CHILTON®

ASIAN
MECHANICAL SERVICE
2006 EDITION
VOLUME II
Infiniti
Mazda
Mitsubishi
Nissan

THOMSON

DELMAR LEARNING

Australia • Canada • Mexico • Singapore • Spain • United Kingdom • United States

THOMSON
™
DELMAR LEARNING

Chilton®

Asian Mechanical Service
2006 Edition
Volume II

Infiniti, Mazda, Mitsubishi, Nissan

**Vice President,
Technology Professional Business Unit:**
Gregory L. Clayton

**Publisher,
Technology Professional Business Unit:**
David Koontz

Director of Marketing:
Beth A. Lutz

Production Director:
Patty Stephan

Editorial Assistant:
Rebecca Rokitowski

Production Manager:
Andrew Crouth

Marketing Manager:
Brian McGrath

Marketing Specialist:
Marissa Maiella

Marketing Coordinator:
Jennifer Stall

Publishing Coordinator:
Paula Baillie

Sr. Content Project Manager:
Elizabeth C. Hough

Managing Editor:
Terry Blomquist

Senior Editor:
Rich Rivele

Editors:
Nick D'Andrea
Matt Frederick
Thomas A. Mellon
Jon Wallace

Graphical Designer:
Melinda Possinger

ISBN: 1-4180-0948-2

ISSN: 1548-0887

NOTICE TO THE READER

Table of Contents

Model Index

USING THIS INFORMATION

Organization

To find where a particular model section or procedure is located, look in the Table of Contents. Main topics are listed with the page number on which they may be found. Following the main topics is an alphabetical listing of all of the procedures within the section and their page numbers.

Manufacturer and Model Coverage

This product covers 2002–2006 Asian models that are produced in sufficient quantities to warrant coverage, and which have technical content available from the vehicle manufacturers before our publication date. Although this information is as complete as possible at the time of publication, some manufacturers may make changes which cannot be included here. While striving for total accuracy, the publisher cannot assume responsibility for any errors, changes, or omissions that may occur in the compilation of this data.

Part Numbers & Special Tools

Part numbers and special tools are recommended by the publisher and vehicle manufacturer to perform specific jobs. Before substituting any part or tool for the one recommended, you must be completely satisfied that neither your personal safety, nor the performance of the vehicle will be endangered.

ACKNOWLEDGEMENT

The publisher would like to express appreciation to the following vehicle manufacturers for their assistance in producing this publication. No further reproduction or distribution of the material in this manual is allowed without the expressed written permission of the vehicle manufacturers and the publisher. Mazda Motor Corporation, Mitsubishi Motors North America, Inc., Nissan North America, including Infiniti and Nissan Divisions.

PRECAUTIONS

Before servicing any vehicle, please be sure to read all of the following precautions, which deal with personal safety, prevention of component damage, and important points to take into consideration when servicing a motor vehicle:

• Always wear safety glasses or goggles when drilling, cutting, grinding or prying.

• Steel-toed work shoes should be worn when working with heavy parts. Pockets should not be used for carrying tools. A slip or fall can drive a screwdriver into your body.

• Work surfaces, including tools and the floor should be kept clean of grease, oil or other slippery material.

• When working around moving parts, don't wear loose clothing. Long hair should be tied back under a hat or cap, or in a hair net.

• Always use tools only for the purpose for which they were designed. Never pry with a screwdriver.

• Keep a fire extinguisher and first aid kit handy.

• Always properly support the vehicle with approved stands or lift.

• Always have adequate ventilation when working with chemicals or hazardous material.

• Carbon monoxide is colorless, odorless and dangerous. If it is necessary to operate the engine with vehicle in a closed area such as a garage, always use an exhaust collector to vent the exhaust gases outside the closed area.

• When draining coolant, keep in mind that small children and some pets are attracted by ethylene glycol antifreeze, and are quite likely to drink any left in an open container, or in puddles on the ground. This will prove fatal in sufficient quantity. Always drain the coolant into a sealable container.

• To avoid personal injury, do not remove the coolant pressure relief cap while the engine is operating or hot. The cooling system is under pressure; steam and hot liquid can come out forcefully when the cap is loosened slightly. Failure to follow these instructions may result in personal injury. The coolant must be recovered in a suitable, clean container for reuse. If the coolant is contaminated it must be recycled or disposed of correctly.

• When carrying out maintenance on the starting system be aware that heavy gauge leads are connected directly to the battery. Make sure the protective caps are in place when maintenance is completed. Failure to follow these instructions may result in personal injury.

• Do not remove any part of the engine emission control system. Operating the engine without the engine emission control system will reduce fuel economy and engine ventilation. This will weaken engine performance and shorten engine life. It is also a violation of Federal law.

• Due to environmental concerns, when the air conditioning system is drained, the refrigerant must be collected using refrigerant recovery/recycling equipment. Federal law requires that refrigerant be recovered into appropriate recovery equipment and the process be conducted by qualified technicians who have been certified by an approved organization, such as MACS, ASI, etc. Use of a recovery machine dedicated to the appropriate refrigerant is necessary to reduce the possibility of oil and refrigerant incompatibility concerns. Refer to the instructions provided by the equipment manufacturer when removing refrigerant from or charging the air conditioning system.

• Always disconnect the battery ground when working on or around the electrical system.

• Batteries contain sulfuric acid. Avoid contact with skin, eyes, or clothing. Also, shield your eyes when working near batteries to protect against possible splashing of the acid solution. In case of acid contact with skin or eyes, flush immediately with water for a minimum of 15 minutes and get prompt medical attention. If acid is swallowed, call a physician immediately. Failure to follow these instructions may result in personal injury.

• Batteries normally produce explosive gases. Therefore, do not allow flames, sparks or lighted substances to come near the battery. When charging or working near a battery, always shield your face and protect your eyes. Always provide ventilation. Failure to follow these instructions may result in personal injury.

• When lifting a battery, excessive pressure on the end walls could cause acid to spew through the vent caps, resulting in personal injury, damage to the vehicle or battery. Lift with a battery carrier or with your hands on opposite corners. Failure to follow these instructions may result in personal injury.

• Observe all applicable safety precautions when working around fuel. Whenever servicing the fuel system, always work in a well-ventilated area. Do not allow fuel spray or vapors to come in contact with a spark, open flame, or excessive heat (a hot drop light, for example). Keep a dry chemical fire extinguisher near the work area. Always keep fuel in a container specifically designed for fuel storage; also, always properly seal fuel containers to avoid the possibility of fire or explosion. Do not smoke or carry lighted tobacco or open flame of any type when working on or near any fuel-related components.

• Fuel injection systems often remain pressurized, even after the engine has been turned OFF. The fuel system pressure must be relieved before disconnecting any fuel lines. Failure to do so may result in fire and/or personal injury.

• The evaporative emissions system contains fuel vapor and condensed fuel vapor. Although not present in large quantities, it still presents the danger of explosion or fire. Disconnect the battery ground cable from the battery to minimize the possibility of an electrical spark occurring, possibly causing a fire or explosion if fuel vapor or liquid fuel is present in the area. Failure to follow these instructions can result in personal injury.

• The EPA warns that prolonged contact with used engine oil may cause a number of skin disorders, including cancer! You should make every effort to minimize your exposure to used engine oil. Protective gloves should be worn when changing oil. Wash your hands and any other exposed skin areas as soon as possible after exposure to used engine oil. Soap and water, or waterless hand cleaner should be used.

• Some vehicles are equipped with an air bag system, often referred to as a Supplemental Restraint System (SRS) or Supplemental Inflatable Restraint (SIR) system. The system must be disabled before performing service on or around system components, steering column, instrument panel components, wiring and sensors. Failure to follow safety and disabling procedures could result in accidental air bag deployment, possible personal injury and unnecessary system repairs.

• Always wear safety goggles when working with, or around, the air bag system. When carrying a non-deployed air bag, be sure the bag and trim cover are pointed away from your body. When placing a non-deployed air bag on a work surface, always face the bag and trim cover upward, away from the surface. This will reduce the motion of the module if it is accidentally deployed.

• Electronic modules are sensitive to electrical charges. The ABS module can be damaged if exposed to these charges.

• Brake pads and shoes may contain asbestos, which has been determined to be a cancer-causing agent. Never clean brake surfaces with compressed air. Avoid inhaling brake dust. Clean all brake surfaces with a commercially available brake cleaning fluid.

• When replacing brake pads, shoes, discs or drums, replace them as complete axle sets.

• When servicing drum brakes, disassemble and assemble one side at a time, leaving the remaining side intact for reference.

• Brake fluid often contains polyglycol ethers and polyglycols. Avoid contact with the eyes and wash your hands thoroughly after handling brake fluid. If you do get brake fluid in your eyes, flush your eyes with clean, running water for 15 minutes. If eye irritation persists, or if you have taken brake fluid internally, immediately seek medical assistance.

• Clean, high quality brake fluid from a sealed container is essential to the safe and proper operation of the brake system. You should always buy the correct type of brake fluid for your vehicle. If the brake fluid becomes contaminated, completely flush the system with new fluid. Never reuse any brake fluid. Any brake fluid that is removed from the system should be discarded. Also, do not allow any brake fluid to come in contact with a painted or plastic surface; it will damage the paint.

• Never operate the engine without the proper amount and type of engine oil; doing so will result in severe engine damage.

• Timing belt maintenance is extremely important! Many models utilize an interference-type, non-freewheeling engine. If the timing belt breaks, the valves in the cylinder head may strike the pistons, causing potentially serious (also time-consuming and expensive) engine damage.

• Disconnecting the negative battery cable on some vehicles may interfere with the functions of the on-board computer system (s) and may require the computer to undergo a relearning process once the negative battery cable is reconnected.

• Steering and suspension fasteners are critical parts because they affect performance of vital components and systems and their failure can result in major service expense. They must be replaced with the same grade or part number or an equivalent part if replacement is necessary. Do not use a replacement part of lesser quality or substitute design. Torque values must be used as specified during reassembly to ensure proper retention of these parts.

INFINITI

G20 • G35 • I35 • M35 • M45 • Q45

SPECIFICATIONS AND MAINTENANCE CHARTS

ENGINE AND VEHICLE IDENTIFICATION

Engine							Model Year	
Code ①	Liters (cc)	Cu. In.	Cyl.	Fuel Sys.	Engine Type	Eng. Mfg.	Code ②	Year
SR20DE	2.0 (1998)	122	4	MFI	DOHC	Nissan	2	2002
VQ35DE	3.5 (3498)	213	6	MFI	DOHC	Nissan	3	2003
VK45DE	4.5 (4494)	274	8	MFI	DOHC	Nissan	4 ③	2004
							5	2005
							6	2006

MFI: Multi-port Fuel Injection

DOHC: Double Overhead Camshaft

① 4th digit of the Vehicle Identification Number (VIN)

② 10th digit of the Vehicle Identification Number (VIN)

③ Serial numbers 800001 and up may be referenced as 2004.5 models

09482_INFI_C0001

GENERAL ENGINE SPECIFICATIONS

Year	Model	Engine Displacement Liters	Engine ID	Net Horsepower @ rpm	Net Torque @ rpm (ft. lbs.)	Bore x Stroke (in.)	Com- pression Ratio	Oil Pressure @ rpm
2002	G20	2.0	SR20DE	145@6000	136@4800	3.39x3.39	9.5:1	46-57@3200
	I35	3.5	VQ35DE	255@5800	246@4400	3.76X3.20	10.3:1	43@2000
	Q45	4.5	VK45DE	340@6400	333@4000	3.66x3.25	10.5:1	43@3000
2003	G35	3.5	VQ35DE	260@6000 ①	260@4800 ②	3.76X3.20	10.3:1	43@2000
	I35	3.5	VQ35DE	255@5800	246@4400	3.76X3.20	10.3:1	43@2000
	M45	4.5	VK45DE	340@6400	333@4000	3.66x3.25	10.5:1	43@3200
	Q45	4.5	VK45DE	340@6400	333@4000	3.66x3.25	10.5:1	43@3000
2004	G35	3.5	VQ35DE	260@6000 ①	260@4800 ②	3.76X3.20	10.3:1	43@2000
	I35	3.5	VQ35DE	255@5800	246@4400	3.76X3.20	10.3:1	43@2000
	M45	4.5	VK45DE	340@6400	333@4000	3.66x3.25	10.5:1	43@3200
	Q45	4.5	VK45DE	340@6400	333@4000	3.66x3.25	10.5:1	43@3000
2005	G35	3.5	VQ35DE	280@6200 ③	270@4800 ④	3.76X3.20	10.3:1	43@2000
	Q45	4.5	VK45DE	340@6400	333@4000	3.66x3.25	10.5:1	43@3000
2006	G35	3.5	VQ35DE	280@6200 ③	270@4800 ④	3.76X3.20	10.3:1	43@2000
	M35	3.5	VQ35DE	280@6200	270@4800	3.76X3.20	10.3:1	43@2000
	M45	4.5	VK45DE	335@6400	340@4000	3.66x3.25	10.5:1	43@3200
	Q45	4.5	VK45DE	340@6400	333@4000	3.66x3.25	10.5:1	43@3000

① Coupe: 280@6200 5-Spd A/T

② Coupe: 270@4800 5-Spd A/T

③ 298@6400 6-Spd M/T

④ 260@6400 6-Spd M/T

09482_INFI_C0002

ENGINE TUNE-UP SPECIFICATIONS

Year	Engine Displacement Liters	Engine ID	Spark Plug Gap (in.)	Ignition Timing (deg.) MT	Ignition Timing (deg.) AT	Fuel Pump (psi) ①	Idle Speed (rpm) MT	Idle Speed (rpm) AT	Valve Clearance Intake	Valve Clearance Exhaust
2002	2.0	SR20DE	0.031-0.035	15B	15B	34	800	800	②	②
	3.5	VQ35DE	0.043	—	15B	34	—	600-700	②	②
	4.5	VK45DE	0.039-0.041	—	15B	34	—	650	②	②
2003	3.5	VQ35DE	0.043	15B	15B	34	600-700	600-700	②	②
	4.5	VK45DE	0.039-0.041	—	15B	34	—	650	②	②
2004	3.5	VQ35DE	0.043	15B	15B	34	600-700	600-700	②	②
	4.5	VK45DE	0.043	—	12B	34	—	650	②	②
2005	3.5	VQ35DE	0.043	15B	15B	34	600-700	600-700	②	②
	4.5	VK45DE	0.043	—	12B	34	—	650	②	②
2006	3.5	VQ35DE	0.043	15B	15B	34	600-700	600-700	②	②
	4.5	VK45DE	0.043	—	12B	34	—	650	②	②

NOTE: The Vehicle Emission Control Information label often reflects specification changes made during production.

The label figures must be used if they differ from those in this chart.

B: Before top dead center

① 43 psi with regulator vacuum hose disconnected

② Hydraulic lash adjuster. See text for procedure.

09482_INFI_C0003

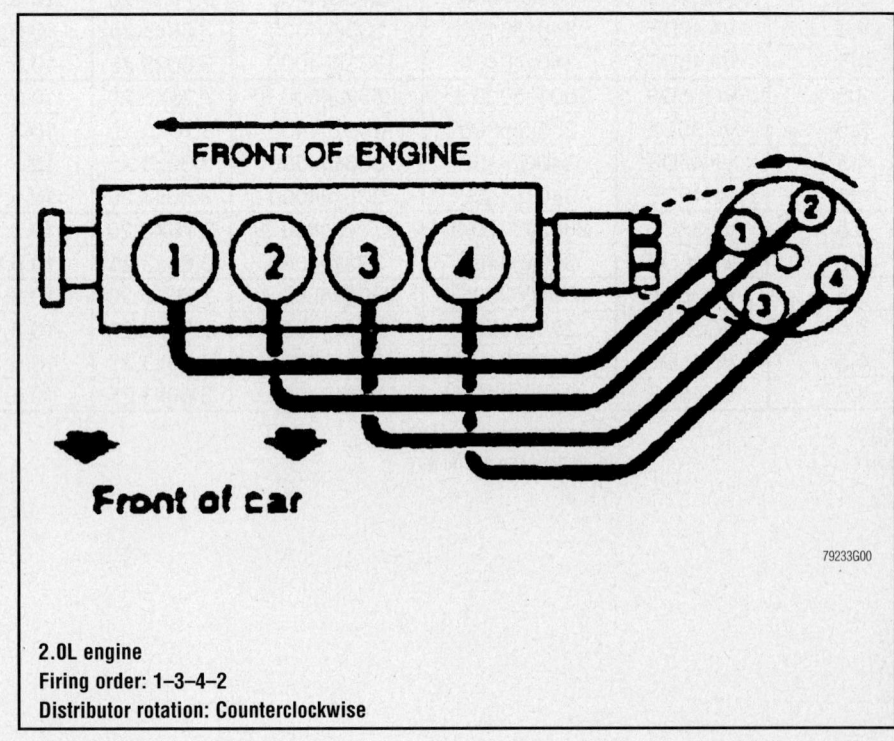

2.0L engine
Firing order: 1–3–4–2
Distributor rotation: Counterclockwise

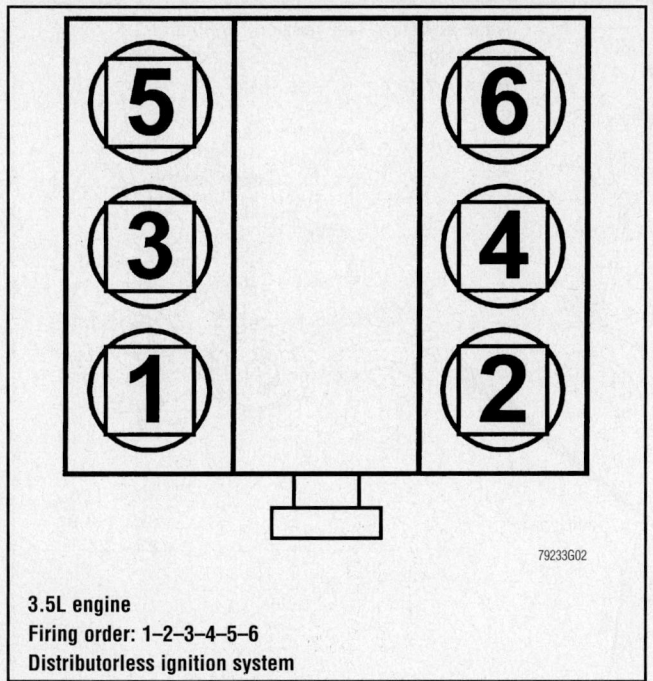

3.5L engine
Firing order: 1–2–3–4–5–6
Distributorless ignition system

79233G02

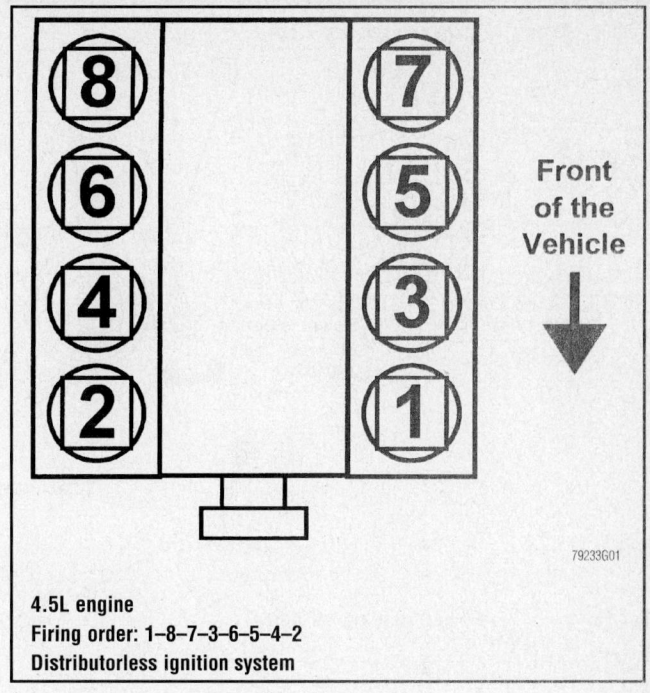

Front of the Vehicle

4.5L engine
Firing order: 1–8–7–3–6–5–4–2
Distributorless ignition system

79233G01

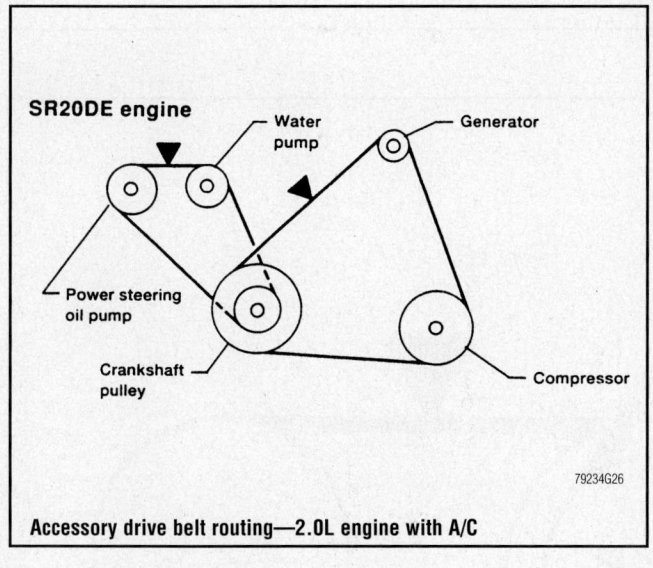

SR20DE engine

Water pump

Generator

Power steering oil pump

Crankshaft pulley

Compressor

79234G26

Accessory drive belt routing—2.0L engine with A/C

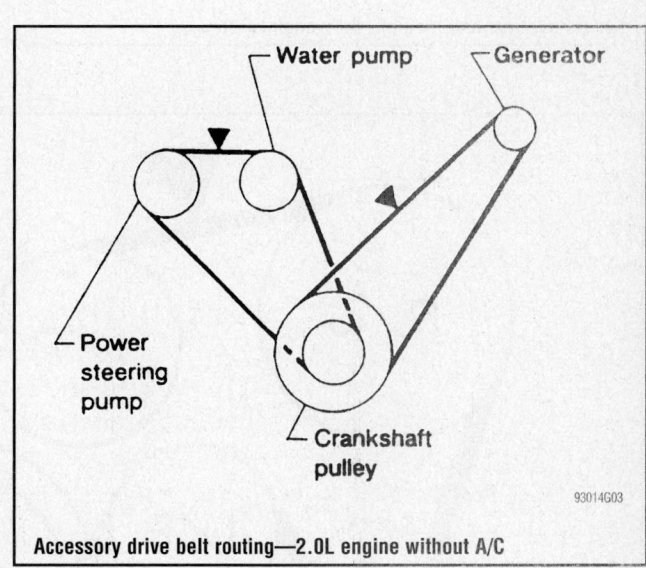

Water pump

Generator

Power steering pump

Crankshaft pulley

93014G03

Accessory drive belt routing—2.0L engine without A/C

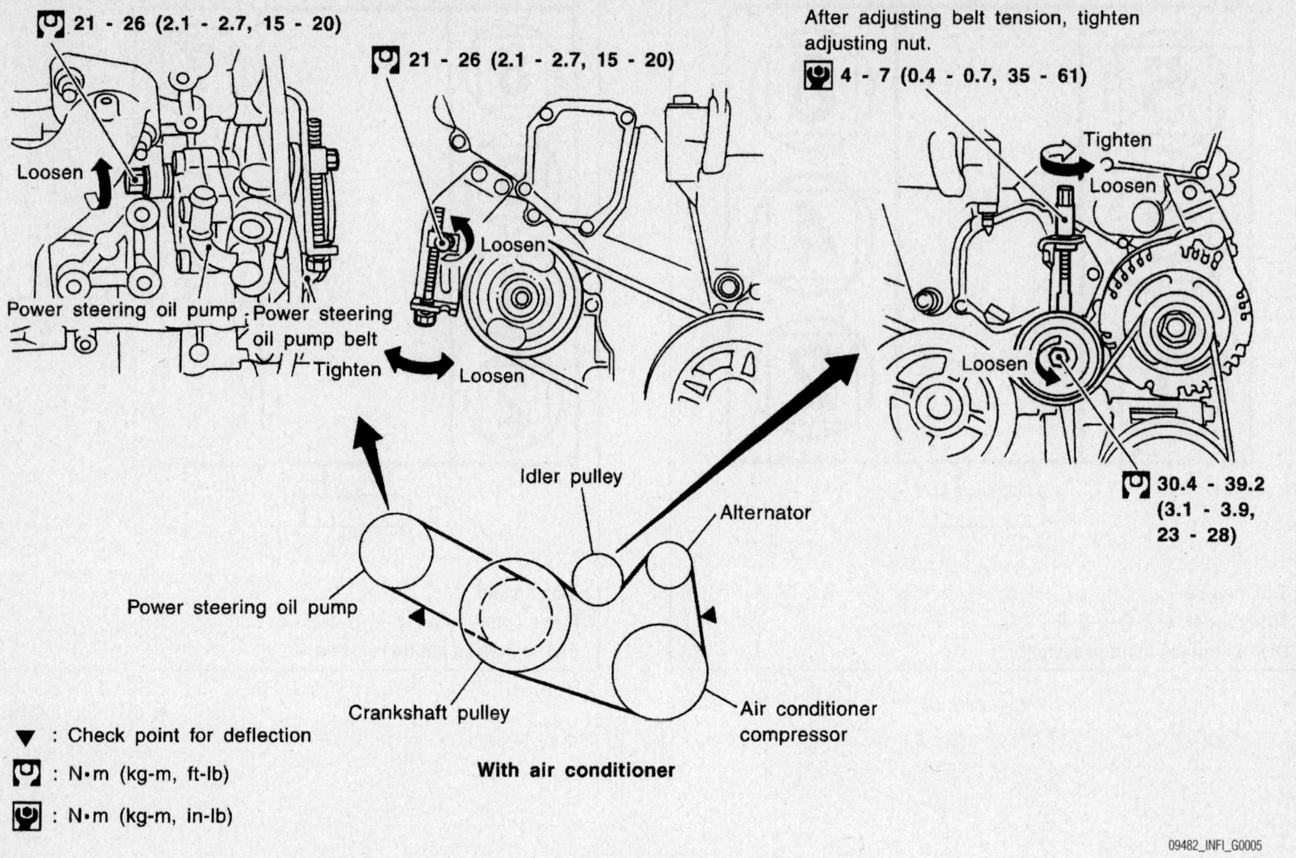

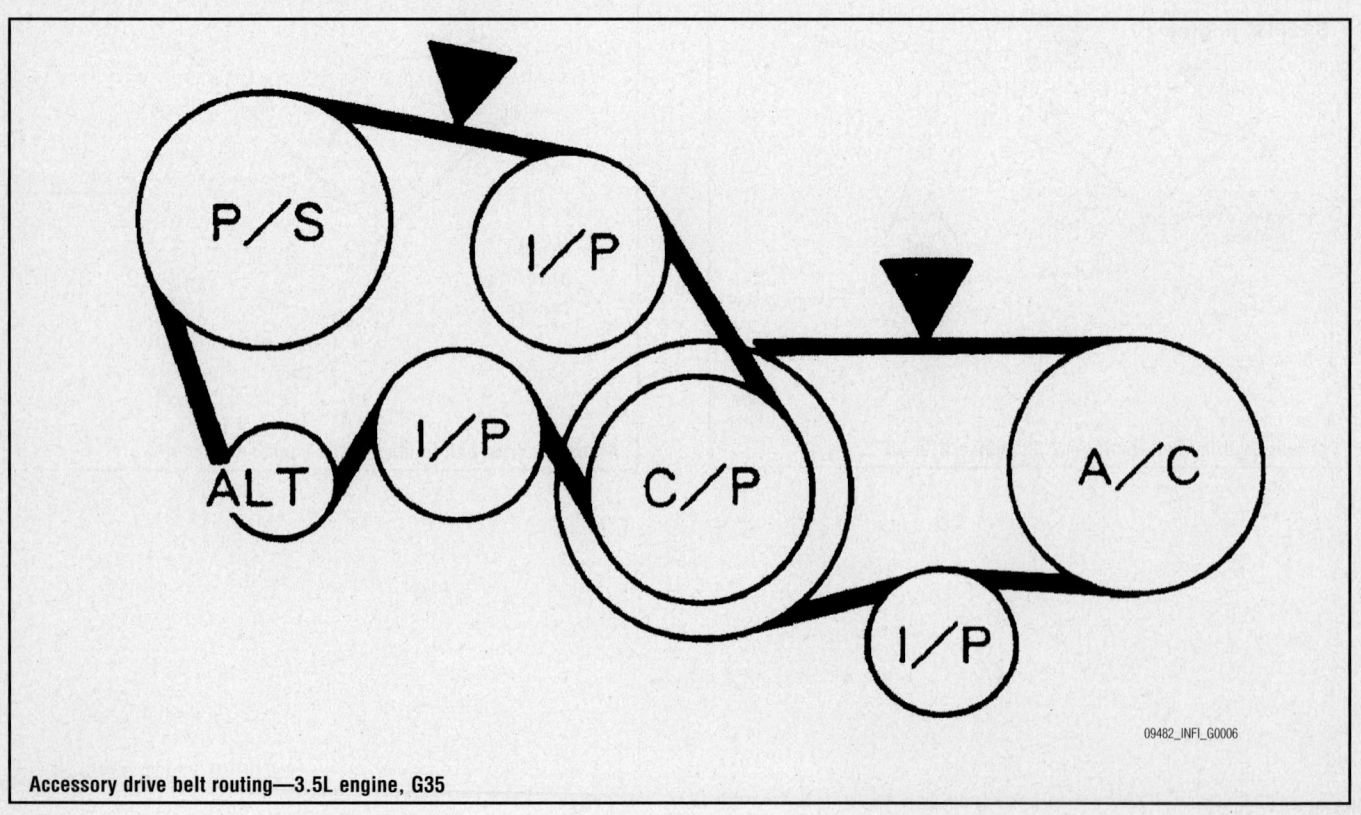

21 - 26 (2.1 - 2.7, 15 - 20)

Loosen

Power steering oil pump

Power steering oil pump belt

Tighten

21 - 26 (2.1 - 2.7, 15 - 20)

Loosen

Loosen

After adjusting belt tension, tighten adjusting nut.

4 - 7 (0.4 - 0.7, 35 - 61)

Tighten

Loosen

Loosen

30.4 - 39.2 (3.1 - 3.9, 23 - 28)

Idler pulley

Alternator

Power steering oil pump

Crankshaft pulley

Air conditioner compressor

▼ : Check point for deflection

: N•m (kg-m, ft-lb)

: N•m (kg-m, in-lb)

With air conditioner

09482_INFI_G0005

Accessory drive belt routing—3.5L engine, I35

P/S

I/P

ALT

I/P

C/P

A/C

I/P

09482_INFI_G0006

Accessory drive belt routing—3.5L engine, G35

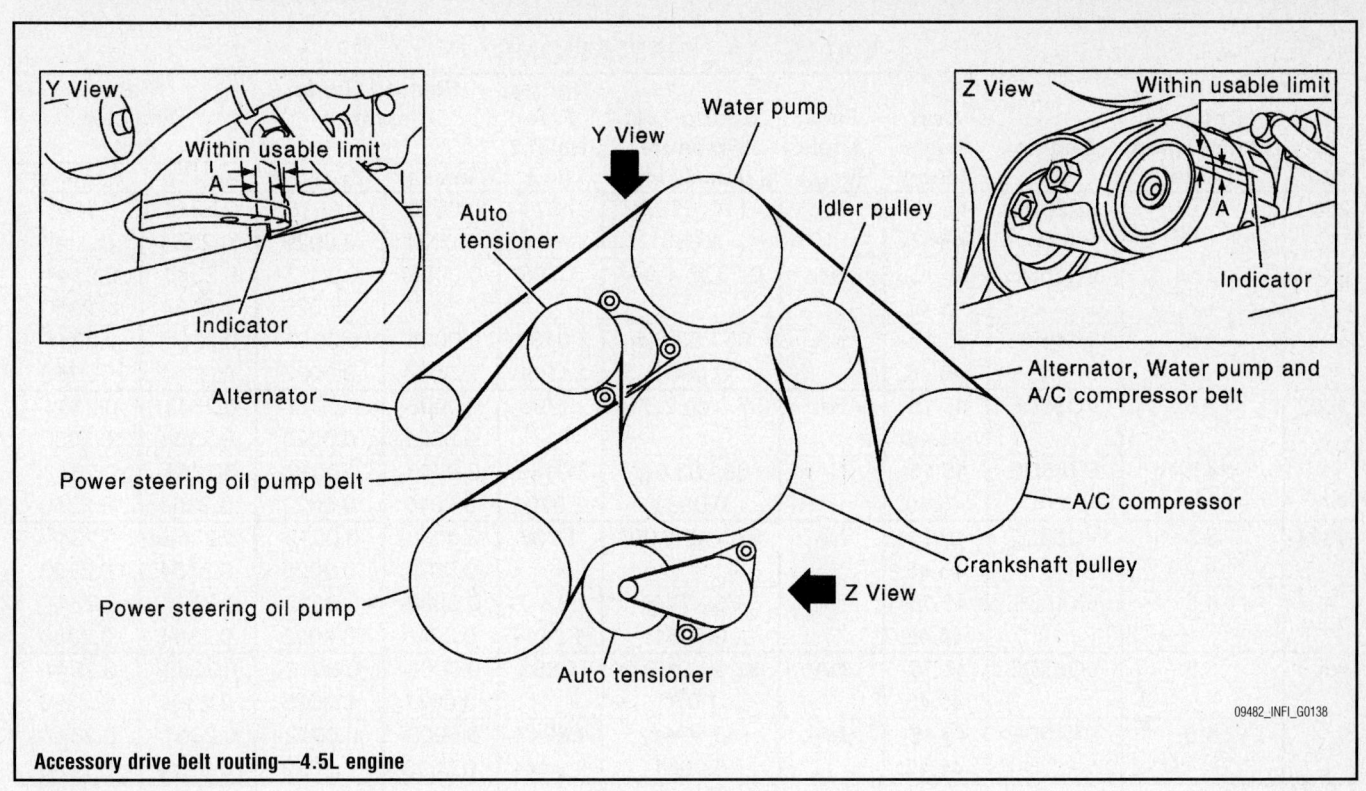

Accessory drive belt routing—4.5L engine

CAPACITIES

Year	Model	Engine Displacement Liters	Engine ID	Engine Oil with Filter	Transmission (pts.) Manual	Transmission (pts.) Auto.	Transfer Case (pts.)	Drive Axle (pts.)	Fuel Tank (gal.)	Cooling System (qts.)
2002	G20	2.0	SR20DE	3.62	①	14.8	—	—	15.9	6.50
	I35	3.5	VQ35DE	5.00	—	21.8	—	—	18.5	9.00
	Q45	4.5	VK45DE	5.63	—	21.3	—	2.75	21.4	10.38
2003	G35	3.5	VQ35DE	5.00	6.25	21.8	—	3.00	18.5	②
	I35	3.5	VQ35DE	5.00	—	21.8	—	—	18.5	9.00
	M45	4.5	VK45DE	5.88	—	21.7	—	3.00	21.1	10.38
	Q45	4.5	VK45DE	5.63	—	21.3	—	2.75	21.4	10.38
2004	G35	3.5	VQ35DE	5.00	6.25	21.8	2.6	③	20.1	②
	I35	3.5	VQ35DE	5.00	—	18.0	—	—	18.5	9.00
	M45	4.5	VK45DE	5.88	—	21.7	—	3.00	21.1	10.38
	Q45	4.5	VK45DE	5.88	—	21.7	—	3.00	21.4	10.38
2005	G35	3.5	VQ35DE	5.00	6.25	21.8	2.6	③	20.1	②
	Q45	4.5	VK45DE	5.91	—	21.7	—	3.00	21.4	10.88
2006	G35	3.5	VQ35DE	5.00	6.25	21.8	2.6	③	20.1	②
	M35	3.5	VQ35DE	5.00	—	21.8	—	③	20.0	9.38
	M45	4.5	VK45DE	5.75	—	21.7	—	3.00	20.0	11.00
	Q45	4.5	VK45DE	5.91	—	21.7	—	3.00	21.4	10.88

NOTE: All capacities are approximate. Add fluid gradually and check to be sure a proper fluid level is obtained.

① RSF50V: 9.13-9.50   ② Man. Trans. 9.25   ③ Rear Axle 3.00
RSF50A: 9.50-10.13   Auto. Trans. 9.00   Front Axle 1.38 (AWD)

VALVE SPECIFICATIONS

Year	Engine Displacement Liters	Engine ID	Seat Angle (deg.)	Face Angle (deg.)	Spring Test Pressure (lbs. @ in.)	Spring Free Height (in.)	Stem-to-Guide Clearance (in.)		Stem Diameter (in.)	
							Intake	Exhaust	Intake	Exhaust
2002	2.0	SR20DE	45.15-45.45	NA	117 -128@ 1.063	1.871	0.0008-0.0021	0.0016-0.0029	0.2348-0.2354	0.2341-0.2346
	3.5	VQ35DE	45.15-45.45	NA	97.3 @ 1.094	1.796	0.0008-0.0021	0.0012-0.0025	0.2348-0.2354	0.2344-0.2350
	4.5	VK45DE	45.15-45.45	NA	65 - 83.6@ 0.0961	2.0189-2.0268	0.0008-0.0018	0.0012-0.0022	0.2351-0.2354	0.2347-0.2350
2003	3.5	VQ35DE	45.15-45.45	NA	97.3 @ 1.094	1.796	0.0008-0.0021	0.0012-0.0025	0.2348-0.2354	0.2344-0.2350
	4.5	VK45DE	45.15-45.45	NA	65 - 83.6@ 0.0961	2.0189-2.0268	0.0008-0.0018	0.0012-0.0022	0.2351-0.2354	0.2347-0.2350
2004	3.5	VQ35DE	45.15-45.45	NA	97.3 @ 1.094	1.796	0.0008-0.0021	0.0012-0.0025	0.2348-0.2354	0.2344-0.2350
	4.5	VK45DE	45.15-45.45	NA	65 - 74@ 0.0961	1.8247-1.8444	0.0008-0.0018	0.0012-0.0022	0.2351-0.2354	0.2347-0.2350
2005	3.5	VQ35DE	45.15-45.45	NA	83.9 - 94.6@ 1.070	1.853	0.0008-0.0021	0.0012-0.0025	0.2348-0.2354	0.2344-0.2350
	4.5	VK45DE	45.15-45.45	NA	65 - 74@ 0.0961	1.8247-1.8444	0.0008-0.0018	0.0012-0.0022	0.2351-0.2354	0.2347-0.2350
2006	3.5	VQ35DE	45.15-45.45	NA	①	②	0.0008-0.0021	0.0012-0.0025	0.2348-0.2354	0.2344-0.2350
	4.5	VK45DE	45.15-45.45	NA	65 - 74@ 0.0961	1.8247-1.8444	0.0008-0.0018	0.0012-0.0022	0.2351-0.2354	0.2347-0.2350

NA: Not Available

09482_INFI_C0005

PISTON AND RING SPECIFICATIONS

All measurements are given in inches.

Year	Engine Displacement Liters	Engine ID	Piston Clearance	Ring Gap			Ring Side Clearance		
				Top Compression	Bottom Compression	Oil Control	Top Compression	Bottom Compression	Oil Control
2002	2.0	SR20DE	0.0004-0.0012	0.0079-0.0154	0.0138-0.0232	0.0079-0.0272	0.0016-0.0031	0.0012-0.0028	0.0026-0.0053
	3.5	VQ35DE	0.0004-0.0012	0.0091-0.0130	0.0130-0.0189	0.0079-0.0197	0.0018-0.0031	0.0012-0.0028	0.0026-0.0053
	4.5	VK45DE	0.0004-0.0012	0.0087-0.0126	0.0126-0.0185	0.0079-0.0236	0.0018-0.0031	0.0012-0.0028	0.0026-0.0053
2003	3.5	VQ35DE	0.0004-0.0012	0.0091-0.0130	0.0130-0.0189	0.0079-0.0197	0.0018-0.0031	0.0012-0.0028	0.0026-0.0053
	4.5	VK45DE	0.0004-0.0007	0.0087-0.0126	0.0087-0.0126	0.0079-0.0236	0.0018-0.0031	0.0012-0.0028	0.0026-0.0053
2004	3.5	VQ35DE	0.0004-0.0012	0.0091-0.0130	0.0130-0.0189	0.0079-0.0236①	0.0018-0.0031	0.0012-0.0028	0.0026-0.0053
	4.5	VK45DE	0.0004-0.0012	0.0087-0.0126	0.0087-0.0126	0.0079-0.0236	0.0018-0.0031	0.0012-0.0028	0.0026-0.0053
2005	3.5	VQ35DE	0.0004-0.0012	0.0091-0.0130	0.0130-0.0189	0.0079-0.0197	0.0018-0.0031	0.0012-0.0028	0.0026-0.0053
	4.5	VK45DE	0.0004-0.0012	0.0087-0.0126	0.0087-0.0126	0.0079-0.0197	0.0018-0.0031	0.0012-0.0028	0.0026-0.0053
2006	3.5	VQ35DE	0.0004-0.0012	0.0091-0.0130	0.0130-0.0189	0.0079-0.0197	0.0018-0.0031	0.0012-0.0028	0.0026-0.0053
	4.5	VK45DE	0.0004-0.0012	0.0087-0.0126	0.0087-0.0126	0.0079-0.0197	0.0018-0.0031	0.0012-0.0028	0.0026-0.0053

① Serial numbers up to 800000; Serial numbers 800001 and above: 0.0079-0.0197

09482_INFI_C0006

CRANKSHAFT AND CONNECTING ROD SPECIFICATIONS
All measurements represent standard values and are given in inches.

Year	Engine Displacement Liters	Engine ID	Crankshaft				Connecting Rod		
			Main Brg. Journal Dia.	Main Brg. Oil Clearance	Shaft End-play	Thrust on No.	Journal Diameter	Oil Clearance	Side Clearance
2002	2.0	SR20DE	2.1643-2.1646	0.0002-0.0009	0.0039-0.0102	3	1.8885-1.8887	0.0008-0.0018	0.0079-0.0138
	3.5	VQ35DE	2.3603-2.3612	0.0014-0.0018	0.0039-0.0098	3	2.0460-2.0462	0.0013-0.0023	0.0079-0.0138
	4.5	VK45DE	2.5173-2.5183	①	0.0039-0.0102	3	2.0460-2.0462	0.0008-0.0018	0.0079-0.0138
2003	3.5	VQ35DE	2.3603-2.3612	0.0014-0.0018	0.0039-0.0098	3	2.0460-2.0462	0.0013-0.0023	0.0079-0.0138
	4.5	VK45DE	2.5173-2.5183	①	0.0039-0.0098	3	2.0460-2.0462	0.0008-0.0018	0.0079-0.0138
2004	3.5	VQ35DE	2.3603-2.3612	0.0014-0.0018	0.0039-0.0098	3	2.0460-2.0462	0.0013-0.0023	0.0079-0.0138
	4.5	VK45DE	2.5173-2.5183	①	0.0039-0.0098	3	2.0460-2.0462	0.0008-0.0018	0.0079-0.0138
2005	3.5	VQ35DE	2.3603-2.3612	0.0014-0.0018	0.0039-0.0098	3	2.0460-2.0462	0.0013-0.0023	0.0079-0.0138
	4.5	VK45DE	2.5173-2.5183	①	0.0039-0.0098	3	2.0460-2.0462	0.0008-0.0018	0.0079-0.0138
2006	3.5	VQ35DE	2.3603-2.3612	0.0014-0.0018	0.0039-0.0098	3	②	0.0013-0.0023	0.0079-0.0138
	4.5	VK45DE	2.5173-2.5183	①	0.0039-0.0098	3	②	0.0008-0.0018	0.0079-0.0138

① Nos. 1 and 5: 0.00004-0.0004 in.
Nos. 2, 3 and 4: 0.0003-0.0007

② Grade 0: 2.0460-2.0462 in.
Grade 1: 2.0457-2.0460 in.
Grade 2: 2.0455-2.0457 in.

09482_INFI_C0007

TORQUE SPECIFICATIONS

All readings in ft. lbs.

Year	Engine Displacement Liters	Engine ID	Cylinder Head Bolts	Main Bearing Bolts	Rod Bearing Bolts	Crankshaft Damper Bolts	Flywheel Bolts	Manifold		Spark Plugs	Oil Pan Drain Plug
								Intake	Exhaust		
2002	2.0	SR20DE	①	②	③	105-112	61-69	13-15	28-35	14-22	NA
	3.5	VQ35DE	④	⑤	⑥	⑦	61-69	⑧	22-24	15-21	22-29
	4.5	VK45DE	⑨	⑩	③	⑪	62-68	18-23	19-22	15-21	22-28
2003	3.5	VQ35DE	④	⑤	⑥	⑦	61-69	⑧	21-23	15-21	22-29
	4.5	VK45DE	⑨	⑩	③	⑪	62-68	21	19-22	18	22-28
2004	3.5	VQ35DE	④	⑤	⑥	⑦	61-69	⑧	21-23	15-21	22-29
	4.5	VK45DE	⑨	⑩	③	⑪	62-68	21	19-22	18	22-28
2005	3.5	VQ35DE	④	⑤	⑥	⑦	61-69	⑧	21-23	15-21	22-29
	4.5	VK45DE	⑨	⑩	③	⑪	62-68	18-23	19-22	15-21	22-28
2006	3.5	VQ35DE	④	⑤	⑥	⑫	61-69	⑧	21-23	18	25
	4.5	VK45DE	⑨	⑩	③	⑪	62-68	21	19-22	18	25

NA: Not Available

① Step 1: 29 ft. lbs.
Step 2: 58 ft. lbs.
Step 3: Loosen bolts completely
Step 4: 27 ft. lbs.
Step 5: Tighten an additional 90-95 degrees
Step 6: Repeat Step 5.

② Step 1: 61-112 ft. lbs.
Step 2: Tighten an additional 70-80 degrees.
Step 3: Loosen bolts completely.
Step 4: 24-28 ft.lbs.
Step 5: Tighten an additional 30-35 degrees.

③ Step 1: 10-12 ft. lbs.
Step 2: Tighten an additional 60 degrees.

④ Step 1: 72 ft. lbs.
Step 2: Loosen bolts completely
Step 3: 25-33 ft. lbs.
Step 4: Tighten an additional 90 degrees
Step 5: Repeat Step 4

⑤ Step 1: Shift crankshaft to align the bearing beam
Step 2: Tighten all bolts to 24-28 ft. lbs.
Step 3: Tighten an additional 90-95 degrees

⑥ Step 1: Tighten to 15 ft. lbs.
Step 2: Tighten an additional 90-95 degrees

⑦ Step 1: 29-36 ft. lbs.
Step 2: Tighten an additional 60-66 degrees

⑧ Step 1: Tighten to 4-7 ft. lbs.
Step 2: Tighten to 20-23 ft. lbs.
Step 3: Tighten, again, to 20-23 ft. lbs.

⑨ Step 1: 72 ft. lbs.
Step 2: Loosen bolts completely
Step 3: 33 ft. lbs.
Step 4: Tighten an additional 60 degrees
Step 5: Repeat step 4.

⑩ Step 1: M12 bolts: 29 ft. lbs.
Step 2: M9 bolts: 22 ft. lbs.
Step 3: M12 bolts: Tighten an additional 40 degrees
Step 4: M9 bolts: Tighten an additional 30 degrees
Step 5: M10 side bolts: 10-12 ft. lbs. (2002-2003) or 36 ft. lbs. (2004-2006)

⑪ Step 1: 65-72 ft. lbs.
Step 2: Tighten an additional 90-96 degrees

⑫ Step 1: 33 ft. lbs.
Step 2: plus 90 degrees

09482_INFI_C0008

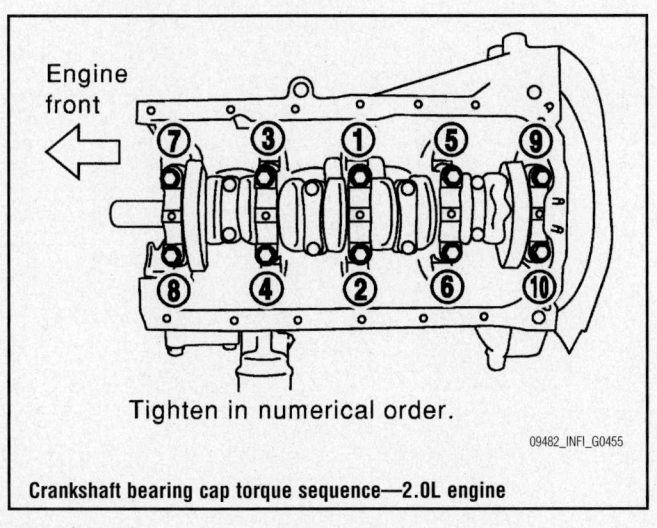

Tighten in numerical order.

09482_INFI_G0455

Crankshaft bearing cap torque sequence—2.0L engine

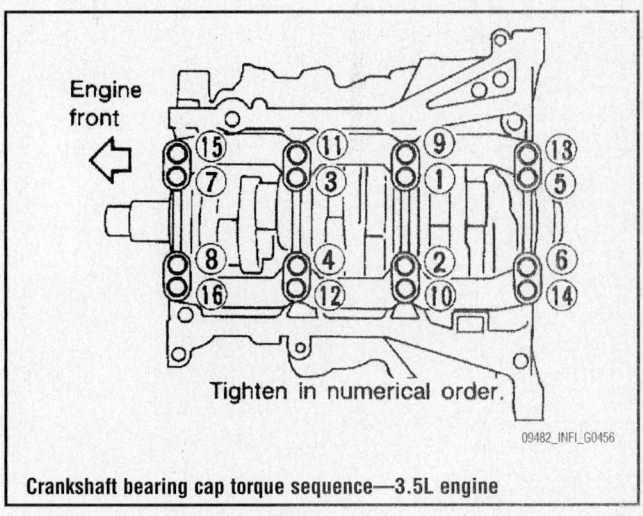

Tighten in numerical order.

09482_INFI_G0456

Crankshaft bearing cap torque sequence—3.5L engine

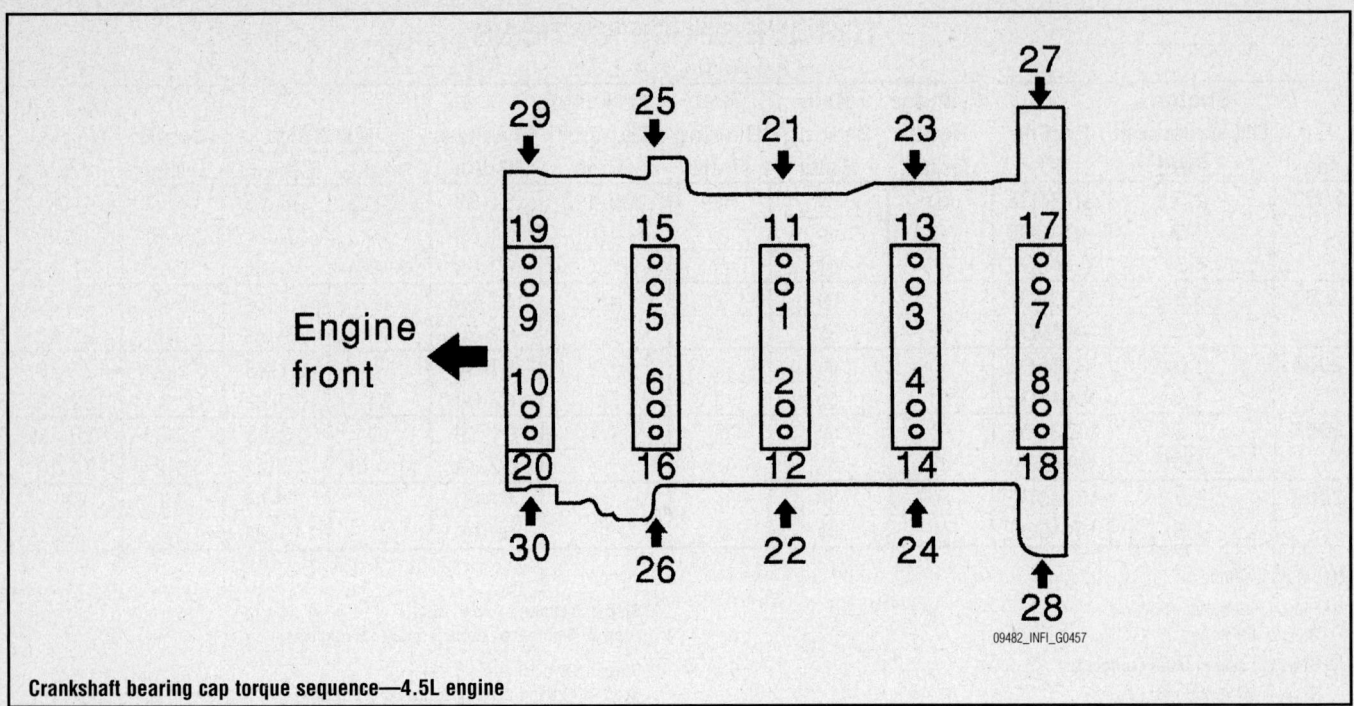

Crankshaft bearing cap torque sequence—4.5L engine

WHEEL ALIGNMENT

Year	Model		Caster Range (+/-Deg.)	Caster Preferred Setting (Deg.)	Camber Range (+/-Deg.)	Camber Preferred Setting (Deg.)	Toe-in (in.)
2002	G20	F	0.75	+1.92	0.75	0.00	0.04 +/- 0.04
		R	—	—	0.75	-1.30	0.16 +/- 0.15
	I35	F	0.75	+2.75	0.75	-0.25	0.04 +/- 0.04
		R	—	—	0.75	-1.00	0.04 +/- 0.16
	Q45	F	0.75	+6.17	0.75	-0.75	0.04 +/- 0.04
		R	—	—	0.50	-0.67	0.09 +/- 0.09
2003	G35 Coupe	F	0.75	①	0.75	-0.50	0.04 +/- 0.04
		R	—	—	0.50	-1.50	0.11 +/- 0.11
	G35 Sedan	F	0.75	7.75	0.75	-0.08	0.04 +/- 0.04
		R	—	—	0.50	-0.58	0.11 +/- 0.11
	I35	F	0.75	2.75	0.75	-0.25	0.04 +/- 0.04
		R	—	—	0.75	-1.00	0.04 +/- 0.16
	M45	F	0.75	6.58	0.75	-0.67	0.08 +/- 0.04
		R	—	—	0.50	-0.67	0.10 +/- 0.10
	Q45	F	0.75	6.17	0.75	-0.75	0.05 +/- 0.05
		R	—	—	0.50	②	③
2004	G35 Coupe	F	0.75	①	0.75	-0.50	0.04 +/- 0.04
		R	—	—	0.50	-1.25	0.11 +/- 0.11
	G35 2WD Sedan	F	0.75	+7.75	0.75	-0.08	0.04 +/- 0.04
		R	—	—	0.50	④	0.11 +/- 0.10
	G35 AWD Sedan	F	0.75	+6.67	0.75	-0.25	0.04 +/- 0.04
		R	—	—	0.50	④	0.11 +/- 0.10
	I35	F	0.75	+2.75	0.75	-0.25	0.04 +/- 0.04
		R	—	—	0.75	-1.00	0.04 +/- 0.16
	M45	F	0.75	+6.58	0.75	-0.67	0.08 +/- 0.04
		R	—	—	0.50	-0.67	0.10 +/- 0.10
	Q45	F	0.75	+6.17	0.75	-0.75	0.04 +/- 0.04
		R	—	—	0.50	②	③
2005	G35 Coupe	F	0.75	① ⑤	0.75	-0.50	0.04 +/- 0.04
		R	—	—	0.50	-1.25	0.11 +/- 0.10
	G35 2WD Sedan	F	0.75	+7.75	0.75	-0.08	0.04 +/- 0.04
		R	—	—	0.50	④	0.11 +/- 0.10
	G35 AWD Sedan	F	0.75	+6.67	0.75	-0.25	0.04 +/- 0.04
		R	—	—	0.50	④	0.11 +/- 0.10
	Q45	F	0.75	+6.17	0.75	-0.75	0.04 +/- 0.04
		R	—	—	0.50	②	③
2006	G35 2WD Sedan	F	0.75	+7.75	0.75	-0.08	0.04 +/- 0.04
		R	—	—	0.50	⑧	0.11 +/- 0.10
	G35 AWD Sedan	F	0.75	+6.67	0.75	-0.25	0.04 +/- 0.04
		R	—	—	0.50	⑧	0.11 +/- 0.10
	M35 2WD Sedan	F	0.65	⑥	0.75	-0.25	0.04 +/- 0.04
		R	—	—	0.50	⑦	0.11 +/- 0.10
	M35 AWD Sedan	F	0.75	+3.83	0.50	-0.25	0.04 +/- 0.04
		R	—	—	0.50	-0.17	0.11 +/- 0.10
	M45	F	0.65	+3.83	0.75	-0.25	0.04 +/- 0.04
		R	—	—	0.50	⑦	0.11 +/- 0.10
	Q45	F	0.75	+6.17	0.75	-0.75	0.04 +/- 0.04
		R	—	—	0.50	②	③

① 17" Wheel: +8.17
   18" Wheel: +8.00
② 17" Wheel: -042
   18" Wheel: -058
③ 17" Wheel: 0.031 +/- 0.05
   18" Wheel: 0.047 +/- 0.10
④ Auto. Trans.: -0.58
   Man. Trans.: -0.67
⑤ 19" Wheel: Not Available
⑥ 18" Wheel: +0.450
   19" Wheel: +0.458
⑦ 18" Wheel: -0.67
   19" Wheel: -0.83
⑧ Exc. 18 in. wheels: -0.58
   18 inch wheels: -0.67

09482_INFI_C0009

TIRE, WHEEL AND BALL JOINT SPECIFICATIONS

Year	Model	OEM Tires		Tire Pressures (psi)		Wheel Size	Ball Joint Inspection	Lug Nut (Ft. Lbs.)
		Standard	Optional	Front	Rear			
2002	G20	P195/65HR15	P205/50R16	35	35	Std: 6.0-JJ Opt: 7.0-JJ	①	NA
	I35	P215/55R17	None	30	30	Std: 7.0-JJ	①	73-94
	Q45	P225/55VR17	P245/45VR18	33	33	Std: 7.5-JJ Opt: 7.5-JJ	①	72-87
2003	G35 Coupe	②	③	30	30	Std: ④ Opt: 8.0-JJ	①	72-87
	G35 Sedan	P205/65R16	P215/55R17	30	30	Std: 6.5-JJ Opt: 7-JJ	①	72-87
	I35	P215/55R17	None	30	30	Std: 7.0-JJ	①	73-94
	M45	P235/45R18	None	33	33	Std: 7.5-JJ	①	72-87
	Q45	P225/55VR17	P245/45VR18	33	33	Std: 7.5-JJ Opt: 7.5-JJ	①	72-87
2004	G35 Coupe	②	③	30	30	Std: ④ Opt: 8-JJ	①	72-87
	G35 Sedan	P205/65R16	P215/55R17	30	30	Std: 6.5-JJ Opt: 7-JJ	①	72-87
	I35	P215/55R17	None	30	30	Std: 7.0-JJ	①	73-94
	M45	P235/45R18	None	33	33	Std: 7.5-JJ	①	72-87
	Q45	P225/55R17	P245/45R18	33	33	Std: 7.5-JJ Opt: 7.5-JJ	①	72-87
2005	G35 Coupe	②	③ ⑤	30	30	Std: 7.0-JJ Opt: ⑥	①	72-87
	G35 Sedan	P215/55R17	P235/45R18⑦	30	30	Std: 7.0-JJ Opt: 7.5-JJ	①	72-87
	Q45	P225/55R17	P245/45R18	33	33	Std: 7.5-JJ Opt: 8.0-JJ	①	72-87
2006	G35 Coupe	②	③ ⑤	30	30	Std: 7.0-JJ Opt: ⑥	①	72-87
	G35 Sedan	P215/55R17	P235/45R18⑦⑧	30	30	Std: 7.0-JJ Opt: 7.5-JJ⑧	①	72-87
	M35	P245/45R18	P245/40R19⑦	33	33	Std: 8.0-JJ Opt: 8.5-JJ	①	72-87
	M45	P245/45R18	P245/40R19	33	33	Std: 8.0-JJ Opt: 8.5-JJ	①	72-87
	Q45	P225/55R17	P245/45R18 P245/40R19	33	33	Std: 7.5-JJ Opt: 8.0-JJ, 8.5-JJ	①	72-87

OEM: Original Equipment Manufacturer
PSI: Pounds Per Square Inch
STD: Standard
OPT: Optional
NA: Not Available
M/T: Manual Transmission
① Replace if any measurable movement is found.

② Front: P225/50R17, Rear: P235/50R17
③ Front: P225/45R18, Rear: P245/45R18
④ Front: 7.5JJ, Rear: 8.0JJ
⑤ Front: P225/40R19, Rear: P245/40R19 (Standard 6-Spd M/T)
⑥ 18" Wheel: 8.0-JJ;   19" Wheel: Front 8.0-JJ, Rear 8.5-JJ
⑦ 2WD only
⑧ Standard 6-Spd M/T

BRAKE SPECIFICATIONS
All measurements in inches unless noted

| Year | Model | Front Brake Disc | | | Rear Brake Disc | | | Minimum Lining Thickness | | Brake Caliper | |
		Original Thickness	Minimum Thickness	Maximum Run-out	Original Thickness	Minimum Thickness	Maximum Run-out	Front	Rear	Bracket Bolts (ft. lbs.)	Mounting Bolts (ft. lbs.)
2002	G20	0.870	0.787	0.003	0.350	0.315	0.0028	0.079	0.059	①	16-23
	I35	0.940	0.866	0.0028	0.350	0.315	0.0028	0.079	0.059	②	16-23
	Q45	1.100	1.020	0.0028	0.630	0.550	0.0039	0.079	0.079	③	④
2003	G35	0.945	0.866	0.0028	0.630	0.551	0.0039	0.079	0.079	⑤	⑥
	I35	0.940	0.866	0.0028	0.350	0.315	0.0028	0.079	0.059	②	16-23
	M45	1.102	1.024	0.0028	0.630	0.551	0.004	0.079	0.079	⑦	④
	Q45	1.100	1.020	0.0028	0.630	0.550	0.0039	0.079	0.079	③	④
2004	G35	0.945	0.866	0.0014	0.630	0.551	0.0039	0.079	0.079	⑤	⑥
	I35	0.940	0.866	0.0028	0.350	0.315	0.0028	0.079	0.059	②	16-23
	M45	1.102	1.024	0.0028	0.630	0.551	0.004	0.079	0.079	⑦	④
	Q45	1.100	1.020	0.002	0.630	0.550	0.0028	0.079	0.079	③	④
2005	G35	1.102	1.024	0.0014	0.630	0.551	0.0039	0.079	0.079	⑤	⑥
	Q45	1.100	1.020	0.0028	0.630	0.550	0.0028	0.079	0.079	⑧	⑨
2006	G35	1.102	1.024	0.0014	0.630	0.551	0.0039	0.079	0.079	⑤	⑥
	M35	1.102	1.024	0.0014	0.630	0.551	0.0022	0.079	0.079	⑤	⑥
	M45	1.102	1.024	0.0014	0.630	0.551	0.0022	0.079	0.079	⑤	⑥
	Q45	1.100	1.020	0.0028	0.630	0.550	0.0028	0.079	0.079	⑧	⑨

① Front: 53-72 Rear: 28-38
② Front: 106-126 Rear: 28-38
③ Front: 102-122 Rear: 29-37
④ Front: 16-23 Rear: 25-30
⑤ Front: 113-114 Rear: 53-71
⑥ Front: 17-22 Rear: 28-36
⑦ Front: 102-122 Rear: 29-37
⑧ Front: 112 Rear: 33
⑨ Front: 20 Rear: 28

09482_INFI_C0011

SCHEDULED MAINTENANCE INTERVALS
Infiniti—G20, G35, I35, M35, M45 & Q45

TO BE SERVICED	TYPE OF SERVICE	VEHICLE MILEAGE INTERVAL (x1000)												
		7.5	15	22.5	30	37.5	45	52.5	60	67.5	75	82.5	90	97.5
Engine oil & filter	R	✓	✓	✓	✓	✓	✓	✓	✓	✓	✓	✓	✓	✓
Automatic transaxle fluid ①	S/I		✓		✓		✓		✓		✓		✓	
Brake lines & cables	S/I		✓		✓		✓		✓		✓		✓	
Brake pads & discs	S/I		✓		✓		✓		✓		✓		✓	
Differential gear oil (G35, M35 & Q45) ①	S/I		✓		✓		✓		✓		✓		✓	
Driveshaft boots (G20 & I35)	S/I		✓		✓		✓		✓		✓		✓	
Active suspension fluid (Q45) ②	S/I		✓		✓		✓		✓		✓		✓	
Manual transaxle oil (G20)	S/I		✓		✓		✓		✓		✓		✓	
In-cabin microfilter	R	✓		✓		✓			✓		✓		✓	
Air cleaner filter ③	R				✓				✓				✓	
Exhaust system	S/I				✓				✓				✓	
Fuel lines	S/I				✓				✓				✓	
Steering gear & linkage, axle & suspension parts	S/I				✓				✓				✓	
SUPER HICAS linkage (Q45)	S/I				✓				✓				✓	
Vapor lines	S/I				✓				✓				✓	
Engine coolant ④	R								✓				✓	
Spark plugs (Conventional) ⑤	R			✓					✓					
Drive belts ⑥	S/I								✓					

R: Replace   S/I: Service or Inspect

① If towing a trailer, using a camper or car-top carrier, or driving on rough or muddy roads, CHANGE oil every 30,000 miles or 24 months.

② Replace at 60,000 miles (if not previously replaced).

③ If operating in dusty conditions, more frequent maintenance may be required.

④ After 60,000 miles or 48 months, replace coolant every 30,000 miles or 24 months.

⑤ Platinum-tipped spark plugs should be changed every 105,000 miles.

⑥ After 60,000 miles or 48 months, inspect every 15,000 miles or 12 months. Replace belts if found damaged.

FREQUENT OPERATION MAINTENANCE (SEVERE SERVICE)

If a vehicle is operated under any of the following conditions it is considered severe service:

- Extremely dusty areas.

- 50% or more of the vehicle operation is in 32°C (90°F) or higher temperatures, or constant operation in temperatures below 0°C (32°F).

- Prolonged idling (vehicle operation in stop and go traffic).

- Frequent short running periods (engine does not warm to normal operating temperatures).

- Police, taxi, delivery or trailer towing usage.

Oil & oil filter: change every 3750 miles.

Brake pads & discs: service or inspect every 7500 miles.

Driveshaft boots (G20 & I35): service or inspect every 7500 miles

Exhaust system: service or inspect every 7500 miles.

Steering gear, linkage, axle & suspension ball joints: service or inspect every 7500 miles.

Steering linkage, ball joints & front suspension ball joints: service or inspect every 7500 miles.

SUPER HICAS linkage (Q45): service or inspect every 7500 miles.

09482_INFI_C0012

GASOLINE ENGINE REPAIR

Distributor

REMOVAL

2.0L Engine

1. Before servicing the vehicle, refer to the precautions in the beginning of this section.
2. Remove or disconnect the following:
   - Negative battery cable
   - Splash shield, if equipped
   - Distributor connections but leave the ignition wires in place
   - Distributor cap hold-down screws and lift off the distributor cap with all ignition wires still connected
3. Matchmark the rotor to the distributor housing, and the distributor housing to the engine.

➡ **Do not crank the engine during this procedure. If the engine is cranked, the matchmark must be disregarded.**

   - Hold-down bolt
   - Distributor from the engine

To install:

ENGINE NOT DISTURBED

1. If the engine was not disturbed, install or connect the following:
   - New distributor housing O-ring
   - Distributor in the engine so the rotor is aligned with the matchmark on the housing and the housing is aligned with the matchmark on the engine. Be sure the distributor is fully seated and the distributor gear is fully engaged.
   - Snug the hold-down bolt
   - Distributor pick-up lead wires
   - Distributor cap and tighten the screws
   - Splash shield
   - Negative battery cable
2. Check and/or adjust the ignition timing and tighten the hold-down bolt to 10–12 ft. lbs. (14–16 Nm).

ENGINE DISTURBED

1. If the engine was disturbed (cranked or turned over with the distributor removed), install or connect the following:
   - New distributor housing O-ring
2. Position the engine so the No. 1 piston is at TDC of its compression stroke and the mark on the vibration damper is aligned with **0** on the timing indicator.

   - Distributor in the engine so the rotor is aligned with the position of the No. 1 ignition wire on the distributor cap. Be sure the distributor is fully seated and that the distributor shaft is fully engaged.

➡ **There are distributor cap runners inside the cap on 2.0L engine. Be sure the rotor is pointing to where the No. 1 runner originates inside the cap.**

   - Snug the hold-down bolt
   - Distributor pick-up lead wires
   - Distributor cap and tighten the screws
   - Splash shield, if equipped
   - Negative battery cable
3. Check and/or adjust the ignition timing and tighten the hold-down bolt to 10–12 ft. lbs. (14–16 Nm).

3.5L and 4.5L Engines

These engines are equipped with a distributorless ignition.

Alternator

REMOVAL

2.0L Engine

1. Before servicing the vehicle, refer to the precautions in the beginning of this section.
2. Remove or disconnect the following:
   - Negative battery cable

   - Drive belt
   - Alternator harness connector
   - Alternator bracket, if necessary
   - Alternator retainers
   - Alternator

To install:

3. Installation is the reverse order of removal.

3.5L Engines

I35 MODELS

1. Before servicing the vehicle, refer to the precautions in the beginning of this section.
2. Remove or disconnect the following:
   - Negative battery cable
   - Right side engine undercover and side inspection cover
   - Radiator
   - Drive belt
   - Alternator and A/C compressor harness connectors
   - Upper and lower alternator bolts
   - Alternator

To install:

3. Installation is the reverse order of removal.

G35 MODEL—AUTOMATIC TRANSMISSION

1. Before servicing the vehicle, refer to the precautions in the beginning of this section.
2. Remove or disconnect the following:
   - Negative battery cable
   - Engine undercover

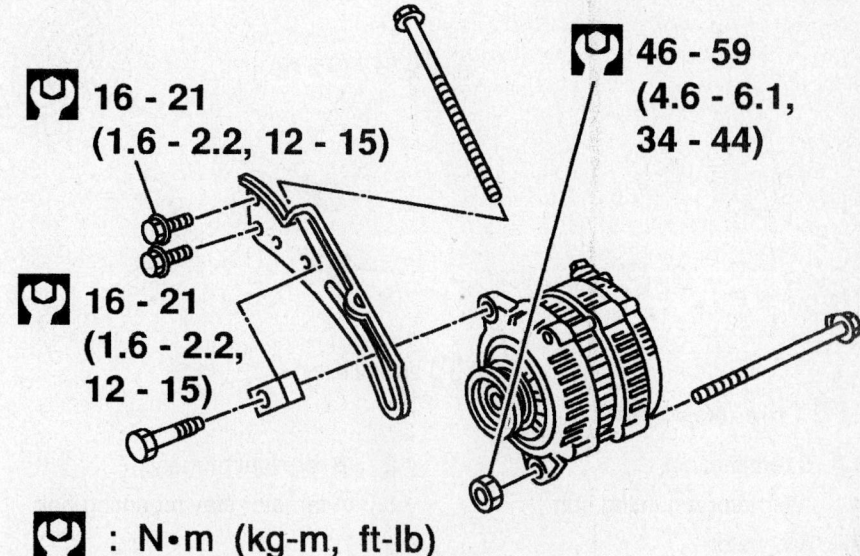

16 - 21 (1.6 - 2.2, 12 - 15)

46 - 59 (4.6 - 6.1, 34 - 44)

16 - 21 (1.6 - 2.2, 12 - 15)

: N•m (kg-m, ft-lb)

09482_INFI_G0139

Alternator and bracket retainer locations and torque specifications—G20 models

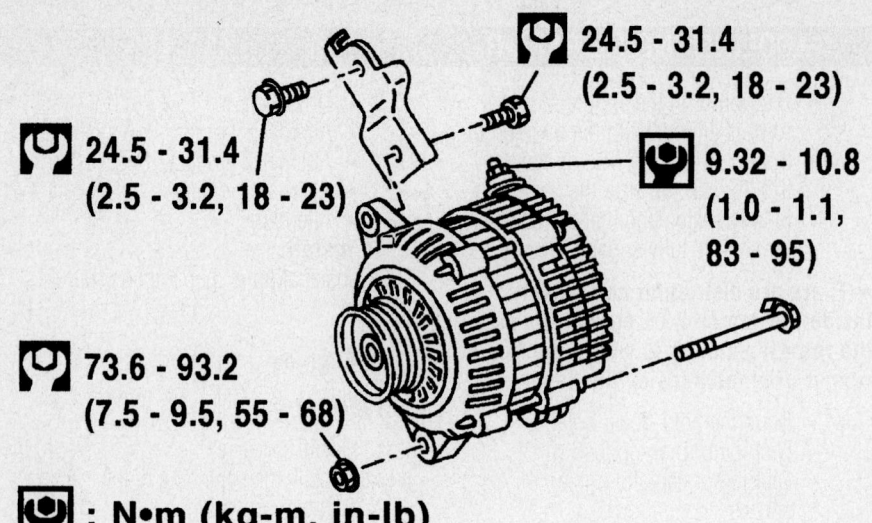

24.5 - 31.4 (2.5 - 3.2, 18 - 23)

24.5 - 31.4 (2.5 - 3.2, 18 - 23)

9.32 - 10.8 (1.0 - 1.1, 83 - 95)

73.6 - 93.2 (7.5 - 9.5, 55 - 68)

: N•m (kg-m, in-lb)

: N•m (kg-m, ft-lb)

09482_INFI_G0024

Alternator and bracket retainer locations and torque specifications—I35 models

- Stabilizer bar clamps, then slide stabilizer downward
- Loosen drive belts
- Alternator electrical connector
- Oil pressure switch harness connector
- "B" terminal mounting nut

- Upper and lower alternator mounting bolt
- Both alternator bracket bolts
- Alternator from the vehicle

To install:
3. Installation is the reverse order of removal.

G35 MODELS—AWD & MANUAL TRANSMISSION

1. Before servicing the vehicle, refer to the precautions in the beginning of this section.
2. Remove or disconnect the following:

- Negative battery cable
- Engine undercover
- Radiator fan assembly
- Remove drive belts
- Oil pressure switch harness connector
- Alternator stay mounting bolts and stay
- Alternator mounting bolt
- Alternator electrical connector
- "B" terminal mounting nut
- Harness clip and water hose bracket
- Alternator from the vehicle

To install:
3. Installation is the reverse order of removal.

M35 MODELS—2WD

1. Before servicing the vehicle, refer to the precautions in the beginning of this section.
2. Remove or disconnect the following:
- Negative battery cable

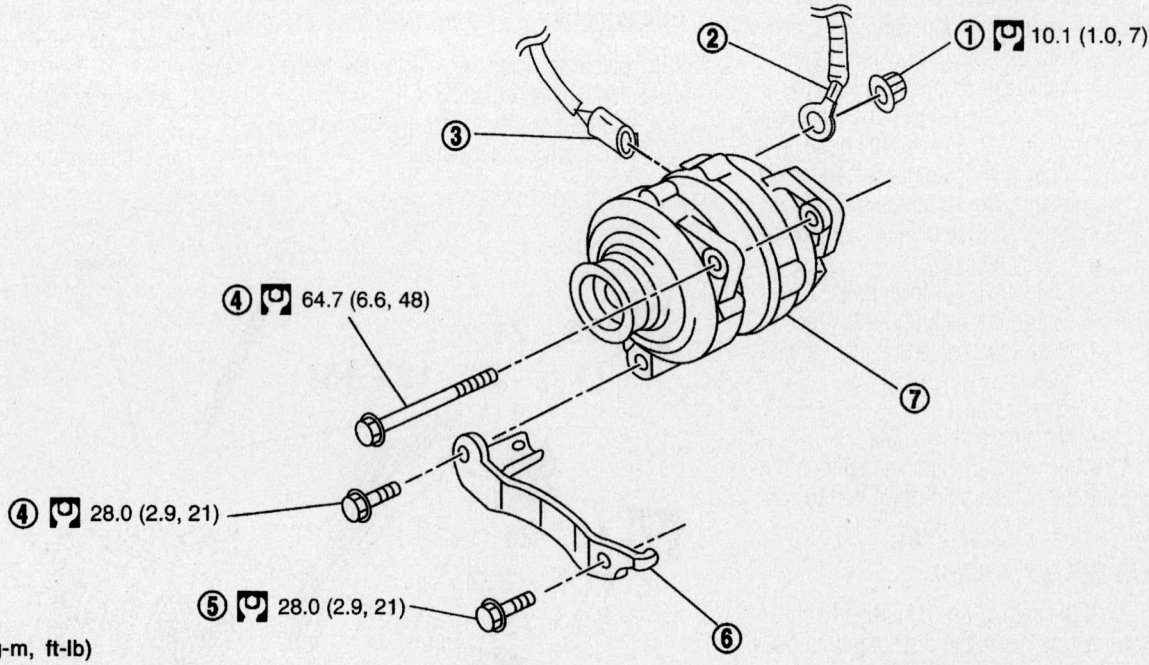

① 10.1 (1.0, 7)

④ 64.7 (6.6, 48)

④ 28.0 (2.9, 21)

⑤ 28.0 (2.9, 21)

: N•m (kg-m, ft-lb)

1. B terminal nut
2. B terminal harness
3. Alternator connector
4. Alternator mounting bolt
5. Alternator stay mounting bolt
6. Alternator stay
7. Alternator

09482_INFI_G0021

Alternator mounting and torque specifications—G35 and M35 models

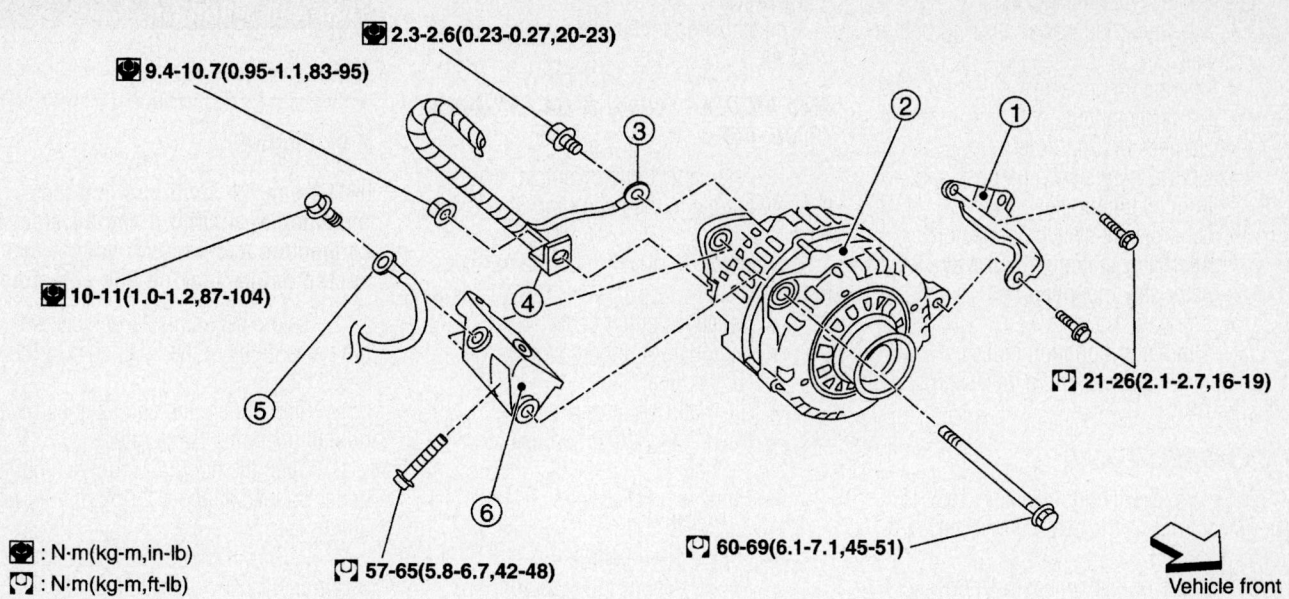

2.3-2.6(0.23-0.27,20-23)

9.4-10.7(0.95-1.1,83-95)

10-11(1.0-1.2,87-104)

21-26(2.1-2.7,16-19)

60-69(6.1-7.1,45-51)

57-65(5.8-6.7,42-48)

: N·m(kg-m,in-lb)
: N·m(kg-m,ft-lb)

Vehicle front

1. Alternator stay
2. Alternator
3. Alternator ground harness
4. Alternator B terminal
5. Battery ground harness
6. Alternator bracket

09482_INFI_G0140

Alternator mounting and torque specifications—2004 M45, 2002–2004 Q45

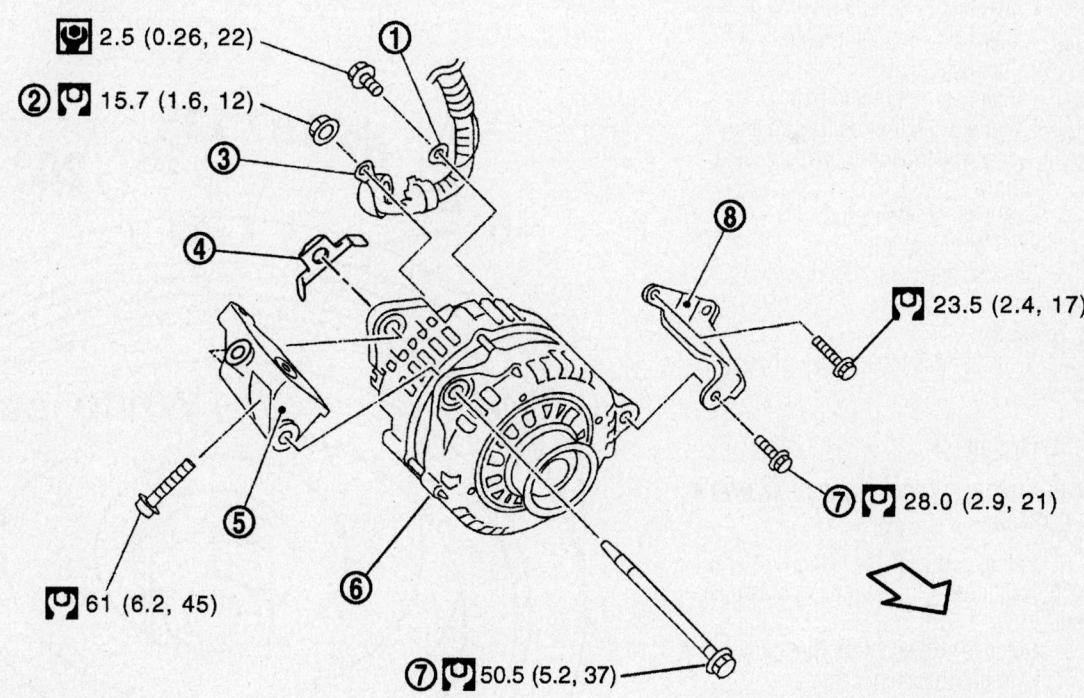

2.5 (0.26, 22)

15.7 (1.6, 12)

23.5 (2.4, 17)

28.0 (2.9, 21)

61 (6.2, 45)

50.5 (5.2, 37)

1. Alternator ground harness
2. "B" terminal nut
3. "B" terminal harness
4. Alternator nut
5. Alternator bracket
6. Alternator
7. Alternator mounting bolt
8. Alternator stay

: N·m (kg-m, in-lb)
: N·m (kg-m, ft-lb)
: Engine front

09482_INFI_G0023

Alternator mounting and torque specifications—2006 M45, 2005–06 Q45

- Engine front undercover
- Alternator and power steering pump belt
- Alternator connector
- "B" terminal nut
- Harness bracket bolts
- Oil pressure switch harness clip from alternator stay
- Oil pressure switch connector
- Alternator mounting bolt and alternator stay mounting bolt
- Alternator stay
- Alternator mounting bolt

3. Remove alternator in a downward direction

M35 MODELS—AWD

1. Before servicing the vehicle, refer to the precautions in the beginning of this section.
2. Remove or disconnect the following:
- Negative battery cable
- Power steering oil reservoir tank from bracket
- Clips and hose clamp from harness bracket
- Engine front undercover
- Alternator and power steering belts
- Alternator mounting bolt and alternator stay mounting bolt
- Alternator stay
- Alternator mounting bolt
- Harness bracket bolts; pull and rotate alternator to gain access to bolts
- Alternator connector
- "B" terminal nut

3. Remove alternator in a downward direction

To install:

4. Installation is the reverse order of removal.

4.5L Engine

M45 MODELS (2004) & Q45 MODELS (2002–04)

1. Before servicing the vehicle, refer to the precautions in the beginning of this section.
2. Remove or disconnect the following:
- Negative battery cable
- Air intake duct
- Engine drive belt
- Power steering hose clamp bolt
- Alternator stay
- Radiator reservoir tank
- Ground harness bolt
- Engine compartment harness clip
- Alternator electrical connector
- Alternator

To install:

3. Installation is the reverse order of removal.

M45 MODELS (2006) & Q45 MODELS (2005–06)

1. Before servicing the vehicle, refer to the precautions in the beginning of this section.
2. Remove or disconnect the following:
- Negative battery cable
- Engine front undercover
- "B" terminal nut
- Alternator connector
- Alternator ground harness mounting bolt
- Harness bracket bolts
- Air intake duct
- Engine drive belts
- Power steering oil reservoir tank from bracket
- Engine coolant reservoir tank
- Vacuum tank
- Harness clips
- Alternator mounting bolts
- Alternator

To install:

3. Installation is the reverse order of removal.

Ignition Timing

ADJUSTMENT

2.0L Engine

➡The engine should be in good mechanical condition and all electrical connectors and vacuum hoses connected before making this adjustment.

1. Before servicing the vehicle, refer to the precautions in the beginning of this section.
2. Start the engine and let it warm up to normal operating temperature.
3. Open the hood and run the engine under no load at about 2,000 rpm for about 2 minutes.
4. Perform Diagnostic Test Mode II and repair any causes of trouble codes as needed.
5. Run the engine under no load at 2,000 rpm for about 2 minutes. Rev the engine 2 or 3 times and let it idle for 1 minute.
6. Turn **OFF** the engine and disconnect the Throttle Position (TP) sensor connector. Connect a timing light to the No. 1 spark plug wire. Start the engine.

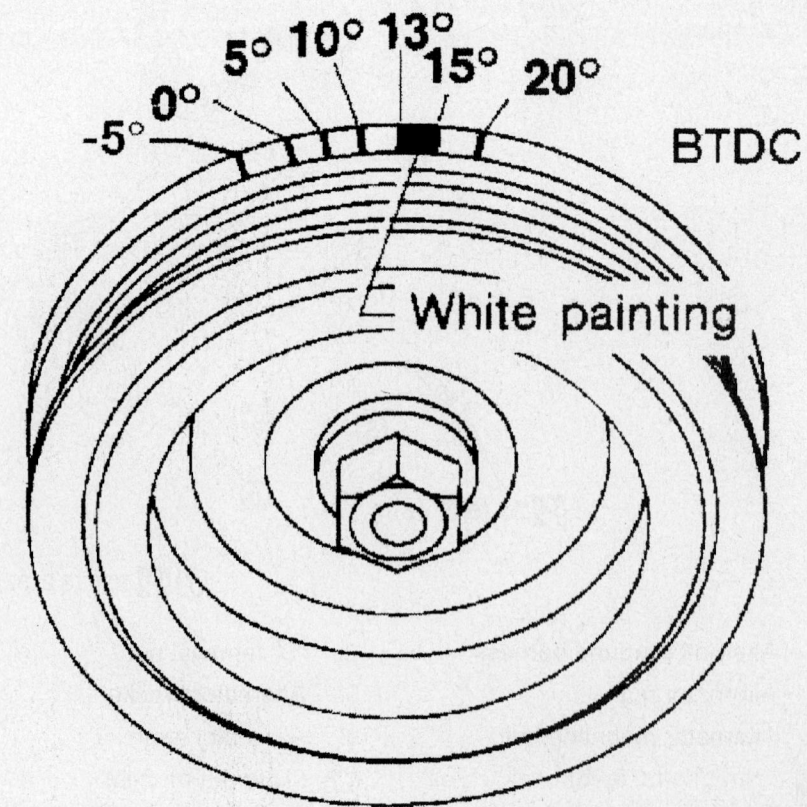

Crankshaft pulley and timing marks—2.0L engine

09482_INFI_G0013

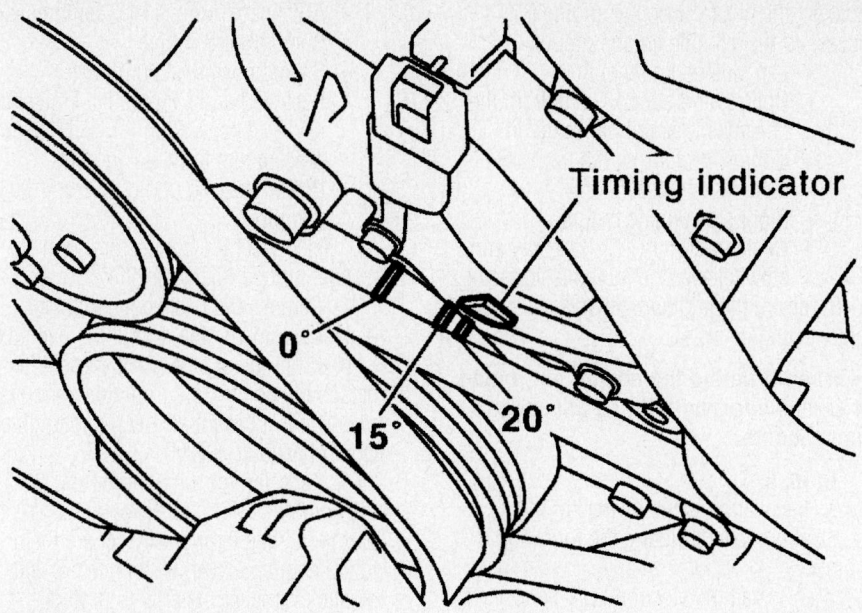

Location of timing marks—3.5L engines

09482_INFI_G0014

7. Adjust the timing to 13–17° BTDC by loosening the distributor mounting bolts and turning the distributor. When the timing is correct, tighten the mounting bolts and turn the engine **OFF**.

8. Reconnect the TP sensor connector. Start the engine and check the ignition timing again.

3.5L Engines

➡**The engine should be in good mechanical condition and all electrical connectors and vacuum hoses attached before making this adjustment.**

1. Before servicing the vehicle, refer to the precautions in the beginning of this section.

2. Start the engine and let it warm up to normal operating temperature.

3. Open the hood and run the engine under no load at about 2,000 rpm for about 2 minutes.

4. Perform Diagnostic Test Mode II and repair any causes of trouble codes as needed.

5. Run the engine under no load at 2,000 rpm for about 2 minutes. Rev the engine 2 or 3 times and let it idle for 1 minute.

6. Turn **OFF** the engine and disconnect the Throttle Position sensor connector. Remove the No. 1 ignition coil. Connect the coil to the spark plug using a spare piece of high-tension wire so you have a place to connect your timing light. Start the engine.

7. Run the engine under no load at

2,000 rpm for about 2 minutes. Rev the engine 2 or 3 times and let it idle.

8. Check the ignition timing and adjust if needed (Refer to Engine Tune-Up Specifications at beginning of chapter for proper settings; always defer to timing settings on tag under hood if different)

9. Adjustment is made by loosening the screws and turning the Camshaft Position (CMP) sensor until the mark on the crankshaft pulley is pointing at 10° BTDC. Tighten the mounting screws and confirm ignition timing has not changed.

10. Turn the engine **OFF** and connect the TP sensor connector.

4.5L Engines

1. The ignition timing is controlled by the ECM and is not adjustable. With all electrical accessories off, wheels in the straight ahead position, the ignition timing is 7–17 degrees BTDC.

Engine Assembly

REMOVAL & INSTALLATION

2.0L Engine

1. Before servicing the vehicle, refer to the precautions in the beginning of this section.

2. Drain the coolant system.

3. Drain the engine oil.

4. Drain the transaxle fluid.

5. Release fuel system pressure and remove fuel line.

6. Remove or disconnect the following:
• Hood and hinges
• Negative battery cable
• Both front wheels
• Engine undercover
• Air cleaner assembly and duct
• Battery and battery tray
• All vacuum lines and wiring harness connectors
• Heater hoses
• Oil cooler lines, if equipped
• Power steering hoses
• Fuel lines
• Throttle cable
• Cruise control cable, if equipped
• A/T control cable, if equipped
• Cooling fans, radiator and recovery tank
• Drive shafts
• Front exhaust pipe
• Starter and intake manifold support
• Drive belts
• Alternator, A/C compressor, and the power steering pump from the engine and lay them aside. Do not disconnect the compressor or power steering pump lines.

7. Support the engine with a hoist and the transaxle with a suitable jack. Raise the engine and transaxle slightly and remove the center member.
• Engine mounting bolts from both sides and slowly lower the hoist and transaxle jack
• Engine and transaxle from beneath the vehicle

To install:

8. Install or connect the following:
• Center member bracket (manual transmission) on the engine, if removed. Ensure that all insulators are correctly positioned on the brackets. Torque the insulator through-bolts to 32–41 ft. lbs. (43–55 Nm).

9. If equipped with manual transaxle, ensure that the distance between the center of the insulator through-bolt and the center member is 2.28–2.36 in. (58–60mm). Torque the through-bolt to 46–58 ft. lbs. (62–78 Nm).
• Engine. Torque the center member-to-frame bolts to 57–72 ft. lbs. (77–98 Nm).
• Alternator, air conditioning compressor, and power steering pump
• Drive belts
• Starter and intake manifold support
• Front exhaust pipe
• Drive shafts
• Cooling fans, radiator and recovery tank
• A/T control cable, if equipped

- Cruise control cable, if equipped
- Throttle cable
- Fuel lines
- Power steering hoses
- Oil cooler lines, if equipped
- Heater hoses
- All vacuum lines and wiring harness connectors
- Engine undercover
- Front wheels
- Battery and battery tray
- Air cleaner assembly and duct
- Negative battery cable
- Hood and hinges

10. Fill and bleed the cooling system.
11. Fill the engine with clean oil.
12. Fill the transaxle to the proper level.
13. Start the vehicle, check for leaks and repair if necessary.

3.5L Engines

I35 MODELS

It is recommended the engine and transaxle be removed as a single unit. If need be, the units may be separated after removal.

➡ **The engine and transaxle assembly must be removed from the underside of the vehicle.**

1. Before servicing the vehicle, refer to the precautions in the beginning of this section.
2. Release the fuel system pressure.
3. Drain the cooling system.
4. Drain the engine oil.
5. Drain the automatic transaxle, if equipped.
6. Remove or disconnect the following:
   - Negative battery cable
   - Hood
   - Engine undercover
   - All vacuum hoses, fuel lines, wires and connectors; tag before disconnecting
   - Front exhaust pipe from the manifold
   - Ball joints from the steering knuckle
   - Halfshafts
   - Radiator and fans
   - Drive belts
   - Alternator
   - A/C compressor. Position it aside with the lines attached. Do NOT disconnect the refrigerant lines.
   - Power steering pump and position aside with the lines attached. Do NOT disconnect the fluid lines.
7. Place a suitable jack under the transaxle. Install engine slingers and a suit-

able engine hoist. Raise the engine for access to the left side engine mount.
   - Left side engine mount
   - Control and support rods from the transaxle, manual transaxle only
   - Control cable from the transaxle, automatic transaxle only
   - Right side engine mount
   - Center member, then carefully and slowly lower the transmission jack
8. Lower the engine/transaxle assembly onto an engine stand.

➡ **When lowering the engine out, guide it carefully to avoid hitting any other components.**

To install:

9. Installation is the reverse of the removal procedure, noting the following points:
   a. Install the electronically controlled engine mount harness to the specifications shown in the accompanying figure.
   b. Make sure to connect all vacuum hoses, lines, and electrical connectors as tagged during removal.
   c. Fill the cooling system.
   d. Fill the engine with clean oil.
   e. Start the vehicle, check for leaks and repair if necessary.

G35 MODELS—2WD

1. Before servicing the vehicle, refer to the precautions in the beginning of this section.
2. Evacuate the A/C system.
3. Release the fuel system pressure.
4. Drain the engine oil.
5. Drain the cooling system.
6. Drain the transaxle fluid.
7. Remove or disconnect the following:
   - Negative battery cable
   - Hood
   - Engine cover
   - Battery cover
   - Engine undercover
   - Front wheels
   - Wiper arm and cowl top cover
   - Air cleaner assembly, including air ducts
   - Fan, radiator shroud, radiator, reservoir tank and hoses
   - Heater hose from the engine and plug to avoid leaks
   - Ground wire from left hand cylinder head
   - Positive battery cable from the vehicle and temporarily fasten it on the engine
   - Battery
   - Engine harness connector; tag before disconnecting

- A/C lines from the A/C compressor
- Body ground cables
- Brake booster vacuum hose
- Fuel feed and Evaporative Emission (EVAP) hoses. Plug the fuel lines to prevent fuel from leaking out.
- Power steering pump reservoir tank and pipes

8. Disconnect the connectors from the passenger compartment as follows:
   a. Remove the passenger side kick plate, dashboard side trim and glove box.
   b. From the engine compartment, detach the connectors from the Transmission Control Module (TCM), and Engine Control Module (ECM).
   c. Unfasten the wire harnesses, pull the harnesses out into the engine compartment, then temporarily secure them to the engine. Cover all connectors with plastic or similar material to protect them.
9. Remove or disconnect the following:
   - Front exhaust pipe from the manifold
   - Steering lower joint and release the steering shaft
   - Propeller shaft from the transmission
   - Shift control linkage from the gear selector. Secure it temporarily to the transmission, so it doesn't drag or catch on any other components.
   - Rear plate cover from upper oil pan
   - Bolts securing the drive plate to the torque converter
   - Bolts securing the transmission to the lower rear side of the oil pan
   - Front stabilizer shaft
   - Left and right tie rod ends from the steering knuckle
   - Disconnect the lower strut from the lower control arms on the left and right sides
   - Left and right control arms from the suspension crossmember
10. Position a suitable engine table or other suitable tool under the engine. Securely support the bottom of the suspension member and transmission before removing:
   - Rear member mounting bolt
   - Suspension member mounting bolt and nut
11. Carefully lower the jack, or raise the lift to remove the engine, transmission and suspension member assembly. Make sure that all lines, hoses, connectors etc. have been disconnected. If necessary, support the rear of the vehicle at the rear jacking point, as its center of gravity has changed with the engine removed.

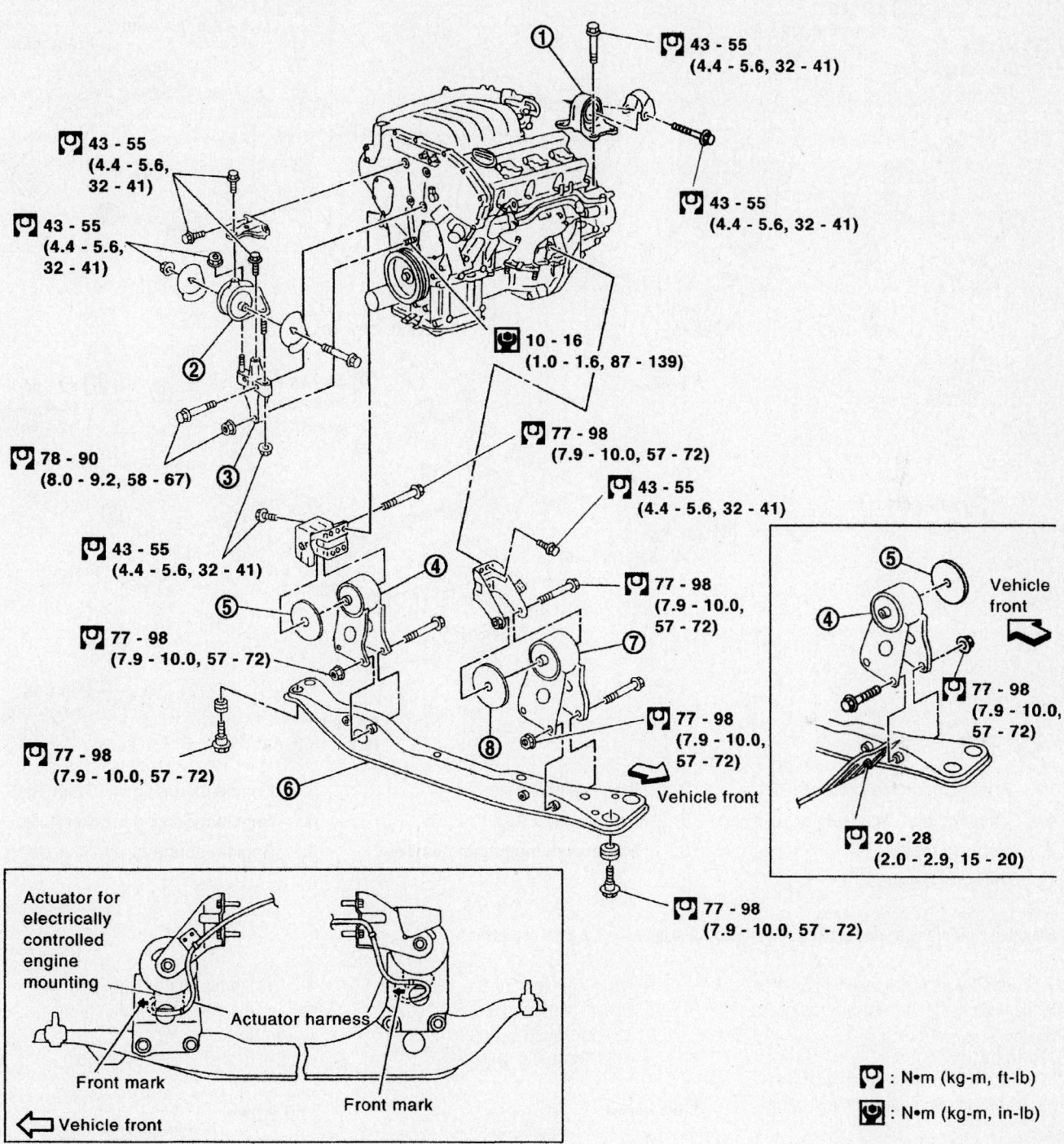

43 - 55
(4.4 - 5.6, 32 - 41)

43 - 55
(4.4 - 5.6, 32 - 41)

43 - 55
(4.4 - 5.6, 32 - 41)

43 - 55
(4.4 - 5.6, 32 - 41)

78 - 90
(8.0 - 9.2, 58 - 67)

43 - 55
(4.4 - 5.6, 32 - 41)

10 - 16
(1.0 - 1.6, 87 - 139)

77 - 98
(7.9 - 10.0, 57 - 72)

43 - 55
(4.4 - 5.6, 32 - 41)

77 - 98
(7.9 - 10.0, 57 - 72)

77 - 98
(7.9 - 10.0, 57 - 72)

77 - 98
(7.9 - 10.0, 57 - 72)

77 - 98
(7.9 - 10.0, 57 - 72)

77 - 98
(7.9 - 10.0, 57 - 72)

Vehicle front

77 - 98
(7.9 - 10.0, 57 - 72)

Vehicle front

20 - 28
(2.0 - 2.9, 15 - 20)

Actuator for electrically controlled engine mounting

Actuator harness

Front mark

Front mark

Vehicle front

: N•m (kg-m, ft-lb)

: N•m (kg-m, in-lb)

1.	LH engine mounting	4.	Rear engine mounting	7.	Front engine mounting
2.	RH engine mounting	5.	Insulator	8.	Insulator
3.	Mounting bracket	6.	Center member		

09482_INFI_G0015

For the electronically controlled engine mount harness, the proper length from A to B is 6.69 in. (170mm) and from C to D is 5.12 in. (130mm)

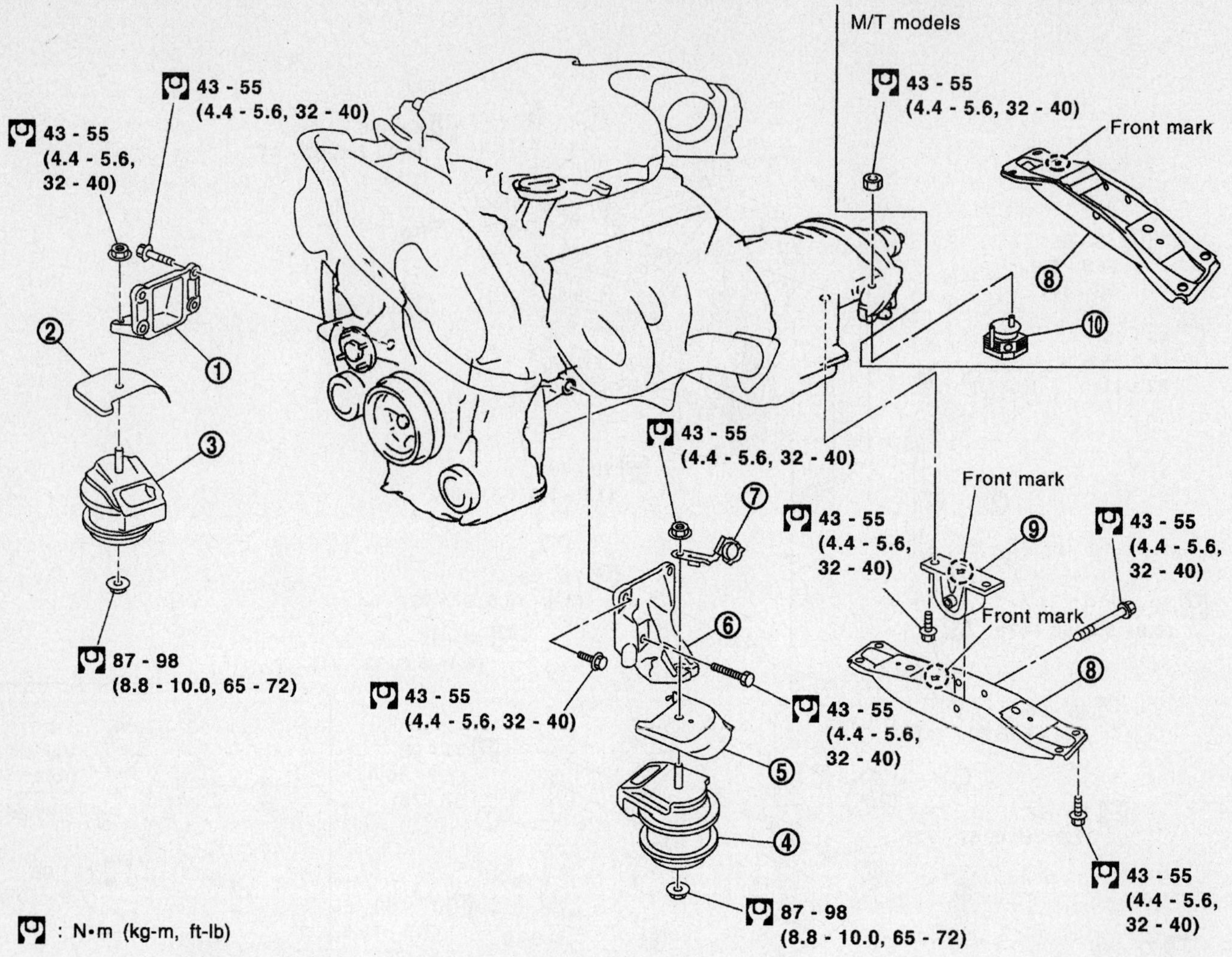

M/T models

🔧 43 - 55
(4.4 - 5.6, 32 - 40)

🔧 43 - 55
(4.4 - 5.6, 32 - 40)

🔧 43 - 55
(4.4 - 5.6, 32 - 40)

Front mark

🔧 43 - 55
(4.4 - 5.6, 32 - 40)

🔧 43 - 55
(4.4 - 5.6, 32 - 40)

🔧 87 - 98
(8.8 - 10.0, 65 - 72)

🔧 43 - 55
(4.4 - 5.6, 32 - 40)

Front mark

🔧 43 - 55
(4.4 - 5.6, 32 - 40)

🔧 43 - 55
(4.4 - 5.6, 32 - 40)

Front mark

🔧 43 - 55
(4.4 - 5.6, 32 - 40)

🔧 87 - 98
(8.8 - 10.0, 65 - 72)

🔧 43 - 55
(4.4 - 5.6, 32 - 40)

🔧 : N•m (kg-m, ft-lb)

1. Engine mounting bracket (RH)	2. Heat insulator (RH)	3. Engine mounting insulator (RH)
4. Engine mounting insulator (LH)	5. Heat insulator (LH)	6. Engine mounting bracket (LH)
7. Harness bracket	8. Rear engine mounting member	9. Engine mounting insulator (rear)
10. Dynamic damper		

09482_INFI_G0025

Exploded view of the engine mounts and related components—G35 2WD models

12. Remove and components and separate the engine and transmission as necessary.

To install:

13. Installation is the reverse of the removal procedure, noting the following points:

- Tighten all engine mounts and related components to the specifications shown in the accompanying figure
- Tighten the rear member mounting bolts in sequence.
- When installing engine mounting brackets on cylinder block, tighten two upper bolts ("A" in the figure) first, then tighten two lower bolts ("B" in the figure).

- Make sure to connect all vacuum hoses, lines, and electrical connectors as tagged during removal.
- Fill the cooling system.

- Fill the engine with clean oil.
- Fill the transmission to the proper level.
- Recharge the A/C system.

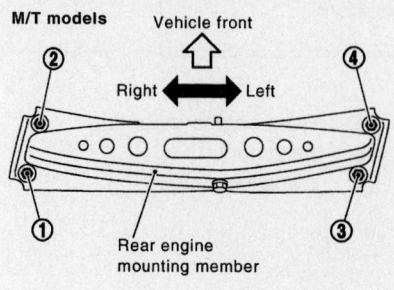

Rear member bolt torque sequence—G35 2WD models w/manual transmission

09482_INFI_0016

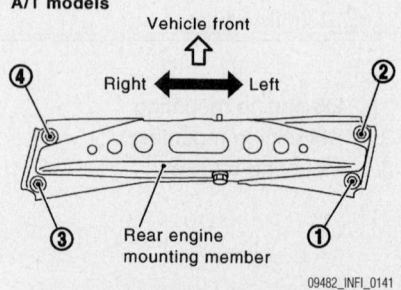

Rear member bolt torque sequence—G35 2WD models w/automatic transmission

09482_INFI_0141

Example: Left

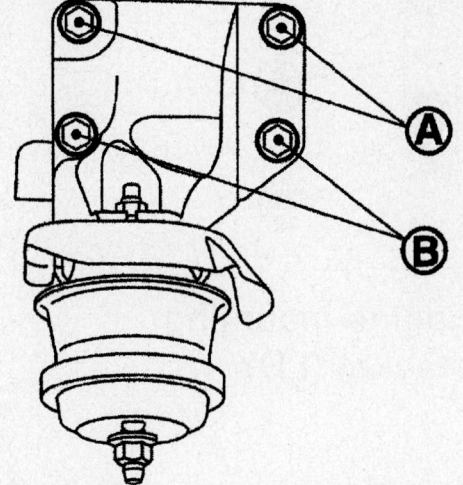

09482_INFI_G0027

Engine mount bolt torque sequence—G35 2WD

14. Start the vehicle, check for leaks and repair if necessary.

G35 MODELS—AWD

1. Before servicing the vehicle, refer to the precautions in the beginning of this section.
2. Evacuate the A/C system.
3. Release the fuel system pressure.
4. Drain the engine oil.
5. Drain the cooling system.
6. Drain the transaxle fluid.
7. Remove or disconnect the following:

- Negative battery cable
- Hood
- Engine cover
- Battery cover
- Engine undercover
- Front wheels
- Wiper arm and cowl top cover

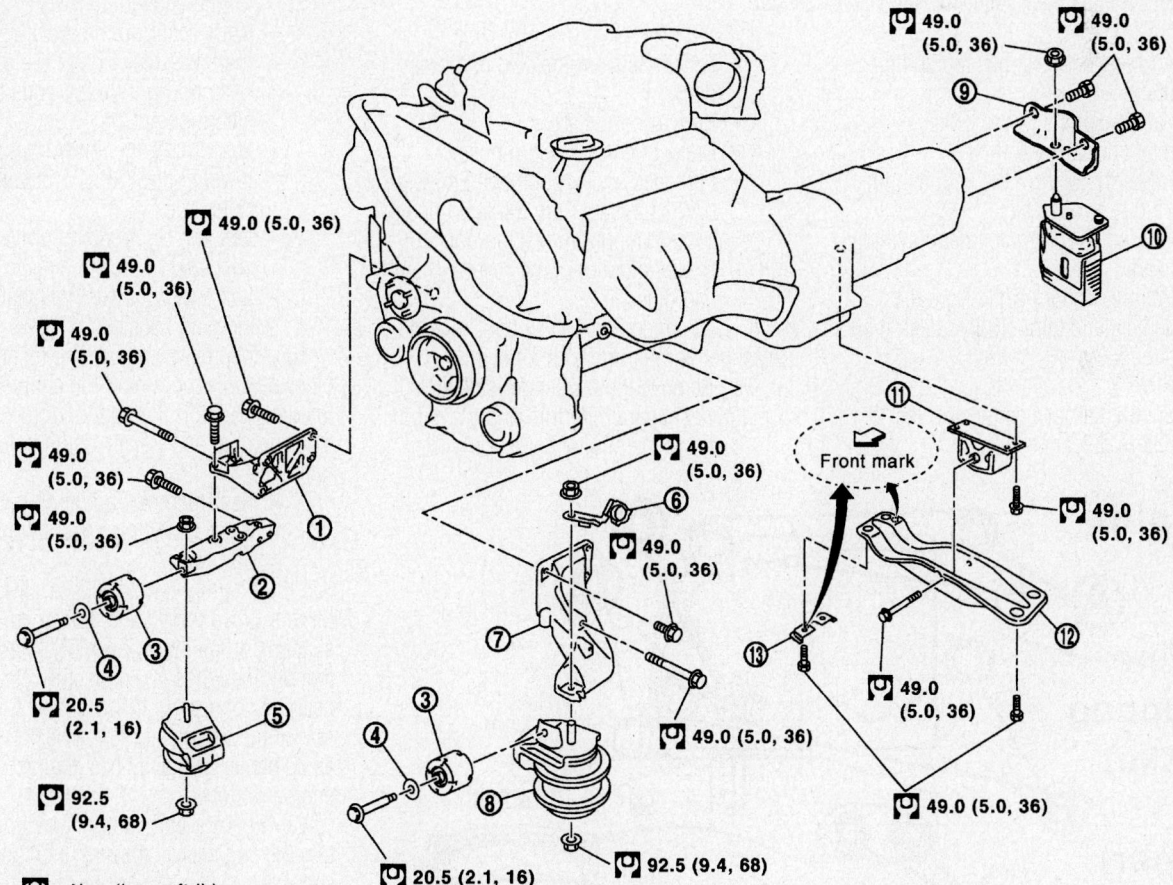

: N•m (kg-m, ft-lb)

1.	Engine mounting bracket (RH) (Upper)	2.	Engine mounting bracket (RH) (Lower)	3.	Dynamic damper
4.	Washer	5.	Engine mounting insulator (RH)	6.	Harness bracket
7.	Engine mounting bracket (LH)	8.	Engine mounting insulator (LH)	9.	Dynamic damper bracket
10.	Dynamic damper	11.	Engine mounting insulator (Rear)	12.	Rear engine mounting member
13.	Heat insulator				

09482_INFI_G0026

Exploded view of the engine mounts and related components—G35 AWD models

LH side

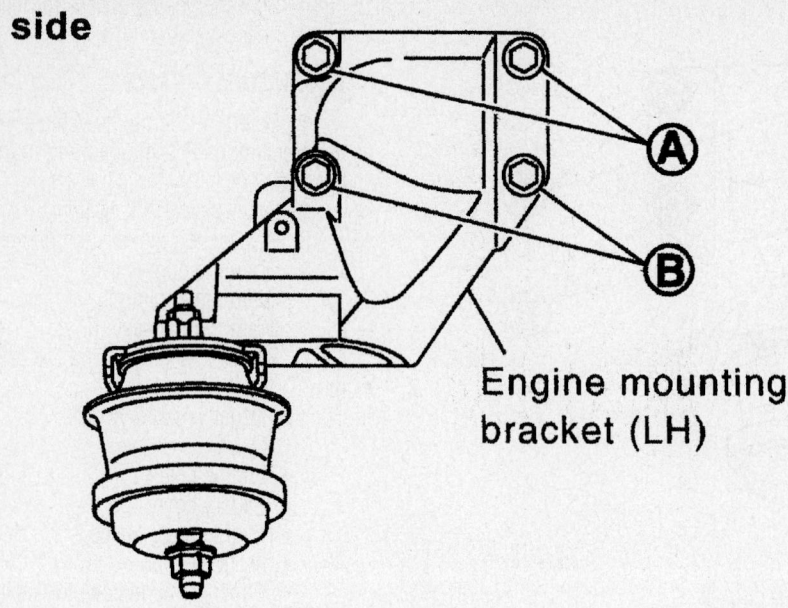

Engine mounting bracket (LH)

09482_INFI_G0028

Engine mount bolt torque sequence, left side—G35 AWD

- Air cleaner assembly, including air ducts
- Fan, radiator shroud, radiator, reservoir tank and hoses
- Heater hose from the engine and plug to avoid leaks
- Ground wire from left hand cylinder head
- Positive battery cable from the vehicle and temporarily fasten it on the engine
- Battery
- Engine harness connector; tag before disconnecting

- A/C lines from the A/C compressor
- Body ground cables
- Brake booster vacuum hose
- Fuel feed and Evaporative Emission (EVAP) hoses. Plug the fuel lines to prevent fuel from leaking out.
- Power steering pump reservoir tank and pipes

8. Disconnect the connectors from the passenger compartment as follows:

a. Remove the passenger side kick plate, dashboard side trim and glove box.

b. From the engine compartment, detach the connectors from the Transmission Control Module (TCM), and Engine Control Module (ECM).

c. Unfasten the wire harnesses, pull the harnesses out into the engine compartment, then temporarily secure them to the engine. Cover all connectors with plastic or similar material to protect them.

9. Remove or disconnect the following:

- Front exhaust pipe from the manifold
- Steering lower joint and release the steering shaft
- Propeller shaft from the transmission
- Shift control linkage from the gear selector. Secure it temporarily to the transmission, so it doesn't drag or catch on any other components.
- Rear plate cover from upper oil pan
- Bolts securing the drive plate to the torque converter
- Bolts securing the transmission to the lower rear side of the oil pan
- Left and right tie rod ends from the steering knuckle
- Disconnect the lower strut from the lower control arms on the left and right sides
- Left and right control arms from the suspension crossmember
- Left and right front halfshafts from steering knuckles

10. Position a suitable engine table or other suitable tool under the engine. Securely support the bottom of the suspension member and transmission before removing:

- Rear member mounting bolt
- Suspension member mounting bolt and nut

11. Carefully lower the jack, or raise the lift to remove the engine, transmission and suspension member assembly. Make sure that all lines, hoses, connectors etc. have been disconnected. If necessary, support the rear of the vehicle at the rear jacking point, as its center of gravity has changed with the engine removed.

12. Remove and components and separate the engine and transmission as necessary.

To install:

13. Installation is the reverse of the removal procedure, noting the following points:

- Tighten all engine mounts and related components to the specifications shown in the accompanying figure
- When installing left engine mount-

RH side

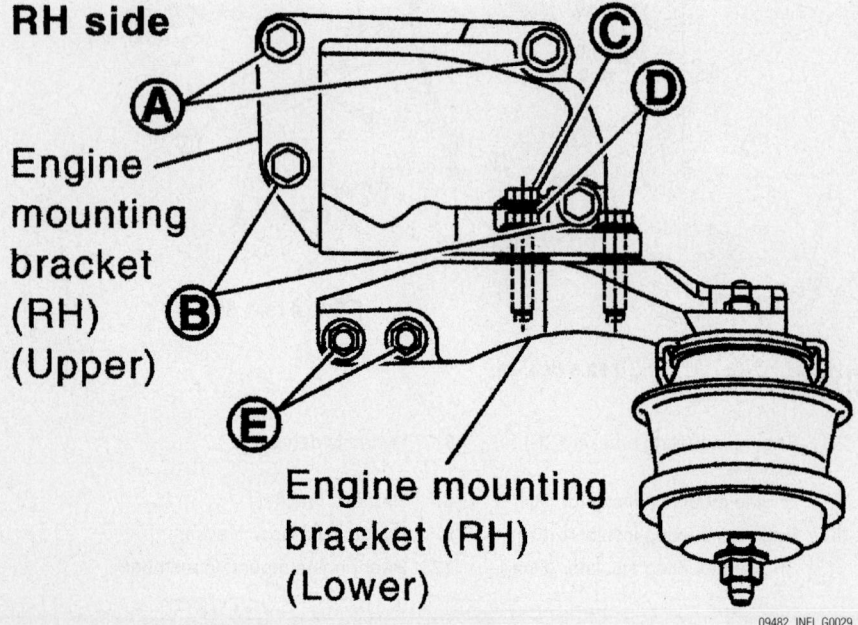

Engine mounting bracket (RH) (Upper)

Engine mounting bracket (RH) (Lower)

09482_INFI_G0029

Engine mount bolt torque sequence, right side—G35 AWD

ing bracket on cylinder block, tighten two upper bolts ("A" in the figure) first, then tighten two lower bolts ("B" in the figure).

- The right engine mounting bracket comprises two pieces. Install right upper engine mounting bracket on cylinder block, tightening two upper bolts ("A" in the figure) first, then tightening two lower bolts ("B" in the figure). Make certain right lower engine mounting bracket is in full contact with the upper mounting bracket and the front final drive assembly before tightening bolts "D" and "E" in sequence.
- Make sure to connect all vacuum hoses, lines, and electrical connectors as tagged during removal.
- Fill the cooling system.
- Fill the engine with clean oil.
- Fill the transmission to the proper level.
- Recharge the A/C system.

14. Start the vehicle, check for leaks and repair if necessary.

4.5L Engines

1. Before servicing the vehicle, refer to the precautions in the beginning of this section.
2. Evacuate the A/C system.
3. Release the fuel system pressure.
4. Drain the engine oil.
5. Drain the cooling system.
6. Drain the transmission fluid.
7. Remove or disconnect the following:

- Battery cables
- Hood
- Engine appearance cover
- Engine undercover
- Tower bar
- Battery
- Air duct and air cleaner
- Drive belts
- Accelerator cable
- Cooling fan and tubes
- Radiator hoses
- Radiator and shroud
- Engine harnesses from both sides
- Heater hoses
- Exhaust manifold cover

- Vacuum hoses to body
- Cooling reservoir tank
- A/C compressor
- Power steering reservoir and wire aside
- Power steering oil pump from the engine compartment
- Automatic transmission cooler line
- Front exhaust tubes
- Steering shaft lower joint
- Front cross bar
- Driveshaft
- Transmission
- All necessary vacuum hoses, fuel hoses and electrical connections that would interfere with engine removal

8. Attach engine slingers to the cylinder head and attach a suitable hoist to the slinger

- Engine mounting bolts from both sides and then slowly raise the engine
- Engine from the engine compartment

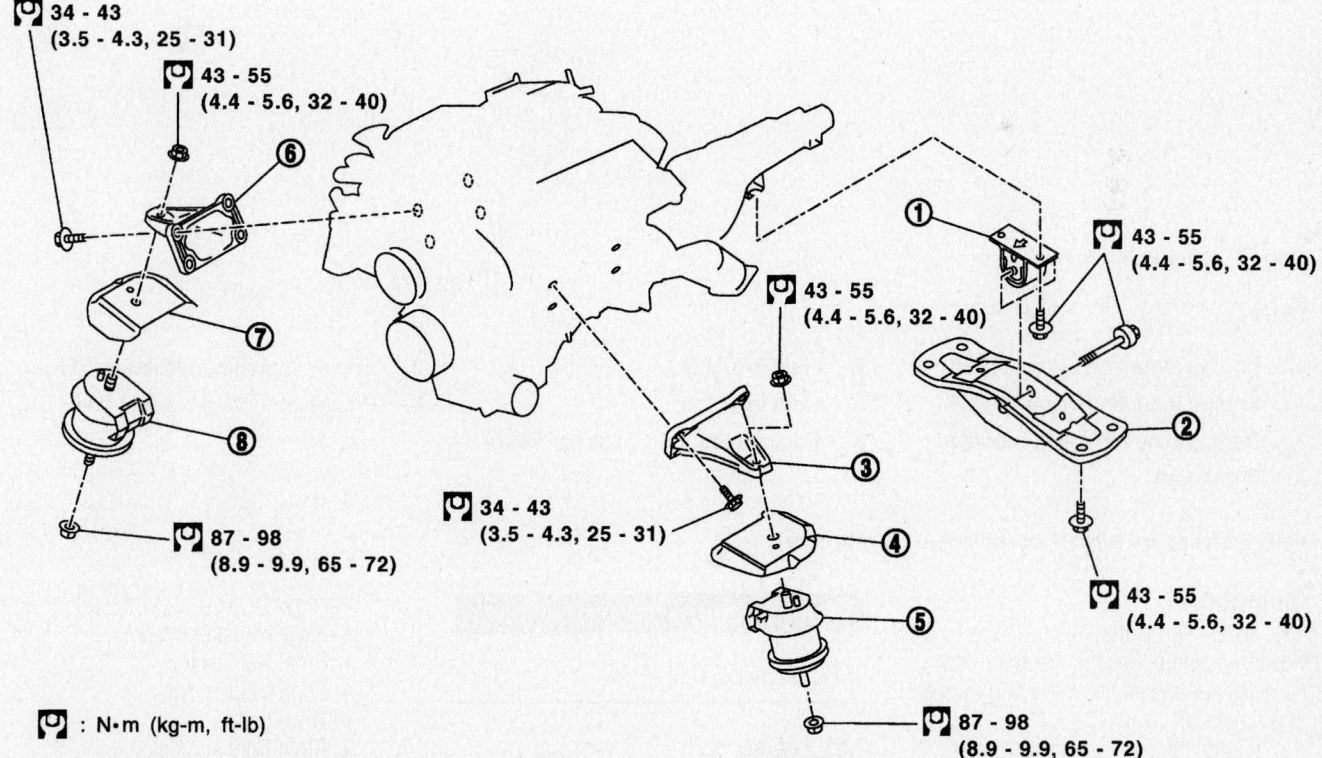

1. Rear engine mounting insulator
2. Rear member
3. Left engine mounting bracket
4. Heat insulator
5. Left engine mounting insulator
6. Right engine mounting bracket
7. Heat insulator
8. Right engine mounting insulator

Engine mounting and torque specifications—2004 M45 and 2002–06 Q45 models

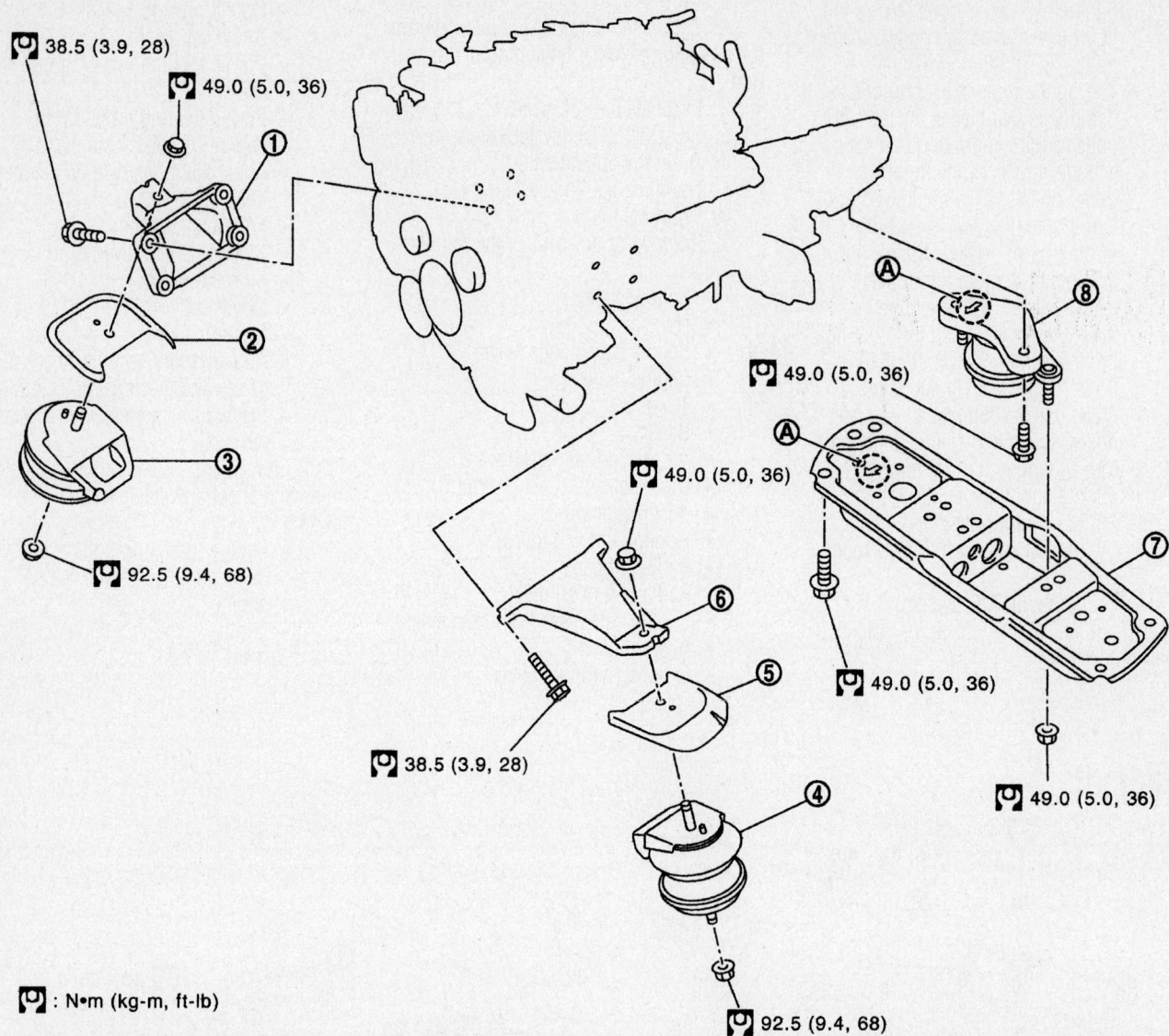

1. Engine mounting bracket (RH)
2. Heat insulator
3. Engine mounting insulator (RH)
4. Engine mounting insulator (LH)
5. Heat insulator
6. Engine mounting bracket (LH)
7. Rear engine mounting member
8. Engine mounting insulator (rear)
A. Front mark

09482_INFI_G0143

Engine mounting and torque specifications—2006 M45 models

To install:

9. Installation is the reverse of the removal procedure. See the illustration for front cross bar and engine mounting torque specifications.

10. Fill the transmission to the proper level.

11. Fill and bleed the cooling system.

12. Fill the engine with clean oil.

13. Recharge the A/C system.

14. Start the vehicle, check for leaks and repair if necessary.

Water Pump

REMOVAL & INSTALLATION

2.0L Engine

1. Before servicing the vehicle, refer to the precautions in the beginning of this section.

2. Drain the coolant from the radiator and engine block. The drain plug in the engine block is located at the left front of the cylinder block.

3. Remove or disconnect the following:
   • Negative battery cable
   • Right front wheel
   • Engine side cover
   • Drive belts
   • Right front engine mount
   • Water pump

To install:

4. Clean all mating surfaces and place a 2–3mm bead of liquid gasket on the water pump mating surface.

5. Install or connect the following:

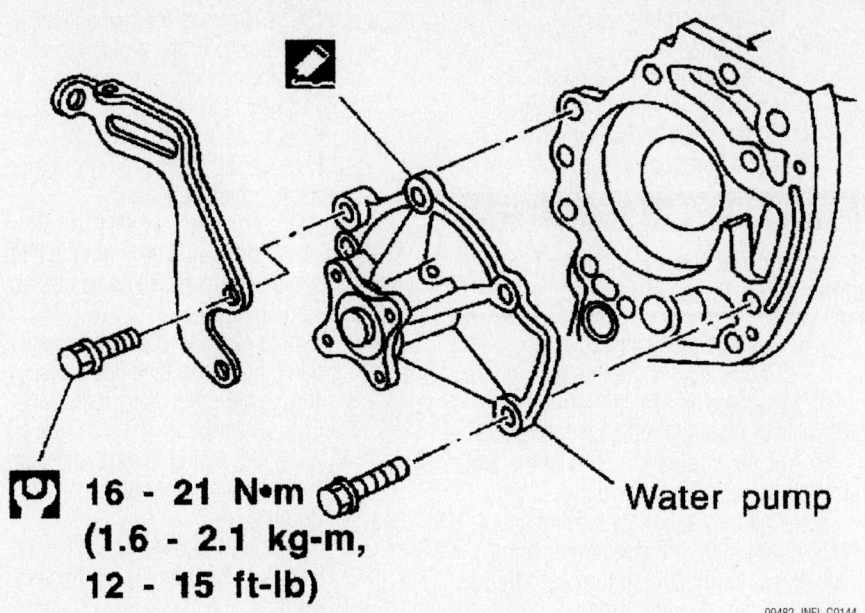

16 - 21 N•m
(1.6 - 2.1 kg-m,
12 - 15 ft-lb)

09482_INFI_G0144

Exploded view of the water pump mounting—2.0L engine

- Water pump. Torque the bolts to 15 ft. lbs. (21 Nm).
- Drive belts
- Front engine mount
- Engine side cover
- Right front wheel
- Negative battery cable

6. Fill and bleed the cooling system.

7. Start the vehicle, check for leaks and repair if necessary.

3.5L Engines

EXCEPT G35 MODELS

1. Before servicing the vehicle, refer to the precautions in the beginning of this section.

2. Drain the cooling system.

3. Remove or disconnect the following:

- Negative battery cable

- Right side engine mount and bracket
- Drive belts
- Idler pulley bracket
- Water pump drain plug, if equipped
- Timing chain tensioner cover
- Water pump cover

4. Push the timing chain tensioner sleeve and apply a stopper pin so it does not return.

- Timing chain tensioner assembly
- 3 bolts that secure the water pump

5. Rotate the crankshaft 20° counter-clockwise to provide timing chain slack.

6. Put the 2 grade M8 bolts in the 2 M8 threaded holes of the water pump.

7. Tighten each bolt by turning alternately ½ turn until they reach the timing chain rear case. Be sure to turn each bolt ½ turn at a time to prevent damage.

8. Lift up the water pump and remove it.

➡**When removing the water pump, do not allow the water pump gear to hit the timing chain.**

9. Remove and discard the O-rings from the water pump.

10. Clean all traces of liquid gasket from the water pump and covers.

To install:

11. Install or connect the following:

- Water pump with new O-rings. Torque the bolts to 89 inch lbs. (10

8.5 - 10.7
(0.86 - 1.10, 75 - 95)

O-ring (Black)

White paint

Water pump

O-ring (Black)
(Apply engine coolant.)

6.9 - 9.3
(0.70 - 0.95, 61 - 82)

❌ : Always replace after every disassembly.

: Lubricate with new engine oil.

: N•m (kg-m, in-lb)

: Apply liquid gasket. (Use Genuine Liquid Gasket or equivalent.

10 - 13
(1.0 - 1.3, 87 - 113)

Drain plug

7.8 - 11.8
(0.80 - 1.20, 69.4 - 104.2)

10 - 13
(1.0 - 1.3, 87 - 113)

09482_INFI_G0145

Water pump and timing cover assembly—I35

Nm) and rotate the crankshaft pulley to its original position by turning it 20° clockwise.

- Timing chain tensioner. Torque the bolts to 89 inch lbs. (10 Nm). Remove the stopper pin from the timing chain tensioner.

12. Apply a continuous 0.091–0.130 in. (2–3mm) bead of liquid sealant to the mating surfaces of the timing chain tensioner and water pump covers.

- Timing chain tensioner and water pump covers to the engine block. Torque the cover bolts to 89 inch lbs. (10 Nm).
- Water pump drain plug, if equipped
- Drive belts
- Idler pulley bracket
- Right side engine mounting bracket and the engine mount
- Negative battery cable

13. Fill and bleed the cooling system.

14. Start the vehicle, check for leaks and repair if necessary.

G35 MODELS

1. Before servicing the vehicle, refer to the precautions in the beginning of this section.

2. Drain the cooling system.

3. Remove or disconnect the following:
- Engine undercover
- Drive belts
- Air duct

- Radiator upper and lower hoses
- Radiator shrouds
- Cooling fan
- Water drain plug from the water pump side of the cylinder block
- Timing chain tensioner cover and water pump cover

✳✳ WARNING

Be careful not the drop the mounting bolts inside the chain case.

- Timing chain tensioner
- 3 bolts that secure the water pump

4. Rotate the crankshaft 20° counterclockwise to provide timing chain slack.

5. Put the 2 grade M8 bolts in the 2 M8 threaded holes of the water pump.

6. Tighten each bolt by turning alternately ½ turn until they reach the timing chain rear case. Be sure to turn each bolt ½ turn at a time to prevent damage.

7. Lift the water pump straight out to remove it.

8. When removing the water pump, do not allow the water pump gear to hit the timing chain.

9. Remove and discard the O-rings from the water pump.

10. Clean all traces of liquid gasket from the water pump and covers.

To install:

11. Install or connect the following:
- Water pump with new O-rings.

Torque the bolts to 75–95 inch lbs. (8–11 Nm) and rotate the crankshaft pulley to its original position by turning it 20° clockwise.

- Timing chain tensioner. Torque the bolts to 89 inch lbs. (10 Nm). Remove the stopper pin from the timing chain tensioner.

12. Apply a continuous 0.091–0.130 in. (2–3mm) bead of liquid sealant to the mating surfaces of the timing chain tensioner and water pump covers.

- Timing chain tensioner and water pump covers to the engine block. Torque the cover bolts to 89 inch lbs. (10 Nm).
- Water drain plug to the water pump side of the cylinder block
- Cooling fan
- Radiator shrouds
- Radiator upper and lower hoses
- Air duct
- Drive belts
- Engine undercover
- Negative battery cable

13. Fill and bleed the cooling system.

14. Start the vehicle, check for leaks and repair it necessary.

4.5L Engines

1. Before servicing the vehicle, refer to the precautions in the beginning of this section.

2. Drain the cooling system.

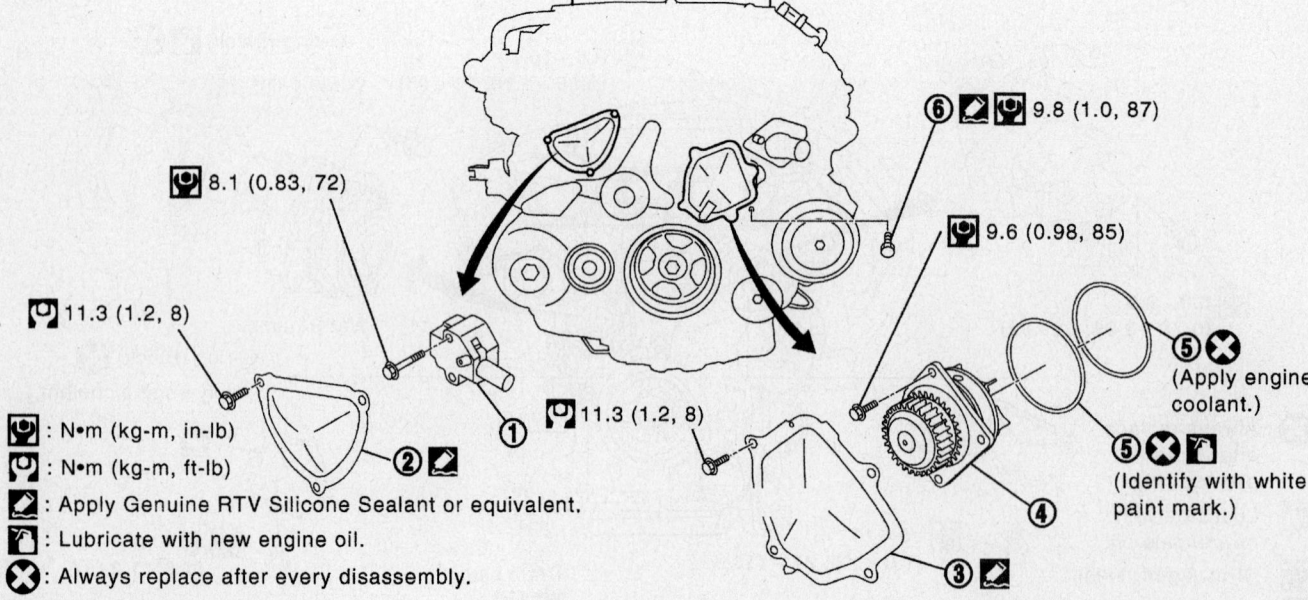

- : N•m (kg-m, in-lb)
- : N•m (kg-m, ft-lb)
- : Apply Genuine RTV Silicone Sealant or equivalent.
- : Lubricate with new engine oil.
- : Always replace after every disassembly.

1. Timing chain tensioner (primary)
2. Chain tensioner cover
3. Water pump cover
4. Water pump
5. O- ring
6. Water drain plug (front)

09482_INFI_G0146

Exploded view of the water pump mounting—G35 models

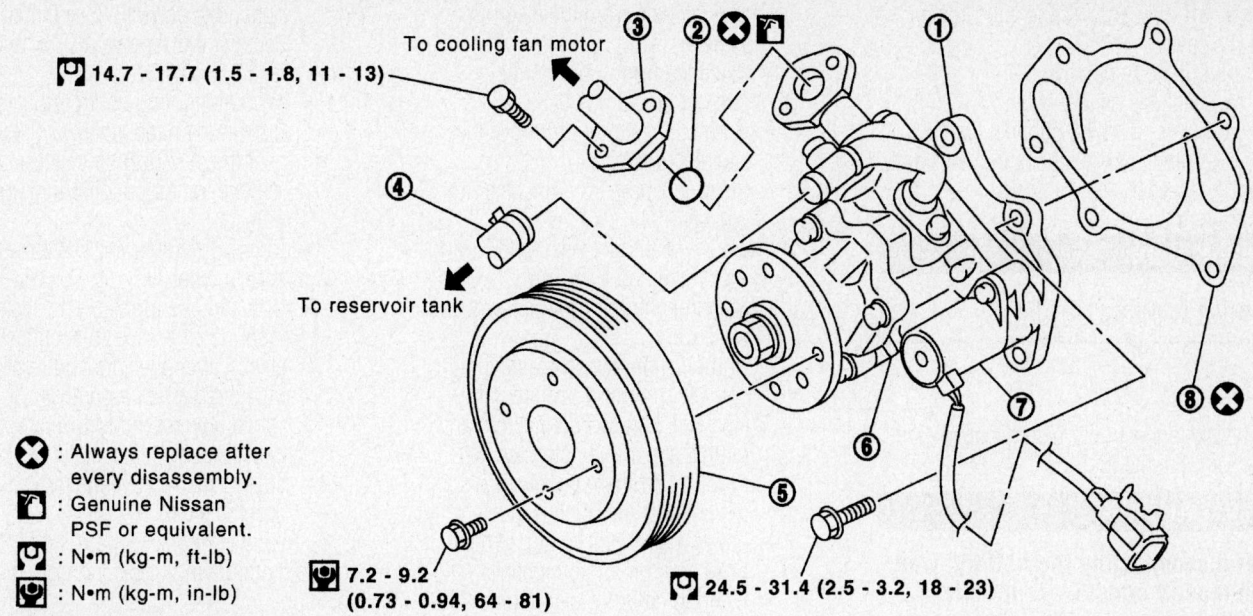

To cooling fan motor

⊡ 14.7 - 17.7 (1.5 - 1.8, 11 - 13)

To reservoir tank

❌ : Always replace after
every disassembly.

🗎 : Genuine Nissan
PSF or equivalent.

⊡ : N•m (kg-m, ft-lb)

⊞ : N•m (kg-m, in-lb)

⊡ 7.2 - 9.2
(0.73 - 0.94, 64 - 81)

⊡ 24.5 - 31.4 (2.5 - 3.2, 18 - 23)

1. Water pump
   (Do not disassemble.)

2. O-ring

3. Fluid hose (feed side)

4. Fluid hose (return side)

5. Water pump pulley

6. Cooling fan pump
   (Do not disassemble.)

7. Cooling fan speed control solenoid
   valve (Do not disassemble.)

8. Gasket

09482_INFI_G0147

Component view of water pump—2002–04 Q45 and 2004 M45

3. Remove or disconnect the following:
   • Negative battery cable
   • Engine undercover
   • Air duct and engine appearance
     cover
   • Drive belt

   • Cooling fan connector
   • Radiator hose
   • Water pump
   To install:
4. Thoroughly clean and dry the mating
   surfaces, bolts and bolt holes.

5. Install or connect the following:
   • Water pump. Torque the bolts to
     18–23 ft. lbs. (25–32 Nm).
   • Radiator hose
   • Cooling fan connector
   • Drive belt

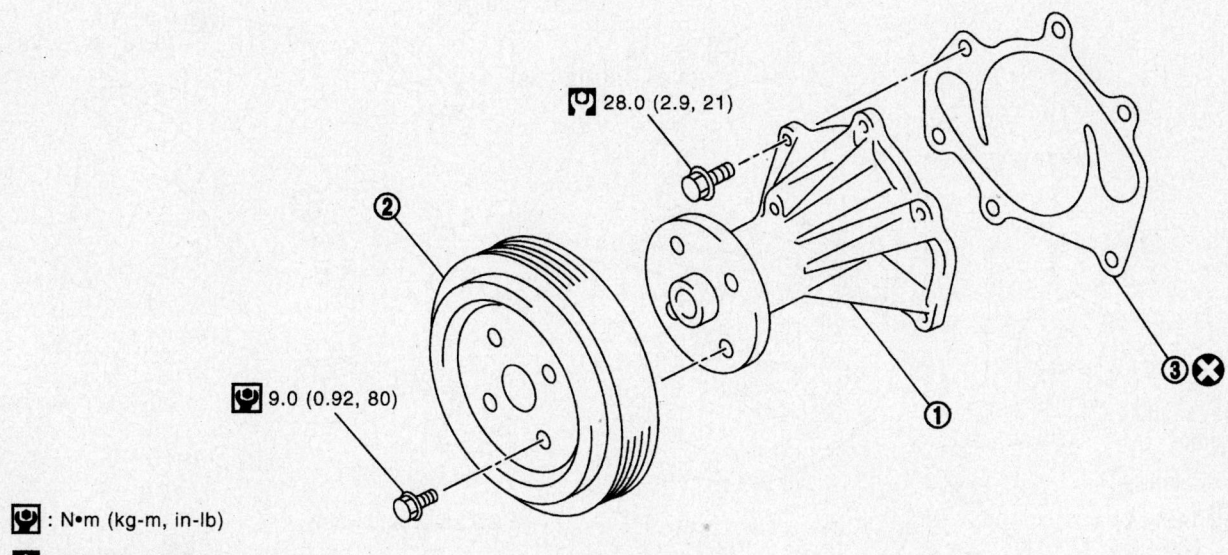

⊡ 28.0 (2.9, 21)

⊞ 9.0 (0.92, 80)

⊞ : N•m (kg-m, in-lb)

⊡ : N•m (kg-m, ft-lb)

1. Water pump

2. Water pump pulley

3. Gasket

09482_INFI_G0030

Component view of water pump—2005–06 Q45 and M45

- Air duct and engine appearance cover
- Engine undercover
- Negative battery cable

6. Fill and bleed the cooling system.

7. Start the vehicle, check for leaks and repair if necessary.

Heater Core

REMOVAL & INSTALLATION

G20

1. Disconnect both battery cables, the negative (–) cable first.

✻✻ CAUTION

After disconnecting the battery, wait for a least 3 minutes for the SRS module to deplete its energy before working on the steering column or instrument panel.

2. Drain the cooling system into a clean container for reuse.

3. Remove the driver's side SRS module and the steering wheel by removing or disconnecting the following:

- Lower cover at the base of the steering wheel
- SRS module electrical connector
- Side covers at both sides of the steering wheel
- SRS module-to-steering wheel bolts using a T50 Torx® wrench

- SRS module from the steering wheel
- Place the front wheel in the straight-ahead position
- Horn connector and remove the steering wheel nut
- Steering wheel from the steering column

4. Remove or disconnect the following:
- Dash side and floor trim
- Steering column cover screws and the covers
- Combination switch-to-steering column screws, disconnect the electrical harness connectors and remove the combination switch
- Lower instrument panel screws and the panel at the driver's side
- Lower instrument reinforcement bolts and the reinforcement
- Steering column-to-instrument panel nuts and lower the steering column
- Cluster lid "C"-to-instrument panel screws, disconnect the electrical connectors and remove the cluster lid
- Cluster lid "A"-to-instrument panel screws, disconnect the electrical connectors and remove the cluster lid
- Combination meter-to-instrument panel screws, disconnect the electrical connectors and remove the combination meter
- Audio center-to-instrument panel

bolts, disconnect the electrical connectors and remove the audio center
- Air conditioning control unit-to-instrument panel screws, disconnect the electrical connectors and remove the air conditioning control unit
- Console finisher clips and the console finisher
- Console box assembly-to-instrument panel screws, disconnect the electrical connectors and remove the console box assembly
- Glove box assembly-to-instrument panel screws, disconnect the lamp socket and remove the glove box assembly

5. Remove the passenger's side SRS module by removing or disconnecting the following:

- Open the glove box and remove the SRS module cover
- SRS module electrical connector
- Glove box assembly
- SRS module-to-instrument panel bolts using a T50 Torx® wrench
- SRS module from the instrument panel

6. Remove or disconnect the following:

- Lower instrument cover clip and the cover
- Lower instrument panel center-to-instrument panel screws and the panel

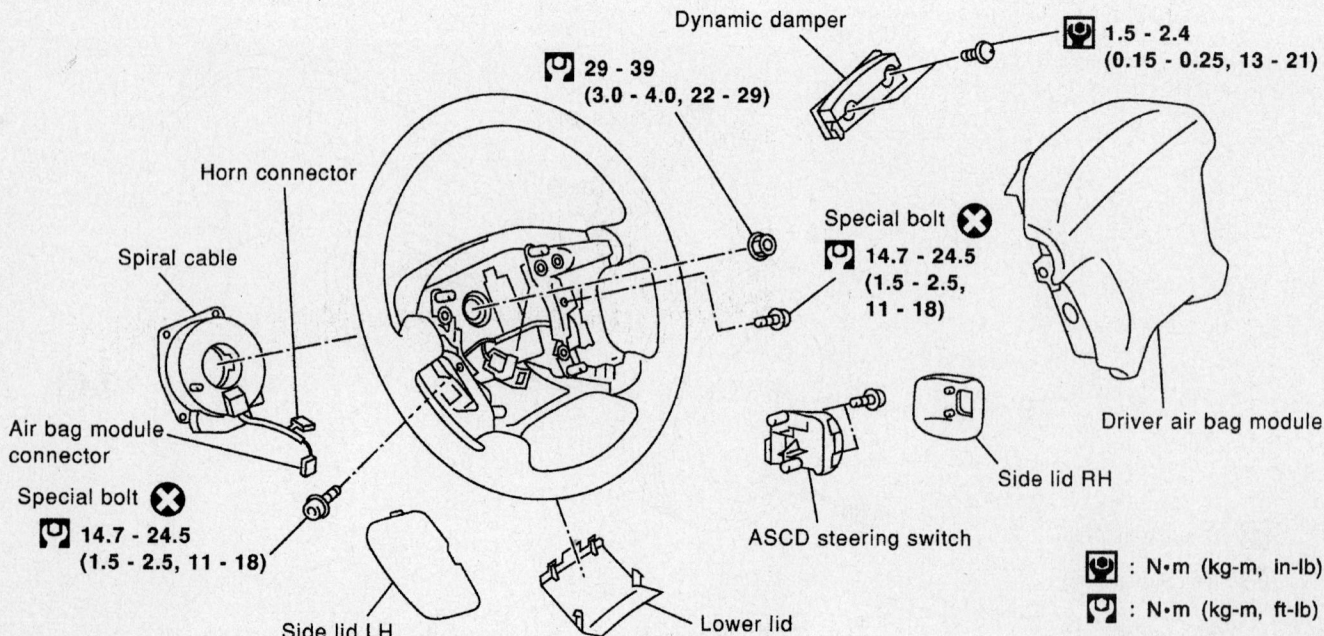

Exploded view of the air bag module, the steering wheel and related components—G20

09482_INFI_G0148

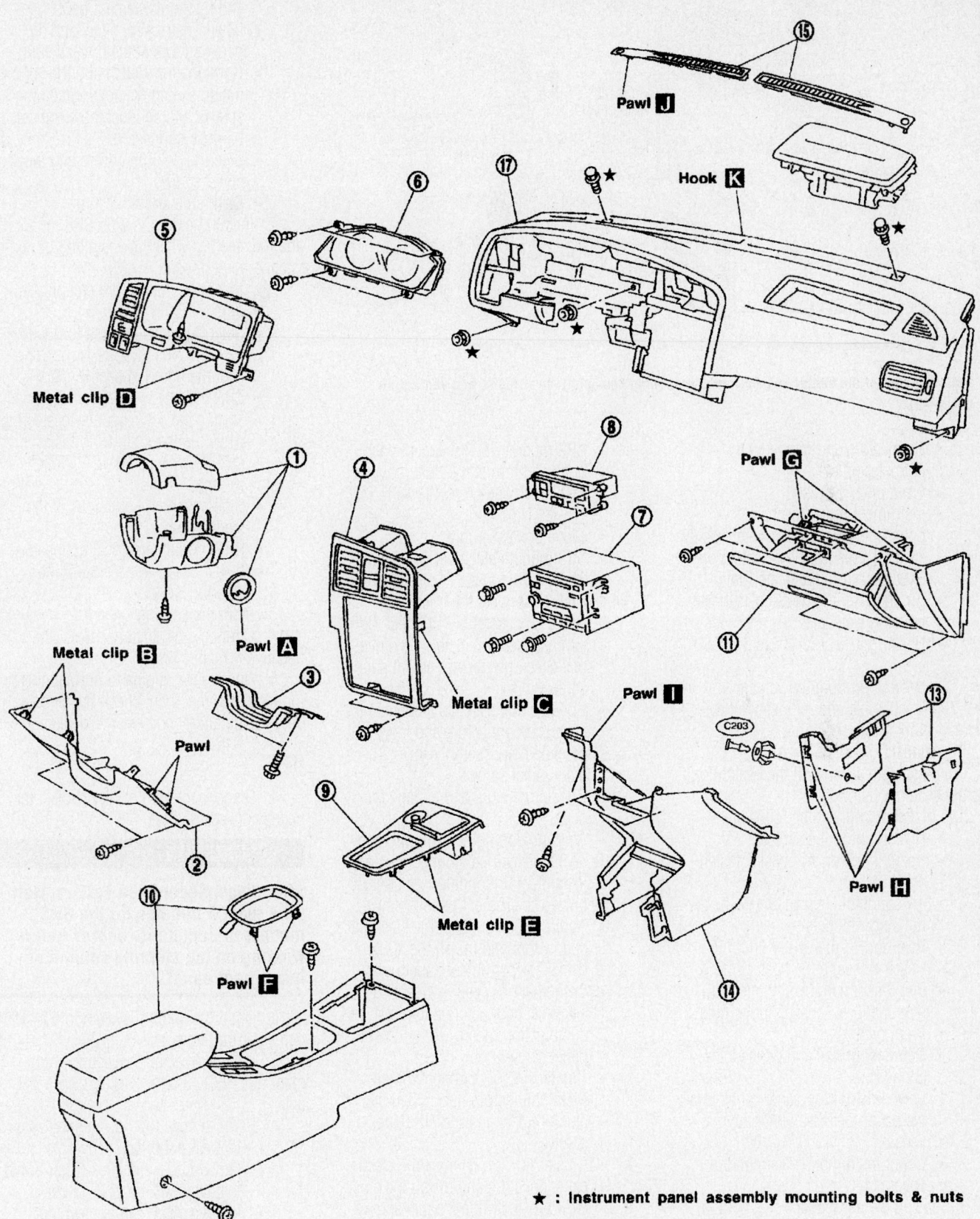

Pawl **J**

Hook **K**

Metal clip **D**

Pawl **A**

Metal clip **B**

Pawl

Metal clip **C**

Pawl **G**

Pawl **I**

C203

Pawl **H**

Metal clip **E**

Pawl **F**

★ : Instrument panel assembly mounting bolts & nuts

09482_INFI_G0149

Exploded view of the instrument panel—G20

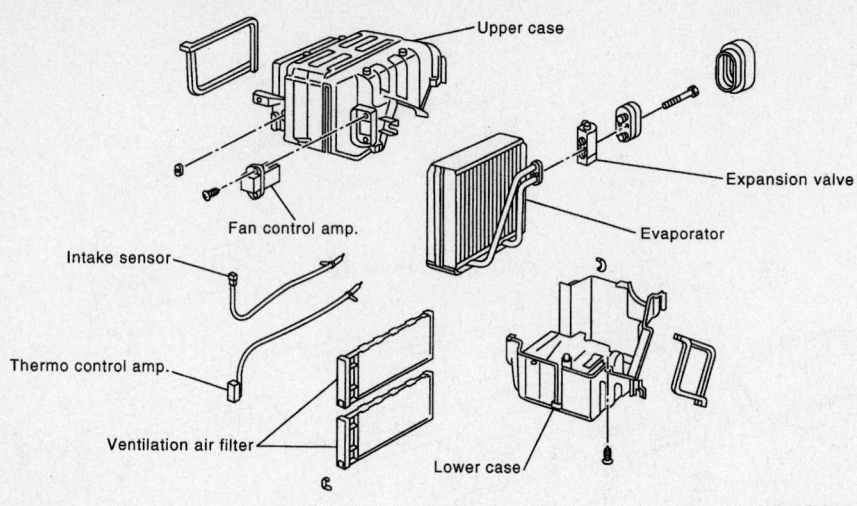

Exploded view of the heater housing, air conditioning housing, blower motor and ventilation ducts—G20

- Connectors and remove the defroster grilles
- Front pillar garnish
- Instrument panel-to-chassis bolts/nuts and the instrument panel
- Heater hoses from the heater core
- Rear duct from the heater unit
- Air conditioning housing-to-heater housing fasteners
- Heater housing-to-chassis fasteners
- Heater unit from the vehicle

7. Disassemble and remove the heater core from the heater housing.

To install:

8. Assemble and install the heater core to the heater housing.

9. Install or connect the following:
- Heater unit to the vehicle
- Heater housing-to-chassis fasteners
- Air conditioning housing-to-heater housing fasteners
- Rear duct to the heater unit
- Heater hoses to the heater core
- Instrument panel and the instrument panel-to-chassis bolts/nuts
- Front pillar garnish
- Defroster grilles and connect the connectors
- Lower instrument panel center and the panel-to-instrument panel screws
- Lower instrument cover and the cover clip

10. Install the passenger's side SRS module by installing or connecting the following:
- SRS module to the instrument panel
- SRS module-to-instrument panel bolts and torque the bolts using a T50 Torx® wrench to 11–18 ft. lbs. (15–25 Nm)
- Glove box assembly
- SRS module electrical connector
- SRS module cover

11. Install or connect the following:
- Glove box assembly, connect the lamp socket and install the glove box assembly-to-instrument panel screws
- Console box assembly, connect the electrical connectors and install the console box assembly-to-instrument panel screws
- Console finisher and engage the clips
- Air conditioning control unit, connect the electrical connectors and install the air conditioning control unit-to-instrument panel screws
- Audio center, connect the electrical connectors and install the audio center-to-instrument panel bolts
- Combination meter, connect the electrical connectors and install the combination meter-to-instrument panel screws
- Cluster lid "A", connect the electrical connectors and install the cluster lid-to-instrument panel screws
- Cluster lid "C", connect the electrical connectors and install the cluster lid-to-instrument panel screws
- Steering column and torque the steering column-to-instrument panel nuts to 11–14 ft. lbs. (15–19 Nm)

- Lower instrument reinforcement and the reinforcement bolts
- Lower instrument panel and the panel screws at the driver's side
- Combination switch and the combination switch-to-steering column screws and connect the electrical harness connectors
- Steering column covers and the cover screws
- Dash side and floor trim

12. Install the driver's side SRS module and the steering wheel by installing or connecting the following:
- Steering wheel to the steering column
- Steering wheel nut and torque to 22–29 ft. lbs. (29–39 Nm). Connect the horn connector
- SRS module to the steering wheel
- Torque the SRS module-to-steering wheel bolts to 11–18 ft. lbs. (15–25 Nm) using a T50 Torx® wrench
- Side covers at both sides of the steering wheel
- SRS module electrical connector
- Lower cover at the base of the steering wheel

13. Refill the cooling system.

14. Connect both battery cables, the negative (–) cable last.

15. Operate the engine to normal operating temperatures; then, check the climate control operation and check for leaks.

G35

1. Disconnect both battery cables, the negative (–) cable first.

✳✳ CAUTION

After disconnecting the battery, wait for a least 3 minutes for the SRS module to deplete its energy before working on the steering column or instrument panel.

2. Drain the cooling system into a clean container for reuse.

3. Use a refrigerant collecting equipment (for HFC-134a) to discharge refrigerant.

4. Remove cowl top cover as follows:
- Remove hood ledge cover.
- Remove both right/left wiper arms.
- Remove cowl top seal rubber.
- Remove clips, cap of cowl top cover and remove cowl top cover (right).
- Remove clips, cap, screws and remove cowl top cover (left).

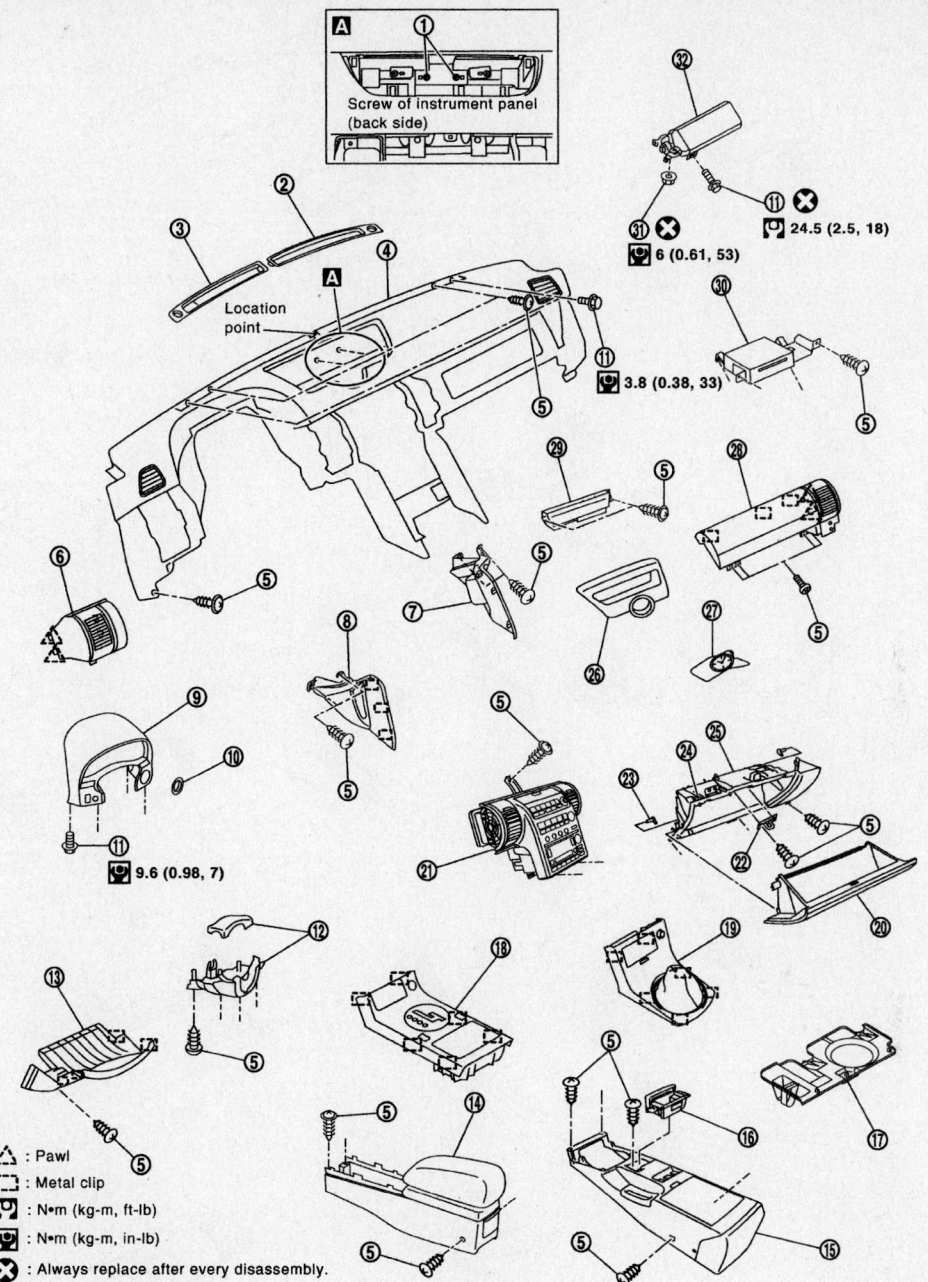

Exploded view of instrument panel assembly—G35 2002–04

1.	Instrument panel and pad mounting screw (View A)	
2.	Front defroster grille (RH)	
3.	Front defroster grille (LH)	

1. Instrument panel and pad mounting screw (View A)
2. Front defroster grille (RH)
3. Front defroster grille (LH)
4. Instrument panel and pad
5. Screw
6. Side ventilator grille (LH)
7. Instrument side panel (RH)
8. Instrument side panel (LH)
9. Cluster lid A
10. Steering lock escutcheon
11. Bolt
12. Steering column cover
13. Instrument lower driver panel
14. Center console (A/T Models)
15. Center console (M/T Models)
16. Ashtray
17. Instrument lower cover
18. Console finisher (A/T Models)
19. Console boot (M/T Models)
20. Glove box
21. Cluster lid C
22. Glove box striker
23. Glove box pin
24. Resin clip
25. Instrument lower passenger panel
26. Cluster lid finisher
27. Clock assembly
28. Center box assembly
29. Display and amplifier assembly
30. NAVI control unit
31. Nut
32. Passenger air bag module

△ : Pawl
▢ : Metal clip
🔧 : N•m (kg-m, ft-lb)
🔧 : N•m (kg-m, in-lb)
✖ : Always replace after every disassembly.

09482_INFI_G0036

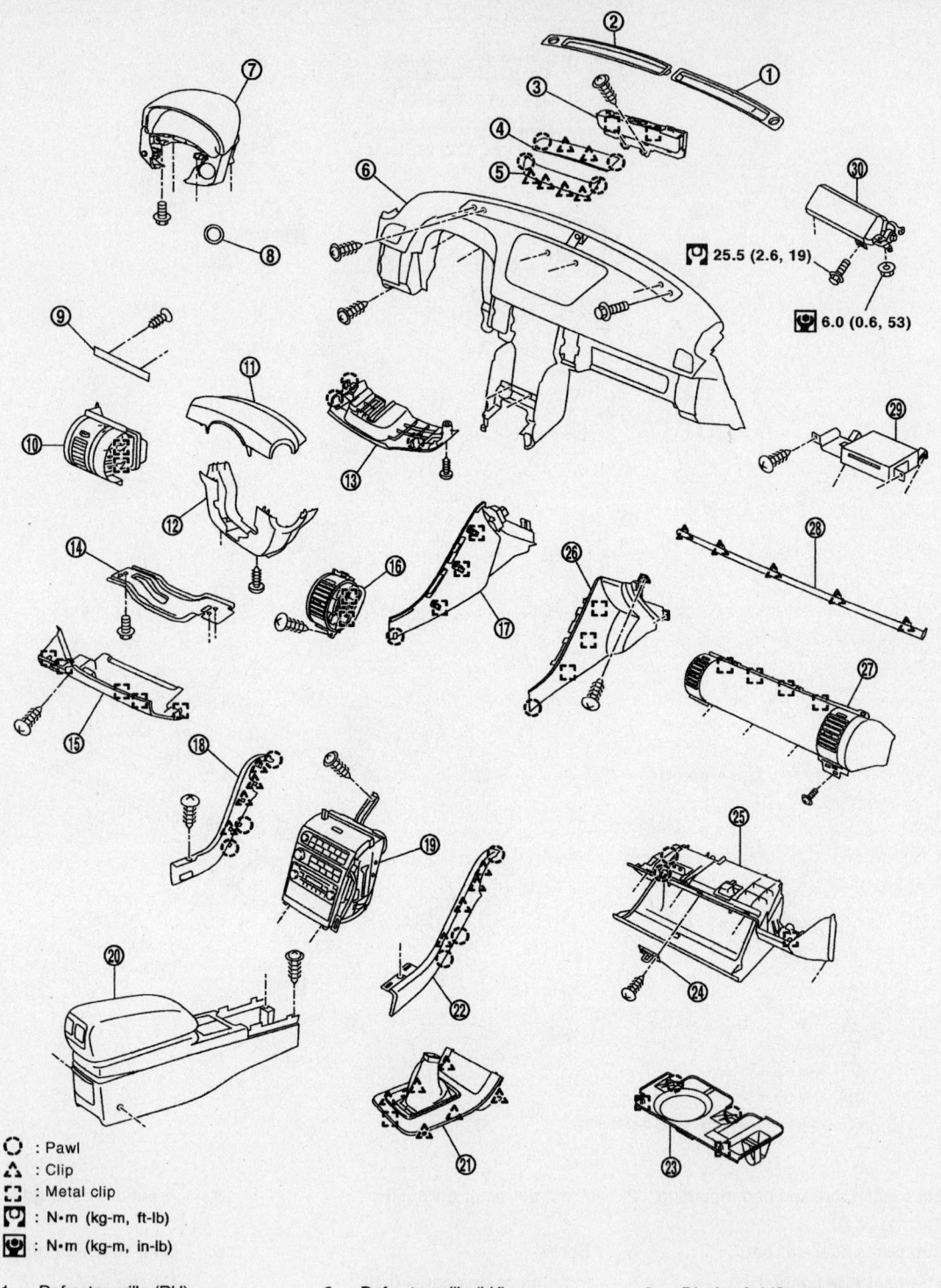

25.5 (2.6, 19)

6.0 (0.6, 53)

○ : Pawl
△ : Clip
⊓ : Metal clip
▣ : N·m (kg-m, ft-lb)
▣ : N·m (kg-m, in-lb)

1. Defroster grille (RH)	2. Defroster grille (LH)	3. Display & A/C auto amp
4. Cluster lid finisher upper	5. Cluster lid finisher lower	6. Instrument panel & pad
7. Cluster lid A assembly	8. Steering lock escutcheon	9. Instrument finisher A
10. Side ventilator grille (LH)	11. Steering column cover upper	12. Steering column cover lower
13. Steering column cover front lower	14. Knee protector lower	15. Instrument driver lower panel
16. Center ventilator grille (LH)	17. Instrument side panel (LH)	18. Cluster lid C side finisher (LH)
19. Cluster lid C assembly	20. Center console assembly	21. Console finisher
22. Cluster lid C side finisher (RH)	23. Instrument lower cover	24. Glove box striker
25. Glove box assembly	26. Instrument side panel (RH)	27. Center box assembly
28. Instrument finisher B	29. NAVI control unit	30. Passenger air bag module

09482_INFI_G0038

Exploded view of instrument panel assembly—G35 2005–06

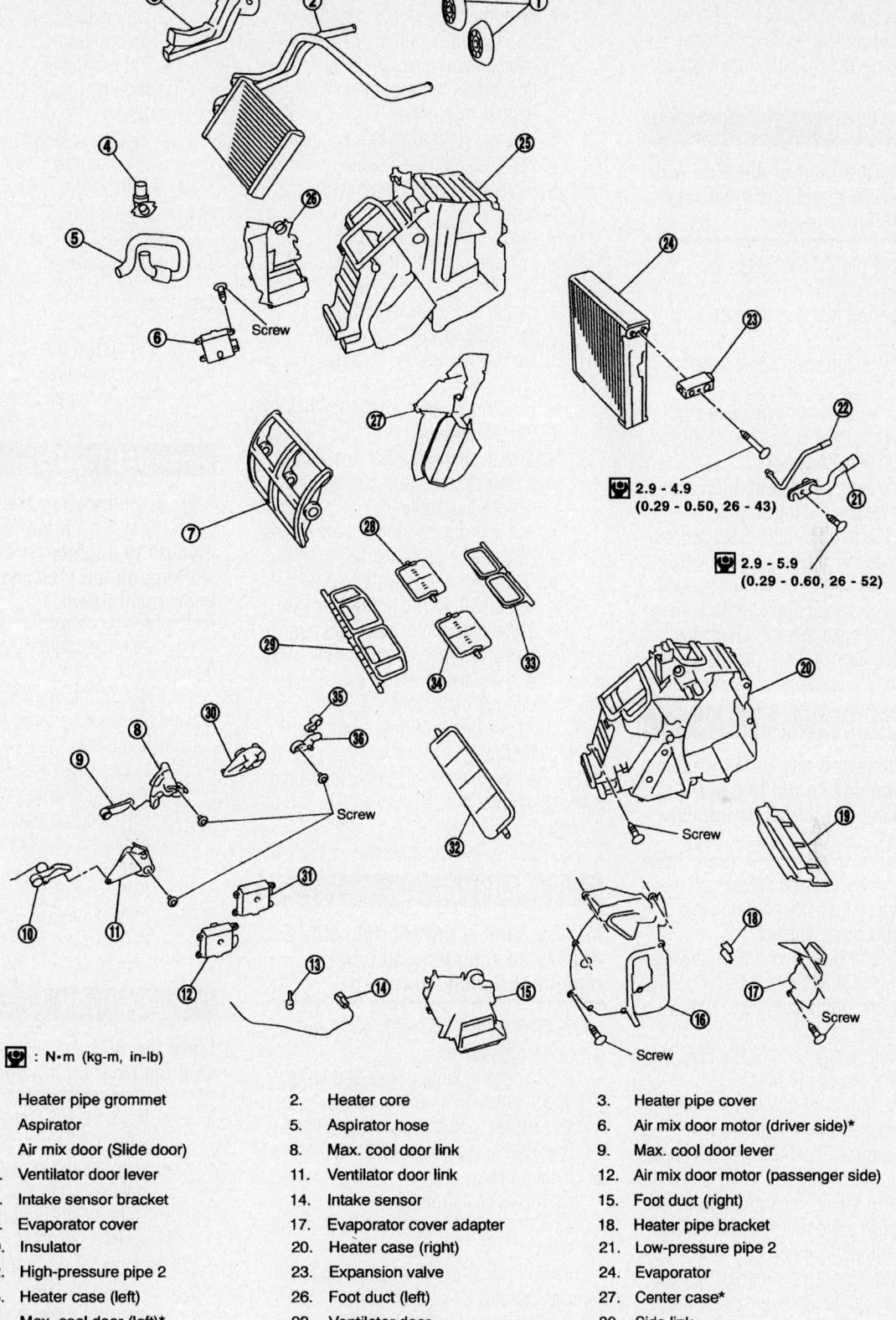

2.9 - 4.9
(0.29 - 0.50, 26 - 43)

2.9 - 5.9
(0.29 - 0.60, 26 - 52)

⬛ : N•m (kg-m, in-lb)

1. Heater pipe grommet	2. Heater core	3. Heater pipe cover
4. Aspirator	5. Aspirator hose	6. Air mix door motor (driver side)*
7. Air mix door (Slide door)	8. Max. cool door link	9. Max. cool door lever
10. Ventilator door lever	11. Ventilator door link	12. Air mix door motor (passenger side)
13. Intake sensor bracket	14. Intake sensor	15. Foot duct (right)
16. Evaporator cover	17. Evaporator cover adapter	18. Heater pipe bracket
19. Insulator	20. Heater case (right)	21. Low-pressure pipe 2
22. High-pressure pipe 2	23. Expansion valve	24. Evaporator
25. Heater case (left)	26. Foot duct (left)	27. Center case*
28. Max. cool door (left)*	29. Ventilator door	30. Side link
31. Mode door motor	32. Max. cool door (Without left and right ventilation temperature separately control system)	33. Defroster door
34. Max. cool door (right)*	35. Defroster door lever	36. Defroster door link

*: With left and right ventilation temperature separately control system.

09482_INFI_G0037

Exploded view of heater and cooling unit—G35

- Remove washer nozzles and hose from cowl top cover.

5. Disconnect low-pressure flexible hose and high-pressure pipe from evaporator.

❄❄ WARNING

Cap or wrap the joint of the pipe with suitable material such as vinyl tape to avoid the entry of air.

6. Remove air hose and electronic control throttle assembly.

7. Disconnect two heater hoses from heater core.

8. Remove instrument panel assembly as follows:

- Remove shift knob (manual transmission)/gear selector knob (automatic transmission)
- Place selector lever in DRIVE position (automatic transmission)
- Insert a thin flat-bladed screwdriver wrapped with tape from behind console boot/console finisher and remove metal clip(s) on back, then remove clips at front (automatic transmission). Then pull up and back to disengage from console.

❄❄ WARNING

Guide pin inserted into automatic transmission device guide can be easily broken; exercise caution during removal.

- Disconnect hazard switch harness connector, and remove console boot/console finisher.

9. Remove or disconnect the following:

- Cluster Lid "C" side finisher screws (left and right) and pawls; remove finisher
- Upper Cluster Lid finisher clips and pawls; remove finisher
- Lower Cluster Lid finisher clips and pawls; remove finisher
- Instrument Panel finisher "B" clips; remove finisher
- Cluster Lid "C" screws; disconnect display/audio unit harness connector and remove unit
- Driver Side Lower Instrument Panel screws and clips; disconnect harness connectors and remove panel
- Hood lock cable and grommet
- Front lower steering column cover
- Upper and lower side steering column covers
- Lighting and turn signal switch
- Wiper and washer switch

- Lower knee protector
- Steering column lock escutcheon
- Cluster Lid "A" bolts; disconnect combination meter and mirror control switch harness connectors and remove cluster lid
- Left and right front kick panels
- Left and right dash side finishers
- Right side lower instrument panel cover
- Glove box assembly
- Left and right Instrument Side Panels
- Center box assembly
- Navigation Control Unit screws and harness connector; remove control unit
- Passenger side Air Bag Module
- Center console
- Display and Amplifier Assembly screws and harness connector; remove assembly
- Left and right front defroster grilles
- Left side ventilator grille
- Center ventilator grille
- Left and right front pillar garnish
- Lower steering column and remove instrument panel bolts and screws; remove instrument panel and pad through passenger-side door

10. Remove Engine Control Module (ECM) with bracket attached

11. Remove blower assembly mounting bolt and screws; disconnect harness connectors

12. Remove blower assembly

❄❄ WARNING

Move assembly toward right side and remove locating pin and joint; remove assembly downward

13. Remove clips of vehicle harness from steering member.

14. Remove mounting nuts and bolts, and then remove instrument stays (driver side and passenger side).

15. Remove mounting bolts from heater & cooling unit.

16. Disconnect drain hose.

17. Remove defroster nozzle and ventilator ducts.

18. Remove steering member, and then remove heater & cooling assembly.

19. Remove heater core

To install:

20. Installation is the reverse order of removal; observe the following:

- Replace air conditioning O-rings with new O-rings
- Exercise caution when reassem-

bling air conditioning piping; the connection points are thin and deform easily.

- Tighten heater & cooling assembly mounting bolts to 61 inch lbs. (6.9 Nm)
- Tighten steering member mounting nut and bolt to 9 ft. lbs. (12 Nm)

21. Refill cooling system and check for leaks.

22. Recharge air conditioning system and check for leaks.

I35

1. Disconnect both battery cables, the negative (–) cable first.

2. Drain the cooling system into a clean container for reuse.

❄❄ CAUTION

After disconnecting the battery, wait for a least 3 minutes for the SRS module to deplete its energy before working on the steering column or instrument panel.

3. Drain the cooling system into a clean container for reuse.

4. Remove the driver's side SRS module and the steering wheel by removing or disconnecting the following:

- Lower cover at the base of the steering wheel
- SRS module electrical connector
- Side covers at both sides of the steering wheel
- SRS module-to-steering wheel bolts using a T50 Torx® wrench
- SRS module from the steering wheel

❄❄ CAUTION

Store the SRS module in a safe place with the front facing upward.

- Place the front wheel in the straight-ahead position
- Horn connector and remove the steering wheel nut
- Steering wheel from the steering column

5. Remove or disconnect the following:

- Instrument panel assembly
- Front seats
- Defroster ducts, the ventilator ducts and the floor ducts from the heater unit
- Vacuum hoses and electrical connectors leading to the heater unit
- Heater unit from the cooling unit.

Take care not to damage the air conditioning tubes
  • Heater unit attaching bolts. Remove the heater unit from the passenger compartment
6. Disassemble the heater unit and remove the heater core.

To install:

7. Install the heater core and assemble the heater unit.
8. Install or connect the following:
  • Heater unit in the passenger compartment and tighten the attaching bolts securely
  • All vacuum hoses and electrical connectors
  • Defroster ducts, the ventilator ducts and the floor ducts to the heater unit
  • Front seats
  • Instrument panel assembly
  • Heater hoses
9. Install the driver's side SRS module and the steering wheel by installing or connecting the following:
  • Steering wheel to the steering column
  • Steering wheel nut and torque to 22–29 ft. lbs. (29–39 Nm). Connect the horn connector
  • SRS module to the steering wheel
  • SRS module-to-steering wheel bolts to 11–18 ft. lbs. (15–25 Nm) using a T50 Torx® wrench
  • Side covers at both sides of the steering wheel
  • SRS module electrical connector

  • Lower cover at the base of the steering wheel
10. Refill the cooling system.
11. Connect both battery cables, the negative (–) cable last.
12. Operate the engine to normal operating temperatures; then, check the climate control operation and check for leaks.

2004 M45

1. Disconnect both battery cables, the negative (–) cable first.

✳✳ CAUTION

After disconnecting the battery, wait for a least 3 minutes for the SRS module to deplete its energy before working on the steering column or instrument panel.

2. Recover the air conditioning refrigerant.
3. Drain the engine coolant.
4. Remove the engine appearance cover.
5. Remove the front suspension tower bar.
6. Remove or disconnect the following:

  • Heater hoses
  • Evaporator lines
  • Steering lock cover
  • Steering column upper and lower covers
  • Driver side lower instrument panel cover
  • Hood opening lever

  • Top instrument cluster cover
  • Combination meter
  • Inner instrument cluster panel
  • Instrument cluster
  • Clock
  • Automatic transmission shift lever cover
  • Cup holder
  • Ashtray and center lower panel
  • Console upper finish panel
  • Audio unit
  • Instrument panel lower cover
  • Glove box
  • Glove box cover
  • CD unit
  • Center ventilation grille
  • Console box assembly
  • Front defroster grille
  • Both A pillar trim panels
  • Ignition key lamp and instrument panel bracket
  • Instrument panel reinforcement bracket
  • Instrument panel and pad
  • Blower motor, intake door motor and amplifier connectors
  • Blower unit
  • Vehicle harness clips from steering member
  • Instrument panel stays at both sides
  • Defroster nozzle
  • Ventilator ducts
  • Heating and cooling unit
  • Foot duct from heating and cooling unit
  • Heater core from heating and cooling unit

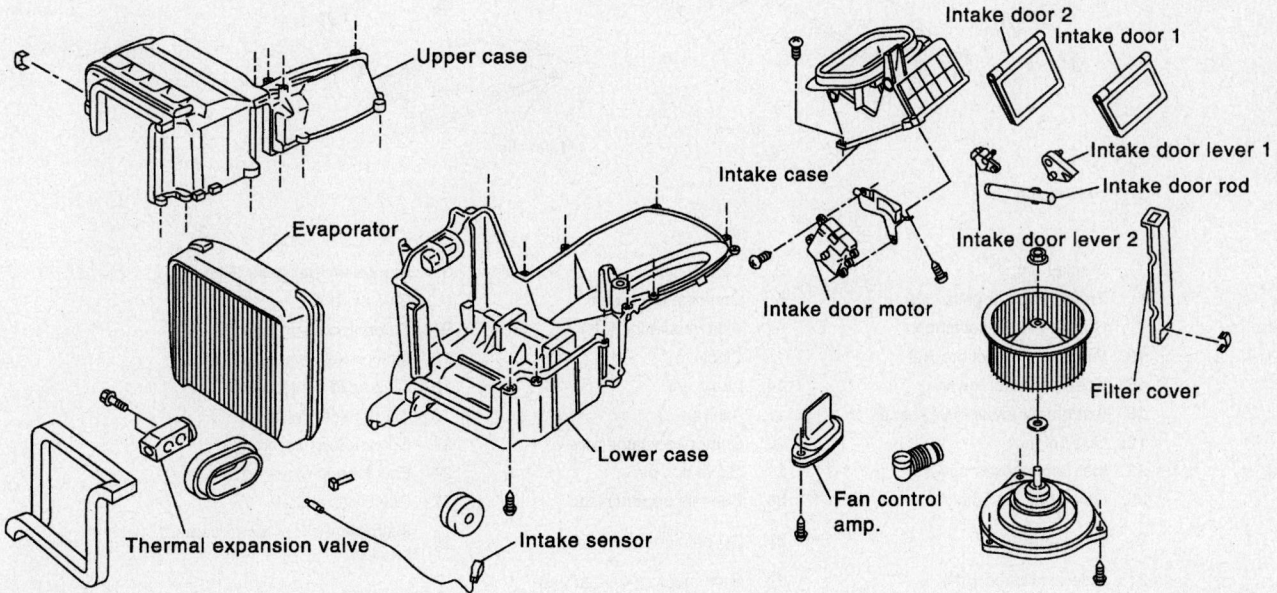

Exploded view of the heating-air conditioning system assemblies and related components—I35

09482_INFI_G0151

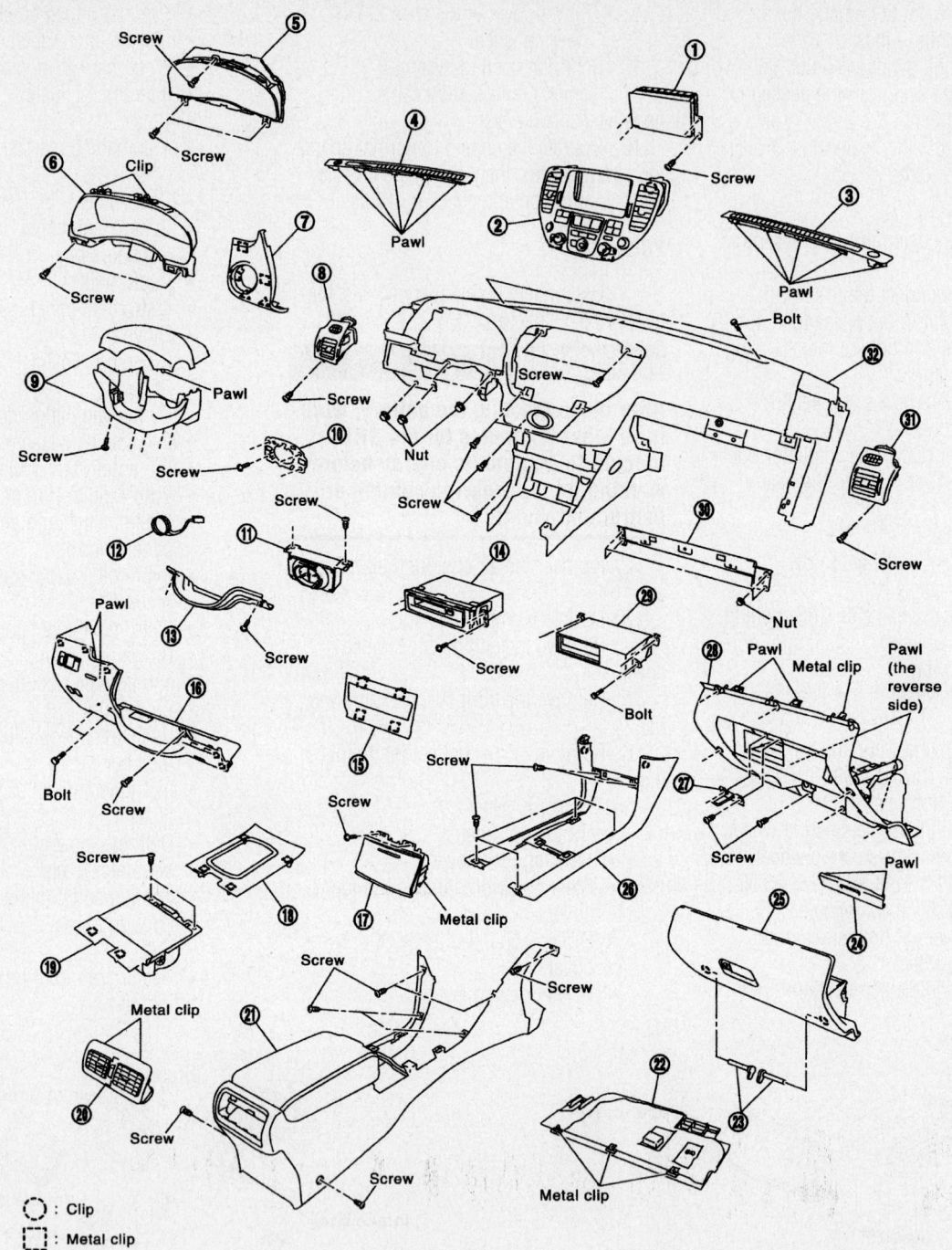

1.	Display unit	2.	Cluster lid C	3.	Front defroster grille (RH)
4.	Front defroster grille (LH)	5.	Combination meter	6.	Cluster lid A
7.	Steering lock escutcheon	8.	Side ventilation (LH)	9.	Steering column cover
10.	Instrument panel bracket	11.	Clock	12.	Ignition key lamp assembly
13.	Knee protector lower	14.	Audio unit	15.	Cluster lid center lower
16.	Instrument lower driver panel	17.	Ashtray	18.	A/T console finisher
19.	Cup holder	20.	Center ventilator grille	21.	Console box assembly
22.	Instrument lower cover	23.	Glove box pin	24.	Escutcheon glove box
25.	Glove box assembly	26.	Console upper finisher	27.	Glove box striker
28.	Glove box cover	29.	CD auto changer	30.	Instrument panel reinforcement bracket
31.	Side ventilation (RH)	32.	Instrument panel and pad		

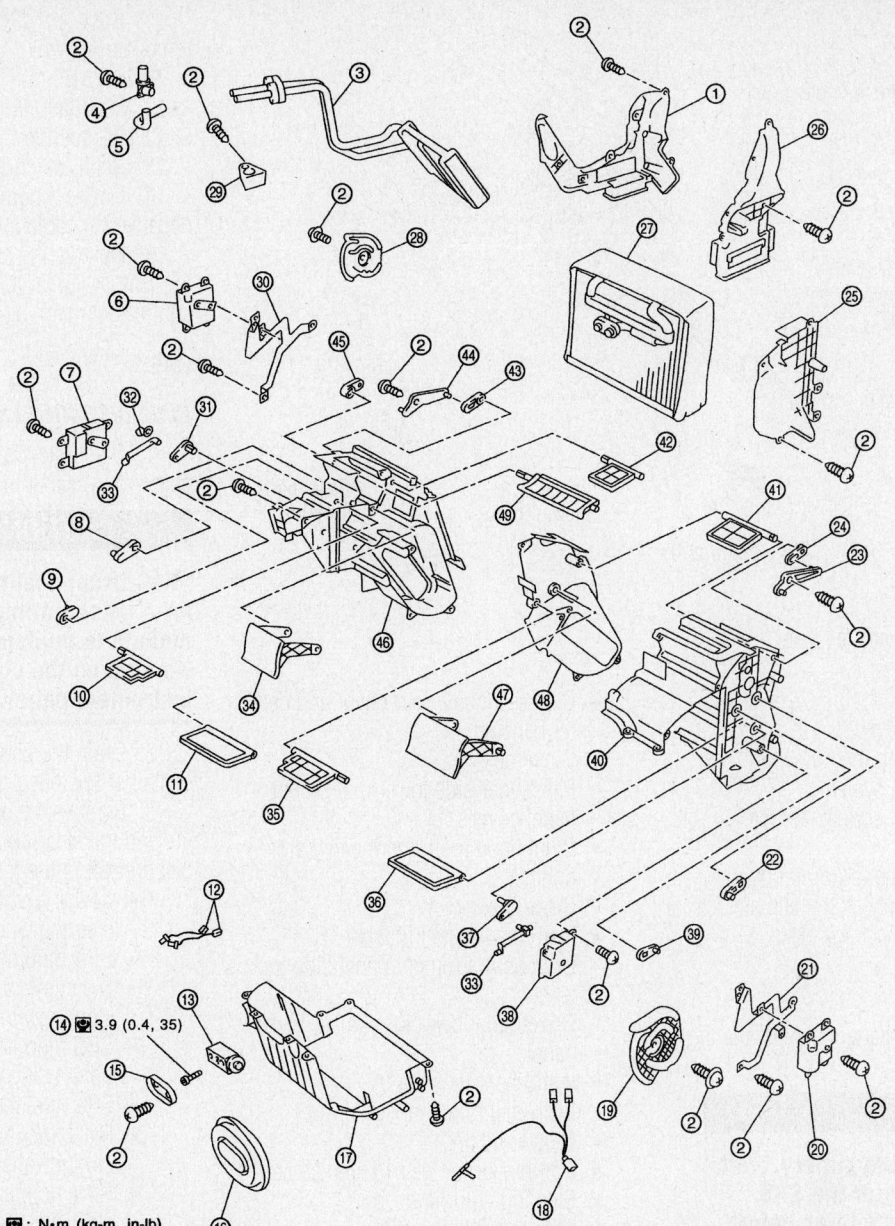

14 ⊡ 3.9 (0.4, 35)

⊡ : N·m (kg-m, in-lb)

1.	Foot duct (right)	2.	Screw	3.	Heater core
4.	Aspirator	5.	Aspirator duct	6.	Mode door motor (passenger side)
7.	Air mix door motor (passenger side)	8.	Ventilator door lever (right)	9.	Bi-level door lever (right)
10.	Bi-level door (right)	11.	Air mix door (right)	12.	Sub harness
13.	Expansion valve	14.	Bolt	15.	Expansion valve cover
16.	Cooler grommet	17.	Heater & cooling unit case (lower)	18.	Intake sensor
19.	Side link (left)	20.	Mode door motor (driver side)	21.	Mode door motor bracket (left)
22.	Ventilator door lever (left)	23.	Foot door link (left)	24.	Foot door lever (left)
25.	Evaporator cover	26.	Foot duct (left)	27.	Evaporator
28.	Side link (right)	29.	Heater pipe bracket	30.	Mode door motor bracket (right)
31.	Air mix door lever (right)	32.	Rod holder	33.	Rod
34.	Ventilator door (right)	35.	Bi-level door (left)	36.	Air mix door (left)
37.	Air mix door lever (left)	38.	Air mix door motor (driver side)	39.	Bi-level door lever (left)
40.	Heater & cooling unit case (driver side)	41.	Foot door (left)	42.	Foot door (right)
43.	Foot door lever (right)	44.	Foot door link (right)	45.	Defroster door lever (right)
46.	Heater & cooling unit case (passenger side)	47.	Ventilator door (left)	48.	Heater & cooling unit case (center)
49.	Defroster door				

09482_INFI_G0153

Heating and cooling unit mounting—2003–04 M45 & Q45

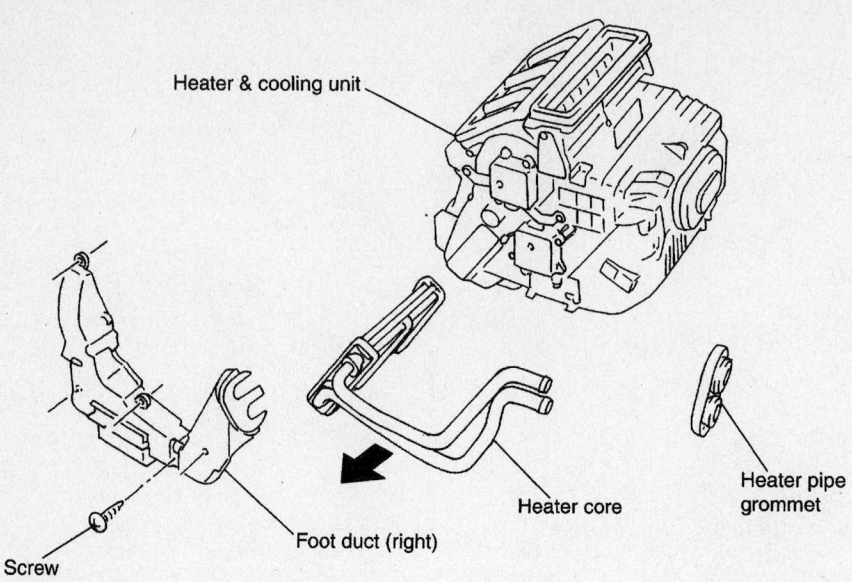

Heater & cooling unit

Heater core

Foot duct (right)

Screw

Heater pipe grommet

09482_INFI_G0154

Heater core removal—2003–04 M45

To install:
7. Installation is the reverse of the removal procedure noting the following:

8. Refill the cooling system.

9. Connect both battery cables, the negative (–) cable last.

10. Operate the engine to normal operating temperatures; then, check the climate control operation and check for leaks.

M35 and 2006 M45
1. Disconnect both battery cables, the negative (–) cable first.

✷✷ CAUTION

After disconnecting the battery, wait for a least 3 minutes for the SRS module to deplete its energy before working on the steering column or instrument panel.

2. Recover the air conditioning refrigerant.

3. Drain the engine coolant.

4. Remove the engine appearance cover.

5. Remove cowl top cover

6. Remove air hose and electronic throttle control assembly

7. Disconnect heater hoses from heater core

8. Disconnect A/C evaporator lines

9. Remove or disconnect the following:

- Cup holder
- Automatic transmission shift lever cover
- Cluster lid "C"

- Center console rear cover and center console
- Glove box
- Passenger side lower instrument panel cover
- Right hand instrument panel side finisher
- Glove box cover
- Instrument panel finisher "B"
- Left had instrument panel side finisher
- Driver side lower instrument lower panel
- Instrument finisher "A"
- Instrument finisher "C"
- Cluster lid "A"
- Combination meter (gauge cluster)
- Steering column cover
- Upper ventilation grille
- Center ventilation grille
- Multifunction switch
- Front display unit
- Audio assembly
- Navigation Control Unit
- Left and right side ventilation grilles
- Front defroster grille
- Instrument panel and pad
- Blower unit
- Wiring harness clips from steering member
- Instrument panel stays
- Drain hose
- Side defroster nozzles
- Steering member
- Heater and cooling unit

10. Remove heater pipe cover and left foot duct from unit

11. Slide heater core to left side to remove from unit

To install:
12. Installation is the reverse of the removal procedure noting the following:

13. Refill the cooling system.

14. Connect both battery cables, the negative (–) cable last.

15. Operate the engine to normal operating temperatures; then, check the climate control operation and check for leaks.

Q45

2002–03 MODELS
1. Disconnect both battery cables, the negative (–) cable first.

✷✷ CAUTION

After disconnecting the battery, wait for a least 3 minutes for the SRS module to deplete its energy before working on the steering column or instrument panel.

2. Drain the cooling system into a clean container for reuse.

3. Remove the driver's side SRS module and the steering wheel by removing or disconnecting the following:

- Lower cover at the base of the steering wheel
- SRS module electrical connector
- Side covers at both sides of the steering wheel
- SRS module-to-steering wheel bolts using a T50 Torx® wrench
- SRS module from the steering wheel
- Place the front wheel in the straight-ahead position
- Horn connector and remove the steering wheel nut
- Steering wheel from the steering column

4. Remove or disconnect the following:

- Dash side lower finishers
- Steering column cover screws and the covers
- Combination switch-to-steering column screws, disconnect the electrical harness connectors and remove the combination switch
- Lower instrument panel screws/bolts and the panel. Disconnect the electrical harness connectors and the in-vehicle sensor at the driver's side
- Lower instrument reinforcement bolts and the reinforcement
- Steering column-to-instrument panel nuts and lower the steering column

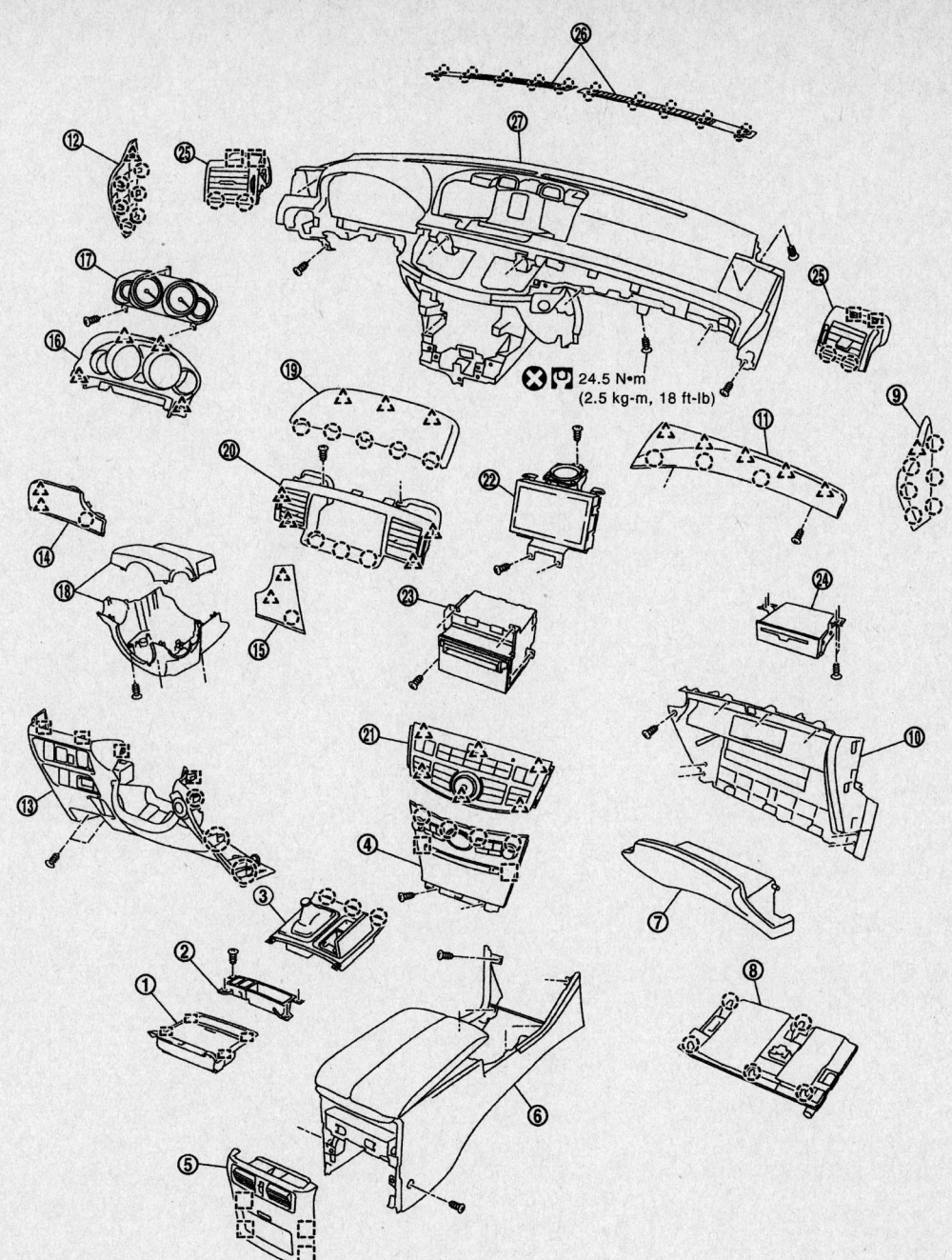

24.5 N•m
(2.5 kg-m, 18 ft-lb)

1.	Cup holder	2.	Switch finisher	3.	A/T console finisher
4.	Cluster lid C	5.	Console rear finisher	6.	Center console
7.	Glove box	8.	Instrument passenger lower cover	9.	Instrument side finisher (RH)
10.	Glove box cover	11.	Instrument finisher B	12.	Instrument side finisher (LH)
13.	Instrument driver lower panel	14.	Instrument finisher A	15.	Instrument finisher C
16.	Cluster lid A	17.	Combination meter	18.	Steering column cover
19.	Upper ventilator gurille	20.	Center ventilator grille	21.	Multifunction switch
22.	Front display unit	23.	Audio assembly	24.	NAVI C/U
25.	Side ventilator grille (RH/LH)	26.	Front defroster grille	27.	Instrument panel & pad
⌒	Pawl	△	Clip	⌐	Metal clip

09482_INFI_G0034

Details of the instrument panel—2006 M35 & M45

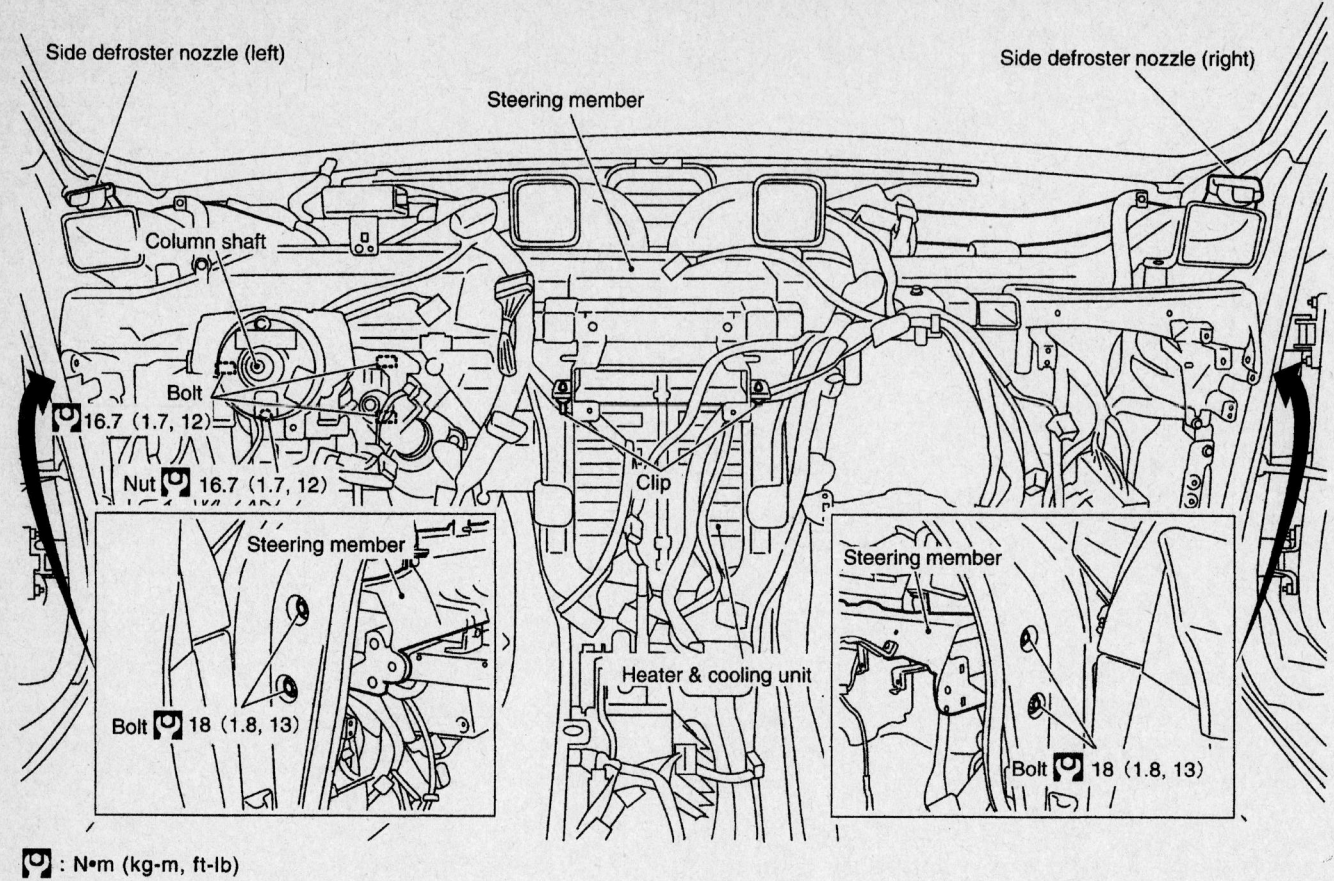

Side defroster nozzle (left)

Side defroster nozzle (right)

Steering member

Column shaft

Bolt

16.7 (1.7, 12)

Nut ⬜ 16.7 (1.7, 12)

Steering member

Clip

Steering member

Bolt ⬜ 18 (1.8, 13)

Heater & cooling unit

Bolt ⬜ 18 (1.8, 13)

⬜ : N•m (kg-m, ft-lb)

09482_INFI_G0033

Heater and cooling unit mounting—2006 M35 & M45

- Cluster lid "A"-to-instrument panel screws and remove the cluster lid
- Steering lock escutcheon screws and the escutcheon
- Cluster lid "D"
- Combination meter-to-instrument panel screws, disconnect the electrical connectors and remove the combination meter
- Glove box assembly-to-instrument panel screws, disconnect the lamp socket and remove the glove box assembly

5. Remove the passenger's side SRS module by removing or disconnecting the following:

- Lower instrument panel cover
- SRS module electrical connector
- SRS module-to-instrument panel bolts using a T50 Torx® wrench
- SRS module from the instrument panel

✻✻ CAUTION

Store the SRS module in a safe place with the front facing upward.

6. Remove or disconnect the following:

- Center ventilation grille using a suitable prytool
- Lock console
- Card pocket assembly screws and the assembly
- Console finisher clips and the console finisher
- Audio center, the cluster lid "C" and the air conditioning control unit-to-instrument panel screws, disconnect the electrical connectors and remove the panel
- Console box assembly-to-instrument panel screws, disconnect the electrical connectors and remove the console box assembly
- Connectors and remove the defroster grilles
- Connectors and remove the sun-load sensor
- Front pillar garnish
- Instrument panel-to-chassis bolts/nuts and the instrument panel
- Heater hoses from the heater core
- Rear duct from the heater unit

- Air conditioning housing-to-heater housing fasteners
- Heater unit from the vehicle

7. Disassemble and remove the heater core from the heater housing.

To install:

8. Assemble and install the heater core to the heater housing.

9. Install or connect the following:

- Heater unit to the vehicle
- Heater housing-to-chassis fasteners
- Air conditioning housing-to-heater housing fasteners
- Rear duct to the heater unit
- Heater hoses to the heater core
- Instrument panel and the instrument panel-to-chassis bolts/nuts
- Front pillar garnish
- Defroster grilles and connect the connectors
- Lower instrument panel center and the panel-to-instrument panel screws
- Lower instrument cover and the cover clip
- Center ventilation grille

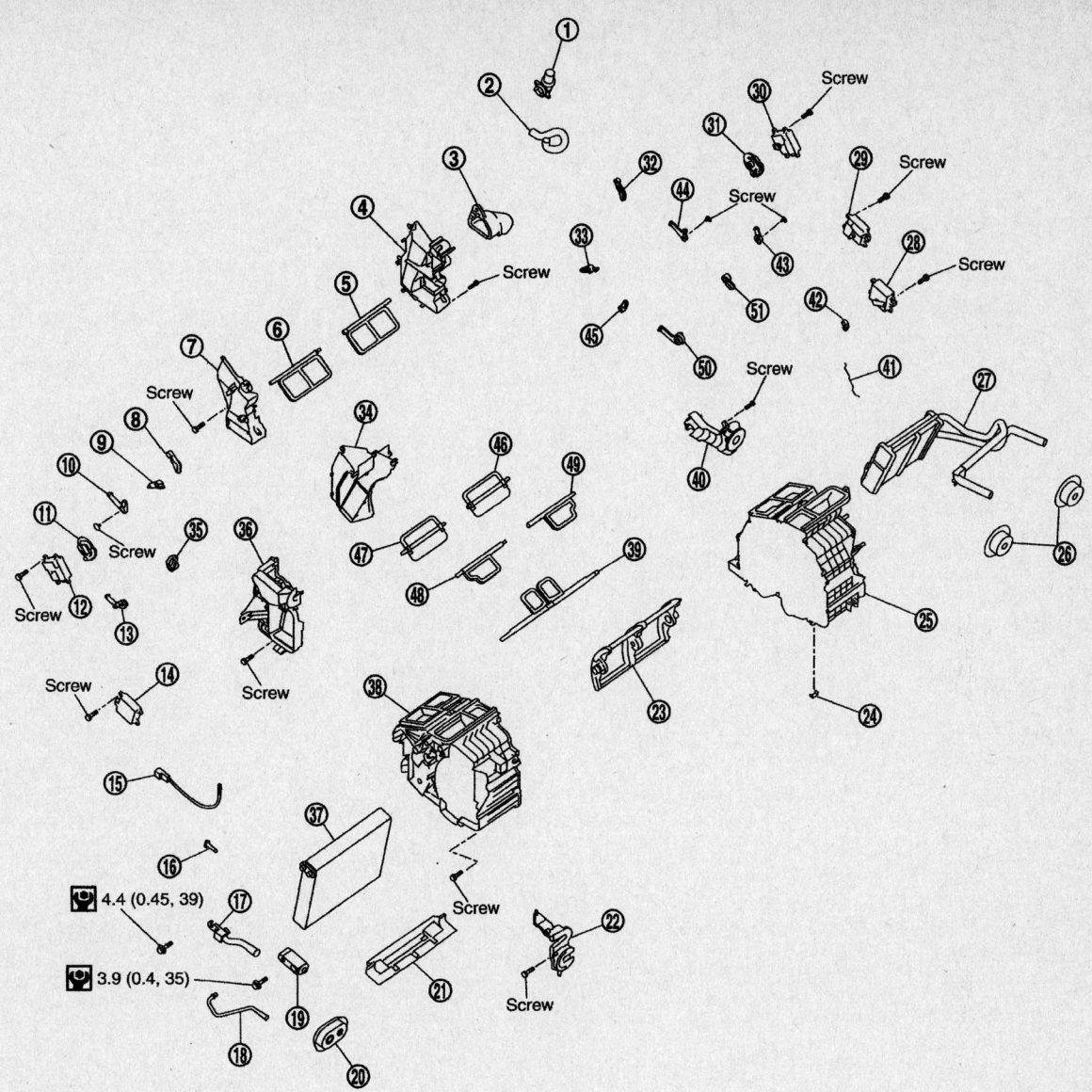

: N•m (kg-m, in-lb)

1.	Aspirator	2.	Aspirator hose	3. Front heater duct (left)

1. Aspirator
2. Aspirator hose
3. Front heater duct (left)
4. Foot duct (left)
5. Ventilator door (left)
6. Ventilator door (right)
7. Foot duct (right)
8. Main link sub (right)
9. Ventilator door lever (right)
10. Ventilator door link (right)
11. Main link (right)
12. Mode door motor (passenger side)
13. Max. cool door link (right)
14. Air mix door motor (passenger side)
15. Intake sensor
16. Intake sensor bracket
17. Low-pressure pipe 1
18. High-pressure pipe 2
19. Expansion valve
20. Cooler pipe grommet
21. Insulator
22. Evaporator cover adapter
23. Air mix door (Slide door)
24. Clip
25. Heater & cooling unit case (left)
26. Heater pipe grommet
27. Heater core
28. Upper ventilator door motor
29. Air mix door motor (driver side)
30. Mode door motor (driver side)
31. Main link (left)
32. Main link sub (left)
33. Ventilator door lever (left)
34. Center case
35. Max. cool door lever (right)
36. Evaporator cover
37. Evaporator
38. Heater & cooling unit case (right)
39. Upper ventilator door
40. Heater pipe cover
41. Upper ventilator door rod
42. Upper ventilator door lever
43. Defroster door link
44. Ventilator door link (left)
45. Max. cool door lever (left)
46. Max. cool door (left)
47. Max. cool door (right)
48. Defroster door (right)
49. Defroster door (left)
50. Max. cool door link (left)
51. Defroster door lever

Details of heater and cooling unit—2006 M35 & M45

09482_INFI_G0035

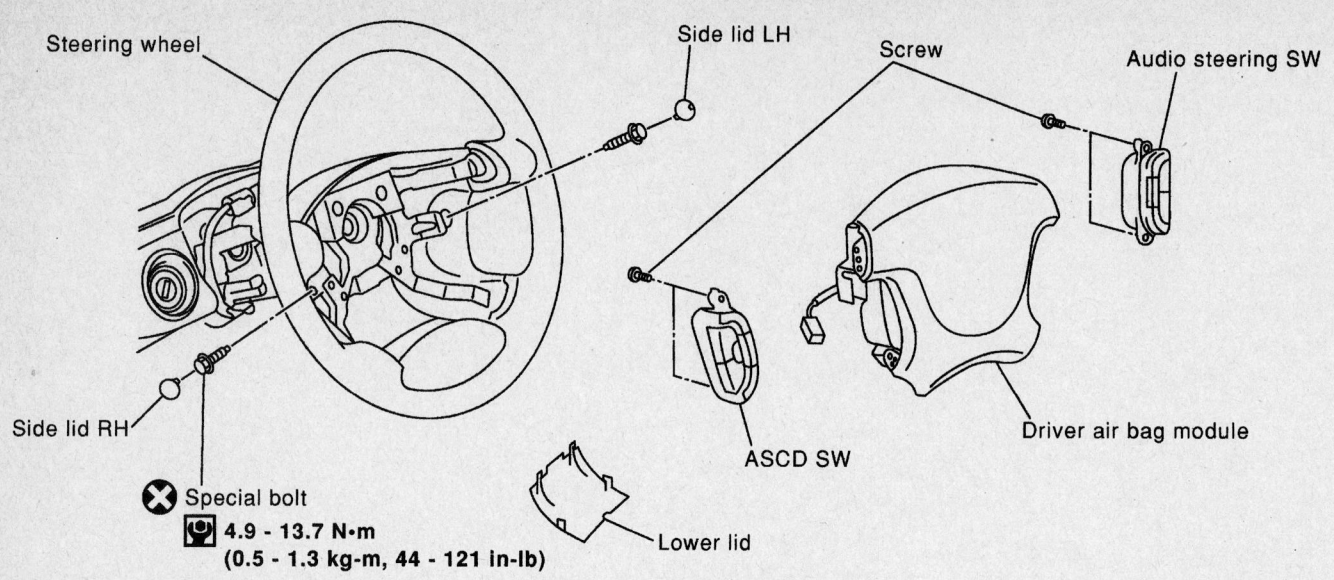

Exploded view of the air bag module, the steering wheel and related components—2002–03 Q45

- Lock console
- Card pocket assembly and the assembly screws
- Console finisher and engage the clips
- Audio center, the cluster lid "C" and the air conditioning control unit panel, connect the electrical connectors and install the panel-to-instrument panel screws
- Console box assembly, connect the electrical connectors and install the console box assembly-to-instrument panel screws
- Defroster grilles and connect the connector
- Sunload sensor and connectors

10. Install the passenger's side SRS module by installing or connecting the following:

- SRS module to the instrument panel
- SRS module-to-instrument panel bolts and torque the bolts to 11–18 ft. lbs. (15–25 Nm) using a T50 Torx® wrench
- SRS module electrical connector
- Lower instrument panel cover
- Glove box assembly, connect the lamp socket and install the glove box assembly-to-instrument panel screws

11. Install or connect the following:

- Combination meter, connect the electrical connectors and install the combination meter-to-instrument panel screws

- Cluster lid "D"
- Steering lock escutcheon and the escutcheon screws
- Cluster lid "A" and the cluster lid-to-instrument panel screws
- Steering column and torque the steering column-to-instrument panel nuts to 11–14 ft. lbs. (15–19 Nm)
- Lower instrument reinforcement and the reinforcement bolts.
- Electrical harness connectors and the in-vehicle sensor; then, install the lower instrument panel and the panel screws/bolts at the driver's side
- Combination switch and the combination switch-to-steering column screws and connect the electrical harness connectors
- Steering column covers and the cover screws
- Dash side lower finishers

12. Install the driver's side SRS module and the steering wheel by installing or connecting the following:

- Steering wheel to the steering column
- Steering wheel nut and torque to 22–29 ft. lbs. (29–39 Nm). Connect the horn connector
- SRS module to the steering wheel
- Torque the SRS module-to-steering wheel bolts to 11–18 ft. lbs. (15–25 Nm) using a T50 Torx® wrench

- Side covers at both sides of the steering wheel
- SRS module electrical connector
- Lower cover at the base of the steering wheel

13. Refill the cooling system.

14. Connect both battery cables, the negative (–) cable last.

15. Operate the engine to normal operating temperatures; then, check the climate control operation and check for leaks.

Q45

2004–06 MODELS

1. Disconnect both battery cables, the negative (–) cable first.

> ✳✳ **CAUTION**
>
> **After disconnecting the battery, wait for a least 3 minutes for the SRS module to deplete its energy before working on the steering column or instrument panel.**

2. Drain the cooling system into a clean container for reuse.

3. Remove the engine appearance cover and the air cleaner cover.

4. Remove the engine compartment rear cross brace.

5. Remove the heater hoses from the heater core.

6. Disconnect the refrigerant lines from the evaporator.

7. Remove the driver's side SRS mod-

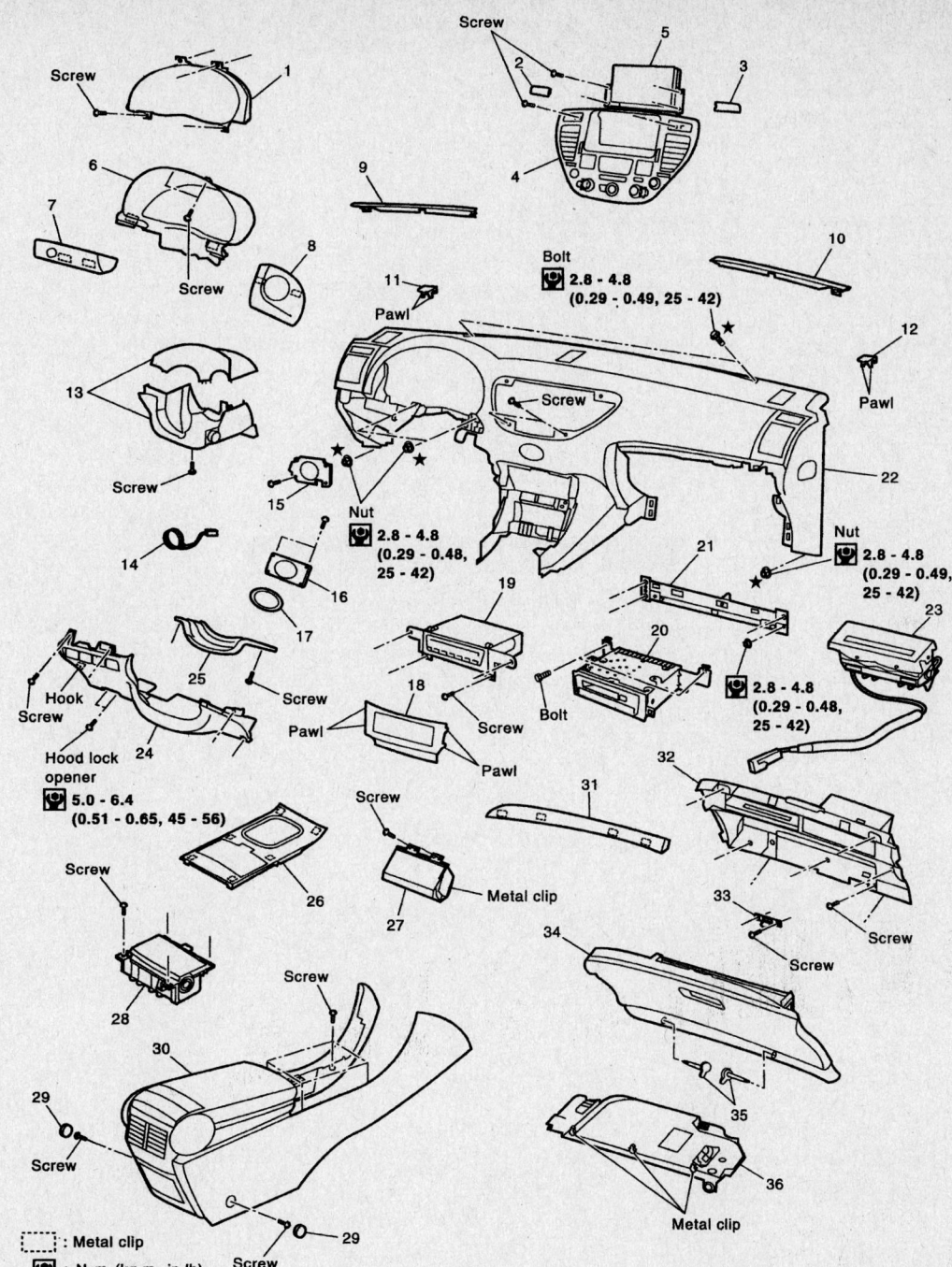

Screw

Screw

Screw

Screw

Screw

1

6

7

8

5

2

3

4

9

10

12

11

Pawl

Pawl

Bolt
🔧 **2.8 - 4.8**
(0.29 - 0.49, 25 - 42)

13

Screw

Screw

15

14

Nut
🔧 **2.8 - 4.8**
(0.29 - 0.48, 25 - 42)

16

17

22

Nut
🔧 **2.8 - 4.8**
(0.29 - 0.49, 25 - 42)

21

19

20

23

25

Screw

Hook
Screw

Hood lock
opener

24

🔧 **5.0 - 6.4**
(0.51 - 0.65, 45 - 56)

18

Pawl

Screw

Bolt

Pawl

🔧 **2.8 - 4.8**
(0.29 - 0.48, 25 - 42)

Screw

26

27

Metal clip

31

32

33

Screw

34

Screw

28

30

29

Screw

29

Screw

35

36

Metal clip

▭ : Metal clip

🔧 : N•m (kg-m, in-lb)

1.	Combination meter	2.	Ventilation mask (LH)	3.	Ventilation mask (RH)
4.	Cluster lid C	5.	Display unit	6.	Cluster lid A
7.	Cluster lid D	8.	Steering lock escutcheon	9.	Front defroster grille (LH)
10.	Front defroster grille (RH)	11.	Instrument panel mask (LH)	12.	Instrument panel mask (RH)
13.	Steering column cover	14.	Ignition key lamp assembly	15.	Instrument panel bracket
16.	Clock	17.	Cluster lid C lower	18.	Cluster lid center lower
19.	CD auto changer	20.	Audio unit	21.	Instrument panel bracket
22.	Instrument panel and pad assembly	23.	Front passenger air bag module	24.	Instrument lower driver panel
25.	Knee protector lower	26.	A/T console finisher	27.	Ashtray
28.	Cup holder	29.	Bolt cap	30.	Console box assembly
31.	Instrument finisher	32.	Glove box cover	33.	Glove box striker
34.	Glove box assembly	35.	Glove box pin	36.	Instrument lower cover

09482_INFI_G0156

Exploded view of the instrument panel—2002–03 Q45

ule and the steering wheel by removing or disconnecting the following:

- Lower cover at the base of the steering wheel
- SRS module electrical connector
- Side covers at both sides of the steering wheel
- SRS module-to-steering wheel bolts using a T30 Torx® wrench
- SRS module from the steering wheel
- Place the front wheel in the straight-ahead position
- Horn connector and remove the steering wheel nut
- Steering wheel from the steering column

8. Remove or disconnect the following:
- Steering column cover screws and the covers
- Steering lock escutcheon screws and the escutcheon
- Cluster lid "D"
- Cluster lid "A"-to-instrument panel screws and remove the cluster lid
- Combination meter-to-instrument panel screws, disconnect the electrical connectors and remove the combination meter
- Cluster lid "C"
- Visual display unit

- Clock
- Ashtray and then the center lower cluster lid
- Automatic transmission finish panel
- Cup holder
- Center console box
- CD changer and audio unit
- Instrument panel lower cover
- Glove box
- Glove box upper finish panel
- Glove box opening cover
- Instrument panel reinforcement
- Passenger SRS module electrical connector
- SRS module-to-instrument panel bolts using a Torx® wrench
- SRS module from the instrument panel

※ CAUTION

Store the SRS module in a safe place with the front facing upward.

- Hood release lever
- Driver side lower instrument panel cover
- Knee protector
- Instrument panel mask
- Front pillar garnishes

- Ignition key lamp and instrument panel bracket
- Remove 2 bolts and lower the steering column
- Instrument panel-to-chassis bolts/nuts and the instrument panel
- Blower motor
- Vehicle harness clips at left of A/C-heater unit
- Instrument panel stays
- Defroster and ventilation ducts
- Air conditioning housing-to-heater housing fasteners
- A/C-heater unit from the vehicle

9. Disassemble and remove the heater core from the A/C-heater housing.

To install:

10. Installation is the reverse of the removal procedure noting the following:

  a. Tighten the driver air bag module screws to 44–121 inch lbs. (4–14 Nm).

  b. Tighten the passenger air bag module bolts to 11–18 ft. lbs. (15–25 Nm).

  c. Tighten the steering column bolts to 11–13 ft. lbs. (15–18 Nm).

  d. Tighten the steering wheel nut to 23–28 ft. lbs. (30–39 Nm).

11. Refill the cooling system.

12. Connect both battery cables, the negative (–) cable last.

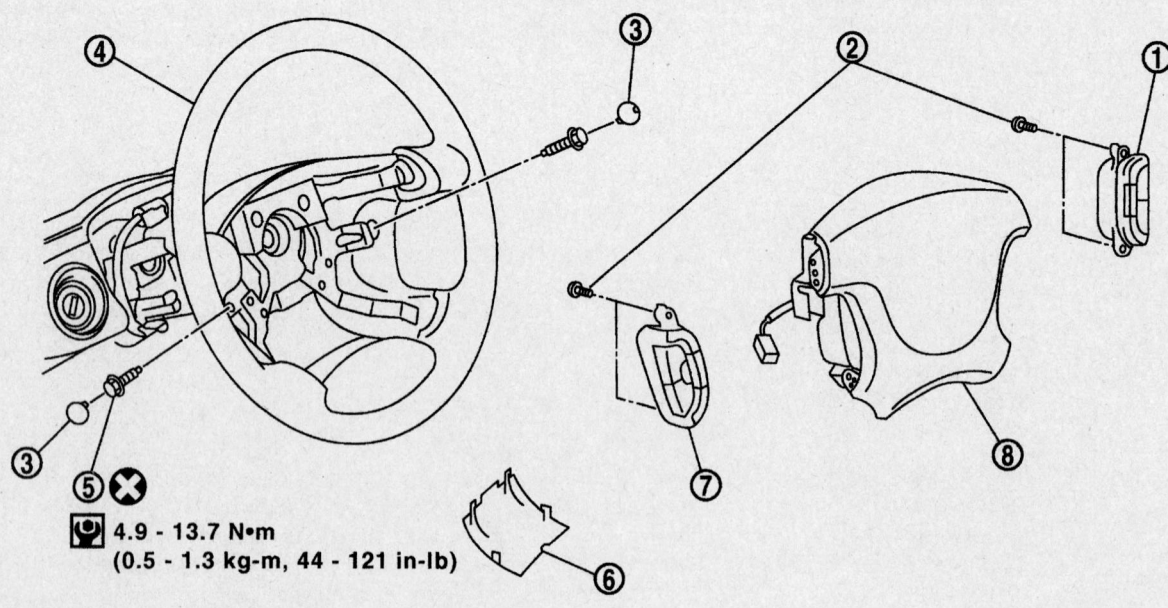

4.9 - 13.7 N•m
(0.5 - 1.3 kg-m, 44 - 121 in-lb)

: Always replace after every disassembly.

1.	Audio steering switch	2.	Screw	3.	Side lid
4.	Steering wheel	5.	TORX bolt (T30)	6.	Lower lid
7.	ASCD switch	8.	Driver air bag module		

Exploded view of the air bag module, the steering wheel and related components—2004–06 Q45

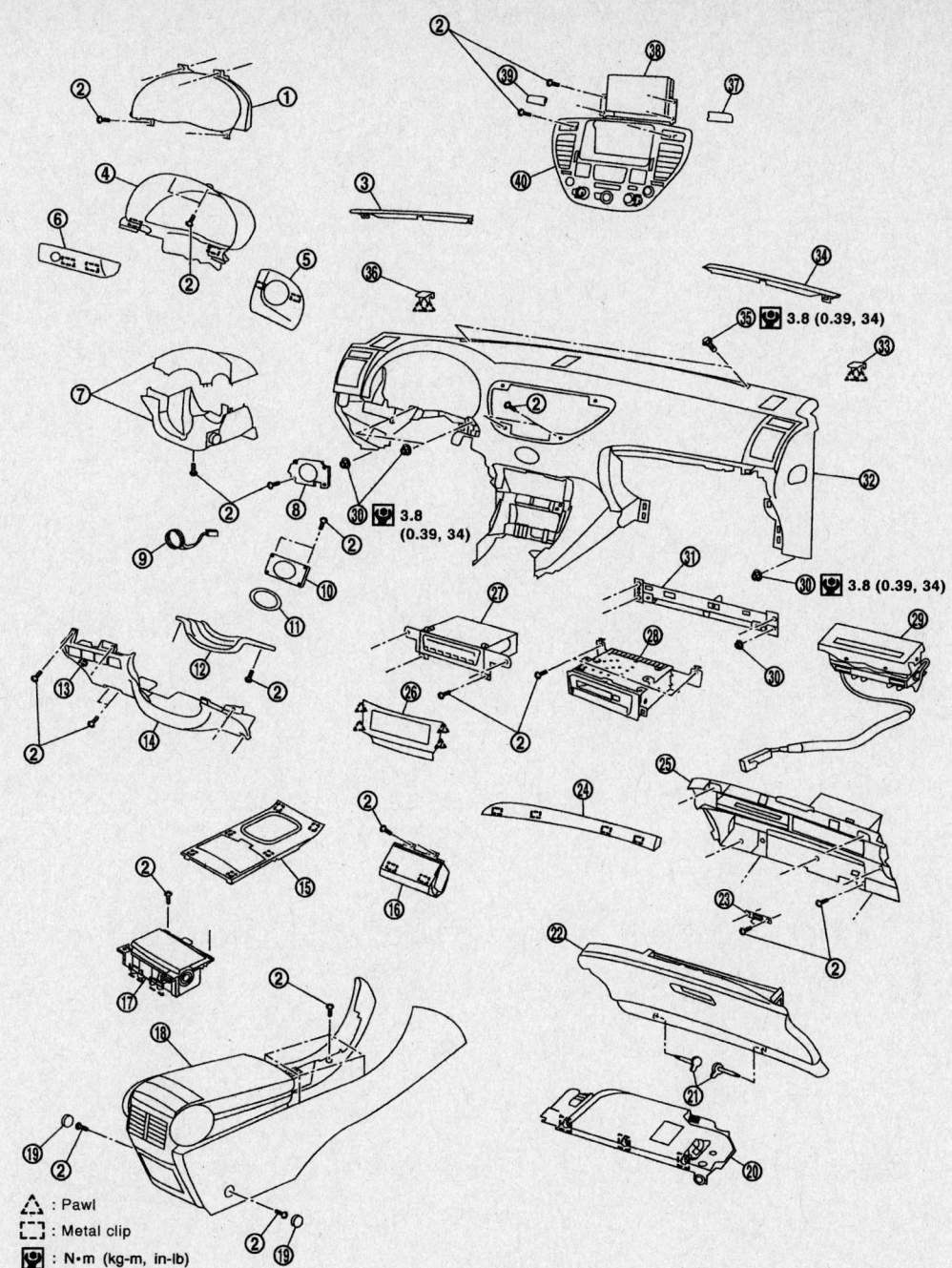

Key:
△ : Pawl
⊏⊐ : Metal clip
⚙ : N•m (kg-m, in-lb)

1.	Combination meter	2.	Screw	3.	Front defroster grille (LH)
4.	Cluster lid A	5.	Steering lock escutcheon	6.	Cluster lid D
7.	Steering column cover	8.	Instrument panel bracket	9.	Ignition key lamp assembly
10.	Clock	11.	Cluster lid C lower	12.	Knee protector lower
13.	Hook	14.	Instrument lower driver panel	15.	A/T console finisher
16.	Ashtray	17.	Cup holder	18.	Console box assembly
19.	Screw cap	20.	Instrument lower cover	21.	Glove box pin
22.	Glove box assembly	23.	Glove box striker	24.	Instrument finisher
25.	Glove box cover	26.	Cluster lid center lower	27.	CD auto changer
28.	Audio unit	29.	Front passenger air bag module	30.	Nut
31.	Instrument panel bracket	32.	Instrument panel and pad assembly	33.	Instrument panel mask (RH)
34.	Front defroster grille (RH)	35.	Bolt	36.	Instrument panel mask (LH)
37.	Ventilation mask (RH)	38.	Display unit	39.	Ventilation mask (LH)
40.	Cluster lid C				

09482_INFI_G0158

Exploded view of the instrument panel—2004–06 Q45

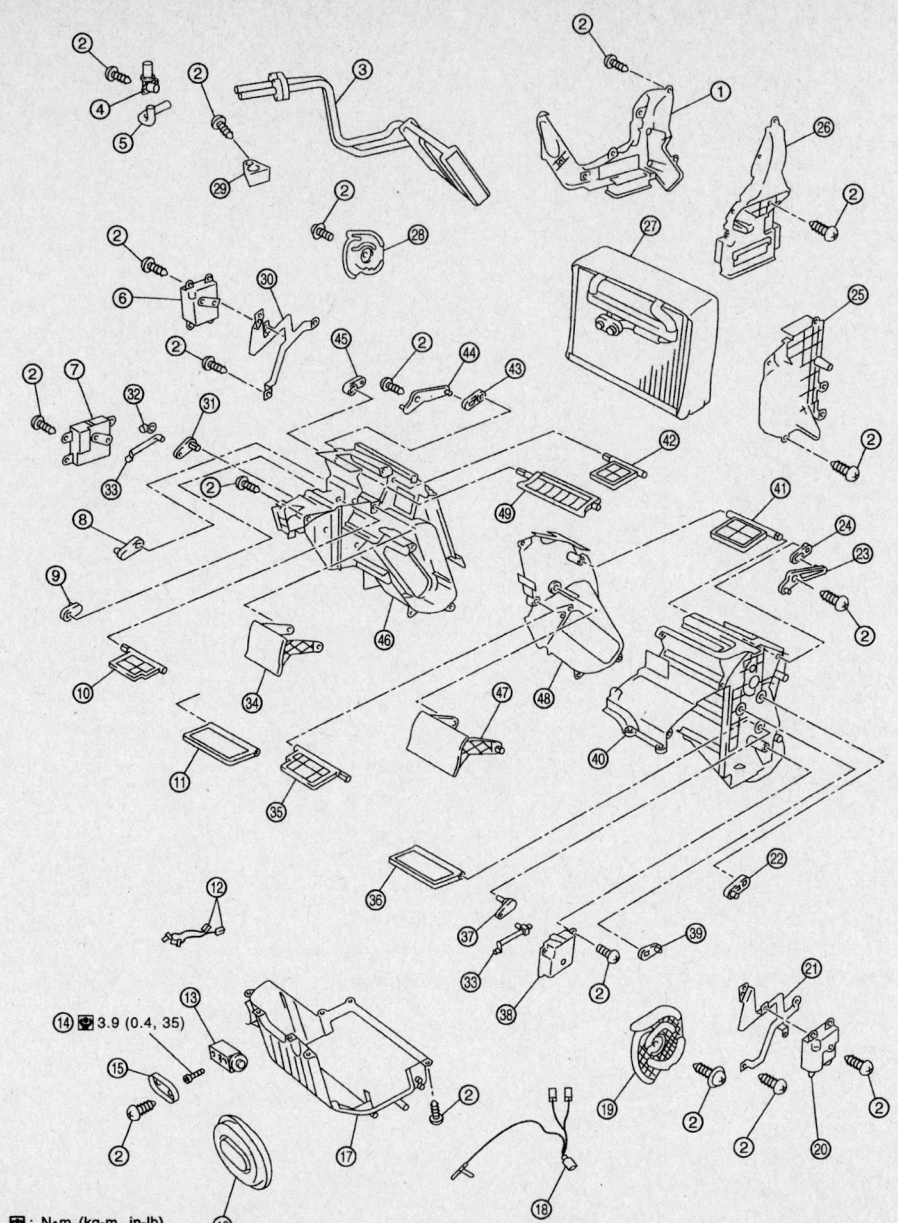

☒ : N•m (kg-m, in-lb)

1.	Foot duct (right)	2.	Screw	3.	Heater core
4.	Aspirator	5.	Aspirator duct	6.	Mode door motor (passenger side)
7.	Air mix door motor (passenger side)	8.	Ventilator door lever (right)	9.	Bi-level door lever (right)
10.	Bi-level door (right)	11.	Air mix door (right)	12.	Sub harness
13.	Expansion valve	14.	Bolt	15.	Expansion valve cover
16.	Cooler grommet	17.	Heater & cooling unit case (lower)	18.	Intake sensor
19.	Side link (left)	20.	Mode door motor (driver side)	21.	Mode door motor bracket (left)
22.	Ventilator door lever (left)	23.	Foot door link (left)	24.	Foot door lever (left)
25.	Evaporator cover	26.	Foot duct (left)	27.	Evaporator
28.	Side link (right)	29.	Heater pipe bracket	30.	Mode door motor bracket (right)
31.	Air mix door lever (right)	32.	Rod holder	33.	Rod
34.	Ventilator door (right)	35.	Bi-level door (left)	36.	Air mix door (left)
37.	Air mix door lever (left)	38.	Air mix door motor (driver side)	39.	Bi-level door lever (left)
40.	Heater & cooling unit case (driver side)	41.	Foot door (left)	42.	Foot door (right)
43.	Foot door lever (right)	44.	Foot door link (right)	45.	Defroster door lever (right)
46.	Heater & cooling unit case (passenger side)	47.	Ventilator door (left)	48.	Heater & cooling unit case (center)
49.	Defroster door				

09482_INFI_G0039

Exploded view of the heating/cooling unit—2004–06 Q45

13. Operate the engine to normal operating temperatures; then, check the climate control operation and check for leaks.

Cylinder Head

REMOVAL & INSTALLATION

2.0L Engine

1. Before servicing the vehicle, refer to the precautions in the beginning of this section.
2. Relieve the fuel system pressure.
3. Drain the cooling system.
4. Remove or disconnect the following:
   - Negative battery cable
   - Right front wheel
   - Engine side cover
   - Radiator
   - Air duct to intake manifold
   - ASCD actuator
   - Vacuum and fuel hoses
   - Electrical connectors and wiring
   - Spark plugs
   - Rocker cover bolts in sequence
   - Rocker cover
   - Steering pump
   - Intake manifold supports
   - Water pipe assembly and set the No. 1 piston at Top Dead Center (TDC) of its compression stroke

➡**Rotate the crankshaft until the mating mark on the camshaft sprocket is properly set.**

   - Timing chain tensioner
   - Distributor. Do not turn the rotor with the distributor removed.

- Camshaft sprockets and brackets
- Starter
- Heater hoses
- Cylinder head bolts in the proper sequence

To install:

5. Apply liquid gasket to the top of the chain cover where it meets the cylinder block before installing the head gasket.
6. Install the gasket and cylinder head on the block.

➡**Cylinder head bolts may be reused providing the dimension from the bottom of the head to the end of the bolt does not exceed 6.228 in. (158.2mm). If the dimension exceeds the specification, replace the cylinder head bolts.**

7. Tighten the cylinder head bolts as follows:
   a. Tighten all bolts to 29 ft. lbs. (39 Nm) using the proper sequence.
   b. Tighten all bolts to 58 ft. lbs. (78 Nm) using the proper sequence.
   c. Loosen all bolts completely.
   d. Tighten all bolts to 25–33 ft. lbs. (34–44 Nm) using the proper sequence.
   e. Tighten all bolts 90–95°.
   f. Tighten all bolts an additional 90–95°.
8. Install or connect the following:
   - Heater hoses
   - Camshafts and brackets. Ensure that the camshaft keys are at 12 o'clock.
9. The procedure for tightening camshaft bolts must be followed exactly to prevent camshaft damage. Tighten the bolts as follows:

   a. Tighten the right camshaft bolts Nos. 9 and 10 to 18 inch lbs. (2 Nm). Tighten bolts 1 through 8 to the same amount.
   b. Tighten the left camshaft bolts No. 11 and No. 12 to 18 inch lbs. (2 Nm). Tighten bolts 1 through 10 to the same amount.
   c. Tighten all bolts in sequence to 54 inch lbs. (6 Nm).
   d. Tighten all bolts in sequence again. Tighten type A, B, and D bolts to 78–102 inch lbs. (9–12 Nm) and type C bolts to 13–19 ft. lbs. (18–25 Nm).
10. Line up the mating marks on the timing chain and camshaft sprockets and install the sprockets. Tighten the sprocket bolts to 101–116 ft. lbs. (137–157 Nm).
11. Install or connect the following:
    - Timing chain guide, distributor, chain tensioner, oil filter bracket and power steering oil pump bracket
    - Intake manifold supports
    - Rocker cover. Torque the bolts, in sequence, to 89 inch lbs. (10 Nm).
    - Spark plugs, power steering pump, alternator, water pump pulley and drive belts, air duct to the intake manifold and the radiator
    - Vacuum and fuel hoses and reconnect all electrical connections
    - Engine undercover
    - Right front wheel
    - Negative battery cable
12. Fill and bleed the cooling system.
13. Start the vehicle, check for leaks and repair if necessary.

3.5L Engine

I35 MODELS

➡**For this procedure, you must remove the engine from the vehicle in order to remove the cylinder head.**

1. Before servicing the vehicle, refer to the precautions in the beginning of this section.
2. Relieve the fuel system pressure.
3. Drain the engine oil.
4. Drain the cooling system.

➡**Before detaching any hoses or connectors, note the locations for reassembly.**

5. Remove or disconnect the following:
   - Negative battery cable
   - Engine assembly
   - Exhaust manifold
6. Place the engine on a suitable work stand.
   - Oil pan

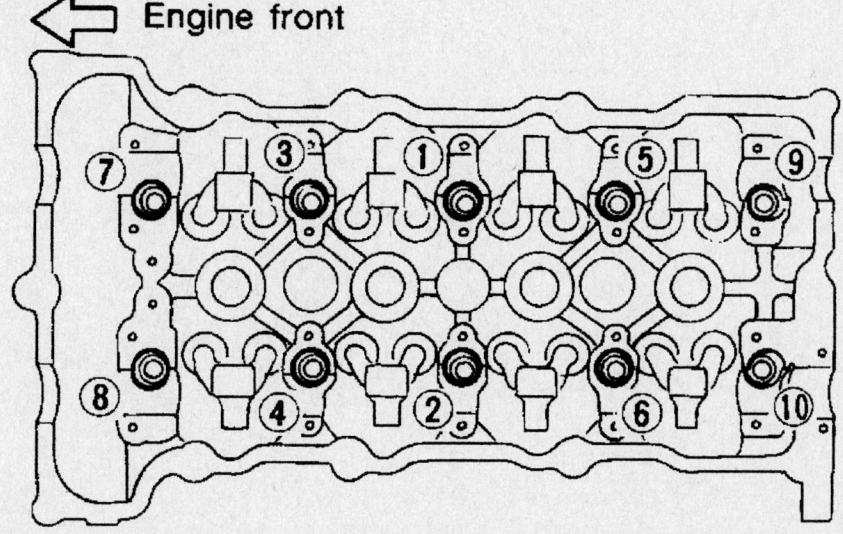

⬅ Engine front

09482_INFI_G0159

Cylinder head torque sequence—2.0L engine

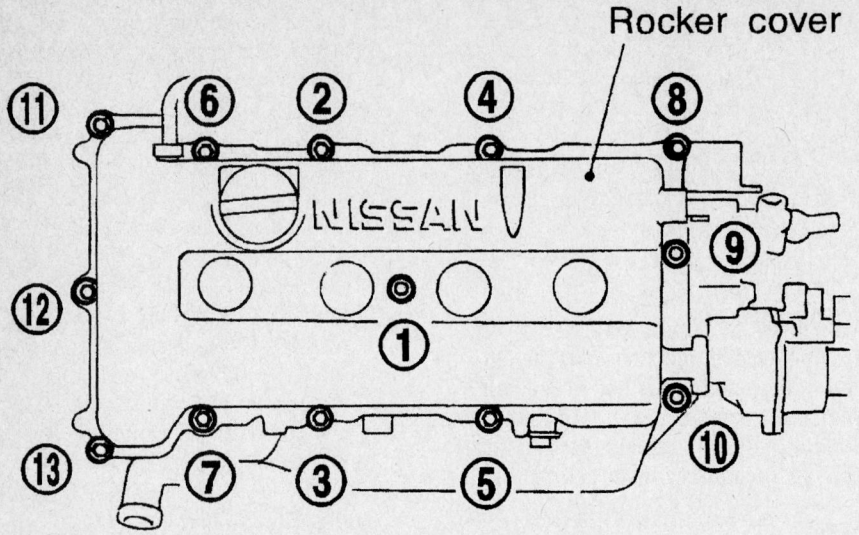

Tighten the rocker cover bolts according to the sequence shown—2.0L engine

- Timing chain
- Intake manifold
- Water outlet
- Rear timing chain case bolts, in the reverse of the sequence shown
- Rear timing chain case
- O-rings from the cylinder head and block
- Intake valve timing control solenoid valves

➡ **For installation purposes, matchmark the camshaft brackets before removing them.**

- Intake and exhaust camshafts and brackets. Loosen the bracket bolts in several steps, in the sequence shown.
- Right and left side cam chain tensioner from the cylinder head
- Cylinder head bolts. Loosen in several steps, in the sequence shown.

➡ **A warped or cracked cylinder head could result from removing the bolts in incorrect order.**

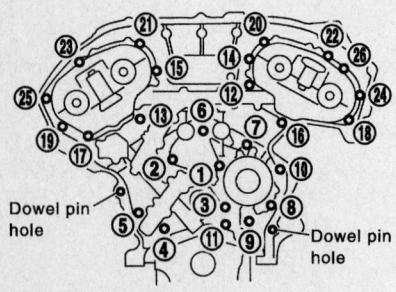

Loosen the rear timing chain case bolts in the reverse of this sequence—I35

- Cylinder heads from the vehicle
- Discard the head gaskets

7. Remove all traces of liquid gasket from the timing chain case and from the water pump covers.

8. Remove all traces of liquid gasket from the engine block.

9. Inspect the timing chain for excessive wear or damage and replace as necessary.

To install:

10. Turn the crankshaft until the No. 1 piston is a Top Dead Center (TDC) on compression stroke. The crankshaft key should face toward the right bank.

11. Using new head gaskets, install the cylinder heads.

➡ **If possible, replacement of the head bolts is suggested.**

12. If replacement of the head bolts is not possible, perform the following bolt measurement:

   a. Measure the diameter of the head bolt 0.43 in. (11mm) from the bottom of the bolt.

   b. Measure the diameter of the head bolt 1.89 in. (48mm) from the bottom of the bolt.

   c. Whenever the size difference between the 2 measurements exceeds 0.0043 in. (0.11mm) the head bolts must be replaced.

13. Install the cylinder head bolts and

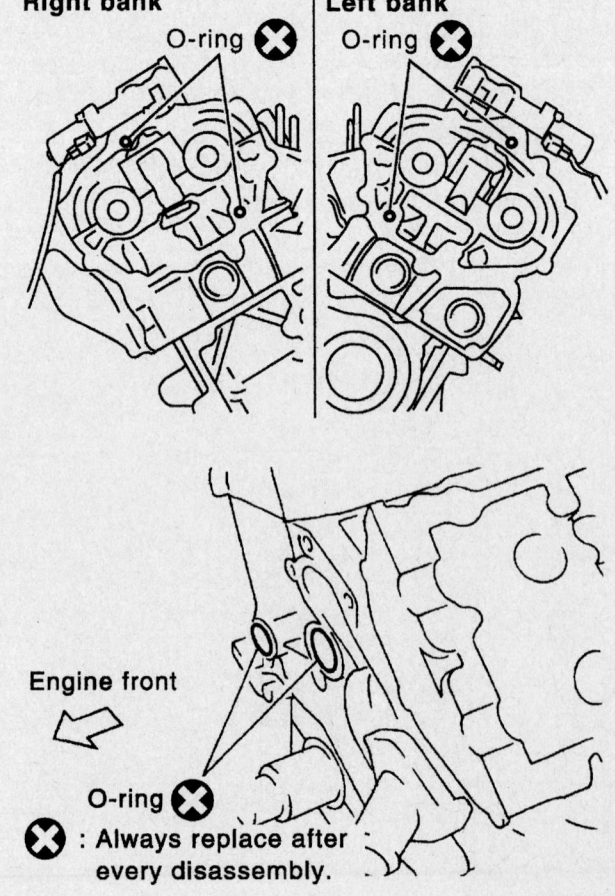

Remove the O-rings from the cylinder head and block— 3.5L engine

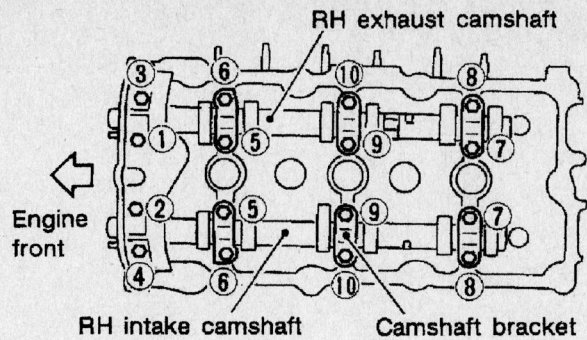

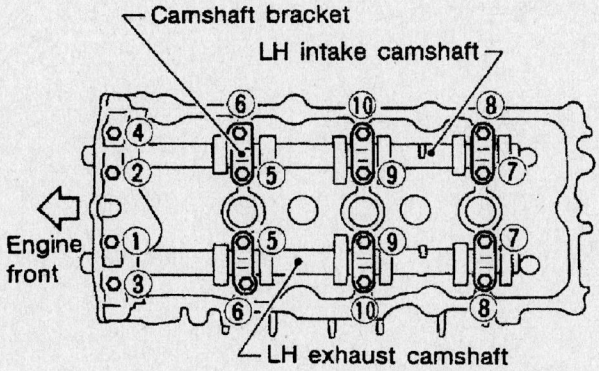

Right and left camshaft bracket bolt loosening sequence—3.5L engine

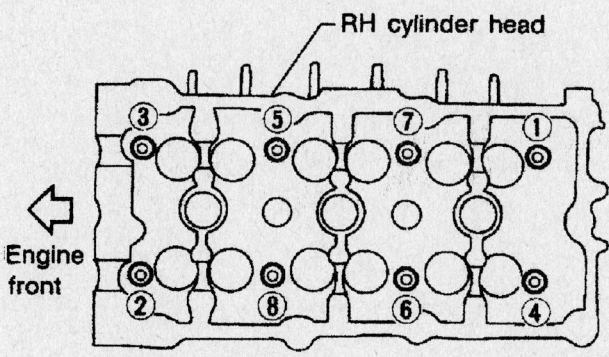

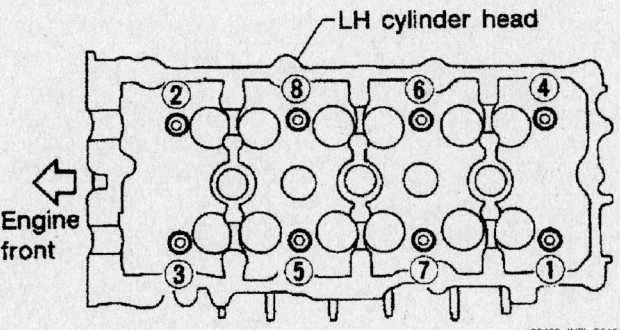

Cylinder head bolt loosening sequence—3.5L engine

torque in sequence (Refer to Torque Specifi-cations Chart in beginning of chapter for torque values and procedures)

14. Install or connect the following:
  • Camshafts and related components
  • Intake valve timing control solenoid valves
  • New O-rings to the front of the engine block and cylinder head

15. Apply sealant to the hatched portion of the of the rear timing chain case.

16. Align the rear timing chain case with the dowel pins and install onto the cylinder heads and engine block.

17. Torque the rear timing chain case mounting bolts in sequence to 105–121 inch lbs. (11.8–13.7 Nm).

18. Install or connect the following:
  • Water outlet
  • Intake manifold
  • Timing chain
  • Oil pan
  • Exhaust manifold
  • Engine assembly into the vehicle
  • Negative battery cable

19. Fill the cooling system.
20. Fill the engine with clean oil.
21. Start the vehicle, check for leaks and repair if necessary.

G35 MODELS

➡For this procedure, you must remove the engine from the vehicle in order to remove the cylinder head.

1. Before servicing the vehicle, refer to the precautions in the beginning of this sec-tion.

2. Properly relieve the fuel system pressure.

3. Drain the cooling system.
4. Drain the engine oil.
5. Remove or disconnect the following:
  • Engine and place on a suitable stand
  • Intake manifold collector
  • Fuel rail and injector assembly
  • Intake and exhaust manifolds
  • Ignition coil
  • Rocker arm (valve) cover
  • Water inlet and thermostat housing
  • Water outlet and hoses
  • Upper and lower oil pans and strainer
  • Front timing chain case, timing chain and rear timing chain case
  • Camshaft
  • Cylinder head bolts. Loosen in sev-eral steps, in the sequence shown.

➡A warped or cracked cylinder head could result from removing the bolts in incorrect order.

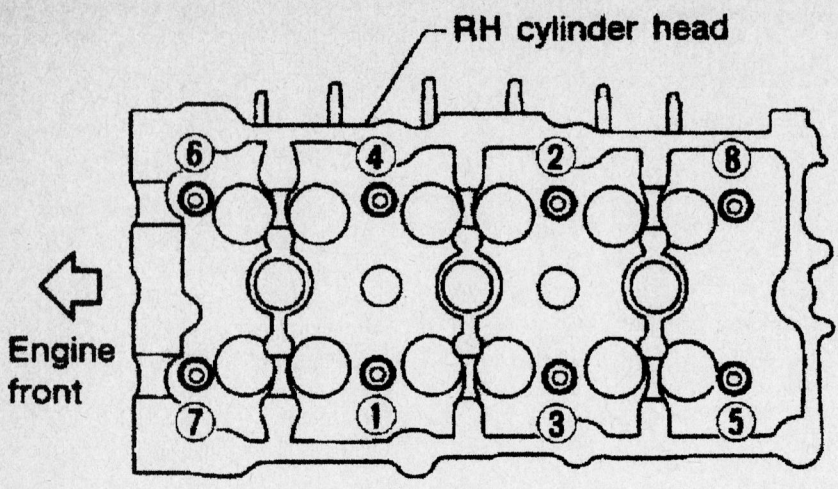

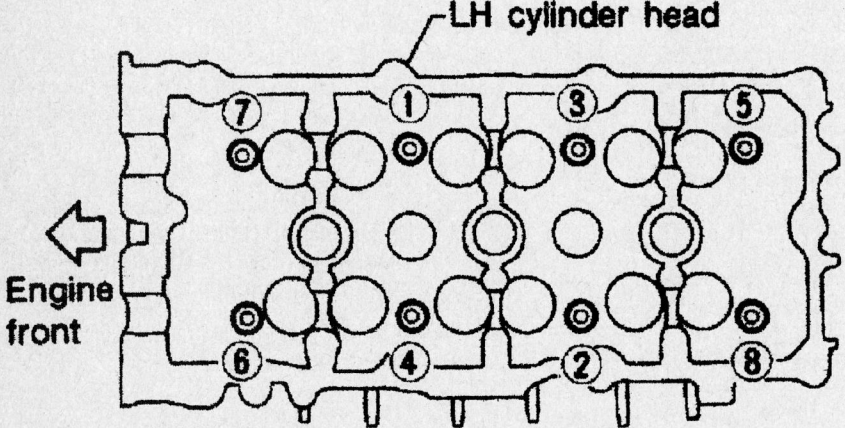

09482_INFI_G0165

Cylinder head bolt torque sequence— I35 and G35

• Cylinder heads from the vehicle
• Discard the head gaskets

6. Remove all traces of liquid gasket from the timing chain case and from the water pump covers.

7. Remove all traces of liquid gasket from the engine block.

8. Inspect the timing chain for excessive wear or damage and replace as necessary.

To install:

9. Turn the crankshaft until the No. 1 piston is a Top Dead Center (TDC) on compression stroke. The crankshaft key should face toward the right bank.

10. Using new head gaskets, install the cylinder heads.

➡**If possible, replacement of the head bolts is suggested.**

11. If replacement of the head bolts is not possible, perform the following bolt measurement:

a. Measure the diameter of the head bolt 0.43 in. (11mm) from the bottom of the bolt.

b. Measure the diameter of the head bolt 1.89 in. (48mm) from the bottom of the bolt.

c. Whenever the size difference between the 2 measurements exceeds 0.0043 in. (0.11mm) the head bolts must be replaced.

12. Install the cylinder head bolts and torque in sequence (Refer to Torque Specifications Chart in beginning of chapter for torque values and procedures)

13. After installing the cylinder head, measure the distance between the front end

faces of the cylinder block and cylinder head. If the specification does not fall within 0.555–0.587 in. (14.1–14.9mm), you must reinstall the cylinder head.

14. Install or connect the following:

• Camshaft
• Front timing chain case, timing chain and rear timing chain case
• Oil strainer and upper and lower oil pans
• Water outlet and hoses
• Thermostat housing and water inlet
• Rocker arm (valve) cover
• Ignition coil
• Intake and exhaust manifolds
• Fuel rail and injector assembly
• Intake manifold collector
• Engine into the vehicle
• Negative battery cable

15. Fill the cooling system.

16. Fill the engine with clean oil.

17. Start the vehicle, check for leaks and repair if necessary.

4.5L Engine

1. Before servicing the vehicle, refer to the precautions in the beginning of this section.

2. Properly relieve the fuel system pressure.

3. Drain the cooling system.

4. Drain the engine oil.

5. Remove or disconnect the following:

• Both battery cables
• Engine assembly from the vehicle and place it on a work stand
• Drive belt and idler pulley
• Thermostat housing
• Oil pan and strainer
• Upper and lower intake manifolds
• Fuel rail and injector assembly
• Ignition coils
• Spark plugs
• Valve cover
• Timing chain
• Camshafts

6. Loosen the cylinder head bolts gradually in reverse of the tightening sequence, then remove the cylinder head.

To install:

7. Be sure all mating surfaces are clean before installation.

8. Check the cylinder head surface for warpage using a feeler gauge and a suitable straightedge. If the cylinder head is warped more than 0.004 in. (0.1mm), it must be resurfaced or replaced. The total amount machined from the head or head and block combined, cannot total more than 0.008 in. (0.2mm).

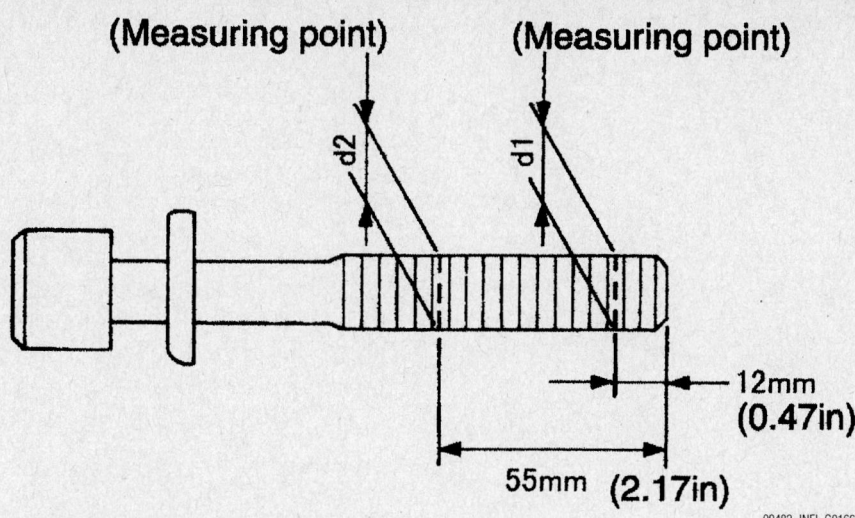

(Measuring point)   (Measuring point)

d2   d1

12mm
(0.47in)

55mm (2.17in)

09482_INFI_G0166

Checking the cylinder bolt threads for distortion—4.5L engine

9. If the cylinder head bolts are to be reused, measure the outside diameter of the bolt threads for distortion. If the dimension between d1 and d2 exceeds 0.0071 inch (0.18 mm), replace the bolts.

10. Place new gaskets on the cylinder block.

11. Apply clean engine oil to the cylinder head bolt threads.

12. Install the cylinder head and tighten the bolts in sequence (Refer to Torque Specifications Chart in beginning of chapter for torque values and procedures)

13. Install or connect the following:
- Camshafts

- Timing chain
- Valve cover
- Spark plugs
- Ignition coils
- Fuel rail and injector assembly
- Upper and lower intake manifolds
- Oil pan and strainer
- Thermostat housing
- Drive belt and idler pulley
- Engine assembly
- Both battery cables

14. Fill and bleed the cooling system.

15. Fill the engine with clean oil.

16. Start the vehicle, check for leaks and repair if necessary.

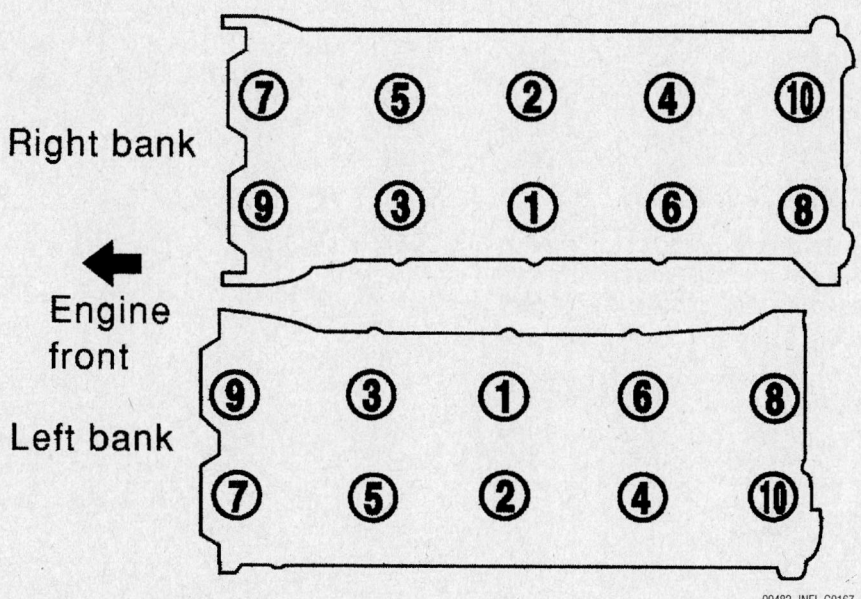

Right bank

Engine front

Left bank

09482_INFI_G0167

Cylinder head bolt tightening sequence—4.5L engine

Rocker Arms/Shafts

REMOVAL & INSTALLATION

2.0L Engine

1. Before servicing the vehicle, refer to the precautions in the beginning of this section.

2. Relieve the fuel system pressure.

3. Drain the cooling system.

4. Remove or disconnect the following:

- Negative battery cable
- Right front wheel
- Engine side cover
- Radiator
- Air duct to the intake manifold
- Drive belts, water pump pulley, alternator and power steering pump
- Vacuum hoses, fuel hoses and wiring harness connectors
- Spark plugs, the Air Intake Valve (AIV) and resonator
- Rocker cover and oil separator. Loosen rocker cover bolts, using 2 to 3 steps, in the opposite sequence of tightening.
- Intake manifold supports
- Oil filter bracket and power steering oil pump bracket

5. Set No. 1 piston at Top Dead Center (TDC) on the compression stroke by rotating the crankshaft.

- Chain tensioner
- Distributor. Do not turn the rotor with the distributor removed.
- Timing chain guide, camshaft sprockets, camshafts, brackets, oil tubes and baffle plate. The camshaft bracket bolts must be loosened in sequence to prevent damage to the camshafts or the head.
- Rocker arm assembly

To install:

6. Check the hydraulic lash adjusters to ensure they did not bleed down during disassembly by trying to compress them. If the lash adjuster can be compressed 0.04 in. (1mm), air has entered and it must be bled.

➡**Air cannot be bled from the lash adjusters by running the engine.**

7. Clean the camshaft end bracket and coat with liquid gasket. Install the camshafts, camshaft brackets, oil tubes and baffle plate. Ensure the left camshaft key is at 12 o'clock and the right camshaft key is at 10 o'clock.

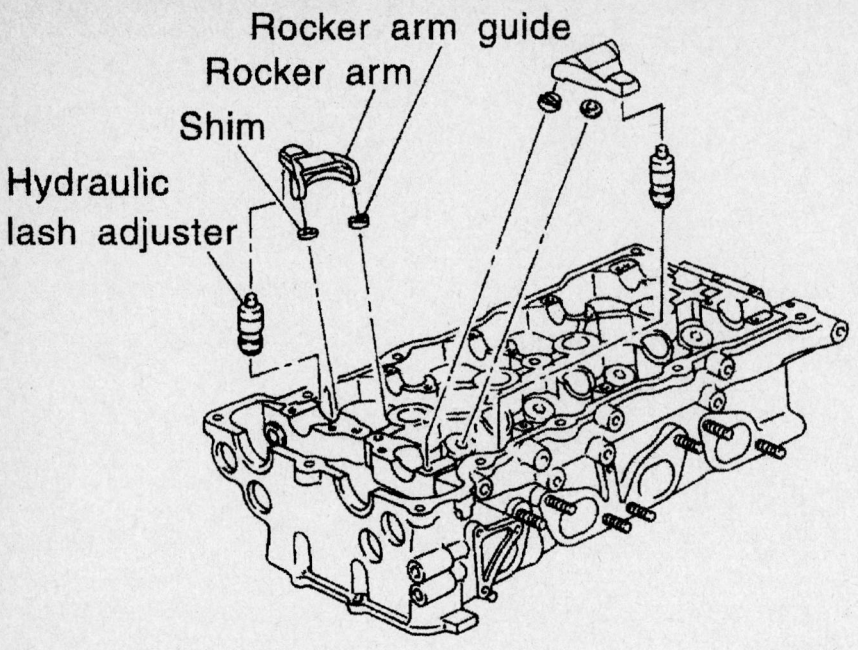

Rocker arm guide
Rocker arm
Shim
Hydraulic lash adjuster

09482_INFI_G0168

Exploded view of the rocker arms and related components—2.0L engine

➡ **The procedure for tightening camshaft bracket bolts must be followed exactly to prevent camshaft damage.**

8. Line up the mating marks on the timing chain and camshaft sprockets and install the sprockets. Torque the sprocket bolts to 101–116 ft. lbs. (137–157 Nm).

9. Install or connect the following:
- Timing chain guide, distributor (ensure that rotor head is at 5 o'clock position) and chain tensioner
- Intake manifold supports. Clean the rocker cover and mating surfaces and apply a continuous bead of liquid gasket to the mating surface.

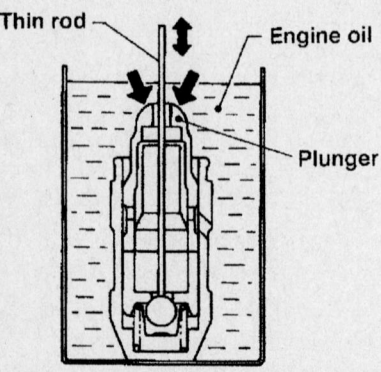

Thin rod — Engine oil
Plunger

09482_INFI_G0169

Submerge the lash adjuster in engine oil, lightly unseat the check ball with a thin rod and push on the plunger to release the air

- Rocker cover and oil separator. Tighten the rocker cover bolts in sequence.
- Oil filter bracket and the power steering pump bracket
- Spark plugs, AIV valve and the resonator
- Fuel lines, vacuum hoses and wiring connectors
- Water pump pulley, alternator and power steering pump
- Drive belts
- Intake manifold air duct, engine side cover and right wheel
- Radiator
- Negative battery cable

10. Fill and bleed the cooling system.

11. Start the vehicle, check for leaks and repair if necessary.

3.5L and 4.5L Engines

➡ **The valves in the 3.5L and 4.5L engines are actuated directly by the camshaft. No rocker arms are used in these engines.**

Intake Manifold

REMOVAL & INSTALLATION

G20 Models

1. Before servicing the vehicle, refer to the precautions in the beginning of this section.

2. Properly relieve the fuel system pressure.

3. Drain the cooling system.

4. Remove or disconnect the following:
- Negative battery cable
- Fuel lines, vacuum hoses and electrical connectors
- Throttle linkage
- Intake manifold supports from the front and rear
- Intake manifold and collector. Loosen the bolts in the sequence illustrated.
- Injector tube assembly
- Power steering oil pump and the oil filter bracket
- Intake manifold and discard the gasket

To install:

5. Be sure all mating surfaces are clean prior to installation.

6. Fit a new gasket and the manifold into place. Start the support bolts to hold the manifold in place.

7. Install or connect the following:
- Intake manifold bolts. Torque the bolts in sequence to 13–15 ft. lbs. (18–21 Nm). Torque the bolts in 2 steps, starting at the center and working towards the ends.
- Injector tube assembly. Torque the bolts first to 84–96 inch lbs. (9–11 Nm), then to 15–20 ft. lbs. (21–26 Nm).
- Power steering oil pump and the oil filter bracket
- Intake manifold collector using a new gasket. Torque the bolts in sequence to 13–15 ft. lbs. (18–21 Nm).
- Fuel lines, vacuum hoses, electrical connectors and the throttle linkage
- Negative battery cable

8. Fill and bleed the cooling system.

9. Start the vehicle, check for leaks and repair if necessary.

I35 Models

1. Before servicing the vehicle, refer to the precautions in the beginning of this section.

2. Release the fuel system pressure.

3. Drain the cooling system.

4. Remove or disconnect the following:
- Negative battery cable
- Throttle body coolant hoses
- Electrical connectors from the Throttle Position (TP) sensor
- Hoses from the throttle body, the Exhaust Gas Recirculation (EGR) valve, intake manifold collector, Idle

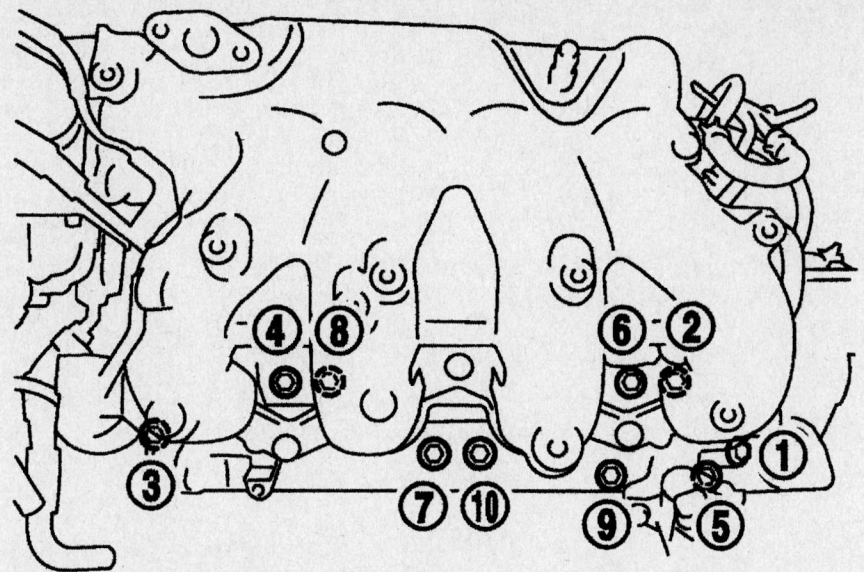

Intake manifold and collector bolt loosening sequence—2.0L engine

09482_INFI_G0170

Air Control (IAC) valve, and the fuel pressure regulator
• Evaporative emissions (EVAP) canister purge hose and blow-by hose
• EGR guide tube
• Accelerator cable from the throttle body
• Intake manifold collector support brackets
• Right side electrical connectors from the ignition coils
• Electrical connector from the crank angle sensor and the power transistor, if necessary

• Intake manifold collector-to-intake manifold bolts/nuts and remove the intake manifold collector
5. Fuel injector assembly by removing or disconnecting the following:
• Electrical connectors from the fuel injectors
• Fuel lines from the fuel injector assembly
• Fuel rail-to-cylinder head bolts
• Fuel rail assembly from the engine
• Intake manifold bolts/nuts in the reverse sequence of the torque procedure

6. Remove the intake manifold from the engine and discard the gaskets
7. Clean all gasket mounting surfaces.
To install:
8. Using new gaskets, install the intake manifold to the engine.
9. Tighten the bolts/nuts in sequence as follows:
a. Step 1: Tighten nuts and bolts to 44–86 inch lbs. (5–10 Nm).
b. Step 2: Tighten nuts and bolts to 20–23 ft. lbs. (26–31 Nm).
c. Step 3: Repeat step 2 at least 5 times until all nuts and bolts have a final torque of 20–23 ft. lbs. (26–31 Nm).
10. Install the fuel injector assembly by installing or connecting the following:
• Fuel rail assembly to the engine
11. Install the fuel rail-to-cylinder head bolts and torque the bolts to 15–20 ft. lbs. (21–26 Nm) in the following sequence:
a. Step 1: Tighten bolts in sequence to 84–96 inch lbs. (9–10 Nm).
b. Step 2: Tighten the bolts in sequence to 15–20 ft. lbs. (21–26 Nm).
• Fuel lines to the fuel injector assembly
• Electrical connectors to the fuel injectors
• Intake manifold collector using a new gasket, and torque the intake manifold collector-to-intake manifold bolts/nuts to 13–16 ft. lbs. (18–22 Nm) in the sequence illustrated
• Intake manifold collector supports. Torque the bolts to 14–18 ft. lbs. (20–25 Nm).
• Electrical connector to the crank angle sensor and the power transistor, if disconnected
• Electrical connectors to the ignition coils and torque the mounting bolts to 27–33 inch lbs. (3–4 Nm)
• Accelerator cable to the throttle body
• EGR guide tube. Torque the bolts to 15–20 ft. lbs. (21–26 Nm) in 2 progressive steps
• EVAP canister purge hose and blow-by hose
• Hoses to the throttle body, EGR valve, intake manifold collector, IAC valve, and the fuel pressure regulator
• Electrical connectors to the TP sensor
• Throttle body coolant hoses
• Negative battery cable
12. Fill and bleed the cooling system.
13. Start the vehicle, check for leaks and repair if necessary.

Intake manifold and collector installation torque sequence—2.0L engine

09482_INFI_G0171

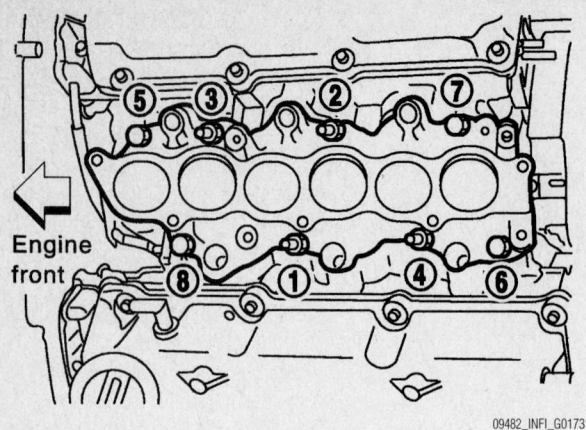

Intake manifold torque sequence—I35

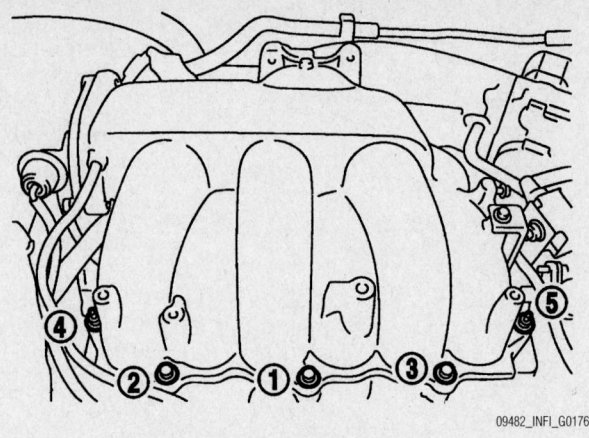

Lower intake manifold collector torque sequence—I35

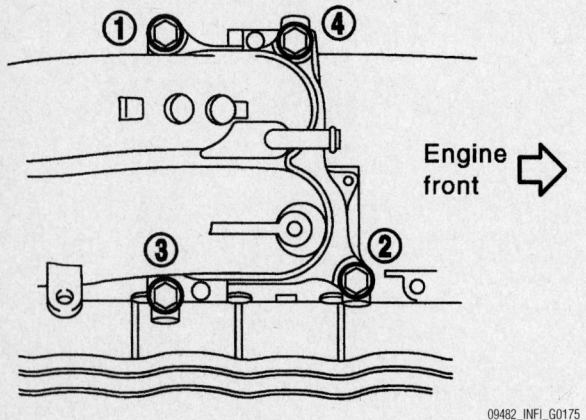

Upper intake manifold collector torque sequence—I35

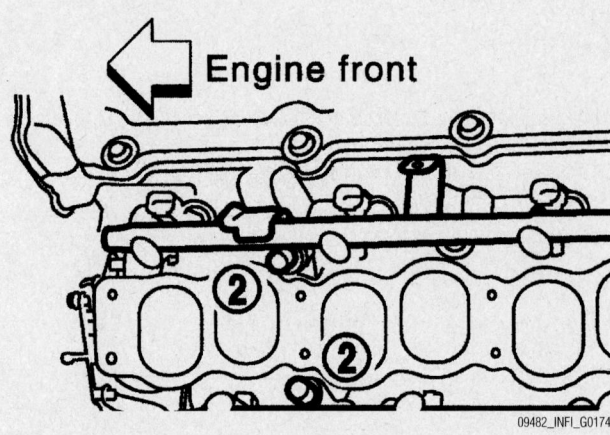

Tighten the fuel rail (tube) bolts in the sequence illustrated—I35

G35 and M35 Models

1. Before servicing the vehicle, refer to the precautions in the beginning of this section.

2. Release the fuel system pressure.

3. Drain the cooling system.

4. Remove or disconnect the following:
- Negative battery cable
- Engine cover
- Air cleaner and duct
- Electric throttle control actuator, loosening the bolts in a criss-cross pattern
- Fuel sub-tube mounting bolt to disconnect it from the rear of the lower intake manifold (collector)
- Vacuum hose and water hose from the intake manifold collector
- Evaporative emissions (EVAP) canister purge volume control solenoid valve bracket mounting bolt from the intake manifold collector
- Intake manifold collector (upper) bolts in the sequence shown, then remove the collector

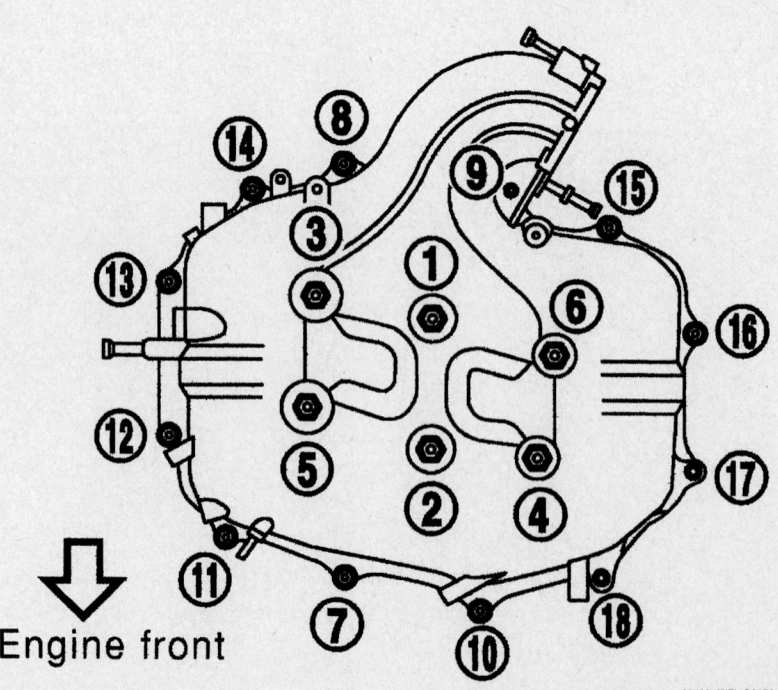

Intake manifold collector (upper) bolt tightening sequence. Use the reverse sequence for bolt removal —G35 models

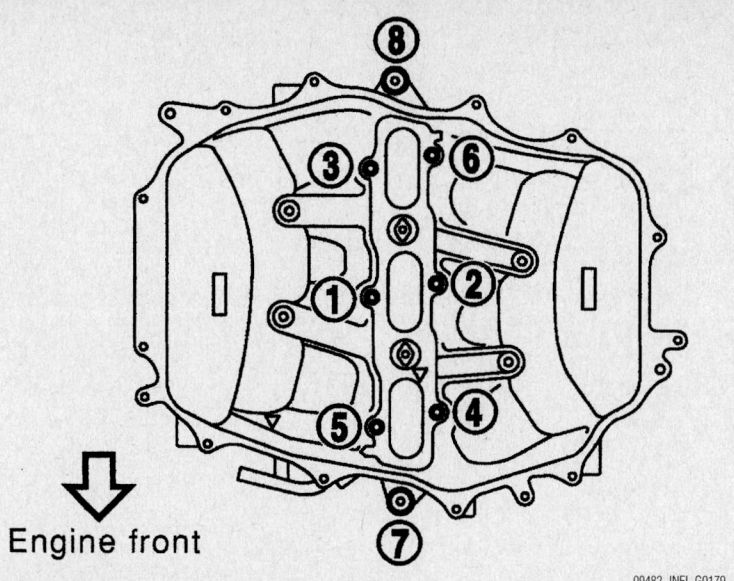

09482_INFI_G0179

Intake manifold collector (lower) bolt tightening sequence. Use the reverse sequence for bolt removal—G35 models

- Positive Crankcase Ventilation (PCV) hose between the intake manifold collector and right side rocker arm cover
- Intake manifold collector (lower) bolts in the reverse of the sequence shown
- Intake manifold collector (lower) cover, gaskets and lower manifold collector
- Fuel rail and injector assembly
- Intake manifold bolts and nuts, in the reverse of the installation sequence

5. Discard all gaskets and clean all gasket mounting surfaces.

To install:

6. Install the intake manifold, with new gaskets. Install the intake manifold nut and bolts, in sequence, and tighten as follows:

a. Stud bolts, if removed: 7.3–8.7 ft. lbs. (9.8–11.8 Nm).

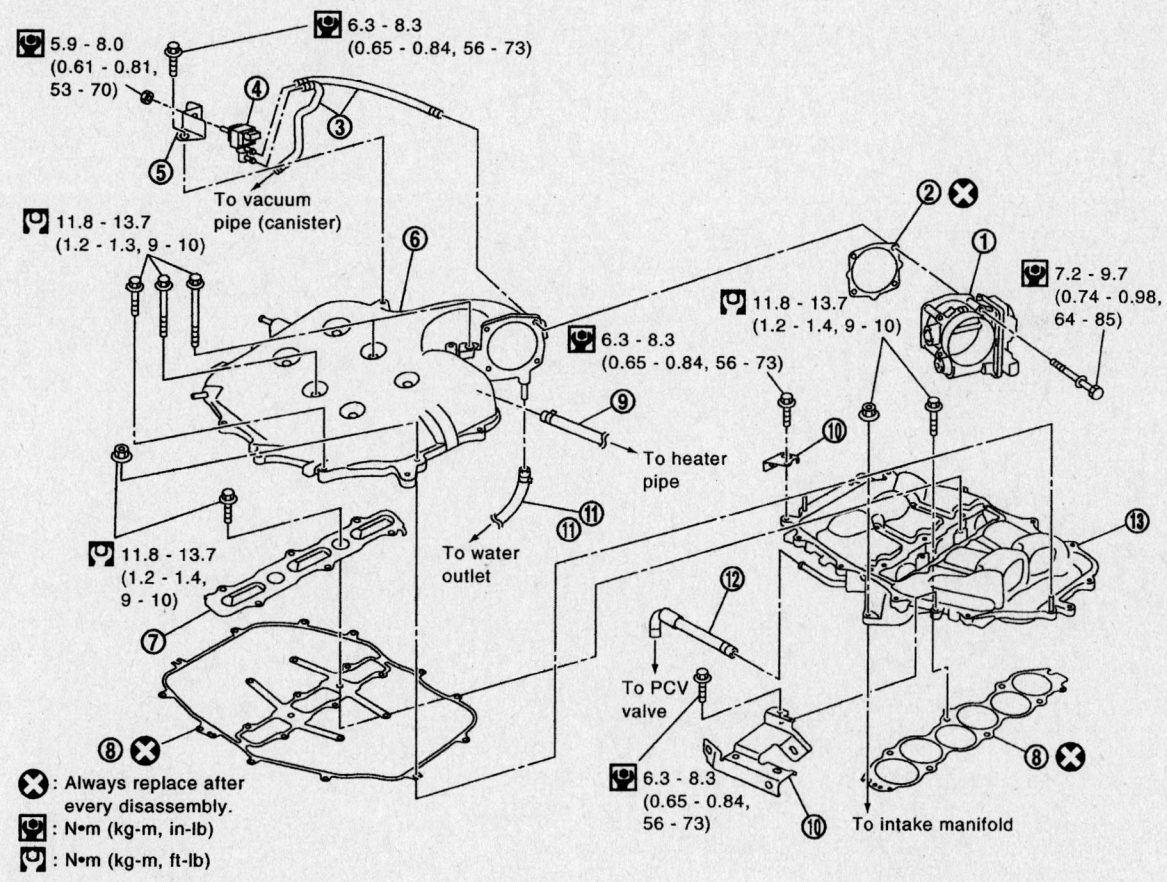

⊗ : Always replace after every disassembly.
🔧 : N•m (kg-m, in-lb)
🔧 : N•m (kg-m, ft-lb)

1. Electric throttle control actuator
2. Gasket
3. Vacuum hose
4. EVAP canister purge volume control solenoid valve
5. Bracket
6. Intake manifold collector (upper)
7. Intake manifold collector cover
8. Gasket
9. Water hose
10. Bracket
11. Water hose
12. PCV hose
13. Intake manifold collector (lower)

09482_INFI_G0177

Exploded view of the upper and lower intake manifold collectors and tightening specifications—G35 models

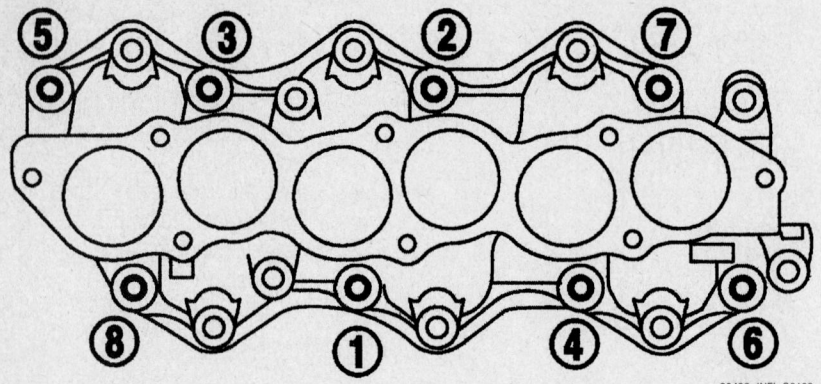

← Engine front

Intake manifold bolt tightening sequence. Use the reverse sequence when removing the bolts—G35 models

b. Step 1: 4–7 ft. lbs. (5–10 Nm).
c. Step 2: 20–23 ft. lbs. (26–31 Nm).
d. Step 3: 20–23 ft. lbs. (26–31 Nm).
7. Install or connect the following:
• Fuel rail and injector assembly
• New gaskets, lower manifold collector, and collector cover, as shown. Tighten the bolts, in sequence to 13–15 ft. lbs. (17.7–21.6 Nm).
• Intake manifold collector (upper). Torque the bolts and nuts, in sequence, to the specifications shown in the illustration.

8. The remainder of installation is the reverse of the removal procedure.
9. Fill and bleed the cooling system.
10. Start the vehicle, check for leaks and repair if necessary.

2006 M45 Models

➡**Refer to Q45 Models for 2004 M45 procedures**

1. Before servicing the vehicle, refer to the precautions in the beginning of this section.
2. Disconnect negative battery cable

3. Remove engine cover
4. Air duct inlet, air cleaner case, duct and resonator
5. Drain the cooling system.
6. Relieve the fuel system pressure.
7. Remove or disconnect the following:
• Fuel feed hose
• Fuel damper and hose assembly

➡**Plug hoses to prevent leaks; do not separate fuel damper and fuel hose**

• Engine wiring harnesses
• Left and right side engine cover brackets
• Vacuum hose, EVAP tube and hose and PCV tube and hose
8. Remove upper intake manifold bolts in reverse of order indicated in diagram and remove upper intake manifold
9. Disconnect electronic throttle control assembly harness
10. Loosen assembly harness bolts diagonally and remove electronic throttle control actuator

➡**Handle actuator with care; do not attempt to disassemble**

11. Remove fuel rail and tube
12. Disconnect water hoses from intake manifold adaptor
13. Remove lower intake manifold bolts in reverse of order indicated in diagram and remove lower intake manifold
14. Remove lower intake manifold adaptor

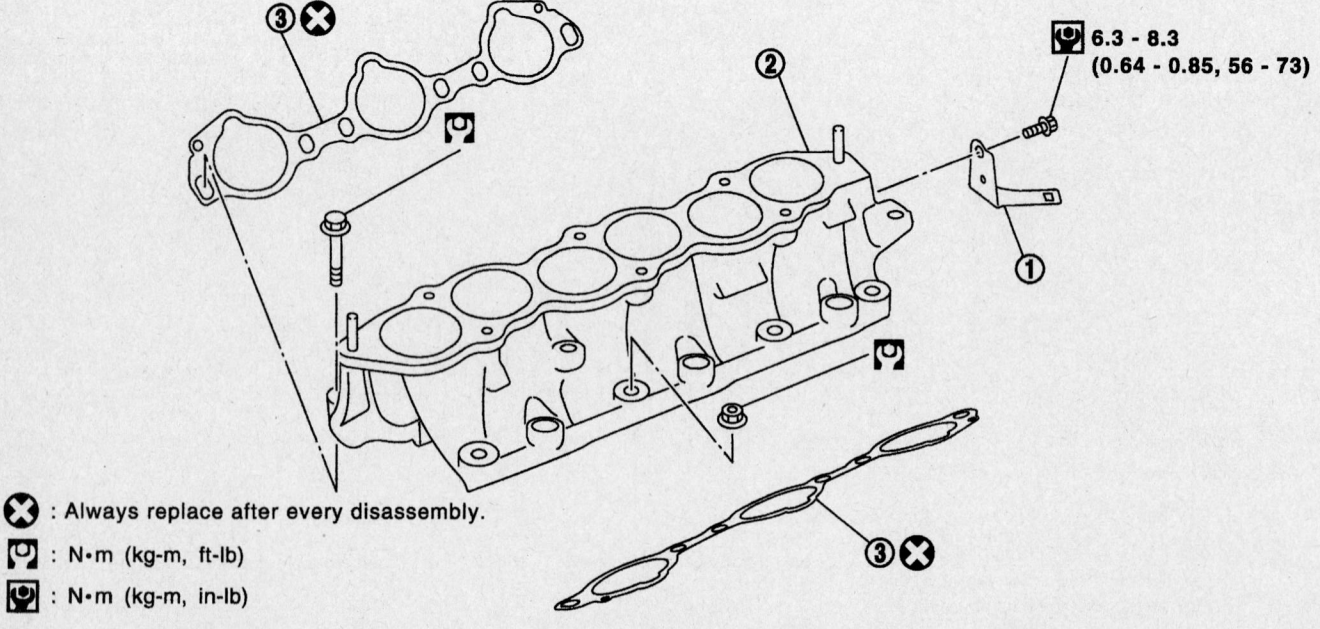

❌ : Always replace after every disassembly.

🔧 : N·m (kg-m, ft-lb)

🔧 : N·m (kg-m, in-lb)

1. Harness bracket
2. Intake manifold
3. Gasket

Component view of the intake manifold—G35 models

09482_INFI_G0181

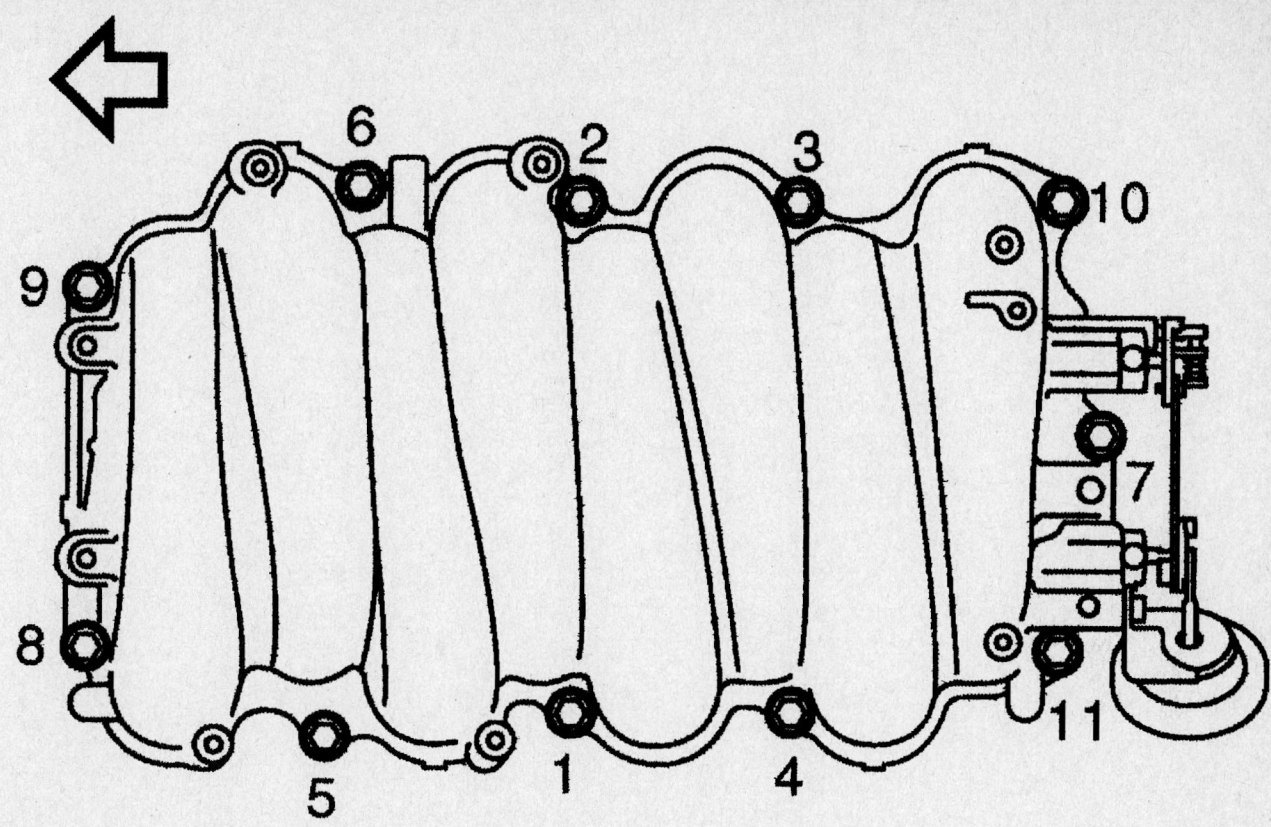

09482_INFI_G0196

Upper intake manifold torque sequence—2006 M45

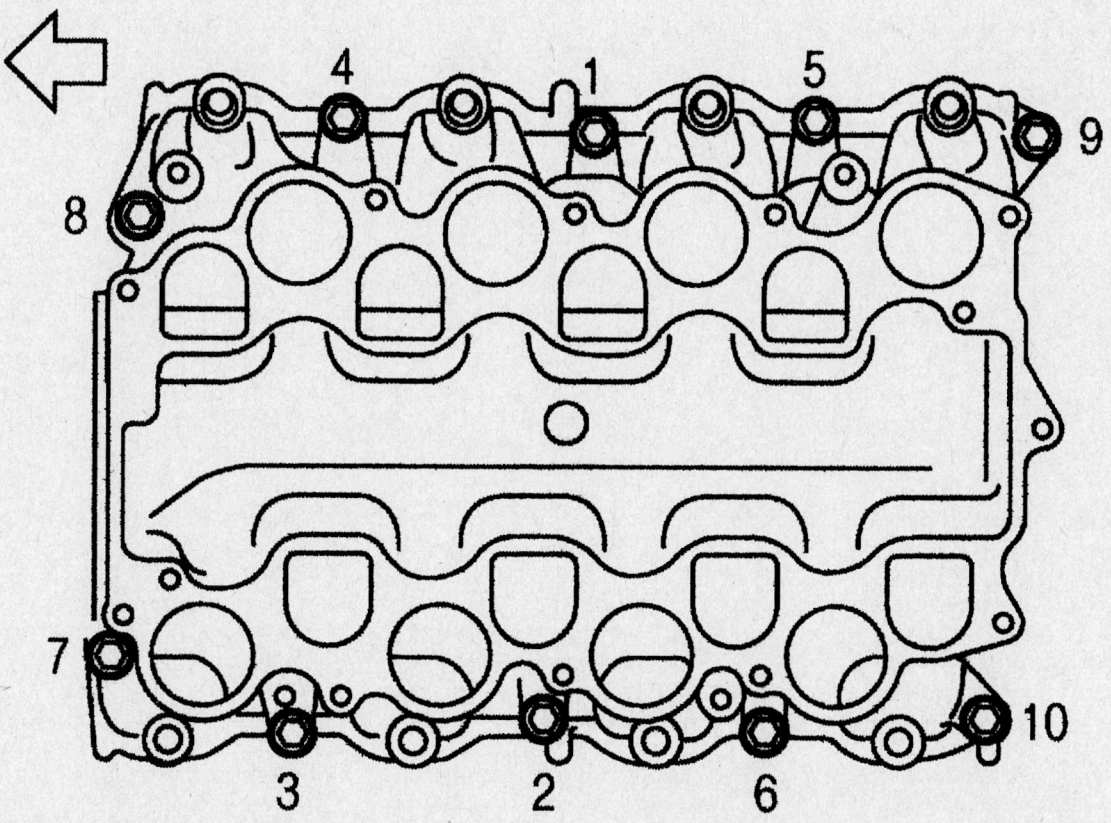

09482_INFI_G0197

Lower intake manifold torque sequence—2006 M45

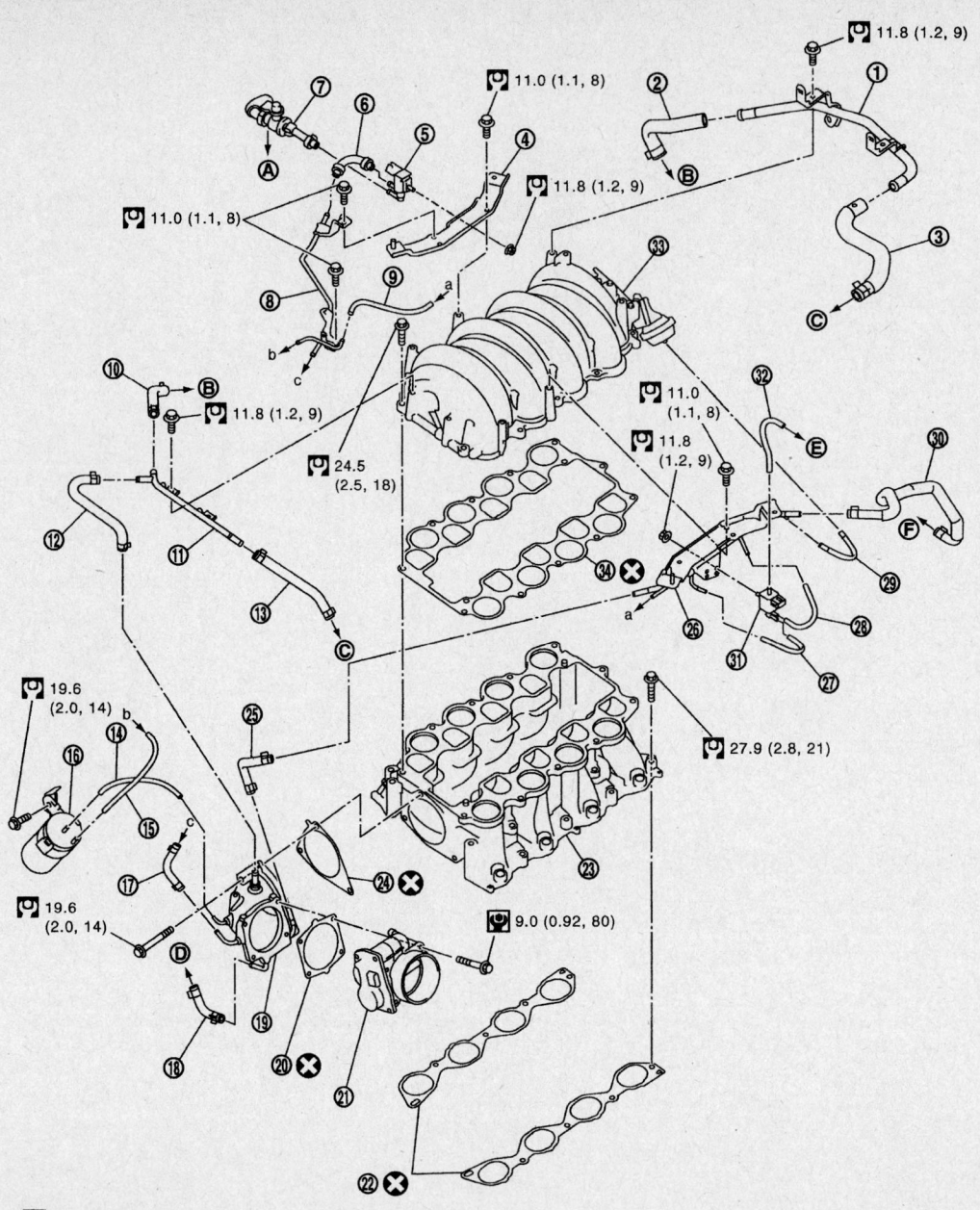

☒ : N•m (kg-m, in-lb)

☒ : N•m (kg-m, ft-lb)

1.	PCV tube	2.	PCV hose	3.	PCV hose
4.	Engine cover bracket (RH)	5.	EVAP canister purge control solenoid valve	6.	EVAP hose
7.	EVAP service port	8.	EVAP tube	9.	Vacuum hose
10.	PCV hose	11.	PCV tube	12.	PCV hose
13.	PCV hose	14.	Vacuum hose	15.	Vacuum hose
16.	Vacuum tank	17.	EVAP hose	18.	Water hose
19.	Intake manifold adapter	20.	Gasket	21.	Electric throttle control actuator
22.	Gasket	23.	Intake manifold (lower)	24.	Gasket
25.	Water hose	26.	Engine cover bracket (LH)	27.	Vacuum hose
28.	Vacuum hose	29.	Vacuum hose	30.	Water hose
31.	VIAS control solenoid valve	32.	Vacuum hose	33.	Intake manifold (upper)
34.	Gasket				
A.	To centralized under-floor piping	B.	To rocker cover (right bank)	C.	To rocker cover (left bank)
D.	To thermostat housing	E.	To air duct and resonator assembly	F.	To heater pipe

09482_INFI_G0188

Exploded view of the intake manifold assembly—2006 M45

15. Remove vacuum tank

16. Remove and discard intake manifold gaskets

To install:

17. Installation is in reverse order of removal

18. Always install new intake manifold gaskets

19. Tighten electronic throttle control actuator bolts evenly and in a diagonal pattern

Q45 and 2004 M45 Models

1. Before servicing the vehicle, refer to the precautions in the beginning of this section.

2. Drain the cooling system.

3. Relieve the fuel system pressure.

4. Remove or disconnect the following:

- Negative battery cable
- Engine appearance cover
- Air cleaner case and air duct
- Fuel tube quick connector
- Accelerator cable
- Wiring harnesses, bracket, vacuum hoses, vacuum gallery and PCV hose and tube from upper manifold
- Electric throttle control actuator
- Intake manifold adapter
- Water hoses

5. Loosen the upper intake manifold bolts in reverse of the tightening sequence and remove the upper manifold.

6. Remove the fuel rail and fuel injectors.

7. Loosen the lower intake manifold bolts in reverse of the tightening sequence and remove the lower manifold.

8. Remove and discard the gaskets.

To install:

9. Clean the intake manifold mounting surfaces

10. Install or connect the following:

- Lower intake manifold using new gasket. Torque the bolts in sequence to 21 ft. lbs. (28 Nm).
- Fuel rail and injectors
- Upper intake manifold using new gasket. Torque the bolts in sequence to 106 inch lbs. (12 Nm).
- Water hoses
- Intake manifold adapter
- Electric throttle control actuator. Torque the bolts to 15 ft. lbs. (20 Nm).
- Wiring harnesses, bracket, vacuum hoses, vacuum gallery and PCV hose and tube from upper manifold
- Accelerator cable
- Fuel tube quick connector
- Air cleaner case and air duct

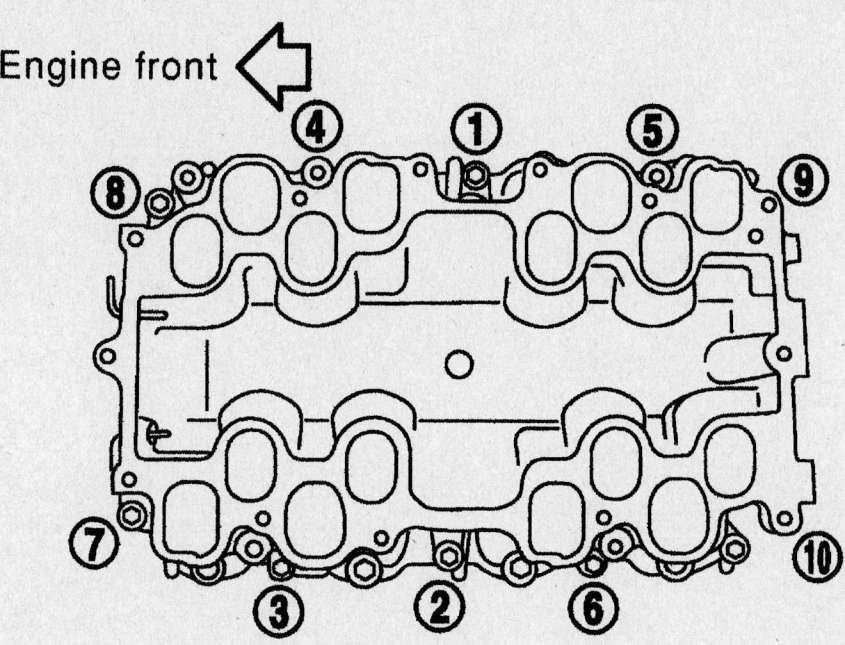

Lower intake manifold torque sequence—Q45 and 2004 M45

09482_INFI_G0184

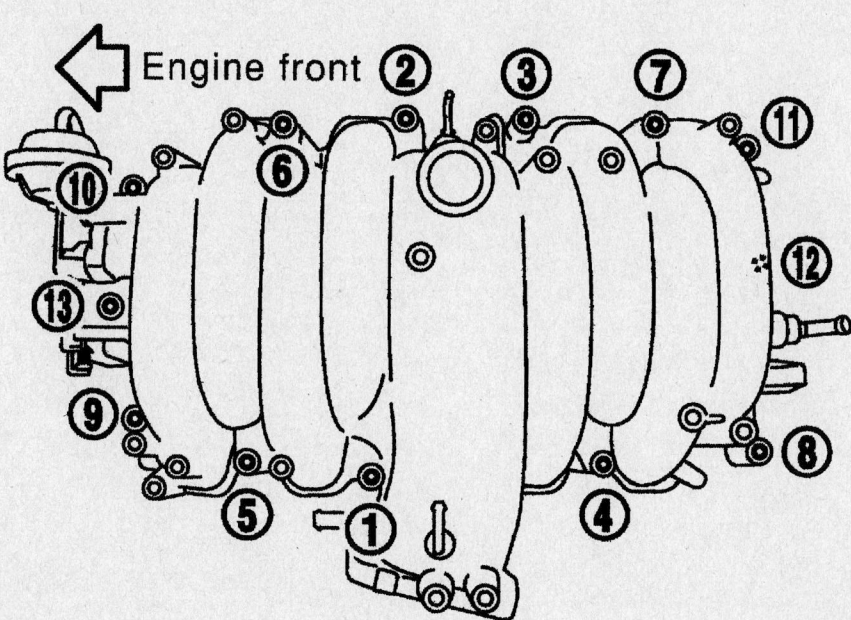

Upper intake manifold torque sequence—Q45 and 2004 M45

09482_INFI_G0185

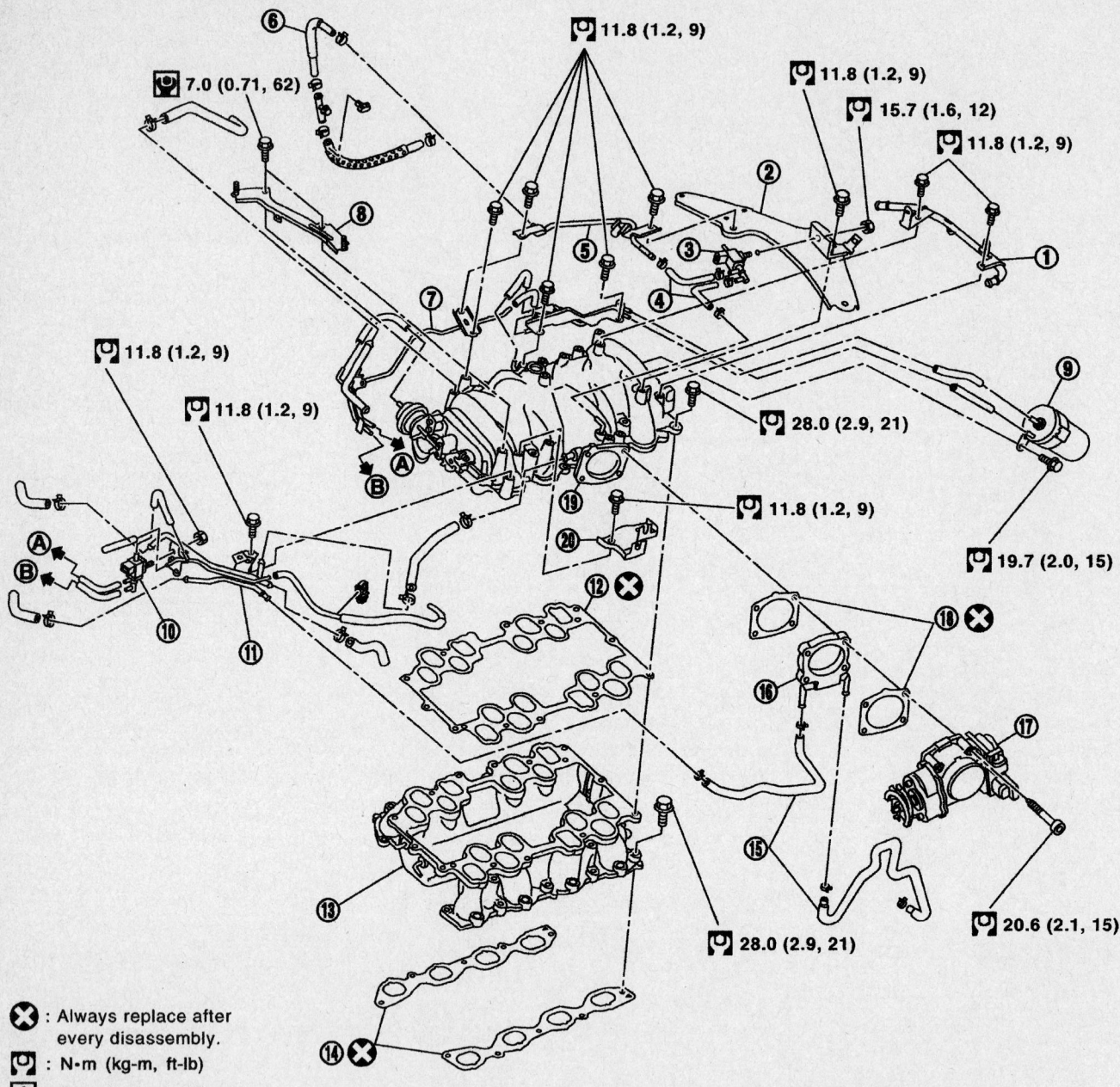

- ☒ : Always replace after every disassembly.
- ◫ : N·m (kg-m, ft-lb)
- ◫ : N·m (kg-m, in-lb)

1. PCV tube	2. Engine cover rear bracket	3. EVAP canister purge volume control solenoid valve
4. EVAP hose	5. EVAP tube	6. EVAP hose
7. Vacuum gallery	8. Engine cover front bracket	9. Vacuum tank
10. VIAS control solenoid valve	11. Water gallery	12. Gasket
13. Intake manifold lower	14. Gasket	15. Water hose
16. Intake manifold adapter	17. Electric throttle control actuator	18. Gasket
19. Intake manifold upper	20. Bracket	

09482_INFI_G0186

Exploded view of the intake manifold assembly—2004 M45

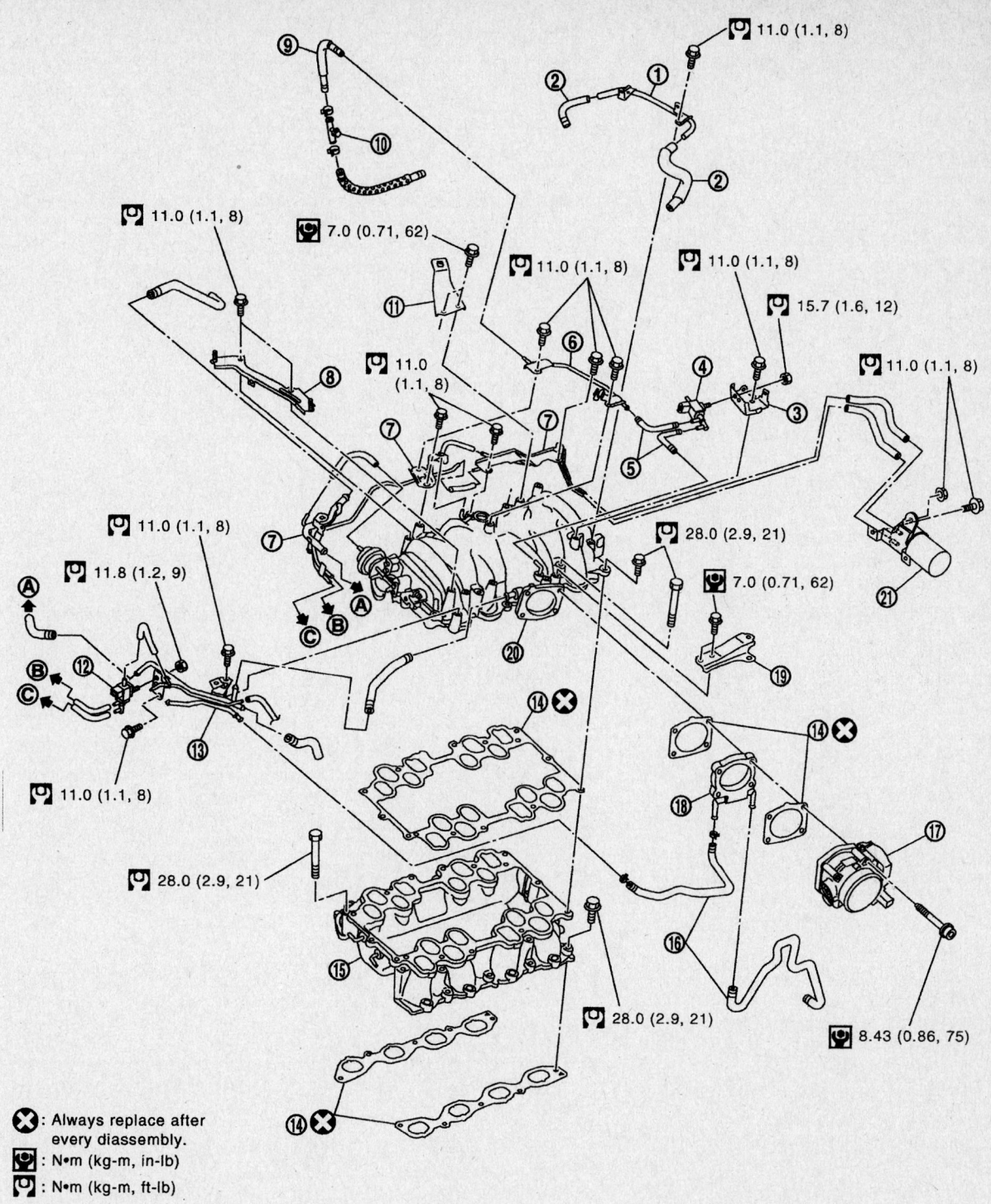

11.0 (1.1, 8)

11.0 (1.1, 8)

7.0 (0.71, 62)

11.0 (1.1, 8)

11.0 (1.1, 8)

15.7 (1.6, 12)

11.0 (1.1, 8)

11.0 (1.1, 8)

11.0 (1.1, 8)

28.0 (2.9, 21)

7.0 (0.71, 62)

11.8 (1.2, 9)

11.0 (1.1, 8)

28.0 (2.9, 21)

28.0 (2.9, 21)

8.43 (0.86, 75)

⊗ : Always replace after every diassembly.

[symbol] : N•m (kg-m, in-lb)

[symbol] : N•m (kg-m, ft-lb)

1. PCV tube	2. PCV hose	3. EVAP canister purge volume control solenoid valve bracket
4. EVAP canister purge volume control solenoid valve	5. EVAP hose	6. EVAP tube
7. Vacuum gallery	8. Engine cover front bracket	9. EVAP hose
10. EVAP service port	11. Engine cover rear bracket	12. VIAS control solenoid valve
13. Water gallery	14. Gasket	15. Intake manifold (lower)
16. Water hose	17. Electric throttle control actuator	18. Intake manifold adapter
19. Engine cover rear bracket	20. Intake manifold (upper)	21. Vacuum tank

09482_INFI_G0187

Exploded view of the intake manifold assembly—Q45

- Engine appearance cover
- Negative battery cable

11. Fill and bleed the cooling system.

12. Start the vehicle, check for leaks and repair if necessary.

Exhaust Manifold

REMOVAL & INSTALLATION

G20 Models

1. Before servicing the vehicle, refer to the precautions in the beginning of this section.

2. Remove or disconnect the following:
- Negative battery cable
- Undercover and dust covers, if equipped
- Exhaust pipe at the manifold flange
- Air Injection Valve (AIV), AIV tube, and the attaching bracket, if equipped
- Exhaust Gas Recirculation (EGR) sensor electrical connection and the sensor
- Exhaust manifold cover
- Exhaust manifold nuts, starting at the outside and working towards the middle
- Exhaust manifold and discard the gasket

To install:

3. Clean the gasket mating surface and install a new exhaust manifold gasket.

4. Install or connect the following:
- Exhaust manifold. Torque the nuts in sequence, in 2 steps, to 27–35 ft. lbs. (37–48 Nm).
- Exhaust manifold cover and EGR sensor
- Exhaust gas sensor electrical connection
- AIV, AIV tube, and the attaching bracket, if equipped
- Exhaust pipe to the manifold flange. Torque the nuts to 30–35 ft. lbs. (41–48 Nm).
- Negative battery cable

5. Start the engine, check for leaks and repair if necessary.

I35, G35 and M35 Models

1. Before servicing the vehicle, refer to the precautions in the beginning of this section.

2. Remove or disconnect the following:
- Negative battery cable

➥**If necessary, soak the exhaust pipe retaining nuts with penetrating oil to loosen them.**

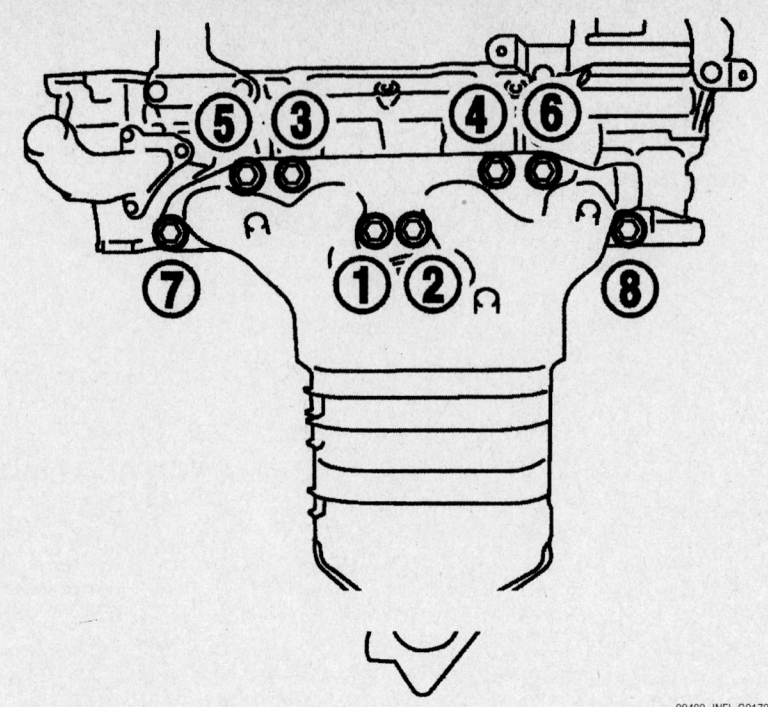

09482_INFI_G0172

Be sure to tighten the exhaust manifold nuts in the proper sequence—2.0L engines

Right bank

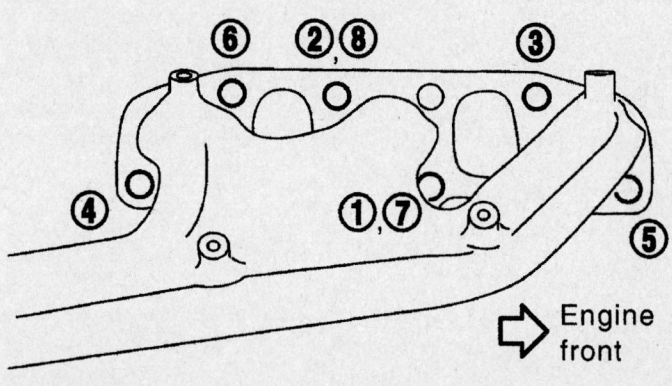

Engine front

Left bank

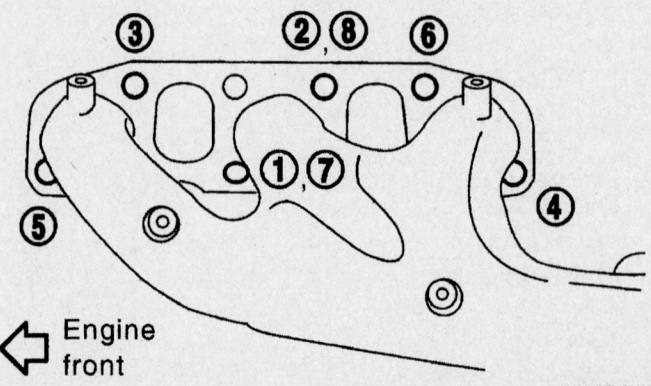

Engine front

09482_INFI_G0189

Exhaust manifold tightening sequence. Use the reverse of the sequence for removal—G35 shown

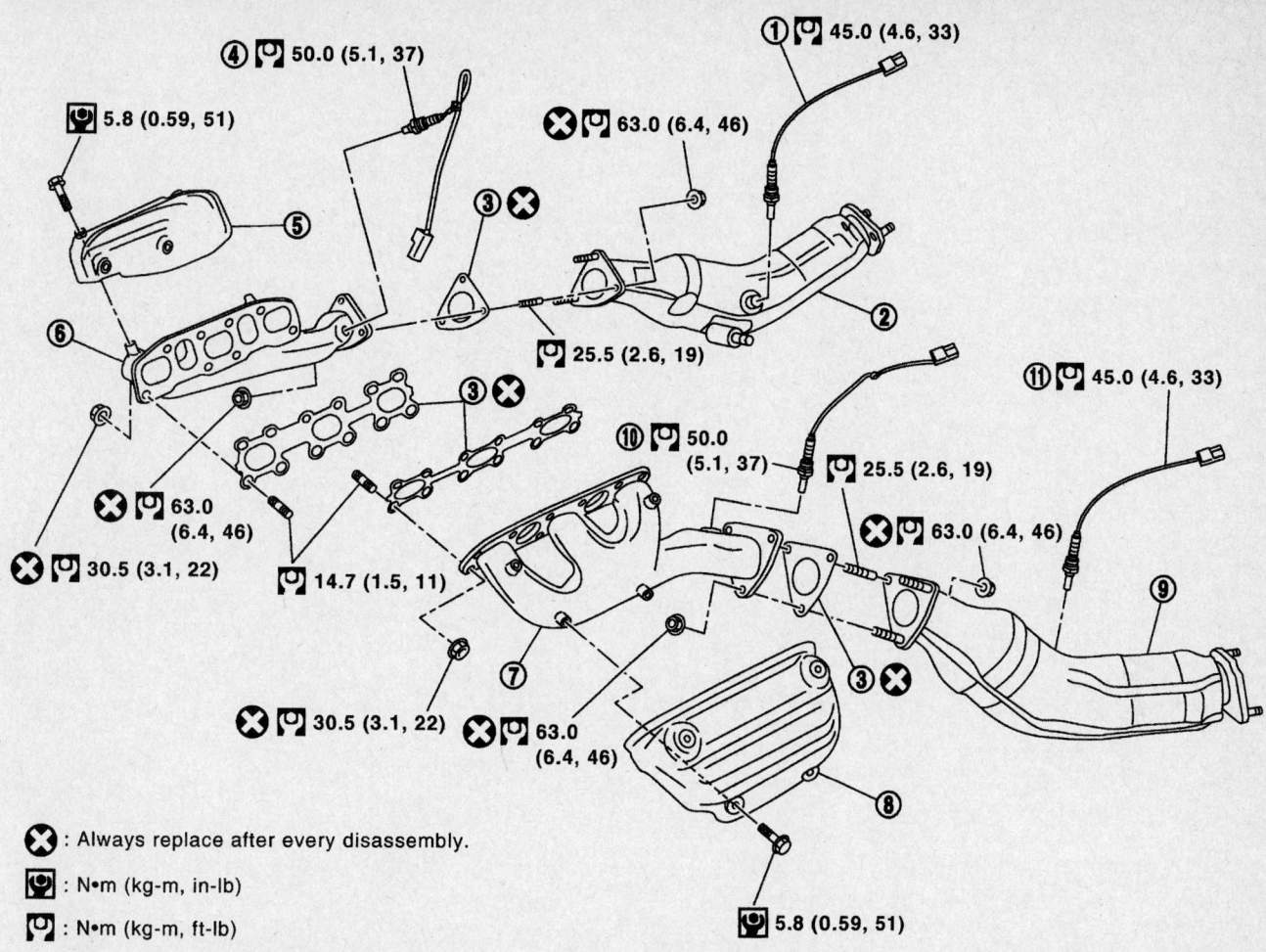

④ 🔧 50.0 (5.1, 37)

🔧 5.8 (0.59, 51)

① 🔧 45.0 (4.6, 33)

❌ 🔧 63.0 (6.4, 46)

⑤

③ ❌

⑥

②

🔧 25.5 (2.6, 19)

⑪ 🔧 45.0 (4.6, 33)

③

⑩ 🔧 50.0 (5.1, 37)

🔧 25.5 (2.6, 19)

❌ 🔧 63.0 (6.4, 46)

❌ 🔧 63.0 (6.4, 46)

❌ 🔧 30.5 (3.1, 22)

🔧 14.7 (1.5, 11)

⑦

⑨

③ ❌

❌ 🔧 30.5 (3.1, 22)

❌ 🔧 63.0 (6.4, 46)

⑧

🔧 5.8 (0.59, 51)

❌ : Always replace after every disassembly.

🔧 : N•m (kg-m, in-lb)

🔧 : N•m (kg-m, ft-lb)

1. Heated oxygen sensor 2 (bank 1)
2. Three way catalyst (right bank)
3. Gasket
4. Air fuel ratio sensor 1 (bank 1)
5. Exhaust manifold cover (right bank)
6. Exhaust manifold (right bank)
7. Exhaust manifold (left bank)
8. Exhaust manifold cover (left bank)
9. Three way catalyst (left bank)
10. Air fuel ratio sensor 1 (bank 2)
11. Heated oxygen sensor 2 (bank 2)

09482_INFI_G0190

Exploded view of the exhaust manifold and related components—G35 shown

- Engine cover, G35 only
- Air cleaner assembly and duct, G35 only
- Front exhaust pipe from the exhaust manifolds
- Heated Oxygen Sensors (HO2S) from the manifold, as necessary
- Protective covers from the manifolds
- Exhaust manifold-to-engine mounting nuts
- Manifold from the engine and discard the gaskets

To install:
3. Clean all gasket mounting surfaces.
4. Install or connect the following:
- Exhaust manifold with new gaskets. Torque the bolts, in sequence, in 2 steps to 21–23 ft. lbs. (28–32 Nm).

- Protective shields. Torque the bolts in 2 steps to 46–57 inch lbs. (5–6 Nm).
- HO2S to the manifold, as necessary
- Exhaust manifolds to the exhaust pipes. Torque the bolts/nuts to 32–37 ft. lbs. (43–50 Nm) for all models except G35. For the G35, torque the nuts to 45–48 ft. lbs. (60–66 Nm).
- Air cleaner assembly and engine cover, G35 only
- Negative battery cable
5. Start the engine, check for exhaust leaks and repair if necessary.

2006 M45 Models

➡ **Refer to Q45 Models for 2004 M45 procedures.**

1. Before servicing the vehicle, refer to the precautions in the beginning of this section.
2. Drain and save engine coolant
3. Disconnect negative battery cable
4. Remove or disconnect the following:
- Engine cover
- Air duct inlet, air cleaner case and resonator assembly
- Front and rear engine undercover
- Radiator assembly
- Drive belts
- Front exhaust tube
- Air/fuel ration sensors and harness (both banks)
- A/C piping from compressor manifold
- A/C compressor
- Lower steering column joint

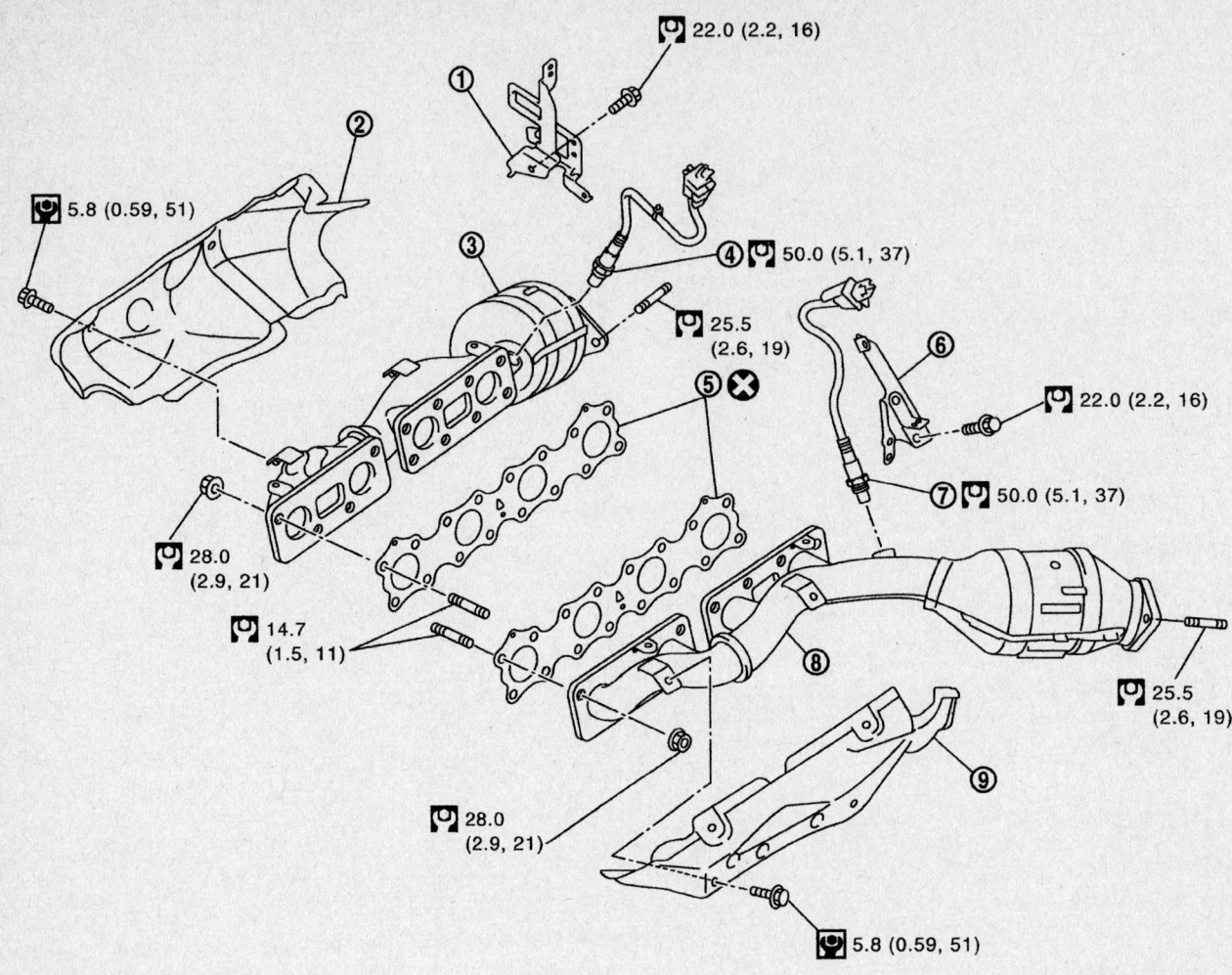

: N•m (kg-m, in-lb)

: N•m (kg-m, ft-lb)

1.	Harness bracket	2.	Exhaust manifold cover (right bank)	3.	Exhaust manifold and three way catalyst (right bank)
4.	Air fuel ratio sensor 1 (bank 2)	5.	Gasket	6.	Harness bracket
7.	Air fuel ratio sensor 1 (bank 1)	8.	Exhaust manifold and three way catalyst (left bank)	9.	Exhaust manifold cover (left bank)

09482_INFI_G0191

Exploded view of the exhaust manifold assembly—2006 M45 models

• Starter

5. Remove fasteners on bottom of left engine mounting insulator and raise left side of engine approximately 2 inches (5 cm) with a suitable jack

6. Remove left bank exhaust manifold cover

7. Remove left exhaust manifold nuts in reverse order shown in diagram; remove left exhaust manifold and three-way catalyst

8. Remove alternator and mounting bracket

9. Remove fasteners on bottom of right engine mounting insulator and raise right

side of engine approximately 2 inches (5 cm) with a suitable jack

10. Remove right bank exhaust manifold cover

11. Remove right exhaust manifold nuts in reverse order shown in diagram; remove right exhaust manifold and three-way catalyst

➡Cover engine openings to prevent entry of foreign material

To install:

12. Installation is in the reverse order of removal

13. Always use new exhaust manifold gaskets and lower studs

14. Observe tightening sequence indicated in diagram and make certain alignment tab on gasket faces upward

15. Start the engine, check for exhaust leaks and repair if necessary.

Q45 and 2004 M45 Models

1. Before servicing the vehicle, refer to the precautions in the beginning of this section.

2. Remove or disconnect the following:
   • Both battery cables

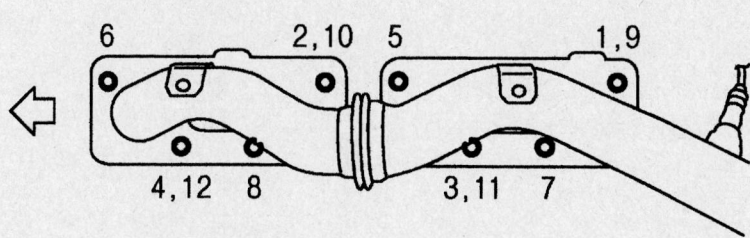

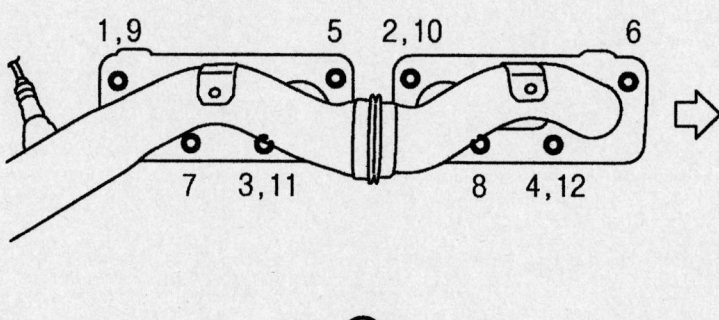

09482_INFI_G0192

Exhaust manifold tightening sequence—2006 M45 models

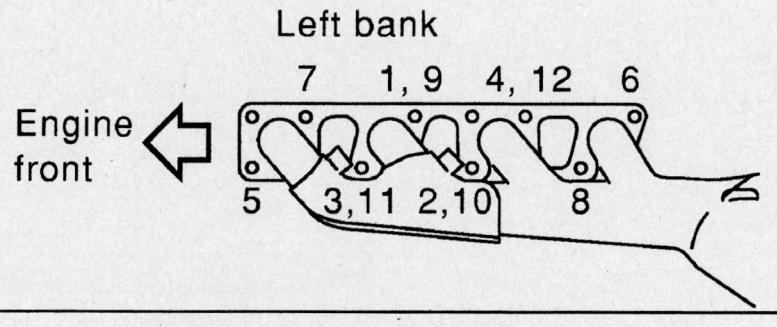

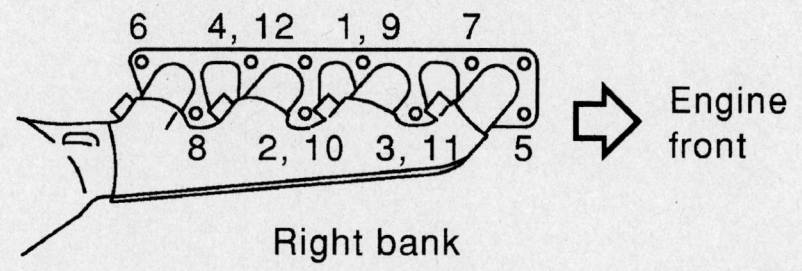

09482_INFI_G0195

Make certain alignment tab is pointed upward—2006 M45 models

- Battery and tray
- Coolant reservoir tank and bracket
- Engine appearance cover
- Engine undercover
- Both oxygen sensor harnesses and sensors
- A/C compressor without disconnecting the refrigerant lines and set aside
- Left side exhaust front tube
- Steering column lower joint

3. Use a jack and raise the bottom of the engine.

4. Remove the left engine mounting insulator and the engine mounting brackets.

5. Remove the left manifold heat shield.

6. Remove the left manifold mounting nuts in the reverse order of the tightening sequence and remove the manifold and catalyst.

7. Remove the right side front exhaust tube.

8. Remove the nuts from the bottom right engine mounting insulator and lift the right side of the engine up about 1 and a half inches.

9. Remove the starter, then remove the right engine mounting insulator and mounting brackets.

10. Remove the right manifold mounting nuts in the reverse order of the tightening sequence and remove the manifold and catalyst.

To install:

11. Clean the gasket mating surfaces.

12. Install or connect the following:
- Exhaust manifold with new gaskets. Torque the nuts to 20–22 ft. lbs. (26–30 Nm) in two steps.

13. Install the starter, then install the right engine mounting insulator and mounting brackets. Tighten the bracket bolts to 25–31 ft. lbs. (35–43 Nm). Tighten the upper insulator nut to 32–40 ft. lbs. (43–553 Nm) and the lower nut to 65–72 ft. lbs. (87–98 Nm).

14. Install the heat shields.

15. Install the right side front exhaust tube.

16. Install the left engine mounting insulator and the engine mounting brackets and tighten the fasteners to the same specification as the right side.

17. Connect the steering column lower joint and tighten the bolt to 20 ft. lbs. (27 Nm).

18. Install the left side front exhaust tube.

19. Install or connect the following:
- A/C compressor
- Both oxygen sensor harnesses and sensors

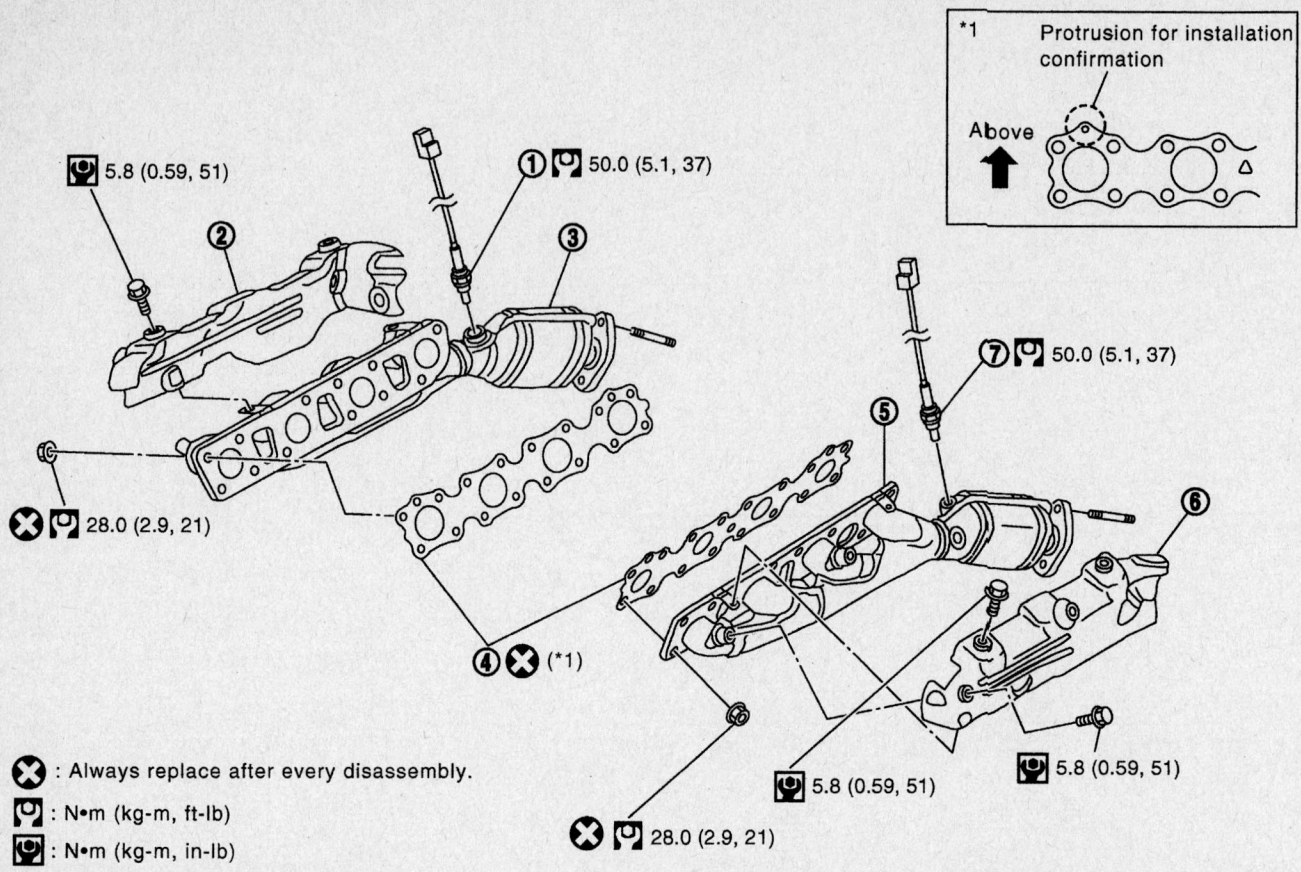

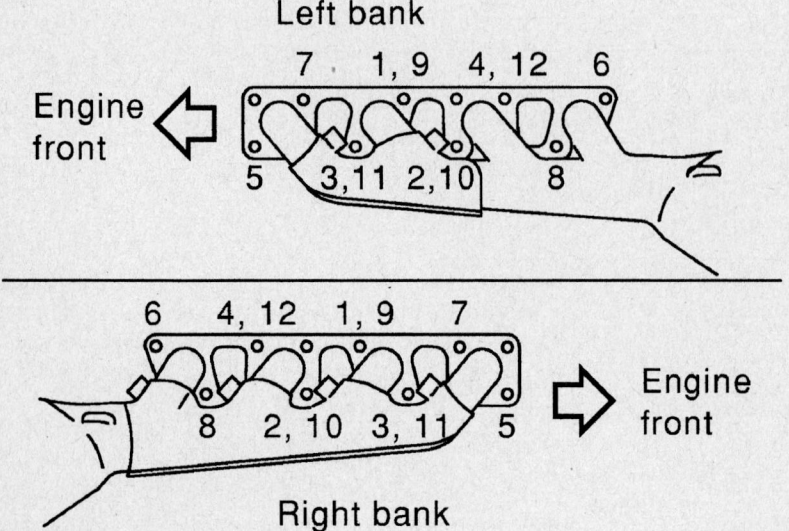

❌ : Always replace after every disassembly.

🔧 : N•m (kg-m, ft-lb)

🔧 : N•m (kg-m, in-lb)

1. Heated oxygen sensor 1 (bank 2)
2. Exhaust manifold cover (right bank)
3. Exhaust manifold and three way catalyst (right bank)
4. Gasket
5. Exhaust manifold and three way catalyst (left bank)
6. Exhaust manifold cover (left bank)
7. Heated oxygen sensor 1 (bank 1)

09482_INFI_G0193

Exploded view of the exhaust manifold assembly—Q45 and 2004 M45 models

Exhaust manifold tightening sequence—Q45 and 2004 M45 models

09482_INFI_G0194

• Engine undercover
• Engine appearance cover
• Coolant reservoir tank and bracket
• Battery and tray
• Both battery cables

20. Start the engine, check for leaks and repair if necessary.

Camshaft and Valve Lifters

REMOVAL & INSTALLATION

2.0L Engine

1. Before servicing the vehicle, refer to the precautions in the beginning of this section.
2. Relieve the fuel system pressure.
3. Remove or disconnect the following:
   • Negative battery cable
   • Rocker arm cover

- Oil separator

4. Rotate the crankshaft until the No. 1 piston is at Top Dead Center (TDC) on the compression stroke and the mating marks on the camshaft sprockets line up with the mating marks on the timing chain.
- Timing chain tensioner
- Distributor
- Timing chain guide
- Camshaft sprockets. Use a wrench to hold the camshaft while loosening the sprocket bolt.

5. Loosen the camshaft bearing cap bolts in sequence.
- Camshaft from the cylinder head

6. When removing the rocker arm, be careful not to drop the valve shims into the cylinder head. After removing the adjuster, set them upright or lay them down in a pan of clean engine oil. Do not lay them down on the bench or the oil will drain out and the adjuster will become air bound. Keep all of these parts in order so they can be installed in the same locations.

To install:

7. Install the adjusters, shims and rockers into their original locations.

8. Clean the left-hand camshaft end bearing cap and coat the mating surface with liquid gasket. Install the camshafts, bearing caps, oil tubes and baffle plate. Ensure the left camshaft key is at 12 o'clock and the right camshaft key is at 10 o'clock.

9. The procedure for tightening bearing cap bolts must be followed exactly to prevent camshaft damage. Torque bolts as follows:

   a. STEP 1: Intake camshaft, tighten bolts 9–10 in that order then tighten bolts 1–8 in numerical order to 2 Nm (17 inch lbs.); Exhaust camshaft, tighten bolts 11–12 in that order then tighten bolts 1–10 in numerical order to 2 Nm (17 inch lbs.)

   b. STEP 2: Tighten all bolts in the specified order to 6 Nm (52 inch lbs.)

   c. STEP 3: Tighten all bolts in the specified order to
   - Bolt type A B: 10–12 Nm (86–104 inch lbs.)
   - Bolt type C: 18–25 Nm (13–19 ft. lbs.)

10. Line up the mating marks on the timing chain and camshaft sprockets and install the sprockets. Torque sprocket bolts to 101–116 ft. lbs. (137–157 Nm).

11. Install or connect the following:
- Timing chain guide and chain tensioner
- Distributor making certain that the rotor head is at the 5 o'clock position

12. Clean the rocker cover and mating surfaces and apply a continuous bead of liquid gasket to the mating surface.

13. Install the rocker cover and oil separator. Tighten the rocker cover bolts as follows:

   a. Torque the nuts 1, 10, 11 and 8, in that order to 36 inch lbs. (4 Nm).

   b. Torque the nuts 1 through 13 as indicated in the figure to 72–84 inch lbs. (8–10 Nm).

14. Connect the negative battery cable.

3.5L Engines

I35

1. Before servicing the vehicle, refer to the precautions in the beginning of this section.
2. Relieve the fuel system pressure.
3. Drain the engine oil.
4. Drain the cooling system.
5. Remove or disconnect the following:
- Negative battery cable
- Left side rocker cover ornament

➡**Before disconnecting any hoses or connectors, note the locations for reassembly.**

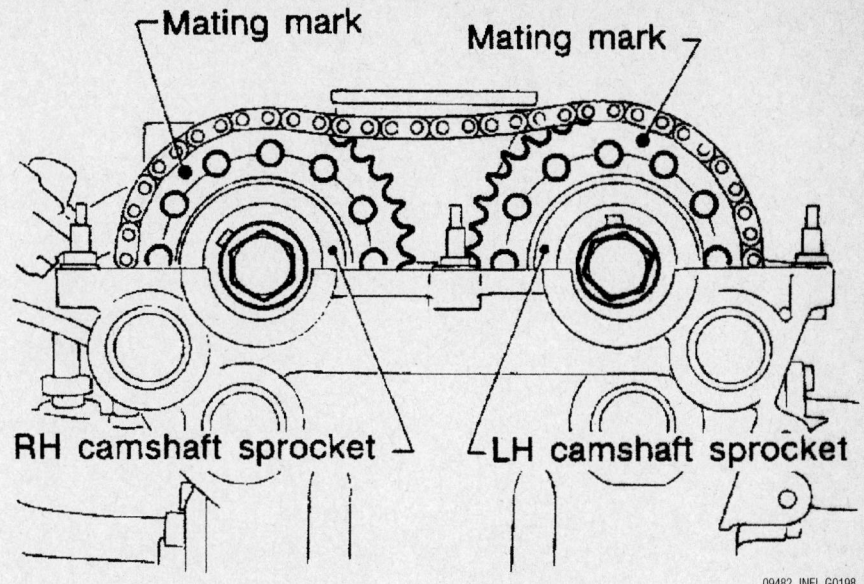

09482_INFI_G0198

Be sure to align the marks on the sprockets with the marks on the chain—2.0L engine

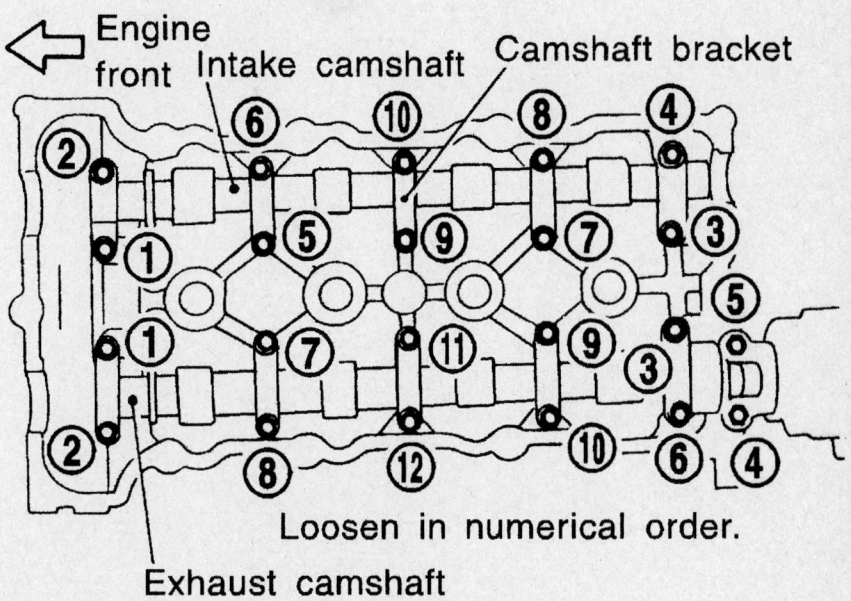

09482_INFI_G0199

Camshaft bearing cap bolt loosening sequence—2.0L engines

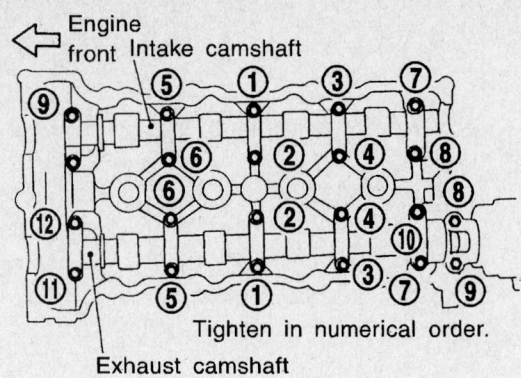

Tighten in numerical order.

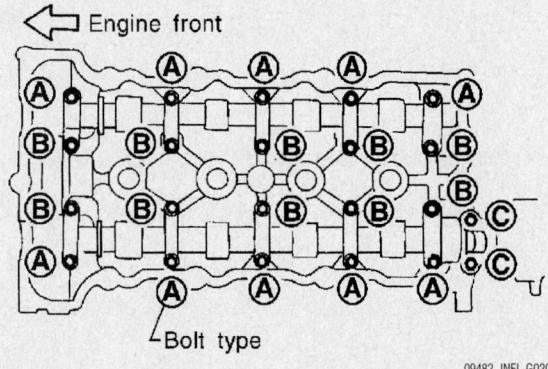

Camshaft bearing cap bolt torque sequence—2.0L engines

09482_INFI_G0200

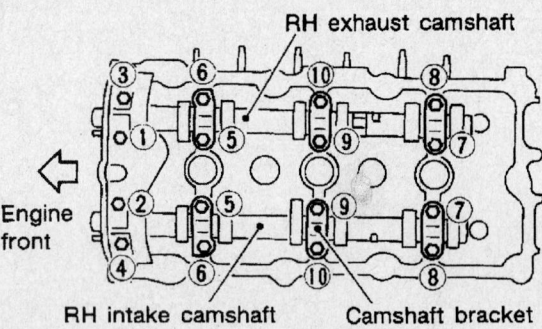

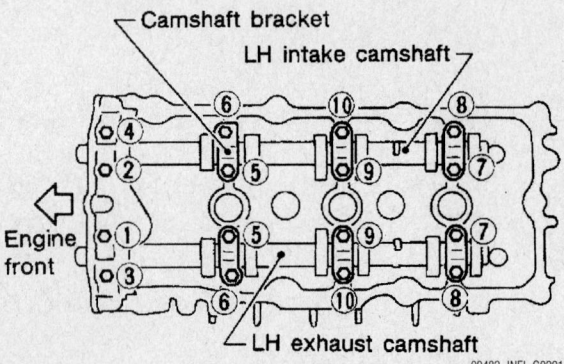

To avoid camshaft damage, loosen the bearing cap bolts in the sequence shown—3.5L engines, except G35

09482_INFI_G0201

- Air duct to intake manifold hose, collector hose, blow-by hose, and vacuum hoses
- Fuel hoses and disconnect the harness connection
- Evaporative emissions (EVAP) canister purge hoses
- Water hoses from the cylinder head and intake manifold
- Ignition coils
- Spark plugs
- Exhaust Gas Recirculation (EGR) tube
- Intake manifold collector supports and the collector
- Fuel tube
- Intake manifold
- Rocker arm covers
- Engine undercover
- Right front wheel
- Engine side covers
- Drive belts and idler pulley
- Power steering oil pump and belt
- Camshaft Position (CMP) sensor (PHASE) and Crankshaft Position (CKP) sensors (REF)/(POS)

6. Set the No. 1 piston to Top Dead center (TDC) of compression stroke by rotating the crankshaft.
- Ring gear cover access plate

7. Loosen the crankshaft pulley bolt while securing the ring gear so the crankshaft cannot rotate.

➡**Use care not to damage the ring gear teeth.**

- Crankshaft pulley, using a suitable puller
- A/C compressor and bracket
- Front exhaust pipe and install engine slingers

8. Support the transaxle with jack.
- Right side engine mounting bracket
- Center crossmember assembly
- Oil pan bolts and oil pans
- Timing chain
- O-rings from the front of the engine block

9. Loosen the camshaft bearing caps in several steps. The bearing caps MUST be loosened in sequence.

➡**Keep all bearing caps and camshafts in proper order for reinstallation.**

- Left-hand and right-hand camshaft tensioners from the cylinder head
- Camshafts from the cylinder heads

➡**The valve adjusters have a replaceable shim on the top of the adjuster. Note the proper locations of each shim to adjuster and remove the shims from the adjusters.**

- Valve adjusting shim from the adjuster, using a magnet
- Adjuster assembly from the bore. Be sure to note the locations from where each adjuster came.

10. Check the diameter of the valve adjuster and the valve adjuster guide bore.

11. The diameter of the adjuster should be 1.3764–1.3770 in. (34.960–34.975mm) and the diameter of the bore should be 1.3780–1.3788 in. (35.000–35.021mm).

- All traces of liquid gasket from the timing chain case and from the water pump covers
- All traces of liquid gasket from the engine block

12. Inspect the camshafts for excessive wear or damage and replace as necessary.

To install:

13. Lubricate the valve adjusters with clean engine oil and install the adjusters into the bore from which they were removed.

14. Lubricate the valve adjuster shims with clean engine oil and install the shims into the adjuster from which they were removed.

15. Turn the crankshaft clockwise until the No. 1 piston is set 240° before TDC on compression stroke.

16. Install or connect the following:

- Camshaft tensioners on both sides of the cylinder heads. Torque the tensioner mounting bolts to 75–96 inch lbs. (8–11 Nm).

➡The camshafts can be identified by the paint marks on the camshaft. The left cylinder head camshafts have a YELLOW paint mark and the right cylinder head camshafts have a WHITE paint mark. When installing the camshafts, position the camshaft dowel pins at the 12 o'clock position in respect to the cylinder head angle.

- Exhaust and intake camshafts and install the bearing caps. Before installing the No. 1 bearing cap, apply liquid gasket to the corners of the cap.

17. Torque the camshaft bearing caps as follows:

    a. Nos. 7 through 10 then, Nos. 1 through 6 to 17 inch lbs. (2 Nm).

    b. All bolts in order to 52 inch lbs. (6 Nm).

    c. Nos. 1 though 6 to 80–104 inch lbs. (9–11 Nm).

    d. Nos. 7 though 10 to 74–91 inch lbs. (9–11 Nm).

- New O-rings to the front of the engine block

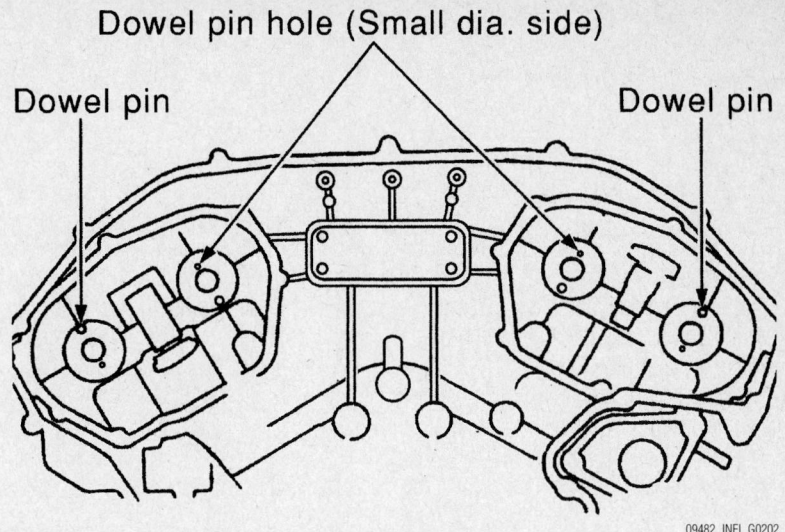

09482_INFI_G0202

Positioning of the camshaft dowel pins during installation—I35

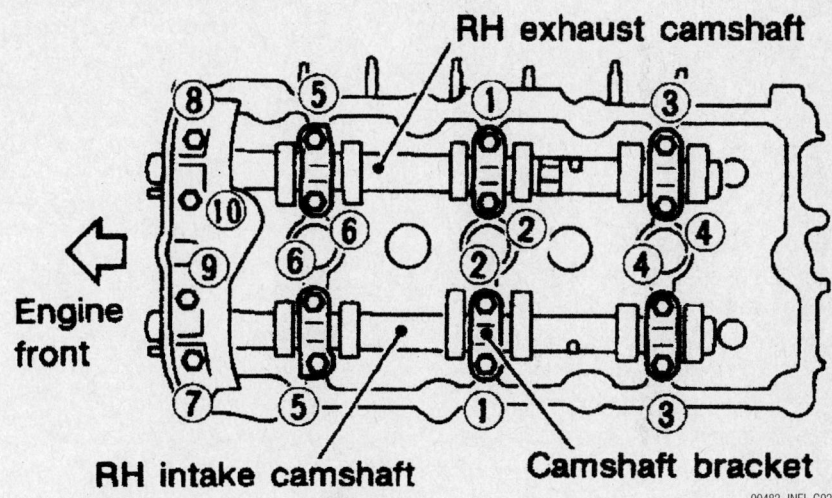

09482_INFI_G0203

Be sure to tighten the camshaft bearing cap bolts in the correct sequence—I35, right cylinder head

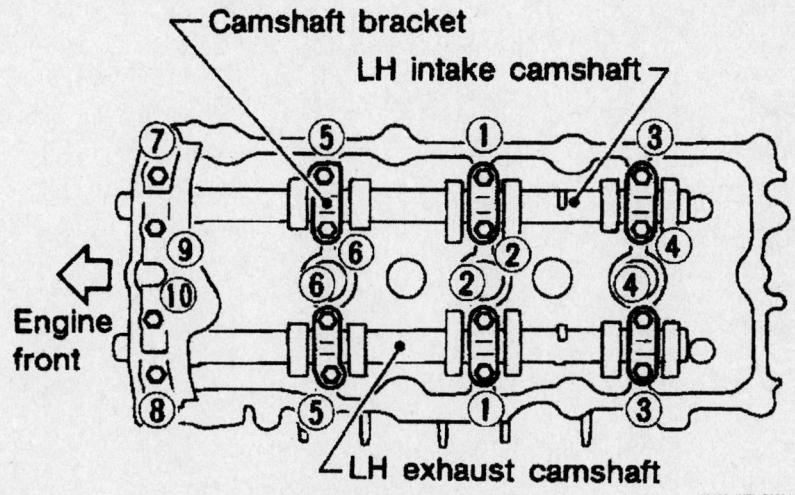

09482_INFI_G0204

Tighten the camshaft bearing cap bolts in the correct sequence—I35, left cylinder head

18. Apply sealant to the hatched portion of the of the rear timing chain case.

19. Align the rear timing chain case with the dowel pins and install onto the cylinder heads and engine block. Torque the rear timing chain case mounting bolts in sequence to 105–121 inch lbs. (11.8–13.7 Nm).

- Crankshaft sprocket with the mating mark facing out

20. Rotate the crankshaft clockwise and position the crankshaft to TDC of compression stroke and align the dowels of the camshaft sprockets to the 12 o'clock position in respect to the cylinder head

- Lower chain guide on the dowel pin with the front mark on the guide facing upward

21. On a workbench, align the marks on the intake and exhaust camshaft sprockets with the marks of the chain.

- Exhaust camshaft sprockets onto the dowel pin. Torque the bolts to 88–95 ft. lbs. (120–129 Nm). Be sure to secure the camshafts while tightening the bolts.
- Align and install the timing chains and sprockets to the camshafts
- Timing cover evenly and gently. Be sure to align the dowel pin holes

➡ **Leave the bolts unattended for 30 minutes or more after tightening.**

22. Apply a 0.091–0.130 in. (2.3–3.3mm) continuous bead of liquid gasket to the water pump cover and install the cover. Tighten the bolts to 84–108 inch lbs. (10–13 Nm).

- Rocker covers

23. Apply sealant to the front and rear seal of the oil pan and install the oil pan.

- Center crossmember assembly
- Right side engine mounting bracket and mount assembly
- Engine slinger assembly
- Front exhaust pipe and its support
- Air conditioning compressor and bracket
- Crankshaft pulley to the crankshaft and install the mounting bolt

24. Torque the mounting bolt to 14–22 ft. lbs. (20–29 Nm). Torque the crankshaft bolt an additional 60–66° clockwise. This is approximately the angle from 1 hexagon bolt head corner to another.

- Ring gear cover plate
- CMP sensor (PHASE) and CKP sensors
- Power steering pump assembly, drive belts and the idler pulley
- Engine side cover
- Right front wheel

- Engine undercover
- Intake manifold, using new gaskets
- Fuel tube assembly using new insulators. Torque the bolts in several steps to 15–20 ft. lbs. (21–26 Nm).
- Intake manifold collector gasket with the arrow facing forward
- Intake manifold collector assembly and support bracket
- EGR tube using new insulators. Torque the bolts in two steps to 15–20 ft. lbs. (21–26 Nm).
- Spark plugs and ignition coils. Torque the bolts to 27–33 inch lbs. (2.9–3.8 Nm).
- Water hoses to the cylinder head and intake manifold
- Fuel hoses and wiring harness connections to the fuel rail
- Air duct to intake manifold hose, collector hose, blow-by hose, and vacuum hoses
- Negative battery cable

25. Fill the engine with clean oil.

26. Fill and bleed the cooling system.

27. Start the engine and run at 3000 RPM under no load to purge the air from the high pressure chamber. The engine may produce a rattling noise. This indicates that air still remains in the chamber and is not a matter of concern.

G35 MODELS

1. Before servicing the vehicle, refer to the precautions in the beginning of this section.

2. Drain the engine oil and cooling system.

3. Relieve the fuel system pressure.

4. Remove or disconnect the following:

- Negative battery cable
- Front timing chain case
- Camshaft sprocket
- Timing chain
- Rear timing chain case
- Camshaft Position (CMP) sensors (PHASE) from the back side of the cylinder heads
- Intake valve timing control solenoid valves from the No. 1 camshaft bracket

➡ **Before removal, matchmark the position of the camshafts, brackets and bolts, so they are reinstalled in their original locations.**

- Intake and exhaust camshaft brackets. Loosen the bolts in several stages, in reverse of the sequence shown.
- Camshafts
- Valve lifters if necessary, noting their installed positions

To install:

5. Install or connect the following:

- Valve lifters, in their original positions
- Camshafts with the dowel pin attached to its front end face on the exhaust side
- Camshaft brackets, as shown in the illustration

Location of the intake valve timing control solenoid valves—G35 models

09482_INFI_G0205

Right bank

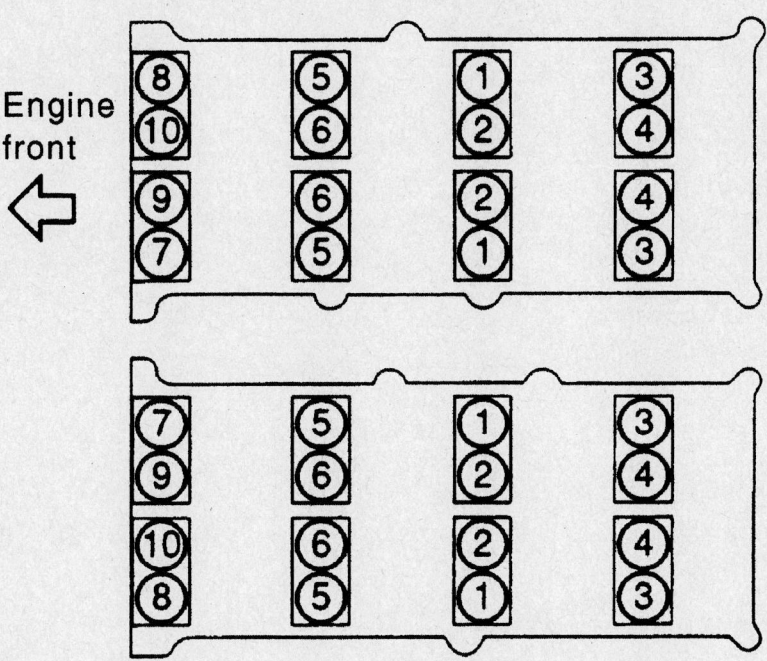

Engine front

Left bank

09482_INFI_G0206

Intake and exhaust camshaft bracket bolts tightening sequence. Use the reverse sequence for removal—G35 models

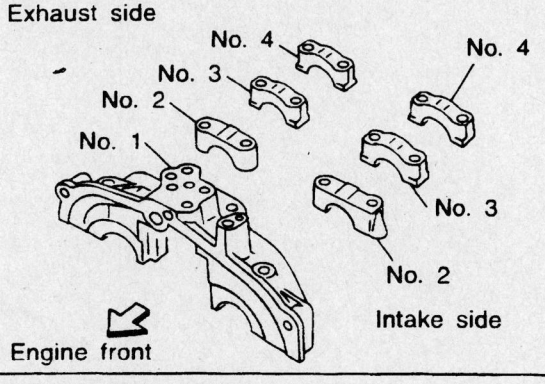

RH camshaft brackets

Exhaust side

Engine front

LH camshaft brackets

Intake side

Engine front

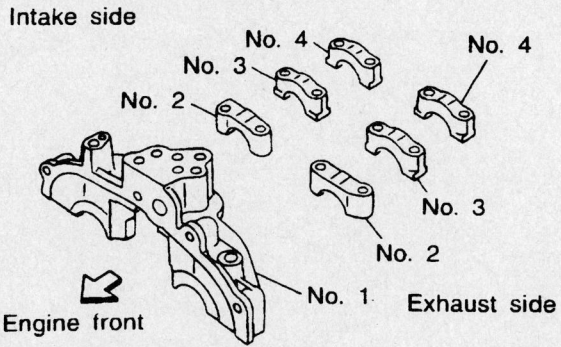

9357MG17

Camshaft identification and installation—G35 models

6. Torque the camshaft bracket bolts, in sequence, as follows:

 a. Step 1: Nos. 7–10, then Nos. 1–6 to 17 inch lbs. (2 Nm).

 b. Step 2: Nos. 1–10 to 52 inch lbs. (5.9 Nm).

 c. Step 3: Nos. 1–6 to 80–104 inch lbs. (9–12 Nm).

 d. Step 4: Nos. 7–10 to 74–91 inch lbs.

7. Measure the difference in levels between the front end faces of the No. 1 camshaft bracket and the cylinder head. If the measurement falls out of the range of -0.0055–0.0055 in. (-0.14–0.14mm), you must reinstall the camshaft and brackets.

8. Check and adjust the valve clearance.

- Intake valve timing control solenoid valves
- CMP sensors (PHASE)
- Rear timing chain case, timing chain, camshaft sprocket and front timing chain case
- Negative battery cable

9. Fill the engine with oil.

10. Fill and bleed the cooling system.

11. Start the vehicle, check for leaks and repair if necessary.

4.5L Engine

1. Before servicing the vehicle, refer to the precautions in the beginning of this section.

2. Drain the cooling system.

3. Relieve the fuel system pressure.

4. Remove or disconnect the following:
- Engine
- Timing chain assemblies
- Camshaft sprockets while holding camshaft locked in place
- Camshaft bracket bolts in the reverse order of the tightening sequence to prevent damage to the camshaft. Keep the brackets in order so they can be installed in the correct locations.
- Camshafts, shims and lifters. Keep all parts in order so they can be installed in their original positions.

To install:

5. Install or connect the following:
- Valve lifters and shims

6. Identify the correct camshaft for proper placement as shown in the illustration.

7. With no 1 cylinder at TDC, install the camshafts so the dowel pins are in the correct locations.

8. Install the camshaft brackets so the installation location can be correctly read

Right camshaft brackets
Exhaust side

No. 4
No. 3
No. 2
No. 1
No. 4
No. 3
No. 2

Engine front   Intake side

Left camshaft brackets
Intake side

No. 4
No. 3
No. 2
No. 4
No. 3
No. 2
No. 1

Engine front   Exhaust side

09482_INFI_G0208

Camshaft bracket installation—G35 models

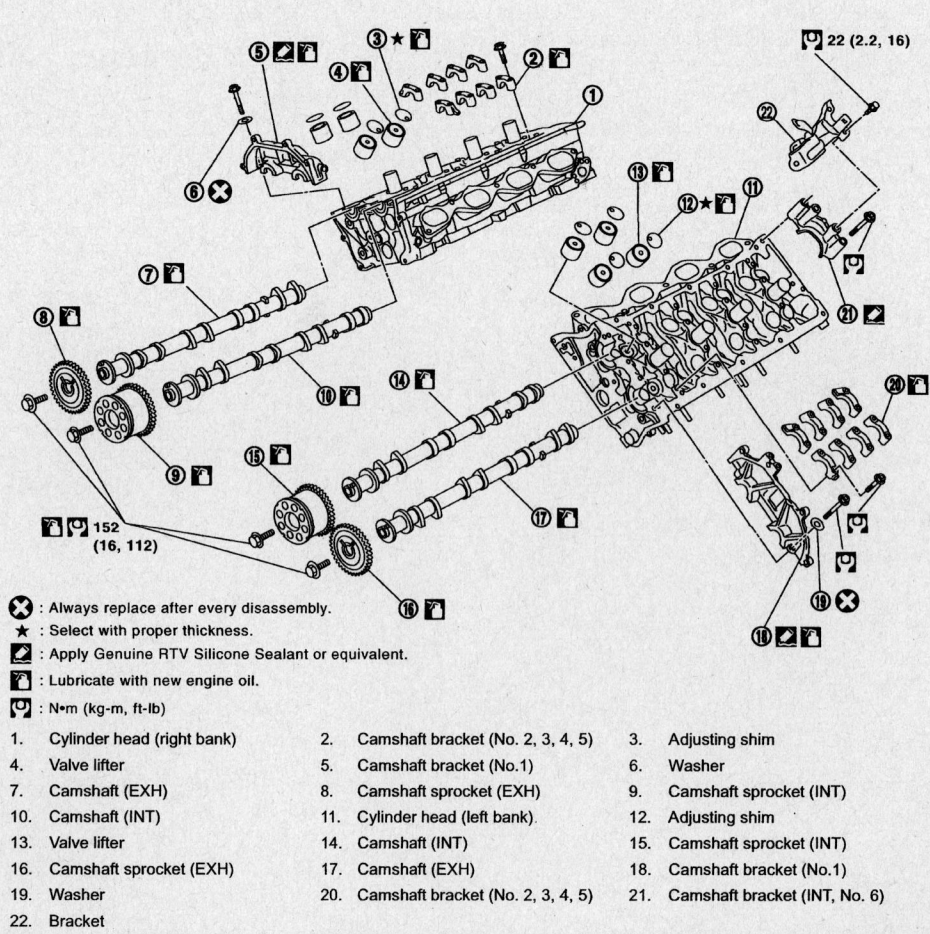

22 (2.2, 16)

152
(16, 112)

⊗ : Always replace after every disassembly.
★ : Select with proper thickness.
▨ : Apply Genuine RTV Silicone Sealant or equivalent.
▩ : Lubricate with new engine oil.
▣ : N•m (kg-m, ft-lb)

1. Cylinder head (right bank)	2. Camshaft bracket (No. 2, 3, 4, 5)	3. Adjusting shim
4. Valve lifter	5. Camshaft bracket (No.1)	6. Washer
7. Camshaft (EXH)	8. Camshaft sprocket (EXH)	9. Camshaft sprocket (INT)
10. Camshaft (INT)	11. Cylinder head (left bank)	12. Adjusting shim
13. Valve lifter	14. Camshaft (INT)	15. Camshaft sprocket (INT)
16. Camshaft sprocket (EXH)	17. Camshaft (EXH)	18. Camshaft bracket (No.1)
19. Washer	20. Camshaft bracket (No. 2, 3, 4, 5)	21. Camshaft bracket (INT, No. 6)
22. Bracket		

67162-INFI-G18

Exploded view of camshaft assemblies—4.5L engine

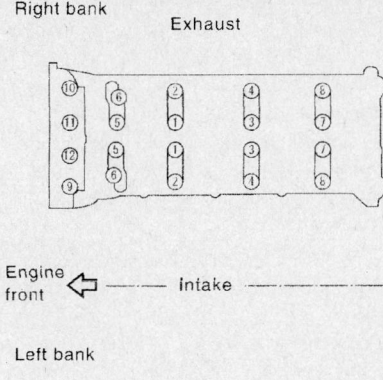

Camshaft bracket tightening sequence—
4.5L engine

67162-INFI-G19

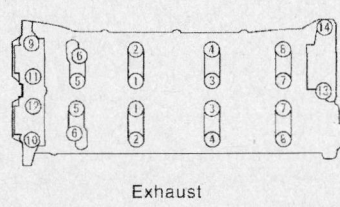

Identifying correct camshaft placement—
4.5L engine

67162-INFI-G20

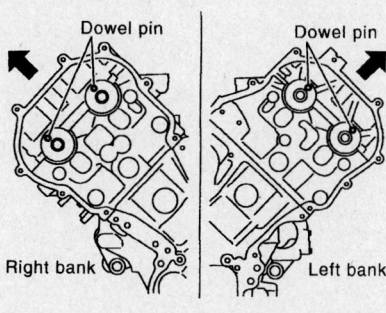

Identifying correct camshaft dowel pin
locations—4.5L engine

67162-INFI-G21

when viewed from the side of the left
exhaust bank.

9. Apply liquid gasket to camshaft
brackets No. 1 and left side no. 6 in the
locations shown.

10. Tighten the camshaft brackets in the
sequence shown using the following steps:

   a. Tighten numbers 9 to 12 to 17 inch
lbs. (1.96 Nm).

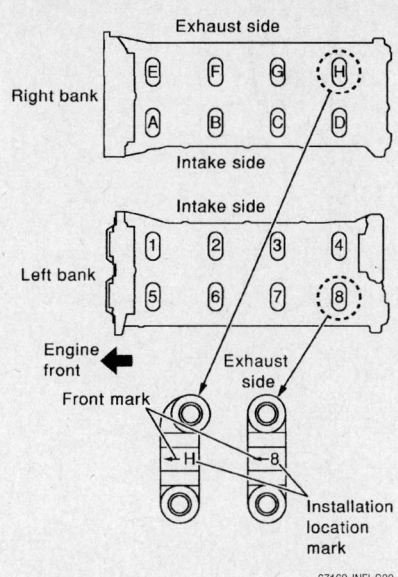

Identifying correct camshaft bracket installation—4.5L engine

67162-INFI-G22

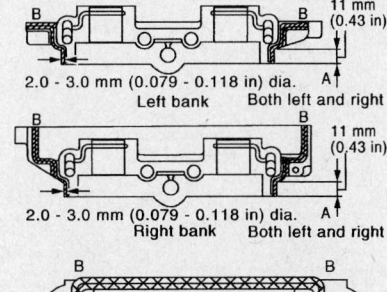

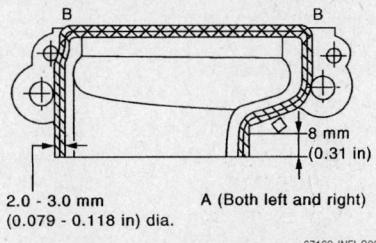

Identifying liquid gasket application areas
to camshaft brackets—4.5L engine

67162-INFI-G23

   b. Tighten numbers 1 to 8 to 17 inch
lbs. (1.96 Nm).

   c. Tighten numbers 13 and 14 to 17
inch lbs. (1.96 Nm).

   d. Tighten all bolts in sequence to 52
inch lbs. (5.88 Nm).

   e. Tighten numbers 1 to 12 to 92 inch
lbs. (10.41 Nm).

   f. Tighten numbers 13 and 14 to 23
ft. lbs. (31 Nm).

11. Install the camshaft sprockets and
while holding the camshaft in place tighten
the bolts to 112 ft. lbs. (152 Nm).

12. Adjust the valve clearances.

13. Install or connect the following:
   • Timing chain assemblies

   • Engine
   • Negative battery cable

14. Fill and bleed the cooling system.

15. Start the vehicle, check for leaks and
repair if necessary.

Valve Lash

ADJUSTMENT

2.0L Engine

➡A special gauge plate and collar will
be needed to complete this procedure.

1. Before servicing the vehicle, refer to
the precautions in the beginning of this section.

2. Remove the camshafts.

3. Install the J38957-1 gauge plate to
the cylinder head. Use the bolts supplied in
the kit to secure the plate to the cam bearing
journals.

4. Install the collar J38957-2 on the
dial indicator. Be sure the dished side of the
collar is toward the gauge and tighten the
setscrew.

5. Place the gauge on the No. 1 intake
valve (shim side). Be sure the shim has
been removed. Place the tip of the dial
gauge on the top of the valve stem and the
collar on the gauge plate. Zero the dial
gauge.

6. Move the dial gauge to the other
intake valve (rocker guide side). Place the
tip of the dial gauge on the rocker guide and
the collar of the gauge plate. Record the
measurement.

7. Select the correct size shim using the
chart. Shims are available in 17 different
sizes ranging from 0.1102 in. (2.800mm) to
0.1260 in. (3.200mm) in increments of
0.001 in. (0.025mm).

3.5L Engines

➡Check and adjust the valve clearances while the engine is cold and not
running.

1. Before servicing the vehicle, refer to
the precautions in the beginning of this section.

2. Remove the intake manifold collector.

3. Remove the left and right rocker covers.

4. Remove the spark plugs.

5. Set the No. 1 cylinder at Top Dead
Center (TDC) on its compression stroke.
Align the pointer with the TDC mark on the
crankshaft pulley. Check that the valve
adjusters on the No. 1 cylinder are loose

Available shim

Thickness mm (in)	Identification mark
2.800 (0.1102)	28 00
2.825 (0.1112)	28 25
2.850 (0.1122)	28 50
2.875 (0.1132)	28 75
2.900 (0.1142)	29 00
2.925 (0.1152)	29 25
2.950 (0.1161)	29 50
2.975 (0.1171)	29 75
3.000 (0.1181)	30 00
3.025 (0.1191)	30 25
3.050 (0.1201)	30 50
3.075 (0.1211)	30 75
3.100 (0.1220)	31 00
3.125 (0.1230)	31 25
3.150 (0.1240)	31 50
3.175 (0.1250)	31 75
3.200 (0.1260)	32 00

7923HG56

Select the correct valve lash adjusting shim using the chart—2.0L engine

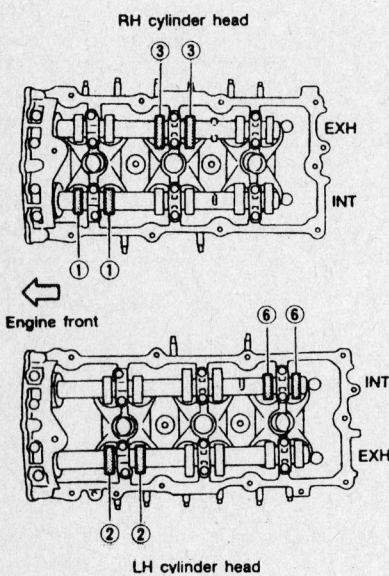

RH cylinder head

EXH

INT

Engine front

INT

EXH

LH cylinder head

7923HG57

Valve lash checking sequence at TDC of cylinder No. 1—3.5L engine

and valve adjusters on the No. 4 cylinder are tight. If not, turn the crankshaft 1 revolution (360°) and align the pointer with the TDC mark on the crankshaft pulley.

6. Check the following valves:
- Both No. 1 intake valves
- Both No. 2 exhaust valves
- Both No. 3 exhaust valves
- Both No. 6 intake valves

7. Using a feeler gauge, measure the clearance between the valve adjuster and the camshaft. Record any valve clearance measurements that are out of specification. Intake valve clearance (cold) is 0.010–0.013 in. (0.26–0.34mm) and exhaust valve clearance (cold) is 0.011–0.015 in. 0.29–0.37mm).

8. Turn the crankshaft 240° and set the No. 3 cylinder to TDC of its compression stroke.

9. Check the following valves:
- Both No. 2 intake valves
- Both No. 3 intake valves
- Both No. 4 exhaust valves

- Both No. 5 exhaust valves

10. Using a feeler gauge, measure the clearance between the valve adjuster and the camshaft. Record any valve clearance measurements that are out of specification. Intake valve clearance (cold) is 0.010–0.013 in. (0.26–0.34mm) and exhaust valve clearance (cold) is 0.011–0.015 in. (0.29–0.37mm).

11. Turn the crankshaft 240° and set the No. 5 cylinder to TDC of its compression stroke.

12. Check the following valves:
- Both No. 1 exhaust valves
- Both No. 4 intake valves
- Both No. 5 intake valves
- Both No. 6 exhaust valves

13. Using a feeler gauge, measure the clearance between the valve adjuster and the camshaft. Record any valve clearance measurements that are out of specification. Intake valve clearance (cold) is 0.010–0.013 in. (0.26–0.34mm) and exhaust valve clearance (cold) is 0.011–0.015 inches (0.29–0.37mm).

14. If all the valve clearances are within specification, install the cylinder head cover, spark plugs, and the intake manifold collector.

15. If an adjustment is necessary, adjust the valve clearance while engine is cold by removing the adjusting shim. The adjusting shim can be removed by using the following procedures:

a. Turn the crankshaft so the camshaft lobe of the valve to be adjusted is pointed straight up.

b. Turn the adjuster so the notch is pointed towards the center of the cylinder head; this will facilitate the shim removal process.

c. Using a depressor tool No. KV10115110 push down on the adjuster and insert a keeper tool on the edge of the adjuster to keep the adjuster in the depressed position.

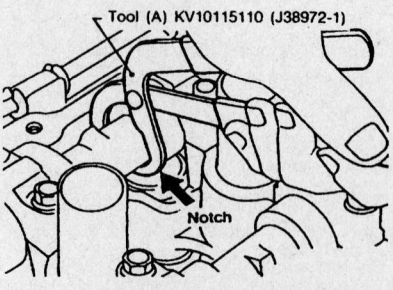

Tool (A) KV10115110 (J38972-1)

Notch

7923HG58

Install the depressor tool around the camshaft being careful not to damage the surfaces—3.5L engine shown

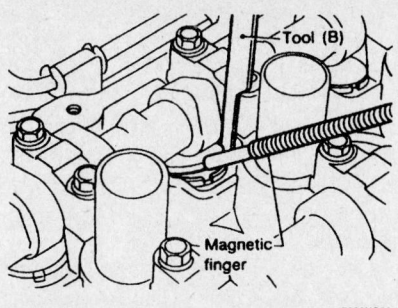

Use a magnet to remove the shim from the adjuster. Sometimes a shot of compressed air can help lift the shim up—3.5L engine shown

d. Remove the depressor tool and remove the shim with a magnet.

➡**Compressed air can be blown into the hole of the adjuster to separate the adjusting shim from the adjuster.**

16. Determine the replacement adjusting shim size by using the following procedures and formula:

    a. Using a micrometer determine thickness of the removed shim.

    b. Calculate the thickness of a new adjusting shim so valve clearance is within the specified values.

- R= thickness of the removed shim
- N= thickness of the new shim
- M= measured valve clearance
- Calculate the Intake Shim as follows: N = R + M−0.0118 in. (0.30mm)
- Calculate the Exhaust Shim as follows: N = R + M−0.0130 in. (0.33mm)

17. Shims are available in 64 sizes from 0.0913–0.1161 in. (2.32–2.95mm) in steps of 0.004 in. (0.01mm). The thickness is stamped on the shim; this side is always installed facing down. Select new shims with thickness as close as possible to calculated valve and install it in the adjuster.

18. Install the new shim onto the adjuster.

19. Depress the adjuster and remove the keeper tool. Remove the depressor tool and recheck the valve clearance. Repeat this procedure for any other valves requiring adjustment.

20. When all valve adjustments are finished, install the cylinder head cover, spark plugs, and the intake manifold collector.

4.5L Engines

➡**Check the valve clearances while the engine is warm and not running. Adjustments must be made when the engine is COLD.**

➡**The 4.5L firing order is 1-8-7-3-6-5-4-2. The left bank has cylinders No. 1, 3, 5 and 7 from front to rear and the right bank has cylinders No. 2, 4, 6 and 8 from front to rear.**

1. Before servicing the vehicle, refer to the precautions in the beginning of this section.

2. Remove the engine appearance cover.

3. Remove the left and right rocker covers.

4. Turn the crankshaft clockwise until the TDC mark without the paint mark aligns with the timing pointer.

5. Check that the camshaft lobes on the number one cylinder are pointing outward.

6. Check the following valves:

- Cylinder numbers 1 and 2 intake valves
- Cylinder number 1 exhaust valves
- Cylinder numbers 4 and 5 intake valves
- Cylinder numbers 7 and 8 exhaust valves

7. Using a feeler gauge, measure the clearance between the valve lifter and the camshaft. Record any valve clearance measurements that are out of specification. Intake valve clearance (hot) is 0.012–0.016 in. (0.30–0.41mm) and exhaust valve clearance (hot) is 0.012–0.017 in. 0.30–0.43mm).

8. Turn the crankshaft 270° and set the No. 3 cylinder to TDC of its compression stroke.

9. Check the following valves:

- Cylinder numbers 3 and 4 exhaust valves
- Cylinder numbers 3 and 7 intake valves
- Cylinder numbers 5 and 8 exhaust valves

10. Using a feeler gauge, measure the clearance between the valve lifter and the camshaft. Record any valve clearance measurements that are out of specification. Intake valve clearance (hot) is 0.012–0.016

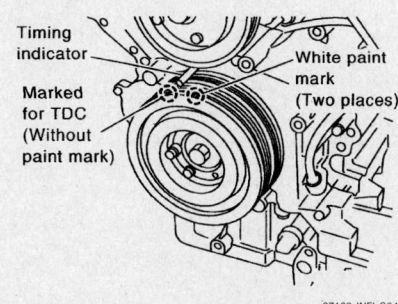

Identifying No. 1 cylinder TDC mark—4.5L engine

in. (0.30–0.41mm) and exhaust valve clearance (hot) is 0.012–0.017 in. 0.30–0.43mm).

11. Turn the crankshaft 90° (360° from No. 1 TDC) and set the No. 6 cylinder to TDC of its compression stroke.

12. Check the following valves:

- Cylinder numbers 2 and 6 exhaust valves
- Cylinder number 6 intake valves

13. If all the valve clearances are within specification, install the cylinder head cover and the engine appearance cover.

14. If an adjustment is necessary, adjust the valve clearance while engine is **COLD** by removing the adjusting shim. Refer to Tune Up Specifications for cold clearance measurements.

15. The adjusting shim can be removed by using the following procedures:

    a. Turn the crankshaft so the camshaft lobe of the valve to be adjusted is pointed straight up.

    b. Using an extra fine screwdriver, turn the round hole of the adjusting shim so it faces toward the center of the cylinder head.

    c. Using a depressor tool No. KV10115110 push down on the adjuster and insert a keeper tool on the edge of the adjuster to keep the adjuster in the depressed position.

    d. Remove the depressor tool and remove the shim with a magnet.

➡**Compressed air can be blown into the hole of the adjuster to separate the adjusting shim from the adjuster.**

16. Determine the replacement adjusting shim size by using the following procedures and formula:

    a. Using a micrometer determine thickness of the removed shim.

    b. Calculate the thickness of a new adjusting shim so valve clearance is within the specified values.

- R= thickness of the removed shim
- N= thickness of the new shim
- M= measured valve clearance
- Calculate the Intake Shim as follows: N = R + M−0.0118 in. (0.30mm)
- Calculate the Exhaust Shim as follows: N = R + M−0.0130 in. (0.33mm)

17. Shims are available in 64 sizes from 0.0913–0.1161 in. (2.32–2.95mm) in steps of 0.004 in. (0.01mm). The thickness is stamped on the shim; this side is always installed facing down. Select new shims with thickness as close as possible to calculated valve and install it in the adjuster.

18. Install the new shim onto the adjuster.

19. Depress the adjuster and remove the keeper tool. Remove the depressor tool and recheck the valve clearance. Repeat this procedure for any other valves requiring adjustment.

20. When all valve adjustments are finished, install the cylinder head cover and the engine appearance cover.

Starter Motor

REMOVAL & INSTALLATION

2.0L Engine

1. Before servicing the vehicle, refer to the precautions in the beginning of this section.
2. Remove or disconnect the following:
   - Negative battery cable
   - Starter insulator
   - Starter harness connector and cable
   - Starter mounting bolt and nut
   - Starter

To install:
3. Install or connect the following:
   - Starter. Torque the bolts to 27 ft. lbs. (36 Nm).
   - Starter harness connector and cable

- Starter insulator
- Negative battery cable

3.5L Engines

EXCEPT G35 & M35 MODELS

1. Before servicing the vehicle, refer to the precautions in the beginning of this section.
2. Remove or disconnect the following:
   - Negative battery cable
   - Air duct assembly
   - Harness protector
   - Starter harness
   - Both starter bolts
   - Starter

To install:
3. Install or connect the following:
   - Starter
   - Both starter bolts. Tighten the long bolt to 57–72 ft. lbs. (77–98 Nm) and the short bolt to 22–30 ft. lbs. (30–41 Nm).
   - Starter harness
   - Harness protector
   - Air duct assembly
   - Negative battery cable

G35 MODELS

1. Before servicing the vehicle, refer to the precautions in the beginning of this section.

2. Remove or disconnect the following:
   - Negative battery cable
   - Engine undercover (2WD); front and rear engine undercover (AWD).
   - S and B terminals from the starter
   - Starter mounting bolts and harness bracket
   - Starter

To install:
3. Install or connect the following:
   - Starter motor
   - Terminals to the starter
   - Engine undercover (2WD); front and rear engine undercover (AWD)
   - Negative battery cable

M35 MODELS

1. Before servicing the vehicle, refer to the precautions in the beginning of this section.

2. Remove or disconnect the following:
   - Negative battery cable
   - Front and rear engine undercover
   - Exhaust mounting bracket (AWD only)
   - S and B terminals from the starter
   - Starter mounting bolts and harness bracket
   - Starter

To install:
3. Install or connect the following:

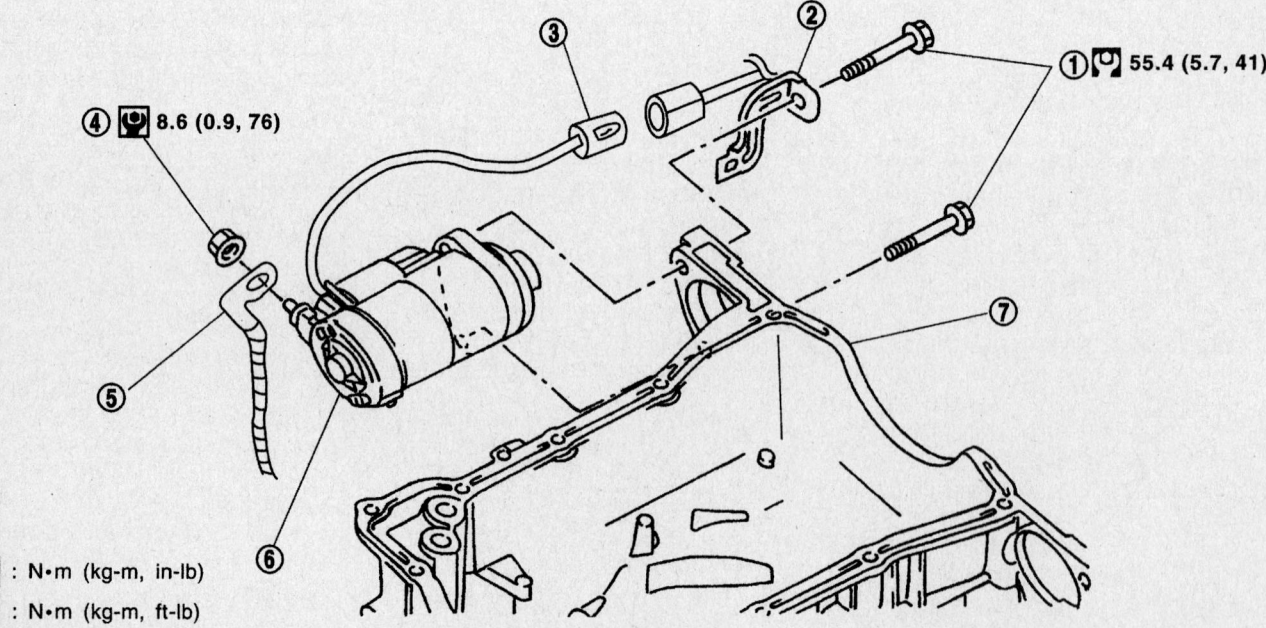

④ 🔧 8.6 (0.9, 76)   ① 🔧 55.4 (5.7, 41)

🔧 : N•m (kg-m, in-lb)

🔧 : N•m (kg-m, ft-lb)

1. Starter motor mounting bolt
2. Harness clip bracket
3. S connector
4. B terminal nut
5. B terminal harness
6. Starter motor
7. Oil pan

09482_INFI_G0040

Starter mounting and torque specifications—G35 & M35 (2WD)

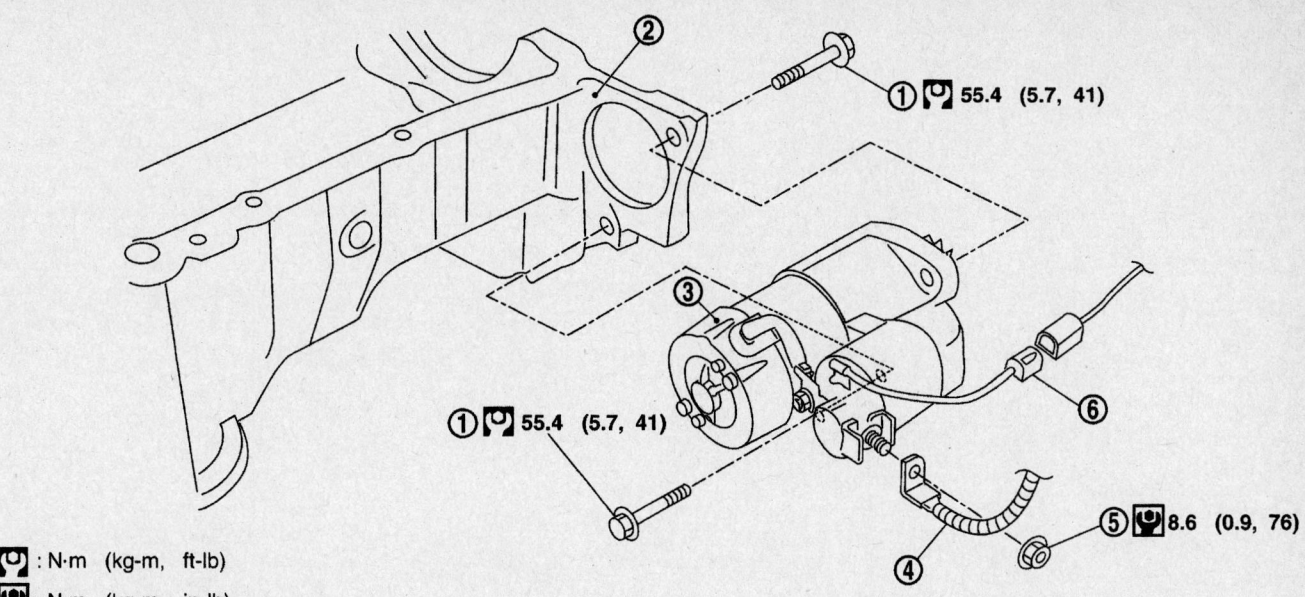

☐ : N·m (kg-m, ft-lb)

☐ : N·m (kg-m, in-lb)

1.	Starter motor mounting bolt	2.	Oil pan	3.	Starter motor
4.	B terminal harness	5.	B terminal nut	6.	S connector

09482_INFI_G0041

Starter mounting and torque specifications—G35 & M35 (AWD)

- Starter motor
- Terminals to the starter
- Exhaust mounting bracket
- Front and rear engine undercover
- Negative battery cable

4.5L Engine

Q45 AND 2004 M45 MODELS

1. Before servicing the vehicle, refer to the precautions in the beginning of this section.

2. Remove or disconnect the following:

- Negative battery cable
- Engine undercover
- Harness connectors

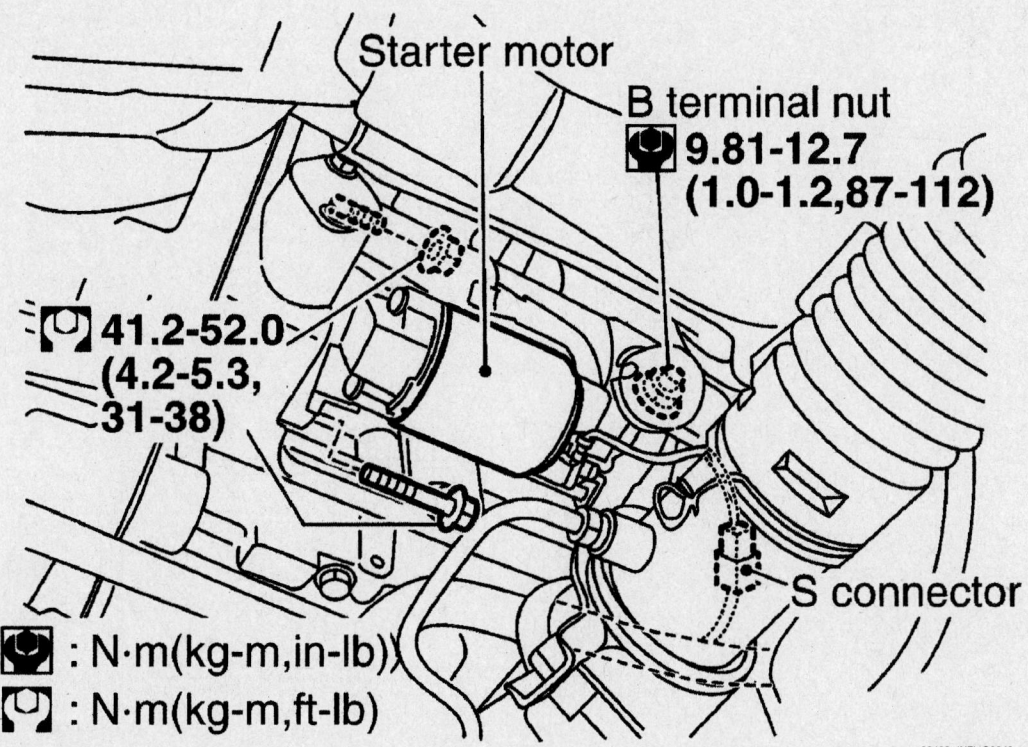

09482_INFI_G0042

Starter motor mounting and torque specifications—Q45 & 2004 M45

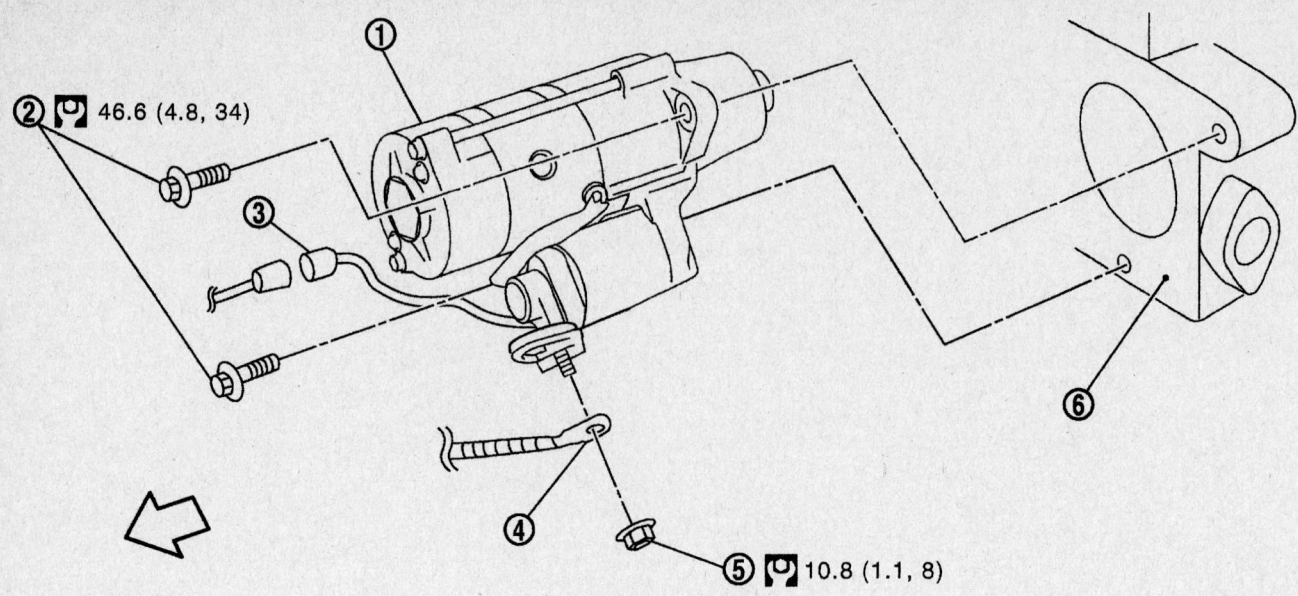

2 🔧 46.6 (4.8, 34)

5 🔧 10.8 (1.1, 8)

1. Starter motor
2. Starter motor mounting bolt
3. "S" connector
4. "B" terminal harness
5. "B" terminal nut
6. Cylinder block
🔧 : N·m (kg-m, ft-lb)
⇦ : Engine front

09482_INFI_G0043

Starter motor mounting and torque specifications—2006 M45

- Starter motor mounting bolt and nut
3. Remove engine mounting insulator bottom nuts, and lift engine approximately 2 inches with a transmission jack
4. Remove starter motor
To install:
5. Installation is in reverse order of removal; observe torque values shown in diagram

M45 MODELS—2006

1. Before servicing the vehicle, refer to the precautions in the beginning of this section.
2. Remove or disconnect the following:

- Negative battery cable
- Engine undercover
- Left engine mounting insulator and bracket
- Harness connectors and harness bracket
- Starter motor mounting bolts
3. Remove starter toward front of vehicle
To install:
4. Installation is in reverse order of removal; observe torque values shown in diagram

Oil Pan

REMOVAL & INSTALLATION

2.0L Engine

1. Before servicing the vehicle, refer to the precautions in the beginning of this section.
2. Raise and support the vehicle safely.
3. Drain the engine oil.
4. Remove or disconnect the following:

- Negative battery cable
- Engine undercover

- Steel oil pan bolts in the proper sequence
- Steel oil pan. Insert tool KV10111100 between steel oil pan and aluminum oil pan to break the seal.
- Front exhaust tube and support the transaxle with a suitable jack and raise the engine with an engine hoist
- Center crossmember
- Transaxle shift control cable, if equipped with an automatic transaxle
- A/C compressor bracket gussets and the rear cover plate
- Aluminum oil pan bolts in sequence

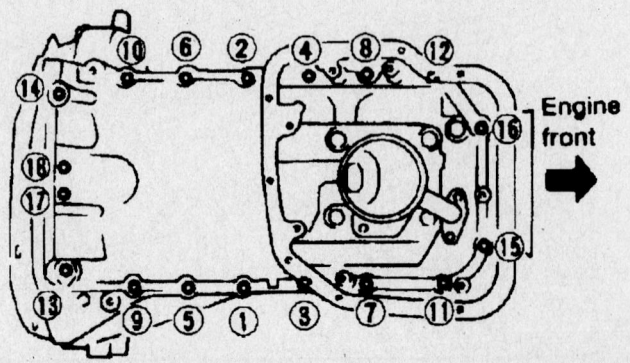

Tighten the aluminum oil pan mounting bolts in the sequence shown—2.0L engine

7923HG60

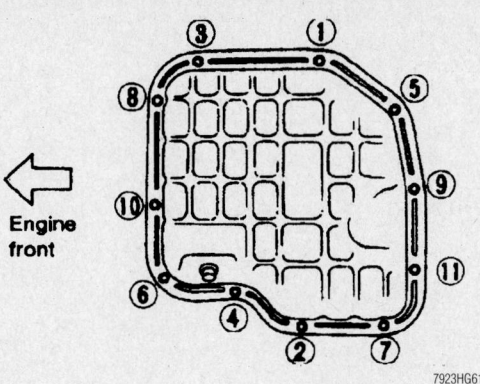

7923HG61

Be sure to tighten the steel oil pan mounting bolts in the proper order to prevent leakage—2.0L engine

- Baffle plate
- 2 engine-to-transaxle bolts and install them into vacant bolt holes on the oil pan. Tighten the bolts to release the oil pan from the cylinder block. Use tool KV10111100 to break the remaining seal.

To install:

5. Remove the 2 bolts previously installed in the oil pan.

6. Clean the oil pan rail of all liquid gasket and apply a new bead of ⅛ inch thickness to the oil pan rail.

7. Install or connect the following:
- Aluminum oil pan.

8. Torque the bolts in the proper sequence as follows:
  a. Bolts 1 through 16 to 12–14 ft. lbs. (16–19 Nm).
  b. Bolts 17 and 18 to 56–66 inch lbs. (6.5–7.5 Nm).
- 2 engine-to-transaxle bolts, rear cover plate, compressor bracket gussets, automatic transmission shift control cable (if equipped), center member, front exhaust tube and baffle plate

9. Clean the oil pan rail of all liquid gasket and apply a new bead of ⅛ inch thickness to the oil pan rail.
- Steel oil pan. Torque the bolts in numbered sequence to 56–66 inch lbs. (6–8 Nm). Wait 30 minutes before refilling engine case with oil.
- Negative battery cable

10. Fill the engine with clean oil.
- Start the vehicle, check for leaks and repair if necessary.

3.5L Engines

I35 MODELS

1. Before servicing the vehicle, refer to the precautions in the beginning of this section.

2. Drain the engine oil.

3. Remove or disconnect the following:
- Negative battery cable
- Engine undercover(s)
- Steel (lower) oil pan bolts in the reverse sequence of the torque sequence

4. Insert a seal cutter between the steel and aluminum oil pan.

5. Tapping the cutter with a hammer, slide it around the entire edge of the oil pan. Be careful not to damage the aluminum mating surface of the upper oil pan.
- Steel oil pan and the oil strainer
- Front exhaust pipe and its support

6. Hang the engine at the right and left side engine slingers with a suitable hoist.

7. Position a suitable jack under the transaxle.

8. Remove or disconnect the following:
- Crankshaft Position (CKP) sensors (REFERENCE and POSITION) from the oil pan
- Front and rear engine mounting nuts and bolts
- Center crossmember assembly
- Engine drive belts
- A/C compressor and bracket
- Rear cover plate
- Aluminum (upper) oil pan bolts in

the reverse sequence of the torque sequence
- 4 engine-to-transaxle bolts

9. Insert a seal cutter between the aluminum oil pan and the engine block.

10. Tapping the cutter with a hammer, slide it around the entire edge of the oil pan. Be careful not to damage the mating surfaces of the oil pan or engine block.
- Oil pan assembly
- O-rings from the cylinder block and oil pump body

To install:

11. Install or connect the following:
- Baffle plate to the oil pan. Torque the bolts to 22–27 inch lbs. (2–3 Nm).

12. Apply sealant to the front and rear seal of the oil pan.
- New O-rings to the cylinder block and the oil pump body

13. Apply a 0.177–0.217 in. (4.5–5.5mm) continuous bead of liquid gasket to the upper oil pan mating surface and install the oil pan. Torque the bolts in sequence to 12–14 ft. lbs. (16–19 Nm).
- Oil pan strainer. Torque the bolts to 12–14 ft. lbs. (16–19 Nm).
- Rear cover plate and the lower transaxle bolts
- A/C compressor and bracket
- Engine drive belts and adjust as necessary

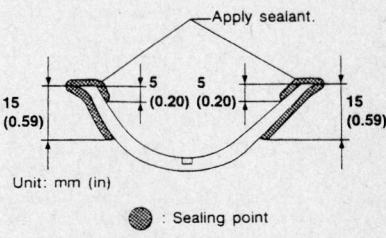

9307HG11

Apply sealant to the front and rear seal of the oil pan as shown—I35

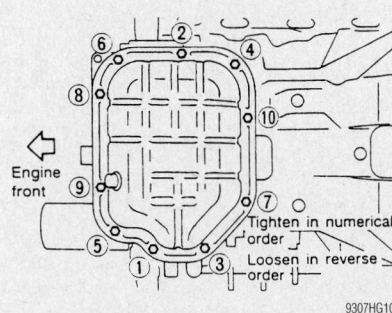

9307HG10

Steel (lower) oil pan loosening and tightening sequence—I35

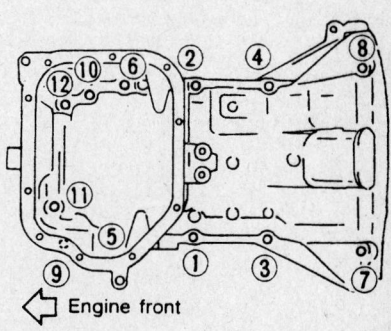

7923HG63

Aluminum oil pan torque sequence (loosen in reverse sequence)—I35

- Center crossmember assembly
- Front and rear engine mounting nuts and bolts and remove the support jack and the engine hoist
- CKP sensors (REFERENCE and POSITION) to the oil pan. Torque the bolts to 75–96 inch lbs. (9–10 Nm).
- Front exhaust pipe and its support
- Oil strainer

14. Apply a 0.177–0.217 in. (4.5–5.5mm) continuous bead of liquid gasket to the lower oil pan mating surface and install the oil pan. Tighten the mounting bolts in sequence to 57–66 inch lbs. (6–8 Nm)

➡ **Wait at least 30 minutes before refilling the engine oil.**

- Engine undercover(s)
- Negative battery cable

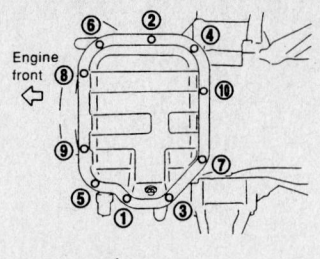

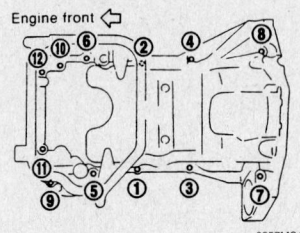

9357MG19

Lower and upper oil pan tightening sequence (use reverse for loosening)— G35 and M35 models (all)

15. Fill the engine with clean oil.
16. Start the engine, check for leaks and repair if necessary.

G35 AND M35— 2WD MODELS

1. Before servicing the vehicle, refer to the precautions in the beginning of this section.
2. Drain the engine oil and coolant.
3. Remove the hood.
4. Install a suitable engine slinger to secure the engine for crossmember removal.
5. Remove or disconnect the following:
   - Front suspension crossmember
   - Drive belts
   - Alternator and starter
   - Idler pulley and bracket
   - Crankshaft Position (CKP) sensor
   - Oil filter and oil cooler, as necessary

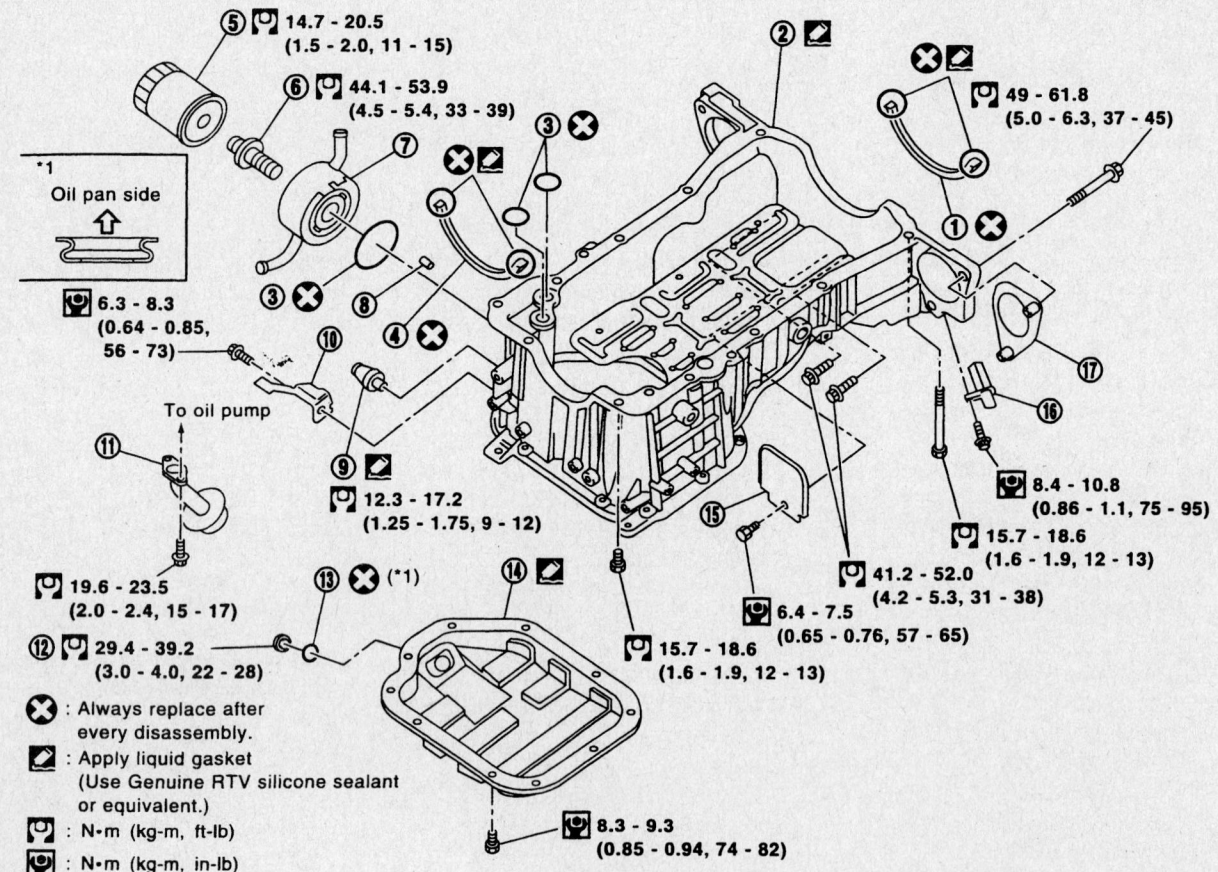

No.	Component	No.	Component	No.	Component
1.	Oil pan gasket	2.	Oil pan (upper)	3.	O-ring
4.	Oil pan gasket	5.	Oil filter	6.	Connector bolt
7.	Oil cooler	8.	Relief valve	9.	Oil pressure switch
10.	Bracket	11.	Oil strainer	12.	Drain plug
13.	Drain plug washer	14.	Oil pan (lower)	15.	Rear plate
16.	Crankshaft position sensor (POS)	17.	Rear cover plate		

9357MG18

Exploded view of the upper and lower oil pans and related components—G35 and M35 2WD models

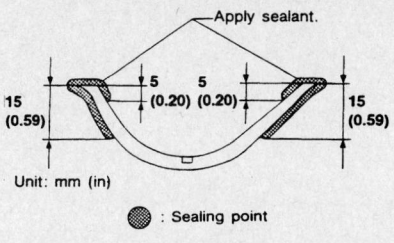

Proper sealant application for upper oil pan gasket—G35 and M35 models (all)

- Lower oil pan bolts in the reverse sequence of the torque sequence

6. Insert a seal cutter between the upper and lower oil pan. Tapping the cutter with a hammer, slide it around the entire edge of the oil pan. Be careful not to damage the aluminum mating surface of the upper oil pan.

- Transmission joint bolts which pierce upper oil pan
- Rear cover plate
- Upper oil pan bolts in the reverse of the torque sequence

7. Insert a seal cutter between the steel and aluminum oil pan. Tapping the cutter with a hammer, slide it around the entire edge of the oil pan. Be careful not to damage the mating surfaces.

- Oil strainer
- O-rings and discard

8. Clean all gasket mating surfaces.

To install:

9. Install or connect the following:

- Oil strainer to the oil pump
- New O-rings to the cylinder block and oil pump side
- Oil pan gasket, applying RTV sealant as shown. Align the protrusion of the oil pan gasket with the notches of the front timing chain case and rear oil seal retainer.

10. Apply sealant as shown in the illustration. Install the upper oil pan and tighten the bolts, in sequence, to 12–13 ft. lbs. (15.7–18.6 Nm).

- Transmission joint bolts

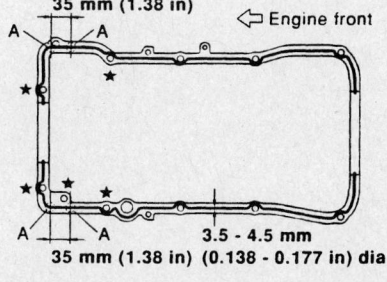

Upper oil pan sealant application—G35 and M35 models (all)

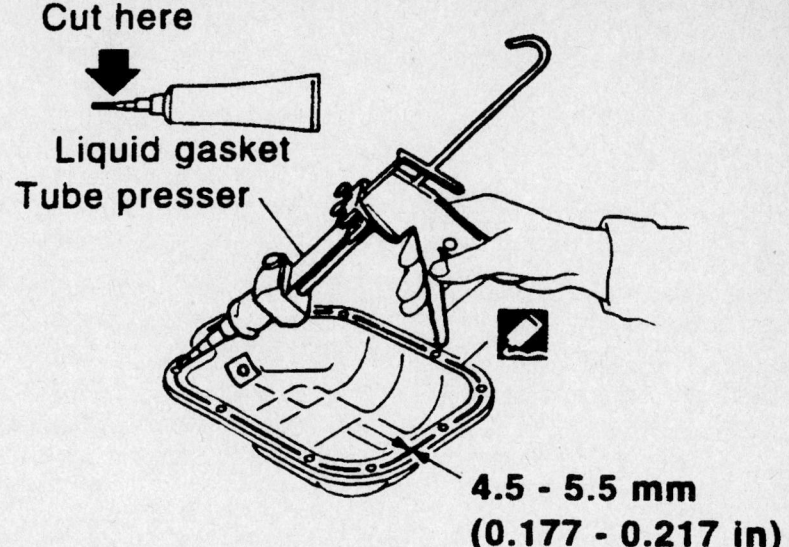

: Apply liquid gasket. Use Genuine RTV silicone sealant or equivalent.

Lower oil pan sealant application—G35 and M35 models (all)

11. For the lower oil pan, apply sealant as shown in the illustrations, then install and tighten the bolts, in sequence to 74–82 inch lbs. (8.3–9.3 Nm).

12. The remainder of installation is the reverse of the removal procedure.

➡ **Wait at least 30 minutes before refilling the engine oil.**

13. Connect the negative battery cable

14. Fill the engine with clean oil and coolant.

15. Start the engine, check for leaks and repair if necessary.

G35 AND M35 AWD MODELS

1. Before servicing the vehicle, refer to the precautions in the beginning of this section.

2. Drain the engine oil and coolant.

3. Remove the hood.

4. Remove engine cover

5. Remove air hose and air duct to mass air flow sensor side and electronic throttle control actuator side

6. Remove front and rear undercover

7. Install a suitable engine sling to secure the engine for crossmember removal.

8. Remove or disconnect the following:

- Drive belts
- Front suspension crossmember
- Left and right halfshafts
- Side (intermediate) shaft
- Left and right engine mounting brackets and insulators
- Front driveshaft
- Alternator and starter
- Idler pulley and bracket
- Oil filter and oil cooler, as necessary
- Automatic transmission oil cooler
- Crankshaft Position (CKP) sensor

➡ **Handle sensor with care; do not allow metal powder to adhere to magnetic tip or place sensor in a location where it might be exposed to magnetism**

- Lower oil pan bolts in the reverse sequence of the torque sequence

9. Insert a seal cutter between the upper and lower oil pan. Tapping the cutter with a hammer, slide it around the entire edge of the oil pan. Be careful not to damage the aluminum mating surface of the upper oil pan.

10. Remove oil strainer

11. Remove transmission joint bolts that pierce upper oil pan

12. Remove rear plate cover

13. Remove upper oil pan bolts in the reverse sequence of the torque sequence

14. Remove O-rings from bottom of cylinder block and oil pump and discard

15. Remove oil pan gaskets and clean all mating surfaces

16. Remove axle pipe from upper oil pan (if necessary) and discard O-rings.

To install:

17. Install axle pipe to upper oil pan, if removed, from axle pipe flange side (left)

➡️**Lubricate O-rings with engine oil and insert pipe with care to prevent O-ring from sliding**

18. Install or connect the following:
• Oil strainer to the oil pump
• New O-rings to the cylinder block and oil pump side
• Oil pan gasket, applying RTV sealant as shown. Align the protrusion of the oil pan gasket with the notches of the front timing chain case and rear oil seal retainer.

19. Apply sealant as shown in the illustration. Install the upper oil pan and tighten the bolts, in sequence, to 12–13 ft. lbs. (15.7–18.6 Nm).

➡️**M8 bolts number 5, 7, 8 and 11 are 3.97 inches (100mm); remaining bolts are 0.98 inches (25mm)**

• Transmission joint bolts

20. For the lower oil pan, apply sealant as shown in the illustrations, then install and tighten the bolts, in sequence to 74–82 inch lbs. (8.3–9.3 Nm).

21. The remainder of installation is the reverse of the removal procedure.

➡️**Wait at least 30 minutes before refilling the engine oil.**

22. Connect the negative battery cable

23. Fill the engine with clean oil and coolant.

24. Start the engine, check for leaks and repair if necessary.

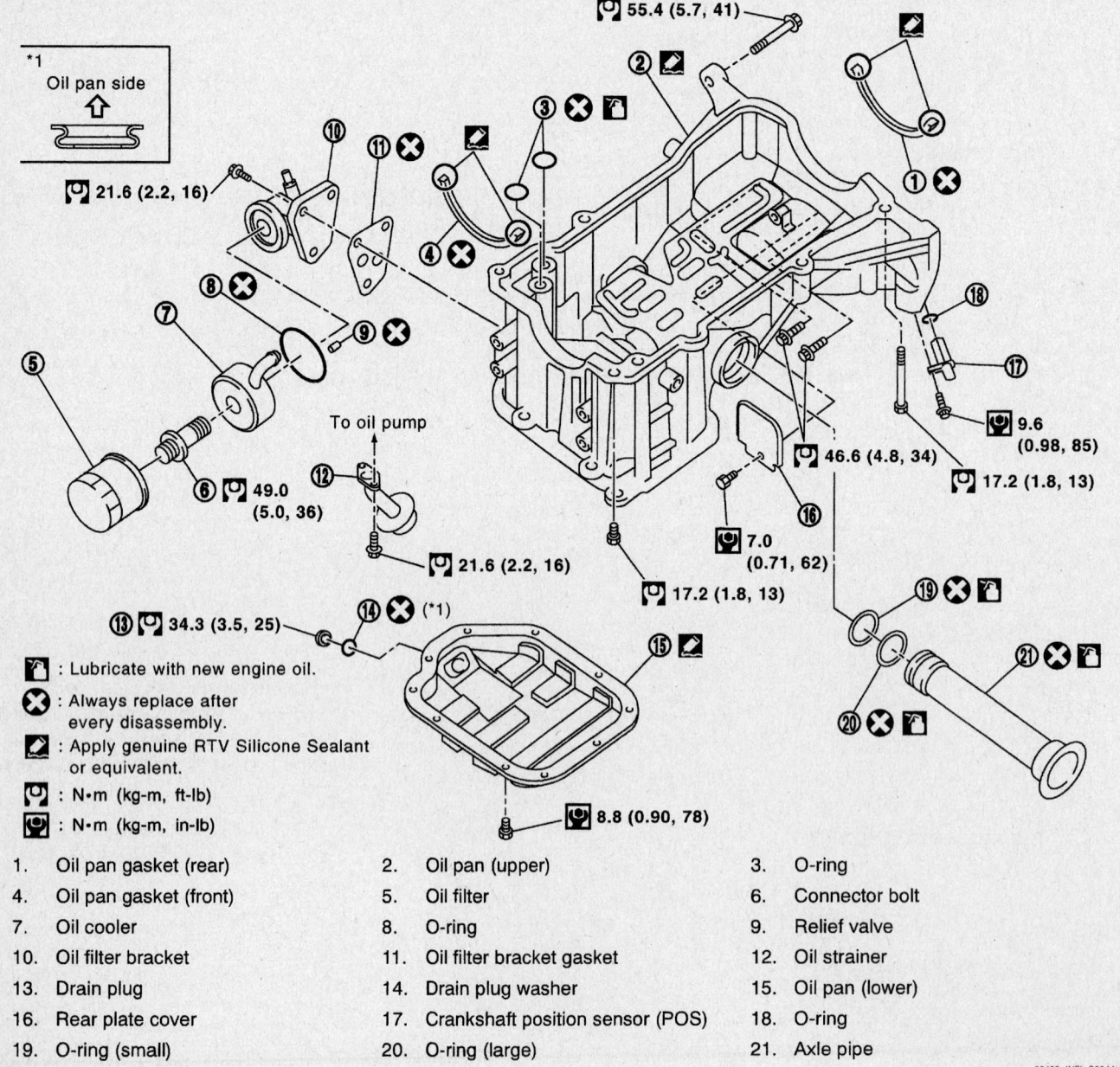

*1 Oil pan side

55.4 (5.7, 41)
21.6 (2.2, 16)
49.0 (5.0, 36)
21.6 (2.2, 16)
34.3 (3.5, 25)
9.6 (0.98, 85)
17.2 (1.8, 13)
46.6 (4.8, 34)
7.0 (0.71, 62)
17.2 (1.8, 13)
8.8 (0.90, 78)

To oil pump

🛢️ : Lubricate with new engine oil.

❌ : Always replace after every disassembly.

🖊️ : Apply genuine RTV Silicone Sealant or equivalent.

🔧 : N•m (kg-m, ft-lb)

🔧 : N•m (kg-m, in-lb)

1. Oil pan gasket (rear)	2. Oil pan (upper)	3. O-ring
4. Oil pan gasket (front)	5. Oil filter	6. Connector bolt
7. Oil cooler	8. O-ring	9. Relief valve
10. Oil filter bracket	11. Oil filter bracket gasket	12. Oil strainer
13. Drain plug	14. Drain plug washer	15. Oil pan (lower)
16. Rear plate cover	17. Crankshaft position sensor (POS)	18. O-ring
19. O-ring (small)	20. O-ring (large)	21. Axle pipe

09482_INFI_G0044

Exploded view of the upper and lower oil pans and related components—G35 and M35 AWD models

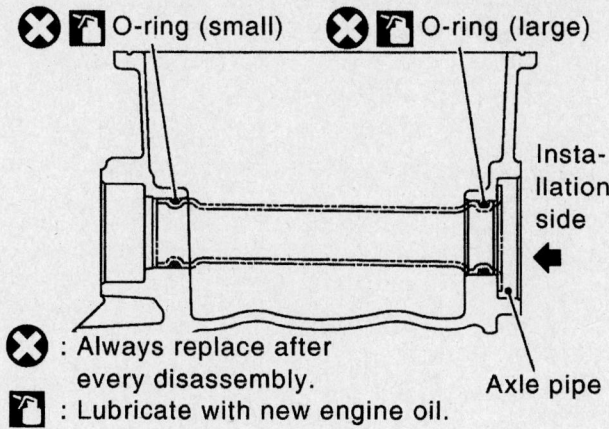

❌ 🛢 O-ring (small) ❌ 🛢 O-ring (large)

Installation side

❌ : Always replace after every disassembly.
🛢 : Lubricate with new engine oil.

Axle pipe

09482_INFI_G0045

Details of axle pipe installation—G35 and M35 AWD models

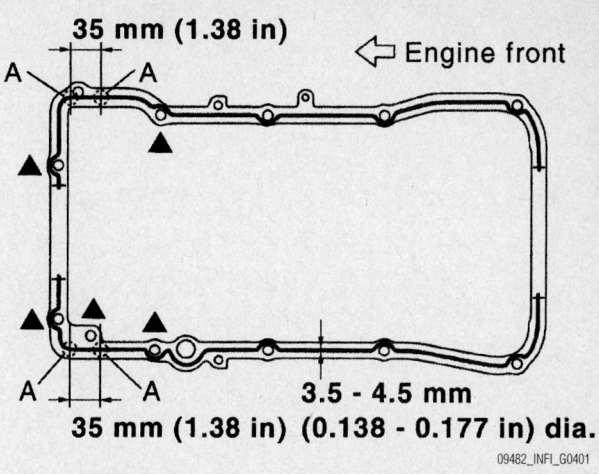

35 mm (1.38 in) ⇐ Engine front

3.5 - 4.5 mm
35 mm (1.38 in) (0.138 - 0.177 in) dia.

09482_INFI_G0401

Sealer application to the upper pan—G35 and M35 AWD models

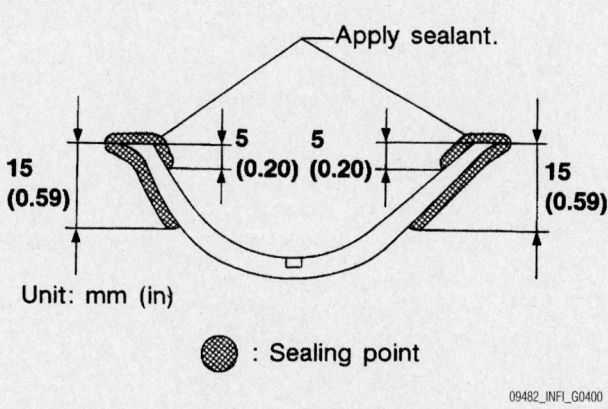

Apply sealant.

15 (0.59) 5 (0.20) 5 (0.20) 15 (0.59)

Unit: mm (in)

⬤ : Sealing point

09482_INFI_G0400

Sealer application to the upper pan gasket—G35 and M35 AWD models

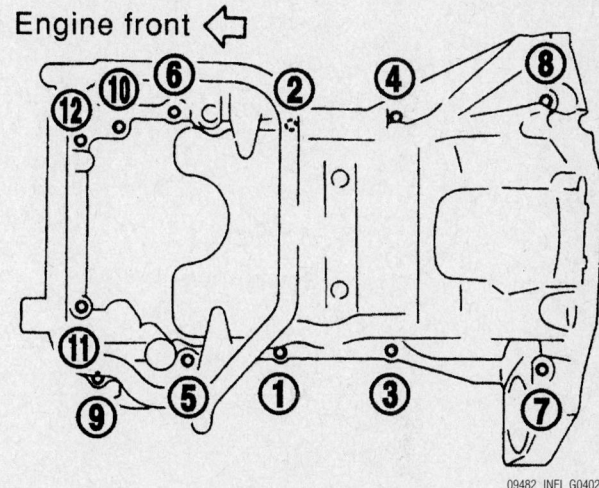

Engine front ⇐

09482_INFI_G0402

Upper pan torque sequence—G35 and M35 AWD models

4.5L Engine

1. Before servicing the vehicle, refer to the precautions in the beginning of this section.
2. Drain the engine oil.
3. Remove or disconnect the following:
   - Negative battery cable
   - Front wheels
   - Hood
   - Engine appearance cover
   - Engine splash shield
   - Drive belts
   - Power steering pump belt tensioner
   - Power steering oil pump and wire aside
   - Oil filter
   - A/C compressor and wire aside
   - Wiring harnesses at oil pan
   - Crankshaft Position Sensor (CKP)
4. Attach an engine lifting device and hold engine in place.
5. Remove the front lower crossmember.
6. Remove the oil pan rear plate cover.

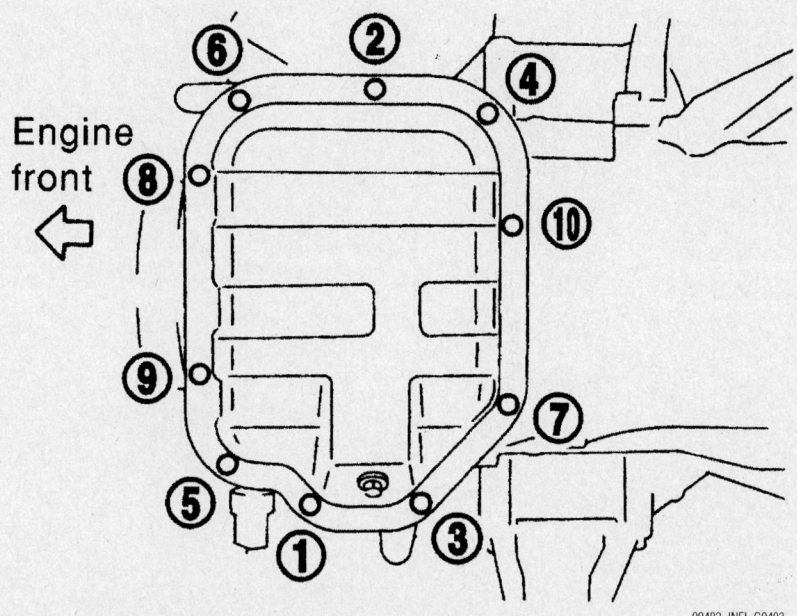

Engine front ⇐

09482_INFI_G0403

Lower pan torque sequence—G35 and M35 AWD models

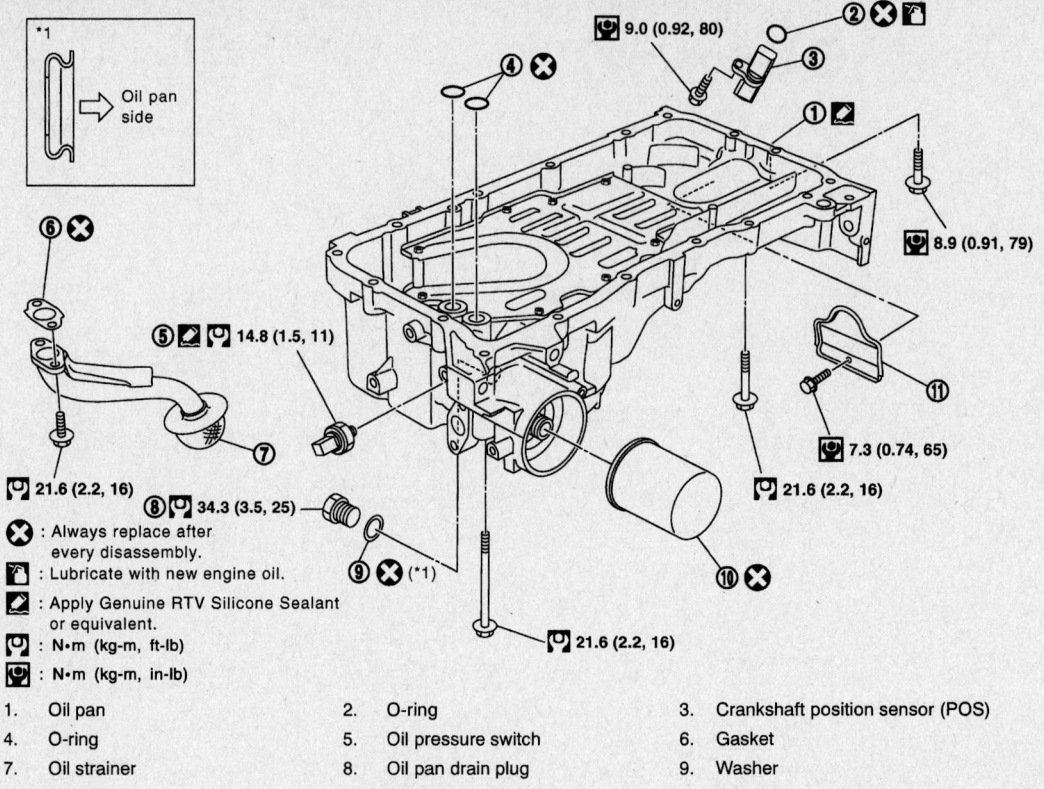

[*1] Oil pan side

⊗ : Always replace after every disassembly.

⊡ : Lubricate with new engine oil.

✎ : Apply Genuine RTV Silicone Sealant or equivalent.

⊡ : N•m (kg-m, ft-lb)

⊡ : N•m (kg-m, in-lb)

9.0 (0.92, 80)
8.9 (0.91, 79)
14.8 (1.5, 11)
21.6 (2.2, 16)
34.3 (3.5, 25)
7.3 (0.74, 65)
21.6 (2.2, 16)
21.6 (2.2, 16)

1. Oil pan
2. O-ring
3. Crankshaft position sensor (POS)
4. O-ring
5. Oil pressure switch
6. Gasket
7. Oil strainer
8. Oil pan drain plug
9. Washer
10. Oil filter
11. Rear plate cover

09482_INFI_G0047

Exploded view of oil pan assembly—4.5L engine, exc. 2006 M45

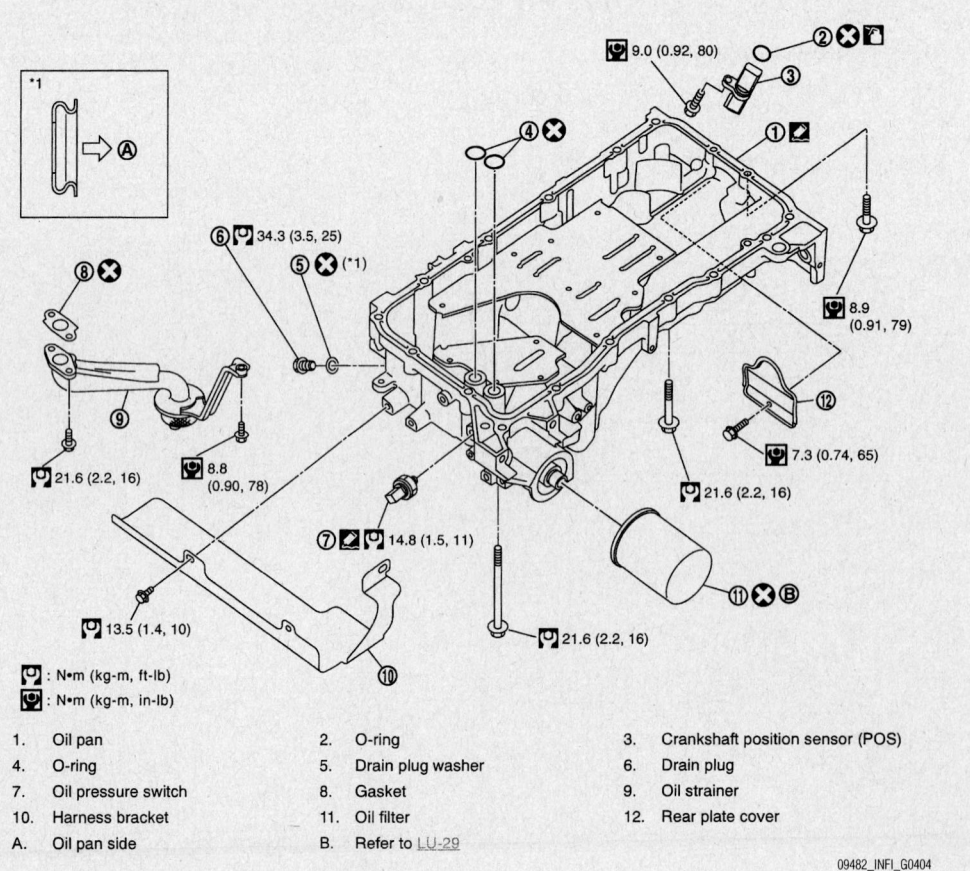

[*1] → Ⓐ

⊡ : N•m (kg-m, ft-lb)

⊡ : N•m (kg-m, in-lb)

9.0 (0.92, 80)
8.9 (0.91, 79)
34.3 (3.5, 25)
21.6 (2.2, 16)
8.8 (0.90, 78)
14.8 (1.5, 11)
7.3 (0.74, 65)
21.6 (2.2, 16)
21.6 (2.2, 16)
13.5 (1.4, 10)

1. Oil pan
2. O-ring
3. Crankshaft position sensor (POS)
4. O-ring
5. Drain plug washer
6. Drain plug
7. Oil pressure switch
8. Gasket
9. Oil strainer
10. Harness bracket
11. Oil filter
12. Rear plate cover
A. Oil pan side
B. Refer to LU-29

09482_INFI_G0404

Exploded view of oil pan assembly—4.5L engine, 2006 M45

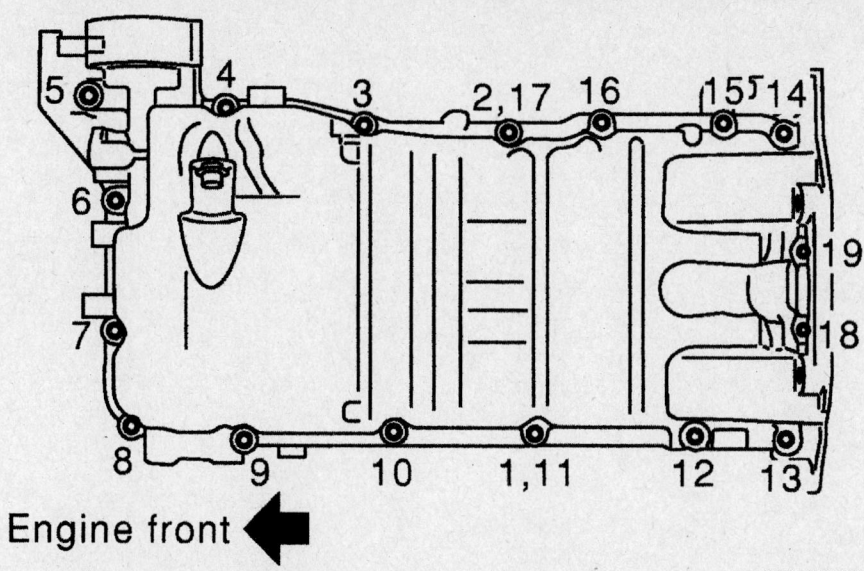

Oil pan bolt tightening sequence—4.5L engine

09482_INFI_G0048

7. Remove 4 transmission-to-oil pan bolts.

8. Remove oil pan bolts in reverse order of the tightening sequence.

9. Remove the oil pan and oil strainer.

To install:

10. Clean all gasket mating surfaces thoroughly.

11. Apply a continuous bead of liquid gasket to the oil pan mating surface. Be sure the bead is ⅛ inch (3mm) wide.

12. Install the oil pan and oil strainer. Tighten the bolts in sequence to 15–17 ft. lbs. (20–23 Nm).

13. Reverse the removal procedure to complete the installation. Tighten the front crossmember bolts to 80–93 ft. lbs. (108–123 Nm).

14. Fill the engine with clean oil.

15. Start the vehicle and check for leaks.

Oil Pump

REMOVAL & INSTALLATION

2.0L Engine

1. Before servicing the vehicle, refer to the precautions in the beginning of this section.

2. Relieve the fuel system pressure.

3. Drain the engine oil.

4. Remove or disconnect the following:
- Negative battery cable
- Drive belts
- Cylinder head with the intake and exhaust manifolds attached
- Oil pans
- Oil strainer and baffle plate
- Crankshaft pulley and the front cover assembly
- Oil pump from the inside of the front cover

To install:

5. Coat the oil pump gears with oil and fit the pump to the cover, using a new oil seal and O-ring.

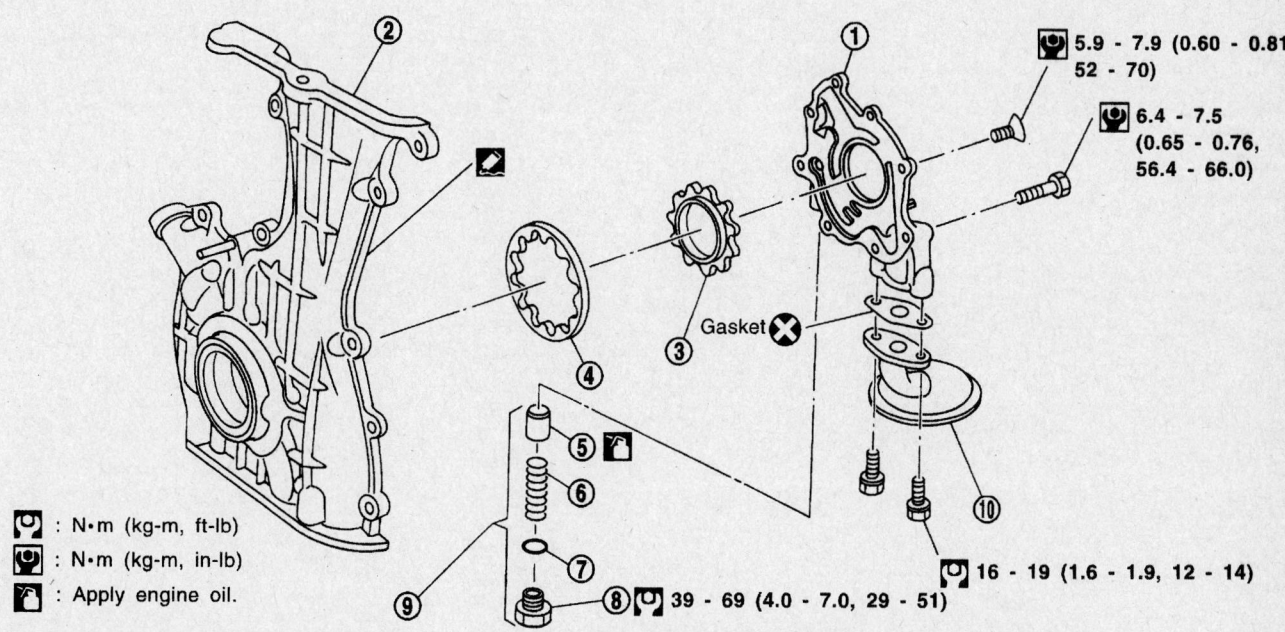

1. Oil pump cover
2. Front cover
3. Inner gear
4. Outer gear
5. Regulator valve
6. Spring
7. Shim
8. Plug
9. Regulator valve assembly
10. Oil strainer

Oil pump exploded view—2.0L engine

09482_INFI_G0405

6. Clean the mating surfaces of liquid gasket and apply a fresh bead of ⅛ inch (3mm) sealer to the sealing surface of the front cover.

7. Install or connect the following:
- Front cover assembly
- Crankshaft pulley
- Oil strainer, baffle plate, oil pans, cylinder head and drive belts
- Negative battery cable

8. Fill the engine with clean oil.

9. Start the vehicle, check for leaks and repair if necessary.

3.5L Engines

➡ **The oil pump bolts to the front of the engine block and is driven by the crankshaft. Removal of the timing cover and chains are necessary for oil pump service.**

1. Before servicing the vehicle, refer to the precautions in the beginning of this section.

2. Drain the engine oil.

3. Rotate the engine and position it to Top Dead Center (TDC) compression stroke of cylinder No. 1.

4. Remove or disconnect the following:

- Negative battery cable
- Drive belts
- Camshaft Position (CMP) sensor (PHASE) and the Crankshaft Position (CKP) sensor (REF/POS)
- Right front wheel and inner fender cover
- Engine undercover
- Crankshaft pulley
- Front exhaust pipe and its support and support the engine at the left and right side slingers with a suitable hoist
- Engine right side mounting insulator and bracket nuts and bolts
- Center crossmember assembly
- A/C compressor and mounting bracket
- Lower and upper oil pans
- Oil strainer from the oil pump
- Water pump cover and the front cover assembly
- Lower timing chain assembly
- Oil pump

To install:

➡ **When installing the oil pump, be sure to apply engine oil to the gears.**

5. Install or connect the following:

- Oil pump. Refer to illustration for torque values
- Lower timing chain assembly
- Front timing cover and water pump covers
- Oil strainer using a new gasket. Torque the bolts to 12–14 ft. lbs. (16–19 Nm).
- Upper and lower oil pans. Be sure to use new O-rings at the oil pump to upper oil pan mating surface.
- A/C compressor and mounting bracket
- Center crossmember assembly
- Engine right side mounting insulator and bracket and remove the engine support hoist
- Front exhaust pipe and its support
- Crankshaft pulley
- Engine undercover and the right side inner fender cover
- Right front wheel
- CMP sensor (PHASE) and the CKP sensor (REF/POS)
- Engine drive belts and adjust as necessary
- Negative battery cable

6. Fill the engine with clean oil.

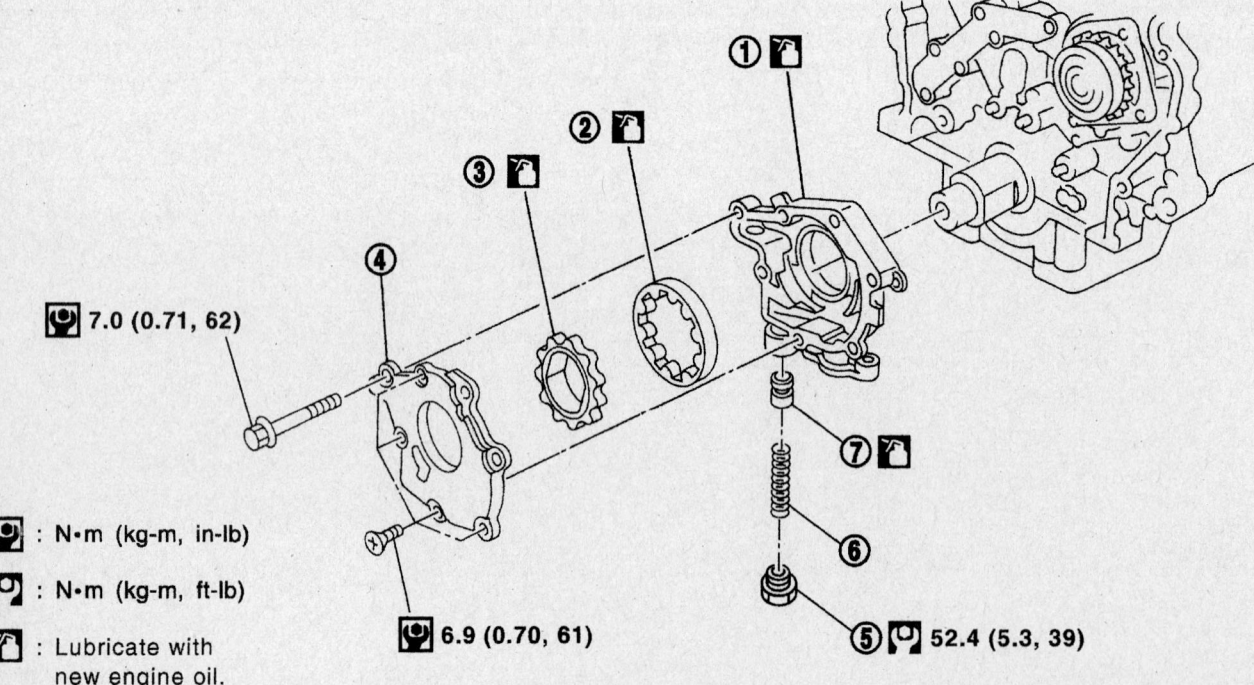

7.0 (0.71, 62)

⬛ : N•m (kg-m, in-lb)

⬛ : N•m (kg-m, ft-lb)

⬛ : Lubricate with new engine oil.

6.9 (0.70, 61)

52.4 (5.3, 39)

1.	Oil pump body	2.	Oil pump outer rotor	3.	Oil pump inner rotor
4.	Oil pump cover	5.	Regulator valve plug	6.	Regulator valve spring
7.	Regulator valve				

Exploded view of the oil pump—3.5L engine

09482_INFI_G0046

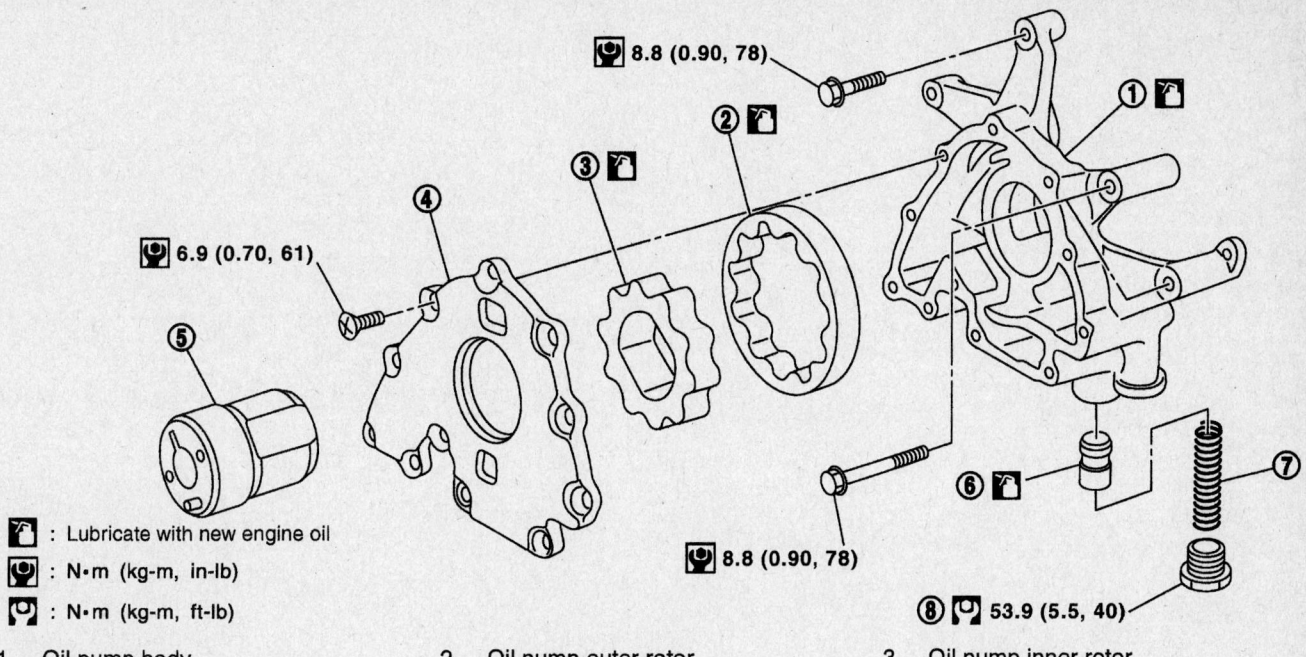

8.8 (0.90, 78)

6.9 (0.70, 61)

8.8 (0.90, 78)

8 53.9 (5.5, 40)

: Lubricate with new engine oil

: N•m (kg-m, in-lb)

: N•m (kg-m, ft-lb)

1. Oil pump body
4. Oil pump cover
7. Regulator valve spring

2. Oil pump outer rotor
5. Oil pump drive spacer
8. Regulator valve plug

3. Oil pump inner rotor
6. Regulator valve

09482_INFI_G0049

Exploded view of the oil pump—4.5L engine

7. Start the engine, check the oil pressure, and check for oil leaks.

4.5L Engine

➡The oil pump is mounted in the cylinder block below the left bank and behind the left timing chain; engine removal is required before servicing. Refer to Engine Assembly Removal and Installation.

1. Before servicing the vehicle, refer to the precautions in the beginning of this section.
2. Drain the engine oil.
3. Remove or disconnect the following:
   - Negative battery cable
   - Engine assembly
   - Front cover and timing chains
   - Oil pump assembly from the front of the engine

To install:
4. Clean the oil pump mounting surface.
5. Install or connect the following:
   - Oil pump using a new gasket
   - Timing chains and front cover
   - Negative battery cable
6. Fill the engine with clean oil.
7. Start the vehicle, check for leaks and repair if necessary.

Rear Main Seal

REMOVAL & INSTALLATION

1. Before servicing the vehicle, refer to the precautions in the beginning of this section.
2. Remove or disconnect the following:
   - Transmission or transaxle
   - Drive plate from the crankshaft
3. Carefully pry the seal out of the retainer without damaging the crankshaft or the seal retainer.

To install:
4. Lubricate the seal with clean engine oil.

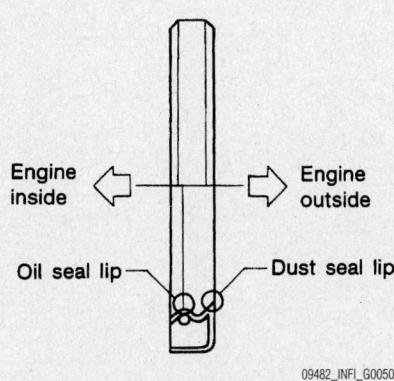

Engine inside ⇐   ⇒ Engine outside

Oil seal lip   Dust seal lip

09482_INFI_G0050

Proper orientation of rear main seal

5. Install or connect the following:
   - Seal into the retainer using the appropriate seal driver

➡Install rear oil seal so that each seal lip is oriented as shown in the figure.

   - Driveplate and transmission or transaxle

Timing Chain, Sprockets, Front Cover and Seal

REMOVAL & INSTALLATION

2.0L Engine

1. Before servicing the vehicle, refer to the precautions in the beginning of this section.
2. Relieve the fuel system pressure.
3. Raise and support the vehicle safely.
4. Drain the cooling system.
5. Remove or disconnect the following:
   - Negative battery cable
   - Engine undercover
   - Right front wheel
   - Engine side cover and lower the vehicle
   - Radiator
   - Intake manifold air duct
   - Drive belts, water pump pulley, alternator and power steering pump

- Vacuum hoses, fuel hoses and wiring harness connectors
- Spark plugs
- Cylinder head cover and oil separator
- Intake manifold supports
- Oil filter bracket and the power steering oil pump bracket

6. Place the No. 1 piston at Top Dead Center (TDC) on the compression stroke.

- Chain tensioner
- Distributor. Do not turn the rotor while the distributor is removed.
- Timing chain guide
- Camshaft sprockets
- Camshafts, camshaft brackets, oil tubes and baffle plate
- Starter
- Heater hoses and the water hoses from the cylinder head
- Knock sensor harness connector
- Cylinder head outside bolts
- Cylinder head with the intake and exhaust manifolds and raise and support the vehicle safely
- Oil pans
- Oil strainer and baffle plate
- Crankshaft pulley and place a transmission jack under the main bearing beam and raise the engine slightly to take the weight off of the front engine mount
- Front engine mount
- Timing chain cover. Tap the seal out of the cover with a suitable seal driver.
- Timing chain sprocket bolts
- Timing chain guides, timing chain and sprockets

To install:

7. Be sure all sealing surfaces are clean and prepared for assembly.

8. Install or connect the following:
- Crankshaft sprocket. Position the crankshaft so No. 1 piston is set at TDC (keyway at 12 o'clock, mating mark at 4 o'clock).

9. Fit the timing chain to crankshaft sprocket with the gold mating mark on the chain aligned with the mark on the sprocket. (The mating marks for the camshaft sprockets are silver).

- Timing chain guides and hang the chain off the left (front) guide. If necessary, secure the chain so it does not disengage from the crankshaft sprocket during assembly.
- New seal in the front cover and apply engine oil to the lip of the seal and apply a bead of liquid gasket to the front cover
- Oil pump drive spacer and front

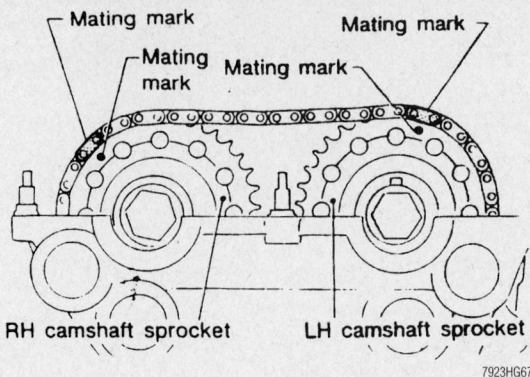

During disassembly, be sure to align the timing chain and camshaft sprocket mating marks—2.0L engine

cover. Torque the bolts evenly to 60 inch lbs. (6.7 Nm) and wipe away any excess liquid gasket.
- Front engine mount
- Crankshaft pulley and temporarily tighten the bolt to hold the sprocket in place. The timing mark should align with the TDC mark.
- Oil strainer, baffle plate and oil pan
- Cylinder head, camshafts, oil tubes and baffles. Position the left camshaft key at 12 o'clock and the right camshaft key at 10 o'clock.
- Camshaft sprockets by lining up the mating marks on the timing chain with the mating marks on the camshaft sprockets. Torque the camshaft bolts to 101–116 ft. lbs. (137–157 Nm) and the crankshaft pulley bolt to 105–112 ft. lbs. (142–152 Nm).

- Upper timing chain guide and distributor. Ensure that the rotor is at the 5 o'clock position.

10. Before installing the chain tensioner, press the cam stopper down and the push in the sleeve until the hook can be engaged on the pin. When tensioner is bolted in position, the hook will release automatically. Ensure the arrow on the outside faces the front of the engine.

- Oil filter bracket and the power steering pump bracket
- Intake manifold supports
- Oil separator and the cylinder head cover
- Spark plugs
- Vacuum hoses, fuel hoses, and wiring harness connectors
- Alternator and power steering pump
- Water pump pulley
- Drive belts

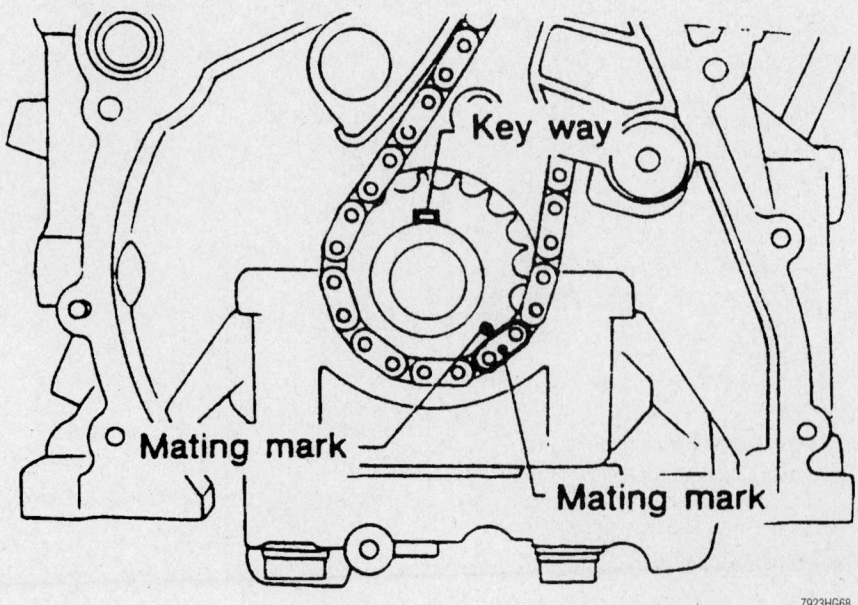

Crankshaft sprocket and timing chain alignment marks—2.0L engine

Right

Left

Dowel pin hole

Dowel pin hole

09482_INFI_G0051

Intake valve timing control cover tightening sequence—3.5L engine

- Radiator
- Engine undercover
- Right front wheel
- Negative battery cable
11. Fill and bleed the cooling system.
12. Start the vehicle, check for leaks and repair if necessary.

3.5L Engines

1. Before servicing the vehicle, refer to the precautions in the beginning of this section.
2. Properly relieve the fuel system pressure.
3. Drain the engine oil and cooling system.
4. Remove or disconnect the following:
- Negative battery cable
- Right and left side rocker covers
- Cooling fan and radiator
- A/C compressor from bracket and position side with the lines attached
- A/C compressor bracket

- Power steering pump from its bracket and position aside with the lines attached
- Power steering pump bracket
- Water bypass hose and cooling fan bracket from the front timing chain case
- Lower and upper oil pans
- Right and left side intake valve timing control covers. Loosen the bolts in the reverse of the tightening sequence. Use a seal cutter to cut the gasket.

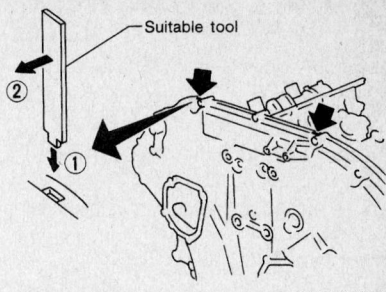

Front timing chain case removal, using a suitable tool—3.5L engine

5. Set the No. 1 piston to Top Dead Center (TDC) on its compression stroke. Align the timing mark (grooved line) on the crankshaft pulley with the timing indicator on the front cover.
6. Make sure the intake and exhaust cam nose for the No. 1 cylinder is positioned as shown in the illustration. If not, turn the crankshaft on revolution (360°) and align.
7. Remove or disconnect the following:
- Crankshaft pulley using a suitable puller
- Front timing chain case. Loosen the bolts in the reverse of the torque sequence.

✳✳ WARNING

Do not use a screwdriver to pry the case off!

- Timing chain case. Insert the proper tool into the notch at the top

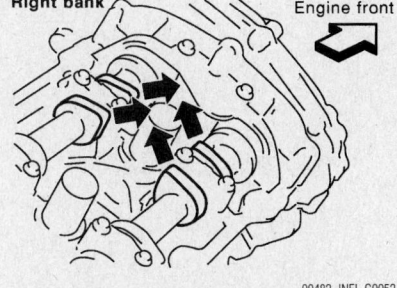

09482_INFI_G0052

Proper orientation of the intake and exhaust cams—3.5L engine

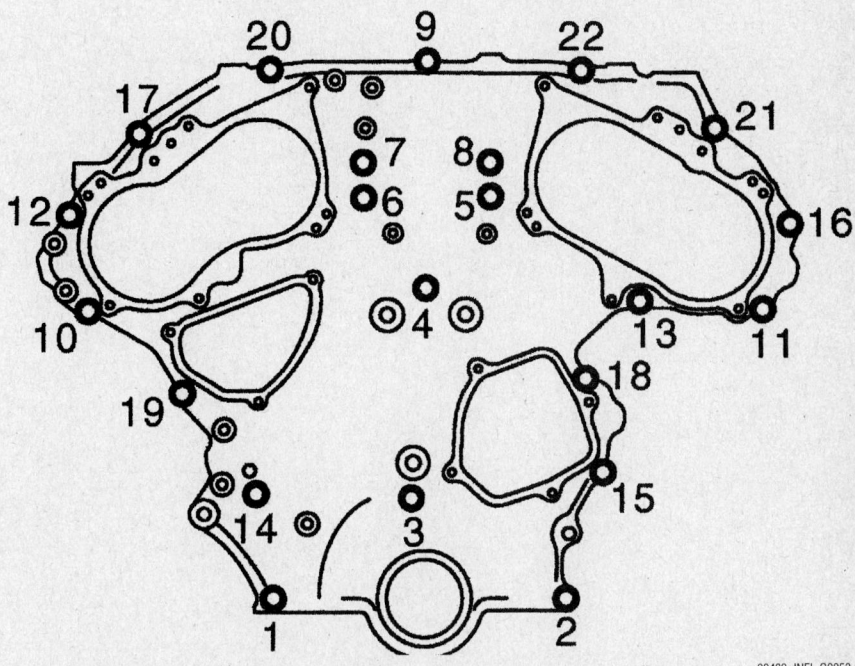

09482_INFI_G0053

Front timing chain case torque sequence (loosen in reverse order)—3.5L engine

of the case as shown. Pry the case off by levering the tool as shown. Use a seat cutter to cut the liquid gasket.

• O-rings from rear timing chain case
• Water pump cover and chain tensioner cover from the case
• Front oil seal from the case using a suitable prytool, being careful not the damage the case
• Internal chain guide, timing chain tensioner and slack guide. Remove the upper chain tensioner by pressing the tensioner in and inserting a 0.098 inch (2.5mm) diameter pin in the pin hole. Once secured remove the bolts and the tensioner.

✳✳ WARNING

After the timing chain is removed, do NOT turn the crankshaft and camshaft separately, or the valve will strike the pistons.

8. Remove the timing chain and camshaft sprocket, as follows:

a. Attach a suitable stopper pin to the right and left camshaft chain tensioners (for secondary timing chains).

b. Hold the hex part of the camshaft secure with a wrench, then remove the camshaft sprocket mounting bolts.

c. Remove the primary and secondary timing chains with the camshaft sprockets.

9. Remove or disconnect the following:

• Tension guide and crankshaft sprocket
• Rear timing chain case. Use the reverse of the tightening sequence, then use a seal cutter to separate the gasket.

To install:

10. Install the rear timing chain case as follows:

a. Install new O-rings onto the cylinder block and head.

b. Apply suitable RTV sealant to the back side of the rear timing chain case as shown in the illustration.

c. Install the rear case and tighten the bolts, in sequence, to 9–10 ft. lbs. (11.7–13.7 Nm). There are 2 bolts lengths: Bolts 1, 2, 3, 6–10 are 0.79 in. (20mm) long and the other bolts are 0.63 in. (16mm) long.

11. Set the No. 1 piston to Top Dead Center (TDC) on its compression stroke.

12. Install the crankshaft sprocket, making sure the mating marks on the sprocket face the front of the engine.

13. Push the plunger of the secondary chain tensioner and keep it pressed in with a stopper pin.

14. Install the secondary timing chains and camshaft sprockets, as follows:

a. Align the matchmarks on the secondary timing chain (gold link) with the stamped marks on the intake and exhaust sprockets, then install them.

➡**Matchmarks for the intake sprocket are on the back of the secondary sprocket. There are 2 kinds of marks, the right bank uses round marks and the left bank uses oval marks.**

b. Align the dowel pin and pin hole on the camshaft with the groove and dowel pin on the sprocket, then install them.

c. On the intake side, align the pin hold on the small diameter side of the camshaft front end with the dowel pin on the on the back side of the camshaft sprocket, and install them.

d. On the exhaust side, align the dowel pin on the camshaft front end with

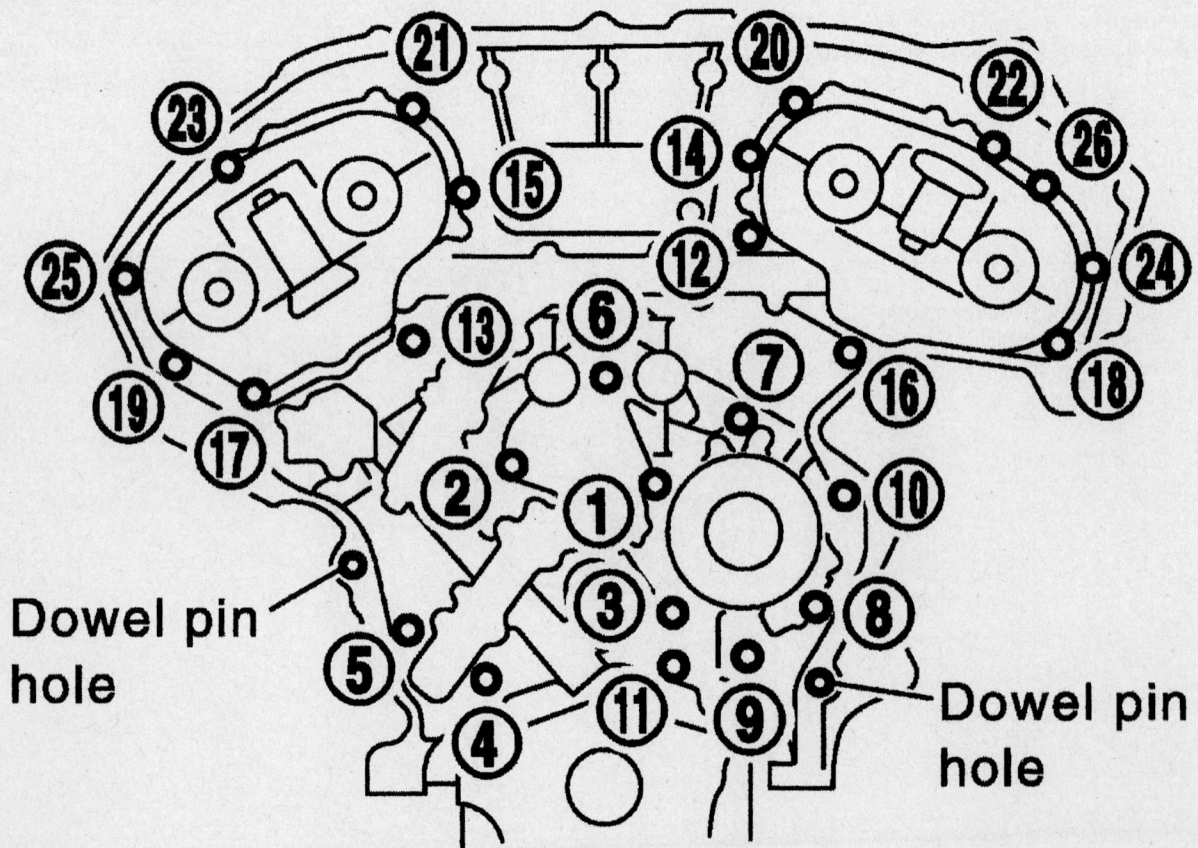

Dowel pin hole

Dowel pin hole

Rear timing chain case torque sequence (loosen in reverse order)—3.5L engines

09482_INFI_G0056

Rear timing chain case: Back side

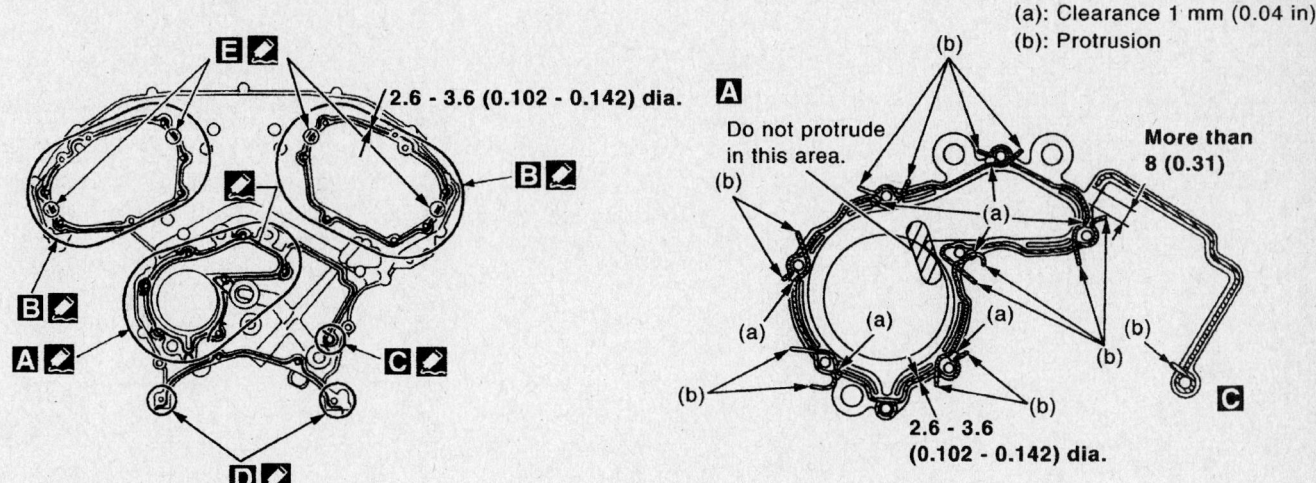

(a): Clearance 1 mm (0.04 in)
(b): Protrusion

B Cross both ends as shown and be sure to minimize the overlapped area.

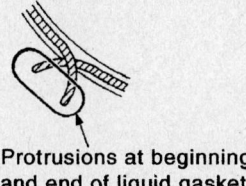

Protrusions at beginning and end of liquid gasket

E Camshaft axis area

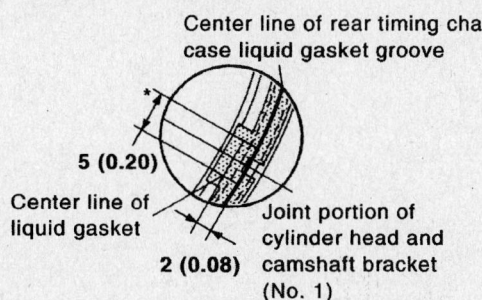

Center line of rear timing chain case liquid gasket groove

5 (0.20)

Center line of liquid gasket

2 (0.08)

Joint portion of cylinder head and camshaft bracket (No. 1)

D

2.6 - 3.6 (0.102 - 0.142) dia.

Protrusions at beginning and end of liquid gasket

◀ : Run along bolt hole outer side

*: Apply liquid gasket to the chamfered surface between camshaft bracket (No. 1) and cylinder head.

✎ : Apply Genuine RTV Silicone Sealant or equivalent.

Unit: mm (in)

09482_INFI_G0055

Rear timing chain case sealant application—3.5L engines

the pin groove on the camshaft sprocket, then install them.

e. Tighten the camshaft sprocket bolts hand-tight to prevent the dowel pins from dislocating.

15. Install the primary timing chain, as follows:

a. Install the primary chain so the punched mating marks on the cam sprockets are aligned with **matching colored** links on the timing chain, while the notched mating mark on the crankshaft sprocket is aligned with a **different color** link on the timing chain. Cam

sprocket mating links must be the same color in order to ensure proper cam timing.

b. Use a wrench on the hex portion of the camshaft to secure it in place, tighten the camshaft sprocket mounting bolts to 73–78 ft. lbs. (98–108 Nm).

c. Remove the stopper pins from secondary chain tensioners.

16. Install the internal chain guide, timing chain tensioner, tension guide and slack guide. Do not overtighten the slack guide mounting bolts. It's normal to have a gap under the bolt seats when the mounting

bolts are properly tightened to 10–13 ft. lbs. (14–18 Nm).

17. Recheck that all matchmarks are still aligned.

18. Install or connect the following:
- New O-rings on the rear timing chain case
- New front oil seal in to the timing chain cover
- Water pump and chain tensioner covers to the front cover
- Suitable liquid gasket to the back side of the timing chain case
- Dowel pin on the rear timing chain

Example: Right bank (Rear view)

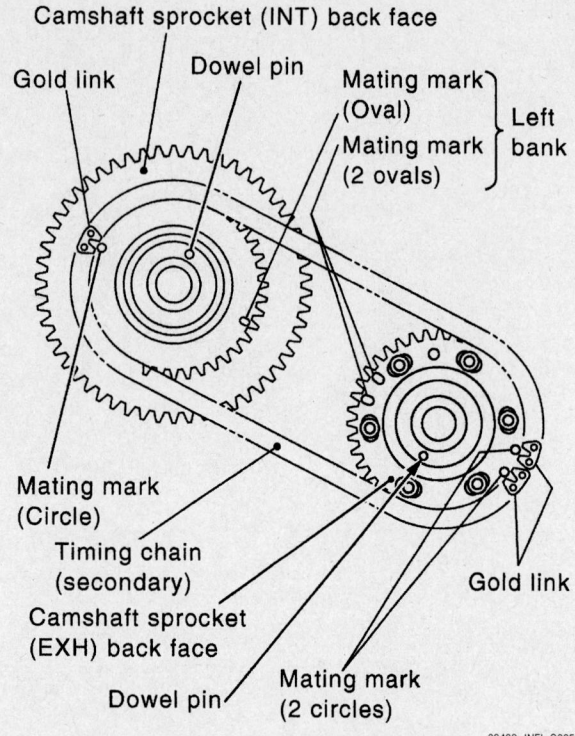

Camshaft sprocket alignment, right side bank shown—3.5L engine

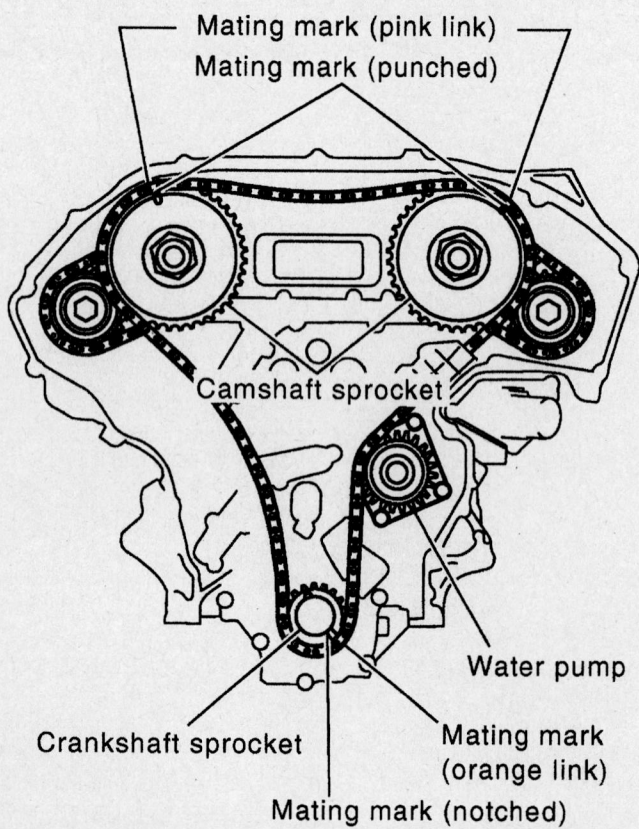

Proper alignment for the primary timing chain—3.5L engine

case into the dowel pin hold on the front chain case

- Front timing chain case bolts, in sequence. Tighten the M6 bolts to 9–10 ft. lbs. (11.7–13.7 Nm) and the M8 bolts to 19–23 ft. lbs. (25–31 Nm).
- Right and left intake valve timing control cover
- Crankshaft pulley to the crankshaft and the mounting bolt. Torque the mounting bolt to 29–36 ft. lbs. (39–49 Nm). Torque the crankshaft bolt an additional 60–66° clockwise. This is approximately the angle from 1 hexagon bolt head corner to another.

19. Installation of the remaining components is the reverse of the removal procedure.

20. Fill the engine with clean oil.

21. Fill and bleed the cooling system.

22. Start the engine and check for proper operation.

4.5L Engine

1. Before servicing the vehicle, refer to the precautions in the beginning of this section.

2. Properly relieve the fuel system pressure.

3. Remove or disconnect the following:
- Negative battery cable
- Engine from the vehicle
- Drive belt tensioner and idler pulley
- Thermostat housing
- Ignition coils
- Cylinder head cover
- Intake valve timing control sensor from both sides
- Camshaft Position sensor (CMP)
- Intake valve timing control solenoid from both sides
- Intake valve timing control covers by loosening bolts in the reverse of the tightening sequence
- Front cover o-rings

4. Turn the crankshaft pulley clockwise until No. 1 cylinder is on the compression stroke. The TDC mark on the pulley without the paint mark should align with the timing pointer.

5. Verify the correct TDC position by ensuring that the lobes of the No. 1 cylinder camshafts are pointing outward. If not rotate the crankshaft 360° until the lobes are correct.

6. Remove the oil pan rear plate cover and install a locking tool into the drive gear.

7. Remove the crankshaft pulley.

8. Remove the oil pan and strainer

Front timing chain case

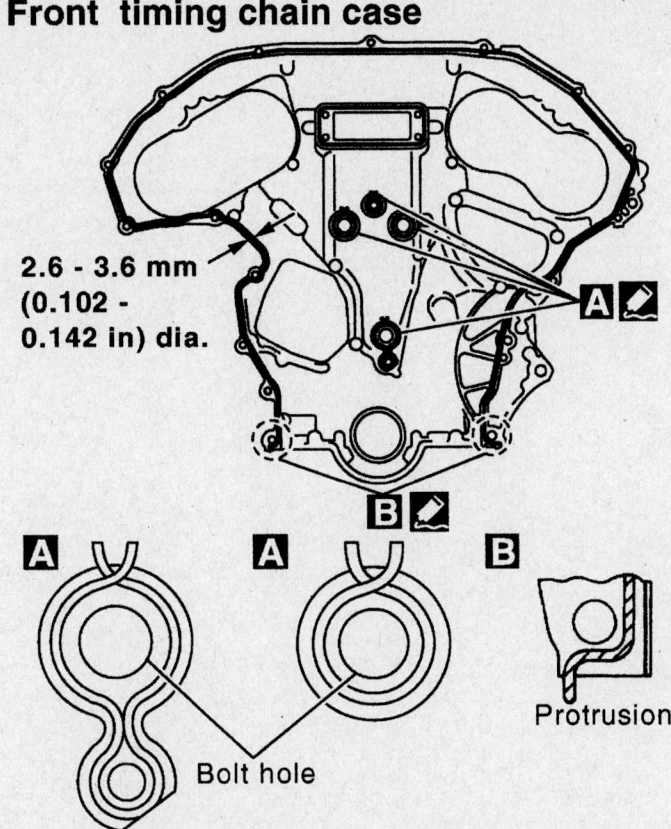

2.6 - 3.6 mm
(0.102 -
0.142 in) dia.

Protrusion

Bolt hole

Liquid gasket protrusion away from bolt hole

: Apply Genuine RTV silicone sealant
or equivalent.

09482_INFI_G0059

Apply proper sealant as shown on the back side of the front timing chain case—3.5L engine

9. Loosen and remove the front cover bolts in the reverse of the tightening sequence.

10. Remove the front cover and pry out the front cover oil seal.

11. Remove the 3 O-rings from the cylinder heads and block.

12. Remove the chain tensioner cover from the front cover.

13. Remove the oil pump drive spacer and oil pump.

14. Compress the left side chain tensioner and install a pin through the hole to secure it, then remove the tensioner.

15. Remove the chain tensioner, tension guide and slack guide.

16. Remove the left side timing chain.

17. While holding the camshaft in place, remove the left side camshaft sprockets.

18. Repeat the procedure on the right side to remove the chain tensioner, guides, timing chain and camshaft sprockets.

To install:

19. Be sure the crankshaft key is pointing toward the center of the left bank. This should be a 45° angle from the center. Check that the camshaft dowel pins are in the correct location as shown.

20. Install or connect the following:
- Camshaft sprockets by holding the camshaft with a wrench and tightening the sprocket bolts to 112 ft. lbs. (152 Nm).
- Crankshaft sprockets and be sure the thick side of the sprocket faces the cylinder block to provide clearance between the block and chain.
- Right bank timing chain by aligning the marks on the chain with the marks on the sprockets
- Right slack and chain guides. Be sure to install the bolts in the correct locations. Torque the bolts to 10–14 ft. lbs. (13–19 Nm).
- Timing chain for the left bank in the same manner as the right one
- Left slack and chain guide. Torque the bolts to 10–14 ft. lbs. (13–19 Nm).
- Chain tensioners with the pins installed using new gaskets. Torque

the bolts to 61 inch lbs. (6.9 Nm) and remove the pins.

21. Confirm that the timing marks on the crankshaft sprockets and chains are still aligned.
- Oil pump and tighten the bolts to 78 inch lbs. (9 Nm).
- Oil pump drive spacer, aligning the spacer key groove with the crankshaft key, and tapping it in with a plastic hammer
- New seal into the front cover until it is flush with the cover face
- Chain tensioner cover after applying liquid gasket to the mating surface. Tighten the bolts to 78 inch lbs. (9 Nm).
- New O-rings into the block and cylinder heads
- Front cover after applying liquid gasket to the mating surface. Be sure to install the bolts in the correct locations. Tighten the bolts in sequence to 78 inch lbs. (9 Nm).
- Timing control cover using new O-

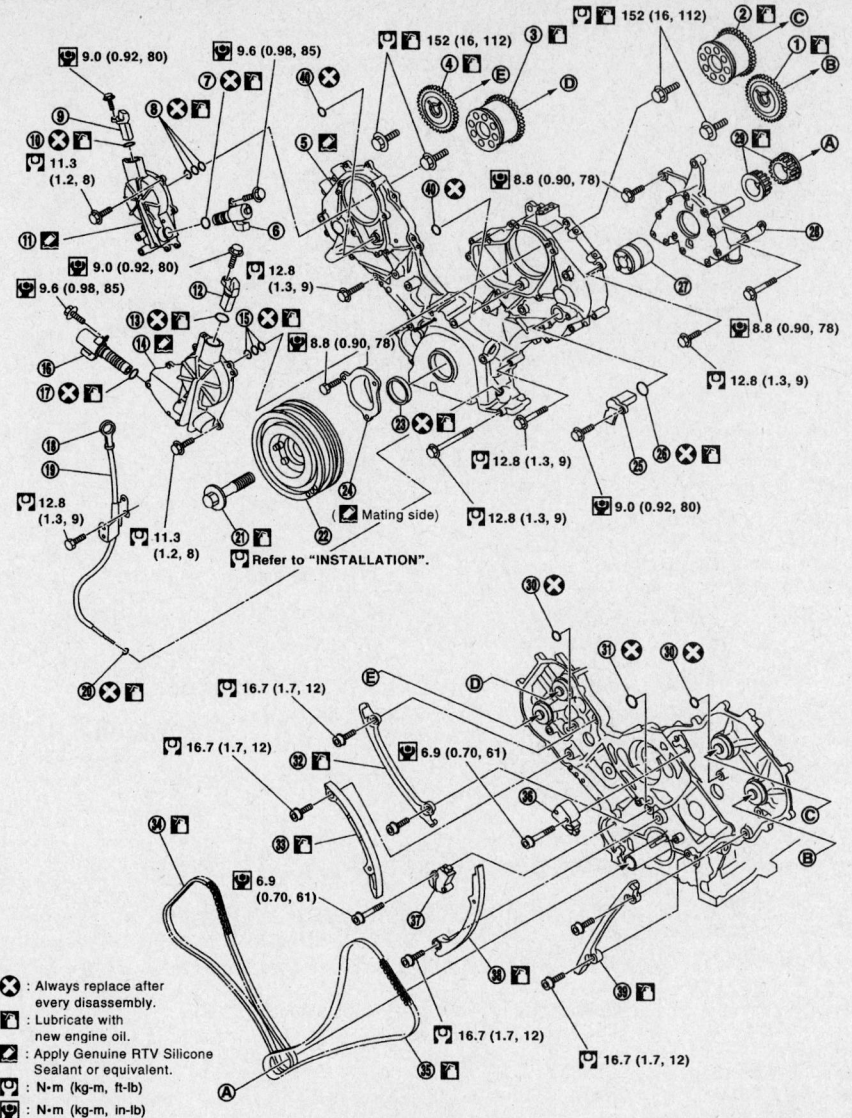

- 9.0 (0.92, 80)
- 9.6 (0.98, 85)
- 152 (16, 112)
- 152 (16, 112)
- 11.3 (1.2, 8)
- 9.0 (0.92, 80)
- 12.8 (1.3, 9)
- 8.8 (0.90, 78)
- 9.6 (0.98, 85)
- 8.8 (0.90, 78)
- 8.8 (0.90, 78)
- 12.8 (1.3, 9)
- 12.8 (1.3, 9)
- 12.8 (1.3, 9)
- 9.0 (0.92, 80)
- 12.8 (1.3, 9)
- 11.3 (1.2, 8)
- (: Mating side)
- : Refer to "INSTALLATION".
- 16.7 (1.7, 12)
- 16.7 (1.7, 12)
- 6.9 (0.70, 61)
- 6.9 (0.70, 61)
- 16.7 (1.7, 12)
- 16.7 (1.7, 12)

- : Always replace after every disassembly.
- : Lubricate with new engine oil.
- : Apply Genuine RTV Silicone Sealant or equivalent.
- : N•m (kg-m, ft-lb)
- : N•m (kg-m, in-lb)

#		#		#	
1.	Camshaft sprocket (EXH)	2.	Camshaft sprocket (INT)	3.	Camshaft sprocket (INT)
4.	Camshaft sprocket (EXH)	5.	Front cover	6.	Intake valve timing control solenoid valve (right bank)
7.	O-ring	8.	Seal ring	9.	Intake valve timing control position sensor (right bank)
10.	O-ring	11.	Intake valve timing control cover (right bank)	12.	Intake valve timing control position sensor (left bank)
13.	O-ring	14.	Intake valve timing control cover (left bank)	15.	Seal ring
16.	Intake valve timing control solenoid valve (left bank)	17.	O-ring	18.	Oil level gauge
19.	Oil level gauge guide	20.	O-ring	21.	Crankshaft pulley bolt
22.	Crankshaft pulley	23.	Front oil seal	24.	Chain tensioner cover
25.	Camshaft position sensor (PHASE)	26.	O-ring	27.	Oil pump drive spacer
28.	Oil pump assembly	29.	Crankshaft sprocket	30.	O-ring
31.	O-ring	32.	Chain tension guide (right bank)	33.	Chain slack guide (right bank)
34.	Timing chain (right bank)	35.	Timing chain (left bank)	36.	Chain tensioner (left bank)
37.	Chain tensioner (right bank)	38.	Chain slack guide (left bank)	39.	Chain tension guide (left bank)
40.	O-ring				

Exploded view of timing chain and front cover components—4.5L engine

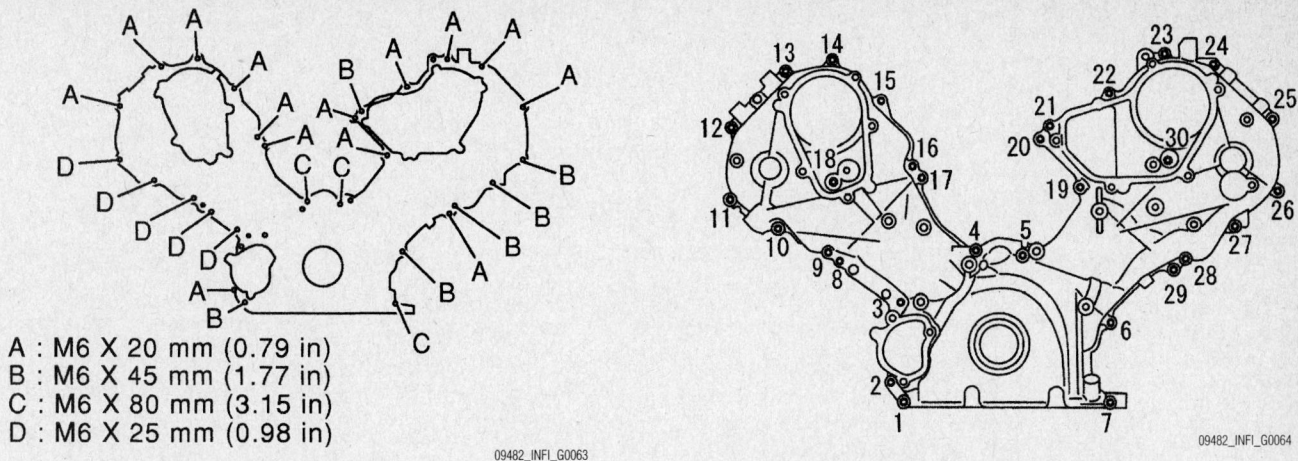

A : M6 X 20 mm (0.79 in)
B : M6 X 45 mm (1.77 in)
C : M6 X 80 mm (3.15 in)
D : M6 X 25 mm (0.98 in)

09482_INFI_G0063

Front cover bolt identification—4.5L engine

09482_INFI_G0064

Front cover bolt tightening sequence—4.5L engine

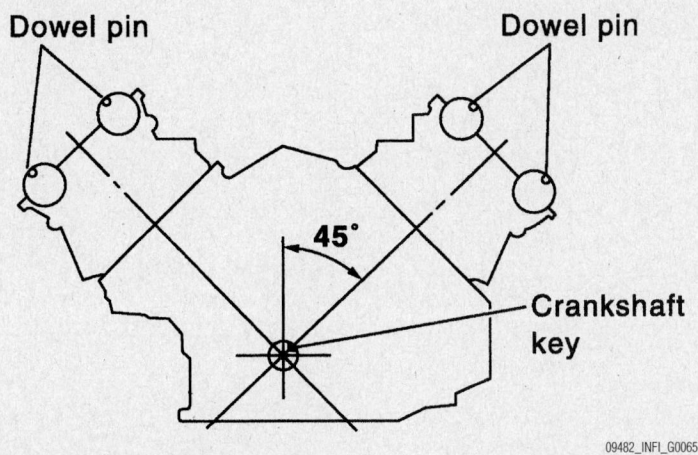

09482_INFI_G0065

Aligning crankshaft key and camshaft dowel pins—4.5L engine

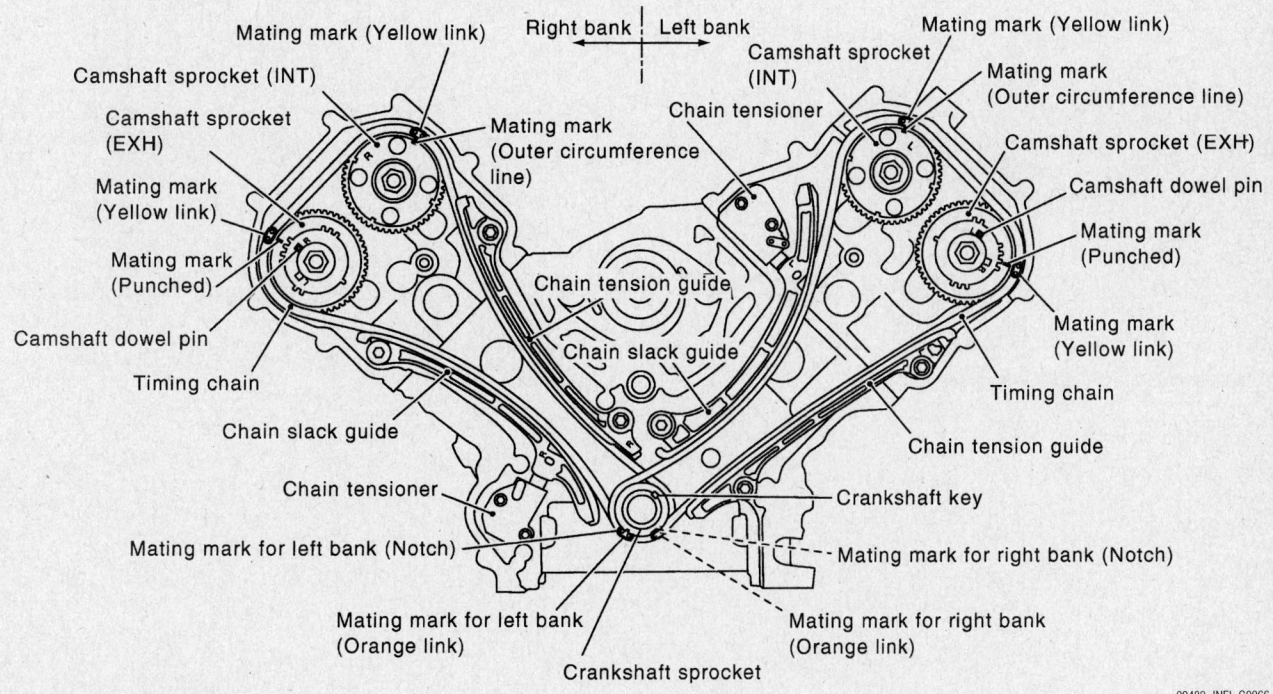

09482_INFI_G0066

Timing chain and sprocket alignment marks—4.5L engine

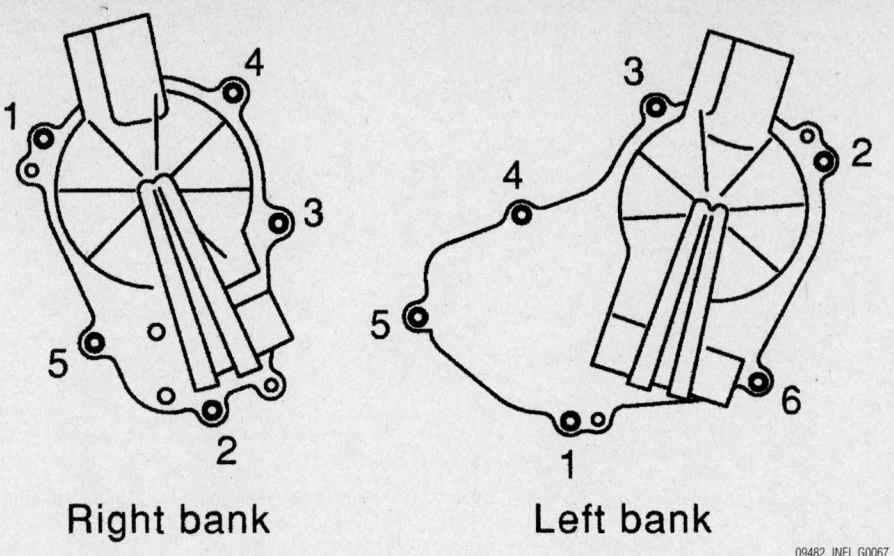

Right bank   Left bank

09482_INFI_G0067

Timing control cover tightening sequence—4.5L engine

rings and after applying liquid gasket to the mating surface. Tighten the bolts in sequence to 97 inch lbs. (11 Nm).
- Intake valve timing control sensor to both sides
- Camshaft Position sensor (CMP)
- Intake valve timing control solenoid to both sides

22. Install the crankshaft pulley so it aligns with the dowel pin of the oil pump drive spacer.

23. Apply clean engine oil the crankshaft pulley bolt and tighten the bolt to 69 ft. lbs. (93 Nm), plus an additional 90°.

24. Installation of the remaining components is the reverse of the removal procedure.

25. Fill the engine with clean oil.
26. Fill and bleed the cooling system.
27. Start the engine and check for proper operation.

Piston and Ring

POSITIONING

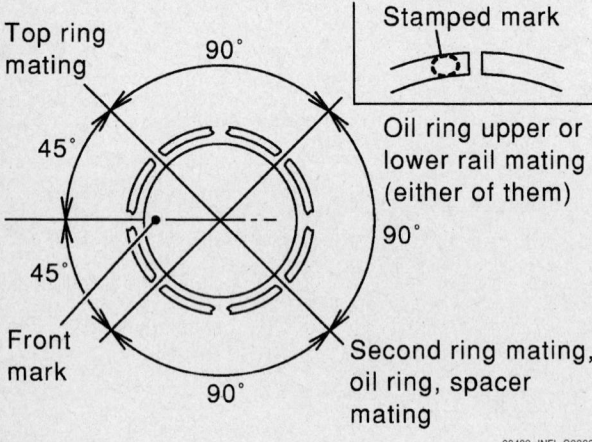

09482_INFI_G0068

Piston ring installation—Infiniti engines

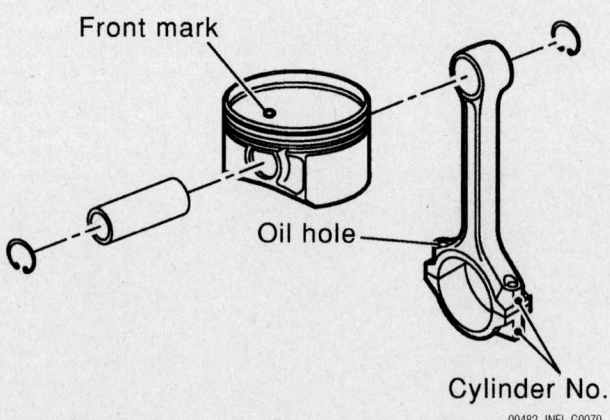

09482_INFI_G0070

Piston/connecting rod assembly-to-engine orientation—Infiniti engines

FUEL SYSTEM

Fuel System Service Precautions

Safety is the most important factor when performing not only fuel system maintenance but any type of maintenance. Failure to conduct maintenance and repairs in a safe manner may result in serious personal injury or death. Maintenance and testing of the vehicle's fuel system components can be accomplished safely and effectively by adhering to the following rules and guidelines.

1. To avoid the possibility of fire and personal injury, always disconnect the negative battery cable unless the repair or test procedure requires that battery voltage be applied.

2. Always relieve the fuel system pressure prior to disconnecting any fuel system component (injector, fuel rail, pressure regulator, etc.), fitting or fuel line connection. Exercise extreme caution whenever relieving fuel system pressure, to avoid exposing skin, face and eyes to fuel spray. Please be advised that fuel under pressure may penetrate the skin or any part of the body that it contacts.

3. Always place a shop towel or cloth around the fitting or connection prior to loosening to absorb any excess fuel due to spillage. Ensure that all fuel spillage (should it occur) is quickly removed from engine surfaces. Ensure that all fuel soaked cloths or towels are deposited into a suitable waste container.

4. Always keep a dry chemical (Class B) fire extinguisher near the work area.

5. Do not allow fuel spray or fuel vapors to come into contact with a spark or open flame.

6. Always use a back-up wrench when loosening and tightening fuel line connection fittings. This will prevent unnecessary stress and torsion to fuel line piping. Always follow the proper torque specifications.

7. Always replace worn fuel fitting O-rings with new. Do not substitute fuel hose where fuel pipe is installed.

Relieving Fuel System Pressure

1. Before servicing the vehicle, refer to the precautions in the beginning of this section.

2. Remove the fuel pump fuse.

3. Start the engine.

4. Allow the engine to run until it stalls.

5. After the engine stalls, crank the engine 2 or 3 times to release the remaining fuel pressure.

6. Turn the ignition switch **OFF**. Reinstall the fuel pump fuse into the fuse block.

➡ **Do not crank the engine or turn the ignition switch ON after the fuel pump fuse has been reinstalled, or the fuel pressure will be reestablished.**

Fuel Filter

REMOVAL & INSTALLATION

2.0L Engine

1. Before servicing the vehicle, refer to the precautions in the beginning of this section.

2. Relieve the fuel system pressure.

3. Remove or disconnect the following:

- Negative battery cable
- Fuel hoses from the fuel filter, located at the right side of the engine compartment
- Filter mounting screws
- Filter from the vehicle

To install:

4. Inspect all hoses and clamps for damage of any type. Replace parts, as required.

➡ **The fuel filters are directional and should be installed with the arrow facing the direction of fuel flow.**

5. Install or connect the following:

- New filter in the bracket and install new hose clamp
- Negative battery cable

6. Start the vehicle, check for fuel leaks and repair if necessary.

➡ **On some vehicles, a code will be set and/or the check engine light will remain on after starting the vehicle. This is because a code was set for an open fuel pump circuit when the fuel pressure was released. If you did not disconnect the negative battery cable during this procedure, do it now so the code will be erased. The negative battery cable should be disconnected for at least 1 minute. Also, remember to reset the clock and radio stations when finished.**

3.5L Engine

1. Before servicing the vehicle, refer to the precautions in the beginning of this section.

2. Properly relieve fuel system pressure.

3. Remove or disconnect the following:

- Negative battery cable
- Rear seat bottom
- Inspection hole cover
- Electrical and quick connectors
- Six screws
- Fuel level sensor unit and fuel pump assembly
- Flange and snap fit portion of the fuel pump

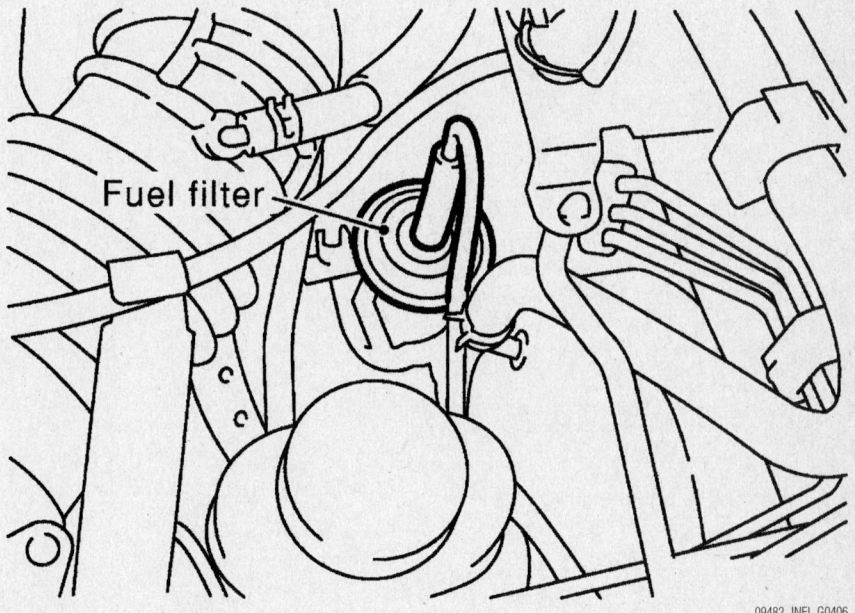

Fuel filter—2.0L engines

09482_INFI_G0406

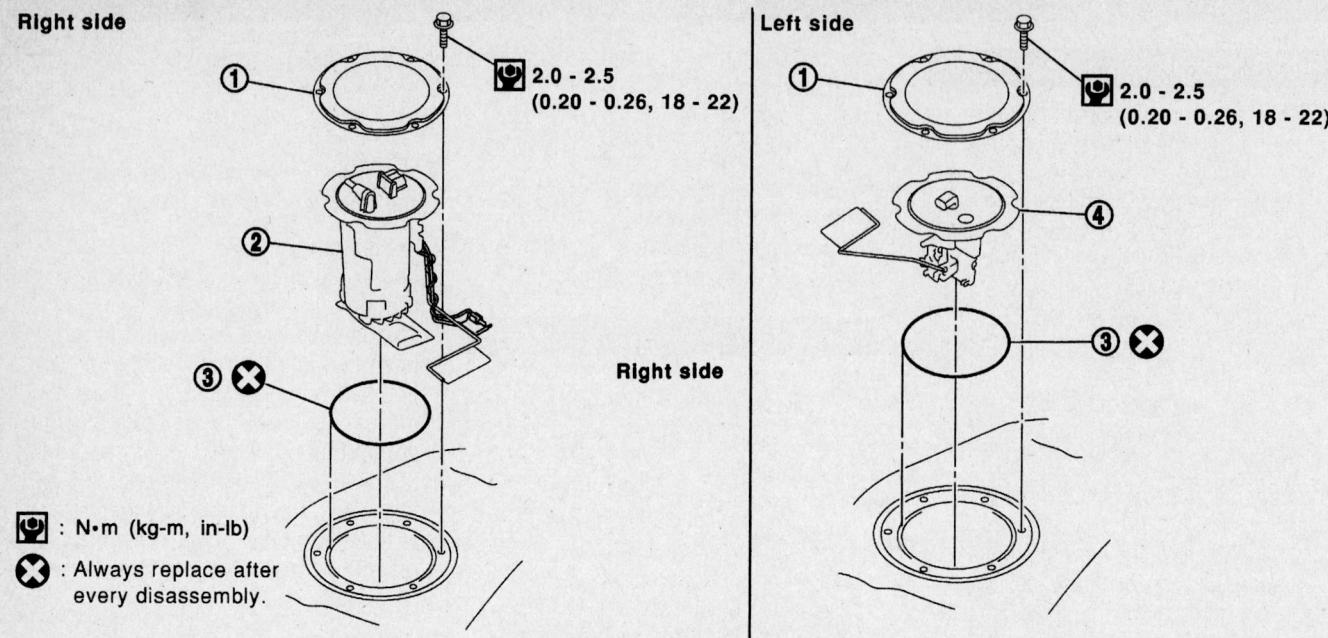

Right side

Left side

Right side

2.0 - 2.5
(0.20 - 0.26, 18 - 22)

: N•m (kg-m, in-lb)

: Always replace after every disassembly.

1. Retainer
2. Main fuel level sensor unit, fuel filter - and fuel pump assembly
3. O-ring
4. Sub fuel level sensor unit

Fuel pump and fuel filter assembly—G35

09482_INFI_G0071

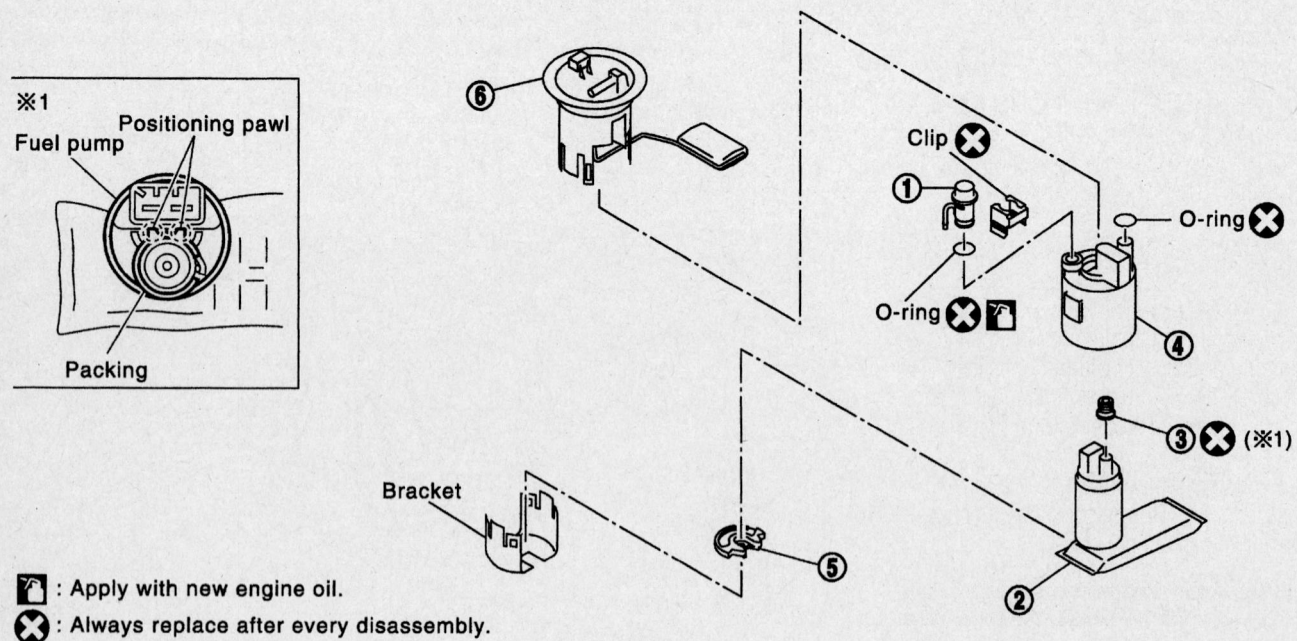

※1

Fuel pump

Positioning pawl

Packing

Clip

O-ring

O-ring

Bracket

③ (※1)

: Apply with new engine oil.

: Always replace after every disassembly.

1. Pressure regulator
2. Fuel pump
3. Packing
4. Fuel filter
5. Rubber
6. Fuel level sensor unit

Fuel pump and fuel filter assembly—I35

09482_INFI_G0407

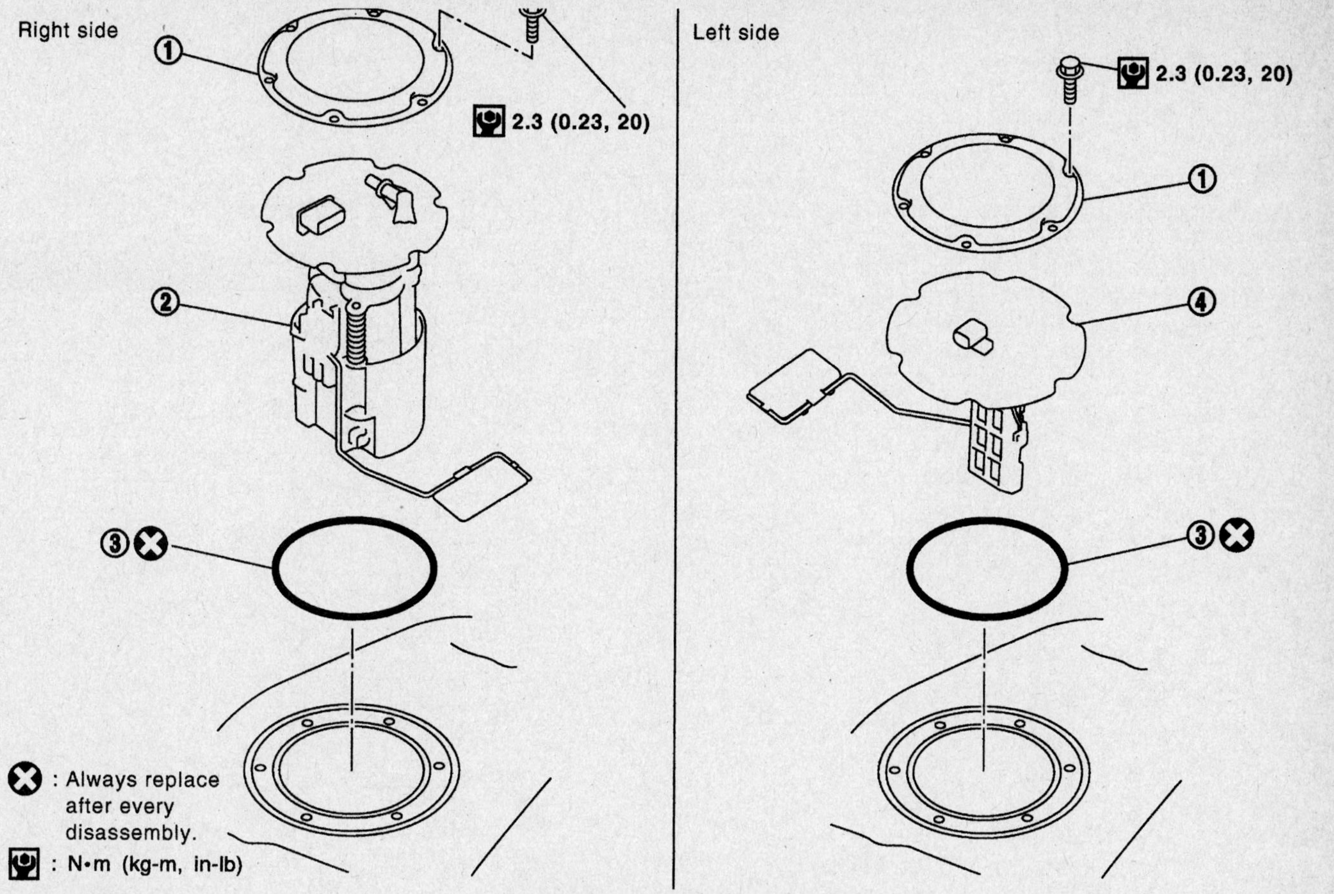

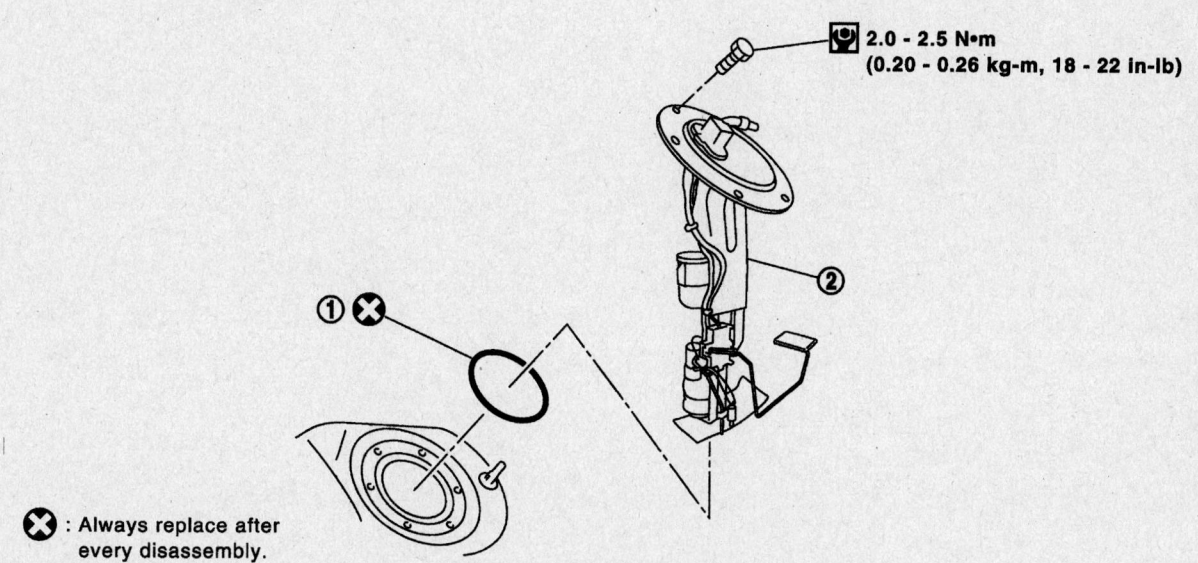

Right side

① 2.3 (0.23, 20)

②

③ ⊗

Left side

2.3 (0.23, 20)

①

④

③ ⊗

⊗ : Always replace after every disassembly.

: N•m (kg-m, in-lb)

1. Retainer

2. Main fuel level sensor unit, fuel filter and fuel pump assembly

3. O-ring

4. Sub fuel level sensor unit

09482_INFI_G0408

Fuel pump and fuel filter assembly—M35 and 2006 M45

2.0 - 2.5 N•m
(0.20 - 0.26 kg-m, 18 - 22 in-lb)

②

① ⊗

⊗ : Always replace after every disassembly.

1. O-ring

2. Fuel level sensor unit, fuel filter and fuel pump assembly

09482_INFI_G0072

Fuel pump and fuel filter assembly—Q45 and 2004 M45

- Fuel tank temperature sensor harness
- Fuel level sensor flange
- Fuel pump connector
- Quick connectors from the fuel level sensor
- Fuel level sensor from the chamber
- Fuel filter from the chamber

To install:

4. Install or connect the following:
- Fuel filter to the chamber
- Fuel level sensor to the chamber
- Quick connectors to the fuel level sensor
- Fuel pump connector
- Fuel level sensor flange
- Fuel tank temperature sensor harness
- Fuel pump assembly to the fuel tank

- Screws and electrical connectors
- Quick connectors
- Negative battery cable

5. Start the vehicle, check for leaks and repair if necessary.
- Inspection hole cover
- Rear seat bottom

4.5L Engine

1. Before servicing the vehicle, refer to the precautions in the beginning of this section.
2. Properly relieve fuel system pressure.
3. Remove or disconnect the following:
- Negative battery cable
- Fuel filler cap
- Truck front finish panel
- Fuel pump connector

- Fuel feed line
- Fuel pump assembly
- Fuel filter

To install:

4. Installation is the reverse of the removal procedure.

5. Start the vehicle, check for leaks and repair if necessary.

Fuel Pump

REMOVAL & INSTALLATION

2.0L Engine

1. Before servicing the vehicle, refer to the precautions in the beginning of this section.

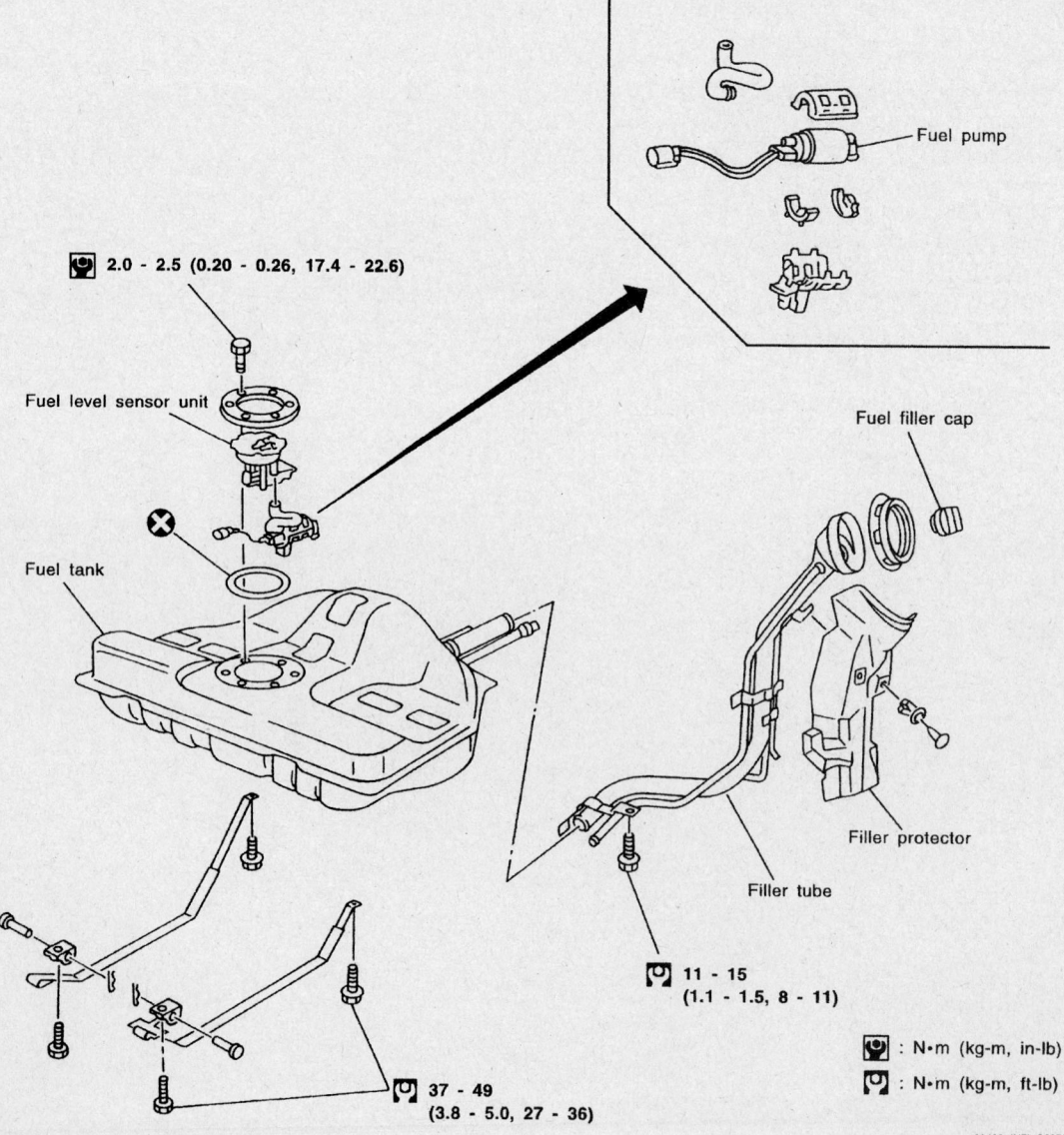

2.0 - 2.5 (0.20 - 0.26, 17.4 - 22.6)

Fuel level sensor unit

Fuel pump

Fuel tank

Fuel filler cap

Filler protector

Filler tube

11 - 15
(1.1 - 1.5, 8 - 11)

37 - 49
(3.8 - 5.0, 27 - 36)

: N•m (kg-m, in-lb)

: N•m (kg-m, ft-lb)

Fuel pump and fuel level sending unit—2.0L engine

09482_INFI_G0073

2. Release the fuel system pressure.
3. Remove or disconnect the following:
- Negative battery cable
- Rear seat back and bottom
- Inspection hole cover located beneath the rear seat
- Connectors and fuel tubes
- Six screws
- Fuel pump/gauge assembly and disconnect the tubes and connector
- Fuel pump by sliding it out on an angle

To install:

4. Install or connect the following:
- Fuel pump/gauge assembly
- All fuel lines and connectors
- Six screws
- Negative battery cable

5. Start the vehicle, check for leaks and repair if necessary.
6. Install the inspection cover.
7. Rear seat back and bottom.

3.5L Engine

1. Before servicing the vehicle, refer to the precautions in the beginning of this section.
2. Relieve the fuel system pressure
3. Remove or disconnect the following:
- Negative battery cable
- Rear seat bottom
- Inspection hole cover from under the rear seat
- Electrical and quick connectors
- Six screws

- Fuel level sensor/fuel pump assembly

➡**If replacement of the fuel filter is required, proceed with the following steps.**

- Fuel level sensor/fuel pump assembly flange
- Fuel tank temperature sensor harness
- Fuel level sensor flange and raise the fuel level sensor
- Fuel pump electrical connector
- Quick connectors from the fuel level sensor
- Fuel level sensor from the assembly
- Snap fit connectors and remove the fuel filter from the fuel pump assembly

To install:

4. Install or connect the following:
- Fuel filter to the fuel pump assembly
- Fuel level sensor to the fuel pump assembly
- Fuel level sensor quick connectors
- Electrical connectors
- Fuel tank temperature sensor harness
- Fuel level sensor unit and fuel pump flanges. Make certain that they snap together.
- Fuel pump assembly to the fuel tank
- Six screws

- Quick and electrical connectors
- Negative battery cable

5. Start the vehicle, check for leaks and repair if necessary.
6. Install the inspection hole cover plate.
7. Install the rear seat bottom.

4.5L Engine

1. Before servicing the vehicle, refer to the precautions in the beginning of this section.
2. Properly relieve fuel system pressure.
3. Remove or disconnect the following:
- Negative battery cable
- Fuel filler cap
- Truck front finish panel
- Fuel pump connector
- Fuel feed line
- Fuel pump assembly
- Fuel pump

To install:

4. Installation is the reverse of the removal procedure.
5. Start the vehicle, check for leaks and repair if necessary.

Fuel Rail & Injectors

REMOVAL & INSTALLATION

Except 4.5L Engine

1. Before servicing the vehicle, refer to the precautions in the beginning of this section.

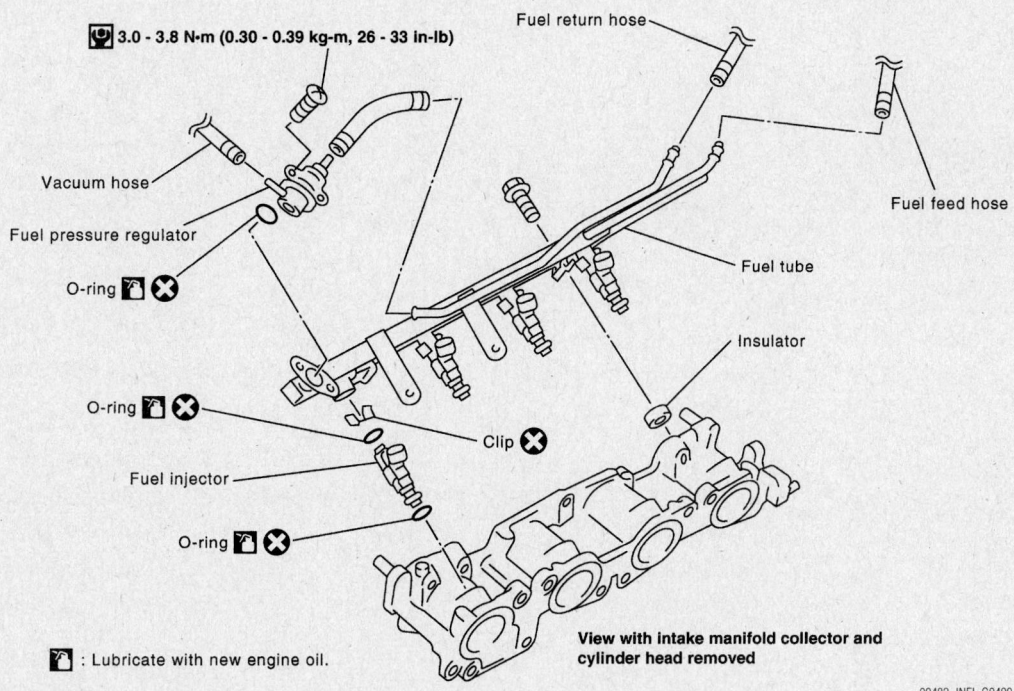

🔧 3.0 - 3.8 N·m (0.30 - 0.39 kg-m, 26 - 33 in-lb)

Fuel return hose

Vacuum hose

Fuel pressure regulator

O-ring 🛢️ ✖️

Fuel feed hose

Fuel tube

O-ring 🛢️ ✖️

Insulator

Clip ✖️

Fuel injector

O-ring 🛢️ ✖️

🛢️ : Lubricate with new engine oil.

View with intake manifold collector and cylinder head removed

09482_INFI_G0409

Fuel rail and injector assembly—2.0L engine

2. Relieve the fuel system pressure
3. Remove or disconnect the following:
   - Negative battery cable
   - Engine cover, as necessary
   - Intake manifold collector
   - Vacuum hose from the fuel pressure regulator
   - Fuel hoses from fuel rail
   - Fuel rail bolts
   - Injector harness connectors
   - Injectors and the fuel rail as an assembly
   - Injector(s) from the fuel rail by pushing them out

4. Remove and discard the fuel injector O-rings

To install:
5. Lubricate the new O-rings with clean engine oil and install the O-rings on the injector(s).
6. Install or connect the following:
   - Fuel injectors to the fuel rail
   - Fuel rail and injectors as an assembly to the intake manifold
7. Tighten the fuel rail bolts in the following sequence;
   a. Step 1: 7–8 ft. lbs. (9–10 Nm).
   b. Step 2: 15–20 ft. lbs. (21–26 Nm).

   - Fuel hoses to the fuel rail
   - Vacuum hose to the fuel pressure regulator
   - Intake manifold collector
   - Engine cover, as necessary
   - Negative battery cable
8. Start the vehicle, check for leaks and repair if necessary.

4.5L Engine

1. Before servicing the vehicle, refer to the precautions in the beginning of this section.

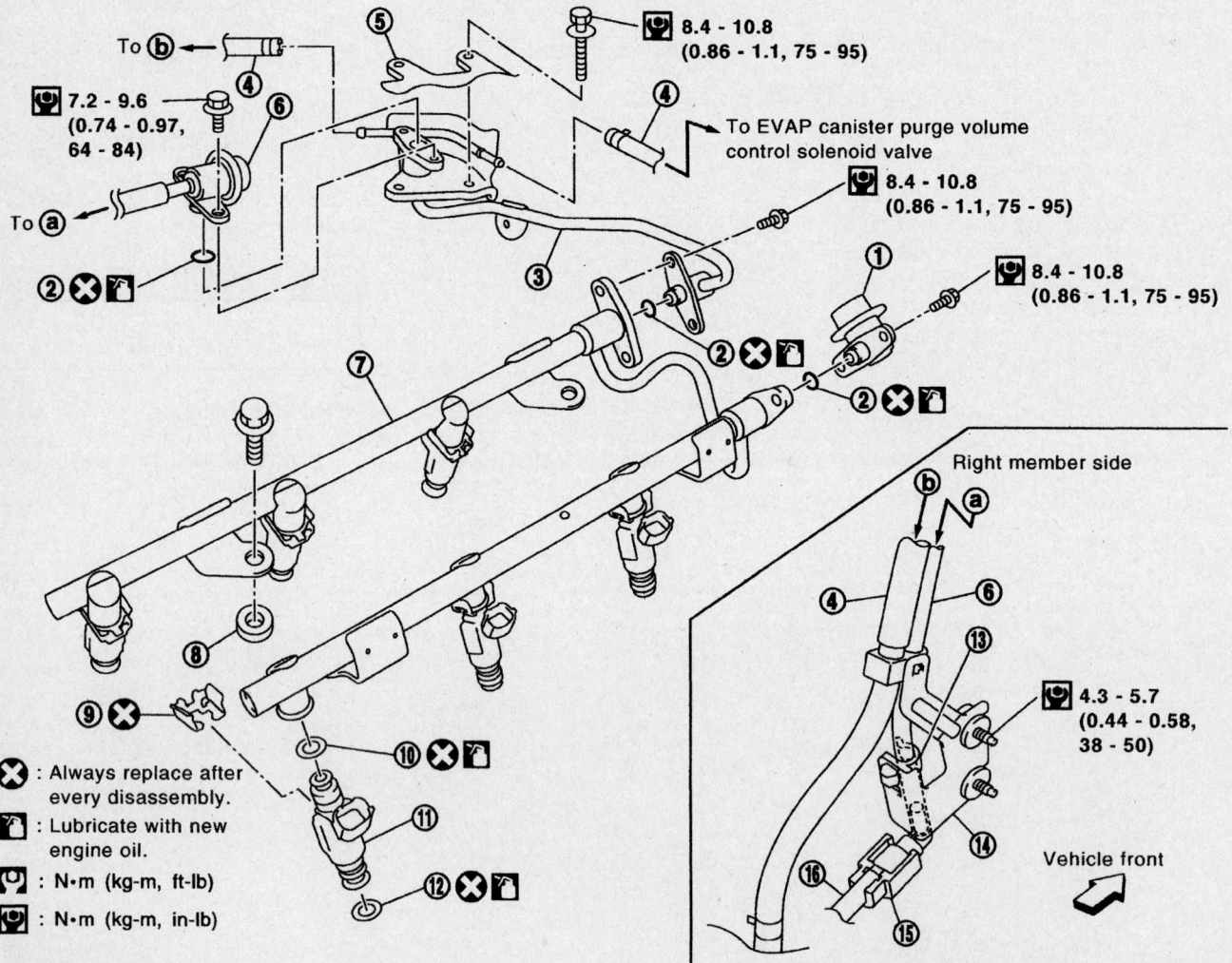

1. Fuel damper
2. O-ring
3. Fuel sub-tube
4. EVAP hose
5. Intake manifold collector (lower) rear right side
6. Fuel feed hose (with damper)
7. Fuel tube
8. Spacer
9. Clip
10. O-ring (blue)
11. Fuel injector
12. O-ring (brown)
13. Hose clamp
14. Bracket
15. Quick connector cap
16. Centralized under-floor piping

Fuel rail and injector assembly—3.5L engine

2. Relieve the fuel system pressure
3. Remove or disconnect the following:
- Negative battery cable
- Engine appearance cover
- Fuel damper and hoses from fuel rail
- Fuel rail bolts
- Injector harness connectors
- Injectors and the fuel rail as an assembly

To install:

4. Lubricate the new O-rings with clean engine oil and install the O-rings on the injector(s). The Green O-ring goes on the bottom and the Black O-ring goes on the top of the injector.
5. Install or connect the following:
- Fuel injectors to the fuel rail
- Fuel rail and injectors as an assembly to the intake manifold

6. Tighten the fuel rail bolts in the following sequence;
  a. Step 1: 7 ft. lbs. (10 Nm).
  b. Step 2: 17 ft. lbs. (23 Nm).
- Fuel hoses to the fuel rail
- Fuel damper
- Engine appearance cover
- Negative battery cable

7. Start the vehicle, check for leaks and repair if necessary.

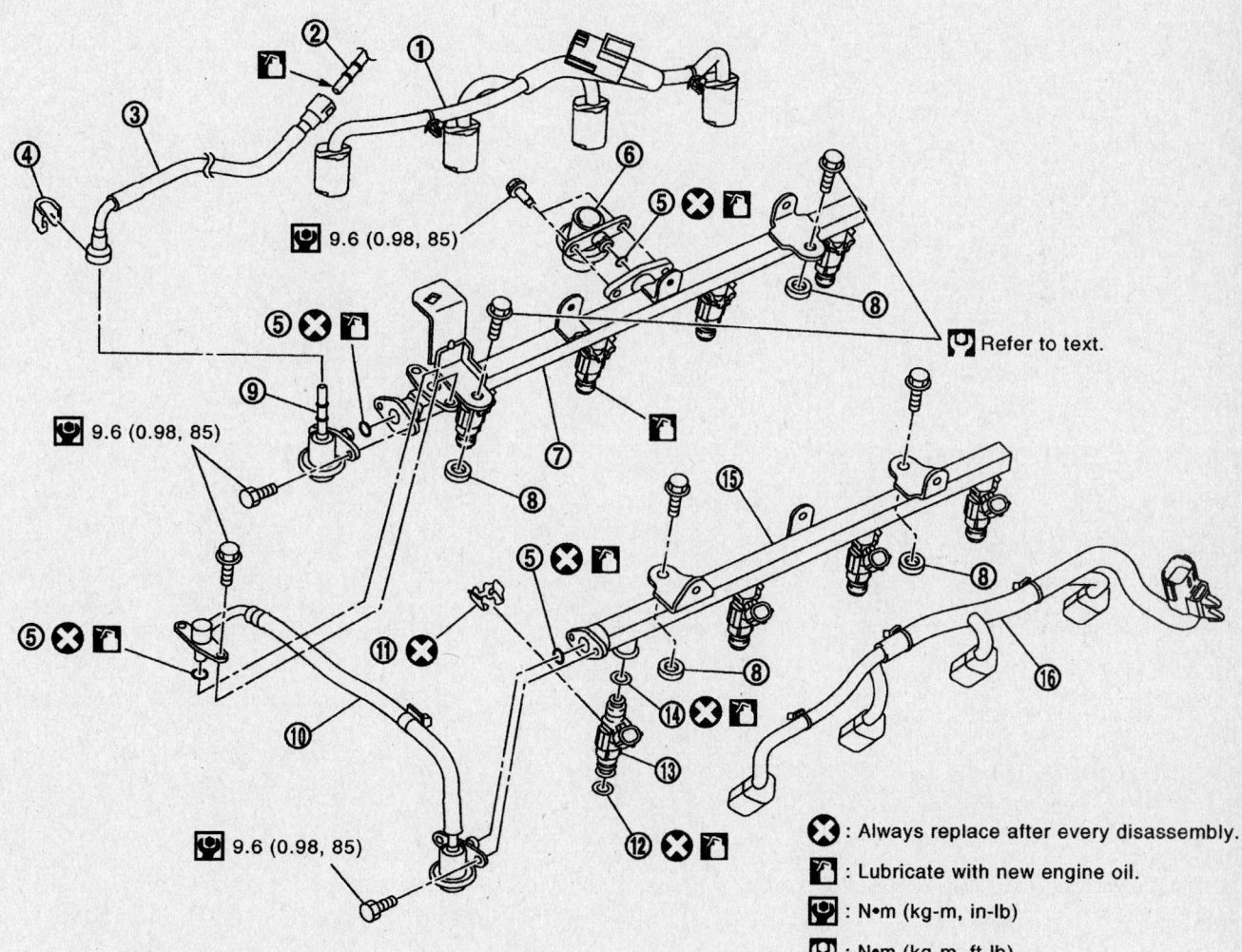

1.	Fuel injector sub harness (right)	2.	Centralized under-floor piping	3.	Fuel feed hose
4.	Quick connector cap	5.	O-ring	6.	Fuel damper (right)
7.	Fuel tube (right)	8.	Spacer	9.	Fuel feed damper
10.	Fuel damper and fuel hose assembly	11.	Clip	12.	O-ring (Green)
13.	Fuel injector	14.	O-ring (Black)	15.	Fuel tube (left)
16.	Fuel injector sub harness (left)				

Fuel rail and fuel injector assembly—4.5L engine, exc. 2006 M45

09482_INFI_G0074

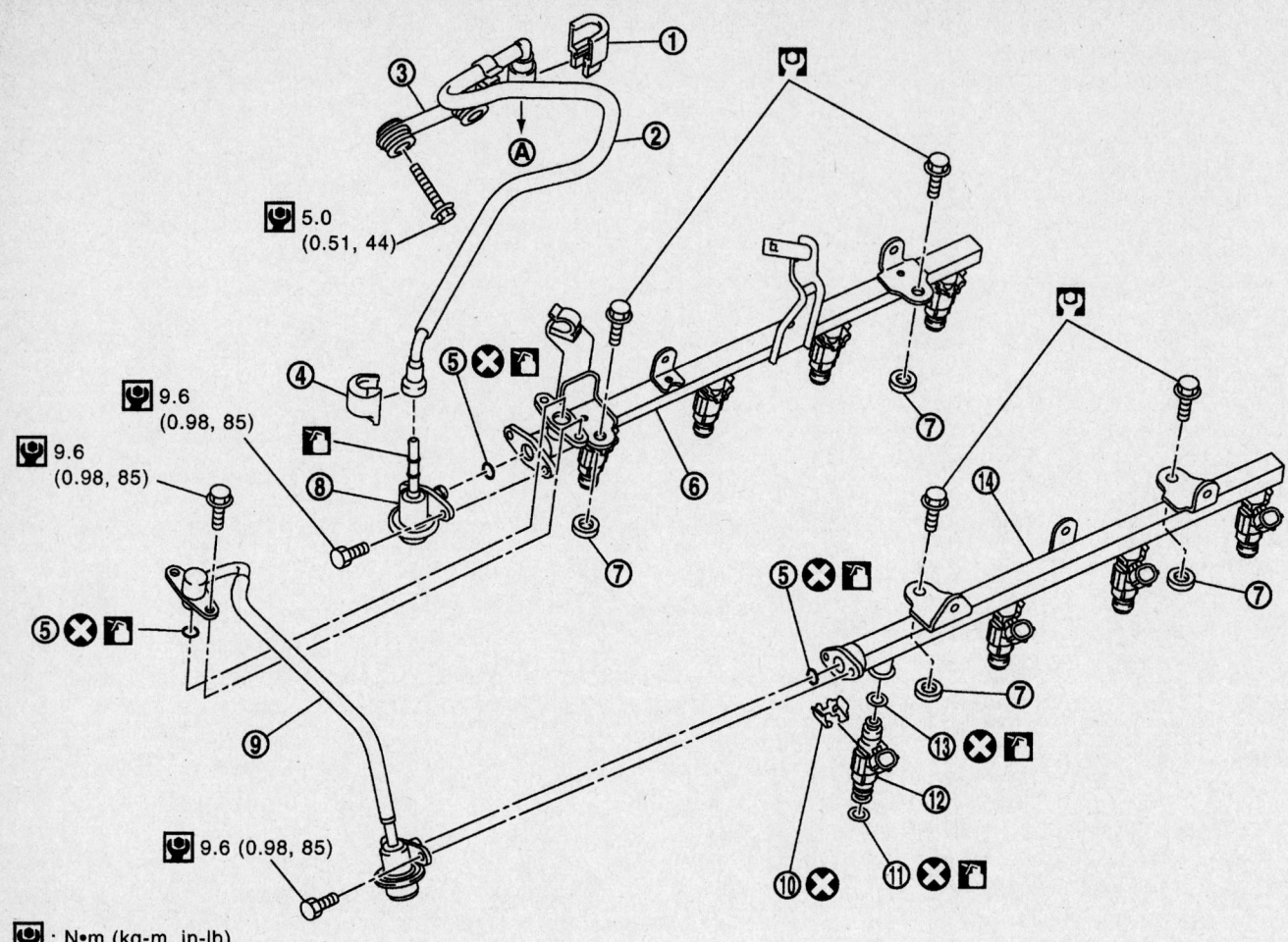

5.0 (0.51, 44)

9.6 (0.98, 85)

9.6 (0.98, 85)

9.6 (0.98, 85)

: N•m (kg-m, in-lb)

: N•m (kg-m, ft-lb)

1.	Quick connector cap	2.	Fuel feed hose	3.	Fuel feed hose bracket
4.	Quick connector cap	5.	O-ring	6.	Fuel tube (RH)
7.	Spacer	8.	Fuel feed damper	9.	Fuel damper and fuel hose assembly
10.	Clip	11.	O-ring (Green)	12.	Fuel injector
13.	O-ring (Black)	14.	Fuel tube (LH)		
A.	To centralized under-floor piping				

09482_INFI_G0410

Fuel rail and fuel injector assembly—4.5L engine, 2006 M45

DRIVE TRAIN

Transaxle Assembly

REMOVAL & INSTALLATION

Manual

G20 MODELS

1. Before servicing the vehicle, refer to the precautions in the beginning of this section.
2. Drain the transaxle fluid.
3. Remove or disconnect the following:

- Negative battery cable
- Air cleaner and air duct assembly
- Clutch operating cylinder from the transaxle
- Back-up light switch, neutral switch and ground harness connectors
- Speedometer sensor
- Starter
- Air bleeder hose
- Shift control and support rods
- Front exhaust tube
- Halfshafts and support the engine with a suitable jack under the oil pan
- Rear and left engine mount

4. Raise the jack and remove the lower transaxle housing bolts. Lower jack and remove the upper housing bolts. Keep the bolts in order as they are different lengths and must be returned to the same position.
5. Lower the transaxle.

To install:

6. Raise the transaxle into place and install the attaching bolts. Torque the shortest bolt to 22–30 ft. lbs. (30–40 Nm) and the remaining bolts to 51–59 ft. lbs. (70–79 Nm).
7. Install or connect the following:

- Rear and left engine mounts
- Driveshafts
- Shift control rods, support rod, bleeder air hose and starter
- Air bleeder hose
- Starter
- Speedometer sensor
- Back-up light switch, neutral switch and ground harness connectors
- Clutch operating cylinder. Torque the bolts to 22–27 ft. lbs. (29–37 Nm).
- Negative battery cable

8. Fill the transaxle to the proper level.
9. Start the vehicle, check for leaks and repair if necessary.

Automatic

G20 MODELS

1. Before servicing the vehicle, refer to the precautions in the beginning of this section.
2. Drain the transaxle fluid.
3. Remove or disconnect the following:

- Battery and bracket
- Air duct assembly
- Transaxle solenoid harness connector, Park Neutral Position (PNP) switch connector and revolution switch connector
- Crankshaft Position (CKP) sensor from the transaxle
- Control cable and transaxle coolant lines
- Halfshafts, the intake manifold support bracket and the starter
- Upper bolts attaching transaxle to the engine

4. Support the engine with a suitable stand and use a suitable jack to support the transaxle.

➡**Bolts are of different lengths, note the locations that the bolts are removed from.**

- Center member
- Rear cover plate and the bolts securing the torque converter to the driveplate. Rotate the crankshaft to gain access to the bolts.
- Transaxle mounts
- Lower transaxle mounting bolts and lower the transaxle

To install:

5. Place a straightedge across the bell housing of the transaxle and measure the distance to the mounting bosses on the torque converter. The distance should be 0.626 in. (16mm). If not, the torque converter is not installed correctly.
6. Check the driveplate run-out with a dial indicator. Maximum allowable run-out is 0.008 in. (0.2mm).
7. Raise the transaxle into position and install the transaxle mounting bolts. Tighten the 50, 55, and 65mm long bolts to 51–59 ft. lbs. (70–79 Nm). Torque the 35 and 45 mm long bolts to 12–15 ft. lbs. (16–21 Nm).
8. Install or connect the following:

- Torque converter bolts. Torque the bolts to 33–43 ft. lbs. (44–59 Nm). Rotate the crankshaft to gain access to the bolts.
- Rear cover and center member

- Transaxle mounts
- Halfshafts, intake manifold support bracket and the starter
- Control cable and transaxle coolant lines
- CKP sensor to the transaxle
- Transaxle solenoid harness connector, PNP switch connector and revolution switch connector
- Air duct assembly
- Battery and bracket

9. Fill the transaxle with fluid.
10. Start the vehicle, check for leaks and repair if necessary.

I35 MODELS

➡**The radio may contain a coded theft protection circuit. Always obtain the code number from the customer before disconnecting the battery.**

1. Before servicing the vehicle, refer to the precautions in the beginning of this section.
2. Drain the transaxle fluid.
3. Remove or disconnect the following:

- Battery and bracket
- Air cleaner and resonator
- Terminal cord assembly harness connector and Park Neutral Position (PNP) switch harness connector
- Revolution and Vehicle Speed Sensor (VSS) electrical connections
- Crankshaft Position (CKP) sensor from the transaxle
- Left-hand mounting bracket from transaxle and body
- Control cable from the transaxle
- Driveshafts
- Oil cooler pipes and cap pipes to avoid contamination
- Starter motor and place a jack under the oil pan to support the engine. Do not place the jack under the oil pan drain plug.
- Crossmember
- Rear cover plate and bolts attaching the torque converter to the drive plate

4. Support the transaxle with a jack.
5. Remove the transaxle-to-engine bolts and lower the transaxle using the jack.

To install:

6. Install or connect the following:

- Torque converter in the transmission. Be sure the torque converter is fully seated in the front pump assembly. The distance from the

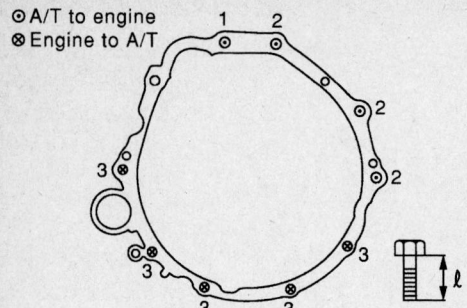

⊙ A/T to engine
⊗ Engine to A/T

Bolt No.	Tightening torque N·m (kg-m, ft-lb)	ℓ mm (in)
1	69.6 - 79.4 (7.1 - 8.0, 52 - 58)	65 (2.56)
2	69.6 - 79.4 (7.1 - 8.0, 52 - 58)	52 (2.05)
3	69.6 - 79.4 (7.1 - 8.0, 52 - 58)	40 (1.57)

09482_INFI_G0076

Transaxle bolts tightening sequence and torque specifications–I35 models with automatic transaxle

front edge of the transmission to the bolt hole of the torque converter should be 0.75 in. (19mm).
- Position the transmission to the engine and install a few bolts to hold the transmission in place. Do not fully tighten the bolts at this time.
- Torque converter-to-flexplate bolts. Torque the bolts to 33–43 ft. lbs. (44–59 Nm).

7. Secure the transmission to the engine. Torque the bolts in the sequence illustrated.

8. Install all remaining components in the reverse order of removal.

9. Fill the transmission with new fluid. Use the same amount of fluid that was drained before removal.

10. Connect negative battery cable and start the engine. Allow the engine to reach normal operating temperature and check the transmission fluid level. Add fluid as needed.

Transmission Assembly

REMOVAL & INSTALLATION

Manual

G35 MODELS

1. Before servicing the vehicle, refer to the precautions in the beginning of this section.
2. Remove or disconnect the following:
   - Negative battery cable

- Exhaust mounting bracket
- Catalytic converter and front exhaust tube
- Rear driveshaft
- Shift lever assembly
- Clutch slave cylinder
- Crankshaft position sensor (POS)

➡ **Handle sensor with care; do not place in an area affected by magnetism.**

- Neutral safety switch and back-up lamp switch
- Oxygen sensor, POS, and back-up lamp harness
- Starter motor

3. Support weight of transmission on a suitable transmission jack

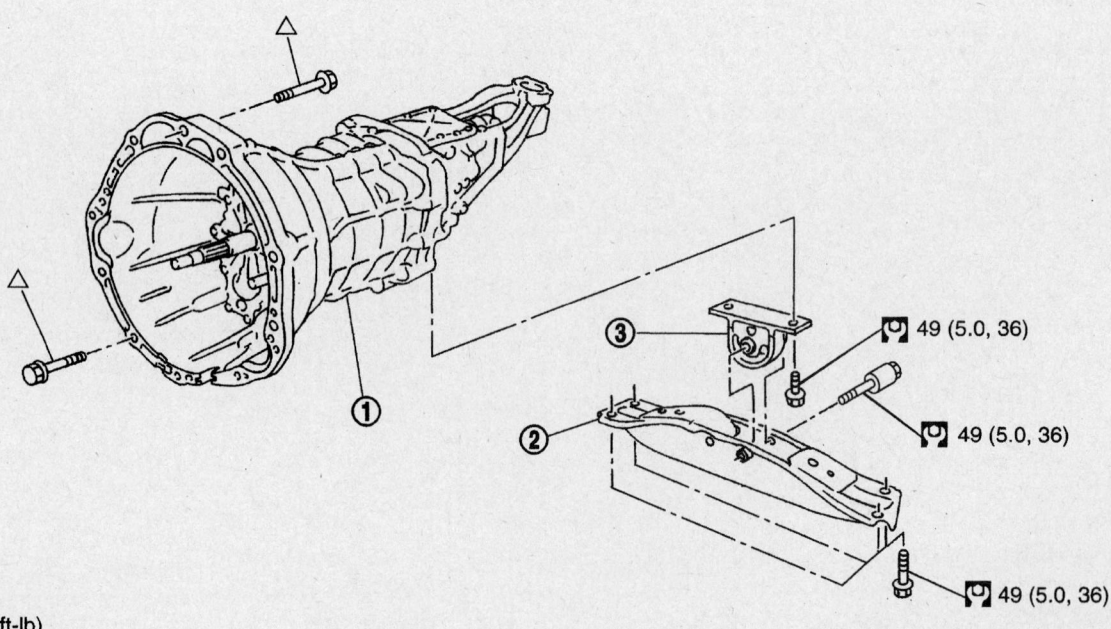

⊡ : N·m (kg-m, ft-lb)

1. Transmission case
2. Rear engine mounting member
3. Insulator

09482_INFI_G0080

Manual transmission and mounting member—G35 models

Bolt No.	1	2	3
Quantity	1	5	2
" ℓ " mm (in)	55 (2.17)	65 (2.56)	35 (1.38)
Tightening torque N·m (kg-m, ft-lb)	75 (7.7, 55)		46.6 (4.8, 34)

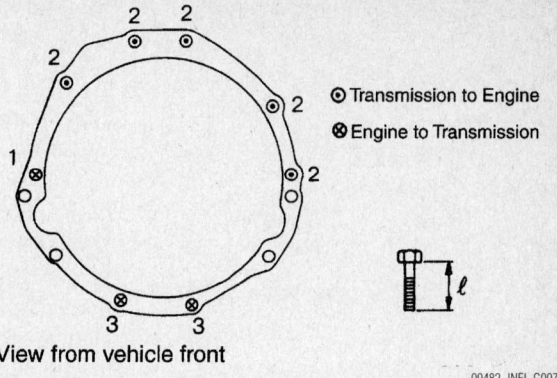

⊙ Transmission to Engine

⊗ Engine to Transmission

View from vehicle front

09482_INFI_G0078

Bolt position and torque values for manual transmission—G35 models

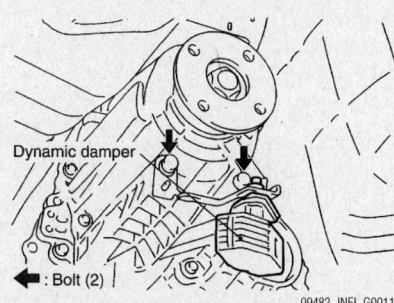

Dynamic damper

← : Bolt (2)

09482_INFI_G0011

Dynamic damper assembly (AWD)

➡ **Make certain transmission does not rest on switch terminals**

4. Remove rear engine mounting member
5. Remove engine and transmission mounting bolts
6. Lower jack and remove transmission from vehicle

To install:

7. Installation is the reverse of the removal procedure, observing mounting bolt position and torque values shown in chart.

Automatic

G35 MODELS

1. Before servicing the vehicle, refer to the precautions in the beginning of this section.
2. Remove or disconnect the following:
   - Negative battery cable
   - Engine cover
   - Positive battery cable and battery
   - Exhaust pipe

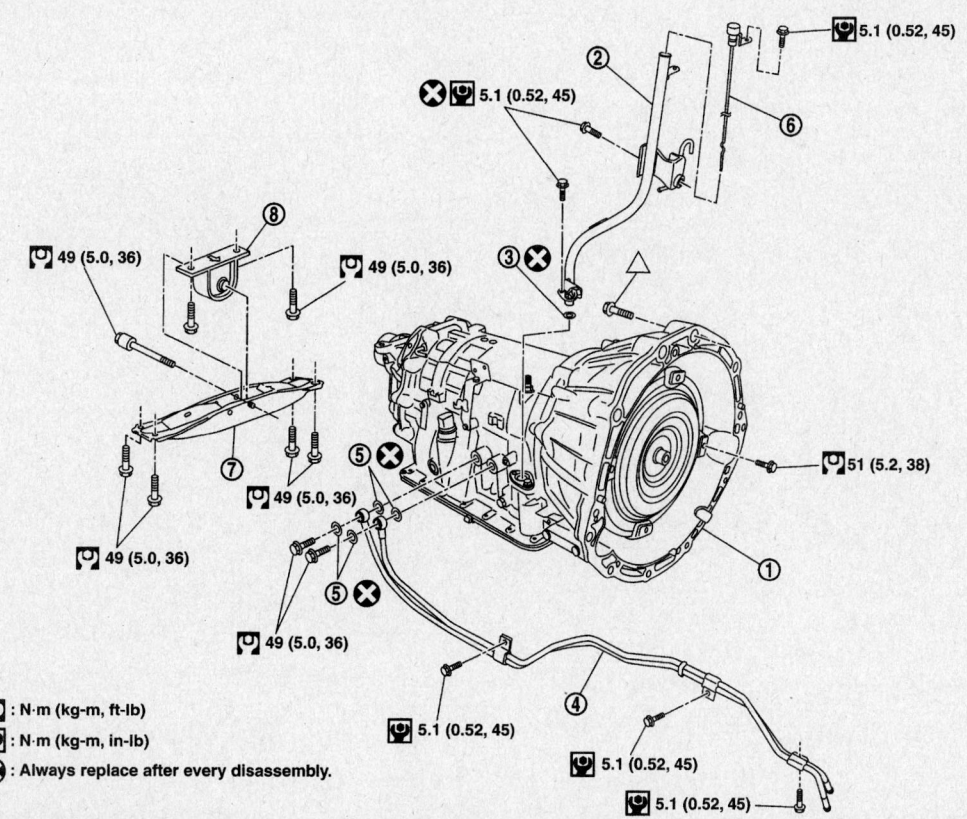

🔧 : N·m (kg-m, ft-lb)

🔧 : N·m (kg-m, in-lb)

⊗ : Always replace after every disassembly.

1. Transmission assembly
2. A/T fluid charging pipe
3. O-ring
4. Fluid cooler tube
5. Copper washer
6. A/T fluid level gauge
7. Rear engine mounting member
8. Engine mounting insulator (rear)

09482_INFI_G0077

Exploded view of the automatic transmission mounting—G35 2WD models

- Rear driveshaft
- Front driveshaft (AWD)
- Automatic Transmission (A/T) control rod and solenoid valve harness connector
- Crankshaft Position (CKP) sensor from the transmission
- Oil cooler and transmission fluid pipes, then plug the lines
- Air breather hose
- Starter motor
- Dust cover from converter housing

→ **When turning the crankshaft, turn it clockwise as viewed from the front of the engine.**

3. Turn the crankshaft, then remove the 4 tightening bolts for the drive plate and torque converter.

4. Remove the dynamic damper (AWD).

5. Support the transmission/transfer (AWD) assembly with a suitable jack. Be careful not the let the jack hit the drain plug.

6. Remove or disconnect the following:

- Rear member
- Engine-to-transmission bolts

✳✳ WARNING

Before removal, secure the transmission to the jack and secure the torque converter to prevent it from falling.

- Transmission from the vehicle by carefully lowering it with the jack

7. Remove transfer assembly mounting bolts and separate transfer assembly from transmission (AWD; Refer to Transfer Case Assembly procedures).

To install:

8. Installation is the reverse of the removal procedure, noting the following:

a. Tighten the transmission-to-engine bolts as shown in the illustration.

b. Align the positions of the tightening bolts for the drive plate with those of the torque converter and hand-tighten. Then, tighten to 33–42 ft. lbs. (44–58 Nm).

c. After the converter is installed, rotate the crankshaft a few times to be sure the transmission rotates without any binding

d. Fill the transmission with fluid.

e. Start the vehicle, check for leaks and repair if necessary.

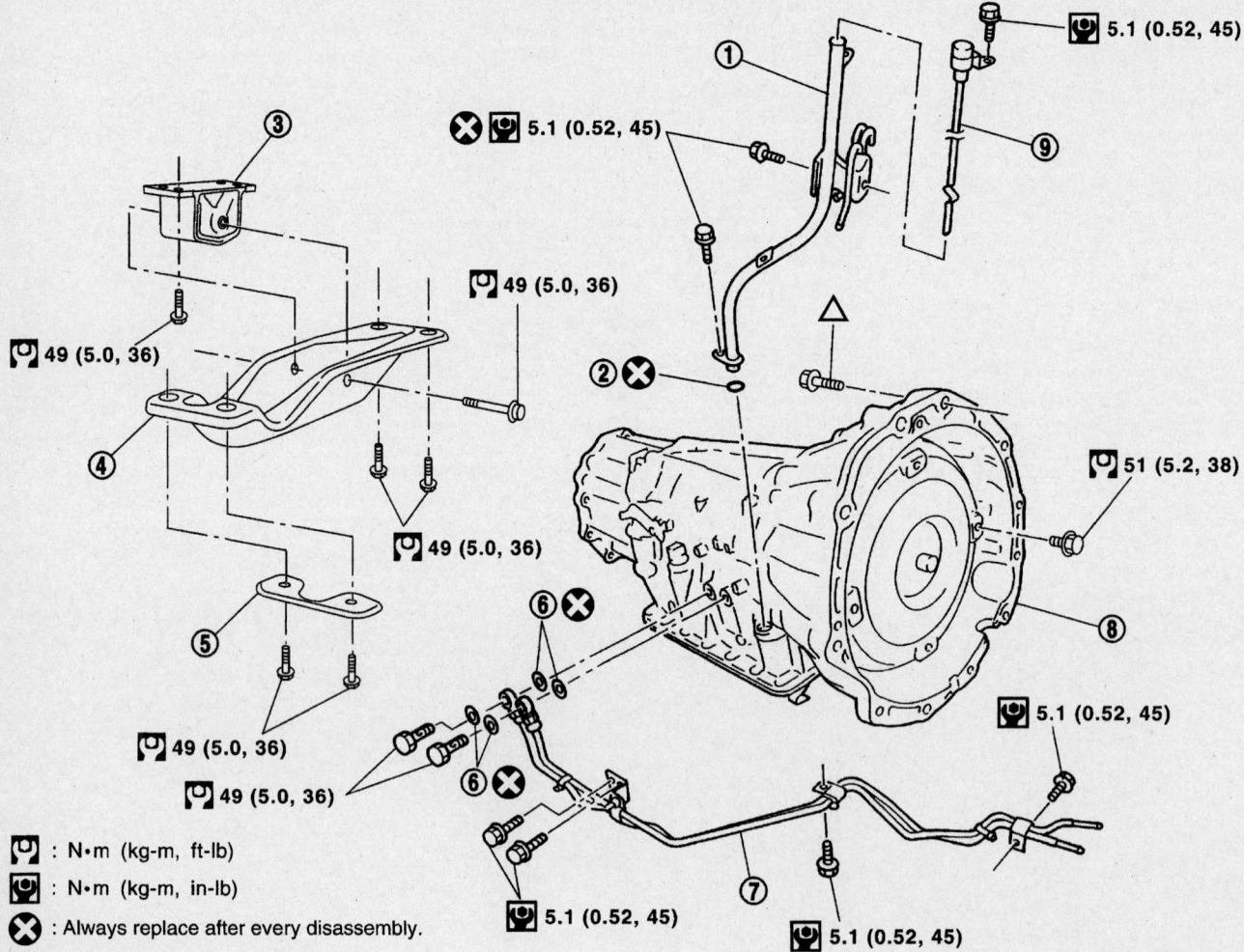

: N•m (kg-m, ft-lb)

: N•m (kg-m, in-lb)

: Always replace after every disassembly.

1.	A/T fluid charging pipe	2.	O-ring	3.	Engine mounting insulator (rear)
4.	Rear engine mounting member	5.	Heat insulator	6.	Copper washer
7.	Fluid cooler tube	8.	Transmission assembly	9.	A/T fluid level gauge

09482_INFI_G0079

Exploded view of the automatic transmission mounting—G35 AWD models

Bolt No.	1	2	3	4
Number of bolts	1	5	2	2
Bolt length " ℓ "mm (in)	55 (2.17)	65 (2.56)	50 (2.20)	35 (1.38)
Tightening torque N·m (kg-m, ft-lb)	75 (7.7, 55)		55 (5.6, 41)	47 (4.8, 35)

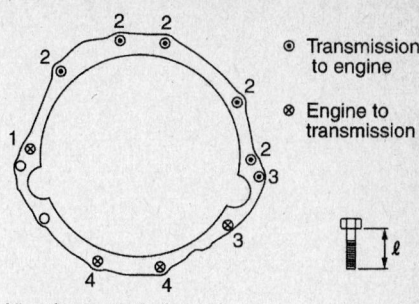

View from vehicle front

09482_INFI_G0411

Automatic transmission-to-engine bolt tightening specifications—G35 2WD models

2002–03 Q45 MODELS

1. Before servicing the vehicle, refer to the precautions in the beginning of this section.
2. Remove or disconnect the following:
   - Negative battery cable
   - Crankshaft Position (CKP) sensor
   - Rear Heated Oxygen Sensor (HO2S) connector
   - Exhaust tubes
   - Fluid charging pipe
   - Oil cooler lines. Plug fluid charging and oil cooler fittings after removing the lines.
   - Control linkage from the selector lever
   - Neutral safety switch and solenoid harness connectors
   - Speed sensor connection
   - Driveshaft (make matchmarks to ease in installation). Insert plug into rear seal opening to prevent loss of fluid.
3. Support the transmission safely.
   - Bolts securing the torque converter to the flexplate
   - Gussets securing the transmission to the engine
   - Bolts attaching the transmission to the engine

➡The bolts securing the transmission to the engine are of different lengths. Note the length of the bolts as they are removed.

4. Support the engine safely. Avoid jacking directly under the oil pan drain plug.
5. Remove the transmission from the vehicle.

To install:

6. Install or connect the following:
   - Transmission in the vehicle and install the torque converter-to-flex-plate bolts. Torque the bolts to 33–43 ft. lbs. (44–59 Nm).
7. Secure the transmission to the engine. Torque the bolts as follows:
   a. 70mm bolts: 80–87 ft. lbs. (108–118 Nm).
   b. 90mm bolts: 51–58 ft. lbs. (69–78 Nm).
   - Torque converter-to-drive plate bolts and tighten in 2 steps to 33–43 ft. lbs. (44–59 Nm)
   - Driveshaft, aligning the matchmarks made before removal
   - Speed sensor connection
   - Neutral safety switch and solenoid harness connectors
   - Control linkage to the selector lever
   - Fluid charging and oil cooler lines
   - Exhaust tubes

8. Lower the vehicle.
9. Fill the transmission with fluid.
10. Connect negative battery cable.
11. Start the vehicle, check for leaks and repair if necessary.

2004–06 Q45 AND 2004 M45 MODELS

1. Before servicing the vehicle, refer to the precautions in the beginning of this section.
2. Remove or disconnect the following:
   - Negative battery cable
   - Engine splash shield
   - Front cross bar (if equipped)
   - Front exhaust tube and center muffler
   - Driveshaft (make matchmarks to ease in installation). Insert plug into rear seal opening to prevent loss of fluid.
   - Transmission control rod
   - Transmission harness connectors
   - Crankshaft Position (CKP) sensor
   - Oil cooler lines
   - Breather hose
   - Starter
   - Torque converter cover
   - Torque converter bolts
3. Place a transmission jack under the transmission
4. Remove or disconnect the following:
   - Rear engine crossmember

Bolt No.	1	2	3	4
Number of bolts	1	5	2	1
Bolt length " ℓ "mm (in)	55 (2.17)	65 (2.56)	35 (1.38)	40 (1.57)
Tightening torque N·m (kg-m, ft-lb)	75 (7.7, 55)		47 (4.8, 35)	34 (3.5, 25)

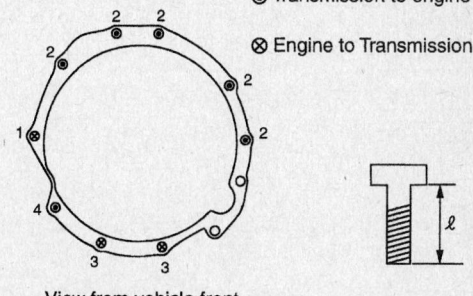

View from vehicle front

09482_INFI_G0082

Automatic transmission-to-engine bolt tightening specifications—G35 AWD models

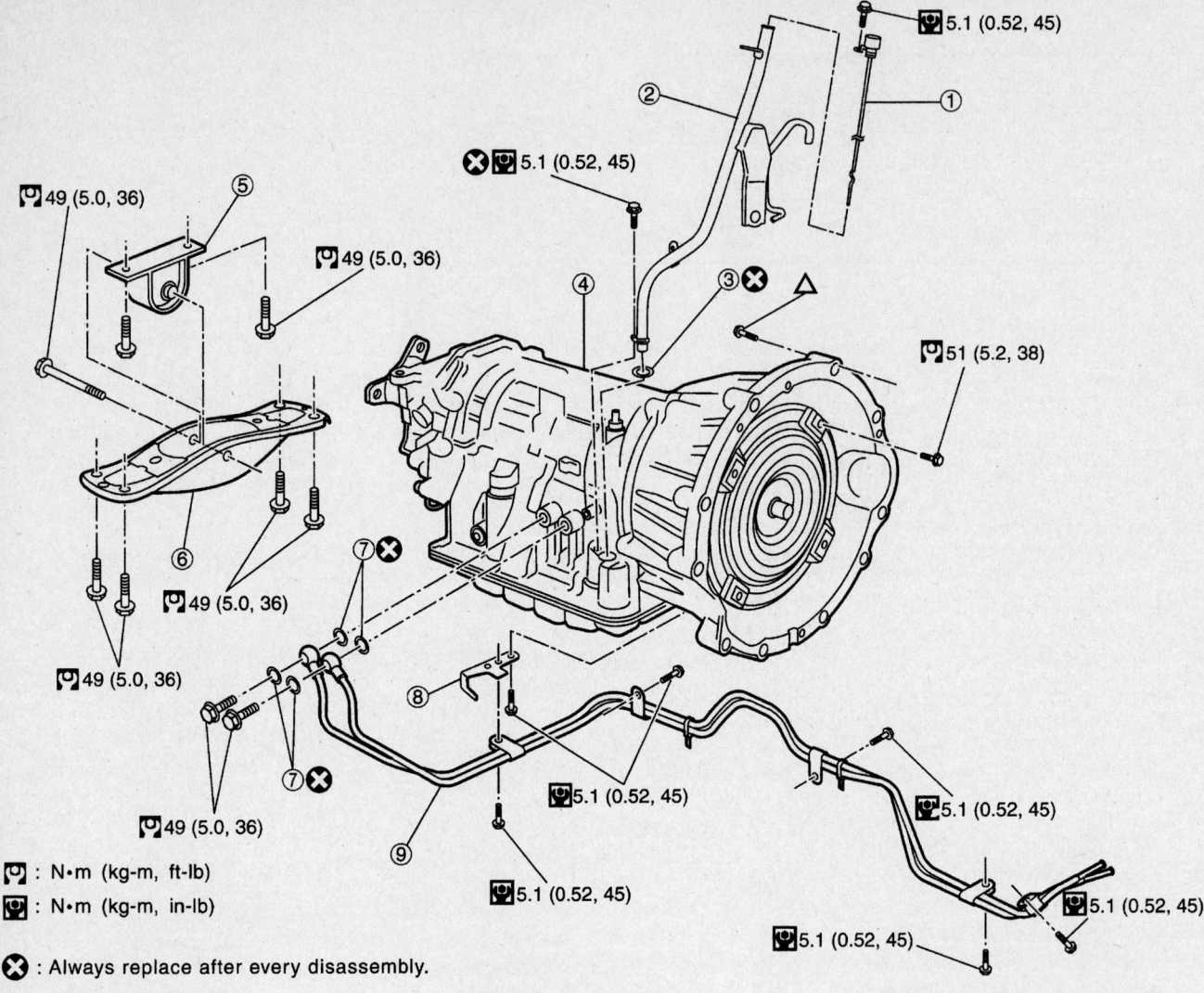

49 (5.0, 36)

5.1 (0.52, 45)

5.1 (0.52, 45)

49 (5.0, 36)

49 (5.0, 36)

51 (5.2, 38)

49 (5.0, 36)

49 (5.0, 36)

49 (5.0, 36)

5.1 (0.52, 45)

5.1 (0.52, 45)

5.1 (0.52, 45)

5.1 (0.52, 45)

5.1 (0.52, 45)

: N•m (kg-m, ft-lb)

: N•m (kg-m, in-lb)

⊗ : Always replace after every disassembly.

1. A/T fluid level gauge	2. A/T fluid charging pipe	3. O-ring
4. A/T assembly	5. Insulator	6. Engine rear member
7. Copper washer	8. Bracket	9. Fluid cooler tube

09482_INFI_G0083

Exploded view of the automatic transmission mounting—Q45 and 2004 M45 models

- Transmission-to-engine bolts
- Transmission

To install:

5. Installation is the reverse of the removal procedure, noting the following:

   a. When installing the transmission, note the correct bolt locations. The 65 mm bolts are the four bolts at the bottom. Tighten the 65 mm bolts to 55 ft. lbs. (74 Nm) and the 70 mm bolts to 84 ft. lbs. (114 Nm).

   b. Tighten the torque converter-to-drive plate bolts in 2 steps to 38 ft. lbs. (51 Nm)

   c. Lower the vehicle.

   d. Fill the transmission with fluid.
   e. Connect negative battery cable.
   f. Start the vehicle, check for leaks and repair if necessary.

2006 M45/M35 MODELS—2WD

➡Refer to 2004-06 Q45 procedures for 2004 M45 models.

1. Before servicing the vehicle, refer to the precautions in the beginning of this section.

2. Remove or disconnect the following:
- Negative battery cable
- Engine undercover
- Transmission dipstick

- Front exhaust tube and center muffler
- Heat insulator
- Rear driveshaft
- Steering rack stay
- Exhaust mounting bracket
- Transmission shifter control rod
- Crankshaft position sensor (POS)

➡Handle sensor with care; do not place in an area affected by magnetism.

- Starter motor
- Torque converter cover plate
- Torque converter to drive plate bolts

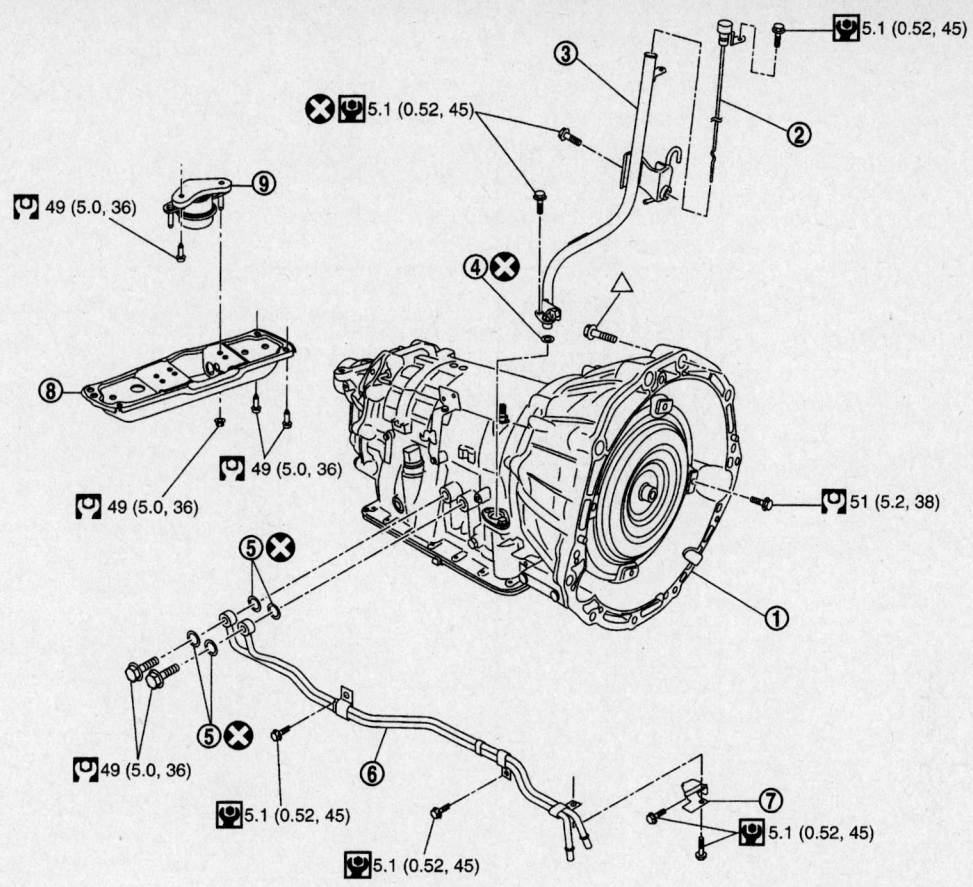

1. A/T assembly
2. A/T fluid level gauge
3. A/T fluid charging pipe
4. O-ring
5. Copper washer
6. Fluid cooler tube
7. Bracket
8. Rear engine mounting member
9. Engine mounting insulator (rear)

09482_INFI_G0084

Exploded view of transmission mounting—2006 M45/M35 2WD models

VQ35DE models

Bolt No.	1	2	3	4
Number of bolts	1	5	2	2
Bolt length "ℓ" mm (in)	55 (2.17)	65 (2.56)	65 (2.56)	35 (1.38)
Tightening torque N·m (kg-m, ft-lb)	75 (7.7, 55)		55 (5.6, 41)	47 (4.8, 35)

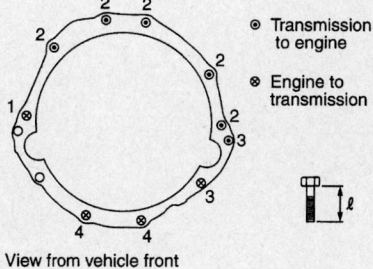

⊙ Transmission to engine
⊗ Engine to transmission

View from vehicle front

VK45DE models

Bolt No.	1	2*	3
Number of bolts	5	1	4
Bolt length "ℓ" mm (in)	70 (2.76)	70 (2.76)	65 (2.56)
Tightening torque N·m (kg-m, ft-lb)	113 (12, 83)		74 (7.5, 55)

*: No.2 bolt also secures A/T fluid charging pipe.

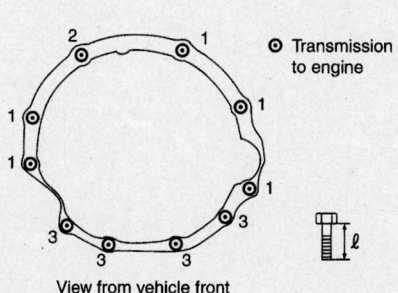

⊙ Transmission to engine

View from vehicle front

09482_INFI_G0087

Bolt positioning and torque values for M35/M45 2WD models

3. Support transmission assembly with a suitable transmission jack

➡ **Make certain weight of transmission does not rest on drain plug.**

4. Remove rear engine mounting member and insulator
5. Disconnect transmission assembly wiring harness and air breather tube
6. Remove transmission filler pipe and O-ring
7. Disconnect transmission oil cooler tubes
8. Remove bolts fastening transmission assembly to engine block

➡ **The bolts securing the transmission to the engine are of different lengths.**

Note the length of the bolts as they are removed.

9. Lower jack and remove transmission from vehicle

✳✳ WARNING

Do not allow torque converter to drop from housing!

To install:

10. Installation is the reverse of the removal procedure, noting the following:

a. Tighten the transmission-to-engine bolts as shown in the illustration.
b. Fill the transmission with fluid.
c. Connect negative battery cable.

d. Start the vehicle, check for leaks and repair if necessary.

2006 M35 AWD MODELS

1. Before servicing the vehicle, refer to the precautions in the beginning of this section.
2. Remove or disconnect the following:
- Negative battery cable
- Engine undercover
- Transmission dipstick
- Front exhaust tube and center muffler
- Heat insulator
- Rear driveshaft
- Front cross bar
- Exhaust mounting bracket and 3-way catalyst

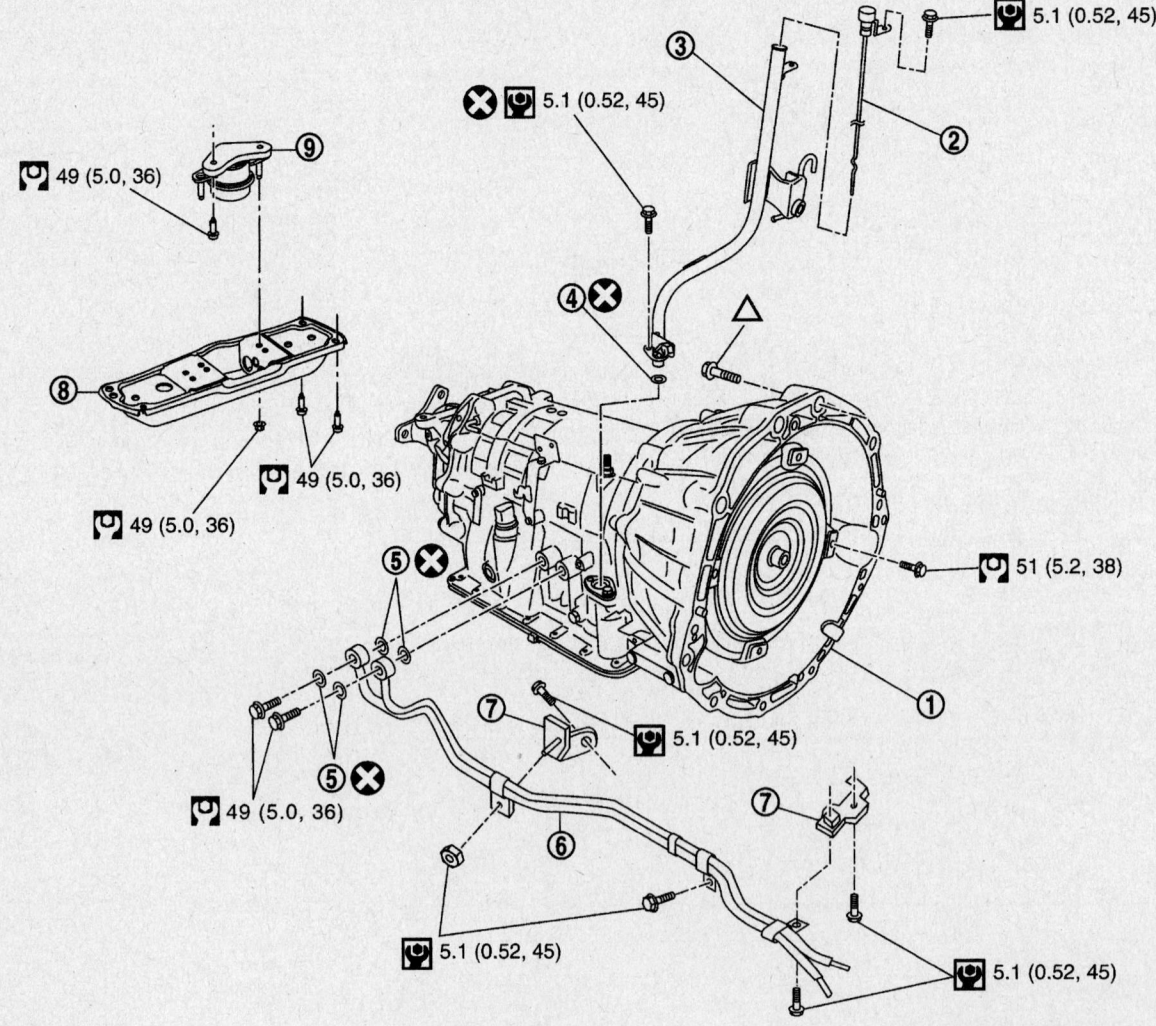

1. A/T assembly
4. O-ring
7. Bracket

2. A/T fluid level gauge
5. Copper washer
8. Rear engine mounting member

3. A/T fluid charging pipe
6. Fluid cooler tube
9. Engine mounting insulator (rear)

09482_INFI_G0088

Exploded view of transmission mounting—2006 M35 AWD models

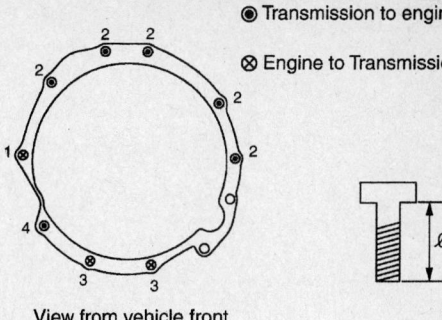

⊙ Transmission to engine

⊗ Engine to Transmission

View from vehicle front

09482_INFI_G0089

Bolt No.	1	2	3	4
Number of bolts	1	5	2	1
Bolt length "ℓ" mm (in)	55 (2.17)	65 (2.56)	35 (1.38)	40 (1.57)
Tightening torque N·m (kg-m, ft-lb)	75 (7.7, 55)		47 (4.8, 35)	34 (3.5, 25)

Bolt positioning and torque values for M35/M45 AWD models

- Front driveshaft
- Transmission shifter control rod
- Crankshaft position sensor (POS)

➡**Handle sensor with care; do not place in an area affected by magnetism.**

- Starter motor
- Torque converter cover plate
- Torque converter to drive plate bolts

3. Support transmission assembly with a suitable transmission jack

➡**Make certain weight of transmission does not rest on drain plug.**

4. Remove rear engine mounting member and insulator
5. Disconnect transmission assembly wiring harness and air breather tube
6. Remove transmission filler pipe and O-ring

7. Disconnect transmission oil cooler tubes
8. Remove bolts fastening transmission assembly to engine block

➡**The bolts securing the transmission to the engine are of different lengths. Note the length of the bolts as they are removed.**

9. Lower jack and remove transmission from vehicle

☀ WARNING

Do not allow torque converter to drop from housing!

To install:

10. Installation is the reverse of the removal procedure, noting the following:
  a. Tighten the transmission-to-engine bolts as shown in the illustration.
  b. Fill the transmission with fluid.

  c. Connect negative battery cable.
  d. Start the vehicle, check for leaks and repair if necessary.

Clutch

ADJUSTMENT

All models are equipped with a hydraulic clutch, which is self-adjusting.

REMOVAL & INSTALLATION

G20 Models

1. Before servicing the vehicle, refer to the precautions in the beginning of this section.
2. Raise and support the vehicle safely.
3. Remove or disconnect the following:
  - Negative battery cable
  - Transaxle

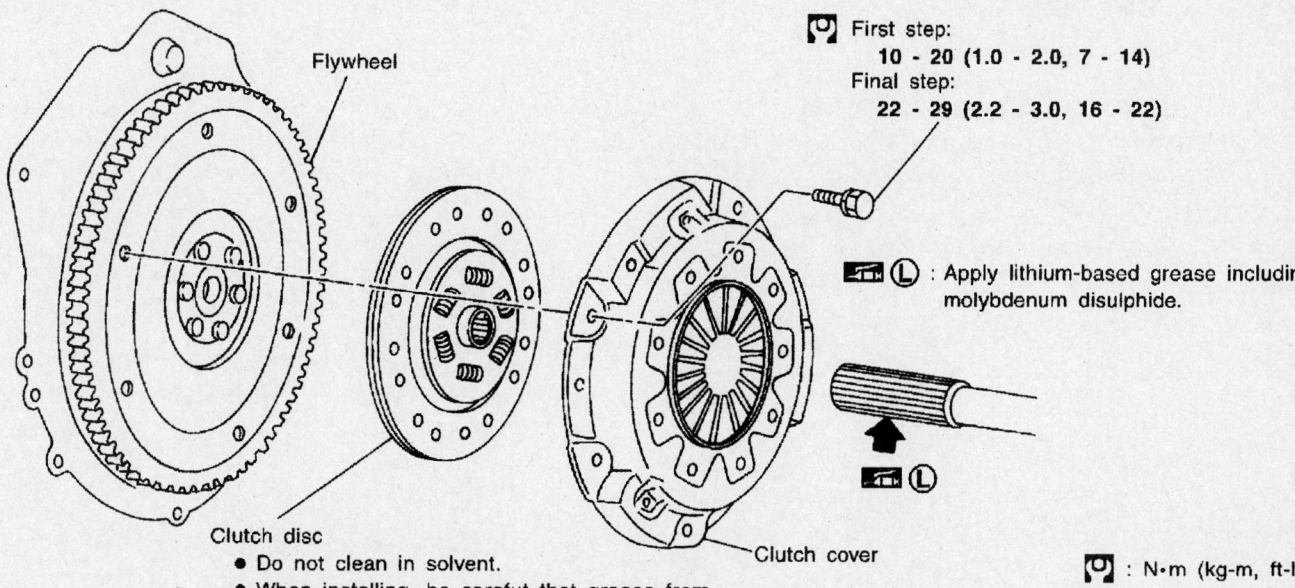

Flywheel

First step:
10 - 20 (1.0 - 2.0, 7 - 14)
Final step:
22 - 29 (2.2 - 3.0, 16 - 22)

🔧 Ⓛ : Apply lithium-based grease including molybdenum disulphide.

Clutch disc
- Do not clean in solvent.
- When installing, be careful that grease from main drive shaft does not adhere to clutch disc.

Clutch cover

🔧 : N·m (kg-m, ft-lb)

09482_INFI_G0086

Clutch disc and pressure plate—G20

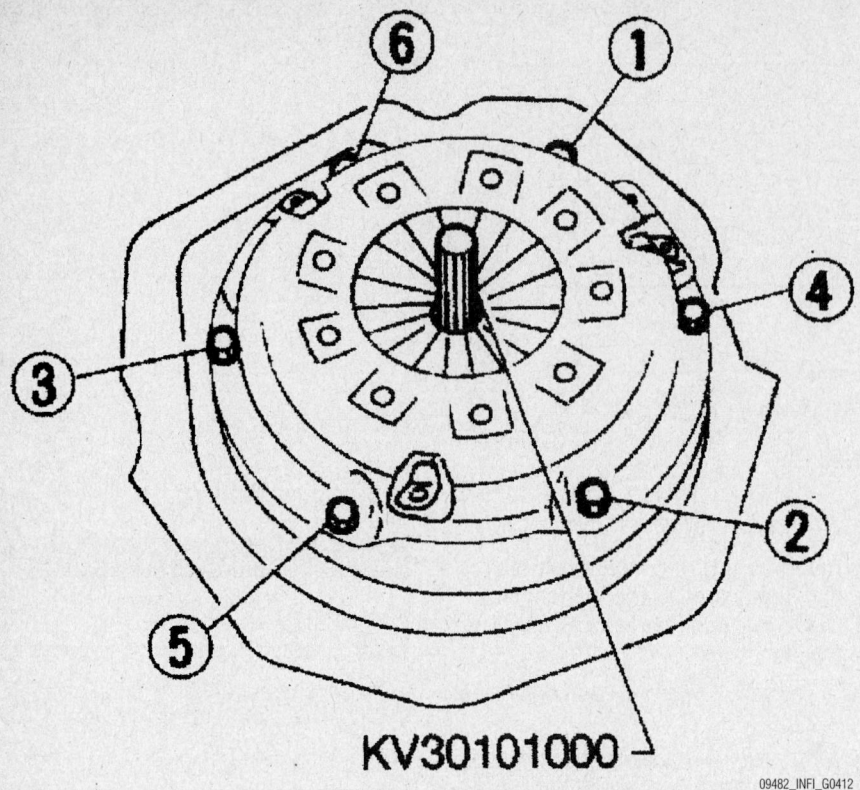

KV30101000

09482_INFI_G0412

Clutch pressure plate torque sequence—G20

4. Insert alignment tool KV30101000 into the clutch disc hub and loosen the pressure plate bolts in small increments using a star-type pattern.

- Pressure plate and clutch disc as an assembly
- Release bearing by pulling the bearing retainers outward from the transaxle case

5. Inspect the clutch disc for surface wear. Measure from the friction surface to the top of the rivets. Wear limit is 0.012 in. (0.3mm). Replace clutch disc as necessary.

6. Inspect the contact surface of the flywheel for burns or discoloration. Check flywheel run-out. Maximum run-out is 0.0059 in. (0.15mm).

7. Using tools ST20050100 and ST20050010, check the pressure plate diaphragm springs. Measure from the pressure plate/flywheel mating surface to the top of the diaphragm spring. Height should be 1.201–1.280 in. (30.5–32.5mm). Replace pressure plate as necessary.

8. Inspect the release bearing for damage. Spin the bearing to see that it rolls freely.

To install:

9. Lightly lubricate the transaxle input shaft, input shaft collar, clutch lever assem-

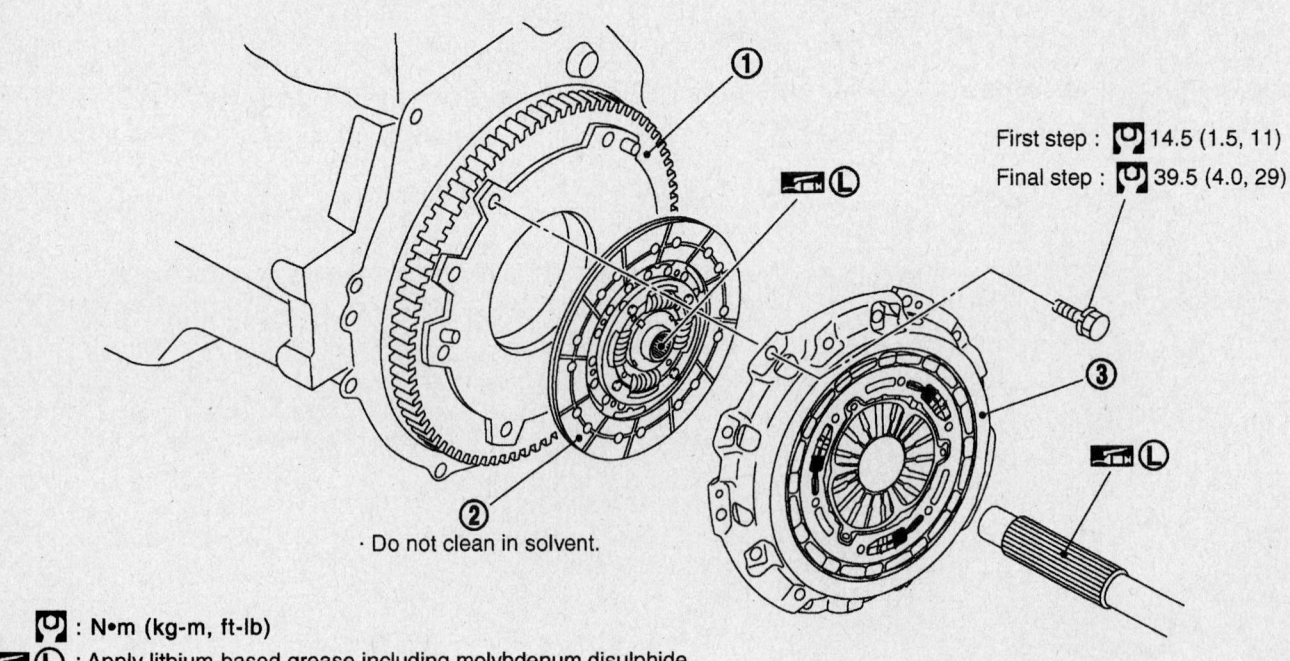

First step : 14.5 (1.5, 11)
Final step : 39.5 (4.0, 29)

· Do not clean in solvent.

: N•m (kg-m, ft-lb)

: Apply lithium-based grease including molybdenum disulphide.

1. Flywheel   2. Clutch disc   3. Clutch cover

09482_INFI_G0413

Clutch disc and pressure plate—G35

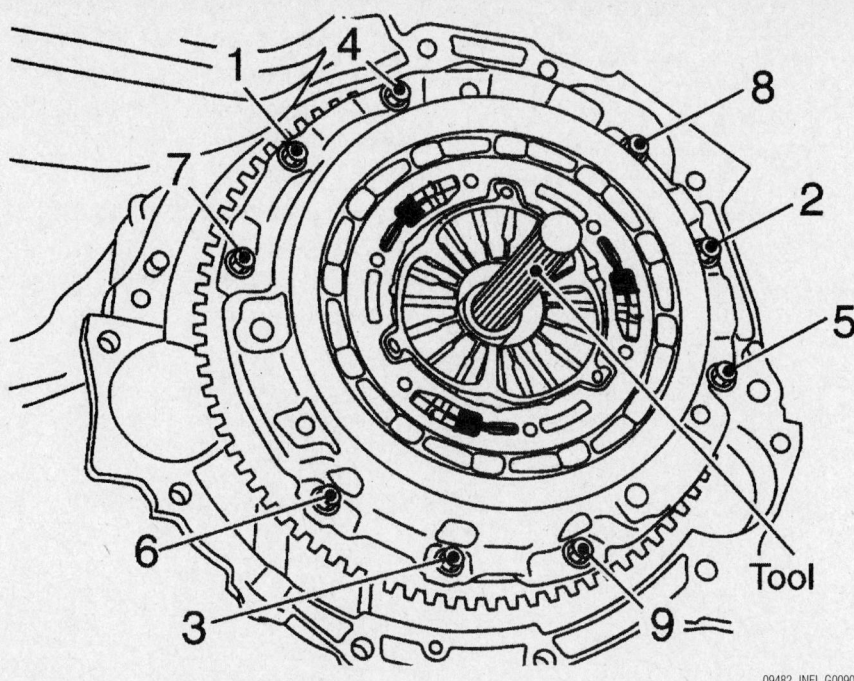

Tighten clutch cover bolts in the order shown—G35 models

09482_INFI_G0090

bly and the clutch release bearing with a lithium based grease.

➡️**Keep clutch disc and all clutch components clean during installation. Do not allow grease to contact the clutch disc.**

10. Insert alignment tool KV30101000 into the clutch disc hub. Install the clutch disc and pressure plate on the tool and torque the pressure plate bolts to 16–22 ft. lbs. (22–29 Nm) in 2–3 steps using a criss-cross pattern. Remove the tool.

11. Install or connect the following:
- Release bearing in the transaxle. Ensure that the bearing retainer clips are fully engaged.
- Transaxle
- Negative battery cable

12. If necessary, adjust clutch pedal height and free-play.

G35 Models

1. Before servicing the vehicle, refer to the precautions in the beginning of this section.

2. Raise and support the vehicle safely.

3. Remove or disconnect the following:
- Negative battery cable
- Manual transmission assembly

4. Loosen clutch cover mounting bolts in a star pattern to avoid warping

5. Remove clutch cover and clutch disc

6. Remove throwout bearing sleeve assembly, holder spring, and throwout lever from inside clutch housing

7. Remove dust cover

8. Remove snapring from throwout lever

9. Remove throwout bearing from bearing sleeve using a puller

10. Inspect the clutch disc for surface wear. Measure from the friction surface to the top of the rivets. Wear limit is 0.012 in. (0.3mm). Replace clutch disc as necessary.

11. Inspect the contact surface of the flywheel for burns or discoloration. Check flywheel run-out; deflection should be less that 0.0177 inches (0.45mm). Replace flywheel if measure is exceeded.

12. Inspect the release bearing for damage. Turn the bearing to see that it spins freely.

To install:

13. Lightly lubricate the transaxle input shaft, input shaft collar, clutch lever assembly and the clutch release bearing with a lithium based grease.

➡️**Keep clutch disc and all clutch components clean during installation. Do not allow grease to contact the clutch disc.**

14. Insert an alignment tool into the clutch disc hub. Install the clutch disc and pressure plate on the tool and torque the pressure plate bolts in two passes in the order shown in diagram:
- First pass to 11 ft. lbs. (14.5 Nm)
- Second pass to 29 ft. lbs. (39.6 Nm)

15. Remove the tool.

16. Install or connect the following:

- Throwout bearing in the transmission housing; ensure that the bearing retainer clips are fully engaged
- Transmission
- Negative battery cable

17. If necessary, adjust clutch pedal height and free-play.

Hydraulic Clutch System

BLEEDING

Bleeding is required to remove air trapped in the hydraulic system. The bleed screw is located on the clutch slave (release) cylinder.

1. Before servicing the vehicle, refer to the precautions in the beginning of this section.

2. Remove the bleed screw dust cap.

3. Attach a transparent vinyl tube to the bleed screw, immersing the free end in a clean container of clean brake fluid.

4. Fill the clutch master cylinder with the proper fluid.

5. Slowly depress the clutch pedal all the way several times and hold it down.

6. Have an assistant open the bleeder valve about ¾ turn to release the air. Then, close the bleeder valve while the pedal is still depressed.

7. Repeat the above procedure until no more air bubbles are seen in the fluid container.

8. Remove the bleed tube.

9. Replace the dust cap and refill the master cylinder.

10. Bleed the clutch damper, if equipped.

➡️**Monitor fluid level in reservoir tank to make sure it does not empty. Do not spill clutch fluid onto painted surfaces; If spills occur, wipe immediately and wash affected area with water.**

➡️**Do not use a vacuum assist or any other type of power bleeder on this system. Use of vacuum assist or power bleeder will not purge all the air from the system.**

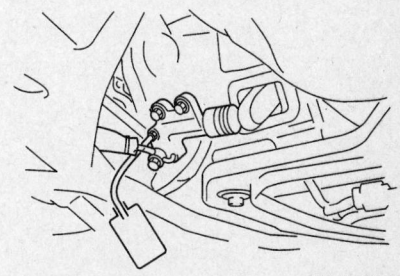

09482_INFI_G0414

Bleeding the clutch hydraulic system

Transfer Case Assembly

REMOVAL & INSTALLATION

1. Before servicing the vehicle, refer to the precautions in the beginning of this section.

2. Raise and support the vehicle safely.

3. Remove the front driveshaft as follows:

- Remove exhaust front pipe.
- Remove engine undercover.
- Remove the right bank catalytic converter.
- Put matchmarks on flanges and separate driveshaft from final drive.

➡ For matchmark, use paint. Do not damage the propeller shaft flange and companion flange on the front final drive.

4. Remove the rear driveshaft as follows:

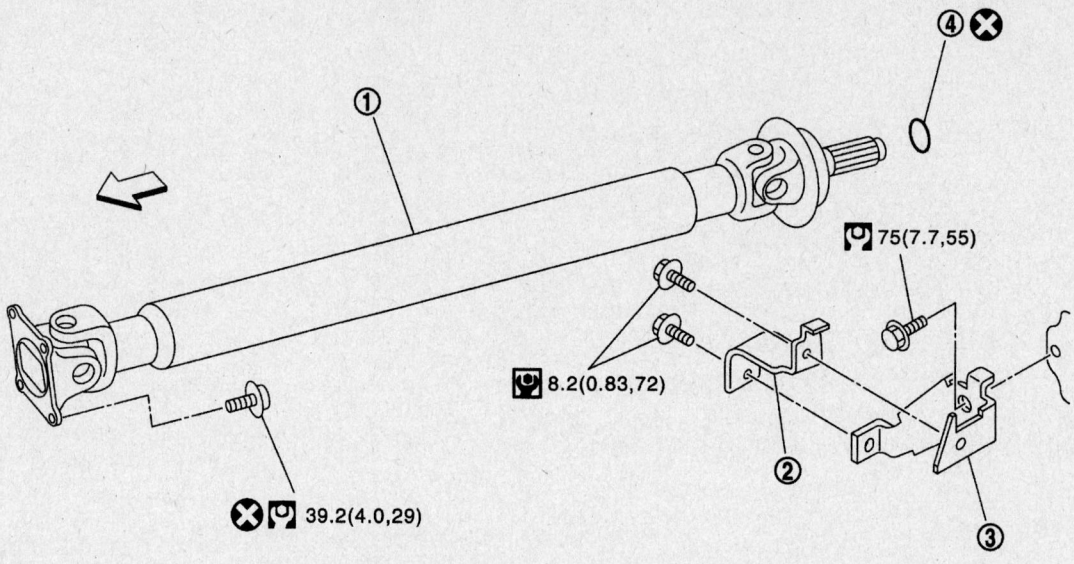

1. Propeller shaft assembly
2. Heat bracket (A)
3. Heat bracket (B)
4. O-ring

09482_INFI_G0004

Front driveshaft components

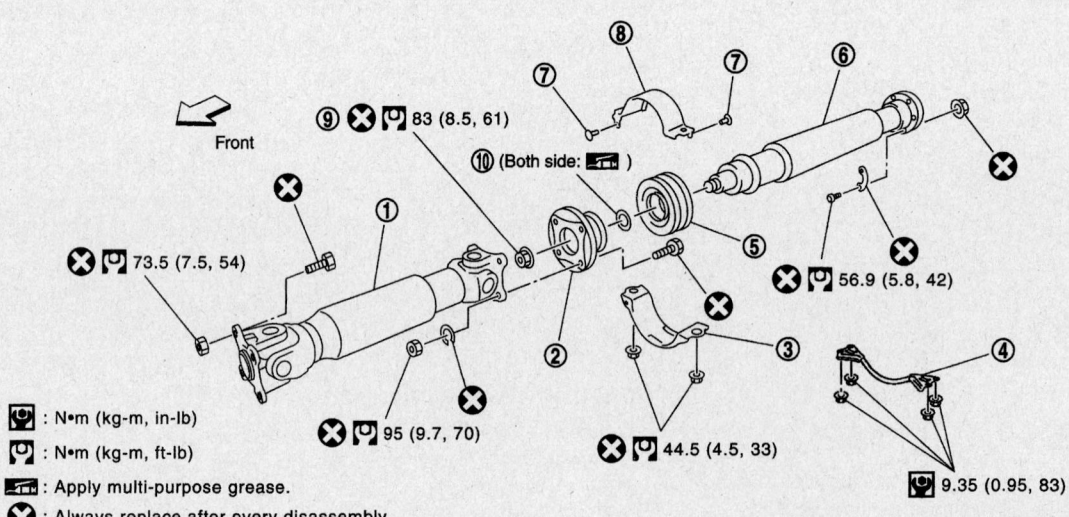

- : N•m (kg-m, in-lb)
- : N•m (kg-m, ft-lb)
- : Apply multi-purpose grease.
- : Always replace after every disassembly.

1. Propeller shaft (1st shaft)
2. Center flange
3. Center bearing mounting bracket (Lower)
4. Floor rain force
5. Center bearing assembly
6. Propeller shaft (2nd shaft)
7. Clip
8. Center bearing mounting bracket (Upper)
9. Lock nut
10. Washer

09482_INFI_G0003

Rear driveshaft components—G35 w/AWD

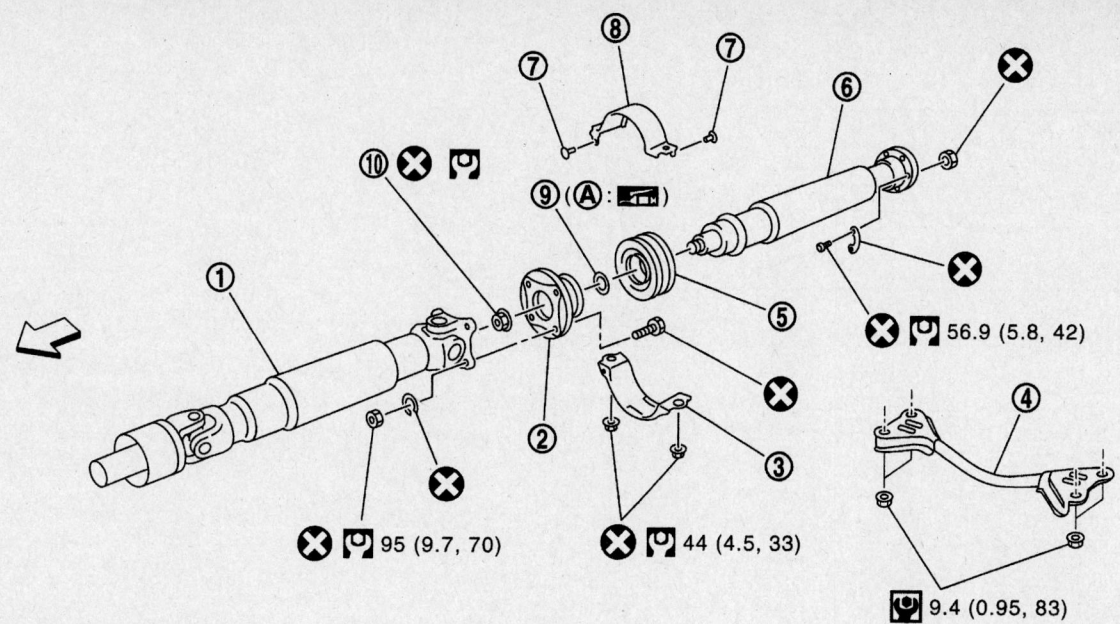

1. Propeller shaft (1st shaft)
2. Center flange
3. Center bearing mounting bracket (Lower)
4. Floor reinforcement
5. Center bearing assembly
6. Propeller shaft (2nd shaft)
7. Clip
8. Center bearing mounting bracket (Upper)
9. Washer
10. Lock nut
A: Both side

09482_INFI_G0415

Rear driveshaft components—M35 w/AWD

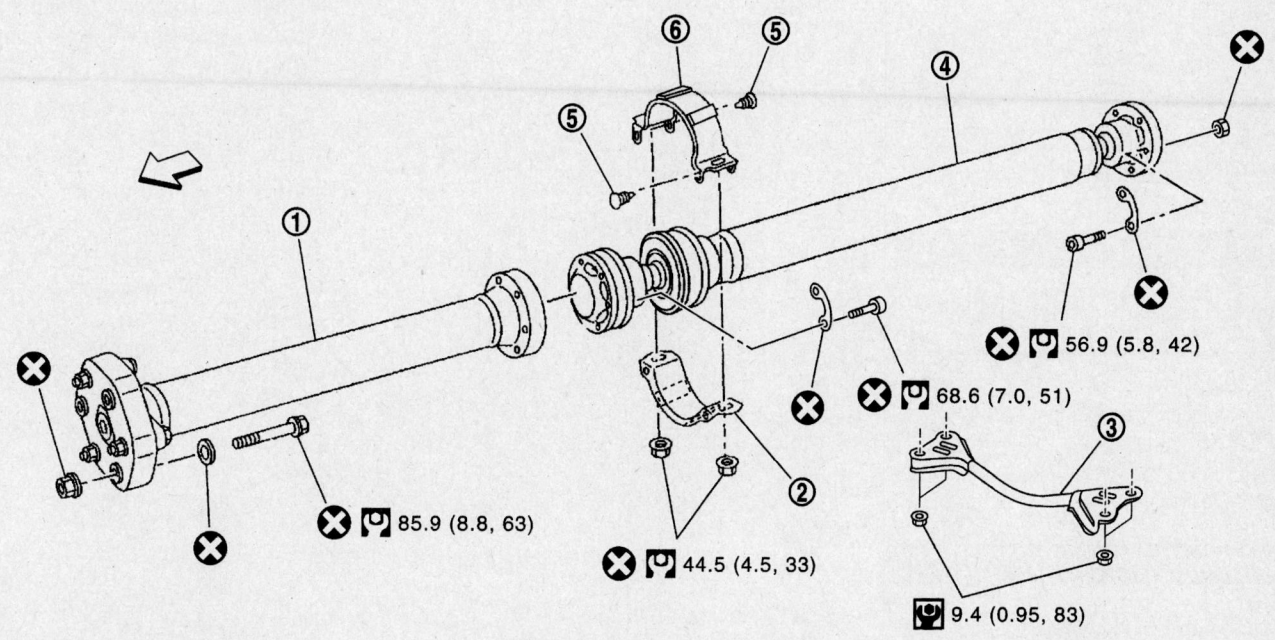

1. Propeller shaft (1st shaft)
2. Center bearing mounting bracket (Lower)
3. Floor reinforcement
4. Propeller shaft (2nd shaft)
5. Clip
6. Center bearing mounting bracket (Upper)

09482_INFI_G0416

Rear driveshaft components—M45 w/AWD

⊙ : Transfer to Transmission
⊗ : Transmission to transfer

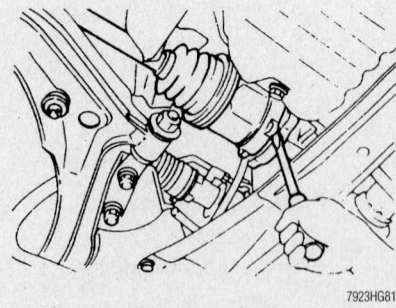

09482_INFI_G0001

Bolt No.	1	2	3	4
Quantity	4	3	2	1
Bolt length "ℓ" mm (in)	75 (2.95)	45 (1.77)	40 (1.57)	30 (1.18)
Tightening torque N·m (kg-m, ft-lb)	37 (3.8, 27)			

Use the accompanying chart for proper bolt placement when installing the transfer case.

- Move the transmission select lever to N position and release the parking brake.
- Remove the center muffler.
- Loosen the center bearing mounting bracket fixing nuts.
- Put matchmarks on flange and rear driveshaft.

➡**For matchmark, use paint. Do not damage the propeller shaft flange and companion flange on the rear final drive.**

- Remove the driveshaft fixing bolts and nuts.
- Remove the center bearing mounting bracket fixing nuts.
- Remove driveshaft from the vehicle.

5. Disconnect transfer assembly harness connector and separate harness from transfer assembly.

6. Remove air breather hose.

7. Support transfer assembly and transmission assembly with a jack.

8. Remove rear engine mounting member.

9. Remove transfer assembly mounting bolts and separate transfer assembly from transmission.

❋❋ WARNING

Secure transfer assembly and transmission assembly to a jack.

To install:

10. Installation is the reverse of the removal procedure.

11. Tighten transfer mounting bolts to 27 ft. lbs. (37 Nm).

Halfshaft

REMOVAL & INSTALLATION

G20 Models

FRONT

1. Before servicing the vehicle, refer to the precautions in the beginning of this section.

2. Raise and support the vehicle safely.

3. Remove or disconnect the following:
- Front wheel
- Wheel bearing locknut
- Brake caliper assembly and rotor. Using a piece of wire, position the caliper so that it is not supported by the brake line.
- Tie-rod from the ball joint
- Kingpin from the knuckle
- Halfshaft from the wheel hub/ knuckle by lightly tapping it with a wood drift. Take care not to damage the CV-boots.
- Halfshaft from the transaxle by prying outward with a suitable tool at the transaxle case

4. On automatic transaxle models, remove the left halfshaft by tapping it out with a drift from the right side of the transaxle case. Take care not to damage the pinion mate shaft and side gear.

To install:

5. Drive a new oil seal into the transaxle. For the right side use tool KV38106800 along the inner circumference of the oil seal. For the left side use tool KV38106700.

6. Install or connect the following:
- Halfshaft into the transaxle. Ensure that the serrations are aligned. Remove the tool.

7. Push the halfshaft inward and install the circular clip in the groove of the side gear. After inserting the clip, pull outward on the flange of the slide joint to ensure the clip is properly meshed with the side gear. If it pulls out, the clip was not installed properly.
- Halfshaft into the wheel hub/ knuckle. Torque the upper knuckle nut to 72–87 ft. lbs. (98–118 Nm) and wheel bearing locknut to 174–231 ft. lbs. (235–314 Nm).

8. Using a dial indicator, check wheel bearing axial end-play. Specification calls for 0.0020 in. (0.05mm) or less.

- Rotor and brake caliper
- Wheel

9. **Always** install new cotter pins.

I35 Models

FRONT—RIGHT SIDE

1. Before servicing the vehicle, refer to the precautions in the beginning of this section.

2. Raise and support the front of the vehicle safely.

3. Remove or disconnect the following:
- Front wheel
- Anti-lock Brake System (ABS) wheel sensor and move it out of the way
- Brake hose from the strut
- Wheel bearing locknut

4. Matchmark and remove the bolts attaching the steering knuckle to the strut

➡**Cover axle boots with waste cloth so as not to damage them when removing halfshaft.**

- Halfshaft from the knuckle by slightly tapping it
- Halfshaft from the transaxle using a suitable flat-bladed tool
- Circlip on the end of the halfshaft and discard circlip
- Seal from the transaxle

To install:

5. Install or connect the following:
- New seal into the transaxle and

Separating the right halfshaft from the transaxle—I35 models

7923HG81

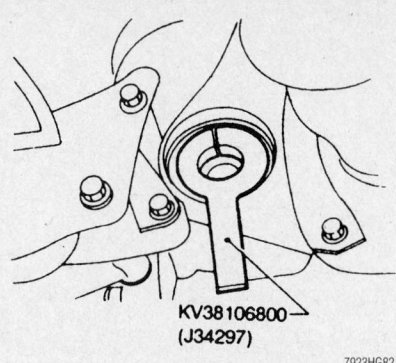

KV38106800 —
(J34297)

7923HG82

Right halfshaft alignment tool—I35 models

install a halfshaft alignment tool KV38106800 into the transaxle seal
- New circlip to the halfshaft, then insert the halfshaft into the transaxle

6. With the serration's aligned remove the alignment tool.

7. Push the halfshaft fully into the transaxle to seat the circlip. Try to pull the halfshaft from the transaxle by hand to verify that the circlip is properly seated.
- Halfshaft into the steering knuckle and install the hub locknut, do not tighten the hub nut at this time
- Steering knuckle to the strut
- Strut mounting bolts and align the matchmarks. Torque the bolts to 103–117 ft. lbs. (140–159 Nm).
- Brake hose to the strut
- ABS wheel sensor. Torque the attaching bolt to 13–17 ft. lbs. (18–24 Nm).
- Front wheels, lower the vehicle and torque the hub locknut to 174–231 ft. lbs. (235–314 Nm)

8. **Always** install new cotter pins.

9. Check and/or adjust the wheel alignment as necessary.

FRONT—LEFT SIDE

1. Before servicing the vehicle, refer to the precautions in the beginning of this section.

2. Raise and support the front of the vehicle safely.

3. Remove or disconnect the following:
- Front wheel
- Anti-lock Brake System (ABS) wheel sensor and move it out of the way
- Brake hose from the strut
- Wheel bearing locknut
- Bolts attaching the steering knuckle to the strut. Matchmark the bolts prior to removal

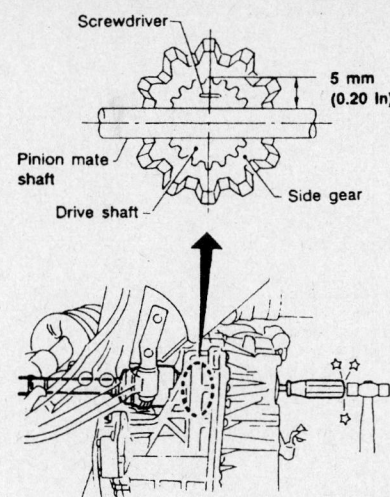

7923HG83

Separating the left halfshaft from an automatic transaxle—I35 models

➡Cover axle boots with waste cloth so as not to damage them when removing halfshaft.

- Halfshaft from the knuckle by slightly tapping it
- Bolts attaching the support bearing to the support bearing bracket
- Halfshaft from the transaxle using a suitable prytool, if equipped with a manual transaxle

4. If equipped with a automatic transaxle perform the following:

a. Remove the right halfshaft from the vehicle.

b. Insert a flat-bladed tool into the transaxle where the right halfshaft was, place the end of the tool on the halfshaft and drive the left shaft from the pinion side gear.
- Support bearing bolts
- Halfshaft from the vehicle
- Circlip on the end of the halfshaft and discard circlip
- Seal from the transaxle

To install:

5. Install or connect the following:
- New seal into the transaxle and install a halfshaft alignment tool KV38106700 into the transaxle seal
- New circlip to the halfshaft, then insert the halfshaft into the transaxle

6. With the serration's aligned remove the alignment tool.
- Halfshaft fully into the transaxle to seat the circlip. Try to pull the halfshaft from the transaxle by hand to verify that the circlip is properly seated.
- Support bearing bolts and torque

the bolts to 10–14 ft. lbs. (13–19 Nm)
- Halfshaft into the steering knuckle and install the hub locknut, do not tighten the hub nut at this time
- Steering knuckle to the strut
- Strut mounting bolts and align the matchmarks. Torque the bolts to 103–117 ft. lbs. (140–159 Nm).
- Brake hose to the strut
- ABS wheel sensor. Torque the attaching bolt to 13–17 ft. lbs. (18–24 Nm).
- Front wheels, lower the vehicle and torque hub locknut to 174–231 ft. lbs. (235–314 Nm)

7. **Always** install new cotter pins.

8. Check and/or adjust the wheel alignment as necessary.

G35 and 2002–03 Q45 Models

FRONT (G35 AWD MODELS)

❋❋ CAUTION

The amount of force need to loosen the front wheel bearing nut is high enough to cause the vehicle to fall off the jack. Remove cotter pin and loosen or tighten this nut with the vehicle on the ground.

1. Before servicing the vehicle, refer to the precautions in the beginning of this section.

2. Remove or disconnect the following:
- Front wheels
- Engine undercover
- Brake calipers; use wire to support calipers where they will not interfere with work
- Brake rotors
- Antilock Brake wheel sensors

➡**Do not pull on sensor harness**

- Brake hose bracket from steering knuckle
- Front axle nut (nut should be loosened while vehicle is on the ground)
- Tie rod end

➡**Do not damage threads or boot on tie rod end**

- Upper link ball joint

3. Remove halfshaft from wheel hub and bearing assembly

4. On left side, remove bolts securing halfshaft from side shaft and remove halfshaft from vehicle

5. On right side, remove halfshaft from splines in final drive

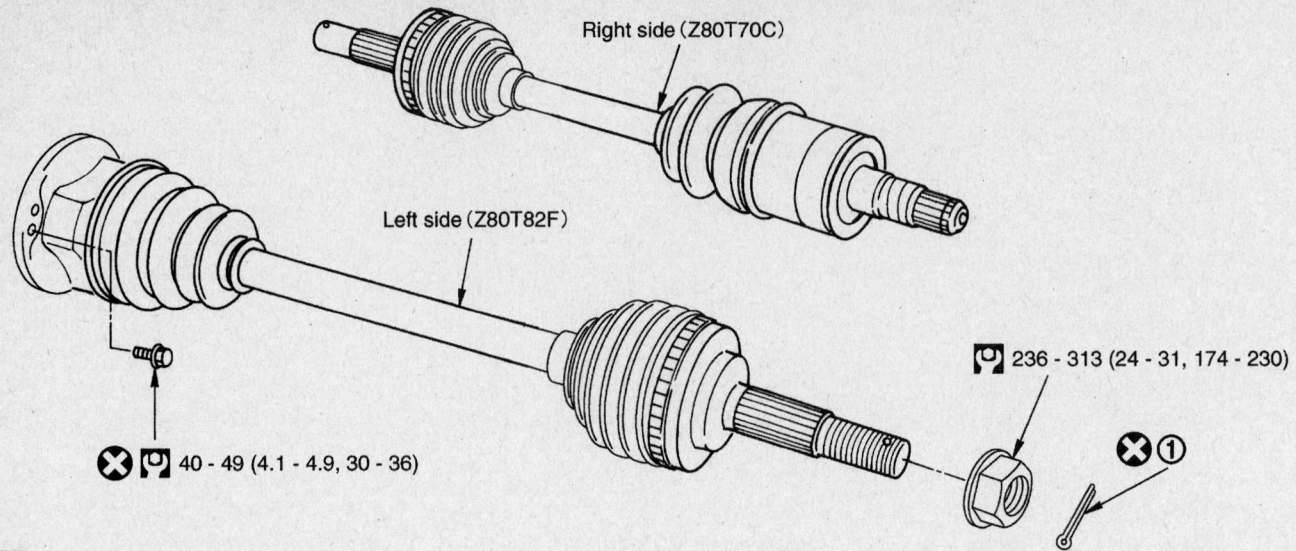

Right side (Z80T70C)

Left side (Z80T82F)

⊡ 236 - 313 (24 - 31, 174 - 230)

❌⊡①

❌⊡ 40 - 49 (4.1 - 4.9, 30 - 36)

⊡ : N•m (kg-m, ft-lb)

❌ : Always replace after every disassembly.

1. Cotter pin

09482_INFI_G0091

Front halfshafts—G35/M35 AWD models

6. Remove halfshaft from vehicle
To install:
7. Installation is in reverse order of removal. Observe the following:
  • Tighten left halfshaft flange to side shaft bolts to 30–36 ft. lbs. (40–49 Nm)
  • Tighten axle nuts to 174–230 ft. lbs. (236–313 Nm)
  • Tighten upper link ball joint to 40–46 ft. lbs. (54–63.7 Nm)
  • Tighten tie rod end ball joint to 22–28 ft. lbs. (29.5–39.2 Nm)
8. **Always** install new cotter pins.
9. Check and perform front end alignment if needed.

REAR

✳✳ CAUTION

The amount of force need to loosen the rear wheel bearing nut is high enough to cause the vehicle to fall off the jack. Loosen and tighten this nut with the vehicle on the ground.

1. Before servicing the vehicle, refer to the precautions in the beginning of this section.
2. Remove or disconnect the following:
  • Rear wheel cotter pin, adjusting cap and insulator. Loosen the wheel

bearing nut with the brakes applied and the vehicle sitting on the ground.
3. Raise the vehicle and support safely.
  • Rear wheel
  • Differential side flange bolts and nuts and separate shaft from the differential
  • Wheel bearing locknut and washer from halfshaft
  • Halfshaft by lightly tapping it with a copper hammer
  • Halfshaft assembly from the vehicle
To install:
4. Install or connect the following:
  • Halfshaft into wheel hub and install

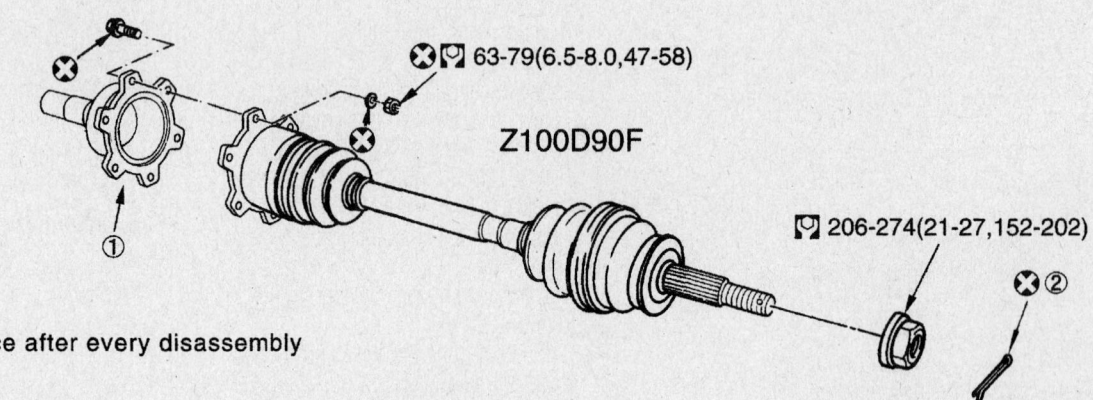

❌⊡ 63-79(6.5-8.0,47-58)

Z100D90F

⊡ 206-274(21-27,152-202)

❌②

❌ : Always replace after every disassembly

⊡ : N·m(kg-m,ft-lb)

1. Side flange   2. Cotter pin

09482_INFI_G0092

Rear halfshaft removal and installation—G35

washer and wheel bearing locknut; temporarily tighten the locknut.
- Halfshaft with the differential side flange. Install the nuts and bolts and tighten to 61–69 ft. lbs. (83–93 Nm) for the Q45 or to 47–57 ft. lbs. (64–78 Nm) for the G35.
- Wheels and lower the vehicle to the ground
- Torque the wheel bearing locknut with the brakes applied to 152–201 ft. lbs. (206–274 Nm)
- Insulator, adjusting cap and a new cotter pin

Q45 and 2004 M45 Models

REAR

⁑ **CAUTION**

The amount of force need to loosen the rear wheel bearing nut is high enough to cause the vehicle to fall off the jack. Loosen and tighten this nut with the vehicle on the ground.

1. Before servicing the vehicle, refer to the precautions in the beginning of this section.
2. Remove or disconnect the following:
- Rear wheel cotter pin, adjusting cap and insulator. Loosen the wheel bearing nut with the brakes applied
- Exhaust center tube
- Final drive-to-axle shaft nuts and bolts
3. Using a puller, separate the axle shaft from the axle.
4. Remove the axle shaft.

To install:
5. Installation is the reverse of the removal procedure noting the following:
6. Tighten the axle flange-to-axle shaft nuts to 47–58 ft. lbs. (63–79 Nm).
7. Tighten the axle shaft nut to 152–202 ft. lbs. (206–274 Nm).
8. **Always** install new cotter pins.

2006 M35/M45 Models

FRONT (M35 AWD MODELS)

⁑ **CAUTION**

The amount of force need to loosen the front wheel bearing nut is high enough to cause the vehicle to fall off the jack. Remove cotter pin and loosen or tighten this nut with the vehicle on the ground.

1. Before servicing the vehicle, refer to the precautions in the beginning of this section.
2. Remove or disconnect the following:
- Front wheels
- Antilock Brake wheel sensors

➡ **Do not pull on sensor harness**

- Brake hose bracket
- Front calipers
- Brake rotors
3. Loosen axle nut and separate wheel hub and bearing assembly from halfshaft by tapping the end with a hammer and a wood block (the nut should have been loosened while the vehicle was on the ground); remove nut.

➡ **Do not allow the axle to hang without support or at extreme angles.**

4. Remove tie rod end from steering knuckle, using an appropriate tool

➡ **Do not damage threads or ball joint on tie rod end**

5. Remove shock absorber arm
6. On left side, remove bolts securing halfshaft to side shaft and remove halfshaft from vehicle
7. On right side, remove axle from splines in final drive, using an appropriate puller if necessary
8. Remove halfshaft from vehicle

To install:
9. Installation is in reverse order of removal. Observe the following:
- Tighten left halfshaft flange to side shaft bolts to 33 ft. lbs. (45 Nm)
- Tighten axle nuts to 92 ft. lbs. (125 Nm)
- Tighten shock absorber mount upper nut to 46 ft. lbs. (63 Nm)
- Tighten shock absorber mount lower nut to 68 ft. lbs. (92 Nm)
- Tighten tie rod end ball joint to 25 ft. lbs. (34.4 Nm)
10. **Always** install new cotter pins
11. Check and perform front end alignment if needed

REAR

⁑ **CAUTION**

The amount of force need to loosen the rear wheel bearing nut is high enough to cause the vehicle to fall off the jack. Loosen and tighten this nut with the vehicle on the ground.

1. Before servicing the vehicle, refer to the precautions in the beginning of this section.

2. Remove or disconnect the following:
- Rear wheel cotter pin, adjusting cap and insulator. Loosen the wheel bearing nut with the brakes applied
- Stabilizer connecting rod mounting bracket bolt; free stabilizer connecting rod.
- Final drive-to-halfshaft nuts and bolts
3. Using a puller if necessary, separate the axle shaft from the axle.
4. Remove the axle shaft from the vehicle

To install:
5. Installation is in reverse order of removal. Observe the following:
- Tighten final drive-to-halfshaft bolts to 54 ft. lbs. (74 Nm) on M35 models
- Tighten final drive-to-halfshaft bolts to 87 ft. lbs. (118 Nm) on M35 models.
- Tighten axle nuts to 130 ft. lbs. (175 Nm) on both models
- Tighten stabilizer connecting rod mounting bracket to 41 ft. lbs. (55 Nm)
6. **Always** install new cotter pins

CV-Joints

OVERHAUL

G20 Models

TRANSAXLE SIDE—DS83 TYPE

1. Before servicing the vehicle, refer to the precautions in the beginning of this section.
2. Disassemble the joint as follows:
   a. Remove the boot bands.
   b. Matchmark the slide joint housing and inner race before separating the assembly
   c. Using a suitable prytool, remove the stopper ring and pull out the slide joint

09482_INFI_G0093

The inner CV joint uses a large C-clip to retain the ball and cage assembly in the outer housing

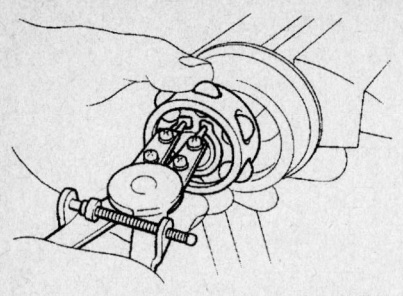

After the outer housing is removed, the ball and cage assembly can slide from the shaft by removing the C-clip

d. Matchmark the inner race and half-shaft

3. Remove or disconnect the following:
- Snapring
- Ball cage, inner race and balls as a unit
- Boot

➡Cover the halfshaft serrations with tape, so as not to damage the boot.

To install:

4. Assemble the joint as follows:

a. Thoroughly clean all parts in solvent and dry with compressed air. Check parts for evidence of damage, and replace as necessary.

b. Install the boot and new boot band on the halfshaft.

c. Install a new inner snapring.

d. Install the ball cage, inner race and balls as a unit. Confirm that the match-marks are aligned.

e. Install a new outer snapring.

f. Pack the CV joint with 5.0–6.0 ounces of grease.

g. Ensure that the boot is properly installed on the halfshaft groove.

h. Set the boot so that it does not swell or deform when its length is 3.82–3.90 in. (97–99mm).

i. Lock the new boot bands securely.

TRANSAXLE SIDE—TS83 TYPE

1. Before servicing the vehicle, refer to the precautions in the beginning of this section.

2. Disassemble the joint as follows:

a. Remove the boot bands.

b. Matchmark the slide joint housing and inner race before separating the assembly.

c. Matchmark the spider assembly and driveshaft.

d. Remove the snapring and the spider assembly.

➡Do not disassemble the spider assembly.

e. Remove the boot.

➡Cover the halfshaft serrations with tape, so as not to damage the boot.

To install:

3. Assemble the joint as follows:

a. Thoroughly clean all parts in solvent and dry with compressed air. Check parts for evidence of damage, and replace as necessary.

b. Install the boot and new boot band on the halfshaft.

c. Install the spider assembly making sure the matchmarks made during removal are properly aligned.

d. Install a new snapring.

e. Pack the joint with 4.5–5.11 ounces (124–145g) of grease.

f. Install the slide joint housing.

g. Ensure that the boot is properly installed on the halfshaft groove.

h. Set the boot so that it does not swell or deform when its length is 3.90 in. (99mm).

4. Lock the new boot bands securely.

WHEEL SIDE

➡The joint on the wheel side cannot be disassembled.

1. Before servicing the vehicle, refer to the precautions in the beginning of this section.

2. Prior to separating the joint assembly, matchmark the halfshaft and joint assembly.

3. Separate the joint using a slide hammer.

4. Remove the boot bands.

To assemble:

5. Thoroughly clean all parts in solvent and dry with compressed air. Check parts for evidence of damage and replace as necessary.

➡Cover the halfshaft serrations with tape, so as not to damage the boot.

6. Install the boot and small boot band on the halfshaft.

7. Set the joint assembly onto the half-shaft and align the matchmarks.

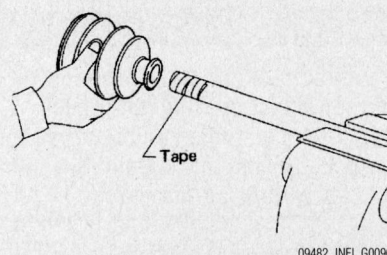

Use vinyl tape and wrap the end of the shaft to protect the boot during installation

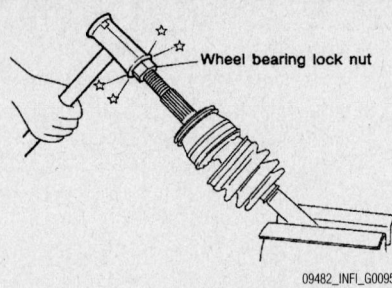

Use an old nut or wooden hammer to protect the threads when tapping the outer CV joint onto the shaft

8. Attach the joint assembly to the half-shaft by lightly tapping the serrated end with a plastic hammer.

➡Using a metal hammer may damage the threads on the end of the joint.

9. Pack the CV joint with 3.5–4.0 ounces of grease.

10. Ensure that the boot is properly installed on the halfshaft groove.

11. Set the boot so that it does not swell or deform when its length is 3.327–3.406 in. (84.5–86.5mm).

12. Lock the new boot bands securely.

G35 Models

FINAL DRIVE SIDE

➡Procedures apply to front and rear halfshafts.

1. Before servicing the vehicle, refer to the precautions in the beginning of this section.

2. Remove or disconnect the following:
- Plug seal from the slide joint by gently tapping around the joint with a hammer
- Boot bands

3. Put matchmarks on the slide joint housing and halfshaft prior to separating the joint assembly.

4. Matchmark the spider assembly and driveshaft.
- Snapring and the spider assembly

➡Do not disassemble the spider assembly.
- Slide joint housing and the boot

➡Cover the halfshaft serrations with tape, so as not to damage the boot.

To assemble:

5. Thoroughly clean all parts in solvent and dry with compressed air. Check parts for evidence of damage and replace as necessary.

a. Install the boot and new boot band on the halfshaft.

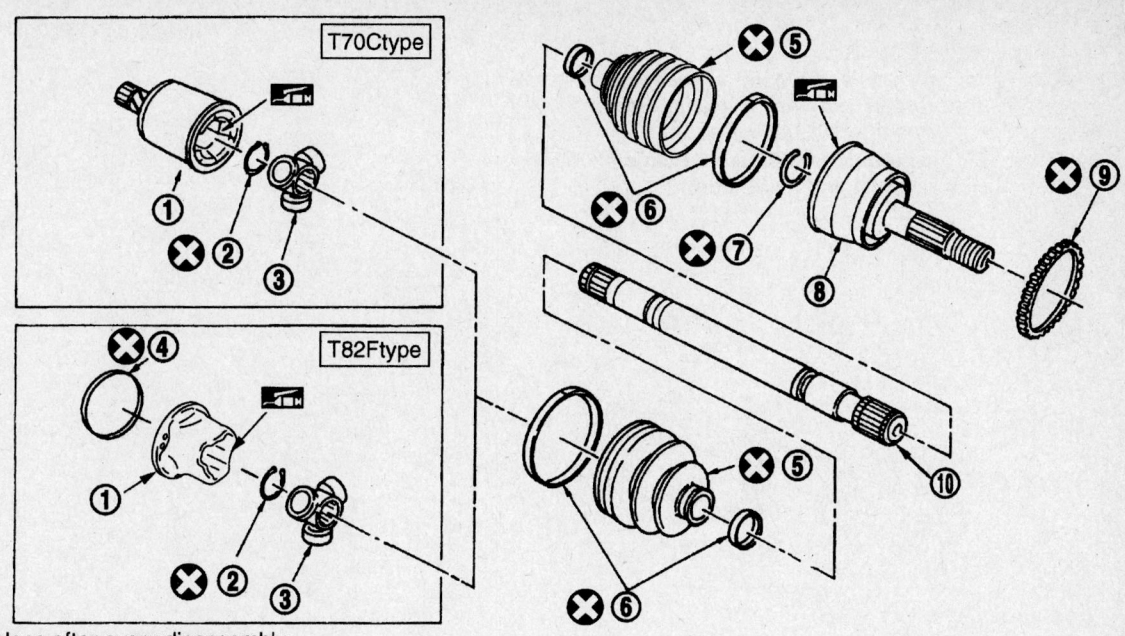

⊗ : Always replace after every disassembly.

🔧 : Nissan genuine grease or equivalent.

1. Housing
2. Snap ring
3. Spider assembly
4. Plug
5. Boot
6. Boot band
7. Circular clip
8. joint sub-assembly
9. sensor rotor
10. Shaft

09482_INFI_G0098

Exploded view of front halfshaft—G35 and M35 AWD models

b. Install the spider assembly making sure the matchmarks made during removal are properly aligned.

➡**The spider is press fit with the serration chamfer facing the shaft.**

c. Install a new snapring.
6. Pack the joint with 4.37–4.73 ounces (124–134g) of grease.
a. Install the slide joint housing and the snapring.
b. Ensure that the boot is properly installed on the halfshaft groove
7. Set the boot so that it does not swell or deform when its length is 3.70 in. (94mm).
8. Lock the new boot bands securely.

WHEEL SIDE

1. Before servicing the vehicle, refer to the precautions in the beginning of this section.
2. Remove or disconnect the following:
a. Remove the boot bands.
b. Thread a slide hammer into the joint sub-assembly and pull the joint out of the shaft.

➡**Do not disassemble the spider assembly.**

c. Remove the boot.

To install:
3. Install the joint as follows:
a. Thoroughly clean all parts in solvent and dry with compressed air. Check parts for evidence of damage, and replace as necessary.
b. Pack the joint with 3.03–3.39 ounces (86–96g) of grease.
c. Install the boot and new boot band on the halfshaft.
d. Install a new snapring.
e. Ensure that the boot is properly installed on the halfshaft groove.
4. Set the boot so that it does not swell or deform when its length is 3.82 in. (97mm).
5. Lock the new boot bands securely.

I35 Models

TRANSAXLE SIDE

1. Before servicing the vehicle, refer to the precautions in the beginning of this section.
2. Disassemble the joint as follows:
a. Remove the boot bands.
b. Matchmark the slide joint housing and inner race before separating the assembly

c. Using a suitable prytool, remove the stopper ring and pull out the slide joint
d. Matchmark the inner race and halfshaft
e. Remove the snapring.
f. Remove the ball cage, inner race and balls as a unit.
g. Remove the boot.

➡**Cover the halfshaft serration's with tape, so as not to damage the boot.**

To install:
3. Assemble the joint as follows:
a. Thoroughly clean all parts in solvent and dry with compressed air. Check parts for evidence of damage, and replace as necessary.
b. Install the boot and new boot band on the halfshaft.
4. Install a new inner snapring.
5. Install the ball cage, inner race and balls as a unit. Confirm that the matchmarks are aligned.
6. Install a new outer snapring.
7. Pack the CV joint with 5.8–6.17 ounces of grease.
8. Ensure that the boot is properly installed on the halfshaft groove.

Circular clip:
　Circular clips should be properly meshed with differential side gear (transaxle side) and with joint assembly (wheel side). Make sure they will not come out.
Be careful not to damage boots. Use suitable protector or cloth during removal and installation.

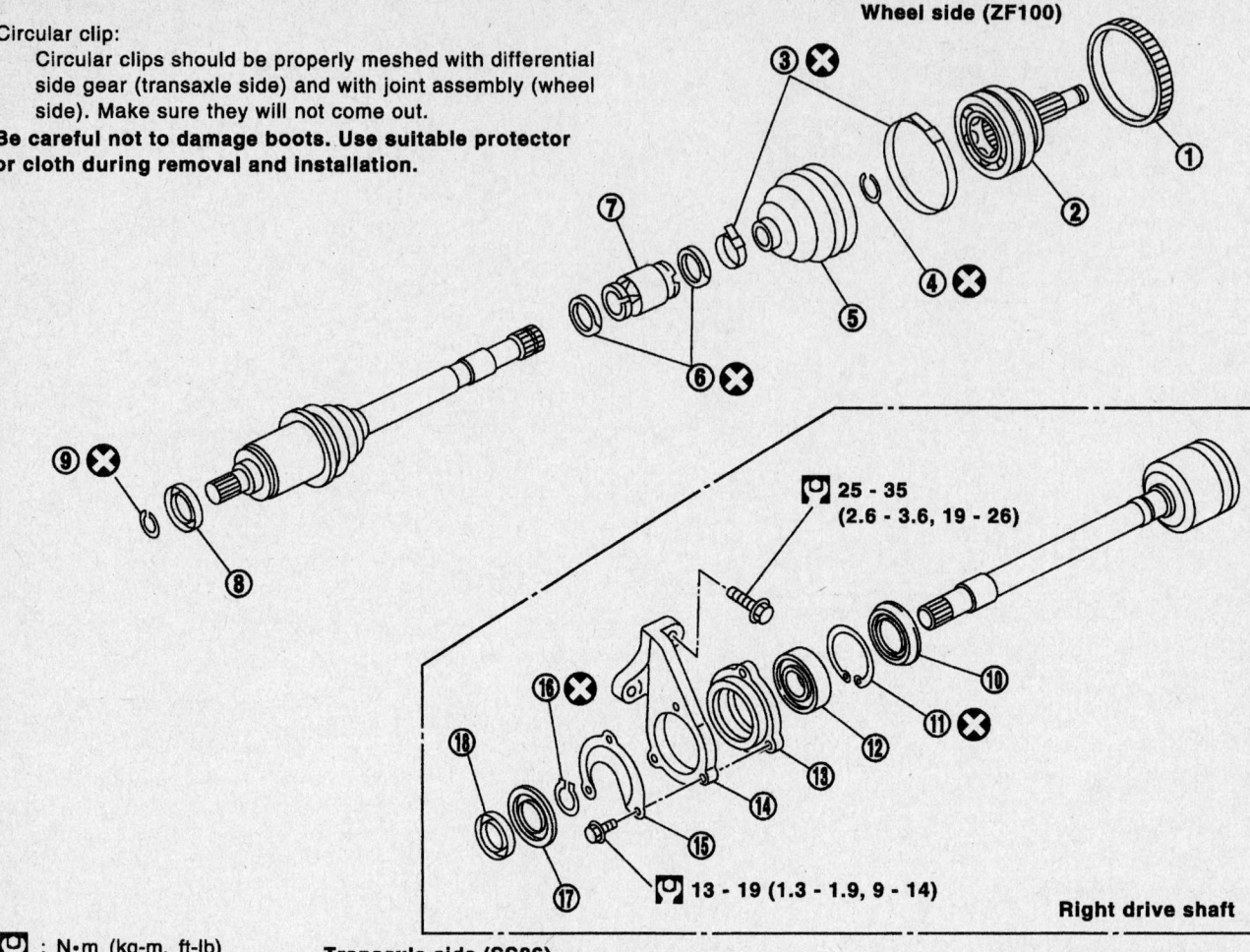

Wheel side (ZF100)

$\boxed{\text{Ω}}$ 25 - 35 (2.6 - 3.6, 19 - 26)

$\boxed{\text{Ω}}$ 13 - 19 (1.3 - 1.9, 9 - 14)

Right drive shaft

$\boxed{\text{Ω}}$: N•m (kg-m, ft-lb)

Transaxle side (SS86)

1. ABS sensor rotor
2. Joint assembly
3. Boot band
4. Circular clip
5. Boot
6. Dynamic damper band
7. Dynamic damper
8. Dust shield
9. Circular clip
10. Support bearing dust shield
11. Snap ring
12. Support bearing
13. Support bearing retainer
14. Support bearing bracket
15. Shield heat plate
16. Snap ring
17. Support bearing dust shield
18. Dust shield

09482_INFI_G0097

Exploded view of the halfshaft—I35

　9. Set the boot so that it does not swell or deform when its length is 3.82–3.90 in. (97–99mm).
　10. Lock the new boot bands securely.

WHEEL SIDE

➡**The joint on the wheel side cannot be disassembled.**

　1. Before servicing the vehicle, refer to the precautions in the beginning of this section.
　2. Prior to separating the joint assembly, matchmark the halfshaft and joint assembly.
　3. Separate the joint using a slide hammer.

　4. Remove the boot bands.

To assemble:
　5. Thoroughly clean all parts in solvent and dry with compressed air. Check parts for evidence of damage and replace as necessary.

➡**Cover the halfshaft serration's with tape, so as not to damage the boot.**

　6. Install the boot and small boot band on the halfshaft.
　7. Set the joint assembly onto the halfshaft and align the matchmarks.
　8. Attach the joint assembly to the halfshaft by lightly tapping the serrated end with a plastic hammer.

➡**Using a metal hammer may damage the threads on the end of the joint.**

　9. Pack the CV joint with 4.7–5.11 ounces of grease.
　10. Ensure that the boot is properly installed on the halfshaft groove.
　11. Set the boot so that it does not swell or deform when its length is 3.78–3.86 in. (96–98mm)
　12. Lock the new boot bands securely

2006 M35/M45 Models

➡**Refer to 2004–06 Q45 procedures for servicing 2004 M45 models**

FINAL DRIVE SIDE

➡**Procedures apply to front and rear halfshafts.**

1. Before servicing the vehicle, refer to the precautions in the beginning of this section.

2. Remove or disconnect the following:
- Plug seal from the slide joint by gently tapping around the joint with a hammer
- Boot bands

3. Put matchmarks on the slide joint housing and halfshaft prior to separating the joint assembly.

4. Matchmark the spider assembly and driveshaft.
- Snapring and the spider assembly

➡**Do not disassemble the spider assembly.**

- Slide joint housing and the boot

➡**Cover the halfshaft serration's with tape, so as not to damage the boot.**

To assemble:

5. Thoroughly clean all parts in solvent and dry with compressed air. Check parts for evidence of damage and replace as necessary.

a. Install the boot and new boot band on the halfshaft.

b. Install the spider assembly making sure the matchmarks made during removal are properly aligned.

➡**The spider is press fit with the serration chamfer facing the shaft.**

c. Install a new snapring.

6. Pack the joint with 4.37–4.73 ounces (124–134g) of grease.

a. Install the slide joint housing and the snapring.

b. Ensure that the boot is properly installed on the halfshaft groove

7. Set the boot so that it does not swell or deform when its length is 3.70 in. (94mm).

8. Lock the new boot bands securely.

WHEEL SIDE

1. Before servicing the vehicle, refer to the precautions in the beginning of this section.

2. Remove or disconnect the following:
a. Remove the boot bands.
b. Thread a slide hammer into the joint sub-assembly and pull the joint out of the shaft.

➡**Do not disassemble the spider assembly.**

c. Remove the boot.

To install:

3. Install the joint as follows:

a. Thoroughly clean all parts in solvent and dry with compressed air. Check parts for evidence of damage, and replace as necessary.

b. Pack the joint with 3.03–3.39 ounces (86–96g) of grease.

c. Install the boot and new boot band on the halfshaft.

d. Install a new snapring.

e. Ensure that the boot is properly installed on the halfshaft groove.

4. Set the boot so that it does not swell or deform when its length is 3.82 in. (97mm).

5. Lock the new boot bands securely.

2002–03 Q45 Models

FINAL DRIVE SIDE

1. Before servicing the vehicle, refer to the precautions in the beginning of this section.

2. Remove or disconnect the following:
- Plug seal from the slide joint by gently tapping around the joint with a hammer
- Boot bands

3. Put matchmarks on the slide joint housing and halfshaft prior to separating the joint assembly

4. Matchmark the spider assembly and driveshaft
- Snapring and the spider assembly

➡**Do not disassemble the spider assembly.**

- Slide joint housing and the boot

➡**Cover the halfshaft serration's with tape, so as not to damage the boot.**

To assemble:

5. Thoroughly clean all parts in solvent and dry with compressed air. Check parts for evidence of damage and replace as necessary

6. Install or connect the following:
- Boot and small boot band on the halfshaft
- Joint housing onto halfshaft
- Spider assembly, making sure the matchmarks are properly aligned

➡**The spider is a press fit with the serration chamfer facing the shaft.**

- Snapring
- Coil spring, spring cap and new plug seal to the slide joint housing.

Apply a suitable sealant to the plug seal prior to installation

➡**Hold the plug seal horizontally when pressing it into place. This will prevent the spring inside from falling down or tilting.**

7. Move the shaft in an axial direction to make sure that the spring is installed properly. If there is a drag or the spring is installed improperly, replace the plug seal with a new one.

8. Pack the halfshaft with 5.82–6.17 ounces (165–175g) of grease.

9. Ensure that the boot is properly installed on the halfshaft groove

10. Set the boot so that it does not swell or deform when its length is 3.66–3.74 in. (93–95mm).

11. Lock the new boot bands securely.

WHEEL SIDE

1. Before servicing the vehicle, refer to the precautions in the beginning of this section.

2. Remove or disconnect the following:

a. Remove the boot bands.

b. Matchmark the housing with the shaft and halfshaft before separating the assembly

c. Matchmark the spider assembly and halfshaft

d. Remove the snapring and the spider assembly.

➡**Do not disassemble the spider assembly.**

e. Remove the boot.

➡**Cover the halfshaft serration's with tape, so as not to damage the boot.**

To install:

3. Install the joint as follows:

a. Thoroughly clean all parts in solvent and dry with compressed air. Check parts for evidence of damage, and replace as necessary.

b. Install the boot and new boot band on the halfshaft.

c. Install the spider assembly making sure the matchmarks made during removal are properly aligned.

➡**The spider is press fit with the serration chamfer facing the shaft.**

d. Install a new snapring.

4. Pack the joint with 4–4.34 ounces (113–123g) of grease.

a. Install the slide joint housing and the snapring.

b. Ensure that the boot is properly installed on the halfshaft groove.

5. Set the boot so that it does not swell or deform when its length is 3.78–3.86 in. (96–98mm).

6. Lock the new boot bands securely.

2004–06 Q45 Model and 2004 M45 Models

FINAL DRIVE SIDE

1. Before servicing the vehicle, refer to the precautions in the beginning of this section.

2. Remove or disconnect the following:
- Plug seal from the slide joint by gently tapping around the joint with a hammer
- Boot bands

3. Put matchmarks on the slide joint housing and halfshaft prior to separating the joint assembly.

4. Matchmark the spider assembly and driveshaft.
- Snapring and the spider assembly

➡**Do not disassemble the spider assembly.**

- Slide joint housing and the boot

➡**Cover the halfshaft serration's with tape, so as not to damage the boot.**

To assemble:

5. Thoroughly clean all parts in solvent and dry with compressed air. Check parts for evidence of damage and replace as necessary.

a. Install the boot and new boot band on the halfshaft.

b. Install the spider assembly making sure the matchmarks made during removal are properly aligned.

➡**The spider is press fit with the serration chamfer facing the shaft.**

c. Install a new snapring.

6. Pack the joint with 4.37–4.73 ounces (124–134g) of grease.

a. Install the slide joint housing and the snapring.

b. Ensure that the boot is properly installed on the halfshaft groove

7. Set the boot so that it does not swell or deform when its length is 3.70 in. (94mm).

8. Lock the new boot bands securely.

WHEEL SIDE

1. Before servicing the vehicle, refer to the precautions in the beginning of this section.

2. Remove or disconnect the following:

a. Remove the boot bands.

b. Thread a slide hammer into the joint sub-assembly and pull the joint out of the shaft.

➡**Do not disassemble the spider assembly.**

c. Remove the boot.

To install:

3. Install the joint as follows:

a. Thoroughly clean all parts in solvent and dry with compressed air. Check parts for evidence of damage, and replace as necessary.

b. Pack the joint with 3.03–3.39 ounces (86–96g) of grease.

c. Install the boot and new boot band on the halfshaft.

d. Install a new snapring.

e. Ensure that the boot is properly installed on the halfshaft groove.

4. Set the boot so that it does not swell or deform when its length is 3.82 in. (97mm).

5. Lock the new boot bands securely.

Front Differential Pinion Seal

REMOVAL & INSTALLATION

1. Raise and safely support the vehicle.

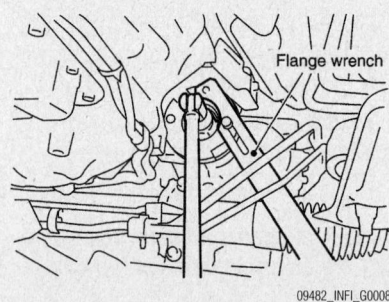

09482_INFI_G0008

A flange tool such as one shown here will aid pinion nut removal and installation

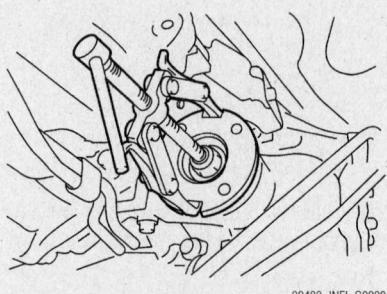

09482_INFI_G0009

Use the appropriate puller to remove companion flange and pinion seal

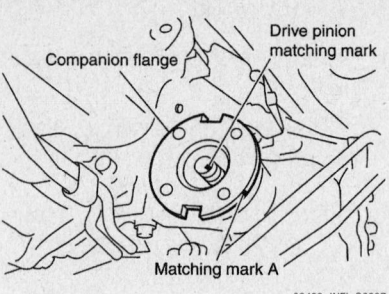

09482_INFI_G0007

Paint match mark in line with mark "A" on companion flange

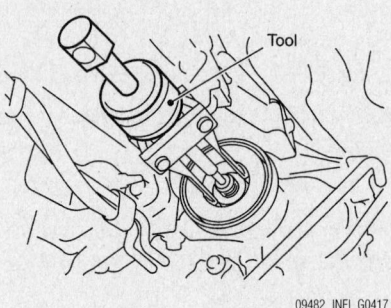

09482_INFI_G0417

Use the appropriate puller to remove pinion seal

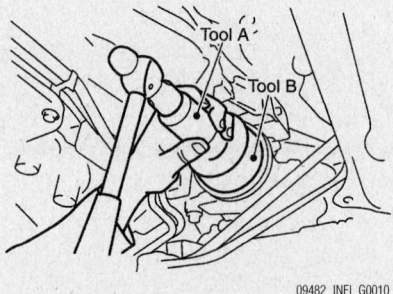

09482_INFI_G0010

Use a drift or seal installation tool when installing seal

2. Remove driveshaft

3. Put a match mark on the end of the drive pinion. The match mark should be in line with match mark "A" on companion flange.

➡**For match mark, use paint. Do not damage drive pinion.**

➡**The matching mark A on the final drive companion flange indicates the maximum vertical runout position.**

4. Remove drive pinion lock nut using flange tool KV40104000 or equivalent.

5. Remove companion flange using a puller.

6. Remove front oil seal using a puller.

To install:

7. Apply multi-purpose grease to front oil seal lips.

8. Using a drift, install front oil seal.

➡ **Do not reuse oil seal. When installing make certain seal is squarely seated.**

9. Align the match mark of drive pinion with the match mark "A" of companion flange, then install the companion flange.

10. Apply gear oil on the screw part of drive pinion and the seating surface of drive pinion lock nut.

11. Install drive pinion lock nut with the flange wrench and tighten to 109–238 ft. lbs. (147–323 Nm).

➡ **Do not reuse drive pinion lock nut.**

12. Install driveshaft.

13. Lower the vehicle.

STEERING

Air Bag

PRECAUTIONS

Several precautions must be observed when handling the inflator module to avoid accidental deployment and possible personal injury.

1. Never carry the inflator module by the wires or connector on the underside of the module.

2. When carrying a live inflator module, hold securely with both hands, and ensure that the bag and trim cover are pointed away.

3. Place the inflator module on a bench or other surface with the bag and trim cover facing up.

4. With the inflator module on the bench, never place anything on or close to the module that may be thrown in the event of an accidental deployment.

DISARMING

➡ **All Air Bag electrical wiring harnesses and connectors are covered with YELLOW outer insulation. Do not use electrical test equipment on any circuit related to the Air Bag sensors. When installing Air Bag components, always install with the arrow marks facing the front of the vehicle.**

1. Before servicing the vehicle, refer to the precautions in the beginning of this section.

2. Turn the ignition switch to the **OFF** position.

3. Disconnect both battery cables starting with the negative cable first and wait at least 10 minutes after the cables are disconnected. Be sure to insulate the battery terminal ends.

REARMING

1. Before servicing the vehicle, refer to the precautions in the beginning of this section.

2. Turn the ignition switch to the **OFF** position.

3. Connect both battery cables starting with the positive cable first.

➡ **The Air Bag or Air Bag system is equipped with a self-diagnostic operation. After turning the ignition key to the ON or START position, the AIR BAG warning lamp will illuminate for 7 seconds. After 7 seconds, the AIR BAG lamp will extinguish if no malfunction is detected. If the AIR BAG lamp does not extinguish after 7 seconds, check the Air Bag self-diagnostic system for a malfunction.**

Power Rack and Pinion Steering Gear

REMOVAL & INSTALLATION

G20 Models

1. Before servicing the vehicle, refer to the precautions in the beginning of this section.

✳✳ CAUTION

The air bag system must be disarmed before removing the steering wheel. Failure to do so may cause accidental deployment, property damage or personal injury.

2. Point the front tires straight ahead and lock the steering in this position.

✳✳ WARNING

Do not turn the steering wheel or column with the lower joint removed from the steering column or the spiral cable may be damaged.

3. Remove the steering wheel.

➡ **The steering wheel must be removed before disconnecting the steering column lower joint to avoid damaging the Supplemental Restraint System (SRS) spiral cable.**

4. Raise and support the vehicle safely and remove the front wheels.

5. Remove or disconnect the following:
- Tie rod ends from the steering knuckles
- Carbon canister and properly support the engine
- Bolts attaching the engine mounts to the engine mounting center member
- Engine mounting center member
- Front stabilizer bar from the vehicle, if necessary
- Nuts attaching the hole cover to the bulkhead

6. Move the hole cover aside and disconnect the lower joint from the rack and pinion. Matchmark the pinion shaft and the pinion housing to record the steering neutral position.
- Power steering fluid pipes from the rack and pinion
- Bolts attaching the mounting brackets and the rack and pinion from the vehicle

To install:

7. Install or connect the following:
- Rack and pinion in the vehicle
- Mounting brackets and tighten the mounting nuts and bolts in the proper sequence
- New O-rings to the power steering fluid pipes and connect them to the rack and pinion. Torque the low pressure line 20–29 ft. lbs. (27–39 Nm) and the high pressure line to 11–18 ft. lbs. (15–25 Nm).

8. Align the lower steering joint to the pinion shaft and install the joint onto the pinion shaft. Torque the bolt to 17–22 ft. lbs. (24–29 Nm).

9. Properly position the hole cover. Torque the nuts to 2.9–3.6 ft. lbs. (4–5 Nm).
- Front stabilizer
- Engine mounting center member and tighten the attaching bolts. Attach the engine mounts to the center member and tighten the bolts. Remove the support from the engine.
- Remaining components in the reverse order of removal

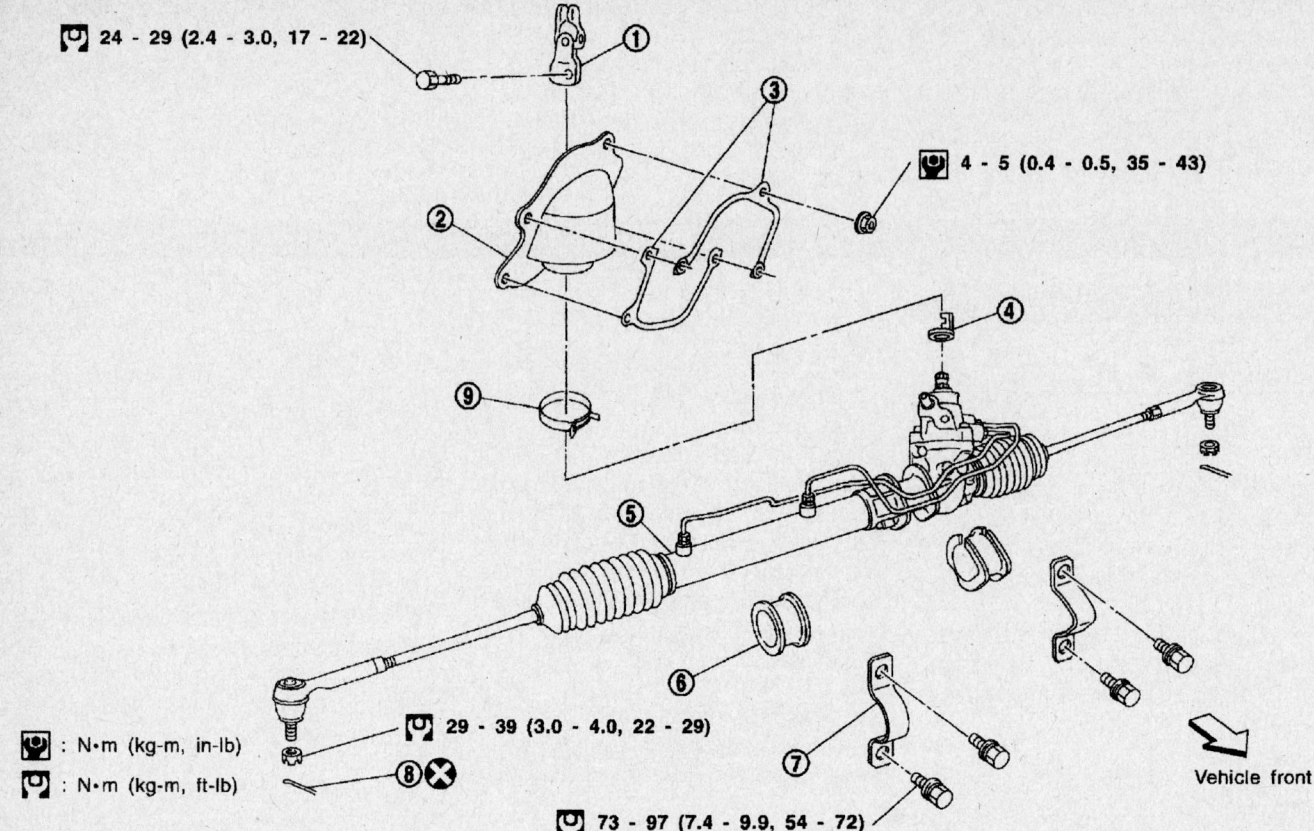

⊡ 24 - 29 (2.4 - 3.0, 17 - 22)

⊡ 4 - 5 (0.4 - 0.5, 35 - 43)

⊡ : N·m (kg-m, in-lb)

⊡ : N·m (kg-m, ft-lb)

⊡ 29 - 39 (3.0 - 4.0, 22 - 29)

⊡ 73 - 97 (7.4 - 9.9, 54 - 72)

Vehicle front

1. Lower joint
2. Hole cover
3. Insulator bracket
4. Rear cover cap
5. Gear and linkage assembly
6. Rack mounting insulator
7. Gear housing mounting bracket
8. Cotter pin
9. Clamp

09482_INFI_G0100

Component view of the steering gear assembly—G20 Models

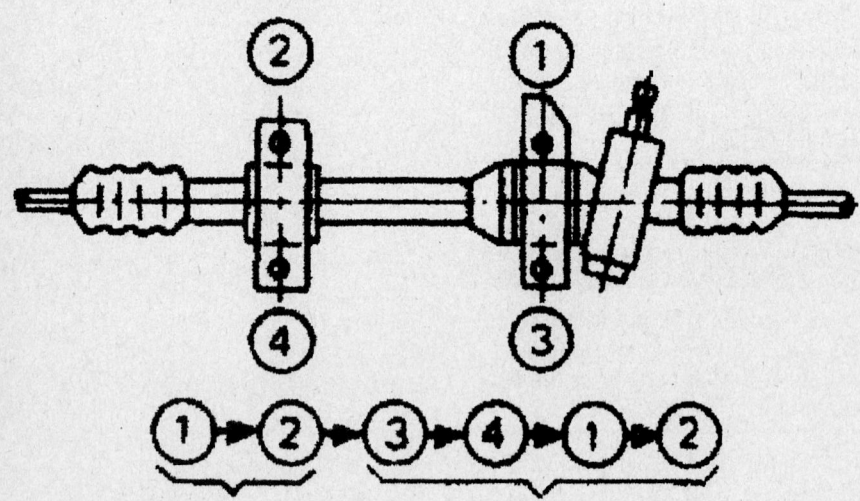

Temporary tightening   Secure tightening

09482_INFI_G0099

Tighten the steering rack fasteners in this order—G20 Models

10. Torque the tie rod end nuts to 22–29 ft. lbs. (29–39 Nm), then install a new cotter pin.

11. Fill the power steering reservoir with fluid and bleed the air from the power steering system.

12. Check the vehicle front end alignment and adjust as necessary.

I35 Models

1. Before servicing the vehicle, refer to the precautions in the beginning of this section.

※※ CAUTION

The air bag system must be disarmed before removing the steering wheel. Failure to do so may cause accidental deployment, property damage or personal injury.

2. Point the front tires straight ahead and lock the steering in this position.

❈❈ WARNING

Do not turn the steering wheel or column with the lower joint removed from the steering column or the spiral cable may be damaged.

3. Remove the steering wheel.

➡**The steering wheel must be removed before disconnecting the steering column lower joint to avoid damaging the Supplemental Restraint System (SRS) spiral cable.**

4. Remove or disconnect the following:
- Front exhaust pipe

5. Place a suitable jack under the transaxle.
- Center member and rear engine mount
- Front stabilizer bar
- Separate the tie rod ends from the steering knuckle
- Power steering lines and plug them

to prevent contaminants from entering
- Steering gear retaining bolts
- Steering gear assembly

To install:

6. Installation is the reverse of the removal procedure, noting the following:

a. Tighten all fasteners as shown in the illustration.

b. Fill the power steering reservoir with fluid and bleed the air from the power steering system.

c. Check the vehicle front end alignment and adjust as necessary.

G35 Models

1. Before servicing the vehicle, refer to the precautions in the beginning of this section.

❈❈ WARNING

Do not turn the steering wheel or column with the steering gear removed.

2. Drain the power steering fluid.
3. Remove or disconnect the following:
- Both front wheels
- Engine undercover
- Tie rod ends from the steering knuckles
- Pinch bolts from upper and lower sides
- Power steering fluid pipes from the steering gear assembly
- Bolt from rack mounting bracket insulator
- Bolts attaching the mounting brackets and the steering gear from the vehicle

To install:

4. Installation is the reverse of the removal procedure, noting the following:

a. Tighten all fasteners as shown in the illustration.

b. Fill the power steering reservoir with fluid and bleed the air from the power steering system.

c. Check the vehicle front end alignment and adjust as necessary.

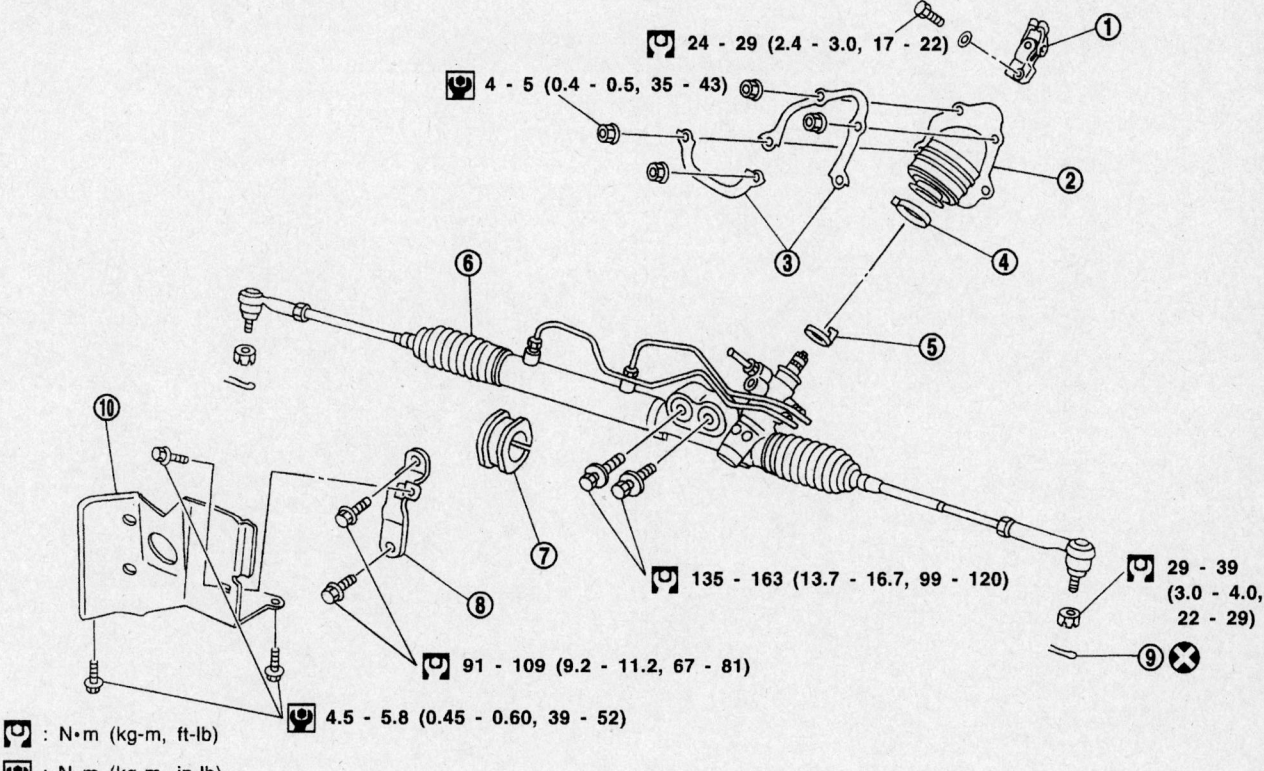

1. Lower joint
2. Hole cover
3. Insulator bracket
4. Clamp
5. Rear cover cap
6. Gear and linkage assembly
7. Rack mounting insulator
8. Gear housing mounting bracket
9. Cotter pin
10. Heat insulator

09482_INFI_G0101

Component view of the power steering gear assembly—I35 models

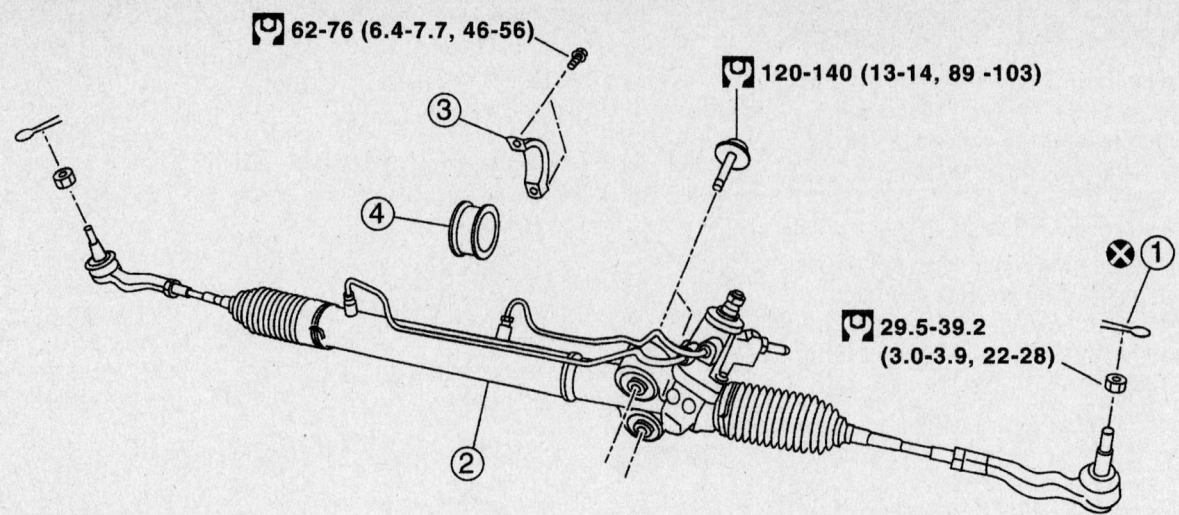

🔧 62-76 (6.4-7.7, 46-56)

🔧 120-140 (13-14, 89 -103)

🔧 29.5-39.2 (3.0-3.9, 22-28)

❌①

❌ : Always replace after every disassembly.

🔧 : N•m (kg-m, ft-lb)

1. Cotter pin
2. Steering gear assembly
3. Rack mounting bracket
4. Rack mounting insulator

09482_INFI_G0102

Exploded view of the power steering gear assembly—G35 models

2006 M35/M45 Models

➡️Refer to 2004–06 Q45 Models for 2004 M45 service procedures.

1. Before servicing the vehicle, refer to the precautions in the beginning of this section.

※※ WARNING

Do not turn the steering wheel or column with the steering gear removed.

2. Remove or disconnect the following:

- Both front wheels
- Tie rod ends from the steering knuckles
- Lower steering column joint bolt
- Solenoid valve harness connector
- Power steering fluid pipes from the rack and pinion

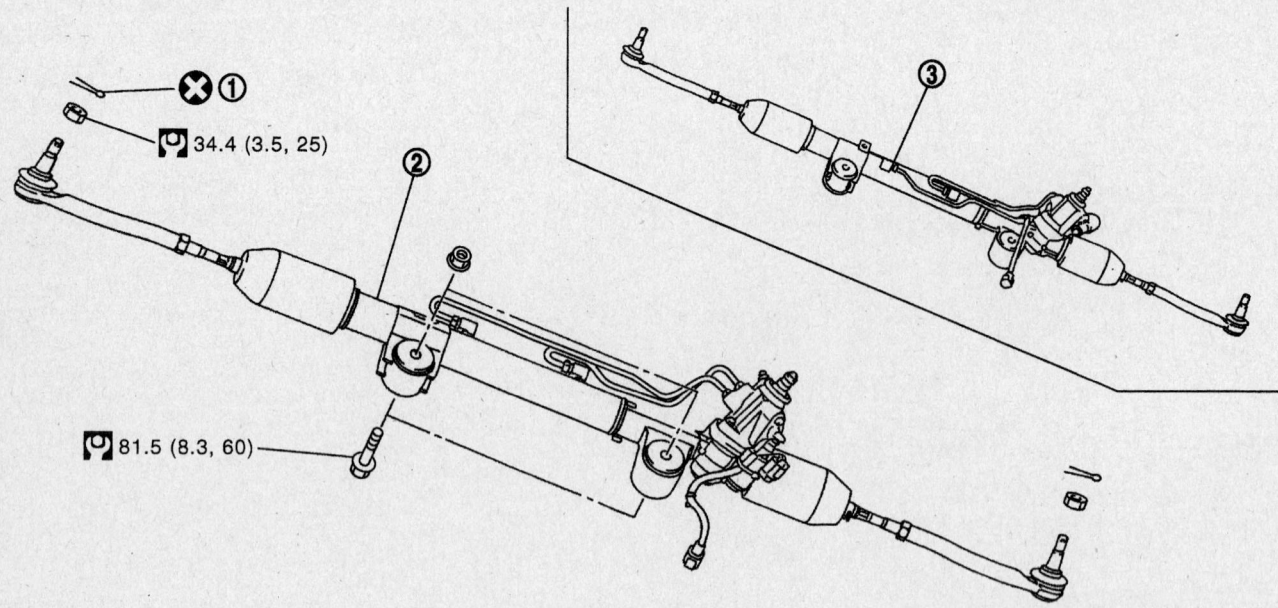

❌①

🔧 34.4 (3.5, 25)

②

③

🔧 81.5 (8.3, 60)

1. Cotter pin
2. Steering gear assembly
3. Steering gear assembly (AWD models)

09482_INFI_G0105

Component view of the steering gear assembly—2006 M35/M45 Models

- Mounting bolts and nuts

3. Remove steering gear assembly from vehicle

To install:

4. Install or connect the following:
- Rack and pinion in the vehicle.
- Mounting bolts; tighten the gear-to-frame bolts as indicated in diagram.

5. Check the vehicle front end alignment and adjust as necessary.

2002–03 Q45 Models

1. Before servicing the vehicle, refer to the precautions in the beginning of this section.

> ✳✳ **WARNING**
>
> **Do not turn the steering wheel or column with the steering gear removed.**

2. Remove or disconnect the following:
- Both front wheels
- Tie rod ends from the steering knuckles
- Carbon canister from the vehicle and properly support the engine
- Bolts attaching the engine mounts to the engine mounting center member
- Front stabilizer bar from the vehicle
- Lower joint bolts

- Power steering fluid pipes from the rack and pinion
- Bolts attaching the mounting brackets and the rack and pinion from the vehicle

To install:

3. Install or connect the following:
- Rack and pinion in the vehicle.
- Mounting brackets. Torque the bolts to 112–127 ft. lbs. (152–172 Nm).
- New O-rings to the power steering fluid pipes and connect them to the rack and pinion. Torque the low pressure line 30–33 ft. lbs. (40–44 Nm) and the high pressure line to 11–18 ft. lbs. (15–25 Nm).

4. Align the lower steering joint to the pinion shaft and install the joint onto the pinion shaft. Install the bolt and torque to 17–22 ft. lbs. (24–29 Nm).
- Front stabilizer, if removed
- Engine mounting center member and tighten the attaching bolts. Attach the engine mounts to the center member and tighten the bolts if removed. Remove the support from the engine.
- Remaining components in the reverse order of removal

5. Torque the tie rod end nuts to 47–80 ft. lbs. (64–108 Nm) and install a new cotter pin.

6. Fill the power steering reservoir with fluid and bleed the air from the power steering system.

7. Check the front end alignment and adjust as necessary.

2004–06 Q45 and 2004 M45 Models

1. Before servicing the vehicle, refer to the precautions in the beginning of this section.

> ✳✳ **WARNING**
>
> **Do not turn the steering wheel or column with the steering gear removed.**

2. Remove or disconnect the following:
- Both front wheels
- Engine splash shield
- Tie rod ends from the steering knuckles
- Lower steering column joint bolt
- Solenoid valve harness connector
- Power steering fluid pipes from the rack and pinion
- Mounting bolts and nuts

3. Remove steering gear assembly from vehicle

To install:

4. Install or connect the following:
- Rack and pinion in the vehicle.

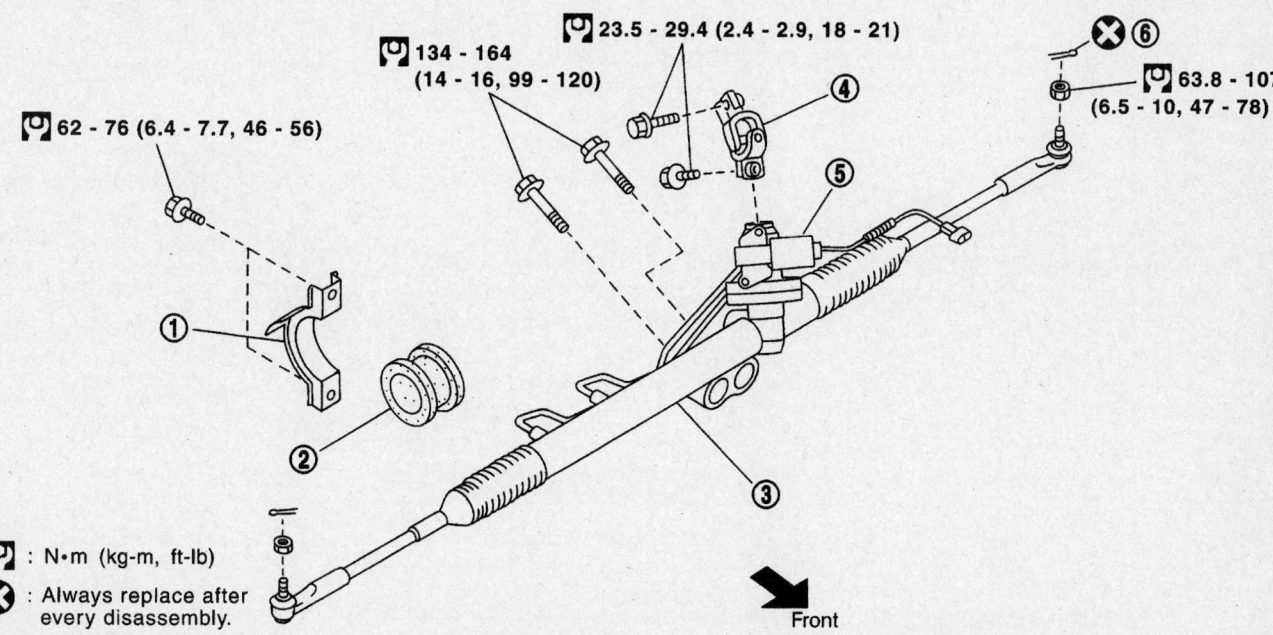

62 - 76 (6.4 - 7.7, 46 - 56)

134 - 164 (14 - 16, 99 - 120)

23.5 - 29.4 (2.4 - 2.9, 18 - 21)

63.8 - 107 (6.5 - 10, 47 - 78)

🔧 : N•m (kg-m, ft-lb)

❌ : Always replace after every disassembly.

Front

1. Rack mounting bracket
2. Rack mounting insulator
3. Steering gear assembly
4. Lower joint
5. EPS solenoid valve
6. Cotter pin

09482_INFI_G0104

Component view of the steering gear assembly—Q45 and 2004 M45 Models

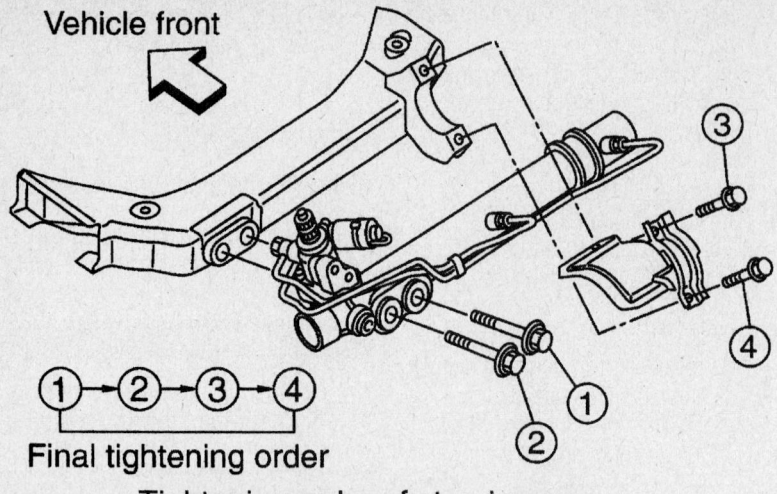

Vehicle front

Final tightening order

Tightening order of steering gear

09482_INFI_G0103

Tightening sequence—Q45 and 2004 M45 models

- Mounting bolts. Torque the bolts in sequence. Tighten the mounting bracket bolts and the gear-to-frame bolts as indicated in diagram.
- New O-rings to the power steering fluid pipes and connect them to the rack and pinion.

5. Align the lower steering joint to the pinion shaft and install the joint onto the pinion shaft. Install the bolt and torque to 18–21 ft. lbs. (24–29 Nm).

6. Torque the tie rod end nuts to 47–76 ft. lbs. (64–107 Nm) and install a new cotter pin.

7. Fill the power steering reservoir with fluid and bleed the air from the power steering system.

8. Check the front end alignment and adjust as necessary.

FRONT SUSPENSION

Strut

REMOVAL & INSTALLATION

G20 Models

1. Before servicing the vehicle, refer to the precautions in the beginning of this section.

2. Raise and support the vehicle safely.

3. Remove or disconnect the following:
- Strut mounting bolt at the lower suspension member and the 3 nuts inside the engine compartment. Do not remove the piston rod locknut
- Strut assembly and place in a suitable holding device

4. Using a prybar to hold the upper spring mount, loosen but do not remove the piston rod locknut.

5. Compress the spring with a spring compressor so the strut mounting insulator can be turned by hand.
- Piston rod locknut
- Coil spring from strut assembly

To install:

6. Inspect all components carefully for damage or wear. Replace as necessary.

7. Install or connect the following:
- Compressed coil spring on the strut. Torque the locknut to 13–17 ft. lbs. (18–24 Nm).
- Strut. Ensure the bend in the lower strut bracket faces rearward on the left side and forward on the right side of the vehicle.
- Upper spring seat with the cutout facing the inside of the vehicle.

Torque the upper mounting bolts to 31–40 ft. lbs. (42–54 Nm) and the lower through-bolt to 82–93 ft. lbs. (112–126 Nm). Final tightening must take place with the suspension loaded (vehicle at normal ride height).

I35 AND G35 2wd Models

1. Before servicing the vehicle, refer to the precautions in the beginning of this section.

2. Remove or disconnect the following:
- Wheel. Matchmark the position of the strut-to-steering knuckle location
- Brake hose from the strut
- Anti-lock Brake System (ABS) wheel sensor and move it out of the way
- Bolts attaching the steering knuckle transverse link to the strut. Matchmark the assembly prior to removing the bolts
- Strut attaching nuts while holding the strut from inside the engine compartment

✳✳ CAUTION

Do not remove the center locknut from the strut assembly until the strut is safely compressed.

- Strut from the vehicle
3. Place the strut assembly in a vise with the special holding tool ST35652000 or in a spring compressor.
4. Loosen the piston rod locknut.

✳✳ CAUTION

Do not remove the piston rod locknut, the spring is under tension and can cause serious personal injury.

5. Compress the spring with the spring compressor, then remove the piston rod locknut.

➡**Before removing the strut from the coil spring, note the positioning of the strut in relationship to the coil spring for reassembly.**

- Strut mounting insulator bracket, strut mounting bearing, upper spring seat, and the upper spring rubber seat
- Strut, leaving the coil spring compressed
- Piston boot and rebound bumper from the strut

To install:
6. Install or connect the following:
- Rebound bumper and the boot to the strut piston
- Strut into the coil spring, be sure the strut and spring are properly positioned
- Upper spring rubber seat, upper spring seat, strut mounting bearing, and the strut mounting insulator bracket. Be sure that the cutout on the upper spring seat is facing the outside of the vehicle.
- Piston rod locknut, then remove the spring compressor. Torque the piston rod locknut to 43–65 ft. lbs. (59–88 Nm) I35 models, or to

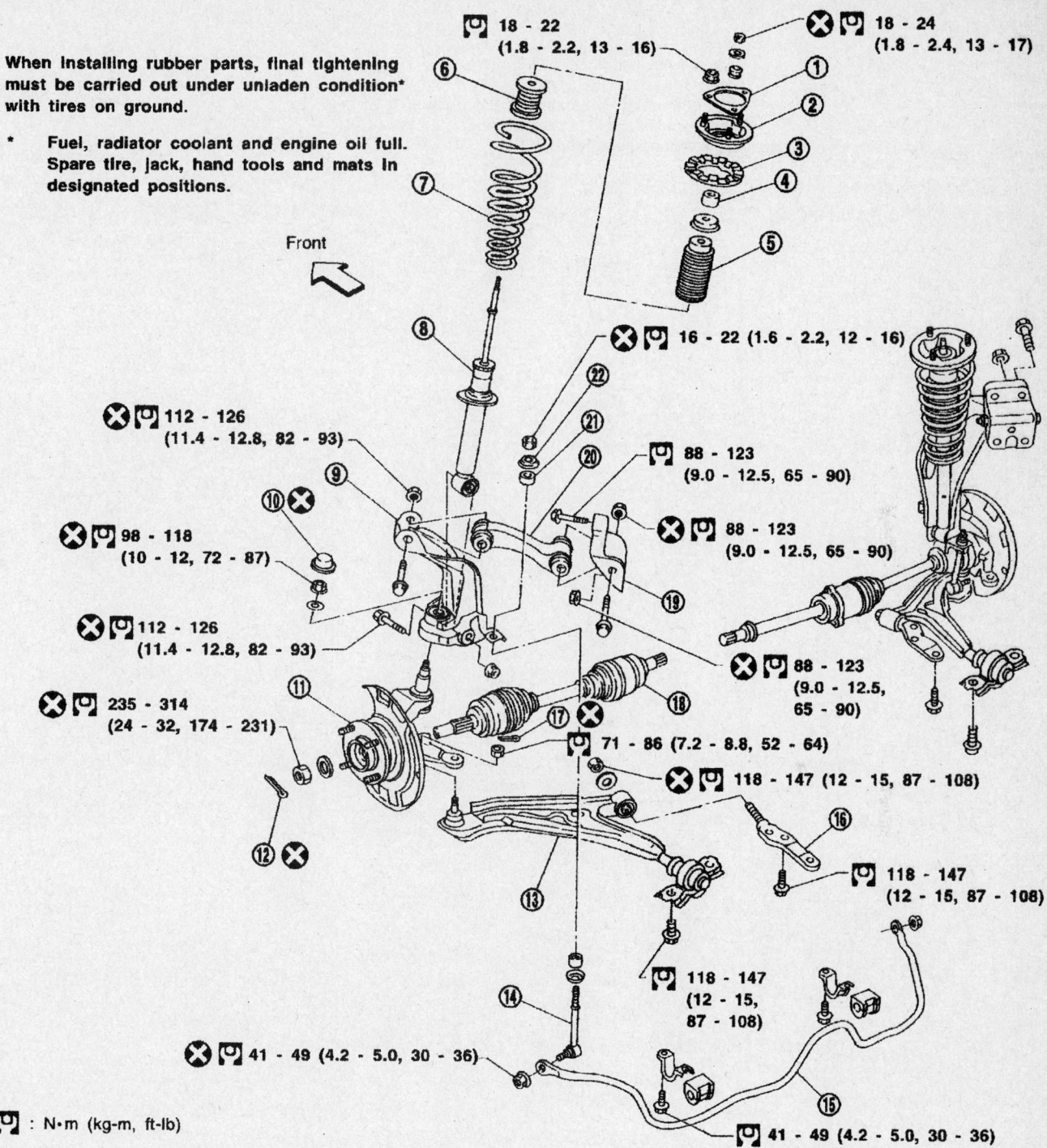

When installing rubber parts, final tightening must be carried out under unladen condition* with tires on ground.

* Fuel, radiator coolant and engine oil full. Spare tire, jack, hand tools and mats in designated positions.

Front

18 - 22 (1.8 - 2.2, 13 - 16)

18 - 24 (1.8 - 2.4, 13 - 17)

16 - 22 (1.6 - 2.2, 12 - 16)

112 - 126 (11.4 - 12.8, 82 - 93)

88 - 123 (9.0 - 12.5, 65 - 90)

88 - 123 (9.0 - 12.5, 65 - 90)

98 - 118 (10 - 12, 72 - 87)

112 - 126 (11.4 - 12.8, 82 - 93)

88 - 123 (9.0 - 12.5, 65 - 90)

235 - 314 (24 - 32, 174 - 231)

71 - 86 (7.2 - 8.8, 52 - 64)

118 - 147 (12 - 15, 87 - 108)

118 - 147 (12 - 15, 87 - 108)

118 - 147 (12 - 15, 87 - 108)

41 - 49 (4.2 - 5.0, 30 - 36)

41 - 49 (4.2 - 5.0, 30 - 36)

[ ] : N·m (kg-m, ft-lb)

1. Gasket
2. Shock absorber mounting insulator
3. Upper rubber seat
4. Shock absorber bushing
5. Dust cover
6. Bound bumper rubber
7. Coil spring
8. Shock absorber

9. Third link
10. Cap
11. Wheel hub and steering knuckle assembly
12. Cotter pin
13. Transverse link
14. Connecting rod
15. Stabilizer bar

16. Gusset pin
17. Cotter pin
18. Drive shaft
19. Upper link bracket
20. Upper link
21. Bushing
22. Washer

09482_INFI_G0418

Front suspension components—G20

When installing rubber parts, final tightening must be carried out under unladen condition* with tires on ground.
* Fuel, radiator coolant and engine oil full. Spare tire, jack, hand tools and mats in designated positions.

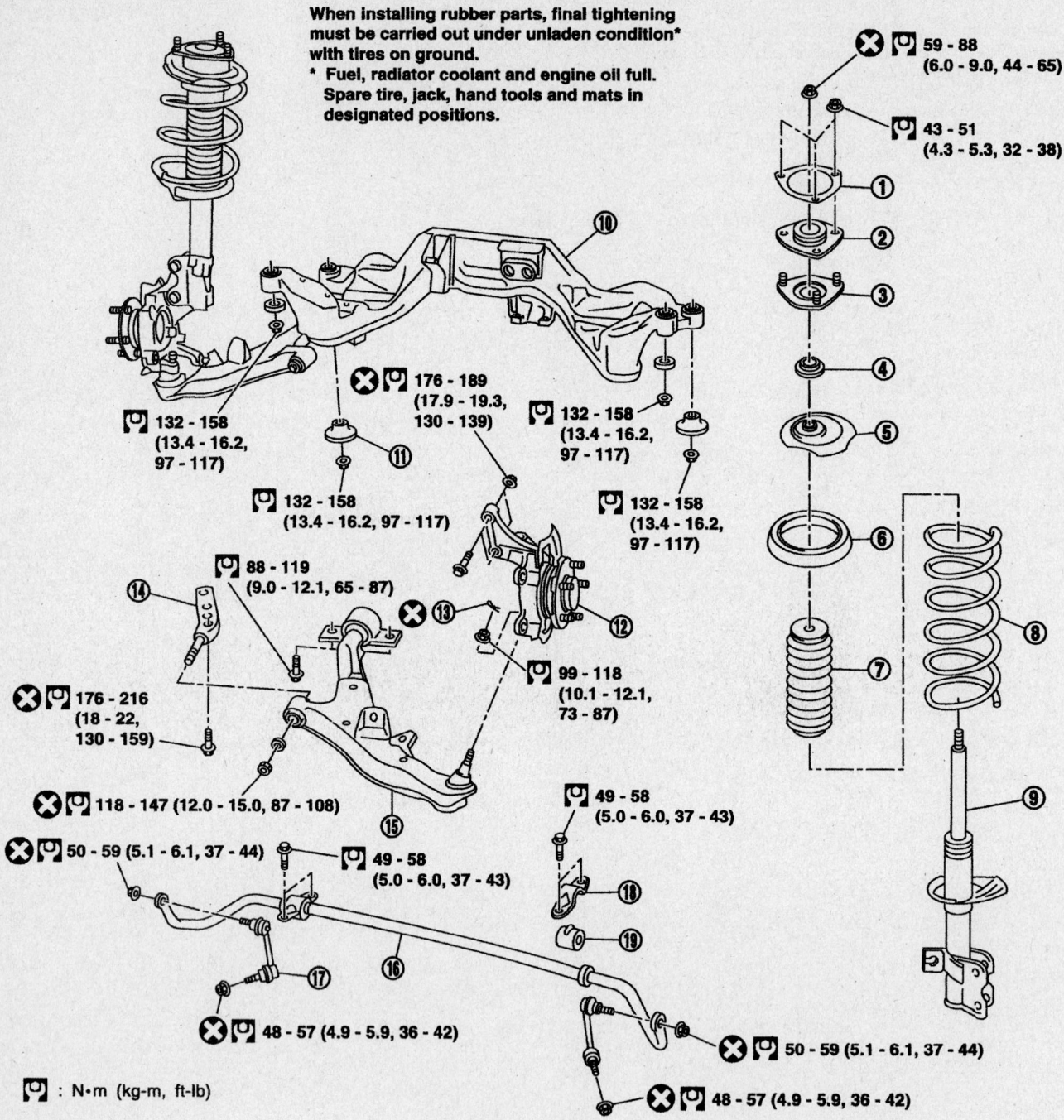

59 - 88 (6.0 - 9.0, 44 - 65)

43 - 51 (4.3 - 5.3, 32 - 38)

176 - 189 (17.9 - 19.3, 130 - 139)

132 - 158 (13.4 - 16.2, 97 - 117)

132 - 158 (13.4 - 16.2, 97 - 117)

132 - 158 (13.4 - 16.2, 97 - 117)

132 - 158 (13.4 - 16.2, 97 - 117)

88 - 119 (9.0 - 12.1, 65 - 87)

99 - 118 (10.1 - 12.1, 73 - 87)

176 - 216 (18 - 22, 130 - 159)

118 - 147 (12.0 - 15.0, 87 - 108)

49 - 58 (5.0 - 6.0, 37 - 43)

50 - 59 (5.1 - 6.1, 37 - 44)

49 - 58 (5.0 - 6.0, 37 - 43)

48 - 57 (4.9 - 5.9, 36 - 42)

50 - 59 (5.1 - 6.1, 37 - 44)

48 - 57 (4.9 - 5.9, 36 - 42)

⟨U⟩ : N·m (kg-m, ft-lb)

1. Strut spacer
2. Strut mount insulator
3. Strut mount bracket
4. Strut mount bearing
5. Spring upper seat
6. Spring rubber seat
7. Bound bumper rubber

8. Coil spring
9. Shock absorber
10. Suspension member
11. Rebound stopper
12. Wheel hub and steering knuckle
13. Cotter pin

14. Bush link pin
15. Transverse link
16. Stabilizer
17. Connecting rod
18. Stabilizer clamp
19. Bushing

09482_INFI_G0419

Front suspension components—I35

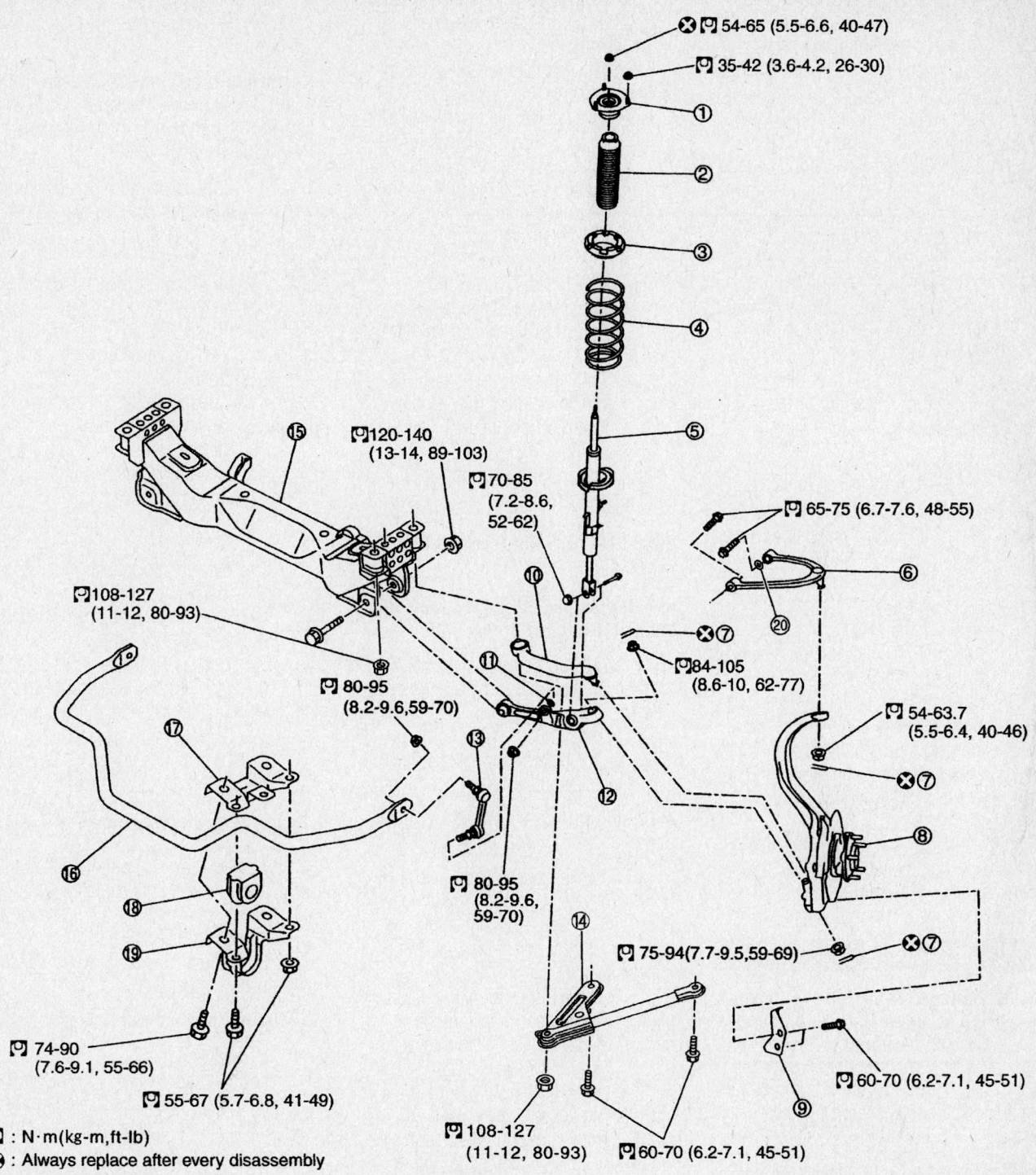

⊗ 🔧 54-65 (5.5-6.6, 40-47)
🔧 35-42 (3.6-4.2, 26-30)
🔧 120-140 (13-14, 89-103)
🔧 70-85 (7.2-8.6, 52-62)
🔧 65-75 (6.7-7.6, 48-55)
🔧 108-127 (11-12, 80-93)
🔧 84-105 (8.6-10, 62-77)
⊗7
🔧 80-95 (8.2-9.6, 59-70)
🔧 54-63.7 (5.5-6.4, 40-46)
⊗7
🔧 80-95 (8.2-9.6, 59-70)
🔧 75-94 (7.7-9.5, 59-69)
⊗7
🔧 74-90 (7.6-9.1, 55-66)
🔧 60-70 (6.2-7.1, 45-51)
🔧 55-67 (5.7-6.8, 41-49)
🔧 108-127 (11-12, 80-93)
🔧 60-70 (6.2-7.1, 45-51)

🔧 : N·m(kg-m,ft-lb)
⊗ : Always replace after every disassembly

1. Mounting insulator
2. Bound bumper
3. Spring rubber seat
4. Coil spring
5. Shock absorber
6. Upper link
7. Cotter pin
8. Front axle
9. Steering stopper bracket
10. Compression rod
11. Washer
12. Transverse link
13. Stabilizer connecting rod
14. Compression rod stay
15. Front suspension member
16. Stabilizer bar
17. Stabilizer clamp bracket
18. Stabilizer bushing
19. Stabilizer clamp
20. Stopper rubber

09482_INFI_G0420

Front suspension components—2WD G35

40–47 ft. lbs. (54–65 Nm) for G35 models.

- Strut into the strut tower and install new attaching nuts. Torque the nuts to 32–38 ft. lbs. (43–51 Nm) for I35 models, or to 28 ft. lbs. (38.5 Nm) for G35 models.
- Bolts attaching the steering knuckle or transverse link to the strut and align the matchmarks. Torque the bolts to 130–139 ft. lbs. (176–189 Nm) for I35 models, or to 79 ft. lbs. (107 Nm) for G35 models.
- ABS wheel sensor. Torque the bolt to 13–17 ft. lbs. (18–24 Nm).
- Brake hose to the strut
- Front wheels and lower the vehicle

7. Check and/or adjust the wheel alignment as necessary.

G35 AWD Models

1. Before servicing the vehicle, refer to the precautions in the beginning of this section.
2. Remove or disconnect the following:
- Front wheels
- Engine undercover
- ABS (antilock brake sensor) sensor harness
- Brake hose mounting bracket
- Strut attachment (shock absorber arm) from lower control arm
- Upper strut mounting nuts
3. Remove strut assembly from vehicle
4. Place the strut assembly in a vise with the special holding tool ST35652000 or in a spring compressor.
5. Loosen the piston rod locknut.

✳✳ CAUTION

Do not remove the piston rod locknut, the spring is under tension and can cause serious personal injury.

6. Compress the spring with the spring compressor, then remove the piston rod locknut.

➡**Before removing the strut from the coil spring, note the positioning of the strut in relationship to the coil spring for reassembly.**

- Strut mounting insulator bracket, strut mounting bearing, upper spring seat, and the upper spring rubber seat
- Strut, leaving the coil spring compressed
- Piston boot and rebound bumper from the strut

- Strut attachment (shock absorber arm) from shock absorber

To install:

7. Installation is in reverse order of removal. Observe the following:
- Install coil spring with large diameter up and small diameter down
- Make certain coil spring is securely seated in spring mounting groove of spring upper seat
- Tighten strut attachment (shock absorber arm) bolt to 42–51 ft. lbs. (56–70 Nm)
- Tighten strut attachment-to-lower control arm nut to 93–107 ft. lbs. (126–146 Nm)
- Tighten shock absorber piston nut to 40–47 ft. lbs. (54–65 Nm)
- Tighten mounting insulator nuts to 26–30 ft. lbs. (35–42 Nm)

8. Check and/or adjust the wheel alignment as necessary.

2006 2wd M35 Models and M45 Models

➡**Refer to 2004–06 Q45 models for 2004 M45 service procedures**

1. Before servicing the vehicle, refer to the precautions in the beginning of this section.
2. Remove front wheels
3. Remove harness of wheel sensor from shock absorber.

➡**Do not pull on wheel sensor harness.**

4. Remove brake hose bracket.
5. Remove the mounting nut on the upper side of stabilizer connecting rod and remove stabilizer connecting rod from transverse link.
6. Remove mounting nut and bolt on the lower side of shock absorber and remove shock absorber from transverse link.
7. Remove cotter pin of transverse link and steering knuckle and loosen nut.
8. Remove transverse link from steering knuckle
9. Remove the mounting nuts of shock absorber mounting bracket and remove shock absorber from vehicle.
10. Place the strut assembly in a vise and attach a suitable spring compressor.
11. Loosen the piston rod locknut.

✳✳ CAUTION

Do not remove the piston rod locknut, the spring is under tension and can cause serious personal injury.

12. Compress the spring with the spring compressor and remove the piston rod locknut.

➡**Before removing the strut from the coil spring, note the positioning of the strut in relationship to the coil spring for reassembly.**

- Strut mounting insulator bracket, strut mounting bearing, upper spring seat, and the upper spring rubber seat
- Strut, leaving the coil spring compressed
- Piston boot and rebound bumper from the strut

To install:

13. Installation is in the reverse order of removal. Observe the following:
- Make certain coil spring is securely seated in spring mounting groove of spring upper seat
- Tighten strut-to-transverse arm nut to 79 ft. lbs. (107 Nm)
- Tighten shock absorber piston nut to 44 ft. lbs. (60 Nm)
- Tighten mounting insulator nuts to 29 ft. lbs. (39 Nm)
- Tighten upper stabilizer connecting rod nut to 66 ft. lbs. (90 Nm)
- Tighten steering knuckle nut to 100 ft. lbs. (136 Nm)

14. Check and/or adjust the wheel alignment as necessary.

2006 M35 AWD Models

1. Before servicing the vehicle, refer to the precautions in the beginning of this section.
2. Remove front wheels
3. Remove harness of wheel sensor from shock absorber

➡**Do not pull on wheel sensor harness**

4. Remove brake hose bracket
5. Remove the mounting nut on the upper side of stabilizer connecting rod and remove stabilizer connecting rod from transverse link
6. Remove mounting nut and bolt on the lower side of shock absorber arm and remove shock absorber from transverse link
7. Remove cotter pin of transverse link and steering knuckle and loosen nut.
8. Remove transverse link from steering knuckle
9. Remove the nuts of shock absorber upper mounting bracket and remove shock absorber from vehicle
10. Remove shock absorber arm from shock absorber

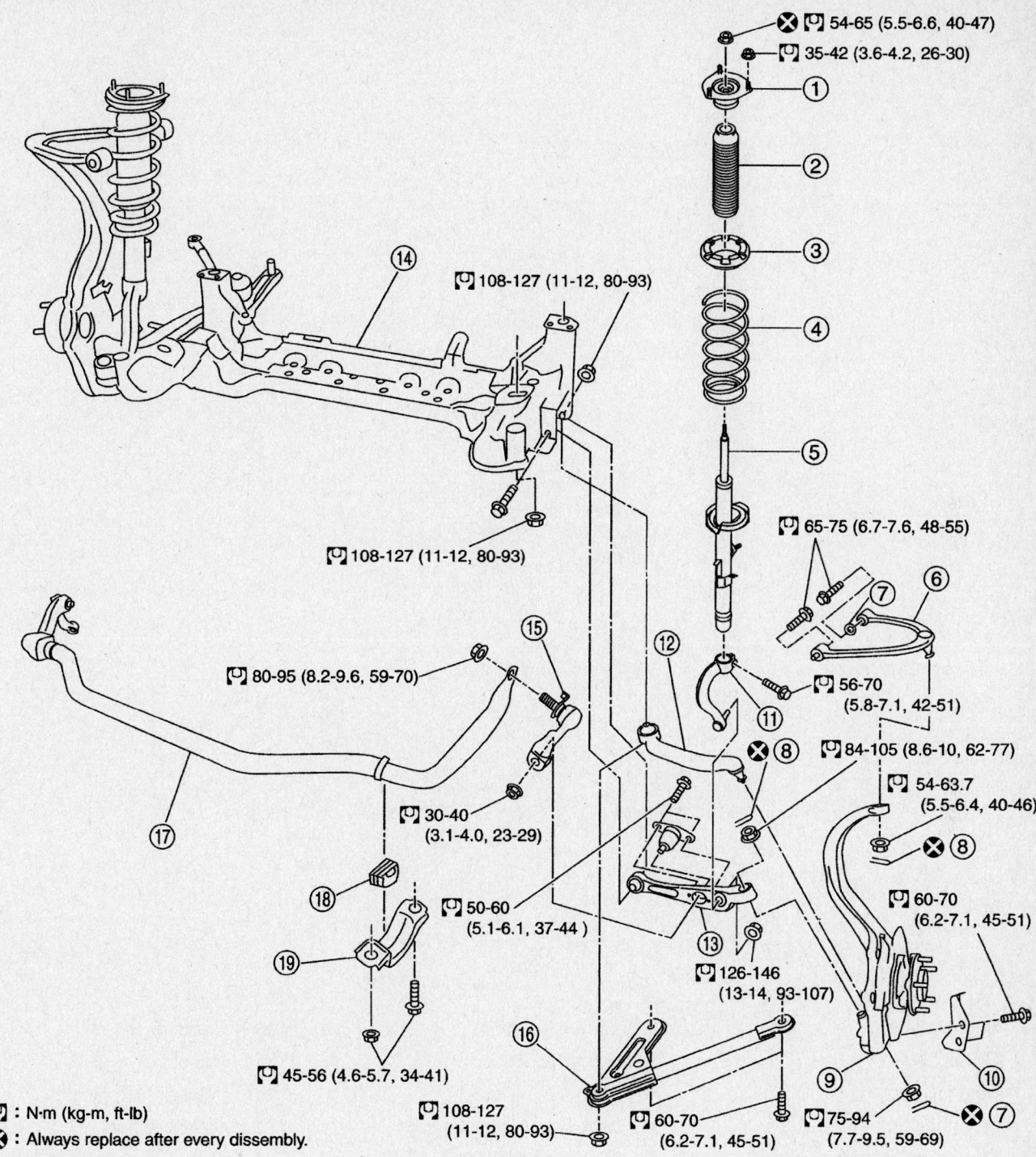

54-65 (5.5-6.6, 40-47)
35-42 (3.6-4.2, 26-30)
108-127 (11-12, 80-93)
108-127 (11-12, 80-93)
65-75 (6.7-7.6, 48-55)
56-70 (5.8-7.1, 42-51)
84-105 (8.6-10, 62-77)
54-63.7 (5.5-6.4, 40-46)
80-95 (8.2-9.6, 59-70)
30-40 (3.1-4.0, 23-29)
50-60 (5.1-6.1, 37-44)
60-70 (6.2-7.1, 45-51)
126-146 (13-14, 93-107)
45-56 (4.6-5.7, 34-41)
108-127 (11-12, 80-93)
60-70 (6.2-7.1, 45-51)
75-94 (7.7-9.5, 59-69)

: N·m (kg-m, ft-lb)
: Always replace after every dissembly.

1. Mounting insulator	2. Bound bumper	3. Spring upper seat
4. Coil spring	5. Shock absorber	6. Upper link
7. Washer	8. Cotter pin	9. Front axle
10. Steering stopper bracket	11. Shock absorber arm	12. Compression rod
13. Transverse link	14. Front suspension member	15. Stabilizer connecting rod
16. Compression rod stay	17. Stabilizer bar	18. Stabilizer bushing
19. Stabilizer clamp		

09482_INFI_G0421

Front suspension components—AWD G35

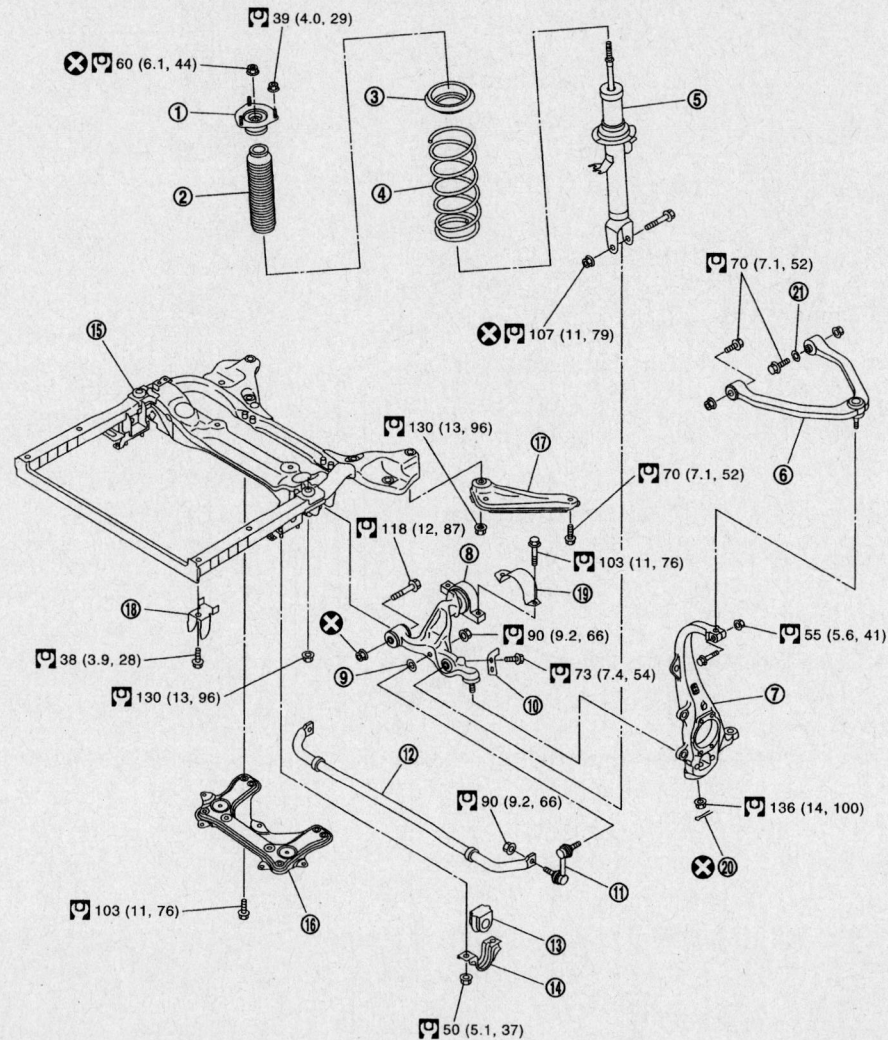

1. Shock absorber mounting bracket
2. Bound bumper
3. Rubber seat
4. Coil spring
5. Shock absorber
6. Upper link
7. Steering knuckle
8. Transverse link
9. Washer
10. Steering stopper bracket
11. Stabilizer connecting rod
12. Stabilizer bar
13. Stabilizer bushing
14. Stabilizer clamp
15. Front suspension member
16. Rack stay
17. Member stay
18. Member bracket
19. Clamp
20. Cotter pin
21. Stopper rubber

09482_INFI_G0470

Front suspension components—2006 2WD M35 & M45 models

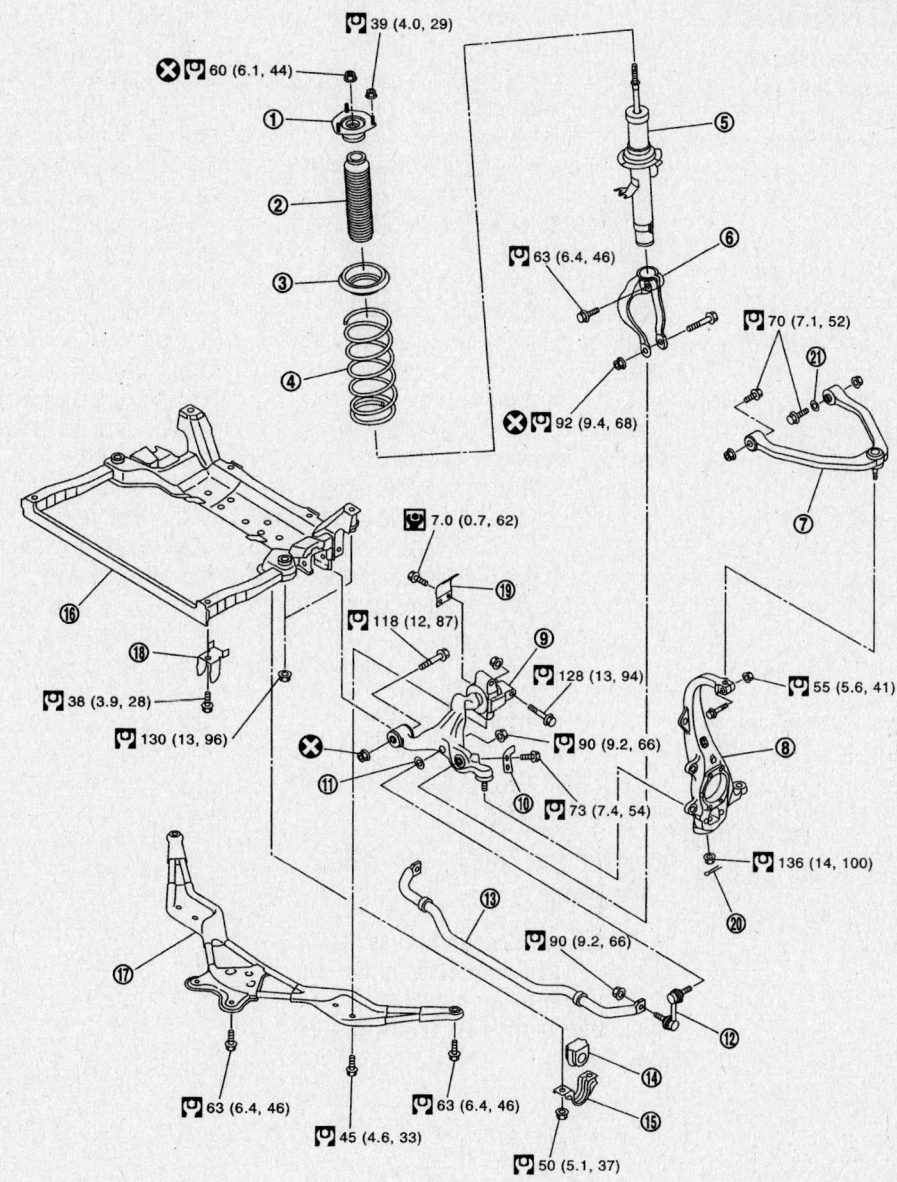

1. Shock absorber mounting bracket
2. Bound bumper
3. Rubber seat
4. Coil spring
5. Shock absorber
6. Shock absorber arm
7. Upper link
8. Steering knuckle
9. Transverse link
10. Steering stopper bracket
11. Washer
12. Stabilizer connecting rod
13. Stabilizer bar
14. Stabilizer bushing
15. Stabilizer clamp
16. Front suspension member
17. Front cross bar
18. Member bracket
19. Clamp
20. Cotter pin
21. Stopper rubber

09482_INFI_G0422

Front suspension components—2006 AWD M35

11. Place the strut assembly in a vise and attach a suitable spring compressor

12. Loosen the piston rod locknut

✳✳ CAUTION

Do not remove the piston rod locknut, the spring is under tension and can cause serious personal injury

13. Compress the spring with the spring compressor and remove the piston rod locknut

➡**Before removing the strut from the coil spring, note the positioning of the strut in relationship to the coil spring for reassembly**

- Strut mounting insulator bracket, strut mounting bearing, upper spring seat, and the upper spring rubber seat
- Strut, leaving the coil spring compressed
- Piston boot and rebound bumper from the strut

To install:

14. Installation is in the reverse order of removal. Observe the following:

- Make certain coil spring is securely seated in spring mounting groove of spring upper seat
- Tighten shock absorber arm bolt to 46 ft. lbs. (63 Nm)
- Tighten shock absorber arm lower nut to 68 ft. lbs. (92 Nm)
- Tighten shock absorber piston nut to 44 ft. lbs. (60 Nm)
- Tighten mounting insulator nuts to 29ft. lbs. (39Nm)
- Tighten upper stabilizer connecting rod nut to 66 ft. lbs. (90 Nm)
- Tighten steering knuckle nut to 100 ft. lbs. (136 Nm)

15. Check and/or adjust the wheel alignment as necessary.

2002–03 Q45 Models with Standard Suspension

1. Before servicing the vehicle, refer to the precautions in the beginning of this section.

2. Remove or disconnect the following:

- Front wheel
- Brake caliper and rotor
- Tie rod ball joint and lower ball joint with tool ST29020001
- Stabilizer connecting rod upper nut and separate the strut from the connecting rod

- Upper mounting insulator bolts
- Strut assembly

3. Secure the strut in a suitable holding fixture.

4. Loosen the piston rod locknut. Do not remove the locknut.

5. Compress the spring with the proper tool so the strut assembly mounting insulator can be turned by hand.

6. Remove the piston rod locknut. Remove the spring assembly, dust cover and rubber seat. Remove the strut insert.

To install:

7. Inspect the rubber parts for deterioration. If the rubber is pulling away from the metal, the mounting insulator should be replaced.

8. Fit the spring into the lower seat, install the dust cover/bumper and upper seat and mounting insulator.

9. Install or connect the following:

- Piston rod locknut and torque to 13–17 ft. lbs. (17–23 Nm).
- Strut into place and torque the upper mounting nuts to 30–35 ft. lbs. (40–47 Nm).

10. Torque the lower mounting bolt to 80–94 ft. lbs. (108–128 Nm).

- Brake rotor and caliper
- Front wheel

11. Check and/or adjust the wheel alignment as necessary.

2002–03 Q45 Models with Active Suspension

➡**The Nissan Consult or a scan tool that can issue commands to the control unit is required for bleeding the hydraulics in the Full Active Suspension system.**

1. Before servicing the vehicle, refer to the precautions in the beginning of this section.

2. Relieve the hydraulic pressure as follows:

a. Raise all 4 wheels off the ground and wait at least 3 minutes for the system to stabilize.

b. Remove both front inner fenders and the rear pressure control unit cover.

c. Loosen the locknut and slowly open the bypass valve on each pressure control unit. Open the valves all the way and leave them open until the job is finished.

3. Remove the flange joint from the top of the actuator.

4. Install 2, 15mm bolts into the actuator in the flange joint mounting bolt holes.

5. Insert a bar between the bolts and loosen the joint adapter. Do not remove it yet.

6. Remove or disconnect the following:

- Upper mount insulator nuts
- Hydraulic lines, then cap the lines to keep the system clean
- Lower actuator mounting nut and remove the assembly

7. Secure the actuator/spring assembly in a suitable holding fixture. Scribe alignment marks on the spring, upper mount insulator and actuator unit.

8. Compress the spring with the proper tool so the joint adapter can be turned by hand. Remove the joint adapter and lift off the mount insulator, spring, and any other components necessary.

To install:

9. If the actuator is being replaced, the rubber bumper should also be replaced. Fit the bumper, dust cover and rubber seat onto the actuator.

10. Fit the spring into the lower seat with the matchmarks aligned. Install the upper seat/mounting insulator with the marks aligned and start the joint adapter. The joint adapter will be tightened after installing the actuator assembly.

11. Fit the strut into place and torque the upper mounting nuts to 30–41 ft. lbs. (40–55 Nm).

12. Torque the lower mounting bolt to 76–94 ft. lbs. (103–128 Nm).

13. Torque the joint adapter to 63–72 ft. lbs. (85–98 Nm).

14. Install the flange adapter and torque the bolts to 11–13 ft. lbs. (15–18 Nm).

15. Close the bypass valves on the pressure control units.

16. Bleed the system as follows:

a. With all 4 wheels about 2 in. (50mm) off the ground, run the engine for about 2 minutes.

b. Connect the Consult scan tool and enter "WORK SUPPORT" mode. Select "4. AIR BLEEDING".

c. Check the fluid level in the reservoir. It should be slightly overfilled.

d. Touch "START" on the scan tool. The display will show a regular rise and fall in system pressure. When the pressure stabilizes, stop the engine.

e. Connect a clear tube to the air bleeder at the actuator and place the other end in a container.

➡**Do not allow the fluid to contact the body or the paint will be damaged.**

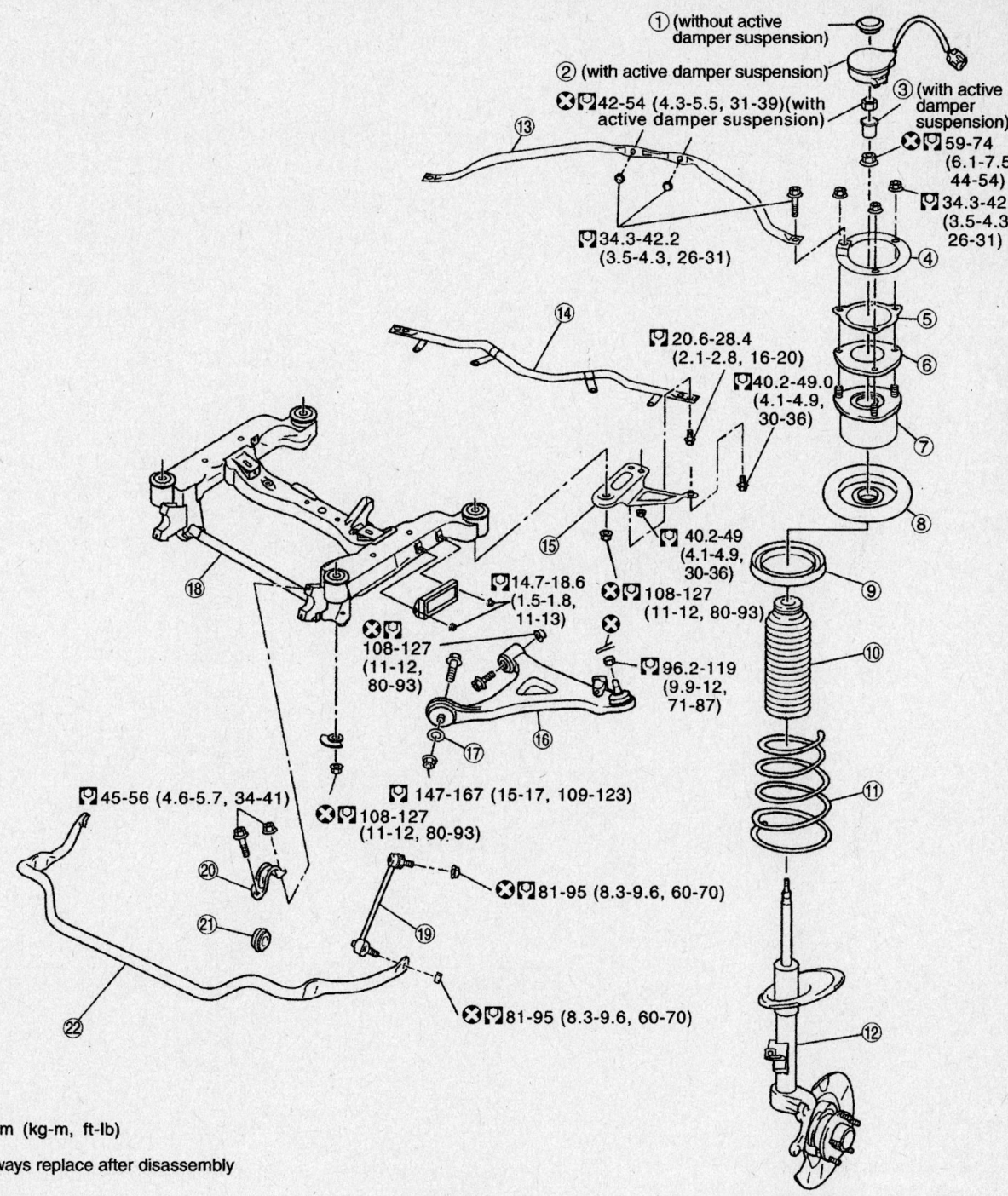

① (without active
damper suspension)

② (with active damper suspension)

⊗ 📷 42-54 (4.3-5.5, 31-39)(with
active damper suspension)

③ (with active
damper
suspension)

⊗ 📷 59-74
(6.1-7.5,
44-54)

📷 34.3-42.2
(3.5-4.3,
26-31)

⑬

📷 34.3-42.2
(3.5-4.3, 26-31)

④

⑤

⑥

⑦

⑭

📷 20.6-28.4
(2.1-2.8, 16-20)

📷 40.2-49.0
(4.1-4.9,
30-36)

⑧

⑨

⑮

📷 40.2-49
(4.1-4.9,
30-36)

📷 14.7-18.6
(1.5-1.8,
11-13)

⊗ 📷 108-127
(11-12, 80-93)

⊗ 📷 108-127
(11-12,
80-93)

📷 96.2-119
(9.9-12,
71-87)

⑯

⑩

⑰

📷 147-167 (15-17, 109-123)

📷 45-56 (4.6-5.7, 34-41)

⊗ 📷 108-127
(11-12, 80-93)

⑪

⑳

㉑

⑲

⊗ 📷 81-95 (8.3-9.6, 60-70)

⑫

㉒

⊗ 📷 81-95 (8.3-9.6, 60-70)

📷 : N•m (kg-m, ft-lb)

⊗ : Always replace after disassembly

1. Cap	2. Actuator assembly	3. Actuator plate
4. Tower bar bracket	5. Strut mounting insulator bracket	6. Strut mounting insulator
7. Strut mounting bearing	8. Spring upper seat	9. Rubber seat
10. Bound bumper	11. Coil spring	12. Strut assembly
13. Tower bar	14. Front cross bar	15. Member stay
16. Suspension arm	17. Washer	18. Front suspension member
19. Stabilizer connecting rod	20. Stabilizer clamp	21. Stabilizer bushing
22. Stabilizer bar		

09482_INFI_G0423

Front suspension components—2002–03 Q45 Models

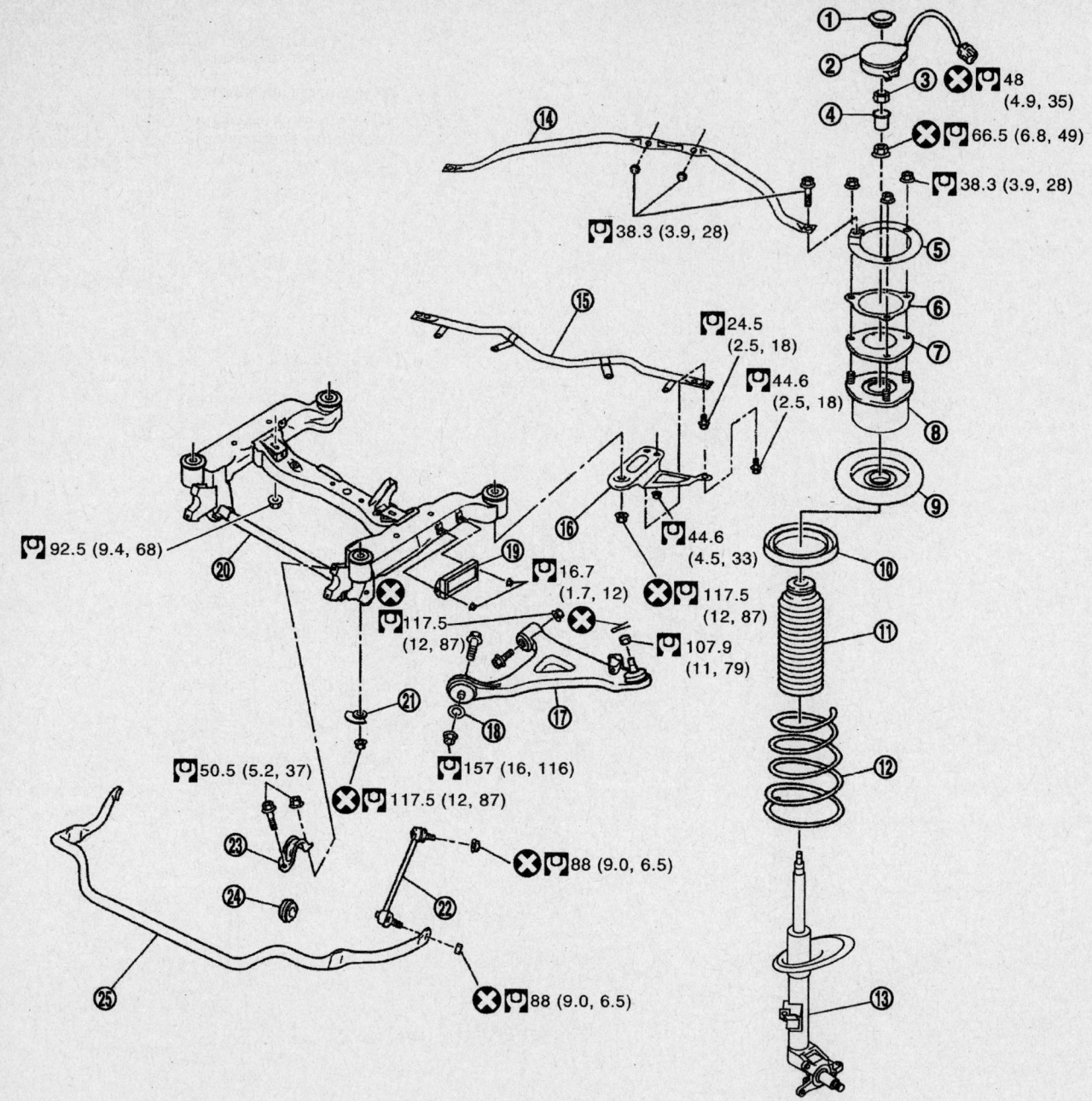

1. Cap (Without active damper suspension)
2. Actuator assembly (With active damper suspension)
3. Nut (With active damper suspension)
4. Actuator plate (With active damper suspension)
5. Tower bar bracket
6. Spacer
7. Strut mounting insulator bracket
8. Strut mounting insulator assembly
9. Spring upper seat
10. Upper rubber seat
11. Bound bumper
12. Coil spring
13. Strut assembly
14. Tower bar
15. Front cross bar
16. Member stay
17. Suspension arm
18. Washer
19. Dynamic damper
20. Front suspension member
21. Rebound stopper
22. Stabilizer connecting rod
23. Stabilizer clamp
24. Stabilizer bushing
25. Stabilizer bar

Front suspension components—2004–06 Q45 and 2004 M45 Models

09482_INFI_G0424

f. Open the bleeder and watch the fluid move through the tube. If there are still air bubbles in the fluid when the flow stops, check the fluid level, pressurize the system again and repeat the process.

17. Check and/or adjust the wheel alignment as necessary.

2004–06 Q45 and 2004 M45 Models

1. Before servicing the vehicle, refer to the precautions in the beginning of this section.

2. Remove or disconnect the following:
- Front wheel
- Brake caliper and rotor
- Wheel speed sensor harness
- Brake hose
- Stabilizer connecting rod
- Steering outer socket from strut
- Lower strut end from control arm
- Suspension actuator assemblies on Q45
- Strut damper and bracket
- Strut tower bar and bracket
- Upper mounting insulator bolts
- Strut assembly

3. Secure the strut in a suitable holding fixture.

4. Loosen the piston rod locknut. Do not remove the locknut.

5. Compress the spring with the proper tool so the strut assembly mounting insulator can be turned by hand.

6. Remove the piston rod locknut. Remove the spring assembly, dust cover and rubber seat. Remove the strut insert.

To install:

7. Inspect the rubber parts for deterioration. If the rubber is pulling away from the metal, the mounting insulator should be replaced.

8. Fit the spring into the lower seat and compress the spring.

9. Install the rubber seat, upper seat and mounting insulator.

10. Install or connect the following:
- Piston rod locknut and torque to 44–54 ft. lbs. (59–47 Nm).
- Strut into place and torque the upper mounting nuts to 26–31 ft. lbs. (34–42 Nm).
- Strut tower bar and bracket and torque the bolts to 26–31 ft. lbs. (34–42 Nm).
- Strut damper and bracket and torque the bolts to 11–13 ft. lbs. (15–19 Nm).
- Lower strut end to control arm
- Steering outer socket to strut
- Brake hose
- Wheel speed sensor harness

- Brake caliper and rotor
- Front wheel

11. Check and/or adjust the wheel alignment as necessary.

Stabilizer (Sway) Bar

REMOVAL & INSTALLATION

G20

1. Matchmark all pieces.
2. Remove the fasteners and remove stabilizer bar.
3. When installing stabilizer bar, make sure that paint mark and clamp face in their correct directions.
4. When removing and installing stabilizer bar, hold portion A.
5. Install stabilizer bar with ball joint socket properly placed. See the exploded view for torque values.

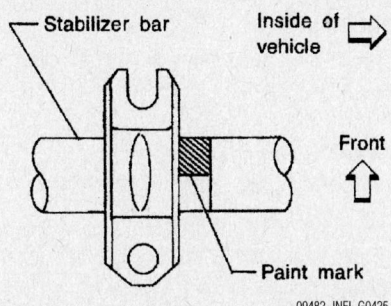

09482_INFI_G0425

Make sure that paint mark and clamp face in their correct directions—G20

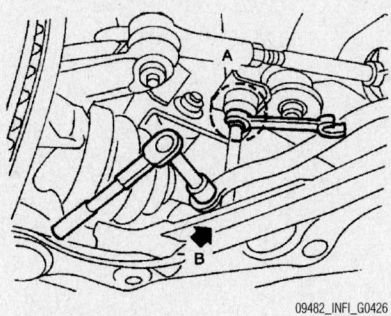

09482_INFI_G0426

Hold portion A—G20

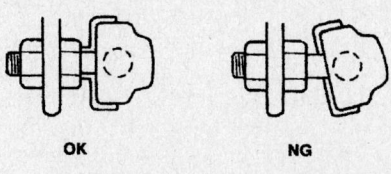

09482_INFI_G0427

Install stabilizer bar with ball joint socket properly placed—G20

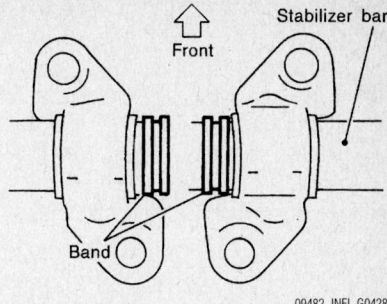

09482_INFI_G0428

Install stabilizer bar with ball joint socket properly placed—I35

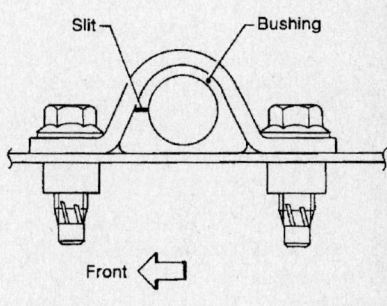

09482_INFI_G0429

Make sure that slit in bushing is in the position shown—I35

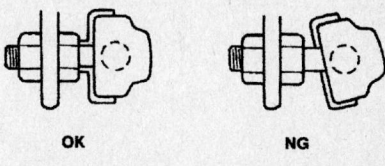

09482_INFI_G0427

Install stabilizer bar with ball joint socket properly placed—I35

I35

1. Remove power steering gear.
2. Remove the fasteners and remove the stabilizer bar.
3. When installing stabilizer, make sure that band and clamp face in their correct directions.
4. Make sure that slit in bushing is in the position shown in the figure.
5. Install stabilizer bar with ball joint socket properly placed. See the exploded view for torque values.

2wd G35

1. Remove the wheel.
2. Remove undercover.
3. Remove mounting nut on upper por-

tion of stabilizer connecting rod with power tool.

4. Remove fixing bolts and nuts, then remove stabilizer clamp, stabilizer bushing, and stabilizer clamp bracket.

5. Remove stabilizer bar from vehicle.

6. Check stabilizer bar, stabilizer connecting rod, stabilizer bushing, stabilizer clamp and stabilizer clamp bracket for deformation, cracks and damage, and replace if necessary.

To install:

7. Refer to the exploded view for tightening torque. Install in the reverse order of removal. Tighten each bolt and nut as shown in the figure.

8. The stabilizer bar uses pillow ball type connecting rod. Position ball joint with case on pillow ball head parallel to stabilizer bar.

AWD G35

1. Remove tire with power tool.

2. Remove undercover with power tool.

3. Remove mounting nut on upper portion of stabilizer connecting rod with power tool.

4. Remove fixing bolt and nut, then remove stabilizer clamp, stabilizer bushing.

5. Remove stabilizer bar from vehicle.

6. Check stabilizer bar, stabilizer connecting rod, stabilizer bushing and clamp for deformation, cracks and damage, and replace if necessary.

To install:

7. Refer to the exploded view for tightening torque. Install in the reverse order of removal.

8. The stabilizer bar uses pillow ball type connecting rod. Position ball joint with case on pillow ball head parallel to stabilizer bar.

Q45 and 2003–04 M45

1. Remove the wheels.

2. Remove undercover.

3. Remove mounting nut on upper position of stabilizer connecting rod with a power tool.

4. Remove stabilizer clamp mounting bolts and nuts with a power tool.

5. Remove stabilizer bar from vehicle.

6. Check stabilizer bar, stabilizer connecting rod, stabilizer bushing and stabilizer clamp for deformation, cracks and damage, and replace if necessary.

To install:

7. Refer to the exploded view for tightening torque. Install in the reverse order of removal.

➡**Refer to component parts location and do not reuse non-reusable parts.**

8. After removing/installing or replacing suspension components, check wheel alignment.

9. After adjusting wheel alignment,

adjust neutral position of steering angle sensor.

10. Stabilizer bar uses pillow ball type connecting rod.

M35 and 2006 M45 2wd

1. Remove tires from vehicle.

2. Remove undercover.

3. Remove the mounting nut on the lower side of stabilizer connecting rod with a power tool, and then remove stabilizer connecting rod from stabilizer bar.

4. If necessary remove the mounting nut on the upper side of stabilizer connecting rod with a power tool, and then remove stabilizer connecting rod from transverse link.

5. Remove the mounting nuts of stabilizer clamp, and then remove stabilizer clamp and stabilizer bushing.

6. Remove stabilizer bar from vehicle.

7. Check stabilizer bar, stabilizer connecting rod, stabilizer bushing and stabilizer clamp for deformation, cracks or damage. Replace it if a malfunction is detected.

8. Installation is the reverse order of removal. For tightening torque, refer to the exploded view.

M35 and 2006 M45 AWD

1. Remove tires from vehicle.

2. Remove undercover.

3. Remove the mounting nut on the lower side of stabilizer connecting rod with a power tool, and then remove stabilizer connecting rod from stabilizer bar.

4. If necessary remove the mounting nut on the upper side of stabilizer connecting rod with a power tool, and then remove stabilizer connecting rod from transverse link.

5. Remove the mounting nuts of stabilizer clamp, and then remove stabilizer clamp and stabilizer bushing.

6. Remove stabilizer bar from vehicle.

7. Check stabilizer bar, stabilizer connecting rod, stabilizer bushing, and stabilizer clamp for deformation, cracks or damage. Replace it if a malfunction is detected.

8. Installation is the reverse order of removal. For tightening torque, refer to the exploded view.

Lower Ball Joint

REMOVAL & INSTALLATION

The lower ball joint assembly is part of the lower control arm/transverse link. If replacement of the ball joint is required, the lower control arm needs to be replaced.

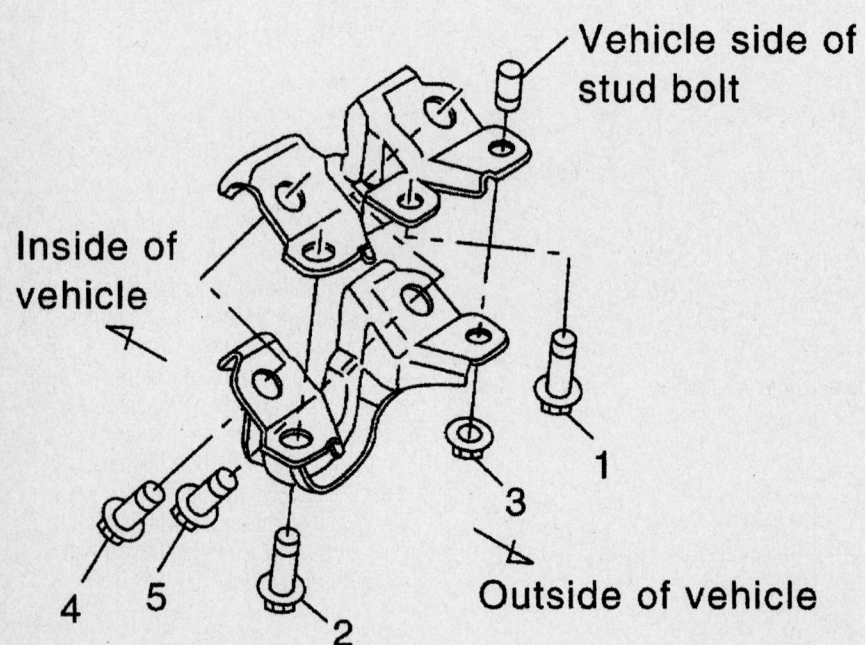

Vehicle side of stud bolt

Inside of vehicle

Outside of vehicle

09482_INFI_G0430

Stabilizer bar clamp torque sequence—2wd G35

Lower Control Arm

REMOVAL AND & INSTALLATION

G20 Models

1. Before servicing the vehicle, refer to the precautions in the beginning of this section.
2. Remove or disconnect the following:
   • Stabilizer bar

➡**Take note of paint mark and clamp position when removing stabilizer bar for correct reinstallation.**

3. Support the steering knuckle with a suitable jack and remove the lower ball joint nut. Separate the ball joint from the knuckle.
   • Bolts attaching the lower control arm to the chassis
   • Lower control arm

To install:

4. If the lower ball joint is worn or damaged, the lower control arm must be replaced. The ball joint is not serviceable separately.
5. Install or connect the following:
   • Lower control arm to the chassis with the attaching bolts and nut
   • Ball joint stud in the knuckle. Torque the nut to 52–64 ft. lbs. (71–86 Nm).
   • Stabilizer bar and wheel
6. Lower the vehicle.

➡**Final tightening must be done with the vehicle at normal ride height, tires on the ground and the chassis loaded.**

7. Torque front control arm bolts to 87–108 ft. lbs. (118–147 Nm) and rear gusset nut to 69–87 ft. lbs. (93–118 Nm).

I35 Models

1. Before servicing the vehicle, refer to the precautions in the beginning of this section.
2. Remove or disconnect the following:
   • Front wheels
   • Anti-lock Brake System (ABS) wheel sensor and move it out of the way
   • Wheel bearing locknut
   • Tie rod from the steering knuckle
   • Bolts attaching the strut to the steering knuckle. Matchmark prior to removal.
   • Halfshaft from the steering knuckle by lightly tapping the end of the shaft
   • Steering knuckle and the lower ball joint

   • Stabilizer bar from the lower control arm
   • Bolts attaching the link bushing pin to the chassis
   • Nut attaching the link to the control arm and remove the link
   • Bolts attaching the compression rod bushing clamp
   • Lower control arm/traverse link

To install:

3. Install or connect the following:
   • Lower control arm and the compression rod bushing clamp into the vehicle
   • Link bushing pin, if removed from the control arm
4. Tighten all bolts and nuts until they are snug enough to support the weight of the vehicle but not fully tight. The bolts should be tightened to specification with the vehicle on the floor.
   • Steering knuckle to the lower control arm and connect the ball joint. Torque the nut to 46–56 ft. lbs. (62–76 Nm).

➡**Always use a new nut when installing the ball joint to the control arm.**

   • Steering knuckle to the strut and to the halfshaft
   • Strut mounting bolts and align the matchmarks. Torque the bolts to 103–117 ft. lbs. (140–159 Nm).
   • Tie rod ball joint and tighten the nut to 46–54 ft. lbs. (63–73 Nm)
   • Wheel bearing locknut
   • ABS wheel sensor. Torque the attaching bolt to 13–17 ft. lbs. (18–24 Nm).
   • Front wheels
5. Lower the vehicle and torque the hub locknut to 174–231 ft. lbs. (235–314 Nm).
6. Torque the bolts attaching the compression rod bushing clamp and the link bushing pin, in the proper sequence to 87–108 ft. lbs. (118–147 Nm).
7. If the link bushing pin was removed from the control arm, torque the attaching nut to 87–108 ft. lbs. (118–147 Nm).
8. Tighten the sway bar attaching nut to 30–35 ft. lbs. (41–47 Nm).
9. Check the vehicle alignment.

G35 and 2002–03 Q45 Models

1. Before servicing the vehicle, refer to the precautions in the beginning of this section.
2. Remove or disconnect the following:
   • Negative battery cable
   • Front wheel

   • Engine undercover, if necessary
   • Nuts securing the tension rod to the transverse link (control arm)
   • Nut and separate the ball joint stud from the knuckle
   • Transverse link from the sub-frame

To install:
3. Install or connect the following:
   • Transverse link on the sub-frame. Temporarily install the bolt and nut.
   • Tension rod on the transverse link. Torque the nuts to 87–94 ft. lbs. (118–127 Nm).
   • Nut on the ball joint stud. Torque the nut to 71–88 ft. lbs. (96–120 Nm) for Q45 models or to 59–69 ft. lbs. (75–94 Nm) for G35 models.
   • Front wheel and lower the vehicle to the floor
   • Transverse link mounting bolt. Torque the bolt to 72–87 ft. lbs. (98–118 Nm).
   • Engine undercover, if necessary
   • Negative battery cable

M45 and 2004 Q45 Models

1. Before servicing the vehicle, refer to the precautions in the beginning of this section.
2. Remove or disconnect the following:
   • Negative battery cable
   • Front wheel
   • Engine undercover, if necessary
   • Nut and separate the ball joint stud from the suspension arm
   • Nuts securing the suspension arm to the front suspension member
   • Suspension arm
3. Installation is the reverse of the removal procedure. Tighten all components, as listed in the rear suspension illustrations.

CONTROL ARM BUSHING REPLACEMENT

The bushing are part of the transverse or suspension arm assembly; if they are defective the whole assembly must be replaced.

Upper Link

REMOVAL & INSTALLATION

See Strut Removal & Installation

Wheel Bearings

ADJUSTMENT

The wheel bearing assemblies on all models are pressed in and are not adjustable. If the bearing assembly does not turn smoothly or has more than 0.002 in. (0.05mm) of axial play, replace the bearing assembly.

REMOVAL & INSTALLATION

G20 Models

1. Before servicing the vehicle, refer to the precautions in the beginning of this section.

2. The axle nut torque is very high and should be loosened and tightened with the vehicle on the ground.

3. Remove the cotter pin, adjusting cap and insulator and loosen the front axle nut.

4. Remove or disconnect the following:
- Brake caliper, carrier, and the rotor. Hang the caliper from the body with wire; do not let it hang by the brake hose.
- Cotter pin and nut and use a ball joint press to disconnect the tie rod end
- Cap and the upper king pin mounting nut and separate the kingpin from the third link

5. Hold a block of wood against the axle stub and strike it with a hammer to

release it from the hub. Withdraw the axle from the hub and fold the steering knuckle down on the ball joint.
- Cotter pin and nut and use a ball joint press to disconnect the ball joint
- Steering knuckle

➡**Wheel bearings must be replaced any time the hub is removed.**

6. Pry the grease seals out of the steering knuckle.

7. Support the steering knuckle and press the hub out of the bearing.

8. Remove the snaprings and press the

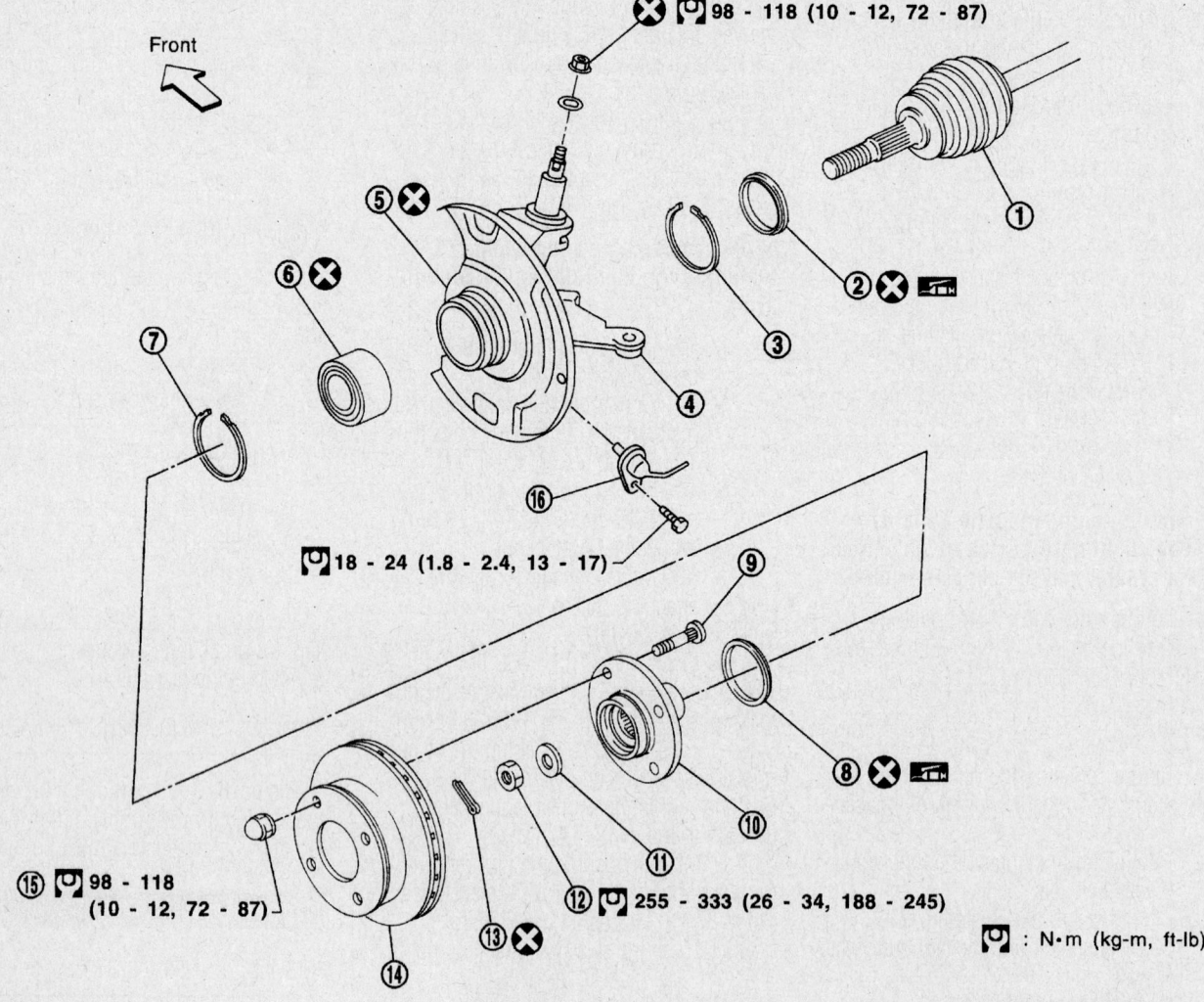

1. Drive shaft	7. Snap ring	12. Wheel bearing lock nut
2. Inner grease seal	8. Outer grease seal	13. Cotter pin
3. Snap ring	9. Hub bolt	14. Brake disc
4. Knuckle	10. Wheel hub	15. Wheel nut
5. Baffle plate	11. Plain washer	16. ABS sensor
6. Wheel bearing assembly		

Front hub and knuckle—G20

09482_INFI_G0431

bearing out towards the inside of the knuckle.

To install:

9. Be sure all parts are clean and dry. The hub and steering knuckle should be inspected for cracks using dye or a magnetic crack detection process.

10. Install or connect the following:

- Inner snapring and carefully press the new bearing into the steering knuckle. Be sure the press tool contacts only the outer bearing race or the bearing will be damaged.
- Outer snapring. Pack the new grease seals with clean grease and install them. If removed, install the splash guard.

11. Support the inner race on the press table and carefully press the hub into the bearing. Be sure the hub turns smoothly in both directions.

- Steering knuckle onto the lower ball joint and start the nut. Fit the axle shaft through the hub and start the nut.

12. Pack the king pin bearing housing with grease and fit the third link into place. Torque the kingpin nut to 72–87 ft. lbs. (98–118 Nm) and install the dust cap.

13. Torque the lower ball joint nut to 52–64 ft. lbs. (71–86 Nm). Install a new cotter pin.

- Tie rod end. Torque the nut to 22–29 ft. lbs. (29–39 Nm). Torque as needed to install a new cotter pin but do not exceed 36 ft. lbs. (49 Nm).
- Brake caliper, carrier, rotor and the wheel

14. Lower the vehicle to the ground.

15. Torque the front axle nut to 174–231

ft. lbs. (235–314 Nm). Install the insulator, adjusting cap and cotter pin.

I35 Models

➡**Whenever the hub or bearing assembly is removed, the wheel bearing assembly must be replaced. Never reuse the old bearing assembly.**

1. Before servicing the vehicle, refer to the precautions in the beginning of this section.

2. Remove the knuckle assembly from the vehicle by separating the ball joint and tie rod end, then removing the retaining hardware securing the knuckle to the strut.

3. Using a shop press and a suitable tool, press the hub with the inner race from the steering knuckle.

4. Using a shop press and a suitable

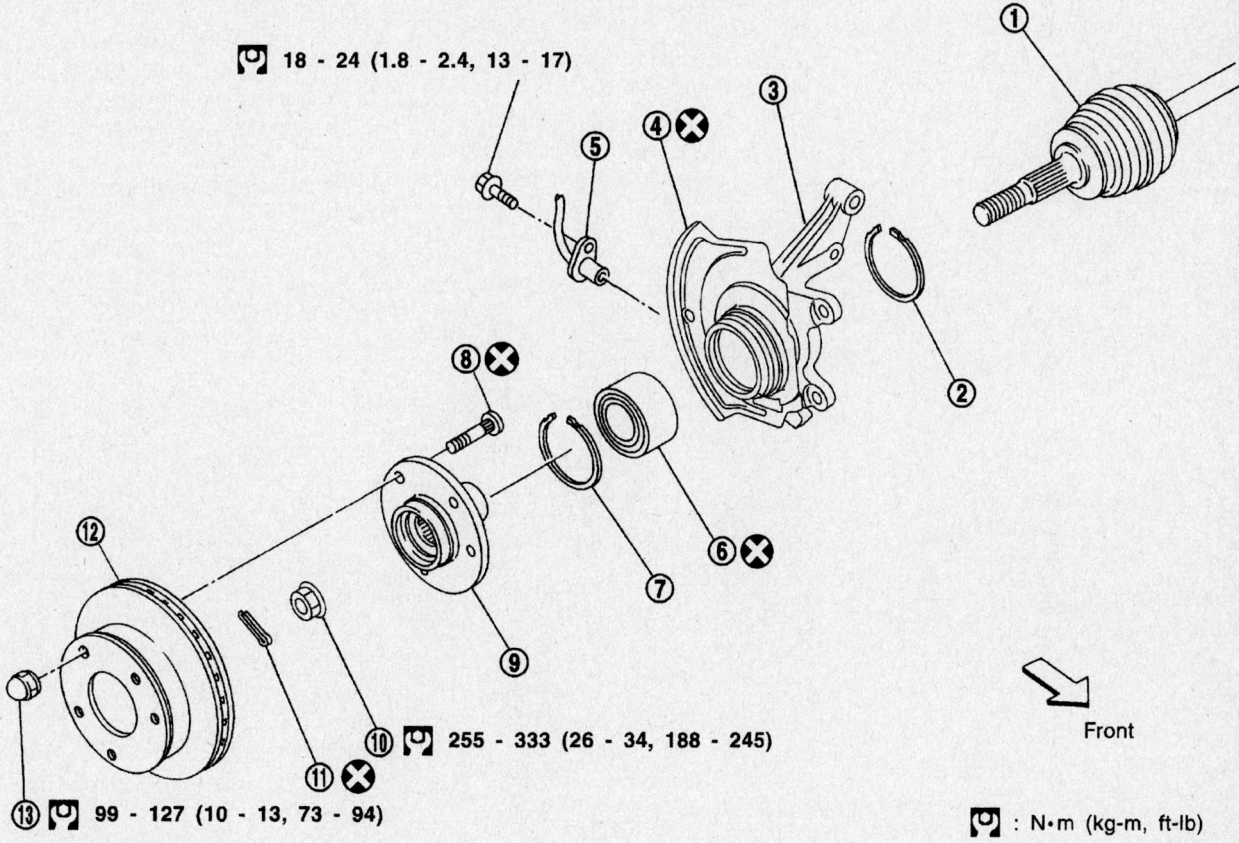

1.	Drive shaft	6.	Wheel bearing assembly	10.	Wheel bearing lock nut
2.	Snap ring	7.	Snap ring	11.	Cotter pin
3.	Knuckle	8.	Hub bolt	12.	Brake disc
4.	Baffle plate	9.	Wheel hub	13.	Wheel nut
5.	ABS sensor				

Component view of the front knuckle assembly—I35

09482_INFI_G0119

tool, press the bearing inner race from the hub and remove the outer grease seal.

5. Use snapring pliers to remove the snaprings from the steering knuckle.

6. Inspect the hub, steering knuckle and snaprings for cracks and/or wear; if necessary, replace the damaged part(s).

To install:

7. Install the inner snapring in the steering knuckle groove.

8. Using a shop press and a suitable tool, press the new wheel bearing assembly into the steering knuckle, until it seats, using a maximum pressure of 3 tons (2722kg).

9. Install the outer snapring.

10. Pack the new grease seal lips with multi-purpose grease.

11. Using a shop press and a suitable tool, press the new outer grease seal into the steering knuckle.

12. Using a shop press and a suitable tool, press the new inner grease seal into the steering knuckle.

13. Using a shop press and a suitable tool, press the hub into the steering knuckle, until it seats, using a maximum pressure of 5.5 tons (4990kg); be careful not to damage the grease seal.

14. To check the bearing operation, perform the following procedures:

   a. Increase the press pressure to 3.5–5.0 tons (3175–4536kg).

   b. Spin the steering knuckle, several turns, in both directions.

   c. Be sure the wheel bearings operate smoothly.

15. If the wheel bearings do not operate smoothly, replace the wheel bearing assembly.

16. Install the knuckle assembly.

17. Install the halfshaft into the hub. Torque the locknut to188–245 ft. lbs. (255–333 Nm) for I35 models.

18. Install the wheel assembly and lower the vehicle.

19. Road test the vehicle and verify proper operation.

G35 and Q45 Models

1. Before servicing the vehicle, refer to the precautions in the beginning of this section.

2. Support the hub assembly with a suitable jack.

3. Remove or disconnect the following:
   - Brake caliper, carrier and rotor. Hang the caliper from the body with wire, do not let it hang by the brake hose.
   - Cotter pins and nuts and use a ball joint press to disconnect the lower ball joint and tie rod end
   - Kingpin lower mounting nut to remove the steering knuckle assembly

➡**Wheel bearings must be replaced any time the hub is removed.**

4. Use a vise or a wheel to hold the hub and remove the hub cap and nut from the back of the hub. Remove the wheel speed sensor rotor.

5. Use a press or large drift pin to press the hub out of the steering knuckle.

6. Remove the snapring and press the bearings and grease seal out of the steering knuckle.

To install:

➡**See the accompanying illustration for torque values.**

7. Be sure all parts are clean. Carefully press the new bearing into the steering knuckle. Be sure the press tool contacts only the outer bearing race or the bearing will be damaged.

8. Install a new grease seal and the

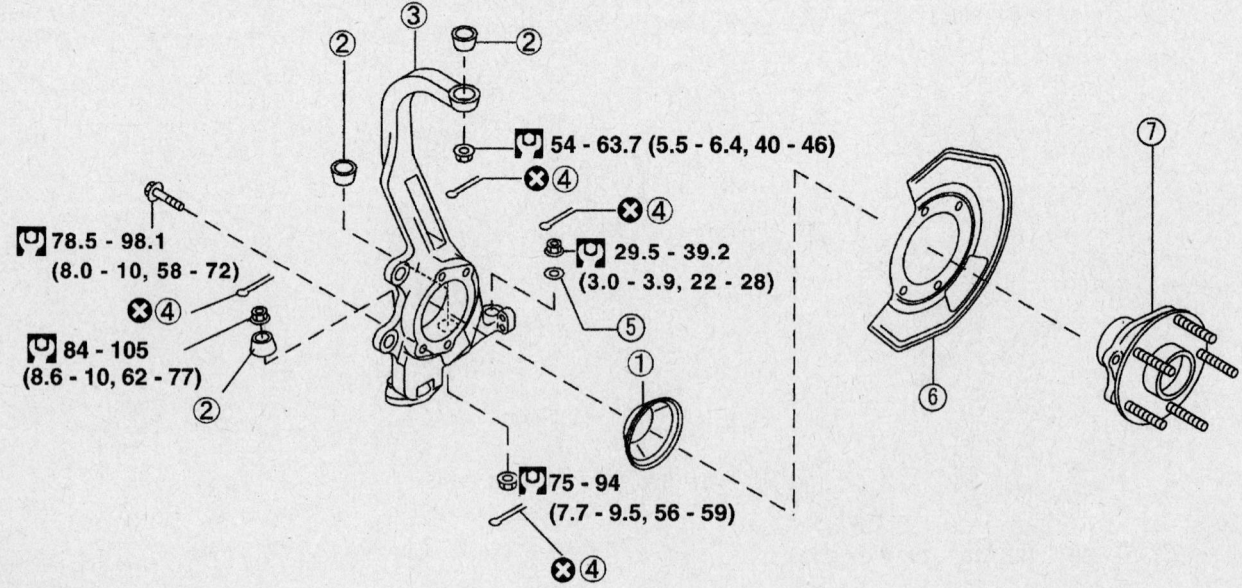

🔧 54 - 63.7 (5.5 - 6.4, 40 - 46)

🔧 78.5 - 98.1 (8.0 - 10, 58 - 72)

🔧 84 - 105 (8.6 - 10, 62 - 77)

🔧 29.5 - 39.2 (3.0 - 3.9, 22 - 28)

🔧 75 - 94 (7.7 - 9.5, 56 - 59)

🔧 : N•m (kg-m, ft-lb)

✖ : Always replace after disassembly

1. Hub cap	2. Ball seat	3. Steering knuckle
4. Cotter pin	5. Washer	6. Splash guard
7. Wheel hub and bearing assembly		

Component view of the front knuckle assembly—2wd G35

09482_INFI_G0432

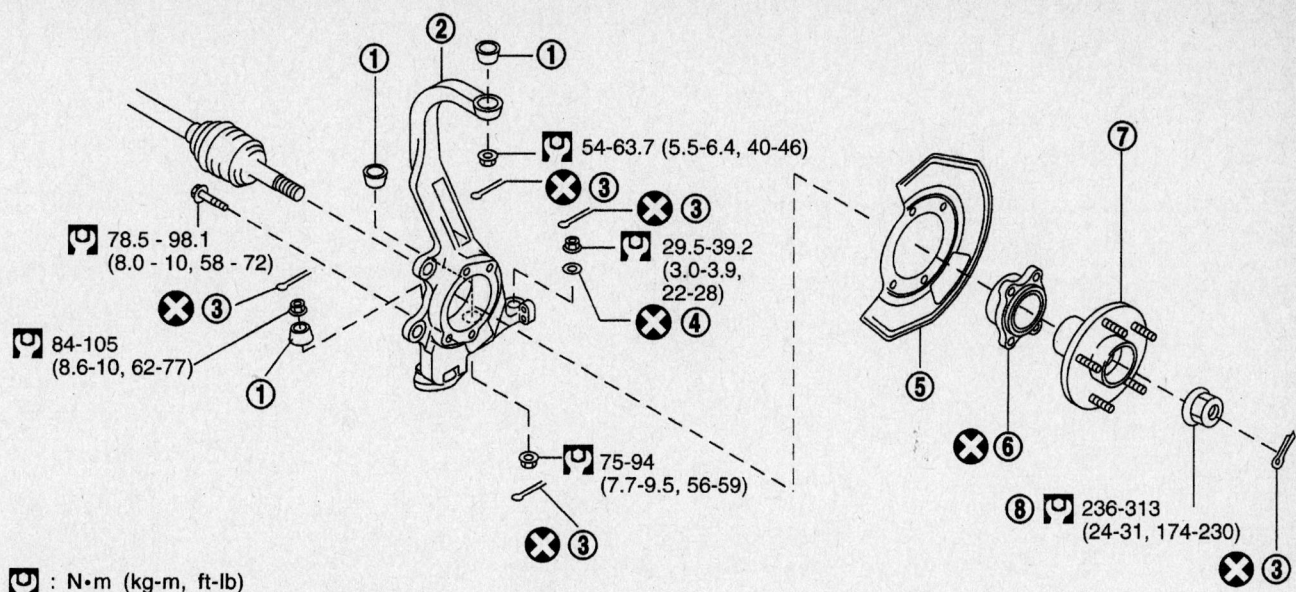

1. Ball seat
4. Washer
7. Wheel hub

2. Steering knuckle
5. Splash guard
8. Wheel bearing lock nut

3. Cotter pin
6. Wheel bearing

⊡ : N•m (kg-m, ft-lb)

✖ : Always replace after every disassembly.

09482_INFI_G0433

Component view of the front knuckle assembly—AWD G35

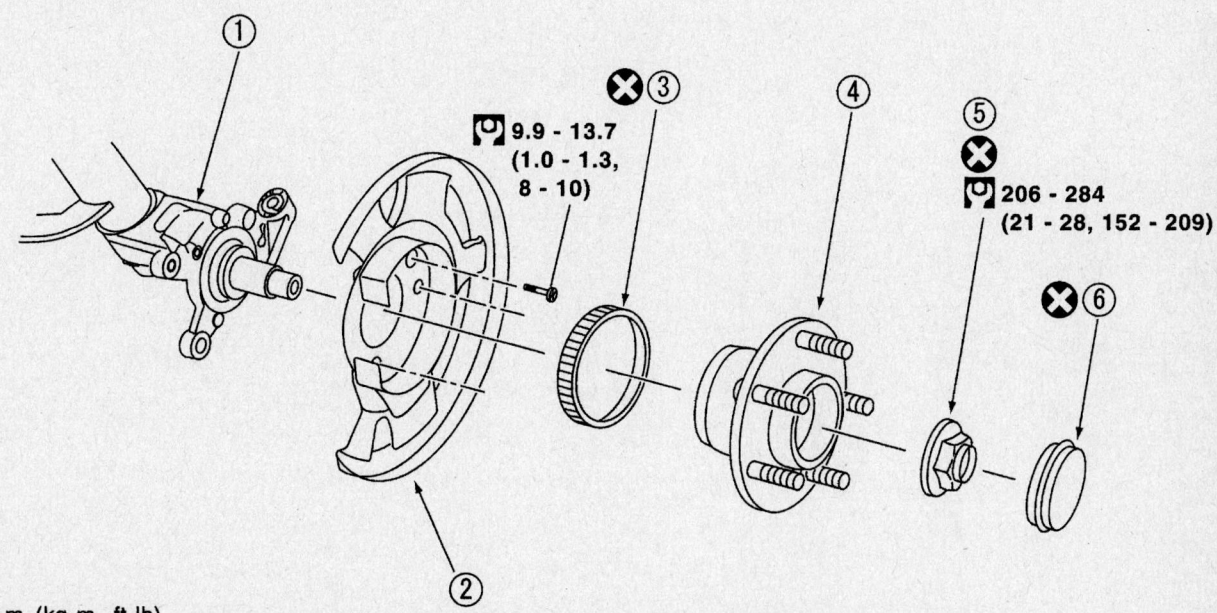

⊡ : N•m (kg-m, ft-lb)

✖ : Always replace after every disassembly

1. Strut assembly
4. Wheel hub and bearing assembly

2. Splash guard
5. Lock nut

3. Sensor rotor
6. Hub cap

09482_INFI_G0434

Component view of the front knuckle assembly—Q45

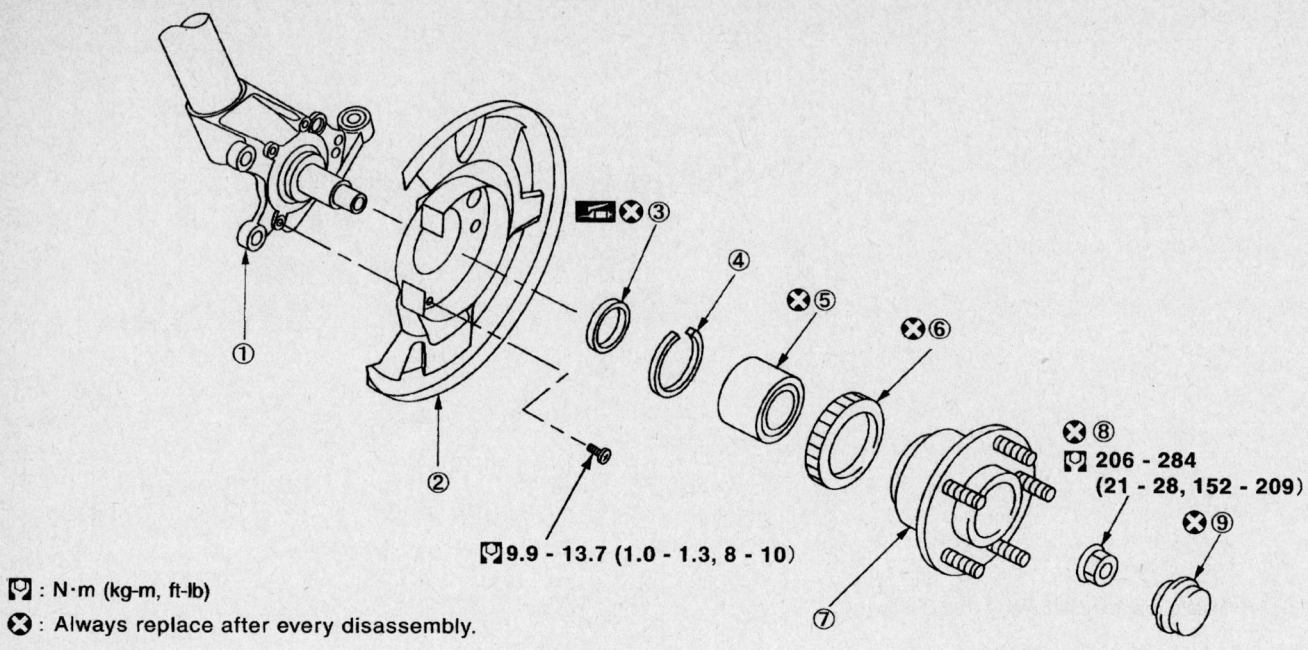

⊡ 9.9 - 13.7 (1.0 - 1.3, 8 - 10)

⊡ 206 - 284
(21 - 28, 152 - 209)

⊡ : N·m (kg-m, ft-lb)

⊗ : Always replace after every disassembly.

1.	Strut assembly	2.	Splash guard	3.	Grease seal
4.	Snap ring	5.	Wheel bearing	6.	ABS sensor rotor
7.	Wheel hub	8.	Lock nut	9.	Hub cap

67162-INFI-G40

Exploded view of the front hub and bearing—2003–04 M45 Models

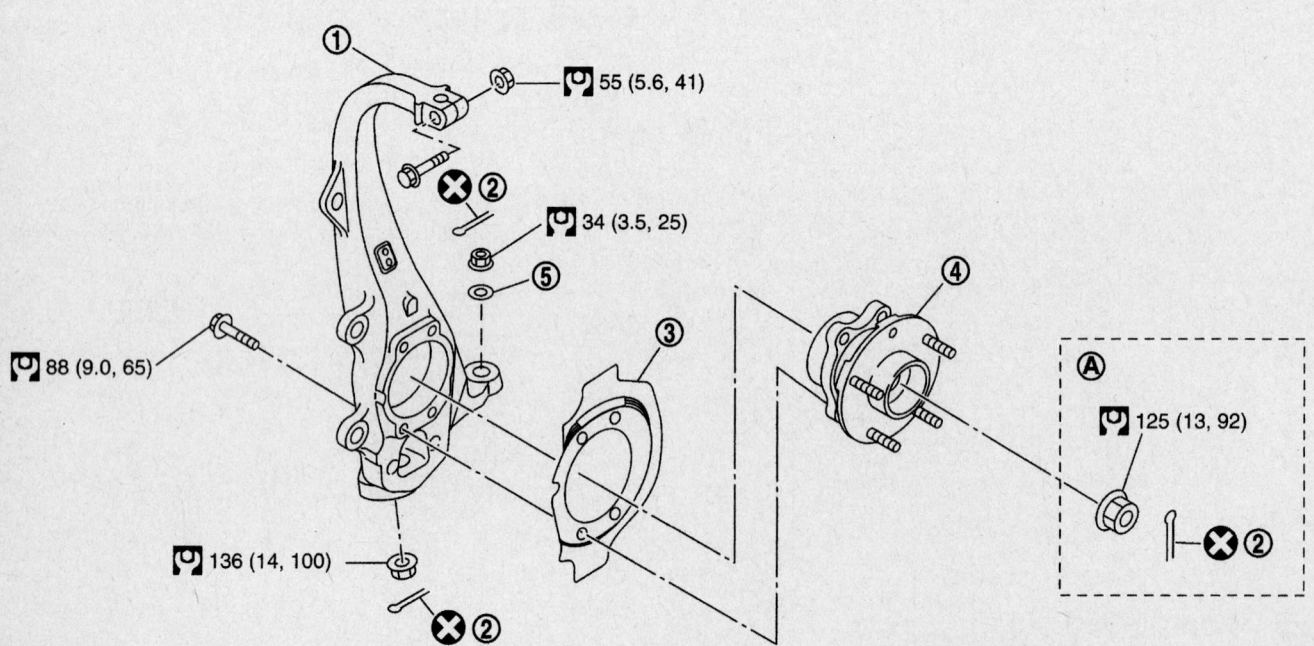

⊡ 55 (5.6, 41)

⊡ 34 (3.5, 25)

⊡ 88 (9.0, 65)

⊡ 125 (13, 92)

⊡ 136 (14, 100)

1.	Steering knuckle	2.	Cotter pin	3.	Splash guard
4.	Wheel hub and bearing assembly	5.	Washer	A.	AWD models

09482_INFI_G0435

Exploded view of the front hub and bearing—M35 and 2006 M45 Models

snapring. If removed, install the splash guard.

9. Lightly lubricate the lips of the seal with clean grease. Be careful not to grease the bearing or hub mating surfaces.

10. Carefully press the hub into the bearing. Support the inner race on the press table or the bearing will be damaged. Do not exceed 3.9 tons (3538kg) pressure.

11. Install or connect the following:
- Speed sensor rotor and nut on the hub. Stake the nut into place.

12. Lightly tap the cap into place and install the bolts.
- Steering knuckle to the king pin.
- Lower ball joint. Install a new cotter pin.
- Tie rod end. Torque as needed to

install a new cotter pin but do not exceed 36 ft. lbs. (49 Nm).
- Rotor and the brake caliper
- Wheel and tire assembly

M35 and M45 Models

1. Before servicing the vehicle, refer to the precautions in the beginning of this section.

2. Support the hub assembly with a suitable jack.

3. Remove or disconnect the following:
- Brake caliper and rotor. Hang the caliper from the body with wire, do not let it hang by the brake hose.
- Grease cap and axle nut
- Wheel hub and bearing assembly
- Pry the grease seal from the hub
- Use a puller, drift and bearing

replacer and press the ABS sensor ring off the hub
- Use a drift and press the wheel bearing from the hub

To install:

4. Be sure all parts are clean. Carefully press the new bearing into the wheel hub. Be sure the press tool does not contact the inner bearing race. Do not exceed 3.6 tons (3000kg) pressure.

5. Press the new ABS sensor ring onto the hub.

6. Install a new snapring.

7. Press the grease seal into the hub. Do not exceed 2.2 tons (1000kg) pressure.

8. The remainder of the installation is the reverse of the removal procedure. Tighten all components, as listed in the wheel hub illustration.

REAR SUSPENSION

Strut (Shock Absorber)

REMOVAL & INSTALLATION

G20 Models

1. Before servicing the vehicle, refer to the precautions in the beginning of this section.

2. Remove or disconnect the following:
- Wheels

※ WARNING

Be sure to disconnect the Anti-lock Brake System (ABS) wheel sensor from the assembly. Failure to do so may result in damage to the sensor wire and the sensor becoming inoperative.

- Brake calipers and suspend them with a piece of wire. Do not let them hang by the hose.

3. Using a transmission jack, raise the torsion beam slightly
- Strut lower mounting bolt

4. Open the trunk, remove the trim and remove the two nuts attaching the strut to the vehicle.
- Strut

※ CAUTION

Do not remove the center locknut from the strut assembly until the strut is safely compressed.

5. Place the strut assembly in a vise with the special holding tool HT71780000 or in a spring compressor.

6. Loosen the piston rod locknut.

※ CAUTION

Do not remove the piston rod locknut, the spring is under tension and can cause serious personal injury.

7. Compress the spring with the spring compressor then remove the piston rod locknut.

➡**Before removing the strut from the coil spring, note the positioning of the strut in relationship to the coil spring for reassembly.**

8. Remove or disconnect the following:
- Bushing, strut mounting bracket, and the upper spring seat rubber
- Strut, leaving the coil spring compressed
- Bushing, bound bumper cover and the bound bumper

To install:

9. Install or connect the following:
- Bound bumper, bound bumper cover and the bushing
- Strut into the coil spring, make sure the strut and spring are properly positioned
- Upper spring seat rubber, strut mounting bracket, and the bushing. Make sure that the mounting bracket is properly positioned.
- Piston rod locknut then remove the

spring compressor. Torque the locknut to 13–17 ft. lbs. (18–24 Nm).
- Strut with new attaching nuts. Torque the nuts to 14–16 ft. lbs. (19–22 Nm).
- Strut on the rear torsion beam. Torque the bolt to 80–94 ft. lbs. (108–127 Nm).

10. Remove the support from the rear torsion beam.

11. Install the rear wheels and lower the vehicle.

12. Check the vehicle's alignment and adjust as necessary.

I35 Models

1. Before servicing the vehicle, refer to the precautions in the beginning of this section.

2. Remove or disconnect the following:
- Rear wheels

3. Support the rear torsion beam assembly with a jack.

※ CAUTION

Do not remove the center locknut from the strut assembly until the strut is safely compressed.

- 2 nuts attaching the strut to the vehicle located inside the trunk
- Bolt attaching the strut to the rear torsion beam assembly and remove the strut

4. Place the strut assembly in a vise

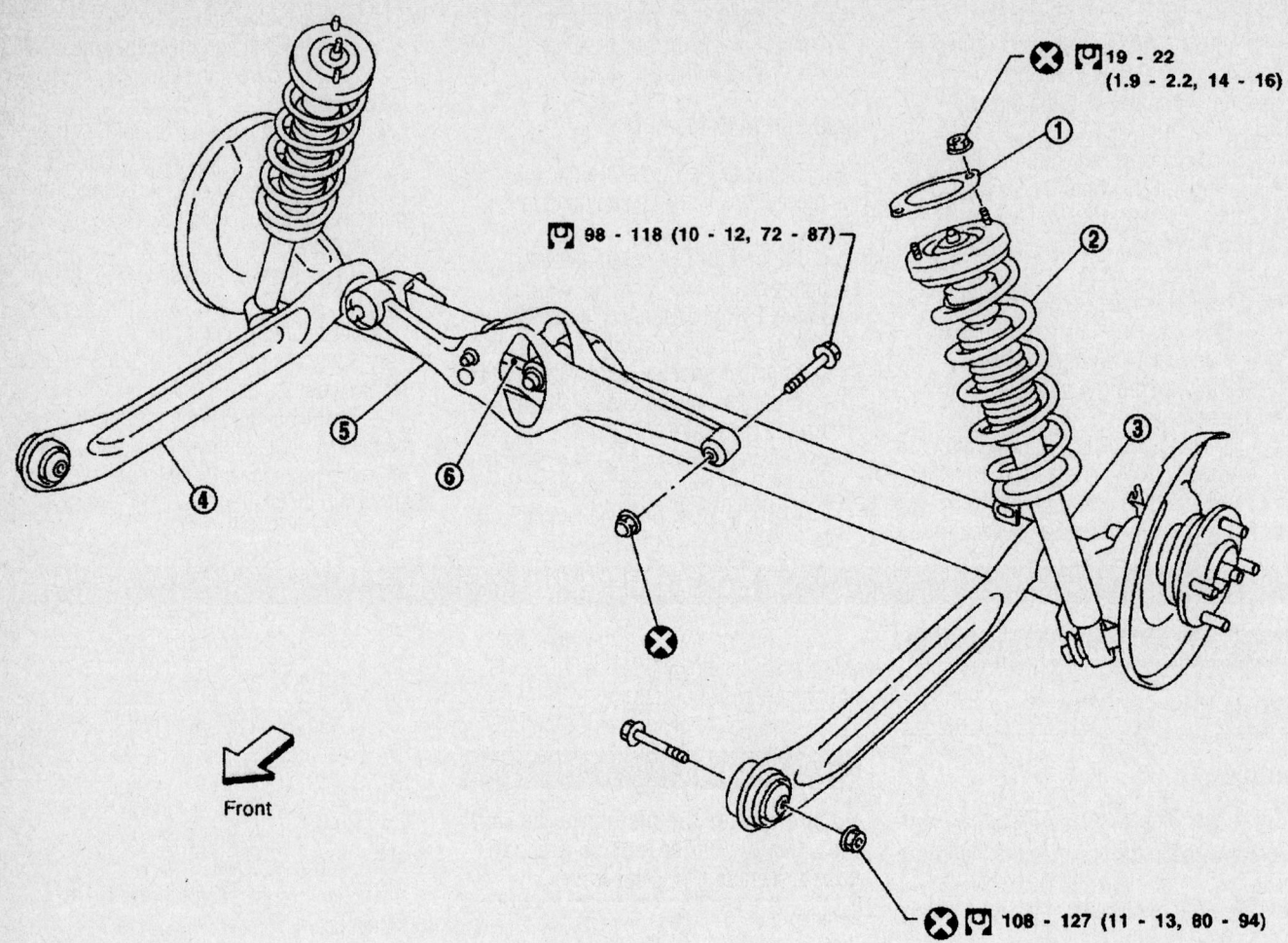

⊗ 🔧 19 - 22
(1.9 - 2.2, 14 - 16)

①

🔧 98 - 118 (10 - 12, 72 - 87)

②

③

⑤

⑥

④

⊗

⊗ 🔧 108 - 127 (11 - 13, 80 - 94)

Front

🔧 : N•m (kg-m, ft-lb)

1.   Gasket
2.   Coil spring

3.   Shock absorber
4.   Torsion beam

5.   Lateral link
6.   Control rod

09482_INFI_G0436

Component view of the rear suspension—G20 models

with the special holding tool HT71780000 or in a spring compressor.
• Piston rod locknut

※※ CAUTION

Do not remove the piston rod locknut, the spring is under tension and can cause serious personal injury.

5.   Compress the spring with the spring compressor, then remove the piston rod locknut.

➡**Before removing the strut from the coil spring, note the positioning of the strut in relationship to the coil spring for reassembly.**

6.   Remove or disconnect the following:

• Bushing, strut mounting bracket, and the upper spring seat rubber
• Strut, leaving the coil spring compressed
• Bushing, bound bumper cover, and the bound bumper

To install:

7.   Install or connect the following:

• Bound bumper, bound bumper cover, and the bushing
• Strut into the coil spring, be sure the strut and spring are properly positioned
• Upper spring seat rubber, strut mounting bracket, and the bushing.

Be sure that the mounting bracket is properly positioned.

• Piston rod locknut, then remove the spring compressor. Torque the locknut to 13–17 ft. lbs. (18–24 Nm)
• Strut with new attaching nuts. Torque the nuts to 12–16 ft. lbs. (16–22 Nm).
• Strut on the rear torsion beam. Torque the bolt to 80–94 ft. lbs. (108–127Nm).
• Support from the rear torsion beam
• Rear wheels and lower the vehicle

8.   Check the vehicle's alignment and adjust as necessary.

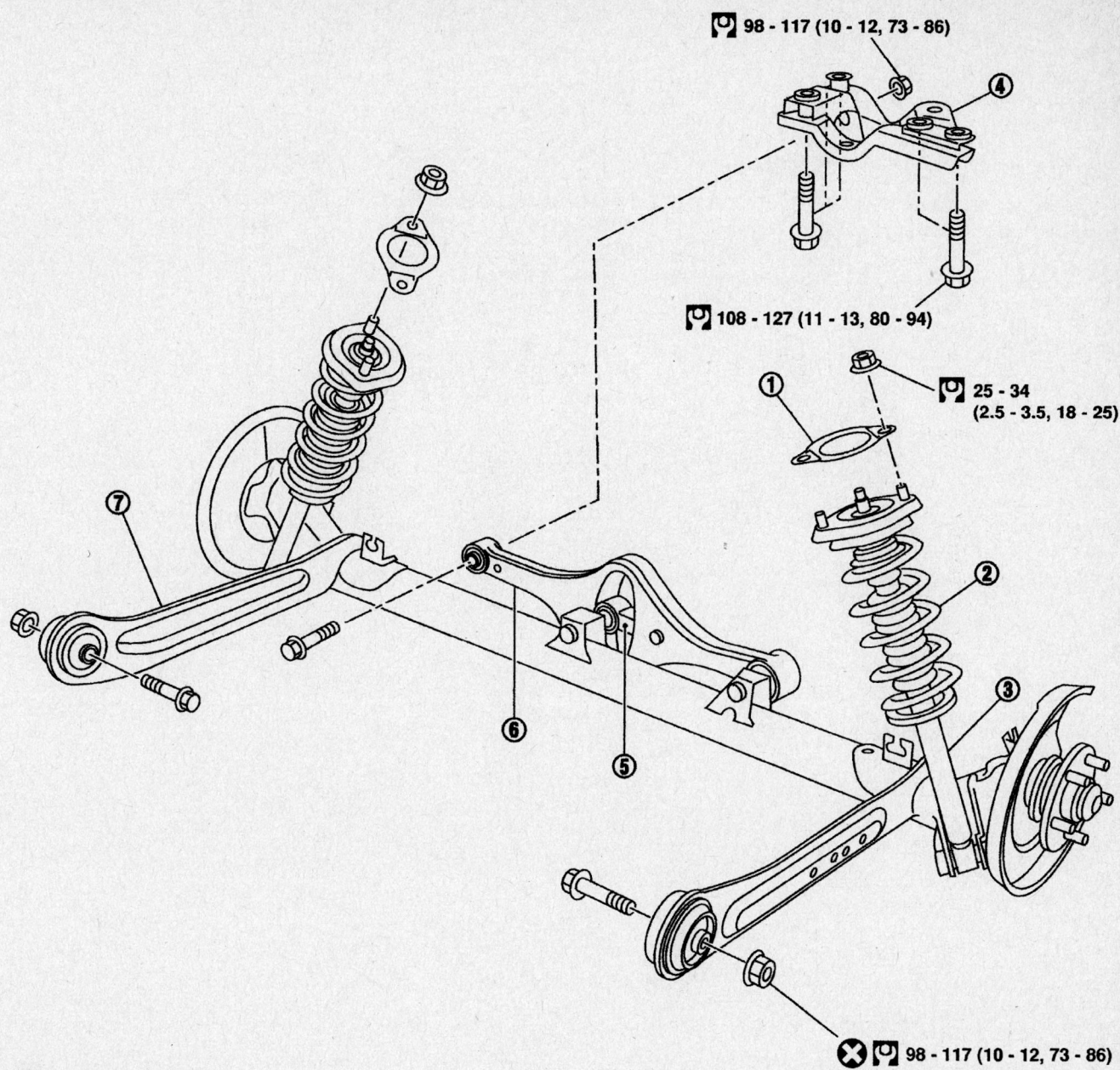

98 - 117 (10 - 12, 73 - 86)

108 - 127 (11 - 13, 80 - 94)

25 - 34
(2.5 - 3.5, 18 - 25)

98 - 117 (10 - 12, 73 - 86)

: N•m (kg-m, ft-lb)

1. Shock absorber mounting seal
2. Coil spring
3. Shock absorber
4. Suspension member
5. Control rod
6. Lateral link
7. Torsion beam

09482_INFI_G0437

Component view of the rear suspension—I35 models

2002–04 Q45 Models with Standard Suspension

✳✳ CAUTION

Do not remove piston rod locknut with the shock absorber on vehicle.

1. Before servicing the vehicle, refer to the precautions in the beginning of this section.

2. Remove the upper strut mounting nuts.
3. Raise and safely support the vehicle and remove the lower mounting bolt. Remove coil spring/strut absorber assembly.
4. Place the assembly into a suitable holding fixture and matchmark the spring, strut and upper seat. Loosen but do not remove the piston rod locknut.
5. Install a spring compressor and com-

press the spring until the upper spring seat can be turned by hand.
6. Remove the locknut, spring seat components, spring, bushings and bumper.
To install:
7. Fit the bumper, spring, upper seat and other components onto the strut with the matchmarks aligned. The top of the spring is flat.

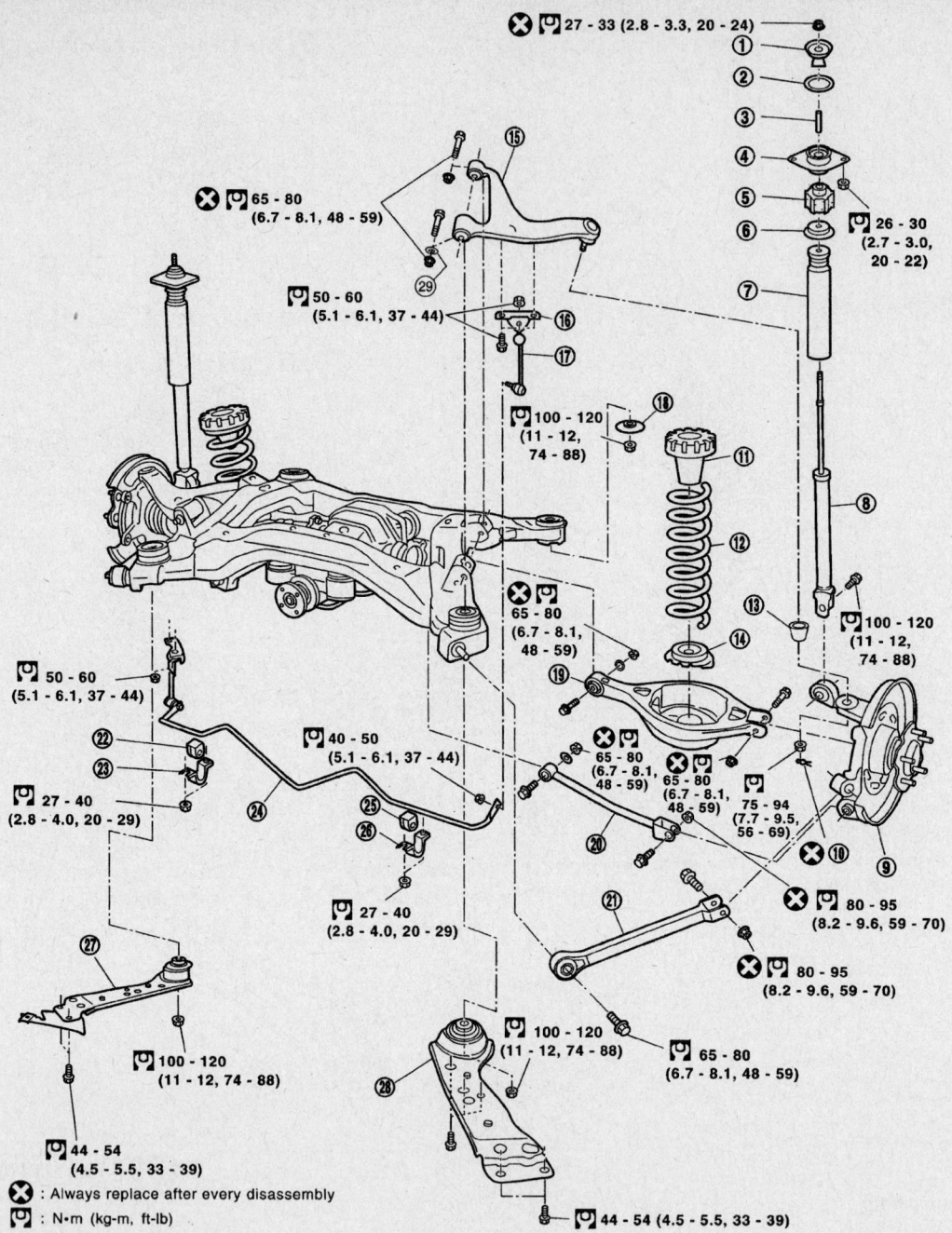

- ⊗ 🔧 27 - 33 (2.8 - 3.3, 20 - 24)
- ①
- ②
- ③
- ④
- ⑤
- ⑥
- 🔧 26 - 30 (2.7 - 3.0, 20 - 22)
- ⑦
- ⊗ 🔧 65 - 80 (6.7 - 8.1, 48 - 59)
- ⑮
- 29
- 🔧 50 - 60 (5.1 - 6.1, 37 - 44)
- ⑯
- ⑰
- 🔧 100 - 120 (11 - 12, 74 - 88)
- ⑱
- ⑪
- ⑫
- ⑧
- ⑬
- ⑭
- 🔧 100 - 120 (11 - 12, 74 - 88)
- ⊗ 🔧 65 - 80 (6.7 - 8.1, 48 - 59)
- 🔧 50 - 60 (5.1 - 6.1, 37 - 44)
- ⑲
- ㉒
- ㉓
- 🔧 40 - 50 (5.1 - 6.1, 37 - 44)
- ⊗ 🔧 65 - 80 (6.7 - 8.1, 48 - 59)
- ⊗ 🔧 65 - 80 (6.7 - 8.1, 48 - 59)
- 🔧 75 - 94 (7.7 - 9.5, 56 - 69)
- ⑨
- ⑩
- 🔧 27 - 40 (2.8 - 4.0, 20 - 29)
- ㉔
- ㉕
- ㉖
- ⑳
- ㉑
- ⊗ 🔧 80 - 95 (8.2 - 9.6, 59 - 70)
- ⊗ 🔧 80 - 95 (8.2 - 9.6, 59 - 70)
- ㉗
- 🔧 27 - 40 (2.8 - 4.0, 20 - 29)
- 🔧 100 - 120 (11 - 12, 74 - 88)
- 🔧 65 - 80 (6.7 - 8.1, 48 - 59)
- ㉘
- 🔧 100 - 120 (11 - 12, 74 - 88)
- 🔧 44 - 54 (4.5 - 5.5, 33 - 39)
- ⊗ : Always replace after every disassembly
- 🔧 : N•m (kg-m, ft-lb)
- 🔧 44 - 54 (4.5 - 5.5, 33 - 39)

1.	Washer	2.	Shock absorber mounting seal	3.	Distance tube
4.	Shock absorber mounting insulator	5.	Bushing	6.	Bound bumper cover
7.	Bound bumper	8.	Shock absorber	9.	Axle assembly
10.	Cotter pin	11.	Upper seat	12.	Coil spring
13.	Ball seat	14.	Rubber seat	15.	Suspension arm
16.	Connecting rod mounting bracket	17.	Connecting rod	18.	Mount stopper
19.	Rear lower link	20.	Front lower link	21.	Radius rod
22.	Bushing	23.	Clamp	24.	Stabilizer bar
25.	Bushing	26.	Clamp	27.	Member stay
28.	Member stay	29.	Stopper rubber		

09482_INFI_G0107

Component view of the rear suspension—G35 models

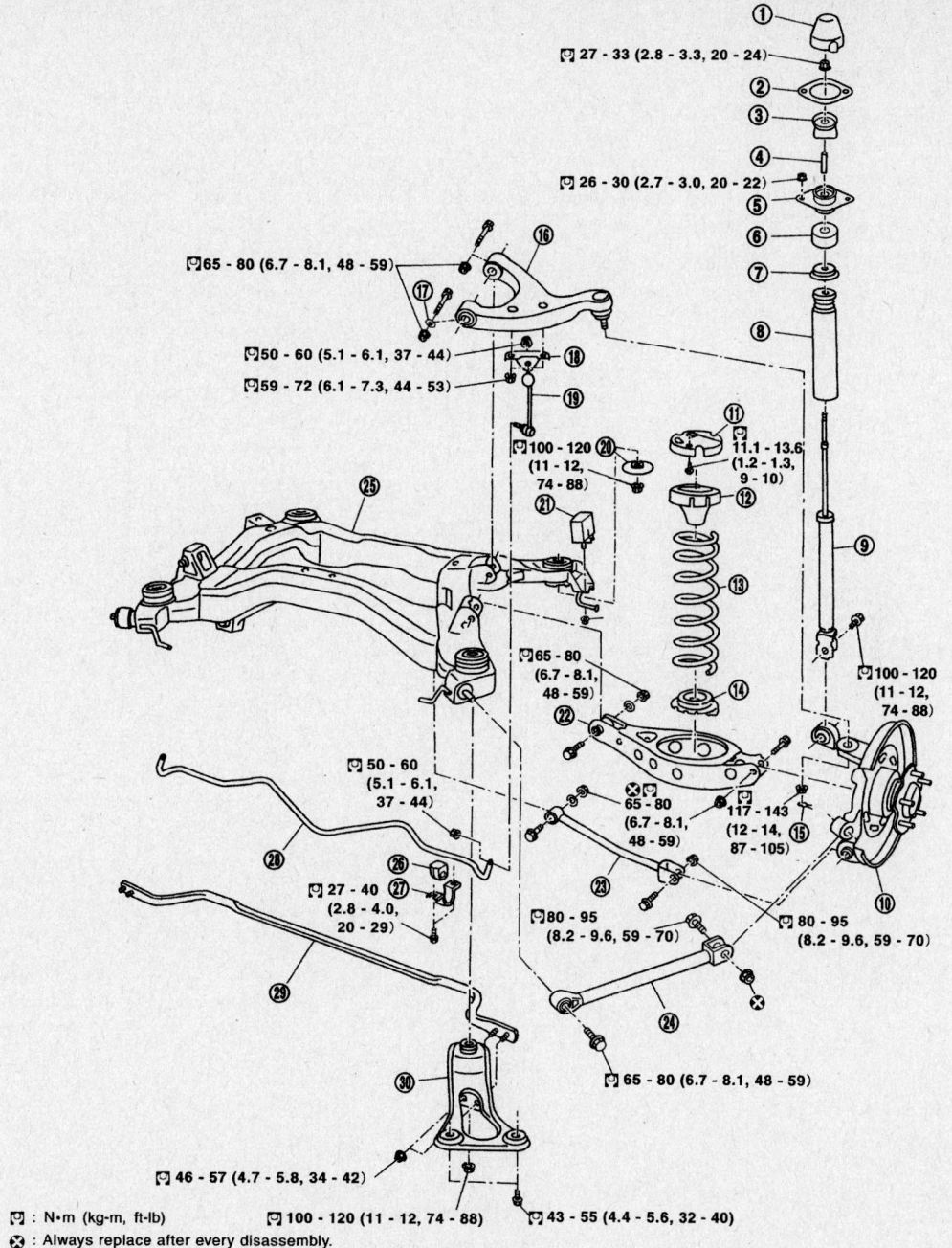

27 - 33 (2.8 - 3.3, 20 - 24)
26 - 30 (2.7 - 3.0, 20 - 22)
65 - 80 (6.7 - 8.1, 48 - 59)
50 - 60 (5.1 - 6.1, 37 - 44)
59 - 72 (6.1 - 7.3, 44 - 53)
100 - 120 (11 - 12, 74 - 88)
11.1 - 13.6 (1.2 - 1.3, 9 - 10)
65 - 80 (6.7 - 8.1, 48 - 59)
100 - 120 (11 - 12, 74 - 88)
50 - 60 (5.1 - 6.1, 37 - 44)
65 - 80 (6.7 - 8.1, 48 - 59)
117 - 143 (12 - 14, 87 - 105)
27 - 40 (2.8 - 4.0, 20 - 29)
80 - 95 (8.2 - 9.6, 59 - 70)
80 - 95 (8.2 - 9.6, 59 - 70)
65 - 80 (6.7 - 8.1, 48 - 59)
46 - 57 (4.7 - 5.8, 34 - 42)

: N·m (kg-m, ft-lb)   100 - 120 (11 - 12, 74 - 88)   43 - 55 (4.4 - 5.6, 32 - 40)
: Always replace after every disassembly.

1. Cap	2. Shock absorber mounting seal	3. Bushing
4. Distance tube	5. Shock absorber mounting bracket	6. Bushing
7. Bound bumper cover	8. Bound bumper	9. Shock absorber
10. Axle assembly	11. Bracket	12. Upper seat
13. Coil spring	14. Rubber seat	15. Cotter pin
16. Suspension arm	17. Stopper rubber	18. Stabilizer connecting rod mounting bracket
19. Stabilizer connecting rod	20. Mount stopper	21. Dynamic dumper
22. Rear lower link	23. Front lower link	24. Radius rod
25. Rear suspension member	26. Stabilizer bushing	27. Stabilizer clamp
28. Stabilizer bar	29. Cross bar	30. Member stay

09482_INFI_G0108

Component view of the rear suspension—2003–04 M45 models

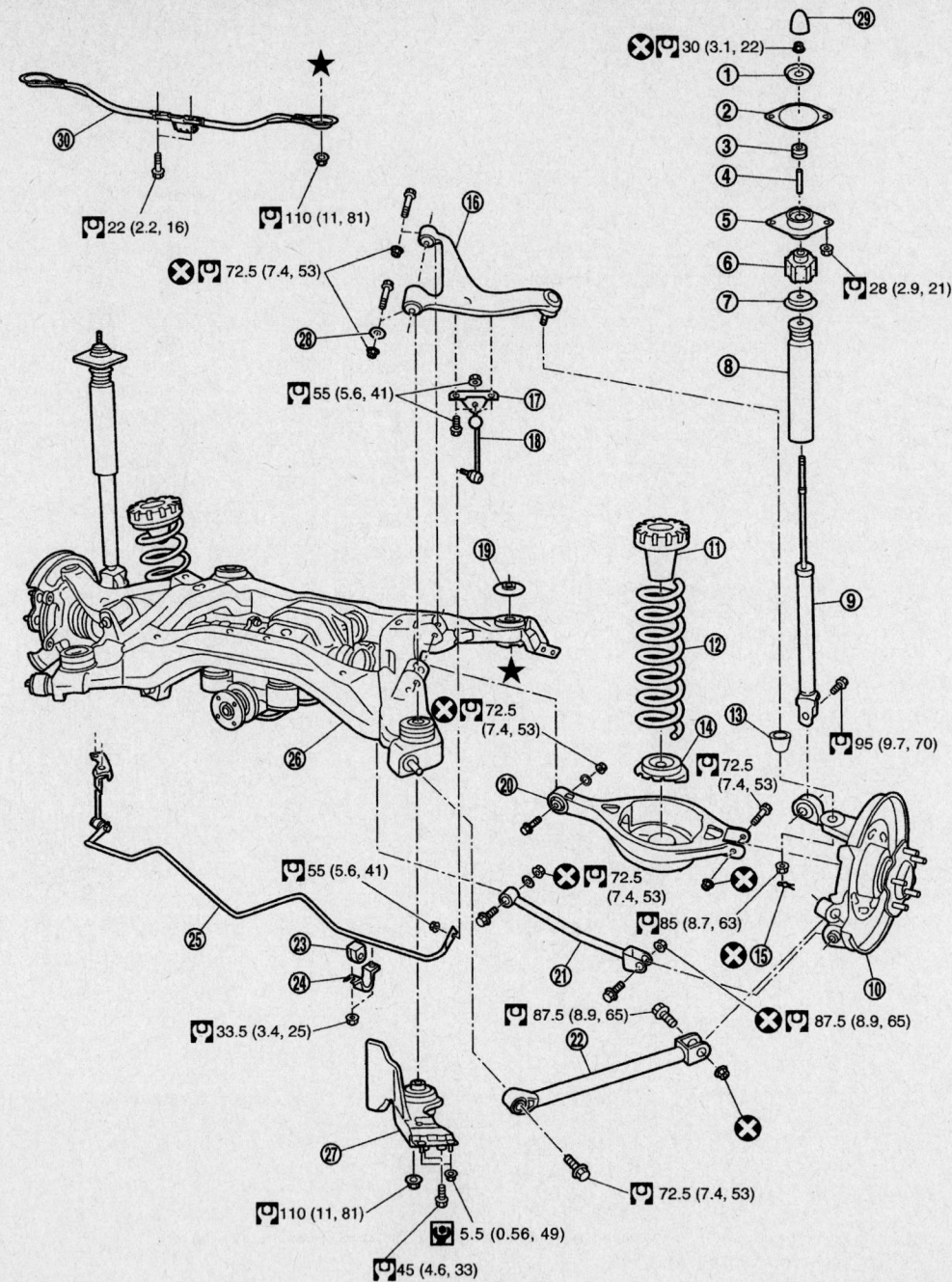

1. Washer	2. Mounting seal	3. Bushing (upper side)
4. Distance tube	5. Mounting bracket	6. Bushing (lower side)
7. Bound bumper cover	8. Bound bumper	9. Shock absorber
10. Axle assembly	11. Upper seat	12. Coil spring
13. Ball seat	14. Rubber seat	15. Cotter pin
16. Suspension arm	17. Connecting rod mounting bracket	18. Connecting rod
19. Mount stopper	20. Rear lower link	21. Front lower link
22. Radius rod	23. Stabilizer Bushing	24. Stabilizer Clamp
25. Stabilizer bar	26. Rear suspension member	27. Member stay
28. Stopper rubber	29. Cap	30. Rear pin stay

09482_INFI_G0109

Component view of the rear suspension—M35 and 2006 M45 models

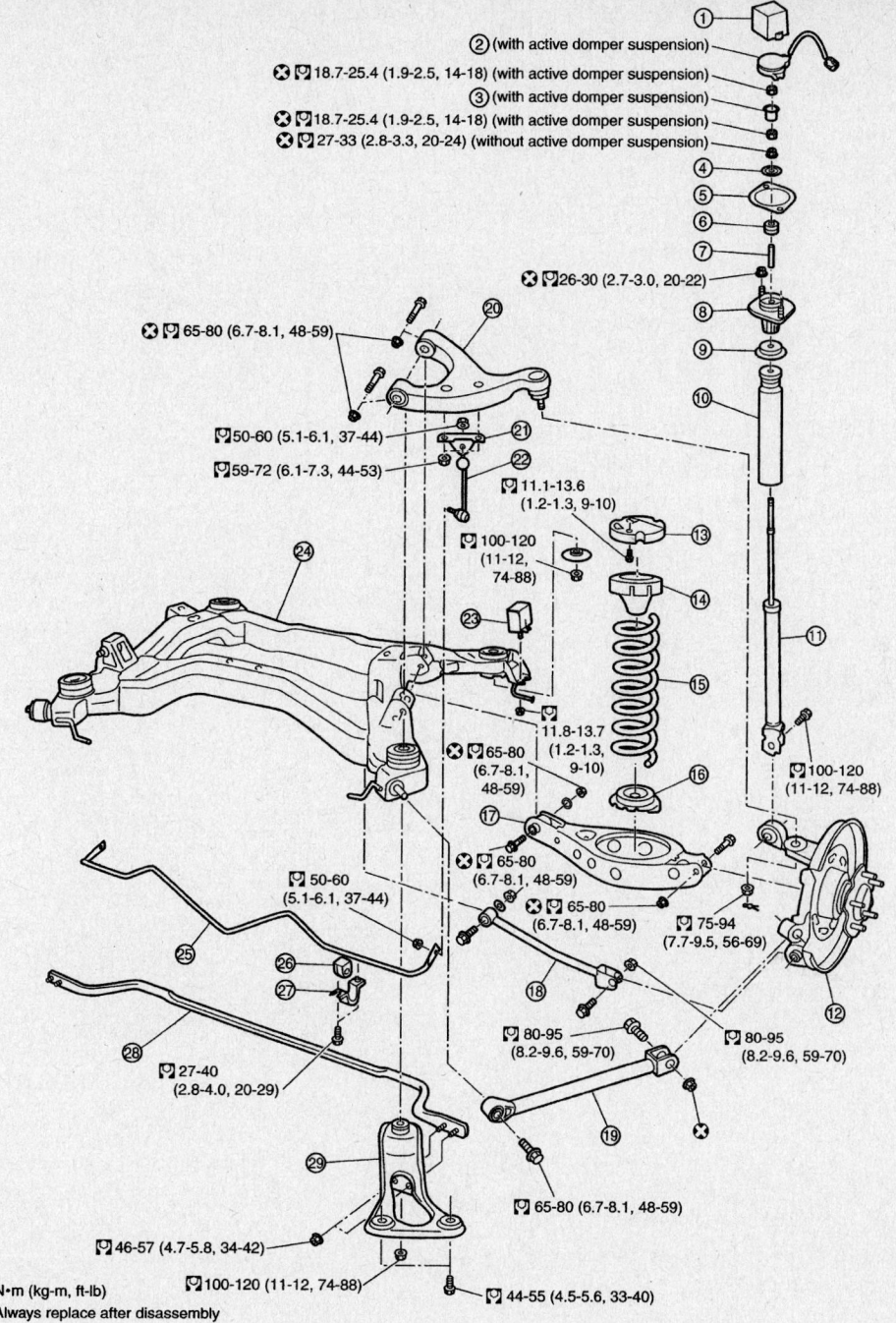

① (with active damper suspension)
② (with active damper suspension)
⊗ ⚲18.7-25.4 (1.9-2.5, 14-18) (with active damper suspension)
③ (with active damper suspension)
⊗ ⚲18.7-25.4 (1.9-2.5, 14-18) (with active damper suspension)
⊗ ⚲27-33 (2.8-3.3, 20-24) (without active damper suspension)
④
⑤
⑥
⑦
⊗ ⚲26-30 (2.7-3.0, 20-22)
⑧
⑨
⑩

⊗ ⚲65-80 (6.7-8.1, 48-59)   ⑳
⚲50-60 (5.1-6.1, 37-44)   ㉑
⚲59-72 (6.1-7.3, 44-53)   ㉒   ⚲11.1-13.6 (1.2-1.3, 9-10)
⑬
⚲100-120 (11-12, 74-88)   ⑭
㉔   ㉓   ⑮
⊗ ⚲65-80 (6.7-8.1, 48-59)   ⚲11.8-13.7 (1.2-1.3, 9-10)
⑯
⑰
⊗ ⚲65-80 (6.7-8.1, 48-59)   ⑪
⚲50-60 (5.1-6.1, 37-44)   ⊗ ⚲65-80 (6.7-8.1, 48-59)   ⚲100-120 (11-12, 74-88)
⑫
⚲75-94 (7.7-9.5, 56-69)
㉕   ㉖   ⑱
㉗
⚲80-95 (8.2-9.6, 59-70)   ⚲80-95 (8.2-9.6, 59-70)
㉘   ⑲   ⊗
⚲27-40 (2.8-4.0, 20-29)
㉙   ⊗ ⚲65-80 (6.7-8.1, 48-59)
⚲46-57 (4.7-5.8, 34-42)
⚲:N•m (kg-m, ft-lb)   ⚲100-120 (11-12, 74-88)   ⚲44-55 (4.5-5.6, 33-40)
⊗:Always replace after disassembly

1.	Cap	2.	Actuator assembly	3.	Actuator plate
4.	Washer	5.	Shock absorber mounting seal	6.	Bushing
7.	Distance tube	8.	Shock absorber mounting bracket	9.	Bound bumper cover
10.	Bound bumper	11.	Shock absorber	12.	Axle
13.	Bracket	14.	Upper seat	15.	Coil spring
16.	Rubber seat	17.	Rear lower link	18.	Front lower link
19.	Radius rod	20.	Suspension arm	21.	Stabilizer connecting rod mounting bracket
22.	Stabilizer connecting rod	23.	Dynamic damper	24.	Rear suspension member
25.	Stabilizer bar	26.	Stabilizer bushing	27.	Stabilizer clamp
28.	Cross bar	29.	Member stay		

09482_INFI_G0110

Exploded view of the rear suspension—2002–04 Q45 models

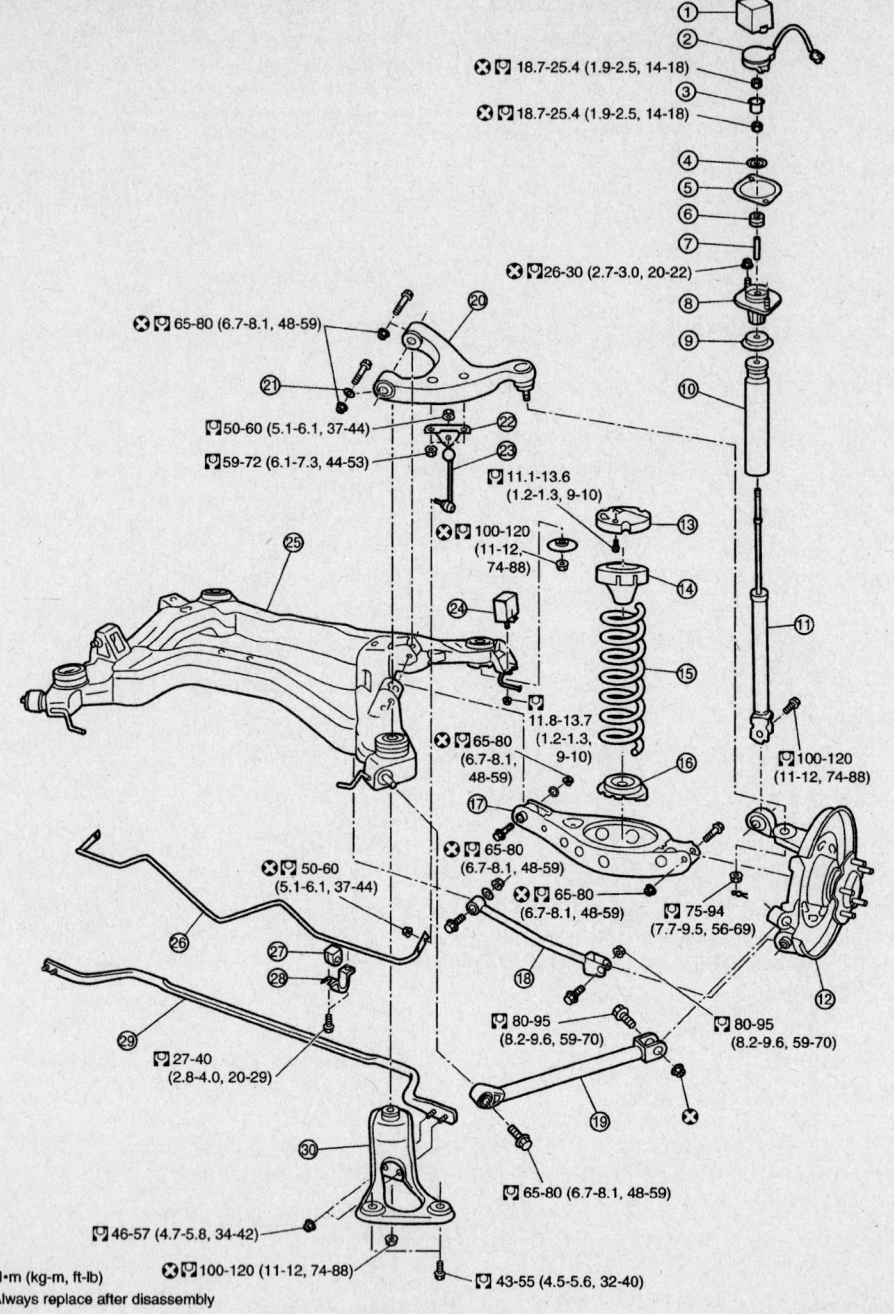

❌ 🔧 18.7-25.4 (1.9-2.5, 14-18)

❌ 🔧 18.7-25.4 (1.9-2.5, 14-18)

❌ 🔧 26-30 (2.7-3.0, 20-22)

❌ 🔧 65-80 (6.7-8.1, 48-59)

🔧 50-60 (5.1-6.1, 37-44)

🔧 59-72 (6.1-7.3, 44-53)

🔧 11.1-13.6 (1.2-1.3, 9-10)

❌ 🔧 100-120 (11-12, 74-88)

🔧 11.8-13.7 (1.2-1.3, 9-10)

❌ 🔧 65-80 (6.7-8.1, 48-59)

❌ 🔧 100-120 (11-12, 74-88)

❌ 🔧 65-80 (6.7-8.1, 48-59)

🔧 75-94 (7.7-9.5, 56-69)

❌ 🔧 50-60 (5.1-6.1, 37-44)

❌ 🔧 65-80 (6.7-8.1, 48-59)

🔧 80-95 (8.2-9.6, 59-70)

🔧 80-95 (8.2-9.6, 59-70)

🔧 27-40 (2.8-4.0, 20-29)

🔧 65-80 (6.7-8.1, 48-59)

🔧 46-57 (4.7-5.8, 34-42)

🔧 :N•m (kg-m, ft-lb)

❌ 🔧 100-120 (11-12, 74-88)

🔧 43-55 (4.5-5.6, 32-40)

❌ :Always replace after disassembly

1.	Cap	2.	Actuator assembly	3.	Actuator plate
4.	Washer	5.	Shock absorber mounting seal	6.	Bushing
7.	Distance tube	8.	Shock absorber mounting bracket	9.	Bound bumper cover
10.	Bound bumper	11.	Shock absorber	12.	Axle
13.	Bracket	14.	Upper seat	15.	Coil spring
16.	Rubber seat	17.	Rear lower link	18.	Front lower link
19.	Radius rod	20.	Suspension arm	21.	Stopper rubber
22.	Stabilizer connecting rod mounting bracket	23.	Stabilizer connecting rod	24.	Dynamic damper
25.	Rear suspension member	26.	Stabilizer bar	27.	Stabilizer bushing
28.	Stabilizer clamp	29.	Cross bar	30.	Member stay

09482_INFI_G0439

Exploded view of the rear suspension—2005 Q45 models

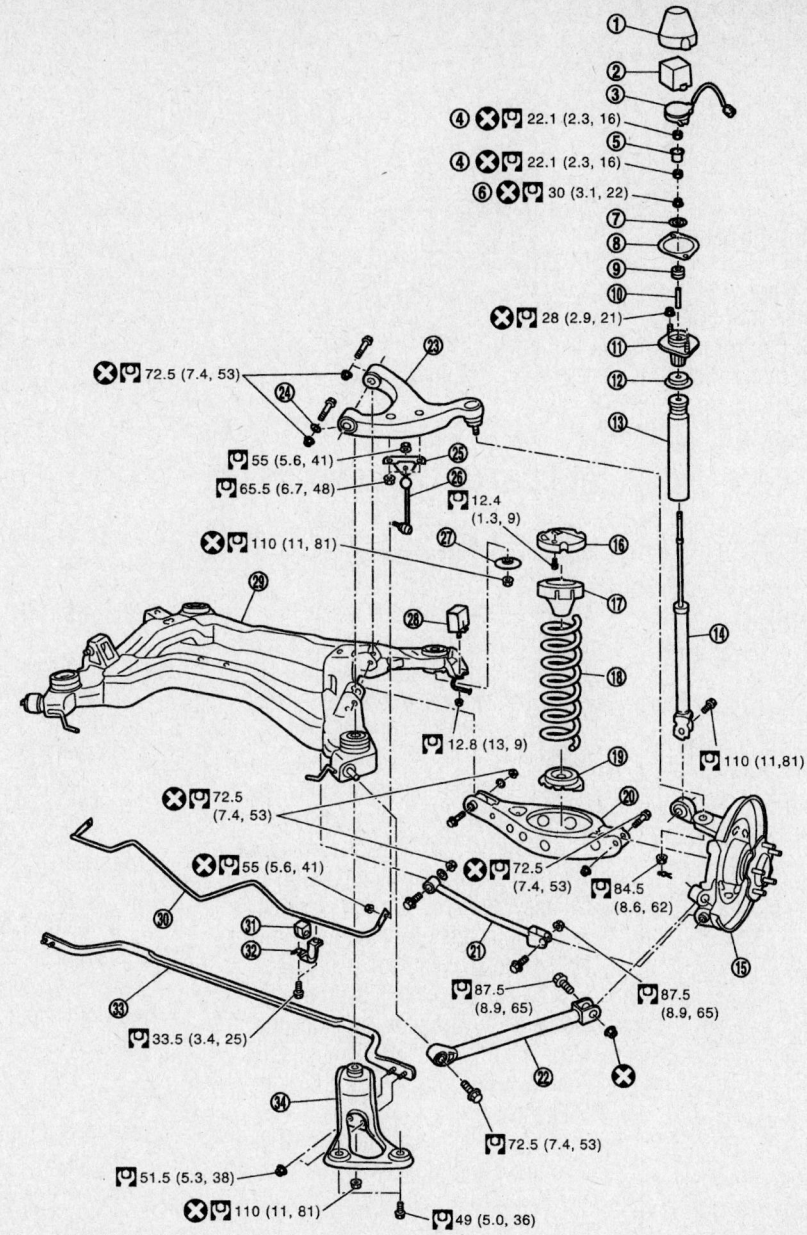

1. Cap (Without active damper suspension)
2. Cap (With active damper suspension)
3. Actuator assembly (With active damper suspension)
4. Nut (With active damper suspension)
5. Actuator plate (With active damper suspension)
6. Nut (Without active damper suspension)
7. Washer
8. Shock absorber mounting seal
9. Bushing
10. Distance tube
11. Shock absorber mounting bracket
12. Bound bumper cover
13. Bound bumper
14. Shock absorber
15. Axle housing assembly
16. Bracket
17. Upper seat
18. Coil spring
19. Rubber seat
20. Rear lower link
21. Front lower link
22. Radius rod
23. Suspension arm
24. Stopper rubber
25. Stabilizer connecting rod mounting bracket
26. Stabilizer connecting rod
27. Mounting stopper
28. Dynamic damper
29. Rear suspension member
30. Stabilizer bar
31. Stabilizer bushing
32. Stabilizer clamp
33. Cross bar
34. Member stay

Exploded view of the rear suspension—2006 Q45 models

09482_INFI_G0438

8. Install the locknut and torque it to 13–17 ft. lbs. (18–24 Nm) and remove the spring compressor.

9. Install strut assembly. Torque the upper shock mounting nuts to 12–14 ft. lbs. (16–19 Nm) and the lower bolt to 57–72 ft. lbs. (77–98 Nm).

2002–04 Q45 Models with Full Active Suspension

➡The Nissan Consult or scan tool that can issue commands to the control unit is required for bleeding the hydraulics in the Full Active Suspension system.

1. Before servicing the vehicle, refer to the precautions in the beginning of this section.

2. Relieve the hydraulic pressure as follows:

a. Raise and safely support the vehicle with all 4 wheels off the ground and wait at least 3 minutes for the system to stabilize.

b. Remove both front inner fenders and the rear pressure control unit cover.

c. Loosen the locknut and slowly open the bypass valve on each pressure control unit. Do not open the bleeder valves.

d. Open the bypass valves all the way, and leave them open, until the job is finished.

3. Remove or disconnect the following:
- Upper mount insulator nuts
- Hydraulic lines. Cap the lines to keep the system clean
- Lower actuator mounting bolt and remove the actuator/spring assembly

4. Secure the actuator/spring assembly in a suitable holding fixture. Scribe alignment marks on the spring, upper mount insulator and actuator unit.

5. Compress the spring with the proper tool. Remove the piston rod locknut lift off the mount insulator, hose joint adapter, spring, and any other components necessary.

To install:

6. Fit the bumper and dust cover onto the actuator.

7. Fit the spring into the lower seat with the matchmarks aligned. Install the upper seat, mounting insulator and other components with the marks aligned.

8. Install the locknut and torque to 43–54 ft. lbs. (59–74 Nm).

9. Fit the assembly onto the vehicle. Torque the upper mounting nuts to 12–24 ft. lbs. (16–19 Nm) and the lower mounting bolt to 58–72 ft. lbs. (78–98 Nm).

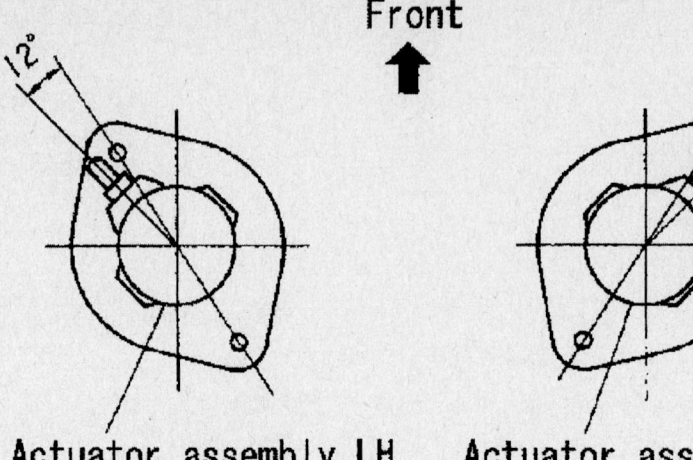

Front

Actuator assembly LH   Actuator assembly RH

09482_INFI_G0106

Rear actuator assembly positioning—Q45 models

10. Bleed the system as follows:

a. With all 4 wheels off the ground, run the engine for about 2 minutes.

b. Connect the scan tool and enter "WORK SUPPORT" mode. Select "4. AIR BLEEDING".

c. Check the fluid level in the reservoir and make it slightly overfilled.

d. Touch "START" on the scan tool. The display will show a regular rise and fall in system pressure that may last for several minutes. When the pressure stabilizes, stop the engine.

e. Connect a clear tube to the air bleeder at the actuator and place the other end in a container. Do not allow fluid to contact the body or the paint will be damaged.

f. Open the bleeder and watch the fluid move through the tube. If there are still air bubbles in the fluid when the flow stops, close the bleeder and check the fluid level. Pressurize the system again and repeat the bleeding process.

G35

1. Before servicing the vehicle, refer to the precautions in the beginning of this section.

2. Place a transmission jack under the rear axle to remove the fitting bolt and nut in the lower side of the shock absorber.

3. Remove or disconnect the following:
- Rear seat cushion, seat back and rear package shelf finish panel
- Fitting nut from the upper side of the shock absorber
- Shock absorber from the vehicle

4. Installation is the reverse of the removal procedure. Refer to the exploded view for tightening values.

2005–06 Q45

1. Remove tire.

2. Set jack under rear lower link to remove fixing bolt in the lower side of shock absorber.

3. Remove jack from rear lower link.

4. Remove rear seat cushion, rear seat back and rear parcel shelf finisher.

5. Remove cap and actuator assembly.

6. Remove fixing nut in the upper side of shock absorber.

7. Check the following: Shock absorber for deformation, cracks, damage, and replace if there are. Piston rod for damage, uneven wear, or distortion, and replace if there are. Welded and sealed areas for oil leakage, and replace if there are.

To install:

8. Installation is the reverse of removal.

➡Refer to the exploded view for tightening torque. Install in the reverse order of the removal. Do not reuse non-reusable parts.

9. Perform the final tightening of shock absorber assembly lower side (rubber bushing) under unladen conditions with tires on level ground. Check wheel alignment.

10. After adjusting wheel alignment, adjust neutral of steering angle sensor.

11. Be sure to install actuator assembly correctly as shown in the figures.

❊❊ WARNING

If a strong shock has been given to actuator assembly or if it has been dropped, replace it with a new one.

2003–04 M45

1. Remove tire with power tool.
2. Set jack under rear lower link to remove fixing bolt and nut in lower side of shock absorber with power tool.
3. Remove jack from rear lower link.
4. Remove rear seat cushion, rear seat back and rear parcel shelf finisher.
5. Remove cap and then remove shock absorber mounting bracket fixing nuts of shock absorber upper side and remove shock absorber from vehicle.
6. Check shock absorber for deformation, cracks, damage, and replace if necessary.
7. Check piston rod for damage, uneven wear, distortion, and replace if necessary.
8. Check welded and sealed areas for oil leakage, and replace if necessary.

To Install:

9. Installation is the reverse of removal.
10. Refer to the exploded view for tightening torque. Tighten in the reverse order of removal.

✻✻ WARNING

Refer to component parts location and do not reuse non-reusable parts.

11. Perform final tightening of shock absorber lower side (rubber bushing) under unladen condition with tires on level ground. Check wheel alignment.

M35 and 2006 M45

1. Remove tires from vehicle with a power tool.
2. Set a jack under rear lower link to relieve the coil spring tension.
3. Remove shock absorber lower end bolt with a power tool.
4. Gradually lower the jack to remove it from rear lower link.
5. Remove shock absorber assembly upper end nuts with a power tool, and then remove shock absorber assembly from vehicle.
6. Check shock absorber assembly for deformation, cracks, damage, and replace if there are.
7. Check welded and sealed areas for oil leakage, and replace if there are.

To install:

8. Installation is the reverse order of removal. For tightening torques, refer to the exploded view. Do not reuse non-reusable parts.
9. Perform final tightening of shock absorber assembly lower side (rubber bushing) under unladen condition with tires on level ground. Check wheel alignment.
10. Adjust neutral position of steering angle sensor after checking the wheel alignment.

Coil Spring

REMOVAL & INSTALLATION

G35 Models

For G20 and I35 models, refer to the Strut Removal and Installation procedure for coil spring replacement.

1. Before servicing the vehicle, refer to the precautions in the beginning of this section.
2. Remove or disconnect the following:
   - Tire and wheel
3. Place a jack under the rear lower link.
4. Loosen the fixing bolt and nut attaching the rear lower line to the side of the suspension member.
   - Fixing bolt and nut from the side of the axle housing
5. Slowly lower the jack, then remove the upper rubber seat, coil spring and rubber sheet from the lower link.
   - Fixing bolt and nut from the side of the suspension member to remove the lower link.
6. Installation is the reverse of the removal procedure. Tighten all components, as listed in the rear suspension illustration.

M45 and Q45 Models

1. Before servicing the vehicle, refer to the precautions in the beginning of this section.
2. Remove or disconnect the following:
   - Tire and wheel
3. Place a jack under the rear lower link.
4. Loosen the fixing bolt and nut attaching the rear lower link to the side of the suspension member.
   - Fixing bolt and nut from the side of the axle housing
5. Slowly lower the jack, then remove the upper rubber seat, coil spring and rubber sheet from the lower link.
   - Fixing bolt and nut from the side of the suspension member to remove the lower link.
6. Installation is the reverse of the removal procedure. Tighten all components, as listed in the rear suspension illustration.

Torsion Bars

REMOVAL & INSTALLATION

G20 and I35 Models

1. Before servicing the vehicle, refer to the precautions in the beginning of this section.
2. Loosen the lug nuts.
3. Remove or disconnect the following:
   - Wheels

✻✻ WARNING

Be sure to disconnect the Anti-lock Brake System (ABS) wheel sensor from the assembly. Failure to do so may result in damage to the sensor wire and the sensor becoming inoperative.

   - Brake calipers and suspend them with a piece of wire. Do not let them hang by the hose
4. Using a transmission jack, raise the torsion beam slightly, then remove the suspension mounting bolts.
5. Lower the jack and remove the suspension assembly.
6. The lateral link and control rod can now be removed.
7. Inspect the torsion beam and control rod for cracks, wear and deformation. The length of the lateral link and control rod is as follows:
   a. A—8.15–8.19 in. (207–208mm)
   b. B—15.51–15.55 in. (394–395mm)
   c. C—23.66–23.74 in. (601–603mm)
   d. D—4.17–4.25 in. (106–108mm)

To install:

8. When installing the control rod, connect the bushing with the smaller inner

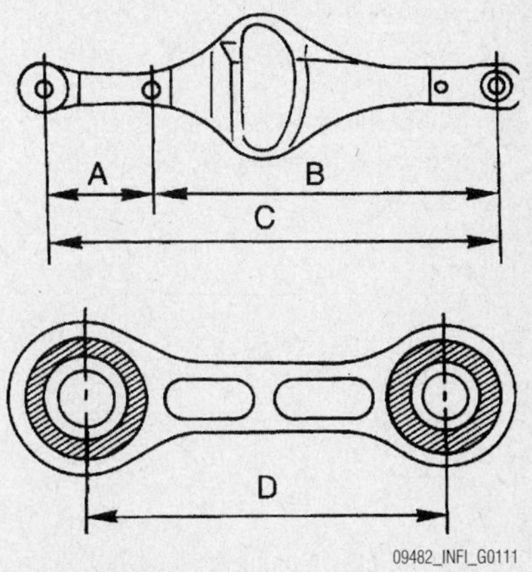
A
B
C
D
09482_INFI_G0111

Measure the control rod and lateral links at these points—G20 and I35 models

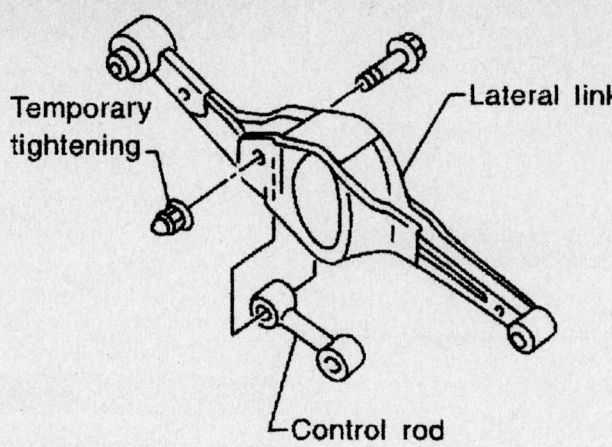

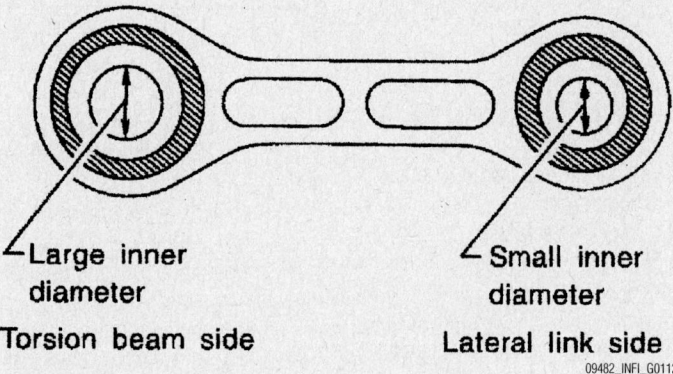

Be sure to install the control rod correctly—G20 and I35 models

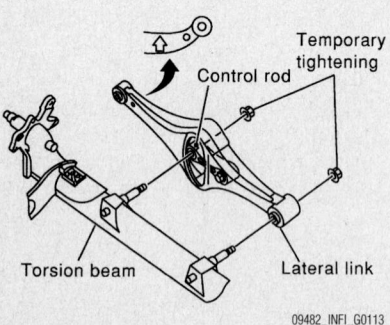

The lateral link must be in the horizontal position when tightening the bolts—G20 and I35 models

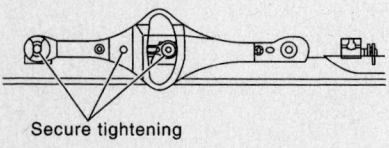

Tighten the torsion beam-to-chassis bolts with the suspension unloaded—G20 and I35 models

diameter to the lateral link. Install the lateral link and the control rod on the torsion beam. Place the lateral link with the arrow topside.

9. Place the lateral link and control rod horizontally against the beam, and tighten the bolts. Refer to the illustration.

10. Secure the torsion beam to the vehicle. Make sure the lateral link is horizontal, then tighten the link to the chassis.

11. Attach the struts to the torsion beam and tighten the fasteners.

12. Tighten the torsion beam-to-chassis bolts.

13. Install the calipers, ABS sensor and wheels. Lower the vehicle to the ground. Final tighten the lug nuts.

Stabilizer (Sway) Bar

REMOVAL & INSTALLATION

G35

1. Remove dynamic damper of exhaust tube.

2. Remove lower side fixing nut on stabilizer connecting rod and remove stabilizer connecting rod from stabilizer bar with power tool.

3. Remove fixing nut on stabilizer clamp and remove stabilizer from vehicle with power tool.

4. Check stabilizer bar, stabilizer bushings, stabilizer clamp, stabilizer connecting rod, stabilizer connecting rod mounting bracket for any deformation, crack or damage. Replace if necessary.

To install:

5. Refer to the exploded view for tightening torque. Install in the reverse order of removal.

➡**Do not reuse non-reusable parts.**

6. The stabilizer bar uses pillow ball type connecting rod, position ball joint with case on pillow ball head parallel to stabilizer bar.

Q45

1. Remove center muffler.

2. Remove stabilizer connecting rod from stabilizer bar with a power tool.

3. Remove mounting bolts of stabilizer clamp and then remove stabilizer clamp and stabilizer bushing from stabilizer bar.

4. Remove stabilizer bar from vehicle behind.

5. Check stabilizer bar, stabilizer bushings, stabilizer clamp, stabilizer connecting rod, stabilizer connecting rod mounting bracket for any deformation, cracks or damage. Replace if there are.

To install:

6. Refer to the exploded view for tightening torque. Install in the reverse order of removal.

➡**Do not reuse non-reusable parts.**

7. The stabilizer bar uses pillow ball type connecting rod.

2003–04 M45

1. Remove dynamic dampener of exhaust tube.

2. Remove stabilizer connecting rod from stabilizer bar with power tool.

3. Remove mounting bolts of stabilizer clamp and then remove stabilizer clamp and stabilizer bushing from stabilizer bar with power tool.

4. Remove stabilizer bar from vehicle behind.

5. Check stabilizer bar, stabilizer bushing, stabilizer clamp, stabilizer connecting rod, stabilizer connecting rod mounting bracket for any deformation, crack or damage. Replace if necessary.

To install:

6. Refer to the exploded view for tightening torques. Install in the reverse order of removal.

→**Refer to component parts location and do not reuse non-reusable parts.**

7. Stabilizer bar uses the pillow ball type connecting rod, position ball joint with case on pillow ball head parallel to stabilizer bar.

M35 and 2006 M45

1. Remove mounting bracket of center muffler and remove mounting rubber of main muffler.
2. Remove lower side mounting nut on stabilizer connecting rod and remove stabilizer connecting rod from stabilizer bar with power tool.
3. Remove mounting nut on stabilizer clamp and remove stabilizer from vehicle with power tool.

4. Check stabilizer bar, stabilizer bushings, stabilizer clamp, stabilizer connecting rod and stabilizer connecting rod mounting bracket for any deformation, crack or damage. Replace if there are.

To install:

5. Installation is the reverse order of removal. For tightening torques, refer to the exploded view.

→**Do not reuse non-reusable parts.**

Wheel Bearings

ADJUSTMENT

The wheel bearing assemblies on all models are pressed in and are not adjustable. If the bearing assembly does not turn smoothly or has more than 0.002 in. (0.05mm) of axial play, replace the bearing assembly.

REMOVAL & INSTALLATION

G20 Models

1. Before servicing the vehicle, refer to the precautions in the beginning of this section.
2. Raise and support the vehicle safely.
3. Remove or disconnect the following:
- Rear caliper and rotor. Hang the caliper from the body with wire, do not let hang by the brake hose.
- Rear wheel hub cap, cotter pin and locknut
- Hub off the stub axle

→**The wheel bearing is integral with the hub and cannot be serviced separately.**

To install:
4. Install or connect the following:
- New hub assembly onto the axle stub

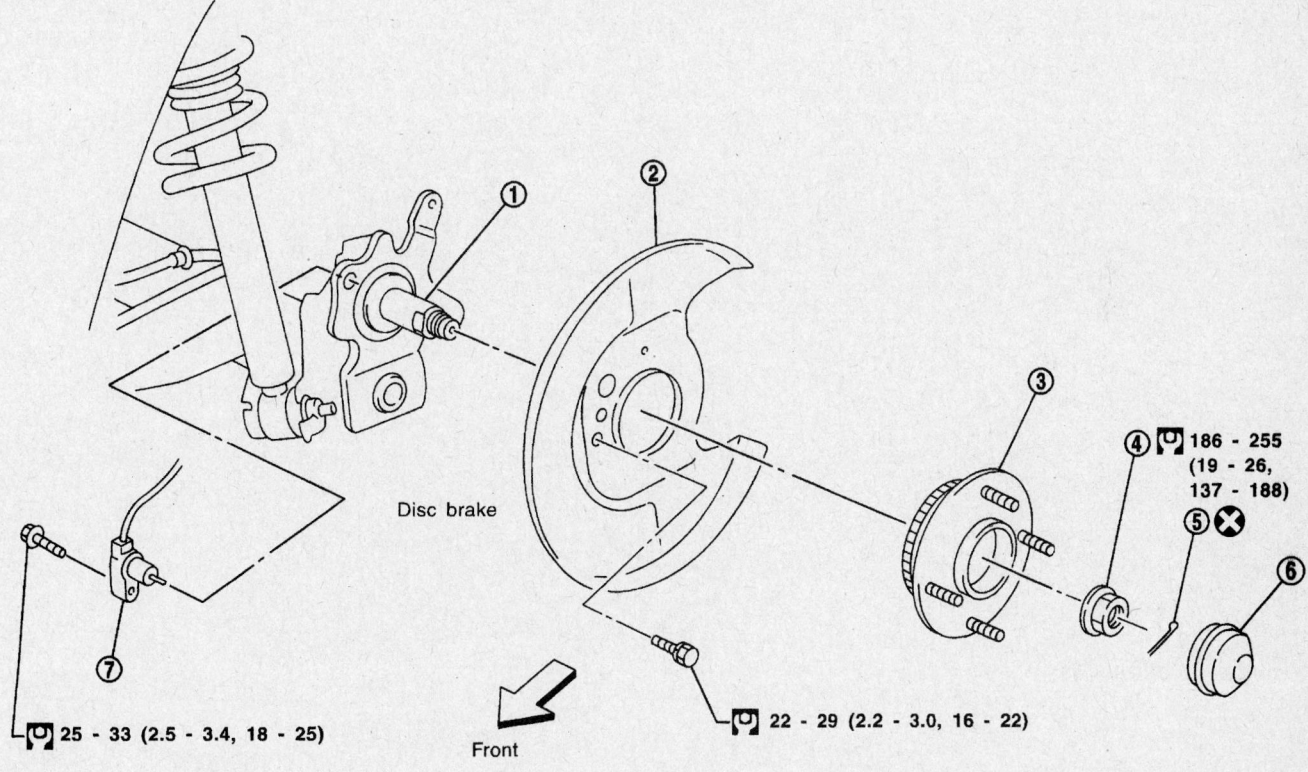

Disc brake

Front

25 - 33 (2.5 - 3.4, 18 - 25)

22 - 29 (2.2 - 3.0, 16 - 22)

186 - 255 (19 - 26, 137 - 188)

: N•m (kg-m, ft-lb)

1. Spindle
2. Baffle plate
3. Wheel hub bearing
4. Wheel bearing lock nut
5. Cotter pin
6. Hub cap
7. ABS sensor

09482_INFI_G0440

Rear hub and related parts—G20 models

5. Replace the washer and wheel bearing locknut. Torque the locknut to 137–174 ft lbs. (186–235 Nm). Install a new cotter pin.
• Brake rotor, caliper and wheel
6. Lower the vehicle to the ground.

I35

Before removing the rear wheel hub assembly, disconnect the ABS wheel sensor from the assembly. Then move it away from the hub assembly. Failure to do so may result in damage to the sensor wires and the sensor becoming inoperative.

Wheel hub bearing does not require maintenance. If any of the following symptoms are noted, replace wheel hub bearing assembly.

  a. Growling noise is emitted from wheel hub bearing during operation.

  b. Wheel hub bearing drags or turns roughly. This occurs when turning hub by hand after bearing lock nut is tightened to specified torque.

1. Remove brake caliper assembly.
2. Remove wheel bearing lock nut.
3. Remove brake rotor.
4. Remove wheel hub bearing from spindle.

➡ **Brake hose does not need to be disconnected from brake caliper. Suspend caliper assembly with wire so as not to stretch brake hose.**

✳✳ WARNING

Be careful not to depress brake pedal, or piston will pop out. Make sure brake hose is not twisted.

5. Remove the sensor rotor using suitable puller, drift and bearing replacer.
To install:
6. With vehicles equipped with ABS, press-fit ABS sensor rotor into wheel hub bearing using a drift.

➡ **Do not reuse ABS sensor rotor. When installing, replace it with a new one.**

7. Press-fit ABS sensor rotor as far as the location shown.
8. Install wheel hub bearing. Before tightening, apply oil to threaded portion of rear spindle.

➡ **Do not reuse wheel bearing lock nut.**

9. Tighten wheel bearing lock nut to 187–254 Nm (138–188 ft. lbs.)
10. Check that wheel bearings operate smoothly.
11. Check wheel hub bearing axial end play. Axial end play should be less than 0.05 mm (0.0020 in.)
12. Stake the lock nut in two places.
13. Install hub cap using a suitable tool.

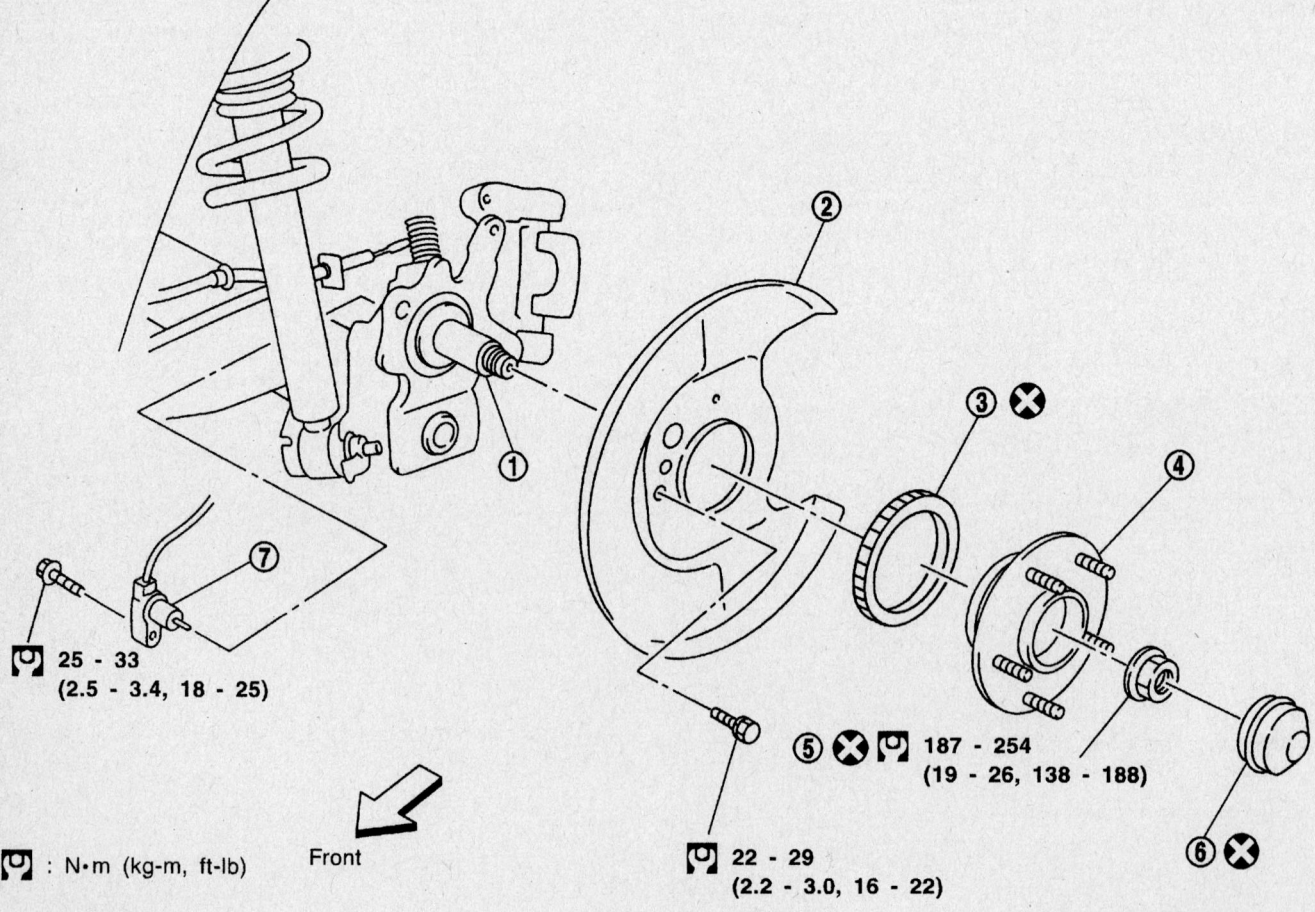

25 - 33
(2.5 - 3.4, 18 - 25)

: N•m (kg-m, ft-lb)

Front

187 - 254
(19 - 26, 138 - 188)

22 - 29
(2.2 - 3.0, 16 - 22)

1. Spindle
2. Baffle plate
3. ABS sensor rotor
4. Wheel hub bearing
5. Wheel bearing lock nut
6. Hub cap
7. ABS sensor

Rear hub and related parts—I35 models

09482_INFI_G0442

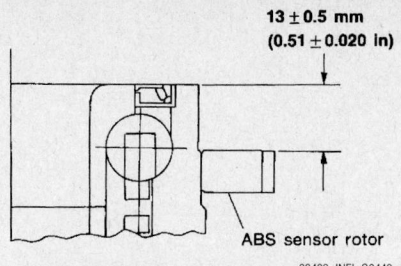

13 ± 0.5 mm
(0.51 ± 0.020 in)

ABS sensor rotor

09482_INFI_G0443

Press-fit ABS sensor rotor—I35 models

➡**Do not reuse hub cap. When installing, replace it with a new one.**

G35 and Q45 Models

1. Before servicing the vehicle, refer to the precautions in the beginning of this section.
2. Remove or disconnect the following:
   • Cotter pin and adjusting cap and loosen the wheel bearing nut. Carefully tap the end of the axle shaft or use a puller to loosen the shaft from the hub.
   • Brake caliper and rotor. Do not let the caliper hang by the brake hose, support it with wire.
   • Parking brake assembly
   • Nuts and through-bolts to remove the axle housing from the suspension. If equipped with rear wheel steering, use a ball joint press to separate the tie rod end.

4 bolts at the back and remove the bearing flange and hub from the bearing housing
3. Press the hub out of the bearing flange and use a puller to remove the bearing from the hub. If it is not damaged, the hub can be used again but the bearing and flange are supplied as a single unit.

To install:

➡**Refer to the exploded view for tightening values.**

➡**The wheel bearing and flange are supplied as an assembly.**

4. Place the hub on a press table and press the new bearing and flange onto the hub. Be sure the press tool contacts only the inner bearing race and take care not to damage the seal.
5. Install or connect the following:
   • Bearing flange onto the axle housing.
   • Axle housing onto the lower ball joint, and install a new cotter pin
   • Torque the tie rod end nut to 33–44 ft. lbs. (45–60 Nm) and install a new cotter pin, if equipped with rear wheel steering
   • Axle shaft into the hub and install the bolts through the suspension bushings. Tighten the bolts temporarily, they will be tightened with the vehicle resting on the wheels.

• Brake components and apply the brake to hold the hub from turning
   • Wheel bearing locknut and torque it to 152–203 ft. lbs. (206–275 Nm)
   • Insulator and adjusting cap and a new cotter pin
   • Wheel and lower the vehicle to the ground. Torque the suspension bushing bolts to 57–72 ft. lbs. (77–98 Nm).

2003–04 M45

1. Before servicing the vehicle, refer to the precautions in the beginning of this section.
2. Raise and support the vehicle.
3. Remove or disconnect the following:

   • Rear tires
   • Axle shaft nut
   • Brake caliper and rotor. Hang the caliper from the body with wire, do not let it hang by the brake hose.
   • Radius rod bolts and nuts
   • Lower link nuts and bolts
   • Coil spring
   • Lower shock bolts
   • Axle shaft
   • Suspension arm ball joint
   • Wheel hub and bearing assembly

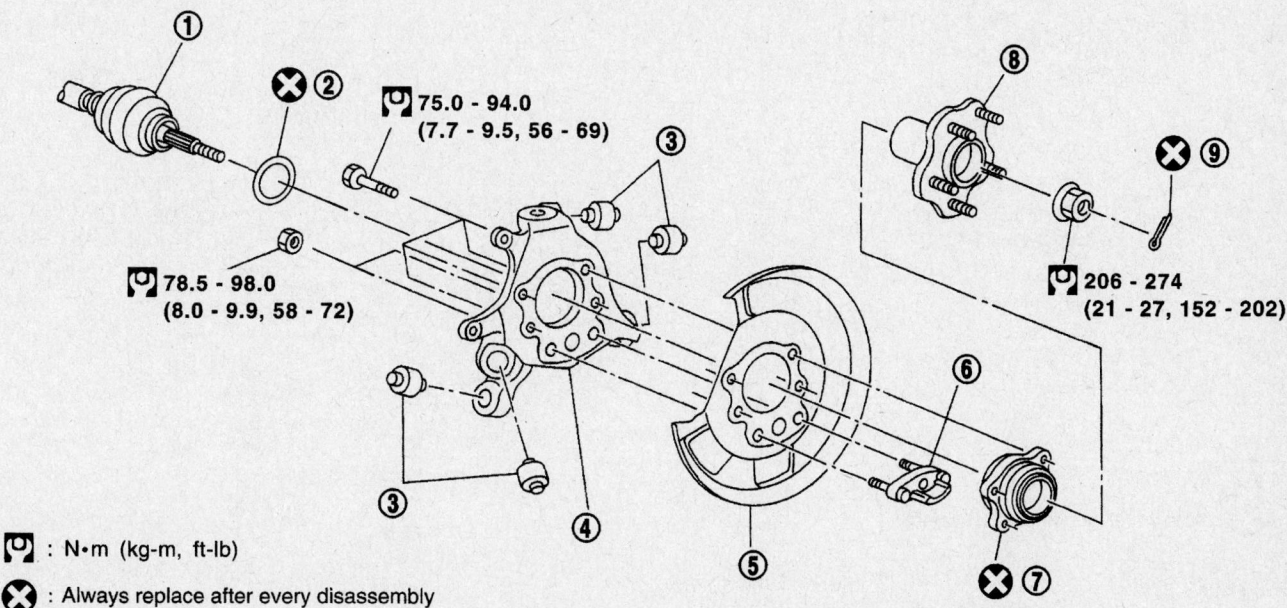

① Drive shaft
❌ ②
🔧 75.0 - 94.0 (7.7 - 9.5, 56 - 69)
③
🔧 78.5 - 98.0 (8.0 - 9.9, 58 - 72)
④
⑤
⑥
🔧 206 - 274 (21 - 27, 152 - 202)
⑧
❌ ⑨
❌ ⑦

🔧 : N•m (kg-m, ft-lb)
❌ : Always replace after every disassembly

1. Drive shaft	2. Dust shield	3. Bushing
4. Axle	5. Back plate	6. Anchor block
7. Wheel bearing	8. Wheel hub	9. Cotter pin

Rear hub and related parts—G35 and Q45 models

09482_INFI_G0441

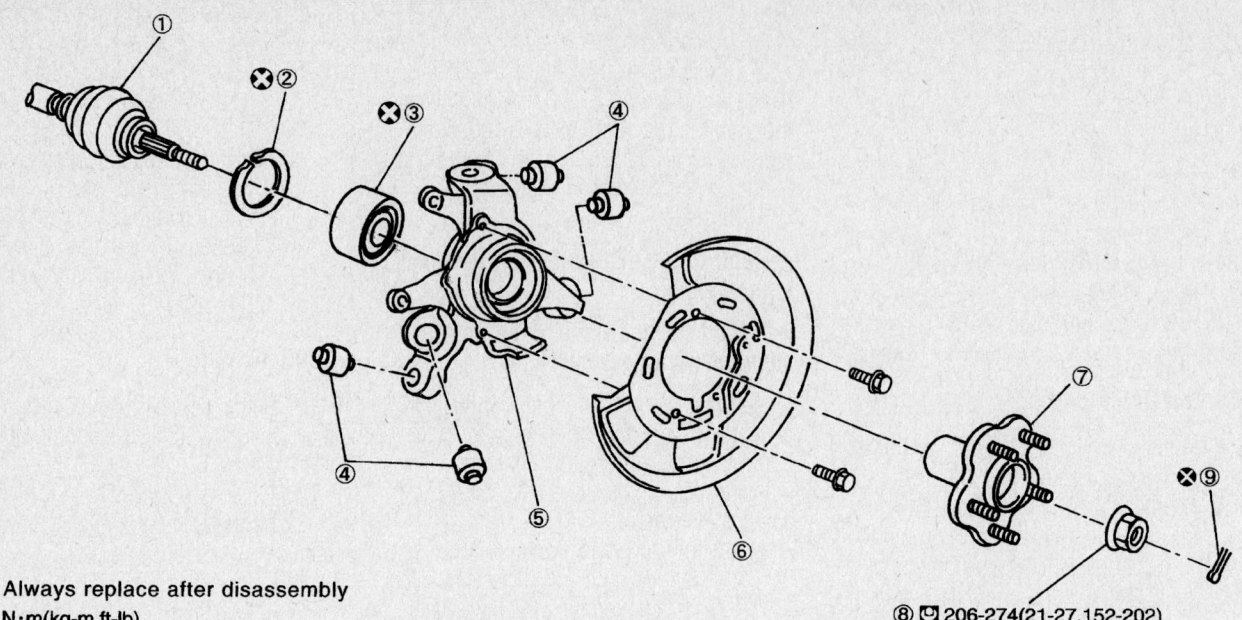

❎ : Always replace after disassembly

🔧 : N·m(kg-m,ft-lb)

⑧ 🔧 206-274(21-27,152-202)

1.	Drive shaft	2.	Snap ring	3.	Wheel bearing
4.	Bushing	5.	Axle	6.	Back plate
7.	Wheel hub	8.	Lock nut	9.	Cotter pin

67162-INFI-G41

Exploded view of the rear hub and bearing—M45 Models

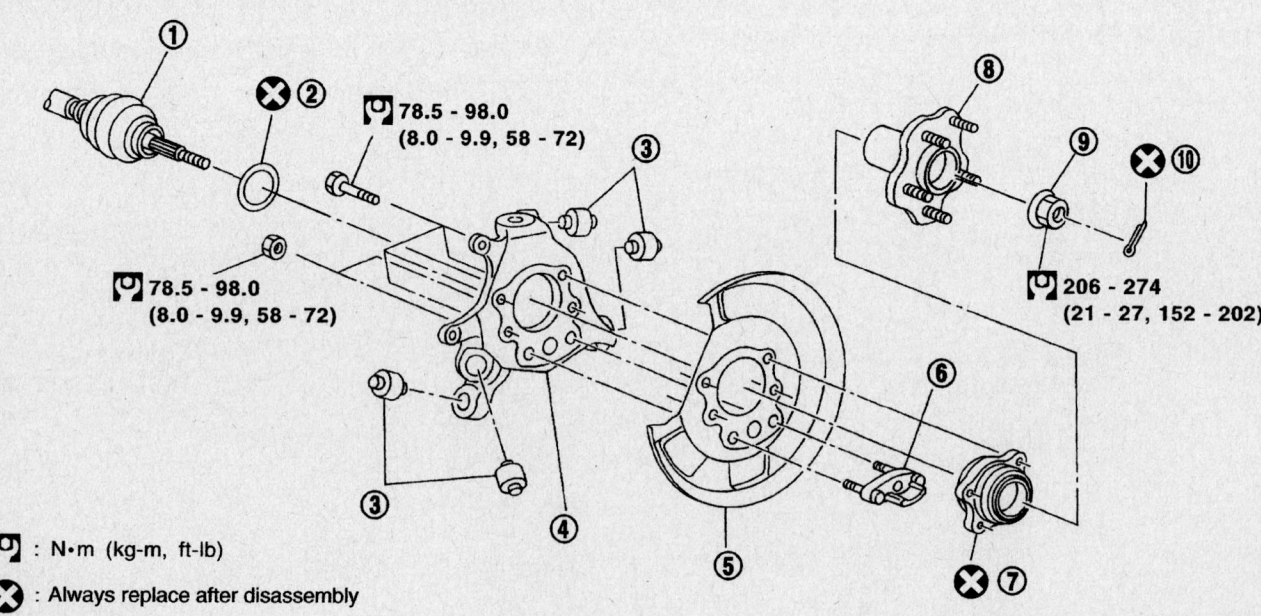

🔧 78.5 - 98.0
(8.0 - 9.9, 58 - 72)

🔧 78.5 - 98.0
(8.0 - 9.9, 58 - 72)

🔧 206 - 274
(21 - 27, 152 - 202)

🔧 : N·m (kg-m, ft-lb)

❎ : Always replace after disassembly

1.	Drive shaft	2.	Dust shield	3.	Bushing
4.	Axle	5.	Back plate	6.	Anchor block
7.	Wheel bearing	8.	Wheel hub	9.	Lock nut
10.	Cotter pin				

67162-INFI-G42

Exploded view of the rear hub and bearing—2004 Q45 Models

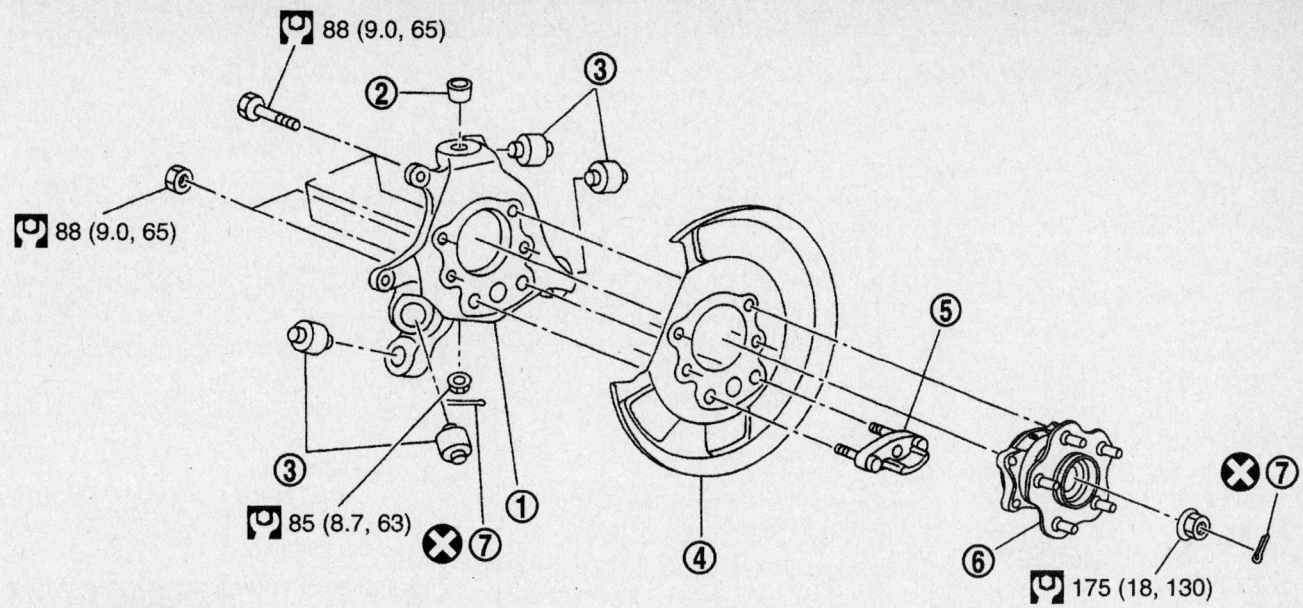

88 (9.0, 65)

88 (9.0, 65)

85 (8.7, 63)

175 (18, 130)

1. Axle housing
2. Ball seat
3. Bushing
4. Back plate
5. Anchor block
6. Wheel hub and bearing assembly
7. Cotter pin

09482_INFI_G0444

Rear hub and related parts—M35 and 2006 M45 models

- Use a puller, drift and bearing replacer and press the wheel bearing outer race off the hub
- Use a drift and remove the wheel bearing from the hub

To install:

4. Be sure all parts are clean. Carefully press the hub into the new bearing. Be sure the press tool does not contact the inner bearing race. Do not exceed 11,000 lbs. (5000kg) pressure.

5. Install backing plate and the hub and bearing assembly.

6. Reverse the remainder of the removal procedure to complete installation.

M35 and 2006 M45

1. Remove tire.
2. Remove rear brake caliper with a power tool. Hang it in a place where it will not interfere with work.

WARNING

Do not depress brake pedal while brake caliper is removed.

3. Put matching mark on disc rotor and the wheel hub and bearing assembly then removing disc rotor.

4. Remove cotter pin, then loosen hub lock nut with a power tool.

5. Separate the wheel hub and bearing assembly from halfshaft by lightly tapping the end with a hammer (suitable tool) and wood block, and then remove hub lock nut.

WARNING

Do not place halfshaft joint at an extreme angle. Also be careful not to overextend slide joint. Do not allow halfshaft to hang down without support for housing (or joint sub-assembly), shaft and other parts.

➡Use a puller (suitable tool), if the wheel hub and bearing assembly and halfshaft cannot be separated even after performing the above procedure.

6. Remove the wheel hub and bearing assembly mounting bolts.

7. Remove the wheel hub and bearing assembly.

8. Check the wheel hub and bearing assembly for wear, cracks, and damage. Replace if there are.

To install:

9. Installation is the reverse order of removal. See the accompanying figure for torque values.

➡Do not reuse non-reusable parts.

10. Assemble disc rotor and the wheel hub and bearing assembly by aligning each matching mark as shown in the figure when installing disc rotor.

FRONT BRAKES

Brake Caliper

REMOVAL & INSTALLATION

G20 Models

1. Before servicing the vehicle, refer to the precautions in the beginning of this section.
2. Remove the wheels.
3. Loosen the brake hose connecting bolt.
4. Remove the bolts connecting the caliper to the torque member.
5. Slide the caliper out from the rotor.
6. Remove the brake hose connecting bolt from the caliper.
7. Remove the caliper from the vehicle.

To install:

8. Fit the caliper onto the torque member and torque the bolts to 16–23 ft. lbs. (22–31 Nm).
9. Using new copper washers, connect the hydraulic hose to the caliper. Torque the union bolt to 12–14 ft. lbs. (17–19 Nm).
10. Bleed the air from the system.

G35 Models

W/STANDARD BRAKES

1. Before servicing the vehicle, refer to the precautions in the beginning of this section.
2. Remove the front wheels.
3. Remove both guide pin bolts securing the caliper to the steering knuckle.
4. Loosen and remove the brake hose connector from the caliper.
5. Remove the caliper assembly from the vehicle.

To install:

➡ **See the accompanying illustration for torque values.**

6. Using new copper washers, install the brake line to the brake caliper.
7. Install the caliper to the steering knuckle using the guide pins bolts.
8. Install the wheels and tighten the lug nuts to the proper specification.
9. Bleed the brake system and top off the master cylinder as necessary.

W/BREMBO® OPTION

1. Before servicing the vehicle, refer to the precautions in the beginning of this section.
2. Remove the front wheels.
3. Remove brake pads (Refer to Disc Brake Pads, Removal and Installation).
4. Remove brake tube flare nut using a flare wrench.
5. Remove brake tube bracket from knuckle.
6. Remove caliper bolts and remove the caliper.

To install:

7. Install caliper assembly to vehicle, and tighten bolts to 113–114 ft. lbs. (152.1–154.9 Nm).

✳✳ WARNING

When attaching the caliper assembly to vehicle, wipe any oil off knuckle spindle washers and caliper assembly attachment surfaces.

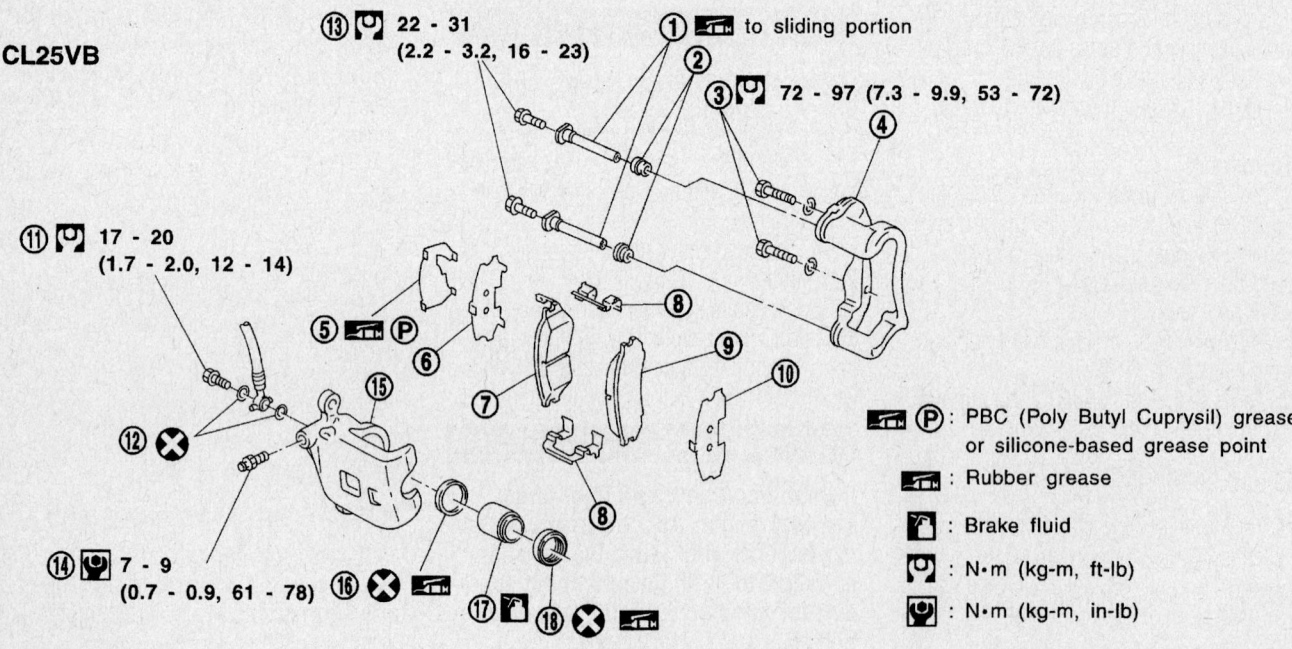

CL25VB

⑬ 22 - 31 (2.2 - 3.2, 16 - 23)
① to sliding portion
③ 72 - 97 (7.3 - 9.9, 53 - 72)
⑪ 17 - 20 (1.7 - 2.0, 12 - 14)
⑭ 7 - 9 (0.7 - 0.9, 61 - 78)

P : PBC (Poly Butyl Cuprysil) grease or silicone-based grease point
: Rubber grease
: Brake fluid
: N·m (kg-m, ft-lb)
: N·m (kg-m, in-lb)

1. Main pin	7. Inner pad	13. Main pin bolt
2. Pin boot	8. Pad retainer	14. Bleed valve
3. Torque member fixing bolt	9. Outer pad	15. Cylinder body
4. Torque member	10. Outer shim	16. Piston seal
5. Shim cover	11. Connecting bolt	17. Piston
6. Inner shim	12. Copper washer	18. Piston boot

09482_INFI_G0127

Component view of front brake assembly—G20

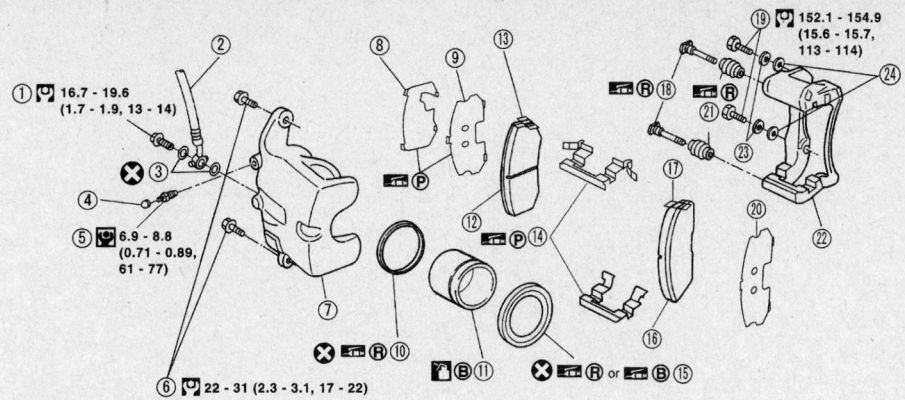

① 16.7 - 19.6
(1.7 - 1.9, 13 - 14)

⑤ 6.9 - 8.8
(0.71 - 0.89,
61 - 77)

⑥ 22 - 31 (2.3 - 3.1, 17 - 22)

⑲ 152.1 - 154.9
(15.6 - 15.7,
113 - 114)

(P) : PBC (Poly Butyl Cuprysil) grease or silicone-based grease point

(R) : Rubber grease point

(B) : Brake fluid point

: N·m (kg-m, ft-lb)

: N·m (kg-m, in-lb)

⊗ : Always replace after every disassembly.

1.	Union bolt	2.	Brake hose	3.	Copper washer
4.	Cap	5.	Air bleeder	6.	Sliding pin bolt
7.	Cylinder body	8.	Inner shim cover	9.	Inner shim
10.	Piston seal	11.	Piston	12.	Inner pad
13.	Pad wear sensor	14.	Pad retainer	15.	Piston boot
16.	Outer pad	17.	Pad wear sensor	18.	Sliding pin bolt
19.	Torque member bolts	20.	Outer shim	21.	Slide pin boot
22.	Torque member	23.	Washer	24.	Decrement shim (Not inserted in some vehicles.)

09482_INFI_G0445

Front caliper and related parts—2003 G35 with standard brakes

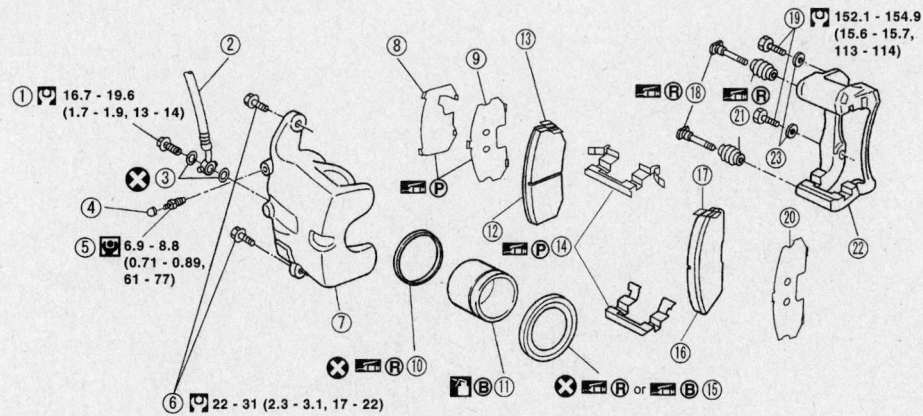

① 16.7 - 19.6
(1.7 - 1.9, 13 - 14)

⑤ 6.9 - 8.8
(0.71 - 0.89,
61 - 77)

⑥ 22 - 31 (2.3 - 3.1, 17 - 22)

⑲ 152.1 - 154.9
(15.6 - 15.7,
113 - 114)

(P) : PBC (Poly Butyl Cuprysil) grease or silicone-based grease point

(R) : Rubber grease point

(B) : Brake fluid point

: N·m (kg-m, ft-lb)

: N·m (kg-m, in-lb)

⊗ : Always replace after every disassembly.

1.	Union bolt	2.	Brake hose	3.	Copper washer
4.	Cap	5.	Bleed valve	6.	Sliding pin bolt
7.	Cylinder body	8.	Inner shim cover	9.	Inner shim
10.	Piston seal	11.	Piston	12.	Inner pad
13.	Pad wear sensor	14.	Pad retainer	15.	Piston boot
16.	Outer pad	17.	Pad wear sensor	18.	Sliding pin
19.	Torque member bolts	20.	Outer shim	21.	Slide pin boot
22.	Torque member	23.	Washer		

09482_INFI_G0446

Front caliper and related parts—2004 and 2004.5 G35

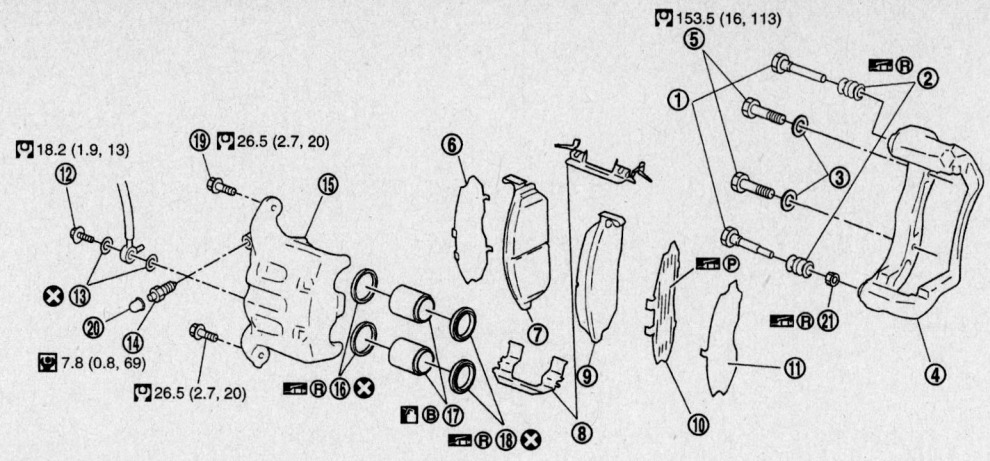

: Always replace after every disassembly.

: PBC(Poly Butyl Cuprysil) grease or silicone-based grease point

: Rubber grease point

: Brake fluid point

: N•m (kg-m, in-lb)

: N•m (kg-m, ft-lb)

1. Sliding pin	2. Sliding pin boot	3. Washer
4. Torque member	5. Torque member mounting bolt	6. inner shim
7. Inner pad	8. Pad retainer	9. Outer pad
10. Outer shim	11. Outer shim cover	12. Union bolt
13. Copper washer	14. Bleed valve	15. Cylinder body
16. Piston seal	17. Piston	18. Piston boot
19. Sliding pin bolt	20. Cap	21. Bushing

09482_INFI_G0447

Front caliper and related parts—2005 G35 (CLZ31VD)

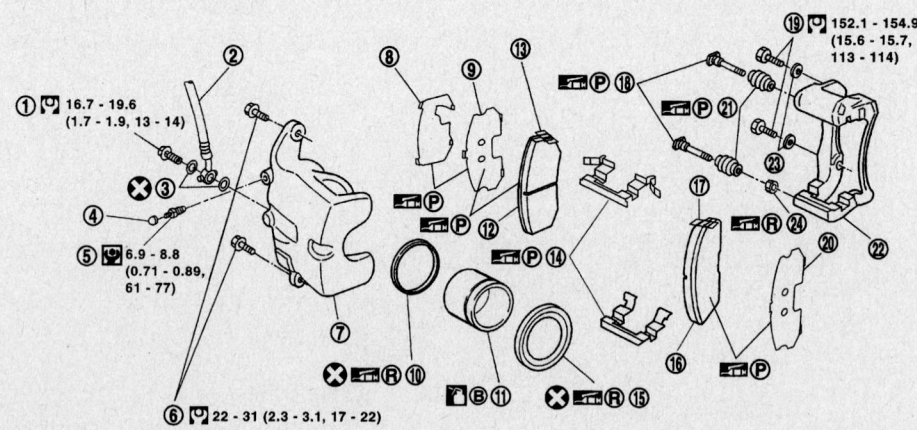

: PBC (Poly Butyl Cuprysil) grease or silicone-based grease point

: Rubber grease point

: Brake fluid point

: N•m (kg-m, ft-lb)

: N•m (kg-m, in-lb)

: Always replace after every disassembly.

1. Union bolt	2. Brake hose	3. Copper washer
4. Cap	5. Bleed valve	6. Sliding pin bolt
7. Cylinder body	8. Inner shim cover	9. Inner shim
10. Piston seal	11. Piston	12. Inner pad
13. Pad wear sensor	14. Pad retainer	15. Piston boot
16. Outer pad	17. Pad wear sensor	18. Sliding pin
19. Torque member bolts	20. Outer shim	21. Slide pin boot
22. Torque member	23. Washer	24. Bushing

09482_INFI_G0448

Front caliper and related parts—2005 G35 (CLZ25VD)

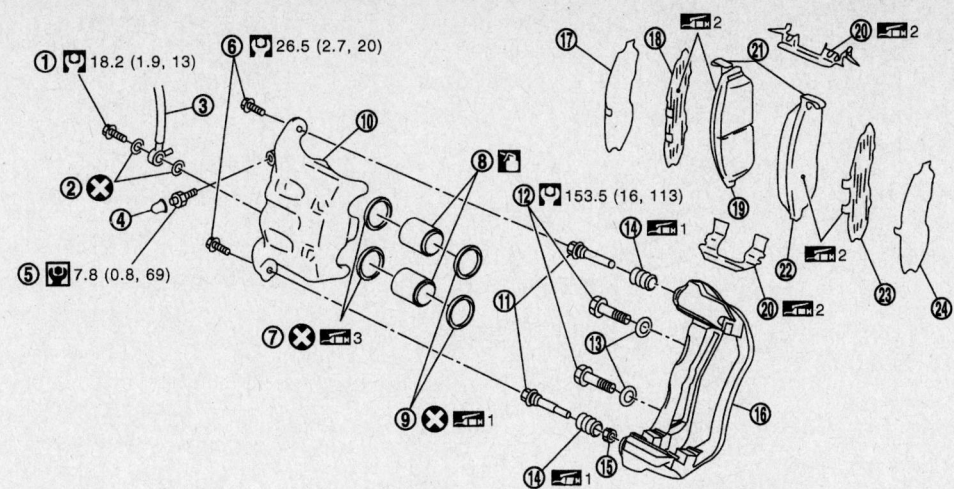

1. Union bolt
2. Copper washer
3. Brake hose
4. Cap
5. Bleed valve
6. Sliding pin bolt
7. Piston seal
8. Piston
9. Piston boot
10. Cylinder body
11. Sliding pin
12. Torque member mounting bolt
13. Washer
14. Sliding pin boot
15. Bushing
16. Torque member
17. inner shim cover
18. inner shim
19. Inner pad
20. Pad retainer
21. Pad wear sensor
22. Outer pad
23. Outer shim
24. Outer shim cover

09482_INFI_G0449

Front caliper and related parts—2006 G35

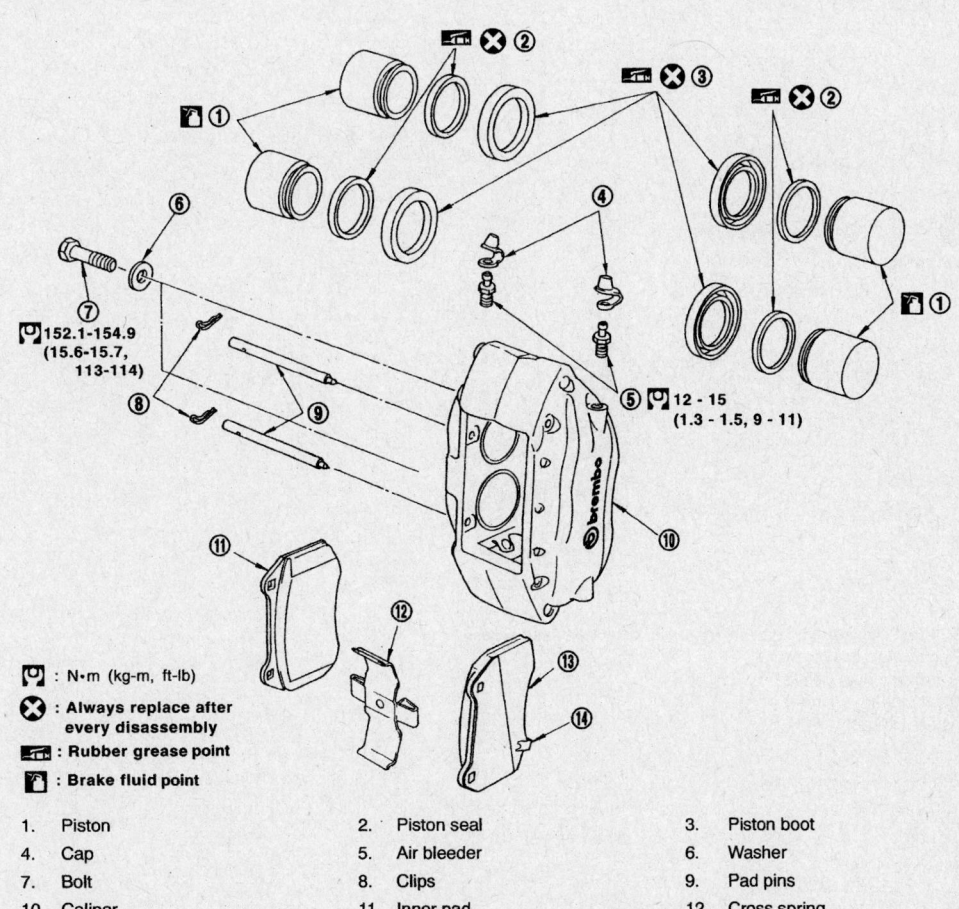

⊙ : N•m (kg-m, ft-lb)

✕ : **Always replace after every disassembly**

: **Rubber grease point**

: **Brake fluid point**

1. Piston
2. Piston seal
3. Piston boot
4. Cap
5. Air bleeder
6. Washer
7. Bolt
8. Clips
9. Pad pins
10. Caliper
11. Inner pad
12. Cross spring
13. Outer pad
14. Pad wear sensor

09482_INFI_G0017

Component view of front caliper assembly—2003 Brembo® option

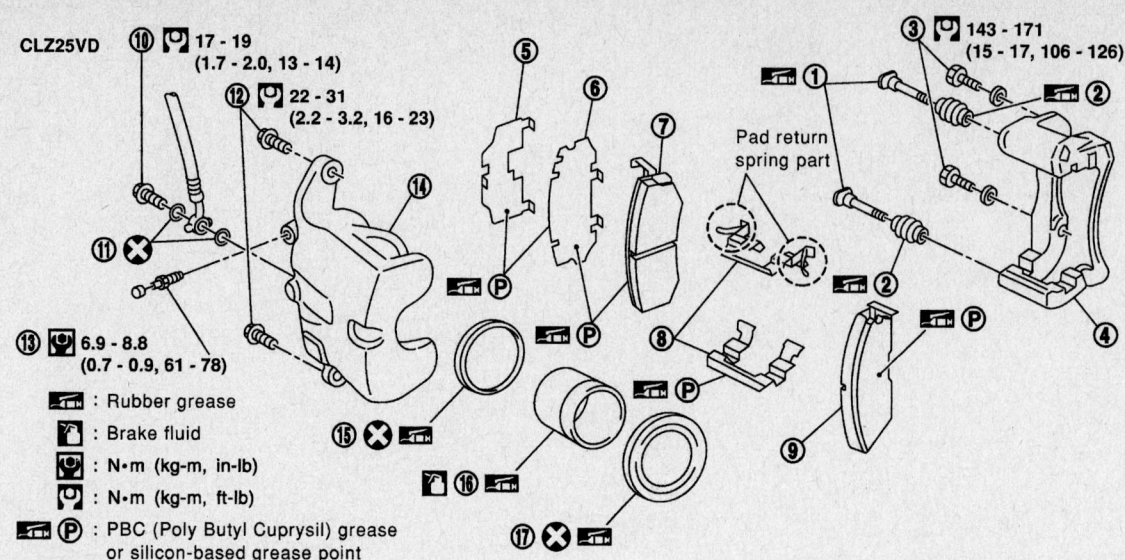

CLZ25VD

① 17 - 19
(1.7 - 2.0, 13 - 14)

② 22 - 31
(2.2 - 3.2, 16 - 23)

③ 143 - 171
(15 - 17, 106 - 126)

⑬ 6.9 - 8.8
(0.7 - 0.9, 61 - 78)

: Rubber grease

: Brake fluid

: N·m (kg-m, in-lb)

: N·m (kg-m, ft-lb)

Ⓟ : PBC (Poly Butyl Cuprysil) grease
or silicon-based grease point

Pad return spring part

1. Main pin
2. Pin boot
3. Torque member fixing bolt
4. Torque member
5. Shim cover
6. Inner shim
7. Inner pad
8. Pad retainer
9. Outer pad
10. Connecting bolt
11. Copper washer
12. Main pin bolt
13. Bleed valve
14. Cylinder body
15. Piston seal
16. Piston
17. Piston boot

09482_INFI_G0136

Component view of front brake assembly—I35

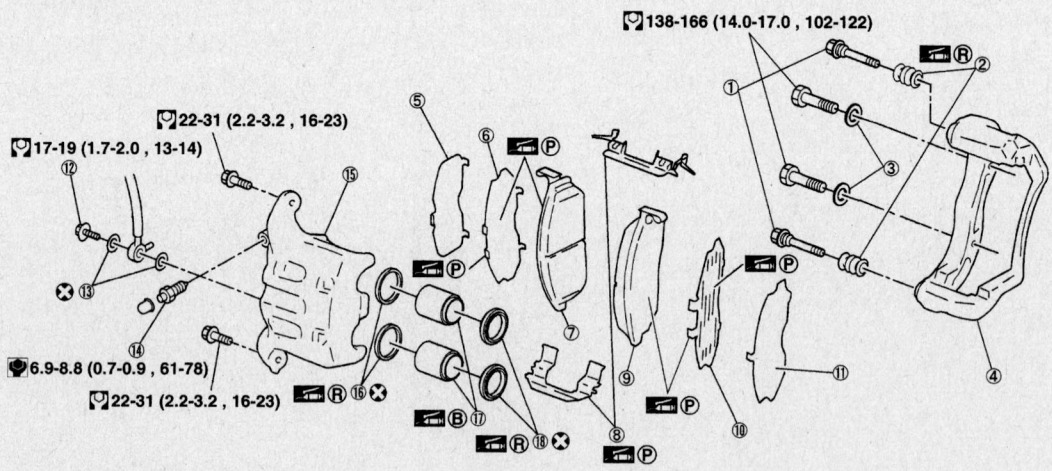

CLZ31VA

138-166 (14.0-17.0 , 102-122)

22-31 (2.2-3.2 , 16-23)

17-19 (1.7-2.0 , 13-14)

6.9-8.8 (0.7-0.9 , 61-78)

22-31 (2.2-3.2 , 16-23)

Ⓧ : Always replace after every disassembly
Ⓟ : PBC(Poly Butyl Cuprysil) grease or silicone-based grease point
Ⓡ : Rubber Grease point
Ⓑ : Brake fluid point
: N·m (kg-m , in-lb)
: N·m (kg-m , ft-lb)

1. Main pin
2. Pin boot
3. Washer
4. Torque member
5. Inner shim cover
6. inner shim
7. Inner pad
8. Pad retainer
9. Outer pad
10. Outer shim
11. Outer shim cover
12. Connecting bolt
13. Copper washer
14. Bleed valve
15. Cylinder body
16. Piston seal
17. Piston
18. Piston boot

09482_INFI_G0452

Front caliper and related parts—2002–05 Q45

8. Install brake tube to the caliper assembly and partially tighten flare nut.

9. Install brake tube bracket to knuckle spindle and tighten to 8–11 ft. lbs. (10.8–15.6 Nm).

10. Using a flare nut torque wrench, tighten the caliper assembly

and brake tube connection flare nut to 11–12 ft. lbs. (14.8–17.6 Nm).

11. Install brake pads (Refer to Disc Brake Pads, Removal and Installation).

12. Install the wheels and tighten the lug nuts to the proper specification.

13. Bleed the brake system and top off the master cylinder as necessary.

I35 Models

1. Before servicing the vehicle, refer to the precautions in the beginning of this section.

2. Remove the front wheels.

3. Remove both guide pin bolts securing the caliper to the steering knuckle.

4. Loosen and remove the brake hose connector from the caliper.

5. Remove the caliper assembly from the vehicle.

To install:

6. Using new copper washers, install the brake line to the brake caliper Refer to the accompanying illustrations for torque values and procedures.

7. Install the caliper to the steering knuckle using the guide pins bolts.

8. Install the wheels and tighten the lug nuts to the proper specification.

9. Bleed the brake system and top off the master cylinder as necessary.

2002–06 Q45 Models

1. Before servicing the vehicle, refer to the precautions in the beginning of this section.

2. Remove the wheels.

3. Remove the brake hose connecting bolt.

4. Remove the torque member mounting bolts and disconnect the brake fluid hose from the caliper.

5. Slide the caliper off of the rotor.

6. Remove the caliper from the vehicle.

To install:

7. Position the torque member on the knuckle assembly and install the mounting bolts. Refer to the accompanying illustrations for torque values and procedures.

8. Using new copper washers, connect the hydraulic hose to the caliper. Torque the union bolt to 12–14 ft. lbs. (17–19 Nm).

9. Bleed the air from the system and fill the master cylinder with clean brake fluid.

M35 and M45 Models

1. Before servicing the vehicle, refer to the precautions in the beginning of this section.

2. Remove the wheels.

3. Remove the brake hose connecting bolt.

4. Remove the torque member mounting bolts and disconnect the brake fluid hose from the caliper.

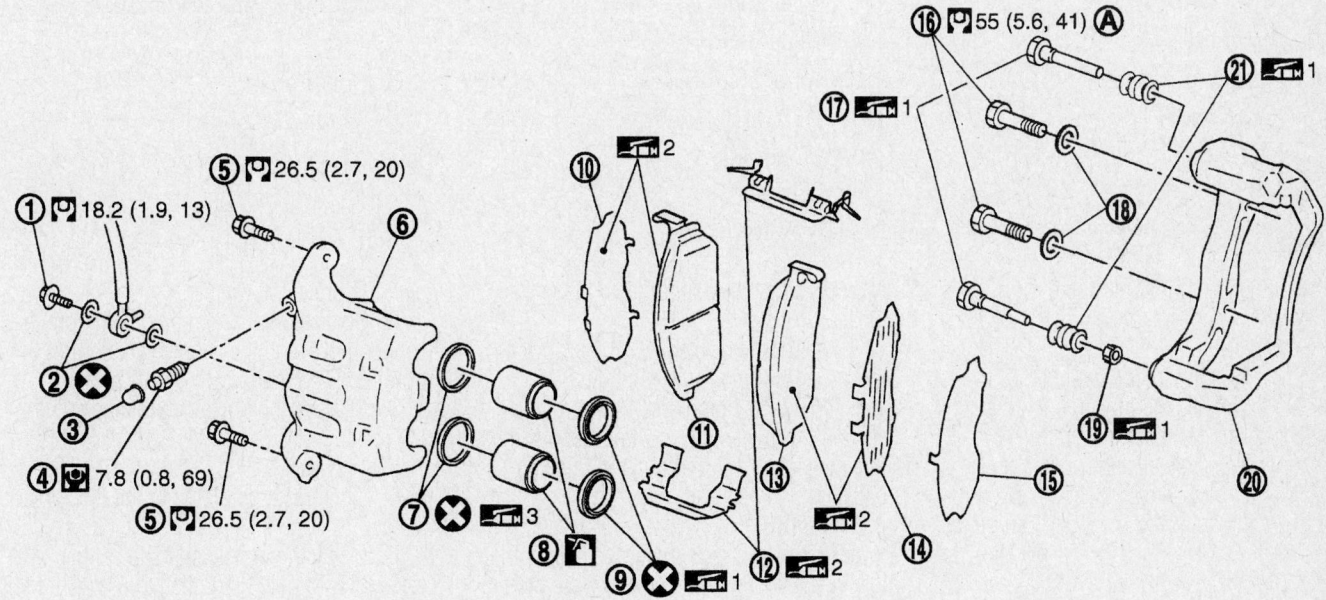

1. Union bolt	2. Copper washer	3. Cap
4. Bleed valve	5. Sliding pin bolt	6. Cylinder body
7. Piston seal	8. Piston	9. Piston boot
10. Inner shim	11. Inner pad	12. Pad retainer
13. Outer pad	14. Outer shim	15. Outer shim cover
16. Torque member mounting bolt	17. Sliding pin	18. Washer
19. Bushing	20. Torque member	21. Sliding pin boot

A: After tightening the bolts to the specified torque, tighten the bolts additionally by turning the bolts 32 to 37 degrees.

09482_INFI_G0453

Front caliper and related parts—2006 Q45

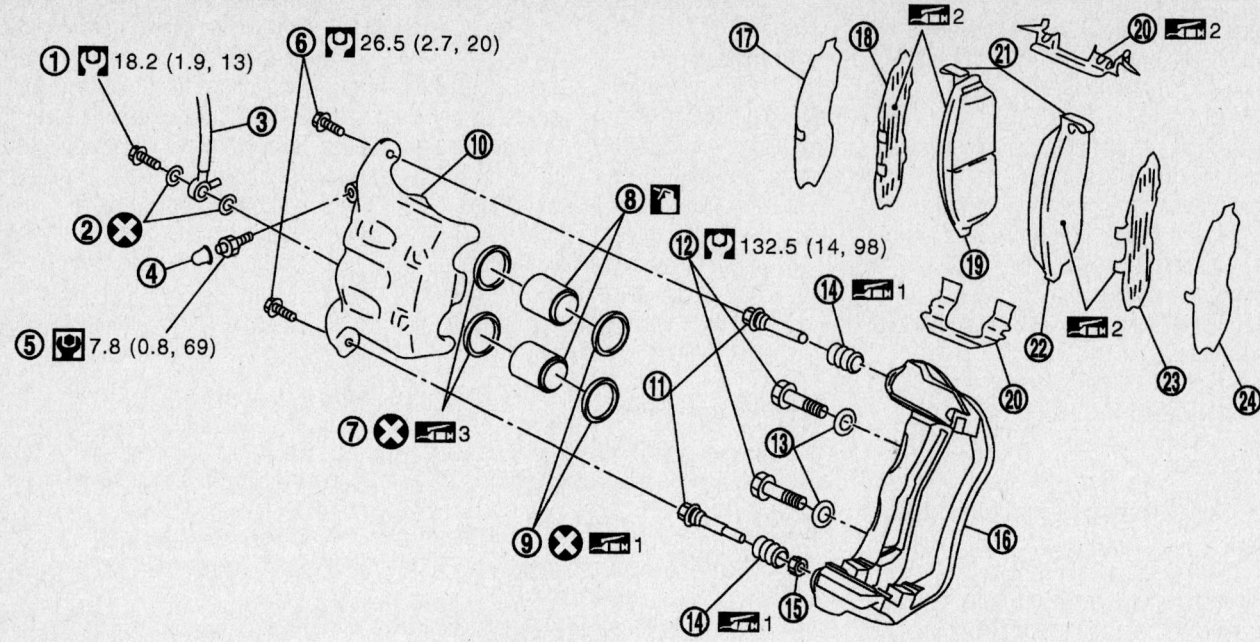

① 🔧 18.2 (1.9, 13) ⑥ 🔧 26.5 (2.7, 20)

⑫ 🔧 132.5 (14, 98)

⑤ 🔧 7.8 (0.8, 69)

1.	Union bolt	2.	Copper washer	3.	Brake hose
4.	Cap	5.	Bleed valve	6.	Sliding pin bolt
7.	Piston seal	8.	Piston	9.	Piston boot
10.	Cylinder body	11.	Sliding pin	12.	Torque member mounting bolt
13.	Washer	14.	Sliding pin boot	15.	Bushing
16.	Torque member	17.	Inner shim cover	18.	Inner shim
19.	Inner pad	20.	Pad retainer	21.	Pad wear sensor
22.	Outer pad	23.	Outer shim	24.	Outer shim cover

09482_INFI_G0129

Component view of the front brake assembly—M35 and M45 Models

5. Slide the caliper off of the rotor.
6. Remove the caliper from the vehicle.
To install:
7. Position the torque member on the knuckle assembly and install the mounting bolts. Refer to the accompanying illustrations for torque values and procedures.
8. Using new copper washers, connect the hydraulic hose to the caliper. Torque the union bolt to 12–14 ft. lbs. (17–19 Nm).
9. Bleed the air from the system and fill the master cylinder with clean brake fluid.

Disc Brake Pads

REMOVAL AND INSTALLATION

G20 Models

1. Before servicing the vehicle, refer to the precautions in the beginning of this section.
2. Remove the cap from the master cylinder reservoir and extract about ⅓ of the brake fluid from the reservoir to prevent

overflow when the caliper piston is compressed.
3. Remove the wheels.
4. Remove the lower pin bolt.
5. Pivot the caliper body upward and secure it with a length of wire. Remove the retainers and inner and outer shims and pads.
To install:
6. Place an old pad over the caliper piston. Use a C-clamp to compress the piston.
7. Install the new pads and shims and rotate caliper down onto rotor. Install the pin bolt and torque it to 16–23 ft. lbs. (22–31 Nm).
8. Install the wheels and lower the vehicle to the floor.
9. Check and then refill the master cylinder if needed.

G35 Models

W/STANDARD BRAKES

1. Before servicing the vehicle, refer to the precautions in the beginning of this section.

2. Remove the wheels.
3. Remove the bottom guide pin from the caliper and swing the caliper cylinder body upward. Support the caliper with a wire.
4. Remove the brake pad retainers and the pads.
To install:
5. Compress the piston of the disc brake caliper.
6. Install the brake pads and caliper assembly. Torque the guide pin according to the illustration.
7. Install the wheels.
8. Check the master cylinder and add fluid if necessary.

W/BREMBO® OPTION

1. Before servicing the vehicle, refer to the precautions in the beginning of this section.
2. Remove the front wheels.
3. Remove clips from pad pin.
4. Remove pad pin while holding down cross spring, then remove cross spring from caliper.

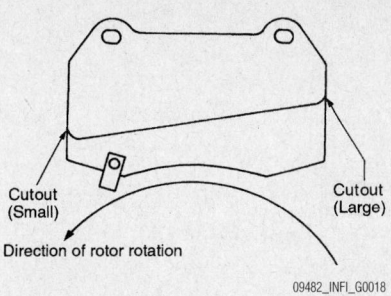

Note position of large cutout in relation to disc rotation — Brembo®

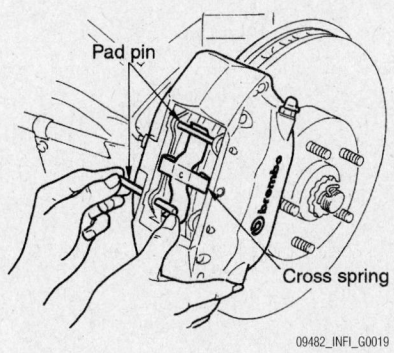

Make certain cross spring passes UNDER pad pins—Brembo®

5. Using pliers, remove pad from caliper.
   To install:
6. Apply grease to brake pad back side.
7. Compress the pistons of the disc brake caliper.
8. Install pads.

➡**Install pad with wear sensor on the outer side.**

➡**The side of pad (shim) with the larger cutouts should be on the entry side of the disc rotor spin.**

9. Insert upper pad pin from the inner piston side and press firmly to the outer piston side through the hole in the top of pads.
10. Place the top of the cross spring under the top pad pin; while pressing bottom of cross spring against pads, insert lower pad pin from the inner piston side and press firmly to the outer piston side through the hole in the top of pads.
11. Insert clips in the small hole at the end of pad pins.

❋❋ CAUTION

If clips are not fully attached, pad pin or the pad could fall out while vehicle is in motion.

12. Install the wheels.
13. Check the master cylinder and add fluid if necessary.

I35

1. Before servicing the vehicle, refer to the precautions in the beginning of this section.
2. Remove the wheels.
3. Remove the bottom guide pin from the caliper and swing the caliper cylinder body upward. Support the caliper with a wire.

4. Remove the brake pad retainers and the pads.
   To install:
5. Compress the piston of the disc brake caliper.
6. Install the brake pads and caliper assembly. Torque the guide pin according to the illustration.
7. Install the wheels.
8. Check the master cylinder and add fluid if necessary.

M45 and Q45 Models

1. Before servicing the vehicle, refer to the precautions in the beginning of this section.
2. Remove the cap from the master cylinder reservoir and extract about ⅓ of the brake fluid from the reservoir to prevent overflow when the caliper piston is compressed.
3. Remove the wheels.
4. Remove the lower pin bolt.
5. Pivot the caliper body upward and secure it with a length of wire. Remove the retainers and inner and outer shims and pads.
   To install:
6. Place an old pad over the caliper piston. Use a C-clamp to compress the piston.
7. Install the new pads and shims and rotate caliper down onto rotor. Install the pin bolt and torque it according to the illustration.
8. Install the wheels.
9. Check and refill master cylinder if needed.

REAR BRAKES

Brake Caliper

REMOVAL & INSTALLATION

G20 Models

1. Before servicing the vehicle, refer to the precautions in the beginning of this section.
2. Remove the rear wheels.
3. Remove the brake cable mounting bolt and lock spring.
4. Disconnect the parking brake cable from the caliper.
5. Disconnect the brake fluid hose.
6. Remove the torque member mounting bolts and remove the caliper assembly.
   To install:
7. Fit the caliper onto the torque member and torque the bolts to 16–23 ft. lbs. (22–31 Nm).

8. Using new copper washers, connect the hydraulic hose to the caliper. Torque the union bolt to 12–14 ft. lbs. (17–19 Nm).
9. Connect the parking brake cable to the rear caliper.
10. Bleed the brake system.
11. Install the rear wheels.

G35 Models

W/STANDARD BRAKES

1. Before servicing the vehicle, refer to the precautions in the beginning of this section.
2. Remove the rear wheels.
3. Remove the parking brake cable stay fixing bolt and the lock spring.
4. Remove the brake fluid hose from the caliper.
5. Remove the guide pin bolts and remove the caliper.

To install:
6. Install the caliper body into position Refer to the accompanying illustrations for torque values and procedures.
7. Reconnect the brake fluid hose and tighten the flare nut to 12–14 ft. lbs. (17–20 Nm).
8. Install the lock spring and the parking brake stay attaching bolt.
9. Bleed the brake system and top off the master cylinder as necessary.
10. Install the wheels.

W/BREMBO® OPTION

1. Before servicing the vehicle, refer to the precautions in the beginning of this section.
2. Remove the rear wheels.
3. Remove brake pads (Refer to Disc Brake Pads, Removal and Installation).

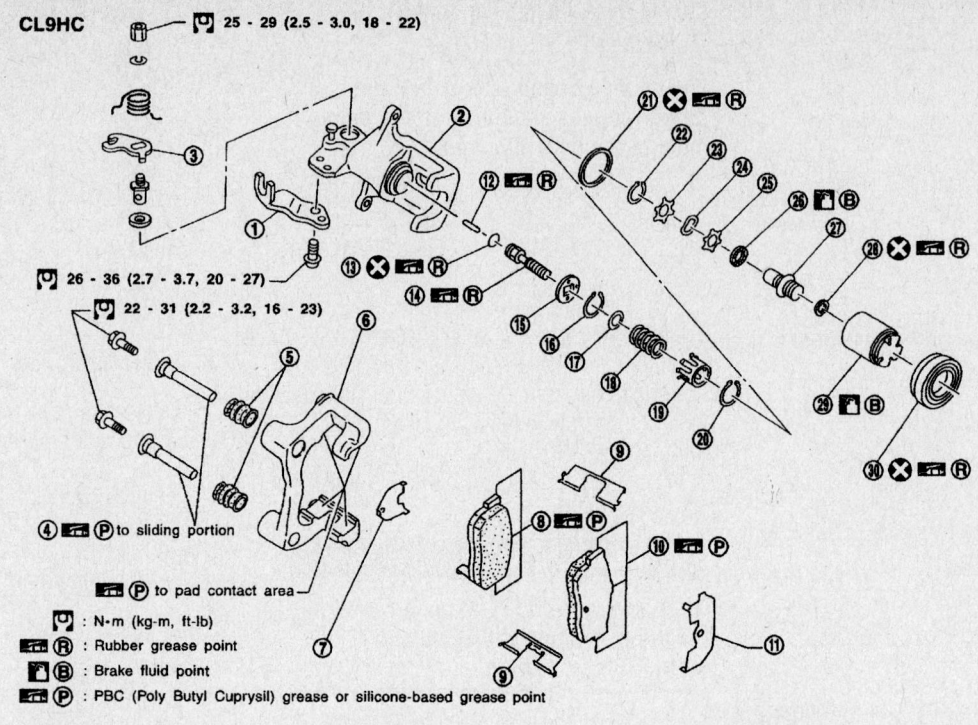

CL9HC

25 - 29 (2.5 - 3.0, 18 - 22)

26 - 36 (2.7 - 3.7, 20 - 27)

22 - 31 (2.2 - 3.2, 16 - 23)

to sliding portion

(P) to pad contact area

: N·m (kg-m, ft-lb)

(R) : Rubber grease point

(B) : Brake fluid point

(P) : PBC (Poly Butyl Cuprysil) grease or silicone-based grease point

1. Cable guide	11. Outer shim	21. Piston seal
2. Cylinder	12. Strut	22. Snap ring A
3. Toggle lever	13. O-ring	23. Spacer
4. Pin	14. Push rod	24. Wave washer
5. Pin boot	15. Key plate	25. Spacer
6. Torque member	16. Snap ring C	26. Bearing
7. Inner shim	17. Seat	27. Adjuster
8. Inner pad	18. Spring	28. Cup
9. Pad retainer	19. Spring cover	29. Piston
10. Outer pad	20. Snap ring B	30. Piston boot

09482_INFI_G0128

Component view of rear brake assembly—G20

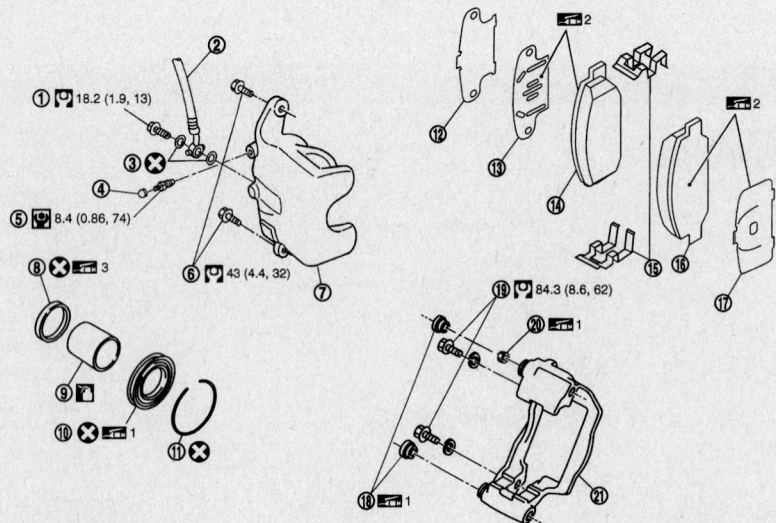

① 18.2 (1.9, 13)

⑤ 8.4 (0.86, 74)

⑥ 43 (4.4, 32)

⑲ 84.3 (8.6, 62)

1. Union bolt	2. Brake hose	3. Copper washer
4. Cap	5. Bleed valve	6. Sliding pin bolt
7. Cylinder body	8. Piston seal	9. Piston
10. Piston boot	11. Retaining ring	12. Inner shim cover
13. Inner shim	14. Inner pad	15. Pad retainer
16. Outer pad	17. Outer shim	18. Slide pin boot
19. Torque member bolt	20. Bushing	21. Torque member

09482_INFI_G0450

Rear brake caliper and related parts—2005–06 G35

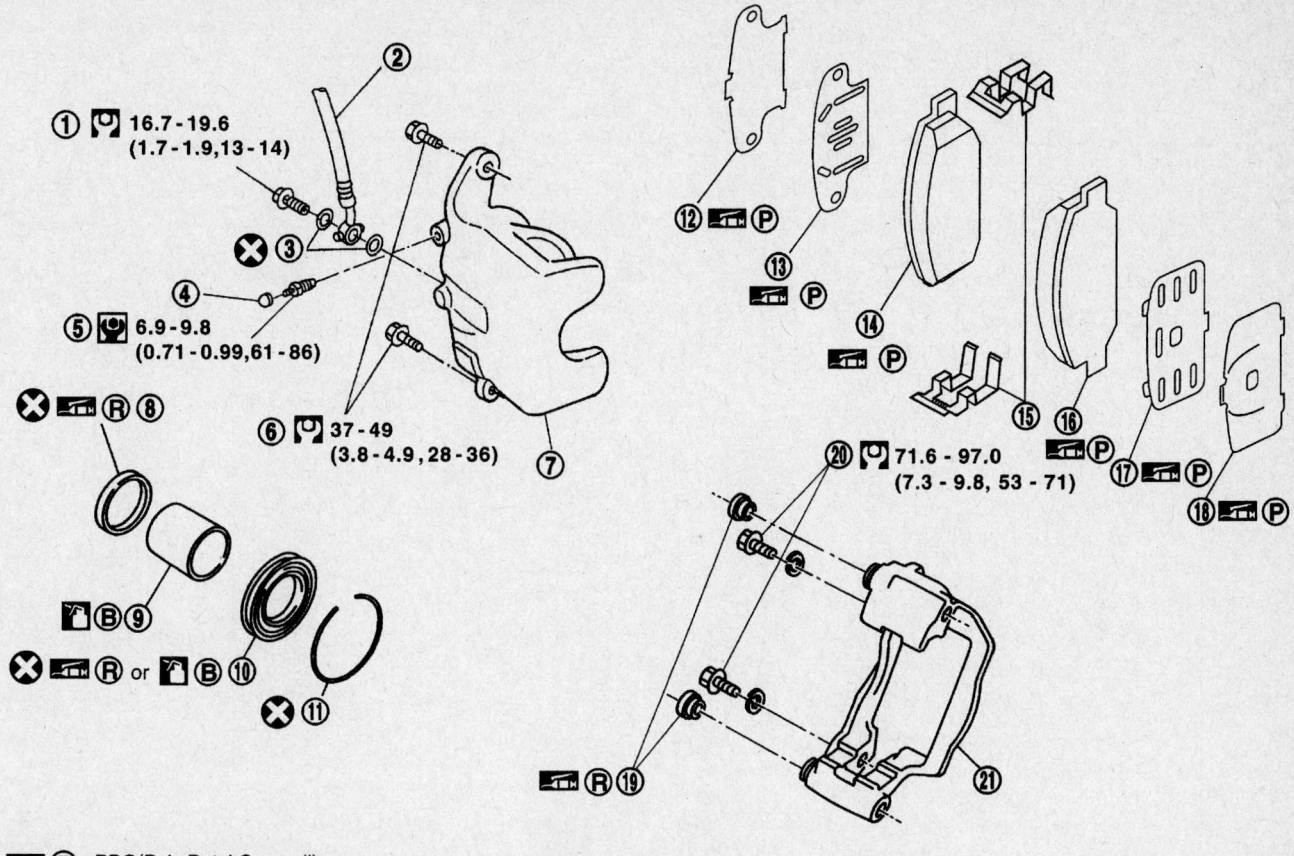

① 🔧 16.7 - 19.6
(1.7 - 1.9,13 - 14)

⑤ 🔧 6.9 - 9.8
(0.71 - 0.99,61 - 86)

⑥ 🔧 37 - 49
(3.8 - 4.9 , 28 - 36)

🔧 71.6 - 97.0
(7.3 - 9.8 , 53 - 71)

🔧 (P) : PBC(Poly Butyl Cuprysil) grease or silicone-based grease point

🔧 (R) : Rubber grease point

🔧 (B) : Brake fluid point

🔧 : N•m (kg-m, ft-lb)

🔧 : N•m (kg-m, in-lb)

❌ : Always replace after every disassembly.

1. Union bolt	2. Brake hose	3. Copper washer
4. Cap	5. Bleed valve	6. Sliding pin bolt
7. Cylinder body	8. Piston seal	9. Piston
10. Piston boot	11. Retaining ring	12. Inner shim cover
13. Inner shim	14. Inner pad	15. Pad retainer
16. Outer pad	17. Outer shim	18. Outer shim cover
19. Slide pin boot	20. Torque member bolts	21. Torque member

09482_INFI_G0451

Rear brake caliper and related parts—2002–2004.5 G35 w/standard brakes

4. Remove brake tube flare nut using a flare wrench.

5. Remove brake hose bolt and move brake hose aside.

6. Remove caliper bolt and remove caliper from vehicle.

To install:

7. Install brake tube to caliper assembly and partially tighten flare nut.

8. Install caliper assembly to vehicle and tighten bolt to 53–71 ft. lbs. (71.6–97 Nm).

9. Tighten flare nut to 11–12 ft. lbs. (14.8–17.6 Nm).

10. Install brake hose bolt.

11. Install brake pad and shim (Refer to Disc Brake Pads, Removal and Installation).

12. Install the wheels and tighten the lug nuts to the proper specification.

13. Bleed the brake system and top off the master cylinder as necessary.

I35 Models

1. Before servicing the vehicle, refer to the precautions in the beginning of this section.

2. Remove the rear wheels.

3. Remove the parking brake cable stay fixing bolt and the lock spring.

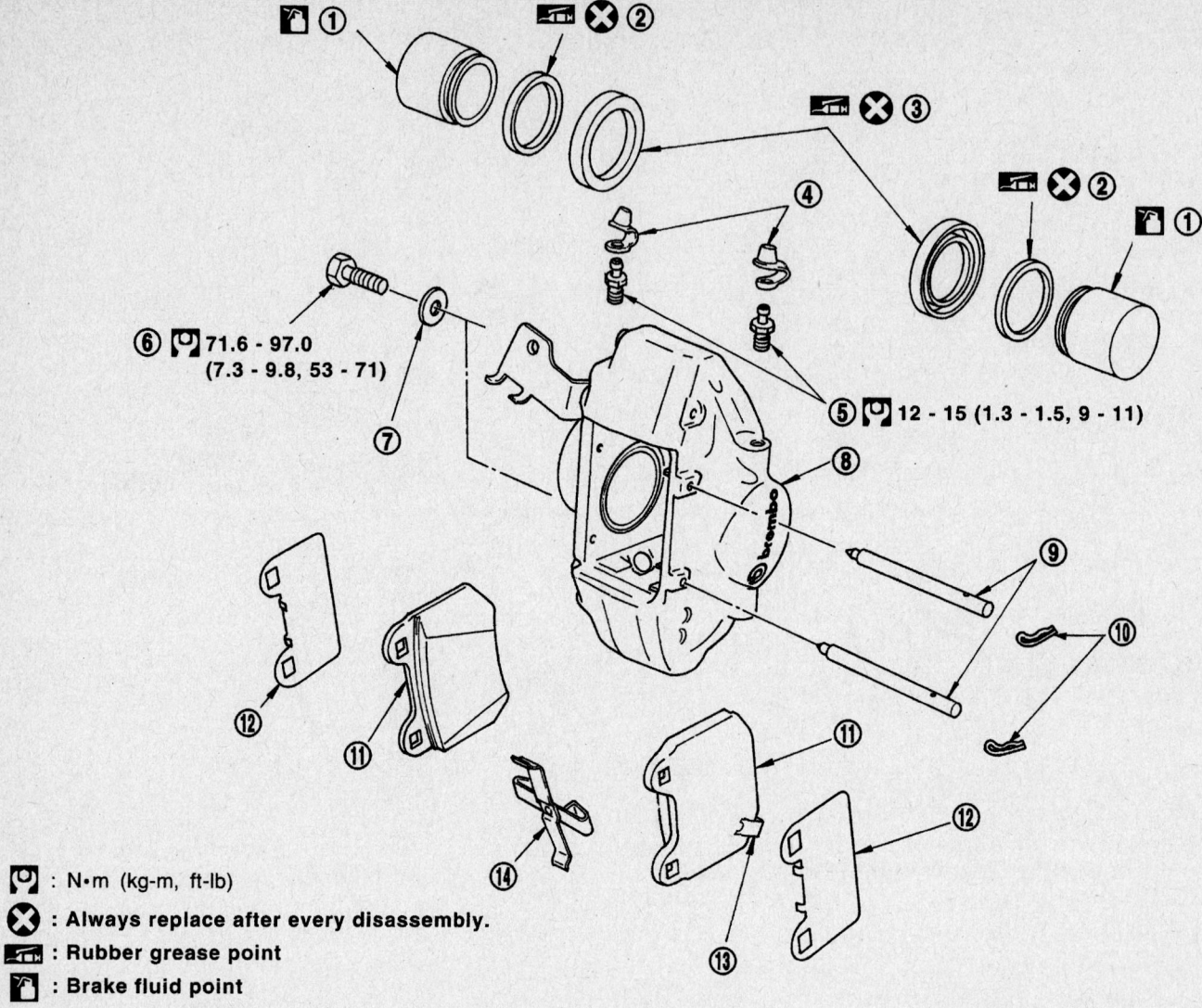

: N•m (kg-m, ft-lb)

: Always replace after every disassembly.

: Rubber grease point

: Brake fluid point

1.	Piston	2.	Piston seal	3.	Piston boot
4.	Cap	5.	Bleed valve	6.	Bolt
7.	Washer	8.	Caliper	9.	Pad pins
10.	Clips	11.	Brake pad	12.	Shim
13.	Pad wear sensor	14.	Cross spring		

09482_INFI_G0020

Component view of rear brake assembly—Brembo®

4. Remove the brake fluid hose from the caliper.

5. Remove the guide pin bolts and remove the caliper.

To install:

6. Install the caliper body into position Refer to the accompanying illustrations for torque values and procedures.

7. Reconnect the brake fluid hose and tighten the flare nut to 12–14 ft. lbs. (17–20 Nm).

8. Install the lock spring and the parking brake stay attaching bolt.

9. Bleed the brake system and top off the master cylinder as necessary.

10. Install the wheels.

2002–06 Q45 Models

1. Before servicing the vehicle, refer to the precautions in the beginning of this section.

2. Remove the rear wheels.

3. Disconnect the brake fluid hose.

4. Remove the torque member mounting bolts and remove the caliper assembly.

To install:

5. Fit the caliper over the rotor and install the mounting bolts. Refer to the accompanying illustrations torque values and procedures.

6. Using new copper washers, connect the hydraulic hose to the caliper. Torque the union bolt to 12–14 ft. lbs. (17–19 Nm).

7. Bleed the brake system.

8. Install the rear wheels and lower the vehicle to the floor.

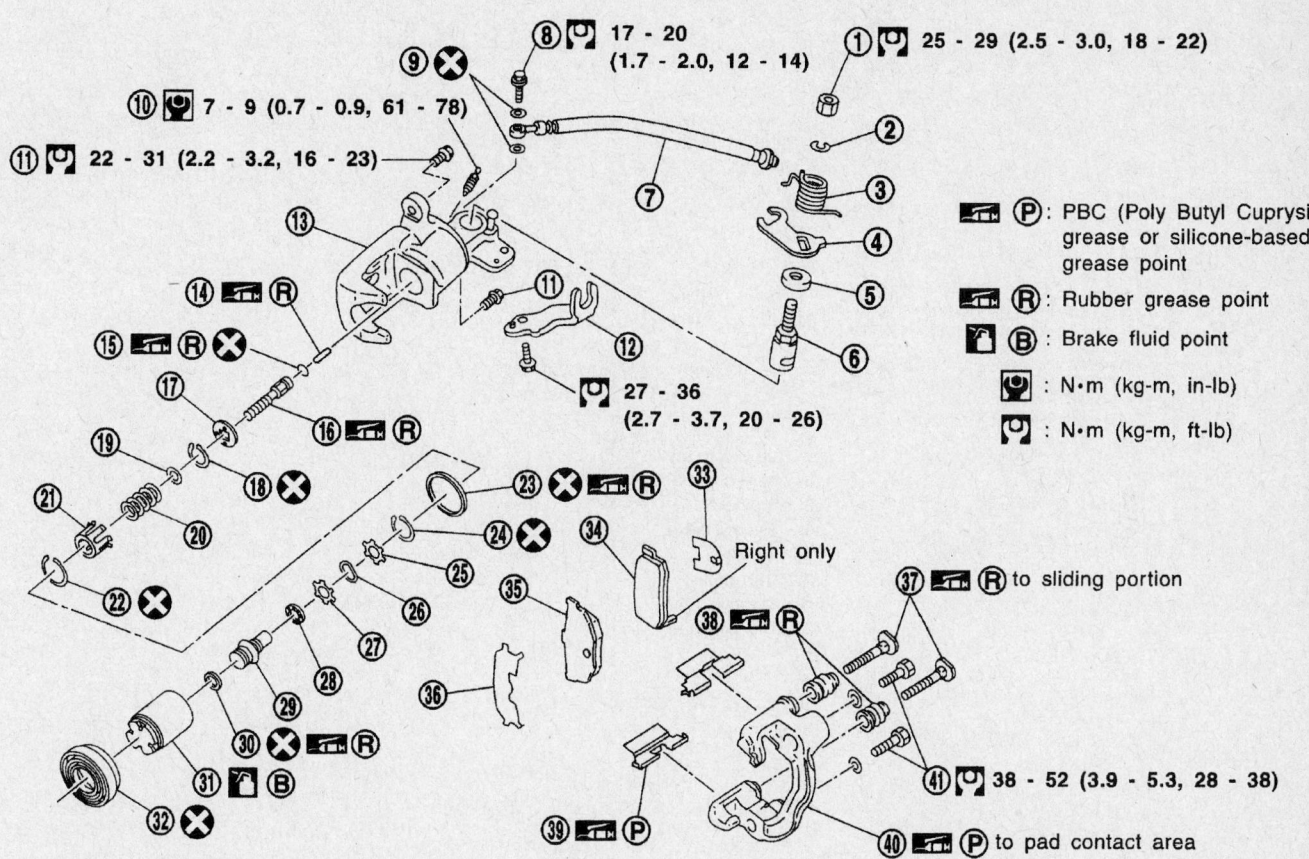

⑧ 17 - 20
(1.7 - 2.0, 12 - 14)

① 25 - 29 (2.5 - 3.0, 18 - 22)

⑩ 7 - 9 (0.7 - 0.9, 61 - 78)

⑪ 22 - 31 (2.2 - 3.2, 16 - 23)

Ⓟ: PBC (Poly Butyl Cuprysil) grease or silicone-based grease point

Ⓡ: Rubber grease point

Ⓑ: Brake fluid point

: N•m (kg-m, in-lb)

: N•m (kg-m, ft-lb)

27 - 36
(2.7 - 3.7, 20 - 26)

Right only

Ⓡ to sliding portion

⑪ 38 - 52 (3.9 - 5.3, 28 - 38)

Ⓟ to pad contact area

1. Nut	15. O-ring	29. Adjust nut
2. Washer	16. Push rod	30. Cup
3. Return spring	17. Key plate	31. Piston
4. Parking brake lever	18. Ring C	32. Dust seal
5. Cam boot	19. Seat	33. Inner shim
6. Cam	20. Spring	34. Inner pad
7. Brake hose	21. Spring cover	35. Outer pad
8. Connecting bolt	22. Ring B	36. Outer shim
9. Copper washer	23. Piston seal	37. Pin
10. Bleed screw	24. Ring A	38. Pin boot
11. Pin bolt	25. Spacer	39. Pad retainer
12. Cable mounting bracket	26. Wave washer	40. Torque member
13. Cylinder	27. Spacer	41. Torque member fixing bolt
14. Strut	28. Ball bearing	

09482_INFI_G0137

Component view of rear brake assembly—I35

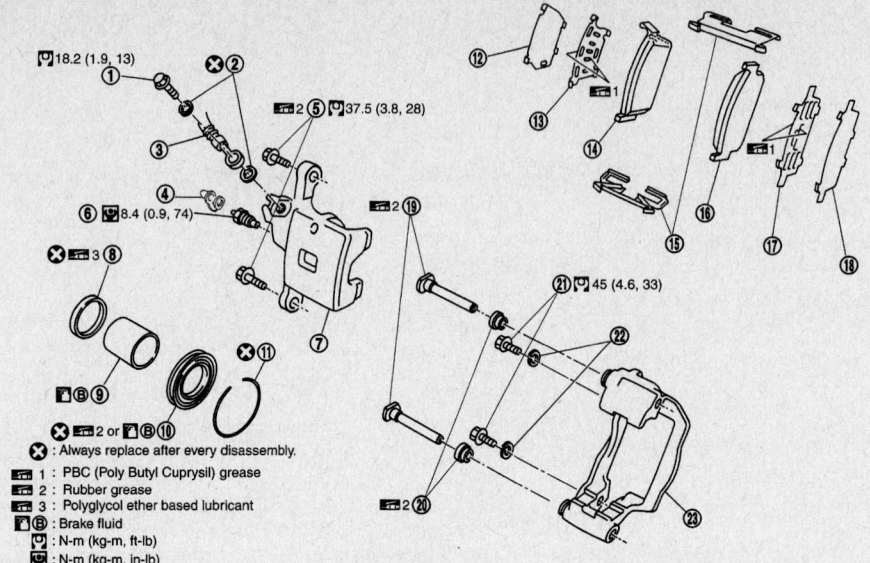

1. Union bolt
2. Copper washer
3. Brake hose
4. Cap
5. Sliding pin bolt
6. Bleed valve
7. Cylinder body
8. Piston seal
9. Piston
10. Piston boot
11. Retaining ring
12. Inner shim cover
13. Inner shim
14. Inner pad
15. Pad retainer
16. Outer pad
17. Outer shim
18. Outer shim cover
19. Sliding pin
20. Sliding pin boot
21. Torque member mounting bolt
22. Washer
23. Torque member

⊗ : Always replace after every disassembly.
⊟ 1 : PBC (Poly Butyl Cuprysil) grease
⊟ 2 : Rubber grease
⊟ 3 : Polyglycol ether based lubricant
⊟ B : Brake fluid
N-m (kg-m, ft-lb)
N-m (kg-m, in-lb)

09482_INFI_G0454

Rear caliper and related parts—2002–06 Q45

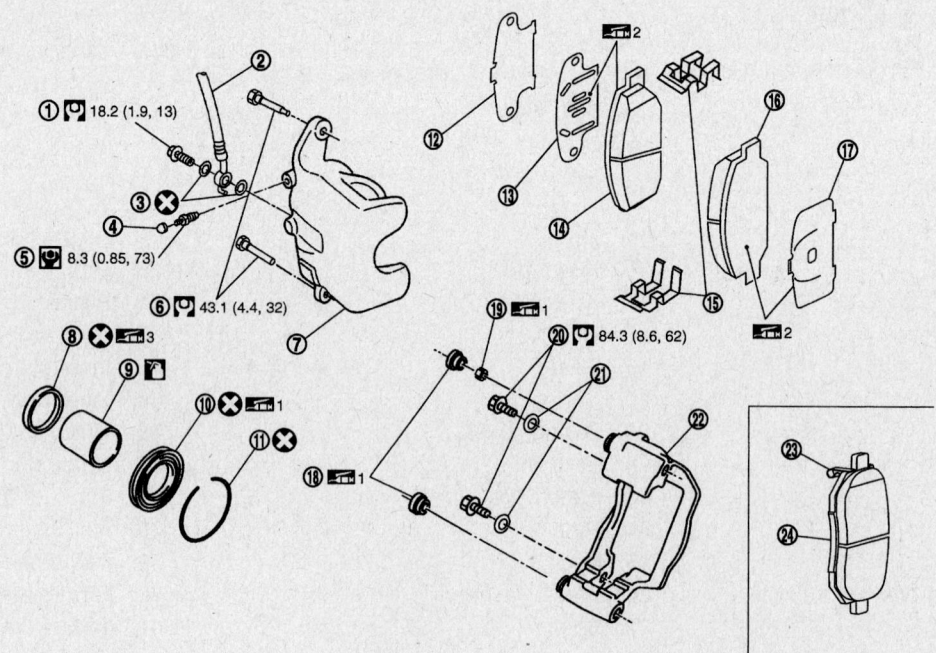

1. Union bolt
2. Brake hose
3. Copper washer
4. Cap
5. Bleed valve
6. Sliding pin bolt
7. Cylinder body
8. Piston seal
9. Piston
10. Piston boot
11. Retaining ring
12. Inner shim cover
13. Inner shim
14. Inner pad
15. Pad retainer
16. Outer pad
17. Outer shim
18. Sliding pin boot
19. Bushing
20. Torque member mounting bolt
21. Washer
22. Torque member
23. Pad wear sensor
24. Inner pad (RH)

09482_INFI_G0130

Component view of the rear brake assembly—M35 and M45 Models

M35 and M45 Models

1. Before servicing the vehicle, refer to the precautions in the beginning of this section.
2. Remove the rear wheels.
3. Disconnect the brake fluid hose.
4. Remove the torque member mounting bolts and remove the caliper assembly.

To install:

5. Fit the caliper over the rotor and install the mounting bolts. Refer to accompanying illustrations for torque values and procedures.
6. Using new copper washers, connect the hydraulic hose to the caliper. Torque the union bolt to 12–14 ft. lbs. (17–19 Nm).
7. Bleed the brake system.
8. Install the rear wheels and lower the vehicle to the floor.

Disc Brake Pads

REMOVAL AND INSTALLATION

G20 Models

1. Before servicing the vehicle, refer to the precautions in the beginning of this section.
2. Remove the cap from the master cylinder reservoir and extract about ⅓ of the brake fluid from the reservoir to prevent overflow when the caliper piston is compressed.
3. Remove the wheels.
4. Remove the brake cable mounting bracket bolt and lock spring.
5. Disconnect the parking brake cable.
6. Remove the lower pin bolt.
7. Pivot the caliper body upward and secure it with a length of wire. Remove the retainers and inner and outer shims and pads.

To install:

8. Push the piston into the cylinder body by turning the piston clockwise.
9. Install the new pads and shims and rotate the caliper down onto rotor. Install the pin bolt and torque it according to the illustration.
10. Connect the parking brake cable and install the bracket.
11. Install the wheels.
12. Check and then refill the master cylinder if needed.

G35 Models

W/STANDARD BRAKES

1. Before servicing the vehicle, refer to the precautions in the beginning of this section.

2. Remove the rear wheels.
3. Remove the parking brake cable mounting bolt and lock spring.
4. Disconnect the cable from the caliper.
5. Remove the upper pin bolt.
6. Pivot the caliper body downward.
7. Pull out the pad springs and remove the pads and shims.

To install:

8. Turn the piston clockwise back into the caliper body. Take care not to damage the piston boot.
9. Coat the pad contact area on the mounting support with grease.
10. Install the pads, shims, and the pad springs.
11. Position the caliper body in the mounting support and tighten the pin bolts according to the illustration.
12. Install the wheels.
13. Check the master cylinder and add fluid if necessary.

W/BREMBO® OPTION

1. Before servicing the vehicle, refer to the precautions in the beginning of this section.
2. Remove the rear wheels.
3. Remove clips from pad pins.
4. Remove pad pins while holding down cross spring, then remove cross spring from caliper.

5. Using pliers, remove pad and shim from caliper.

To install:

6. Apply grease to brake pad back side.
7. Compress the pistons of the disc brake caliper.
8. Install pads.

➡ **Install pad with wear sensor on the outer side.**

➡ **The side of pad (shim) with the larger cutouts should be on the entry side of the disc rotor spin.**

9. Insert upper pad pin from the inner piston side and press firmly to the outer piston side through the hole in the top of pads.
10. Place the top of the cross spring under the top pad pin; while pressing bottom of cross spring against pads, insert lower pad pin from the inner piston side and press firmly to the outer piston side through the hole in the top of pads.
11. Insert clips in the small hole at the end of pad pins.

✳✳ CAUTION

If clips are not fully attached, pad pin or the pad could fall out while vehicle is in motion.

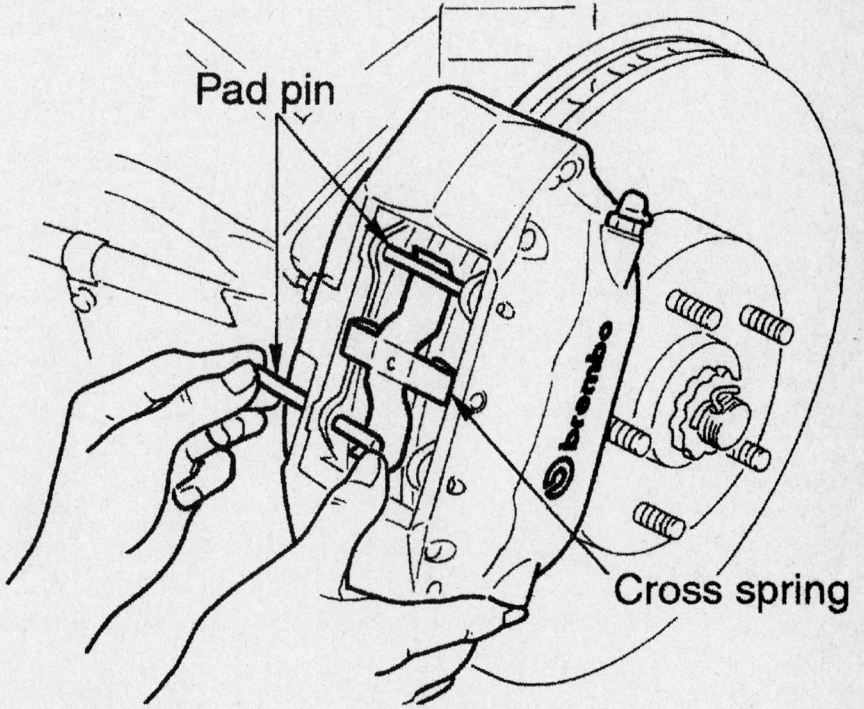

Make certain cross spring passes UNDER pad pins.

09482_INFI_G0019

12. Install the wheels.

13. Check the master cylinder and add fluid if necessary.

I35

1. Before servicing the vehicle, refer to the precautions in the beginning of this section.

2. Remove the rear wheels.

3. Remove the parking brake cable mounting bolt and lock spring.

4. Disconnect the cable from the caliper.

5. Remove the upper pin bolt.

6. Pivot the caliper body downward.

7. Pull out the pad springs and then remove the pads and shims.

To install:

8. Turn the piston clockwise back into the caliper body. Take care not to damage the piston boot.

9. Coat the pad contact area on the mounting support with grease.

10. Install the pads, shims, and the pad springs.

11. Position the caliper body in the mounting support and tighten the pin bolts according to the illustration.

12. Install the wheels.

13. Check the master cylinder and add fluid if necessary.

M45 and Q45 Models

1. Before servicing the vehicle, refer to the precautions in the beginning of this section.

2. Remove the cap from the master cylinder reservoir and extract about ⅓ of the brake fluid from the reservoir to prevent overflow when the caliper piston is compressed.

3. Remove the wheels.

4. Remove the lower pin bolt.

5. Pivot the caliper body upward and secure it with a length of wire. Remove the retainers and inner and outer shims and pads.

To install:

6. Place an old pad over the caliper piston. Use a C-clamp to compress the piston.

7. Install the new pads and shims and rotate caliper down onto rotor. Install the pin bolt and torque it according to the illustration.

8. Install the wheels.

9. Check and refill master cylinder if needed.

INFINITI

FX35 • FX45

SPECIFICATIONS AND MAINTENANCE CHARTS

ENGINE AND VEHICLE IDENTIFICATION

Code ①	Liters (cc)	Cu. In.	Cyl.	Fuel Sys.	Engine Type	Eng. Mfg.
VQ35DE	3.5 (3498)	213.5	6	SFI	DOHC	Nissan
VK45DE	4.5 (4494)	274	8	SFI	DOHC	Nissan

Code ②	Year
3	2003
4	2004
5	2005
6	2006

SFI: Sequential Fuel Injection

DOHC: Double Overhead Camshaft

① Stamped on the upper rear of the engine block, just behind a cylinder head.

② 10th digit of the Vehicle Identification Number (VIN)

09482_FX35_C0001

GENERAL ENGINE SPECIFICATIONS

Year	Model	Engine Displacement Liters	Engine Series ID	Net Horsepower @ rpm	Net Torque @ rpm (ft. lbs.)	Bore x Stroke (in.)	Com-pression Ratio	Oil Pressure @ rpm
2003	FX35	3.5	VQ35DE	280@6200	270@4800	3.76 x 3.21	10.3:1	43@2000
	FX45	4.5	VK45DE	315@6400	329@4000	3.66 x 3.26	10.5:1	43@2000
2004	FX35	3.5	VQ35DE	280@6200	270@4800	3.76 x 3.21	10.3:1	43@2000
	FX45	4.5	VK45DE	315@6400	329@4000	3.66 x 3.26	10.5:1	43@2000
2005	FX35	3.5	VQ35DE	280@6200	270@4800	3.76 x 3.21	10.3:1	43@2000
	FX45	4.5	VK45DE	315@6400	329@4000	3.66 x 3.26	10.5:1	43@2000
2006	FX35	3.5	VQ35DE	280@6200	270@4800	3.76 x 3.21	10.3:1	43@2000
	FX45	4.5	VK45DE	320@6000	335@4000	3.66 x 3.26	10.5:1	43@2000

09482_FX35_C0002

ENGINE TUNE-UP SPECIFICATIONS

Year	Engine Displacement Liters	Engine ID	Spark Plug Gap (in.)	Ignition Timing (deg.) ①	Fuel Pump (psi)	Idle Speed (rpm)	Valve Clearance ② Intake	Valve Clearance ② Exhaust
2003	3.5	VQ35DE	0.043	15B	51	600-700	0.010-0.013	0.011-0.015
	4.5	VK45DE	0.043	12B	51	600-700	0.010-0.013	0.011-0.015
2004	3.5	VQ35DE	0.043	15B	51	600-700	0.010-0.013	0.011-0.015
	4.5	VK45DE	0.043	12B	51	600-700	0.010-0.013	0.011-0.015
2005	3.5	VQ35DE	0.043	15B	51	600-700	0.010-0.013	0.011-0.015
	4.5	VK45DE	0.043	12B	51	600-700	0.010-0.013	0.011-0.015
2006	3.5	VQ35DE	0.043	15B	51	600-700	0.010-0.013	0.011-0.015
	4.5	VK45DE	0.043	12B	51	600-700	0.010-0.013	0.011-0.015

NOTE: The Vehicle Emission Control Information label often reflects specification changes made during production.

The label figures must be used if they differ from those in this chart.

B: Before top dead center

① With terminals TC and CG of DLC3 connected

② Engine cold - approximately 68°F (20°C)

09482_FX35_C0003

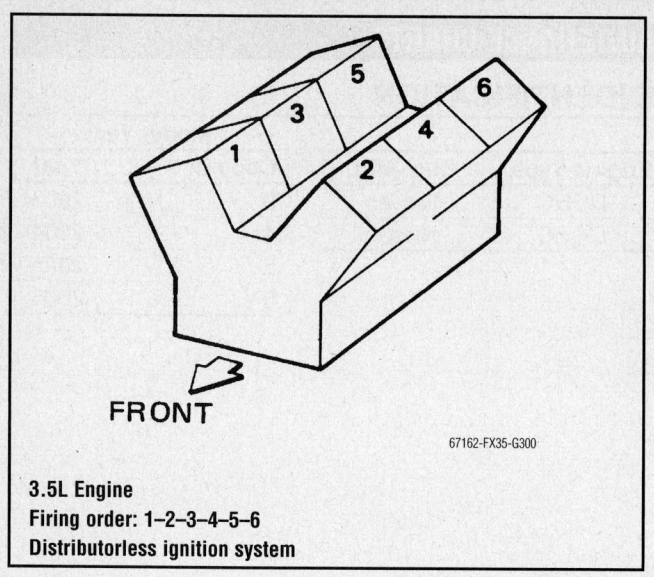

FRONT

67162-FX35-G300

3.5L Engine
Firing order: 1–2–3–4–5–6
Distributorless ignition system

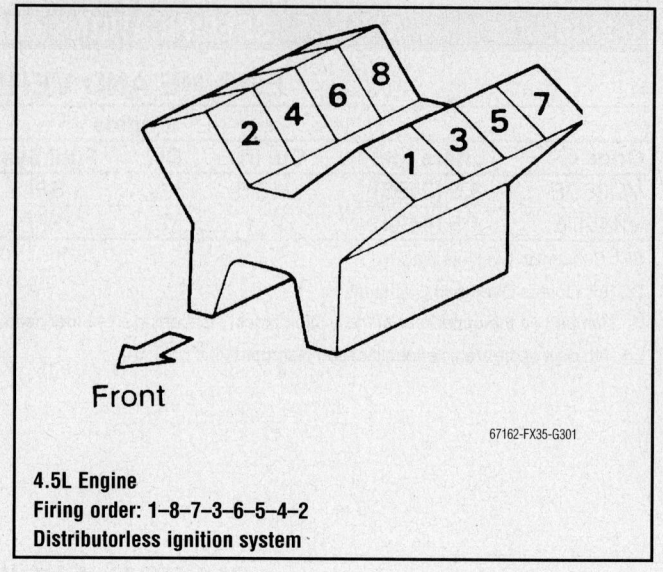

Front

67162-FX35-G301

4.5L Engine
Firing order: 1–8–7–3–6–5–4–2
Distributorless ignition system

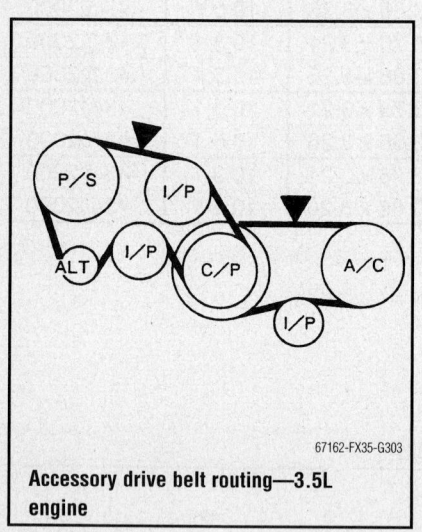

67162-FX35-G303

Accessory drive belt routing—3.5L engine

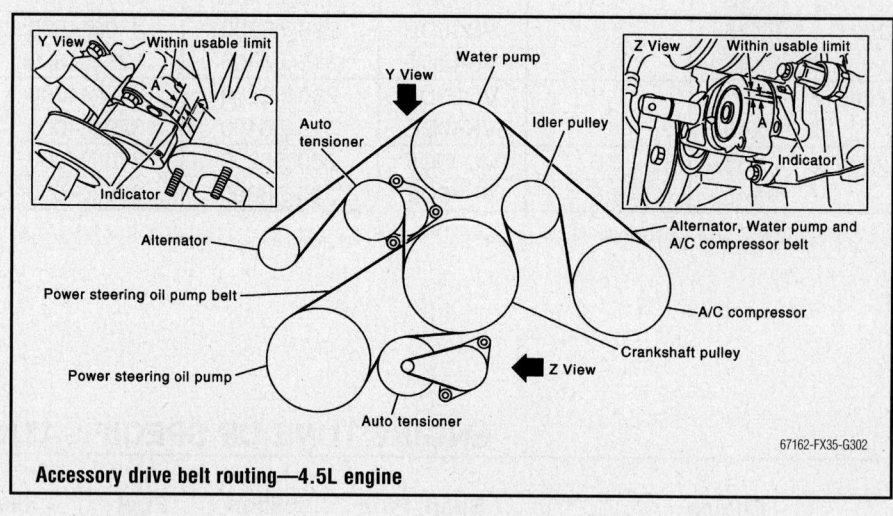

Y View — Within usable limit
Indicator
Alternator
Power steering oil pump belt
Power steering oil pump
Auto tensioner
Auto tensioner
Water pump
Y View
Idler pulley
Z View — Within usable limit
Indicator
Alternator, Water pump and A/C compressor belt
A/C compressor
Crankshaft pulley
Z View

67162-FX35-G302

Accessory drive belt routing—4.5L engine

CAPACITIES

Year	Model	Engine Displacement Liters	Engine ID	Engine Oil with Filter (qts.)	Transmission (pts)	Transfer Case (pts.)	Front Drive Axle (pts.)	Rear Drive Axle (pts.)	Fuel Tank (gal.)	Cooling System (qts.)
2003	FX35	3.5	VQ35DE	5.0	21.7	2.6	1.4	3.0	23.8	9.1
	FX45	4.5	VK45DE	7.0	21.7	2.6	1.4	3.0	23.8	10.6
2004	FX35	3.5	VQ35DE	5.0	21.7	2.6	1.4	3.0	23.8	9.1
	FX45	4.5	VK45DE	7.0	21.7	2.6	1.4	3.0	23.8	10.0
2005	FX35	3.5	VQ35DE	5.0	21.7	2.6	1.4	3.0	23.8	9.1
	FX45	4.5	VK45DE	6.75	21.7	2.6	1.4	3.0	23.8	10.6
2006	FX35	3.5	VQ35DE	5.0	21.7	2.6	1.4	3.0	23.8	9.1
	FX45	4.5	VK45DE	6.75	21.7	2.6	1.4	3.0	23.8	10.6

09482_FX35_C0004

VALVE SPECIFICATIONS

Year	Engine Displacement Liters	Engine ID	Seat Angle (deg.)	Face Angle (deg.)	Spring Test Pressure (lbs. @ in.)	Spring Installed Height (in.)	Stem-to-Guide Clearance (in.) Intake	Stem-to-Guide Clearance (in.) Exhaust	Stem Diameter (in.) Intake	Stem Diameter (in.) Exhaust
2003	3.5	VQ35DE	45	45°15'-45°45'	84-95@1.071	1.457	0.0008-0.0021	0.0012-0.0025	0.2348-0.2354	0.2344-0.2350
	4.5	VK45DE	44°23'-44°67'	45°15'-45°45'	65-74@0.961	1.331	0.0008-0.0018	0.0012-0.0022	0.2351-0.2354	0.2347-0.2350
2004	3.5	VQ35DE	45	45°15'-45°45'	84-95@1.071	1.457	0.0008-0.0021	0.0012-0.0025	0.2348-0.2354	0.2344-0.2350
	4.5	VK45DE	44°23'-44°67'	45°15'-45°45'	65-74@0.961	1.331	0.0008-0.0018	0.0012-0.0022	0.2351-0.2354	0.2347-0.2350
2005	3.5	VQ35DE	45	45°15'-45°45'	84-95@1.071	1.457	0.0008-0.0021	0.0012-0.0025	0.2348-0.2354	0.2344-0.2350
	4.5	VK45DE	44°23'-44°67'	45°15'-45°45'	65-74@0.961	1.331	0.0008-0.0018	0.0012-0.0022	0.2351-0.2354	0.2347-0.2350
2006	3.5	VQ35DE	45	45°15'-45°45'	84-95@1.071	1.457	0.0008-0.0021	0.0012-0.0025	0.2348-0.2354	0.2344-0.2350
	4.5	VK45DE	44°23'-44°67'	45°15'-45°45'	65-74@0.961	1.331	0.0008-0.0018	0.0012-0.0022	0.2351-0.2354	0.2347-0.2350

09482_FX35_C0005

CAMSHAFT SPECIFICATIONS CHART

All measurements are given in inches.

Year	Engine Displ. Liters	Engine ID	Journal Dia.	Brg. Oil Clearance	Shaft End-play	Runout	Lobe Height	
							Intake	Exhaust
2003	3.5	VQ35DE	①	②	0.0045-0.0074	0.0020 max.	1.7633-1.7738	1.7633-1.7738
	4.5	VK45DE	③	④	0.0045-0.0074	0.0020 max.	1.7633-1.7738	1.7293-1.7368
2004	3.5	VQ35DE	①	②	0.0045-0.0074	0.0020 max.	1.7633-1.7738	1.7633-1.7738
	4.5	VK45DE	③	④	0.0045-0.0074	0.0020 max.	1.7633-1.7738	1.7293-1.7368
2005	3.5	VQ35DE	①	②	0.0045-0.0074	0.0020 max.	1.7633-1.7738	1.7633-1.7738
	4.5	VK45DE	③	④	0.0045-0.0074	0.0020 max.	1.7633-1.7738	1.7293-1.7368
2006	3.5	VQ35DE	①	②	0.0045-0.0074	0.0020 max.	1.7633-1.7738	1.7633-1.7738
	4.5	VK45DE	③	④	0.0045-0.0074	0.0020 max.	1.7633-1.7738	1.7293-1.7368

① No. 1: 1.0211-1.0218 in.
  Nos. 2, 3, 4: 0.9230-0.9238 in.

② No. 1: 0.0018-0.0034 in.
  Nos. 2, 3, 4: 0.0014-0.0030 in.

③ No. 1: 1.0212-1.0218 in.
  Nos. 2, 3, 4, 5: 1.0218-1.0224 in.

④ No. 1: 0.0018-0.0033 in.
  Nos. 2, 3, 4, 5: 0.0012-0.0027 in.

09482_FX35_G0006

CRANKSHAFT AND CONNECTING ROD SPECIFICATIONS

All measurements are given in inches.

Year	Engine Displacement Liters	Engine ID	Crankshaft				Connecting Rod		
			Main Brg. Journal Dia.	Main Brg. Oil Clearance	Shaft End-play	Thrust on No.	Journal Diameter	Oil Clearance	Side Clearance
2003	3.5	VQ35DE	①	0.0014-0.0018	0.0039-0.0098	3	④	0.0013-0.0023	0.0079-0.0138
	4.5	VK45DE	②	③	0.0039-0.0098	3	④	0.0008-0.0018	0.0079-0.0138
2004	3.5	VQ35DE	①	0.0014-0.0018	0.0039-0.0098	3	④	0.0013-0.0023	0.0079-0.0138
	4.5	VK45DE	②	③	0.0039-0.0098	3	④	0.0008-0.0018	0.0079-0.0138
2005	3.5	VQ35DE	①	0.0014-0.0018	0.0039-0.0098	3	④	0.0013-0.0023	0.0079-0.0138
	4.5	VK45DE	②	③	0.0039-0.0098	3	④	0.0008-0.0018	0.0079-0.0138
2006	3.5	VQ35DE	①	0.0014-0.0018	0.0039-0.0098	3	④	0.0013-0.0023	0.0079-0.0138
	4.5	VK45DE	②	③	0.0039-0.0098	3	④	0.0008-0.0018	0.0079-0.0138

① Depends on the grade of the crankshaft. The nominal range is 2.3603-2.3612 in.

② Depends on the grade of the crankshaft. The nominal range is: 2.5173-2.5183 in.

③ Journals 1 and 5: 0.00004-0.0004 in.

Journals 2, 3 & 4: 0.0003-0.0007 in.

④ Grade 0: 2.0460-2.0462 in.

Grade 1: 2.0457-2.0460 in.

Grade 2: 2.0455-2.0457 in.

09482_FX35_C0007

PISTON AND RING SPECIFICATIONS

All measurements are given in inches.

Year	Engine Displ. Liters	Engine ID	Piston Clearance	Ring Gap			Ring Side Clearance		
				Top Comp.	Bottom Comp.	Oil Control	Top Comp.	Bottom Comp.	Oil Control
2003	3.5	VQ35DE	0.0004-0.0012	0.0091-0.0130	0.0130-0.0189	0.0079-0.0197	0.0018-0.0031	0.0012-0.0028	0.0026-0.0053
	4.5	VK45DE	0.0004-0.0012	0.0087-0.0126	0.0087-0.0126	0.0079-0.0236	0.0018-0.0031	0.0012-0.0028	0.0026-0.0053
2004	3.5	VQ35DE	0.0004-0.0012	0.0091-0.0130	0.0130-0.0189	0.0079-0.0197	0.0018-0.0031	0.0012-0.0028	0.0026-0.0053
	4.5	VK45DE	0.0004-0.0012	0.0087-0.0126	0.0087-0.0126	①	0.0018-0.0031	0.0012-0.0028	0.0026-0.0053
2005	3.5	VQ35DE	0.0004-0.0012	0.0091-0.0130	0.0130-0.0189	0.0079-0.0197	0.0018-0.0031	0.0012-0.0028	0.0026-0.0053
	4.5	VK45DE	0.0004-0.0012	0.0087-0.0126	0.0087-0.0126	0.0079-0.0197	0.0018-0.0031	0.0012-0.0028	0.0026-0.0053
2006	3.5	VQ35DE	0.0004-0.0012	0.0091-0.0130	0.0130-0.0189	0.0079-0.0197	0.0018-0.0031	0.0012-0.0028	0.0026-0.0053
	4.5	VK45DE	0.0004-0.0012	0.0087-0.0126	0.0087-0.0126	0.0079-0.0197	0.0018-0.0031	0.0012-0.0028	0.0026-0.0053

① Up to VINs U4X105000 and W4X21000: 0.0079-0.0236 in.

From VINs 4X105001, W4X21001 and T4X000001: 0.0079-0.0197 in.

09482_FX35_C0008

TORQUE SPECIFICATIONS
All readings in ft. lbs.

Year	Engine Displacement Liters	Engine ID	Cylinder Head Bolts	Main Bearing Bolts	Rod Bearing Bolts	Crankshaft Damper Bolts	Flywheel Bolts	Manifold*		Spark Plugs	Oil Pan Drain Plug
								Intake	Exhaust		
2003	3.5	VQ35DE	①	②	③	④	65	⑤	22	18	25
	4.5	VK45DE	⑥	⑦	⑧	⑨	65	21	21	18	25
2004	3.5	VQ35DE	①	②	③	④	65	⑤	22	18	25
	4.5	VK45DE	⑥	⑦	⑧	⑨	65	21	21	18	25
2005	3.5	VQ35DE	①	②	③	④	65	⑤	22	18	25
	4.5	VK45DE	⑥	⑦	⑧	⑨	65	21	21	18	25
2006	3.5	VQ35DE	①	⑩	③	⑪	65	⑤	22	18	25
	4.5	VK45DE	⑥	⑦	⑧	⑨	65	21	21	18	25

* Lower manifold only

① Step 1: Tighten in sequence to 72 ft. lbs.
   Step 2: Completely loosen all in reverse sequence
   Step 3: Tighten in sequence to 29 ft. lbs.
   Step 4: Tighten 90 degrees.
   Step 5: Tighten another 90 degrees.

② Step 1: Tighten in sequence to 10 ft. lbs.
   Step 2: Tighten in sequence to 26 ft. lbs.
   Step 4: Tighten 90 degrees.

③ Step 1: 14 ft. lbs.
   Step 2: Plus 90 degrees

④ Step 1: 33 ft. lbs.
   Step 2: Plus 60 degrees

⑤ Step 1: 60 inch lbs.
   Step 2: 21 ft. lbs.

⑥ Step 1: Tighten in sequence to 72 ft. lbs.
   Step 2: Completely loosen all in reverse sequence
   Step 3: Tighten in sequence to 33 ft. lbs.
   Step 4: Tighten 60 degrees.
   Step 5: Tighten another 60 degrees.

⑦ Step 1: Tighten M12 bolts in sequence 1-10 to 29 ft. lbs.
   Step 2: Tighten M9 bolts in sequence 11-20 to 22 ft. lbs.
   Step 3: Tighten M12 bolts in sequence 1-10 another 40 degrees.
   Step 4: Tighten M9 bolts in sequence 11-20 another 30 degrees.
   Step 5: Tighten M10 bolts in sequence 21-30 to 36 ft. lbs.

⑧ Step 1: 11 ft. lbs.
   Step 2: Plus 60 degrees

⑨ Step 1: 69 ft. lbs.
   Step 2: Plus 90 degrees

⑩ Step 1: 26 ft. lbs.
   Step 2: plus 90 degrees

⑪ Step 1: 33 ft. lbs.
   Step 2: plus 90 degrees

09482_FX35_C0009

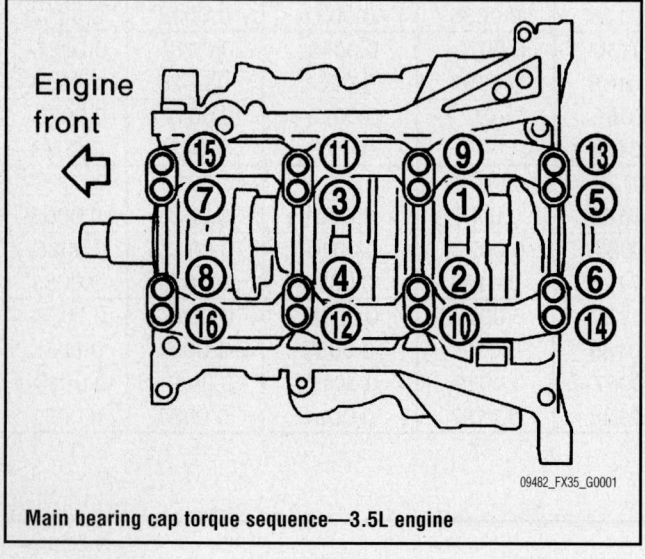

Main bearing cap torque sequence—3.5L engine

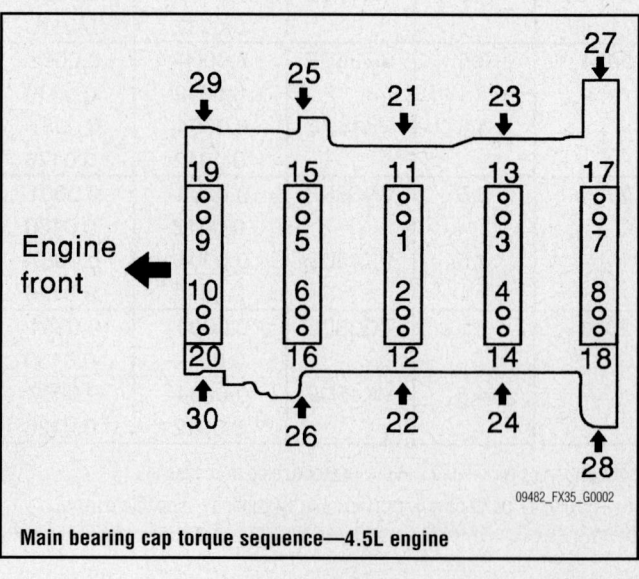

Main bearing cap torque sequence—4.5L engine

WHEEL ALIGNMENT

Year	Model		Caster Range (+/-Deg.)	Caster Preferred Setting (Deg.)	Camber Range (+/-Deg.)	Camber Preferred Setting (Deg.)	Toe-in (in.)
2003	FX35/FX45	F	0.75	+3.78	0.75	-0.73	0.063+/-0.039
		R	—	—	0.50	-0.80	0.185+/-0.091
2004	FX35/FX45	F	0.75	+3.78	0.75	-0.73	0.063+/-0.039
		R	—	—	0.50	-0.80	0.185+/-0.091
2005	FX35/FX45	F	0.75	+3.78	0.75	-0.73	0.063+/-0.039
		R	—	—	0.50	-0.80	0.17+/-0.08
2006	FX35/FX45	F	0.75	+3.78	0.75	-0.73	0.063+/-0.039
		R	—	—	0.50	-0.80	0.17+/-0.08

F: Front

R: Rear

09482_FX35_C0010

TIRE, WHEEL AND BALL JOINT SPECIFICATIONS

Year	Model	OEM Tires Standard	OEM Tires Optional	Tire Pressures (psi) Front	Tire Pressures (psi) Rear	Wheel Bead Width (in.)	Ball Joint Inspection	Lugnut Torque (ft. lbs.)
2003	FX35	P265/60R18	P265/50R20	32	32	8	①	80
	FX45	P265/50R20	none	32	32	8	①	80
2004	FX35	P265/60R18	P265/50R20	32	32	8	①	80
	FX45	P265/50R20	none	32	32	8	①	80
2005	FX35	P265/60R18	P265/50R20	32	32	8	①	80
	FX45	P265/50R20	none	32	32	8	①	80
2006	FX35	P265/60R18	P265/50R20	32	32	8	①	80
	FX45	P265/50R20	none	32	32	8	①	80

OEM: Original Equipment Manufacturer

PSI: Pounds Per Square Inch

STD: Standard

OPT: Optional

① Replace if any measurable movement is found.

09482_FX35_C0011

BRAKE SPECIFICATIONS
All measurements in inches unless noted

Year	Model		Brake Disc Original Thickness	Brake Disc Minimum Thickness	Brake Disc Maximum Runout	Minimum Lining Thickness	Brake Caliper Bracket Bolts (ft. lbs.)	Brake Caliper Mounting Bolts (ft. lbs.)
2003	FX35/FX45	F	1.102	1.024	0.0016	0.079	116	20
		R	0.630	0.551	0.0020	0.079	62	32
2004	FX35/FX45	F	1.102	1.024	0.0016	0.079	①	20
		R	0.630	0.551	0.0020	0.079	62	32
2005	FX35/FX45	F	1.102	1.024	0.0016	0.079	①	20
		R	0.630	0.551	0.0020	0.079	62	32
2006	FX35/FX45	F	1.339	1.260	0.0016	0.079	122	34
		R	0.630	0.551	0.0020	0.079	62	32

F: Front

R: Rear

① Flange bolt: 122 ft. lbs.

  Bolt with washer: 116 ft. lbs.

09482_FX35_C0012

SCHEDULED MAINTENANCE INTERVALS

INFINITI FX35/FX45 — NORMAL MAINTENANCE SCHEDULE
Follow this Maintenance Schedule if none of the conditions listed under Severe Maintenance apply.

TO BE SERVICED	TYPE OF SERVICE	7.5	15	23	30	37.5	45	52.5	60
Accessory drive belt ①	S/I								✓
Air cleaner filter	R				✓				✓
EVAP vapor lines	S/I				✓				✓
Fuel lines & connections	S/I				✓				✓
Engine coolant	R	At 60,000 miles (48 mos.), then every 30,000 (24 mos.) miles thereafter							
Engine oil	R	✓	✓	✓	✓	✓	✓	✓	✓
Engine oil filter	R	✓	✓	✓	✓	✓	✓	✓	✓
Spark plugs	R	Every 105,000 miles							
Valve clearance	Adj.	Whenever valve noise increases							
Brake lines & cables	S/I		✓		✓		✓		✓
Brake pads & rotors	S/I		✓		✓		✓		✓
Auto. Trans. fluid	S/I		✓		✓		✓		✓
Transfer case fluid	S/I		✓		✓		✓		✓
Differential gear oil	S/I		✓		✓		✓		✓
Steering system	S/I				✓				✓
Suspension components	S/I				✓				✓
CV-Joint boots & driveshafts	S/I		✓		✓		✓		✓
Exhaust system	S/I				✓				✓
In-cabin microfilter	R		✓		✓		✓		✓

① Replace if worn or damaged or if the auto tensioner has reached its limit (V8)

09482_FX35_C0013

SCHEDULED MAINTENANCE INTERVALS
INFINITI FX35/FX45 — SEVERE MAINTENANCE SCHEDULE
Follow this Maintenance Schedule if driving habits frequently include one or more of the following:
Repeated short trips of less than 5 miles (8 km).
Repeated short trips of less than 10 miles (16 km) with outside temperatures remaining below freezing.
Operating in hot weather in stop-and-go "rush hour" traffic.
Extensive idling and/or low speed driving for long distances, such as police, taxi or delivery.
Driving in dusty conditions.
Driving on rough, muddy, or salt spread roads.
Towing a trailer, using a camper or a car-top carrier.

TO BE SERVICED	TYPE OF SERVICE	VEHICLE MILEAGE INTERVAL (x1000)							
		3.75	7.5	11.25	15	17.5	22.5	26.25	30
Accessory drive belt	S/I	At 60,000 miles (48 mos.), then every 15,000 (12 mos.) miles thereafter							
Air cleaner filter	R								✓
EVAP vapor lines	S/I								✓
Fuel lines & connections	S/I								✓
Engine coolant	R	At 60,000 miles (48 mos.), then every 30,000 (24 mos.) miles thereafter							
Engine oil	R	✓	✓	✓	✓	✓	✓	✓	✓
Engine oil filter	R	✓	✓	✓	✓	✓	✓	✓	✓
Spark plugs	R	Every 105,000 miles							
Valve clearance	Adj.	Whenever valve noise increases							
Brake lines & cables	S/I				✓				✓
Brake pads & rotors	S/I		✓		✓		✓		✓
Auto. Trans. fluid ①	S/I				✓				✓
Transfer case fluid ①	S/I				✓				✓
Differential gear oil ①	S/I				✓				✓
Steering system	S/I		✓		✓		✓		✓
Suspension components	S/I		✓		✓		✓		✓
CV-Joint boots & driveshafts	S/I		✓		✓		✓		✓
Exhaust system	S/I		✓		✓		✓		✓
In-cabin microfilter	R				✓		✓		

R: Replace   S/I: Service or Inspect

① If towing a trailer or using a roof-top carrier, or regular driving on rough or muddy roads, change every 30,000 miles or 24 mos.

09482_FX35_C0014

ENGINE REPAIR

Ignition Coil

REMOVAL & INSTALLATION

1. Before servicing the vehicle, refer to the Precautions Section.
2. Remove the engine cover.
3. Remove the air duct (for ignition coil of left bank side).
4. Move aside the wiring harness, wiring harness bracket, and hoses located above ignition coil.
5. Disconnect the wiring harness connector from the ignition coil.
6. Remove the ignition coil.

✳✳ CAUTION

Do not subject the ignition coils to excessive shock or vibration.

To install:

7. Install the ignition coil on the engine.
8. Reconnect the wiring harness to the coil.
9. Reposition the wiring harness, bracket and hoses.
10. Install the air duct and the engine cover.

Alternator

REMOVAL & INSTALLATION

3.5L Engine

1. Before servicing the vehicle, refer to the Precautions Section.
2. Disconnect the negative battery cable.

3. Remove the front engine undercover.
4. Remove the lower cooling fan shroud.
5. Remove the alternator/power steering belt, as follows:
   a. Loosen idler pulley lock nut (A) and loosen belt tension by turning the adjusting bolt (B).
   b. Remove the belt from the pulleys.
6. Remove the alternator mounting bolts.
7. Disconnect the alternator wiring connector.
8. Remove the water hose bracket, the oil pressure switch wiring harness clip (2WD models), and the oil pressure switch connector (2WD models).
9. Remove the alternator assembly, by lowering it out of the bottom of the engine compartment.

To install:

10. Reposition the alternator in place on the engine and tighten the mounting bolts. Tighten the long alternator bolt 48 ft. lbs. (65 Nm) and the short alternator bracket bolts to 21 ft. lbs. (28 Nm).
11. Tighten the B terminal nut carefully to 7 ft. lbs. (10 Nm).
12. Reattach the oil pressure switch connector, the oil pressure switch wiring harness clip, and the water hose bracket.
13. Reconnect the alternator wiring.
14. Install the accessory drive belt and adjust the belt tension, as follows:
   a. Inspect drive belt deflection or tension at a point on the belt midway between the power steering and idler pulleys (power steering/alternator belt), and between the crankshaft and idler pulleys (A/C belt).
   - Inspection should be done only when engine is cold, or after the engine has been off for at least 30 minutes.
   - Measure the belt tension with a tension gauge at points marked shown in the figure.
   - When measuring deflection, apply 98 N (10 kg, 22 lb) at the marked point.
   - Adjust the belts if deflection exceeds the limit or if the tension is not within specifications.

➡**When checking belt deflection or tension after installation, first adjust it to the specified value. Then, turn the crankshaft at least two complete revolutions and re-adjust the tension to the specified value. Tighten the idler pul-**

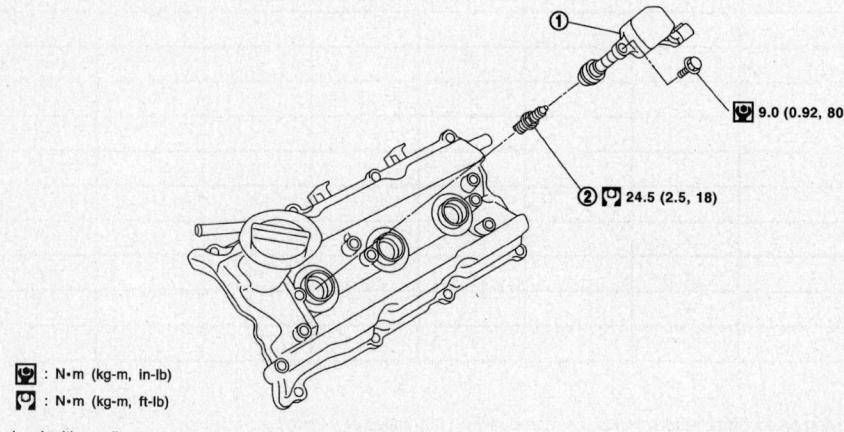

🔧 : N•m (kg-m, in-lb)
🔧 : N•m (kg-m, ft-lb)

1. Ignition coil   2. Spark plug

67162-FX35-G98

Exploded view of ignition coil mounting—3.5L engine

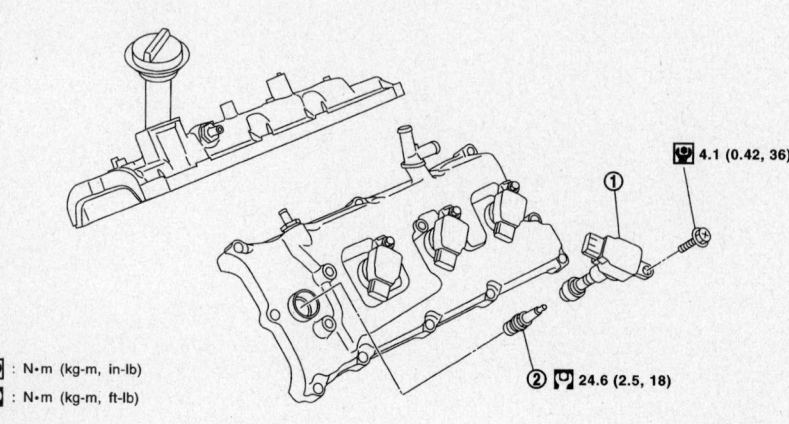

🔧 : N•m (kg-m, in-lb)
🔧 : N•m (kg-m, ft-lb)

1. Ignition coil   2. Spark plug

67162-FX35-G99

Exploded view of ignition coil mounting—4.5L engine

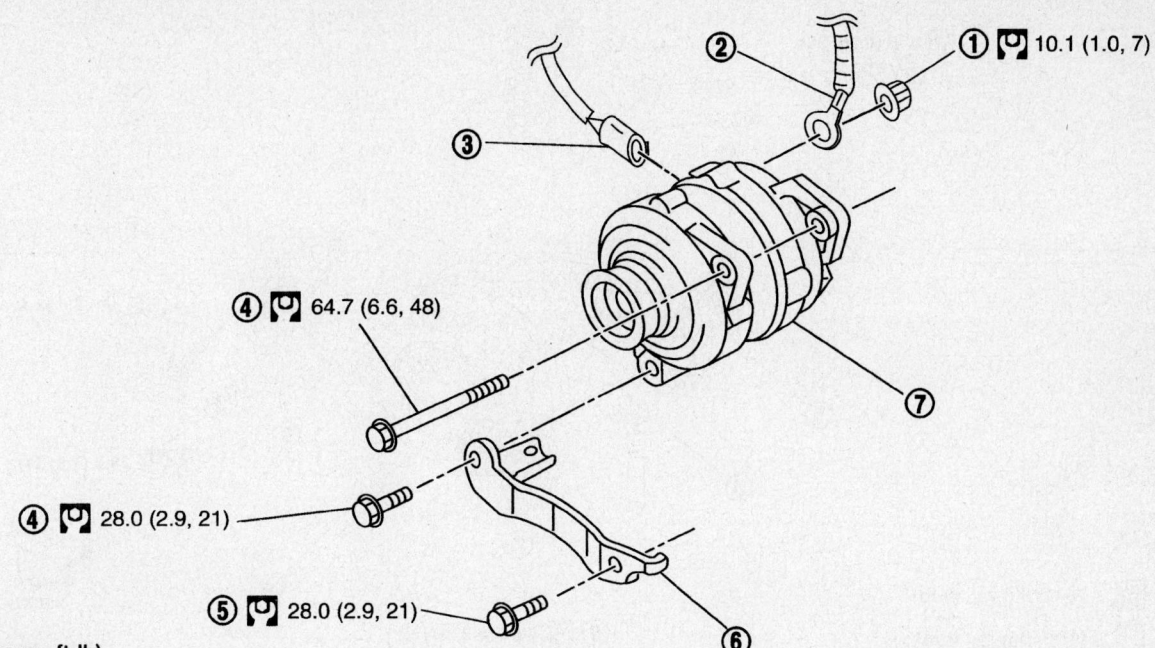

① 🔧 10.1 (1.0, 7)

④ 🔧 64.7 (6.6, 48)

④ 🔧 28.0 (2.9, 21)

⑤ 🔧 28.0 (2.9, 21)

🔧 : N•m (kg-m, ft-lb)

1. B terminal nut	2. Alternator B terminal harness	3. Alternator connector
4. Alternator mounting bolt	5. Alternator stay mounting bolt	6. Alternator stay
7. Alternator		

67162-FX35-G100

Exploded view of alternator mounting—3.5L engine

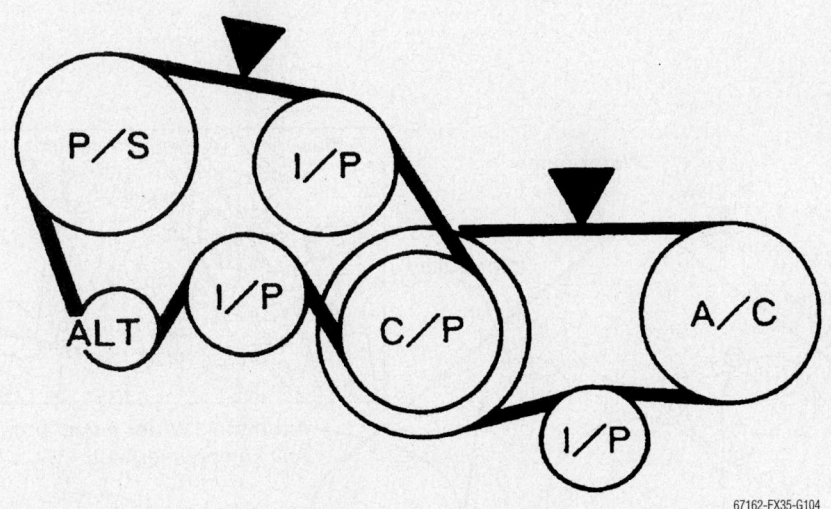

67162-FX35-G104

Accessory drive belt routing—3.5L engine

ley locknut by hand, then measure deflection.

15. Install the lower cooling fan shroud.
16. Install the front engine undercover.
17. Reconnect the negative battery cable.

4.5L Engine

1. Before servicing the vehicle, refer to the Precautions Section.

2. Disconnect the negative battery cable.
3. Remove the front engine undercover.
4. Remove the lower cooling fan shroud.
5. Remove the alternator/water pump/A/C belt as follows:
   a. Remove air duct (inlet).
   b. While securely holding the hexagonal part in the pulley center of the auto-tensioner with a wrench, move wrench handle in the direction of the arrow (loosening direction of the tensioner).
   c. Under the above condition, insert a metal bar of approximately 6mm (0.24 in) in diameter through the holding boss to lock the auto-tensioner pulley arm in position.

➡**Leave the auto-tensioner pulley arm locked until the belt is reinstalled.**

6. Remove the alternator mounting bolts.
7. Disconnect the alternator wiring connector.
8. Remove the alternator assembly, by lowering it out of the bottom of the engine compartment.

To install:

9. Reposition the alternator in place on the engine and tighten the mounting bolts. Tighten the long alternator bolt 37 ft. lbs. (50 Nm), the alternator-to-bracket bolt to 21 ft. lbs. (28 Nm), and the bracket-to-engine bolts to 17 ft. lbs. (24 Nm).
10. Tighten the B terminal nut carefully to 7 ft. lbs. (10 Nm).
11. Reconnect the alternator wiring.
12. Install the alternator/water pump/A/C drive belt. The accessory drive belt uses an auto-tensioner; no manual tensioning is necessary.

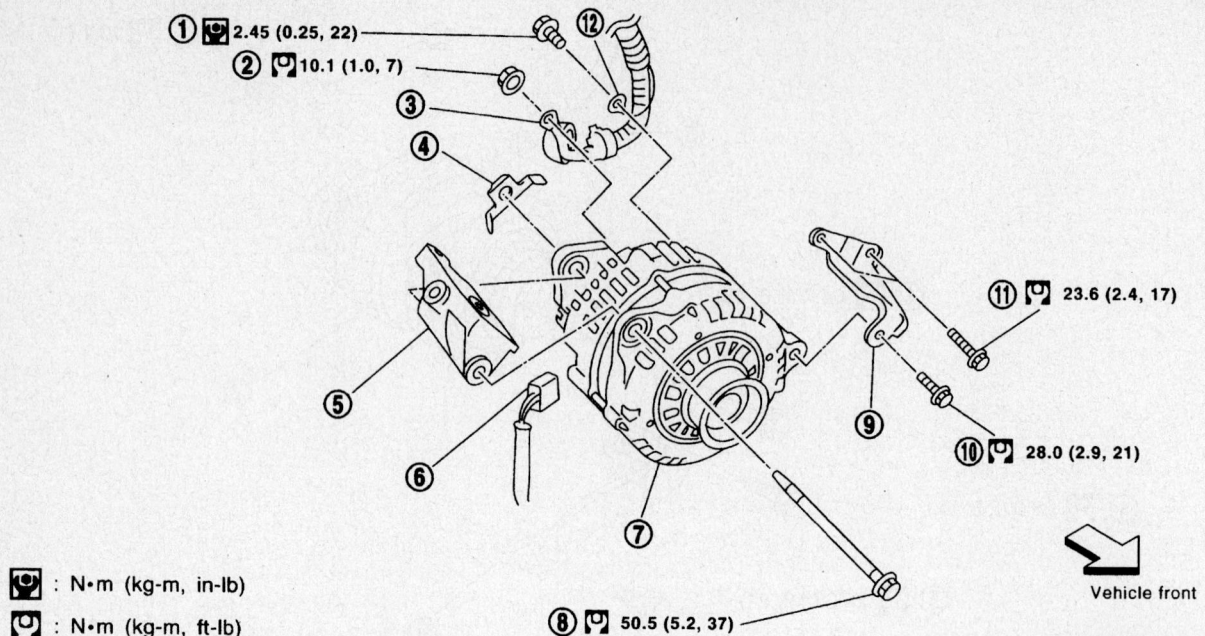

1. Alternator ground harness mounting bolt
2. B terminal nut
3. Alternator B terminal harness
4. Alternator Nut
5. Alternator bracket
6. Alternator connector
7. Alternator
8. Alternator mounting bolt
9. Alternator stay
10. Alternator mounting bolt
11. Alternator stay mounting bolt
12. Alternator ground harness

67162-FX35-G102

Exploded view of alternator mounting—4.5L engine

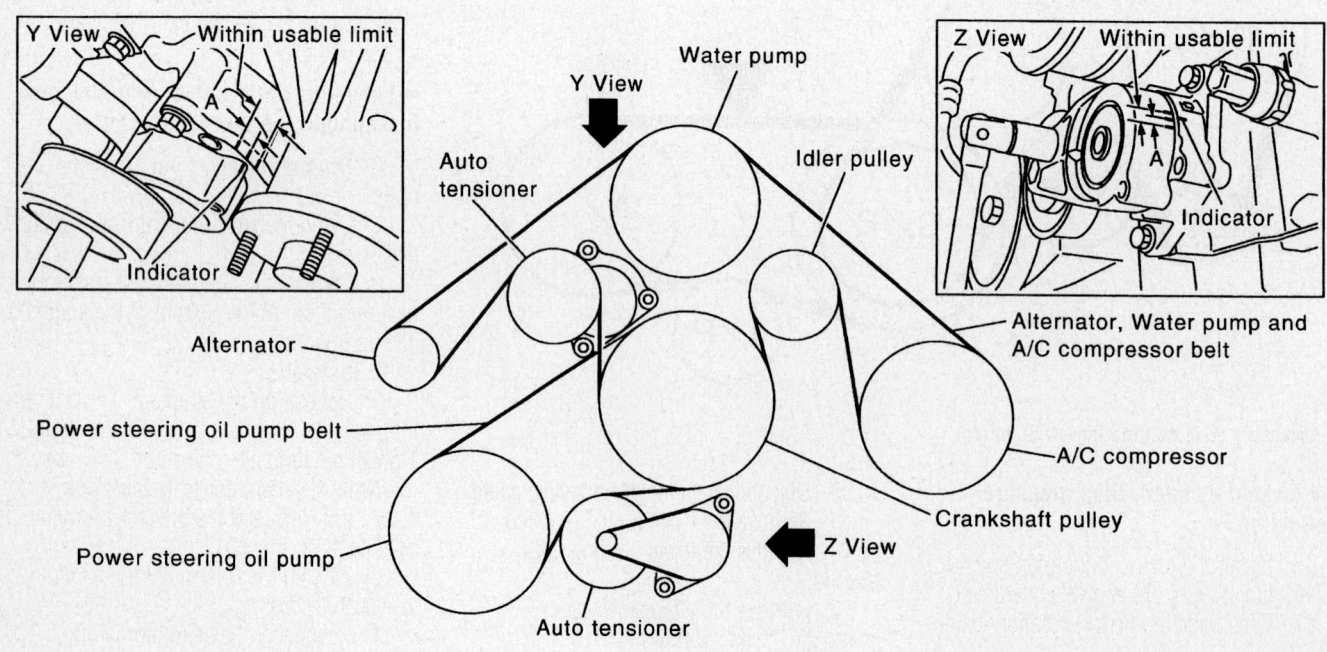

67162-FX35-G103

Accessory drive belt routing—4.5L engine

13. Install the lower cooling fan shroud.
14. Install the front engine undercover.
15. Reconnect the negative battery cable.

Ignition Timing

ADJUSTMENT

Ignition timing is controlled by the electronic fuel injection system and is not adjustable. If ignition timing is incorrect, there could be an issue with the ECM, Crankshaft Position Sensor or Camshaft Position Sensor.

Engine Assembly

REMOVAL & INSTALLATION

1. Before servicing the vehicle, refer to the Precautions Section.

2. Situate vehicle on a flat and solid surface.

3. Place chocks at front and back of rear wheels.

4. For engines not equipped with engine slingers, attach proper slingers and bolts. Tighten the slinger bolts to 21 ft. lbs. (28 Nm).

> ❋❋ **CAUTION**
>
> **Always use the support point specified for lifting. Use either a 2-point lift type or a separate type lift as best you can. If a board-on type is used for unavoidable reasons, support at the rear axle jacking point with a transmission jack or similar tool before starting work. The rear axle jacking point needs to be supported since the center of gravity will shift rearward once the engine assembly is removed.**

The engine and transmission assembly are removed from the vehicle with the suspension member from the underside of the vehicle. After removal, separate the engine from the transmission.

5. Release the fuel system pressure.

6. Disconnect both battery terminals.

7. Remove the engine cover, battery cover and both front wheel/tire assemblies.

8. Remove the front and rear engine under covers and front cross bar.

9. Drain the engine coolant from the radiator.

10. On AWD models, remove the clips of the hood ledge cover and remove the hood ledge cover.

11. On AWD models, remove both wiper arms. Operate the wiper motor, and stop it at the auto stop position. Then, remove the washer tube from the washer tube joint. Remove the wiper arm mounting nuts and wiper arms from the vehicle.

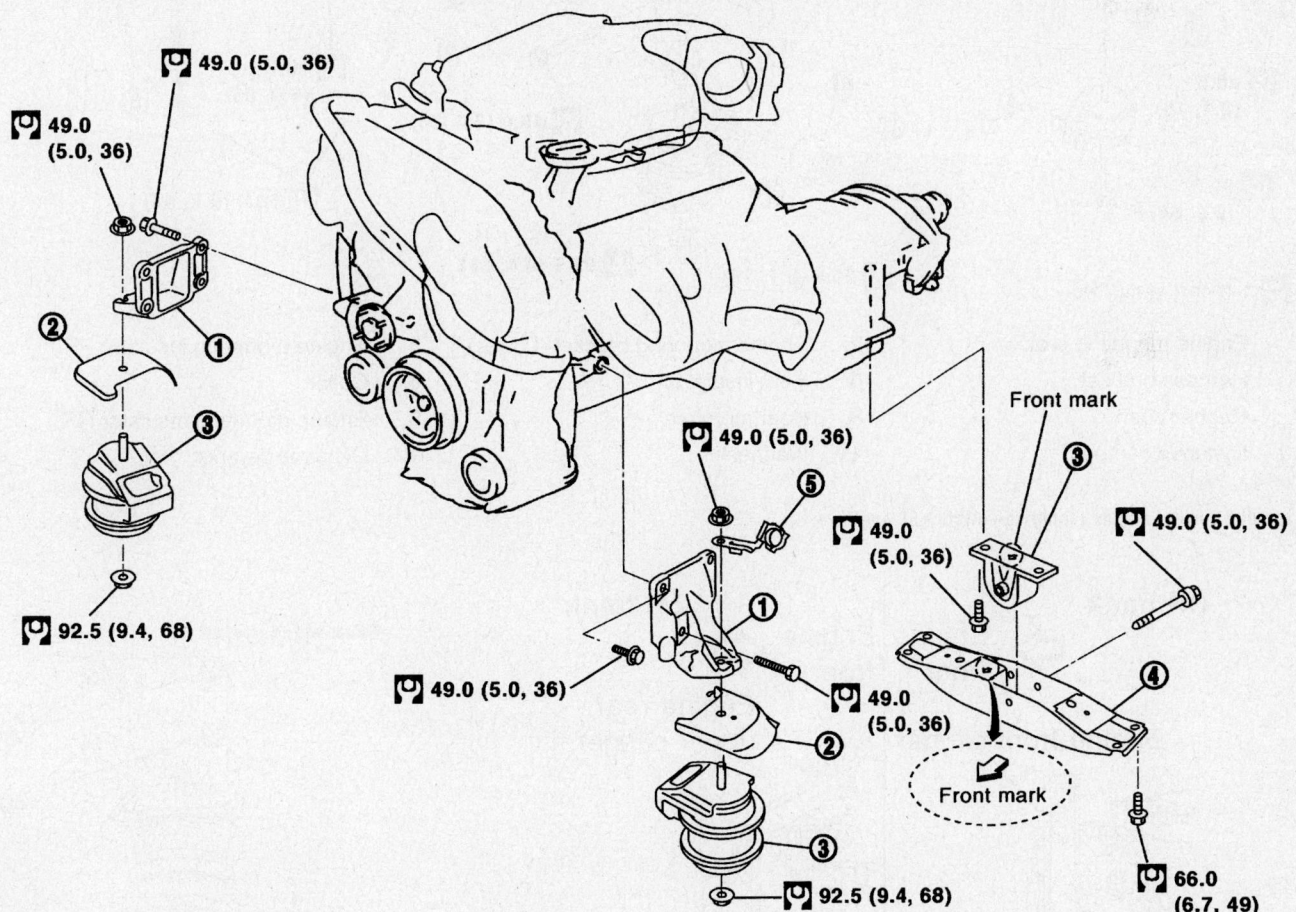

1. **Engine mounting bracket**
2. **Heat insulator**
3. **Engine mounting insulator**
4. **Rear member**
5. **Harness bracket**

67162-FX35-G105

Exploded view of engine mounting—2WD 3.5L engine

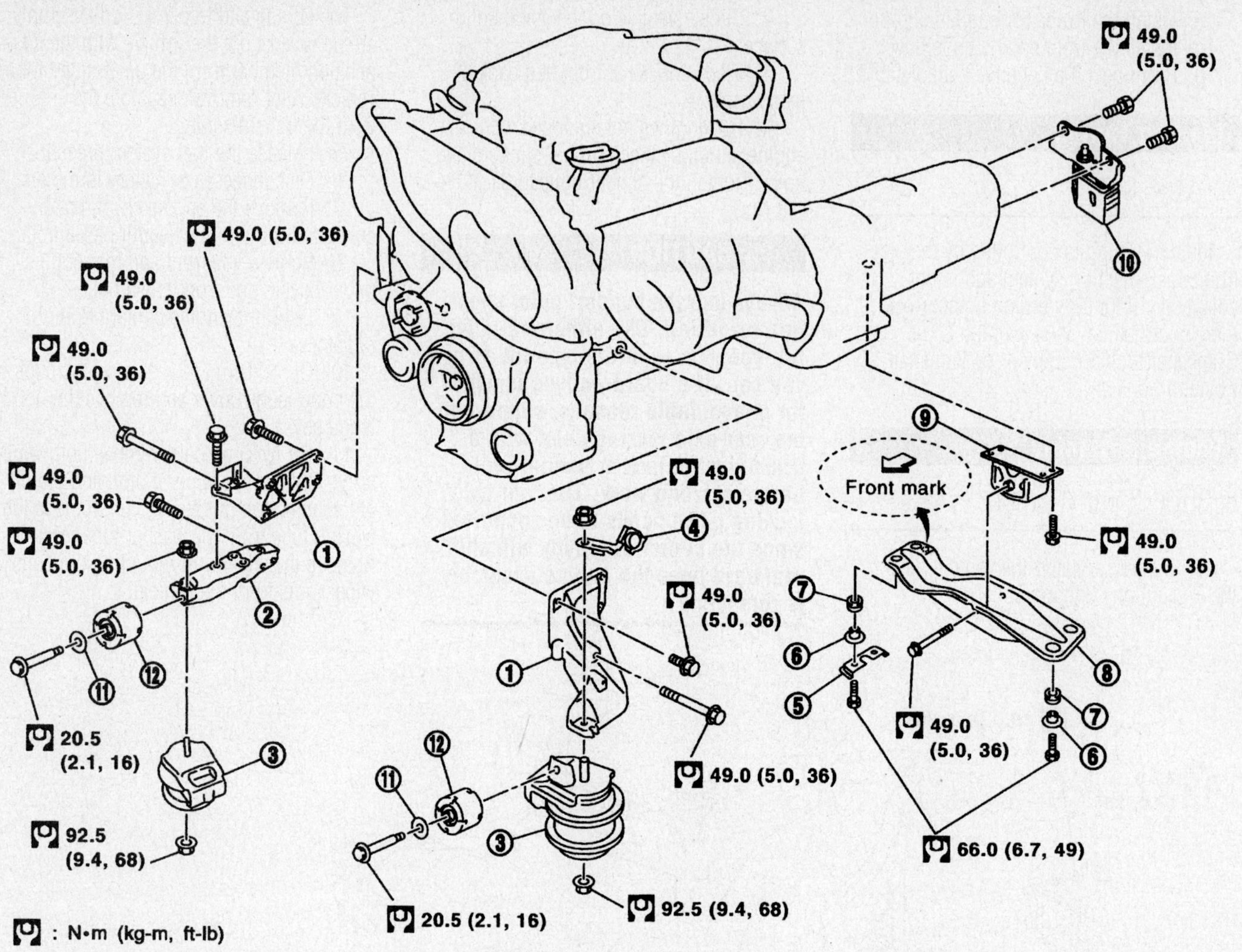

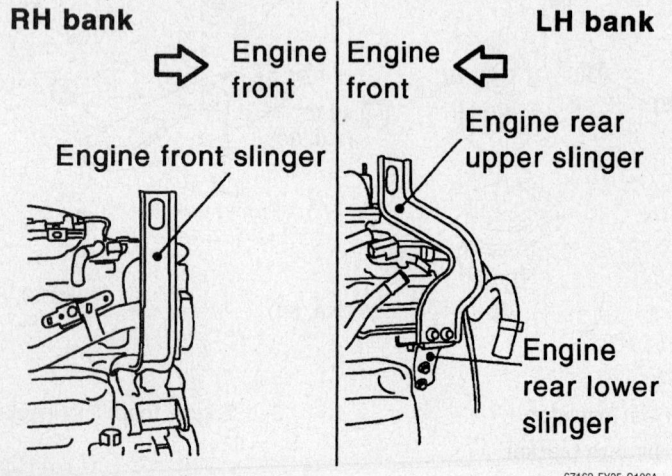

🔧 49.0 (5.0, 36)

🔧 49.0 (5.0, 36)

🔧 49.0 (5.0, 36)

🔧 49.0 (5.0, 36)

🔧 49.0 (5.0, 36)

🔧 49.0 (5.0, 36)

🔧 49.0 (5.0, 36)

🔧 49.0 (5.0, 36)

🔧 49.0 (5.0, 36)

🔧 49.0 (5.0, 36)

🔧 20.5 (2.1, 16)

🔧 92.5 (9.4, 68)

Front mark

🔧 66.0 (6.7, 49)

🔧 20.5 (2.1, 16)   🔧 92.5 (9.4, 68)

🔧 : N•m (kg-m, ft-lb)

1. Engine mounting bracket
4. Harness bracket
7. Rubber bush
10. Dynamic damper

2. Engine mounting bracket (Lower)
5. Heat insulator
8. Rear member
11. Washer

3. Engine mounting insulator FR
6. Caller
9. Engine mounting insulator RR
12. Dynamic damper

67162-FX35-G109

Exploded view of engine mounting—AWD 3.5L engine

RH bank   **LH bank**

⇨ Engine | Engine ⇦
front | front

Example: LH side

Engine front slinger

Engine rear
upper slinger

Engine
rear lower
slinger

67162-FX35-G106A

Engine slinger installation positions—2WD 3.5L engine

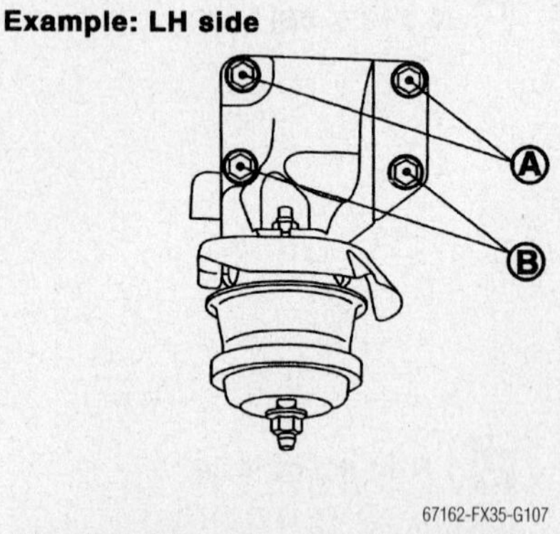

Ⓐ

Ⓑ

67162-FX35-G107

When installing engine mounting bracket
on cylinder block, tighten two upper bolts
(A), then tighten two lower bolts (B)—
2WD 3.5L engine

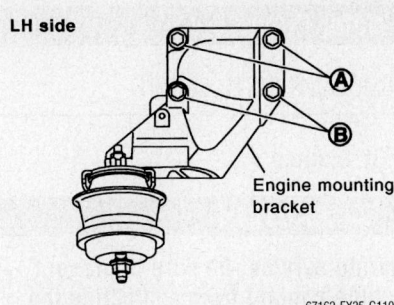

LH side

Engine mounting bracket

67162-FX35-G110

Left-hand engine mounting bracket bolt identification—AWD 3.5L engine

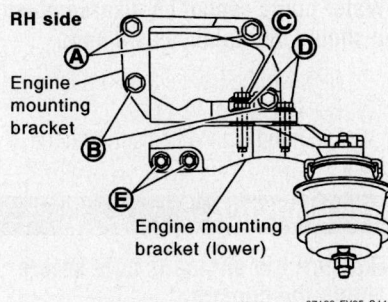

RH side

Engine mounting bracket

Engine mounting bracket (lower)

67162-FX35-G111

Right-hand engine mounting bracket bolt identification—AWD 3.5L engine

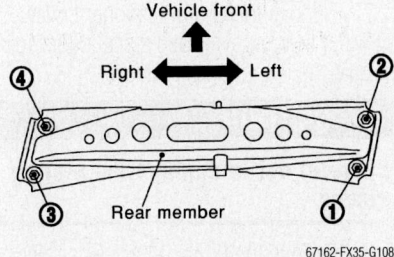

Vehicle front

Right ←→ Left

Rear member

67162-FX35-G108

Rear member mounting bolt tightening sequence—2WD 3.5L engine

12. Remove the air duct and air cleaner assembly.

13. Remove the radiator hoses.

14. Disconnect and plug the chassis-end of the heater hose.

15. Disconnect the chassis/left bank cylinder head ground wire.

16. Disconnect and tag all engine wiring harness connectors.

17. Disconnect the A/C tubing from the A/C compressor, and fasten it to the inside of the engine compartment with rope or strong cord.

18. Disconnect the two chassis ground cables.

19. Disconnect the brake booster vacuum hose.

20. Disconnect and plug the fuel feed hose and EVAP hose.

21. Remove the power steering oil pump reservoir tank and tubing from the chassis, and fasten them onto the engine.

➡ **When securing the power steering tubing to the engine, situate it so that the open end is pointed up to avoid a fluid leak.**

22. Disconnect the engine compartment wiring harness connectors from the passenger compartment, as follows:

 a. Remove the passenger-side kick plate, dashboard side trim, and glove box.

 b. Disconnect the engine compartment wiring harness connectors from the ECM.

 c. Disengage the intermediate fixing point. Pull the engine compartment wiring harnesses through to the engine compartment and fasten them to the engine.

✳ CAUTION

Be careful not to damage the wiring harness when pulling it out of the passenger compartment. Also, cover the wiring harness connectors to prevent dirt or other material from contaminating the connector openings.

23. Remove the A/T fluid cooler hoses and power steering oil pump oil cooler hoses.

24. Remove the front exhaust pipe.

25. Disconnect the lower steering joint, and disengage the steering shaft.

26. Disconnect the driveshaft from the transmission.

27. Disconnect the shift control linkage from selector lever, then secure it onto the transmission so that it doesn't hang free.

28. Remove the rear plate cover from the upper oil pan. Remove the bolts holding the flexplate to the torque converter.

29. Remove the mounting bolts from the transmission to the lower rear side of the upper oil pan.

30. On 2WD models, remove the front stabilizer.

31. Detach the left-hand and right-hand side tie-rod ends from the steering knuckles.

32. Remove the lower ends of the left-hand and right-hand struts from the lower arms.

33. Disconnect the left-hand and right-hand lower arms from the suspension member.

34. On AWD models, disconnect both front halfshafts from the knuckles.

35. Use a manual lift table caddy (commercial service tool) or equivalently rigid tool such as a transmission jack. Securely support the bottom of the suspension member and transmission.

✳ CAUTION

Put a piece of wood or something similar as the supporting surface, secure in a completely stable condition.

36. Remove the rear member mounting bolt.

37. Remove the suspension member mounting bolt and nut.

38. Carefully lower the jack to remove engine, transmission, and suspension member assembly. While performing this, observe the following:

 • Confirm there is no interference with vehicle.
 • Make sure all connection points have been disconnected.
 • Keep in mind the center of vehicle gravity changes.
 • If necessary, use jackstands to support the vehicle at the rear jacking point(s) to prevent it from falling off the lift.

39. Install engine slingers/supports into the front of the right bank cylinder head and the rear of the left bank cylinder head.

40. Remove the power steering oil pump from the engine.

41. Remove the engine mounting insulator bottom nut.

42. Lifting with a hoist, separate the engine and transmission assembly from the suspension member.

✳ CAUTION

Before and during lifting, always check whether any wiring harnesses are still connected. Also, avoid damage to and oil/grease contacting the engine mounting insulator.

43. On AWD models, remove both front halfshafts.

44. Remove the alternator.

45. Remove the starter motor.

46. On AWD models, detach the front driveshaft from the front final drive assembly.

47. Separate the engine from the transmission assembly.

48. Remove the engine mounting insulator and bracket.

49. On AWD models, remove front final drive assembly from oil pan (upper).

To install:

50. On AWD models, install the front final drive assembly onto the upper oil pan.

➡**When installing the engine mounting bracket on the cylinder block, tighten the two upper bolts (shown as A) first. Then tighten the two lower bolts (shown as B).**

51. Install the engine mounting insulator and bracket.
52. Join the engine and transmission assemblies.
53. On AWD models, reattach the front driveshaft to the front final drive assembly.
54. Install the starter motor.
55. Install the alternator.
56. On AWD models, install both front halfshafts.
57. Join the engine and transmission assembly to the suspension member.
58. Install the engine mounting insulator bottom nut.
59. Install the power steering oil pump.
60. Remove the engine slingers/supports.
61. Carefully raise the jack to install the engine, transmission, and suspension member assembly.
62. Install the suspension member mounting bolt and nut.
63. Install the rear member mounting bolts in the order shown.
64. On AWD models, reconnect both front halfshafts to the knuckles.
65. Connect the left-hand and right-hand lower arms to the suspension member.
66. Reinstall the lower ends of the left-hand and right-hand struts from the lower arms.
67. Reattach the left-hand and right-hand side tie-rod ends to the steering knuckles.
68. On 2WD models, install the front stabilizer.
69. Install the mounting bolts from the transmission to the lower rear side of the upper oil pan.
70. Install the bolts holding the flexplate to the torque converter.
71. Install the rear plate cover onto the upper oil pan.
72. Reconnect the shift control linkage to the selector lever.
73. Install the driveshaft to the transmission.
74. Connect the lower steering joint, and engage the steering shaft.
75. Install the front exhaust pipe.
76. Reattach the A/T fluid cooler hoses and power steering oil pump oil cooler hoses.
77. Route the engine compartment

wiring harnesses through to the passenger compartment.
78. Connect the engine compartment wiring harness connectors to the ECM.
79. Install the passenger-side kick plate, dashboard side trim, and glove box.
80. Install the power steering oil pump reservoir tank and tubing.
81. Reconnect the fuel feed and EVAP hoses.
82. Reconnect the brake booster vacuum hose.
83. Reattach the two chassis ground cables.
84. Reconnect the A/C tubing to the A/C compressor.
85. Reconnect all engine wiring harness connectors.
86. Reconnect the chassis/left bank cylinder head ground wire.
87. Reconnect the heater hose.
88. Install the radiator hoses.
89. Install the air duct and air cleaner assembly.
90. On AWD models, install the wiper arms, as follows:
    a. Clean the pivot area, which will help reduce the possibility of wiper arm looseness.
    b. Prior to wiper arm instillation, turn on the wiper switch to operate the wiper motor and then turn it "OFF" (auto stop).
    c. Push the wiper arm onto pivot shaft, paying attention to the blind spline.
    d. Attach the washer tube to the washer tube joint.
    e. Lift the blade up and then set it down onto the glass surface to set the blade center to clearance "L1" & "L2" immediately before tightening the nut.
    f. Operate the washer fluid pump. Turn on the wiper switch to operate the wiper motor and then turn it "OFF".
    g. Ensure that the wiper blades stop within the clearance "L1" & "L2".
    h. Tighten the wiper arm nuts to 18 ft. lbs. (24 Nm).
    i. Install the cowl top seal rubber.
    j. Install the clips of the cowl top cover (right) and the cowl top cover (right).
    k. Install the clips, cap and bolts and cowl top cover (left).
    l. Install the washer hose from cowl top cover.
91. Install the top right-hand cowl cover.
92. Fill the engine cooling system.
93. Install the front crossbar and the front and rear lower engine covers.
94. Install the engine cover, the battery cover and both front wheel assemblies.
95. Connect both battery cables.

Water Pump

REMOVAL & INSTALLATION

3.5L Engine

❊❊ CAUTION

During service, be sure to prevent engine coolant from contacting the drive belt.

1. Before servicing the vehicle, refer to the Precautions Section.

➡**Water pump cannot be disassembled and should be replaced as a unit.**

2. Remove the front engine undercover.
3. Remove the drive belts.
4. Drain the engine coolant from the radiator.

❊❊ WARNING

Make sure the engine is cold before draining the coolant.

5. Remove the air duct (inlet), power duct and air cleaner case assembly.
6. Remove the cylinder block drain plug (front side) of engine.
7. Remove the chain tensioner cover and water pump cover. Use a seal cutter to separate the two mating surfaces.

❊❊ CAUTION

Be careful not to damage the mating surfaces.

8. Remove the timing chain tensioner (primary), as follows:
    a. Pull the lever down and release the plunger stopper tab.

➡**The plunger stopper tab can be pushed up to release (coaxial structure with lever).**

    b. Insert a stopper pin into the tensioner body hole to hold the lever and keep the plunger stopper tab released.

➡**An Allen wrench can be used for a stopper pin.**

    c. Insert the plunger into the tensioner body by pressing the timing chain slack guide.
    d. Keep the slack guide depressed and hold the plunger in by pushing the stopper pin deeper through the lever and into the tensioner body hole.
    e. Turn the crankshaft pulley approximately 20 degrees clockwise, so that the

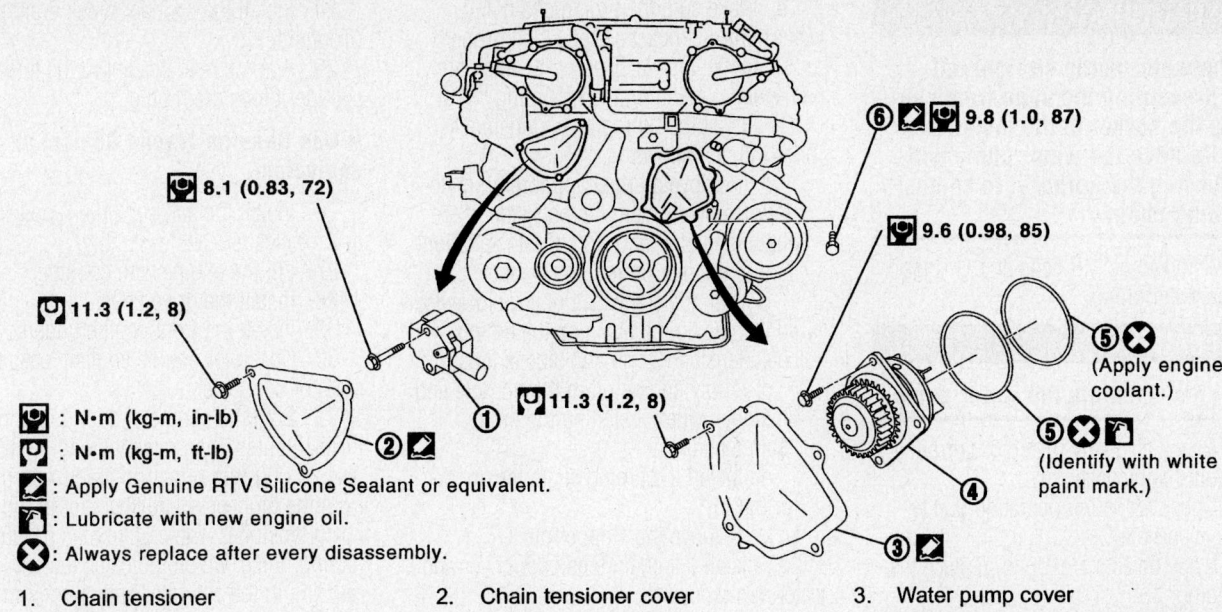

$\boxed{}$: N•m (kg-m, in-lb)

$\boxed{}$: N•m (kg-m, ft-lb)

$\boxed{}$: Apply Genuine RTV Silicone Sealant or equivalent.

$\boxed{}$: Lubricate with new engine oil.

$\boxtimes$: Always replace after every disassembly.

1. Chain tensioner
2. Chain tensioner cover
3. Water pump cover
4. Water pump
5. O-rings
6. Cylinder block drain plug (front)

67162-FX35-G114

Exploded view of water pump and related components—3.5L engine

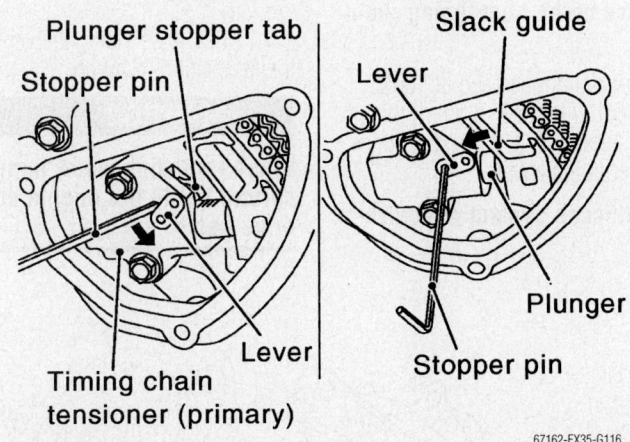

Primary timing chain tensioner details—3.5L engine

67162-FX35-G116

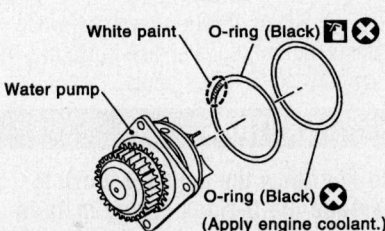

$\boxtimes$: Always replace after every disassembly.

$\boxed{}$: Lubricate with new engine oil.

67162-FX35-G119

Water pump O-ring positioning for installation—3.5L engine

timing chain on the timing chain tensioner (primary) side is loose.

f. Remove the mounting bolts and timing chain tensioner (primary).

✳✳ CAUTION

Be careful not to drop mounting bolts inside the chain case.

9. Remove the three water pump fixing bolts. Secure a gap between the water pump gear and timing chain, by turning the crankshaft pulley counterclockwise until timing chain slack on the water pump sprocket is at its maximum.

10. Insert M8x1.25 bolts, approx. 50mm long, into the water pump upper and lower mounting bolt holes until they contact the timing chain case. Then, alternately tighten each bolt for a half turn, and pull out water pump.

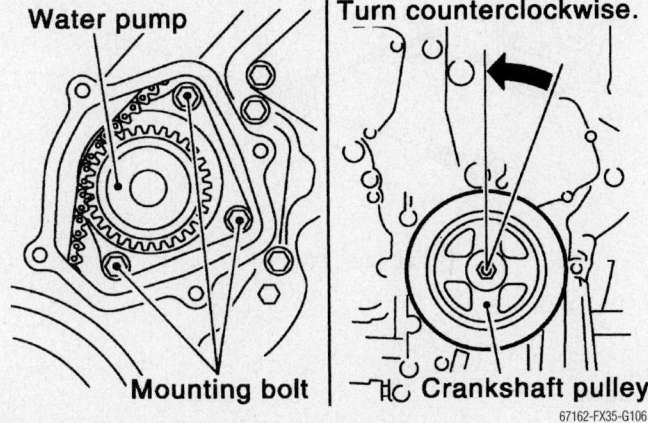

Water pump mounting bolt locations—3.5L engine

67162-FX35-G106

✳✳ CAUTION

Pull the water pump straight out while preventing the vane from contacting the socket in the installation area. Remove the water pump without allowing the sprocket to contact the timing chain.

11. Remove the M8 bolts and O-rings from the water pump.

✥ CAUTION

Do not disassemble the water pump.

12. Check for badly rusted or corroded water pump body assembly.
13. Check for rough operation due to excessive end play.
14. If any defects are found, replace the water pump.

To install:
15. Install new O-rings on the water pump.
16. Apply engine oil and engine coolant to the water pump O-rings.
17. Position the O-ring with the white paint mark toward the engine front side.
18. Install the water pump.

✥ CAUTION

Do not allow the cylinder block to damage the O-rings during installation.

19. Make sure the timing chain and water pump sprocket are engaged properly.
20. Install the water pump by alternately and evenly tightening the mounting bolts.
21. Install the timing chain tensioner (primary), as follows:
   a. Remove all dust and foreign material completely from the backside of the chain tensioner and from the installation area of the rear timing chain case.
   b. Turn the crankshaft pulley clockwise so that the timing chain on the timing chain tensioner (primary) side is loose.
   c. Apply engine oil to the oil hole and tensioner, when installing the timing chain tensioner.
   d. Install the timing chain tensioner (primary).
   e. Remove the stopper pin.
22. Install the chain tensioner cover and water pump cover.

➡ Before installing, remove all traces of liquid gasket from the mating surface of the water pump cover and chain tensioner cover using a scraper. Also, remove traces of liquid gasket from the mating surface of the front timing chain case.

23. Apply a continuous bead of liquid gasket 0.091–0.130 in. (2.3–3.3mm) thick to the mating surface of the chain tensioner cover and water pump cover.

➡ Use RTV Silicone Sealant or equivalent.

24. Install the cylinder block drain plug (front side).
25. Apply thread sealant to the thread of cylinder block drain plug.

➡ Use Genuine Thread Sealant or equivalent.

26. Install the air duct (inlet), power duct and air cleaner case assembly.
27. Fill the engine with coolant.
28. Install the drive belts.
29. Install the front engine undercover.
30. Check for engine coolant leaks using radiator cap tester.
31. Start the engine and let it idle for three minutes, then raise the engine RPM up to 3,000 rpm under no load to purge air from the high-pressure chamber of the chain tensioner. The engine may produce a rattling noise, which indicates that air remains in the chamber and is not a matter of concern.
32. With the engine idling, visually make sure that there are no leaks of engine coolant or A/T fluid.

4.5L Engine

1. Before servicing the vehicle, refer to the Precautions Section.

✳✳ CAUTION

When removing water pump, be careful not to get engine coolant on drive belt.

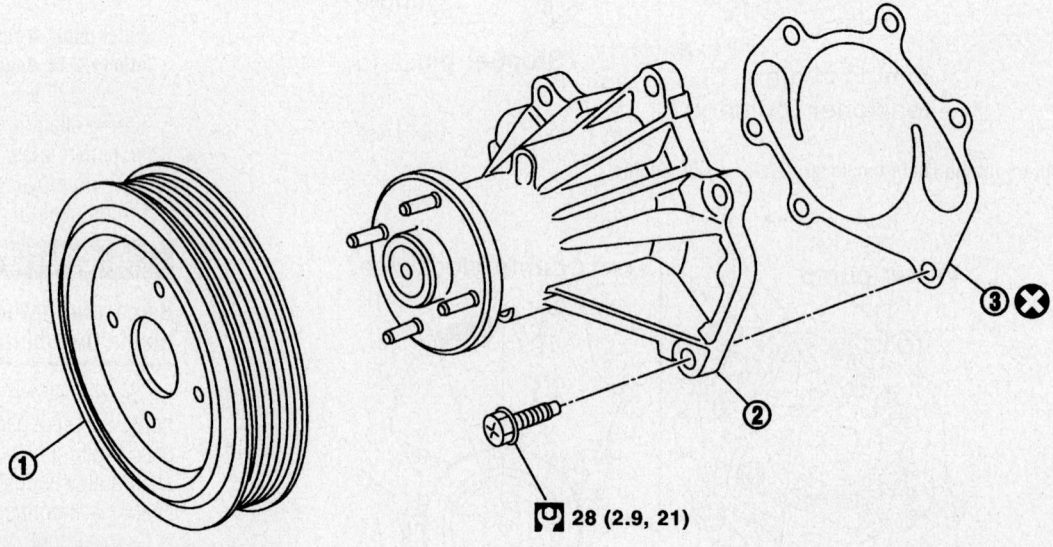

❌ : Always replace after every disassembly.

⚙ : N•m (kg-m, ft-lb)

1.   Water pump pulley
2.   Water pump
3.   Gasket

⚙ 28 (2.9, 21)

67162-FX35-G121

Exploded view of the water pump mounting—4.5L engine

➡**Water pump can not be disassembled and should be replaced as a unit.**

2. Drain the engine coolant from the drain plugs on radiator and both sides of the engine block.

❈❈ WARNING

Make sure the engine is cold before draining the coolant.

3. Remove the engine front undercover.
4. Remove the air duct (inlet).
5. Remove the alternator, water pump and A/C compressor belt.
6. Remove the fan coupling with the cooling fan, and then the water pump pulley.
7. Remove the water pump.

➡**Engine coolant will leak from the cylinder block, so have a receptacle ready under vehicle.**

❈❈ CAUTION

Handle the water pump vane so that it does not contact any other parts.

8. Inspect the water pump after removal for the following:
   • Significant dirt or rusting on water pump body and vane.
   • looseness in vane shaft, and that it turns smoothly when rotated by hand.
If anything is found, replace the water pump.
To install:
9. Install the water pump.
10. Install the water pump pulley, then the fan coupling with the cooling fan.
11. Install the alternator, water pump and A/C compressor belt.
12. Install the air duct (inlet).
13. Install the engine front undercover.
14. Tighten the drain plugs on the radiator and both sides of the engine block.
15. Fill the cooling system with coolant.
16. Check for leaks of engine coolant using a radiator cap tester.
17. Start and warm up the engine. Visually make sure that there are no leaks of engine coolant.

Heater Core

REMOVAL & INSTALLATION

1. Before servicing the vehicle, refer to the Precautions Section.
2. Use approved refrigerant collecting equipment to discharge refrigerant.

❈❈ WARNING

Make sure the engine is cold before draining the coolant.

3. Drain the coolant from the cooling system.
4. Remove the cowl top cover.
5. Remove the two high-pressure pipe mounting clips.
6. Remove the low-pressure flexible hose bracket mounting bolts.
7. Disconnect the evaporator-side one touch joint, as follows:
   a. Set a disconnect tool (High-pressure side: 92530–89908, Low pressure side: 92530–89916) on the A/C piping.

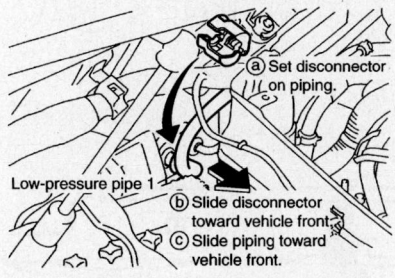

Heater core quick-connect coupling locations

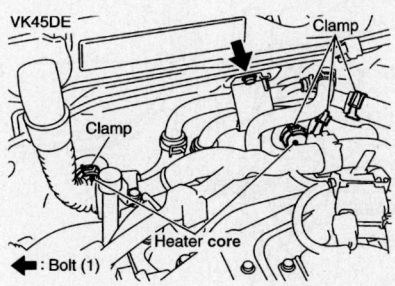

Heater hose clamp locations for heater core service

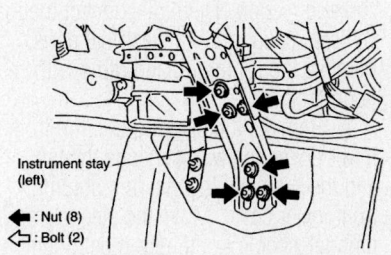

Instrument panel stay mounting bolt locations

b. Slide the disconnect tool toward the front of the vehicle until it clicks.
c. Slide the A/C pipe toward the front of the vehicle front and disconnect it.

❈❈ CAUTION

Seal the connection opening of the pipe with a cap or vinyl tape to avoid exposure to atmosphere.

8. On 3.5L engines, remove the electronic control throttle assembly.
9. Disconnect the two heater hoses from the heater core.
10. Remove the instrument panel assembly, as follows:
   a. Remove the front kicking plate on both sides of the vehicle.
   b. Remove dash side finisher plastic nuts, then remove the dash side finisher.
   c. Pull to the inside of the vehicle, disengage the metal clips and remove the front pillar garnish.
   d. To remove the A/T Select Lever Knob, pull down the knob cover. Remove the lock-pin of the select lever knob. Then, lift up the select lever knob and remove it.
   e. Insert a remover into the side between the gaps of the instrument clock finisher and pull back to your side.
   f. Disconnect the clips and wiring harness connector, then remove instrument clock finisher.
   g. Insert a remover into the side between the gaps of the A/T console finisher and remove it by lifting the A/T console finisher.
   h. Disconnect the wiring harness connector.
   i. Remove the console finisher screws.
   j. Remove the console finishers.
   k. To remove the center console, remove the mounting screws, then remove the console sub-wiring harness.

❈❈ CAUTION

When removing console, be careful not to pull the wiring harness.

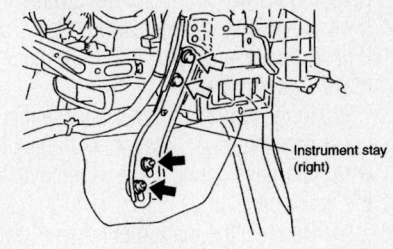

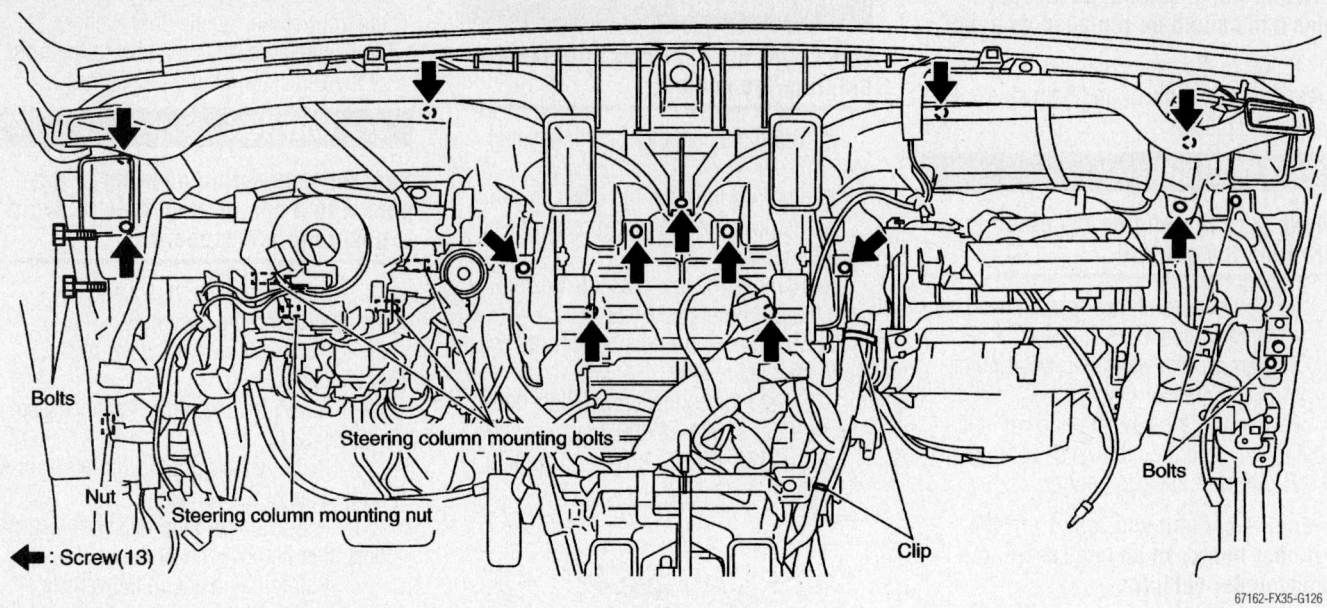

Bolts

Steering column mounting bolts

Nut

Steering column mounting nut

← : Screw(13)

Clip

Bolts

67162-FX35-G126

Instrument panel mounting bolt locations

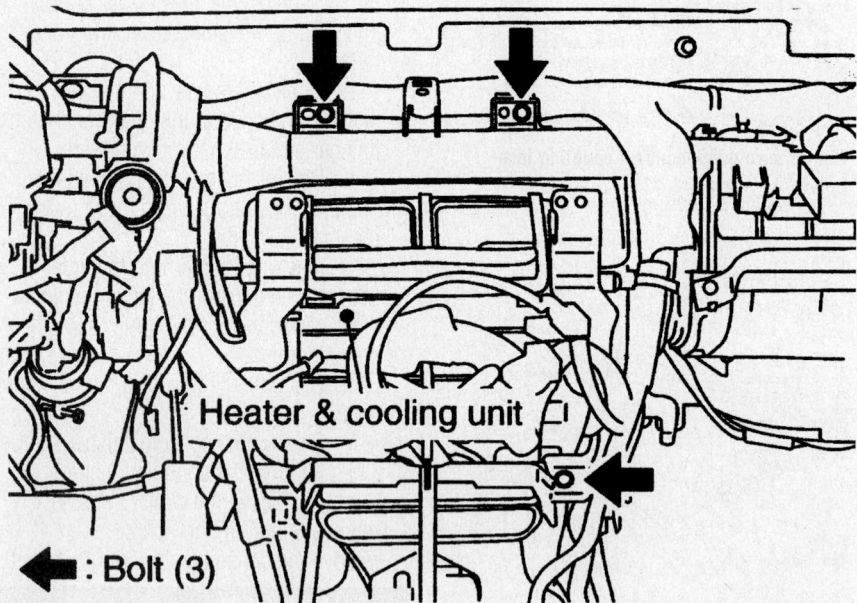

Heater & cooling unit

← : Bolt (3)

67162-FX35-G125

Heater core assembly mounting bolt locations

l. Remove the instrument lower cover by pulling down on the front instrument lower cover and disconnecting clips. Pull it horizontally, and remove it from the lower cover pawls.

m. Remove the instrument passenger lower panel screws, disconnect the wiring harness connector, and remove the lower panel.

n. Remove the instrument driver lower panel bolt and screws, detach the data link connector, pull to disengage the clip

and pawl by removing panel in a horizontal direction. Then, disconnect the in-vehicle sensor and all electrical parts. Remove the grommet, and remove the hood lock cable.

o. Remove the steering column front lower cover screw, disengage the tab, and then remove the steering column front lower cover. Move the steering column telescopic to the rear most position, and move the steering column tilt to the top position.

p. Remove the steering column lower cover screws, then disengage the tab and remove steering column lower cover.

q. Remove the steering column upper cover.

r. Remove the wiper and washer switch

s. Remove the lighting and turn signal switch.

t. Pull the steering lock escutcheon back to your side, and remove it.

u. Remove the Combination Meter Assembly by removing the bolts and disconnecting the connector bracket. Remove the bolts and disconnect the wiring harness connector.

❄ CAUTION

To prevent it from damaged by interference with the combination meter assembly, protect the combination meter assembly with cloths.

v. Remove the instrument panel side panel screws, then pull the panels to the side, disconnect the clip and pawls, and remove the instrument side panels. Perform for both right-hand and left-hand panels.

w. To remove the cluster lid C, insert a pry tool into the gap between the instrument panel and pad, pull back towards you, and disconnect the metal clips. Then, disconnect the wiring harness connectors, and remove the cluster lid C.

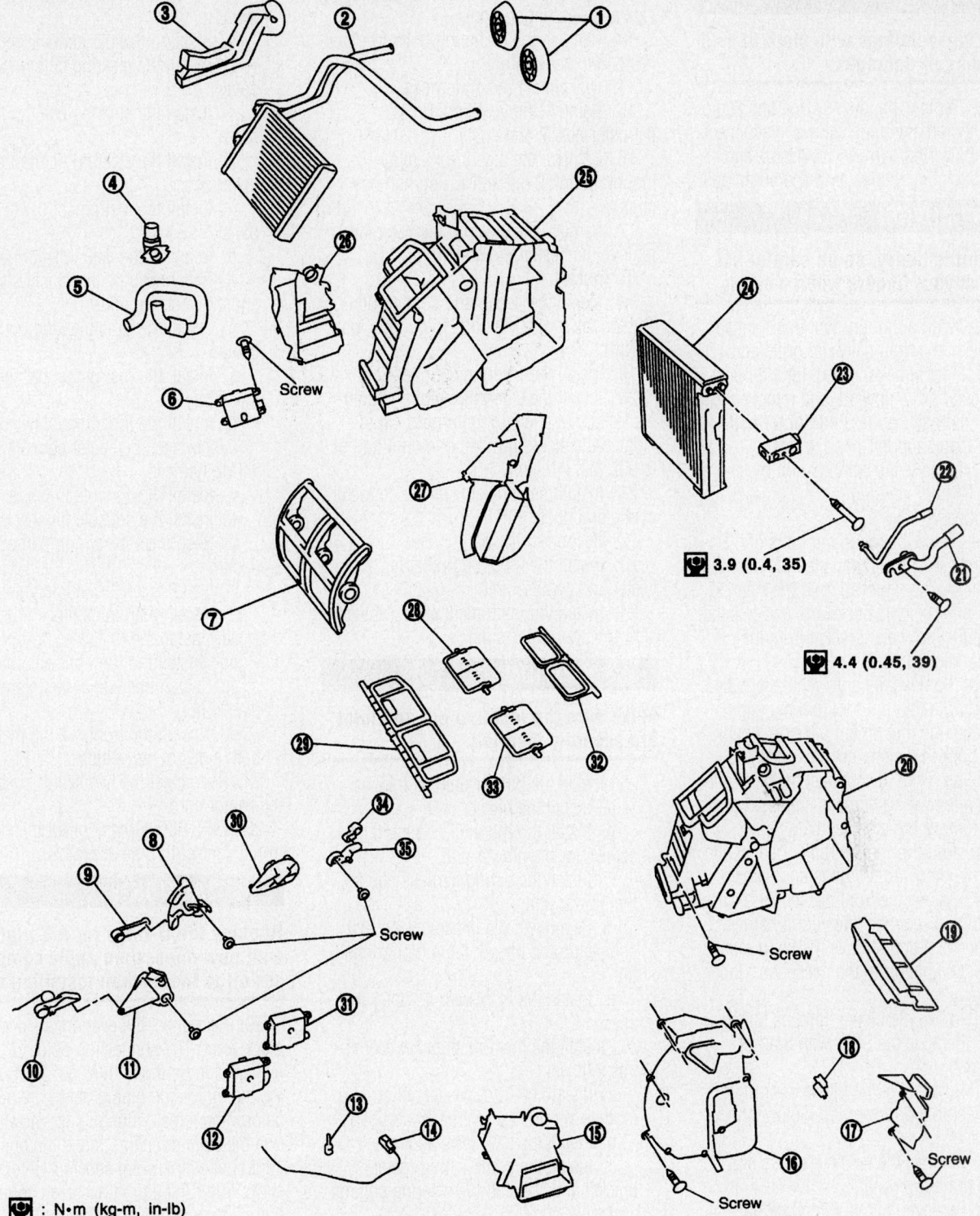

3.9 (0.4, 35)

4.4 (0.45, 39)

Screw

Screw

Screw

Screw

Screw

: N•m (kg-m, in-lb)

1.	Heater pipe grommet	2.	Heater core	3.	Heater pipe cover
4.	Aspirator	5.	Aspirator hose	6.	Air mix door motor (driver side)
7.	Air mix door (slide door)	8.	Max. cool door link	9.	Max. cool door lever
10.	Ventilator door lever	11.	Ventilator door link	12.	Air mix door motor (passenger side)
13.	Intake sensor bracket	14.	Intake sensor	15.	Foot duct (right)
16.	Evaporator cover	17.	Evaporator cover adaptor	18.	Heater pipe bracket

67162-FX35-G127

Exploded view of the Heater & Cooling Unit, which contains the heater core

Cover surroundings with cloth to avid scratches or damages.

x. Remove the display unit and audio unit by removing the screws, disconnecting the wiring harness connector, and removing the display unit and audio unit.

The Unit is heavy, so be careful not to pinch your fingers when working.

y. Insert a thin pry tool into the gaps between the front defroster grille and instrument panel and pad, lift the front defroster grille upward, and remove the front defroster grille. Perform for both right-hand and left-hand grilles.

z. Remove the combination meter bracket bolts and remove the bracket from the vehicle.

aa. Once the mounting bolts of the wiring harness clip and steering column assembly are removed, pull the steering column assembly backward, and free the combination meter bracket from the instrument panel and pad.

bb. Remove the side ventilations by inserting a thin pry tool into the gaps between the instrument panel and pad, pull back to disconnect the metal retaining clips. Then, disconnect the door mirror switch wiring harness connectors, and remove the side ventilations.

cc. Remove the instrument panel and pad by removing the bolts and screws. Then, remove the front passenger air bag module, disconnect the wiring harness connectors, and remove the instrument panel and pad from the passenger door opening.

11. Remove the blower unit, as follows:

a. Remove the ECM with bracket attached.

b. Disconnect the intake door motor connector and blower fan motor connector.

c. Remove the wiring harness clip from the blower unit.

d. Remove the mounting bolt and screws from the blower unit.

Move blower unit rightward, and remove the locating pin and joint. Then, remove the blower unit downward.

e. Remove the blower unit.

12. Remove the instrument stays (driver-side and passenger-side).

13. Remove the mounting bolts from the heater and cooling unit.

14. Disconnect the drain hose.

15. Remove the ventilator ducts, defroster nozzle and ducts.

16. Remove the steering member mounting bolts, nut and wiring harness clips.

17. Remove the steering member, then remove the heater and cooling unit.

To install:

18. Install the heater and cooling unit. Tighten the mounting bolts to 60 inch lbs. (7 Nm).

19. Install the steering member.

20. Install the steering member mounting bolts, nut and wiring harness clips. Tighten the steering member mounting bolts to 9 ft. lbs. (12 Nm).

21. Install the ventilator ducts, defroster nozzle and ducts.

22. Reconnect the drain hose.

23. Install the mounting bolts for the heater and cooling unit.

24. Install the instrument stays (driver-side and passenger-side).

Make sure the locating pin and joint are securely inserted.

25. Install the blower unit, as follows:

a. Install the blower unit.

b. Install the mounting bolt and screws for the blower unit.

c. Install the wiring harness clip for the blower unit.

d. Reconnect the intake door motor connector and blower fan motor connector.

e. Install the ECM with its bracket attached.

26. Install the instrument panel assembly, as follows:

a. Install the front passenger air bag module and instrument panel and pad.

b. Install the side ventilations.

c. Install the combination meter bracket and reinstall the steering column assembly.

d. Install the front defroster grilles.

e. Install the display unit and audio unit.

f. Install the cluster lid C.

g. Install the instrument side panels..

h. Install the combination meter assembly.

i. Install the steering lock escutcheon.

j. Install the lighting and turn signal switch.

k. Install the wiper and washer switch

l. Install the steering column upper cover.

m. Install the steering column lower cover.

n. Install the steering column front lower cover.

o. Install the instrument driver lower panel.

p. Reattach the data link connector.

q. Reconnect the in-vehicle sensor and all electrical parts.

r. Install the grommet and hood lock cable.

s. Install the instrument passenger lower panel.

t. Install the instrument lower cover.

u. Connect the center console sub-wiring harness.

v. Install the console finishers.

w. Install the console finisher screws.

x. Reconnect the wiring harness connector.

y. Install the A/T console finisher.

z. Install instrument clock finisher.

aa. Install the A/T select lever knob.

bb. Install the front pillar garnish.

cc. Install the dash side finisher and plastic nuts.

dd. Install the front kicking plate on both sides of the vehicle.

27. Reconnect the two heater hoses to the heater core.

28. On 3.5L engines, install the electronic control throttle assembly.

Replace the O-rings for A/C piping with new ones, then apply compressor oil to them when installing them.

29. Reconnect the evaporator-side one touch joint. The connection point for the female-side piping is thin. So, when inserting the male-side piping, take care not to deform the female-side piping. Slowly insert it in the axial direction. Insert the one-touch joint connection point securely until it clicks. After the piping has been connected, pull on the male-side piping by hand to make sure the piping does not come off.

30. Install the low-pressure flexible hose bracket mounting bolts.

31. Install the two high-pressure pipe mounting clips.

32. Install the cowl top cover.

33. Fill the cooling system.

34. Recharge the vehicle A/C system and check for leaks.

Cylinder Head

REMOVAL & INSTALLATION

3.5L Engine

1. Before servicing the vehicle, refer to the Precautions Section.
2. Remove the camshaft.

➡ **It is also possible to perform the following steps 2 and 3 just before removing camshaft.**

3. Temporarily support the front suspension member to support the engine.

✳✳ CAUTION

Temporary support means that the engine is adequately stable although the weight supported by the hoist may be released.

➡ **At the time of the start of this procedure the front suspension member is removed, and the cylinder head is hanged by the hoist with engine slinger installed.**

Right bank

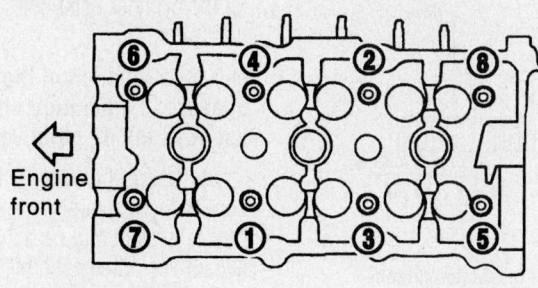

Left bank

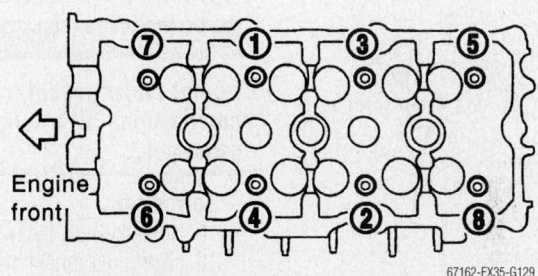

67162-FX35-G129

Cylinder head mounting bolt tightening sequence—3.5L engine

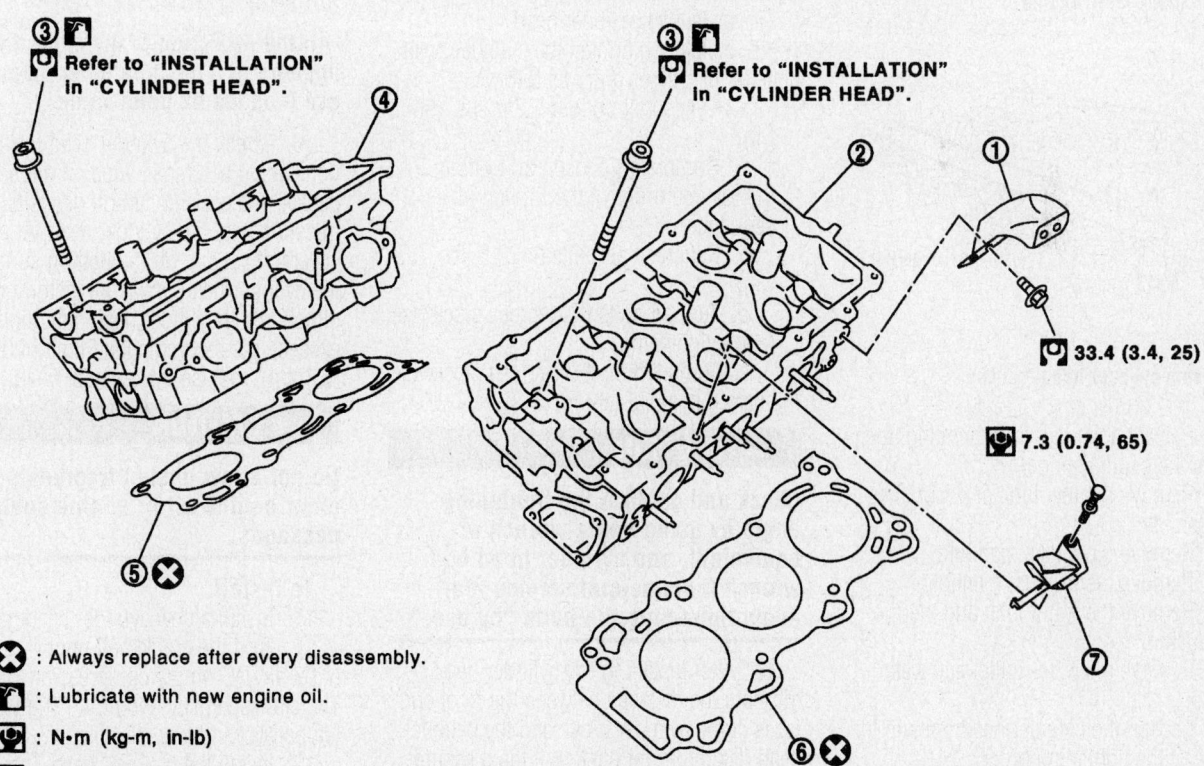

⊗ : Always replace after every disassembly.

⊡ : Lubricate with new engine oil.

⬓ : N•m (kg-m, in-lb)

⬓ : N•m (kg-m, ft-lb)

1. Engine rear lower slinger
2. Cylinder head (left bank)
3. Cylinder head bolt
4. Cylinder head (right bank)
5. Cylinder head gasket (right bank)
6. Cylinder head gasket (left bank)
7. Oil level gauge guide

67162-FX35-G128

Exploded view of cylinder head mounting and related components—3.5L engine

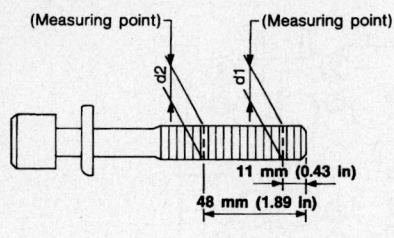

Cylinder head bolt

(Measuring point)   (Measuring point)

d_2 d_1

11 mm (0.43 in)

48 mm (1.89 in)

67162-FX35-G130

Cylinder head bolt inspection dimensions—3.5L engine

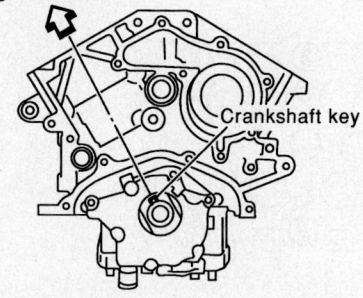

Right bank side

Crankshaft key

67162-FX35-G131

Crankshaft positioning for cylinder head installation—3.5L engine

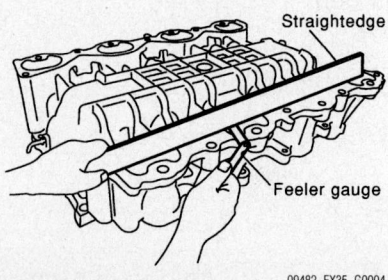

Straightedge

Feeler gauge

09482_FX35_G0004

Checking cylinder head flatness

4. Release the hoist from hanging, then remove the engine slinger.

5. Remove the fuel tube and fuel injector assembly.

6. Remove the intake manifold.

7. Remove the exhaust manifold.

8. Remove the water inlet and thermostat housing.

9. Remove the water outlet and water piping.

10. Loosen the cylinder head bolts in the reverse of the tightening order.

11. Remove the cylinder head.

12. Remove the cylinder head gaskets.

13. Inspect the cylinder head bolt diameters. The cylinder head bolts are tightened by the plastic zone tightening method. Whenever the size difference between d1

and d2 exceeds the limit, replace the bolt with a new one. The specification for d1-d2 for the cylinder head bolts is 0.0043 in. (0.11mm).

➡**If the reduction of the outer diameter appears in a position other than at d2, use it as the d2 point value.**

14. Check the cylinder head for distortion. Using a scraper, wipe off oil, scale, gasket, sealant and carbon deposits from the surface of the cylinder head. At each of several locations on the bottom surface of the cylinder head, measure distortion in six directions. If cylinder head distortion exceeds the recommended limit of 0.004 in. (0.1mm), replace the cylinder head.

❋❋ **CAUTION**

Do not allow gasket fragments to enter engine oil or engine coolant passages.

To install:

15. Install a new cylinder head gasket.

16. Turn the crankshaft until No. 1 piston is set at TDC on the compression stroke. The crankshaft key should line up with the right bank cylinder center line.

17. Install the cylinder head.

18. Install and tighten the cylinder head bolts in the proper order as follows:

    a. Tighten all bolts to 72 ft. lbs. (98 Nm).

    b. Completely loosen all bolts in the reverse order of the tightening sequence.

    c. Retighten all bolts to 29 ft. lbs. (39 Nm)

    d. Turn all bolts 90 degrees clockwise (angle tightening).

    e. Turn all bolts 90 degrees clockwise again (angle tightening).

❋❋ **CAUTION**

Check and confirm the tightening angle by using angle wrench or equivalent, and cylinder head bolt wrench (commercial service tool). Avoid tightening the bolts "by eye."

19. After installing the cylinder head, measure the distance between the front end faces of the cylinder block and the cylinder head (left and right banks). If the measurement is outside the specified range, re-install the cylinder head.

20. Install the water outlet and water piping.

21. Install the water inlet and thermostat housing.

22. Install the exhaust manifold.

23. Install the intake manifold.

24. Install the fuel tube and fuel injector assembly.

25. Install the camshaft.

4.5L Engine

1. Before servicing the vehicle, refer to the Precautions Section.

➡**According to Nissan, cylinder head removal requires removal of the engine assembly from the vehicle.**

2. Remove the engine assembly from vehicle.

3. Remove the exhaust manifold.

4. Remove the camshaft.

5. Loosen the cylinder head bolts in the reverse of the tightening order.

6. Remove the cylinder head.

7. Remove the cylinder head gaskets.

8. Inspect the cylinder head bolt diameters. The cylinder head bolts are tightened by the plastic zone tightening method. Whenever the size difference between d1 and d2 exceeds the limit, replace the bolt with a new one. The specification for d1-d2 for the cylinder head bolts is 0.0071 in. (0.18mm)

➡**If the reduction of the outer diameter appears in a position other than at d2, use it as the d2 point value.**

9. Check the cylinder head for distortion. Using a scraper, wipe off oil, scale, gasket, sealant and carbon deposits from the surface of the cylinder head. At each of several locations on the bottom surface of the cylinder head, measure distortion in six directions. If cylinder head distortion exceeds the recommended limit of 0.004 in. (0.1mm), replace the cylinder head.

❋❋ **CAUTION**

Do not allow gasket fragments to enter engine oil or engine coolant passages.

To install:

10. Install a new cylinder head gasket.

11. Turn the crankshaft until No. 1 piston is set at TDC on the compression stroke. The crankshaft key should line up with the left bank cylinder center line.

12. Install the cylinder head.

❋❋ **CAUTION**

If cylinder head bolts are to be reused, check their outer diameters before installation.

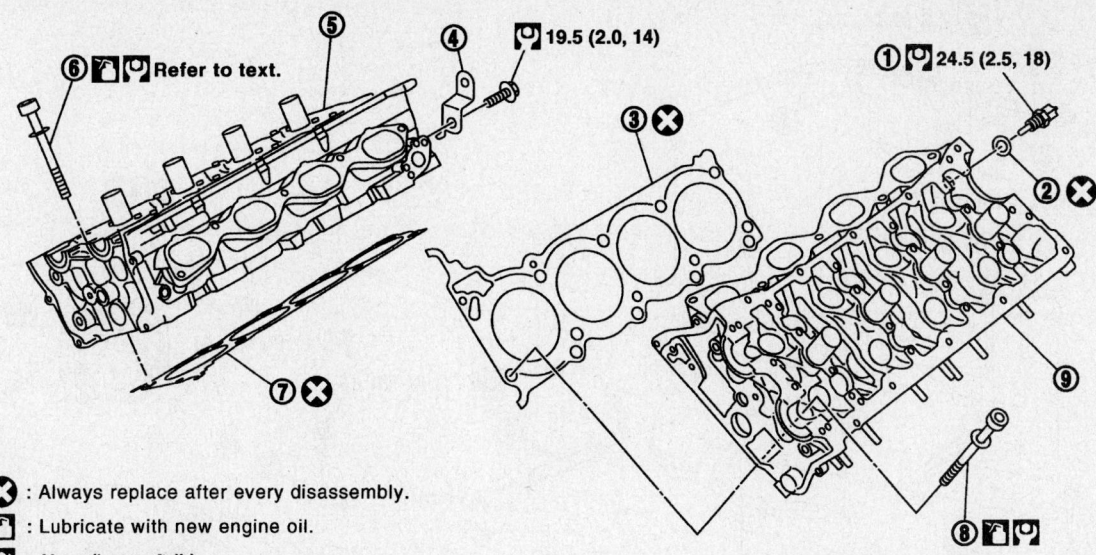

- ⊗ : Always replace after every disassembly.
- ▢ : Lubricate with new engine oil.
- ▢ : N•m (kg-m, ft-lb)

1. Engine coolant temperature sensor	2. Washer	3. Cylinder head gasket (left bank)
4. Harness bracket	5. Cylinder head (right bank)	6. Cylinder head bolt
7. Cylinder head gasket (right bank)	8. Cylinder head bolt	9. Cylinder head (left bank)

67162-FX35-G132

Exploded view of the cylinder head mounting and related components—4.5L engine

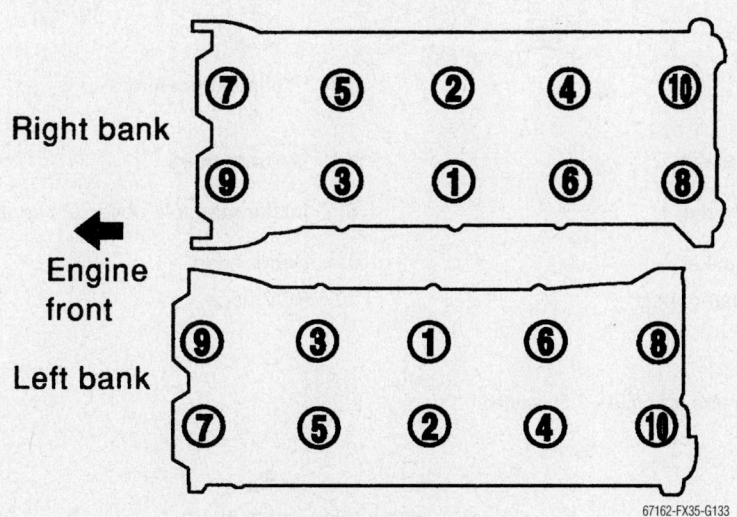

Right bank

← Engine front

Left bank

67162-FX35-G133

Cylinder head mounting bolt tightening sequence—4.5L engine

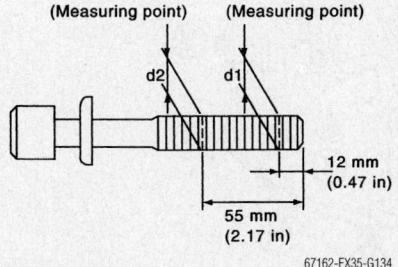

67162-FX35-G134

Cylinder head mounting bolt inspection dimensions—4.5L engine

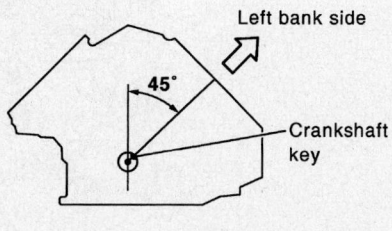

67162-FX35-G135

Crankshaft positioning for cylinder head installation—4.5L engine

13. Install and tighten the cylinder head bolts in the proper order as follows:

a. Tighten all bolts to 72 ft. lbs. (98 Nm).

b. Completely loosen all bolts in the reverse order of the tightening sequence.

c. Retighten all bolts to 33 ft. lbs. (44 Nm)

d. Turn all bolts 60 degrees clockwise (angle tightening).

e. Turn all bolts 60 degrees clockwise again (angle tightening).

✳✳ CAUTION

Check and confirm the tightening angle by using angle wrench or equivalent.

14. Install the camshaft.
15. Install the exhaust manifold.
16. Install the engine assembly into the vehicle.

Intake Manifold

REMOVAL & INSTALLATION

3.5L Engine

UPPER INTAKE MANIFOLD

1. Before servicing the vehicle, refer to the Precautions Section.

The upper intake manifold is constructed of two halves: upper intake manifold collec-

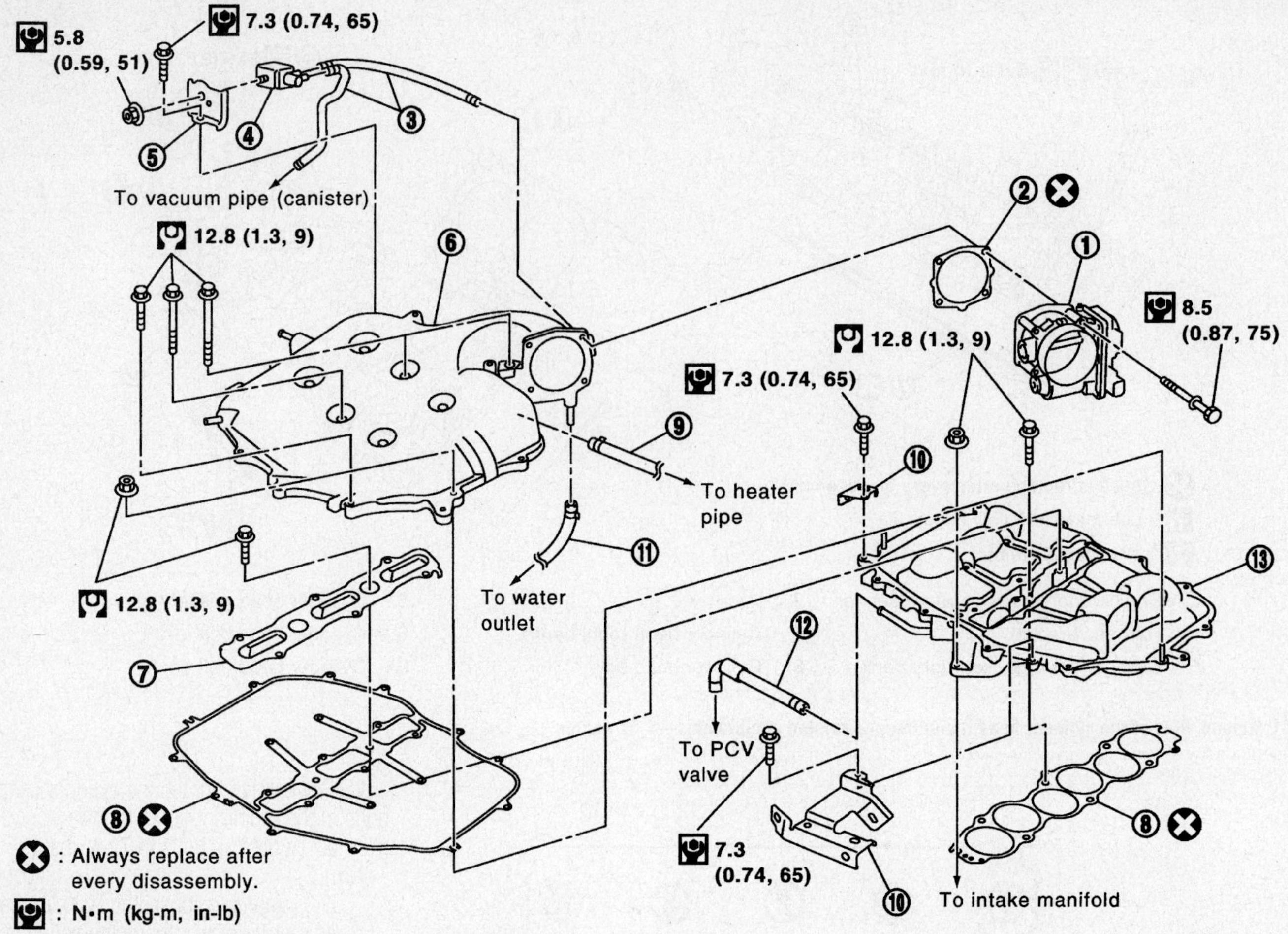

5.8
(0.59, 51)

7.3 (0.74, 65)

To vacuum pipe (canister)

12.8 (1.3, 9)

7.3 (0.74, 65)

8.5
(0.87, 75)

12.8 (1.3, 9)

To heater
pipe

To water
outlet

12.8 (1.3, 9)

To PCV
valve

7.3
(0.74, 65)

To intake manifold

❌ : Always replace after
    every disassembly.

🔧 : N•m (kg-m, in-lb)

1.	Electric throttle control actuator	2.	Gasket	3.	Vacuum hose
4.	EVAP canister purge volume control solenoid valve	5.	Bracket	6.	Intake manifold collector (upper)
7.	Intake manifold collector cover	8.	Gasket	9.	Water hose
10.	Bracket	11.	Water hose	12.	PCV hose
13.	Intake manifold collector (lower)				

67162-FX35-G136

Exploded view of the upper and lower halves of the upper intake manifold—3.5L engine

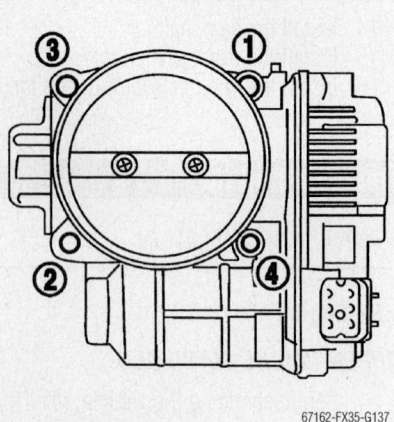

67162-FX35-G137

Throttle body mounting bolt tightening sequence—3.5L engine

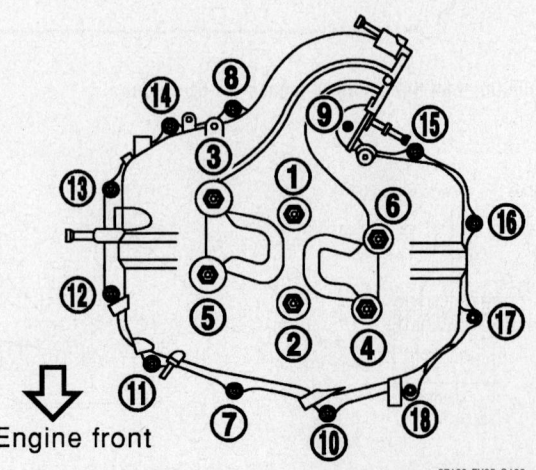

Engine front

67162-FX35-G138

Tightening sequence for the upper half of the upper intake manifold—3.5L engine

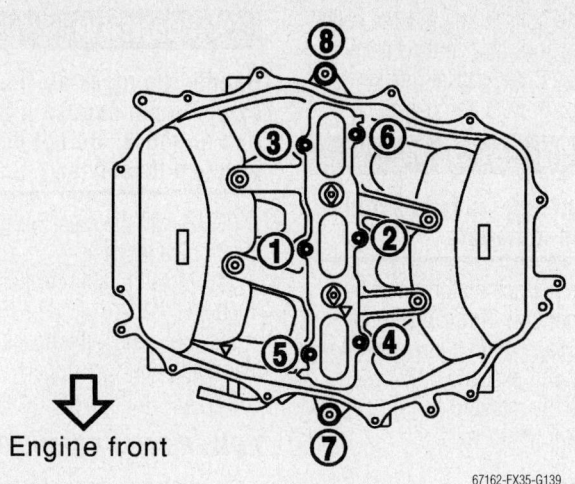

Engine front

67162-FX35-G139

Tightening sequence for the lower half of the upper intake manifold—3.5L engine

tor and lower intake manifold collector. Together these two components create the upper intake manifold. The 3.5L engine also uses a lower intake manifold plenum.

✳✳ WARNING

To avoid the danger of being scalded, never drain engine coolant when engine is hot.

➡The gasket for the intake manifold collector (upper) is secured together with the mounting bolt for the intake manifold collector (lower). Thus, even when only the gasket for upper side is replaced, the gasket for lower side must also be replaced.

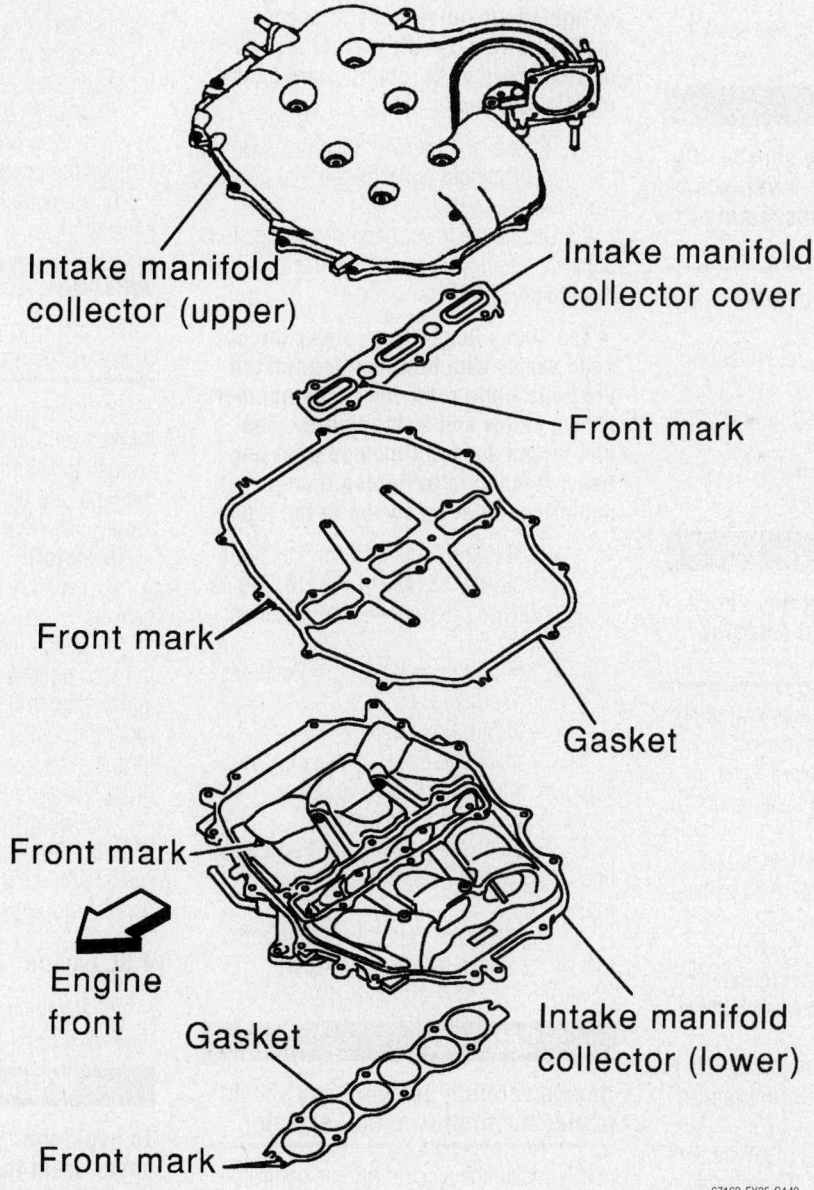

Intake manifold collector (upper)

Intake manifold collector cover

Front mark

Front mark

Gasket

Front mark

Engine front

Gasket

Intake manifold collector (lower)

Front mark

67162-FX35-G140

Upper intake manifold gasket positioning—3.5L engine

2. Remove the engine cover.

3. Disconnect and plug the water hoses from intake manifold collector (upper).

✶ CAUTION

Do not spill engine coolant on the drive belts.

4. Remove the air cleaner case and air duct, as follows:

   a. Remove the air duct (inlet).

   b. Disconnect the mass air flow sensor wiring harness connector.

   c. Remove the air cleaner case/mass air flow sensor assembly and the air duct/resonator assembly disconnecting their joints.

➡**Add marks as necessary for easier installation.**

   d. Remove the mass air flow sensor from air cleaner case.

✶ CAUTION

Handle the mass air flow sensor with care. Do not expose it to harsh vibration or shock. Do not disassemble it or touch its sensor.

   e. Remove the resonator in the fender, lifting the left fender protector.

5. Remove the electric throttle control actuator, as follows:

   a. Disconnect the wiring harness connector.

   b. Loosen the bolts in the reverse order as shown in the figure.

✶✶ CAUTION

Handle carefully to avoid any shock to electric throttle control actuator. Do not disassemble.

6. Remove the fuel sub-tube mounting bolt to disconnect it from the rear of the intake manifold collector (lower). Refer to the fuel injector removal and installation procedure.

7. Disconnect the vacuum hose and water hose from the intake manifold collector (upper).

8. Disconnect the EVAP canister purge volume control solenoid valve bracket mounting bolt from the intake manifold collector (upper).

9. Loosen the bolts in the reverse order of the illustration to remove the intake manifold collector (upper).

10. Remove the PCV hose (between the intake manifold collector and the right-hand rocker cover).

11. Loosen the bolts in the reverse order of the illustration, and remove the intake manifold collector cover, gasket, intake manifold collector (lower) and gasket.

✶✶ CAUTION

Cover all engine openings to avoid entry of foreign materials.

12. Check the surface distortion of both the intake manifold collector (upper and lower) mating surfaces with a straightedge and feeler gauge. If it exceeds 0.004 in. (0.1mm), replace the intake manifold collector (upper and/or lower).

To install:

13. Install the intake manifold collector (lower). Tighten the mounting bolts in numerical order as shown in the figure.

➡**Tighten mounting bolts to secure gasket (lower), intake manifold collector (lower), gasket (upper), and intake manifold collector cover.**

14. Reconnect the PCV hose (between the intake manifold collector and the right-hand rocker cover).

15. Install the lower manifold. If the stud bolts were removed, install them and tighten them to the specified torque.

➡**The shank length from under the bolt head varies with bolt location. Install the bolts while referring to the numbers shown below and in the figure. (The bolt length does not include pilot portion.) Make sure to tighten them in numerical order as shown in the figure.**

- M6 bolt (length 25mm): Positions 7, 8, 10, 11, 13, 14, 15, 16, and 18
- M6 bolt (length 45mm): Positions 2, 4, 5
- M6 bolt (length 60mm): Positions 1, 3, 6, and 9
- M6 nut: Positions 12, 17

16. Reattach the EVAP canister purge volume control solenoid valve bracket to the intake manifold collector (upper).

17. Reattach the vacuum hose and water hose to the intake manifold collector (upper).

18. Reattach the fuel sub-tube to the rear of the intake manifold collector (lower).

✶✶ CAUTION

Handle carefully to avoid any shock to electric throttle control actuator.

19. Install the electric throttle control actuator.

✶✶ CAUTION

Handle the mass air flow sensor with care. Do not expose it to harsh vibration or shock. Do not disassemble it or touch its sensor.

20. Install the mass air flow sensor in the air cleaner case.

21. Install the air cleaner case and air duct.

22. Reconnect the hoses to the intake manifold collector (upper).

23. Install the engine cover.

LOWER INTAKE MANIFOLD

1. Before servicing the vehicle, refer to the Precautions Section.

2. Release the fuel pressure.

3. Remove the intake manifold collector (upper) and (lower).

4. Remove the fuel tube and fuel injector assembly.

5. Loosen the mounting bolts and nuts in the reverse order of the illustration to remove the lower intake manifold.

6. Remove the intake manifold gaskets.

✶✶ CAUTION

Cover all engine openings to avoid entry of foreign materials.

7. Check the surface distortion of the intake manifold mating surface with a straightedge and feeler gauge. If it exceeds the limit of 0.04 in. (0.1mm), replace the lower intake manifold.

To install:

8. Install new lower intake manifold gaskets.

9. Install the lower intake manifold. If the stud bolts were removed, install them and tighten to 8 ft. lbs. (11 Nm). Tighten all mounting bolts and nuts to the specified torque in two or more steps in numerical order shown in the figure, as follows:

   a. Step 1: 60 inch lbs. (7 Nm)

   b. Step 2: 21 ft. lbs. (29 Nm)

10. The remainder of installation is the reverse of removal.

4.5L Engine

1. Before servicing the vehicle, refer to the Precautions Section.

✶✶ WARNING

To avoid the danger of being scalded, never drain the engine coolant when the engine is hot.

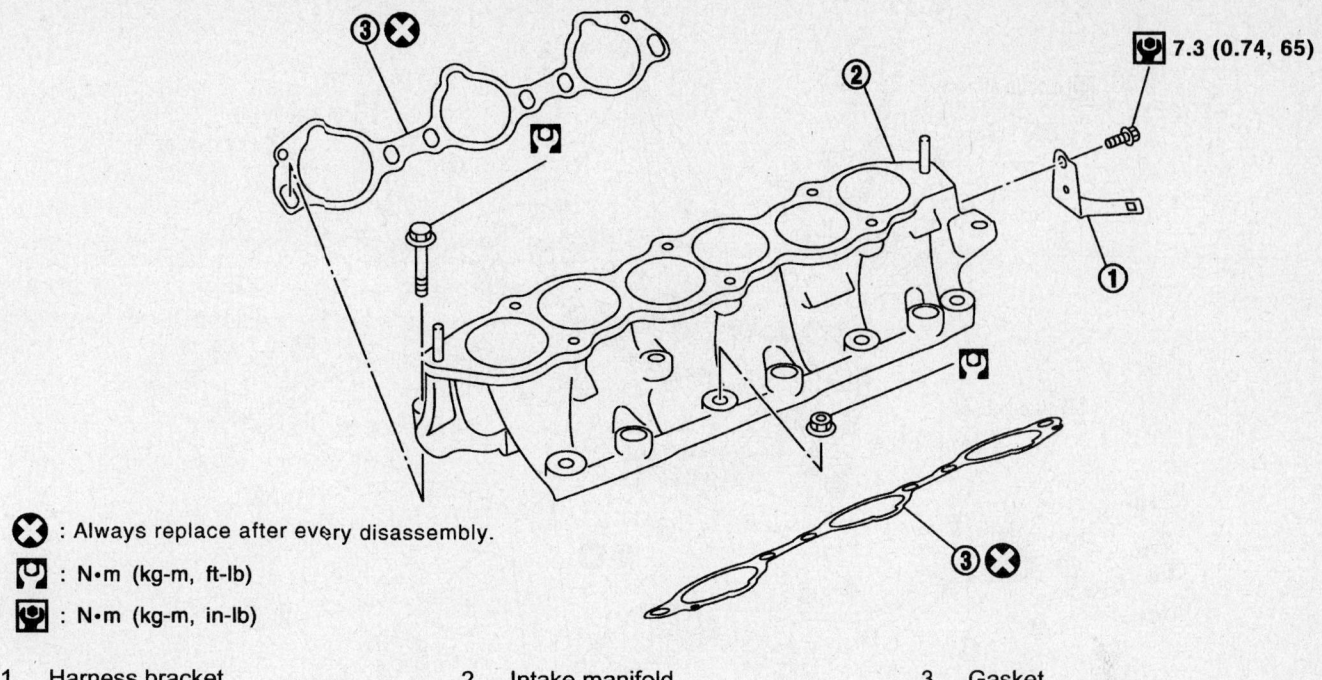

7.3 (0.74, 65)

❌ : Always replace after every disassembly.

🔧 : N•m (kg-m, ft-lb)

🔧 : N•m (kg-m, in-lb)

1. Harness bracket

2. Intake manifold

3. Gasket

67162-FX35-G141

Exploded view of the lower intake manifold—3.5L engine

← **Engine front**

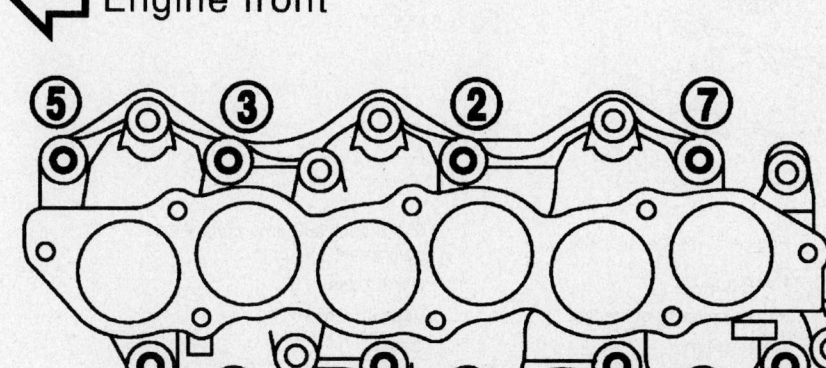

67162-FX35-G142

Lower intake manifold mounting bolt tightening sequence—3.5L engine

2. Remove the engine cover.
3. Release the fuel pressure.
4. Drain the engine coolant.
5. Remove the air duct (inlet), air cleaner case and mass air flow sensor assembly, air duct and resonator assembly.
6. Disconnect the fuel feed hose quick connector on the engine side, the fuel damper and fuel hose assembly. Refer to the fuel injector removal and installation procedure.

✳✳ CAUTION

While hoses are disconnected, plug them to prevent fuel from draining. Do not separate fuel damper and fuel hose.

7. Remove or disconnect the wiring harnesses, brackets, vacuum hose, vacuum gallery and PCV hose and tube from the intake manifold (upper).

8. Remove the electric throttle control actuator as follows:
a. Disconnect the wiring harness connector.
b. Loosen the mounting bolts diagonally.

✳✳ CAUTION

Handle carefully to avoid any shock to the electric throttle control actuator. Do not disassemble the electric throttle control actuator.

9. Disconnect the water hoses from the water gallery.
10. Remove the intake manifold adaptor and water gallery.
11. Loosen the bolts in the reverse order as shown in the figure to remove the intake manifold (upper).
12. Remove the vacuum tank from the intake manifold (lower).
13. Remove the fuel injector and fuel tube assembly. Refer to the fuel injector removal and installation procedure.
14. Loosen the bolts in the reverse order as shown in the figure to remove the intake manifold (lower).
15. Remove the intake manifold gaskets.

✳✳ CAUTION

Cover all engine openings to avoid entry of foreign materials.

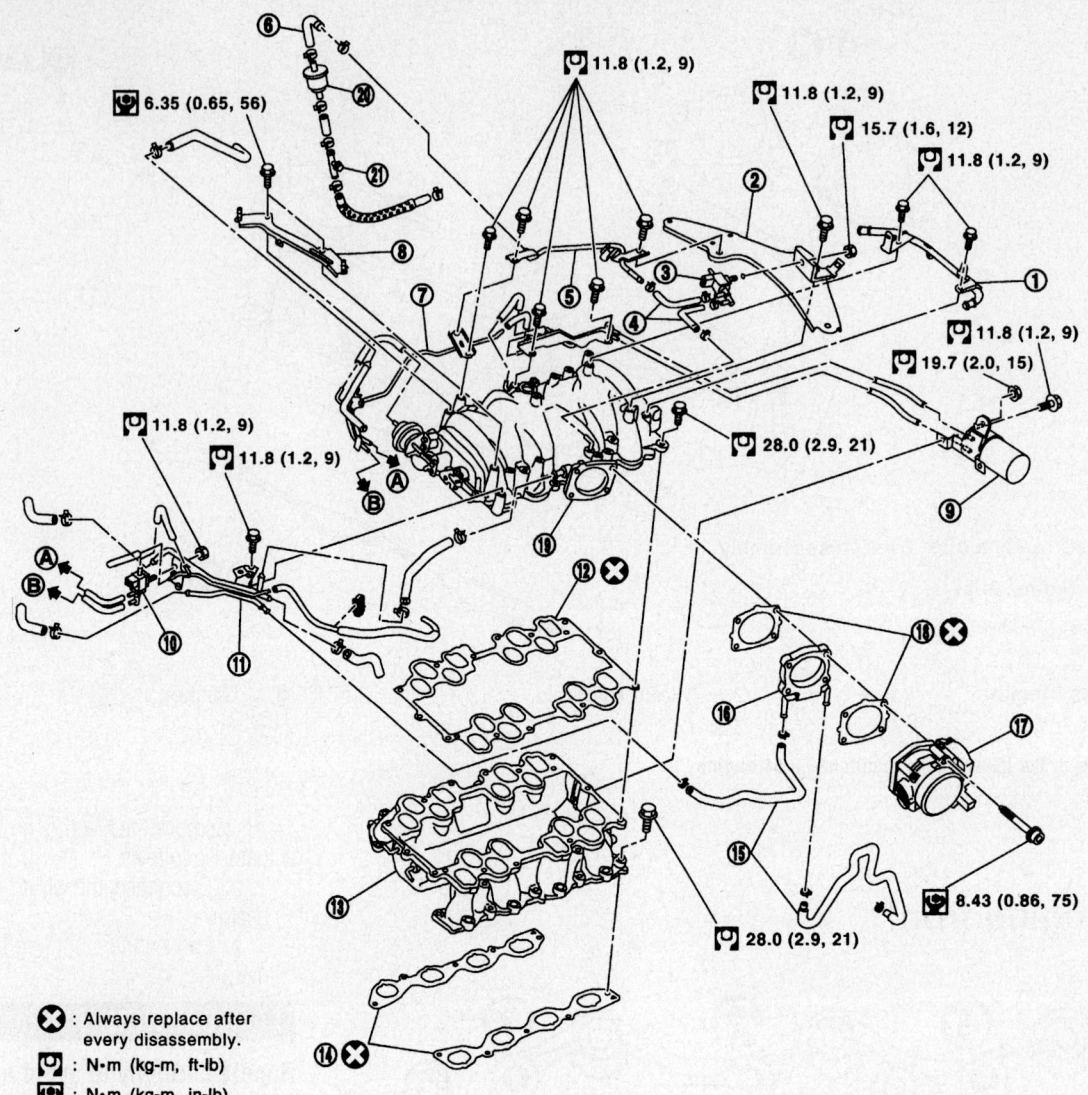

6.35 (0.65, 56)

11.8 (1.2, 9)

11.8 (1.2, 9)

15.7 (1.6, 12)

11.8 (1.2, 9)

11.8 (1.2, 9)

19.7 (2.0, 15)

28.0 (2.9, 21)

11.8 (1.2, 9)

11.8 (1.2, 9)

8.43 (0.86, 75)

28.0 (2.9, 21)

❌ : Always replace after every disassembly.

⊡ : N•m (kg-m, ft-lb)

⊡ : N•m (kg-m, in-lb)

1. PCV tube	2. Engine cover rear bracket	3. EVAP canister purge control solenoid valve
4. EVAP hose	5. EVAP tube	6. EVAP hose
7. Vacuum gallery	8. Engine cover front bracket	9. Vacuum tank
10. VIAS control solenoid valve	11. Water gallery	12. Gasket
13. Intake manifold (lower)	14. Gasket	15. Water hose
16. Intake manifold adapter	17. Electric throttle control actuator	18. Gasket
19. Intake manifold (upper)	20. Resonator	21. EVAP service port

67162-FX35-G143

Exploded view of the intake manifold mounting and related components—2003–05 4.5L engine

16. Check the surface distortion of both the intake manifold (upper and lower) mating surfaces with a straightedge and feeler gauge. If it exceeds the limit of 0.004 in (0.1mm), replace the intake manifolds (lower and/or upper).

To install:

17. Install new intake manifold gaskets.

18. Install the intake manifold and tighten the mounting bolts in numerical order as shown in the figure. There are two types of mounting bolts. Refer to the following for locating bolts:
- M8 bolts (length 90mm): Positions 7, 8
- M8 bolts (length 35mm): Positions except 7 and 8

19. Install the fuel injector and fuel tube assembly.

20. Install the vacuum tank onto the intake manifold (lower).

21. Install the intake manifold (upper). Tighten the mounting bolts in the numerical order shown in the figure. There are two types of mounting bolts. Refer to the following for locating bolts:
- M8 bolts (length 80mm): Positions 4, 5, 6, and 7
- M8 bolts (length 25mm): Positions except 4, 5, 6, and 7

22. Install the intake manifold adaptor and water gallery.

23. Connect the water hoses to the water gallery.

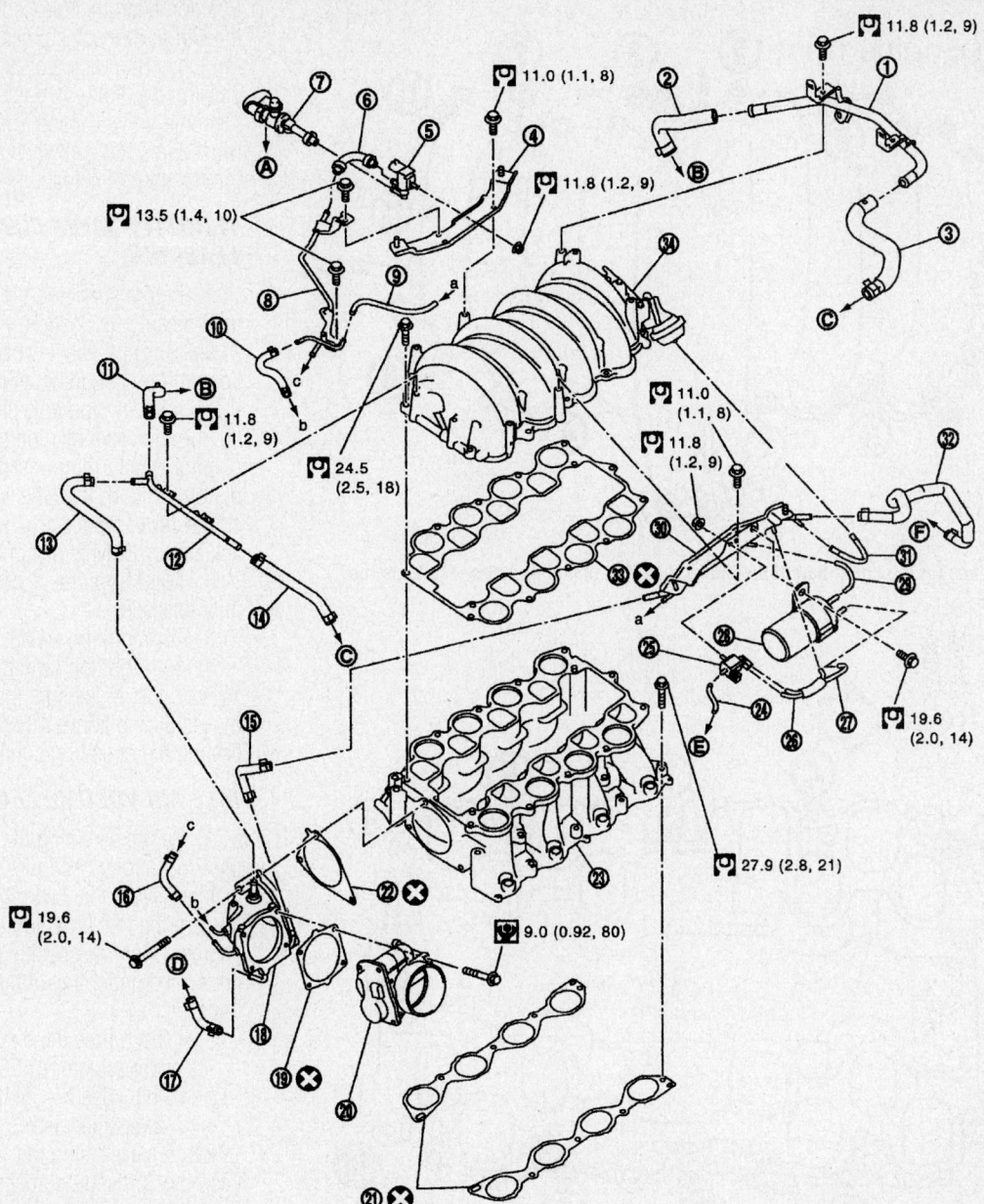

□ : N•m (kg-m, in-lb)
□ : N•m (kg-m, ft-lb)

1.	PCV tube	2.	PCV hose	3.	PCV hose
4.	Engine cover bracket (RH)	5.	EVAP canister purge control solenoid valve	6.	EVAP hose
7.	EVAP service port	8.	EVAP tube	9.	Vacuum hose
10.	Vacuum hose	11.	PCV hose	12.	PCV tube
13.	PCV hose	14.	PCV hose	15.	Water hose
16.	EVAP hose	17.	Water hose	18.	Intake manifold adapter
19.	Gasket	20.	Electric throttle control actuator	21.	Gasket
22.	Gasket	23.	Intake manifold (lower)	24.	Vacuum hose
25.	VIAS control solenoid valve	26.	Vacuum hose	27.	Vacuum hose
28.	Vacuum tank	29.	Vacuum hose	30.	Engine cover bracket (LH)
31.	Vacuum hose	32.	Water hose	33.	Gasket
34.	Intake manifold (upper)				
A.	To centralized under-floor piping	B.	To rocker cover (right bank)	C.	To rocker cover (left bank)
D.	To thermostat housing	E.	To air duct and resonator assembly	F.	To heater pipe

09482_FX35_G0003

Exploded view of the intake manifold—2006 4.5L engine

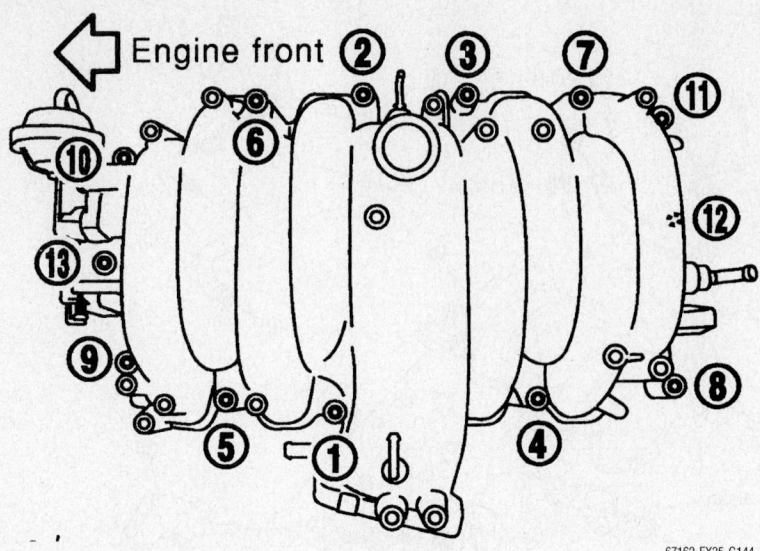

Tightening sequence for the upper half of the intake manifold mounting bolts—4.5L engine

67162-FX35-G144

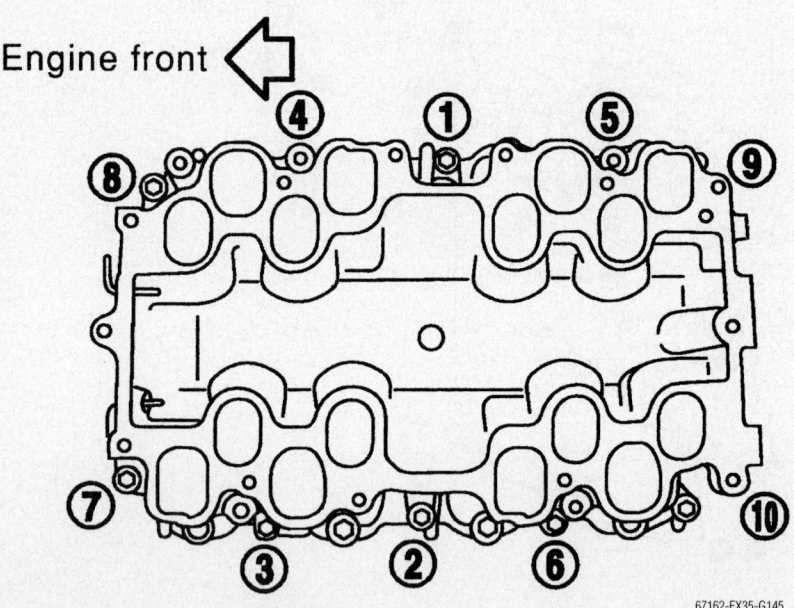

Tightening sequence for the lower half of the intake manifold mounting bolts—4.5L engine

67162-FX35-G145

❋❋ CAUTION

Handle carefully to avoid any shock to the electric throttle control actuator.

24. Install the electric throttle control actuator. Install the intake manifold adapter gasket and electric throttle control actuator gasket so that the three protrusions for installation do not face downward.

25. Tighten the mounting bolts of the electric throttle control actuator equally and diagonally in several steps.

26. Reconnect the wiring harnesses, brackets, vacuum hose, vacuum gallery and PCV hose and tube to the intake manifold (upper).

27. Connect the fuel feed hose quick connector on the engine side, the fuel damper and fuel hose assembly.

28. Install the air duct (inlet), air cleaner case and mass air flow sensor assembly, air duct and resonator assembly.

29. Fill the engine cooling system with engine coolant.

30. Install the engine cover.

31. Perform the following drivability adjustments:

32. Perform the "Throttle Valve Closed Position Learning" procedure (below) when the wiring harness connector of the electric throttle control actuator is disconnected, or perform the "Idle Air Volume Learning" and "Throttle Valve Closed Position Learning" procedures (below) when the electric throttle control actuator is replaced.

THROTTLE VALVE CLOSED POSITION LEARNING

1. Before servicing the vehicle, refer to the Precautions Section.

The Throttle Valve Closed Position Learning procedure is an operation for the ECM to relearn the fully closed position of the throttle valve by monitoring the throttle position sensor output signal. It must be performed each time the wiring harness connector of the electric throttle control actuator or ECM is disconnected.

2. Make sure that accelerator pedal is fully released.

3. Turn ignition switch ON.

4. Turn ignition switch OFF wait at least 10 seconds. Make sure that throttle valve moves during above 10 seconds by confirming the operating sound.

IDLE AIR VOLUME LEARNING

1. Before servicing the vehicle, refer to the Precautions Section.

Idle Air Volume Learning is an operation to learn the idle air volume that keeps each engine within the specific range. It must be performed under any of the following conditions:

- Each time the electric throttle control actuator or ECM is replaced.
- Idle speed or ignition timing is out of specification.

Before performing the "Idle Air Volume Learning" procedure, make sure that all of the following conditions are satisfied. Learning will be cancelled if any of the following conditions are missed for even a moment.

- Battery voltage: More than 12.9V (At idle)
- Engine coolant temperature: 70–100°C (158–212°F)
- PNP switch: ON
- Electric load switch: OFF (air conditioner, headlamp, and rear window defogger)

➡**On vehicles equipped with daytime light systems, if the parking brake is applied before the engine is started, the headlamp will not be illuminated.**

- Steering wheel: Neutral (Straight-ahead position)
- Vehicle speed: Stopped

- Transmission: Warmed-up
- For models with CONSULT-II, drive vehicle until "FLUID TEMP SE" in "DATA MONITOR" mode of "A/T" system indicates less than 0.9V.
- For models without CONSULT-II, drive vehicle for 10 minutes.

2. If using the CONSULT-II tool, perform the following:

a. Perform the "Accelerator Pedal Released Position Learning" procedure.

b. Perform the "Throttle Valve Closed Position Learning" procedure.

c. Start the engine and warm it up to normal operating temperature.

d. Check that all items listed above are properly set.

e. Select "IDLE AIR VOL LEARN" in "WORK SUPPORT" mode.

f. Touch "START" and wait 20 seconds.

g. Make sure that "CMPLT" is displayed on CONSULT-II screen. If "CMPLT" is not displayed, the Idle Air Volume Learning procedure will not be carried out successfully.

h. Rev up the engine two or three times and make sure that idle speed and ignition timing are within specifications.

3. If NOT using the CONSULT-II tool, perform the following:

➡It is best to keep track of time accurately with a clock.

➡It is impossible to switch the diagnostic mode when an accelerator pedal position sensor circuit has a malfunction.

a. Perform the "Accelerator Pedal Released Position Learning" procedure.

b. Perform the "Throttle Valve Closed Position Learning" procedure.

c. Start the engine and warm it up to normal operating temperature.

d. Check that all items listed above are properly set.

e. Turn the ignition switch OFF and wait at least 10 seconds.

f. Confirm that the accelerator pedal is fully released, turn the ignition switch ON and wait 3 seconds.

➡Repeat the following two steps quickly five times within 5 seconds.

g. Fully depress the accelerator pedal.

h. Fully release the accelerator pedal.

i. Wait 7 seconds, fully depress the accelerator pedal and keep it for approx. 20 seconds until the MIL stops blinking and remains ON.

j. Fully release the accelerator pedal

within 3 seconds after the MIL turned ON.

k. Start the engine and let it idle.

l. Wait 20 seconds.

m. Rev up the engine two or three times and make sure that idle speed and ignition timing are within specifications.

n. If idle speed and ignition timing are not within specification, the Idle Air Volume Learning procedure will not be successful.

ACCELERATOR PEDAL RELEASED POSITION LEARNING

1. Before servicing the vehicle, refer to the Precautions Section.

The "Accelerator Pedal Released Position Learning" procedure is an operation for the ECM to relearn the fully released position of the accelerator pedal by monitoring the accelerator pedal position sensor output signal. It must be performed each time the wiring harness connector of the accelerator pedal position sensor or ECM is disconnected.

2. Make sure that the accelerator pedal is fully released.

3. Turn the ignition switch ON and wait at least 2 seconds.

4. Turn the ignition switch OFF wait at least 10 seconds.

5. Turn the ignition switch ON and wait at least 2 seconds.

6. Turn the ignition switch OFF wait at least 10 seconds.

Exhaust Manifold

REMOVAL & INSTALLATION

3.5L Engine

1. Before servicing the vehicle, refer to the Precautions Section.

✳✳ WARNING

Perform the work when the exhaust and cooling system have completely cooled down.

2. Remove the engine cover.

3. Remove the air cleaner case and air duct.

4. Remove the front and rear engine undercover and front cross bar.

5. Disconnect the heated oxygen sensors 2 (bank 1 and bank 2) wiring harness connectors.

6. Using a heated oxygen sensor wrench, remove heated oxygen sensors 2 (bank 1 and bank 2).

✳✳ CAUTION

Be careful not to damage heated oxygen sensor. Discard any heated oxygen sensor which has been dropped from a height of more than 20 in. (0.5m) onto a hard surface such as a concrete floor; replace with a new sensor.

7. Remove the exhaust mounting bracket between the right/left catalytic converter and transmission.

8. Remove the three way catalyst (right and left bank).

9. Disconnect the heated oxygen sensor 1 (bank 1 and bank 2) wiring harness connectors and remove the wiring harness clip.

10. Using the heated oxygen sensor wrench, remove the heated oxygen sensor 1 (bank 1 and bank 2).

✳✳ CAUTION

Be careful not to damage heated oxygen sensor. Discard any heated oxygen sensor which has been dropped from a height of more than 20 in. (0.5m) onto a hard surface such as a concrete floor; replace with a new sensor.

11. Remove water pipes on both the right and left side.

12. Remove the exhaust manifold cover (right and left bank).

13. Loosen the mounting nuts in the reverse order shown in the illustration to remove the exhaust manifold.

➡Disregard Nos. 7 and No. 8 in removal.

14. Remove the exhaust manifold gaskets.

✳✳ CAUTION

Cover all engine openings to avoid entry of foreign materials.

15. Check the surface distortion of the exhaust manifold mating surface with a straightedge and feeler gauge. If it exceeds the limit, replace the exhaust manifold.

To install:

16. Install new exhaust manifold gaskets. On installation, locate the thick side of the port connecting part on the right-hand side (from your viewpoint). Locate the round press in the thick side of the port connecting part above the center level line of port.

17. Install the manifold and tighten the mounting nuts in the order shown. If the

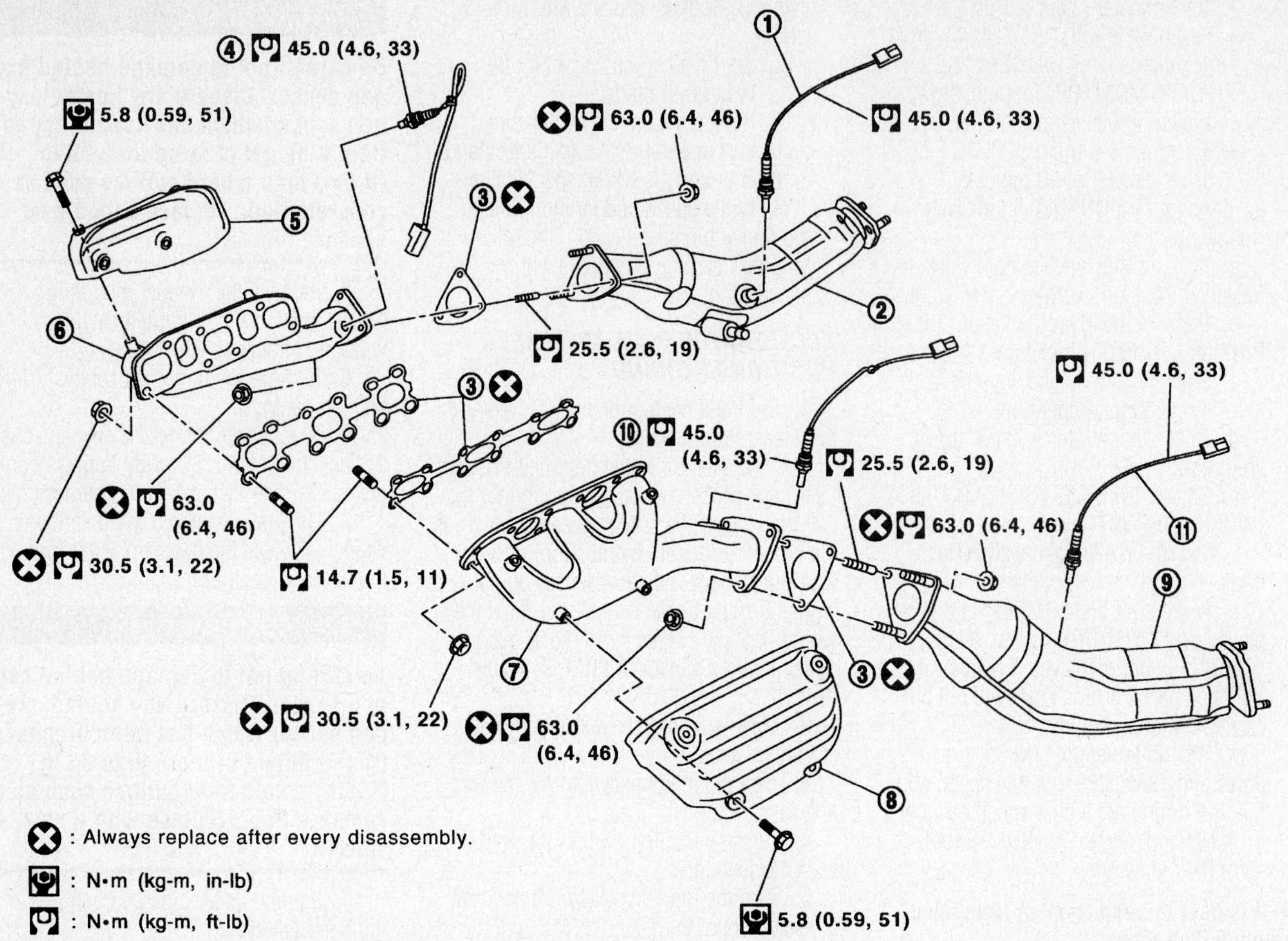

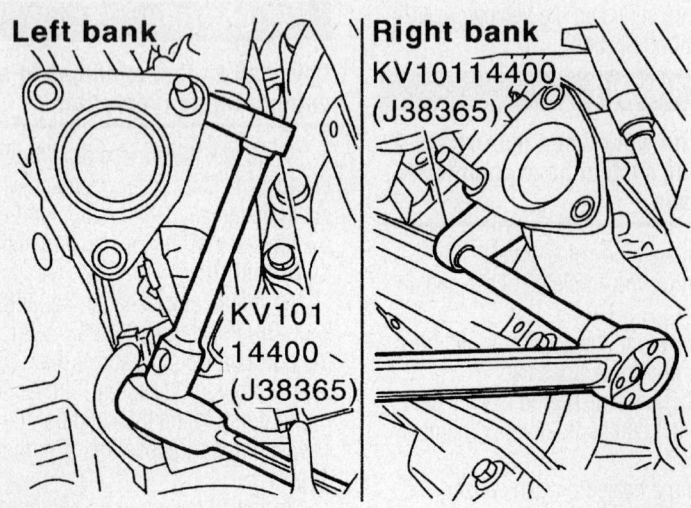

4 🔧 45.0 (4.6, 33)

🔧 5.8 (0.59, 51)

❌ 🔧 63.0 (6.4, 46)

1

🔧 45.0 (4.6, 33)

5

6

3 ❌

🔧 25.5 (2.6, 19)

2

10 🔧 45.0 (4.6, 33)

🔧 25.5 (2.6, 19)

🔧 45.0 (4.6, 33)

❌ 🔧 63.0
(6.4, 46)

❌ 🔧 63.0 (6.4, 46)

11

❌ 🔧 30.5 (3.1, 22)

🔧 14.7 (1.5, 11)

9

7

3 ❌

❌ 🔧 30.5 (3.1, 22)

❌ 🔧 63.0
(6.4, 46)

8

🔧 5.8 (0.59, 51)

❌ : Always replace after every disassembly.

🔧 : N•m (kg-m, in-lb)

🔧 : N•m (kg-m, ft-lb)

1. Heated oxygen sensor 2 (bank 1)
2. Three way catalyst (right bank)
3. Gasket
4. heated oxygen sensor 1 (bank 1)
5. Exhaust manifold cover (right bank)
6. Exhaust manifold (right bank)
7. Exhaust manifold (left bank)
8. Exhaust manifold cover (left bank)
9. Three way catalyst (left bank)
10. heated oxygen sensor 1 (bank 2)
11. Heated oxygen sensor 2 (bank 2)

67162-FX35-G146

Exploded view of the exhaust manifold mounting and related components—3.5L engine

Left bank

Right bank
KV10114400
(J38365)

KV101
14400
(J38365)

67162-FX35-G147

Heated oxygen sensor locations—3.5L engine

stud bolts were removed, install them and tighten them to the torque specified. Tighten nuts No. 1 and No. 2 in two steps. The numerical order No. 7 and No. 8 shows the second step.

18. Install the exhaust manifold cover (right and left bank).

19. Install the water pipes on both the right and left side.

20. Install the heated oxygen sensor 1 (bank 1 and bank 2).

21. Install the three way catalyst (right and left bank).

22. Install the exhaust mounting bracket between the right/left catalytic converter and transmission.

23. Install the heated oxygen sensors 2 (bank 1 and bank 2).

24. Reconnect the heated oxygen sensor wiring harness connectors.

Right bank

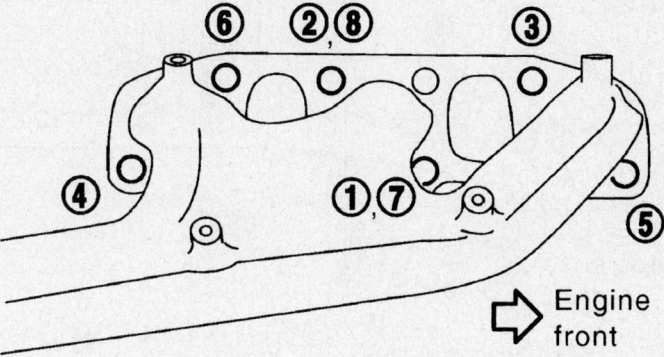

Left bank

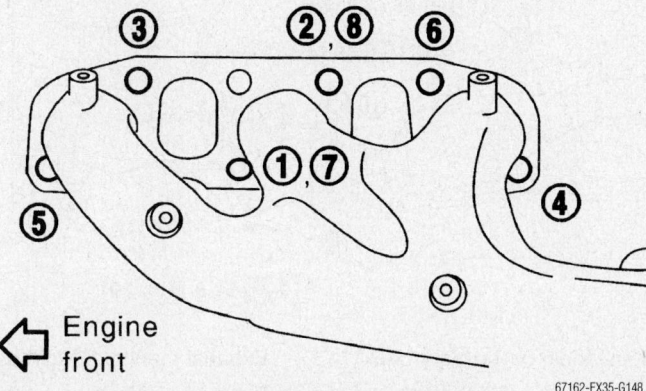

67162-FX35-G148

Exhaust manifold mounting bolt tightening sequence—3.5L engine

Right bank

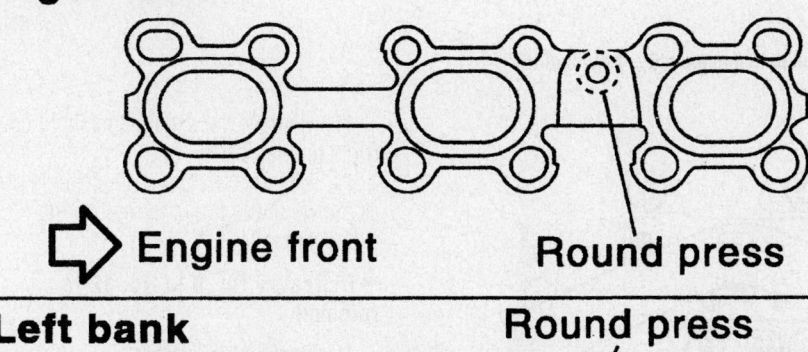

➡ Engine front   Round press

Left bank   Round press

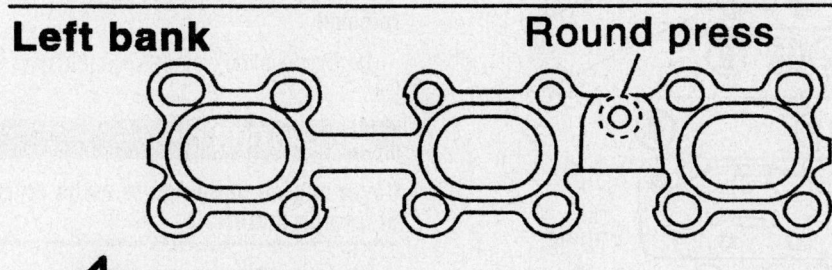

➡ Engine front

67162-FX35-G149

Exhaust manifold gasket positioning—3.5L engine

25. Install the front and rear engine undercover and front cross bar.
26. Install the air cleaner case and air duct.
27. Install the engine cover.

4.5L Engine

1. Before servicing the vehicle, refer to the Precautions Section.

⁂ WARNING

Perform the work, when the exhaust and cooling system have completely cooled down.

2. Remove the engine cover.
3. Remove the front and rear engine under covers.
4. Remove the air duct (inlet), air cleaner case and mass air flow sensor assembly, air duct and resonator assembly.
5. Remove the front cross bar.

⁂ CAUTION

Do not spill engine coolant on drive belts.

6. Drain the engine coolant from the radiator.
7. Remove the radiator.
8. Remove the drive belts.
9. Remove the heated oxygen sensors, as follows:
   a. Disconnect the wiring harness connector of each heated oxygen sensors.
   b. Remove the heated oxygen sensor 1 and 2 on both banks with a heated oxygen sensor wrench.

⁂ CAUTION

Be careful not to damage heated oxygen sensor. Discard any heated oxygen sensor which has been dropped from a height of more than 20 in. (0.5m) onto a hard surface such as a concrete floor; replace with a new sensor.

10. Remove the exhaust mounting bracket between the three-way catalysts (right and left bank) and the transmission.
11. Evacuate the A/C system. Disconnect the A/C piping from the A/C compressor, then remove the A/C compressor.
12. Remove the alternator and bracket.
13. Remove the exhaust front tube.
14. Remove the steering lower joint at the power steering gear assembly side, and release the steering lower shaft.
15. Remove the three-way catalysts (right and left bank).

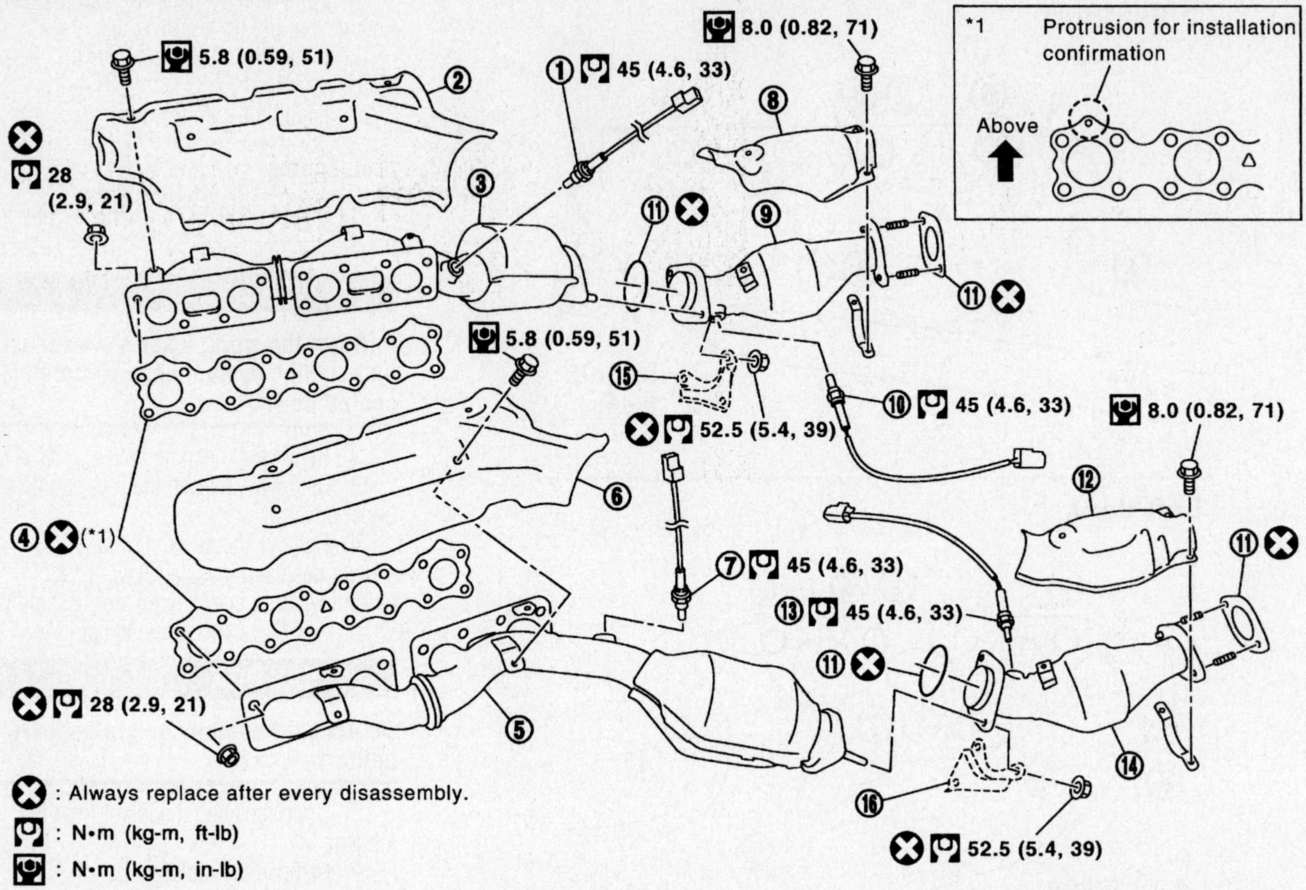

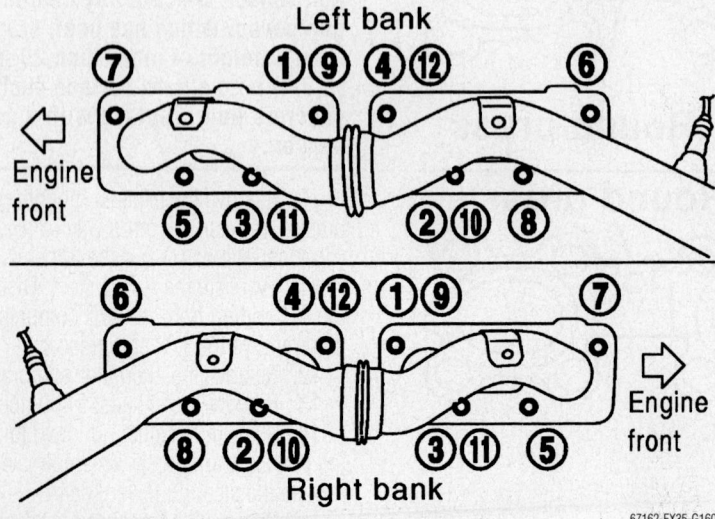

8.0 (0.82, 71)

5.8 (0.59, 51)

① 45 (4.6, 33)

28 (2.9, 21)

*1 Protrusion for installation confirmation

Above

5.8 (0.59, 51)

52.5 (5.4, 39)

(*1)

10 45 (4.6, 33)

8.0 (0.82, 71)

28 (2.9, 21)

⑦ 45 (4.6, 33)

13 45 (4.6, 33)

52.5 (5.4, 39)

❌ : Always replace after every disassembly.

⬚ : N•m (kg-m, ft-lb)

⬚ : N•m (kg-m, in-lb)

1. Heated oxygen sensor 1 (bank 2)
2. Exhaust manifold cover (right bank)
3. Exhaust manifold (right bank)
4. Gasket
5. Exhaust manifold (left bank)
6. Exhaust manifold cover (left bank)
7. Heated oxygen sensor 1 (bank 1)
8. Three way catalyst cover (right bank)
9. Three way catalyst (right bank)
10. Heated oxygen sensor 2 (bank 2)
11. Gasket
12. Three way catalyst cover (left bank)
13. Heated oxygen sensor 2 (bank 1)
14. Three way catalyst (left bank)
15. Mounting bracket
16. Mounting bracket

67162-FX35-G150

Exploded view of the exhaust manifold mounting and related components—4.5L engine

Left bank

Engine front

Engine front

Right bank

67162-FX35-G160

Exhaust manifold mounting bolt tightening sequence—4.5L engine

16. Remove the exhaust manifold covers. (right and left bank)

17. Loosen the nuts in the reverse order as shown in the figure to remove the exhaust manifold.

➡Disregard No. 9 to No. 12 in removal.

18. Remove the exhaust manifold gaskets.

⁕⁕ CAUTION

Cover engine openings to avoid entry of foreign materials.

19. Check the surface distortion of each exhaust manifold flange mating surface with a straightedge and feeler gauge. If it exceeds the limit, replace exhaust manifold.

To install:

20. Install new exhaust manifold gaskets. Install each exhaust manifold gasket with its directional protrusion set upward. Refer to the illustration.

21. Install the exhaust manifold. Install the exhaust manifold mounting nuts in numerical order as shown in the figure. Tighten nuts No. 1 to No. 4 in two steps. The numerical order No. 9 to No. 12 shown second steps.

22. Install the exhaust manifold covers. (right and left bank)

23. Install the three-way catalysts (right and left bank).

24. Install the steering lower joint at the power steering gear assembly side.

25. Install the exhaust front tube.

26. Install the alternator and bracket.

27. Install the A/C compressor, and reconnect the A/C piping to the A/C compressor.

28. Install the exhaust mounting bracket between the three-way catalysts (right and left bank) and the transmission.

✳✳ CAUTION

Be careful not to damage heated oxygen sensor. Discard any heated oxygen sensor which has been dropped from a height of more than 20 in. (0.5m) onto a hard surface such as a concrete floor; replace with a new sensor.

29. Install the heated oxygen sensors, as follows:

30. Install the drive belts.

31. Install the radiator.

32. Install the front cross bar.

33. Install the air duct (inlet), air cleaner case and mass air flow sensor assembly, air duct and resonator assembly.

34. Install the front and rear engine under covers.

35. Install the engine cover.

36. Fill the engine cooling system with coolant.

37. Recharge the A/C system.

Camshaft and Valve Lifters

REMOVAL & INSTALLATION

3.5L Engine

1. Before servicing the vehicle, refer to the Precautions Section.

2. Remove the front timing chain case, camshaft sprocket, timing chain and rear timing chain case.

3. If necessary, remove the camshaft position sensor (PHASE) (right and left banks) from the cylinder head back side.

✳✳ CAUTION

Handle the camshaft position sensor carefully to avoid dropping it and exposing it to abrupt or severe shocks. Do not disassemble it, and do not allow metal powder to adhere to magnetic part at the sensor tip. Do not place sensors in a location where they are exposed to magnetism.

4. Remove the intake valve timing control solenoid valve from the No.1 camshaft bracket.

5. Remove the intake and exhaust camshaft brackets. Mark the camshafts, camshaft brackets, and bolts so they are reinstalled in the same position and direction for installation. Loosen the camshaft bracket bolts equally in several steps in the reverse order as shown.

6. Remove the camshaft.

7. Remove the valve lifters. Identify the installation positions, and store them without mixing them up.

8. Remove the secondary timing chain tensioner from the cylinder head.

9. Remove the chain tensioner with its stopper pin attached.

➡**The stopper pin was attached when the secondary timing chain was removed.**

10. Inspect camshaft run out, as follows:
    a. Put a V block on a precise flat bed, and support the No. 2 and No. 4 journals of the camshaft.

✳✳ CAUTION

Do not support journal No. 1 (on the side of camshaft sprocket) because it has a different diameter than the other three locations.

    b. Set a dial gauge vertically to the No. 3 journal.
    c. Turn the camshaft in one direction by hand, and measure camshaft run out on the dial gauge. (Total indicator reading)
    d. If run out exceeds specification, replace the camshaft.

➡**Camshaft run out limit: 0.0020 in (0.05mm)**

11. Inspect the camshaft cam height, as follows:

    a. Measure the camshaft cam height.

➡**Standard cam height (intake and exhaust): 1.7663–1.7738 in. (44.865–45.055mm). Allowable cam wear limit: 0.008 in. (0.2mm).**

    b. If wear is beyond specification, replace the camshaft.

12. Measure the outer diameter of the camshaft journals.

➡**Standard outer camshaft journal diameters—No. 1: 1.0211–1.0218 in. (25.935–25.955mm), Nos. 2, 3 & 4: 0.9230–0.9238 in. (23.445–23.465mm)**

13. Measure the inner diameter of the camshaft bracket journals, as follows:

    a. Tighten the camshaft bracket bolts with the specified torque.
    b. Using an inside micrometer, measure the inner diameter "A" of the camshaft bracket.
    c. If the inner camshaft journals exceed allowable specification, replace the cylinder head. The camshaft brackets cannot be replaced as a single part, because it is machined together with the cylinder head. If necessary, replace the whole cylinder head assembly.

➡**Standard inner diameter—No. 1: 1.0236–1.0244 in. (26.000–26.021mm), Nos. 2, 3 & 4: 0.9252–0.9260 in. (23.500–23.521mm)**

14. Calculate the camshaft journal oil clearance, using the following equation: journal oil clearance = (inner diameter of camshaft bracket) − (outer diameter of camshaft journal).

    a. Subtract the inner diameter of the camshaft bracket from the outer diameter of the camshaft journal.
    b. If the oil clearance exceeds the allowable limit, replace either the cylinder head or camshaft or both, if necessary.

➡**Standard oil clearance—No. 1: 0.0018–0.0034 in. (0.045–0.086mm), Nos. 2, 3 & 4: 0.0014–0.0030 in. (0.035–0.076mm), Maximum limit: 0.0059 in. (0.15mm)**

15. Inspect camshaft end-play, as follows:

    a. Install a dial indicator in the thrust direction on the front end of the camshaft.
    b. Measure the end play on the dial indicator when the camshaft is moved forward and backward (in direction to axis).

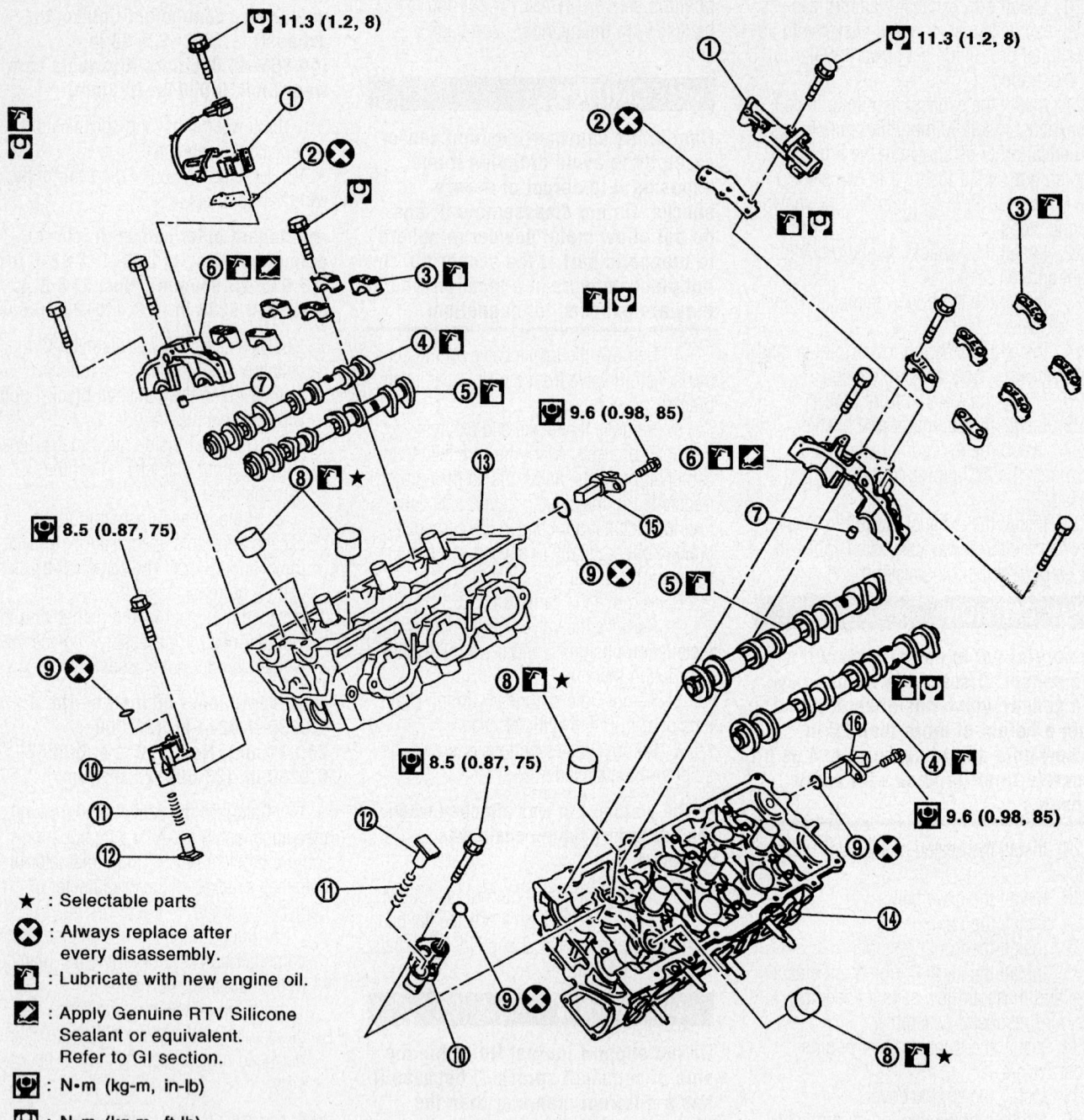

★ : Selectable parts

✕ : Always replace after every disassembly.

🛢 : Lubricate with new engine oil.

✏ : Apply Genuine RTV Silicone Sealant or equivalent. Refer to GI section.

🔧 : N•m (kg-m, in-lb)

🔩 : N•m (kg-m, ft-lb)

1.	Intake valve timing control solenoid valve	2.	Gasket	3.	Camshaft bracket (No. 2 to No. 4)
4.	Camshaft (EXH)	5.	Camshaft (INT)	6.	Camshaft bracket (No. 1)
7.	Dowel pin	8.	Valve lifter	9.	O-ring
10.	Chain tensioner	11.	Spring	12.	Plunger
13.	Cylinder head (right bank)	14.	Cylinder head (left bank)	15.	Camshaft position sensor (PHASE) (right bank)
16.	Camshaft position sensor (PHASE) (left bank)				

Exploded view of camshaft and related components—3.5L engine

67162-FX35-G11

Right bank

Left bank

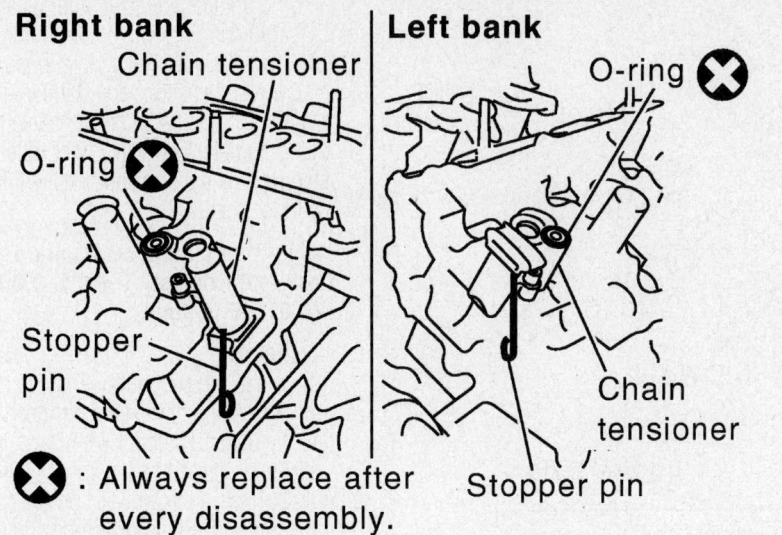

Right bank: Chain tensioner, O-ring, Stopper pin

Left bank: O-ring, Chain tensioner, Stopper pin

✖ : Always replace after every disassembly.

67162-FX35-G12

O-ring positions—3.5L engine

➡**Camshaft end-play standard: 0.0045–0.0074 in. (0.115–0.188mm), limit: 0.0094 in. (0.24mm).**

c. Measure the following parts if the camshaft end-play exceeds specifications, and replace the camshaft and/or cylinder if the following measurements are out of specification:
• Dimension "A" for camshaft No. 1

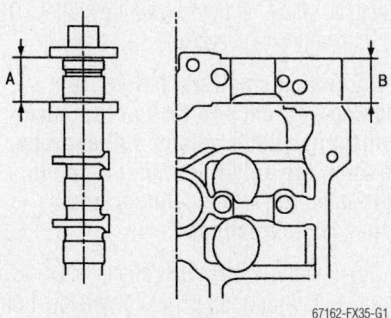

67162-FX35-G1

If the measurements A and B are out of range, replace the cylinder head or camshaft—3.5L engine

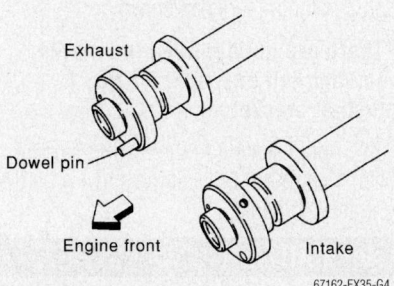

67162-FX35-G4

Exhaust and intake camshaft differences—3.5L engine

journal—Standard: (1.0827–1.0846 in. (27.500–27.548mm).
• Dimension "B" for cylinder head No. 1 journal—Standard: 1.0772–1.0781 in. (27.360–27.385mm).
16. Inspect camshaft sprocket run out, as follows:
a. Put a V-block on precise flat table, and support the No. 2 and No. 4 journals of the camshaft.

➡**Do not support journal No. 1 (on the side of camshaft sprocket) because it has a different diameter than the other three journals.**

b. Measure camshaft sprocket run out with a dial indicator. (Total indicator

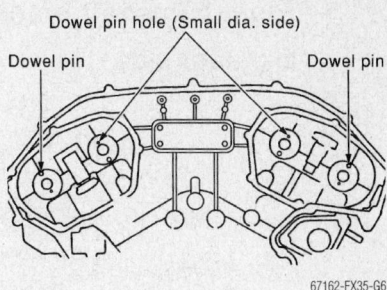

67162-FX35-G6

Camshaft positioning for installation—3.5L engine

reading), and replace the camshaft sprocket it the run out measurement exceeds specification.

➡**Camshaft sprocket run out limit: 0.0059 in. (0.15mm).**

17. Inspect the valve lifters, as follows:
a. Check the surface of valve lifter for wear or cracks. If any defects are found, replace the valve lifter.
18. Inspect valve lifter clearance, as follows:
a. Measure the outer diameter of each valve lifter.

➡**Valve lifter outer diameter (Intake and exhaust): 1.3377–1.3381 in. (33.977–33.987mm).**

b. Using an inside micrometer, measure the diameter of valve lifter hole of the cylinder head.

➡**Valve lifter bore standard (Intake and exhaust): 1.3386–1.3392 in. (34.000–34.016mm).**

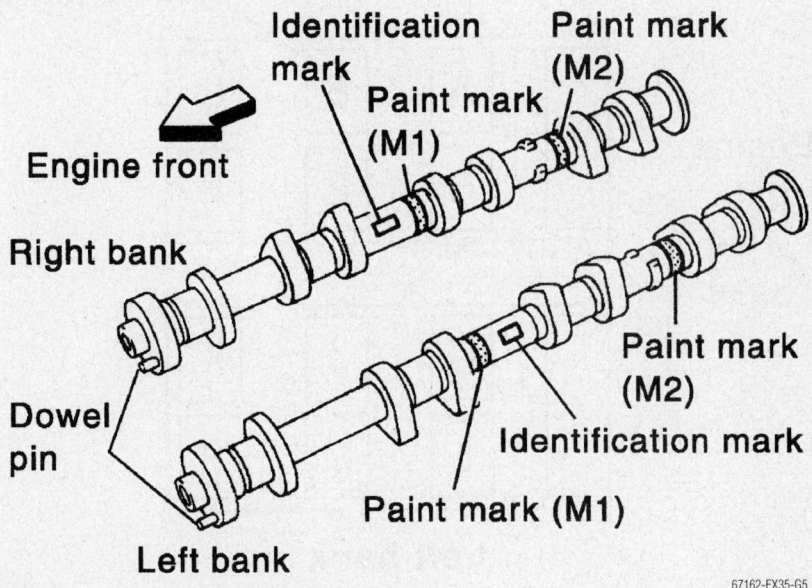

67162-FX35-G5

Camshaft identification mark locations—3.5L engine

Right camshaft brackets

Exhaust side

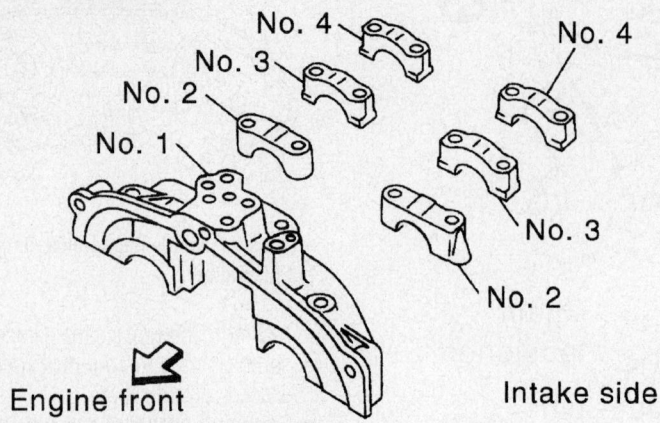

Camshaft bracket positions—3.5L engine

Left camshaft brackets

Intake side

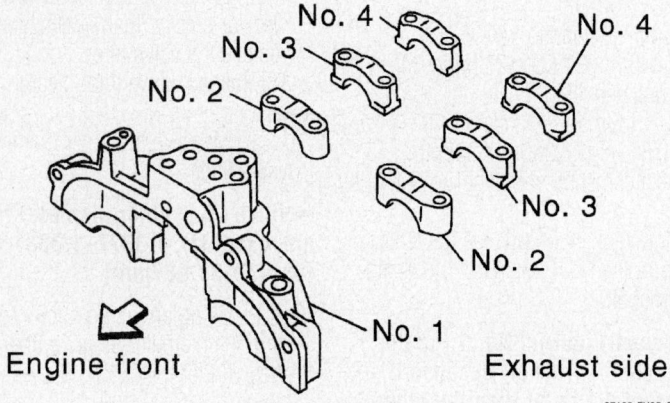

67162-FX35-G8

Right bank **Stamp mark**

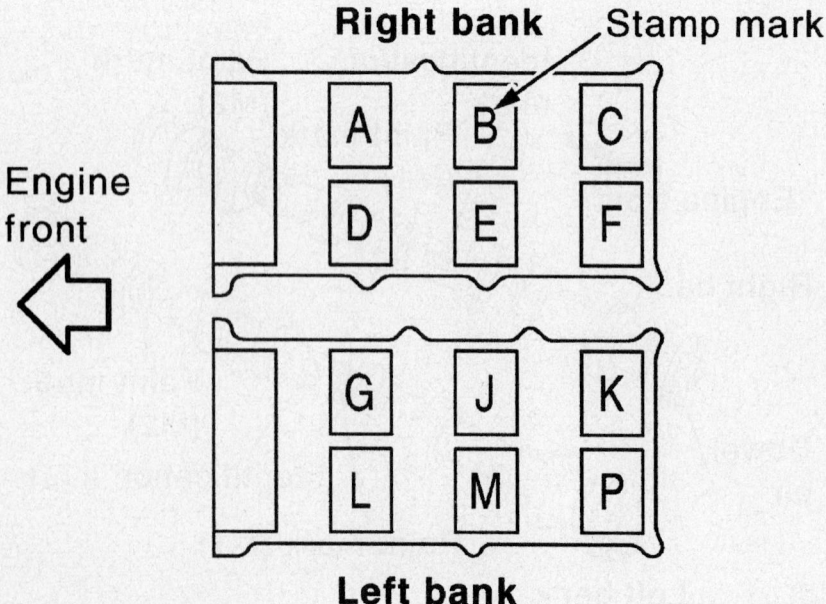

Camshaft bracket identification mark positions—3.5L engine

67162-FX35-G9

c. Calculate the valve lifter clearance by subtracting the valve lifter outer diameter from the valve lifter hole inner diameter. If result exceeds the standard, referring to each standard of valve lifter outer diameter and valve lifter hole diameter, replace either or both valve lifter and cylinder head.

➡️**Valve lifter clearance standard (Intake and exhaust): 0.0005–0.0015 in. (0.013–0.039mm).**

To install:

19. Install the secondary chain tensioners on both sides of the cylinder head. Install the chain tensioner with its stopper pin attached. Install the tensioner with the sliding part facing downward on the right-hand cylinder head, and with the sliding part facing upward on the left-hand cylinder head.

20. Install new O-rings as shown.

21. Install the valve lifters. If using the original valve lifters, make sure to install them in their original positions.

22. Install the camshafts. Install camshaft with dowel pin attached to its front end face on the exhaust side. Follow your identification marks made during removal, or follow the identification marks that are present on new camshafts for proper placement and direction. Install the camshaft so that the dowel pin hole and dowel pin on the front end face are positioned as shown in figure (No. 1 cylinder TDC on its compression stroke).

➡️**Large and small pin holes are located on front end face of the intake camshaft, at intervals of 180 degrees. Position the small diameter side pin hole upward (in the cylinder head upper face direction).**

23. Install the camshaft brackets. Remove all foreign material completely from the camshaft bracket backside and from the cylinder head installation face. Install the camshaft brackets in their original positions and directions as shown in illustration. Install the No. 2 to 4 camshaft brackets aligning stamp marks as shown.

➡️**There are no identification marks indicating left and right for No. 1 camshaft bracket.**

24. Apply sealant to the mating surface of No. 1 camshaft bracket (as shown) on the right and left banks.

✳✳ CAUTION

Use Genuine Thread Sealant or equivalent.

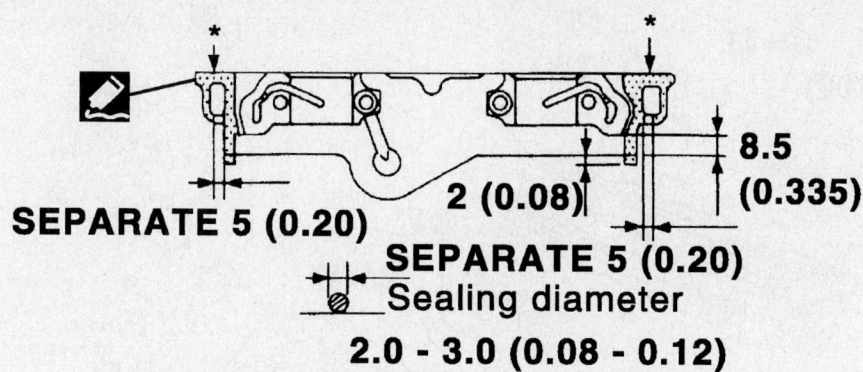

SEPARATE 5 (0.20)

2 (0.08)

8.5 (0.335)

SEPARATE 5 (0.20)
Sealing diameter

2.0 - 3.0 (0.08 - 0.12)

* : Remove the protruding sealant from front face. (Remove the hardended sealant from surface only.)

67162-FX35-G10

Liquid gasket positioning for camshaft bracket No. 1 installation—3.5L engine

Right bank

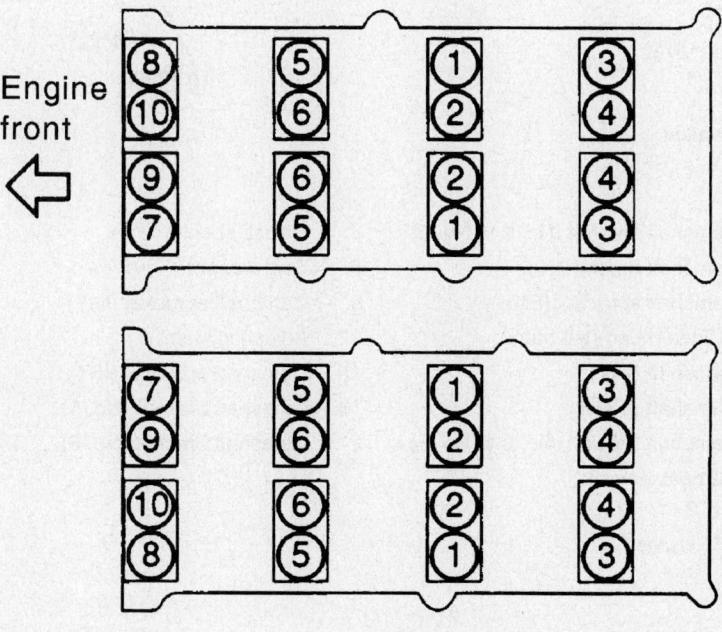

Engine front

Left bank

67162-FX35-G7

Camshaft bearing cap mounting bolt tightening sequence—3.5L engine

25. Tighten the camshaft brackets in the following steps, in numerical order as shown:

a. Tighten No. 7 to 10, then tighten No. 1 to 6 in order as shown to 1 ft. lbs. (2 Nm)

b. Tighten all bolts in numerical order to 4 ft. lbs. (6 Nm).

c. Tighten all bolts in order to 8 ft. lbs. (10 Nm).

➡**After tightening the mounting bolts of camshaft brackets (No. 1), be sure to wipe off excessive liquid gasket from these parts: mating surface of the rocker cover, mating surface of the rear timing chain case**

26. Measure the difference in levels between the front end faces of the No. 1 camshaft bracket and cylinder head. If the measurement is outside the specified range, re-install the camshaft and camshaft brackets.

➡**Camshaft bracket and cylinder head standard: -0.0055 to 0.0055 in. (-0.14 to 0.14 mm)**

27. Inspect and adjust the valve clearance.
28. The remainder of installation is the reverse of removal.

4.5L Engine

1. Before servicing the vehicle, refer to the Precautions Section.
2. Remove the engine assembly from vehicle.
3. Remove the timing chain.
4. With the hexagonal part of the camshaft locked with a wrench, loosen the bolts securing the camshaft sprocket, then remove the camshaft sprocket.

❊❊ **CAUTION**

After removing timing chain, do not turn the crankshaft or camshaft separately, otherwise the valves will strike the piston head.

5. Remove the intake and exhaust camshaft brackets. Mark the camshafts, camshaft brackets and bolts so that they can be installed in their original positions and direction. Equally loosen the camshaft brackets and bolts in several steps in the reverse order of the tightening sequence. Lightly tapping with a plastic hammer, remove camshaft brackets Nos. 1 & 6.

➡**The bottom surface of each bracket will be stuck to the cylinder head because of liquid gasket.**

6. Remove the camshaft.
7. Remove the adjusting shims and valve lifters, if necessary.
8. Identify installation positions, and store all parts without mixing them up.
9. Inspect the camshaft run out, as follows:

a. Situate a V-block on a precise flat table, and support the camshaft by journals No. 2 and No. 5.

➡**Do not support the camshaft by journal No. 1 (on the side of camshaft sprocket) since it has a different diameter from the other three locations.**

b. Set the dial indicator vertically to measure against journal No. 3.

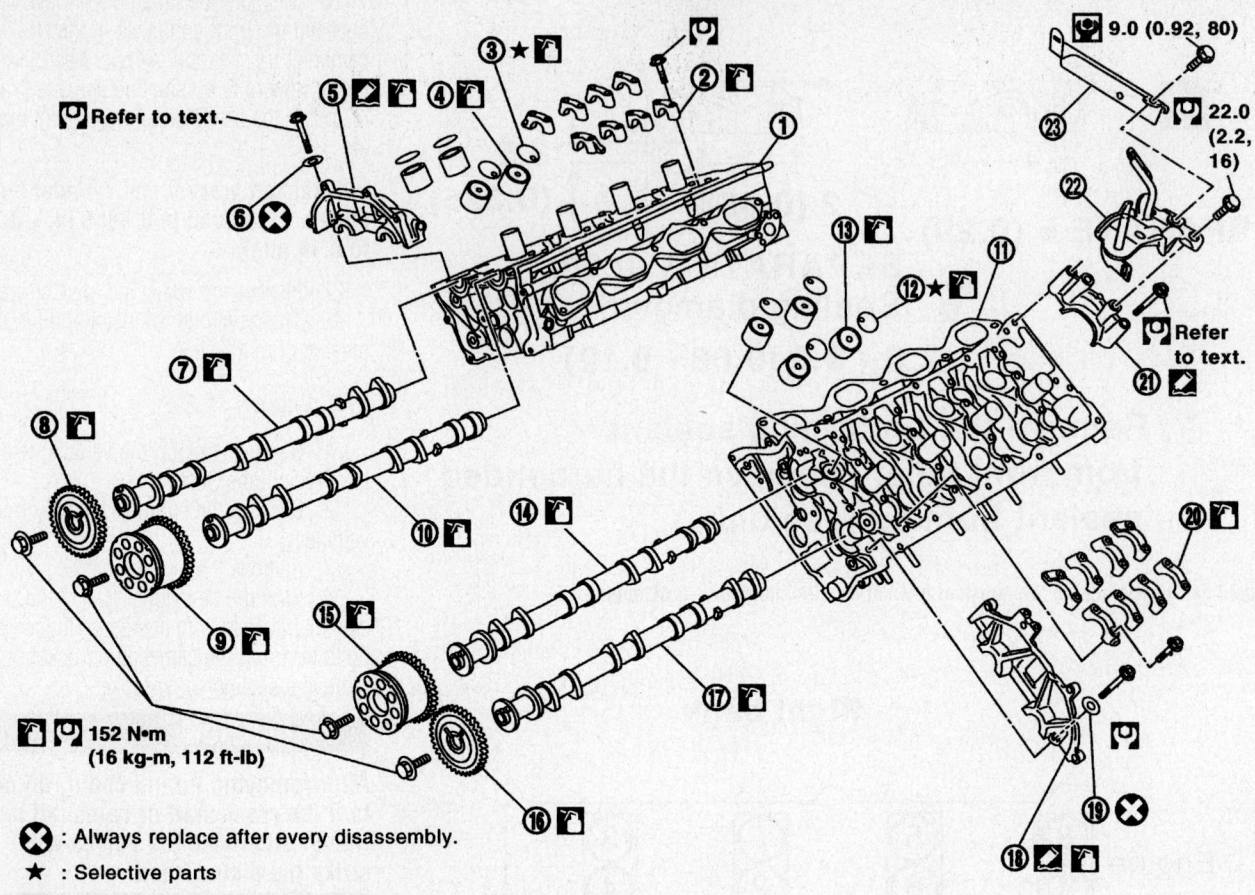

☒ Refer to text.

9.0 (0.92, 80)

22.0 (2.2, 16)

Refer to text.

152 N•m (16 kg-m, 112 ft-lb)

⊗ : Always replace after every disassembly.

★ : Selective parts

☒ : Apply Genuine RTV Silicone Sealant or equivalent. Refer to GI section.

☒ : Lubricate with new engine oil.

1. Cylinder head (right bank)	2. Camshaft bracket (No. 2 to No. 5)	3. Adjusting shim
4. Valve lifter	5. Camshaft bracket (No. 1)	6. Seal washer
7. Camshaft (EXH)	8. Camshaft sprocket (EXH)	9. Camshaft sprocket (INT)
10. Camshaft (INT)	11. Cylinder head (left bank)	12. Adjusting shim
13. Valve lifter	14. Camshaft (INT)	15. Camshaft sprocket (INT)
16. Camshaft sprocket (EXH)	17. Camshaft (EXH)	18. Camshaft bracket (No. 1)
19. Seal washer	20. Camshaft bracket (No. 2 to No. 5)	21. Camshaft bracket (No. 6)
22. Harness bracket	23. Harness bracket	

67162-FX35-G13

Exploded view of the camshaft and related components—4.5L engine

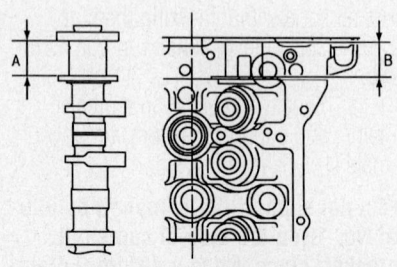

67162-FX35-G16

If the following measurements are out of specifications, replace the camshaft and/or cylinder head—4.5L engine

Bank	INT/EXH	Identification rib	Paint marks		Identification mark
			M1	M2	
RH	INT	Yes	Blue	No	RH
	EXH	Yes	No	Orange	RH
LH	INT	No	Blue	No	LH
	EXH	No	No	Orange	LH

67162-FX35-G21

New camshaft identification marks—4.5L engine

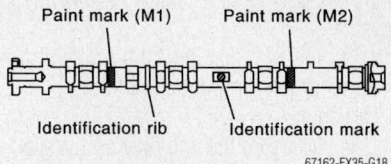

Identification mark locations on the camshaft—4.5L engine

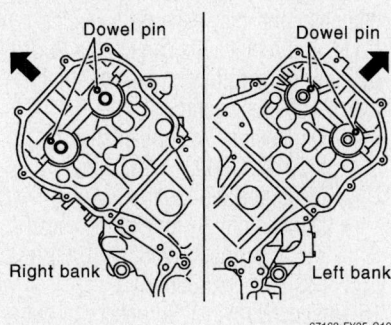

Camshaft dowel location for installation—4.5L engine

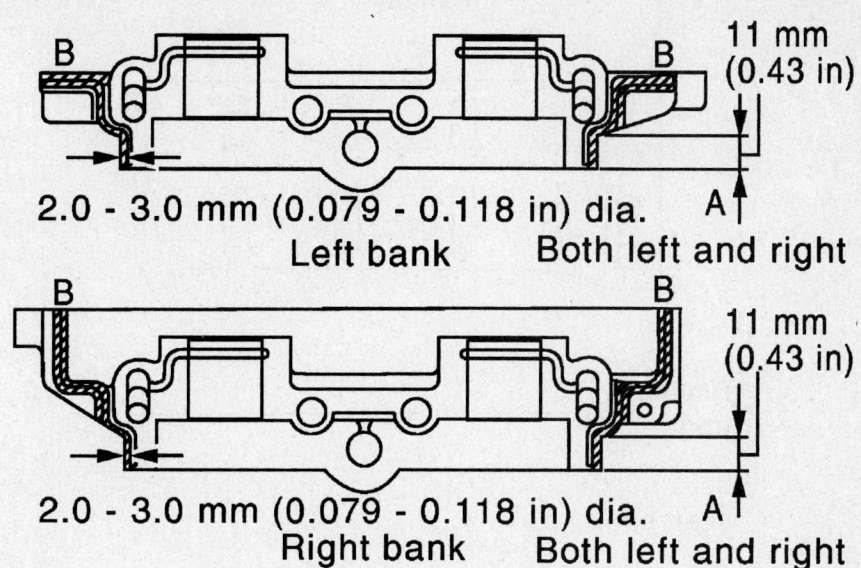

2.0 - 3.0 mm (0.079 - 0.118 in) dia.
Left bank
Both left and right

11 mm (0.43 in)

2.0 - 3.0 mm (0.079 - 0.118 in) dia.
Right bank
Both left and right

11 mm (0.43 in)

Liquid gasket positioning for No. 1 camshaft bracket installation—4.5L engine

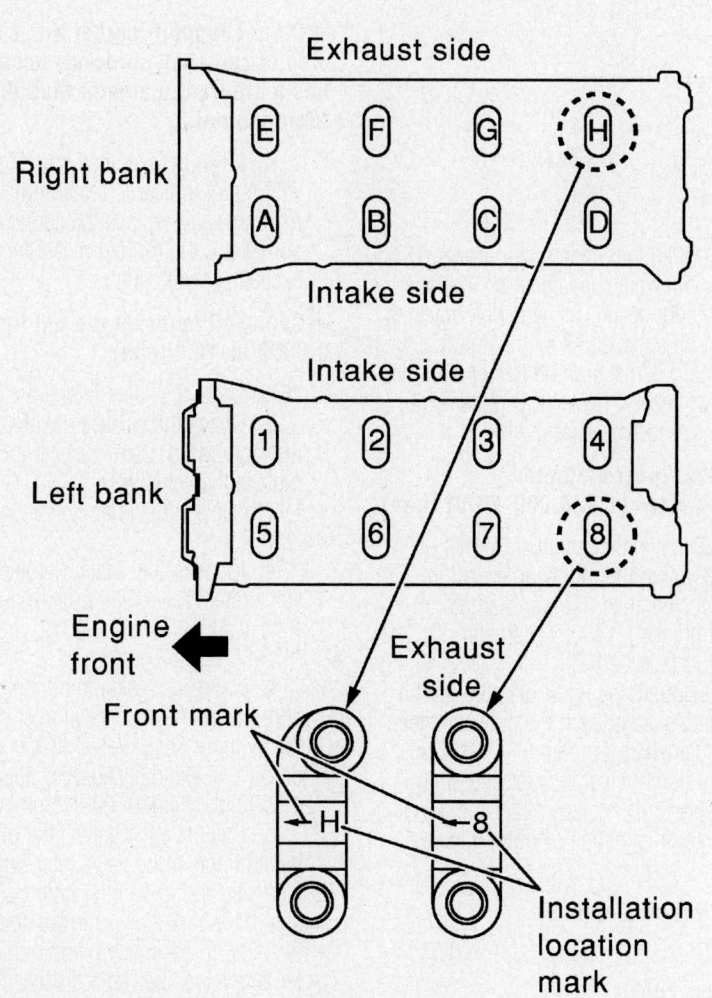

Camshaft bearing cap identification—4.5L engine

Liquid gasket positioning for No. 6 camshaft bracket installation—4.5L engine

2.0 - 3.0 mm (0.079 - 0.118 in) dia.

8 mm (0.31 in)

A (Both left and right)

c. Turn the camshaft in one direction by hand, and measure camshaft the run out on the dial indicator.

d. If run out exceeds the limit, replace the camshaft.

➡**Camshaft run out limit: 0.0008 in. (0.02mm).**

10. Inspect the camshaft cam height, as follows:

a. Measure the camshaft cam height.

➡**Standard cam height:**

- Intake: 1.7663–1.7738 in. (44.865–45.055mm)
- Exhaust: 1.7293–1.7368 in. (43.925–44.115mm)
- Allowable cam wear limit: 0.008 in. (0.2mm).

b. If wear is beyond specification, replace the camshaft.

11. Measure the outer diameter of the camshaft journals with a micrometer.

Right bank

Exhaust

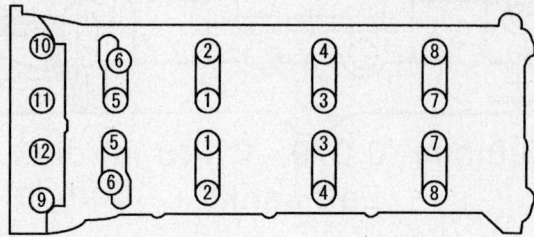

Engine front ⟸ —— Intake ———

Left bank

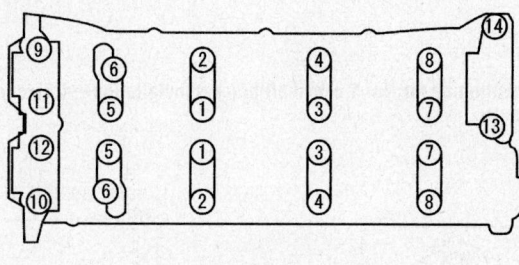

Exhaust

67162-FX35-G20

Tightening sequence for camshaft bearing caps—4.5L engine

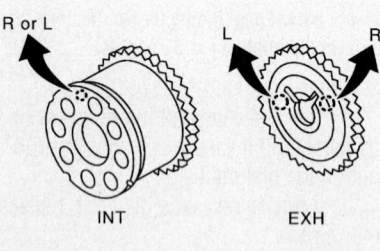

INT   EXH

67162-FX35-G22

Intake and Exhaust camshaft sprocket identification—4.5L engine

➡**Standard outer camshaft journal diameters:**

- No. 1: 1.0212–1.0218 in (25.938–25.955mm)
- Nos. 2, 3 & 4: 1.0218–1.0224 in. (25.953–25.970mm)

12. Measure the inner diameter of the camshaft bracket journals, as follows:

a. Tighten the camshaft bracket bolts with the specified torque.

b. Using an inside micrometer, measure the inner diameter of the camshaft bracket.

c. If the inner camshaft journals exceed allowable specification, replace the cylinder head. The camshaft brackets cannot be replaced as a single part, because it is machined together with the cylinder head. If necessary, replace the whole cylinder head assembly.

➡**Standard inner diameter: 1.0236–1.0244 in. (26.000–26.021mm)**

13. Calculate the camshaft journal oil clearance, using the following equation: journal oil clearance = (inner diameter of camshaft bracket) – (outer diameter of camshaft journal).

a. Subtract the inner diameter of the camshaft bracket from the outer diameter of the camshaft journal.

b. If the oil clearance exceeds the allowable limit, replace either the cylinder head or camshaft or both, if necessary.

➡**Standard oil clearance**

- No. 1: 0.0018–0.0033 in. (0.045–0.083mm)
- Nos. 2, 3 & 4: 0.0012–0.0027 in. (0.030–0.068mm)

14. Inspect camshaft end-play, as follows:

a. Install a dial indicator in the thrust direction on the front end of the camshaft.

b. Measure the end play on the dial indicator when the camshaft is moved forward and backward (in direction to axis). Camshaft end-play standard: 0.0045–0.0074 in. (0.115–0.188mm).

c. Measure the following parts if the camshaft end-play exceeds specifications, and replace the camshaft and/or cylinder if the following measurements are out of specification:

- Dimension "A" for camshaft No. 1 journal: 1.2008–1.2027 in. (30.500–30.548mm)
- Dimension "B" for cylinder head No. 1 journal: 1.1953–1.1963 in. (30.360–30.385mm)

15. Inspect camshaft sprocket run out, as follows:

a. Put a V-block on precise flat table, and support the No. 2 and No. 5 journals of the camshaft.

➡**Do not support journal No. 1 (on the side of camshaft sprocket) because it has a different diameter than the other three journals.**

b. Measure camshaft sprocket run out with a dial indicator. (Total indicator reading), and replace the camshaft sprocket it the run out measurement exceeds specification.

➡**Camshaft sprocket run out limit: 0.0059 in. (0.15mm)**

16. Inspect the valve lifters, as follows:

a. Check the surface of valve lifter for wear or cracks. If any defects are found, replace the valve lifter.

17. Inspect valve lifter clearance, as follows:

a. Measure the outer diameter of each valve lifter. Valve lifter outer diameter (Intake and exhaust): 1.3372–1.3376 in. (33.965–33.975mm)

b. Using an inside micrometer, measure the diameter of valve lifter hole of the cylinder head. Valve lifter bore standard (Intake and exhaust): 1.3386–1.3392 in. (34.000–34.016mm)

c. Calculate the valve lifter clearance by subtracting the valve lifter outer diameter from the valve lifter hole inner diameter. If the result exceeds the standard, referring to each valve lifter outer diameter and valve lifter hole diameter, replace either or both valve lifter and cylinder head.

➡Valve lifter clearance standard (Intake and exhaust): 0.0010–0.0020 in. (0.025–0.51mm)

To install:

18. Install valve lifters and adjusting shims if removed. Install them in their original positions.

19. Install the camshafts. Follow your identification marks made during removal, or follow the identification marks that are present on new camshafts for proper placement and direction. Install the camshafts so that the dowel pins on the front end faces are positioned as shown (No. 1 cylinder TDC on its compression stroke).

20. Install the camshaft brackets. Remove all foreign material completely from the camshaft bracket backsides and from the cylinder head installation faces. Install by referring to the installation location mark on the upper surface and front mark. Install so that the installation location mark can be correctly read when viewed from the side of the left exhaust bank.

21. Apply liquid gasket to the mating surface of the camshaft bracket (No. 1) as shown.

➡Use Genuine RTV Silicone Sealant or equivalent.

✴✴ CAUTION

After installation, be sure to wipe off any excessive liquid gasket leaking from parts "A" and "B" (both on right and left sides). Remove completely any excess of liquid gasket inside the bracket.

22. Apply liquid gasket to the mating surface of camshaft bracket (No. 6) on left bank intake as shown.

23. Tighten the camshaft bracket bolts in the following steps, in the order shown.
- No. 9 to 12: 12 inch lbs. (2 Nm).
- No. 1 to 8: 12 inch lbs. (2 Nm).
- No. 13 and 14: 12 inch lbs. (2 Nm). (Left bank only)
- All bolts: 48 inch lbs. (6 Nm).
- No. 1 to 12: 96 inch lbs. (10.5 Nm).
- No. 13 and 14: 23 ft. lbs. (31 Nm). (Left bank only)

✴✴ CAUTION

After tightening mounting bolts of camshaft brackets, be sure to wipe off excessive liquid gasket from these parts: mating surface of the rocker cover, mating surface of the front cover.

24. Install the camshaft sprockets. Make sure to position the sprockets as shown. Install the camshaft sprocket (EXH) by selectively using the groove of dowel pin according to the bank. Lock the hexagonal part of camshaft in the same way as for removal, and tighten the mounting bolts.

25. Check and adjust valve clearance.

26. The remainder of installation is the reverse of removal.

Valve Lash

ADJUSTMENT

3.5L Engine

1. Before servicing the vehicle, refer to the Precautions Section.

Perform inspection after removal, installation or replacement of camshaft or valve-related parts, or if there is unusual engine condition regarding valve clearance.

2. Remove the right and left rocker covers.

3. Set the No. 1 cylinder at TDC of its compression stroke, as follows:

a. Rotate the crankshaft pulley clockwise until the timing mark (grooved line without color) is aligned with the timing indicator.

b. Make sure the No. 1 cylinder intake and exhaust cam noses are facing inward and upward from the cylinder head, as shown. If they are not positioned as shown, rotate the crankshaft pulley 360 degrees clockwise (when viewed from the front).

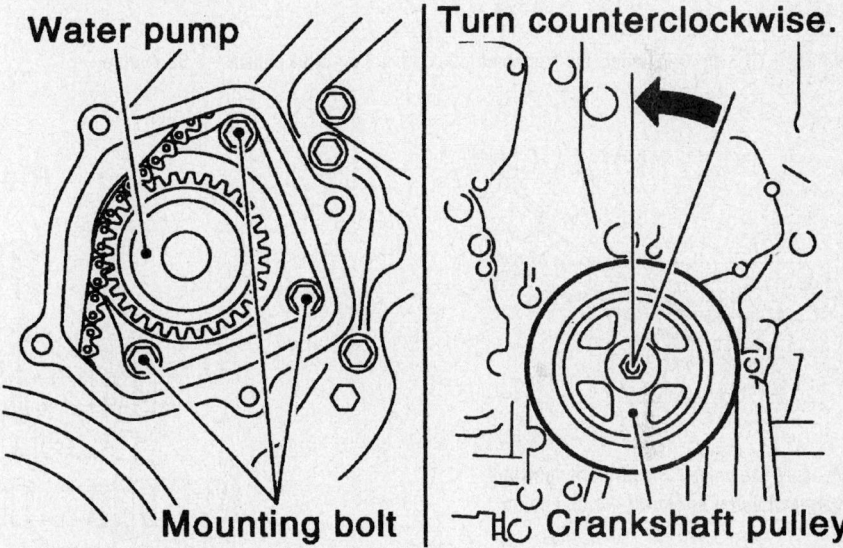

Engine cylinder number identification—3.5L engine

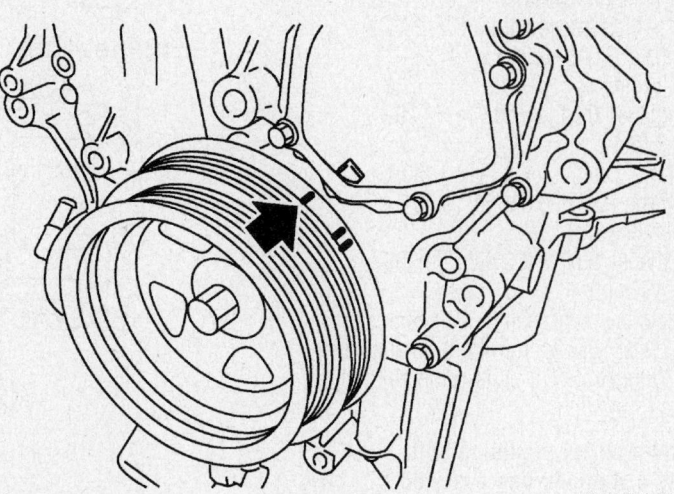

Position the crankshaft at TDC No. 1—3.5L engine

Right bank

Engine front

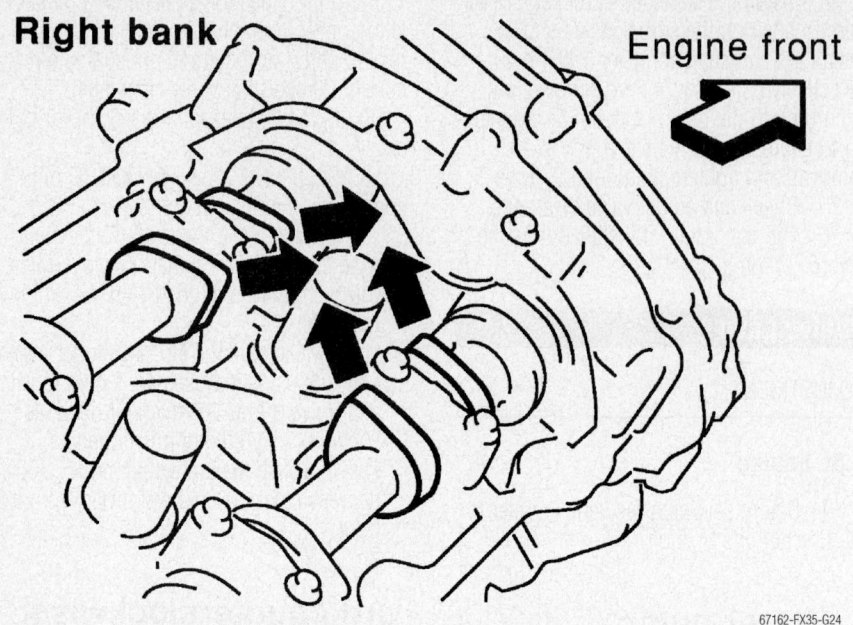

When at TDC No. 1 cylinder, the camshaft lobes should point as shown—3.5L engine

67162-FX35-G24

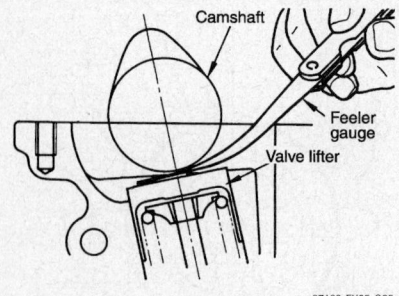

67162-FX35-G25

Measure the valve lash with the camshaft lobe positioned as shown—3.5L engine

4. Using a feeler gauge, measure the valve clearance for the following cylinders: Cylinder 1 Intake, Cylinder 2 Exhaust, Cylinder 3 Exhaust, Cylinder 6 Intake valves

    a. Valve clearance cold (68°F/20°C):
- Intake: 0.010–0.013 in. (0.26–0.34mm)
- Exhaust: 0.011–0.015 in. (0.29–0.37mm)

    b. Valve clearance hot (176°F/80°C):
- Intake: 0.012–0.016 in. (0.304–0.416mm)
- Exhaust: 0.012–0.016 in. (0.308–0.432mm)

5. Rotate the crankshaft by 240 degrees clockwise (when viewed from front) to align the No. 3 cylinder at TDC of its compression stroke.

➡**Crankshaft pulley mounting bolt flange has a stamped line every 60 degrees. They can be used as a guide to rotation angle.**

6. Using a feeler gauge, measure the valve clearance for the following cylinders: Cylinder 2 Intake valves, Cylinder 3 Intake valves, Cylinder 4 Exhaust valves, Cylinder 5 Exhaust valves.

    a. Valve clearance standard cold (68°F/20°C):
- Intake: 0.010–0.013 in. (0.26–0.34mm)
- Exhaust: 0.011–0.015 in. (0.29–0.37mm)

    b. Valve clearance hot (176°F/80°C):
- Intake: 0.012–0.016 in. (0.304–0.416mm)
- Exhaust: 0.012–0.016 in. (0.308–0.432mm)

7. Rotate the crankshaft by 240 degrees clockwise (when viewed from front) to align the No. 5 cylinder at TDC of its compression stroke.

➡**Crankshaft pulley mounting bolt flange has a stamped line every 60 degrees. They can be used as a guide to rotation angle.**

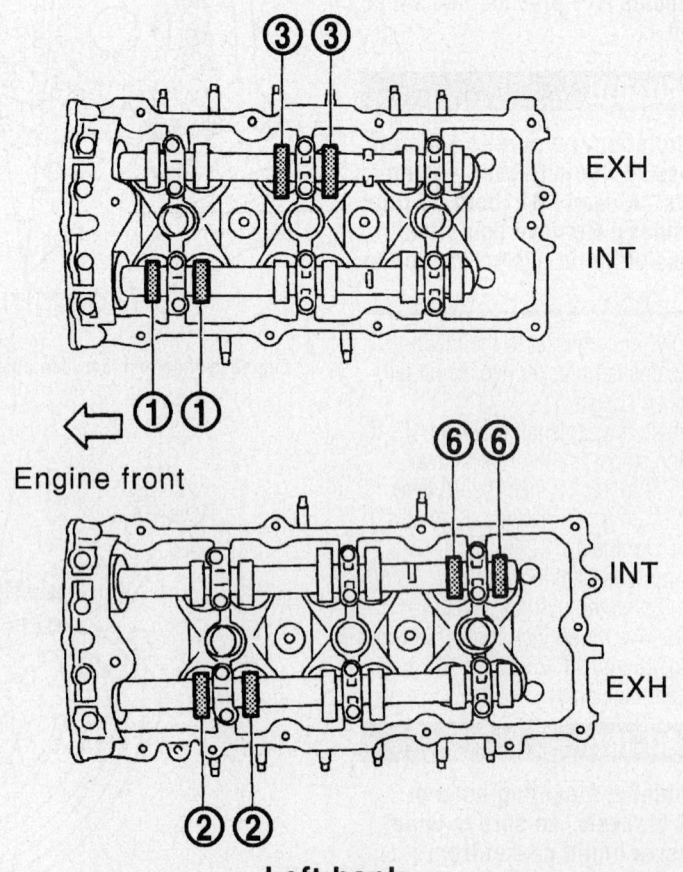

Right bank

EXH

INT

Engine front

INT

EXH

Left bank

67162-FX35-G26

Measure the valve lash for the valves shown with the engine at No. 1 cylinder TDC—3.5L engine

Right bank

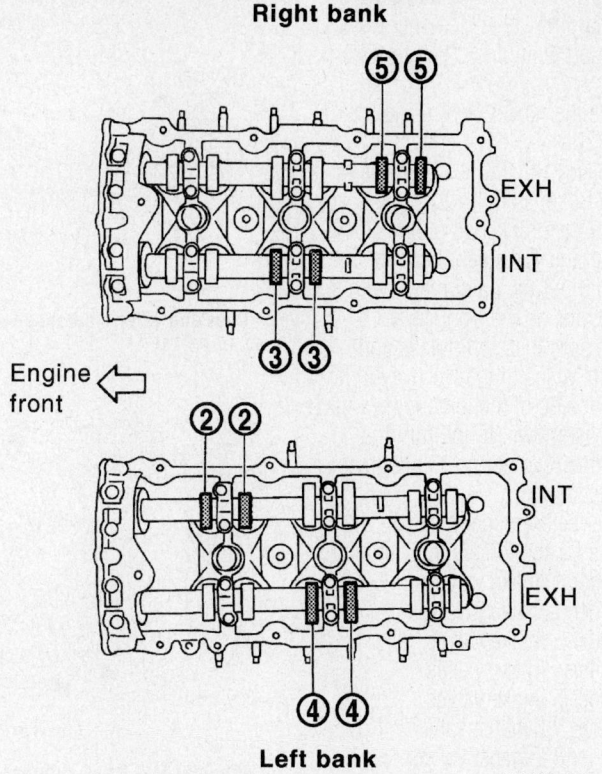

Engine ← front

Left bank

67162-FX35-G28

Measure the valve lash for the valves shown with the engine at No. 3 cylinder TDC—3.5L engine

Right bank

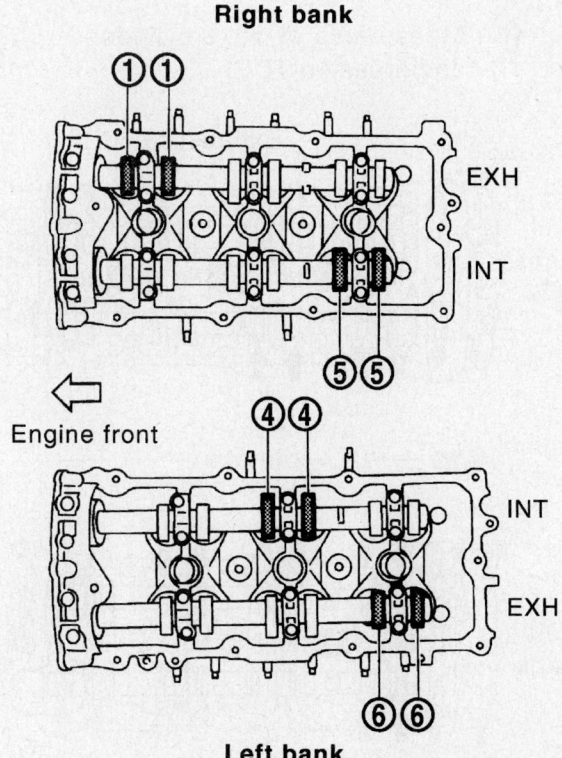

Engine front ←

Left bank

67162-FX35-G29

Measure the valve lash for the valves shown with the engine at No. 5 cylinder TDC—3.5L engine

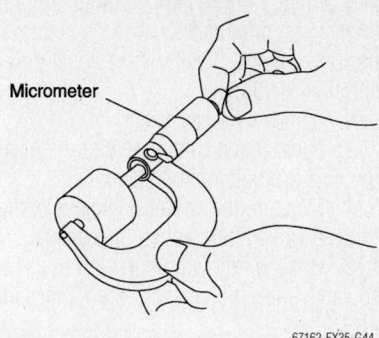

67162-FX35-G44

Measure the valve lifter height as shown—3.5L engine

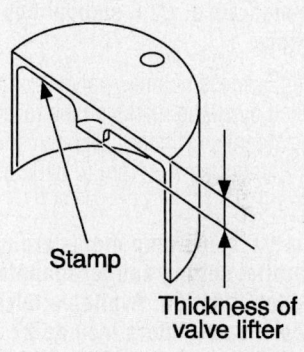

67162-FX35-G45

Valve lifter identification stamp location—3.5L engine

8. Using a feeler gauge, measure the valve clearance for the following cylinders: Cylinder 1 Exhaust valves, Cylinder 4 Intake valves, Cylinder 5 Intake valves, Cylinder 6 Exhaust valves.

   a. Valve clearance standard cold (68°F/20°C):
   - Intake: 0.010–0.013 in. (0.26–0.34mm)
   - Exhaust: 0.011–0.015 in. (0.29–0.37mm)
   b. Valve clearance hot (176°F/80°C):
   - Intake: 0.012–0.016 in. (0.304–0.416mm)
   - Exhaust: 0.012–0.016 in. (0.308–0.432mm)

✳✳ CAUTION

If the inspection was carried out with cold engine, make sure values with fully warmed up engine are still within specifications.

9. For all valve lifters that are found to be outside the specified range, perform the following steps.

Perform adjustment depending on selected head thickness of valve lifter. The specified valve lifter thickness is the dimension at normal temperatures. Ignore dimensional differences caused by temperature. Use the specifications for hot engine condition to adjust.

10. Remove the camshaft.

11. Remove the valve lifters at the locations that are outside the standard.

12. Measure the center thickness of the removed valve lifters with a micrometer.

13. Use the following equation to calculate valve lifter thickness for the replacement lifters.

➡**Valve lifter thickness calculation: thickness of replacement valve lifter = t1 + (C1 − C2). t1 = Thickness of removed valve lifter, C1 = measured valve clearance, C2 = standard valve clearance.**

The thickness of a new valve lifter can be identified by stamp marks on the reverse side (inside the cylinder). Stamp mark 788U or 788R indicates 7.88 mm (0.3102 in) in thickness.

➡**Two types of stamp marks are used for parallel setting and for manufacturer identification. Available thicknesses of valve lifters include 27 sizes covering a range of 0.3102–0.3307 in. (7.88–8.40mm) in steps of 0.0008 in. (0.02mm).**

14. Install the selected valve lifter(s).

15. Install the camshaft.

16. Manually turn crankshaft pulley a few turns.

17. Make sure the valve clearances for the cold engine are within specifications by referring to the specified values.

18. After completing the repair, check valve clearances again with the specifications for a warmed engine. Make sure the values are within specifications.

➡**Valve clearance specifications:**

- Cold Intake (68 degrees F/20 degrees C): 0.010–0.013 in. (0.26–0.34mm)
- Cold Exhaust: 0.011–0.015 in. (0.29–0.37mm)
- Hot Intake (176 degrees F/80 degrees C): 0.012–0.016 in. (0.304–0.416mm)
- Hot Exhaust: 0.012–0.016 in. (0.308–0.432mm)

4.5L Engine

1. Before servicing the vehicle, refer to the Precautions Section.

Perform inspection after removal, installation or replacement of camshaft or valve-related parts, or if there is unusual engine conditions regarding valve clearance.

2. Warm up the engine. Then turn it OFF.

3. Remove the rocker covers (right and left bank).

4. Set the No. 1 cylinder at TDC of its compression stroke, as follows:

a. Rotate the crankshaft pulley clockwise to align the TDC identification notch (without the paint mark) with the timing indicator on the front cover.

b. Make sure that both intake and exhaust cam noses of the No. 1 cylinder (engine front side of the left engine bank) are located as shown. If not, turn the crankshaft one revolution (360 degrees) and align as shown.

5. Using a feeler gauge, measure the valve clearance for the following cylinders:

- Cylinder 1 Intake valves
- Cylinder 1 Exhaust valves
- Cylinder 2 Intake valves
- Cylinder 4 Intake valves
- Cylinder 5 Intake valves
- Cylinder 7 Exhaust valves
- Cylinder 8 Exhaust valves

a. Valve clearance standard cold (68°F/20°C):

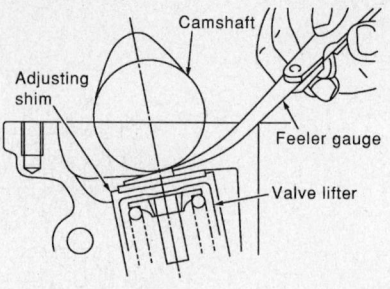

Checking valve clearance—4.5L engine

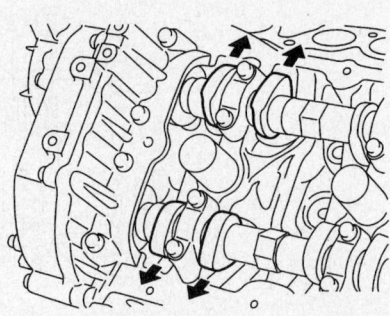

When at TDC No. 1 cylinder, the camshaft lobes should point as shown—3.5L engine

⬆ : Measurable at No. 1 cylinder compression TDC

⇧ : Measurable at No. 3 cylinder compression TDC

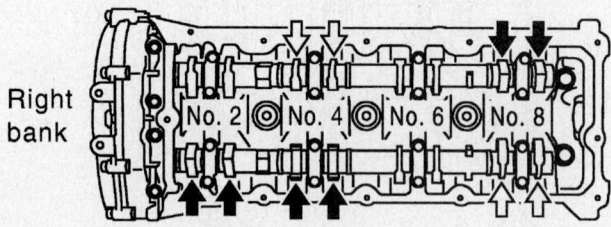

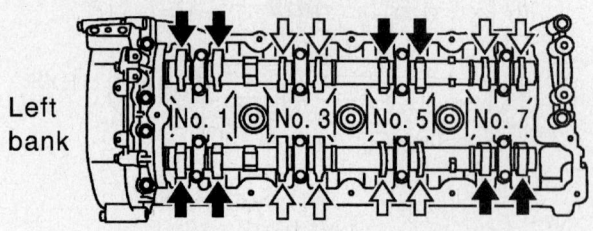

Valve lash inspection positions at No. 1 TDC and No. 3 TDC—4.5L engine

Exhaust

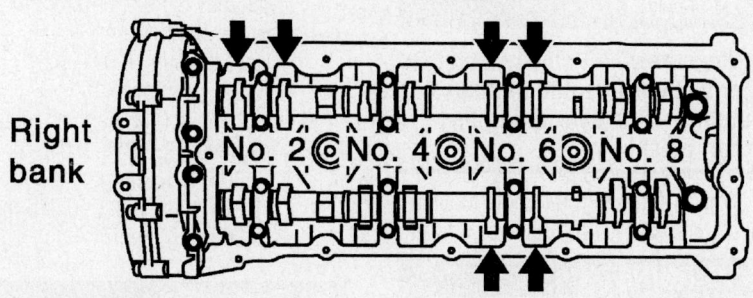

Right bank

No. 2   No. 4   No. 6   No. 8

Engine front

Intake

67162-FX35-G37

Valve lash inspection positions at No. 6 TDC—4.5L engine

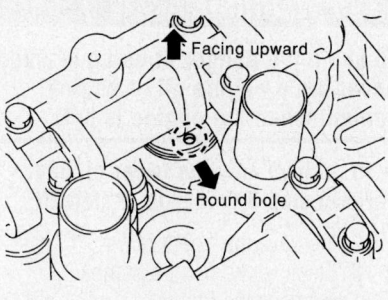

Facing upward

Round hole

67162-FX35-G38

Position the camshaft lobe as shown for shim replacement—4.5L engine

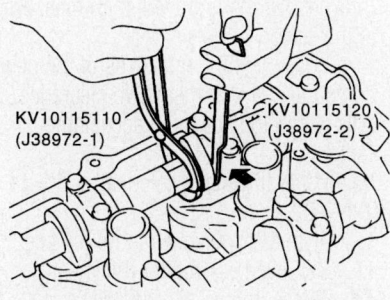

KV10115110 (J38972-1)   KV10115120 (J38972-2)

67162-FX35-G40

Use the tools shown to depress the valve spring for shim replacement—4.5L engine

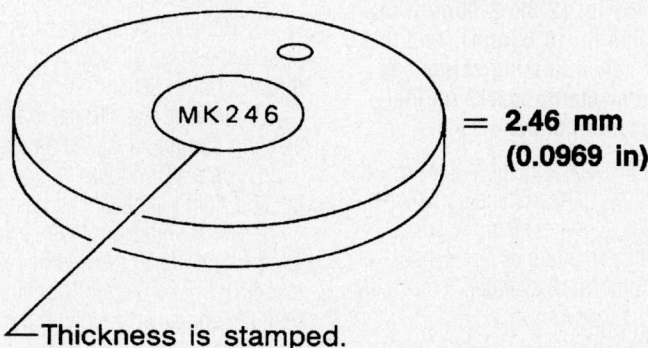

MK246   = 2.46 mm (0.0969 in)

Thickness is stamped.

67162-FX35-G42

The shims are stamped with their thickness—4.5L engine

- Intake: 0.010–0.013 in. (0.26–0.34mm)
- Exhaust: 0.011–0.015 in. (0.29–0.37mm)
- b. Valve clearance hot (176°F/80°C):
- Intake: 0.012–0.016 in. (0.304–0.416mm)
- Exhaust: 0.012–0.016 in. (0.308–0.432mm)

※※ CAUTION

If inspection was carried out with cold engine, make sure values with fully warmed up engine are still within specifications. Valve clearance standard:

6. Rotate the crankshaft pulley clockwise (when viewed from engine front) by

270 degrees from the position of No. 1 cylinder compression TDC to align No. 3 cylinder at TDC of its compression stroke.

➡**The crankshaft pulley mounting bolt flange has an angle mark every 90 degrees. They can be used as a guide to rotation angle.**

7. Using a feeler gauge, measure the valve clearance for the following cylinders:
- Cylinder 3 Intake valves
- Cylinder 3 Exhaust valves
- Cylinder 4 Exhaust valves
- Cylinder 5 Exhaust valves
- Cylinder 7 Intake valves
- Cylinder 8 Intake valves
  a. Valve clearance standard cold (68°F/20°C):
- Intake: 0.010–0.013 in. (0.26–0.34mm)
- Exhaust: 0.011–0.015 in. (0.29–0.37mm)
  b. Valve clearance hot (176°F/80°C):
- Intake: 0.012–0.016 in. (0.304–0.416mm)
- Exhaust: 0.012–0.016 in. (0.308–0.432mm)

8. Rotate the crankshaft pulley clockwise (when viewed from engine front) by 90 degrees from the position of No. 3 cylinder compression TDC to align No. 6 cylinder at TDC of its compression stroke.

9. Using a feeler gauge, measure the valve clearance for the following cylinders:
- Cylinder 2 Exhaust valves
- Cylinder 6 Exhaust valves
- Cylinder 6 Intake valves
  a. Valve clearance standard cold (68°F/20°C):
- Intake: 0.010–0.013 in. (0.26–0.34mm)
- Exhaust: 0.011–0.015 in. (0.29–0.37mm)
  b. Valve clearance hot (176°F/80°C):
- Intake: 0.012–0.016 in. (0.304–0.416mm)
- Exhaust: 0.012–0.016 in. (0.308–0.432mm)

10. For any valves that measure out of the standard allowable range, perform the following adjustment steps.

➡**Adjust valve clearance while the engine is cold. After adjustment, make sure that the valve clearance is within the standard range while the engine is hot.**

11. Thoroughly wipe off all engine oil around the adjusting shim.

12. Rotate the crankshaft to position the cam nose upward for the camshaft for the valve that must be adjusted.

13. Using small screwdriver or pick, turn the round hole of adjusting shim in the direction of the arrow (toward the center of the cylinder head).

14. For all valves except the exhaust side of No. 7 and No. 8 cylinders, install the lifter stopper SST: 10115120 (J38972-2) or equivalent, as follows:

 a. Place the camshaft pliers around camshaft as shown in the figure.

 b. Rotate the camshaft pliers so that the valve lifter is pushed down.

✳ CAUTION

Be careful not to damage the cam surface, valve lifter or cylinder head with the camshaft pliers.

 c. Place the lifter stopper between the camshaft and the edge of the valve lifter to retain the valve lifter.

✳ CAUTION

The lifter stopper must be placed as close to the camshaft bracket as possible. Be careful not to damage the cam surface, valve lifter or cylinder head with the lifter stopper.

 d. Remove camshaft pliers.

✳ CAUTION

The camshaft pliers should be removed by rotating it slowly because the lifter stopper hits and damages the journal portion when rotating the camshaft pliers quickly.

15. For the exhaust side of No. 7 and No. 8 cylinders, perform the following:

➡**Exhaust side of No. 7 and No. 8 cylinders do not have space for installing with camshaft pliers SST: KV10115110 (J38972-1). Install lifter stopper SST: KV10115120 (J38972-2)] or equivalent according to the following instructions:**

 a. Rotate the crankshaft to press the cam nose onto the adjusting part of the valve lifter.

 b. Place the lifter stopper between the camshaft and the edge of the valve lifter to retain the valve lifter.

✳ CAUTION

The lifter stopper must be placed as close to the camshaft bracket as possible. Be careful not to damage the cam surface, valve lifter or cylinder head with the lifter stopper.

 c. Rotate the crankshaft slowly 180 degrees clockwise.

✳ CAUTION

Rotate the crankshaft slowly because the lifter stopper hits and damages the journal portion by rotating the crankshaft quickly.

16. Blow air into the round hole to separate the adjusting shim from the valve lifter.

✳ WARNING

When blowing, use goggles to protect your eye.

17. Remove the adjusting shim with a magnetic tool.

18. Use the following to calculate the adjusting shim thickness for replacement:

 a. Use a micrometer to determine the thickness of the removed shim measured at the center.

 b. Calculate the thickness of the new adjusting shim so that valve clearance falls within the specified acceptable range.

Valve lifter thickness calculation: $t = t1 + (C1 - C2)$

t = Valve lifter thickness to be replaced
$t1$ = Removed valve lifter thickness
$C1$ = Measured valve clearance
$C2$ = Standard valve clearance (Intake: 0.012 in./0.30mm, Exhaust: 0.013 in./ 0.33mm @ 68 degrees F/20 degrees C)

➡**Shims are available in 64 sizes from 0.0913–0.1161 in. (2.32–2.95mm) in steps of 0.0004 in. (0.01mm). And the thickness of new adjusting shims can be identified by stamp marks on the underside (inside the cylinder).**

19. Install the new adjusting shim using a suitable tool, with the surface on which the thickness is stamped facing down.

20. For all valve lifters except exhaust side of No. 7 and No. 8 cylinder, remove the lifter stopper as follows:

 a. Perform the same procedure as described for removal using the camshaft pliers.

 b. Remove the lifter stopper.

 c. Remove the camshaft pliers.

21. For exhaust side of No. 7 and No. 8 cylinder valve lifters, rotate the crankshaft slowly 180 degrees clockwise, then remove the lifter stopper.

22. Manually rotate the crankshaft pulley a few turns.

23. Make sure that the valve clearance is within specifications for cold and hot settings.

➡**Valve clearance specifications:**

- Cold Intake (68 degrees F/20 degrees C): 0.010–0.013 in. (0.26–0.34mm)
- Cold Exhaust: 0.011–0.015 in. (0.29–0.37mm)
- Hot Intake (176 degrees F/80 degrees C): 0.012–0.016 in. (0.304–0.416mm)
- Hot Exhaust: 0.012–0.016 in. (0.308–0.432mm)

Oil Pan

REMOVAL & INSTALLATION

3.5L Engine

1. Before servicing the vehicle, refer to the Precautions Section.

✳ WARNING

To avoid the danger of being scalded, never drain engine oil or engine coolant when the engine is hot.

➡**To remove only the lower oil pan, perform step 5 then skip to step 25.**

2. Remove the front tire.
3. Remove the hood assembly.
4. Remove the front and rear engine undercover.
5. Remove the front cross bar.
6. Drain the engine oil.
7. Drain the engine coolant.
8. Remove the engine cover.
9. Remove the air hose from the air duct to the mass air flow and the electric throttle control actuator side.
10. Remove the alternator and power steering pump and A/C compressor belt.
11. On AWD models, remove the front left and right halfshafts, and side shaft.
12. Remove the engine rear lower slinger, and install the engine rear slinger tool SST: 10006 31U00 (or equivalent) to hold the engine assembly in position. Tighten the engine rear slinger tool mounting bolts to 21 ft. lbs. (28 Nm).
13. Remove the front suspension member.
14. On AWD models, remove the engine mounting bracket, lower engine mounting bracket and insulator.
15. On AWD models, remove the front driveshaft.
16. On AWD models, remove the oil filter and oil filter bracket.
17. Remove the alternator stay.

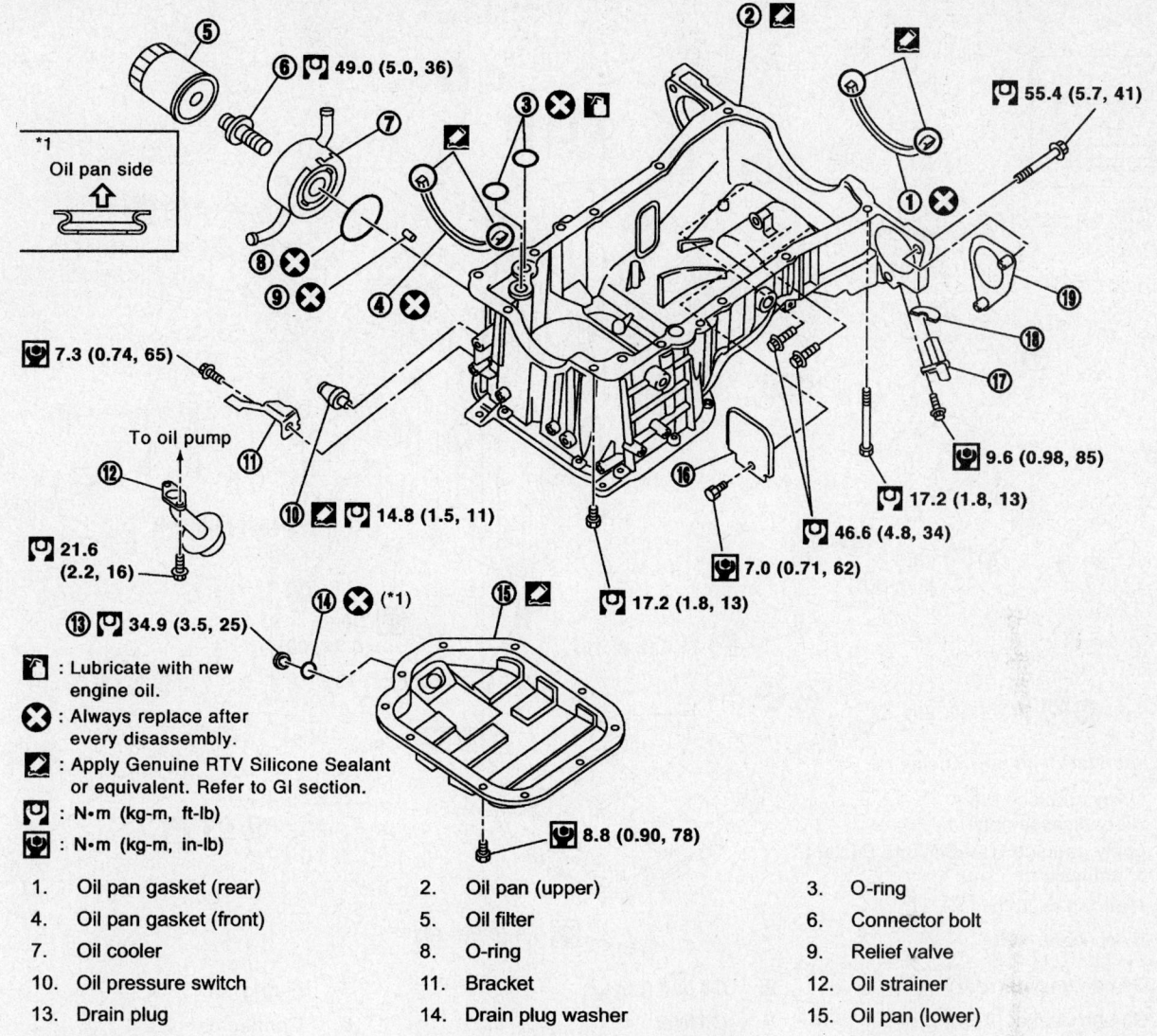

49.0 (5.0, 36)

*1
Oil pan side

55.4 (5.7, 41)

7.3 (0.74, 65)

To oil pump

21.6 (2.2, 16)

14.8 (1.5, 11)

9.6 (0.98, 85)

17.2 (1.8, 13)

46.6 (4.8, 34)

7.0 (0.71, 62)

17.2 (1.8, 13)

34.9 (3.5, 25)

(*1)

: Lubricate with new engine oil.

: Always replace after every disassembly.

: Apply Genuine RTV Silicone Sealant or equivalent. Refer to GI section.

: N•m (kg-m, ft-lb)

: N•m (kg-m, in-lb)

8.8 (0.90, 78)

1. Oil pan gasket (rear)	2. Oil pan (upper)	3. O-ring
4. Oil pan gasket (front)	5. Oil filter	6. Connector bolt
7. Oil cooler	8. O-ring	9. Relief valve
10. Oil pressure switch	11. Bracket	12. Oil strainer
13. Drain plug	14. Drain plug washer	15. Oil pan (lower)
16. Rear plate	17. Crankshaft position sensor (POS)	18. Seal rubber
19. Rear cover plate		

67162-FX35-G46

Exploded view of oil pan and related components—3.5L engine 2WD

18. Remove the starter motor.

19. Remove the alternator and power steering pump and A/C compressor idler pulley and bracket assembly.

20. Disconnect the A/T fluid cooler hoses, and remove the oil cooler water pipe mounting bolt.

21. Disconnect the A/T fluid cooler tube.

22. On AWD models, remove the front final drive assembly.

23. Remove the crankshaft position sensor (POS).

✳✳ CAUTION

Handle the POS carefully to avoid dropping it and exposing it to abrupt shocks. Do not disassemble it, do not allow metal powder to adhere to

magnetic part at the sensor tip, and do not place sensor in a location where it may be exposed to magnetism.

24. Remove the oil filter, as necessary.

25. Remove the oil cooler, as necessary.

26. Remove the oil pan (lower), as follows:

a. Loosen the mounting bolts in the reverse order of the tightening sequence.

b. Insert seal cutter SST: KV10111100 (J37228) or equivalent, between the upper oil pan and lower oil pan.

c. Slide the seal cutter by tapping on the side of the tool with a hammer.

d. Remove the lower oil pan.

✳✳ CAUTION

Be careful not to damage the mating surface. Do not use a flat-bladed screwdriver—this will damage the mating surfaces.

27. Remove the oil strainer.

28. Remove the transmission joint bolts which pass through the upper oil pan.

29. On 2WD models, remove the rear cover plate.

30. Loosen the upper oil pan bolts in the reverse order of the tightening sequence.

31. Insert seal cutter SST: KV10111100 (J37228) between the upper oil pan and cylinder block. Slide the seal cutter by tapping on the side of the tool with a hammer.

32. Remove the upper oil pan.

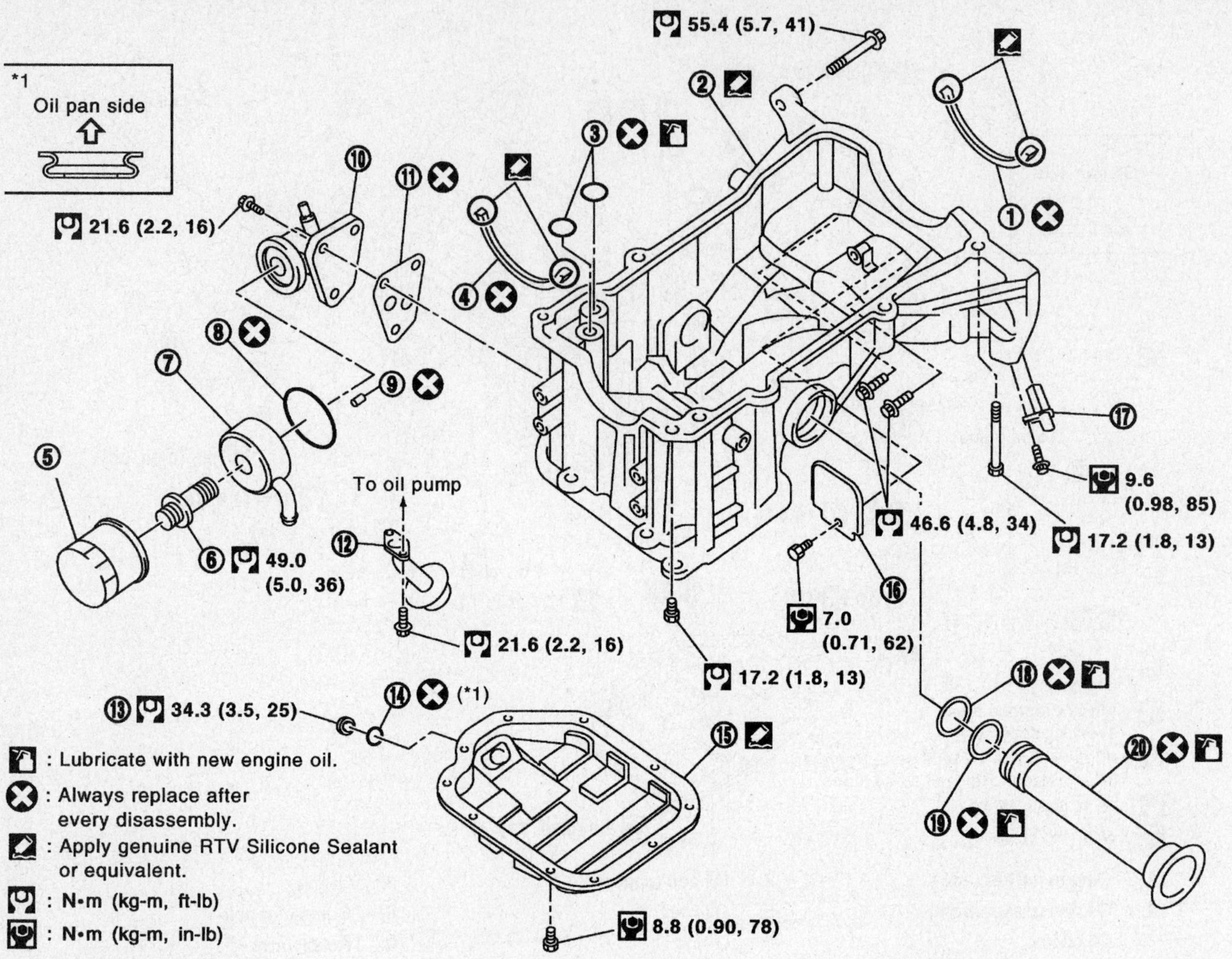

*1
Oil pan side
⬆

📐 21.6 (2.2, 16)

📐 55.4 (5.7, 41)

📐 49.0 (5.0, 36)

To oil pump

📐 21.6 (2.2, 16)

📐 34.3 (3.5, 25)

(*1)

📐 17.2 (1.8, 13)

📐 46.6 (4.8, 34)

📐 7.0 (0.71, 62)

📐 17.2 (1.8, 13)

📐 9.6 (0.98, 85)

📐 17.2 (1.8, 13)

📐 8.8 (0.90, 78)

🛢 : Lubricate with new engine oil.

❌ : Always replace after every disassembly.

📐 : Apply genuine RTV Silicone Sealant or equivalent.

📐 : N•m (kg-m, ft-lb)

📐 : N•m (kg-m, in-lb)

1.	Oil pan gasket (rear)		2.	Oil pan (upper)		3.	O-ring	
4.	Oil pan gasket (front)		5.	Oil filter		6.	Connector bolt	
7.	Oil cooler		8.	O-ring		9.	Relief valve	
10.	Oil filter bracket		11.	Oil filter bracket gasket		12.	Oil strainer	
13.	Drain plug		14.	Drain plug washer		15.	Oil pan (lower)	
16.	Rear plate		17.	Crankshaft position sensor (POS)		18	O-ring (small)	
19.	O-ring (large)		20.	Axle pipe				

67162-FX35-G47

Exploded view of oil pan and related components—3.5L engine AWD

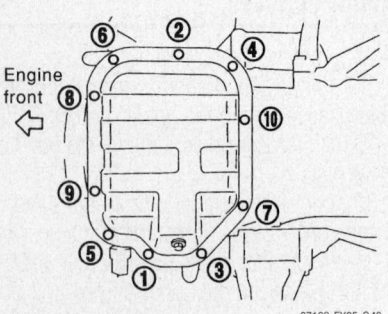

67162-FX35-G48

Lower oil pan mounting bolt tightening sequence—3.5L engine

✳✳ CAUTION

Be careful not to damage the mating surface. Do not use a flat-bladed screwdriver—this will damage the mating surfaces.

33. Remove the O-rings from the bottom of the cylinder block and oil pump.

34. Remove the oil pan gaskets.

35. For AWD models, remove the axle pipe from the upper oil pan using a suitable drift, if necessary.

36. Clean the oil strainer, if necessary.

To install:

37. Install the upper oil pan, as follows:

a. Use a scraper to remove the old liquid gasket from all mating surfaces. Remove old liquid gasket from the mating surface of the cylinder block, and the bolt holes and threads.

✳✳ CAUTION

Do not scratch or damage the mating surfaces when cleaning off old liquid gasket.

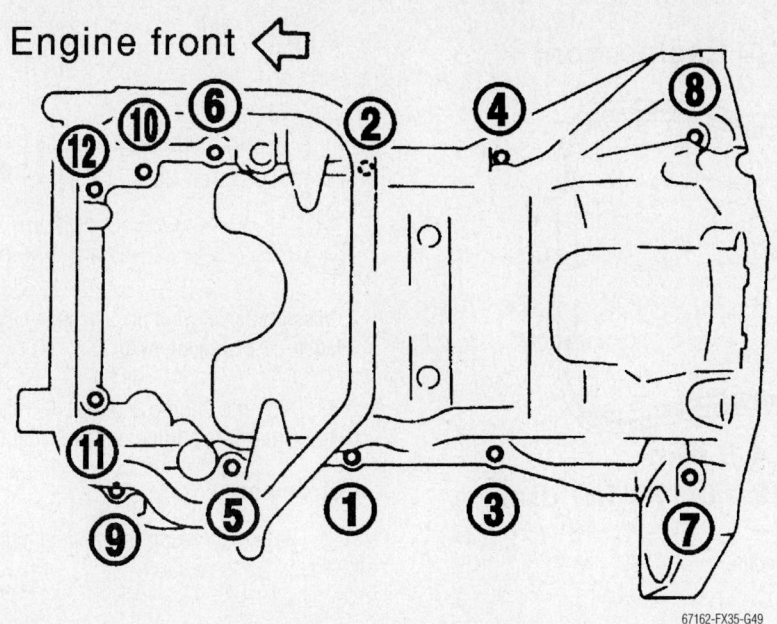

Engine front ⇐

Upper oil pan mounting bolt tightening sequence—3.5L engine

67162-FX35-G49

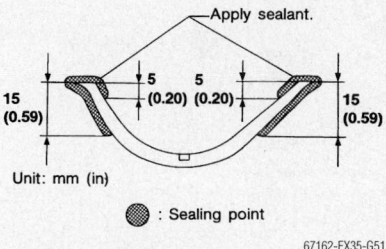

Apply sealant.

15 (0.59)   5 (0.20)   5 (0.20)   15 (0.59)

Unit: mm (in)

⬤ : Sealing point

67162-FX35-G51

Sealant positioning for front oil pan seal installation—3.5L engine

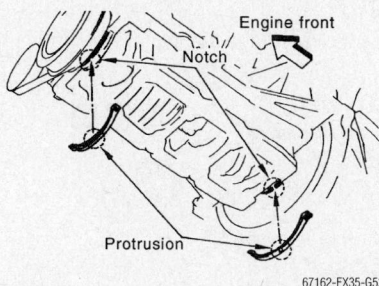

Engine front

Notch

Protrusion

67162-FX35-G52

Oil pan seal positioning—3.5L engine

b. Apply liquid gasket to the oil pan gaskets as shown.

➡**Use Genuine RTV Silicone Sealant or equivalent.**

c. Install the new gasket. Align the protrusion of the oil pan gasket with the notches of the front timing chain case and rear oil seal retainer.

d. Install the oil pan gasket with the smaller arc to the front timing chain case side.

e. Install new O-rings on the cylinder block and oil pump.

f. Apply a continuous bead of liquid gasket to the cylinder block mating surface of the upper oil pan to a limited portion as shown.

➡**Use Genuine RTV Silicone Sealant or equivalent.**

- For bolt holes with marks (5 locations), apply liquid gasket outside the holes.
- Apply a bead of 0.177–0.217 in.

(4.5–5.5mm) in diameter to area "A".

- Installation should be done within 5 minutes after coating.

g. Install the upper oil pan. Tighten the mounting bolts in the order shown. There are two types of mounting bolts. Refer to the following for locating the bolt positions:

- M8 x 100 mm (3.97 in): positions 5, 7, 8 & 11
- M8 x 25 mm (0.98 in) : positions except 5, 7, 8 & 11

h. Tighten the transmission joint bolts.

38. Install the oil strainer onto the oil pump.

39. Install the lower oil pan, as follows:

a. Use a scraper to remove all old liquid gasket from the mating surfaces. Also remove old liquid gasket from the mating surface of the upper oil pan.

b. Apply new liquid gasket.

➡**Use Genuine RTV Silicone Sealant or equivalent. Attaching should be done within 5 minutes after coating.**

c. Tighten the mounting bolts in numerical order as shown.

40. Install the oil pan drain plug.

41. The remainder of installation is the reverse of removal.

➡**Wait at least 30 minutes after the oil pan is installed, fill the engine with new oil.**

42. Start the engine, and check there is no leak of engine oil.

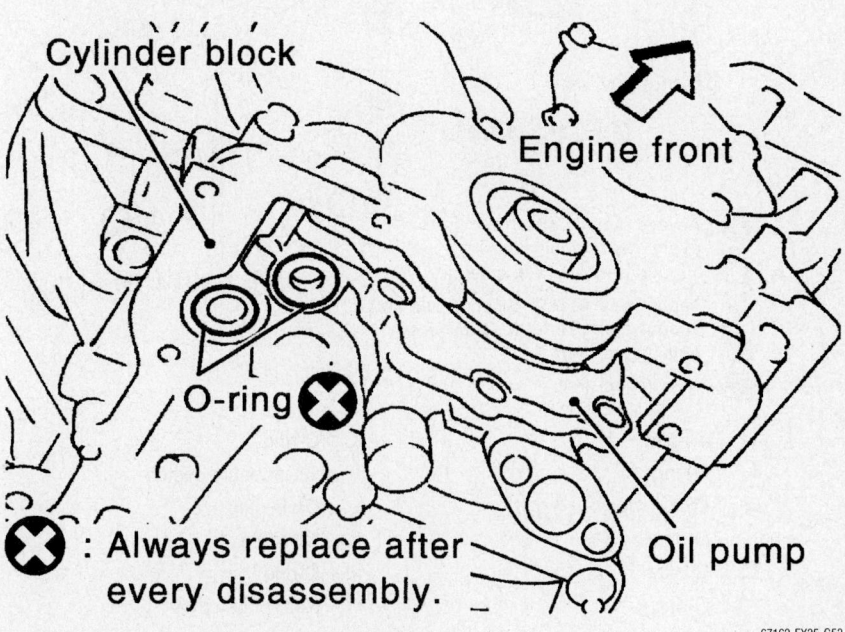

Cylinder block

Engine front

O-ring ⊗

⊗ : Always replace after every disassembly.

Oil pump

67162-FX35-G53

O-ring locations for oil pan service—3.5L engine

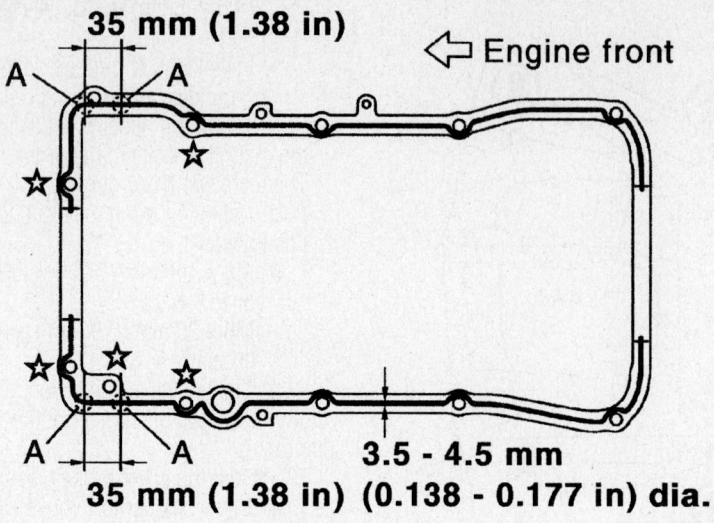

35 mm (1.38 in)

⟵ Engine front

A

A

A

A

3.5 - 4.5 mm

35 mm (1.38 in)   (0.138 - 0.177 in) dia.

67162-FX35-G54

Liquid gasket positioning for oil pan installation—3.5L engine

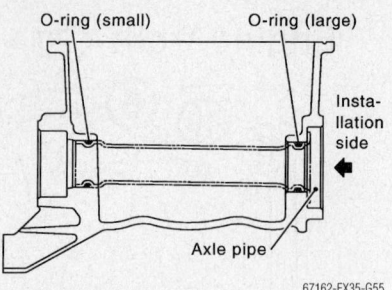

O-ring (small)   O-ring (large)

Installation side

Axle pipe

67162-FX35-G55

Cross-section view of the axle pipe installation—3.5L engine AWD

43. Stop the engine and wait 10 minutes.
44. Check the engine oil level again.

4.5L Engine

    1. Before servicing the vehicle, refer to the Precautions Section.

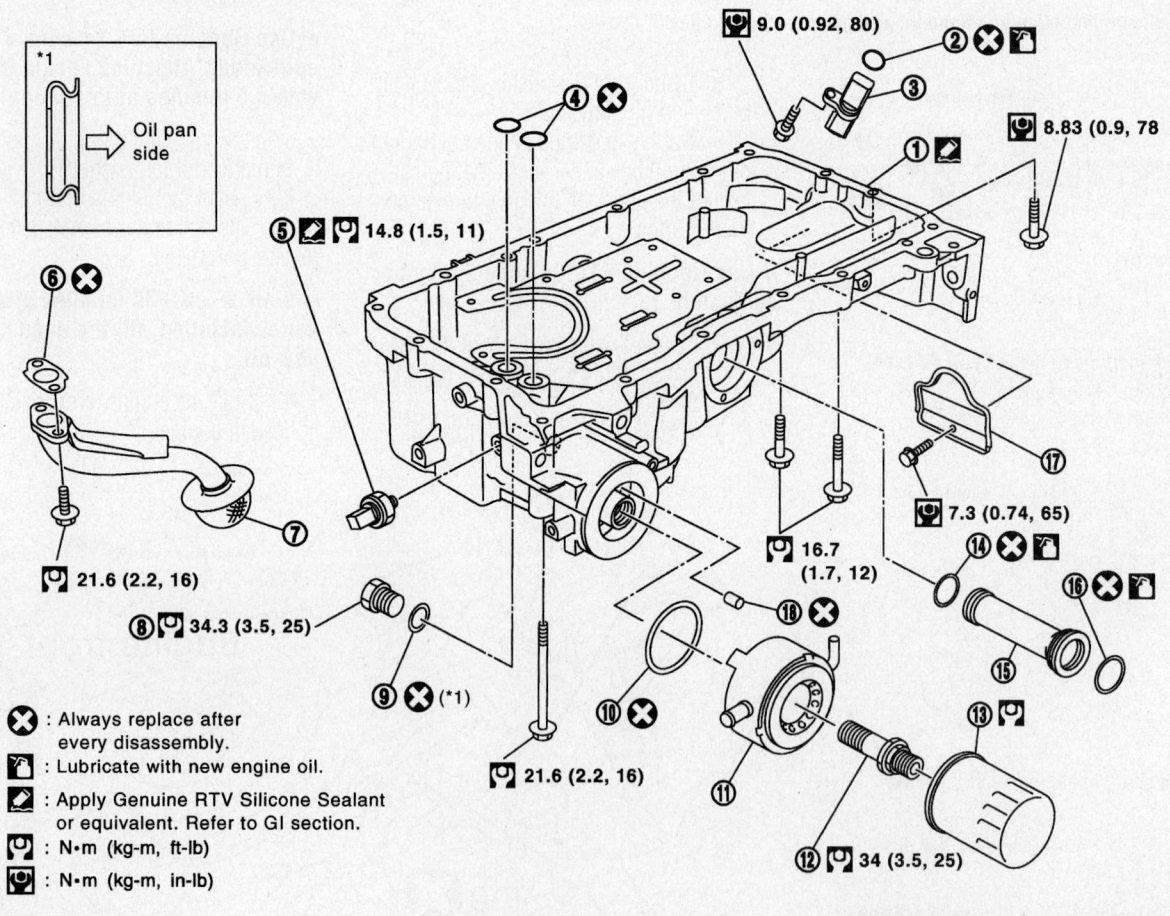

*1
→ Oil pan side

9.0 (0.92, 80)

8.83 (0.9, 78)

14.8 (1.5, 11)

21.6 (2.2, 16)

34.3 (3.5, 25)

7.3 (0.74, 65)

16.7 (1.7, 12)

(*1)

21.6 (2.2, 16)

34 (3.5, 25)

❌ : Always replace after every disassembly.
🛢 : Lubricate with new engine oil.
🖊 : Apply Genuine RTV Silicone Sealant or equivalent. Refer to GI section.
🔧 : N·m (kg-m, ft-lb)
🔧 : N·m (kg-m, in-lb)

1.	Oil pan	2.	O-ring	3.	Crankshaft position sensor (POS)
4.	O-ring	5.	Oil pressure switch	6.	Gasket
7.	Oil strainer	8.	Drain plug	9.	Drain plug washer
10.	O-ring	11.	Oil cooler	12.	Connector bolt
13.	Oil filter	14.	O-ring	15.	Axle pipe
16.	O-ring	17.	Rear plate cover	18.	Relief valve

67162-FX35-G56

Exploded view of the oil pan and related components—4.5L engine

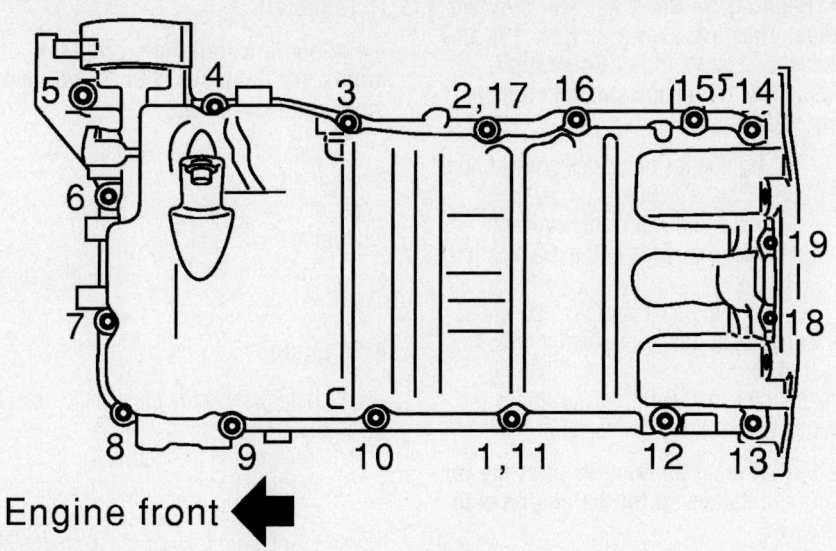

Oil pan mounting bolt tightening sequence—4.5L engine

67162-FX35-G57

Engine front ←

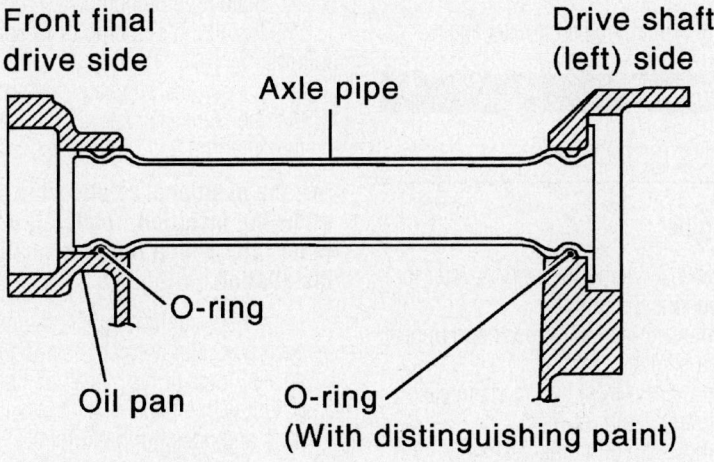

Axle pipe installation—4.5L engine

67162-FX35-G58

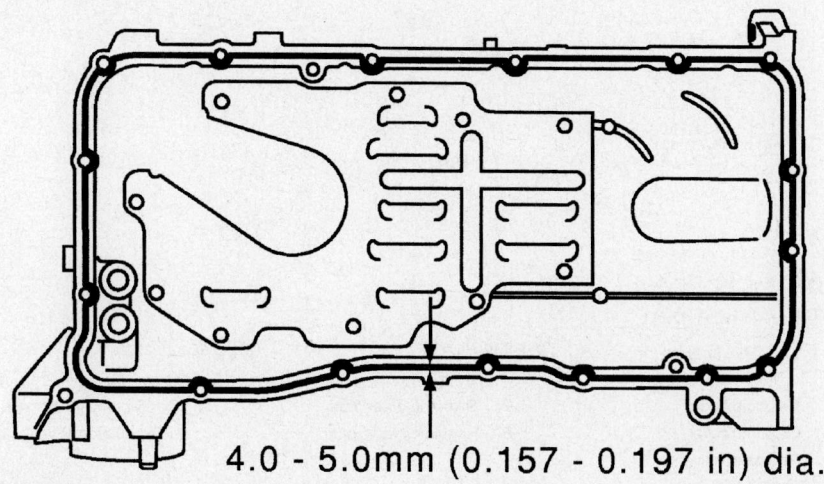

4.0 - 5.0mm (0.157 - 0.197 in) dia.

67162-FX35-G59

Liquid gasket positioning for oil pan installation—4.5L engine

To avoid the danger of being scalded, do not drain engine oil or coolant when the engine is hot.

2. Remove the front tire.
3. Remove the hood assembly.
4. Remove the engine cover.
5. Remove the front and rear engine under covers.
6. Drain the engine oil and engine coolant.
7. Remove the accessory drive belts.
8. Remove the auto tensioner for the power steering oil pump belt.
9. Remove the power steering oil pump with the piping connected, and temporarily secure it aside with ropes or equivalent.
10. Remove the A/C compressor with the piping connected, and temporarily secure it aside with ropes or equivalent.
11. Remove the A/C compressor fitting bolts, and install the A/C compressor temporarily on the vehicle side with ropes or equivalent.
12. Remove the wiring harness of the lower side of oil pan.
13. Remove the crankshaft position sensor (POS) from the transmission.

Handle the POS carefully to avoid dropping it and exposing it to abrupt shocks. Do not disassemble it, do not allow metal powder to adhere to magnetic part at the sensor tip, and do not place sensor in a location where it may be exposed to magnetism.

14. Install the engine slinger to hold the engine assembly in a secure position. Tighten the slinger mounting bolts to 25 ft. lbs. (33.5 Nm).
15. Remove the front suspension member.
16. Remove the front final drive assembly.
17. Remove the oil filter.
18. Disconnect the oil cooler water hoses, and remove the oil cooler water pipe and oil cooler.
19. Remove the oil pan as follows:
    a. Remove the rear plate cover.
    b. Remove the transmission joint bolts which pass through the oil pan.
    c. Loosen the oil pan bolts in the reverse order of the tightening sequence.

➡Disregard the tightening sequence numbers No. 11 and No. 17 during removal.

d. Insert seal cutter between the oil pan and cylinder block. Slide the seal cutter by tapping on the side of seal cutter with hammer.

e. Remove the oil pan.

> ※※ **CAUTION**
>
> **Be careful not to damage the mating surfaces. Do not use a flat-bladed screwdriver—this will damage the mating surfaces.**

f. Remove the O-rings from the bottom of the oil pump and front cover.

20. As necessary, pull the axle pipe from the oil pan. Hold the pipes and pull them out to the left side.

21. Remove the oil strainer.

22. Clean the oil strainer, if necessary.

To install:

23. Install the oil strainer.

24. Install the axle pipe to the oil pan, if removed.

25. Lubricate the O-ring groove of the axle pipe, O-ring, and O-ring joint of the oil pan with new engine oil.

➡**The right and left O-ring diameters differ from each other. The O-ring with an identification paint mark must be installed on the left front halfshaft side.**

26. Install the axle pipe to the oil pan on the left side.

> ※※ **CAUTION**
>
> **Insert the axle pipe with care to prevent the O-ring from sliding.**

27. Install the oil pan as follows:

a. Install new O-rings onto the oil pump and the side of the front cover.

b. Apply a continuous bead of RTV liquid gasket to the cylinder block mating surface of the oil pan as shown.

➡**Use Genuine RTV Silicone Sealant or equivalent.**

> ※※ **CAUTION**
>
> **Installation should be done within 5 minutes after coating, otherwise the liquid gasket may not seal properly.**

c. Install the oil pan and tighten the mounting bolts in order as shown. There are three types of mounting bolts. Refer to the following for locating bolts.

• M6 x 30mm: positions 18 & 19
• M8 x 100mm: positions 5 & 9
• M8 x 45mm: positions except 5, 9, 18 & 19

➡**Tighten bolts No. 1 and No. 2 in two steps. They are shown as Nos. 1 & 11 and Nos. 2 & 17 in the illustration. Nos. 11 & 17 are the second steps for Nos. 1 & 2.**

d. Tighten the transmission joint bolts to 55 ft. lbs. (74 Nm).

e. Install the rear plate cover and tighten the mounting bolt to 65 inch lbs. (7.3 Nm).

28. Install the oil pan drain plug with a new drain plug washer. Tighten it to 25 ft. lbs. (34 Nm).

29. The remainder of installation is the reverse of the removal procedure.

➡**Wait at least 30 minutes after the oil pan is installed, to fill the engine with new oil.**

30. Add engine oil.

31. Start the engine, and check for leakage of engine oil.

32. Stop the engine and wait for 15 minutes.

33. Check the engine oil level again.

Oil Pump

REMOVAL & INSTALLATION

3.5L Engine

1. Before servicing the vehicle, refer to the Precautions Section.

2. Remove the oil pan (lower and upper) and the oil strainer.

3. Remove the front timing chain case and the timing chain (primary).

4. Remove the oil pump assembly.

To install:

➡**For pump installation, align the crankshaft flat faces with the oil pump inner rotor flat faces.**

5. Installation is the reverse of the removal procedure.

6. After warming up the engine, check for engine oil leakage.

7. Check the engine oil level and add engine oil, as needed.

4.5L Engine

1. Before servicing the vehicle, refer to the Precautions Section.

2. Remove the front cover.

3. Remove the oil pump drive spacer.

4. Set bolts in the two bolt holes (M6 x 1.0mm) on the front surface. Using a suitable puller, pull oil pump drive spacer off the crankshaft.

5. Remove the oil pump.

To install:

6. Install the oil pump.

7. Install the oil pump drive spacer as follows:

a. Insert the oil pump drive spacer so that the crankshaft key and flat surfaces of the oil pump inner rotor mesh properly.

➡**If the positional relationship does not allow the insertion, rotate oil pump inner rotor with a finger to facilitate installation.**

b. After confirming that the position of each part is in correct position for the spacer, force fit the spacer by lightly tapping it with a plastic hammer until it contacts and does not go further.

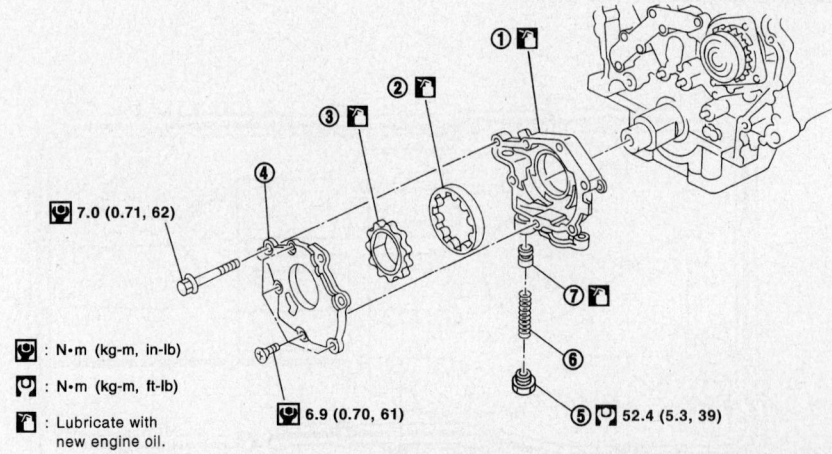

⊡ : N·m (kg-m, in-lb)
⊡ : N·m (kg-m, ft-lb)
⊡ : Lubricate with new engine oil.

⊡ 7.0 (0.71, 62)
⊡ 6.9 (0.70, 61)
⊡ 52.4 (5.3, 39)

1. Oil pump body	2. Oil pump outer rotor	3. Oil pump inner rotor
4. Oil pump cover	5. Regulator valve plug	6. Regulator valve spring
7. Regulator valve		

67162-FX35-G60

Exploded view of the oil pump assembly—3.5L engine

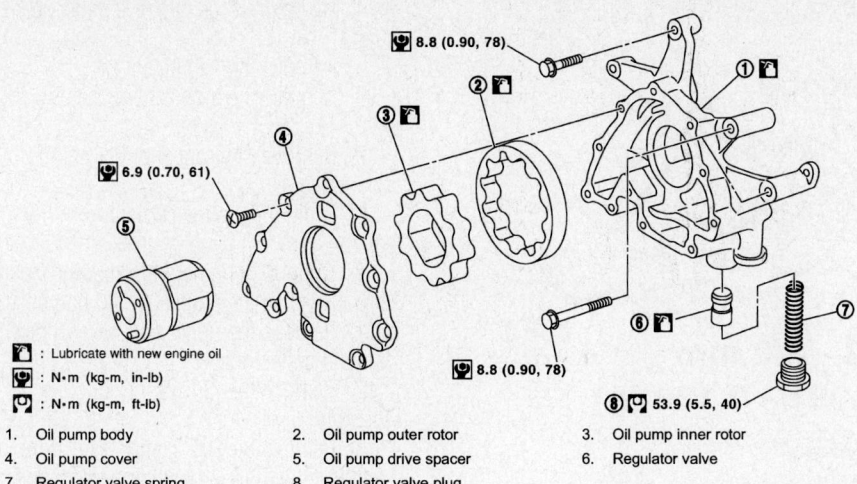

8.8 (0.90, 78)

6.9 (0.70, 61)

8.8 (0.90, 78)

8 53.9 (5.5, 40)

: Lubricate with new engine oil

: N·m (kg-m, in-lb)

: N·m (kg-m, ft-lb)

1. Oil pump body
2. Oil pump outer rotor
3. Oil pump inner rotor
4. Oil pump cover
5. Oil pump drive spacer
6. Regulator valve
7. Regulator valve spring
8. Regulator valve plug

67162-FX35-G61

Exploded view of the oil pump—4.5L engine

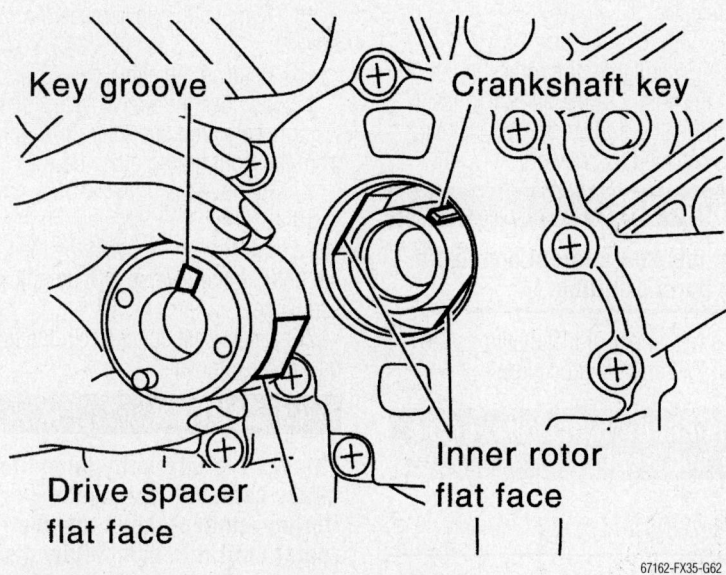

Key groove

Crankshaft key

Drive spacer flat face

Inner rotor flat face

67162-FX35-G62

Oil pump drive spacer installation orientation—4.5L engine

8. Installation is the reverse of the removal procedure.

9. After warming up the engine, check for engine oil leakage.

10. Check the engine oil level, and add more oil, if necessary.

Rear Main Seal

REMOVAL & INSTALLATION

3.5L Engine

1. Before servicing the vehicle, refer to the Precautions Section.

2. Remove the upper oil pan.

3. Remove the transmission assembly.

4. Remove the drive plate. Install a ring

gear stopper (SST: KV1011770 (J44716) or equivalent) on the crankshaft and remove the mounting bolts in a diagonal order.

✳✳ CAUTION

Do not disassemble the drive plate. Never place the drive plate with the signal plate facing down. When handling the signal plate, take care not to damage or scratch it. Handle the signal plate in a manner that prevents it from becoming magnetized.

5. Use a seal cutter to cut away the old liquid gasket and remove the rear oil seal retainer.

✳✳ CAUTION

Be careful not to damage the mounting surfaces.

➡The rear oil seal and retainer form a single part and are handled as one assembly.

To install:

6. Remove the old liquid gasket from the mating surface of the cylinder block and oil pan using a scraper.

7. Apply new engine oil to the oil and dust seal lips.

8. Apply liquid gasket to the rear oil seal retainer.

➡Use Genuine Thread Sealant or equivalent.

✳✳ CAUTION

Installation should be done within 5 minutes after coating, otherwise the liquid gasket may not seal properly.

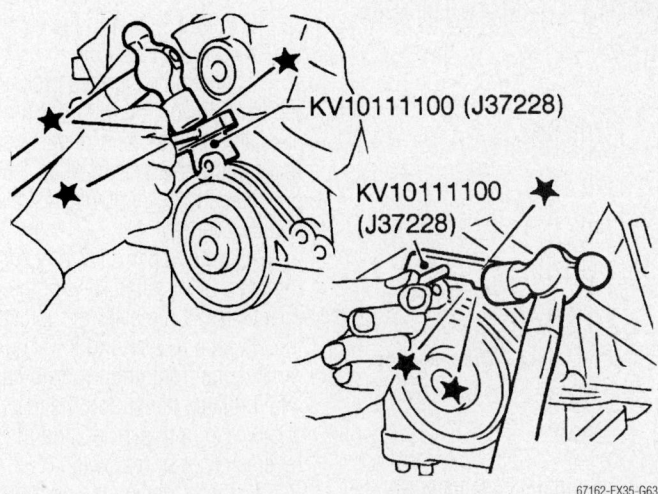

KV10111100 (J37228)

KV10111100 (J37228)

67162-FX35-G63

Liquid gasket seal cutting tools—3.5L engine

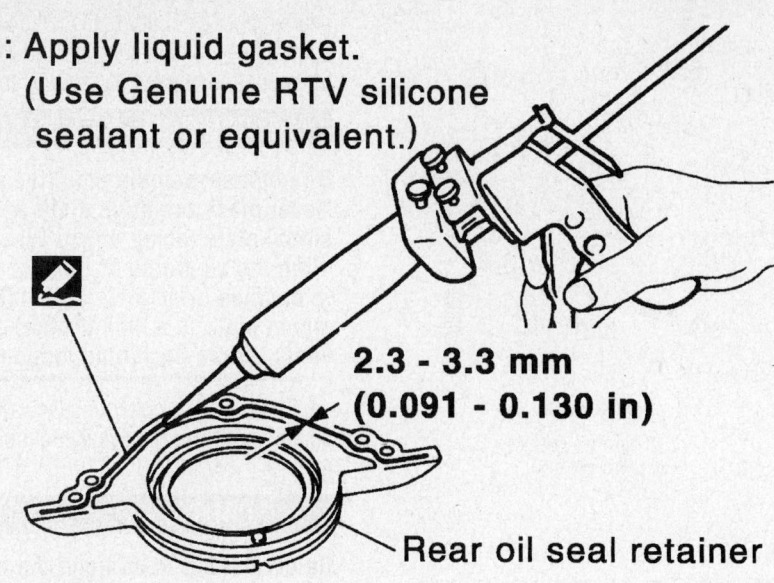

: Apply liquid gasket.
(Use Genuine RTV silicone
sealant or equivalent.)

2.3 - 3.3 mm
(0.091 - 0.130 in)

Rear oil seal retainer

67162-FX35-G64

Rear main seal liquid gasket installation positioning—3.5L engine

9. Install the rear oil seal retainer onto the cylinder block.

10. The remainder of installation is the reverse of the removal procedure.

4.5L Engine

1. Before servicing the vehicle, refer to the Precautions Section.

2. Remove the transmission and transfer assembly.

3. Remove the drive plate.

4. Remove the engine rear plate.

5. Remove the rear oil seal using a suitable tool.

❊❊ CAUTION

Be careful not to damage the crankshaft and oil seal retainer surface.

To install:

6. Apply engine oil to both the oil seal lip and dust seal lip.

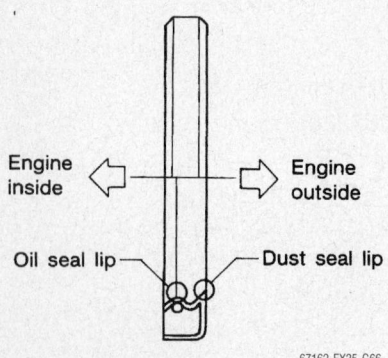

Engine inside ⇦   ⇨ Engine outside

Oil seal lip   Dust seal lip

67162-FX35-G66

**Rear main seal installation orientation—
4.5L engine**

7. Using a suitable drift, press fit the oil seal until the height of the oil seal is level with the mounting surface and so that each seal lip is oriented as shown.

❊❊ CAUTION

Press fit the seal straight and avoid causing burrs or tilting.

8. The remainder of installation is the reverse of the removal procedure.

Timing Chain, Sprockets, Front Cover and Seal

REMOVAL & INSTALLATION

3.5L Engine

WITH OIL PAN REMOVAL

1. Before servicing the vehicle, refer to the Precautions Section.

This section describes procedures for removing/installing the front timing chain case and timing chain related parts, and the rear timing chain case, when the upper oil pan needs to be removed/installed for engine overhaul, etc.

When the upper oil pan needs to be removed or installed, or when rear timing chain case is removed or installed, remove the oil pans (upper and lower) first. Then remove the front timing chain case, timing chain related parts, and the rear timing chain case in this order, and install in the reverse order of removal.

2. Place the vehicle on a lift.

3. Remove the front tire.

4. Disconnect the negative battery terminal.

5. Remove the engine cover.

6. Remove the air cleaner case assembly.

7. Remove the front and rear engine under covers.

8. Drain the engine coolant from the radiator.

9. Drain the engine oil from the oil pan.

10. Remove the engine wiring harnesses.

11. Remove the upper and lower intake manifold collectors.

12. Remove the radiator cooling fan assembly.

13. Remove the A/C compressor from the bracket with its piping connected, and temporarily secure it aside.

14. Remove the power steering oil pump from the bracket with its piping connected, and temporarily secure it aside.

15. Remove the power steering oil pump bracket.

16. Remove the alternator.

17. Remove the water bypass hose, water hose clamp and idler pulley bracket from the front timing chain case.

18. Remove the upper and lower oil pan.

19. Remove the right and left intake valve timing control covers, by loosening the bolts in the reverse or the tightening sequence.

20. Use a seal cutter to cut the liquid gasket for removal.

❊❊ CAUTION

The shaft is internally joined to the intake camshaft sprocket center hole. During removal, keep the shaft horizontal until it is completely disconnected.

21. Remove the collared O-ring from the front timing chain case (left and right side).

22. Remove the right and left rocker covers.

23. Position the engine at compression TDC of No. 1 cylinder as follows:

a. Rotate the crankshaft pulley clockwise to align the timing mark (grooved line without color) with the timing indicator.

b. Make sure the intake and exhaust cam noses on the No. 1 cylinder (engine front side of right bank) are located so that they point inward and upward compared to the cylinder head.

c. If the cam lobes are not positioned pointing inward and upward, rotate the crankshaft one full revolution (360 degrees).

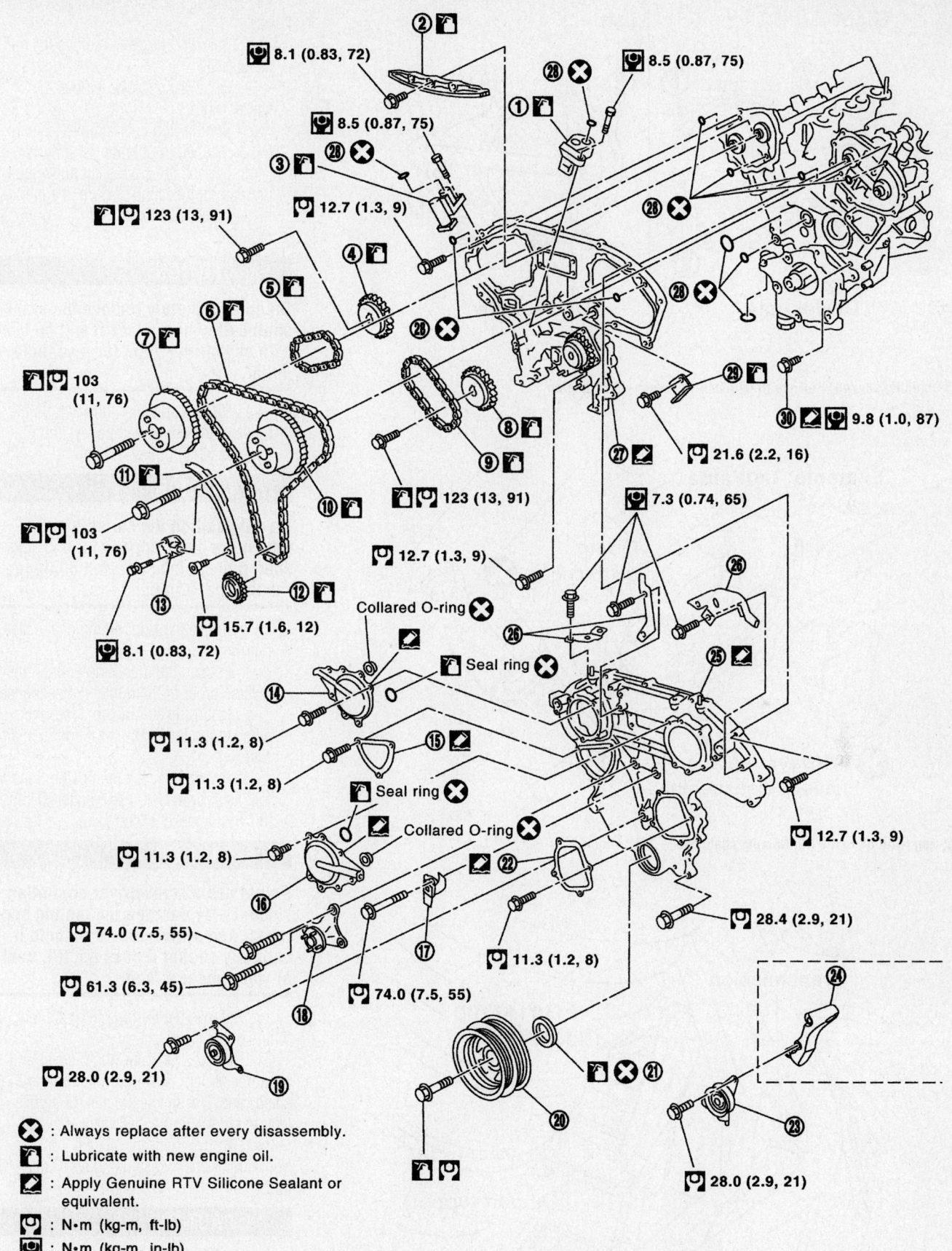

8.1 (0.83, 72)

8.5 (0.87, 75)

8.5 (0.87, 75)

12.7 (1.3, 9)

123 (13, 91)

103 (11, 76)

103 (11, 76)

8.1 (0.83, 72)

15.7 (1.6, 12)

123 (13, 91)

12.7 (1.3, 9)

Collared O-ring

Seal ring

11.3 (1.2, 8)

11.3 (1.2, 8)

Seal ring

Collared O-ring

11.3 (1.2, 8)

74.0 (7.5, 55)

61.3 (6.3, 45)

74.0 (7.5, 55)

11.3 (1.2, 8)

28.0 (2.9, 21)

9.8 (1.0, 87)

21.6 (2.2, 16)

7.3 (0.74, 65)

12.7 (1.3, 9)

28.4 (2.9, 21)

28.0 (2.9, 21)

: Always replace after every disassembly.

: Lubricate with new engine oil.

: Apply Genuine RTV Silicone Sealant or equivalent.

: N•m (kg-m, ft-lb)

: N•m (kg-m, in-lb)

67162-FX35-G67

Exploded view of timing chain and related components—3.5L engine

Right | Left

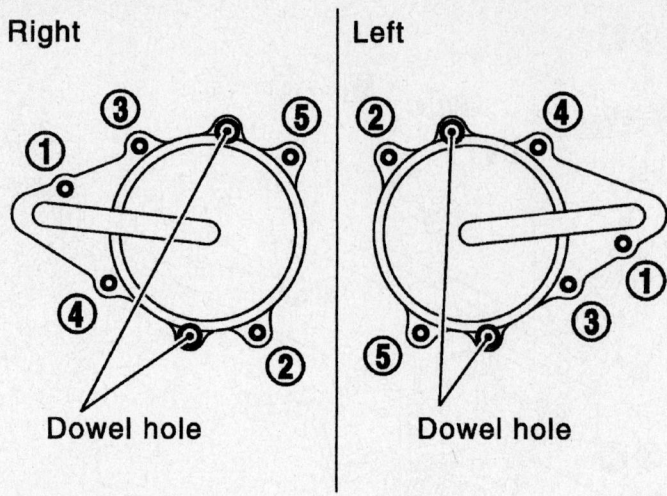

Dowel hole   **Dowel hole**

67162-FX35-G68

Tightening sequence for the timing control covers—3.5L engine

Example: Left side

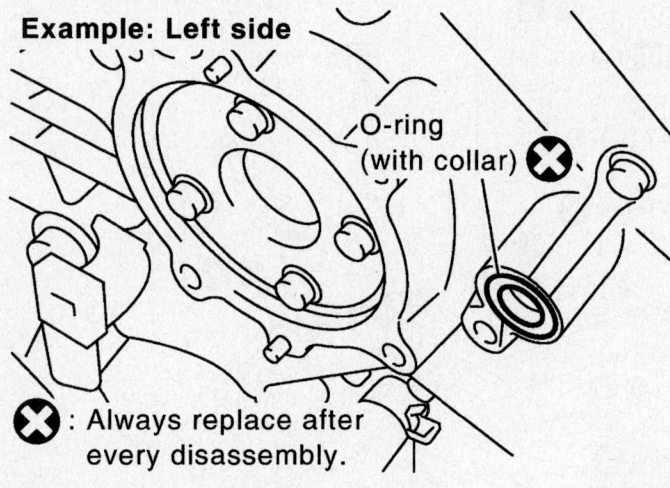

O-ring
(with collar) ✖

✖ : Always replace after
every disassembly.

67162-FX35-G69

O-ring location in the front timing chain case—3.5L engine

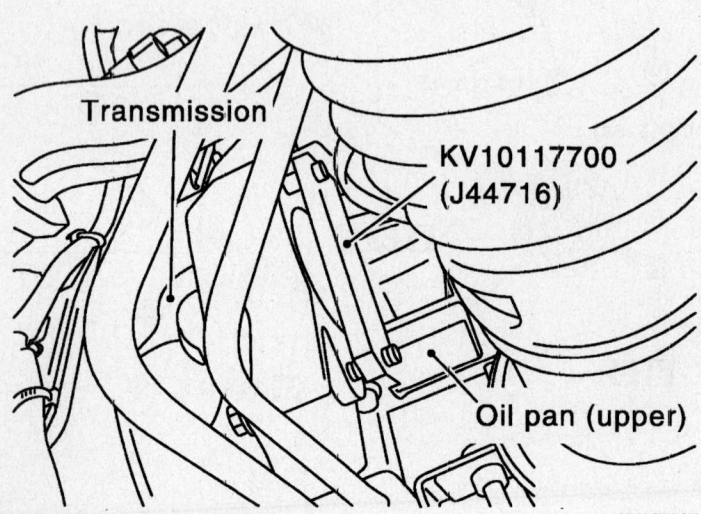

Transmission

KV10117700
(J44716)

Oil pan (upper)

67162-FX35-G70

Ring gear stopper positioning—3.5L engine

24. Remove the crankshaft pulley, as follows:
   a. For 2WD models, remove the rear cover plate.
   b. For AWD models, remove the starter motor.
   c. Set the ring gear stopper to hold the crankshaft in position.
   d. Loosen the crankshaft pulley bolt until the bolt seating surface is approximately 0.39 in. (10mm) from its original position.

> ✳✳ **CAUTION**
>
> **Do not completely remove the crankshaft pulley bolt, since it will be used as a supporting point for a suitable puller.**

   e. Place a suitable puller tab on the holes of the crankshaft pulley, and pull the crankshaft pulley through.

> ✳✳ **CAUTION**
>
> **Do not position the suitable puller tab on the outer edges of the crankshaft pulley since this can damage the internal damper.**

25. Remove the front timing chain case, as follows:
   a. Loosen the mounting bolts in the reverse order of the tightening sequence.
   b. Insert the suitable tool into the notch at the top of the front timing chain case.
   c. Pry off the case by moving the pry tool as shown. Use a seal cutter to cut the liquid gasket for removal.

> ✳✳ **CAUTION**
>
> **Do not use a screwdriver or similar, since it may damage the mating surfaces. And after removal, handle it carefully so that it does not tilt, cant, or warp under a load.**

26. Remove the O-rings from the rear timing chain case.
27. Remove the water pump cover and chain tensioner cover from the front timing chain case. Use the seal cutter, or equivalent, to cut the liquid gasket for removal.
28. Remove the front oil seal from the front timing chain case using a suitable tool.

> ✳✳ **CAUTION**
>
> **Be careful not to damage the front timing chain case.**

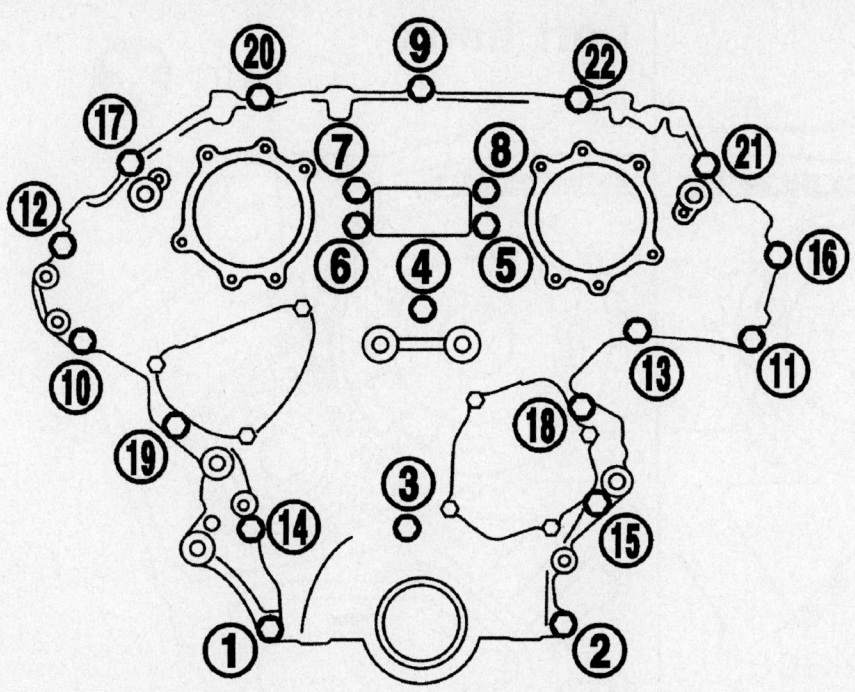

Tightening sequence for the front timing chain case mounting bolts—3.5L engine

67162-FX35-G71

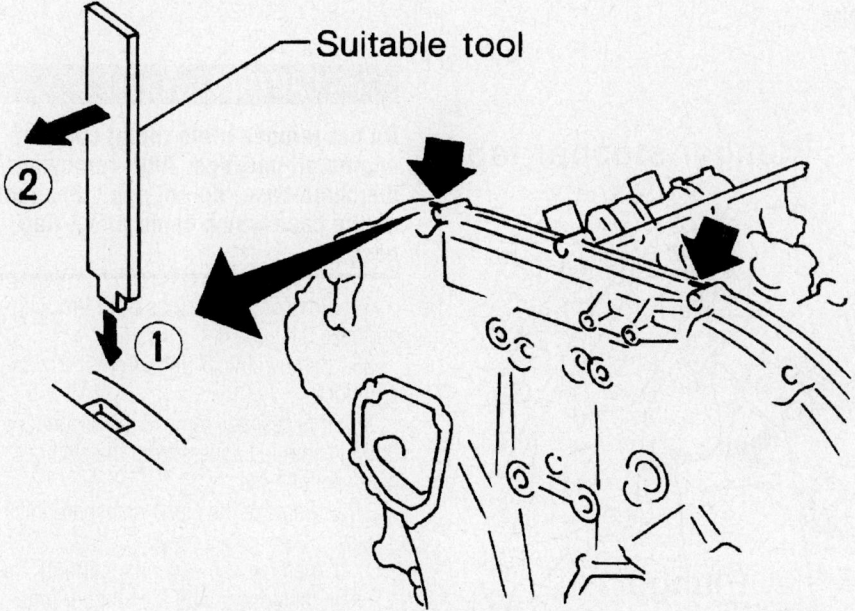

— Suitable tool

67162-FX35-G72

Pry off the front timing chain case as shown—3.5L engine

29. Remove the primary timing chain tensioner, as follows:

a. Pull the lever down and release the plunger stopper tab. The plunger stopper tab can be pushed up to release.

b. Insert a stopper pin into the tensioner body hole to hold the lever, and keep the tab released.

➡A 0.098 in. (2.5mm) Allen wrench can be used for a stopper pin.

c. Insert the plunger into the tensioner body by pressing on the slack guide.

d. Keep the slack guide depressed and hold it by pushing the stopper pin through the lever hole and body hole.

e. Remove the mounting bolts and remove the primary timing chain tensioner.

30. Remove the internal chain guide, tension guide and slack guide.

➡The tension guide can be removed after removing the primary timing chain.

31. Remove the primary timing chain, tension guide and crankshaft sprocket.

※ CAUTION

After removing the timing chain, do not turn the crankshaft and camshaft separately, or the valves will strike piston heads.

32. Remove the secondary timing chain and camshaft sprockets, as follows:

a. Attach a suitable stopper pin to the right and left secondary timing chain camshaft chain tensioners.

b. Remove the intake and exhaust camshaft sprocket bolts. Apply paint to the timing chain and camshaft sprockets for alignment during installation. Secure the hexagonal portion of the camshaft using an open-end wrench to hold the camshaft steady while loosening the mounting bolts.

c. Remove the secondary timing chain together with camshaft sprockets. Turn the camshaft slightly to create slack in the timing chain on the secondary timing chain tensioner side. Insert a 0.020 in. (0.5mm) thick metal or resin plate between the timing chain and timing chain tensioner plunger guide. Remove the secondary timing chain together with the camshaft sprockets with the timing chain loose from the guide groove.

※ CAUTION

Be careful of the plunger coming-off when removing the secondary timingchain. This is because the plunger of the secondary timing chain tensioner moves during operation, which can result in the fixed stopper pin coming off.

➡Camshaft sprocket (INT) is a two-for-one assembly of the primary and secondary sprockets.

※ CAUTION

When handling the camshaft sprocket (INT), be careful of the following:

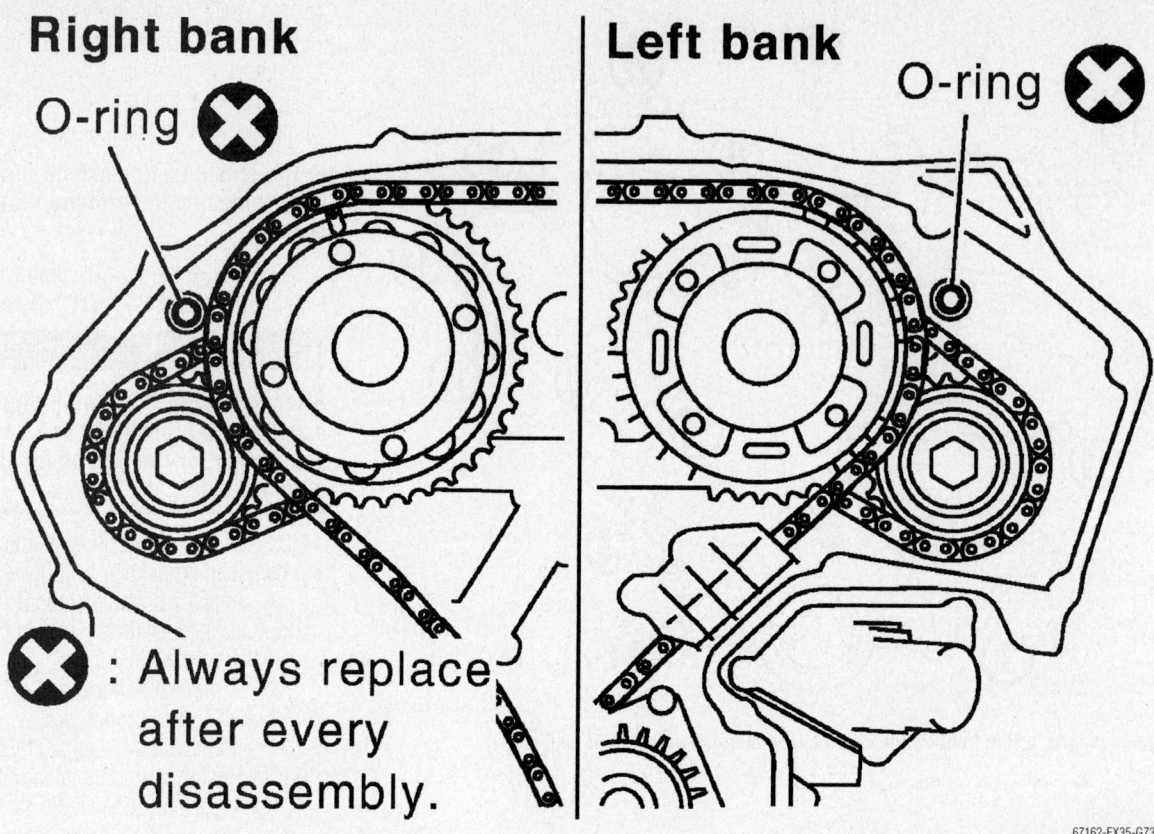

Right bank

O-ring ⊗

Left bank

O-ring ⊗

⊗ : Always replace after every disassembly.

67162-FX35-G73

O-ring positions in rear timing chain case—3.5L engine

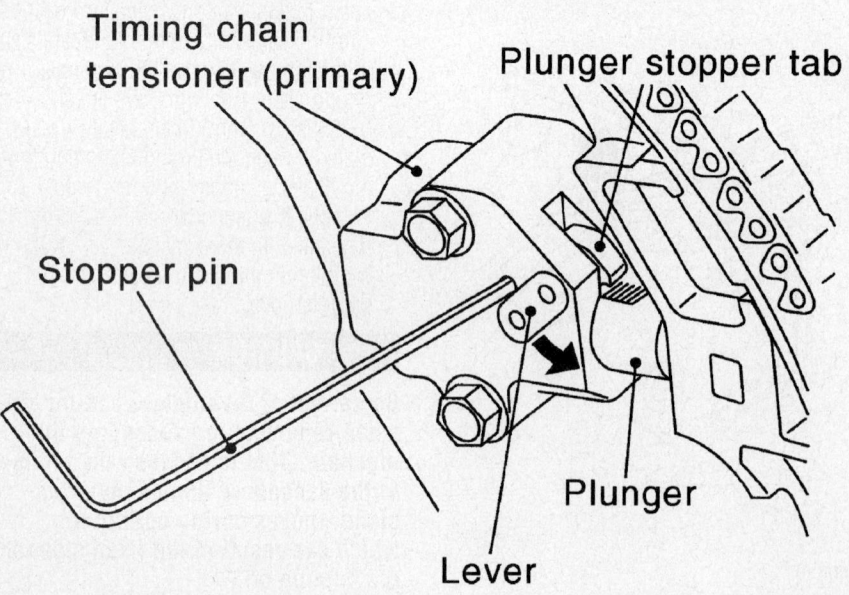

Timing chain tensioner (primary)

Plunger stopper tab

Stopper pin

Plunger

Lever

67162-FX35-G74

Timing chain tensioner detail—3.5L engine

- Handle it carefully to avoid any shock to the camshaft sprocket.
- Do not disassemble. Do not loosen bolts "A" and "B".

33. Remove the rear timing chain case, as follows:

a. Loosen and remove the mounting bolts in the reverse order of the tightening sequence.

b. Cut the sealant using a seal cutter and remove the rear timing chain case.

> ✳✳ **CAUTION**
>
> **Do not remove plate metal cover of engine oil passage. After removing the chain case, do not apply any load on the case which could affect flatness.**

34. Remove the O-rings from the cylinder head.

35. Remove the O-rings from the cylinder block.

36. If necessary, remove the secondary timing chain tensioners from the cylinder head, as follows:

a. Remove the No. 1 camshaft brackets.

b. Remove the secondary timing chain tensioners with the stopper pins attached.

37. Use a scraper to remove all traces of liquid gasket from the front and rear timing chain cases, and opposite mating surfaces.

> ✳✳ **CAUTION**
>
> **Be careful not to allow gasket fragments to enter the oil pan.**

38. Remove the old liquid gasket from the bolt holes and threads.

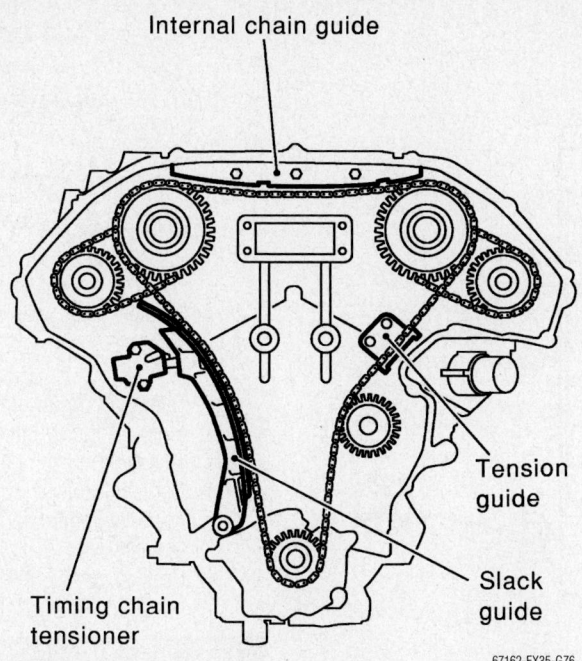

Internal chain guide

Tension guide

Slack guide

Timing chain tensioner

67162-FX35-G76

Internal chain guide, tension guide and slack guide positions—3.5L engine

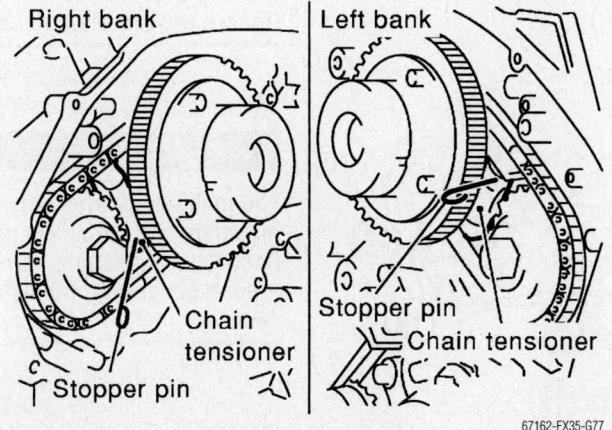

Right bank

Left bank

Chain tensioner

Stopper pin

Stopper pin

Chain tensioner

67162-FX35-G77

Secondary timing chain tensioner positions—3.5L engine

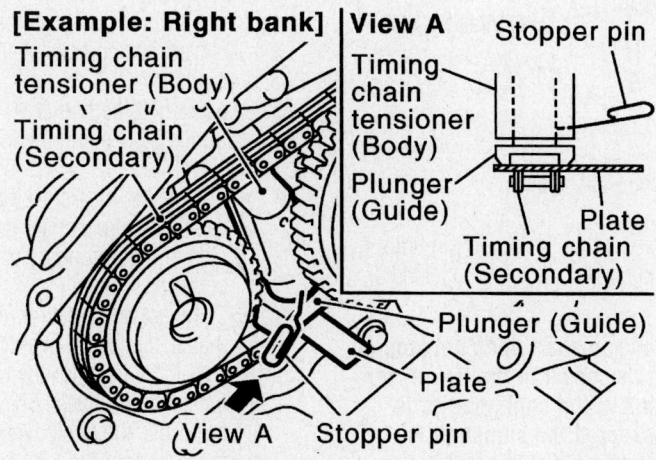

[Example: Right bank]

Timing chain tensioner (Body)

Timing chain (Secondary)

View A

Stopper pin

Timing chain tensioner (Body)

Plunger (Guide)

Plate

Timing chain (Secondary)

Plunger (Guide)

Plate

View A

Stopper pin

67162-FX35-G78

Secondary timing chain plunger setting for removal—3.5L engine

39. Use a scraper to remove all traces of liquid gasket from the water pump cover, chain tensioner cover and intake valve timing control covers.

40. Inspect the timing chain for cracks and any excessive wear at link plates and roller links of the timing chain. Replace the timing chain, as necessary.

To install:

→**The accompanying illustration shows the relationship between the mating mark on each timing chain and that on the corresponding sprocket, with the components installed.**

41. If removed, install the secondary timing chain tensioners on the cylinder head, as follows:

a. Install the chain tensioners with stopper pins attached and new O-rings.

b. Install the No. 1 camshaft brackets. Refer to the camshaft removal and installation procedure.

42. Install new O-rings onto the cylinder block.

43. Install new O-rings on the cylinder head.

44. Apply liquid gasket to the rear timing chain case backside, as shown.

✳✳ CAUTION

Use Genuine RTV Silicone Sealant or equivalent.

45. For "A" in the illustration, completely wipe out the liquid gasket extended on a portion touching at engine coolant.

46. Apply liquid gasket on the installation position of the water pump and cylinder head completely.

47. Align the rear timing chain case and water pump assembly with right and left dowel pins on cylinder block and install the case.

→**Make sure the O-rings stay in place during installation on the cylinder block and cylinder head.**

48. Tighten the mounting bolts in the sequence shown. After all bolts are temporarily tightened, retighten them to 9 ft. lbs. (13 Nm) in the tightening sequence.

→**There are two bolt lengths used for the timing chain case:**

- 20mm length: Pos. 1, 2, 3, 6, 7, 8, 9, 10
- 16mm length: Except Pos. 1, 2, 3, 6, 7, 8, 9, 10

→**If RTV Silicone Sealant protrudes, wipe it off immediately.**

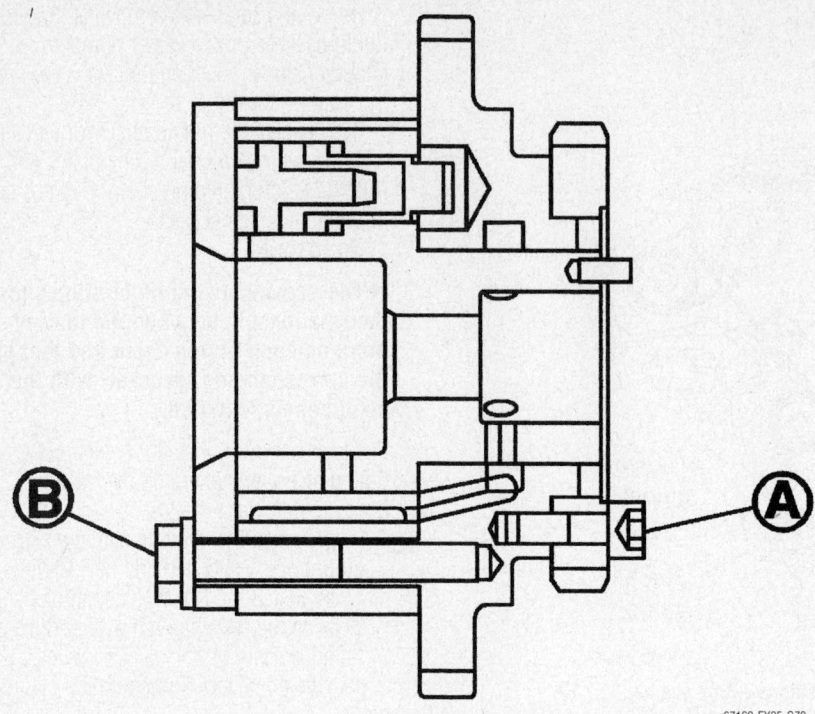

Cross-section of intake camshaft sprocket—Do NOT loosen bolts A or B—3.5L engine

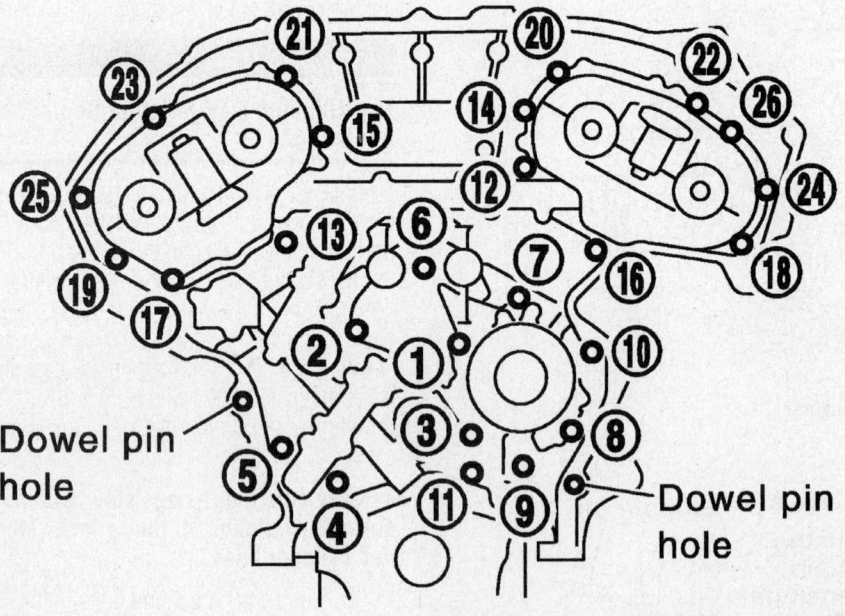

Tightening sequence for the rear timing chain case mounting bolts—3.5L engine

49. After installing the rear timing chain case, check the surface height difference between the rear timing chain case to the cylinder block at the oil pan mounting surface. If the difference is not within -0.0094–0.0055 in. (-0.24–0.14 mm), repeat the installation procedure.

50. Position the crankshaft so the No. 1 piston is set at TDC on the compression stroke. Make sure that the dowel pin hole, dowel pin and crankshaft key are located as shown.

➡Though the camshaft does not stop at the position as shown in the figure, for the placement of the cam nose, it is generally accepted the camshaft is placed in the same direction of the figure and as follows:

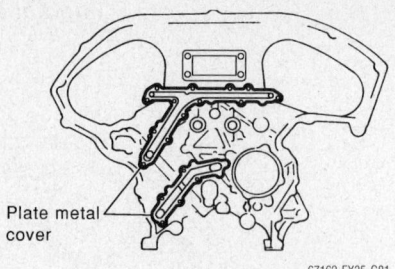

Do not remove these plate metal covers— 3.5L engine

- Camshaft dowel pin hole (intake side): At the cylinder head upper face side in each bank.
- Camshaft dowel pin (exhaust side): At the cylinder head upper face side in each bank.
- Crankshaft key: At the cylinder head side of the right bank.

✳✳ CAUTION

The hole on the small diameter side must be used for the intake side dowel pin hole. Do not misidentify (ignore the big diameter side).

51. Install the secondary timing chains and camshaft sprockets, as follows:

✳✳ CAUTION

The matching marks between the timing chain and sprockets slip easily. Confirm all matching mark positions repeatedly during the installation process.

a. Push the plunger of the secondary chain tensioner and keep it pressed in with a stopper pin.

b. Install the secondary timing chains and camshaft sprockets. Align the mating marks on the secondary timing chain (gold link) with the ones on the intake and exhaust camshaft sprockets (stamped), and install them.

52. Ensure that the sprockets are properly positioned by heeding the following items:

- The mating marks for the intake camshaft sprocket are on the back side of the secondary camshaft sprocket.
- There are two types of mating marks, circle and oval types. They should be used for the right and left banks, respectively. For the right bank, use the circle type of mating mark, and for the left bank use the oval type of mating mark.

Right bank

O-ring ⊗

Left bank

O-ring ⊗

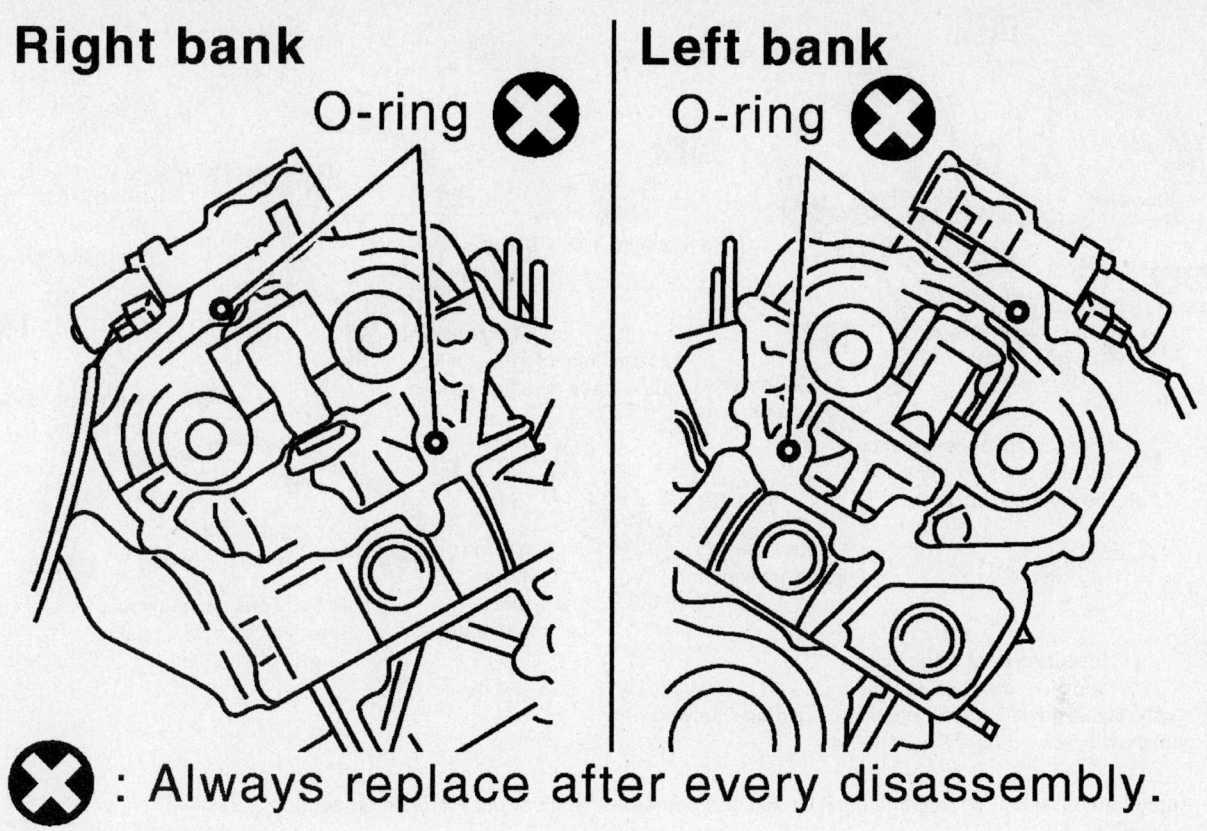

⊗ : Always replace after every disassembly.

67162-FX35-G82

Cylinder head O-ring positions—3.5L engine

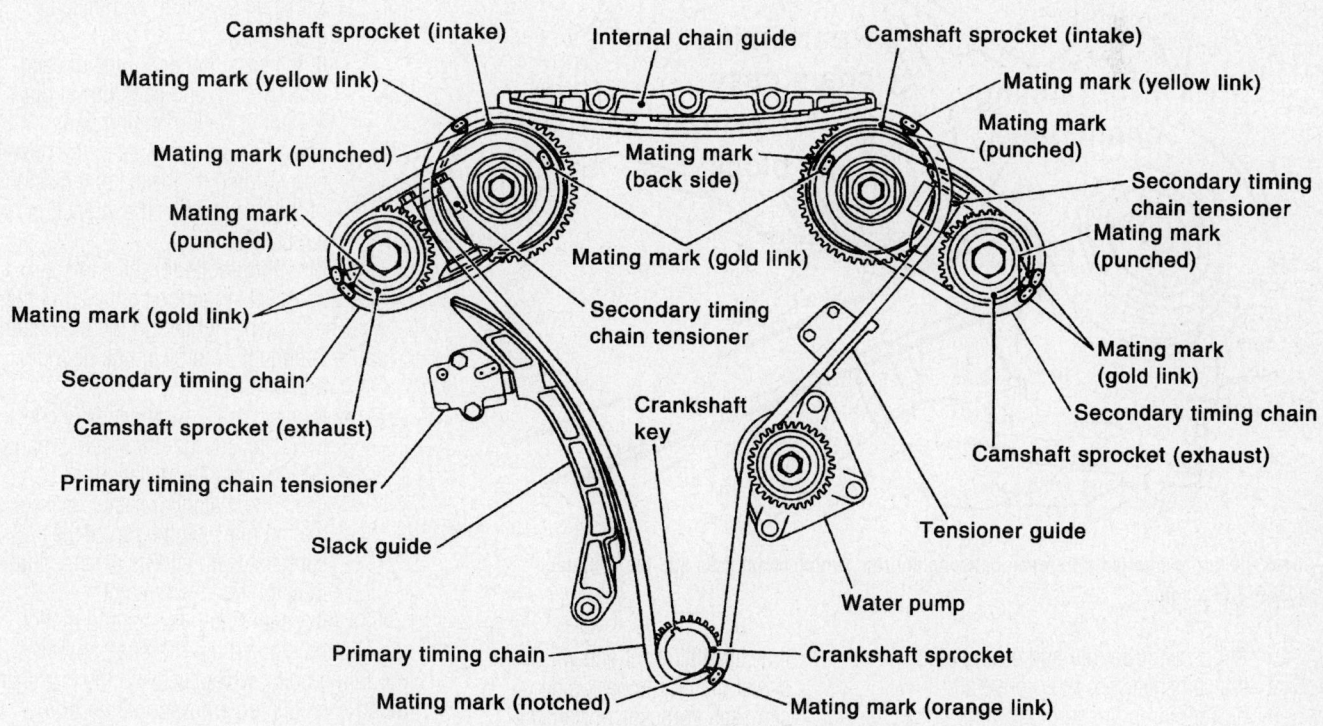

Camshaft sprocket (intake)
Mating mark (yellow link)
Mating mark (punched)
Mating mark (punched)
Mating mark (gold link)
Secondary timing chain
Camshaft sprocket (exhaust)
Primary timing chain tensioner
Slack guide
Primary timing chain
Mating mark (notched)

Internal chain guide
Mating mark (back side)
Mating mark (gold link)
Secondary timing chain tensioner
Crankshaft key
Tensioner guide
Water pump
Crankshaft sprocket
Mating mark (orange link)

Camshaft sprocket (intake)
Mating mark (yellow link)
Mating mark (punched)
Secondary timing chain tensioner
Mating mark (punched)
Mating mark (gold link)
Secondary timing chain
Camshaft sprocket (exhaust)

67162-FX35-G84

Alignment marks for timing chain installation—3.5L engine

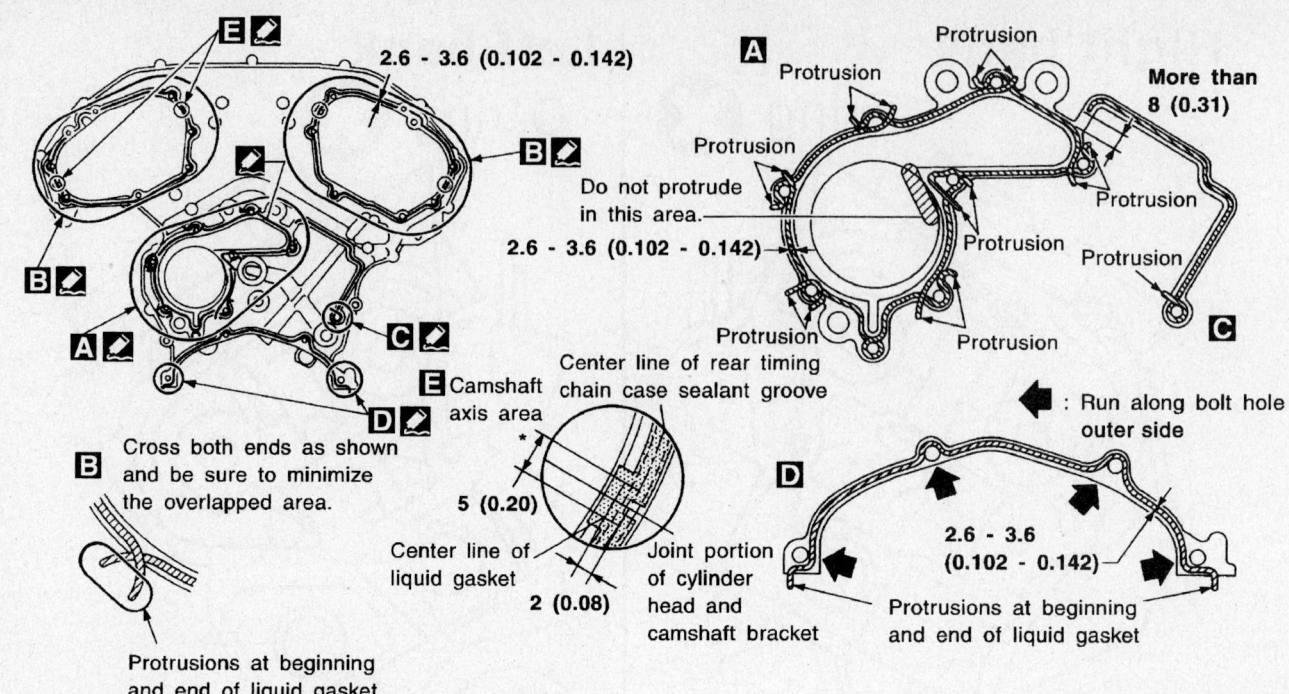

E 2.6 - 3.6 (0.102 - 0.142)

B

B

A

C

D

A Protrusion Protrusion

Protrusion

Protrusion

More than 8 (0.31)

Do not protrude in this area.

2.6 - 3.6 (0.102 - 0.142)

Protrusion

Protrusion

Protrusion

Protrusion

Protrusion

C

E Camshaft axis area

Center line of rear timing chain case sealant groove

5 (0.20)

Center line of liquid gasket

2 (0.08)

Joint portion of cylinder head and camshaft bracket

: Run along bolt hole outer side

D

2.6 - 3.6 (0.102 - 0.142)

Protrusions at beginning and end of liquid gasket

B Cross both ends as shown and be sure to minimize the overlapped area.

Protrusions at beginning and end of liquid gasket

* : Apply liquid gasket to the chamfered surface between camshaft bracket and cylinder head.

: Apply liquid gasket. (Use Genuine RTV silicone sealant or equivalent. Refer to GI section.)

Unit: mm (in)

67162-FX35-G85

Liquid gasket positions for timing chain case installation—3.5L engine

---

Front timing chain case

Rear timing chain case

Cylinder block

67162-FX35-G86

Check the surface height difference between the rear timing chain case and the cylinder block—3.5L engine

- Align the dowel pin and pin hole on the camshaft with the groove and dowel pin on the sprocket, and install them.
- On the intake side, align the pin hole on the small diameter side of

the camshaft front end with the dowel pin on the back side of camshaft sprocket, and install them.

- On the exhaust side, align the dowel pin on the camshaft front

end with the pin groove on the camshaft sprocket, and install them.

- In the case that positions of each mating mark and each dowel pin are not fit on the mating parts, make fine adjustments to the position holding the hexagonal portion of the camshaft with a wrench or equivalent.
- Mounting bolts for the camshaft sprockets must be tightened in the next step. Tightening them by hand is enough to prevent the dislocation of dowel pins.
- It may be difficult to visually check the dislocation of the mating marks during and after installation. To make the matching easier, make a mating mark on the top of the sprocket teeth and its extended line in beforehand with paint.

53. After confirming the mating marks are aligned, tighten the camshaft sprocket mounting bolts, while securing the camshaft using a wrench on a hexagonal portion of the camshaft.

54. Pull the stopper pins out from the secondary timing chain tensioners.

Dowel pin hole (Small dia. side)

Dowel pin   Dowel pin

Crankshaft key

67162-FX35-G87

Camshaft positioning for timing chain installation—3.5L engine

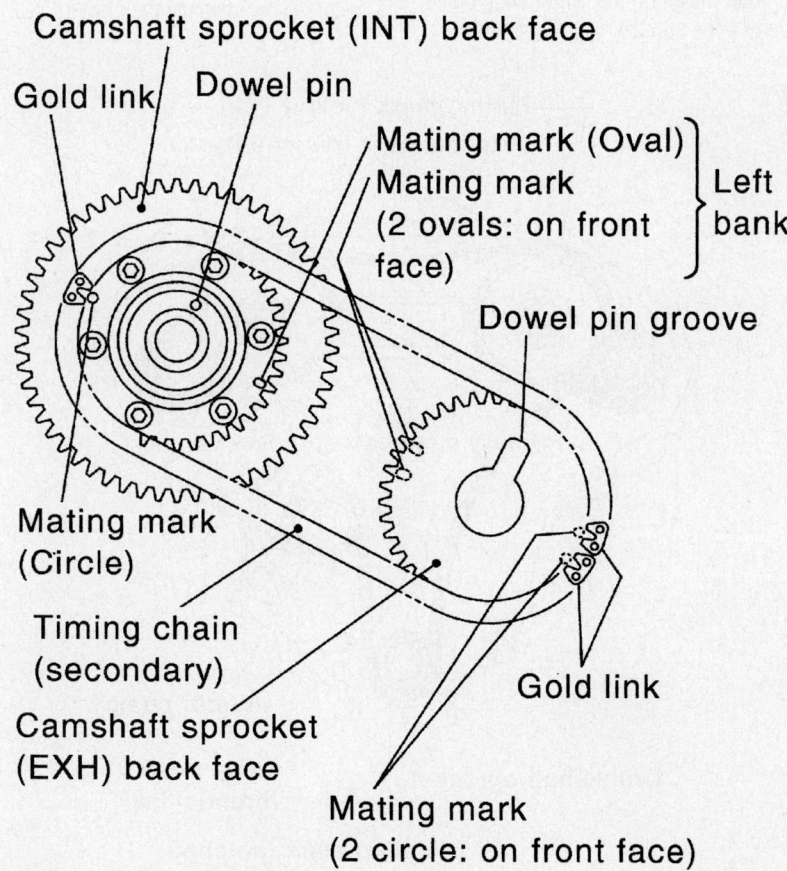

Camshaft sprocket (INT) back face

Gold link   Dowel pin

Mating mark (Oval)
Mating mark
(2 ovals: on front   } Left
face)   bank

Dowel pin groove

Mating mark
(Circle)

Timing chain
(secondary)

Camshaft sprocket
(EXH) back face

Gold link

Mating mark
(2 circle: on front face)

67162-FX35-G89

Secondary timing chain installation alignment—3.5L engine

55. Install the primary timing chain, as follows:

➡**During alignment, be careful to prevent dislocation of the mating mark alignments of the secondary timing chains.**

　a. Install the crankshaft sprocket, making sure the mating marks on the crankshaft sprocket face the front of the engine.

　b. Install the primary timing chain, so that the mating mark (punched) on the camshaft sprocket is aligned with the yellow link on the timing chain, while the mating mark (notched) on the crankshaft sprocket is aligned with the orange one on the timing chain, as shown.

➡**If it is difficult to align mating marks of the primary timing chain with each sprocket, gradually turn the camshaft using a wrench on the hexagonal portion of the camshaft to align it with the mating marks.**

56. Install the internal chain guide and primary timing chain tensioner.

57. Install the slack guide.

✳✳ CAUTION

Do not over tighten the slack guide mounting bolts. It is normal for a gap to exist under the bolt seats when the mounting bolts are tightened to specification.

58. Remove all dirt and foreign materials completely from the back and mounting surfaces of the chain tensioner.

59. Install chain tensioner for slack guide. When installing the chain tensioner, push in the sleeve and keep it depressed with a stopper pin.

60. After chain tensioner installation, pull out the stopper pin by pressing in the slack guide.

61. Reconfirm that the mating marks on sprockets and timing chains have not slipped out of alignment.

62. Install new O-rings on the rear timing chain case.

63. Install the front oil seal on the front timing chain case, as follows:

　a. Apply new engine oil to the oil seal edges.

　b. Make sure the garter spring is in position and the seal lip is not inverted, and so that each seal lip is oriented as shown in the figure.

　c. Using a suitable drift, press-fit the oil seal until it is flush with the front timing chain case end face.

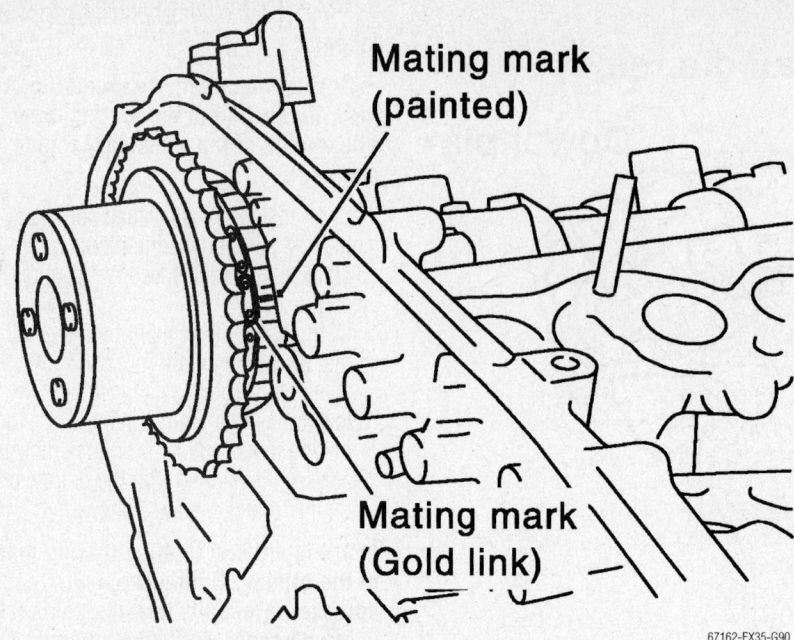

Mating mark (painted)

Mating mark (Gold link)

67162-FX35-G90

Intake camshaft sprocket and secondary timing chain alignment—3.5L engine

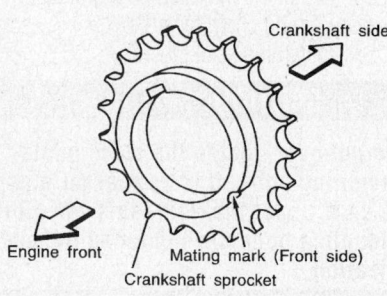

Crankshaft side

Engine front

Mating mark (Front side)
Crankshaft sprocket

67162-FX35-G92

Crankshaft sprocket installation position—3.5L engine

64. Install the water pump cover and chain tensioner cover to the front timing chain case. Apply liquid gasket to the front timing chain case front side as shown.

➡**Use Genuine RTV Silicone Sealant or equivalent.**

65. Install the front timing chain case as follows:

a. Apply liquid gasket to front timing chain case back side as shown.

➡**Use Genuine RTV Silicone Sealant or equivalent.**

b. Install the dowel pin on the rear timing chain case into the dowel pin hole on the front timing chain case.

c. Tighten the bolts to the specified torque in the order shown. Refer to the following for locating the bolts.

• 8mm bolts (positions 1 & 2): 21 ft. lbs. (28 Nm)

• 6mm bolts (except positions 1 & 2): 9 ft. lbs. (13 Nm)

d. After tightening, retighten them to the specified torque in the numerical order shown.

66. After installing the front timing chain case, check the surface height difference between the front timing chain case to rear timing chain case on the oil pan mounting surface. The allowable height difference is -0.005–0.0055 in. (-0.14–0.14mm). If not within specification, repeat the installation procedure.

67. Install the right and left intake valve timing control covers, as follows:

a. Install the seal rings in the shaft grooves.

b. Apply liquid gasket to the intake valve timing control covers.

➡**Use Genuine RTV Silicone Sealant or equivalent.**

c. Install the collared O-ring in the front cover engine oil hole (left and right sides).

d. Being careful not to move the seal ring from the installation groove, align the dowel pins on the chain case with the holes to install the intake valve timing control covers. Tighten the bolts in the order shown.

68. Install the crankshaft pulley, as follows:

a. Fix the crankshaft in position using ring gear stopper SST: KV10117700 (J44716) or equivalent.

b. Install the crankshaft pulley, taking care not to damage the front oil seal.

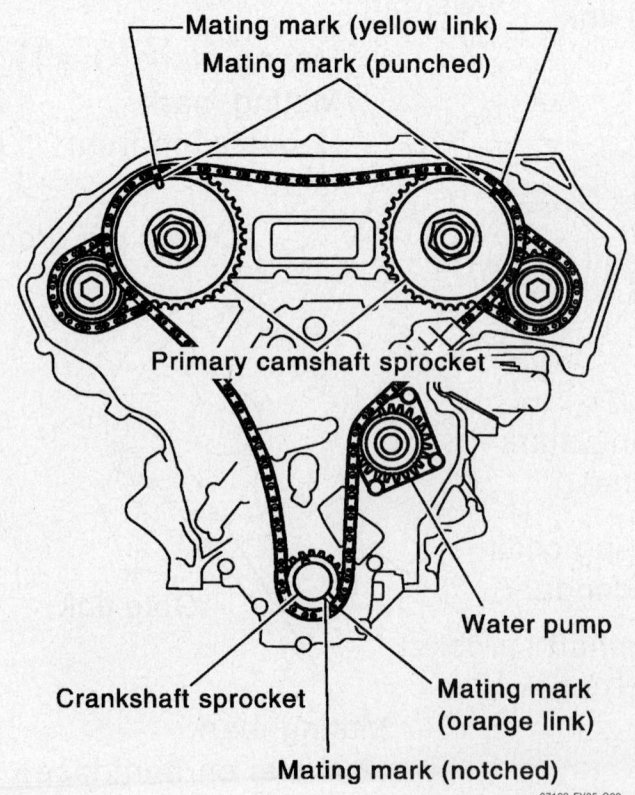

Mating mark (yellow link)
Mating mark (punched)

Primary camshaft sprocket

Water pump

Crankshaft sprocket

Mating mark (orange link)

Mating mark (notched)

67162-FX35-G93

Primary and secondary timing chain installation positioning—3.5L engine

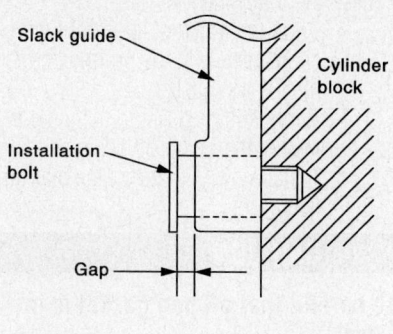

A gap between the slack guide bolt head and the slack guide is normal—3.5L engine

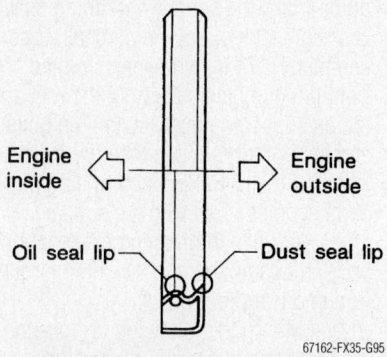

Crankshaft oil seal installation positioning—3.5L engine

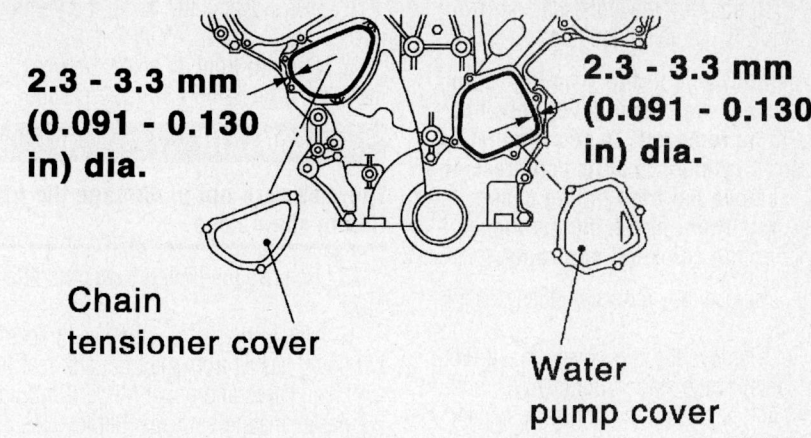

Chain tensioner cover liquid gasket positioning—3.5L engine

❊❊ CAUTION

When press-fitting the crankshaft pulley with a plastic hammer, tap on its center portion, not on the circumference.

c. Tighten the crankshaft bolt to 33 ft. lbs. (44 Nm).

d. Put a paint mark on the crankshaft pulley aligned with the angle mark on the crankshaft pulley bolt. Then, further tighten the bolt 60 degrees (equivalent to one graduation) on 2003–05 engines, or 90 degrees on 2006 engines.

69. Rotate the crankshaft pulley in the normal direction (clockwise when viewed from front) to confirm that it turns smoothly.

70. The remainder of installation is the reverse of removal.

➡**If the hydraulic pressure inside the chain tensioner drops after removal/ installation, the slack in the guide may generate a pounding noise during and just after engine start. However, this does not indicate a problem. The noise will stop after hydraulic pressure rises.**

71. Perform the following once installation is complete:

a. Before starting the engine, check the levels of the engine coolant, lubrications and working fluid. If less than required quantity, fill to the specified level.

b. Run the engine to check for unusual noise and vibration.

c. Warm up the engine thoroughly to make sure there is no leakage of engine coolant, engine oil and working fluid, fuel and exhaust gas.

d. Bleed air from passages in pipes and tubes of applicable lines, such as in the cooling system.

e. After cooling down the engine, again check the amounts of the engine coolant, engine oil and working fluid. Refill to the specified level, if necessary.

WITHOUT OIL PAN REMOVAL

1. Before servicing the vehicle, refer to the Precautions Section.

This section describes the removal/ installation procedure of the front timing chain case and timing chain related parts without removing the upper oil pan.

2. Position the vehicle onto a lift, or support it in a safe manner so that work can be performed on the underside of the vehicle.

3. Disconnect the negative battery terminal.

4. Remove the engine cover.

5. Remove the air cleaner case assembly.

6. Remove the front and rear engine undercover.

7. Drain the engine coolant from the radiator.

8. Drain the engine oil from oil pan.

9. Remove and label the engine wiring harnesses.

10. Remove the upper and lower intake manifold collectors.

11. Remove the power steering oil pump from the bracket with the piping connected, then temporarily secure it aside.

12. Remove the power steering oil pump bracket.

13. Remove the alternator.

14. Remove the water bypass hose, water hose clamp, idler pulley bracket, and accessory drive belt tensioner from the front timing chain case.

15. Remove the right and left intake valve timing control covers, by loosening the bolts in the reverse order as shown.

16. Use a seal cutter to cut the liquid gasket.

❊❊ CAUTION

A shaft is internally jointed with the intake camshaft sprocket center hole. When removing it, keep it horizontal until it is completely disconnected.

17. Remove the collared O-ring from the front timing chain case left and right sides.

18. Remove the right and left rocker covers.

➡**When the secondary timing chain is not removed/installed, the following step and associated sub-steps are not required.**

19. Position the engine with cylinder No. 1 on TDC of its compression stroke, as follows:

a. Rotate the crankshaft pulley clockwise to align the timing mark (grooved line without color) with the timing indicator.

b. Make sure the intake and exhaust camshaft lobes noses on the No. 1 cylinder (engine front side of right bank) are located as shown.

c. If not, turn the crankshaft one revolution (360 degrees) and align as shown.

→**When only the primary timing chain is removed, the rocker cover does not need to be removed. To confirm that the No. 1 cylinder is at its compression TDC, remove the front timing chain case first. Then, check the mating marks on the camshaft sprockets.**

20. Remove the crankshaft pulley, as follows:

a. Remove the rear cover plate (2WD models) or the starter motor (AWD models) and set install the ring gear stopper or equivalent as shown.

b. Loosen the crankshaft pulley bolt until the bolt seating surface is 0.39 in. (10mm) from its original position.

✳✳ CAUTION

Do not remove the crankshaft pulley bolt, since it will be used as a supporting point for the pulley puller.

c. Place a suitable puller tab in the holes of the crankshaft pulley, and remove the crankshaft pulley.

✳✳ CAUTION

Do not situate the puller tab on crankshaft pulley periphery, as this will damage the internal damper.

21. Remove the lower oil pan.
22. Loosen the two mounting bolts in the front of the upper oil pan in the reverse order shown.
23. Remove the front timing chain case, as follows:

a. Loosen the mounting bolts in the reverse order as shown.

b. Insert a suitable tool into the notch at the top of the front timing chain case, as shown.

c. Use a seal cutter to cut the liquid gasket for removal.

✳✳ CAUTION

Use only an approved pry tool to remove the case.

d. Pry off the case.

e. After removal, handle the case carefully so it does not tilt, cant, or warp under a load.

24. Remove the O-rings from the rear timing chain case.
25. Remove the water pump cover and the chain tensioner cover from the front timing chain case.

26. Use a seal cutter to cut the liquid gasket.
27. Remove front oil seal from front timing chain case using a suitable pry tool.

✳✳ CAUTION

Exercise care not to damage the front timing chain case.

28. Remove the timing chain and related parts.
29. Use a scraper to remove all traces of old liquid gasket from the front and rear timing chain cases and upper oil pan, and liquid gasket mating surfaces. Remove the old liquid gasket from bolt holes and threads.

✳✳ CAUTION

Be careful not to allow gasket fragments to enter the oil pan.

30. Use a scraper to remove all traces of liquid gasket from the water pump cover, chain tensioner cover and intake valve timing control covers.

To install:

→**Throughout the installation procedure, whenever liquid gasket is to be used make sure to use Genuine RTV Silicone Sealant or equivalent.**

31. Install the timing chain and related parts.
32. Hammer the dowel pins (left and left) into the front timing chain case up to a point close to the taper in order to shorten the protrusion length.
33. Install the front oil seal in the front timing chain case. During seal installation, heed the following:

- Apply new engine oil to the oil seal edges.
- Install the seal so that each seal lip is oriented as shown.
- Using a suitable drift, press-fit the oil seal until it becomes flush with front timing chain case end face.
- Make sure the retaining coil spring is in position and seal lip is not inverted.

34. Apply liquid gasket to the front timing chain case front side as shown.
35. Install the water pump cover and chain tensioner cover to the front timing chain case.
36. Install the front timing chain case, as follows:

a. Apply liquid gasket to the front timing chain case back side as shown.

b. Apply liquid gasket to the oil pan gasket as shown.

c. Install the new oil pan gasket, and heed the following:

- Align the notch of the front timing chain case with the protrusion of the oil pan gasket.
- Apply liquid gasket to the top surface of the upper oil pan as shown.

d. Install new O-rings on the rear timing chain case.

✳✳ CAUTION

Be careful that oil pan gasket is in place.

e. Assemble the front timing chain case by fitting the lower end of the front timing chain case tightly onto the top face of the upper oil pan. From the fitting point, make the entire front timing chain case contact the rear timing chain case completely. Then, while pressing the front timing chain case from its front and top as shown in figure, install the bolts and temporarily tighten them by hand. Hammer the dowel pin until the outer end becomes flush with the surface.

f. After hand tightening the mounting bolts, retighten them to specified torque in numerical order shown.

g. Refer to the following for locating the bolt positions and torque values:

- 8mm bolts: Positions 1 & 2: 21 ft. lbs. (28 Nm)
- 6mm bolts: Exc. positions 1 & 2: 9 ft. lbs. (13 Nm)

37. Install two mounting bolts in the front of the upper oil pan in numerical order shown to 13 ft. lbs. (17 Nm).
38. Install the lower oil pan.
39. Install the right and left intake valve timing control covers, as follows:

a. Install seal rings in the shaft grooves.

b. Apply liquid gasket to the intake valve timing control covers.

c. Install the collared O-ring in the front timing chain case oil hole (left and right sides).

d. Being careful not to move the seal ring from the installation groove, align the dowel pins on the chain case with the holes in the intake valve timing control covers.

e. Tighten the bolts in the numerical order shown to 8 ft. lbs. (11 Nm).

40. Install the crankshaft pulley, as follows:

a. Fix the crankshaft in position using ring gear stopper SST: KV10117700 (J44716) or equivalent.

b. Install the crankshaft pulley, taking

care not to damage front oil seal. When press-fitting crankshaft pulley with a plastic hammer, tap on its center portion (not circumference).

c. Tighten the crankshaft pulley bolt to 33 ft. lbs. (44 Nm).

d. Put a paint mark on the crankshaft pulley aligned with the angle mark on the crankshaft pulley bolt. Then, further tighten the bolt by 60 degrees (equivalent to one graduation) for 2003–05 models, or 90 degrees for 2006 models.

41. Rotate the crankshaft pulley in the normal direction (clockwise when viewed from the front of the engine) to confirm it turns smoothly.

42. The remainder of installation is the reverse of removal.

➡ **If hydraulic pressure inside the chain tensioner drops after removal/installation, slack in the guide may generate a pounding noise during and just after engine start. However, this is normal and the noise will stop after hydraulic pressure rises.**

43. Perform the following once installation is complete:

- Before starting the engine, check the levels of the engine coolant, lubrications and working fluids. If less than required quantity, fill to the specified level.
- Run the engine to check for unusual noise and vibration.
- Warm up the engine thoroughly to make sure there is no leakage of engine coolant, engine oil, working fluid, fuel and exhaust gas.
- Bleed air from passages in pipes and tubes of applicable lines, such as in the cooling system.
- After cooling down the engine, again check the amounts of engine coolant, engine oil and working fluid. Refill to specified level, if necessary.

4.5L Engine

1. Before servicing the vehicle, refer to the Precautions Section.

2. Remove the engine assembly from the vehicle.

3. Remove the drive belt auto tensioner and idler pulley, as follows:

a. Remove the front engine undercover.

b. Remove the drive belts.

4. Remove the thermostat housing and hoses.

5. Remove the ignition coil.

6. Remove the rocker cover, as follows:

a. Remove the engine cover.

b. Remove the inlet air duct, air cleaner case and mass air flow sensor assembly, air duct and resonator assembly.

c. Move the wiring harness on the upper rocker cover and its peripheral aside.

d. Remove the wiring harness brackets from the No. 6 camshaft bracket.

e. Remove the electric throttle control actuator.

f. Remove the ignition coil.

g. Remove the PCV hose from the PCV valve.

h. Move the wiring harness on the upper rocker cover.

i. Remove the ignition coil.

j. Remove the PCV hose from the PCV valve.

k. Remove the grommets from the right and left cowl top panel. For the right side grommet, remove the battery, battery tray, then the grommet.

l. Loosen the rocker cover bolts in reverse order as shown.

✳✳ CAUTION

Do not hold the oil filler neck on the right bank so that it is not damaged.

➡ **Loosen No. 10 bolt of the right bank and No. 10 and No. 12 bolts at the left bank from cowl top panel hole.**

m. Use a scraper to remove all traces of the liquid gasket from the cylinder head and camshaft bracket.

7. If necessary, remove the intake valve timing control position sensor (right and left banks) and the camshaft position sensor (PHASE) from the intake valve timing control cover and front cover.

✳✳ CAUTION

Handle the sensors carefully to avoid dropping and/or shocking them. Also, do not disassemble them. Do not allow metal powder to adhere to the magnetic part at the sensor tip. Do not place sensors in a location where they are exposed to magnetism.

8. If necessary, remove the intake valve timing control solenoid valve from the intake valve timing control cover.

9. Remove the intake valve timing control cover, as follows:

a. Loosen and remove the mounting bolts in the reverse order shown.

b. Use a seal cutter to cut the liquid gasket for removal.

✳✳ CAUTION

Exercise care not to damage the mating surfaces. Pull out the cover keeping it level, since an inner part of the cover is engaged with the center of the camshaft sprocket (INT).

10. Remove the O-rings from the front cover.

11. Position the crankshaft with No. 1 cylinder at TDC of its compression stroke, as follows:

a. Rotate the crankshaft pulley clockwise to align the TDC identification notch (without paint mark) with the timing indicator on the front cover.

b. Make sure that both intake and exhaust camshaft lobes of the No. 1 cylinder (engine front side of left bank) are located as shown in the figure.

c. If the camshaft lobes are not positioned appropriately, turn the crankshaft pulley one revolution (360 degrees) and align it as shown.

12. Remove the crankshaft pulley, as follows:

a. Remove rear plate cover, and install the ring gear stopper or equivalent.

b. Loosen the crankshaft pulley bolt, and then pull the crankshaft pulley with both hands to remove it.

✳✳ CAUTION

Do not completely remove the crankshaft pulley bolt. Keep the loosened crankshaft pulley bolt in place to protect the removed crankshaft pulley from dropping. Do not remove the balance weight (inner hexagonal bolt) at the front of the crankshaft pulley.

13. Remove the oil pan and oil strainer.

14. Remove the front cover, as follows:

a. Loosen the mounting bolts in the reverse order shown.

b. Use a seal cutter to cut the liquid gasket.

✳✳ CAUTION

Exercise care not to damage mating surfaces. After removal, handle the front cover carefully so it does not tilt, cant, or warp under load.

15. Remove the front oil seal from the front cover using a suitable thin-bladed pry tool.

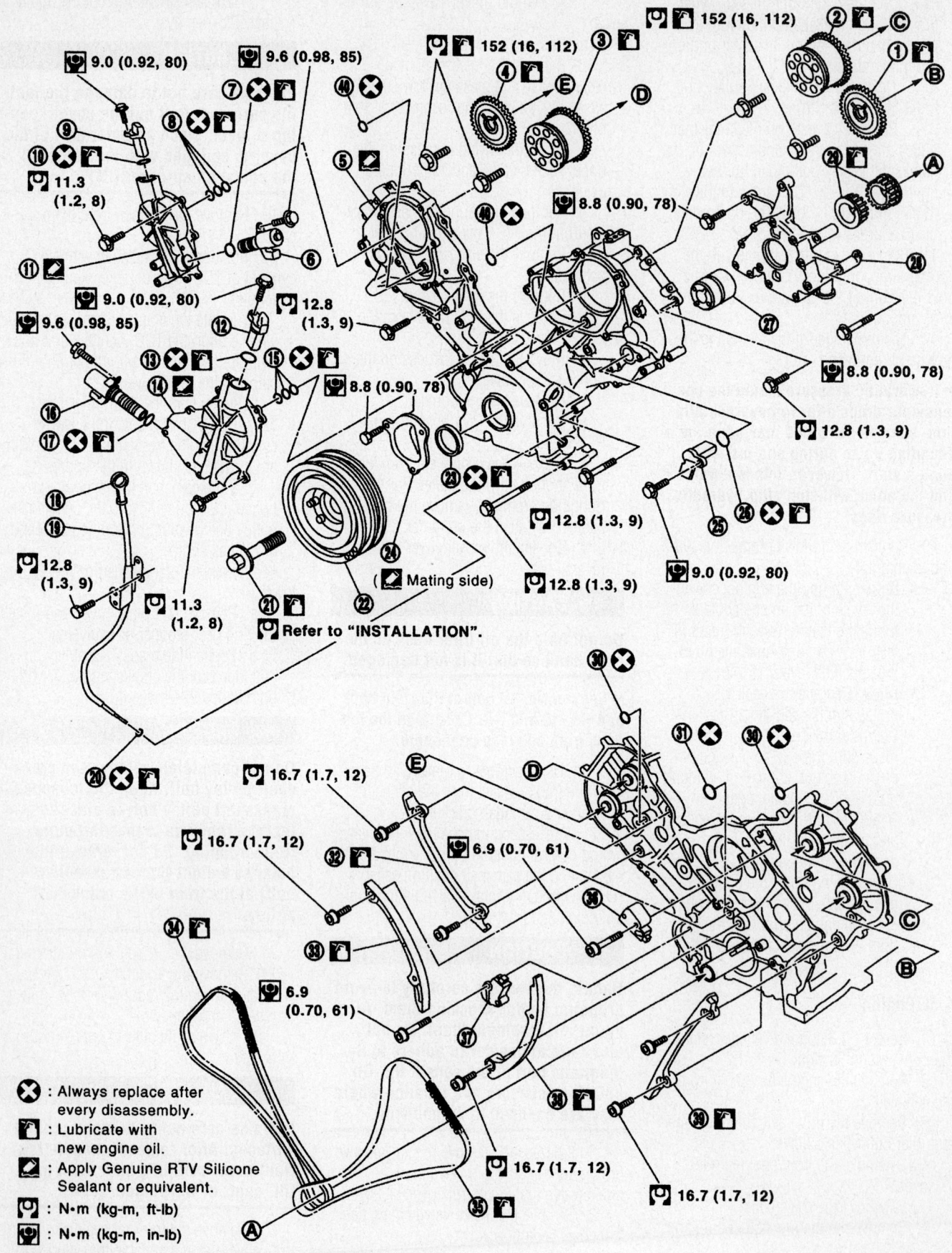

Exploded view of the timing chain and related components—4.5L engine

67162-FX35-G162A

1. Camshaft sprocket (EXH)
2. Camshaft sprocket (INT)
3. Camshaft sprocket (INT)
4. Camshaft sprocket (EXH)
5. Front cover
6. Intake valve timing control solenoid valve (right bank)
7. O-ring
8. Seal ring
9. Intake valve timing control position sensor (right bank)
10. O-ring
11. Intake valve timing control cover (right bank)
12. Intake valve timing control position sensor (left bank)
13. O-ring
14. Intake valve timing control cover (left bank)
15. Seal ring
16. Intake valve timing control solenoid valve (left bank)
17. O-ring
18. Oil level gauge
19. Oil level gauge guide
20. O-ring
21. Crankshaft pulley bolt
22. Crankshaft pulley
23. Front oil seal
24. Chain tensioner cover
25. Camshaft position sensor (PHASE)
26. O-ring
27. Oil pump drive spacer
28. Oil pump assembly
29. Crankshaft sprocket
30. O-ring
31. O-ring
32. Timing chain tension guide (right bank)
33. Timing chain slack guide (right bank)
34. Timing chain (right bank)
35. Timing chain (left bank)
36. Chain tensioner (left bank)
37. Chain tensioner (right bank)
38. Timing chain slack guide (left bank)
39. Timing chain tension guide (right bank)
40. O-ring

67162-FX35-G162B

❈❈ CAUTION

Be careful not to damage the front cover.

16. Remove the O-rings from both cylinder heads and cylinder block.

17. Remove the chain tensioner cover from the front cover. Use a seal cutter to cut the liquid gasket for removal.

18. Remove the oil pump drive spacer, by setting bolts in the two bolt holes (M6

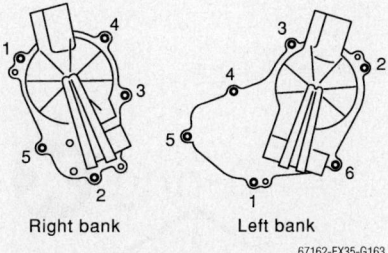

Right bank   Left bank

67162-FX35-G163

Intake valve timing control cover tightening sequence—4.5L engine

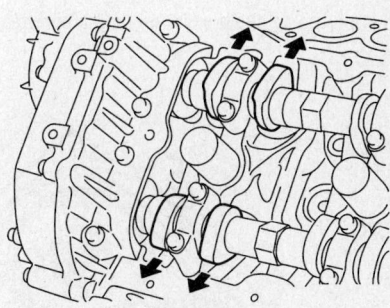

67162-FX35-G166

With the engine positioned at TDC on No. 1 cylinder, the camshaft lobes should be positioned as shown—4.5L engine

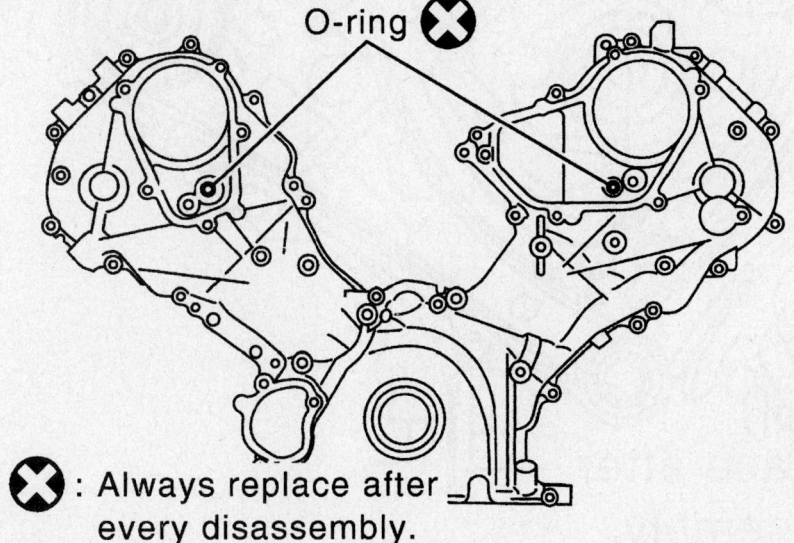

O-ring ⊗

⊗ : Always replace after every disassembly.

67162-FX35-G164

Timing chain front cover O-ring locations—4.5L engine

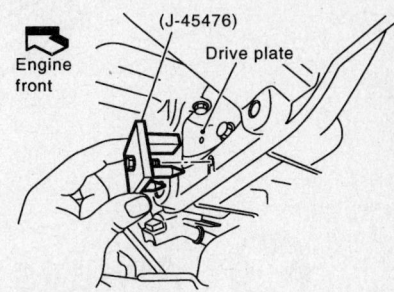

67162-FX35-G167

Use a ring gear holder to fix the crankshaft to loosen the crankshaft pulley bolt—4.5L engine

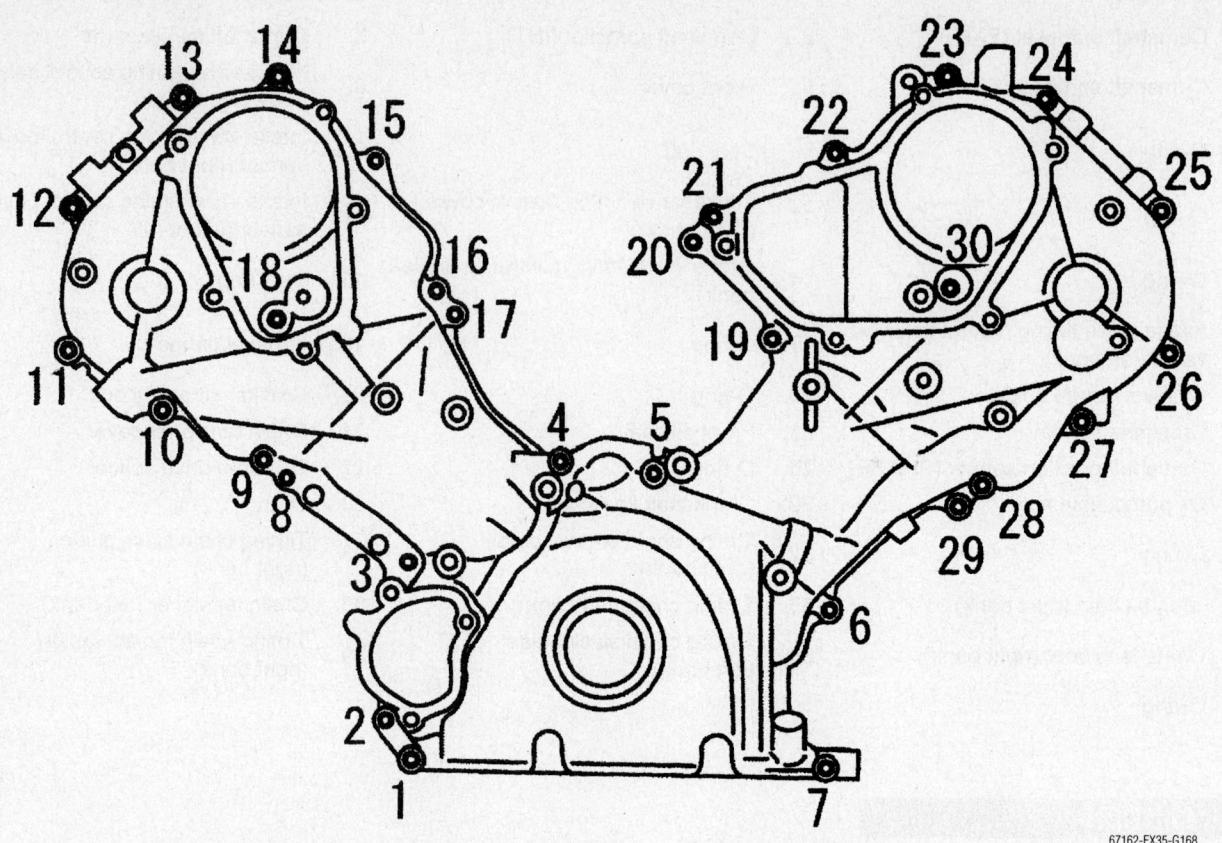

Front timing chain cover mounting bolt tightening sequence—4.5L engine

67162-FX35-G168

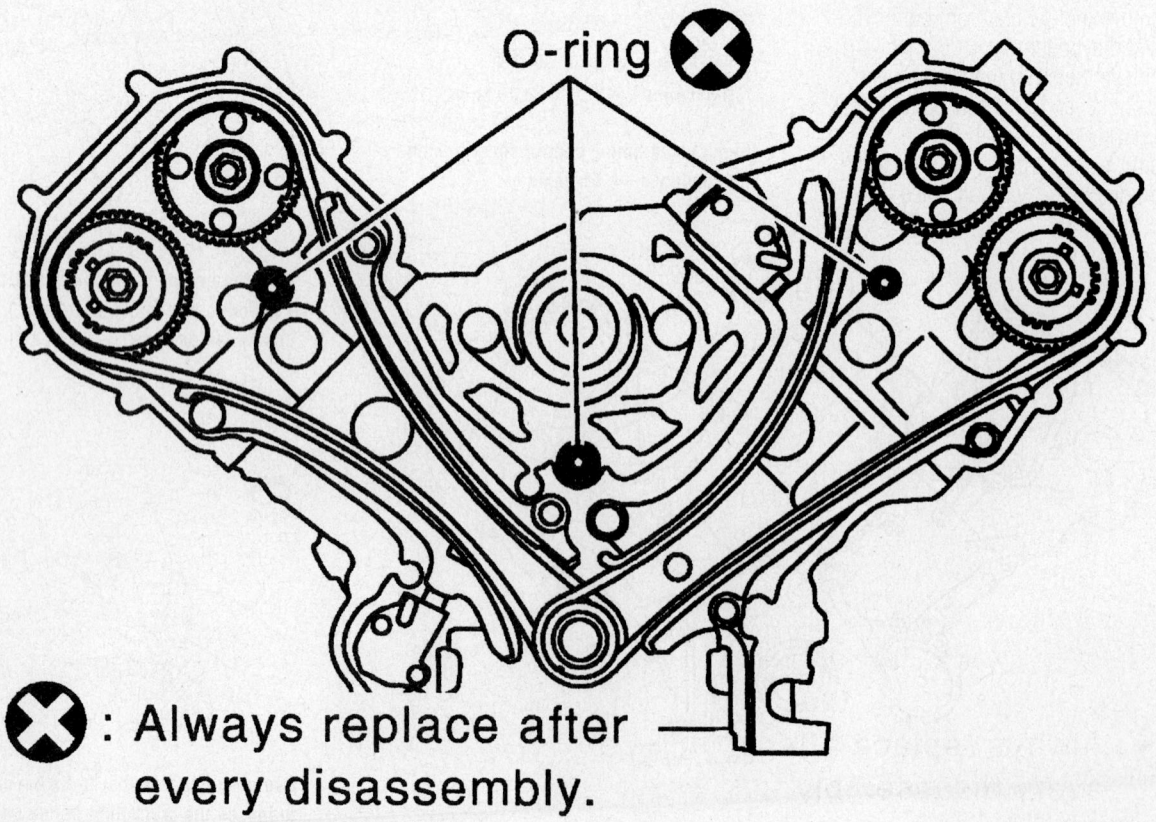

⊗ : Always replace after every disassembly.

O-ring positions on engine block and cylinder heads—4.5L engine

67162-FX35-G169

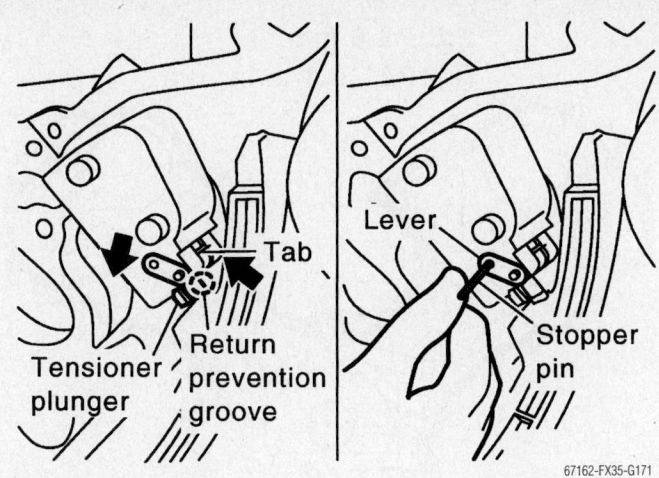

Timing chain tensioner details—4.5L engine

pitch) on the front surface. Using a suitable puller, pull on the oil pump drive spacer off the crankshaft.

➡ **The dimension between the centers of the two bolt holes is 1.30 in. (33mm).**

19. Remove the oil pump.
20. Remove the chain tensioner from the left bank, as follows:

➡ **To remove timing chain and related parts, start with those on left bank. The procedure for removing parts on right bank is omitted because it is the same as that for the left bank.**

a. While lightly pressing the tensioner plunger, depress the tensioner tab in (or turn the lever in the direction of the arrow in the accompanying illustration) to unlock the mechanism that stops the tensioner plunger.

b. Push in the tensioner plunger to align the hole on the lever and that on the pump main body. If you push in the tensioner too far, the holes will not align. Therefore, push in the plunger to the degree at which the start of stopper groove and tab engages.

c. Insert a stopper pin (hard wire approximately 0.020 in./0.5mm thick or

similar tool) to hold the plunger in position. With the plunger held, remove the chain tensioner.

21. Remove the chain tension guide and timing chain slack guide.
22. Remove the timing chain and crankshaft sprocket.

❋❋ CAUTION

After removing the timing chain, do not turn the crankshaft and camshaft separately, or the valves will strike the piston head.

23. With the hexagonal part of the camshaft held with a wrench, loosen the bolts securing the camshaft sprocket.
24. Perform the same procedure on the right side.
25. Use a scraper to remove all traces of old liquid gasket from the front cover and opposite mating surfaces. Remove the oil liquid gasket from the bolt holes and threads.
26. Use scraper to remove all trace of liquid gasket from the chain tensioner cover and the intake valve timing control covers.
27. Check the timing chain for cracks and any excessive wear at the roller links and link plates. Replace the timing chain as necessary.

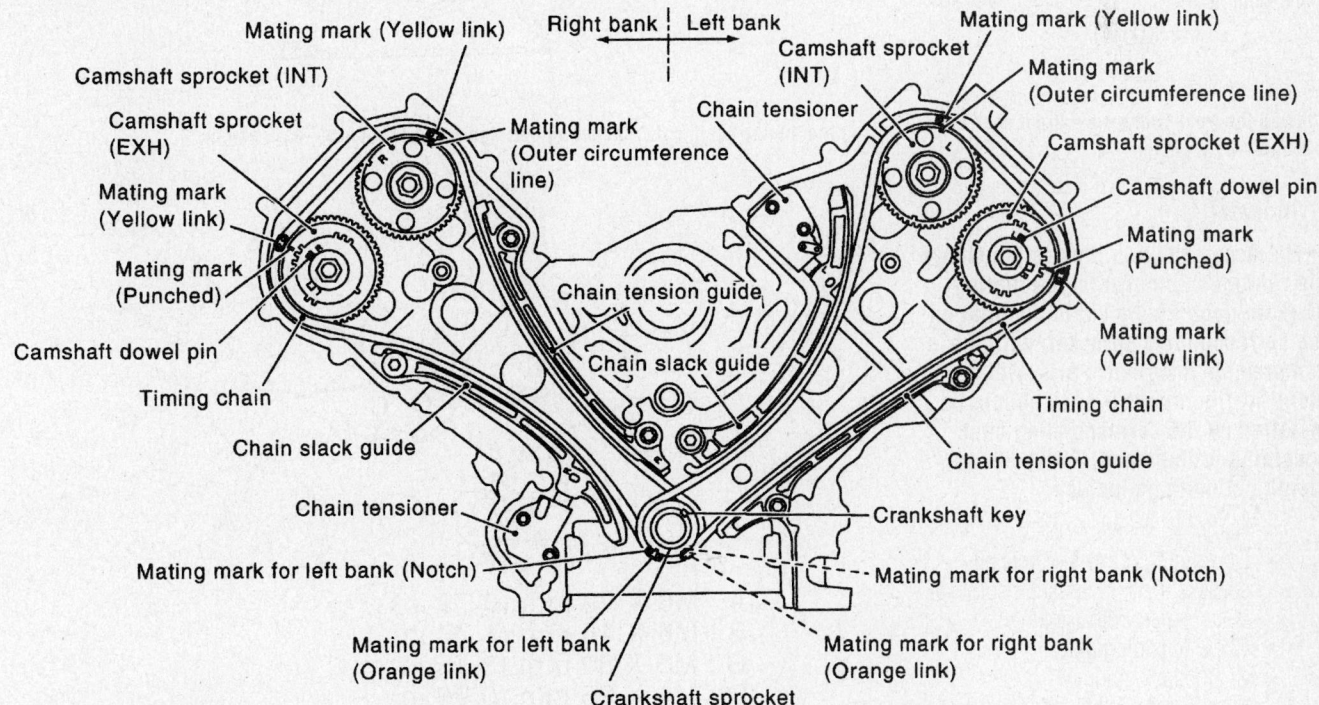

Timing chain positioning for installation—4.5L engine

67162-FX35-G172

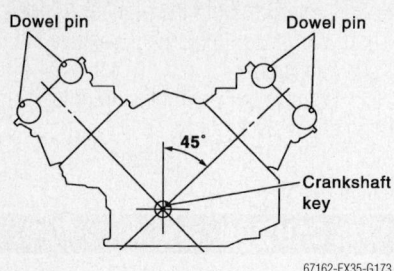

Crankshaft positioning for timing chain installation—4.5L engine

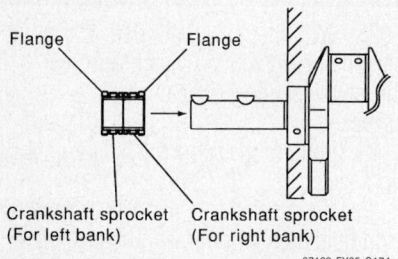

Crankshaft sprocket orientation for installation on the crankshaft—4.5L engine

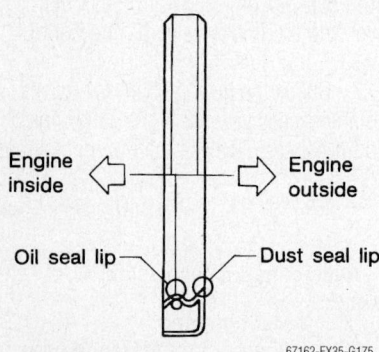

Install the front timing cover front seal as shown—4.5L engine

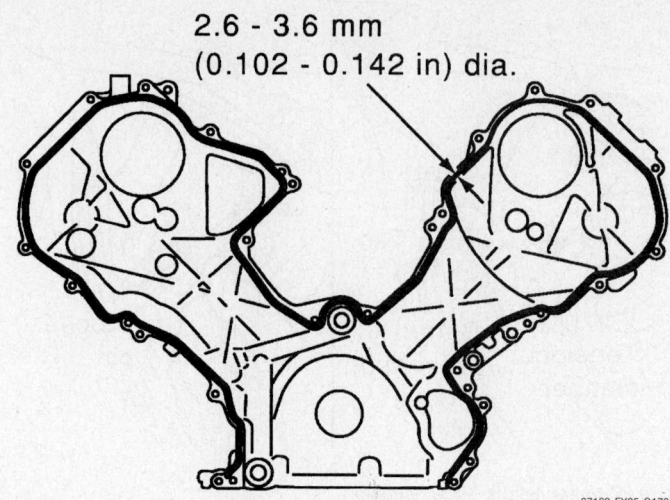

Liquid gasket positioning for timing chain cover installation—4.5L engine

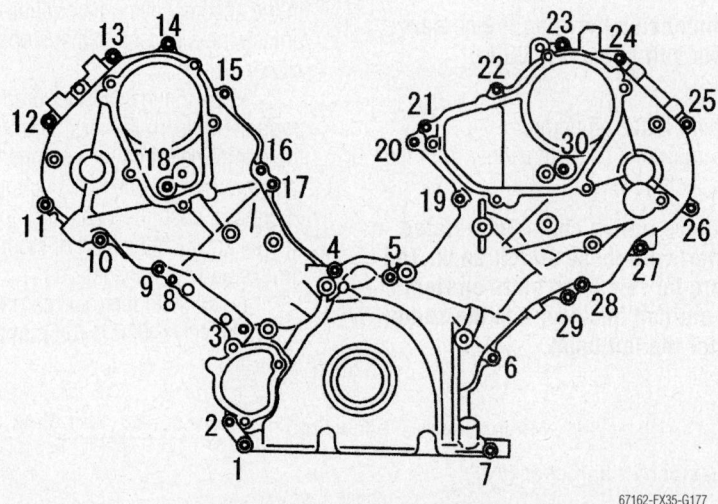

Front timing chain cover mounting bolt tightening sequence—4.5L engine

To install:

➡The accompanying illustration shows the relationship between the mating mark on each timing chain and that on the corresponding sprocket, with the components installed. Parts with an identification mark (R or L) should be installed on the corresponding bank according to the mark. Parts with an identification mark include:

- Camshaft sprocket (INT)
- Dowel pin groove of camshaft sprocket (EXH) (camshaft sprocket is same part both banks)
- Chain tension guide
- Chain slack guide
- Because of parallel manufacture, there are two types of mark (link colors) for timing chains.

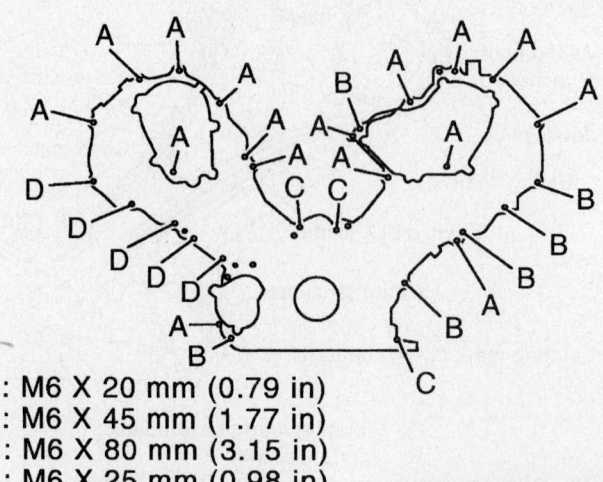

A : M6 X 20 mm (0.79 in)
B : M6 X 45 mm (1.77 in)
C : M6 X 80 mm (3.15 in)
D : M6 X 25 mm (0.98 in)

Front timing chain cover mounting bolt identification—4.5L engine

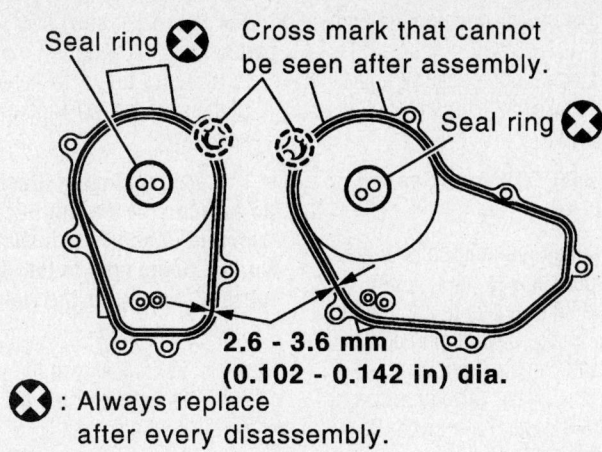

X: Always replace after every disassembly.

67162-FX35-G179

Intake valve timing control cover liquid gasket positioning for installation—4.5L engine

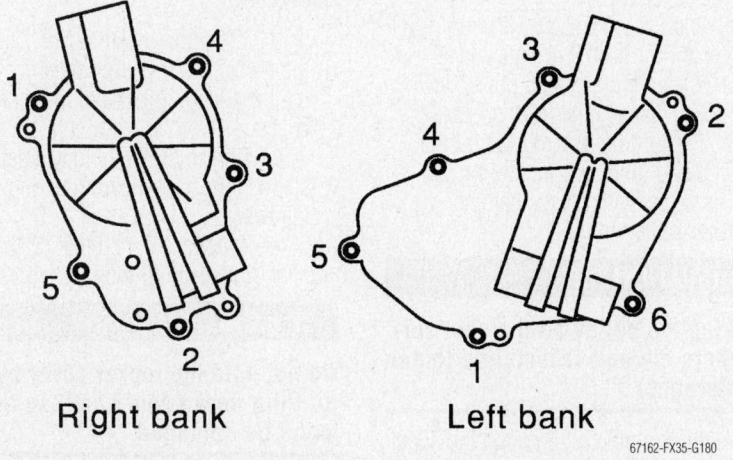

Right bank   Left bank

67162-FX35-G180

Intake valve timing control cover mounting bolt tightening sequence—4.5L engine.

➡ **To install the timing chain and related parts, start with those on right bank. The procedure for installing parts on left bank is omitted because it is the same as that for the right bank.**

28. Make sure that the crankshaft key and dowel pin of each camshaft are located as shown in the figure with the crankshaft at No. 1 cylinder at compression TDC. The positioning is as follows:
- Camshaft dowel pin—At the cylinder head upper face side in each bank
- Crankshaft key—At the cylinder head side of left bank

➡ **Though the camshaft may not stop at the position as shown in the figure, for the placement of camshaft lobe nose, it is generally accepted that the camshaft is placed for the same direction of the figure.**

29. Install the camshaft sprockets, while heeding the following items:

- Install the sprockets onto the correct side by checking with the identification mark on surface.
- Install the camshaft sprocket (EXH) by selectively using the groove of the dowel pin according to the bank. (Common part used for both banks.)
- Lock the hexagonal part of the camshaft in the same procedure as removal, and tighten the mounting bolts.

30. Install the crankshaft sprockets for both banks. Install each crankshaft sprocket so that its flange side (the larger diameter side without teeth) faces in the direction shown.

➡ **The same parts are used but facing directions are different.**

31. Install the timing chains and related parts, as follows:
    a. Align the mating mark on each sprocket and timing chain for installation.

b. After the mating marks are aligned, keep them aligned by holding them in position by hand.

➡ **Before installing the chain tensioner, it is possible to change the position of the mating mark on the timing chain for that on each sprocket for alignment.**

c. Install the slack guides and tension guides onto the correct side by checking with the identification mark on the surface.

d. Install the chain tensioner with the plunger fixed as described in its removal.

✳ CAUTION

Before and after the installation of the chain tensioner, make sure that the mating mark on the timing chain is not out of alignment.

e. After installing the chain tensioner, remove the stopper pin to release the tensioner. Make sure the tensioner is released.

f. To avoid chain-link skipping of timing chain teeth, do not move the crankshaft or camshafts until the front cover is installed.

32. Perform the same procedure on the right bank, installing the timing chain and related parts on the left side.

33. Install the oil pump.

34. Install the oil pump drive spacer, as follows:
    a. Insert the oil pump drive spacer according to the directions of the crankshaft key and the two flat surfaces of the oil pump inner rotor.

➡ **If the positional relationship does not allow the insertion, rotate the oil pump inner rotor by finger to allow spacer.**

b. After confirming that the position of each part is in correct condition to allow for spacer, force fit the spacer by lightly tapping it with a plastic hammer until it contacts and does not go further.

35. Install the front oil seal in the front cover, as follows:
    a. Apply new engine oil to both the oil seal lip and the dust seal lip.
    b. Position the seal so that each seal is oriented as shown, then, using a suitable drift, press the front oil seal into the front cover until the face of the oil seal is flush with the mounting surface. The drift used to seat the oil seat should have an outside diameter of 2.20 in. (56mm) and inner diameter of 1.93 in. (49mm).

✳✳ CAUTION

Be careful not to scratch or make burrs on the circumference of the oil seal.

c. Make sure the garter spring is in position and the seal lips not inverted.

36. Install the chain tensioner cover on the front cover.

37. Apply a continuous bead of liquid gasket to the front cover as shown.

➡**Use Genuine RTV Silicone Sealant or equivalent.**

38. Install the front cover, as follows:

a. Install new O-rings onto the right and left cylinder heads and cylinder block.

39. Apply a continuous bead of liquid gasket to the front cover as shown.

➡**Use Genuine RTV Silicone Sealant or equivalent.**

a. Make sure again that the mating marks on the timing chain and that on each sprocket are aligned. Then, install the front cover.

↻ CAUTION

Be careful to avoid interference with the front end of the oil pump drive spacer. Such interference may damage the front oil seal.

b. Tighten the mounting bolts in numerical order shown. There are four type mounting bolts. The bolts are as follows:

- Position A: M6 x 20mm
- Position B: M6 x 45mm
- Position C: M6 x 80mm
- Position D: M6 x 25mm

c. After all bolts are tightened, retighten them in numerical order shown.

↻ CAUTION

Be sure to wipe off any excessive liquid gasket leaking onto the surface mating with the oil pan.

40. Install the intake valve timing control cover, as follows:

a. At the back of the intake valve timing control cover, install new seal rings (three for each bank) to the area to be inserted into the camshaft sprocket (INT).

✳✳ CAUTION

Do not spread the seal ring excessively to avoid breaks and deformation.

b. Install the new O-rings on the front cover.

c. Apply a continuous bead of liquid gasket to the intake valve timing control covers as shown.

➡**Use Genuine RTV Silicone Sealant or equivalent.**

d. Tighten the mounting bolts in the numerical order shown.

41. Install the intake valve timing control position sensor, intake valve timing control solenoid valve and camshaft position sensor (PHASE) on the intake valve timing control cover and front cover, if removed. Be sure to tighten the bolts with flanges completely seated.

42. Install the oil pan and oil strainer.

43. Install the crankshaft pulley, as follows:

a. Hold the crankshaft with ring gear stopper SST: J-45476 or equivalent.

b. Install the crankshaft pulley, taking care not to damage the front oil seal. Install it according to the dowel pin of the oil pump drive spacer. Lightly tapping its center with plastic hammer, insert the pulley.

✳✳ CAUTION

Do not tap the pulley on the side surface where the belt is installed (outer circumference).

c. Apply engine oil onto the threaded parts of crankshaft pulley bolt and seating area.

d. Tighten the crankshaft pulley bolt to 69 ft. lbs. (93 Nm).

e. Mark the crankshaft pulley to align it with the angle mark on the crankshaft pulley bolt.

f. Further tighten the pulley bolt 90 degrees. (Angle tightening) Check the tightening angle by referencing it to the notches. The angle between two notches is 90 degrees.

44. Rotate the crankshaft pulley in the normal direction of rotation (clockwise when viewed from the front of the engine) to confirm it turns smoothly.

45. The remainder of installation is the reverse of removal.

➡**If hydraulic pressure inside the chain tensioner drops after removal/installation, slack in the guide may generate a pounding noise during and just after engine start. However, this is normal and the noise will stop after hydraulic pressure rises.**

46. For rocker arm cover installation, perform the following:

a. Apply liquid gasket to the joint part of the cylinder head and camshaft bracket as shown.

➡**The accompanying illustration shows an example of the left bank side (zoomed in shows camshaft bracket No. 1). Apply only to the camshaft bracket (No. 1) for the right bank side.**

b. Refer to "a" in the illustration to apply liquid gasket to the joint part of the camshaft bracket (both No. 1 and No. 6) and the cylinder head.

c. Refer to "b" in the illustration to apply liquid gasket in 90 degrees to "a".

➡**Use Genuine RTV Silicone Sealant or equivalent.**

d. Install the rocker cover. Check if the rocker cover gasket does not fall from the installation groove of the rocker cover.

e. Tighten the bolts in two separate steps in the numerical order shown, as follows:

- 1st Step: 18 inch lbs. (2 Nm)
- 2nd step: 73 inch lbs. (8.3 Nm)

✳✳ CAUTION

Do not hold the rocker cover by the oil filler neck (right bank) so that it won't be damaged.

➡**Tighten No. 10 bolt of the right bank and No. 10 and No. 12 bolts of the left bank from cowl top panel hole.**

47. Perform the following once installation is complete:

- Before starting the engine, check the levels of engine coolant, lubrications and working fluid. If less than at the required level, fill them to the specified level.
- Run the engine to check for unusual noise and vibration.
- Warm up the engine thoroughly to make sure there is no leakage of engine coolant, engine oil, working fluid, fuel and exhaust gas.
- Bleed air from passages in pipes and tubes of applicable lines, such as in the cooling system.
- After cooling down the engine, again check the amounts of engine coolant, engine oil and working fluid. Refill them to the specified level, if necessary.

Piston and Ring

POSITIONING

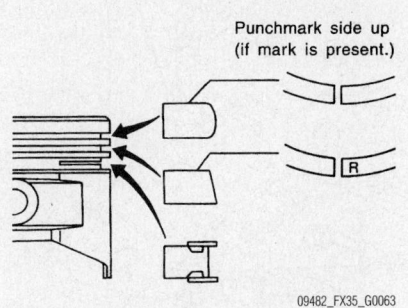

Piston ring installation—3.5L and 4.5L engines

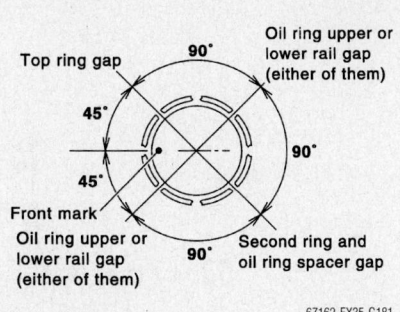

67162-FX35-G181

Piston ring gap positioning—3.5L and 4.5L engines

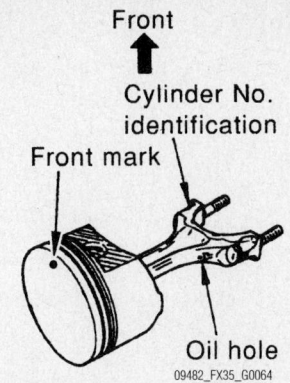

09482_FX35_G0064

Piston and connecting rod installation—3.5L and 4.5L engines

EXHAUST SYSTEM

Components

REMOVAL & INSTALLATION

1. Disconnect each joint and mounting as necessary.

2. When removing main muffler, remove main muffler mounting bracket bolts.

3. Remove exhaust mounting bracket as follows:

   a. Remove engine rear undercover.

   b. Remove front cross bar.

   c. Remove exhaust mounting bracket.

4. Installation is the reverse of removal. Check the accompanying illustration for tightening torques.

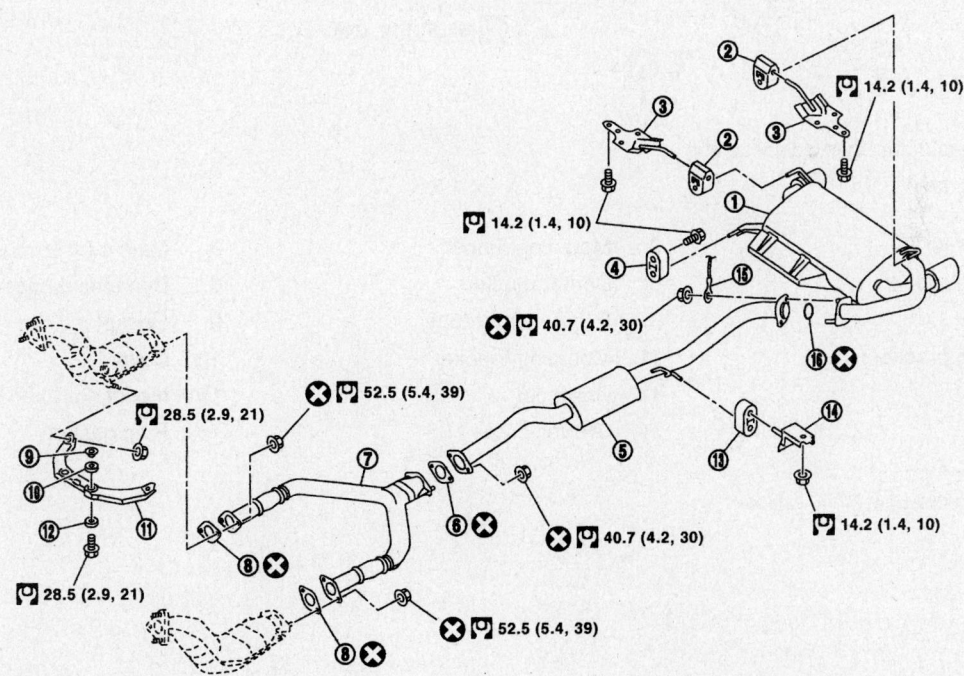

❌ : Always replace after every disassembly.

🔧 : N•m (kg-m, ft-lb)

1. Main muffler	2. Mounting rubber	3. Main muffler mounting bracket
4. Mounting rubber	5. Center muffler	6. Gasket
7. Exhaust front tube	8. Gasket	9. Collar
10. Grommet	11. Mounting bracket	12. Grommet
13. Mounting rubber	14. Bracket	15. Ground cable
16. Ring gasket		

09482_FX35_G0065

Exhaust system exploded view—3.5L engine

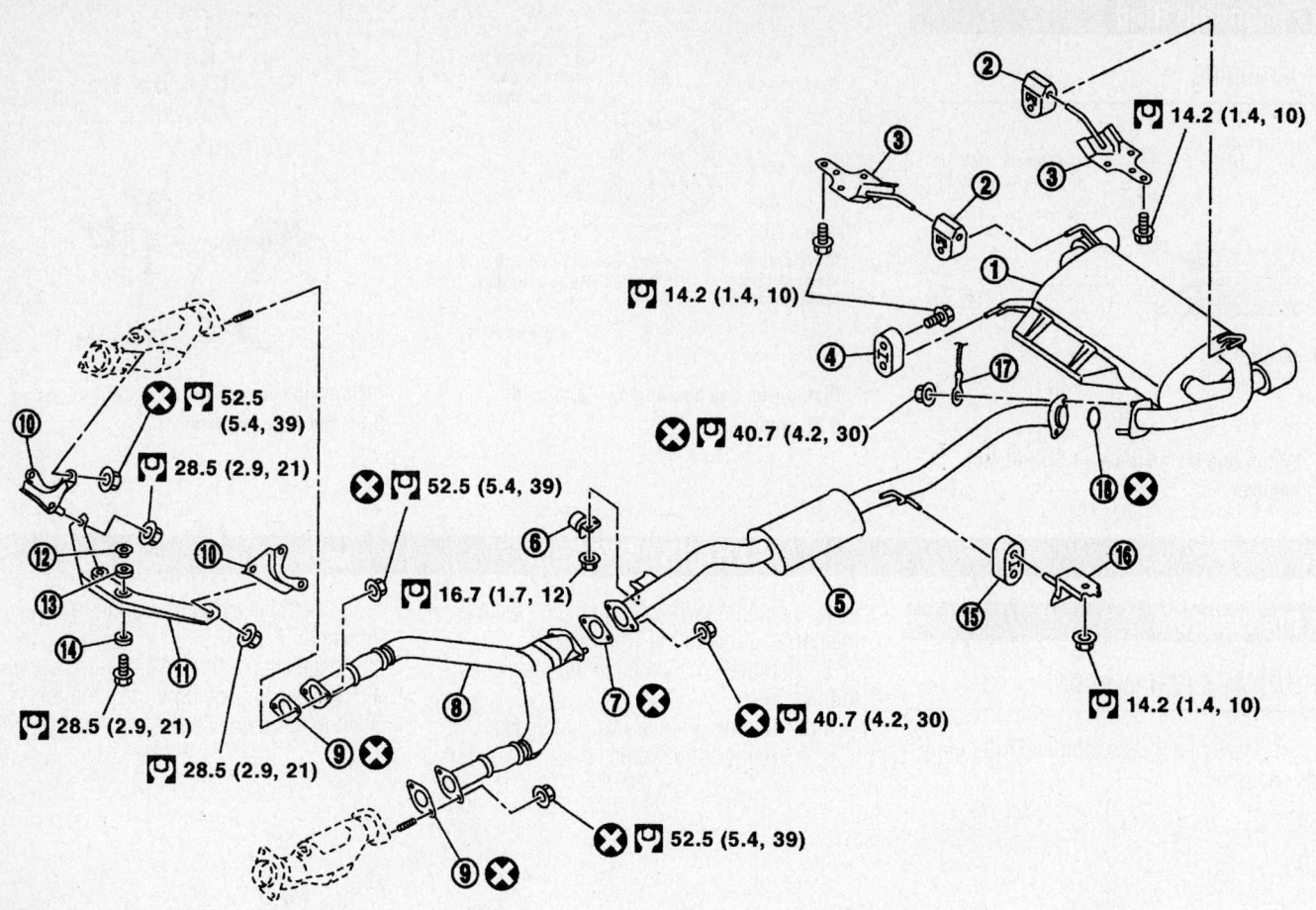

X : Always replace after every disassembly.

⚙ : N•m (kg-m, ft-lb)

1. Main muffler	2. Mounting rubber	3. Main muffler mounting bracket
4. Mounting rubber	5. Center muffler	6. Dynamic damper
7. Gasket	8. Exhaust front tube	9. Gasket
10. Mounting bracket	11. Mounting bracket	12. Collar
13. Grommet	14. Grommet	15. Mounting rubber
16. Bracket	17. Ground cable	18. Ring gasket

Exhaust system exploded view—4.5L engine

09482_FX35_G0066

FUEL SYSTEM

Fuel System Service Precautions

Safety is the most important factor when performing not only fuel system maintenance but any type of maintenance. Failure to conduct maintenance and repairs in a safe manner may result in serious personal injury or death. Maintenance and testing of the vehicle's fuel system components can be accomplished safely and effectively by adhering to the following rules and guidelines.

• To avoid the possibility of fire and personal injury, always disconnect the negative battery cable unless the repair or test procedure requires that battery voltage be applied.

• Always relieve the fuel system pressure prior to disconnecting any fuel system component (injector, fuel rail, pressure regulator, etc.), fitting or fuel line connection. Exercise extreme caution whenever relieving fuel system pressure, to avoid exposing skin, face and eyes to fuel spray. Please be advised that fuel under pressure may penetrate the skin or any part of the body that it contacts.

• Always place a shop towel or cloth around the fitting or connection prior to loosening to absorb any excess fuel due to spillage. Ensure that all fuel spillage (should it occur) is quickly removed from engine surfaces. Ensure that all fuel soaked cloths or towels are deposited into a suitable waste container.

• Always keep a dry chemical (Class B) fire extinguisher near the work area.

• Do not allow fuel spray or fuel vapors to come into contact with a spark or open flame.

• Always use a back-up wrench when loosening and tightening fuel line connection fittings. This will prevent unnecessary stress and torsion to fuel line piping.

• Always replace worn fuel fitting O-rings with new. Do not substitute fuel hose or equivalent, where fuel pipe is installed.

Relieving Fuel System Pressure

WITH THE CONSULT-II TOOL

1. Before servicing the vehicle, refer to the Precautions Section.
2. Turn the ignition switch ON.
3. Perform the "FUEL PRESSURE RELEASE" in "WORK SUPPORT" mode with the CONSULT-II.

FUEL PRESSURE RELEASE

FUEL PUMP WILL STOP BY TOUCHING START IN IDLING. CRANK A FEW TIMES AFTER ENGINE STALL.

67162-FX35-G112

Follow the directions on the CONSULT-II tool for fuel pressure release

4. Start the engine.
5. After engine stalls, crank it over two or three times to release all fuel pressure.
6. Turn the ignition switch OFF.

WITHOUT THE CONSULT-II TOOL

1. Before servicing the vehicle, refer to the Precautions Section.
2. Remove the fuel pump fuse located in IPDM E/R.
3. Start the engine.
4. After the engine stalls, crank it over two or three times to release all fuel pressure.
5. Turn the ignition switch OFF.
6. Reinstall the fuel pump fuse after servicing the fuel system.

Fuel Filter

REMOVAL & INSTALLATION

The fuel filter is an integral component of the fuel pump. Please refer to the fuel pump removal and installation procedure.

Fuel Pump

REMOVAL & INSTALLATION

1. Before servicing the vehicle, refer to the Precautions Section.
2. Check the fuel level on the fuel gauge. If the fuel gauge indicates more than the level as shown in the figure (full or almost full), drain the fuel from the fuel tank until the gauge indicates a level as shown in the figure.

✳✳ CAUTION

Fuel will be spilled when removing the main and sub fuel level sensor units if level of the fuel in the tank is higher than the installation positions of the sensor units.

3. In the case that the fuel pump does not operate, perform the following procedure:
   a. Insert a hose of less than 1 in. (25mm) in diameter into the fuel filler tube through the fuel filler opening to draw fuel from the fuel filler tube.

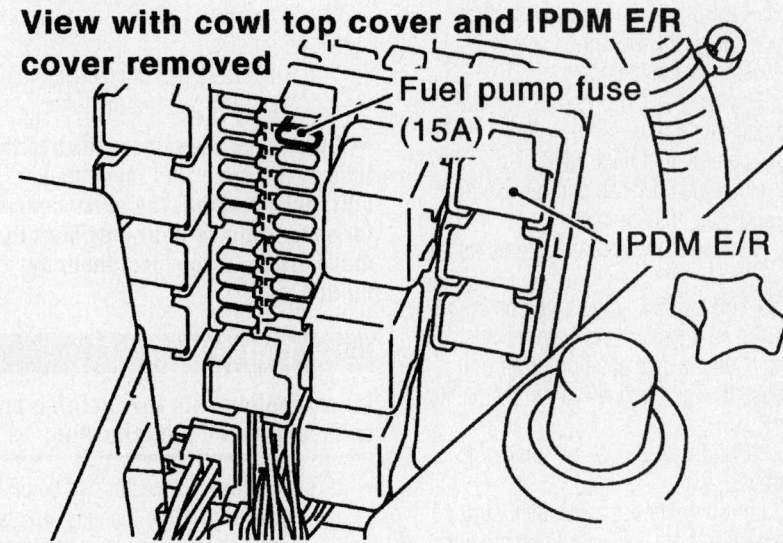

View with cowl top cover and IPDM E/R cover removed

Fuel pump fuse (15A)

IPDM E/R

67162-FX35-G113

Fuel pump fuse location for fuel pressure release

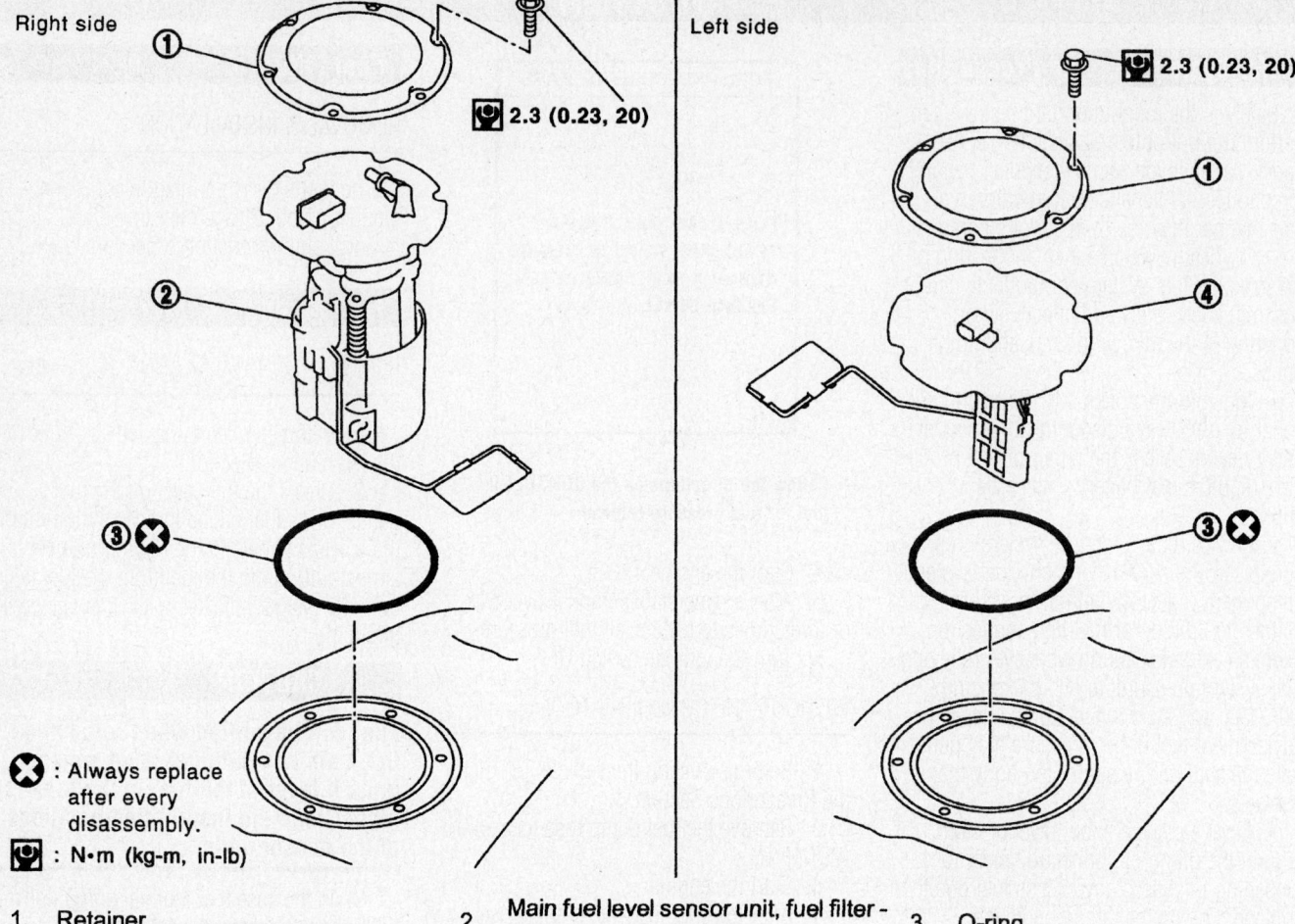

Right side

Left side

2.3 (0.23, 20)

2.3 (0.23, 20)

❌ : Always replace after every disassembly.

⬛ : N•m (kg-m, in-lb)

1. Retainer
2. Main fuel level sensor unit, fuel filter - and fuel pump assembly
3. O-ring
4. Sub fuel level sensor unit

67162-FX35-G182

Exploded view of fuel pump and filter assembly and related components

b. Disconnect the fuel filler hose from the fuel filler tube.

c. Insert the fuel tube into the fuel tank through the fuel filler hose to draw the fuel from the fuel tank.

4. Release the fuel pressure from the fuel lines.

5. Open the fuel filler lid.

6. Open the filler cap and release the pressure inside the fuel tank.

7. Remove the rear seat cushion, as follows:

a. Pull the lock at the front bottom of the seat cushion forward (1 for each side).

b. Pull the seat cushion upward to release the retaining wire from the plastic hook.

c. Pull the seat cushion forward to remove.

8. Lift up the floor carpet, then remove the inspection hole cover for the main and sub fuel level sensor units by turning the retaining clips clockwise by 90 degrees.

9. Disconnect the wiring harness connector and fuel feed tube.

10. Disconnect the fuel line quick connector, as follows:

a. Hold the sides of connector, push in the tabs and pull out the tube.

➡ **If the quick connector sticks to the tube of the main fuel level sensor unit, push and pull the quick connector several times until they start to move. Then, disconnect them by pulling.**

✳✳ CAUTION

When dealing with the fuel line quick connector, heed the following:

- The quick connector can be disconnected when the tabs are completely depressed. Do not twist it more than necessary.

- Do not use any tools to disconnect the quick connector.

- Keep the resin tube away from heat. Be especially careful when welding near the resin tube.

- Prevent acidic liquid such as battery electrolyte, etc. from getting on the resin tube.

- Do not bend or twist the resin tube during connection and disconnection.

- Do not remove the remaining retainer on the hard tube (or the equivalent) except when the resin tube or retainer is replaced.

- When the resin tube or hard tube (or the equivalent) is replaced, also replace the retainer with new one.

- To keep the connecting portion clean and to avoid damage and foreign materials, cover them completely with plastic bags or something similar.

※ CAUTION

Make sure to not bend the float arm during removal, and avoid impacts, such as falling, when handling components.

11. Remove the main fuel level sensor unit, fuel filter and fuel pump assembly, and sub fuel level sensor unit, as follows:
    a. Remove the main fuel sensor unit retainer.
    b. Raise the main fuel level sensor unit, fuel filter and fuel pump assembly,

and using snapring pliers, remove fuel hose connector.

※ CAUTION

Be careful not to damage the fuel hose connector by expanding them excessively.

    c. Removal of sub fuel level sensor unit:
    d. Remove the sub fuel level sensor unit retainer.
    e. Raise and release the sub fuel level sensor unit.

To install:
12. Installation is the reverse of removal.

Fuel Injectors

REMOVAL & INSTALLATION

3.5L Engine

1. Before servicing the vehicle, refer to the Precautions Section.
2. Remove the engine cover.
3. Release fuel pressure.

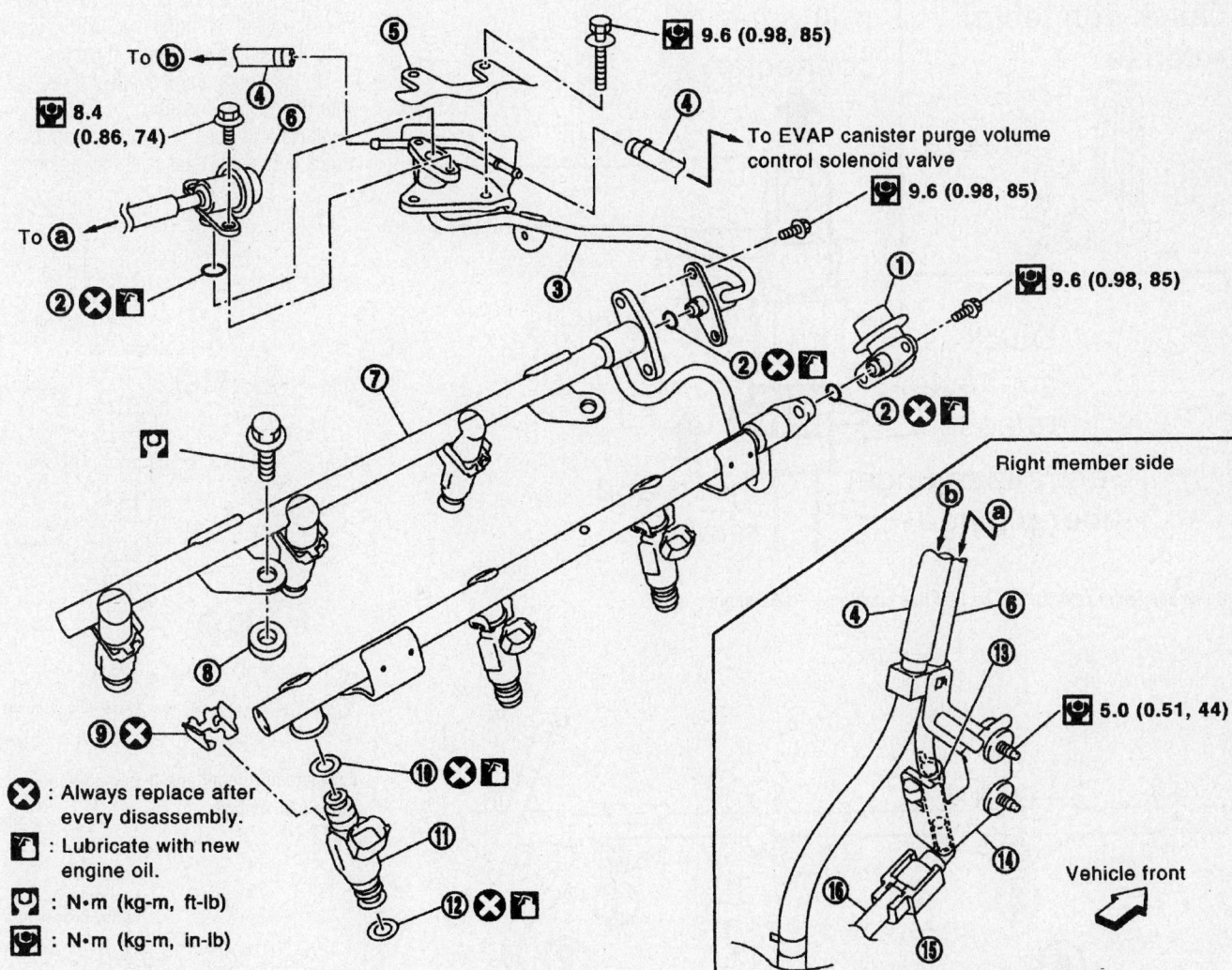

X : Always replace after every disassembly.
🛢 : Lubricate with new engine oil.
🔧 : N•m (kg-m, ft-lb)
🔧 : N•m (kg-m, in-lb)

1.	Fuel damper	2.	O-ring	3.	Fuel sub-tube
4.	EVAP hose	5.	Intake manifold collector (lower)	6.	Fuel feed hose (with damper)
7.	Fuel tube	8.	Spacer	9.	Clip
10.	O-ring (blue)	11.	Fuel injector	12.	O-ring (brown)
13.	Hose clamp	14.	Bracket	15.	Quick connector cap
16.	Centralized under-floor piping				

67162-FX35-G183

Exploded view of the fuel injector and rail assembly—3.5L engine

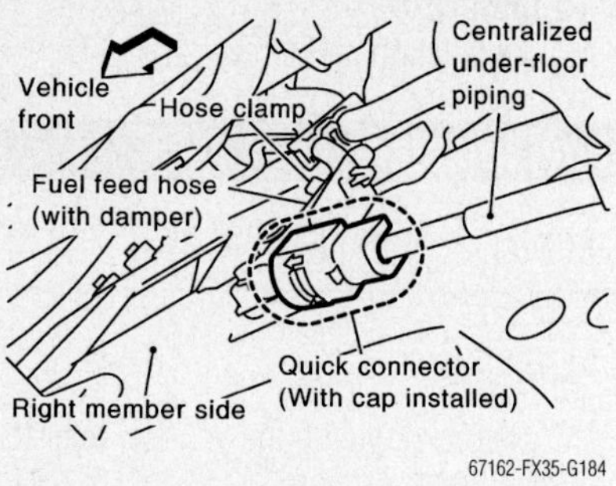
Vehicle front
Hose clamp
Centralized under-floor piping
Fuel feed hose (with damper)
Quick connector (With cap installed)
Right member side
67162-FX35-G184

Quick connect coupling detail for underfloor piping—3.5L engine

4. Remove the fuel feed hose (with damper) from the fuel sub-tube.

➡There is no fuel return route.

✲✲ CAUTION

While the hoses are disconnected, plug them to prevent fuel from draining. Also, do not separate the damper and hose.

5. When separating the fuel feed hose (with damper) and the centralized underfloor piping connection, disconnect the quick connector, as follows:

a. Remove the quick connector cap from the quick connector connection on the right member side.

b. Disconnect the fuel feed hose (with damper) from the bracket hose clamp.

✲✲ CAUTION

Disconnect the quick connector by using quick connector release too SST: (J-45488) or equivalent, not by picking out the retainer tabs.

c. With the sleeve side of the quick connector release facing the quick connector, install the quick connector release onto the centralized under-floor piping.

d. Insert the quick connector release into the quick connector until the sleeve contacts and goes no further. Hold the quick connector release at that position.

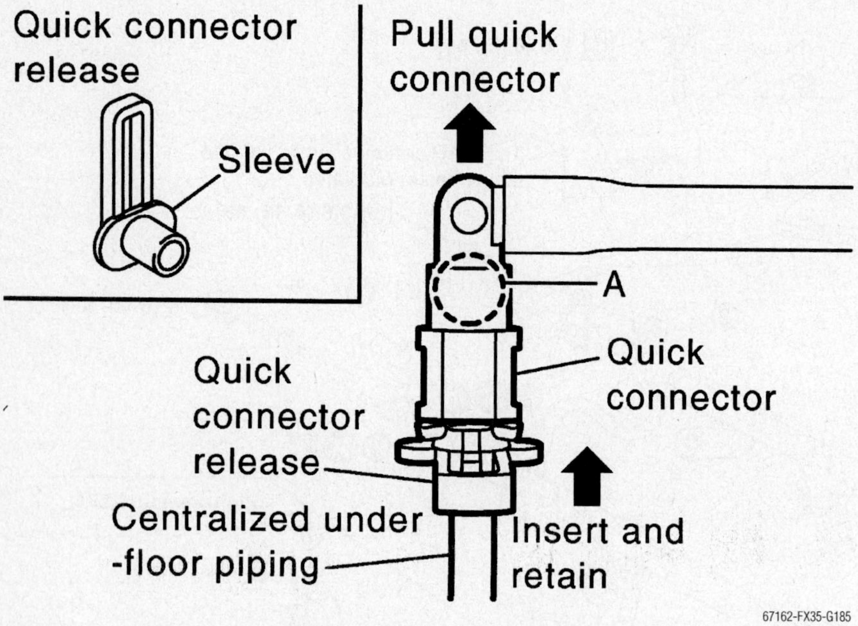
Quick connector release
Sleeve
Pull quick connector
A
Quick connector
Quick connector release
Centralized under-floor piping
Insert and retain
67162-FX35-G185

Quick connector release use on fuel line couplings—3.5L engine

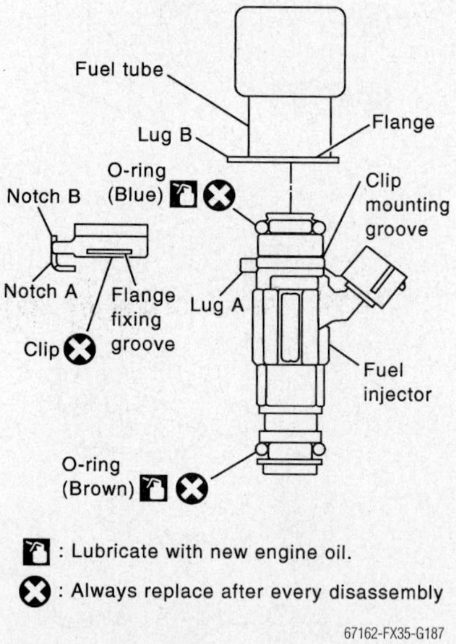
Fuel tube
Flange
Lug B
O-ring (Blue)
Notch B
Clip mounting groove
Notch A
Flange fixing groove
Lug A
Clip
Fuel injector
O-ring (Brown)
: Lubricate with new engine oil.
: Always replace after every disassembly
67162-FX35-G187

Fuel injector detail and O-ring positioning—3.5L engine

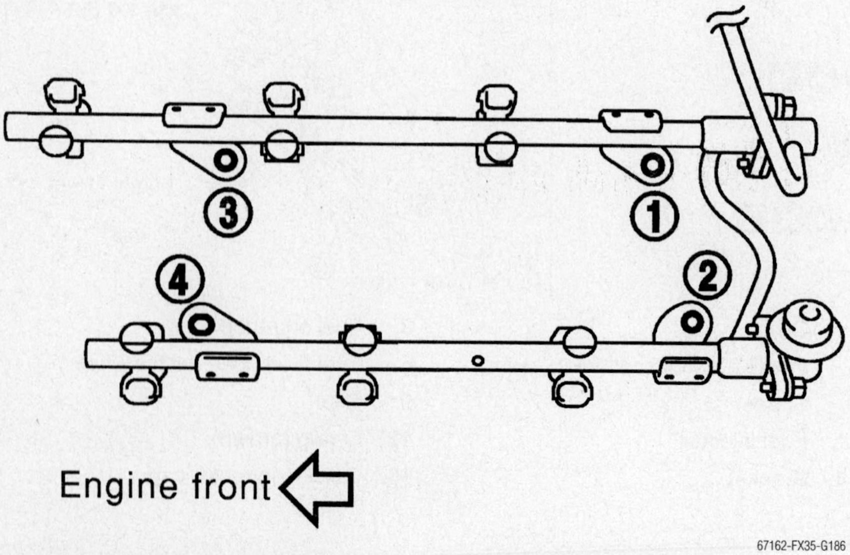
3
1
4
2
Engine front
67162-FX35-G186

Fuel rail assembly mounting bolt tightening sequence—3.5L engine

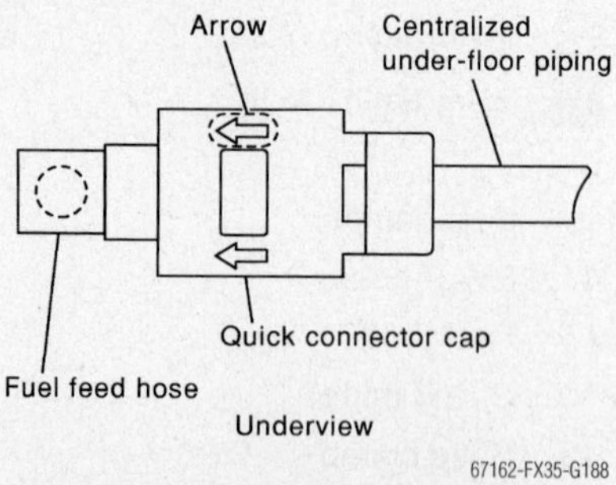
Arrow
Centralized under-floor piping
Quick connector cap
Fuel feed hose
Underview
67162-FX35-G188

Install the quick connector so that the arrows face the fuel feed tube—3.5L engine

> **✲✲ CAUTION**
>
> **Inserting the quick connector release hard will not disconnect the quick connector. Hold the quick connector release where it contacts and goes no further.**

e. Draw and pull out the quick connector straight from the centralized under-floor piping.

> **✲✲ CAUTION**
>
> **When disconnecting the fuel line, heed the following:**

- Pull the quick connector holding "A" position as shown in the figure. Do not pull it with lateral force applied. The O-ring inside the quick connector may be damaged.
- Prepare a container and cloth beforehand as fuel will leak out.
- Avoid fire and sparks.
- Keep parts away from all heat sources. Especially, be careful when welding is performed around them.
- Do not expose the parts to battery electrolyte or other acids.
- Do not bend or twist the connection between the quick connector and the fuel feed hose (with damper) during installation/removal.
- To keep the connecting portion clean and to avoid damage from foreign materials, cover them completely with plastic bags or something similar.

6. Remove the upper and lower intake manifold collectors.

7. Disconnect the wiring harness connector from the fuel injector.

8. Loosen the mounting bolts in reverse order shown, and remove the fuel tube and fuel injector assembly.

> **✲✲ CAUTION**
>
> **Do not tilt the assembly, or the remaining fuel may leak.**

9. Remove the fuel injectors from fuel the tube with the following procedure:
a. Open and remove the clip.
b. Remove the fuel injector from the fuel tube by pulling straight.

> **✲✲ CAUTION**
>
> **During injector removal, heed the following items:**

- Be careful with the remaining fuel that leaks from the fuel tube.
- Be careful not to damage the injector nozzles during removal.
- Do not bump or drop the fuel injectors.
- Do not disassemble the fuel injectors.

10. Remove the fuel sub-tube and fuel damper.
To install:

> **✲✲ CAUTION**
>
> **When handling all O-rings in this procedure, heed the following:**

- Handle the O-ring with bare hands. Never wear gloves.
- Lubricate the O-ring with new engine oil.
- Do not clean the O-ring with solvent.
- Make sure that the O-ring and its mating part are free of foreign material.
- When installing the O-ring, be careful not to scratch it with a tool or fingernails.
- Be careful not to twist or stretch the O-ring. If the O-ring was stretched while it was being attached, do not insert it quickly into the fuel tube.
- Insert the O-ring straight into the fuel tube. Do not de-center or twist it.

11. Install the fuel damper and fuel sub-tube.

12. Insert the fuel damper and fuel sub-tube straight into the fuel tube.

13. Tighten the mounting bolts evenly in turn.

14. After tightening the mounting bolts, make sure that there is no gap between the flange and fuel tube.

15. Install O-rings onto fuel injector. The upper and lower O-ring are different. Be careful not to confuse them. The O-rings are identified as follows:
- Fuel tube side O-ring: Blue
- Nozzle side O-ring: Brown

16. Install each fuel injector onto the fuel tube, as follows:
a. Insert the clip into the clip mounting groove on the fuel injector.
b. Insert the clip so that lug "A" of fuel injector matches notch "A" of the clip.

> **✲✲ CAUTION**
>
> **Do not reuse old clips. Replace them with new ones. Be careful to keep the clip from interfering with the O-ring.**

If interference occurs, replace the O-ring with a new one.

c. Insert the fuel injector into the fuel tube, matching it to the axial center, with the clip attached. Insert the fuel injector so that lug "B" of fuel tube matches notch "B" of the clip. Make sure that the fuel tube flange is securely fixed in the groove on the clip.
d. Make sure that installation is complete by checking that the fuel injector does not rotate or come off.

17. Install the fuel tube and fuel injector assembly onto the intake manifold, and tighten the mounting bolts in two steps as shown.
- 1st Step: 7 ft. lbs. (10 Nm)
- 2nd Step: 17 ft. lbs. (24 Nm)

> **✲✲ CAUTION**
>
> **Be careful not to let the tips of the injector nozzles come into contact with other parts.**

18. Connect the injector sub-wiring harness.

19. Install the upper and lower intake manifold collectors.

20. Install the fuel sub-tube on the rear end of the lower intake manifold collector.

21. Connect the fuel feed hose and damper. After tightening the mounting bolts, make sure that there is no gap between the flange and the fuel sub-tube.

22. Connect the quick connector between the fuel feed hose and centralized under-floor piping connection with the following procedure:
a. Check the connection for damage and foreign materials.
b. Align the connector with the tube, then insert the connector straight into the tube until a click is heard.
c. After connecting the quick connector, visually confirm that the two retainer tabs are connected to connector, then pull the tube and connector to make sure they are securely connected.
d. Install quick connector cap to quick connector connection.
e. Install the quick connector cap with the arrow on the surface facing in the direction of the quick connector (fuel feed hose side).

> **✲✲ CAUTION**
>
> **If the cap cannot be installed smoothly, the quick connector may not have been installed correctly. Check the connection again.**

f. Secure the fuel feed hose to the clamp.

23. The remainder of installation is the reverse of removal.

24. Perform the following once installation is complete:

25. After installing the fuel tubes, make sure there is no fuel leakage at the connections in the following steps:

a. Apply fuel pressure to the fuel lines by turning the ignition switch "ON" (with the engine stopped). Then check for fuel leaks at all connections.

b. Start the engine and while holding it at a high rpm check for fuel leaks at all connections.

→Use mirrors to check hard-to-see connections.

4.5L Engine

1. Before servicing the vehicle, refer to the Precautions Section.

⁕⁕ CAUTION

Do not remove or disassemble parts unless instructed as shown in the figure.

2. Remove the engine cover.
3. Release fuel pressure.
4. Disconnect the fuel feed hose on the engine side, as follows:

→Except to confirm whether or not there is a quick connector cap, perform the same procedure for the side of centralized under-floor piping as well.

a. Remove the quick connector cap from the quick connector connection (engine side only).

⁕⁕ CAUTION

Disconnect the quick connector by using quick connector release tool SST: J-45488 or equivalent; not by picking out the retainer tabs. Inserting the quick connector release tool too hard will not disconnect the quick

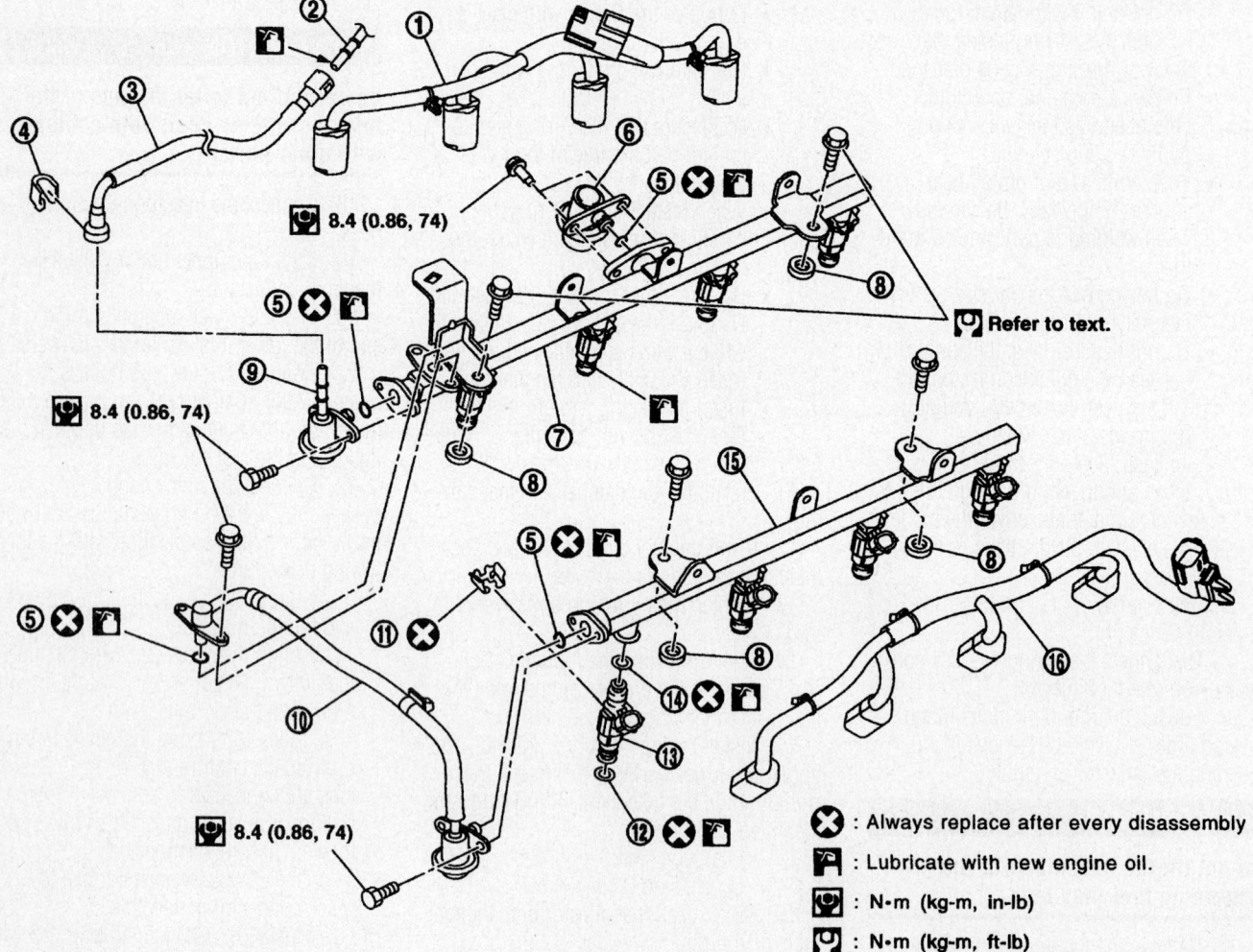

❌ : Always replace after every disassembly
🛢 : Lubricate with new engine oil.
🔧 : N·m (kg-m, in-lb)
🔧 : N·m (kg-m, ft-lb)

1. Fuel injector sub harness (RH)
2. Centralized under-floor piping
3. Fuel feed hose
4. Quick connector cap
5. O-ring
6. Fuel damper (RH)
7. Fuel tube (RH)
8. Spacer
9. Fuel feed damper
10. Fuel damper and fuel hose assembly
11. Clip
12. O-ring (Green)
13. Fuel injector
14. O-ring (Black)
15. Fuel tube (LH)
16. Fuel injector sub harness (LH)

67162-FX35-G189

Exploded view of fuel rail and injector mounting and related components—4.5L engine

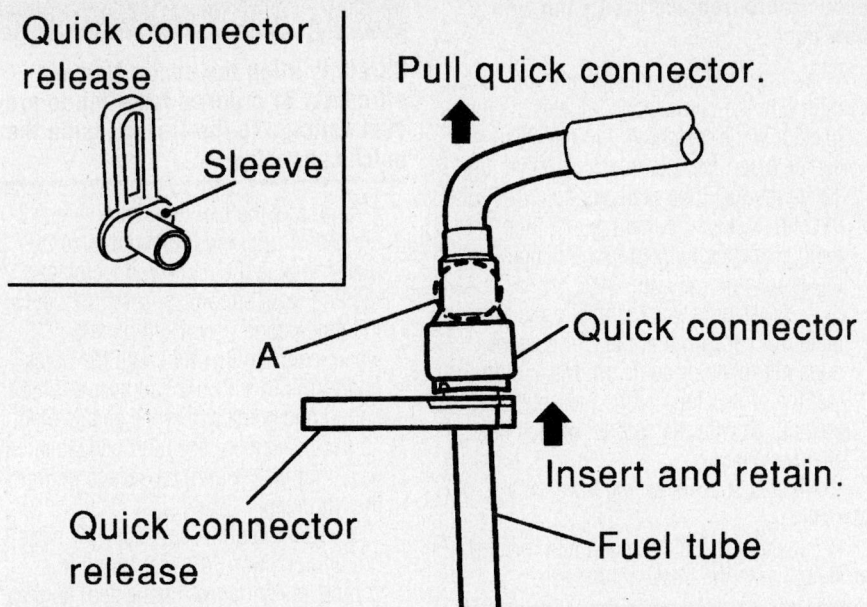

Quick connector release tool used to disconnect the fuel line quick connector—4.5L engine

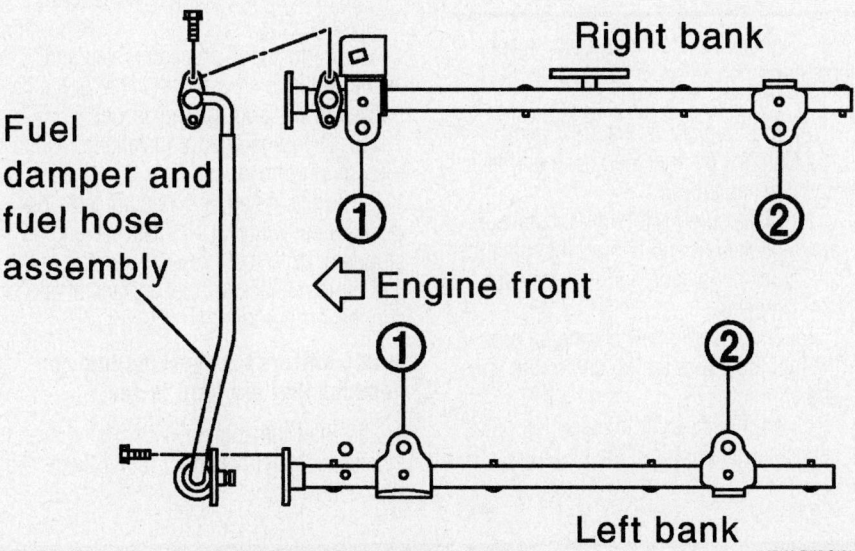

Fuel rail mounting bolt tightening sequence—4.5L engine

connector. Hold quick connector release where it contacts and goes no further. Do not pull with lateral force applied. O-ring inside quick connector may be damaged.

b. Disconnect the quick connector from the fuel feed damper. With the sleeve side of the quick connector release tool facing the quick connector, install the quick connector release tool onto the fuel tube. Insert the quick connector release tool into the quick connector until the sleeve contacts and goes no

further. Hold the quick connector release tool in that position. Draw and pull out the quick connector, holding position "A", straight from the fuel damper.

✳✳ CAUTION

Heed the following when working with the fuel system tubing:

- Prepare a container and cloth beforehand since fuel will leak out.
- Avoid fire and sparks.
- Keep parts away from a heat source. Especially, be careful when

welding is performed around them.
- Do not expose parts to battery electrolyte or other acids.
- Do not bend or twist the connection between the quick connector and the fuel feed hose during installation/removal.
- To keep the connecting portion clean and to avoid damage and foreign materials, cover them completely with plastic bags or something similar.

5. Disconnect the fuel damper and fuel hose assembly from the left-hand and right-hand fuel tubes.

✳✳ CAUTION

While the hoses are disconnected, plug them to prevent fuel from leaking and do not separate the fuel damper and fuel hose.

6. Remove the upper intake manifold.
7. Disconnect the wiring harness connector from the fuel injector.
8. Loosen the mounting bolts in reverse order shown, and remove the fuel tube and fuel injector assembly.

✳✳ CAUTION

Do not tilt the fuel injector assembly, or remaining fuel in the pipes may flow out from the pipes.

9. Remove the spacers on the lower intake manifold.
10. Remove the fuel injector from the fuel tube, as follows:
    a. Open and remove the clip.
    b. Remove the fuel injector from the fuel tube by pulling it straight out.
    c. During injector removal, heed the following items:
    - Be careful with the remaining fuel that leaks from the fuel tube.
    - Be careful not to damage the injector nozzles during removal.
    - Do not bump or drop the fuel injectors.
    - Do not disassemble the fuel injectors.
11. Remove the right-hand fuel damper and fuel feed damper.
To install:

✳✳ CAUTION

When handling all O-rings in this procedure, heed the following:

- Handle the O-ring with bare hands. Never wear gloves.

- Lubricate the O-ring with new engine oil.
- Do not clean the O-ring with solvent.
- Make sure that the O-ring and its mating part are free of foreign material.
- When installing the O-ring, be careful not to scratch it with a tool or fingernails.
- Be careful not to twist or stretch the O-ring. If the O-ring was stretched while it was being attached, do not insert it quickly into the fuel tube.
- Insert the O-ring straight into the fuel tube. Do not de-center or twist it.

12. Install the right-hand fuel damper and fuel feed damper, as follows:

a. Insert the right-hand fuel damper and fuel feed damper straight into the right-hand fuel tube.

b. Tighten the mounting bolts evenly in turn.

c. After tightening the mounting bolts, make sure that there is no gap between the flange and the right-hand fuel tube.

13. Install new O-rings on the fuel injector, paying attention to the fact that the upper and lower O-rings are different. Be careful not to confuse them. The fuel tube side O-ring is black, whereas the nozzle side O-ring is green.

14. Install the fuel injector onto the fuel tube, as follows:

a. Insert the clip into clip mounting groove on the fuel injector. Insert the clip so that lug "A" of the fuel injector matches notch "A" of the clip. Do not reuse clips; replace them with new ones.

➡ Be careful to keep the clip from interfering with the O-ring. If interference occurs, replace the O-ring with a new one.

b. Insert the fuel injector into the fuel tube with the clip attached. Insert the fuel injector while matching it to the axial center. Insert the fuel injector so that lug "B" of the fuel tube matches notch "B" of the clip. Make sure that the fuel tube flange is securely fixed in the flange fixing groove on the clip.

c. Make sure that installation is complete by checking that the fuel injector does not rotate or come off. Make sure that the protrusions of the fuel injectors are aligned with the cutouts of the clips after installation.

15. Install spacers on the lower intake manifold.

16. Install the fuel tube and fuel injector assembly onto the intake manifold.

✳✳ CAUTION

Be careful not to let the tip of the injector nozzle come in contact with other parts.

17. Tighten the mounting bolts in two steps, in numerical order shown:
- 1st Step: 7 ft. lbs. (10 Nm)
- 2nd Step: 17 ft. lbs. (23.5 Nm)

18. Connect the fuel feed hose on the engine side, as follows

a. Make sure no foreign substances are deposited in and around the fuel tube and quick connector, and that they are not damaged.

b. Thinly apply new engine oil around the fuel tube from the tip end to the spool end.

c. Align the center to insert the quick connector straight into the fuel tube.

✳✳ CAUTION

Carefully align the center to avoid off-center or crooked insertion to prevent damage to the O-ring inside the quick connector.

d. Insert the fuel tube into the quick connector until the top spool is completely inside the quick connector, and the 2nd level spool exposes right below the quick connector. Hold position "A" when inserting fuel tube into the quick connector. Insert until you hear a "click" sound and actually feel the engagement. To avoid misidentification of engagement with a similar sound, be sure to perform the next step.

e. Pull the quick connector by hand holding position "A". Make sure it is completely engaged (connected) so that it does not disconnect from fuel tube.

f. Install the quick connector cap on quick connector connection. (on the engine side only).

g. Install the fuel feed hose to the hose clamps.

19. Perform the preceding step and associated sub steps for the side of the centralized under-floor piping as well.

20. The remainder of installation is the reverse of removal.

21. Once installation is complete, turn the ignition switch "ON" (with the engine stopped). With fuel pressure applied to the fuel piping, check for fuel leakage at all connection points.

➡ Use mirrors for checking fuel connections that are hard to see.

22. Start the engine. With engine speed increased, check again for fuel leakage at all connection points.

DRIVE TRAIN

Automatic Transmission

REMOVAL & INSTALLATION

1. Before servicing the vehicle, refer to the Precautions Section.

✳✳ CAUTION

When removing the automatic transmission assembly from the engine, first remove the crankshaft position sensor (POS) from the A/T assembly. Be careful not to damage the sensor edge.

2. Disconnect the negative battery terminal.

3. Remove the engine cover.

4. Remove the A/T fluid level gauge.

5. Raise and safely support the vehicle.

6. Remove the engine under cover.

7. Remove the exhaust front tube and center muffler.

8. Remove the driveshaft.

9. Detach the transmission shifter control rod by removing the snap pin retaining the control rod bracket to the control device assembly lever.

10. Remove the crankshaft position sensor (POS) from A/T assembly.

11. When handling the POS, heed the following:
- Do not subject it to impact by dropping or hitting it.
- Do not disassemble it.
- Do not allow metal filings, etc., to get on the sensor's front edge magnetic area.
- Do not place in an area affected by magnetism.

12. Remove the starter motor, as follows:

a. Disconnect the negative battery terminal.

b. Remove the engine rear undercover.

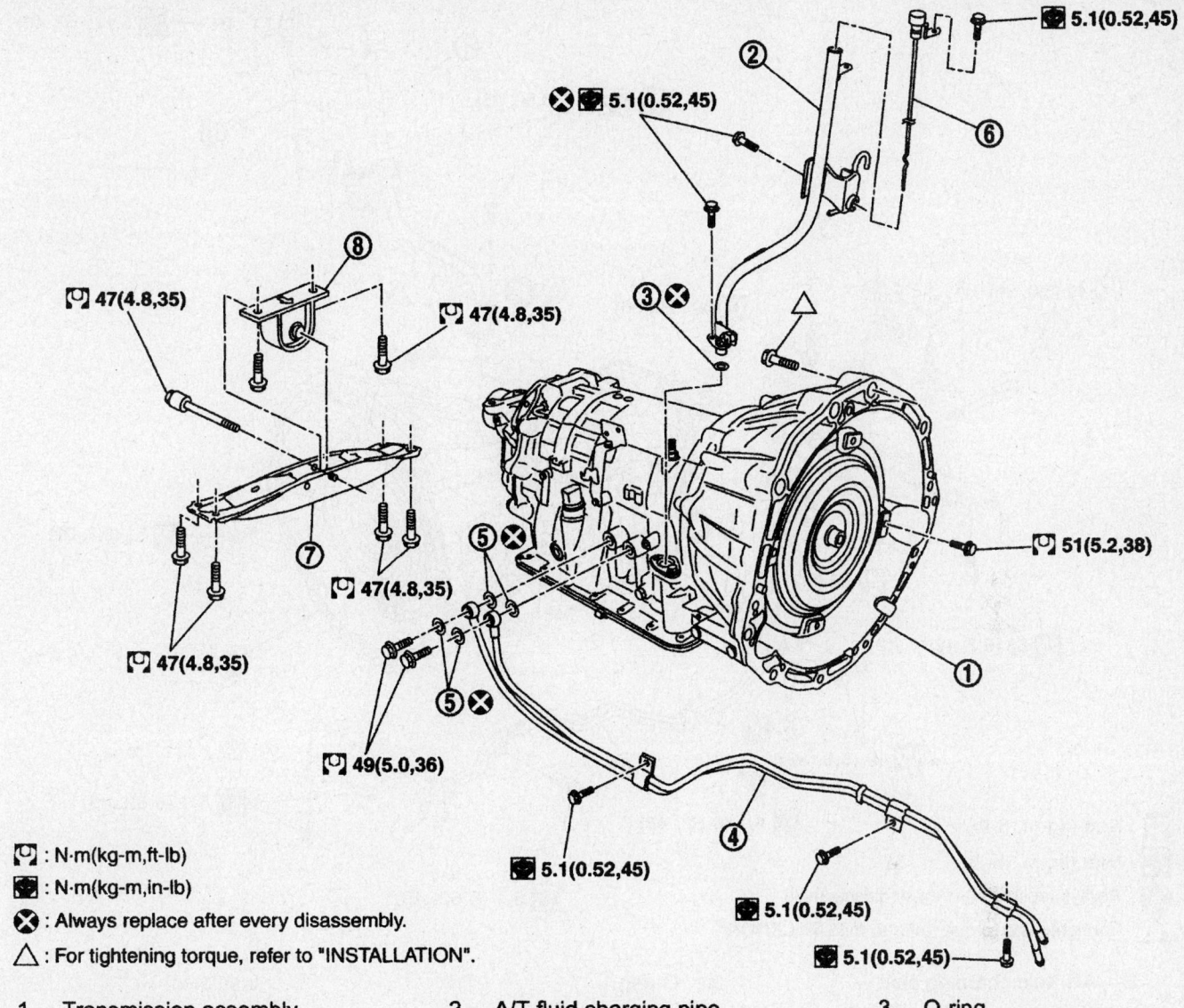

⊠ 🔀 5.1(0.52,45)

🔀 5.1(0.52,45)

⊠ 🔀 5.1(0.52,45)

🔧 47(4.8,35)

🔧 47(4.8,35)

🔧 47(4.8,35)

🔧 47(4.8,35)

🔧 49(5.0,36)

🔧 51(5.2,38)

🔀 5.1(0.52,45)

🔀 5.1(0.52,45)

🔀 5.1(0.52,45)

🔧 : N·m(kg-m,ft-lb)

🔀 : N·m(kg-m,in-lb)

⊠ : Always replace after every disassembly.

△ : For tightening torque, refer to "INSTALLATION".

1. Transmission assembly	2. A/T fluid charging pipe	3. O-ring
4. Fluid cooler tube	5. Copper washer	6. A/T fluid level gauge
7. Engine rear member	8. Insulator	

67162-FX35-G194

Exploded view of transmission mounting—2003 2WD Models

c. Disconnect the S connector.
d. Remove the B terminal nut.
e. Remove the starter motor mounting bolts and wiring harness clip bracket.
f. Remove the starter motor toward the underside of the vehicle.
13. Disconnect the A/T fluid cooler tube from the transmission assembly.
14. Remove the dust cover from the converter housing part.
15. Turn the crankshaft, and remove the four tightening bolts for the driveplate and torque converter.

※※ **CAUTION**

When turning the crankshaft, turn it clockwise as viewed from the front of the engine.

16. For AWD models with the 3.5L engine, remove the dynamic damper.
17. Support the transmission assembly with a transmission jack.

※※ **CAUTION**

When setting the transmission jack, be careful not to allow it to collide against the drain plug.

18. Remove the engine rear member.
19. Tilt the transmission slightly to keep the clearance between the body and the transmission, and then disconnect the air breather hose from the charging pipe.
20. Disconnect the A/T assembly connector.
21. Remove the A/T fluid charging pipe from the A/T assembly.
22. Plug up the openings such as the fluid charge pipe hole, etc.
23. Remove the bolts fixing the transmission assembly to the engine.
24. Secure the torque converter to prevent it from dropping.

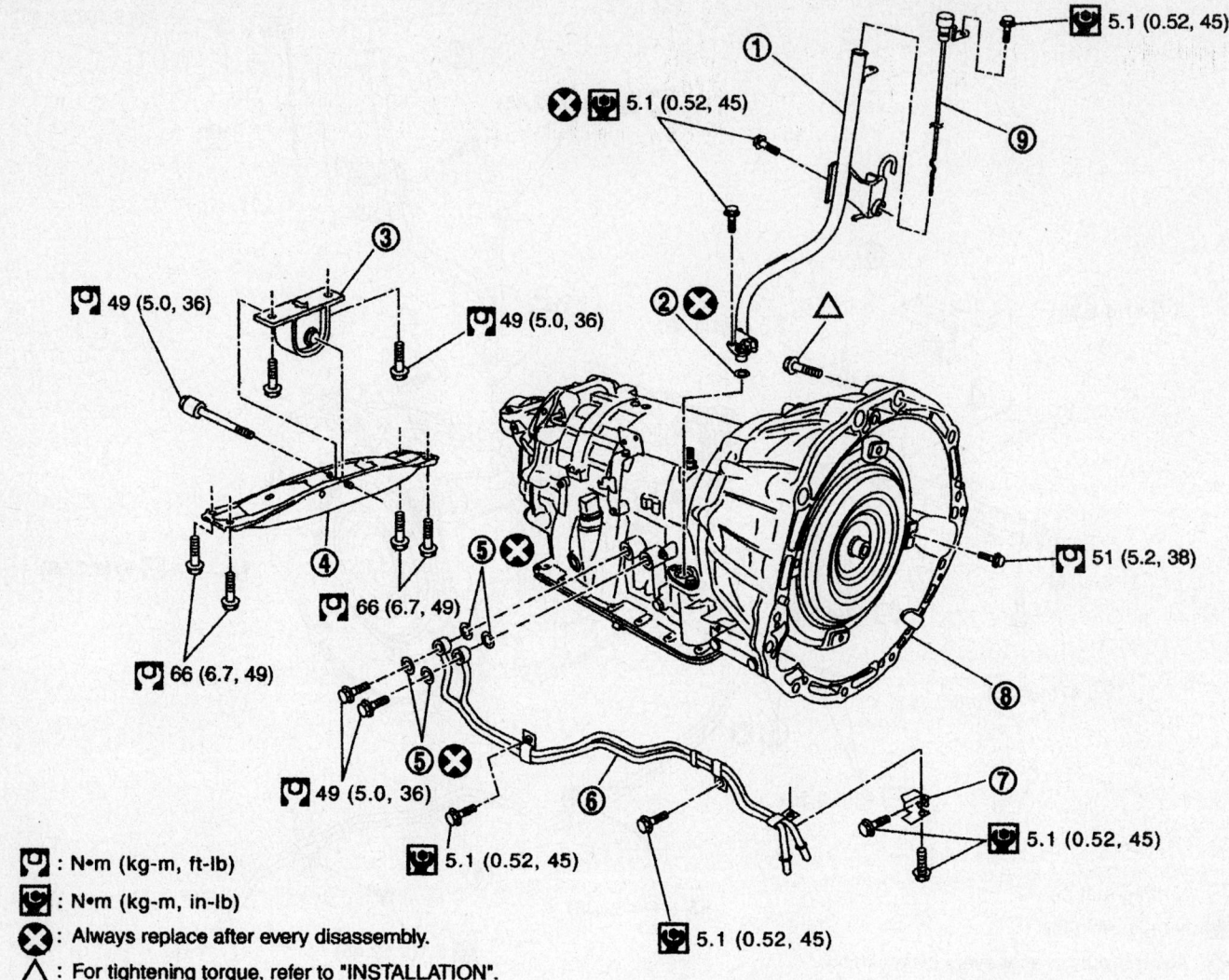

Legend:

⚙ : N•m (kg-m, ft-lb)

⚙ : N•m (kg-m, in-lb)

❌ : Always replace after every disassembly.

△ : For tightening torque, refer to "INSTALLATION".

1. A/T fluid charging pipe
4. Rear member
7. Bracket

2. O-ring
5. Copper washer
8. Transmission assembly

3. Insulator
6. Fluid cooler tube
9. A/T fluid level gauge

09482_FX35_G0005

Exploded view of transmission mounting—2004–06 2WD Models

25. Secure the transmission assembly to the jack.

26. For 2WD models, remove the transmission assembly from the vehicle with the transmission jack.

27. For AWD models, remove the transmission assembly with the transfer case from the vehicle with the transmission jack. Then remove the transfer mounting bolts and separate the transfer from the transmission.

To install:

28. After inserting the torque converter to the transmission, be sure to check dimension A to ensure it is within the reference value limit of 0.98 in. (25.0mm).

✳✳ CAUTION

When setting the transmission jack, be careful not to allow it to collide against the drain plug.

29. For AWD models, install the transfer case to the transmission assembly.

30. Install the transmission assembly into the vehicle with the transmission jack.

31. Install the bolts fixing the transmission assembly to the engine. Use the accompanying illustrations for installation torques.

32. Unplug the fluid holes.

33. Install the A/T fluid charging pipe onto the A/T assembly.

34. Connect the A/T assembly connector.

35. Install the air breather hose.

36. Install the engine rear member.

37. Carefully, remove the transmission jack.

✳✳ CAUTION

When turning the crankshaft, turn it clockwise as viewed from the front of the engine.

38. For AWD models with the 3.5L engine, install the dynamic damper.

39. Align the positions of tightening bolts for the driveplate with those of the torque converter, and temporarily tighten the

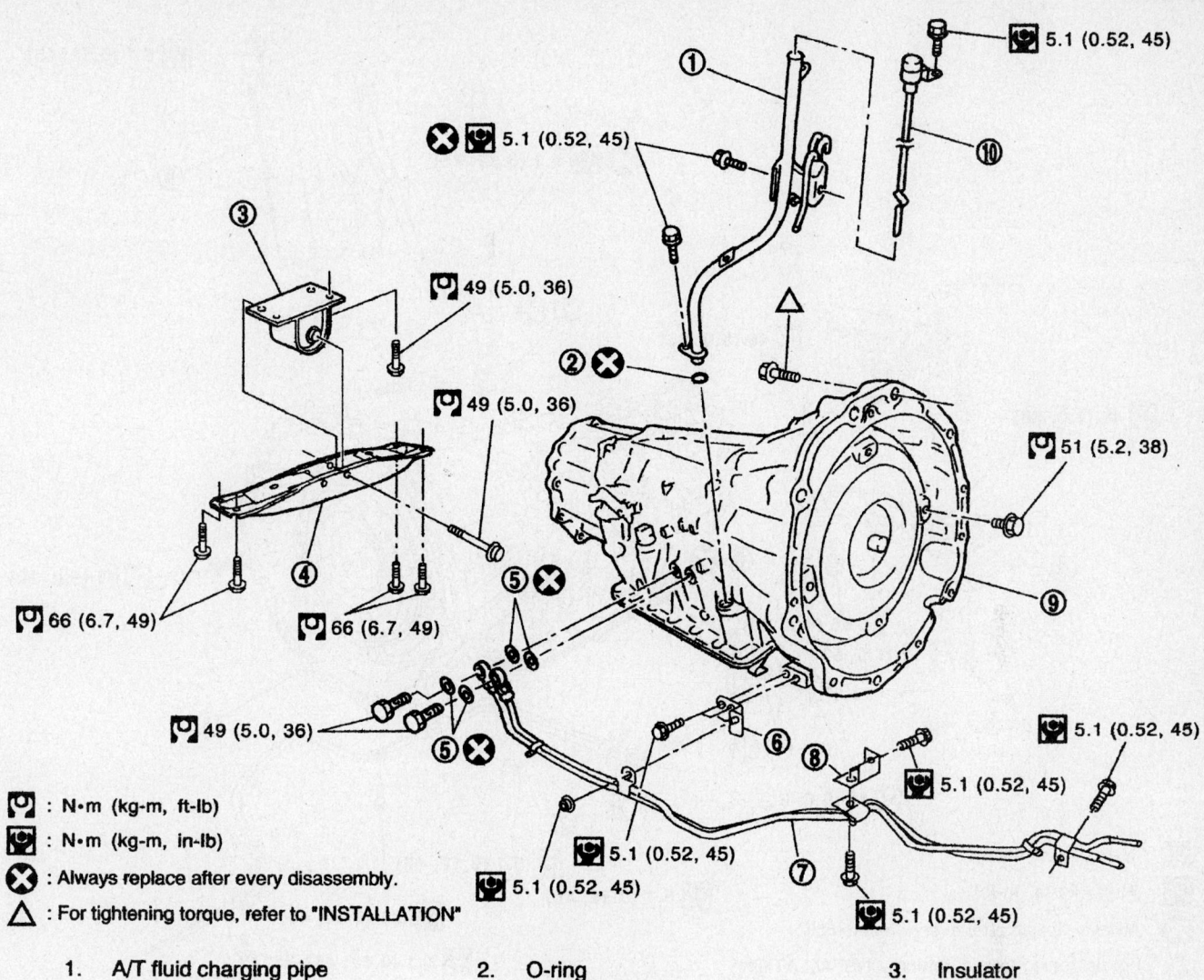

5.1 (0.52, 45)

5.1 (0.52, 45)

49 (5.0, 36)

49 (5.0, 36)

51 (5.2, 38)

66 (6.7, 49)

66 (6.7, 49)

49 (5.0, 36)

5.1 (0.52, 45)

5.1 (0.52, 45)

5.1 (0.52, 45)

5.1 (0.52, 45)

5.1 (0.52, 45)

: N•m (kg-m, ft-lb)

: N•m (kg-m, in-lb)

: Always replace after every disassembly.

: For tightening torque, refer to "INSTALLATION"

1. A/T fluid charging pipe	2. O-ring	3. Insulator
4. Rear member	5. Copper washer	6. Bracket
7. Fluid cooler tube	8. Bracket	9. Transmission assembly
10. A/T fluid level gauge		

09482_FX35_G0006

Exploded view of transmission mounting—2003–06 AWD Models w/3.5L engine

bolts. Then, tighten the bolts with the specified torque. When tightening the tightening bolts for the torque converter, after fixing the crankshaft pulley bolts, be sure to confirm the tightening torque of the crankshaft pulley mounting bolts. After the converter is installed to the driveplate, rotate the crankshaft several turns and check to be sure that transmission rotates freely without binding.

40. Install the dust cover onto the converter housing part.

41. Install the fluid cooler tube.

42. Install the starter motor, as follows:

a. Install the starter motor on the engine.

b. Install the starter motor mounting bolts and wiring harness clip bracket.

c. Install the B terminal nut.

d. Connect the S connector.

e. Install the engine rear undercover.

f. Connect the negative battery terminal.

43. Install the crankshaft position sensor (POS) on the A/T assembly.

44. Reattach the transmission shifter control rod by installing the snap pin retaining the control rod bracket to the control device assembly lever.

45. Install the driveshaft.

46. Install the exhaust front tube and center muffler.

47. Install the engine under cover.

48. Install the A/T fluid level gauge.

49. Install the engine cover.

50. Connect the negative battery terminal.

51. Check the adjustment of the A/T position, as follows:

a. Loosen the nut of the control rod.

b. Place the PNP switch and selector lever in the "P" position.

c. While pressing the lower lever toward the rear of the vehicle (in P position direction), tighten the nut to the specified torque.

❈❈ **CAUTION**

Do not push the bracket.

52. Check the A/T position, as follows:

a. Place the selector lever in "P"

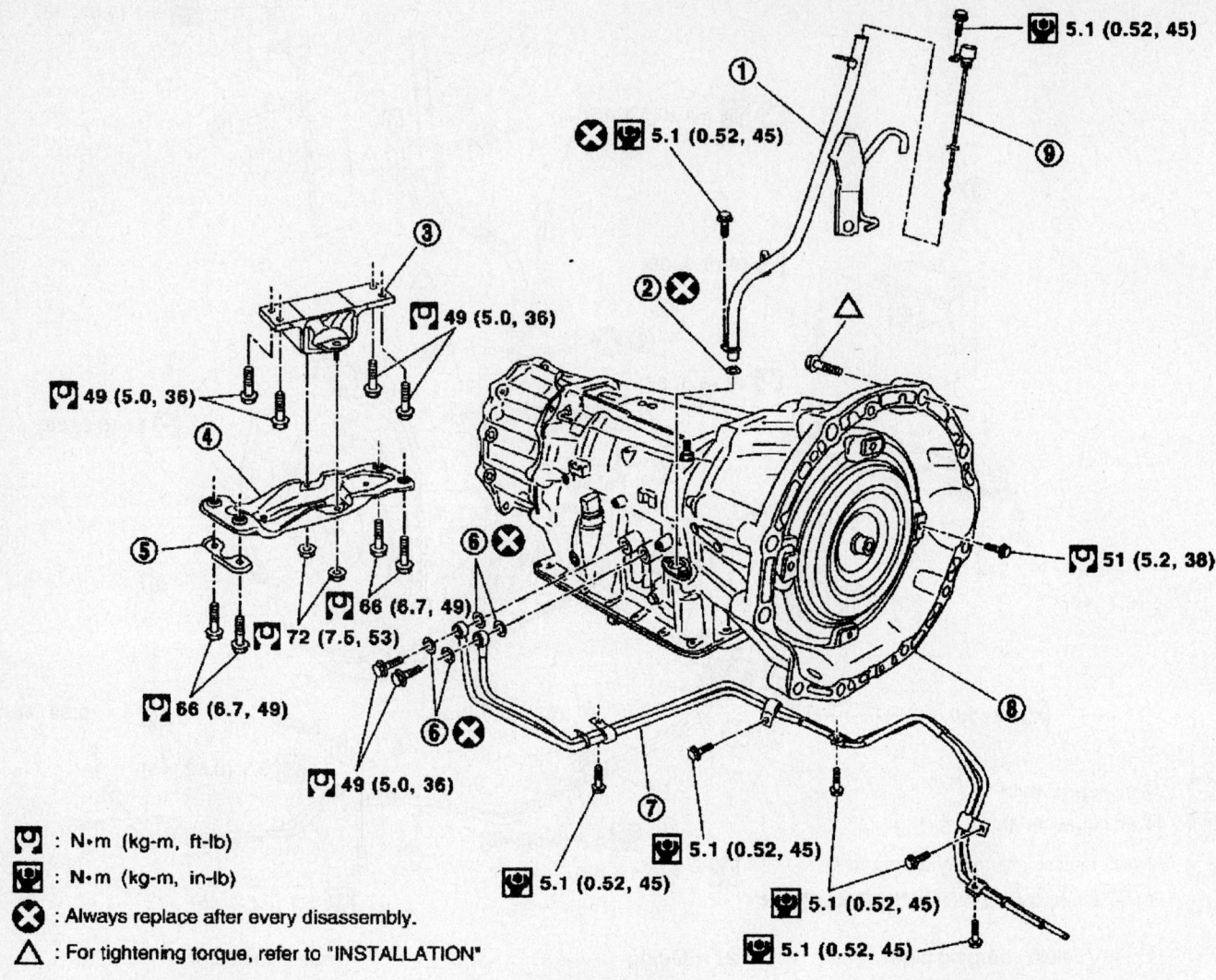

5.1 (0.52, 45)

5.1 (0.52, 45)

49 (5.0, 36)

49 (5.0, 36)

66 (6.7, 49)

72 (7.5, 53)

66 (6.7, 49)

49 (5.0, 36)

51 (5.2, 38)

5.1 (0.52, 45)

5.1 (0.52, 45)

5.1 (0.52, 45)

: N•m (kg-m, ft-lb)

: N•m (kg-m, in-lb)

: Always replace after every disassembly.

: For tightening torque, refer to "INSTALLATION"

1. Transmission assembly
2. A/T fluid charging pipe
3. O-ring
4. Fluid cooler tube
5. Copper washer
6. A/T fluid level gauge
7. Engine rear member
8. Insulator

09482_FX35_G0007

Exploded view of transmission mounting—2003 AWD Models w/4.5L engine

position, and turn ignition switch ON (engine OFF).

b. Make sure the selector lever can be shifted to positions other than "P" when the brake pedal is depressed. Also, make sure the selector lever can be shifted from "P" position only when the brake pedal is depressed.

c. Move the selector lever and check for excessive effort, sticking, noise or rattle.

d. Confirm the selector lever stops at each position with the feel of engagement when it is moved through all of the positions. Check whether or not the

actual position the selector lever is in matches the position shown by the shift position indicator and the transmission body.

e. The method of operating the lever to individual positions correctly should be as shown in the figure.

f. When selector button is pressed in "P", "R", or "N" position without applying forward/backward force to the selector lever, check the button operation for sticking.

g. Confirm the back-up lamps illuminate only when the lever is placed in the "R" position. Confirm the back-up lamps

do not illuminate when the selector lever is pushed against the "R" position while in the "P" or "N" positions.

h. Confirm the engine can only be started with the selector lever in the "P" and "N" positions.

i. Make sure the transmission is locked completely in the "P" position.

j. When the selector lever is set to manual shift gate, make sure the manual mode is displayed on the combination meter. Shift the selector lever to "+" and "-" sides, and make sure set shift position changes.

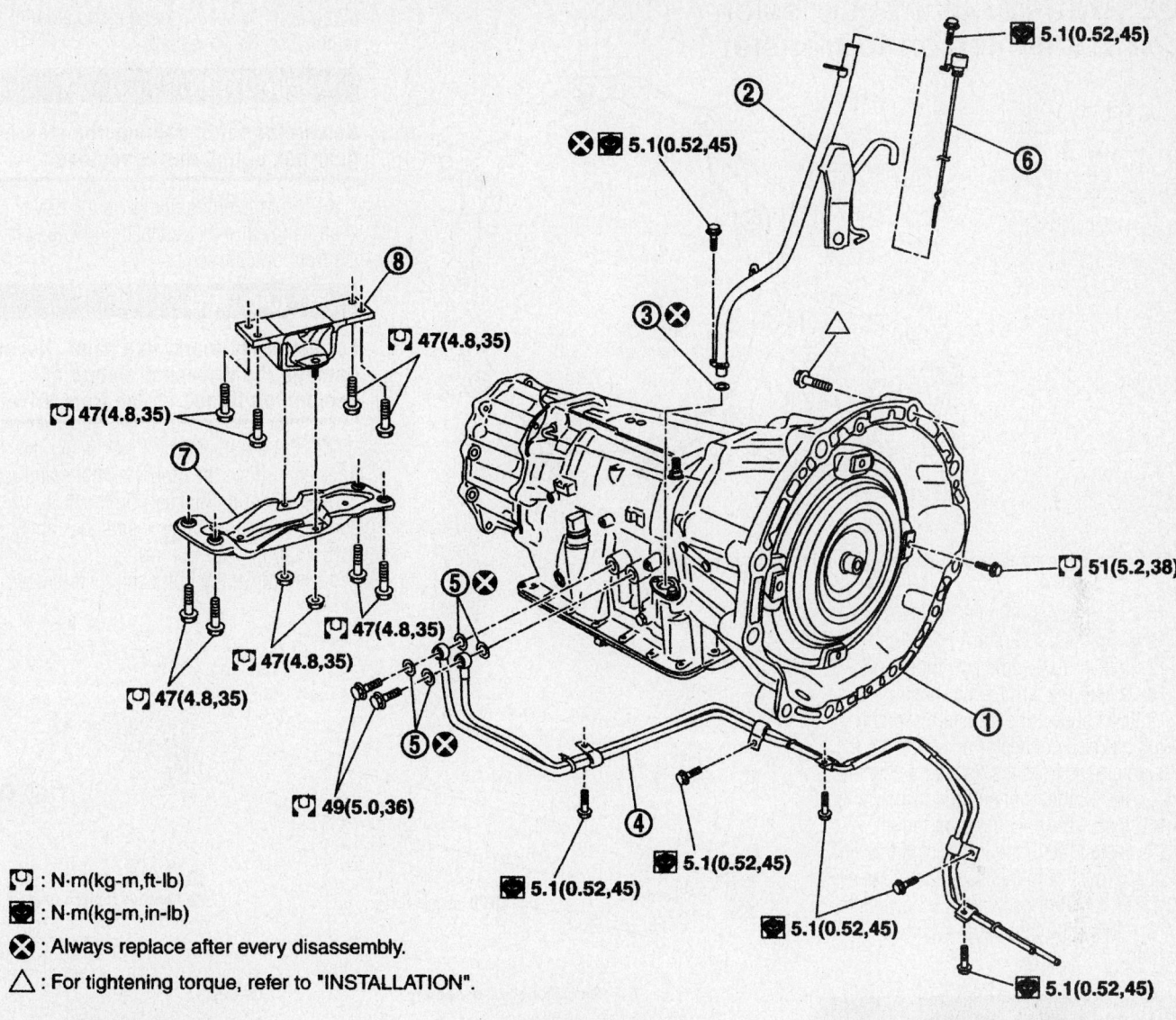

Legend:
- ⊙ : N·m(kg-m,ft-lb)
- ◼ : N·m(kg-m,in-lb)
- ✕ : Always replace after every disassembly.
- △ : For tightening torque, refer to "INSTALLATION".

1. Transmission assembly
2. A/T fluid charging pipe
3. O-ring
4. Fluid cooler tube
5. Copper washer
6. A/T fluid level gauge
7. Engine rear member
8. Insulator

67162-FX35-G195

Exploded view of transmission mounting—2004-06 AWD Models w/4.5L engine

Transfer Case Assembly

REMOVAL & INSTALLATION

1. Before servicing the vehicle, refer to the Precautions Section.
2. Raise and safely support the vehicle.
3. Remove the tunnel stay. Remove fixing bolts and nuts of tunnel stay and member stay, then remove those parts from vehicle.
4. Remove the exhaust front tube.

5. Remove the front and rear drive-shafts.
6. Disconnect the transfer assembly wiring harness connector and separate the wiring harness from the transfer assembly.
7. Remove the air breather hose.
8. Support the transfer assembly with a jack.
9. Remove the engine rear mounting.
10. Remove the transfer mounting bolts and separate the transfer from the transmission.

✷✷ CAUTION

Secure the transfer assembly to a jack or similar support.

To install:
11. Install the transfer mounting bolts and mount the transfer onto the transmission. Tighten the bolts to 27 ft. lbs. (37 Nm) in the sequence shown:
- Bolt No. 1 (75mm)
- Bolt No. 2 (45mm)

⊙ : Transfer to Transmission
⊗ : Transmission to transfer

Transfer case-to-transmission mounting bolt tightening sequence

67162-FX35-G196

- Bolt No. 3 (40mm)
- Bolt No. 4 (30mm)

12. Install the engine rear mounting.

13. Install the air breather hose.

14. Disconnect the transfer assembly wiring harness connector and separate the wiring harness from the transfer assembly.

15. Install the front and rear driveshafts.

16. Install the exhaust front tube.

17. Install the tunnel stay. Install fixing bolts and nuts of tunnel stay and member stay, then install those parts in the vehicle.

18. Check the fluid level and for fluid leakage.

Driveshaft

REMOVAL & INSTALLATION

Front

1. Before servicing the vehicle, refer to the Precautions Section.

2. Raise and safely support the vehicle.

3. Remove the front and rear engine undercover.

4. Remove the front cross bar.

5. Remove the exhaust front tube bracket.

6. Disconnect the heated oxygen sensor wiring harness connector.

7. Remove the exhaust front tube mounting nuts.

8. Remove the right bank catalytic converter.

9. Remove the power steering piping mounting bolts.

10. Remove the power steering gear box fixing bolts to secure the working area for removal of the driveshaft.

※※ CAUTION

Be careful not to damage the steering gear box piping during removal.

11. Put matching marks on the driveshaft flange and the companion flange on the front drive assembly.

※※ CAUTION

For matching mark, use paint. Never damage the driveshaft flange and companion flange on the front drive.

12. Set a transmission jack under the transfer, remove the mounting bolts and rear engine mounting bracket. On the 4.5L engine, lower the transmission jack about 0.16–0.20 in. (40–50mm).

13. Remove the bolts and then remove

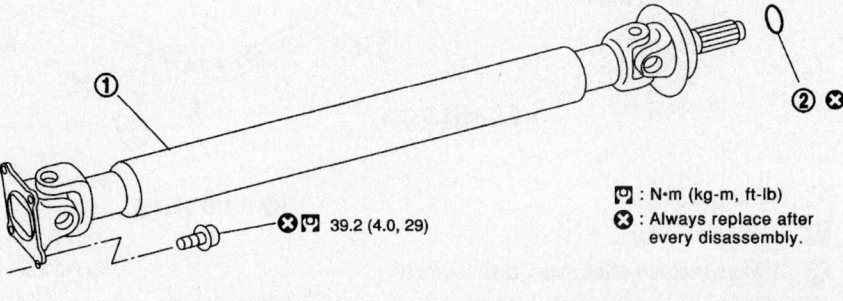

1. Propeller shaft assembly
2. O-ring

📐 : N·m (kg-m, ft-lb)
❌ : Always replace after every disassembly.

❌ 📐 39.2 (4.0, 29)

67162-FX35-G197

Front driveshaft—AWD Models

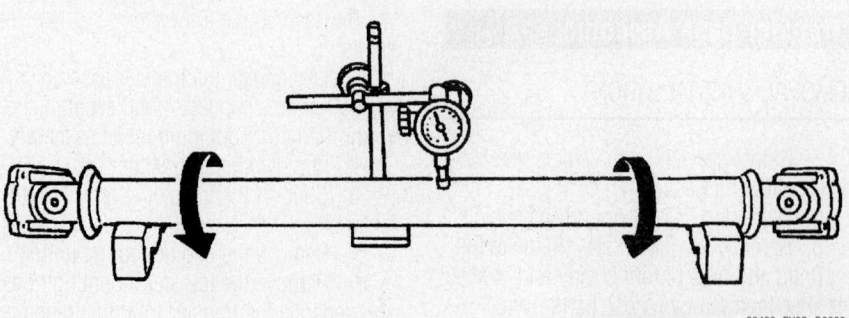

09482_FX35_G0008

Checking driveshaft runout

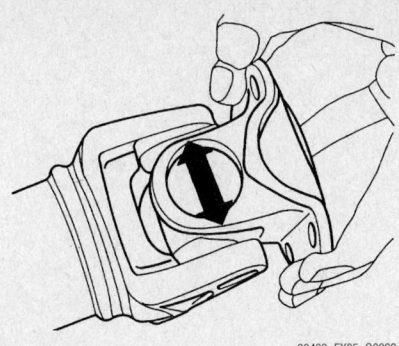

Checking driveshaft U-joint axial play

09482_FX35_G0009

the driveshaft from the front final drive and transfer.

14. Inspect the driveshaft runout as shown. If runout exceeds specifications, replace the driveshaft assembly. Runout limit: 0.8 mm (0.031 in.).

15. As shown in the accompanying illustration, fix the yoke on one side and check the axial play of the joint. The standard is **0**, therefore if there is any play, replace the driveshaft assembly.

16. Check the driveshaft for bending and damage. If damage is detected, replace the driveshaft assembly.

To install:

17. Install the driveshaft into the transfer, then install the driveshaft onto the front final drive flange while the matching marks are aligned.

18. Tighten the drive flange bolts to 29 ft. lbs. (39 Nm).

※※ CAUTION

Do not reuse the bolts and nuts. Always replace them with new ones.

19. The remainder of installation is the reverse of removal.

20. After installation, check the vibration by driving the vehicle. If vibration is present, remove the driveshaft from the final drive companion flange and turn the driveshaft 90, 180 or 270 degrees. Then, reinstall the driveshaft onto the companion flange.

21. Recheck vibration by driving the vehicle.

Rear

2003–04 MODELS

1. Before servicing the vehicle, refer to the Precautions Section.

2. Move the A/T select lever to "N" position and release the parking brake.

3. Raise and safely support the vehicle.

4. Remove the tunnel stay.

5. Remove the center muffler.

6. Loosen the center bearing lower mounting bracket fixing nuts.

7. Put matching marks on the flange and the rear driveshaft.

※※ CAUTION

For matching marks, use paint. Never damage the driveshaft flange and companion flange on the rear final drive.

8. Remove the driveshaft fixing bolts and nuts.

9. Remove the center bearing lower mounting bracket fixing nuts, remove the driveshaft from the vehicle.

10. Inspect driveshaft run out. Measure run out at the points shown below. If run out exceeds 0 in. (0mm), replace the driveshaft assembly.

2WD models:
- Position A: 192mm
- Position B: 190mm
- Position C: 185mm

2WD models:
- Position A: 162mm
- Position B: 245mm
- Position C: 185mm

To install:

➡ **During installation, refer to the following torque specifications:**

2WD models with 3.5L engine:
- Center flange attaching bolts and nuts: 70 ft. lbs. (95 Nm)
- First (front-most) flange-to-flange attaching nut: 61 ft. lbs. (83 Nm)
- Center bearing lower mounting bracket nuts: 32 ft. lbs. (44 Nm)
- Second (rear-most) rear-to-rear drive unit fasteners: 42 ft. lbs. (57 Nm)

AWD models with 3.5L engine:
- First (front-most) transmission-to-transmission attaching nuts and bolts: 54 ft. lbs. (74 Nm)
- Center flange-to-first driveshaft attaching bolts and nuts: 70 ft. lbs. (95 Nm)
- Second (rear-most) flange-to-flange center attaching nut: 61 ft. lbs. (83 Nm)
- Center bearing lower mounting bracket nuts: 32 ft. lbs. (44 Nm)
- Second (rear-most) rear-to-rear drive unit fasteners: 42 ft. lbs. (57 Nm)

AWD models with 4.5L engine:
- First (front-most) transmission-to-transmission attaching nuts and bolts: 54 ft. lbs. (74 Nm)
- First (front-most) flange-to-flange center attaching nut: 144 ft. lbs. (195 Nm)
- Center bearing lower mounting bracket nuts: 32 ft. lbs. (44 Nm)

- Center flange-to-second driveshaft attaching bolts and nuts: 70 ft. lbs. (95 Nm)
- Second (rear-most) rear-to-rear drive unit fasteners: 42 ft. lbs. (57 Nm)

11. Install the driveshaft onto the rear final drive companion flange while aligning the matching marks that were made during removal.

12. Adjust the position of the bearing cushion so as not to apply thrust play to the center bearing insulator.

13. Position the bearing cushion overlap as shown in the accompanying illustration.

14. Install the center bearing upper bracket with its arrow mark facing forward.

15. Tighten the center bearing upper mounting bracket fixing nuts to specified torque.

※※ CAUTION

Do not reuse the nuts. Always replace the nuts with a new ones.

16. If the companion flange has been removed, put a new alignment matching mark "C" on it. Then, reassemble using the following procedure:

➡ **Also, perform these steps when either of final drive and driveshaft is replaced with a new one.**

a. Erase the original mark C from companion flange with suitable solvent.

b. Measure the companion flange vertical run-out.

c. Determine the position where maximum run out is read on the dial indicator. Put a mark (shown by C in figure) on the flange perimeter corresponding to the maximum run out position.

17. If the driveshaft or final drive has been replaced, connect the driveshaft and final drive as follows:

➡ **Avoid damaging the joint boot. Protect it with a shop towel or equivalent.**

a. Install the driveshaft while aligning its matching mark A with the mark C on the joint as close as possible.

b. Tighten the joint bolts/nuts to the specified torque.

※※ CAUTION

Do not reuse the bolts, and washers. Always replace the them with new ones.

18. After installation, check driveline vibration by driving the vehicle. If vibration is present, remove the driveshaft from the final drive companion flange.

VQ35DE 2WD MODEL
3S80A-1VL107

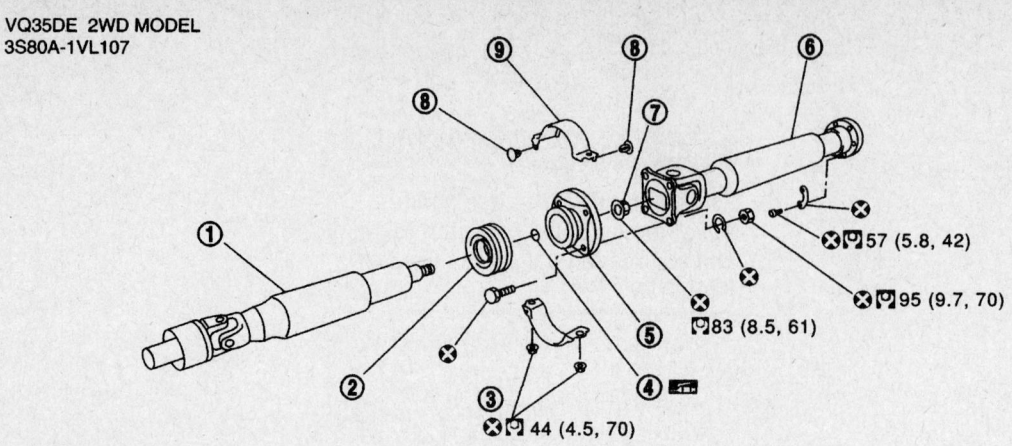

⊗🔧 57 (5.8, 42)

⊗🔧 95 (9.7, 70)

🔧 83 (8.5, 61)

⊗🔧 44 (4.5, 70)

VQ35DE AWD MODEL
3F80A-1VL107

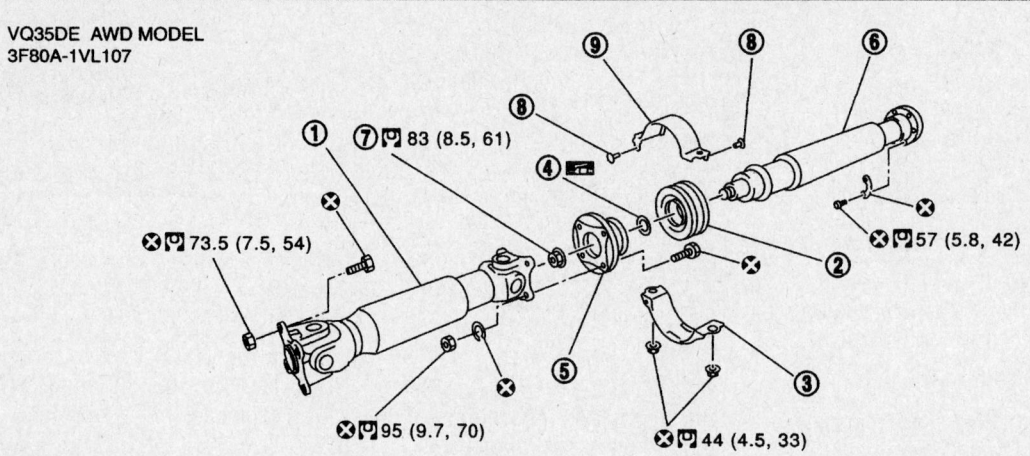

⑦ 🔧 83 (8.5, 61)

⊗🔧 73.5 (7.5, 54)

⊗🔧 57 (5.8, 42)

⊗🔧 95 (9.7, 70)

⊗🔧 44 (4.5, 33)

VK45DE AWD MODEL
3F80A-1VL107

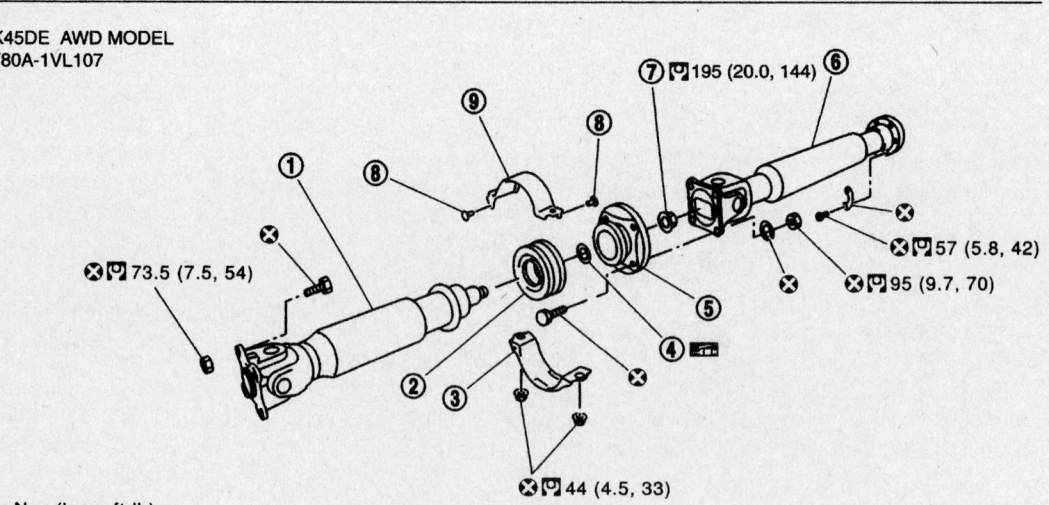

⑦🔧195 (20.0, 144)

⊗🔧 73.5 (7.5, 54)

⊗🔧 57 (5.8, 42)

⊗🔧 95 (9.7, 70)

⊗🔧 44 (4.5, 33)

🔧 : N•m (kg-m, ft-lb)

📇 : N•m (kg-m, in-lb)

⊗ : Always replase after every disassembly.

▨ : Apply multi-purpose grease.

1. Propeller shaft (1st shaft)

2. Center bearing assembly

3. Center bearing mounting bracket (lower)

4. Washer

5. Center flange

6. Propeller shaft (2nd shaft)

7. Lock nut

8. Clip

9. Center bearing mounting bracket (upper)

67162-FX35-G198

Exploded view of the three rear driveshaft configurations—2003–04 models

19. Turn the driveshaft 60, 120, 180, 240 or 300 degrees and reinstall the driveshaft to the companion flange, then check for vibration again by driving the vehicle on each angle position.

20. The remainder of installation is the reverse of removal.

2005–06 MODELS

1. Move the A/T select lever to N position and release the parking brake.
2. Remove the tunnel stay.
3. Before servicing the vehicle, refer to the Precautions Section.

4. Raise and safely support the vehicle.
5. Remove the center muffler.
6. Loosen mounting nuts of center bearing mounting brackets.

➡ **Tighten mounting nuts temporarily.**

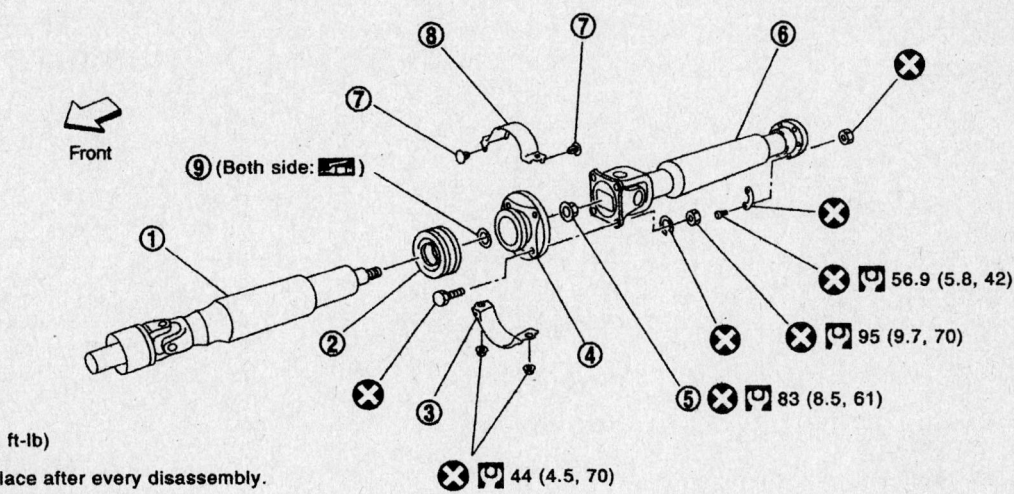

◘ : N•m (kg-m, ft-lb)

❌ : Always replace after every disassembly.

▨ : Apply multi-purpose grease.

1. Propeller shaft (1st shaft)	2. Center bearing assembly	3. Center bearing mounting bracket (Lower)
4. Center flange	5. Lock nut	6. Propeller shaft (2nd shaft)
7. Clip	8. Center bearing mounting bracket (Upper)	9. Washer

09482_FX35_G0010

Rear driveshaft—2005–06 2wd models

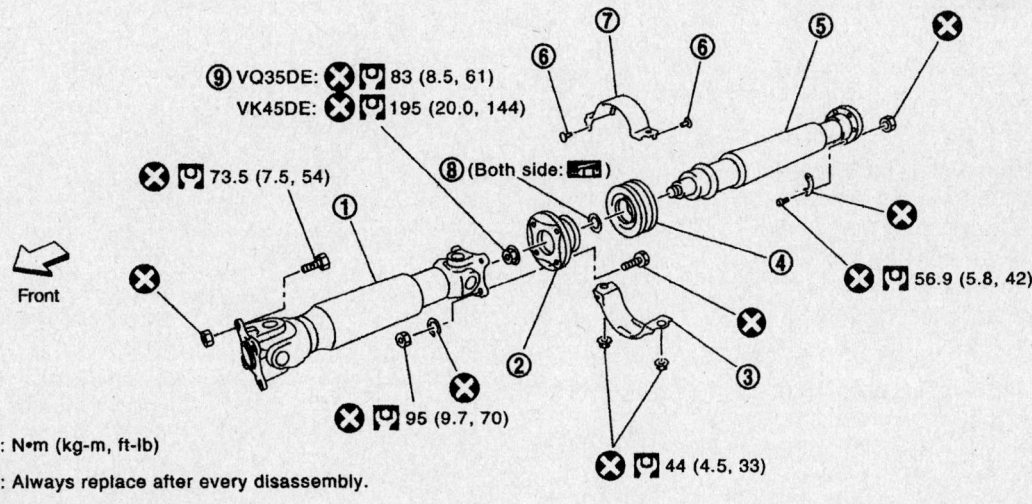

◘ : N•m (kg-m, ft-lb)

❌ : Always replace after every disassembly.

▨ : Apply multi-purpose grease.

1. Propeller shaft (1st shaft)	2. Center flange	3. Center bearing mounting bracket (Lower)
4. Center bearing assembly	5. Propeller shaft (2nd shaft)	6. Clip
7. Center bearing mounting bracket (Upper)	8. Washer	9. Lock nut

09482_FX35_G0011

Rear driveshaft—2005–06 4wd models

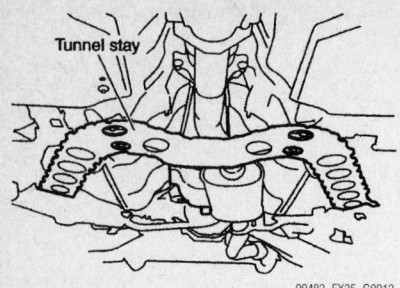

Remove the tunnel stay

09482_FX35_G0012

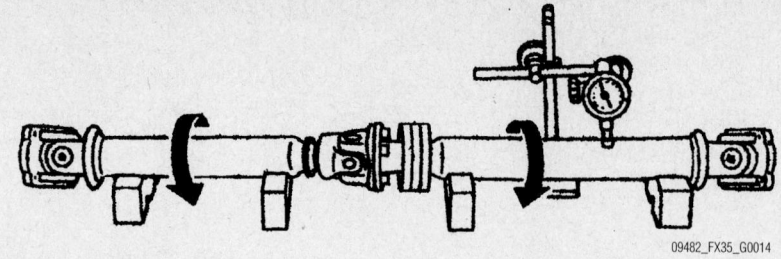

09482_FX35_G0014

Checking rear driveshaft runout

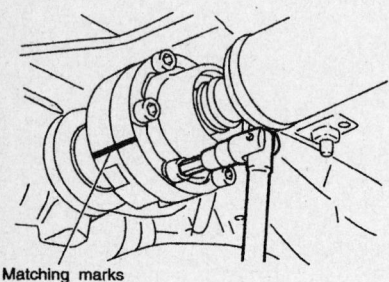

Matching marks

09482_GX35_G0013

Matchmark the joints

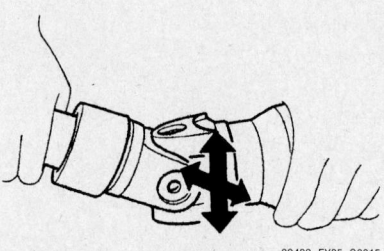

09482_FX35_G0015

Checking rear driveshaft U-joint axial play

7. For 2WD models, matchmark the driveshaft Rebro joint and final drive companion flange.

➡Do not damage the driveshaft and companion flange.

8. For AWD models, matchmark the driveshaft flange yoke and transfer companion flange, and the Rebro joint with the final drive companion flange.

9. Remove the driveshaft fixing bolts and nuts.

10. Remove the center bearing mounting bracket fixing nuts.

11. Remove the driveshaft.

➡If the constant velocity joint was bent during driveshaft assembly removal, installation, or transportation, its boot may be damaged. Wrap the boot interference area to metal part with shop cloth or rubber to protect boot from breakage.

12. Inspect the driveshaft runout as shown. If runout exceeds specifications, replace the driveshaft assembly. Driveshaft runout limit: 0.8 mm (0.031 in.).

13. Check the axial play of the U-joint. If any play is noted, replace the U-joint. Check the driveshaft for bend and damage. If damage is detected, replace the driveshaft.

➡Do not disassemble joints.

14. Check the center bearing for noise and damage. If noise or damage is detected, replace the center bearing.

To install:

15. Installation is the reverse of removal. Note the following:

➡Avoid damaging the Rebro joint boot, protect it with a shop towel or equivalent.

a. Align matching marks to install driveshaft to final drive and transfer (AWD models only) companion flanges,

and then tighten to the torques shown in the accompanying illustrations.

b. Install center bearing mounting bracket (upper) with its arrow mark facing forward.

c. Adjust position of mounting bracket sliding back and forth to prevent play in thrust direction of center bearing insulator. Install bracket to vehicle.

d. After assembly, perform a driving test to check driveshaft vibration. If vibration occurred, separate driveshaft from final drive. Reinstall companion flange after rotating it by 60, 120, 180, 240, 300 degrees. Then perform driving test and check driveshaft vibration again at each point.

e. If driveshaft or final drive has been replaced, connect them as follows: Install the driveshaft while aligning its matching mark A with the matching mark B on the joint as close as possible. Tighten the joint bolts to the specified torque.

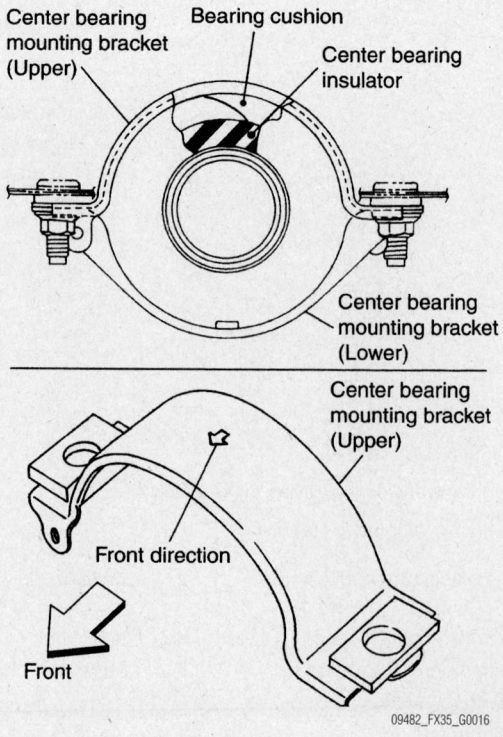

09482_FX35_G0016

Correct center bearing installation—2005–06 models

➡**Do not reuse the bolts, nuts and washers.**

Front Halfshaft

REMOVAL & INSTALLATION

Front

1. Before servicing the vehicle, refer to the Precautions Section.
2. Raise and safely support the vehicle.
3. Remove the wheels.
4. Remove the undercover.
5. Remove the cotter pin. Then remove the locknut from the driveshaft.
6. Remove the wheel sensor wiring harness from the strut assembly.

✳✳ CAUTION

Do not pull on the wheel sensor wiring harness.

7. Remove the brake hose lockplate. Then remove the brake hose from the strut assembly.
8. Remove the mounting bolts and nuts between the strut assembly and the steering knuckle.
9. Separate the halfshaft from the steering knuckle.

✳✳ CAUTION

When removing the halfshaft, do not apply an excessive angle to the half-shaft joint. Also, be careful not to excessively extend the slide joint.

10. For the left-hand halfshaft, remove the fixing bolts of the front final drive side assembly halfshaft, then remove the half-shaft from the vehicle.
11. For the right-hand halfshaft, pry off the halfshaft from the front final drive assembly.
12. Inspect the halfshaft, as follows:
    a. Move the joint up/down, left /right, and in the axial direction. Check for any rough movement or significant looseness.
    b. Check the boot for cracks or other damage, and also for grease leakage.
    c. If damage is found, disassemble the halfshaft and replace the defective part(s) with new one(s).

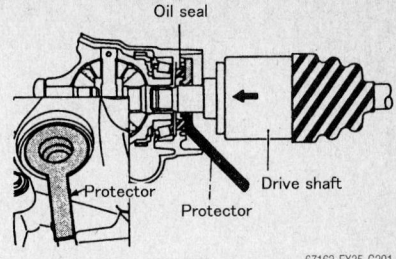
67162-FX35-G201

Use a protector to prevent damage to the side oil seal during installation

To install:

➡**Refer to component parts location and do not reuse non-reusable parts.**

13. Installation is the reverse of removal. During installation, tighten the fasteners to the following specifications:
    - Halfshaft-to-hub locknut: 203 ft. lbs. (275 Nm)
    - Left-hand side halfshaft-to-drive unit bolts: 33 ft. lbs. (44.5 Nm)
    - Right-hand side, in order to prevent damage to the front final drive assembly side oil seal, first fit a protector onto the oil seal before inserting the halfshaft. Slide the halfshaft into the slide joint and tap it with a hammer to install it securely.

✳✳ CAUTION

Be sure to check that circular clip is securely fastened.

14. Check the condition of the wheel sensor wiring harness.

Rear

1. Before servicing the vehicle, refer to the Precautions Section.
2. Raise and safely support the vehicle.
3. Remove the wheels.
4. Remove the cotter pin. Then, remove the locknut from the outer end of the half-shaft.
5. Remove the fixing nuts and bolts between the side flange and halfshaft.
6. Separate the halfshaft from the wheel hub and bearing assembly by lightly tapping the end with a suitable hammer and wood block. If it is hard to separate, use a suitable puller.
7. Remove the halfshaft from the axle.

✳✳ CAUTION

When removing the halfshaft, do not apply an excessive angle to the half-shaft joint. Also be careful not to excessively extend the slide joint.

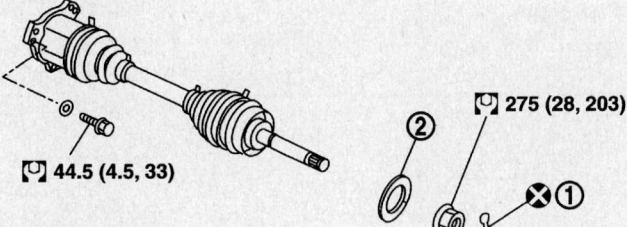

44.5 (4.5, 33)   **275 (28, 203)**

❌ : Always replace after every disassembly

🔧 : N·m(kg-m,ft-lb)

1.   Cotter pin   2.   Washer

67162-FX35-G199

Exploded view of left-hand front halfshaft mounting

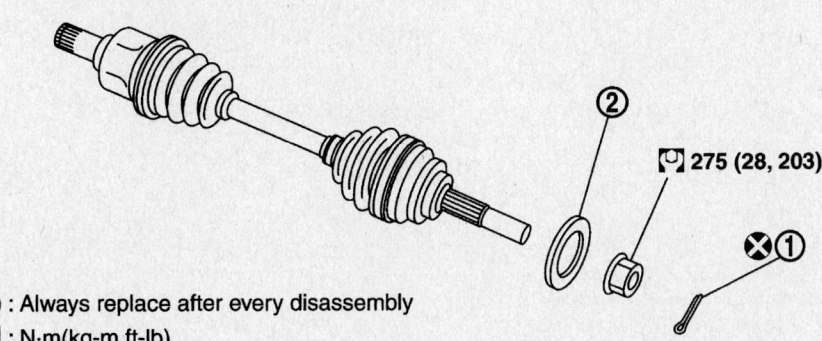

275 (28, 203)

❌ : Always replace after every disassembly

🔧 : N·m(kg-m,ft-lb)

1.   Cotter pin   2.   Washer

67162-FX35-G200

Exploded view of right-hand front halfshaft mounting

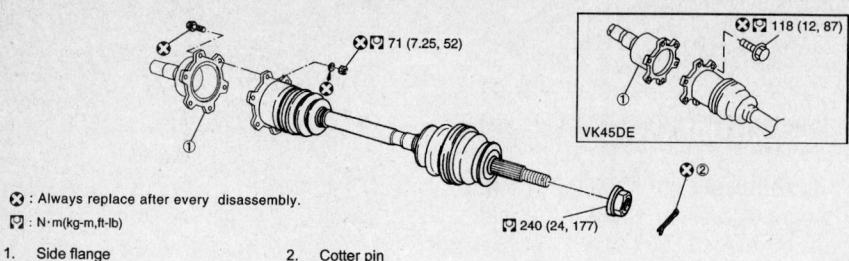

⊗ : Always replace after every disassembly.

⏷ : N·m(kg-m,ft-lb)

1. Side flange   2. Cotter pin

67162-FX35-G207

Exploded view of the rear halfshaft mounting

8. Inspect the halfshaft, as follows:

a. Move the joint up/down, left /right, and in the axial direction. Check for any rough movement or significant looseness.

b. Check the boot for cracks or other damage, and also for grease leakage.

c. If damage is found, disassemble the halfshaft and replace the defective part(s) with new one(s).

To install:

➡Refer to component parts location and do not reuse non-reusable parts.

9. Installation is the reverse of removal. During installation, tighten the fasteners to the following specifications:

- Halfshaft-to-hub locknut: 177 ft. lbs. (240 Nm)
- Halfshaft-to-side flange bolts: 52 ft. lbs. (71 Nm)

CV-Joints

OVERHAUL

Front

1. Before servicing the vehicle, refer to the Precautions Section.

2. Raise and safely support the vehicle.

3. For the front final drive assembly side, perform the following:

a. Press the halfshaft in a vice.

⁕⁕ CAUTION

When retaining the shaft in a vice, always use copper or aluminum plates between the vise and shaft.

b. Remove the boot bands.

c. If the plug needs to be removed, move the boot toward the wheel side, and drive it out with a plastic hammer.

d. Put matching marks on the spider assembly and shaft.

⁕⁕ CAUTION

Use paint for the matching mark, but don't damage the spider assembly and halfshaft.

e. Remove the snapring, then remove the spider assembly from the shaft.

f. Remove the boot from the shaft.

g. Remove the old grease on the slide joint assembly with paper towels.

4. For the wheel side, perform the following:

a. Place the halfshaft in a vice.

⁕⁕ CAUTION

When retaining the halfshaft in a vice, always use copper or aluminum plates between a vise and halfshaft.

b. Remove the boot bands.

c. Remove the boot from the joint sub-assembly.

d. Screw a halfshaft puller or equivalent 1.18 in. (30 mm) or more into the threaded part of the joint sub-assembly. Pull the joint sub-assembly out of the halfshaft.

⁕⁕ CAUTION

If the joint sub-assembly cannot be removed after five or more unsuccessful attempts, replace the shaft and joint subassembly as a set. Use a sliding hammer on the halfshaft and remove the halfshaft.

e. Remove the boot from the halfshaft.

f. Remove the circular clip from the halfshaft.

g. While rotating the ball cage, remove the old grease from the joint sub-assembly with paper towels.

5. Replace the halfshaft if there is any run out, cracking, or other damage.

6. Make sure there is no rough rotation or unusual axial looseness of the joint sub-assembly.

7. Make sure there is no foreign material inside joint sub-assembly.

8. Check joint sub-assembly for compression scar, cracks or fractures.

⁕⁕ CAUTION

If there are any irregular conditions of joint sub-assembly components, replace the entire joint sub-assembly.

9. Inspect the housing and spider assembly. If the roller or roller surface of the spider assembly has scratches or other

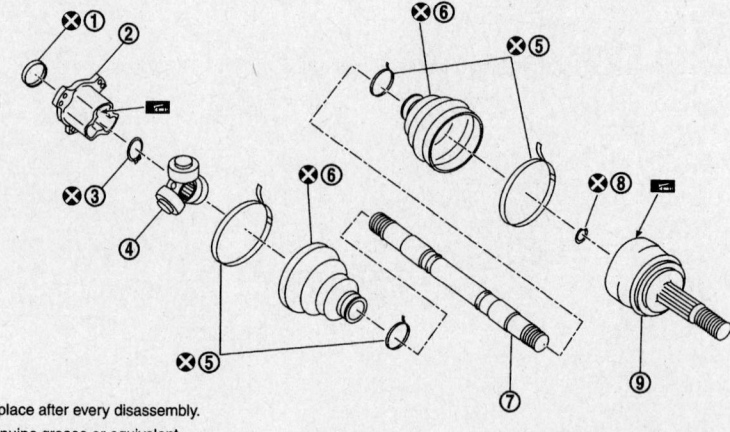

⊗ : Always replace after every disassembly.

▱ : Nissan genuine grease or equivalent.

1. Plug	2. Housing	3. Snap ring
4. Spider assembly	5. Boot band	6. Boot
7. Shaft	8. Circular clip	9. Joint sub-assembly

67162-FX35-G202

Exploded view of the left-hand halfshaft and CV-joint

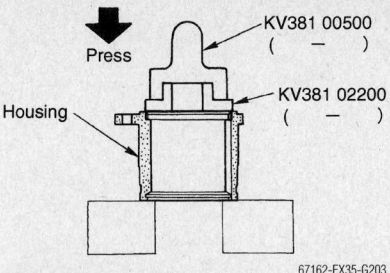

If the plug has been removed, use a press and tools to drive in a new one

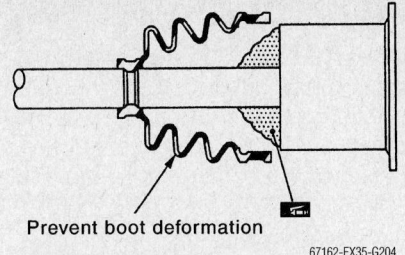

Install grease where indicated

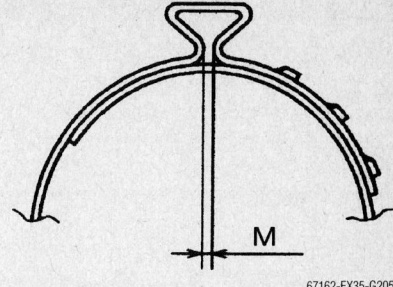

Install the band so that the gap M matches specification

wear, replace the housing and the spider assembly.

➡The housing and spider assembly are components which are used as a set.

To assemble:
10. To assemble the front final drive assembly side, perform the following:
a. If the plug has been removed, use a drift to press in a new one.

➡Discard the old plug, and replace it with a new one.

b. Wind the serrated part of the shaft with tape. Install the boot band and boot on the shaft. Be careful not to damage the boot.

➡Discard the old boot band and boot, and replace them with new ones.

c. Remove the protective tape wound around the serrated part of the shaft.
d. Align the alignment marks, which were made when the spider assembly was removed. Install the spider assembly with the serration chamfer facing shaft.
e. Secure the spider assembly with a snapring.

➡Discard the old snapring, and replace it with a new one.

f. Apply multi-purpose grease to the spider assembly and sliding surface.
g. Install the housing onto the spider assembly. Apply multi-purpose grease to the housing.

h. Install the boot securely into the grooves (indicated by * marks shown in the figure).

✳✳ CAUTION

If there is grease on the boot mounting surfaces (indicated by * marks) of the shaft and housing, the boot may come off. Remove all grease from surfaces.

i. Make sure the boot installation length "L" is 3.74–3.82 in. (95–97mm). Insert a flat-bladed pry tool or similar tool into the smaller side of the boot. Bleed air from the boot to prevent boot deformation.

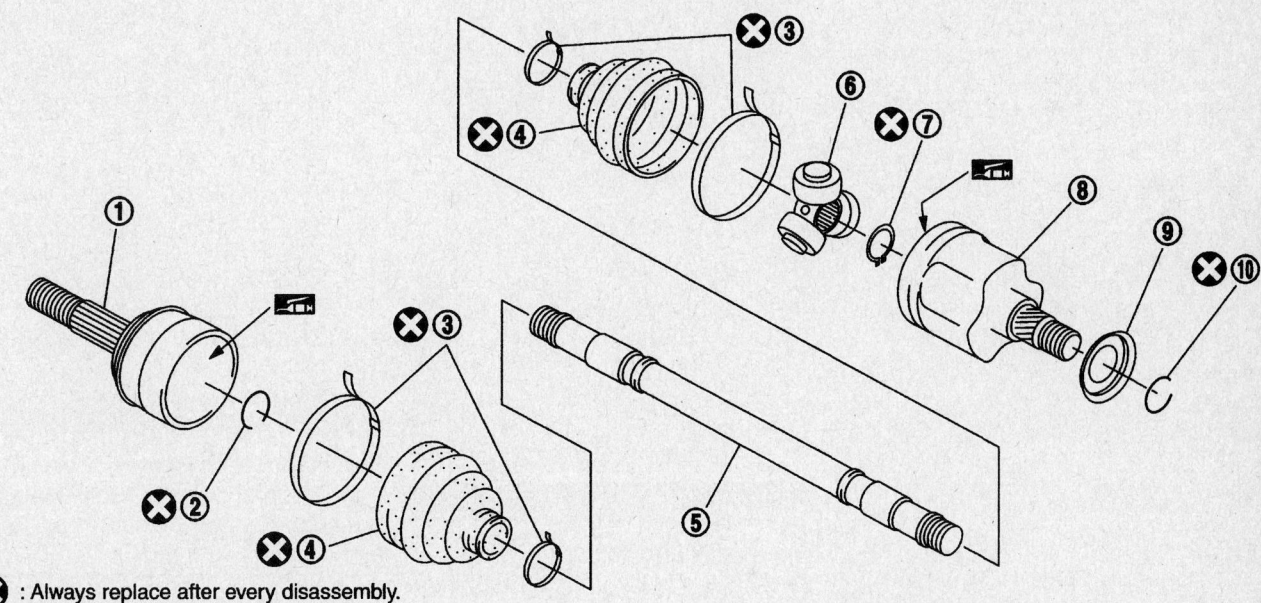

❌ : Always replace after every disassembly.

▦ : Nissan genuine grease or equivalent.

1. Joint sub-assembly	2. Circular clip	3. Boot band
4. Boot	5. Shaft	6. Spider assembly
7. Snap ring	8. Housing	9. Dust shield
10. Circular clip		

Exploded view of the right-hand halfshaft and CV-joint

✳✳ CAUTION

The boot may break if the boot installation length is less than the standard value. Take care not to touch the tip of the pry tool to the inside surface of the boot.

j. Install the new larger and smaller boot bands securely with a suitable tool.

➡**Discard the old boot bands, and replace them with new ones.**

k. Secure the boot band so that dimension "M" shown is:
- Large diameter side: 0.118 in. (3.0mm)
- Small diameter side: 0.079 in. (2.0mm)

l. After installing the housing and shaft, rotate the boot to check whether or not the actual position is correct. If the boot position is not correct, secure the boot with new boot bands again.

11. Assemble the wheel side of the halfshaft as follows:

a. Insert grease (Nissan genuine grease or equivalent) into the joint subassembly serration hole until grease begins to ooze from the ball groove and serration hole. After inserting the grease, use a shop cloth to wipe off the old grease that has oozed out.

b. Wind the serrated part of the shaft with tape for protection of the boot, then install the boot band and boot on the shaft. Be careful not to damage the boot.

➡**Discard the old boot band and boot, and replace them with new ones.**

c. Remove the protective tape from the shaft.

d. Attach the circular clip to the shaft. At this time, the circular clip must fit securely into the shaft groove. Attach the nut to the joint sub-assembly. Use a wooden hammer to press-fit it in place.

➡**Discard the old circular clip, and replace it with a new one.**

e. Insert the specified amount of grease (Nissan genuine grease or equivalent) into the boot from the large end of the boot.

➡**Grease amount: 3.35–4.06 oz. (95-115 g) for the left-hand side or 3.99–4.34 oz. (113-123 g) for the right-hand side.**

f. Install the boot securely into the grooves (indicated by * marks) shown.

✳✳ CAUTION

If there is grease on the boot mounting surfaces (indicated by * marks) of the shaft and housing of the joint sub assembly, the boot may come off. Remove all grease from the surfaces.

g. Make sure the boot installation length "L" is 5.35 in. (136mm) for the left-hand side or 6.21–6.29 in. (157.8–159.8mm) for the right-hand side or. Insert a flat-bladed pry tool or similar tool into the smaller side of boot. Bleed air from the boot to prevent boot deformation.

✳✳ CAUTION

The boot may brake if the boot installation length is less than the standard value. Be careful that the pry tool tip does not contact the inside surface of the boot.

h. Install the new larger and smaller boot bands securely with a suitable tool.

➡**Discard the old boot bands, and replace them with new ones. Secure the boot band so that dimension "M" shown is:**

- Large diameter side: 0.118 in. (3.0mm)
- Small diameter side: 0.079 in. (2.0mm)

i. After installing the joint subassembly and shaft, rotate the boot to check whether or not the actual position is correct. If the boot position is not correct, secure the boot with new boot bands again.

Rear

1. Before servicing the vehicle, refer to the Precautions Section.
2. Raise and safely support the vehicle.
3. Disassemble the final drive side, as follows:

a. Press shaft in a vice.

✳✳ CAUTION

When retaining the halfshaft in a vice, always use copper or aluminum plates between a vise and halfshaft.

b. Remove the boot bands.
c. If the plug needs to be removed, move the boot toward the wheel side, and drive it out with a plastic hammer.
d. Remove the stopper ring with a flat-bladed pry tool, and pull out the housing.

e. Remove the snaping, then remove the ball cage/steel ball/inner race assembly from the shaft.
f. Remove the boot from the shaft.
g. Remove the old grease on the housing with paper towels.

4. Disassemble the wheel side, as follows:

a. Place the shaft in a vice.

✳✳ CAUTION

When retaining the halfshaft in a vice, always use copper or aluminum plates between a vise and halfshaft.

b. Remove the boot bands. Then remove the boot from the joint subassembly.
c. Thread a halfshaft puller 30 mm (1.18 in) or more into threaded part of joint sub-assembly. Pull joint sub-assembly out of shaft.

✳✳ CAUTION

If the joint sub-assembly cannot be removed after five or more unsuccessful attempts, replace the shaft and joint subassembly as a set. Align the sliding hammer and halfshaft and remove them by pulling directly out.

d. Remove the boot from shaft.
e. Remove the circular clip from the shaft.
f. While rotating the ball cage, remove the old grease on the joint sub-assembly with paper towels.

5. Replace the shaft if there is any run out, cracking, or other damage.

6. Inspect the joint sub-assembly. Make sure there is no rough rotation or unusual axial looseness. Make sure there is no foreign material inside the joint. Check the joint sub-assembly for compression scars, cracks, or fractures.

✳✳ CAUTION

If there are any irregular conditions of the joint sub-assembly components, replace the entire joint sub-assembly.

7. Inspect the sliding joint side housing. Make sure there are no compression scars, cracks, fractures or unusual wear on the ball rolling surface. Make sure there is no damage to the shaft screws. Make sure there is no deformation of the boot installation parts.

8. Inspect the ball cage. Make sure there are no compression scars, cracks, fractures of the sliding surface.

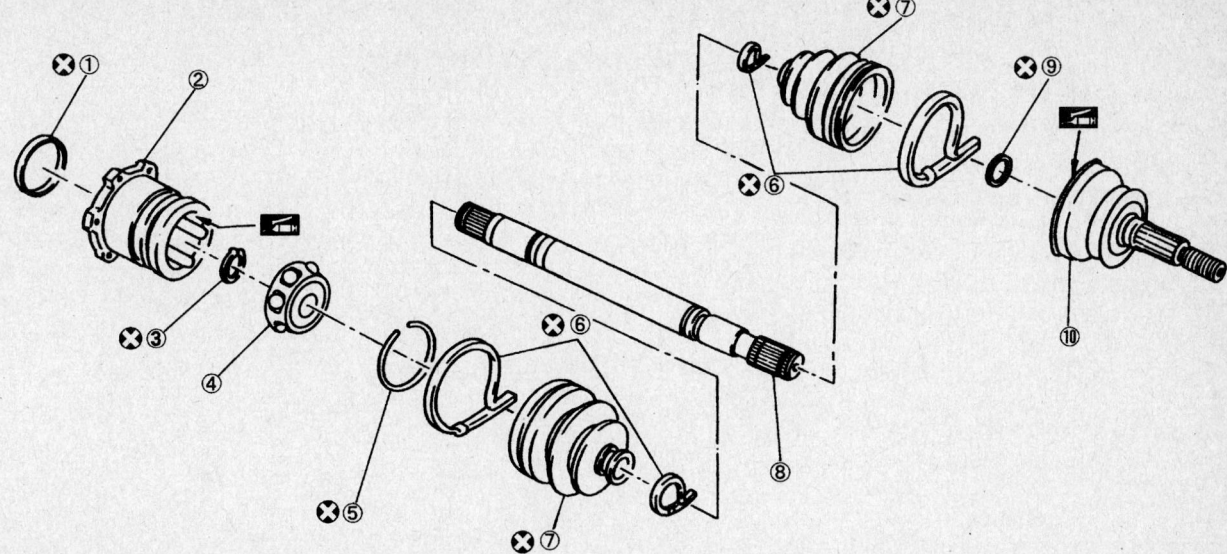

67162-FX35-G208

⊗ : Always replace after every disassembly.

1. Plug	2. Housing	3. Snap ring
4. Ball cage/Steel ball/Inner race assembly	5. Stopper ring	6. Boot band
7. Boot	8. Shaft	9. Circular clip
10. Joint sub-assembly		

Exploded view of the rear halfshaft and CV-joint

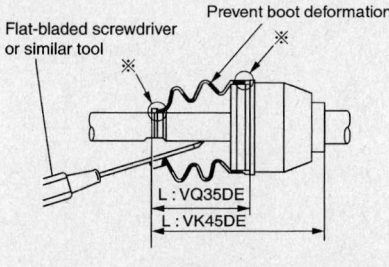

67162-FX35-G209

When installing the CV-joint boot make sure installation length L meets specification

9. Inspect the steel balls. Make sure there are no compression scars, cracks, fractures or unusual wear.

10. Inspect the inner race. Check ball sliding surface for compression scars, cracks, or fractures. Make sure there is no damage to the serrated part.

✳✳ CAUTION

If there are any irregular conditions on the components, replace with a new set of housing, ball cage, steel ball and inner race.

To assemble:

11. Assemble the final drive side as follows:

a. If the plug has been removed, use a drift to press in a new one.

➡**Discard old plug, and replace with a new one.**

b. Wind the serrated part of the shaft with tape to protect the boot. Install the boot band and boot on the shaft. Be careful not to damage the boot.

➡**Discard the old boot band and boot, and replace it with new ones.**

c. Remove the protective tape wound around the serrated part of shaft.

d. Install the ball cage/steel ball/inner race assembly onto the shaft, and secure them tightly with a snapring.

➡**Discard the old snapring, and replace it with a new one.**

e. Insert the proper amount of grease (NISSAN genuine grease or equivalent) into the housing (* point), and install it to shaft. Grease amount:
- 4.5L models: 6.17–6.88 oz. (175–195g)
- 3.5L models: 4.37–4.73 oz. (124–134g)

f. Install the stopper ring onto the housing.

g. After installation, pull the shaft to check engagement between the joint sub-assembly and the stopper ring.

h. Install the boot securely into the grooves (indicated by * marks) shown.

✳✳ CAUTION

If there is grease on the boot mounting surfaces (indicated by * marks) of the shaft and housing, the boot may come off. Remove all grease from the surfaces.

i. Make sure the boot installation length "L" is the length indicated. Insert a flat-bladed pry tool or similar tool into the smaller side of boot. Bleed air from the boot to prevent boot deformation. Boot installation length "L":
- 4.5L models: 5.82 in. (147.9mm)
- 3.5L models: 3.697 in. (93.9mm)

✳✳ CAUTION

The boot may break if the boot installation length is less than the standard value. Take care not to touch the tip of the pry tool to the inside of the boot.

j. Secure the big and small ends of the boot with the new boot bands as shown.

➡**Discard the old boot bands, and replace with new ones.**

k. After installing the housing and shaft, rotate the boot to check whether or not the actual position is correct. If the boot position is not correct, secure the boot with new boot band again.

12. Assemble the wheel side as follows:

a. Insert the proper amount of grease (NISSAN genuine grease or equivalent) into the joint sub-assembly serration hole until grease begins to ooze from the ball groove and the serration hole. Afterward, use a shop cloth to wipe off the old grease that has oozed out.

b. Wind the serrated part of the shaft with tape. Install the boot band and boot onto the shaft. Be careful not to damage the boot.

➡**Discard the old boot band and boot, replace each with new ones.**

c. Remove the protective tape wound around the serrated part of the shaft.

➡**Discard the old circular clip, and replace with a new one.**

d. Attach the circular clip to the shaft. At this time, the circular clip must fit securely into the shaft groove. Attach the nut to the joint sub-assembly. Use a wooden hammer to press-fit.

e. Insert the proper amount of grease (NISSAN genuine grease or equivalent) into the housing from the large end of the boot.

Grease amount:
- 4.5L models: 4.93–5.64 in. (140–160g)
- 3.5L models: 3.03–3.39 in. (86–96g)

f. Install the boot securely into grooves (indicated by * marks) shown.

※※ CAUTION

If there is grease on the boot mounting surfaces (indicated by * marks) of the shaft and the housing, the boot may come off. Remove all grease from the surfaces.

g. Make sure the boot installation length "L" is the length indicated below. Insert a flat-bladed pry tool or similar tool into the smaller side of the boot. Bleed air from the boot to prevent boot deformation.

Boot installation length "L":

- 4.5L models: 5.57 in. (141.5mm)
- 3.5L models: 3.82 in. (97mm)

※※ CAUTION

The boot may break if the boot installation length is less than the standard value. Be careful that the pry tool tip does not contact the inside surface of the boot.

h. Secure the big and small ends of the boot with new boot bands as shown.

➡**Discard the old boot bands, and replace with new ones.**

i. After installing the joint sub-assembly and shaft, rotate the boot to check whether or not the actual position is correct. If the boot position is not correct, secure the boot with new boot bands again.

Front Differential Seals

REMOVAL & INSTALLATION

Pinion Seal

2003–05 MODELS

1. Before servicing the vehicle, refer to the Precautions Section.
2. Raise and safely support the vehicle.
3. Remove the front driveshaft.
4. Put a matching mark on the end of the drive pinion corresponding to the B position matching mark on the final drive companion flange.

※※ CAUTION

Use paint for the matching mark; never damage drive pinion. The matching mark B on the final drive companion flange indicates the maximum vertical run out position.

5. Using the drive pinion flange wrench. Remove the drive pinion locknut with tool number: KV40104000 or equivalent.
6. Remove the companion flange using a puller (commercial service tool).
7. Remove the front oil seal using outer race puller ST33290001 (J34286) or equivalent.

To install:
8. Apply multi-purpose grease to the sealing lips of the oil seal. Drive the oil seal into the differential case using special service tools ST33400001 (J26082) and KV38102510 or equivalents so that the oil seal is flush with the gear carrier end.

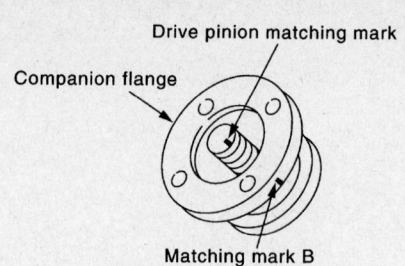

During installation the matching marks should align—2003–05 models

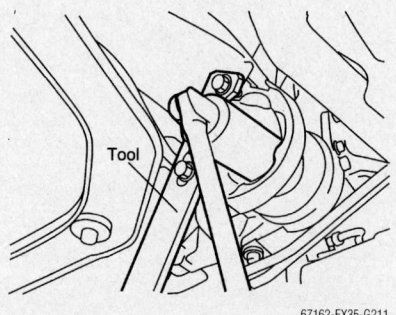

To remove the flange bolt, a spanner wrench will need to be used—2003–05 models

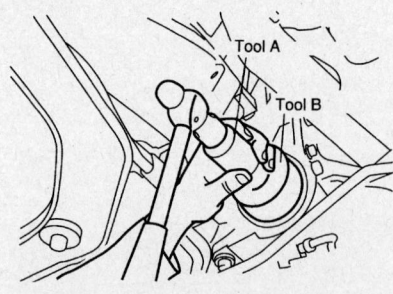

Drive the new seal into the differential housing as shown—2003–05 models

➡**When installing the front oil seal, be careful not to get it inclined.**

9. Discard the old front oil seal. Always replace it with a new one.
10. Install the companion flange while aligning the matching mark of the drive pinion with the matching mark B of the companion flange.
11. Apply oil to the drive pinion threads and the seating surface of the drive pinion locknut.
12. Using the drive pinion flange wrench KV40104000 or equivalent. Install the drive pinion locknut with the tool. Tighten to 94–181 ft. lbs. (127–245 Nm).

✳✳ **CAUTION**

Do not reuse the drive pinion locknut. Always replace it with a new one.

13. Install the front driveshaft.

2006 MODELS

1. Before servicing the vehicle, refer to the Precautions Section.
2. Raise and safely support the vehicle.
3. Drain the gear oil.
4. Remove the front driveshaft.
5. Remove the both front halfshafts.
6. Remove the side shaft assembly.
7. Remove the drive pinion lock nut using a flange wrench.
8. Match (B) the end of the drive pinion. The matchmark (B) should be in line with the matching mark (A) on the companion flange (1).

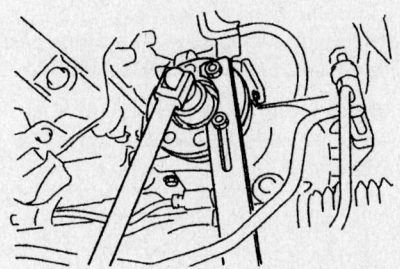

Remove the drive pinion lock nut using a flange wrench—2006 models

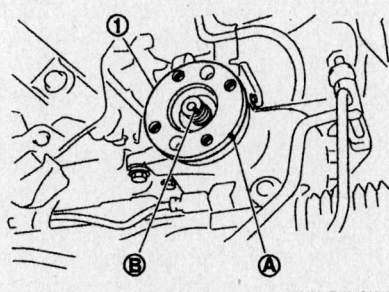

Matchmark positioning—2006 models

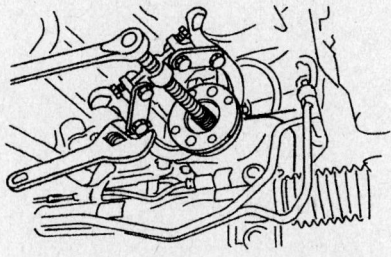

Remove the companion flange using a puller—2006 models

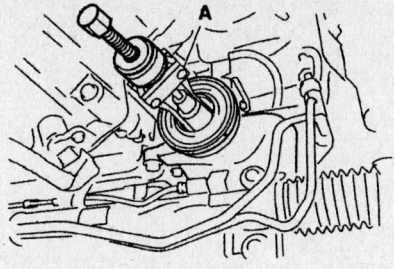

Remove the pinion seal using a puller—2006 models

➥ **For matching mark, use paint. Do not damage companion flange and drive pinion.**

➥ **The matching mark (A) on the final drive companion flange (1) indicates the maximum vertical runout position.**

9. Remove the companion flange using a puller.
10. Remove the front oil seal using a puller.

➥ **Be careful not to damage the gear carrier.**

To install:

11. Apply multi-purpose grease to the front oil seal lips.
12. Install the front oil seal as shown.

➥ **Do not reuse the oil seal. When installing, do not cock the oil seal.**

13. Align the matchmark (B) of drive pinion with the matchmark (A) of the companion flange, and then install the companion flange (1).
14. Apply anti-corrosion oil to the threads and seat of the new drive pinion lock nut, and temporarily tighten the drive pinion lock nut to drive pinion.

➥ **Do not reuse the drive pinion lock nut.**

15. Tighten the drive pinion lock nut, while adjusting total preload torque. Drive

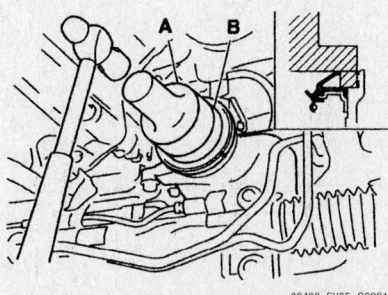

Installing the pinion seal—2006 models

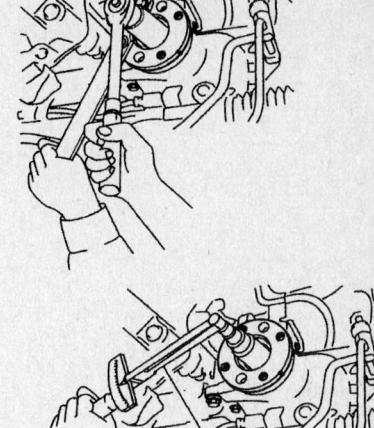

Adjusting the preload—2006 models

pinion lock nut tightening torque: 127.4–245.0 Nm (94–181 ft. lbs.). Total preload torque: 1.56–2.65 Nm (14–23 inch lbs.)

16. Adjust the lower limit of the drive pinion lock nut tightening torque first. After adjustment, rotate the drive pinion back and forth 2 to 3 times to check for unusual noise, rotation malfunction, and other malfunctions.
17. If the measured value is out of specification, remove the final drive assembly and disassemble drive pinion parts to check and adjust each part.
18. Install the front driveshaft.
19. Install the side shaft assembly.
20. Install both front halfshafts.
21. Refill the gear oil and the check oil level.
22. Check the final drive for oil leakage.

Side Seals

LEFT SIDE

1. Before servicing the vehicle, refer to the Precautions Section.
2. Raise and safely support the vehicle.

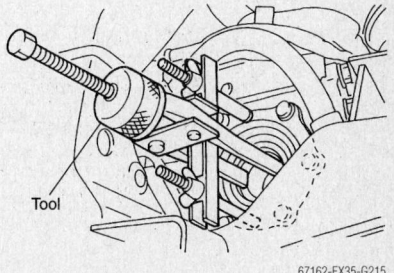

Remove the old side seal using a puller as shown—2003–05 models

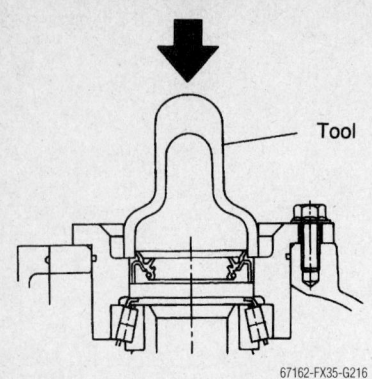

Drive the new seal into the housing as shown—2003–05 models and 2006 models right side

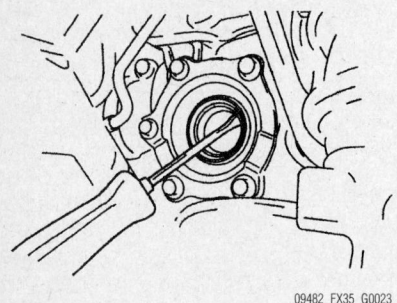

Remove the side oil seal using a flat-bladed screwdriver—2006 models

3. Remove the front final drive assembly from vehicle.
4. Remove the side oil seal using a flat-bladed screwdriver.

➥Be careful not to damage gear carrier.

To install:

5. Apply multi-purpose grease to sealing lips of side oil seal.
6. Using a driver (A), install the side oil

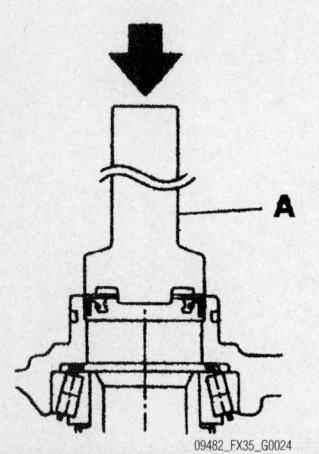

Left side seal installation—2006 models

seal so that its surface comes even the end surface of the gear carrier.

➥Do not reuse oil seal. When installing, do not incline oil seal.

7. Install the front final drive assembly on vehicle.
8. Install the front halfshaft.

RIGHT SIDE

1. Before servicing the vehicle, refer to the Precautions Section.
2. Raise and safely support the vehicle.
3. Remove the halfshaft.
4. Remove the side oil seal using a flat-bladed screwdriver.

➥Be careful not to damage gear carrier.

To install:

5. Apply multi-purpose grease to sealing lips of side oil seal.
6. Using a driver, install the side oil seal so that its surface is even with the end surface of the side retainer.

➥Do not reuse oil seal. When installing, do not incline oil seal.

7. Install the front drive shaft.

Rear Differential Seals

REMOVAL & INSTALLATION

Pinion Seal

2003–05 MODELS

1. Before servicing the vehicle, refer to the Precautions Section.
2. Raise and safely support the vehicle.
3. Remove the rear driveshaft.
4. Put a matching mark on the end of the drive pinion corresponding to the C position matching mark on the final drive companion flange.

⁂ CAUTION

Use paint for the matching mark; never damage drive pinion. The matching mark B on the final drive companion flange indicates the maximum vertical run out position.

5. Using the drive pinion flange wrench KV40104000 or equivalent, remove the drive pinion locknut.
6. Remove the companion flange using the puller.
7. Remove the front oil seal using the side bearing outer race puller ST33290001 (J34286) or equivalent.

To install:

8. Apply multi-purpose grease to sealing lips of oil seal. Press the front oil seal into the carrier with tool ST30720000 (J25405) or equivalent.

⁂ CAUTION

When installing the side oil seal, be careful not to install it crooked. Do not reuse the old side oil seal. Always replace the oil seal with a new one.

9. Align the matching mark of the drive pinion with the matching mark C of the companion flange, then install the companion flange.
10. Apply oil on the threaded part of the drive pinion and the seating surface of the drive pinion locknut.
11. Install the drive pinion nut with tool KV40104000 or equivalent. Tighten to 109–238 ft. lbs. (147–323 Nm).

⁂ CAUTION

The drive pinion locknut is not reusable. Never reuse the drive pinion nut.

12. Install the rear driveshaft.

2006 MODELS

1. Before servicing the vehicle, refer to the Precautions Section.
2. Raise and safely support the vehicle.
3. Check the identification stamp on the differential to determine if replacement of the collapsible spacer is necessary when replacing the front oil seal. If collapsible spacer replacement is necessary, remove the final drive assembly and disassemble it to replace front oil seal and collapsible spacer.

➥The reuse of a collapsible spacer is prohibited in principle. However, it is reusable on a one-time basis only in cases when replacing front oil seal.

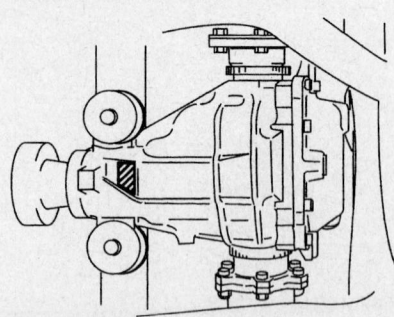

Stamping location

- No stamp: Not required
- **0** on the far right of the stamp: Required
- **1 or 01** on the far right of the stamp: Not required

➡**Be sure to make a stamping after replacing front oil seal.**

4. After replacing the front oil seal, make a stamping on the stamping point.

➡**Stamping should be made from left to right.**

5. Drain the gear oil.
6. Determine if a collapsible spacer replacement is required.
7. Remove the center muffler.
8. Remove the rear wheel sensor.
9. Remove the halfshaft from the final drive. Then suspend it by wire. Install an adapter to the side flange, and then pull out the side flange with a slide hammer.

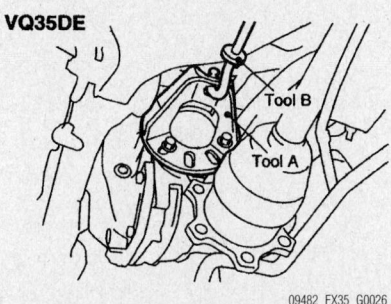

Slide flange removal—2006 3.5L models

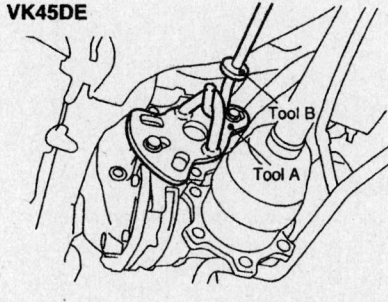

Slide flange removal—2006 4.5L models

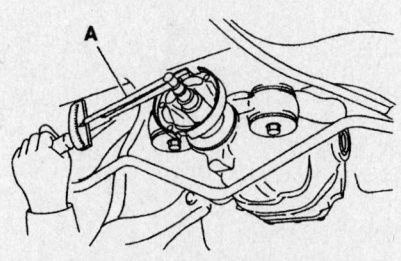

Preload measurement—2006 models

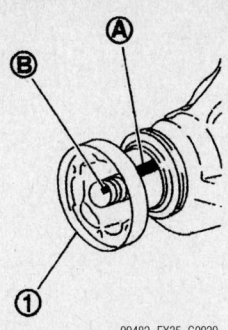

Matchmarking the drive pinion and flange—2006 models

10. Remove the driveshaft.
11. Measure the total preload with an inch-pound torque wrench gauge.

➡**Record the preload measurement.**

12. Put a matchmark (B) on the end of the drive pinion. The matching mark (B) should be in line with the matching mark (A) on companion flange (1).

➡**For matchmarks, use paint. Do not damage the companion flange and drive pinion.**

➡**The matchmark (A) on the final drive companion flange (1) indicates the maximum vertical runout position.**

13. Remove the drive pinion lock nut using the flange wrench.
14. Remove companion flange using a puller.

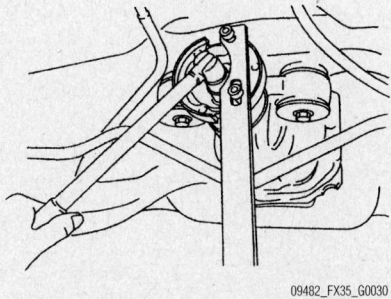

Removing the pinion nut—2006 models

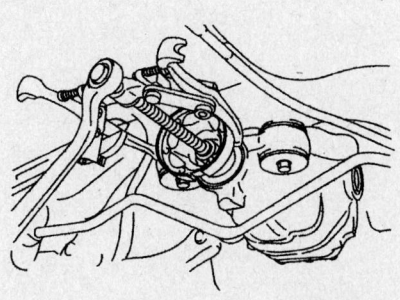

Companion flange removal—2006 models

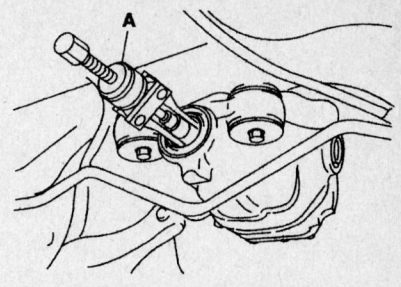

Pinion seal removal—2006 models

15. Remove front oil seal using the puller.

To install:

16. Apply multi-purpose grease to the front oil seal lips.
17. Install the front oil seal using the drift as shown.

➡**Do not reuse the oil seal. Do not cock the oil seal when installing. Align the matchmark (B) of the drive pinion with the matching mark (A) of companion flange (1), and then install the companion flange (1).**

18. Apply anti-corrosion oil to the thread and seat of the new drive pinion lock nut, and temporarily tighten the drive pinion lock nut.
19. Align the matchmark (B) of drive pinion with the matchmark (A) of the companion flange (1), and then install the companion flange (1).
20. Apply anti-corrosion oil to the threads and seat of the new drive pinion lock nut, and temporarily tighten the drive pinion lock nut.

➡**Do not reuse the drive pinion lock nut.**

21. Tighten the drive pinion lock nut, while adjust total preload torque. Drive pinion lock nut tightening torque: 147–323 Nm (109–238 ft. lbs.). Total preload torque should equal the measurement taken during removal plus an additional 0.1–0.4 Nm (1–3 inch lbs.).

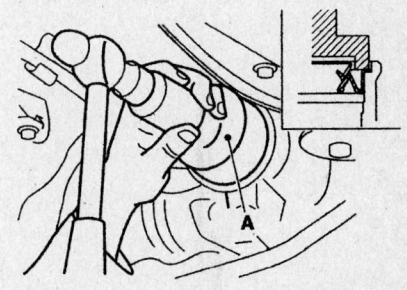

Pinion seal installation—2006 models

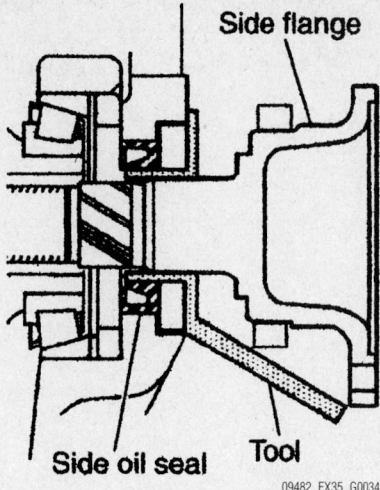

Side flange

Side oil seal   Tool

09482_FX35_G0034

After the side flange is inserted and the serrated part of side gear has engaged the serrated part of flange, remove the protector—2006 models

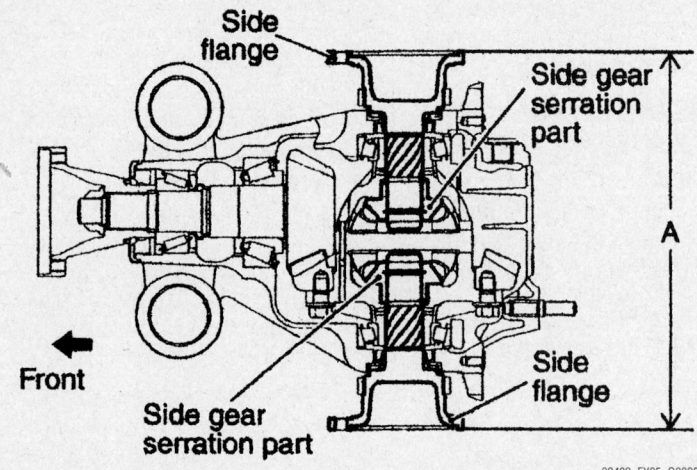

Side flange installation dimensions—2006 models

a. Adjust the lower limit of the drive pinion lock nut tightening torque first.

b. If the preload torque exceeds the specified value, replace the collapsible spacer and tighten it again to adjust. Do not loosen the drive pinion lock nut to adjust the preload torque.

22. Make a stamping for identification of front oil seal replacement frequency.

➡ Be sure to make a stamping after replacing front oil seal.

23. Install the driveshaft.
24. Install the side flange with the following procedure:

a. Attach the protector to the side oil seal.

b. After the side flange is inserted and the serrated part of side gear has engaged the serrated part of flange, remove the protector.

c. Put a suitable drift on the center of side flange, then drive it until it bottoms (the sound changes).

d. Confirm the dimension of the side flange installation. Measurement A: 326–328mm (12.83–12.91 in.).

25. Install the halfshaft.
26. Install the rear wheel sensor.
27. Install the center muffler.
28. Refill the gear oil.
29. Check the final drive for oil leakage.

Side Seals

1. Before servicing the vehicle, refer to the Precautions Section.
2. Raise and safely support the vehicle.
3. Remove the center muffler.
4. Remove the side flange, as follows:

a. Remove ABS rear wheel sensor.

b. Remove the halfshaft and axle assemblies.

c. Install the axle stand onto the side

VQ35DE

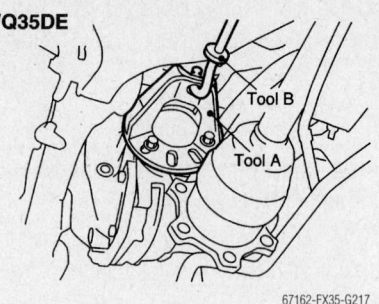

67162-FX35-G217

Remove the side seal using an adapter and puller—3.5L engine

VK45DE

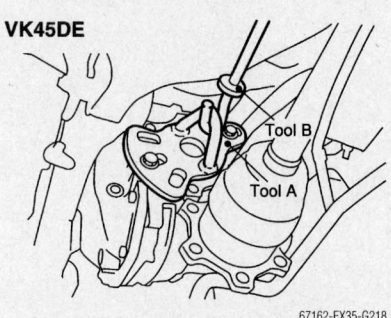

67162-FX35-G218

Remove the side seal using an adapter and puller—4.5L engine

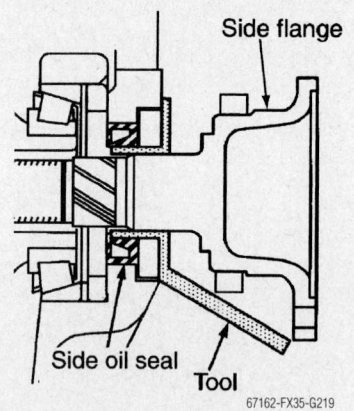

67162-FX35-G219

Use a protector when installing the side flange, as shown

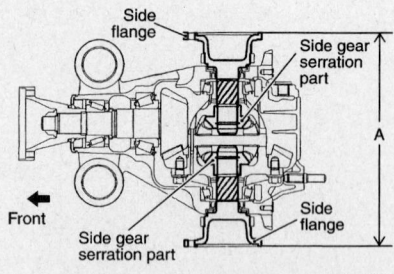

67162-FX35-G220

After installation, confirm that the side flange installed dimension A is approximately 12.83–12.91 in. (326–328mm)

flange, then pull out the side flange with a slidehammer.

5. Remove side oil seal using a flat-bladed pry tool.

To install:

6. Apply multi-purpose grease to the sealing lips of the side oil seal.

7. Using drift KV38100200 (J26233) or equivalent, press-fit the side oil seal so that its surface is flush with the end surface of the case.

✵✵ CAUTION

When installing the side oil seal be careful not to press it in crooked. Discard the old side oil seal. Always replace them with new ones.

8. Install the side flange, as follows:

a. Attach protector KV38107900 (J39352) or equivalent to the side oil seal.

b. After the side flange is inserted and the serrated part of the side gear has engaged the serrated part of the flange, remove the protector.

c. Put a suitable drift on the center of the side flange, then drive it in until it clicks or snaps in place.

➡**When installation is completed, the driving sound of the side flange turns into a sound which seems to affect the whole final drive.**

9. Confirm that the dimension of the side flange installation (Measurement A) in

the illustration falls within the range of 12.83–12.91 in. (326–328mm).

10. Install the halfshaft and axle assembly.

11. Install the center muffler.

12. Align the installation position of the ABS rear wheel sensor.

Front Differential

REMOVAL & INSTALLATION

With the 3.5L Engine

1. Before servicing the vehicle, refer to the Precautions Section.

2. Raise and safely support the vehicle.

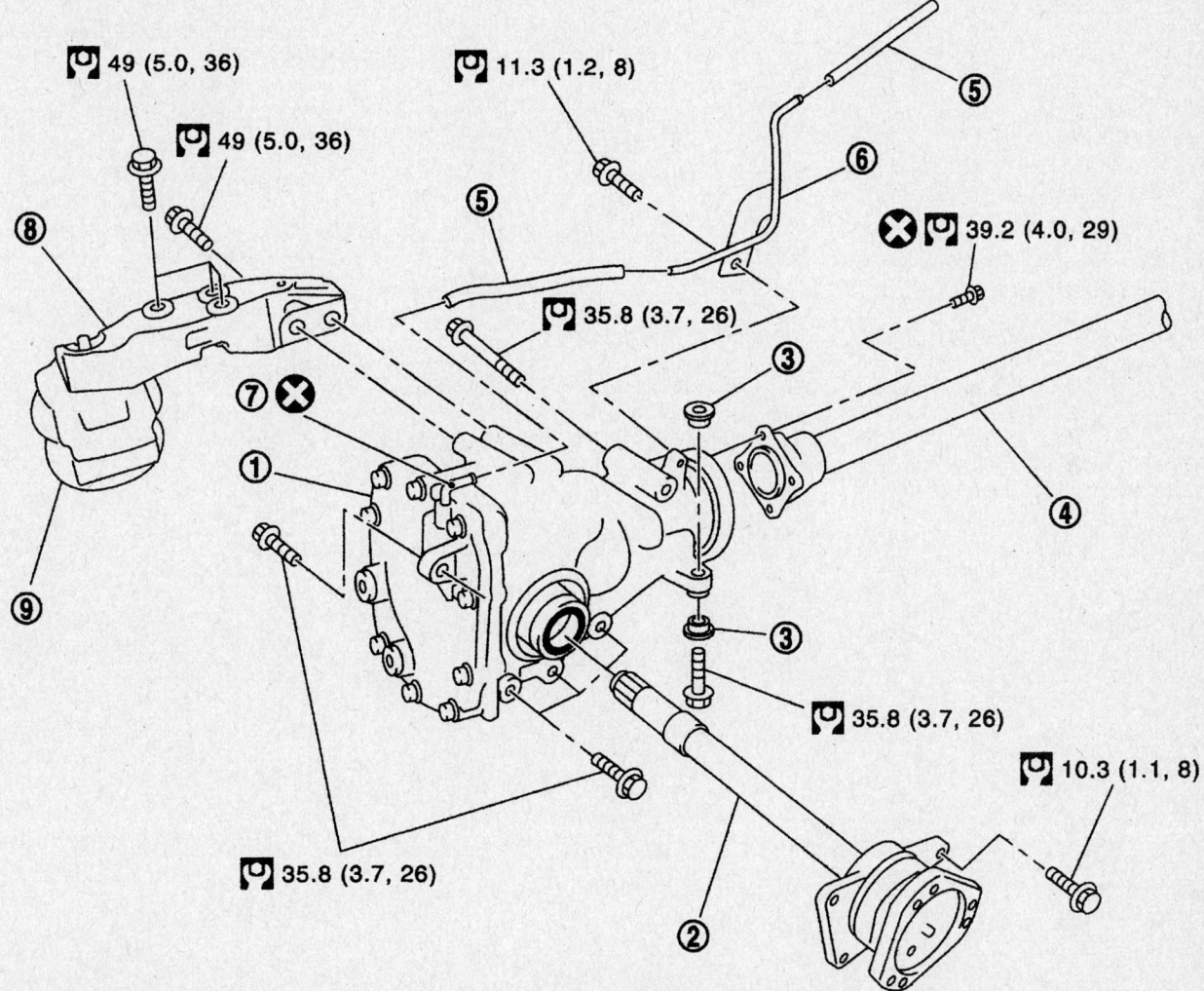

1. Front final drive assembly
2. Side shaft
3. Bushing
4. Front propeller shaft
5. Breather hose
6. Breather tube
7. Breather connector
8. Engine mounting bracket
9. Insulator

Front differential mounting—with the 3.5L engine

09482_FX35_G0036

3. Remove the three engine mounting bracket upper bolts.

4. Remove the three way catalyst (right bank).

5. Remove the stabilizer assembly.

6. Remove the steering gearbox mounting bolts.

7. Remove both front halfshafts.

8. Remove the side shaft assembly.

9. Remove the front driveshaft.

10. Remove the front suspension member.

11. Remove the breather hose and tube.

12. Remove the mounting bolts and remove the front final drive assembly from the vehicle.

To install:

13. Installation is the reverse of removal. Refer to the accompanying illustration for tightening torques. Note the following:

- When installing the side shaft, apply multi-purpose grease to contact surface of side shaft and side shaft oil seal.
- When installing the front final drive assembly, refer to the accompanying illustration for the proper tightening sequence.

➡**Align the mating faces of gear carrier and oil pan for installation.**

- When installing the breather hoses (1) and tube (2), refer to the illustration. Make sure the paint mark facing up.

➡**Make sure there are no pinched or restricted areas on the breather hose caused by bending or winding when installing it.**

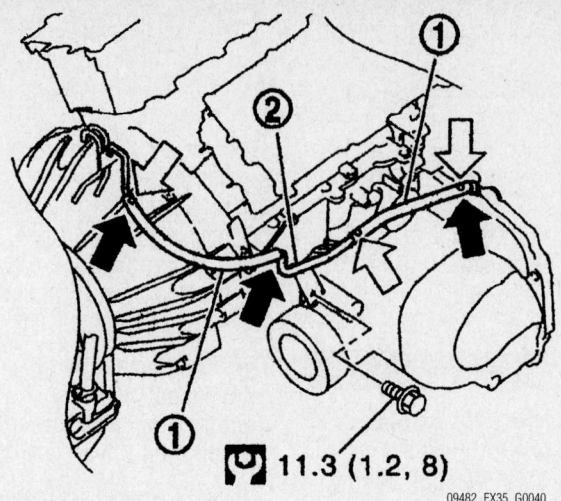

⬡ 11.3 (1.2, 8)

09482_FX35_G0040

When installing the breather hoses (1) and tube (2), refer to the illustration. Make sure the paint mark facing up

- Install the breather hose until it seats the rounded portion of the tube.
- Install the breather connector as shown in the accompanying illustration. Angle "A": 0–30°.
- Seat the breather tube bracket end (A) to the machined face (B) of gear carrier boss.
- Check the oil level after installation.

With the 4.5L Engine

1. Before servicing the vehicle, refer to the Precautions Section.

2. Raise and safely support the vehicle.

3. Remove the three way catalyst (right bank).

4. Remove the stabilizer assembly.

5. Remove the steering gearbox mounting bolts.

6. Remove both front halfshafts.

7. Remove the side shaft assembly.

8. Remove the front driveshaft.

9. Remove the front suspension member.

10. Remove the engine wire harness clamp bolts from the front final drive.

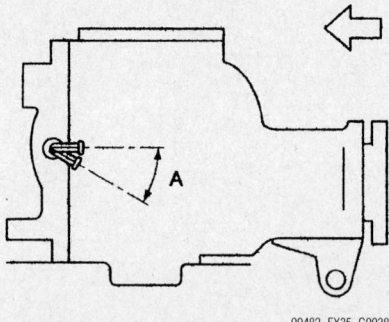

09482_FX35_G0038

Breather connector installation angle— with the 3.5L engine

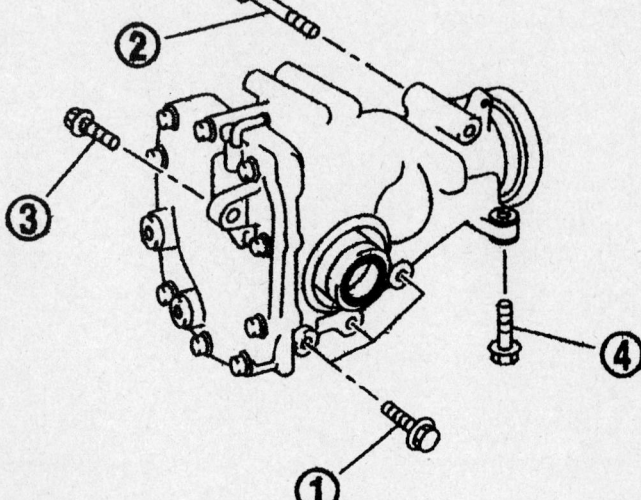

09482_FX35_G0037

Final drive bolt torque sequence—with the 3.5L engine

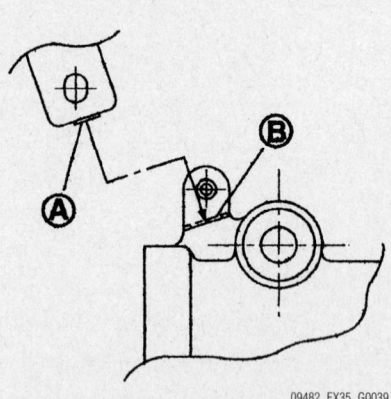

09482_FX35_G0039

Breather bracket installation—with the 3.5L engine

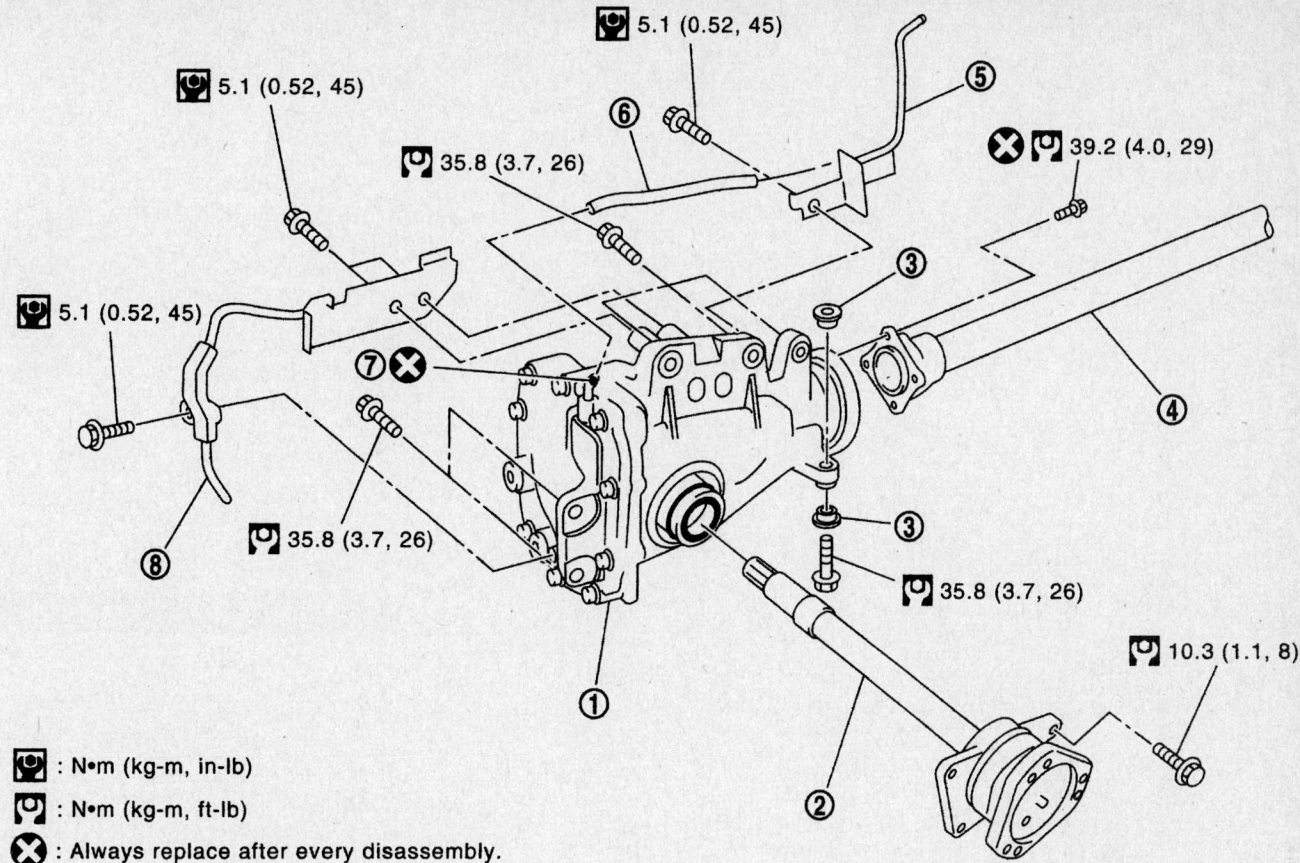

5.1 (0.52, 45)

5.1 (0.52, 45)

35.8 (3.7, 26)

39.2 (4.0, 29)

5.1 (0.52, 45)

35.8 (3.7, 26)

35.8 (3.7, 26)

10.3 (1.1, 8)

🔧 : N•m (kg-m, in-lb)

🔧 : N•m (kg-m, ft-lb)

✕ : Always replace after every disassembly.

1. Front final drive assembly
4. Front propeller shaft
7. Breather connector

2. Side shaft
5. Breather tube
8. Harness bracket

3. Bushing
6. Breather hose

09482_FX35_G0041

Front final drive mounting—with the 4.5L engine

09482_FX35_G0042

Mounting bolt torque sequence—with the 4.5L engine

11. Remove the breather hose and tube.
12. Remove the mounting bolts and remove the front final drive assembly from the vehicle.
   To install:
13. Installation is the reverse of removal. Refer to the accompanying illustration for tightening torques. Note the following:
   • When installing the side shaft, apply multi-purpose grease to the contact surface of the side shaft and side shaft oil seal.
   • Tighten the mounting bolts in the order shown in the accompanying illustration.

➡**Align the mating faces of the gear carrier and oil pan for installation.**

   • When installing the breather hoses and tube, refer to the illustration.

➡**Make sure there are no pinched or restricted areas on the breather hose caused by bending or winding when installing it.**

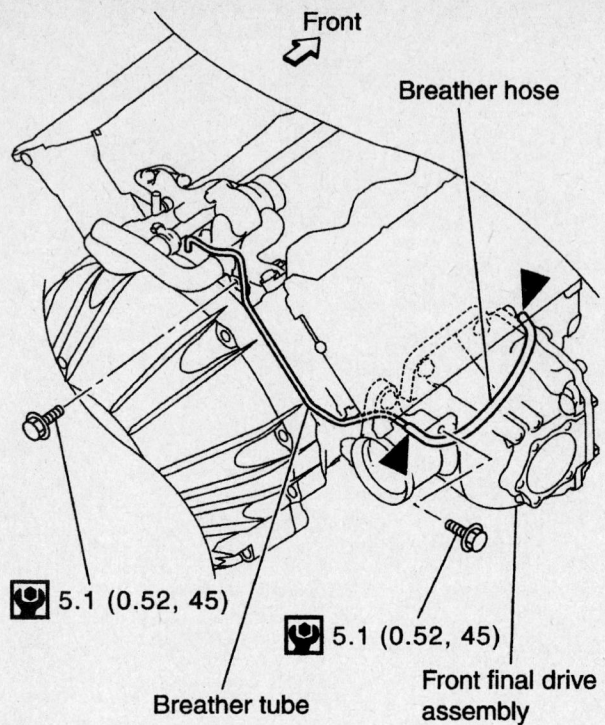

Breather tube installation—with the 4.5L engine

09482_FX35_G0043

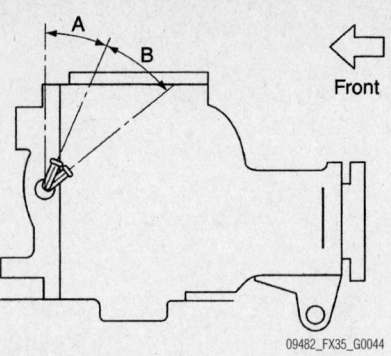

09482_FX35_G0044

Breather connector installation—with the 4.5L engine

- Install the breather connector as shown. Angle "A": 22.5°; Angle "B": 22.5—45°
- Check the oil level.

Rear Differential (Final Drive)

REMOVAL & INSTALLATION

1. Before servicing the vehicle, refer to the Precautions Section.
2. Raise and safely support the vehicle.
3. Remove the center muffler.
4. Remove the rear stabilizer bar.
5. Remove the driveshaft from the final drive.

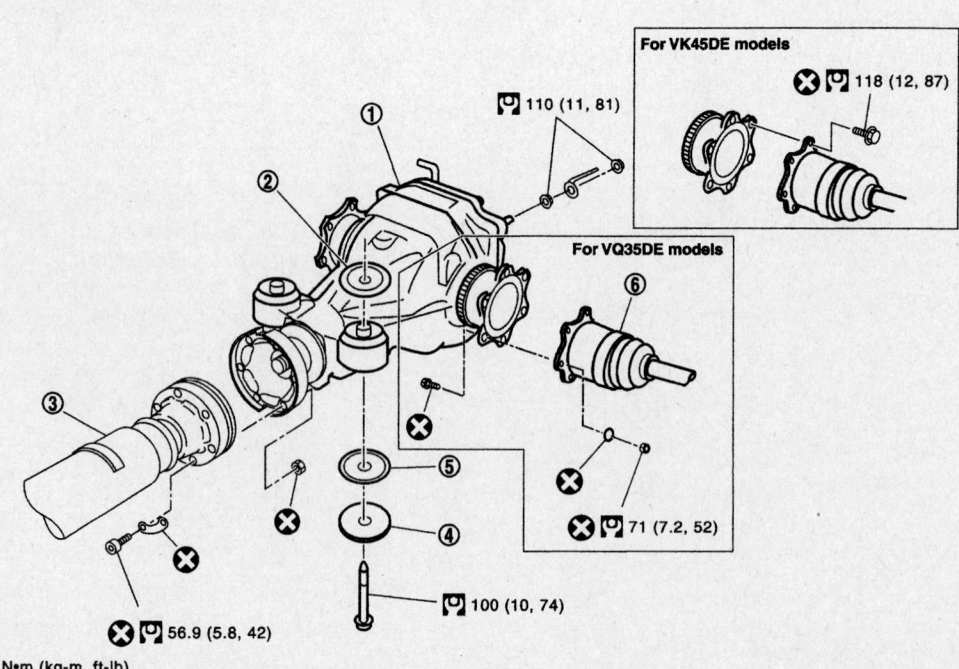

09482_FX35_G0045

| 1. | Rear final drive assembly | 2. | Upper stopper | 3. | Propeller shaft |
| 4. | Washer | 5. | Lower stopper | 6. | Drive shaft |

Rear final drive mounting—2005–06 models

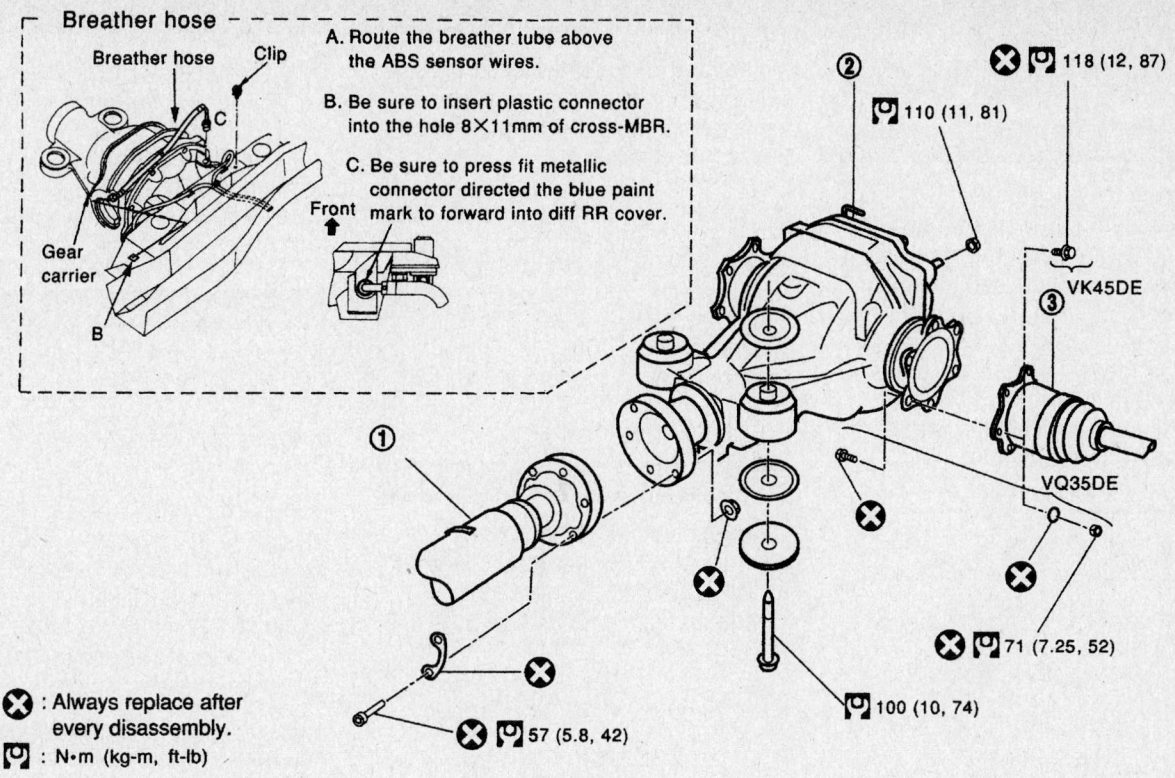

Breather hose

Breather hose   Clip

A. Route the breather tube above the ABS sensor wires.

B. Be sure to insert plastic connector into the hole 8×11mm of cross-MBR.

C. Be sure to press fit metallic connector directed the blue paint mark to forward into diff RR cover.

Front

Gear carrier

B

② ⊗ 🔧 118 (12, 87)

🔧 110 (11, 81)

VK45DE

③

VQ35DE

①

⊗ 🔧 71 (7.25, 52)

🔧 100 (10, 74)

⊗ 🔧 57 (5.8, 42)

⊗ : Always replace after every disassembly.

🔧 : N•m (kg-m, ft-lb)

1.   Rear propeller shaft

2.   Rear final drive assembly

3.   Drive shaft

09482_FX35_G0047

Rear final drive mounting—2003–04 models

6. Remove the halfshaft from final drive. Then suspend it by wire.

7. Remove the breather hose from the final drive.

8. Remove the rear wheel sensor.

9. Place a suitable jack under the rear final drive assembly and secure it.

➡**Do not place the jack on the rear cover (aluminum case).**

10. Remove the mounting bolts and nuts connecting the final drive assembly to the suspension member, and remove the rear final drive assembly.

To install:

11. Installation is the reverse of removal. Refer to the accompanying illustration for tightening torques.

12. When installing the breather hoses (1), refer to the illustration. The vehicle side end should be inserted in the suspension member (2). Install the metal connector (3) side of this hose to the rear cover by inserting it with the painted marking facing the front of vehicle.

➡**Make sure there are no pinched or restricted areas on the breather hose caused by bending or winding when installing it.**

13. Check the oil level.

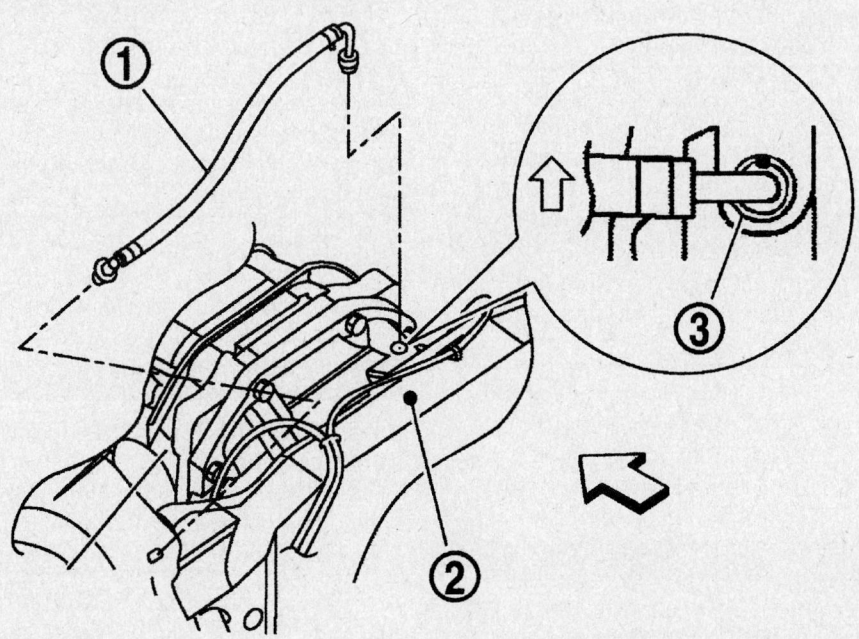

09482_FX35_G0046

Breather hose installation—rear final drive

STEERING

Air Bag

> ✳✳ CAUTION
>
> Some vehicles are equipped with an air bag system. The system must be disarmed before performing service on, or around, system components, the steering column, instrument panel components, wiring and sensors. Failure to follow the safety precautions and the disarming procedure could result in accidental air bag deployment, possible injury and unnecessary system repairs.

PRECAUTIONS

Several precautions must be observed when handling the inflator module to avoid accidental deployment and possible personal injury.
• Never carry the inflator module by the wires or connector on the underside of the module.
• When carrying a live inflator module, hold securely with both hands and ensure that the bag and trim cover are pointed away.
• Place the inflator module on a bench or other surface with the bag and trim cover facing up.
• With the inflator module on the bench, never place anything on or close to the module which may be thrown in the event of an accidental deployment.

DISARMING THE SYSTEM

1. Before servicing the vehicle, refer to the Precautions Section.
2. Turn the ignition switch to OFF.
3. Disconnect and isolate both battery cables.
4. Wait at least 3 minutes prior to servicing the SRS system.

ARMING THE SYSTEM

1. Before servicing the vehicle, refer to the Precautions Section.
2. Reconnect the battery cables.

Steering Angle Sensor Neutral

ADJUSTMENT

1. Before servicing the vehicle, refer to the Precautions Section.

➡ After any of the following conditions, check the adjustment of the steering angle sensor neutral position before driving the vehicle:

• Replacing ABS actuator and electric unit (control unit)
• Removing/installing steering angle sensor
• Removing/installing steering components
• Removing/installing suspension components
• Adjusting wheel alignment

> ✳✳ CAUTION
>
> To adjust the neutral position of steering angle sensor, make sure to use the CONSULT-II. (Adjustment cannot be done without CONSULT-II.)

2. Stop the vehicle with the front wheels in the straight-ahead position.
3. Connect the CONSULT-II and CONSULT-II CONVERTER to the data link connector on the vehicle, and turn the ignition switch ON (do not start engine).

> ✳✳ CAUTION
>
> If the CONSULT-II is used with no connection of the CONSULT-II CONVERTER, malfunctions might be detected in self-diagnosis depending on the control unit which carry out CAN communication.

4. Touch "ABS", "WORK SUPPORT" and "ST ANGLE SENSOR ADJUSTMENT" on the CONSULT-II screen in this order.
5. Touch "START".

> ✳✳ CAUTION
>
> Do not touch the steering wheel while adjusting the steering angle sensor.

6. After approximately 10 seconds, touch "END". (After approximately 60 seconds, it ends automatically.)
7. Turn the ignition switch OFF, then turn it ON again.

> ✳✳ CAUTION
>
> Be sure to carry out above operation.

8. Run the vehicle with the front wheels in the straight-ahead position, then stop.
9. Select "DATA MONITOR", "SELECTION FROM MENU", and "STR ANGLE SIG" on the CONSULT-II screen. Then make sure

"STR ANGLE SIG" is within 0 Ò3.5 deg. If the value is more than specified, repeat steps 3 to 7.
10. Erase memory of the ABS actuator and electric unit (control unit) and the ECM.
11. Turn the ignition switch OFF.

Decel G-Sensor (AWD Models)

CALIBRATION

1. Before servicing the vehicle, refer to the Precautions Section.

➡ After removing/installing or replacing the yaw rate/side/decel G-sensor, ABS actuator and electric unit (control unit), suspension components, or after adjusting the wheel alignment, make sure to calibrate the decel G-sensor before running the vehicle.

> ✳✳ CAUTION
>
> To calibrate the decel G-sensor, make sure to use the CONSULT-II. (Adjustment cannot be done without the CONSULT-II.)

2. Stop the vehicle with the front wheels in the straight-ahead position.

> ✳✳ CAUTION
>
> The work should be done on a flat area when the vehicle is in the unloaded vehicle condition. Keep all tires inflated to the correct pressures. Adjust the tire pressure to the specified value.

3. Connect the CONSULT-II and CONSULT-II CONVERTER to data link connector on vehicle, and turn the ignition switch ON (do not start engine).

> ✳✳ CAUTION
>
> If the CONSULT-II is used without the CONSULT-II CONVERTER, malfunctions might be detected in self-diagnosis, depending on control unit which carry out CAN communication.

4. Touch "ABS", "WORK SUPPORT" and "DECEL G-SEN CALIBRATION" on the CONSULT-II screen in this order.
5. Touch "START".
6. After approximately 10 seconds, touch "END". (After approximately 60 seconds, it ends automatically.)

7. Turn the ignition switch OFF, then turn it ON again.

⁂ CAUTION

Be sure to carry out above operation.

8. Run the vehicle with the front wheels in the straight-ahead position, then stop.

9. Select "DATA MONITOR", "SELECTION FROM MENU", and "DECEL G-SEN" on the CONSULT-II screen. Then make sure "DECEL G-SEN" is within 0 ±0.08 G. If value is more than specified, repeat steps 3 to 7.

10. Erase the memory of the ABS actuator and electric unit (control unit) and ECM.

11. Turn the ignition switch OFF.

Power Steering Gear

REMOVAL & INSTALLATION

1. Before servicing the vehicle, refer to the Precautions Section.

⁂ CAUTION

The spiral cable may snap due to steering operation if the steering column is separated from the steering gear assembly. Therefore fix the steering wheel with a string to avoid turning too far.

2. Set the wheels in the straight-ahead position.

3. Raise and safely support the vehicle.

4. Remove the tires from vehicle.

5. Remove the undercover.

6. Confirm the slit of the lower joint fits with the projection on the rear cover cap, furthermore marking position on steering gear assembly nearly fits with the projection on rear cover cap.

7. Remove the cotter pin at steering outer socket, then loosen the mounting nut.

8. Use a ball joint remover to remove the steering outer socket from the steering knuckle. Be careful not to damage the ball joint boot.

⁂ CAUTION

Temporarily tighten the mounting nut to prevent damage to the threads and to prevent the ball joint remover from coming off.

9. Remove the high pressure side and low pressure side oil pipes from the steering gear assembly, then drain the fluid from the pipes.

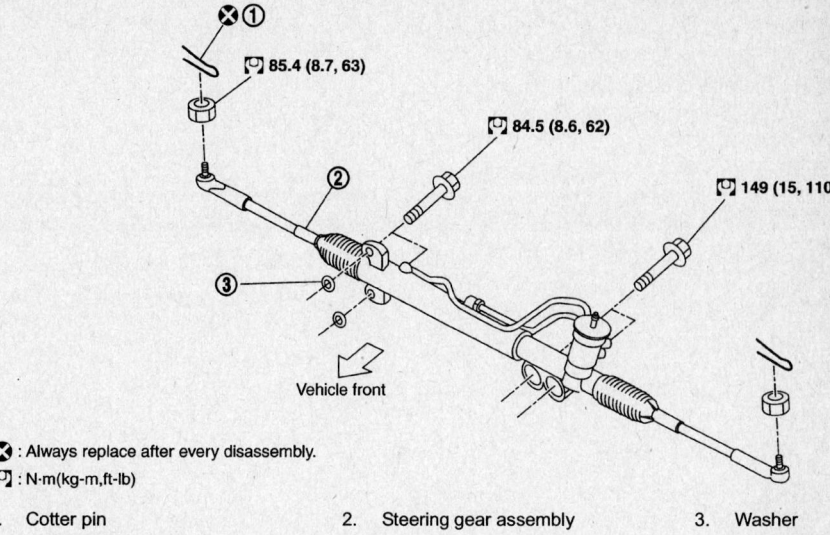

⊗ : Always replace after every disassembly.
🔧 : N·m(kg-m,ft-lb)

1. Cotter pin	2. Steering gear assembly	3. Washer

67162-FX35-G221

Exploded view of the power steering gear mounting

10. Remove the mounting bolt of the steering hydraulic pipe bracket from the steering gear assembly.

11. Remove the lower side mounting bolt of the lower joint.

12. Remove the mounting bolts of the steering gear assembly, then remove the steering gear assembly from the vehicle.

To install:

13. Installation is the reverse of the removal procedure.

➡**Refer to component parts location and do not reuse non-reusable parts.**

14. After installation, check wheel alignment.

15. After adjusting wheel alignment, adjust the neutral position of the steering angle sensor.

16. When steering wheel is set in the straight ahead direction, confirm the slit of the lower joint fits with the projection on the rear cover cap, and that the marking position on steering gear assembly nearly fits with the projection on rear cover cap.

17. Bleed all air from the power steering system, as follows:

Incomplete air bleeding causes the following. When this happens, bleed the system again:

- Generation of air bubbles in the reservoir tank.
- Generation of clicking noise in the oil pump.
- Excessive buzzing in the oil pump.

➡**When the vehicle is stationary or while the steering wheel is being**

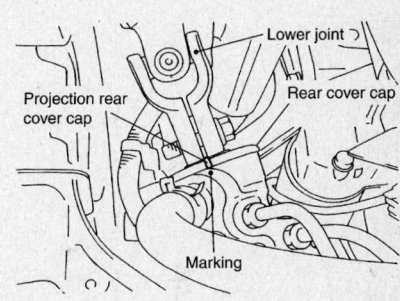

67162-FX35-G222

Steering gear projection position

turned slowly, some noise may be heard from the oil pump or gear. This noise is normal and does not affect any system.

a. Stop the engine, then turn the steering wheel fully to the right and left several times.

⁂ CAUTION

Do not allow the steering fluid reservoir tank level to drop below the low-level line. Check the tank frequently and add fluid as needed.

b. Run the engine at idle speed. Turn the steering wheel fully to the right and then fully to the left, and keep hold for about three seconds. Then check whether any fluid leaks have occurred.

c. Repeat sub-step b several times at about three second intervals.

✻✻ CAUTION

Do not hold the steering wheel in the locked position for more than 10 seconds. (There is the possibility that the oil pump may be damaged.)

d. Check for air bubbles and cloudiness in the fluid.

e. If air bubbles and/or cloudiness don't fade, stop the engine, and stop air bleeding until the air bubbles and cloudiness fade.

f. Perform until all bubbles and cloudiness are gone.

g. Stop the engine and check the fluid level.

18. Check that the steering wheel turns smoothly when it is turned several times fully to the end of the left and right.

FRONT SUSPENSION

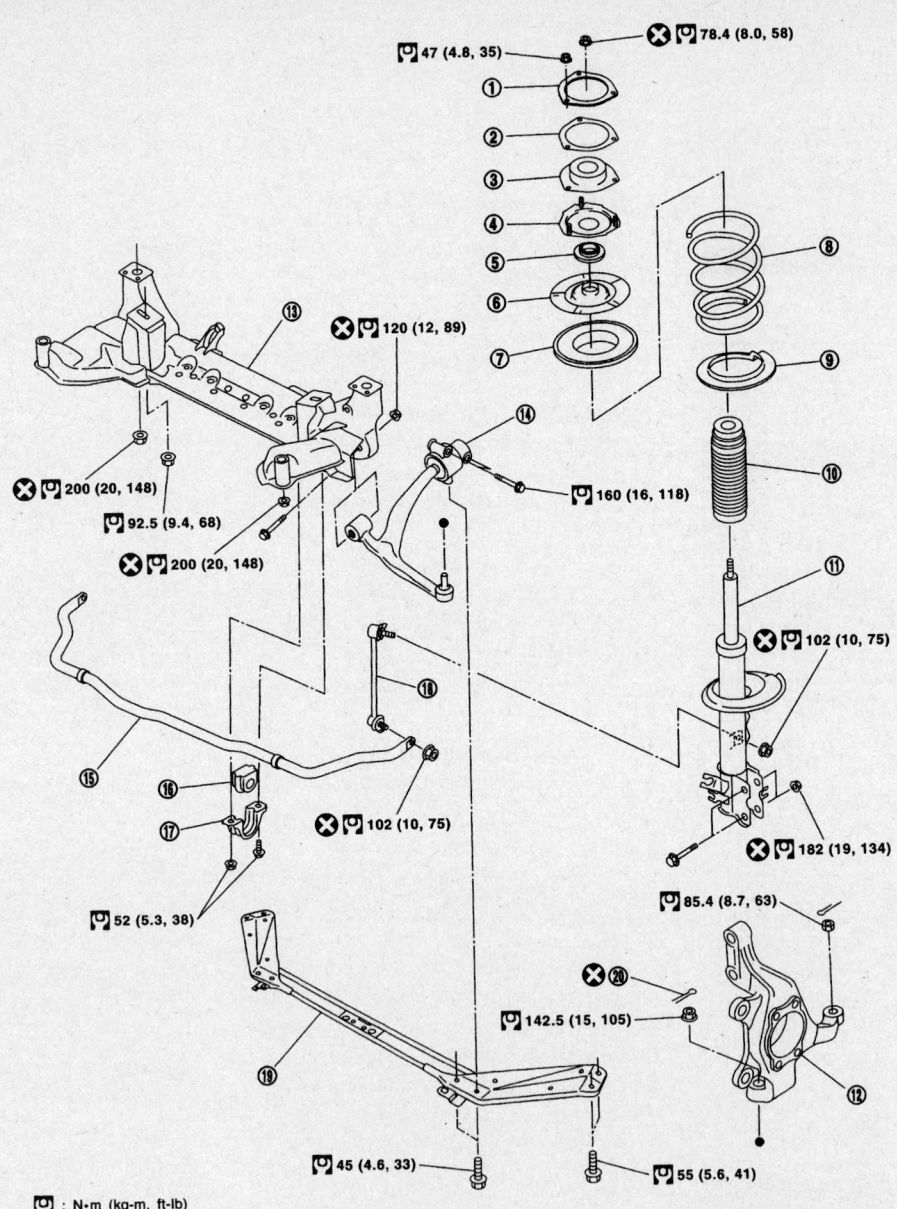

☒ : Always replace after every disassembly.

⊡ : N•m (kg-m, ft-lb)

1.	Strut upper plate	2.	Strut spacer	3.	Mounting insulator
4.	Mounting insulator bracket	5.	Mounting bearing	6.	Spring upper seat
7.	Spring upper rubber seat	8.	Coil spring	9.	Spring lower rubber seat
10.	Bound bumper	11.	Strut	12.	Steering knuckle
13.	Front suspension member	14.	Transverse link	15.	Stabilizer bar
16.	Stabilizer bushing	17.	Stabilizer clamp	18.	Stabilizer connecting rod
19.	Front cross bar	20.	Cotter pin		

67162-FX35-G223

Exploded view of the front suspension components

MacPherson Strut

REMOVAL & INSTALLATION

1. Before servicing the vehicle, refer to the Precautions Section.
2. Raise and safely support the vehicle.
3. Remove the wheels from the vehicle.
4. Remove the brake hose lockplate. Then, remove the brake hose from the strut assembly.
5. Remove the wheel sensor wiring harness from the strut assembly.

✳✳ CAUTION

Do not pull the wheel sensor wiring harness.

6. Remove the stabilizer connecting rod upper nut, separate the stabilizer connecting rod and strut assembly.
7. Remove the attaching bolts and nuts between the strut assembly and the steering knuckle.
8. Remove the mounting nuts on the mounting insulator bracket, then remove the strut upper plate, strut spacer and the strut from the vehicle.

To install:

✳✳ CAUTION

Attach strut upper plate as shown in the accompanying illustration.

9. Install the strut upper plate, strut spacer and the strut onto the vehicle.
10. Install and tighten the mounting nuts on the mounting insulator bracket to 35 ft. lbs. (47 Nm).
11. Install and tighten the attaching bolts and nuts between the strut assembly and the steering knuckle to 134 ft. lbs. (182 Nm).
12. Reattach the stabilizer connecting rod to the strut and tighten the upper nut to 75 ft. lbs. (102 Nm).
13. Install the wheel sensor wiring harness onto the strut assembly.
14. Install the brake hose and the hose lockplate.
15. Install the tire from the vehicle.

➡**Refer to component parts location and do not reuse non-reusable parts.**

16. After installation, check wheel alignment.
17. After adjusting the wheel alignment, adjust the neutral position of steering angle sensor.
18. Double-check to ensure that the wheel sensor wiring harness is properly routed.

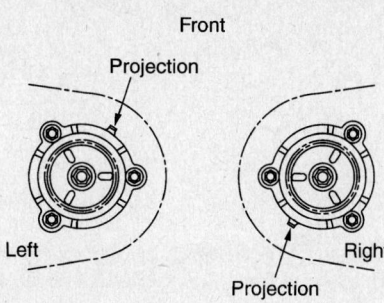

MacPherson strut projection positioning

DISASSEMBLY

1. Before servicing the vehicle, refer to the Precautions Section.

➡**Make sure the piston rod on the strut is not damaged when removing the components from the strut assembly.**

2. Install the strut attachment SST: ST35652000 or equivalent to the strut and secure it in a vise.

✳✳ CAUTION

When installing the strut attachment to the strut, wrap a shop towel around the strut to protect it from damage.

3. Using a spring compressor, compress the coil spring between the spring upper seat and spring lower seat (on the strut) until the coil spring is free.

✳✳ CAUTION

Be sure the spring compressor is securely attached to the coil spring before compressing the spring.

4. After making sure the coil spring is free between the spring upper seat and spring lower seat of the strut, remove the piston rod locknut.
5. Remove the mounting insulator, mounting insulator bracket, mounting bearing, spring upper seat, spring upper rubber seat, and bound bumper. Then, remove the coil spring and spring lower rubber seat from the strut.
6. Gradually release the spring compressor, and remove the coil spring.

✳✳ CAUTION

Loosen the spring compressor while making sure the coil spring attachment position does not move.

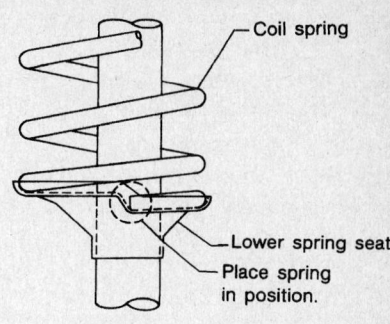

Proper coil spring placement for installation.

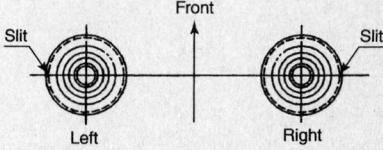

Upper spring seat positioning for installation.

7. Remove the strut attachment from the strut.
8. Check the strut for deformation, cracks, damage, and replace if necessary.
9. Check the piston rod for damage, uneven wear or distortion, and replace if necessary.
10. Check the welded and sealed areas for oil leakage, and replace if necessary.
11. Check the mounting insulator for cracks and rubber parts for wear. Replace them if necessary.
12. Check the coil spring for cracks, wear or damage, and replace if necessary.

ASSEMBLY

1. Before servicing the vehicle, refer to the Precautions Section.

➡**Make sure the piston rod on the strut is not damaged when attaching the components to the strut.**

2. Install the strut attachment to the strut and fix it in a vise.

✳✳ CAUTION

When installing strut attachment to strut, wrap a shop cloth around strut to protect it from damage.

3. Compress coil spring using a spring compressor (commercial service tool), and install it onto the strut.

✳✳ CAUTION

Face the tube side of the coil spring downward. Align the lower end to the spring rubber seat as shown. Be sure the spring compressor is securely attached to coil spring before compressing the coil spring.

4. Apply soapy water to the bound bumper and insert into the mounting insulator.

✳✳ CAUTION

Do not use machine oil.

5. Install the mounting insulator bracket, mounting bearing, bound bumper, spring upper seat, spring upper rubber seat and spring lower rubber seat.

➡ **The installation position of spring upper seat is as shown.**

6. Fix the mounting insulator, then tighten the piston rod locknut to 58 ft. lbs. (78 Nm).

✳✳ CAUTION

Be sure not to deform the mounting insulator bracket.

7. Gradually the release spring compressor, and remove the coil spring.

✳✳ CAUTION

Loosen the spring compressor while making sure coil spring attachment position does not move.

8. Remove the strut attachment from the strut.

Lower Ball Joint

REMOVAL & INSTALLATION

The front lower control arm ball joints are not separately replaceable from the control arms themselves. If the joints are found to be defective, the entire assembly must be replaced.

Lower Control Arm

REMOVAL & INSTALLATION

1. Before servicing the vehicle, refer to the Precautions Section.

2. Raise and safely support the vehicle.
3. Remove the wheels from vehicle.
4. Remove the undercover.
5. Remove the front cross bar.
6. Remove the cotter pin at the transverse link, then loosen the mounting nut.
7. Use a ball joint remover to remove the transverse link from the steering knuckle. Be careful not to damage the ball joint boot.

✳✳ CAUTION

Temporarily tighten the mounting nut to prevent damage to the threads and to prevent the ball joint remover from coming off.

8. Remove the mounting bolts which are at the back of the transverse link (mounting part with body), then separate the transverse link.
9. Remove the mounting bolts which are at the front of the transverse link (mounting part with the front suspension member), then separate the transverse link.
10. Remove the transverse link from the vehicle.
11. Check transverse link and bushing for deformation, cracks, or damage. If any non-standard condition is found, replace it.
12. Check the boot of the ball joint for cracks, or other damage, and also for grease leakage. If any non-standard condition is found, replace it.
13. Manually move ball stud to confirm it moves smoothly with no binding.

➡ **Before measurement, move ball joint at least ten times by hand to check for smooth movement.**

14. Hook a spring scale onto the ball stud tip. Confirm that the spring scale measurement value is within specifications when the ball stud begins to move. If it is outside the specified range, replace the transverse link assembly.

➡ **Swing torque specification: Less than 5–43 inch. lbs. (0.5–4.9 Nm), measure value of spring scale: less than 5–43 inch. lbs. (0.5–4.9 Nm).**

15. Attach the mounting nut onto the ball stud. Check that the rotating torque is within specifications with a preload gauge . If it is outside the specified range, replace transverse link assembly.

➡ **Rotating torque specification: Less than 5–43 inch lbs. (0.5–4.9 Nm).**

16. Move the tip of ball joint in axial direction to check for looseness. If it is outside the specified range, replace transverse link assembly.

➡ **Axial end play specification: 0.004 in. (0.1mm).**

To install:

17. Install the transverse link on the vehicle.
18. Install and tighten the mounting bolts at the front of the transverse link (mounting part with the front suspension member) to 89 ft. lbs. (120 Nm), then install and tighten the mounting bolts which are at the back of the transverse link (mounting part with body) to 118 ft. lbs. (160 Nm).
19. Reattach the transverse link to the steering knuckle, and tighten the ball joint nut to 105 ft. lbs. (142 Nm). Be careful not to damage the ball joint boot.
20. Install a new cotter pin.
21. Install the front cross bar. Tighten the inner two bolts on each end to 33 ft. lbs. (45 Nm) and the outer two bolts on each end to 41 ft. lbs. (55 Nm).
22. Install the undercover.
23. Install the tire.
24. Check the wheel alignment.
25. After adjusting wheel alignment, adjust the neutral position of the steering angle sensor.

LOWER CONTROL ARM BUSHING REPLACEMENT

The front lower control arm bushings are not separately replaceable from the control arms themselves. If the bushings are found to be defective, the entire assembly must be replaced.

Stabilizer Bar

REMOVAL & INSTALLATION

1. Before servicing the vehicle, refer to the Precautions Section.
2. Raise and safely support the vehicle.
3. Remove the wheels.
4. Remove the undercover.
5. Remove the stabilizer connecting rod lower nut, and separate the stabilizer bar and stabilizer connecting rod.
6. Remove the stabilizer clamp mounting bolts and nuts.
7. Remove the stabilizer bar, stabilizer clamp, and stabilizer bushing from the vehicle.
8. Remove the stabilizer connecting rod upper nut and separate the stabilizer connecting rod and strut.
9. Check the stabilizer bar, stabilizer connecting rod, stabilizer bushing and stabilizer clamp for deformation, cracks and damage, and replace if necessary.

To install:

10. Installation is the reverse of removal. Refer to the front suspension exploded view for tightening torques.

➡**Refer to the component parts location and do not reuse non-reusable parts.**

11. After removing/installing or replacing suspension components, check the wheel alignment.

12. After adjusting the wheel alignment, adjust the neutral position of the steering angle sensor.

13. The stabilizer bar uses a pillow ball-type connecting rod. Position the ball joint with the case on the pillow ball head parallel to the stabilizer bar.

Transverse Link

REMOVAL

1. Before servicing the vehicle, refer to the Precautions Section.
2. Raise and safely support the vehicle.
3. Remove the wheels.
4. Remove the undercover.
5. Remove the front cross bar.
6. Remove the cotter pin at the transverse link, then loosen the mounting nut.
7. Use a ball joint remover to remove the transverse link from the steering knuckle. Be careful not to damage the ball joint boot.

✽✽ WARNING

Tighten the mounting nut temporarily to prevent damage to the threads and to prevent the ball joint remover from coming off.

8. Remove the mounting bolts at the back of transverse link and separate the transverse link.

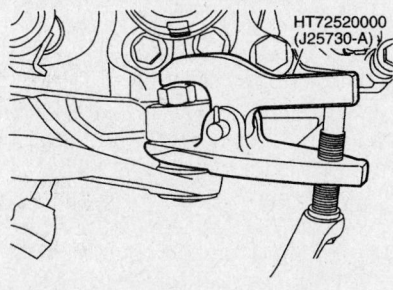

09482_FX35_G0048

Use a ball joint remover (SST) to remove the transverse link from the steering knuckle

9. Remove the mounting bolts which at the front of transverse link and separate the transverse link.
10. Remove the transverse link from vehicle.
11. Check the transverse link and bushing for deformation, cracks, or damage. If any non-standard condition is found, replace it.
12. Check the boot of ball joint for cracks, or other damage, and also for grease leakage. If any non-standard condition is found, replace it.
13. Manually move the ball stud to confirm it moves smoothly with no binding.

SWING TORQUE INSPECTION

➡**Before measurement, move the ball joint at least ten times by hand to check for smooth movement.**

1. Hook the spring scale at the ball stud. Confirm that the spring scale measurement value is within specifications when the ball stud begins moving. Swing torque: Less than 0.5–4.9 Nm (5–43 inch lbs.).
2. If it is outside the specified range, replace the transverse link assembly.

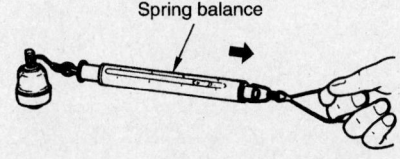

Spring balance

09482_FX35_G0049

Transverse link swing torque measurement

ROTATING TORQUE INSPECTION

1. Attach a mounting nut to the ball stud. Check that the rotating torque is within the specifications with a preload gauge. Rotating Torque: Less than 0.5–4.9 Nm (5–43 inch lbs.).

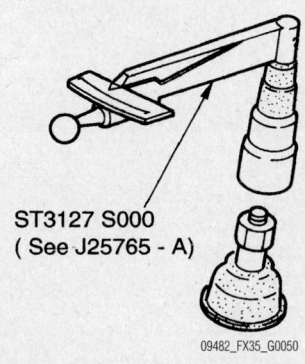

ST3127 S000
(See J25765 - A)

09482_FX35_G0050

Transverse link ball stud rotating torque check

2. If it is outside the specified range, replace the transverse link assembly.

AXIAL END PLAY INSPECTION

1. Move the tip of the ball joint in an axial direction to check for looseness. Axial end play: 0.1 mm (0.004 in.).
2. If it is outside the specified range, replace the transverse link assembly.

INSTALLATION

1. Installation is the reverse of removal. Refer to the front suspension exploded view for tightening torques.

➡**Refer to exploded view and do not reuse non-reusable parts.**

2. After removing/installing or replacing suspension components, check the wheel alignment.

3. After adjusting the wheel alignment, adjust the neutral position of the steering angle sensor.

Front Suspension Member

REMOVAL & INSTALLATION

1. Before servicing the vehicle, refer to the Precautions Section.
2. Take up the weight of the engine with a shop crane.
3. Raise and safely support the vehicle.
4. Remove the wheels.
5. Remove the undercover.
6. Remove the front cross bar.
7. Remove the mounting bolts at the back of transverse link and separate the transverse link.
8. Remove the mounting bolts at the front of transverse link and separate the transverse link.
9. Remove the steering hydraulic piping bracket from front suspension member.
10. Remove the steering gear mounting bolts, then hang steering gear out of the way.
11. Remove the stabilizer bar from the front suspension member and stabilizer connecting rod lower side.
12. Remove the mounting nuts between engine mounting insulator and front suspension member.
13. Remove the mounting nuts between front suspension member and body. Move the crane down slowly to remove the front suspension member from the vehicle.
14. Check the front suspension member for deformation, cracks, or any other damage. Replace if necessary.

To install:

15. Refer to the front suspension exploded view for tightening torques. Installation is the reverse of removal.

➡**Refer to the exploded view and do not reuse non-reusable parts.**

16. After removing/installing or replacing suspension components and steering components, check the wheel alignment.

17. After adjusting the wheel alignment, adjust the neutral position of the steering angle sensor.

Wheel Hub and Knuckle

REMOVAL & INSTALLATION

1. Before servicing the vehicle, refer to the Precautions Section.
2. Raise and safely support the vehicle.
3. Remove the appropriate wheel.
4. Remove the brake caliper. Support it in a place where it will not interfere with work.

➡**Avoid depressing brake pedal while brake caliper is removed.**

5. Remove the disc rotor.
6. Remove the wheel sensor from the wheel hub and bearing assembly.

❈ CAUTION

Do not pull on wheel sensor wiring harness.

7. Remove the cotter pin from the steering outer socket, then loosen the mounting nut.
8. Use a ball joint remover to separate the steering outer socket from the steering knuckle. Be careful not to damage the ball joint boot.

❈❈ CAUTION

Temporarily tighten the mounting nut to prevent damage to the threads and to prevent the ball joint remover from coming off.

9. Remove the cotter pin at the transverse link, then loosen the mounting nut.
10. Use a ball joint remover to separate the transverse link from the steering knuckle. Be careful not to damage ball joint boot.

❈❈ CAUTION

Temporarily tighten the mounting nut to prevent damage to the threads and

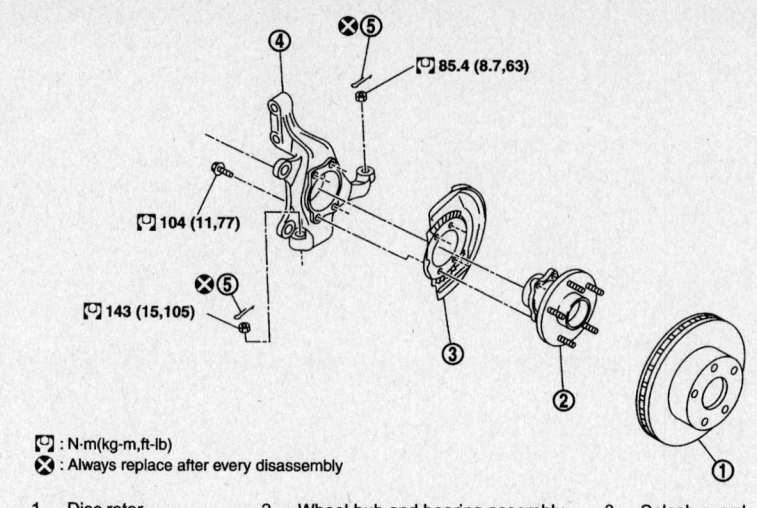

⊡ : N·m(kg-m,ft-lb)
⊗ : Always replace after every disassembly

1. Disc rotor
2. Wheel hub and bearing assembly
3. Splash guard
4. Steering knuckle
5. Cotter pin

67162-FX35-G228

Exploded view of front wheel hub mounting—2WD models

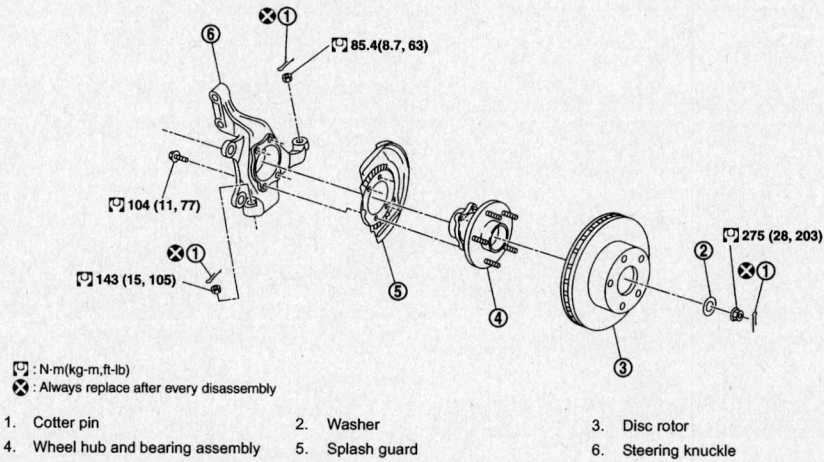

⊡ : N·m(kg-m,ft-lb)
⊗ : Always replace after every disassembly

1. Cotter pin
2. Washer
3. Disc rotor
4. Wheel hub and bearing assembly
5. Splash guard
6. Steering knuckle

67162-FX35-G229

Exploded view of front wheel hub mounting—AWD models

to prevent the ball joint remover from coming off.

11. On AWD models, perform the following:
   a. Remove the cotter pin, then remove the lock nut from the halfshaft.
   b. Remove the steering knuckle from the halfshaft.

❈❈ CAUTION

When removing steering knuckle, do not apply an excessive angle to the halfshaft joint. Also be careful not to excessively extend the slide joint. Do not hang over the halfshaft without proper support.

12. Remove the mounting bolts and nuts between the strut assembly and the steering knuckle.

13. Remove the steering knuckle from the vehicle.

14. Remove the mounting bolts between the steering knuckle and the wheel hub/bearing assembly.

15. Remove the splash guard and wheel hub/bearing assembly from the steering knuckle.

16. Check for deformities, cracks and damage on all parts and replace if necessary.

17. Inspect the ball joint for boot breakage, axial looseness, and torque of transverse link and steering outer socket ball joint. Maximum of allowable axial end play is 0.002 in. (0.05mm) or less.

To install:

18. Install the splash guard and wheel hub/bearing assembly onto the steering knuckle.

19. Install the mounting bolts between the steering knuckle and the wheel

hub/bearing assembly. Tighten the bolts to 77 ft. lbs. (104 Nm).

20. Install the steering knuckle on the vehicle.

21. Install the mounting bolts and nuts to the strut assembly and the steering knuckle. Tighten the bolts to 134 ft. lbs. (182 Nm).

22. On AWD models, perform the following:

   a. Install the steering knuckle onto the halfshaft.

   b. Install and tighten the lock nut on the halfshaft to 203 ft. lbs. (275 Nm). Install a new cotter pin.

23. Reattach the transverse link to the steering knuckle. Tighten the ball joint nut to 105 ft. lbs. (143 Nm).

24. Install a new cotter pin.

25. Reconnect the steering outer socket to the steering knuckle. Tighten the ball joint nut to 63 ft. lbs. (86 Nm).

26. Install a new cotter pin.

27. Install the wheel sensor onto the wheel hub and bearing assembly.

28. Install the disc rotor.

29. Install the brake caliper.

30. Install the wheel.

31. Check wheel alignment.

32. After adjusting wheel alignment, adjust the neutral position of the steering angle sensor.

33. Check the installation condition of the wheel sensor wiring harness.

REAR SUSPENSION

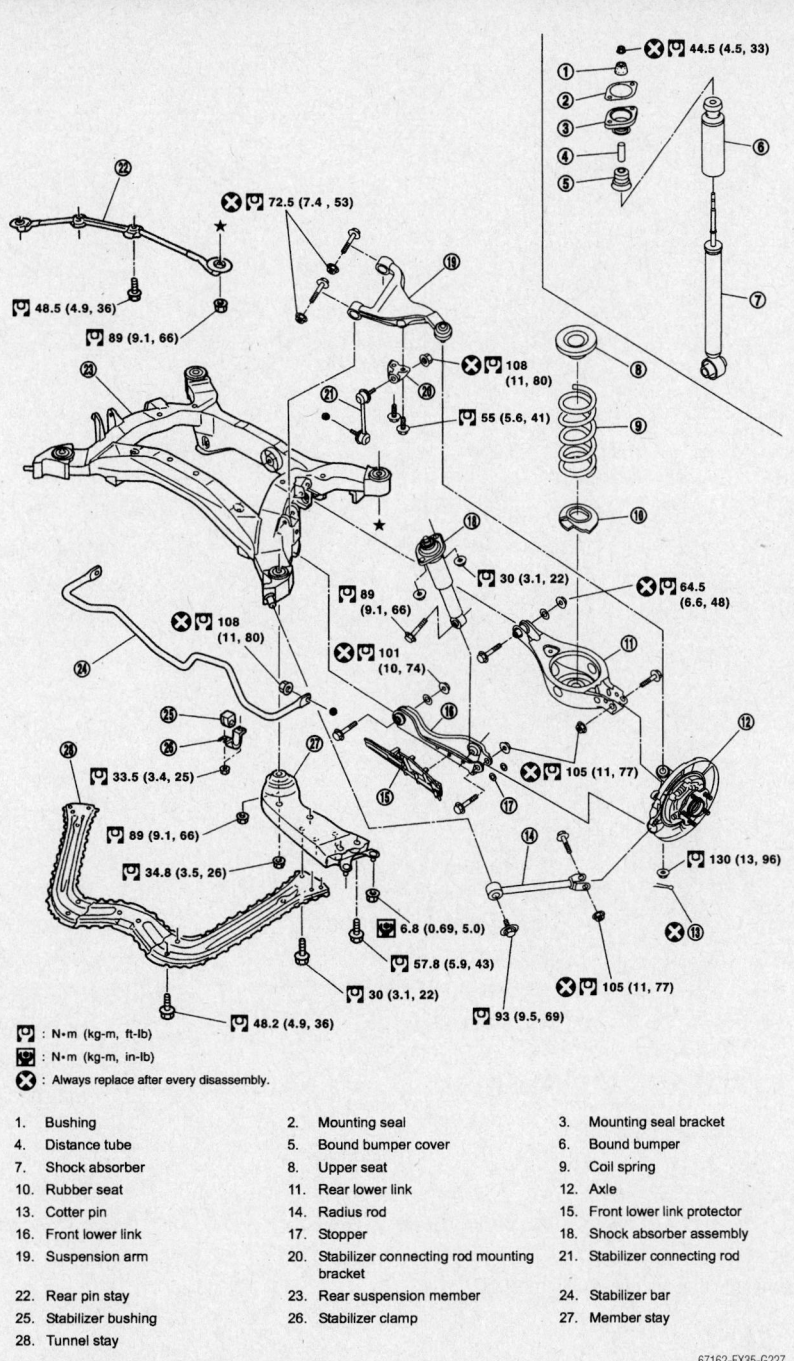

1. Bushing
2. Mounting seal
3. Mounting seal bracket
4. Distance tube
5. Bound bumper cover
6. Bound bumper
7. Shock absorber
8. Upper seat
9. Coil spring
10. Rubber seat
11. Rear lower link
12. Axle
13. Cotter pin
14. Radius rod
15. Front lower link protector
16. Front lower link
17. Stopper
18. Shock absorber assembly
19. Suspension arm
20. Stabilizer connecting rod mounting bracket
21. Stabilizer connecting rod
22. Rear pin stay
23. Rear suspension member
24. Stabilizer bar
25. Stabilizer bushing
26. Stabilizer clamp
27. Member stay
28. Tunnel stay

67162-FX35-G227

Exploded view of the rear suspension components—2003–05 models

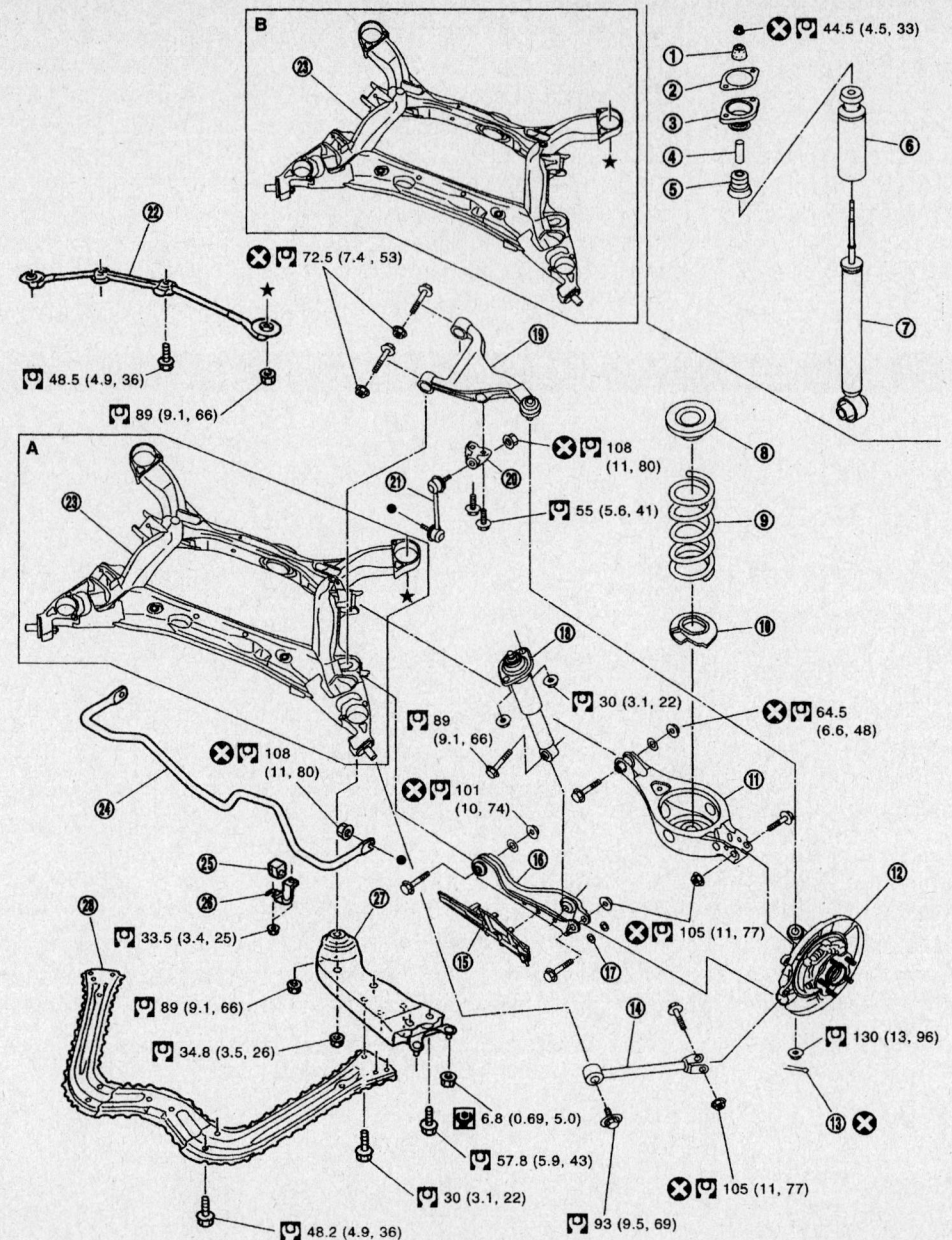

1. Bushing
2. Mounting seal
3. Mounting seal bracket
4. Distance tube
5. Bound bumper cover
6. Bound bumper
7. Shock absorber
8. Upper seat
9. Coil spring
10. Rubber seat
11. Rear lower link
12. Axle assembly
13. Cotter pin
14. Radius rod
15. Front lower link protector
16. Front lower link
17. Stopper rubber
18. Shock absorber assembly
19. Suspension arm
20. Stabilizer connecting rod mounting bracket
21. Stabilizer connecting rod
22. Rear pin stay
23. Rear suspension member
24. Stabilizer bar
25. Stabilizer bushing
26. Stabilizer clamp
27. Member stay
28. Tunnel stay
A: With height sensor
B: Without height sensor

09482_FX35_G0051

Exploded view of the rear suspension components—2006 models

Shock Absorber

REMOVAL & INSTALLATION

1. Before servicing the vehicle, refer to the Precautions Section.
2. Raise and safely support the vehicle.
3. Remove the rear tire.
4. Position a jack or equivalent support under the rear lower link.
5. Remove the fixing bolt in the lower side of the shock absorber assembly.
6. Remove the attaching nuts in the upper side of the shock absorber assembly and remove the shock absorber assembly from the vehicle.
7. Check the shock absorber assembly for deformation, cracks, or damage, and replace if necessary.
8. Check the piston rod for damage, uneven wear, or distortion, and replace if necessary.
9. Check the welded and sealed areas for oil leakage, and replace if necessary.

To install:
10. Position the shock absorber assembly in the vehicle.
11. Install and tighten the attaching nuts in the upper side of the shock absorber assembly to 22 ft. lbs. (30 Nm).
12. Install the fixing bolt in the lower side of the shock absorber assembly, and tighten until snug.
13. Remove the jack or equivalent support from under the rear lower link.
14. Install the tire.
15. With the weight of the vehicle resting on the suspension (empty vehicle), tighten the shock absorber lower fixing bolt to 66 ft. lbs. (89 Nm).

➡**Refer to component parts location and do not reuse non-reusable parts.**

16. Check wheel alignment.
17. After adjusting wheel alignment, adjust the neutral position of the steering angle sensor.

Coil Spring

REMOVAL & INSTALLATION

Rear

Refer to the rear lower link procedure for the rear lower control arm.

Lower Ball Joint

REMOVAL & INSTALLATION

Rear

The rear lower control arms (lower links) do not use replaceable bushings or ball joints. Replace the entire arm assembly if any damage is noticed, including cracks, dents or deformations.

Upper Control Arm

REMOVAL & INSTALLATION

1. Before servicing the vehicle, refer to the Precautions Section.
2. Raise and safely support the vehicle.
3. Remove the rear tire.
4. Remove the stabilizer connecting rod mounting bracket from the suspension arm.
5. Remove the halfshaft from the vehicle.
6. Remove the cotter pin of the suspension arm ball joint, and loosen the nut.
7. Use a ball joint remover or suitable tool to remove the suspension arm from the axle. Be careful not to damage the ball joint boot.

✳✳ CAUTION

Temporarily tighten the mounting nut to prevent damage to the threads and to prevent the ball joint remover from coming off.

8. Remove the fixing nuts and bolts between the suspension arm and the rear suspension member.
9. Remove the suspension arm from vehicle.
10. Check the suspension arm and bushing for deformation, cracks, or damage. If any non-standard condition is found, replace it.
11. Check the boot of the ball joint for cracks, or damage, and also for grease leakage.
12. Manually move the ball stud to confirm it moves smoothly with no binding.

➡**Before measuring, move ball joint at least ten times by hand to check for smooth movement.**

13. Hook a spring scale at the cotter pin mounting hole. Confirm the spring scale measurement value is within 2.18–14.8 lbs.

(9.7–66.0 N) when the ball joint stud begins moving. If it is outside the specified range, replace the suspension arm assembly.
14. Attach the mounting nut to the ball stud. Make sure the rotating torque is within 5–30 inch lbs. (0.5–3.4 Nm) with a preload gauge . If it is outside the specified range, replace the suspension arm assembly.
15. Move the tip of the ball joint in the axial direction to check for looseness. If it is outside the specified range of 0 in. (0mm), replace the suspension arm assembly.

To install:
16. Install the suspension arm onto the vehicle.
17. Install the fixing nuts and bolts between the suspension arm and the rear suspension member. Tighten them to 53 ft. lbs. (73 Nm).
18. Reattach the suspension arm ball joint to the axle. Tighten the ball joint nut to 96 ft. lbs. (130 Nm).
19. Install a new cotter pin.
20. Install the halfshaft.
21. Install the stabilizer connecting rod mounting bracket onto the suspension arm, and tighten the two mounting bolts to 41 ft. lbs. (55 Nm).
22. Install the rear tire.

➡**Refer to component parts location and do not reuse non-reusable parts.**

- Perform final tightening of the rear suspension member installation position (rubber bushing) under unladen conditions with the tires on level ground.
23. Check wheel alignment.
24. After adjusting wheel alignment, adjust the neutral position of the steering angle sensor.

Lower Control Arm

REMOVAL & INSTALLATION

Front Lower Link

1. Before servicing the vehicle, refer to the Precautions Section.
2. Raise and safely support the vehicle.
3. Remove the rear tire.
4. Position a jack under the rear lower link for support.
5. Remove the front lower link protector.
6. Remove the shock absorber assembly from the vehicle.

7. Remove the mounting nut and bolt between the front lower link and the axle.

8. Remove the mounting nut and bolt between the front lower link and the rear suspension member.

9. Remove the front lower link from the vehicle.

10. Check the front lower link and bushing for any deformation, cracks, or damage. Replace it if necessary.

To install:

11. Position the front lower link on the vehicle.

12. Install and tighten the mounting nut and bolt between the front lower link and the rear suspension member to 74 ft. lbs. (101 Nm).

13. Install and tighten the mounting nut and bolt between the front lower link and the axle to 77 ft. lbs. (105 Nm).

14. Install the shock absorber assembly.

15. Install the front lower link protector.

16. Slowly lower the jack from under the rear lower link.

17. Install the rear tire.

✳✳ CAUTION

Perform final tightening of the rear suspension member and axle installation position (rubber bushing) under unladen conditions with the tires on level ground.

18. Check the wheel alignment.

19. After adjusting wheel alignment, adjust the neutral position of the steering angle sensor.

Rear Lower Link

1. Before servicing the vehicle, refer to the Precautions Section.

2. Raise and safely support the vehicle.

3. Remove the rear tire.

4. Position a jack under the rear lower link for support.

5. Loosen the fixing bolt and nut of the rear lower link in the side of the suspension member, and then remove the fixing bolt and nut in the side of the axle.

6. Slowly lower the jack, then remove the upper seat, coil spring and rubber sheet from the rear lower link.

7. Remove the fixing bolt and nut in the side of the rear suspension member to remove the rear lower link.

8. Check rear the lower link, bushing and coil spring for deformation, cracks, and damage. Replace the rear lower link and coil spring, if necessary.

To install:

9. Position the rear lower link on the vehicle.

10. Install and tighten the fixing bolt and nut in the side of the rear suspension member to 48 ft. lbs. (65 Nm).

➡**Check that the upper seat is attached as shown. Insert the bracket into the upper seat with the setting three tabs of the upper seat to the projecting part of the bracket beforehand as shown.**

11. Position the upper seat, coil spring and rubber sheet in place, and then slowly raise the jack under the rear lower link.

➡**Match up the rubber seat indentions and rear lower link grooves. Also, make sure spring is not upside down. The top and bottom are indicated by paint color.**

12. Install and tighten the fixing bolt and nut in the side of the axle to 77 ft. lbs. (105 Nm).

13. Slowly lower the jack from under the rear lower link.

14. Install the rear tire.

✳✳ CAUTION

Perform final tightening of the rear suspension member and axle installation position (rubber bushing) under unladen conditions with the tires on level ground.

15. Check the wheel alignment.

16. After adjusting wheel alignment, adjust the neutral position of the steering angle sensor.

LOWER CONTROL ARM BUSHING REPLACEMENT

The rear lower control arms (lower links) do not use bushings or ball joints. Replace the entire arm assembly if any damage is noticed, including cracks, dents or deformations.

Suspension Arm

REMOVAL

1. Before servicing the vehicle, refer to the Precautions Section.

2. Raise and safely support the vehicle.

3. Remove the wheels.

4. Remove the stabilizer connecting rod mounting bracket from the suspension arm.

5. Remove the driveshaft from vehicle (VK45DE models).

6. Remove the cotter pin from the suspension arm ball joint, and loosen the nut.

7. Use a ball joint remover to remove the suspension arm from the axle assembly. Be careful not to damage the ball joint boot.

✳✳ WARNING

Temporarily tighten the mounting nut to prevent damage to the threads and to prevent the ball joint remover from coming off.

8. Remove the fixing nuts and bolts between the suspension arm and rear suspension member.

9. Remove the suspension arm from the vehicle.

VISUAL INSPECTION

1. Check the suspension arm and bushing for deformation, cracks, or damage. If any non-standard condition is found, replace it.

2. Check the boot of the ball joint for cracks, or damage, and also for grease leakage.

BALL JOINT INSPECTION

Manually move the ball stud to confirm it moves smoothly with no binding.

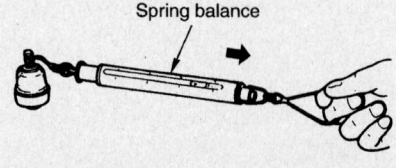

09482_FX35_G0049

Rear suspension arm ball stud swing torque check

SWING TORQUE INSPECTION

➡**Before measuring, move the ball joint at least ten times by hand to check for smooth movement.**

1. Hook a spring scale at the cotter pin mounting hole. Confirm that the measurement value is within specifications when the ball stud begins moving. Swing torque: 0.5–3.4 Nm (5–30 inch lbs.). Measured value of spring scale: 9.7–66.0 N (2.18–14.8 lbs.).

2. If it is outside the specified range, replace suspension arm assembly.

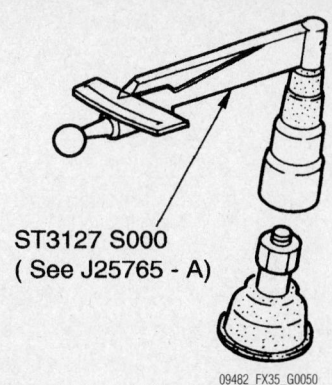

ST3127 S000
(See J25765 - A)

09482 FX35 G0050

Rear suspension arm ball stud rotating torque check

ROTATING TORQUE INSPECTION

1. Attach mounting nut to ball stud. Make sure rotating torque is within the specifications with a preload gauge (SST). Rotating torque: 0.5–3.4 Nm (5–30 inch lbs.).
2. If it is outside the specified range, replace suspension arm assembly.

AXIAL END PLAY INSPECTION

1. Move tip of ball joint in axial direction to check for looseness. Axial end play: 0 mm (0 in.).
2. If it is outside the specified range, replace suspension arm assembly.

INSTALLATION

1. Refer to the rear suspension exploded view for tightening torques. Installation is the reverse of removal.

➡**Refer to exploded view and do not reuse non-reusable parts.**

2. Perform final tightening of rear suspension member installation position (rubber bushing) under unladen conditions with tires on level ground. Check wheel alignment.
3. After adjusting wheel alignment, adjust neutral position of steering angle sensor.

Radius Rod

REMOVAL & INSTALLATION

1. Before servicing the vehicle, refer to the Precautions Section.
2. Raise and safely support the vehicle.
3. Remove the wheels.
4. Support the rear lower link with a jack.

5. Remove the fixing bolt and nut in the axle side of the radius rod.
6. Remove the fixing bolt in the rear suspension member side of the radius rod, then remove the radius rod from the vehicle.
7. Check the radius rod and bushing for any deformation, cracks, or damage. Replace if necessary.

To install:

8. Refer to the exploded view for tightening torques. Installation is the reverse of removal.

➡**Refer to the exploded view and do not reuse non-reusable parts.**

9. Perform the final tightening of the rear suspension member and axle installation position (rubber bushing) with the tires on level ground. Check the wheel alignment.
10. After adjusting the wheel alignment, adjust the neutral position of the steering angle sensor.

Stabilizer Bar

REMOVAL & INSTALLATION

1. Before servicing the vehicle, refer to the Precautions Section.
2. Raise and safely support the vehicle.
3. Remove the wheels.
4. Remove the center muffler from vehicle.
5. Remove the fixing bolts and remove the stabilizer connecting rod mount bracket from the suspension arm.
6. Remove the lower side fixing nut on the stabilizer connecting rod and remove the stabilizer connecting rod from the stabilizer bar.
7. Remove the fixing nuts on the stabilizer clamps and remove the stabilizer from the vehicle.
8. Check the stabilizer bar, stabilizer bushings, stabilizer clamps, stabilizer connecting rod, and stabilizer connecting rod mounting bracket for any deformation, cracks or damage. Replace if necessary.

To install:

9. Refer to the exploded view for tightening torques. Installation is the reverse of removal.

➡**Refer to exploded view and do not reuse non-reusable parts.**

10. The stabilizer bar uses pillow ball type connecting rod, position the ball joint with the case on pillow ball head parallel to the stabilizer bar.
11. When the bushing and clamp are

installed to the stabilizer bar, position the bushing and clamp inside of the side slip prevention clamp.

Rear Suspension Member

REMOVAL & INSTALLATION

1. Before servicing the vehicle, refer to the Precautions Section.
2. Raise and safely support the vehicle.
3. Remove the wheels.
4. Remove the brake caliper. Hang it in a place where it will not interfere with work.

➡**Avoid depressing the brake pedal while the brake caliper is removed.**

5. Remove the wheel sensor from the rear final drive, then remove the wheel sensor harness from the rear suspension member.
6. Remove the height sensor harness from the rear suspension member (if equipped).
7. Remove the center muffler and main muffler.
8. Remove the stabilizer bar.
9. Remove the rear halfshafts.
10. Remove the driveshaft.
11. Remove the rear final drive.
12. Separate the attachments between the parking brake cable and vehicle and the rear suspension member.
13. Remove the rear lower link and coil spring.
14. Remove the fixing bolt in the lower side of the shock absorber.
15. Support the rear suspension member with a jack.
16. Remove the fixing bolts and nuts of the tunnel stay and member, then remove those parts from the vehicle and rear suspension member.
17. Remove the fixing bolts and nuts of the rear pin stay then remove the rear pin stay from the vehicle and rear suspension member.
18. Slowly lower the jack, then remove the rear suspension member, suspension arm, radius rod, front lower link and axle from the vehicle as a unit.
19. Remove the fixing bolts and nuts, then remove the suspension arm, front lower link, and radius rod from the rear suspension member.
20. Check the rear suspension member for deformation, cracks, and other damage and replace if necessary.

To install:

21. Refer to the exploded view for tightening torques. Installation in the reverse order of removal.

➡**Refer to the exploded view and do not reuse non-reusable parts.**

22. Perform the final tightening of the links (rubber bushing) with the tires on level ground. Check the wheel alignment.

23. After adjusting the wheel alignment, adjust the neutral position of the steering angle sensor.

Wheel Hub and Knuckle

REMOVAL & INSTALLATION

1. Before servicing the vehicle, refer to the Precautions Section.
2. Raise and safely support the vehicle.
3. Remove the rear wheel.
4. Remove the brake caliper. Hang it in a place where it will not interfere with work.

➡**Avoid depressing brake pedal while brake caliper is removed.**

5. Remove the disc rotor.
6. Remove the wheel sensor from the axle.

✳ CAUTION

Do not pull on the wheel sensor wiring harness.

7. Remove the cotter pin, then remove the lock nut from the halfshaft.
8. Separate the halfshaft from the wheel hub and bearing assembly by lightly tapping the end with a suitable hammer and block of wood. If it is hard to separate, use a suitable puller.
9. Remove the mounting bolts of the wheel hub/bearing assembly, then remove the wheel hub/bearing assembly from the axle.
10. Remove the parking brake cable and parking brake shoe from the brake backing plate.
11. Remove the mounting nuts of the anchor block, then remove the anchor block and backing plate from the axle.
12. Loosen mounting bolts and nuts of the front lower link, radius rod, and rear lower link on the side of the suspension member.
13. Set a jack under the rear lower link, then remove the mounting bolt in the front lower link side of the shock absorber.
14. Remove the bolt and nut in the axle side of the rear lower link, then remove the coil spring.

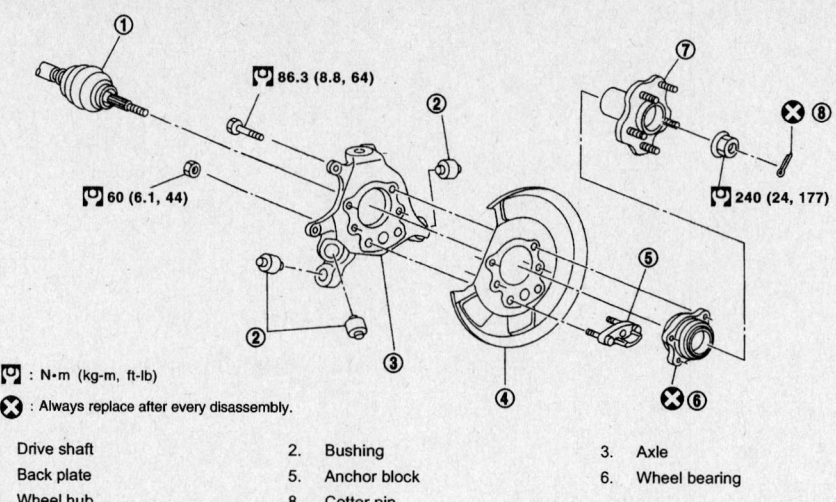

Exploded view of rear wheel hub mounting

🔧 86.3 (8.8, 64)

🔧 60 (6.1, 44)

🔧 240 (24, 177)

🔧 : N·m (kg-m, ft-lb)

❌ : Always replace after every disassembly.

1.	Drive shaft	2.	Bushing	3.	Axle
4.	Back plate	5.	Anchor block	6.	Wheel bearing
7.	Wheel hub	8.	Cotter pin		

67162-FX35-G230

15. Remove the mounting bolts and nuts in the axle side of the front lower link and radius rod.
16. Remove the suspension arm and cotter pin at the axle, then loosen the mounting nut.
17. Use a ball joint remover (or equivalent) to remove the suspension arm from the axle. Be careful not to damage the ball joint boot.

✳✳ CAUTION

Temporarily tighten the mounting nut to prevent damage to the threads and to prevent the ball joint remover from coming off.

18. Remove the axle from the vehicle.
19. Inspect the Ball Joint for boot breakage, axial looseness, and torque of suspension arm ball joint. Maximum of allowable axial end play is 0.00 in. (0.0mm).

To install:

20. Install the axle in the vehicle.
21. Reattach the upper control arm to the axle.
22. Tighten the upper control arm mounting nut to 96 ft. lbs. (130 Nm), and install a new cotter pin at the axle.
23. Install the mounting bolts and nuts in the axle side of the front lower link and radius rod.
24. Install the coil spring, then install the bolt and nut in the axle side of the rear lower link.
25. Install the mounting bolt in the front lower link side of the shock absorber.

26. Tighten the mounting bolts and nuts of the front lower link, radius rod, and rear lower link on the side of the suspension member.
27. Install the anchor block and backing plate onto the axle.
28. Install and tighten the mounting nuts of the anchor block to 44 ft. lbs. (60 Nm).
29. Install the parking brake cable and parking brake shoe from the brake backing plate.
30. Install the wheel hub/bearing assembly onto the axle, then install the mounting bolts of the wheel hub/bearing assembly and tighten to 64 ft. lbs. (86 Nm).
31. Install the halfshaft to the wheel hub/bearing assembly.
32. Install and tighten the lock nut on the halfshaft to 177 ft. lbs. (240 Nm), then install a new cotter pin.
33. Install the wheel sensor on the axle.
34. Install the disc rotor.
35. Install the brake caliper.
36. Install the appropriate wheel.

✳✳ CAUTION

Perform final tightening of the rear suspension fasteners under unladen conditions with the tires on level ground.

37. Check the wheel alignment.
38. After adjusting wheel alignment, adjust the neutral position of the steering angle sensor.

FRONT DISC BRAKES

Brake Caliper

REMOVAL & INSTALLATION

2003–05 Models

1. Before servicing the vehicle, refer to the Precautions Section.
2. Raise and safely support the vehicle.
3. Remove the front wheels from the vehicle.
4. Drain the brake fluid.
5. Remove the union bolts and torque member bolts, and remove the brake caliper assembly from the vehicle.
6. If necessary, remove the rotor.

To install:

✳✳ CAUTION

Use Only new "DOT3" brake fluid. Do not reuse drained brake fluid.

7. If removed, install the rotor.
8. Install the caliper assembly onto the vehicle. Tighten the sliding pin bolts to 20 ft. lbs. (27 Nm) and the torque member bolts to 116 ft. lbs. (157 Nm) for 2003 models. 2004–05 models have one of 2 bolt types, either a flange bolt or a bolt with a washer. Torque the flange bolts to 122 ft. lbs. (165 Nm) or the bolt/washers to 116 ft. lbs. (157 Nm).

✳✳ CAUTION

When attaching the caliper assembly to the vehicle, wipe any oil off the knuckle spindle, washers and caliper assembly attachment surfaces.

9. Reattach the brake hose to the brake caliper assembly, and tighten the union bolt to 13 ft. lbs. (18 Nm).

✳✳ CAUTION

Do not reuse the old copper washers for the union bolt. Attach the brake hose to the caliper assembly together only using the specified union bolt and washers.

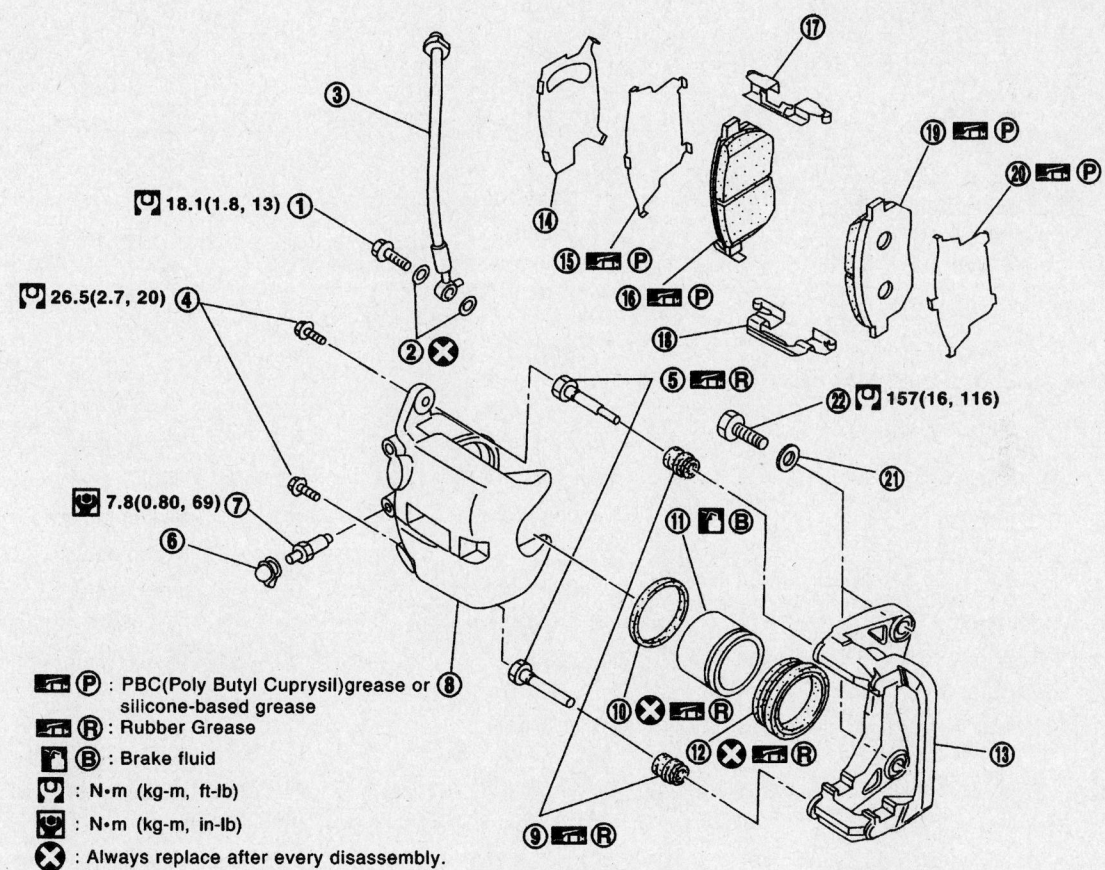

- ⬛ ℗ : PBC(Poly Butyl Cuprysil)grease or silicone-based grease
- ⬛ ℝ : Rubber Grease
- ⬛ Ⓑ : Brake fluid
- 🔧 : N•m (kg-m, ft-lb)
- 🔧 : N•m (kg-m, in-lb)
- ✖ : Always replace after every disassembly.

1. Union bolt	2. Copper washer	3. Brake hose
4. Sliding pin bolt	5. Sliding pin	6. Cap
7. Bleed valve	8. Cylinder body	9. Sliding pin boot
10. Piston seal	11. Piston	12. Piston boot
13. Torque member	14. Inner shim cover	15. Inner shim
16. Inner pad	17. Pad retainer (Upper)	18. Pad retainer (Lower)
19. Outer pad	20. Outer shim	21. Washer
22. Torque member bolt		

67162-FX35-G231

Exploded view of the front disc brake mounting and components—2003 models

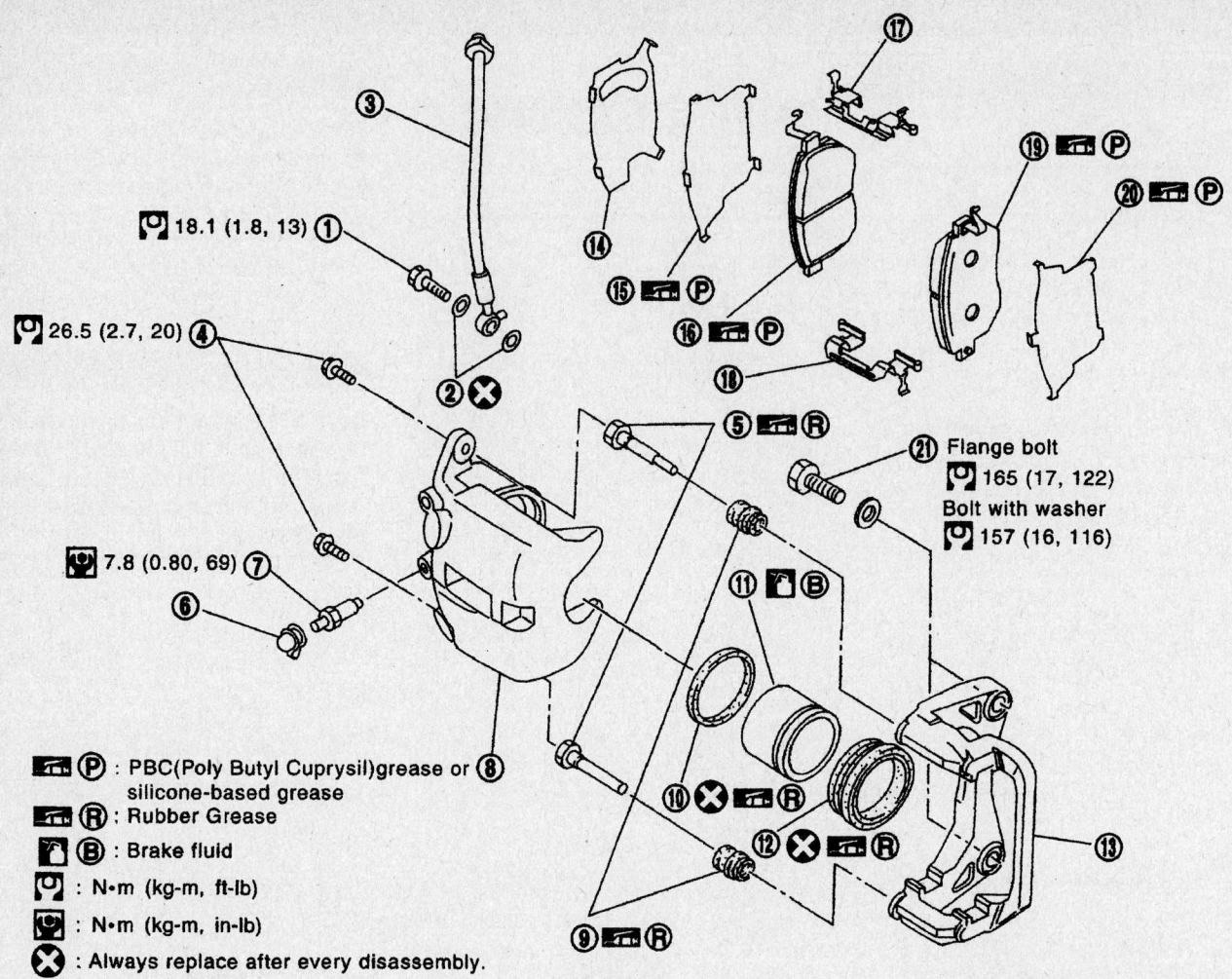

18.1 (1.8, 13) ①

26.5 (2.7, 20) ④

7.8 (0.80, 69) ⑦

⑥

⑰

⑲ P

⑳ P

⑭

⑮ P

⑯ P

⑱

⑤ R

㉑ Flange bolt
165 (17, 122)
Bolt with washer
157 (16, 116)

⑪ B

⑧

⑩ R

⑫ R

⑨ R

⑬

P : PBC(Poly Butyl Cuprysil)grease or silicone-based grease
R : Rubber Grease
B : Brake fluid
: N•m (kg-m, ft-lb)
: N•m (kg-m, in-lb)
: Always replace after every disassembly.

1.	Union bolt	2.	Copper washer	3.	Brake hose
4.	Sliding pin bolt	5.	Sliding pin	6.	Cap
7.	Bleed valve	8.	Cylinder body	9.	Sliding pin boot
10.	Piston seal	11.	Piston	12.	Piston boot
13.	Torque member	14.	Inner shim cover	15.	Inner shim
16.	Inner pad	17.	Pad retainer (Upper)	18.	Pad retainer (Lower)
19.	Outer pad	20.	Outer shim	21.	Torque member bolt

09482_FX35_G0052

Exploded view of the front disc brake mounting and components—2004–05 models

10. Refill the brake system with new brake fluid, then bleed the system.
11. Install the wheels.

2006 Models

1. Before servicing the vehicle, refer to the Precautions Section.
2. Raise and safely support the vehicle.
3. Remove the wheels.

➡**Use a lug nut to hold the rotor, if necessary.**

4. Remove the union bolt and disconnect the brake line from the caliper. Discard the copper washers. Plug the brake line to prevent fluid loss.
5. Remove the torque member bolts, and remove the brake caliper assembly from the vehicle.
6. Remove the sliding pin bolts and remove the caliper from the torque member.

➡**Put matchmarks on both the rotor and the wheel hub if removing rotor.**

7. Remove the rotor if necessary.
To install:
8. Install the rotor.

➡**If using the same rotor, align the matchmarks.**

9. Attach the caliper to the torque member. Torque the bolts to 34 ft. lbs. (46 Nm).

➡**Before installing the torque member, wipe any oil and grease from the washer seats on the steering knuckle and mounting surface of the torque member.**

10. Install the caliper assembly and tighten the mounting bolts to 122 ft. lbs. (165 Nm).
11. Connect the brake hose and tighten the union bolt to 13 ft. lbs. (18 Nm).

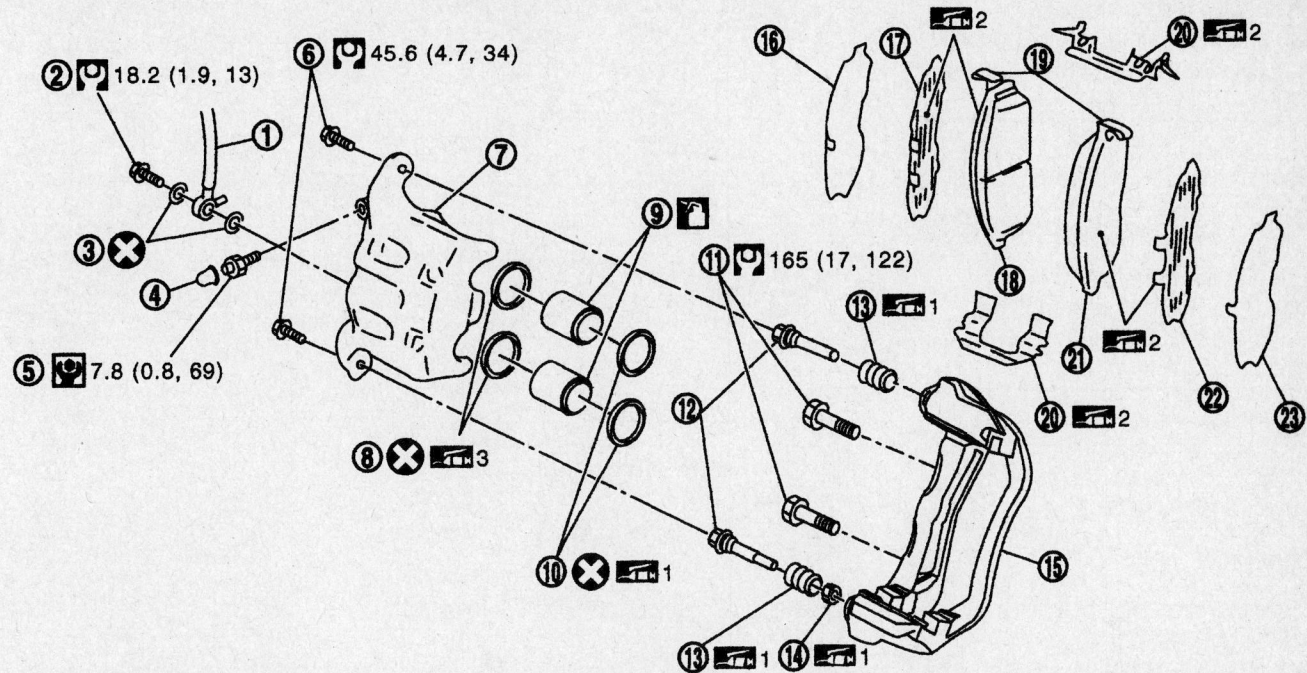

1. Brake hose
4. Cap
7. Cylinder body
10. Piston boot
13. Sliding pin boot
16. Inner shim cover
19. Pad wear sensor
22. Outer shim

2. Union bolt
5. Bleed valve
8. Piston seal
11. Torque member mounting bolt
14. Bushing
17. Inner shim
20. Pad retainer
23. Outer shim cover

3. Copper washer
6. Sliding pin bolt
9. Piston
12. Sliding pin
15. Torque member
18. Inner pad
21. Outer pad

1: Apply rubber grease.

2: Apply PBC (Poly Butyl Cuprysil) grease or silicone-based grease.

3: Apply polyglycol ether based lubricant.

: Apply brake fluid.

09482_FX35_G0053

Exploded view of the front disc brake mounting and components—2006 models

✳✳ CAUTION

Use new copper washers.

12. Refill with new brake fluid and bleed the air.

✳✳ CAUTION

Refill with new "DOT3" brake fluid. Do not reuse brake fluid.

13. Install the wheels.

Disc Brake Pads

REMOVAL & INSTALLATION

2003–05 Models

1. Before servicing the vehicle, refer to the Precautions Section.
2. Raise and safely support the vehicle.
3. Remove the front wheels.
4. Remove the lower sliding pin bolt.
5. Suspend the cylinder body with strong cord or wire, then remove the pad and shim from the torque member.

To install:

6. Position the inner shim and shim cover onto the inner pad, and outer shim onto the outer pad.
7. Push the caliper piston in so that the brake pad is firmly installed.
8. Position the cylinder body on the torque member.

➡**Using a disc brake piston tool, makes it easier to push in the piston.**

✳✳ CAUTION

By pushing in the piston, brake fluid returns to the master cylinder reservoir tank. Watch the level of the surface of reservoir tank.

9. Reattach the pad retainer to the torque member.

✳✳ CAUTION

When attaching the pad retainer, attach it firmly so that it does not float up higher than the torque member.

10. Install the lower sliding pin bolt and tighten it to 20 ft. lbs. (27 Nm).
11. Check the brake assembly for drag.
12. Install the wheels

2006 Models

1. Before servicing the vehicle, refer to the Precautions Section.
2. Raise and safely support the vehicle.
3. Remove the wheel.
4. Remove about ⅓ of the fluid from the master cylinder.
5. Remove the lower sliding pin bolt, pivot the caliper up and retain it, if necessary, with a length of wire.
6. Remove the pads, pad retainers, and shims from the torque member.

⬌ WARNING

When removing the pad retainer from the torque member, lift the pad retainer in the direction shown by the arrow in the illustration to avoid deforming it.

To install:

7. Apply silicone brake grease to both sides of the inner shim and outer shim. Install the inner shim on the inner pad, and the outer shim and outer shim cover on the outer pad.
8. Apply silicone brake grease to the pad contact surface on the pad retainer, and install the pad retainers and pads on the torque member.

⬌ WARNING

The inner pad and outer pad have a pad-return mechanism on the upper

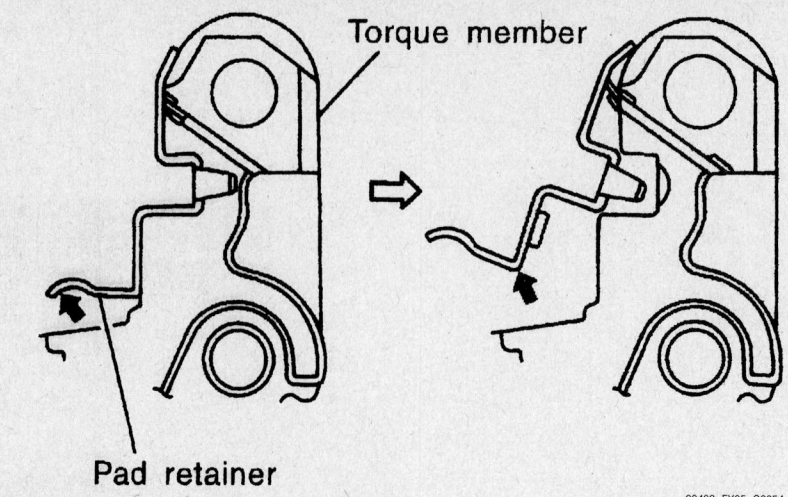

Pad retainer removal direction—2006 models

09482_FX35_G0054

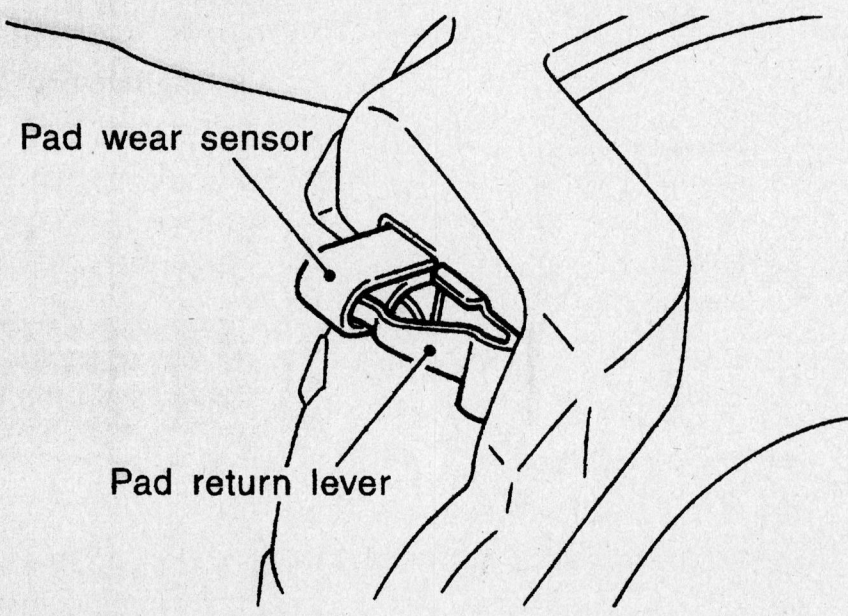

Pad return mechanism installation–2006 models

09482_FX35_G0055

side of the pad retainer. When installing the pad on the torque member, be sure to install the pad return lever on the pad wear sensor securely.

9. Depress the piston until the pads can be installed. Carefully monitor the brake fluid level in the reservoir tank.

10. Install the caliper on the torque member.
11. Install the lower sliding pin bolt, and tighten it to 34 ft. lbs. (46 Nm).
12. Secure the rotor with lug nuts. Depress brake pedal a few times until it gets a responsive touch.
13. Check the brakes for drag.
14. Install the wheels.

REAR DISC BRAKES

Brake Caliper

REMOVAL & INSTALLATION

2003–05 Models

1. Before servicing the vehicle, refer to the Precautions Section.
2. Raise and safely support the vehicle.
3. Remove the wheels.
4. Drain the brake fluid.
5. Remove the union bolts and torque member bolts, and remove the brake caliper assembly from the vehicle.
6. If necessary, remove the disc rotor.

To install:

✳✳ CAUTION

Only use new "DOT3" brake fluid. Do not reuse drained brake fluid.

7. If removed, install the disc rotor.
8. Install the caliper assembly onto the vehicle. Tighten the sliding pin bolts to 32 ft. lbs. (43 Nm) and the torque member bolts to 62 ft. lbs. (84 Nm).

✳✳ CAUTION

When attaching the caliper assembly to the vehicle, wipe any oil off the knuckle spindle, washers and caliper assembly attachment surfaces.

9. Reattach the brake hose to the brake caliper assembly, and tighten the union bolt to 13 ft. lbs. (18 Nm).

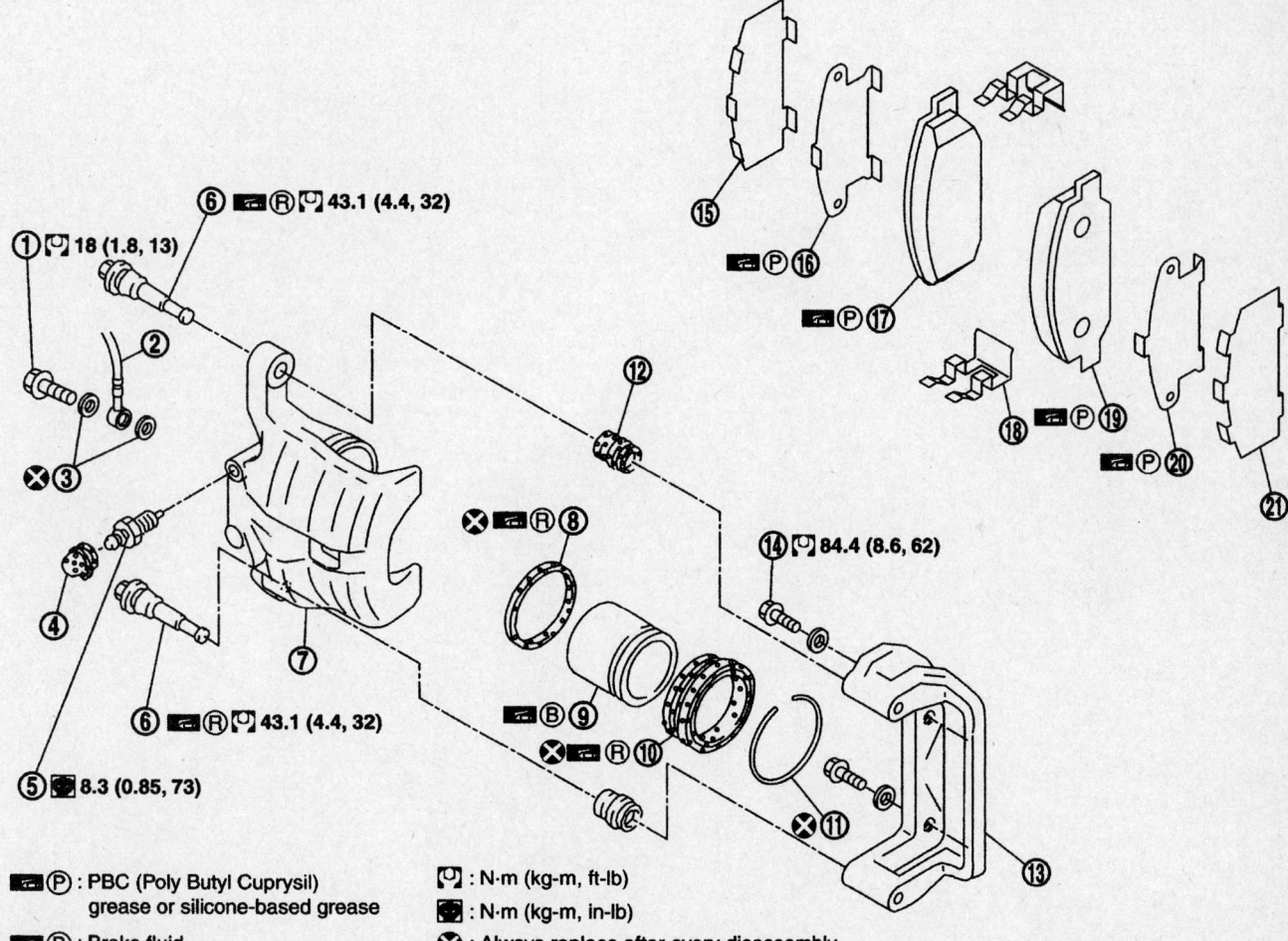

📖ℙ : PBC (Poly Butyl Cuprysil) grease or silicone-based grease

📖Ⓑ : Brake fluid

📖Ⓡ : Rubber grease

🔧 : N·m (kg-m, ft-lb)

⊙ : N·m (kg-m, in-lb)

❌ : Always replace after every disassembly.

1. Union bolt	2. Brake hose	3. Copper washer
4. Cap	5. Bleed valve	6. Sliding pin bolt
7. Cylinder body	8. Piston seal	9. Piston
10. Piston boot	11. Retaining ring	12. Sliding pin boot
13. Torque member	14. Torque member bolt	15. Inner shim cover
16. Inner shim	17. Inner pad	18. Pad retainer
19. Outer pad	20. Outer shim	21. Outer shim cover

67162-FX35-G232

Exploded view of the rear disc brake mounting and components—2003–05 models

✱✱ CAUTION

Do not reuse the old copper washers for the union bolt. Attach the brake hose to the caliper assembly together only using the specified union bolt and washers.

10. Refill the brake system with new brake fluid, then bleed the system.

11. Install the wheels.

2006 Models

1. Before servicing the vehicle, refer to the Precautions Section.

2. Raise and safely support the vehicle.

3. Remove the wheels.

4. If necessary, secure the rotor with a lug nut.

5. Remove the union bolt and disconnect the brake line from the caliper. Discard the copper washers. Plug the brake line to prevent fluid loss.

6. Remove the torque member bolts, and remove the brake caliper assembly from the vehicle.

7. Remove the slide pin bolts and separate the caliper from the torque member.

➡**If removing the rotor, put matchmarks on both the rotor and the wheel hub.**

8. Remove the rotor if necessary.

To install:

9. Install the rotor, aligning the matchmarks.

10. Attach the caliper to the torque member. Torque the bolts to 32 ft. lbs. (43 Nm).

➡**Before installing the caliper assembly, wipe off any oil and grease from the washer seats on the axle assembly and mounting surface of the caliper assembly.**

11. Install the caliper assembly to the vehicle, and tighten the bolts to 62 ft. lbs. (84 Nm).

12. Using NEW copper washers, install the brake hose to caliper assembly and tighten the union bolt to 13 ft. lbs. (18 Nm).

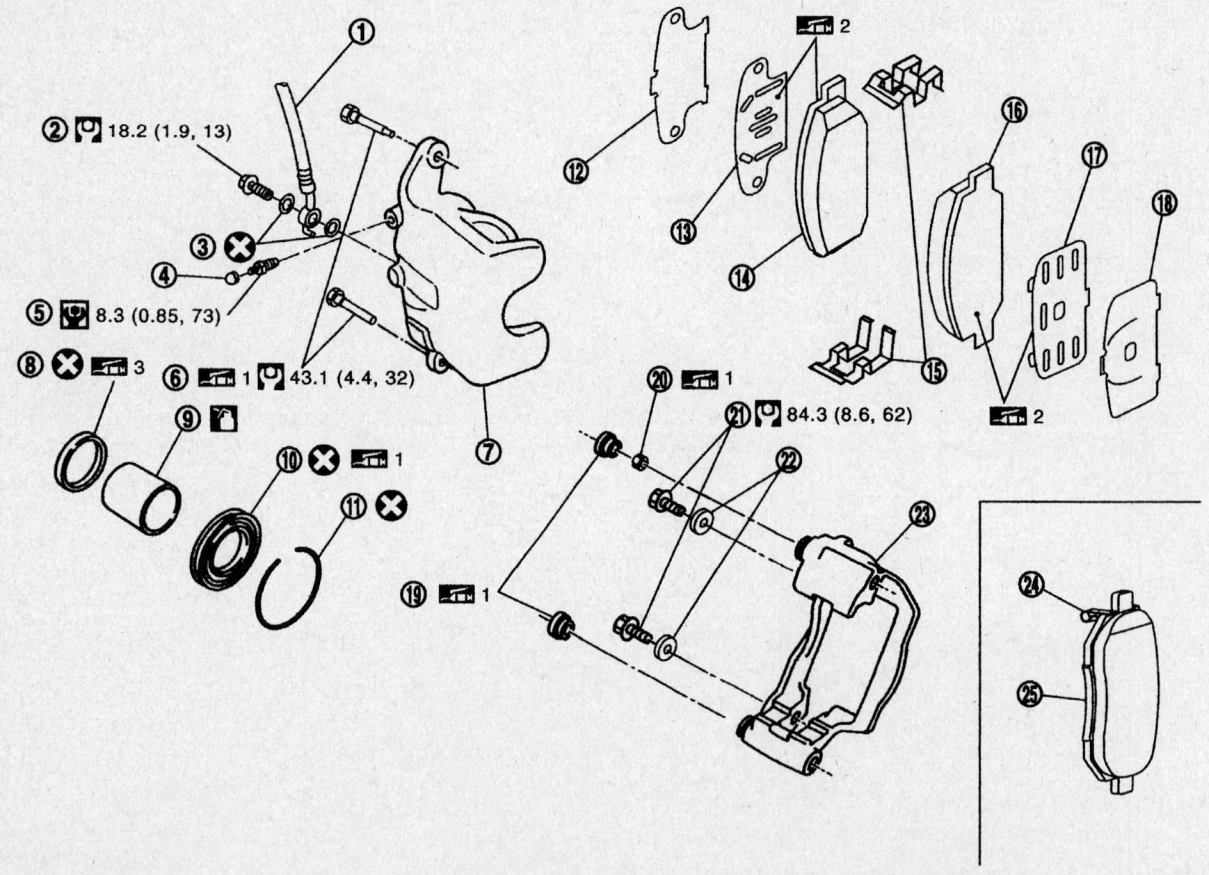

1.	Brake hose	2.	Union bolt	3.	Copper washer
4.	Cap	5.	Bleed valve	6.	Sliding pin bolt
7.	Cylinder body	8.	Piston seal	9.	Piston
10.	Piston boot	11.	Retaining ring	12.	Inner shim cover
13.	Inner shim	14.	Inner pad	15.	Pad retainer
16.	Outer pad	17.	Outer shim	18.	Outer shim cover
19.	Sliding pin boot	20.	Bushing	21.	Torque member mounting bolt
22.	Washer	23.	Torque member	24.	Pad wear sensor
25.	Inner pad (RH)				

Exploded view of the rear disc brake mounting and components—2006 models

09482_FX35_G0056

✳✳ CAUTION

Use new copper washers.

13. After installing the caliper assembly, refill with new brake fluid and bleed the air.

➡**Refill with new "DOT 3" fluid. Do not reuse drained brake fluid.**

14. Install the wheels.

Disc Brake Pads

REMOVAL & INSTALLATION

1. Before servicing the vehicle, refer to the Precautions Section.

2. Raise and safely support the vehicle.
3. Remove the wheels.
4. Remove the upper sliding pin bolt.
5. Suspend the cylinder body with strong cord or wire, then remove the pad and shim from the torque member.

To install:

6. Apply silicon-based grease to the backside of the pad and to both sides of the shim, then attach the inner shim and shim cover to the inner pad. Attach the outer shim and outer shim cover to the outer pad.
7. Install the pad retainer and mount the pad onto the torque member.
8. Position the cylinder body on the torque member.

➡**Using a disc brake piston tool (commercial service tool), makes it easier to push in the piston.**

✳✳ CAUTION

By pushing in the piston, brake fluid returns to the master cylinder reservoir tank. Watch the level of the surface of the reservoir tank.

9. Install the top sliding pin bolt and tighten it to 32 ft. lbs. (43 Nm).
10. Check the brake assembly for drag.
11. Install the wheels

PARKING BRAKE

Shoes

REMOVAL

1. Before servicing the vehicle, refer to the Precautions Section.
2. Raise and safely support the vehicle.
3. Release the parking brake.
4. Remove the rear wheels.
5. Remove the rotor. If rotor cannot be removed:
   a. Secure the rotor in place with lug nuts and remove adjuster hole plug.
   b. Using a flat-bladed screwdriver, rotate the adjuster in direction "B" to retract and loosen the brake shoes.

6. Remove the anti-rattle pins, retainers, anti-rattle springs, and return springs.
7. Remove the parking brake shoes, adjuster assembly, adjuster spring and toggle lever.

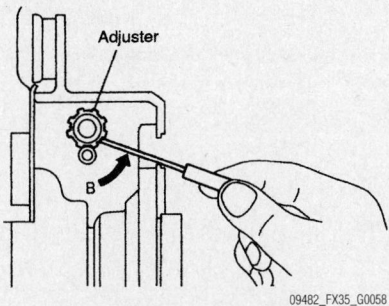

09482_FX35_G0058

Backing off the brake shoe adjuster

INSPECTION

1. Check the thickness of the lining. Standard thickness "A": 3.2mm (0.126 in.). Repair limit thickness "A": 1.5 mm (0.059 in.).
2. Check the drum inner diameter. Standard inner diameter: 190mm (7.48 in.). Maximum inner diameter: 191mm (7.52 in.).
3. Check the following:
   - Shoes for excessive wear, damage, and peeling.
   - Shoe sliding surfaces for excessive wear and damage.
   - Anti-rattle pins for excessive wear and corrosion.
   - Return springs for sagging.

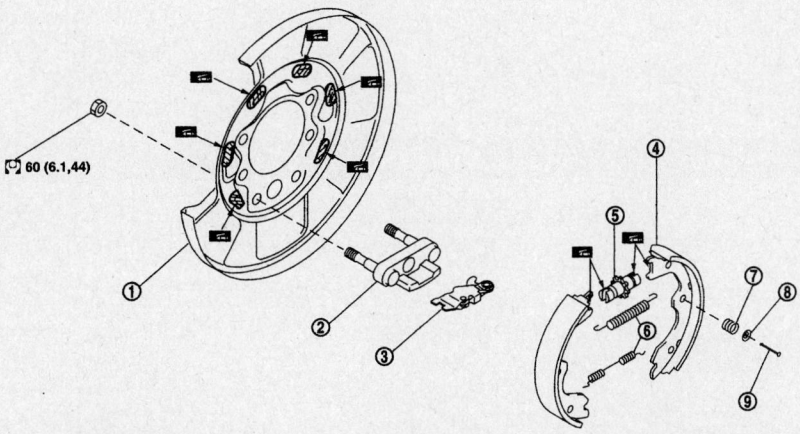

▪ : PBC (Poly Butyl Cuprysil) grease or silicone-based grease point
🔧 : N·m (kg-m, ft-lb)

1.	Back plate	2.	Anchor block	3.	Toggle lever
4.	Shoe	5.	Adjuster	6.	Return spring
7.	Anti-rattle spring	8.	Retainer	9.	Anti-rattle pin

09482_FX35_G0057

Parking brake shoes and related components

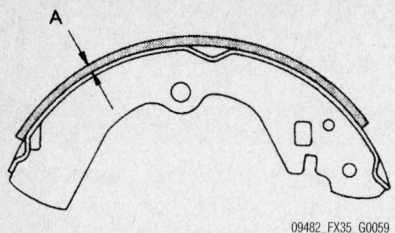

Checking the lining thickness

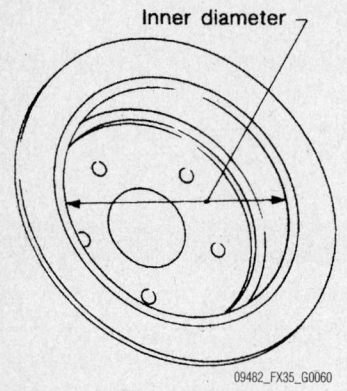

Checking the drum inner diameter

- Check that adjuster moves smoothly.
- Visually check the inside of drum for excessive wear, cracks, and damage. Check the inside diameter of the drum using a pair of Vernier calipers.
4. Replace any suspect parts.

➡**When disassembling the adjuster, apply silicone grease or equivalent to the threads.**

INSTALLATION

1. Installation is the reverse of removal. Note the following:
- Apply brake grease to the specified points during assembly. See the accompanying illustration.

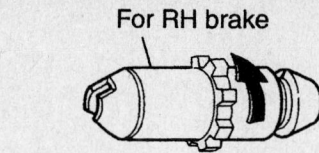

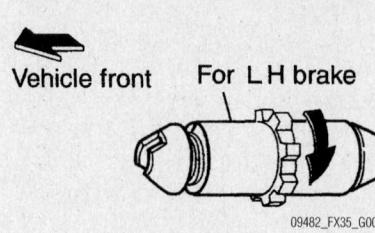

Adjuster expansion rotation

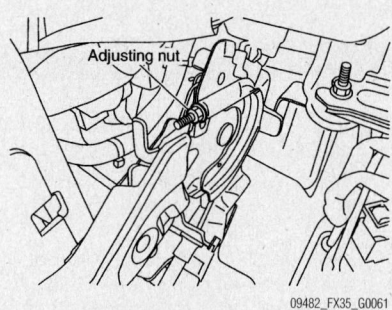

Parking brake cable adjusting nut

- Assemble the adjuster so that the threaded part expands when rotating it in the direction shown by the arrow.
- When disassembling the adjuster, apply silicone based grease to the threads.
2. Perform the final adjustment:
   a. Adjust the parking brake pedal stroke to 4–5 clicks fully depressed. Insert a deep socket wrench to rotate the adjusting nut and loosen the cable sufficiently. Then, return pedal.

b. Remove the wheels.
c. Using a lug nut, secure the rotor to hub to prevent it from tilting.
d. Remove the adjusting hole plug. Using a flat-bladed screwdriver, turn the adjuster clockwise until the rotor is locked. After locking, turn the adjuster in the opposite direction by 5 or 6 notches.
e. Rotate the rotor to make sure that there is no drag. Then install the adjusting hole plug.
3. After adjusting the clearance of rear shoes, with no drag on rear brake, adjust the cable as follows:
   a. Operate the pedal 10 or more times with a force of 490 N (110 lbs.).
   b. Depress the pedal until a deep socket can be inserted.
   Insert the deep socket, and rotate the adjusting nut to adjust the pedal stroke.

➡**Do not reuse the adjusting nut.**

   c. When the parking brake pedal is operated with a force of 200 N (45 lbs.), make sure the stroke is 4–5 notches. (Check it by listening and counting the ratchet clicks.)
   d. With the parking brake pedal completely returned, make sure there is no drag on the rear brake.
4. Perform the parking brake break-in operation as follows: Safely, drive forward at approximately 40 km/h (25 mph) with the parking brake set with a force of approx. 200 N (45 lbs.) for about 30 seconds.
5. After the break-in operation, check the pedal stroke of parking brake. Readjust if necessary.

➡**To prevent lining from getting too hot, allow a cool off period of approximately 5 minutes after every break-in operation.**

BRAKE HYDRAULIC SYSTEM

BRAKE SYSTEM BLEEDING

1. Before servicing the vehicle, refer to the Precautions Section.

✳✳ CAUTION

While bleeding the brake system, pay attention to the master cylinder fluid level.

2. Turn the ignition switch OFF and disconnect the ABS actuator and electric unit (control unit) connector or negative battery cable.
3. Raise and safely support the vehicle.
4. Attach a vinyl tube to the right, rear bleeder valve.
5. Depress the brake pedal fully 4 or 5 times.
6. With the brake pedal depressed, loosen the bleeder valve to let the air out, then tighten it immediately.
7. Repeat steps 3 and 4 until no more air comes out.
8. Tighten the bleeder valve for front calipers to 69 inch lbs. (7.8 Nm) and rear calipers to 73 inch lbs. (8.3 Nm)
9. Fill the master cylinder reservoir.
10. Repeat steps 2 through 7 for the left front, left rear, and the right front calipers, in that order.

SPECIFICATIONS AND MAINTENANCE CHARTS

ENGINE AND VEHICLE IDENTIFICATION

Code ①	Liters (cc)	Cu. In.	Cyl.	Fuel Sys.	Engine	Eng. Mfg.
VQ35DE	3.5 (3498)	213	6	MFI	DOHC	Nissan
VQ40DE	4.0 (3954)	241	6	MFI	DOHC	Nissan

MFI: Multi-port Fuel Injection

DOHC: Double Overhead Camshafts

① Located on the timing belt cover

② 10th digit of the Vehicle Identification Number (VIN)

Code ②	Year
2	2002
3	2003
4	2004
5	2005
6	2006

09482_PATH_C0001

GENERAL ENGINE SPECIFICATIONS

Year	Model	Engine Displacement Liters (cc)	Engine ID/VIN	Fuel System Type	Net Horsepower @ rpm	Net Torque @ rpm (ft. lbs.)	Bore x Stroke (in.)	Compression Ratio	Oil Pressure @ rpm
2002	QX4	3.5 (3498)	VQ35DE	MFI	240@6000	265@3200	3.76X3.20	10.0:1	43@2000
	Pathfinder	3.5 (3498)	VQ35DE	MFI	240@6000	265@3200	3.76X3.20	10.0:1	43@2000
2003	QX4	3.5 (3498)	VQ35DE	MFI	240@6000	265@3200	3.76X3.20	10.0:1	43@2000
	Pathfinder	3.5 (3498)	VQ35DE	MFI	240@6000	265@3200	3.76X3.20	10.0:1	43@2000
2004	Pathfinder	3.5 (3498)	VQ35DE	MFI	240@6000	265@3200	3.76X3.20	10.0:1	43@2000
2005	Pathfinder	4.0 (3954)	VQ40DE	MFI	270@5600	291@4000	3.76X3.62	9.7:1	43@2000

MFI: Multi-port Fuel Injection

09482_PATH_C0002

ENGINE TUNE-UP SPECIFICATIONS

Year	Engine Displacement Liters (cc)	Engine ID/VIN	Spark Plug Gap (in.)	Ignition Timing (deg.) MT	Ignition Timing (deg.) AT	Fuel Pump (psi)	Idle Speed (rpm) MT	Idle Speed (rpm) AT ②	Valve Clearance (in.) In.	Valve Clearance (in.) Ex.
2002	3.5 (3498)	VQ35DE	0.044	15B	15B	35 ①	700-800	700-800	HYD	HYD
2003	3.5 (3498)	VQ35DE	0.044	15B	15B	35 ①	700-800	700-800	HYD	HYD
2004	3.5 (3498)	VQ35DE	0.044	15B	15B	35 ①	700-800	700-800	HYD	HYD
2005	4.0 (3954)	VQ40DE	0.043	15B	15B	51	575-675	575-675	HYD	HYD

NOTE: The Vehicle Emission Control Information label often reflects specification changes made during production. The label figures must be used if they differ from those in this chart.

B: Before top dead center

HYD: Hydraulic

① System pressure at idle with vacuum hose connected
Should increase to 43 psi when disconnected

② Automatic transmission in Neutral or Park

09482_PATH_C0003

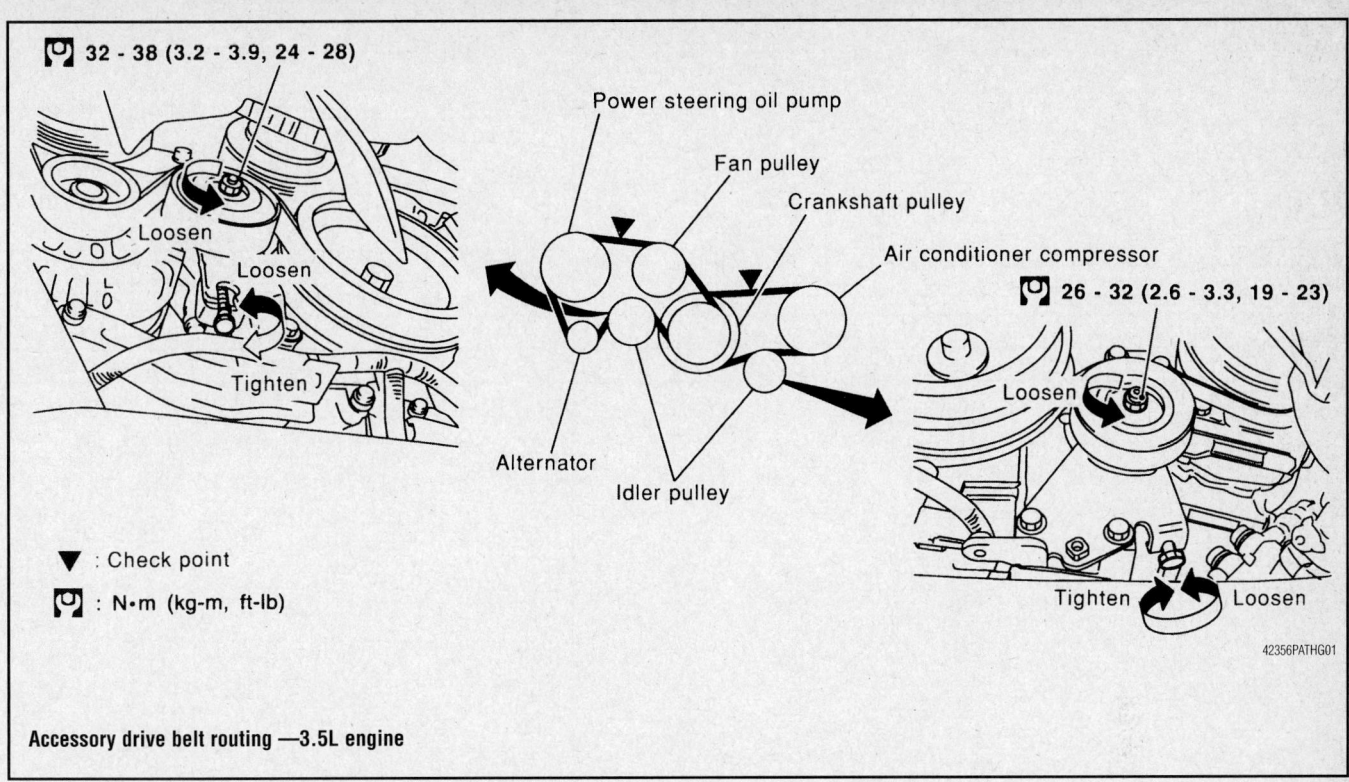

Accessory drive belt routing —3.5L engine

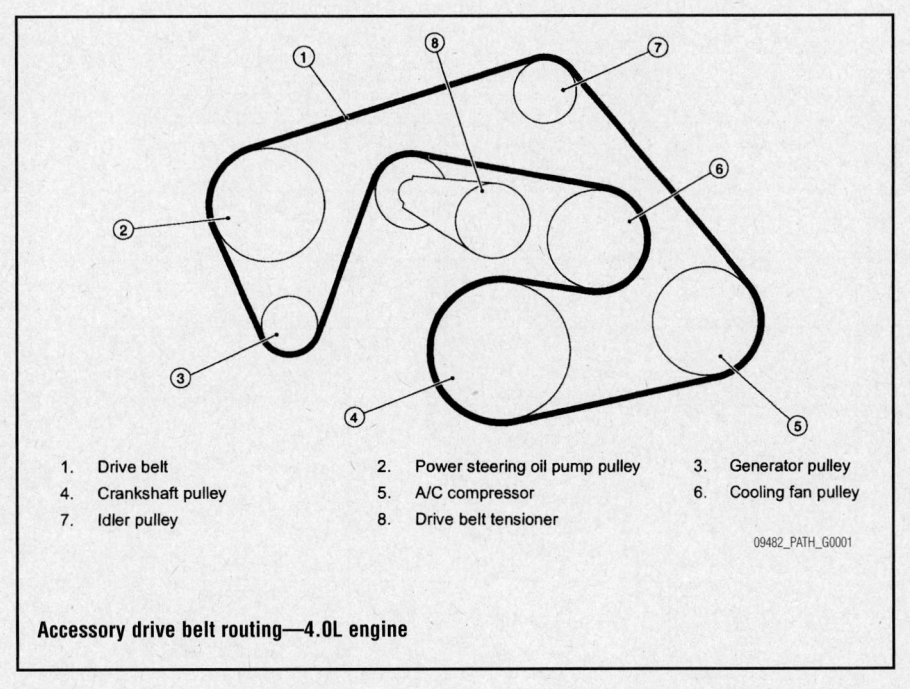

1.	Drive belt	2.	Power steering oil pump pulley	3.	Generator pulley
4.	Crankshaft pulley	5.	A/C compressor	6.	Cooling fan pulley
7.	Idler pulley	8.	Drive belt tensioner		

Accessory drive belt routing—4.0L engine

CAPACITIES

Year	Model	Engine Displacement Liters (cc)	Engine ID/VIN	Engine Oil with Filter (qts.)	Transmission (pts.) 5-Spd	Transmission (pts.) Auto.	Transfer Case (pts.)	Drive Axle Front (pts.)	Drive Axle Rear (pts.)	Fuel Tank (gal.)	Cooling System (qts.)
2002	QX4	3.5 (3498)	VQ35DE	5.25	—	18.0	5.3	3.9	5.9	21.1	9.75
	Pathfinder	3.5 (3498)	VQ35DE	5.25	—	18.0	①	3.9	5.9	21.1	9.75
2003	QX4	3.5 (3498)	VQ35DE	5.25	—	18.0	5.3	3.9	5.9	21.1	9.75
	Pathfinder	3.5 (3498)	VQ35DE	5.25	—	18.0	①	3.9	5.9	21.1	9.75
2004	Pathfinder	3.5 (3498)	VQ35DE	5.25	—	18.0	①	3.9	5.9	21.1	9.75
2005	Pathfinder	4.0 (3954)	VQ40DE	5.375	—	21.8	②	1.75	3	21.1	③

NOTE: All capacities are approximate. Add fluid gradually and check to be sure a proper fluid level is obtained.

① Part time: 2.375; full time: 2.625 pts.

② Part time: 4.25 pts; full time: 6.25 pts.

③ Without rear A/C: 11 qts.; with rear A/C: 14 qts.

09482_PATH_C0004

CRANKSHAFT AND CONNECTING ROD SPECIFICATIONS

All measurements are given in inches.

Year	Engine Displacement Liters (cc)	Engine ID/VIN	Crankshaft Main Brg. Journal Dia.	Crankshaft Main Brg. Oil Clearance	Crankshaft Shaft End-play	Crankshaft Thrust on No.	Connecting Rod Journal Diameter	Connecting Rod Oil Clearance	Connecting Rod Side Clearance
2002	3.5 (3498)	VQ35DE	①	0.0014-0.0018	0.0118	4	②	0.0013-0.0023	0.0079-0.0138
2003	3.5 (3498)	VQ35DE	①	0.0014-0.0018	0.0118	4	②	0.0013-0.0023	0.0079-0.0138
2004	3.5 (3498)	VQ35DE	①	0.0014-0.0018	0.0118	4	②	0.0013-0.0023	0.0079-0.0138
2005	4.0 (3954)	VQ40DE	③	0.0014-0.0018	0.0039-0.0098	4	NA	0.0013-0.0023	NA

NA - Not Available

① There are 24 different grades, ranging from A (2.3612) to 7 (2.3603)

② Grade 0: 2.0460-2.0462

   Grade 1: 2.0457-2.0460

   Grade 2: 2.0445-2.0457

③ There are 24 different grades, ranging from A (2.7549) to 7 (2.7540)

09482_PATH_C0005

VALVE SPECIFICATIONS

Year	Engine Displacement Liters (cc)	Engine ID/VIN	Seat Angle (deg.)	Face Angle (deg.)	Spring Test Pressure (lbs. @ in.)	Spring Installed Height (in.)	Stem-to-Guide Clearance (in.)		Stem Diameter (in.)	
							Intake	Exhaust	Intake	Exhaust
2002	3.5 (3498)	VQ35DE	45.15-45.45	45	45.4@1.457	1.457	0.0008-0.0021	0.0016-0.0029	0.2348-0.2354	0.2341-0.2346
2003	3.5 (3498)	VQ35DE	45.15-45.45	45	45.4@1.457	1.457	0.0008-0.0021	0.0016-0.0029	0.2348-0.2354	0.2341-0.2346
2004	3.5 (3498)	VQ35DE	45.15-45.45	45	45.4@1.457	1.457	0.0008-0.0021	0.0016-0.0029	0.2348-0.2354	0.2341-0.2346
2005	4.0 (3954)	VQ40DE	45.15-45.45	45	37-42 @1.457	1.457	0.0008-0.0021	0.0012-0.0025	0.2348-0.2354	0.2344-0.2350

09482_PATH_C0006

CAMSHAFT AND BEARING SPECIFICATIONS CHART

All measurements are given in inches.

Year	Engine Displ. Liters	Engine ID/VIN	Journal Diameter	Brg. Oil Clearance	Shaft End-play	Runout	Journal Bore	Lobe Height	
								Intake	Exhaust
2002	3.5	VQ35DE	①	②	0.0045-0.0074	0.0008	③	1.7900-1.7921	1.7746-1.7821
2003	3.5	VQ35DE	①	②	0.0045-0.0074	0.0008	③	1.7900-1.7921	1.7746-1.7821
2004	3.5	VQ35DE	①	②	0.0045-0.0074	0.0008	③	1.7900-1.7921	1.7746-1.7821
2005	4.0	VQ40DE	①	②	0.0045-0.0074	0.0008	③	1.7900-1.7921	1.7746-1.7821

① No. 1: 1.0211-1.0218
  Nos. 2-4: 0.9230-0.9238
② No. 1: 0.0018-0.0034
  Nos. 2-4: 0.0014-0.0030

③ No. 1: 1.0236-1.0244
  Nos. 2-4: 0.9252-0.9260

09482_PATH_C0007

PISTON AND RING SPECIFICATIONS

All measurements are given in inches.

Year	Engine Displacement Liters (cc)	Engine ID/VIN	Piston Clearance	Ring Gap Top Comp.	Ring Gap Bottom Comp.	Ring Gap Oil Control	Ring Side Clearance Top Comp.	Ring Side Clearance Bottom Comp.	Ring Side Clearance Oil Control
2002	3.5 (3498)	VQ35DE	0.0004-0.0012	0.0091-0.0130	0.0130-0.0189	0.0079-0.0236	0.0016-0.0031	0.0012-0.0028	0.0006-0.0020
2003	3.5 (3498)	VQ35DE	0.0004-0.0012	0.0091-0.0130	0.0130-0.0189	0.0079-0.0236	0.0016-0.0031	0.0012-0.0028	0.0006-0.0020
2004	3.5 (3498)	VQ35DE	0.0004-0.0012	0.0091-0.0130	0.0130-0.0189	0.0079-0.0236	0.0016-0.0031	0.0012-0.0028	0.0006-0.0020
2005	4.0 (3954)	VQ40DE	0.0004-0.0012	0.0091-0.0130	0.0130-0.0189	0.0079-0.0197	0.0018-0.0031	0.0012-0.0028	0.0026-0.0053

09482_PATH_C0008

TORQUE SPECIFICATIONS

All readings in ft. lbs.

Year	Engine Displacement Liters (cc)	Engine ID/VIN	Cylinder Head Bolts	Main Bearing Bolts	Rod Bearing Bolts	Crankshaft Damper Bolts	Flywheel Bolts	Manifold Intake	Manifold Exhaust	Spark Plugs	Lug Nuts
2002	3.5 (3498)	VQ35DE	①	②	③	④	61-69	⑤	21-24	14-22	87-108
2003	3.5 (3498)	VQ35DE	①	②	③	④	61-69	⑤	21-24	14-22	87-108
2004	3.5 (3498)	VQ35DE	①	②	③	④	61-69	⑤	21-24	14-22	87-108
2005	4.0 (3954)	VQ40DE	①	②	③	④	61-69	⑤	11	18	98

① Step 1: 72 ft. lbs.
Step 2: Loosen all bolts completely
Step 3: 25-33 ft. lbs.
Step 4: +90 degrees
Step 5: +90 degrees
② Step 1: 24-28 ft. lbs.
Step 2: +90 degrees

④ Step 1: 15 ft. lbs.
Step 2: +90 degrees
⑤ Step 1: 44-86 inch lbs.
Step 2: 20-23 ft. lbs.

09482_PATH_C0009

WHEEL ALIGNMENT

Year	Model	Caster Range (+/-Deg.)	Caster Preferred Setting (Deg.)	Camber Range (+/-Deg.)	Camber Preferred Setting (Deg.)	Toe-in (in.)
2002	Pathfinder	0.75	+3.00	0.75	+0.17	0.08+/-0.04
	QX4 ①	0.75	+3.00	0.75	+0.17	0.08+/-0.04
2003	Pathfinder	0.75	+3.00	0.75	+0.17	0.08+/-0.04
	QX4 ①	0.75	+3.00	0.75	+0.17	0.08+/-0.04
2004	Pathfinder	0.75	+3.00	0.75	+0.17	0.08+/-0.04
2005	Pathfinder	0.75	+2.00	0.5	-0.25	0.08+/-0.04

① Assumes P245/65R17 tire

09482_PATH_C0010

TIRE, WHEEL AND BALL JOINT SPECIFICATIONS

Year	Model	OEM Tires Standard	Optional	Tire Pressures (psi) Front	Rear	Wheel Size	Ball Joint Inspection
2002	Pathfinder LE	P245/65SR17	none	①	①	8J	②
	Pathfinder SE	P255/65SR16	none	①	①	7JJ	②
	QX4	P245/70R16	P245/65R17	①	①	Std: 7J/Opt: 8J	②
2003	Pathfinder LE	P245/65SR17	none	①	①	8J	②
	Pathfinder SE	P255/65SR16	none	①	①	7JJ	②
	QX4	P245/65R17	none	①	①	Std: 7J/Opt: 8J	②
2004	Pathfinder LE	P245/65SR17	none	①	①	8J	②
	Pathfinder SE	P255/65SR16	none	①	①	7JJ	②
2005	Pathfinder S	P245/75R16	none	35	35	N/A	②
	Pathfinder SE	P265/70R16	P265/75R16	35	35	N/A	②
	Pathfinder LE	P265/75R17	none	35	35	N/A	②

OEM: Original Equipment Manufacturer
PSI: Pounds Per Square Inch
STD: Standard
OPT: Optional
N/A: Not Available

① See placard on vehicle
② Axial play
Upper: 0
Lower: 0.008 in.

09482_PATH_C0011

BRAKE SPECIFICATIONS
All measurements in inches unless noted

Year	Model	Brake Disc Original Thickness	Minimum Thickness	Maximum Runout	Brake Drum Diameter Original Inside Diameter	Max. Wear Limit	Maximum Machine Diameter	Minimum Lining Thickness Front	Rear	Brake Caliper Bracket Bolts (ft. lbs.)	Mounting Bolts (ft. lbs.)
2002	Pathfinder	1.100	1.024	0.003	11.61	NA	11.67	0.079	0.059	①	①
	QX4	1.100	1.024	0.003	11.61	NA	11.67	0.079	0.059	②	②
2003	Pathfinder	1.100	1.024	0.003	11.61	NA	11.67	0.079	0.059	①	①
	QX4	1.100	1.024	0.003	11.61	NA	11.67	0.079	0.059	②	②
2004	Pathfinder	1.100	1.024	0.003	11.61	NA	11.67	0.079	0.059	①	①
2005	Pathfinder	1.102	1.024	0.0006	NA	NA	NA	0.079	0.079	①	①

NA: Not Available
① Torque member mounting bolt: 127-134
  Main pin bolt: 24-31

② 2WD: 0.870
4WD: 1.020

09482_PATH_C0012

SCHEDULED MAINTENANCE INTERVALS
2002-05 Nissan—Pathfinder; 2002-03 Infiniti—QX4

TO BE SERVICED	TYPE OF SERVICE	VEHICLE MILEAGE INTERVAL (x1000)											
		7.5	15	22.5	30	37.5	45	52.5	60	67.5	75	82.5	90
Engine oil & filter	R	✓	✓	✓	✓	✓	✓	✓	✓	✓	✓	✓	✓
Brake lines & cables	S/I		✓		✓		✓		✓		✓		✓
Brake pads& rotors	I		✓		✓		✓		✓		✓		✓
Driveshaft boots & propeller shaft (4x4)	L/I		✓		✓		✓		✓		✓		✓
Front wheel bearings (4x2)	I				✓				✓				
Automatic & manual transmission & transfer case	I		✓		✓		✓		✓				
LSD gear oil	I		✓		✓		✓		✓		✓		✓
Front wheel bearing grease (4x4)	R				✓				✓				
Timing belt ②	R	Replace every 105,000 miles											
Air cleaner filter	R			✓					✓				✓
Engine coolant	R								✓				✓
Spark plugs	R	Replace every 105,000 miles											
Drive belt(s)	S/I								✓		✓		✓
Cabin air filter	R		✓		✓		✓		✓		✓		✓
Exhaust system	I				✓				✓				✓
Fuel lines	S/I				✓				✓				
Steering gear (box) & linkage, axle & suspension parts	I				✓				✓				✓
Vapor lines	S/I				✓				✓				✓

R: Replace   S/I: Service or Inspect   L: Lubricate

FREQUENT OPERATION MAINTENANCE (SEVERE SERVICE)

If a vehicle is operated under any of the following conditions it is considered severe service:

- Extremely dusty areas.

- 50% or more of the vehicle operation is in 32°C (90°F) or higher temperatures, or constant operation in temperatures below 0°C (32°F).

- Prolonged idling (vehicle operation in stop and go traffic).

- Frequent short running periods (engine does not warm to normal operating temperatures).

- Police, taxi, delivery usage or trailer towing usage.

Oil & oil filter: replace every 3750 miles.

Brake pads, discs, drums & linings: service or inspect every 7500 miles.

Driveshaft boots & propeller shaft: service or inspect every 7500 miles.

Exhaust system: service or inspect every 7500 miles.

Steering gear (box) & linkage, (steering damper-4x4), axle & suspension parts: service or inspect every 7500 miles.

Steering linkage ball joints & front suspension ball joints: service or inspect every 7500 miles.

09482_PATH_C0013

ENGINE REPAIR

➡**Disconnecting the negative battery cable on some vehicles may interfere with the functions of the on board computer system. The computer may undergo a relearning process once the negative battery cable is reconnected.**

Distributor

The Nissan 3.5L and 4.0L engines are equipped with a Distributorless Ignition System (DIS).

Alternator

REMOVAL & INSTALLATOIN

3.5L Engines

1. Before servicing the vehicle, refer to the Precautions Section.
2. Remove or disconnect the following:
   - Negative battery cable
   - Alternator harness connectors
   - Engine under cover
   - Alternator belt
   - Alternator

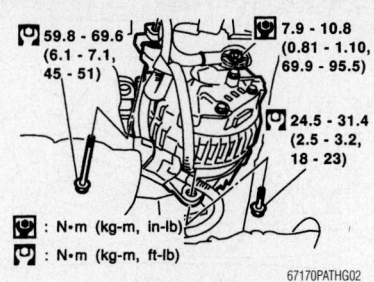

59.8 - 69.6 (6.1 - 7.1, 45 - 51)
7.9 - 10.8 (0.81 - 1.10, 69.9 - 95.5)
24.5 - 31.4 (2.5 - 3.2, 18 - 23)

: N•m (kg-m, in-lb)
: N•m (kg-m, ft-lb)

67170PATHG02

Alternator mounting—3.5L Engine

To install:

3. Install or connect the following:
   - Alternator
   - Alternator belt. Tighten the adjustment bolts as shown in the accompanying illustration.
   - Engine under cover
   - Alternator harness connectors
   - Negative battery cable

4.0L Engines

1. Before servicing the vehicle, refer to the Precautions Section.
2. Remove or disconnect the following:
   - Negative battery cable

- Fan shroud
- Accessory drive belt
- Alternator support brace
- Alternator upper bolt
- Alternator harness connectors
- Alternator

To install:

3. Install or connect the following:
   - Alternator harness connections. Tighten nut to 8 ft. lbs. (11 Nm).
   - Alternator upper bolt. Tighten to 48 ft. lbs. (65 Nm).
   - Alternator support brace. Tighten bolts to 21 ft. lbs. (29 Nm).
   - Accessory drive belt
   - Fan shroud
   - Negative battery cable

Engine Assembly

REMOVAL & INSTALLATION

3.5L Engine

1. Release fuel pressure.
2. Remove engine hood and front RH and LH wheels.

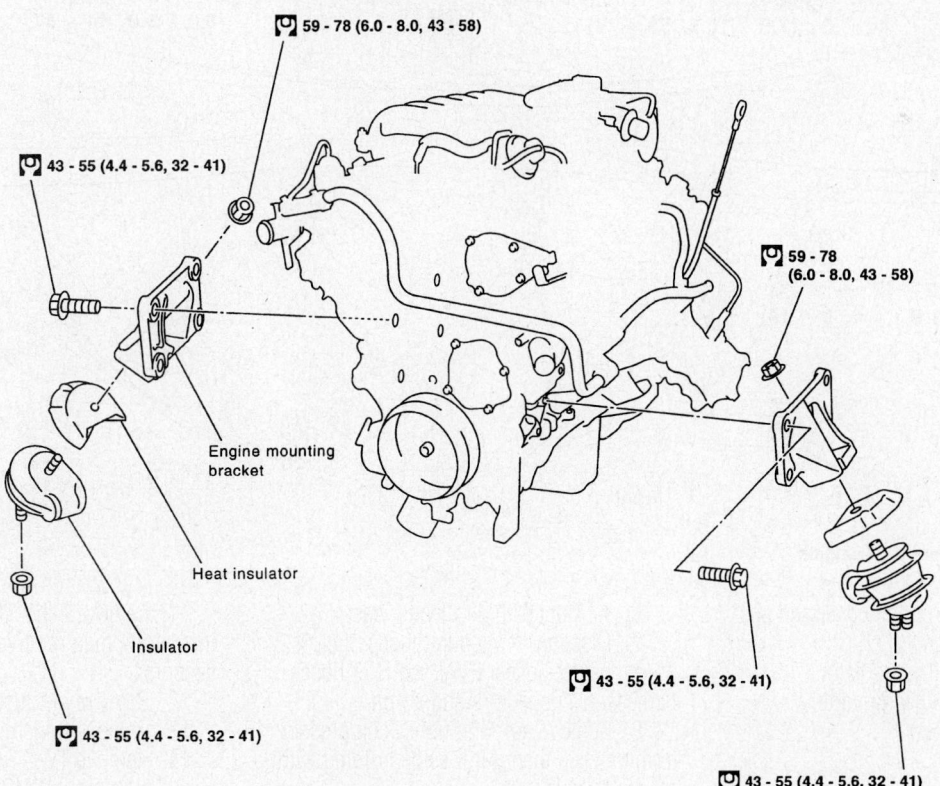

59 - 78 (6.0 - 8.0, 43 - 58)

43 - 55 (4.4 - 5.6, 32 - 41)

59 - 78 (6.0 - 8.0, 43 - 58)

Engine mounting bracket

Heat insulator

Insulator

43 - 55 (4.4 - 5.6, 32 - 41)

43 - 55 (4.4 - 5.6, 32 - 41)

43 - 55 (4.4 - 5.6, 32 - 41)

: N•m (kg-m, ft-lb)

9359VG01

Front engine mounting—3.5L engine

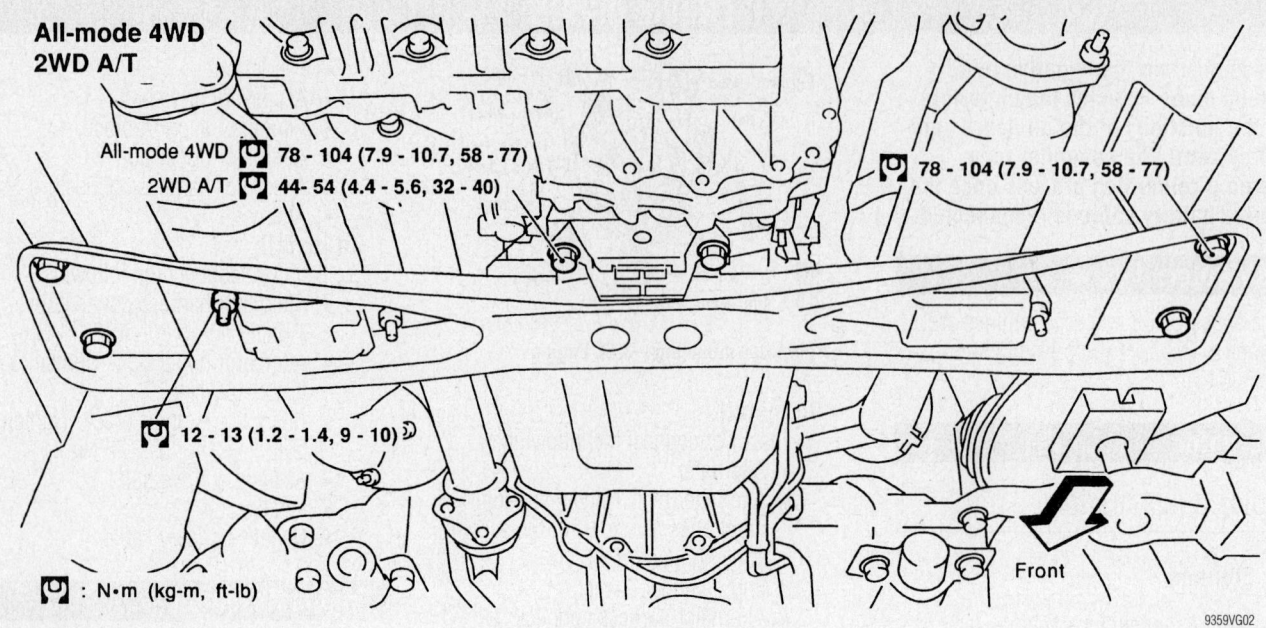

All-mode 4WD
2WD A/T

All-mode 4WD 78 - 104 (7.9 - 10.7, 58 - 77)
2WD A/T 44 - 54 (4.4 - 5.6, 32 - 40)

78 - 104 (7.9 - 10.7, 58 - 77)

12 - 13 (1.2 - 1.4, 9 - 10)

: N•m (kg-m, ft-lb)

Front

9359VG02

Rear engine mounting—3.5L engine

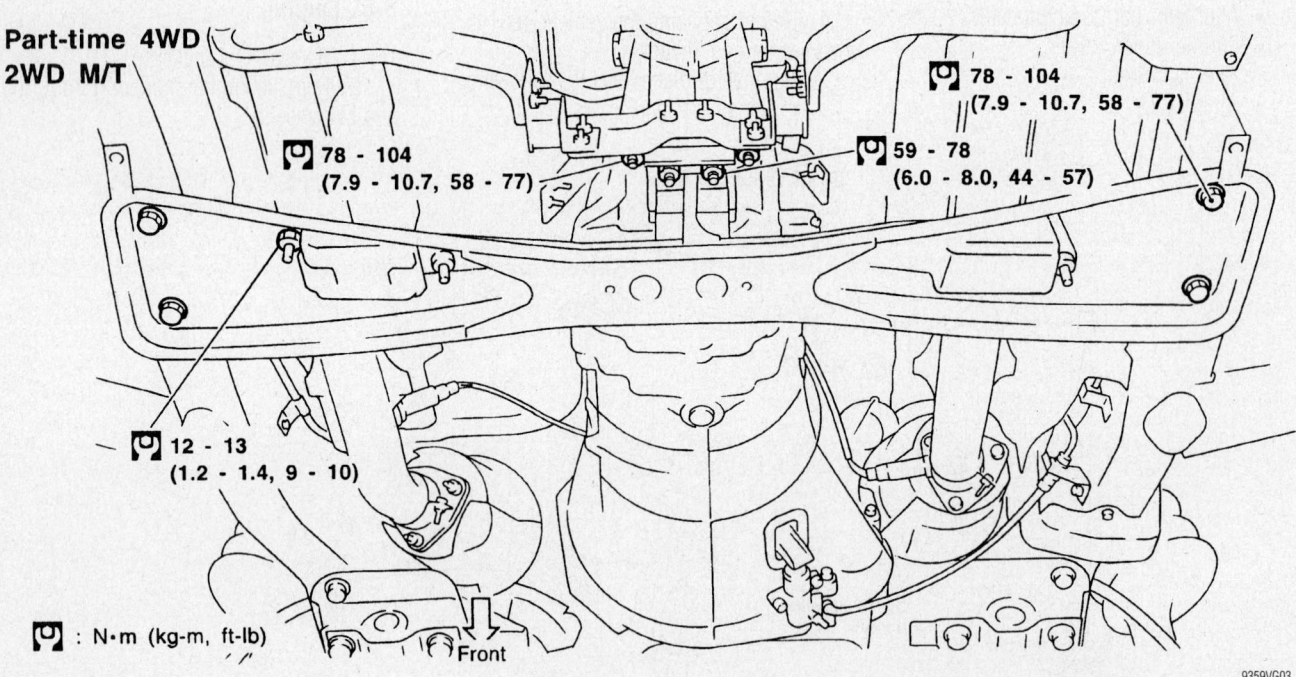

Part-time 4WD
2WD M/T

78 - 104
(7.9 - 10.7, 58 - 77)

59 - 78
(6.0 - 8.0, 44 - 57)

78 - 104
(7.9 - 10.7, 58 - 77)

12 - 13
(1.2 - 1.4, 9 - 10)

: N•m (kg-m, ft-lb)

Front

9359VG03

Rear engine mounting—3.5L engine

3. Remove engine undercover and suspension member stay.
4. Drain coolant from radiator.
5. Remove the following parts.
- Radiator shroud
- Radiator
- Cooling fan
- Drive belts
- Battery
- Engine cover
- Throttle wires

6. Air duct with air cleaner case.
7. Disconnect vacuum hoses, fuel hoses, heater hoses, EVAP canister hoses, harnesses, connectors and so on.
8. Remove air conditioner compressor from bracket, then put it aside holding with a suitable wire.
9. Remove power steering oil pump and reservoir tank with bracket, then put it aside holding with a suitable wire.
10. Remove alternator.

11. Remove exhaust front tube heat insulators, then remove rear heated oxygen sensors.
12. Remove exhaust front and rear tubes.
13. Remove transmission.
14. Remove TWC (manifold) heat insulators, then remove TWC (manifold).
15. Install engine slingers.
16. Hoist engine with engine slingers and remove front engine mounting nuts.
17. Remove engine from vehicle.

To install:

Installation is in the reverse order of removal. Observe the following torques:

- Front engine mount-to-bracket: 43–58 ft. lbs.
- Front mount-to-frame: 32–41 ft. lbs.
- Front bracket-to-block: 32–41 ft. lbs.
- Rear engine mount-to-bracket: all exc. 2wd with AT: 58–77 ft. lbs.; 2wd with AT: 32–40 ft. lbs.
- Crossmember-to-frame: 58–77 ft. lbs.

4.0L Engine

1. Before servicing the vehicle, refer to the Precautions Section.
2. Drain the cooling system.
3. Drain the engine oil.
4. Partially drain the automatic transmission fluid.
5. Remove or disconnect the following:
- Negative battery cable
- Hood
- Engine cover
- Air intake assembly
- Vacuum hose between vehicle and engine
- Radiator hose
- Radiator
- Cooling fan
- Engine wiring harnesses
- Power steering reservoir tank
- Power steering pump
- A/C compressor
- Brake booster vacuum line
- EVAP line
- Fuel supply hose
- Heater hoses
- A/T fluid level indicator and tube
- Front differential assembly, if equipped with 4WD
- Catalytic converter
6. Install engine slingers and tighten to 21 ft. lbs. (28 Nm).
7. Attach a hoist to the engine slingers and secure into position.
8. Remove the transmission.
9. Remove the engine assembly from the vehicle.

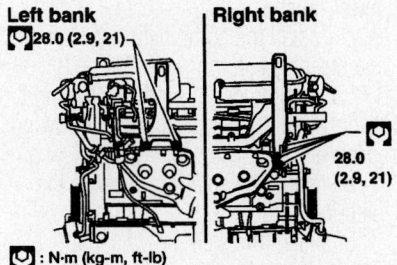

Left bank **Right bank**

28.0 (2.9, 21)

28.0 (2.9, 21)

: N·m (kg-m, ft-lb)

09482_PATH_G0002

Engine mounting bracket torques—4.0L engine

10. Installation is the reverse of removal. Note the torque of the engine mounting brackets in the accompanying illustration.

Heater Core

REMOVAL & INSTALLATION

2002–2004 Models

1. Disconnect the negative battery cable.

✳ CAUTION

After disconnecting the negative battery cable, wait for at least 3 minutes before working on the steering column or instrument panel.

2. Drain the cooling system into a clean container for reuse.
3. Disconnect the heater hoses from the heater core.
4. Remove the driver's side air bag and steering wheel by performing the following procedure:
   a. Place the front wheels in the straight-ahead position.
   b. Remove the lower lid from the steering wheel and disconnect the air bag module connector.
   c. Remove the side lids from both sides of the steering wheel.
   d. Using the Tamper Resistant Torx® tool T50, remove the left and right Torx® bolts.
   e. Carefully, remove the air bag module.

✳ CAUTION

Place the air bag module in safe place with the front facing upward.

   f. Remove the steering wheel nut.
   g. Using a steering wheel puller, press the steering wheel from the steering column.
5. Remove the passenger's side air bag by performing the following procedure:
   a. Remove the glove box clips and disconnect the passenger's side air bag module connector.
   b. Remove the lower panel screws; then, disconnect the harness connector and remove the air bag module bracket.
   c. Using the Tamper Resistant Torx® tool T50, remove the passenger's side air bag module bolts.
   d. Carefully, remove the air bag module.

✳ CAUTION

Place the air bag module in safe place with the front facing upward.

6. Remove the instrument panel by performing the following procedure:
   a. Remove the steering column cover and the combination switch.
   b. Remove the instrument panel side lower finisher.
   c. At the driver's side, remove the lower panel screws, disconnect the electrical harness connectors and remove the panel.
   d. Remove the cluster lid "A" screws and the cluster lid "A".
   e. Remove the combination meter screws, disconnect the electrical harness connectors and remove the combination meter.
   f. Remove the cluster lid "C" screws, disconnect the electrical harness connectors and remove the cluster lid "C".
   g. Remove the audio assembly screws and the audio assembly.
   h. Remove the air conditioning control unit screws, disconnect the electrical harness connectors and the air conditioning control unit.
   i. Remove the ashtray.
   j. Remove the shifter (automatic transmission) or shift lever boot (manual transmission); then, remove the screw and disconnect the harness connector.
   k. Remove the console box; then, remove the screw and disconnect the harness connector.
   l. Remove the lower instrument center panel screws and the lower instrument center panel.
   m. Remove the defroster grille.
   n. At both sides, remove the pillar garnishes.
   o. Remove the instrument panel and pads nuts and bolts.
   p. Using an assistant, remove the instrument panel.
7. Remove the defroster nozzle and the heater nozzle from the heater housing.
8. Disconnect the electrical connector and/or control cable from the heater housing.
9. Remove the heater housing-to-chassis fasteners and remove the heater housing.
10. Separate the heater core from the heater housing and remove the heater core.

To install:

11. Install the heater core and assemble the heater housing.

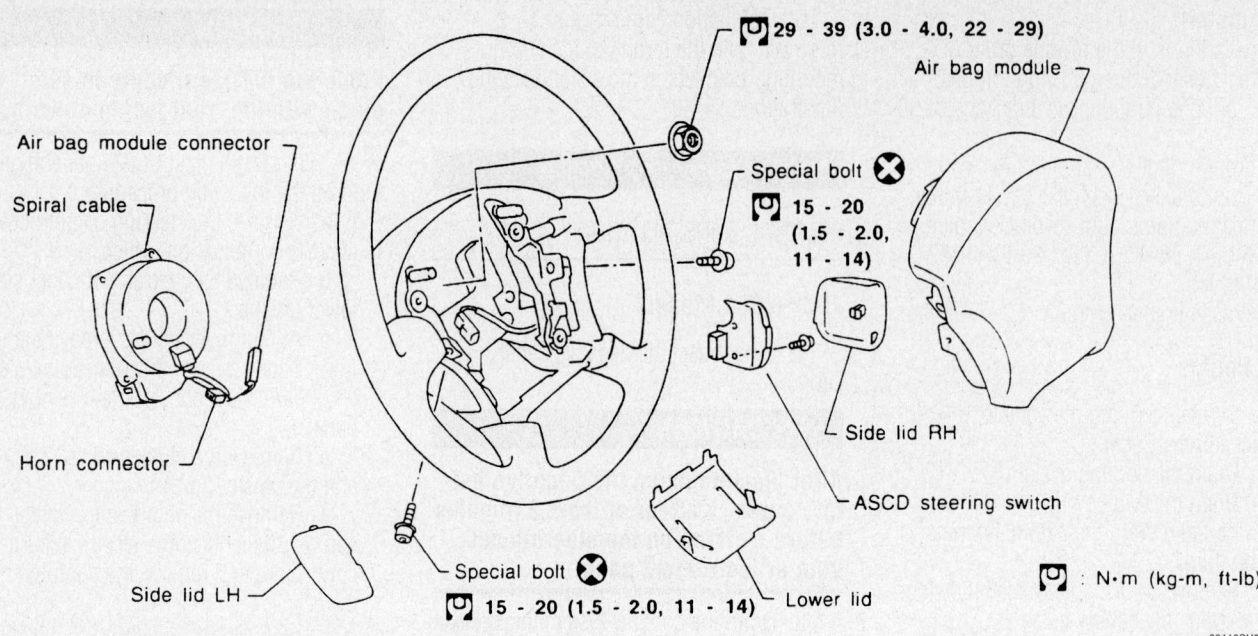

Exploded view of the driver's side air bag module and steering wheel

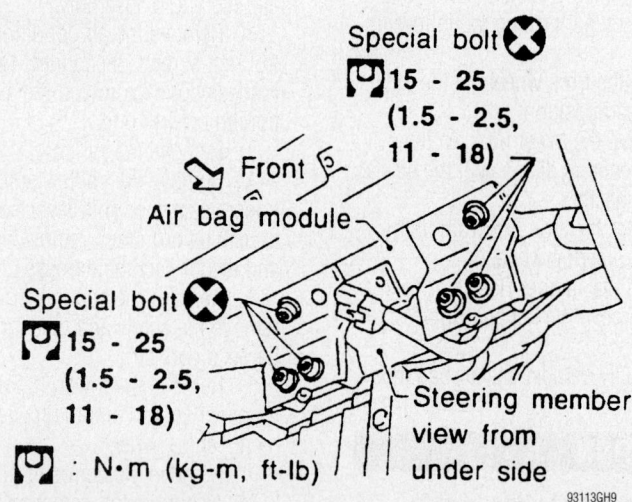

Exploded view of the passenger's side air bag module

12. Install the heater housing and the heater housing-to-chassis fasteners.

13. Connect the electrical connector and/or control cable to the heater housing.

14. Install the defroster nozzle and the heater nozzle to the heater housing.

15. Install the passenger's side air bag by performing the following procedure:

   a. Carefully, install the air bag module.

   b. Using the Tamper Resistant Torx® tool T50, install the passenger's side air bag module bolts. Torque the bolts to 11–18 ft. lbs. (15–25 Nm).

   c. Connect the harness connector and install the air bag module bracket; then, install the lower panel screws.

   d. Connect the passenger's side air

bag module connector and install the glove box clips.

16. Install the instrument panel by performing the following procedure:

   a. Using an assistant, position the instrument panel.

   b. Install the instrument pads, nuts and bolts.

   c. At both sides, install the pillar garnishes.

   d. Install the defroster grille.

   e. Install the lower instrument center panel and the lower instrument center panel screws.

   f. Install the console box; then, install the screw and connect the harness connector.

   g. Connect the harness connector and install the screw; then, install the shifter (automatic transmission) or shift lever boot (manual transmission).

   h. Install the ashtray.

   i. Install the air conditioning control unit, connect the electrical harness connectors and the air conditioning control unit screws.

   j. Install the audio assembly and the audio assembly screws.

   k. Install the cluster lid "C", connect the electrical harness connectors and install the cluster lid "C" screws.

   l. Install the combination meter, connect the electrical harness connectors and install the combination meter screws.

   m. Install the cluster lid "A" and the cluster lid "A" screws.

   n. At the driver's side, install the lower panel, connect the electrical harness connectors and install the panel screws.

   o. Install the instrument panel side lower finisher.

   p. Install the combination switch and the steering column cover.

17. Install the driver's side air bag and steering wheel by performing the following procedure:

   a. Install the steering wheel to the steering column.

   b. Install the steering wheel nut. Torque the nut to 22–29 ft. lbs. (29–39 Nm).

   c. Carefully, install the air bag module.

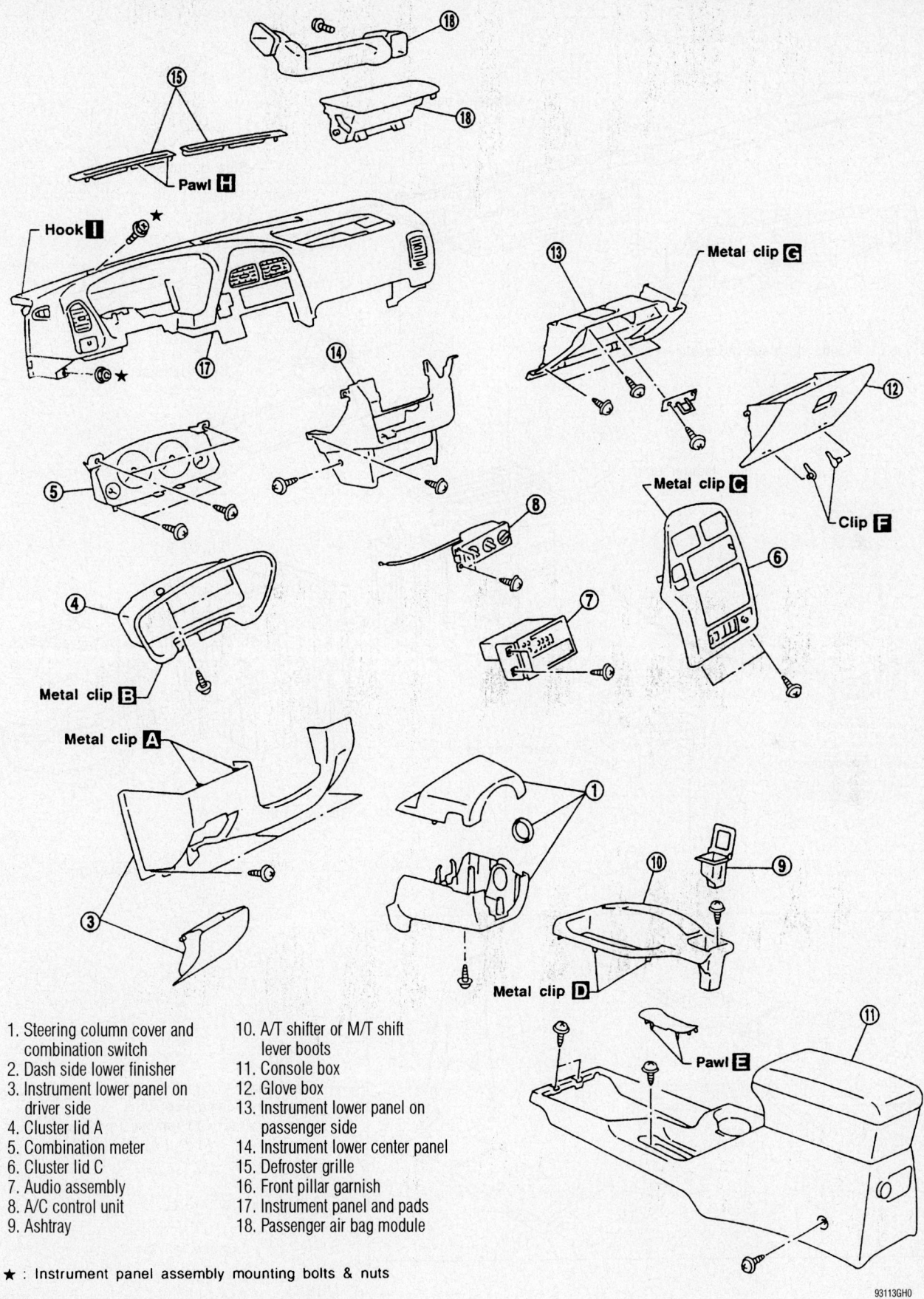

1. Steering column cover and combination switch
2. Dash side lower finisher
3. Instrument lower panel on driver side
4. Cluster lid A
5. Combination meter
6. Cluster lid C
7. Audio assembly
8. A/C control unit
9. Ashtray
10. A/T shifter or M/T shift lever boots
11. Console box
12. Glove box
13. Instrument lower panel on passenger side
14. Instrument lower center panel
15. Defroster grille
16. Front pillar garnish
17. Instrument panel and pads
18. Passenger air bag module

★ : Instrument panel assembly mounting bolts & nuts

93113GH0

Exploded view of the instrument panel and related accessories

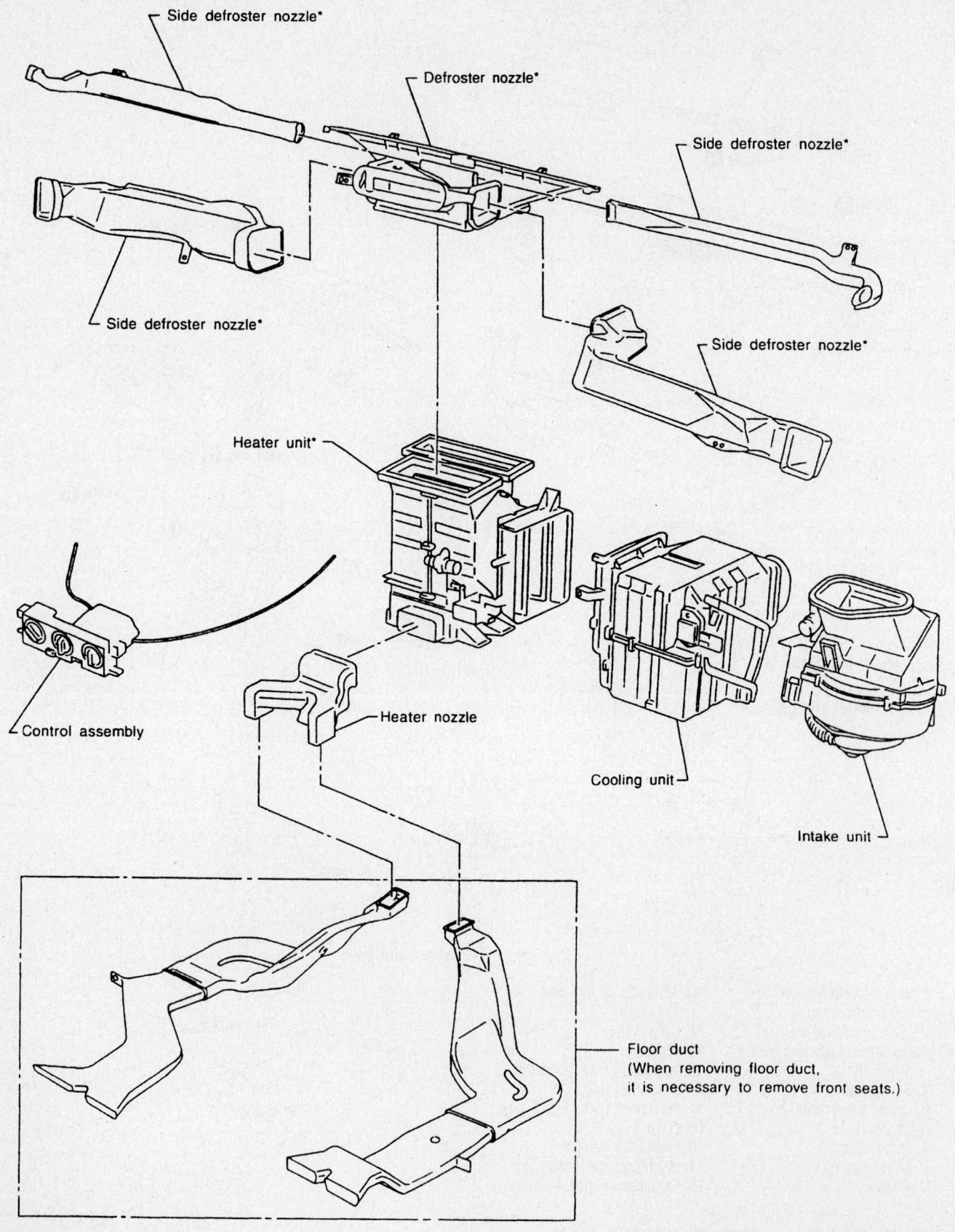

Exploded view of the heater housing, the evaporator housing, the ventilation dusts and related accessories

d. Using the Tamper Resistant Torx® tool T50, install the left and right Torx® bolts. Torque the bolts to 11–14 ft. lbs. (15–20 Nm).

e. Install the side lids to both sides of the steering wheel.

f. Connect the air bag module connector and install the lower lid to the steering wheel.

18. Connect the heater hoses to the heater core.

19. Refill the cooling system.
20. Connect the negative battery cable.
21. Run the engine to normal operating temperatures; then, check the climate control operation and check for leaks.

2005 Model

1. Before servicing the vehicle, refer to the Precautions Section.
2. Drain the cooling system.

3. Discharge the refrigerant from the A/C system.

4. Remove the right hand front heater core pipes nut.

5. Disconnect the front heater core hoses from the front heater core.

6. Disconnect the high and low pressure A/C pipes from the front expansion valve.

7. Move the front seats to the rearmost position.

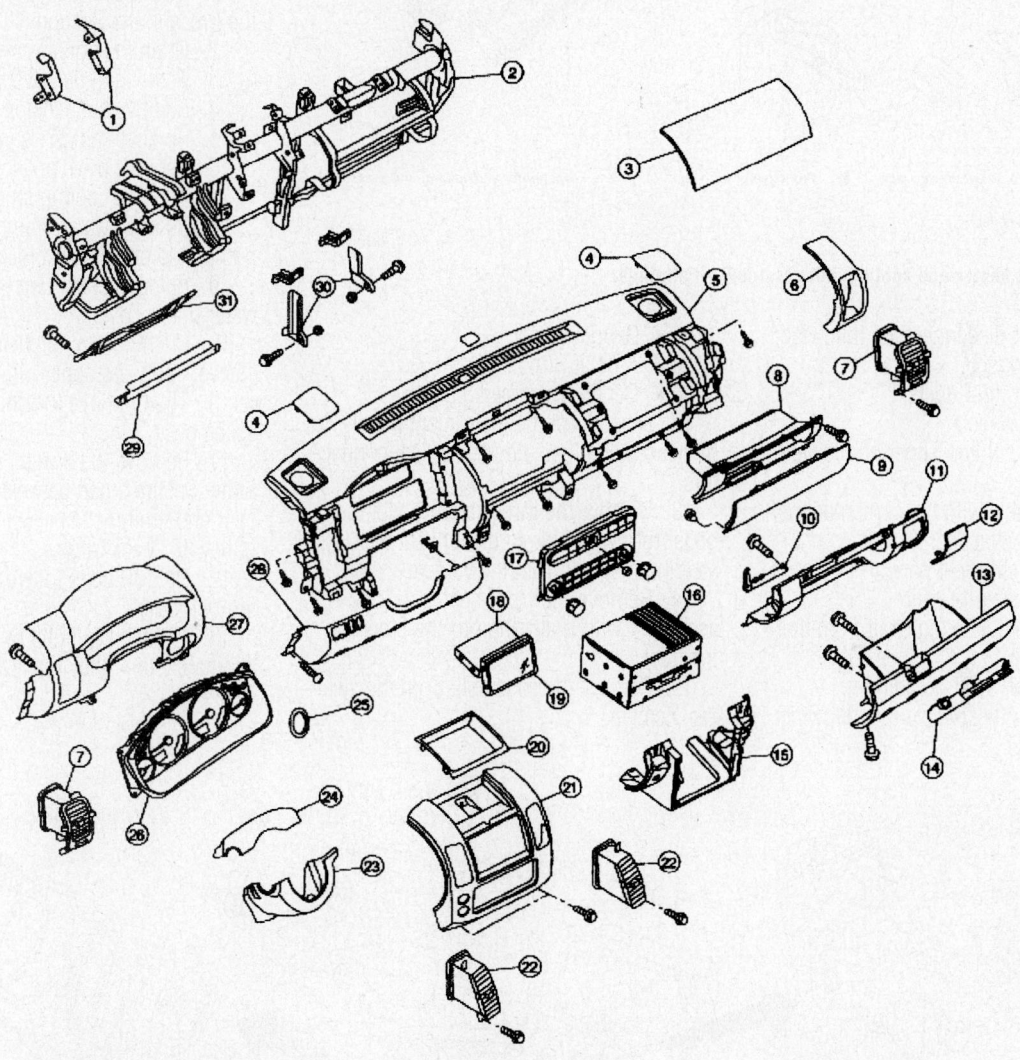

1.	Display unit bracket RH/LH	2.	Steering member assembly	3.	Passenger air bag module cover
4.	Speaker grille RH/LH	5.	Instrument panel and pad assembly	6.	Instrument side finisher
7.	Side ventilator grille RH/LH	8.	Upper glove box bin	9.	Upper glove box door
10.	Lower glove box damper assembly	11.	Lower instrument panel RH	12.	Fuse block cover
13.	Lower glove box assembly	14.	Lower glove box latch assembly	15.	Cluster lid D
16.	Audio unit	17.	NAVI/Audio control panel	18.	NAVI display controller
19.	NAVI display	20.	Storage tray	21.	Cluster lid C
22.	Center ventilator grill RH/LH	23.	Steering column cover lower	24.	Steering column cover upper
25.	Steering lock escutcheon	26.	Combination meter	27.	Cluster lid A
28.	Lower instrument panel LH	29.	Knee protector brace	30.	Instrument stay RH/LH
31.	Knee protector				

09482_PATH_G0003

Exploded view of the instrument panel assembly—2005 Pathfinder

16. Remove the front heater core cover.
17. Remove the heater core and evaporator pipe bracket.
18. Separate the heater core from the pipe bracket.
19. Installation is the reverse of removal.

Water Pump

REMOVAL & INSTALLATION

1. Before servicing the vehicle, refer to the Precautions Section.
2. Drain the cooling system
3. Remove undercover.
4. Remove suspension member stay.
5. Remove radiator shrouds.
6. Remove drive belts.
7. Remove cooling fan.
8. Remove water drain plug on water pump side of cylinder block.
9. Remove chain tensioner cover and water pump cover.
10. Pushing timing chain tensioner sleeve, apply a stopper pin so it does not return. Then remove the chain tensioner assembly.
11. Remove the 3 water pump fixing bolts. Secure a gap between water pump gear and timing chain, by turning crankshaft pulley 20° backwards.
12. Put M8 bolts to two water pump fixing bolt holes.
13. Tighten M8 bolts by turning half turn alternately until they reach timing chain rear case.

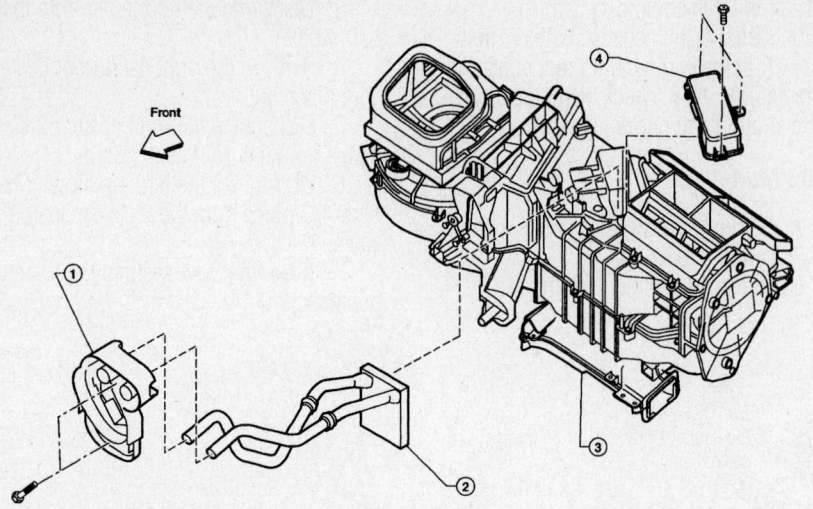

1. Front heater core and evaporator pipe bracket
2. Front heater core
3. Front heater and cooling unit assembly
4. Front heater core cover

09482_PATH_G0004

Exploded view of heater and cooling unit assembly—Pathfinder

8. Remove or disconnect the following:
   • Center console
   • Two front floor ducts
   • Steering wheel
   • Steering column upper and lower covers
   • Spiral cable with the combinations switches attached.
   • Lower instrument panel
   • Lower knee protector
9. Separate the steering shaft from the upper joint.
10. Remove the steering column.
11. Remove or disconnect the following:
   • Gauge cluster
   • A pillar trim
   • Center instrument panel
12. Disconnect the instrument panel wire harnesses and fuse block electrical connectors.
13. Remove the three steering member bolts from each side to disconnect the steering member from the vehicle.
14. Remove the front heater unit assembly with it attached to the steering member.
15. Remove the assembly from the steering member.

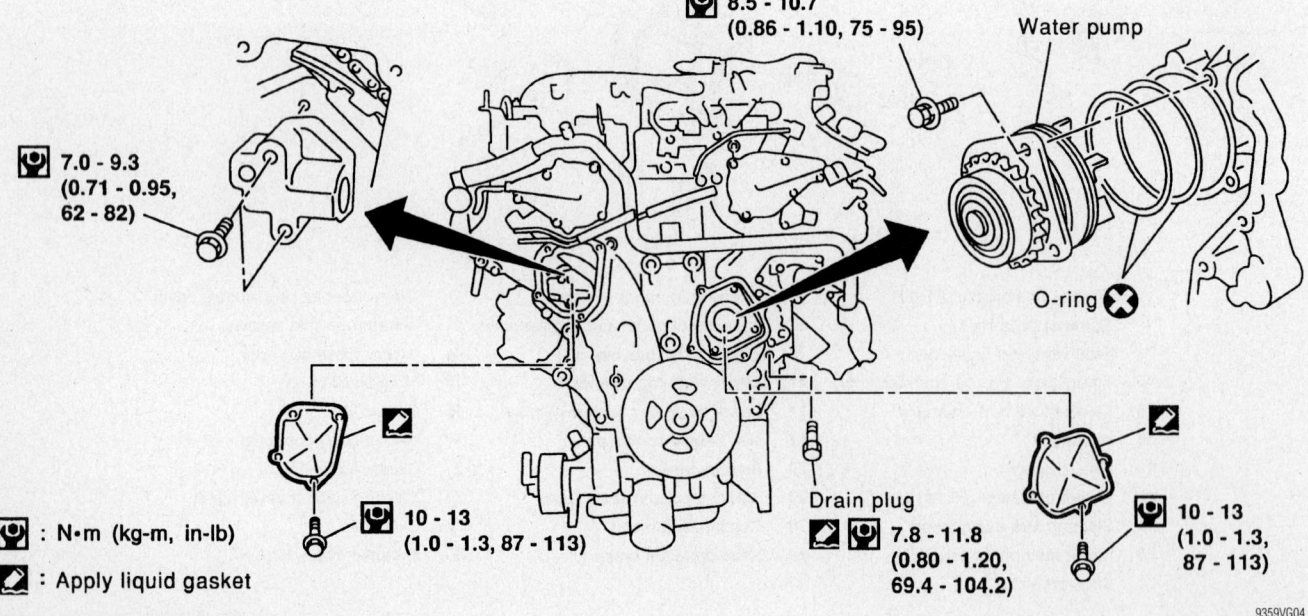

Exploded view of the water pump assembly

9359VG04

➡ In order to prevent damages to water pump or timing chain rear case, do not tighten one bolt continuously. Always turn each bolt half turn each time.

14. Lift up water pump and remove it.

➡ When lifting up water pump, do not allow water pump gear to hit timing chain.

To install:

15. Apply engine oil and coolant to O-rings as shown in the figure.
16. Install water pump.

➡ Do not allow cylinder block to nip O-rings when installing water pump.

17. Before installing, remove all traces of liquid gasket from mating surface of water pump cover and chain tensioner cover using a scraper. Also remove traces of liquid gasket from mating surface of front cover.

18. Apply a continuous bead of liquid gasket to mating surface of chain tensioner cover and water pump cover. Use Genuine RTV silicone sealant or equivalent.

19. Return the crankshaft pulley to its original position by turning it 20° forward.

20. Install timing chain tensioner, then remove the stopper pin.

➡ When installing the timing chain tensioner, engine oil should be applied to the oil hole and tensioner.

➡ After starting engine, let idle for three minutes, then rev engine up to 3,000 rpm under no load to purge air from the high-pressure chamber of the chain tensioners. The engine may produce a rattling noise. This indicates that air still remains in the chamber and is not a matter of concern.

21. Reinstall any parts removed in reverse order of removal.

Cylinder Head

REMOVAL & INSTALLATION

3.5L Engine

1. Remove engine from vehicle.
2. Remove exhaust manifolds in reverse order of installation.
3. Place engine on a work stand.
4. Remove aluminum oil pan
5. Remove timing chain.
6. Remove intake manifold in reverse order of installation.
7. Remove water outlet.
8. Remove rear timing chain case bolts.

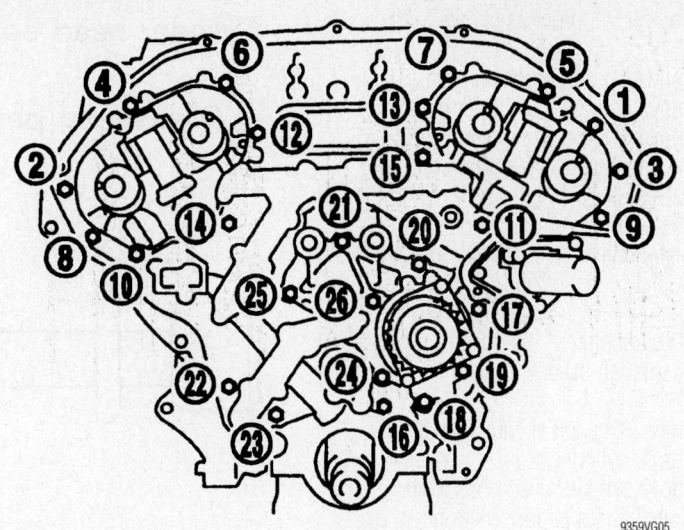

Rear timing case loosening sequence—3.5L engine

9359VG05

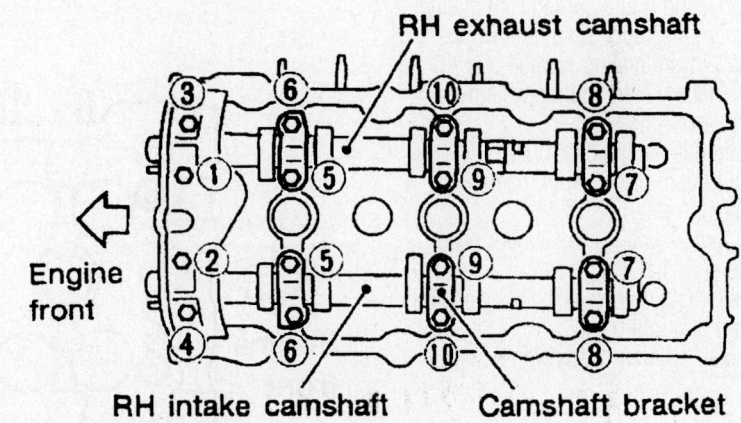

RH exhaust camshaft

Engine front

RH intake camshaft · Camshaft bracket

Loosen in numerical order.

9359VG06

Right camshaft loosening sequence—3.5L engine

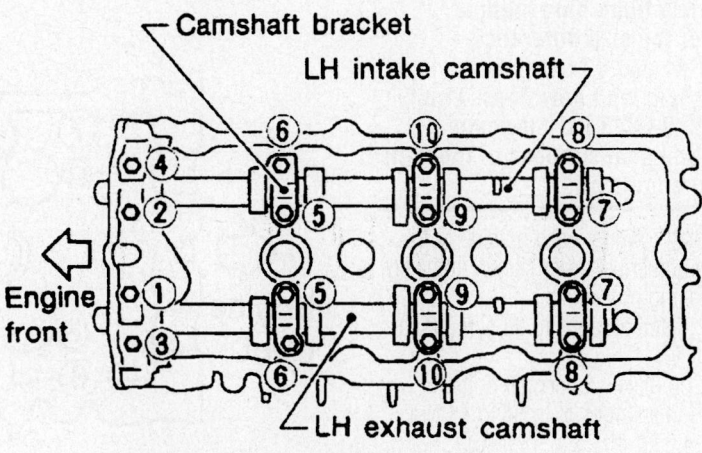

Camshaft bracket

LH intake camshaft

Engine front

LH exhaust camshaft

Loosen in numerical order.

9359VG07

Left camshaft loosening sequence—3.5L engine

Loosen in numerical order as shown in the figure.

9. Remove rear timing chain case.

10. Remove O-rings to cylinder head.

11. Remove O-rings to cylinder block.

12. Remove intake valve timing control solenoid valves.

13. Remove intake and exhaust camshafts and camshaft brackets. Equally loosen camshaft bracket bolts in several steps in the numerical order shown in the figure. For reinstallation, be sure to put marks on camshaft bracket before removal.

14. Remove RH and LH camshaft chain tensioners from cylinder head.

15. Remove cylinder head bolts. Cylinder head bolts should be loosened in two or three steps.

16. Remove cylinder head.

To install:

17. Before installing rear timing chain case, remove old liquid gasket from mating surface using a scraper. Also remove old liquid gasket from mating surface of cylinder block. Remove old liquid gasket from the bolt hole and thread.

18. Before installing cam bracket, remove old liquid gasket from mating surface using a scraper.

19. Before installing the cylinder head gasket, be sure that No. 1 cylinder is at TDC. At this time, the crankshaft key should face toward the right bank.

20. Install cylinder heads with new gaskets.

➡ Do not rotate crankshaft and camshaft separately, or valves will strike piston heads.

✳✳ CAUTION

Cylinder head bolts are tightened by plastic zone tightening method. Whenever the size difference between d1 and d2 exceeds the limit, replace them with new ones. Limit (d1 – d2): 0.0043 in. Lubricate threads and seat surfaces of the bolts with new engine oil.

21. Install cylinder head outside bolts Tighten in numerical order shown in the figure. Tightening procedure:

   a. Tighten all bolts to 98 Nm (10 kg-m, 72 ft-lb).

   b. Completely loosen all bolts.

   c. Tighten all bolts to 34 to 44 Nm (3.5 to 4.5 kg-m, 25 to 33 ft-lb).

   d. Turn all bolts 90 to 95 degrees clockwise.

Cylinder head bolt

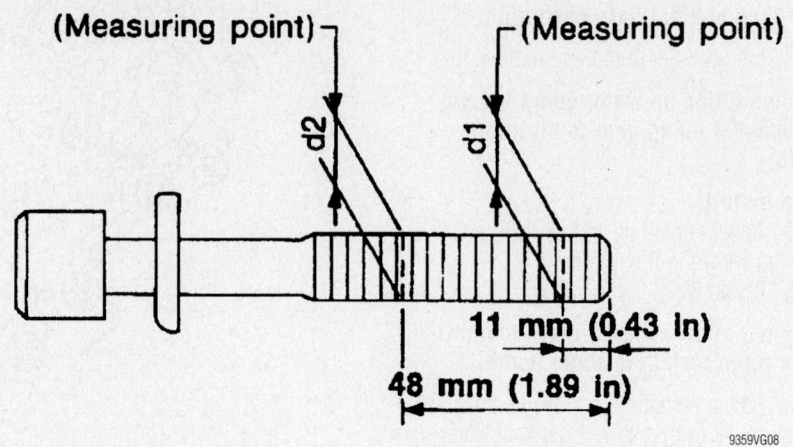

Head bolt checking—3.5L engine

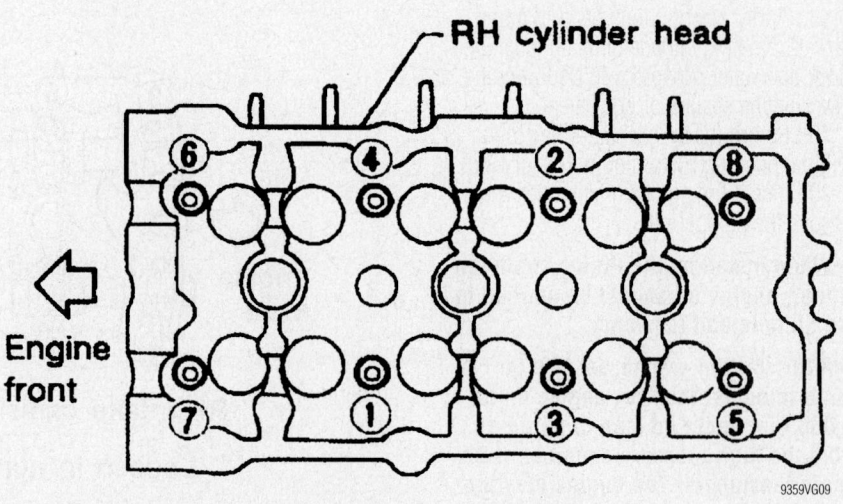

Right cylinder head bolt torque sequence—3.5L and 4.0L engines

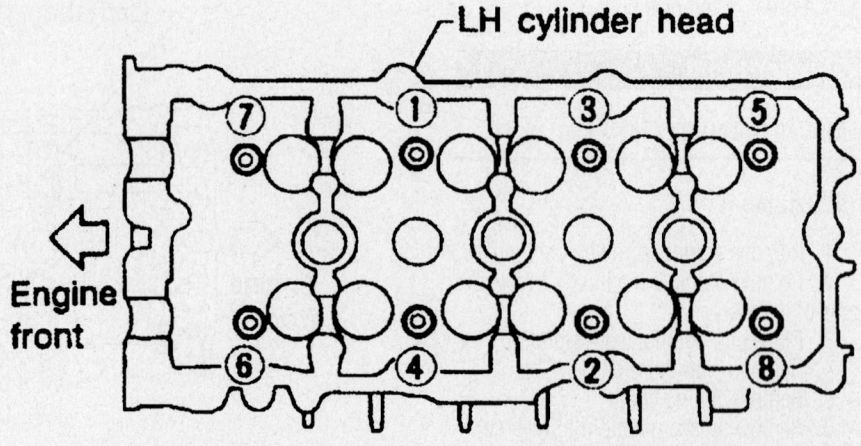

Left cylinder head bolt torque sequence—3.5L and 4.0L engines

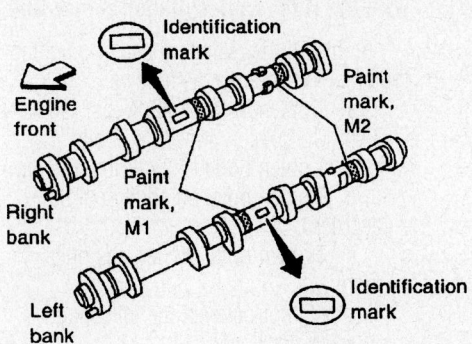

Camshaft identification—3.5L engine

● **Identification marks are present on camshafts.**

Bank	INT/EXH	ID mark	Drill mark	Paint mark	
				M1	M2
RH	INT	R3	Yes	Yes	No
	EXH	R3	No	No	Yes
LH	INT	L3	Yes	Yes	No
	EXH	L3	No	No	Yes

9359VG11

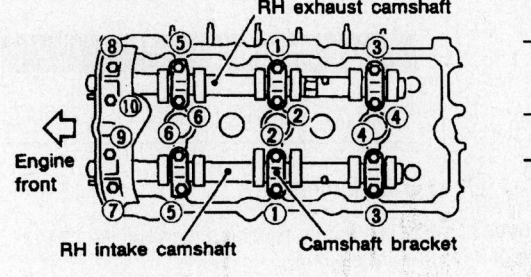

Right camshaft bolt torque sequence—3.5L engine

● Tighten the camshaft brackets in the following steps.

Step	Tightening torque	Tightening order
1	1.96 N·m (0.2 kg-m, 17 in-lb)	Tighten in the order of 7 to 10, then tighten 1 to 6.
2	5.88 N·m (0.6 kg-m, 52 in-lb)	Tighten in the numerical order.
3	9.02 - 11.8 N·m (0.92 - 1.20 kg-m, 79.9 - 104.2 in-lb)	Tighten in the order of 1 to 6.
	8.3 - 10.3 N·m (0.9 - 1.0 kg-m, 74 - 91 in-lb)	Tighten in the order of 7 to 10.

9359VG12

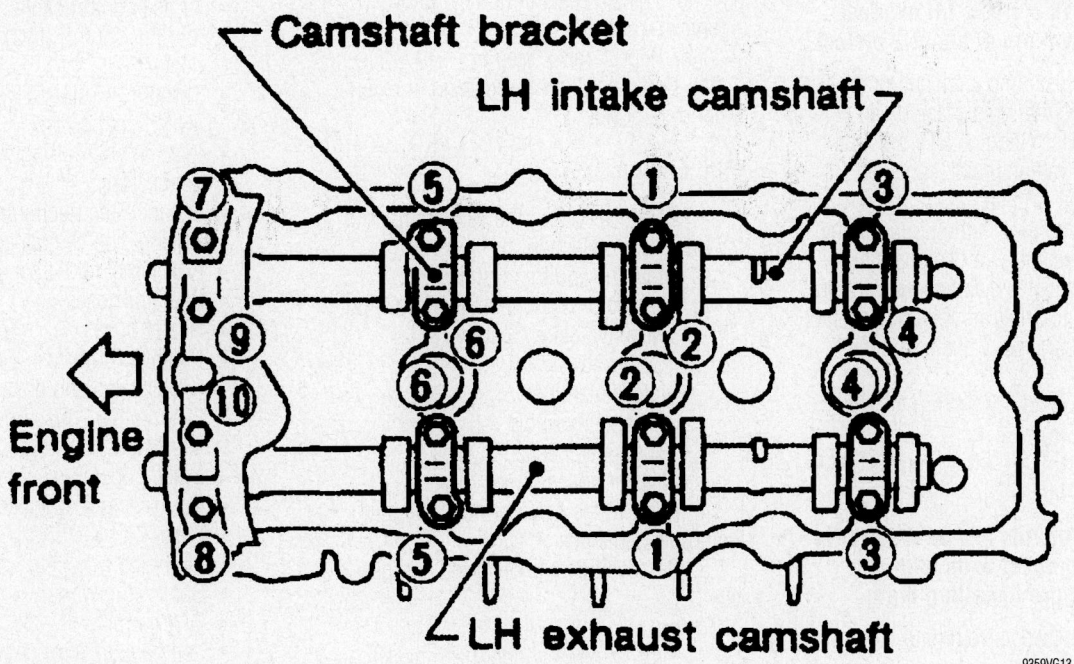

Left camshaft bolt torque sequence—3.5L engine

9359VG13

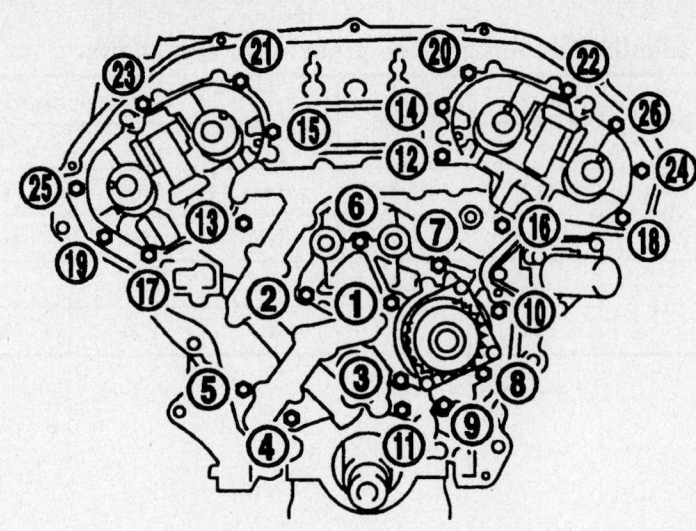

⚙ **12 - 13 N•m**
(1.2 - 1.4 kg-m, 9 - 10 ft-lb)

9359VG14

Rear timing case bolt torque sequence—3.5L engine

e. Turn all bolts 90 to 95 degrees clockwise.

22. Install camshaft chain tensioners on both sides of cylinder head.

23. Install exhaust and intake camshafts and camshaft brackets.

➡ **Intake camshaft has a drill mark on camshaft sprocket mounting flange. Install it on the intake side. Position camshaft. RH exhaust camshaft dowel pin at about 10 o'clock; LH exhaust camshaft dowel pin at about 2 o'clock**

24. Before installing camshaft brackets, apply sealant to mating surface of No. 1 journal head. Use Genuine RTV silicone sealant or equivalent. Install camshaft brackets in their original positions. Align stamp mark as shown in the figure. If any part of valve assembly or camshaft is replaced, check valve clearance according to reference data. After completing assembly check valve clearance. Valve clearance (Cold):

- Intake 0.26 – 0.34 mm (0.010 – 0.013 in)
- Exhaust 0.29 – 0.37 mm (0.011 – 0.015 in)

➡ **Lubricate threads and seat surfaces of camshaft bracket bolts with new engine oil before installing them.**

25. Install intake valve timing control solenoid valves.

26. Install O-rings to cylinder block.

27. Install O-rings to cylinder head.

28. Apply sealant to the hatched portion of rear timing chain case. Apply continuous bead of liquid gasket to mating surface of rear timing chain case. Before installation, wipe off the protruding sealant.

29. Align rear timing chain case with dowel pins, then install on cylinder head and block.

30. Tighten rear chain case bolts.

a. Tighten bolts in numerical order shown in the figure.

b. Repeat above step a.

31. Reinstall all removed parts in reverse order of removal.

4.0L Engine

1. Before servicing the vehicle, refer to the Precautions Section.

2. Drain the cooling system.

3. Remove or disconnect the following:

- Negative battery cable
- Camshaft
- Intake manifold
- Exhaust manifold
- Thermostat assembly
- Heater hoses

4. Remove the cylinder head bolts in the proper sequence.

5. Remove the cylinder head and gaskets.

To install:

6. Install a new cylinder head gasket.

7. Turn the crankshaft until the No. 1 piston is set at Top Dead Center (TDC).

➡ **The crankshaft key should line up with the right bank cylinder center line.**

8. Install the cylinder head and tighten the bolts in sequence as follows:

a. Tighten bolts to 72 ft. lbs. (98 Nm).

b. Completely loosen all bolts.

c. Tighten bolts to 29 ft. lbs. (39 Nm).

d. Tighten bolts an additional 90 degrees.

e. Tighten bolts 90 degrees once more.

9. Install or connect the following:

- Heater hoses
- Thermostat assembly
- Exhaust manifold
- Intake manifold
- Camshaft
- Negative battery cable

10. Refill the cooling system to the correct level.

11. Start the engine and check for leaks.

Intake Manifold

REMOVAL & INSTALLATION

3.5L Engines

1. Before servicing the vehicle, refer to the Precautions Section.

2. Drain the cooling system.

3. Relieve the fuel system pressure.

4. Remove or disconnect the following:

- Negative battery cable
- Air intake duct
- Accelerator cable
- Cruise control cable
- Idle Air Control (IAC) valve connector
- Throttle Position (TP) sensor and switch connectors
- Ignition coil and power transistor connectors
- Exhaust Gas Recirculation (EGR) Solenoid valve connector
- EGR temperature sensor connector
- Radiator hoses
- Heater hoses
- Positive Crankcase Ventilation (PCV) valve and hose
- Evaporative Emissions (EVAP) canister vacuum and purge hoses
- Brake booster vacuum hose
- Fuel pressure regulator vacuum hose
- EGR tube
- Spark plug wires
- Distributor
- Left bank injector connectors
- Thermal transmitter
- Upper intake manifold ground cable

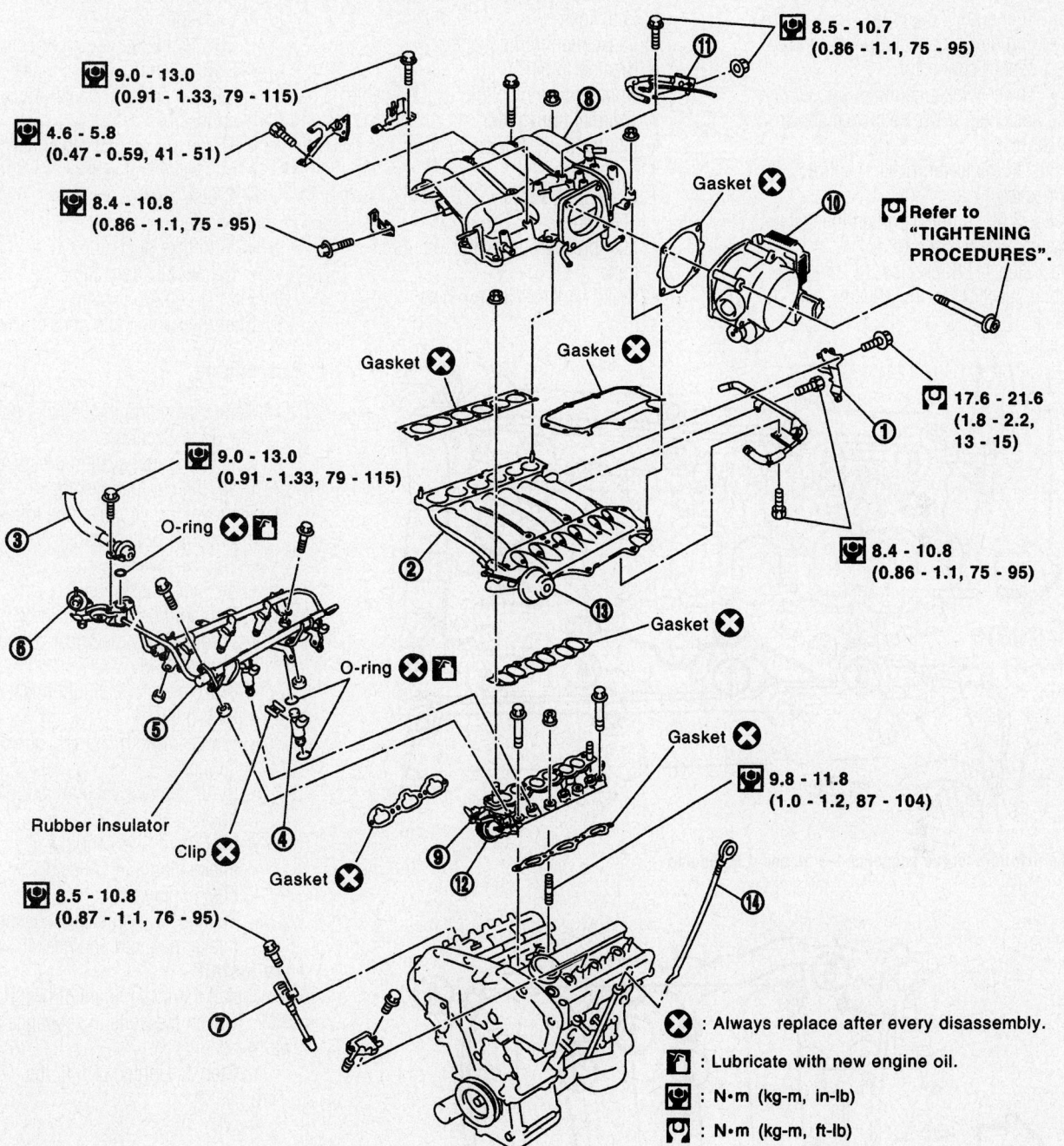

9.0 - 13.0
(0.91 - 1.33, 79 - 115)

4.6 - 5.8
(0.47 - 0.59, 41 - 51)

8.4 - 10.8
(0.86 - 1.1, 75 - 95)

8.5 - 10.7
(0.86 - 1.1, 75 - 95)

Gasket

Refer to "TIGHTENING PROCEDURES".

Gasket

Gasket

17.6 - 21.6
(1.8 - 2.2, 13 - 15)

9.0 - 13.0
(0.91 - 1.33, 79 - 115)

O-ring

8.4 - 10.8
(0.86 - 1.1, 75 - 95)

O-ring

Gasket

Rubber insulator

Clip

Gasket

Gasket

9.8 - 11.8
(1.0 - 1.2, 87 - 104)

8.5 - 10.8
(0.87 - 1.1, 76 - 95)

Gasket

: Always replace after every disassembly.

: Lubricate with new engine oil.

: N•m (kg-m, in-lb)

: N•m (kg-m, ft-lb)

1. Intake manifold collector support
2. Lower intake manifold collector
3. Fuel damper and fuel feed hose assembly
4. Fuel injector
5.
6. Fuel pressure regulator
7. Ignition coil with power transistor
8. Upper intake manifold collector
9. Intake manifold
10. Electric throttle control actuator
11. EVAP canister purge volume control solenoid valve
12. Swirl control valve actuator
13. Power valve actuator (A/T)
14. Oil level gauge

Intake manifold exploded view—3.5L engine

67170PATHG03

- Breather pipe
- Upper intake manifold
- Fuel lines
- Right bank injector connectors
- Fuel supply manifold
- Engine Coolant Temperature (ECT) sensor connector
- Lower intake manifold. Loosen the fasteners in the sequence shown.

To install:

5. Install the lower intake manifold with a new gasket.

6. For 3.5L engines, tighten the fasteners in sequence as follows:
  a. Step 1: 86 inch lbs. (4 Nm)
  b. Step 2: 23 ft. lbs. (9 Nm)

7. Install or connect the following:
- ECT sensor connector
- Fuel supply manifold
- Right bank injector connectors
- Fuel lines
- Intake manifold collector
- Breather pipe
- Upper intake manifold ground cable
- Thermal transmitter
- Left bank injector connectors
- Distributor
- Spark plug wires
- EGR tube
- Fuel pressure regulator vacuum hose
- Brake booster vacuum hose

- EVAP canister vacuum and purge hoses
- PCV valve and hose
- Heater hoses
- Radiator hoses
- EGR temperature sensor connector
- EGR Solenoid valve connector
- Ignition coil and power transistor connectors
- TP sensor and switch connectors
- IAC valve connector
- Cruise control cable
- Accelerator cable
- Air intake duct
- Negative battery cable

8. Fill the cooling system.
9. Start the engine and check for leaks.

4.0L Engine

1. Before servicing the vehicle, refer to the Precautions Section.
2. Relieve the fuel system pressure.
3. Drain the cooling system.
4. Remove or disconnect the following:
- Negative battery cable
- Engine cover
- Air intake assembly
- Coolant hoses from throttle body
- Throttle body electrical connections
- Throttle body
- Brake booster vacuum hose
- PCV hose
- Lower Intake manifold collector support
- EVAP canister vacuum and purge hoses
- VIAS control solenoid valve
- Intake manifold collector
- Fuel supply hose
- Fuel rail with injectors attached
- Intake manifold and gasket

To install:

5. Install the intake manifold with a new gasket. Tighten the bolts in sequence as follows:
  a. Step 1: Tighten to 5 ft. lbs. (7.4 Nm).

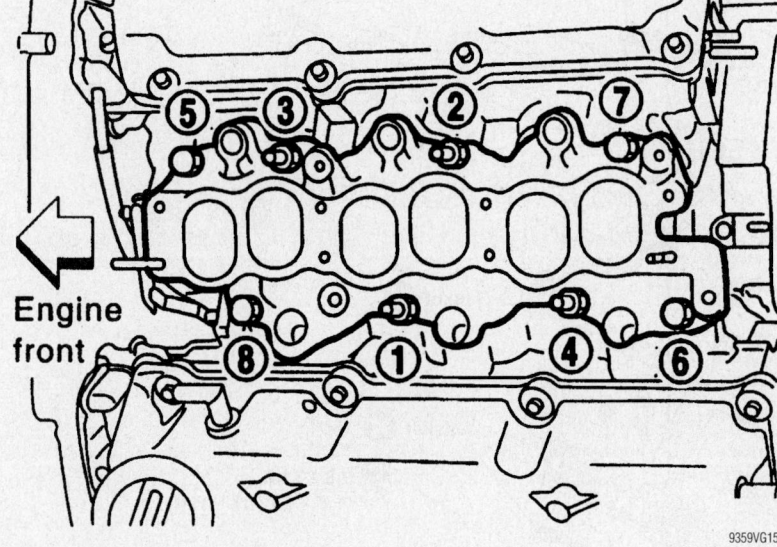

Intake manifold torque sequence—3.5L and 4.0L engine

9359VG15

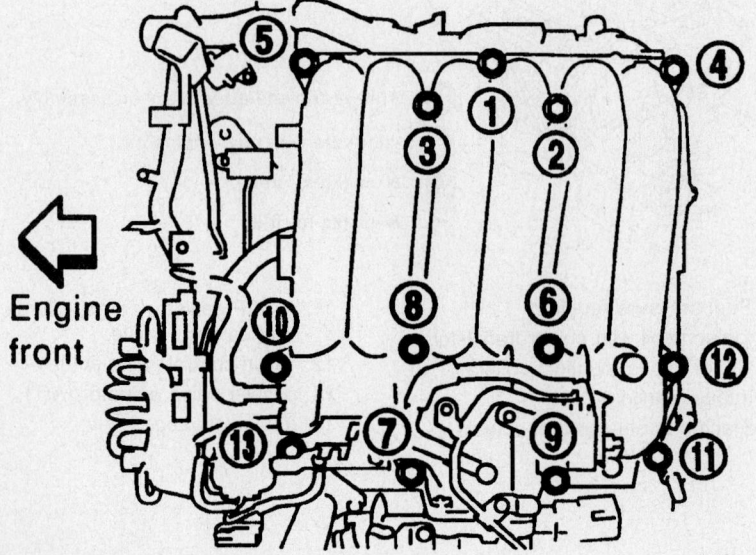

Intake manifold collector torque sequence—3.5L engine

9359VG16

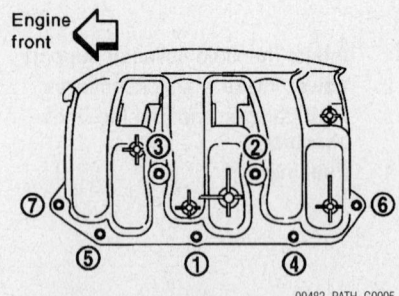

Intake manifold collector torque sequence—4.0L engine

09482_PATH_G0005

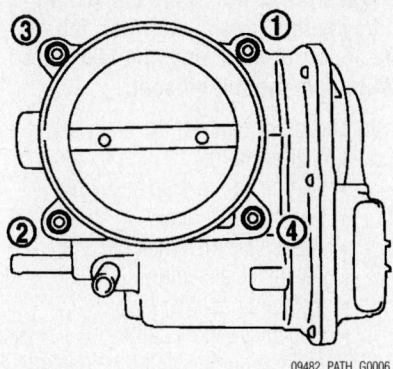

Throttle body torque sequence—4.0L engine

b. Step 2: Tighten to 21 ft. lbs. (29 Nm).
6. Install or connect the following:
- Fuel rail
- Fuel supply hose
- Intake manifold collector. Tighten the bolts in sequence to 8 ft. lbs. (11 Nm).
- VIAS control solenoid valve
- EVAP canister vacuum and purge hoses
- Intake manifold collector support
- PCV hose
- Brake booster vacuum hose
- Throttle body and tighten the bolts in sequence
- Throttle body electrical connections
- Coolant hoses to the throttle body
- Air intake assembly
- Engine cover
- Negative battery cable
7. Fill the cooling system to the correct level.
8. Start the engine and check for leaks.

Exhaust Manifold

REMOVAL & INSTALLATION

3.5L Engines

1. Before servicing the vehicle, refer to the Precautions Section.
2. Remove or disconnect the following:
- Negative battery cable
- Exhaust manifold heat shields
- Exhaust Gas Recirculation (EGR) tube
- Heated Oxygen (HO_2S) sensor connectors
- Exhaust front pipes
- Exhaust manifolds with catalytic converters attached. Loosen the nuts in the reverse of the torque sequence.

Right bank

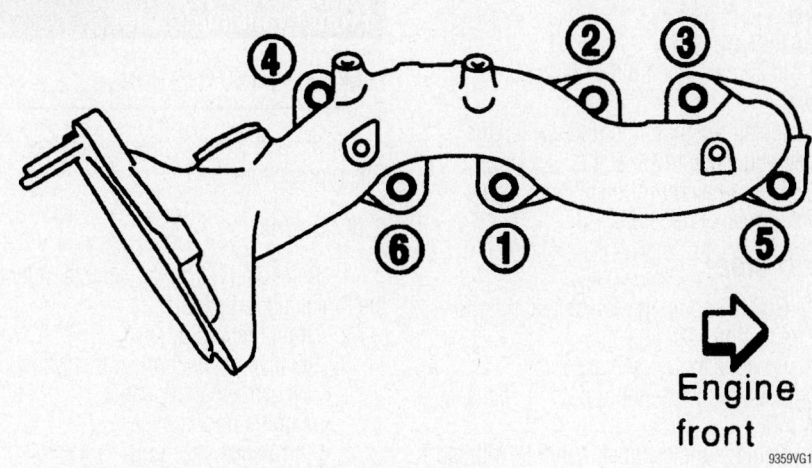

Engine front

Right exhaust manifold torque sequence—3.5L engine

Left bank

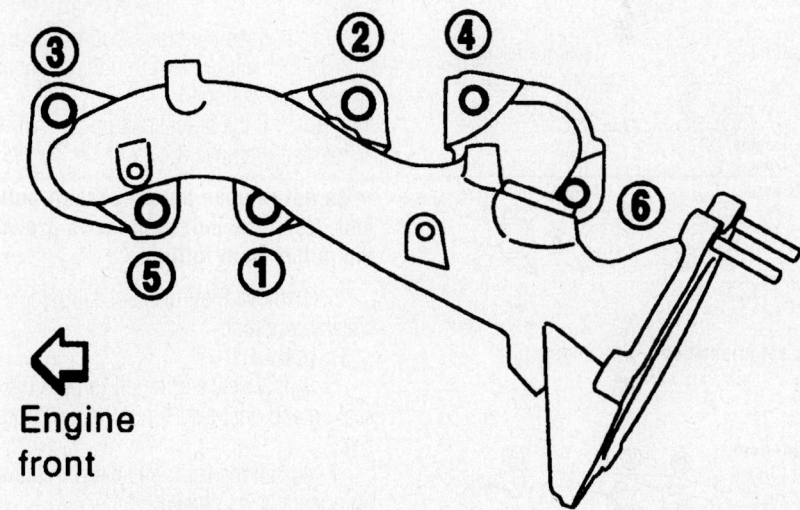

Engine front

Left exhaust manifold torque sequence—3.5L engine

To install:
3. Install or connect the following:
- Exhaust manifolds with catalytic converters attached. Tighten the nuts in sequence to 21–25 ft. lbs. (28–33 Nm).
- Exhaust front pipes. Tighten the bolts to 21–25 ft. lbs. (28–33 Nm).
- Heated Oxygen (HO_2S) sensor connectors
- EGR tube. Tighten the flange fittings to 29–36 ft. lbs. (39–49 Nm).
- Exhaust manifold heat shields. Tighten the bolts to 84–96 inch lbs. (9–11 Nm)
- Negative battery cable
4. Start the engine and check for leaks.

4.0L Engine

LEFT SIDE

1. Before servicing the vehicle, refer to the Precautions Section.
2. Remove or disconnect the following:
- Negative battery cable
- Engine undercover
- Heated Oxygen (HO_2S) sensor connectors
- Center exhaust pipe
- Muffler
- Left front exhaust downpipe
- Exhaust manifold heat shield
- Air-fuel ratio sensor
- Catalytic converter
3. Loosen the manifold nuts in the

reverse order of the installation sequence shown.

4. Remove the exhaust manifold with the gasket.

To install:

5. Install the manifold in direction as shown.

6. Install the manifold and tighten the nuts in sequence to 11 ft. lbs. (15 Nm).

7. The remainder of installation is the reverse order of removal.

RIGHT SIDE

1. Before servicing the vehicle, refer to the Precautions Section.

2. Remove the engine assembly.

3. Loosen the manifold nuts in sequence as shown.

4. Remove the exhaust manifold with the gasket.

5. Installation is the reverse order of removal. Note the following:

    a. Install the gasket in the direction as shown.

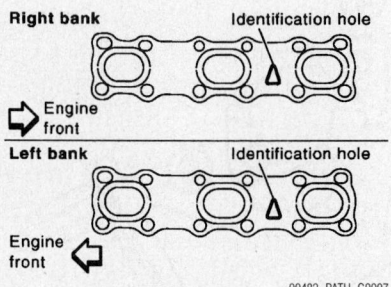

Gasket orientation—4.0L engine

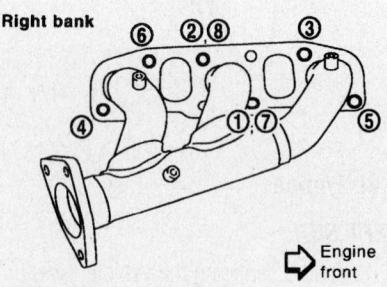

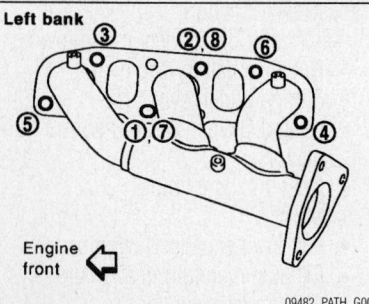

Exhaust manifold torque sequence—4.0L engine

b. Tighten the manifold nuts in sequence to 11 ft. lbs. (15 Nm).

Front Crankshaft Seal

REMOVAL & INSTALLATION

For the 3.5L engine, see the Timing Chain procedure.

4.0L Engine

1. Before servicing the vehicle, refer to the Precautions Section.

2. Drain the engine oil.

3. Remove or disconnect the following:

- Negative battery cable
- Engine undercover
- Air intake assembly
- Accessory drive belt
- Upper and lower radiator shrouds
- Engine cooling fan assembly
- Starter motor

4. Remove the crankshaft pulley as follows:

    a. Loosen the crankshaft pulley bolt 0.39 inches (10 mm) from its original seating position.

    b. Pull the crankshaft pulley with both hands to remove it.

➡ **Do not remove the crankshaft pulley bolt. Keep the bolt in place to prevent the pulley from falling.**

5. Remove the front oil seal using a suitable pry tool.

To install:

6. Apply new engine oil to both the oil seal lip and dust seal lip of the new front seal.

7. Install the front oil seal with the seal lip orientated as shown.

8. Press-fit the seal with a suitable tool until it is flush with the mounting surface.

9. Install the crankshaft pulley and tighten the bolt to 33 ft. lbs. (44 Nm).

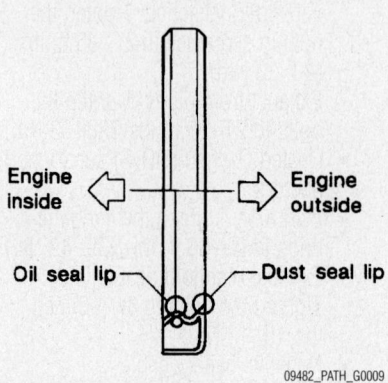

Front oil seal orientation—4.0L engine

➡ **When press-fitting the crankshaft pulley with a plastic hammer, tap on the center portion only and take care not to damage the oil seal.**

10. Install or connect the following:

- Starter motor
- Engine cooling fan assembly
- Radiator shrouds
- Accessory drive belt
- Air intake assembly
- Engine undercover
- Negative battery cable

11. Refill the engine oil to the correct level.

12. Start the engine and check for leaks.

Camshaft and Valve Lifters

REMOVAL & INSTALLATION

3.5L Engines

See the Cylinder Head Removal and Installation procedure.

4.0L Engines

1. Before servicing the vehicle, refer to the Precautions Section.

2. Drain the engine oil.

3. Drain the cooling system.

4. Remove or disconnect the following:

- Front cover
- Camshaft sprockets
- Timing chain
- Rear timing chain case
- Camshaft position sensor from cylinder head back side
- Intake valve timing control solenoid valves

5. Remove the camshaft bearing caps in several steps in the reverse order of the installation sequence shown.

➡ **Keep all camshaft components in order so they are reinstalled in the same position and direction.**

6. Remove the camshafts.

7. Remove the valve lifters.

To install:

8. Install the valve lifters in their original position.

9. Ensure the No. 1 cylinder is at Top Dead Center.

10. Install the camshaft with the dowel pin attached to its front end face on the exhaust side.

➡ **Large and small pin holes are located on the front end face of the intake camshaft at intervals of 180 degrees. Face the small diameter side**

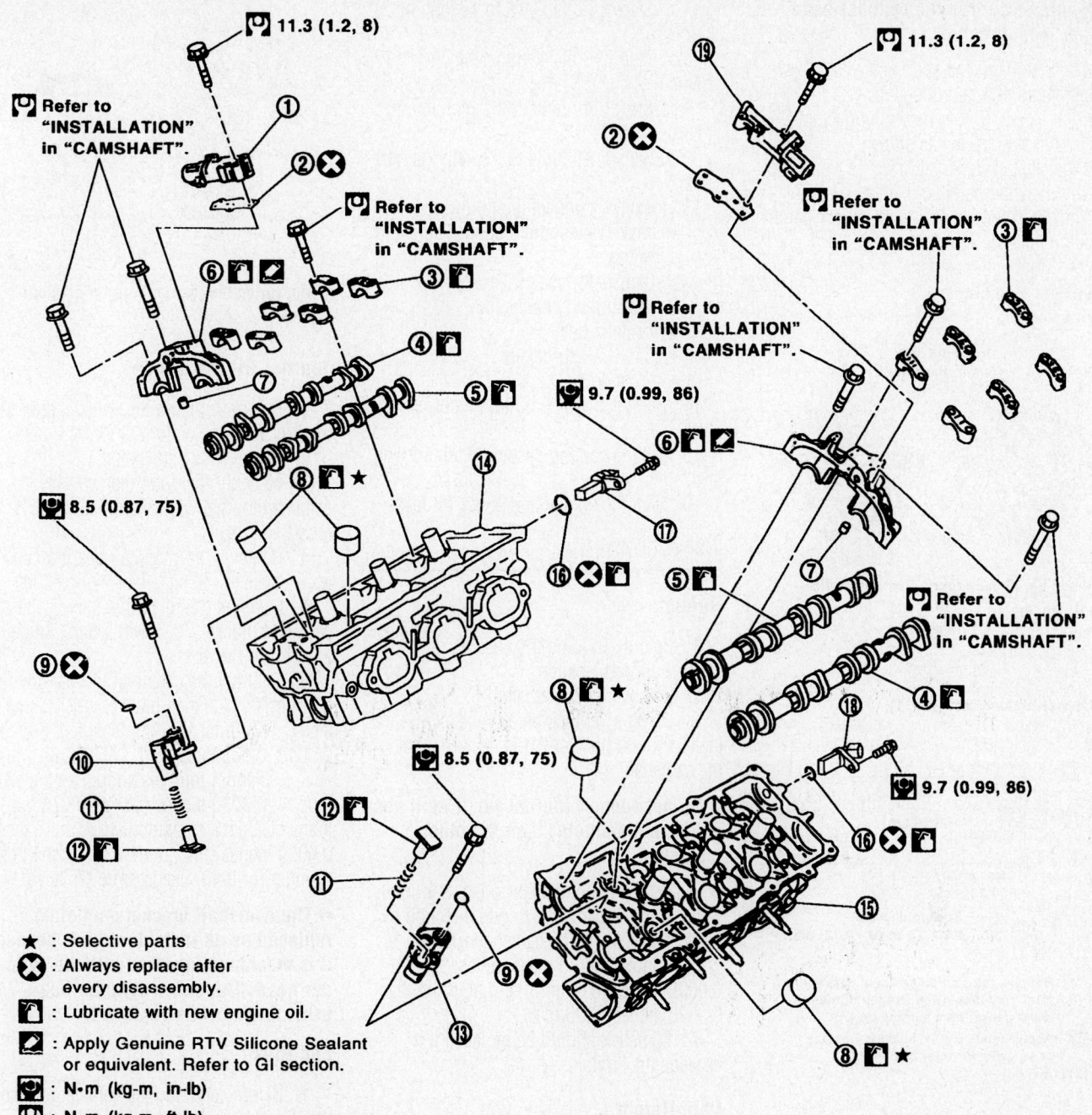

★ : Selective parts

Ⓧ : Always replace after every disassembly.

: Lubricate with new engine oil.

: Apply Genuine RTV Silicone Sealant or equivalent. Refer to GI section.

: N•m (kg-m, in-lb)

: N•m (kg-m, ft-lb)

1. Intake valve timing control solenoid valve (right bank)
2. Gasket
3. Camshaft bracket (No. 2 to 4)
4. Camshaft (EXH)
5. Camshaft (INT)
6. Camshaft bracket (No. 1)
7. Dowel pin
8. Valve lifter
9. O-ring
10. Timing chain tensioner (secondary) (right bank)
11. Spring
12. Plunger
13. Timing chain tensioner (secondary) (left bank)
14. Cylinder head (right bank)
15. Cylinder head (left bank)
16. O-ring
17. Camshaft position sensor (PHASE) (right bank)
18. Camshaft position sensor (PHASE) (left bank)
19. Intake valve timing control solenoid valve (left bank)

Exploded view of the camshaft assembly—4.0L engine

09482_PATH_G0010

pin hole upward in the cylinder head upper face direction.

11. Install the camshaft brackets in their original position and direction:

a. Apply RTV to mating surface of No 1 camshaft bracket as shown.

b. Install camshaft brackets Nos. 2-4 with the stamped marks aligned as shown. There are no identification identifying left and right for bracket No. 1.

12. Tighten the camshaft bearing caps in sequence as shown.

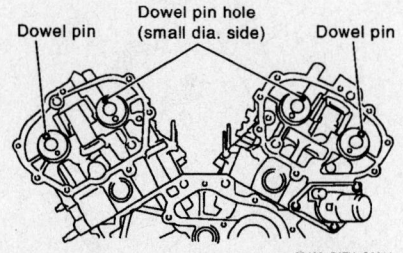

Dowel pin placement on the camshafts— 4.0L engine

Camshaft bracket (No. 1)

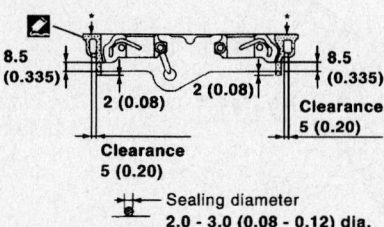

* : Remove the protruding liquid gasket from front face. (Remove the hardened liquid gasket from surface only.)

▨ : Apply Genuine RTV Silicone Sealant or equivalent. Refer to GI section.

Unit: mm (in)

No. 1 camshaft bracket liquid gasket application—4.0L engine

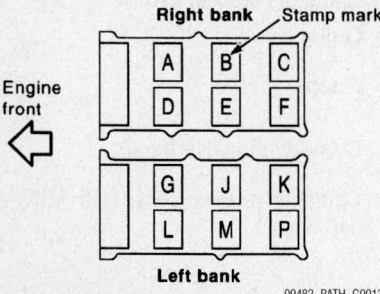

Camshaft stamped identifying marks— 4.0L engine

a. Step 1: Bolts 7–10 to 17 inch lbs. (2 Nm).

b. Step 2: Bolts 1–6 to 17 inch lbs. (2 Nm).

c. Step 3: All bolts to 52 inch lbs. (6 Nm).

d. Step 4: All bolts to 92 inch lbs. (10 Nm).

13. Install or connect the following:
- Intake valve timing control solenoid valves
- Camshaft position sensors
- Rear timing chain case
- Timing chain
- Camshaft sprockets
- Front cover

14. Refill the engine oil to the correct level.

15. Fill the cooling system to the correct level.

16. Start the engine and check for leaks.

INSPECTION

Runout

1. Before servicing the vehicle, refer to the Precautions Section.

2. Remove the camshafts.

3. Using a V-block on a precise flat table, support the No. 2 and 4 journals of the camshaft.

➡ **Do not support journal No. 1 as it has a different diameter than the other locations.**

4. Set the dial indicator to No. 3 journal.

5. Turn the camshaft to one direction by hand and measure the camshaft runout.

6. Runout should measure less than 0.0008 inches (0.02 mm) and no more than 0.0020 inches (0.05 mm).

7. Camshaft should be replaced if it exceeds the limit.

Cam Height

1. Before servicing the vehicle, refer to the Precautions Section.

2. Remove the camshafts.

3. Measure the cam height with a micrometer.

4. The intake camshaft should measure between 1.7900–1.7974 inches (45.465–45.655 mm) and not less than 1.7821 inches (45.265 mm).

5. The exhaust camshaft should measure between 1.7746–1.7821 inches (45.075–45.265 mm) and not less than 1.7667 inches (44.875 mm).

6. Camshaft should be replaced if it exceeds the limit.

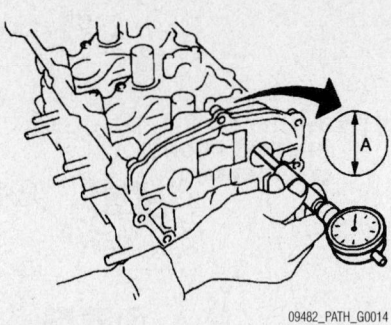

Measuring the journal bore for oil clearance

Journal Oil Clearance

1. Before servicing the vehicle, refer to the Precautions Section.

2. Remove the camshafts.

3. Measure the outer diameter of camshaft journal with micrometer and record the result.

4. Reinstall the camshaft bearing caps in accordance to the installation procedure.

5. Measure the inner diameter of the camshaft bracket ("A") with a bore gauge and record the result.

6. Subtract the camshaft journal diameter from the camshaft bracket inner diameter. No. 1 journal should measure between 0.0018–0.0034 inches (0.045–0.086 mm). Nos. 2, 3 and 4 journals should be between 0.0014–0.0030 inches (0.035–0.076 mm). If any clearance measurement is more than 0.0059 inches (0.15 mm), the camshaft or cylinder (or both) needs to be replaced.

➡ **The camshaft bracket cannot be replaced as an individual part, because it is machined together with the cylinder head. The entire cylinder head assembly must be replaced.**

End Play

1. Before servicing the vehicle, refer to the Precautions Section.

2. Install a dial indicator in the thrust direction on the front end of the camshaft. Measure the end play of the dial indicator when the camshaft is moved back and forth. The dial indicator should measure between 0.0045–0.0074 inches (0.115–0.188 mm) and not exceed 0.0094 inches (0.24 mm).

3. Measure the No. 1 journal as shown. The distance ("A") should be 1.0827–1.0846 inches (27.500–27.548 mm).

4. Measure the No. 1 journal bearing as shown. The distance ("B") should be 1.0772–1.0781 inches (27.360–27.385 mm).

5. Replace either the camshaft or cylin-

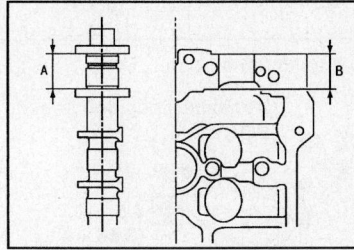

09482_PATH_G0015

Measuring for camshaft endplay

der head assembly if the measurement is exceeded.

Valve Lash

ADJUSTMENT

3.5L Engines

➡ **Adjust valve clearance while engine is cold.**

1. Turn crankshaft, to position cam lobe on camshaft of valve that must be adjusted upward.
2. Thoroughly wipe off engine oil around adjusting shim using a rag.
3. Using an extra-fine screwdriver, turn the round hole of the adjusting shim in the direction of the arrow.
4. Place Tool (A) around camshaft as shown in figure.
Before placing Tool (A), rotate notch toward center of cylinder head (See figure.), to simplify shim removal later.

❄❄ **CAUTION**

Be careful not to damage cam surface with Tool (A).

5. Rotate Tool (A) (See figure.) so that valve lifter is pushed down.
6. Place Tool (B) between camshaft and the edge of the valve lifter to retain valve lifter.

❄❄ **CAUTION**

Tool (B) must be placed as close to camshaft bracket as possible. Be

careful not to damage cam surface with Tool (B).

7. Remove Tool (A).
8. Blow air into the hole to separate adjusting shim from valve lifter.
9. Remove adjusting shim using a small screwdriver and a magnetic finger.
10. Determine replacement adjusting shim size following formula. Using a micrometer determine thickness of removed shim. Calculate thickness of new adjusting shim so valve clearance comes within specified values.

• R = Thickness of removed shim
• N = Thickness of new shim
• M = Measured valve clearance
• Intake: N = R + [M − 0.30 mm (0.0118 in)]
• Exhaust: N = R + [M − 0.33 mm (0.0130 in)]

Shims are available in 64 sizes from 2.32 mm (0.0913 in) to 2.95 mm (0.1161 in), in steps of 0.01 mm (0.0004 in). Select new shim with thickness as close as possible to calculated value.

11. Install new shim using a suitable tool. Install with the surface on which the thickness is stamped facing down.
12. Place Tool (A) as mentioned in steps 2 and 3.
13. Remove Tool (B).
14. Remove Tool (A).
15. Recheck valve clearance.
Valve clearance (Cold)
• Intake: 0.010 − 0.013
• Exhaust: 0.011 − 0.015

Starter Motor

REMOVAL & INSTALLATION

1. Before servicing the vehicle, refer to the Precautions Section.
2. Remove or disconnect the following:

• Negative battery cable
• Engine under cover
• Starter harness connectors
• Starter motor

To install:
3. Install or connect the following:
• Starter motor. Tighten the bolts to 37–45 ft. lbs. (61-69 Nm) on the 3.5L and 33 ft. lbs. (45 Nm) on the 4.0L.
• Starter harness connectors
• Engine under cover
• Negative battery cable

Oil Pan

REMOVAL & INSTALLATION

3.5L Engines

1. Remove front RH and LH wheels.
2. Remove battery.
3. Remove oil level gauge.
4. Remove engine undercover.
5. Remove suspension member stay.
6. Drain engine coolant from radiator drain plug.
7. Disconnect A/T oil cooler hoses. (A/T)
8. Drain engine oil.
9. Remove the crankshaft position sensors (REF and POS).
10. Remove drive belts and idler pulley with bracket.
11. Remove power steering oil pump, then put it aside holding with a suitable wire.
12. Remove alternator.
13. Install engine slingers.
14. Remove front propeller shaft. (4WD)
15. Remove exhaust front tube heat insulators, then remove rear heat oxygen sensors.
16. Remove exhaust front tube from both sides.
17. Remove front final drive. (4WD)
18. Remove starter motor.
19. Disconnect oil pressure switch harness connector.
20. Loosen and disconnect the bolts fixing the steering column assembly lower joint and the power steering gear.
21. Set a suitable transmission jack under the front suspension member and hoist engine with engine slingers.
22. Remove front engine mounting nuts from both sides.
23. Remove front suspension member bolts.
24. Lower the transmission jack carefully to secure clearance between the oil pan and suspension member.
25. Remove A/T oil cooler tube. (A/T)
26. Remove water hose and tube. (A/T)
27. Remove the four engine-to-transmission bolts.
28. Remove aluminum oil pan bolts in numerical order.
29. Remove aluminum oil pan.
  a. Insert tool between aluminum oil pan and cylinder block.

➡ **Be careful not to damage aluminum mating surface. I Do not insert screwdriver, or oil pan flange will be deformed.**

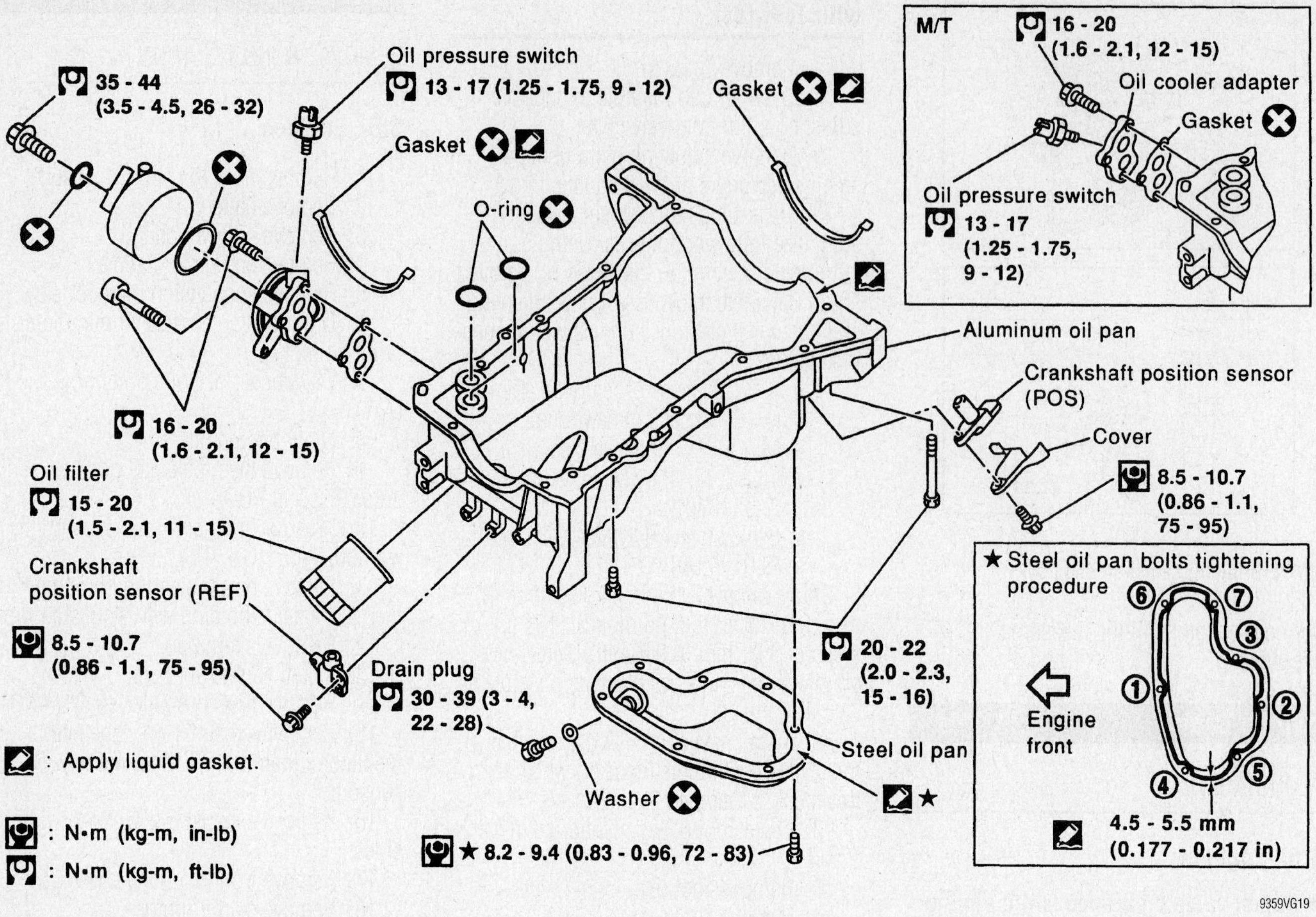

35 - 44
(3.5 - 4.5, 26 - 32)

Oil pressure switch
13 - 17 (1.25 - 1.75, 9 - 12)

Gasket

M/T   16 - 20
(1.6 - 2.1, 12 - 15)

Oil cooler adapter

Gasket

Gasket

O-ring

Oil pressure switch
13 - 17
(1.25 - 1.75,
9 - 12)

Aluminum oil pan

16 - 20
(1.6 - 2.1, 12 - 15)

Crankshaft position sensor
(POS)

Cover

Oil filter
15 - 20
(1.5 - 2.1, 11 - 15)

8.5 - 10.7
(0.86 - 1.1,
75 - 95)

Crankshaft
position sensor (REF)

8.5 - 10.7
(0.86 - 1.1, 75 - 95)

★ Steel oil pan bolts tightening
procedure

20 - 22
(2.0 - 2.3,
15 - 16)

Drain plug
30 - 39 (3 - 4,
22 - 28)

Steel oil pan

Engine
front

: Apply liquid gasket.

: N•m (kg-m, in-lb)

: N•m (kg-m, ft-lb)

Washer

★ 8.2 - 9.4 (0.83 - 0.96, 72 - 83)

4.5 - 5.5 mm
(0.177 - 0.217 in)

9359VG19

Oil pan exploded view—3.5L engine

b. Slide tool by tapping its side with a hammer.

30. Remove O-rings from cylinder block and oil pump body.

31. Remove front cover gasket and rear oil seal retainer gasket.

To install:

32. Before installing oil pan, remove old liquid gasket from mating surface using a scraper. Also remove old liquid gasket from mating surface of cylinder block. Remove old liquid gasket from the bolt hole and thread.

33. Apply sealant to front cover gasket and rear oil seal retainer gasket.

34. Install front cover gasket and rear oil seal retainer gasket.

35. Apply a continuous bead of liquid gasket to mating surface of aluminum oil pan. Use RTV silicone sealant or equivalent.

36. Apply liquid gasket to inner sealing surface as shown in figure. Be sure liquid gasket is 4.0 to 5.0 mm (0.157 to 0.197 in) or 4.5 to 5.5 mm (0.177 to 0.217 in) wide. Attaching should be done within 5 minutes after coating.

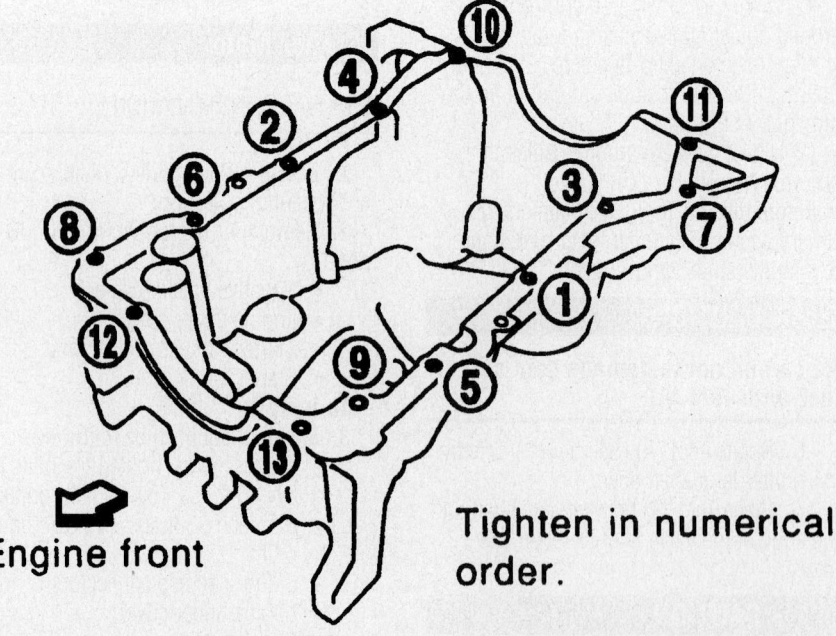

Engine front

Tighten in numerical order.

9359VG20

Oil pan bolt torque sequence—3.5L engine

37. Install O-rings, cylinder block and oil pump body.

38. Install aluminum oil pan. Tighten bolts in numerical order. Wait at least 30 minutes before refilling engine oil.

39. Install the four engine-to-transmission bolts.

40. The remainder of installation is the reverse order of removal.

4.0L Engine

LOWER OIL PAN

1. Before servicing the vehicle, refer to the Precautions Section.

2. Drain the engine oil.

3. Remove the lower oil pan bolts in the reverse order of the installation sequence.

4. Use a seal cutter if necessary to remove the oil pan.

❊❊ WARNING

Do not use a screwdriver to pry the pan off, this may damage the mating surfaces.

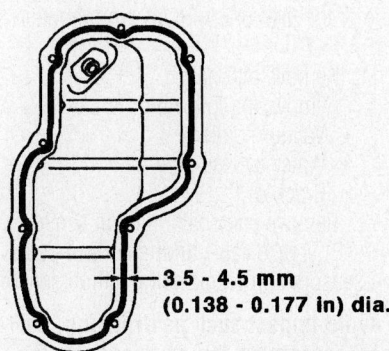

3.5 - 4.5 mm
(0.138 - 0.177 in) dia.

09482_PATH_G0016

RTV application for the lower pan—4.0L engine

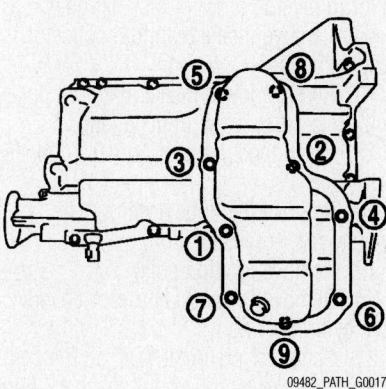

09482_PATH_G0017

Lower oil pan torque sequence—4.0L engine

To install:

5. Use a scraper to completely remove any old sealant.

6. Apply a continuous bead of RTV to the lower oil pan as shown.

7. Install the lower oil pan and tighten the bolts in sequence to 80 inch lbs. (9 Nm).

➥**Wait at least 30 minutes for RTV to dry before filling with oil.**

8. Refill the engine oil to the correct level.

9. Start the engine and check for leaks.

UPPER OIL PAN

1. Before servicing the vehicle, refer to the Precautions Section.

2. Drain the engine oil.

3. Drain the engine coolant.

4. Remove or disconnect the following:
   - Front final drive, if equipped with 4WD
   - Steering gear mounting nuts and shaft bolt and position out of the way
   - Starter motor
   - Transmission cooler hoses
   - Oil filter
   - Oil cooler
   - Lower oil pan
   - Oil strainer
   - Transmission mounting bolts that bolt into the upper oil pan
   - Rear cover plate

5. Loosen the upper oil pan bolts in the reverse order of the installation sequence.

6. Use a seal cutter if necessary to remove the oil pan.

❊❊ WARNING

Do not use a screwdriver to pry the pan off, this may damage the mating surfaces.

To install:

7. Use a scraper to completely remove any old sealant.

8. Apply a continuous bead of RTV to the upper oil pan as shown.

9. Install the upper oil pan bolts and tighten is sequence to 16 ft. lbs. (22 Nm).

10. Tighten any transmission mounting bolts removed to 55 ft. lbs. (74 Nm).

11. Install or connect the following:
   - Oil strainer
   - Lower oil pan
   - Oil cooler
   - Oil filter
   - Transmission cooler hoses
   - Starter motor

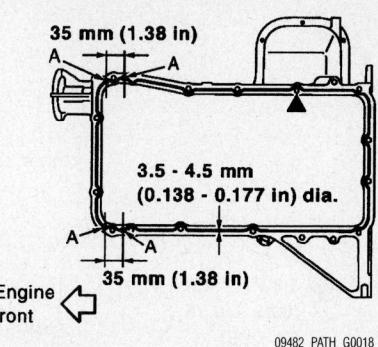

35 mm (1.38 in)

3.5 - 4.5 mm
(0.138 - 0.177 in) dia.

35 mm (1.38 in)

Engine front

09482_PATH_G0018

RTV application for the upper oil pan—4.0L engine

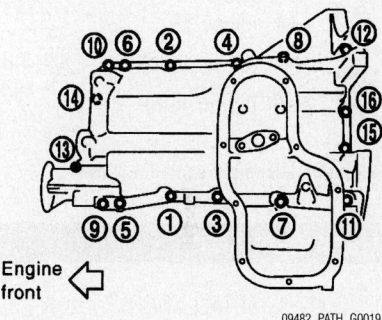

Engine front

09482_PATH_G0019

Upper oil pan torque sequence—4.0L engine

   - Steering gear
   - Front final drive, if equipped with 4WD

➥**Wait at least 30 minutes for RTV to dry before filling with oil.**

12. Refill the engine oil to the correct level.

13. Fill the cooling system to the correct level.

14. Start then engine and check for leaks.

Oil Pump

REMOVAL & INSTALLATION

1. Before servicing the vehicle, refer to the Precautions Section.

2. Drain the engine oil.

3. Drain the cooling system.

4. Remove or disconnect the following:
   - Oil pans (upper and lower)
   - Timing chain
   - Oil pump assembly

5. Installation is the reverse order of removal.

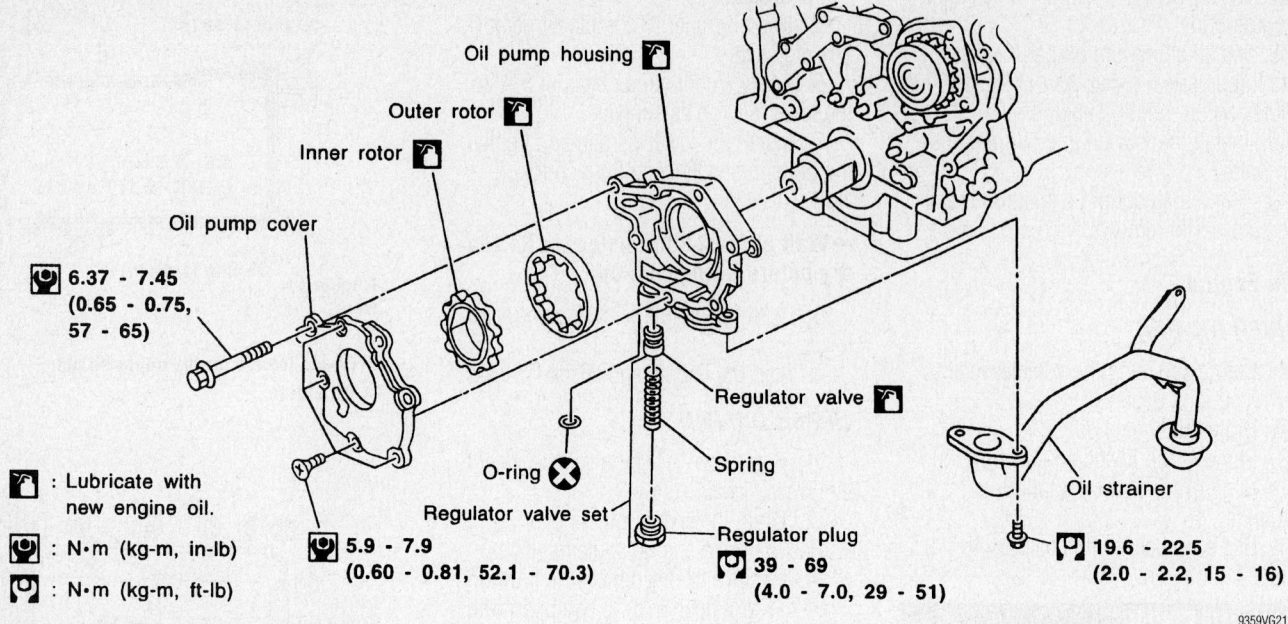

Oil pump assembly exploded view

Oil pump housing

Outer rotor

Inner rotor

Oil pump cover

6.37 - 7.45
(0.65 - 0.75,
57 - 65)

: Lubricate with
new engine oil.

: N·m (kg-m, in-lb)

: N·m (kg-m, ft-lb)

O-ring

Regulator valve set

5.9 - 7.9
(0.60 - 0.81, 52.1 - 70.3)

Regulator valve

Spring

Regulator plug

39 - 69
(4.0 - 7.0, 29 - 51)

Oil strainer

19.6 - 22.5
(2.0 - 2.2, 15 - 16)

9359VG21

Rear Main Seal

REMOVAL & INSTALLATION

3.5L Engine

1. Remove transmission.
2. Remove flywheel or drive plate.
3. Remove oil pan.
4. Remove rear oil seal retainer.
5. Remove the oil seal from the retainer.

To install:

6. Remove old liquid gasket using scraper. Remove old liquid gasket from the bolt hole and thread.
7. Press fit the oil seal into the retainer.
8. Apply liquid gasket to rear oil seal retainer.
9. The remainder of installation is the reverse order of removal.

4.0L Engine

1. Before servicing the vehicle, refer to the Precautions Section.
2. Remove the transmission assembly.
3. Remove the rear oil seal with a suitable pry tool.

To install:

4. Install the rear oil seal so each seal lip is oriented as shown.
5. Press in the seal with a suitable seal installer tool.
6. Install the transmission assembly.
7. Refill the engine oil to the correct level.
8. Start the engine and check for leaks.

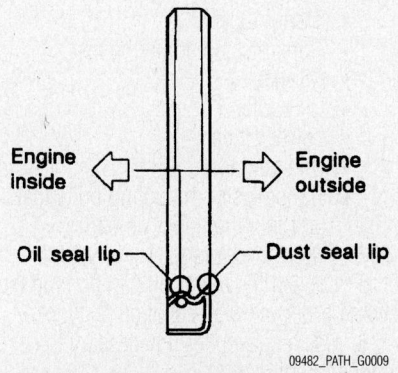

Engine inside

Engine outside

Oil seal lip

Dust seal lip

09482_PATH_G0009

Rear oil seal lip orientation

Timing Chain, Sprockets and Front Cover

REMOVAL & INSTALLATION

3.5L Engine

1. Release fuel pressure.
2. Remove battery.
3. Remove radiator.
4. Drain engine oil.
5. Remove drive belts and idler pulley with brackets.
6. Remove cooling fan with bracket.
7. Remove engine cover.
8. Remove air duct with air cleaner case, collector, blow-by hose, vacuum hoses, fuel hoses, water hoses, wires, harnesses, connectors and so on.
9. Remove the air compressor, and tie it

down using rope or the like to keep it from interfering.

10. Remove the power steering oil pump and reservoir tank. Tie them down using rope or the like to keep them from interfering.
11. Remove alternator.
12. Remove the following.
- Vacuum gallery
- Water bypass pipe
- Brackets

13. Remove camshaft position sensor (PHASE), intake valve timing control position sensors and crankshaft position sensor.

➡**Avoid impact such as dropping. Do not disassemble the components. Do not place them on areas where iron powder may adhere. Keep away from the objects susceptible to magnetism.**

14. Remove upper intake manifold collector in reverse order of installation.
15. Remove intake manifold collector support bolts.
16. Remove lower intake manifold collector in reverse order of installation.
17. Disconnect injector harness connectors.
18. Remove fuel tube assembly in reverse order of installation.
19. Remove ignition coils.
20. Remove RH and LH rocker covers from cylinder head.
21. Set No. 1 piston at TDC on the compression stroke by rotating crankshaft. Align pointer with TDC mark on crankshaft pulley. Check that intake and exhaust cam nose on

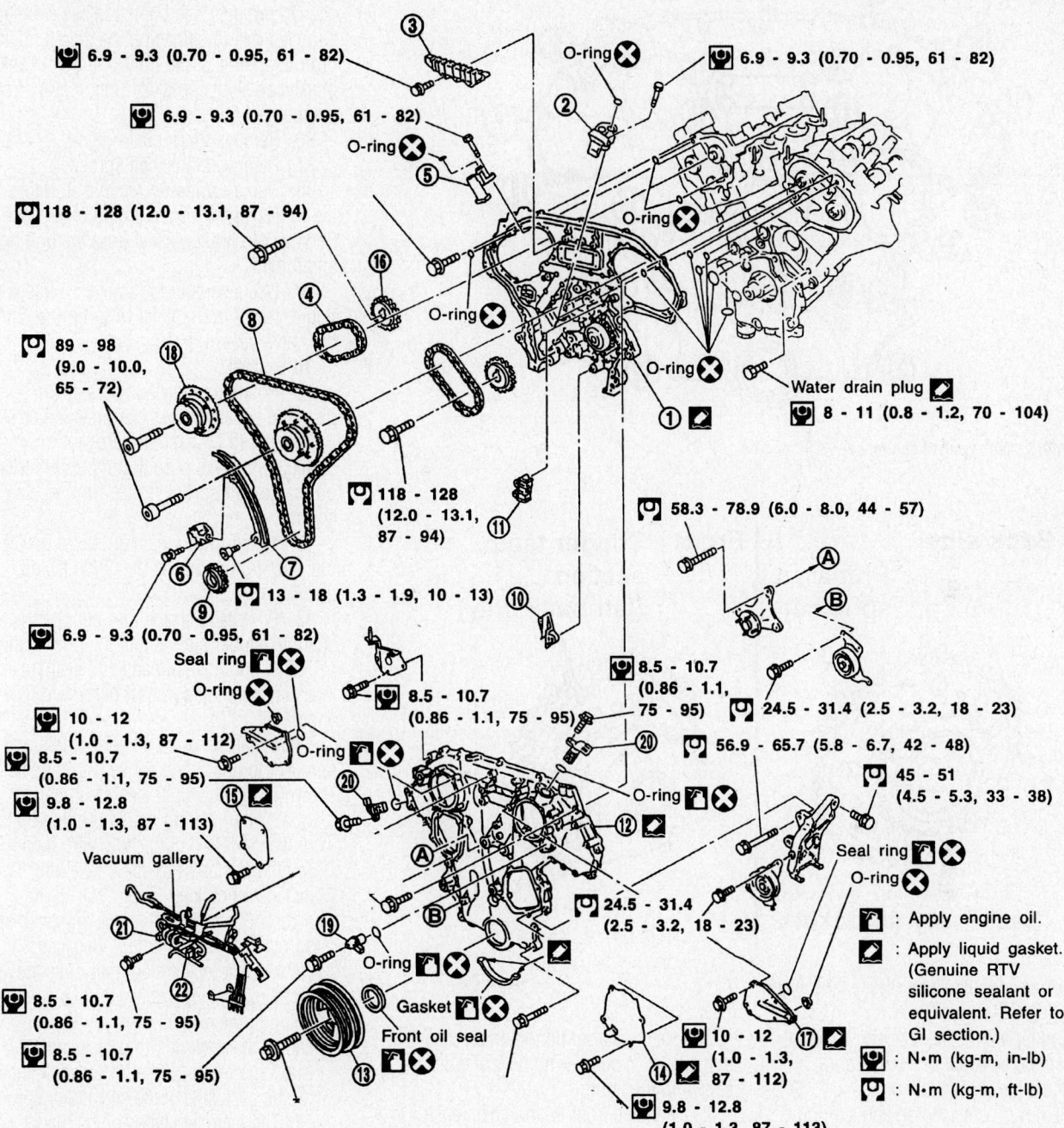

6.9 - 9.3 (0.70 - 0.95, 61 - 82)

6.9 - 9.3 (0.70 - 0.95, 61 - 82)

O-ring

6.9 - 9.3 (0.70 - 0.95, 61 - 82)

O-ring

O-ring

O-ring

118 - 128 (12.0 - 13.1, 87 - 94)

O-ring

89 - 98 (9.0 - 10.0, 65 - 72)

O-ring

Water drain plug

8 - 11 (0.8 - 1.2, 70 - 104)

118 - 128 (12.0 - 13.1, 87 - 94)

58.3 - 78.9 (6.0 - 8.0, 44 - 57)

13 - 18 (1.3 - 1.9, 10 - 13)

6.9 - 9.3 (0.70 - 0.95, 61 - 82)

Seal ring

O-ring

8.5 - 10.7 (0.86 - 1.1, 75 - 95)

8.5 - 10.7 (0.86 - 1.1, 75 - 95)

24.5 - 31.4 (2.5 - 3.2, 18 - 23)

10 - 12 (1.0 - 1.3, 87 - 112)

O-ring

56.9 - 65.7 (5.8 - 6.7, 42 - 48)

8.5 - 10.7 (0.86 - 1.1, 75 - 95)

45 - 51 (4.5 - 5.3, 33 - 38)

9.8 - 12.8 (1.0 - 1.3, 87 - 113)

O-ring

Vacuum gallery

Seal ring

O-ring

24.5 - 31.4 (2.5 - 3.2, 18 - 23)

8.5 - 10.7 (0.86 - 1.1, 75 - 95)

O-ring

Gasket

8.5 - 10.7 (0.86 - 1.1, 75 - 95)

Front oil seal

10 - 12 (1.0 - 1.3, 87 - 112)

9.8 - 12.8 (1.0 - 1.3, 87 - 113)

: Apply engine oil.

: Apply liquid gasket. (Genuine RTV silicone sealant or equivalent. Refer to GI section.)

: N•m (kg-m, in-lb)

: N•m (kg-m, ft-lb)

1. Rear timing chain case
2. Left camshaft chain tensioner
3. Internal guide
4. Timing chain (Secondary)
5. Right camshaft chain tensioner
6. Timing chain tensioner
7. Slack guide
8. Timing chain (Primary)
9. Crankshaft sprocket

10. Lower tension guide
11. Upper tension guide
12. Front timing chain case
13. Crankshaft pulley
14. Water pump cover
15. Chain tensioner cover
16. Exhaust camshaft sprocket
17. Intake valve timing control valve cover

18. Intake camshaft sprocket
19. Camshaft position sensor (PHASE)
20. Intake valve timing control position sensor
21. Power valve actuator (A/T)
22. Swirl control valve control solenoid valve

Timing chain components—3.5L engine

9359VG30

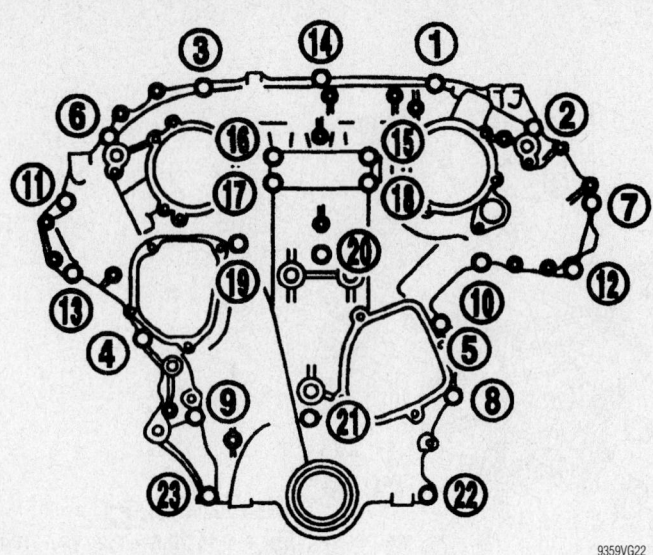

Rear timing case removal sequence—3.5L engine

9359VG22

Back side

Primary sprocket

Secondary sprocket

Front Trigger teeth section (left bank only)

9359VG23

Primary and secondary sprockets—3.5L engine

No. 1 cylinder are installed as shown left. If not, turn the crankshaft one revolution (360°) and align as above.

22. Remove starter motor, and set ring gear stopper using the mounting bolt hole. Be careful not to damage the signal plate teeth.

23. Loosen the crankshaft pulley bolt.

24. Remove crankshaft pulley with a suitable puller.

25. Remove aluminum oil pan.

26. Temporarily install the suspension member bolts and engine mounting nuts.

27. Remove intake valve timing control valve covers. Loosen bolts in numerical order as shown in the figure. In the cover, the shaft is engaged with the center hole of the intake cam sprocket. Remove it straight out until the engagement comes off.

28. Remove front timing chain case bolts. Loosen bolts in numerical order as shown in the figure.

29. Remove front timing chain case. Do not scratch sealing surfaces.

30. Remove internal chain guide.

31. Remove upper tension guide.

32. Remove timing chain tensioner and slack guide. Remove timing chain tensioner. (Push piston and insert a suitable pin into pinhole.)

33. Attach a suitable stopper pin to RH and LH camshaft chain tensioners.

34. Remove intake and exhaust camshaft sprocket bolts. I Apply paint to timing chain and camshaft sprockets for alignment during installation. Secure the hexagonal head of the camshaft using a spanner to loosen mounting bolts.

35. Remove primary and secondary timing chains along with the camshaft sprockets. Do not disassemble the intake camshaft sprocket. Avoid damaging the signal mark protrusion area at the front of the left bank intake camshaft sprocket. Keep it away from magnetized objects.

36. Remove lower chain guide.

37. Remove crankshaft sprocket.

38. Use a scraper to remove all traces of liquid gasket from front timing chain case. Remove old liquid gasket from the bolt hole and thread.

39. Use a scraper to remove all traces of liquid gasket from intake valve timing control valve cover.

To install:

40. Position crankshaft so that No. 1 piston is set at TDC on compression stroke.

41. Install crankshaft sprocket on crankshaft. Make sure that mating marks on crankshaft sprocket face front of engine.

42. Install lower chain guide on dowel pin, with front mark on the guide facing upside.

43. Press and shrink the secondary chain tensioner sleeve, and fix it using stopper pins. Lubricate threads and seat surfaces of camshaft sprocket bolts with new engine oil.

44. Install secondary timing chain and sprocket to one of the banks (Right bank shown in the figure) as described below.

a. Align mating marks (golden links) on secondary timing chain with those (punched marks) on the intake and exhaust sprockets.

b. Align camshaft knock pins with the sprocket groove and hole. Because camshaft sprocket mounting bolts are tightened in step 7, perform manual tightening to the extent necessary to keep camshaft knock pin from dislocating. Matching marks of the intake sprocket are on the back side of the secondary sprockets. There are two types of the marks; round and oval types, which should be used for right and left banks respectively.

• Right bank: Round
• Left bank: Oval

It may be difficult to visually check the dislocation of mating marks during and after installation. To make the matching easier, make a mating mark on the sprocket teeth in advance using paint.

45. Install secondary timing chain and sprocket to the other bank. Install primary timing chain at the same time. Installation of the secondary timing chain follows the procedure described in step 5.

46. Install primary timing chain so that

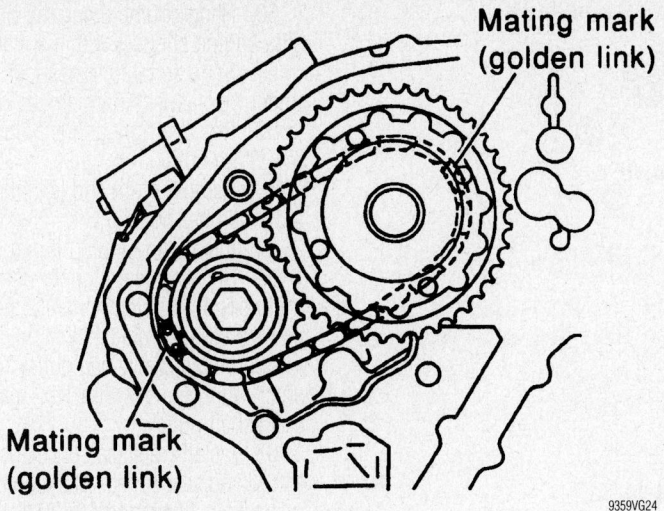

Secondary timing chain installed—3.5L engine

9359VG24

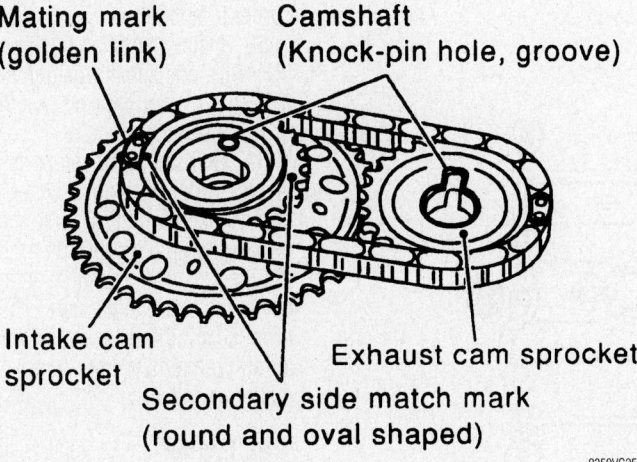

Intake sprocket mating marks—3.5L engine

9359VG25

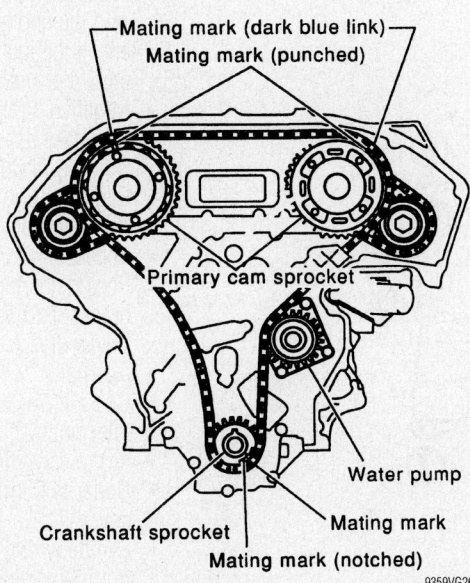

Primary timing chain installation—3.5L engine

9359VG26

mating mark (punched) on camshaft sprocket is aligned with that (dark blue link) on the timing chain, and mating mark (notched) on crankshaft sprocket is aligned with that on the timing chain, respectively.

47. When it is difficult to align mating marks of the primary timing chain with each sprocket, gradually turn the camshaft hexagonal head using a spanner so it is aligned with the mating mark.

48. During alignment, be careful to prevent dislocation of mating marks on the secondary timing chain.

49. After confirming the mating marks are aligned, tighten the camshaft sprocket mounting bolts. Secure the camshaft hexagonal head using a spanner to tighten mounting bolts.

50. Pull out the stopper pin from the secondary timing chain tensioner.

51. Install internal guide.

52. Install upper tension guide and slack guide.

53. Install timing chain tensioner, then remove the stopper pin. When installing the timing chain tensioner, engine oil should be applied to the oil hole and tensioner.

54. Install O-rings on rear timing chain case.

55. Apply liquid gasket to front timing chain case. Before installation, wipe off the protruding sealant.

56. Install rear case pin into dowel pin hole on front timing chain case.

57. Tighten bolts to the specified torque in order shown in the figure. Leave the bolts unattended for 30 minutes or more after tightening.

58. Install intake valve timing control valve cover.

　a. Install O-rings at front timing chain case.

　b. Install seal ring at intake valve timing control valve covers.

　c. Apply liquid gasket to intake valve timing control valve covers. Use RTV silicone sealant or equivalent. I Being careful not to move the seal ring from the installation groove, align the dowel pins on the chain case with the holes to install the intake valve timing control valve cover. Tighten in numerical order as shown in the figure.

59. Install RH and LH rocker covers. Rocker cover tightening procedure:

- Tighten in numerical order as shown in the figure.
- Tighten bolts 1 to 10 in that order to 61–78 inch lbs. (7–9 Nm).
- Then tighten bolts 1 to 10 as indicated in figure to 61–78 inch lbs. (7–9 Nm).

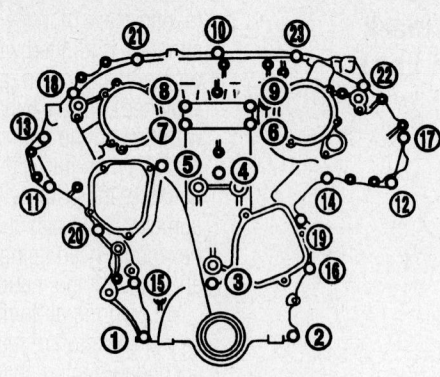

🔧 ① - ② 8 mm dia. bolts
25.5 - 31.4 N•m
(2.6 - 3.2 kg-m, 18.8 - 23.1 ft-lb)
③ - ㉔ 6 mm dia. bolts
11.8 - 13.7 N•m
(1.2 - 1.4 kg-m, 8.7 - 10.1 ft-lb)

9359VG27

Rear timing case installation—3.5L engine

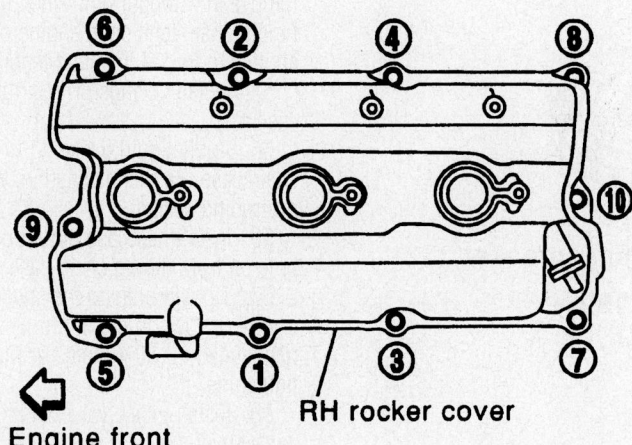

RH rocker cover

Engine front

9359VG28

Right rocker cover installation—3.5L engine

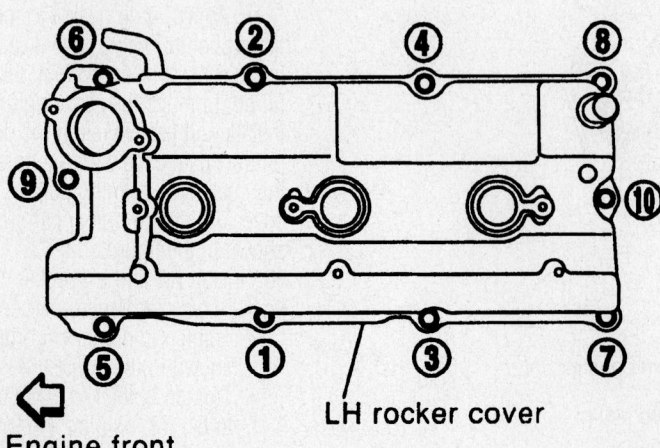

LH rocker cover

Engine front

9359VG29

Left rocker cover installation—3.5L engine

60. Hang engine using the right and left side engine slingers with a suitable hoist.

61. Set a suitable transmission jack under the suspension member.

62. Remove right and left side engine mounting nuts.

63. Remove right and left side suspension member bolts.

64. Install aluminum oil pan.

65. Set ring gear stopper using the mounting bolt hole. Be careful not to damage the signal plate teeth.

66. Install crankshaft pulley to crankshaft. Align pointer with TDC mark on crankshaft pulley.

67. Install crankshaft pulley bolt. Lubricate thread and seat surface of the bolt with new engine oil. Tighten to 29–36 ft. lbs. (39–49 Nm). Put a paint mark on the crankshaft pulley. Again tighten by turning 60° to 66°, about the angle from one hexagon bolt head corner to another.

68. Install camshaft position sensor (PHASE), crankshaft position sensors (REF)/(POS) and intake valve timing control position sensors.

69. Reinstall removed parts in the reverse order of removal. After starting engine, keep idling for three minutes. Then rev engine up to 3,000 rpm under no load to purge air from the high-pressure chamber of the chain tensioners. The engine may produce a rattling noise. This indicates that air still remains in the chamber and is not a matter of concern.

4.0L Engine

1. Before servicing the vehicle, refer to the Precautions Section.
2. Drain the engine oil.
3. Drain the cooling system.
4. Relieve the fuel system pressure.
5. Remove or disconnect the following:
- Negative battery cable
- Air intake assembly
- Cooling fan assembly
- Engine wiring harness brackets from the front cover
- Accessory drive belts
- Power steering oil pump from the bracket and hang securely
- Power steering oil pump bracket
- Alternator
- Coolant bypass hose and bracket
- Idler pulley bracket
- Intake valve timing control covers
- Collared O-rings from the front cover
- Engine wiring harness brackets from the cylinder head
- Ignition coil

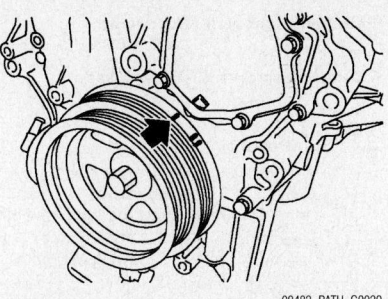

Crankshaft pulley timing marks—4.0L engine

09482_PATH_G0020

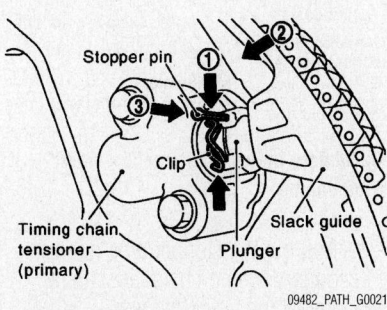

Exploded view of the timing chain tensioner—4.0L engine

09482_PATH_G0021

- PCV hoses from valve covers
- Valve covers

6. Rotate the crankshaft pulley and align the timing mark with the timing indicator to put the No. 1 cylinder at Top Dead Center (TDC).

7. Remove the starter motor.

8. Remove the crankshaft pulley.

9. Loosen the two front bolts of the upper oil pan.

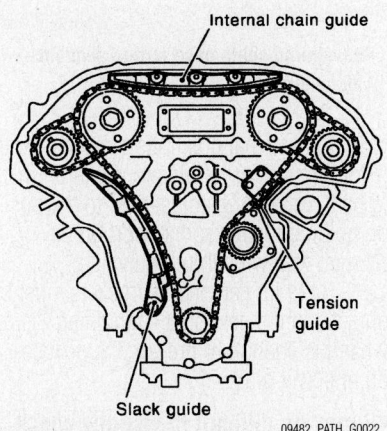

Timing chain guide locations—4.0L engine

09482_PATH_G0022

10. Remove the front cover as follows:

a. Loosen the bolts in the reverse order of the installation sequence.

b. Insert a suitable pry tool into the notch at the top of the front cover and pry off the front cover.

11. Remove the O-rings from the rear timing chain case.

12. Remove the primary timing chain tensioner as follows:

a. Loosen the clip of timing chain tensioner (primary), and release plunger stopper.

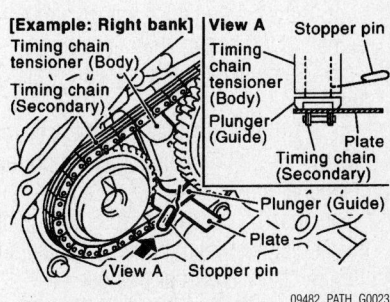

Insert a metal plate between the timing chain and tensioner plunger to remove—4.0L engine

09482_PATH_G0023

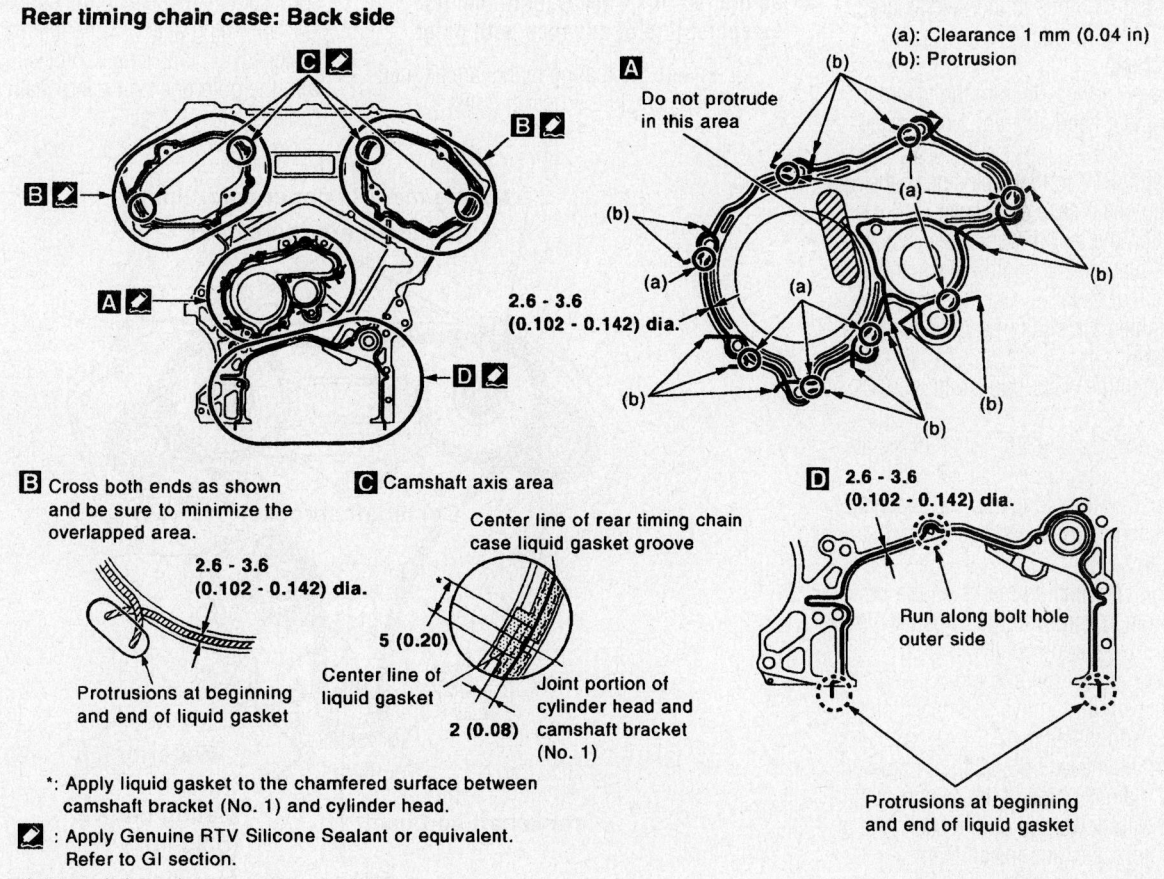

B Cross both ends as shown and be sure to minimize the overlapped area.

2.6 - 3.6 (0.102 - 0.142) dia.

Protrusions at beginning and end of liquid gasket

C Camshaft axis area

Center line of rear timing chain case liquid gasket groove

5 (0.20)

Center line of liquid gasket

2 (0.08)

Joint portion of cylinder head and camshaft bracket (No. 1)

D 2.6 - 3.6 (0.102 - 0.142) dia.

Run along bolt hole outer side

Protrusions at beginning and end of liquid gasket

*: Apply liquid gasket to the chamfered surface between camshaft bracket (No. 1) and cylinder head.

▨ : Apply Genuine RTV Silicone Sealant or equivalent. Refer to GI section.

Unit: mm (in)

Rear timing chain case RTV application—4.0L engine

09482_PATH_G0024

b. Insert the plunger into tensioner body by pressing slack guide.

c. Keep the slack guide pressed and hold plunger in by pushing the stopper pin through the tensioner body hole and plunger groove.

d. Remove bolts and remove timing chain tensioner (primary).

13. Remove the internal chain guide, tension guide and slack guide.

14. Remove the primary timing chain.

15. Remove the crankshaft sprocket.

16. Remove the secondary timing chain and camshaft sprockets as follows:

a. Attach a suitable stopper pin to the right and left timing chain tensioners.

b. Remove the camshaft sprocket bolts.

c. Turn the camshaft slight to secure the slackness of the timing chain on the tensioner side.

d. Insert a 0.020 inch (0.5 mm) thick metal plate between the timing chain and timing chain tensioner plunger.

e. Remove the secondary timing chain together with the camshaft sprockets.

17. Remove the water pump.

18. Remove the rear timing chain case by loosening the bolts in the reverse order of the installation sequence.

To install:

19. Install new O-rings to the cylinder block, cylinder head and No. 1 camshaft bracket.

20. Apply RTV to the back side of the rear timing chain case as shown.

21. Align the rear timing chain case with the dowel pins on the engine block and install the rear case.

22. Tighten the bolts in sequence to 9 ft. lbs. (13 Nm).

23. Install the water pump with new O-rings.

24. Ensure that the No. 1 cylinder is at TDC.

25. Push in the plunger of the secondary timing chain tensioners and secure with a stopper pin.

26. Align the mating marks on the secondary timing chain with the camshaft sprockets marks and install.

27. Align dowel pin and pin hole on camshafts with the groove and dowel pin on sprockets, and install them.

28. On the intake side, align pin hole on the small diameter side of the camshaft front end with dowel pin on the back side of camshaft sprocket, and install them.

29. On the exhaust side, align dowel pin on camshaft front end with pin groove on camshaft sprocket, and install them.

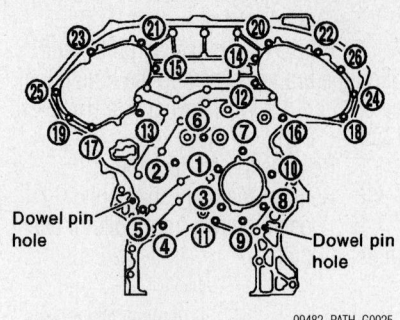

09482_PATH_G0025

Rear timing chain case torque sequence— 4.0L engine

30. In case that positions of each mating mark and each dowel pin are not fit on mating parts, make fine adjustment to the position holding the hexagonal portion on camshaft with wrench or equivalent.

31. Bolts for camshaft sprockets must be tightened in the next step. Tightening them by hand is enough to prevent the dislocation of dowel pins.

➡**It may be difficult to visually check the dislocation of mating marks during and after installation. To make the matching easier, make a mating mark on the top of sprocket teeth and its extended line in advance with paint.**

32. Ensure the mating marks are aligned and tighten camshaft sprocket bolts.

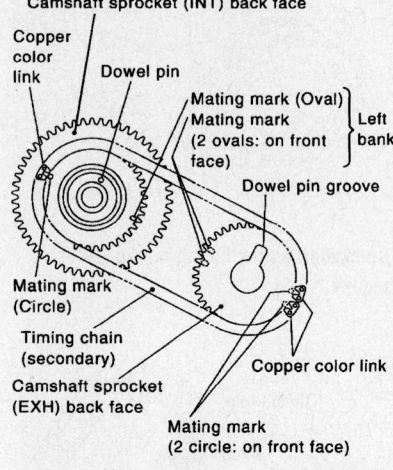

09482_PATH_G0026

Secondary timing chain mating marks— 4.0L engine

33. Pull out the stopper pins out from the secondary timing chain tensioners.

34. Install the tension guide.

35. Install the crankshaft sprocket.

36. Install the primary timing chain so the mating mark (punched) on camshaft sprocket is aligned with the copper color link on timing chain, while the mating mark (notched) on crankshaft sprocket is aligned with the gold one on timing chain, as shown.

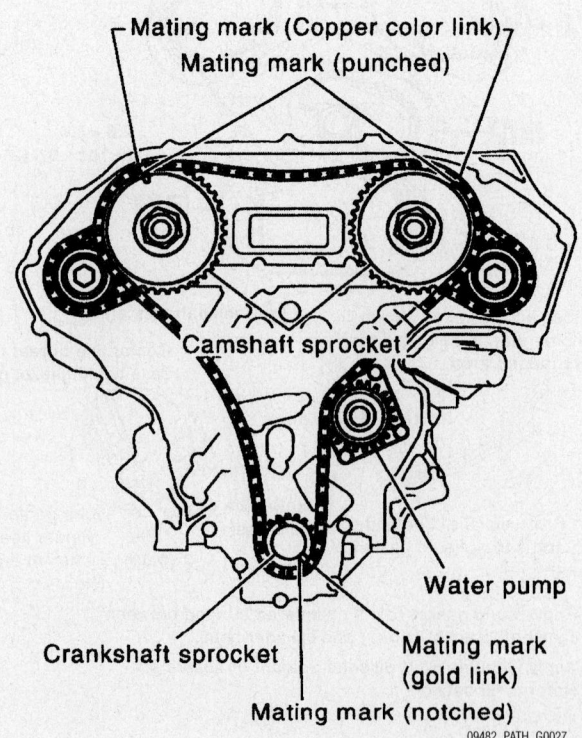

09482_PATH_G0027

Primary timing chain mating marks—4.0L engine

➡**When it is difficult to align mating marks of the primary timing chain with each sprocket, gradually turn camshaft using wrench on the hexagonal portion to align it with the mating marks.**

❊❊ WARNING

During alignment, be careful to prevent dislocation of mating mark alignments of the secondary timing chains.

37. Install internal chain guide, slack guide and primary timing chain tensioner.

❊❊ CAUTION

Do not overtighten slack guide bolts. It is normal for a gap to exist under the bolt seats when bolts are tightened to specification.

38. Install new O-rings on the rear timing case.
39. Install or connect the following:
   • Front oil seal
   • Water pump cover
   • Chain tensioner cover
40. Apply a continuous bead of RTV to the back side of the front timing chain case as shown.
41. Fit lower end of front cover tightly onto top face of the upper oil pan. From the fitting point, make entire front timing chain case contact rear timing chain case completely.
42. Tighten the front cover bolts in sequence shown as follows:
   a. Bolts 1–5: 41 ft. lbs. (55 Nm)
   b. Bolts 6–25: 9 ft. lbs. (13 Nm)
43. Install the two front bolts of the upper oil pan and tighten to 16 ft. lbs. (22 Nm).
44. Install new seal rings in the shaft grooves of the intake valve timing control covers.
45. Apply a bead of RTV to the back side of the timing control covers.
46. Install the valve timing control covers by aligning the dowel pins on the front cover with the holes on the timing control covers. Tighten the bolts in sequence as shown.
47. Install or connect the following:
   • Crankshaft pulley and tighten to 33 ft. lbs. (44 Nm) plus 60 degrees
   • Oil pans
   • Valve covers
   • PCV hoses from valve covers
   • Ignition coil
   • Engine wiring harness brackets from the cylinder head
   • Collared O-rings from the front cover

Front timing chain case

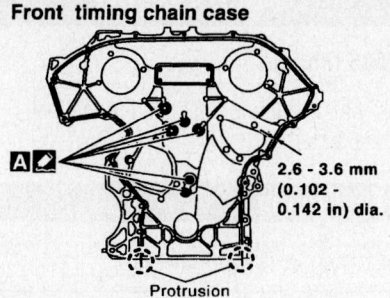

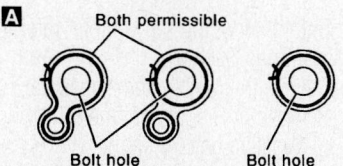

Liquid gasket protrusion away from bolt hole

✎ : Apply Genuine RTV silicone sealant or equivalent. Refer to GI section.

09482_PATH_G0028

Front timing chain case RTV sealant location—4.0L engine

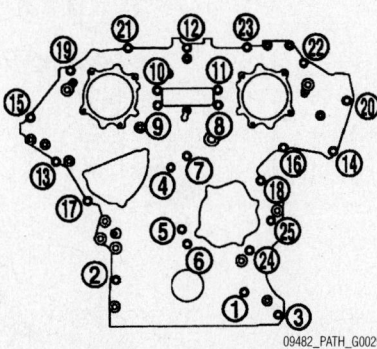

09482_PATH_G0029

Front cover torque sequence—4.0L engine

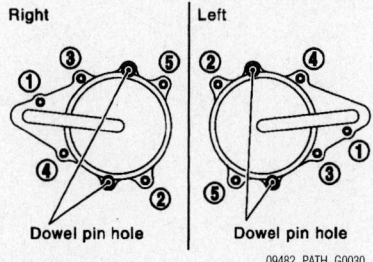

09482_PATH_G0030

Intake valve timing control cover torque sequence—4.0L engine

• Intake valve timing control covers
• Idler pulley bracket
• Coolant bypass hose and bracket
• Alternator
• Power steering oil pump bracket
• Power steering oil pump
• Accessory drive belts
• Engine wiring harness brackets from the front cover
• Cooling fan assembly
• Air intake assembly
• Negative battery cable

48. Refill the engine oil to the correct level.
49. Fill the cooling system to the correct level.
50. Start the engine and check for leaks.

Piston and Ring

POSITIONING

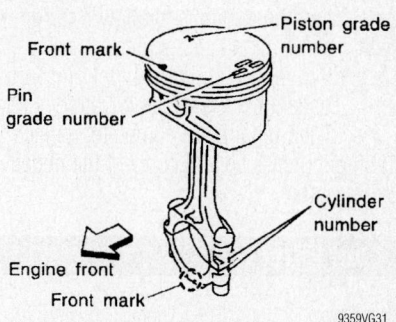

9359VG31

Piston and connecting rod positioning–3.5L

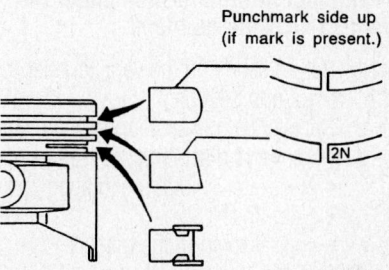

9359VG32

Piston ring positioning–3.5L, 4.0L second ring is marked with a "R"

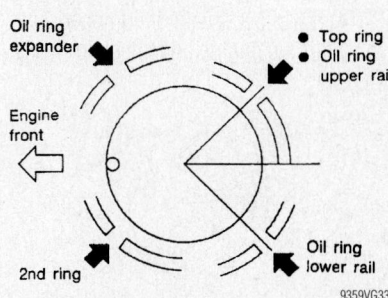

9359VG33

Piston ring positioning–3.5L

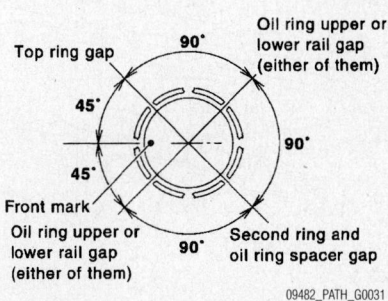

09482_PATH_G0031

Piston ring positioning—4.0L

FUEL SYSTEM

Fuel System Pressure

RELIEVING

1. Before servicing the vehicle, refer to the Precautions Section.
2. Remove the fuel pump fuse from the panel.
3. Start the engine and allow it to run until it stalls. Crank the engine for a few seconds to relieve additional fuel pressure.
4. Disconnect the negative battery cable.
5. When repairs are complete, replace the fuel pump fuse and connect the negative battery cable.

Fuel Filter

REMOVAL & INSTALLATION

2002–2004 Models

➡ **The fuel filter is located under the vehicle near the fuel tank.**

1. Before servicing the vehicle, refer to the Precautions Section.
2. Relieve the fuel system pressure.
3. Remove or disconnect the following:
   • Fuel filter shield, if equipped
   • Fuel lines
   • Fuel filter from the bracket

To install:

4. Install or connect the following:
   • Fuel filter to the bracket
   • Fuel lines
   • Fuel filter shield, if equipped
5. Start the engine and check for leaks.

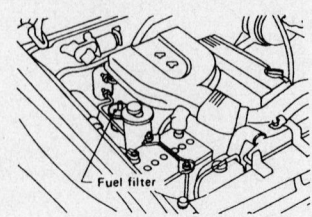

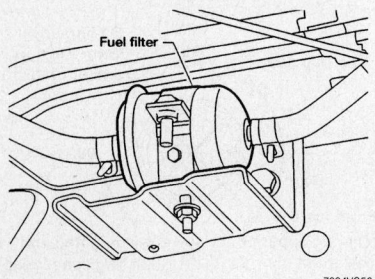

7924VG56

Typical fuel filter locations

2005 Models

The fuel filter is integrated in the fuel pump assembly.

Fuel Pump

REMOVAL & INSTALLATION

1. Before servicing the vehicle, refer to the Precautions Section.
2. Relieve the fuel system pressure.
3. Remove or disconnect the following:
   • Negative battery cable
   • Access panel behind the rear seat
   • Fuel lines
   • Fuel pump and gauge harness connectors
   • Fuel gauge sender
   • Fuel pump

To install:

4. Install or connect the following:
   • Fuel pump
   • Fuel gauge sender. Tighten the screws to 17–23 inch lbs. (2.0–2.5 Nm).
   • Fuel pump and gauge harness connectors
   • Fuel lines
   • Access panel
   • Negative battery cable
5. Start the engine and check for leaks.

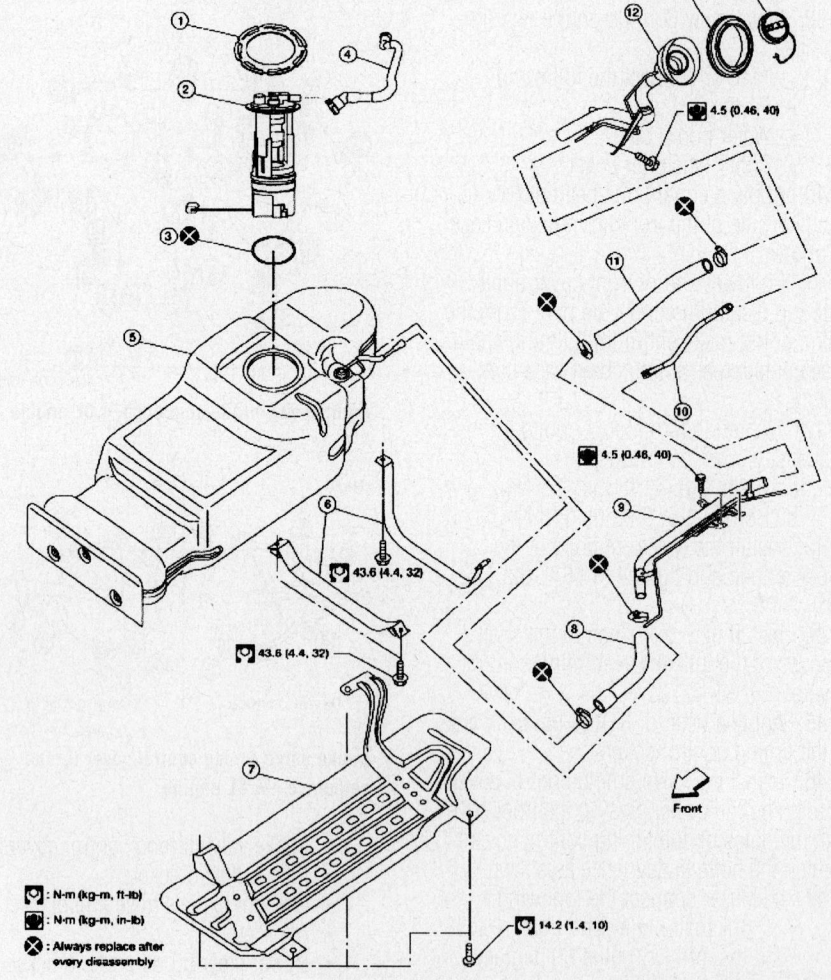

| : N·m (kg-m, ft-lb) |
| : N·m (kg-m, in-lb) |
| ✕ : Always replace after every disassembly |

1.	Lock ring	2.	Fuel level sensor, fuel filter, and fuel pump assembly	3.	Fuel level sensor, fuel filter, and fuel pump assembly O-ring
4.	EVAP hose	5.	Fuel tank	6.	Fuel tank straps
7.	Fuel tank shield	8.	Lower fuel filler hose	9.	Fuel filler pipe and vent pipe
10.	Vent hose	11.	Upper fuel filler hose	12.	Fuel filler pipe and cup
13.	Fuel filler hose grommet	14.	Fuel filler cap		

09482_PATH_G0032

Exploded view of the fuel system components—2005 Models

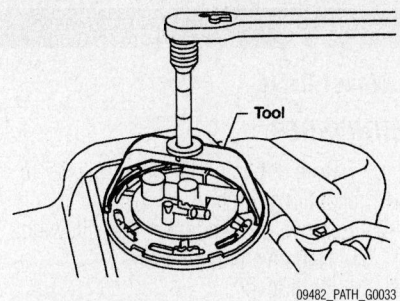

09482_PATH_G0033

Removing the fuel pump assembly—4.0L engine

2005 Models

1. Before servicing the vehicle, refer to the Precautions Section.
2. Relieve the fuel system pressure.
3. Remove or disconnect the following:
   - Negative battery cable
   - Fuel filler hose from the fuel tank
   - EVAP hose
   - Vent pipe
   - Fuel tank shield
   - Driveshaft
4. Support the fuel tank with a suitable jack.
5. Remove fuel tank strap bolts and lower the tank.

※※ WARNING

Take care when lowering the fuel tank not to damage the fuel supply hose, level sensor and pump assembly connector.

6. Disconnect the fuel pump assembly electrical connector and fuel supply hose.
7. Remove the fuel pump assembly lock ring using Special Tool J-45722.
8. Installation is the reverse order of removal.
9. Turn the ignition on, but do not start the engine, to check for leaks.

Fuel Injectors

REMOVAL & INSTALLATION

1. Before servicing the vehicle, refer to the Precautions Section.
2. Release fuel pressure to zero.
3. Remove intake manifold collector.

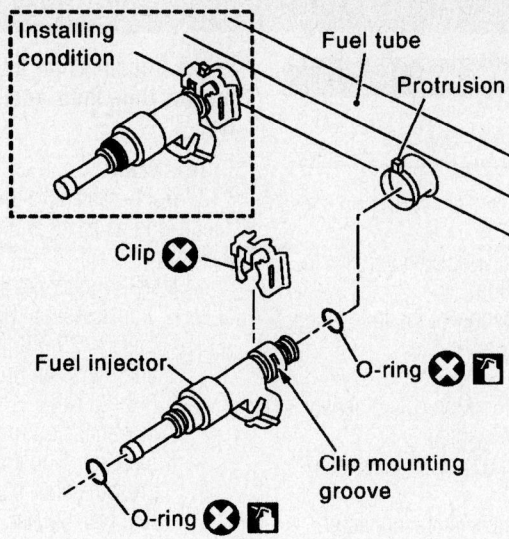

❌ : Always replace after every disassembly.

🔧 : Lubricate with new engine oil.

09482_PATH_G0034

Exploded view of the fuel injector installation

4. Remove fuel tube assemblies in numerical sequence as shown in the figure at left.
5. Expand and remove clips securing fuel injectors.
6. Extract fuel injectors straight from fuel tubes.

➡**Be careful not to damage injector nozzles during removal. Do not bump or drop fuel injectors.**

7. Carefully install O-rings, including the one used with the pressure regulator. Lubricate O-rings with a smear of engine oil.

➡**Be careful not to damage O-rings with service tools, finger nails or clips. Do not expand or twist O-rings. Discard old clips; replace with new ones.**

8. Position clips in grooves on fuel injectors. Make sure that protrusions of fuel injectors are aligned with cutouts of clips after installation.
9. Align protrusions of fuel tubes with those of fuel injectors. Insert fuel injectors straight into fuel tubes.

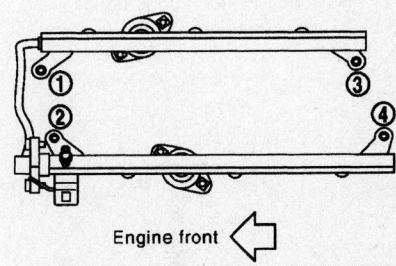

Engine front

09482_PATH_G0035

Fuel rail torque sequence

10. After properly inserting fuel injectors, check to make sure that fuel tube protrusions are engaged with those of fuel injectors, and that flanges of fuel tubes are engaged with clips.
11. Tighten fuel tube assembly mounting nuts in sequence in two stages:
    a. Step 1: 84–96 inch lbs.
    b. Step 2: 16–19 ft. lbs.
12. Install all parts removed in reverse order of removal.
13. Turn the ignition to the **ON**, but do not start, to check for leaks.

DRIVE TRAIN

Manual Transmission

REMOVAL & INSTALLATION

2 Wheel Drive

1. Before servicing the vehicle, refer to the Precautions Section.
2. Remove or disconnect the following:
   - Negative battery cable
   - Shift lever
   - Crankshaft Position (CKP) sensor
   - Clutch slave cylinder
   - Vehicle Speed (VSS) sensor connector
   - Back-up lamp switch connector
   - Park/Neutral Position (PNP) switch connector
   - Rear Heated Oxygen (HO2S) sensor connector
   - Starter motor
   - Driveshaft
   - Exhaust mounting bracket
   - Transmission mount and crossmember. Support the transmission.
   - Transmission flange bolts
   - Transmission

➡ The transmission flange bolts vary in length. Note their positions for assembly.

To install:

3. Apply sealant to the transmission flange, engine block and engine rear plate as shown.
4. Install or connect the following:
   - Transmission. Tighten the large bolts to 29–36 ft. lbs. (39–49 Nm) and the small bolts to 12–16 ft. lbs. (16–22 Nm).
   - Transmission mount and crossmember. Tighten the mount and crossmember fasteners to 30–38 ft. lbs. (41–52 Nm).
   - Exhaust mounting bracket
   - Driveshaft
   - Starter motor
   - HO2S sensor connector
   - PNP switch connector
   - Back-up lamp switch connector
   - VSS sensor connector
   - Clutch slave cylinder
   - CKP sensor
   - Shift lever
   - Negative battery cable

4 Wheel Drive

PATHFINDER

1. Before servicing the vehicle, refer to the Precautions Section.
2. Remove or disconnect the following:
   - Negative battery cable
   - Shift lever
   - Transfer case select lever
   - Crankshaft Position (CKP) sensor
   - Clutch slave cylinder
   - Vehicle Speed (VSS) sensor connector
   - Back-up lamp switch connector
   - Park/Neutral Position (PNP) switch connector
   - Rear Heated Oxygen (HO2S) sensor connector
   - Starter motor
   - Front and rear driveshafts
   - Exhaust front pipes
   - Exhaust center pipe
   - Transmission mount and crossmember. Support the transmission.
   - Transmission flange bolts
   - Transmission

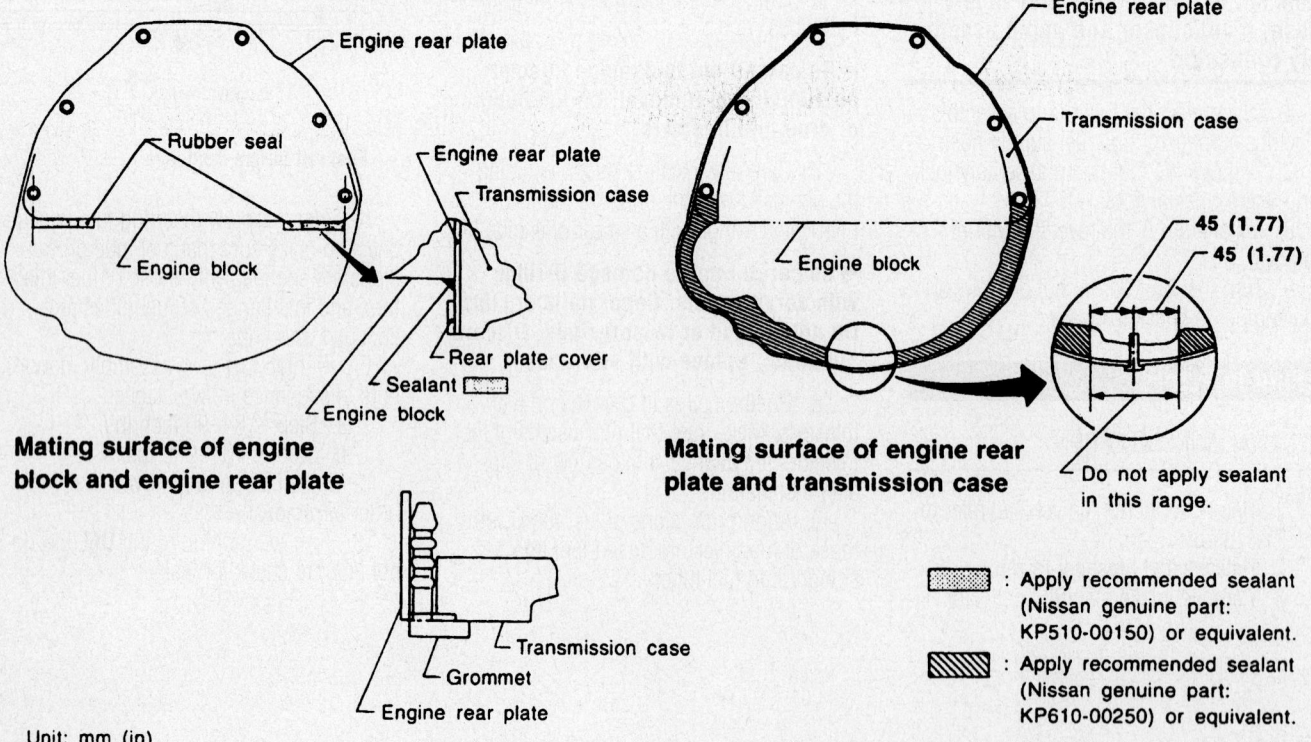

Mating surface of engine block and engine rear plate

Mating surface of engine rear plate and transmission case

45 (1.77)
45 (1.77)

Do not apply sealant in this range.

⬚ : Apply recommended sealant (Nissan genuine part: KP510-00150) or equivalent.

▨ : Apply recommended sealant (Nissan genuine part: KP610-00250) or equivalent.

Unit: mm (in)

7924VG61

Apply sealant to the indicated areas between the engine block, transmission and engine rear plate—4 Wheel Drive shown

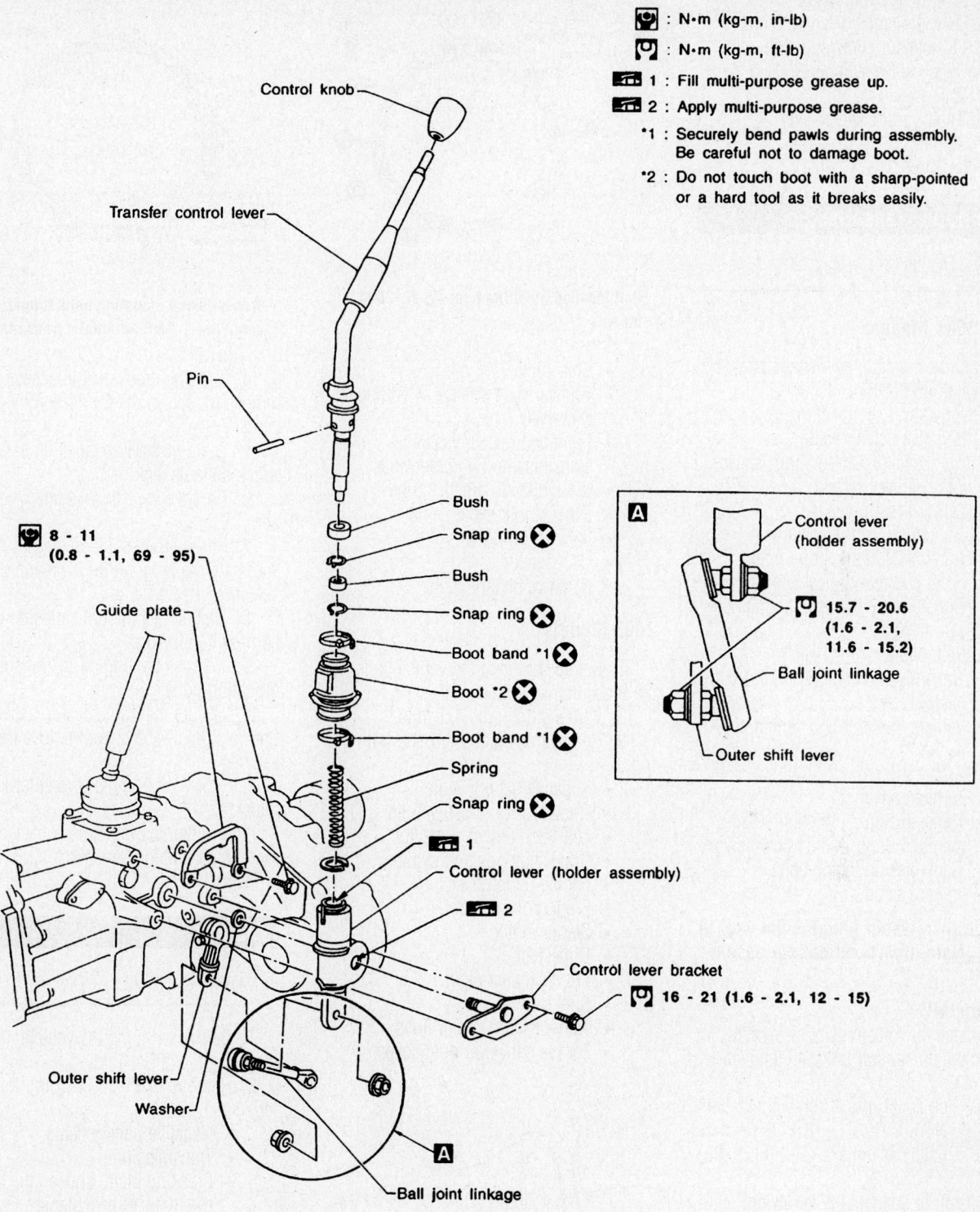

: N•m (kg-m, in-lb)

: N•m (kg-m, ft-lb)

1 : Fill multi-purpose grease up.

2 : Apply multi-purpose grease.

*1 : Securely bend pawls during assembly.
     Be careful not to damage boot.

*2 : Do not touch boot with a sharp-pointed
     or a hard tool as it breaks easily.

Control knob
Transfer control lever
Pin
8 - 11 (0.8 - 1.1, 69 - 95)
Guide plate
Bush
Snap ring
Bush
Snap ring
Boot band *1
Boot *2
Boot band *1
Spring
Snap ring
1
Control lever (holder assembly)
2
Control lever bracket
16 - 21 (1.6 - 2.1, 12 - 15)
Outer shift lever
Washer
Ball joint linkage

Control lever (holder assembly)
15.7 - 20.6 (1.6 - 2.1, 11.6 - 15.2)
Ball joint linkage
Outer shift lever

Exploded view of the transfer case shifter lever and related components—Pathfinder 4WD

➡**The transmission flange bolts vary in length. Note their positions for assembly.**

To install:

3. Apply sealant to the transmission flange, engine block, and engine rear plate as shown.

4. Install or connect the following:
• Transmission. Tighten the large bolts to 29–36 ft. lbs. (39–49 Nm) and the small bolts to 22–29 ft. lbs. (29–39 Nm).
• Transmission mount and crossmember. Tighten the mount and crossmember fasteners to 30–38 ft. lbs. (41–52 Nm).
• Exhaust center pipe
• Exhaust front pipes
• Front and rear driveshafts
• Starter motor
• HO2S sensor connector

- PNP switch connector
- Back-up lamp switch connector
- VSS sensor connector
- Clutch slave cylinder
- CKP sensor
- Transfer case select lever
- Shift lever
- Negative battery cable

Automatic Transmission

REMOVAL & INSTALLATION

2002–2004 Models

1. Before servicing the vehicle, refer to the Precautions Section.
2. Remove or disconnect the following:
   - Negative battery cable
   - Crankshaft Position (CKP) sensor
   - Exhaust front pipes
   - Exhaust rear pipes
   - Transmission dipstick tube
   - Transmission oil cooler lines
   - Front and rear driveshafts
   - Transfer case linkage, if equipped with 4WD
   - Shift cable
   - Transmission control harness connectors
   - Vehicle Speed (VSS) sensor connector
   - Starter motor
   - Torque converter
   - Transmission mount and crossmember. Support the transmission.
   - Transmission flange bolts
   - Transmission

➡ **The transmission flange bolts vary in length. Note their positions for assembly.**

To install:

3. Install the transmission assembly. Refer to the illustration for bolt identification.
   a. Tighten #1 and #3 bolts to 52–59 (70–80 Nm).
   b. Tighten #2 bolts to 22–29 (29–39 Nm).
4. Install or connect the following:
   - Transmission mount and crossmember. Tighten the mount and crossmember fasteners to 30–38 ft. lbs. (41–52 Nm).
   - Torque converter. Tighten the bolts to 33–43 ft. lbs. (44–59 Nm).
   - Starter motor
   - VSS sensor connector
   - Transmission control harness connectors

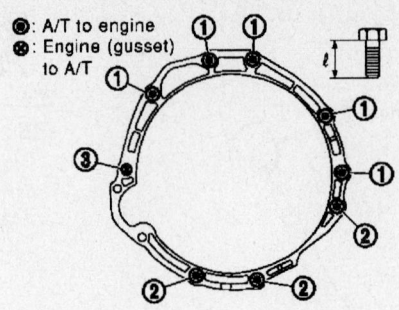

Bolt location identification—2002–2004 models

- Shift cable
- Transfer case linkage, if equipped with 4WD
- Front and rear driveshafts
- Transmission oil cooler lines
- Transmission dipstick tube
- Exhaust rear pipes
- Exhaust front pipes
- CKP sensor
- Negative battery cable

2005 Models

1. Before servicing the vehicle, refer to the Precautions Section.
2. Partially drain the A/T fluid.
3. Remove or disconnect the following:

- Negative battery cable
- Transmission dipstick tube
- Left-hand fender protector
- Crankshaft position sensor
- Engine undercover
- Front crossmember
- Starter motor
- Driveshaft
- Front exhaust pipes
- Shift cable
- Transmission cooler lines
- Torque converter dust cover

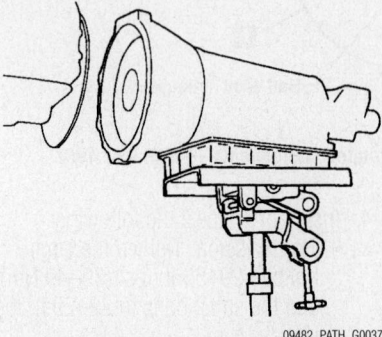

Secure the transmission to the Special Tool to remove—2005 Model shown

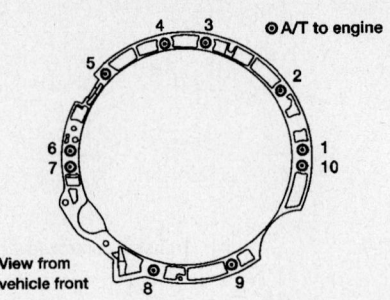

Transmission mounting bolts torque sequence—2005 Automatic transmission

4. Turn the crankshaft to access and remove the four bolts for the torque converter.
5. Support the transmission assembly with a suitable jack.
6. Remove the transmission crossmember.
7. Tilt the transmission slightly to gain access to the air breather hose and disconnect it.
8. Disconnect the transmission wiring harness connectors.
9. Remove the engine-to-transmission mounting bolts.
10. Secure the transmission assembly to Special Tool J-47002 and remove from the vehicle.
11. Installation is the reverse order of removal.
12. Tighten the transmission-to-engine mounting bolts in sequence to 55 ft. lbs. (74 Nm).

Clutch

REMOVAL & INSTALLATION

1. Before servicing the vehicle, refer to the Precautions Section.
2. Remove or disconnect the following:

- Negative battery cable
- Transmission
- Pressure plate. Loosen the bolts evenly in ½ turn steps.
- Clutch disc

To install:

3. Install or connect the following:
- Clutch disc and pressure plate. Tighten the pressure plate bolts evenly in ½ turns to 16–22 ft. lbs. (22–29 Nm).
- Transmission
- Negative battery cable

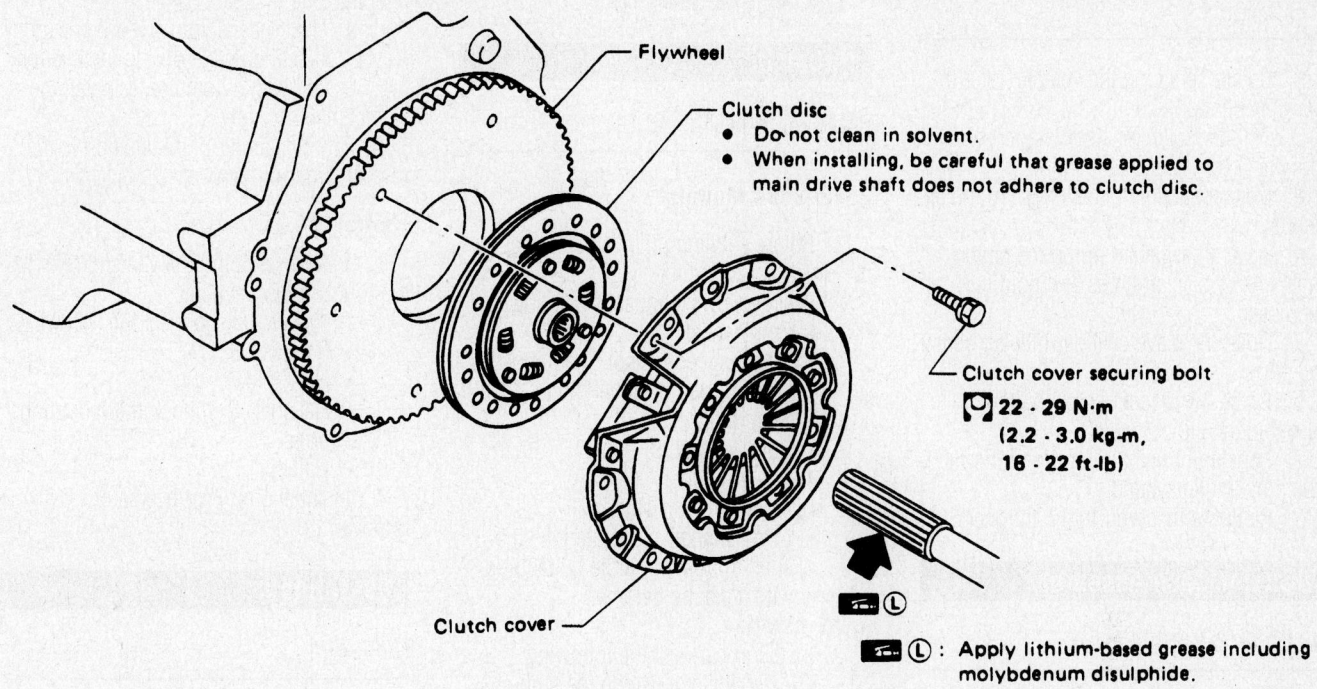

Flywheel

Clutch disc
- Do not clean in solvent.
- When installing, be careful that grease applied to main drive shaft does not adhere to clutch disc.

Clutch cover securing bolt
22 - 29 N·m
(2.2 - 3.0 kg-m,
16 - 22 ft-lb)

Clutch cover

: Apply lithium-based grease including molybdenum disulphide.

7924VG63

Exploded view of the pressure plate and clutch disc and related components—all models

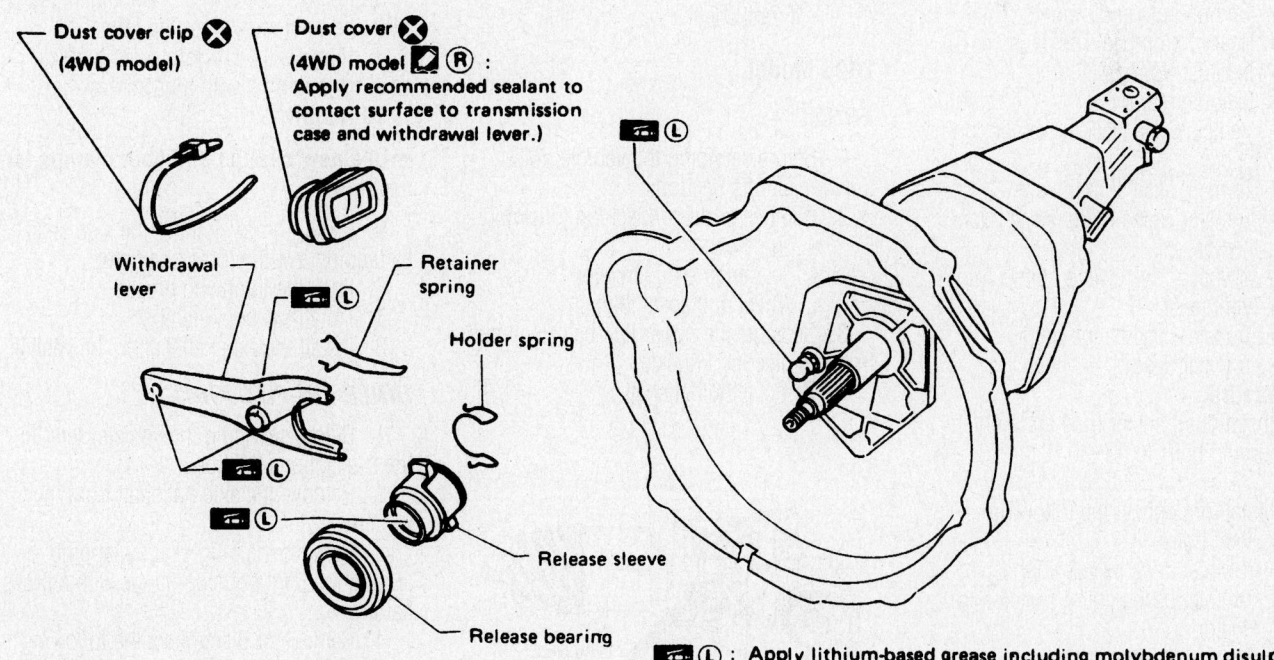

Dust cover clip ⊗
(4WD model)

Dust cover ⊗
(4WD model) ®:
Apply recommended sealant to contact surface to transmission case and withdrawal lever.)

Withdrawal lever

Retainer spring

Holder spring

Release sleeve

Release bearing

: Apply lithium-based grease including molybdenum disulphide

7924VG64

Clutch release mechanism exploded view—all models

Hydraulic Clutch System

BLEEDING

1. Before servicing the vehicle, refer to the Precautions Section.
2. Fill the clutch master cylinder reservoir with fresh clean brake fluid.
3. Connect a clear plastic hose to the air bleeder.
4. Have an assistant pump the clutch pedal slowly several times and hold it depressed.
5. Open the slave cylinder bleeder screw and allow air to escape.
6. Close the bleeder screw before releasing the clutch pedal.
7. Repeat until all air is purged from the clutch hydraulic system.
8. Refill the reservoir to the full mark.

Transfer Case Assembly

REMOVAL & INSTALLATION

1. Before servicing the vehicle, refer to the Precautions Section.
2. Drain the transfer case fluid.
3. Support the transfer case and transmission assembly with suitable jacks.
4. Remove or disconnect the following:
   - Negative battery cable
   - Front and rear driveshafts
   - Torsion bars and mounts
   - Rear torsion bar crossmember
   - Exhaust front pipes
   - Exhaust rear pipes
   - Vehicle Speed (VSS) sensor connector
   - Transfer case shift linkage
   - Transfer case neutral switch connector
   - 4 wheel drive switch connector
   - Vent hose
   - Transfer case flange bolts
   - Transfer case

To install:
5. Install the transfer case and tighten the mounting bolts to 23–30 ft. lbs. (31–41 Nm).
6. Install or connect the following:
   - Vent hose
   - 4 wheel drive switch connector
   - Transfer case neutral switch connector
   - Transfer case shift linkage
   - VSS sensor connector
   - Exhaust rear pipes
   - Exhaust front pipes
   - Rear torsion bar crossmember
   - Torsion bars and mounts
   - Front and rear driveshafts
   - Negative battery cable

Halfshaft

REMOVAL & INSTALLATION

2002–2004 Models

1. Before servicing the vehicle, refer to the Precautions Section.
2. Remove or disconnect the following:
   - Front wheel
   - Wheel speed sensor, if equipped
   - Locking hub or drive flange
   - Snapring
   - Spindle washer
   - Thrust washer
   - Inner CV-joint bolts
   - Axle halfshaft. Separate the stub shaft from the spindle by tapping with a plastic hammer.

To install:
3. Install or connect the following:
   - Axle halfshaft. Guide the stub shaft into the spindle and tighten the inner CV-joint bolts to 25–33 ft. lbs. (34–44 Nm).
   - Thrust washer
   - Spindle washer
   - Snapring
   - Locking hub or drive flange
   - Wheel speed sensor, if equipped
   - Front wheel

2005 Model

FRONT

1. Before servicing the vehicle, refer to the Precautions Section.
2. Remove or disconnect the following:
   - Front wheel
   - Rear engine undercover
   - Wheel hub assembly
3. Separate the upper link ball joint stud from the steering knuckle.
4. Remove the halfshaft.

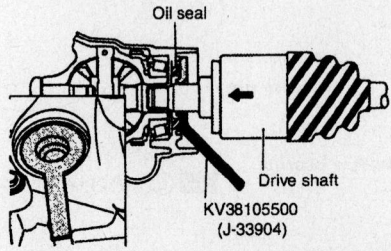

KV38105500
(J-33904)

09482_PATH_G0039

Use the special tool to prevent damage to the oil seal

To install:
5. Install the halfshaft using a new differential side oil seal.
   a. Use Special Tool J-33904 to prevent oil seal damage during installation.
   b. Tap with a hammer to install the halfshaft securely.
6. The remainder of installation is the reverse order of removal.

REAR

1. Before servicing the vehicle, refer to the Precautions Section.
2. Remove or disconnect the following:
   - Rear wheel
   - Hub spindle nut
   - Halfshaft-to-differential mounting bolts
   - Halfshaft
3. Installation is the reverse order of removal.

CV-Joints

OVERHAUL

2002–2004 Models

OUTER CV-JOINT

1. Before servicing the vehicle, refer to the Precautions Section.
2. Remove the axle halfshaft from the vehicle.
3. Remove the CV-joint boot clamps and push the boot away from the joint.
4. Remove the CV-joint from the axle shaft by tapping it with a brass hammer.

To install:

➡ **Use new circlips and boot clamps for assembly.**

5. Install the CV-joint to the axle shaft by tapping it with a brass hammer.
6. Pack the joint with grease.
7. Install the boot clamps.
8. Install the axle halfshaft to the vehicle.

INNER TRI-POT JOINT

1. Before servicing the vehicle, refer to the Precautions Section.
2. Remove the axle halfshaft from the vehicle.
3. Remove the plug seal by tapping around the joint housing flange with a brass hammer.
4. Remove or disconnect the following:
   - CV-joint boot clamps
   - Snapring
   - Spider assembly
   - CV-joint housing
   - CV-joint boot

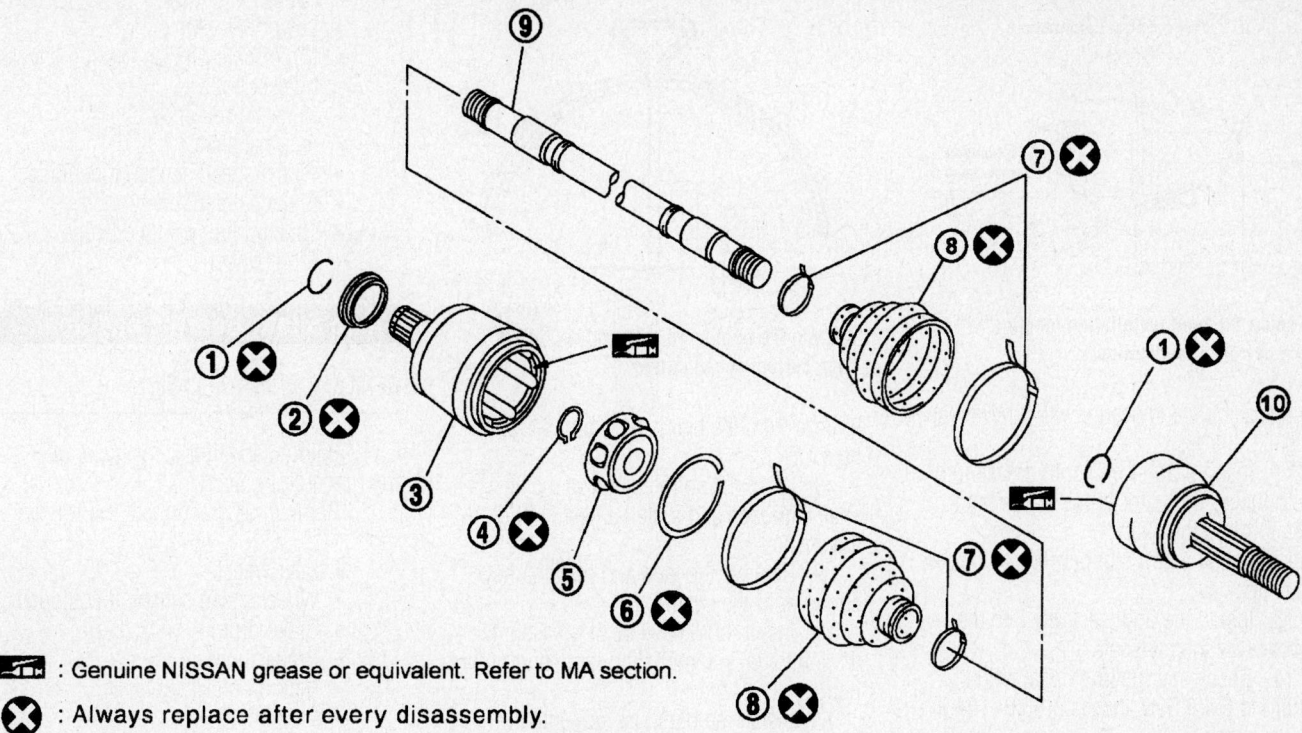

Exploded view of the halfshaft assembly—2005 Model shown

: Genuine NISSAN grease or equivalent. Refer to MA section.

❌ : Always replace after every disassembly.

1. Circlip
2. Dust cover
3. Housing
4. Snap ring
5. Ball cage, steel ball and inner race assembly
6. Stopper ring
7. Boot band
8. Boot
9. Shaft
10. Joint sub-assembly

09482_PATH_G0040

To install:

➡ **Use new snaprings and plug seals for assembly.**

5. Install or connect the following:
 • CV-joint boot
 • CV-joint housing
 • Spider assembly
 • Snapring. Pack the joint with grease.
 • CV-joint boot clamps
 • Plug seal
6. Install the axle halfshaft to the vehicle.

2005 Model

INNER TRI-POT JOINT

1. Before servicing the vehicle, refer to the Precautions Section.
2. Remove the halfshaft and mount in a vise.
3. Remove the boot bands and slide the boot back.
4. Matchmark the joint housing and halfshaft before separating.

Remove the stopper ring with a screwdriver as shown—2005 Pathfinder shown

5. Remove the stopper ring with a flat-bladed screwdriver and pull off the housing.
6. Remove the snap ring, then inner race assembly from the halfshaft.
7. Remove the boot from the shaft.
8. Remove the circlip and dust cover from the housing.

To install:
9. Wrap the serrated part of the shaft with tape and install the boot band and boot to the shaft.
10. Remove the tape and install the inner

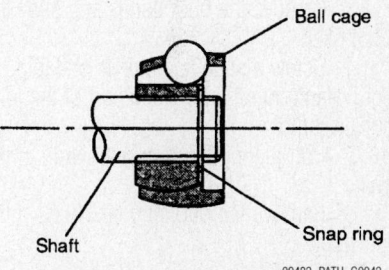

Inner race assembly—2005 Pathfinder shown

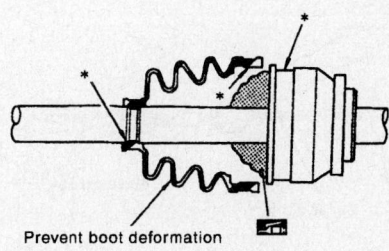

Prevent boot deformation

Joint housing grease locations

09482_PATH_G0043

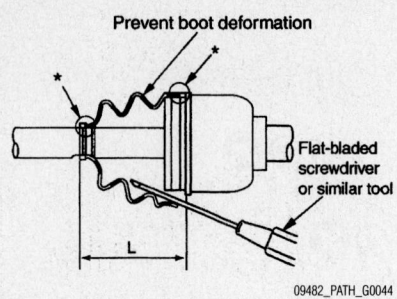

Ensure the boot installation length ("L") is the correct specification.

race assembly and secure with a new snap ring.

11. Insert 4.23–4.94 oz. of grease into the housing in the locations shown and install it into the shaft.

12. Install a new stopper ring onto the housing.

13. Install the boot securely into the grooves on the halfshaft

14. Check that the boot installation length is 6.45–6.47 inches (163.9–164.3 mm).

15. Secure the boot with new boot bands.

16. Install a new circlip and dust cover.

OUTER CV-JOINT

1. Before servicing the vehicle, refer to the Precautions Section.

2. Remove the halfshaft and mount in a vise.

3. Remove the boot bands and slide the boot back.

4. Screw a slide hammer or suitable equivalent onto the threaded part of the joint sub-assembly.

5. Pull the joint sub-assembly off of the shaft.

6. Remove the boot and circlip from the halfshaft.

To install:

7. Install grease into the joint sub-assembly serration hole until grease begins

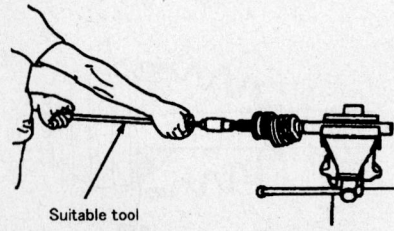

Use a slide hammer to remove to remove joint sub-assembly.

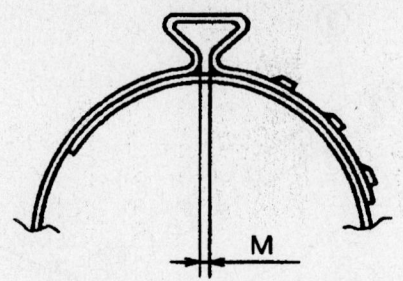

Crimp the boot band so distance "M" is the correct specification.

to ooze from the ball groove and serration hole.

8. Wrap the serrated end of the half-shaft with tape and install a new boot onto the shaft.

9. Remove the tape and install a new circlip onto the shaft.

10. Insert 4.01–4.76 oz of grease into the joint sub-assembly and large end of the boot.

11. Install the boot securely into the shaft grooves. Ensure the boot installation length is 5.32 inches (135.1 mm).

12. Secure both ends of the boot with new bands. Ensure boot band crimp is 0.39–1.57 (1–4 mm).

13. After installation, rotate the boot to ensure proper positioning.

Spindle Bearings

REMOVAL, PACKING AND INSTALLATION

1. Before servicing the vehicle, refer to the Precautions Section.

2. Remove or disconnect the following:

- Front wheel
- Locking hub or drive flange
- Brake caliper and support
- Wheel speed sensor, if equipped
- Axle halfshaft
- Outer tie rod ends
- Upper ball joint or steering knuckle bracket bolts
- Lower ball joint
- Steering knuckle
- Inner seal
- Thrust washer
- Spindle bearing

To install:

3. Install or connect the following:
- Spindle bearing. Coat the bearing with multi-purpose grease.
- Thrust washer
- Inner seal

- Steering knuckle
- Lower ball joint
- Upper ball joint or steering knuckle bracket bolts
- Outer tie rod ends
- Axle halfshaft
- Wheel speed sensor, if equipped
- Brake caliper and support
- Locking hub or drive flange
- Front wheel

Axle Shaft, Bearing and Seal

REMOVAL & INSTALLATION

1. Before servicing the vehicle, refer to the Precautions Section.

2. Remove or disconnect the following:

- Rear wheel
- Wheel speed sensor, if equipped
- Brake drum
- Brake shoes
- Parking brake cable
- Brake fluid line
- Bearing cage and backing plate bolts
- Axle shaft assembly
- Axle seal
- Wheel speed sensor rotor, if equipped
- Lockwasher
- Bearing locknut
- Flat washer
- Wheel bearing
- Wheel bearing cage grease seal

To install:

➡ **Use new lockwashers, seals and bearings for assembly.**

3. Install or connect the following:
- Wheel bearing cage grease seal
- Wheel bearing
- Flat washer
- Bearing locknut
- Lockwasher
- Wheel speed sensor rotor, if equipped
- Axle seal
- Axle shaft assembly
- Bearing cage and backing plate bolts
- Brake fluid line
- Parking brake cable
- Brake shoes
- Brake drum
- Wheel speed sensor, if equipped
- Rear wheel

4. Bleed the rear brakes and check the rear axle lubricant level.

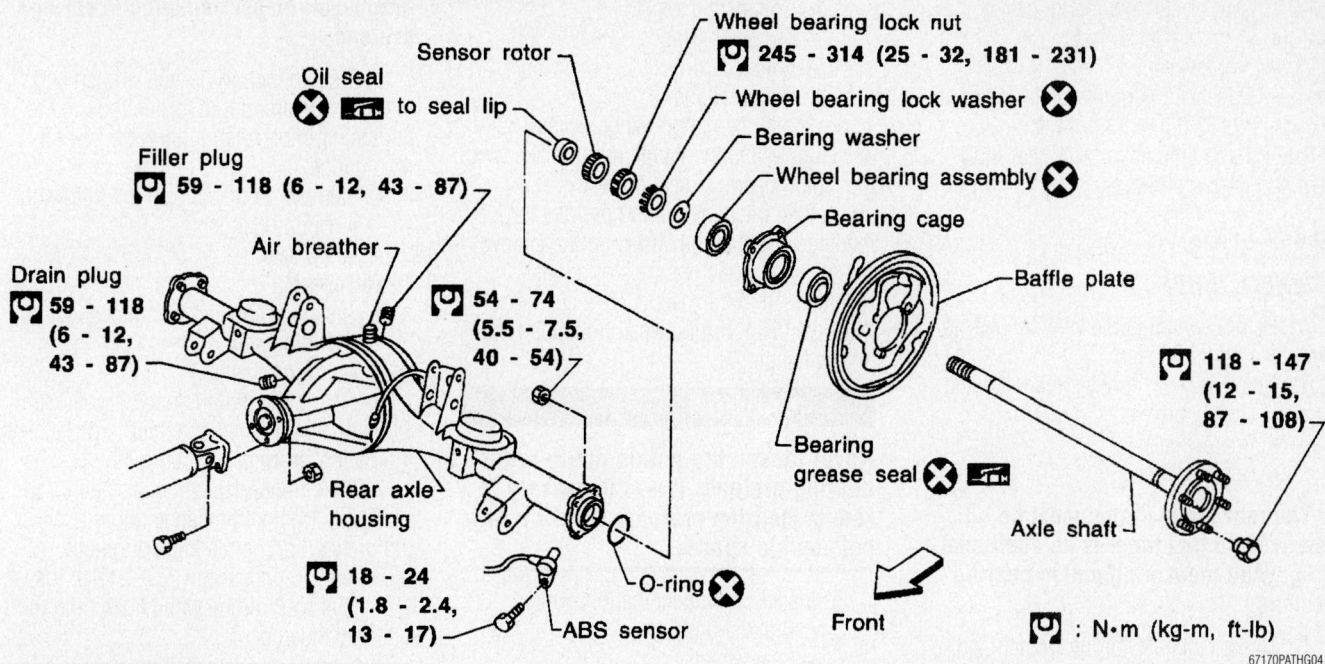

Sensor rotor

Wheel bearing lock nut
🔧 245 - 314 (25 - 32, 181 - 231)

Oil seal ❌ 🔧 to seal lip

Wheel bearing lock washer ❌

Bearing washer

Filler plug
🔧 59 - 118 (6 - 12, 43 - 87)

Wheel bearing assembly ❌

Air breather

Bearing cage

Baffle plate

Drain plug
🔧 59 - 118 (6 - 12, 43 - 87)

🔧 54 - 74 (5.5 - 7.5, 40 - 54)

🔧 118 - 147 (12 - 15, 87 - 108)

Rear axle housing

Bearing grease seal ❌ 🔧

Axle shaft

🔧 18 - 24 (1.8 - 2.4, 13 - 17)

ABS sensor

O-ring ❌

Front

🔧 : N•m (kg-m, ft-lb)

67170PATHG04

Rear axle components

Pinion Seal

Front

1. Before servicing the vehicle, refer to the Precautions Section.
2. Remove or disconnect the following:
   - Driveshaft
   - Front wheels
   - Front brake calipers

➡ **The front brake calipers must be removed so that there is no additional drag when measuring pinion bearing preload.**

3. Use an inch lb. torque wrench and measure the amount of torque required to maintain pinion rotation through several revolutions.

4. Matchmark the pinion flange to the pinion.

5. Remove or disconnect the following:
   - Pinion flange
   - Oil seal

To install:

6. Install or connect the following:
   - Pinion seal
   - Pinion flange

7. Rotate the pinion flange occasionally while tightening the flange nut to make sure the pinion bearings seat correctly.

8. Take frequent bearing preload torque readings. Tighten the flange nut to achieve the preload torque readings originally recorded. Do not exceed 137–217 ft. lbs. (186–294 Nm) torque when tightening the pinion flange nut.

⁑ CAUTION

If the bearing preload can not be achieved at the specified torque, remove the pinion bearing and install a new adjustment spacer.

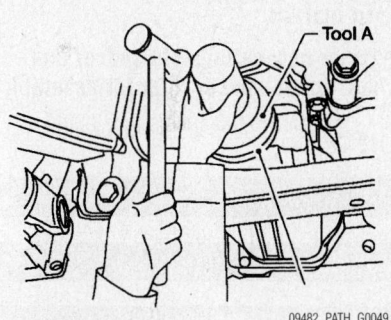

09482_PATH_G0049

Use Special Tool J-25405 to drive in the front pinion oil seal

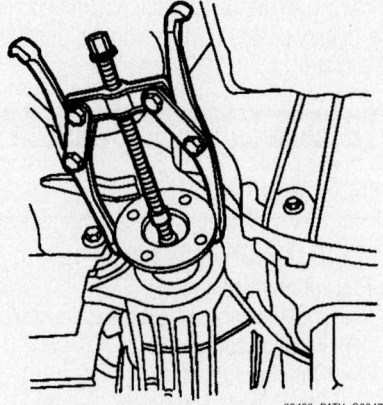

09482_PATH_G0047

Use a puller to remove the front pinion flange.

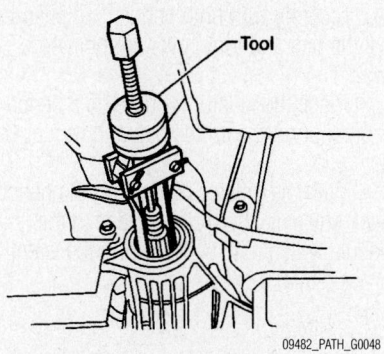

09482_PATH_G0048

Use Special Tool J-34286 to remove the front oil seal.

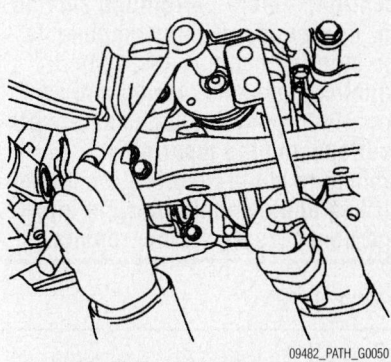

09482_PATH_G0050

Use Special Tool J-44195 to hold the pinion flange when installing the flange nut.

9. Install or connect the following:
- Front brake calipers
- Front wheels
- Driveshaft. Tighten the fasteners to 29–33 ft. lbs. (39–44 Nm).
10. Fill the differential with gear lubricant and check for leaks.

Rear

2 WHEEL DRIVE

1. Before servicing the vehicle, refer to the Precautions Section.
2. Remove or disconnect the following:
- Driveshaft
- Rear wheels
- Brake drums

➡**The rear brake drums must be removed so that there is no additional drag when measuring pinion bearing preload.**

3. Use an inch lb. torque wrench and measure the amount of torque required to maintain pinion rotation through several revolutions.
4. Remove or disconnect the following:
- Pinion flange
- Wheel speed sensor and rotor, if equipped
- Oil seal
- Pinion bearing
- Collapsible spacer

To install:

➡**Use a new collapsible spacer and wheel speed sensor rotor for assembly.**

5. Install or connect the following:
- Collapsible spacer
- Pinion bearing
- Pinion seal
- Pinion flange
6. Rotate the pinion flange occasionally while tightening the flange nut to make sure the pinion bearings seat correctly.
7. Take frequent bearing preload torque readings. Tighten the flange nut to achieve the preload torque readings originally recorded. Do not exceed 137–217 ft. lbs. (186–294 Nm) torque when tightening the pinion flange nut.

✳✳ CAUTION

Never loosen the pinion nut to reduce bearing preload. If it is necessary to reduce bearing preload, install a new collapsible spacer.

8. Install or connect the following:
- Brake drums
- Rear wheels
- Driveshaft. Tighten the fasteners to 58–65 ft. lbs. (78–88 Nm).
9. Fill the differential with gear lubricant and check for leaks.

4 WHEEL DRIVE

1. Before servicing the vehicle, refer to the Precautions Section.
2. Remove or disconnect the following:
- Driveshaft
- Rear wheels
- Brake drums

➡**The rear brake drums must be removed so that there is no additional drag when measuring pinion bearing preload.**

3. Use an inch lb. torque wrench and measure the amount of torque required to maintain pinion rotation through several revolutions.
4. Remove or disconnect the following:
- Pinion flange
- Oil seal

To install:

5. Install or connect the following:
- Pinion seal
- Pinion flange
6. Rotate the pinion flange occasionally while tightening the flange nut to make sure the pinion bearings seat correctly.
7. Take frequent bearing preload torque readings. Tighten the flange nut to achieve the preload torque readings originally recorded. Do not exceed 137–217 ft. lbs. (186–294 Nm) torque when tightening the pinion flange nut.

✳✳ CAUTION

If the bearing preload can not be achieved at the specified torque, remove the pinion bearing and install a new adjustment spacer.

8. Install or connect the following:
- Brake drums
- Rear wheels
- Driveshaft. Tighten the fasteners to 58–65 ft. lbs. (78–88 Nm).
9. Fill the differential with gear lubricant and check for leaks.

STEERING AND SUSPENSION

Air Bag

✳✳ CAUTION

Some vehicles are equipped with an air bag system. The system must be disarmed before performing service on, or around, system components, the steering column, instrument panel components, wiring and sensors. Failure to follow the safety precautions and the disarming procedure could result in accidental air bag deployment, possible injury and unnecessary system repairs.

PRECAUTIONS

Several precautions must be observed when handling the inflator module to avoid accidental deployment and possible personal injury.
- Never carry the inflator module by the wires or connector on the underside of the module.
- When carrying a live inflator module, hold securely with both hands, and ensure that the bag and trim cover are pointed away.
- Place the inflator module on a bench or other surface with the bag and trim cover facing up.
- With the inflator module on the bench, never place anything on or close to the module which may be thrown in the event of an accidental deployment.

DISARMING

To disarm the **SRS** system turn the ignition switch to the **OFF** position. Then, disconnect both battery cables starting with the negative cable first and wait at least 3 minutes after the cables are disconnected.

To rearm the **SRS** system, turn the ignition switch to the **OFF** position. Connect both battery cables starting with the positive cable first.

Rack and Pinion Steering Gear

REMOVAL & INSTALLATION

1. Before servicing the vehicle, refer to the Precautions Section.
2. Remove or disconnect the following:
- Front wheels
- Outer tie rod ends
- Steering shaft coupler
- Power steering hoses
- Steering gear

To install:

3. Install or connect the following:
- Steering gear. Tighten the bolts to 101 ft. lbs. (137 Nm) for 2001–02; 116–137 ft. lbs. (157–186 Nm) for 2003–04 models.
- Power steering hoses. Tighten the fittings to 25 ft. lbs. (35 Nm).
- Steering shaft coupler. Tighten the bolt to 22 ft. lbs. (29 Nm).
- Outer tie rod ends. Tighten the nuts to 65 ft. lbs. (88 Nm).
- Front wheels

2005 Pathfinder

1. Before servicing the vehicle, refer to the Precautions Section.

2. Remove or disconnect the following:
- Front wheels
- Front differential assembly, if equipped with 4WD
- Stabilizer bar brackets
- Steering outer socket cotter pins and nuts

3. Separate the outer sockets from the wheel knuckles.

4. Remove the power steering hoses from the steering gear assembly.

5. Remove the bolt from the lower joint steering shaft to separate the shaft from the steering gear assembly.

6. Remove the steering gear mounting nuts.

7. Remove the steering gear assembly.

8. Installation is the reverse order of removal.

Front Strut

REMOVAL & INSTALLATION

1. Before servicing the vehicle, refer to the Precautions Section.

2. Remove or disconnect the following:
- Front wheel
- Stabilizer bar link
- Steering knuckle bracket bolts
- Upper strut mount nuts
- Strut

When installing rubber parts, final tightening must be carried out under unladen condition* with tires on ground.
Fuel, radiator coolant and engine oil full.
Spare tire, jack, hand tools and mats in designated positions.

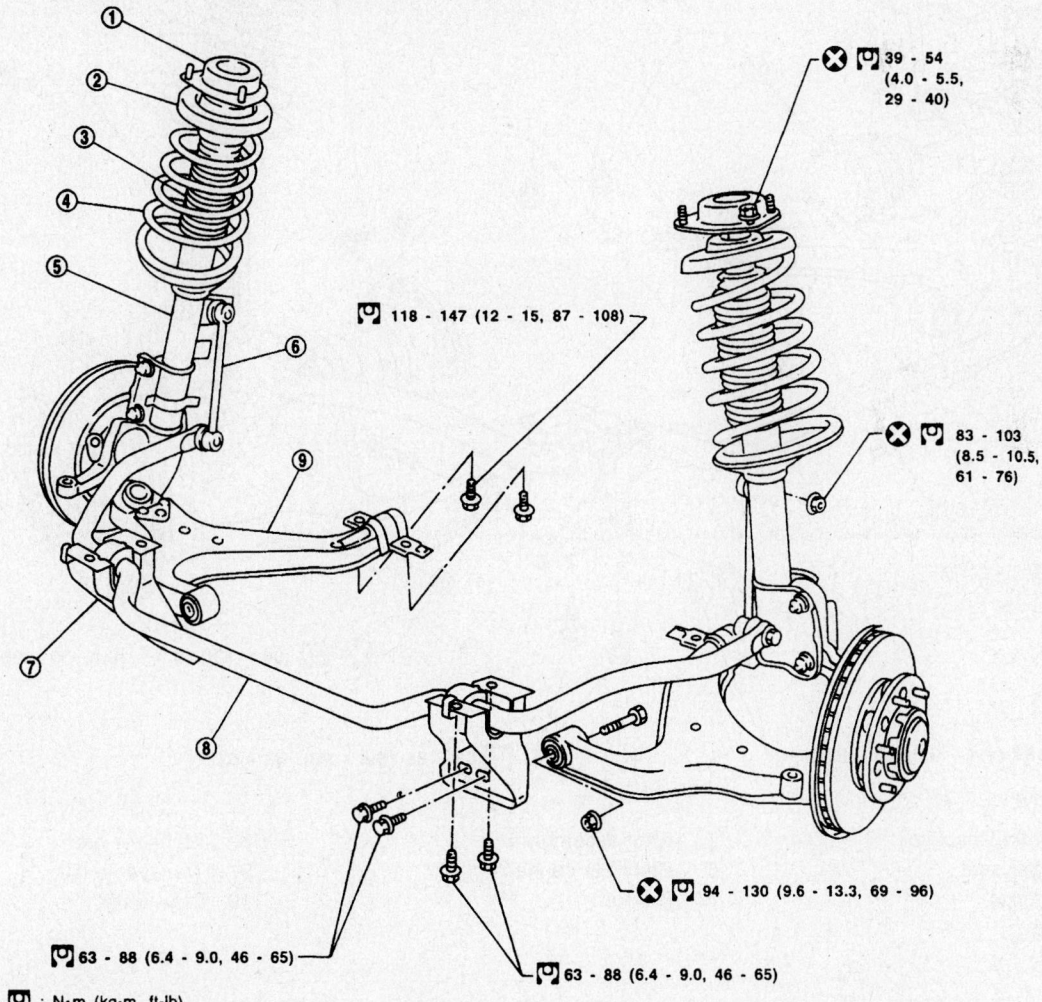

1. Strut mounting insulator
2. Spring upper seat
3. Bound bumper
4. Coil spring
5. Strut assembly
6. Stabilizer connecting rod
7. Bracket
8. Stabilizer bar
9. Transverse link

39 - 54 (4.0 - 5.5, 29 - 40)

118 - 147 (12 - 15, 87 - 108)

83 - 103 (8.5 - 10.5, 61 - 76)

94 - 130 (9.6 - 13.3, 69 - 96)

63 - 88 (6.4 - 9.0, 46 - 65)

63 - 88 (6.4 - 9.0, 46 - 65)

: N·m (kg-m, ft-lb)

Exploded view of the front suspension—2WD shown

7924VG66

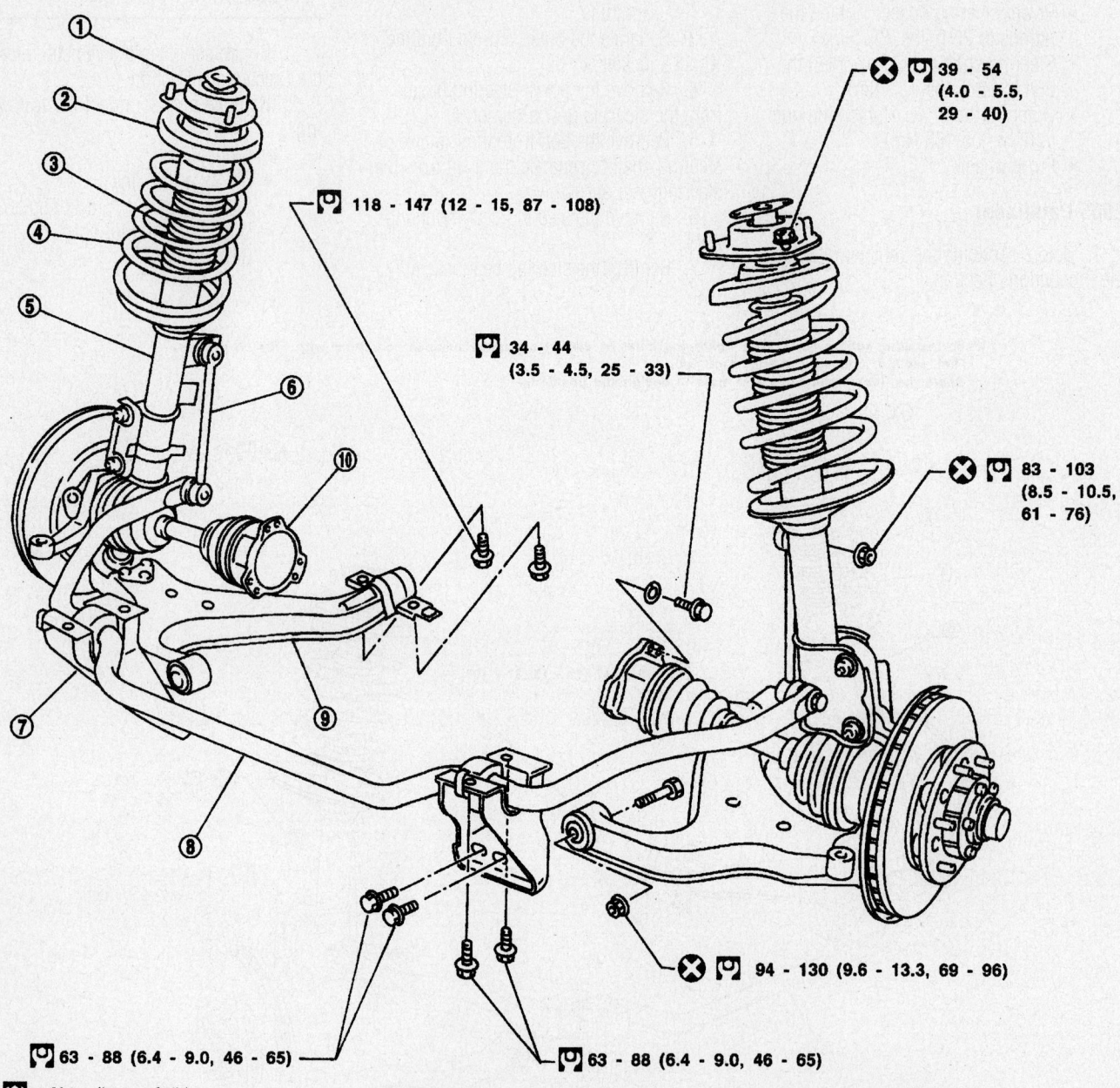

118 - 147 (12 - 15, 87 - 108)

39 - 54 (4.0 - 5.5, 29 - 40)

34 - 44 (3.5 - 4.5, 25 - 33)

83 - 103 (8.5 - 10.5, 61 - 76)

94 - 130 (9.6 - 13.3, 69 - 96)

63 - 88 (6.4 - 9.0, 46 - 65)

63 - 88 (6.4 - 9.0, 46 - 65)

: N•m (kg-m, ft-lb)

1. Strut mounting insulator
2. Spring upper seat
3. Bound bumper
4. Coil spring
5. Strut assembly
6. Stabilizer connecting rod
7. Bracket
8. Stabilizer bar
9. Transverse link
10. Drive shaft

67170PATHG05

Exploded view of the front suspension—4WD shown

NASU0007

NASU0007S01

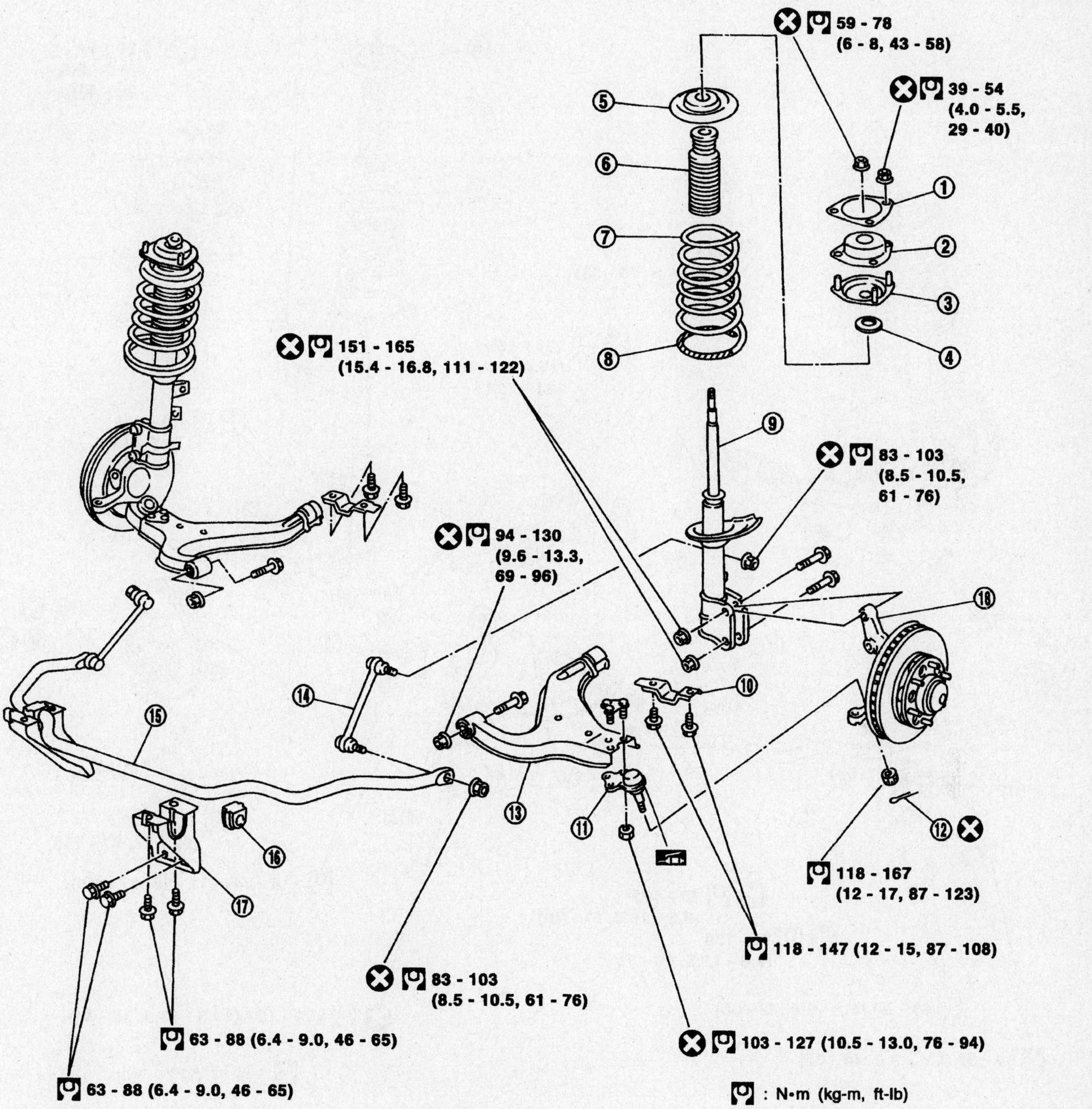

✗ 🔧 59 - 78
(6 - 8, 43 - 58)

✗ 🔧 39 - 54
(4.0 - 5.5,
29 - 40)

① Spacer
②
③
④

⑤
⑥
⑦
⑧

✗ 🔧 151 - 165
(15.4 - 16.8, 111 - 122)

⑨

✗ 🔧 83 - 103
(8.5 - 10.5,
61 - 76)

✗ 🔧 94 - 130
(9.6 - 13.3,
69 - 96)

⑱

⑩

⑭

⑮

⑬

⑪

⑫ ✗

🔧 118 - 167
(12 - 17, 87 - 123)

⑯

🔧 118 - 147 (12 - 15, 87 - 108)

⑰

✗ 🔧 83 - 103
(8.5 - 10.5, 61 - 76)

🔧 63 - 88 (6.4 - 9.0, 46 - 65)

✗ 🔧 103 - 127 (10.5 - 13.0, 76 - 94)

🔧 63 - 88 (6.4 - 9.0, 46 - 65)

🔧 : N•m (kg-m, ft-lb)

1. Spacer	7. Coil spring	13. Transverse link
2. Strut mounting insulator	8. (Polyurethane tube)	14. Stabilizer connecting rod
3. Bracket	9. Strut assembly	15. Stabilizer bar
4. Strut mounting bearing	10. Bracket	16. Bushing
5. Spring upper seat	11. Lower ball joint assembly	17. Bracket
6. Bound bumper	12. Cotter pin	18. Knuckle spindle

67170PATHG06

Front strut and related parts—2002-04 2WD shown

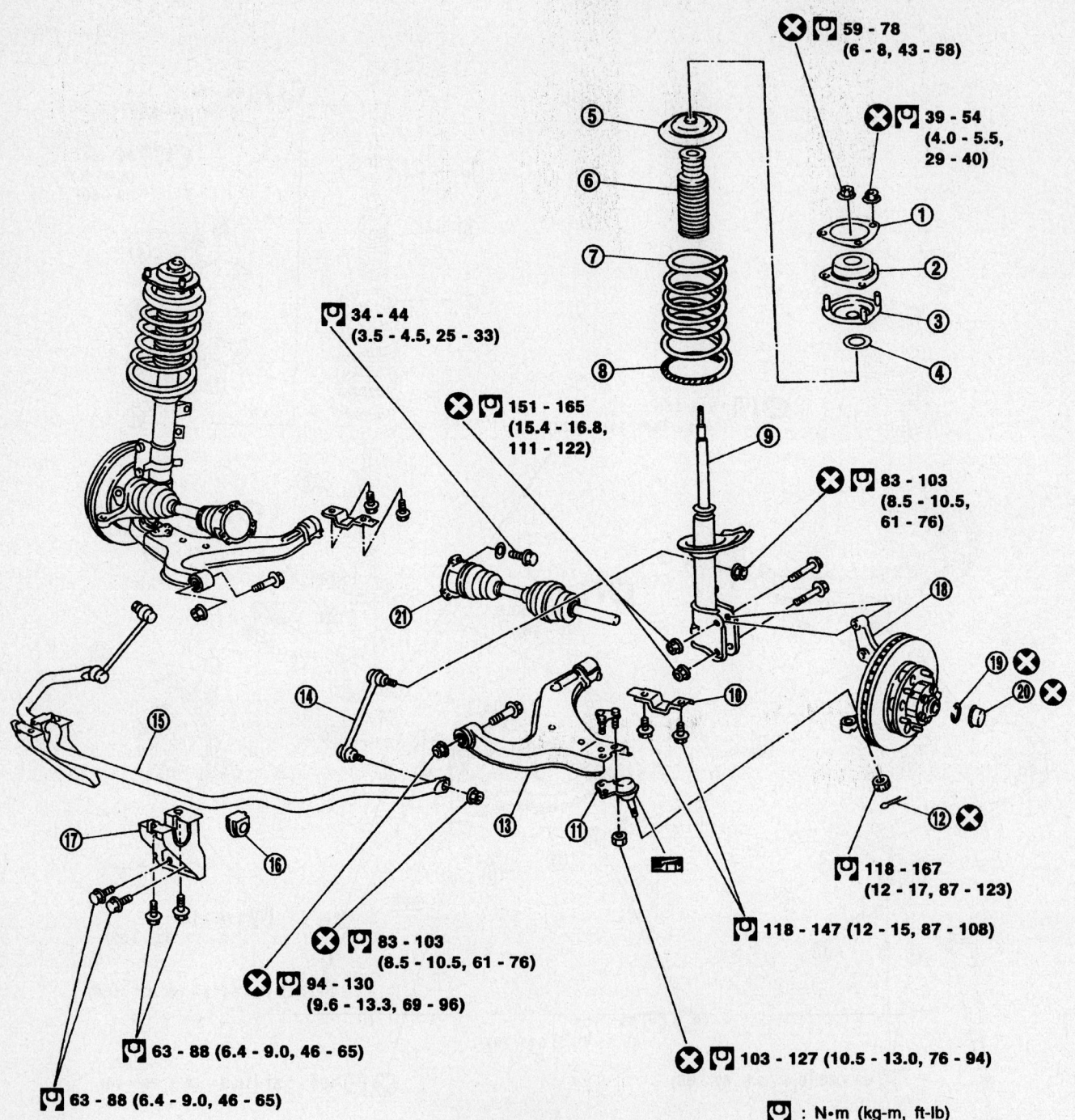

59 - 78
(6 - 8, 43 - 58)

39 - 54
(4.0 - 5.5,
29 - 40)

34 - 44
(3.5 - 4.5, 25 - 33)

151 - 165
(15.4 - 16.8,
111 - 122)

83 - 103
(8.5 - 10.5,
61 - 76)

118 - 167
(12 - 17, 87 - 123)

118 - 147 (12 - 15, 87 - 108)

83 - 103
(8.5 - 10.5, 61 - 76)

94 - 130
(9.6 - 13.3, 69 - 96)

63 - 88 (6.4 - 9.0, 46 - 65)

63 - 88 (6.4 - 9.0, 46 - 65)

103 - 127 (10.5 - 13.0, 76 - 94)

: N•m (kg-m, ft-lb)

1. Spacer
2. Strut mounting insulator
3. Bracket
4. Strut mounting bearing
5. Spring upper seat
6. Bound bumper
7. Coil spring
8. (Polyurethane tube)
9. Strut assembly
10. Bracket
11. Lower ball joint assembly
12. Cotter pin
13. Transverse link
14. Stabilizer connecting rod
15. Stabilizer bar
16. Bushing
17. Bracket
18. Knuckle spindle
19. Snap ring
20. Hub cap
21. Drive shaft

67170PATHG07

Front strut and related parts—2002–04 4WD shown

To install:

➡**Use new nuts and bolts for assembly.**

3. Install or connect the following:
- Strut. Tighten the upper strut mount nuts to 22–40 ft. lbs. (30–54 Nm) and the knuckle bracket bolts to 111–122 ft. lbs. (151–165 Nm) for 2002–2004 models. To 162 ft. lbs. (220 Nm) for 2005 models.
- Stabilizer bar link. Tighten the nut to 61–76 ft. lbs. (83–103 Nm).

- Front wheel
4. Check the wheel alignment and adjust, as necessary.

Rear Shock Absorber

REMOVAL & INSTALLATION

1. Before servicing the vehicle, refer to the Precautions Section.
2. Support the rear axle.

3. Remove or disconnect the following:
- Lower shock absorber bolt
- Upper shock absorber bolt
- Shock absorber

To install:

➡**Use new fasteners for assembly.**

4. Install the shock absorber and tighten the upper bolt to 49–65 ft. lbs. (67–88 Nm); the lower bolt to 44–57 ft. lbs. (59–78 Nm).

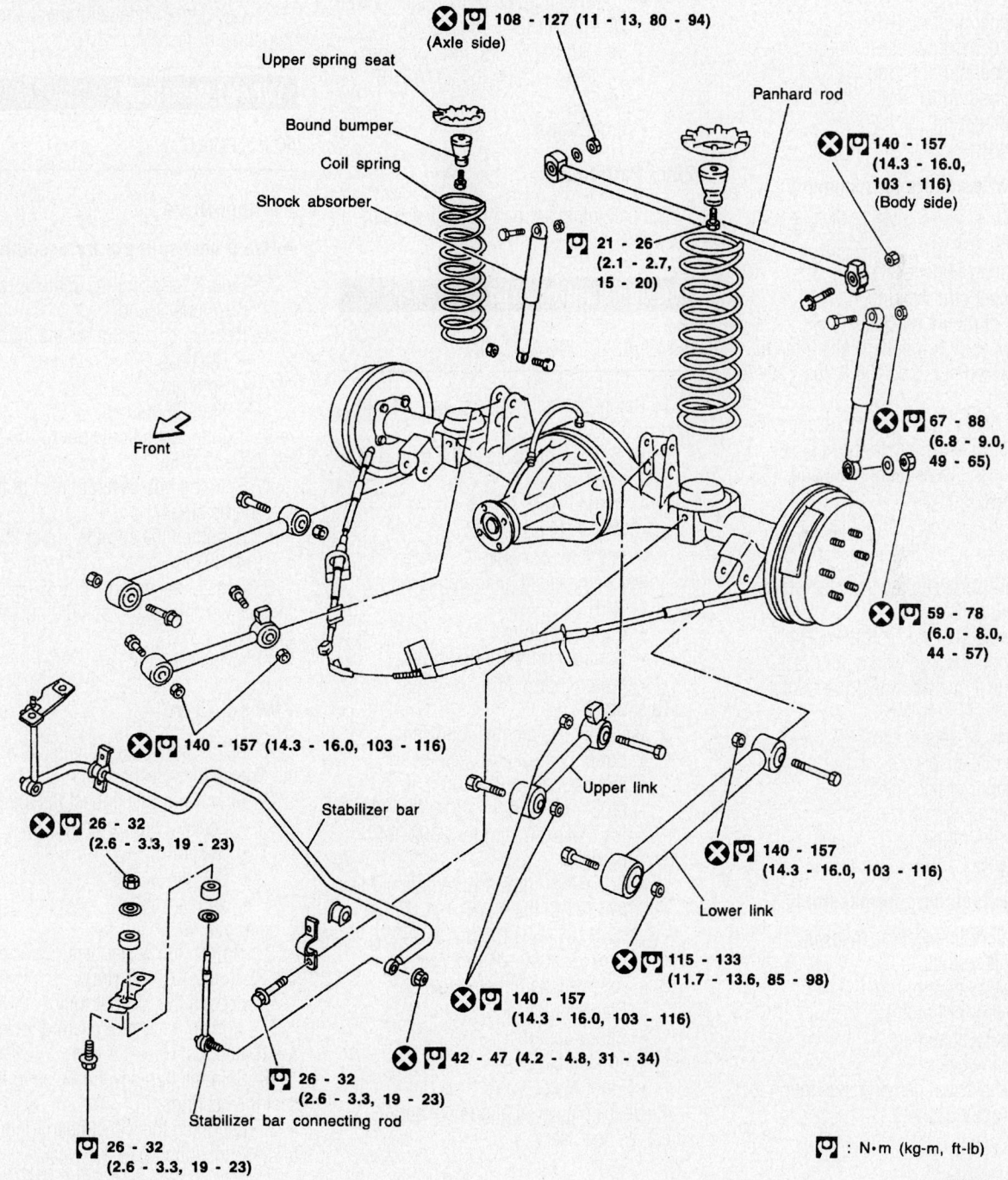

Rear suspension components

67170PATHG08

Coil Spring

REMOVAL & INSTALLATION

Front

1. Before servicing the vehicle, refer to the Precautions Section.
2. Remove the strut assembly.
3. Compress the coil spring and remove the piston rod nut.
4. Remove or disconnect the following:
   - Upper strut mount
   - Strut mount bracket
   - Upper strut bearing
   - Spring upper seat
   - Coil spring

To install:

➡**Use new fasteners for assembly.**

5. Install or connect the following:
   - Coil spring
   - Spring upper seat
   - Upper strut bearing
   - Strut mount bracket
   - Upper strut mount. Tighten the piston rod nut to 43–58 ft. lbs. (59–78 Nm).
6. Remove the spring compressor and install the strut assembly to the vehicle.
7. Check the wheel alignment and adjust, as necessary.

Rear

1. Before servicing the vehicle, refer to the Precautions Section.
2. Support the vehicle at the frame.
3. Support the axle with a floor jack.
4. Remove or disconnect the following:
   - Rear wheels
   - Shock absorbers
   - Stabilizer bar links
   - Lateral control rod
   - Coil springs

To install:

➡**Use new fasteners for assembly.**

5. Install or connect the following:
   - Coil springs
   - Lateral control rod.
   - Stabilizer bar links.
   - Shock absorbers
   - Rear wheels
6. See the accompanying illustration for relevant torques.

Lower Ball Joint

REMOVAL & INSTALLATION

1. Before servicing the vehicle, refer to the Precautions Section.
2. Support the lower control arm.
3. Remove or disconnect the following:
   - Front wheel
   - Lower ball joint

To install:

4. Install or connect the following:
   - Lower ball joint. Tighten the control arm-to-ball joint bolts to 76–94 ft. lbs. (103–127 Nm) and the stud nut to 87–123 ft. lbs. (118–167 Nm).
   - Front wheel

2005 Pathfinder

The ball joints are part of the lower control arms.

Lower Control Arm

REMOVAL & INSTALLATION

1. Before servicing the vehicle, refer to the Precautions Section.
2. Remove or disconnect the following:
   - Front wheel
   - Torsion bar
   - Shock absorber
   - Stabilizer bar link
   - Axle halfshaft, if equipped with 4WD
   - Lower ball joint
   - Control arm mounting bolts
   - Lower control arm

To install:

3. Install or connect the following:
   - Lower control arm. Tighten the mount bolts to 80–105 ft. lbs. (108–142 Nm) for 2001; 69–96 ft. lbs. (94–130 Nm) for 2002–04 models.
   - Lower ball joint. Tighten the nut to 87–141 ft. lbs. (118–191 Nm) for 2001; 87–123 ft. lbs. (118–167 Nm) for 2002–04 models.
   - Axle halfshaft, if equipped
   - Stabilizer bar link
   - Shock absorber
   - Torsion bar
   - Front wheel
4. Check the wheel alignment and adjust, as necessary.

CONTROL ARM BUSHING REPLACEMENT

1. Before servicing the vehicle, refer to the Precautions Section.
2. Remove the control arm from the vehicle.
3. Remove the control arm bushing with a press.

To install:

4. Lubricate the control arm bushings with liquid soap.
5. Install the bushings with a press.
6. Install the control arm to the vehicle.
7. Check the wheel alignment and adjust, as necessary.

Wheel Bearings

ADJUSTMENT

2 Wheel Drive

➡**Use a new split pin for assembly.**

1. Before servicing the vehicle, refer to the Precautions Section.
2. Remove or disconnect the following:
   - Dust cap
   - Split pin
   - Spindle nut cap
3. Tighten the spindle nut to 25–29 ft. lbs. (34–39 Nm).
4. Spin the hub several times to fully seat the bearings.
5. Retighten the spindle nut to 25–29 ft. lbs. (34–39 Nm).
6. Loosen the spindle nut 45–60 degrees and install the spindle nut cap and split pin.
7. Install the dust cap.

4 Wheel Drive

1. Before servicing the vehicle, refer to the Precautions Section.
2. Remove or disconnect the following:
   - Locking hub or driveplate
   - Snapring
   - Spindle washer
   - Thrust washer
   - Lockwasher
3. Tighten the wheel bearing locknut to 58–72 ft. lbs. (78–98 Nm).
4. Loosen the locknut fully.
5. Tighten the wheel bearing locknut to 4–13 inch lbs. (0.5–1.5 Nm).
6. Spin the hub several times to fully seat the bearings.
7. Retighten the wheel bearing locknut to 4–13 inch lbs. (0.5–1.5 Nm).

8. Install or connect the following:
- Lockwasher. Tighten the retaining screw to 10–16 inch lbs. (1–2 Nm).
- Thrust washer
- Spindle washer
- Snapring
- Locking hub or driveplate

2005 Model

The wheel hub and bearing assembly is sealed unit. If there is any axial play or irregular conditions, the hub assembly must be replaced.

REMOVAL & INSTALLATION

2002–2004

2WD

1. Before servicing the vehicle, refer to the Precautions Section.
2. Remove or disconnect the following:
- Front wheel

- Brake caliper and support
- Dust cap
- Split pin
- Spindle nut cap
- Spindle nut
- Bearing washer
- Outer bearing
- Hub and brake rotor assembly
- Inner grease seal
- Inner wheel bearing

To install:
3. Install or connect the following:
- Inner wheel bearing
- Inner grease seal
- Hub and brake rotor assembly
- Outer bearing
- Bearing washer
- Spindle nut. Adjust the wheel bearings.
- Spindle nut cap
- Split pin
- Dust cap
- Brake caliper and support
- Front wheel

4WD

1. Before servicing the vehicle, refer to the Precautions Section.
2. Remove or disconnect the following:
- Front wheel
- Brake caliper and support
- Locking hub or driveplate
- Snapring
- Spindle washer
- Thrust washer
- Lockwasher
- Wheel bearing locknut
- Outer bearing
- Hub and brake rotor assembly
- Inner grease seal
- Inner wheel bearing

To install:
3. Install or connect the following:
- Inner wheel bearing
- Inner grease seal
- Hub and brake rotor assembly
- Outer bearing
- Wheel bearing locknut. Adjust the wheel bearings.
- Lockwasher

NAAX0007S01

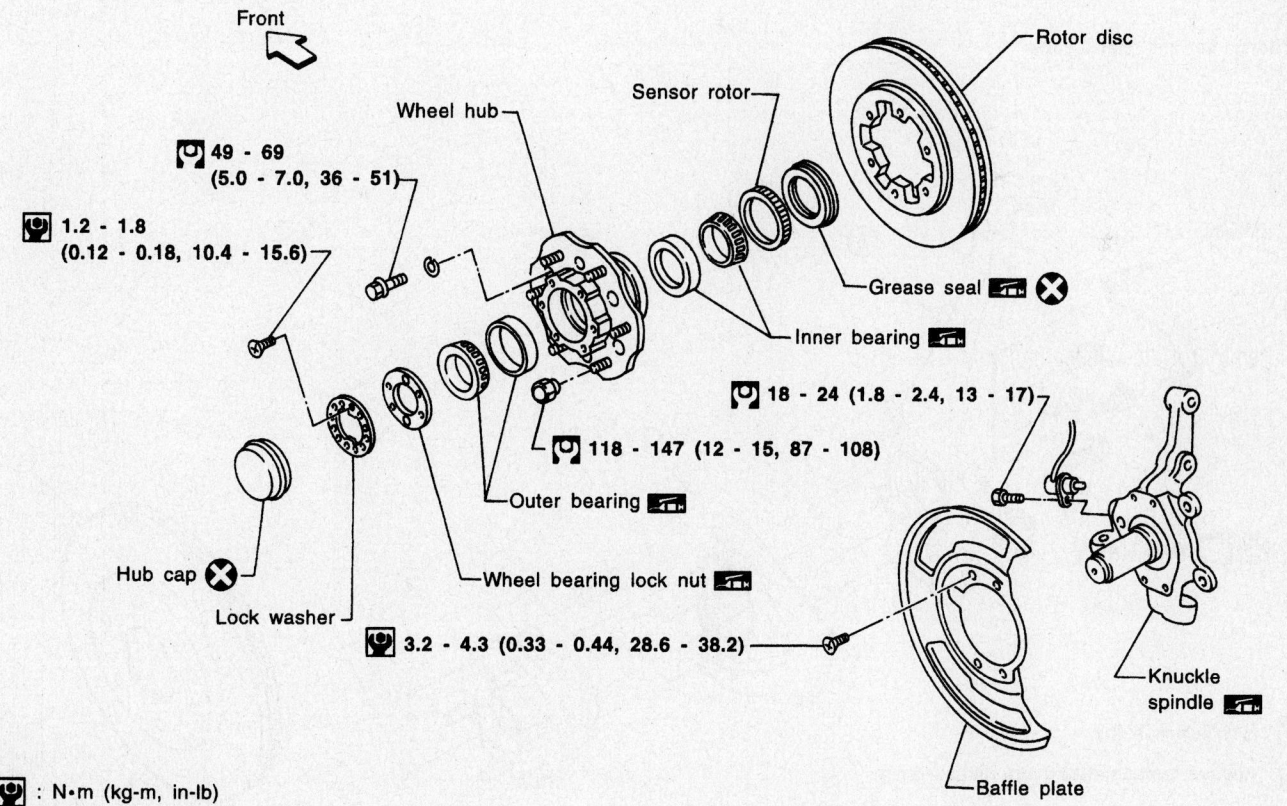

2WD front hub and related parts

NAAX0007S02

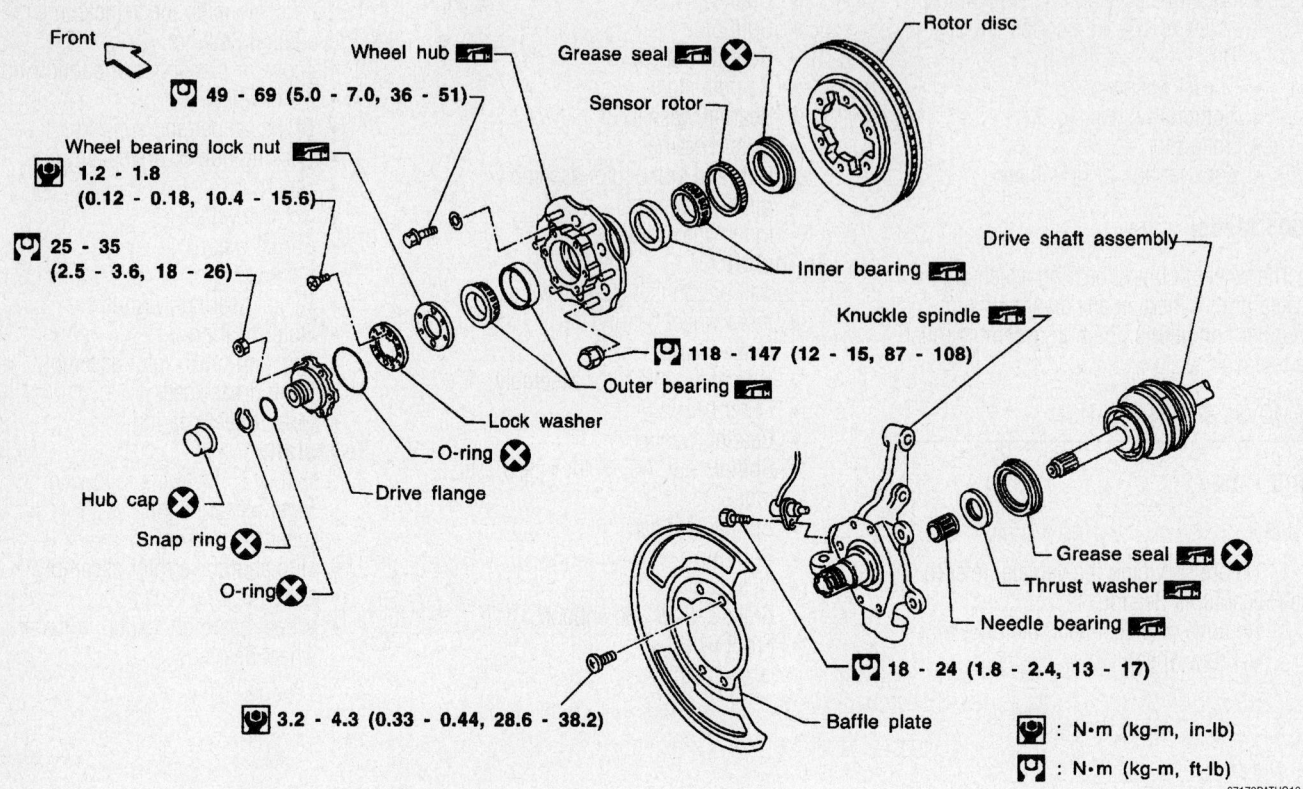

4WD front hub and related parts

67170PATHG10

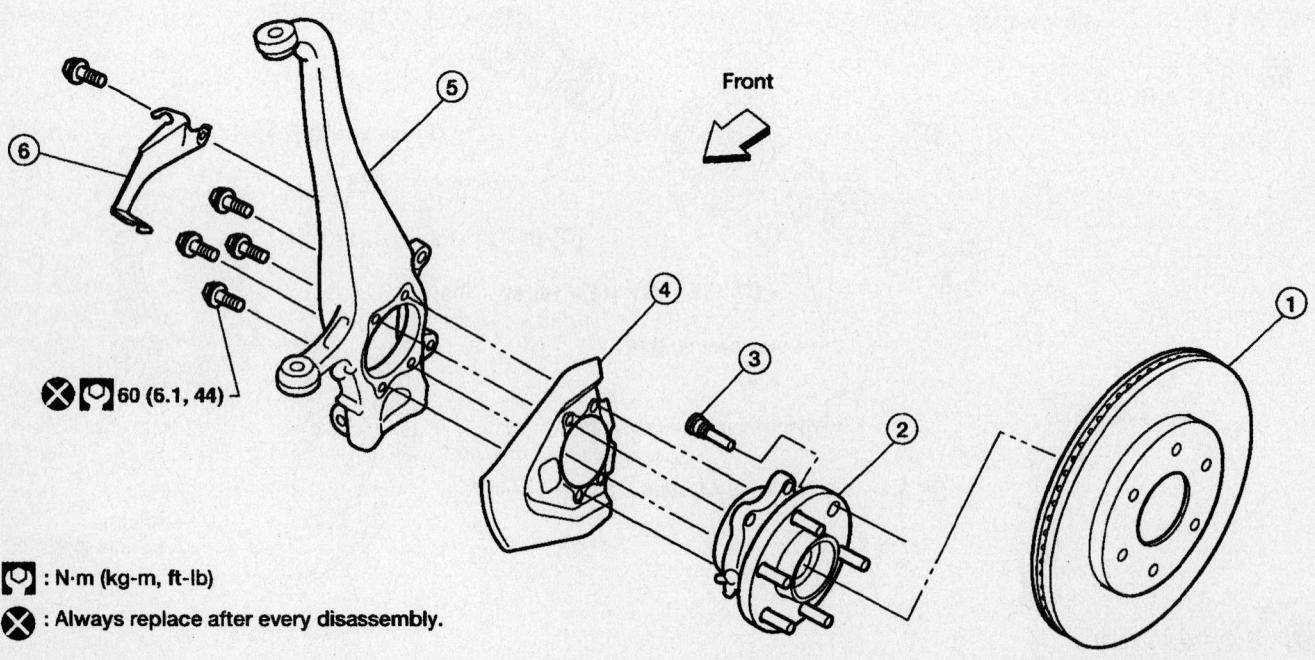

[X] : N·m (kg-m, ft-lb)

[X] : Always replace after every disassembly.

1. Disc rotor
2. Wheel hub and bearing assembly
3. Wheel stud
4. Splash guard
5. Steering knuckle
6. Wheel sensor bracket

Exploded view of the wheel hub assembly—2005 Pathfinder shown

09482_PATH_G0051

- Thrust washer
- Spindle washer
- Snapring
- Locking hub or driveplate
- Brake caliper and support
- Front wheel

2005 Pathfinder

1. Before servicing the vehicle, refer to the Precautions Section.

2. Remove the wheel.
3. Remove the caliper assembly and hang securely with wire.
4. Matchmark the brake rotor and hub assembly and remove the rotor.
5. Remove or disconnect the following:
   - Hub spindle nut
   - Halfshaft, if equipped with 4WD
   - Wheel speed sensor
   - Wheel hub assembly

6. Installation is the reverse order of removal. Note the following during installation:
   a. Use new bolts for the wheel hub assembly.
   b. Tighten the wheel hub assembly bolts to 44 ft. lbs. (60 Nm).

BRAKES

Brake Caliper

REMOVAL AND INSTALLATION

1. Raise the vehicle and support safely.
2. Remove the appropriate tire and wheel assembly.
3. Remove the bolt attaching the brake hose to the caliper. Plug the brake hose to prevent brake fluid loss.
4. Remove the caliper support mounting bolts and lift the caliper/support assembly from the knuckle.

To install

5. Position the caliper assembly onto the knuckle and install the bolts. Make sure

the rotor fits between the brake pads. Torque the bolts to:
- 2001: 53–72 ft. lbs. (72–97 Nm).
- 2002–03: 124–137 ft. lbs. (169–186 Nm)
- 2004: 107–136 ft. lbs. (145–185 Nm)
- 2005: 136 ft. lbs. (185 Nm)

6. Using new copper washers, connect the brake hose to the caliper. Torque the brake hose attaching bolt to 12–14 ft. lbs. (17–20 Nm).
7. Bleed the brake system.
8. Apply the brake pedal and inspect the system. Ensure proper operation and no leakage.
9. Install tire and wheel assembly. Lower the vehicle and road-test.

Disc Brake Pads

REMOVAL AND INSTALLATION

➡ **Both the front and rear disc brake pads can be serviced using the same procedure.**

1. Using a syringe, siphon brake fluid from the reservoir, leaving reservoir approximately ½ full.
2. Raise and properly support the vehicle.
3. Remove the wheel assemblies.
4. Remove the lower pin bolt from the brake caliper.
5. Swivel the caliper up and away from the torque member. Tie the caliper to a

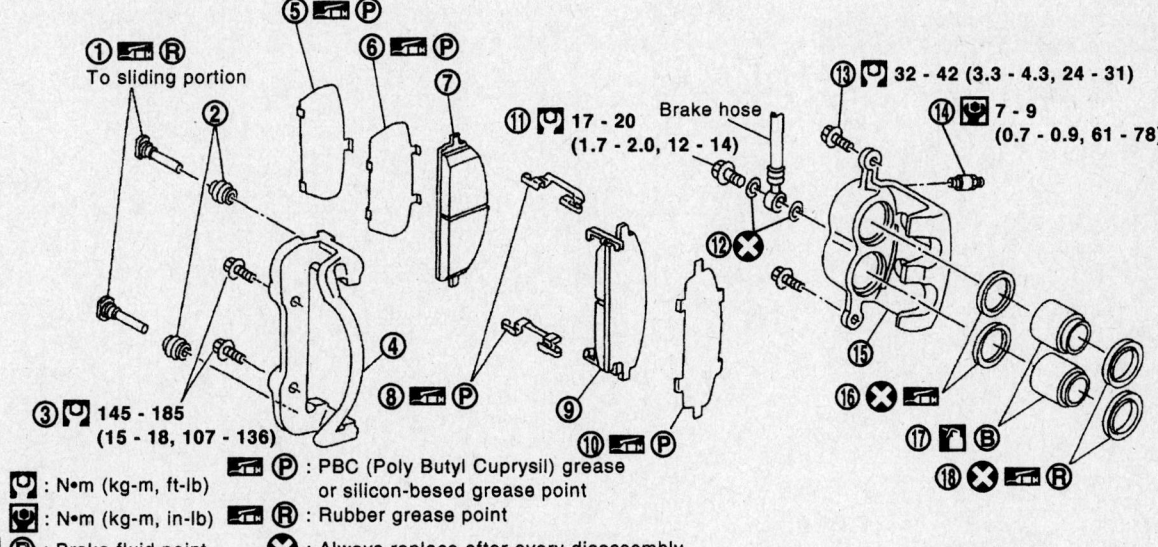

: N•m (kg-m, ft-lb)
: N•m (kg-m, in-lb)
: Brake fluid point

: PBC (Poly Butyl Cuprysil) grease or silicon-besed grease point
: Rubber grease point
: Always replace after every disassembly

1. Main pin	7. Inner pad	13. Main pin bolt
2. Pin boot	8. Pad retainer	14. Bleed valve
3. Torque member fixing bolt	9. Outer pad	15. Cylinder body
4. Torque member	10. Outer shim	16. Piston seal
5. Shim cover	11. Connecting bolt	17. Piston
6. Inner shim	12. Copper washer	18. Piston boot

67170PATHG01

Exploded view of the front brake components—2004 model shown

suspension member so that it is out of the way.

6. Lift the 2 brake pads out of the torque member.

7. Remove the inner and outer shims. Remove the 2 pad retainers if they are not attached to the pads.

8. Check the pad thickness and replace the pads if they are less than 0.079 in. (2mm) thick.

To install:

9. Install the inner and outer shims into the torque member.

10. Install a pad retainer to the bottom of each pad.

11. Install the pads into the torque member.

12. Use a C-clamp or hammer handle and press the caliper piston(s) back into the housing.

13. Untie the caliper and swivel it back into position over the torque plate so that the dust boot is not pinched. Install the pin bolt and torque it to 16–23 ft. lbs. (22–31 Nm) for 2001 models; 24–31 ft. lbs. (32–42 Nm) for 2002–04 models.

14. Check the condition of the pin boot. Gently pull on it to expel any trapped air.

15. Install the wheel and lower the vehicle.

16. Pump the brakes until the pedal is firm and check the level of brake fluid. Road-test the vehicle.

Brake Drums

REMOVAL AND INSTALLATION

1. Remove the hub cap and loosen the lug nuts.

2. Raise the rear of the vehicle and support it on jackstands.

3. Remove the lug nuts, tire and wheel.

4. Release the parking brake.

5. Pull the brake drum from the hub. If difficult to remove try the following:

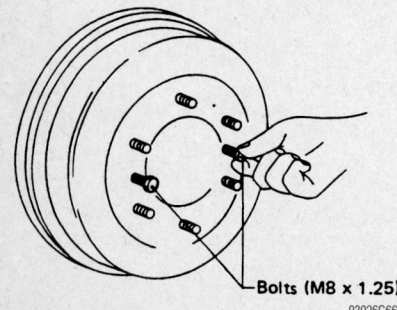

Install and tighten 2 bolts to remove a stubborn brake drum

a. Strike the face of the drum with a plastic or rubber mallet. This will break free any rust that may develop between the drum and the hub.

b. Install 2, M8x1.25mm bolts into the holes in the drum and gradually tighten them to pull the drum off the hub.

To install:

6. Install the brake drum to the hub.

7. Install the wheel.

8. Remove the jackstands and lower the vehicle.

9. Road-test the vehicle to ensure that the brakes are working properly.

Brake Shoes

REMOVAL AND INSTALLATION

Pathfinder

1. Release the parking brake.

2. Safely raise and support the vehicle.

3. Remove the rear wheel and drum.

4. Remove the hold-down pin retainers.

5. Remove the leading shoe and then the trailing shoe.

6. Remove the adjuster.

7. Disconnect the parking brake cable from the toggle lever on the rear shoe.

To install:

8. Transfer the toggle lever to the new rear shoe.

9. Apply a small amount of brake grease to the tips of the shoes and the 6 pads on the backing plate that contact the brake shoe.

10. Shorten the adjuster by turning it.

11. Connect the parking brake cable to the toggle lever on the rear shoe.

12. Install the lower return spring to both shoes and install the shoes on the backing plate with the hold down pins and retainers.

13. Install the adjuster and the remaining springs. Pay attention to the direction of the adjuster assembly.

14. Inspect the complete assembly and install the brake drum.

15. Adjust the shoe to drum clearance.

16. Install the wheel assembly and lower the vehicle to the floor.

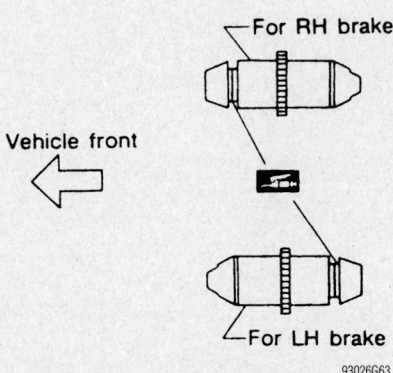

Correct direction of brake shoe adjuster

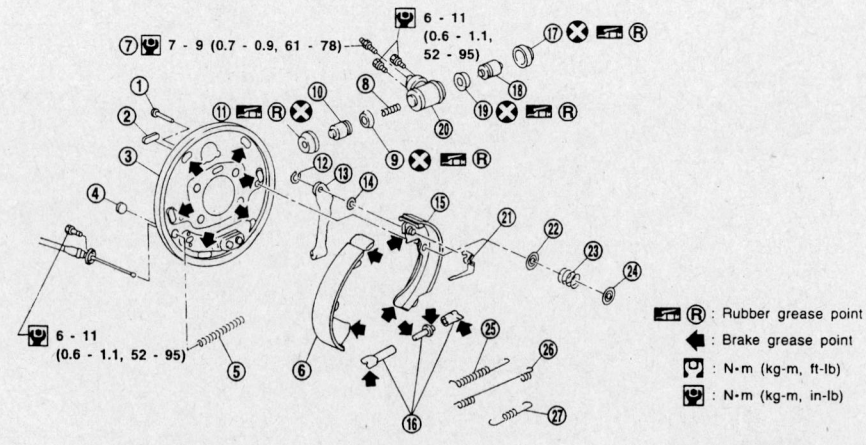

1.	Shoe hold pin	10.	Piston	19.	Piston cup
2.	Plug	11.	Boot	20.	Wheel cylinder
3.	Back plate	12.	Retainer ring	21.	Adjuster lever
4.	Check plug	13.	Toggle lever	22.	Spring seat
5.	Spring	14.	Wave washer	23.	Shoe hold spring
6.	Shoe (leading side)	15.	Shoe (trailing side)	24.	Retainer
7.	Air bleeder	16.	Adjuster	25.	Adjuster spring
8.	Spring	17.	Boot	26.	Return spring (upper)
9.	Piston cup	18.	Piston	27.	Return spring (lower)

Drum brake assembly exploded view

INFINITI & NISSAN

QX56 • Armada

SPECIFICATIONS AND MAINTENANCE CHARTS

ENGINE AND VEHICLE IDENTIFICATION

Engine							Model Year	
Code ①	Liters (cc)	Cu. In.	Cyl.	Fuel Sys.	Engine	Eng. Mfg.	Code ②	Year
VK56DE	5.6 (5552)	338.8	8	MFI	DOHC	Nissan	4	2004
							5	2005
							6	2006

MFI: Multi-port Fuel Injection

DOHC: Double Overhead Camshafts

① Located on the timing belt cover

② 10th digit of the Vehicle Identification Number (VIN)

09482_ARMA_C0001

GENERAL ENGINE SPECIFICATIONS

Year	Model	Engine Displacement Liters	Engine ID	Net Horsepower @ rpm	Net Torque @ rpm (ft. lbs.)	Bore x Stroke (in.)	Com-pression Ratio	Oil Pressure @ rpm
2004	Armada	5.6	VK56DE	305@4900	385@3600	3.86X3.62	9.8:1	43@2000
	QX56	5.6	VK56DE	305@4900	385@3600	3.86X3.62	9.8:1	43@2000
2005	Armada	5.6	VK56DE	305@4900	385@3600	3.86X3.62	9.8:1	43@2000
	QX56	5.6	VK56DE	305@4900	385@3600	3.86X3.62	9.8:1	43@2000

09482_ARMA_C0002

ENGINE TUNE-UP SPECIFICATIONS

Year	Engine Displacement Liters	Engine ID	Spark Plug Gap (in.)	Ignition Timing	Fuel Pump (psi) ①	Idle Speed ②	Valve Clearance (in.) In.	Valve Clearance (in.) Ex.
2004	5.6	VK56DE	0.043	15B	51	600-700	0.010-0.013	0.011-0.016
2005	5.6	VK56DE	0.043	15B	51	600-700	0.010-0.013	0.011-0.016

NOTE: The Vehicle Emission Control Information label often reflects specification changes made during production. The label figures

must be used if they differ from those in this chart.

B: Before top dead center

① System pressure at idle with vacuum hose connected

   Should increase to 43 psi when disconnected

② Automatic transmission in Neutral

09482_ARMA_C0003

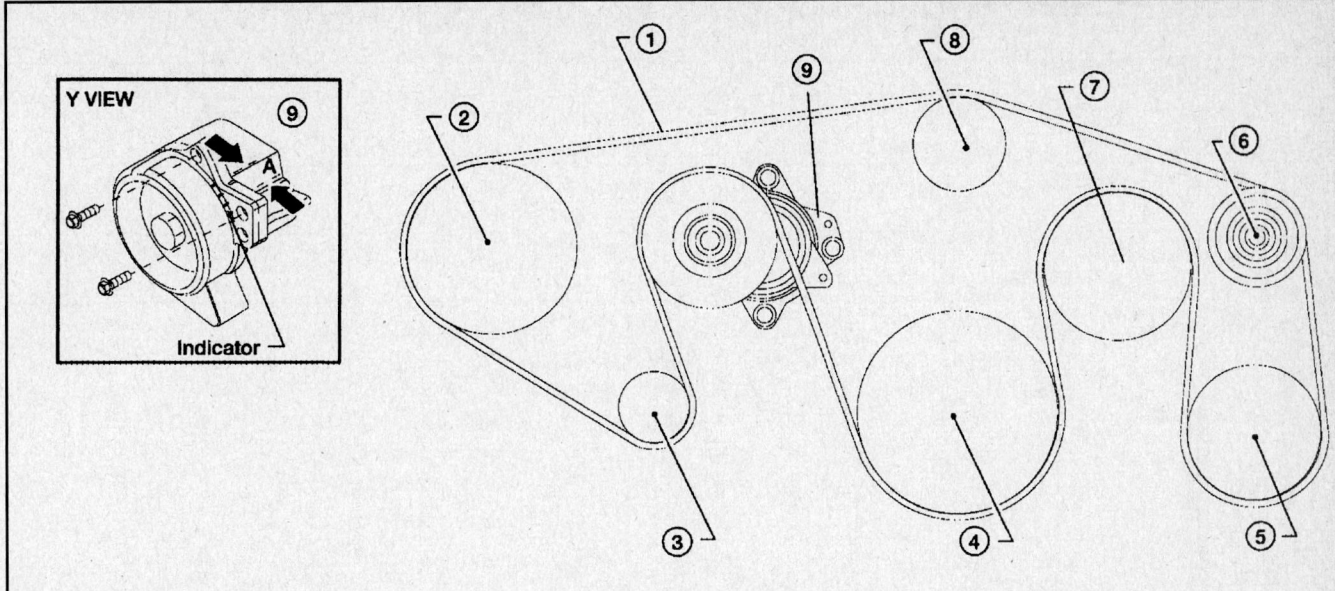

1. Drive Belt
2. Power Steering Pump Pulley
3. Generator pulley
4. Crankshaft Pulley
5. A/C Compressor
6. Idler Pulley
7. Cooling Fan Pulley
8. Water Pump Pulley
9. Drive Belt Tensioner

67162-QX56-G47

Accessory drive belt routing

CAPACITIES

Year	Model	Engine Displacement Liters	Engine ID	Engine Oil with Filter (qts.)	Transmission (pts.)	Transfer Case (pts.)	Drive Axle		Fuel Tank (gal.)	Cooling System (qts.)
							Front (pts.)	Rear (pts.)		
2004	Armada	5.6	VK56DE	6.5	22.5	6.25	3.375	3.75	28.0	15
	QX56	5.6	VK56DE	6.5	22.5	6.25	3.375	3.75	28.0	15
2005	Armada	5.6	VK56DE	6.5	22.5	6.25	3.375	3.75	28.0	15
	QX56	5.6	VK56DE	6.5	22.5	6.25	3.375	3.75	28.0	15

NOTE: All capacities are approximate. Add fluid gradually and check to be sure a proper fluid level is obtained.

09482_ARMA_C0004

CRANKSHAFT AND CONNECTING ROD SPECIFICATIONS

All measurements are given in inches.

Year	Engine Displ. Liters	Engine ID	Crankshaft				Connecting Rod		
			Main Brg. Journal Dia.	Main Brg. Oil Clearance	Shaft End-play	Thrust on No.	Journal Diameter	Oil Clearance	Side Clearance
2004	5.6	VK56DE	①	②	0.0118	3	③	0.0002-0.0007	0.0079-0.0157
2005	5.6	VK56DE	①	②	0.0118	3	③	0.0002-0.0007	0.0079-0.0157

① There are 17 different grades, ranging from 0 (2.483) to 78 (2.510)

② No. 1 and 5: 0.00004-0.0004

   No. 2, 3 and 4: 0.0003-0.0007

③ Grade 0: 2.0441-2.0441

  Grade 1: 2.0441-2.0442

  Grade 2: 2.0442-2.0442

  Grade 3: 2.0442-2.0443

  Grade 4: 2.0443-2.0443

  Grade 5: 2.0443-2.0443

  Grade 6: 2.0443-2.0444

  Grade 7: 2.0444-2.0444

  Grade 8: 2.0444-2.0444

  Grade 9: 2.0444-2.0445

  Grade A: 2.0445-2.0445

  Grade B: 2.0445-2.0446

  Grade C: 2.0446-2.0446

09482_ARMA_C0005

VALVE SPECIFICATIONS

Year	Engine Displacement Liters	Engine ID	Seat Angle (deg.)	Face Angle (deg.)	Spring Test Pressure (lbs. @ in.)	Spring Installed Height (in.)	Stem-to-Guide Clearance (in.)		Stem Diameter (in.)	
							Intake	Exhaust	Intake	Exhaust
2004	5.6	VK56DE	45.15-45.45	45	37.0@1.457	1.991	0.0008-0.0021	0.0012-0.0025	0.2348-0.2354	0.2344-0.2350
2005	5.6	VK56DE	45.15-45.45	45	37.0@1.457	1.991	0.0008-0.0021	0.0012-0.0025	0.2348-0.2354	0.2344-0.2350

09482_ARMA_C0006

CAMSHAFT AND BEARING SPECIFICATIONS CHART

All measurements are given in inches.

Year	Engine Displ. Liters	Engine ID/VIN	Journal Dia.	Brg. Oil Clearance	Shaft End-play	Runout	Journal Bore	Lobe Height	
								Intake	Exhaust
2004	5.6	VK56DE	1.0218-1.0224	0.0012-0.0027	0.0045-0.0074	0.0008	1.0236-1.0244	1.7506-1.7581	1.7506-1.7581
2005	5.6	VK56DE	1.0218-1.0224	0.0012-0.0027	0.0045-0.0074	0.0008	1.0236-1.0244	1.7506-1.7581	1.7506-1.7581

09482_ARMA_C0007

PISTON AND RING SPECIFICATIONS

All measurements are given in inches.

Year	Engine Displacement Liters	Engine ID	Piston Clearance	Ring Gap			Ring Side Clearance		
				Top Comp.	Bottom Comp.	Oil Control	Top Comp.	Bottom Comp.	Oil Control
2004	5.6	VK56DE	0.0004-0.0012	0.0091-0.0110	0.0189-0.0217	0.0079-0.0197	0.0014-0.0033	0.0012-0.0028	0.0006-0.0073
2005	5.6	VK56DE	0.0004-0.0012	0.0091-0.0110	0.0189-0.0217	0.0079-0.0197	0.0014-0.0033	0.0012-0.0028	0.0006-0.0073

09482_ARMA_C0008

TORQUE SPECIFICATIONS

All readings in ft. lbs.

Year	Engine Displacement Liters	Engine ID	Cylinder Head Bolts	Main Bearing Bolts	Rod Bearing Bolts	Crankshaft Damper Bolts	Flywheel Bolts	Manifold		Spark Plugs	Oil Pan Drain Plug
								Intake	Exhaust		
2004	5.6	VK56DE	①	②	③	④	65	6	25	18	25
2005	5.6	VK56DE	①	②	③	④	65	6	25	18	25

① Step 1: 72 ft. lbs

   Step 2: Loosen all bolts completely

   Step 3: 33 ft. lbs.

   Step 4: +60 degrees

   Step 5: +60 degrees

② Step 1: Main Bolts to 29 ft. lbs.

   Step 2: Sub-bolts to 22 ft. lbs.

   Step 3: Main Bolts +40 degrees

   Step 4: Sub-Bolts +30 degrees

   Step 5: Side Bolts to 36 ft. lbs.

③ Step 1: 11 ft. lbs.

   Step 2: +90 degrees

④ Step 1: 65 ft. lbs.

   Step 2: +90 degrees

09482_ARMA_C0009

WHEEL ALIGNMENT

Year	Model	Caster Range (+/-Deg.)	Caster Preferred Setting (Deg.)	Camber Range (+/-Deg.)	Camber Preferred Setting (Deg.)	Toe-in (in.)
2004	Armada ①	0.75	③	0.75	⑤	0.08+/-0.11
	QX56 ②	0.75	④	0.75	⑤	0.08+/-0.03
2005	Armada ①	0.75	③	0.75	⑤	0.11+/-0.04
	QX56 ①	0.75	④	0.75	⑤	0.08+/-0.03

① Assumes P285/70R17 tire
② Assumes P265/70R18 tire
③ 4x2: +3.47
   4x4: +3.05
④ 4x2: +3.87
   4x4: +3.43
⑤ 4x2: -0.10
   4x4: +0.18

09482_ARMA_C0010

TIRE, WHEEL AND BALL JOINT SPECIFICATIONS

Year	Model	OEM Tires Standard	OEM Tires Optional	Tire Pressures (psi) Front	Tire Pressures (psi) Rear	Wheel Size	Ball Joint Inspection	Lugnut Torque (ft. lbs.)
2004	Armada	P285/70R17	P265/70R18	35	35	std.: 17, opt: 18	①	98
	QX56	P265/70R18	None	35	35	18	①	98
2005	Armada	P285/70R17	P265/70R18	35	35	std.: 17, opt: 18	①	98
	QX56	P265/70R18	None	35	35	18	①	98

OEM: Original Equipment Manufacturer

PSI: Pounds Per Square Inch

① Axial play
  Upper: 0

09482_ARMA_C0011

BRAKE SPECIFICATIONS
All measurements in inches unless noted

Year	Model		Brake Disc Original Thickness	Brake Disc Minimum Thickness	Brake Disc Maximum Runout	Minimum Pad Thickness	Brake Caliper Bracket Bolts (ft. lbs.)	Brake Caliper Mounting Bolts (ft. lbs.)
2004	Armada	F	1.024	0.965	0.0016	0.039	155	20
		R	0.551	0.472	0.0020	0.039	—	32
	QX56	F	1.024	0.965	0.0016	0.039	155	32
		R	0.551	0.472	0.0020	0.039	—	24
2005	Armada	F	1.024	0.965	0.0016	0.039	155	32
		R	0.551	0.472	0.0020	0.039	—	24
	QX56	F	1.024	0.965	0.0016	0.039	155	32
		R	0.551	0.472	0.0020	0.039	—	24

09482_ARMA_C0012

SCHEDULED MAINTENANCE INTERVALS
Nissan Armada - Infiniti QX56

TO BE SERVICED	TYPE OF SERVICE	7.5	15	22.5	30	37.5	45	52.5	60
Engine oil & filter	R	✓	✓	✓	✓	✓	✓	✓	✓
Brake lines & cables	S/I		✓		✓		✓		✓
Brake pads and rotors	I	✓	✓	✓	✓	✓	✓	✓	✓
Driveshaft boots & propeller shaft (4x4)	L/I		✓		✓		✓		✓
Transmission, transfer & differential gear oil	I		✓		✓		✓		✓
Air cleaner filter	R				✓				✓
Engine coolant ①	R								✓
Spark plugs (Platinum)	R	Replace every 105,000 miles							
Drive belt(s) ②	S/I								✓
Cabin air filter	R		✓		✓		✓		✓
Exhaust system	I				✓				✓
Fuel lines	S/I				✓				✓
Fuel Filter ③									
Steering gear (box) & linkage, axle & suspension parts	I				✓				✓
Vapor lines	S/I				✓				✓

R: Replace   S/I: Service or Inspect   L: Lubricate

① Coolant: After 60,000 miles, inspect every 30,000 miles.

② Drive Belts: After 60,000 miles, inspect every 15,000 miles. Replace belts if damaged.

③ Fuel Filter: Maintenance free item.

FREQUENT OPERATION MAINTENANCE (SEVERE SERVICE)

 If a vehicle is operated under any of the following conditions it is considered severe service:

- Extremely dusty areas.

- Rough, muddy, or salt spread roads.

- 50% or more of the vehicle constant operation is in 32°C (90°F) or higher temperatures, or temperatures below 0°C (32°F).

- Prolonged idling (vehicle operation in stop and go traffic).

- Frequent short running periods (engine does not warm to normal operating temperatures).

- Police, taxi, delivery usage or trailer towing usage.

Oil & oil filter: replace every 3750 miles.

Brake pads, discs, drums & linings: service or inspect every 7500 miles.

Driveshaft boots & propeller shaft: service or inspect every 7500 miles.

Exhaust system: service or inspect every 7500 miles.

Steering gear (box) & linkage, (steering damper-4x4), axle & suspension parts: service or inspect every 7500 miles.

Steering linkage ball joints & front suspension ball joints: service or inspect every 7500 miles.

09482_ARMA_C0013

ENGINE REPAIR

Distributor

REMOVAL & INSTALLATION

These models use a Distributorless Ignition System (DIS) controlled by the Powertrain Control Module (PCM).

Alternator

REMOVAL & INSTALLATION

1. Before servicing the vehicle, refer to the Precautions Section.
2. Remove or disconnect the following:
   - Negative battery cable
   - Fan shroud
   - Drive belt
   - Lower alternator bracket
   - Alternator upper bolt
   - Alternator harness connectors
   - Alternator

To install:

3. Install or connect the following:
   - Alternator
   - Alternator harness connectors
   - Upper bolt, tighten to 48 ft. lbs. (65 Nm)

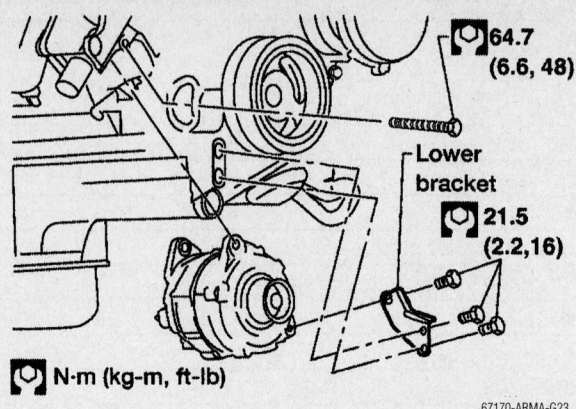

N·m (kg-m, ft-lb)

67170-ARMA-G23

Alternator mounting

   - Lower bracket, tighten to 16 ft. lbs (22 Nm)
   - Drive belt
   - Fan shroud
   - Negative battery cable

Ignition Timing

ADJUSTMENT

Ignition timing is controlled by the ECM and manual adjustment is not possible.

Engine Assembly

REMOVAL & INSTALLATION

1. Before servicing the vehicle, refer to the Precautions Section.
2. Drain the cooling system.
3. Partially drain the automatic transmission fluid.
4. Relieve the fuel system pressure.
5. Remove or disconnect the following:

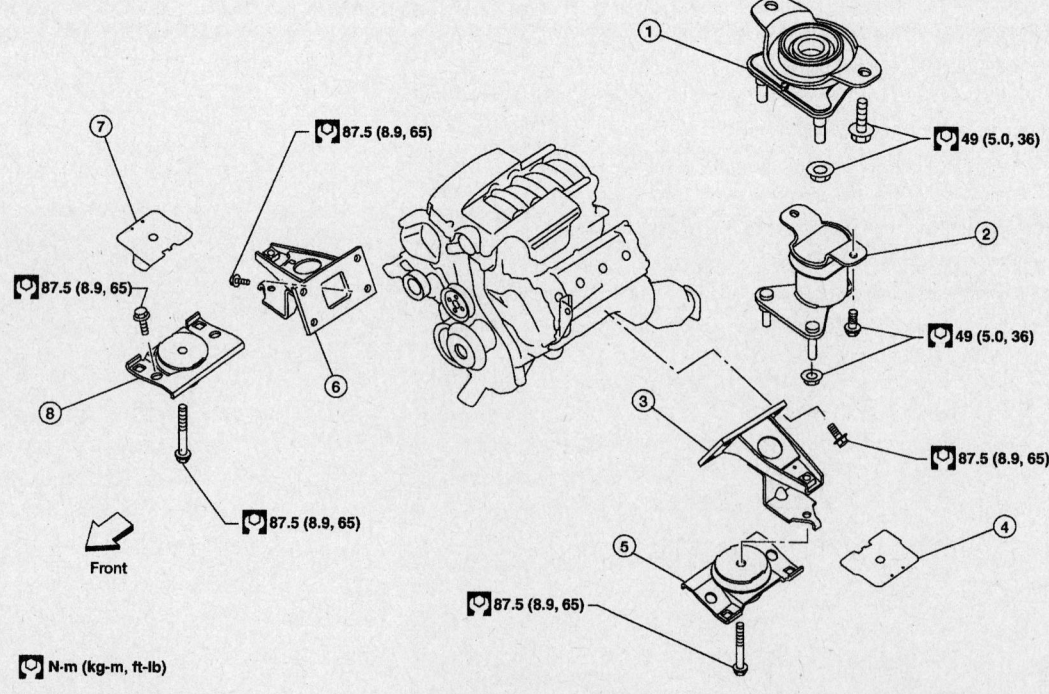

N·m (kg-m, ft-lb)

1. Rear engine mounting insulator 4x4
2. Rear engine mounting insulator 4x2
3. LH engine mounting bracket
4. LH Heat shield plate
5. LH engine mounting insulator
6. RH engine mounting bracket
7. RH Heat shield plate
8. RH engine mounting insulator

Engine mounts—Armada

67170-ARMA-G24

- Hood
- Cowl extension
- Engine cover
- Air intake assembly
- Vacuum hose between vehicle and engine
- Radiator hoses
- Radiator
- Drive belts
- Engine fan
- Wiring harness
- ECM
- Power steering reservoir tank and oil pump
- A/C compressor
- Brake booster vacuum line
- EVAP line
- Fuel hose
- Heater hoses
- Exhaust manifolds-Armada only
- Front final drive assembly-Armada only
- Automatic transmission dipstick tube assembly
- Automatic transmission

6. Install engine slings onto the left and right cylinder heads and tighten to 33 ft. lbs. (45 Nm).

7. Attach an engine hoist to slings and lift engine out of the vehicle

To install:

8. Lower engine into the vehicle

9. Install or connect the following:
- Automatic transmission
- Automatic transmission dipstick tube assembly

- Front final drive assembly-Armada only
- Exhaust manifolds-Armada only
- Heater hoses
- Fuel hose
- EVAP line
- Brake booster vacuum line
- A/C compressor
- Power steering reservoir tank and oil pump
- ECM
- Wiring harness
- Engine fan
- Drive belts
- Radiator and radiator hoses
- Vacuum hose between vehicle and engine
- Air intake assembly
- Engine cover
- Cowl extension
- Hood

10. Refill the automatic transmission fluid.

11. Refill the cooling system.

12. Start the engine and check for leaks.

Water Pump

REMOVAL & INSTALLATION

1. Before servicing the vehicle, refer to the Precautions Section.

2. Drain the cooling system.

3. Remove or disconnect the following:
- Engine splash guard

- Air intake assembly
- Accessory drive belt

➡**Leave tensioner pulley in its fixed position.**

- Water pump pulley
- Water pump

To install:

4. Install or connect the following:
- Water pump with a new gasket. Tighten bolts to 18 ft. lbs. (25 Nm).
- Water pump pulley and tighten bolts to 87 in. lbs. (10 Nm).
- Accessory drive belt
- Air intake assembly
- Engine splash guard

5. Refill the cooling system.

6. Start the engine and check for leaks.

Heater Core

REMOVAL & INSTALLATION

Front Heater Assembly

QX56

1. Before servicing the vehicle, refer to the Precautions Section.

2. Discharge the refrigerant from the A/C system.

3. Drain the cooling system.

4. Remove or disconnect the following:
- Negative battery cable
- Front heater hoses from the front heater core

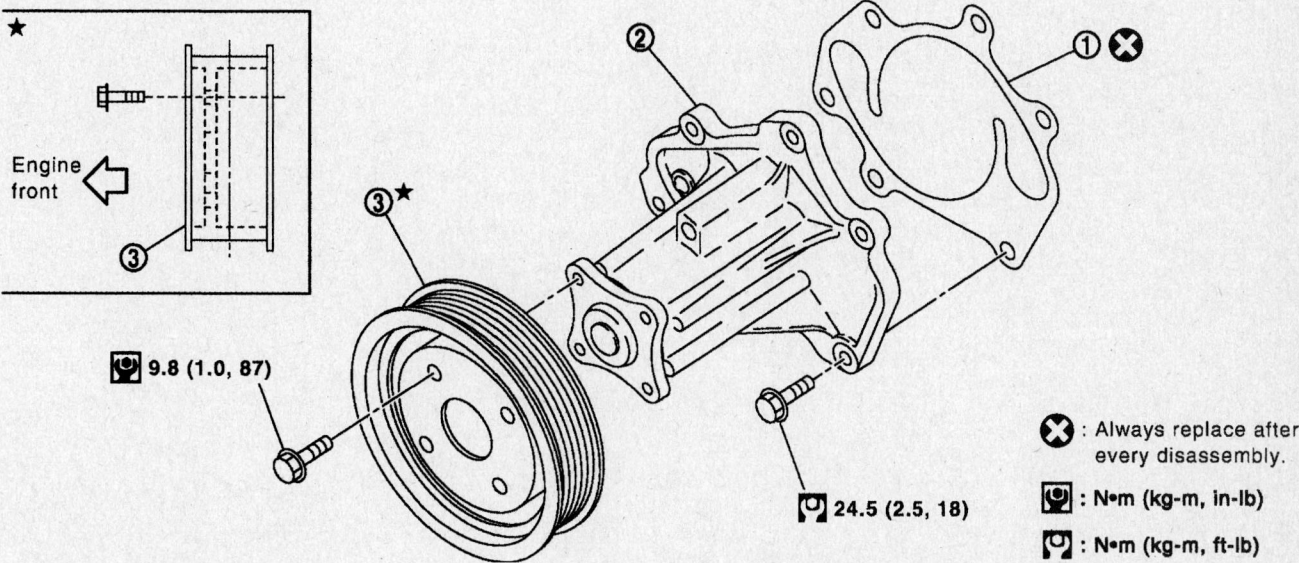

9.8 (1.0, 87)

24.5 (2.5, 18)

❌ : Always replace after every disassembly.

: N•m (kg-m, in-lb)

: N•m (kg-m, ft-lb)

Engine front

1. Gasket   2. Water pump   3. Water pump pulley

Water pump mounting

67170-ARMA-G25

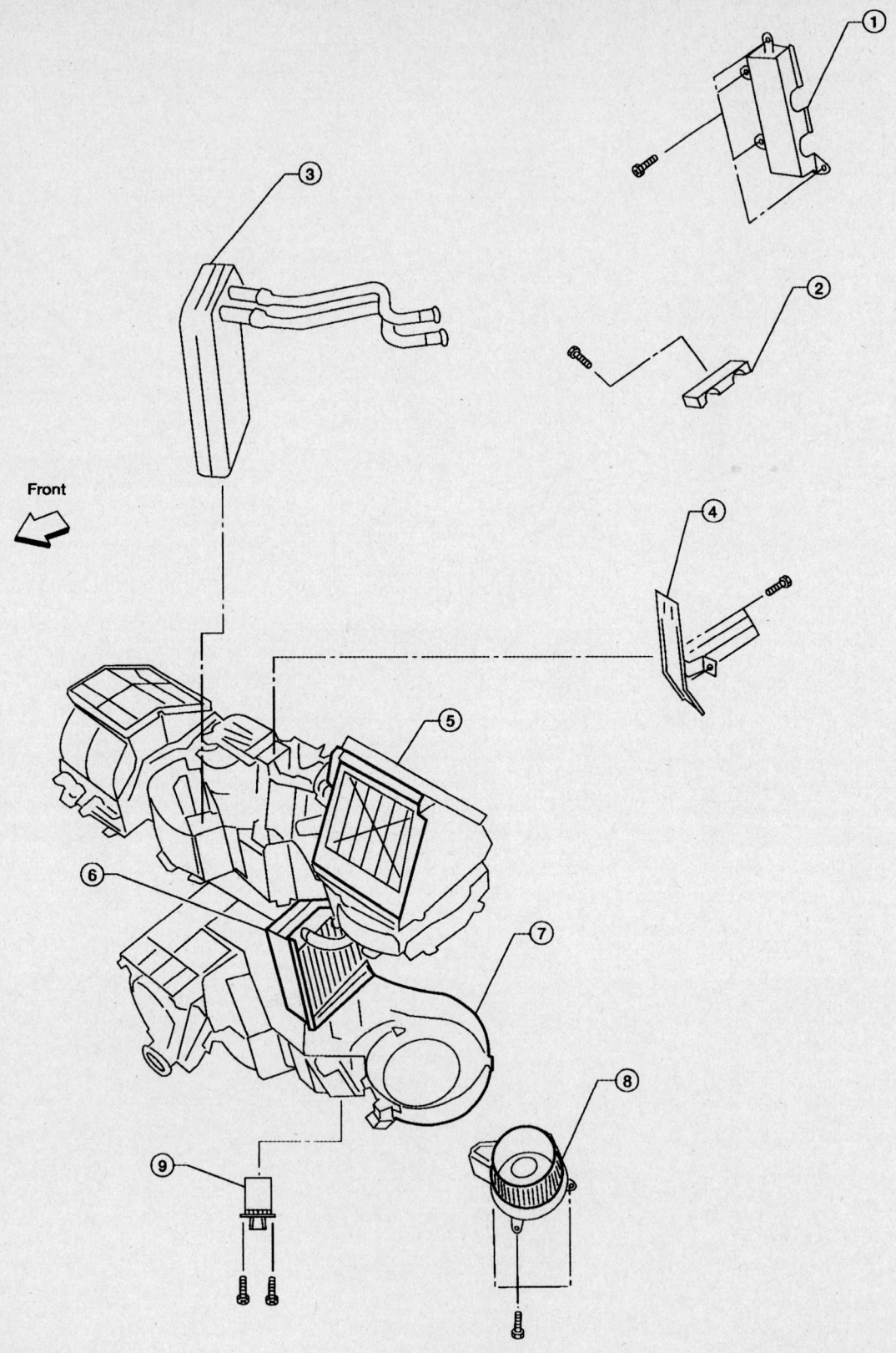

Front

1. Heater core cover
2. Heater core pipe bracket
3. Heater core
4. Upper bracket
5. Upper heater and cooling unit case
6. A/C evaporator
7. Lower heater and cooling unit case
8. Blower motor
9. Variable blower control

09482_ARMA_G0001

Exploded view front heater core assembly—QX56

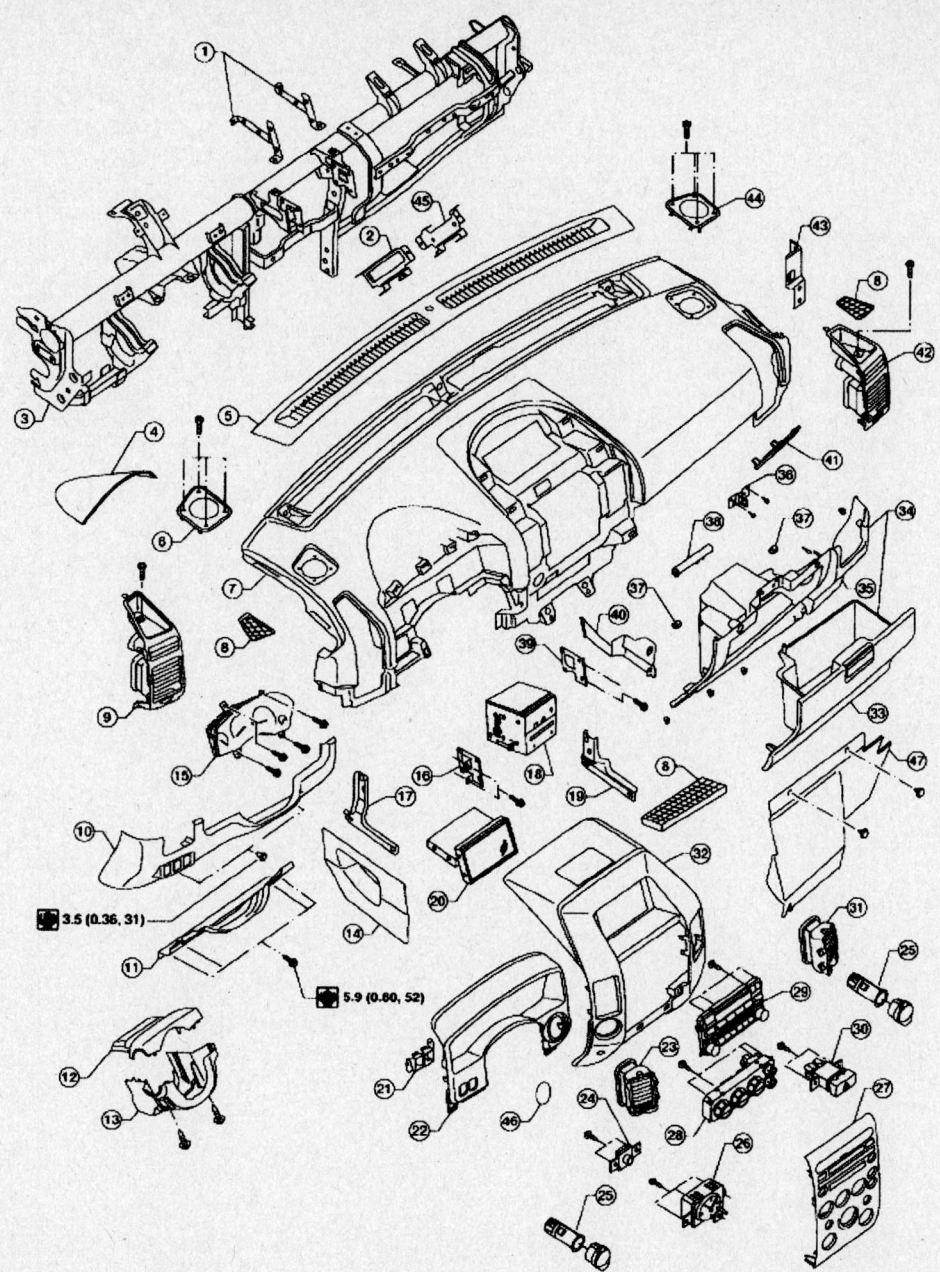

3.5 (0.36, 31)

5.9 (0.60, 52)

1. Steering member brackets	2. Bracket assembly LH	3. Steering member assembly
4. Combination meter cover	5. Defrost grille	6. Instrument panel speaker LH
7. Instrument panel	8. Bin mat	9. Side ventilator assembly LH
10. Instrument lower cover LH	11. Lower knee protector LH	12. Steering column cover upper
13. Steering column cover lower	14. Center console lower cover LH	15. Combination meter
16. Audio unit bracket LH	17. Driver instrument stay	18. Audio unit
19. Passenger instrument stay	20. Display assembly	21. Switch assembly
22. Cluster lid A	23. Cluster lid D ventilator LH	24. Front passenger air bag status light
25. Power socket LH and power socket RH	26. Clock	27. Cluster lid C
28. Front air control	29. Audio control	30. Hazard switch
31. Cluster lid D ventilator RH	32. Cluster lid D	33. Glove box door
34. Glove box assembly	35. Lower instrument panel RH	36. Glove box striker
37. Rubber bumpers	38. Glove box damper	39. Audio unit bracket RH
40. Center console lower cover RH	41. Fuse block cover	42. Side ventilator assembly RH
43. Instrument panel bracket	44. Instrument panel speaker RH	45. Bracket assembly RH
46. Key cylinder escutcheon	47. Instrument lower cover RH	

09482_ARMA_G0002

Exploded view of the Instrument Panel assembly—QX56 (IP-10)

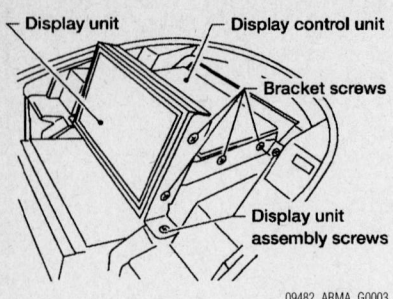

09482_ARMA_G0003

Exploded view of the display unit—QX56

- Pressure pipes from the front expansion valve
- Left-hand instrument panel lower cover
- Steering column
- Right-hand instrument panel lower cover
- Glove box assembly
- Center console lower covers

- Center console electrical connectors
- Center console

5. Remove the power sockets (cigarette lighter) as follows:

   a. Remove inner socket from the ring, while pressing the hook on the ring out from square hole.

   b. Disconnect power socket connector.

   c. Remove ring from power socket finisher while pressing pawls.

6. Remove or disconnect the following:
- Radio cluster cover
- Radio unit
- Display unit cluster lid
- Display unit and center speaker connectors
- Display unit
- Left-hand lower knee protector
- Defroster grille
- GPS antenna

- Optical sensor
- Side ventilator assemblies
- A-pillar trim
- Passenger side airbag module
- Instrument panel electrical connectors

7. Instrument panel

8. Disconnect the steering member from each side of the body

9. Remove the front heater assembly with it attached to the steering member.

10. Remove the upper bracket from the heater assembly.

11. Remove the heater core cover.

12. Remove the heater core.

13. Installation is the reverse orders of removal.

ARMADA

1. Before servicing the vehicle, refer to the Precautions Section.

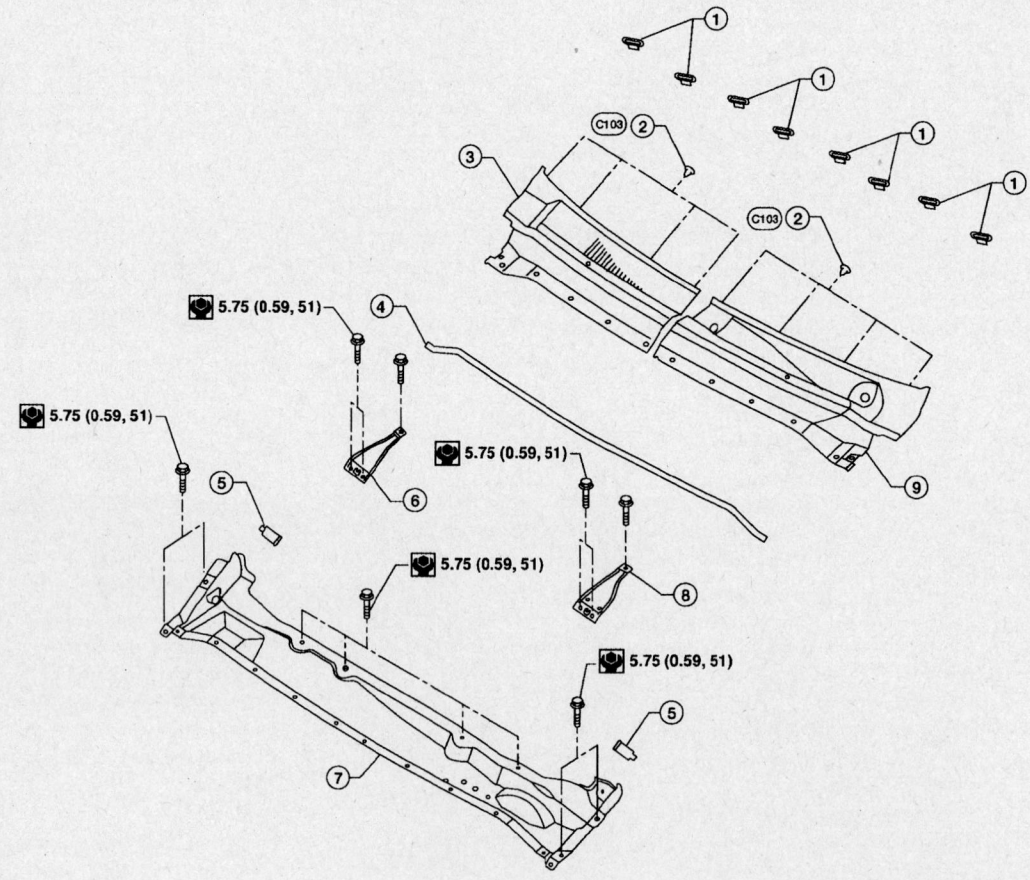

: N·m (kg-m, in-lb)

1.	Grommets	2.	Plastic clips	3.	Cowl top RH
4.	Cowl top seal	5.	Drain tubes	6.	Cowl top extension bracket RH
7.	Cowl top extension	8.	Cowl top extension bracket LH	9.	Cowl top LH

09482_ARMA_G0004

Exploded view of the cowl top—Armada

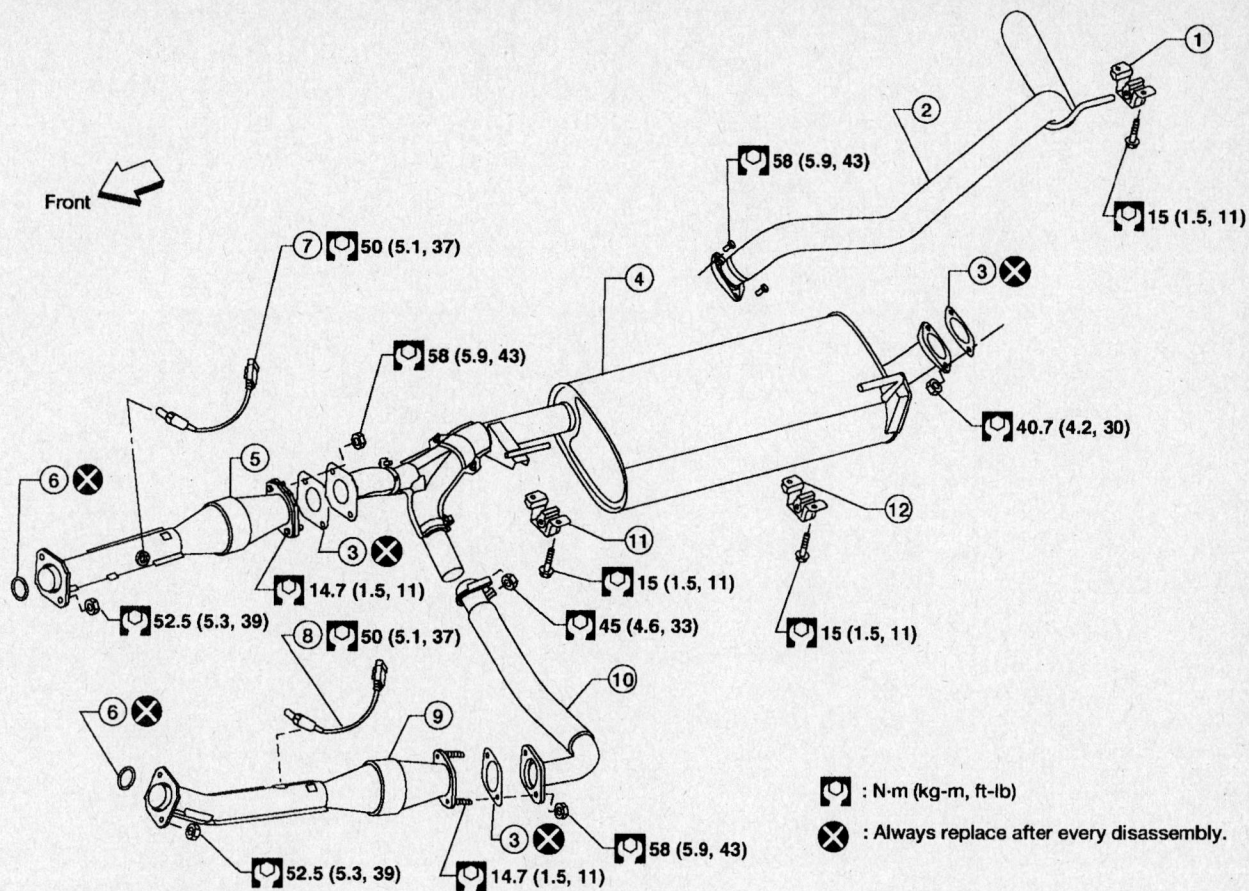

1. Tailpipe hanger bracket
2. Tailpipe
3. Gasket
4. Main muffler
5. Right front exhaust tube
6. Ring gasket
7. Heated oxygen sensor 2 (bank 2)
8. Heated oxygen sensor 2 (bank 1)
9. Left front exhaust tube
10. Center exhaust tube
11. Muffler hanger bracket front
12. Muffler hanger bracket rear

09482_ARMA_G0005

Exploded view of the exhaust system—Armada

2. Discharge the refrigerant from the A/C system.
3. Drain the cooling system.
4. Remove or disconnect the following:
- Negative battery cable
- Wiper arms
- Cowl top seal
- Cowl top extension brackets
- Wiper motor and connecting rod linkage
- Windshield washer hose
- Cowl top extension
- Exhaust system
- Front heater hoses from front heater core
- Pressure pipes from the front expansion valve
- Right-hand lower instrument panel
- Transmission shifter
- Center console
- Steering column
- Left-hand lower instrument panel
- Gauge cluster front and top cover
- Instrument gauge cluster
- Defroster grille
- Optical sensor
- Side ventilator assemblies
- A-pillar trim
- Instrument panel electrical connections
- Instrument panel

5. Disconnect the steering member from each side of the vehicle body.
6. Remove the front heater and cooling unit assembly with it attached to the steering member.
7. Remove the heater unit assembly from the steering member.
8. Remove the upper bracket from the heater assembly.
9. Remove the heater core cover.
10. Remove the heater core.
11. Installation is the reverse orders of removal.

Rear Heater Assembly

1. Before servicing the vehicle, refer to the Precautions Section.
2. Drain the cooling system.
3. Remove or disconnect the following:
- Second and third row seatbelts
- Third row seat belt buckles
- Back door weatherstripping
- Rear cargo area lamp
- Rear upper finishing trim
- Rear door kick plate
- Rear lower finishing trim
- Door open/close link
- Cargo net hooks
- Right-hand lower luggage side trim
- Rear heater hoses from the heater core
- Rear heater core bracket
- Heater
4. Installation is the reverse of removal.

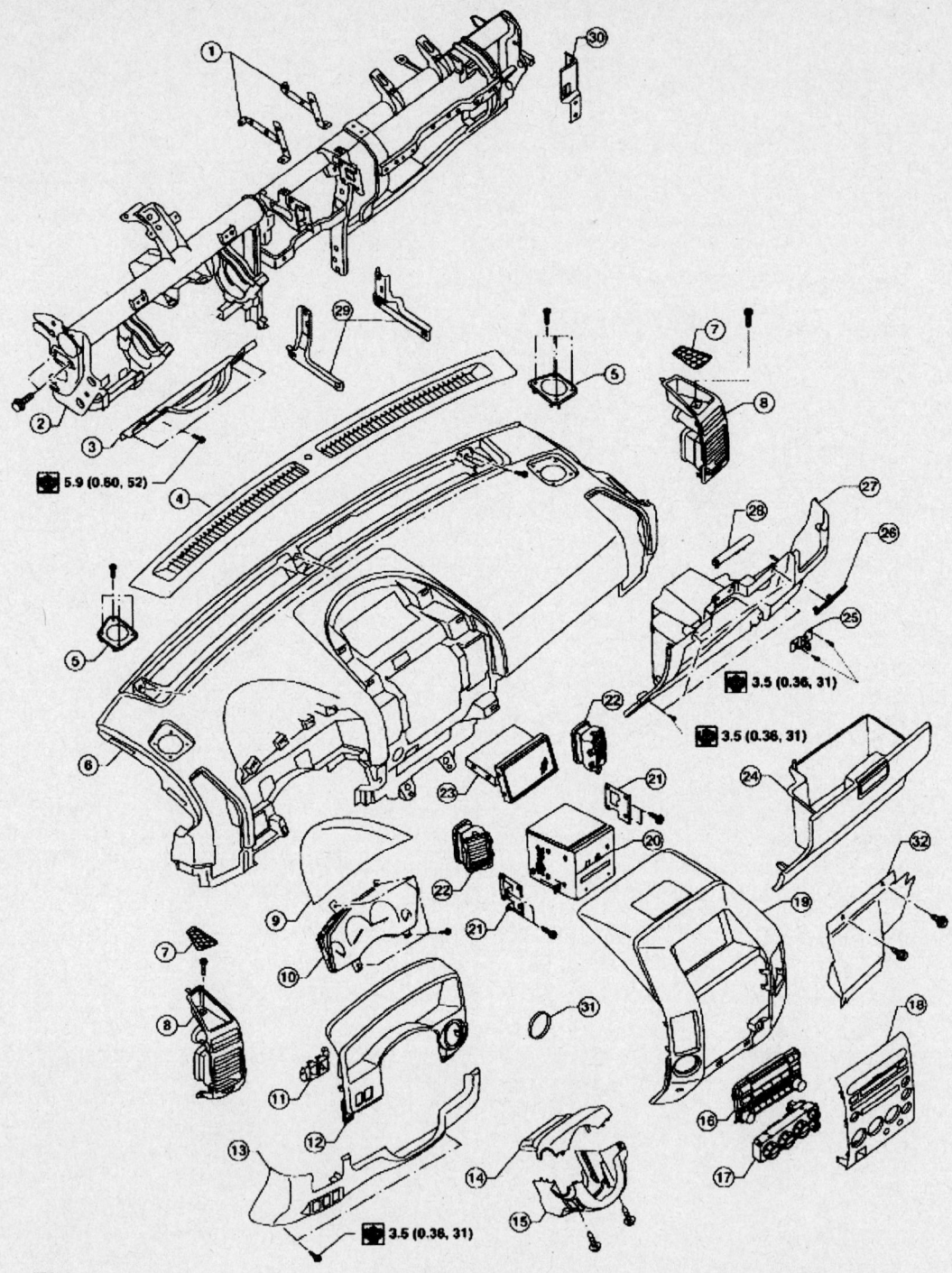

5.9 (0.60, 52)

3.5 (0.36, 31)

3.5 (0.36, 31)

3.5 (0.36, 31)

N·m (kg-m, in-lb)

1. Display unit bracket RH/LH	2. Steering member assembly	3. Lower knee protector
4. Defroster grille	5. Speaker grille RH/LH	6. Instrument panel and pad assembly
7. Deck pocket mat RH/LH	8. Side ventilator assembly RH/LH	9. Meter cover
10. Combination meter	11. Switch assembly	12. Cluster lid A
13. Lower instrument panel LH	14. Upper steering column cover	15. Lower steering column cover
16. Audio display switch assembly	17. Front air control	18. Cluster lid C
19. Cluster lid D	20. Audio unit	21. Radio Bracket RH/LH
22. Center ventilator assembly RH/LH	23. Display assembly	24. Glove box
25. Glove box lid striker	26. Fuse block cover	27. Lower instrument panel RH
28. Glove box damper	29. Instrument stay RH/LH	30. Instrument side bracket
31. Key cylinder escutcheon	32. Lower instrument panel RH	

09482_ARMA_G0006

Exploded view of the instrument panel—Armada

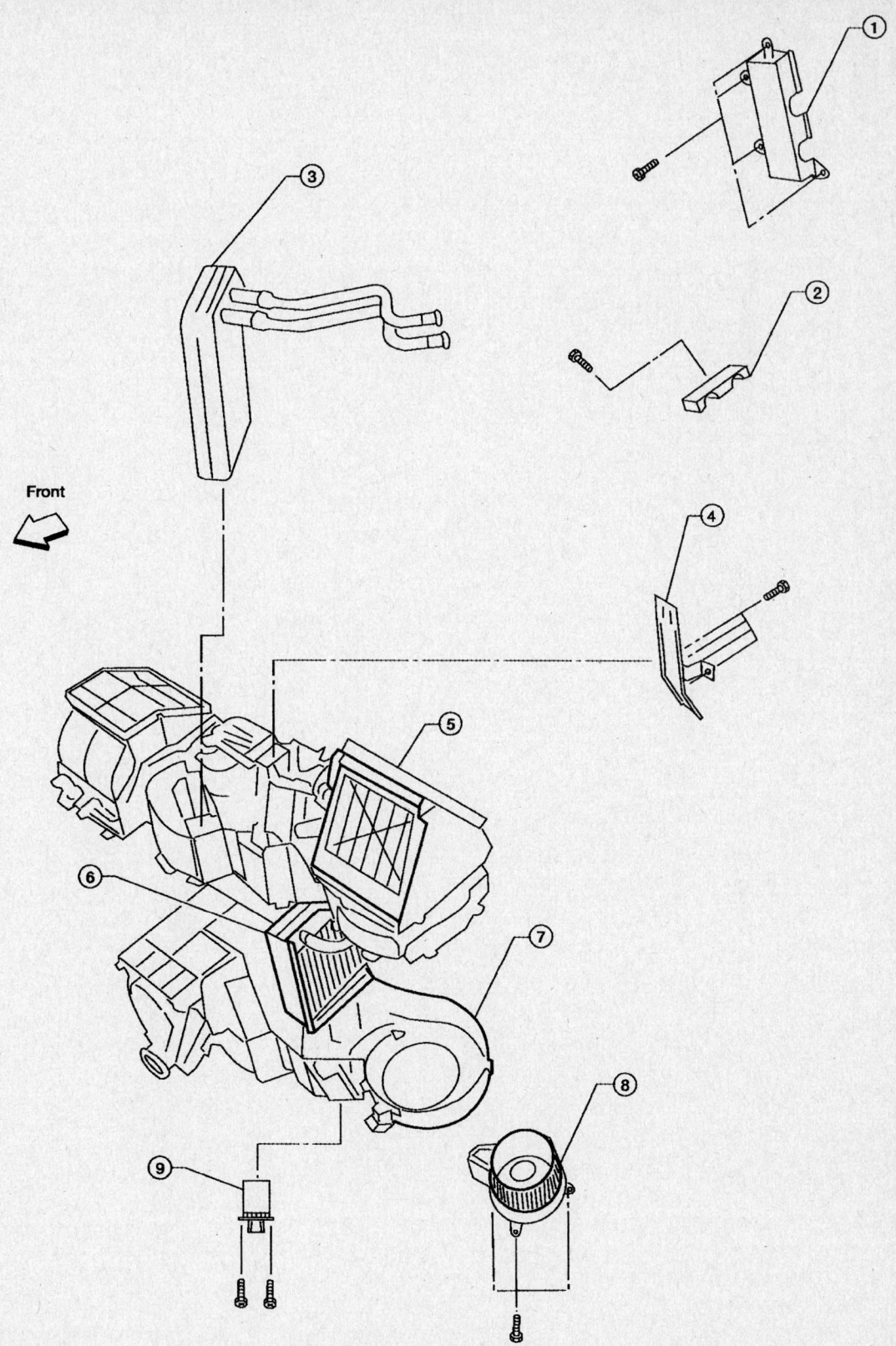

Front

1. Heater core cover	2. Heater core pipe bracket	3. Heater core
4. Upper bracket	5. Upper heater and cooling unit case	6. A/C evaporator
7. Lower heater and cooling unit case	8. Blower motor	9. Variable blower control

67170-ARMA-G26

Front heater/AC assembly—Armada

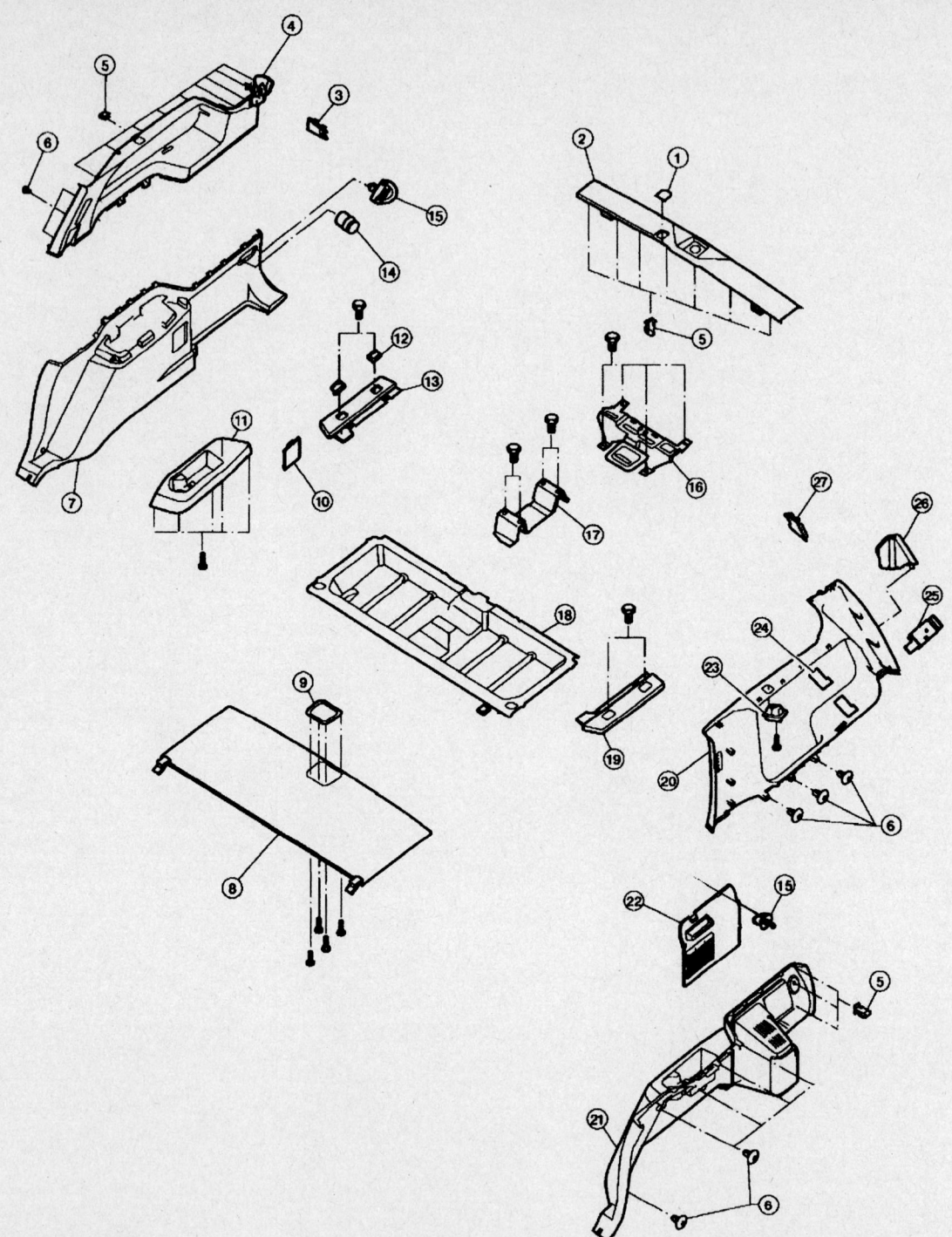

1. Child seat anchor cover	2. Tail gate kicking plate	3. Luggage trim cap RH
4. Luggage side upper finisher RH	5. Metal clips	6. Plastic clips
7. Luggage side lower finisher RH	8. Luggage floor board assembly	9. Luggage floor board latch
10. Leak check cap	11. Cup holder	12. Cargo net hook
13. Floor side finisher RH	14. Power point assembly	15. Trunk net hook
16. Rear luggage box bracket	17. Front luggage floor bracket	18. Storage box
19. Floor side finisher LH	20. Luggage side upper finisher LH	21. Luggage side lower finisher LH
22. Luggage side lid	23. Coat hook	24. Cap lower seat belt finisher
25. Back door open/close switch	26. Luggage side upper cap	27. Luggage trim cap RH

09482_ARMA_G0007

Exploded view of the rear interior trim

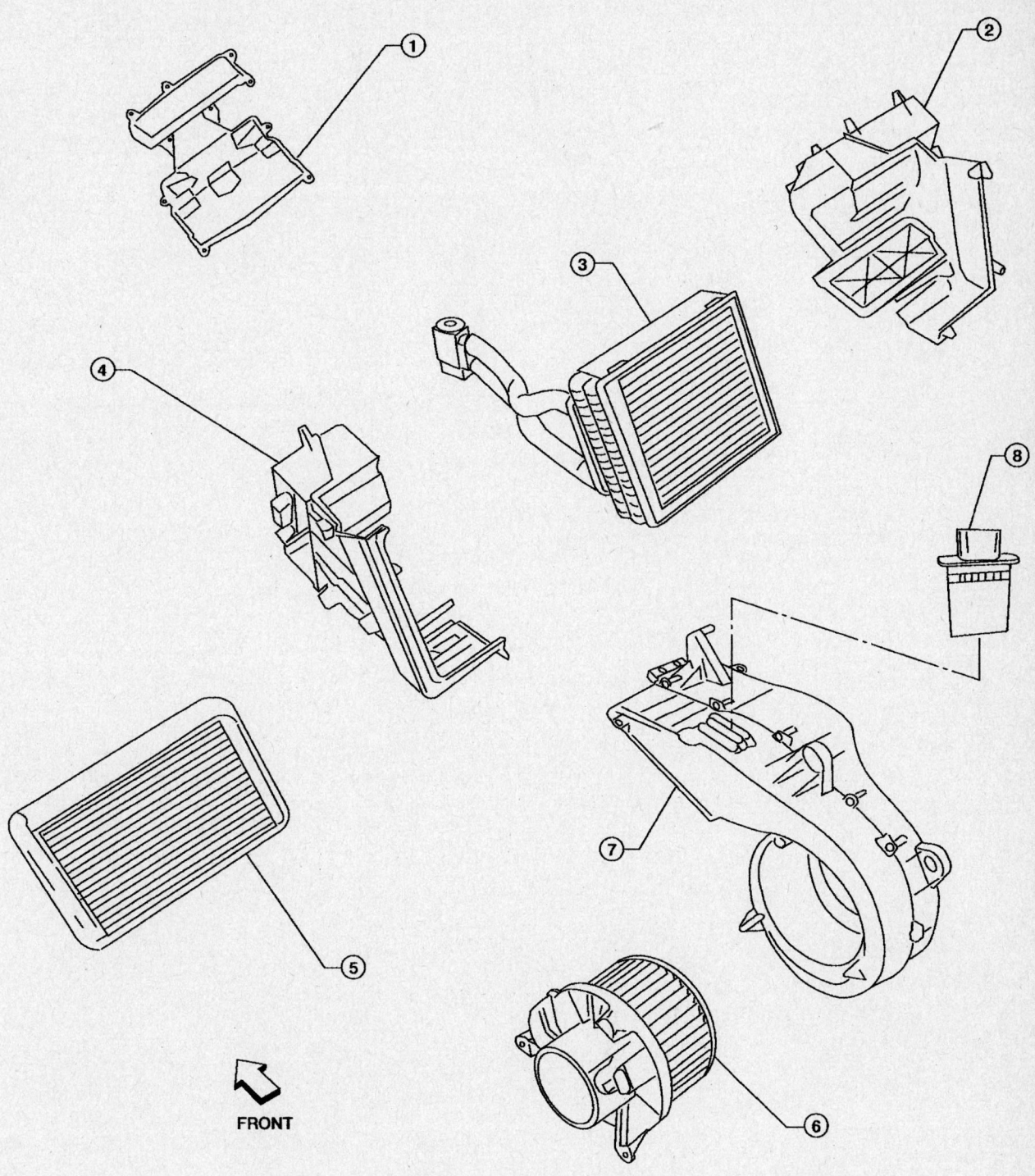

FRONT

1. Front cover
4. Side cover
7. Blower motor case

2. Evaporator and heater core case
5. Heater core
8. Blower motor resistor

3. Evaporator
6. Blower motor

67170-ARMA-G27

Rear heater/AC assembly—Armada

Cylinder Head

REMOVAL & INSTALLATION

1. Before servicing the vehicle, refer to the Precautions Section.
2. Remove or disconnect the following:
   - Engine assembly
   - Belt tensioner
   - Idler pulley
   - Thermostat housing and hose
   - Oil pan and strainer
   - Fuel rail and injector assembly
   - Intake manifold
   - Ignition coil
   - Rocker cover
   - Crankshaft pulley
   - Front engine cover
   - Oil pump
   - Timing chain
   - Camshaft sprockets
   - Camshafts
   - Cylinder head, removing bolts in reverse order shown in figure
3. Install the cylinder head with a new gasket. Tighten the bolts in sequence as follows:
   a. Step 1: 72 ft. lbs. (98 Nm)
   b. Step 2: Loosen all bolts completely
   c. Step 3: 33 ft. lbs. (44 Nm)
   d. Step 4: Plus 60 degrees
   e. Step 5: Plus 60 degrees

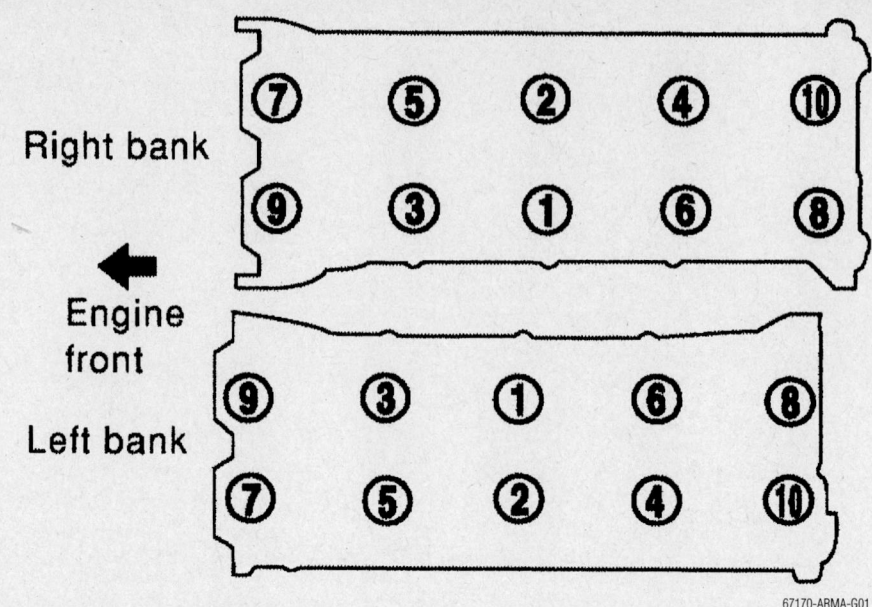

67170-ARMA-G01

Cylinder head torque sequence

4. Install or connect the following:
   - Camshaft
   - Camshaft sprockets
   - Timing chain
   - Oil pump
   - Front engine cover
   - Crankshaft pulley
   - Rocker cover
   - Ignition coil
   - Intake manifold
   - Fuel tube and injector assembly
   - Oil pain and strainer
   - Thermostat housing and hose
   - Idler pulley
   - Belt tensioner
   - Engine assembly
5. Start the engine and check for leaks

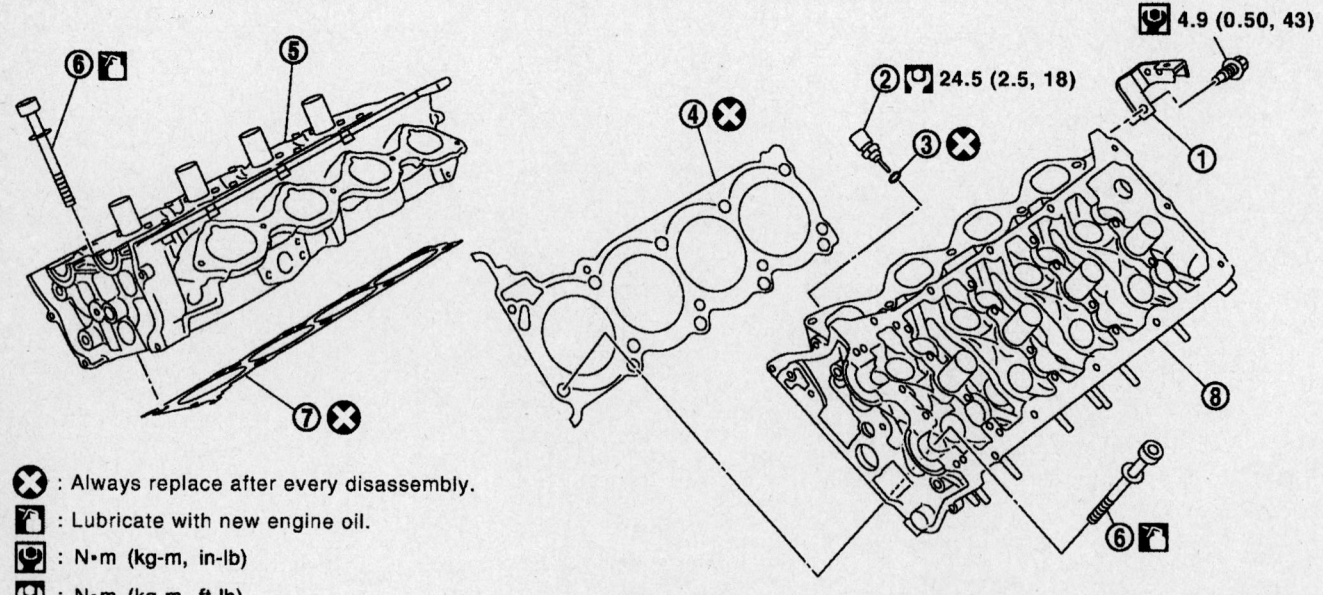

- ⊗ : Always replace after every disassembly.
- 🛢 : Lubricate with new engine oil.
- 🔧 : N•m (kg-m, in-lb)
- 🔧 : N•m (kg-m, ft-lb)

1. Harness bracket	2. Engine coolant temperature sensor	3. Washer
4. Cylinder head gasket (left bank)	5. Cylinder head (right bank)	6. Cylinder head bolt
7. Cylinder head gasket (right bank)	8. Cylinder head (left bank)	

Exploded view of the cylinder head assembly

67170-ARMA-G28

Intake Manifold

REMOVAL & INSTALLATION

1. Before servicing the vehicle, refer to the Precautions Section.
2. Drain the cooling system.
3. Relieve the fuel system pressure.
4. Remove or disconnect the following:
   - Engine cover
   - Air intake assembly
   - Fuel supply hose quick connector using special tool J-45488
   - Wiring harnesses and brackets from manifold
   - Vacuum hoses
   - PCV hose and tube
   - Electric throttle control actuator, loosening bolts diagonally
   - Fuel injectors
   - Fuel rail assembly

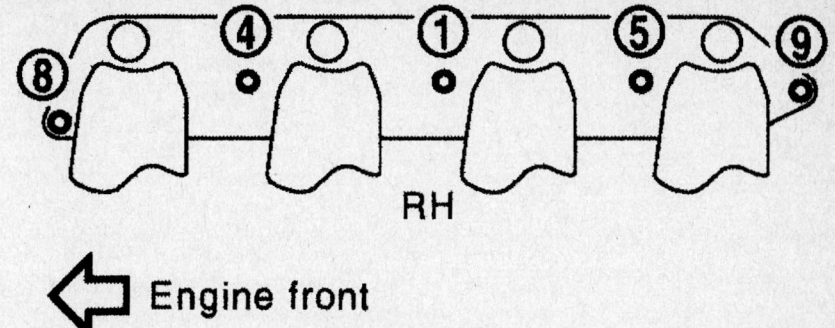

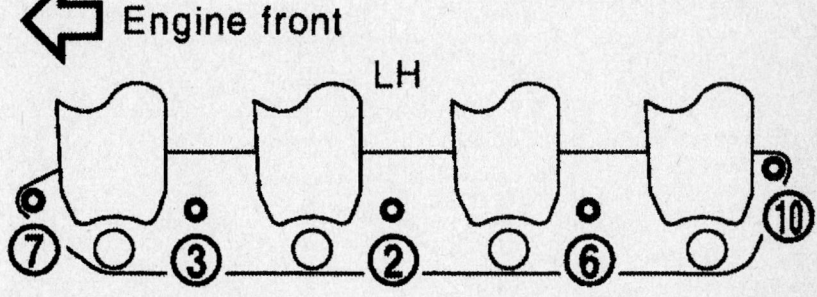

67170-ARMA-G02

Intake manifold torque sequence

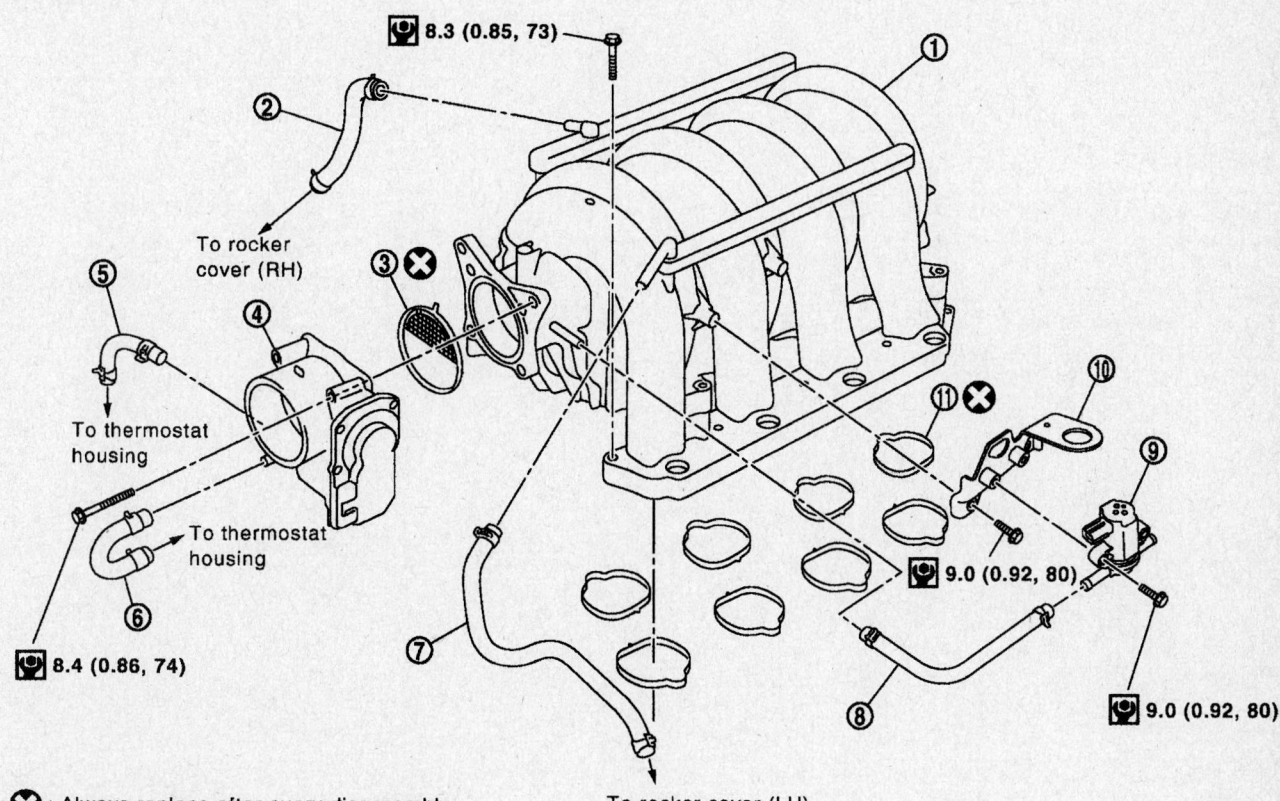

⊗ : Always replace after every disassembly.

🔧 : N•m (kg-m, in-lb)

1.	Intake manifold	2.	PCV hose	3.	Gasket
4.	Electric throttle control actuator	5.	Water hose	6.	Water hose
7.	PCV hose	8.	EVAP hose	9.	EVAP canister purge control solenoid valve
10.	Bracket	11.	Gasket		

Intake manifold and related parts

67170-ARMA-G29

- Intake manifold, removing bolts in reverse order shown in figure

To install:

5. Install the intake manifold with new gaskets. Tighten the bolts in order as shown.

6. Install or connect the following:
- Fuel rail assembly
- Fuel injectors
- Electronic throttle control actuator, tightening the bolts in several steps
- PCV hose
- Vacuum hoses
- Wiring harnesses

7. Connect the fuel supply hose as follows:

a. Apply a thin layer of engine oil on the tube from tip end to spool end.

b. Insert tube into quick connector past the white identification mark

c. Insert tube into quick connector until top spool is completely inside the connector and 2nd level spool is exposed right below the connector.

d. Pull slightly on the quick connector to ensure it is fully engaged.

e. Install quick connector cap on quick connector joint.

8. Install or connect the following:
- Air intake assembly
- Engine cover

9. Refill the cooling system.

10. Start engine and check for leaks.

Exhaust Manifold

REMOVAL & INSTALLATION

1. Before servicing the vehicle, refer to the Precautions Section.

2. Drain the cooling system.

3. Remove or disconnect the following:
- Air intake assembly
- Engine splash guard
- Radiator and radiator hoses
- Accessory drive belt

4. Remove the air fuel ratio sensors as follows:

- Engine cover
- Wiring harness from each sensor
- Sensors, using special tool J-38356
- Front cross bar

5. Remove the left exhaust manifold as follows:

a. Remove the exhaust front tube.

b. Remove the exhaust manifold cover.

c. Loosen the nuts in reverse order shown in figure.

d. Remove studs from position 2, 4, 6, and 8 and remove manifold.

6. Remove right exhaust manifold as follows:

a. Remove the exhaust front tube.

b. Remove the oil level gauge guide.

c. Remove the exhaust manifold cover.

d. Loosen the nuts in reverse order shown in figure.

e. Remove studs from position 2, 4, 6, and 8 and remove manifold.

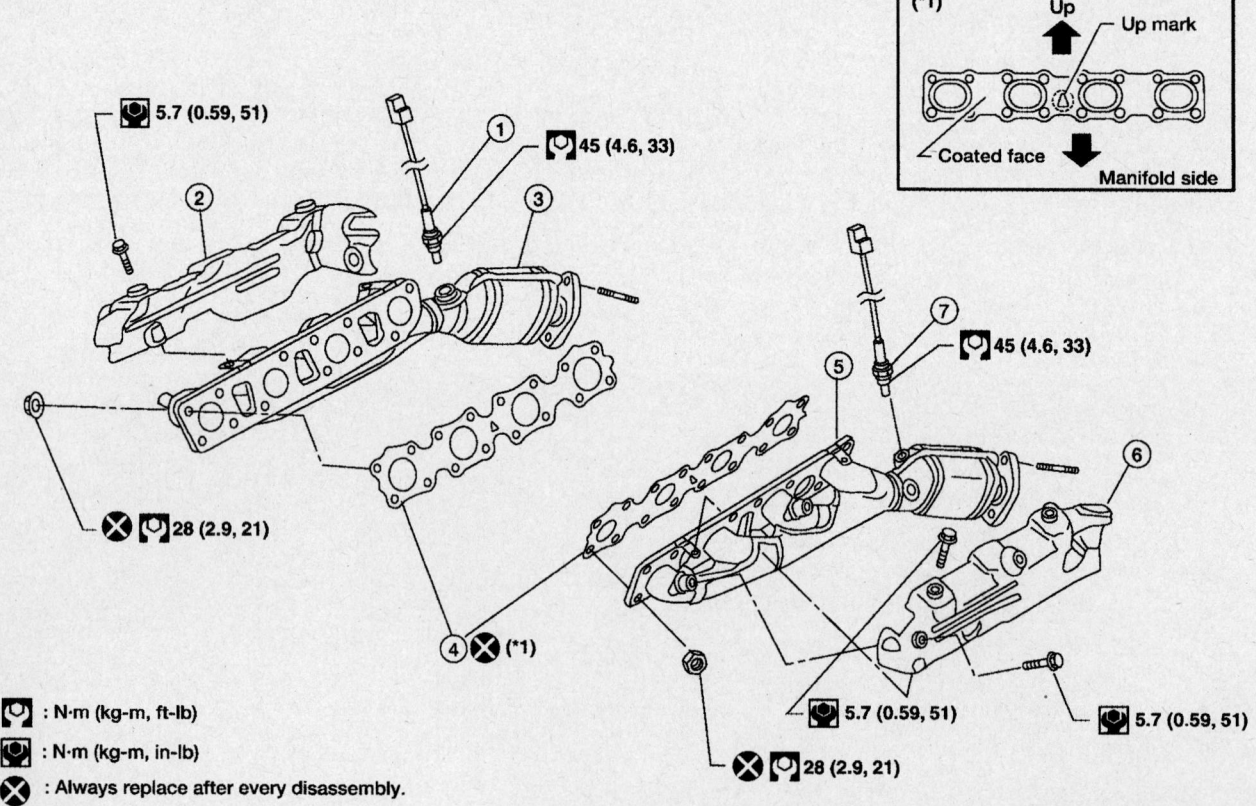

: N·m (kg-m, ft-lb)

: N·m (kg-m, in-lb)

: Always replace after every disassembly.

1. Air fuel ratio (A/F) sensor 1 (bank 2)
2. Exhaust manifold cover (right bank)
3. Exhaust manifold (right bank)
4. Gaskets
5. Exhaust manifold (left bank)
6. Exhaust manifold cover (left bank)
7. Air fuel ratio (A/F) sensor 1 (bank 1)

Exhaust manifolds and related parts

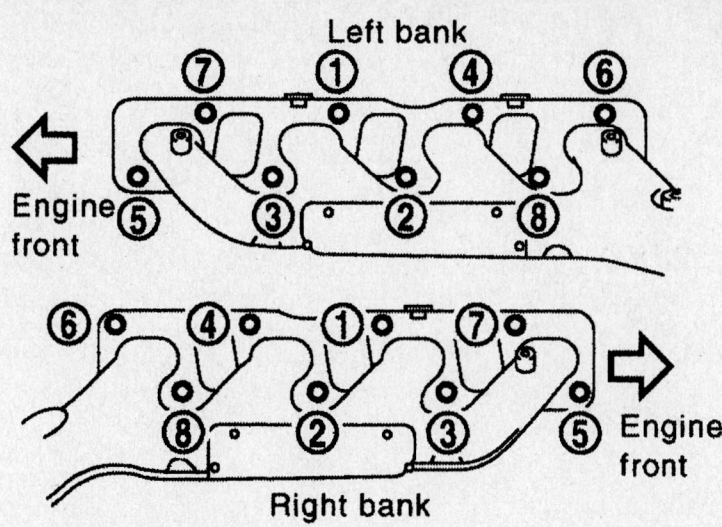

Exhaust manifold torque sequence

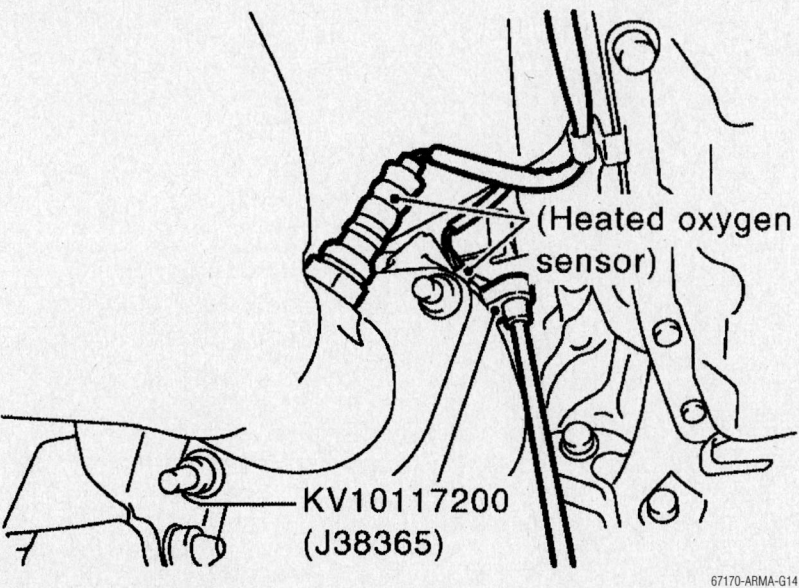

Removing the air-fuel ratio sensors

To install:

7. Install or connect the following:
- Exhaust manifold gasket with triangle mark facing up and coated (gray) face toward exhaust manifold.
- Exhaust manifold, tightening the nuts as shown in figure
- Exhaust manifold cover
- Oil level gauge guide (right side only)
- Exhaust front tube
- Front cross bar
- Air fuel ratio sensors, with anti-seize lubricant
- Engine cover
- Drive belts
- Radiator and radiator hoses
- Engine splash guard
- Air intake assembly
8. Refill the cooling system
9. Start engine and check for leaks.

Camshaft and Valve Lifters

REMOVAL & INSTALLATION

1. Before servicing the vehicle, refer to the Precautions Section.
2. Remove or disconnect the following:
- Negative battery cable
- Engine cover
- Air intake assembly
- Engine wiring harnesses on rocker cover
- Throttle control actuator
- Ignition coil
- PCV hose from PCV valve
3. Remove rocker cover, loosening the bolts in the reverse order.
4. Turn the crankshaft until the No. 1 cylinder is set at Top Dead Center (TDC).
5. Remove timing chain case covers.
6. Matchmark the timing chain, aligning with the camshaft sprocket marks.
7. Remove chain tensioner from left bank as follows:
   a. Squeeze end clips and push plunger into tensioner body.
   b. Secure plunger using stopper pin.
   c. Remove the chain tensioner.
8. Remove the chain tensioner from right bank as follows:
   a. Remove the chain tensioner cover using special tool J-37228.
   b. Squeeze end clips and push plunger into tensioner body.
   c. Secure plunger using stopper pin.
   d. Remove the chain tensioner.
9. With camshaft locked with a wrench, loosen bolts to remove camshaft sprocket.
10. Remove front cover bolts.
11. Remove camshaft brackets, removing bolts in reverse order shown in figure.
12. Remove camshaft.
13. Remove valve lifters.

➡Matchmark the drivetrain components so each part can be reinstalled in its original position.

To install:

14. Install valve lifters.
15. Install camshaft, refer to table for correct placement.
16. Install camshaft brackets as follows:
   a. Refer to location mark on upper surface of bracket.
   b. Installation mark should be correctly read when viewed from intake side.
17. Install camshaft bracket #1 as follows:
   a. Apply liquid gasket to bracket and backside of front cover as shown in figure.
   b. Carefully position and mount camshaft bracket #1.
   c. Temporarily tighten front cover bolts
18. Tighten fixing bolts for camshaft brackets as follows:
   a. Step 1: Bolts 9-12: 17 in. lbs. (1.9 Nm)
   b. Step 2: Bolts 1-8: 17 in. lbs. (1.9 Nm)
   c. Step 3: All bolts: 52 in. lbs. (5.9 Nm)

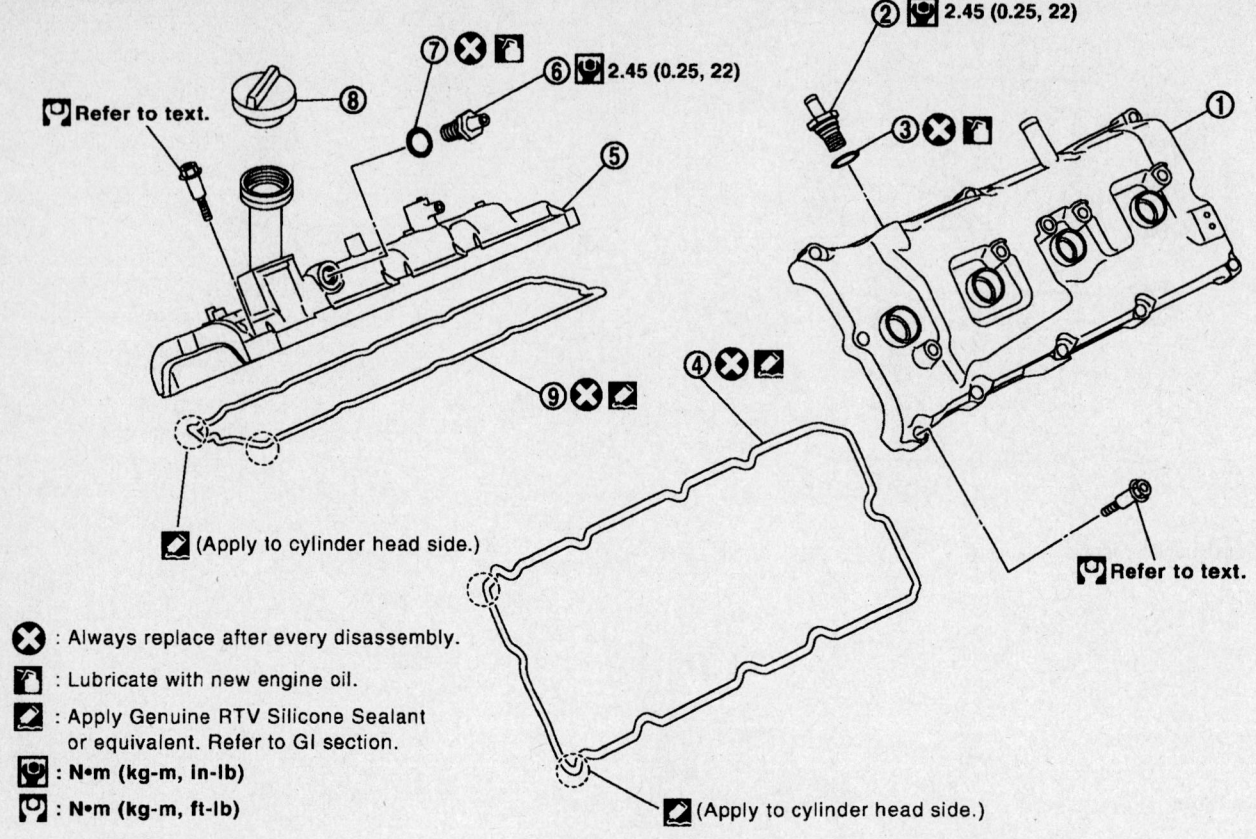

2.45 (0.25, 22)

2.45 (0.25, 22)

Refer to text.

Refer to text.

(Apply to cylinder head side.)

(Apply to cylinder head side.)

❌ : Always replace after every disassembly.

🛢 : Lubricate with new engine oil.

✏ : Apply Genuine RTV Silicone Sealant or equivalent. Refer to GI section.

🔧 : N•m (kg-m, in-lb)

🔧 : N•m (kg-m, ft-lb)

1. Rocker cover (LH)	2. PCV control valve	3. O-ring
4. Rocker cover gasket (LH)	5. Rocker cover (RH)	6. PCV control valve
7. O-ring	8. Oil filler cap	9. Rocker cover gasket (RH)

09482_ARMA_G0008

Exploded view of the rocker cover assembly

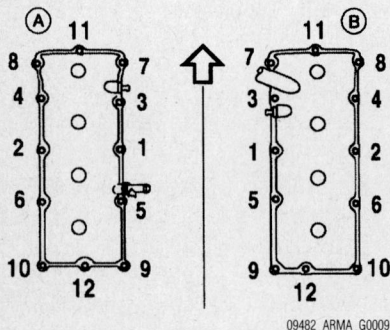

09482_ARMA_G0009

Rocker cover torque sequence

d. Step 4: All bolts: 92 in. lbs. (10 Nm)

19. Tighten front cover bolts to 8 ft. lbs. (11 Nm)

20. Install camshaft sprocket as follows:

a. Install camshaft sprocket aligning matchmarks with timing chain. Align camshaft sprocket key groove with dowel pin on camshaft front edge.

b. Temporarily tighten bolts.

c. Lock the camshaft with a wrench and tighten the bolts.

21. Install chain tensioner as shown:

a. Install chain tensioner, compress plunger and hold with stopper pin.

b. Tighten chain tensioner bolts to 61 in. lbs. (7 Nm)

c. Remove stopper pin, release plunger and apply tension to timing chain.

d. Install chain tensioner front cover (Right-hand bank only) and tighten bolts to 80 in. lbs. (9 Nm).

22. Install or connect the following:

• Timing chain cover
• Rocker cover.
• PCV hose
• Ignition coil
• Throttle control actuator
• Engine wiring harnesses on rocker cover
• Air intake assembly
• Engine cover
• Negative battery cable

23. Start the engine and check for leaks.

INSPECTION

Runout

1. Before servicing the vehicle, refer to the Precautions Section.

2. Remove the camshafts.

3. Using a V-block on a precise flat table, support the No. 2 and 4 journals of the camshaft.

➡ **Do not support journal No. 1 as it has a different diameter than the other locations.**

4. Set the dial indicator to No. 3 journal.

5. Turn the camshaft to one direction by hand and measure the camshaft runout.

6. Runout should measure less than 0.0008 inches (0.02 mm).

7. Camshaft should be replaced if it exceeds the limit.

Cam Height

1. Before servicing the vehicle, refer to the Precautions Section.

Bank	INT EXH	Identification paint (front)	Identification paint (rear)	Identification rib
RH	INT	White	—	Yes.
	EXH	—	Light blue	Yes.
LH	INT	White	—	No.
	EXH	—	Light blue	No.

67162-QX56-G16

Camshaft installation markings

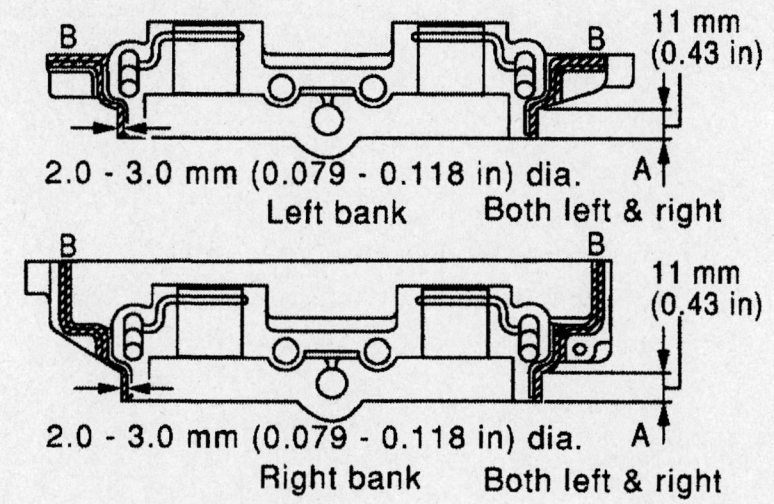

67170-ARMA-G05

Gasket application for camshaft bracket

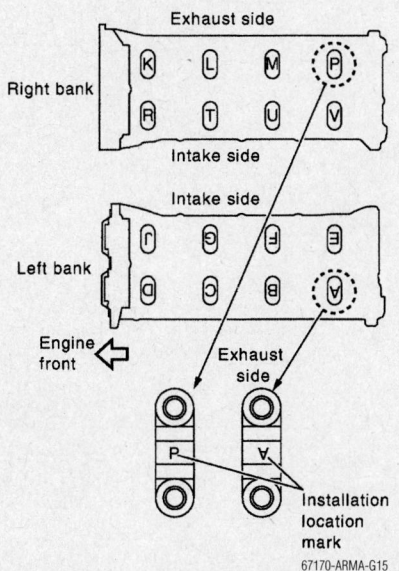

67170-ARMA-G15

Camshaft bracket installation markings

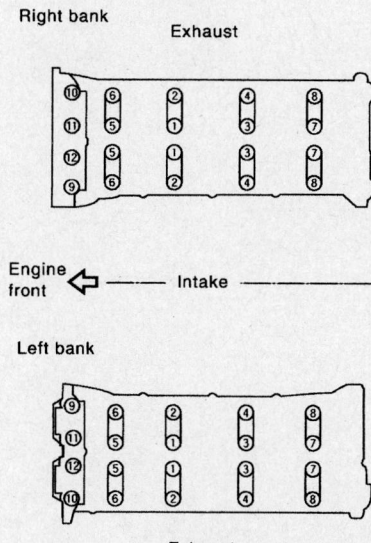

67170-ARMA-G04

Camshaft bracket torque sequence

2. Remove the camshafts.

3. Measure the cam height with a micrometer.

4. The intake and exhaust camshaft should measure between 1.7506–1.7581 inches (44.465–44.655 mm).

5. Camshaft should be replaced if it exceeds the limit.

Journal Oil Clearance

1. Before servicing the vehicle, refer to the Precautions Section.

2. Remove the camshafts.

3. Measure the outer diameter of camshaft journal with micrometer and record the result.

4. Reinstall the camshaft bearing caps in accordance to the installation procedure.

5. Measure the inner diameter of the camshaft bracket ("A") with a bore gauge and record the result.

6. Subtract the camshaft journal diameter from the camshaft bracket inner diameter. The difference should measure between 0.0012–0.0027 inches (0.030–0.068 mm).

7. The camshaft or camshaft bracket should be replaced if it exceeds the limit.

➡**The camshaft bracket cannot be replaced as an individual part, because it is machined together with the cylinder head. The entire cylinder head assembly must be replaced.**

End Play

1. Before servicing the vehicle, refer to the Precautions Section.

2. Install a dial indicator in the thrust direction on the front end of the camshaft. Measure the end play of the dial indicator when the camshaft is moved back and forth. The dial indicator should measure between 0.0045–0.0074 inches (0.115–0.188 mm).

R: For right bank
L: For left bank

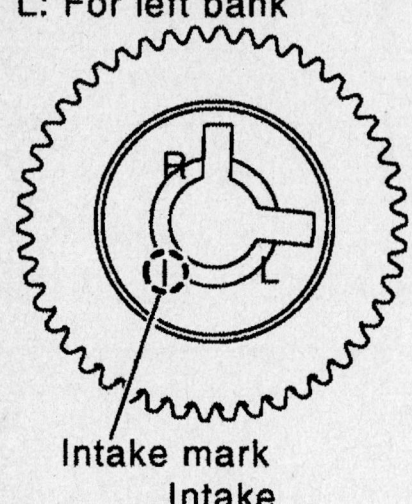

Intake mark
Intake

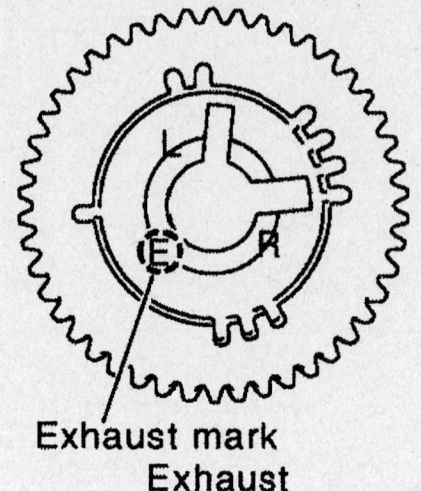

Exhaust mark
Exhaust

67170-ARMA-G16

Camshaft installation markings

3. Measure the No. 1 journal as shown. The distance ("A") should be 1.2008–1.2027 inches (30.500–30.548 mm).

4. Measure the No. 1 journal bearing as shown. The distance ("B") should be 1.1953–1.1963 inches (30.360–30.385 mm).

5. Replace either the camshaft or cylinder head assembly if the measurement is exceeded.

Valve Lash

INSPECTION

➡ Perform the following inspection after removal, installation or replacement of camshaft or valve-related parts, or if there are unusual engine conditions due to changes in valve clearance over time (starting, idling, and/or noise).

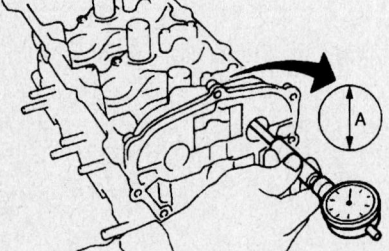

09482_PATH_G0014

Measuring the journal bore for oil clearance

SEM864E

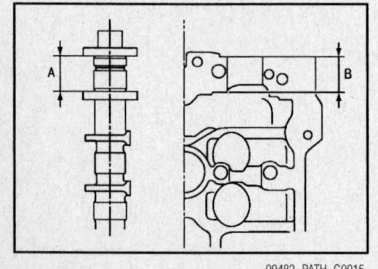

09482_PATH_G0015

Measuring for camshaft endplay

⬆ : Measurable at No. 1 cylinder compression top dead center

⇧ : Measurable at No. 3 cylinder compression top dead center

Exhaust

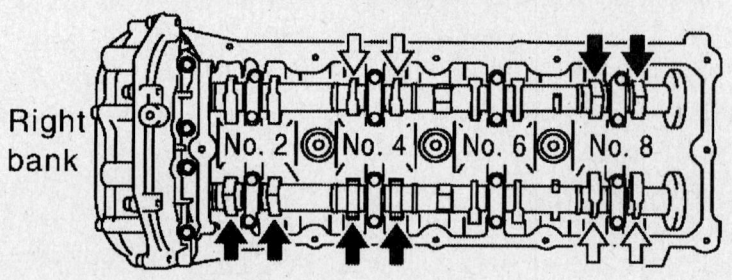

Right bank

No. 2 No. 4 No. 6 No. 8

Engine front ⟵ — — Intake — —

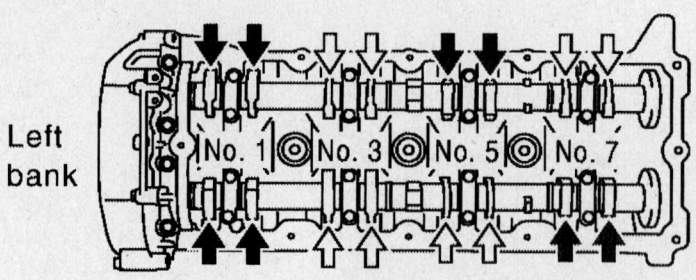

Left bank

No. 1 No. 3 No. 5 No. 7

Exhaust

67170-ARMA-G06

Locations to measure clearance with No. 1 cylinder at TDC

1. Run engine to operating temperature.

2. Remove or disconnect the following:
   - Engine cover
   - Battery cover
   - Air intake assembly
   - Left and right rocker covers

3. Turn the crankshaft pulley clockwise to Top Dead Center (TDC) identification notch with timing indicator.

4. Ensure that both the intake and exhaust cam noses of the No. 1 cylinder face outside.

5. Measure the valve clearances at locations shown in figure.

6. Turn the crankshaft pulley clockwise 270 degrees from the position of No. 1 cylinder compression to obtain No. 3 cylinder compression TDC.

7. Measure the valve clearances at locations shown in the figure.

8. Turn crankshaft pulley clockwise 90 degrees and measure the intake and exhaust valve clearance of No. 6 cylinder and exhaust valve clearance of No. 2 cylinder.

ADJUSTMENT

1. Remove camshaft and valve lifter(s) out of specification.

2. Install replacement valve lifter(s).

3. Install the camshaft.

4. Manually turn the crankshaft pulley several turns.

5. Recheck valve clearances with engine at operating temperature.

Oil Pan

REMOVAL & INSTALLATION

1. Before servicing the vehicle, refer to the Precautions Section.

2. Remove engine assembly.

↑ : Measurable at No. 1 cylinder compression top dead center

⇧ : Measurable at No. 3 cylinder compression top dead center

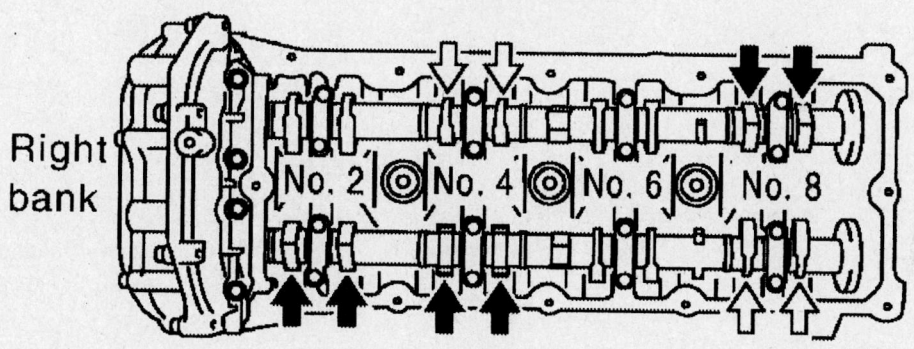

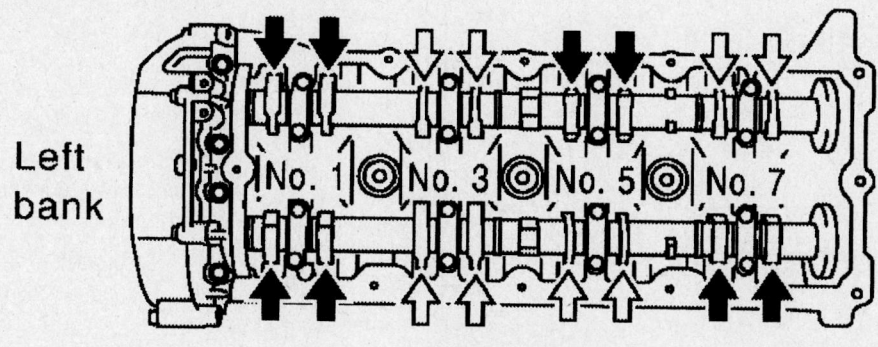

Locations to measure clearance with No. 3 cylinder at TDC

67170-ARMA-G07

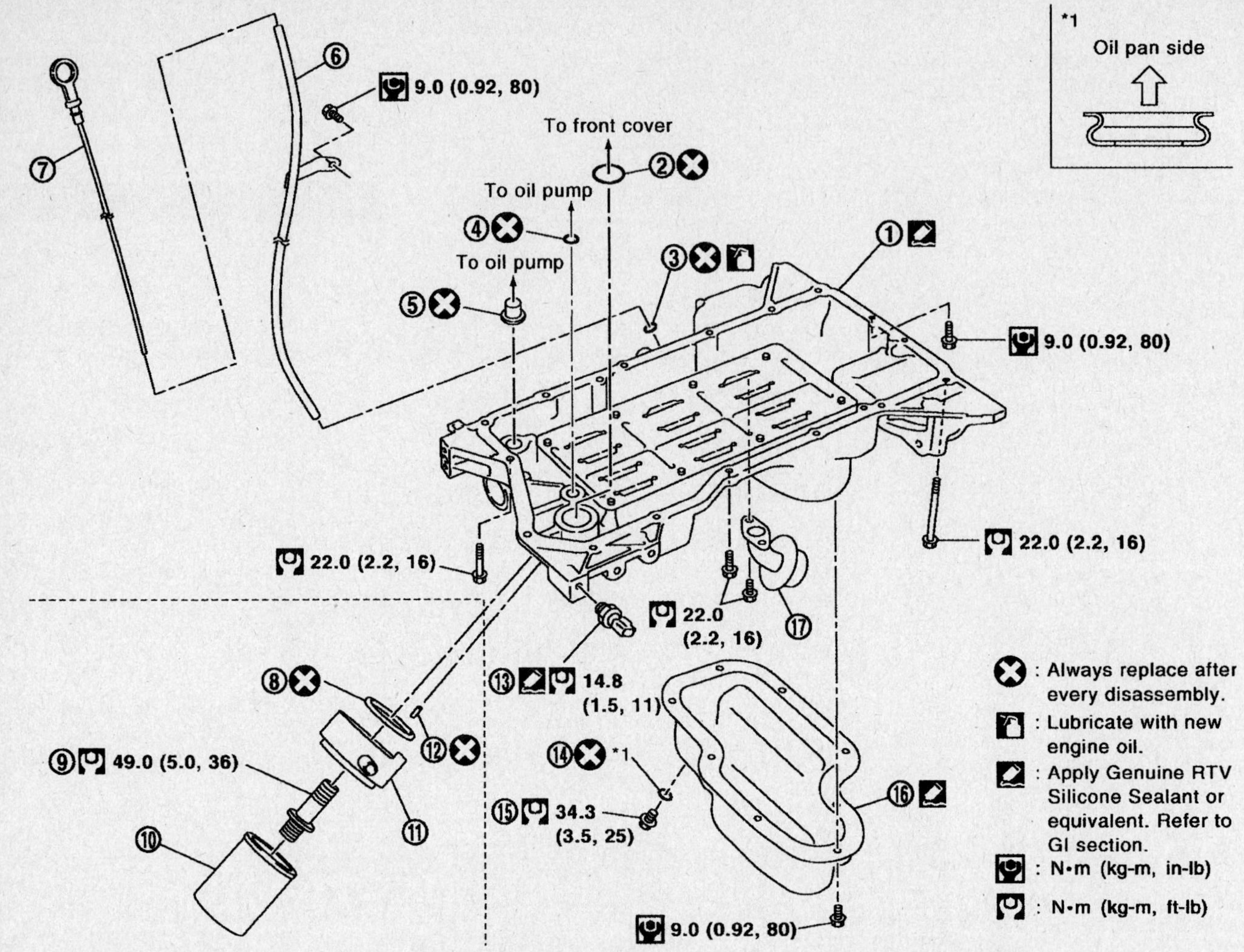

*1
Oil pan side

9.0 (0.92, 80)

To front cover

②⊗
To oil pump

④⊗
To oil pump

⑤⊗

③⊗

①

9.0 (0.92, 80)

22.0 (2.2, 16)

22.0 (2.2, 16)

22.0 (2.2, 16)

⑰

⑧⊗

⑬ 14.8 (1.5, 11)

⑨ 49.0 (5.0, 36)

⑫⊗

⑭⊗ *1

⑮ 34.3 (3.5, 25)

⑩

⑪

⑯

9.0 (0.92, 80)

⊗ : Always replace after every disassembly.

: Lubricate with new engine oil.

: Apply Genuine RTV Silicone Sealant or equivalent. Refer to GI section.

: N•m (kg-m, in-lb)

: N•m (kg-m, ft-lb)

1. Oil pan (Upper)	2. O-ring	3. O-ring
4. O-ring	5. O-ring (with collar)	6. Oil level gauge guide
7. Oil level gauge	8. O-ring	9. Connector bolt
10. Oil filter	11. Oil cooler	12. Relief valve
13. Oil pressure switch	14. Gasket	15. Drain plug
16. Oil pan (Lower)	17. Oil strainer	

67170-ARMA-G31

Oil pan and related parts

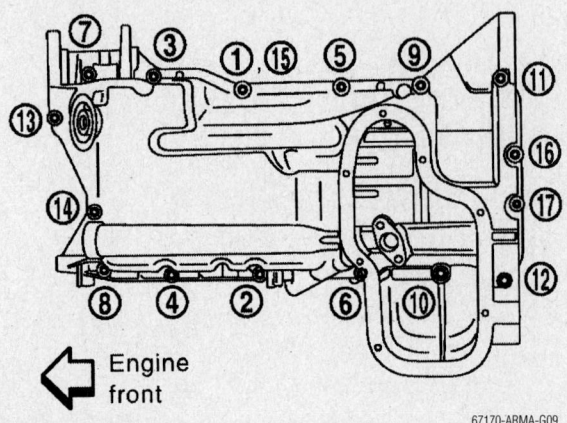

Engine
front

67170-ARMA-G09

Upper oil pan bolt id

3. Remove lower oil pan, loosening bolts in reverse order shown in figure.

4. Remove oil strainer from upper oil pan.

5. Gently pry and remove upper oil pan from engine block.

➡**Bolts are different sizes and should be keep in the correct order for reinstallation.**

To install:

6. Apply liquid gasket to upper oil pan mating surfaces.

7. Install new O-rings to oil pump and front cover side.

8. Tighten upper oil pan bolts to 16 ft. lbs. (22 Nm) in order shown in figure:

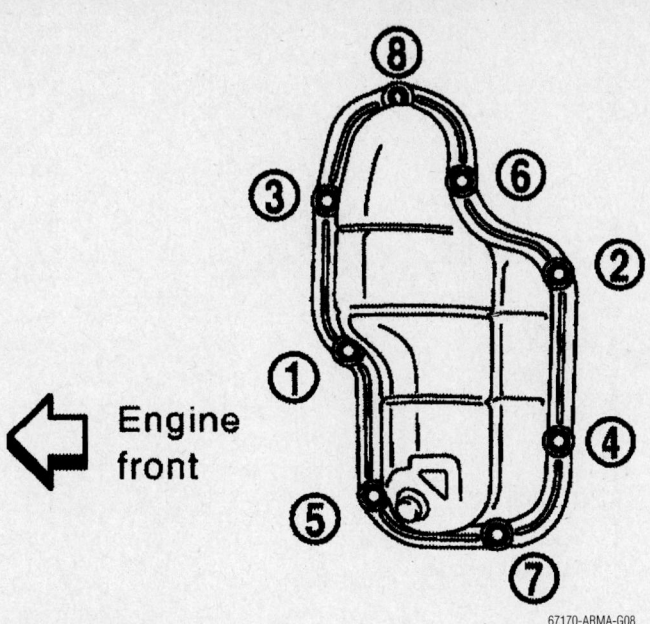

Lower oil pan torque sequence

← Engine front

67170-ARMA-G08

9. Install or connect the following:
   • Rear plate cover
   • Oil strainer to upper oil pan
   • Lower oil pan, tightening bolts to 25 ft. lbs. (34 Nm) in order shown in figure

Oil Pump

REMOVAL & INSTALLATION

1. Before servicing the vehicle, refer to the Precautions Section.
2. Remove or disconnect the following:
   • Timing chain cover
   • Oil pump drive spacer
   • Oil pump
To install:
3. Install or connect the following:
   • Oil pump
   • Oil pump drive spacer
   • Timing chain cover

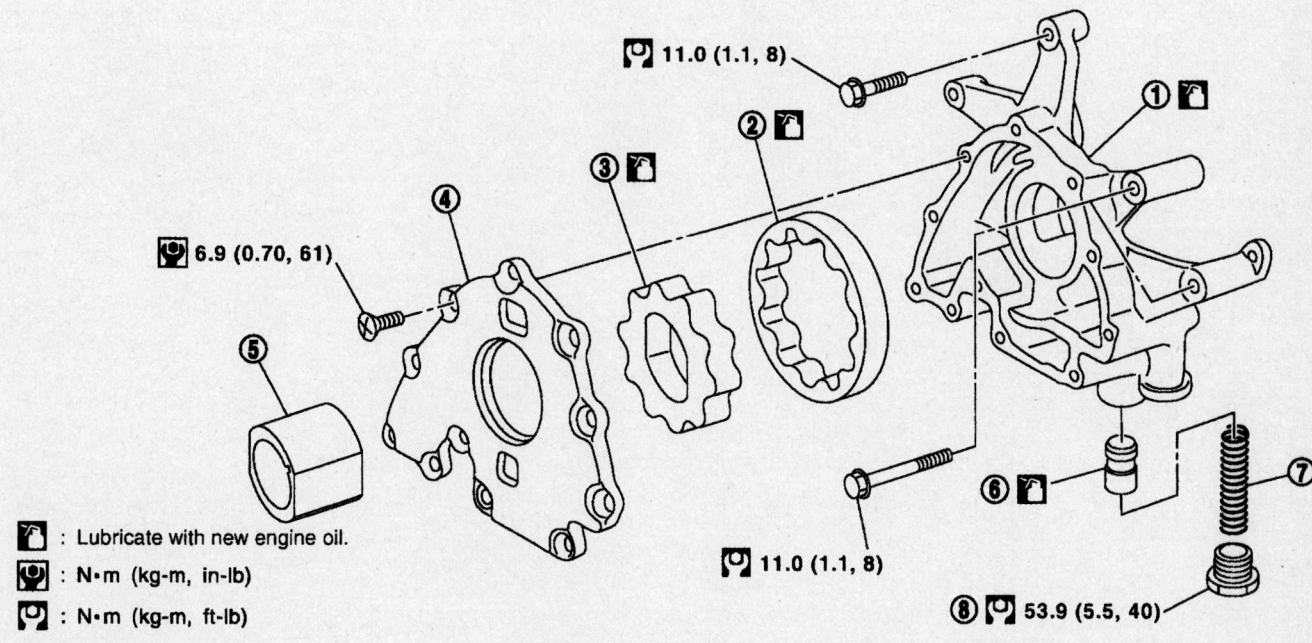

11.0 (1.1, 8)

6.9 (0.70, 61)

11.0 (1.1, 8)

53.9 (5.5, 40)

⬛ : Lubricate with new engine oil.

🔧 : N·m (kg-m, in-lb)

🔧 : N·m (kg-m, ft-lb)

1. Oil pump body	2. Outer rotor	3. Inner rotor
4. Oil pump cover	5. Oil pump drive spacer	6. Regulator valve
7. Regulator spring	8. Regulator plug	

67170-ARMA-G32

Oil pump exploded view

Rear Main Seal

REMOVAL & INSTALLATION

1. Before servicing the vehicle, refer to the Precautions Section.
2. Remove or disconnect the following:
   - Transmission assembly
   - Drive plate
   - Engine rear plate
   - Rear main seal using suitable tool

To install:

3. Install or connect the following:
   - Rear main seal using suitable tool
   - Engine rear plate
   - Drive plate
   - Transmission assembly

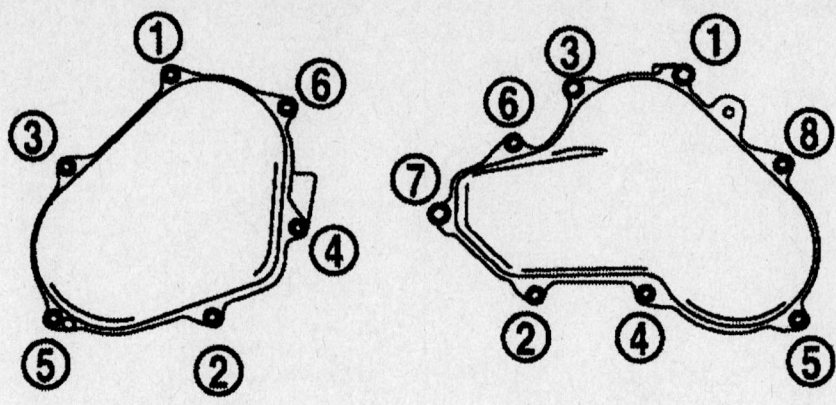

Right bank   **Left bank**

67170-ARMA-G10

Timing case covers torque sequence

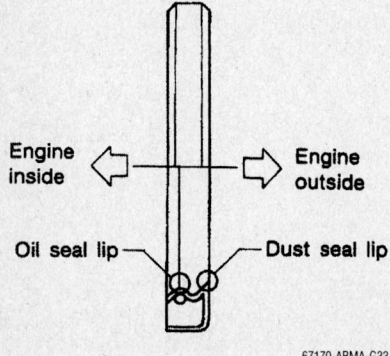

67170-ARMA-G33

Proper seal installation direction

Timing Chain, Sprockets, Front Cover and Seal

REMOVAL & INSTALLATION

1. Before servicing the vehicle, refer to the Precautions Section.
2. Remove or disconnect the following:
   - Engine assembly
   - Drive belt auto tensioner
   - Idler pulley
   - Thermostat housing and water hose
   - Power steering pump bracket
   - Oil pan (upper and lower)
   - Oil strainer
   - Ignition coil
   - Rocker cover
   - Timing chain case cover, loosening bolts in reverse order shown in figure
3. Obtain compression TDC of No. 1 cylinder as follows:
   a. Turn crankshaft pulley to align the

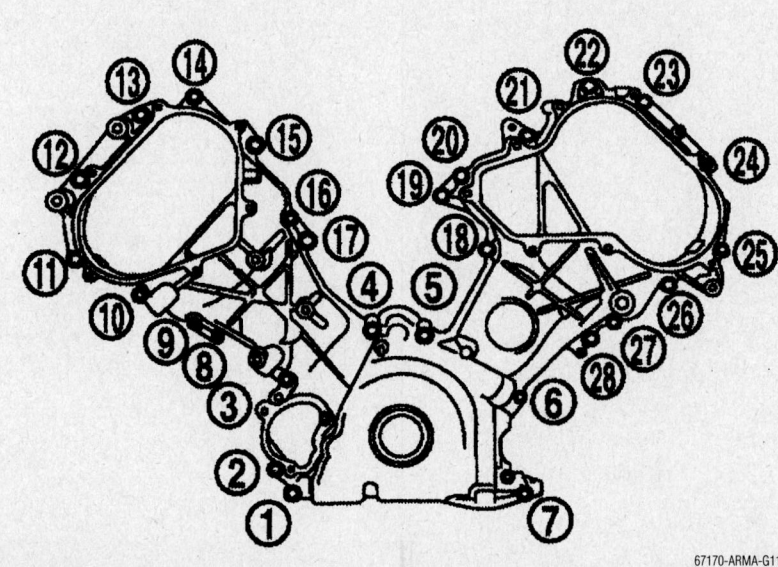

67170-ARMA-G11

Front cover torque sequence

TDC identification notch with timing indicator on front cover.

b. Ensure intake and exhaust cam lobes of No. 1 cylinder point outside.

4. Remove or disconnect the following:
   - Crankshaft pulley from crankshaft using a suitable puller
   - Front cover, loosening bolts in reverse order shown in figure
   - Front oil seal
   - Oil pump drive spacer
   - Oil pump

5. Remove the timing chain tensioner as follows:

a. Squeeze the return-proof clip ends using suitable tool and push the plunger into the tensioner body.

b. Secure the plunger using stopper pin.

➡**Stopper pin is made from hard wire approximately 0.04 in (1mm) in diameter.**

c. Remove the bolts and chain tensioner.

6. Remove the following:
   - Chain tension guide and slack guide
   - Timing chain
7. Using a wrench to hold the hexagon part of the camshaft, loosen the camshaft sprocket bolts.
8. Remove the camshaft sprockets.

To install:

9. Ensure that the crankshaft key and

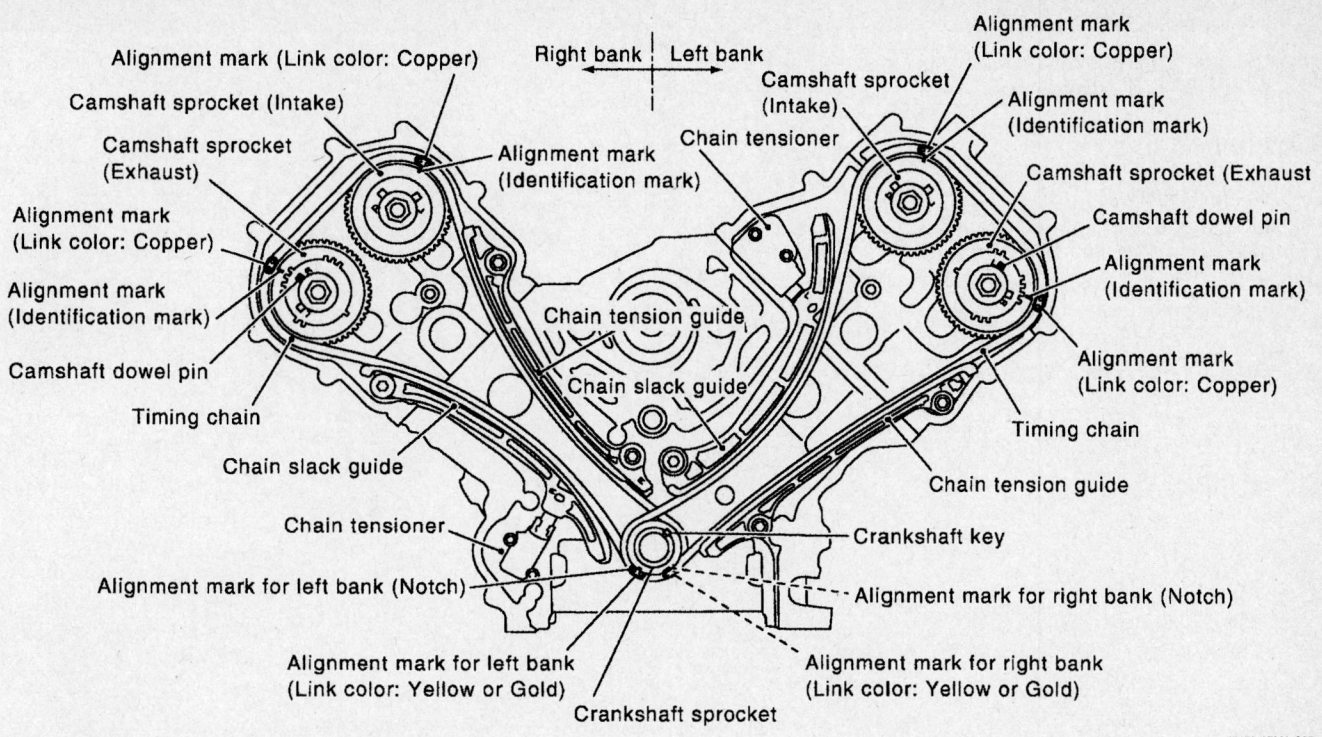

Timing chain orientation and timing mark alignment

dowel pin of each camshaft are facing the same direction.

10. Install or connect the following:
- Camshaft sprockets and tighten to 112 ft. lbs. (152 Nm).
- Timing chain
- Chain tension guide and slack guide
- Oil pump
- Oil pump drive spacer
- Front oil seal, using suitable tool
- Front cover, using new O-rings and tighten bolts in order shown in figure
- Chain case cover, and tighten bolts in order shown in figure
- Crankshaft pulley and tighten bolt to 69 ft. lbs. (93 Nm) plus 90 degrees
- Ignition coil
- Oil strainer
- Lower and upper oil pan
- Power steering pump bracket
- Thermostat housing and water hose
- Idler pulley
- Drive belt auto tensioner
- Engine assembly

Piston and Ring

POSITIONING

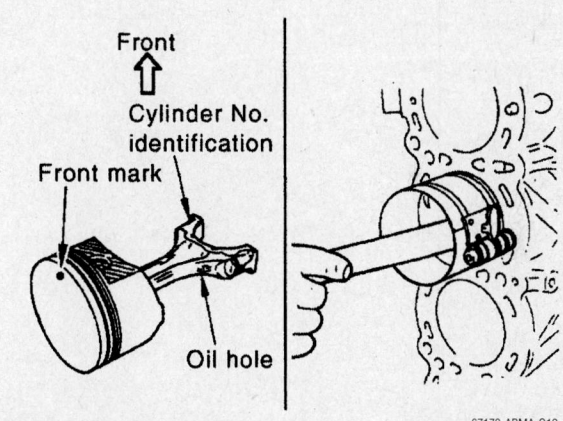

Piston and rod positioning and identification

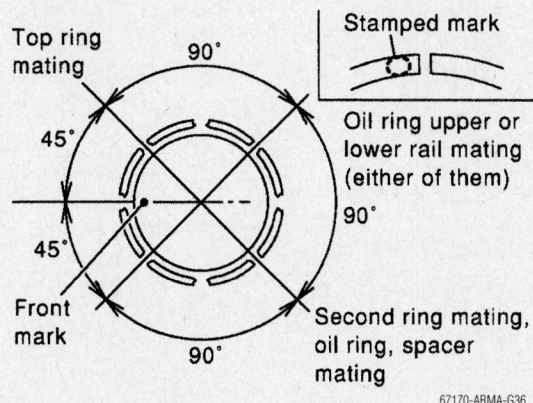

Piston ring installation

FUEL SYSTEM

Relieving Fuel System Pressure

With CONSULT-II

1. Turn ignition switch **ON**.
2. Perform "FUEL PRESSURE RELEASE" in "WORK SUPPORT" mode with CONSULT-II.
3. Start engine.
4. After engine stalls, turn over the engine two or three times to release all fuel pressure.
5. Turn ignition switch **OFF**.

Without CONSULT-II

1. Remove fuel pump fuse located in IPDM E/R.
2. Start engine.
3. After engine stalls, turn over engine two or three times to release all fuel pressure.
4. Turn ignition switch **OFF**.
5. Reinstall fuel pump fuse after servicing fuel system.

Fuel Filter and Fuel Pump

REMOVAL & INSTALLATION

1. Before servicing the vehicle, refer to the Precautions Section.
2. Relieve the fuel system pressure.
3. Remove fuel filler cap to release pressure from inside tank.
4. Remove left hand rear inner fender liner.
5. Disconnect fuel filler hose from fuel filler pipe.
6. Drain fuel tank through the fuel filler hose using a suitable hose.
7. Remove or disconnect the following:
   • Second row left hand seat
   • Third row seat
   • Second and third row seat belt buckles mounted on floor
   • Left hand center pillar trim
   • Left hand rear trim panel
   • Left hand rear side door kick plate and weather stripping
   • Second row rear center console and base, if equipped
   • Inspection hole cover under carpet by turning retainers 90 degrees
   • Electrical connectors
   • EVAP hose
   • Fuel supply hose
   • Lock ring using special tool J-46214

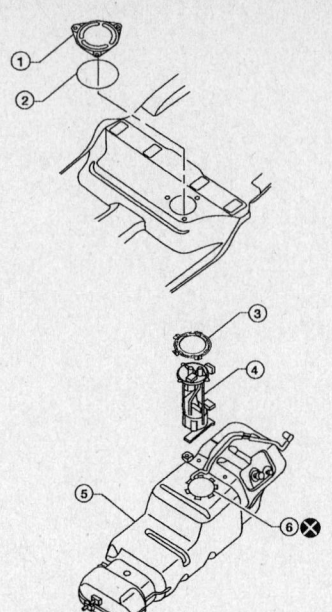

☒ : Always replace after every disassembly

1. Inspection hole cover	2. Inspection hole cover O-ring
4. Fuel level sensor, fuel filter, and fuel pump assembly	5. Fuel tank

3. Lock ring
6. Fuel level sensor, fuel filter, and fuel pump assembly O-ring

67170-ARMA-G37

Fuel pump and related parts

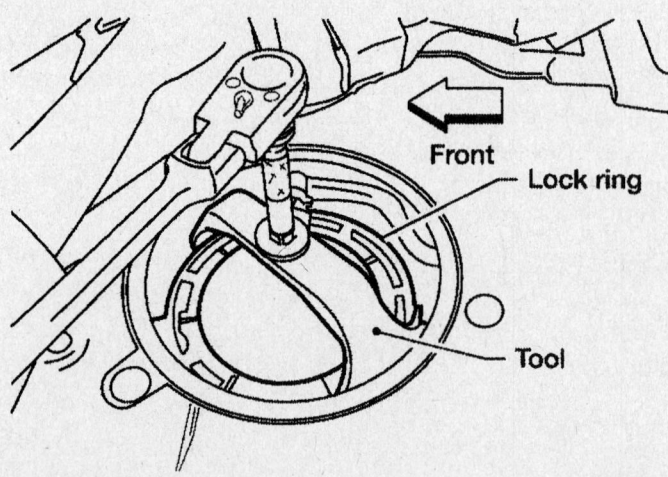

67170-ARMA-G17

Removing fuel assembly lock ring

   • Fuel level sensor
   • Fuel filter
   • Fuel pump assembly
8. Install or connect the following:
   • Fuel pump assembly
   • Fuel filter
   • Fuel level sensor
   • Lock ring using special tool J-46214
   • Fuel supply hose
   • EVAP hose
   • Electrical connectors
   • Inspection hole cover
   • Second row rear center console and base, if equipped
   • Left hand rear side door kick plate and weather stripping
   • Left hand rear trim panel
   • Left hand center pillar trim
   • Second and third row seat belt buckles
   • Third row seat
   • Second row left hand seat

- Fuel filler hose to fuel filler pipe
- Left hand rear inner fender liner

9. Start the engine and check for leaks.

Fuel Injectors

REMOVAL & INSTALLATION

1. Before servicing the vehicle, refer to the Precautions Section.
2. Remove engine cover
3. Relieve fuel system pressure.

4. Remove or disconnect the following:
- Negative battery cable
- Fuel injector harness connectors
- Fuel hose assembly from right and left fuel rails
- Fuel injectors with fuel rail as an assembly
- Fuel injector from fuel rail

To install:
5. Install or connect the following:

➡**Always use a new O-ring when reinstalling the fuel injector to the fuel rail.**

- New clip onto the fuel injector
- Fuel injector to fuel rail
- Fuel injectors and fuel rail as an assembly to the intake manifold. Tighten the bolts to 8 ft. lbs. (11 Nm).
- Fuel hose assembly
- Fuel injector harness connectors
- Negative battery cable
- Engine cover

6. Start engine and check for leaks.

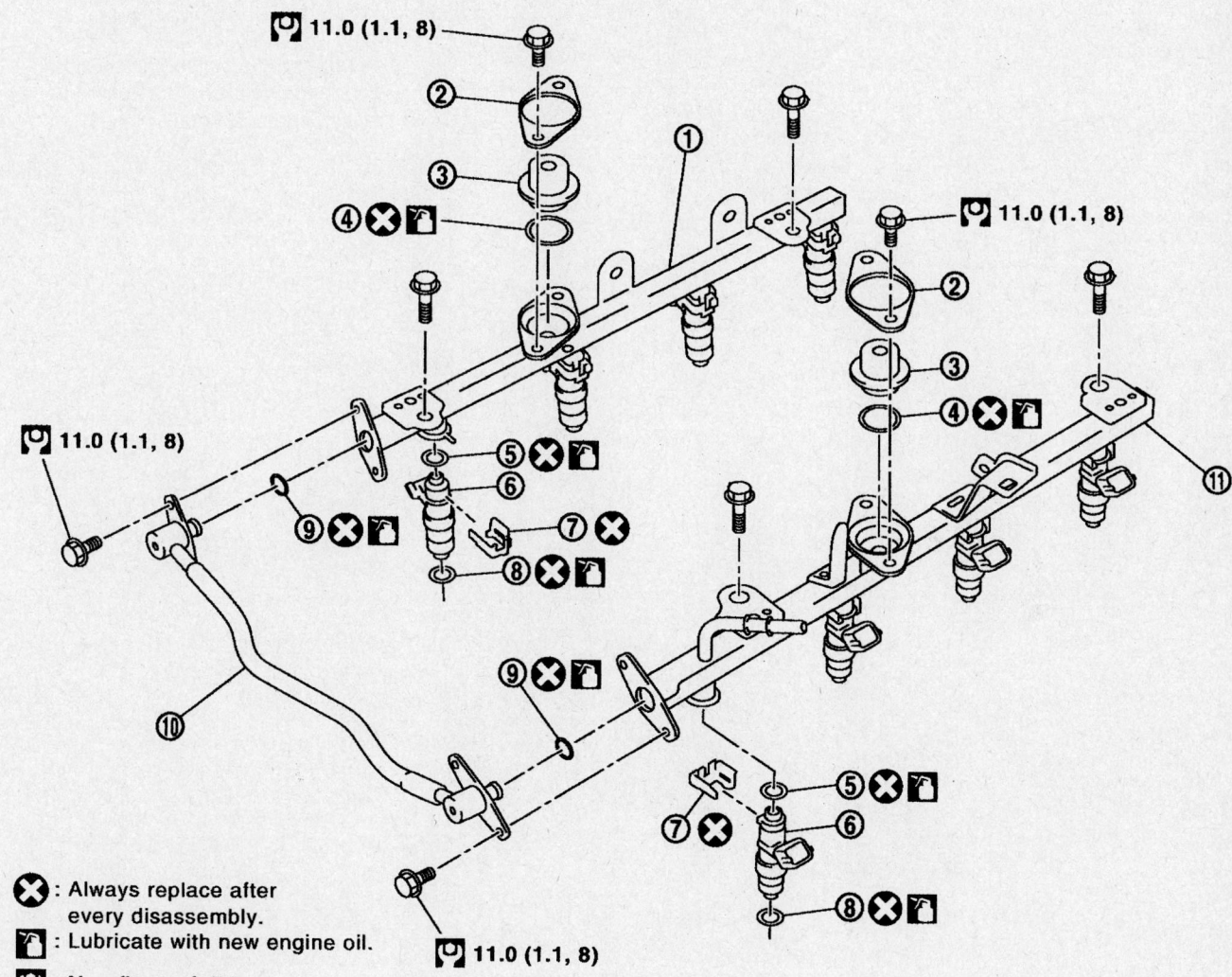

❌ : Always replace after every disassembly.

▣ : Lubricate with new engine oil.

🔧 : N•m (kg-m, ft-lb)

1. Fuel tube (right bank)	2. Cap	3. Fuel damper
4. O-ring	5. O-ring (Blue)	6. Fuel injector
7. Clip	8. O-ring (Brown)	9. O-ring
10. Fuel hose assembly	11. Fuel tube (left bank)	

Fuel injectors and related parts

67170-ARMA-G38

DRIVE TRAIN

Transmission Assembly

REMOVAL & INSTALLATION

2-Wheel Drive

1. Before servicing the vehicle, refer to the Precautions Section.
2. Remove or disconnect the following:
   - Negative battery cable
   - Engine cover
   - Transmission fluid indicator gauge
   - Engine splash guard
   - Exhaust front pipe
   - Center muffler
   - Rear drive shaft
   - Transmission control cable
   - Crankshaft position sensor
   - Transmission cooler tube
   - Dust cover from converter housing
3. Turning crankshaft clockwise, remove the four tightening bolts for drive plate and torque converter
4. Support the transmission with a suitable jack.
5. Remove or disconnect the following:
   - Transmission cross member
   - Air breather hose
   - Transmission assembly connector
   - Fluid indicator tube from transmission assembly
   - Transmission assembly to engine bolts
   - Transmission assembly from vehicle

To install:

6. Install or connect the following:
   - Transmission assembly into vehicle
   - Transmission assembly to engine bolts tightening to 83 ft. lbs. (113 Nm)
   - Fluid indicator tube to transmission assembly
   - Transmission assembly connector
   - Air breather hose
   - Transmission cross member
7. Turning crankshaft clockwise, install the torque converter to drive plate.

➡**After torque converter is installed, rotate the crankshaft to ensure transmission rotates freely.**

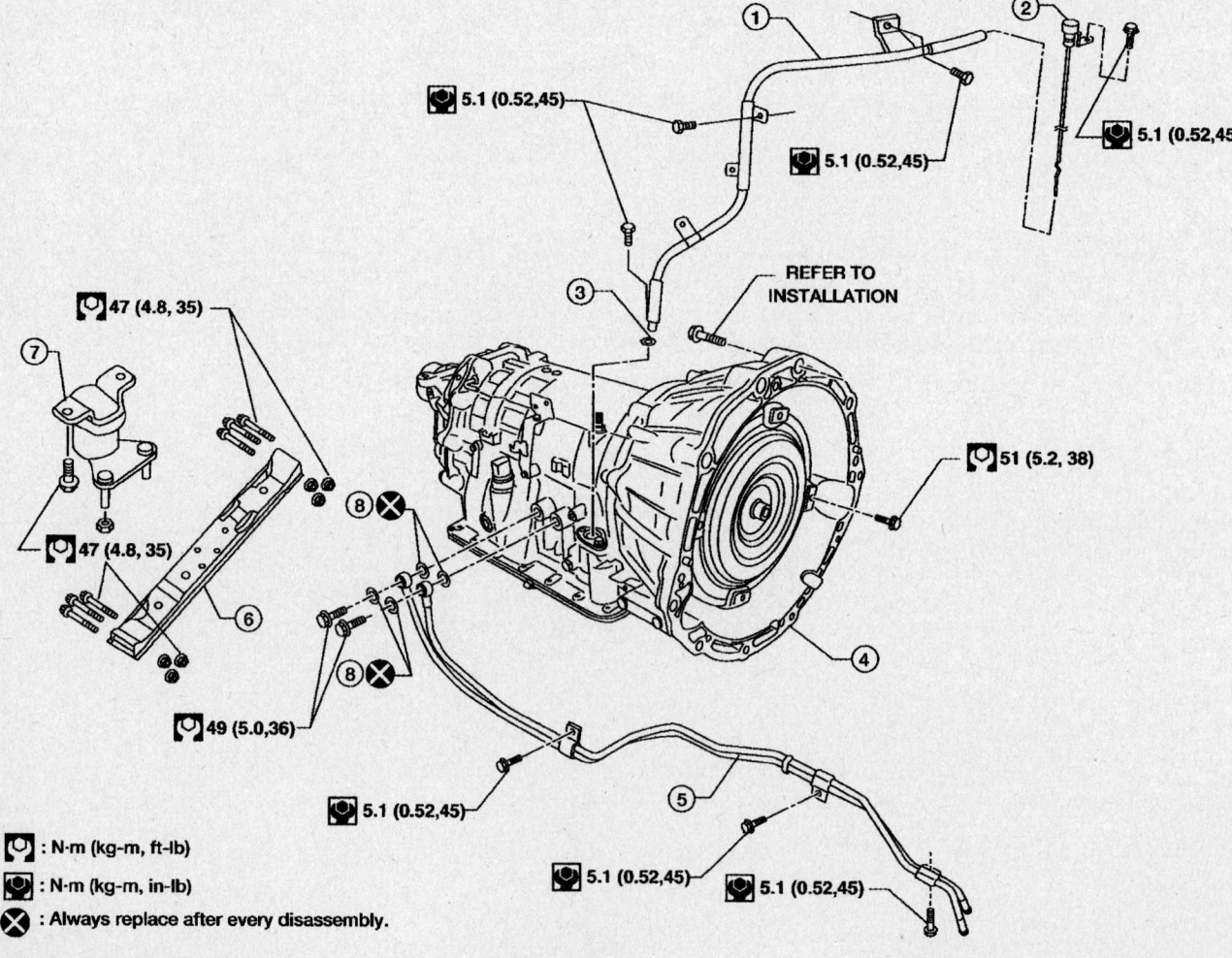

: N·m (kg-m, ft-lb)

: N·m (kg-m, in-lb)

✕ : Always replace after every disassembly.

1. A/T fluid indicator pipe	2. A/T fluid indicator	3. O-ring
4. Transmission assembly	5. A/T fluid cooler tube	6. A/T crossmember
7. Insulator	8. Copper washers	

Transmission and related parts—2-wheel drive

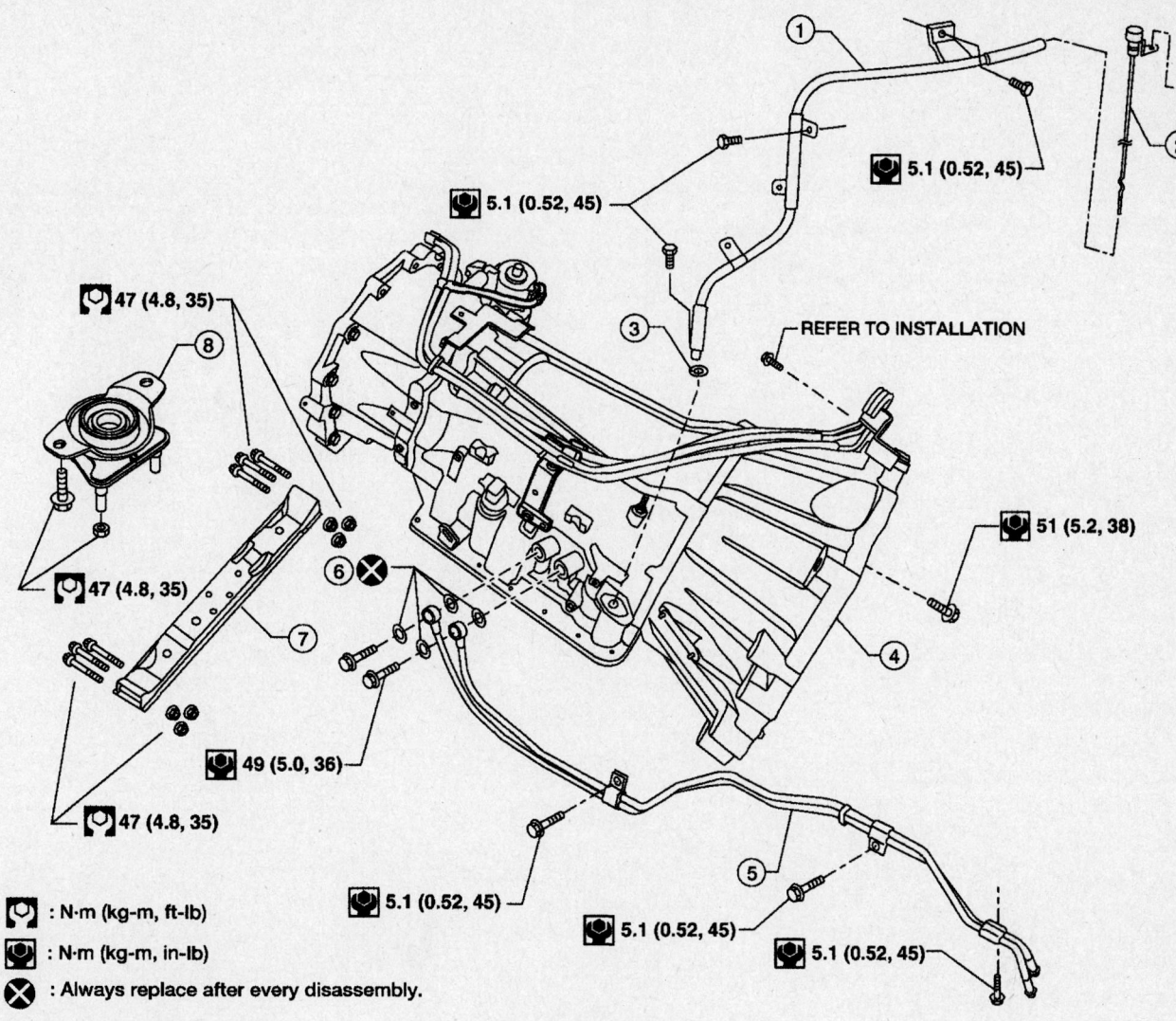

① 5.1 (0.52, 45)

②

③

REFER TO INSTALLATION

47 (4.8, 35)

⑧

47 (4.8, 35)

⑥

⑦

51 (5.2, 38)

④

49 (5.0, 36)

47 (4.8, 35)

⑤

5.1 (0.52, 45)

5.1 (0.52, 45)

5.1 (0.52, 45)

5.1 (0.52, 45)

5.1 (0.52, 45)

: N·m (kg-m, ft-lb)

: N·m (kg-m, in-lb)

: Always replace after every disassembly.

1. A/T fluid indicator pipe
2. A/T fluid indicator
3. O-ring
4. Transmission assembly
5. Fluid cooler tube
6. Copper washer
7. A/T crossmember
8. Insulator

67170-ARMA-G40

Transmission and related parts—with 4-wheel drive

8. Install or connect the following:
   - Dust cover for converter housing
   - Fluid cooler tube
   - Crankshaft position sensor
   - Transmission control cable
   - Rear drive shaft
   - Center muffler
   - Exhaust front pipe
   - Engine splash guard
   - Transmission fluid indicator gauge
   - Engine cover
   - Negative battery cable
9. Start engine and check for leaks.

4-Wheel Drive

1. Before servicing the vehicle, refer to the Precautions Section.
2. Remove or disconnect the following:
   - Negative battery cable
   - Engine cover
   - Transmission fluid indicator gauge
   - Engine splash guard
   - Exhaust front pipe
   - Center muffler
   - Drive shaft
   - Transmission control cable
   - Crankshaft position sensor

- Fluid cooler tube
- Dust housing for torque converter
3. Turning the crankshaft clockwise, remove the four tightening bolts for drive plate and torque converter.
4. Support the transmission assembly with a suitable jack.
5. Remove transmission cross member.
6. Tilt the transmission slightly to keep clearance between the body and the transmission assembly, then disconnect the air breather hose.
7. Remove or disconnect the following:

- Transmission assembly connector and transfer case connector
- Fluid indicator pipe
- Transmission assembly to engine bolts
- Transmission assembly, with transfer case attached, from vehicle
- Transmission assembly from transfer case

To install:

8. Install or connect the following:
   - Transfer case to transmission assembly
   - Transmission assembly into vehicle
   - Transmission assembly to engine bolts tightening to 83 ft. lbs. (113 Nm)

9. With the transmission slightly tilted to allow clearance between body and transmission, connect the air breather hose.

10. Install the transmission cross member.

11. Turning crankshaft clockwise, install the torque converter to drive plate.

➡**After torque converter is installed, rotate the crankshaft to ensure transmission rotates freely.**

12. Install or connect the following:
   - Dust housing for torque converter
   - Fluid cooler tube
   - Crankshaft position sensor
   - Transmission control cable
   - Drive shaft
   - Center muffler
   - Front exhaust pipe
   - Engine splash guard
   - Transmission fluid indicator gauge
   - Engine cover
   - Negative battery cable

13. Start engine and check for leaks.

Transfer Case Assembly

REMOVAL & INSTALLATION

1. Before servicing the vehicle, refer to the Precautions Section.

2. Remove or disconnect the following:
   - Transmission splash guard
   - Center exhaust pipe and muffler
   - Front and rear drive shafts

➡**Plug rear oil seal after removing rear drive shaft.**

   - Transmission assembly mounting bolts

3. Support the transmission assembly with a suitable jack and remove the crossmember.

4. Remove or disconnect the following:

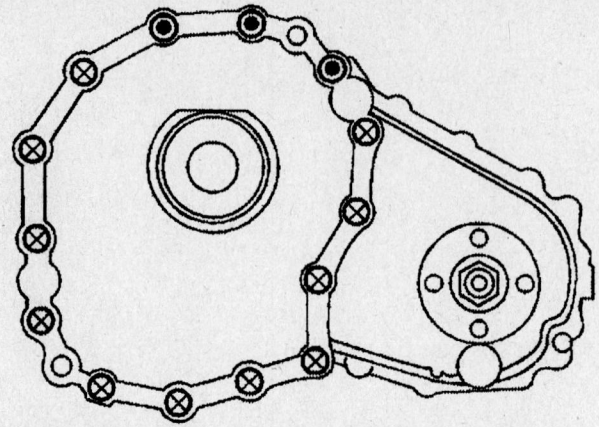

⊙ : Transfer ➞ Automatic transmission
⊗ : Automatic transmission ➞ Transfer

67170-ARMA-G41

Transfer case mounting bolt locations

- ATP switch, neutral 4LO switch, wait detection switch, transfer motor and transfer control device electrical connectors
- Breather hoses
- Shift actuator from the extension housing
- Transfer case to transmission assembly bolts
- Transfer case assembly

To install:

5. Install or connect the following:
   - Transfer case to transmission assembly bolts tightening to 26 ft. lbs. (36 Nm)
   - Shift actuator
   - Breather hoses
   - ATP switch, neutral 4LO switch, wait detection switch, transfer motor and transfer control device electrical connectors
   - Support crossmember
   - Transmission mounting bolts
   - Drive shafts
   - Muffler and center exhaust pipe
   - Transmission splash guard

Halfshaft

REMOVAL & INSTALLATION

Front

1. Before servicing the vehicle, refer to the Precautions Section.

2. Remove or disconnect the following:
   - Wheel
   - Engine splash guard
   - Cotter pin and half shaft nut
   - Half shaft from front differential

- Half shaft from hub and bearing assembly

To install:

3. Install or connect the following:
   - Half shaft into hub
   - Half shaft into front differential
   - Half shaft nut and tighten to 101 ft. lbs. and replace cotter pin
   - Engine splash guard
   - Wheel

Rear

1. Before servicing the vehicle, refer to the Precautions Section.

2. Remove or disconnect the following:
   - Wheel
   - Stabilizer bar clamp
   - Cotter pin and drive shaft nut
   - Bolts from the inside flange of the drive shaft

3. Separate the drive shaft from the wheel hub by lightly tapping the end with suitable hammer and wood block.

4. Remove the half shaft.

❋❋ CAUTION

Do not excessively extend the slide joint.

To install:

5. Install or connect the following:
   - Half shaft
   - Bolts for the inside flange and tighten to 87 ft. lbs. (118 Nm)
   - Drive shaft nut and tighten nut to 101 ft. lbs. (137 Nm) and replace cotter pin
   - Stabilizer bar clamp
   - Wheel

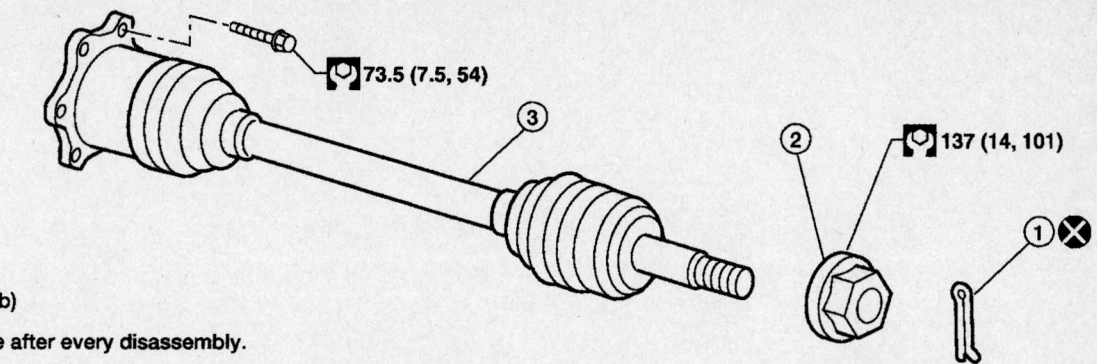

$\boxed{\Omega}$: N·m (kg-m, ft-lb)

$\otimes$: Always replace after every disassembly.

1. Cotter pin
2. Drive shaft nut
3. Drive shaft

67170-ARMA-G42

Front halfshaft

CV-Joints

OVERHAUL

Inner

1. Before servicing the vehicle, refer to the Precautions Section.
2. Remove the halfshaft from the vehicle.
3. Mount halfshaft in a vise.
4. Remove the dust boot bands.

5. Remove the stopper ring with a flat-bladed screwdriver or suitable tool.
6. Remove the snap ring.
7. Disassemble the cage, ball and inner race assembly and dust boot for cleaning and inspection.

To install:

➡**Discard old dust boot, dust boot bands and snap ring and use new ones for assembly.**

8. Wrap the serrated part of the half-shaft with tape.

9. Install new dust boot and band onto halfshaft.
10. Remove tape from serrated part of halfshaft.
11. Install the cage, ball and inner race assembly.
12. Install new snap ring.
13. Insert 4.50-5.3 oz of genuine NISSAN grease or equivalent onto the housing and install onto halfshaft.
14. Install the stopper ring onto the housing.

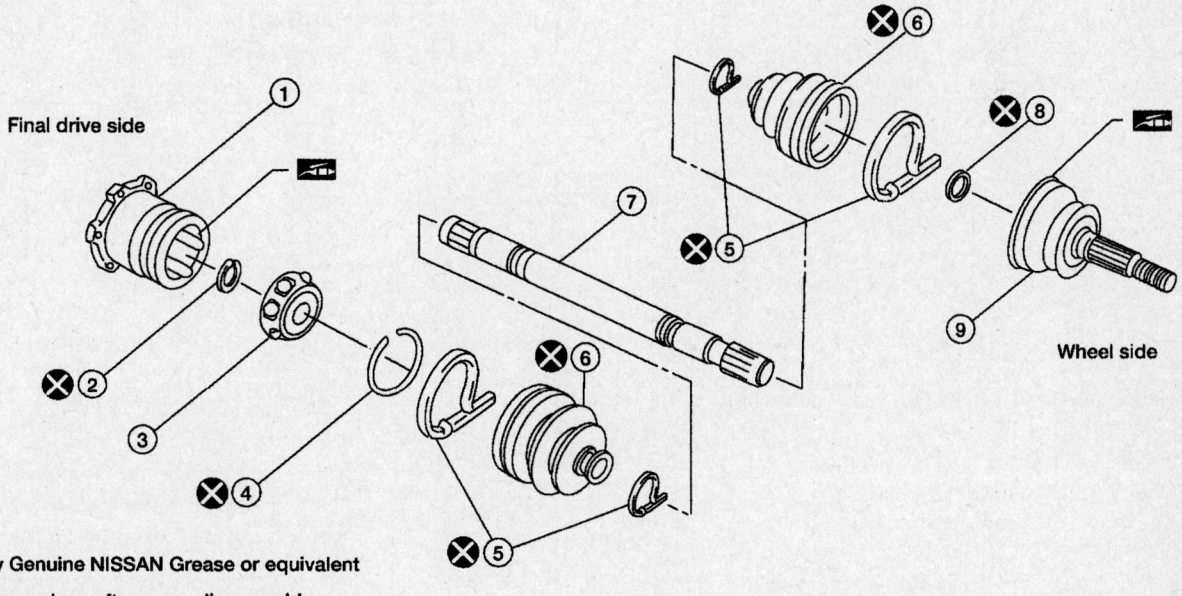

$\boxminus$: Apply Genuine NISSAN Grease or equivalent

$\otimes$: Always replace after every disassembly.

1. Housing
2. Snap ring
3. Ball cage, steel ball, iiner race assembly
4. Stopper ring
5. Boot band
6. Boot
7. Shaft
8. Circlip
9. Joint sub-assembly

67170-ARMA-G43

Front halfshaft—exploded view

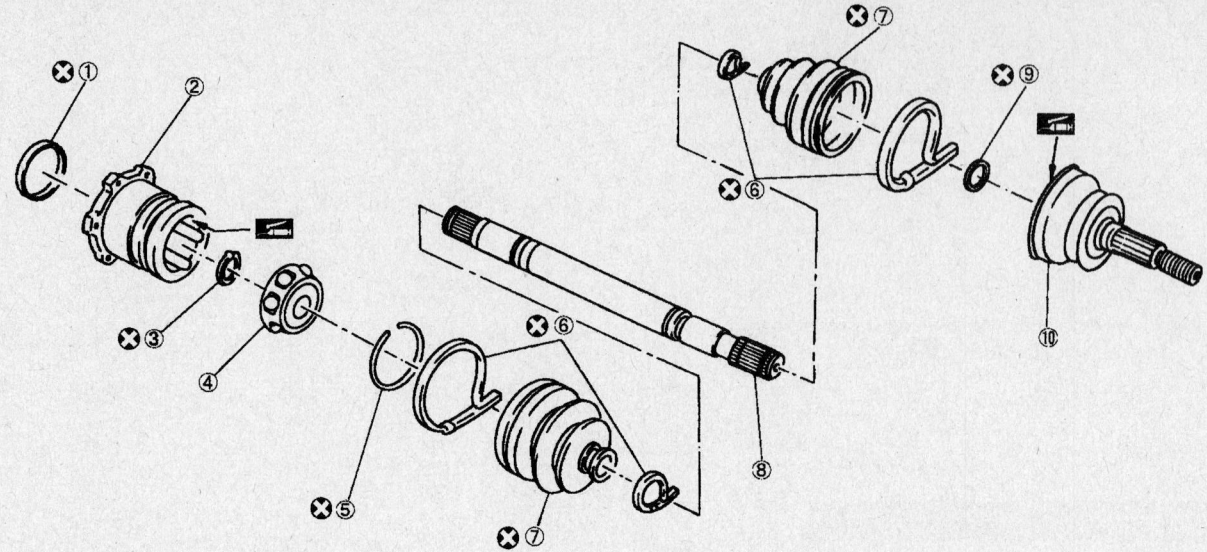

❌ : Always replace after every disassembly.

1. Plug
2. Housing
3. Snap ring
4. Ball cage, steel ball, liner race assembly
5. Stopper ring
6. Boot band
7. Boot
8. Shaft
9. Circlip
10. Joint sub-assembly

67170-ARMA-G44

Rear halfshaft—exploded view

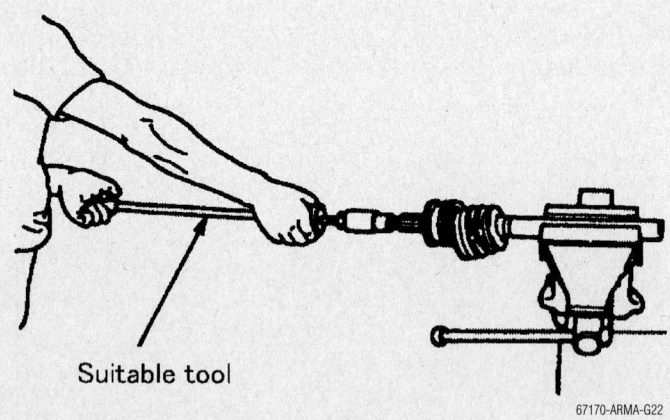

Suitable tool

67170-ARMA-G22

Using a suitable puller to remove joint sub-assembly.

15. Install the dust boot into the grooves on joint sub-assembly.

16. Secure the big and small ends of the dust boot using new boot bands.

Outer

1. Before servicing the vehicle, refer to the Precautions Section.

2. Remove the halfshaft from the vehicle.

3. Mount halfshaft in a vise.

4. Remove the dust boot bands and dust boot from joint sub-assembly.

5. Insert a suitable puller into the threaded part of the halfshaft. Pull the joint sub-assembly off of the halfshaft as shown in figure.

6. Remove dust boot and circlip from halfshaft for cleaning and inspection.

To install:

➡**Discard old dust boot, dust boot bands and circlip and use new ones for assembly.**

7. Insert genuine NISSAN grease or equivalent into the joint sub-assembly until

grease oozes from the ball groove and serration hole.

8. Wrap the serrated part of the halfshaft with tape.

9. Install new dust boot and band onto halfshaft.

10. Remove tape from serrated part of the halfshaft.

11. Press-fit the new circlip to the halfshaft.

12. Insert 5.1-5.8 oz of genuine NISSAN grease or equivalent into the joint sub-assembly and large end of boot.

13. Install the dust boot into the grooves on the joint sub-assembly.

14. Secure the big and small ends of the dust boot using new boot bands.

Front Differential Pinion Seal

REMOVAL & INSTALLATION

1. Before servicing the vehicle, refer to the Precautions Section.

2. Remove or disconnect the following:
   • Front drive shaft
   • Halfshafts

3. Measure and record the pinion bearing preload using special tool J-25765-A.

4. Loosen the pinion nut while holding

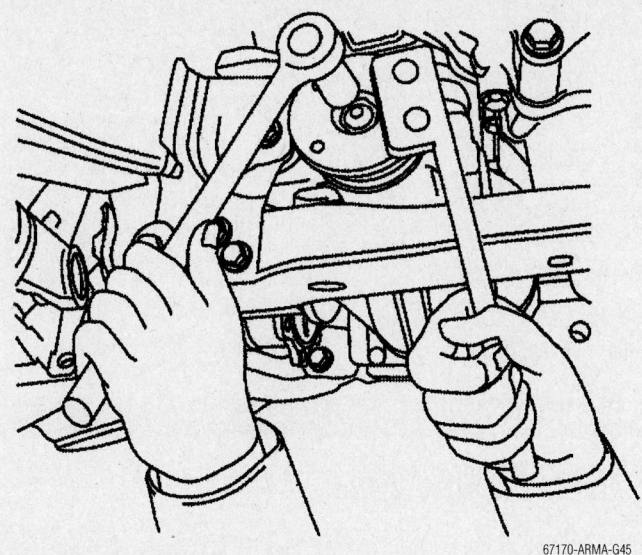

Removing the companion flange using Special Tool J-44195

67170-ARMA-G45

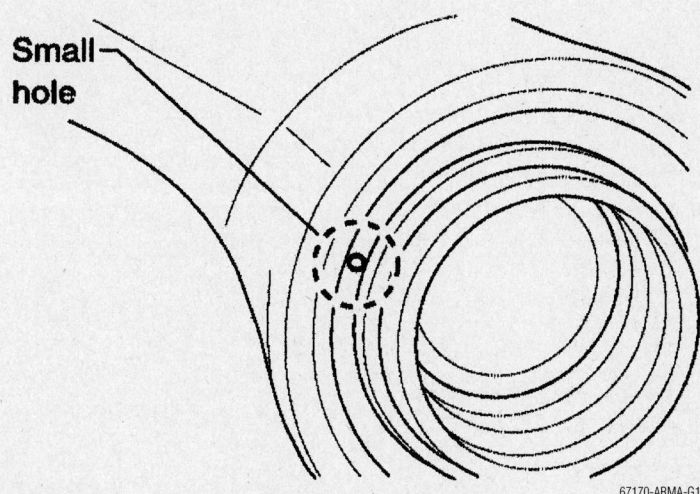

Small
hole

Small hole in casing

67170-ARMA-G18

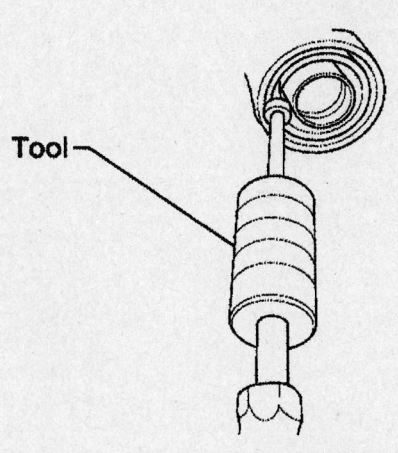

Tool

67170-ARMA-G19

Removing pinion seal using Special Tool
SP8P

the companion flange using special tool J-44195.

5. Remove the companion flange using a suitable tool.

6. Using a punch or drill, place a small hole in the case.

7. Remove the seal using special tool SP8P or equivalent.

To install:

8. Press front seal into carrier using a suitable tool.

9. Install companion flange and new pinion nut. Tighten pinion nut until there is no end play and until recorded pinion bearing preload is met plus an additional 5 inch lbs. (0.5 Nm).

10. Install or connect the following:
- Halfshafts
- Front drive shaft

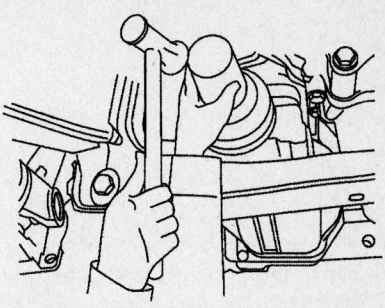

67170-ARMA-G46

Front pinion seal installation

Rear Differential Pinion Seal

REMOVAL & INSTALLATION

1. Before servicing the vehicle, refer to the Precautions Section.

2. Remove the rear drive shaft.

3. Measure and record the total pre-load.

4. Matchmark the drive pinion to position 'B' on the companion flange.

5. Remove the drive pinion nut using suitable tool.

6. Remove the companion flange using suitable tool.

7. Remove the rear pinion seal using special tool J-34286.

To install:

8. Press the rear pinion seal into the carrier using suitable tool.

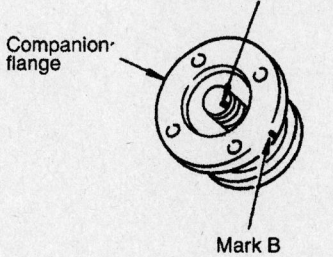

Drive pinion matching mark

Companion flange

Mark B

67170-ARMA-G20

Companion flange marking

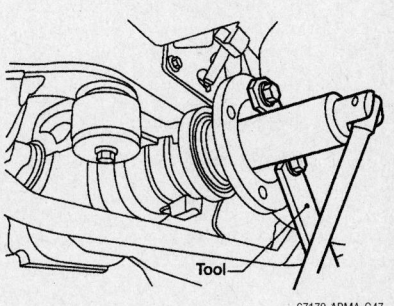

Tool

67170-ARMA-G47

Loosening the flange nut

9. Align the matchmark on the companion flange to the drive pinion and install the companion flange.

10. Lubricate the drive pinion threads and seating surfaces of the drive pinion nut with grease.

11. Using a new drive pinion nut, tighten to 124-274 ft. lbs. (167-372 Nm).

→**Final torque is determined when adjusting total preload using special tool J-25765-A.**

12. Install rear drive shaft.

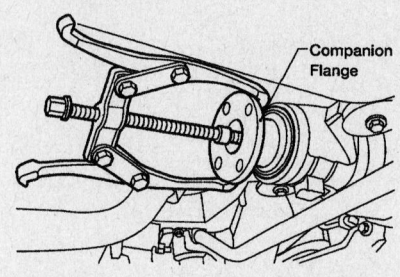

67170-ARMA-G48

Removing the companion flange

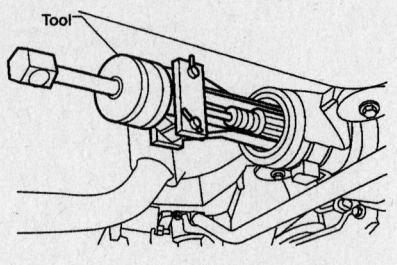

67170-ARMA-G21

Removing the pinion seal

STEERING AND SUSPENSION

Air Bag

DISARMING THE SYSTEM

1. Before servicing the vehicle, refer to the Precautions Section.

2. Disconnect both battery cables.

3. Wait at least 3 minutes before working on the vehicle. The air bag system is designed to retain enough power to deploy the air bag for a short time after the battery has been disconnected.

4. After repairs are complete, connect the negative battery cable. Turn the ignition switch to the **ON** position and check the air bag warning light blinks for proper operation.

Power Steering Gear

REMOVAL & INSTALLATION

1. Before servicing the vehicle, refer to the Precautions Section.

2. Ensure the wheels are in the straight-ahead position.

3. Remove or disconnect the following:
   • Wheels
   • Engine splash guard

4. On 4-wheel drive models only, remove front final drive and support the drive shafts.

5. Remove cotter pin at steering outer socket and loosen mounting nut.

6. Remove steering outer socket from steering knuckle using special tool J-25730-A.

7. On 2-wheel drive models only, remove stabilizer bar mounting bolts and secure the stabilizer bar.

8. Remove or disconnect the following:
   • Oil pipes from steering gear assembly
   • Lower joint mounting bolt from lower shaft
   • Mounting bolts and nuts from steering gear assembly
   • Steering gear assembly

To install:

9. Install or connect the following:
   • Steering gear assembly, tighten nuts to 133 ft. lbs. (180 Nm)
   • Lower joint mounting bolt

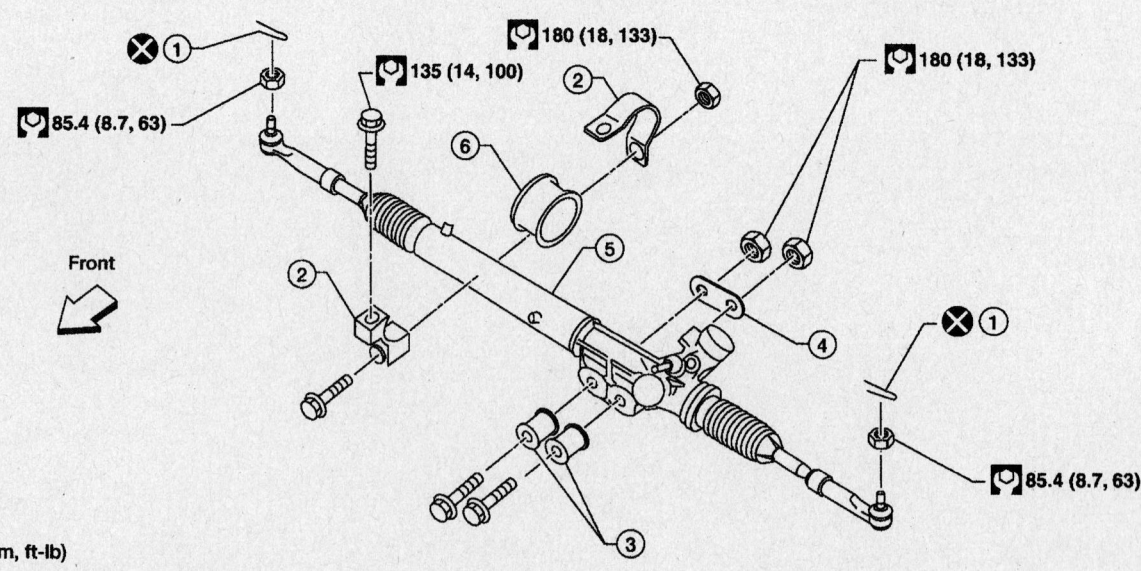

⊗ 1
85.4 (8.7, 63)
135 (14, 100)
180 (18, 133)
2
6
180 (18, 133)
5
Front
2
4
⊗ 1
3
85.4 (8.7, 63)

🔧 : N·m (kg-m, ft-lb)

⊗ : Always replace after every disassembly.

1. Cotter pin	2. Mounting bracket	3. Bushing
4. Washer	5. Steering gear assembly	6. Mounting insulator

67170-ARMA-G49

Steering gear assembly

- Oil pipes to steering gear assembly
- Stabilizer bar, 2 wheel-drive models only
- Steering outer socket to steering knuckle, tighten nut to 63 ft. lbs. (86 Nm)
- Front final drive, 4-wheel drive models only
- Engine splash guard
- Wheels

10. Check the wheel alignment and adjust as necessary.

Shock Absorber

REMOVAL & INSTALLATION

Front

1. Before servicing the vehicle, refer to the Precautions Section.

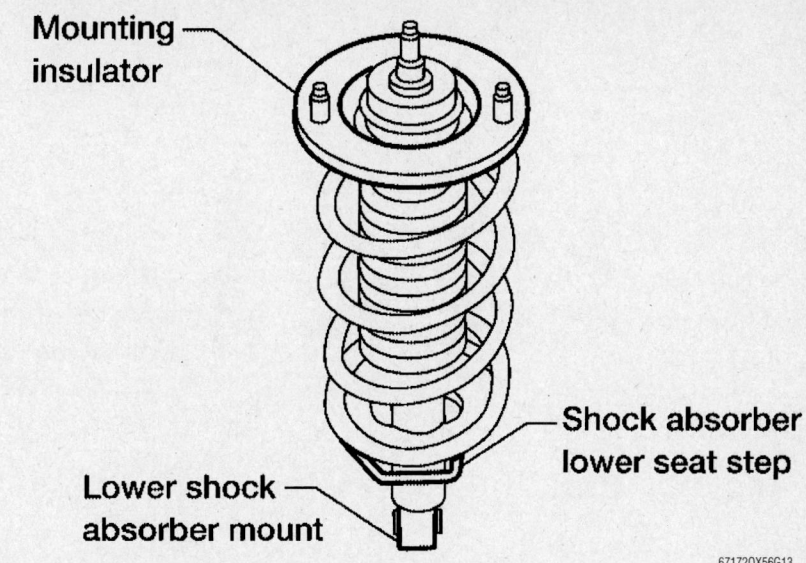

67172QX56G13

Shock absorber installation

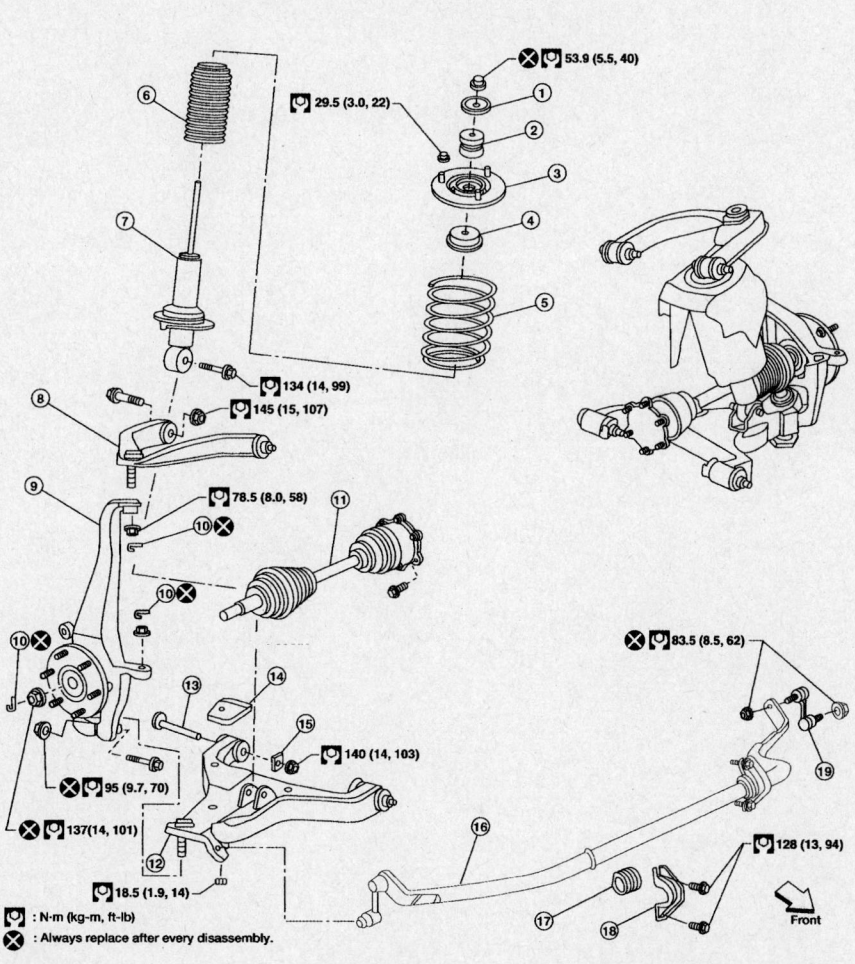

🔧 : N·m (kg-m, ft-lb)
❌ : Always replace after every disassembly.

1. Washer	2. Shock absorber bushing	3. Shock absorber mounting insulator
4. Upper seat	5. Coil spring	6. Dust cover
7. Shock absorber	8. Upper link	9. Steering knuckle
10. Cotter pin	11. Drive shaft	12. Lower link
13. Cam bolt	14. Jounce bumper	15. Cam washer
16. Stabilizer bar	17. Stabilizer bar bushing	18. Stabilizer bar mounting bracket
19. Connecting rod		

67170-ARMA-G13

Front suspension

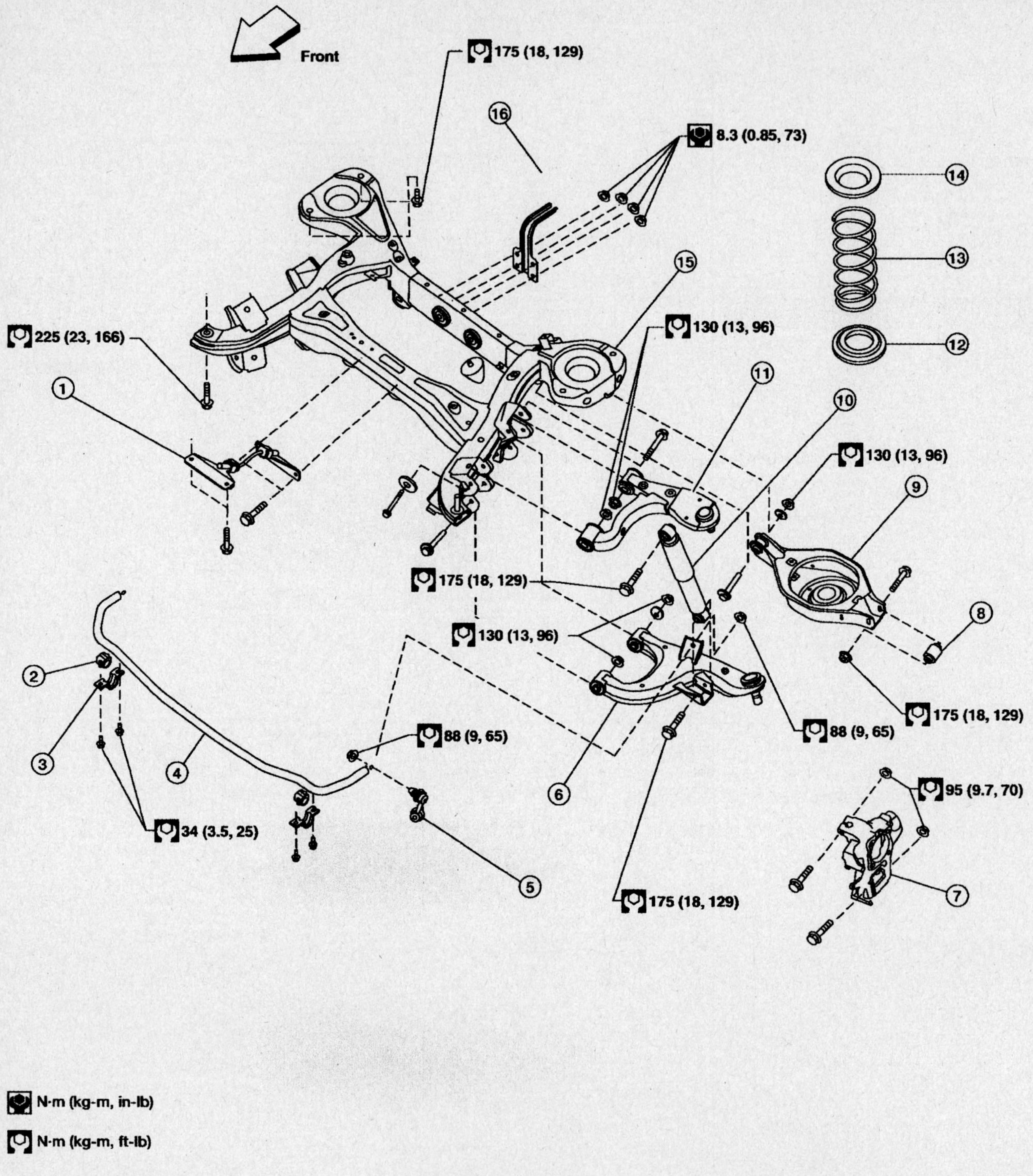

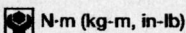

Front

175 (18, 129)

16

8.3 (0.85, 73)

14

15

13

130 (13, 96)

225 (23, 166)

11

12

1

10

130 (13, 96)

9

175 (18, 129)

2

8

130 (13, 96)

175 (18, 129)

3

88 (9, 65)

88 (9, 65)

4

95 (9.7, 70)

5

6

34 (3.5, 25)

7

175 (18, 129)

N·m (kg-m, in-lb)

N·m (kg-m, ft-lb)

1. Seat belt latch anchor	2. Stabilizer bar bushing	3. Stabilizer bar clamp
4. Stabilizer bar	5. Connecting rod	6. Front lower link
7. Wheel hub and spindle assembly	8. Bushing	9. Rear lower link
10. Shock absorber	11. Suspension arm	12. Lower rubber seat
13. Coil spring	14. Upper rubber seat	15. Rear suspension member
16. Spare tire bracket		

67170-ARMA-G50

Standard rear suspension

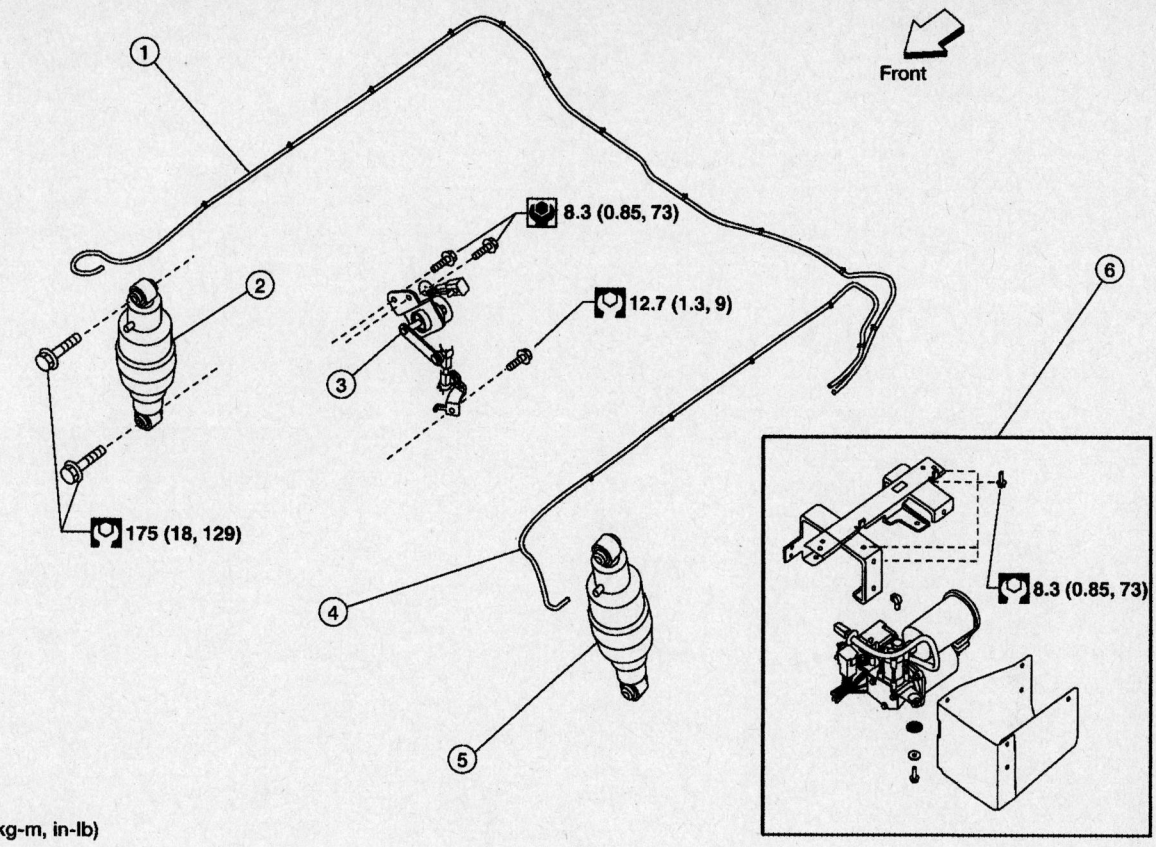

Front

8.3 (0.85, 73)

12.7 (1.3, 9)

175 (18, 129)

8.3 (0.85, 73)

N·m (kg-m, in-lb)

N·m (kg-m, ft-lb)

1. Rear load leveling air suspension hose, RH

2. Shock absorber, RH

3. Height sensor

4. Rear load leveling air suspension hose, LH

5. Shock absorber, LH

6. Rear load leveling air suspension compressor assembly

67170-ARMA-G51

Rear load leveling air suspension

2. Remove or disconnect the following:
   - Wheel
   - Lower shock absorber bolt
   - Upper shock absorber bolts
   - Coil spring and shock absorber assembly

3. Secure the shock absorber in a vice and loosen (without removing) the piston rod lock nut.

4. Install a spring compressor and tighten until the shock absorber mounting insulator can be turned by hand.

5. Remove piston rod lock nut and remove shock absorber.

To install:

6. Install upper mounting insulator in line with the lower shock absorber mount and step in shock absorber lower seat as shown in figure.

7. Tighten the new piston rod lock nut to 40 ft. lbs. (54 Nm).

8. Install or connect the following:

- Coil spring and shock absorber assembly
- Upper shock absorber bolts and tighten to 22 ft. lbs (30 Nm)
- Lower shock absorber bolt and tighten to 99 ft. lbs. (134 Nm)
- Wheel

9. Check wheel alignment and adjust as necessary.

Rear

1. Before servicing the vehicle, refer to the Precautions Section.

2. Remove the rear wheel.

3. Release the air pressure from the rear load leveling air suspension system using the CONSULT-II "EXHAUST SOLENOID" active test.

4. Remove or disconnect the following:
   - Rear fender protector
   - Rear load leveling air suspension hose from the shock absorber

- Shock absorber upper and lower end bolts
- Shock absorber

To install:

5. Install or connect the following:
   - Shock absorber and tighten end bolts to 129 ft. lbs. (175 Nm)
   - Rear load leveling air suspension hose
   - Rear fender protector
   - Rear wheel

Coil Spring

REMOVAL & INSTALLATION

Front

1. Before servicing the vehicle, refer to the Precautions Section.

2. Remove or disconnect the following:
   - Wheel

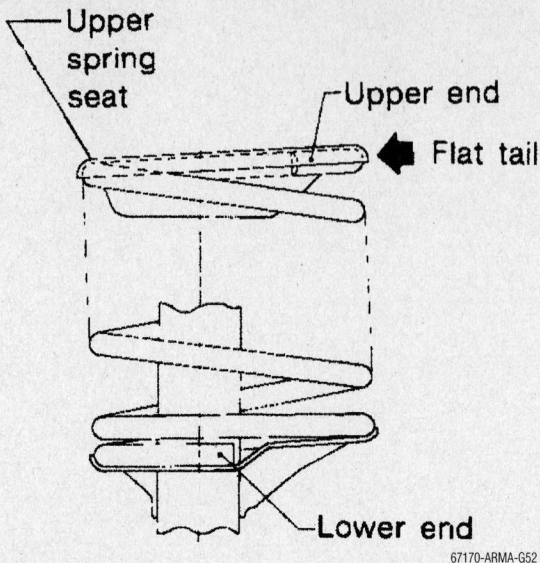

Front coil spring positioning

67170-ARMA-G52

- Lower shock absorber bolt
- Upper shock absorber bolts
- Coil spring and shock absorber assembly

3. Secure the shock absorber in a vice and loosen (without removing) the piston rod lock nut.

4. Install a spring compressor and tighten until the shock absorber mounting insulator can be turned by hand.

5. Remove piston rod lock nut and remove shock absorber from the coil spring.

To install:

6. Install upper mounting insulator in line with the lower shock absorber mount and step in shock absorber lower seat as shown in figure.

7. Tighten the new piston rod lock nut to 40 ft. lbs. (54 Nm).

8. Install or connect the following:
- Coil spring and shock absorber assembly
- Upper shock absorber bolts and tighten to 22 ft. lbs (30 Nm)
- Lower shock absorber bolt and tighten to 99 ft. lbs. (134 Nm)
- Wheel

9. Check wheel alignment and adjust as necessary.

Rear

1. Before servicing the vehicle, refer to the Precautions Section.

2. Remove the rear wheel.

3. Release the air pressure from the rear load leveling air suspension system using the CONSULT-II "EXHAUST SOLE-NOID" active test.

4. Remove the height sensor arm

bracket bolt from the left-hand rear lower link.

5. Place a suitable jack under the rear lower link and relieve the coil spring tension.

6. Loosen the rear lower link adjusting bolt and nut connected to the rear suspension member.

7. Remove the rear lower link bolt and nut from the knuckle.

8. Slowly lower the jack to relieve the coil spring tension.

9. Remove the coil spring.

To install:

10. Install or connect the following:
- Coil spring

➡ When installing the rubber seats for the coil spring, ensure the embossed arrows point outward towards the wheel.

- Rear lower link bolt to knuckle and tighten nut to 70 ft. lbs. (95 Nm)
- Rear lower link adjusting bolt to rear suspension member and tighten nut to 101 ft. lbs. (137 Nm)
- Height sensor arm bracket bolt to left-head rear lower link and tighten to 9 ft. lbs. (12 Nm)
- Rear wheel

Upper Ball Joint

REMOVAL & INSTALLATION

1. Before servicing the vehicle, refer to the Precautions Section.

2. Remove or disconnect the following:
- Wheel

- Coil spring and shock absorber assembly
- Cotter pin and nut from upper ball joint

3. Separate upper ball joint from steering knuckle using special tool J-24319-01

To install:

4. Install or connect the following:
- Upper ball joint
- New cotter pin and tighten nut to 58 ft. lbs. (79 Nm)
- Coil spring and shock absorber assembly
- Wheel

Lower Ball Joint

REMOVAL & INSTALLATION

1. Before servicing the vehicle, refer to the Precautions Section.

2. Remove or disconnect the following:
- Wheel
- Lower shock absorber bolt
- Stabilizer bar connecting rod
- Drive shaft, if equipped with 4WD
- Pinch bolt from steering knuckle

3. Separate lower ball joint from steering knuckle

To install:

4. Install or connect the following:
- Lower ball joint
- Pinch bolt to steering knuckle
- Drive shaft, if equipped with 4WD
- Stabilizer bar connecting rod
- Lower shock absorber bolt
- Wheel

Upper Control Arm

REMOVAL & INSTALLATION

1. Before servicing the vehicle, refer to the Precautions Section.

2. Remove or disconnect the following:
- Wheel
- Coil spring and shock absorber assembly
- Cotter pin and nut from upper ball joint

3. Separate upper ball joint stud from steering knuckle using special tool J-24319-01.

4. Remove the following:
- Upper control arm mounting bolts
- Upper control arm

To install:

5. Install or connect the following:
- Upper control arm and tighten bolts to 107 ft. lbs. (145 Nm)

- Upper ball joint with new cotter pin and tighten nut to 58 ft. lbs. (79 Nm)
- Coil spring and shock absorber assembly
- Wheel

Lower Control Arm

REMOVAL & INSTALLATION

1. Before servicing the vehicle, refer to the Precautions Section.
2. Remove or disconnect the following:
   - Wheel
   - Lower shock absorber bolt
   - Stabilizer bar connecting rod
   - Drive shaft, if equipped with 4WD
   - Pinch bolt from steering knuckle
3. Separate the lower ball joint from the steering knuckle.
4. Remove the following:
   - Lower link adjusting bolts
   - Lower link

To install:

5. Install or connect the following:
   - Lower link and tighten adjusting bolts to 98 ft. lbs. (133 Nm)

- Lower ball joint
- Pinch bolt
- Drive shaft, if equipped with 4WD
- Stabilizer bar connected rod
- Lower shock absorber bolt
- Wheel

Wheel Bearings

REMOVAL & INSTALLATION

Front

1. Before servicing the vehicle, refer to the Precautions Section.
2. Remove or disconnect the following:
   - Wheel
   - Engine splash guard
   - Brake caliper without disconnecting the hydraulic lines, and reposition aside with wire
3. Matchmark the brake rotor to the wheel hub and remove the brake rotor.
4. Remove or disconnect the following:
   - Cotter pin and lock nut from drive shaft
   - Drive shaft from wheel hub and bearing assembly
   - ABS sensor

- Wheel hub and bearing assembly bolts
- Wheel hub and bearing assembly

To install:

5. Install or connect the following:
   - Wheel hub and bearing assembly, using new bolts and tighten to 155 ft. lbs. (210 Nm)
   - ABS sensor
   - Drive shaft to wheel hub and bearing assembly
   - Cotter pin and lock nut and tighten to 101 ft. lbs. (137 Nm)
   - Brake rotor
   - Brake caliper
   - Engine splash guard
   - Wheel

Rear

1. Before servicing the vehicle, refer to the Precautions Section.
2. Remove or disconnect the following:
   - Wheel
   - Brake caliper without disconnecting the hydraulic lines, and reposition aside with wire
   - Brake rotor
   - Cotter pin and nut from drive shaft
   - Drive shaft

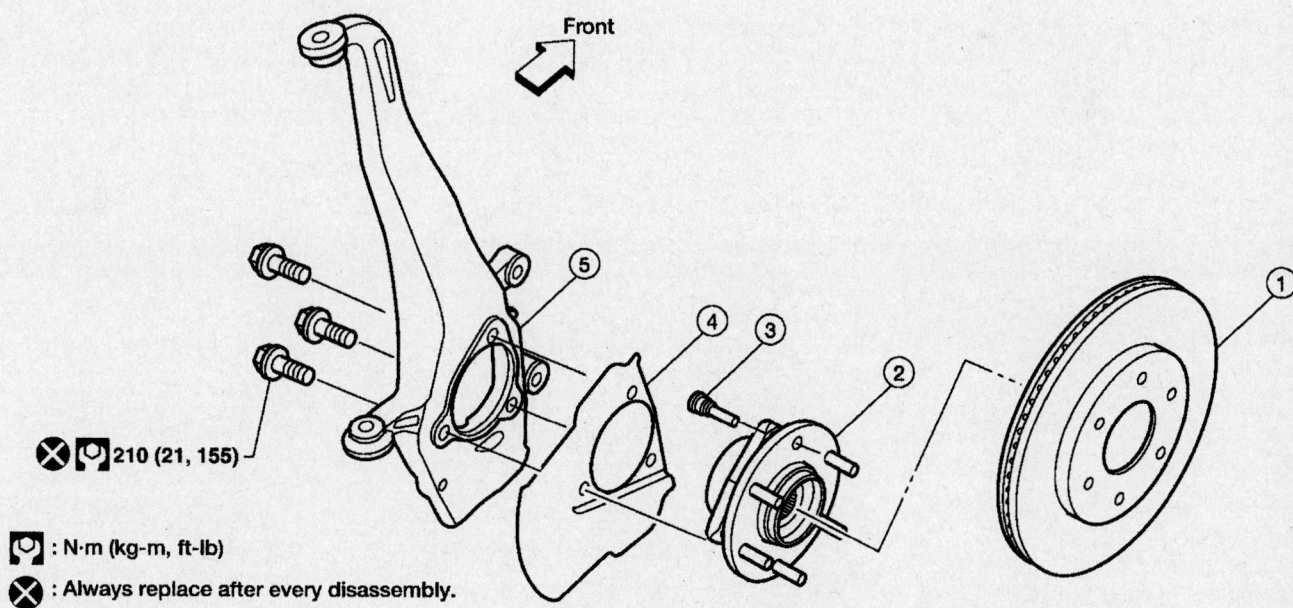

210 (21, 155)

⊡ : N·m (kg-m, ft-lb)

❌ : Always replace after every disassembly.

1. Disc rotor
4. Splash guard
2. Wheel hub and bearing assembly
5. Steering knuckle
3. Wheel stud

Front hub/bearing assembly

67170-ARMA-G53

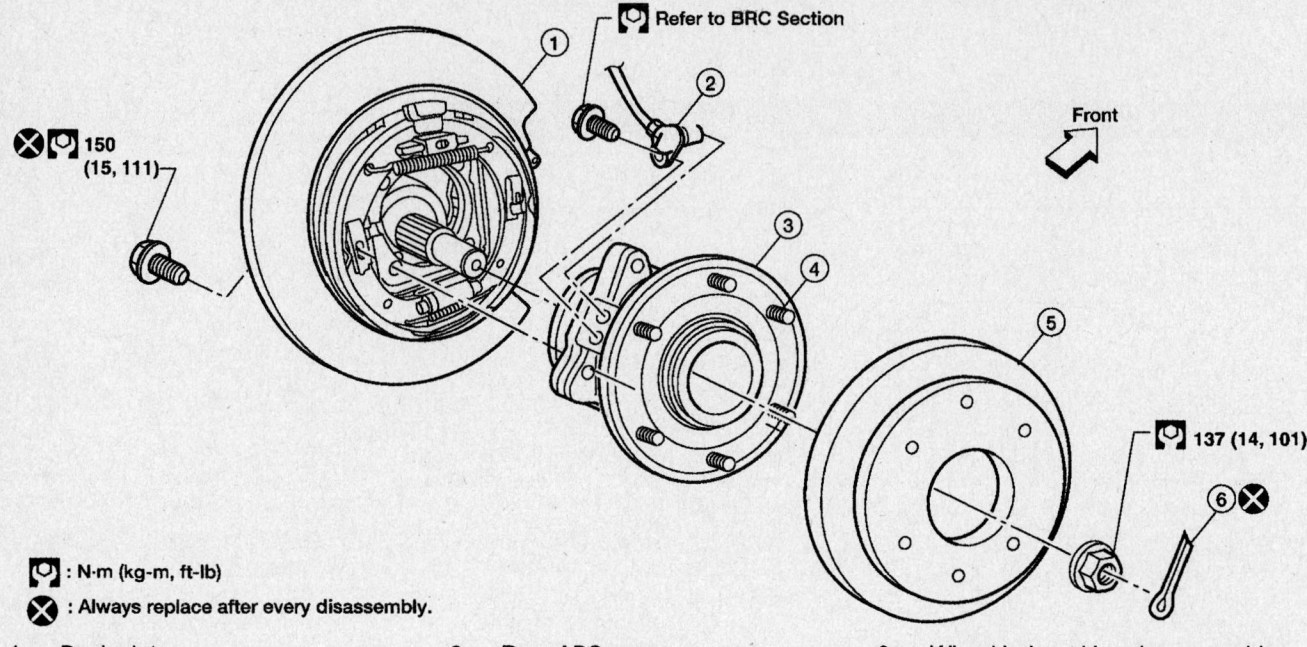

Refer to BRC Section

Front

150
(15, 111)

137 (14, 101)

: N·m (kg-m, ft-lb)

: Always replace after every disassembly.

1. Back plate
2. Rear ABS sensor
3. Wheel hub and bearing assembly
4. Wheel stud
5. Rear disc rotor
6. Cotter pin

67170-ARMA-G54

Rear hub/bearing assembly

- Wheel hub and bearing assembly bolts
3. Pulling out the wheel hub and bearing assembly slightly, remove the ABS sensor.
4. Remove the wheel hub and bearing assembly.

To install:
5. Install or connect the following:
- ABS sensor
- Wheel hub and bearing assembly, using new bolts and tighten to 111 ft. lbs. (150 Nm)

- Drive shaft
- Lock nut and tighten to 101 ft. lbs. (137 Nm) and new cotter pin
- Brake rotor
- Brake caliper
- Wheel

BRAKES

Brake Caliper

REMOVAL & INSTALLATION

Front

1. Before servicing the vehicle, refer to the Precautions Section.
2. Drain brake fluid as necessary.
3. Remove or disconnect the following:
- Wheel
- Union bolt
- Caliper-to-torque member slide pins, or remove the caliper and torque member as an assembly.
- Brake caliper

To install:
4. Install or connect the following:
- Brake caliper, tighten torque member bolts to 155 ft. lbs. (210 Nm); the caliper slide pins to 20 ft. lbs. (27 Nm)
- Union bolt and tighten to 13 ft. lbs. (18 Nm)
5. Fill the master cylinder and bleed the brake system.
6. Install the wheels.

Rear

1. Before servicing the vehicle, refer to the Precautions Section.
2. Drain brake fluid as necessary.

3. Remove or disconnect the following:

- Wheel
- Union bolt
- Mounting bolts
- Brake caliper assembly

To install:
4. Install or connect the following:
- Brake caliper assembly and tighten mounting bolts to 23ft. lbs. (44 Nm)
- Union bolt and tighten to 13 ft. lbs. (18 Nm)
5. Fill the master cylinder and bleed the brake system.
6. Install the wheels.

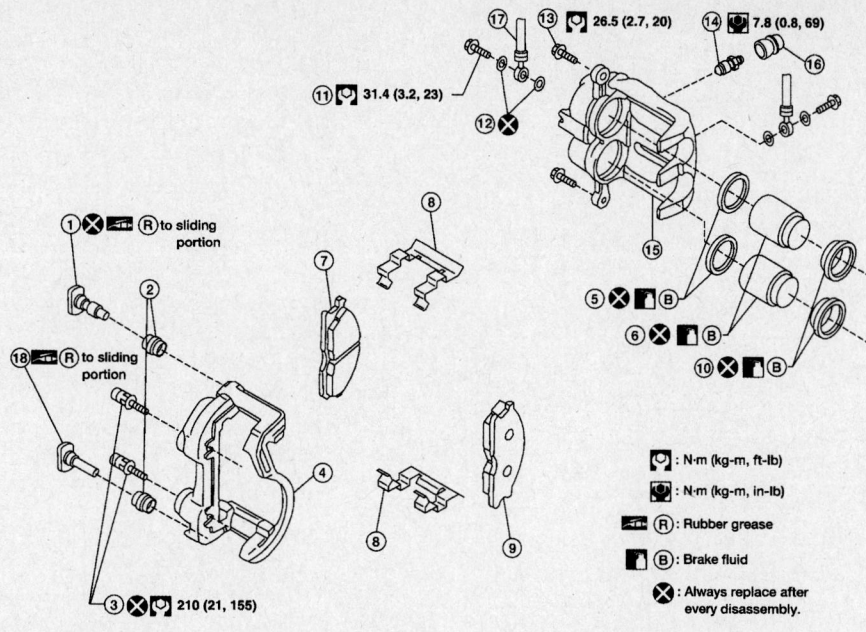

- ①❌ ⇐Ⓡ to sliding portion
- ⑱ ⇐Ⓡ to sliding portion
- ⑪ 🔧 31.4 (3.2, 23)
- ⑰
- ⑬ 🔧 26.5 (2.7, 20)
- ⑭ 🔧 7.8 (0.8, 69)
- ⑯
- ⑫❌
- ⑮
- ⑤❌🔧Ⓑ
- ⑥❌🔧Ⓑ
- ⑩❌🔧Ⓑ
- ③❌ 🔧 210 (21, 155)

- 🔧 : N·m (kg-m, ft-lb)
- 🔧 : N·m (kg-m, in-lb)
- ⇐Ⓡ : Rubber grease
- 🔧Ⓑ : Brake fluid
- ❌ : Always replace after every disassembly.

1.	Upper sliding pin	2.	Sliding pin boot	3.	Torque member bolt
4.	Torque member	5.	Piston seal	6.	Piston
7.	Inner pad	8.	Pad retainer	9.	Outer pad
10.	Piston boot	11.	Union bolt	12.	Copper washer
13.	Sliding pin bolt	14.	Bleed valve	15.	Cylinder body
16.	Cap	17.	Brake hose	18.	Lower sliding pin

67170-ARMA-G55

Front brake components

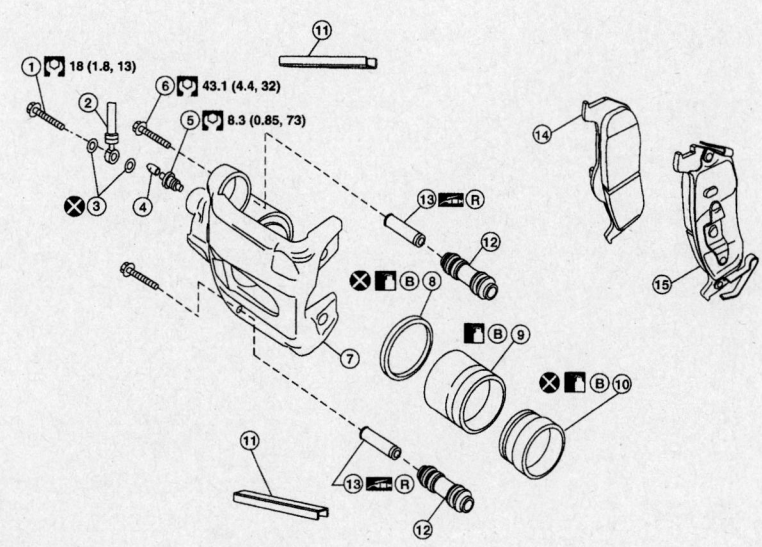

- ① 🔧 18 (1.8, 13)
- ⑥ 🔧 43.1 (4.4, 32)
- ⑤ 🔧 8.3 (0.85, 73)
- ⑪
- ⑫
- ⑭
- ⑮
- ⑬ ⇐Ⓡ
- ③❌
- ④
- ❌🔧Ⓑ⑧
- 🔧Ⓑ⑨
- ❌🔧Ⓑ⑩
- ⑦
- ⑪
- ⑬ ⇐Ⓡ
- ⑫

- 🔧Ⓑ : Brake fluid
- ⇐Ⓡ : Rubber grease
- 🔧 : N·m (kg-m, ft-lb)
- 🔧 : N·m (kg-m, in-lb)
- ❌ : Always replace after every disassembly.

1.	Union bolt	2.	Brake hose	3.	Copper washer
4.	Cap	5.	Bleed valve	6.	Mounting bolt
7.	Cylinder body	8.	Piston seal	9.	Piston
10.	Piston boot	11.	Knuckle slide	12.	Sliding sleeve boot
13.	Sliding sleeve	14.	Inner pad	15.	Outer pad

67170-ARMA-G56

Rear brake components

Disc Brake Pads

REMOVAL AND INSTALLATION

Front

1. Before servicing the vehicle, refer to the Precautions Section.
2. Remove the wheel.
3. Remove lower sliding pin bolt.
4. Suspend brake caliper with a remove and remove brake pad and shim from torque member.

To install:

5. Push pistons in so that the pad is firmly installed, using a suitable tool.
6. Mount the brake caliper to torque member.
7. Attach pad retainer to torque member.
8. Lubricate lower sliding pin bolt with a thin layer of silicone grease and install. Torque to 20 ft. lbs. (27 Nm).
9. Install the wheel.

Rear

1. Before servicing the vehicle, refer to the Precautions Section.

2. Remove the wheel.
3. Remove mounting bolt from the top mount.
4. Swing brake caliper open and remove the brake pads.

To install:

5. Push pistons in so that the pad is firmly installed, using a suitable tool.
6. Install pads to the brake caliper.
7. Install top mounting bolt and tighten to 32 ft. lbs. (44 Nm).
8. Install the wheel.

MAZDA

5

626 • 2002–04 Miata • Millenia • Protégé

SPECIFICATION AND MAINTENANCE CHARTS

ENGINE AND VEHICLE IDENTIFICATION

Code ①	Liters (cc)	Cu. In.	Cyl.	Fuel Sys.	Engine Type	Eng. Mfg.	Code ②	Year
ZM	1.6 (1597)	97.4	4	MPFI	DOHC	Mazda	2	2002
BP	1.8 (1839)	112.2	4	MPFI	DOHC	Mazda	3	2003
FS	2.0 (1991)	121.5	4	MPFI	DOHC	Mazda	4	2004
KJ	2.3 (2254)	137.2	6	MPFI	DOHC	Mazda		
KL	2.5 (2496)	152.3	6	MPFI	DOHC	Mazda		

MPFI: Multi-Point Fuel Injection

DOHC: Double Over Head Cam

① Located above the starter

② 10th digit of the Vehicle Identification Number (VIN)

09482_MAZC1_C0001

GENERAL ENGINE SPECIFICATIONS

Year	Model	Engine Displacement Liters (VIN)	Net Horsepower @ rpm	Net Torque @ rpm (ft. lbs.)	Bore x Stroke (in.)	Compression Ratio	Oil Pressure @ rpm
2002	626 ES	2.0 (FS)	125@5500	127@3300	3.27x3.62	9.0:1	57-71@3000
	626 ES-V6	2.5 (KL)	165@6000	161@5000	3.33x2.92	9.5:1	49-71@3000
	626 LX	2.0 (FS)	125@5500	127@3300	3.27x3.62	9.0:1	57-71@3000
	626 LX-V6	2.5 (KL)	165@6000	161@5000	3.33x2.92	9.5:1	49-71@3000
	Miata	1.8 (BP)	142@7000	125@5500	3.3x3.4	10.0:1	43-56@3000
	Millenia	2.5 (KL)	170@5800	160@4800	3.33x2.92	9.2:1	49-71@3000
	Millenia S	2.3 (KJ)	210@5300	210@3500	3.16x2.92	10.0:1	44-66@3000
	Protege DX	1.6 (ZM)	①	②	3.07x3.29	9.0:1	43-57@3000
	Protege LX	1.6 (ZM)	①	②	3.07x3.29	9.0:1	43-57@3000
	Protege5	2.0 (FS)	125@5500	127@3300	3.27x3.62	9.1:1	57-71@3000
2003-04	626 ES	2.0 (FS)	125@5500	127@3300	3.27x3.62	9.0:1	57-71@3000
	626 ES-V6	2.5 (KL)	165@6000	161@5000	3.33x2.92	9.5:1	49-71@3000
	626 LX	2.0 (FS)	125@5500	127@3300	3.27x3.62	9.0:1	57-71@3000
	626 LX-V6	2.5 (KL)	165@6000	161@5000	3.33x2.92	9.5:1	49-71@3000
	Miata	1.8 (BP)	142@7000	125@5500	3.3x3.4	10.0:1	43-56@3000
	Millenia	2.5 (KL)	170@5800	160@4800	3.33x2.92	9.2:1	49-71@3000
	Millenia S	2.3 (KJ)	210@5300	210@3500	3.16x2.92	10.0:1	44-66@3000
	Protege DX	1.6 (ZM)	①	②	3.07x3.29	9.0:1	43-57@3000
	Protege LX	1.6 (ZM)	①	②	3.07x3.29	9.0:1	43-57@3000
	Protege5	2.0 (FS)	125@5500	127@3300	3.27x3.62	9.1:1	57-71@3000

EFI: Electronic Fuel Injection

① LEV states: 103@5500

Except LEV states: 105@5500

② LEV states: 106@4000

Except LEV states: 107@4000

09482_MAZC1_C0002

ENGINE TUNE-UP SPECIFICATIONS

Year	Engine Displacement Liters (VIN)	Spark Plug Gap (in.)	Ignition Timing (deg.)		Fuel Pump (psi)	Idle Speed (rpm)		Valve Clearance	
			MT	AT		MT	AT	In.	Ex.
2002	1.6 (ZM)	0.040-0.043	6-18B	6-18B	39-45	650-750	650-750	0.010-0.011	0.010-0.011
	1.8 (BP)	0.040-0.043	6-18B	6-18B	53-61	750-850	750-850	0.008-0.0011	0.012-0.013
	2.0 (FS)	0.040-0.043	6-18B	6-18B	37-45	550-850	500-800	0.008-0.0011	0.008-0.0011
	2.3 (KJ)	28-31	6-8B	6-8B	41-48	600-700	600-700	0.011-0.012	0.011-0.012
	2.5 (KL)	①	②	②	39-45	600-700	600-700	③	④
2003-04	1.6 (ZM)	0.040-0.043	6-18B	6-18B	39-45	650-750	650-750	0.010-0.011	0.010-0.011
	1.8 (BP)	0.040-0.043	6-18B	6-18B	53-61	750-850	750-850	0.008-0.0011	0.012-0.013
	2.0 (FS)	0.040-0.043	6-18B	6-18B	37-45	550-850	500-800	0.008-0.0011	0.008-0.0011
	2.3 (KJ)	28-31	8B	8B	41-48	600-700	600-700	0.011-0.012	0.011-0.012
	2.5 (KL)	①	②	②	39-45	600-700	600-700	③	④

NOTE: The Vehicle Emission Control Information label often reflects specification changes made during production. The label figures must be used if they differ from those in this chart.

NA: Not Available

B: Before top dead center

HYD: Hydraulic

① 626 models: 0.028-0.031
Millenia models: 0.039-0.043

② 626 models: 4-16
Millenia models: 9-11

③ 626 models: 0.0097-0.012
Millenia models: maintenance free

④ 626 models: 0.010-0.013
Millenia models: maintenance free

09482_MAZC1_C0003

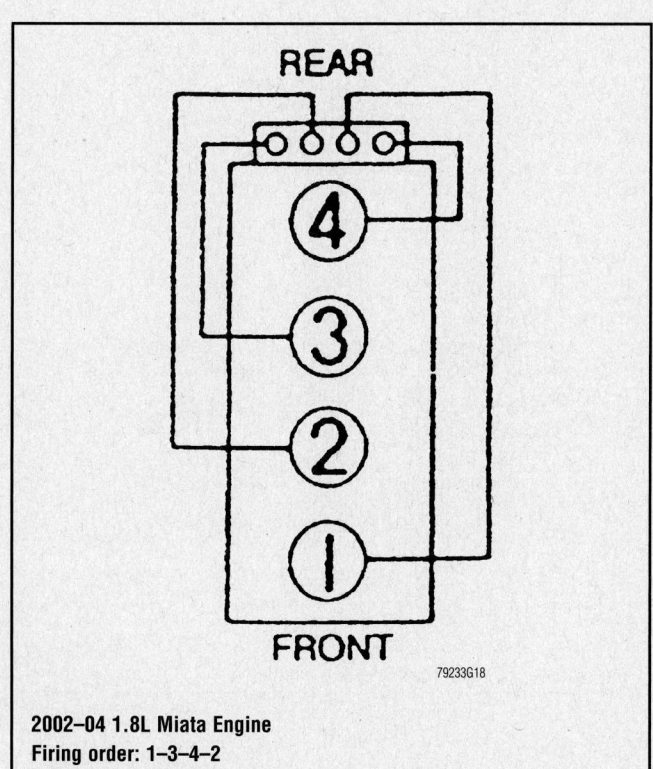

2002–04 1.8L Miata Engine
Firing order: 1–3–4–2
Distributorless ignition system

79233G18

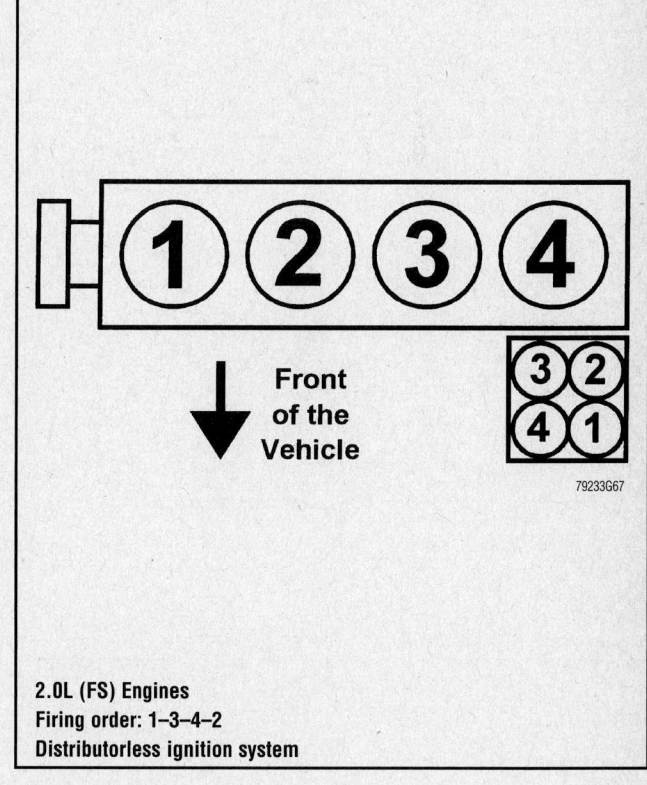

2.0L (FS) Engines
Firing order: 1–3–4–2
Distributorless ignition system

79233G67

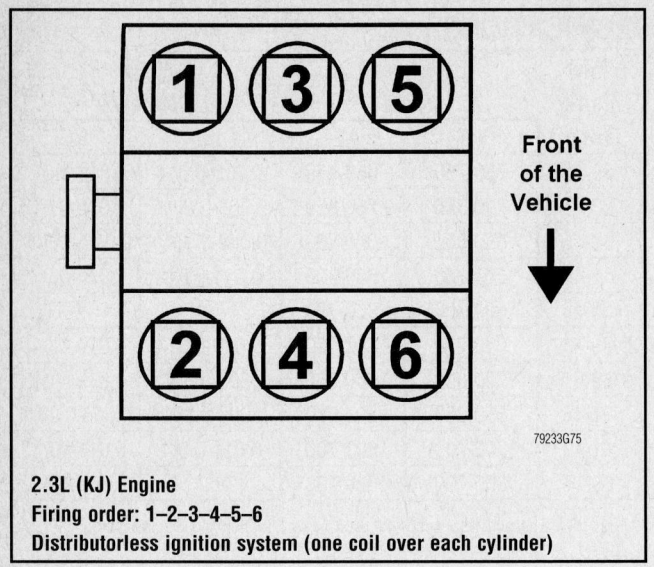

2.3L (KJ) Engine
Firing order: 1–2–3–4–5–6
Distributorless ignition system (one coil over each cylinder)

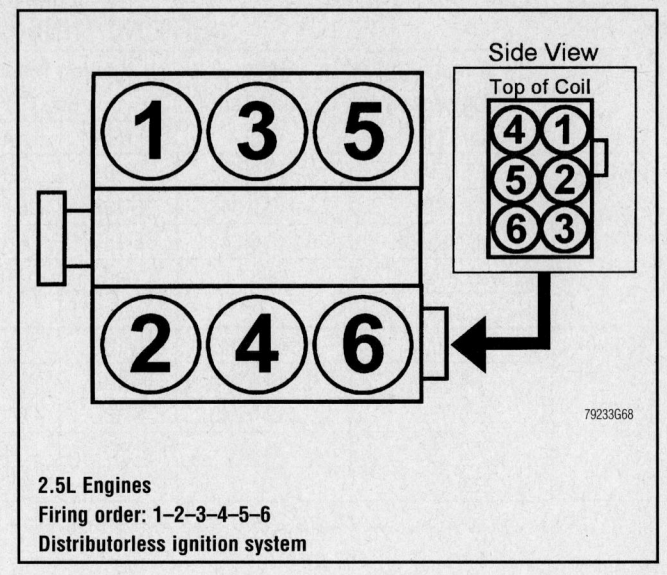

2.5L Engines
Firing order: 1–2–3–4–5–6
Distributorless ignition system

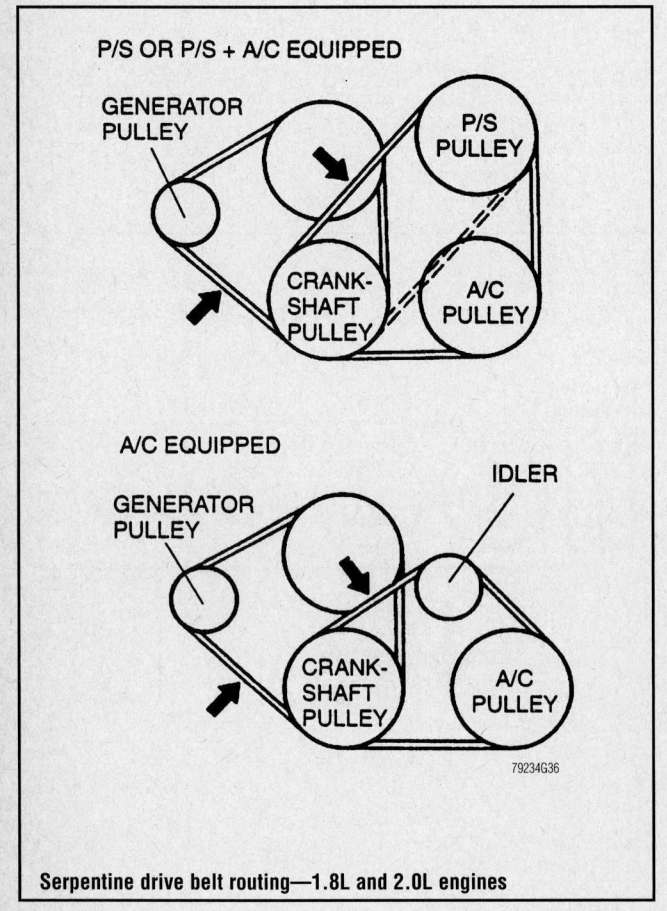

Serpentine drive belt routing—1.8L and 2.0L engines

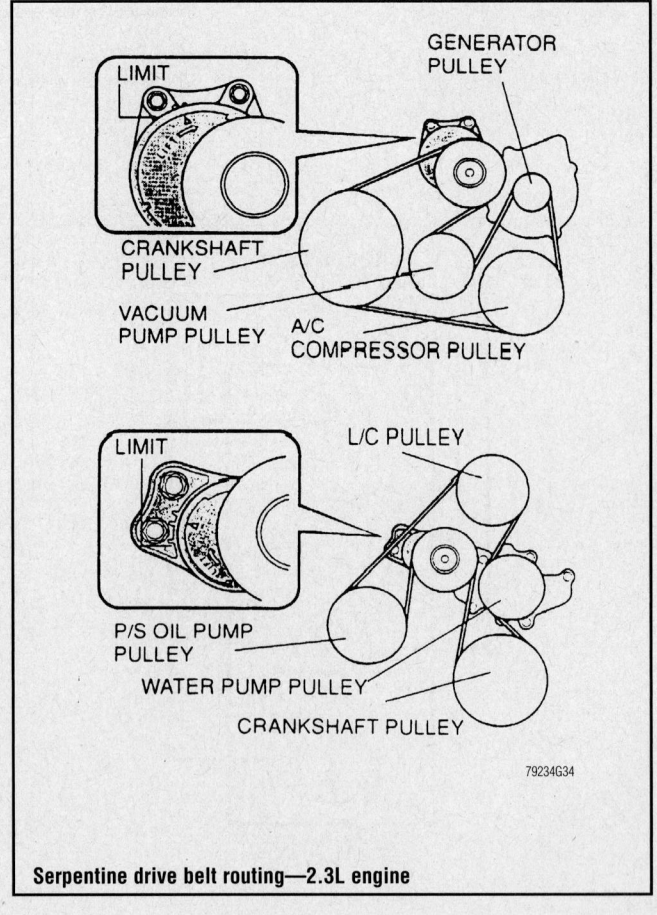

Serpentine drive belt routing—2.3L engine

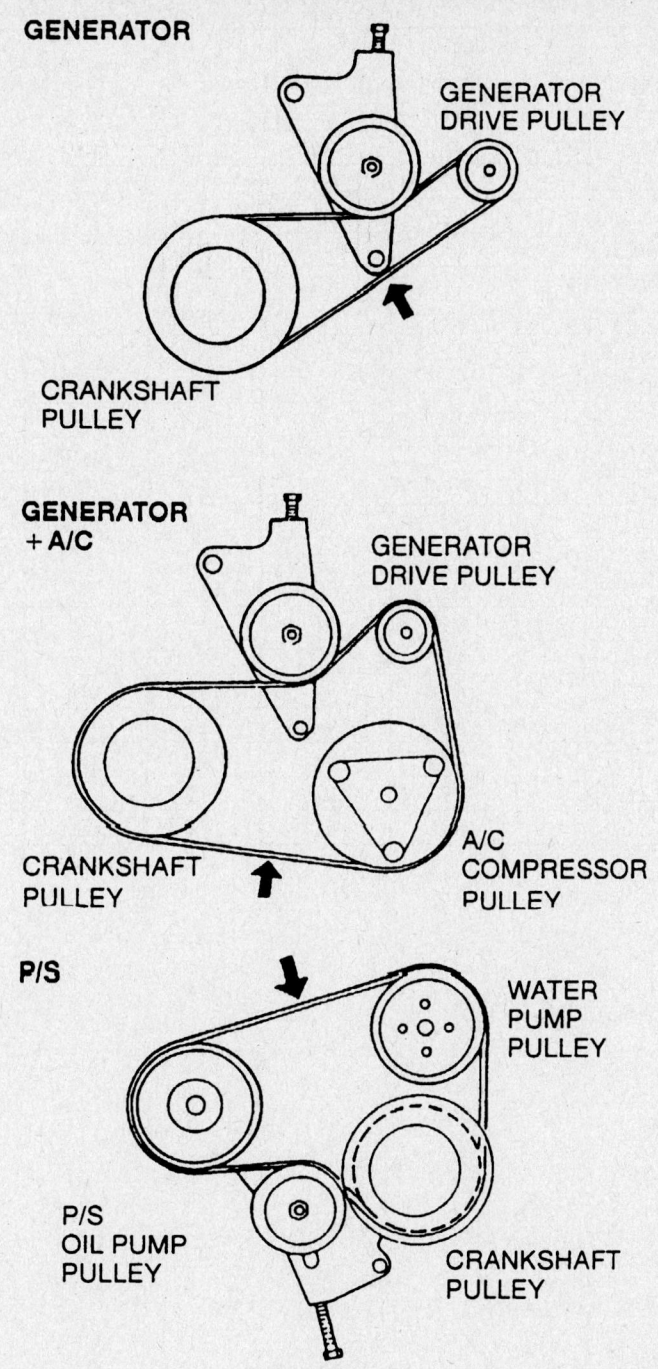

GENERATOR

GENERATOR
DRIVE PULLEY

CRANKSHAFT
PULLEY

GENERATOR
+ A/C

GENERATOR
DRIVE PULLEY

CRANKSHAFT
PULLEY

A/C
COMPRESSOR
PULLEY

P/S

WATER
PUMP
PULLEY

P/S
OIL PUMP
PULLEY

CRANKSHAFT
PULLEY

79234G35

Serpentine drive belt routing—2.5L engines

CAPACITIES

Year	Model	Engine Displacement Liters (VIN)	Engine Oil with Filter (qts.)	Transmission (pts.)		Drive Axle		Fuel Tank (gal.)	Cooling System (qts.)
				5-Spd	Auto.	Front (pts.)	Rear (pts.)		
2002	626 ES	2.0 (FS)	3.7	5.8	18.6	①	—	16.9	7.9
	626 ES-V6	2.5 (KL)	4.2	6.0	17.0	①	—	16.9	7.9
	626 LX	2.0 (FS)	3.7	5.8	18.6	①	—	16.9	7.9
	626 LX-V6	2.5 (KL)	4.2	6.0	17.0	①	—	16.9	7.9
	Miata	1.8 (BP)	4.0	②	14.2	—	2.1	12.7	6.3
	Millenia	2.5 (KL)	4.2	—	16.9	①	—	18.0	7.9
	Millenia S	2.3 (KJ)	4.3	—	16.9	①	—	18.0	7.9
	Protege DX	1.6 (ZM)	3.4	5.8	15.2	①	—	13.2	7.9
	Protege LX	2.0 (FS)	3.7	5.8	15.2	①	—	14.5	7.9
	Protege5	2.0 (FS)	3.7	6.0	17.6	①	—	14.5	7.9
2003-04	626 ES	2.0 (FS)	3.7	5.8	18.6	①	—	16.9	7.9
	626 ES-V6	2.5 (KL)	4.2	6.0	17.0	①	—	16.9	7.9
	626 LX	2.0 (FS)	3.7	5.8	18.6	①	—	16.9	7.9
	626 LX-V6	2.5 (KL)	4.2	6.0	17.0	①	—	16.9	7.9
	Miata	1.8 (BP)	4.0	②	14.2	—	2.1	12.7	6.3
	Millenia	2.5 (KL)	4.2	—	16.9	①	—	18.0	7.9
	Millenia S	2.3 (KJ)	4.3	—	16.9	①	—	18.0	7.9
	Protege DX	1.6 (ZM)	3.4	5.8	15.2	①	—	13.2	7.9
	Protege LX	2.0 (FS)	3.7	5.8	15.2	①	—	14.5	7.9
	Protege5	2.0 (FS)	3.7	6.0	17.6	①	—	14.5	7.9

NOTE: All capacities are approximate. Add fluid gradually and check to be sure a proper fluid level is obtained.

① Included in transaxle

② 5-speed: 4.2 qts.Included in transaxle

　6-speed: 3.8 qts

09482_MAZC1_C0004

VALVE SPECIFICATIONS

Year	Engine Displacement Liters (VIN)	Seat Angle (deg.)	Face Angle (deg.)	Maximum out of Square (in.)	Spring Free Length (in.)	Stem-to-Guide Clearance (in.)		Stem Diameter (in.)	
						Intake	Exhaust	Intake	Exhaust
2002	1.6 (ZM)	NA	NA	NA	NA	NA	NA	NA	NA
	1.8 (BP)	45	45	0.062	①	0.0010-0.0023	0.0012-0.0025	0.2351-0.2356	0.2349-0.2354
	2.0 (FS)	45	45	0.061	1.732	0.0010-0.0023	0.0012-0.0025	0.2351-0.2356	0.2349-0.2354
	2.3 (KJ)	NA	45	0.062	1.413	0.0010-0.0023	0.0012-0.0025	0.2351-0.2356	0.2349-0.2354
	2.5 (KL)	45	45	0.064	1.847	0.0010-0.0023	0.0012-0.0025	0.2351-0.2356	0.2349-0.2354
2003-04	1.6 (ZM)	NA	NA	NA	NA	NA	NA	NA	NA
	1.8 (BP)	45	45	0.062	①	0.0010-0.0023	0.0012-0.0025	0.2351-0.2356	0.2349-0.2354
	2.0 (FS)	45	45	0.061	1.732	0.0010-0.0023	0.0012-0.0025	0.2351-0.2356	0.2349-0.2354
	2.3 (KJ)	NA	45	0.062	1.413	0.0010-0.0023	0.0012-0.0025	0.2351-0.2356	0.2349-0.2354
	2.5 (KL)	45	45	0.064	1.847	0.0010-0.0023	0.0012-0.0025	0.2351-0.2356	0.2349-0.2354

NA: Not Available

① Intake: 1.80 in.
   Exhaust: 1.903 in.

09482_MAZC1_C0005

CRANKSHAFT AND CONNECTING ROD SPECIFICATIONS
All measurements are given in inches.

Year	Engine Displacement Liters (VIN)	Crankshaft				Connecting Rod		
		Main Brg. Journal Dia.	Main Brg. Oil Clearance	Shaft End-play	Thrust on No.	Journal Diameter	Oil Clearance	Side Clearance
2002	1.6 (ZM)	NA	NA	NA	NA	NA	NA	NA
	1.8 (BP)	1.9661-1.9667	0.0008-0.0014	0.0032-0.0111	4	1.7693-1.7699	0.0008-0.0017	0.0044-0.0103
	2.0 (FS)	2.2022-2.2029	①	0.0031-0.0111	4	1.8874-1.8880	0.0005-0.0015	0.0043-0.0103
	2.3 (KJ)	2.4385-2.4391	0.0015-0.0022	0.0032-0.0111	4	2.0843-2.0848	0.0010-0.0016	0.0071-0.0157
	2.5 (KL)	2.4385-2.4391	0.0015-0.0022	0.0032-0.0111	4	2.0843-2.0848	0.0010-0.0016	0.0071-0.0157
2003-04	1.6 (ZM)	NA	NA	NA	NA	NA	NA	NA
	1.8 (BP)	1.9661-1.9667	0.0008-0.0014	0.0032-0.0111	4	1.7693-1.7699	0.0008-0.0017	0.0044-0.0103
	2.0 (FS)	2.2022-2.2029	①	0.0031-0.0111	4	1.8874-1.8880	0.0005-0.0015	0.0043-0.0103
	2.3 (KJ)	2.4385-2.4391	0.0015-0.0022	0.0032-0.0111	4	2.0843-2.0848	0.0010-0.0016	0.0071-0.0157
	2.5 (KL)	2.4385-2.4391	0.0015-0.0022	0.0032-0.0111	4	2.0843-2.0848	0.0010-0.0016	0.0071-0.0157

NA: Not Avilable

① No. 1, 2, 4 & 5: 0.0009-0.0020 in.
 No. 3: 0.0012-0.0022 in.

09482_MAZC1_C0006

PISTON AND RING SPECIFICATIONS
All measurements are given in inches.

Year	Engine Displacement Liters (VIN)	Piston Clearance	Ring Gap			Ring Side Clearance		
			Top Compression	Bottom Compression	Oil Control	Top Compression	Bottom Compression	Oil Control
2002	1.6 (ZM)	NA	NA	NA	NA	NA	NA	NA
	1.8 (BP)	0.0010-0.0014	0.006-0.011	0.006-0.011	0.008-0.027	0.0012-0.0026	0.0012-0.0027	0.0030-0.0060
	2.0 (FS)	0.0015-0.0020	0.006-0.012	0.006-0.012	0.008-0.028	0.0014-0.0026	0.0014-0.0026	SNUG
	2.3 (KJ)	0.0004-0.0014	0.006-0.010	0.010-0.014	0.008-0.030	0.0014-0.0025	0.0012-0.0025	0.0028-0.0062
	2.5 (KL)	0.0012-0.0022	0.006-0.012	0.010-0.016	0.008-0.028	0.0008-0.0025	0.0012-0.0025	0.0008-0.0020
2003-04	1.6 (ZM)	NA	NA	NA	NA	NA	NA	NA
	1.8 (BP)	0.0010-0.0014	0.006-0.011	0.006-0.011	0.008-0.027	0.0012-0.0026	0.0012-0.0027	0.0030-0.0060
	2.0 (FS)	0.0015-0.0020	0.006-0.012	0.006-0.012	0.008-0.028	0.0014-0.0026	0.0014-0.0026	SNUG
	2.3 (KJ)	0.0004-0.0014	0.006-0.010	0.010-0.014	0.008-0.030	0.0014-0.0025	0.0012-0.0025	0.0028-0.0062
	2.5 (KL)	0.0012-0.0022	0.006-0.012	0.010-0.016	0.008-0.028	0.0008-0.0025	0.0012-0.0025	0.0008-0.0020

NA: Not Available

09482_MAZC1_C0007

TORQUE SPECIFICATIONS
All readings in ft. lbs.

Year	Engine Displacement Liters (VIN)	Cylinder Head Bolts	Main Bearing Bolts	Rod Bearing Bolts	Crankshaft Damper Bolts	Flywheel Bolts	Manifold Intake	Manifold Exhaust	Spark Plugs	Oil Pan Drain Plug
2002	1.6 (ZM)	①	NA	NA	116-122	71-76	14-18	15-20	11-16	65-87
	1.8 (BP)	56-60	40-43	35-36	116-130	71-76	17-21	29-33	11-16	65-87
	2.0 (FS)	①	②	③	116-122	71-76	14-18	④	11-16	65-87
	2.3 (KJ)	⑤	⑥	③	116-122	45-49	14-18	14-18	11-16	65-87
	2.5 (KL)	⑤	⑦	③	116-122	45-49	14-18	12-22	11-16	65-87
2003-04	1.6 (ZM)	①	NA	NA	116-122	71-76	14-18	15-20	11-16	65-87
	1.8 (BP)	56-60	40-43	35-36	116-130	71-76	17-21	29-33	11-16	65-87
	2.0 (FS)	①	②	③	116-122	71-76	14-18	④	11-16	65-87
	2.3 (KJ)	⑤	⑥	③	116-122	45-49	14-18	14-18	11-16	65-87
	2.5 (KL)	⑤	⑦	③	116-122	45-49	14-18	12-22	11-16	65-87

NA: Not Available

① Step 1: 13-16 ft. lbs.
Step 2: Tighten 85-95 degees
Step 3: Repeat step 2

② Step 1: 16 ft. lbs.
Step 2: Tighten each bolt 90 degrees

③ Nuts: 15-20 ft. lbs
Bolts: 12-16 ft. lbs.

④ Nuts: 20 ft. lbs
Bolts: 22 ft. lbs.

⑤ Step 1: 17-19 ft. lbs.
Step 2: Tighten 85-95 degees
Step 3: Repeat step 2

⑥ Step 1: Inner bolts: 17-19 ft. lbs.
Step 2: Outer bolts: 13.5-15.5 ft. lbs.
Step 3: Inner bolt Nos. 1-3: Tighten each bolt 70 degrees
Step 4: Inner bolt No. 4: Tighten each bolt 80 degrees
Step 5: Outer bolts: Tighten each bolt 60 degrees
Step 6: Repeat Step 3-5

⑦ Step 1: Inner bolts: 17-18 ft. lbs.; Outer bolts: 13-15 ft. lbs.
Step 2: Inner bolt Nos. 1-3: Tighten each bolt 70 degrees
Step 3: Inner bolt No. 4: Tighten each bolt 80 degrees
Step 4: Outer bolts: Tighten each bolt 60 degrees
Step 5: Repeat Step 2

09482_MAZC1_C0008

WHEEL ALIGNMENT

Year	Model		Caster Range (+/-Deg.)	Caster Preferred Setting (Deg.)	Camber Range (+/-Deg.)	Camber Preferred Setting (Deg.)	Toe-in (in.)
2002	626 ①	F	1.00	+2.13	1.00	-0.70	0.12 +/- 0.16
		R	—	—	1.00	-0.10	0.12 +/- 0.16
	626 ②	F	1.00	+2.15	1.00	-0.70	0.12 +/- 0.16
		R	—	—	1.00	-0.10	0.12 +/- 0.16
	626 ③	F	1.00	+2.05	1.00	-0.70	0.12 +/- 0.16
		R	—	—	1.00	-0.10	0.12 +/- 0.16
	Miata	F	1.00	+5.75	1.00	+0.05	0.12 +/- 0.16
		R	—	—	1.00	-0.75	0.12 +/- 0.16
	Millenia	F	1.00	+2.23	0.75	-0.19	0.12 +/- 0.16
		R	—	—	1.00	-0.31	0.12 +/- 0.16
	Protégé	F	1.00	+1.88	1.00	-0.75	0.08 +/- 0.16
		R	—	—	1.00	-0.52	0.08 +/- 0.16
2003-04	626 ①	F	1.00	+2.13	1.00	-0.70	0.12 +/- 0.16
		R	—	—	1.00	-0.10	0.12 +/- 0.16
	626 ②	F	1.00	+2.15	1.00	-0.70	0.12 +/- 0.16
		R	—	—	1.00	-0.10	0.12 +/- 0.16
	626 ③	F	1.00	+2.05	1.00	-0.70	0.12 +/- 0.16
		R	—	—	1.00	-0.10	0.12 +/- 0.16
	Miata	F	1.00	+5.75	1.00	+0.05	0.12 +/- 0.16
		R	—	—	1.00	-0.75	0.12 +/- 0.16
	Millenia	F	1.00	+2.23	0.75	-0.19	0.12 +/- 0.16
		R	—	—	1.00	-0.31	0.12 +/- 0.16
	Protégé	F	1.00	+1.88	1.00	-0.75	0.08 +/- 0.16
		R	—	—	1.00	-0.52	0.08 +/- 0.16

① With 14 in. wheels
② With 15 in. wheels
③ With 16 in. wheels

09482_MAZC1_C0009

TIRE, WHEEL AND BALL JOINT SPECIFICATIONS

Year	Model	OEM Tires		Tire Pressures (psi)		Wheel Size	Ball Joint Inspection	Lug Nut
		Standard	Optional	Front	Rear			
2002	Millenia	P215/55R16	P215/50R17	32	29	Std: 6 1/2-JJ Opt:7-JJ	④	65-87
	Miata	P195/50R14	P205/45VR16	26	26	6-JJ	① ②	65-87
	Protégé 1.6L	P185/65R14	None	32	32	6-JJ	8-43 in. ①	65-87
	Protégé 2.0L	P185/65R14	P195/50VR15	32	32	Std: 5.5-JJ	8-43 in. ①	65-87
	626 2.0L	P205/60R15	P205/55R16	32	36	6-JJ	③	65-87
	626 2.5L	P205/60R15	P205/55R16	32	36	6-JJ	③	65-87
2003-04	Millenia	P215/55R16	P215/50R17	32	29	Std: 6 1/2-JJ Opt:7-JJ	④	65-87
	Miata	P195/50R14	P205/45VR16	26	26	6-JJ	① ②	65-87
	Protégé 1.6L	P185/65R14	None	32	32	6-JJ	8-43 in. ①	65-87
	Protégé 2.0L	P185/65R14	P195/50VR15	32	32	Std: 5.5-JJ	8-43 in. ①	65-87

OEM: Original Equipment Manufacturer

PSI: Pounds Per Square Inch

STD: Standard

OPT: Optional

① Torque required in ft. lbs. to rotate ball joint using a pull scale

② Lower arm ball rotation torque: 0.78-4.29

  Upper arm ball rotation torque: 0.7-5.0

③ Lower arm ball rotation torque: 0.3-11

④ Lower arm ball rotation torque: 0.7-7.7

  Upper arm ball rotation torque: 0.7-8.8

  Upper leading link rotation torque: 0.7-4.4

09482_MAZC1_C0010

BRAKE SPECIFICATIONS
All measurements in inches unless noted

| Year | Model | | Brake Disc | | | Brake Drum | | | Minimum Lining Thickness | Brake Caliper | |
			Original Thickness	Minimum Thickness	Maximum Runout	Original Inside Diameter	Max. Wear Limit	Maximum Machine Diameter		Bracket Bolts (ft. lbs.)	Mounting Bolts (ft. lbs.)
2002	626	F	0.940	0.870	0.002	—	—	—	0.080	58-75	22-28
		R	0.390	0.310	0.002	9.00	NA	9.059	0.040	34-49	26-28
	Miata	F	0.790	①	0.002	—	—	—	0.080	37-50	58-65
		R	0.350	0.310	0.002	—	—	—	0.040	37-50	26-28
	Millenia	F	1.100	1.000	0.002	—	—	—	0.080	47-62	47-62
		R	0.370	0.300	0.002	—	—	—	0.080	37-50	28-36
	Protege	F	NA	②	0.002	—	—	—	③	58-75	④
		R	NA	0.310	0.002	7.87	7.993	NA	0.040	26-28	26-28
2003-04	626	F	0.940	0.870	0.002	—	—	—	0.080	58-75	22-28
		R	0.390	0.310	0.002	9.00	NA	9.059	0.040	34-49	26-28
	Miata	F	0.790	①	0.002	—	—	—	0.080	37-50	58-65
		R	0.350	0.310	0.002	—	—	—	0.040	37-50	26-28
	Millenia	F	1.100	1.000	0.002	—	—	—	0.080	47-62	47-62
		R	0.370	0.300	0.002	—	—	—	0.080	37-50	28-36
	Protege	F	NA	②	0.002	—	—	—	③	58-75	④
		R	NA	0.310	0.002	7.87	7.993	NA	0.040	26-28	26-28

NA: Not Avilable

F: Front

R: Rear

① With 15 inch wheel: 0.071 in.
   With 16 inch wheel: 0.079 in.

② With 1.6L engine: 0.780 in.
   With 1.8L engine: 0.870 in.
   With 2.0L engine: 0.870 in.

③ With 1.6L engine: 0.059 in.
   With 1.8L engine: 0.080 in.
   With 2.0L engine: 0.079 in.

④ Type A: 37-39
   Type B: 16-23

09482_MAZC1_C0011

SCHEDULED MAINTENANCE INTERVALS
Mazda Car

TO BE SERVICED	TYPE OF SERVICE	7.5	15	22.5	30	37.5	45	52.5	60	67.5	75	82.5	90	97.5
		VEHICLE MILEAGE INTERVAL (x1000)												
Engine oil & filter	R	✓	✓	✓	✓	✓	✓	✓	✓	✓	✓	✓	✓	✓
Air cleaner element	R				✓				✓				✓	
Engine coolant ①	R				✓				✓				✓	
Spark plugs	R				✓				✓				✓	
Automatic transaxle fluid	S/I				✓				✓				✓	
Bolts & nuts on chassis & body	S/I				✓				✓				✓	
Brake lines, hoses & connections	S/I				✓				✓				✓	
Cooling system	S/I				✓				✓				✓	
Disc brakes	S/I				✓				✓				✓	
Drive belts (Millenia ②)	S/I				✓				✓				✓	
Drive shaft dust boots	S/I				✓				✓				✓	
Exhaust system heat shield	S/I				✓				✓				✓	
Front & rear suspension ball joints	S/I				✓				✓				✓	
Fuel lines & hoses	S/I				✓				✓				✓	
Idle speed	S/I				✓				✓				✓	
Steering operation & linkages	S/I				✓				✓				✓	
Engine timing belt ③ ④	R								✓					
Fuel filter	R								✓					
Valve Clearance	I								✓					
Manual transmission	R								✓					
Hose & tube for emission	S/I								✓					

R: Replace   S/I: Service or Inspect

① Replace initially at 45,000 miles & every 30,000 miles thereafter except on 626. On 626, replace at 105,000 miles & every 30, 000 miles therafter

② (Millenia KJ engine): replace every 105,000 miles

③ Except 2002-04 Miata and Millenia KJ engine; inspect every 60,000 miles & replace at 105,000 miles (if not replaced previously).

③ 2002-04 Miata and Millenia KJ engine replace at 60,000 miles (if not replaced previously).

FREQUENT OPERATION MAINTENANCE (SEVERE SERVICE)

If a vehicle is operated under any of the following conditions it is considered severe service

- Extremely dusty areas.

- 50% or more of the vehicle operation is in 32°C (90°F) or higher temperatures, or constant operation in temperatures below 0°C (32°F).

- Prolonged idling (vehicle operation in stop and go traffic).

- Frequent short running periods (engine does not warm to normal operating temperatures).

- Police, taxi, delivery usage or trailer towing usage.

Oil & oil filter: change every 5000 miles.

Oil & oil filter (Puerto Rico): change every 3000 miles.

Air cleaner element: service or inspect every 15,000 miles

Automatic transaxle fluid: service or inspect every 15,000 miles.

Bolts & nuts on chassis & body: tighten every 15,000 miles.

Disc brakes: service or inspect every 15,000 miles.

09482_MAZC1_C0012

ENGINE REPAIR

Distributor

REMOVAL

1. Before servicing the vehicle, refer to the precautions section.
2. Remove or disconnect the following:
   - Negative battery cable
   - Distributor cap and position it aside, leaving the ignition wires connected
   - Distributor electrical connector(s) from the side of the distributor
3. Using a wrench on the crankshaft pulley, rotate the crankshaft to position the No. 1 piston on Top Dead Center (TDC) of the compression stroke; the crankshaft pulley mark should align with the timing indicator and the distributor rotor should point towards the No. 1 spark plug wire tower position of the cap.
4. Using chalk or paint, mark the position of the distributor housing on the cylinder head. Also mark the position of the distributor rotor in relation to the distributor housing.
5. Remove distributor hold-down bolt(s).
6. On distributors attached to the end of the cylinder head (or inline with the camshaft), remove it by pulling it straight outward.

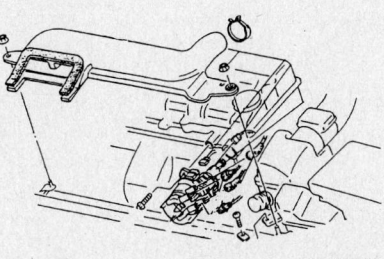

7923MG01

Exploded view of a typical side mounted distributor

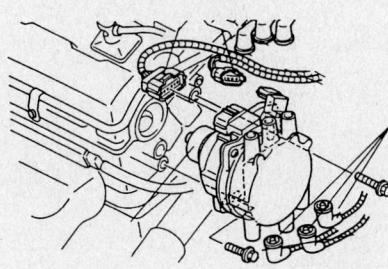

7923MG02

Exploded view of a typical end or inline mounted distributor

7. On distributors attached to the side of the cylinder head (or perpendicular with the camshaft), slowly pull it outward while watching the rotor. These distributors are gear driven and as you remove it, the gears will disengage inside the engine, causing the rotor to rotate. when the rotor stops moving, stop pulling outward. Re-align the distributor body-to-cylinder head match-mark (do not push it back in to do this, simply rotate the body to align the marks). Place a third mark indicating the new rotor position-to-distributor body relation. When installing the distributor, align this mark and the body-to-head mark to properly position the distributor.
8. Inspect the O-ring on the distributor housing and replace it, if it is damaged or worn.

INSTALLATION

Engine Not Disturbed

1. Using engine oil, lubricate the O-ring.
2. Install or connect the following:
   - Distributor

➡**Be sure to engage the distributor drive gear or tangs with the camshaft gear or slot. Align the mark that was made on the distributor housing with the mark that was made on the cylinder head.**

   - Distributor hold-down bolt(s)
   - Electrical connector(s)
   - Distributor cap
   - Negative battery cable
3. Start the engine and check or adjust the ignition timing.

Engine Disturbed

1. Remove or disconnect the following:
   - Spark plug wire from the No. 1 cylinder spark plug
   - Spark plug from the No. 1 cylinder
2. Press a thumb over the spark plug hole.
3. Using a wrench on the crankshaft pulley, rotate the crankshaft until pressure is felt at the spark plug hole, indicating the piston is approaching TDC on the compression stroke. Continue rotating the crankshaft until the crankshaft pulley mark aligns with the timing cover indicator.
4. Place the distributor rotor in position so that it aligns with the No. 1 spark plug wire tower on the distributor cap.
5. Using engine oil, lubricate the O-ring.

6. Install or connect the following:
   - Distributor

➡**Be sure to engage the distributor drive gear or tangs with the camshaft gear or slot. Align the mark that was made on the distributor housing with the mark that was made on the cylinder head.**

   - Distributor hold-down bolt(s)
   - Electrical connector(s)
   - Distributor cap
   - Spark plug in the No. 1 cylinder and connect the spark plug wire
   - Negative battery cable
7. Start the engine and check or adjust the ignition timing.

Alternator

REMOVAL

2002–04 Miata

1. Before servicing the vehicle, refer to the precautions section.
2. Remove or disconnect the following:
   - Negative battery cable
   - Intake manifold bracket
   - Electrical connectors from the alternator
   - Alternator bolts
   - Alternator

Protégé

1. Before servicing the vehicle, refer to the precautions section.
2. Remove or disconnect the following:
   - Negative battery cable
   - Electrical connectors from the alternator
   - Alternator drive belt
   - Alternator pivot and adjusting bar bolts
   - Alternator

626

2.0L (FS) ENGINES

1. Before servicing the vehicle, refer to the precautions section.
2. Remove or disconnect the following:
   - Negative battery cable
   - Alternator upper mounting bolt
   - Alternator adjusting bolt
   - Drive belt from the alternator pulley
   - Transverse member
   - Electrical connectors from the alternator

- Front exhaust pipe at the catalytic converter and suspend it on a piece of wire
- Oxygen Sensor (O_2S)
- 3 exhaust manifold flange nuts and the hold-down bracket clamp
- Exhaust pipe
- Alternator lower through-bolt
- Alternator

2.5L (KL) ENGINES

1. Before servicing the vehicle, refer to the precautions section.
2. Remove or disconnect the following:
   - Negative battery cable
   - Fresh air duct
   - Radiator upper bracket
   - Condenser fan, if equipped
   - Electrical connectors from the alternator
   - Loosen the belt tensioner locknut and tension adjusting bolt
   - Alternator upper mounting bolt
   - Right splash shield
   - Drive belt from the alternator pulley
   - A/C compressor and support it aside, leaving the refrigerant lines connected, if necessary
   - Alternator through-bolt
   - Alternator

Millenia

1. Before servicing the vehicle, refer to the precautions section.
2. Disconnect the negative battery cable.
3. If equipped with the 2.3L (KJ) engine, remove the front charge air cooler, radiator upper seal board and the condenser fan assembly.
4. Remove or disconnect the following:
   - Intake air system on 2.4 (KL) engine
   - Electrical connectors from the alternator
   - Right splash shield
   - Drive belt from the alternator pulley
   - A/C compressor and support it aside, leaving the refrigerant lines connected
   - Upper and lower alternator mounting bolts
   - Alternator

INSTALLATION

2002–04 Miata

1. Install or connect the following:
   - Alternator
   - Alternator bolts. Tighten the lower bolts to 38 ft. lbs. (51 Nm) and the upper bolt 18 ft. lbs. (25 Nm).

- Electrical connectors to the alternator
- Intake manifold bracket
- Negative battery cable

Protégé

1. Install or connect the following:
   - Alternator with the pivot bolt
   - Alternator electrical connectors
   - Drive belt
   - Upper mounting bolt
2. Adjust the belt tension. Torque the lower through bolt to 28–38 ft. lbs. (38–51 Nm) and the upper mounting bolt to 14–18 ft. lbs. (19–26 Nm).
   - Negative battery cable

626

2.0L (FS) ENGINES

Install or connect the following:
- Alternator
- Alternator bolts. Tighten the lower bolt to 28–38 ft. lbs. (38–51 Nm) and the upper bolt to 14–18 ft. lbs. (19–25 Nm).
- Drive belt, check the belt deflection by applying moderate pressure 22 lbs. (98 N) between the alternator and the water pump pulley. The deflection on a new belt should be 6.5–7 inch (0.26–0.27mm) or on a used belt, 7.0–9.0 inch (0.28–0.35mm). Loosen the alternator mounting bolts and use the adjusting bolt to adjust the belt tension to the correct specification.

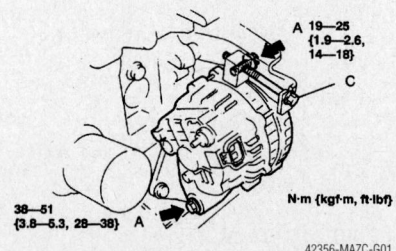

Adjust the alternator belt tension using the adjustment bolt C—626 4 cylinder models

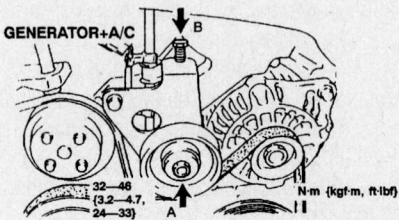

Adjust the alternator belt tension using the adjustment bolt B—626 6 cylinder models

- Electrical connector and terminal wire
- Front exhaust pipe
- Transverse pipe
- Negative battery cable

2.5L (KL) ENGINES

1. Install or connect the following:
   - Alternator. Torque the lower bolt to 28–38 ft. lbs. (38–51 Nm) and the upper bolt to 14–18 ft. lbs. (19–25 Nm).
   - A/C compressor, if removed. Torque the bolts to 11–15 ft. lbs. (15–21 Nm).
   - Drive belt, check the belt deflection by applying moderate pressure 22 lbf. (98 N) between the alternator and the crankshaft pulley to check the alternator belt or midway between the A/C compressor and the crankshaft pulley to check the Air conditioning/alternator belt. The deflection on a new belt should be 6.0–7 inch (0.24–0.27mm) or on a used belt, 7.0–8.0 inch (0.28–0.31mm) on the alternator belt. On models with a air conditioning/alternator belt; the deflection on a new belt should be 5.5–6.5 inch (0.22–0.25mm) or on a used belt, 6.5–7.5 inch (0.26–0.29mm) Loosen the alternator mounting bolts and use the adjusting bolt to adjust the belt tension to the correct specification.
   - Negative battery cable

Millenia

1. Install or connect the following:
   - Alternator. Torque the upper bolt to 14–18 ft. lbs. (19–25 Nm) and the lower bolt to 28–38 ft. lbs. (38–51 Nm).
   - A/C compressor. Torque the bolts to 12–16 ft. lbs. (16–22 Nm).
   - Right splash shield

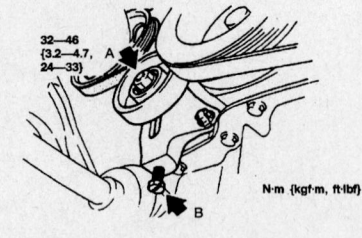

- Drive belt
- Electrical connectors to the alternator
- Intake air system on 2.4 (KL) engine

2. If equipped with the 2.3L engine, install the condenser fan assembly, radiator upper seal board and the front charge air cooler using new O-rings. Torque the mounting bolts to 12–16 ft. lbs. (16–22 Nm).

3. Connect the negative battery cable.

Ignition Timing

ADJUSTMENT

Except Millenia 2.5L (KL) Engines

1. Before servicing the vehicle, refer to the precautions section.

➡**If the information given in the following procedures differs from that on the emission information label located in the engine compartment, follow the directions given on the label. The label often reflects production changes made during the model year.**

The timing is controlled by the computer. Ignition timing adjustment is not possible or necessary.

2. If the timing is still not within specification. the following components may be defective:

- Camshaft position (CMP) sensor
- Crankshaft Position (CKP) sensor
- Throttle Position (TP) sensor
- Engine Coolant Temperature (ECT) sensor
- Neutral switch if equipped with a manual transaxle
- Clutch switch if equipped with a manual transaxle
- Transaxle range switch if equipped with an automatic transaxle

3. If the above components are normal, replace the Powertrain Control Module (PCM).

Millenia 2.5L (KL) Engine

1. Before servicing the vehicle, refer to the precautions section.

2. Let the engine warm to normal operating temperature.

3. Apply the parking brake. If equipped with a manual transaxle, place the shifter in the neutral position. If equipped with an automatic transaxle, place the shift lever in **P**.

4. Start the engine and allow it to come

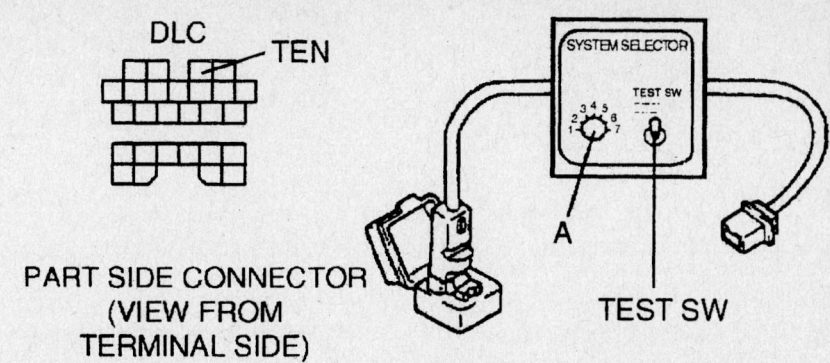

PART SIDE CONNECTOR (VIEW FROM TERMINAL SIDE)

42356-MAZC-GFA

Jumper the connections shown on the data link connector and system selector tool

to normal operating temperature. Be sure all accessories are **OFF**.

5. Wait until the cooling fan stops, then, connect the scan tool to the Data Link Connector 2 (DLC2) and access the RPM PID.

6. Locate the timing marks on the crankshaft pulley and timing belt lower cover. The engine may have to be cranked slightly to see the mark on the crankshaft pulley.

7. Check the idle speed and adjust, if necessary.

8. Connect a jumper wire between the

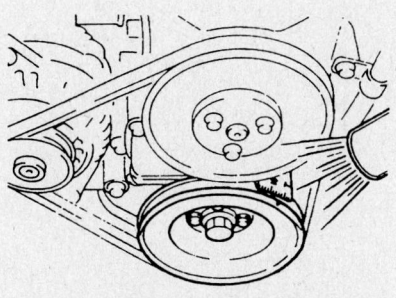

7923MG05

Connect an inductive timing light and aim it at the crankshaft pulley. Read the pulley mark against the scale

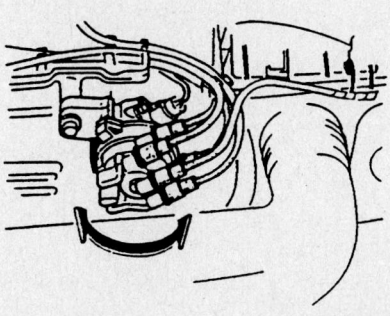

7923MG06

If adjustment is necessary, loosen the distributor lockbolts and rotate it until the mark is aligned

TEN terminal and the GND terminal at the underhood diagnosis connector.

9. Connect the system selector tool to the DLC and set switch **A** to position **1**.

10. Set the test switch to **SELF TEST**.

11. Connect an inductive timing light according to the manufacturer's instructions.

12. Start the engine and allow the idle to stabilize. Aim the timing light at the timing marks. The timing should be 9–11 degrees Before Top Dead Center (TDC).

13. Loosen the distributor lockbolts just enough to turn the distributor. While aiming the timing light at the timing marks, turn the distributor until the marks are aligned. Tighten the distributor lockbolts to 14–18 ft. lbs. (19–25 Nm) and recheck the timing.

14. The ignition timing is now set. Disconnect the jumper wire from the DLC.

15. Remove all test equipment.

Engine Assembly

REMOVAL & INSTALLATION

626

2.0L (FS) ENGINE

➡**The procedure for pulling the engine requires removing the transaxle along with it. As a result, when the halfshafts are pulled from the transaxle, a special plug/side gear holding tool is recommended.**

1. Before servicing the vehicle, refer to the precautions section.

2. Properly relieve the fuel system pressure.

3. Drain the engine oil.

4. Drain the transaxle fluid.

5. Drain the cooling system.

6. Remove or disconnect the following:

- Negative battery cable
- Radiator
- Air cleaner assembly

- Accelerator cable
- Fuel hoses
- Front exhaust pipe
- Any rods, pipes or cables related to the transaxle that would hinder removal
- Battery
- Fuse box
- Power steering pump with the lines still attached and set aside
- A/C compressor with the lines still attached and position aside
- Cruise actuator connector, actuator retainers and the actuator

- Halfshafts
- Number 5 engine mount retainers. Refer to the accompanying illustration for location.
- Engine mount member (2) retainers. Refer to the accompanying illustration for location.
- Number 1 engine mount stay bracket retainers. Refer to the accompanying illustration for location.
- Number 3 engine mount rubber retainers. Refer to the accompanying illustration for location.

- Number 3 engine bracket retainers. Refer to the accompanying illustration for location.
- Number 1 engine mount nut and bolt. Refer to the accompanying illustration for location.
- Number 4 engine mount rubber retainers. Refer to the accompanying illustration for location.
- Number 4 engine bracket retainers. Refer to the accompanying illustration for location.
- Engine and transaxle assembly from the vehicle.

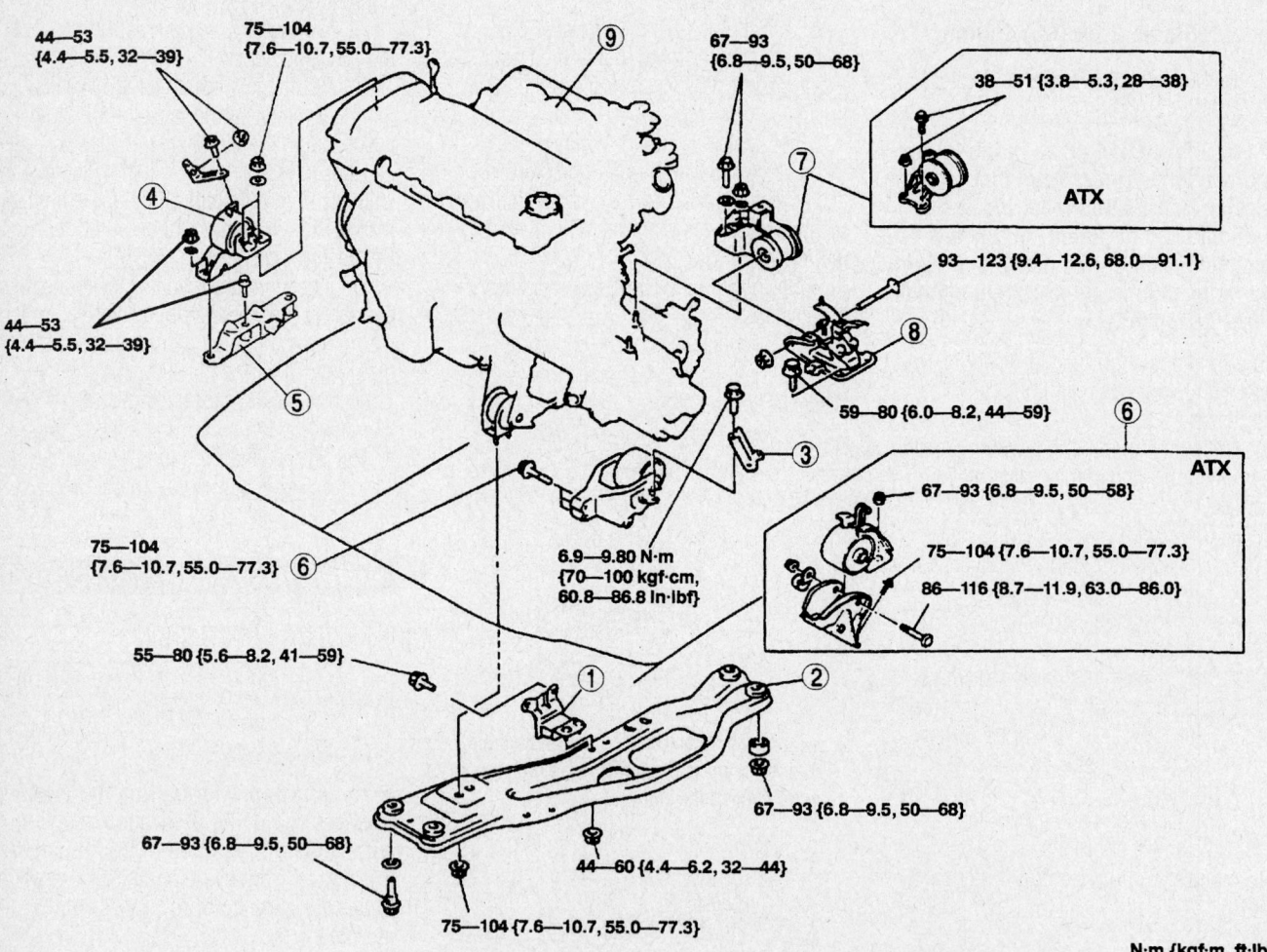

1 No.5 engine mount rubber	6 No.1 engine mount bolt and nut
2 Engine mount member	7 No.4 engine mount rubber
3 No.1 engine mount stay bracket	8 No.4 engine mount bracket
4 No.3 engine mount rubber	9 Engine, transaxle
5 No.3 engine bracket	

42356-MAZC-G03

Location of the engine mounting components and their torque specifications—626 models equipped with the 2.0L (FS) engine

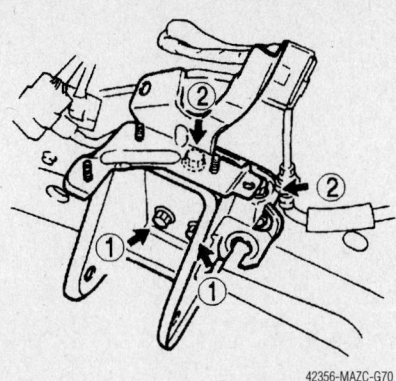

42356-MAZC-G70

Tighten the number 4 engine bracket retainers in the sequence shown—626 models equipped with the 2.0L (FS) engine

To install:

7. Installation is the reverse of removal. Tighten the fasteners to the specifications shown in the accompanying illustration.

8. When possible, leave the engine mounting nuts/bolts loose (hand tight) until all mounts are aligned and bolted. This may help in aligning the engine and transmission assembly in the vehicle.

9. Install or connect the following:
- Number 4 engine bracket retainers. Tighten the fasteners to the specifications shown in the accompanying illustration.
- Number 4 engine mount rubber retainers. Tighten the fasteners to the specifications shown in the accompanying illustration.
- Number 1 engine mount nut and bolt. Tighten the fasteners to the specifications shown in the accompanying illustration.
- Number 3 engine bracket retainers. Tighten the fasteners to the specifications shown in the accompanying illustration.
- Number 3 engine mount rubber retainers. Tighten the fasteners to the specifications shown in the accompanying illustration.
- Number 1 engine mount stay bracket retainers. Tighten the fasteners to the specifications shown in the accompanying illustration.
- Engine mount member (2) retainers. Tighten the fasteners to the specifications shown in the accompanying illustration.
- Number 5 engine mount retainers. Tighten the fasteners to the specifications shown in the accompanying illustration.

10. When connecting the accelerator cable, perform the following adjustment:

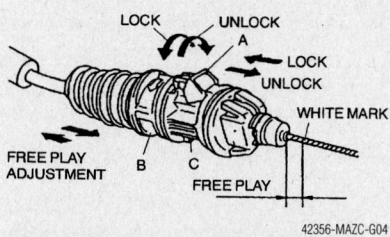

Free play 1.5—4.0 mm {0.06—0.15 in}

42356-MAZC-G04

Accelerator cable adjustment components—626 models equipped with the 2.0L (FS) engine

a. Move the white locking tab to the unlock position.
b. Turn stopper B to the to the unlock position.

→**If stopper B will not unlock, it may be necessary to carefully bend back tab C out a little using a suitable pry tool.**

c. Push or pull the cable housing directly behind the spring.
d. Turn the stopper B to the lock position.
e. Measure the free play which should be 0.06–0.15 inch (1.5–4mm) and make sure the cable free play is within specification.
f. Move the white locking tab to the lock position and check for proper accelerator operation.

11. Fill the engine and the transaxle with the proper type and amount of fluids. Fill the cooling system.

12. Connect the negative battery cable, start the engine and check for leaks.

13. Check the ignition timing and the idle speed.

14. Check all fluid levels.

2.5L (KL) ENGINE

→**The procedure for pulling the engine requires removing the transaxle along with it. As a result, when the halfshafts are pulled from the transaxle, a special plug/side gear holding tool is recommended.**

1. Before servicing the vehicle, refer to the precautions section.

2. Properly relieve the fuel system pressure.

3. Drain the engine oil.

4. Drain the transaxle fluid.

5. Drain the cooling system.

6. Remove or disconnect the following:
- Negative battery cable
- Radiator
- Air cleaner assembly
- Accelerator cable
- Fuel hoses
- Front exhaust pipe
- Any rods, pipes or cables related to the transaxle that would hinder removal
- Battery
- Power steering pump with the lines still attached and set aside
- A/C compressor with the lines still attached and position aside
- Cruise actuator connector, actuator retainers and the actuator
- Halfshafts
- Number 5 engine mount retainers. Refer to the accompanying illustration for location.
- Engine mount member (2) retainers. Refer to the accompanying illustration for location.
- Number 1 engine mount stay bracket retainers. Refer to the accompanying illustration for location.
- Number 3 engine mount rubber retainers. Refer to the accompanying illustration for location.
- Number 1 engine mount nut and bolt. Refer to the accompanying illustration for location.
- Number 4 engine mount rubber retainers. Refer to the accompanying illustration for location.
- Number 4 engine bracket retainers. Refer to the accompanying illustration for location.
- Engine and transaxle assembly from the vehicle.

To install:

7. Installation is the reverse of removal. Tighten the fasteners to the specifications shown in the accompanying illustration.

8. When possible, leave the engine mounting nuts/bolts loose (hand tight) until all mounts are aligned and bolted. This may help in aligning the engine and transmission assembly in the vehicle.

9. Install or connect the following:
- Number 4 engine bracket retainers. Tighten the fasteners to the specifications shown in the accompanying illustration.
- Number 4 engine mount rubber retainers. Tighten the fasteners to the specifications shown in the accompanying illustration.
- Number 1 engine mount nut and bolt. Tighten the fasteners to the specifications shown in the accompanying illustration.
- Number 3 engine mount rubber retainers. Tighten the fasteners to

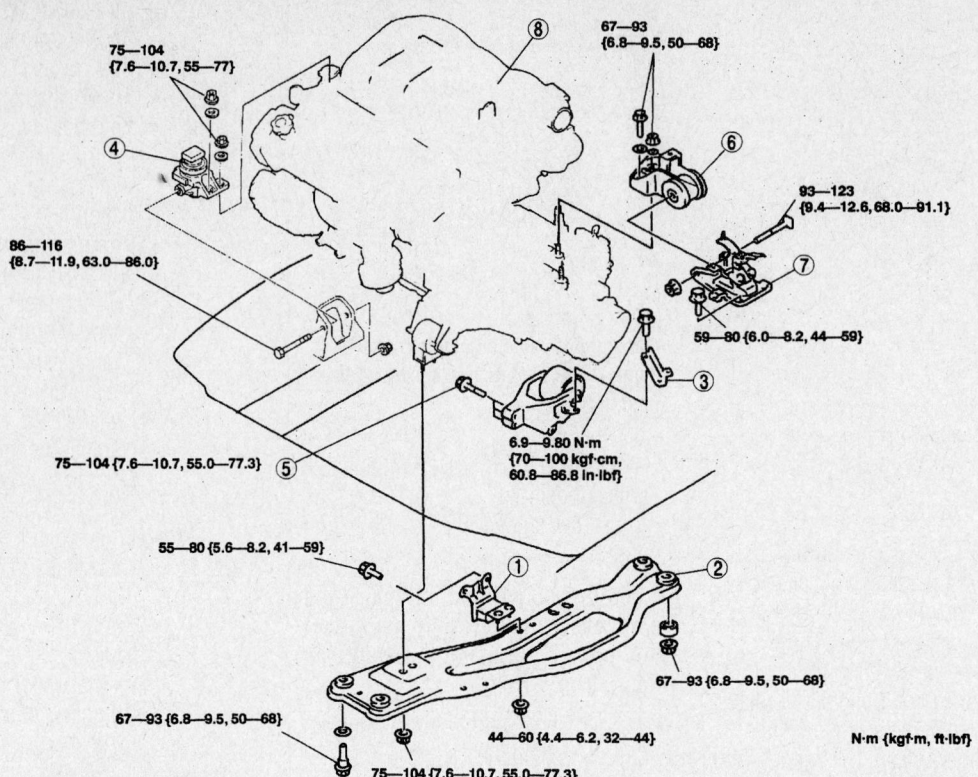

75—104
{7.6—10.7, 55—77}

67—93
{6.8—9.5, 50—68}

86—116
{8.7—11.9, 63.0—86.0}

93—123
{9.4—12.6, 68.0—91.1}

59—80 {6.0—8.2, 44—59}

75—104 {7.6—10.7, 55.0—77.3}

6.9—9.80 N·m
{70—100 kgf·cm,
60.8—86.8 in·lbf}

55—80 {5.6—8.2, 41—59}

67—93 {6.8—9.5, 50—68}

67—93 {6.8—9.5, 50—68}

44—60 {4.4—6.2, 32—44}

75—104 {7.6—10.7, 55.0—77.3}

N·m {kgf·m, ft·lbf}

1 No.5 engine mount rubber
2 Engine mount member
3 No.1 engine mount stay bracket
4 No.3 engine mount rubber
5 No.1 engine mount bolt and nut

6 No.4 engine mount rubber
7 No.4 engine mount bracket
8 Engine, transaxle

No.4 Engine Mount Bracket Installation Note
• Tighten the bolt in the order shown.

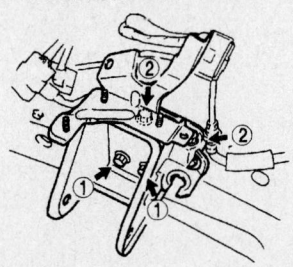

42356-MAZC-G05

Location of the engine mounting components and their torque specifications—626 models equipped with the 2.5L (KL) engine

the specifications shown in the accompanying illustration.
• Number 1 engine mount stay bracket retainers. Tighten the fasteners to the specifications shown in the accompanying illustration.
• Engine mount member (2) retainers. Tighten the fasteners to the specifications shown in the accompanying illustration.
• Number 5 engine mount retainers. Tighten the fasteners to the specifications shown in the accompanying illustration.

10. When connecting the accelerator cable, perform the following adjustment:
  a. Move the white locking tab to the unlock position.
  b. Turn stopper B to the to the unlock position.

➡ **If stopper B will not unlock, it may be necessary to carefully bend back tab C out a little using a suitable pry tool.**

  c. Push or pull the cable housing directly behind the spring.
  d. Turn the stopper B to the lock position.

Free play
1.5—4.0 mm {0.06—0.15 in}

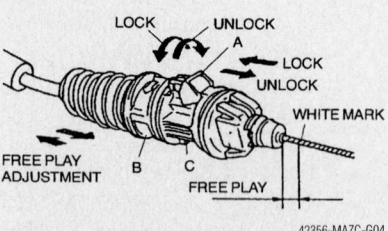

42356-MAZC-G04

Accelerator cable adjustment components—626 models equipped with the 2.5L (KL) engine

e. Measure the free play which should be 0.06–0.15 inch (1.5–4mm) and make sure the cable free play is within specification.

f. Move the white locking tab to the lock position and check for proper accelerator operation.

11. Fill the engine and the transaxle with the proper type and amount of fluids. Fill the cooling system.

12. Connect the negative battery cable, start the engine and check for leaks.

13. Check the ignition timing and the idle speed.

14. Check all fluid levels.

Millenia

2.5L (KL) ENGINE

➡ **The procedure for pulling the engine requires removing the transaxle along with it. As a result, when the halfshafts are pulled from the transaxle, a special plug/side gear holding tool is recommended.**

1. Before servicing the vehicle, refer to the precautions section.

2. Properly relieve the fuel system pressure.

3. Drain the engine oil.

4. Drain the transaxle fluid.

5. Drain the cooling system.

6. Remove or disconnect the following:
- Hood
- Front wheels
- Splash shield
- Battery clamp, cover, battery, battery carrier and battery air duct
- Air cleaner assembly
- Upper seal board
- Radiator reservoir hose and the reservoir
- Condenser fan motor connector and the fan
- Cooling fan connector and fan

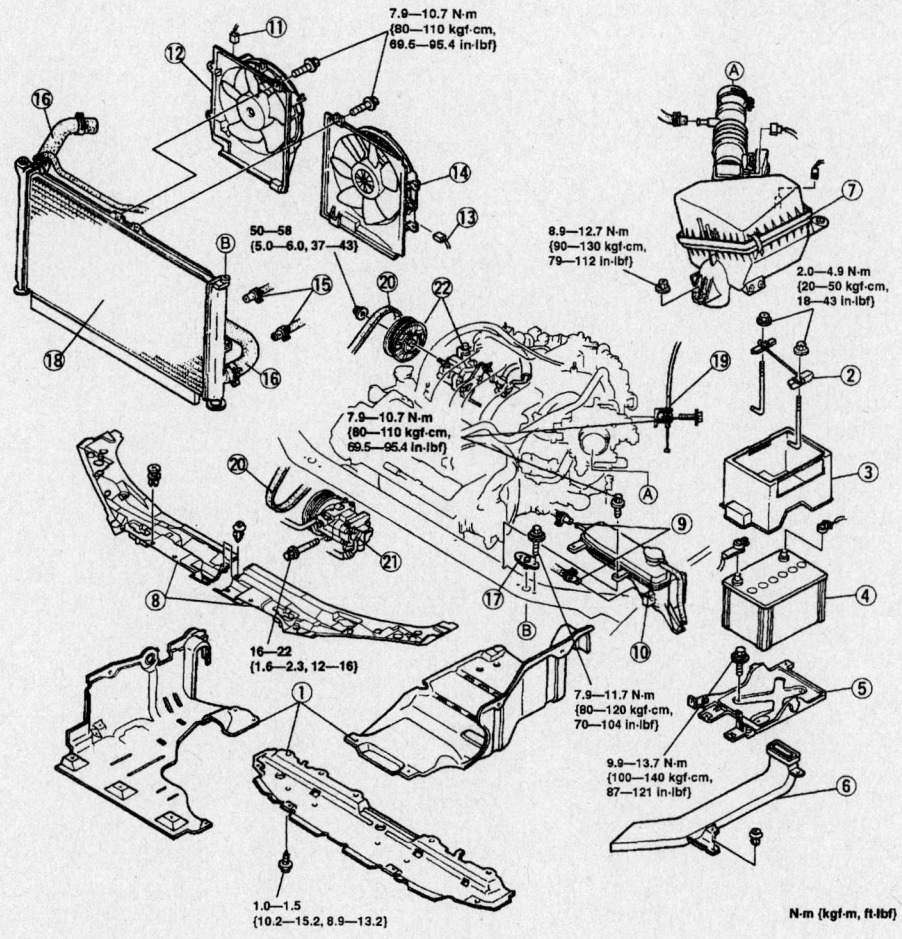

1	Splash shield	12	Condenser fan component
2	Battery clamp	13	Cooling fan motor connector
3	Battery cover	14	Cooling fan component
4	Battery	15	Oil cooler hose
5	Battery carrier	16	Radiator hose
6	Battery air duct	17	Radiator bracket
7	Air cleaner component	18	Radiator
8	Upper seal board	19	Accelerator cable
9	Radiator reservoir hose	20	Drive belt
10	Radiator reservoir	21	A/C compressor
11	Condenser fan motor connector	22	P/S oil pump

42356-MAZC-GDA

Location of the engine mounting components and their torque specifications (part 1 of 3)—Millenia models equipped with the 2.5L (KL) engine

- Oil cooler hose
- Radiator hose, bracket and the radiator
- Accelerator cable
- Drive belt
- A/C compressor with the lines still attached and position aside
- Power steering pump with the lines still attached and position aside
- Selector cable
- Front exhaust pipe
- Tie rod end ball joint
- Upper and lower ball joints
- Halfshafts

- Joint shaft

7. Support the engine using a suitable support device.
- Number 1 engine mount stay. Refer to the accompanying illustration for location.
- Number 1 engine mount bracket. Refer to the accompanying illustration for location.
- Engine mount member retainers. Remove the bolts **A** first and then the bolts **B** and remove the member. Refer to the accompanying illustration for location.

✵✵ CAUTION

Engine load can damage the number 4 mount bolts holes when removing the bolts so make sure all weight is off the mount.

8. Attach a hoist to the engine, take the weight of the engine with the hoist and remove the engine support device, Lift up the engine/transaxle assembly slightly until the number 3 and 4 mounts are free from engine weight.
- Number 4 engine mount bracket.

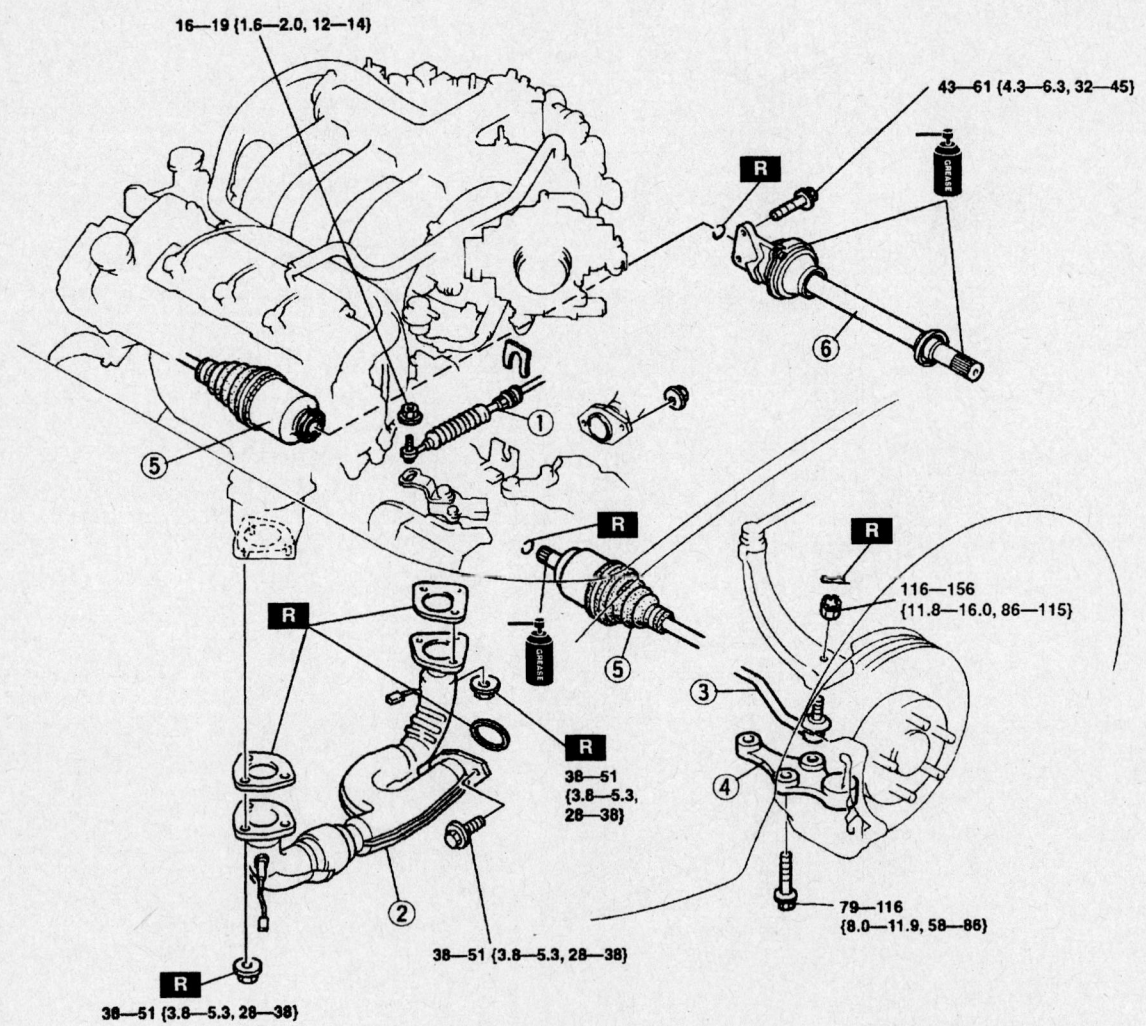

16—19 {1.6—2.0, 12—14}

43—61 {4.3—6.3, 32—45}

116—156 {11.8—16.0, 86—115}

38—51 {3.8—5.3, 28—38}

38—51 {3.8—5.3, 28—38}

79—116 {8.0—11.9, 58—86}

38—51 {3.8—5.3, 28—38}

N·m {kgf·m, ft·lbf}

1	Selector cable	5	Drive shaft
2	Front pipe	6	Joint shaft
3	Upper lateral link ball joint		
4	Lower ball joint		

42356-MAZC-GDB

Location of the engine mounting components and their torque specifications (part 2 of 3)—Millenia models equipped with the 2.5L (KL) engine

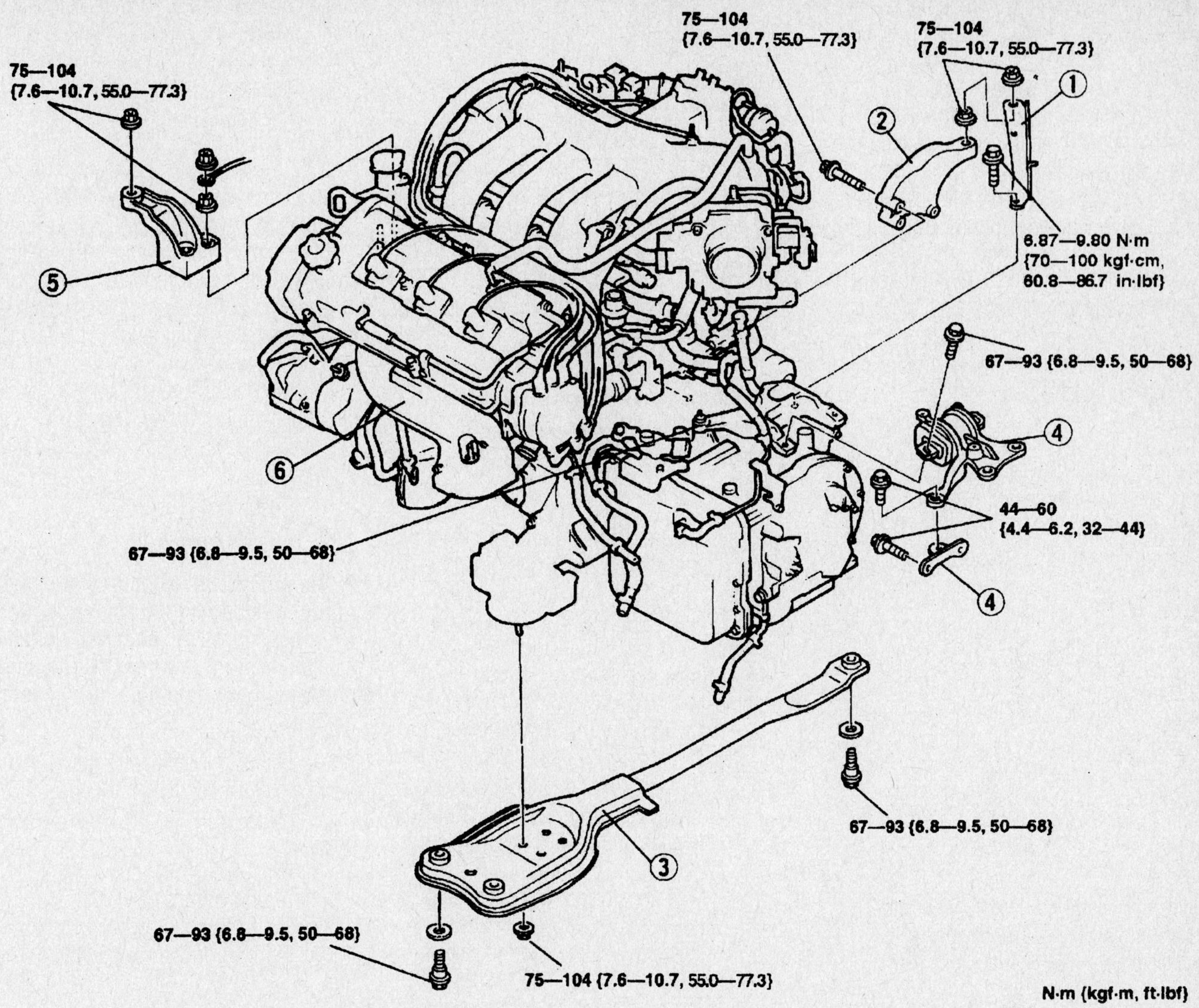

75—104
{7.6—10.7, 55.0—77.3}

75—104
{7.6—10.7, 55.0—77.3}

75—104
{7.6—10.7, 55.0—77.3}

①

②

6.87—9.80 N·m
{70—100 kgf·cm,
60.8—86.7 in·lbf}

67—93 {6.8—9.5, 50—68}

④

44—60
{4.4—6.2, 32—44}

④

67—93 {6.8—9.5, 50—68}

⑤

⑥

67—93 {6.8—9.5, 50—68}

③

67—93 {6.8—9.5, 50—68}

75—104 {7.6—10.7, 55.0—77.3}

N·m {kgf·m, ft·lbf}

1	No.1 engine mount stay	4	No.4 engine mount bracket
2	No.1 engine mount bracket	5	No.3 engine mount sub bracket
3	Engine mount member	6	Engine and transaxle

42356-MAZC-GDC

Location of the engine mounting components and their torque specifications (part 3 of 3)—Millenia models equipped with the 2.5L (KL) engine

Refer to the accompanying illustration for location.
- Number 3 engine mount sub bracket. Refer to the accompanying illustration for location.
- Engine and transaxle assembly from the vehicle. Be careful the powertrain assembly does not swing and strike the vehicle causing damage

To install:
9. Installation is the reverse of removal. Tighten the fasteners to the specifications shown in the accompanying illustration.
10. When possible, leave the engine

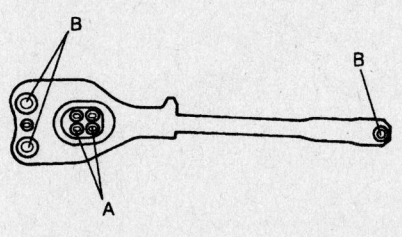

42356-MAZC-GDD

Remove the bolts A first and then the bolts B and remove the engine mount member—Millenia models equipped with the 2.5L (KL) engine

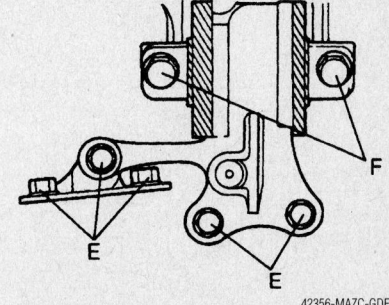

42356-MAZC-GDE

Tighten the number 4 engine bracket retainers E to specification and hand tighten bolts F—Millenia models equipped with the 2.5L (KL) engine

mounting nuts/bolts loose (hand tight) until all mounts are aligned and bolted. This may help in aligning the engine and transmission assembly in the vehicle.

11. Install or connect the following:
  • Number 4 engine bracket retainers and hand tighten the retainers

12. Using the hoist lower the powertrain assembly and hand tighten the number 3 mount sub bracket retainers
  • Number 4 engine bracket retainers and tighten the bolts **E** to 44 ft. lbs. (60 Nm) and hand tighten bolts **F**.

13. Remove the engine hoist and install the engine support device

14. Install the engine mount member and tighten bolts **B** to 68 ft. lbs. (93 Nm) and hand tighten bolts **A**.

15. Install the number 1 bracket and tighten the retainers to 77 ft. lbs. (104 Nm) and remove the engine support.

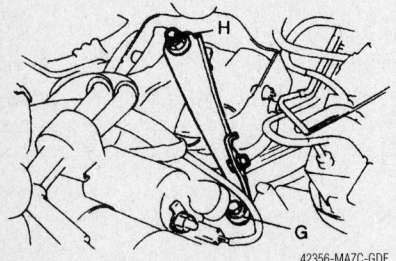

42356-MAZC-GDF

Tighten the number 1 engine mount stay retainers G and H to specification—Millenia models equipped with the 2.5L (KL) engine

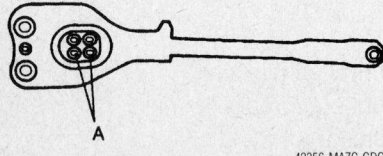

42356-MAZC-GDG

Tighten the number 2 engine mount bolts A to specification—Millenia models equipped with the 2.5L (KL) engine

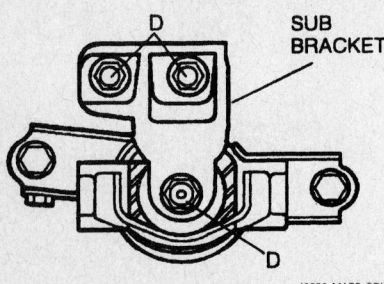

42356-MAZC-GDH

Tighten the number 3 engine mount sub bracket nuts D to specification—Millenia models equipped with the 2.5L (KL) engine

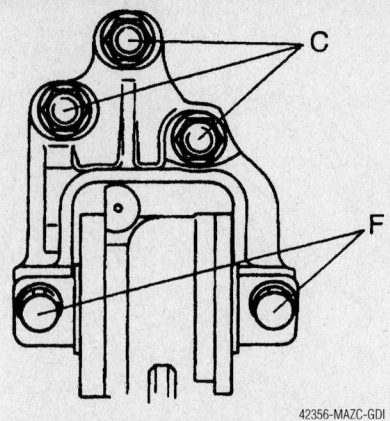

42356-MAZC-GDI

Tighten the number 4 engine mount nuts and bolts C and F to specification—Millenia models equipped with the 2.5L (KL) engine

16. Install the number 1 engine mount stay and tighten retainers **G** to 86 inch lbs. (10 Nm) and **H** to 77 ft. lbs. (104 Nm).

17. Tighten the number 2 engine mount bolts **A** to 77 ft. lbs. (104 Nm).

18. Tighten the number 3 engine mount sub bracket nuts **D** to 77 ft. lbs. (104 Nm).

19. Tighten the number 4 engine mount nuts and bolts **C** and **F** to 68 ft. lbs. (93 Nm)
  • Joint shaft. Install a new circlip with the opening facing up. Tighten the bolts to 32–45 ft. lbs. (43–61 Nm).
  • Halfshafts. Install new circlips on the inner CV-joint stub shafts, if equipped, and the intermediate shaft. Grease the shaft splines before installing the halfshaft/intermediate shaft into the transaxle.
  • Ball joints
  • Front pipe
  • Selector cable
  • Power steering pump
  • A/C Compressor
  • Drive belt

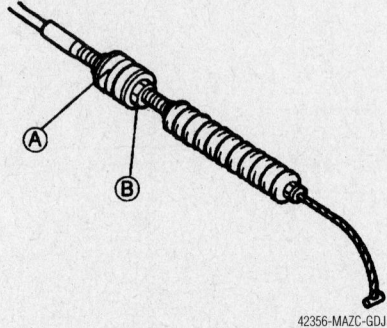

42356-MAZC-GDJ

Accelerator cable A and throttle cable B adjustment nuts—Millenia models equipped with the 2.5L (KL) engine

20. When connecting the accelerator cable, perform the following adjustment:
  a. Measure the free play of the cable, it should be 0.04–0.11 inch (1–3mm), if not turn adjustment nut **A** until the specification is reached.
  b. Fully depress the accelerator pedal and check the throttle is wide open, if not adjust using nut **B**.

21. Install the remaining components in the reverse of removal.

22. Fill the engine and the transaxle with the proper type and amount of fluids. Fill the cooling system.

23. Connect the negative battery cable, start the engine and check for leaks.

24. Check the ignition timing and the idle speed.

25. Check all fluid levels.

2.3L (KJ) ENGINE

➡The procedure for pulling the engine requires removing the transaxle along with it. As a result, when the halfshafts are pulled from the transaxle, a special plug/side gear holding tool is recommended.

1. Before servicing the vehicle, refer to the precautions section.

2. Properly relieve the fuel system pressure.

3. Drain the engine oil.

4. Drain the transaxle fluid.

5. Drain the cooling system.

6. Remove or disconnect the following:
  • Hood
  • Front wheels
  • Air cleaner assembly
  • Intake manifold cover
  • Splash shield
  • Charge air cooler duct
  • Battery clamp, cover, battery, battery carrier and battery air duct
  • Air duct
  • Accelerator cable
  • Dust cover
  • Drive belt
  • Crankshaft and vacuum pump pulleys
  • A/C compressor with the lines still attached and position aside
  • Power steering pump with the lines still attached and position aside
  • Radiator grille
  • Upper seal board
  • Radiator reservoir
  • Cooling fan connector and fan
  • Condenser fan motor connector and the fan
  • Oil cooler hose

- Radiator hose, bracket and the radiator
- Selector cable
- Front exhaust pipe
- Upper and lower ball joints
- Halfshafts
- Joint shaft

7. Support the engine using a suitable support device.

- Number 1 engine mount bracket. Refer to the accompanying illustration for location.
- Engine mount member retainers.

Remove the bolts **A** first and then the bolts **B** and remove the member. Refer to the accompanying illustration for location.

✳✳ CAUTION

Engine load can damage the number 4 mount bolts holes when removing the bolts so make sure all weight is off the mount.

8. Attach a hoist to the engine, take the weight of the engine with the hoist and

remove the engine support device, Lift up the engine/transaxle assembly slightly until the number 3 and 4 mounts are free from engine weight.

- Number 4 engine mount bracket. Refer to the accompanying illustration for location.
- Number 3 engine mount sub bracket. Refer to the accompanying illustration for location.
- Engine and transaxle assembly from the vehicle. Be careful the powertrain assembly does not

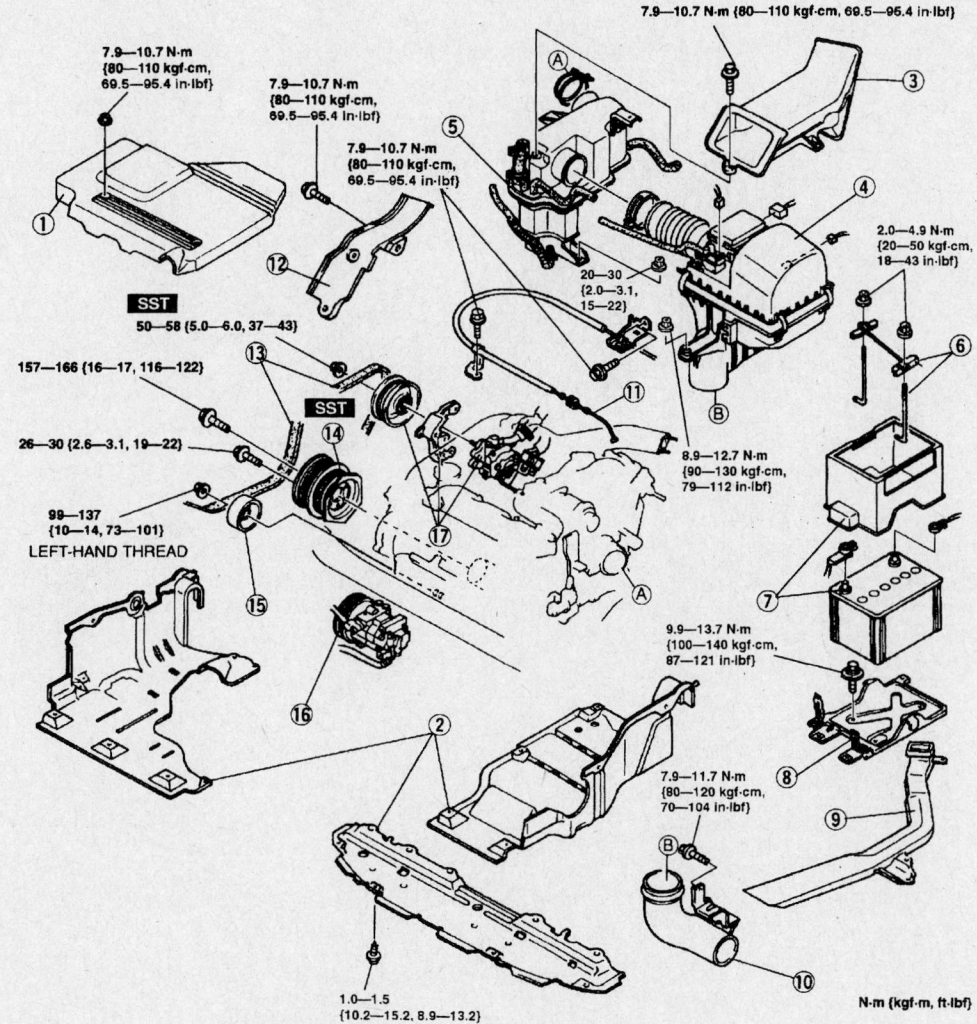

1 Dynamic chamber cover	10 Air duct
2 Splash shield	11 Accelerator cable
3 Charge air cooler air duct	12 Dust cover
4 Air cleaner component	13 Drive belt
5 Resonator	14 Crankshaft pulley
6 Battery clamp	15 Vacuum pump pulley
7 Battery and battery cover	16 A/C compressor
8 Battery carrier	17 P/S oil pump
9 Battery air duct	

42356-MAZC-GDK

Location of the engine mounting components and their torque specifications (part 1 of 4)—Millenia models equipped with the 2.3L (KJ) engine

swing and strike the vehicle causing damage

To install:

9. Installation is the reverse of removal. Tighten the fasteners to the specifications shown in the accompanying illustration.

10. When possible, leave the engine mounting nuts/bolts loose (hand tight) until all mounts are aligned and bolted. This may help in aligning the engine and transmission assembly in the vehicle.

11. Install or connect the following:
- Number 4 engine bracket retainers and hand tighten the retainers

12. Using the hoist lower the powertrain assembly and hand tighten the number 3 mount sub bracket retainers
- Number 4 engine bracket retainers and tighten the bolts **E** to 44 ft. lbs. (60 Nm) and hand tighten bolts **F**.

13. Remove the engine hoist and install the engine support device

14. Install the engine mount member and tighten bolts **B** to 68 ft. lbs. (93 Nm) and hand tighten bolts **A**.

15. Install the number 1 bracket and tighten the retainers to 77 ft. lbs. (104 Nm) and remove the engine support.

16. Install the number 1 engine mount stay and tighten retainers **G** to 68 inch lbs. (10 Nm) and **H** to 77 ft. lbs. (104 Nm).

17. Tighten the number 2 engine mount bolts **A** to 77 ft. lbs. (104 Nm).

18. Tighten the number 3 engine mount sub bracket nuts **D** to 77 ft. lbs. (104 Nm).

19. Tighten the number 4 engine mount nuts and bolts **C** and **F** to 68 ft. lbs. (93 Nm)

20. Install or connect the following:
- Joint shaft. Install a new circlip with the opening facing up. Tighten the bolts to 32–45 ft. lbs. (43–61 Nm).

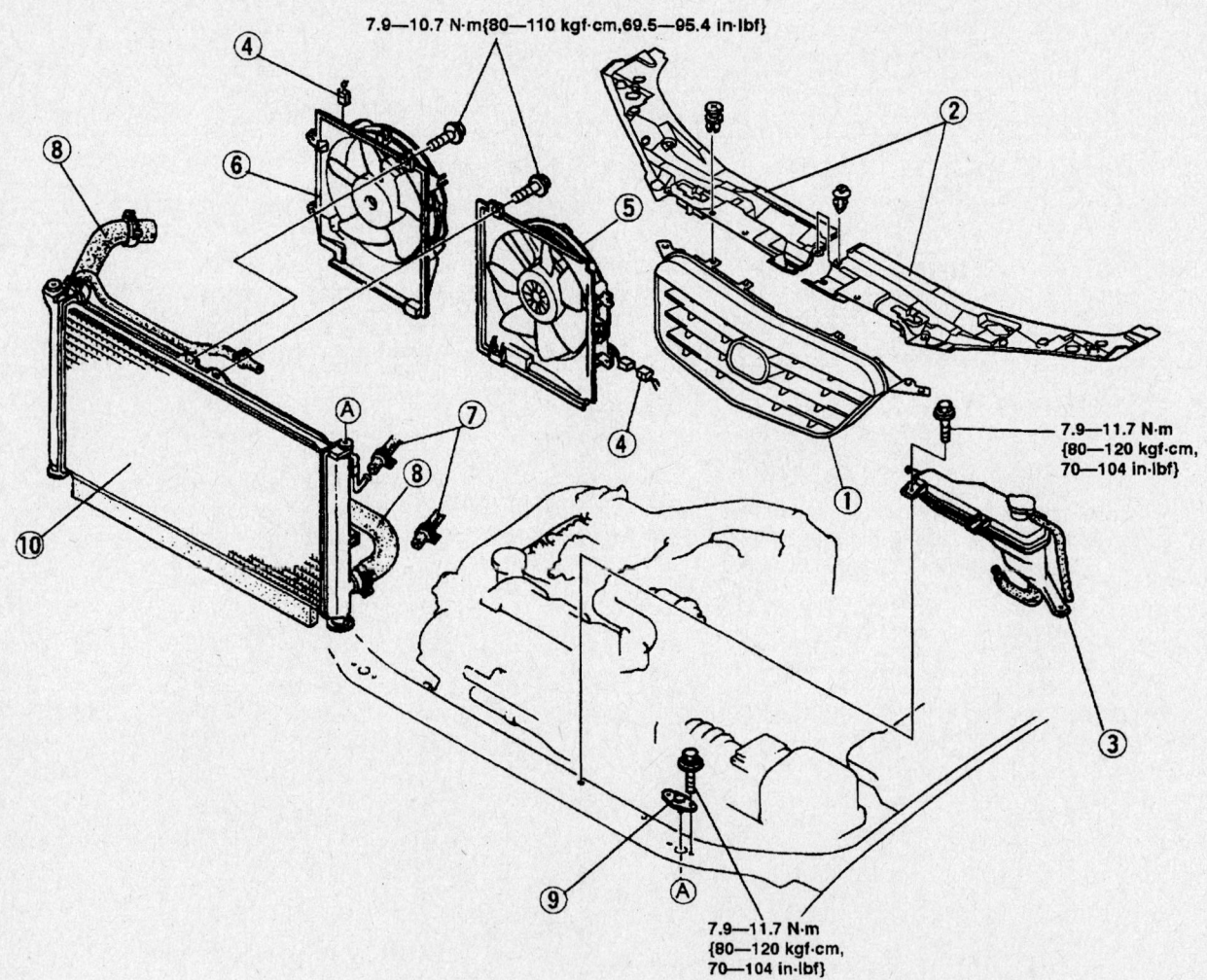

1	Radiator grille	6	Condenser fan component
2	Upper seal board	7	Oil cooler hose
3	Radiator reservoir	8	Radiator hose
4	Fan motor connector	9	Radiator bracket
5	Cooling fan component	10	Radiator

Location of the engine mounting components and their torque specifications (part 2 of 4)—Millenia models equipped with the 2.3L (KJ) engine

42356-MAZC-GDL

- Halfshafts. Install new circlips on the inner CV-joint stub shafts, if equipped, and the intermediate shaft. Grease the shaft splines before installing the halfshaft/intermediate shaft into the transaxle.
- Ball joints
- Front pipe
- Selector cable

21. Power steering pump
- A/C Compressor
- Drive belt

22. When connecting the accelerator cable, perform the following adjustment:

a. Measure the free play of the cable, it should be 0.04–0.11 inch (1–3mm), if not turn adjustment nut **A** until the specification is reached.

b. Fully depress the accelerator pedal and check the throttle is wide open, if not adjust using nut **B**.

23. Install the remaining components in the reverse of removal.

24. Fill the engine and the transaxle with the proper type and amount of fluids. Fill the cooling system.

25. Connect the negative battery cable, start the engine and check for leaks.

26. Check the ignition timing and the idle speed.

27. Check all fluid levels.

2002–04 Miata

1. Before servicing the vehicle, refer to the precautions section.

2. Properly relieve the fuel system pressure.

3. Drain the engine oil.

4. Drain the cooling system.

5. Remove or disconnect the following:
- Negative battery cable
- Air cleaner assembly

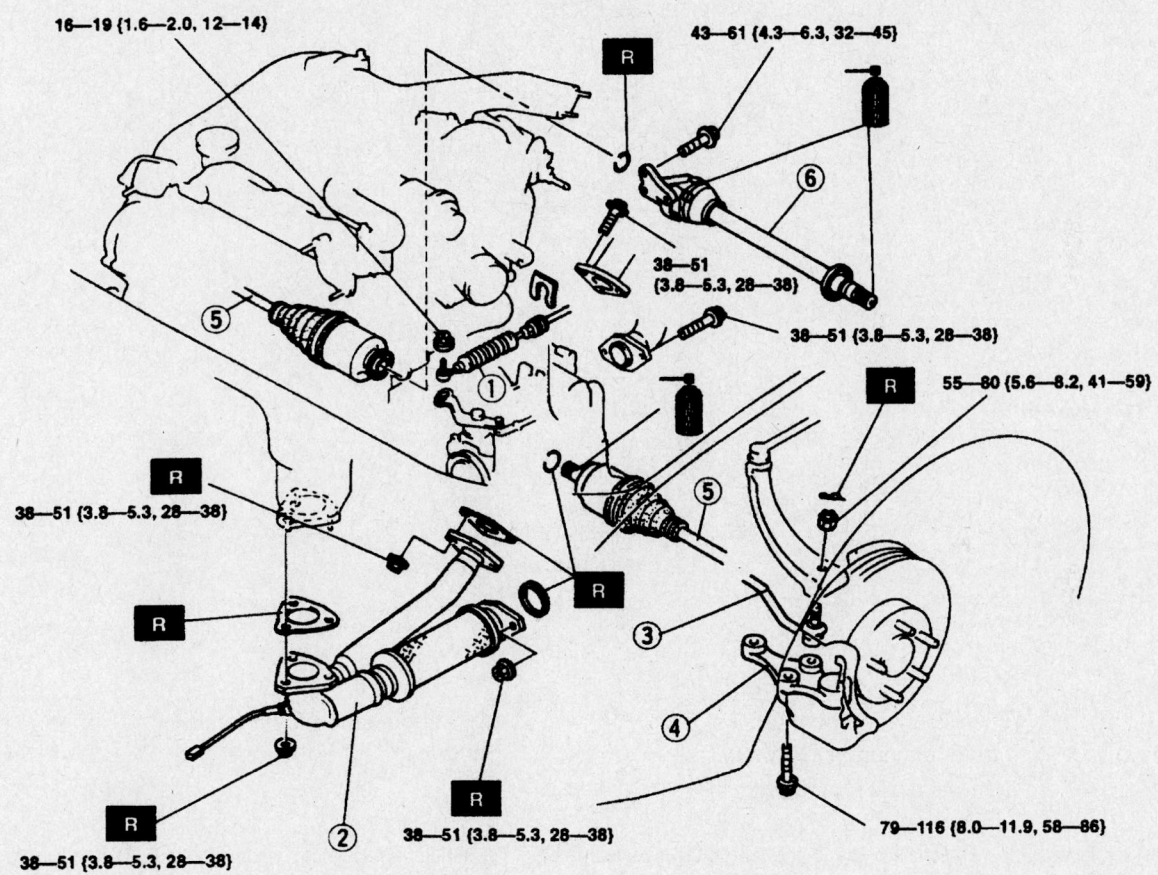

N·m {kgf·m, ft·lbf}

1	Selector cable	4	Lower ball joint
2	Front pipe	5	Drive shaft
3	Upper lateral link ball joint	6	Joint shaft

42356-MAZC-GDM

Location of the engine mounting components and their torque specifications (part 3 of 4)—Millenia models equipped with the 2.3L (KJ) engine

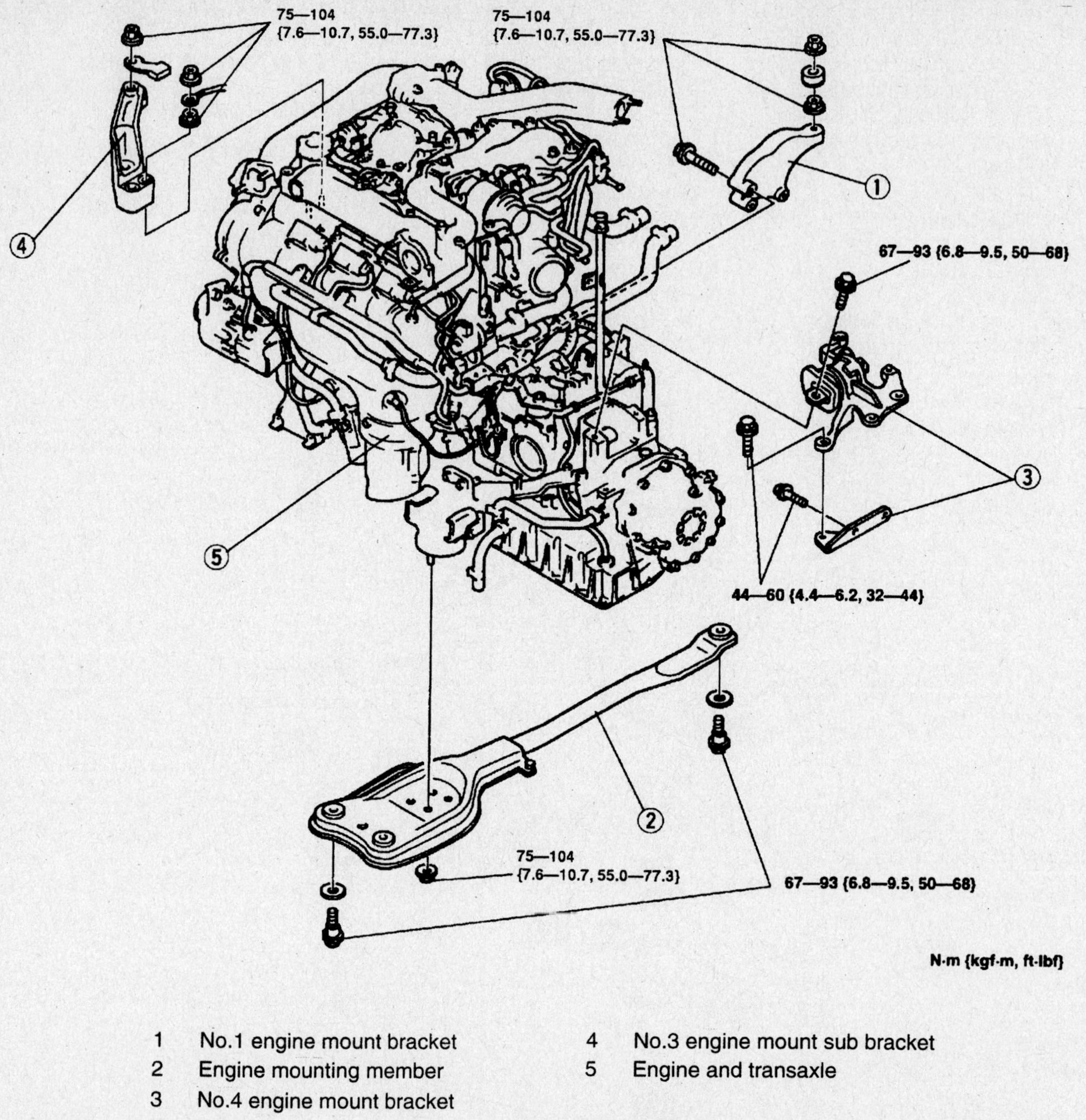

75—104
{7.6—10.7, 55.0—77.3}

75—104
{7.6—10.7, 55.0—77.3}

④

①

67—93 {6.8—9.5, 50—68}

③

⑤

44—60 {4.4—6.2, 32—44}

②

75—104
{7.6—10.7, 55.0—77.3}

67—93 {6.8—9.5, 50—68}

N·m {kgf·m, ft-lbf}

1	No.1 engine mount bracket		4	No.3 engine mount sub bracket
2	Engine mounting member		5	Engine and transaxle
3	No.4 engine mount bracket			

42356-MAZC-GDN

Location of the engine mounting components and their torque specifications (part 4 of 4)—Millenia models equipped with the 2.3L (KJ) engine

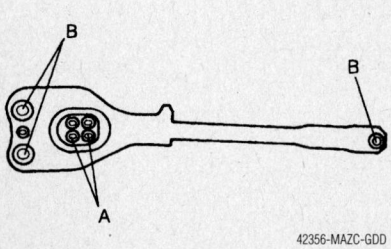

42356-MAZC-GDD

Remove the bolts A first and then the bolts B and remove the engine mount member—Millenia models equipped with the 2.3L (KJ) engine

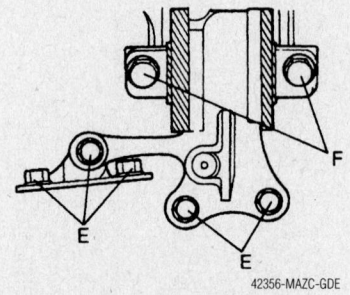

42356-MAZC-GDE

Tighten the number 4 engine bracket retainers E to specification and hand tighten bolts F—Millenia models equipped with the 2.3L (KJ) engine

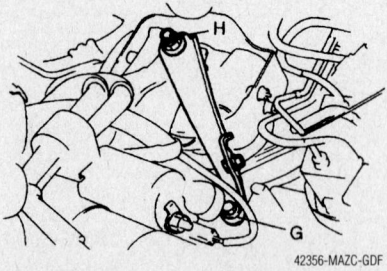

42356-MAZC-GDF

Tighten the number 1 engine mount stay retainers G and H to specification—Millenia models equipped with the 2.3L (KJ) engine

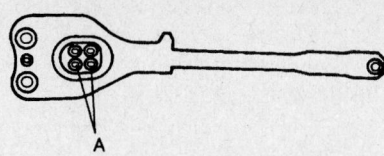

Tighten the number 2 engine mount bolts A to specification—Millenia models equipped with the 2.3L (KJ) engine

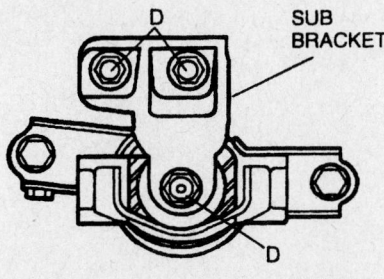

Tighten the number 3 engine mount sub bracket nuts D to specification—Millenia models equipped with the 2.3L (KJ) engine

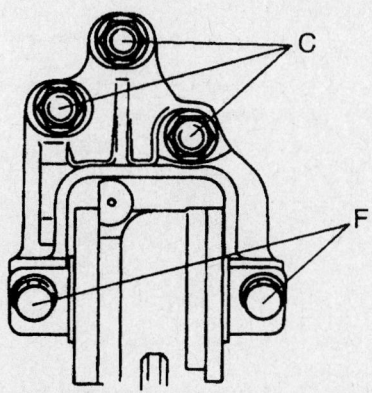

Tighten the number 4 engine mount nuts and bolts C and F to specification—Millenia models equipped with the 2.3L (KJ) engine

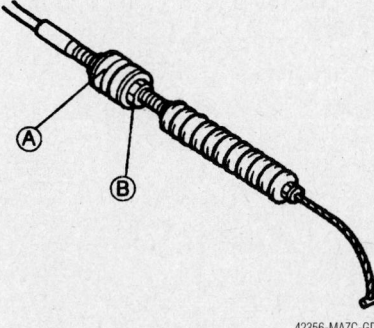

Accelerator cable A and throttle cable B adjustment nuts—Millenia models equipped with the 2.3L (KJ) engine

- Radiator
- Accelerator cable and bracket from the throttle body
- Fuel hose
- Vacuum hoses and engine harness connectors
- Heater hose
- Accessory drive belts
- Power steering pump and move it aside without disconnecting the hydraulic hoses
- Air conditioner compressor and move it aside without disconnecting the refrigerant lines
- Transmission

6. Disconnect the following electrical connectors, if equipped:
- Steering pressure sensor electrical connector
- Throttle Position (TP) sensor electrical connector
- Idle Air Control (IAC) valve electrical connector
- Heated Oxygen (HO_2S) sensor electrical connector
- Ignition coil electrical connectors
- Crankshaft Position (CKP) sensor electrical connector
- Ground electrical connectors
- Fuel injector electrical connectors
- Alternator electrical connectors
- Oil pressure sensor electrical connector
- Starter electrical connectors

7. Remove or disconnect the following:
- Exhaust pipe from the exhaust manifold and install suitable lifting equipment onto the engine.
- Engine from the vehicle

To install:

8. Install or connect the following:
- Engine assembly by tilting the engine downward and aligning the engine mounts with the crossmember holes. Torque the nuts to 42–57 ft. lbs. (57–78 Nm).
- Exhaust pipe to the manifold using a new gasket. Torque the nuts to 34 ft. lbs. (46 Nm).
- All vacuum hoses

9. Connect the following electrical connectors, if equipped:
- Steering pressure sensor electrical connector
- TP sensor electrical connector
- IAC valve electrical connector
- HO_2S electrical connector
- Ignition coil electrical connectors
- CKP sensor electrical connector
- Ground electrical connectors
- Fuel injector electrical connectors

- Alternator electrical connectors
- Oil pressure sensor electrical connector
- Starter electrical connectors

10. Install or connect the following:
- Transmission
- Air conditioner compressor and power steering pump
- Drive belt(s)
- Any remaining vacuum hoses
- Radiator and fans and all cooling system hoses
- Accelerator cable and bracket
- Air cleaner assembly
- Negative battery cable

11. Fill and bleed the cooling system.
12. Fill the engine and transmission.
13. Start the engine and check for leaks.
14. Check the ignition timing and idle speed.

Protégé

2.0L (FS) ENGINE

➡The procedure for pulling the engine requires removing the transaxle along with it. As a result, when the halfshafts are pulled from the transaxle, a special plug/side gear holding tool is recommended.

1. Before servicing the vehicle, refer to the precautions section.
2. Properly relieve the fuel system pressure.
3. Drain the engine oil.
4. Drain the transaxle fluid.
5. Drain the cooling system.
6. Remove or disconnect the following:
- Negative battery cable
- Radiator
- Air cleaner assembly
- Accelerator cable
- Fuel hoses
- Front exhaust pipe
- Any rods, pipes or cables related to the transaxle that would hinder removal
- Battery
- Fuse box
- Power steering pump with the lines still attached and set aside
- A/C compressor with the lines still attached and position aside
- Halfshafts
- Roll dampener

7. Support the engine with a suitable engine support device.
- Engine mount member retainers. Refer to the accompanying illustration for location.
- Number 3 engine mount. Refer to

the accompanying illustration for location.
- Number 1 engine mount. Refer to the accompanying illustration for location.
- Number 2 engine mount. Refer to the accompanying illustration for location.
- Number 4 engine mount. Refer to the accompanying illustration for location.
- Engine and transaxle assembly from the vehicle.

To install:

8. Installation is the reverse of removal. Tighten the fasteners to the specifications shown in the accompanying illustration.

9. When possible, leave the engine mounting nuts/bolts loose (hand tight) until all mounts are aligned and bolted. This may help in aligning the engine and transmission assembly in the vehicle.

10. Install the number 4 engine mount. Tighten the fasteners to the specifications shown in the accompanying illustration as follows:

a. Hand tighten the number 3 and 4 engine mount bolts and nuts **A–M**.

b. Tighten the number 4 engine mount bolts and nuts **A–H**.

c. Tighten the number 3 engine mount bolts and nuts **I–N**.

d. Measure the number 4 mount clearance which should be 0.12–0.15 inch (3–4mm). If not within specification, loosen the nuts and bolts and retorque using the same procedure.

11. Install or connect the following:
- Number 1 engine mount. Tighten

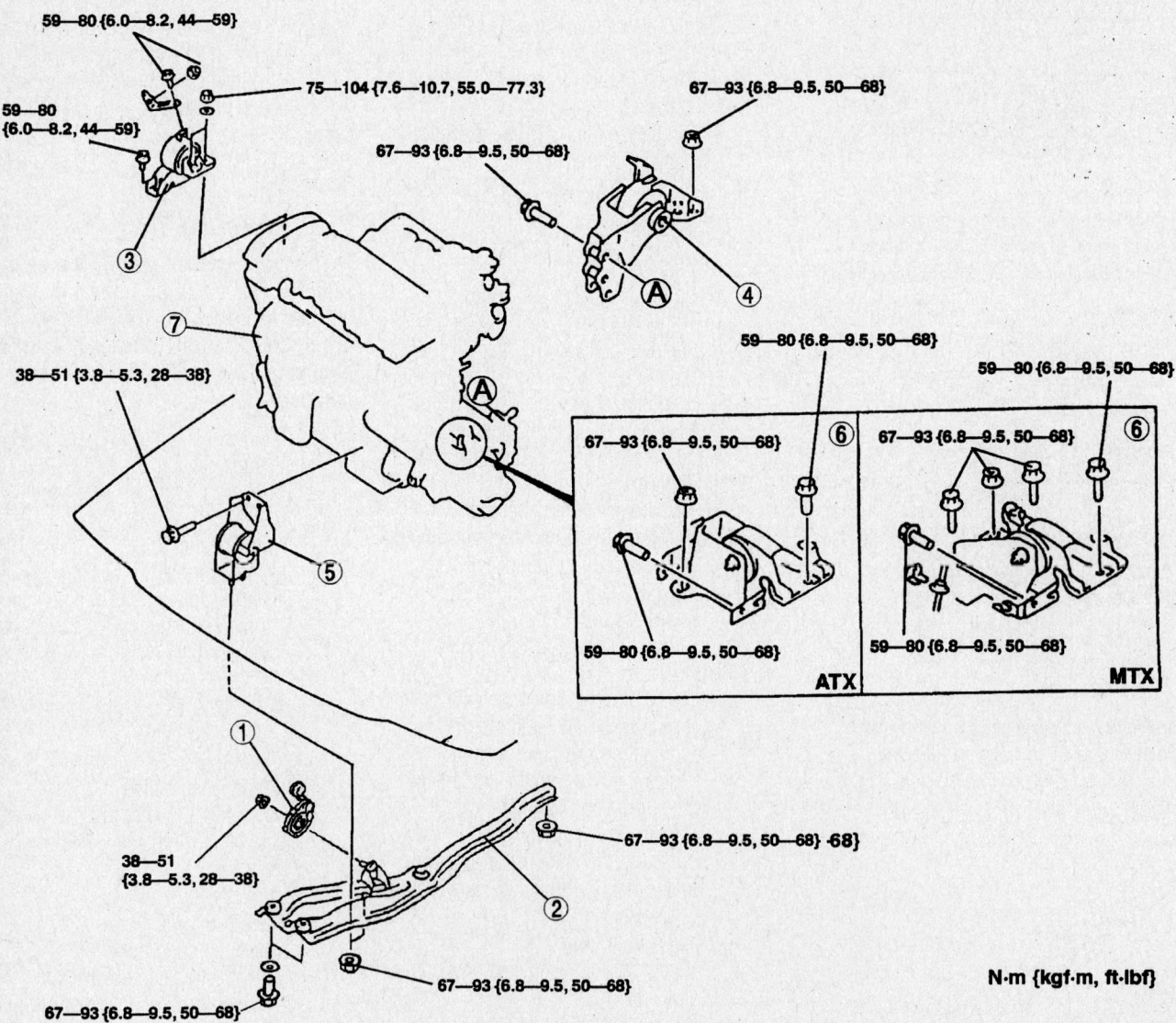

1	Roll damper	5	No.2 Engine mount
2	Engine mount member	6	No.4 Engine mount
3	No.3 Engine mount	7	Engine, transaxle
4	No.1 Engine mount		

Location of the engine mounting components and their torque specifications—Protégé models equipped with the 2.0L (FS) engine

42356-MAZC-GGA

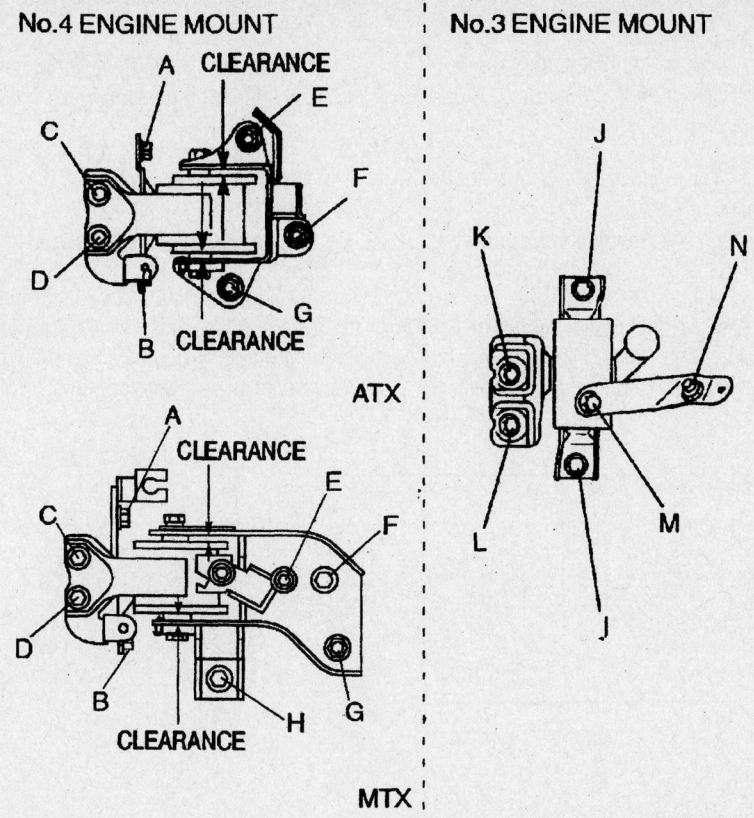

No.4 ENGINE MOUNT
ATX

MTX

42356-MAZC-GGB

Tighten the number 4 engine bracket retainers in the sequence shown—Protégé models equipped with the 2.0L (FS) engine

No.3 ENGINE MOUNT

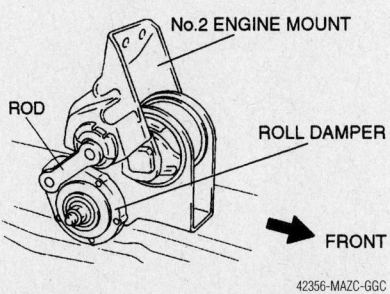

No.2 ENGINE MOUNT

ROD

ROLL DAMPER

FRONT

42356-MAZC-GGC

Assemble the roll damper as illustrated—Protégé models equipped with the 2.0L (FS) engine

the fasteners to the specifications shown in the accompanying illustration.
- Number 3 engine mount, refer to the number 4 mount tightening sequence. Tighten the fasteners to the specifications shown in the accompanying illustration.
- Engine mount member. Tighten the fasteners to the specifications shown in the accompanying illustration.
- Roll damper, refer to the illustration or proper assembly.

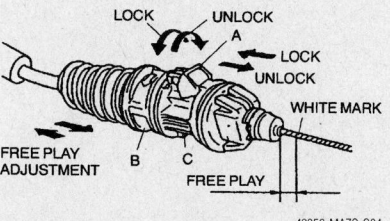

Free play
1.5—4.0 mm {0.06—0.15 in}

42356-MAZC-G04

Accelerator cable adjustment components—Protégé models equipped with the 2.0L (FS) engine

- Halfshafts. Install new circlips on the inner CV-joint stub shafts, if equipped, and the intermediate shaft. Grease the shaft splines before installing the halfshaft/intermediate shaft into the transaxle.
- A/C compressor
- Power steering pump
- Fuse box
- Battery
- Any rods, pipes or cables related to the transaxle
- Front exhaust pipe. Always install

new gaskets and/or O-rings. Use new self-locking nuts, especially on the exhaust.
- Fuel hoses

12. When connecting the accelerator cable, perform the following adjustment:
  a. Move the white locking tab to the unlock position.
  b. Turn stopper B to the to the unlock position.

➡**If stopper B will not unlock, it may be necessary to carefully bend back tab C out a little using a suitable pry tool.**

  c. Push or pull the cable housing directly behind the spring.
  d. Turn the stopper B to the lock position.
  e. Measure the free play which should be 0.04–0.11 inch (1–3mm) and make sure the cable free play is within specification.
  f. Move the white locking tab to the lock position and check for proper accelerator operation.

13. Install or connect the following:
- Air cleaner assembly
- Radiator
- Negative battery cable

14. Fill the engine and the transaxle with the proper type and amount of fluids. Fill the cooling system.

15. Connect the negative battery cable, start the engine and check for leaks.

16. Check the ignition timing and the idle speed.

17. Check all fluid levels.

1.6L (ZM) ENGINE

➡**The procedure for pulling the engine requires removing the transaxle along with it. As a result, when the halfshafts are pulled from the transaxle, a special plug/side gear holding tool is recommended.**

1. Before servicing the vehicle, refer to the precautions section.

2. Properly relieve the fuel system pressure.

3. Drain the engine oil.

4. Drain the transaxle fluid.

5. Drain the cooling system.

6. Remove or disconnect the following:
- Battery
- Air cleaner, air hose and resonance chamber
- Front exhaust pipe
- Accelerator cable and bracket
- Heater and vacuum hoses
- Radiator
- Drive belt

- Fuel hoses
- Any rods, pipes or cables related to the transaxle that would hinder removal
- Halfshafts
- Power steering pump with the lines still attached and set aside
- A/C compressor with the lines still attached and position aside
- Air cleaner bracket
- Battery carrier bracket

7. Support the engine with a suitable engine support device.
- Number 1 engine mount. Refer to the accompanying illustration for location.
- Number 2 engine mount nut **A**. Refer to the accompanying illustration for location.
- Engine mount member retainers. Refer to the accompanying illustration for location.
- Number 2 engine mount. Refer to the accompanying illustration for location.

8. Remove the engine support device and attach a hoist and chain to the engine.
- Number 4 engine mount. Refer to the accompanying illustration for location.
- Number 3 engine mount. Refer to the accompanying illustration for location.
- Engine and transaxle assembly from the vehicle.

To install:

9. Installation is the reverse of removal. Tighten the fasteners to the specifications shown in the accompanying illustration.

10. When possible, leave the engine mounting nuts/bolts loose (hand tight) until all mounts are aligned and bolted. This may

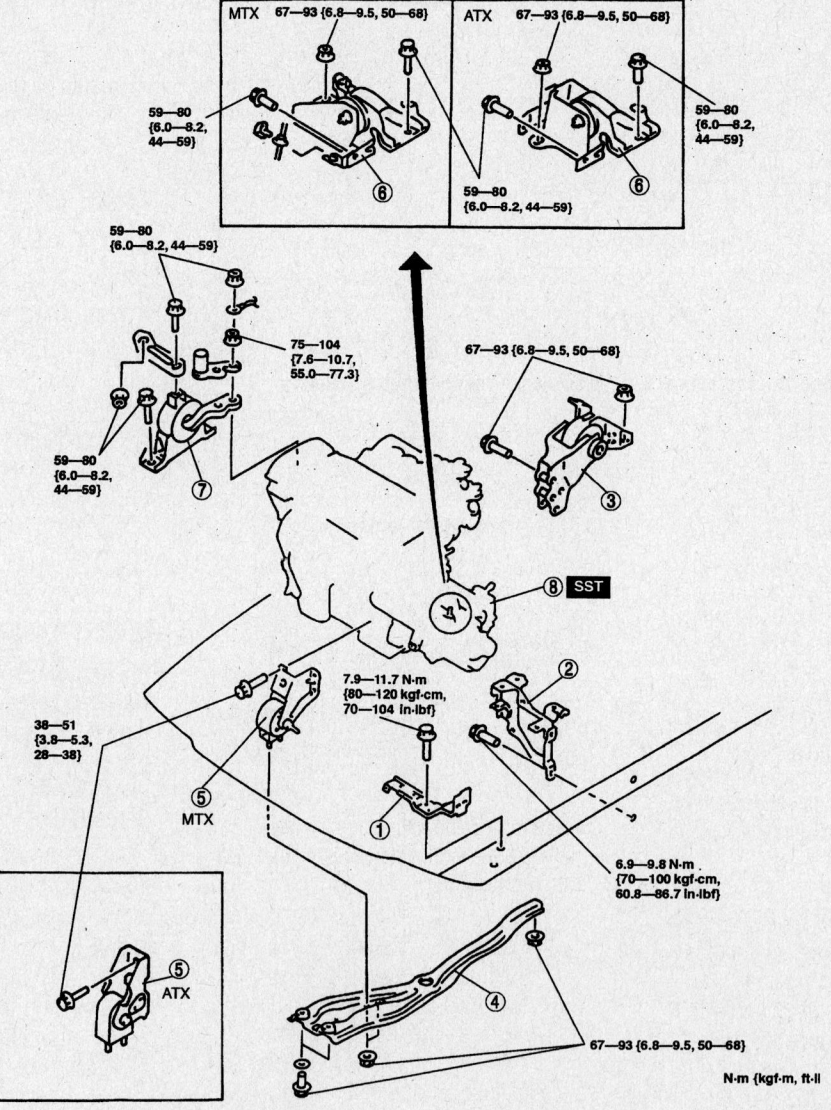

1	Air cleaner bracket	5	No.2 engine mount
2	Battery carrier bracket	6	No.4 engine mount
3	No.1 engine mount	7	No.3 engine mount
4	Engine mount member	8	Engine, transaxle

42356-MAZC-GHD

Location of the engine mounting components and their torque specifications—Protégé models equipped with the 1.6L (ZM) engine

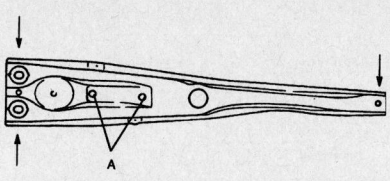

42356-MAZC-GHE

Remove the number 2 engine mount nut retainers A, then the engine mount retainers—Protégé models equipped with the 1.6L (ZM) engine

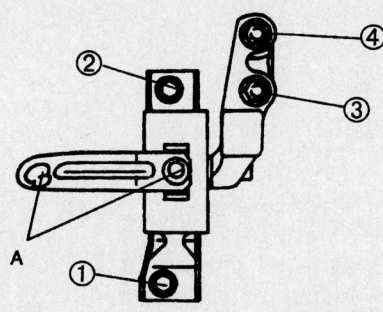

42356-MAZC-GHF

Tighten the number 3 engine mount bolt and nut in this sequence and the mount stay bolt A—Protégé models equipped with the 1.6L (ZM) engine

MTX

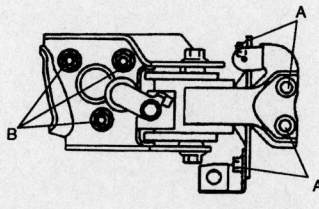

42356-MAZC-GHG

Number 4 engine mount retainer torque sequence on models with a manual transaxle—Protégé models equipped with the 1.6L (ZM) engine

help in aligning the engine and transmission assembly in the vehicle.

11. Install or connect the following:
- Tighten the number 3 engine mount bolt and nut in the sequence illustrated, then tighten the mount stay bolt and nut **A**. Tighten the fasteners to the specifications shown in the exploded view of the engine mounting assembly illustration.
- Tighten the number 4 engine mount bolts **A**, then tighten bolts **B**. Tighten the fasteners to the specifications shown in the exploded view of the engine mounting assembly illustration.

ATX

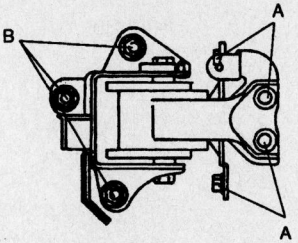

42356-MAZC-GHH

Number 4 engine mount retainer torque sequence on models with a automatic transaxle—Protégé models equipped with the 1.6L (ZM) engine

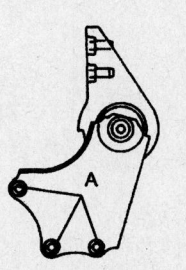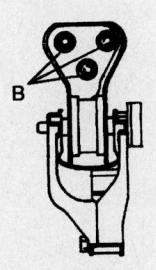

42356-MAZC-GHI

Number 1 engine mount retainer torque sequence—Protégé models equipped with the 1.6L (ZM) engine

12. Remove the hoist and chain and install the engine support device.
- Number 2 engine mount nut **A**, then tighten the mount member bolt and nut. Tighten the fasteners to the specifications shown in the exploded view of the engine mounting assembly illustration.
- Number 1 engine mount bolts **A** and then the bolt **B**. Tighten the fasteners to the specifications shown in the accompanying illustration.
- Battery carrier bracket
- Air cleaner bracket
- A/C compressor
- Power steering pump
- Halfshafts
- Any rods, pipes or cables related to the transaxle that were removal
- Fuel hoses
- Drive belt
- Radiator
- Heater and vacuum hoses
- Accelerator cable and bracket
- Front exhaust pipe
- Air cleaner, air hose and resonance chamber
- Battery
13. Fill the engine and the transaxle with

the proper type and amount of fluids. Fill the cooling system.
14. Connect the negative battery cable, start the engine and check for leaks.
15. Check the ignition timing and the idle speed.
16. Check all fluid levels.

Water Pump

REMOVAL & INSTALLATION

1.6L (ZM) Engines

1. Before servicing the vehicle, refer to the precautions section.
2. Drain the cooling system.
3. Remove or disconnect the following:
- Negative battery cable
- Fresh air duct
- Exhaust manifold insulator
- Timing belt
- Power steering oil pump with the lines attached and set aside
- A/C compressor with the lines attached and set aside
- Water inlet pipe
- Water pump mounting bolts
- Water pump

To install:
4. Clean all gasket mating surfaces.
5. Install or connect the following:
- Water pump using a new gasket. Torque the main bolts to 12–17 ft. lbs. (16–23 Nm) and the upper left hand bolt to 28–38 ft. lbs. (38–51 Nm).
- Water inlet pipe with new gasket and O–ring and tighten the bolts to 12–17 ft. lbs. (16–23 Nm)
- A/C compressor
- Power steering oil pump

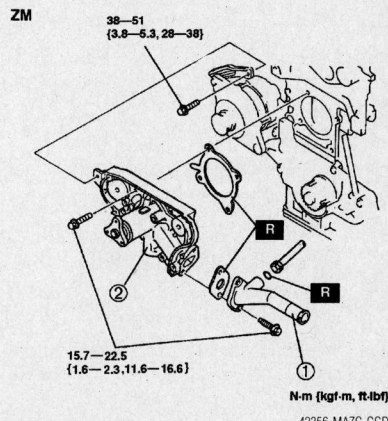

42356-MAZC-GGD

Exploded view of the water pump assembly—1.6L (ZM) engines

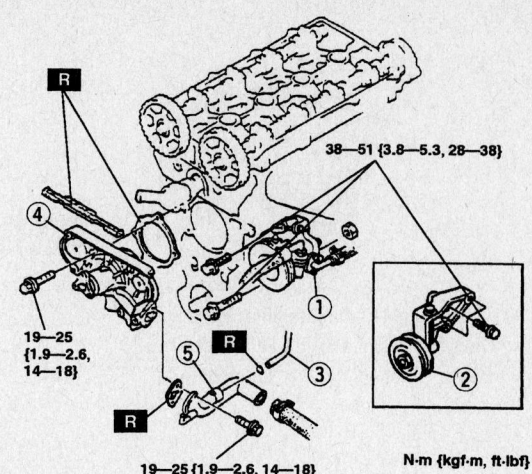

1. P/S oil pump
2. Idler (without P/S oil pump)
3. Water hose
4. Water pump
5. Water inlet pipe

Exploded view of the water pump assembly—1.8L (BP) engines

- Timing belt
- Exhaust manifold insulator
- Fresh air duct
- Negative battery cable
6. Fill and bleed the cooling system.
7. Start the engine, check for leaks and repair if necessary.

1.8L (BP) Engines

1. Before servicing the vehicle, refer to the precautions section.
2. Disconnect the negative battery cable.
3. Drain the engine coolant.
4. Remove or disconnect the following:
   - Timing belt covers and timing belt
   - Power steering pump, leaving the lines attached and set aside
   - Idler, on models not equipped with power steering
   - Coolant inlet pipe and gasket
   - Water pump

To install:
5. Clean all gasket mating surfaces.
6. Using a new gasket, install the water pump on the engine. Torque the mounting bolts to 14–18 ft. lbs. (19–25 Nm). Torque the bolt from the water pump to the alternator bracket to 28–38 ft. lbs. (38–51 Nm).
7. Install or connect the following:
   - Coolant inlet pipe with a new gasket. Torque the coolant inlet pipe bolts to 14–18 ft. lbs. (19–25 Nm).
   - Idler, on models not equipped with power steering
   - Power steering pump
   - Timing belt and the timing belt covers
   - Negative battery cable
8. Fill and bleed the cooling system.
9. Start the engine and bring to normal operating temperature. Check for leaks.

2.0L (FS) Engines

1. Before servicing the vehicle, refer to the precautions section.
2. Drain the cooling system.
3. Remove or disconnect the following:
   - Negative battery cable
   - Timing belt
   - Power steering oil pump adjuster

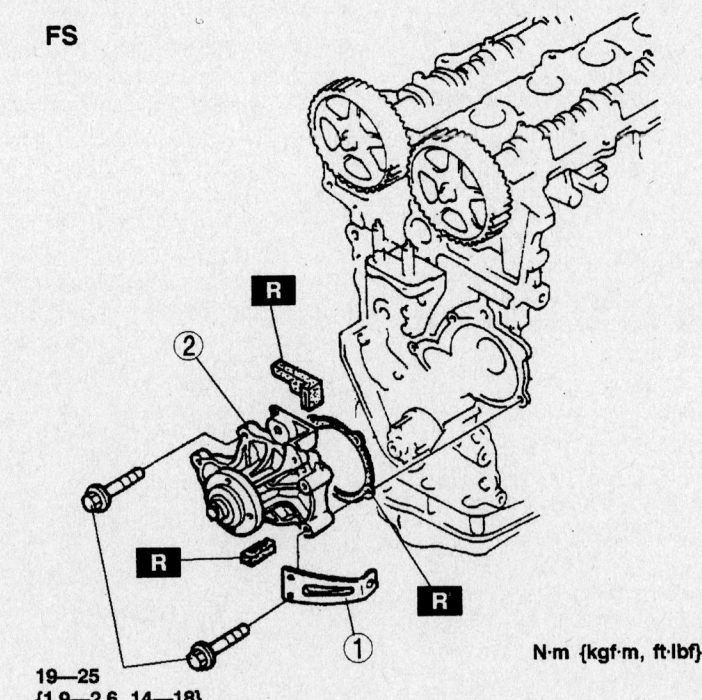

1   P/S oil pump adjuster
2   Water pump

Exploded view of the water pump assembly—2.0L (FS) engines

- Water pump mounting bolts
- Water pump

To install:
4. Clean all gasket mating surfaces.
5. Install or connect the following:
   - Water pump using a new gasket. Torque the bolts to 14–18 ft. lbs. (19–25 Nm).
   - Power steering oil pump adjuster. Torque the bolts to 12–16 ft. lbs. (16–22 Nm).
   - Timing belt
   - Negative battery cable
6. Fill and bleed the cooling system.
7. Start the engine, check for leaks and repair if necessary.

2.3L (KJ) Engine

1. Before servicing the vehicle, refer to the precautions section.
2. Drain the cooling system.
3. Remove or disconnect the following:
   - Negative battery cable
   - Timing belt and water pump together

To install:
4. Clean the mating surfaces.
5. Install or connect the following:
   - Water pump using a new gasket

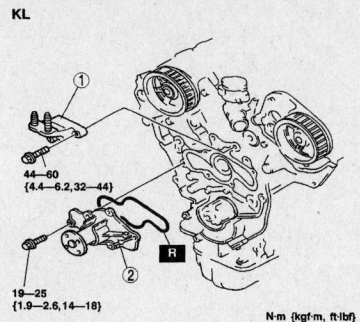

1 No.3 engine mount bracket
2 Water pump

42356-MAZC-G06

Exploded view of the water pump assembly—2.5L (KL) engines

along with the timing belt. Torque the bolts to 14–18 ft. lbs. (19–25 Nm).
• Negative battery cable
6. Fill the cooling system.
7. Start the engine, check for leaks and repair if necessary.

2.5L (KL) Engine

1. Before servicing the vehicle, refer to the precautions section.
2. Disconnect the negative battery cable.
3. Drain the cooling system.
• Timing belt
• Number 3 engine mount bracket
• Water pump bolts
• Water pump
To install:
4. Clean the mating surfaces.
5. Install or connect the following:
• Water pump using a new gasket. Tighten the bolts to 14–18 ft. lbs. (19–25 Nm).
• Number 3 engine mount bracket and tighten the bolts to 32–44 ft. lbs. (44–60 Nm)
• Timing belt
• Negative battery cable
6. Fill the cooling system.
7. Start the engine and check for leaks.

Heater Core

REMOVAL AND INSTALLATION

2002–04 Miata

1. Before servicing the vehicle, refer to the precautions section.
2. Disconnect the negative battery cable.

After disconnecting the battery, wait for more than 1 minute for the SAS to deplete its stored energy.

3. Drain the cooling system into a clean container for reuse.
4. Disconnect the heater hoses from the heater core.
5. Discharge and recover the air conditioning system refrigerant.
6. At the driver's side, remove the SAS module and the steering wheel by removing or disconnecting the following:
• Place the wheel in the straight-ahead position and turn the ignition switch to LOCK
• Cover clips at both sides of the steering wheel
• Steering wheel-to-SAS module bolts
• SAS module from the steering wheel and disconnect the electrical connector
• Steering wheel-to-column nut
• Steering wheel from the steering column using a suitable puller
7. At the passenger's side, remove the

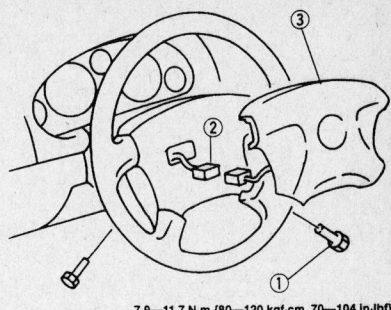

7.9—11.7 N·m {80—120 kgf-cm, 70—104 in-lbf}
93112GG0

View of the SAS module and the steering wheel—2002–04 Miata

SAS module by removing or disconnecting the following:
• Glove compartment and the glove compartment cover
• SAS module-to-dash bolts
• SAS module and disconnect the electrical connector
• Console
8. Remove the instrument cluster by removing or disconnecting the following:
• A-pillar trim at both sides
• Lower panel
• Instrument cluster hood

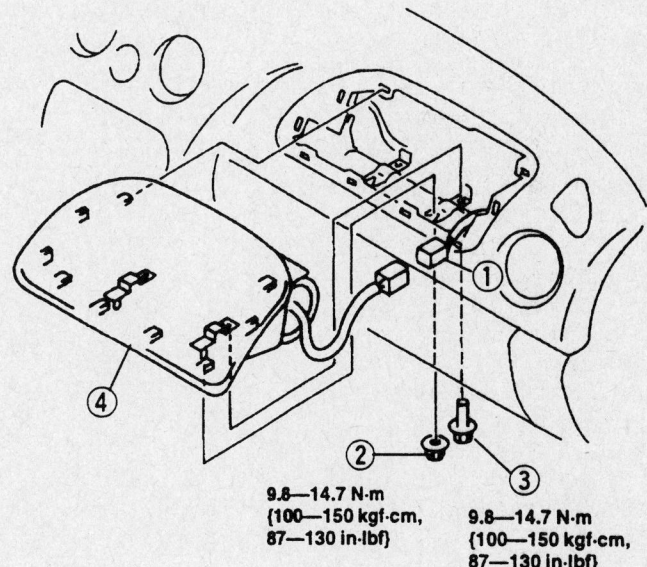

9.8—14.7 N·m {100—150 kgf-cm, 87—130 in·lbf}   9.8—14.7 N·m {100—150 kgf-cm, 87—130 in·lbf}

1 Connector
2 Nut
3 Bolt
4 Passenger-side air bag module

93112GH1

View of the passenger's side SAS module—2002–04 Miata

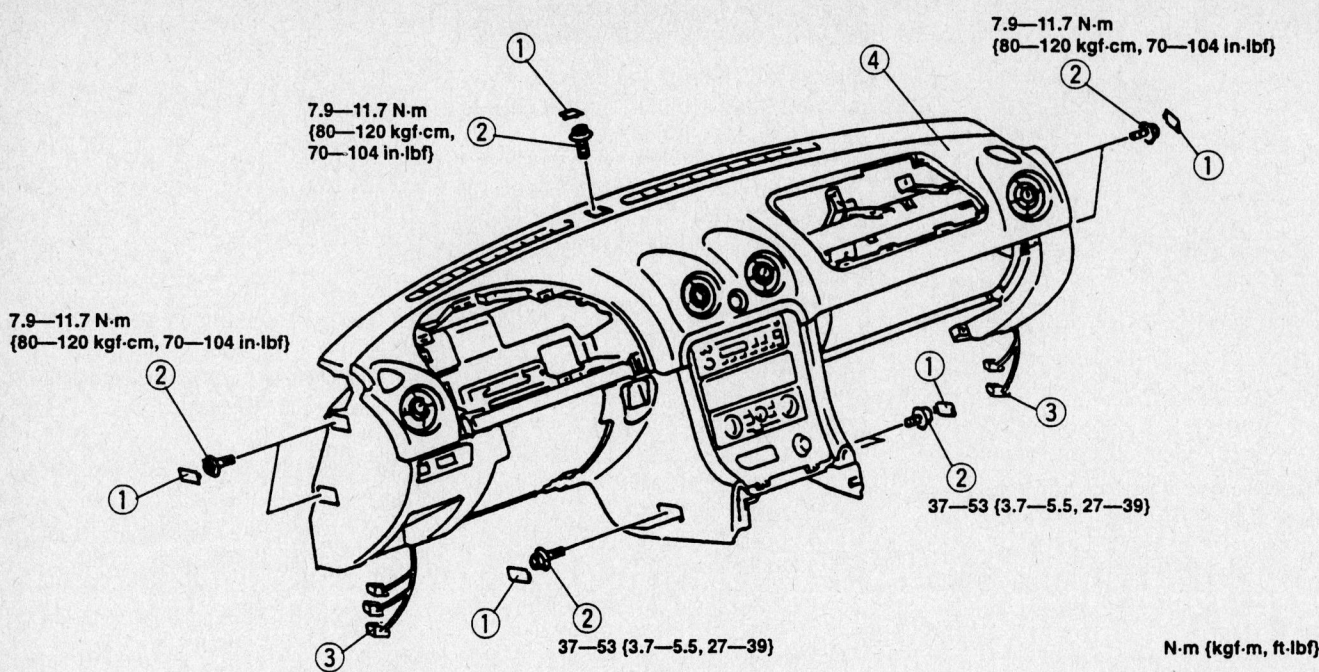

7.9—11.7 N·m
{80—120 kgf·cm, 70—104 in·lbf}

7.9—11.7 N·m
{80—120 kgf·cm,
70—104 in·lbf}

7.9—11.7 N·m
{80—120 kgf·cm, 70—104 in·lbf}

37—53 {3.7—5.5, 27—39}

37—53 {3.7—5.5, 27—39}

N·m {kgf·m, ft·lbf}

1	Cover	3	Connector
2	Bolt	4	Dashboard

93112GH2

View of the instrument panel—2002–04 Miata

- Instrument cluster-to-dash screws and the instrument cluster
- Hood release lever
- Control wire from the heater unit and the blower unit
- Steering column-to-instrument panel bolts and lower the steering column
- Instrument panel-to-chassis bolt covers and the bolts
- Instrument panel with the help of an assistant

9. Remove or disconnect the following:

- Heater unit-to-evaporator housing seal plate
- Heater unit-to-chassis nuts and the heater unit

10. Separate the heater unit cases and remove the heater core.

To install:

11. Install the heater core and assemble the heater unit cases.

12. Install or connect the following:

- Heater unit-to-chassis nuts and the heater unit
- Heater unit-to-evaporator housing seal plate

13. Install the instrument cluster by installing or connecting the following:

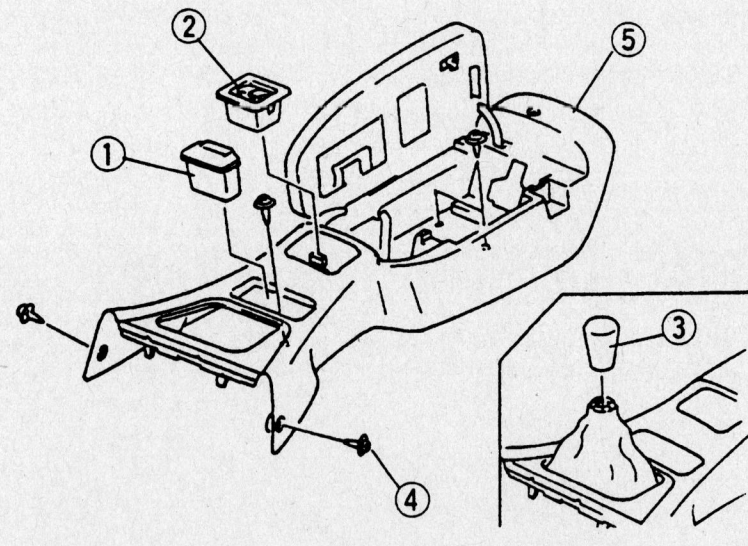

1	Ashtray
2	Power window switch
3	Shift lever knob (MT)
4	Screw
5	Console

93112GH3

View of the console—2002–04 Miata

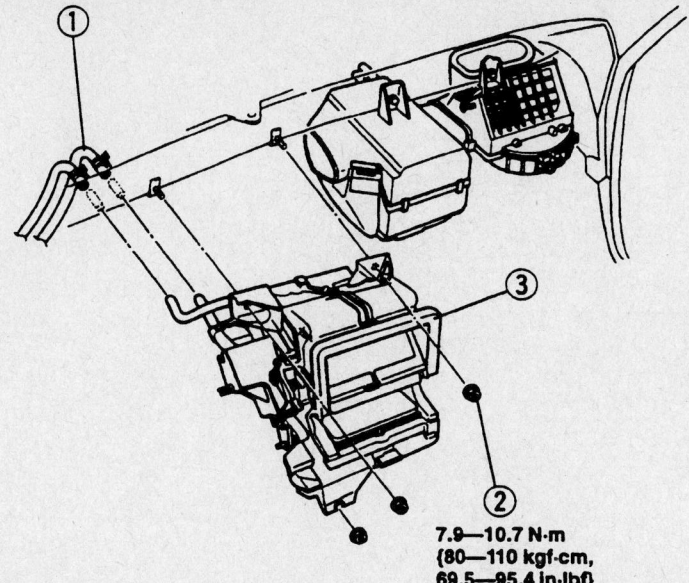

7.9—10.7 N·m
{80—110 kgf·cm,
69.5—95.4 in·lbf}

1	Heater hose
2	Nut
3	Heater unit

93112GH4

View of the heater unit and the heater unit-to-evaporator unit seal plate—2002–04 Miata

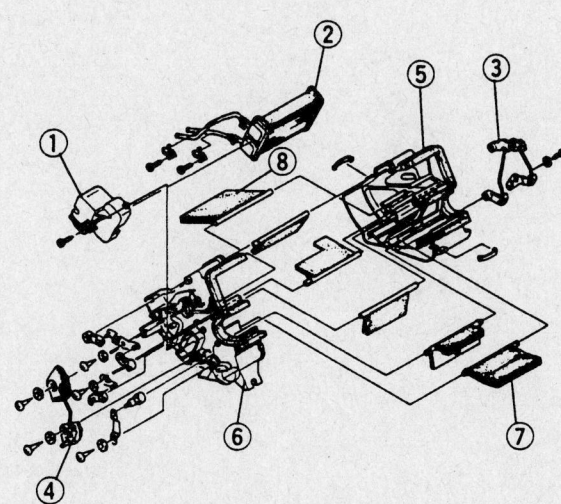

1	Cover
2	Heater core
3	Air mix link
4	Airflow mode link
5	Case (RH)
6	Case (LH)
7	Air mix door
8	Airflow mode door

93112GH5

Exploded view of the heater core and heater unit assembly—2002–04 Miata

- Instrument panel with the help of an assistant
- Instrument panel-to-chassis bolt covers and the bolts
- Steering column-to-instrument panel bolts
- Control wire to the heater unit and the blower unit
- Hood release lever
- Instrument cluster and the instrument cluster-to-dash screws
- Instrument cluster hood
- Lower panel
- A-pillar trim at both sides
- Console

14. At the passenger's side, install the SAS module by installing or connecting the following:
- SAS module and connect the electrical connector
- SAS module-to-dash bolts
- Glove compartment cover and the glove compartment

15. At the driver's side, install the SAS module and the steering wheel by installing or connecting the following:
- Steering wheel-to-column nut. Torque the steering wheel nut to 29–36 ft. lbs. (40–49 Nm).
- SAS module to the steering wheel and connect the electrical connector
- Steering wheel-to-SAS module bolts. Torque the steering column-to-SAS module bolts to 70–104 inch lbs. (8–12 Nm).
- Cover clips at both sides of the steering wheel

16. Connect the heater hoses to the heater core.
17. Refill the cooling system.
18. Connect the negative battery cable.
19. Evacuate, charge and leak test the air conditioning system refrigerant.
20. Operate the engine to normal operating temperatures; then, check the climate control operation and check for leaks.

Millenia

1. Before servicing the vehicle, refer to the precautions section.
2. Disconnect the negative battery cable.

✳✳ CAUTION

After disconnecting the battery, wait for more than 1 minute for the SAS to deplete its stored energy.

3. Drain the cooling system into a clean container for reuse.

4. Discharge and recover the air conditioning system refrigerant.

5. At the driver's side, remove the SAS module and the steering wheel by removing or disconnecting the following:

- Place the wheel in the straight-ahead position and turn the ignition switch to LOCK
- Cover clips at both sides of the steering wheel
- Steering wheel-to-SAS module clips
- SAS module from the steering wheel and disconnect the electrical connector
- Steering wheel-to-column nut
- Steering wheel from the steering column using a suitable puller

6. At the passenger's side, remove the SAS module by removing or disconnecting the following:

- Glove compartment and the glove compartment cover
- SAS module-to-dash bolts
- SAS module and disconnect the electrical connector

7. Remove the rear console by removing or disconnecting the following:

- Rear console's box
- Bake boot
- Center panel
- Rear console

8. Remove or disconnect the following:

- A-pillar trim at both sides
- Undercover at the passenger's side
- Upper and lower steering column covers
- Electrical connectors and remove the combination switch from the steering column
- Meter hood
- Electrical connectors and remove the instrument cluster
- Steering column-to-chassis bolts and the steering column
- Hood release lever
- Both side panels
- Instrument panel with the help of an assistant

9. Remove the evaporator housing by removing or disconnecting the following:

- Air conditioning system refrigerant lines and discard the gaskets
- Aspirator hose
- Power transistor connector
- MAX-HI connector
- Evaporator temperature connector
- Evaporator housing assembly

10. Disconnect and remove the heater unit assembly.

11. Separate the heater unit cases and remove the heater core.

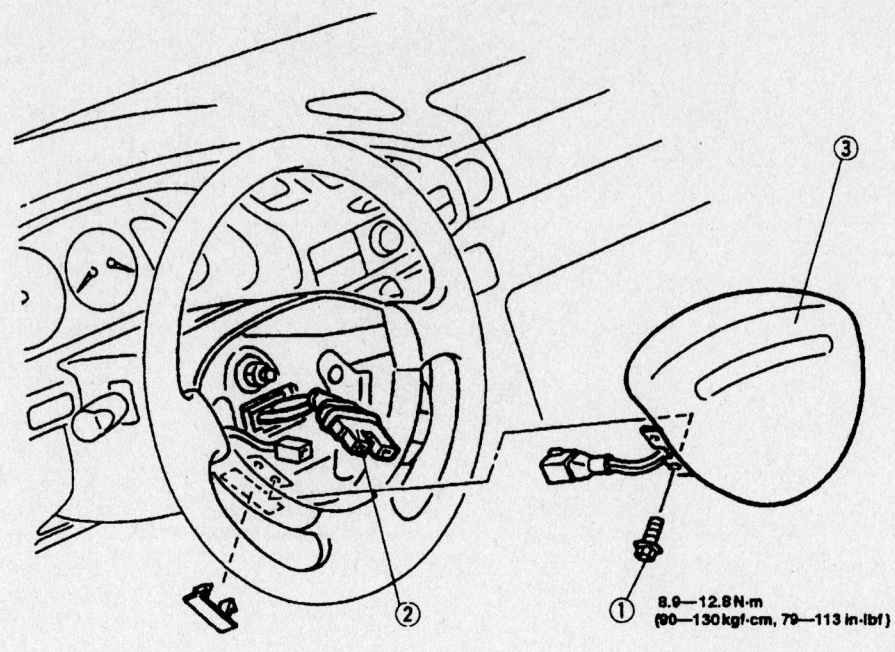

8.9—12.8 N·m
(90—130 kgf·cm, 79—113 in·lbf)

1 Bolt
2 Connector
3 Driver-side air bag module

93112GE5

Exploded view of the SAS module and steering wheel assembly—Millenia

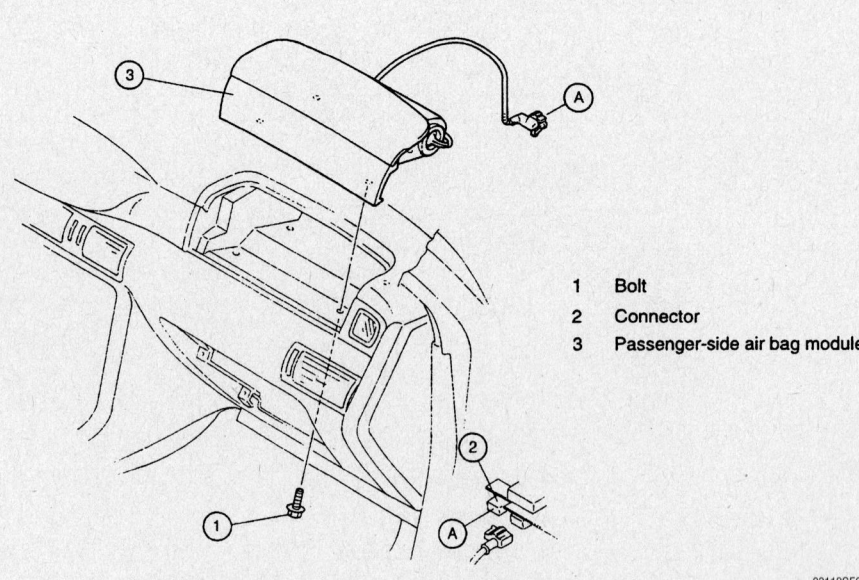

1 Bolt
2 Connector
3 Passenger-side air bag module

93112GE6

Exploded view of the passenger's side SAS module—Millenia

To install:

12. Install the heater core and assemble the heater unit cases.

13. Connect and install the heater unit assembly.

14. Install the evaporator housing by installing or connecting the following:
- Evaporator housing assembly
- Evaporator temperature connector
- MAX-HI connector
- Power transistor connector
- Aspirator hose

- Air conditioning system refrigerant lines using new gaskets

15. Install or connect the following:
- Instrument panel with the help of an assistant
- Both side panels
- Hood release lever
- Steering column and the steering column-to-chassis bolts
- Instrument cluster and connect the electrical connectors
- Meter hood

- Combination switch to the steering column and connect the electrical connectors
- Upper and lower steering column covers
- Undercover at the passenger's side
- A-pillar trim at both sides

16. Install the rear console by installing or connecting the following:
- Rear console
- Center panel
- Brake boot

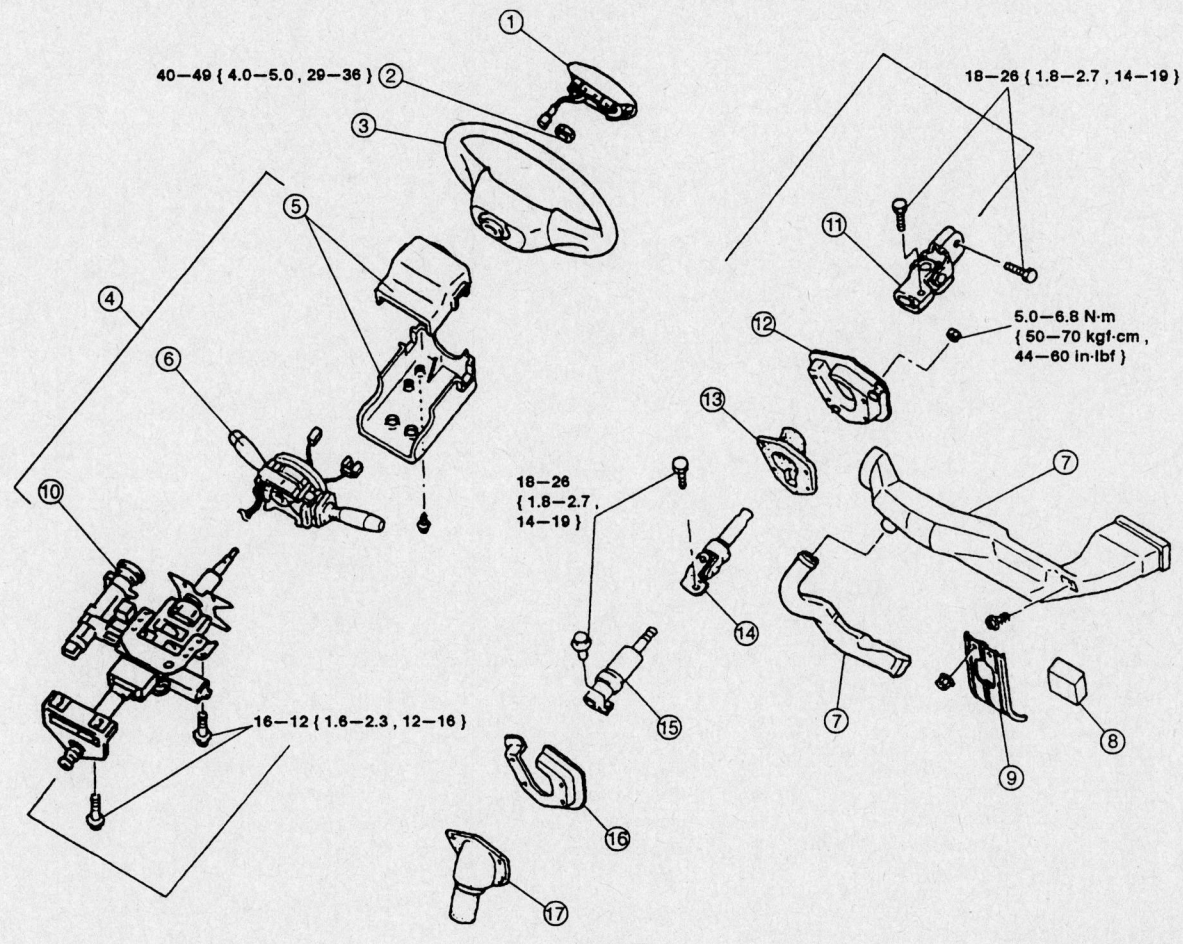

1	Air bag module	10	Steering shaft component
2	Locknut	11	Universal joint (intermediate shaft)
3	Steering wheel	12	Cover
4	Dashboard, console, and steering shaft component	13	Shaft seal
5	Column cover	14	Intermediate shaft
6	Combination switch	15	Collapsible shaft
7	Air duct	16	Set plate
8	Flasher unit	17	Dust cover
9	Bracket		

N·m { kgf·m , ft·lbf }

93112GE9

Exploded view of the steering wheel and steering column assembly—Millenia

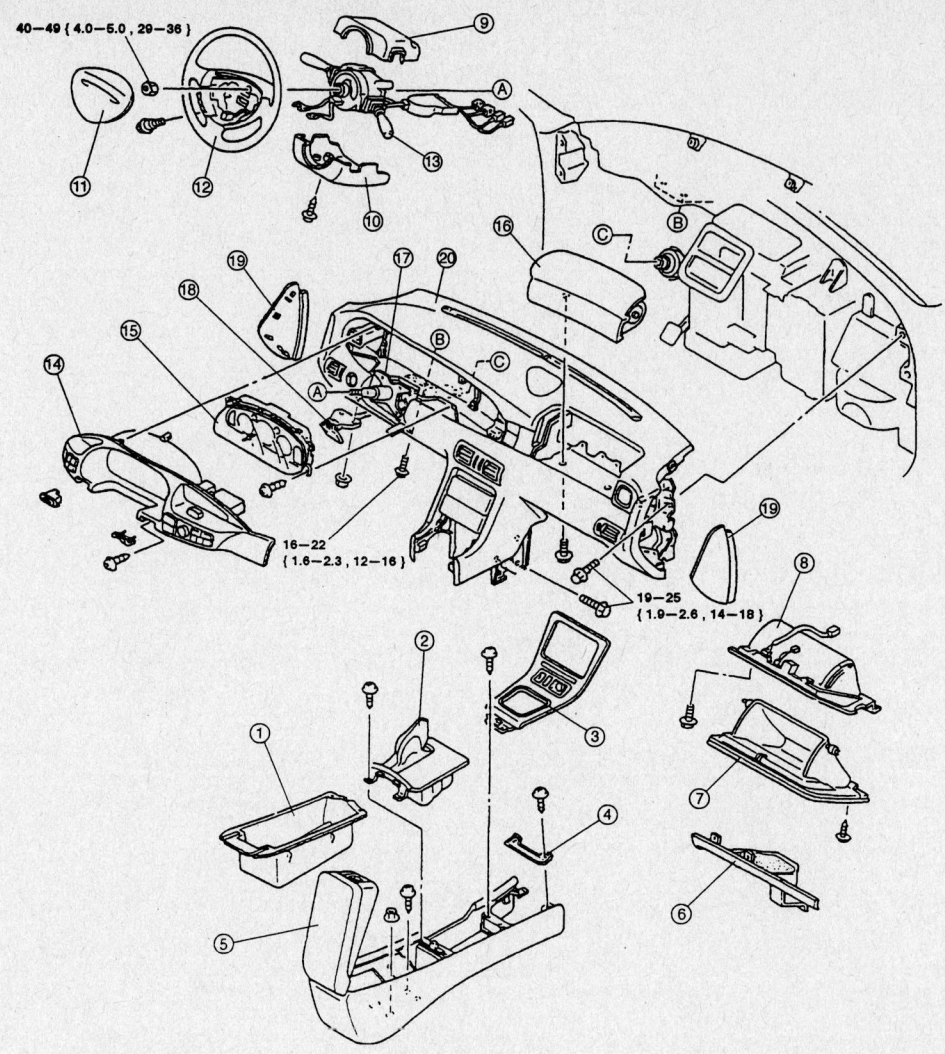

1	Rear console box	8	Glove compartment cover	15	Instrument cluster
2	Brake boot	9	Upper column cover	16	Passenger-side air bag module
3	Center panel	10	Lower column cover	17	Steering shaft
4	Bracket	11	Driver-side air bag module	18	Hood release lever
5	Rear console	12	Steering wheel	19	Side panel
6	Under cover	13	Combination switch	20	Dashboard
7	Glove compartment	14	Meter hood		

93112GE8

Exploded view of the instrument panel and rear console assemblies—Millenia

- Rear console's box
17. At the passenger's side, install the SAS module by installing or connecting the following:
    - SAS module and connect the electrical connector
    - SAS module-to-dash bolts
    - Glove compartment cover and the glove compartment
18. At the driver's side, install the SAS module and the steering wheel by installing or connecting the following:
    - Steering wheel to the steering column
    - Steering wheel-to-column nut.

Torque the nut to 29–36 ft. lbs. (40–49 Nm).
- SAS module to the steering wheel and connect the electrical connector. Torque the bolts to 79–113 inch lbs. (9–13 Nm).
- Steering wheel-to-SAS module clips
- Cover clips at both sides of the steering wheel
19. Refill the cooling system.
20. Connect the negative battery cable.
21. Evacuate, charge and leak test the air conditioning system refrigerant.
22. Operate the engine to normal operat-

ing temperatures; then, check the climate control operation and check for leaks.

Protégé

1. Before servicing the vehicle, refer to the precautions section.
2. Disconnect the negative battery cable.

❊❊ CAUTION

After disconnecting the battery, wait for more than 1 minute for the SAS to deplete its stored energy.

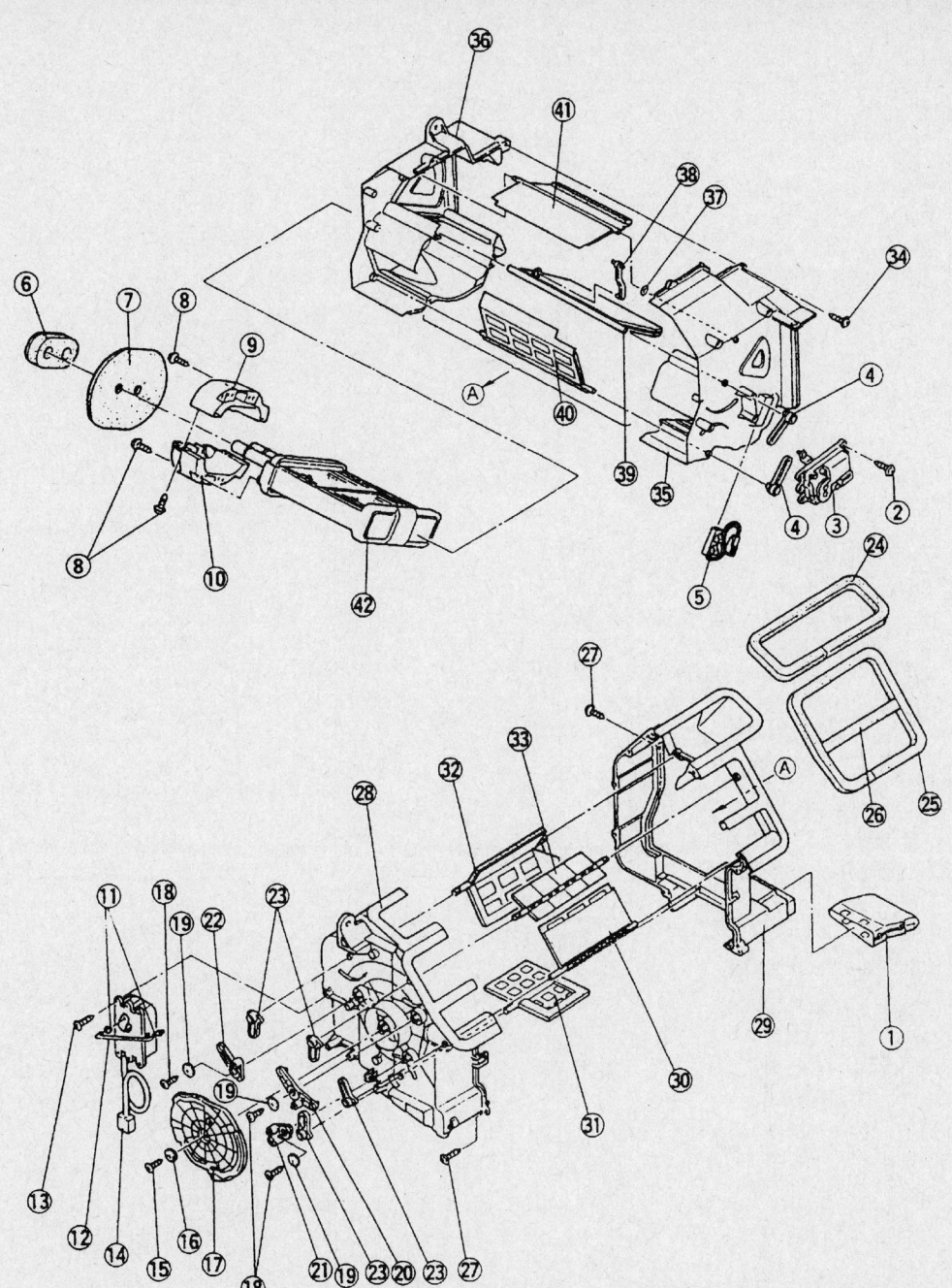

1	Heat duct	16	Link collar	30	Vent door
2	Tapping screw	17	Airflow mode main link	31	Heat door
3	Air mix actuator	18	Tapping screw	32	Defroster door
4	Air mix crank	19	Link collar	33	Side vent door
5	Water temperature sensor	20	Airflow mode sub link (VENT)	34	Tapping screw
6	Polyurethane foam (thick)	21	Airflow mode sub link (HEAT)	35	Heater case (1)
7	Polyurethane foam (thin)	22	Airflow mode sub link (DEFROSTER)	36	Heater case (2)
8	Tapping screw	23	Airflow mode crank	37	Collar
9	Heater core bracket (1)	24	Polyurethane protector (DEFROSTER)	38	Air mix rod
10	Heater core bracket (2)	25	Polyurethane protector (VENT)	39	Air mix main door
11	Rod stopper	26	Polyurethane protector (SIDE VENT)	40	Air mix sub door
12	Airflow mode rod	27	Tapping screw	41	Air mix guide door
13	Tapping screw	28	Heater case (4)	42	Heater core
14	Airflow mode actuator	29	Heater case (3)		
15	Tapping screw				

93112GE7

Exploded view of the heater core and heater case assembly—Millenia

3. Drain the cooling system into a clean container for reuse.

4. Disconnect the heater hoses from the heater core.

5. Discharge and recover the air conditioning system refrigerant.

6. Place the wheel in the straight-ahead position and turn the ignition switch to LOCK.

7. At the driver's side, remove the SAS module and the steering wheel by removing or disconnecting the following:
- Cover clips at both sides of the steering wheel
- Steering wheel-to-SAS module bolts
- SAS module from the steering wheel and disconnect the electrical connector
- Steering wheel-to-column nut
- Steering wheel from the steering column using a suitable puller

8. At the passenger's side, remove the SAS module by removing or disconnecting the following:
- Glove compartment and the glove compartment cover
- SAS module-to-dash bolts
- SAS module and disconnect the electrical connector

9. Remove the console by removing or disconnecting the following:
- Shift lever knob, if equipped with a manual transmission
- Console's cover
- Console-to-chassis screws and console

10. Remove the combination switch by removing or disconnecting the following:
- Steering column cover
- Electrical connectors and remove the combination switch-to-steering column bolts and the combination switch

11. Remove the instrument cluster by:
- Meter hood
- Instrument cluster-to-dash panel screws
- Electrical connectors and remove the instrument cluster

12. Remove or disconnect the following:
- Lower panel
- Hood release cable installation nut
- Side wall trim
- "A" pillar trim at both sides
- Side panel
- Antenna connector
- Blower motor and heater unit electrical connectors, if equipped with the wire-type climate control unit
- Electrical connectors and the bolts
- Dashboard-to-chassis bolts

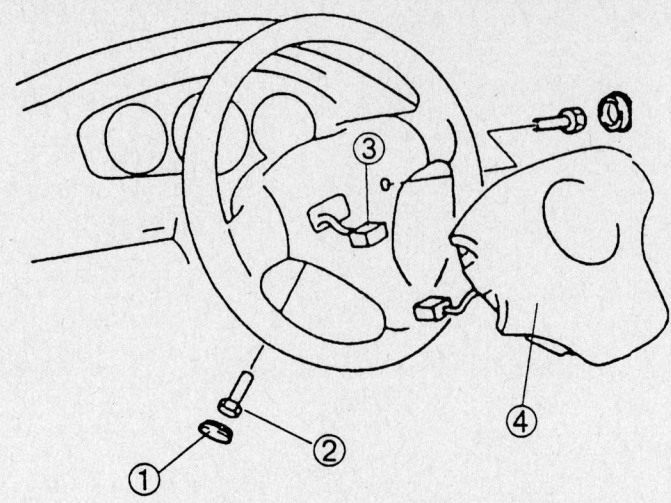

7.9—11.7 N·m {80—120 kgf·cm, 70—104 in·lbf}

1	Cap
2	Bolt
3	Connector
4	Driver-side air bag module

93112GG4

Exploded view of the SAS module and the steering wheel assembly—Protégé

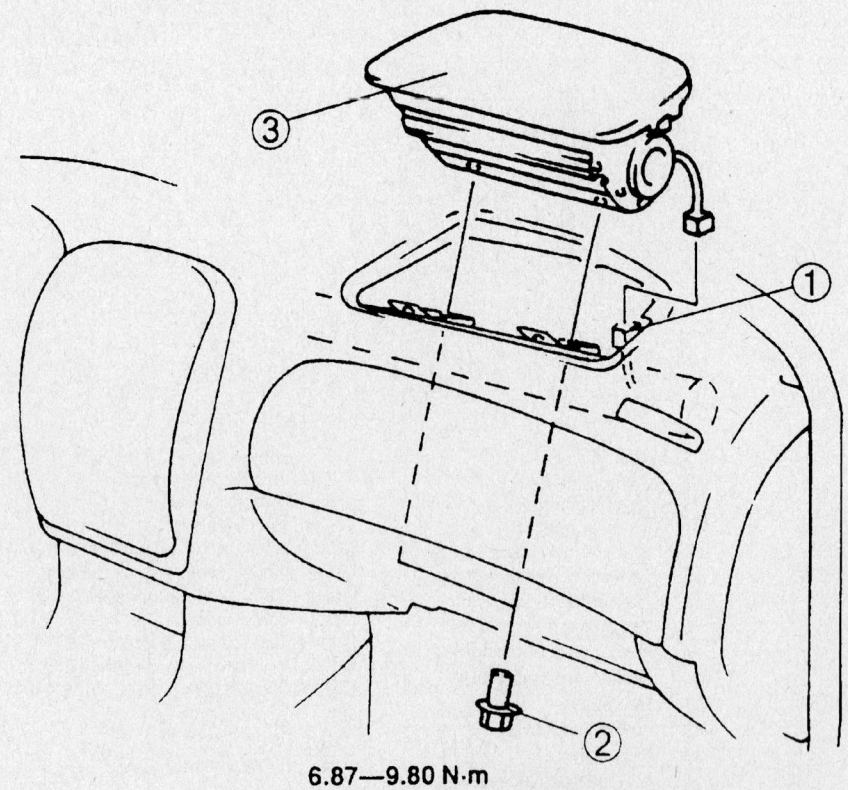

6.87—9.80 N·m
{70—100 kgf·cm, 60.8—86.7 in·lbf}

93112GG5

Exploded view of the passenger's side SAS module—Protégé

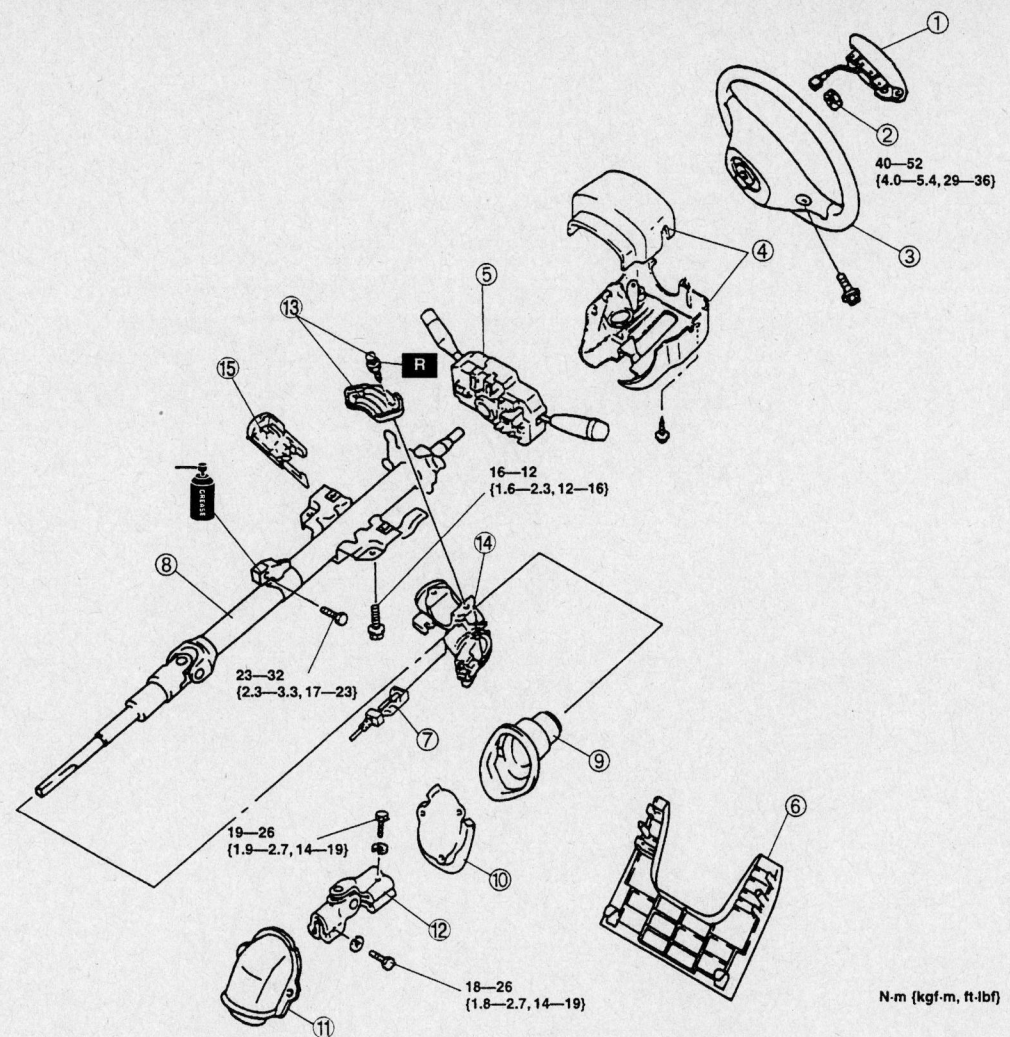

40—52
{4.0—5.4, 29—36}

16—12
{1.6—2.3, 12—16}

23—32
{2.3—3.3, 17—23}

19—26
{1.9—2.7, 14—19}

18—26
{1.8—2.7, 14—19}

N·m {kgf·m, ft·lbf}

1	Air bag module	9	Shaft seal
2	Locknut	10	Set plate
3	Steering wheel	11	Dust cover
4	Column cover	12	Universal joint
5	Combination switch	13	Steering lock mounting bolts and bracket
6	Lower panel	14	Steering lock component
7	Key interlock cable	15	Cylinder outer component
8	Steering shaft		

93112GG6

Exploded view of the steering column assembly—Protégé

- Dashboard from the vehicle with the help of an assistant
- Passenger's side lower panel
- Air intake wire from the climate control unit
- Air conditioning refrigerant lines from the evaporator and discard the O-rings
- Evaporator electrical connector(s)
- Evaporator housing

13. Disassemble the heater housing and remove the heater core.

To install:

14. Install the heater core and assemble the heater housing.

15. Install or connect the following:
  - Evaporator housing
  - Evaporator electrical connector(s)
  - Air conditioning refrigerant lines to the evaporator using new O-rings
  - Air intake wire to the climate control unit
  - Passenger's side lower panel
  - Dashboard to the vehicle with the help of an assistant
  - Dashboard-to-chassis bolts
  - Electrical connectors and the bolts
  - Blower motor and heater unit electrical connectors, if equipped with the wire-type climate control unit

- Antenna connector
- Side panel
- A-pillar trim at both sides
- Side wall trim
- Hood release cable installation nut
- Lower panel

16. Install the instrument cluster by installing or connecting the following:
  - Instrument cluster and connect the electrical connectors
  - Instrument cluster-to-dash panel screws
  - Meter hood

17. Install the combination switch by installing or connecting the following:

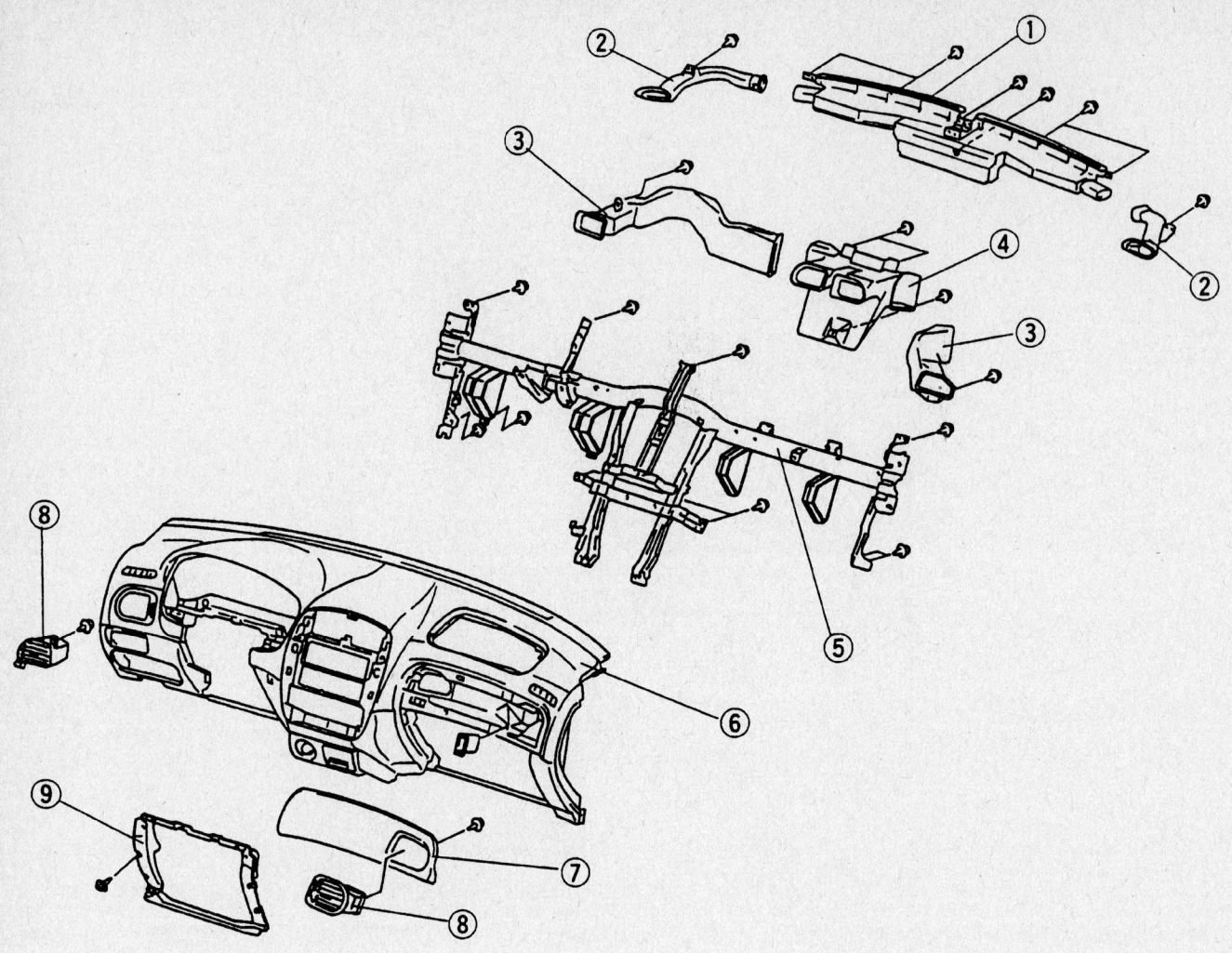

1 Defroster nozzle	6 Crush pad
2 Side demister duct	7 Pad
3 Duct	8 Ventilator grille
4 Center duct	9 Passenger-side lower panel
5 Dashboard member	

93112GG7

Exploded view of the dashboard assembly—Protégé

- Electrical connectors and install the combination switch-to-steering column bolts and the combination switch
- Steering column cover
18. Install the console by installing or connecting the following:
  - Console and the console-to-chassis screws
  - Console's cover
  - Shift lever knob, if equipped with a manual transmission
19. At the passenger's side, install the SAS module by installing or connecting the following:

- SAS module and connect the electrical connector
- SAS module-to-dash bolts
- Glove compartment and the glove compartment cover
20. At the driver's side, install the SAS module and the steering wheel by installing or connecting the following:
  - Steering wheel-to-column nut
  - SAS module from the steering wheel and connect the electrical connector
  - Steering wheel-to-SAS module bolts
  - Cover clips at both sides of the steering wheel

21. Connect the heater hoses to the heater core.
22. Refill the cooling system.
23. Connect the negative battery cable.
24. Evacuate, charge and leak test the air conditioning system.
25. Operate the engine to normal operating temperatures; then, check the climate control operation and check for leaks.

626

1. Before servicing the vehicle, refer to the precautions section.
2. Disconnect the negative battery cable.

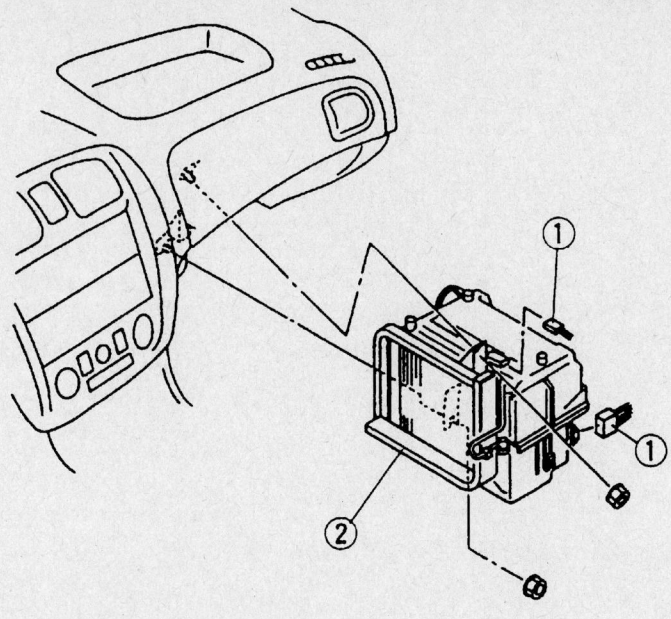

1   Connector
2   Cooling unit

93112GG8

View of the evaporator housing assembly—Protégé

After disconnecting the battery, wait for more than 1 minute for the SAS to deplete its stored energy.

3. Drain the cooling system into a clean container for reuse.

4. Discharge and recover the air conditioning system refrigerant.

5. Place the wheel in the straight-ahead position and turn the ignition switch to LOCK.

6. At the driver's side, remove the SAS module and the steering wheel removing or disconnecting the following:
- Cover clips at both sides of the steering wheel
- Steering wheel-to-SAS module bolts
- SAS module from the steering wheel and disconnect the electrical connector
- Steering wheel-to-column nut
- Steering wheel from the steering column using a suitable puller

7. At the passenger's side, remove the

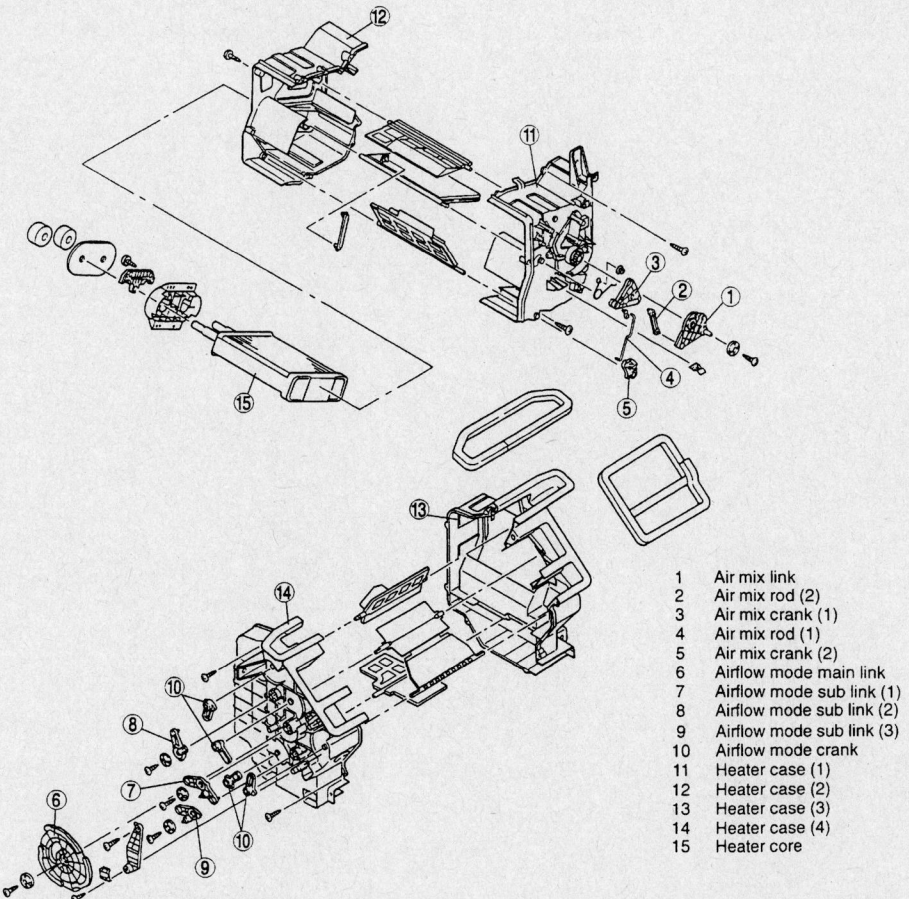

1	Air mix link
2	Air mix rod (2)
3	Air mix crank (1)
4	Air mix rod (1)
5	Air mix crank (2)
6	Airflow mode main link
7	Airflow mode sub link (1)
8	Airflow mode sub link (2)
9	Airflow mode sub link (3)
10	Airflow mode crank
11	Heater case (1)
12	Heater case (2)
13	Heater case (3)
14	Heater case (4)
15	Heater core

93112GG9

Exploded view of the heater housing assembly—Protégé

SAS module by removing or disconnecting the following:
- Glove compartment and the glove compartment cover
- SAS module-to-dash bolts
- SAS module and disconnect the electrical connector

8. Remove the instrument cluster by removing or disconnecting the following:
- Instrument cluster meter hood
- Instrument cluster-to-dash screws, the instrument cluster and disconnect the electrical connectors

9. Remove the climate control assembly by removing or disconnecting the following:
- Climate control meter hood
- Climate control assembly screws
- Climate control assembly and disconnect the electrical connector and the assembly

10. Remove the audio unit by removing or disconnecting the following:
- Hole covers by inserting a small tape-wrapped flathead screwdriver into the slot and carefully pry off the hole covers
- Using 2 removal tools 49 UN01 050 or equivalent, insert them into sides of the audio unit.
- Slide audio unit outward and forward
- Electrical connectors and the antenna jack

11. Remove the instrument panel by removing or disconnecting the following:
- Upper center panel cover
- Dash panel-to-chassis bolts
- Dash panel with the help of an assistant

12. Remove the evaporator housing by removing or disconnecting the following:
- Center lower panel
- Refrigerant lines from the air conditioning evaporator and discard the gaskets
- Blower motor assembly, if necessary
- Evaporator assembly fasteners and remove the evaporator assembly

13. Remove or disconnect the following:
- Rear heater duct
- Heater housing fasteners and the heater housing
- Airflow mode actuator

14. Separate the heater housing and remove the heater core.

To install:

15. Install the heater core and assemble the heater housing.

16. Install or connect the following:
- Airflow mode actuator

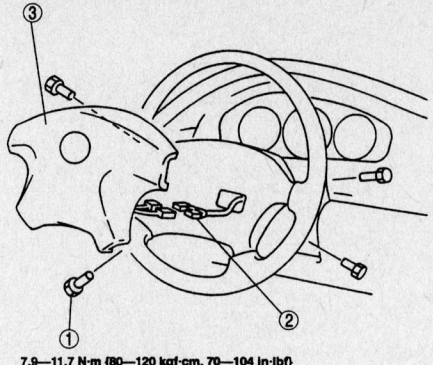

7.9—11.7 N·m {80—120 kgf·cm, 70—104 in·lbf}

1	Bolt
2	Connector
3	Driver-side air bag module

93112GH6

Exploded view of the steering wheel and SAS module—626

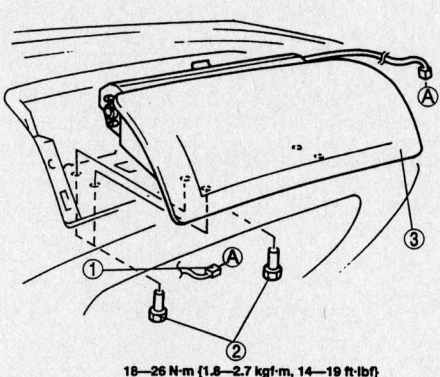

18—26 N·m {1.8—2.7 kgf·m, 14—19 ft·lbf}

1	Connector
2	Bolt
3	Passenger-side air bag module

93112GH7

Exploded view of the passenger's side SAS module—626

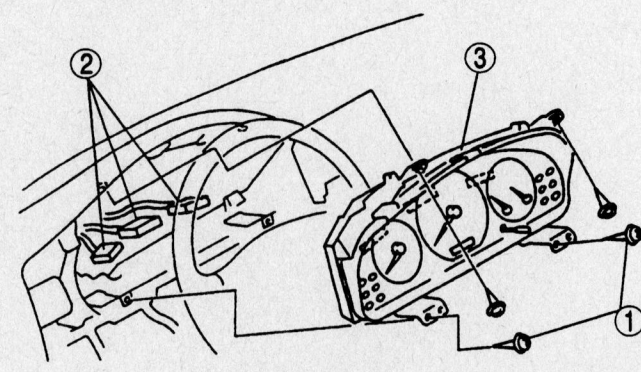

1	Screw
2	Connector
3	Instrument cluster

93112GH8

View of the instrument cluster assembly—626

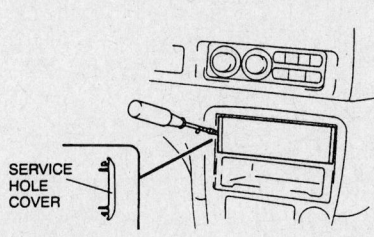

Removing the audio unit—626

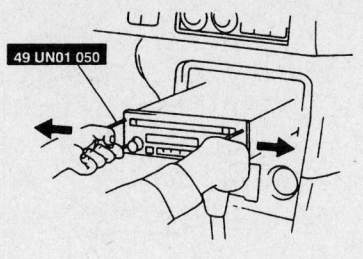

49 UN01 050

93112GH9

- Heater housing and the heater housing fasteners
- Rear heater duct

17. Install the evaporator housing by installing or connecting the following:
- Center lower panel
- Refrigerant lines to the air conditioning evaporator using new gaskets
- Blower motor assembly, if necessary
- Evaporator assembly and the evaporator assembly fasteners

18. Install the instrument panel by installing or connecting the following:
- Upper center panel cover
- Dash panel-to-chassis bolts
- Dash panel with the help of an assistant

19. Install the audio unit by installing or connecting the following:
- Hole covers
- Audio unit
- Electrical connectors and the antenna jack

20. Install the climate control assembly by installing or connecting the following:
- Climate control meter hood
- Climate control assembly screws
- Climate control assembly and connect the electrical connector

21. Install the instrument cluster by installing or connecting the following:
- Instrument cluster meter hood
- Instrument cluster-to-dash screws, the instrument cluster and connect the electrical connectors

22. At the passenger's side, install the SAS module by installing or connecting the following:
- Glove compartment and the glove compartment cover
- SAS module-to-dash bolts
- SAS module and connect the electrical connector

23. At the driver's side, install the SAS module and the steering wheel by installing or connecting the following:
- Steering wheel and the steering

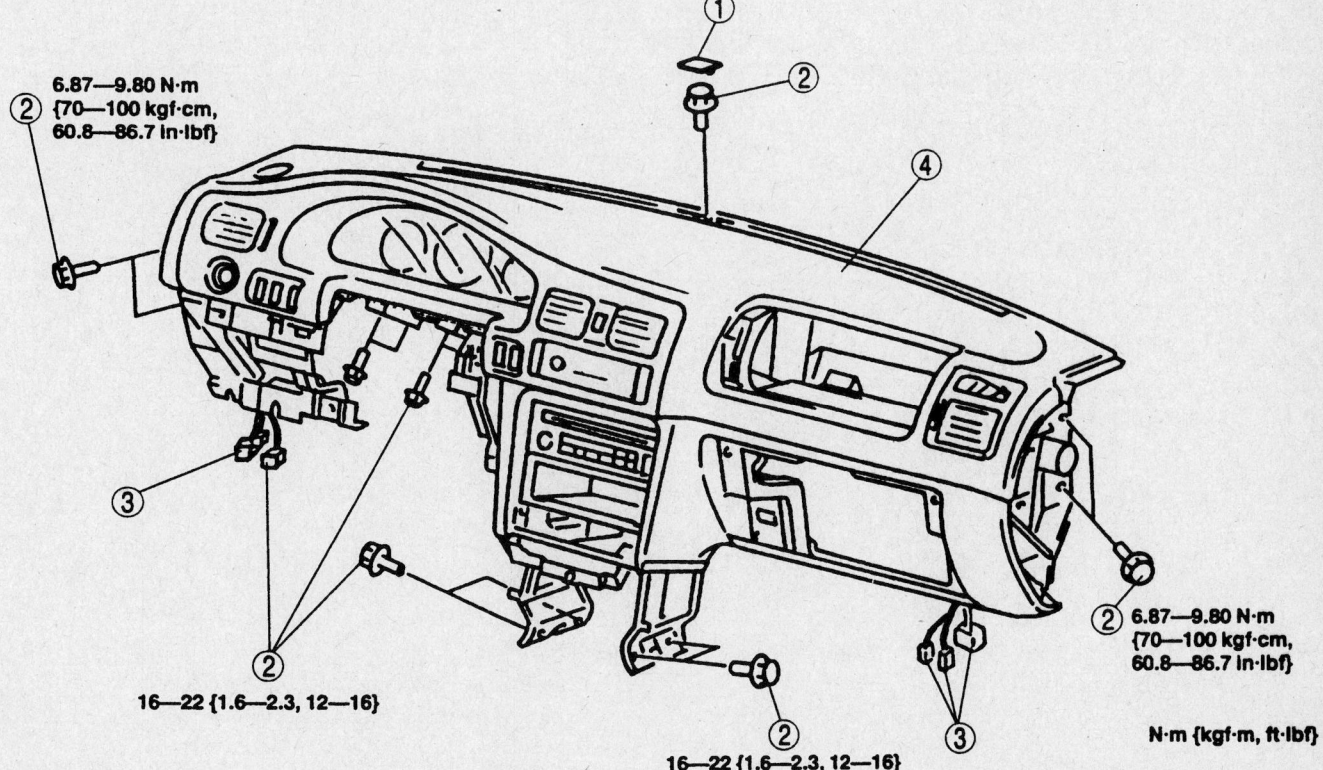

6.87—9.80 N·m {70—100 kgf·cm, 60.8—86.7 in·lbf}

16—22 {1.6—2.3, 12—16}

16—22 {1.6—2.3, 12—16}

6.87—9.80 N·m {70—100 kgf·cm, 60.8—86.7 in·lbf}

N·m {kgf·m, ft·lbf}

| 1 | Cover | 3 | Connector |
| 2 | Bolt | 4 | Dashboard |

View of the instrument panel assembly—626

93112GH0

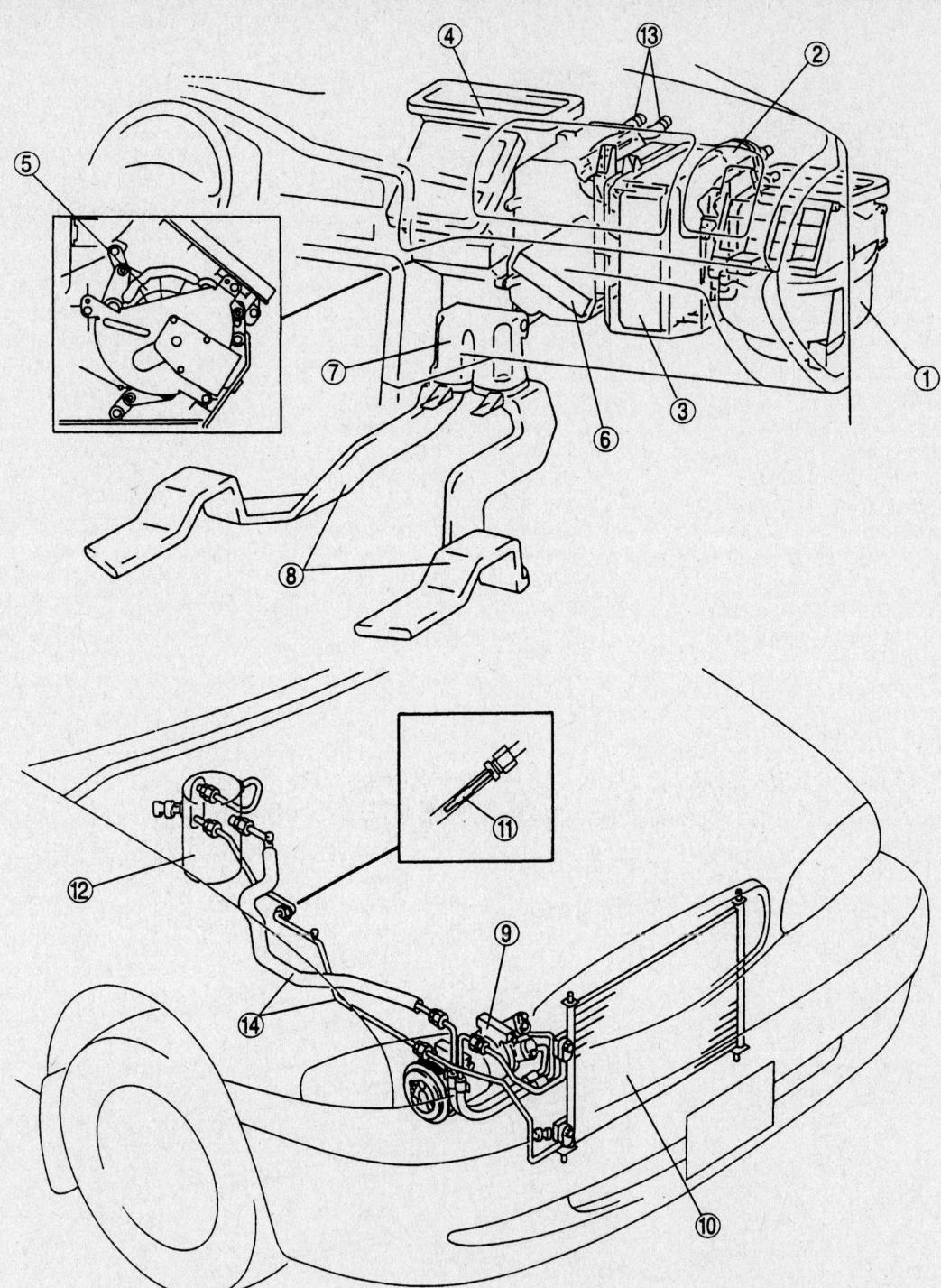

1	Blower unit	8	Rear heat duct (CANADA only)
2	Cooling unit	9	A/C compressor
3	Evaporator	10	Condenser
4	Heater unit	11	Orifice tube
5	Airflow mode main link	12	Accumulator tank
6	Heater core	13	Heater hose
7	Rear duct (CANADA only)	14	Refrigerant lines

93112GI1

View of the heater and air conditioning housing assemblies—626

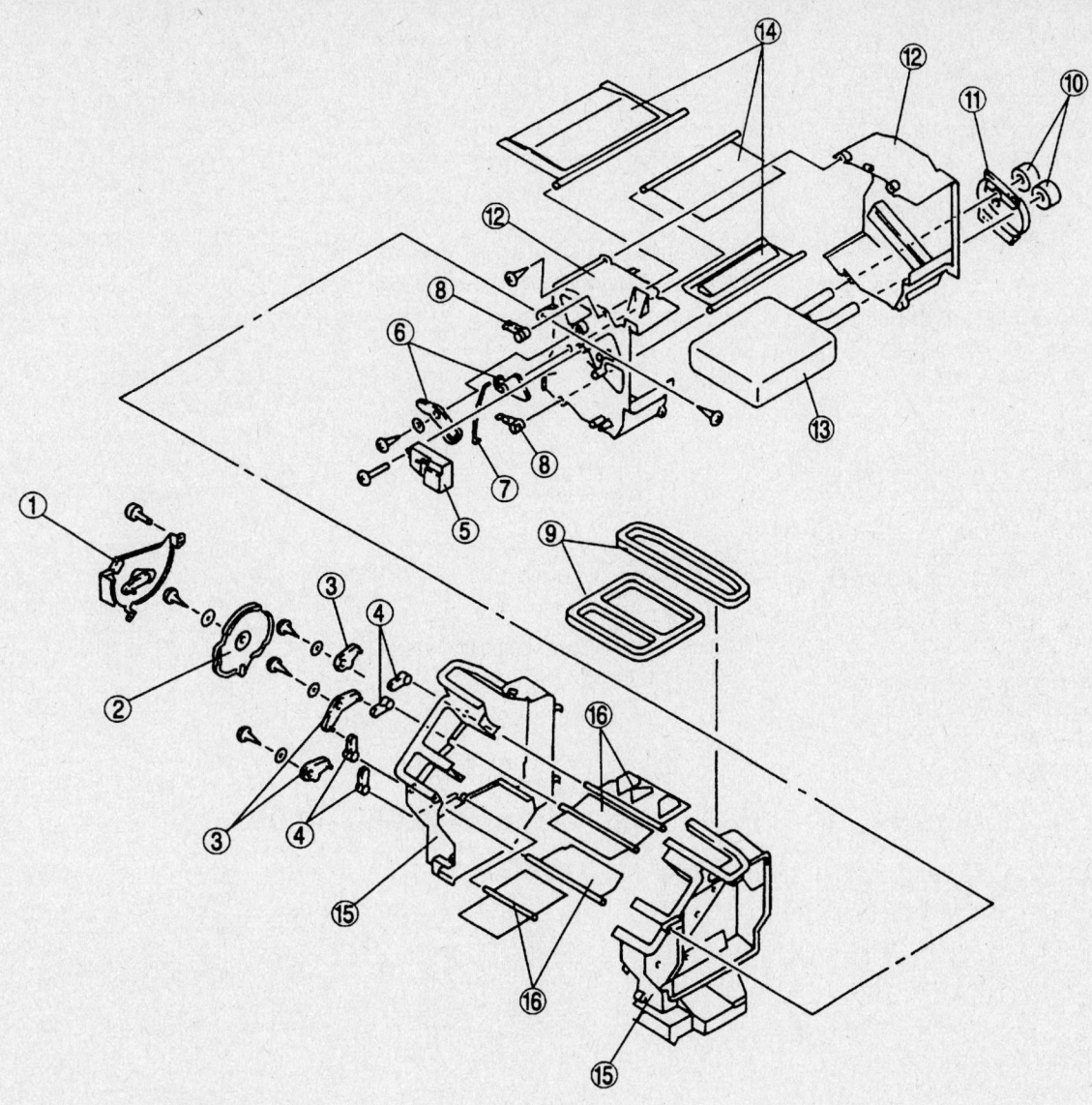

1	Airflow mode actuator	9	Polyurethane protector
2	Airflow mode main link	10	Seal
3	Airflow mode sub link	11	Cover
4	Airflow mode crank	12	Case
5	Air mix actuator	13	Heater core
6	Air mix link	14	Air mix door
7	Air mix rod	15	Case
8	Air mix crank	16	Airflow mode door

93112Gl2

Exploded view of the heater core and heater housing assembly—626

wheel-to-column nut. Torque the steering wheel nut to 29–36 ft. lbs. (40–49 Nm).
- SAS module to the steering wheel and connect the electrical connector. Torque the steering column-to-SAS module bolts to 70–104 inch lbs. (8–12 Nm).
- Steering wheel-to-SAS module clips

- Cover clips, at both sides of the steering wheel
24. Refill the cooling system.
25. Connect the negative battery cable.
26. Evacuate, charge and leak test the air conditioning system refrigerant.
27. Operate the engine to normal operating temperatures; then, check the climate control system and check for leaks.

Rocker Arm (Valve) Cover

REMOVAL & INSTALLATION

626 and Protégé

2.0L (FS) ENGINE

1. Before servicing the vehicle, refer to the precautions section.

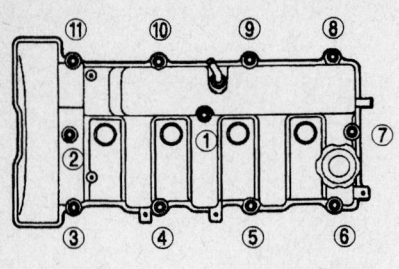

42356-MAZC-G06A

Remove the rocker arm cover bolts in the sequence shown–2.0L (FS) engine

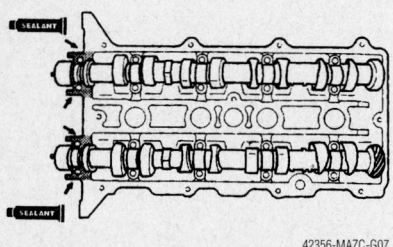

42356-MAZC-G07

Apply silicone sealant to cylinder head at the areas illustrated–2.0L (FS) engine

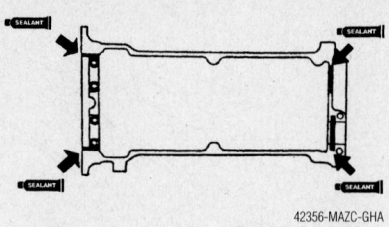

42356-MAZC-GHA

Apply silicone sealant to cylinder head at the areas illustrated–1.6L (ZM) engine

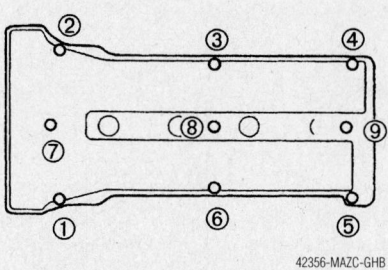

42356-MAZC-GHB

Tighten the rocker arm cover bolts in the sequence shown–1.6L (ZM) engine

1.6L (ZM) Engines

1. Before servicing the vehicle, refer to the precautions section.
2. Remove or disconnect the following:
   - Negative battery cable
   - Spark plug wires
   - Vent hose
   - Positive Crankcase Ventilation (PCV) hose
   - Rocker arm cover and discard the gasket
3. Clean all mating surfaces of any residual gasket material.

To install:

4. Install or connect the following:
5. Apply a 0.12–0.15 inch (3–4mm) bead of silicone sealant to cylinder head at the areas illustrated.
6. Install or connect the following:
   - Rocker arm cover with a new gasket. Torque the bolts in the sequence illustrated to 95 inch lbs. (11 Nm).
   - Remaining components removed to

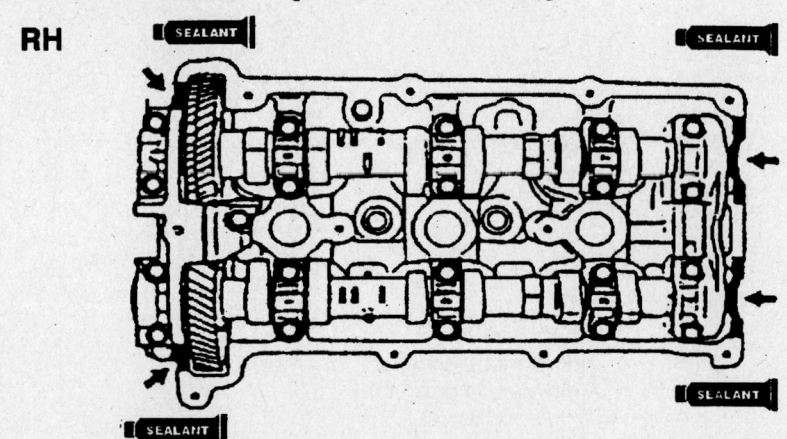

42356-MAZC-G08

Tighten the rocker arm cover bolts in the sequence shown–2.0L (FS) engine

2. Remove or disconnect the following:
   - Negative battery cable
   - Any components that would interfere with cover removal
   - Rocker arm cover bolts in the sequence illustrated
   - Rocker arm cover and discard the gasket
3. Clean all mating surfaces of any residual gasket material.

To install:

4. Apply silicone sealant to cylinder head at the areas illustrated.
5. Install or connect the following:
   - Rocker arm cover with a new gasket. Torque the bolts in the sequence illustrated to 95 inch lbs. (11 Nm).
   - Remaining components removed to facilitate the rocker arm cover removal
   - Negative battery cable

Thickness
1.5—2.5 mm {0.060—0.098 in}

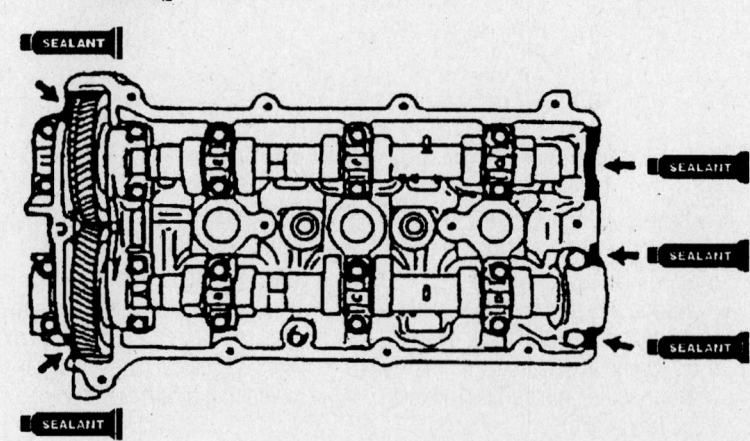

42356-MAZC-G09

Apply silicone sealant to cylinder head at the areas illustrated–2.5L (KL) engine

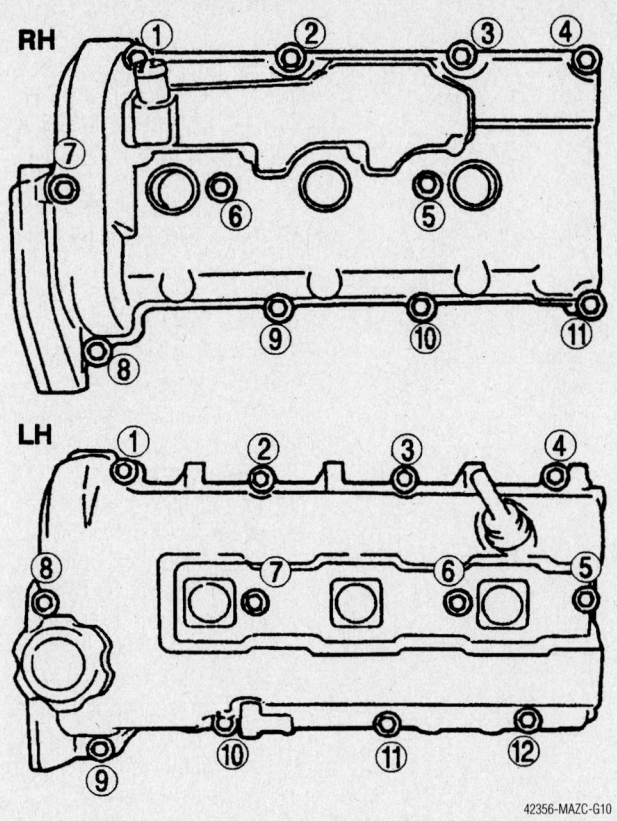

Tighten the rocker arm cover bolts in the sequence shown—2.5L (KL) engine

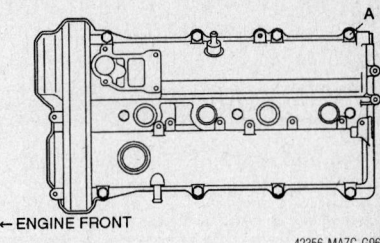

← ENGINE FRONT

42356-MAZC-G96

Temporarily tighten the cover bolts A–1.8L (BP) engine

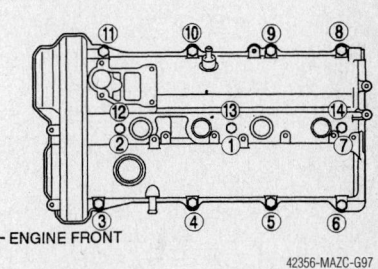

← ENGINE FRONT

42356-MAZC-G97

Tighten the rocker arm cover bolts in the sequence shown–1.8L (BP) engine

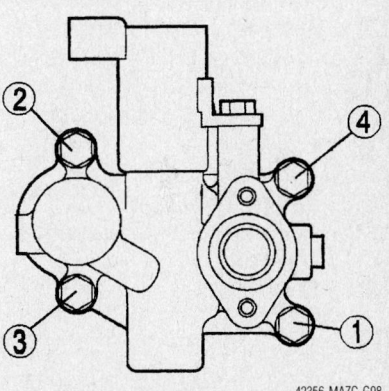

42356-MAZC-G98

Tighten the OCV bolts in the sequence shown–1.8L (BP) engine

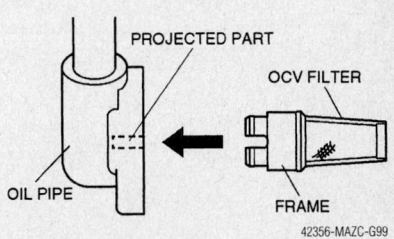

PROJECTED PART

OCV FILTER

OIL PIPE

FRAME

42356-MAZC-G99

Hold the frame of the OCV valve filter and install the OCV so it is aligned with the projected part on the flange end of the oil pipe–1.8L (BP) engine

facilitate the rocker arm cover removal
- Negative battery cable

2.5L (KL) ENGINE

1. Before servicing the vehicle, refer to the precautions section.
2. Remove or disconnect the following:
- Negative battery cable
- Any components that would interfere with cover removal
- Rocker arm cover bolts
- Rocker arm cover and discard the gasket
3. Clean all mating surfaces of any residual gasket material.

To install:

4. Apply silicone sealant to cylinder head at the areas illustrated.
5. Install or connect the following:
- Rocker arm cover with a new gasket. Torque the bolts in the sequence illustrated to 95 inch lbs. (11 Nm) using 2 or 3 steps. Retighten the right hand cover number 5 and 6 bolts and the left hand cover number 6 and 7 bolts.
- Remaining components removed to facilitate the rocker arm cover removal
- Negative battery cable

2002–04 Miata

1.8L (BP) ENGINES

1. Before servicing the vehicle, refer to the precautions section.
2. Remove or disconnect the following:
- Negative battery cable
- Upper radiator hose
- Water hose
- Oil pipe
- Oil Control Valve (OCV)
- Rocker arm cover and discard the gasket
3. Clean all mating surfaces of any residual gasket material.

To install:

4. Install or connect the following:
- Rocker arm cover with a new gasket.

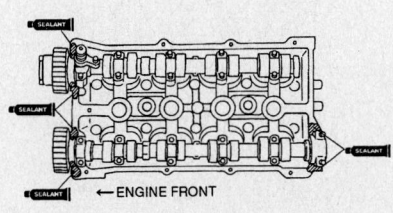

SEALANT

← ENGINE FRONT

42356-MAZC-G95

Apply silicone sealant to cylinder head at the areas illustrated–1.8L (BP) engine

- Temporarily tighten the cover bolts **A**, refer to the illustration for location. Torque the bolts in sequence to 80 inch lbs. (9 Nm).

☀☀ CAUTION

When installing the OCV valve, be careful not to damage the O-ring, if damaged it may cause leaking.

- OCV and tighten the bolts in the sequence illustrated

5. Install the oil pipe as follows:

a. Oil pipe. Hold the frame of the OCV valve filter and install the OCV so it is aligned with the projected part on the flange end of the oil pipe. Coat the new washer with clean engine oil and temporarily install the upper and side oil pipe and position the pipe.

b. **A** using several passes to 95 inch lbs. (11 Nm).

c. Tighten oil pipe bolts **B** using several passes to 61 inch lbs. (7 Nm).

d. Tighten oil pipe bolts **C** using several passes to 13 ft. lbs. (17 Nm).

e. Tighten oil pipe bolts 1, 2, 3, 4 and 7 using several passes to 95 inch lbs. (11 Nm).

f. Tighten oil pipe bolt 5 using several passes to 17 ft. lbs. (23 Nm).

g. Tighten oil pipe bolt 6 using several passes to 34 ft. lbs. (47 Nm).

6. Install or connect the following:
- Water hose
- Upper radiator hose
- Spark plug wires
- Power steering hose bracket, if removed
- Negative battery cable

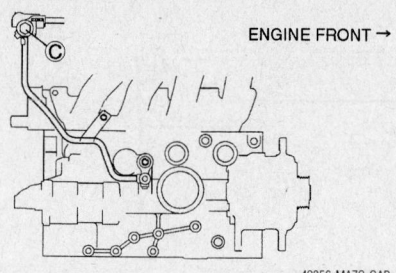

Tighten oil pipe bolts C using several passes to the specification in the text–1.8L (BP) engine

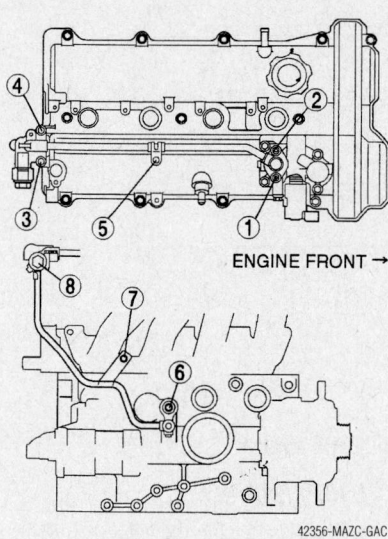

Tighten oil pipe bolts using several passes to the specification in the text–1.8L (BP) engine

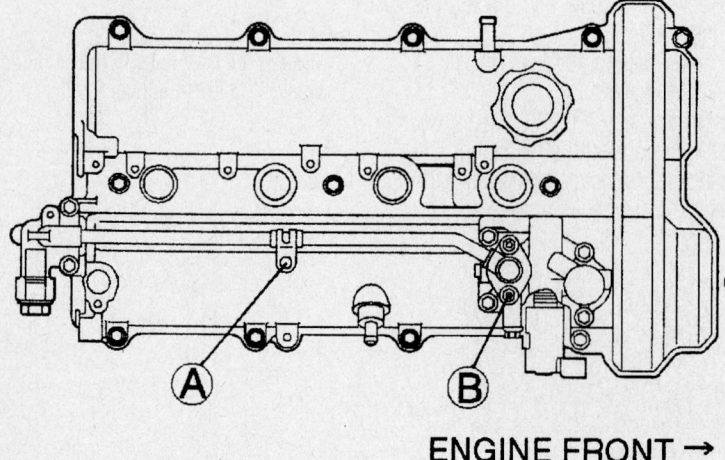

Tighten oil pipe bolts A and B using several passes to the specification in the text–1.8L (BP) engine

Millenia

2.3L (KJ) Engine

1. Before servicing the vehicle, refer to the precautions section.

2. Remove or disconnect the following:
- Negative battery cable
- Ignition coils
- Vent hose
- Positive Crankcase Ventilation (PCV) hose
- Rocker arm cover and discard the gasket

3. Clean all mating surfaces of any residual gasket material.

To install:

4. Install or connect the following:

5. Apply a 0.004–0.07 inch (1–2mm) bead of sealant to rocker arm cover.
- New rocker arm gasket

6. Apply a 0.059–0.098 inch (1.5–2.5mm) to the areas illustrated
- Rocker arm cover. Torque the bolts in sequence to 86 inch lbs. (10 Nm), then retighten in sequence to 86 inch lbs. (10 Nm).
- PCV hose
- Vent hose
- Ignition coils
- Negative battery cable

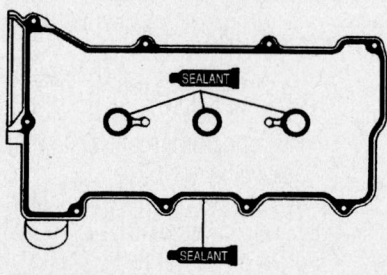

Apply a 0.004–0.07 inch (1–2mm) bead of sealant to rocker arm cover—Millenia models equipped with the 2.3L (KJ) engine

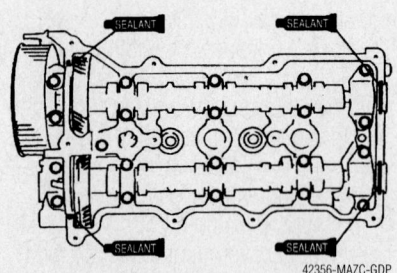

Apply a 0.059–0.098 inch (1.5–2.5mm) to the areas shown—Millenia models equipped with the 2.3L (KJ) engine

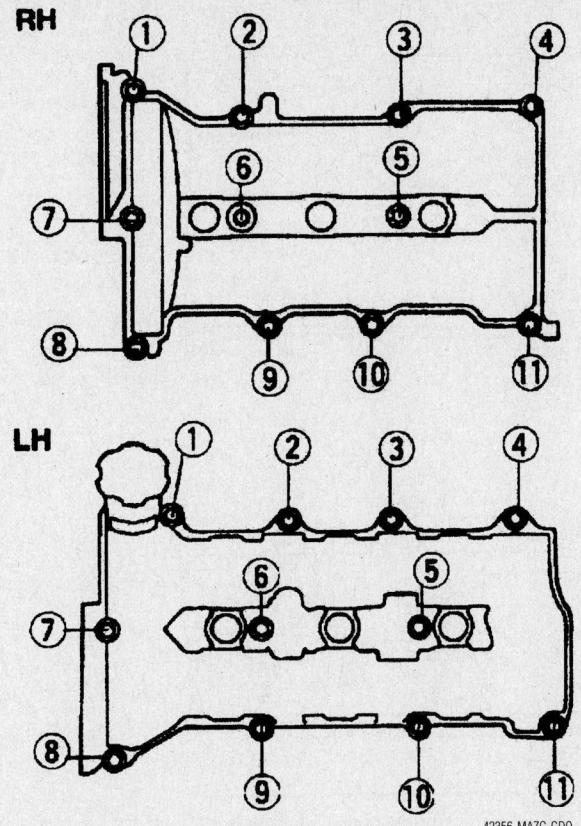

Rocker arm cover bolt sequence—Millenia models equipped with the 2.3L (KJ) engine

Thickness
1.5—2.5 mm {0.060—0.098 in}

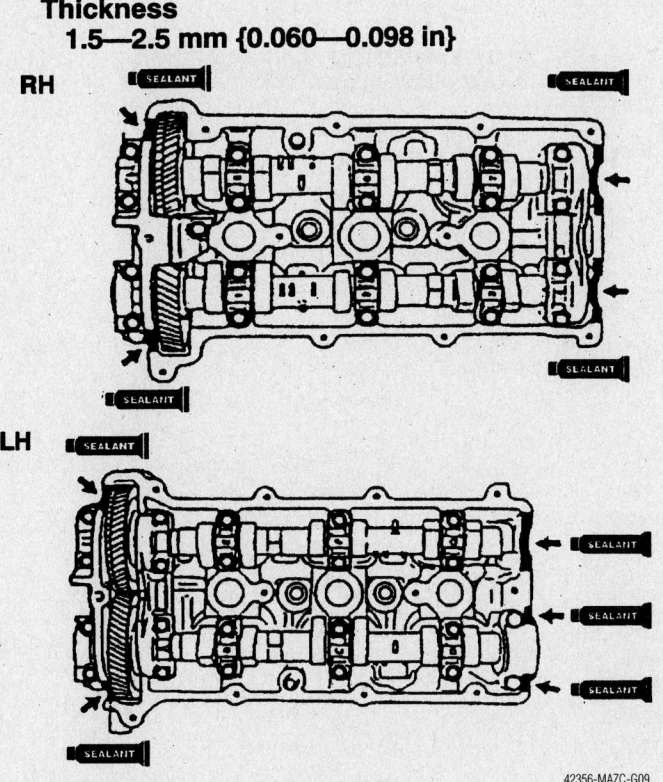

Apply silicone sealant to cylinder head at the areas illustrated–2.5L (KL) engine

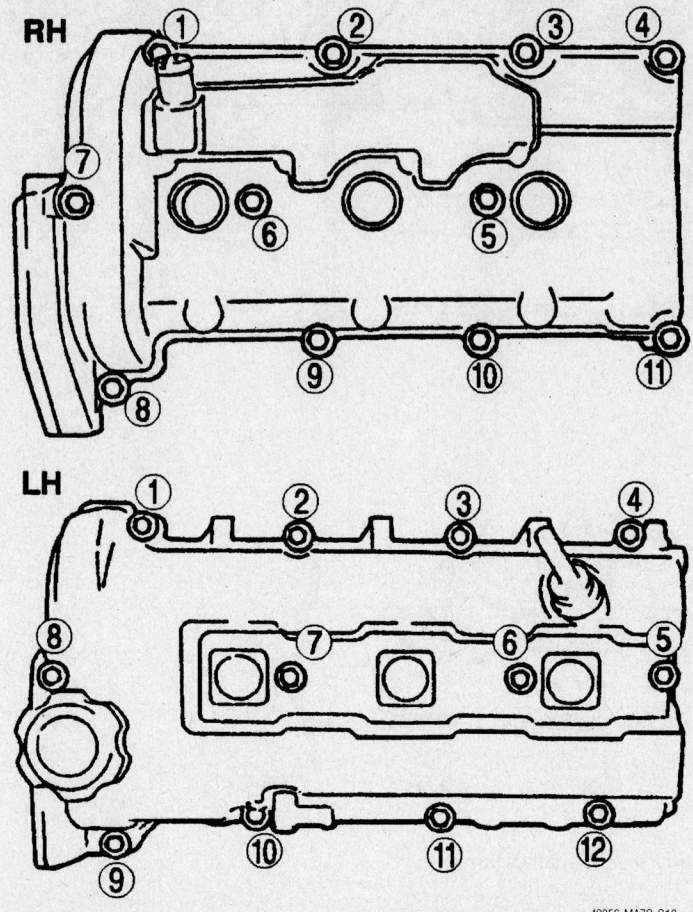

Tighten the rocker arm cover bolts in the sequence shown–2.5L (KL) engine

2.5L (KL) ENGINE

1. Before servicing the vehicle, refer to the precautions section.
2. Remove or disconnect the following:
   - Negative battery cable
   - Any components that would interfere with cover removal.
   - Rocker arm cover bolts
   - Rocker arm cover and discard the gasket
3. Clean all mating surfaces of any residual gasket material.

To install:

4. Apply silicone sealant to cylinder head at the areas illustrated.
5. Install or connect the following:
   - Rocker arm cover with a new gasket. Torque the bolts in the sequence illustrated to 95 inch lbs. (11 Nm) using 2 or 3 steps. Retighten the right hand cover number 5 and 6 bolts and the left hand cover number 6 and 7 bolts.
   - Remaining components removed to facilitate the rocker arm cover removal
   - Negative battery cable

Cylinder Head

REMOVAL & INSTALLATION

1.8L (BP) Engines

1. Before servicing the vehicle, refer to the precautions section.

2. Relieve the fuel system pressure.
3. Drain the cooling system.
4. Remove or disconnect the following:
   - Negative battery cable
   - Timing belt
   - Air cleaner assembly and front pipe
   - Exhaust manifold
   - Vacuum hoses
   - All engine harness connectors necessary to access the cylinder head
   - Fuel hose
   - Intake manifold bracket
   - Accelerator cable bracket
   - Cylinder head bolts, in 2–3 steps, in sequence
   - Cylinder head

To install:

5. Thoroughly, clean the cylinder head and the block contact surfaces. Examine the head gasket and check the cylinder head for cracks. Check the cylinder head for warpage using a feeler gauge and straightedge. The maximum allowable distortion is 0.004 inch (0.10mm).
6. Clean the cylinder head bolts and the threads in the block. Be sure the bolts turn freely in the block.
7. Install new head gasket on the engine block.
8. Install the cylinder head.
9. Lubricate the bolt threads and seat surfaces with clean engine oil and install them as follows:
   a. Torque the bolts in 2–3 steps to 56–60 ft. lbs. (75–81 Nm) in the proper sequence.
10. Install or connect the following:
   - Accelerator cable bracket
   - Intake manifold bracket
   - Fuel hose
   - All engine harness connectors removed to access the cylinder head

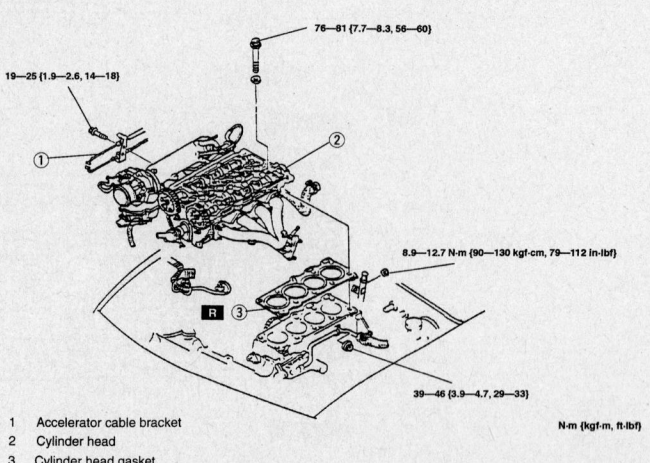

1 Accelerator cable bracket
2 Cylinder head
3 Cylinder head gasket

N·m {kgf·m, ft·lbf}

Exploded view of the cylinder head assembly—1.8L (BP) engines

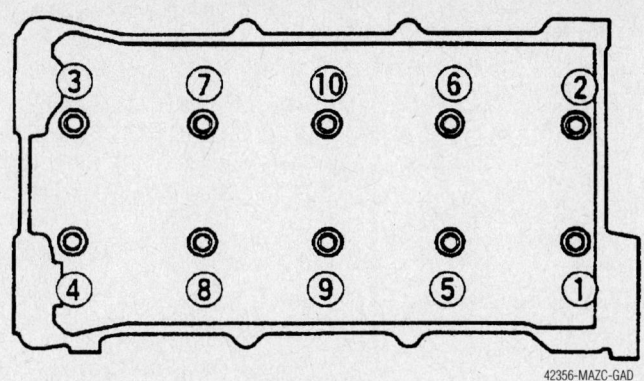

Cylinder head loosening sequence—1.8L (BP) engines

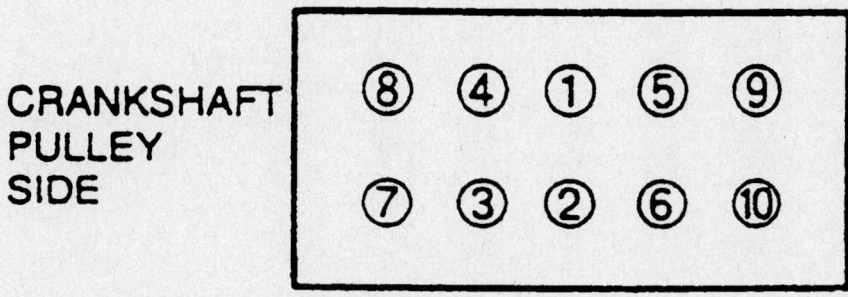

CRANKSHAFT PULLEY SIDE

Cylinder head bolt tightening sequence—1.8L (BP) engines

- Vacuum hoses
- Exhaust manifold using a new gasket
- Front pipe and air cleaner assembly
- Timing belt
- Negative battery cable
11. Fill and bleed the cooling system.
12. Change the oil and filter.
13. Run the engine and check for proper operation.

626 and Protégé

2.0L (FS) ENGINE

1. Before servicing the vehicle, refer to the precautions section.
2. Drain the cooling system.
3. Relieve the fuel system pressure.
4. Drain the cooling system.
5. Remove or disconnect the following:

- Negative battery cable
- Timing belt
- Front exhaust pipe
- Air cleaner assembly
- Power steering pump with the lines attached and set aside, if necessary on Protégé models
- Accelerator cable
- Fuel hose
- Ignition coil
- Camshaft pulley. Hold the camshaft

pulley with a back–up wrench while loosening the pulley bolt.
- Camshaft
- Intake manifold bracket
6. Temporarily install the Number 3 engine mount to support the engine, on 626 models.

- Cylinder head bolts by loosening them, in 2–3 steps, in the sequence illustrated
- Cylinder head

To install:
7. Install or connect the following:
- New cylinder head gasket
- Cylinder head
8. Apply clean engine oil to the bolt threads and seating faces.
9. Install new cylinder head bolts and torque in 2–3 steps, in sequence, to 13–16 ft. lbs. (17–22 Nm). If reusing old bolts, which is not recommended, make sure the maximum length of the bolt is 4.154 inch (105.5mm). The standard length of the bolt should be 4.103–4.125 inch (104.2–104.8mm).
10. Paint a mark on the edge of each cylinder head bolt to use as a reference. Turn each bolt, in sequence, 85–95 degrees. Again, turn each bolt, in sequence, an additional 85–95 degrees.

ENGINE FRONT

Cylinder head bolt removal sequence—2.0L (FS) engines

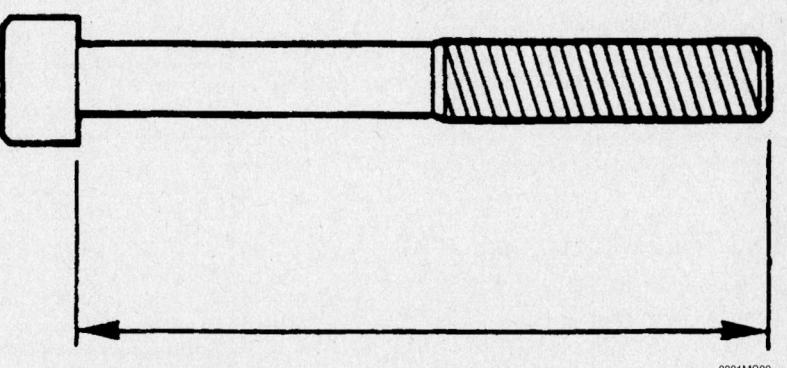

Replace any bolts that exceed the maximum length of 4.154 inch (105.5mm)—2.0L (FS)engines

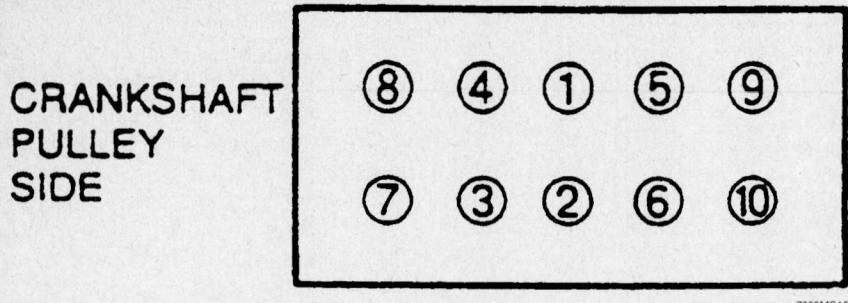

Cylinder head bolt tightening sequence—2.0L (FS) engines

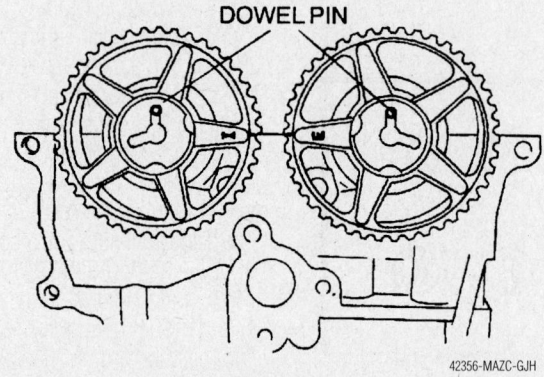

Make sure the camshaft sprocket pulleys are positioned with the pins facing up—2.0L (FS) engine

- Negative battery cable
- Timing belt
- Front exhaust pipe
- Exhaust manifold insulator and the Exhaust Gas Recirculation (EGR) pipe
- Fresh air duct and air cleaner assembly
- Accelerator cable and bracket
- All vacuum hoses
- Engine wiring harness connectors
- Fuel supply and return hoses
- Intake manifold support bracket
- Heater hoses
- Camshaft pulleys by holding the them with a wrench
- Camshafts
- Cylinder head bolts in sequence
- Cylinder head

To install:

5. Thoroughly, clean the cylinder head and the block contact surfaces.

6. Clean the cylinder head bolts and the threads in the block. Be sure the bolts turn freely in the block.

7. Measure the length of the cylinder head bolts, as shown, maximum bolt length is 3.956 inch (100.5mm).

8. Install or connect the following:

11. Support the engine assembly with a suitable lifting device and remove the number 3 mount.

12. Install or connect the following:
- Intake manifold bracket
- Camshafts
- Camshaft pulley, make sure the camshaft sprocket pulleys are positioned with the pins facing up. Hold the camshaft pulley with a back–up wrench while tightening the pulley bolt to 37–44 ft. lbs. (50–60 Nm).
- Ignition coil
- Fuel hose
- Accelerator cable
- Power steering pump, if removed
- Air cleaner assembly
- Front exhaust pipe
- Timing belt
- Negative battery cable

13. Fill and bleed the cooling system.

14. Change the oil and filter.

15. Run the engine and check for proper operation.

1.6L (ZM) ENGINES

1. Before servicing the vehicle, refer to the precautions section.

2. Properly relieve the fuel system pressure.

3. Drain the engine coolant.

4. Remove or disconnect the following:

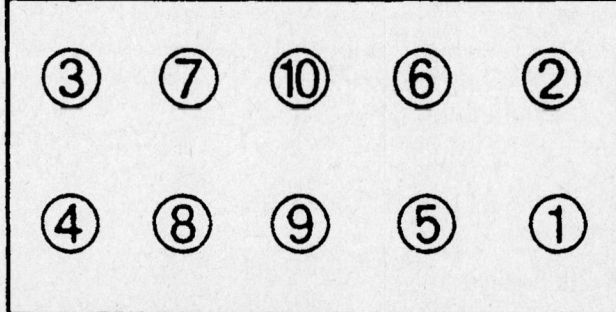

Cylinder head bolt removal sequence—1.6L (ZM) engines

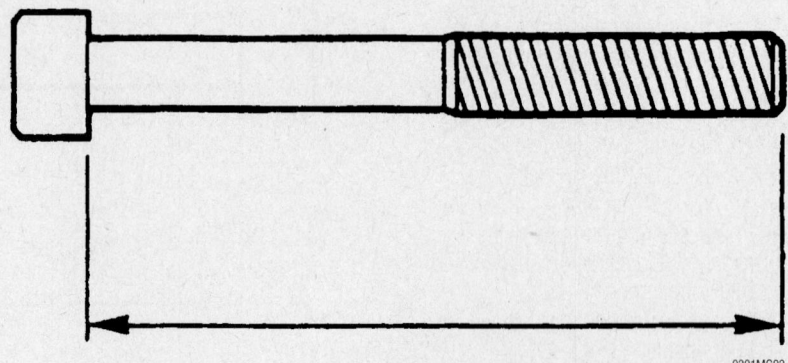

Replace any bolts that exceed the maximum length—1.6L (ZM) engines

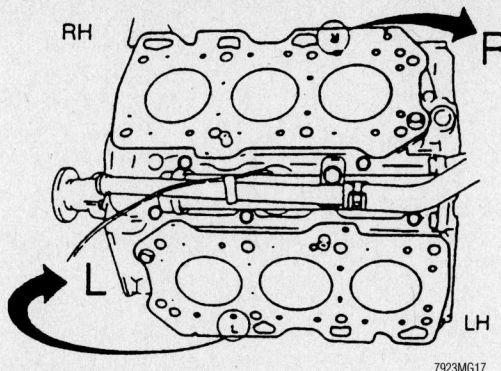

Cylinder head gasket positioning—6-cylinder engines

CRANKSHAFT PULLEY SIDE

Cylinder head bolt tightening sequence—1.6L (ZM) engines

- New head gasket
- Cylinder head

9. Torque the cylinder head bolts, in sequence, as follows.

   a. Step 1: 13–16 ft. lbs. (17–22 Nm).
   b. Step 2: Turn 85–95 degrees.
   c. Step 3: Turn an additional 85–95 degrees.

10. Install or connect the following:
   - Camshafts
   - Camshaft pulleys, install the pulleys so that the **I** mark on the intake side or **E** mark on the exhaust side are facing up. Torque the bolts to 37–44 ft. lbs. (50–60 Nm).
   - Heater hoses
   - Intake manifold support bracket
   - Fuel hoses
   - Engine wiring harness connector
   - Any vacuum hoses that were removed
   - Accelerator cable and bracket
   - Air cleaner assembly and fresh air duct
   - EGR system
   - Exhaust manifold insulator and front exhaust pipe
   - Timing belt
   - Negative battery cable

11. Fill the cooling system.

12. Start the vehicle, check for leaks and repair if necessary.

2.5L (KL) ENGINE

1. Before servicing the vehicle, refer to the precautions section.
2. Drain the cooling system.
3. Relieve the fuel system pressure.
4. Drain the cooling system.
5. Remove or disconnect the following:

   - Negative battery cable
   - Timing belt
   - Front exhaust pipe
   - Air cleaner assembly
   - Intake manifold
   - Fuel hose
   - Cylinder head cover
   - Camshaft pulley. Hold the camshaft

pulley with a back–up wrench while loosening the pulley bolt.
   - Upper radiator hose
   - Number 3 engine mount bracket
   - Seal plate
   - Water outlet
   - Camshaft
   - Alternator stay

6. Temporarily install the Number 3 engine mount to support the engine.
   - Cylinder head bolts by loosening them, in 2–3 steps, in the sequence illustrated
   - Cylinder head

To install:

7. Install or connect the following:
   - New cylinder head gasket
   - Cylinder head

8. Apply clean engine oil to the bolt threads and seating faces.

9. Install new cylinder head bolts and torque in 2–3 steps, in sequence, to 17–19 ft. lbs. (23–26 Nm). If reusing old bolts, which is not recommended; make sure the maximum length of the bolt is 5.217 inch (132.5mm). The standard length of the bolt should be 5.166–5.188 inch (131.2–131.8mm).

10. Paint a mark on the edge of each cylinder head bolt to use as a reference. Turn each bolt, in sequence, 85–95 degrees. Again, turn each bolt, in sequence, an additional 85–95 degrees.

11. Support the engine assembly with a suitable lifting device and remove the number 3 mount.

12. Install or connect the following:
   - Alternator stay
   - Camshaft
   - Water outlet
   - Seal plate
   - Number 3 engine mount bracket
   - Upper radiator hose
   - Camshaft pulley. Hold the camshaft pulley with a back–up wrench while tightening the pulley bolt to 90–97 ft. lbs. (123–132 Nm).

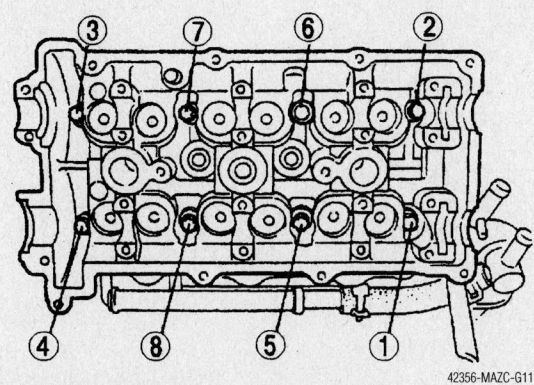

Cylinder head bolt removal sequence—2.5L (KL) engines

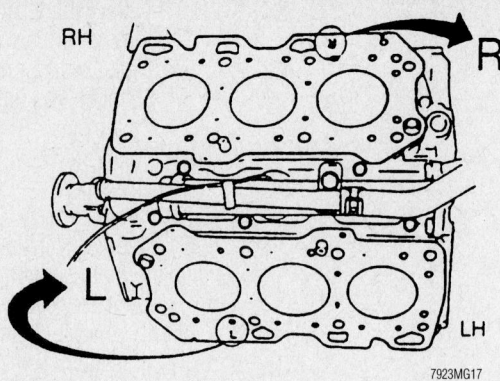

Cylinder head gasket positioning—2.5L (KL) engines

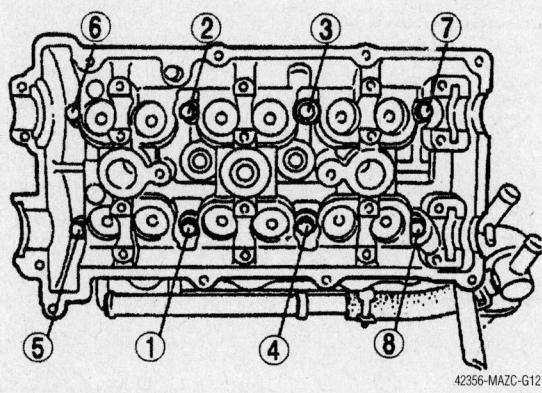

Cylinder head bolt tightening sequence—2.5L (KL) engines

- Cylinder head cover
- Fuel hose
- Intake manifold
- Air cleaner assembly
- Front exhaust pipe
- Timing belt
- Negative battery cable

13. Fill and bleed the cooling system.
14. Change the oil and filter.
15. Run the engine and check for proper operation.

Millenia

2.3L (KJ) ENGINES

1. Before servicing the vehicle, refer to the precautions section.
2. Relieve the fuel system pressure.
3. Drain the engine coolant.
4. Remove or disconnect the following:

- Negative battery cable
- Oxygen (O2S) sensor connectors
- Exhaust pipe-to-manifold nuts and lower the exhaust pipes
- Right-hand 3-way catalytic converter
- Compressor (supercharger)
- Intake manifold
- Timing belt covers and timing belt

- Spacer and O-ring from the front of the camshaft
- Ignition coils
- Cylinder head cover mounting bolts, in 5–6 steps, using the reverse of the tightening sequence
- Cylinder head cover
- Camshaft sprockets

5. Turn the camshafts so the knock pins are aligned with the marks on the camshaft caps. This will reduce the pressure on the adjustment shims.

6. Note the markings on the camshaft caps prior to removal, so they can be reinstalled in the same positions. The right-hand (rear) caps are marked with numbers and the left-hand (front) caps are marked with letters.

7. Loosen the front camshaft cap bolts in sequence, in 5–6 steps. Remove the front camshaft caps. Remove the remaining camshaft cap bolts in the proper sequence. Remove the caps, being sure to remove the thrust caps last. Do not damage the cylinder head thrust bearing support.

8. Remove or disconnect the following:

- Camshafts and oil seals
- Lifters and adjustment shims

9. Identify and mark each lifter as it is removed so it can be reinstalled in the same position.

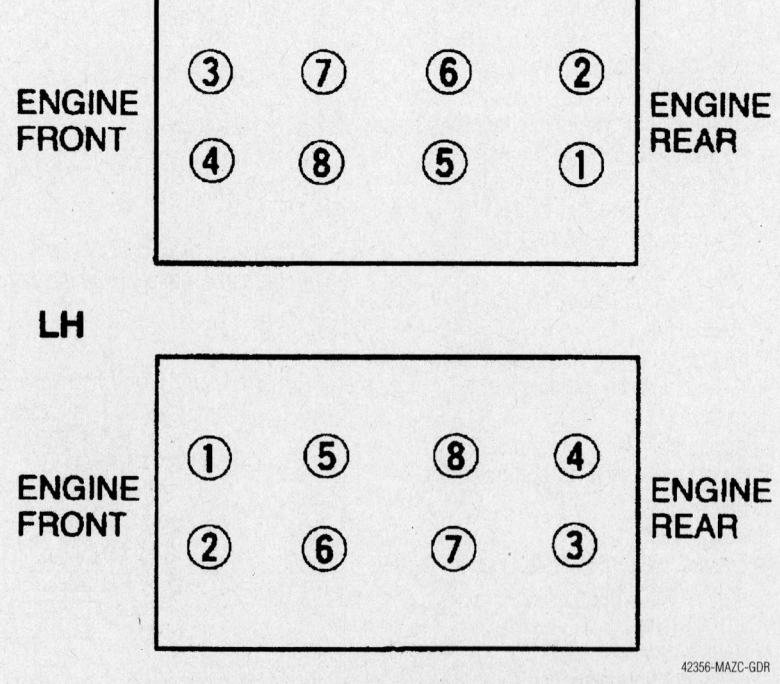

Loosen the cylinder head bolts in this sequence using several passes—Millenia models equipped with the 2.3L (KJ) engine

10. Remove or disconnect the following:
- Lower radiator hose
- Water inlet pipe
- Compressor bracket
- Alternator bracket bolt to gain additional clearance
- Rubber insulator from the left-hand cylinder head

11. Temporarily install the No. 3 engine mount, which was removed with the timing belt, to support the engine.

12. Remove or disconnect the following:
- Engine support device
- Cylinder head bolts, in 2–3 steps, in the sequence illustrated
- Cylinder heads
- Oil control plug O-rings

13. Clean all gasket mating surfaces. Inspect the cylinder head for damage, cracks, and water and oil leakage. Check the head gasket surface for distortion using a straightedge and feeler gauge. Maximum allowable distortion is 0.004 inch (0.10mm).

To install:

14. Apply clean engine oil to the O-rings, and install them onto the oil control plugs.

15. Position new head gaskets on the cylinder block. The gaskets cannot be interchanged between sides and are marked **R** and **L** for right and left side.

16. Install the cylinder heads.

17. Apply clean engine oil to the threads of new cylinder head bolts and install. Torque the bolts in 2–3 steps, in sequence, to 17–19 ft. lbs. (23–26 Nm).

18. Paint a mark on the edge of each cylinder head bolt to use as a reference. Turn each bolt, in sequence, 85–95 degrees. Again, turn each bolt, in sequence, an additional 85–95 degrees.

19. Install the rubber insulator onto the left-hand cylinder head.

20. Fit the knock sensor harness into the drill hole on the cylinder block. Pass the harness under the rubber insulator.

21. Install or connect the following:
- Engine support device and remove the No. 3 engine mount
- Alternator bracket bolt. Torque the bolt to 12–16 ft. lbs. (16–22 Nm).
- Compressor bracket. Torque the bolts to 14–18 ft. lbs. (19–25 Nm).
- Water inlet pipe. Torque the bolts to 14–18 ft. lbs. (19–25 Nm).
- Lower radiator hose
- Lifters in their original positions by lubricating them with engine oil.

Verify that they move smoothly in their bore
- New oil seals on the camshafts

22. Apply clean engine oil to the camshaft lobes, journals and supports.

23. Install or connect the following:
- Camshafts so the gear marks align
- Thrust caps. Torque the bolts, in 5–6 steps, until they are fully seated on the cylinder head

24. Apply silicone sealant, at a thickness of 0.06–0.09 inch (1.5–2.5mm), to the cylinder head surface in the area forward of the camshaft gear cavity.

25. Install or connect the following:
- Remaining camshaft caps in their original positions. Torque the caps, in sequence, in 5 equal steps, with the final step being 100–125 inch lbs. (11–14 Nm).
- New camshaft oil seal lubricated with engine oil. Tap the seal in evenly with a Seal Installer tool 49 F401 337A with a final protrusion of 0–0.02 inch (0–0.5mm). Tap in a new blind cap
- Camshaft sprockets. Torque the bolts to 91–103 ft. lbs. (123–140 Nm).

26. Measure and adjust the valve clearances.

27. Remove any sealant and gasket material from the cylinder head cover contact surfaces.

28. Apply silicone sealant to the cylinder head in the area adjacent to the front and rear camshaft caps.

29. Install or connect the following:
- Cylinder head cover using a new gasket. Torque the bolts in 5–6 steps, in sequence, to 44–78 inch lbs. (5–9 Nm).
- Distributor using a new O-ring
- Ignition coils
- Spacer using a new O-ring. Torque the bolt to 14–18 ft. lbs. (19–25 Nm).
- Timing belt and timing belt cover
- Intake manifold
- Compressor (supercharger)
- Right-hand 3-way catalytic converter
- Exhaust pipes to the manifolds. Torque the nuts to 28–38 ft. lbs. (38–51 Nm).
- O2S connectors
- Negative battery cable

30. Fill and bleed the coolant system.

31. Run the engine and check for leaks.

2.5L (KL) ENGINE

1. Before servicing the vehicle, refer to the precautions section.

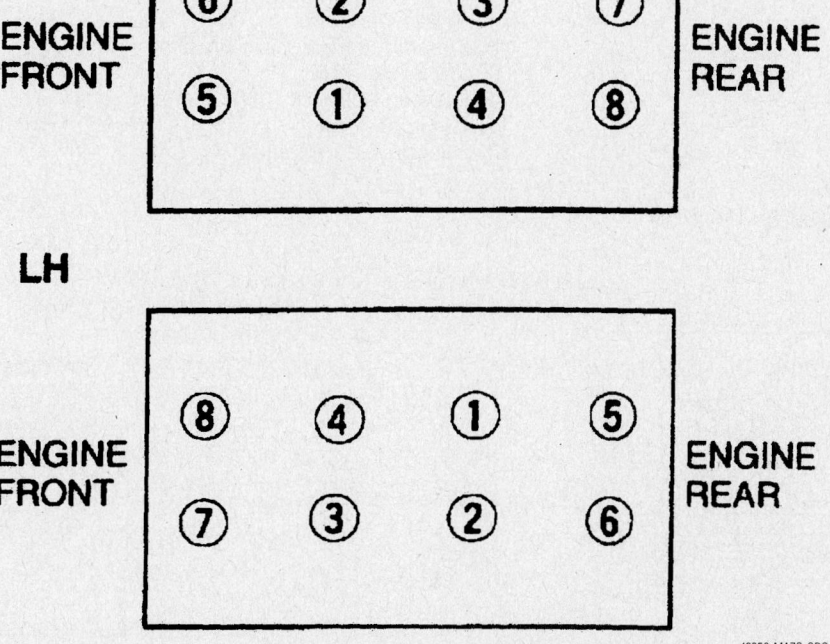

RH

ENGINE FRONT · ENGINE REAR

LH

ENGINE FRONT · ENGINE REAR

42356-MAZC-GDS

Cylinder head bolt torque sequence—Millenia models equipped with the 2.3L (KJ) engine

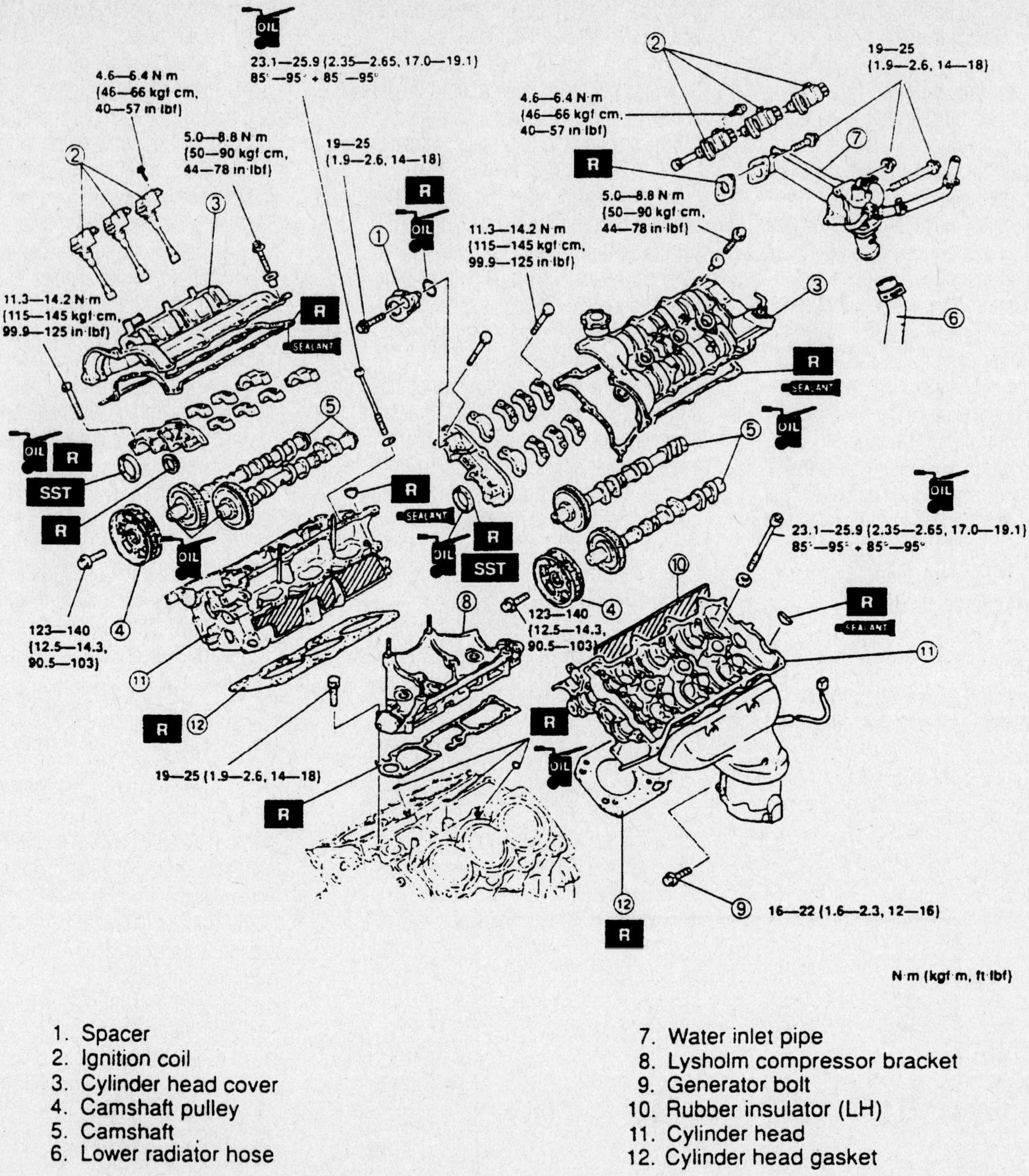

23.1—25.9 (2.35—2.65, 17.0—19.1)
85°—95° + 85 —95°

4.6—6.4 N·m
(46—66 kgf cm,
40—57 in lbf)

5.0—8.8 N·m
(50—90 kgf cm,
44—78 in lbf)

19—25
(1.9—2.6, 14—18)

11.3—14.2 N·m
(115—145 kgf cm,
99.9—125 in lbf)

19—25 (1.9—2.6, 14—18)

4.6—6.4 N·m
(46—66 kgf cm,
40—57 in lbf)

5.0—8.8 N·m
(50—90 kgf cm,
44—78 in lbf)

11.3—14.2 N·m
(115—145 kgf cm,
99.9—125 in lbf)

123—140
(12.5—14.3,
90.5—103)

23.1—25.9 (2.35—2.65, 17.0—19.1)
85°—95° + 85°—95°

123—140
(12.5—14.3,
90.5—103)

19—25 (1.9—2.6, 14—18)

16—22 (1.6—2.3, 12—16)

N·m (kgf·m, ft·lbf)

1. Spacer
2. Ignition coil
3. Cylinder head cover
4. Camshaft pulley
5. Camshaft
6. Lower radiator hose
7. Water inlet pipe
8. Lysholm compressor bracket
9. Generator bolt
10. Rubber insulator (LH)
11. Cylinder head
12. Cylinder head gasket

7923MG24

Exploded view of the cylinder head and related components—2.3L engine

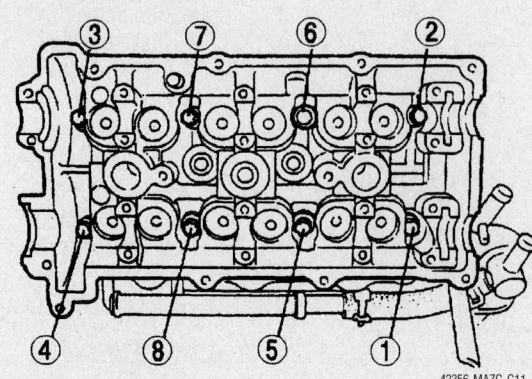

42356-MAZC-G11

Cylinder head bolt removal sequence—2.5L (KL) engines

2. Drain the cooling system.
3. Relieve the fuel system pressure.
4. Drain the cooling system.
5. Remove or disconnect the following:
   • Negative battery cable
   • Timing belt
   • Air cleaner assembly
   • Spark plug wires
   • Distributor
   • Intake manifold
   • Upper radiator hose
   • Water outlet
   • Number 3 engine mount bracket
   • Seal plate
   • Front exhaust pipe

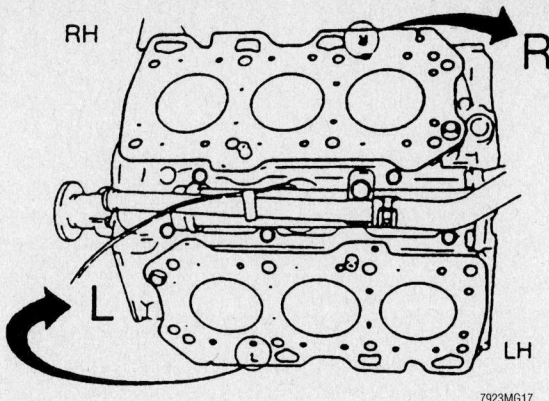

Cylinder head gasket positioning—2.5L (KL) engines

7923MG17

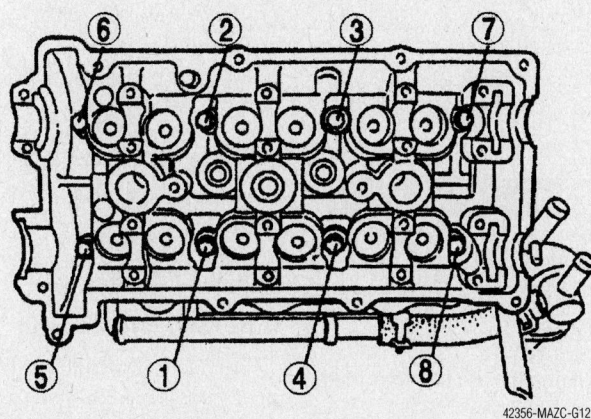

Cylinder head bolt tightening sequence—2.5L (KL) engines

42356-MAZC-G12

- Alternator stay
- Cylinder head cover
- Camshaft pulley. Hold the camshaft pulley with a back–up wrench while loosening the pulley bolt.
- Camshaft

6. Temporarily install the Number 3 engine mount to support the engine.

- Cylinder head bolts by loosening them, in 2–3 steps, in the sequence illustrated
- Cylinder head

To install:

7. Install or connect the following:
- New cylinder head gasket
- Cylinder head

8. Apply clean engine oil to the bolt threads and seating faces.

9. Install new cylinder head bolts and torque in 2–3 steps, in sequence, to 17–19 ft. lbs. (23–26 Nm). If reusing old bolts, which is not recommended; make sure the maximum length of the bolt is 5.217 inch (132.5mm). The standard length of the bolt should be 5.166–5.188 inch (131.2–131.8mm).

10. Paint a mark on the edge of each cylinder head bolt to use as a reference.

Turn each bolt, in sequence, 85–95 degrees. Again, turn each bolt, in sequence, an additional 85–95 degrees.

11. Support the engine assembly with a suitable lifting device and remove the number 3 mount.

12. Install or connect the following:
- Camshaft
- Camshaft pulley. Hold the camshaft pulley with a back–up wrench while tightening the pulley bolt to 90–97 ft. lbs. (123–132 Nm).
- Cylinder head cover
- Alternator stay
- Front exhaust pipe
- Seal plate
- Number 3 engine mount bracket
- Water outlet
- Upper radiator hose
- Intake manifold
- Distributor and spark plug wires
- Air cleaner assembly
- Timing belt
- Negative battery cable

13. Fill and bleed the cooling system.

14. Change the oil and filter.

15. Run the engine and check for proper operation.

REMOVAL & INSTALLATION

All Mazda engines covered in this manual are not equipped with rocker arms/shafts, the camshafts directly actuate the valves through a bucket type cam follower.

Supercharger

REMOVAL & INSTALLATION

2.3L Engine

1. Before servicing the vehicle, refer to the precautions section.

2. Relieve the fuel system pressure.

3. Drain the cooling system.

4. Remove or disconnect the following:
- Negative battery cable
- Dynamic chamber cover
- Charge air cooler air duct
- Vacuum hoses and electrical connectors from the air cleaner housing
- Air cleaner assembly
- Fresh air ducts
- Mass Air Flow (MAF) sensor and the air intake hose from the throttle body
- Resonator
- Right-hand charge air cooler
- Left-hand charge air cooler
- Accelerator cable
- Vacuum hoses from the rear of the intake manifold and Exhaust Gas Recirculation (EGR) valve
- EGR valve
- Air intake pipe assembly
- Charge air cooler pipe
- Fuel supply line at the fuel rails and discard the copper washers
- Fuel and vacuum lines from the fuel pressure regulator
- Coolant hoses
- Wiring harness from the intake manifold
- Intake manifold mounting nuts and bolts in 2–3 steps
- Intake manifold
- Fuel hoses and electrical connectors from the throttle body
- Throttle body
- Drive belt from the compressor (supercharger)
- Mounting bolts from the compressor
- Compressor

To install:

5. Clean all gasket mating surfaces.

6. Position the rubber shield for the

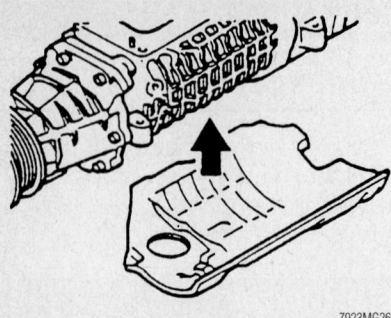

7923MG26

When installing the compressor, ensure that the rubber insulating pad is temporarily affixed to the compressor

compressor onto the compressor using double sided adhesive tape.

7. Install or connect the following:
- Compressor. Torque the nuts to 14–18 ft. lbs. (19–25 Nm).
- Compressor drive belt
- Throttle body. Torque the bolts to 14–18 ft. lbs. (19–25 Nm).
- Fuel hoses and electrical connectors
- Intake manifold using a new gasket. Torque the bolts in 2–3 steps, from the center to the ends, to 14–18 ft. lbs. (19–25 Nm).
- Wiring harness onto the intake manifold
- Coolant hoses
- Fuel and vacuum lines to the fuel pressure regulator
- Fuel supply line to the fuel rail using new copper crush washers
- Charge air cooler pipe
- Position the air intake pipe assembly using new gaskets

8. Hand-tighten the nuts/bolts in the order shown until the air intake pipe contacts the intake manifold. Verify that the rubber gaskets are not twisted or distorted. Torque the bolts marked **A** to 70–95 inch lbs. (8–11 Nm) and all others, in sequence, to 14–18 ft. lbs. (19–25 Nm).

9. Install or connect the following:
- EGR valve using a new gasket
- Vacuum hoses to the intake manifold and EGR valve
- Accelerator cable and adjust it
- Left and right-hand charge air coolers using new gaskets. Hand-tighten the nuts/bolts in the order shown until the air intake pipes and charge air coolers contact the intake manifold. Verify that the rubber gaskets are not twisted or distorted.

10. Torque the charge air cooler bolts to:
   a. Marked **A**: 44–78 inch lbs. (5–9 Nm).

b. Marked **B**: to 70–95 inch lbs. (8–11 Nm).
   c. All others, in sequence: 14–18 ft. lbs. (19–25 Nm).
11. Install or connect the following:
- Resonator. Torque the bolts to 12–16 ft. lbs. (16–22 Nm).
- Air intake hose onto the throttle body
- MAF sensor
- Fresh air ducts
- Air cleaner assembly
- Vacuum hoses and electrical connectors to the air cleaner housing
- Charge air cooler air duct. Torque the bolts to 70–95 inch lbs. (8–11 Nm).
- Dynamic chamber cover
- Negative battery cable
12. Fill the cooling system.
13. Start the vehicle, check for leaks and repair if necessary.

Intake Manifold

REMOVAL & INSTALLATION

1.8L (BP) Engines

1. Before servicing the vehicle, refer to the precautions section.
2. Relieve the fuel system pressure.
3. Drain the cooling system.
4. Remove or disconnect the following:
- Negative battery cable
- Intake Air Temperature (IAT) sensor
- Air cleaner assembly, Mass Air Flow (MAF) sensor and resonance chamber
- Throttle (automatic transaxle only) and accelerator cables
- Idle Air Control (IAC) valve and the Throttle Position Sensor (TPS) electrical connectors
- Throttle body
- Dynamic chamber (upper intake manifold) bracket
- Dynamic chamber (upper intake manifold) and gasket
- Exhaust Gas Recirculation (EGR) pipe
- Intake manifold and discard the gasket

To install:
5. Clean all gasket mating surfaces.
6. Install or connect the following:
- Intake manifold using a new gasket, make sure the convex side of the gasket is facing up. Torque the bolts to 17–21 ft. lbs. (19–25 Nm).
- EGR pipe to the manifold

- Dynamic chamber (upper intake manifold) using new gaskets, make sure the convex side of the gasket is facing up. Tighten the bolts to 14–18 ft. lbs. (19–25 Nm).
- Dynamic chamber (upper intake manifold) bracket and tighten the bolts 95 inch lbs. (11 Nm). Tighten bolts firmly, then tighten the chamber side bolt before tightening the fuel rail side bolt.
- Throttle body using a new gasket. Tighten the bolts to 14–18 ft. lbs. (19–25 Nm).
- Electrical connectors for the IAC valve and the TPS
- Connect and adjust the throttle and accelerator cables
- IAT sensor
- Air cleaner assembly, MAF sensor and ducts
- Negative battery cable
7. Fill the cooling system.
8. Run the engine and check for leaks.

626 and Protégé

1.6L (ZM) ENGINES

1. Before servicing the vehicle, refer to the precautions section.
2. Relieve the fuel system pressure. Drain the cooling system.
3. Remove or disconnect the following:
- Negative battery cable
- Fuel hoses from the fuel rail
- Fuel injector electrical connectors
- Fuel rail with the injectors connected
- Components in the order illustrated

To install:
4. Clean all gasket mating surfaces.

➡**Be sure that the convex side of the intake manifold gasket is facing the manifold side, as shown.**

5. Install or connect the following:
- Intake manifold using a new gasket, make sure the convex side of the gasket faces the manifold. Torque the bolts to 14–18 ft. lbs. (19–25 Nm).
- Components in the reverse order of the removal sequence
- Fuel rail. Tighten the bolts to 14–18 ft. lbs. (19–25 Nm).
- Fuel lines and electrical connectors to the fuel rail
- Negative battery cable
6. Fill the cooling system.
7. Run the engine and check for leaks.

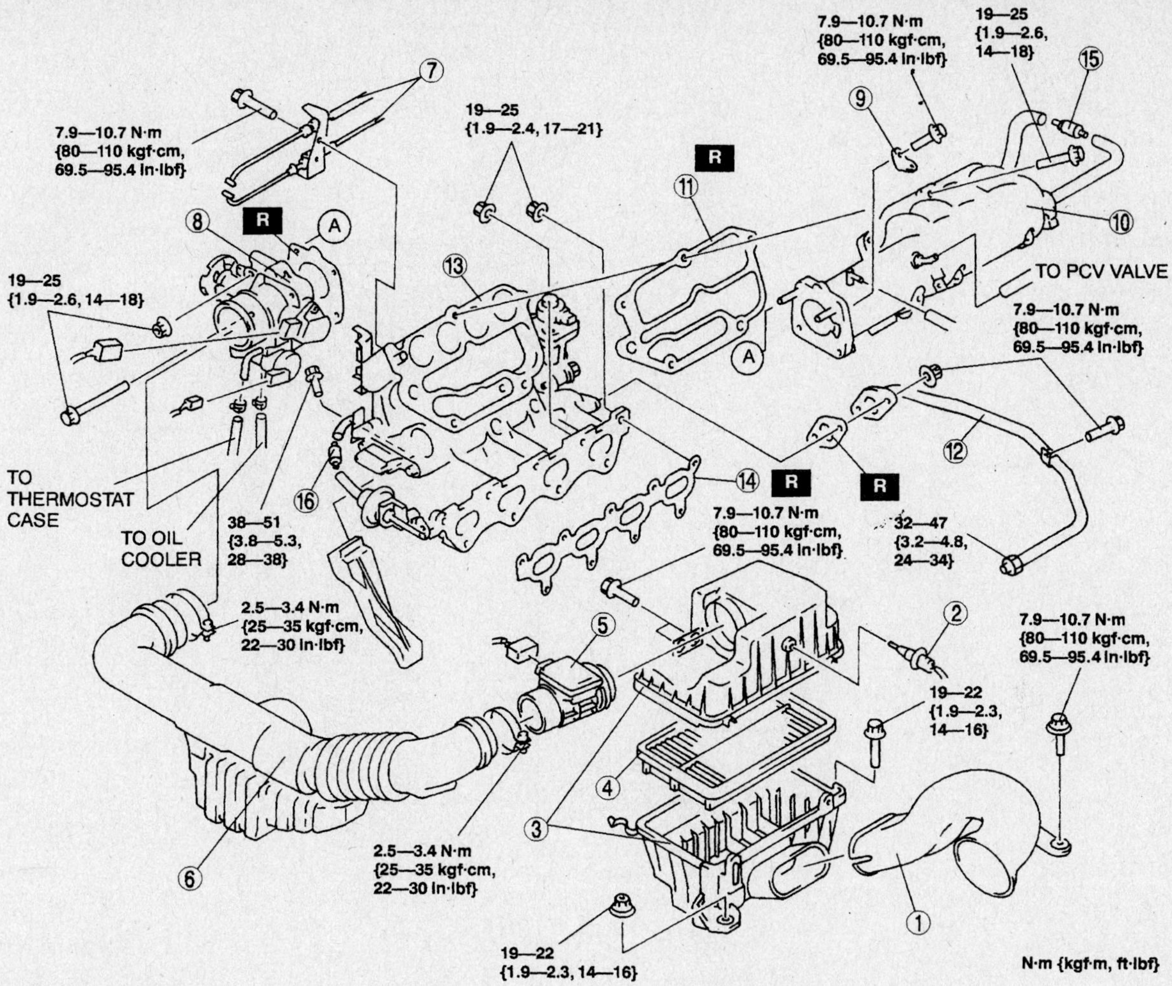

7.9—10.7 N·m
{80—110 kgf·cm,
69.5—95.4 in·lbf}

19—25
{1.9—2.6,
14—18}

7.9—10.7 N·m
{80—110 kgf·cm,
69.5—95.4 in·lbf}

19—25
{1.9—2.4, 17—21}

TO PCV VALVE

7.9—10.7 N·m
{80—110 kgf·cm,
69.5—95.4 in·lbf}

19—25
{1.9—2.6, 14—18}

TO
THERMOSTAT
CASE

TO OIL
COOLER

38—51
{3.8—5.3,
28—38}

2.5—3.4 N·m
{25—35 kgf·cm,
22—30 in·lbf}

7.9—10.7 N·m
{80—110 kgf·cm,
69.5—95.4 in·lbf}

32—47
{3.2—4.8,
24—34}

7.9—10.7 N·m
{80—110 kgf·cm,
69.5—95.4 in·lbf}

19—22
{1.9—2.3,
14—16}

2.5—3.4 N·m
{25—35 kgf·cm,
22—30 in·lbf}

19—22
{1.9—2.3, 14—16}

N·m {kgf·m, ft·lbf}

1	Fresh-air duct	9	Dynamic chamber bracket
2	IAT sensor	10	Dynamic chamber
3	Air cleaner (ACL)	11	Dynamic chamber gasket
4	Air cleaner (ACL) element	12	EGR pipe
5	MAF sensor	13	Intake manifold
6	Air hose	14	Intake manifold gasket
7	Accelerator cable (and throttle cable (AT only))	15	VTCS check valve (one-way)
8	Throttle body (TB)	16	Delay valve

42356-MAZC-GAF

Exploded view of the intake manifold and related components—1.8L (BP) engine

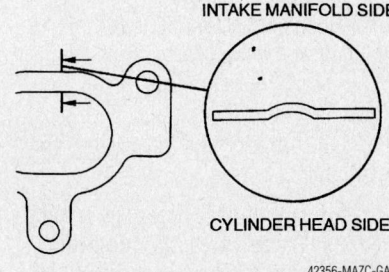

INTAKE MANIFOLD SIDE

CYLINDER HEAD SIDE

42356-MAZC-GAG

Make sure the convex side of the gasket is facing up when installing the intake manifold gasket—1.8L (BP) engine

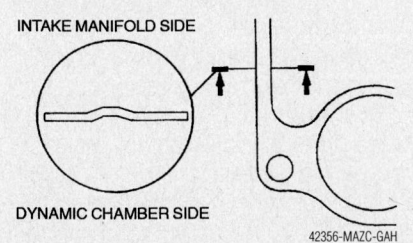

INTAKE MANIFOLD SIDE

DYNAMIC CHAMBER SIDE

42356-MAZC-GAH

Make sure the convex side of the gasket is facing up when installing the upper intake manifold gasket—1.8L (BP) engine

626 WITH 2.0L (FS) ENGINE

1. Before servicing the vehicle, refer to the precautions section.
2. Relieve the fuel system pressure.
3. Drain the cooling system.
4. Remove or disconnect the following:
   - Negative battery cable
   - Fresh air duct and air cleaner assembly
   - Mass Air Flow (MAF) sensor
   - Air hose
   - Accelerator cable
   - Throttle body assembly
   - Idle Air Control (IAC) valve

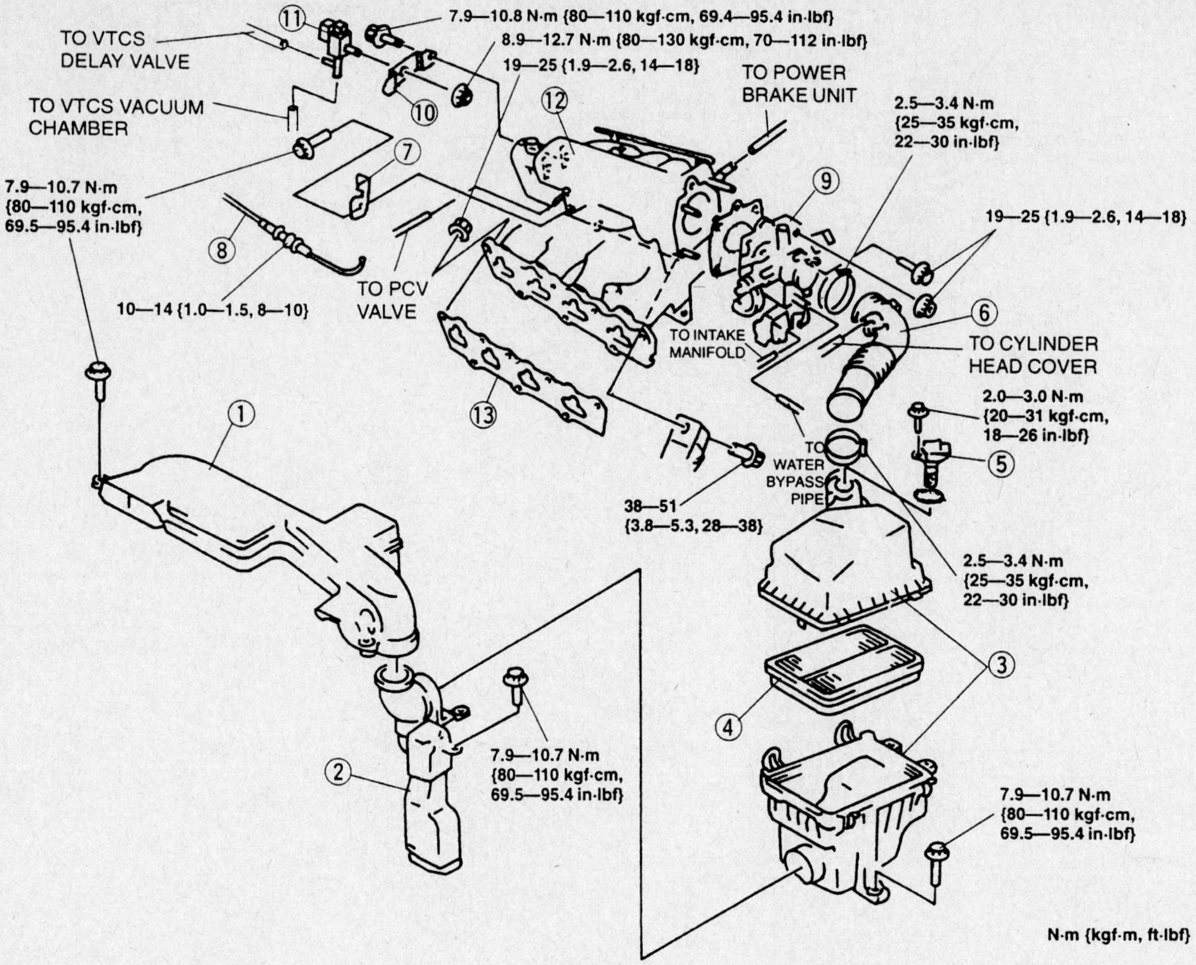

TO VTCS
DELAY VALVE

TO VTCS VACUUM
CHAMBER

7.9—10.7 N·m
{80—110 kgf·cm,
69.5—95.4 in·lbf}

7.9—10.8 N·m {80—110 kgf·cm, 69.4—95.4 in·lbf}

8.9—12.7 N·m {80—130 kgf·cm, 70—112 in·lbf}

19—25 {1.9—2.6, 14—18}

TO POWER
BRAKE UNIT

2.5—3.4 N·m
{25—35 kgf·cm,
22—30 in·lbf}

19—25 {1.9—2.6, 14—18}

10—14 {1.0—1.5, 8—10}

TO PCV
VALVE

TO INTAKE
MANIFOLD

TO CYLINDER
HEAD COVER

TO
WATER
BYPASS
PIPE

2.0—3.0 N·m
{20—31 kgf·cm,
18—26 in·lbf}

38—51
{3.8—5.3, 28—38}

2.5—3.4 N·m
{25—35 kgf·cm,
22—30 in·lbf}

7.9—10.7 N·m
{80—110 kgf·cm,
69.5—95.4 in·lbf}

7.9—10.7 N·m
{80—110 kgf·cm,
69.5—95.4 in·lbf}

N·m {kgf·m, ft·lbf}

1	Fresh-air duct	8	Accelerator cable
2	Resonance chamber	9	Throttle body
3	Air cleaner	10	VTCS solenoid valve bracket
4	Air cleaner element	11	VTCS solenoid valve
5	MAF sensor (Integrated with IAT sensor)	12	Intake manifold
6	Air hose	13	Intake manifold gasket
7	Accelerator cable bracket		

9301MG03

Exploded view of the intake manifold, illustrating the removal and installation components with tightening values—1.6L engine

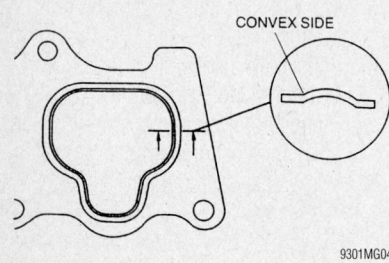

CONVEX SIDE

9301MG04

**Cross-sectional view of the intake mani-
fold gasket—1.6L (ZM) engines**

- Intake manifold stay
- Intake manifold and discard the gasket
5. Clean the mating surfaces of any gasket material
To install:
6. Install or connect the following:
- Intake manifold with a new gasket. Make certain that the convex side of the new gasket faces the intake manifold. Torque the bolts to 18 ft. lbs. (25 Nm).
- Intake manifold stay
- IAC valve. Torque the bolt to 39–57 inch lbs. (4–6 Nm).

- Throttle body. Torque the nuts to 18 ft. lbs. (25 Nm).
7. When connecting the accelerator cable, perform the following adjustment:
   a. Move the white locking tab to the unlock position.
   b. Turn stopper B to the to the unlock position.

➡**If stopper B will not unlock, it may be necessary to carefully bend back tab C out a little using a suitable pry tool.**

   c. Push or pull the cable housing directly behind the spring.

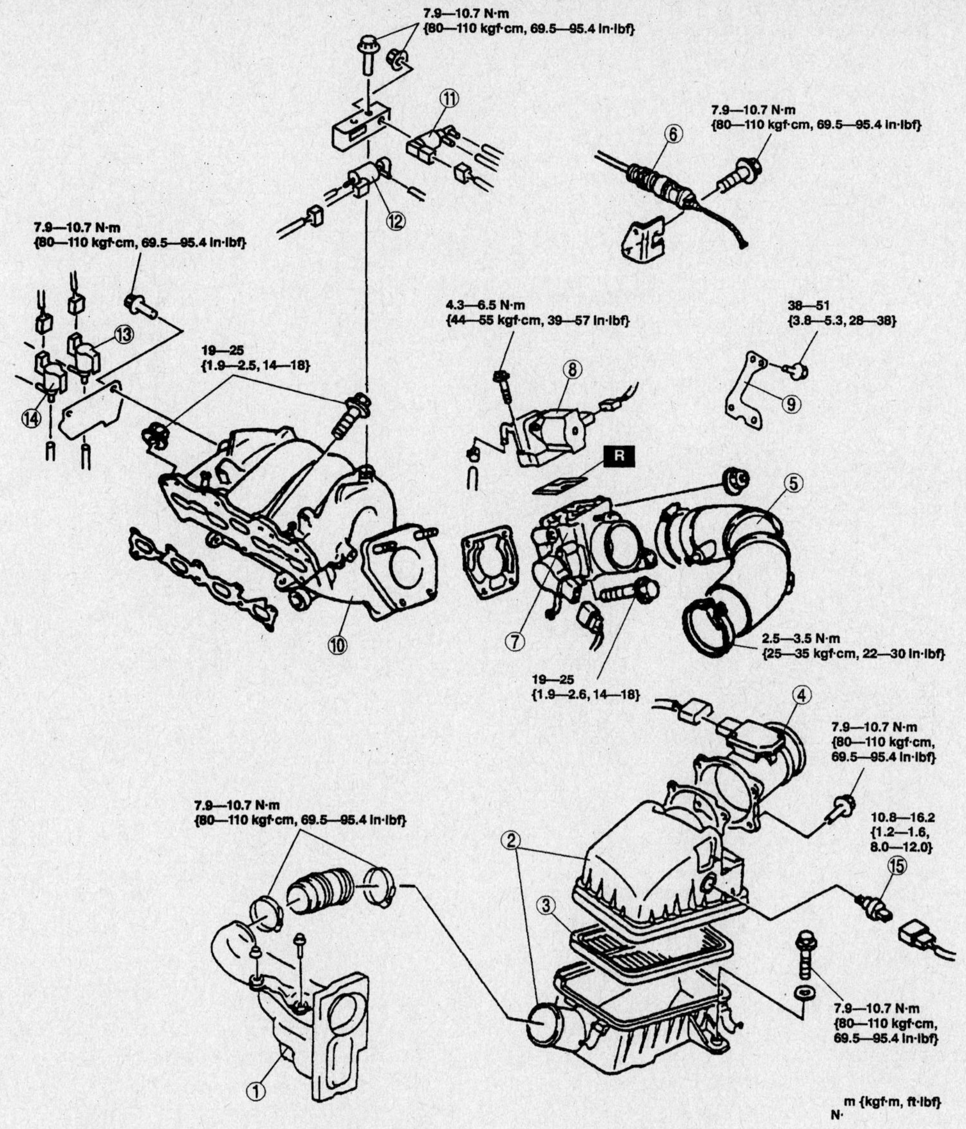

7.9—10.7 N·m
{80—110 kgf·cm, 69.5—95.4 In·lbf}

7.9—10.7 N·m
{80—110 kgf·cm, 69.5—95.4 In·lbf}

7.9—10.7 N·m
{80—110 kgf·cm, 69.5—95.4 In·lbf}

4.3—6.5 N·m
{44—55 kgf·cm, 39—57 In·lbf}

38—51
{3.8—5.3, 28—38}

19—25
{1.9—2.5, 14—18}

2.5—3.5 N·m
{25—35 kgf·cm, 22—30 In·lbf}

19—25
{1.9—2.6, 14—18}

7.9—10.7 N·m
{80—110 kgf·cm,
69.5—95.4 In·lbf}

7.9—10.7 N·m
{80—110 kgf·cm, 69.5—95.4 In·lbf}

10.8—16.2
{1.2—1.6,
8.0—12.0}

7.9—10.7 N·m
{80—110 kgf·cm,
69.5—95.4 In·lbf}

m {kgf·m, ft·lbf}
N·

1	Fresh–air duct	6	Accelerator cable	11	Purge solenoid valve
2	Air cleaner	7	Throttle body	12	PRC solenoid valve
3	Air cleaner element	8	IAC valve	13	EGR boost solenoid valve
4	Mass air flow sensor	9	Intake manifold stay	14	VTCS solenoid valve
5	Air hose	10	Intake manifold	15	Intake air temperature sensor

42356-MAZC-G13

Exploded view of the intake manifold assembly–626 with 2.0L (FS) engine

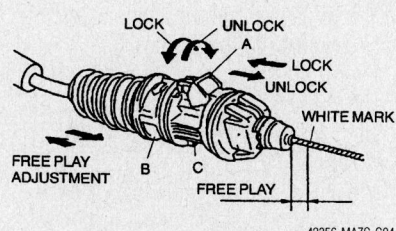

Free play
1.5—4.0 mm {0.06—0.15 in}

LOCK   UNLOCK
LOCK
UNLOCK
WHITE MARK
FREE PLAY
ADJUSTMENT   B   C
FREE PLAY

42356-MAZC-G04

Accelerator cable adjustment components—626 and Protégé models equipped with the 2.0L (FS) engine

d. Turn the stopper B to the lock position.

e. Measure the free play which should be 0.06—0.15 inch (1.5—4mm) and make sure the cable free play is within specification.

f. Move the white locking tab to the lock position and check for proper accelerator operation.

8. Install or connect the following:
• Air hose
• MAF sensor. Torque the bolt to 95 inch lbs. (11 Nm).
• Air cleaner/air duct assembly
• Negative battery cable

9. Fill the cooling system.
10. Start the vehicle, check for leaks and repair if necessary.

PROTÉGÉ WITH 2.0L (FS) ENGINE

1. Before servicing the vehicle, refer to the precautions section.
2. Relieve the fuel system pressure.
3. Drain the cooling system.
4. Remove or disconnect the following:
• Negative battery cable
• Fresh air duct and resonance assembly
• Intake Air Temperature (IAT) sensor
• Air cleaner and filter

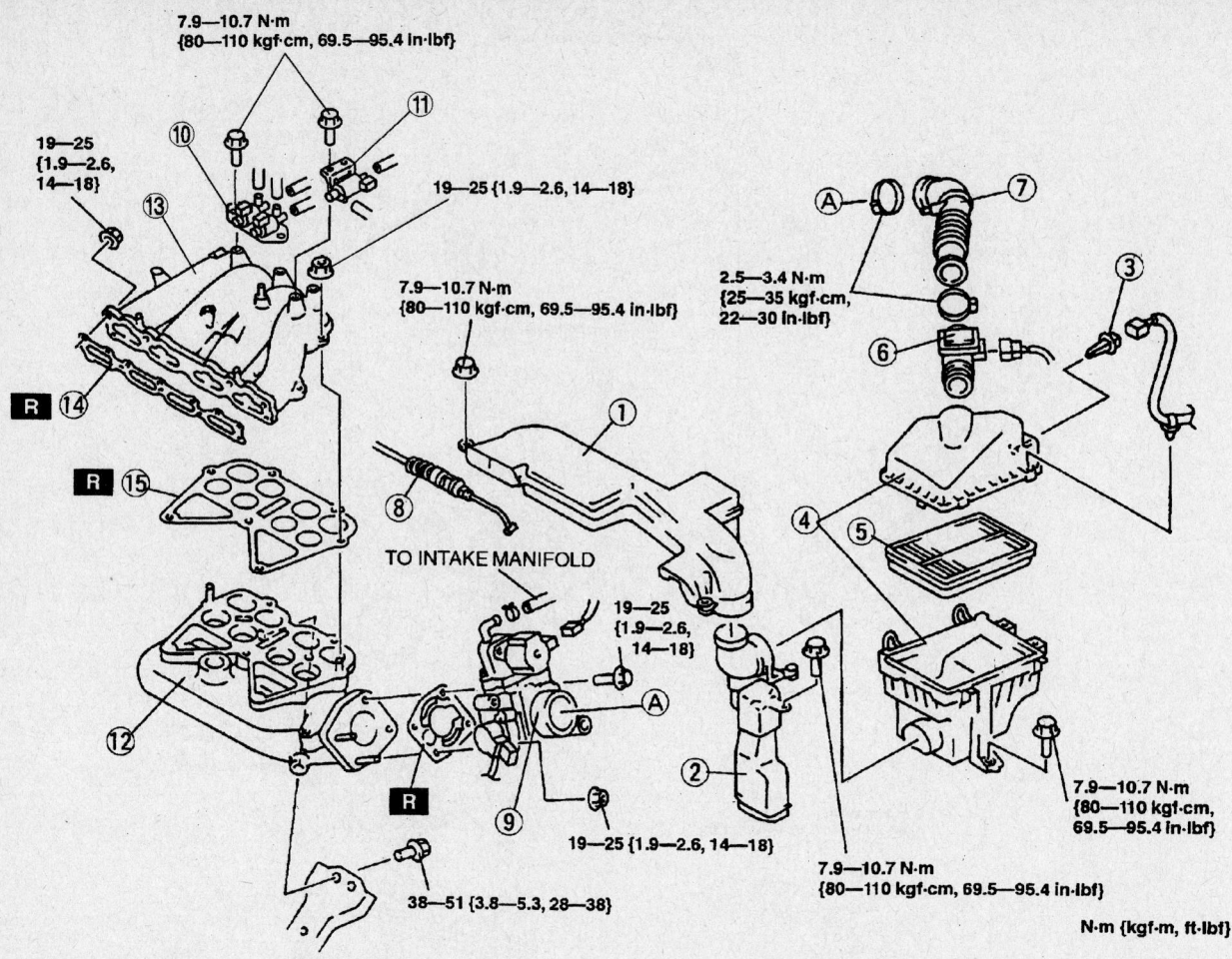

7.9—10.7 N·m
{80—110 kgf·cm, 69.5—95.4 in·lbf}

19—25
{1.9—2.6,
14—18}

19—25 {1.9—2.6, 14—18}

7.9—10.7 N·m
{80—110 kgf·cm, 69.5—95.4 in·lbf}

2.5—3.4 N·m
{25—35 kgf·cm,
22—30 in·lbf}

TO INTAKE MANIFOLD

19—25
{1.9—2.6,
14—18}

19—25 {1.9—2.6, 14—18}

38—51 {3.8—5.3, 28—38}

7.9—10.7 N·m
{80—110 kgf·cm, 69.5—95.4 in·lbf}

7.9—10.7 N·m
{80—110 kgf·cm,
69.5—95.4 in·lbf}

N·m {kgf·m, ft·lbf}

1	Fresh-air duct	9	Throttle body
2	Resonance chamber	10	Solenoid valve bracket
3	IAT sensor	11	PRC solenoid valve
4	Air cleaner	12	Dynamic chamber
5	Air cleaner element	13	Intake manifold
6	MAF sensor	14	Intake manifold gasket
7	Air hose	15	Dynamic chamber gasket
8	Accelerator cable		

42356-MAZC-GHC

Exploded view of the intake manifold assembly–Protégé with 2.0L (FS) engine

- Mass Air Flow (MAF) sensor
- Air hose
- Accelerator cable
- Throttle body assembly
- Solenoid valve bracket
- PRC solenoid valve
- Dynamic chamber
- Intake manifold and discard the gasket

5. Clean the mating surfaces of any gasket material

To install:

6. Install or connect the following:
- Intake manifold with a new gasket. Make certain that the convex side

of the new gasket faces the intake manifold. Torque the bolts to 18 ft. lbs. (25 Nm).
- Dynamic chamber
- PRC solenoid valve
- Solenoid valve bracket
- Throttle body assembly. Torque the bolts and nuts to 18 ft. lbs. (25 Nm).

7. When connecting the accelerator cable, perform the following adjustment:
   a. Move the white locking tab to the unlock position.
   b. Turn stopper B to the to the unlock position.

➡**If stopper B will not unlock, it may be necessary to carefully bend back tab C out a little using a suitable pry tool.**

   c. Push or pull the cable housing directly behind the spring.
   d. Turn the stopper B to the lock position.
   e. Measure the free play which should be 0.06–0.15 inch (1.5–4mm) and make sure the cable free play is within specification.
   f. Move the white locking tab to the lock position and check for proper accelerator operation.

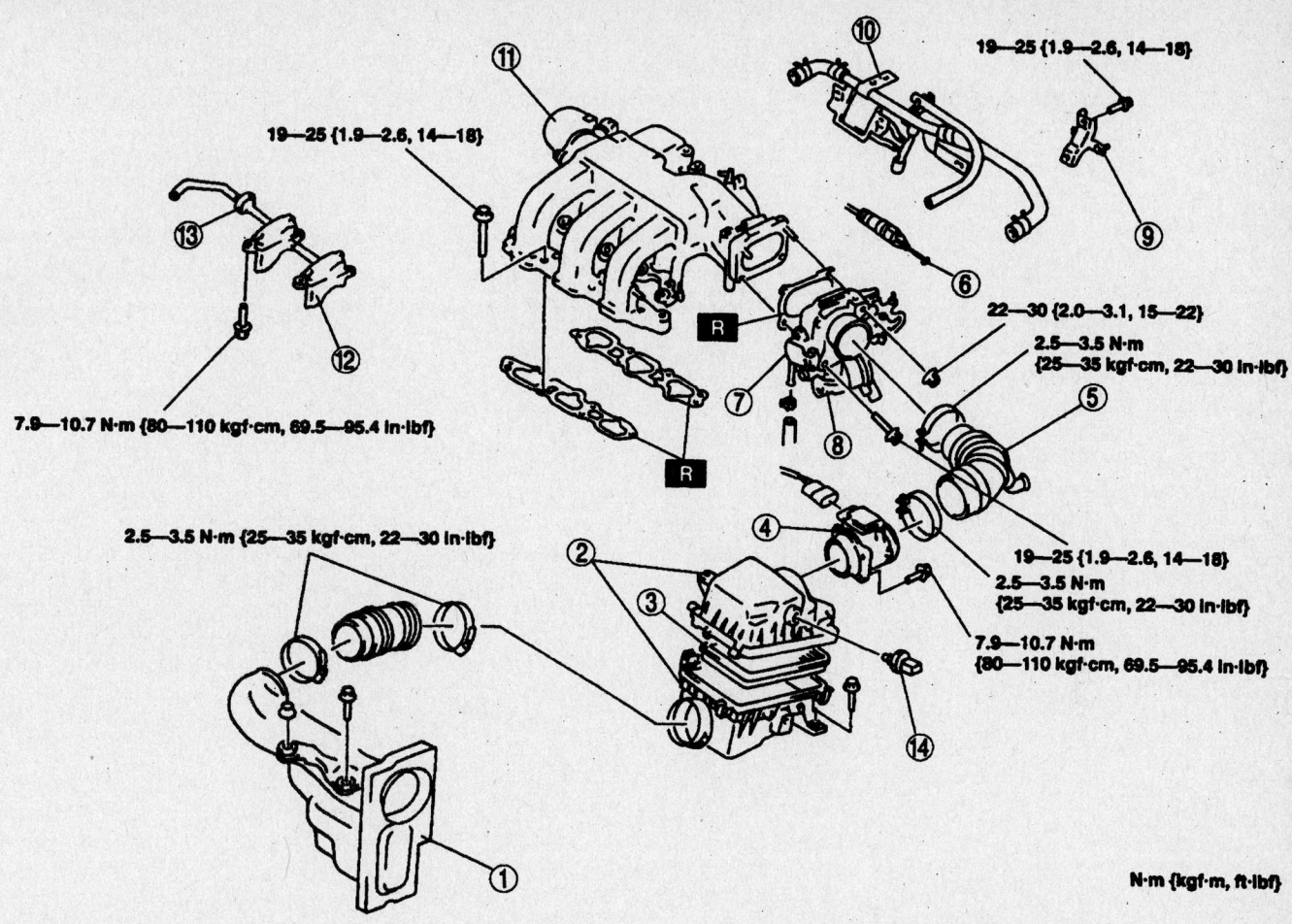

19—25 {1.9—2.6, 14—18}

19—25 {1.9—2.6, 14—18}

22—30 {2.0—3.1, 15—22}

2.5—3.5 N·m
{25—35 kgf·cm, 22—30 in·lbf}

7.9—10.7 N·m {80—110 kgf·cm, 69.5—95.4 in·lbf}

2.5—3.5 N·m {25—35 kgf·cm, 22—30 in·lbf}

19—25 {1.9—2.6, 14—18}

2.5—3.5 N·m
{25—35 kgf·cm, 22—30 in·lbf}

7.9—10.7 N·m
{80—110 kgf·cm, 69.5—95.4 in·lbf}

N·m {kgf·m, ft·lbf}

1	Fresh–air duct	8	IAC valve
2	Air cleaner	9	Intake manifold stay
3	Air cleaner element	10	Ventilation pipe
4	Mass air flow sensor	11	Intake manifold component
5	Air hose	12	Vacuum chamber
6	Accelerator cable	13	VRIS check valve (one–way)
7	Throttle body	14	Intake air temperature sensor

42356-MAZC-G14

Exploded view of the intake manifold assembly–626 with 2.5L (KL) engine

8. Install or connect the following:
   - Air hose
   - AF sensor
   - Air filter and cleaner
   - IAT sensor
   - Fresh air duct and resonance assembly
   - Negative battery cable
9. Fill the cooling system.
10. Start the vehicle, check for leaks and repair if necessary.

2.5L (KL) ENGINE

1. Before servicing the vehicle, refer to the precautions section.

2. Relieve the fuel system pressure.
3. Drain the cooling system.
4. Remove or disconnect the following:
   - Negative battery cable
   - Fresh air duct and air cleaner assembly
   - Mass Air Flow (MAF) sensor
   - Air hose
   - Accelerator cable
   - Throttle body assembly
   - Idle Air Control (IAC) valve
   - Intake manifold stay
   - Intake manifold and discard the gasket

5. Clean the mating surfaces of any gasket material
To install:
6. Install or connect the following:
   - Intake manifold with a new gasket. Torque the bolts to 18 ft. lbs. (25 Nm).
   - Intake manifold stay
   - IAC valve.
   - Throttle body. Torque the nuts to 18 ft. lbs. (25 Nm).
   - Accelerator cable
   - Air hose
   - MAF sensor. Torque the bolt to 95 inch lbs. (11 Nm).

- Air cleaner/air duct assembly
- Negative battery cable
7. Fill the cooling system.
8. Start the vehicle, check for leaks and repair if necessary.

Millenia

2.3L ENGINE

1. Before servicing the vehicle, refer to the precautions section.
2. Relieve the fuel system pressure.
3. Drain the cooling system.
4. Remove or disconnect the following:
   - Negative battery cable
   - Dynamic chamber cover
   - Charge air cooler air duct
   - Vacuum hoses and electrical connectors from the air cleaner housing
   - Air cleaner assembly
   - Fresh air ducts
   - Mass Air Flow (MAF) sensor and the air intake hose from the throttle body
   - Resonator

- Right-hand charge air cooler
- Left-hand charge air cooler
- Accelerator cable
- Vacuum hoses from the rear of the intake manifold and Exhaust Gas Recirculation (EGR) valve
- EGR valve
- Air intake pipe assembly
- Charge air cooler pipe
- Fuel supply line at the fuel rails and discard the copper washers
- Fuel and vacuum lines from the fuel pressure regulator
- Coolant hoses
- Wiring harness from the intake manifold
- Intake manifold mounting nuts and bolts in 2–3 steps
- Intake manifold

5. If necessary, label and disconnect the fuel hoses and electrical connectors from the throttle body. Remove the throttle body.

To install:

6. Clean all gasket mating surfaces.
7. Install or connect the following:

- Throttle body, if removed. Tighten the nuts/bolts to 14–18 ft. lbs. (19–25 Nm).
- Fuel hoses and electrical connectors.
- Intake manifold using new gaskets. Tighten the nuts/bolts in 2–3 steps, from the center to the ends, to 14–18 ft. lbs. (19–25 Nm).
- Wiring harness onto the intake manifold
- Coolant hoses
- Fuel and vacuum lines to the fuel pressure regulator
- Fuel supply line to the fuel rail using new copper washers
- Charge air cooler pipe
- Air intake pipe assembly using new gaskets. Verify that the rubber gaskets are not twisted or distorted. Tighten the bolts marked **A** to 70–95 inch lbs. (8–11 Nm) and all other bolts, in sequence, to 14–18 ft. lbs. (19–25 Nm).
- EGR valve using a new gasket

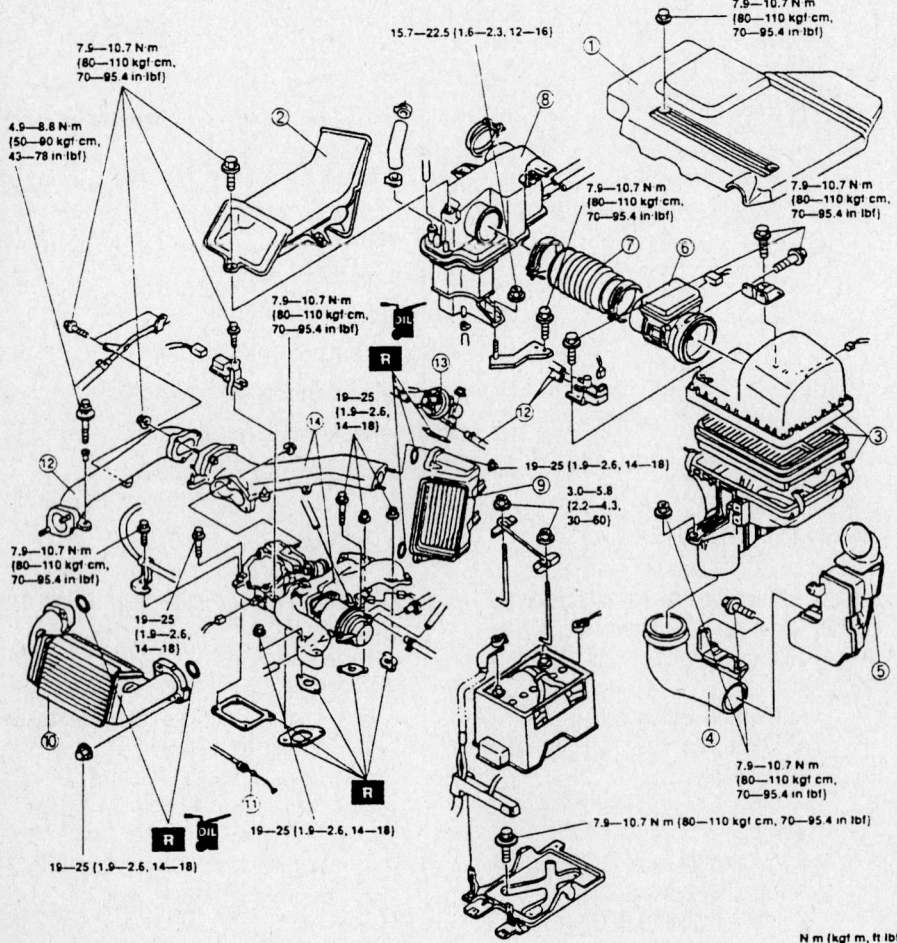

1. Dynamic chamber cover
2. Charge air cooler air duct
3. Air cleaner assembly
4. Air duct
5. Fresh air duct
6. Mass air flow sensor
7. Air intake hose
8. Resonator
9. Charge air cooler (RH)
10. Charge air cooler (LH)
11. Accelerator cable
12. Vacuum hose assembly
13. EGR control valve
14. Air intake pipe assembly

Exploded view of the intake manifold assembly (1 of 2)—2.3L engine

7923MG31

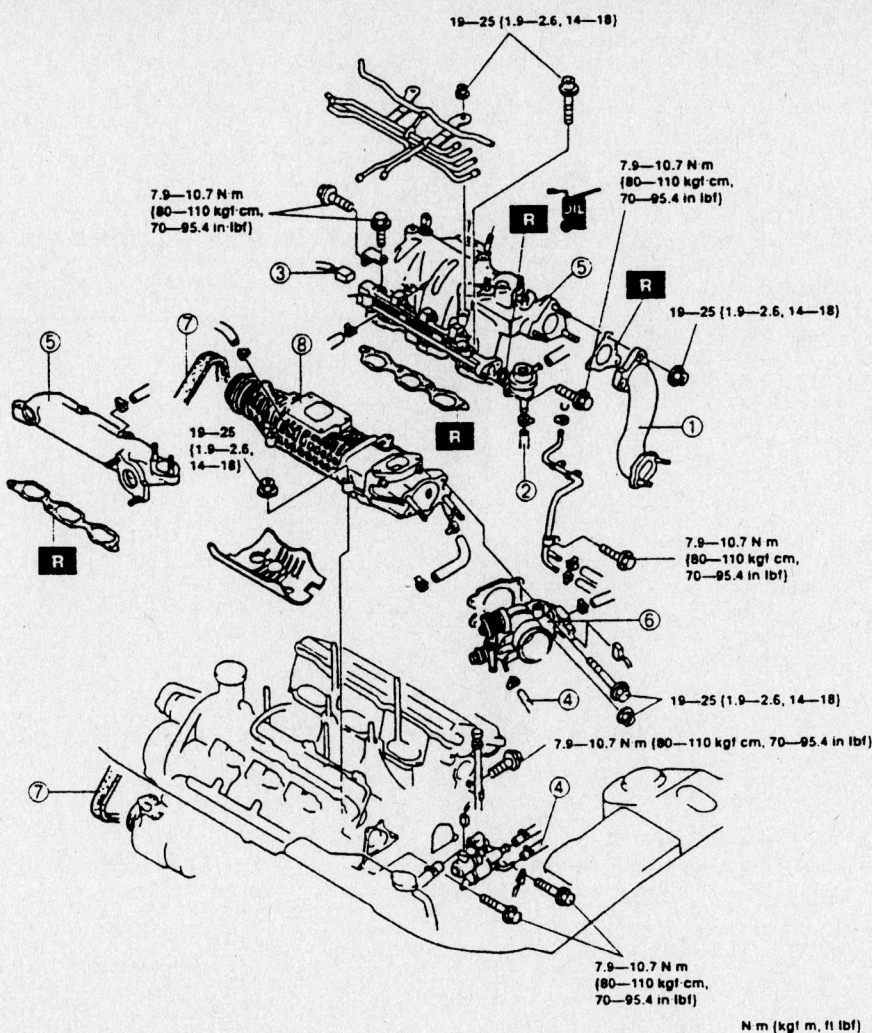

19—25 (1.9—2.6, 14—18)

7.9—10.7 N·m
(80—110 kgf·cm,
70—95.4 in·lbf)

7.9—10.7 N·m
(80—110 kgf·cm,
70—95.4 in·lbf)

19—25 (1.9—2.6, 14—18)

19—25
(1.9—2.6,
14—18)

7.9—10.7 N·m
(80—110 kgf·cm,
70—95.4 in·lbf)

19—25 (1.9—2.6, 14—18)

7.9—10.7 N·m (80—110 kgf·cm, 70—95.4 in·lbf)

7.9—10.7 N·m
(80—110 kgf·cm,
70—95.4 in·lbf)

N·m (kgf·m, ft·lbf)

7923MG32

1. Charge air cooler pipe
2. Fuel hose
3. Fuel distributor connector
4. Coolant hose
5. Intake manifold assembly
6. Throttle body assembly
7. Drive belt
8. Lysholm compressor

Exploded view of the intake manifold assembly (2 of 2)—2.3L engine

- Vacuum hoses to the intake manifold and EGR valve
- Accelerator cable, adjust as necessary
- Left and right-hand charge air coolers using new gaskets. Verify that the rubber gaskets are not twisted or distorted. Tighten the bolts marked **A** to 44–78 inch lbs. (5–9 Nm). Tighten the bolts marked **B** to 70–95 inch lbs. (8–11 Nm) and all other bolts, in sequence, to 14–18 ft. lbs. (19–25 Nm).
- Resonator. Tighten the nuts/bolts to 12–16 ft. lbs. (16–22 Nm).
- Air intake hose onto the throttle body
- MAF sensor
- Fresh air and air ducts
- Air cleaner assembly
- Vacuum hoses and electrical connectors to the air cleaner housing

- Charge air cooler air duct. Tighten the bolts to 70–95 inch lbs. (8–11 Nm).
- Dynamic chamber cover
- Negative battery cable
8. Fill the cooling system.
9. Run the engine and check for leaks.

2.5L (KL) ENGINE

1. Before servicing the vehicle, refer to the precautions section.
2. Relieve the fuel system pressure.
3. Drain the cooling system.
4. Remove or disconnect the following:

- Negative battery cable
- Fresh air duct and air cleaner assembly
- Air hose
- Mass Air Flow (MAF) sensor
- Water hoses from the throttle body
- Throttle body assembly

- Accelerator cable
- Fuel hoses from the rail
- Pipe and harness, refer to the illustration for component location
- Exhaust Gas Recirculation (EGR) Valve
- EGR pipe
- Intake manifold and discard the gasket

5. Clean the mating surfaces of any gasket material

To install:

6. Install or connect the following:

- Intake manifold with a new gasket. Torque the bolts to 18 ft. lbs. (25 Nm).
- EGR pipe and valve
- Pipe and harness, refer to the illustration for component location
- Fuel hoses to the rail
- Accelerator cable
- Throttle body assembly

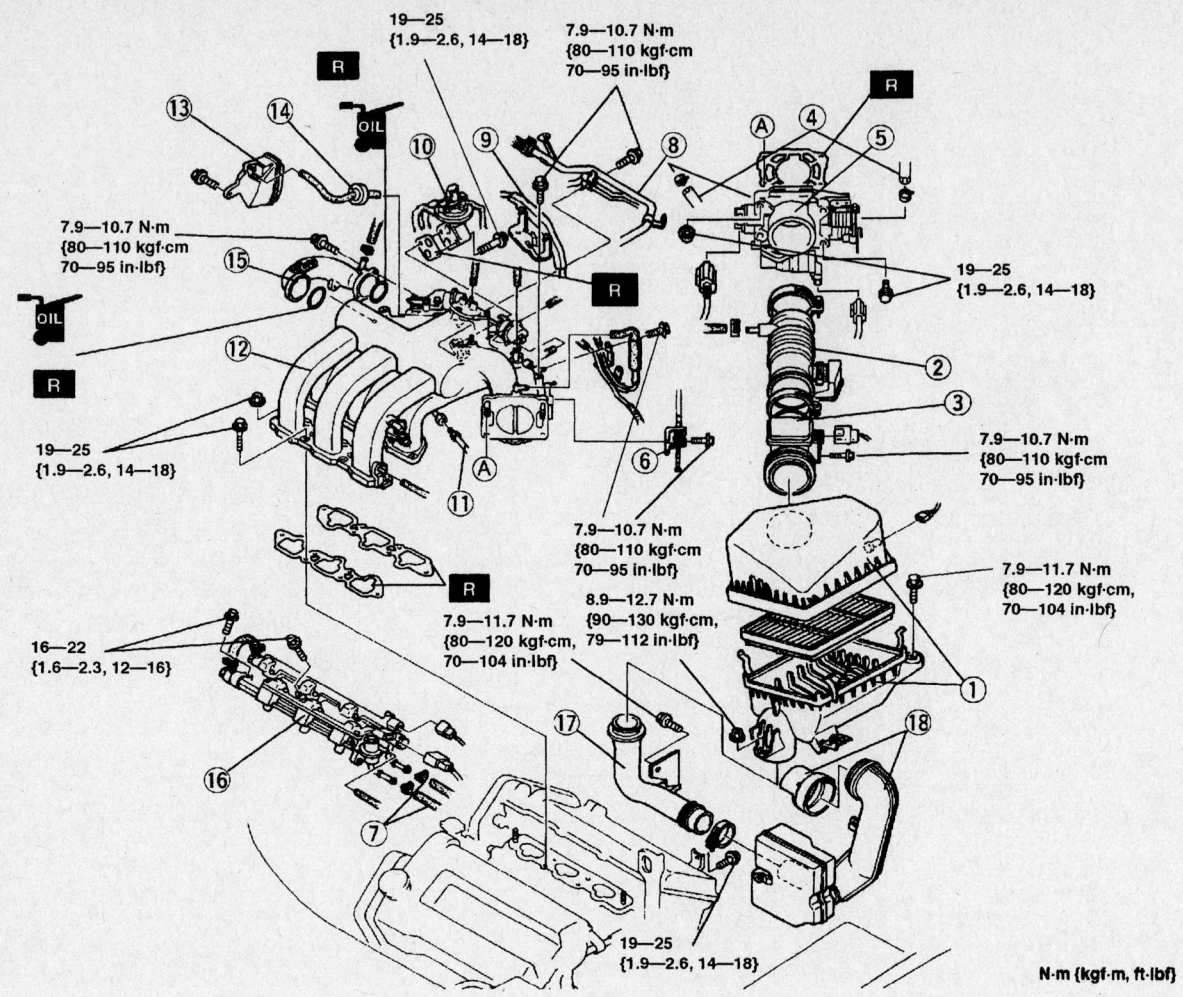

Exploded view of the intake manifold assembly–Millenia with 2.5L (KL) engine

1	Air cleaner	10	EGR valve
2	Air intake hose	11	EGR pipe
3	MAF sensor	12	Intake manifold
4	Water hose	13	Vacuum chamber
5	Throttle body component	14	Check valve
6	Accelerator cable	15	Air intake pipe
7	Fuel hose	16	Fuel distributor component
8	Pipe	17	Air duct
9	Harness	18	Fresh-air duct

42356-MAZC-GDT

- Water hoses to the throttle body. Torque the nuts to 18 ft. lbs. (25 Nm).
- MAF sensor. Torque the bolt to 95 inch lbs. (11 Nm).
- Air hose
- Fresh air duct and air cleaner assembly
- Negative battery cable

7. Fill the cooling system.

8. Start the vehicle, check for leaks and repair if necessary.

Exhaust Manifold

REMOVAL & INSTALLATION

2002–04 Miata

1.8L (BP) ENGINES

1. Before servicing the vehicle, refer to the precautions section.

2. Remove or disconnect the following:
   - Negative battery cable

- Air cleaner and air hose
- Exhaust manifold heat shield bolts and the heat shield
- Oxygen (O_2S) sensor electrical connector
- Exhaust pipe-to-exhaust manifold nuts and discard them
- Exhaust Gas Recirculation (EGR) pipe from the exhaust manifold
- Exhaust manifold nuts and bolts and discard the nuts
- Exhaust manifold

To install:

3. Clean all gasket mating surfaces.
4. Install or connect the following:
    *   Exhaust manifold. Torque the bolts to 29–33 ft. lbs. (39–46 Nm).
    *   Exhaust pipe. Torque the new nuts to 38 ft. lbs. (52 Nm).
    *   O2S connector
    *   EGR pipe. Torque the pipe to 34 ft. lbs. (47 Nm).
    *   Heat shield. Torque the bolts to 95 inch lbs. (11 Nm).
    *   Air hose and air cleaner
    *   Negative battery cable

626 and Protégé

1.6L (ZM) ENGINES

1. Before servicing the vehicle, refer to the precautions section.
2. Remove or disconnect the following:
    *   Negative battery cable
    *   Air cleaner and hose assembly

*   Water bypass pipe-to-engine block bolt, if equipped
*   Exhaust Gas Recirculation (EGR) pipe
*   Oxygen Sensor (O2S) from the exhaust system
*   Front exhaust pipe from the Warm-Up Three Way Catalytic (WU-TWC) converter
*   Exhaust manifold insulator
*   WU-TWC converter from the manifold
*   Exhaust manifold

To install:

3. Be sure all gasket mating surfaces are clean prior to assembly.
4. Tighten the components following the illustration.
5. Install or connect the following:
    *   Exhaust manifold
    *   WU-TWC converter to the manifold
    *   Exhaust manifold insulator

*   Front exhaust pipe from the WU-TWC converter
*   O2S to the exhaust system
*   EGR pipe
*   Water bypass pipe-to-engine block bolt
*   Air cleaner and hose assembly
*   Negative battery cable

2.0L (FS) ENGINE

1. Before servicing the vehicle, refer to the precautions section.
2. Remove or disconnect the following:
    *   Negative battery cable
    *   Exhaust manifold insulator
    *   Oxygen Sensor (O2S) electrical connector
    *   O2S, if necessary
    *   Exhaust Gas Recirculation (EGR) pipe, if equipped
    *   Exhaust pipe flange nuts
    *   Exhaust pipe from the manifold
    *   Exhaust manifold

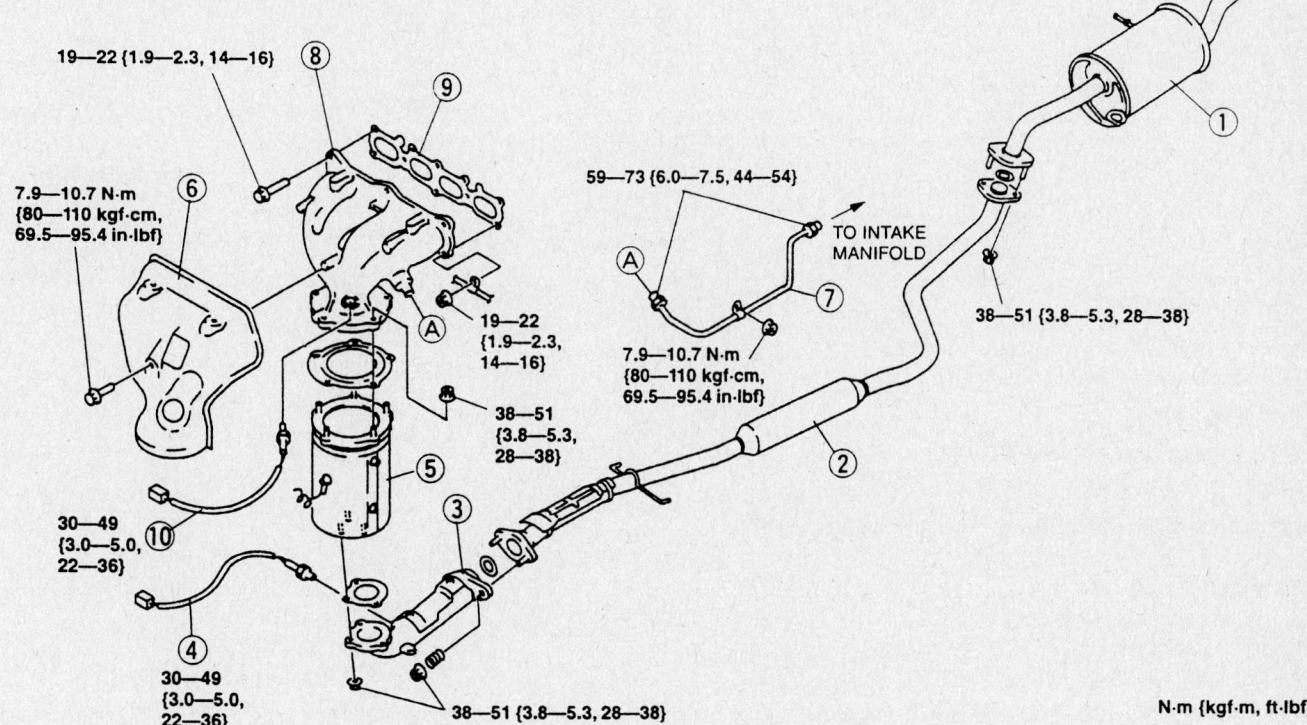

1	Main silencer	6	Exhaust manifold insulator
2	Presilencer	7	EGR Pipe
3	Front pipe	8	Exhaust manifold
4	HO2S (Rear)	9	Exhaust manifold gasket
5	WU-TWC	10	HO2S (Front)

9301MG06

Exploded view of the exhaust system—1.6L (ZM) engines

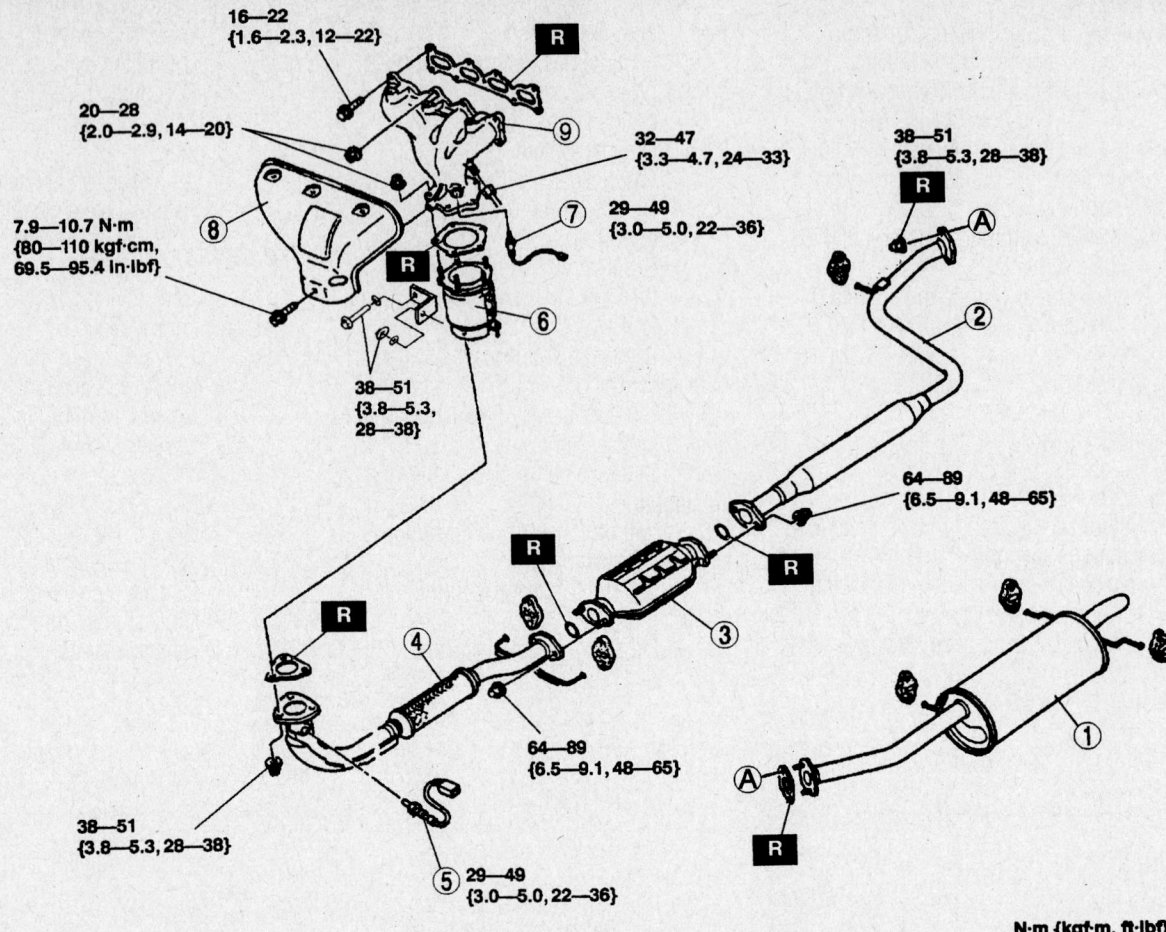

16—22
{1.6—2.3, 12—22}

20—28
{2.0—2.9, 14—20}

7.9—10.7 N·m
{80—110 kgf·cm,
69.5—95.4 in·lbf}

32—47
{3.3—4.7, 24—33}

29—49
{3.0—5.0, 22—36}

38—51
{3.8—5.3, 28—38}

38—51
{3.8—5.3,
28—38}

64—89
{6.5—9.1, 48—65}

38—51
{3.8—5.3, 28—38}

64—89
{6.5—9.1, 48—65}

29—49
{3.0—5.0, 22—36}

N·m {kgf·m, ft·lbf}

42356-MAZC-G15

1	Main silencer	6	Warm up three way catalytic converter
2	Presilencer	7	Heated oxygen sensor (Front)
3	Three way catalytic converter	8	Exhaust manifold insulator
4	Front pipe	9	Exhaust manifold
5	Heated oxygen sensor (Rear)		

Exploded view of the exhaust system—626 shown, Protégé similar

To install:
3. Be sure all gasket mating surfaces are clean prior to assembly.
4. Install or connect the following:
- Exhaust manifold using a new gasket. Torque the bolts to 22 ft. lbs. (30 Nm) and the nuts to 20 ft. lbs. (28 Nm).
- Exhaust pipe flange using a new gasket. Torque the nuts to 38 ft. lbs. (51 Nm).
- EGR pipe, if equipped
- O_2S, if necessary
- O_2S electrical connector
- Exhaust manifold insulator
- Negative battery cable

2.5L ENGINE

1. Before servicing the vehicle, refer to the precautions section.
2. Remove or disconnect the following:
- Negative battery cable
- Oxygen (O_2S) sensor connectors
- Front and rear exhaust pipe nuts and lower the exhaust system

➡**Both pipes must be disconnected, even if only one manifold is to be removed.**

- Exhaust Gas Recirculation (EGR) pipe, if equipped; if removing the rear (right side) manifold
- 3Heat shield

- 2 nuts and 5 bolts and the exhaust manifold
To install:
3. Clean all gasket mating surfaces.
4. Install or connect the following:
- Exhaust manifold using a new gasket. Torque the nuts and bolts to 12–22 ft. lbs. (16–22 Nm).
- Heat shield. Torque the bolts to 95 inch lbs. (11 Nm).
- EGR pipe, if equipped; if installing the rear (right side) manifold
- Exhaust pipes using new gaskets. Torque the nuts to 38 ft. lbs. (51 Nm).
- O_2S connectors
- Negative battery cable

Millenia

2.3L (KJ) Engine

1. Before servicing the vehicle, refer to the precautions section.
2. Remove or disconnect the following:
  - Negative battery cable
  - Front and rear exhaust pipe nuts and lower the exhaust system

➡ **Both pipes must be disconnected, even if only one manifold is to be removed.**

  - Exhaust Gas Recirculation (EGR) pipe, if removing the rear (right side) manifold
  - Charge air cooler and coolant/condenser fans, If removing the front (left side) manifold
  - Front and rear Oxygen Sensor (O_2S) connectors
  - 3 heat shield bolts and the heat shield
  - Exhaust manifold

To install:

3. Clean all gasket mating surfaces.
4. Install or connect the following:
  - Exhaust manifold using a new gasket. Torque the bolts to 14–18 ft. lbs. (18–24 Nm).
  - Heat shield. Torque the bolts to 70–95 inch lbs. (8–11 Nm).
  - O_2S connectors
  - Coolant/condenser fans and the charge air cooler, if installing the front (left side) manifold. Torque the bolts to 14–18 ft. lbs. (19–25 Nm).
  - EGR pipe, if installing the rear (right side) manifold
  - Exhaust pipes using new gaskets. Torque the nuts to 28–38 ft. lbs. (38–51 Nm).
  - Negative battery cable

2.5L (KL) Engine

1. Before servicing the vehicle, refer to the precautions section.
2. Remove or disconnect the following:
  - Negative battery cable
  - Oxygen Sensor (O_2S) connectors
  - Front and rear exhaust pipe nuts and lower the exhaust system

➡ **Both pipes must be disconnected, even if only one manifold is to be removed.**

  - Exhaust Gas Recirculation (EGR) pipe, if equipped; if removing the rear (right side) manifold

  - Heat shield
  - Nuts and bolts and the exhaust manifold

To install:

3. Clean all gasket mating surfaces.
4. Install or connect the following:
  - Exhaust manifold using a new gasket. Torque the nuts and bolts to 12–16 ft. lbs. (16–22 Nm).
  - Heat shield. Torque the bolts to 95 inch lbs. (11 Nm).
  - EGR pipe, if equipped; if installing the rear (right side) manifold
  - Exhaust pipes using new gaskets. Torque the nuts to 38 ft. lbs. (51 Nm).

  - O_2S connectors
  - Negative battery cable

Front Crankshaft Seal

REMOVAL & INSTALLATION

1. Before servicing the vehicle, refer to the precautions section.
2. Remove or disconnect the following:
  - Negative battery cable
  - Timing belt
  - Crankshaft damper bolt and damper
  - Timing belt sprocket
  - Sprocket key from the crankshaft

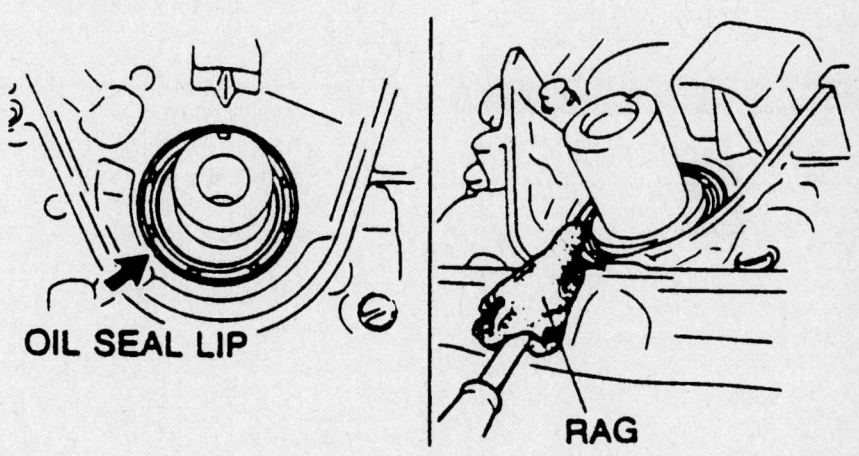

Remove the front engine seal by cutting the seal lip, then, so as not to damage the crankshaft, carefully pry the seal out with a prybar

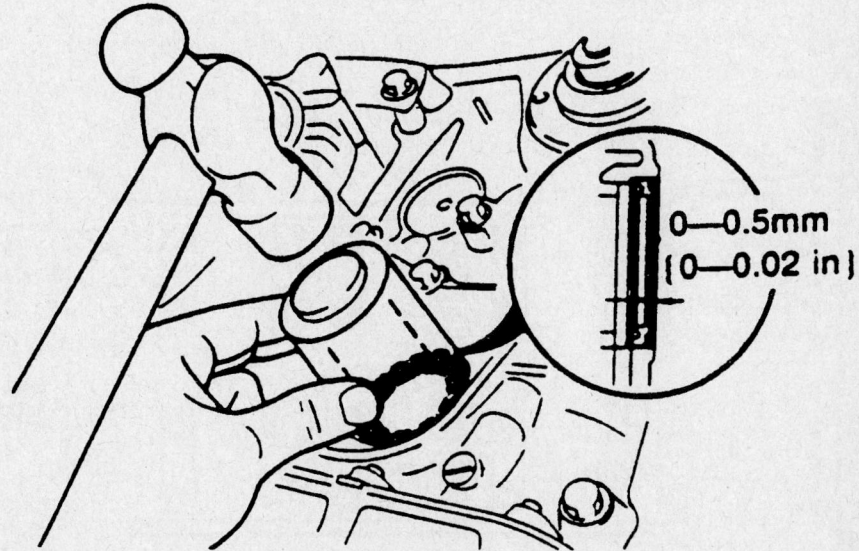

Install the seal using an appropriate driver, which fits over the crankshaft snout and presses on the outside edge of the seal

- Oil seal from the engine block using a prybar

✳ WARNING

Be careful not to score the crankshaft or the seal seat.

3. Clean the seal bore.

To install:

4. Install or connect the following:
   - New oil seal lubricated with clean engine oil, drive it into the engine using an installation tool until it seats
   - Sprocket key onto the crankshaft
   - Timing belt sprocket
   - Crankshaft damper
   - Timing belt
   - Negative battery cable

Camshaft

REMOVAL & INSTALLATION

1.6L (ZM) Engines

1. Before servicing the vehicle, refer to the precautions section.
2. Remove or disconnect the following:
   - Negative battery cable
   - Spark plug wires
   - Spark plugs
   - Cylinder head cover hoses, if equipped
   - Cylinder head cover bolts
   - Cylinder head cover
   - Timing belt
   - Camshaft by holding it with a wrench on the hexagon cast into the camshaft

- Sprocket bolts
- Sprockets

➡ Label the caps so they can be reinstalled in their original positions.

   - Camshaft cap bolts by loosening in 2–3 steps in the sequence shown
   - Camshaft caps

To install:

✳✳ CAUTION

Because there is little thrust clearance, the camshaft must be held in the horizontal position during installation. If not excessive force will be applied to the thrust area causing a burr on the receiving area of the cylinder head journal. Make sure to use the following procedure to avoid damage.

3. Install or connect the following:
   - Lubricate the camshaft journals and lobes with clean engine oil
   - Camshafts onto the head facing the cam noses at the no. 1 and 3 cylinders as illustrated. Make sure the camshaft sliding surface is free of sealant.
4. Apply a 0.04 inch (1mm) bead of silicone sealant to the areas illustrated in the illustration. Do not allow any sealant on the camshaft journals.
5. Install or connect the following:
   - Camshaft caps in their original positions
   - Cap bolts and hand tighten bolts 5, 7, 2 and 4
   - Cap bolts a few turns in the sequence illustrated
6. Make sure the camshaft settles hori-

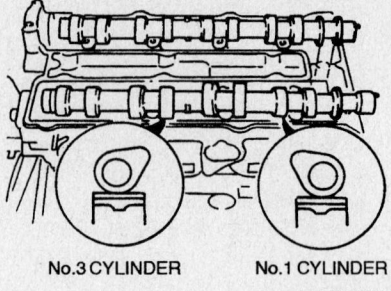

No.3 CYLINDER   No.1 CYLINDER

42356-MAZC-GHK

Install the camshafts onto the head facing the cam noses at the no. 1 and 3 cylinders as shown—1.6L (ZM) engines

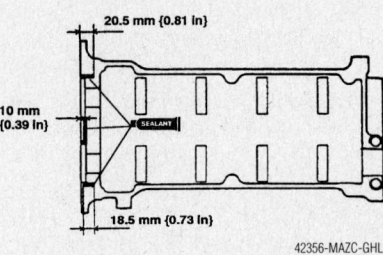

42356-MAZC-GHL

Apply silicone sealant to the cylinder head in the positions shown—1.6L (ZM) engines

zontally when the bearing cap bolts at the number 3 journals are tightened. Torque the bolts in 2–3 steps to 125 inch lbs. (14 Nm) in the proper sequence.
   - New camshaft seal by lubricating it with engine oil. Tap the seal into position, using a seal installer, until it is recessed into the cylinder head 0.01 inch (0.4mm.
7. Turn the camshafts until the dowel pins face straight up.
8. Install or connect the following:
   - Camshaft sprockets. Torque the bolts to 44 ft. lbs. (60 Nm) by holding the camshaft with the wrench on the cast hexagon.
   - Remaining components

2.0L (FS) Engine

1. Before servicing the vehicle, refer to the precautions section.
2. Drain the cooling system.
3. Relieve the fuel system pressure.
4. Drain the cooling system.
5. Remove or disconnect the following:
   - Negative battery cable
   - Timing belt
   - Front exhaust pipe
   - Air cleaner assembly
   - Accelerator cable
   - Fuel hose
   - Ignition coil

42356-MAZC-GHJ

Camshaft cap bolt loosening sequence—1.6L (ZM) engines

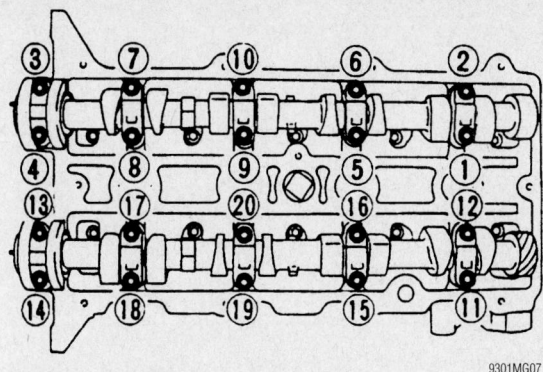

9301MG07

Camshaft cap bolt loosening sequence—2.0L (FS) engines

- Camshaft pulley. Hold the camshaft pulley with a back–up wrench while loosening the pulley bolt.

➥**Label the caps so they can be reinstalled in their original positions.**

- Camshaft cap bolts by loosening in 2–3 steps in the sequence shown
- Camshaft caps

To install:

6. Install or connect the following:
- Camshafts making sure to lubricate the journals and lobes with clean engine oil. Place the camshafts onto the cylinder head facing the cam noses at the No. 1 and No. 3 cylinders as illustrated.

7. Apply silicone sealant to the cylinder head on the front camshaft cap mating surfaces. Do not allow any sealant on the camshaft journals.

8. Install or connect the following:
- Camshaft caps in their original positions
- Hand tighten cap bolts marked 5, 5, 2 and 4, refer to the torque sequence illustration for location
- Cap bolts. Torque the bolts in 2–3 steps to 125 inch lbs. (14 Nm) in the proper sequence.

9. Make sure the camshaft is settled horizontally when the 2 bearing cap bolts at the number 3 journal are tightened.
- New camshaft seal by lubricating it with engine oil. Tap the seal into position, using a seal installer, until it is recessed into the cylinder head 0.012–0.027 inch (0.3–0.7mm).

10. Turn the camshafts until the dowel pins face straight up.

11. Install or connect the following:
- Camshaft sprockets. Torque the bolts to 44 ft. lbs. (60 Nm) by holding the camshaft with the wrench on the cast hexagon.
- Ignition coil

- Fuel hose
- Accelerator cable
- Air cleaner assembly
- Front exhaust pipe
- Timing belt
- Negative battery cable

12. Fill and bleed the cooling system.

13. Change the oil and filter.

14. Run the engine and check for proper operation.

1.8L (BP) Engines

1. Before servicing the vehicle, refer to the precautions section.

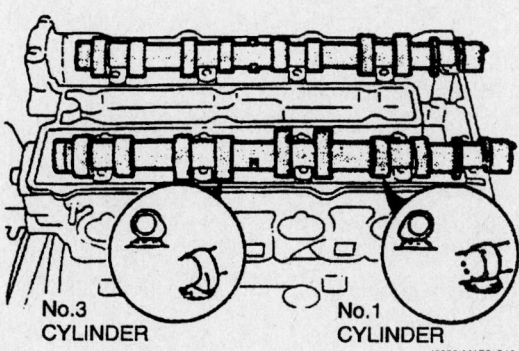

42356-MAZC-G16

Install the camshafts so the cam noses are positioned at the No. 1 and No. 3 cylinders—2.0L (FS) engine

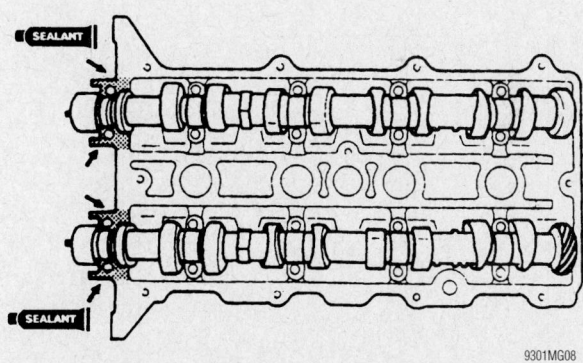

9301MG08

Apply silicone sealant to the cylinder head in the positions shown—2.0L (FS) engines

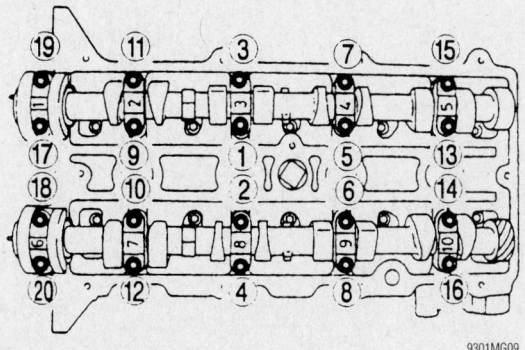

9301MG09

Camshaft cap bolt tightening sequence—2.0L (FS) engines

2. Remove or disconnect the following:
- Negative battery cable
- Timing belt
- Variable valve timing actuator and camshaft pulley. Hold the hexagonal part of the camshaft with a wrench to prevent the camshaft from turning and remove the actuator and pulley bolt.

➡**Label the caps so they can be reinstalled in their original positions.**

- Camshaft cap bolts by loosening in 2–3 steps in the sequence shown
- Camshaft caps
- Lifters and adjustment shims, if necessary

➡**Identify and mark each lifter as it is removed so it can be reinstalled in the same position.**

To install:

3. Install or connect the following:
- Lifters in their original positions by lubricating them with engine oil

➡**Verify that they move smoothly in their bore.**

4. Lubricate the camshaft journals and lobes with clean engine oil.
5. Install the camshafts so the cam projections of cylinders 1 and 3 face in the direction as illustrated.
6. Apply silicone sealant to the cylinder head on the front camshaft cap mating surfaces. Do not allow any sealant on the camshaft journals.
7. Install or connect the following:
- Camshaft caps in their original positions
- Cap bolts. Torque the bolts in 2–3 steps to 125 inch lbs. (14 Nm) in the proper sequence.
8. Make sure the camshaft is settled horizontally when the 2 bearing cap bolts at the number 3 journal are tightened.
- New camshaft seal by lubricating it with engine oil. Tap the seal into position, using a seal installer, until it is flush with the edge of the camshaft.
9. Install the variable valve timing actuator and camshaft pulley as follows:
    a. Rotate the camshaft and face the knock pin to the position illustrated.
    b. Install the camshaft so that the alignment mark of the pulley faces as illustrated.
    c. Install the camshaft so the knock pin of the camshaft is connected to the camshaft knock pin hole of the actuator.

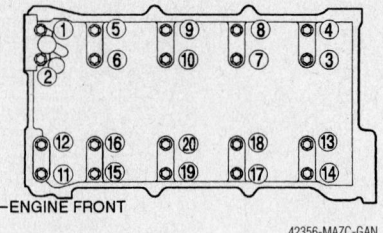

Camshaft cap bolt loosening sequence——1.8L (BP) engine

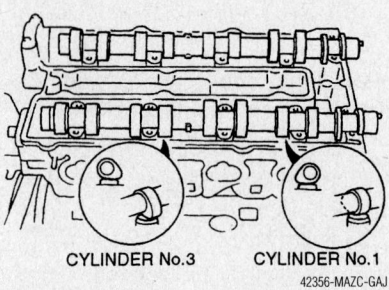

Install the camshafts so the cam projections of cylinders 1 and 3 face in the direction shown—1.8L (BP) engine

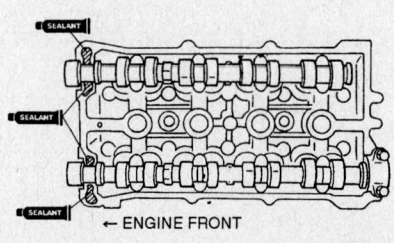

Apply silicone sealant to the cylinder head in the positions shown—1.8L (BP) engine

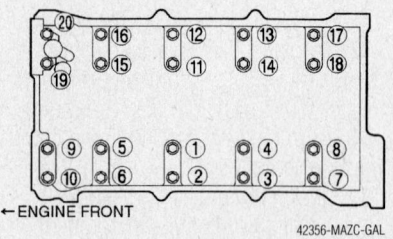

Camshaft cap bolt tightening sequence—1.8L (BP) engine

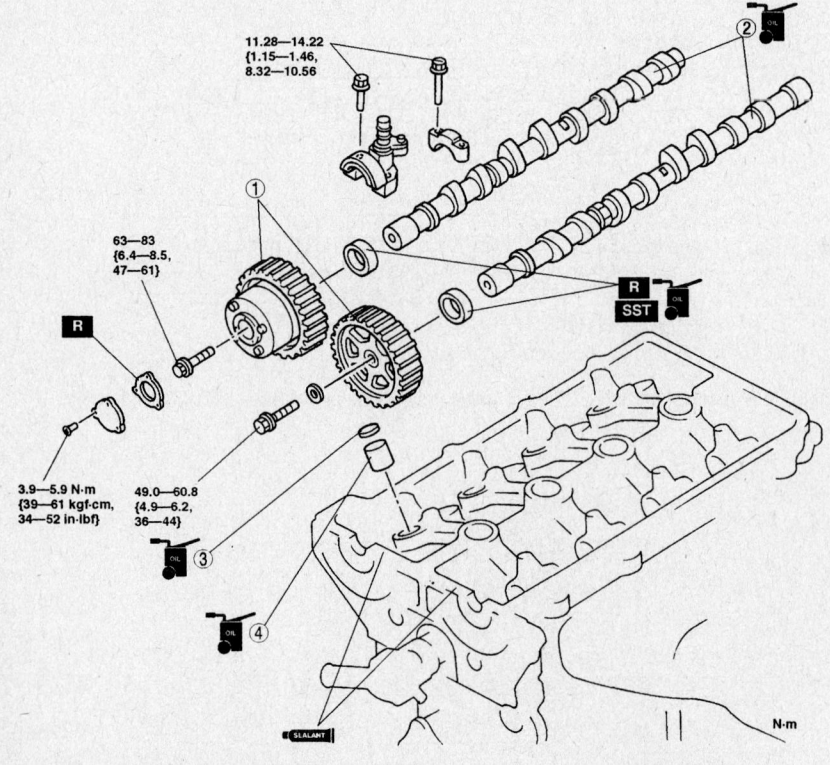

| 1 | Variable valve timing actuator and Camshaft Pulley | 3 | Adjustment shim |
| 2 | Camshaft | 4 | Tappet |

Exploded view of the camshaft assembly and related components—1.8L (BP) engine

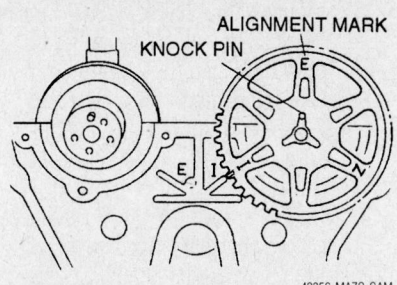

When installing the variable valve timing actuator and camshaft pulley, make sure the knock pin and alignment marks are aligned as shown—1.8L (BP) engine

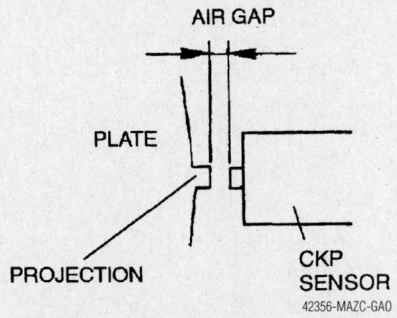

Check the Crankshaft position (CKP) sensor gap—1.8L (BP) engine

d. Hold the hexagonal part of the camshaft with a wrench to prevent the camshaft from turning and tighten the actuator and pulley bolt. Refer to the illustration for torque specifications.

10. Install the remaining components in the reverse order of removal and inspect the Crankshaft position (CKP) sensor gap. Measure the gap between each 4 projections of the plate behind crankshaft pulley. The gap should be 0.020–0.059 inch (0.5–1.5mm). If not as specified, adjust.

2.3L (KJ) Engine

1. Before servicing the vehicle, refer to the precautions section.
2. Relieve the fuel system pressure.
3. Drain the engine coolant.
4. Remove or disconnect the following:
   - Negative battery cable
   - Oxygen (O_2S) sensor connectors
   - Exhaust pipe-to-manifold nuts and lower the exhaust pipes
   - Right-hand 3-way catalytic converter
   - Compressor (supercharger)
   - Intake manifold
   - Timing belt covers and timing belt
   - Spacer and O-ring from the front of the camshaft
   - Ignition coils

- Cylinder head cover mounting bolts, in 5–6 steps, using the reverse of the tightening sequence
- Cylinder head cover
- Camshaft sprockets

5. Turn the camshafts so the knock pins are aligned with the marks on the camshaft caps. This will reduce the pressure on the adjustment shims.

6. Note the markings on the camshaft caps prior to removal, so they can be reinstalled in the same positions. The right-hand (rear) caps are marked with numbers and the left-hand (front) caps are marked with letters.

7. Loosen the front camshaft cap bolts in sequence, in 5–6 steps. Remove the front camshaft caps. Remove the remaining camshaft cap bolts in the proper sequence. Remove the caps, being sure to remove the thrust caps last. Do not damage the cylinder head thrust bearing support.

8. Remove or disconnect the following:
   - Camshafts and oil seals
   - Lifters and adjustment shims

9. Identify and mark each lifter as it is removed so it can be reinstalled in the same position.

To install:

10. Install or connect the following:
   - Lifters in their original positions, lubricated with engine oil. Verify that they move smoothly in their bore.
   - New oil seals on the camshafts
   - Camshafts with the camshaft lobes, journals and supports lubricated with engine oil and the gear marks aligned

11. Install the camshafts so the intake and exhaust camshaft gear marks align. Adjust

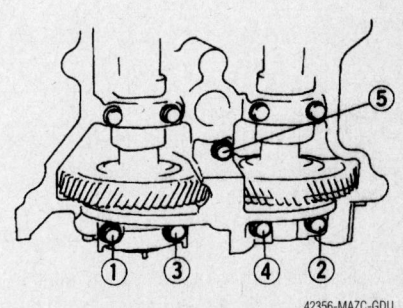

Loosen the front camshaft cap bolts using this sequence—2.3L (KJ) engine

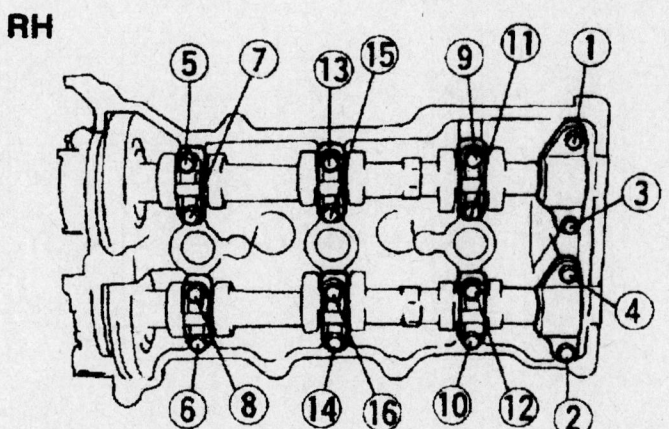

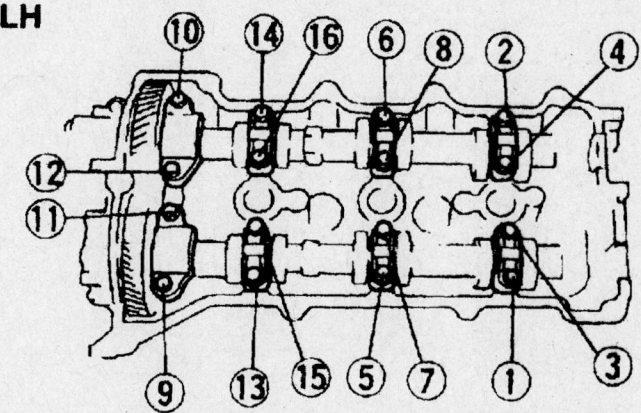

Loosen the remaining camshaft cap bolts using this sequence—2.3L (KJ) engine

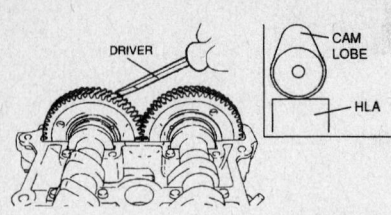

The intake cam lobes of the number 1 cylinder on the right hand side and the number 2 cylinder on the left hand side face straight up—2.3L (KJ) engine

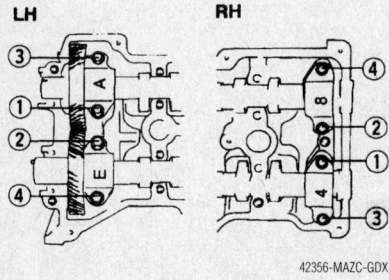

Tighten the front camshaft cap bolts in this sequence—2.3L (KJ) engine

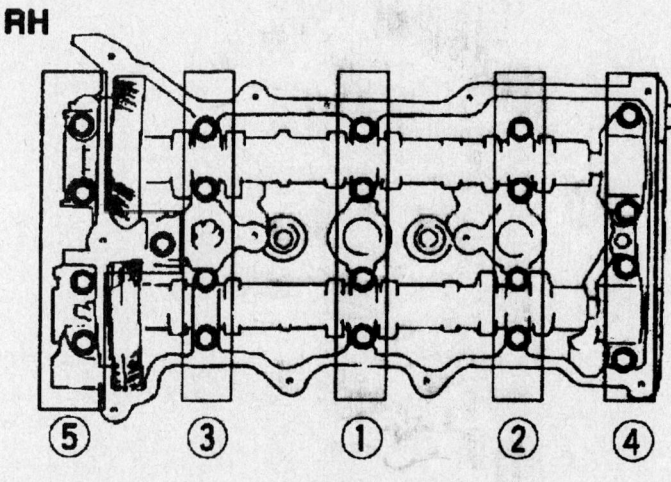

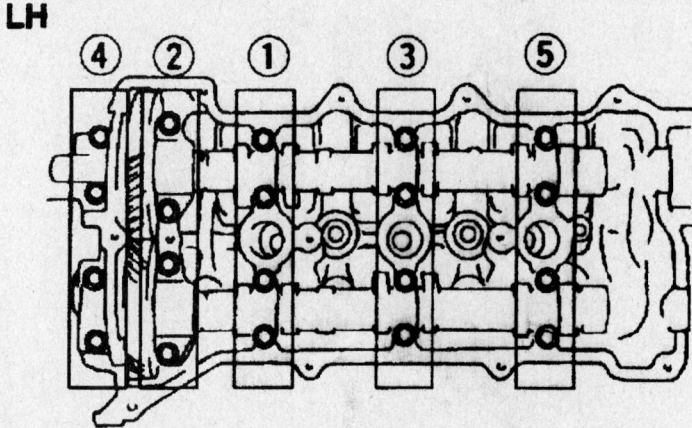

Tighten the camshaft cap bolts using this sequence—2.3L (KJ) engine

the friction gear position so the tappets are not lifted using the cam lobes. The intake cam lobes of the number 1 cylinder on the right hand side and the number 2 cylinder on the left hand side face straight up.

- Front trust caps. Torque the bolts, in 5–6 steps, until the caps are fully seated on the cylinder head.

12. Apply silicone sealant, at a thickness of 0.06–0.09 inch (1.5–2.5mm), to the cylinder head surface in the area forward of the camshaft gear cavity.

13. Install or connect the following:

- Remaining camshaft caps in their original positions. Make sure the camshaft remains horizontal as the camshaft bolts marked **1** in the illustration are tightened. Torque the bolts, in sequence, in 5 equal steps, with the final step being 100–125 inch lbs. (11–14 Nm). Then retighten the bolts in the same sequence to 100–125 inch lbs. (11–14 Nm).
- New camshaft oil seal lubricated with engine oil. Tap the seal in

evenly with a Seal Installer tool 49 F401 337A with a final protrusion of 0–0.02 inch (0–0.5mm).

- New blind cap by tapping it in
- Camshaft sprockets. Torque the bolts to 91–103 ft. lbs. (123–140 Nm).

14. Measure and adjust valve clearances.

15. Remove any sealant and gasket material from the cylinder head cover contact surfaces.

16. Apply silicone sealant to the cylinder head in the area adjacent to the front and rear camshaft caps.

17. Install or connect the following:

- New gasket on the cylinder head
- Cylinder head cover. Torque the cover bolts in 5–6 steps, in sequence, to 44–78 inch lbs. (5–9 Nm).
- Distributor using a new O-ring
- Ignition coils
- Intake manifold
- Spacer using a new O-ring. Torque the bolt to 14–18 ft. lbs. (19–25 Nm).
- Timing belt and timing belt cover
- Negative battery cable

18. Start the engine, check for leaks and repair if necessary.

2.5L (KL) Engine

626

1. Before servicing the vehicle, refer to the precautions section.

2. Drain the cooling system.

3. Relieve the fuel system pressure.

4. Drain the cooling system.

5. Remove or disconnect the following:

- Negative battery cable
- Timing belt
- Front exhaust pipe
- Air cleaner assembly
- Intake manifold
- Fuel hose
- Cylinder head cover
- Camshaft sprocket bolt by holding the camshaft with a wrench on the hexagon cast into the camshaft
- Upper radiator hose
- Number 3 engine mount bracket
- Seal plate
- Water outlet

6. Turn the camshaft, using a wrench on the cast hexagon, until the camshaft knock pin is aligned with the cylinder head marks.

➡ **Do not remove the camshaft caps when the camshaft lobe is pressing on a lifter, as the thrust journal support may become damaged.**

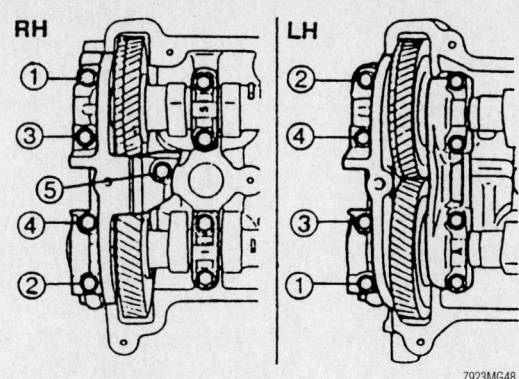

Front camshaft cap bolt loosening sequence—626 with 2.5L engines

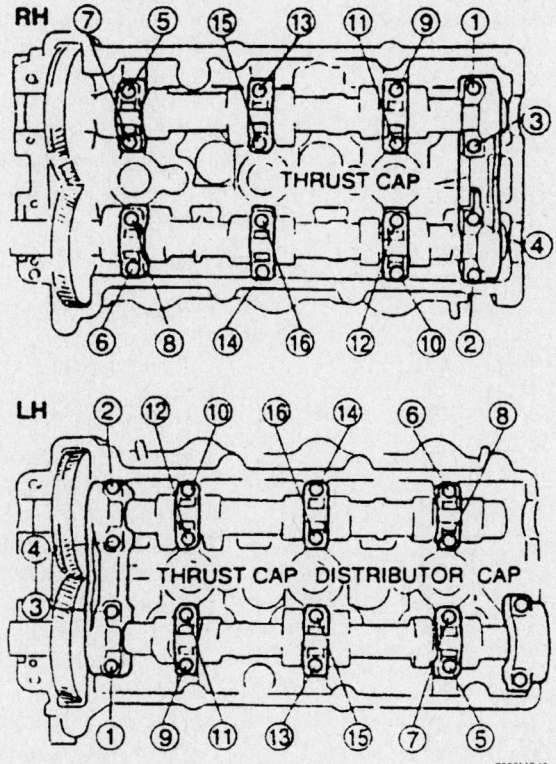

Camshaft cap bolt loosening sequence—626 with 2.5L (KL) engines

7. Loosen the front camshaft cap bolts in 5–6 steps, in the proper sequence. Bolt **A** is only on the right cylinder head. Remove the front camshaft cap.

8. Mark the position of the camshaft caps so they can be reinstalled in their original locations. Loosen the remaining camshaft cap bolts in 5–6 steps, in the proper sequence, then remove the caps.

9. Remove or disconnect the following:
   • Camshafts
   • Lifters and adjustment shims, if necessary

➡Identify and mark each lifter as it is removed so it can be reinstalled in the same position.

To install:

10. Install or connect the following:
   • Lifters in their original positions lubricated with engine oil. Verify that they move smoothly in their bore.
   • Camshafts by lubricating the camshaft journals, lobes and gears with clean engine oil and aligning the timing marks

➡**The thrust plate positions for the right and left cylinder head camshafts are different.**

11. Be sure the camshaft cap and cylinder head surfaces are clean. Apply a small amount of sealant to the mating surface of the front camshaft cap on both cylinder heads and the rear exhaust camshaft cap on the left cylinder head. Do not get any sealant on the camshaft rotating surfaces.

12. Install or connect the following:
   • Front camshaft caps and thrust plate caps. Torque the bolts until the cap seats fully to the cylinder head.
   • Remaining caps in their original locations. Torque the bolts in 5–6 steps to 126 inch lbs. (14 Nm), in the proper sequence.
   • New oil seals in the cylinder head using an installer
   • New blind cap coated with sealant, tap it in place using a plastic hammer
   • Camshaft sprockets

➡**On the right cylinder head, install the sprocket so theRmark can be seen and the timing mark aligns with the camshaft knock pin. On the left cylinder head, install the sprocket so theLmark can be seen and the timing mark aligns with the camshaft knock pin.**

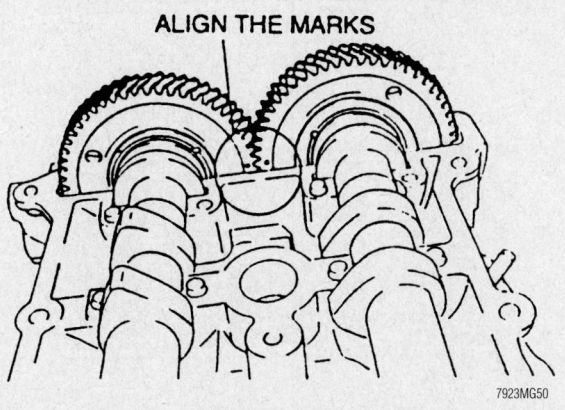

When installing the camshafts, ensure that the marks on the cam gears are aligned—2.3L and 2.5L engines

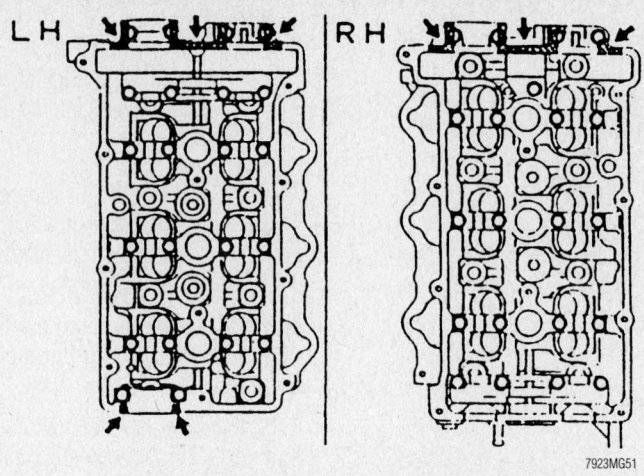

Put silicone sealant on the cylinder head at the positions shown—626 with 2.5L (KL) engines

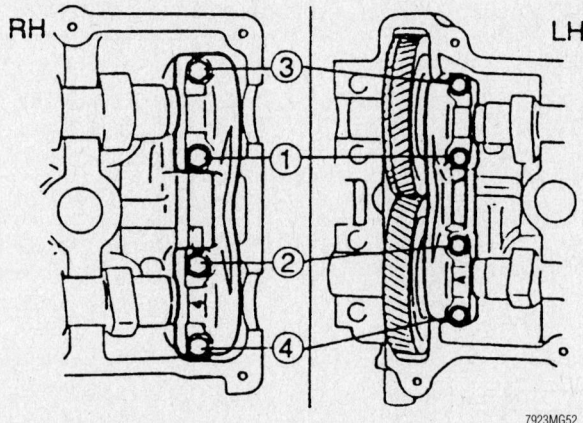

Front camshaft cap bolt tightening sequence—626 with 2.5L (KL) engines

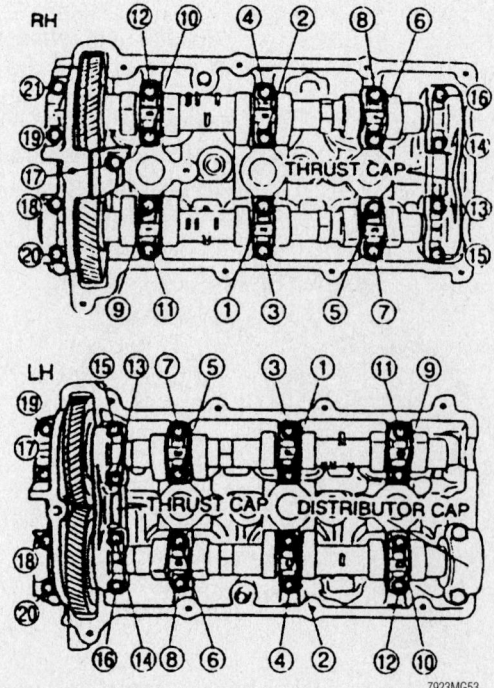

Camshaft cap bolt tightening sequence—626 with 2.5L (KL) engines

- Camshaft sprocket bolts lubricated with engine oil, by holding the camshaft with a wrench on the cast hexagon. Torque the bolt to 97 ft. lbs. (132 Nm).
- Water outlet
- Seal plate
- Number 3 engine mount bracket
- Upper radiator hose
- Camshaft pulley. Hold the camshaft pulley with a back–up wrench while tightening the pulley bolt to 90–97 ft. lbs. (123–132 Nm).
- Cylinder head cover
- Fuel hose
- Intake manifold
- Air cleaner assembly
- Front exhaust pipe
- Timing belt
- Negative battery cable

13. Fill and bleed the cooling system.
14. Change the oil and filter.
15. Run the engine and check for proper operation.

MILLENIA

1. Before servicing the vehicle, refer to the precautions section.
2. Drain the cooling system.
3. Relieve the fuel system pressure.
4. Drain the cooling system.
5. Remove or disconnect the following:
- Negative battery cable
- Timing belt
- Air cleaner assembly
- Spark plug wires
- Distributor
- Intake manifold
- Upper radiator hose
- Water outlet
- Number 3 engine mount bracket
- Seal plate
- Front exhaust pipe
- Alternator stay
- Cylinder head cover
- Camshaft pulley. Hold the camshaft pulley with a back–up wrench while loosening the pulley bolt.

6. Turn the camshaft, using a wrench on the cast hexagon, until the camshaft knock pin is aligned with the cylinder head marks.

➡**Do not remove the camshaft caps when the camshaft lobe is pressing on a lifter, as the thrust journal support may become damaged.**

7. Loosen the front camshaft cap bolts in 5–6 steps, in the proper sequence. Bolt **A** is only on the right cylinder head. Remove the front camshaft cap.

8. Mark the position of the camshaft

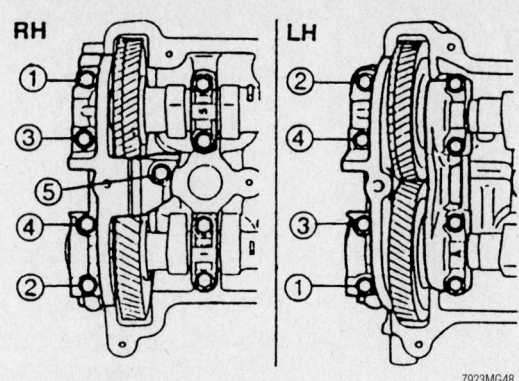

Front camshaft cap bolt loosening sequence—2.5L engines

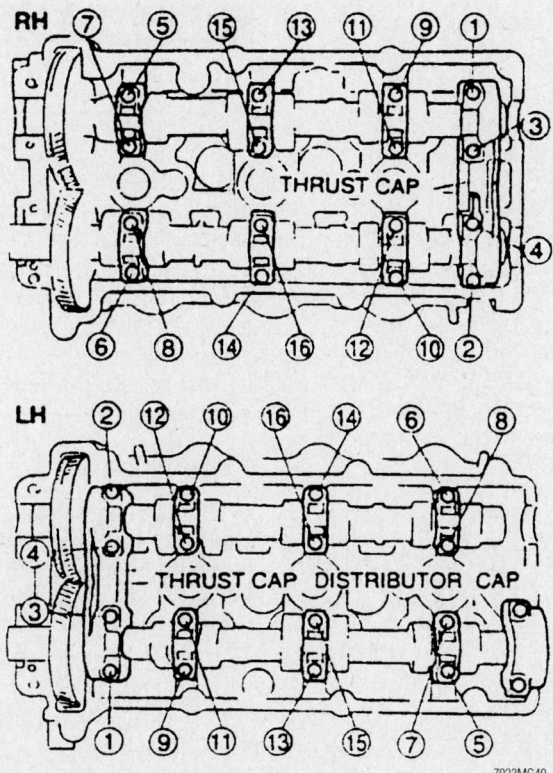

Camshaft cap bolt loosening sequence—2.5L (KL) engines

caps so they can be reinstalled in their original locations. Loosen the remaining camshaft cap bolts in 5–6 steps, in the proper sequence, then remove the caps.

9. Remove or disconnect the following:
   • Camshafts
   • Lifters and adjustment shims, if necessary

➡Identify and mark each lifter as it is removed so it can be reinstalled in the same position.

To install:

10. Install or connect the following:
   • Lifters in their original positions lubricated with engine oil. Verify that they move smoothly in their bore.

   • Camshafts by lubricating the camshaft journals, lobes and gears with clean engine oil and aligning the timing marks

➡**The thrust plate positions for the right and left cylinder head camshafts are different.**

11. Be sure the camshaft cap and cylinder head surfaces are clean. Apply a small amount of sealant to the mating surface of the front camshaft cap on both cylinder heads and the rear exhaust camshaft cap on the left cylinder head. Do not get any sealant on the camshaft rotating surfaces.

12. Install or connect the following:
   • Front camshaft caps and thrust plate caps. Torque the bolts until the cap seats fully to the cylinder head.
   • Remaining caps in their original locations. Torque the bolts in 5–6 steps to 126 inch lbs. (14 Nm), in the proper sequence.
   • New oil seals in the cylinder head using an installer
   • New blind cap coated with sealant, tap it in place using a plastic hammer
   • Camshaft sprockets

➡**On the right cylinder head, install the sprocket so theRmark can be seen and the timing mark aligns with the camshaft knock pin. On the left cylinder head, install the sprocket so theLmark can be seen and the timing mark aligns with the camshaft knock pin.**

   • Camshaft sprocket bolts lubricated with engine oil, by holding the camshaft with a wrench on the cast hexagon. Torque the bolt to 97 ft. lbs. (132 Nm).
   • Cylinder head cover

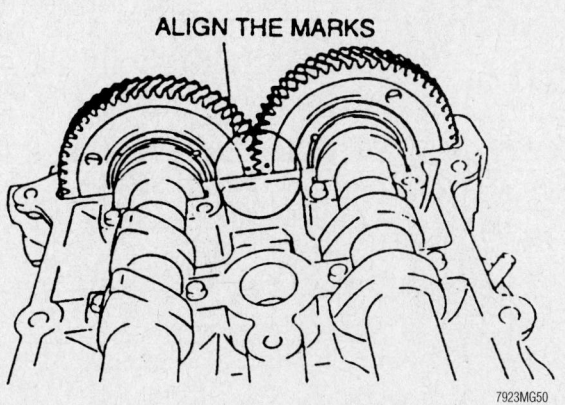

When installing the camshafts, ensure that the marks on the cam gears are aligned—2.5L (KL) engines

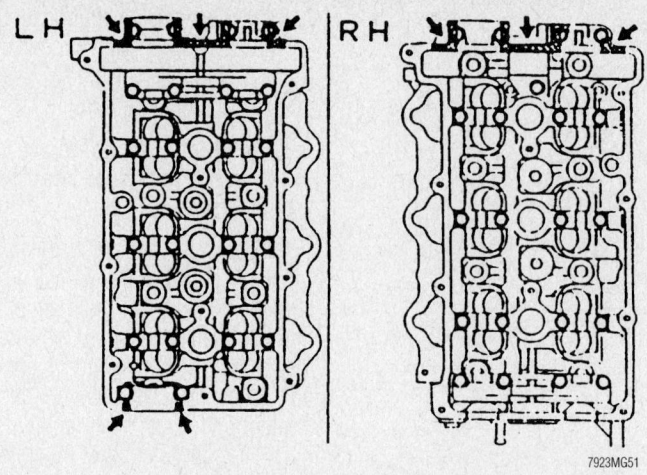

Put silicone sealant on the cylinder head at the positions shown—2.5L (KL) engines

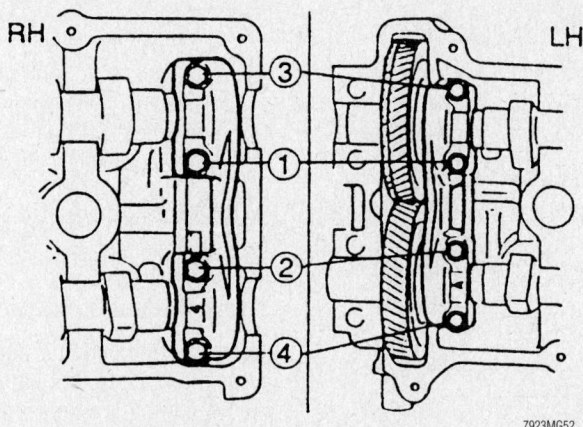

Front camshaft cap bolt tightening sequence—2.5L (KL) engines

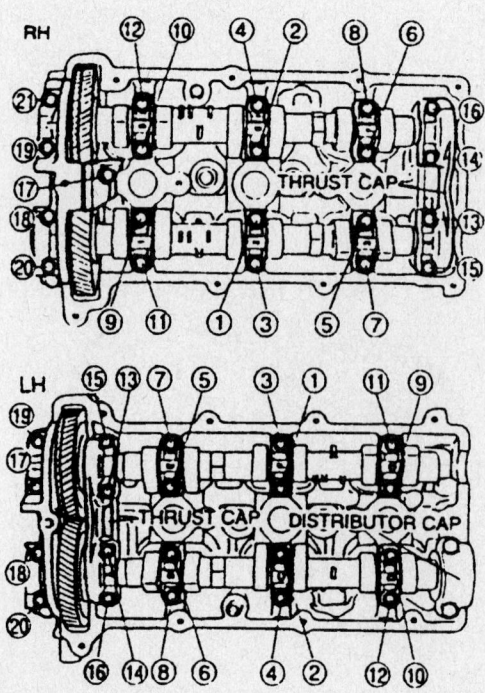

Camshaft cap bolt tightening sequence—2.5L (KL) engines

- Alternator stay
- Front exhaust pipe
- Seal plate
- Number 3 engine mount bracket
- Water outlet
- Upper radiator hose
- Intake manifold
- Distributor and spark plug wires
- Air cleaner assembly
- Timing belt
- Negative battery cable

13. Fill and bleed the cooling system.
14. Change the oil and filter.
15. Run the engine and check for proper operation.

Valve Lash

ADJUSTMENT

These engines use solid cam followers with a removable adjustment shim. The valve lash clearance is measured with the original shim installed and checked against the specification. If adjustment is necessary, the original shim is removed, and a thicker or thinner shim is installed to obtain the proper clearance. Special tools are required in order to adjust the shim without removing the camshaft.

1.8L (BP) Engine

➡With the engine cold, standard valve clearance is 0.08–0.009 inch (0.18–0.24mm) on intake side and 0.012–0.013 inch (0.28–0.34mm) on the exhaust side.

1. Before servicing the vehicle, refer to the precautions section.
2. Remove the cylinder head cover.
3. Measure the valve clearance by turning the crankshaft clockwise until the No. 1 piston is at Top Dead Center (TDC).
4. Measure the valve clearance at **A**. If the clearance exceeds specifications, replace the adjustment shim.
5. Turn the crankshaft clockwise until the cam on the camshaft requiring the adjustment is positioned straight up.
6. Turn the crankshaft clockwise 360 degrees until the No. 4 piston is at TDC. Measure the valve clearance at **B**. If the clearance exceeds specifications, replace the adjustment shim.
7. Repeat this procedure for all the camshafts.
8. Turn the crankshaft clockwise until the cam on the camshaft requiring the adjustment is positioned straight up.
9. Remove the camshaft cap bolts one pair at a time as follows:

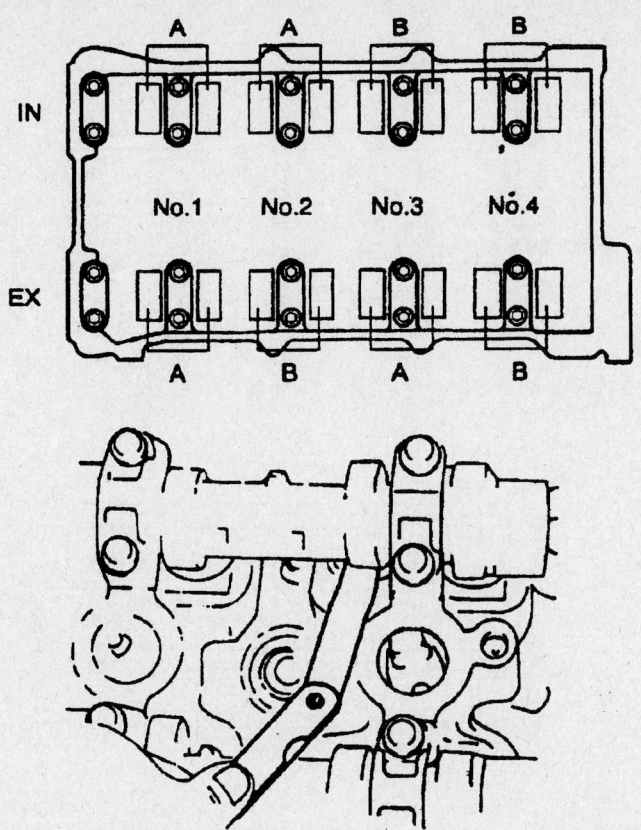

Valve clearance checking positions—1.8L (BP) engines

42356-MAZC-GAP

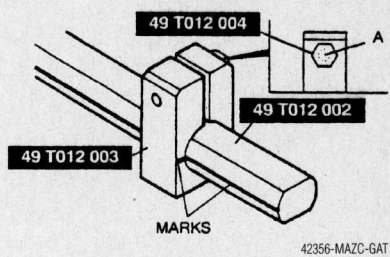

42356-MAZC-GAT

Align the mark on the 49-T012-002 (shaft) with the mark on the 49-T012-003 (clamp—1.8L (BP) engines

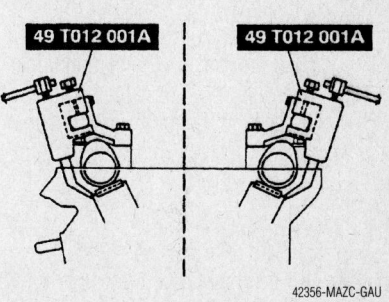

42356-MAZC-GAU

Position special tool 49-T012-001A toward the center of the cylinder head and mount it on the shaft where the adjustment shim needs replacement—1.8L (BP) engines

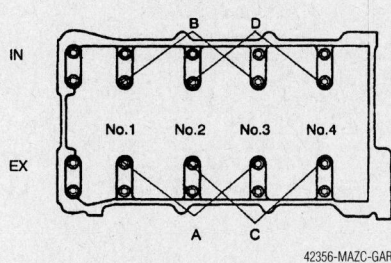

42356-MAZC-GAR

Cam bearing cap bolt removal positions—1.8L (BP) engines

a. For exhaust side No. 1, 2 and 3 cylinder adjustment shim removal use **A**.
b. For intake side No. 1, 2 and 3 cylinder adjustment shim removal use **B**.
c. For exhaust side No. 2, 3 and 4 cylinder adjustment shim removal use **C**.
d. For intake side No. 2, 3 and 4 cylinder adjustment shim removal use **D**.

➡ **For exhaust side No's. 2 and 3, cylinder adjustment shim removal; remove bolts A or C. For intake side No's 2 and 3 cylinder adjustment shim removal, remove bolts B or D.**

10. Install special tools 49-T012-002 and 003, using the camshaft cap bolt holes.

Tighten the bolts to 100–125 inch lbs. (11–14 Nm).

11. Align the mark on the 49-T012-002 (shaft) with the mark on the 49-T012-003 (clamp). Tighten special tool 49-T012-004 (bolt) to secure the shaft.

12. Position special tool 49-T012-001A toward the center of the cylinder head and

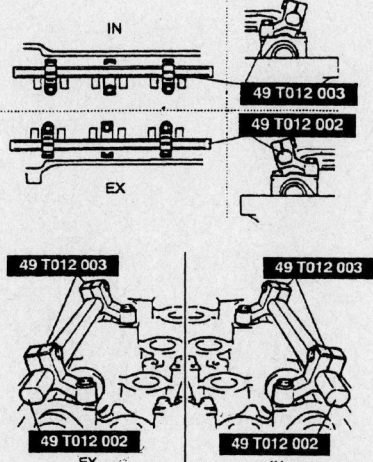

42356-MAZC-GAS

Install special tools 49-T012-002 and 003, using the camshaft cap bolt holes—1.8L (BP) engines

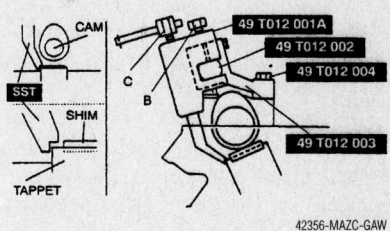

42356-MAZC-GAW

Tighten the mounting bolt B securing it on the shaft. Tighten bolt C and press down the tappet—1.8L (BP) engines

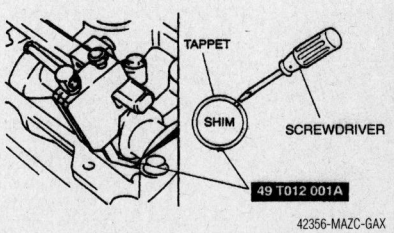

42356-MAZC-GAX

Using a small prytool, pry the adjustment shim upwards through the notch on the tappet—1.8L (BP) engines mount it on the shaft where the adjustment shim needs replacement.

13. Position the notch of the tappet to allow a small prytool to be inserted.

14. Set the special tool on the tappet by its notch. Tighten the mounting bolt **B** securing it on the shaft.

15. Tighten bolt **C**, and press down the tappet.

16. Using a small prytool, pry the adjustment shim upwards through the notch on the tappet. Remove the shim with a magnet.

17. Select and install the proper adjustment shim. Loosen bolt **C** to allow the tappet to move up, and loosen bolt **B** to remove special tool 49-T012-001A.

18. Remove special tools 49-T012-002, 003 and 004, and torque the camshaft cap bolts to 100–125 inch lbs. (11–14 Nm).

19. Repeat the procedure for all necessary adjustment shims. Check the valve clearance.

2.0L (FS) Engine

➡**With the engine cold, standard valve clearance is 0.089–0.0116 inch (0.225–0.295mm) on intake and exhaust sides.**

1. Before servicing the vehicle, refer to the precautions section.

2. Remove the cylinder head cover.

3. Measure the valve clearance by turning the crankshaft clockwise until the No. 1 piston is at Top Dead Center (TDC).

4. Measure the valve clearance at **A**. If the clearance exceeds specifications, replace the adjustment shim.

5. Turn the crankshaft clockwise 360 degrees until the No. 4 piston is at TDC. Measure the valve clearance at **B**. If the clearance exceeds specifications, replace the adjustment shim.

6. Repeat this procedure for all the camshafts.

7. Turn the crankshaft clockwise until

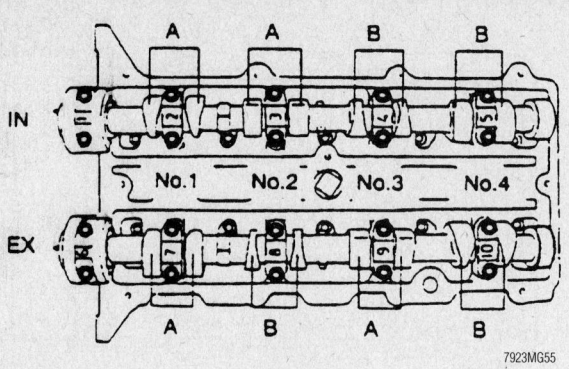

Valve clearance checking positions—2.0L (FS) engines

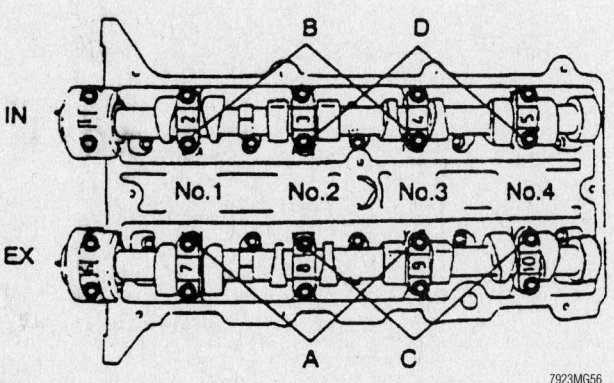

Cam bearing cap bolt removal positions—2.0L (FS) engines

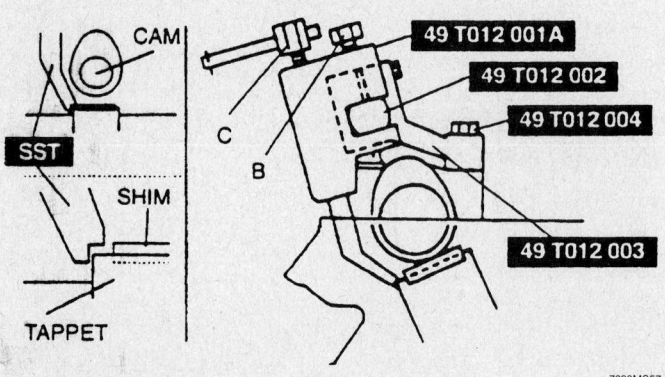

Mount the tappet depressor tool onto the shaft above the tappet that needs adjustment—2.0L (FS) engines

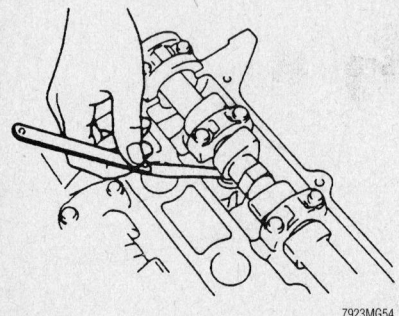

Ensure that the cam lobe faces away from the follower when checking the valve clearance

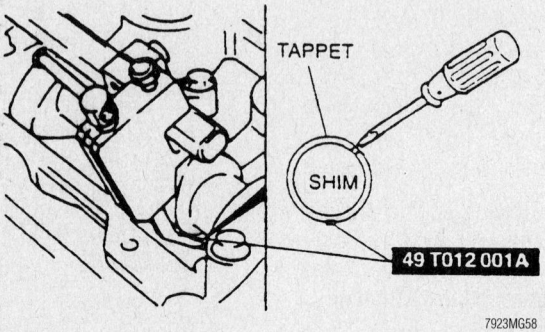

With the tappet depressed, use a small prytool to remove the adjustment shim—2.0L (FS) engines

the cam on the camshaft requiring the adjustment is positioned straight up.

8. Remove the camshaft cap bolts as follows:

    a. For exhaust side No. 1, 2 and 3 cylinder adjustment shim removal use **A**.

    b. For intake side No. 1, 2 and 3 cylinder adjustment shim removal use **B**.

    c. For exhaust side No. 2, 3 and 4 cylinder adjustment shim removal use **C**.

    d. For intake side No. 2, 3 and 4 cylinder adjustment shim removal use **D**.

9. Install special tools 49-T012-002 and 003, using the camshaft cap bolt holes. Tighten the bolts to 100–125 inch lbs. (11–14 Nm).

10. Align the mark on the 49-T012-002 (shaft) with the mark on the 49-T012-003 (clamp). Tighten special tool 49-T012-004 (bolt) to secure the shaft.

11. Position special tool 49-T012-001A toward the center of the cylinder head and mount it on the shaft where the adjustment shim needs replacement.

12. Position the notch of the tappet to allow a small prytool to be inserted.

13. Set the special tool on the tappet by its notch. Tighten the mounting bolt **B** securing it on the shaft.

14. Tighten bolt **C**, and press down the tappet.

15. Using a small prytool, pry the adjustment shim upwards through the notch on the tappet. Remove the shim with a magnet.

16. Select and install the proper adjustment shim. Loosen bolt **C** to allow the tappet to move up, and loosen bolt **B** to remove special tool 49-T012-001A.

17. Remove special tools 49-T012-002, 003 and 004, and torque the camshaft cap bolts to 100–125 inch lbs. (11–14 Nm).

18. Repeat the procedure for all necessary adjustment shims. Check the valve clearance.

1.6L (ZM) Engines

➡With the engine cold, standard valve clearance is 0.010–0.012 inch (0.25–0.31mm) on both the intake and exhaust sides.

1. Before servicing the vehicle, refer to the precautions section.

2. Remove the cylinder head cover.

3. Measure the valve clearance by turning the crankshaft clockwise until the No. 1 piston is at Top Dead Center (TDC).

4. Measure the valve clearance at **A**. If the clearance exceeds specifications, replace the adjustment shim.

5. Turn the crankshaft clockwise 360

degrees until the No. 4 piston is at TDC. Measure the valve clearance at **B**. If the clearance exceeds specifications, replace the adjustment shim.

6. Repeat this procedure for all the camshafts.

7. Turn the crankshaft clockwise until the cam on the camshaft requiring the adjustment is positioned straight up.

8. Remove the camshaft cap bolts as follows:

    a. For exhaust side No. 1, 2 and 3 cylinder adjustment shim removal use **A**.

    b. For intake side No. 1, 2 and 3 cylinder adjustment shim removal use **B**.

    c. For exhaust side No. 2, 3 and 4 cylinder adjustment shim removal use **C**.

    d. For intake side No. 2, 3 and 4 cylinder adjustment shim removal use **D**.

9. Install special tools 49-T012-002 and 003, using the camshaft cap bolt holes. Tighten the bolts to 100–125 inch lbs. (11–14 Nm).

10. Align the mark on the 49-T012-002 (shaft) with the mark on the 49-T012-003 (clamp). Tighten special tool 49-T012-004 (bolt) to secure the shaft.

11. Position special tool 49-T012-001A toward the center of the cylinder head and

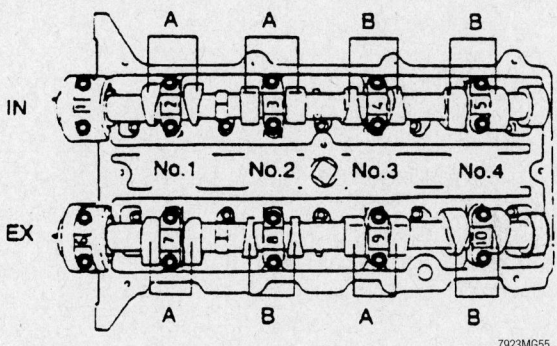

Valve clearance checking positions—4-cylinder engine

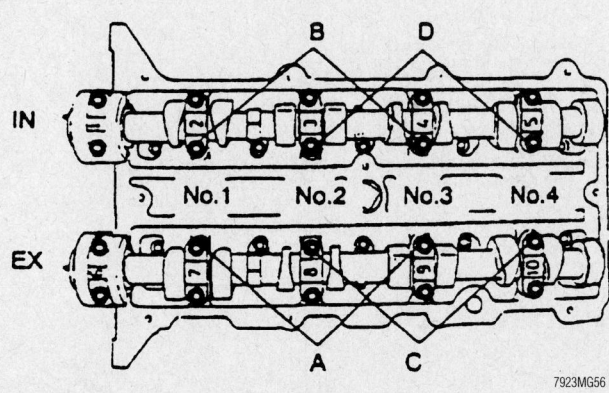

Cam bearing cap bolt removal positions—4-cylinder engine

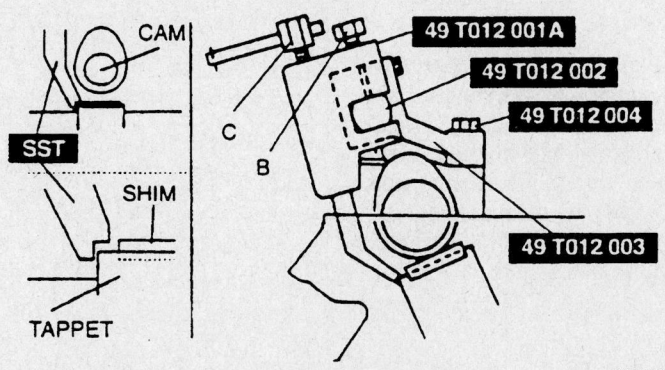

Mount the tappet depressor tool onto the shaft above the tappet that needs adjustment

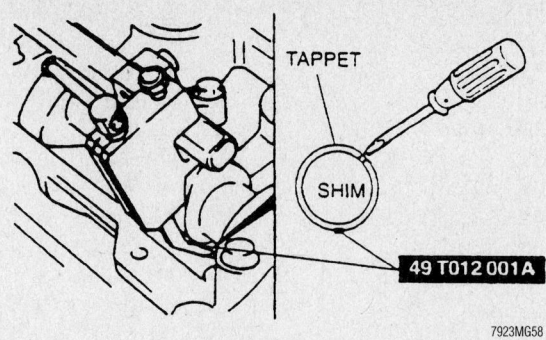

TAPPET

SHIM

49 T012 001A

7923MG58

With the tappet depressed, use a small prytool to remove the adjustment shim

mount it on the shaft where the adjustment shim needs replacement.

12. Position the notch of the tappet to allow a small prytool to be inserted.

13. Set the special tool on the tappet by its notch. Tighten the mounting bolt **B** securing it on the shaft.

14. Tighten bolt **C**, and press down the tappet.

15. Using a small prytool, pry the adjustment shim upwards through the notch on the tappet. Remove the shim with a magnet.

16. Select and install the proper adjustment shim. Loosen bolt **C** to allow the tappet to move up, and loosen bolt **B** to remove special tool 49-T012-001A.

17. Remove special tools 49-T012-002, 003 and 004, and torque the camshaft cap bolts to 100–125 inch lbs. (11–14 Nm).

18. Repeat the procedure for all necessary adjustment shims. Check the valve clearance.

2.3L (KJ) and 2.5L (KL) Engines

This procedure for the 2.5L (KL) engine valve adjustment is for 626 models only. The Millenia model equipped with the 2.5L engine is equipped with hydraulic lash adjusters and no adjustment is possible or necessary.

1. Before servicing the vehicle, refer to the precautions section.

➡**With the engine cold, standard valve clearance on the 2.5L (KL) engine is 0.0097–0.0124 inch (0.245–0.311mm) on intake side and 0.0105–0.0131 inch (0.265–0.335mm) on the exhaust side. On the 2.3L (KJ) the measurement is 0.011–0.012 inch (0.27–0.33mm) on both the intake and exhaust sides.**

2. Measure the valve clearance by turning the crankshaft clockwise until the No. 1 piston is at Top Dead Center (TDC).

3. Measure the valve clearance at **A**. Turn the crankshaft clockwise 240 degrees

until the No. 3 piston is at TDC. Measure the valve clearance at **B**. Turn the crankshaft clockwise 240 degrees until the No. 5 piston is at TDC. Measure the valve clearance at **C**.

➡**If the valve clearance exceeds the standard, replace the adjustment shim.**

4. Turn the crankshaft clockwise until the cam, on the camshaft requiring the adjustment shim replacement, is positioned straight up.

5. Camshaft cap bolts as follows:

a. For right-hand exhaust side shim removal use **A**.

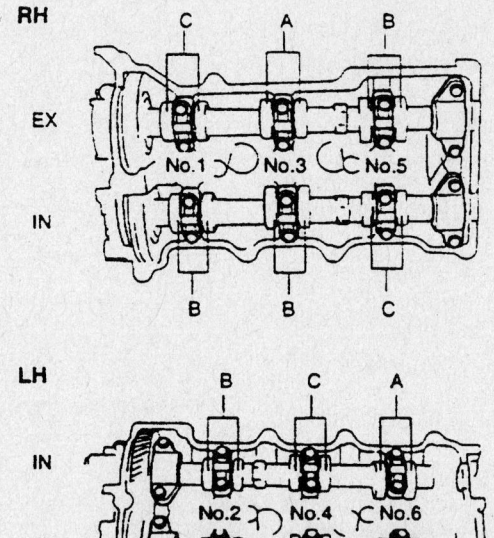

7923MG59

Valve clearance checking positions—6-cylinder engine

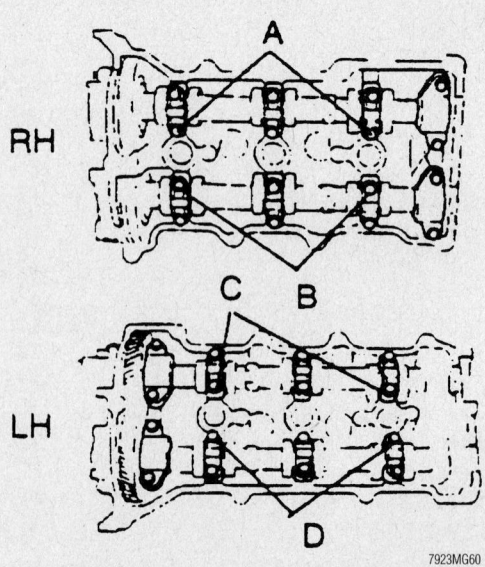

7923MG60

Camshaft cap bolt removal positions—6-cylinder engine—refer to text

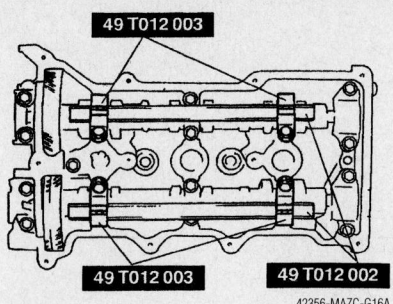

Install special tools 49-T012-002 and 003, using the camshaft cap bolt holes—6-cylinder engine

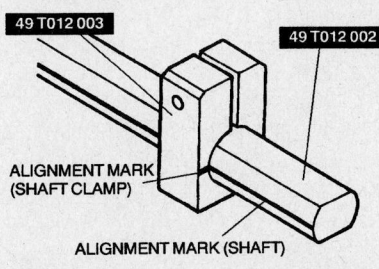

Align the mark on the 49-T012-002 (shaft) with the mark on the 49-T012-003 (clamp)—6-cylinder engine

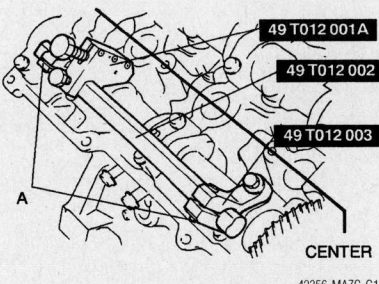

Position special tool 49-T012-001 toward the center of the cylinder head and mount it on the shaft where the adjustment shim needs replacement—6-cylinder engine

b. For right-hand intake side shim removal use **B**.

c. For left-hand intake side shim removal use **C**.

d. For left-hand exhaust side shim removal use **D**.

6. Install special tools 49-T012-002 and 003, using the camshaft cap bolt holes.

7. Align the mark on the 49-T012-002 (shaft) with the mark on the 49-T012-003 (clamp).

8. Position special tool 49-T012-001 toward the center of the cylinder head and mount it on the shaft where the adjustment shim needs replacement.

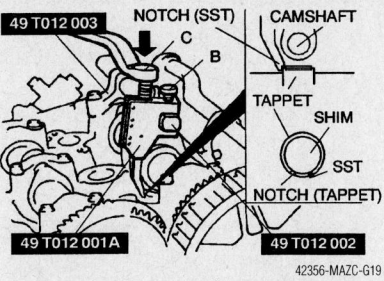

Tighten the mounting bolt B securing it on the shaft. Tighten bolt C and press down the tappet—6-cylinder engine

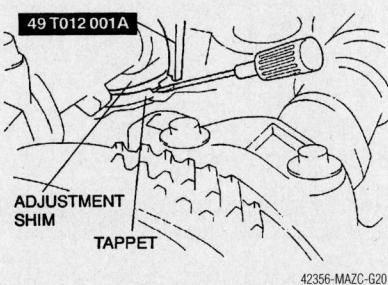

Using a small prytool, pry the adjustment shim upwards through the notch on the tappet—6-cylinder engine

9. Position the notch of the tappet to allow a small prytool to be inserted.

10. Set the special tool on the tappet by its notch. Tighten the mounting bolt **B** securing it on the shaft.

11. Tighten bolt **C** and press down the tappet.

12. Using a small prytool, pry the adjustment shim upwards through the notch on the tappet. Remove the shim with a magnet.

13. Select and install the proper adjustment shim. Loosen bolt **C** to allow the tappet to move up and loosen bolt **B** to remove special tool 49-T012-001.

14. Remove special tools 49-T012-002, 003 and 004 and tighten the camshaft cap bolts to 100–125 inch lbs. (11–14 Nm).

15. Repeat the procedure for all necessary adjustment shims. Check the valve clearance.

Starter Motor

REMOVAL & INSTALLATION

Protégé

1. Remove or disconnect the following:
   - Negative battery cable
   - Air cleaner
   - Intake manifold support bracket bolts
   - Starter electrical connectors
   - Starter

To install:

2. Install or connect the following:
   - Starter and loosely tighten the lower starter mounting bolt
   - Starter electrical connectors
   - Intake manifold support bracket. Torque the bolts to 28–38 ft. lbs. (38–51 Nm).
   - Starter bolts. Torque the bolts 28–38 ft. lbs. (38–51 Nm). The upper mounting bolts must be tightened first.
   - Air cleaner
   - Negative battery cable

626

2.0L (FS) ENGINE

1. Remove or disconnect the following:
   - Battery
   - Air cleaner assembly
   - Transverse member
   - Intake manifold bracket
   - Wiring at the starter
   - Starter

To install:

2. Install or connect the following:
   - Starter. Torque the bolts to 38 ft. lbs. (51 Nm).
   - Wiring at the starter
   - Intake manifold bracket
   - Transverse member
   - Air cleaner assembly
   - Battery

2.5L (KL) ENGINE

1. Remove or disconnect the following:
   - Battery
   - Air cleaner assembly
   - Transverse member
   - Fuel filter with the hose still attached and set aside
   - Oil filler pipe
   - Transaxle selector cable from the automatic transaxle and remove the cable bracket
   - Wiring at the starter
   - Starter

To install:

2. Install or connect the following:
   - Starter. Torque the bolts to 38 ft. lbs. (51 Nm).
   - Wiring at the starter
   - Cable bracket and connect the transaxle selector cable to the transaxle
   - Oil filler pipe
   - Fuel filter
   - Transverse member

- Air cleaner assembly
- Battery

2002–04 Miata

1. Remove or disconnect the following:
   - Negative battery cable
   - Air cleaner assembly
   - Dipstick tube
   - Intake manifold bracket
   - Wiring at the starter
   - Starter

To install:
2. Install or connect the following:
   - Starter. Torque the bolts to 33 ft. lbs. (46 Nm).
   - Wiring at the starter
   - Intake manifold bracket. Torque the bolts to 38 ft. lbs. (51 Nm).
   - Dipstick tube
   - Air cleaner assembly
   - Negative battery cable

Millenia

2.3L (KJ) ENGINE

1. Remove or disconnect the following:
   - Negative battery cable
   - Charge air cooler duct
   - Battery clamp, box and battery
   - Battery tray
   - Rear charge air cooler
   - Pipe bracket
   - Starter electrical connectors
   - Starter

To install:
2. Install or connect the following:
   - Starter. Torque the bolts to 28–38 ft. lbs. (38–51 Nm).
   - B-terminal wire
   - S-terminal wire
   - Pipe bracket. Torque the bolt to 14–18 ft. lbs. (19–25 Nm).
   - Rear charge air cooler using new O-rings. Torque the nuts to 14–18 ft. lbs. (19–25 Nm).
   - Battery tray
   - Battery, box and clamp
   - Charge air cooler duct
   - Negative battery cable

2.5L (KL) ENGINE

1. Remove or disconnect the following:
   - Negative battery cable
   - Battery clamp, box and battery
   - Battery tray
   - Shift cable from the selector lever
   - Cable from the bracket by squeeze the lock tabs
   - Electrical connectors from the starter solenoid

- 2 selector cable bracket bolts and the bracket
- 2 nuts and the bolt from the starter bracket and the bracket
- Starter electrical connectors
- Starter

To install:
2. Install or connect the following:
   - Starter. Torque the bolts to 28–38 ft. lbs. (38–51 Nm).
   - B-terminal wire.
   - S-terminal wire to the solenoid
   - Starter bracket
   - Selector cable bracket. Torque the bolts to 5–7 ft. lbs. (7–9 Nm).
   - Starter solenoid electrical connectors
   - Shift cable into the cable bracket and into the selector lever
   - Battery tray
   - Battery, box and clamp
   - Negative battery cable

Oil Pan

REMOVAL & INSTALLATION

1.8L (BP) Engines

1. Before servicing the vehicle, refer to the precautions section.
2. Drain the engine oil.
3. Remove or disconnect the following:
   - Negative battery cable
   - Air cleaner assembly
   - Wheel speed sensor
   - Dipstick tube
   - Intermediate shaft
4. Attach an engine to the hoist, loosen the oil pan bolts, remove the engine mounting bolts and raise the engine slightly using a hoist.
   - Stabilizer link nut
   - Shock absorber-to-knuckle nut and bolt

✳✳ CAUTION

When removing the crossmember, be careful do to damage the brake hoses, A/C pipes and power steering pipes when lowering the crossmember.

5. Support the crossmember with a transmission jack, remove the crossmember retainers. Separate the steering intermediate shaft from the pinion shaft and lower the crossmember until the clearance between the oil pan and the steering gear exceeds 5.12 inch (130mm).
6. Remove or disconnect the following:
   - Engine mount

- Oil pan bolts and the oil pan using a seal cutter, then insert a flat pry tool into the locations illustrated.

To install:
7. Clean the oil pan. Clean all dirt, oil, gasket and old sealant from the oil pan and cylinder block contact surfaces.
8. Apply a continuous bead of silicone sealant to the contact surfaces of the new oil pan gaskets as illustrated.
9. Install new gaskets into the oil pump body and rear cover facing the notches as illustrated.
10. Apply a 0.079 inch (2mm) continuous bead of silicone sealant on the oil pan as illustrated.
11. Apply a continuous bead of silicone sealant on the oil pan-to-block areas as illustrated.
12. Apply a 0.099–0.137 inch (2.5–3.5mm) continuous bead of silicone sealant on the oil pan along the inside of the bolt holes and overlap the ends.
13. Install or connect the following:
    - New oil pan gaskets
    - Oil pan. Torque the vertical bolts to 70–95 inch lbs. (8–11 Nm) and the horizontal bolts to 48–65 ft. lbs. (64–86 Nm).
    - Engine mount
14. Raise the crossmember into position and tighten the bolts to 15 ft. lbs. (21 Nm) and the nuts to 101 ft lbs. (137 Nm).
    - Shock absorber lower nut and bolt and tighten to 86 ft. lbs. (116 Nm)
    - Stabilizer link nut and tighten to 44 ft. lbs. (60 Nm)
    - Engine mount nut and tighten to 57 ft. lbs. (78 Nm)
    - Intermediate shaft and tighten the bolt to 19 ft. lbs. (26 Nm)
    - Dipstick tube
    - Wheel speed sensor
    - Air cleaner assembly
    - Negative battery cable
15. Fill the engine with clean oil.
16. Start the vehicle, check for leaks and repair if necessary.

1.6L (ZM) Engines

1. Before servicing the vehicle, refer to the precautions section.
2. Drain the engine oil.
3. Remove or disconnect the following:
   - Negative battery cable
   - Oxygen (O_2S) sensors
   - Front exhaust pipe
   - Integrated stiffener (1)
   - Oil pan (2) and oil strainer (3) and the Main Bearing Support Plate (MBSP) (4)

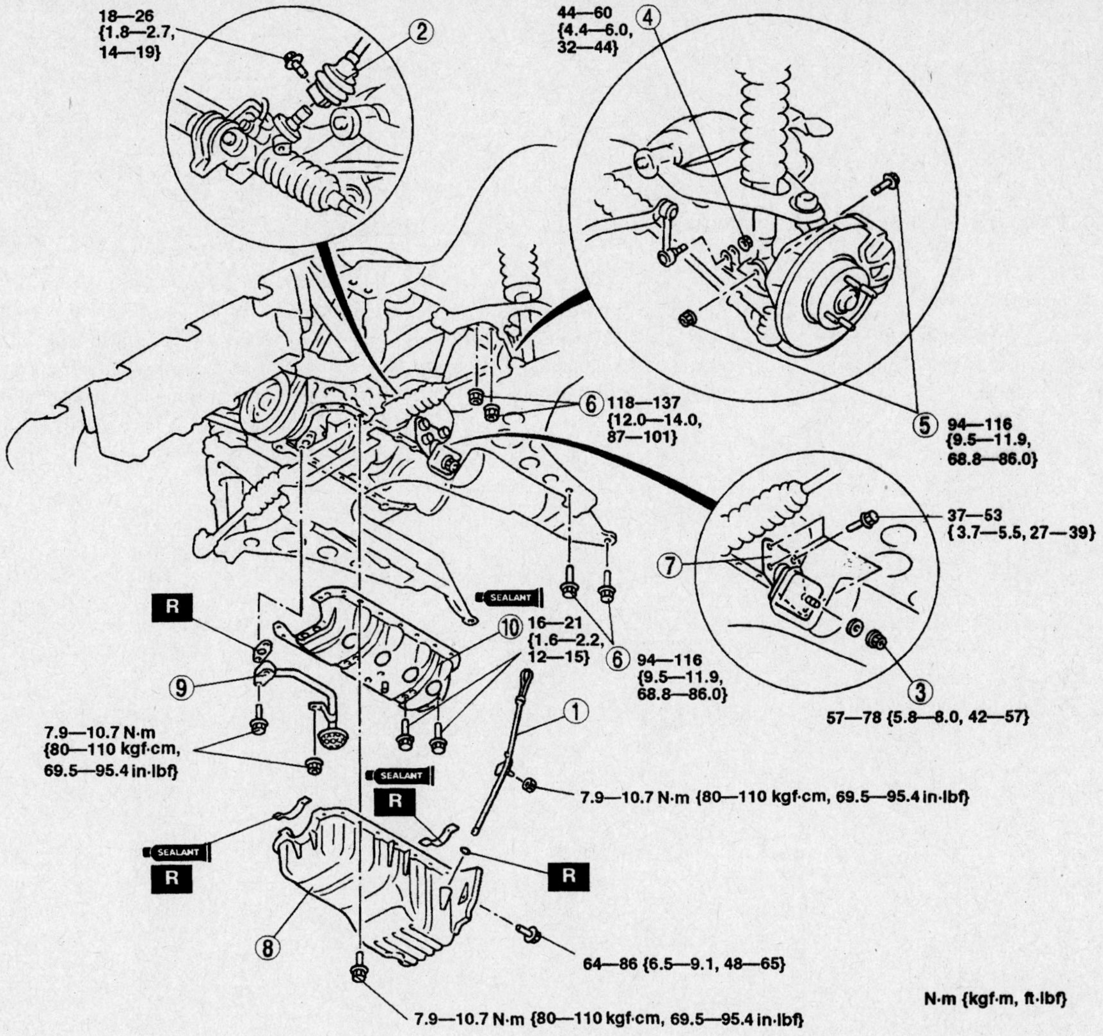

18—26
{1.8—2.7,
14—19}
2
44—60
{4.4—6.0,
32—44}
4
6 118—137
{12.0—14.0,
87—101}
5 94—116
{9.5—11.9,
68.8—86.0}
37—53
{3.7—5.5, 27—39}
7
SEALANT
R
10 16—21
{1.6—2.2,
12—15}
6 94—116
{9.5—11.9,
68.8—86.0}
9
3
1
57—78 {5.8—8.0, 42—57}
7.9—10.7 N·m
{80—110 kgf·cm,
69.5—95.4 in·lbf}
7.9—10.7 N·m {80—110 kgf·cm, 69.5—95.4 in·lbf}
8
64—86 {6.5—9.1, 48—65}
N·m {kgf·m, ft·lbf}
7.9—10.7 N·m {80—110 kgf·cm, 69.5—95.4 in·lbf}

1	Dipstick and pipe	6	Crossmember bolt and nut
2	Intermediate shaft	7	Engine mount
3	Engine mount nut	8	Oil pan
4	Stabilizer control link nut	9	Oil strainer
5	Shock absorber bolt and nut	10	MBSP

42356-MAZC-GAY

Exploded view of the oil pan and related components—1.8L (BP) engine

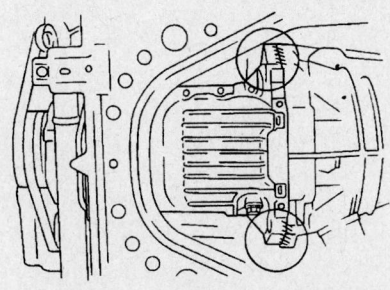

42356-MAZC-GAZ

Remove the oil pan by inserting a flat pry tool into the locations shown—1.8L (BP) engine

SEALANT

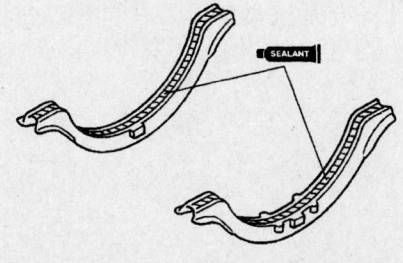

42356-MAZC-GBA

Apply a continuous bead of silicone sealant to the contact surfaces of the new oil pan gaskets as shown—1.8L (BP) engine

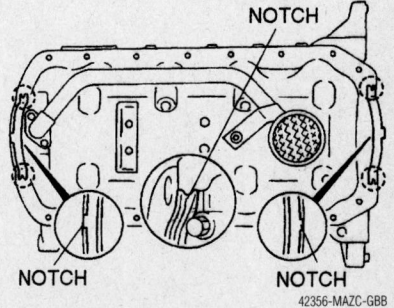

NOTCH
NOTCH
NOTCH

42356-MAZC-GBB

Install new gaskets into the oil pump body and rear cover facing the notches as shown—1.8L (BP) engine

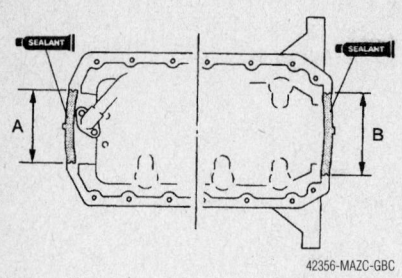

SEALANT
SEALANT
A
B
42356-MAZC-GBC

Apply a 0.079 inch (2mm) continuous bead of silicone sealant on the oil pan as shown—1.8L (BP) engine

SEALANT
SEALANT
SEALANT
SEALANT
42356-MAZC-GBD

Apply a continuous bead of silicone sealant on the oil pan-to-block areas as shown—1.8L (BP) engine

SEALANT
42356-MAZC-GBE

Apply a 0.099–0.137 inch (2.5–3.5mm) continuous bead of silicone sealant on the oil pan along the inside of the bolt holes and overlap the ends—1.8L (BP) engine

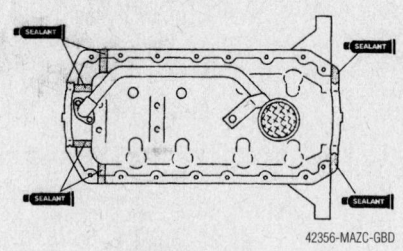

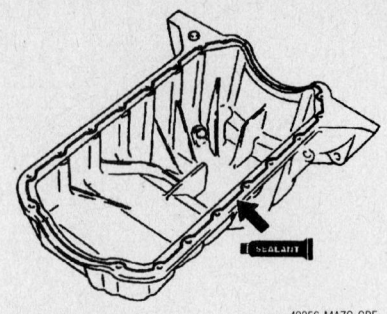

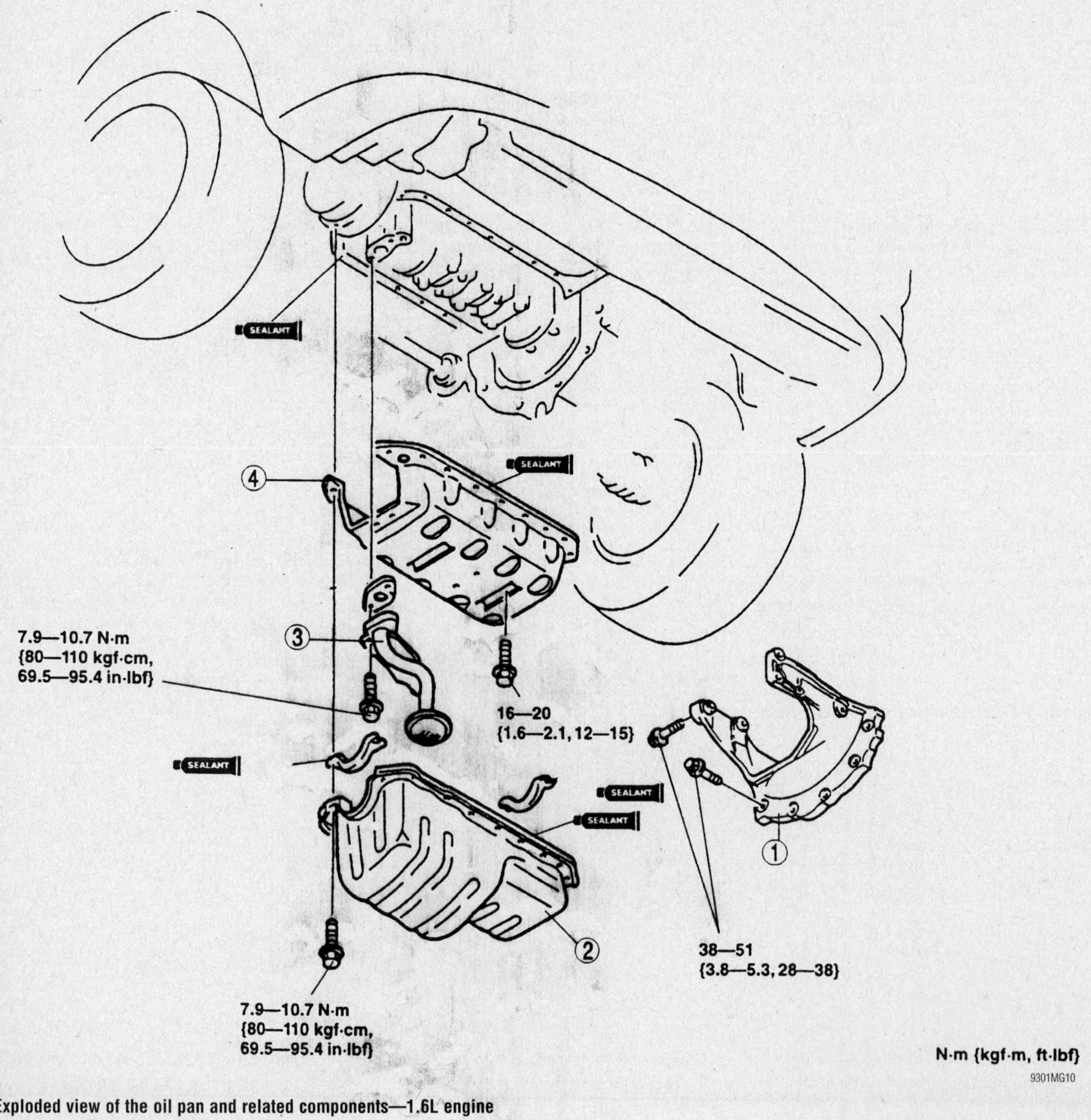

SEALANT
SEALANT
4
3
7.9—10.7 N·m
{80—110 kgf·cm,
69.5—95.4 in·lbf}
16—20
{1.6—2.1, 12—15}
SEALANT
SEALANT
SEALANT
1
2
38—51
{3.8—5.3, 28—38}
7.9—10.7 N·m
{80—110 kgf·cm,
69.5—95.4 in·lbf}
N·m {kgf·m, ft·lbf}
9301MG10

Exploded view of the oil pan and related components—1.6L engine

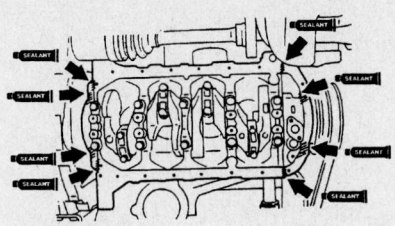

Apply a bead of silicone sealant to area as illustrated—1.6L (ZM) engines

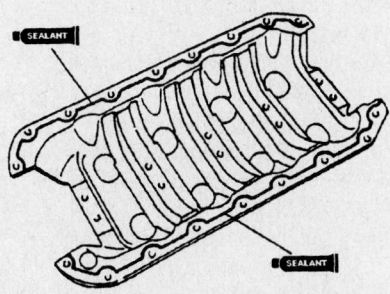

Apply a 0.099–0.137 inch (2.5–3.5mm) bead of silicone sealant to the MBSP and along the inside of the bolt holes—1.6L (ZM) engines

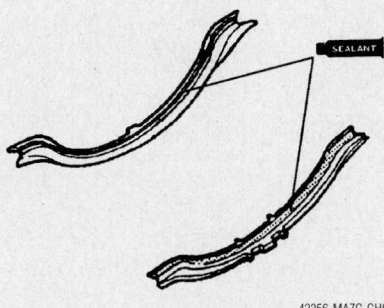

Apply a bead of silicone sealant to new oil pan gaskets . . .

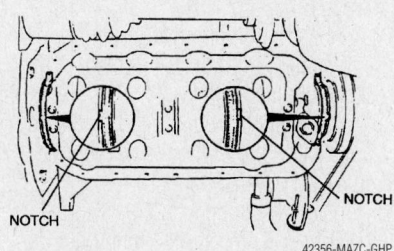

. . . . and install the gaskets onto the oil pump body and rear cover with the projections in the notches as illustrated—1.6L (ZM) engines

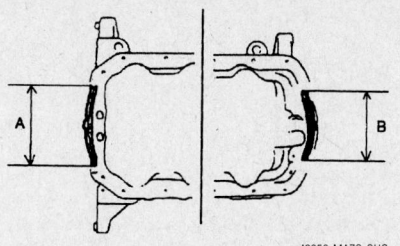

Apply a 0.079 inch (2mm) bead of silicone sealant to area of the oil pan gaskets marked by A and B—1.6L (ZM) engines

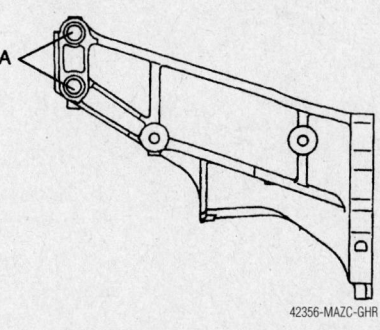

Location of the integrated stiffener bolt A—1.6L (ZM) engines

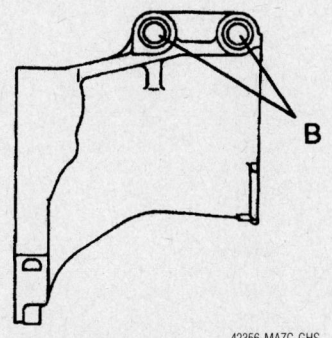

Location of the integrated stiffener bolt B—1.6L (ZM) engines

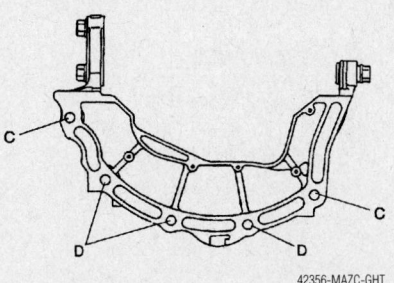
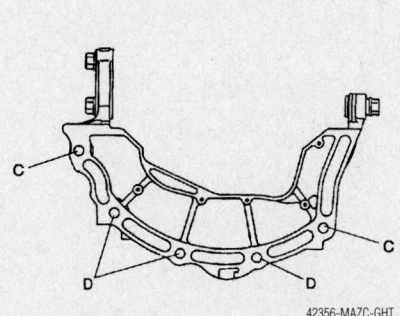

Location of the integrated stiffener bolts C and D—1.6L (ZM) engines

4. Clean the oil pan. Clean all dirt, oil, gasket and old sealant from the oil pan and cylinder block contact surfaces.

To install:

5. Apply a 0.099–0.137 inch (2.5–3.5mm) bead of silicone sealant to the MBSP and along the inside of the bolt holes.

6. Apply a bead of silicone sealant to new oil pan gaskets. Install the gaskets onto the oil pump body and rear cover with the projections in the notches as illustrated.

7. Apply a 0.079 inch (2mm) bead of silicone sealant to area of the oil pan gaskets marked by **A and B** in the illustration. Install the gaskets onto the oil pump body and rear cover with the projections in the notches as illustrated.

8. Apply a 0.99–0.137 inch (2.5–3.5mm) bead of silicone sealant to the oil pan along the inside of the bolt holes and overlap the ends.

9. Install or connect the following:
- Oil pan. Torque the bolts to 69.5–95.4 inch lbs. (7.9–10.7 Nm).

10. Install the integrated stiffener and hand tighten bolt **A**, then bolt **B**.

11. Tighten bolt **C** to 38 ft. lbs. (52 Nm).

12. Tighten bolt **D** to 38 ft. lbs. (52 Nm).

13. Tighten bolt **A** to 38 ft. lbs. (52 Nm).

14. Tighten bolt **B** to 38 ft. lbs. (52 Nm).
- Front exhaust pipe
- O$_2$S
- Negative battery cable

15. Fill the engine with clean oil.

16. Start the vehicle, check for leaks and repair if necessary.

2.0L (FS) Engine

1. Before servicing the vehicle, refer to the precautions section.

2. Drain the engine oil.

3. Remove or disconnect the following:
- Negative battery cable
- Oxygen (O$_2$S) sensor
- Front exhaust pipe
- Oil pan bolts and the oil pan

To install:

4. Clean the oil pan. Clean all dirt, oil and old sealant from the oil pan and cylinder block contact surfaces.

5. Apply a continuous bead of silicone sealant around the oil pan, going on the inside of the bolt holes.

6. Install or connect the following:
- Oil pan. Torque the bolts to 14–18 ft. lbs. (19–25 Nm).
- Front pipe. Torque the nuts to 28–38 ft. lbs. (38–51 Nm).
- O$_2$S connector
- Negative battery cable

FS

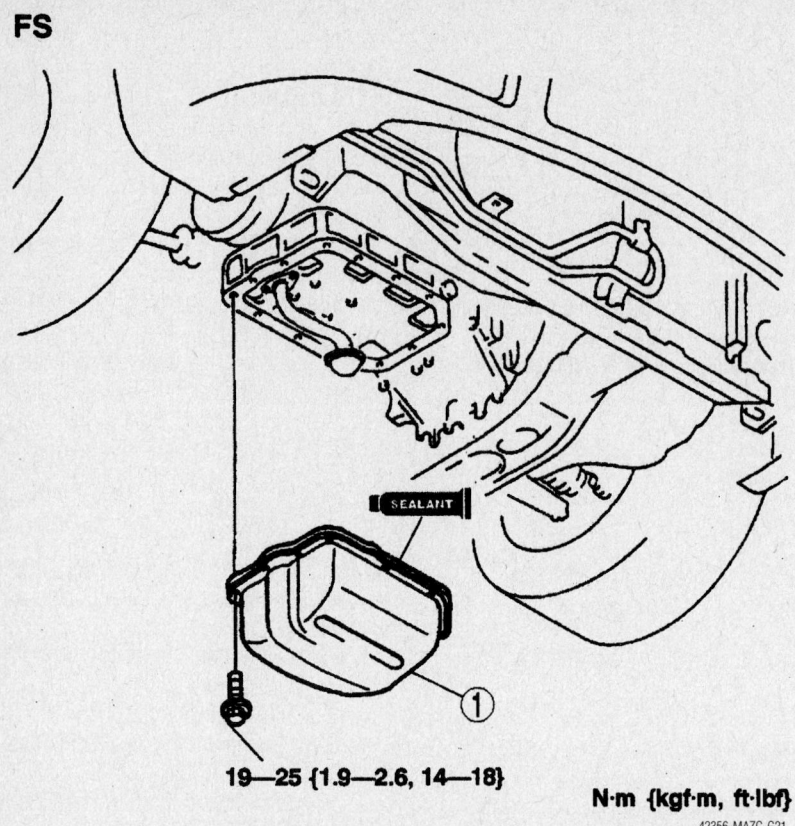

19—25 {1.9—2.6, 14—18}

N·m {kgf·m, ft·lbf}

42356-MAZC-G21

Exploded view of the oil pan and related components for the 2.0L (FS) engines

KL

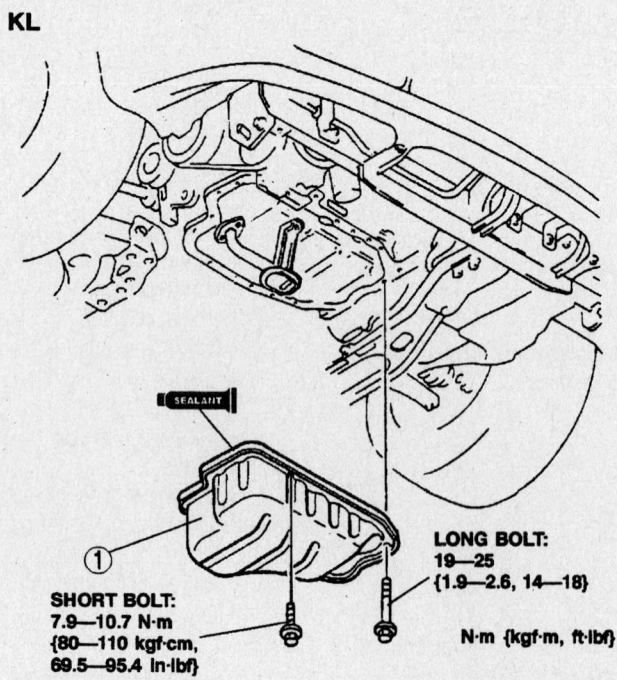

SHORT BOLT:
7.9—10.7 N·m
{80—110 kgf·cm,
69.5—95.4 in·lbf}

LONG BOLT:
19—25
{1.9—2.6, 14—18}

N·m {kgf·m, ft·lbf}

1   Oil pan

42356-MAZC-G72

Exploded view of the oil pan and related components—2.5L (KL) engines, 626 model shown
Millenia similar

7. Fill the engine with clean oil.
8. Start the engine, check for leaks and repair if necessary.

2.5L (KL) Engine

1. Before servicing the vehicle, refer to the precautions section.
2. Drain the engine oil.
3. Remove or disconnect the following:
   - Negative battery cable
   - Right hand splash shield on Millenia models
   - Oxygen (O_2S) sensor
   - Front exhaust pipe
   - Oil pan bolts and the oil pan

To install:

4. Clean the oil pan. Clean all dirt, oil and old sealant from the oil pan and cylinder block contact surfaces.
5. Apply a continuous bead of silicone sealant around the oil pan, going on the inside of the bolt holes.
6. Install or connect the following:
   - Oil pan. Torque the long bolts to 14–18 ft. lbs. (19–25 Nm) and the short bolts to 70–95 inch lbs. (8–11 Nm).
   - Front pipe
   - O_2S connector
   - Right hand splash shield on Millenia models
   - Negative battery cable
7. Fill the engine with clean oil.
8. Start the engine, check for leaks and repair if necessary.

2.3L (KJ) Engines

1. Before servicing the vehicle, refer to the precautions section.
2. Drain the engine oil.
3. Remove or disconnect the following:
   - Negative battery cable
   - Passenger side splash shield
   - Oxygen (O_2S) sensor
   - Front exhaust pipe
   - Oil pan bolts and the oil pan

To install:

4. Clean the oil pan. Clean all dirt, oil and old sealant from the oil pan and cylinder block contact surfaces.
5. Apply a continuous bead of silicone sealant around the oil pan, going on the inside of the bolt holes.
6. Install or connect the following:
   - Oil pan. Torque the long bolts to 14–18 ft. lbs. (19–25 Nm) and the short bolts to 70–95 inch lbs. (8–11 Nm).
   - Front pipe. Torque the nuts to 28–38 ft. lbs. (38–51 Nm).

- O₂S connector
- Splash shield
- Negative battery cable

7. Fill the engine with clean oil.

8. Start the engine, check for leaks and repair if necessary.

Oil Pump

REMOVAL & INSTALLATION

626, Protégé and Millenia

1.6L (ZM) ENGINE

1. Before servicing the vehicle, refer to the precautions section.

2. Remove or disconnect the following:
- Negative battery cable
- Crankshaft pulley
- Timing belt cover
- Timing belt
- Timing belt pulley
- Oil pan
- A/C compressor and move it aside, leaving the refrigerant lines attached
- A/C compressor mounting bracket
- Alternator
- Oil pump attaching bolts
- Front crankshaft seal from the oil pump, if the pump is being replaced

To install:

3. Clean the oil, dirt and old sealant from all contact surfaces.

4. If the oil seal was removed from the oil pump, apply clean engine oil to the lip of the seal. Push the seal in lightly be hand. Press the seal, with a protrusion of 0.02–0.04 inch (0.5–1.0mm) into the oil pump with a Seal Installer tool 49 B014 001.

5. Apply a bead of silicone to the oil pump at the cylinder block contact surface, going inside the bolt holes.

6. Install or connect the following:
- New O-rings on the oil pump
- Oil pump. Torque the bolts to 14–18 ft. lbs. (19–25 Nm).
- Alternator
- Air conditioning compressor bracket
- Air conditioning compressor
- Oil pan
- Timing belt pulley
- Timing belt
- Timing belt cover
- Crankshaft pulley
- Negative battery cable

7. Fill the engine with clean oil.

8. Start the vehicle, check for leaks and repair if necessary.

2.0L (FS) ENGINE

1. Before servicing the vehicle, refer to the precautions section.

2. Remove or disconnect the following:
- Negative battery cable
- Crankshaft pulley
- Timing belt cover
- Timing belt
- Timing belt pulley
- Oil pan
- A/C compressor and move it aside, leaving the refrigerant lines attached
- A/C compressor mounting bracket
- Alternator
- Transaxle
- Two bolts at the rear of the oil pan upper block
- Rubber caps at the bottom surface, then remove the two bolts from the upper block through the holes uncovered by removing the caps, if equipped an automatic transaxle, on 2.0L (FS) engines
- Oil pan upper block bolts using 2–3 steps in the sequence illustrated
- Oil pan upper block using a rubber mallet and a suitable separator tool
- Oil pump body bolts and the oil pump body
- Seal from the oil pump, if the pump is being replaced

To install:

3. Clean the oil, dirt and old sealant from all contact surfaces.

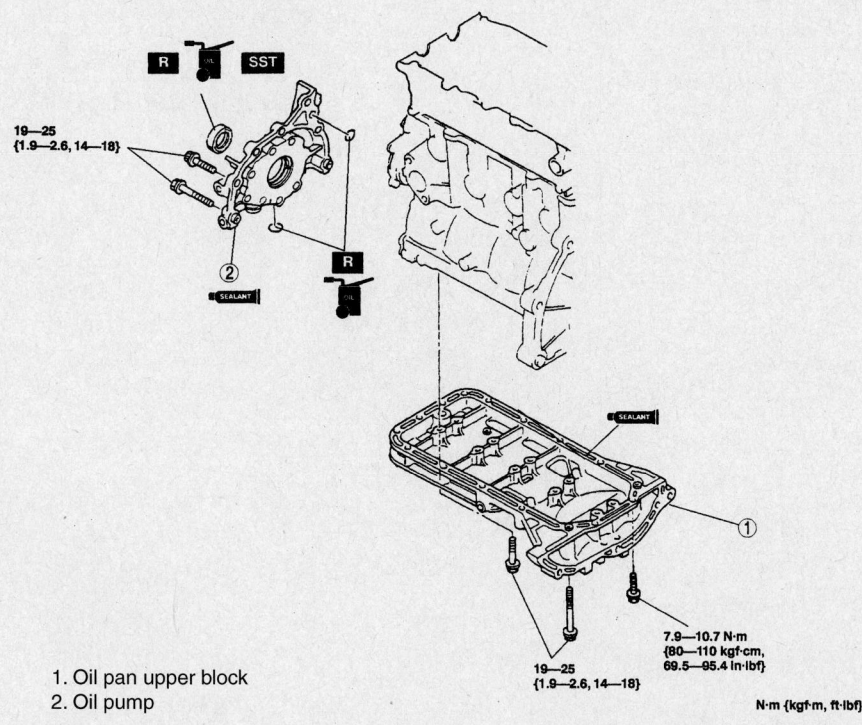

1. Oil pan upper block
2. Oil pump

Exploded view of the oil pump and related components for the 2.0L (FS) engines

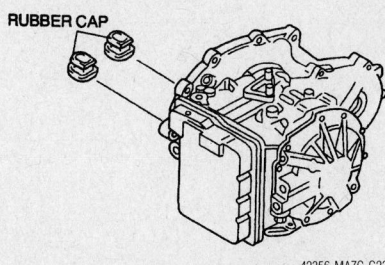

Remove the rubber caps at the bottom surface of the upper block, if equipped an automatic transaxle–2.0L (FS) engines

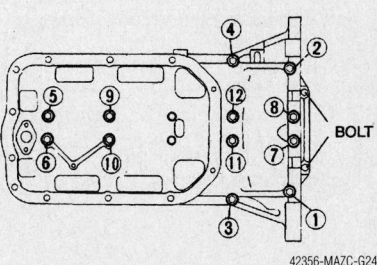

Remove the oil pan upper block bolts using 2–3 steps in the sequence illustrated–2.0L (FS) engines

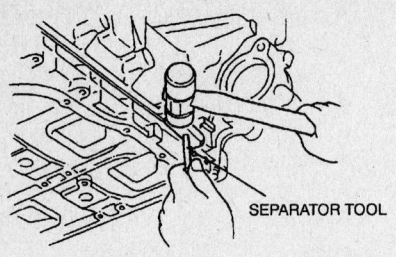

42356-MAZC-G25

Remove the oil pan upper block using a rubber mallet and a suitable separator tool–2.0L (FS) engines

4. If the oil seal was removed from the oil pump, apply clean engine oil to the lip of the seal. Push the seal in lightly be hand. Press the seal, with a protrusion of 0–0.019 inch (0–0.5mm) into the oil pump.

5. Apply 0.04–0.07 inch (1–2mm) bead of sealant to the oil pump at the locations illustrated

6. Install or connect the following:
- Oil pump. Torque the bolts to 14–18 ft. lbs. (19–25 Nm).
- Oil pan

7. Apply a bead of sealant 0.08–0.11 inch (2–3mm) the mating surface of the upper block as illustrated, place the block assembly into position and tighten the bolts in sequence illustrated to 18 ft. lbs. (25 Nm) using 2–3 steps.

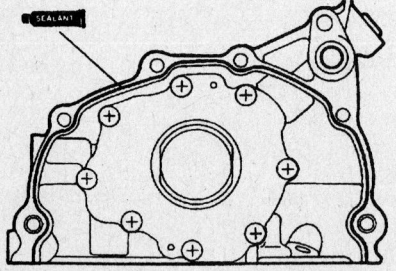

42356-MAZC-G26

Apply 0.04–0.07 inch (1–2mm) bead of sealant to the oil pump–2.0L (FS) engines

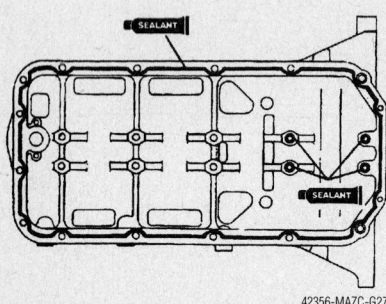

42356-MAZC-G27

Apply a bead of sealant 0.08–0.11 inch (2–3mm) the mating surface of the upper block–2.0L (FS) engines

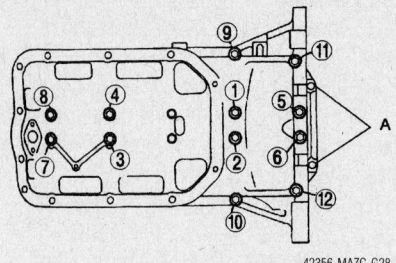

42356-MAZC-G28

Tighten the upper block bolts in this sequence–2.0L (FS) engines

- Two bolts at the rear of the oil pan upper block and torque to 95 inch lbs. (11 Nm) and install the caps
- Two bolts to the upper block through the holes uncovered by removing the caps, if equipped an automatic transaxle, tighten the bolts to 95 inch lbs. (11 Nm)
- Transaxle, if equipped with a manual transaxle
- Air conditioning compressor bracket
- Air conditioning compressor
- Oil pan
- Timing belt pulley
- Timing belt
- Timing belt cover
- Crankshaft pulley
- Negative battery cable

8. Fill the engine with clean oil.

9. Start the vehicle, check for leaks and repair if necessary.

2.5L (KL) ENGINE

1. Before servicing the vehicle, refer to the precautions section.

2. Remove or disconnect the following:
- Negative battery cable
- Crankshaft pulley
- Timing belt cover
- Timing belt
- Timing belt pulley
- Oil pan
- A/C compressor and move it aside, leaving the refrigerant lines attached
- A/C compressor mounting bracket, if necessary
- Alternator
- Oil baffle
- Alternator bracket
- Oil pump body bolts and the oil pump body
- Seal from the oil pump, if the pump is being replaced

To install:

3. Clean the oil, dirt and old sealant from all contact surfaces.

4. If the oil seal was removed from the oil pump, apply clean engine oil to the lip of the seal. Push the seal in lightly be hand. Press the seal, with a protrusion of 0–0.019 inch (0–0.5mm) into the oil pump.

5. Apply clean engine oil to the NEW O–ring and install it in the pump

6. Install or connect the following:
- Oil pump bolts in the locations illustrated noting the location and length of the bolts. The long bolts **A** are 1.6 inch (40mm) long and the short bolts **B** are 1 inch (25mm) long. Torque the bolts to 14–18 ft. lbs. (19–25 Nm).
- Oil baffle and torque the bolts in sequence to 14–18 ft. lbs. (19–25 Nm)

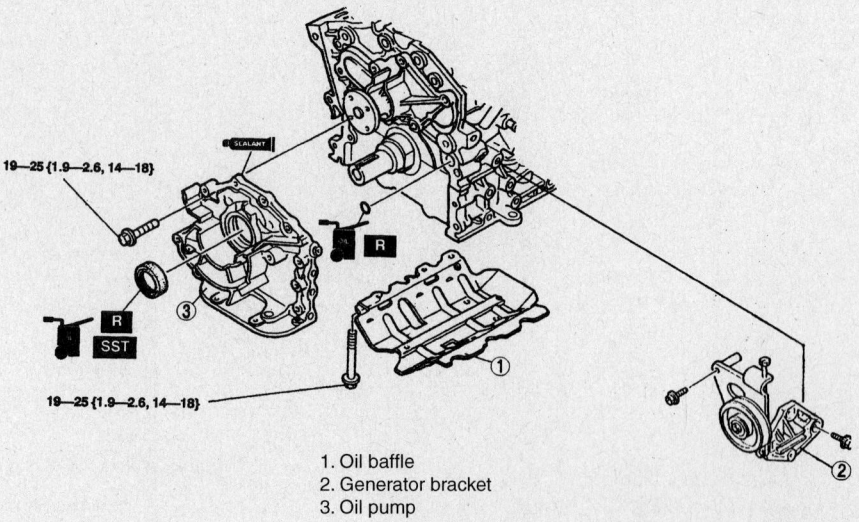

1. Oil baffle
2. Generator bracket
3. Oil pump

42356-MAZC-G29

Exploded view of the oil pump and related components for the 2.5L (KL) engines

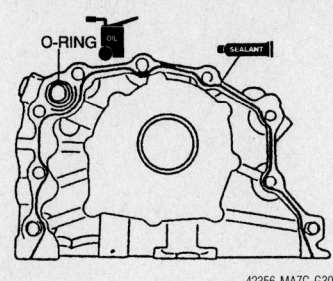

Apply clean engine oil to the NEW O-ring and install it in the pump—2.5L (KL) engines

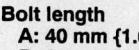

Bolt length
A: 40 mm {1.6 in}
B: 25 mm {1.0 in}

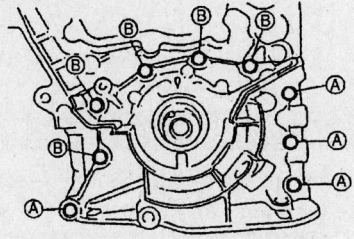

The oil pump bolts are different lengths, make sure to install them where indicated. The long bolts A are 1.6 inch (40mm) long and the short boltsBare 1 inch (25mm) long—2.5L (KL) engines

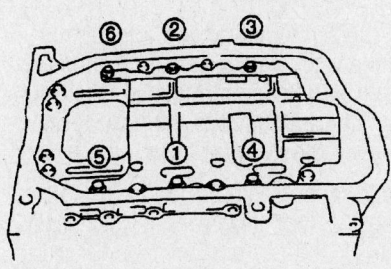

Tighten the oil baffle and torque the bolts in sequence—2.5L (KL) engines

- Alternator bracket
- Air conditioning compressor bracket
- Air conditioning compressor
- Oil pan
- Timing belt pulley
- Timing belt
- Timing belt cover
- Crankshaft pulley
- Negative battery cable
7. Fill the engine with clean oil.
8. Start the vehicle, check for leaks and repair if necessary.

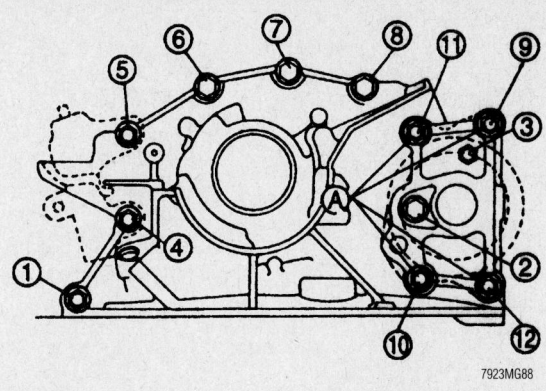

Tightening the oil pump mounting bolts in sequence—2.3L (KJ) engines

2.3L (KJ) ENGINE

1. Before servicing the vehicle, refer to the precautions section.
2. Remove or disconnect the following:
- Negative battery cable
- Crankshaft pulley
- Timing belt cover
- Timing belt
- Timing belt pulley
- Oil pan
- Alternator
- Power steering pump and bracket, move the pump aside leaving the refrigerant lines attached
- Oil baffle
- A/C compressor and move it aside, leaving the refrigerant lines attached
- A/C compressor mounting bracket, if necessary
- Vacuum pump
- Oil pump body bolts and the oil pump body
- Seal from the oil pump, if the pump is being replaced

To install:
3. Clean the oil, dirt and old sealant from all contact surfaces.
4. If the oil seal was removed from the oil pump, apply clean engine oil to the lip of the seal. Push the seal in lightly be hand. Press the seal, with a protrusion of 0–0.02 inch (0–0.7mm) into the oil pump.
5. Apply a 0.04–0.07 inch (1–2mm) bead of silicone sealant to the contact surface of the pump.
6. Apply clean engine oil to the NEW O-ring and install it in the pump
7. Install or connect the following:
- Oil pump. Torque the bolts, in sequence, to 15–22 ft. lbs. (20–30 Nm) for the **A** bolts and to 14–18 ft. lbs. (19–25 Nm) for all other bolts.
- Vacuum pump
- Air conditioning compressor bracket

- Oil baffle and torque the bolts in sequence to 14–18 ft. lbs. (19–25 Nm)
- Power steering pump and bracket
- Alternator
- Air conditioning compressor
- Oil pan
- Timing belt pulley
- Timing belt
- Timing belt cover
- Crankshaft pulley
- Negative battery cable
8. Fill the engine with clean oil.
9. Start the vehicle, check for leaks and repair if necessary.

2002–04 Miata

1.8L (BP) ENGINES

1. Before servicing the vehicle, refer to the precautions section.
2. Remove or disconnect the following:
- Negative battery cable
- Crankshaft pulley
- Timing belt cover
- Timing belt
- Crankshaft sprocket
- A/C compressor and move it aside, leaving the refrigerant lines attached
- A/C compressor mounting bracket
- Alternator
- Oil pump bolts and the oil pump
- Front crankshaft seal from the oil pump, if the pump is being replaced

To install:
3. Clean the oil, dirt and old sealant from all contact surfaces.
4. If the oil seal was removed from the oil pump, apply clean engine oil to the lip of the seal. Push the seal in lightly be hand. Press the seal, with a protrusion of 0–0.02 inch (0–0.5mm) into the oil pump.
5. Apply a bead of silicone to the oil pump body-to-cylinder block contact surface, going inside the bolt holes.

6. Install or connect the following:
- Oil pump. Torque the bolts to 14–18 ft. lbs. (19–25 Nm).
- Air conditioning compressor bracket
- Air conditioning compressor
- Crankshaft sprocket
- Timing belt
- Timing belt cover
- Crankshaft pulley
- Negative battery cable

7. Fill the engine with clean oil.
8. Start the vehicle, check for leaks and repair if necessary.

Rear Main Seal

REMOVAL & INSTALLATION

1. Before servicing the vehicle, refer to the precautions section.
2. Remove or disconnect the following:
- Negative battery cable
- Transaxle/transmission assembly
- Clutch/flywheel assembly, if equipped with a manual transaxle/transmission
- Flexplate/shim plates, if equipped with an automatic transaxle/transmission

3. Cut the oil seal lip with a knife. Install a rag to the housing and using a prytool, carefully pry the oil seal from the oil seal housing.
4. Clean the gasket mounting surfaces.

To install:

5. Clean the oil seal housing. Coat the oil seal and the housing with clean engine oil.
6. Install or connect the following:
- New oil seal into the housing by tapping it evenly into place with a hammer and a seal installer until it is flush with the edge of the rear cover
- Clutch/flywheel assembly or the flexplate, as applicable
- Transaxle/transmission
- Negative battery cable

Timing Belt

REMOVAL & INSTALLATION

626 and Protégé

1.6L (ZM) ENGINE

1. Before servicing the vehicle, refer to the precautions section.
2. Drain the cooling system.
3. Remove or disconnect the following:

- Negative battery cable
- Camshaft Position (CMP) sensor
- Ignition coils
- Drive belt
- Crankshaft pulley and plate
- Water pump pulley
- Rocker arm cover and discard the gasket

4. Support the engine with a suitable support device and remove the number 3 engine mount.
- Timing belt cover
- Pulley using tool 49 D011 102 to prevent crankshaft rotation and remove the pulley boss

5. Install the pulley boss on the crankshaft and tighten the bolt.
6. Turn the crankshaft until the timing marks on the crankshaft and camshaft sprockets are aligned. Face the camshaft pulley marks **I** and **E** straight up, then align the timing marks with the horizontal surface on the cylinder head. The pin on the pulley boss must face upward. Hold the crankshaft pulley boss with a suitable tool and remove the pulley lockbolt, being careful not to rotate the crankshaft. Remove the crankshaft pulley boss.

➡**Protect the tensioner with a shop towel before prying on it. Do not rotate the crankshaft after the timing belt has been removed.**

7. Mark the direction of rotation on the timing belt. Loosen the tensioner lockbolt and pry the tensioner outward. Tighten the lockbolt with the tensioner spring fully extended. Remove the timing belt.
8. Remove the tensioner and spring. If necessary, remove the idler pulley.
9. Inspect the belt for wear, peeling, cracking, hardening or signs of oil contamination. Inspect the tensioner for free and smooth rotation. Check the tensioner spring free length; it should not exceed 2.43 inch (68mm). Inspect the sprocket teeth for wear or damage. Replace parts, as necessary.

To install:

10. Install the crankshaft sprocket bolt and temporarily tighten.
11. Install the tensioner and tensioner spring. Install the spring with the damper rubber closing face on the right side. Temporarily tighten the tensioner lockbolt with the tensioner spring fully extended.
12. Face the **I** and **E** marks of the camshaft pulley marks straight up, then align the timing marks with the horizontal surface on the cylinder head. Refer to the illustration for timing mark alignment.
13. Install the timing belt so there is no

looseness, refer to the illustration for location as follows:
- Crankshaft pulley
- Idler pulley
- Exhaust camshaft pulley
- Intake camshaft pulley
- Tensioner

14. Make sure no pressure other than the tensioner spring is applied to the belt. If reusing the old belt, be sure it is installed in the same direction of rotation.
15. Temporarily install the pulley boss and lockbolt.
16. Turn the crankshaft 1⅚ turns clockwise and align the crankshaft sprocket timing mark with the tension set mark for proper belt tension adjustment. Remove the lockbolt and pulley boss.
- Remove the pulley bolt and boss. and verify the timing belt pulley mark is still aligned with the tensioner set mark.

✳✳ CAUTION

Do not let the tensioner move with the tensioner lock bolt as it is turned.

17. Tighten the tensioner lock bolt.
18. Temporarily install the pulley boss and lockbolt.
19. Turn the crankshaft 2⅙ turns clockwise and face the pin on the pulley boss upright. Be sure the camshaft sprocket timing marks are aligned. If they are not, repeat the alignment steps.

➡**The timing marks are aligned normally if the camshaft pulley marks I and E are facing straight up, the timing marks are aligned to the horizontal surface on the cylinder head.**

20. Apply approximately 22 lbs. (10kg) pressure to the timing belt at a point midway between the camshaft sprockets. The belt should deflect 0.24–0.29 inch (6–7.5mm). If the deflection is not correct, repeat the alignment and tensioning procedure.
21. Install the pulley boss and lockbolt. Tighten the bolt to 116–122 ft. lbs. (157–166 Nm).
22. Install the timing belt covers and tighten the bolts to 95 inch lbs. (11 Nm).
23. Install the number 3 mount and remove the engine support device.
24. Install or connect the following:
- Rocker arm cover with a new gasket
- Water pump pulley
- Plate
- Crankshaft pulley and tighten the bolts to 109–151 inch lbs. (12–17 Nm)

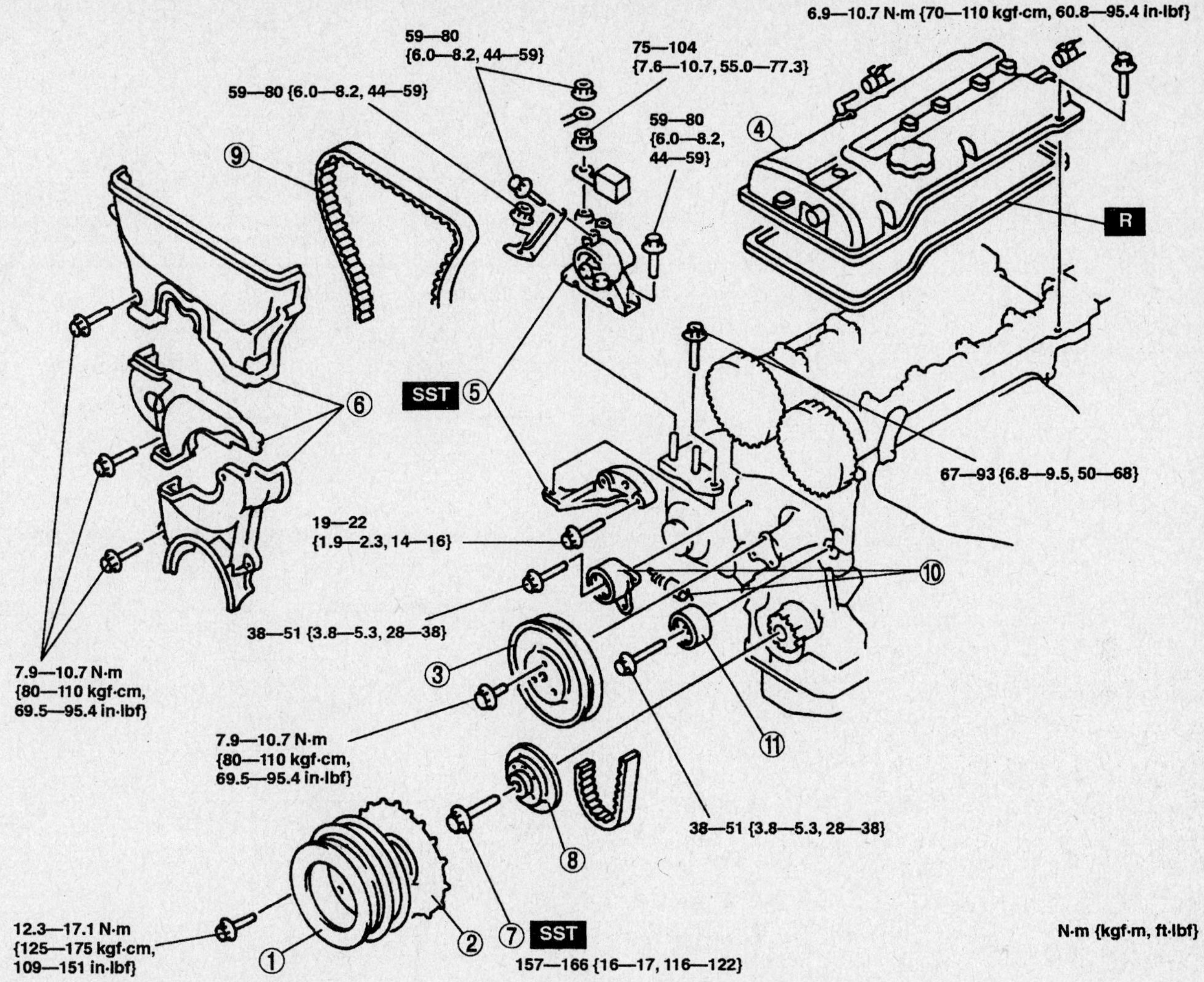

6.9—10.7 N·m {70—110 kgf·cm, 60.8—95.4 in·lbf}

59—80 {6.0—8.2, 44—59}

59—80 {6.0—8.2, 44—59}

75—104 {7.6—10.7, 55.0—77.3}

59—80 {6.0—8.2, 44—59}

⑨

④

R

SST ⑤

67—93 {6.8—9.5, 50—68}

⑥

19—22 {1.9—2.3, 14—16}

⑩

38—51 {3.8—5.3, 28—38}

③

7.9—10.7 N·m {80—110 kgf·cm, 69.5—95.4 in·lbf}

⑪

7.9—10.7 N·m {80—110 kgf·cm, 69.5—95.4 in·lbf}

38—51 {3.8—5.3, 28—38}

⑧

N·m {kgf·m, ft·lbf}

12.3—17.1 N·m {125—175 kgf·cm, 109—151 in·lbf}

①

②

⑦

SST

157—166 {16—17, 116—122}

1	Crankshaft pulley	7	Pulley lock bolt
2	Plate	8	Pulley boss
3	Water pump pulley	9	Timing belt
4	Cylinder head cover	10	Tensioner, tensioner spring (See 01-10A-11 Tensioner, Tensioner Spring Installation Note)
5	No.3 engine mount		
6	Timing belt cover	11	Idler

42356-MAZC-GIA

Exploded view of the timing belt assembly—1.6L (ZM) engines

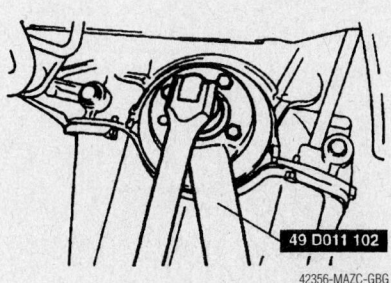

42356-MAZC-GBG

Remove pulley bolt and boss by using tool 49 D011 102 to prevent crankshaft rotation—1.6L (ZM) engines

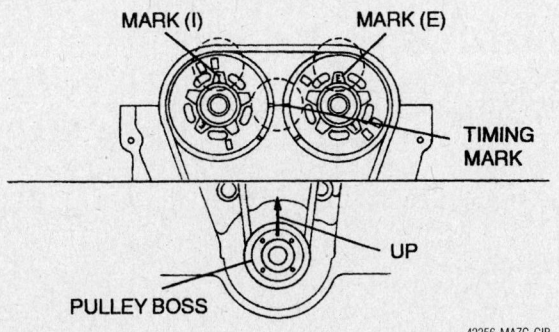

42356-MAZC-GIB

Turn the crankshaft until the timing marks on the crankshaft and camshaft sprockets are aligned. The pin on the pulley boss must face upward—1.6L (ZM) engines

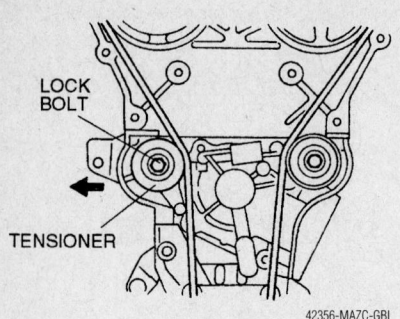

42356-MAZC-GBI

Loosen the tensioner lockbolt and pry the tensioner outward—1.6L (ZM) engines

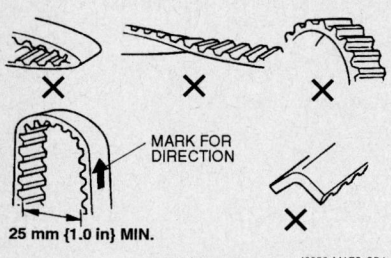

42356-MAZC-GBJ

Do not to bend, twist or get oil or other contaminates on the belt as this will damage the belt, if reusing the belt; mark the direction of rotation prior to removal—1.6L (ZM) engines

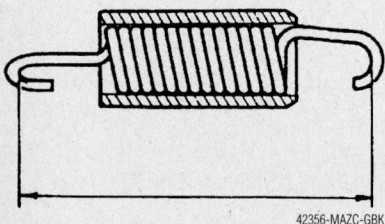

42356-MAZC-GBK

Check the tensioner spring free length; it should not exceed 2.43 inch (68mm)—1.6L (ZM) engines

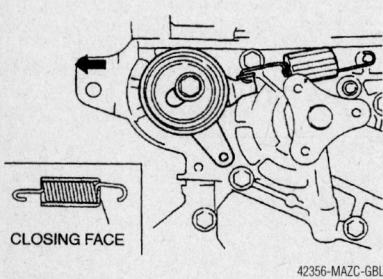

42356-MAZC-GBL

Install the tensioner and tensioner spring—1.6L (ZM) engines

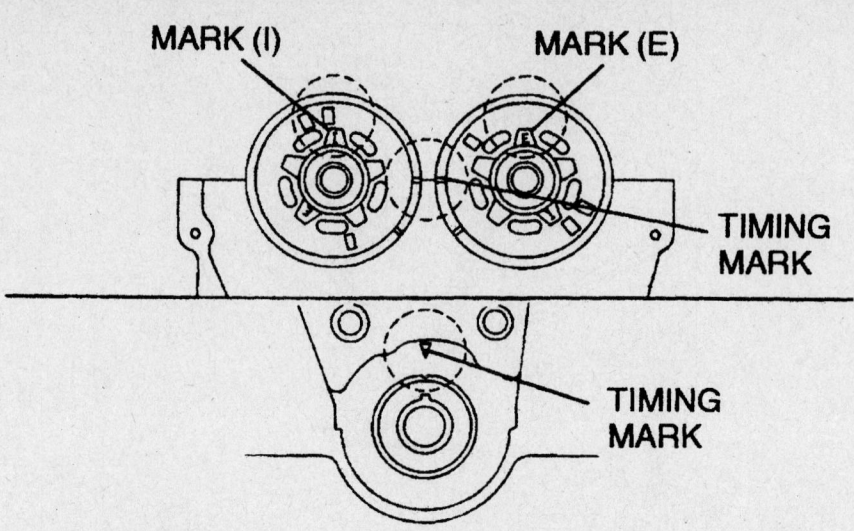

42356-MAZC-GIC

Verify the timing marks are aligned as shown before installing the belt—1.6L (ZM) engines

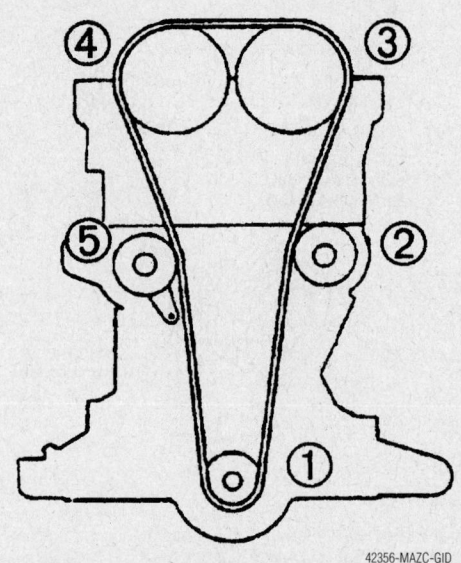

42356-MAZC-GID

Install the timing belt in this order—1.6L (ZM) engines

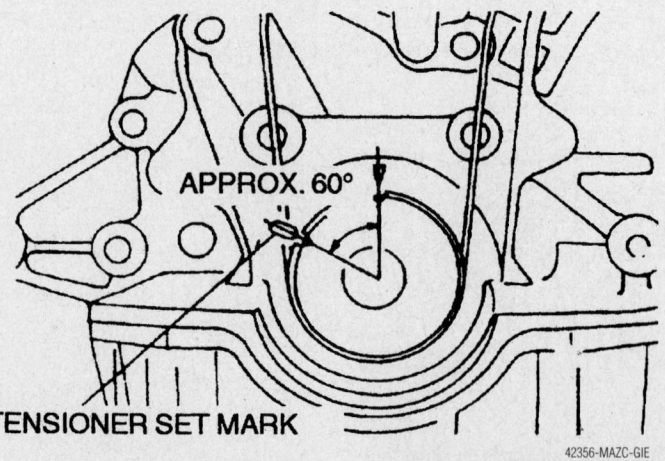

42356-MAZC-GIE

Turn the crankshaft 1⅚ turns clockwise and align the crankshaft sprocket timing mark with the tension set mark for proper belt tension adjustment—1.6L (ZM) engines

MEASURING POINT

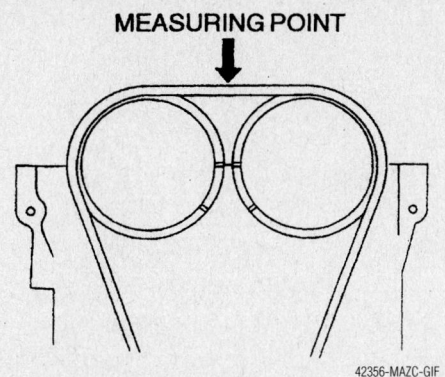

42356-MAZC-GIF

The belt should deflect 0.24–0.29 inch (6–7.5mm)—1.6L (ZM) engines

- Drive belt
- Ignition coils
- CMP sensor and check the air gap between the sensor and the plate teeth using a brass feeler gauge. If the gap is not 0.020–0.059 inch (0.5–1.5mm) adjust the sensor until the proper gap is reached.
- Negative battery cable

2.0L (FS) ENGINES

1. Before servicing the vehicle, refer to the precautions section.
2. Refer to the illustration of the exploded

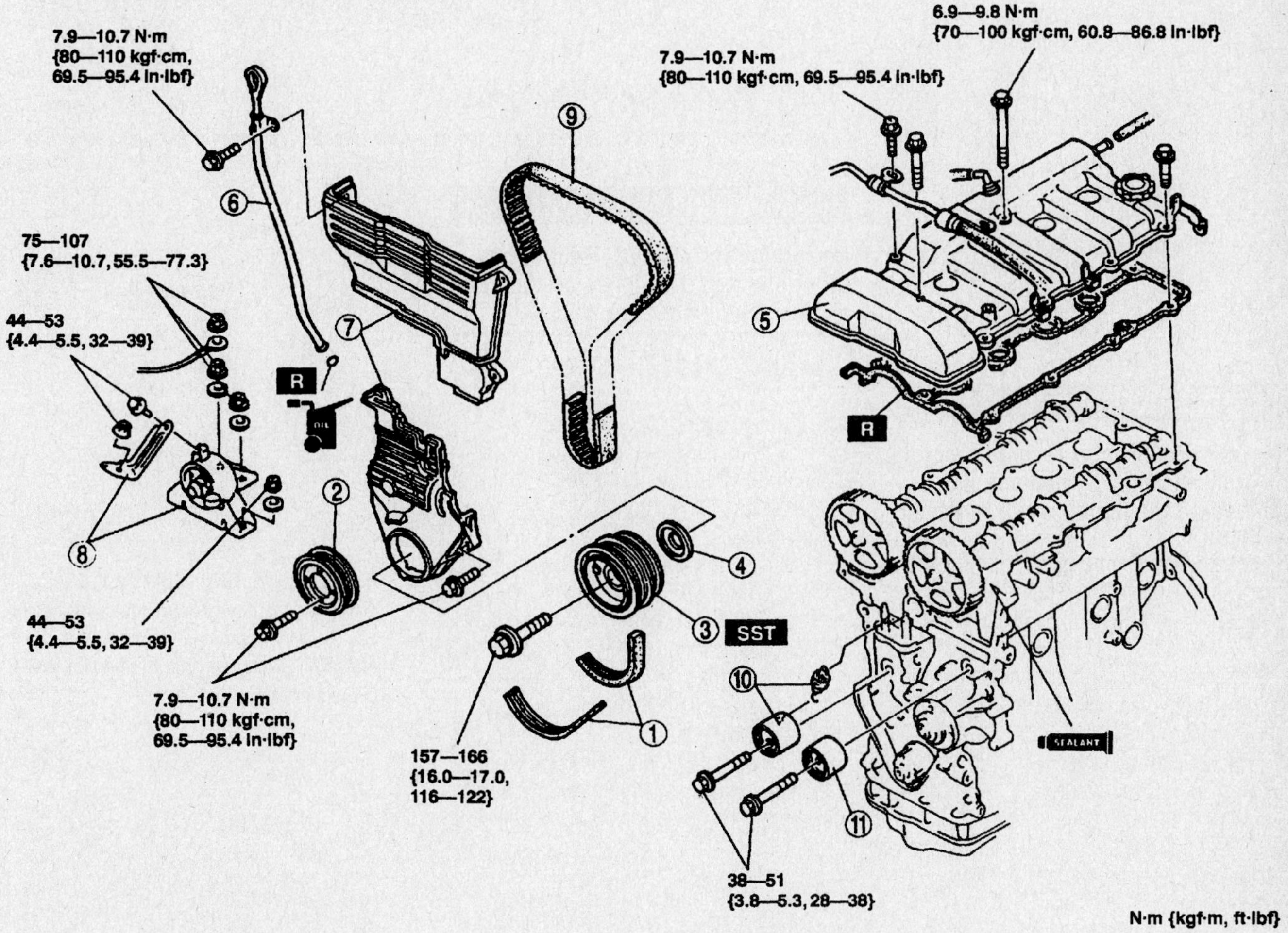

7.9—10.7 N·m {80—110 kgf·cm, 69.5—95.4 in·lbf}

6.9—9.8 N·m {70—100 kgf·cm, 60.8—86.8 in·lbf}

7.9—10.7 N·m {80—110 kgf·cm, 69.5—95.4 in·lbf}

75—107 {7.6—10.7, 55.5—77.3}

44—53 {4.4—5.5, 32—39}

R

OIL

R

44—53 {4.4—5.5, 32—39}

7.9—10.7 N·m {80—110 kgf·cm, 69.5—95.4 in·lbf}

157—166 {16.0—17.0, 116—122}

SST

SEALANT

38—51 {3.8—5.3, 28—38}

N·m {kgf·m, ft·lbf}

1	Drive belt	7	Timing belt cover
2	Water pump pulley	8	No.3 engine mount rubber
3	Crankshaft pulley	9	Timing belt
4	Guide plate	10	Tensioner, tensioner spring
5	Cylinder head cover	11	Idler
6	Dipstick and pipe		

42356-MAZC-G33

Exploded view of the timing belt assembly–2.0L (FS) engines shown

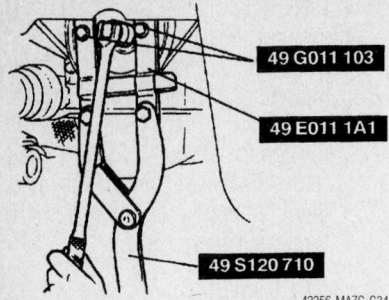

Remove the crankshaft pulley using special tools 49 G011 v103, 49 E011 1A1 and 49 S120 710–2.0L (FS) engines

42356-MAZC-G34

view of the timing belt assembly for component location and torque specifications.

3. Remove or disconnect the following:
- Negative battery cable
- Crankshaft Position (CKP) sensor
- Camshaft Position (CMP) sensor on Protégé models
- Spark plugs
- Power steering pump and bracket, leave the lines attached and position the pump aside on 2.0L (FS) engines
- Drive belt
- Water pump pulley

✳✳ CAUTION

The CKP sensor rotor is on the rear of the crankshaft pulley and can be damaged if the pulley is not removed carefully.

- Crankshaft pulley using special tools 49 G011 103, 49 E011 1A1 and 49 S120 710. Refer to the illustration for tool positioning.
- Guide plate from behind the crankshaft pulley
- Rocker arm cover
- Oil dipstick and tube
- Timing belt cover

4. Support the engine assembly with a hoist and remove the number 3 engine mount rubber.

5. Turn the crankshaft until the timing mark on the crankshaft sprocket aligns with the timing mark on the oil pump and the camshaft sprocket timing marks **E** and **I** align on the camshaft sprockets. Refer to the illustration for proper timing mark alignment.

6. Lower the vehicle. Insert a camshaft sprocket holding tool between the camshaft sprockets.

7. Turn the timing belt tensioner with an Allen wrench and remove the tensioner spring from the hook pinch

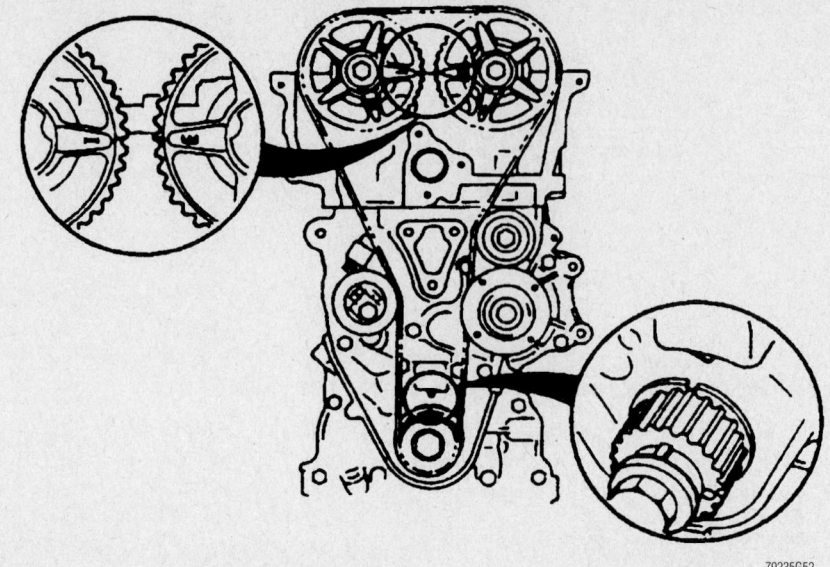

79235G52

When properly aligned for belt removal, the cam gear marks should face each other–2.0L (FS)

✳✳ CAUTION

Be careful not to bend, twist or get oil or other contaminates on the belt as this will damage the belt.

8. If the timing belt is to be reused, mark the direction of rotation on the timing belt. Remove the timing belt.

9. Rotate the tensioner, if the tension rotates with no resistance or does not rotate; replace the tensioner

To install:

10. Install the crankshaft sprocket bolt. Install the flywheel locking tool, if equipped with automatic transaxle, or place the shift lever in **4th** gear and apply the parking brake, if equipped with manual transaxle. Tighten the bolt to 116–122 ft. lbs. (157–166 Nm).

11. Be sure the timing marks on the camshaft and crankshaft sprockets are still aligned.

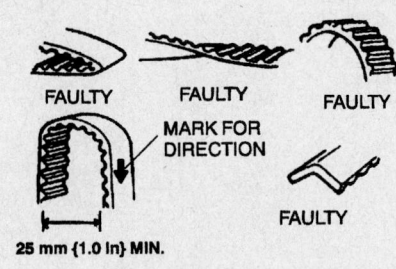

42356-MAZC-G36

Do not to bend, twist or get oil or other contaminates on the belt as this will damage the belt, if reusing the belt; mark the direction of rotation prior to removal–2.0L (FS) engines

42356-MAZC-G37

Rotate the tensioner, if the tension rotates with no resistance or does not rotate; replace the tensioner–2.0L (FS) engines

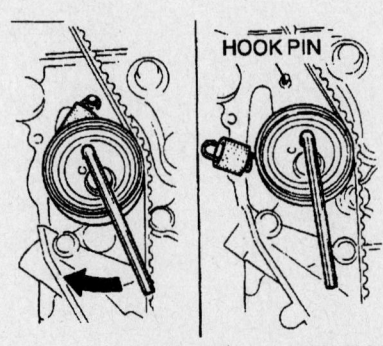

42356-MAZC-G35

Remove the tensioner spring from the hook pin–2.0L (FS) engines

12. Install the timing belt. If reusing the original timing belt, be sure it is installed in the same direction of rotation. Make sure there is no looseness on the idler side.

13. Rotate the crankshaft 2 turns in the normal direction of rotation and align the timing marks. Be sure all marks are still correctly aligned. If the marks are not aligned, remove the belt and then reinstall it and make sure the marks are properly aligned.

14. Check the tensioner spring length, if the free length is not 1.44 inch (36.5mm), replace the spring.

✸✸ CAUTION

Do not use tension other than that supplied by the tension spring or damage could occur

15. Turn the tensioner clockwise with an Allen wrench and install the tensioner spring. Remove the holding tool from between the camshaft sprockets.

16. Rotate the crankshaft 2 turns in the normal direction of rotation and align the timing marks. Be sure all marks are still correctly aligned.

17. Install the number 3 engine mount rubber and tighten the fasteners to the specifications shown in the illustration. Remove the engine hoist.

18. Install or connect the following:
- Timing belt cover. Refer to the exploded view of the timing belt components for torque specifications.
- Oil dipstick and tube. Refer to the exploded view of the timing belt components for torque specifications.
- Rocker arm cover
- Guide plate

✸✸ CAUTION

The CKP sensor rotor is on the rear of the crankshaft pulley and can be damaged if the pulley is not installed carefully.

- Crankshaft pulley using special tools 49 G011 v103, 49 E011 1A1 and 49 S120 710 and tighten the bolt to 116–122 ft. lbs. (157–166 Nm).
- Water pump pulley. Refer to the exploded view of the timing belt components for torque specifications.
- Drive belt

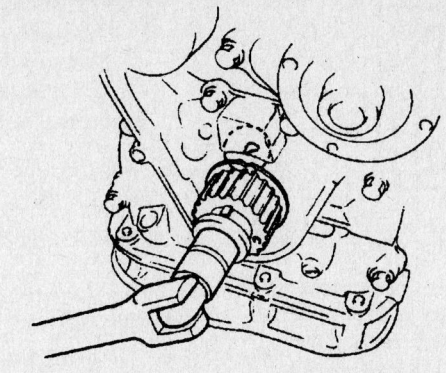

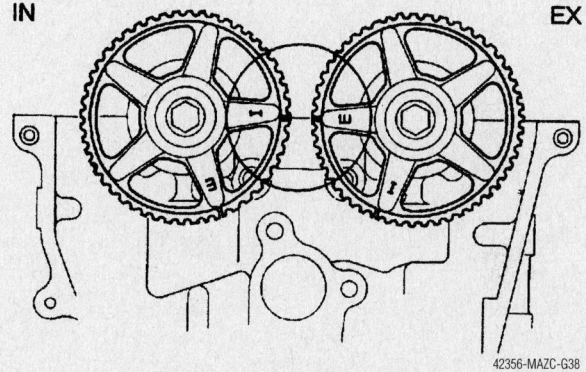

IN   EX

42356-MAZC-G38

Be sure the timing marks on the camshaft and crankshaft sprockets are still aligned—2.0L (FS) engines

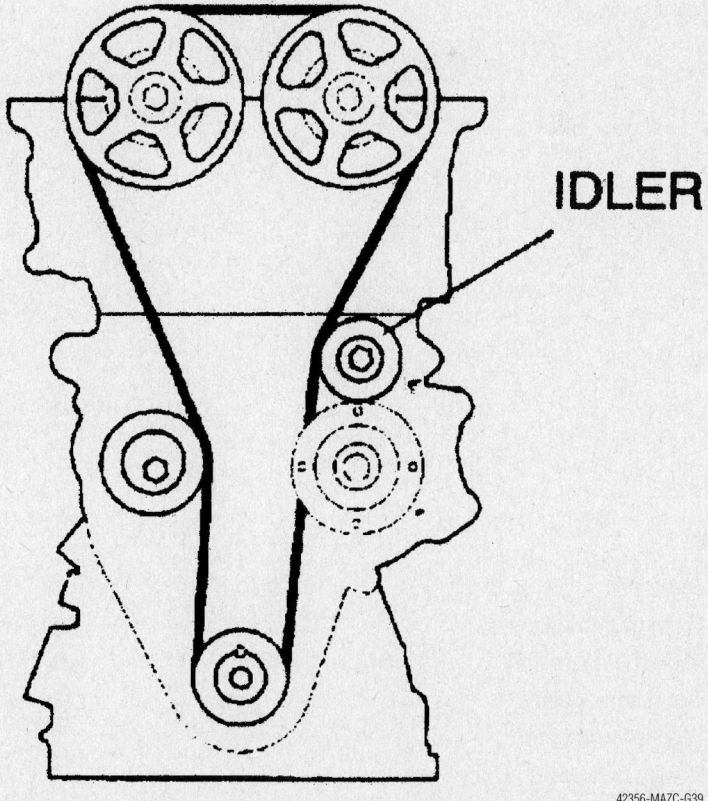

IDLER

42356-MAZC-G39

Make sure there is no looseness on the idler side when the belt is installed—2.0L (FS) engines

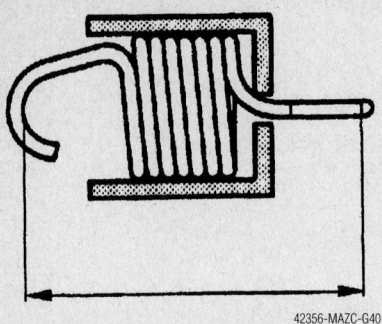

42356-MAZC-G40

Check the tensioner spring length, if the free length is not 1.44 inch (36.5mm); replace the spring–2.0L (FS) engines

- Power steering bracket and pump, if removed
- Spark plugs
- CKP sensor
- Negative battery cable

2.5L (KL) ENGINE

1. Before servicing the vehicle, refer to the precautions section.
2. Refer to the illustration of the exploded view of the timing belt assembly for component location and torque specifications.
3. Remove or disconnect the following:
   - Negative battery cable

- Crankshaft Position (CKP) sensor
- Spark plugs
- Power steering pump, leave the lines attached and position the pump aside
- Drive belt
- Water pump pulley
- Idler pulley bracket

✴✴ CAUTION

The CKP sensor rotor is on the rear of the crankshaft pulley and can be damaged if the pulley is not removed carefully.

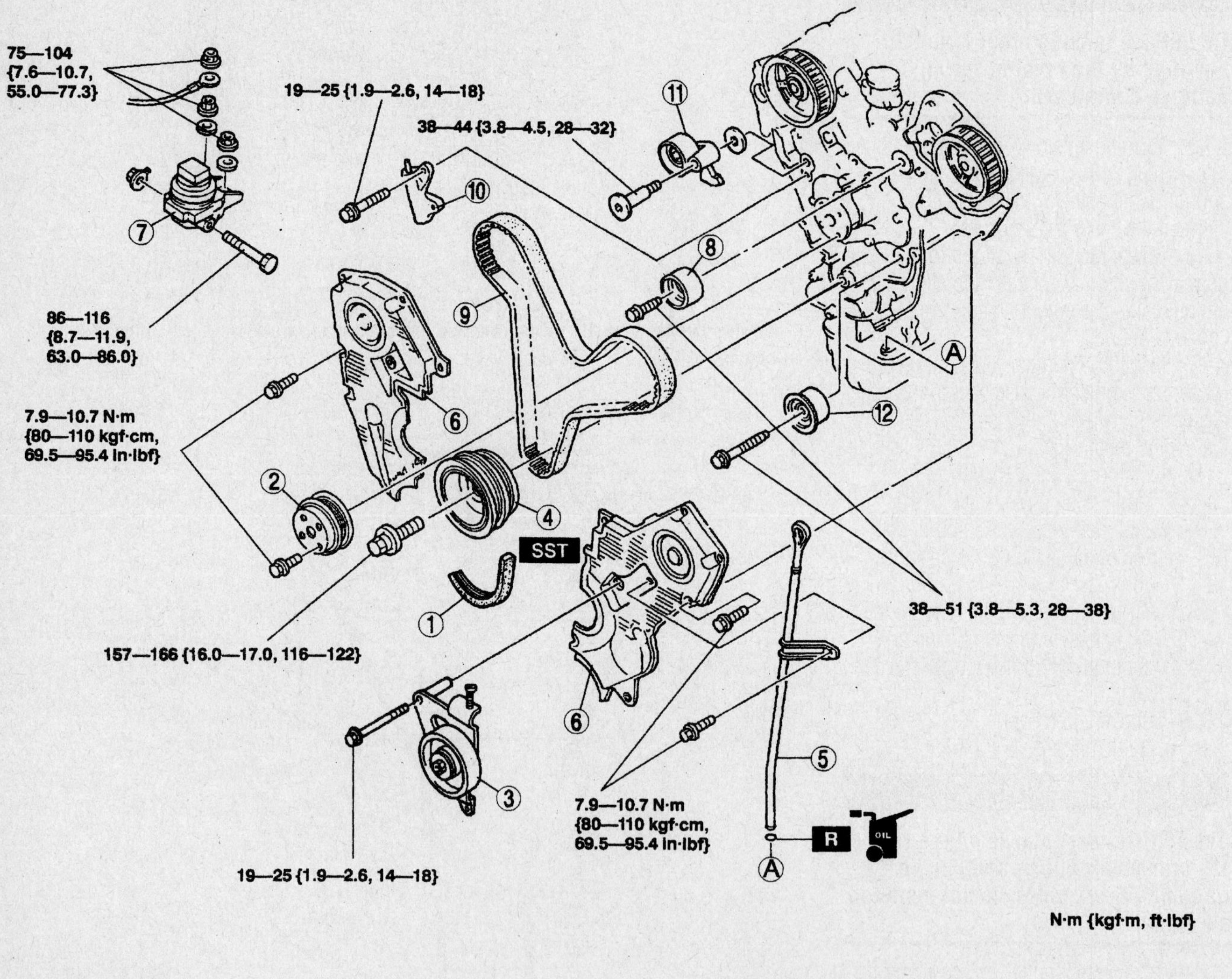

75—104 {7.6—10.7, 55.0—77.3}

19—25 {1.9—2.6, 14—18}

38—44 {3.8—4.5, 28—32}

86—116 {8.7—11.9, 63.0—86.0}

7.9—10.7 N·m {80—110 kgf·cm, 69.5—95.4 In·lbf}

157—166 {16.0—17.0, 116—122}

19—25 {1.9—2.6, 14—18}

SST

38—51 {3.8—5.3, 28—38}

7.9—10.7 N·m {80—110 kgf·cm, 69.5—95.4 In·lbf}

R OIL

N·m {kgf·m, ft·lbf}

1	Drive belt	7	No.3 engine mount rubber
2	Water pump pulley	8	No.1 idler
3	Idler pulley bracket	9	Timing belt
4	Crankshaft pulley	10	Timing belt auto tensioner
5	Dipstick and pipe	11	Tensioner
6	Timing belt cover	12	No.2 idler

42356-MAZC-G41

Exploded view of the timing belt assembly–2.5L (KL) engines—626 models

- Crankshaft pulley using special tools 49 E011 1A1 and 49 S120 710. Refer to the illustration for tool positioning.
- Rocker arm cover
- Oil dipstick and tube
- Timing belt cover

4. Support the engine assembly with a hoist and remove the number 3 engine mount rubber.

5. Install the crankshaft pulley bolt and turn the crankshaft clockwise and align the timing marks as illustrated. The number 1 piston should be at Top Dead Center (TDC) of the compression stroke.

✳✳ CAUTION

When removing the number 1 idler pulley bolt, hold the pulley so that the threads are aligned or the threads can be damaged.

6. Remove the number 1 idler pulley

✳✳ CAUTION

When removing the auto tensioner bolts, hold the tensioner so that the threads are aligned or the threads can be damaged.

7. Loosen the auto tensioner bolts and remove the lower bolt.

✳✳ CAUTION

Be careful not to bend, twist or get oil or other contaminates on the belt as this will damage the belt.

8. If the timing belt is to be reused, mark the direction of rotation on the timing belt. Remove the timing belt.

9. Remove the belt.

10. If necessary, remove the auto tensioner.

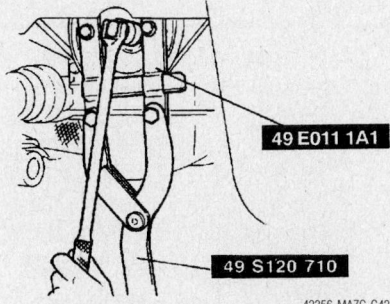

Remove the crankshaft pulley using special tools 49 E011 1A1 and 49 S120 710–2.5L (KL) engines

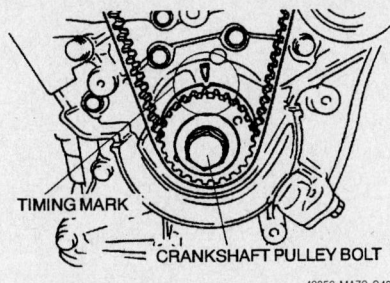

Turn the crankshaft clockwise and align the timing marks as illustrated. The number 1 piston should be at Top Dead Center (TDC) of the compression stroke.–2.5L (KL) engines

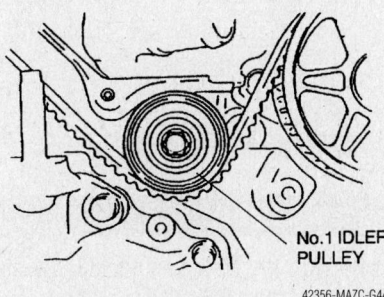

Remove the number 1 idler pulley.–2.5L (KL) engines

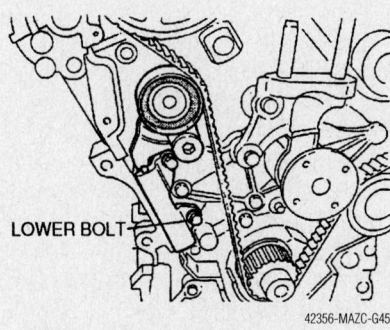

Loosen the auto tensioner bolts and remove the lower bolt–2.5L (KL) engines

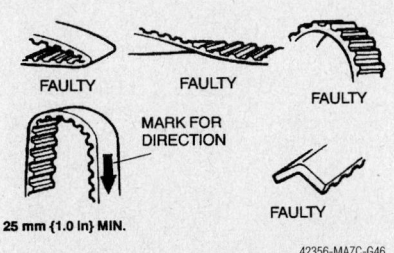

Do not to bend, twist or get oil or other contaminates on the belt as this will damage the belt, if reusing the belt; mark the direction of rotation prior to removal–2.5L (KL) engines

To install:

✳✳ CAUTION

There are two type of auto tensioners and they are interchangeable.

11. If removed install the auto tensioner as follows:

a. Measure the tensioner rod projection length. If the length exceeds 0.563–0.594 inch (14.3–15.1mm) on type **A** tensioners or 0.473–0.511 inch (12–13mm) on type **B** tensioners replace the tensioner. Refer to the illustration to distinguish tensioner type.

b. Inspect the tensioner for leakage and replace if defective.

✳✳ CAUTION

Do not apply pressure of more than 2,200 lbf. (9.8 Kn) to the tensioner or damage will occur.

c. Using a press, slowly press in the tensioner rod.

d. Insert a 0.055 inch (1.4mm) diameter pin to hold the rod in position.

e. Install the tensioner and hand tighten the bolt.

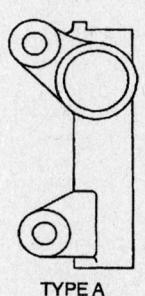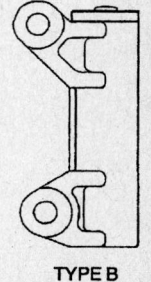

There are two type of auto tensioners and they are interchangeable–2.5L (KL) engines

Projection (Free length)
    Type A: 14.3—15.1 mm {0.563—0.594 in}
    Type B: 12.0—13.0 mm {0.473—0.511 in}

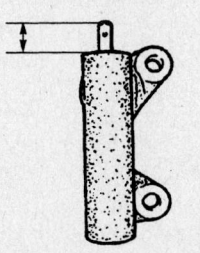

Measure the tensioner rod projection length and replace if it exceeds specification–2.5L (KL) engines

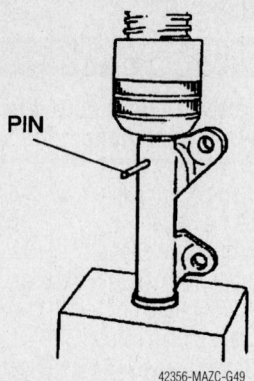

Using a press, slowly press in the tensioner rod and insert a 0.055 inch (1.4mm) diameter pin to hold the rod in position–2.5L (KL) engines

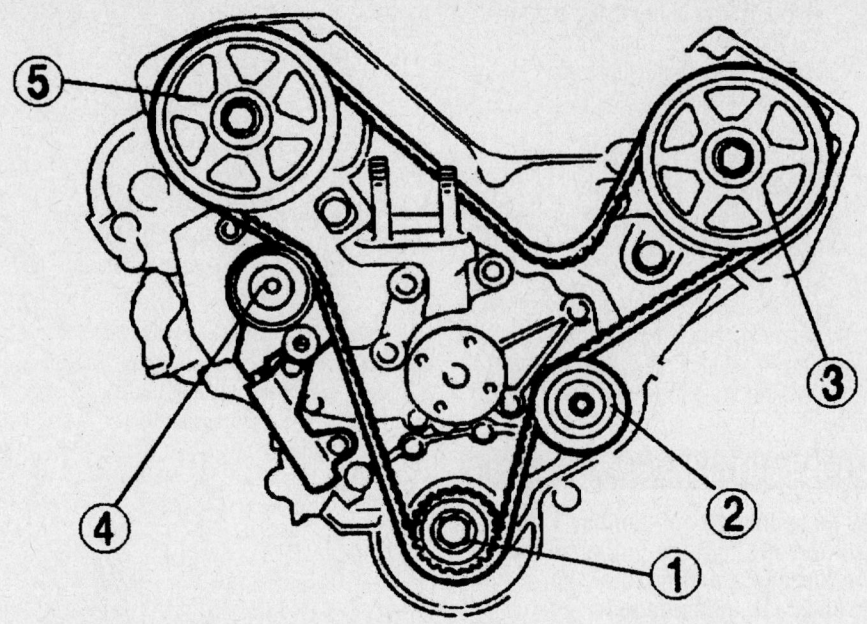

Make sure there is tension between pulleys (3) and (1) and pulleys (1) and (5)–2.5L (KL)

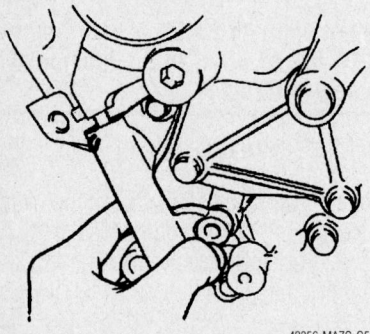

Install the tensioner and hand tighten the bolt–2.5L (KL) engines

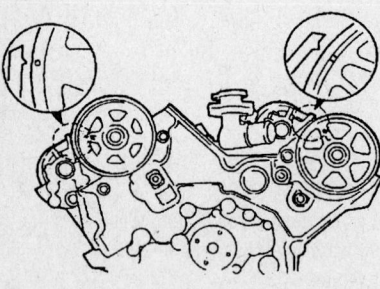

Turn the camshafts clockwise and align the timing marks–2.5L (KL) engines

12. Turn the camshafts clockwise and align the timing marks as illustrated.

13. Install the crankshaft pulley bolt and turn the crankshaft clockwise and align the marks as illustrated.

14. With the number 1 idler pulley removed, install the belt on the pulleys in this order:
- Timing belt pulley
- Number 2 idler pulley
- Left hand camshaft pulley
- Tensioner pulley
- Right hand camshaft pulley

⁕⁕ CAUTION

Make sure the belt has no looseness at the tension side.

15. Make sure there is tension between pulleys (3) and (1) and pulleys (1) and (5). Refer to the illustration for location of the pulleys.

16. Install the number 1 idler pulley while applying the pressure on the timing belt. Tighten the bolt to 28–38 ft. lbs. (38–51 Nm).

17. Push the auto tensioner in the direction of arrow (refer to illustration) and hand tighten the lower bolt, then tighten the bolts. Refer to the exploded view of the timing belt components for torque specifications. Then remove the retaining pin.

18. Turn the crankshaft clockwise and align the timing marks. Make sure the tim-

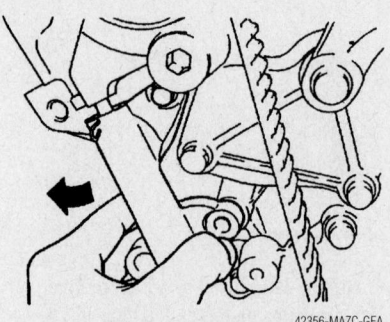

Push the auto tensioner in the direction of arrow and hand tighten the lower bolt–2.5L (KL) engine

ing marks are aligned, if not remove the belt and repeat the installation process.

19. Install or connect the following:
- Number 3 engine mount rubber. Refer to the exploded view of the timing belt components for torque specifications. Remove the engine hoist.
- Timing belt cover. Refer to the exploded view of the timing belt components for torque specifications.
- Oil dipstick and tube. Refer to the exploded view of the timing belt components for torque specifications.
- Rocker arm cover

⁕⁕ CAUTION

The CKP sensor rotor is on the rear of the crankshaft pulley and can be damaged if the pulley is not removed carefully.

- Crankshaft pulley using special tools 49 E011 1A1 and 49 S120 710. Refer to the illustration for tool positioning. Tighten the bolt to 16–122 ft. lbs. (157–166 Nm).
- Idler pulley bracket. Refer to the exploded view of the timing belt components for torque specifications.
- Water pump pulley. Refer to the exploded view of the timing belt components for torque specifications.

- Drive belt
- Power steering pump
- Spark plugs
- CKP sensor
- Negative battery cable

Millenia

2.3L (KJ) ENGINE

1. Before servicing the vehicle, refer to the precautions section.

2. Refer to the illustration of the exploded view of the timing belt assembly for component location and torque specifications.

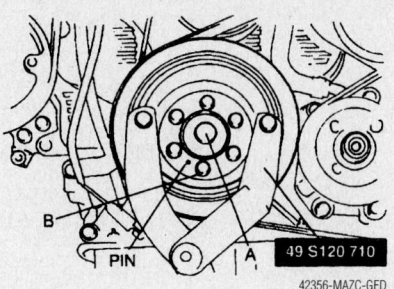

Remove the crankshaft pulley using special tools 49 E120 710–2.3L (KJ) engines

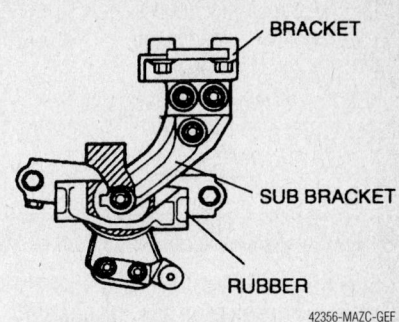

Remove the number 3 engine mount sub bracket, rubber and bracket–2.3L (KJ) engines

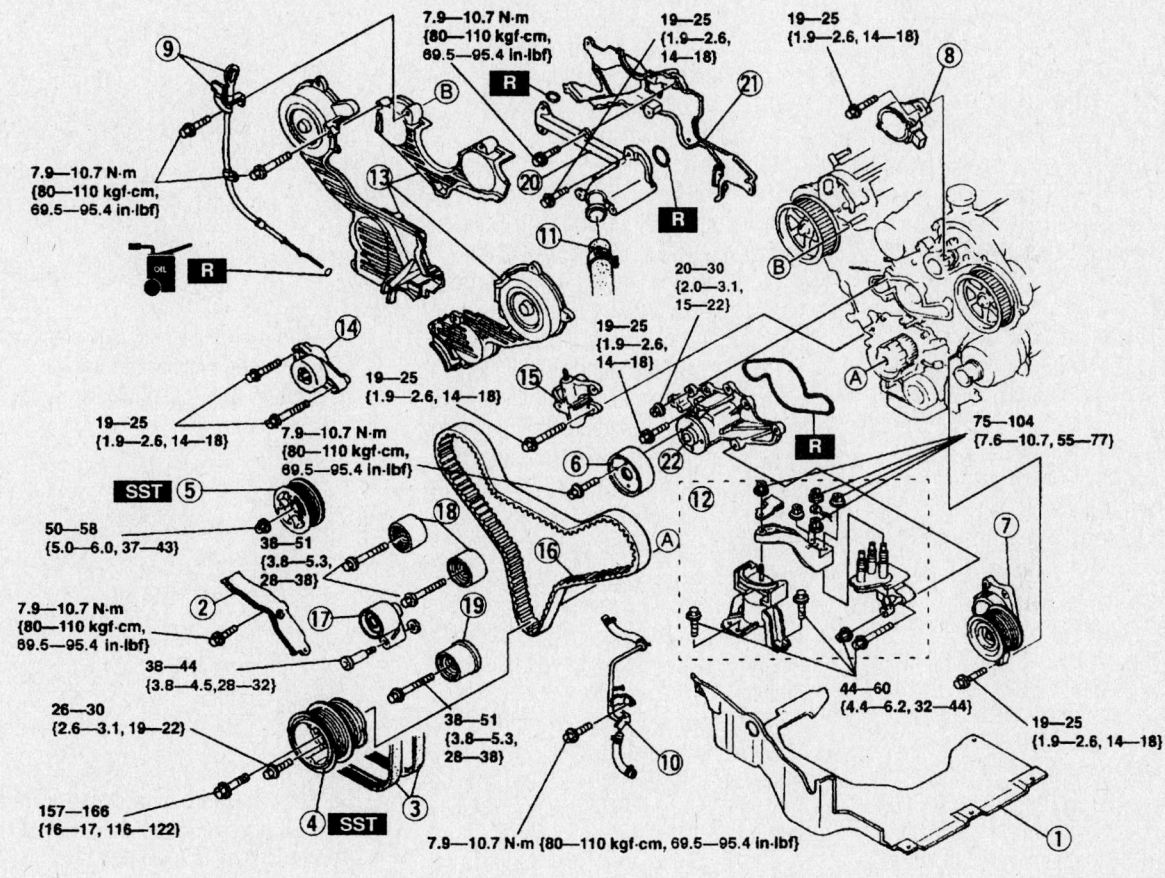

N·m {kgf·m, ft·lbf}

1	Splash shield (RH)	12	No.3 engine mount
2	Dust cover	13	Timing belt cover
3	Drive belt	14	Drive belt auto tensioner (P/S)
4	Crankshaft pulley	15	Timing belt auto tensioner
5	P/S oil pump pulley	16	Timing belt
6	Water pump pulley	17	Tensioner pulley
7	Drive belt auto tensioner (Generator)	18	No.1 idler pulley
8	CMP sensor	19	No.2 idler pulley
9	Dipstick and pipe	20	Water outlet pipe
10	Vacuum pipe	21	Seal plate
11	Upper radiator hose	22	Water pump

Exploded view of the timing belt assembly–2.3L (KJ) engines

3. Remove or disconnect the following:
- Negative battery cable
- Right front wheel
- Splash shield from the right hand side
- Dust cover
- Drive belt

✳✳ CAUTION

The Crankshaft Position (CKP) sensor rotor is on the rear of the crankshaft pulley and can be damaged if the pulley is not removed carefully.

4. Turn the crankshaft pulley clockwise until the pin on the pulley is facing down.
- Crankshaft pulley using special tools 49 E120 710. Refer to the illustration for tool positioning.
- Power steering pump, leave the lines attached and position the pump aside
- Water pump pulley
- Alternator drive belt tensioner
- Camshaft Position (CMP) sensor
- Oil dipstick and tube
- Vacuum pipe
- Upper radiator hose

5. Support the engine assembly with a hoist and remove the number 3 engine mount rubber.
- Number 3 engine mount sub bracket, rubber and bracket
- Timing belt cover
- Power steering pump drive belt tensioner

6. Turn the crankshaft until the timing mark on the crankshaft sprocket aligns with the timing mark on the oil pump and the camshaft sprocket timing marks align with the marks on the cylinder head. The No. 1 piston should be at Top Dead Center (TDC) of the compression stroke.

7. Remove the two bolts from the automatic tensioner, removing the lower one first. Keep the bolt holes aligned by holding

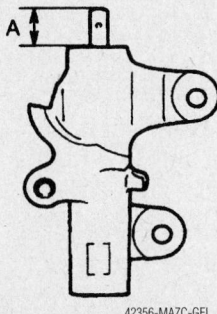

Check the tension rod projection, it should be 0.563–0.594 inch (14.3–15.1mm)–2.3L (KJ) engines

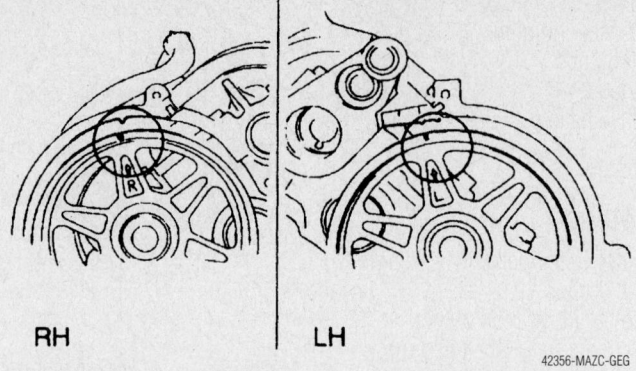

RH   LH

42356-MAZC-GEG

Be sure the camshaft sprocket timing marks are still aligned–2.3L (KJ) engines

the tensioner to reduce the chance of stripping the threads on the bolts.

8. If the timing belt is to be reused, mark the direction of rotation on the timing belt.

9. Remove the timing belt.

To install:

10. Install the crankshaft sprocket bolt. Install the flywheel locking tool. Tighten the bolt to 116–122 ft. lbs. (157–166 Nm).

11. Check the tension rod projection, it should be 0.563–0.594 inch (14.3–15.1mm), if not, replace the tensioner

12. Position the automatic tensioner in a press. Set a flat washer under the tensioner body to prevent damage to the body plug.

13. Compress the tensioner until the hole in the piston is aligned with the 2nd hole in the tensioner case. Insert a 0.055 inch (1.4mm) diameter wire or pin through the 2nd hole to keep the piston compressed.

14. Be sure the camshaft sprocket timing marks are still aligned. Turn the crankshaft clockwise until the timing sprocket is aligned.

15. Install the timing belt. If the original belt is being reused, be sure it is installed in the same direction of rotation. The order of installation is: timing belt (crankshaft) sprocket, No. 2 idler pulley, left-hand camshaft sprocket, both No. 1 idler pulleys, right-hand camshaft sprocket and the tensioner pulley.

16. Install the automatic belt tensioner and tighten the bolts to 14–18 ft. lbs. (19–25 Nm). Remove the wire or pin from the tensioner.

17. Turn the crankshaft clockwise, until the crankshaft sprocket timing mark is again at TDC. This should place all of the belt slack in the automatic tensioner portion of the belt.

18. Rotate the crankshaft two turns in the normal direction of rotation and align the timing marks. Be sure all marks are still correctly aligned.

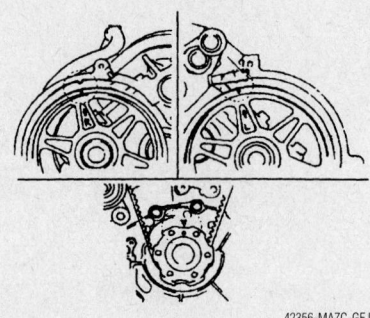

42356-MAZC-GEJ

Check the timing belt mark alignment once all the components are installed–2.3L (KJ) engines

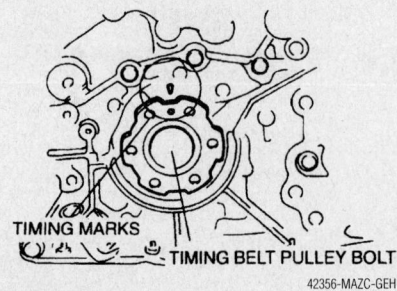

TIMING MARKS

TIMING BELT PULLEY BOLT

42356-MAZC-GEH

Turn the crankshaft clockwise, until the crankshaft sprocket timing mark is again at TDC–2.3L (KJ) engines

19. Inspect timing belt deflection, 0.24–0.31 inch (6–8mm), between the crankshaft sprocket and the tensioner pulley. If it is out of specification, replace the auto-tensioner.

20. Install or connect the following:
- Power steering pump drive belt tensioner
- Upper timing belt cover and tighten the bolts to 95 inch lbs. (11 Nm) in the sequence illustrated
- Right and left hand timing belt cover and tighten the bolts to 95 inch lbs. (11 Nm) in the sequence illustrated

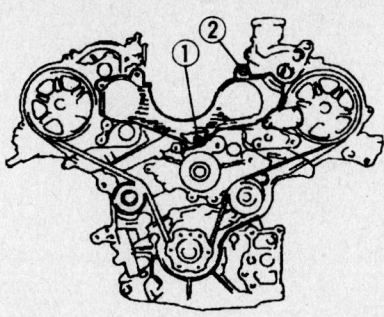

Tighten the upper timing belt cover in this sequence–2.3L (KJ) engine

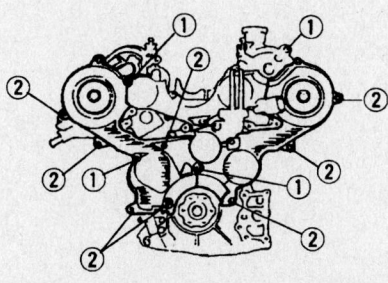

Tighten the right and left hand timing belt cover in this sequence–2.3L (KJ) engine

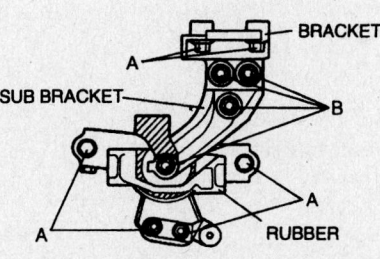

Install the number 3 engine mount rubber, bracket and sub bracket and tighten to specification–2.3L (KJ) engine

- Number 3 engine mount rubber, bracket and sub bracket. Tighten bolts **A** to 44 ft. lbs. (60 Nm) and bolts **B** to 77 ft. lbs. (104 Nm).
- Upper radiator hose
- Vacuum pipe
- Oil dipstick and tube
- CMP sensor
- Alternator drive belt tensioner
- Water pump pulley
- Power steering pump

21. Remove the timing belt pulley **A**, install the crankshaft pulley. Hand tighten bolts **A** and **B**. Tighten the bolt **A** to 116–122 ft. lbs. (157–166 Nm) and bolt **B** to 22 ft. lbs. (30 Nm).
- Drive belt

- Dust cover
- Splash shield from the right hand side
- Right front wheel
- Negative battery cable

2.5L (KL) ENGINE

1. Before servicing the vehicle, refer to the precautions section.
2. Refer to the illustration of the exploded view of the timing belt assembly for component location and torque specifications.
3. Remove or disconnect the following:
- Negative battery cable
- Splash shield from the right hand side
- Drive belt
- Water pump pulley
- Idler pulley and bracket
- Power steering pump, leave the lines attached and position the pump aside

❋❋ CAUTION

The Crankshaft Position (CKP) sensor rotor is on the rear of the crankshaft pulley and can be damaged if the pulley is not removed carefully.

- Crankshaft pulley using special tools 49 E011 1A1 and 49 S120 710. Refer to the illustration for tool positioning.
- Rocker arm cover
- Oil dipstick and tube
- CKP sensor
- Timing belt cover

4. Support the engine assembly with a hoist and remove the number 3 engine mount rubber.
5. Install the crankshaft pulley bolt and turn the crankshaft clockwise and align the timing marks as illustrated. The number 1 piston should be at Top Dead Center (TDC) of the compression stroke.

❋❋ CAUTION

When removing the auto tensioner bolts, hold the tensioner so that the threads are aligned or the threads can be damaged.

6. Loosen the auto tensioner bolts and remove the lower bolt.

❋❋ CAUTION

When removing the number 1 idler pulley bolt, hold the pulley so that the threads are aligned or the threads can be damaged.

7. Remove the number 1 idler pulley

❋❋ CAUTION

Be careful not to bend, twist or get oil or other contaminates on the belt as this will damage the belt.

8. If the timing belt is to be reused, mark the direction of rotation on the timing belt. Remove the timing belt.
9. Remove the belt.
10. If necessary, remove the auto tensioner.
To install:

❋❋ CAUTION

There are two type of auto tensioners and they are interchangeable.

11. If removed install the auto tensioner as follows:
   a. Measure the tensioner rod projection length. If the length exceeds 0.0.512–0.551 inch (13–14mm) on type **A** tensioners or 0.563–0.594 inch (14.3–15.1mm) on type **B** tensioners replace the tensioner. Refer to the illustration to distinguish tensioner type.
   b. Inspect the tensioner for leakage and replace if defective.

❋❋ CAUTION

Do not apply pressure of more than 2,200 lbf. (9.8 Kn) to the tensioner or damage will occur.

   c. Using a press, slowly press in the tensioner rod.
   d. Insert a pin whose diameter is 0.055 inch (1.4mm) on type **A** tensioners or 0.079 inch (2mm) on type **B** tensioners to hold the rod in position.
   e. Install the tensioner and hand tighten the bolt.
12. Turn the camshafts clockwise and align the timing marks as illustrated.
13. Install the crankshaft pulley bolt and turn the crankshaft clockwise and align the marks as illustrated.
14. With the number 1 idler pulley removed, install the belt on the pulleys in this order:
- Timing belt pulley
- Number 2 idler pulley
- Left hand camshaft pulley
- Tensioner pulley
- Right hand camshaft pulley

❋❋ CAUTION

Make sure the belt has no looseness at the tension side.

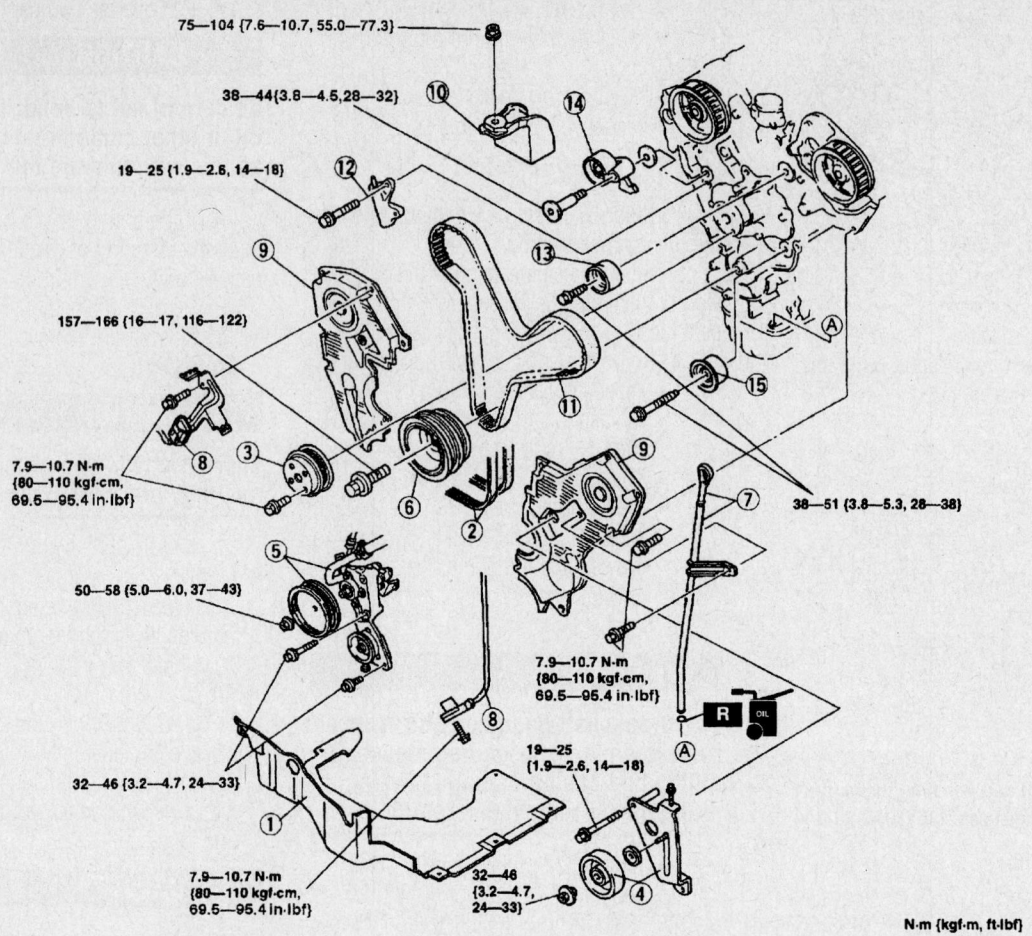

75—104 {7.6—10.7, 55.0—77.3}

38—44 {3.8—4.5, 28—32}

19—25 {1.9—2.6, 14—18}

157—166 {16—17, 116—122}

7.9—10.7 N·m
{80—110 kgf·cm,
69.5—95.4 in·lbf}

50—58 {5.0—6.0, 37—43}

32—46 {3.2—4.7, 24—33}

7.9—10.7 N·m
{80—110 kgf·cm,
69.5—95.4 in·lbf}

32—46
{3.2—4.7,
24—33}

38—51 {3.8—5.3, 28—38}

7.9—10.7 N·m
{80—110 kgf·cm,
69.5—95.4 in·lbf}

19—25
{1.9—2.6, 14—18}

N·m {kgf·m, ft·lbf}

1	Splash shield (RH)	9	Timing belt cover
2	Drive belt	10	No.3 engine mount sub bracket
3	Water pump pulley	11	Timing belt
4	Idler pulley and bracket	12	Timing belt auto tensioner
5	P/S oil pump	13	No.1 idler pulley
6	Crankshaft pulley	14	Tensioner pulley
7	Dipstick and pipe	15	No.2 idler pulley
8	CKP sensor		

42356-MAZC-GEB

Exploded view of the timing belt assembly–2.5L (KL) engines—Millenia models

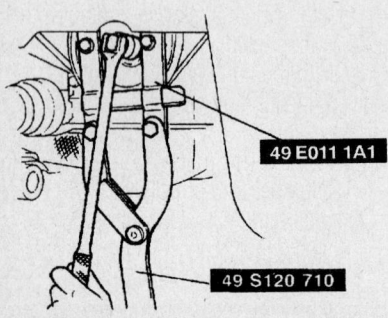

49 E011 1A1

49 S120 710

42356-MAZC-G42

Remove the crankshaft pulley using special tools 49 E011 1A1 and 49 S120 710–2.5L (KL) engines

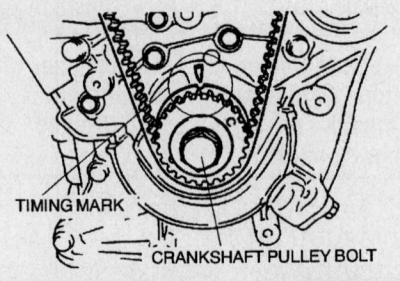

TIMING MARK

CRANKSHAFT PULLEY BOLT

42356-MAZC-G43

Turn the crankshaft clockwise and align the timing marks as illustrated. The number 1 piston should be at Top Dead Center (TDC) of the compression stroke.–2.5L (KL) engines

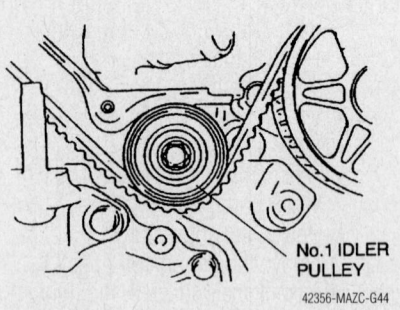

No.1 IDLER
PULLEY

42356-MAZC-G44

Remove the number 1 idler pulley.–2.5L (KL) engines

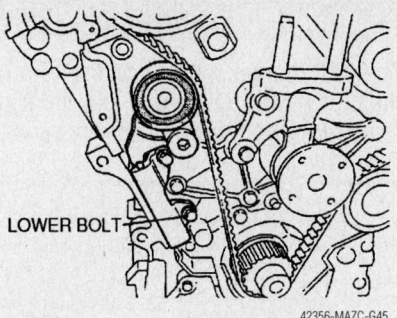

Loosen the auto tensioner bolts and remove the lower bolt–2.5L (KL) engines

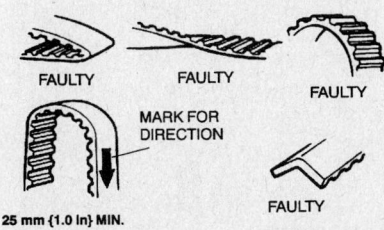

Do not to bend, twist or get oil or other contaminates on the belt as this will damage the belt, if reusing the belt; mark the direction of rotation prior to removal–2.5L (KL) engines

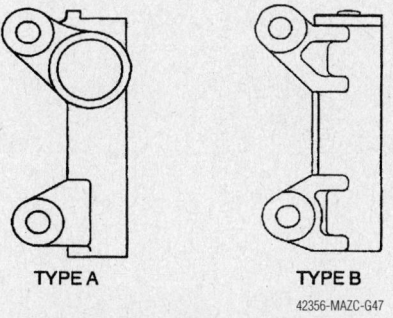

There are two type of auto tensioners and they are interchangeable–2.5L (KL) engines

Projection (Free length)
  Type A: 14.3—15.1 mm {0.563—0.594 in}
  Type B: 12.0—13.0 mm {0.473—0.511 in}

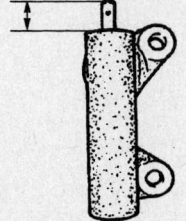

Measure the tensioner rod projection length and replace if it exceeds specification–2.5L (KL) engines

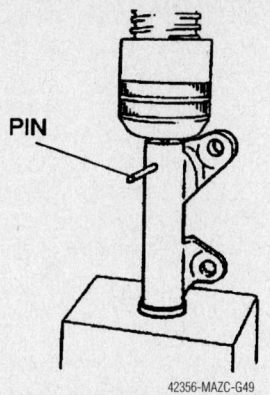

Using a press, slowly press in the tensioner rod and insert a 0.055 inch (1.4mm) diameter pin to hold the rod in position–2.5L (KL) engines

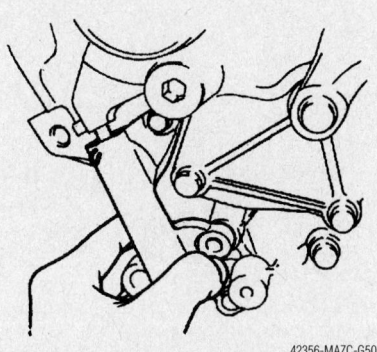

Install the tensioner and hand tighten the bolt–2.5L (KL) engines

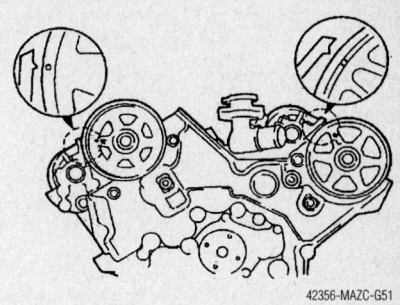

Turn the camshafts clockwise and align the timing marks–2.5L (KL) engines

15. Make sure there is tension between pulleys (3) and (1) and pulleys (1) and (5). Refer to the illustration for location of the pulleys.

16. Install the number 1 idler pulley while applying the pressure on the timing belt. Tighten the bolt to 28–38 ft. lbs. (38–51 Nm).

17. Push the auto tensioner in the direction of arrow (refer to illustration) and hand tighten the lower bolt, then tighten the bolts. Refer to the exploded view of the timing belt components for torque specifications. Then remove the retaining pin.

18. Turn the crankshaft clockwise and align the timing marks. Make sure the timing marks are aligned, if not remove the belt and repeat the installation process.

19. Install or connect the following:

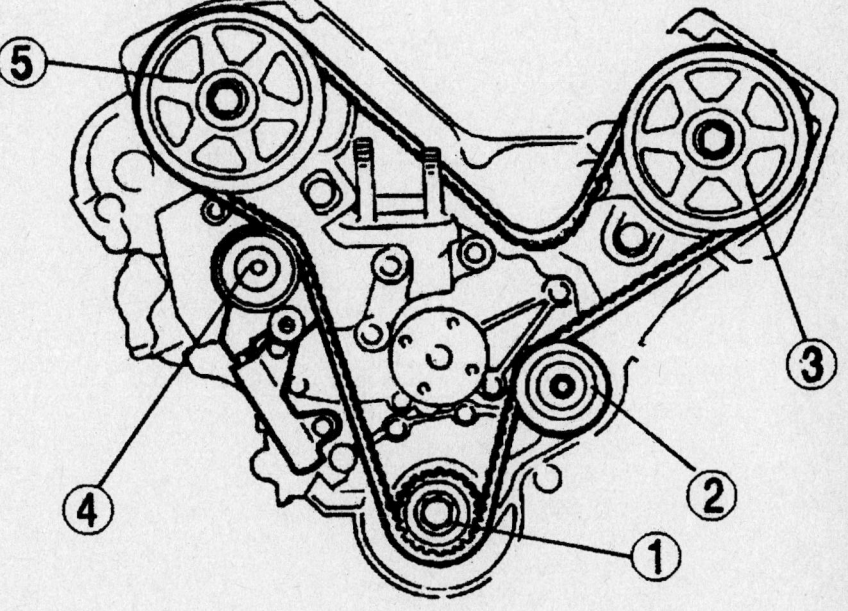

Make sure there is tension between pulleys (3) and (1) and pulleys (1) and (5)–2.5L (KL) engines

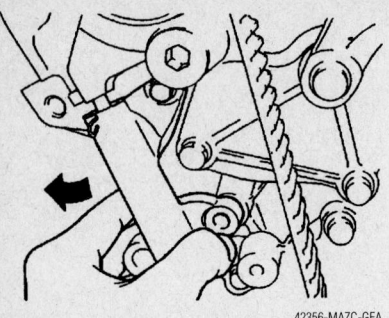

42356-MAZC-GEA

Push the auto tensioner in the direction of arrow and hand tighten the lower bolt–2.5L (KL) engine

- Number 3 engine mount rubber. Refer to the exploded view of the timing belt components for torque specifications. Remove the engine hoist.
- Timing belt cover. Refer to the exploded view of the timing belt components for torque specifications.
- Oil dipstick and tube. Refer to the exploded view of the timing belt components for torque specifications.
- CKP sensor
- Rocker arm cover

❊❊ CAUTION

The CKP sensor rotor is on the rear of the crankshaft pulley and can be damaged if the pulley is not removed carefully.

- Crankshaft pulley using special tools 49 E011 1A1 and 49 S120 710. Refer to the illustration for tool positioning. Tighten the bolt to 16–122 ft. lbs. (157–166 Nm).
- Power steering pump
- Idler pulley bracket. Refer to the

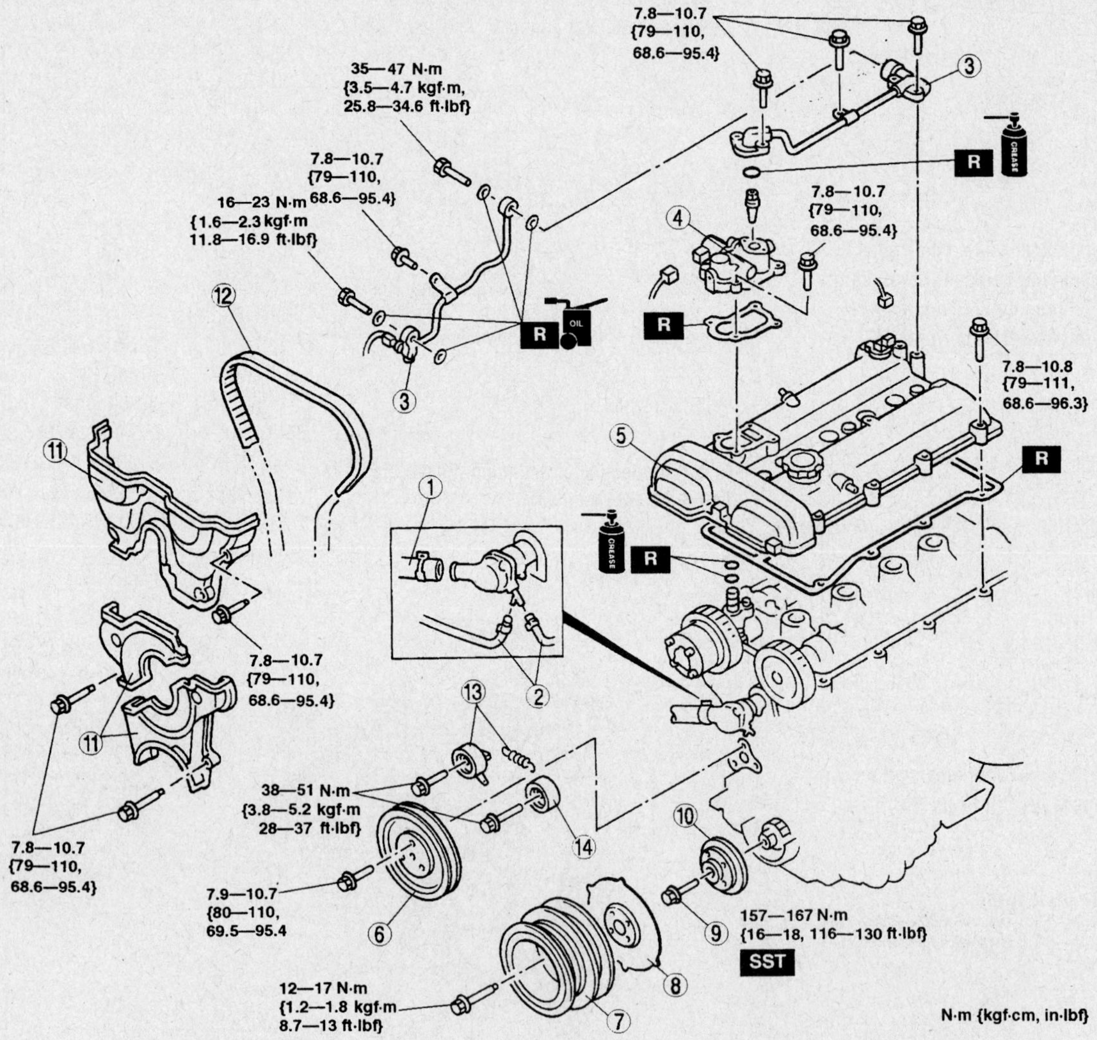

7.8—10.7 {79—110, 68.6—95.4}

35—47 N·m {3.5—4.7 kgf·m, 25.8—34.6 ft·lbf}

7.8—10.7 {79—110, 68.6—95.4}

16—23 N·m {1.6—2.3 kgf·m 11.8—16.9 ft·lbf}

7.8—10.7 {79—110, 68.6—95.4}

7.8—10.8 {79—111, 68.6—96.3}

7.8—10.7 {79—110, 68.6—95.4}

7.8—10.7 {79—110, 68.6—95.4}

38—51 N·m {3.8—5.2 kgf·m 28—37 ft·lbf}

7.9—10.7 {80—110, 69.5—95.4}

157—167 N·m {16—18, 116—130 ft·lbf}

SST

12—17 N·m {1.2—1.8 kgf·m 8.7—13 ft·lbf}

N·m {kgf·cm, in·lbf}

1	Upper radiator hose	8	Plate
2	Water hose	9	Pulley lock bolt
3	Oil pipe	10	Pulley boss
4	Oil control valve (OCV) case	11	Timing belt cover
5	Cylinder head cover	12	Timing belt
6	Water pump pulley	13	Tensioner and tensioner spring
7	Crankshaft pulley	14	Idler

42356-MAZC-GBF

Exploded view of the timing belt assembly—1.8L (BP) engines

exploded view of the timing belt components for torque specifications.
- Water pump pulley. Refer to the exploded view of the timing belt components for torque specifications.
- Drive belt
- Splash shield
- Negative battery cable

2002–04 Miata

1.8L (BP) ENGINE

1. Before servicing the vehicle, refer to the precautions section.
2. Drain the cooling system.
3. Remove or disconnect the following:
- Negative battery cable
- Front suspension lower bar
- Air pipe
- Drive belt
- Crankshaft Position (CKP) sensor
- High tension lead and the ignition coil
- Spark plugs
- Upper radiator hose
- Water hose
- Oil pipe
- Oil Control Valve (OCV)
- Rocker arm cover and discard the gasket
- Water pump pulley
- Crankshaft pulley and plate
- Pulley bolt and boss by using tool 49 D011 102 to prevent crankshaft rotation
- Timing belt cover

4. Install the pulley boss on the crankshaft and tighten the bolt.
5. Turn the crankshaft until the timing marks on the crankshaft and camshaft sprockets are aligned. The pin on the pulley boss must face upward. Hold the crankshaft pulley boss with a suitable tool and remove the pulley lockbolt, being careful not to rotate the crankshaft. Remove the crankshaft pulley boss.

➡Protect the tensioner with a shop towel before prying on it. Do not rotate the crankshaft after the timing belt has been removed.

6. Mark the direction of rotation on the timing belt. Loosen the tensioner lockbolt and pry the tensioner outward. Tighten the lockbolt with the tensioner spring fully extended. Remove the timing belt.
7. Remove the tensioner and spring. If necessary, remove the idler pulley.
8. Inspect the belt for wear, peeling, cracking, hardening or signs of oil contamination. Inspect the tensioner for free and smooth rotation. Check the tensioner spring free length; it should not exceed 2.31 inch

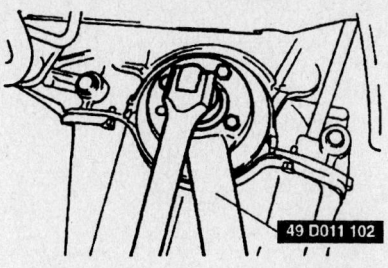

Remove pulley bolt and boss by using tool 49 D011 102 to prevent crankshaft rotation—1.8L (BP) engines

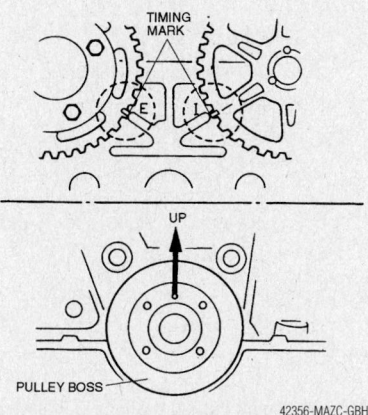

Turn the crankshaft until the timing marks on the crankshaft and camshaft sprockets are aligned. The pin on the pulley boss must face upward—1.8L (BP) engines

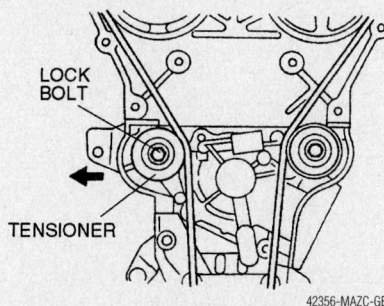

Loosen the tensioner lockbolt and pry the tensioner outward—1.8L (BP) engines

(59.2mm). Inspect the sprocket teeth for wear or damage. Replace parts, as necessary.

To install:
9. Install the crankshaft sprocket bolt and temporarily tighten.
10. If removed, install the idler pulley and tighten the bolt to 37 ft. lbs. 51 Nm).
11. Install the tensioner and tensioner spring. Pry the tensioner outward and temporarily tighten the tensioner lockbolt with the tensioner spring fully extended.
12. Be sure the crankshaft sprocket tim-

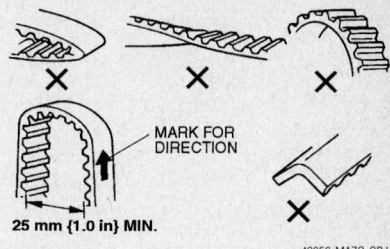

Do not to bend, twist or get oil or other contaminates on the belt as this will damage the belt, if reusing the belt; mark the direction of rotation prior to removal—1.8L (BP) engines

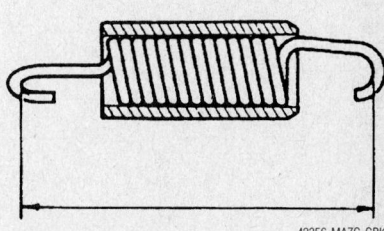

Check the tensioner spring free length; it should not exceed 2.31 inch (59.2mm)—1.8L (BP) engines

ing mark is aligned with the mark on the oil pump housing. Be sure the camshaft sprocket timing marks are aligned with the marks on the seal plate. Refer to the illustration for timing mark alignment.
13. Install the timing belt so there is no looseness at the idler pulley side or between the camshaft sprockets. If reusing the old belt, be sure it is installed in the same direction of rotation.
14. Temporarily install the pulley boss and lockbolt.
15. Turn the crankshaft 2 turns clockwise and align the crankshaft sprocket timing mark. Face the pin on the pulley boss upright. Be sure the camshaft sprocket timing marks are aligned. If they are not, repeat the alignment steps.
16. Turn the crankshaft 1⅚ turns clockwise and align the crankshaft sprocket timing mark with the tension set mark for proper belt tension adjustment. Remove the lockbolt and pulley boss.
17. Be sure the crankshaft sprocket timing mark is aligned with the tension set mark. Loosen the tensioner lockbolt, and allow the spring to apply tension to the belt. Tighten the tensioner lockbolt to 37 ft. lbs. (51 Nm).
18. Install the pulley boss and lockbolt.
19. Turn the crankshaft 2⅙ turns clockwise and be sure the timing marks are cor-

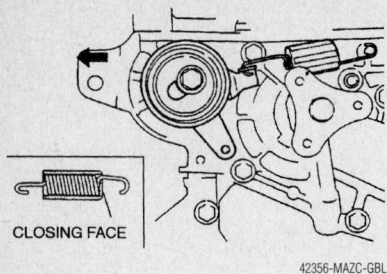

42356-MAZC-GBL

Install the tensioner and tensioner spring—1.8L (BP) engines

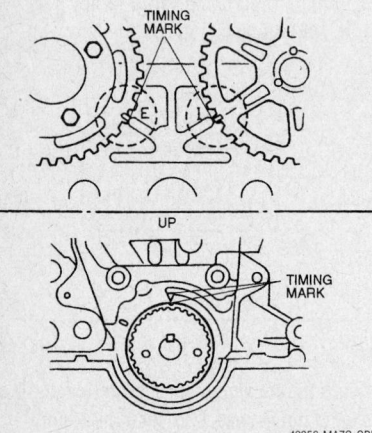

42356-MAZC-GBM

Be sure the crankshaft sprocket timing mark is aligned with the mark on the oil pump housing. Be sure the camshaft sprocket timing marks are aligned with the marks on the seal plate—1.8L (BP) engines

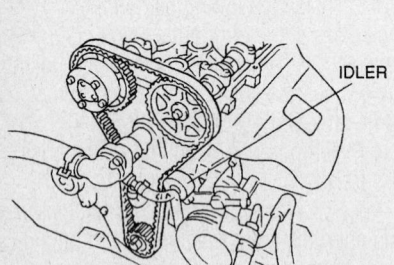

42356-MAZC-GBN

Install the timing belt so there is no looseness at the idler pulley side or between the camshaft sprockets—1.8L (BP) engines

rectly aligned. Make sure the pin on the pulley boss is straight up.

20. Apply approximately 22 lbs. (10kg) pressure to the timing belt at a point midway between the camshaft sprockets. The belt should deflect 0.35–0.45 inch (9.0–11.5mm). If the deflection is not correct, repeat the alignment and tensioning procedure.

21. Install the timing belt covers and tighten the bolts in sequence to 95 inch lbs. (11 Nm).

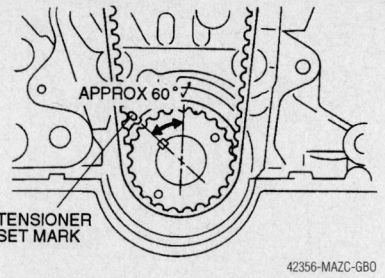

42356-MAZC-GBO

Turn the crankshaft 1⅚ turns clockwise and align the crankshaft sprocket timing mark with the tension set mark for proper belt tension adjustment—1.8L (BP) engines

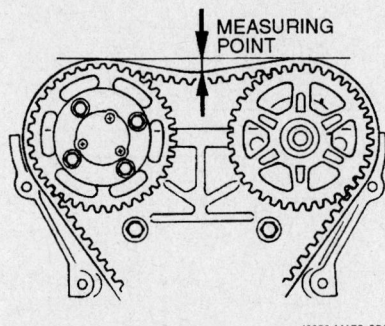

42356-MAZC-GBQ

The belt should deflect 0.35–0.45 inch (9.0–11.5mm)—1.8L (BP) engines

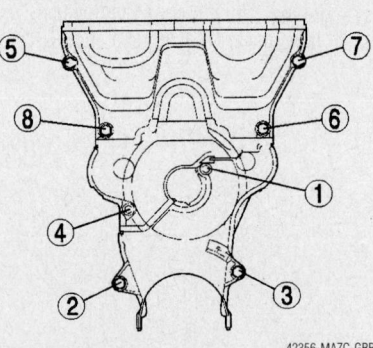

42356-MAZC-GBR

Tighten the timing belt cover using this sequence)—1.8L (BP) engines

22. Install the valve cover and spark plugs, along with all other applicable components.

23. Hold the pulley boss with a suitable tool, and tighten the lockbolt to 116–130 ft. lbs. (157–167 Nm).

24. Install or connect the following:
- Crankshaft plate and pulley
- Water pump pulley
- Rocker arm cover with a new gasket
- Temporarily tighten the cover bolts **A**, refer to the illustration for location. Torque the bolts in sequence to 80 inch lbs. (9 Nm).

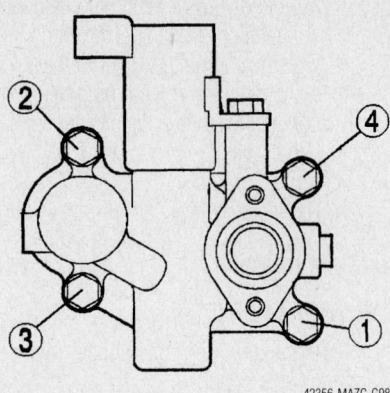

42356-MAZC-G98

Tighten the OCV bolts in the sequence shown—1.8L (BP) engine

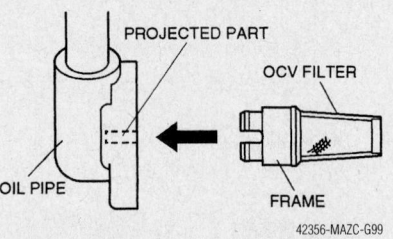

42356-MAZC-G99

Hold the frame of the OCV valve filter and install the OCV so it is aligned with the projected part on the flange end of the oil pipe—1.8L (BP) engine

✳✳ CAUTION

When installing the OCV valve, be careful not to damage the O–ring, if damaged it may cause leaking.

- OCV and tighten the bolts in the sequence illustrated

25. Install the oil pipe as follows:

a. Oil pipe. Hold the frame of the OCV valve filter and install the OCV so it is aligned with the projected part on the flange end of the oil pipe. Coat the new washer with clean engine oil and temporarily install the upper and side oil pipe and position the pipe.

b. **A** using several passes to 95 inch lbs. (11 Nm).

c. Tighten oil pipe bolts **B** using several passes to 61 inch lbs. (7 Nm).

d. Tighten oil pipe bolts **C** using several passes to 13 ft. lbs. (17 Nm).

e. Tighten oil pipe bolts 1, 2, 3, 4 and 7 using several passes to 95 inch lbs. (11 Nm).

f. Tighten oil pipe bolt 5 using several passes to 17 ft. lbs. (23 Nm).

g. Tighten oil pipe bolt 6 using several passes to 34 ft. lbs. (47 Nm).

26. Install or connect the following:

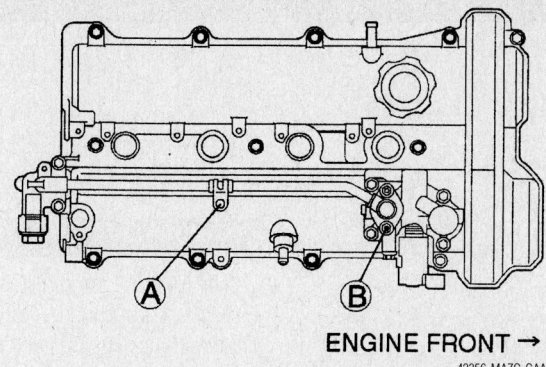

ENGINE FRONT →

42356-MAZC-GAA

Tighten oil pipe bolts A and B using several passes to the specification in the text–1.8L (BP) engine

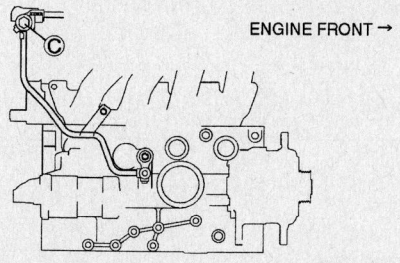

ENGINE FRONT →

42356-MAZC-GAB

Tighten oil pipe bolts C using several passes to the specification in the text–1.8L (BP) engine

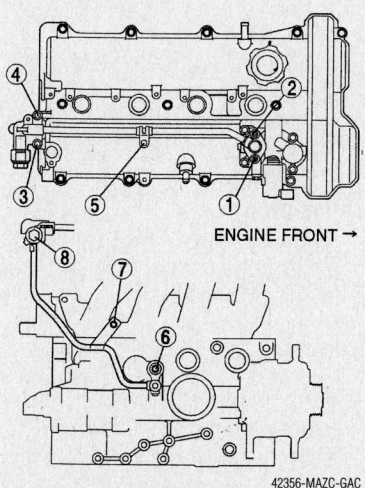

ENGINE FRONT →

42356-MAZC-GAC

Tighten oil pipe bolts using several passes to the specification in the text–1.8L (BP) engine

- Water hose
- Upper radiator hose
- Spark plugs
- High tension lead and the ignition coil
- CKP sensor
- Drive belt
- Air pipe
- Front suspension lower bar
- Negative battery cable

Piston and Ring

POSITIONING

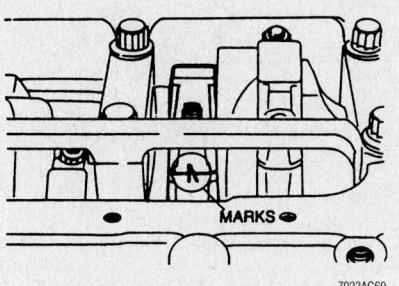

MARKS

7923AG60

Before removing the caps from the connecting rods, be sure to matchmark them—Mazda engines

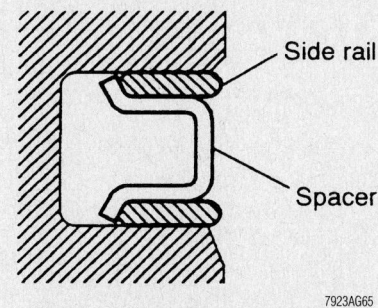

Side rail

Spacer

7923AG65

Compression ring identification and positioning—Mazda engines

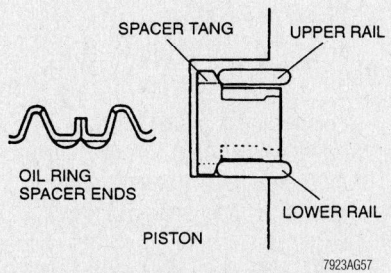

SPACER TANG   UPPER RAIL

OIL RING
SPACER ENDS

PISTON   LOWER RAIL

7923AG57

Upper, spacer and lower oil ring identification and positioning—Mazda engines

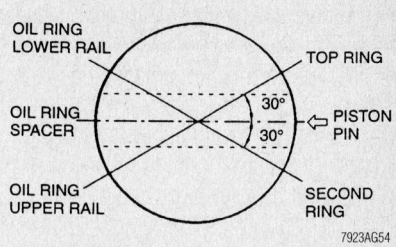

OIL RING
LOWER RAIL   TOP RING

OIL RING
SPACER   30°   PISTON
         30°   PIN

OIL RING
UPPER RAIL   SECOND
              RING

7923AG54

Piston ring end-gap spacing—Mazda engines

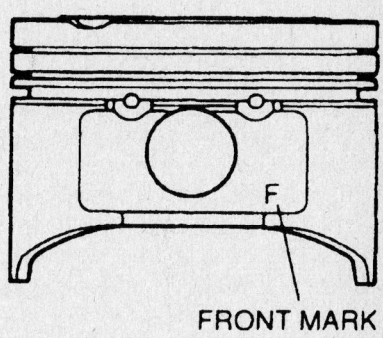

F

FRONT MARK

7923AG61

Piston-to-engine block mark location on the piston—Mazda 1.8L (BP) engines

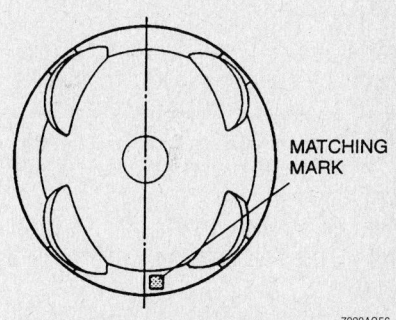

MATCHING
MARK

7923AG56

Piston-to-engine block mark location on the piston face—Mazda 2.0L (FS) and 2.5L (KL) engines

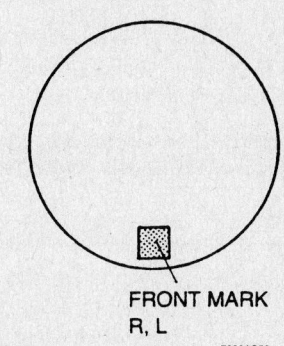

FRONT MARK
R, L

7923AG58

Piston-to-engine positioning mark location—Mazda 2.3L (KJ) engine

FUEL SYSTEM

Fuel System Service Precautions

Safety is the most important factor when performing not only fuel system maintenance but any type of maintenance. Failure to conduct maintenance and repairs in a safe manner may result in serious personal injury or death. Maintenance and testing of the vehicle's fuel system components can be accomplished safely and effectively by adhering to the following rules and guidelines.

1. To avoid the possibility of fire and personal injury, always disconnect the negative battery cable unless the repair or test procedure requires that battery voltage be applied.

2. Always relieve the fuel system pressure prior to disconnecting any fuel system component (injector, fuel rail, pressure regulator, etc.), fitting or fuel line connection. Exercise extreme caution whenever relieving fuel system pressure, to avoid exposing skin, face and eyes to fuel spray. Please be advised that fuel under pressure may penetrate the skin or any part of the body that it contacts.

3. Always place a shop towel or cloth around the fitting or connection prior to loosening to absorb any excess fuel due to spillage. Ensure that all fuel spillage (should it occur) is quickly removed from engine surfaces. Ensure that all fuel soaked cloths or towels are deposited into a suitable waste container.

4. Always keep a dry chemical (Class B) fire extinguisher near the work area.

5. Do not allow fuel spray or fuel vapors to come into contact with a spark or open flame.

6. Always use a back-up wrench when loosening and tightening fuel line connection fittings. This will prevent unnecessary stress and torsion to fuel line piping. Always follow the proper torque specifications.

7. Always replace worn fuel fitting O-rings with new. Do not substitute fuel hose where fuel pipe is installed.

Fuel System Pressure

RELIEVING

626 and Protégé

1. Before servicing the vehicle, refer to the precautions section.
2. Remove the filler cap.
3. Remove the fuel pump relay from the

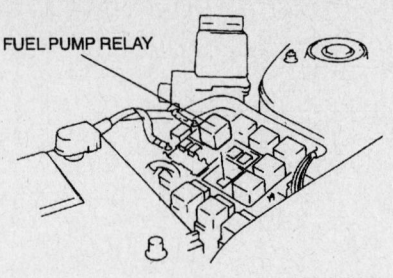

Fuel pump relay location—626 and with 2.0L and 2.5L engines

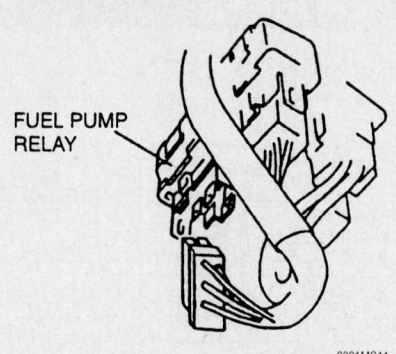

Fuel pump relay connector location—Miata

relay box, located in the engine compartment.
4. Start the engine.
5. After the engine stalls, turn the ignition switch **OFF**.
6. After servicing the vehicle, reinstall the relay.

2002–04 Miata

1. Before servicing the vehicle, refer to the precautions section.
2. Loosen the fuel filler cap to release the pressure in the tank.
3. Remove the fuel pump relay connector, located above the accelerator pedal.
4. Start the engine.
5. After the engine stalls, turn the ignition switch **OFF**.
6. After servicing the vehicle, reinstall the relay and tighten the fuel filler cap.

Millenia

1. Before servicing the vehicle, refer to the precautions section.
2. If necessary for clearance, remove the cruise control actuator and position aside on 2.3L (KJ) engines.
3. Remove the fuel pump relay from the relay box.
4. Start the engine.

5. After the engine stalls, turn the ignition switch **OFF**.
6. After servicing the vehicle, reinstall the relay and the cruise actuator, if necessary.

Fuel Filter

The fuel filter on most Mazda cars can be located on a bracket in the left rear of the engine compartment, next to or beneath the brake master cylinder fluid reservoir or as part of the fuel pump assembly.

On the Millenia, the fuel filter is located beneath an access cover in the trunk. Access to the cover is achieved by removing the trunk mat to expose the cover.

REMOVAL & INSTALLATION

626 and Protégé

1. Before servicing the vehicle, refer to the precautions section.
2. Properly relieve the fuel system pressure.
3. Remove or disconnect the following:
   • Negative battery cable
   • Cruise control actuator with the harness and cable connected on 2.0L (FS) engines
   • Air cleaner assembly, 2.5L (KL) engines
   • Harness from the filter
   • Fuel hoses from the filter
   • Fuel filter

To install:
4. Installation is the reverse of removal.
5. Pressurize the fuel system and check all connections for leaks.

2002–04 Miata

1. Before servicing the vehicle, refer to the precautions section.
2. Properly relieve the fuel system pressure.
3. Remove or disconnect the following:
   • Negative battery cable
4. Raise and safely support the rear of the vehicle.
   • Fuel filter protector
   • Fuel lines by squeezing the tabs. Cover the lines to prevent leakage and contamination.
   • Filter bracket and the filter. Mark the filter and bracket to aid for correct installation

To install:
5. Install or connect the following:
   • Filter and bracket aligning the

marks made during removal. Tighten the bracket bolts to 52 inch lbs. (6 Nm).
- Fuel lines making sure the tabs on the fittings are firmly engaged by pulling lightly on the lines
- Fuel filter protector
- Negative battery cable

6. Start the vehicle and check for leaks, then lower the vehicle.

Millenia

1. Before servicing the vehicle, refer to the precautions section.
2. Insure the ignition switch is **OFF**.
3. Relieve the fuel system pressure.
4. Remove or disconnect the following:
   - Negative battery cable
   - Trunk mat
   - Service hole cover
   - Fuel lines from both ends of the fuel filter
   - Fuel filter and bracket
   - Fuel filter from the mounting bracket

To install:
5. Install or connect the following:
   - Filter in the bracket. Torque the nut to 70–95 inch lbs. (8–11 Nm).
   - Fuel lines to the filter
   - Service hole cover
   - Trunk mat
   - Negative battery cable
6. Run the engine and check for any fuel leaks.

Fuel Pump

REMOVAL & INSTALLATION

Protégé

1. Before servicing the vehicle, refer to the precautions section.
2. Relieve the fuel pressure.
3. Remove or disconnect the following:
   - Negative battery cable
   - Rear seat cushion
   - Fuel pump/sending unit electrical connector
   - Fuel pump/sending unit access cover
   - Fuel supply and return hoses from the fuel pump/sending unit
   - Fuel pump/sending unit from the fuel tank
   - Sending unit electrical connector
   - Sending unit from the fuel pump assembly

To install:
4. Install or connect the following:

- Sending unit to the fuel pump assembly
- Sending unit electrical connector
- Fuel pump/sending unit into the fuel tank with a new gasket
- Fuel supply and return lines
- Access cover
- Sending unit electrical connector
- Rear seat cushion
- Negative battery cable

5. Start the engine and check fuel leaks.

2002–04 Miata

1. Before servicing the vehicle, refer to the precautions section.
2. Properly relieve the fuel pressure.
3. Remove or disconnect the following:
   - Negative battery cable
   - Rear package trim
   - Service hole cover
   - Fuel pump cover
   - Fuel pump connector
   - Fuel hoses
   - Fuel pump and gauge sender unit as an assembly
   - Fuel pump from the sender bracket

To install:
- New O-ring set
- Fuel pump to the sender bracket

➡ **Pull the fuel pump down so that it is tight against the bracket.**

- Fuel pump and gauge sender unit as an assembly
- Fuel hoses
- Fuel pump connector
- Fuel pump cover
- Service hole cover
- Rear package trim
- Negative battery cable

626

1. Before servicing the vehicle, refer to the precautions section.
2. Relieve the fuel system pressure.
3. Drain the fuel from the tank.
4. Remove or disconnect the following:
   - Negative battery cable
   - Fuel pump electrical connector
   - Hoses from the fuel tank
   - Pressure control valve
   - Fuel tank pressure sensor
   - Fuel pipe
   - Presilencer insulator
   - Fuel tank strap while supporting the fuel tank
   - Fuel tank
   - All fuel hoses from the fuel pump unit
   - Fuel pump ring retainers and the ring

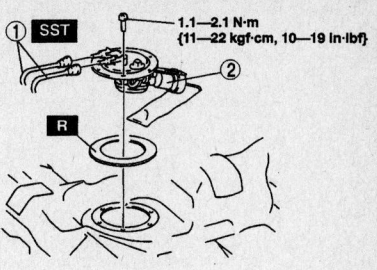

1.1—2.1 N·m {11—22 kgf·cm, 10—19 in·lbf}

1   Fuel pipe
2   Fuel pump

42356-MAZC-G53

Exploded view of the fuel pump assembly—626

- Fuel pump and gaskets from the fuel tank

To install:
5. Install or connect the following:
   - Fuel pump using a new gasket
   - Fuel pump ring and tighten the fasteners to 19 inch lbs. (2 Nm)
   - Fuel hoses to the fuel pump
   - Fuel tank
   - Fuel tank strap while supporting the fuel tank. Refer to the illustration for torque specifications.
   - Presilencer insulator
   - Fuel pipe
   - Fuel tank pressure sensor
   - Pressure control valve
   - Hoses to the fuel tank
   - Fuel pump electrical connector
   - Negative battery cable
6. Add a minimum of 10 gallons of fuel to the tank and check for leaks.

Millenia

1. Before servicing the vehicle, refer to the precautions section.
2. Relieve the fuel system pressure.
3. Remove or disconnect the following:
   - Negative battery cable
   - Rear seat cushion
4. Drain the fuel from the tank.
   - Service hole cover
   - Fuel pump electrical connector
   - Hoses from the fuel tank
   - Fuel tank strap while supporting the fuel tank
   - Evaporative hose
   - Fuel tank
   - All fuel hoses from the fuel pump unit
   - Fuel pump ring using tool 49 T042 011
   - Fuel pump and gaskets from the fuel tank

To install:
5. Install or connect the following:
   - Fuel pump using a new gasket

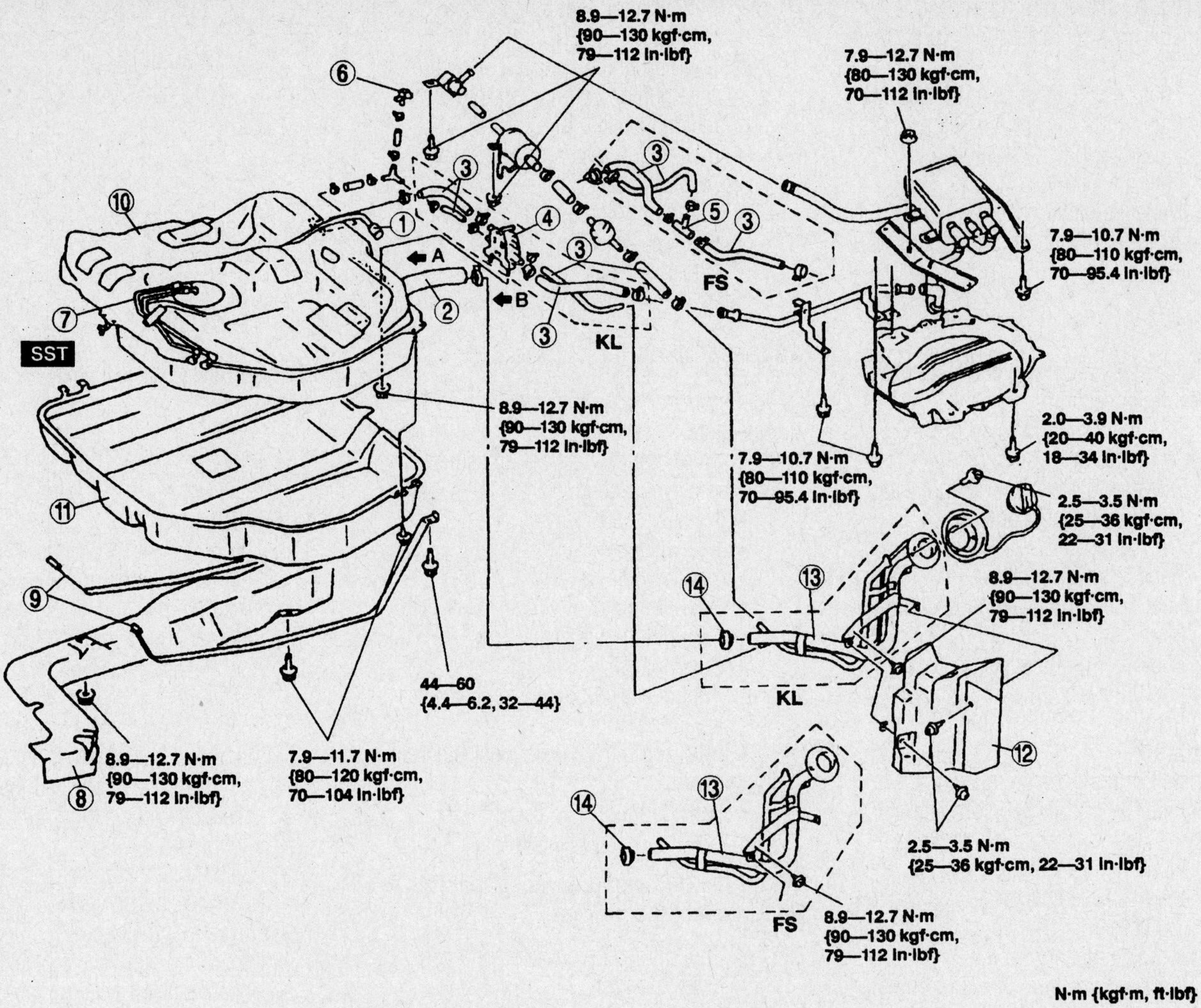

8.9—12.7 N·m
{90—130 kgf·cm,
79—112 in·lbf}
7.9—12.7 N·m
{80—130 kgf·cm,
70—112 in·lbf}
7.9—10.7 N·m
{80—110 kgf·cm,
70—95.4 in·lbf}
A
B
FS
KL
SST
8.9—12.7 N·m
{90—130 kgf·cm,
79—112 in·lbf}
7.9—10.7 N·m
{80—110 kgf·cm,
70—95.4 in·lbf}
2.0—3.9 N·m
{20—40 kgf·cm,
18—34 in·lbf}
2.5—3.5 N·m
{25—36 kgf·cm,
22—31 in·lbf}
8.9—12.7 N·m
{90—130 kgf·cm,
79—112 in·lbf}
44—60
{4.4—6.2, 32—44}
8.9—12.7 N·m
{90—130 kgf·cm,
79—112 in·lbf}
7.9—11.7 N·m
{80—120 kgf·cm,
70—104 in·lbf}
KL
2.5—3.5 N·m
{25—36 kgf·cm, 22—31 in·lbf}
FS
8.9—12.7 N·m
{90—130 kgf·cm,
79—112 in·lbf}
N·m {kgf·m, ft·lbf}

1	Fuel pump connector	8	Presilencer insulator
2	Joint hose	9	Fuel tank strap
3	Evaporative hose	10	Fuel tank
4	Pressure control valve	11	Fuel tank protector
5	Hose joint	12	Fuel-filler pipe insulator
6	Fuel tank pressure sensor	13	Fuel-filler pipe
7	Fuel pipe	14	Nonreturn valve

42356-MAZC-G54

Exploded view of the fuel tank assembly—626

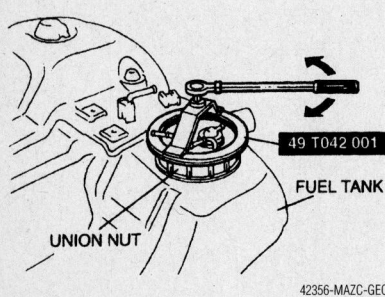

49 T042 001
FUEL TANK
UNION NUT
42356-MAZC-GEO

Remove the fuel pump ring–Millenia

- Fuel pump ring and tighten the fasteners to 75 ft. lbs. (102 Nm)
- Fuel hoses to the fuel pump
- Fuel tank
- Evaporative hose
- Fuel tank strap while supporting the fuel tank. Tighten the strap bolts to 44 ft. lbs. (60 Nm).
- Hoses to the fuel tank
- Fuel pump electrical connector
- Service hole cover
- Rear seat cushion
- Negative battery cable

6. Add a minimum of 10 gallons of fuel to the tank and check for leaks.

Fuel Injector

REMOVAL & INSTALLATION

✲✲ CAUTION

Fuel injection systems remain under pressure after the engine has been turned OFF. Properly relieve fuel pressure before disconnecting any fuel lines. Failure to do so may result in fire or personal injury. Do not allow fuel spray or fuel vapors to come in contact with a spark or open

flame. Keep a dry chemical fire extinguisher nearby. Never store fuel in an open container due to risk of fire or explosion.

626

2.0L (FS) ENGINE

1. Before servicing the vehicle, refer to the precautions section.
2. Refer to the illustration for component location and torque specifications.
3. Relieve the fuel system pressure.
4. Remove or disconnect the following:
   - Negative battery cable
   - Fuel injector wiring harness
   - Fuel lines at the fuel rail
   - Hose from the pressure regulator
   - Fuel rail with the injectors attached
   - Fuel injectors, grommets and O-rings from the fuel rail
   - O-rings from the fuel injectors

To install:

5. Install or connect the following:
   - New O-rings and grommets lubricated with engine oil on the fuel injectors.
   - Insulators and injectors on the intake manifold
   - Grommets and the fuel rail onto the injectors. Torque the bolts to 14–18 ft. lbs. (19–25 Nm).
   - Fuel lines to the fuel rail
   - Fuel injector wiring harness
   - Negative battery cable
6. Turn the ignition switch **ON** to pressurize the fuel system.
7. Check for leaks and correct as necessary, before starting the engine.

2.5L (KL) ENGINE

1. Before servicing the vehicle, refer to the precautions section.
2. Relieve the fuel system pressure.
3. Remove or disconnect the following:
   - Negative battery cable
   - Fuel injector wiring harness
   - Fuel lines at the fuel rail
   - Fuel rail
   - Accumulated connector
   - Spacer
   - Fuel injectors, grommets and O-rings from the fuel rail
   - O-rings from the fuel injectors

To install:

4. Install or connect the following:
   - New O-rings and grommets lubricated with engine oil on the fuel injectors. Fit the injector squarely into the rail while using a twisting motion. Fit the injector tab into the notch on the rail.

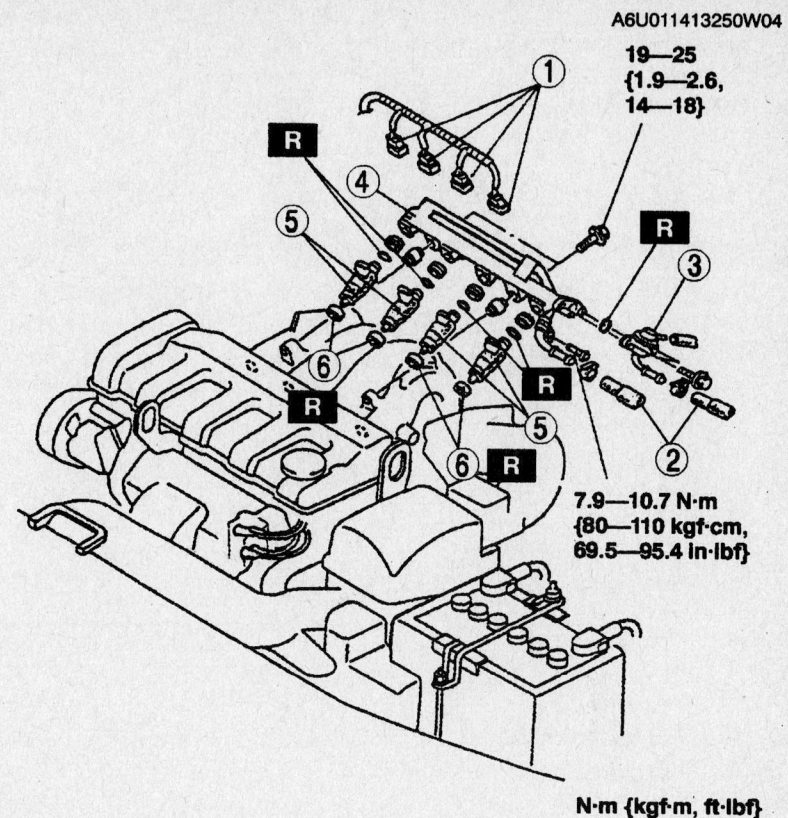

A6U011413250W04

19—25
{1.9—2.6,
14—18}

7.9—10.7 N·m
{80—110 kgf·cm,
69.5—95.4 in·lbf}

N·m {kgf·m, ft·lbf}

42356-MAZC-G56

Exploded view of the fuel rail and injector assembly—626 2.0L (FS)

   - Insulators and injectors on the intake manifold
   - Grommets and the fuel rail onto the injectors. Torque the bolts to 14–18 ft. lbs. (19–25 Nm).
   - Fuel lines to the fuel rail
   - Fuel injector wiring harness
   - Negative battery cable
5. Turn the ignition switch **ON** to pressurize the fuel system.
6. Check for leaks and correct as necessary, before starting the engine.

2002–04 Miata

1. Before servicing the vehicle, refer to the precautions section.
2. Relieve the fuel system pressure.
3. Remove or disconnect the following:
   - Negative battery cable
   - Upper intake manifold
   - Fuel lines at the fuel rail
   - Vacuum hose from the fuel pressure regulator
   - Fuel rail mounting bracket
   - Fuel rail mounting bolts, spacers and insulators
   - Fuel rail, with the injectors attached
   - Fuel injectors, grommets and O-rings from the fuel rail
   - O-rings from the fuel injectors

To install:

4. Install or connect the following:
   - New O-rings and grommets lubricated with engine oil on the fuel injectors
   - Insulators and injectors on the intake manifold
   - Grommets and the fuel rail onto the injectors. Torque the bolts to 14–18 ft. lbs. (19–25 Nm).
   - Fuel line bracket. Torque the bolts to 70–95 inch lbs. (8–11 Nm).
   - Fuel lines to the fuel rail
   - Fuel injector wiring harness
   - Upper intake manifold
   - Negative battery cable
5. Turn the ignition switch **ON** to pressurize the fuel system.
6. Check for leaks and correct as necessary, before starting the engine.

Protégé

1. Before servicing the vehicle, refer to the precautions section.
2. Relieve the fuel system pressure.
3. Remove or disconnect the following:
   - Negative battery cable
   - Accelerator cables and the cable bracket
   - Fuel injector connectors and wiring harness

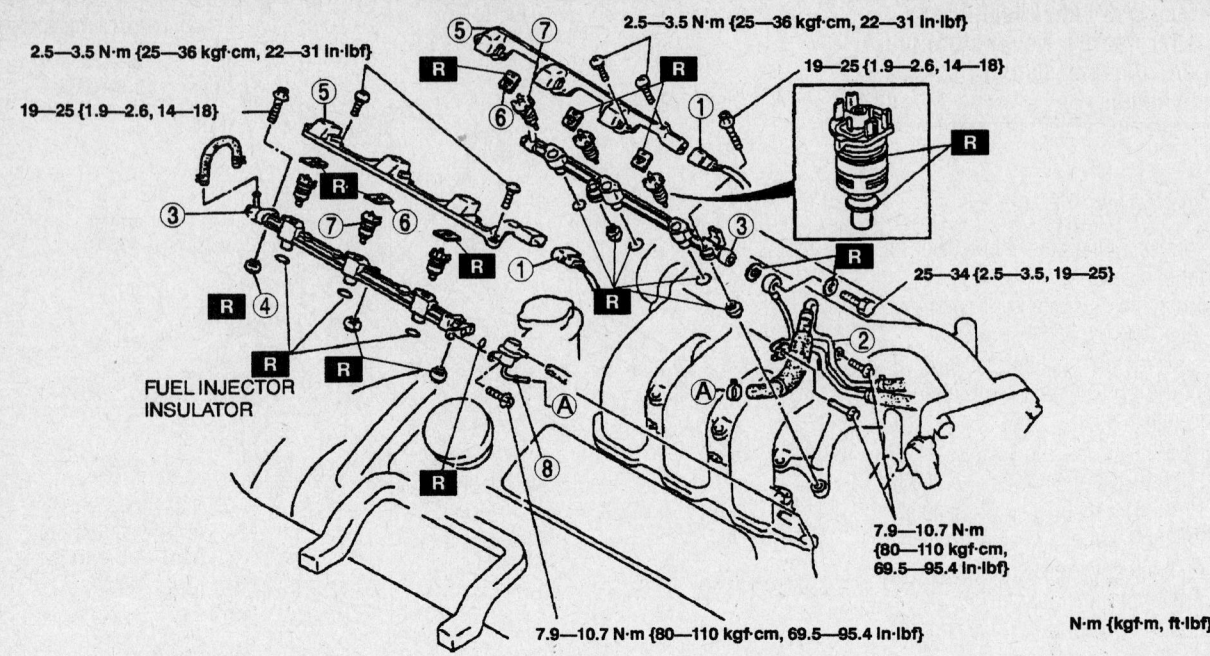

2.5—3.5 N·m {25—36 kgf·cm, 22—31 In·lbf}

19—25 {1.9—2.6, 14—18}

2.5—3.5 N·m {25—36 kgf·cm, 22—31 In·lbf}

19—25 {1.9—2.6, 14—18}

25—34 {2.5—3.5, 19—25}

FUEL INJECTOR INSULATOR

7.9—10.7 N·m {80—110 kgf·cm, 69.5—95.4 In·lbf}

7.9—10.7 N·m {80—110 kgf·cm, 69.5—95.4 In·lbf}

N·m {kgf·m, ft·lbf}

1	Connector	5	Accumulated connector
2	Fuel hose and fuel pipe	6	Spacer
3	Fuel distributor	7	Fuel injector
4	Insulator	8	Pressure regulator

42356-MAZC-G55

Exploded view of the fuel rail and injector assembly—626 2.5L (KL) engine

- Fuel lines at the fuel rail
- Vacuum hose from the fuel pressure regulator
- Fuel rail mounting bolts, spacers and insulators
- Fuel rail, with the injectors attached
- Fuel injectors, grommets and O-rings from the fuel rail
- O-rings from the fuel injectors

To install:
4. Install or connect the following:
- New O-rings and grommets lubricated with engine oil on the fuel injectors
- Insulators and injectors on the intake manifold
- Grommets and the fuel rail onto the injectors. Torque the bolts to 14–18 ft. lbs. (19–25 Nm).
- Vacuum hose to the fuel pressure regulator
- Fuel lines to the fuel rail
- Fuel injector wiring harness
- Cable bracket. Torque the bolt to 70–95 inch lbs. (8–11 Nm).
- Accelerator cables, if removed, adjust as necessary
- Negative battery cable
5. Turn the ignition switch **ON** to pressurize the fuel system.

6. Check for leaks and correct as necessary, before starting the engine.

Millenia

2.5L (KL) ENGINE
1. Before servicing the vehicle, refer to the precautions section.
2. Relieve the fuel system pressure.
3. Remove or disconnect the following:
- Negative battery cable
- Air intake hose
- Fuel injector wiring harness
- Fuel lines at the fuel rail
- Fuel rail
- Insulator
- Accumulated connector
- Spacer
- Fuel injectors, grommets and O-rings from the fuel rail
- O-rings from the fuel injectors

To install:
4. Install or connect the following:
- New O-rings and grommets lubricated with engine oil on the fuel injectors. Fit the injector squarely into the rail while using a twisting motion. Fit the injector tab into the notch on the rail.

- Insulators and injectors on the intake manifold
- Grommets and the fuel rail onto the injectors. Torque the bolts to 14–18 ft. lbs. (19–25 Nm).
- Fuel lines to the fuel rail
- Fuel injector wiring harness
- Negative battery cable
5. Turn the ignition switch **ON** to pressurize the fuel system.
6. Check for leaks and correct as necessary, before starting the engine.

2.3L (KJ) ENGINE
1. Before servicing the vehicle, refer to the precautions section.
2. Relieve the fuel system pressure.
3. Remove or disconnect the following:
- Negative battery cable
- Intake manifold cover
- Charge air cooler air duct
- Air cleaner assembly, as necessary, for clearance
- Resonator
- Left and right-hand charge air coolers
- Accelerator cable
- Air intake pipe assembly
- Vacuum hose assembly
- Fuel injector electrical connectors

- Fuel supply/return lines and discard the copper washers
- Fuel rail
- Insulators
- Distribution harness (accumulated connector) from the fuel rails
- Spacer from the top of each fuel injector and discard it
- Fuel injectors from the fuel rails by rotating back and forth
- Fuel pressure regulator, if necessary

To install:

4. Install or connect the following:
- Fuel pressure regulator, if removed. Torque the bolts to 61–86 inch lbs. (7–10 Nm).
- New O-rings lubricated with engine oil on the injectors

- Fuel injectors into the fuel rails
- New spacers on the injectors
- Distribution harness. Torque the screws to 22–31 inch lbs. (2.5–3.5 Nm).
- 6 insulators and the fuel rails. Torque the bolts to 14–18 ft. lbs. (19–25 Nm).
- Fuel supply and return lines using new copper washers
- Fuel injector electrical connectors
- Vacuum hose assembly. Torque the nuts to 70–95 inch lbs. (8–11 Nm).
- Air intake pipe assembly. Torque the nuts to 70–95 inch lbs. (8–11 Nm) and the bolts to 44–78 inch lbs. (5–9 Nm).

- Accelerator cable and adjust it. Torque the bolt to 70–95 inch lbs. (8–11 Nm).
- Both charge air coolers with new O-rings. Torque the bolts to 14–18 ft. lbs. (19–25 Nm).
- Resonator. Torque the nuts to 70–95 inch lbs. (8–11 Nm).
- Air cleaner assembly. Torque the nuts to 70–95 inch lbs. (8–11 Nm).
- Charge air cooler air duct. Torque the nuts to 70–95 inch lbs. (8–11 Nm).
- Negative battery cable

5. Turn the ignition switch **ON** to pressurize the fuel system.

6. Check for fuel leaks and correct as necessary before starting the engine.

DRIVE TRAIN

Manual Transaxle/Transmission Assembly

REMOVAL & INSTALLATION

2002–04 Miata

1. Before servicing the vehicle, refer to the precautions section.

2. Refer to the illustration for component location and torque specifications.

3. Drain the transaxle oil.

4. Remove or disconnect the following:

- Crossmember and bracket, on models equipped with 16 inch wheels
- Rear crossbar, on models equipped with 16 inch wheels
- Under cover
- Starter
- Front and middle exhaust pipes
- Shifter knob
- Rear console and insulator
- Shift lever assembly
- Dust boot
- Front crossbar
- Drive shaft
- Clutch release cylinder
- Back up light switch connector
- Neutral safety switch connector
- Speedometer sensor connector
- Wiring harness from the Power Plant Frame (PPF)

5. Support the transmission with a jack.

- PPF bracket
- Differential side bolts and pry out the spacer

✳✳ CAUTION

Removing the PPF spacers will reduce the performance of the PPF. If the spacers are removed, replace the PPF as an assembly.

- Differential mounting spacer
- Transmission side bolts and the PPF
- Transmission bolts and the transmission

To install:

6. Tilt the engine by pushing up on the front oil pan with a wooden block placed on a transmission jack.

7. Support the transmission with a jack.

8. Install or connect the following:
- Transmission into position and tighten the bolts to 48–65 ft. lbs. (64–89 Nm) on models equipped with the M15M–D transmission or 76–91 ft. lbs. (104–123 Nm) on models equipped with the Y16M–D transmission.
- Differential mounting spacer and tighten the bolts to 28–38 ft. lbs. (38–51 Nm).

9. Support the transmission with a jack until it is level.

- PPF in position and if removed install the sleeve
- Spacer and bolts and tighten the reamer bolt making sure the threading is aligned correctly. The reamer bolt should be installed in the forward hole. Tighten the outer bolts making sure the threads are properly aligned. Tighten all bolts

in the sequence illustrated to 91 ft. lbs. (123 (Nm).

- PPF bracket and tighten bolts **A** to 91 ft. lbs. (123 Nm) and bolts **B** to 39 ft. lbs. (53 Nm).

10. Remove the jack and connect the wiring harness.

11. Using a straightedge and Vernier caliper measure the distance **A** which should be 2.37–2.83 inch (60–72mm). If the distance is not as specified, reposition the PPF to the transmission.

12. Install or connect the following:
- Speedometer sensor connector
- Neutral safety switch connector
- Back up light switch connector
- Clutch release cylinder
- Drive shaft
- Front crossbar
- Dust boot

➡**The change control assembly must be filled with 4.9–5.8 cubic inch (80–95cc) of GL–4 or 5 oil if the extension housing was removed or the transmission overhauled.**

13. Fill the change control case with the specified amount of the proper oil.

14. Apply grease to the shift lever components as illustrated.

15. Apply sealant to the contact surfaces of the shift lever and change control case.

16. Install the shift lever.

17. Install or connect the following:
- Rear console and insulator
- Shifter knob
- Starter
- Under cover
- Rear crossbar, on models equipped with 16 inch wheels

- Crossmember and bracket, on models equipped with 16 inch wheels
- Front and middle exhaust pipes

18. Fill the transaxle with fluid. Road test the vehicle and check for leaks. Top off all fluids as needed.

626

2.0L (FS) ENGINE

1. Before servicing the vehicle, refer to the precautions section.
2. Refer to the illustration for component location and torque specifications.

3. Drain the transaxle oil.
4. Remove or disconnect the following:
   - Battery and battery box
   - Air cleaner assembly and fresh air duct
   - Wheels
   - Splash shields
   - Starter
   - Neutral safety switch connector
   - Back up light switch connector
   - Vehicle Speed Sensor (VSS) connector
   - Clutch release cylinder
   - Transverse member
   - Extension bar

   - Change control rod
   - Tie rod ends from the knuckle
   - Stabilizer bar link
   - Lower control arm ball joint
   - Halfshafts
   - Joint shaft, install tool 49 G030 455 to hold the side gears after removal

5. Support the engine assembly with a engine support assembly such as 49 E017 5A0.
   - Number 5 engine mount bolt
   - Engine mount member
   - Number 2 engine mount
   - Number 4 engine mount rubber

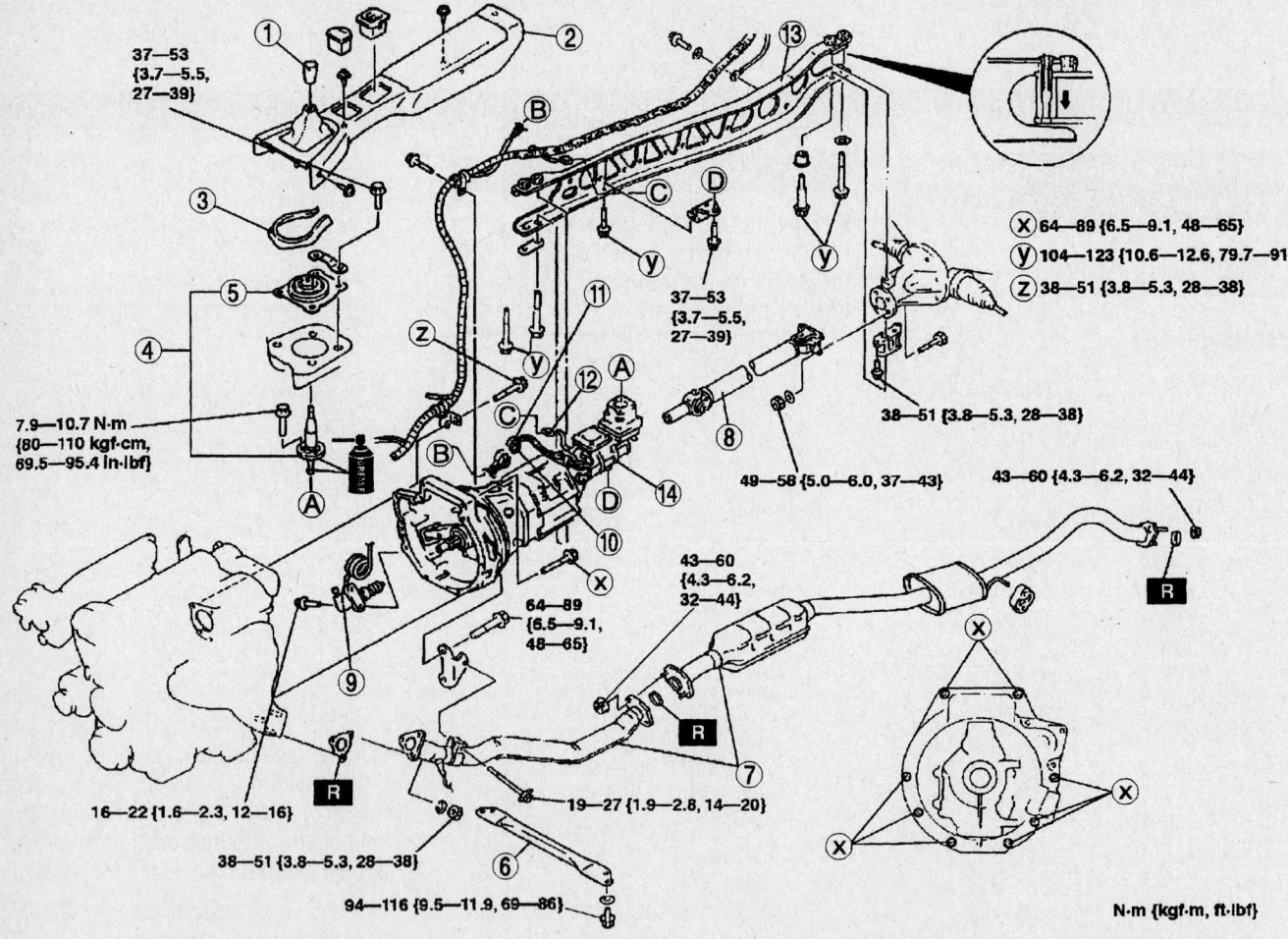

1	Shift lever knob	8	Propeller shaft
2	Rear console	9	Clutch release cylinder
3	Insulation	10	Back-up light switch connector
4	Shift lever component	11	Neutral switch connector
5	Dust boot	12	Speedometer sensor connector
6	Front crossbar	13	Power plant frame (PPF)
7	Front pipe and middle pipe	14	Transmission

Exploded view of the M15M–D manual transmission assembly—2002–04 Miata

42356-MAZC-GBS

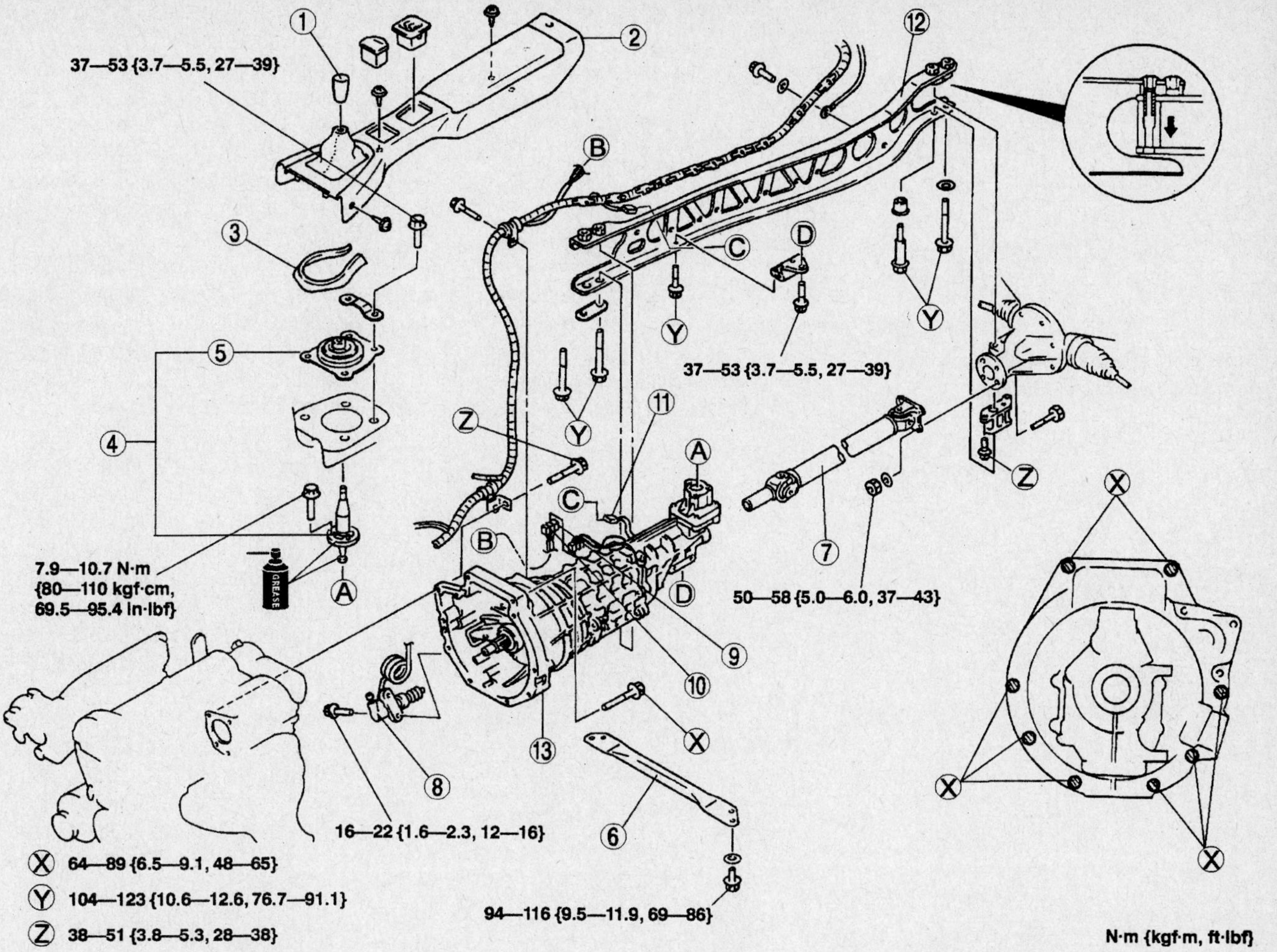

37—53 {3.7—5.5, 27—39}

37—53 {3.7—5.5, 27—39}

50—58 {5.0—6.0, 37—43}

7.9—10.7 N·m
{80—110 kgf·cm,
69.5—95.4 In·lbf}

16—22 {1.6—2.3, 12—16}

Ⓧ 64—89 {6.5—9.1, 48—65}

Ⓨ 104—123 {10.6—12.6, 76.7—91.1}

Ⓩ 38—51 {3.8—5.3, 28—38}

94—116 {9.5—11.9, 69—86}

N·m {kgf·m, ft·lbf}

1	Shift lever knob	8	Clutch release cylinder
2	Rear console	9	Back-up light switch connector
3	Insulator	10	Neutral switch connector
4	Shift lever component	11	Speedometer sensor connector
5	Dust boot	12	Power plant frame (PPF)
6	Front crossbar	13	Transmission
7	Propeller shaft		

42356-MAZC-GCD

Exploded view of the Y16M–D manual transmission assembly—2002–04 Miata

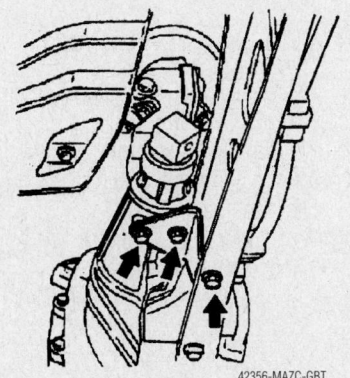

42356-MAZC-GBT

Remove the PPF bracket—2002–04 Miata

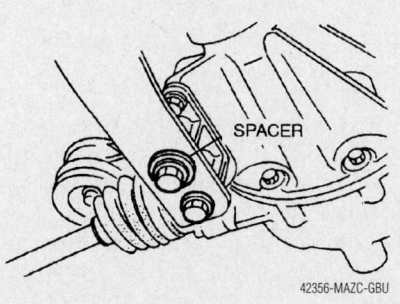

SPACER

42356-MAZC-GBU

Remove the differential side bolts and pry out the spacer—2002–04 Miata

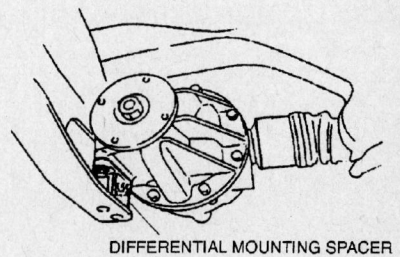

DIFFERENTIAL MOUNTING SPACER

42356-MAZC-GBV

Remove the differential mounting spacer—2002–04 Miata

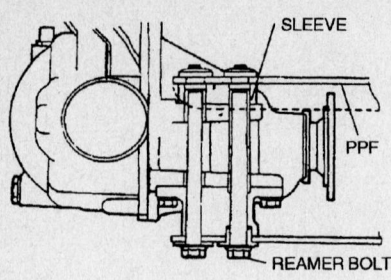

The reamer bolt should be installed in the forward hole—2002–04 Miata

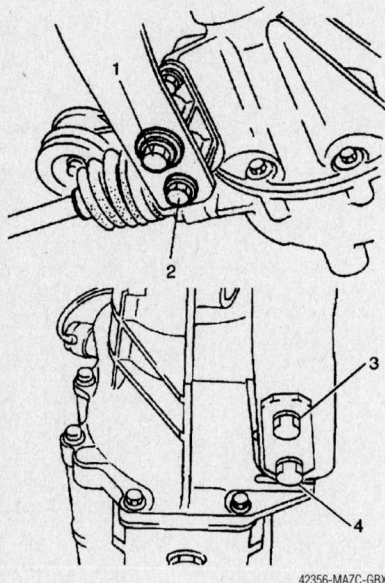

Tighten the PPF bolts using this sequence—2002–04 Miata

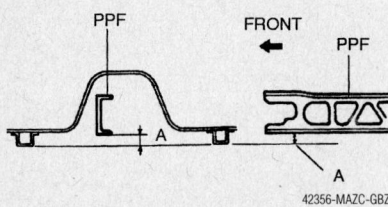

Measure the distance A which should be 2.37–2.83 inch (60–72mm)—2002–04 Miata

- Number 1 engine mount bracket

6. Loosen the engine support assembly and lean the engine towards the transaxle.

7. Support the transaxle with a jack, remove the transaxle bolts and the transaxle.

To install:

8. Place the transaxle onto a jack and raise into position.

9. Install the transaxle bolts and tighten

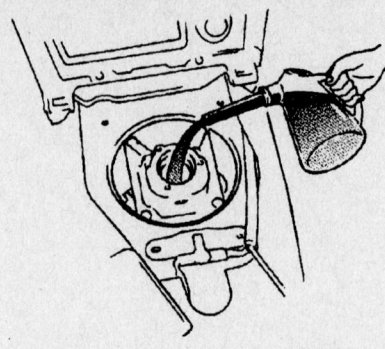

Fill the change control case with the specified amount of the proper oil—2002–04 Miata

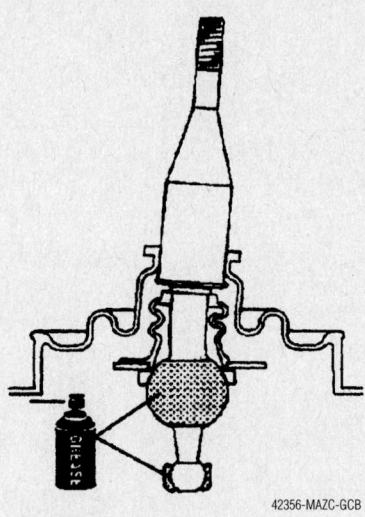

Apply grease to the shift lever components—2002–04 Miata

the upper bolts to 66–86 ft. lbs. (90–116 Nm), the lower bolts (except very bottom bolt) to 28–38 ft. lbs. (38–51 Nm) and the very bottom bolt to 14–18 ft. lbs. (19–25 Nm). Refer to the exploded view illustration for bolt locations.

10. Lean the engine towards its normal position and tighten the support assembly.

11. Install or connect the following:
- Number 1 engine mount bracket. Refer to the illustration for component location and torque specifications.
- Number 4 engine mount rubber. Refer to the illustration for component location and torque specifications.
- Number 2 engine mount. Refer to the illustration for component location and torque specifications.

12. Install the engine mount member as follows while referring to the illustration for bolt locations:

a. Position the direction indicator on the mount member bushings facing towards the front side and install the bushings onto the mount.

b. Put the number 2 engine mount stud bolts through the mount member installation holes. Install mount member bolts **A** and nuts **A** and tighten to 50–68 ft. lbs. (67–93 Nm).

c. Loosely tighten the number 2 engine mount nutsand remove the engine support assembly.

d. Tighten the number 2 engine mount nuts **B** to 55–77 ft. lbs. (75–104 Nm).

13. Install or connect the following:
- Number 5 engine mount bolt. Refer to the illustration for component location and torque specifications.
- Joint shaft. Install a new circlip with the opening facing up. Hand tighten the bolts **A**, then tighten the bolts to 32–45 ft. lbs. (43–61 Nm). Refer to the illustration for bolt location.
- Halfshafts
- Lower control arm ball joint
- Stabilizer bar link
- Tie rod ends to the knuckle and tighten the nut to 24–32 ft. lbs. (32–44 Nm) and install a new cotter pinch
- Change control rod
- Extension bar
- Transverse member
- Clutch release cylinder
- VSS connector
- Back up light switch connector
- Neutral safety switch connector
- Starter
- Splash shields
- Wheels
- Air cleaner assembly and fresh air duct
- Battery and battery box

14. Fill the transaxle with fluid. Road test the vehicle and check for leaks. Top off all fluids as needed.

2.5L (KL) ENGINE

1. Before servicing the vehicle, refer to the precautions section.

2. Refer to the illustration for component location and torque specifications.

3. Drain the transaxle oil.

4. Remove or disconnect the following:
- Battery and battery box
- Air cleaner assembly and fresh air duct
- Wheels
- Splash shields
- Neutral safety switch connector

FS ENGINE

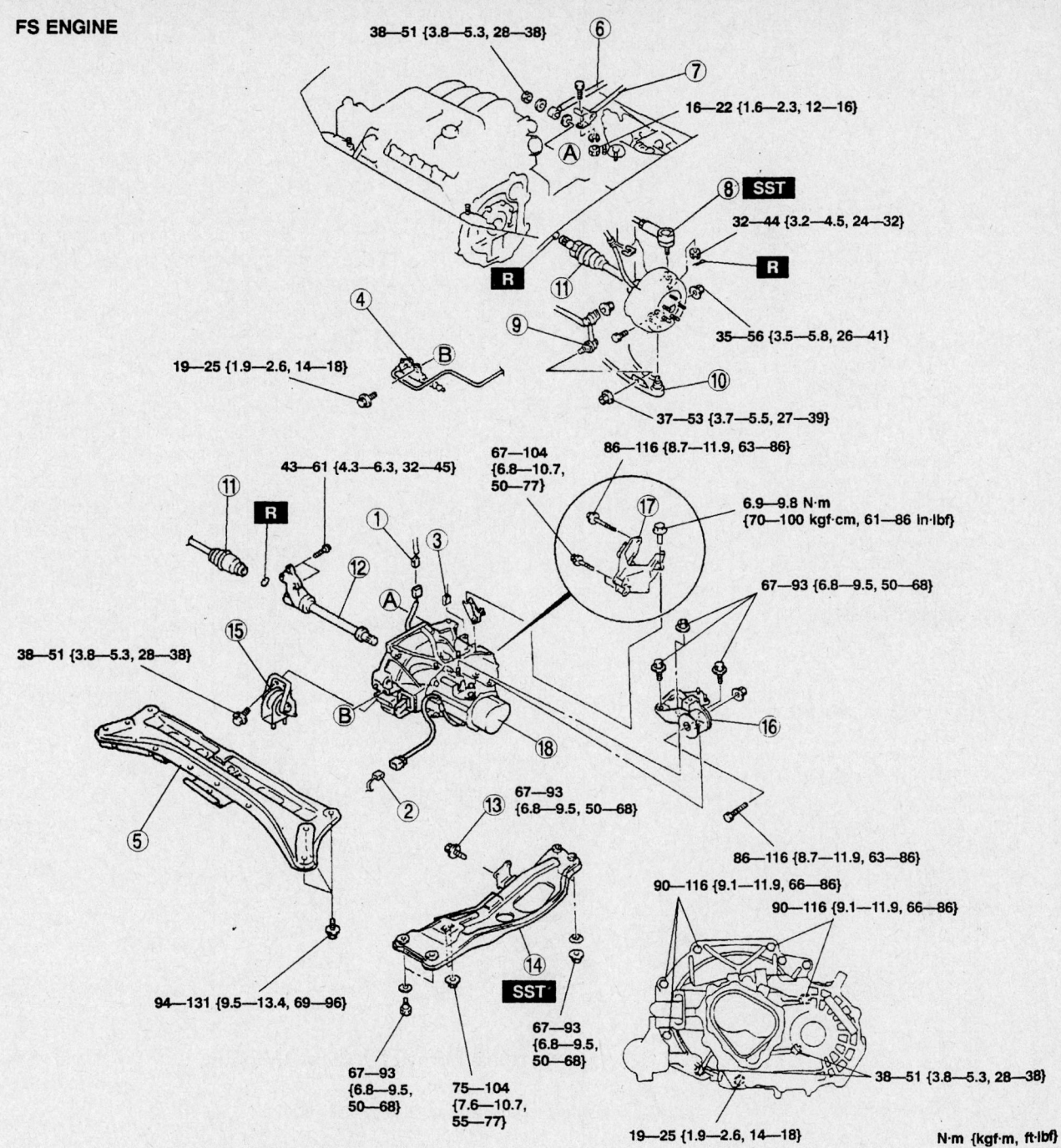

38—51 {3.8—5.3, 28—38}

16—22 {1.6—2.3, 12—16}

32—44 {3.2—4.5, 24—32}

35—56 {3.5—5.8, 26—41}

37—53 {3.7—5.5, 27—39}

19—25 {1.9—2.6, 14—18}

43—61 {4.3—6.3, 32—45}

67—104 {6.8—10.7, 50—77}

86—116 {8.7—11.9, 63—86}

6.9—9.8 N·m {70—100 kgf·cm, 61—86 in·lbf}

67—93 {6.8—9.5, 50—68}

38—51 {3.8—5.3, 28—38}

67—93 {6.8—9.5, 50—68}

86—116 {8.7—11.9, 63—86}

90—116 {9.1—11.9, 66—86}

90—116 {9.1—11.9, 66—86}

94—131 {9.5—13.4, 69—96}

67—93 {6.8—9.5, 50—68}

75—104 {7.6—10.7, 55—77}

67—93 {6.8—9.5, 50—68}

38—51 {3.8—5.3, 28—38}

19—25 {1.9—2.6, 14—18}

N·m {kgf·m, ft·lbf}

1	Neutral switch connector	10	Lower arm ball joint
2	Back-up light switch connector	11	Drive shaft
3	Vehicle speedometer sensor connector	12	Joint shaft
4	Clutch release cylinder	13	No.5 engine mount bolt
5	Transverse member	14	Engine mount member
6	Extension bar	15	No.2 engine mount
7	Change control rod	16	No.4 engine mount rubber
8	Tie-rod end ball joint	17	No.1 engine mount bracket
9	Stabilizer control link	18	Transaxle

42356-MAZC-G57

Exploded view of the manual transaxle assembly mounting—626 with 2.0L (FS) engine

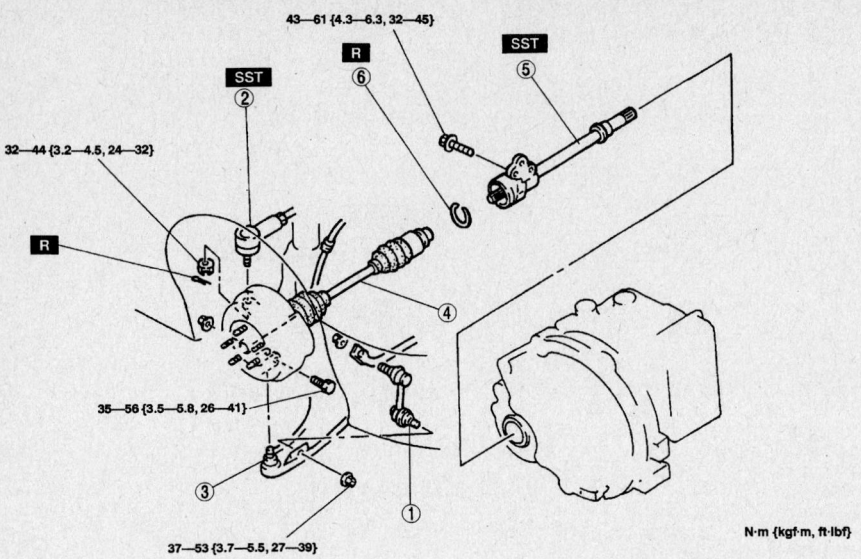

43—61 {4.3—6.3, 32—45}

32—44 {3.2—4.5, 24—32}

35—56 {3.5—5.8, 26—41}

37—53 {3.7—5.5, 27—39}

N·m {kgf·m, ft·lbf}

1	Stabilizer control link	4	Right drive shaft and axle
2	Tie-rod end ball joint	5	Joint shaft
3	Lower arm ball joint	6	Clip

42356-MAZC-G58

Exploded view joint shaft assembly—626

- Extension bar
- Change control rod
- Tie rod ends from the knuckle
- Stabilizer bar link
- Lower control arm ball joint
- Halfshafts
- Joint shaft, install tool 49 G030 455 to hold the side gears after removal

5. Support the engine assembly with a engine support assembly such as 49 E017 5A0.

- Engine mount member
- Number 2 engine mount
- Number 4 engine mount rubber
- Under cover
- Number 1 engine mount bracket

6. Loosen the engine support assembly and lean the engine towards the transaxle.

7. Support the transaxle with a jack, remove the transaxle bolts and the transaxle.

8. Remove the number 5 mount

To install:

9. Install the number 5 mount. Refer to the illustration for component location and torque specifications.

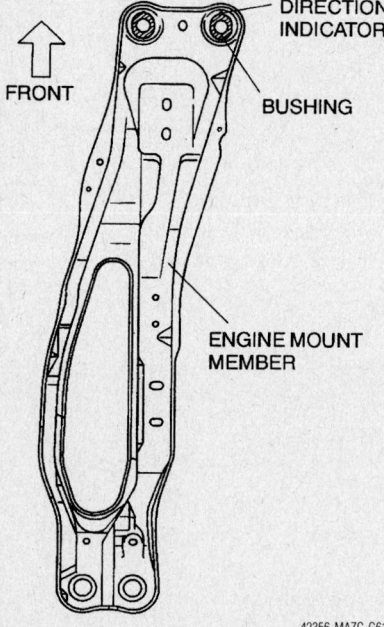

42356-MAZC-G61

Position the direction indicator on the mount member bushings facing towards the front side and install the bushings onto the mount—626

- Back up light switch connector
- Vehicle Speed Sensor (VSS) connector
- Starter
- Clutch release cylinder
- Transverse member

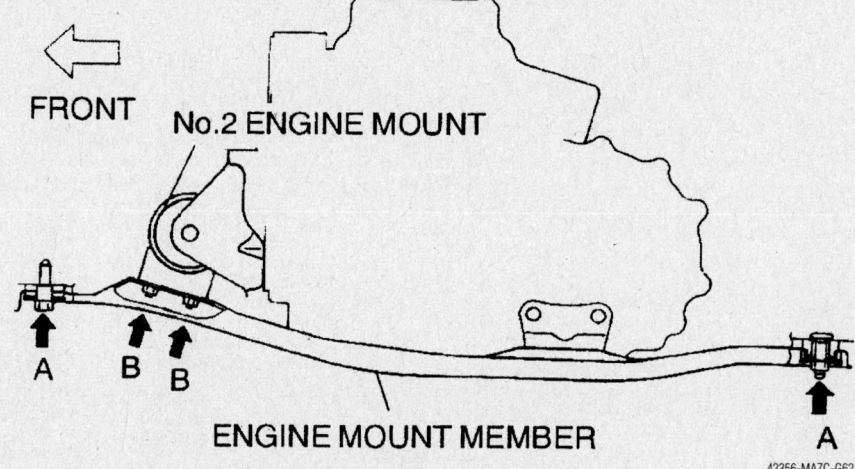

42356-MAZC-G62

Location of the front mount member bolt locations—626

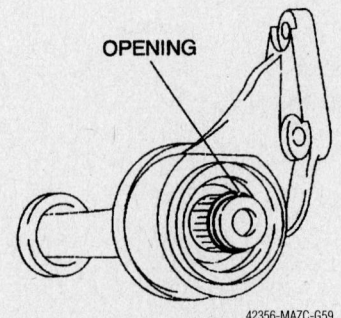

42356-MAZC-G59

When installing the joint shaft make sure to install a new clip with the opening facing up—626

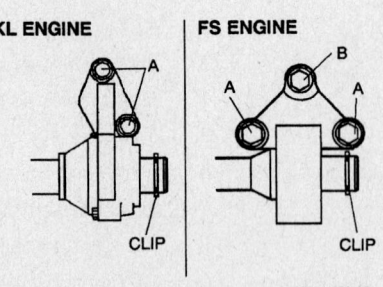

42356-MAZC-G60

Tighten the joint shaft bolts as shown—626

KL ENGINE

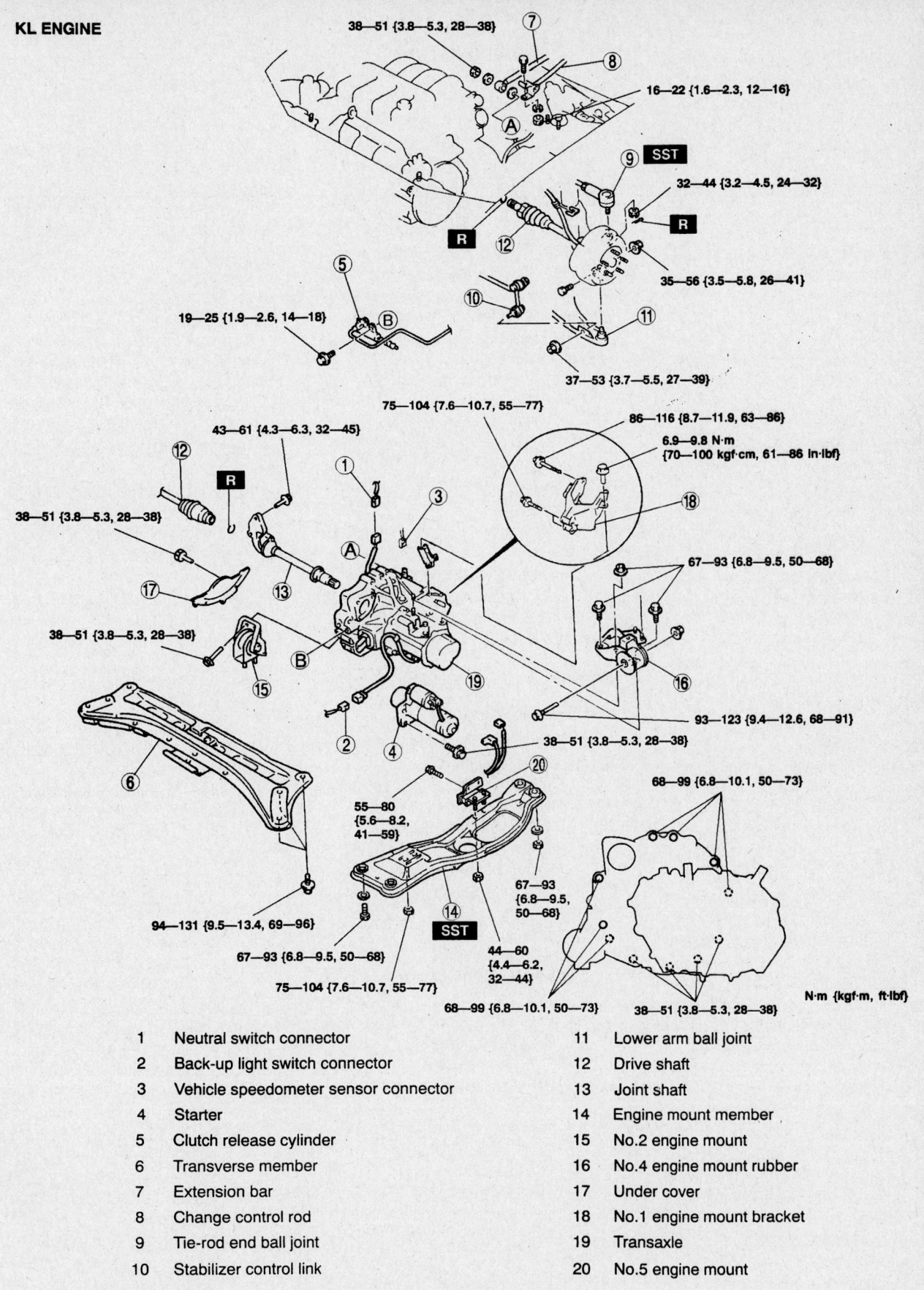

1	Neutral switch connector
2	Back-up light switch connector
3	Vehicle speedometer sensor connector
4	Starter
5	Clutch release cylinder
6	Transverse member
7	Extension bar
8	Change control rod
9	Tie-rod end ball joint
10	Stabilizer control link
11	Lower arm ball joint
12	Drive shaft
13	Joint shaft
14	Engine mount member
15	No.2 engine mount
16	No.4 engine mount rubber
17	Under cover
18	No.1 engine mount bracket
19	Transaxle
20	No.5 engine mount

Exploded view of the manual transaxle assembly mounting—626 with 2.5L (KL) engine

42356-MAZC-G63

10. Place the transaxle onto a jack and raise into position.

11. Install the transaxle bolts and tighten the upper 4 bolts and the 3 side bolts to 50–73 ft. lbs. (68–99 Nm) and the lower 4 bolts to 28–38 ft. lbs. (38–51 Nm). Refer to the exploded view illustration for bolt locations.

12. Lean the engine towards its normal position and tighten the support assembly.

13. Install or connect the following:
- Number 1 engine mount bracket. Refer to the illustration for component location and torque specifications.
- Under cover
- Number 4 engine mount rubber. Refer to the illustration for component location and torque specifications.
- Number 2 engine mount. Refer to the illustration for component location and torque specifications.

14. Install the engine mount member as follows while referring to the illustration for bolt locations:

a. Position the direction indicator on the mount member bushings facing towards the front side and install the bushings onto the mount.

b. Put the number 2 engine mount stud bolts through the mount member installation holes. Install mount member bolts **A** and nuts **A** and tighten to 50–68 ft. lbs. (67–93 Nm).

c. Loosely tighten the number 2 engine mount nutsand remove the engine support assembly.

d. Tighten the number 2 engine mount nuts **B** to 55–77 ft. lbs. (75–104 Nm).

15. Install or connect the following:
- Number 5 engine mount bolt. Refer to the illustration for component location and torque specifications.
- Joint shaft. Install a new circlip with the opening facing up. Hand tighten the bolts **A**, then tighten the bolts to 32–45 ft. lbs. (43–61 Nm). Refer to the illustration for bolt location.
- Halfshafts
- Lower control arm ball joint
- Stabilizer bar link
- Tie rod ends to the knuckle and tighten the nut to 24–32 ft. lbs. (32–44 Nm) and install a new cotter pinch
- Change control rod
- Extension bar
- Transverse member
- Clutch release cylinder
- Starter

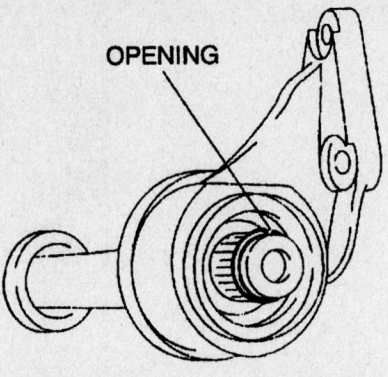

OPENING

42356-MAZC-G59

When installing the joint shaft make sure to install a new clip with the opening facing up—626

- VSS connector
- Back up light switch connector
- Neutral safety switch connector
- Starter
- Splash shields
- Wheels
- Air cleaner assembly and fresh air duct
- Battery and battery box

16. Fill the transaxle with fluid. Road test the vehicle and check for leaks. Top off all fluids as needed.

Protégé

G15M–R TRANSAXLE

1. Before servicing the vehicle, refer to the precautions section.

2. Refer to the illustration for component location and torque specifications.

3. Drain the transaxle oil.

4. Remove or disconnect the following:
- Battery and battery box
- Air cleaner assembly and fresh air duct
- Wheels
- Splash shields
- Exhaust Gas Recirculation (EGR) pipe
- Front exhaust pipe and three way catalytic converter
- Starter
- Speedometer sensor connector
- Neutral safety switch connector
- Back up light switch connector
- Clutch release cylinder
- Extension bar
- Change control rod
- Tie rod ends from the knuckle
- Lower control arm ball joint
- Halfshafts
- Joint shaft, install tool 49 B027 006 to hold the side gears after removal

5. Support the engine assembly with a engine support assembly such as 49 E017 5A0.
- Number 4 engine mount
- Engine mount member
- Number 2 engine mount
- Number 1 engine mount bracket

6. Loosen the engine support assembly and lean the engine towards the transaxle.

7. Support the transaxle with a jack, remove the transaxle bolts and the transaxle.

To install:

8. Place the transaxle onto a jack and raise into position.

9. Install the transaxle bolts and tighten the upper bolts to 65–86 ft. lbs. (89–116 Nm), the lower bolts except the lowest bolt to 28–38 ft. lbs. (38–51 Nm). Tighten the lowest bolt to 13–18 ft. lbs. (18–25 Nm). Refer to the exploded view illustration for bolt locations.

10. Install or connect the following:
- Number 1 engine mount bracket. Refer to the illustration for component location and torque specifications.
- Number 2 engine mount. Refer to the illustration for component location and torque specifications.

11. Install the engine mount member as follows while referring to the illustration for component locations:

a. Position the direction indicator on the mount member bushings as illustrated.

b. Put the number 2 engine mount stud bolts through the mount member installation holes. Install mount member bolts **A** and nuts **B** and tighten to 50–68 ft. lbs. (67–93 Nm).

c. Tighten bolts **C** to 50–68 ft. lbs. (67–93 Nm).

d. By aligning the holes on the stud bolts, install the number 4 mount bracket. Align hole of the bracket with the rubber and hand tighten bolt **D**, then nuts **E** and bolt **D** to 50–68 ft. lbs. (67–93 Nm).

12. Install or connect the following:
- Joint shaft. Install a new circlip with the opening facing up. Hand tighten the bolts **A**, then tighten the bolts to 32–45 ft. lbs. (43–61 Nm). Refer to the illustration for bolt location.
- Halfshafts
- Lower control arm ball joint
- Tie rod ends to the knuckle
- Change control rod
- Extension bar
- Clutch release cylinder

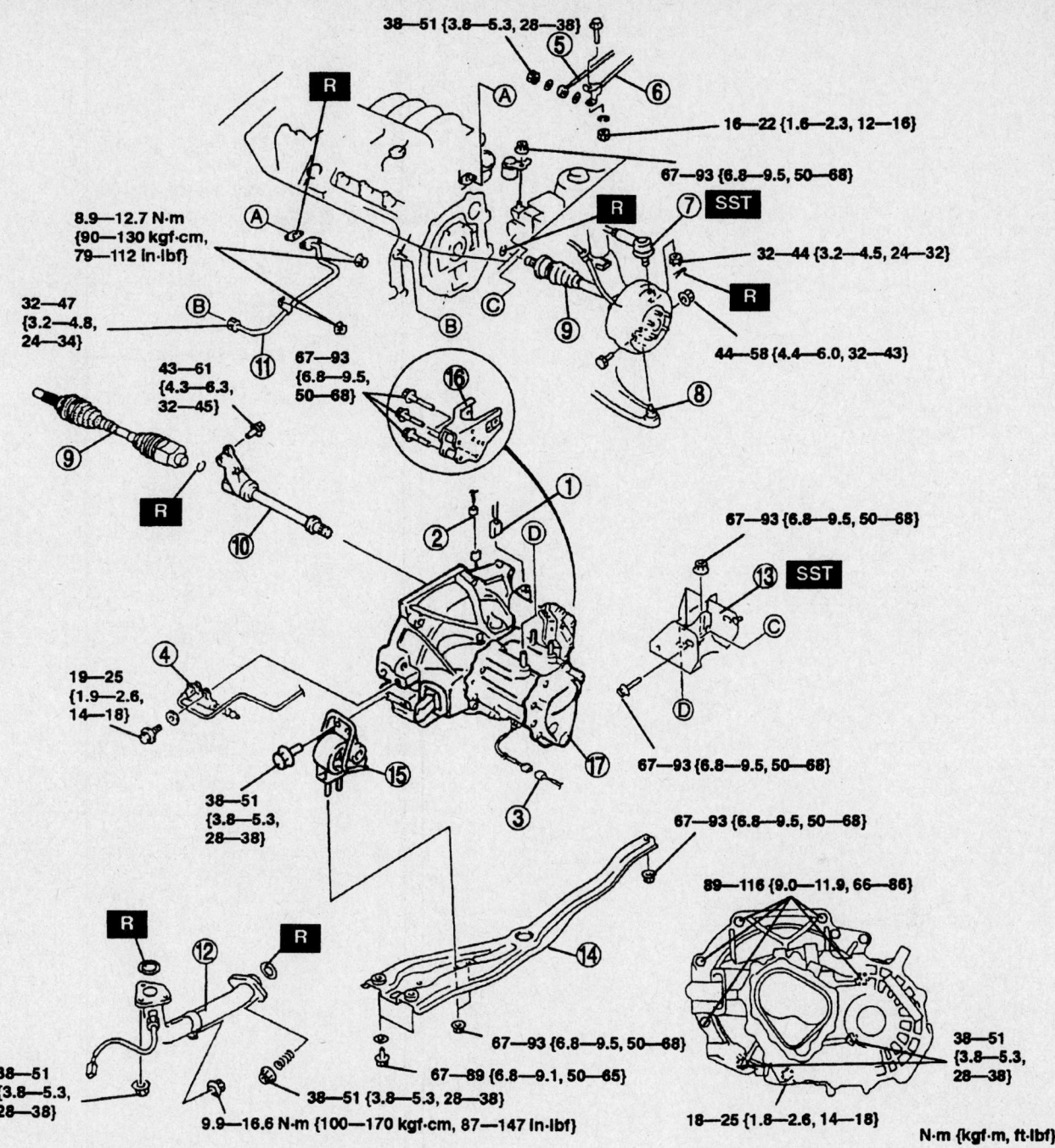

38—51 {3.8—5.3, 28—38}

16—22 {1.6—2.3, 12—16}

67—93 {6.8—9.5, 50—68}

8.9—12.7 N·m {90—130 kgf·cm, 79—112 in·lbf}

32—44 {3.2—4.5, 24—32}

32—47 {3.2—4.8, 24—34}

44—58 {4.4—6.0, 32—43}

43—61 {4.3—6.3, 32—45}

67—93 {6.8—9.5, 50—68}

67—93 {6.8—9.5, 50—68}

19—25 {1.9—2.6, 14—18}

67—93 {6.8—9.5, 50—68}

67—93 {6.8—9.5, 50—68}

89—116 {9.0—11.9, 66—86}

38—51 {3.8—5.3, 28—38}

38—51 {3.8—5.3, 28—38}

67—93 {6.8—9.5, 50—68}

67—89 {6.8—9.1, 50—65}

38—51 {3.8—5.3, 28—38}

18—25 {1.8—2.6, 14—18}

38—51 {3.8—5.3, 28—38}

9.9—16.6 N·m {100—170 kgf·cm, 87—147 in·lbf}

N·m {kgf·m, ft·lbf}

1	Speedometer sensor connector	10	Joint shaft
2	Neutral switch connector	11	EGR pipe
3	Back–up light switch connector	12	Front pipe
4	Clutch release cylinder	13	No.4 engine mount bracket
5	Extension bar	14	Engine mount member
6	Change control rod	15	No.2 engine mount
7	Tie-rod end ball joint	16	No.1
8	Lower arm ball joint	17	MTX
9	Drive shaft		

42356-MAZC-GQQ

Exploded view of the manual transaxle assembly mounting—2000–01 Protégé with G15M–R transaxle

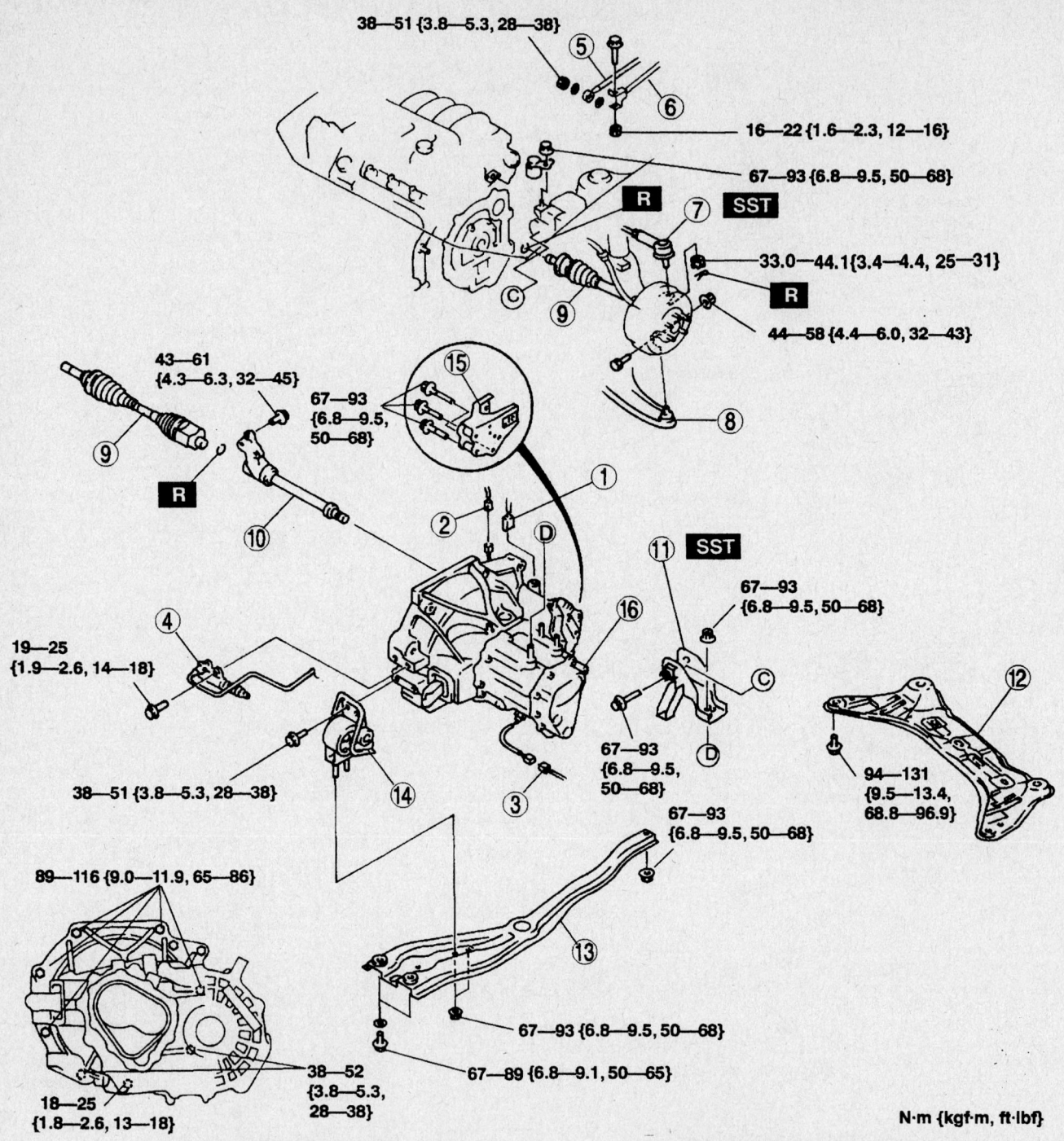

N·m {kgf·m, ft·lbf}

1	Speedometer sensor connector	9	Drive shaft
2	Neutral switch connector	10	Joint shaft
3	Back-up light switch connector	11	No.4 engine mount bracket
4	Clutch release cylinder	12	Transverse member
5	Extension bar	13	Engine mount member
6	Change control rod	14	No.2 engine mount
7	Tie-rod end ball joint	15	No.1 engine mount bracket
8	Lower arm ball joint	16	MTX

Exploded view of the manual transaxle assembly mounting—2002 Protégé with G15M–R transaxle

42356-MAZC-GIM

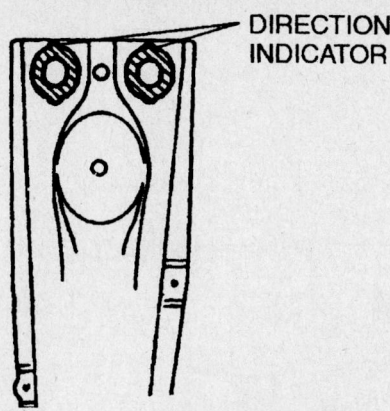

Position the direction indicator on the mount member and bushings—Protégé with both manual and automatic transaxles

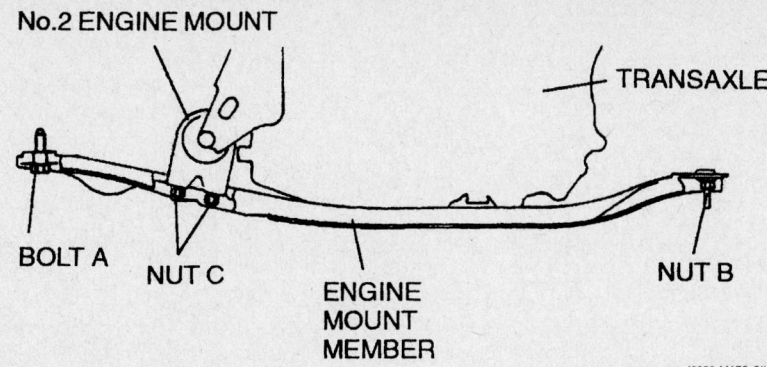

Location of the front mount member bolt locations—Protégé with both manual and automatic transaxles

- Neutral safety switch connector
- Back up light switch connector
- Speedometer connector
- Starter
- EGR pipe
- Splash shields
- Wheels
- Air cleaner assembly and fresh air duct
- Battery and battery box

13. Fill the transaxle with fluid. Road test the vehicle and check for leaks. Top off all fluids as needed.

F25M–R TRANSAXLE

1. Before servicing the vehicle, refer to the precautions section.

2. Refer to the illustration for component location and torque specifications.

3. Drain the transaxle oil.

4. Remove or disconnect the following:
- Battery and battery box
- Air cleaner assembly and fresh air duct
- Wheels
- Splash shields
- Exhaust Gas Recirculation (EGR) pipe
- Front exhaust pipe and three way catalytic converter
- Starter
- Speedometer sensor connector
- Neutral safety switch connector
- Back up light switch connector
- Clutch release cylinder
- Extension bar
- Change control rod
- Tie rod ends from the knuckle
- Lower control arm ball joint
- Halfshafts
- Joint shaft, install tool 49 B027 006 to hold the side gears after removal

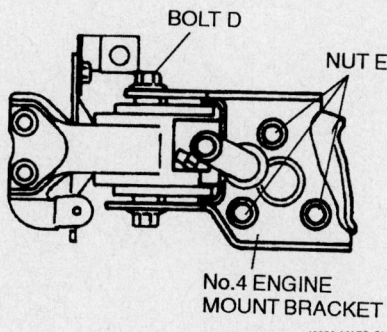

Number 4 mount fastener locations—2000—01 Protégé with G15M–R transaxle

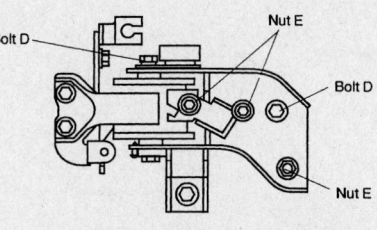

Number 4 mount fastener locations—2002 Protégé with G15M–R transaxle

5. Support the engine assembly with a engine support assembly such as 49 E017 5A0.
- Number 4 engine mount
- Engine mount member
- Number 2 engine mount
- Number 1 engine mount bracket

6. Loosen the engine support assembly and lean the engine towards the transaxle.

7. Support the transaxle with a jack, remove the transaxle bolts and the transaxle.

To install:

8. Place the transaxle onto a jack and raise into position.

9. Install the transaxle bolts and tighten

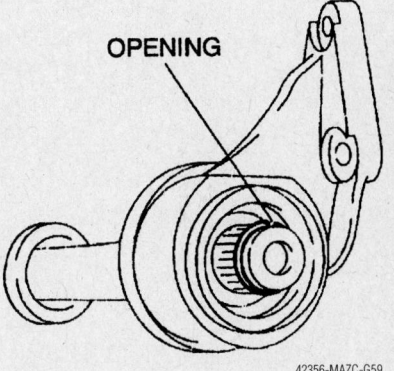

When installing the joint shaft make sure to install a new clip with the opening facing up—Protégé with both manual and automatic transaxles

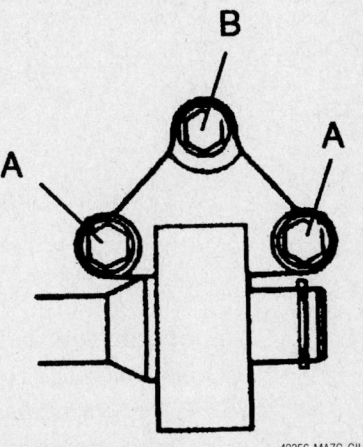

Tighten the joint shaft bolts as shown—Protégé with both manual and automatic transaxles

the upper bolts to 48–65 ft. lbs. (64–89 Nm), the lower bolts to 28–38 ft. lbs. (38–51 Nm). Refer to the exploded view illustration for bolt locations.

10. Install or connect the following:

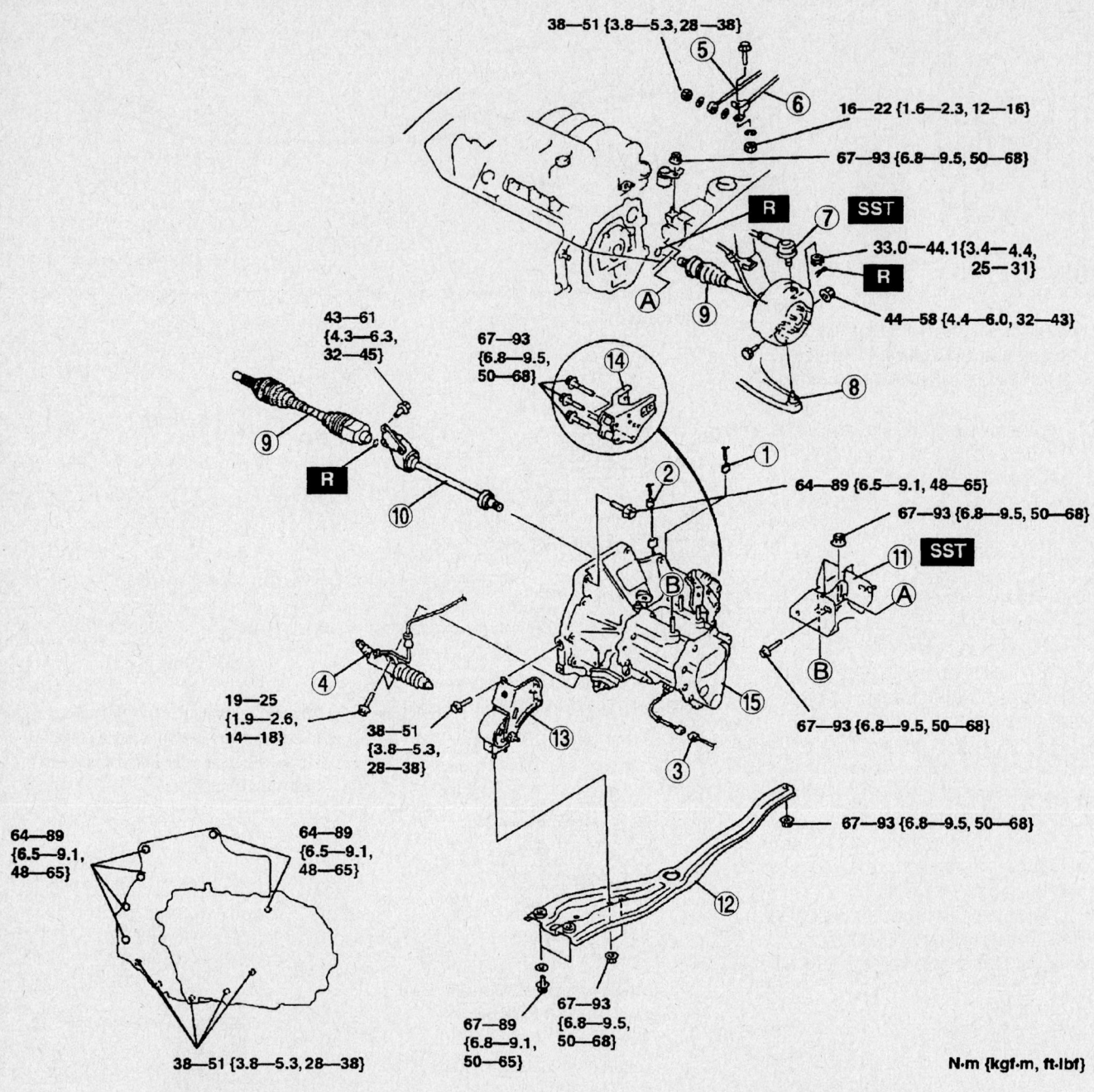

38—51 {3.8—5.3, 28—38}

16—22 {1.6—2.3, 12—16}

67—93 {6.8—9.5, 50—68}

33.0—44.1{3.4—4.4, 25—31}

44—58 {4.4—6.0, 32—43}

43—61 {4.3—6.3, 32—45}

67—93 {6.8—9.5, 50—68}

64—89 {6.5—9.1, 48—65}

67—93 {6.8—9.5, 50—68}

19—25 {1.9—2.6, 14—18}

38—51 {3.8—5.3, 28—38}

67—93 {6.8—9.5, 50—68}

67—93 {6.8—9.5, 50—68}

64—89 {6.5—9.1, 48—65}

64—89 {6.5—9.1, 48—65}

38—51 {3.8—5.3, 28—38}

67—89 {6.8—9.1, 50—65}

67—93 {6.8—9.5, 50—68}

N·m {kgf·m, ft·lbf}

1	Speedometer sensor connector	9	Drive shaft
2	Neutral switch connector	10	Joint shaft
3	Back-up light switch connector	11	No.4 engine mount bracket
4	Clutch release cylinder	12	Engine mount member
5	Extension bar	13	No.2 engine mount
6	Change control rod	14	No.1 engine mount bracket
7	Tie-rod end ball joint	15	MTX
8	Lower arm ball joint		

42356-MAZC-GIG

Exploded view of the manual transaxle assembly mounting—Protégé with F25M–R transaxle

- Number 1 engine mount bracket. Refer to the illustration for component location and torque specifications.
- Number 2 engine mount. Refer to the illustration for component location and torque specifications.

11. Install the engine mount member as follows while referring to the illustration for bolt locations:

a. Position the direction indicator on the mount member bushings as illustrated.

b. Put the number 2 engine mount stud bolts through the mount member installation holes. Install mount member bolts **A** and nuts **B** and tighten to 50–68 ft. lbs. (67–93 Nm).

c. Tighten bolts **C** to 50–68 ft. lbs. (67–93 Nm).

d. By aligning the holes on the stud bolts, install the number 4 mount bracket. Align hole of the bracket with the rubber and hand tighten bolt **D**, then nuts **E** and bolt **D** to 50–68 ft. lbs. (67–93 Nm).

12. Install or connect the following:

- Joint shaft. Install a new circlip with the opening facing up. Hand tighten the bolts **A**, then tighten the bolts to 32–45 ft. lbs. (43–61 Nm). Refer to the illustration for bolt location.
- Halfshafts
- Lower control arm ball joint
- Tie rod ends to the knuckle
- Change control rod
- Extension bar
- Clutch release cylinder
- Neutral safety switch connector
- Back up light switch connector
- Speedometer connector
- Starter
- EGR pipe
- Splash shields
- Wheels
- Air cleaner assembly and fresh air duct
- Battery and battery box

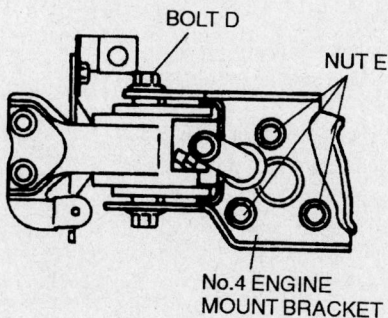

BOLT D

NUT E

No.4 ENGINE MOUNT BRACKET

42356-MAZC-GIJ

Number 4 mount fastener locations—Protégé with F25M–R transaxle

13. Fill the transaxle with fluid. Road test the vehicle and check for leaks. Top off all fluids as needed.

Automatic Transaxle/Transaxle Assembly

REMOVAL & INSTALLATION

Protégé

1. Before servicing the vehicle, refer to the precautions section.

2. Refer to the illustration for component location and torque specifications.

3. Drain the transaxle oil.

4. Remove or disconnect the following:

- Wheels
- Splash shields
- Battery and battery box
- Air cleaner assembly and fresh air duct
- Exhaust Gas Recirculation (EGR) pipe
- Front exhaust pipe and three way catalytic converter
- Speedometer sensor connector
- Transmission range switch connector
- Input/turbine speed sensor connector
- Transaxle connector
- Harness bracket
- Battery tray bracket
- Oil dipstick tube
- Oil hose
- Brake hose clip and ABS sensor bracket
- Tie rod ends from the knuckle
- Lower arm bolt
- Stabilizer link nut
- Lower arm
- Transverse member
- Halfshafts
- Joint shaft, install tool 49 G030 455 to hold the side gears after removal
- Selector cable
- Intake manifold stay
- Starter
- Torque converter nuts

5. Support the engine assembly with a engine support assembly such as 49 E017 5A0.

- Number 4 engine mount
- Number 1 engine mount bracket
- Roll damper on models with the 2.0L (FS) engine
- Engine mount member
- Number 2 engine mount

6. Loosen the engine support assembly and lean the engine towards the transaxle.

7. Support the transaxle with a jack, remove the transaxle bolts and the transaxle.

To install:

8. Place the transaxle onto a jack and raise into position.

9. Install the transaxle bolts and tighten the bolts as follows:

a. Bolts **A** to 65 ft. lbs. (89 Nm).

b. Bolts **B** to 86 ft. lbs. (116 Nm).

c. Bolts **C** to 38 ft. lbs. (51 Nm).

d. Bolts **D** to 18 ft. lbs. (25 Nm).

10. Install or connect the following:

- Number 2 engine mount. Refer to the illustration for component location and torque specifications.

11. Install the engine mount member as follows while referring to the illustration for bolt locations:

a. Position the direction indicator on the mount member bushings as illustrated.

b. Put the number 2 engine mount stud bolts through the mount member installation holes. Install mount member bolts **A** and tighten to 50–68 ft. lbs. (67–93 Nm).

c. Tighten the nuts **B** to 50–68 ft. lbs. (67–93 Nm) on 1.6L (ZM) engines or 63–86 ft. lbs. (86–116 Nm) on 2.0L (FS) engines.

d. By aligning the holes on the stud bolts, install the number 4 mount bracket. Align hole of the bracket with the rubber and hand tighten bolt **A**. Tighten nut **B** and tighten bolt **A** to 50–68 ft. lbs. (67–93 Nm).

12. Install or connect the following:

- Roll damper, on 2.0L (FS) engines
- Number 1 engine mount
- Torque converter nuts
- Starter
- Intake manifold stay
- Selector cable
- Joint shaft. Install a new circlip with the opening facing up. Hand tighten the bolts **A**, then tighten the bolts to 32–45 ft. lbs. (43–61 Nm). Refer to the illustration for bolt location.
- Halfshafts
- Transverse member
- Lower arm
- Stabilizer link nut
- Lower arm bolt
- Tie rod ends to the knuckle
- ABS sensor bracket and brake hose clip
- Oil hose
- Oil dipstick tube
- Battery tray bracket
- Harness bracket

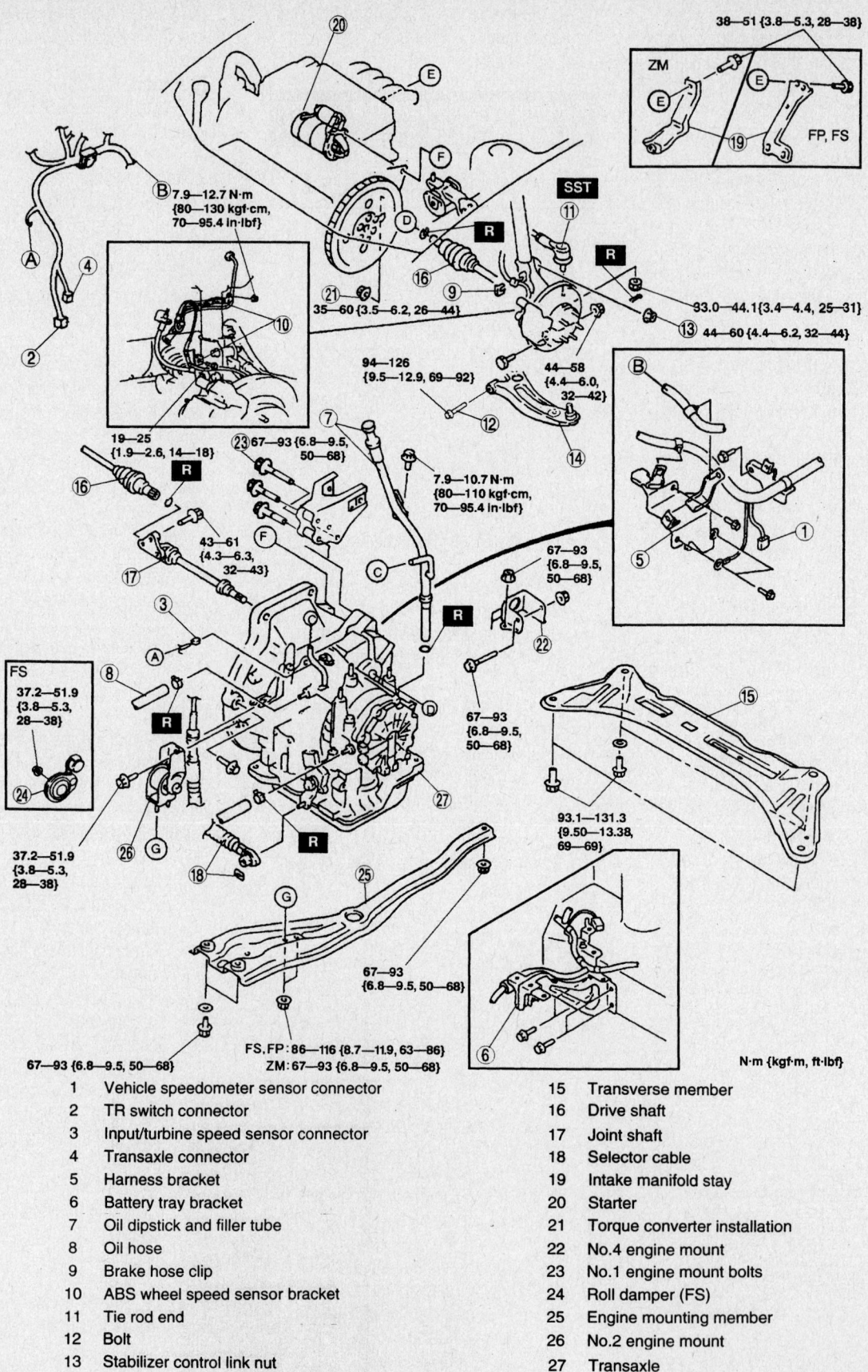

N·m {kgf·m, ft·lbf}

1	Vehicle speedometer sensor connector	15	Transverse member
2	TR switch connector	16	Drive shaft
3	Input/turbine speed sensor connector	17	Joint shaft
4	Transaxle connector	18	Selector cable
5	Harness bracket	19	Intake manifold stay
6	Battery tray bracket	20	Starter
7	Oil dipstick and filler tube	21	Torque converter installation
8	Oil hose	22	No.4 engine mount
9	Brake hose clip	23	No.1 engine mount bolts
10	ABS wheel speed sensor bracket	24	Roll damper (FS)
11	Tie rod end	25	Engine mounting member
12	Bolt	26	No.2 engine mount
13	Stabilizer control link nut	27	Transaxle
14	Lower arm		

42356-MAZC-GIP

Exploded view of the automatic transaxle assembly mounting—Protégé

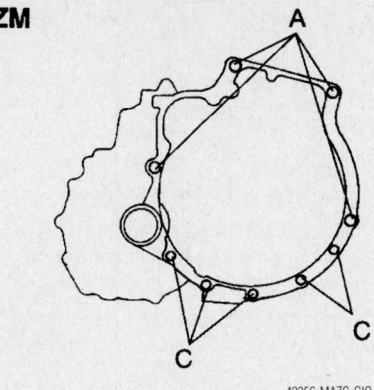

ZM

Automatic transaxle assembly bolt torque sequence—1.6L (ZM) engine

42356-MAZC-GIQ

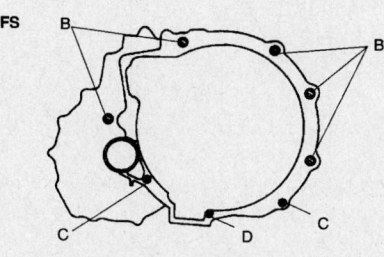

FS

Automatic transaxle assembly bolt torque sequence—2.0L (FS) engines

42356-MAZC-GIR

- Transaxle connector
- Input/turbine speed sensor connector
- Transmission range switch connector
- Speedometer sensor connector
- Front exhaust pipe and three way catalytic converter
- EGR pipe
- Air cleaner assembly and fresh air duct

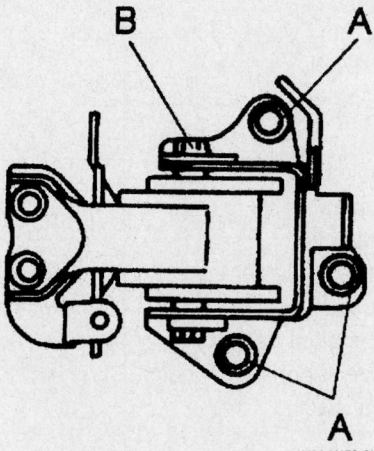

Number 4 mount fastener locations—Protégé with automatic transaxle

42356-MAZC-GIS

- Battery and battery box
- Splash shields
- Wheels

13. Fill the transaxle with fluid. Road test the vehicle and check for leaks. Top off all fluids as needed.

2002–04 Miata

1. Refer to the illustration for component location and torque specifications.
2. Before servicing the vehicle, refer to the precautions section.
3. Refer to the illustration for component location and torque specifications.
4. Drain the transaxle oil, coolant and engine oil.
5. Remove or disconnect the following:
- Negative battery cable
- Crossmember and bracket, on models equipped with 16 inch wheels
- Rear crossbar, on models equipped with 16 inch wheels
- Exhaust system
- Drive shaft
- Throttle cable
- Dipstick tube
- Performance rod
- Exhaust bracket
- Shift rod
- Transmission range selector switch connector
- Output speed sensor connector
- Solenoid connector
- Input/turbine speed sensor
- Harness bracket
- Intake manifold bracket
- Oil filter
- Water hose and oil cooler
- Under cover
- Torque converter bolts
- Harness
- Wiring harness from the Power Plant Frame (PPF)
6. Support the transmission with a jack.
- PPF front bolts
- Differential side bolts and pry out the spacer

❊❊ CAUTION

Removing the PPF spacers will reduce the performance of the PPF. If the spacers are removed, replace the PPF as an assembly.

- Differential mounting spacer
- Transmission side bolts and the PPF
- Transmission bolts and the transmission; keep the torque converter end slightly elevated during removal

To install:

7. Support the transmission with a jack.
8. Install or connect the following:
- Transmission into position, make sure the torque converter side is slightly elevated, once aligned, install and tighten the bolts to 48–65 ft. lbs.
- Differential mounting spacer and tighten the bolts to 28–38 ft. lbs. (38–51 Nm).
9. Support the transmission with a jack until it is level.
- PPF in position and if removed install the sleeve
- Spacer and bolts and tighten the reamer bolt making sure the threading is aligned correctly. The reamer bolt should be installed in the forward hole. Tighten the outer bolts making sure the threads are properly aligned. Tighten all bolts in the sequence illustrated to 91 ft. lbs. (123 (Nm).
- PPF front bolts and tighten to 91 ft. lbs. (123 Nm)
10. Remove the jack and connect the wiring harness.
11. Using a straightedge and Vernier caliper measure the distance **A** which should be 1.99–2.46 inch (50.5–62.5mm). If the distance is not as specified, reposition the PPF to the transmission.
12. Install or connect the following:
- Torque converter bolts. Align the holes while turning the converter, use a suitable tool to lock the flywheel and hand tighten the bolts in a criss–cross pattern. Once all the bolts are hand tight, tighten the bolts to 36 ft. lbs. (49 Nm).
- Under cover
- Oil cooler, tighten the oil cooler bolts to 28 ft. lbs. (39 Nm)
- Water hose
- Oil filter
- Intake manifold bracket
- Harness bracket
- Input/turbine speed sensor
- Solenoid connector
- Output speed sensor connector
- Transmission range selector switch connector
- Shift rod
- Exhaust bracket
- Performance rod
- Dipstick tube
- Throttle cable. Measure the free-play on the cable, if the free-play is not 0.04–0.11 inch (1–3mm), then adjust by turning the lock nuts until the desired specification is reached

and tighten the lock nuts to 10 ft. lbs. (14 Nm).
- Drive shaft
- Exhaust system
- Rear crossbar, on models equipped with 16 inch wheels
- Crossmember and bracket, on models equipped with 16 inch wheels
- Negative battery cable

13. Fill the transmission with fluid. Road test the vehicle and check for leaks. Top off all fluids as needed.

626

LA4A–EL TRANSXLE

1. Refer to the illustration for component location and torque specifications.
2. Before servicing the vehicle, refer to the precautions section.
3. Refer to the illustration for component location and torque specifications.
4. Drain the transaxle oil.
5. Remove or disconnect the following:

- Negative battery cable
- Air cleaner assembly and fresh air duct
- Wheels
- Splash shields
- Ground cable from the transaxle
- Bracket
- Solenoid body connector
- Transaxle range switch connector
- Turbine speed shaft connector
- Oil filler tube
- Fuel filter nut

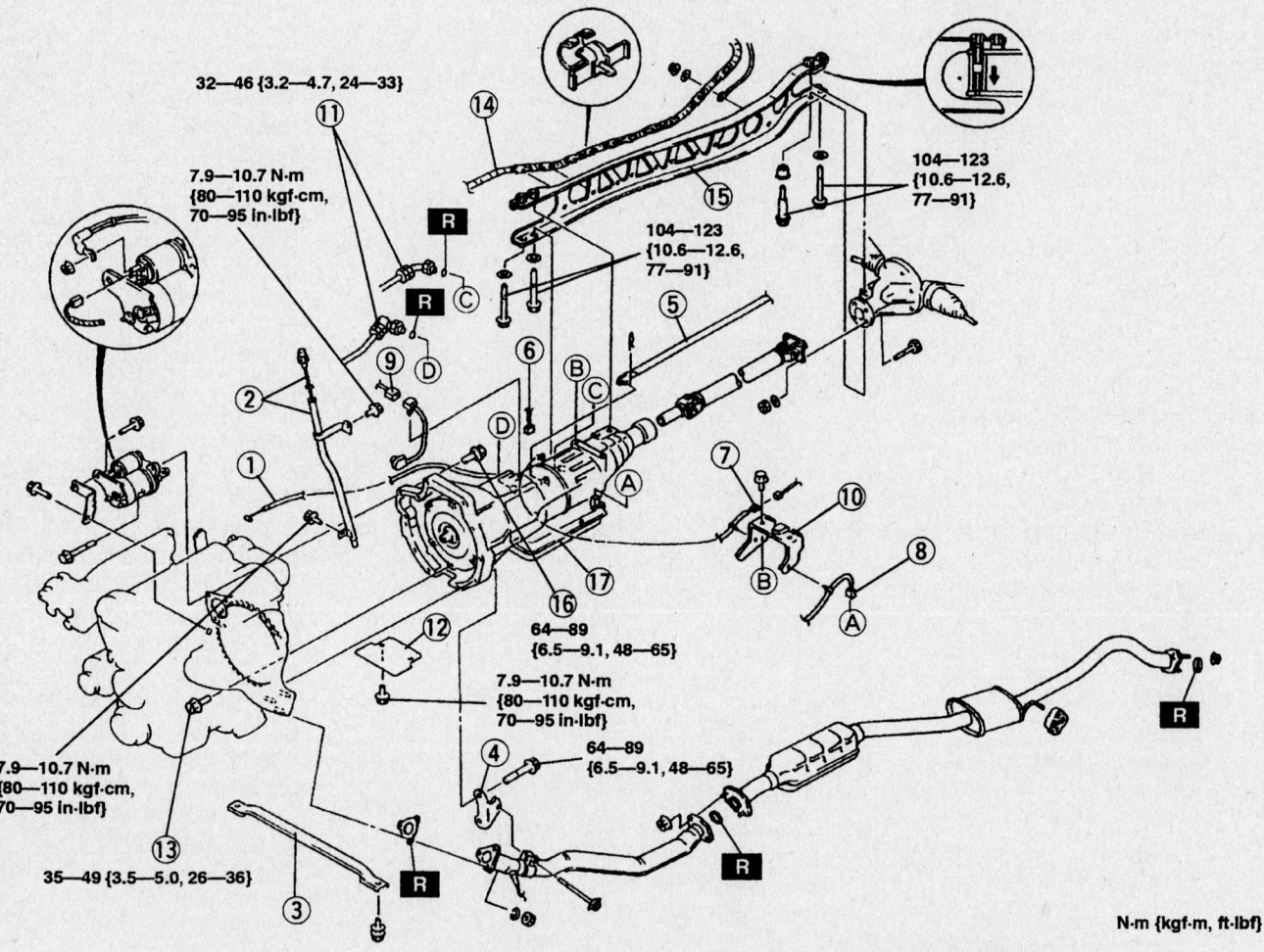

1	Throttle cable
2	Filler tube, dipstick
3	Performance rod
4	Exhaust bracket
5	Shift rod
6	TR switch connector
7	Output speed sensor connector
8	Solenoid connector
9	Input/turbine speed sensor
10	Harness bracket
11	Oil pipe
12	Undercover
13	Torque converter bolts
14	Harness
15	Power plant frame
16	Transmission mount bolts
17	Transmission

42356-MAZC-GXX

Exploded view of the automatic transmission assembly—2002–04 Miata

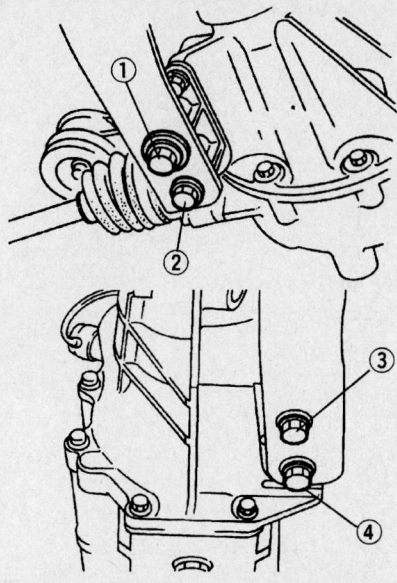

42356-MAZC-GCE

Tighten the differential bolts in the sequence illustrated—2002–04 Miata with an automatic transmission

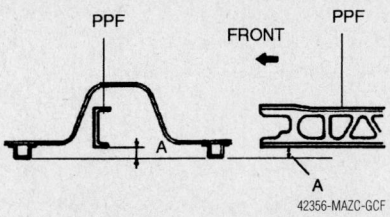

42356-MAZC-GCF

Measure the distance A which should be 1.99–2.46 inch (50.5–62.5mm)—2002–04 Miata with an automatic transmission

- Selector cable
- Number 1 engine mount nut, rubber and bracket

6. Support the engine assembly with a engine support assembly such as 49 E017 5A0.

- Number 4 mount
- Transverse member
- Vehicle Speed Sensor (VSS) connector
- Engine mount member
- Tie rod ends from the knuckle
- Stabilizer bar link

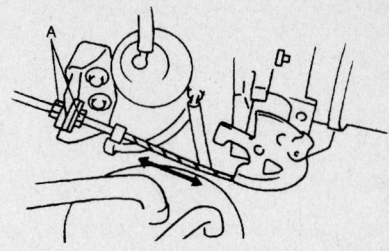

42356-MAZC-GCG

Adjust the free-play on the throttle and adjust using the lock nuts A—Miata with an automatic transmission)—2002–04 Miata with an automatic transmission

- Lower control arm
- Halfshafts
- Joint shaft, install tool 49 G030 455 to hold the side gears after removal
- Intake manifold bracket
- Starter
- Torque converter nuts
- Number 5 and 2 engine mounts
- Oil hose

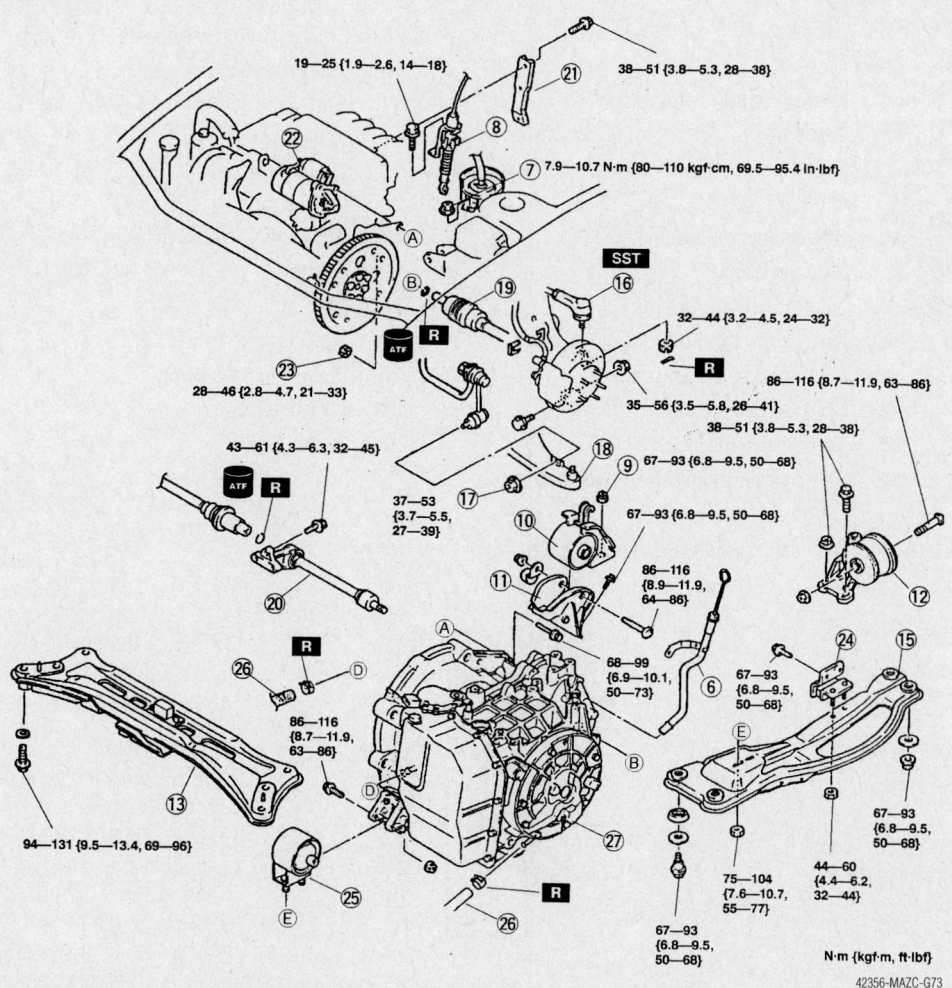

42356-MAZC-G73

Exploded view of the LA4A–EL automatic transaxle assembly (1 OF 2)—626

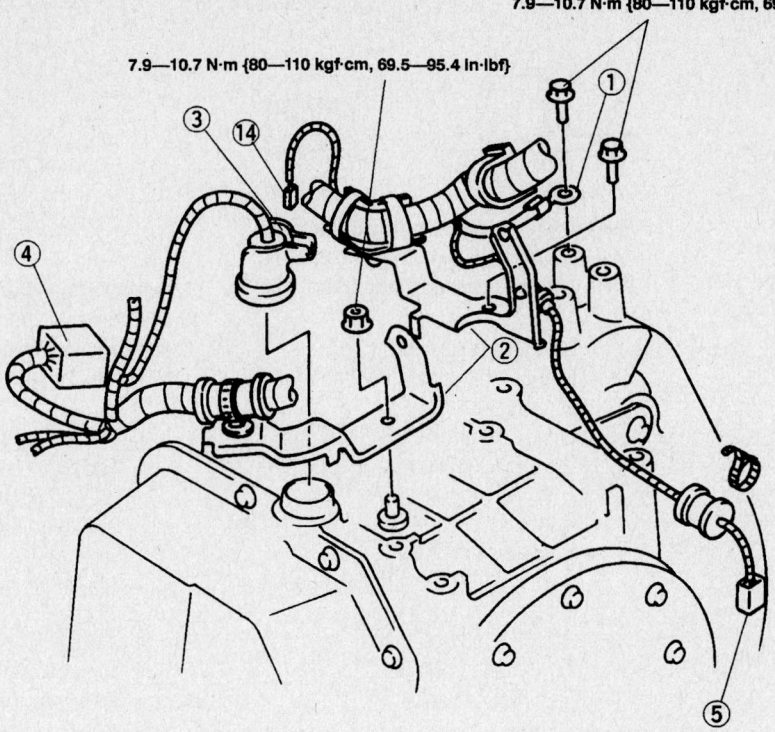

7.9—10.7 N·m {80—110 kgf·cm, 69.5—95.4 in·lbf}

7.9—10.7 N·m {80—110 kgf·cm, 69.5—95.4 in·lbf}

1	Ground	15	Engine mount member
2	Bracket	16	Tie-rod end
3	Solenoid body connector	17	Stabilizer control link
4	Transaxle range switch connector	18	Lower arm
5	Turbine shaft speed sensor connector	19	Drive shaft
6	Oil filler tube	20	Joint shaft
7	Fuel filter mounting nut	21	Intake manifold bracket
8	Selector cable	22	Starter
9	No.1 engine mount nut	23	Torque converter nut
10	No.1 engine mount rubber	24	No.5 engine mount
11	No.1 engine mount bracket	25	No.2 engine mount
12	No.4 engine mount	26	Oil hose
13	Transverse member	27	Transaxle
14	Vehicle speedometer sensor connector		

42356-MAZC-G74

Exploded view of the LA4A–EL automatic transaxle assembly (2 OF 2)—626

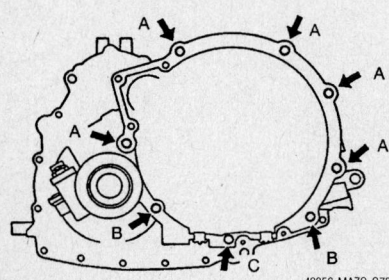

42356-MAZC-G75

LA4A–EL automatic transaxle assembly bolt locations—626

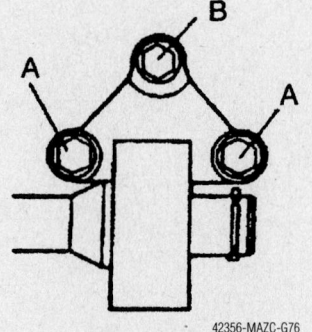

42356-MAZC-G76

Bolt locations on the joint shaft—EL automatic transaxle assembly

7. Loosen the engine support assembly and lean the engine towards the transaxle.

8. Support the transaxle with a jack, remove the transaxle bolts and the transaxle.

To install:

9. Place the transaxle onto a jack and raise into position.

10. Install the transaxle bolts and tighten the bolts **A** to 66–86 ft. lbs. (90–116 Nm), bolts **B** to 28–38 ft. lbs. (38–51 Nm) and bolts **C** to 14–18 ft. lbs. (19–25 Nm). Refer to the illustration for bolt locations.

• Hand tighten the number 4 mount retainers

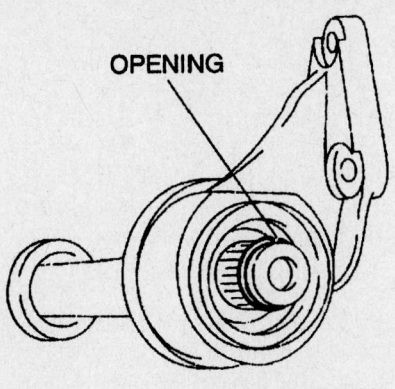

42356-MAZC-G59

When installing the joint shaft make sure to install a new clip with the opening facing up—626

11. Lean the engine towards its normal position and tighten the support assembly.

12. Install or connect the following:
- Oil hose
- Number 5 and 2 engine mounts. Refer to the illustration for component location and torque specifications.
- Torque converter nuts and tighten to 21–33 ft. lbs. (28–46 Nm)
- Starter
- Intake manifold bracket. Refer to the illustration for component location and torque specifications.
- Joint shaft. Install a new circlip with the opening facing up. Hand tighten the bolts **A**, then tighten the bolts to 32–45 ft. lbs. (43–61 Nm). Refer to the illustration for bolt location.
- Halfshafts
- Lower control arm
- Stabilizer bar link. Refer to the illustration for component location and torque specifications.
- Tie rod ends to the knuckle. Refer to the illustration for component location and torque specifications.

13. Install the engine mount member as follows while referring to the illustration for bolt locations:

a. Position the direction indicator on the mount member bushings facing towards the front side and install the bushings onto the mount.

b. Put the number 2 engine mount stud bolts through the mount member installation holes. Install mount member bolts **A** and nuts **A** and tighten to 50–68 ft. lbs. (67–93 Nm).

c. Loosely tighten the number 2 engine mount nutsand remove the engine support assembly.

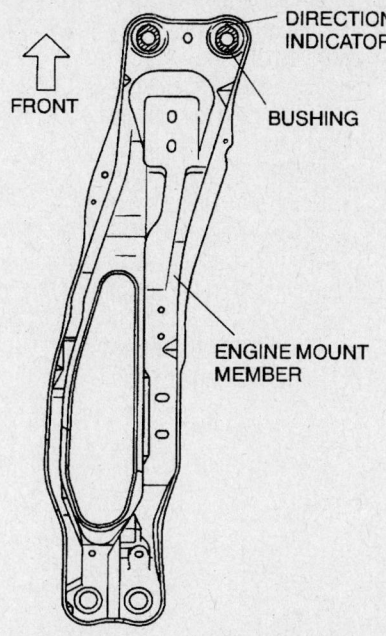

42356-MAZC-G61

Position the direction indicator on the mount member bushings facing towards the front side and install the bushings onto the mount—626

d. Tighten the number 2 engine mount nuts **B** to 55–77 ft. lbs. (75–104 Nm).

14. Install or connect the following:
- VSS connector
- Transverse member. Refer to the illustration for component location and torque specifications.
- Number 4 mount. Refer to the illustration for component location and torque specifications.
- Number 1 engine mount nut, rubber and bracket. Refer to the illustration for component location and torque specifications.
- Selector cable
- Fuel filter nut

- Oil filler tube
- Turbine speed shaft connector
- Transaxle range switch connector
- Solenoid body connector
- Bracket
- Ground cable from the transaxle
- Splash shields
- Wheels
- Air cleaner assembly and fresh air duct
- Negative battery cable

15. Fill the transmission with fluid.

16. Start the engine and check for leaks and proper operation.

GFA4A–EL TRANSAXLE

1. Refer to the illustration for component location and torque specifications.

2. Before servicing the vehicle, refer to the precautions section.

3. Refer to the illustration for component location and torque specifications.

4. Drain the transaxle oil.

5. Remove or disconnect the following:
- Battery
- Air cleaner assembly and fresh air duct
- Wheels
- Splash shields
- Selector cable clip
- Selector cable from the manual shaft lever, pull out the cable from the bracket and remove the cable
- Transaxle range switch connector
- Solenoid valve connector
- Oxygen (O₂S) connector
- Vehicle Speed Sensor (VSS) connector
- Input/Turbine speed shaft connector
- Starter harness
- Engine mount stay
- Fuel filter bolts
- Oil hose and breather hose
- Lower control arm
- Tie rod ends from the knuckle

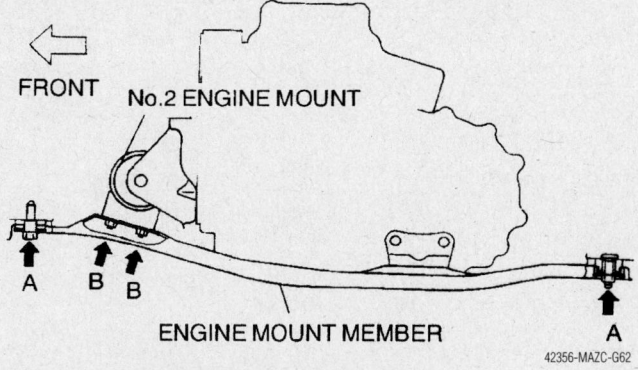

42356-MAZC-G62

Location of the front mount member bolt locations—626

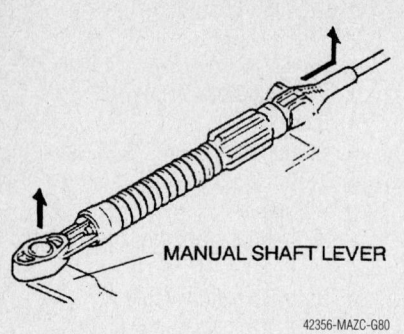

MANUAL SHAFT LEVER

42356-MAZC-G80

Remove the selector cable GF4A–EL automatic transaxle assembly (1 OF 2)—626

- Stabilizer bar link
- Transverse member
- Halfshafts
- Joint shaft, install tool 49 G030 455 to hold the side gears after removal
- Number 5 mount rubber
- Number 1 engine mount nut, rubber and bracket

6. Support the engine assembly with a engine support assembly such as 49 E017 5A0.

- Engine mount member
- Number 2 engine mount

- Under cover
- Torque converter nuts
- Number 4 engine mount

7. Loosen the engine support assembly and lean the engine towards the transaxle.

8. Support the transaxle with a jack, remove the transaxle bolts and the transaxle.

To install:

9. Place the transaxle onto a jack and raise into position.

10. Install the transaxle bolts and tighten the bolts to 50–73 ft. lbs. (68–99 Nm).

- Hand tighten the number 4 mount

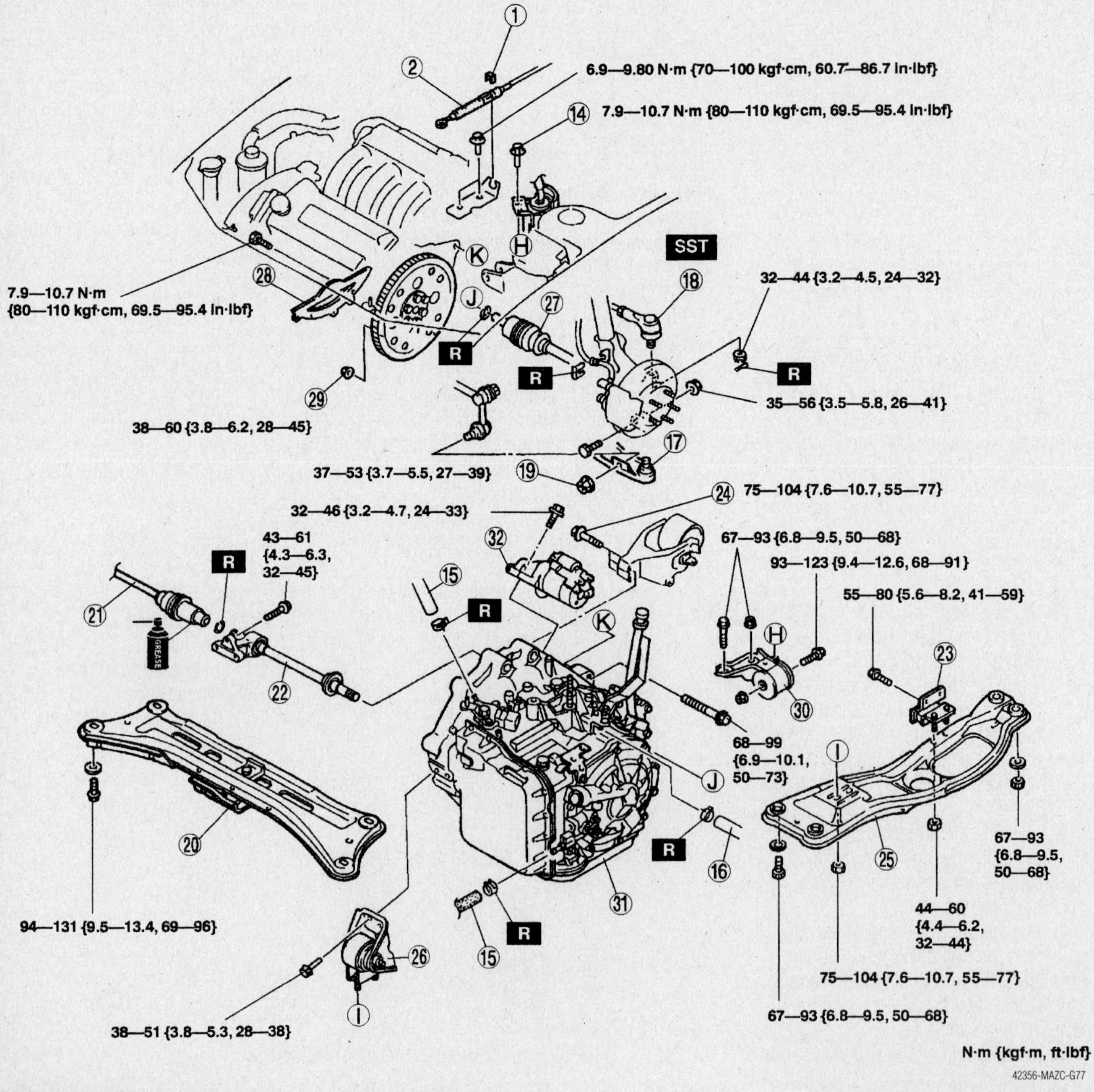

6.9—9.80 N·m {70—100 kgf·cm, 60.7—86.7 in·lbf}

7.9—10.7 N·m {80—110 kgf·cm, 69.5—95.4 in·lbf}

7.9—10.7 N·m {80—110 kgf·cm, 69.5—95.4 in·lbf}

32—44 {3.2—4.5, 24—32}

35—56 {3.5—5.8, 26—41}

38—60 {3.8—6.2, 28—45}

37—53 {3.7—5.5, 27—39}

32—46 {3.2—4.7, 24—33}

43—61 {4.3—6.3, 32—45}

75—104 {7.6—10.7, 55—77}

67—93 {6.8—9.5, 50—68}

93—123 {9.4—12.6, 68—91}

55—80 {5.6—8.2, 41—59}

68—99 {6.9—10.1, 50—73}

67—93 {6.8—9.5, 50—68}

44—60 {4.4—6.2, 32—44}

94—131 {9.5—13.4, 69—96}

38—51 {3.8—5.3, 28—38}

75—104 {7.6—10.7, 55—77}

67—93 {6.8—9.5, 50—68}

N·m {kgf·m, ft·lbf}

42356-MAZC-G77

Exploded view of the GF4A–EL automatic transaxle assembly (1 OF 2)—626

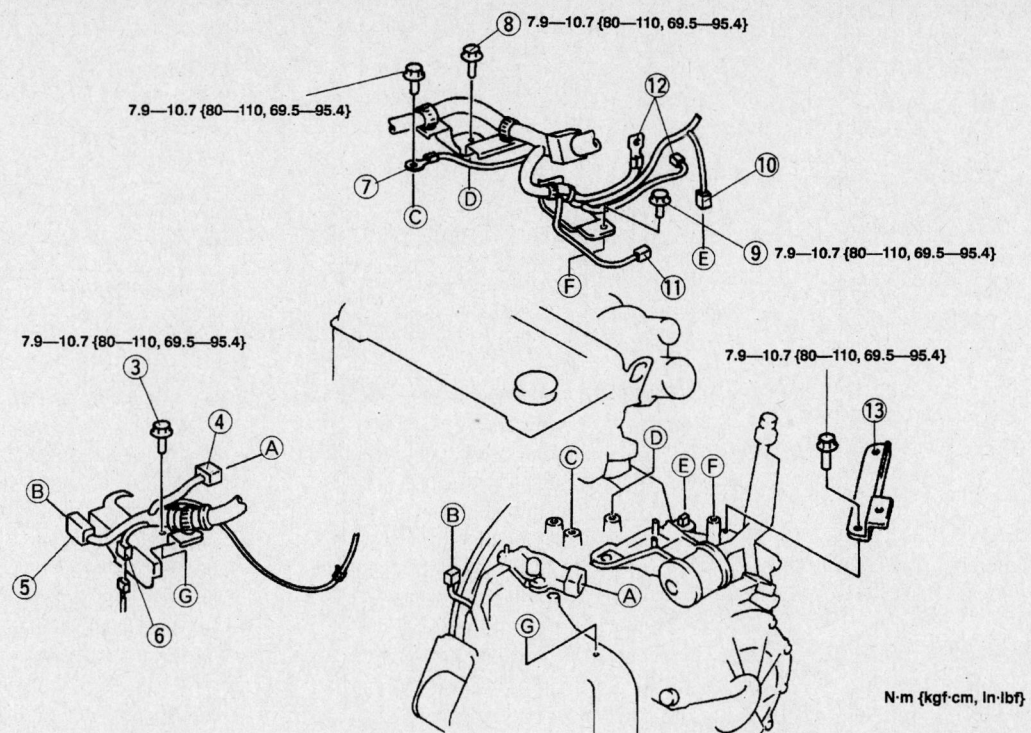

7.9—10.7 {80—110, 69.5—95.4}

7.9—10.7 {80—110, 69.5—95.4}

7.9—10.7 {80—110, 69.5—95.4}

7.9—10.7 {80—110, 69.5—95.4}

7.9—10.7 {80—110, 69.5—95.4}

N·m {kgf·cm, in·lbf}

1	Clip	17	Lower arm
2	Selector cable	18	Tie-rod end
3	Bolt	19	Stabilizer control link nut
4	Transaxle range switch connector	20	Transverse member
5	Solenoid valve connector	21	Drive shaft
6	Oxygen sensor connector	22	Joint shaft
7	Ground	23	No.5 engine mount rubber
8	Bolt	24	No.1 engine mount bolts
9	Bolt	25	Engine mount member
10	Vehicle speedometer sensor connector	26	No.2 engine mount
11	Input/turbine speed sensor connector	27	Drive shaft
12	Starter harness	28	Undercover
13	Engine mount stay	29	Torque converter nuts
14	Fuel filter mounting bolts	30	No.4 engine mount
15	Oil hose	31	Transaxle
16	Breather hose	32	Starter

42356-MAZC-G78

Exploded view of the GF4A–EL automatic transaxle assembly (2 OF 2)—626

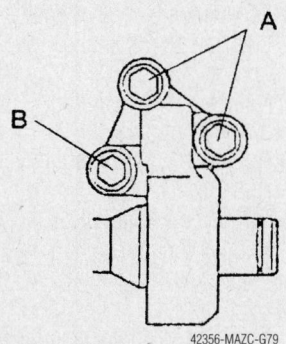

42356-MAZC-G79

Bolt locations on the joint shaft—GF4A–EL automatic transaxle assembly

retainers, then tighten to 55–77 ft. lbs. (75–104 Nm)

11. Lean the engine towards its normal position and tighten the support assembly.

12. Install or connect the following:
- Torque converter nuts
- Under cover
- Number 2 engine mount

13. Install the engine mount member as follows while referring to the illustration for bolt locations:

a. Position the direction indicator on the mount member bushings facing towards the front side and install the bushings onto the mount.

b. Put the number 2 engine mount stud bolts through the mount member installation holes. Install mount member bolts **A** and nuts **A** and tighten to 50–68 ft. lbs. (67–93 Nm).

c. Loosely tighten the number 2 engine mount nutsand remove the engine support assembly.

d. Tighten the number 2 engine mount nuts **B** to 55–77 ft. lbs. (75–104 Nm).

14. Install or connect the following:
- Number 1 engine mount nut, rubber and bracket. Refer to the illustration for component location and torque specifications.

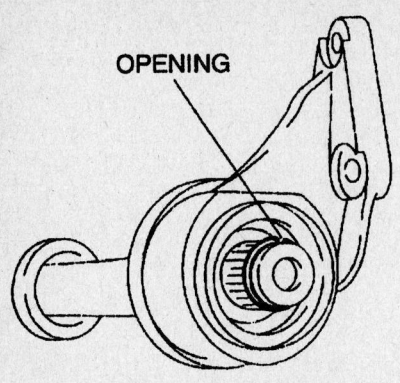

42356-MAZC-G59

When installing the joint shaft make sure to install a new clip with the opening facing up—626

- Number 5 mount rubber. Refer to the illustration for component location and torque specifications.
- Halfshafts
- Joint shaft. Install a new circlip with the opening facing up. Hand tighten the bolts **A**, then tighten the bolts to 32–45 ft. lbs. (43–61 Nm). Refer to the illustration for bolt location.
- Transverse member. Refer to the illustration for component location and torque specifications.
- Stabilizer bar link. Refer to the

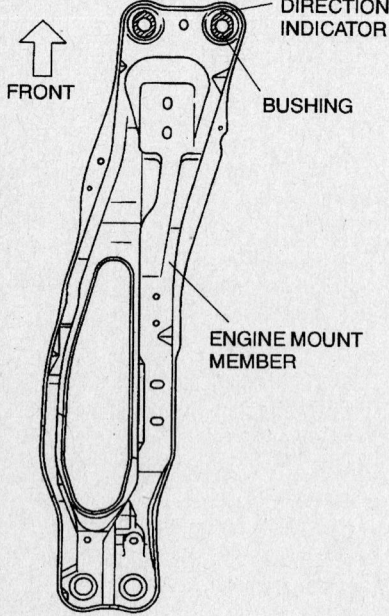

42356-MAZC-G61

Position the direction indicator on the mount member bushings facing towards the front side and install the bushings onto the mount—626

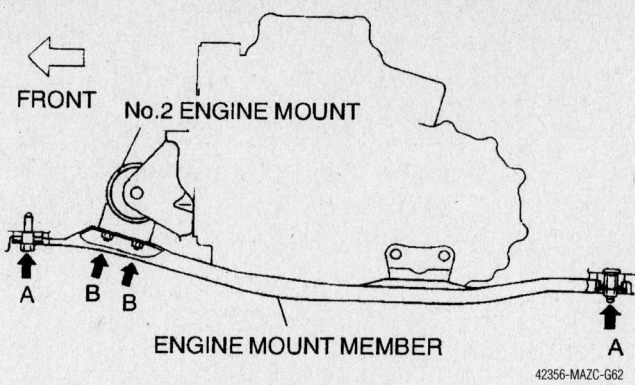

42356-MAZC-G62

Location of the front mount member bolt locations—626

illustration for component location and torque specifications.
- Tie rod ends to the knuckle. Refer to the illustration for component location and torque specifications.
- Lower control arm
- Oil hose and breather hose
- Fuel filter bolts
- Engine mount stay. Refer to the illustration for component location and torque specifications.
- Starter harness
- Input/Turbine speed shaft connector
- VSS connector
- O$_2$S connector
- Solenoid valve connector
- Transaxle range switch connector
- Selector cable to the bracket, install the clip and attach the cable to the manual shaft lever
- Splash shields
- Wheels
- Air cleaner assembly and fresh air duct
- Battery

15. Fill the transmission with fluid.
16. Start the engine and check for leaks and proper operation.

Millenia

LJ4A–EL TRANSAXLE

1. Refer to the illustration for component location and torque specifications.
2. Before servicing the vehicle, refer to the precautions section.
3. Refer to the illustration for component location and torque specifications.
4. Drain the transaxle oil.
5. Remove or disconnect the following:
- Battery cover, battery and tray
- Rear air charge cooler duct
- Air cleaner assembly
- Selector cable clip and nut and the cable

- Output speed sensor connector
- Transaxle range switch connector
- Solenoid valve connector
- Harness bracket
- Selector cable bracket
- Rear intercooler
- Bracket and starter
- Front intercooler
- Electric fan
- Wheels
- Halfshaft locknut
- Splash shields
- Fresh air duct
- Exhaust pipe
- Lower ball joint
- Left side upper lateral link
- Lower arm
- Right side halfshaft
- Joint shaft
- Stabilizer bar link
- Left side halfshaft
- Timing belt
- Drive belts
- Power steering pump leaving the lines attached and set aside
- A/C compressor leaving the lines attached and set aside
- Selector rod
- Transmission dipstick tube

6. Support the engine assembly with a engine support assembly such as 49 E017 5A0.

- Loosen the number 3 engine mount bolt (refer to illustration) and remove the number 4 mount
- Number 1 engine mount nut, rubber and bracket
- Number 2 engine mount nut
- Under cover
- Torque converter bolts
- Number 1 engine mount bolt
- Engine mount member
- Oil hose

7. Loosen the engine support assembly and lean the engine towards the transaxle.
8. Support the transaxle with a jack,

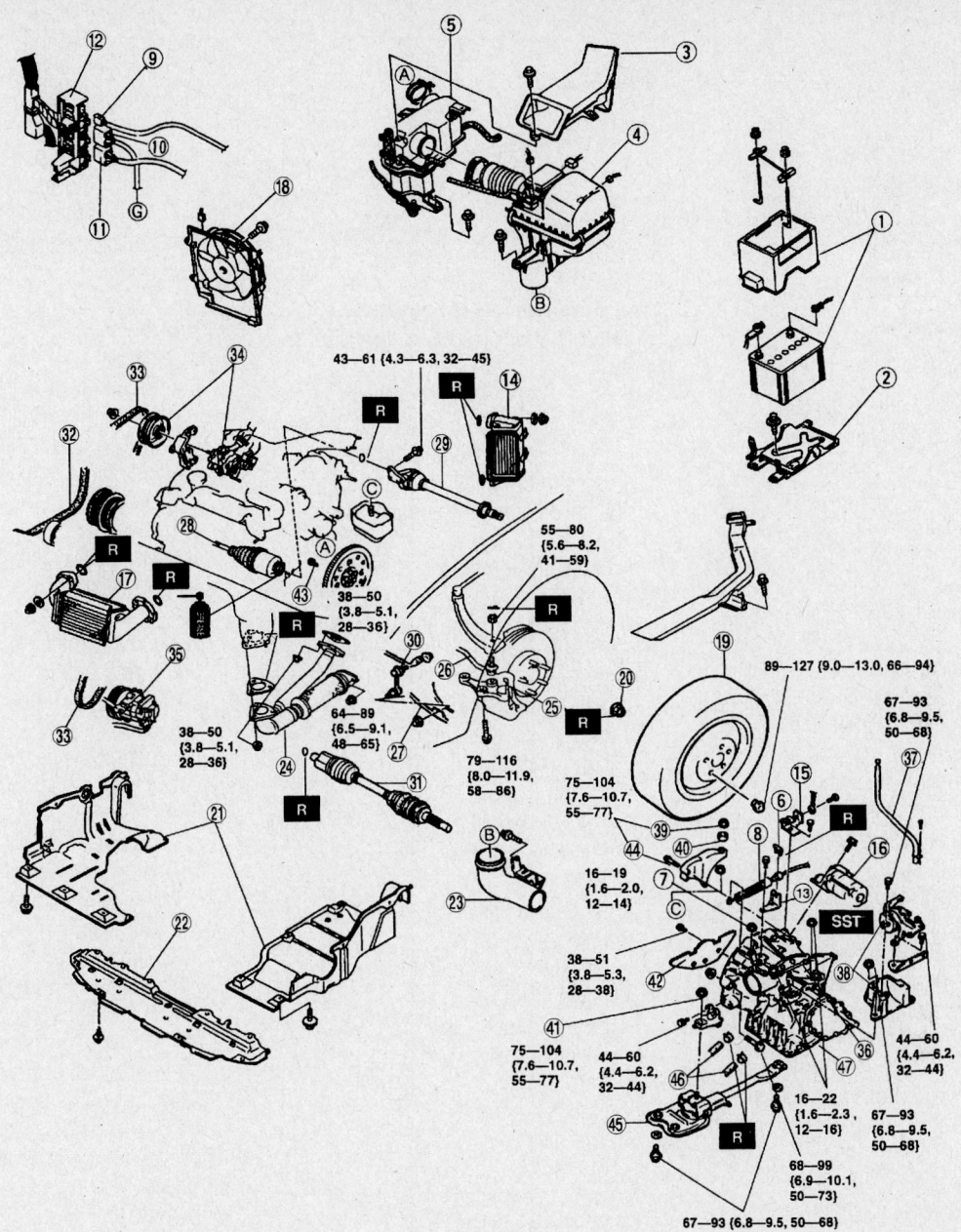

43—61 {4.3—6.3, 32—45}

55—80 {5.6—8.2, 41—59}

38—50 {3.8—5.1, 28—36}

38—50 {3.8—5.1, 28—36}

64—89 {6.5—9.1, 48—65}

79—116 {8.0—11.9, 58—86}

75—104 {7.6—10.7, 55—77}

16—19 {1.6—2.0, 12—14}

38—51 {3.8—5.3, 28—38}

75—104 {7.6—10.7, 55—77}

44—60 {4.4—6.2, 32—44}

16—22 {1.6—2.3, 12—16}

89—127 {9.0—13.0, 66—94}

67—93 {6.8—9.5, 50—68}

44—60 {4.4—6.2, 32—44}

67—93 {6.8—9.5, 50—68}

68—99 {6.9—10.1, 50—73}

67—93 {6.8—9.5, 50—68}

N·m {kgf·m, ft-lbf}

1	Battery and battery cover	17	Front intercooler	33	P/S, A/C Drive belt
2	Battery carrier	18	Electric coolant fan component	34	P/S oil pump
3	Rear change air cooler air duct	19	Wheel and tires	35	A/C compressor
4	Air cleaner component	20	Locknut	36	Selector rod
5	Resonance chamber	21	Splash shield	37	ATF filler tube
6	Clip	22	Splash shield	38	No.4 engine mount
7	Nut	23	Fresh air duct	39	No.1 engine mount nut
8	Selector cable	24	Exhaust pipe	40	No.1 engine mount damper
9	Output speed sensor connector	25	Lower ball joint	41	No.2 engine mount nut
10	Transaxle range switch connector	26	Upper lateral link (left side)	42	Under cover
11	Solenoid valve connector	27	Lower arm	43	Torque converter bolts
12	Harness bracket	28	Drive shaft (right side)	44	No.1 engine mount bolt
13	Selector cable bracket	29	Joint shaft	45	Engine mounting member
14	Rear intercooler	30	Stabilizer control link	46	Oil hose
15	Bracket	31	Drive shaft (left side)	47	Transaxle
16	Starter	32	Timing belt		

Exploded view of the LJ4A–EL automatic transaxle assembly—Millenia

42356-MAZC-GEPQ

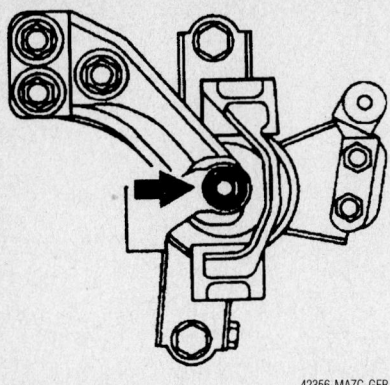

Loosen the number 3 engine mount bolt
LJ4A–EL automatic transaxle assembly—
Millenia

remove the transaxle bolts and the
transaxle.

To install:

9. Place the transaxle onto a jack and
raise into position.

10. Install the transaxle bolts and tighten
the bolts to 50–73 ft. lbs. (68–99 Nm).

11. Install or connect the following:
- Oil hose
- Engine mount member and number
  2 engine mount, make sure the
  number 2 mount stud bolt passes
  through the number 2 mount
  bracket installation hole and tighten
  the bolt **A** to 68 ft. lbs. (93 Nm).
  Tighten bolt **B** to 77 ft. lbs. (104
  Nm). Refer to the illustration for
  component location.
- Number 3 engine mount bolt and
  tighten to 77 ft. lbs. (104 Nm)
- Torque converter bolts and tighten
  to 28–38 ft. lbs. (38–51 Nm)
- Under cover
- Number 2 engine mount nut
- Number 1 engine mount nut, rubber and bracket. Tighten the bolt to
  77 ft. lbs. (104 Nm).
- Hand tighten the number 4 mount
  retainers. Tighten the bolts to 44 ft.
  lbs. (60 Nm), use the jack to

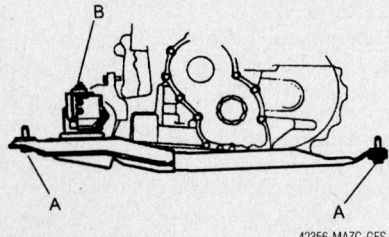

Install the engine mount member bolts
and tighten to specification—LJ4A–EL
automatic transaxle assembly—Millenia
with LJ4A–EL automatic transaxle

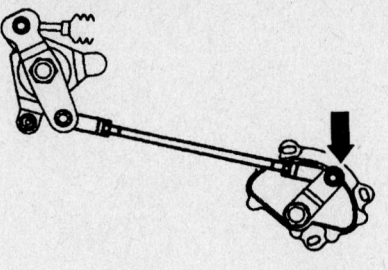

Install the selector rod and tighten the
nut—Millenia with LJ4A–EL automatic
transaxle

ensure the mount holes and bracket
holes are aligned and tighten the
bolts to 68 ft. lbs. (93 Nm).

12. Remove the engine support assembly.
- Transmission dipstick tube

13. Move the selector lever to **P** and the
manual shaft to **P** position. Install the selec-
tor rod and tighten the nut illustrated to 16
ft. lbs. (22 Nm).
- A/C compressor
- Power steering pump
- Drive belts
- Timing belt
- Left side halfshaft
- Stabilizer bar link
- Joint shaft
- Right side halfshaft
- Lower arm
- Left side upper lateral link
- Lower ball joint
- Exhaust pipe
- Fresh air duct
- Splash shields
- Halfshaft locknut
- Wheels
- Electric fan
- Front intercooler
- Bracket and starter
- Rear intercooler
- Selector cable bracket
- Harness bracket
- Solenoid valve connector
- Transaxle range switch connector
- Output speed sensor connector
- Selector cable clip and nut and the
  cable
- Air cleaner assembly
- Rear air charge cooler duct
- Battery tray, battery and cover

14. Fill the transmission with fluid.

15. Start the engine and check for leaks
and proper operation.

GF4A–EL TRANSAXLE

1. Refer to the illustration for component location and torque specifications.

2. Before servicing the vehicle, refer to
the precautions section.

3. Refer to the illustration for component location and torque specifications.

4. Drain the transaxle oil.

5. Remove or disconnect the following:
- Battery cover, battery and tray
- Air cleaner assembly
- Selector cable clip and nut and the
  cable
- Transaxle range switch connector
- Input/Turbine speed sensor con-
  nector
- Solenoid valve connector
- Harness bracket
- Selector cable bracket
- Number 1 engine mount stay
  bracket
- Starter
- Heated Oxygen (HO$_2$S) connector
- Wheels
- Splash shields
- Exhaust pipe
- Lower ball joint
- Right side halfshaft
- Joint shaft

6. Support the engine assembly with a
engine support assembly such as 49 E017
5A0.
- Number 4 mount
- Number 1 engine mount nut
- Engine mount member
- Oil hose
- Under covers
- Torque converter bolts
- Left side halfshaft
- Number 1 engine mount bolt

7. Loosen the engine support assembly
and lean the engine towards the transaxle.

8. Support the transaxle with a jack,
remove the transaxle bolts and the
transaxle.

To install:

9. Place the transaxle onto a jack and
raise into position.

10. Install the transaxle bolts and tighten
the bolts to 50–73 ft. lbs. (68–99 Nm).

11. Install or connect the following:
- Number 1 engine mount bolt. Make
  sure the holes align before tightening. Tighten the bolt to 77 ft. lbs.
  (104 Nm).
- Left side halfshaft
- Torque converter bolts and tighten
  to 32–44 ft. lbs. (44–60 Nm)
- Under covers
- Oil hose
- Engine mount member and number
  2 engine mount, make sure the
  number 2 mount stud bolt passes
  through the number 2 mount
  bracket installation hole and tighten

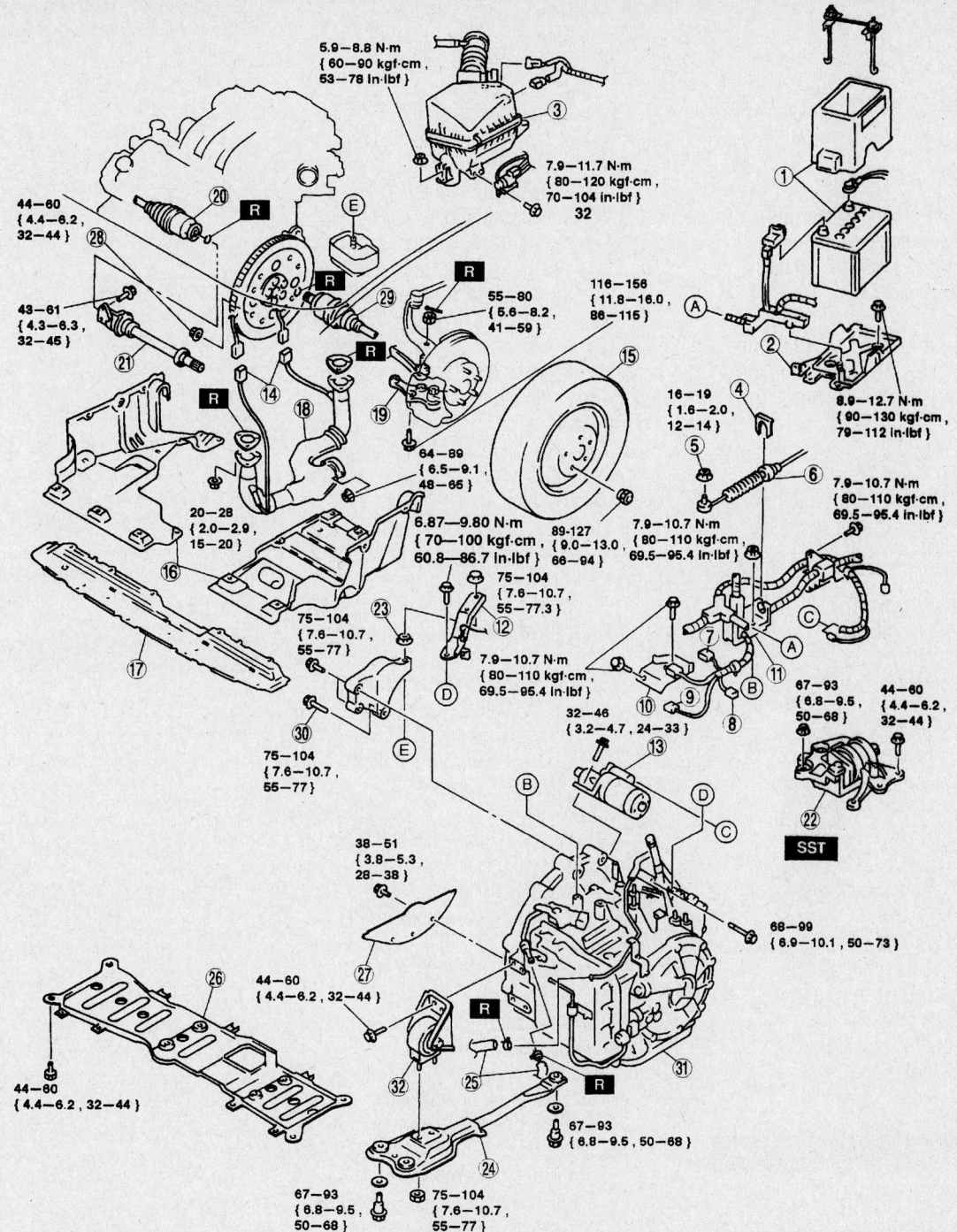

5.9–8.8 N·m
{ 60–90 kgf·cm,
53–78 in·lbf }

7.9–11.7 N·m
{ 80–120 kgf·cm,
70–104 in·lbf }
32

44–60
{ 4.4–6.2,
32–44 }

43–61
{ 4.3–6.3,
32–45 }

55–80
{ 5.6–8.2,
41–59 }

116–156
{ 11.8–16.0,
86–115 }

16–19
{ 1.6–2.0,
12–14 }

8.9–12.7 N·m
{ 90–130 kgf·cm,
79–112 in·lbf }

7.9–10.7 N·m
{ 80–110 kgf·cm,
69.5–95.4 in·lbf }

64–89
{ 6.5–9.1,
48–65 }

20–28
{ 2.0–2.9,
15–20 }

6.87–9.80 N·m
{ 70–100 kgf·cm,
60.8–86.7 in·lbf }

89–127
{ 9.0–13.0,
66–94 }

7.9–10.7 N·m
{ 80–110 kgf·cm,
69.5–95.4 in·lbf }

75–104
{ 7.6–10.7,
55–77.3 }

75–104
{ 7.6–10.7,
55–77 }

7.9–10.7 N·m
{ 80–110 kgf·cm,
69.5–95.4 in·lbf }

67–93
{ 6.8–9.5,
50–68 }

44–60
{ 4.4–6.2,
32–44 }

32–46
{ 3.2–4.7, 24–33 }

SST

75–104
{ 7.6–10.7,
55–77 }

68–99
{ 6.9–10.1, 50–73 }

38–51
{ 3.8–5.3,
28–38 }

44–60
{ 4.4–6.2, 32–44 }

44–60
{ 4.4–6.2, 32–44 }

67–93
{ 6.8–9.5, 50–68 }

67–93
{ 6.8–9.5,
50–68 }

75–104
{ 7.6–10.7,
55–77 }

1	Battery and battery cover	12	No.1 engine mount bracket stay	23	No.1 engine mount nut
2	Battery carrier	13	Starter	24	Engine mounting member
3	Air cleaner component	14	Heated oxygen sensor connector	25	Oil hose
4	Clip	15	Wheel and tire	26	Undercover
5	Nut	16	Splash shield	27	Undercover
6	Selector cable	17	Splash shield	28	Torque converter nut
7	Transaxle range switch connector	18	Front exhaust pipe	29	Drive shaft (left)
8	Input/turbine speed sensor connector	19	Lower ball joint	30	No.1 engine mount bolt
9	Solenoid valve connector	20	Drive shaft (right)	31	Transaxle
10	Harness bracket	21	Joint shaft	32	No.2 engine mount
11	Selector cable bracket	22	No.4 engine mount		

Exploded view of the GF4A–EL automatic transaxle assembly—Millenia

42356-MAZC-GEUV

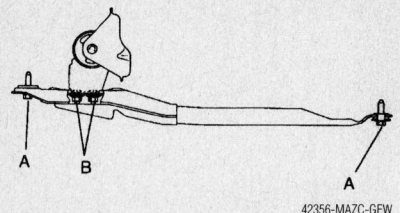

42356-MAZC-GEW

Install the engine mount member bolts and tighten to specification—LJ4A–EL automatic transaxle assembly—Millenia with GF4A–EL automatic transaxle

the bolt **A** to 68 ft. lbs. (93 Nm). Tighten bolt **B** to 77 ft. lbs. (104 Nm). Refer to the illustration for component location.
- Number 1 engine mount nut
- Number 4 mount retainers. Refer to the illustration for component location and torque specifications.
- Joint shaft
- Right side halfshaft
- Lower ball joint
- Exhaust pipe
- Splash shields
- Wheels
- HO2S connector
- Starter
- Number 1 engine mount stay bracket
- Harness bracket
- Solenoid valve connector
- Input/Turbine speed sensor connector
- Transaxle range switch connector
- Selector cable nut and clip
- Air cleaner assembly
- Battery tray, battery and cover

12. Fill the transmission with fluid.
13. Start the engine and check for leaks and proper operation.

Clutch

REMOVAL & INSTALLATION

1. Before servicing the vehicle, refer to the precautions section.
2. Remove or disconnect the following:
- Negative battery cable
- Clutch release cylinder
- Transaxle
- Rubber boot
- Clutch release collar
- Clutch release fork
- Pressure plate loosening the bolts one turn each in a criss–cross pattern
- Clutch disc

3. Inspect the pilot bearing. If it is worn or damaged and does not turn easily by hand, remove it using a puller/slide hammer.
4. Check the flywheel surface for scoring, cracks or burning and machine or replace, as necessary.
5. Install Holder tool 49 E011 1A0 to keep the flywheel from turning. Loosen the flywheel bolts evenly and gradually in a crisscross pattern. Remove the flywheel.
6. Inspect the clutch release bearing for wear. Replace it if it sticks or does not turn easily.
7. Inspect the release fork for wear or damage and replace as necessary.

To install:

8. Lubricate the release fork fingers and pivot with molybdenum grease and install in the release fork boot.
9. Install or connect the following:
- Clutch release bearing on the release fork
- New pilot bearing in the flywheel, if removed, using an installation tool

10. Be sure the flywheel mounting surface and the crankshaft or eccentric shaft mounting surfaces are clean. Remove any old sealant from the flywheel bolt hole threads and the flywheel bolts.
- Flywheel
- Sealant to the flywheel bolt threads and install them hand tight
- Flywheel holding tool. Tighten the bolts, in a crisscross pattern as follows:
  a. Except 2.5L (KL) engines to 71–75 ft. lbs. (97–102 Nm).
  b. On the 2.5L (KL) engines, tighten to 45–49 ft. lbs. (61–66 Nm).

11. Install or connect the following:
- Small amount of molybdenum grease to the clutch disc splines
- Clutch disc on the flywheel with the spring side toward the transaxle
- An alignment tool in the pilot bearing to position the clutch disc
- Clutch pressure plate by aligning the dowel holes with the flywheel dowels
- Pressure plate. Gradually, torque the bolts, in a crisscross pattern, to 19 ft. lbs. (26 Nm).

12. Remove the alignment tool.
- Clutch release fork
- Clutch release collar
- Rubber boot
- Transaxle
- Clutch release cylinder
- Negative battery cable

Hydraulic Clutch System

BLEEDING

1. Before servicing the vehicle, refer to the precautions section.
2. Remove the rubber cap from the bleeder screw on the release cylinder.
3. Place a bleeder tube over the end of the bleeder screw.
4. Submerge the other end of the tube in a jar half filled with hydraulic brake fluid.
5. Slowly pump the clutch pedal fully and allow it to return slowly, several times.
6. While pressing the clutch pedal to the floor, loosen the bleeder screw until the fluid starts to run out. Then, close the bleeder screw. Keep repeating this Step, while watching the hydraulic fluid in the jar. As soon as the air bubbles disappear, close the bleeder screw.
7. During the bleeding procedure the reservoir must be kept at least ¾ full.

Halfshafts

REMOVAL & INSTALLATION

1. Before servicing the vehicle, refer to the precautions section.
2. Drain the transaxle oil.
3. Remove or disconnect the following:
- Wheels
- ABS sensor, if equipped
- Splash shield, if equipped

4. Raise the staked portion of the hub locknut with a hammer and chisel.
5. Lock the hub by applying the brakes and remove the nut.
6. Remove or disconnect the following:
- Stabilizer bar from the lower control arm
- Cotter pin and nut from the tie rod end ball stud
- Tie rod end from the knuckle
- Transverse member, on the 626
- Lower ball joint pinch bolt and nut
- Lower ball joint from the knuckle

7. If removing the left side shaft on 626 with automatic transaxle, proceed as follows:
   a. Suspend the engine using engine support tool 49 G017 5A0.
   b. Remove the engine mount member.
8. Position a prybar between the inner CV-joint and transaxle case. Carefully pry the halfshaft from the transaxle being careful not to damage the oil seal. If equipped with a right side intermediate shaft, insert the prybar between the halfshaft and intermedi-

ate shaft and tap on the bar to uncouple them.

9. Pull outward on the hub/knuckle assembly, push the outer CV-joint stub shaft through the hub and remove the half-shaft. If the halfshaft is stuck in the hub, install the old hub nut to protect the stub shaft threads. Tap on the nut, using only a soft mallet, to remove the halfshaft.

➡**Install plug tool 49 G030 455 into the transaxle after removing the halfshaft, to keep the differential side gear in position. If the gear becomes positioned incorrectly, the differential may have to be removed to realign the gear.**

10. Remove the intermediate shaft, if necessary, by removing the support bearing bolts and pulling the shaft from the transaxle.

To install:

11. Install or connect the following:
   • New circlip on the end of the intermediate shaft, if removed, with the end gap facing upward.
   • Intermediate shaft in the transaxle, being careful not to damage the oil seals
   • Intermediate shaft support bearing bolts. Torque in sequence, to 45 ft. lbs. (61 Nm).

   • New circlip on the end of the halfshaft, with the end gap facing upward
   • Halfshaft into the transaxle, being careful not to damage the oil seal

➡**If equipped, push the halfshaft into the intermediate shaft.**

   • Other end of the halfshaft through the hub. Loosely install a new lock-nut

12. If installing the left side shaft on 626 with automatic transaxle, proceed as follows:
   a. Install the engine mount member.

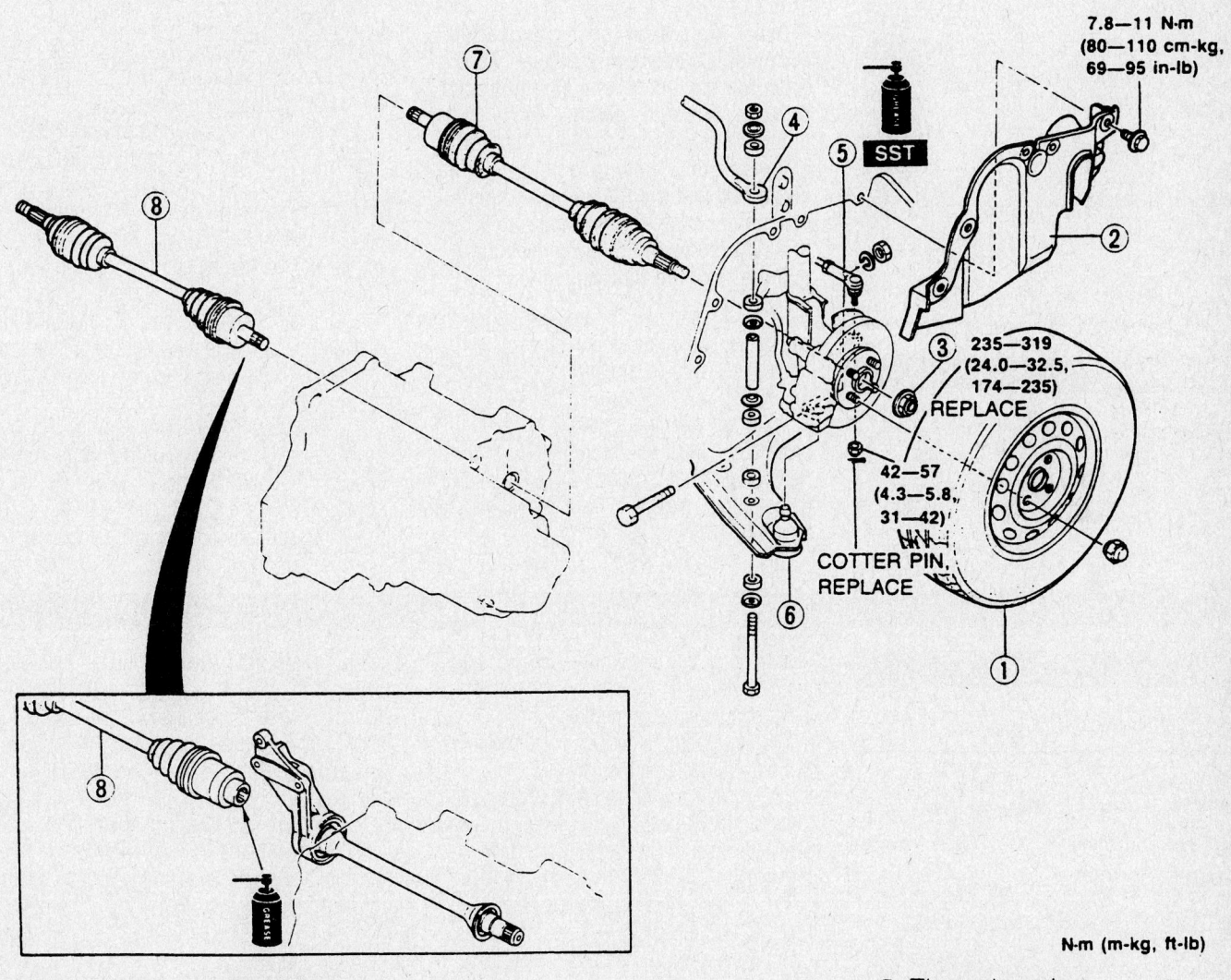

1. Wheel and tire
2. Splash shield
3. Locknut
4. Stabilizer
5. Tie-rod end
6. Lower ball joint
7. Left driveshaft
8. Right driveshaft

7.8—11 N·m (80—110 cm-kg, 69—95 in-lb)

235—319 (24.0—32.5, 174—235) REPLACE

42—57 (4.3—5.8, 31—42)

COTTER PIN, REPLACE

N-m (m-kg, ft-lb)

Exploded view of a typical halfshaft mounting

7923MG76

Torque the mount member-to-body bolts to 66 ft. lbs. (89 Nm).

b. Torque the front mount-to-mount member nuts to 77 ft. lbs. (104 Nm) and the side mount bolts to 44 ft. lbs. (60 Nm).

c. Remove the engine support tool.

13. Install or connect the following:
- Lower ball joint into the knuckle. Torque the pinch bolt to 40 ft. lbs. (54 Nm).
- Transverse member, on 626. Torque the bolts to 96 ft. lbs. (132 Nm).
- Tie rod end to the steering knuckle. Torque the nut to 42 ft. lbs. (57 Nm) on all except 626 and Millenia or to 32 ft. lbs. (44 Nm) for 626 and Millenia.
- New cotter pin. Tighten the nut, if necessary, to align the ball stud hole with the nut castellation.
- Stabilizer bar to the lower control arm
- Splash shield
- Wheels
- New hub nut. Torque it to 174–235 ft. lbs. (235–318 Nm). After tightening, stake the locknut using a hammer and dull bladed chisel.

14. Fill the transaxle.

CV-Joints

OVERHAUL

1. Before servicing the vehicle, refer to the precautions section.

Two types of CV-joints are used. The inboard CV-joints are the tri-Pot type. All outboard CV-joints are Birfield type. The Birfield CV-joint cannot be disassembled; if an outboard CV-joint boot needs replacement, the inboard CV-joint must be removed. If the outboard CV-joint needs to be replaced, replace the entire halfshaft as an assembly.

2. Remove the halfshaft from the vehicle and clamp it in a vise equipped with jaw caps, to prevent damage to the machined surfaces. Do not allow the vise to contact the boot or its clamps.

3. Remove the large boot clamp from the inboard CV-joint, using side cutters. After removing the clamp, roll the boot back over the shaft.

➡Check the grease for contamination by rubbing it between 2 fingers. Any gritty feeling indicates a contaminated CV-joint, in which case the entire CV-joint must be disassembled, cleaned and inspected. If the grease is not contaminated and the CV-joint has been operating satisfactorily, continue with the boot replacement procedure and add the required lubricant.

4. Paint alignment marks on the outer race and shaft for assembly reference. Remove the wire ring bearing retainer and remove the outer race.

5. Paint alignment marks on the tri-pot bearing and shaft for assembly reference. Remove the tri-pot bearing snapring and, using a brass drift and hammer, remove the tri-pot bearing from the shaft.

6. Remove the small clamp and remove the inner boot from the halfshaft. If the boot is to be reused, wrap the shaft splines with tape before removing.

7. If the outer CV-joint boot is to be replaced, remove the clamps and slide the boot off the shaft from the inboard side.

To install:

8. If the outboard boot was removed, slide the boot onto the shaft from the inboard side. Wrap tape on the splines before installing to protect the boot.

9. Install the inboard boot and remove the tape from the shaft.

10. Install the tri-pot assembly on the halfshaft. Tap the assembly onto the shaft using a hammer and brass drift. Install the tri-pot assembly retaining ring.

11. Fill the CV-joint outer race with high temperature CV-joint grease. Install the outer race over the tri-pot joint and install the wire ring bearing retainer.

12. Position the CV-joint boot(s). Make sure the boot is fully seated in the grooves in the shaft and outer race.

13. Insert a small prybar with rounded edges between the boot and the outer bearing race to allow trapped air to escape from the boot. Install new boot clamps.

14. Wrap the clamps around the boots in a clockwise direction, pull tight with pliers and bend the locking tabs to secure in position.

15. Work the CV-joint through its full range of travel at various angles. The joint should flex, extend and compress smoothly.

16. Install the halfshaft into the vehicle.

STEERING AND SUSPENSION

Air Bag

PRECAUTIONS

Several precautions must be observed when handling the inflator module to avoid accidental deployment and possible personal injury.

1. Never carry the inflator module by the wires or connector on the underside of the module.

2. When carrying a live inflator module, hold securely with both hands, and ensure that the bag and trim cover are pointed away.

3. Place the inflator module on a bench or other surface with the bag and trim cover facing up.

4. With the inflator module on the bench, never place anything on or close to the module which may be thrown in the event of an accidental deployment.

5. An air bag is an explosive device. Handle with extreme caution.

6. Always disconnect the battery and the air bag connector before removing the steering wheel or beginning work on the air bag system.

7. Air bag components must not be repaired or opened. Always use new parts, including the wiring harness.

8. Always place a removed air bag unit with the horn pad facing up. Put it in a safe place where it will not be disturbed.

9. The air bag unit must not be exposed to grease, fluids, or cleaning agents.

10. The air bag unit must not be exposed to temperatures above 194°F

(90°C) at any time. Even the heat of a soldering iron can damage or ignite the charge.

11. Storage and transport of air bags is subject to rules governing explosive devices and should be done only in the original package.

12. Failure to follow proper safety precautions may result in personal injury through accidental firing of the air bag, or through failure of the air bag in an accident.

DISARMING

1. Before servicing the vehicle, refer to the precautions section.

2. If equipped, deactivate the audio anti-theft system.

3. Turn the ignition switch to LOCK.

4. Disconnect and isolate the negative battery cable and wait for more than 1

minute to allow the backup power supply to deplete its stored power.

ARMING

1. Before servicing the vehicle, refer to the precautions section.
2. Connect the negative battery cable, turn the ignition switch **ON** and verify the air bag warning light cones on for 6 seconds. If the light does not illuminate there are problems with the system.
3. If equipped, activate the audio anti-theft system.

Rack and Pinion Steering Gear

REMOVAL & INSTALLATION

Manual

2002–04 MIATA

1. Before servicing the vehicle, refer to the precautions section.
2. Remove or disconnect the following:
   - Negative battery cable
   - Front wheels
   - Cotter pins and nuts from both steering tie rod ends
3. Press the tie rod out from the knuckle arm.
   - Intermediate shaft to steering gear pinion shaft bolt. Mark the shaft-to-gear location.
   - Shaft from the steering gear
   - Steering gear mounting nuts
   - Steering gear and linkage from the vehicle

To install:

4. Install or connect the following:
   - Steering gear and linkage to the vehicle. Torque the bolts in sequence to 55–77 ft. lbs. (75–104 Nm).
   - Steering shaft to the steering gear pinion shaft, align the marks made during removal and tighten the bolt to 19 ft. lbs. (26 Nm)
   - Tie rod ends to the knuckle arm. Torque the nuts to 32–41 ft. lbs. (43–56 Nm).
   - New cotter pins
   - Wheels
   - Negative battery cable
5. Check and/or adjust the front end alignment.

Power

2002–04 MIATA

1. Before servicing the vehicle, refer to the precautions section.

2. Remove or disconnect the following:
   - Negative battery cable
   - Front wheels
   - Cotter pins and nuts from both steering tie rod ends
3. Press the tie rod out from the knuckle arm.
   - Intermediate shaft to steering gear pinion shaft bolt. Mark the shaft-to-gear location.
   - Shaft from the steering gear
   - Pressure and return lines
   - Steering gear mounting nuts
   - Steering gear and linkage from the vehicle

To install:

4. Install or connect the following:
   - Steering gear and linkage to the vehicle. Torque the bolts in sequence to 55–77 ft. lbs. (75–104 Nm).
   - Steering shaft to the steering gear pinion shaft, align the marks made during removal and tighten the bolt to 19 ft. lbs. (26 Nm)
   - Pressure and return lines. Tighten the pressure line fitting to 34 ft. lbs. (47 Nm).
   - Tie rod ends to the knuckle arm. Torque the nuts to 32–41 ft. lbs. (43–56 Nm).
   - New cotter pins
   - Wheels
   - Negative battery cable
5. Check and/or adjust the front end alignment.

626

1. Before servicing the vehicle, refer to the precautions section.
2. Remove or disconnect the following:
   - Negative battery cable
   - Front wheels
   - Cotter pins and nuts from both steering tie rod ends
3. Press the tie rod out from the knuckle arm.
4. Remove or disconnect the following:
   - Transverse member
   - Front exhaust pipe on 2.5L (KL) engines
5. Support the engine assembly with a engine support assembly such as 49 E017 5A0.
   - Number 1 engine mount
   - Pressure line and return pipe from the steering gear
   - Intermediate shaft to steering gear pinion shaft bolt. Mark the shaft-to-gear location.
   - Shaft from the steering gear
   - Steering gear mounting nuts

- Steering gear from the right of the vehicle

To install:

6. Install or connect the following:
   - Steering gear to the vehicle. Torque the nuts/bolts in sequence to 28–38 ft. lbs. (37–52 Nm).
   - Steering shaft to the steering gear pinion shaft, align the marks made during removal and tighten the bolt to 19 ft. lbs. (26 Nm)
   - Pressure line and return hose to the steering gear
   - Number 1 engine mount and remove the engine support
   - Front exhaust pipe on 2.5L (KL) engines
   - Transverse member
   - Tie rod ends to the knuckle arm. Torque the nuts to 24–32 ft. lbs. (32–44 Nm).
   - New cotter pins and check the power steering fluid level
   - Wheels
   - Negative battery cable
7. Check and/or adjust the front end alignment.

PROTÉGÉ

1. Before servicing the vehicle, refer to the precautions section.
2. Remove or disconnect the following:
   - Negative battery cable
   - Front wheels
   - Intermediate shaft to steering gear pinion shaft bolt. Mark the shaft-to-gear location.
   - Cotter pins and nuts from both steering tie rod ends
3. Press the tie rod out from the knuckle arm.
   - Transverse member, if necessary on some models to facilitate removal
4. Support the engine assembly with a suitable engine support assembly.
   - Engine mount member
   - Pressure line, return pipe and clamp from the steering gear
   - Oil pipe
   - Shaft from the steering gear
   - Steering gear mounting nuts
   - Steering gear from the right of the vehicle

To install:

5. Install or connect the following:
   - Steering gear to the vehicle. Torque the nuts/bolts 55–77 ft. lbs. (75–104 Nm).
   - Steering shaft to the steering gear pinion shaft, align the marks made during removal
   - Oil pipe

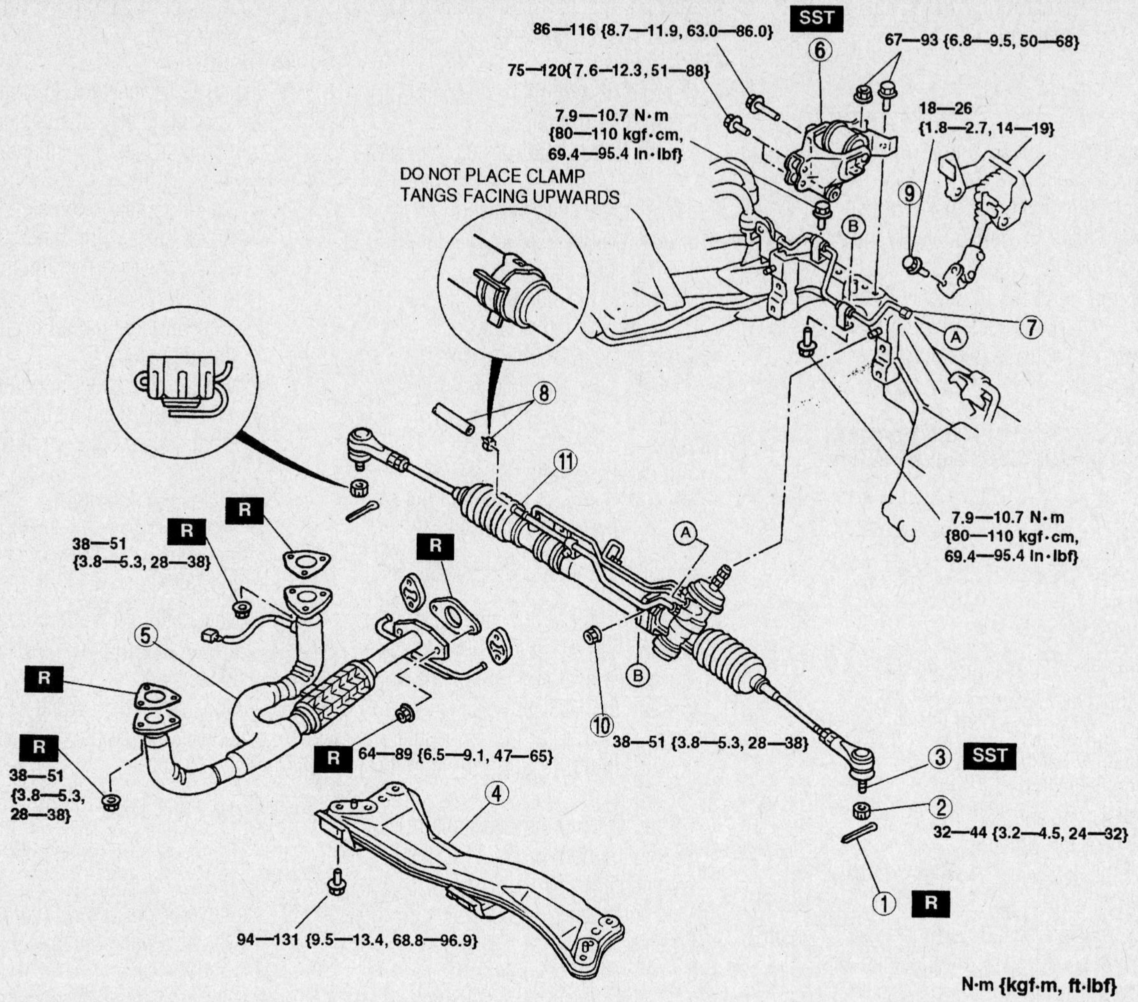

86—116 {8.7—11.9, 63.0—86.0}

75—120{ 7.6—12.3, 51—88}

67—93 {6.8—9.5, 50—68}

7.9—10.7 N·m
{80—110 kgf·cm,
69.4—95.4 in·lbf}

18—26
{1.8—2.7, 14—19}

DO NOT PLACE CLAMP
TANGS FACING UPWARDS

38—51
{3.8—5.3, 28—38}

7.9—10.7 N·m
{80—110 kgf·cm,
69.4—95.4 in·lbf}

38—51
{3.8—5.3,
28—38}

64—89 {6.5—9.1, 47—65}

38—51 {3.8—5.3, 28—38}

32—44 {3.2—4.5, 24—32}

94—131 {9.5—13.4, 68.8—96.9}

N·m {kgf·m, ft·lbf}

1	Cotter pin	7	Pressure pipe
2	Nut	8	Return hose and clamp
3	Tie-rod end ball joint	9	Bolt (intermediate shaft)
4	Transverse member	10	Mounting bracket nut and bolt
5	Front exhaust pipe (KL engine)	11	Steering gear and linkage
6	No.1 engine mount component		

42356-MAZC-G81

Exploded view of the power steering gear assembly–626

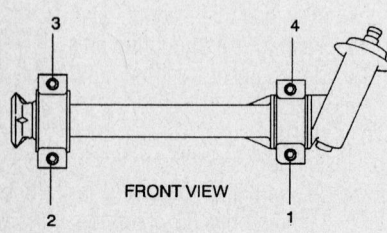

FRONT VIEW

42356-MAZC-G82

Tighten the power steering gear assembly bracket retainers in this sequence–626

- Pressure line, return pipe and clamp to the steering gear
- Engine mount member and remove the engine support
- Transverse member, if removed and tighten the bolts to 68–97 ft. lbs. (94–131 Nm)
- Tie rod ends to the knuckle arm. Torque the nuts to 24–32 ft. lbs. (32–44 Nm).
- New cotter pins
- Tighten the intermediate shaft bolt to 19 ft. lbs. (26 Nm). Check the power steering fluid level.
- Wheels

- Negative battery cable
6. Check and/or adjust the front end alignment.

MILLENIA

1. Before servicing the vehicle, refer to the precautions section.
2. Remove or disconnect the following:
  - Negative battery cable
  - Front wheels
  - Cotter pins and nuts from both steering tie rod ends
3. Press the tie rod out from the knuckle arm.
4. Remove or disconnect the following:

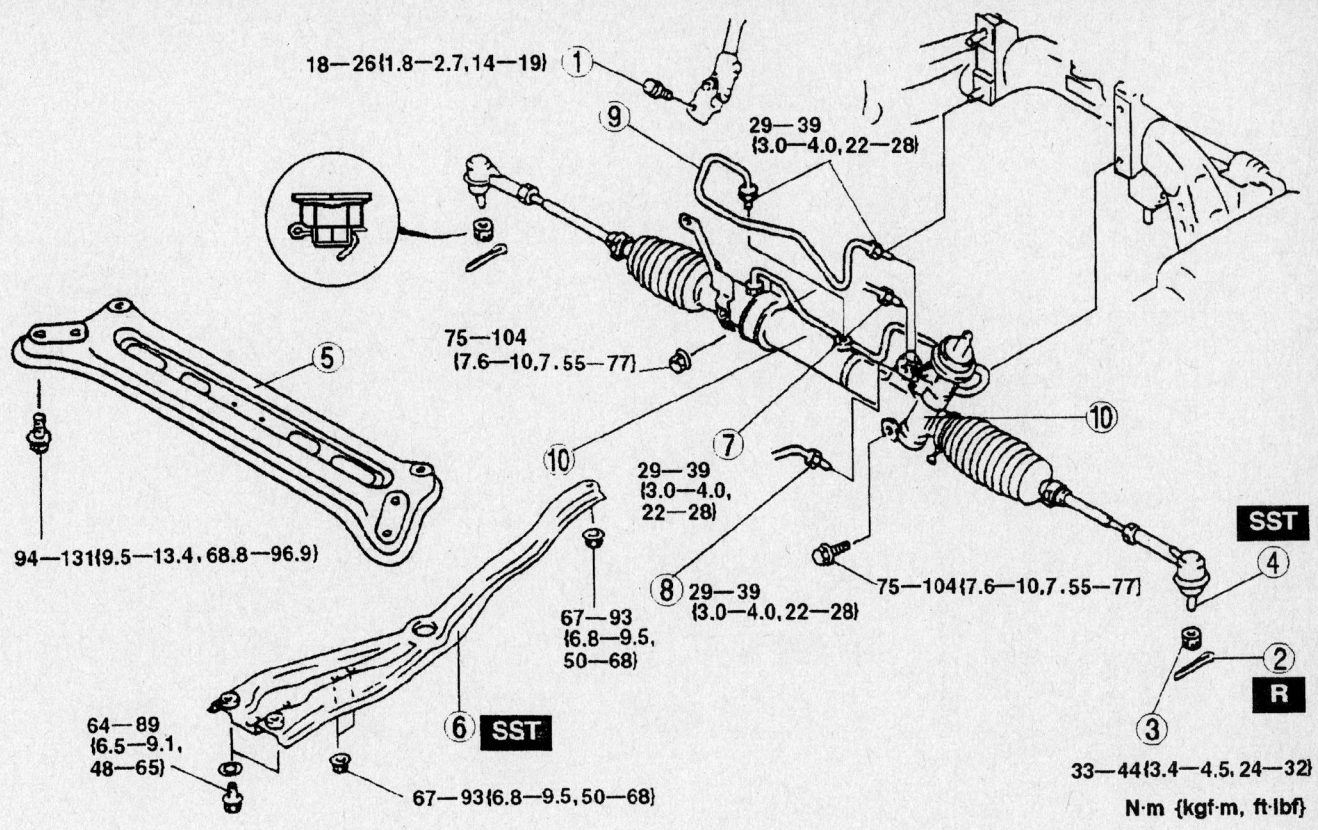

18—26{1.8—2.7,14—19}

29—39 {3.0—4.0, 22—28}

75—104 {7.6—10.7, 55—77}

94—131{9.5—13.4, 68.8—96.9}

29—39 {3.0—4.0, 22—28}

67—93 {6.8—9.5, 50—68}

29—39 {3.0—4.0, 22—28}

75—104{7.6—10.7, 55—77}

SST

R

64—89 {6.5—9.1, 48—65}

67—93{6.8—9.5, 50—68}

SST

33—44{3.4—4.5, 24—32}

N·m {kgf·m, ft·lbf}

1	Bolt (intermediate shaft)	6	Engine mount member
2	Cotter pin	7	Pressure pipe
3	Nut	8	Return pipe and clamp
4	Tie-rod end ball joint	9	Oil pipe
5	Transverse member (ZM (ATX), FS)	10	Steering gear and linkage

42356-MAZC-GKA

Exploded view of the power steering gear assembly–Protégé

- Transverse member
- Pressure line and return pipe from the steering gear
- Intermediate shaft to steering gear pinion shaft bolt. Mark the shaft-to-gear location.

5. Support the engine assembly with a engine support assembly such as 49 E017 5A0.

- Engine mount member
- Number 1 engine mount bolts
- Upper lateral link bolt and nut
- Lower arm bolt on the crossmember side
- Stabilizer bar link
- Crossmember. Support with a jack before removing the bolts and nuts.
- Shaft from the steering gear
- Steering gear mounting nuts
- Steering gear

To install:

6. Install or connect the following:
   - Steering gear to the vehicle. Torque the nuts/bolts in sequence to 23–52 ft. lbs. (44–70 Nm).
   - Steering shaft to the steering gear pinion shaft, align the marks made during removal and hand tighten the bolt
   - Crossmember. Support with a jack before and tighten bolts and nuts to the specifications listed in the illustration.
   - Stabilizer bar link
   - Lower arm bolt on the crossmember side
   - Upper lateral link bolt and nut
   - Number 1 engine mount bolts
   - Engine mount member

7. Remove the engine support assembly.

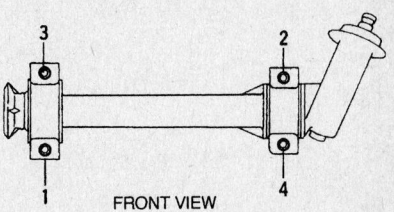

FRONT VIEW

42356-MAZC-GEZ

Tighten the power steering gear assembly bracket retainers in this sequence–Millenia

- Align the marks made during removal of the steering shaft to the steering gear pinion shaft bolt and tighten the bolt to 19 ft. lbs. (26 Nm)
- Pressure line and return pipe to the steering gear
- Transverse member

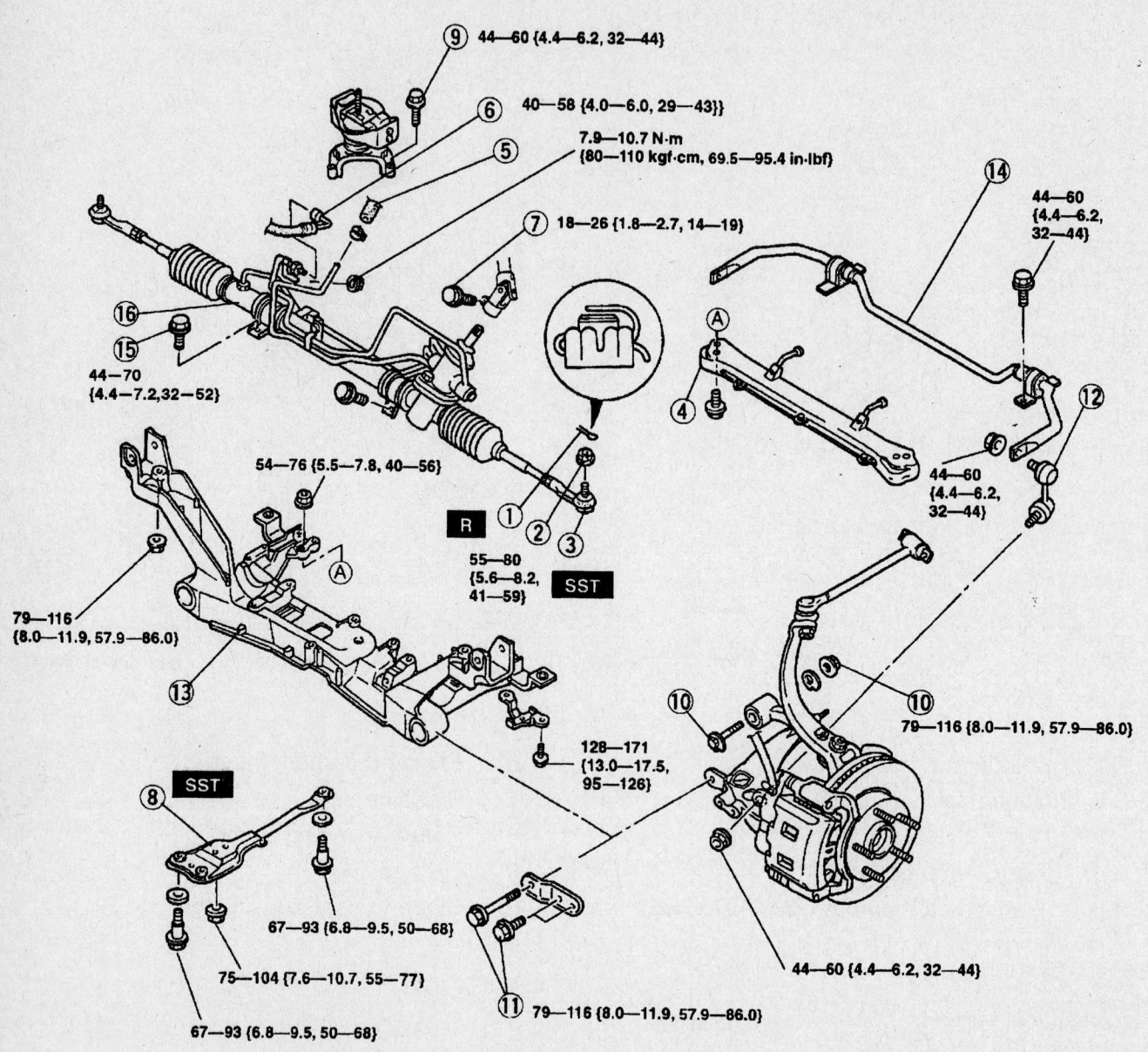

44—60 {4.4—6.2, 32—44}

40—58 {4.0—6.0, 29—43}}

7.9—10.7 N·m
{80—110 kgf·cm, 69.5—95.4 in·lbf}

18—26 {1.8—2.7, 14—19}

44—60
{4.4—6.2,
32—44}

44—70
{4.4—7.2, 32—52}

54—76 {5.5—7.8, 40—56}

79—116
{8.0—11.9, 57.9—86.0}

R

55—80
{5.6—8.2,
41—59}

SST

44—60
{4.4—6.2,
32—44}

79—116 {8.0—11.9, 57.9—86.0}

128—171
{13.0—17.5,
95—126}

SST

67—93 {6.8—9.5, 50—68}

75—104 {7.6—10.7, 55—77}

67—93 {6.8—9.5, 50—68}

79—116 {8.0—11.9, 57.9—86.0}

44—60 {4.4—6.2, 32—44}

N·m {kgf·m, ft·lbf}

1	Cotter pin	9	Bolts (engine mount No.1)
2	Nut	10	Bolt and nut (upper lateral link)
3	Tie-rod end ball joint	11	Lower arm bolt (crossmember side)
4	Transverse member	12	Stabilizer control link
5	Return hose	13	Crossmember
6	Pressure pipe	14	Stabilizer
7	Bolt (intermediate shaft)	15	Mounting bracket bolts
8	Engine mount member	16	Steering gear and linkage

42356-MAZC-GEY

Exploded view of the power steering gear assembly-Millenia

- Tie rods to the knuckle arm. Install the nuts and tighten to 59 ft. lbs. (80 Nm) and install new cotter pins.
- Front wheels
- Negative battery cable

8. Check and/or adjust the front end alignment.

Strut

REMOVAL & INSTALLATION

Front

2002–04 MIATA

1. Before servicing the vehicle, refer to the precautions section.
2. Support the lower control arm with a jack.
3. Remove or disconnect the following:

- Front wheel
- Stabilizer bar nut
- Lower arm ball joint from the knuckle
- Lower arm bolts
- Lower the lower arm and remove the shock. Be careful not to lower the arms too much or it may cause damage.

To install:
4. Installation is the reverse of removal. Tighten the upper shock nuts to 26 ft. lbs. (36 Nm) and the lower shock nut and bolt to 86 ft. lbs. (116 Nm).
5. Check and/or adjust the front end alignment.

626

1. Before servicing the vehicle, refer to the precautions section.
2. Support the lower control arm with a jack.
3. Remove or disconnect the following:

- Front wheel
- Brake hose and/or ABS sensor harness to the strut Bolts or clips
- Actuator from the top of the strut, if equipped with Automatic Adjusting Suspension (AAS)

4. Paint alignment marks on the upper strut mounting block and strut tower, and on the lower strut mount-to-steering knuckle so the strut can be reinstalled in the same position.
5. Remove or disconnect the following:

- Upper strut mounting nuts
- Strut-to-knuckle bolts
- Strut assembly

To install:
6. Install or connect the following:

- Strut into the strut tower, aligning the paint marks made during removal
- Upper mounting nuts and tighten to 34–46 ft. lbs. (47–62 Nm)
- Strut-to-knuckle bolts and tighten to 55–61 ft. lbs. (74–83 Nm)
- Actuator and engage the electrical connector, if equipped with AAS
- Brake hose and/or ABS sensor harness clips or bolts
- Wheel

7. Check and/or adjust the front end alignment.

PROTÉGÉ

1. Before servicing the vehicle, refer to the precautions section.
2. Support the lower control arm with a jack.
3. Remove or disconnect the following:

- Front wheel
- Brake hose and/or ABS sensor harness to the strut Bolts or clips
- Stabilizer link nut
- Strut-to-knuckle bolt
- Upper strut mounting nuts
- Stiffener, if equipped
- Sheet
- Strut assembly

To install:
4. Install or connect the following:

- Face the mounting block direction indicator towards the rear outboard position and install the shock assembly.
- Sheet
- Stiffener, if equipped
- Upper mounting nuts and tighten to 34–46 ft. lbs. (47–62 Nm)
- Strut-to-knuckle bolts and tighten to 68–93 ft. lbs. (94–126 Nm)
- Stabilizer link nut and tighten to 32–44 ft. lbs. (47–60 Nm)
- Brake hose and/or ABS sensor harness clips or bolts
- Wheel

5. Check and/or adjust the front end alignment.

MILLENIA

1. Before servicing the vehicle, refer to the precautions section.
2. Support the lower control arm with a jack.
3. Remove or disconnect the following:

- Front wheel
- Brake hose and/or ABS sensor harness to the strut bolts or clips

4. Paint alignment marks on the upper strut mounting block and strut tower, and on the lower strut mount-to-steering knuckle so the strut can be reinstalled in the same position.
5. Remove or disconnect the following:

- Upper strut mounting nuts
- Stiffener
- Strut-to-knuckle bolts
- Strut assembly

To install:
6. Install or connect the following:

- Stiffener. Make sure the word **LH or RH** faces up and the parting area faces the inside of the vehicle.
- Strut into the strut tower, aligning the paint marks made during removal
- Upper mounting nuts and tighten to 38–49 ft. lbs. (51–67 Nm)
- Strut-to-knuckle bolts and tighten to 73–101 ft. lbs. (99–137 Nm)
- Brake hose and/or ABS sensor harness clips or bolts
- Wheel

7. Check and/or adjust the front end alignment.

Rear

2002–04 MIATA

1. Before servicing the vehicle, refer to the precautions section.
2. Support the lower control arm with a jack.
3. Remove or disconnect the following:

- Rear wheel
- Stabilizer bar nut
- Lower arm ball joint from the knuckle
- Lower arm bolts
- Lower the lower and upper arms and remove the shock. Be careful not to lower the arms too much or it may cause damage.

To install:
4. Installation is the reverse of removal. Tighten the upper shock nuts to 26 ft. lbs. (36 Nm) and the lower shock nut and bolt to 70 ft. lbs. (95 Nm).
5. Check and/or adjust the front end alignment.

626

1. Before servicing the vehicle, refer to the precautions section.
2. Remove or disconnect the following:

- Rear wheel(s)
- Rear package trim
- Speed sensor harness
- Top strut nuts

➡**The suspension will drop when the weight lifts off the wheels.**

- Bottom strut mount bolt(s)
- Strut assembly

To install:

3. Install or connect the following:
- Strut assembly
- Bottom strut mount bolt(s) and tighten to 55–61 ft. lbs. (74–83 Nm)
- Upper mounting nuts and tighten to 33–46 ft. lbs. (47–62 Nm)
- Speed sensor harness
- Rear package trim
- Rear wheel(s)

PROTÉGÉ

1. Before servicing the vehicle, refer to the precautions section.
2. Remove or disconnect the following:
- Rear wheel(s)
- Speed sensor harness
- Rear seat belt on 4SD models or trunk side trim on the 5HB.
- Clip and brake hose
- Stabilizer link nut
- Bottom strut mount bolt
- Cap
- Top strut nuts

➡**The suspension will drop when the weight lifts off the wheels.**

- Strut assembly

To install:

3. Install or connect the following:
- Strut assembly
- Upper mounting nuts and tighten to 34–46 ft. lbs. (47–62 Nm)
- Cap
- Bottom strut mount bolt and tighten to 68–93 ft. lbs. (94–126 Nm)
- Stabilizer link nut and tighten to 32–44 ft. lbs. (44–60 Nm)
- Brake hose and clip
- Rear seat belt on 4SD models or trunk side trim on the 5HB.
- Speed sensor harness
- Rear wheel(s)

MILLENIA

1. Before servicing the vehicle, refer to the precautions section.
2. Remove or disconnect the following:
- Rear wheel(s)
- Rear package front trim
- Speed sensor harness
- Lower lateral link ball joint
- Top strut nuts
- Stiffener

➡**The suspension will drop when the weight lifts off the wheels.**

- Bottom strut mount bolt(s)
- Strut assembly

To install:

3. Install or connect the following:
- Stiffener. Make sure the word **OUT** faces up and the parting area faces the inside of the vehicle.
- Strut assembly
- Bottom strut mount bolt(s) and tighten to 76–101 ft. lbs. (102–137 Nm)
- Upper mounting nuts and tighten to 34–46 ft. lbs. (47–62 Nm)
- Lower lateral link ball joint, tighten the nut to 115 ft. lbs. (156 Nm) and install a new cotter pin
- Speed sensor harness
- Rear package front trim
- Rear wheel(s)

Coil Spring

REMOVAL & INSTALLATION

626 and Millenia

1. Before servicing the vehicle, refer to the precautions section.

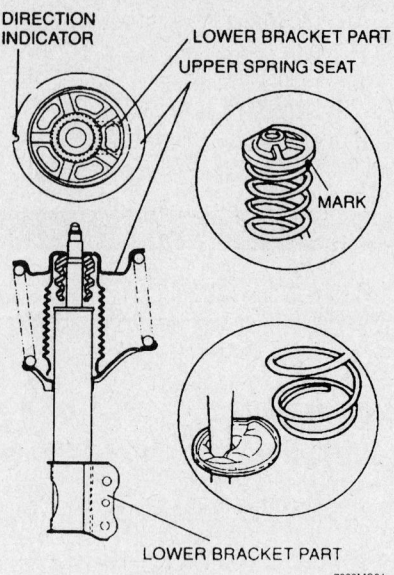

Be sure the end of the coil spring is in the step of the lower seat—626

1. Cap
2. Piston rod nut
3. Mounting rubber
4. Thrust bearing
5. Upper spring seat
6. Upper rubber spring seat
7. Dust cover
8. Bound stopper

9. Coil spring
10. Lower rubber spring seat
11. Shock absorber

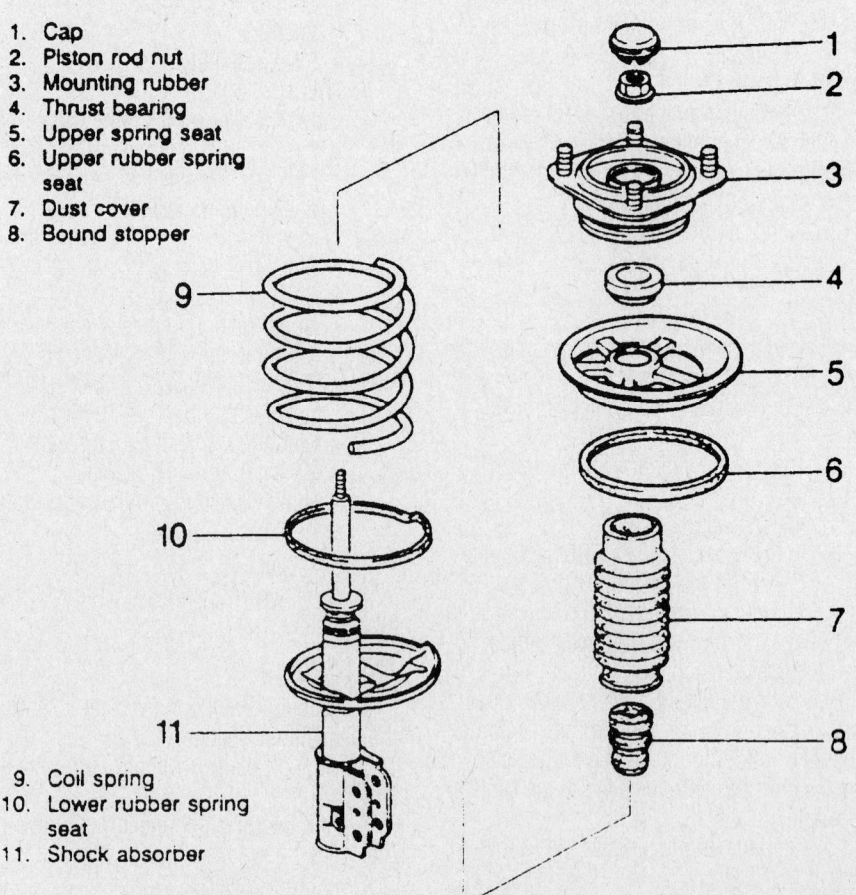

Exploded view of the front strut assembly—rear strut is similar

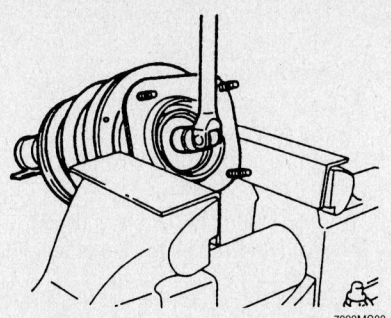

Secure the upper strut mount in a vise and loosen the piston rod nut one turn but do not remove it

2. Remove or disconnect the following:
- Strut from the vehicle
- Cap from the top of the strut, if not equipped with Automatic Adjusting Suspension (AAS)
- Piston rod upper nut 1 turn but do not remove it

3. Place the lower end of the strut in the vise.

4. Install a coil spring compressor and compress the coil spring.

✳✳ CAUTION

Failure to fully compress the spring and hold it securely can be extremely dangerous.

5. Remove or disconnect the following:
- Upper strut nut
- Slowly release the coil spring tension
- Suspension support, dust seal, spring seat, spring insulators, coil spring and bumper

6. While pushing on the piston rod, be sure that the pull stroke is even and that there is no unusual noise or resistance.

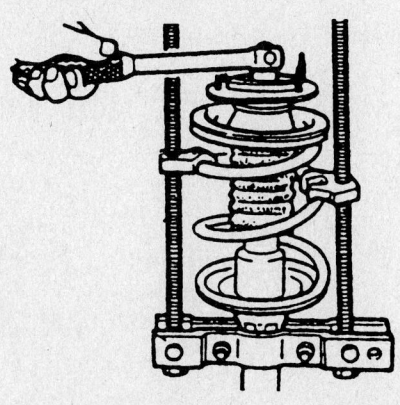

Use a coil spring compressor and relieve the spring tension from the upper mount, then remove the piston rod nut

Also inspect for any oil leakage around the piston rod.

7. Push the piston rod in, then release it. Be sure that the return rate is constant.

8. If the shock absorber does not operate as described, replace it.

To assemble:

9. Install or connect the following:
- Strut assembly into a vise
- Bump stopper and dust boot onto the piston rod
- Temporarily install the upper spring seat, seat rubber and spring. Mark the seat, shock and spring assembly as illustrated for reassembly. Align the marks of the upper seat and coil spring. Protect the assembly with cloth and install the spring compressor.
- Coil spring

10. Compress the coil spring with the spring compressor

11. Install or connect the following:
- Rubber seat, the spring upper seat, the bearing and the mounting block
- Piston rod upper nut

12. Be sure that the spring upper seat notched portion is facing inward and tighten the piston rod upper nut.

13. Remove the spring compressor from the strut. Secure the upper mounting block in the vise and tighten the nut to 40–59 ft. lbs. (55–80 Nm) on 626 models and 24–33 ft. lbs. (32–46 ft. lbs. on Millenia models.

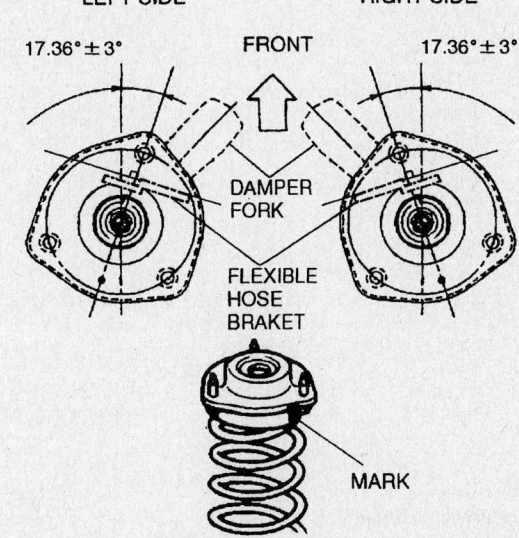

Mark the front seat, shock and spring assembly as illustrated for reassembly–Millenia

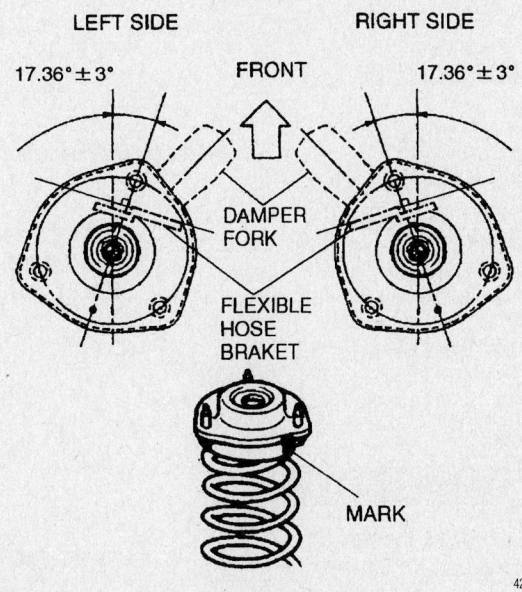

Mark the rear seat, shock and spring assembly as illustrated for reassembly–Millenia

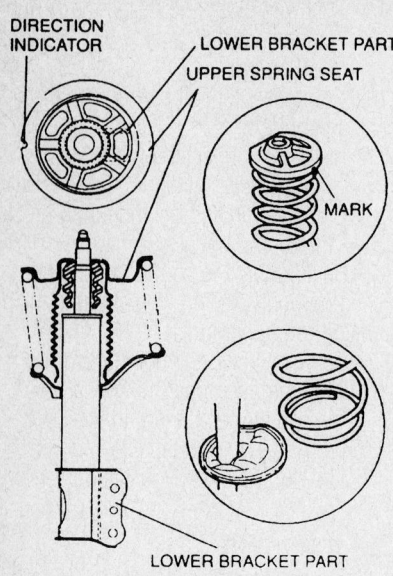

Be sure the end of the coil spring is in the step of the lower seat—Protégé

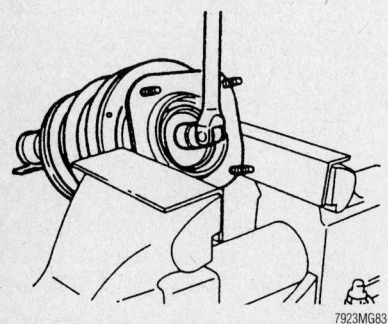

Secure the upper strut mount in a vise and loosen the piston rod nut one turn but do not remove it—Protégé

14. Be sure that the spring is well seated in the upper seats.

15. Install the strut to the vehicle.

Protégé

1. Before servicing the vehicle, refer to the precautions section.

2. Remove or disconnect the following:
   • Strut from the vehicle
   • Piston rod upper nut 1 turn but do not remove it

3. Place the lower end of the strut in the vise.

4. Install a coil spring compressor and compress the coil spring.

✳✳ CAUTION

Failure to fully compress the spring and hold it securely can be extremely dangerous.

5. Remove or disconnect the following:
   • Upper strut nut
   • Slowly release the coil spring tension
   • Suspension support, dust seal, spring seat, spring insulators, coil spring and bumper

6. While pushing on the piston rod, be sure that the pull stroke is even and that there is no unusual noise or resistance. Also inspect for any oil leakage around the piston rod.

7. Push the piston rod in, then release it. Be sure that the return rate is constant.

8. If the shock absorber does not operate as described, replace it.

To assemble:

9. Install or connect the following:
   • Strut assembly into a vise
   • Bump stopper and dust boot onto the piston rod
   • Temporarily install the upper spring seat, seat rubber and spring so that the lower end of the spring is seated on the step of the lower spring seat. Mark the seat, shock and spring assembly as illustrated for reassembly. Align the marks of

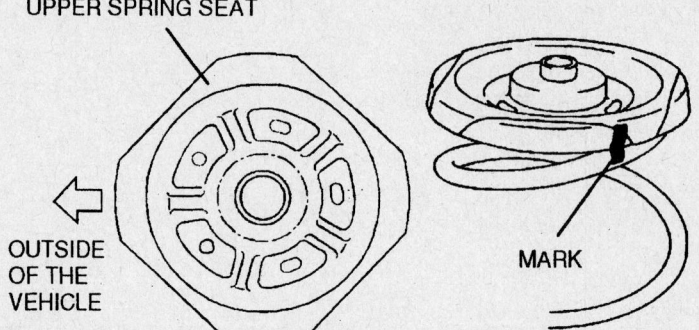

Mark the front seat, shock and spring assembly as illustrated for reassembly–Protégé

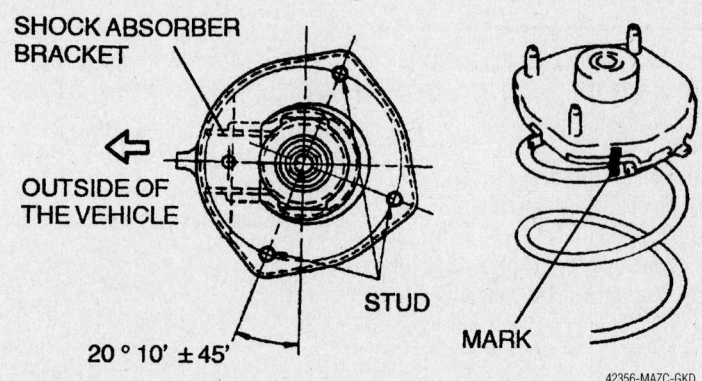

Mark the rear seat, shock and spring assembly as illustrated for reassembly–Protégé

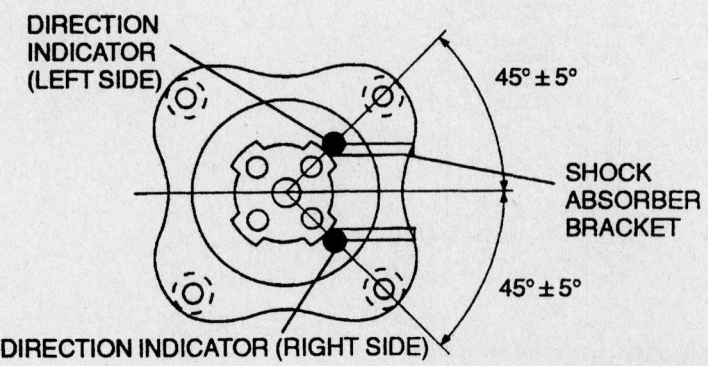

Make sure the bearing, rubber and nut are aligned as shown on the front suspension.—Protégé

the upper seat and coil spring. Protect the assembly with cloth and install the spring compressor.
- Coil spring

10. Compress the coil spring with the spring compressor

11. Install or connect the following:
- Rubber seat, the spring upper seat, the bearing and the mounting block
- Piston rod upper nut

12. Make sure that the marks on the shock absorber and upper spring seat are aligned. Tighten the piston rod upper nut.

13. Make sure the bearing, rubber and nut are aligned as shown on the front suspension.

14. Remove the spring compressor from the strut. Secure the upper mounting block in the vise and tighten the nut to 58–81 ft. lbs. (79–109 Nm) on the front strut and 41–49 ft. lbs. (55–67 ft. lbs. on the rear strut.

15. Be sure that the spring is well seated in the upper seats.

16. Install the strut to the vehicle.

2002–04 Miata

1. Before servicing the vehicle, refer to the precautions section.

2. Remove or disconnect the following:
- Strut from the vehicle
- Cap from the top of the strut, if equipped
- Piston rod upper nut 1 turn but do not remove it

3. Place the lower end of the strut in the vise.

4. Install a coil spring compressor and compress the coil spring.

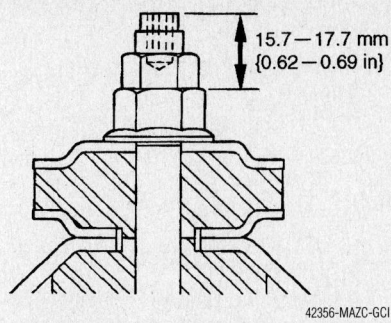

42356-MAZC-GCI

Tighten the lower nut so that 0.62–0.69–17.7mm) of the rod is exposed–2002–04 Miata

✳ CAUTION

Failure to fully compress the spring and hold it securely can be extremely dangerous.

5. Remove or disconnect the following:
- Upper strut nut
- Slowly release the coil spring tension
- Suspension support, dust seal, spring seat, spring insulators, coil spring and bumper

6. While pushing on the piston rod, be sure that the pull stroke is even and that there is no unusual noise or resistance. Also inspect for any oil leakage around the piston rod.

7. Push the piston rod in, then release it. Be sure that the return rate is constant.

8. If the shock absorber does not operate as described, replace it.

To assemble:

9. Install or connect the following:
- Strut assembly into a vise
- Bump stopper and dust boot onto the piston rod. Make sure the lower end of the stopper does not contact the cylinder.
- Temporarily install the upper spring seat, seat rubber and spring. Mark the seat, shock and spring assembly as illustrated for reassembly. Align the marks of the upper seat and coil spring. Protect the assembly with cloth and install the spring compressor.
- Coil spring

10. Compress the coil spring with the spring compressor

11. Install or connect the following:
- Rubber seat, the spring upper seat, the bearing and the mounting block
- Piston rod upper nut. Apply an anti–rust compound on the piston rod thread and tighten the lower nut so that 0.62–0.69–17.7mm) of the rod is exposed. Then, tighten the upper nut to 17 ft. lbs. (23 Nm)

12. Install the strut to the vehicle.

Lower Ball Joint

REMOVAL & INSTALLATION

Except 2002–04 Miata and Millenia

The lower ball joint is an integral part of the lower control and cannot be replaced

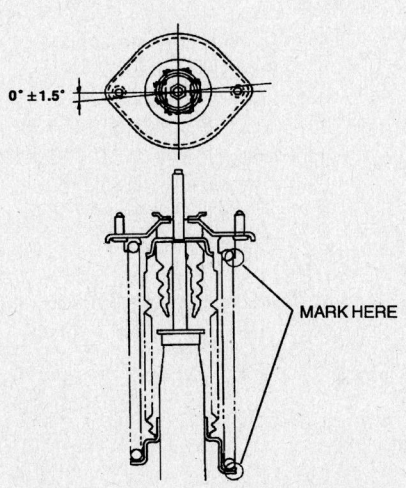

0° ±1.5°

MARK HERE

42356-MAZC-GCH

Mark the seat, shock and spring assembly as illustrated for reassembly–2002–04 Miata

1. Stabilizer nut
2. Retainer, bushing and spacer
3. Stabilizer bolt
4. Bolt, washer
5. Bolt
6. Bolt, nut
7. Nut
8. Washer
9. Lower control arm bushing (rear)
10. Nut
11. Bolt
12. Lower arm ball joint
13. Ball joint dust boot
14. Lower arm bushing (front)
15. Lower arm

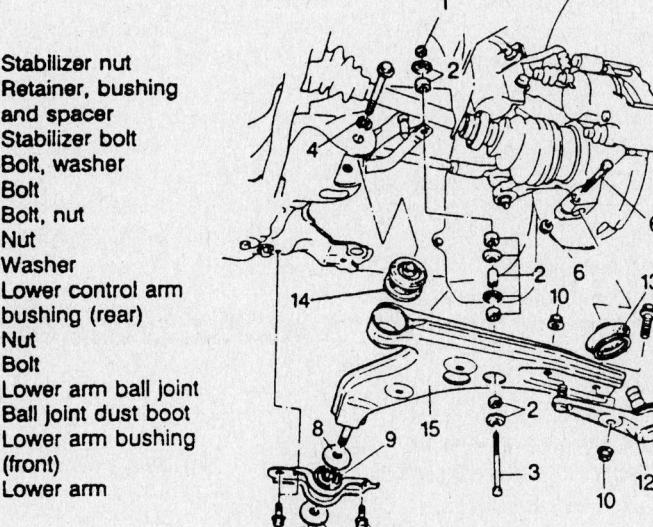

7923MG85

Exploded view of a common lower control arm with replaceable ball joint

separately. If the lower ball joint is defective, the entire lower control arm must be replaced. Refer to the lower control arm procedure.

2002–04 Miata and Millenia

1. Before servicing the vehicle, refer to the precautions section.
2. Remove or disconnect the following:
   • Wheel
   • Ball joint stud pinch bolt and nut from the steering knuckle
   • Ball joint by prying it from the knuckle
   • Ball joint to lower control arm bolt and nut

To install:

3. Install or connect the following:
   • Ball joint to lower control arm. Torque the bolt to 86 ft. lbs. (116 Nm).
   • Ball joint to the knuckle
   • Ball joint to the steering knuckle. Torque the bolt to 57 ft. lbs. (77 Nm).
   • Wheel
4. Check and/or adjust the front wheel alignment.

Upper Control Arm

REMOVAL & INSTALLATION

2002–04 Miata

FRONT

1. Before servicing the vehicle, refer to the precautions section.
2. Remove or disconnect the following:
   • Wheel
3. Support the lower control arm with a jack.
4. Remove or disconnect the following:
   • Cotter pin
   • Upper ball joint nut by loosening it
   • Ball joint by pressing it from the knuckle
   • Upper ball joint nut
   • Lower strut mounting bolt
   • Upper control arm bolt and nut
   • Upper control arm

To install:

5. Install the upper control arm.
6. Loosely tighten the bolt and nut.
7. Loosely install the lower strut mounting bolt.
   • Ball joint to the knuckle. Torque nut to 47–60 ft. lbs. (63–82 Nm).
   • New cotter pin to ball joint nut
   • Wheel

8. Torque upper control arm bolt to 87–101 ft. lbs. (118–137 Nm) and the lower strut mounting bolt to 69–86 ft. lbs. (94–116 Nm).
9. Check and/or adjust the front wheel alignment.

REAR

1. Before servicing the vehicle, refer to the precautions section.
2. Remove or disconnect the following:
   • Wheel
3. Support the lower control arm with a jack.
4. Remove or disconnect the following:
   • Upper control arm bolts and nuts
   • Upper control arm

To install:

5. Install the upper control arm.
6. Loosely tighten the bolt and nut.
   • Wheel
7. Torque upper control arm nuts and bolts to 40–56 ft. lbs. (54–76 Nm).
8. Check and/or adjust the front wheel alignment.

CONTROL ARM BUSHING REPLACEMENT

All Mazda's use a pressed in control arm bushing, and the pressing can be done using two appropriately sized sockets (a press socket and a catch socket) and a large bench vise.

1. Position the control arm and the 2 sockets into a vise.
2. Position the press socket onto the control arm bushing.
3. Position the catch socket onto the control arm, opposite of the press socket.
4. Tighten the bench vise slowly and press the bushing into the catch socket.

To install:

5. Apply soapy water to the new control arm bushing.
6. Position the bushing against the control arm.
7. Using the same sockets, in the same positions, press the new bushing into the control arm.

Lower Control Arm

REMOVAL & INSTALLATION

626

FRONT

1. Before servicing the vehicle, refer to the precautions section.
2. Remove or disconnect the following:
   • Wheel

• Lower ball joint pinch bolt from the steering knuckle
• Stabilizer bar link from the lower control arm
• Lower control arm bolts and nuts
• Lower control arm with the lower ball joint

To install:

3. Install or connect the following:
   • Lower control arm and loosely tighten the mounting nuts and bolts
   • Stabilizer link to the lower control arm. Torque the nut to 27–39 ft. lbs. (37–53 Nm).
   • Lower ball joint to the steering knuckle. Torque the bolt to 26–41 ft. lbs. (35–56 Nm).
   • Wheel
4. With the vehicle at normal rode height, tighten the lower control arm mounting bolts. Torque the front bushing through-bolt to 40–59 ft. lbs. (55–80 Nm) and the rear bushing strap bolts to 69–96 ft. lbs. (94–131 Nm).
5. Check and/or adjust the front wheel alignment.

Protégé

FRONT

1. Before servicing the vehicle, refer to the precautions section.
2. Remove or disconnect the following:
   • Wheel
   • Rear lower arm bolt
   • Lower ball joint pinch bolt from the steering knuckle
   • Bracket
   • Lower control arm with the lower ball joint

To install:

3. Install or connect the following:
   • Bracket and tighten the bolts to 68–97 ft. lbs. (94–126 Nm).
   • Lower control arm and loosely tighten the mounting nuts and bolts
   • Lower ball joint to the steering knuckle. Torque the bolt to 32–43 ft. lbs. (44–58 Nm).
   • Wheel
4. With the vehicle at normal rode height, tighten the lower control arm rear mounting bolts to 69–94 ft. lbs. (94–126 Nm).
5. Check and/or adjust the front wheel alignment.

2002–04 Miata

FRONT

1. Before servicing the vehicle, refer to the precautions section.

2. Remove or disconnect the following:
- Wheel
- Undercover
- Cotter pin and nut from the tie-rod end
- Tie-rod end from the steering knuckle
- Stabilizer bar link bolt and the lower strut mounting bolt
- Cotter pin
- Lower ball joint by loosening it

3. With the nut protecting the ball joint stud, separate the stud from the knuckle. Remove the nut.
- Dust boot, if necessary
- Lower control arm bolts, nuts and adjusting cams
- Lower control arm

To install:

4. Install or connect the following:
- Dust boot, if removed using a press. Always fill the inside of the new boot with grease prior to installation.
- Lower control arm by loosely tightening the bolts and nuts
- Lower ball joint to the knuckle. Torque the nut to 42–57 ft. lbs. (57–77 Nm).
- New cotter pin
- Lower strut and stabilizer link bolts by loosely tightening them
- Tie-rod end to the steering knuckle. Torque the nut to 22–32 ft. lbs. (30–44 Nm).
- New cotter pin
- Wheel and undercover

5. With the vehicle at normal rode height, torque the lower control arm bolts to 69–83 ft. lbs. (94–113 Nm) for the inner and 69–86 ft. lbs. (94–112 Nm) for the outer.

6. Torque the lower strut mounting bolt to 69–86 ft. lbs. (94–116 Nm) and the stabilizer link bolt to 32–44 lbs. (44–60 Nm).

7. Check and/or adjust the front wheel alignment.

REAR

1. Before servicing the vehicle, refer to the precautions section.

2. Remove or disconnect the following:
- Wheel

3. Support the lower control arm with a jack.

4. Remove or disconnect the following:
- Stabilizer bar link nut
- Lower shock absorber bolt
- Lower control arm bolts and nuts
- Lower control arm

To install:

5. Install or connect the following:
- Lower control arm and loosely tighten the bolts and nuts

- Lower shock absorber bolt and tighten to 70 ft. lbs. (85 Nm)
- Stabilizer bar link nut and tighten to 32–44 ft. lbs. (44–60 Nm)
- Wheel

6. Torque lower control arm inner nuts and bolts to 54–70 ft. lbs. (73–95 Nm) and the outer nuts and bolts to 47–54 ft. lbs. (63–74 Nm).

7. Check and/or adjust the front wheel alignment.

Millenia

1. Before servicing the vehicle, refer to the precautions section.

2. Remove or disconnect the following:
- Transverse member
- Power steering return hose and pressure pipe
- Intermediate steering shaft bolt

3. Support the engine from the top.
- Engine mount member
- Bolts for engine mount No. 1
- Lower strut mounting bolt
- Tie-rod end from the steering knuckle
- Upper lateral link ball joint
- Lower ball joint
- Stabilizer control link nut
- Gusset

4. Support the crossmember using a jack.

- Crossmember mounting nuts and lower the crossmember to gain clearance
- Lower arm assembly

To install:

5. Install or connect the following:
- Lower arm assembly to the vehicle
- Crossmember mounting bolts
- Gusset. Torque the bolts to 58–86 ft. lbs. (79–116 Nm).
- Stabilizer control link. Torque the nut to 32–44 ft. lbs. (44–60 Nm).
- Lower ball joint. Torque the bolts to 58–86 ft. lbs. (79–116 Nm) and the nut to 86–115 ft. lbs. (116–156 Nm).
- Upper lateral link nut and bolt. Torque to 58–86 ft. lbs. (79–116 Nm).
- Tie-rod end. Torque the nut to 41–59 ft. lbs. (55–80 Nm).
- Strut lower mounting bolt. Torque to 73–101 ft. lbs. (98–137 Nm).
- No. 1 engine mount. Torque the bolts to 32–44 ft. lbs. (44–60 Nm).
- Engine mount member. Torque the bolts to 50–68 ft. lbs. (67–93 Nm) and the nuts to 55–77.3 ft. lbs. (75–104 Nm).

6. Remove the engine support tool.
- Intermediate steering shaft. Torque the bolt to 14–19 ft. lbs. (18–26 Nm).
- Power steering pressure pipe and return hose
- Transverse member

7. Check the power steering fluid and fill to proper level, bleed if necessary.

8. Check and/or adjust the front end alignment.

CONTROL ARM BUSHING REPLACEMENT

Except Protégé Models

1. Before servicing the vehicle, refer to the precautions section.

All Mazda's use a pressed in control arm bushing, and the pressing can usually be done using 2 appropriately sized sockets (a press socket and a catch socket) and a large vise.

2. Position the control arm and the 2 sockets into a vise.

3. Position the press socket onto the control arm bushing.

4. Position the catch socket onto the control arm, opposite of the press socket.

5. Tighten the vise slowly and press the bushing into the catch socket.

To install:

6. Apply soapy water to the new control arm bushing.

7. Position the bushing against the control arm.

8. Using the same sockets, in the same positions, press the new bushing into the control arm.

Protégé

1. Before servicing the vehicle, refer to the precautions section.

2. Position the control arm into a vise. Cut away the projecting rubber from the front bushing. Use tool 49 B034 2A2 to remove the bushing.

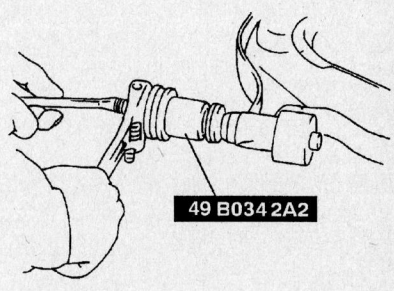

49 B034 2A2

42356-MAZC-GKF

Use tool 49 B034 2A2 to remove and install the front bushing—Protégé

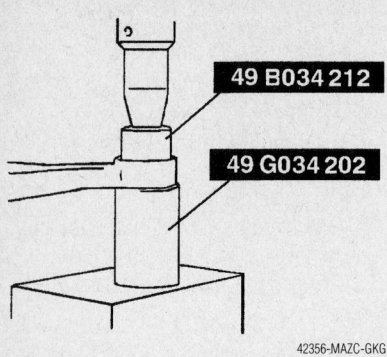

Use tools 49 B034 212 and 49 G034 202 to remove the rear bushing—Protégé

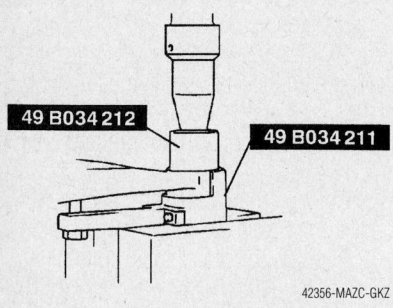

Use tools 49 B034 212 and 49 B034 211 to install the rear bushing—Protégé

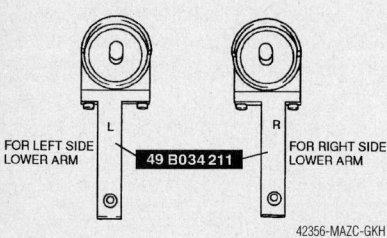

Align the mark of the lower arm and the small projection of the lower arm bushing (rear) when installing the rear bushing—Protégé

3. Remove the rear bushing using a press and tools 49 B034 212 and 49 G034 202.

To install:

4. When installing the rear bushing on the lower arm, align the mark of the lower arm and the small projection of the lower arm bushing (rear) as illustrated, set the arm onto the press and press the bushings into position.

5. Apply soapy water to the new control arm bushing.

6. Install the rear bushing using a press and tools 49 B034 212 and 49 B034 211.

7. Position the control arm into a vise. Use tool 49 B034 2A2 to pull the bushing into the arm bore.

Wheel Bearings

ADJUSTMENT

The front and rear wheel bearings are not adjustable. If the bearings become loose or make noise, they must be replaced.

REMOVAL & INSTALLATION

626

FRONT

1. Before servicing the vehicle, refer to the precautions section.

2. Refer to the illustration for component location and torque specifications.

3. Remove or disconnect the following:
   • Wheels
   • Halfshaft axle nut, unstake the nut prior to removal
   • Brake caliper and rotor
   • Tie rod end from the knuckle
   • Control arm ball joint from the knuckle
   • Knuckle assembly
   • Inner oil seal from the knuckle
   • Hub using a press and Mazda tools 49 G033 103 and 49 G033 105. If the bearing inner race remains in the hub, grind a section of the bearing inner race until 0.02 inch (0.5mm) remains and use a chisel to remove it.
   • Bearing from the hub using a press and Mazda tools 49 G033 102 and 49 G033 106
   • Brake dust shield, if it is being replaced. Mark the cover and knuckle for replacement purposes and use a chisel to remove the shield.

4. Clean and inspect all parts but do not wash or clean the wheel bearing. The bearing must be replaced.

To install:

5. Using Mazda Press tools 49 G033 107a and 49 F027 009, install a new dust shield cover assembly to the knuckle, if removed.

6. Using press tools 49 G033 797, 49 F027 004 and 49 F027 003; press a new wheel bearing into the knuckle assembly.

7. Install or connect the following:
   • Wheel bearing retaining ring

8. Using press tools 49 G033 105, 49 F027 009; press in the hub assembly.
   • New oil seal using installation tool 49 V001 795
   • Knuckle assembly, tighten the upper bolt to 61 ft. lbs. (83 Nm)

and the lower bolt to 41 ft. lbs. (56 Nm)
   • Control arm ball joint to the knuckle
   • Tie rod end to the knuckle
   • Brake rotor and caliper
   • Halfshaft axle nut, tighten the nut to 174–235 ft. lbs. (236–318 Nm) and stake the nut
   • Wheels

REAR

1. Before servicing the vehicle, refer to the precautions section.

2. Refer to the illustration for component location and torque specifications.

➡**The wheel bearings are not service-able. If the bearings are bad, a new hub/bearing assembly must be installed.**

3. Remove or disconnect the following:
   • Rear wheels
   • Brake drum, if equipped
   • Rear caliper and rotor assembly from the hub, if equipped
   • Hub dust cover

4. Raise the staked portion of the hub retaining nut with a hammer and chisel.
   • Hub retaining nut and discard it
   • Hub and bearing assembly from the spindle

To install:

5. Install or connect the following:
   • Bearing assembly on the spindle. Torque the new nut to 131–173 ft. lbs. (177–235 Nm).
   • Stake the nut into the groove in the spindle
   • Dust cover
   • Assemble the brakes
   • Rear wheel

Protégé

FRONT

1. Before servicing the vehicle, refer to the precautions section.

2. Refer to the illustration for component location and torque specifications.

3. Remove or disconnect the following:
   • Wheels
   • Halfshaft axle nut, Unstake the nut prior to removal
   • Brake caliper and rotor
   • Tie rod end from the knuckle
   • Control arm ball joint from the knuckle
   • Knuckle assembly
   • Inner oil seal from the knuckle
   • Hub using a press and Mazda tools

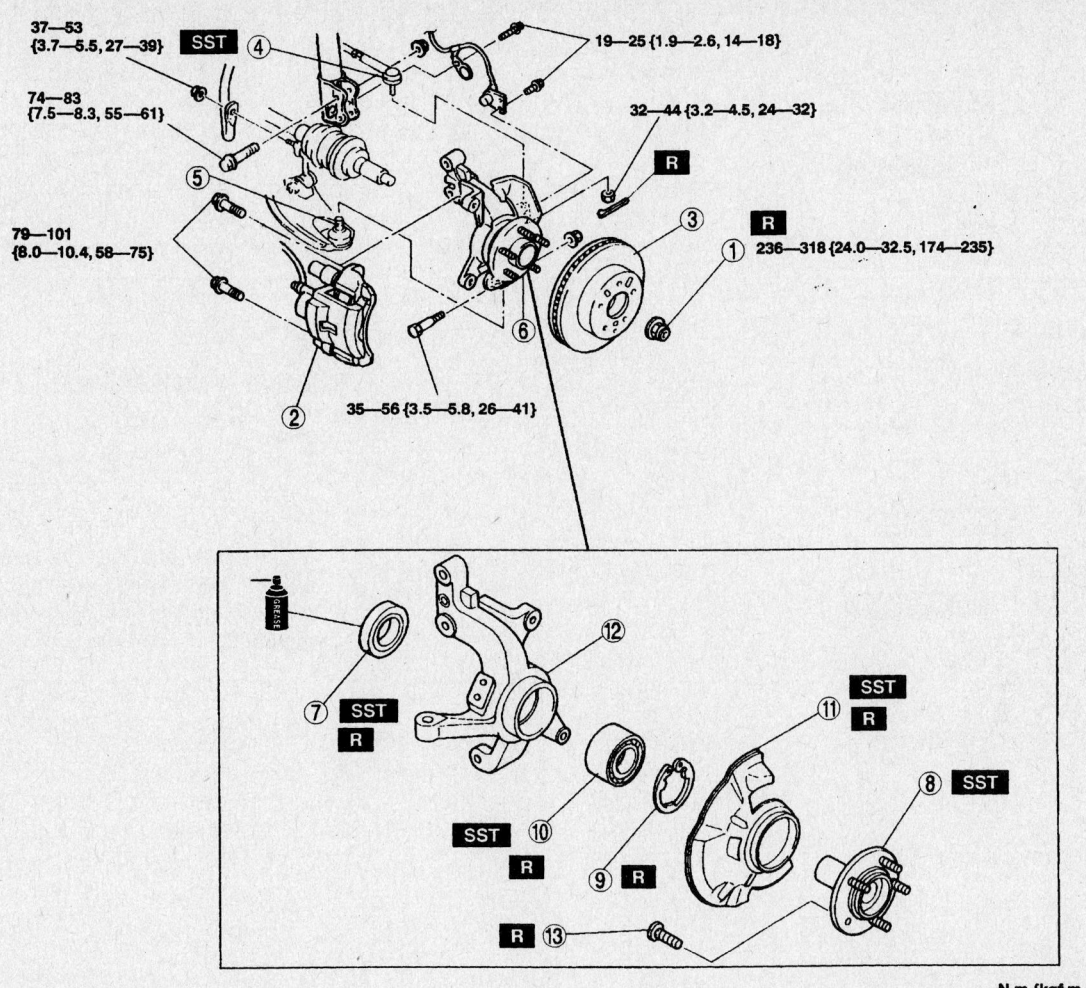

1 Locknut
2 Brake caliper component
3 Disc plate
4 Tie-rod end ball joint
5 Lower arm ball joint
6 Wheel hub, steering knuckle, dust cover
7 Oil seal

8 Wheel hub component
9 Retaining ring
10 Wheel bearing
11 Dust cover
12 Steering knuckle
13 Hub bolt

42356-MAZC-G83

Exploded view of the front wheel bearing and knuckle assembly—626

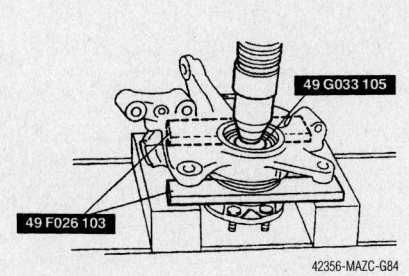

42356-MAZC-G84

Use a press to remove the hub—626

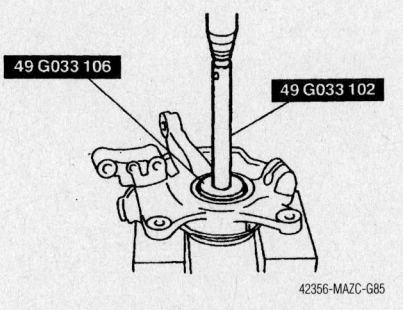

42356-MAZC-G85

Use a press to remove the wheel bearing—626

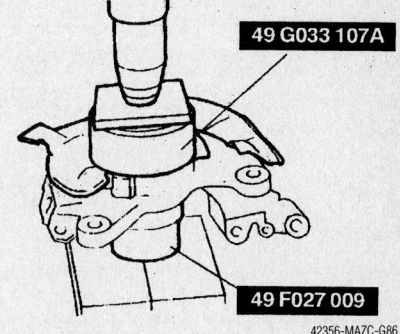

42356-MAZC-G86

Install a new dust cover, if removed—626

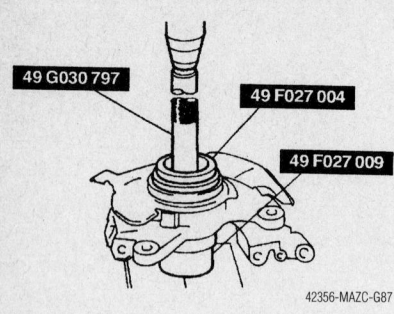

Install a new wheel bearing using a press–626

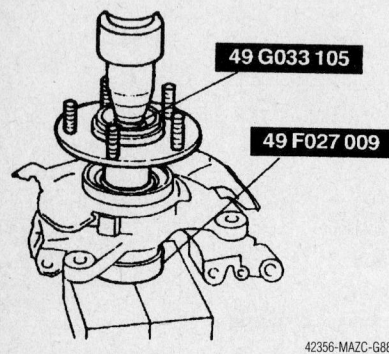

Install a wheel hub using a press–626

49 G030 727, 49 G033 102 and 49 F026 103. If the bearing inner race remains in the hub, grind a section of the bearing inner race until 0.02 inch (0.5mm) remains and use a chisel to remove it.

• Bearing from the hub using a press and Mazda tools 49 F027 005, 49 F027 003 and 49 F026 103

• Brake dust shield, if it is being replaced. Mark the cover and knuckle for replacement purposes and use a chisel to remove the shield.

4. Clean and inspect all parts but do not wash or clean the wheel bearing. The bearing must be replaced.

To install:

5. Using Mazda Press tools 49 E033 101 and 49 F027 009, install a new dust shield cover assembly to the knuckle, if removed.

6. Using press tools 49 F027 003, 49 F027 007 and 49 F027 009; press a new wheel bearing into the knuckle assembly.

7. Install or connect the following:

• Wheel bearing retaining ring

8. Using press tool 49 F027 009 press in the hub assembly.

• New oil seal using installation tool 49 V001 795

• Knuckle assembly, tighten the upper bolt to 93 ft. lbs. (126 Nm) and the lower bolt to 43 ft. lbs. (58 Nm)

• Control arm ball joint to the knuckle

• Tie rod end to the knuckle

• Brake rotor and caliper

• Halfshaft axle nut, tighten the nut to 174–235 ft. lbs. (236–318 Nm) and stake the nut

• Wheels

REAR—DRUM BRAKES

1. Before servicing the vehicle, refer to the precautions section.

2. Refer to the illustration for component location and torque specifications.

➡**The wheel bearings are not service-able. If the bearings are bad, a new hub/bearing assembly must be installed.**

3. Remove or disconnect the following:

• Rear wheels

• Wheel speed sensor

4. Raise the staked portion of the hub retaining nut with a hammer and chisel.

• Hub retaining nut and discard it

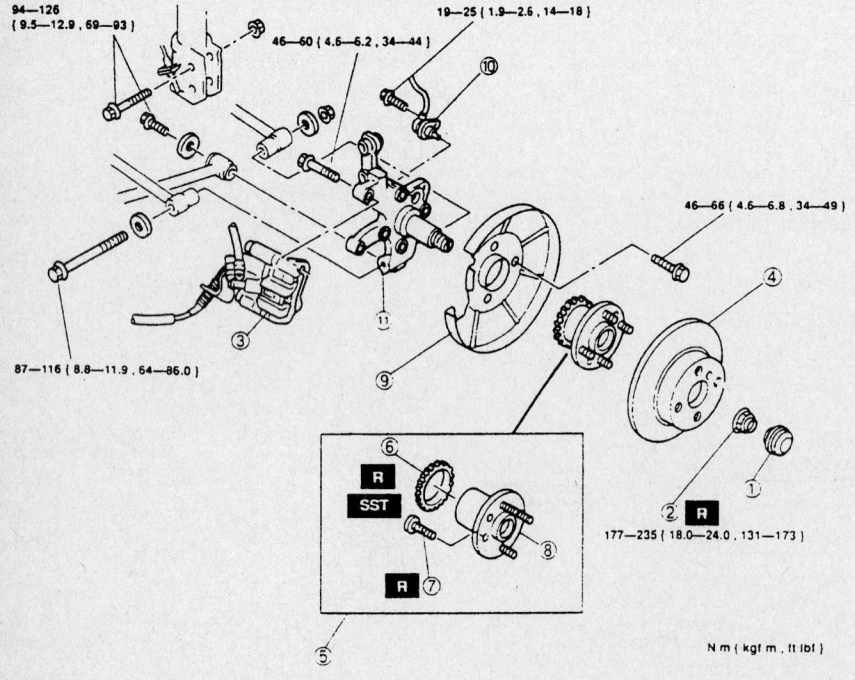

1. Hub cap
2. Locknut
3. Brake caliper assembly
4. Disc plate
5. Wheel hub assembly
   Inspect for damage
   Inspect bearing for damage and rough rotation
6. ABS sensor rotor
7. Hub bolt
8. Wheel hub
9. Dust cover
   Inspect for damage and cracks
10. ABS wheel-speed sensor
11. Hub spindle
    Inspect for damage and cracks

N m (kgf m . ft lbf)

Exploded view of the rear wheel hub and bearing assembly (disc brake model shown, drum is similar)–626

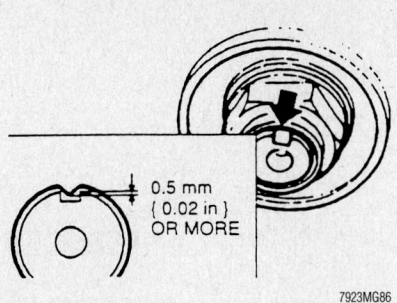

7923MG86

When installing the new hub nut, be sure to stake it into the notch on the spindle

- Brake drum
- Hub and bearing assembly from the spindle with the ABS sensor rotor, if equipped with ABS
- ABS sensor rotor using a chisel

To install:

5. Position the ABS sensor rotor on the hub as illustrated.

6. Position tool 49 B026 103 so that the marking **A** faces the bottom.

7. Press the rotor onto the hub.

8. Install or connect the following:

- Bearing assembly on the spindle. Torque the new nut to 131–173 ft. lbs. (177–235 Nm).

- Stake the nut into the groove in the spindle
- Dust cover
- Assemble the brakes
- Rear wheel

REAR—DISC BRAKES

1. Before servicing the vehicle, refer to the precautions section.

2. Refer to the illustration for component location and torque specifications.

➡**The wheel bearings are not serviceable. If the bearings are bad, a new hub/bearing assembly must be installed.**

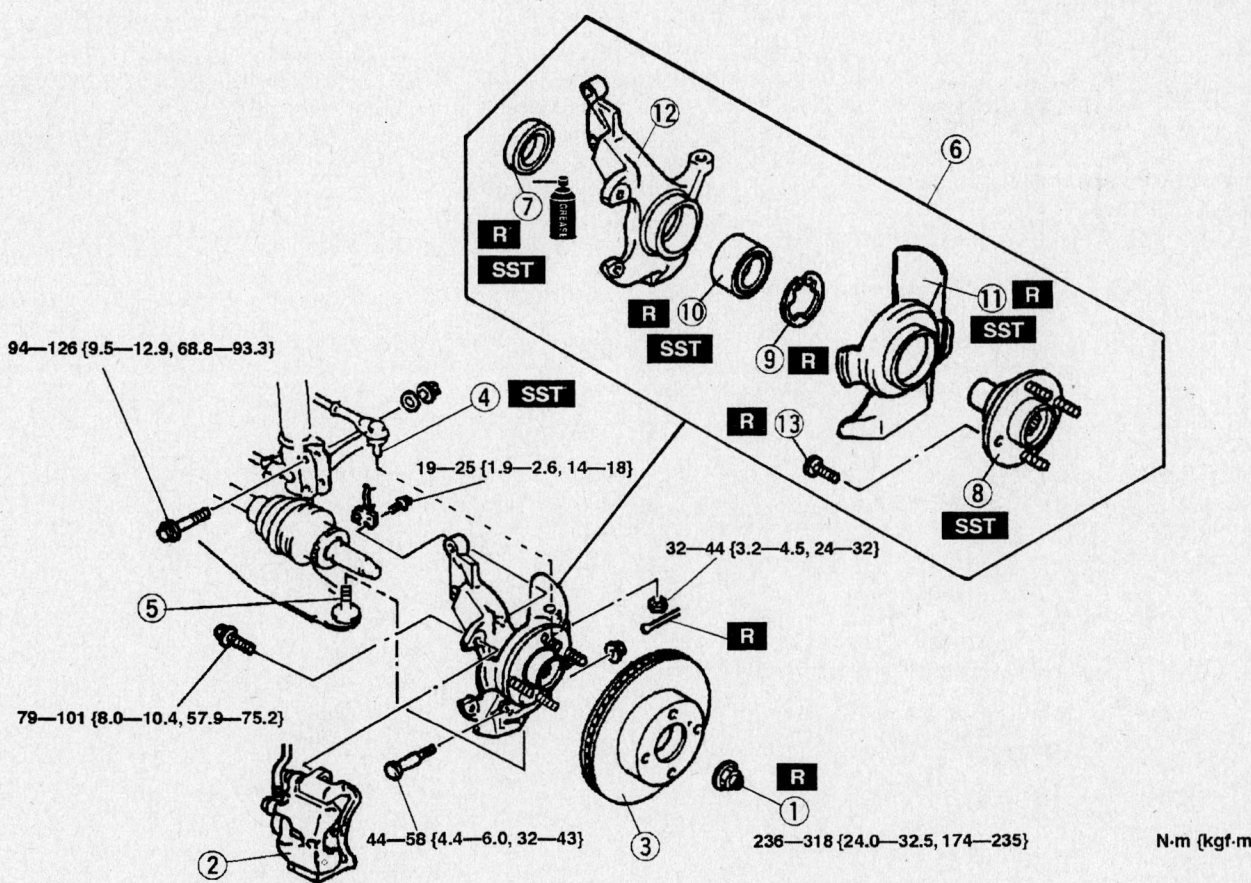

1	Locknut
2	Brake caliper component
3	Disc plate
4	Tie-rod end
5	Lower arm ball joint
6	Wheel hub, steering knuckle, dust cover
7	Oil seal

8	Wheel hub component
9	Retaining ring
10	Wheel bearing
11	Dust cover
12	Steering knuckle
13	Hub bolt

Exploded view of the front wheel bearing and knuckle assembly–Protégé

42356-MAZC-GKI

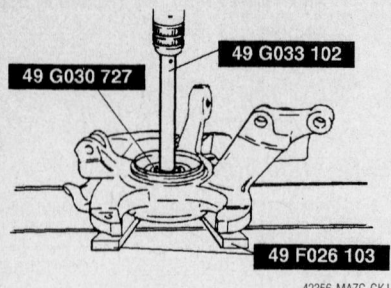

Use a press to remove the hub–Protégé

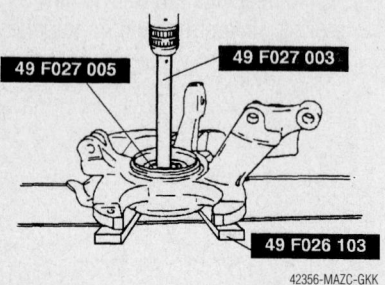

Use a press to remove the wheel bearing–Protégé

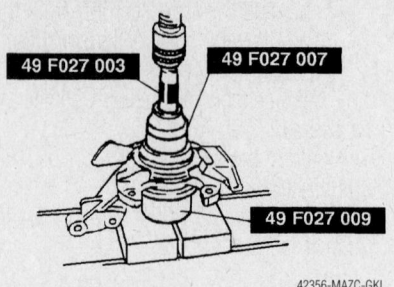

Install a new wheel bearing using a press–Protégé

3. Remove or disconnect the following:
- Rear wheels
- Wheel speed sensor

4. Raise the staked portion of the hub retaining nut with a hammer and chisel.
- Hub retaining nut and discard it
- Brake caliper and rotor
- Hub and bearing assembly from the

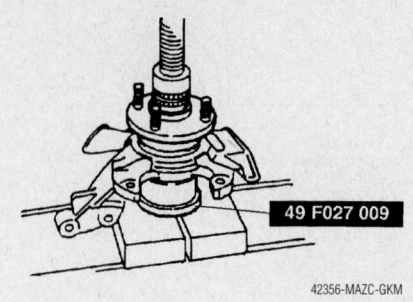

Install a wheel hub using a press–Protégé

spindle with the ABS sensor rotor, if equipped with ABS
- ABS sensor rotor using a chisel

To install:

5. Position the ABS sensor rotor on the hub as illustrated.

6. Position tool 49 B026 105 so that the marking **B** faces the bottom.

7. Press the rotor onto the hub.

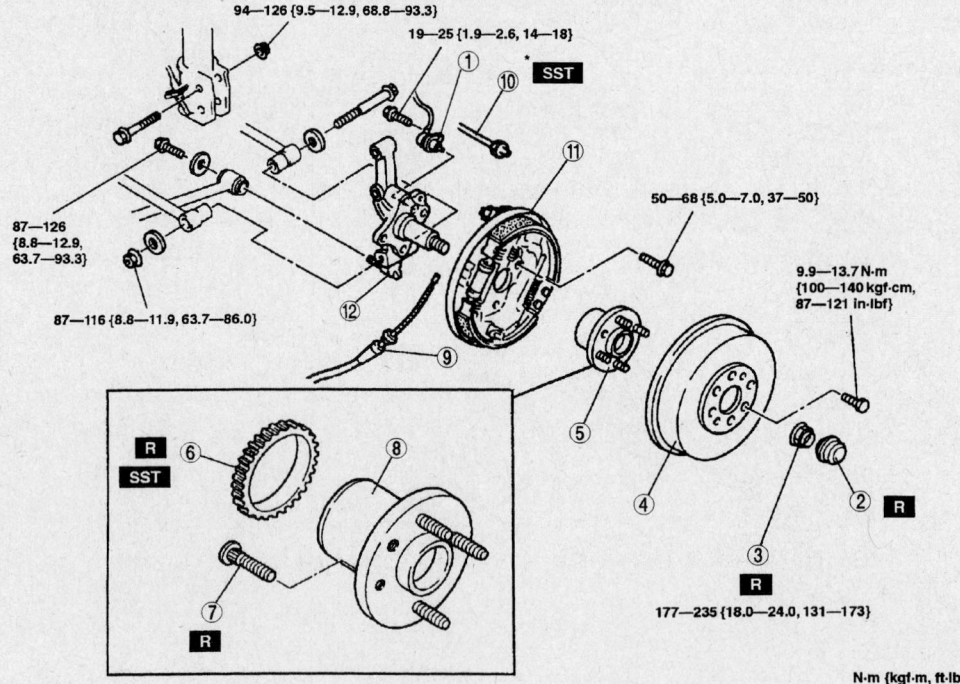

* 49 0259 770B (Flare nut wrench)

N·m {kgf·m, ft·lbf}

1	ABS-wheel speed sensor	
2	Hub cap	
3	Locknut	
4	Brake drum	
5	Wheel hub and ABS sensor rotor (with ABS)	
6	ABS sensor rotor (with ABS)	
7	Hub bolt	
8	Wheel hub	
9	Parking brake cable	
10	Brake pipe	
11	Rear brake component	
12	Hub spindle	

Exploded view of the rear wheel hub and bearing assembly–Protégé with drum brakes

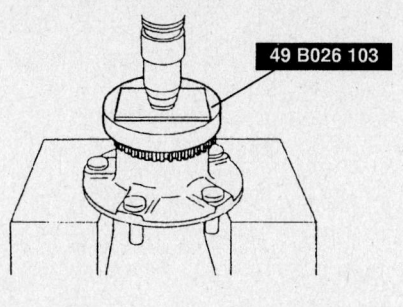

49 B026 103
42356-MAZC-GK0

Position the ABS sensor rotor on the hub–Protégé with drum brakes

WHEEL HUB
49 B026 103
3 mm {0.1 in}
A
ABS SENSOR ROTOR
42356-MAZC-GKP

Position tool 49 B026 103 so that the marking A faces the bottom–Protégé with drum brakes

8. Install or connect the following:
   - Bearing assembly on the spindle. Torque the new nut to 131–173 ft. lbs. (177–235 Nm).
   - Stake the nut into the groove in the spindle
   - Dust cover
   - Assemble the brakes
   - Rear wheel

2002–04 Miata

FRONT

1. Before servicing the vehicle, refer to the precautions section.

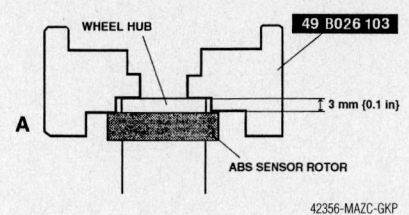

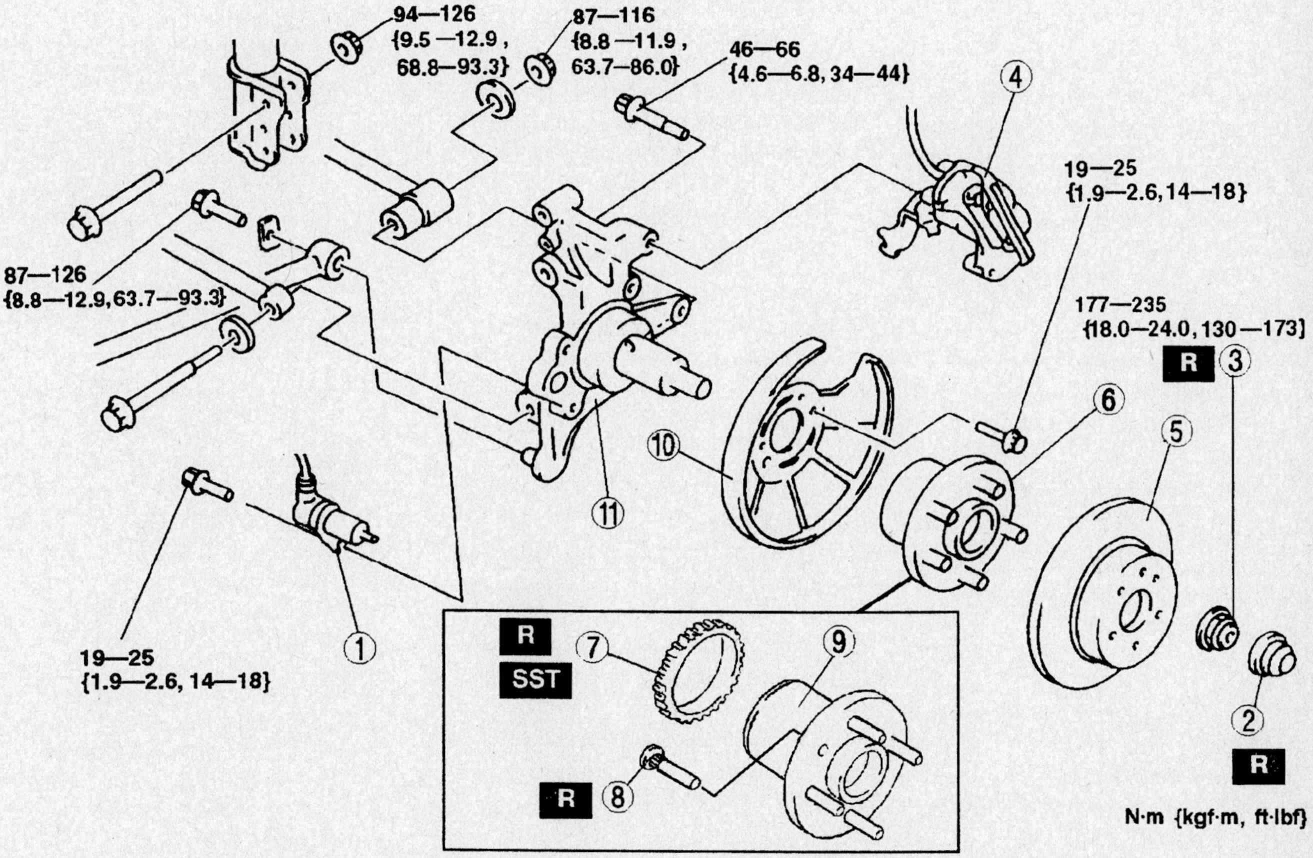

94—126
{9.5—12.9,
68.8—93.3}
87—116
{8.8—11.9,
63.7—86.0}
46—66
{4.6—6.8, 34—44}
4
19—25
{1.9—2.6, 14—18}
87—126
{8.8—12.9, 63.7—93.3}
177—235
{18.0—24.0, 130—173}
R 3
6
5
10
11
19—25
{1.9—2.6, 14—18}
1
R
SST
7
9
R 8
2
R
N·m {kgf·m, ft·lbf}

1 ABS wheel-speed sensor
2 Hub cap
3 Locknut
4 Brake caliper component
5 Disc plate
6 Wheel hub and ABS sensor rotor
7 ABS sensor rotor
8 Hub bolt
9 Wheel hub component
10 Dust cover
11 Hub spindle

42356-MAZC-GKN

Exploded view of the rear wheel hub and bearing assembly–Protégé with disc brakes

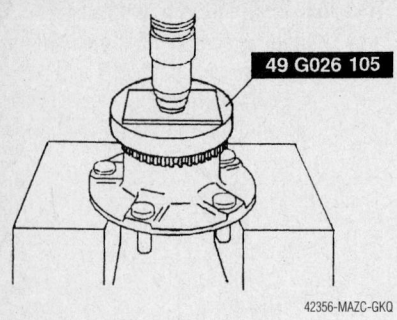

42356-MAZC-GKQ

Position the ABS sensor rotor on the hub–Protégé with disc brakes

➡ **The wheel bearings are not serviceable. If the bearings are bad, a new hub/bearing assembly must be installed.**

2. Remove or disconnect the following:
 • Front wheels
 • Hub center dust cap

3. Raise the staked portion of the hub locknut with a hammer and chisel. Lock the hub by applying the brakes and loosen the nut.
 • Brake caliper
 • Brake rotor
 • Hub locknut and discard it
 • Hub from the spindle

To install:

4. Install or connect the following:
 • Hub to the spindle
 • Hub locknut. Torque the nut to 123–159 ft. lbs. (167–215 Nm).
 • Stake the new nut into the spindle groove using a dull chisel
 • Brake rotor
 • Brake caliper
 • Hub center dust cap
 • Front wheels

REAR

1. Before servicing the vehicle, refer to the precautions section.
2. Remove the rear wheels.
3. Raise the staked portion of the hub locknut with a hammer and chisel. Lock the hub by applying the brakes and loosen the nut.
4. Remove or disconnect the following:
 • Brake caliper and position it aside
 • Brake rotor
 • Hub locknut and discard it
 • Anti-lock Brake System (ABS) wheel speed sensor, if equipped
 • Speed sensor bracket from the rear knuckle
 • Upper and lower knuckle mounting through bolts
 • Knuckle assembly from vehicle
 • Rear bearing oil seal by prying it from the knuckle

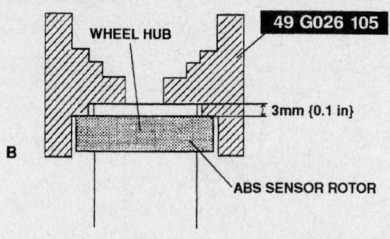

42356-MAZC-GKR

Position tool 49 B026 105 so that the marking B faces the bottom–Protégé with disc brakes

 • Wheel hub by pressing it from the knuckle
 • Retaining snapring from within the knuckle
 • Bearing assembly from the knuckle, once the wheel hub is removed

5. The inner race of the bearing may remain on the hub. Use a chisel and move the bearing race away from the rear hub flange. Once there is enough clearance, press the race from the hub.

To install:

6. Press the new bearing into the rear knuckle assembly.
7. Apply some grease to the wheel bearing inner race and press the rear hub into the bearing. Make sure to position a suitable support on the backside of the bearing.
8. Install or connect the following:
 • Retaining snapring to the knuckle
 • New rear oil seal by lubricating it with grease and pressing it into the rear knuckle
 • Knuckle assembly to the vehicle
 • Upper and lower knuckle mounting through bolts. Torque the upper bolt to 34–49 ft. lbs. (47–66 Nm) and the lower bolt to 47–54 ft. lbs. (63–74 Nm).
 • Speed sensor bracket to the rear knuckle
 • ABS wheel speed sensor, if equipped
 • New hub locknut. Torque the nut to 174–235 ft. lbs. (236–318 Nm).
 • Stake the new nut into the spindle groove using a dull chisel
 • Brake rotor
 • Brake caliper
 • Rear wheels

Millenia

FRONT

1. Before servicing the vehicle, refer to the precautions section.
2. Refer to the illustration for component location and torque specifications.

3. Remove or disconnect the following:
 • Wheels
 • Halfshaft axle nut, Unstake the nut prior to removal.
 • Brake caliper and rotor
 • Tie rod end from the knuckle
 • Upper leading link ball joint
 • Upper lateral link ball joint
 • Lower ball joint
 • Knuckle assembly
 • Snap ring
 • Hub using a press and Mazda tools 49 F026 102 and 49 W017 101. If the bearing inner race remains in the hub, grind a section of the bearing inner race until 0.02 inch (0.5mm) remains and use a chisel to remove it.
 • Bearing from the hub using a press and Mazda tools 49 G0797, 49 G026 102, 49 E033 101 and 49 W017 101
 • Brake dust shield, if it is being replaced. Mark the cover and knuckle for replacement purposes and use a chisel to remove the shield.

4. Clean and inspect all parts but do not wash or clean the wheel bearing. The bearing must be replaced.

To install:

5. Using Mazda Press tools 49 G033 107a and 49 G026 103, install a new dust shield cover assembly to the knuckle, if removed.
6. Using press tools 49 G026 103 and 49 F026 102, press a new wheel bearing into the knuckle assembly.
7. Install or connect the following:
 • Wheel bearing retaining ring

8. Using press tools 49 G033 105, 49 F027 009; press in the hub assembly.
 • Snap ring
 • Knuckle assembly, tighten the upper bolt to the specifications shown in the illustration
 • Lower ball joint
 • Upper lateral link ball joint
 • Upper leading link ball joint
 • Tie rod end to the knuckle
 • Brake caliper and rotor
 • Halfshaft axle nut, tighten the nut to 174–235 ft. lbs. (236–318 Nm) and stake the nut
 • Wheels

REAR

1. Before servicing the vehicle, refer to the precautions section.
2. Refer to the illustration for component location and torque specifications.

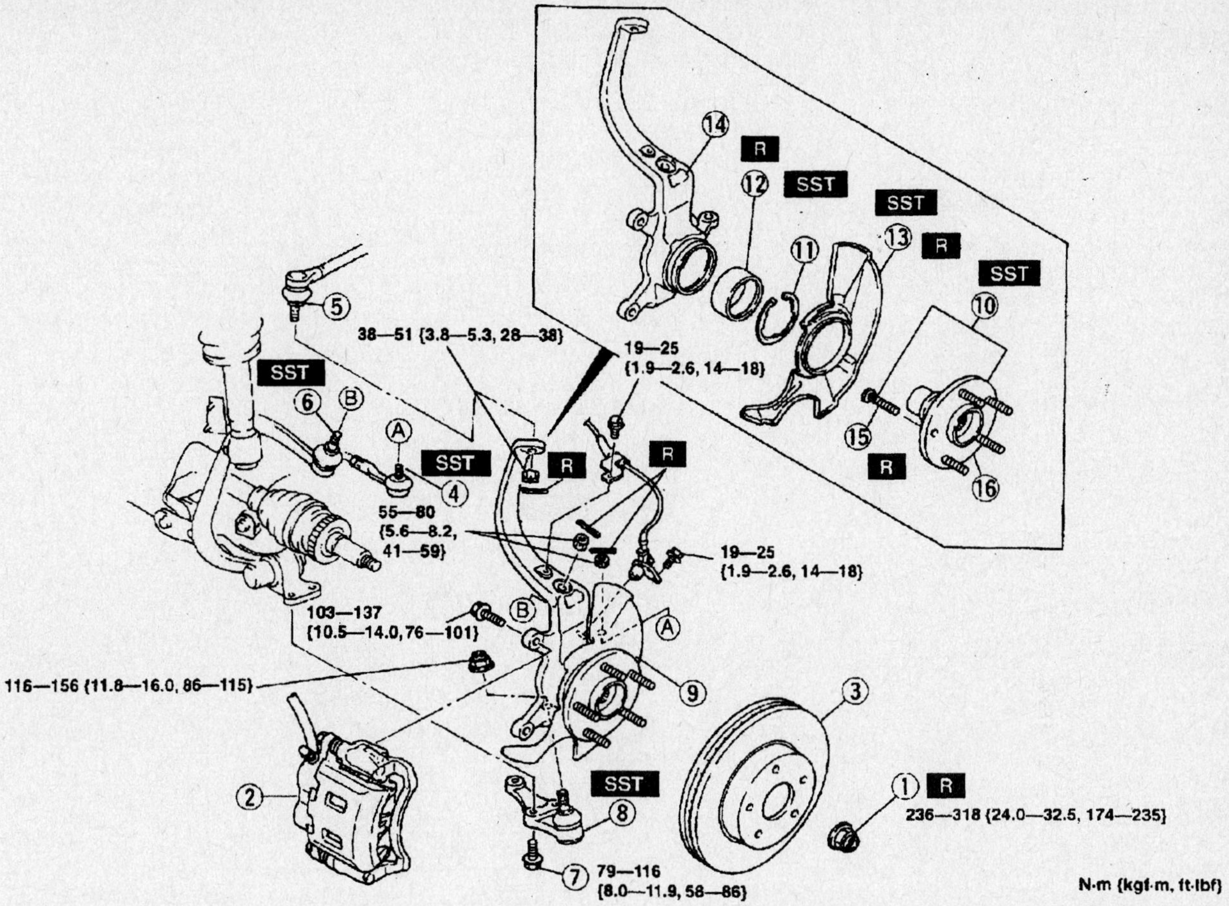

38—51 {3.8—5.3, 28—38}

19—25 {1.9—2.6, 14—18}

55—80 {5.6—8.2, 41—59}

19—25 {1.9—2.6, 14—18}

103—137 {10.5—14.0, 76—101}

116—156 {11.8—16.0, 86—115}

236—318 {24.0—32.5, 174—235}

79—116 {8.0—11.9, 58—86}

N·m {kgf·m, ft·lbf}

1	Locknut	9	Front wheel hub, steering knuckle
2	Brake caliper component	10	Wheel hub component
3	Disc plate	11	Snap ring
4	Tie-rod end ball joint	12	Wheel bearing
5	Upper leading link ball joint	13	Dust cover
6	Upper lateral link ball joint	14	Steering knuckle
7	Lower ball joint bolt	15	Hub bolt
8	Lower ball joint	16	Wheel hub

42356-MAZC-GFC

Exploded view of the front wheel bearing and knuckle assembly–Millenia

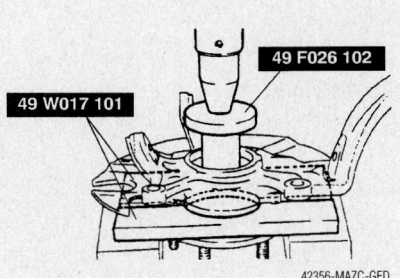

49 W017 101

49 F026 102

42356-MAZC-GFD

Use a press to remove the hub–Millenia

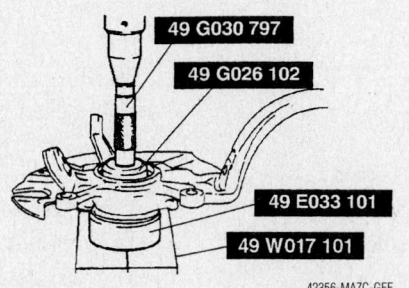

49 G030 797

49 G026 102

49 E033 101

49 W017 101

42356-MAZC-GFE

Use a press to remove the wheel bearing–Millenia

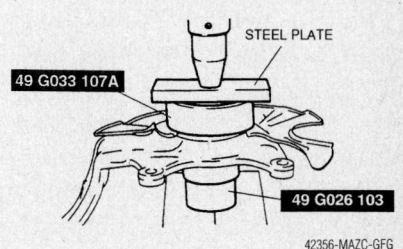

STEEL PLATE

49 G033 107A

49 G026 103

42356-MAZC-GFG

Install a new wheel bearing using a press–Millenia

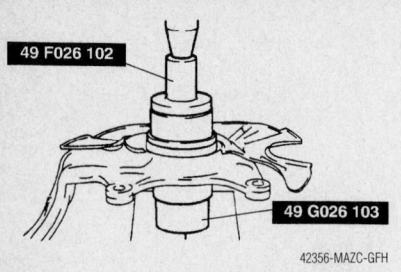

49 F026 102

49 G026 103

42356-MAZC-GFH

Install a wheel hub using a press–Millenia

➥The wheel bearings are not service-able. If the bearings are bad, a new hub/bearing assembly must be installed.

3. Remove or disconnect the following:
- Rear wheels
- Rear caliper and rotor assembly from the hub
- Hub dust cover

4. Raise the staked portion of the hub retaining nut with a hammer and chisel.
- Hub retaining nut and discard it

- Parking brake shoe assembly
- Parking brake cable
- Backing plate
- Rear lower lateral link ball joint
- Lower trailing link ball joint
- Upper trailing link ball joint
- Lower lateral link ball joint
- Upper lateral link ball joint
- Hub spindle
- ABS sensor rotor using a chisel
- Hub bolts

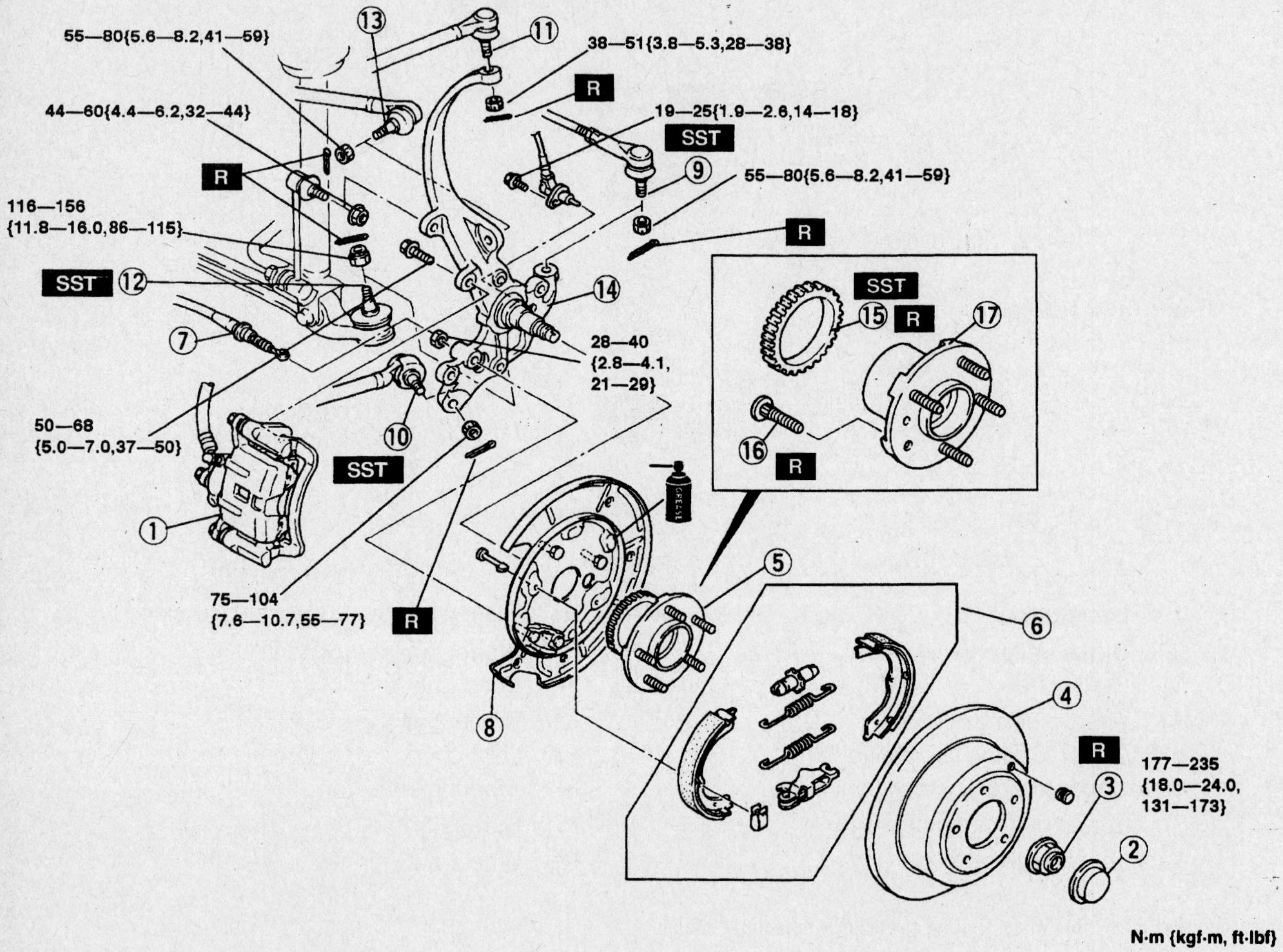

55—80{5.6—8.2,41—59}

38—51{3.8—5.3,28—38}

44—60{4.4—6.2,32—44}

R

19—25{1.9—2.6,14—18}

SST

R

55—80{5.6—8.2,41—59}

116—156 {11.8—16.0,86—115}

R

SST

R

SST

12

7

28—40 {2.8—4.1, 21—29}

15

R

17

16

R

50—68 {5.0—7.0,37—50}

10

SST

1

5

6

75—104 {7.6—10.7,55—77}

R

8

4

R

177—235 {18.0—24.0, 131—173}

3

2

N·m {kgf·m, ft·lbf}

1	Brake caliper component	10	Lower trailing link ball joint
2	Hub cap	11	Upper trailing link ball joint
3	Locknut	12	Lower lateral link ball joint
4	Disc plate	13	Upper lateral link ball joint
5	Wheel hub component	14	Hub spindle
6	Parking brake shoe component	15	ABS sensor rotor
7	Parking brake cable	16	Hub bolt
8	Back plate	17	Wheel hub
9	Rear lower lateral link outer ball joint		

Exploded view of the rear hub and spindle assembly–Millenia

42356-MAZC-GFI

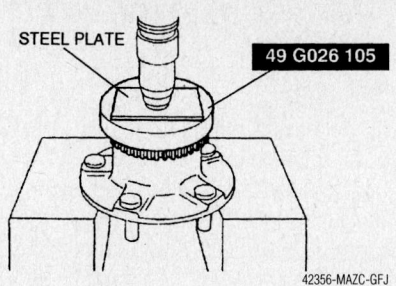

Install the ABS rotor–Millenia

42356-MAZC-GFJ

- Hub and bearing assembly from the spindle

To install:
- Hub and bearing assembly on the spindle
- Hub bolts and tighten to the specifications shown in the illustration

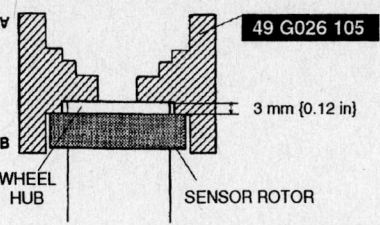

42356-MAZC-GFK

When using the installation tool for the ABS sensor rotor, face the carved side B to the rotor–Millenia

✳✳ CAUTION

When using the installation tool for the ABS sensor rotor, face the carved side B to the rotor.

- ABS sensor rotor using a press, steel plate and tool 49 G02610 until it is flush with the hub.
- Hub spindle
- Upper lateral link ball joint
- Lower lateral link ball joint
- Upper trailing link ball joint
- Lower trailing link ball joint
- Rear lower lateral link ball joint
- Backing plate
- Parking brake cable
- Parking brake shoe assembly
- Hub retaining nut and discard it
- Hub dust cover
- Rear caliper and rotor assembly from the hub, if equipped
- Rear wheels

BRAKES

Brake Caliper

REMOVAL & INSTALLATION

Protégé

FRONT

1. Before servicing the vehicle, refer to the precautions section.
2. Remove or disconnect the following:
   - Wheels
   - Flexible brake hose from the caliper
   - Cap on type **A** brakes or bolt on typebrakes
   - Spring on type **A** brakes
   - Brake pads
   - Upper and lower caliper bolts
   - Caliper

To install:
3. Install or connect the following:
   - Caliper on the brake disc
   - Caliper mounting bolts and tighten the bolts on **A** type brakes to 37–39 ft. lbs. (50–53 Nm) or 16–23 ft. lbs. (22–31 Nm) on type **B** brakes
   - Brake pads
   - Spring on type **A** brakes
4. Replace the washers for the brake line.
   - Brake hose to the caliper and tighten the hose nut to 16–21 ft. lbs. (22–29 Nm)
5. Bleed the brake system.
6. Install the wheels.

REAR

1. Before servicing the vehicle, refer to the precautions section.

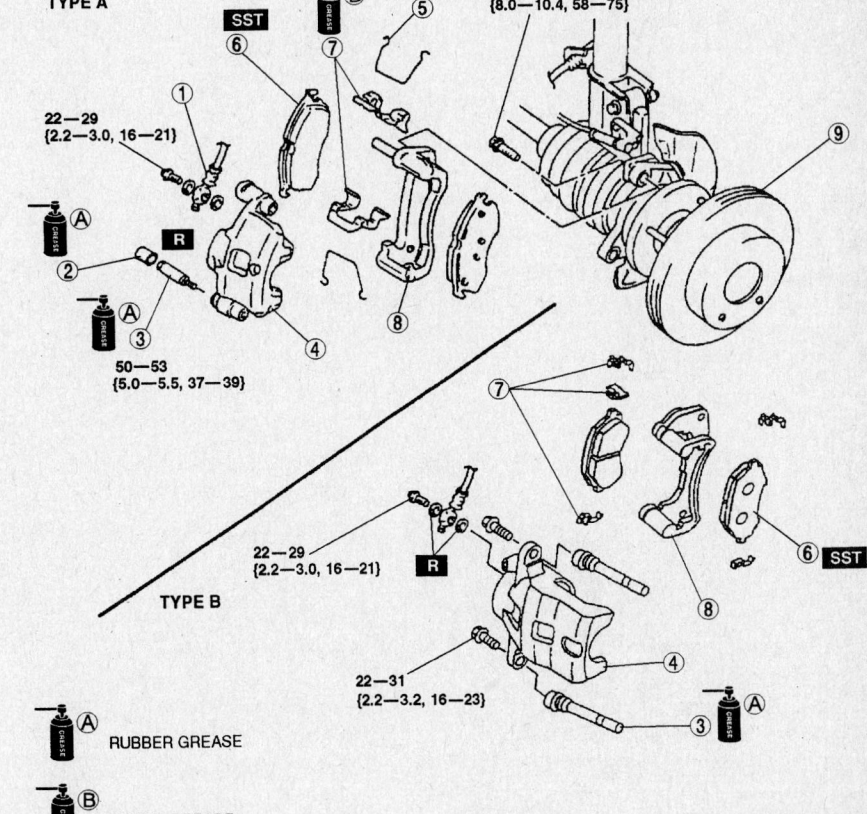

1	Flexible hose	6	Disc pad
2	Cap (type A only)	7	Guide plate
3	Guide pin	8	Mounting support
4	Caliper	9	Disc plate
5	M-spring (type A only)		

N·m {kgf·m, ft·lbf}

42356-MAZC-GKT

Exploded view of the type A and B brake systems–Protégé with disc brakes

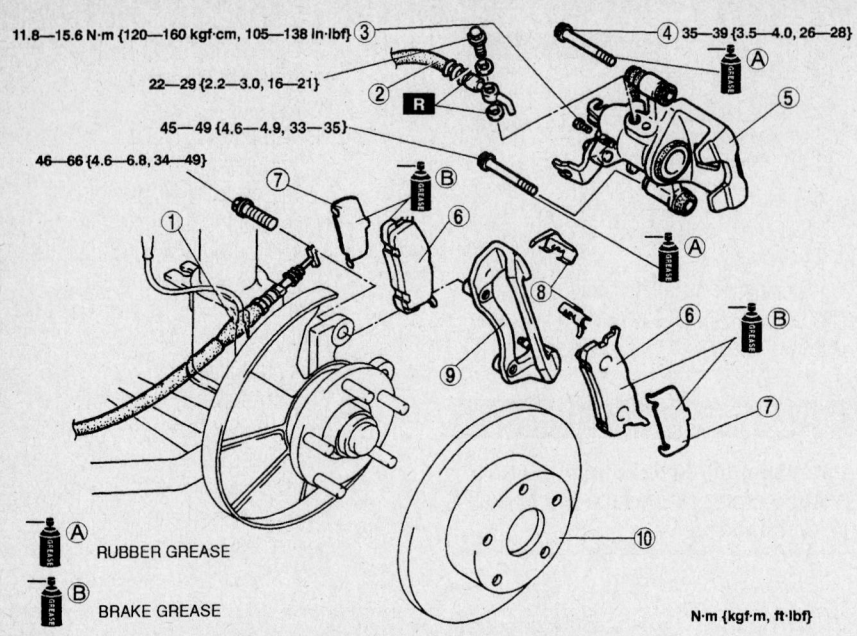

11.8—15.6 N·m {120—160 kgf·cm, 105—138 in·lbf} ③

22—29 {2.2—3.0, 16—21} ②

45—49 {4.6—4.9, 33—35}

46—66 {4.6—6.8, 34—49}

④ 35—39 {3.5—4.0, 26—28}

N·m {kgf·m, ft·lbf}

Ⓐ RUBBER GREASE

Ⓑ BRAKE GREASE

1	Parking brake cable, clip	6	Disc pad
2	Flexible hose	7	Shim
3	Screw plug	8	Guide plate
4	Lock bolt	9	Mounting support
5	Caliper	10	Disc plate

42356-MAZC-GWW

Exploded view of rear brake systems–Protégé with disc brakes

2. Remove or disconnect the following:
- Wheels
- Parking brake cable from the cable bracket and the operating lever
- Flexible brake line from the caliper assembly

3. Turn the manual adjustment gear counterclockwise with an Allen wrench to pull the caliper piston inward (turn until it stops).
- Caliper mounting bolts
- Caliper

To install:

4. Install or connect the following:
- Caliper. Torque the caliper mount bolts to 26–28 ft. lbs. (35–39 Nm).
- Brake hose. Torque the line bolt to 16–22 ft. lbs. (22–30 Nm).
- Parking brake cable

5. Install the wheels and bleed the brake system.

626

FRONT

1. Before servicing the vehicle, refer to the precautions section.

2. Remove or disconnect the following:
- Wheels
- Flexible brake hose from the caliper

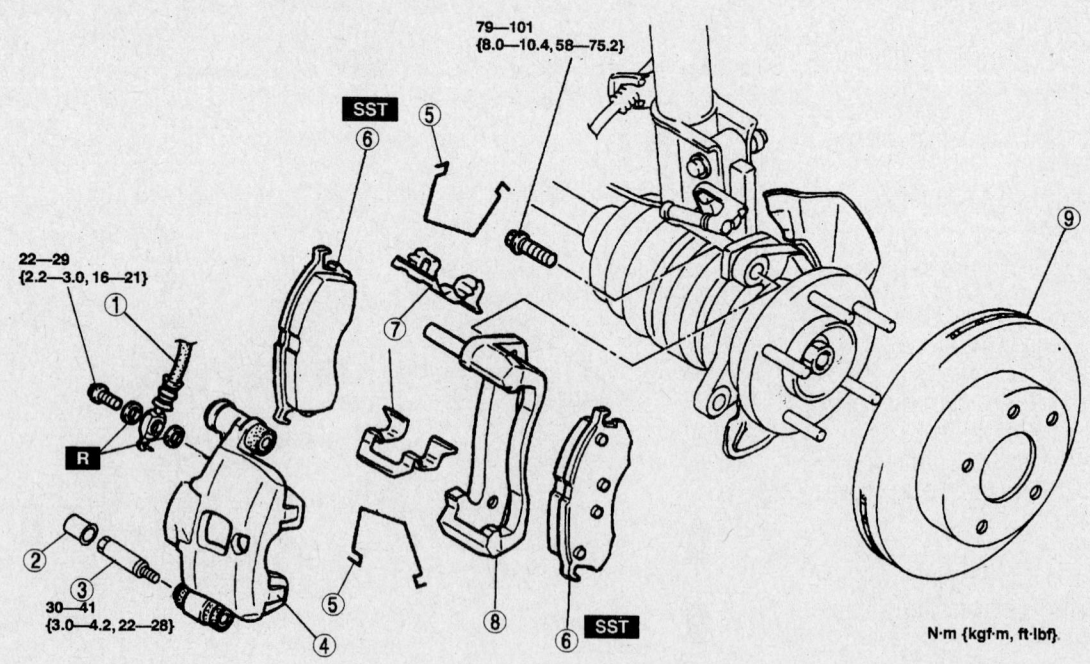

79—101 {8.0—10.4, 58—75.2}

SST

22—29 {2.2—3.0, 16—21}

R

30—41 {3.0—4.2, 22—28}

SST

N·m {kgf·m, ft·lbf}

1	Flexible hose	6	Disc pad
2	Cap	7	Guide plate
3	Lock bolt	8	Mounting support
4	Caliper	9	Disc plate
5	M-spring		

42356-MAZC-G89

Front disc brakes–626

- Lower caliper bolt and pivot the caliper upward. Slide the top of the caliper off of the top pin and remove it from the vehicle.

To install:

3. Lubricate the caliper pin and slide the caliper onto the guide pinch Pivot the caliper over the brake pads.

4. Install or connect the following:
- Brake hose to the caliper and tighten the hose nut to 16–21 ft. lbs. (22–29 Nm)
- Caliper mounting bolt and tighten the bolt to 22–28 ft. lbs. (30–41 Nm)

5. Bleed the brake system and inspect the brake system for proper operation.

6. Install the wheels.

REAR

1. Before servicing the vehicle, refer to the precautions section.

2. Remove the wheels. Loosen the parking brake cable adjustment from inside the vehicle.

3. Remove or disconnect the following:
- Parking brake cable from the cable bracket and the operating lever
- Flexible brake line from the caliper assembly

- Caliper upper mounting bolt and pivot the caliper downward. Slide the caliper off of the guide pinch
- Caliper

To install:

4. Lubricate the caliper pin and slide the caliper onto the guide pinch Pivot the caliper over the brake pads.

5. Install or connect the following:
- Brake hose to the caliper and tighten the hose nut to 16–21 ft. lbs. (22–29 Nm)
- Upper caliper mounting bolt and tighten the bolt to 26–28 ft. lbs. (35–39 Nm)

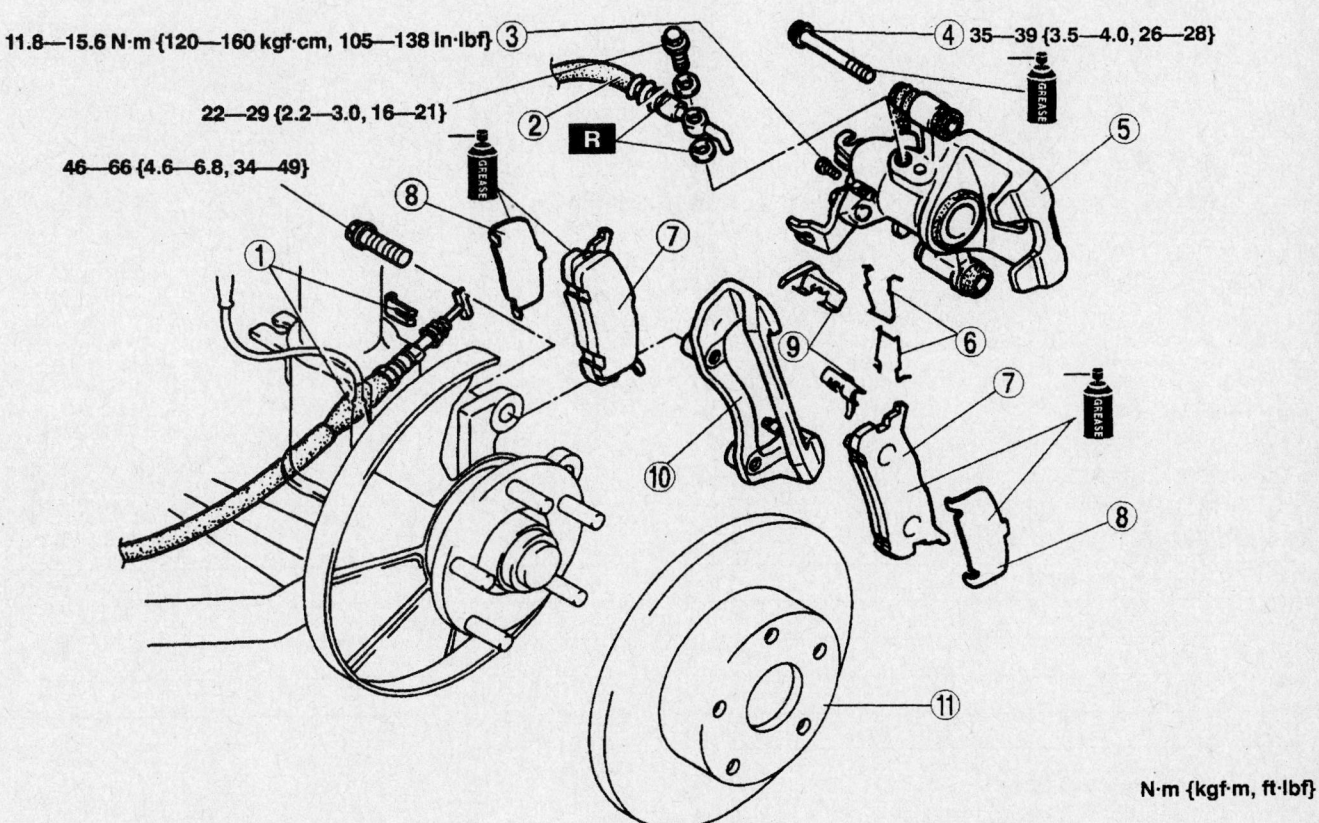

11.8—15.6 N·m {120—160 kgf·cm, 105—138 in·lbf}

22—29 {2.2—3.0, 16—21}

46—66 {4.6—6.8, 34—49}

35—39 {3.5—4.0, 26—28}

N·m {kgf·m, ft·lbf}

1	Parking brake cable, clip	8	Shim
2	Flexible hose	9	Guide plate
3	Screw plug	10	Mounting support
4	Lock bolt	11	Disc plate
5	Caliper		
6	Spring		
7	Disc pad		

42356-MAZC-G91

Rear disc brakes—626

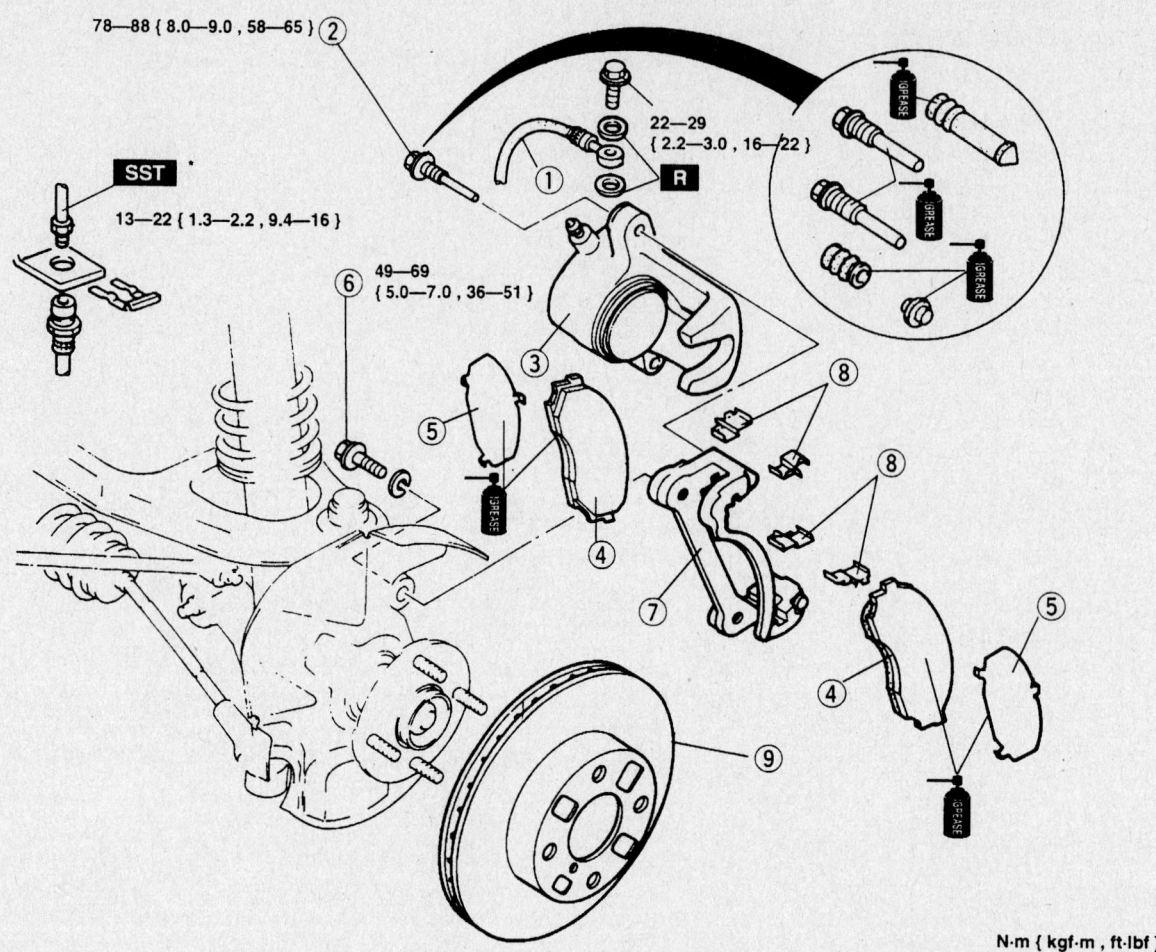

78—88 { 8.0—9.0 , 58—65 }

22—29
{ 2.2—3.0 , 16—22 }

SST

13—22 { 1.3—2.2 , 9.4—16 }

49—69
{ 5.0—7.0 , 36—51 }

N·m { kgf·m , ft·lbf }

1	Brake hose
2	Connecting bolt
3	Caliper
4	Disc pad
5	Shim

6	Bolt
7	Mounting support
8	Guide plate
9	Disc plate

Disc Plate Removal Note
- Mark the wheel hub bolt and disc plate before removal for reference during installation.

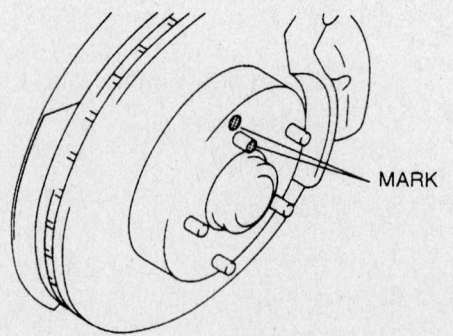

MARK

93016G30

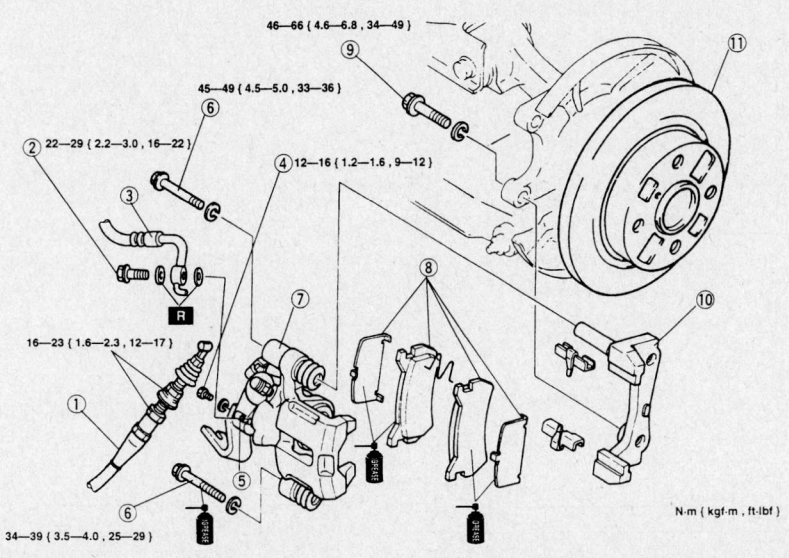

46—66 (4.6—6.8 , 34—49)

45—49 (4.5—5.0 , 33—36)

22—29 (2.2—3.0 , 16—22)

12—16 (1.2—1.6 , 9—12)

16—23 (1.6—2.3 , 12—17)

34—39 (3.5—4.0 , 25—29)

N·m (kgf·m , ft·lbf)

1	Parking brake cable		5	Manual adjustment gear ☞ Removal Note ☞ Installation Note
2	Connecting bolt		6	Lock bolt
3	Brake hose			
4	Plug			

93016G31

Rear disc brakes—2002–04 Miata

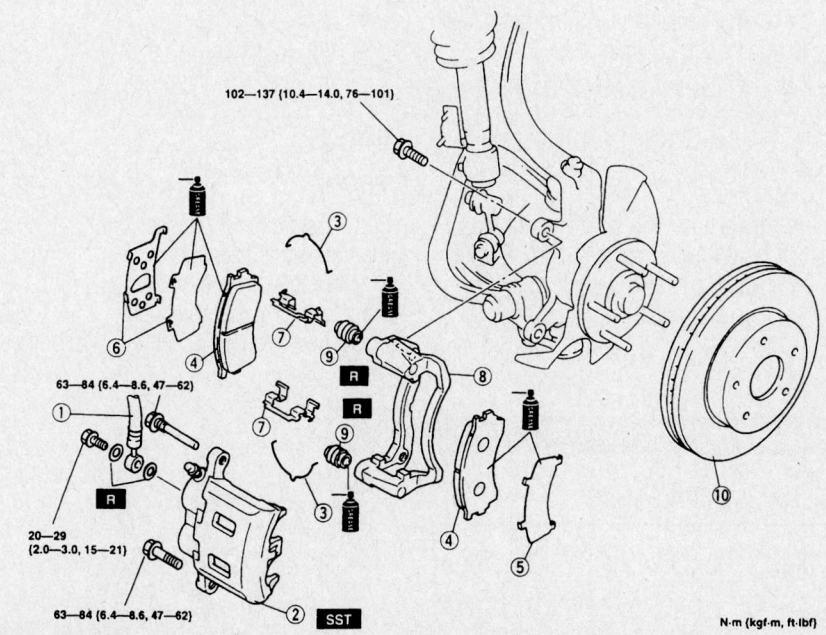

102—137 (10.4—14.0, 76—101)

63—84 (6.4—8.6, 47—62)

20—29 (2.0—3.0, 15—21)

63—84 (6.4—8.6, 47—62)

SST

N·m (kgf·m, ft·lbf)

1	Flexible hose		7	Guide plate
2	Caliper		8	Mounting support
3	V-spring		9	Boot
4	Disc pad		10	Disc plate
5	Outer shim			
6	Inner shim			

93016G32

Front disc brakes—Millenia

- Parking brake cable to the cable bracket and the operating lever
6. Bleed the brake system.
7. Install the wheels.

2002–04 Miata

FRONT

1. Before servicing the vehicle, refer to the precautions section.
2. Remove or disconnect the following:
- Wheels
- Flexible brake hose from the caliper
- Upper and lower caliper bolts
- Caliper from the vehicle

To install:
3. Install or connect the following:
- Caliper on the brake disc
- Caliper mounting bolts and tighten the bolts to 58–65 ft. lbs. (79–88 Nm)
4. Replace the washers for the brake line.
- Brake hose to the caliper and tighten the hose nut to 16–21 ft. lbs. (22–29 Nm)
5. Bleed the brake system and inspect the brake system for proper operation.
6. Install the wheels.

REAR

1. Before servicing the vehicle, refer to the precautions section.
2. Loosen the parking brake cable adjustment from inside the vehicle.
3. Remove or disconnect the following:
- Wheels
- Parking brake cable from the cable bracket and the operating lever
- Flexible brake line from the caliper assembly
- Cover for the manual adjustment gear. Insert an Allen wrench and turn counterclockwise to retract the caliper piston.
- Caliper mounting bolts and remove the caliper from the vehicle

To install:
4. Install or connect the following:
- Caliper and install the mounting bolts. Tighten the upper bolt to 33–36 ft. lbs. (45–49 Nm), and the lower bolt to 26–28 ft. lbs. (35–39 Nm).
5. Turn the manual adjustment gear clockwise to return the caliper until the brake pads just touch the disc, then turn counterclockwise ⅓ of a turn. Replace the cover.
6. After replacing the washers, connect the brake hose to the caliper. Tighten the hose nut to 16–21 ft. lbs. (22–29 Nm).

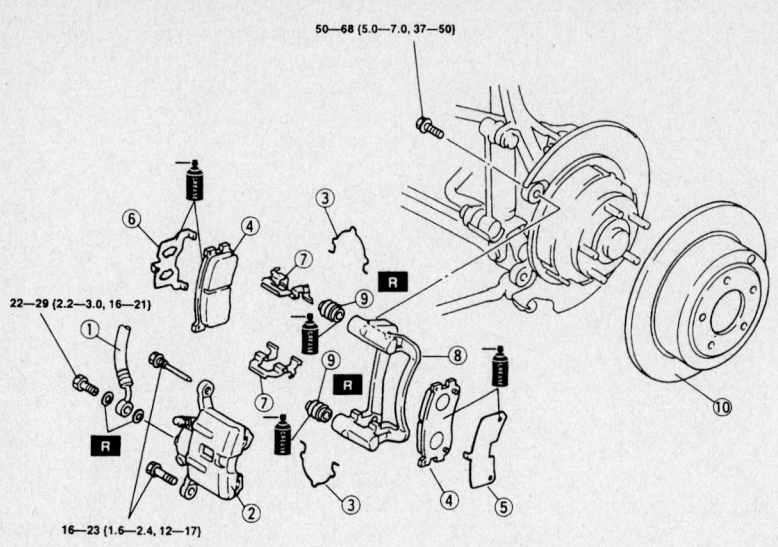

50—68 (5.0—7.0, 37—50)

22—29 (2.2—3.0, 16—21)

16—23 (1.6—2.4, 12—17)

N·m (kgf·m, ft-lbf)

1	Flexible hose
2	Caliper
3	V-spring
4	Disc pad
5	Outer shim
6	Inner shim
7	Guide plate

8	Mounting support
9	Boot
10	Disc plate

93016G33

Rear disc brakes—Millenia

7. Connect the parking brake cable to the cable bracket and the operating lever.

8. Bleed the brake system and adjust the parking brake.

9. Install the wheels.

Millenia

FRONT

1. Before servicing the vehicle, refer to the precautions section.

2. Remove or disconnect the following:
   - Wheels
   - Brake hose and brake pipe.
   - Caliper mounting bolts and the caliper

To install:

3. Install or connect the following:
   - Caliper. Torque the caliper mounting bolts to 47–62 ft. lbs. (63–84 Nm).
   - Brake hose and pipe and tighten to 15–21 ft. lbs. (22–29 Nm). Fill the master cylinder with clean brake fluid and bleed the hydraulic system.
   - Wheels

REAR

1. Before servicing the vehicle, refer to the precautions section.

2. Remove or disconnect the following:
   - Wheels
   - Brake hose
   - Caliper bracket mounting bolts and remove the caliper

To install:

3. Install or connect the following:
   - Caliper. Torque the caliper mounting bolts to 12–17 ft. lbs. (16–23 Nm).
   - Brake hose and pipe and tighten to 16–21 ft. lbs. (22–29 Nm). Fill the master cylinder with clean brake fluid and bleed the hydraulic system.
   - Wheels

Disc Brake Pads

REMOVAL & INSTALLATION

Protégé

FRONT

1. Before servicing the vehicle, refer to the precautions section.

2. Remove or disconnect the following:
   - Wheels
   - Cap on type **A** brakes or bolt on typebrakes
   - Caliper

- Spring on type **A** brakes
- Brake pads

To install:

3. Install or connect the following:
   - Caliper on the brake disc
   - Caliper mounting bolts and tighten the bolts on **A** type brakes to 37–39 ft. lbs. (50–53 Nm) or 16–23 ft. lbs. (22–31 Nm) on type **B** brakes
   - Brake pads
   - Spring on type **A** b rakes

4. Install the wheels.

REAR

1. Before servicing the vehicle, refer to the precautions section.

2. Remove or disconnect the following:
   - Wheels
   - Parking brake cable from the cable bracket and the operating lever

3. Turn the manual adjustment gear counterclockwise with an Allen wrench to pull the caliper piston inward (turn until it stops).
   - Caliper
   - Pads

To install:

4. Install or connect the following:
   - Pads
   - Caliper. Torque the caliper mount bolts to 26–28 ft. lbs. (35–39 Nm).
   - Parking brake cable

5. Install the wheels and bleed the brake system.

626

FRONT

1. Before servicing the vehicle, refer to the precautions section.

2. Remove some of the brake fluid from the master cylinder reservoir.

3. Remove or disconnect the following:
   - Wheels

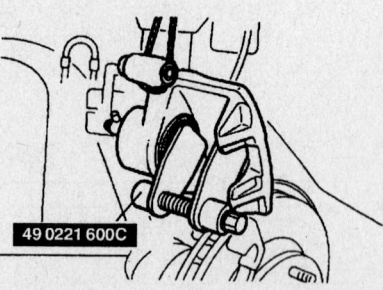

49 0221 600C

42356-MAZC-G90

Using Mazda tool 49-0221-600C and the old inner brake pad, push the caliper piston into the caliper bore—626

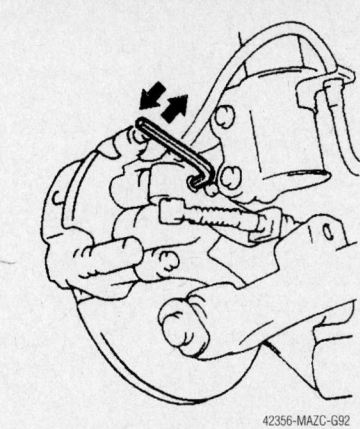

**Location of the manual adjustment gear
on the rear caliper—626**

- Caliper lower mounting bolt and pivot the caliper up and support it
- Brake pads, shims and pin

4. Using Mazda tool 49-0221-600C and the old inner brake pad, push the caliper piston into the caliper bore.

To install:

5. Install or connect the following:
- Brake pads and shims to the caliper support
- Caliper over the brake pads
- Caliper mounting bolt and torque to 22–28 ft. lbs. (30–41 Nm)
- Wheels

6. Test the brakes for proper operation.

REAR

1. Before servicing the vehicle, refer to the precautions section.
2. Remove some of the brake fluid from the master cylinder reservoir.
3. Loosen the parking brake cable adjustment from inside the vehicle.
4. Remove or disconnect the following:
- Wheels.
- Parking brake cable from the cable bracket and the operating lever
- Upper caliper mounting bolt and pivot the caliper downward off of the pads
- Brake pads and spring clips from the caliper support

To install:

5. Turn the manual adjustment gear counterclockwise using an Allen wrench to retract the caliper until it stops.
6. Install or connect the following:
- Brake pads, shims and spring clips to the caliper support. Pivot the caliper over the brake pads.

7. Turn the adjustment gear clockwise until the pads start to touch the rotor and back of the gear ⅓ turn.
8. Lubricate the top caliper mounting

bolt, then tighten the bolt to 26–28 ft. lbs. (35–39 Nm). Attach the parking brake cable to the operating lever.

9. Adjust the parking brake cable, as required.
10. Install the wheels.
11. Test the brakes for proper operation.

2002–04 Miata

FRONT

1. Before servicing the vehicle, refer to the precautions section.
2. Remove some of the brake fluid from the master cylinder reservoir.
3. Remove or disconnect the following:
- Wheels
- Lower lockbolt, and pivot the brake caliper upwards
- Spring, if equipped with 15 inch wheels
- Brake pads and shim

4. Using Mazda piston compressor tool 49-0221-600C and the old inner brake pad, push the caliper piston into the caliper bore.

To install:

- Brake pads and shims onto the caliper
- Caliper
- Spring, if equipped with 15 inch wheels
- Lockbolt and tighten to 58–65 ft. lbs. (78–88 Nm)
- Wheels

5. Test the brakes for proper operation.

REAR

1. Before servicing the vehicle, refer to the precautions section.
2. Remove some of the brake fluid from the master cylinder reservoir.
3. Remove or disconnect the following:
- Wheels
- Plug for the manual adjustment gear. Using an Allen wrench turn the gear counterclockwise to retract the caliper piston.
- Lower caliper mounting bolt and pivot the caliper upward off of the pads
- Brake pads and spring clips from the caliper support

To install:

4. Install or connect the following:
- Brake pads, shims and spring clips to the caliper support. Pivot the caliper over the brake pads.
- Lubricate and install the lower caliper mounting bolt. Tighten the bolt to 26–28 ft. lbs. (35–39 Nm).

5. Turn the manual adjusting gear clock-

wise until the piston contacts the brake disc, then turn it clockwise ⅓ turn. Replace the plug.

6. Install the wheels.
7. Test the brakes for proper operation.

Millenia

FRONT

1. Before servicing the vehicle, refer to the precautions section.
2. Remove or disconnect the following:
- Wheels
- Bottom caliper lock pin and swing the caliper upwards
- V-springs the pads and shims

To install:

3. Install or connect the following:
- Brake pads, shims and V-springs

4. Press the caliper pistons back into their cylinders.
- Calipers. Torque the lock pin to 47–62 ft. lbs. (63–84 Nm).
- Wheels

REAR

1. Before servicing the vehicle, refer to the precautions section.
2. Remove or disconnect the following:
- Wheels
- Lock pin and rotate the caliper upwards
- V-springs, pads and the shims from the pads

To install:

3. Press the caliper piston back into the cylinder.
4. Install or connect the following:
- Pads, shims, and V-springs
- Caliper and torque the lock pin to 37–50 ft. lbs. (50–68 Nm)
- Wheels

Brake Drums

REMOVAL & INSTALLATION

All Models

1. Before servicing the vehicle, refer to the precautions section.
2. Remove or disconnect the following:
- Rear wheel
- Screw securing the rear brake drum, and pull the brake drum outward to remove

To install:

3. Install or connect the following:
- Drum and retaining screw
- Wheels and adjust the rear brakes as necessary

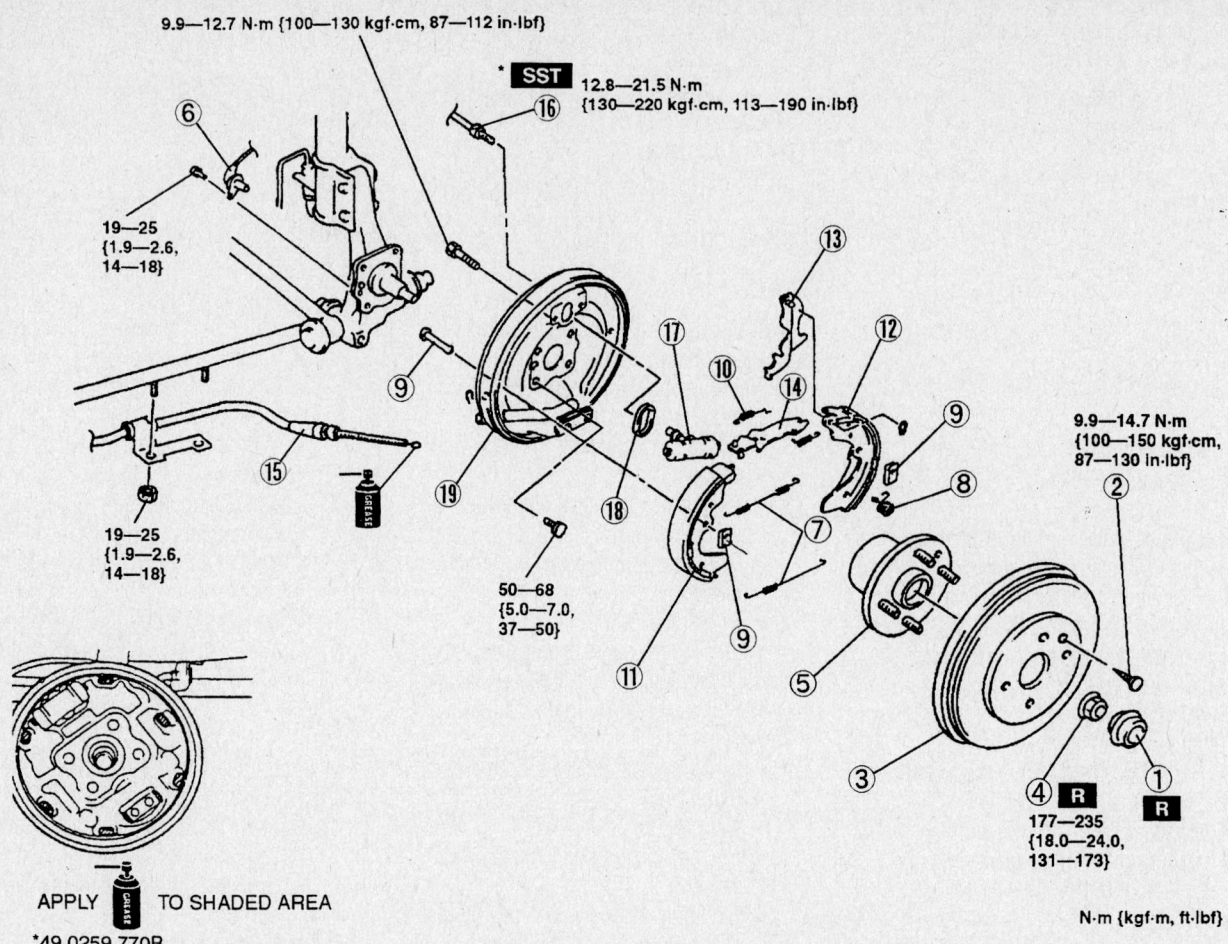

9.9—12.7 N·m {100—130 kgf·cm, 87—112 in·lbf}

* SST 12.8—21.5 N·m {130—220 kgf·cm, 113—190 in·lbf}

19—25 {1.9—2.6, 14—18}

9.9—14.7 N·m {100—150 kgf·cm, 87—130 in·lbf}

GREASE

19—25 {1.9—2.6, 14—18}

50—68 {5.0—7.0, 37—50}

177—235 {18.0—24.0, 131—173}

N·m {kgf·m, ft·lbf}

APPLY GREASE TO SHADED AREA
*49 0259 770B

1	Hub cap	11	Leading shoe
2	Screw	12	Trailing shoe
3	Brake drum	13	Operating lever
4	Locknut	14	Adjuster
5	Wheel hub	15	Parking brake cable
6	ABS wheel-speed sensor (if equipped)	16	Brake pipe
7	Return spring	17	Wheel cylinder
8	Lever spring	18	O-ring
9	Hold pin and hold spring	19	Backing plate
10	Anti-rattle spring		

42356-MAZC-GKU

Typical rear drum brake assembly, Protégé shown others similar

Brake Shoes

REMOVAL & INSTALLATION

1. Before servicing the vehicle, refer to the precautions section.
2. Remove or disconnect the following:
   - Wheel and the brake drum
   - ABS speed sensor, if equipped
   - Parking brake cable from the back-side of the brake backing plate
   - Upper return spring
   - Hold pin and the spring from the lower shoe (leading side)
   - Lower return spring and the anti-rattle spring and shoe from the lower shoe (leading side)
   - Hold pin and spring and remove the upper brake shoe (trailing side)

To install:

3. Install or connect the following:
   - Upper (trailing side) brake shoe to the operating lever and then to the wheel cylinder and backing plate
   - Brake shoe hold spring and hold pin
   - Anti-rattle spring
   - Lower return spring to both brake shoes
   - Leading side brake shoe to the operating lever and then to the wheel cylinder and anchor plate
   - Hold spring and hold pin to the leading side brake shoe
   - Upper return spring
   - Brake drum
   - ABS speed sensor, if equipped
   - Wheels and adjust the brakes

SPECIFICATION AND MAINTENANCE CHARTS

ENGINE AND VEHICLE IDENTIFICATION

Code ①	Liters (cc)	Cu. In.	Cyl.	Fuel Sys.	Engine Type	Eng. Mfg.	Code ②	Year
LF ③	2.0 (1999)	121.9	4	MPFI	DOHC	Mazda	4	2004
LF ④	2.0 (1999)	121.9	4	MPFI	DOHC	Mazda	5	2005
L3 ⑤	2.3 (2261)	137.9	4	MPFI	DOHC	Mazda	6	2006
L3 ⑥	2.3 (2261)	137.9	4	MPFI	DOHC	Mazda		
AJ	3.0 (2999)	181.1	6	MPFI	DOHC	Mazda		

MPFI: Multi-Point Fuel Injection

DOHC: Double Over Head Cam

① Located above the starter

② 10th digit of the Vehicle Identification Number (VIN)

③ Mazda3

④ MX-5 Miata

⑤ Mazda6

⑥ Mazdaspeed6 is a 2.3L engine with a turbocharger

09482_MAZC2_C0001

GENERAL ENGINE SPECIFICATIONS

Year	Model	Engine Displacement Liters (VIN)	Net Horsepower @ rpm	Net Torque @ rpm (ft. lbs.)	Bore x Stroke (in.)	Compression Ratio	Oil Pressure @ rpm
2004	Mazda 3 (I)	2.0 (LF)	①	②	344x3.27	10.0:1	33.9-75.5@3000
	Mazda 3 (S)	2.3 (L3)	160@6500	150@4500	344x3.70	9.7:1	57-94@3000
	Mazda 6 (I)	2.3 (L3)	160@6000	155@4000	344x3.70	9.7:1	57-94@3000
	Mazda 6 (S)	3.0 (AJ)	220@6300	192@5000	3.50x3.13	10.0:1	20@45@1500
2006-06	Mazda 3 (I)	2.0 (LF)	①	②	344x3.27	10.0:1	33.9-75.5@3000
	Mazda 3 (S)	2.3 (L3)	160@6500	150@4500	344x3.70	9.7:1	57-94@3000
	Mazda 6 (I)	2.3 (L3)	160@6000	155@4000	344x3.70	9.7:1	57-94@3000
	Mazda 6 (S)	3.0 (AJ)	220@6300	192@5000	3.50x3.13	10.0:1	20@45@1500
	Mazdaspeed6	2.3 (L3)	274@5500	280@3000	344x3.70	NA	43-79@3000
	MX5-Miata	2.0 (LF)	170@6700	140@5000	344x3.27	10.0:1	49-85@3000

NA: Not Available

EFI: Electronic Fuel Injection

① Partial Zero Emission Vehicle (PZEV): 144@6500

  Except PZEV states: 148@6500

② Partial Zero Emission Vehicle (PZEV): 132@4500

  Except PZEV states: 135@4500

09482_MAZC2_C0002

ENGINE TUNE-UP SPECIFICATIONS

Year	Engine Displacement Liters (VIN)	Spark Plug Gap (in.)	Ignition Timing (deg.)		Fuel Pump (psi)	Idle Speed (rpm)		Valve Clearance	
			MT	AT		MT	AT	In.	Ex.
2004	2.0 (LF)	0.049-0.053	8B	8B	36	600-700	650-750	0.008-0.0011	0.008-0.011
	2.3 (L3)	0.049-0.053	8B	8B	29	600-700	650-750	0.008-0.0011	0.008-0.011
	3.0 (AJ)	0.049-0.053	10B	10B	36	700-800	500-800	NA	NA
2005-06	2.0 (LF) ①	0.049-0.053	8B	8B	36	600-700	650-750	0.008-0.0011	0.008-0.011
	2.0 (LF) ②	0.050-0.053	8B	8B	36	700-800	670-770	0.008-0.0011	0.010-0.012
	2.3 (L3) ③	0.049-0.053	8B	8B	29	600-700	650-750	0.008-0.0011	0.008-0.011
	2.3 (L3) ④	28-31	8B	8B	60-71	650-750	—	0.008-0.011	0.010-0.012
	3.0 (AJ)	0.049-0.053	10B	10B	36	700-800	500-800	NA	NA

NOTE: The Vehicle Emission Control Information label often reflects specification changes made during production. The label figures must be used if they differ from those in this chart.

NA: Not Available

B: Before top dead center

HYD: Hydraulic

① Mazda3

② MX-5 Miata

③ Mazda6

④ Mazdaspeed6

09482_MAZC2_C0003

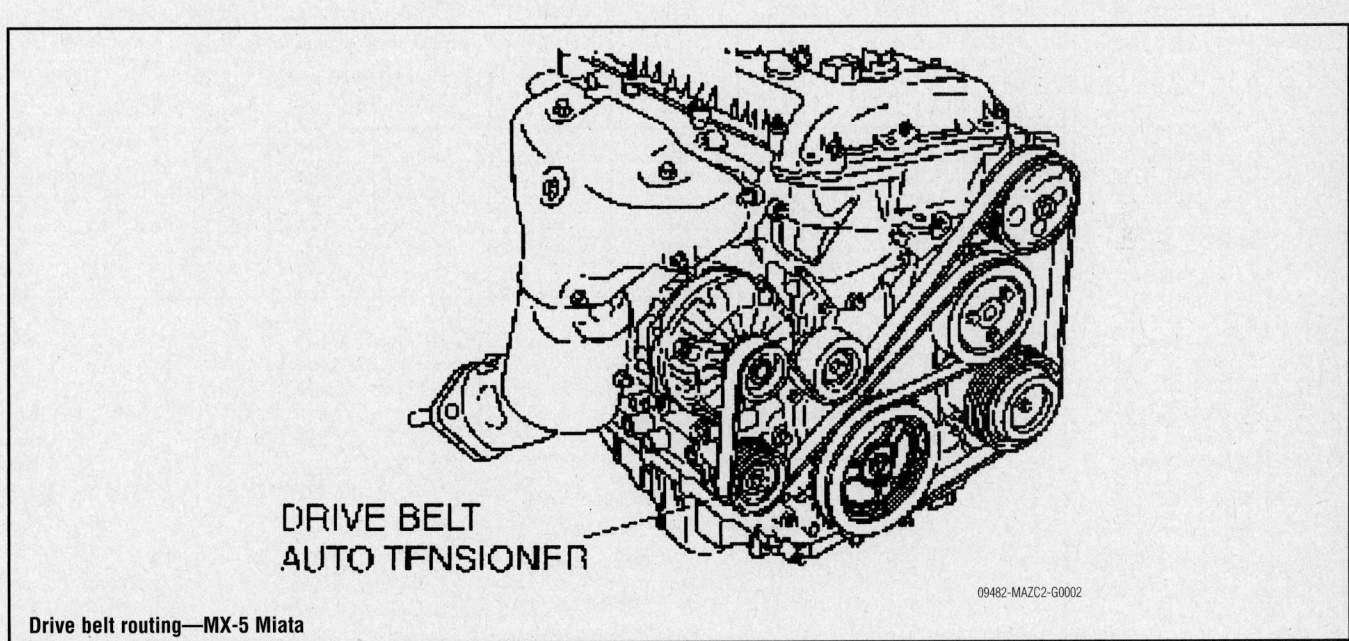

DRIVE BELT
AUTO TENSIONER

09482-MAZC2-G0002

Drive belt routing—MX-5 Miata

CAPACITIES

Year	Model	Engine Displacement Liters (VIN)	Engine Oil with Filter (qts.)	Transmission (pts.) Man	Transmission (pts.) Auto.	Drive Axle Front (pts.)	Drive Axle Rear (pts.)	Fuel Tank (gal.)	Cooling System (qts.)
2004	Mazda 3 (I)	2.0 (LF)	4.5	6.06	15.2	①	—	14.5	7.9
	Mazda 3 (S)	2.3 (L3)	4.5	6.06	15.2	①	—	14.5	7.9
	Mazda 6 (I)	2.3 (L3)	4.5	6.06	15.2	①	—	14.5	7.9
	Mazda 6 (S)	3.0 (AJ)	6.0	4.8	19.4	①	—	14.5	7.9
2005-06	Mazda 3 (I)	2.0 (LF)	4.5	6.06	15.2	①	—	14.5	7.9
	Mazda 3 (S)	2.3 (L3)	4.5	6.06	15.2	①	—	14.5	7.9
	Mazda 6 (I)	2.3 (L3)	4.5	6.06	15.2	①	—	14.5	7.9
	Mazdaspeed6 (S)	2.3 (L3)	6.0	5.38	—	①	—	15.9	8.5
	Mazda 6 (S)	3.0 (AJ)	6.0	4.8	19.4	①	—	14.5	7.9
	MX5-Miata	2.0 (LF)	4.5	4.4	15.6	①	—	12.7	7.9

NOTE: All capacities are approximate. Add fluid gradually and check to be sure a proper fluid level is obtained.

① Included in transaxle

09482_MAZC2_C0004

TORQUE SPECIFICATIONS
All readings in ft. lbs.

Year	Engine Displacement Liters (VIN)	Cylinder Head Bolts	Main Bearing Bolts	Rod Bearing Bolts	Crankshaft Damper Bolts	Flywheel Bolts	Manifold Intake	Manifold Exhaust	Spark Plugs	Oil Pan Drain Plug
2004	2.0 (LF)	①	NA	NA	②	③	④	32-47	8-10	22-30
	2.3 (L3)	①	NA	NA	②	③	④	32-47	8-10	19-22
	3.0 (AJ)	⑤	NA	NA	⑥	⑦	⑧	14-18	⑨	19-22
2005-06	2.0 (LF) ⑩	①	NA	NA	②	③	④	32-47	8-10	22-30
	2.0 (LF) ⑪	⑭	NA	NA	⑮	88	⑯	32-47	8-10	22-30
	2.3 (L3) ⑫	①	NA	NA	②	③	④	32-47	8-10	19-22
	2.3 (L3) ⑬	①	NA	NA	②	③	13-16	32-47	8-10	22-30
	3.0 (AJ)	⑤	NA	NA	⑥	⑦	⑧	14-18	⑨	19-22

NA: Not Available

① Step 1: 97 inch lbs.
  Step 2: Tighten 12.5 ft. lbs
  Step 3: Tighten 34.6 ft. lbs.
  Step 4: Tighten 88 degrees
  Step 5: Tighten 88 degrees

② 71-77 ft. lbs. plus 87-93 degrees

③ Manual transmission: 80-85 ft. lbs.
  Automatic transmission: 71-76 ft. lbs.

④ Mazda 3 with the 2.0L (LF) and 2.3L (L3) engines: 12-14 ft. lbs.
  Mazda 6 with the 2.3L (L3) engine: 71-101 inch lbs.

⑤ Step 1: 17-19 ft. lbs.
  Step 2: Tighten 85-95 degees
  Step 3: Repeat step 2

⑥ Step 1: 88 ft. lbs.
  Step 2: loosen one full turn
  Step 3: Tighten 35-39 ft. lbs.
  Step 4: Tighten 85-95 degrees

⑦ G35M-R A26MX-R manual or FN4A-EL automatic transmission: 80-85 ft. lbs.
  A65M-R manual or JA5A-EL automatic transmission: 54-64 ft. lbs.

⑧ Lower intake: 72-106 inch lbs.
  Upper intake plenum: 72-106 inch lbs.

⑨ 79-177 inch lbs.

⑩ Mazda3

⑪ MX-5 Miata

⑫ Mazda6

⑬ Mazdaspeed6

⑭ Step 1:447 inch lbs.
  Step 2: Tighten to 10-12.5 ft. lbs
  Step 3: Tighten 32-34 ft. lbs.
  Step 4: Tighten 88-92 degrees
  Step 5: Tighten 88-92 degrees

⑮ Step 71-77 ft. lbs.
  Step 4: Tighten 87-93 degrees

⑯ Lower manifold 12-18 ft. lbs.
  Dynamic Chamber 12-14 ft. lbs.

09482_MAZC2_C0008

WHEEL ALIGNMENT

Year	Model		Caster Range (+/-Deg.)	Caster Preferred Setting (Deg.)	Camber Range (+/-Deg.)	Camber Preferred Setting (Deg.)	Toe-in (in.)
2004	Mazda 3	F	1.00	+2.58	1.00	-0.40	0.04 +/- 0.12
		R	—	—	1.00	0	0.08 +/- 0.16
	Mazda 6	F	1.00	①	1.00	②	0.12 +/- 0.16
		R	—	—	1.00	③	0.08 +/- 0.16
	Mazda 3	F	1.00	+2.58	1.00	-0.40	0.04 +/- 0.12
		R	—	—	1.00	0	0.08 +/- 0.16
	Mazda 6	F	1.00	①	1.00	②	0.12 +/- 0.16
		R	—	—	1.00	③	0.08 +/- 0.16
	Mazdaspeed6	F	1.00	+2.58	1.00	-0 17' +/- 1.00	0 +/- 0 22'
		R	—	—	1.00	-0 09' +/- 1.00	0 +/- 0 22'
	MX5-Miata	F	1.00	④	1.00	⑤	⑥
		R	—	—	1.00	⑦	⑧

① Except wagon: 0 degrees 11', plus or minus 0 degrees 22'
Wagon: 3 degrees 33' plus or minus 1 degree

② Except wagon: -0 degrees 16', plus or minus 1 degree
Wagon: -0 degrees 16', plus or minus 1 degree

③ Except wagon: -1 degree 09', plus or minus 1 degree
Wagon: -1 degree 02', plus or minus 1 degree

④ 16 inch wheel: 5 degrees 59', plus or minus 1 degree
17 inch wheel: 6 degrees 06', plus or minus 1 degree

⑤ 16 inch wheel: -0 degrees 06', plus or minus 1 degree
17 inch wheel: -0 degrees 14', plus or minus 1 degree

⑥ 16 inch wheel: 0.05 +/- 0.09
17 inch wheel: 0.06 +/- 0.11

⑦ 16 inch wheel: -1 degrees 04', plus or minus 1 degree
17 inch wheel: -1 degrees 11', plus or minus 1 degree

⑧ 16 inch wheel: 0.071 +/- 0.094
17 inch wheel: 0.083 +/- 0.11

09482_MAZC2_C0009

TIRE, WHEEL AND BALL JOINT SPECIFICATIONS

Year	Model	OEM Tires Standard	OEM Tires Optional	Tire Pressures (psi) Front	Tire Pressures (psi) Rear	Wheel Size	Ball Joint Inspection	Lug Nut
2004	Mazda 3	①	①	32	32	②	③	65-87
	Mazda 6	P205/60R16	P215/50R17	32	32	④	⑤	65-87
2005-06	Mazda 3	①	①	32	32	②	③	65-87
	Mazda 6	P205/60R16	P215/50R17	32	32	④	⑤	65-87
	Mazdaspeed6	P205/50R16	P215/50R17 ⑧	32	32	⑨	⑤	65-87
	MX5-Miata	P205/50R16	P205/45R17	29	29	⑥	⑦	65-87

OEM: Original Equipment Manufacturer

PSI: Pounds Per Square Inch

STD: Standard

OPT: Optional

① Standard Steel rim: P195/65R15
Alluminum alloy: P205/55R16

② Standard Steel rim: 15 x 6 J
Alluminum alloy: 16 x 6 1/2 J or 17 x 6 1/2 J

③ Lower arm ball rotation torque, front: 3-10 ft. lbs.

④ Standard Steel rim: 16 x 6 1/2JJ
Alluminum alloy: 16 x 7 JJ or 17 x 7 JJ

⑤ Lower arm ball rotation torque, front: 10.5-19.7 inch lbs.
Lower arm ball rotation torque, rear: 8.86-19.6 inch lbs.
Upper arm ball rotation torque: 13.2 inch lbs. (max)

⑥ Standard rim: 16 x 6 1/2 J
Optional rim: 17 x 7 J

⑦ Lower arm ball rotation torque, front: 4-25 inch lbs.
Upper arm ball rotation torque: 3-19 inch lbs. (max)

⑧ Also avaialable with 215/45/R18

⑨ Standard Steel rim: 16 x 6 1/2JJ
Alluminum alloy: 17 x 7 JJ or 18 x 7 JJ

09482_MAZC2_C0010

BRAKE SPECIFICATIONS

All measurements in inches unless noted

Year	Model			Brake Disc			Brake Drum			Minimum Lining Thickness	Brake Caliper	
			Original Thickness	Minimum Thickness	Maximum Runout	Original Inside Diameter	Max. Wear Limit	Maximum Machine Diameter		Bracket Bolts (ft. lbs.)	Mounting Bolts (ft. lbs.)
2004	Mazda 3	F	NA	0.910	0.002	—	—	—	0.079	75-87	19-22
		R	NA	0.350	0.002	—	—	—	0.079	43-56	19-22
	Mazda 6	F	NA	0.910	0.002	—	—	—	0.079	58-75	36-39
		R	NA	0.310	0.002	—	—	—	0.079	37-49	27-36
2005-06	Mazda 3	F	NA	0.910	0.002	—	—	—	0.079	75-87	19-22
		R	NA	0.350	0.002	—	—	—	0.079	43-56	19-22
	Mazda 6	F	NA	0.910	0.002	—	—	—	0.079	58-75	36-39
		R	NA	0.310	0.002	—	—	—	0.079	37-49	27-36
	Mazdaspeed 6	F	NA	0.910	0.002	—	—	—	0.059	58-75	16-23
		R	NA	0.310	0.002	—	—	—	0.039	37-49	16-23
	MX5-Miata	F	NA	0.790	0.002	—	—	—	0.079	58-75	15-23
		R	NA	0.310	0.002	—	—	—	0.079	36-50	14-18

NA: Not Avilable

F: Front

R: Rear

09482_MAZC2_C0011

SCHEDULED MAINTENANCE INTERVALS
Mazda Car—Mazda 3, Mazda 6, Mazdaspeed 6 and MX5-Miata

TO BE SERVICED	TYPE OF SERVICE	VEHICLE MILEAGE INTERVAL (x1000)												
		7.5	15	22.5	30	37.5	45	52.5	60	67.5	75	82.5	90	97.5
Engine oil & filter	R	✓	✓	✓	✓	✓	✓	✓	✓	✓	✓	✓	✓	✓
Air cleaner element	R					✓					✓			
Engine coolant ① ②	R													
Spark plugs	R										✓			
Bolts & nuts on chassis & body	S/I				✓				✓				✓	
Brake lines, hoses & connections	S/I				✓				✓				✓	
Cooling system	S/I				✓				✓				✓	
Disc brakes	S/I				✓				✓				✓	
Drive belts ③	S/I				✓				✓				✓	
Drive shaft dust boots	S/I				✓				✓				✓	
Exhaust system heat shield	S/I				✓				✓				✓	
Front & rear suspension ball joints	S/I				✓				✓				✓	
Fuel lines & hoses	S/I				✓				✓				✓	
Steering operation & linkages	S/I				✓				✓				✓	
Fuel filter	R								✓					
Valve Clearance ④	I								✓					
Hose & tube for emission	S/I								✓					

R: Replace
S/I: Service or Inspect

① Mazda 6: replace initially at 105,000 miles or 6 months and every 30, 000 miles or 24 months thereafter

② Mazda 3: replace initially at 60,000 miles or 4 years and every 24 months thereafter

③ 2.0L LF engine and 2.3L L3 engine: inspect every 37, 500 miles. 3.0L AJ engine: inspect every 30,000 miles

④ 2.0L LF and 2.3L L3 engine: Audible inspect every 75, 000 miles and if noisy adjust.

FREQUENT OPERATION MAINTENANCE (SEVERE SERVICE)

If a vehicle is operated under any of the following conditions it is considered severe service

- Extremely dusty areas.

- 50% or more of the vehicle operation is in 32°C (90°F) or higher temperatures, or constant operation in temperatures below 0°C (32°F).

- Prolonged idling (vehicle operation in stop and go traffic).

- Frequent short running periods (engine does not warm to normal operating temperatures).

- Police, taxi, delivery usage or trailer towing usage.

Oil & oil filter: change every 5000 miles.

Oil & oil filter (Puerto Rico): change every 3000 miles.

Air cleaner element: service or inspect every 15,000 miles

Automatic transaxle fluid: service or inspect every 15,000 miles.

Bolts & nuts on chassis & body: tighten every 15,000 miles.

Disc brakes: service or inspect every 15,000 miles.

09482_MAZC2_C0012

ENGINE REPAIR

Alternator

REMOVAL

MX-5 Miata

1. Before servicing the vehicle, refer to the precautions section.

> ❊❊ **CAUTION**
>
> **Remove and install all parts when the engine is cold, otherwise they can cause severe burns or serious injury.**

> ❊❊ **CAUTION**
>
> **The alternator can be damaged by the heat from the exhaust manifold. Make sure the alternator duct is installed securely.**

2. Remove the battery and battery tray.
3. Remove the drive belt.
4. P/S pressure hose bracket.
5. Alternator bracket.
6. Alternator duct.
7. B terminal cable Alternator connector.
8. Alternator.

Mazda3

1. Before servicing the vehicle, refer to the precautions section.
2. Remove the battery cover and disconnect the negative battery cable.
3. Remove the under cover and splash shield as an assembly.
4. Remove the plug hole plate.
5. Remove the drive belt.
6. Disconnect the alternator electrical connections.

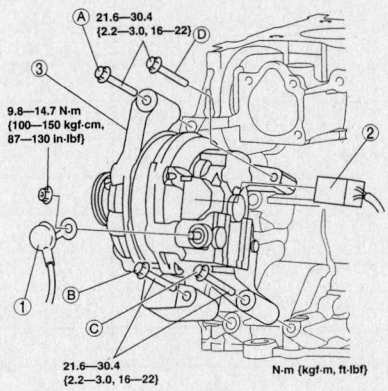

Location of alternator mounting bolts A, B, C, and D—Mazda3 models

7. Position the coolant overflow tank to one side to facilitate alternator removal.
8. Remove the bolts A, B, C, and D in order and the alternator. Refer to the illustration for bolt location.

Mazda6

2.3L (L3) ENGINE

1. Before servicing the vehicle, refer to the precautions section.
2. Disconnect the negative battery cable.
3. Remove the under cover.
4. Remove the drive belt.
5. Remove the alternator duct and heat insulator.
6. Disconnect the alternator electrical connections.
7. Remove the bolts A, B and C. Refer to the illustration for bolt location.

Mazdaspeed6

1. Before servicing the vehicle, refer to the precautions section.
2. Remove the battery and battery tray.
3. Remove the under cover.
4. Remove the drive belt.
5. Remove the charge air cooler cover.
6. Remove the air cleaner.
7. Remove the left side front mudguard, then the resonance chamber.
8. Disconnect the MAF/IAT sensor.
9. Remove the charge air cooler.
10. Remove the charge air cooler bracket.
11. Remove the alternator duct and heat insulator.
12. Disconnect the alternator electrical connections.
13. Remove the bolts A, B and C. Refer to the illustration for bolt location.

3.0L (AJ) ENGINE

1. Before servicing the vehicle, refer to the precautions section.
2. Disconnect the negative battery cable.
3. Remove the under cover.
4. Remove the drive belt.
5. Remove the right hand three way catalytic converter and front pipe.
6. Disconnect the alternator electrical connections.
7. Remove the bolts and the alternator.

INSTALLATION

MX5-Miata

1. Before servicing the vehicle, refer to the precautions section.

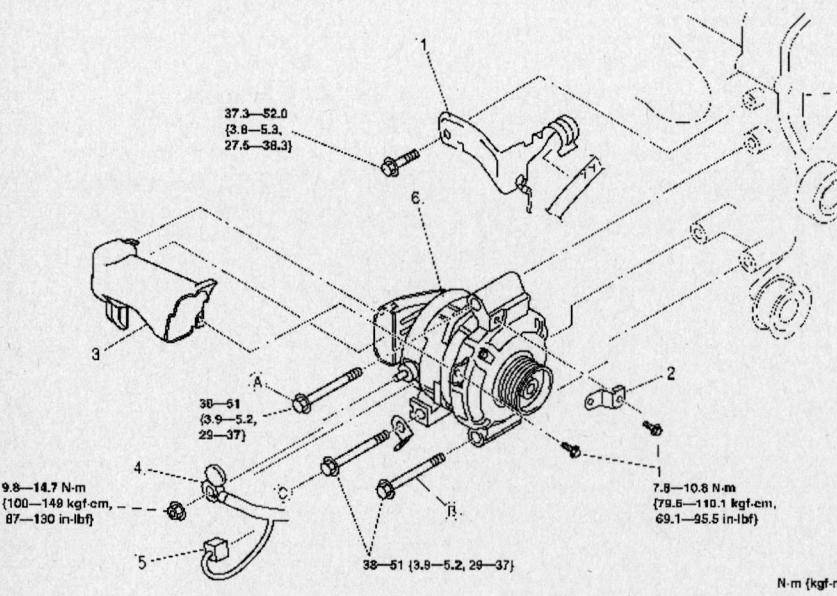

1	P/S pressure hose bracket	4	B terminal cable
2	Generator bracket	5	Generator connector
3	Generator duct	6	Generator

Alternator mounting—MX-5 Miata

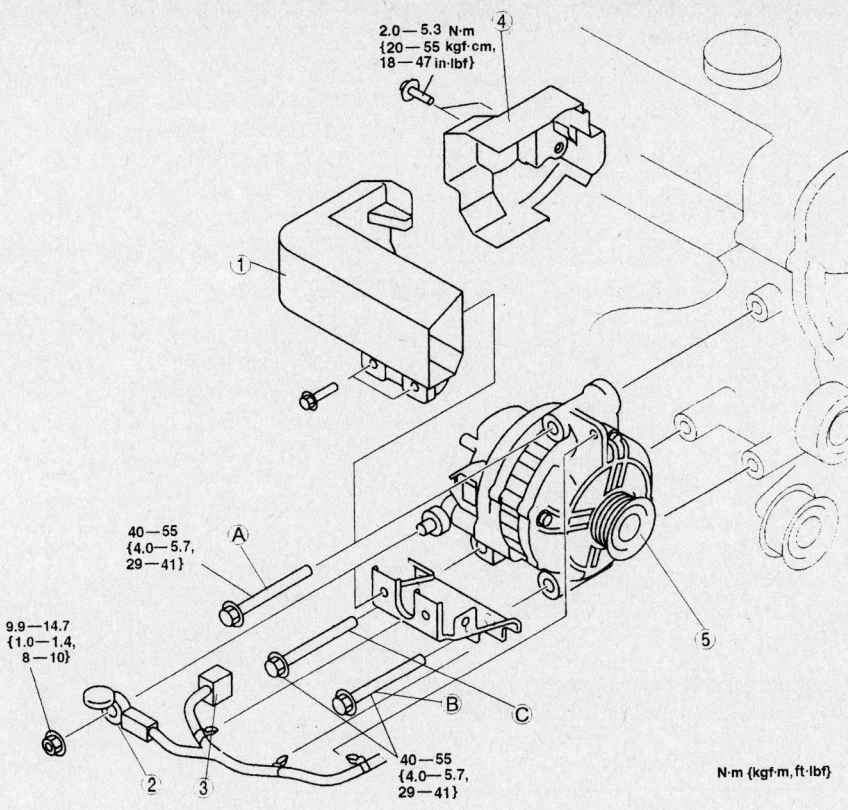

1 Generator duct
2 Terminal B wire
3 Generator connector
4 Generator heat insulator
5 Generator

67162-MAZC-G110

Location of alternator mounting bolts A, B and C—Mazda6 and Mazdaspeed6 models with 2.3L (L3) engine

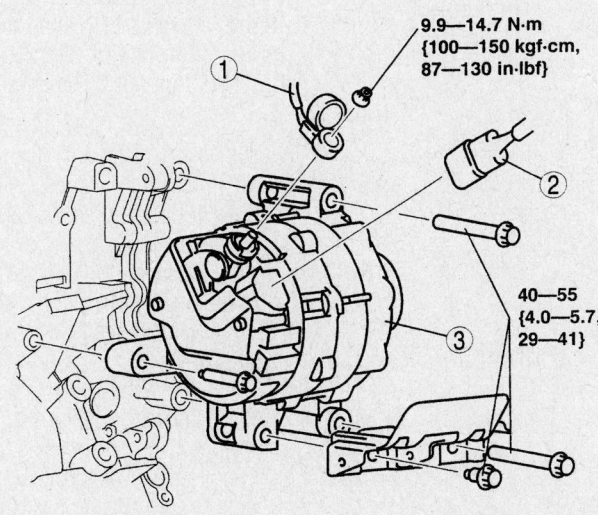

1 Terminal B wire
2 Connector
3 Generator

67162-MAZC-G111

Alternator mounting—Mazda6 models with 3.0L (AJ) engine

2. Installation is the reverse of removal. Please note the following
    a. Tighten bolt A temporarily.
    b. Tighten bolt B, C to 29–37 ft. lbs. (38–51 Nm).
    c. Tighten bolt A to 29–37 ft. lbs. (38–51 Nm).

Mazda3

1. Before servicing the vehicle, refer to the precautions section.
2. Place the alternator in position matching the fixing and engine side holes and hand tighten the bolts A, B, C and D in alphabetical order. Refer to the illustration for bolt locations.
3. Attach the alternator electrical connections. Tighten bolts A, B, C and D in alphabetical order to 16–22 ft. lbs. (21–30 Nm). Refer to the illustration for bolt locations.
4. Install the drive belt.
5. Install the under cover and splash shield.
6. Install the plug hole plate.
7. Connect the negative battery cable and install the battery cover.

Mazda6

2.3L (L3) ENGINE

1. Before servicing the vehicle, refer to the precautions section.
2. Install the alternator, tighten bolt A temporarily, then tighten the bolts in the following order; B, A and C to 29–41 ft. lbs. (40–55 Nm).
3. Connect the alternator electrical connections.
4. Install the heat insulator and tighten the bolt to 18–47 inch lbs. (2–5 Nm).
5. Install the alternator duct
6. Install the under cover.
7. Install the drive belt.
8. Connect the negative battery cable.

Mazdaspeed6

1. Before servicing the vehicle, refer to the precautions section.
2. Install the alternator, tighten bolt A temporarily, then tighten the bolts in the following order; B, A and C to 29–41 ft. lbs. (40–55 Nm).
3. Connect the alternator electrical connections.
4. Install the heat insulator and tighten the bolt to 18–47 inch lbs. (2–5 Nm).
5. Install the alternator duct.
6. Install the charge air cooler bracket
7. Install the charge air cooler.
8. Connect the MAF/IAT sensor.

9. Install the resonance chamber and mudguard.

10. Install the air cleaner.

11. Install the charge air cooler cover.

12. Install the under cover.

13. Install the drive belt.

14. Connect the negative battery cable.

3.0L (AJ) ENGINE

1. Before servicing the vehicle, refer to the precautions section.

2. Install the alternator and tighten the bolts to 29–41 ft. lbs. (40–55 Nm).

3. Connect the alternator electrical connections.

4. Install the right hand three way catalytic converter and front pipe.

5. Install the drive belt.

6. Install the under cover.

7. Connect the negative battery cable.

Ignition Timing

ADJUSTMENT

1. Before servicing the vehicle, refer to the precautions section.

➡ If the information given in the following procedures differs from that on the emission information label located in the engine compartment, follow the directions given on the label. The label often reflects production changes made during the model year.

The timing is controlled by the computer. Ignition timing adjustment is not possible or necessary.

2. If the timing is still not within specification. the following components may be defective:

- Camshaft position (CMP) sensor
- Crankshaft Position (CKP) sensor
- Throttle Position (TP) sensor
- Engine Coolant Temperature (ECT) sensor
- Neutral switch if equipped with a manual transaxle
- Clutch switch if equipped with a manual transaxle
- Transaxle range switch if equipped with an automatic transaxle

3. If the above components are normal, replace the Powertrain Control Module (PCM).

Engine Assembly

REMOVAL & INSTALLATION

MX-5 Miata

➡ The engine, transmission, and crossmember assembly is removed as a unit from under the vehicle.

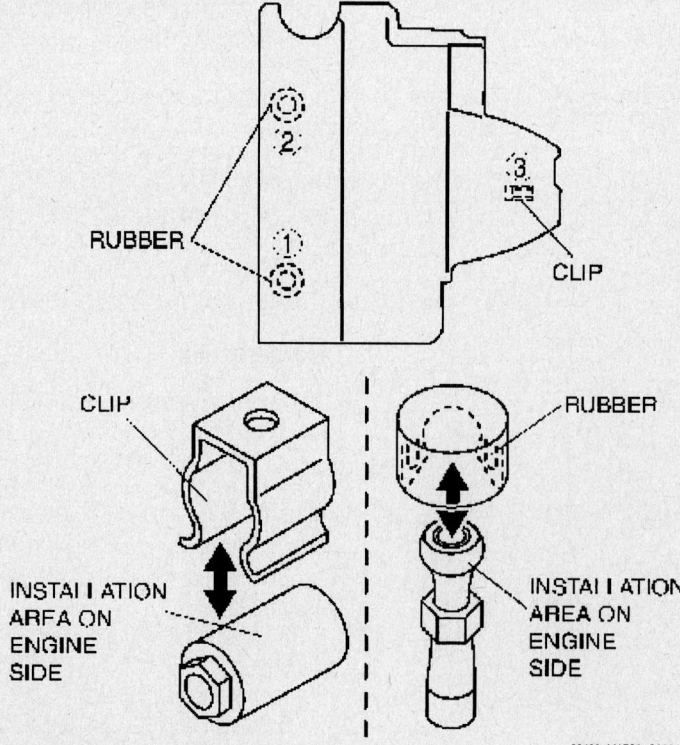

Remove the plug hole plate—MX-5 Miata

09482_MAZC2_G0004

1. Before servicing the vehicle, refer to the precautions section.

2. Remove the fuel-filler cap to release the pressure inside the fuel tank.

3. Remove the fuel pump relay.

4. Start the engine.

5. After the engine stalls, crank the engine several times .

6. Turn the ignition switch to the LOCK position.

7. Install the fuel pump relay.

8. Drain the engine coolant.

9. Remove front tires.

10. Remove the plug hole plate by lifting off and removing the plug hole plate from the installation areas (rubber and clips) as shown in the illustration.

11. Remove the battery cover, battery, battery box, battery tray and battery duct

12. Remove the air cleaner.

13. Remove the throttle body.

14. Remove the PCM, PCM duct and air cleaner insulator, please perform the following to remove the PCM:

a. Connect the WDS or equivalent to the DLC-2.

b. Set up the WDS or equivalent (including the vehicle recognition).

c. Select "Module Programming".

d. Select "Programmable Module Installation".

e. Select "PCM" and perform procedures according to directions on the WDS or equivalent screen.

➡ If the PCM is replaced with a new one, the PCM stores DTC P0602 and illuminates the MIL even though no malfunction is detected. This means the PCM has not been configured yet.

f. Retrieve DTC's by the WDS or equivalent, then verify that there in no DTC present.

g. If DTC is present, perform an applicable DTC inspection.

h. Move the water hose from the PCM cover slightly out of the way.

i. Remove the PCM cover, connector and PCM.

15. Remove the coolant reserve tank.

16. Remove the center console by removing the components as follows:

a. Selector lever knob if equipped with an automatic transmission or the shift lever knob if equipped with a manual transmission.

b. Front cover.

c. Rear cover.

d. Hole cover.

e. Parking brake lever boot.

f. Parking brake lever boot plate.

g. Console.

h. Power window main switch connector.

i. Indicator compartment connector if equipped with an automatic transmission.

17. Disconnect the P/S oil pump hoses and drain the P/S fluid reservoir.

18. Remove the splash shield, under cover and mud guards.

19. Remove the alternator duct.

20. Drain the transmission fluid.

21. Disconnect the brake vacuum hose.

22. Disconnect the quick release connector from the dynamic chamber.

23. Disconnect the quick release connector from the fuel distributor.

24. Remove the drive belt.

25. Remove the A/C compressor with the pipes connected and attach the A/C compressor using wire so that it is out of the way.

26. Disconnect the water hose and heater hose.

27. Secure the front caliper using wire so that it is out of the way.

28. Disconnect the wiring harness.

29. Disconnect front ABS wheel-speed sensor connector.

30. Remove the radiator.

31. If equipped with an automatic transmission, disconnect the manual shaft lever component as follows:

a. Loosen the starter installation bolts only enough that the starter is loose, but not removed.

b. Mark the manual shaft lever component as shown in the illustration.

c. Remove the manual shaft lever component installation nut.

32. If equipped with a manual transmission, Remove the clutch release cylinder with the pipes connected and secure the clutch release cylinder using wire or rope so that it is out of the way. Remove the shift lever knob.

33. Remove the tunnel member.

34. Remove the member bracket.

35. Remove the transverse member.

36. Remove the exhaust middle pipe.

37. Remove the power plant frame as follows:

a. Support the transmission using a transmission jack.

b. Remove the power plant frame.

38. Remove the propeller shaft as follows:

➡When replacing with a new propeller shaft, mark the companion flange to match the position of the tag on the propeller shaft.

a. Before removing the propeller shaft, make alignment marks on the yoke and differential companion flange.

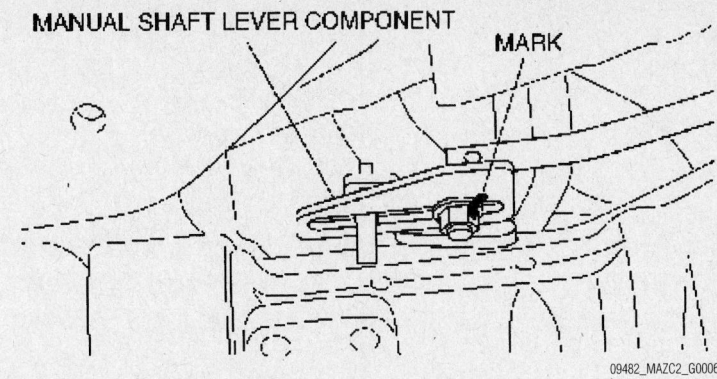

Mark the manual shaft lever component as shown—MX-5 Miata

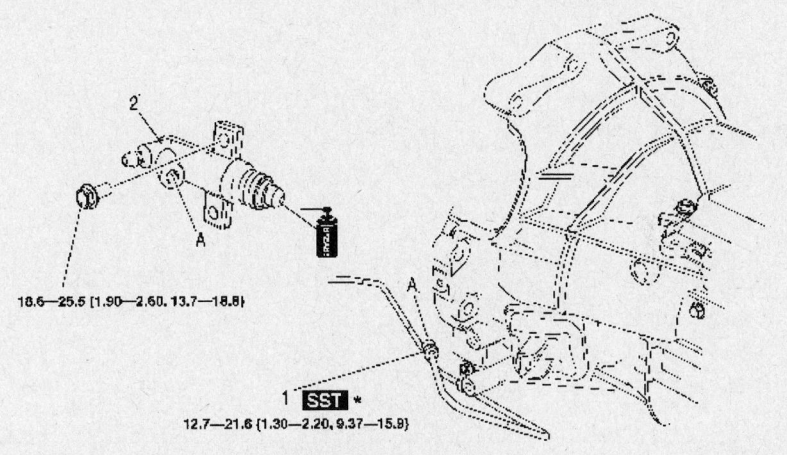

18.6—25.5 [1.90—2.60, 13.7—18.8}

12.7—21.6 {1.30—2.20, 9.37—15.9}

N·m {kgf·m

1 Clutch pipe
2 Clutch release cylinder

Exploded view of the clutch release cylinder components—MX-5 Miata

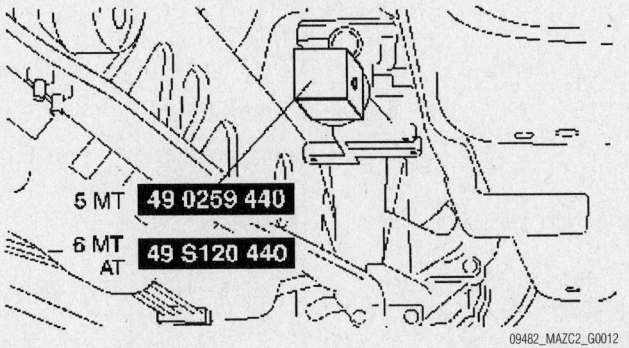

5 MT 49 0259 440
6 MT AT 49 S120 440

Install the tools shown depending on transmission type, onto the to the extension housing—MX-5 Miata

b. Remove the retainers and the shaft.

c. Install the tools illustrated onto the to the extension housing.

39. Remove the intermediate shaft bolt as follows:

a. Mark the pinion shaft and gear housing for proper installation.

b. Remove the bolt.

✳✳ WARNING

Remove the engine, transmission and crossmember carefully, holding it steady. If the transmission falls it could be damaged or cause injury.

40. Secure the engine, transmission, and crossmember using an engine lifter

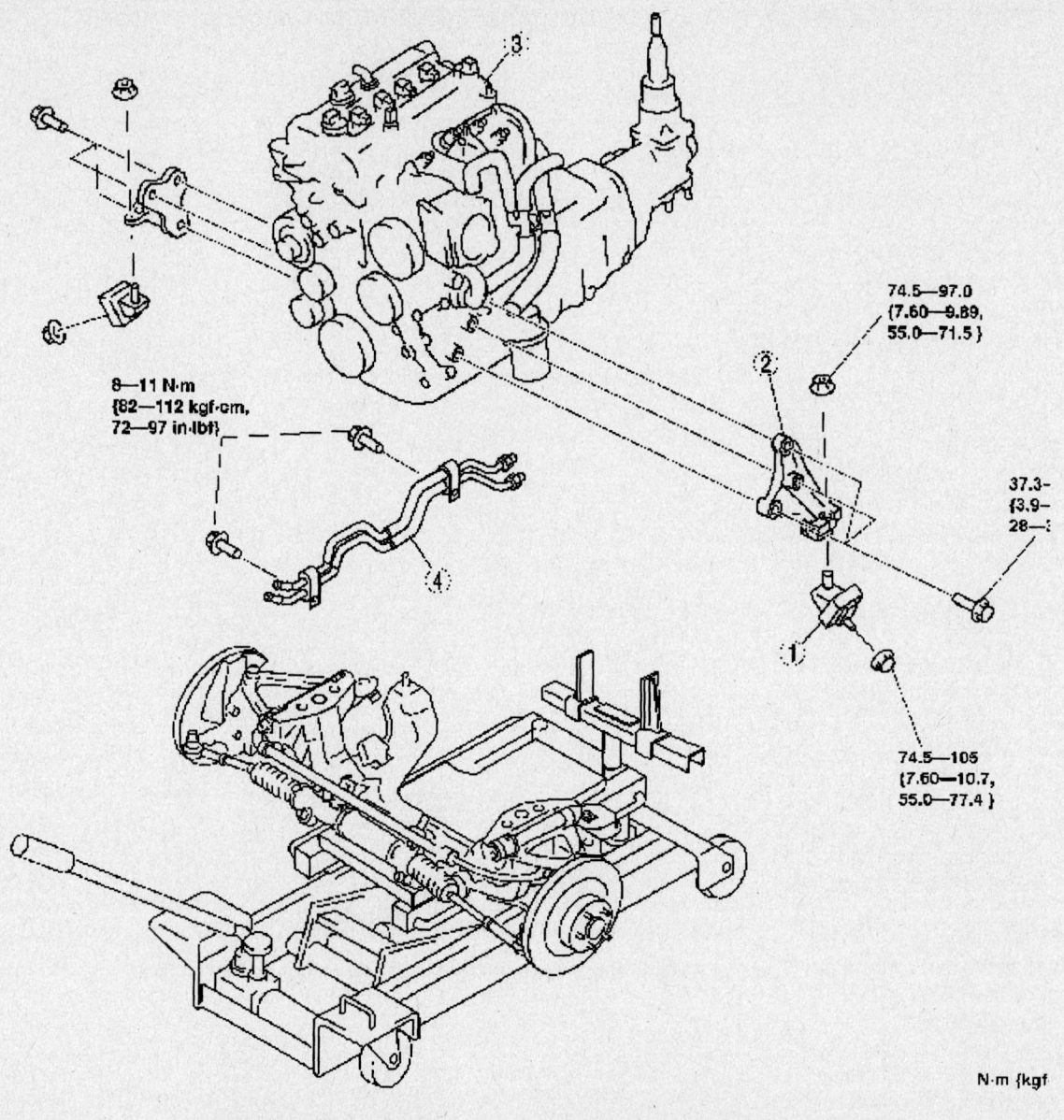

8—11 N·m
{82—112 kgf·cm,
72—97 in·lbf}

74.5—97.0
{7.60—9.89,
55.0—71.5}

37.3—
{3.9—
28—

74.5—105
{7.60—10.7,
55.0—77.4}

N·m {kgf

1 Engine mount rubber
2 Engine mount bracket
3 Engine, transmission
4 Oil pipe, oil hose

09482_MAZC2_G0007

Engine mounts and oil pipe and hose locations—MX-5 Miata

41. Remove the engine and transmission from the crossmember lifter as follows by suspending them with a crane:
   a. Engine mount rubber.
   b. Engine mount bracket.
   c. Engine from the transmission.
   d. Oil pipe and oil hose.
To install:
42. Installation is the reverse of removal. Please refer to the exploded views for component placement and torque specifications.
43. Please note the following special steps:
   a. When installing the oil Pipe, Hose clamp and Oil hose:

• Apply compressed air to cooler-side opening, and blow any remaining grime and foreign material from the cooler pipes. Compressed air should be applied for more than 1 min.
• Be sure to install the oil hose between the power steering pipe and the radiator cowl as shown in the illustration.
• Align the marks, and slide the oil hose onto the oil pipe until it is fully seated as illustrated. Install the hose clamp onto the hose.

➡**If reusing the hose, install the new hose clamp exactly on the mark left by the previous hose clamp. Apply force to the hose clamp in the direction of the arrow in order to fit the clamp in place.**

• Verify that the hose clamp does not interfere with any other components.
   b. When installing the propeller shaft:
• Align the marks and install the propeller shaft.
• When installing a new propeller shaft, align the differential companion flange mark with the tag on the propeller shaft and assemble.

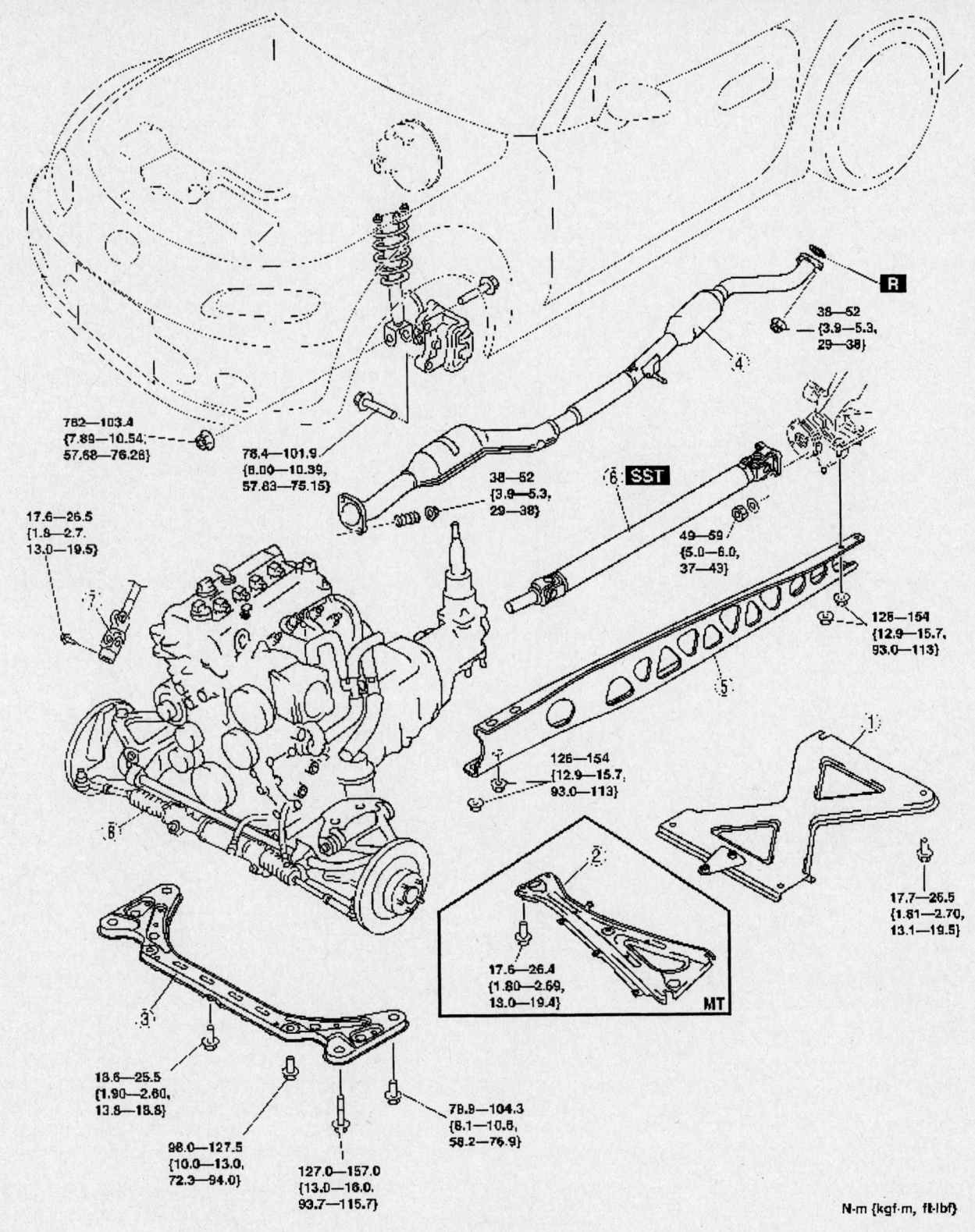

R

38—52
{3.9—5.3,
29—38}

4

782—103.4
{7.99—10.54;
57.68—76.26}

76.4—101.9
{8.00—10.39,
57.83—75.15}

17.6—26.5
{1.9—2.7,
13.0—19.5}

38—52
{3.9—5.3,
29—38}

7

6 SST

49—59
{5.0—6.0,
37—43}

126—154
{12.9—15.7,
93.0—113}

5

126—154
{12.9—15.7,
93.0—113}

8

1

17.7—26.5
{1.81—2.70,
13.1—19.5}

2

17.6—26.4
{1.80—2.69,
13.0—19.4}

MT

3

13.6—25.5
{1.90—2.60,
13.8—18.8}

98.0—127.5
{10.0—13.0,
72.3—94.0}

127.0—157.0
{13.0—16.0,
93.7—115.7}

79.9—104.3
{8.1—10.6,
58.2—76.9}

N·m {kgf·m, ft·lbf}

1	Tunnel member	5	Power plant frame
2	Member bracket	6	Propeller shaft
3	Transverse member	7	Bolt (Intermediate Shaft)
4	Middle pipe	8	Engine, transmission, crossmember component

Exploded view of the engine/transmission assemblies and related components—MX-5 Miata

09482_MAZC2_G0003

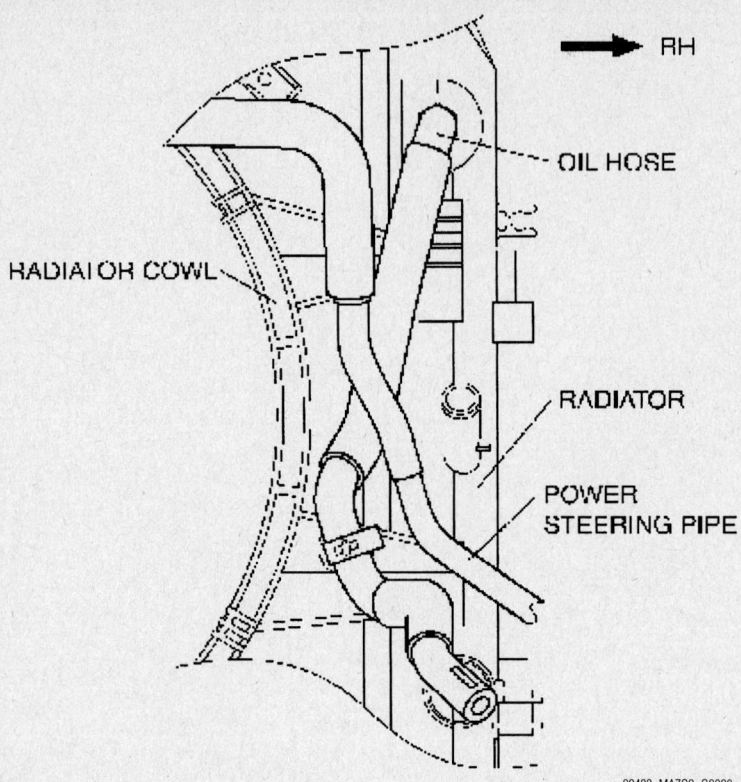

RH

OIL HOSE

RADIATOR COWL

RADIATOR

POWER STEERING PIPE

09482_MAZC2_G0008

Be sure to install the oil hose between the power steering pipe and the radiator cowl—MX-5 Miata

- Raise the front end of the power plant frame (transmission side) or the transmission with the transmission jack, and adjust dimension A to 26.7–34.7 mm (1.06–1.36 in) (lower surface of power plant frame-upper surface of the tunnel member) as shown in the illustration.
- Tighten the power plant frame installation nuts to 126–154 Nm (93–113.5 ft. lbs)
- Verify that dimension A is within the specification with the transmission jack and the adjustment bolt removed.
- If not within the specification, adjust dimension A again.

d. Connect the PCM connector fully into the PCM and push the lever until a click is heard.

e. To position the plug hole plate, grasp rubber 1 and 2, as shown in the illustration, with your hands and press them in.

44. Start the engine and inspect and/or adjust the following:

a. Pulley and belt for runout, tension, and contact

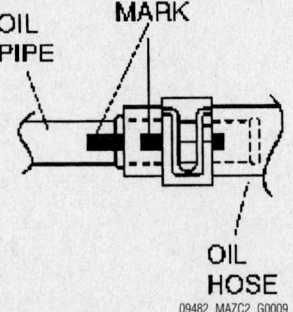

OIL PIPE

MARK

OIL HOSE

09482_MAZC2_G0009

Align the marks, and slide the oil hose onto the oil pipe until it is fully seated—MX-5 Miata

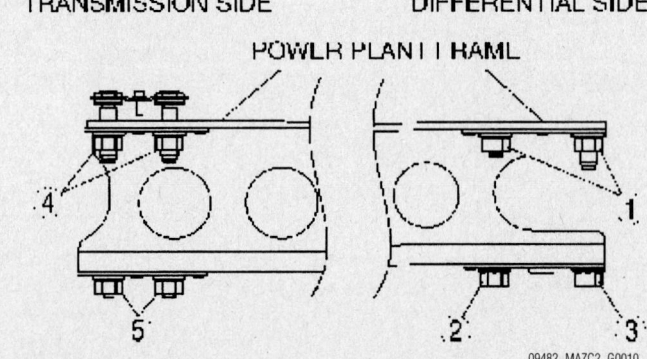

TRANSMISSION SIDE   DIFFERENTIAL SIDE

POWER PLANT FRAME

09482_MAZC2_G0010

Temporarily tighten the nuts 1, 2, 3 in the order shown—MX-5 Miata

c. When installing the power plant frame:
- Support the transmission and differential so that they are level using a transmission jack.
- Install the power plant frame.
- Temporarily tighten the nuts 1, 2, 3 in the order shown in the illustration.
- Tighten nut 2 until the power plant frame is seated in the rear differential.
- Temporarily tighten the nuts 4, 5 in order shown in the illustration.
- Install the middle pipe and tunnel member.

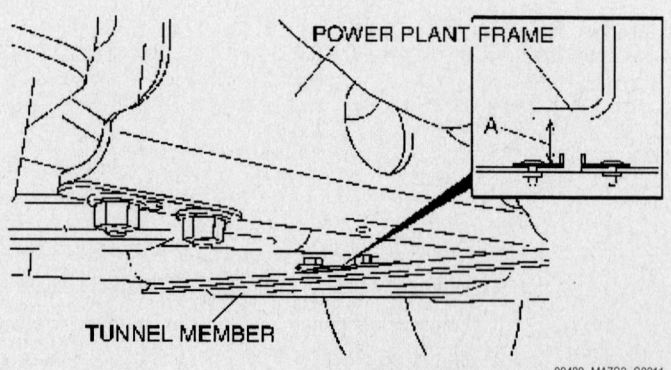

POWER PLANT FRAME

A

TUNNEL MEMBER

09482_MAZC2_G0011

Raise the front end of the power plant frame (transmission side) or the transmission with the transmission jack, and adjust dimension A to 26.7–34.7 mm (1.06–1.36 in—MX-5 Miata

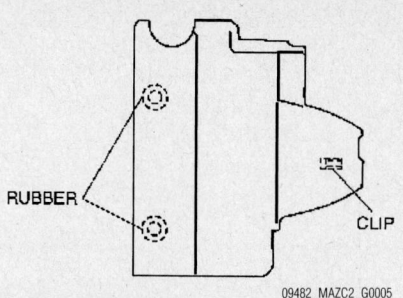

09482_MAZC2_G0005

Install the plug hole plate by grasping rubber and pressing them in—MX-5 Miata

b. Leakage of engine oil, coolant, ATF or MT fluid, and check for fuel leaks

c. Ignition timing, idle speed, and idle.

d. Front wheel alignment.

e. Perform a road test and verify that there is no vibration or noise.

Mazda3

1. Before servicing the vehicle, refer to the precautions section.

2. Remove the plug hole plate by lifting off and removing the plug hole plate from the areas shown in the accompanying illustration.

3. Remove the air hose and air cleaner assembly.

4. Remove the battery cover, duct, clamp, battery and battery tray.

5. Relieve the fuel system pressure and disconnect the fuel hoses.

6. Remove the accelerator cable and bracket.

7. Remove the front wheels, under cover and splash shields.

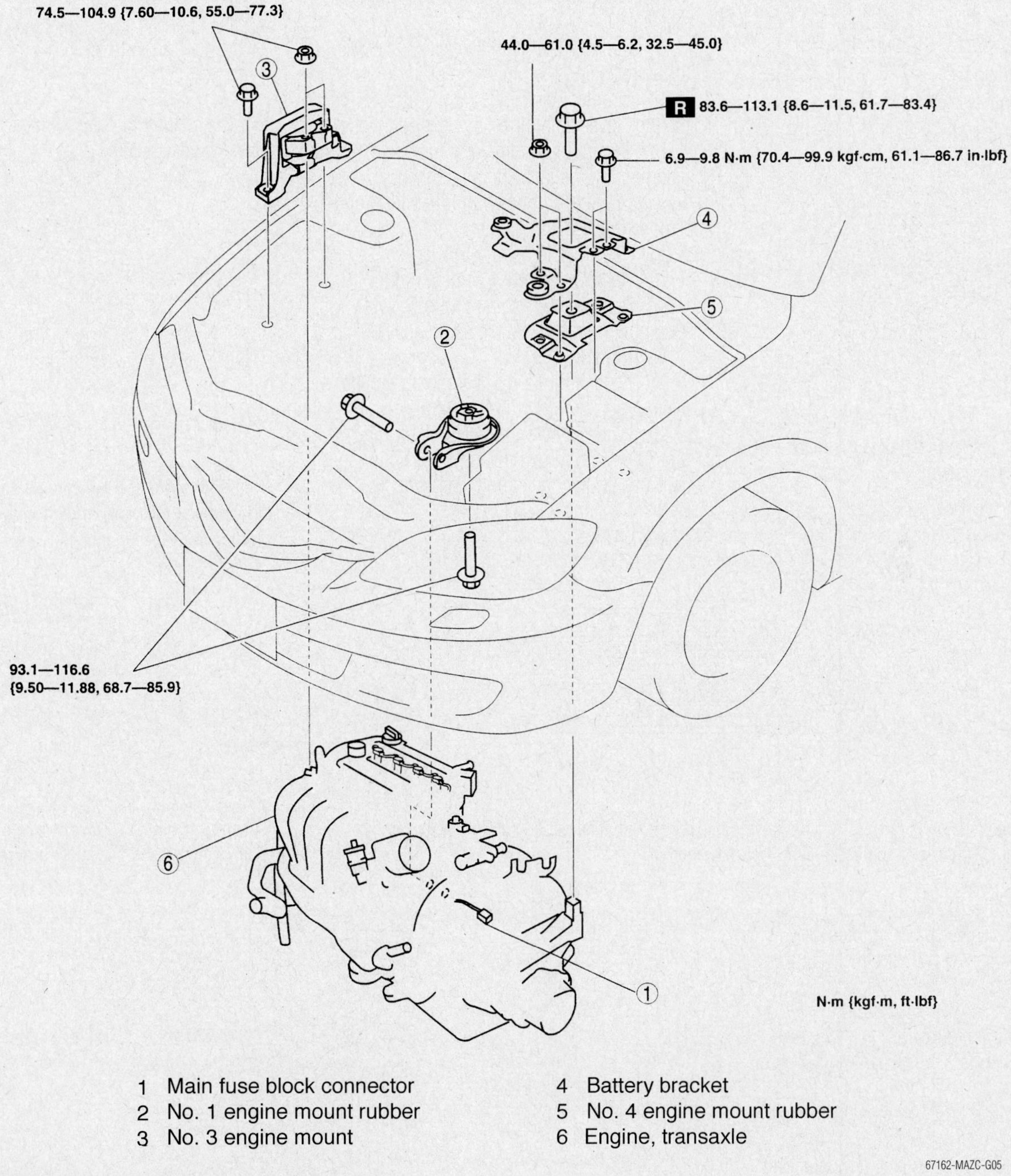

74.5—104.9 {7.60—10.6, 55.0—77.3}

44.0—61.0 {4.5—6.2, 32.5—45.0}

[R] 83.6—113.1 {8.6—11.5, 61.7—83.4}

6.9—9.8 N·m {70.4—99.9 kgf·cm, 61.1—86.7 in·lbf}

93.1—116.6 {9.50—11.88, 68.7—85.9}

N·m {kgf·m, ft·lbf}

1	Main fuse block connector	4	Battery bracket
2	No. 1 engine mount rubber	5	No. 4 engine mount rubber
3	No. 3 engine mount	6	Engine, transaxle

Exploded view of the engine mounting—Mazda3 models

67162-MAZC-G05

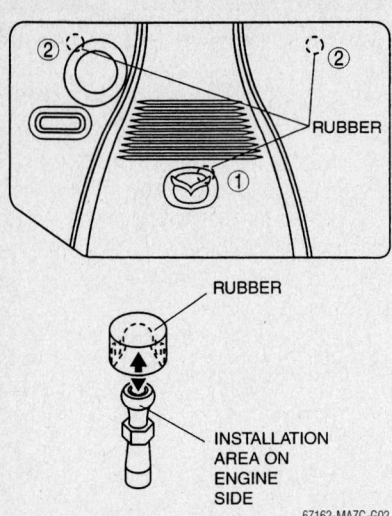

Plug hole plate locations—Mazda3 models

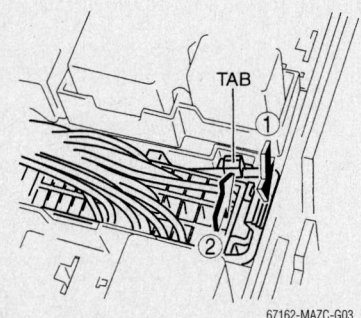

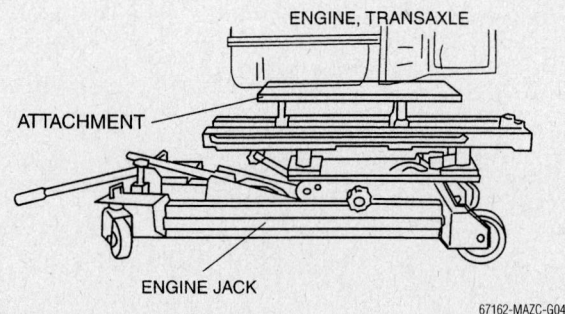

Secure the engine and transaxle using an engine jack and attachment as shown before removing the No. 3 and No. 4 engine mounts—Mazda3 models

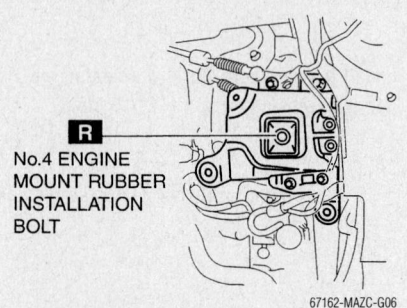

Removing the main fuse block connector—Mazda3 models

15. Remove the drive shafts.

16. Remove the coolant over flow tank with the hose attached and wire it to one side.

17. Remove the cooling fan assembly.

18. If equipped with an automatic transaxle, disconnect the transaxle fluid lines, selector cable and wiring harness.

19. If equipped with a manual transaxle, remove the shift cable and the clutch release cylinder with the line still attached.

20. Disconnect the heater hoses.

21. Remove the radiator hoses.

22. Remove the exhaust system main silencer.

23. Remove the main fuse block connector by releasing the tab in the order shown in the accompanying illustration and pulling the lock lever up and remove the connector.

24. Secure the engine and transaxle using an engine jack and attachment as shown in the accompanying illustration and remove the No. 3 engine mount.

25. Remove the battery bracket.

26. Remove the No. 4 engine mount.

27. Remove the engine and transaxle assembly from the vehicle as an assembly.

To install:

28. Installation is the reverse of removal, please keep in mind the following:

29. When installing the No. 4 engine mount, secure the engine and transaxle using an engine jack and attachment as shown in the accompanying illustration. Install the No. 1 and No. 4 engine mounts, do not tighten the retainers at this time.

30. Use a new No. 4 mount bolt and tighten it to 61–87 ft. lbs. (83–113 Nm).

31. Tighten the No. 4 engine mount and battery bracket nuts and bolts in the

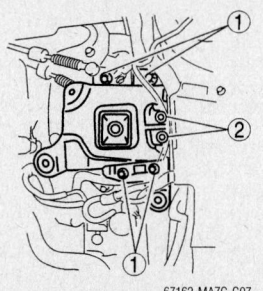

Location of the No. 4 engine mount rubber bolt—Mazda3 models

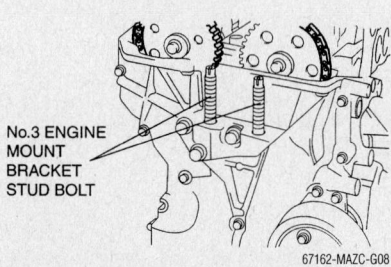

Tighten the No. 4 engine mount and battery bracket nuts and bolts in this sequence—Mazda3 models

8. Remove the A/C drive belt.

9. Remove the A/C compressor with the lines still attached and wire the compressor out of the way.

10. Drain the transaxle fluid.

11. Drain the coolant.

12. Disconnect the brake booster vacuum hose.

13. Remove the exhaust system member.

14. Remove the front crossmember, front stabilizer bar, lower control arm, steering gear and the No. 1 engine mount.

Location of the No. 3 engine mount bracket stud bolts—Mazda3 models

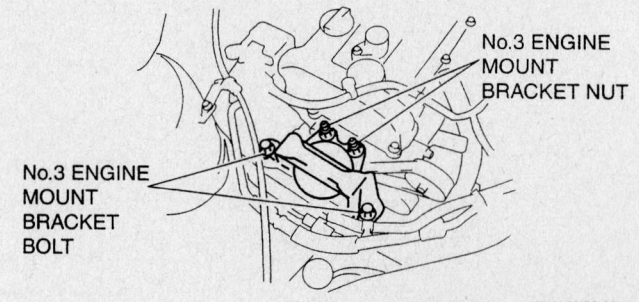

Location of the No. 3 engine joint bracket nuts and bolts—Mazda3 models

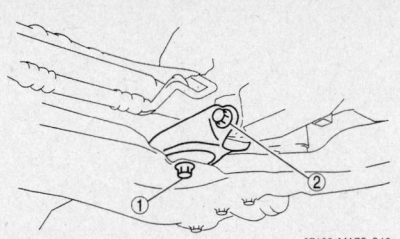

Tighten the No. 1 mount bolts in the sequence illustrated—Mazda3 models

sequence shown in the accompanying illustration. Tighten the retainers (1) to 32–45 ft. lbs. (44–61 Nm) and the retainers (2) to 61–86 inch lbs. (7–10 Nm).

32. Tighten the No. 3 mount bracket stud bolts to 62–115 inch lbs. (7–13 Nm).

33. Tighten the No. 3 joint bracket nuts and bolts to 55–77 ft. lbs. (74–105 Nm).

34. Remove the jack and the attachment and tighten the No. 1 mount bolts in the sequence illustrated to 69–86 ft. lbs. (93–116 Nm).

35. Fill the engine and the transaxle with the proper type and amount of fluids. Fill the cooling system.

36. Connect the negative battery cable, start the engine and check for leaks.

37. Check the ignition timing and the idle speed.

38. Check all fluid levels.

39. Check the vehicle alignment.

Mazda6

2.3L (L3) ENGINE

1. Before servicing the vehicle, refer to the precautions section.

2. Remove the battery and battery tray.

3. Remove the shroud panel.

4. Remove the radiator.

5. Drain the transaxle fluid.

6. Remove the plug hole plate.

7. Remove the power steering pump with the lines attached and set aside.

8. Remove the A/C compressor with the lines still attached and wire the compressor out of the way.

9. Remove the joint shaft as follows:

  a. Remove the ABS sensor.

  b. Separate the tie rod ends from the knuckle.

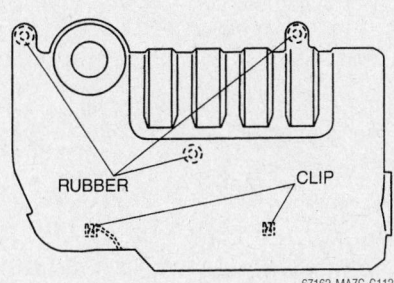

Plug hole plate assembly—Mazda6 shown, Mazdaspeed6 similar with 2.3L (L3) engine

  c. Remove the damper fork bolt.

  d. Separate the lower control arm front and rear ball joints.

  e. Remove the stabilizer bar link nut and the link from the damper fork.

  f. Remove the joint shaft bracket bolt.

  g. Remove the halfshafts.

  h. Disconnect the right halfshaft from the joint shaft by tapping the transaxle side outer ring with a brass bar and a hammer. Disconnect the joint shaft bracket from the block and remove the joint shaft.

  i. Install tool 49 G030 455 to hold the side gears after removal.

10. Remove the air cleaner assembly, air intake duct, bracket and vacuum hoses.

11. On models with a automatic transaxle, disconnect the transmission fluid hose and the selector cable.

12. Remove the vacuum and heater hoses

13. If equipped with a manual transaxle, remove the control cable and the clutch release cylinder with the line still attached.

14. Relieve the fuel system pressure and disconnect the fuel hoses.

15. Disconnect the wring harness at the engine side.

16. Remove the front exhaust pipe.

17. Remove the No. 1 mount rubber bolt A on the engine mount bracket side, loosen

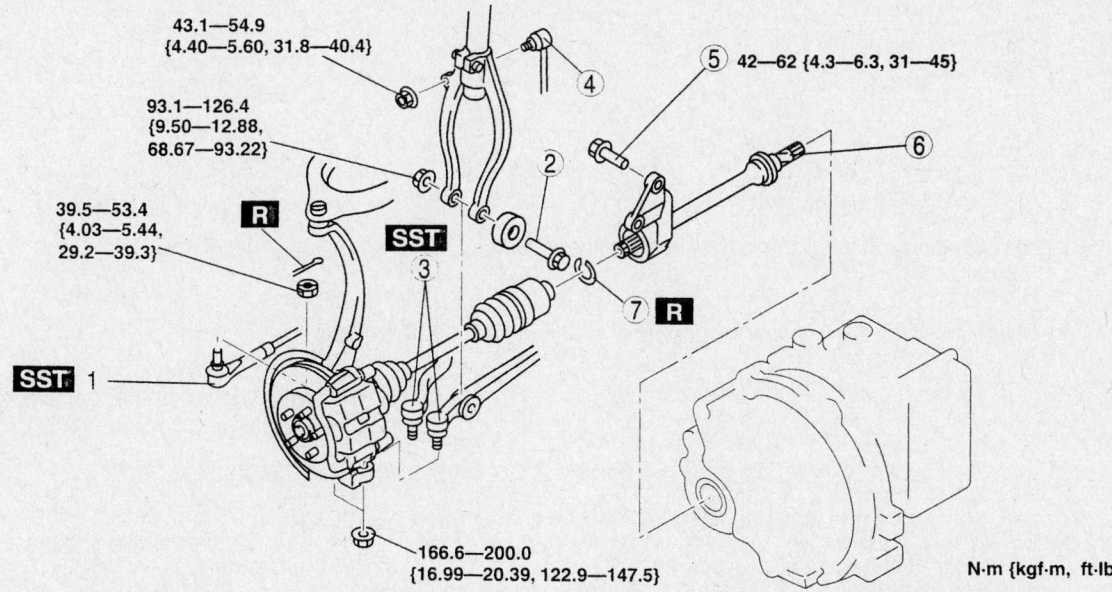

1	Tie-rod end ball joint	5	Joint shaft bracket bolt
2	Bolt	6	Joint shaft
3	Lower arm (front, rear) ball joint	7	Clip
4	Stabilizer control link		

Exploded view of the joint shaft assembly—Mazda6 shown, Mazdaspeed6 similar with the 2.3L (L3) engine

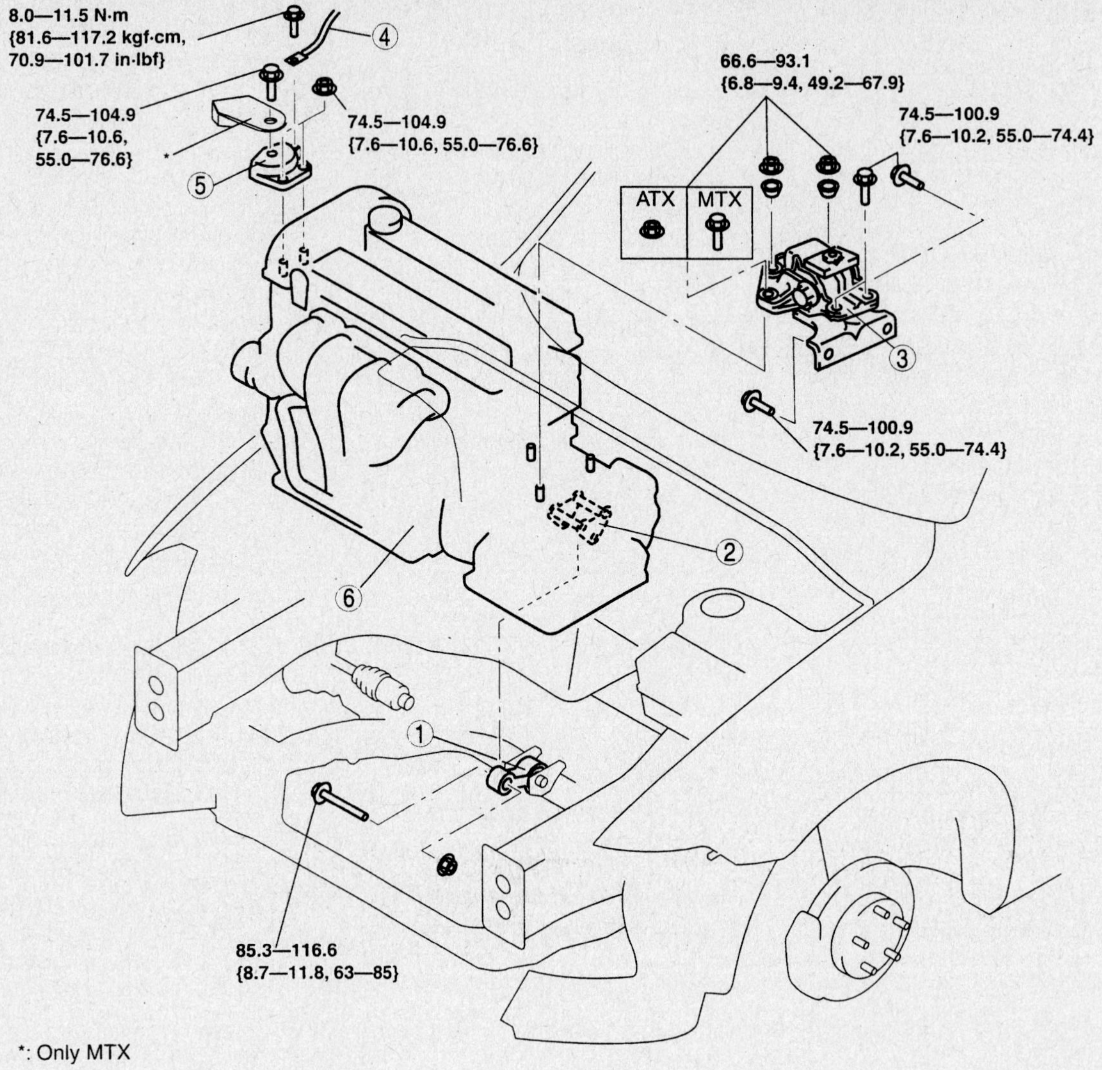

8.0—11.5 N·m
{81.6—117.2 kgf·cm,
70.9—101.7 in·lbf}

74.5—104.9
{7.6—10.6,
55.0—76.6}

74.5—104.9
{7.6—10.6, 55.0—76.6}

66.6—93.1
{6.8—9.4, 49.2—67.9}

74.5—100.9
{7.6—10.2, 55.0—74.4}

ATX MTX

74.5—100.9
{7.6—10.2, 55.0—74.4}

85.3—116.6
{8.7—11.8, 63—85}

*: Only MTX

1	No.1 Engine mount rubber	4	Engine ground
2	No.1 Engine mount bracket	5	No.3 Engine joint bracket
3	No.4 Engine mount bracket and No.4 Engine mount rubber	6	Engine, transaxle

67162-MAZC-G113

Exploded view of the engine mounting—Mazda6 shown, Mazdaspeed6 similar with the 2.3L (L3) engine

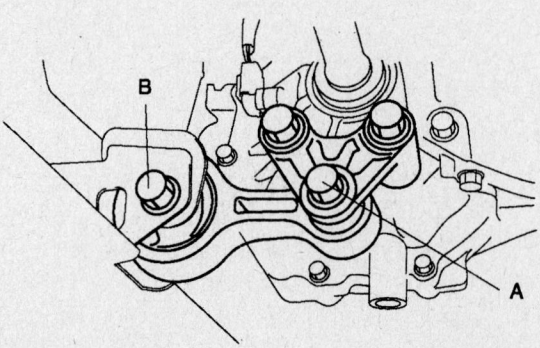

67162-MAZC-G114

Locations of bolts A and B on the No. 1 mount rubber—Mazda6 shown, Mazdaspeed6 similar with the 2.3L (L3) engine

bolt B on the crossmember side until about 3 pitches are showing. DO NOT remove the N0, 1 mount rubber from the vehicle.

18. Remove the No. 1 engine mount bracket.

19. Secure the engine and transaxle using a hoist.

20. Remove the No. 4 mount bracket and rubber as a unit.

21. Remove the engine ground.

22. Remove the No. 3 engine joint bracket.

23. Remove the engine and transaxle from the vehicle as an assembly.

24. Remove the engine–to–transaxle

bolts and separate the engine from the transaxle.

To install:

25. Installation is the reverse of removal, please keep in mind the following:

26. On models with an automatic transaxle, tighten the engine–to–transaxle bolts to 28–38 ft. lbs. (37–52 Nm)

27. On models with a manual transaxle, tighten the engine–to–transaxle bolts to 28–36 ft. lbs. (37–50 Nm)

28. When installing the No. 3 engine joint bracket, tighten the mount bracket stud bolt to 62–115 inch lbs. (7–13 Nm) and the joint bracket bolt and nut in the order illustrated to 55–76 ft. lbs. (74–105 Nm).

29. Install the No. 4 engine mount bracket and rubber. Tighten the No.4 mount bracket and rubber bolts and nuts in the order illustrated as follows:

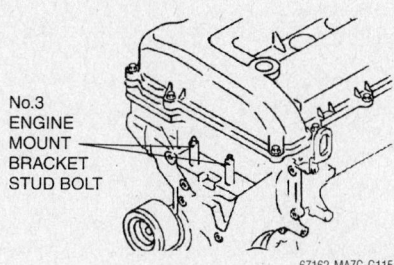

No. 3 engine mount bracket stud bolt locations—Mazda6 shown, Mazdaspeed6 similar with the 2.3L (L3) engine

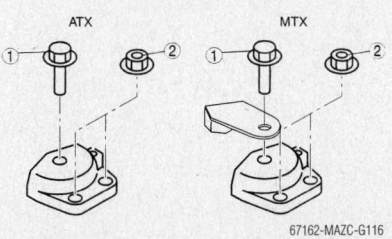

Tighten the No. 3 joint bracket bolt and nut in the order shown—Mazda6 shown, Mazdaspeed6 similar with the 2.3L (L3) engine

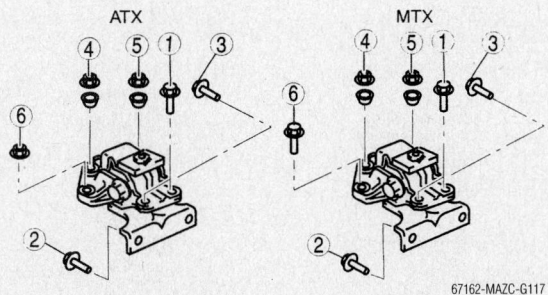

Tighten the No. 4 mount bracket and rubber bolts and nuts in the order shown—Mazda6 shown, Mazdaspeed6 similar with the 2.3L (L3) engine

a. Bolts numbered 1, 2 and 3: 43–58 ft. lbs. (58–80 Nm).

b. Bolts numbered 4, 5 and 6: 49–68 ft. lbs. (66–93 Nm).

30. Install the No. 1 engine mount bracket as follows:

a. Install and tighten bolt A to 68–86 ft. lbs. (93–116 Nm). refer to the illustration for bolt location.

b. Install and tighten bolt B to 68–86 ft. lbs. (93–116 Nm). refer to the illustration for bolt location.

31. Install the No. 1 engine mount rubber as follows:

a. Install and tighten bolt A to 63–85 ft. lbs. (85–116 Nm). refer to the illustration for bolt location.

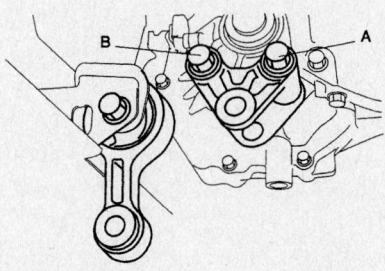

Tighten the No. 1 mount bracket bolts in the order shown—Mazda6 shown, Mazdaspeed6 similar with the 2.3L (L3) engine

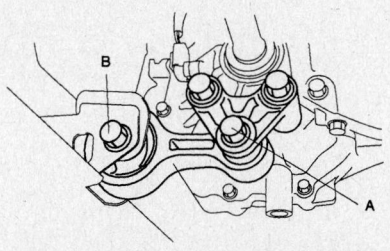

Tighten the No. 1 mount rubber bolts in the order shown—Mazda6 shown, Mazdaspeed6 similar with the 2.3L (L3) engine

b. Install and tighten bolt B to 68–86 ft. lbs. (93–116 Nm). refer to the illustration for bolt location.

32. Fill the engine and the transaxle with the proper type and amount of fluids. Fill the cooling system.

33. Connect the negative battery cable, start the engine and check for leaks.

34. Check the ignition timing and the idle speed.

35. Check all fluid levels.

36. Check the vehicle alignment.

3.0L (AJ) ENGINE

1. Before servicing the vehicle, refer to the precautions section.

2. Remove the under cover.

3. Drain and recycle the engine coolant.

4. Drain the transaxle fluid.

5. Remove the battery and battery tray.

6. Remove the air cleaner assembly, air intake duct, bracket and vacuum hoses.

7. On models with a automatic transaxle, disconnect the transmission fluid hose and the selector cable.

8. If equipped with a manual transaxle, remove the shift and selector cables and the clutch release cylinder with the line still attached.

9. Relieve the fuel system pressure and disconnect the fuel hoses.

10. Disconnect the engine wiring and powertrain Control Module (PCM) harness connections.

11. Remove the joint shaft as follows:

a. Remove the ABS sensor.

b. Remove the halfshaft lock nut.

c. Separate the tie rod ends from the knuckle.

d. Remove the damper fork.

e. Separate the lower control arm front and rear ball joints.

f. Remove the stabilizer bar link nut and the link from the damper fork.

g. Remove the joint shaft bracket bolt.

h. Remove the halfshafts.

i. Disconnect the left halfshaft from the joint shaft by inserting a prybar between the transaxle and the halfshaft outer ring. Disconnect the joint shaft bracket from the block and remove the joint shaft.

j. Install tool 49 G030 455 to hold the side gears after removal

12. Remove the transverse member.

13. Remove the steering gear and linkage assembly from the front crossmember and use mechanics wire to support the gear and linkage away from the crossmember.

14. Lower front shock absorber bolt.

15. No. 1 engine mount center bolt.

16. Support the crossmember with a jack

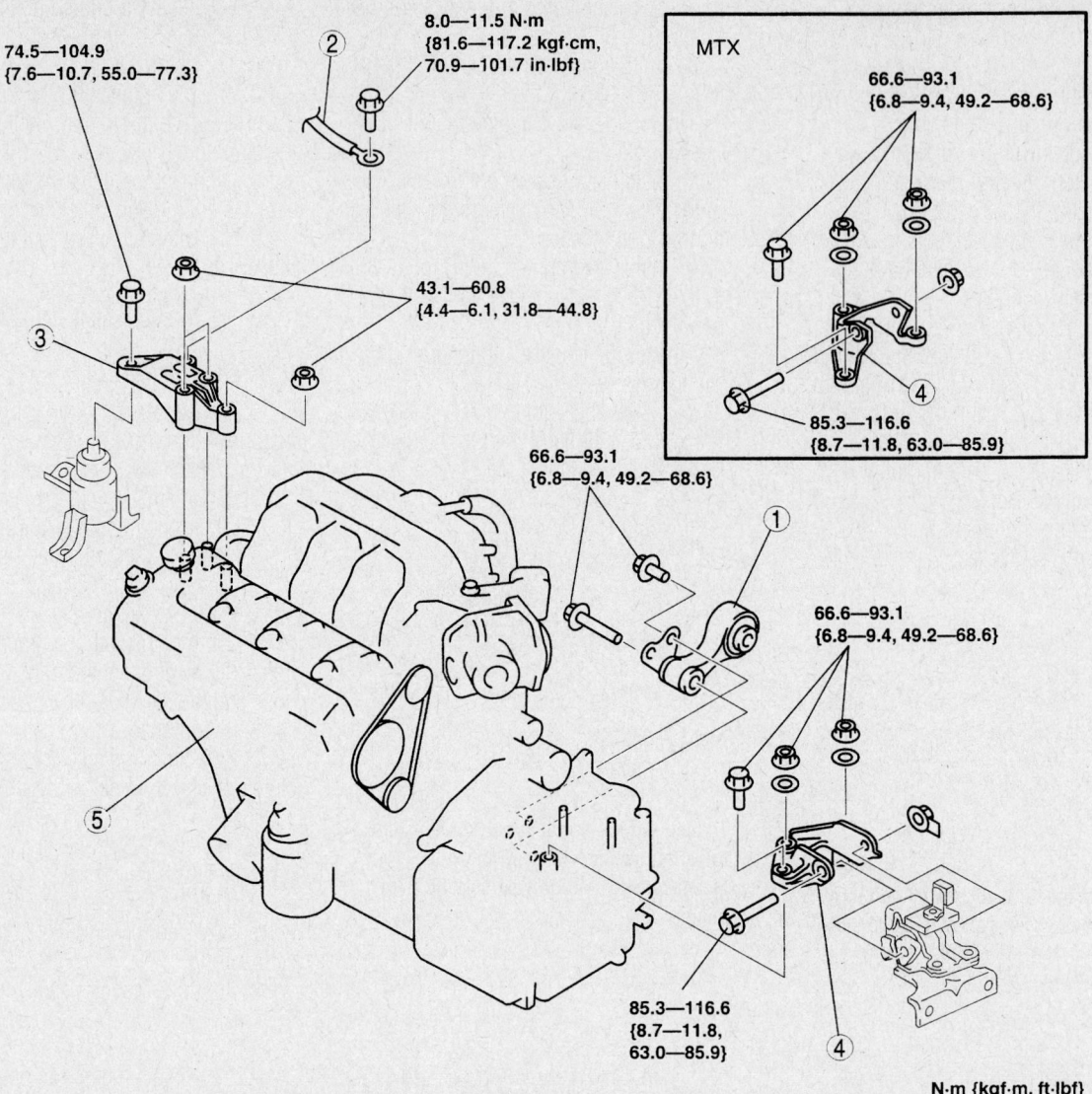

74.5—104.9
{7.6—10.7, 55.0—77.3}

8.0—11.5 N·m
{81.6—117.2 kgf·cm,
70.9—101.7 in·lbf}

MTX

66.6—93.1
{6.8—9.4, 49.2—68.6}

43.1—60.8
{4.4—6.1, 31.8—44.8}

85.3—116.6
{8.7—11.8, 63.0—85.9}

66.6—93.1
{6.8—9.4, 49.2—68.6}

66.6—93.1
{6.8—9.4, 49.2—68.6}

85.3—116.6
{8.7—11.8,
63.0—85.9}

N·m {kgf·m, ft·lbf}

1	No.1 engine mount rubber	4	No.4 engine mount bracket
2	Engine ground	5	Engine, transaxle
3	No.3 engine joint bracket		

67162-MAZC-G120

Exploded view of the engine mounting—Mazda6 models with the 3.0L (AJ) engine

and remove the nuts and the crossmember bracket.

17. Remove the crossmember.

18. Remove the drive belt and the dipstick tube.

19. Disconnect the power steering hoses and drain the power steering reservoir.

20. Properly evacuate the A/C system using approved equipment.

21. Remove the A/c compressor.

22. Remove the left hand three way catalytic converter.

23. Remove the front exhaust pipe.

24. Remove the radiator and heater hoses.

25. Remove the No. 1 engine mount rubber.

26. Disconnect the engine ground.

27. Support the engine using an engine jack and attachment as shown in the illustration.

28. Remove the No. 3 joint bracket and the No. 4 engine mount bracket.

29. Remove the No. 4 engine mount bracket.

30. Remove the engine and transaxle from the vehicle as an assembly.

31. Remove the engine–to–transaxle bolts and separate the engine from the transaxle.

To install:

32. Installation is the reverse of removal, please keep in mind the following:

33. Tighten the engine–to–transaxle bolts to 28–38 ft. lbs. (37–52 Nm)

34. Install the No. 4 engine mount bracket. Tighten the No.4 mount bracket bolts and nuts in the order illustrated as follows:

   a. Bolts numbered 1, 2 and 3: 49–68 ft. lbs. (66–93 Nm).

   b. Bolt 4: 63–86 ft. lbs. (85–116 Nm).

35. When installing the No. 3 engine joint bracket, tighten the No.3 mount bracket bolts and nuts in the order illustrated as follows:

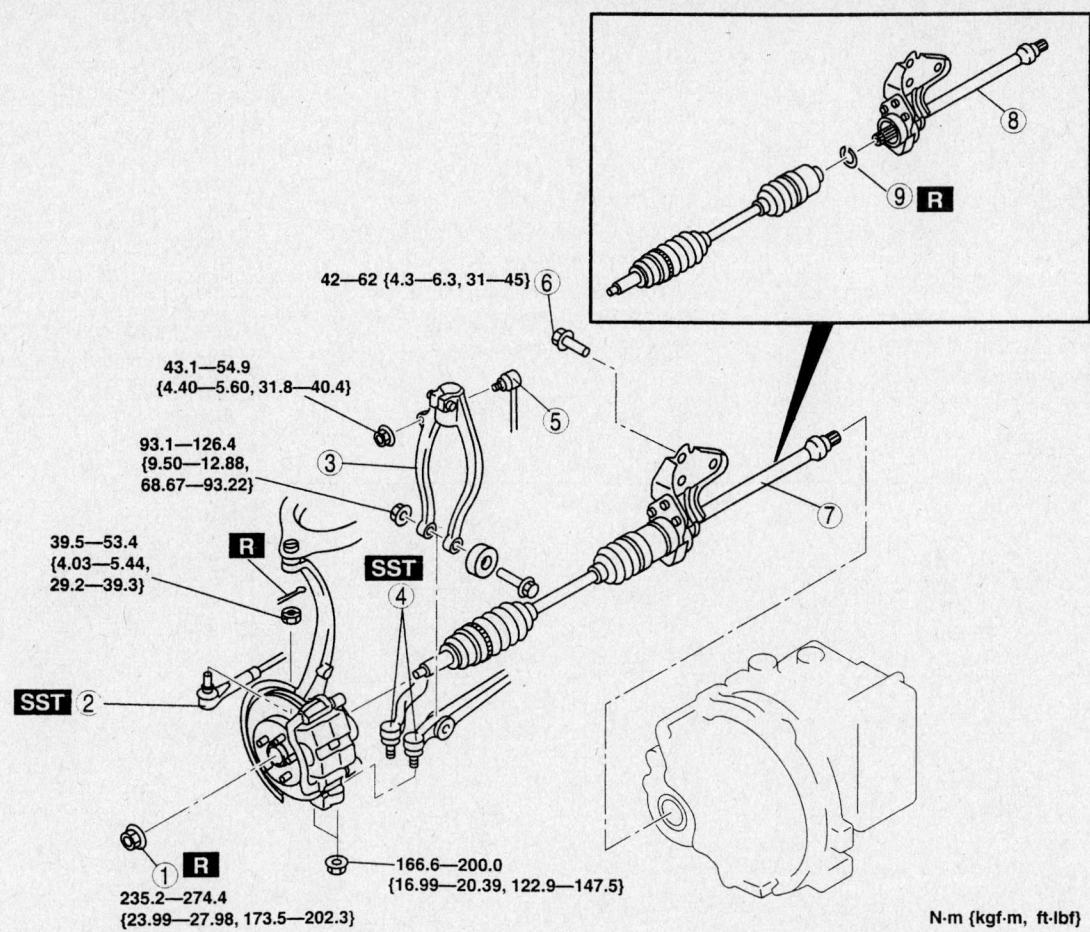

42—62 {4.3—6.3, 31—45}

43.1—54.9
{4.40—5.60, 31.8—40.4}

93.1—126.4
{9.50—12.88, 68.67—93.22}

39.5—53.4
{4.03—5.44, 29.2—39.3}

166.6—200.0
{16.99—20.39, 122.9—147.5}

235.2—274.4
{23.99—27.98, 173.5—202.3}

N·m {kgf·m, ft·lbf}

1 Locknut	5 Stabilizer control link
2 Tie-rod end ball joint	6 Joint shaft bracket bolt
3 Damper fork	7 Drive shaft and joint shaft
4 Lower arm (front, rear) ball joint	8 Joint shaft
	9 Clip

67162-MAZC-G121

Exploded view of the joint shaft assembly—Mazda6 models with the 3.0L (AJ) engine

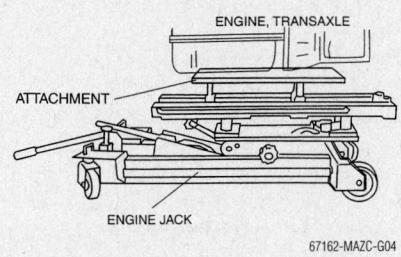

67162-MAZC-G04

Support the engine using an engine jack and attachment as shown—Mazda6 models with the 3.0L (AJ) engine

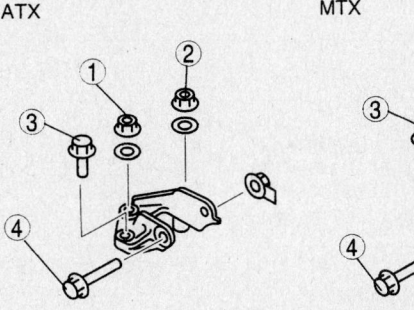

ATX

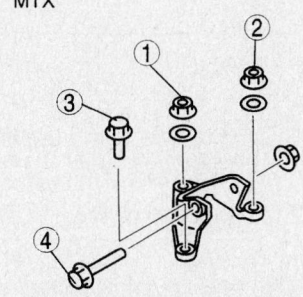

MTX

67162-MAZC-G122

Tighten the No. 4 engine joint bracket as shown—Mazda6 models with the 3.0L (AJ) engine

    a. Bolt 1: 55–77 ft. lbs. (74–105 Nm).
    b. Bolts 2, 3 and 4: 31–44 ft. lbs. (43–60 Nm).
36. Install the No. 1 engine mount bracket and tighten the bolts to 49–68 ft. lbs. (66–93 Nm).

37. Install all remaining components in the reverse of removal.
38. Fill the engine and the transaxle with the proper type and amount of fluids. Fill the cooling system.
39. Recharge the A/C system.

40. Connect the negative battery cable, start the engine and check for leaks.
41. Check the ignition timing and the idle speed.
42. Check all fluid levels.
43. Check the vehicle alignment.

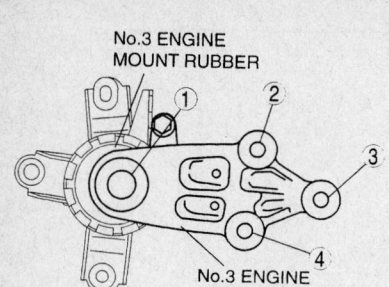

No.3 ENGINE
MOUNT RUBBER

No.3 ENGINE
JOINT BRACKET

67162-MAZC-G123

Tighten the No. 3 engine joint bracket as shown—Mazda6 models with the 3.0L (AJ) engine

Mazdaspeed6

1. Before servicing the vehicle, refer to the precautions section.

2. Remove the battery, battery tray and bracket.

3. Remove the charge air cooler cover.

4. Remove the air cleaner.

5. Remove the left side front mudguard, then the resonance chamber.

6. Disconnect the MAF/IAT sensor.

7. Remove the charge air cooler.

8. Remove the charge air cooler bracket

9. Remove the shroud panel.

10. Remove the radiator.

11. Drain the transaxle fluid.

12. Remove the power steering pump with the lines attached and set aside.

13. Remove the A/C compressor with the lines still attached and wire the compressor out of the way.

14. Disconnect the right hand front drive shaft from the joint shaft side.

15. Disconnect the left hand front drive shaft from the transaxle side.

16. Remove the propeller shaft.

17. Remove the air cleaner assembly, air intake duct, bracket and vacuum hoses.

18. Remove the control cable and the

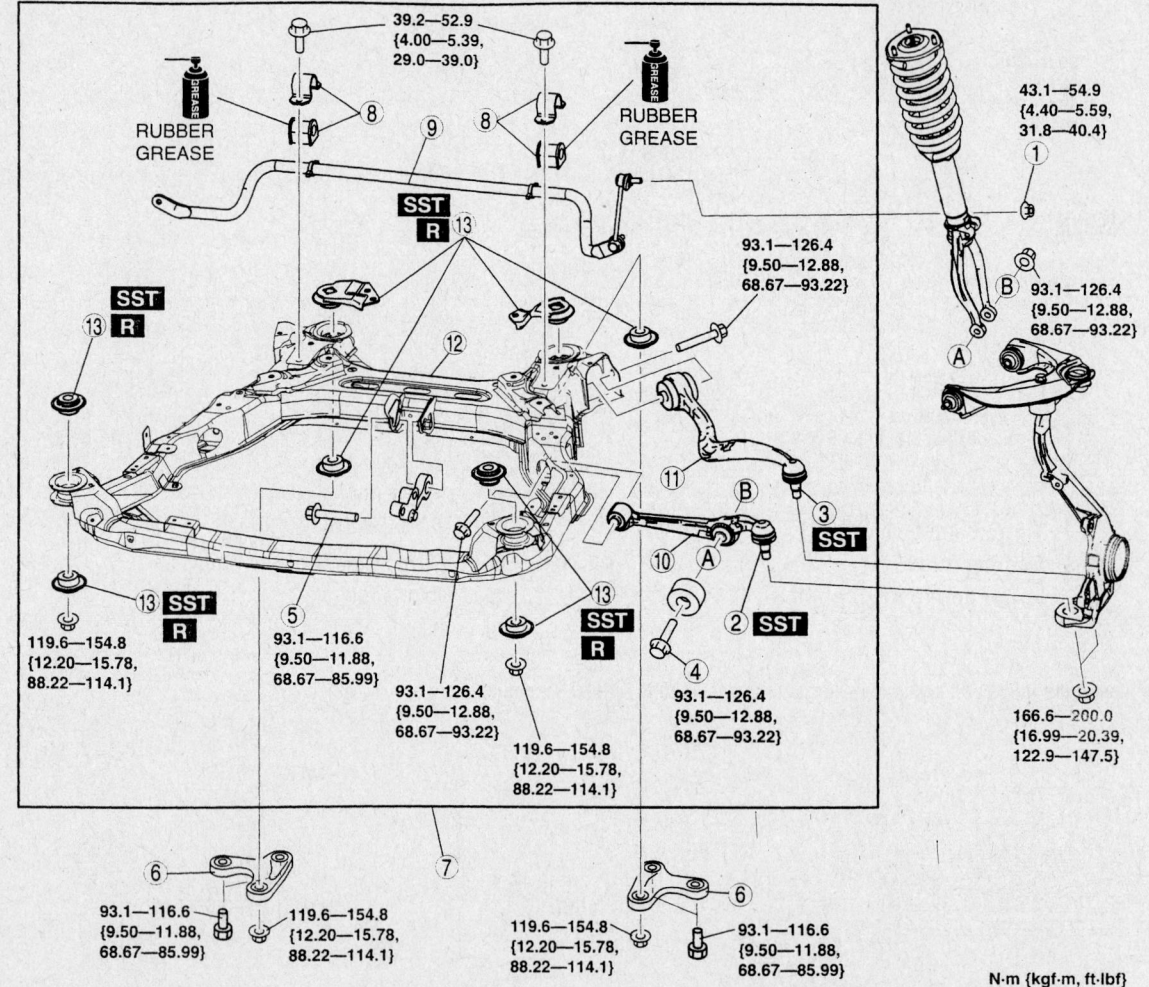

N·m {kgf·m, ft·lbf}

1	Nut (stabilizer control link)	8	Stabilizer bracket and bushing
2	Front lower arm (front) ball joint	9	Front Stabilizer
3	Front lower arm (rear) ball joint	10	Front lower arm (front)
4	Bolt (front shock absorber lower side)	11	Front lower arm (rear)
5	No.1 engine mount center bolt	12	Front crossmember
6	Crossmember bracket	13	Front crossmember bushing
7	Crossmember component		

67162-MAZC-G124

Exploded view of the front crossmember, related components and their torque specifications—Mazda6 models with the 3.0L (AJ) engine

clutch release cylinder with the line still attached.

19. Remove the vacuum and heater hoses.

20. Disconnect the wring harness at the engine side.

21. Remove the three way catalytic converter.

22. Remove the front exhaust pipe.

23. Relieve the fuel system pressure and disconnect the fuel hoses.

24. Remove the No. 1 mount rubber bolt A on the engine mount bracket side, loosen bolt B on the crossmember side until about 3 pitches are showing. DO NOT remove the N0, 1 mount rubber from the vehicle.

25. Remove the No. 1 engine mount bracket.

26. Secure the engine and transaxle using a hoist.

27. Remove the No. 4 mount bracket and rubber as a unit.

28. Remove the engine ground.

29. Remove the No. 3 engine joint bracket.

30. Remove the engine and transaxle from the vehicle as an assembly.

31. Remove the engine–to–transaxle bolts and separate the engine from the transaxle.

To install:

32. Installation is the reverse of removal, please keep in mind the following:

33. Tighten the engine–to–transaxle bolts to 28–36 ft. lbs. (37–50 Nm)

34. When installing the No. 3 engine joint bracket, tighten the mount bracket stud bolt to 62–115 inch lbs. (7–13 Nm) and the joint bracket bolt and nut in the order illustrated to 55–76 ft. lbs. (74–105 Nm).

35. Install the No. 4 engine mount bracket and rubber. Tighten the No.4 mount bracket and rubber bolts and nuts in the order illustrated as follows:

   a. Bolts numbered 1, 2 and 3: 43–58 ft. lbs. (58–80 Nm).

   b. Bolts numbered 4, 5 and 6: 49–68 ft. lbs. (66–93 Nm).

36. Install the No. 1 engine mount bracket as follows:

   a. Install and tighten bolt A to 68–86 ft. lbs. (93–116 Nm). refer to the illustration for bolt location.

   b. Install and tighten bolt B to 68–86 ft. lbs. (93–116 Nm). refer to the illustration for bolt location.

37. Install the No. 1 engine mount rubber as follows:

   a. Install and tighten bolt A to 63–85 ft. lbs. (85–116 Nm). refer to the illustration for bolt location.

   b. Install and tighten bolt B to 68–86

ft. lbs. (93–116 Nm). refer to the illustration for bolt location.

38. Fill the engine and the transaxle with the proper type and amount of fluids. Fill the cooling system.

39. Connect the negative battery cable, start the engine and check for leaks.

40. Check the ignition timing and the idle speed.

41. Check all fluid levels.

42. Check the vehicle alignment.

Water Pump

REMOVAL & INSTALLATION

MX-5 Miata

1. Before servicing the vehicle, refer to the precautions section.

✳✳ CAUTION

Never remove the cooling system cap or loosen the radiator drain plug while the engine is running, or when the engine and radiator are hot. Scalding engine coolant and steam may shoot out and cause serious injury. It may also damage the engine and cooling system.

2. Remove the battery cover.

3. Disconnect the negative battery cable.

4. Drain the engine coolant.

5. Remove the air cleaner.

6. Loosen the water pump pulley bolt and remove the drive belt.

7. Remove the water pump pulley, pump and the o–ring

8. Installation is the reverse of removal, refer to the illustration for component location and torque specifications.

9. Refill the engine coolant, start the vehicle and inspect for leaks.

2.0L (LF) and 2.3L (L3) Engines– Mazda3 and Mazda6 Models

1. Before servicing the vehicle, refer to the precautions section.

2. Remove the battery cover and disconnect the negative battery cable.

3. Remove the under cover and splash shield as an assembly.

4. Drain the cooling system.

5. Position the coolant reservoir tank aside with the hose still attached.

6. Remove the plug hole plate.

7. Loosen the water pump pulley bolt and position the drive belt aside.

8. Remove the water pump pulley.

9. Remove the water pump bolts, the pump and the O–ring.

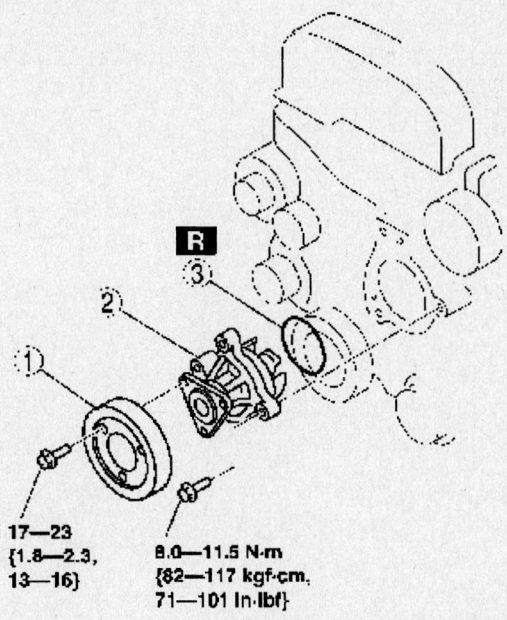

1 Water pump pulley
2 Water pump
3 O-ring

Exploded view of the water pump assembly—MX-5 Miata

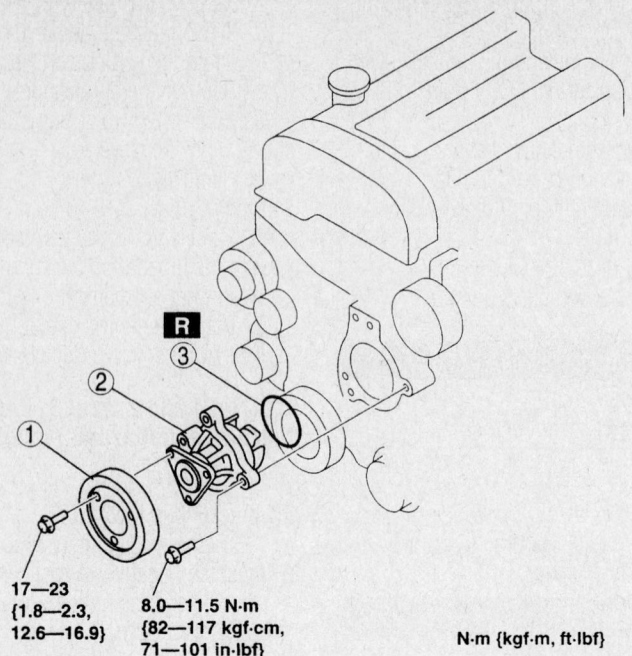

17—23
{1.8—2.3,
12.6—16.9}

8.0—11.5 N·m
{82—117 kgf·cm,
71—101 in·lbf}

N·m {kgf·m, ft·lbf}

1 Water pump pulley
2 Water pump
3 O-ring

67162-MAZC-G11

Water pump mounting and related components—Mazda3 and Mazda6 models equipped with the 2.0L (LF) and 2.3L (L3) engines

To install:
10. Clean the water pump mating surfaces.
11. Install a new O-ring and the water pump. Tighten the bolts to 71–101 inch lbs. (8–11 Nm).
12. Install the water pump pulley and tighten the tighten the bolts to 12–17 ft. lbs. (17–23 Nm).
13. Install the drive belt.
14. Install the plug hole plate.
15. Install the coolant reservoir tank.
16. Install the splash shield and under cover.
17. Connect the battery cable and install the cover.
18. Fill the cooling system.
19. Start the engine, check for leaks and repair if necessary.

3.0L (AJ) Engine
1. Before servicing the vehicle, refer to the precautions section.
2. Disconnect the negative battery cable.
3. Remove the under cover.
4. Drain the cooling system.
5. Remove the air cleaner.

6. Remove the water pump drive belt pulley as follows:
   a. Replace part of tool 49 UN30 3009 with tool 49 UN30 3457.
   b. Install the tools as illustrated and remove the water pump drive pulley.
7. Remove the thermostat case, heater hose and the water outlet pipe.

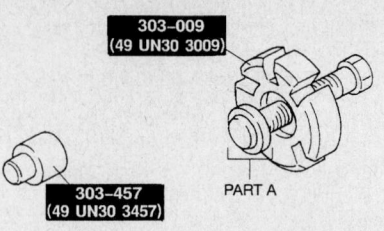

67162-MAZC-G127

Replace part of tool 49 UN30 3009 with tool 49 UN30 3457—Mazda6 models with the 3.0L (AJ) engine

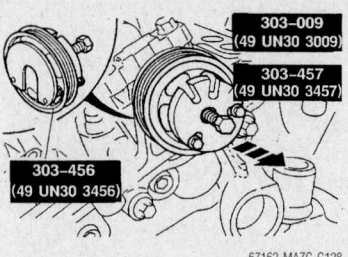

67162-MAZC-G128

Install the tools as illustrated and remove the water pump drive pulley—Mazda6 models with the 3.0L (AJ) engine

8. Remove the water pump bolts and the pump.

To install:
9. Install the water pump and tighten the bolts 89 inch lbs. (10 Nm), plus an additional 85–95 degrees.
10. Install the water outlet pipe, heater hose and thermostat case.
11. Install a new water pump drive belt pulley using tool 49 UN21 1185.
12. Install the air cleaner.
13. Fill the cooling system and connect the negative battery cable.

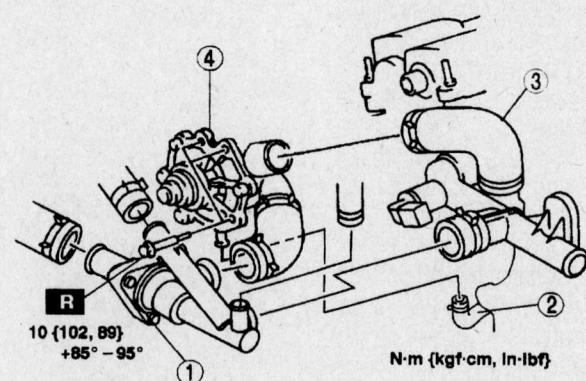

10 {102, 89}
+85°—95°

N·m {kgf·cm, in·lbf}

1 Thermostat case
2 Heater hose
3 Water outlet pipe
4 Water pump

67162-MAZC-G126

Exploded view of the water pump and related components—Mazda6 models with the 3.0L (AJ) engine

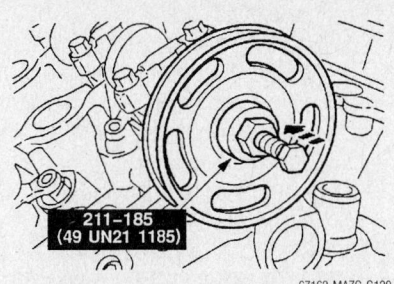

Install a new water pump drive belt pulley using tool 49 UN21 1185—Mazda6 models with the 3.0L (AJ) engine

14. Start the vehicle and check for leaks, repair if necessary.

15. Install the under cover if no leaks are found.

Mazdaspeed6

1. Before servicing the vehicle, refer to the precautions section.

2. Remove the battery cover and disconnect the negative battery cable.

3. Drain the cooling system.

4. Loosen the water pump pulley bolt and position the drive belt aside.

5. Remove the water pump pulley.

6. Remove the water pump bolts, the pump and the O–ring.

To install:

7. Clean the water pump mating surfaces.

8. Install a new O–ring and the water pump. Tighten the bolts to 71–101 inch lbs. (8–11 Nm).

9. Install the water pump pulley and tighten the tighten the bolts to 12–17 ft. lbs. (17–23 Nm).

10. Install the drive belt.

11. Connect the battery cable and install the cover.

12. Fill the cooling system.

13. Start the engine, check for leaks and repair if necessary.

Heater Core

REMOVAL AND INSTALLATION

MX5-Miata

1. Before servicing the vehicle, refer to the precautions section.

2. Disconnect the negative battery cable.

❋❋ CAUTION

After disconnecting the battery, wait for more than 1 minute for the SAS to deplete its stored energy.

3. Drain the cooling system into a clean container for reuse.

4. Disconnect the heater hoses from the heater core.

5. Discharge and recover the air conditioning system refrigerant.

➡**Refer the exploded view illustration for component locations and if applicable, their retainer torque specifications.**

6. Remove the battery cover.

7. Disconnect the negative battery cable.

❋❋ CAUTION

If moisture or foreign material enters the refrigeration cycle, cooling ability will be lowered and abnormal noise or other malfunction could occur. Always plug open fittings immediately after removing any refrigeration cycle parts.

8. To disconnect cooler pipe # 3, disconnect the block joint type pipes by grasping the female side of the block with pliers or similar tool and holding firmly, and then remove the connection bolt or nut.

9. Disconnect the heater hose at the firewall.

10. Remove the console as follows:

a. Remove the selector lever knob if equipped with an automatic transmission, or the shift lever knob if equipped with a manual transmission.

b. Remove the front cover.

c. Remove the rear cover.

d. Remove the hole cover.

e. Remove the parking brake lever boot and boot plate.

f. Remove the console.

g. Disconnect the power window main switch connector.

h. Remove the indicator compartment connector, if equipped with an automatic transmission.

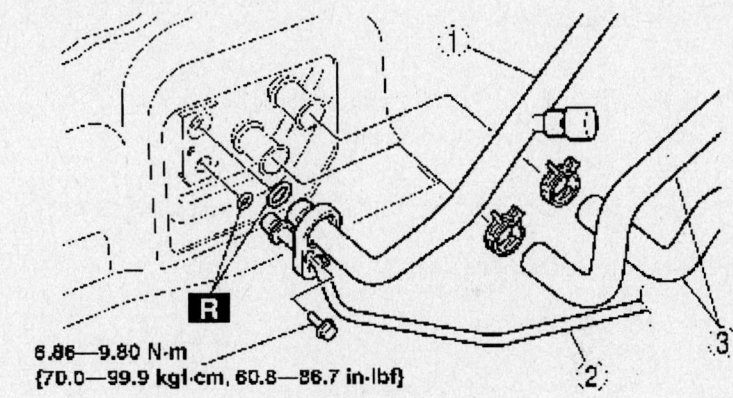

6.86—9.80 N·m
{70.0—99.9 kgf·cm, 60.8—86.7 in·lbf}

1 Cooler pipe No.3 (LO)

2 Cooler pipe No.2 (HI)

3 Heater hose

Location of the cooler and heater pipes at the firewall—MX-5 Miata

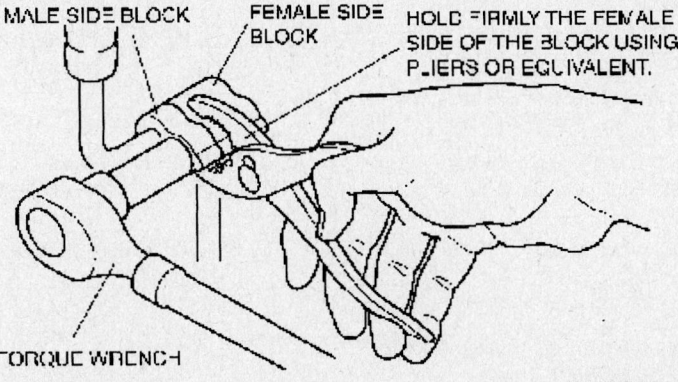

MALE SIDE BLOCK FEMALE SIDE BLOCK HOLD FIRMLY THE FEMALE SIDE OF THE BLOCK USING PLIERS OR EQUIVALENT.

TORQUE WRENCH

Disconnect the block joint type pipes by grasping the female side of the block with pliers or similar tool and holding firmly, and then remove the connection bolt or nut—MX-5 Miata

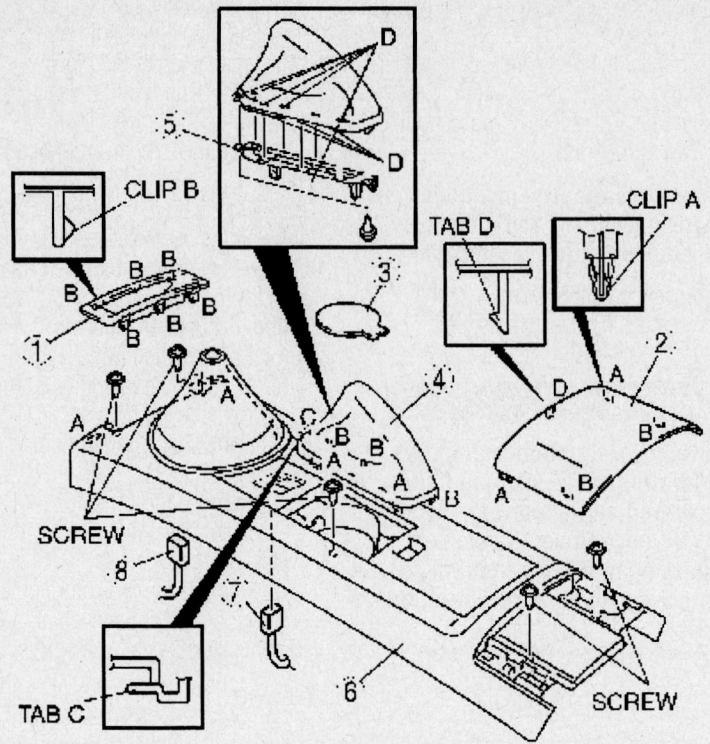

1	Front cover	5	Parking brake lever boot plate
2	Rear cover	6	Console
3	Hole cover	7	Power window main switch connector
4	Parking brake lever boot	8	Indicator compartment connector (AT)

09482_MAZC2_G0133

Exploded view of the console assembly—MX-5 Miata

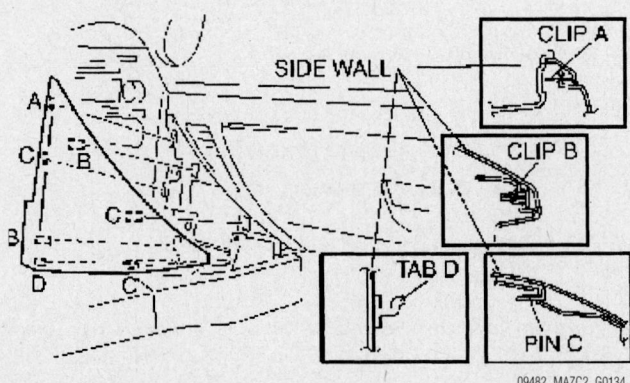

09482_MAZC2_G0134

Exploded view of the side wall assembly—MX-5 Miata

11. Remove the dumper clip.

12. Bend the stoppers inward, then remove.

13. Turn the glove compartment downward and pull the pins.

14. Remove the glove compartment.

➡The side wall removal procedure is the same for the both sides.

15. Pull the side wall rearward and detach clips A, B pins C and tab D.

16. Remove the side wall.

17. Install in the reverse order of removal.

18. Remove the console panel as follows:
    a. Remove the screws.
    b. Pull the console panel outward and detach tabs A and pins B.
    c. Disconnect the seat warmer switch connector, if equipped.
    d. Disconnect the accessory socket connector.

e. Remove the console panel.

19. Remove the center panel as follows:
    a. Pull the lower panel outward and detach clips A and tab B.
    b. Turn the lower panel downward and remove the tabs C.
    c. Remove the lower panel.
    d. Remove the knee bolster.
    e. Remove the bolt and screw.
    f. Pull the center panel unit outward, detach clip A from the dashboard, and then remove the center panel unit.

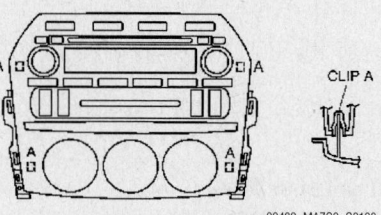

09482_MAZC2_G0138

Pull the center panel unit outward, detach clip A from the dashboard, and then remove the center panel unit—MX-5 Miata

➡Verify that bolt has been removed when pulling the center panel unit outward. If the center panel unit is pulled with bolt installed, stress will be applied to the connectors inside the panel which may cause a malfunction. When removing the center panel unit, disconnect the audio unit connector (24-pin) first to aid in disconnecting other connectors and the antenna plug.

g. Disconnect the connector and antenna feeder plug.

20. Remove the column cover as follows:
    a. Remove the tab A, then remove the upper column cover.
    b. Remove the screws, then remove the lower column cover.
    c. Remove the ignition key illumination.

21. Remove the drivers side air bag as follows:

❊❊ WARNING

Handling the air bag module improperly can accidentally deploy the air bag module, which may seriously injure you.

a. Turn the ignition switch to the LOCK position.

b. Remove the cover, and bolt.

c. Using a flathead screwdriver, pry out the connector stopper plate.

d. Disconnect the connector.

e. Remove the air bag.

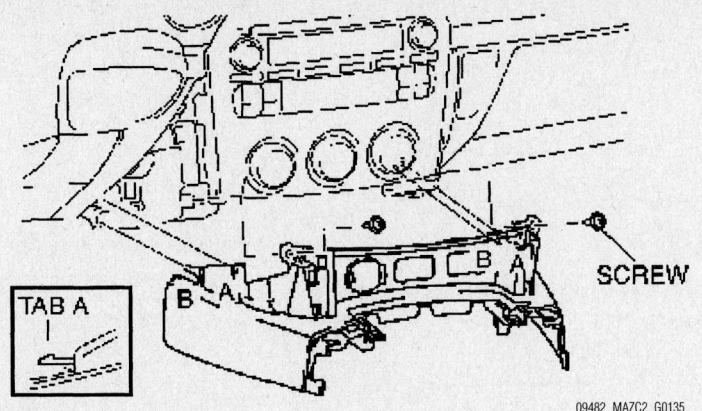

Exploded view of the console panel assembly—MX-5 Miata

09482_MAZC2_G0135

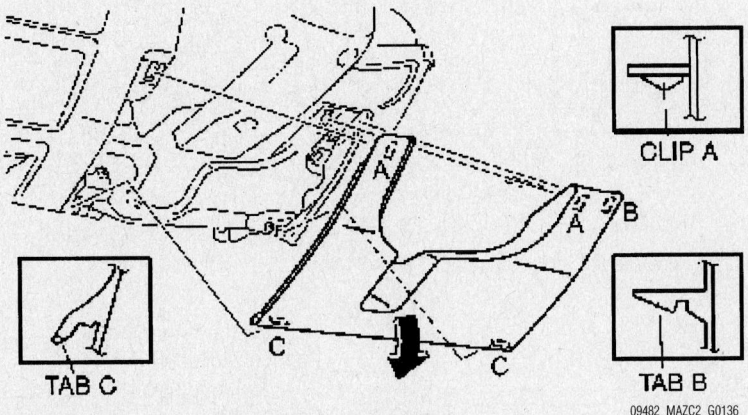

Exploded view of the lower panel assembly—MX-5 Miata

09482_MAZC2_G0136

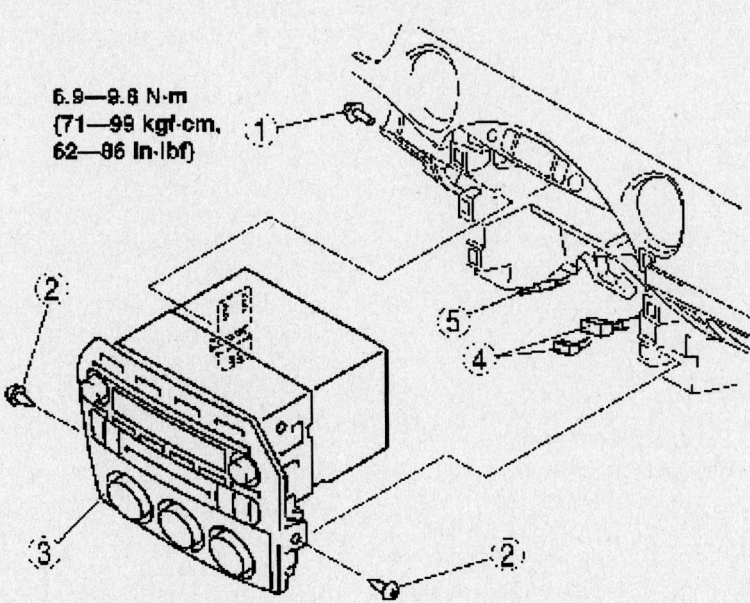

1 Bolt
2 Screw
3 Center panel unit
4 Connector
5 Antenna feeder plug

09482_MAZC2_G0137

Exploded view of the center panel unit assembly—MX-5 Miata

22. Remove the steering wheel as follows:

※※ WARNING

Handling the air bag module improperly can accidentally deploy the air bag module, which may seriously injure you.

   a. Remove the locknut.

※※ CAUTION

Do not try to remove the steering wheel by hitting the shaft with a hammer. The column will collapse.

   b. Set the vehicle in the straight-ahead position.
   c. Remove the steering wheel using a suitable puller.
   d. Remove the column cover.
   e. Remove the clock spring screw, connector and clock spring.

➡**For vehicles with DSC, if the negative battery cable or the steering angle sensor connector or ROOM 15 A fuse is disconnected, the stored initial position of the steering angle sensor will be cleared and the DSC will not operate properly, making the vehicle unsafe to drive. Perform the steering angle sensor initialization procedure after connecting the negative battery cable.**

   f. Disconnect the combination switch connector.
   g. Disconnect the steering angle sensor connector, on vehicles with DSC.
   h. Remove the screws and then remove the combination switch.
23. Remove the steering shaft.
24. Pull the meter hood upward and detach clips A.
25. Remove the meter hood.
26. Remove the instrument cluster as Follows;
   a. Remove the screw.
   b. Remove the cluster by rotating it upwards.
   c. Remove clip A and the connector
   d. Remove clip B by rotating it 90 degrees.

➡**Place the cluster with the display side facing up after removal.**

27. Pull the side panel outward and detach clips A.
28. Pull the side panel rearward and detach tab B from the dashboard.
29. Remove the side panel.
30. Remove the hood release lever as follows:

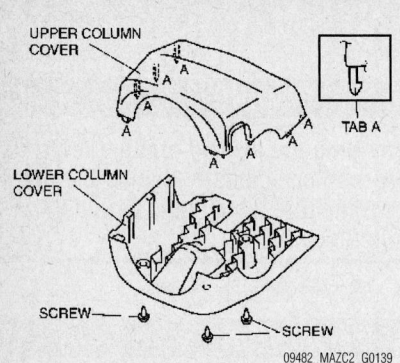

Exploded view of the column cover assembly—MX-5 Miata

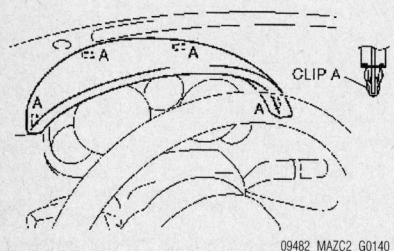

Exploded view of the meter hood assembly—MX-5 Miata

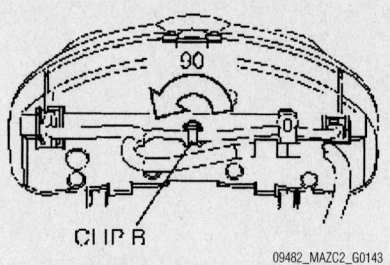

Remove clip B by rotating it 90 degrees—MX-5 Miata

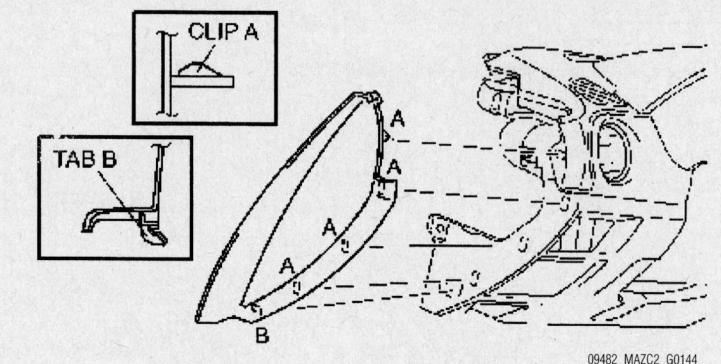

Exploded view of the side panel assembly—MX-5 Miata

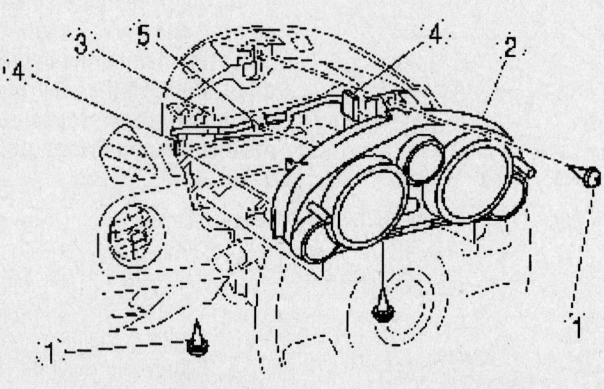

1 Screw	4 Connector
2 Instrument cluster	5 Clip B
3 Clip A	

Exploded view of the instrument cluster assembly—MX-5 Miata

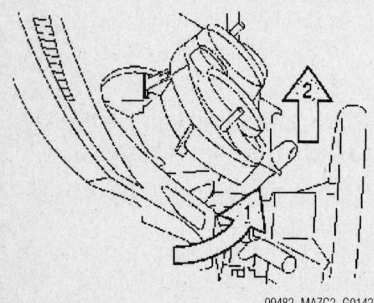

Remove the cluster by rotating it upwards—MX-5 Miata

a. Disconnect the connector.
b. Remove the hood latch.
c. Pull the hood release lever.
d. While pushing the tab in the direction of the arrow using a tape-wrapped, small flathead screwdriver, detach it from the dashboard.

➡**Remove the hood release lever while taking care not to damage the hood release cable with the flathead screwdriver.**

e. Remove the hood release cable.

31. Remove the female bracket as follows:
a. Remove the top lock lever cover.
b. Remove the top lock cover.
c. Remove the top lock.
d. Remove the male wedge cover.
e. Remove the male wedge.
f. Remove the set plate.
g. Remove the slider.
h. Remove the weatherstrip.
i. Mark around the retainer installation screws with paint before removing them.

j. Remove the top fabric from the front bow retainer.
k. Pull out the top fabric from the front header.
l. Remove the covers.
m. Remove the cable installation rivet using a drill.
n. Remove the cables from the cable guide.
o. Remove the band installation rivet using a drill.
p. Remove the nut, then remove the cable end bracket.
q. Remove the rain rail assembly by removing the rivets from the top fabric using a drill.
r. Remove spring A by first loosening the bolt, then remove the spacers, spring and collars from the link.
s. Remove spring B by first loosening the bolt, then pull the link in the direction of the arrow to remove the spacer and spring.
t. Remove the striker, connector and the female wedge.
u. After installing the female bracket, keep in mind the following:
v. Degrease the rain rail using white gasoline.
w. Install the insulation tape to the rivet installation hole of the rain rail.
x. Secure the top fabric and rain rail with the rivet.
y. Flatten the stem using hammer.

z. Place the link onto the top fabric.
aa. Align the link with the set plate installation hole of the top fabric, and install the top fabric to the front header.
bb. Thread the cable into the cable guide.

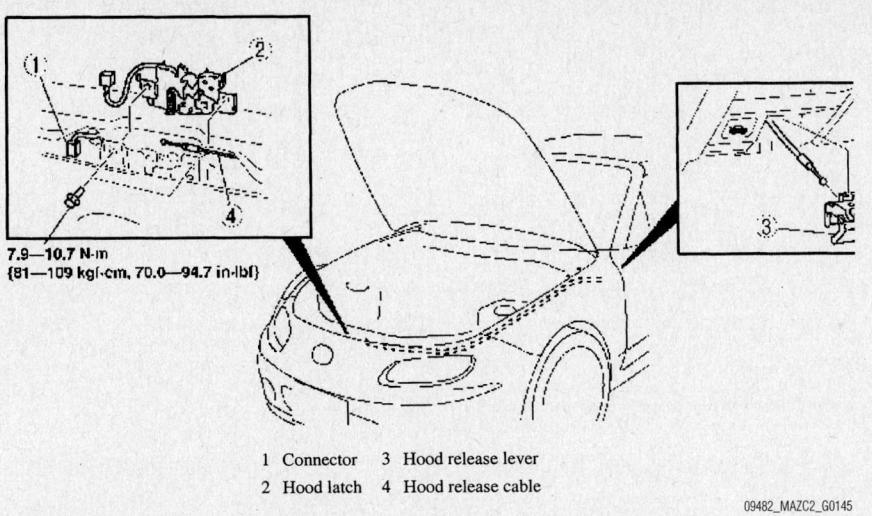

7.9—10.7 N·m
{81—109 kgf·cm, 70.0—94.7 in·lbf}

1	Connector	3	Hood release lever
2	Hood latch	4	Hood release cable

09482_MAZC2_G0145

Exploded view of the hood latch and release lever assembly—MX-5 Miata

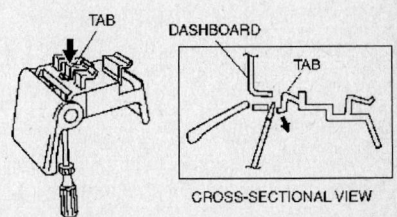

09482_MAZC2_G0146

While pushing the tab in the direction of the arrow using a tape-wrapped, small flathead screwdriver, detach it from the dashboard—MX-5 Miata

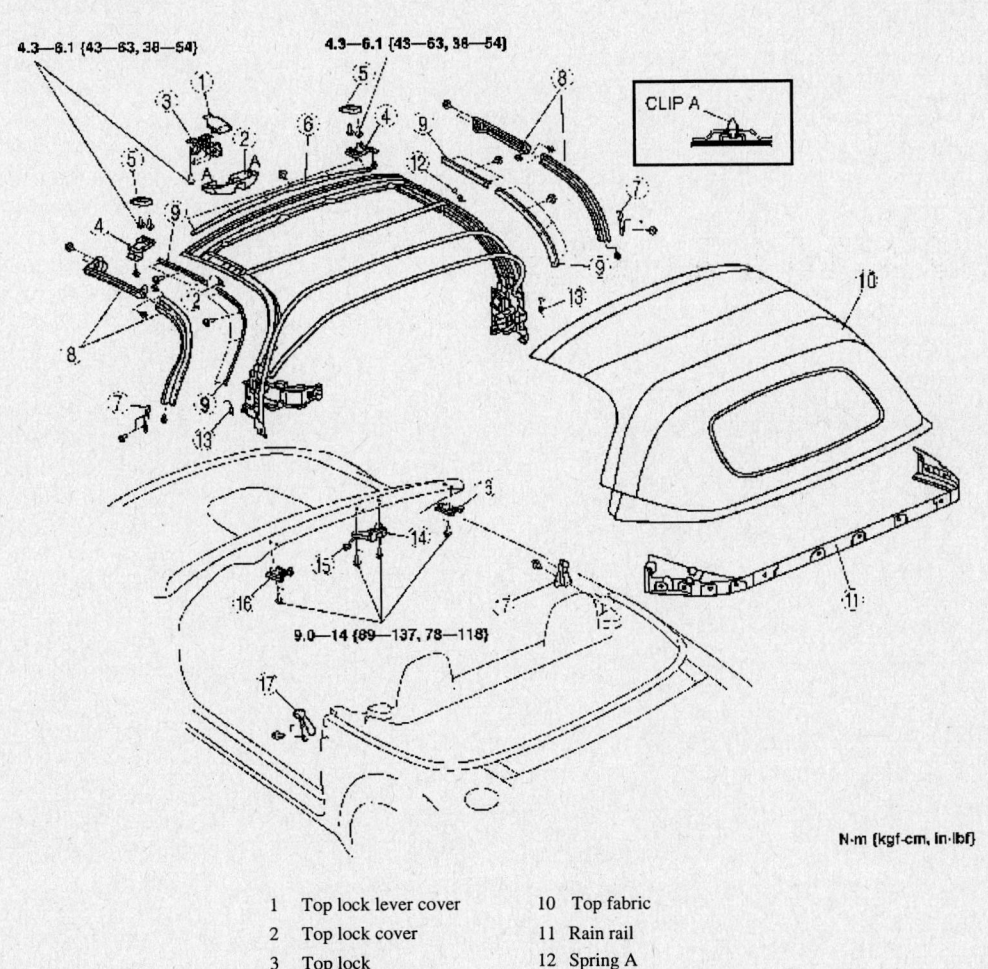

N·m {kgf·cm, in·lbf}

1	Top lock lever cover	10	Top fabric
2	Top lock cover	11	Rain rail
3	Top lock	12	Spring A
4	Male wedge cover	13	Spring B
5	Male wedge	14	Striker
6	Set plate	15	Connector
7	Slider	16	Female wedge
8	Weatherstrip	17	Cab-side weatherstrip
9	Retainer		

09482_MAZC2_G0147

Exploded view of the convertible top assembly—MX-5 Miata

cc. Set the aluminum rivet to the riveter, and then secure the cable to the link with the rivet.

dd. Install the covers.

ee. Install the top fabric to the front bow.

ff. Secure the top fabric to the front bow retainer using a rubber hammer.

gg. Set the aluminum rivet to the riveter, and then secure the band to the link with the rivet.

Install the cable end bracket.

When installing the retainer assembly, install the retainers to the link, aligning the retainer marks with the retainer installation screws.

32. Pull the A-pillar trim, then disengage clips A and B.

33. Pull the A-pillar trim upward, then disengage tabs C from the body.

34. Remove the A-pillar trim.

35. Pull the scuff plate upward while detaching tabs C, and then detach clips A and D, and pins B (4), E (1) from the body, and remove the scuff plate.

36. Install in the reverse order of removal.

37. Remove the seaming welt.

38. Remove the fastener.

39. Pull the front side trim toward you,

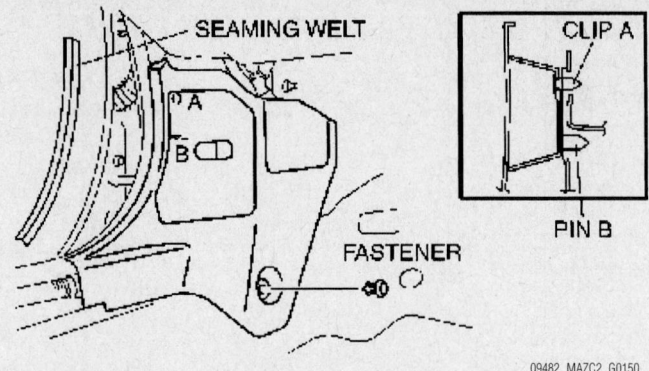

Exploded view of the front side trim assembly—MX-5 Miata

then disengage clip A and pin B from the body.

40. Remove the front side trim.

41. Fuse box No.1

42. Remove the dashboard as follows:

a. Disconnect the dashboard harness connectors.

b. Disconnect the evaporator temperature sensor connector.

c. Disconnect the short code connector.

d. Disconnect the power MOS FET connector.

e. Disconnect the blower motor connector.

f. Disconnect the airflow mode actuator connector.

g. Remove the nut.

h. Remove the cover.

i. Remove bolts A through D.

j. Remove the dashboard.

43. Remove the air dist unit and A/C unit.

44. Remove the heater core as follows:

a. Remove the blower motor.

b. Remove the power MOS FET.

c. Remove the polyurethane foam 1 and 2.

d. Remove the cover.

e. Remove the adhesive polyurethane.

f. Remove the air intake box.

g. Remove the air intake actuator.

h. Remove the air intake crank.

i. Remove the air intake door.

j. Remove the harness.

k. Remove the air mix actuator.

l. Remove the heater core.

To install:

45. Installation is the reverse of removal, please note the following:

a. When replacing the instrument cluster of vehicles with the immobilizer system, perform immobilizer system component replacement/key addition and clearing, if equipped with advanced keyless system or the immobilizer system component replacement/key addition and clearing, if not equipped advanced keyless system .

b. When installing the steering shaft, do not apply a shock in the axial direction of the shaft.

Lock the tilt lever. Tighten nut A then nut B and lastly bolt C.

c. Make sure the wheels in the straight-ahead position, before installing the steering wheel.

d. After installing the combination switch, perform the steering angle sensor initialization procedure:

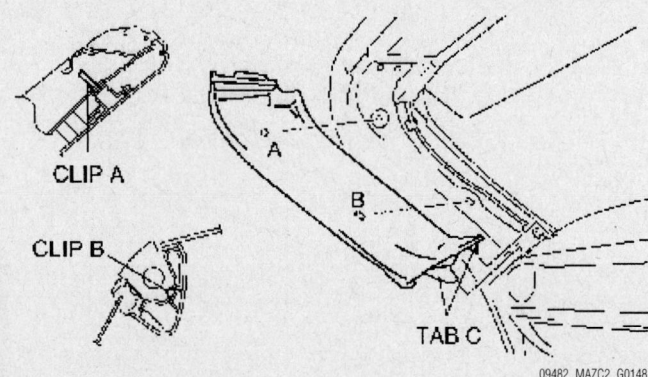

Exploded view of the A-pillar assembly—MX-5 Miata

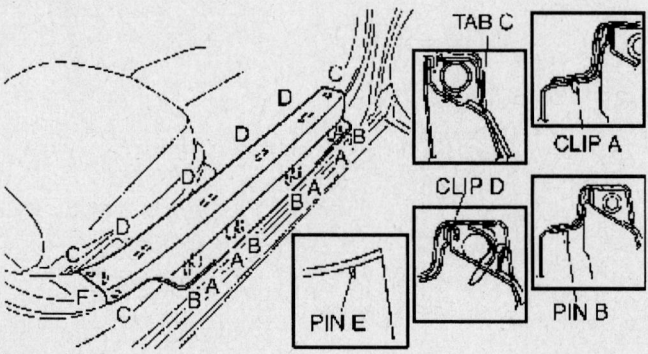

Exploded view of the scuff plate assembly—MX-5 Miata

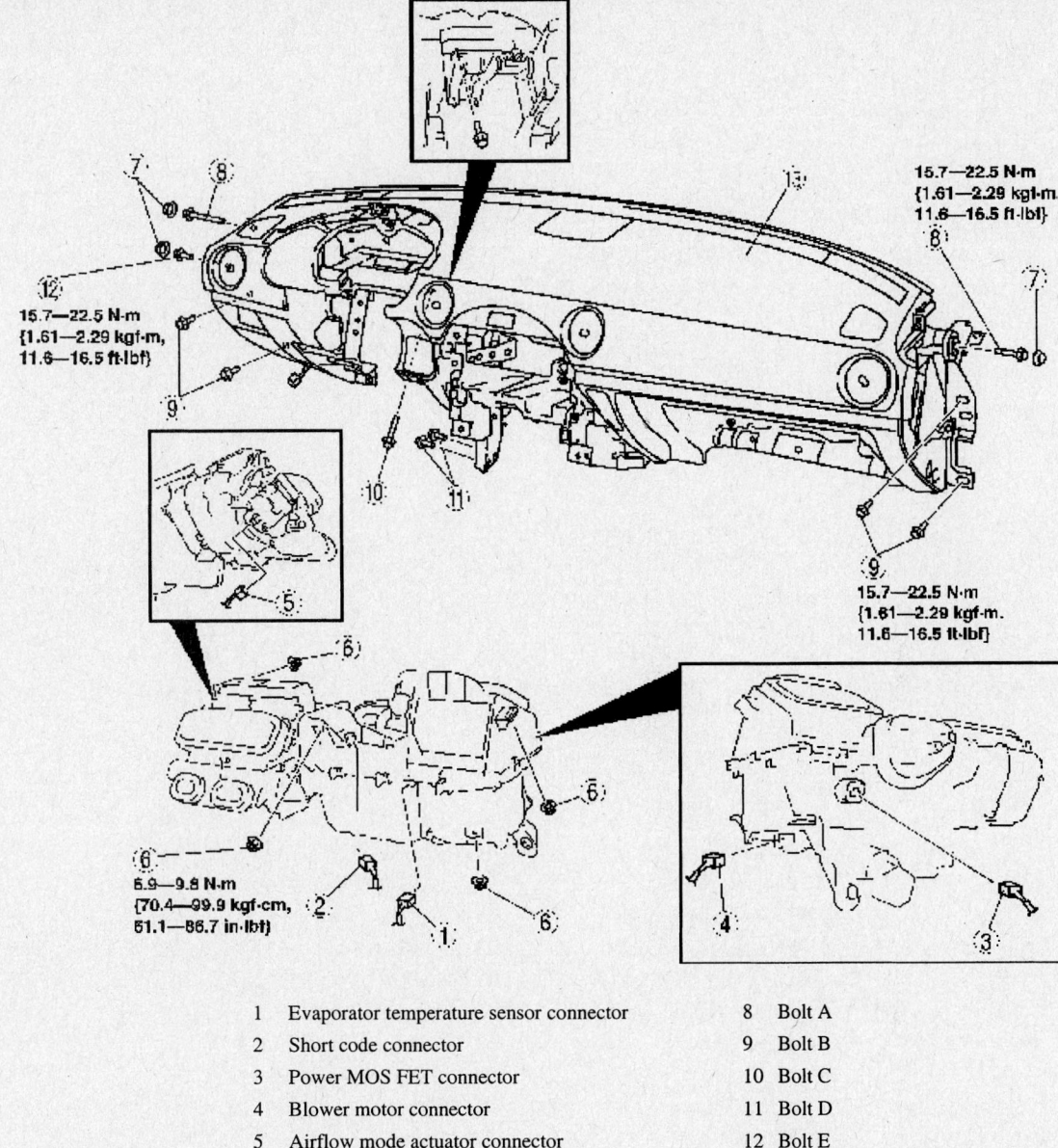

1	Evaporator temperature sensor connector	8	Bolt A
2	Short code connector	9	Bolt B
3	Power MOS FET connector	10	Bolt C
4	Blower motor connector	11	Bolt D
5	Airflow mode actuator connector	12	Bolt E
6	Nut	13	Dashboard
7	Cover		

09482_MAZC2_G0151

Exploded view of the dashboard mounting—MX-5 Miata

✳✳ WARNING

Unless the initialization procedure of the steering angle sensor is completed, the DSC will not operate, causing an unexpected accident. Therefore, always perform the initialization procedure to ensure DSC operation if the power supply to the steering angle sensor has been cut off due to disconnection of the steering angle sensor connector or negative battery cable, or any other cause.

➡The initialization value of the steering angle sensor is stored using the battery power supply. Therefore, the battery power supply of the steering angle sensor is cut and the stored initialization value is cleared when any of the following items are performed.

- Negative battery cable disconnection
- Steering angle sensor connector disconnection
- Fuse (ROOM 15A) removal
- Wiring harness disconnection between battery and steering angle sensor connector

- Inspect the wheel alignment, inflation pressure, and the installation condition of the steering wheel
- If there is any malfunction, adjust the applicable part
- Connect the negative battery cable
- Turn the ignition switch to the ON position
- Confirm that the DSC indicator light illuminates and that the DSC OFF light flashes
- Turn the steering wheel to full right lock, then turn it to full left lock
- Confirm that the DSC OFF light goes out

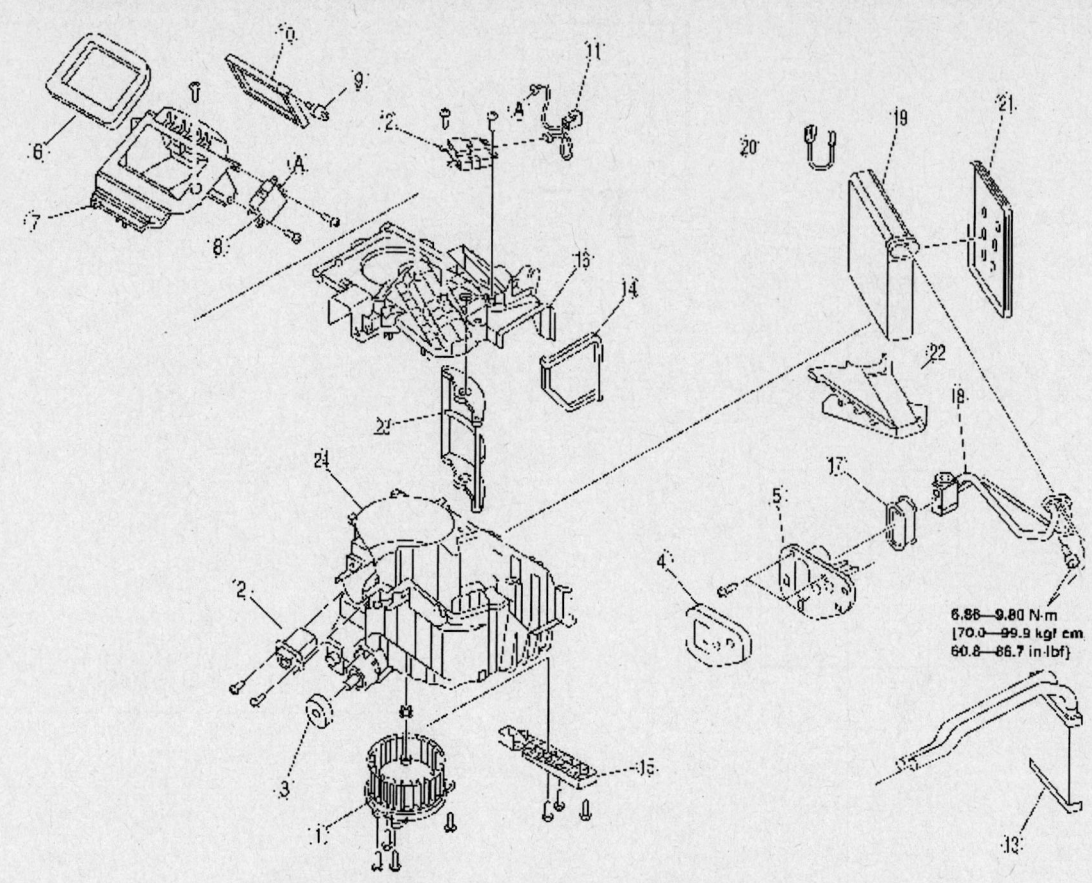

1	Blower motor	13	Heater core
2	Power MOS FET	14	Adhesive polyurethane (2)
3	Polyurethane foam (1)	15	Cover (2)
4	Polyurethane foam (2)	16	A/C case (1)
5	Cover (1)	17	Cover (3)
6	Adhesive polyurethane (1)	18	Expansion valve
7	Air intake box	19	Evaporator
8	Air intake actuator	20	Evaporator temperature sensor
9	Air intake crank	21	Cover (4)
10	Air intake door	22	Insulator
11	Harness	23	Air mix damper
12	Air mix actuator	24	A/C case (2)

09482_MAZC2_G0152

Exploded view of the air distribution and A/C units—MX-5 Miata

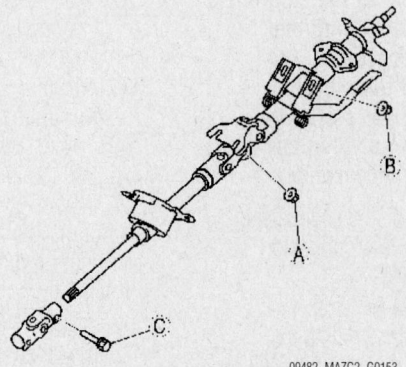

09482_MAZC2_G0153

Exploded view of the steering shaft mounting—MX-5 Miata

- Turn the ignition switch off
- Turn the ignition switch to the ON position again, and confirm that the DSC indicator light goes out
- If the DSC indicator light does not go out, disconnect the negative battery cable, and perform the procedure again
- Drive the vehicle for about 10 minutes and confirm that the ABS warning and DSC indicator lights do not illuminate

e. If the initialization procedure of the steering angle signal is not completed, the DSC will not operate properly and may cause an accident. Therefore, always perform initialization of the DSC HU/CM steering angle signal to ensure proper DSC operation when any of the following items are performed.

- Steering angle sensor replacement DSC HU/CM replacement
- Inspect the wheel alignment and inflation pressure.
- Park the vehicle on level ground
- Turn the ignition switch off
- Connect the WDS or equivalent to the DLC-2
- Access the active command mode, select the following commands, and then follow the indication on the monitor
- Drive the vehicle forward
- After 5 min of driving, verify that the DSC system is normal

f. When installing the center panel unit, make sure that the wiring harness and antenna feeder are not caught between the unit and dashboard. If the wiring harness or the antenna feeder is caught between the unit and dashboard, it may cause malfunctions.

g. Turn the ignition switch to the ON position.

h. Verify that the air bag system warning light illuminates for approx. 6 seconds and goes out.

i. If the air bag system warning light does not operate normally, inspect the system.

Mazda3

1. Before servicing the vehicle, refer to the precautions section.

➡**Refer the exploded view illustration for component locations and if applicable, their retainer torque specifications.**

2. Remove the battery cover.
3. Disconnect the negative battery cable.
4. Discharge the refrigerant from the system.
5. Drain and recycle the engine coolant.
6. Remove the front doors.

✳✳ CAUTION

If moisture or foreign material enters the refrigeration cycle, cooling ability will be lowered and abnormal noise or other malfunction could occur. Always plug open fittings immediately after removing any refrigeration cycle parts.

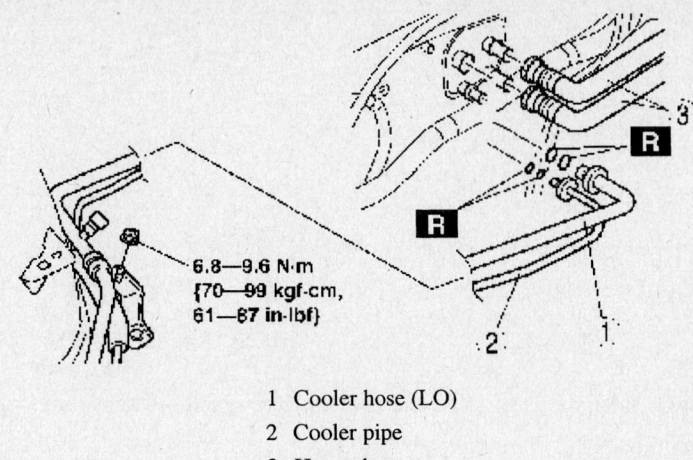

6.8—9.6 N·m
{70—99 kgf-cm,
61—87 in-lbf}

1 Cooler hose (LO)
2 Cooler pipe
3 Heater hose

09482_MAZC2_G0154

Cooler and heater hose connections at the firewall—Mazda3

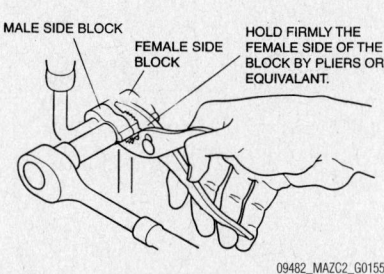

MALE SIDE BLOCK
FEMALE SIDE BLOCK
HOLD FIRMLY THE FEMALE SIDE OF THE BLOCK BY PLIERS OR EQUIVALANT.

09482_MAZC2_G0155

Block type refrigerant joint—Mazda3

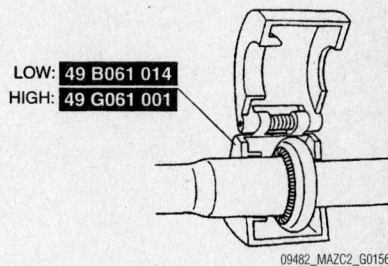

LOW: 49 B061 014
HIGH: 49 G061 001

09482_MAZC2_G0156

Spring lock type refrigerant joint—Mazda3

7. To disconnect the different types of refrigerant line connectors, perform the following

a. On block joint types. disconnect the block joint type pipes by grasping female side of the block with pliers or similar tool and holding firmly, then remove the connection bolt or nut.

b. On spring-lock coupling type , set the service tool shown in the illustration and while looking through the inspection hole of the tool , insert the protruding part of the tool until it makes contact with

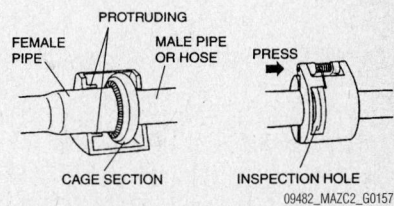

PROTRUDING
FEMALE PIPE
MALE PIPE OR HOSE
PRESS
CAGE SECTION
INSPECTION HOLE

09482_MAZC2_G0157

Looking through the inspection hole of the tool, insert the protruding part of the tool until it makes contact with the cage on spring lock type refrigerant joints—Mazda3

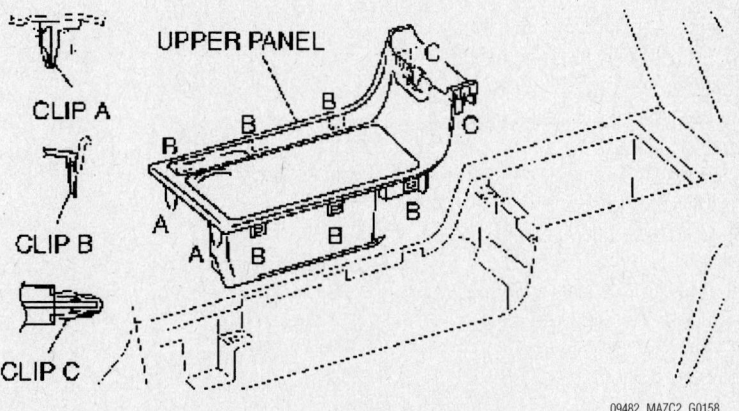

UPPER PANEL
CLIP A
CLIP B
CLIP C

09482_MAZC2_G0158

Detach clips A, B and C and remove the console upper panel—Mazda3

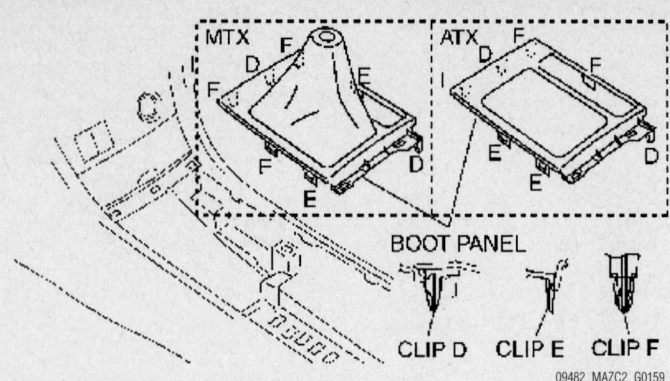

Detach clips D, E and F and remove the boot panel from the console—Mazda3

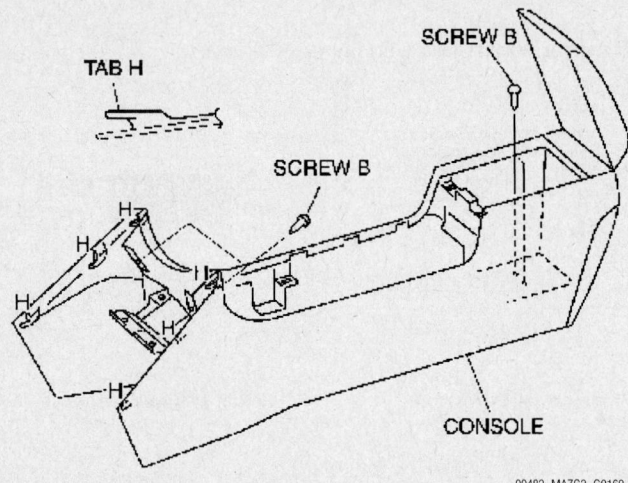

Remove the remaining screws and clips from the console—Mazda3

the cage section. Use the tool to disconnect the male pipe or hose from the female by pulling the male pipe or hose.

➡ **The male pipe or hose can be disconnected easily from the female pipe by pulling from the male pipe or hose while maintaining the pressure of the protruding part of the tool.**

8. Disconnect the heater hose at the firewall.

9. Remove the console as follows:

  a. Detach clips A, B and C and remove the upper panel.

  b. Detach clips D, E and F and remove the boot panel.

  c. Remove screws.

  d. Detach clips G.

  e. Disconnect the cigarette lighter connector.

  f. Remove the ashtray illumination, then remove the ashtray panel.

  g. Remove the screws B.

  h. Detach tabs H and remove the console.

10. Remove the shift lever component on

models with a manual transmission as follows:

  a. Remove the battery and battery tray

  b. Remove the center console.

  c. Remove the front heat insulator.

  d. Remove the shift lever knob.

  e. Remove the boot panel.

  f. Remove the nuts.

  g. Remove the nut.

  h. Remove the bracket.

  i. Remove the seal plate.

  j. Remove both the shift cable end and select cable end using a fastener remover.

  k. Remove the nuts.

  l. Remove the shift lever component.

11. Remove the shift lever component on models with an automatic transmission as follows:

  a. Remove the battery cover.

  b. Disconnect the negative battery cable.

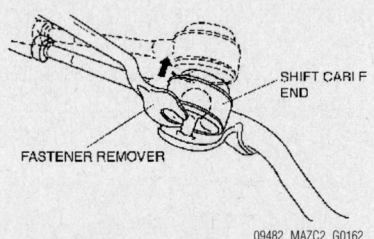

Remove both the shift cable end and select cable end using a fastener remover—Mazda3

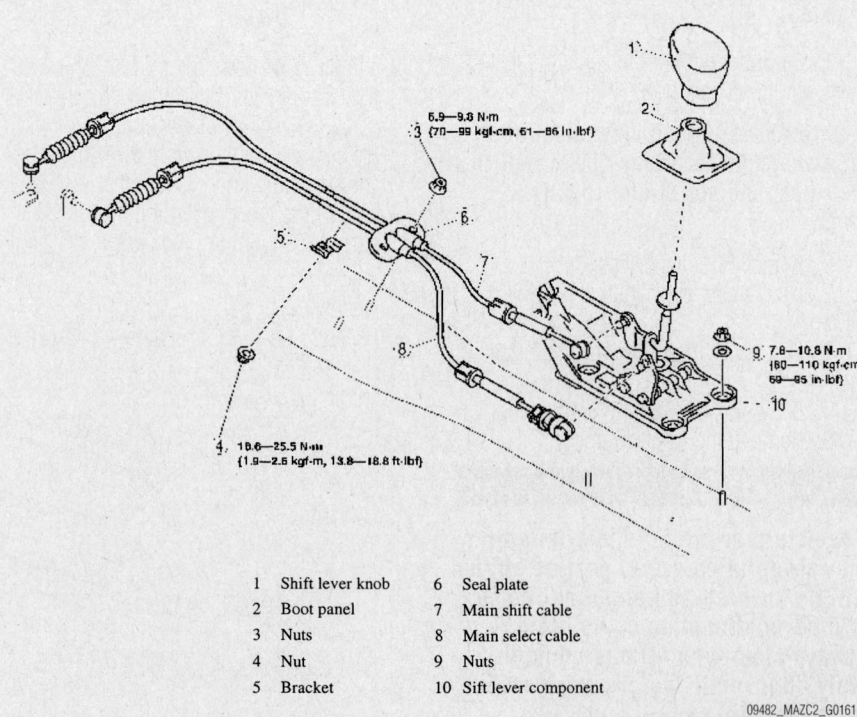

1	Shift lever knob	6	Seal plate
2	Boot panel	7	Main shift cable
3	Nuts	8	Main select cable
4	Nut	9	Nuts
5	Bracket	10	Sift lever component

Exploded view of the shift lever component on models with a manual transmission—Mazda3

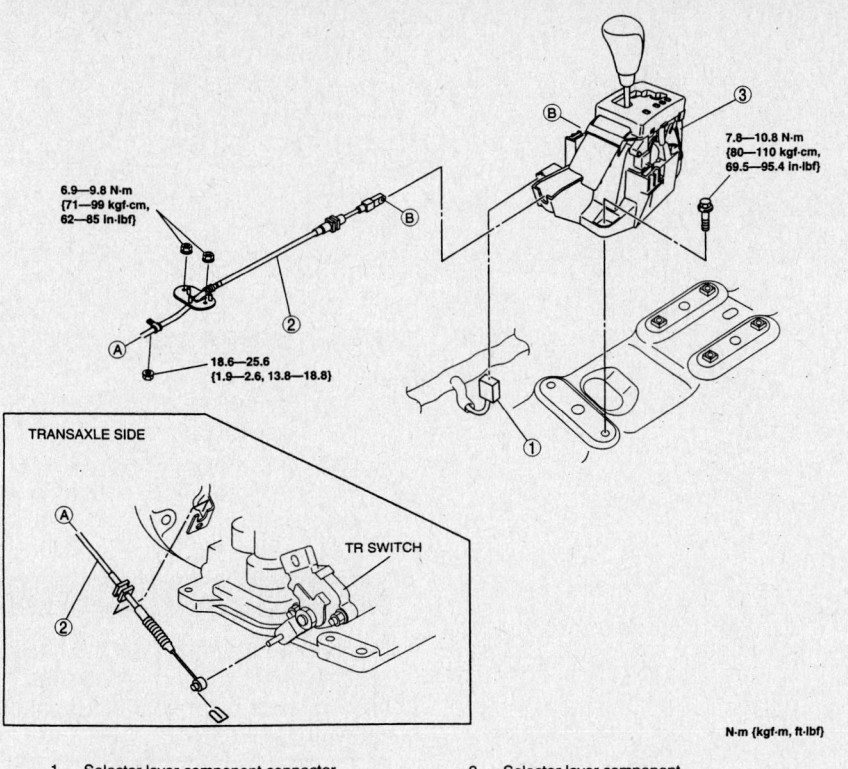

6.9—9.8 N·m
{71—99 kgf-cm,
62—85 in-lbf}

7.8—10.8 N·m
{80—110 kgf-cm,
69.5—95.4 in-lbf}

18.6—25.6
{1.9—2.6, 13.8—18.8}

TRANSAXLE SIDE

TR SWITCH

| 1 | Selector lever component connector | 3 | Selector lever component |
| 2 | Selector cable | | |

N·m {kgf-m, ft-lbf}

09482_MAZC2_G0165

Exploded view of the shift lever component on models with an automatic transmission—Mazda3

c. Remove the battery, battery box and battery tray.

d. Remove the air cleaner component.

e. Remove the console.

f. Remove the front and center heat insulator.

g. Remove the selector lever component connector.

h. Remove the clip and selector cable.

i. Remove the selector lever component.

12. Pull the decoration panel outward and detach the clips A and tab B in the order shown in the illustration.

13. Remove the decoration panel.

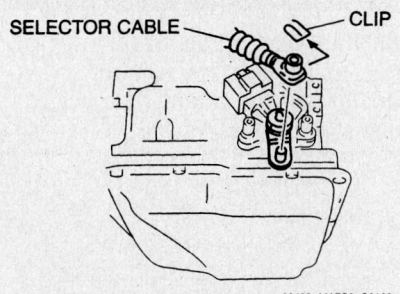

SELECTOR CABLE — CLIP

09482_MAZC2_G0166

Remove the clip and selector cable on models with an automatic transmission—Mazda3

14. Pull the front scuff plate upward, detach clips A, locator pins B and C from the body, and then remove the front scuff plate.

15. Remove the front side trim fastener.

16. Pull the front side trim in the direction of the arrow and detach clip A and locator pin B.

17. Remove the glove box screws.

18. Pull the glove compartment outward and detach clips A and tab B.

19. Remove the glove compartment.

20. Remove the shower ducts.

21. Remove the Passenger Junction Box (PJB) as follows:

a. Remove the cover.

b. On connector A, push the release tab in the direction of the arrow, then rotate the lever in the direction of the arrow and remove connector A.

c. On connector B, rotate the lever in the direction of the arrow and remove connector B.

d. Turn the screws counterclockwise to remove the PJB.

➡ **The screws cannot be removed from the PJB.**

e. Remove the PJB as shown in the illustration.

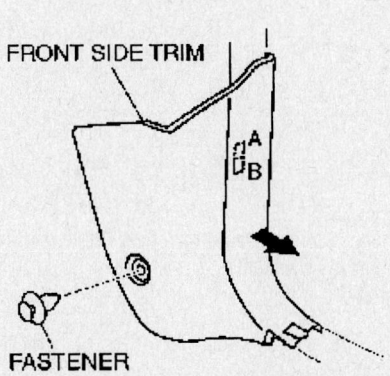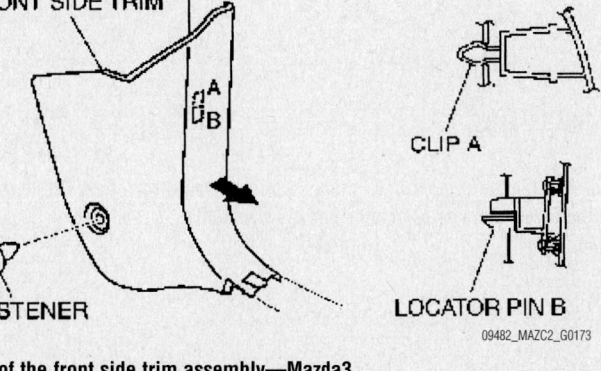

FRONT SIDE TRIM

CLIP A

LOCATOR PIN B

FASTENER

09482_MAZC2_G0173

Exploded view of the front side trim assembly—Mazda3

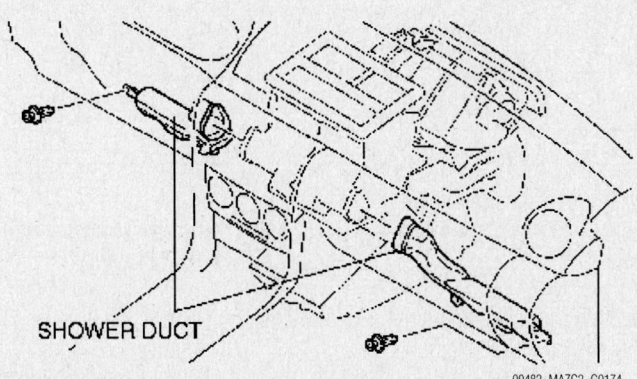

SHOWER DUCT

09482_MAZC2_G0174

Exploded view of the shower duct assembly—Mazda3

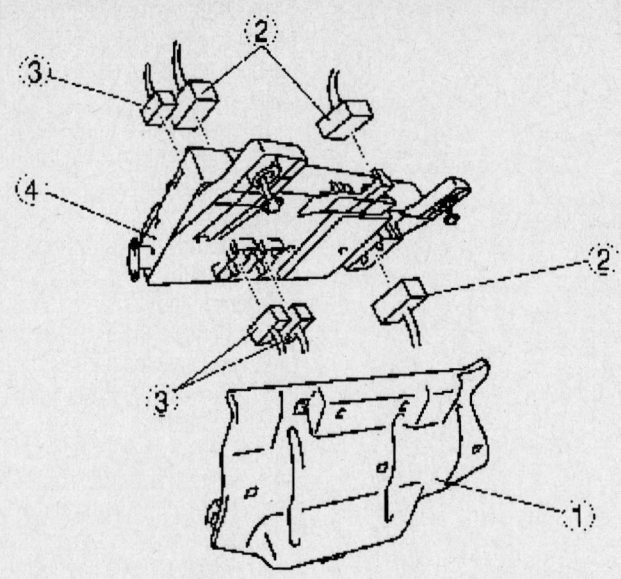

1	Cover	3	Connector B
2	Connector A	4	PJB

09482_MAZC2_G0175

Exploded view of the passenger junction box assembly—Mazda3

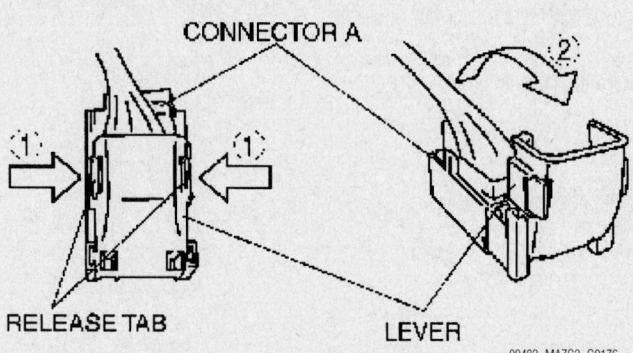

09482_MAZC2_G0176

On connector A, push the release tab in the direction of the arrow, then rotate the lever in the direction of the arrow and remove connector A—Mazda3

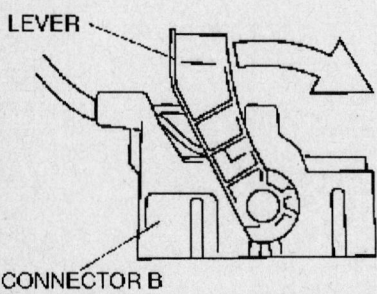

09482_MAZC2_G0177

On connector B, rotate the lever in the direction of the arrow and remove connector B—Mazda3

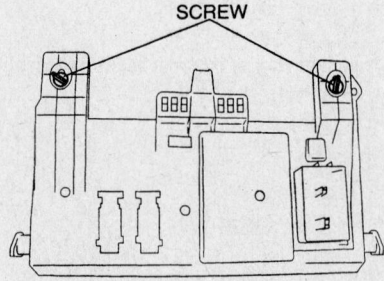

09482_MAZC2_G0178

Turn the screws counterclockwise to remove the PJB—Mazda3

22. Remove the car navigation unit.
23. Remove the lower panel as follows:
    a. Detach the hood release lever from the lower panel.

    b. Remove the lever on 5 door models.
    c. Remove the hood latch.
    d. Remove the hood latch switch connector.

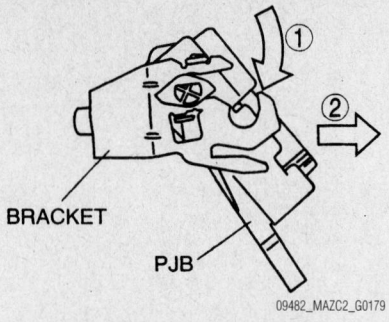

09482_MAZC2_G0179

Remove the PJB as shown—Mazda3

    e. Pull the hood release lever. While pushing the tab in the direction of the arrow using a tape-wrapped, small flathead screwdriver, detach it from the lower panel.
    f. Remove the hood release cable.

✳✳ CAUTION

Be careful not to damage the hood release cable when removing the hood release lever with the flathead screwdriver.

    g. Pull the hood release lever outward, then remove it from the lower panel.
    h. Remove the front scuff plate.
    i. Remove the front side trim.
    j. Remove the screw.
    k. Pull the lower panel outward, and detach clips A and tab.
    l. Disconnect the panel light control switch connector and the headlight leveling switch connector.
    m. Remove the lower panel.
24. Remove the column cover as follows:
    a. Detach the fit of the upper column cover from the meter hood rubber.
    b. Remove the upper column cover.
    c. Remove the ignition key illumination.
    d. Remove the screws.
    e. Remove the lower column cover.
25. Remove the steering shaft as follows:

✳✳ WARNING

Handling the air bag module improperly can accidentally operate (deploy) the air bag module, which may seriously injure you.

    a. Remove the air bag cover.
    b. Remove the bolt.
    c. Using a flathead screwdriver, pry out the connector stopper plate and unplug the connector.
    d. Remove the drivers side air bag module.
    e. Remove the lockbolt.

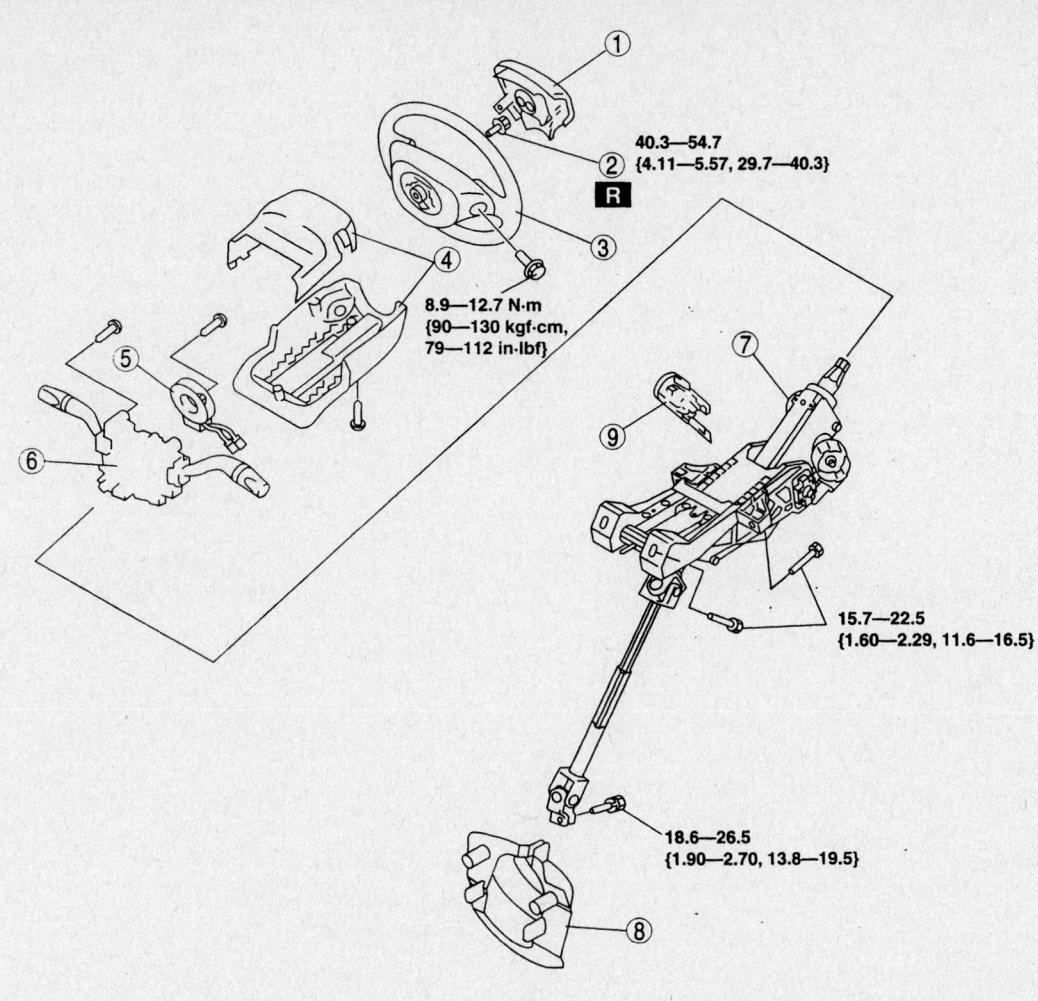

40.3—54.7
{4.11—5.57, 29.7—40.3}
R

8.9—12.7 N·m
{90—130 kgf·cm,
79—112 in·lbf}

15.7—22.5
{1.60—2.29, 11.6—16.5}

18.6—26.5
{1.90—2.70, 13.8—19.5}

N·m {kgf·m, ft·lbf}

1	Air bag module	6	Combination switch
2	Lockbolt	7	Steering shaft
3	Steering wheel	8	Dust cover
4	Column cover	9	Key cylinder
5	Clock spring		

09482_MAZC2_G0182

Exploded view of the steering shaft—Mazda3

Handling the air bag module improperly can accidentally deploy the air bag module, which may seriously injure you.

f. Remove the locknut.

Do not try to remove the steering wheel by hitting the shaft with a hammer. The column will collapse.

g. Set the vehicle in the straight-ahead position.

h. Remove the steering wheel using a suitable puller.

i. Remove the column cover.

j. Remove the clock spring connector.

k. Remove the steering angle sensor connector, if equipped with a steering angle sensor.

l. Remove the screw.

m. Remove the clock spring.

n. Remove the steering angle sensor by detaching the tabs at the four locations shown in the figure and remove the steering angle sensor.

o. Remove the combination switch.

p. Remove the dust cover.

q. Remove the steering shaft.

26. Remove the A-pillar trim shaft as follows:

a. Partially peel back the seaming welt.

b. Detach clips A using a fastener remover.

c. Pull the A-pillar trim and detach clip B (1).

d. Pull the A-pillar trim upward and remove clip B from the A-pillar trim (2).

e. Pull clip B out and rotate it 45 degrees.

f. Remove clip B from the grommet by pulling it upward.

27. Remove the center panel module as follows:

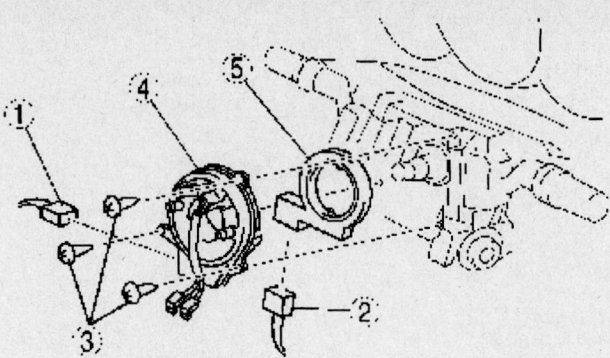

1 Clock spring connector

2 Steering angle sensor connector (With steering angle sensor)

3 Screw

4 Clock spring

5 Steering angle sensor (With steering angle sensor)

09482_MAZC2_G0192

Exploded view of clock spring assembly—Mazda3

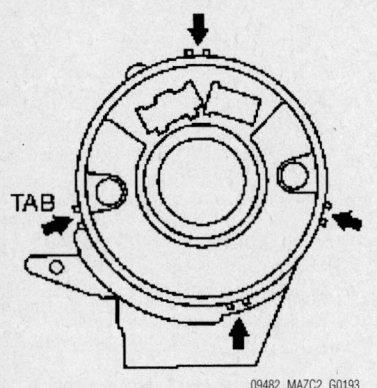

09482_MAZC2_G0193

Remove the steering angle sensor by detaching the tabs at the four locations shown—Mazda3

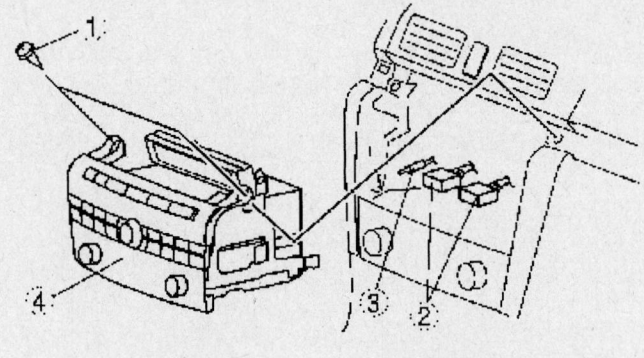

| 1 Screw | 3 Antenna feeder |
| 2 Connector | 4 Center panel module |

09482_MAZC2_G0186

Exploded view of the center panel module—Mazda3

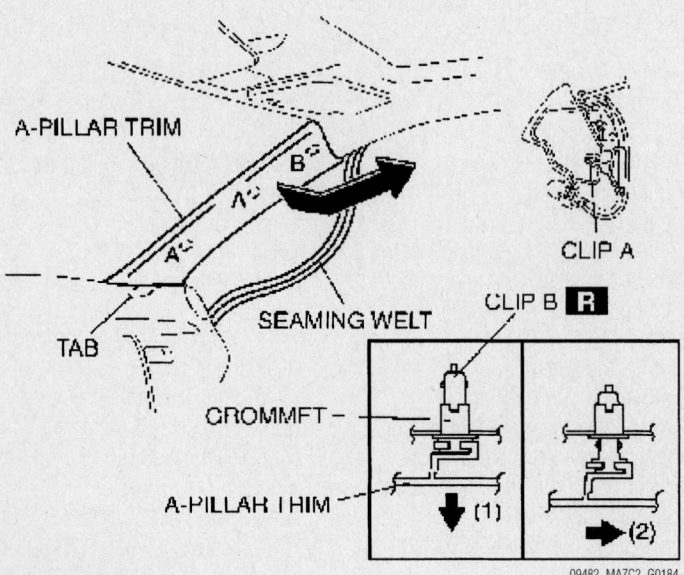

CLIP B

09482_MAZC2_G0184

Exploded view of the A-pillar trim—Mazda3

a. Remove the screw and connector.

b. Remove the antenna feeder and the center panel module. Pull the center panel module outward, detach clip A from the dashboard, and then remove the center panel module.

28. Remove the wiper arm and blade.

29. Remove the cowl grille.

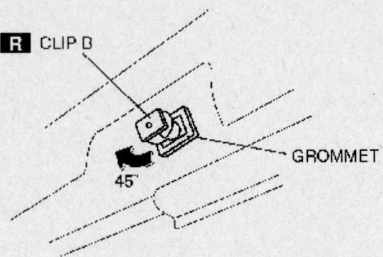

09482_MAZC2_G0185

Pull clip B out and rotate it 45 degrees to remove the A-pillar trim—Mazda3

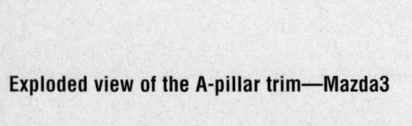

09482_MAZC2_G0187

Pull the center panel module outward, detach clip A from the dashboard, and then remove the center panel module—Mazda3

30. Remove the cowl panel.

31. Remove the wiper motor bolts.

32. Move the wiper motor in the direction of the arrow, remove the securing rubber from the stud pin (for connecting the motor securing rubber), and then remove the windshield wiper motor.

33. Disconnect the windshield wiper motor connector.

34. Remove the A/C unit installation nut

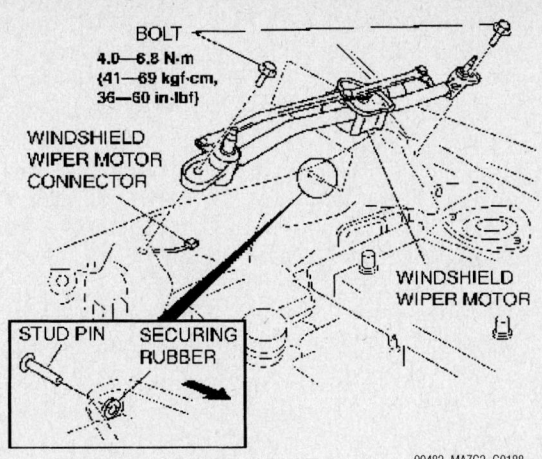

BOLT
4.0—6.8 N·m
{41—69 kgf·cm,
36—60 in·lbf}

WINDSHIELD
WIPER MOTOR
CONNECTOR

WINDSHIELD
WIPER MOTOR

STUD PIN   SECURING
RUBBER

09482_MAZC2_G0188

Exploded view of the wiper motor mounting—Mazda3

from the engine compartment, then remove the A/C unit.

35. Remove the rear heat duct.

36. Disconnect the drain hose connected to the A/C unit.

37. Remove the nuts and bolts for installing the dashboard to the body.

38. Remove the climate control unit.

39. Remove the LCD unit.

✲✲ CAUTION

Handling the air bag module improperly can accidentally deploy the air bag module, which may seriously injure you. Due to the adoption of 2-step deployment control in the passenger-side air bag module, depending on the impact force, it is possible that inflator No.2 might not deploy. In such cases, before disposing of the air bag module, make sure to follow the inflator deployment procedures and verify complete deployment of inflators No.1 and 2.

40. Turn the ignition switch to the LOCK position.

41. Remove the dashboard garnish.

42. Remove the bolt, the passenger side air bag module, the nut and bracket.

43. Remove the heater case and duct.

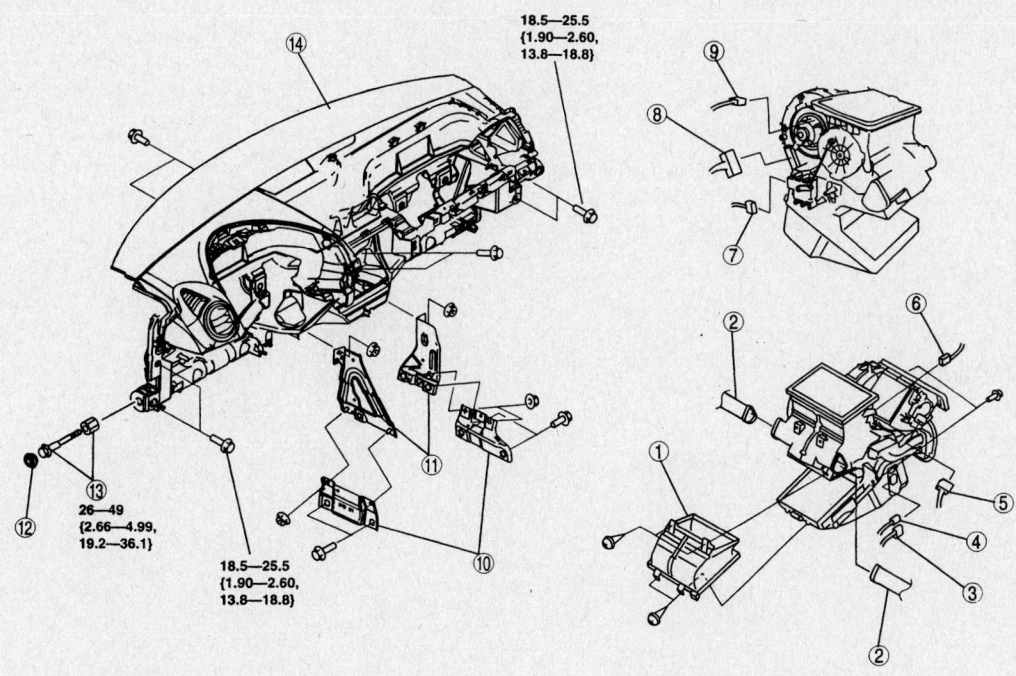

18.5—25.5
{1.90—2.60,
13.8—18.8}

26—49
{2.66—4.99,
19.2—36.1}

18.5—25.5
{1.90—2.60,
13.8—18.8}

N·m {kgf·m, ft·lbf}

1	Heater case	8	Resistor connector (with manual air conditioner system)
2	Duct	9	Blower motor connector
3	Power MOS FET connector (with full-auto air conditioner system)	10	Dashboard bracket (lower)
4	Evaporator temperature sensor connector	11	Dashboard bracket (upper)
5	Air intake actuator connector	12	Cap
6	Air mix actuator connector (with full-auto air conditioner system)	13	Bolt
7	Airflow mode actuator connector (with full-auto air conditioner system)	14	Dashboard

09482_MAZC2_G0189

Exploded view of the dashboard mounting—Mazda3

44. Remove the power MOS FET connector.
45. Remove the evaporator temperature sensor connector.
46. Remove the air intake actuator connector.
47. Remove the air mix actuator connector.
48. Remove the airflow mode actuator connector.
49. Remove the resistor connector on models with manual air conditioner system.

50. Remove the blower motor connector.
51. Remove the upper and lower dashboard brackets.
52. Remove the cap and bolt.
53. Remove the dashboard.
54. Remove the heater core as follows:
  a. Remove the adhesive polyurethane (1).
  b. Remove the blower case 1 and 2.
  c. Remove the air intake actuator.
  d. Remove the air intake link set.
  e. Remove the blower motor.

  f. Remove the power MOS FET.
  g. Remove the air mix link set.
  h. Remove the air mix actuator.
  i. Remove the airflow mode link set.
  j. Remove the airflow mode main link.
  k. Remove the airflow mode actuator.
  l. Remove the polyurethane foam.
  m. Remove the adhesive polyurethane (2 and 3).
  n. Remove the evaporator pipe.
  o. Remove the expansion valve.
  p. Remove the heater core.

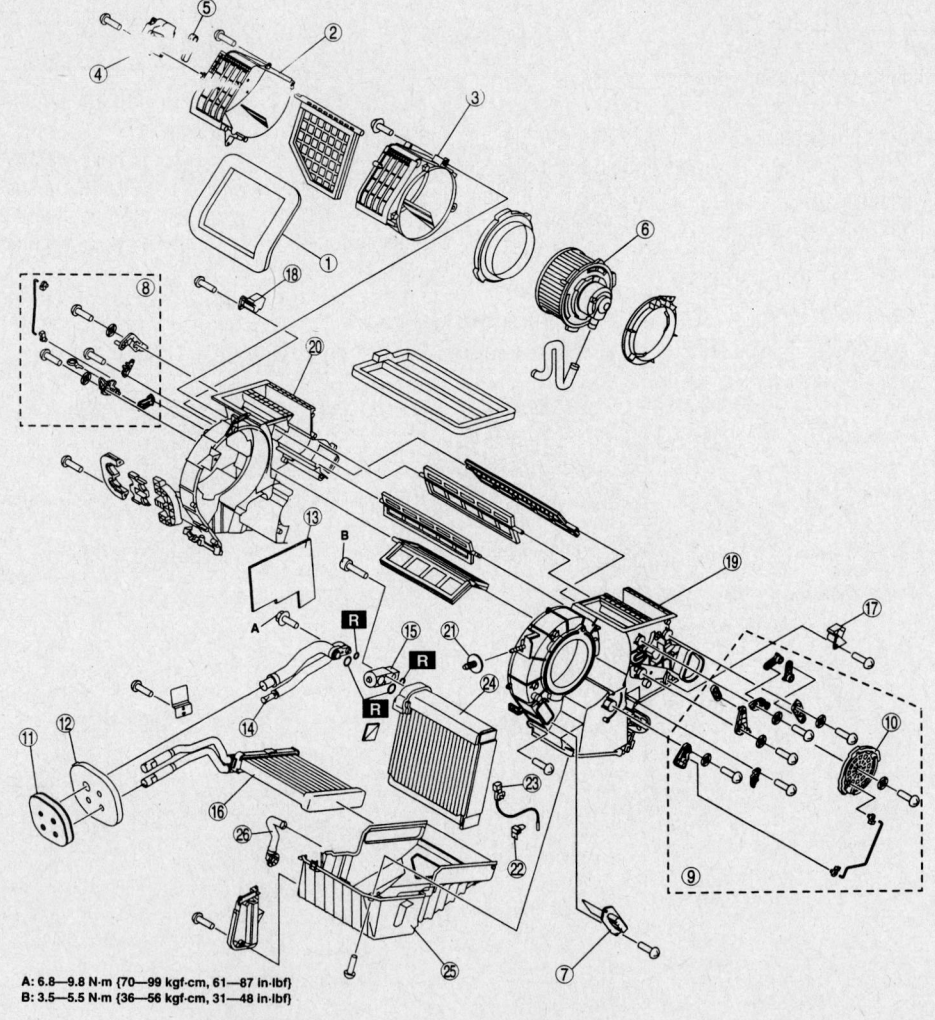

A: 6.8—9.8 N·m {70—99 kgf·cm, 61—87 in·lbf}
B: 3.5—5.5 N·m {36—56 kgf·cm, 31—48 in·lbf}

1	Adhesive polyurethane (1)	14	Evaporator pipe
2	Blower case (1)	15	Expansion valve
3	Blower case (2)	16	Heater core
4	Air intake actuator	17	Bracket (1)
5	Air intake link set	18	Bracket (2)
6	Blower motor	19	A/C case (1)
7	Resistor	20	A/C case (2)
8	Air mix link set	21	Bolt
9	Airflow mode link set	22	Sensor clamp
10	Airflow mode main link	23	Evaporator temperature sensor
11	Polyurethane foam	24	Evaporator
12	Adhesive polyurethane (2)	25	A/C case (3)
13	Adhesive polyurethane (3)	26	Drain hose

09482_MAZC2_G0190

Exploded view of Type (A) A/C unit assembly—Mazda3

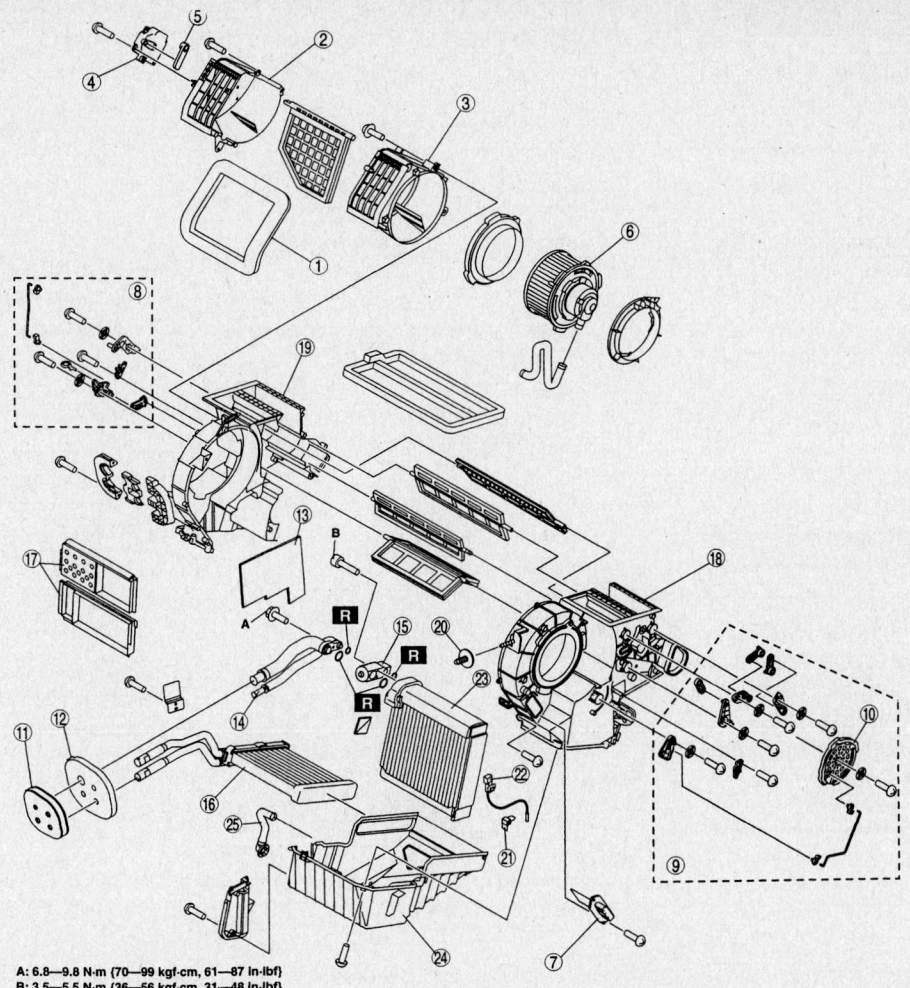

A: 6.8—9.8 N·m (70—99 kgf·cm, 61—87 in-lbf)
B: 3.5—5.5 N·m (36—56 kgf·cm, 31—48 in-lbf)

1	Adhesive polyurethane (1)	14	Evaporator pipe
2	Blower case (1)	15	Expansion valve
3	Blower case (2)	16	Heater core
4	Air intake actuator	17	Air filter cover
5	Air intake link set	18	A/C case (1)
6	Blower motor	19	A/C case (2)
7	Resistor	20	Bolt
8	Air mix link set	21	Sensor clamp
9	Airflow mode link set	22	Evaporator temperature sensor
10	Airflow mode main link	23	Evaporator
11	Polyurethane foam	24	A/C case (3)
12	Adhesive polyurethane (2)	25	Drain hose
13	Adhesive polyurethane (3)		

09482_MAZC2_G0191

Exploded view of Type (B) A/C unit assembly—Mazda3

To install:
55. Installation is the reverse of removal, please note the following:
    a. Adjust the clock spring as follows:

✳✳ CAUTION

If the clock spring is not adjusted, the spring wire in the clock spring could over-wind and break when the steering wheel is turned. Always adjust the clock spring after installing it.

• Set the front tires straight-ahead

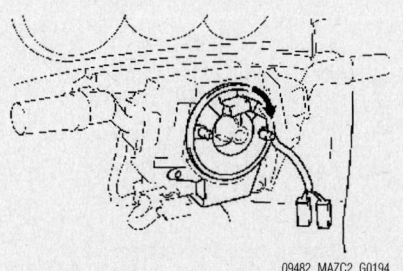

09482_MAZC2_G0194

Turn the clock spring clockwise until it stops—Mazda3

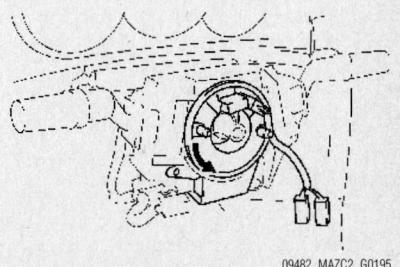

09482_MAZC2_G0195

From the stopped position, turn the clock spring counterclockwise 2 ¾ turns—Mazda3

⚡ CAUTION

The clock spring will break if over-wound. Do not forcibly turn the clock spring.

- Turn the clock spring clockwise until it stops
- From the stopped position, turn the clock spring counterclockwise 2 ¾ turns
- Align the marks
- Verify that the tilt / telescope lever is in the LOCK position

b. Tighten the steering shaft bolts in alphabetical order.

c. Set the wheels in the straight-ahead position and install the steering wheel

⚡ CAUTION

When installing the center panel module, make sure that the wiring harness and antenna feeder are not caught between the unit and dashboard. If the wiring harness or the antenna feeder is caught between the unit and dashboard, it may cause malfunctions.

d. Select Cable for models with an manual transmission should be installed as follows:

- Make sure that the shift lever (transaxle side) is in neutral

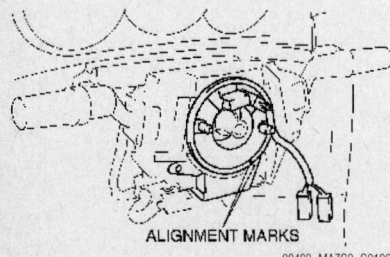

Align the marks on the clock spring—Mazda3

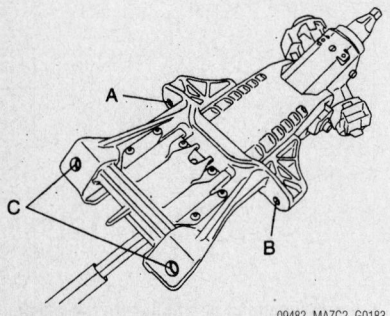

Tighten the steering shaft bolts in alphabetical order—Mazda3

- Push the safety lock, then unlock the lock piece of the select cable in the order shown in the illustration.
- Shift the sift lever to neutral.
- Lock the lock piece of the selector cable in the illustration.
- Shift the shift lever from neutral to other position, and make sure that there are no other components in that area to interfere with the lever.

e. Selector lever for models with an automatic transmission should be installed as follows:

- Insert the location pin of selector lever component to the hole of the floor
- Tighten the selector lever component installation bolts to 7.8–10.8 Nm (69.5–95.4 in. lbs.).

f. Selector Cable for models with an automatic transmission should be installed as follows:

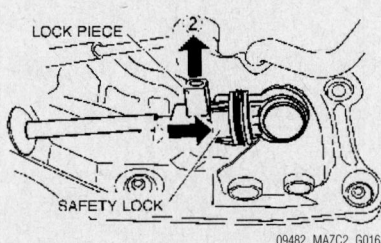

Push the safety lock, then unlock the lock piece of the select cable in the order shown, on manual transmissions—Mazda3

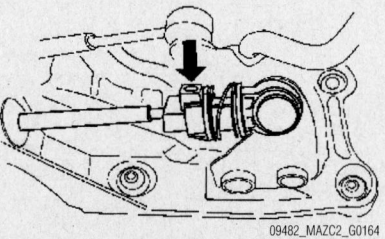

Lock the lock piece of the selector cable, on manual transmissions—Mazda3

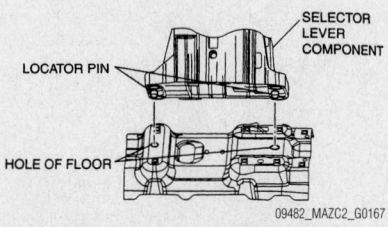

Insert the location pin of selector lever component to the hole of the floor on models with an automatic transmission—Mazda3

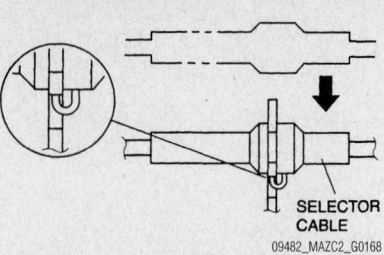

Verify the selector cable is in this position on models with an automatic transmission—Mazda3

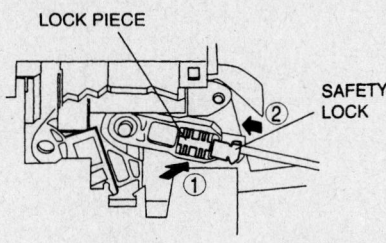

Lock the lock piece and safety lock of the selector cable in this order on models with an automatic transmission—Mazda3

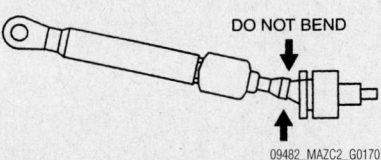

Bending the selector cable in the manner shown in the illustration will damage the cable and it may become loose when shifted on models with an automatic transmission—Mazda3

- Install the selector cable to the selector lever securely.
- Install the selector cable to the bracket securely.

✳✳ CAUTION

Bending the selector cable in the manner shown in the illustration will damage the cable and it may become loose when shifted. When installing the selector cable, hold it straight.

- Install the clip as shown in the illustration.

➡ Install the selector cable to the manual shaft cable with the clip side of the selector cable end facing the front of the vehicle.

- Install the selector cable to the manual shaft lever in such a way that the selector cable does not bear a load.

- Confirm that the end of the manual shift lever projects from the end of the selector cable.
- Install the selector cable to the selector cable bracket securely.

g. Install the connector B as shown in the illustration.

h. After connecting connector A, rotate the lever in the direction of the arrow to install connector A.

i. Turn the ignition switch to the ON position.

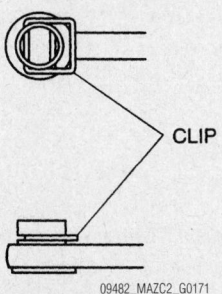

09482_MAZC2_G0171

Install the selector cable to the manual shaft cable with the clip side of the selector cable end facing the front of the vehicle on models with an automatic transmission—Mazda3

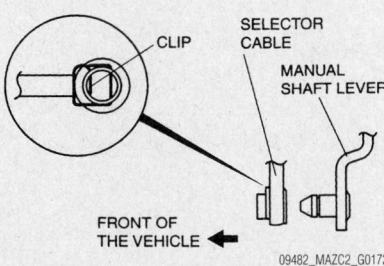

09482_MAZC2_G0172

Install the selector cable to the manual shaft lever in such a way that the selector cable does not bear a load on models with an automatic transmission—Mazda3

j. Verify that the air bag system warning light goes out.

k. If the air bag system warning light does not operate normally, inspect of the system.

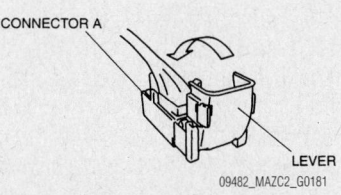

09482_MAZC2_G0181

After connecting connector A, rotate the lever in the direction of the arrow—Mazda3

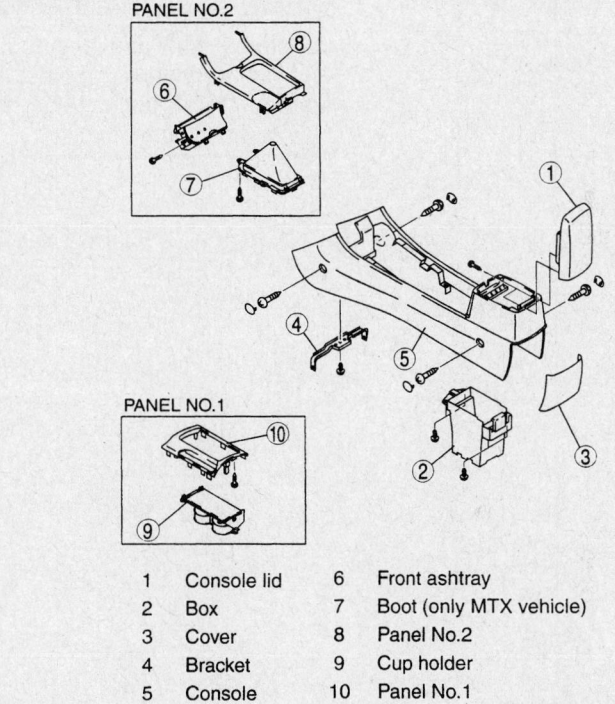

1	Console lid	6	Front ashtray
2	Box	7	Boot (only MTX vehicle)
3	Cover	8	Panel No.2
4	Bracket	9	Cup holder
5	Console	10	Panel No.1

09482_MAZC2_G0197

Exploded view of the console assembly—Mazda6 and Mazdaspeed6 models

Mazda6 and Mazdaspeed6 Models

1. Before servicing the vehicle, refer to the precautions section.

➡**Refer the exploded view illustration for component locations and if applicable, their retainer torque specifications.**

2. Disconnect the negative battery cable.
3. Discharge the refrigerant from the system.
4. Drain and recycle the engine coolant.
5. Remove the dynamic chamber.
6. Remove the glove box as follows:
   a. Remove the damper clip.

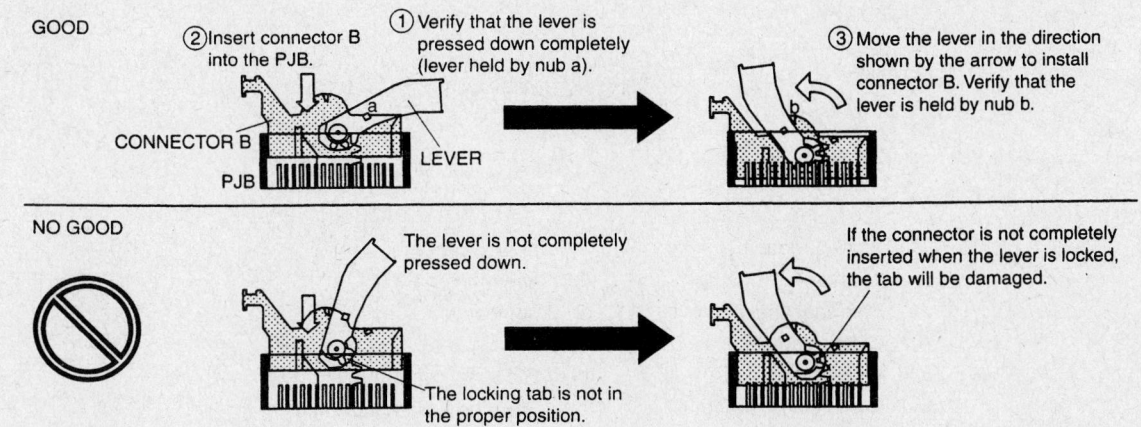

09482_MAZC2_G0180

Install the connector B as shown—Mazda3

b. Bend the stoppers in to disengage them.

c. Lower the glove box door and remove the clips by pulling straight out and remove the glove box.

7. Remove the console as follows:

a. If equipped with a manual transmission, remove the shifter knob.

b. Remove the panels (1 and 2) with a tape wrapped prytool, make sure to disconnect the cigarette lighter and ashtray illumination connections on panel 2.

c. Remove the bolts, disconnect the socket connector and remove the console.

8. Remove the meter hood as follows:

a. Remove the screws, pull the meter hood outwards and detach the column cover.

9. Remove the instrument cluster as follows:

a. Lower the tilt steering wheel to its lowest point.

b. Pull the steering wheel towards you.

c. Remove the screws, connectors and cluster.

10. Remove the column cover.

11. Remove the lower panel as follows:

a. Pull the hood release lever.

b. While pushing the tab in the direction illustrated, using a tape wrapped small flat prytool to pull the release lever outwards from the lower panel.

c. Remove the lower panel screws.

d. Pull the panel out to disengage the clips A from the tabs B and remove the panel.

12. Remove the steering shaft as follows:

❋❋ WARNING

Handling the air bag module improperly can accidentally operate (deploy) the air bag module, which may seriously injure you.

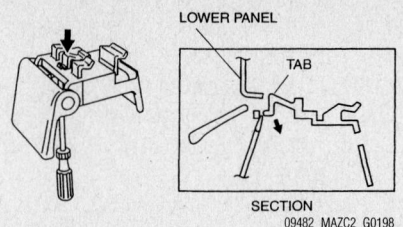

Hood release lever removal—Mazda6 and Mazdaspeed6 models

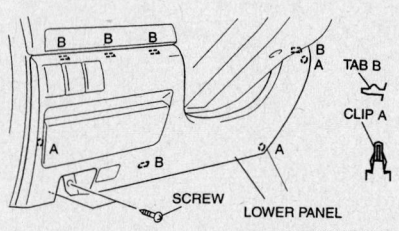

Lower panel mounting—Mazda6 and Mazdaspeed6 models

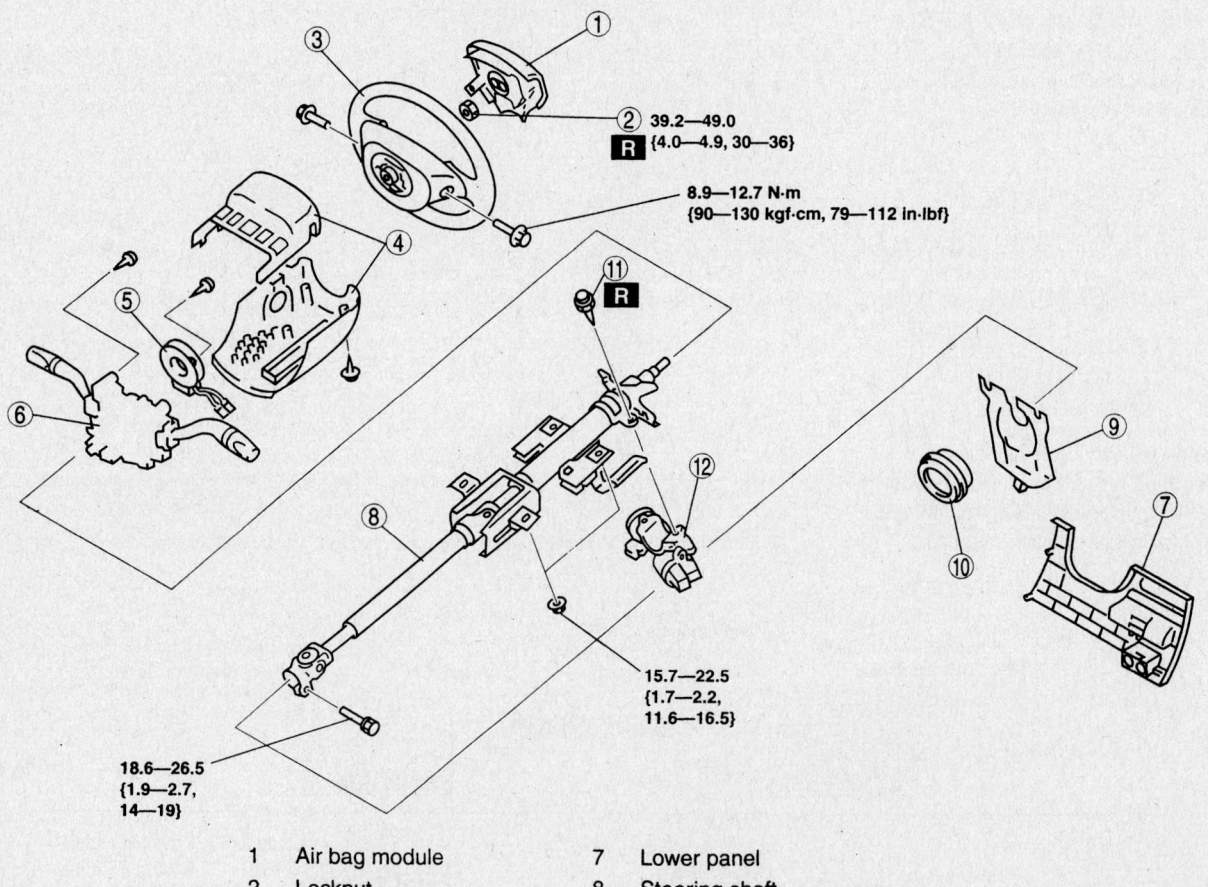

1	Air bag module	7	Lower panel
2	Locknut	8	Steering shaft
3	Steering wheel	9	Joint cover
4	Column cover	10	Dust cover
5	Clock spring	11	Steering lock mounting bolts
6	Combination switch	12	Steering lock component

Exploded view of the steering column and shaft assembly—Mazda6 and Mazdaspeed6 models

a. Remove the air bag cover.

b. Remove the bolt.

c. Using a flathead screwdriver, pry out the connector stopper plate and unplug the connector.

d. Remove the drivers side air bag module.

e. Remove the lockbolt.

✳✳ WARNING

Handling the air bag module improperly can accidentally deploy the air bag module, which may seriously injure you.

f. Remove the locknut.

✳✳ CAUTION

Do not try to remove the steering wheel by hitting the shaft with a hammer. The column will collapse.

g. Set the vehicle in the straight-ahead position.

h. Remove the steering wheel using a suitable puller.

i. Remove the column cover.

j. Remove the clock spring connector.

k. Remove the screw.

l. Remove the clock spring.

m. Remove the combination switch.

n. Remove the steering shaft.

13. Remove the A-pillar trim as follows:

a. Partially pull back the seaming welt.

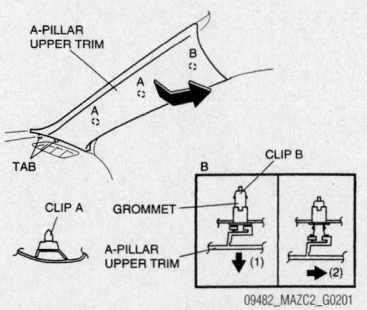

Exploded view of the A-trim mounting—Mazda6 and Mazdaspeed6 models

b. Gently pull the trim the disengage the clips A and then B (1).

c. Pull gently upwards on the trim to disengage clip B (2).

d. Disengage the tabs and remove the trim.

14. Remove the front scuff plates as follows:

a. Pull the scuff plate upwards and disengage clips (B), pins (C and D) and tabs (E).

b. Remove the scuff plate.

15. Remove the front side trims as follows:

a. Partially pull back the seaming welt.

b. Pull the trim outwards to disengage clip (A) and the stud bolt, then remove the trim.

16. Remove the side panels using a flat bladed pry tool to disengage the clips.

17. If equipped with a wire type climate control unit, disconnect the wires for the front A/C unit as follows:

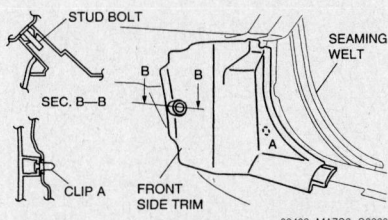

Exploded view of the front side trim mounting—Mazda6 and Mazdaspeed6 models

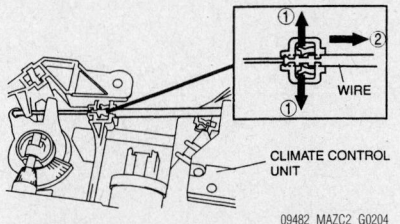

If equipped with a wire type climate control unit, disconnect the wires for the front A/C unit as shown—Mazda6 and Mazdaspeed6 models

a. Remove the illumination bulb.

b. Remove the unit.

c. Remove the fan switch.

d. Disassemble the wire as illustrated.

18. Disconnect the dashboard wiring harness connectors.

19. Remove the dashboard bolts and pull the dashboard pins out from the body.

✳✳ WARNING

When removing the dashboard, make sure to support the dashboard is properly supported to avoid injury and also to avoid damage to the dashboard.

20. Remove the dashboard through the drivers side door.

21. Remove the theft control module bolt, connector and the module.

22. Disconnect the heater hose at the firewall.

23. Disconnect the block joint type pipes by grasping female side of the block with pliers or similar tool and holding firmly, then remove the connection bolt or nut.

24. Remove the heater core on Mazda6 models by removing or disconnecting the following:

- Duct
- Blower motor.
- A/C case 3 and case 4
- Harness
- Air intake actuator
- Air intake crank
- Air filter cover
- Air filter
- Resistor if equipped with a manual A/C system
- PMW unit if equipped with a fully automated A/C system
- Harness if equipped with a fully automated A/C system
- Air flow mode actuator
- Air flow mode bracket, main link, sub link and crank
- Air mix link and crank if equipped with a manual A/C system

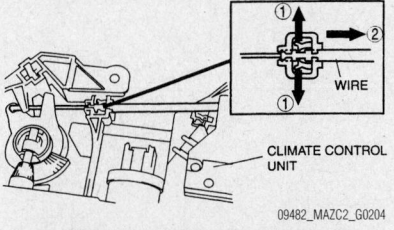

Exploded view of the front scuff plate mounting—Mazda6 and Mazdaspeed6 models

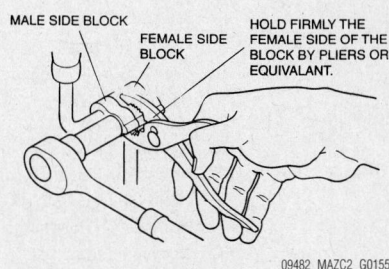

Block type refrigerant joint—Mazda6 and Mazdaspeed6 models

- Air mix actuator, link and crank if equipped with a fully automated A/C system
- Water temperature sensor if equipped with a fully automated A/C system
- Heater core cover and the heater core

25. Remove the heater core on Mazd-speed6 models by removing or disconnecting the following:
- Duct (1)
- Polyurethane protector 1, 2 and 3

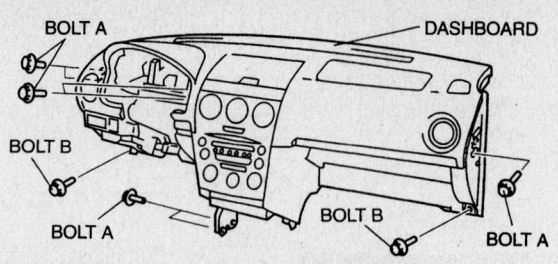

BOLT A: 18.6—25.5 N·m {1.9—2.6 kgf·m, 13.2—18.8 ft·lbf}
BOLT B: 2.1—2.9 N·m {21—30 kgf·cm, 19—26 in·lbf}

09482_MAZC2_G0208

Exploded view of the dashboard mounting—Mazda6 and Mazdaspeed6

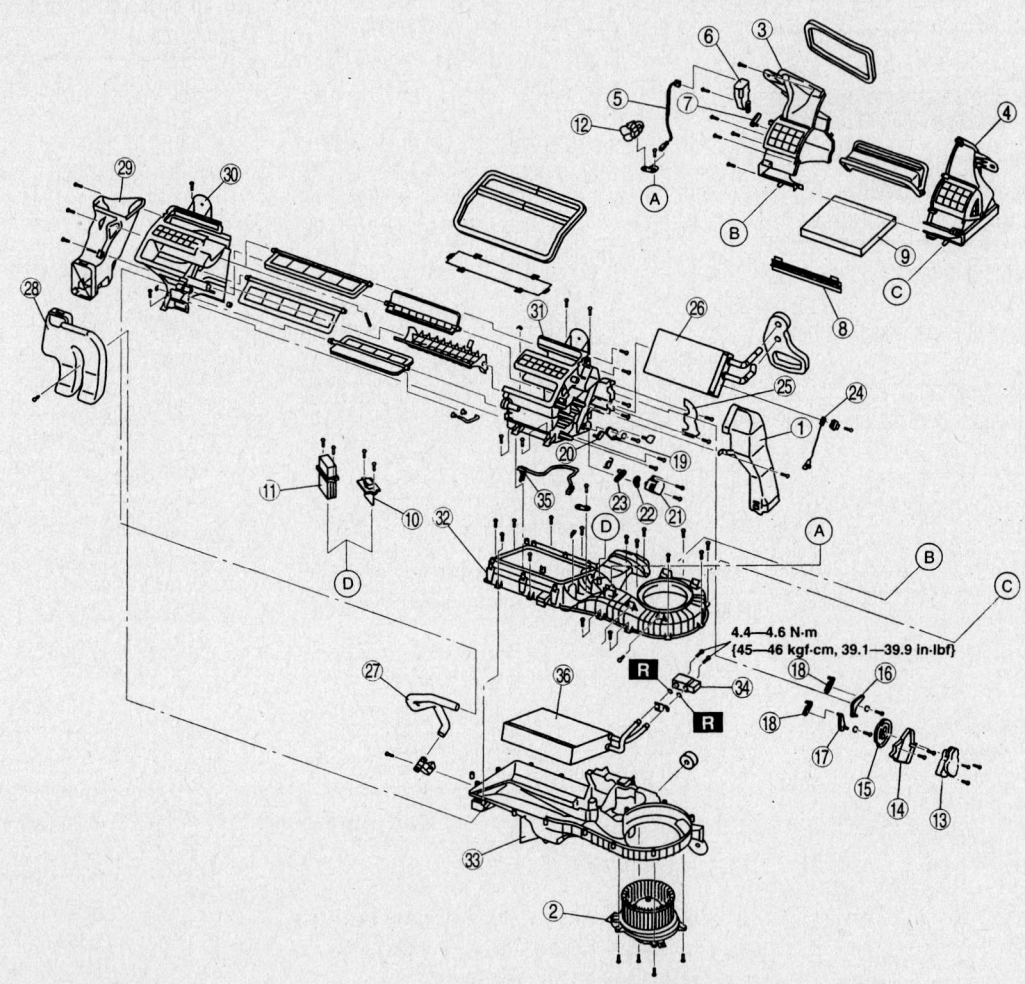

1	Duct (1)	14	Airflow mode bracket
2	Blower motor	15	Airflow mode main link
3	A/C case (3)	16	Airflow mode sub link (1)
4	A/C case (4)	17	Airflow mode sub link (2)
5	Harness (1)	18	Airflow mode crank
6	Air intake actuator	19	Air mix link (1) (manual air conditioner)
7	Air intake crank	20	Air mix crank (1) (manual air conditioner)
8	Air filter cover	21	Air mix actuator (full-auto air conditioner)
9	Air filter	22	Air mix link (2) (full-auto air conditioner)
10	Resistor (manual air conditioner)	23	Air mix crank (2) (full-auto air conditioner)
11	PWM unit (full-auto air conditioner)	24	Water temperature sensor (full-auto air conditioner)
12	Harness (2) (full-auto air conditioner)	25	Heater core cover
13	Airflow mode actuator	26	Heater core

09482_MAZC2_G0206

Exploded view of the A/C unit assembly—Mazda6

- Airflow mode actuator, main link, sublink (2) and sub link (3) and mode crank
- Power MOS FET
- Duct (2)
- Air mix actuator, rod, crank (1 and 2) and air mix rod holder
- Heater core

To install:

26. Installation is the reverse of removal, keep in mind the following
    a. Adjust the clock spring as follows:

❋❋ CAUTION

If the clock spring is not adjusted, the spring wire in the clock spring could

over-wind and break when the steering wheel is turned. Always adjust the clock spring after installing it.

- Set the front tires straight-ahead

❋❋ CAUTION

The clock spring will break if over-wound. Do not forcibly turn the clock spring.

- Turn the clock spring clockwise until it stops
- From the stopped position, turn the clock spring counterclockwise 2 ¾ turns
- Align the marks

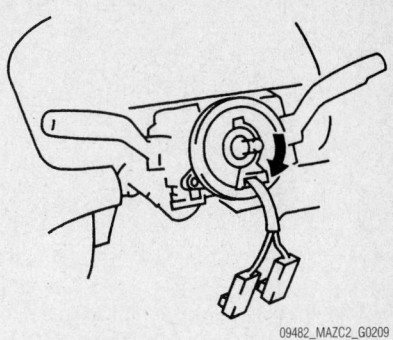

09482_MAZC2_G0209

Turn the clock spring clockwise until it stops—Mazda6 and Mazdaspeed6

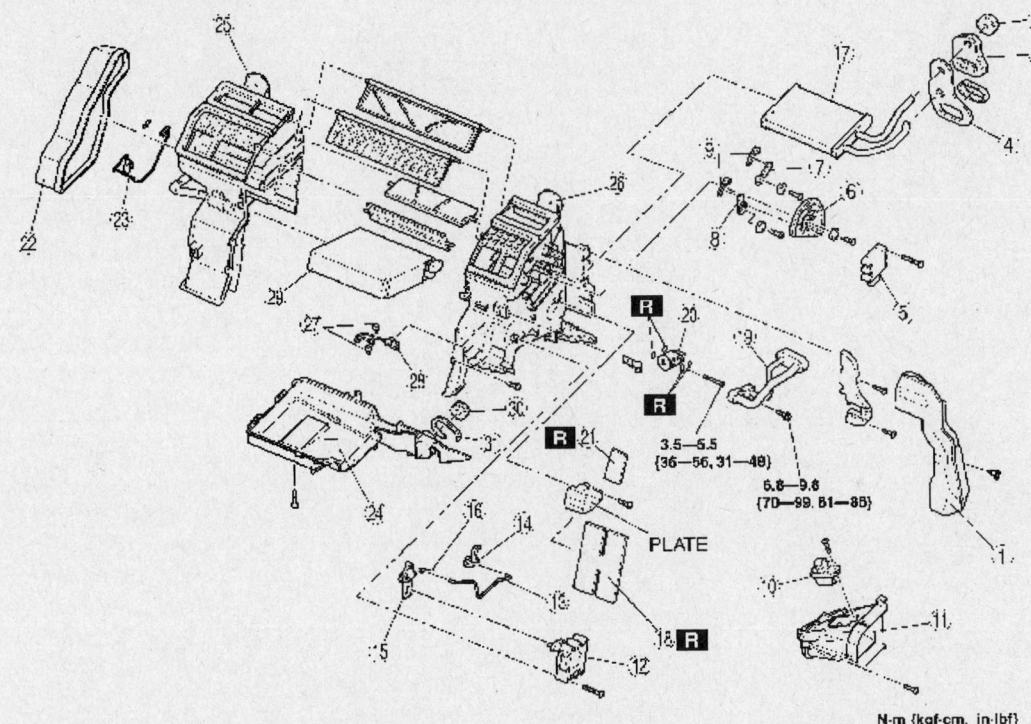

3.5—5.5 {36—56, 31—48}

6.8—9.8 {70—99, 61—85}

PLATE

N·m {kgf·cm, in·lbf}

1	Duct (1)	17	Heater core
2	Polyurethane protector (1)	18	Adhesive polyurethane (1)
3	Polyurethane protector (2)	19	Outlet pipe
4	Polyurethane protector (3)	20	Expansion valve
5	Airflow mode actuator	21	Adhesive polyurethane (2)
6	Airflow mode main link	22	Duct (3)
7	Airflow mode sub link (2)	23	A/C water temperature sensor
8	Airflow mode sub link (3)	24	A/C case (3)
9	Airflow mode crank	25	A/C case (1)
10	Power MOS FET	26	A/C case (2)
11	Duct (2)	27	Sensor clamp
12	Air mix actuator	28	Evaporator temperature sensor
13	Air mix rod	29	Evaporator
14	Air mix crank (1)	30	Polyurethane protector (4)
15	Air mix crank (2)	31	Adhesive polyurethane (3)

09482_MAZC2_G0207

Exploded view of the A/C unit assembly—Mazdaspeed6

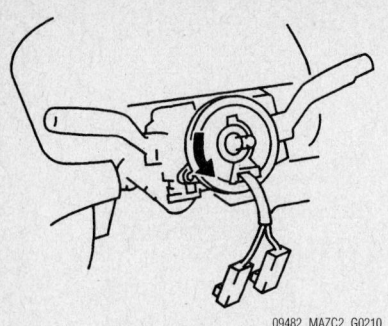

From the stopped position, turn the clock spring counterclockwise 2 ¾ turns—Mazda6 and Mazdaspeed6

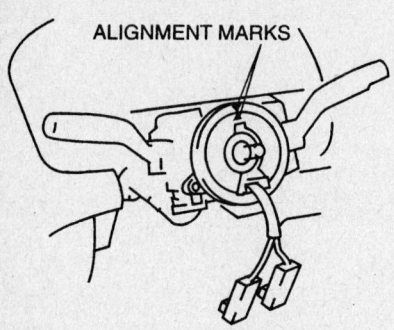

ALIGNMENT MARKS

Align the marks on the clock spring—Mazda6 and Mazdaspeed6

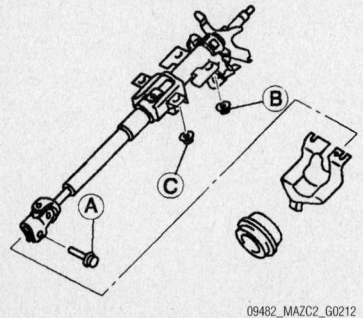

Tighten the steering shaft bolts A, nut B and nut C in that order—Mazda6 and Mazdaspeed6

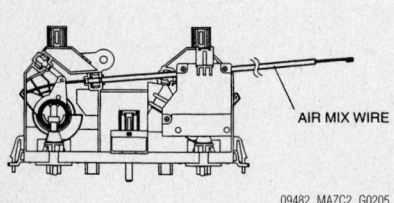

AIR MIX WIRE

If equipped with a wire type climate control unit, connect the wires for the front A/C unit as shown—Mazda6 and Mazdaspeed6 models

---

- Verify that the tilt / telescope lever is in the LOCK position
  b. Tighten the steering shaft bolts A, nut B and nut C in that order.
  c. If equipped with a wire type climate control unit, connect the wires for the front A/C unit as illustrated.

Rocker Arm (Valve) Cover

REMOVAL & INSTALLATION

Mazda3, Mazda6 and Mazdaspeed6 Models

2.0L (LF) AND 2.3L (L3) ENGINES

1. Before servicing the vehicle, refer to the precautions section.
2. Remove or disconnect the following:
   - Negative battery cable
   - Spark plug wires
   - Vent hose
   - Positive Crankcase Ventilation (PCV) hose
   - Rocker arm cover and discard the gasket
3. Clean all mating surfaces of any residual gasket material.

To install:

4. Install or connect the following:
5. Apply a 0.16–0.24 inch (4–7mm) bead of silicone sealant to cylinder head at the areas illustrated. Make sure to install the cover within 10 minutes of applying the sealant.
6. Install or connect the following:
   - Rocker arm cover with a new gasket. Torque the bolts in the sequence illustrated to 71–92 inch lbs. (8–10 Nm) on all Mazda3 engines, 71–101 inch lbs. (8–11.5 Nm) on Mazda6 models, or 71–93 inch lbs. (8–10.5 Nm) on Mazdaspeed6 equipped with the 2.3L (L3) engine.
   - Remaining components removed to

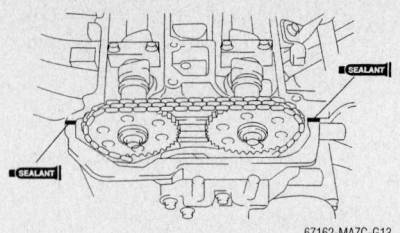

Apply silicone sealant to cylinder head at the areas illustrated–All Mazda3 models, Mazda6 and Mazdaspeed6 models with the 2.3L (L3) engine models

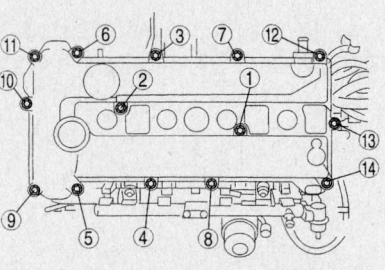

Tighten the rocker arm cover bolts in the sequence shown–All Mazda3 models, Mazda6 and Mazdaspeed6 models with the 2.3L (L3) engine models

facilitate the rocker arm cover removal
   - Negative battery cable

3.0L (AJ) ENGINE

1. Before servicing the vehicle, refer to the precautions in the beginning of this section.
2. Remove or disconnect the following:
   - Negative battery cable
   - Any components that would interfere with cover removal
   - Rocker arm cover bolts in the sequence illustrated
   - Rocker arm cover and discard the gasket
3. Clean all mating surfaces of any residual gasket material.

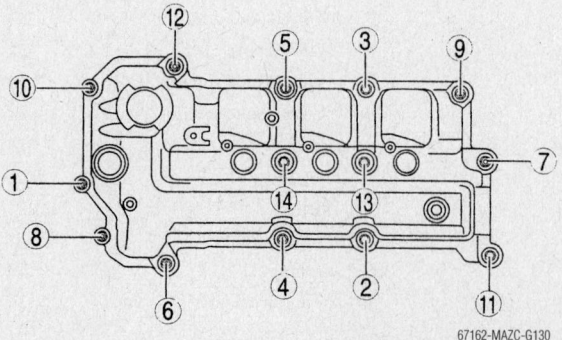

Remove the left hand rocker arm cover bolts in the sequence shown—Mazda6 models with the 3.0L (AJ) engine

To install:

4. Apply silicone sealant to cylinder head at the areas illustrated.

5. Install or connect the following:
- Rocker arm cover with a new gasket.

- Oil control valve with the cylinder head cover raised as shown in the accompanying illustration, being careful not to let the valve retaining bolt slip into the timing chain cover

when installing and tighten the valve bolt to 71–106 INCH LBS. (8–12 Nm).
- Torque the bolts in the sequence illustrated to 71–106 inch lbs. (8–12 Nm).
- Remaining components removed to facilitate the rocker arm cover removal
- Negative battery cable

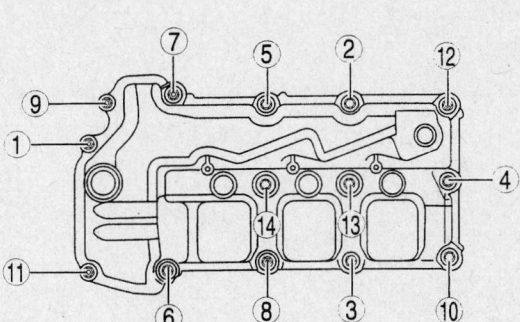

67162-MAZC-G131

Remove the right hand rocker arm cover bolts in the sequence shown—Mazda6 models with the 3.0L (AJ) engine

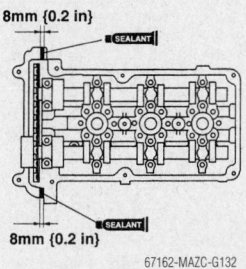

67162-MAZC-G132

Apply silicone sealant to right hand cylinder head at the areas illustrated—Mazda6 models with the 3.0L (AJ) engine

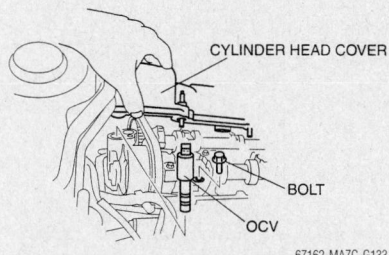

67162-MAZC-G133

Install the oil control valve on the right hand side—Mazda6 models with the 3.0L (AJ) engine

Cylinder Head

REMOVAL & INSTALLATION

MX-5 Miata

1. Before servicing the vehicle, refer to the precautions section.
2. Properly relieve the fuel system pressure.
3. Remove the battery and battery tray.
4. Drain the engine coolant.
5. Remove the front suspension tower bar.
6. Remove the air cleaner.
7. Remove the dynamic chamber.
8. Remove the ignition coil.
9. Remove the drive belt.
10. Remove the CKP sensor.
11. Remove the P/S oil pump with the oil hose still connected and position the P/S oil pump so that it is out of the way.
12. Remove the timing chain.
13. Remove the wiper arm.
14. Remove the cowl grille.
15. Remove the side cowl grille.
16. Remove the service hole cover.
17. Disconnect the alternator, but do not remove it from the vehicle.
18. Remove the exhaust manifold.
19. Remove the OCV sensor, if equipped with variable valve timing.
20. Mark the camshaft cap locations so they may be reinstalled in their original positions.
21. Remove camshafts.
22. Mark the cylinder head bolt locations

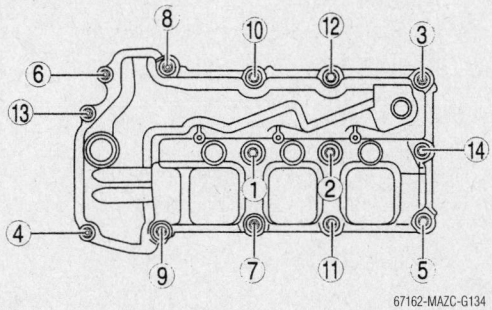

67162-MAZC-G134

Right hand cylinder head cover torque sequence—Mazda6 models with the 3.0L (AJ) engine

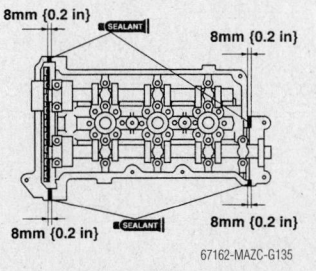

67162-MAZC-G135

Apply silicone sealant to left hand cylinder head at the areas illustrated—Mazda6 models with the 3.0L (AJ) engine

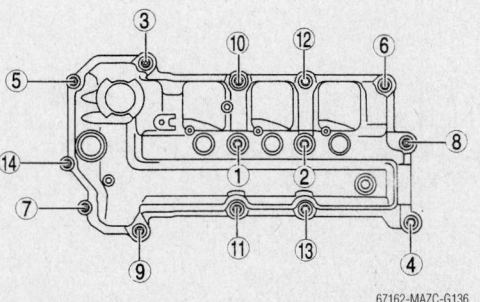

67162-MAZC-G136

Left hand cylinder head cover torque sequence—Mazda6 models with the 3.0L (AJ) engine

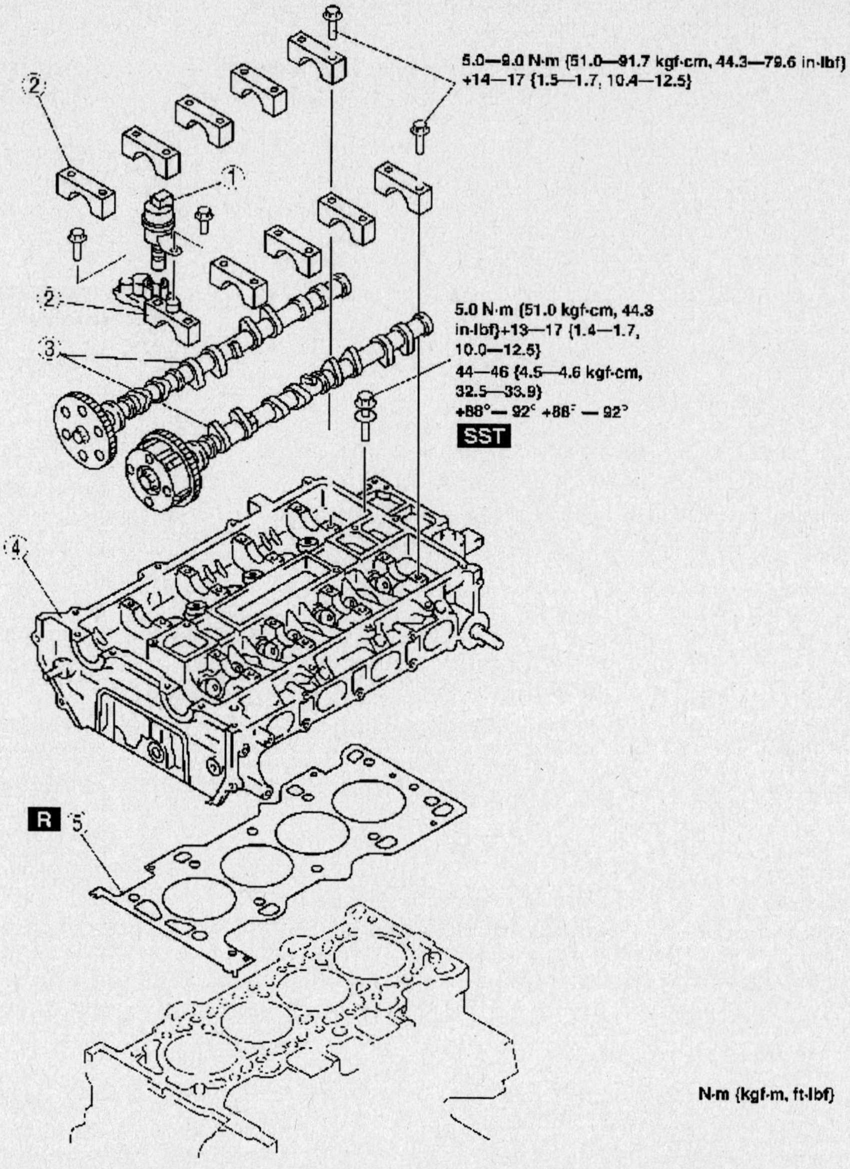

5.0—9.0 N·m {51.0—91.7 kgf·cm, 44.3—79.6 in·lbf}
+14—17 {1.5—1.7, 10.4—12.5}

5.0 N·m {51.0 kgf·cm, 44.3
in·lbf}+13—17 {1.4—1.7,
10.0—12.5}
44—46 {4.5—4.6 kgf·cm,
32.5—33.9}
+88°— 92° +88°— 92°
SST

N·m {kgf·m, ft·lbf}

1 OCV (With variable valve timing mechanism.)
2 Camshaft cap
3 Camshaft

4 Cylinder head
5 Cylinder head gasket

09482_MAZC2_G0015

Exploded view of the cylinder head and camshaft assemblies—MX-5 Miata

so they may be reinstalled in their original positions.

23. Loosen the cylinder head bolts in the sequence illustrated using two or three passes.

24. Remove the cylinder head.

25. Measure the length of the cylinder head bolts and replace any bolts that exceed the maximum length specification of 146 mm (5.77 in.).

To install:

26. Install the cylinder heads.

27. Install the cylinder head bolts in their original locations.

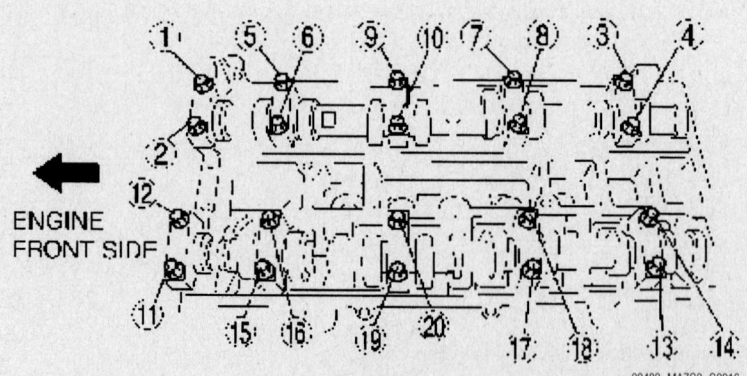

ENGINE FRONT SIDE

09482_MAZC2_G0016

Camshaft cap bolt loosening sequence—MX-5 Miata

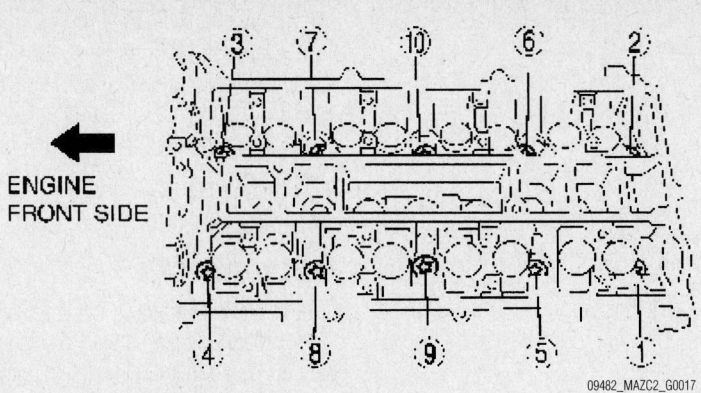

Cylinder head bolt loosening sequence—MX-5 Miata

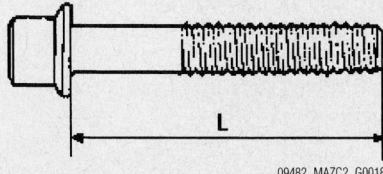

Measure the cylinder head bolt length— MX-5 Miata

- Upper radiator hose
- Water and heater hose
- Wiring harness

6. Support the engine using an engine jack and attachment as shown in the illustration.

7. Remove the camshafts.

➡ **The cylinder head and camshaft caps are numbered and must be reassembled in their original locations. When removing a when removed, keep the caps with the cylinder head they were removed from, do not switch the caps.**

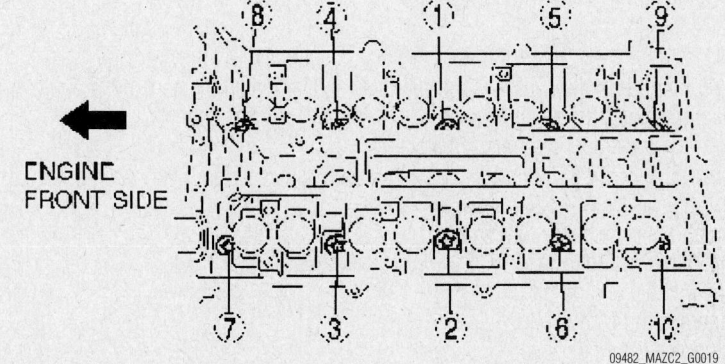

Cylinder head bolt torque sequence—MX-5 Miata

28. Tighten the cylinder head bolts in the sequence illustrated using 5 passes as follows:

    a. Step 1: Tighten to 5 Nm (44.3 in. lbs.)

    b. Step 2: Tighten 13–17 Nm (10–12.5 ft. lbs.)

    c. Step 3: Tighten 44–46 Nm (32.5–33.9 ft. lbs.)

    d. Step 4: Tighten and additional 88–92 degrees.

    e. Step 5: Tighten and additional 88–92 degrees.

29. Set the cam position of the number 1 cylinder to top dead center, install the camshafts, making sure to install the caps in their original locations.

30. Tighten the camshaft cap bolts using two to three steps in the sequence illustrated to 5–9 Nm (44.3–79.6 in. lbs.) and then to 14–17 Nm (10.4–12.5 ft. lbs.)

31. Install the remaining components in the reverse order of removal

32. Start the engine and inspect for leaks.

33. Check the ignition timing, idle speed and idle mixture.

Mazda3

1. Before servicing the vehicle, refer to the precautions section.

2. Disconnect the negative battery cable.

3. Remove the timing chain.

4. Remove the intake manifold.

5. Disconnect the following components:

- Warm Up–Three Way Converter (WU–TWC)

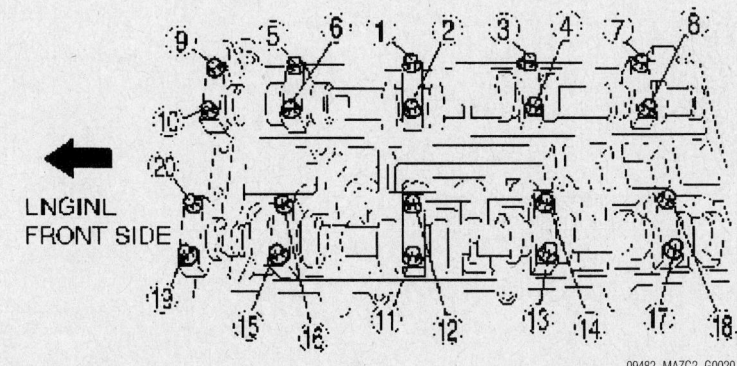

Camshaft cap bolt torque sequence—MX-5 Miata

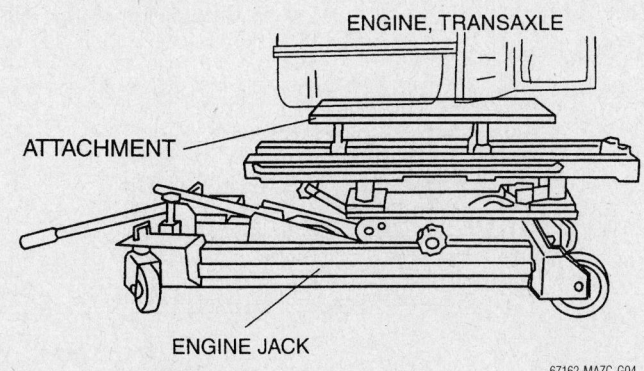

Support the engine using an engine jack and attachment as shown—Mazda3 models

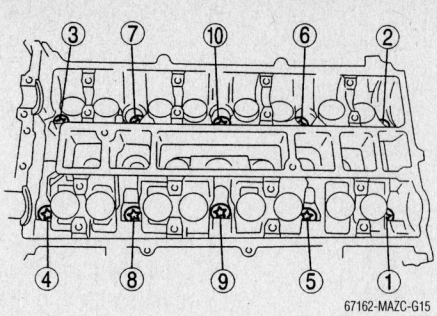

Cylinder head bolt removal sequence—Mazda3 models

8. Loosen the cylinder head bolts using 2–3 passes in the sequence illustrated.

9. Once all the bolts have been removed, remove the cylinder head and gasket.

To install:

10. If reusing old bolts, which is not recommended; measure the length of each head bolt at the locations illustrated. If any bolt exceeds 5.717–5.740 inch (145.2–145.8mm) replace the bolt.

11. Install a new cylinder head gasket.

12. Install the cylinder head.

13. Apply clean engine oil to the bolt threads and seating faces.

14. Install new cylinder head bolts and torque in 2–3 steps, in sequence as follows:

   a. Step 1: Tighten the bolts in sequence to 27–97 inch lbs. (3–11 Nm).

   b. Step 2: Tighten the bolts in sequence to 9.5–12.5 ft. lbs. (13–17 Nm).

   c. Step 3: Tighten the bolts in sequence to 32–34.5 ft. lbs. (43–47 Nm).

   d. Step 4: Paint a mark on the edge of each cylinder head bolt to use as a reference. Turn each bolt, in sequence, 85–92 degrees.

   e. Step 5: Turn each bolt, in sequence, 85–92 degrees.

15. Install the camshafts and caps in their original positions.

16. Install the remaining components in the reverse order of removal.

17. Fill and bleed the cooling system.

18. Change the oil and filter.

19. Run the engine and check for proper operation.

Mazda6

2.3L (L3) ENGINE

1. Before servicing the vehicle, refer to the precautions section.

2. Disconnect the negative battery cable.

3. Remove the timing chain.

4. Remove the ignition coil.

5. Unbolt the alternator and move it aside with the connectors still attached.

6. Remove the front exhaust pipe.

7. Remove the intake manifold.

8. Disconnect the following components:

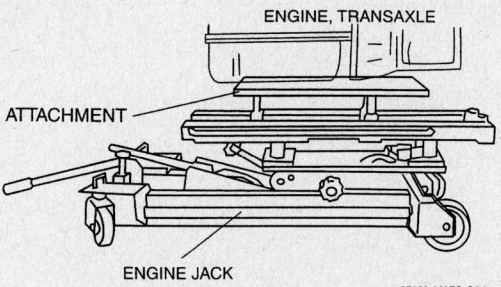

Support the engine using an engine jack and attachment as shown—Mazda6 and Mazdaspeed6 models

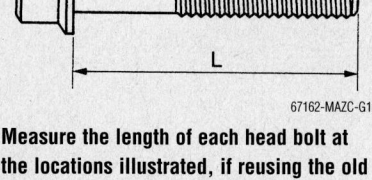

Measure the length of each head bolt at the locations illustrated, if reusing the old bolts. If any bolt exceeds 5.717–5.740 inch (145.2–145.8mm) replace the bolt— Mazda3 models

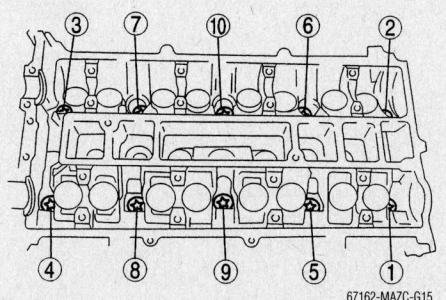

Cylinder head bolt removal sequence—Mazda6 and Mazdaspeed6 models

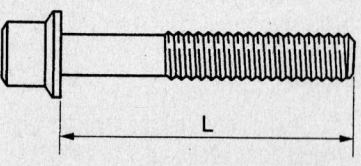

Measure the length of each head bolt at the locations illustrated, if reusing the old bolts. If any bolt exceeds 5.717–5.740 inch (145.2–145.8mm) replace the bolt— Mazda6 models

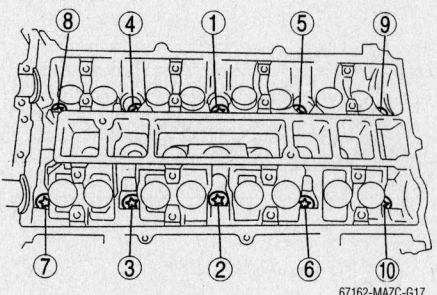

Cylinder head bolt tightening sequence—Mazda3 models

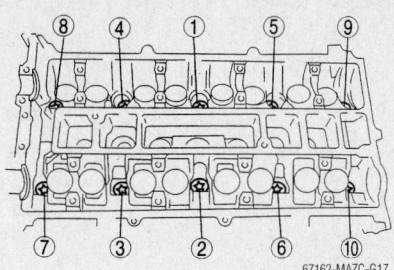

Cylinder head bolt tightening sequence—Mazda6 and Mazdaspeed6 with the 2.3L (L3) engine

- Upper radiator hose
- Water and heater hose

9. Support the engine using an engine jack and attachment as shown in the illustration.

10. Remove the oil control valve.

11. Remove the camshafts.

➡ **The cylinder head and camshaft caps are numbered and must be reassembled in their original locations. When removing a when removed, keep the caps with the cylinder head they were removed from, do not switch the caps.**

12. Loosen the cylinder head bolts using 2–3 passes in the sequence illustrated.

13. Once all the bolts have been removed, remove the cylinder head and gasket.

To install:

14. If reusing old bolts, which is not recommended; measure the length of each head bolt at the locations illustrated. If any bolt exceeds 5.717–5.740 inch (145.2–145.8mm) replace the bolt.

15. Install a new cylinder head gasket.

16. Install the cylinder head.

17. Apply clean engine oil to the bolt threads and seating faces.

18. Install new cylinder head bolts and torque in 2–3 steps, in sequence as follows:

a. Step 1: Tighten the bolts in sequence to 27–97 inch lbs. (3–11 Nm).

b. Step 2: Tighten the bolts in sequence to 9.5–12.5 ft. lbs. (13–17 Nm).

c. Step 3: Tighten the bolts in sequence to 32–34.5 ft. lbs. (43–47 Nm).

d. Step 4: Paint a mark on the edge of each cylinder head bolt to use as a reference. Turn each bolt, in sequence, 85–92 degrees.

e. Step 5: Turn each bolt, in sequence, 85–92 degrees.

19. Install the camshafts and caps in their original positions.

20. Install the remaining components in the reverse order of removal.

21. Fill and bleed the cooling system.

22. Change the oil and filter.

23. Run the engine and check for proper operation.

3.0L (AJ) ENGINE

1. Before servicing the vehicle, refer to the precautions section.

2. Properly relieve the fuel system pressure.

3. Drain the cooling system.

4. Disconnect the negative battery cable.

5. Remove the water pump.

6. Remove the timing chain.

7. Support the engine using an engine jack and attachment.

8. Remove the water bypass tube stud bolt and bolt and disconnect the tube from the cylinder head.

9. Remove the camshaft(s).

10. Remove the rocker arm(s).

11. Remove the camshafts.

➡ **If removing the right hand cylinder head and camshaft caps, remove thrust caps 1R and 5R first. Do not loosen any of the other cap bolts until the thrust caps are removed or you could damage the thrust caps. If removing the left hand cylinder head and camshaft caps, remove thrust caps 1L and 6L first. Do not loosen any of the other cap bolts until the thrust caps are removed or you could damage the thrust caps.**

12. Loosen the camshaft caps using 7–8 passes in sequence after first removing the thrust caps to allow the camshafts to be slowly raised. Keep the caps in the order they were removed so they are re-installed in their original positions.

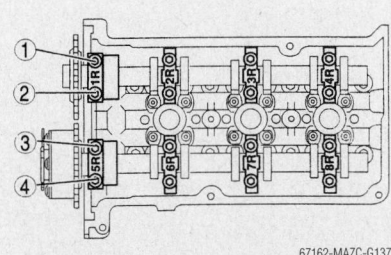

If removing the right hand camshaft caps, remove thrust caps 1R and 5R first—Mazda6 with the 3.0L (AJ) engine

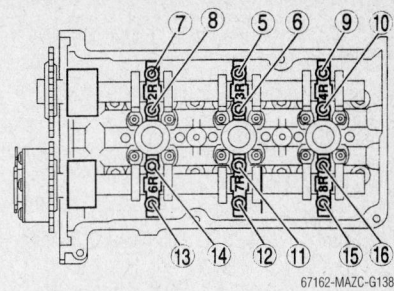

Loosen the right hand camshaft caps using 7–8 passes in this sequence after first removing the thrust caps—Mazda6 with the 3.0L (AJ) engine

13. Loosen the cylinder head bolts using 2–3 passes in the sequence illustrated.

14. Once all the bolts have been removed, remove the cylinder head and gasket.

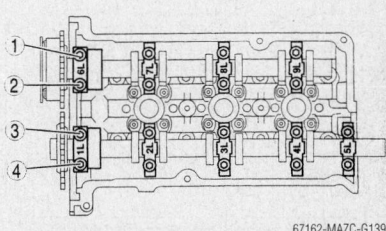

If removing the left hand camshaft caps, remove thrust caps 1L and 6L first—Mazda6 with the 3.0L (AJ) engine

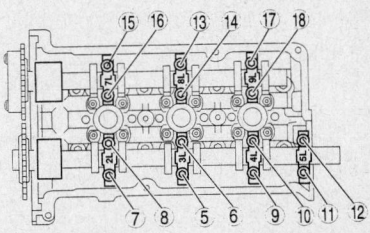

Loosen the left hand camshaft caps using 7–8 passes in this sequence after first removing the thrust caps—Mazda6 with the 3.0L (AJ) engine

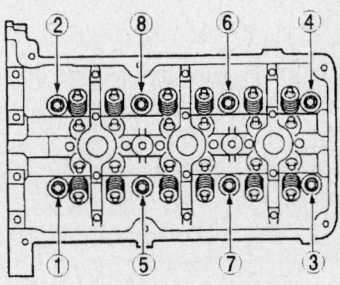

Cylinder head bolt loosening sequence—Mazda6 with the 3.0L (AJ) engine

To install:

15. Install a new head gasket and the cylinder head.

16. Lubricate the cylinder head bolt threads.

17. Torque the cylinder head bolts in the proper sequence as follows:

    a. Step 1: Tighten the bolts in sequence 24–28 ft. lbs. (32–38 Nm).

    b. Step 2: Tighten the bolts in sequence an additional 85–95 degrees.

    c. Step 3: Loosen the bolts in sequence one full turn.

    d. Step 4: Tighten the bolts in sequence 24–28 ft. lbs. (32–38 Nm).

    e. Step 5: Tighten the bolts in sequence an additional 85–95 degrees.

    f. Step 6: Tighten the bolts in sequence an additional 85–95 degrees.

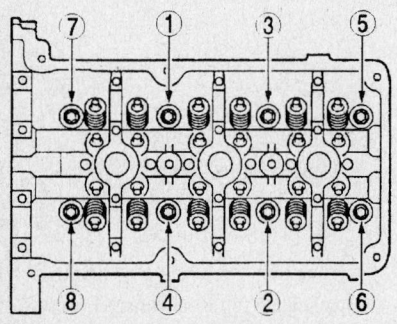

Cylinder head bolt torque sequence—Mazda6 with the 3.0L (AJ) engine

18. Install the camshafts in their original positions.

19. Install the left hand camshaft as follows:

    a. Place the crankshaft keyway at the 11 O'clock position by rotating the crankshaft clockwise.

    b. Install the camshaft and the caps in their original positions and hand tighten the bolts.

    c. Position the mark on the intake camshaft to the 9 O'clock position as illustrated.

    d. Position the mark on the exhaust camshaft to the 12 O'clock position.

    e. Install the timing chain.

➡️**Tighten the camshaft journal thrust caps last to avoid damaging the thrust caps.**

    f. Tighten the camshaft caps using several passes to 71–106 inch lbs. (8–12 Nm). After adjust the camshaft endplay using the thrust caps 1L and 6L, tighten the remaining caps.

20. Install the right hand camshaft as follows:

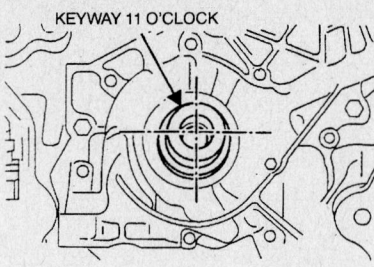

When installing the left hand camshaft place the crankshaft keyway at the 11 O'clock position—Mazda6 with the 3.0L (AJ) engine

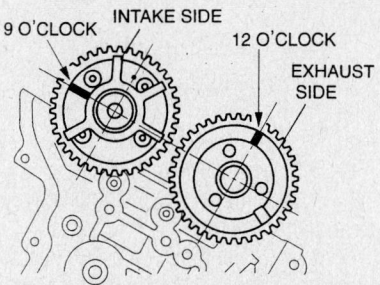

Install the left hand camshaft and position the mark on the intake camshaft to the 9 O'clock and the exhaust camshaft to the 12 O'clock position—Mazda6 with the 3.0L (AJ) engine

    a. Install the camshaft and the caps in their original positions and hand tighten the bolts.

    b. Place the crankshaft keyway at the 3 O'clock position.

    c. Position the mark on the exhaust camshaft to the 12 O'clock position by rotating the crankshaft clockwise.

    d. Position the mark on the intake camshaft to the 3 O'clock position as illustrated.

    e. Install the timing chain.

➡️**Tighten the camshaft journal thrust caps last to avoid damaging the thrust caps.**

When installing the right hand camshaft place the crankshaft keyway at the 3 O'clock position—Mazda6 with the 3.0L (AJ) engine

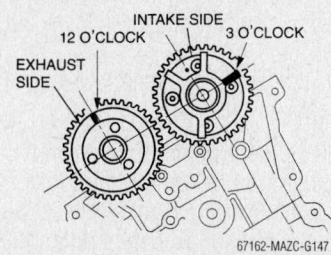

Install the right hand camshaft and position the mark on the intake camshaft to the 3 O'clock and the exhaust camshaft to the 12 O'clock position—Mazda6 with the 3.0L (AJ) engine

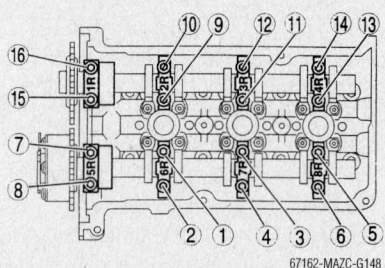

Right side camshaft cap torque sequence—Mazda6 with the 3.0L (AJ) engine

    f. Tighten the camshaft caps using several passes to 71–106 inch lbs. (8–12 Nm). After adjust the camshaft endplay using the thrust caps 1R and 5R, tighten the remaining caps.

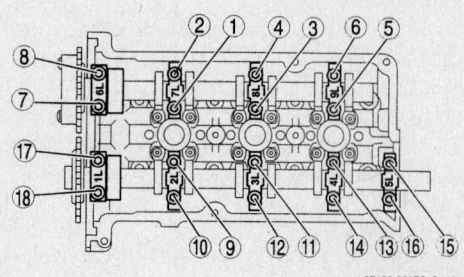

Left side camshaft cap torque sequence—Mazda6 with the 3.0L (AJ) engine

21. Install the water bypass tube.
22. Remove the engine jack and attachment.
23. Install the water pump.
24. Fill the cooling system.
25. Fill and bleed the cooling system.
26. Change the oil and filter.
27. Run the engine and check for proper operation.

Mazdaspeed6

1. Before servicing the vehicle, refer to the precautions section.
2. Disconnect the negative battery cable.
3. Remove the timing chain.
4. Unbolt the alternator and move it aside with the connectors still attached.
5. Remove the front exhaust pipe.
6. Remove the intake manifold.
7. Disconnect the following components:
   • Upper radiator hose
   • Water and heater hose
8. Support the engine using an engine jack and attachment as shown in the illustration.
9. Remove the oil control valve.
10. Remove the camshafts.

➡**The cylinder head and camshaft caps are numbered and must be reassembled in their original locations. When removing a when removed, keep the caps with the cylinder head they were removed from, do not switch the caps.**

11. Loosen the cylinder head bolts using 2–3 passes in the sequence illustrated.
12. Once all the bolts have been removed, remove the cylinder head and gasket.

To install:

13. If reusing old bolts, which is not recommended; measure the length of each head bolt at the locations illustrated. If any bolt exceeds 5.700–5.720 inch (144.7–145.3mm) replace the bolt.
14. Install a new cylinder head gasket.
15. Install the cylinder head.
16. Apply clean engine oil to the bolt threads and seating faces.
17. Install new cylinder head bolts and torque in 2–3 steps, in sequence as follows:
   a. Step 1: Tighten the bolts in sequence to 27–97 inch lbs. (3–11 Nm).
   b. Step 2: Tighten the bolts in sequence to 9.6–12.5 ft. lbs. (13–17 Nm).
   c. Step 3: Tighten the bolts in sequence to 32–34.6 ft. lbs. (43–47 Nm).

   d. Step 4: Paint a mark on the edge of each cylinder head bolt to use as a reference. Turn each bolt, in sequence, 88–92 degrees.
   e. Step 5: Turn each bolt, in sequence, 88–92 degrees.
18. Install the camshafts and caps in their original positions.
19. Install the remaining components in the reverse order of removal.
20. Fill and bleed the cooling system.
21. Change the oil and filter.
22. Run the engine and check for proper operation.

Rocker Arms/Shafts

REMOVAL & INSTALLATION

All Mazda engines covered in this manual are not equipped with rocker arms/shafts, the camshafts directly actuate the valves through a bucket type cam follower.

Turbocharger

REMOVAL & INSTALLATION

Mazdaspeed6

1. Disconnect the negative battery cable.
2. Before servicing the vehicle, refer to the precautions section.
3. Remove the charge air cooler cover.
4. Remove the battery and tray.
5. Remove the air cleaner.
6. Remove the resonance chamber, remove the left hand front mudguard before removing the resonance chamber.
7. Disconnect the MAF/IAT sensor.
8. Remove the duct.
9. Remove the charge air cooler cover.
10. Remove the charge air cooler.
11. Remove the charge air cooler bracket.
12. Remove the air bypass valve.
13. Drain and recycle the engine coolant.
14. Remove the throttle body.
15. Disconnect the EGR valve connector.
16. Remove the air hose.
17. Remove the high pressure fuel pump bracket and the EGR pipe bracket.
18. Remove the fuel delivery pipe cover.
19. Remove the oil level gauge pipe.
20. Remove the drive belt.
21. Remove the power steering pump out of the way with the lines still attached.
22. Remove the vacuum hose.
23. Remove the intake manifold.

24. Remove the three way catalytic converter.
25. Disconnect the rear oxygen sensor connector.
26. Remove the front exhaust pipe.

❋❋ CAUTION

When removing the cowl, a part or tool may hit the edge of the windshield and could damage it. Protect the windshield by covering it with a clean rag to prevent damage to the windshield.

27. Remove the cowl and the insulator under the cowl.
28. Remove the exhaust manifold upper side insulator.
29. Remove the alternator.
30. Remove the exhaust manifold lower side insulator.
31. Disconnect the front oxygen sensor connector.
32. Remove the warm up three-way catalytic converter top side insulator.
33. Remove the brake master back side vacuum hose.
34. Remove the oil pipes.
35. Remove the front oxygen sensor.
36. Remove the warm up three-way catalytic converter.
37. Remove the turbocharger.

To install:

38. Installation is the reverse of removal, note the following:
   a. Tighten the turbocharger bolts to 29–37 ft. lbs. (38–51 Nm).
   b. Tighten the oil pipe installation bolt while the stopper of the oil pipe is faced to the turbocharger.
   c. Tighten the wastegate Control Solenoid Valve bolts to 70–95 inch lbs. (8–11 Nm).
   d. Tighten the intake manifold bolts to 13–16 ft. lbs. (17–23 Nm).
   e. Tighten the fuel delivery pipe cover bolts to 13–16 ft. lbs. (17–23 Nm).
   f. Tighten the charge air cooler bracket to 31–35 ft. lbs. (42–48 Nm).
   g. Tighten the charge air cooler to 14–19 ft. lbs. (42–48 Nm).

❋❋ CAUTION

When installing the charge air cooler cover, be careful not to damage the charge air cooler cover clips.

   h. Tighten the throttle body installation bolts to 71–101 inch lbs. (8–11.5 Nm).

i. Perform the following when installing the air cleaner case:
- Verify that the rubber mounts are set in the air cleaner bracket
- Install the projections on the frame side.
- Verify that the projections on the frame side are installed securely.
- Install the projection on the engine side.
- Verify that the projection on the engine side installed securely.

Intake Manifold

REMOVAL & INSTALLATION

MX-5 Miata

1. Before servicing the vehicle, refer to the precautions section.
2. Properly relieve the fuel system pressure.
3. Remove the battery cover.
4. Disconnect the negative battery cable.
5. Remove the air cleaner cover, this involves first removing the MAF/IAT sensor.

6. Remove the air cleaner element.
    a. Remove the air cleaner case.
7. Disconnect the quick connect fitting from the air cleaner hose.
8. Remove the air hose by moving the purge solenoid valve slightly out of the way.
9. Remove the fresh-air duct by first removing the bumper.
10. Drain the cooling system.
11. Remove the throttle body.
12. Disconnect the quick connect fitting from the dynamic chamber.
13. Remove the plug hole plate.
14. Remove the service hole cover.
15. Remove the front suspension tower bar.
16. Remove the wiper arm.
17. Remove the cowl grille.
18. Remove the side cowl grille.
19. Move the cooler pipe No.3 and heater pipe slightly out of the way.
20. Remove the service hole cover.
21. Disconnect the heater hose and move the heater pipe slightly out of the way.
22. Disconnect the heater hose and move the heater pipe slightly out of the way.
23. Remove the harness bracket.

24. Remove the under cover.
25. Disconnect the variable intake air solenoid valve, EGR valve, CMP sensor and PSP switch connectors.
26. Disconnect the ignition coil and fuel injector connectors and move the harness aside.
27. Disconnect the quick release connector from the fuel rail.
28. Remove the fuel rail.
29. Disconnect the water hose from the EGR valve.
30. Disconnect two water hoses from the thermostat.
31. Remove the heater hose and heater pipe from the dynamic chamber.
32. Remove the variable intake air solenoid valve.
33. Remove the dynamic chamber installation bolts.
34. Remove the EGR pipe.
35. Disconnect the connector from the A/C compressor.
36. Disconnect the knock sensor connector.
37. Move the vacuum hose between the purge solenoid valve and the charcoal canister aside.

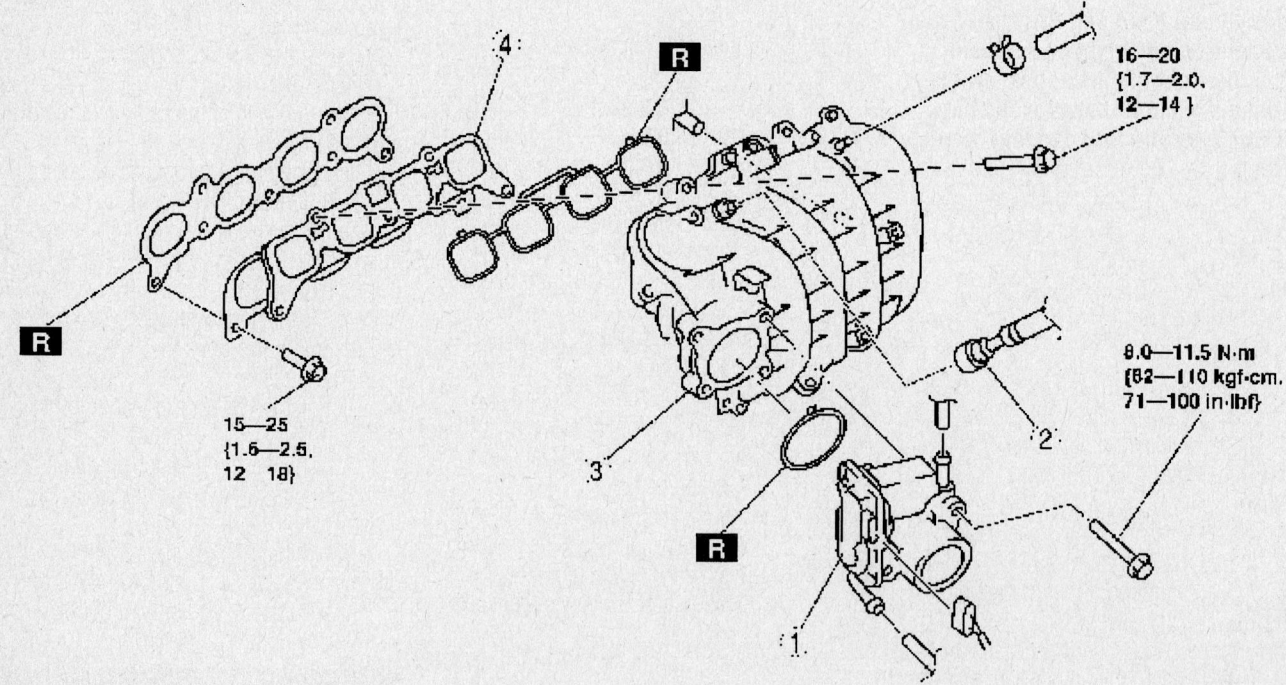

1 Throttle body
2 Quick release connector (Type A)
3 Dynamic chamber
4 Intake manifold

09482_MAZC2_G0022

Exploded view of the dynamic chamber and intake manifold—MX-5 Miata

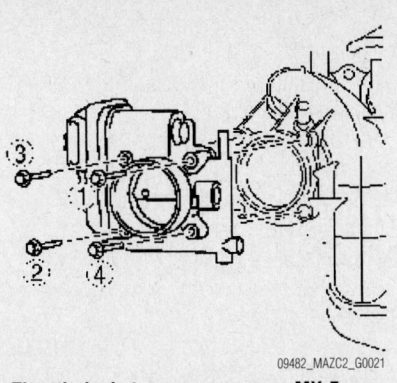

09482_MAZC2_G0021

Throttle body torque sequence—MX-5 Miata

38. Move the clutch release cylinder aside.

39. Disconnect the evaporative hose with the dynamic chamber raised.

40. Remove the dynamic chamber.

41. Remove the intake manifold.

To install:

42. Install the intake manifold and tighten the bolts to 15–25 Nm (12–18 ft. lbs.).

43. Install the dynamic chamber and tighten the bolts to 16–20 Nm (12–14 ft. lbs.)

44. Installation is the reverse order of

removal, make sure to tighten the throttle body bolts in the sequence shown to 8–11 Nm (71–100 in. lbs.).

45. Check and adjust all fluid levels, start the vehicle and check for proper operation.

Mazda3

1. Before servicing the vehicle, refer to the precautions section.

2. Drain the cooling system.

3. Relieve the fuel system pressure.

4. Remove the plug hole plate by lifting off and removing the plug hole plate from

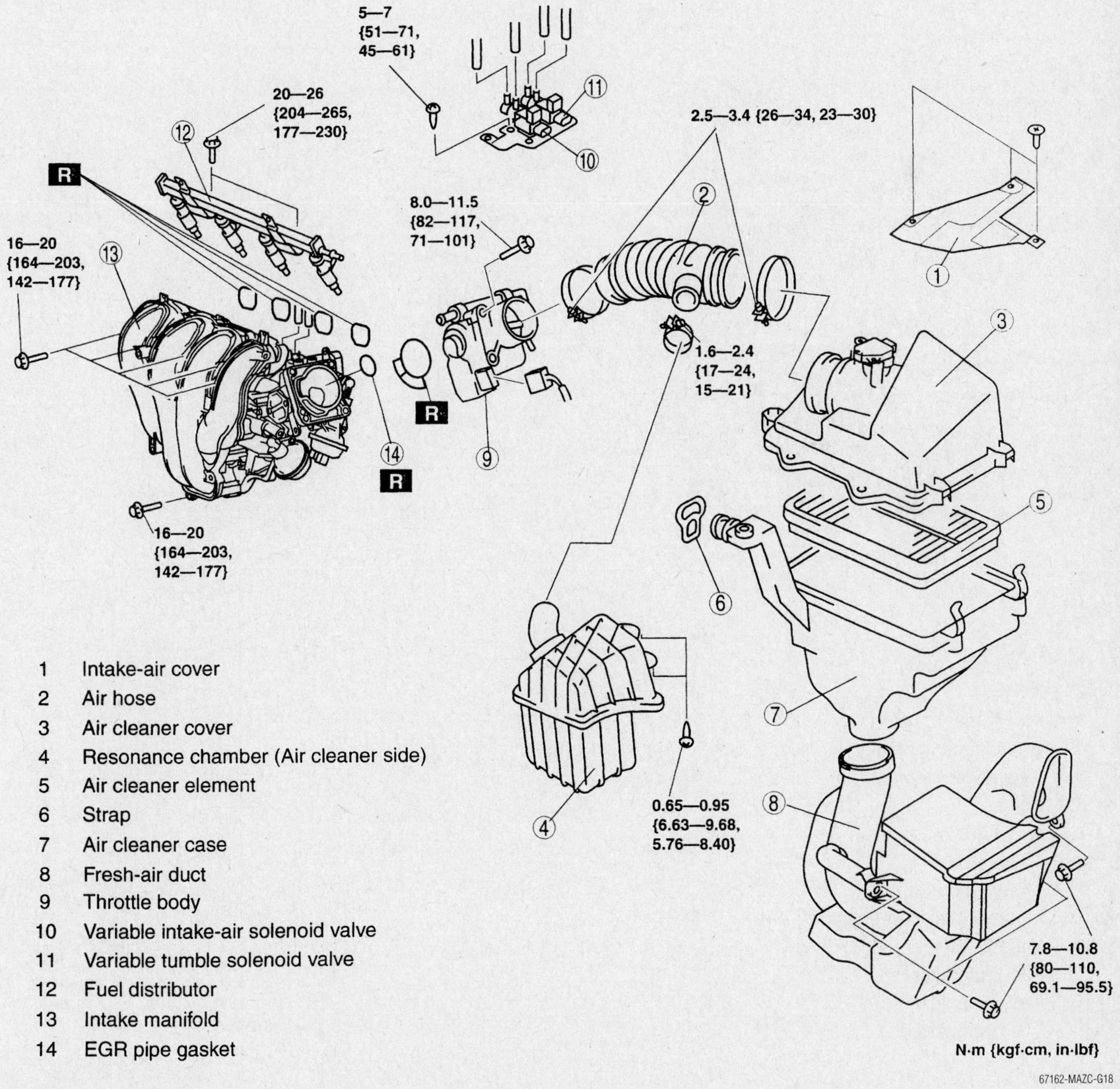

5—7
{51—71,
45—61}

20—26
{204—265,
177—230}

2.5—3.4 {26—34, 23—30}

8.0—11.5
{82—117,
71—101}

16—20
{164—203,
142—177}

1.6—2.4
{17—24,
15—21}

16—20
{164—203,
142—177}

0.65—0.95
{6.63—9.68,
5.76—8.40}

7.8—10.8
{80—110,
69.1—95.5}

1   Intake-air cover
2   Air hose
3   Air cleaner cover
4   Resonance chamber (Air cleaner side)
5   Air cleaner element
6   Strap
7   Air cleaner case
8   Fresh-air duct
9   Throttle body
10 Variable intake-air solenoid valve
11 Variable tumble solenoid valve
12 Fuel distributor
13 Intake manifold
14 EGR pipe gasket

N·m {kgf·cm, in·lbf}

67162-MAZC-G18

Exploded view of the intake manifold assembly and related components—2.0L (LF) engines

the areas shown in the accompanying illustration.

5. Remove the battery cover and battery duct.

6. Remove the under cover and disconnect the negative battery cable.

7. Remove the intake air cover, the air hose and the air cleaner element.

8. Remove the resonance chamber and air filter element.

9. Remove the strap and the air cleaner case.

10. Remove the fresh air duct.

11. Remove the throttle body.

12. Remove the variable intake air solenoid valve and the variable tumble solenoid valve.

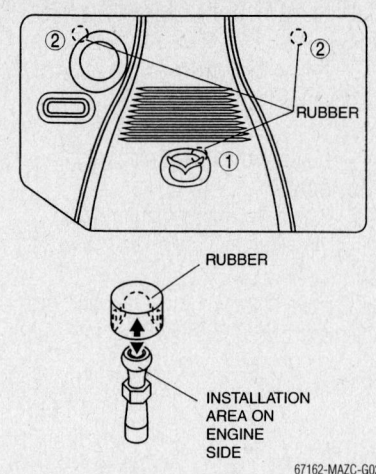

67162-MAZC-G02

Plug hole plate locations—Mazda3 models

13. Remove the fuel rail and injectors as an assembly.

14. Disconnect the vacuum hose from the intake manifold.

15. Remove the engine oil dipstick tube.

16. Remove the intake manifold bolts, the manifold and gasket.

17. Discard the gasket.

18. Clean the mating surfaces of any gasket material.

19. If necessary, remove the Exhaust Gas Recirculation (EGR) pipe gasket.

To install:

20. If removed, install the EGR pipe gasket.

21. Install a new gasket and the mani-

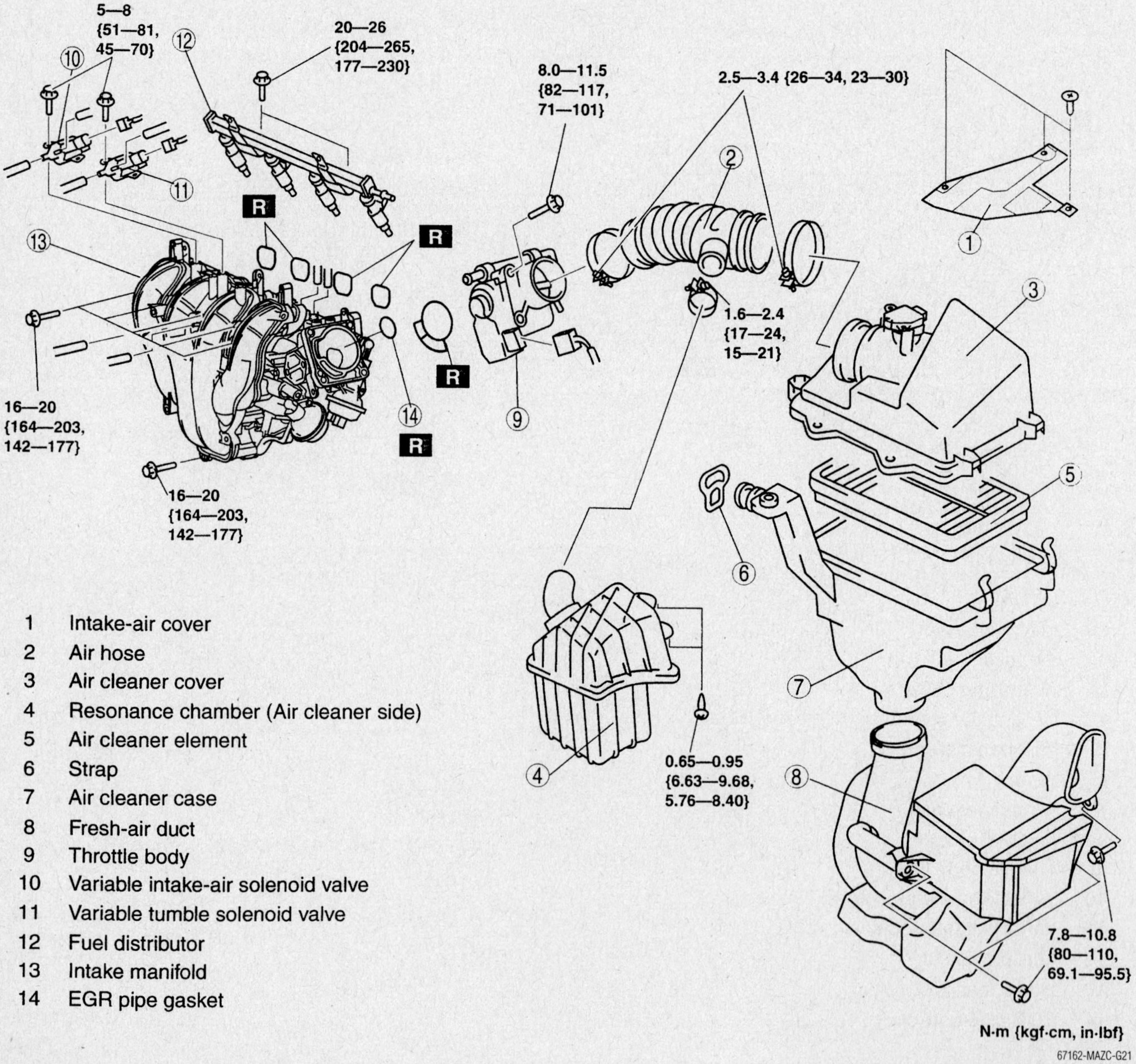

1	Intake-air cover
2	Air hose
3	Air cleaner cover
4	Resonance chamber (Air cleaner side)
5	Air cleaner element
6	Strap
7	Air cleaner case
8	Fresh-air duct
9	Throttle body
10	Variable intake-air solenoid valve
11	Variable tumble solenoid valve
12	Fuel distributor
13	Intake manifold
14	EGR pipe gasket

N·m {kgf·cm, in·lbf}

67162-MAZC-G21

Exploded view of the intake manifold assembly and related components—2.3L (L3) engines

fold. Tighten the manifold bolts to 142–177 inch lbs. (16–20 Nm).

22. Install the engine oil dipstick tube.

23. Connect the vacuum hose to the intake manifold.

24. Install the fuel rail and injectors.

25. Install the variable intake air solenoid valve and the variable tumble solenoid valve.

26. Install the throttle body and tighten the bolts to 71–101 inch lbs. (8–11.5 Nm).

27. Install the fresh air duct.

28. Verify the rubber mounts on the battery support bracket are still in place. Insert the air cleaner case into the rubber mounts, using soapy water if necessary to ease installation.

29. Use the strap to secure the shroud panel and the air cleaner case as shown in the accompanying illustration.

30. Install the air cleaner element and the resonance chamber.

31. Install the air cleaner cover.

32. Install the air hose, make sure to align the alignment marks on the throttle body and the air hose.

33. Install the intake air cover.

34. Install the under cover.

35. Connect the negative battery cable.

36. Install the battery duct and cover.

37. Install the plug hole plate.

38. Fill the cooling system.

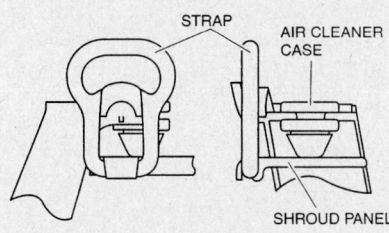

STRAP AIR CLEANER CASE

SHROUD PANEL

67162-MAZC-G19

Use the strap to secure the shroud panel and the air cleaner case—Mazda3 models

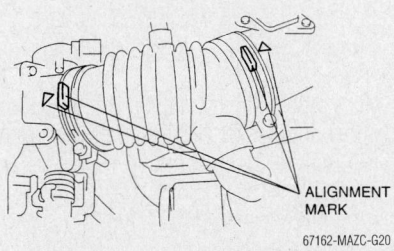

ALIGNMENT MARK

67162-MAZC-G20

Make sure to align the alignment marks on the throttle body and the air hose— Mazda3 models

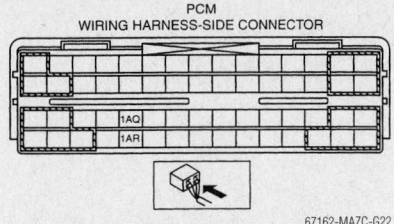

PCM
WIRING HARNESS-SIDE CONNECTOR

1AQ
1AR

67162-MAZC-G22

If equipped with an immobilizer, ground PCM terminal 1AR, if not equipped with an imobilizer, ground PCM terminal 1AQ— Mazda3 models

39. Start the vehicle and check for leaks as follows:

a. Using a jumper wire, ground the Powertrain Control Module (PCM) terminals. If equipped with an immobilizer, ground terminal 1AR, if not equipped with an imobilizer, ground terminal 1AQ. Refer to the illustration for terminal location.

b. Turn the ignition switch the ON position to activate the fuel pump.

c. Check the hoses, clips and other fuel system components for leaks.

d. If there are any leaks, replace the fuel hoses and clips. If the is damage to the seal on the fuel pipe side, replace the pipe.

e. The system must be leak free for five minutes with the terminal grounded. If any component is replaced because of a system, leak, turn the ignition key OFF; remove the jumper wire from the terminal. Reapply the jumper wire, turn the ignition On and check for leaks.

Mazda6

2.3L (L3) ENGINE

1. Before servicing the vehicle, refer to the precautions section.

2. Drain the cooling system.

3. Relieve the fuel system pressure.

4. Disconnect the negative battery cable.

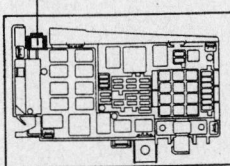

CHECK CONNECTOR

MAIN FUSE BLOCK

Using a jumper wire, short the check connector to terminal F/P to a body ground—Mazda6 models

5. Remove the intake air cover and element.

6. Remove the air cleaner case.

7. Remove the VAD check valve.

8. Remove the left front mud guard and resonance chamber.

9. Remove the Mass Airflow (MAF) and Intake Air Temperature (IAT) sensors.

10. Remove the air hose.

11. Remove the water hose.

12. Remove the throttle body.

13. Remove the variable intake air solenoid valve and the variable tumble solenoid valve.

14. Remove the fuel rail and injectors as an assembly.

15. Disconnect the evaporative hose and vacuum hose from the intake manifold.

16. Remove the intake manifold bolts, the manifold and gasket.

17. Discard the gasket.

18. Clean the mating surfaces of any gasket material.

To install:

19. Install a new gasket and the manifold. Tighten the manifold bolts to 171–101 inch lbs. (8–11.5 Nm).

20. Install the engine oil dipstick tube.

21. Connect the vacuum hose and evaporative hose to the intake manifold.

22. Install the fuel rail and injectors.

23. Install the variable intake air solenoid valve and the variable tumble solenoid valve.

24. Install the throttle body and tighten the bolts to 71–101 inch lbs. (8–11.5 Nm).

25. Install the water hose.

26. Install the air hose.

27. Verify the rubber mounts on the air cleaner bracket are still in place. Insert the air cleaner case into the rubber mounts, using soapy water if necessary to ease installation.

28. Install the remaining components in the reverse order of removal.

29. Fill the cooling system.

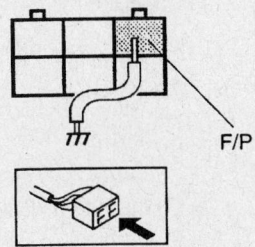

CHECK CONNECTOR

F/P

67162-MAZC-G151

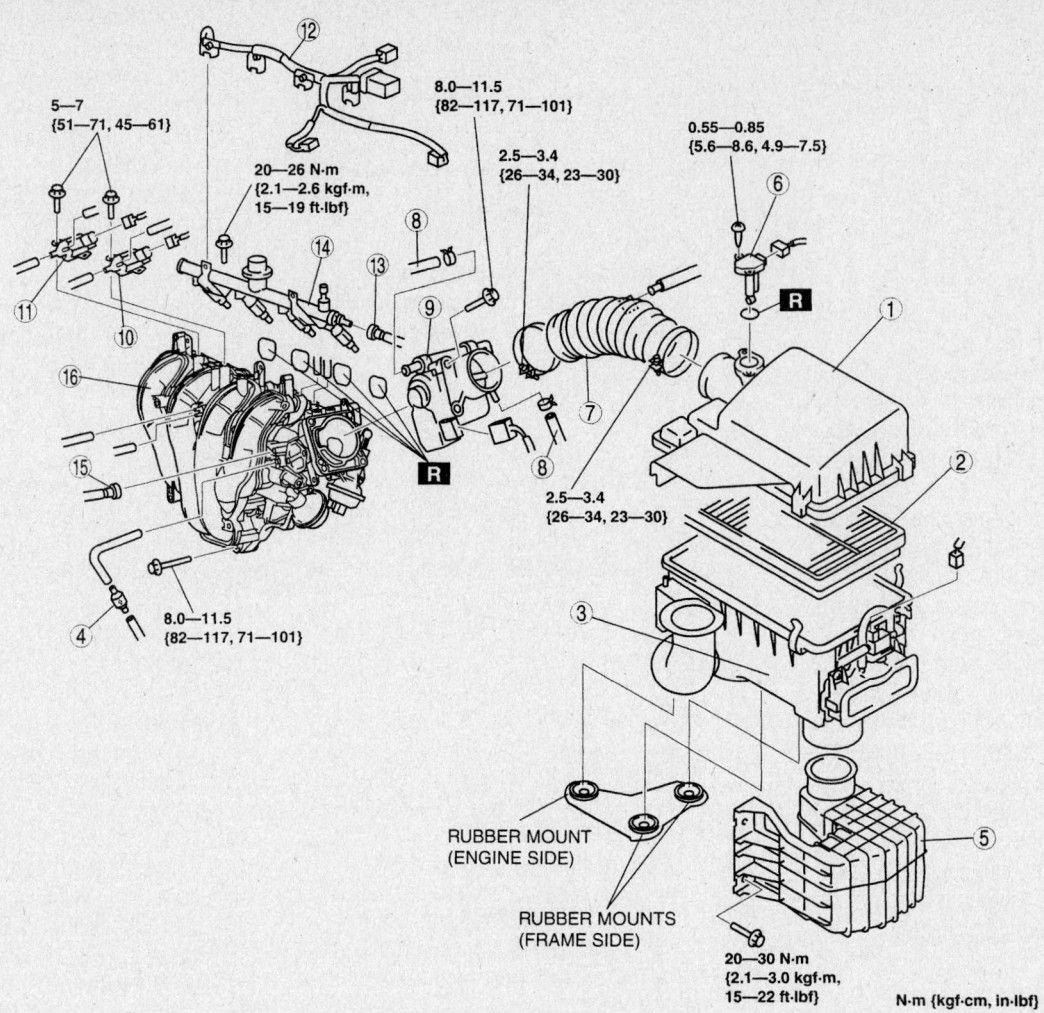

#	Component	#	Component
1	Air cleaner cover	9	Throttle body
2	Air cleaner element	10	Variable tumble control solenoid valve
3	Air cleaner case	11	VIS control solenoid valve
4	VAD check valve (one-way)	12	Fuel injector connector
5	Resonance chamber	13	Plastic fuel hose
6	MAF/IAT sensor	14	Fuel distributor
7	Air hose	15	Evaporative hose
8	Water hose	16	Intake manifold

67162-MAZC-G150

Exploded view of the intake manifold assembly and related components—Mazda6 with the 2.3L (L3) engine

30. Start the vehicle and check for leaks as follows:

a. Using a jumper wire, short the check connector to terminal F/P to a body ground.

b. Turn the ignition switch the ON position to activate the fuel pump.

c. Check the hoses, clips and other fuel system components for leaks.

d. If there are any leaks, replace the fuel hoses and clips. If the is damage to the seal on the fuel pipe side, replace the pipe.

e. The system must be leak free for five minutes with the terminal grounded. If any component is replaced because of a system, leak, turn the ignition key OFF; remove the jumper wire from the terminal. reapply the jumper wire, turn the ignition On and check for leaks.

3.0L (AJ) ENGINE

1. Remove the left front mud guard and resonance chamber.

2. Remove the Mass Airflow (MAF) and Intake Air Temperature (IAT) sensors.

3. Remove the purge solenoid valve vacuum hose.

4. Remove the air hose.

5. Remove the water hose.

6. Remove the throttle body.

7. remove the upper radiator hose, disconnect the Exhaust Gas Recirculation (EGR) pipe, EGR electrical connector and valve.

8. Remove the dynamic chamber bolts, chamber and gasket and discard the gasket.

9. Remove the fuel rail and injectors as an assembly.

10. Remove the intake manifold bolts, the manifold and gasket.

11. Discard the gasket.

12. Clean the mating surfaces of any gasket material.

To install:

13. Installation is the reverse of removal, please note the following:

   a. Install a new gasket and the manifold. Tighten the manifold bolts in the sequence illustrated to 72–106 inch lbs. (8–12 Nm).

   b. Install a new dynamic chamber gasket and the chamber. Tighten the bolts in the sequence illustrated to 72–106 inch lbs. (8–12 Nm).

   c. Install the EGR valve and tighten the retainers to the specifications shown in the accompanying illustration.

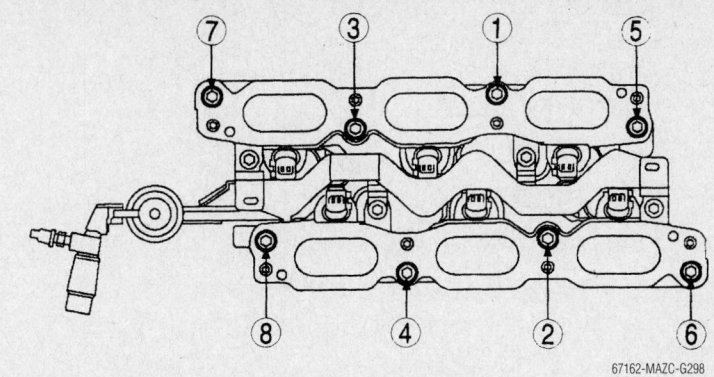

67162-MAZC-G298

Tighten the lower intake manifold bolts in this sequence—Mazda6 with the 3.0L (AJ) engine

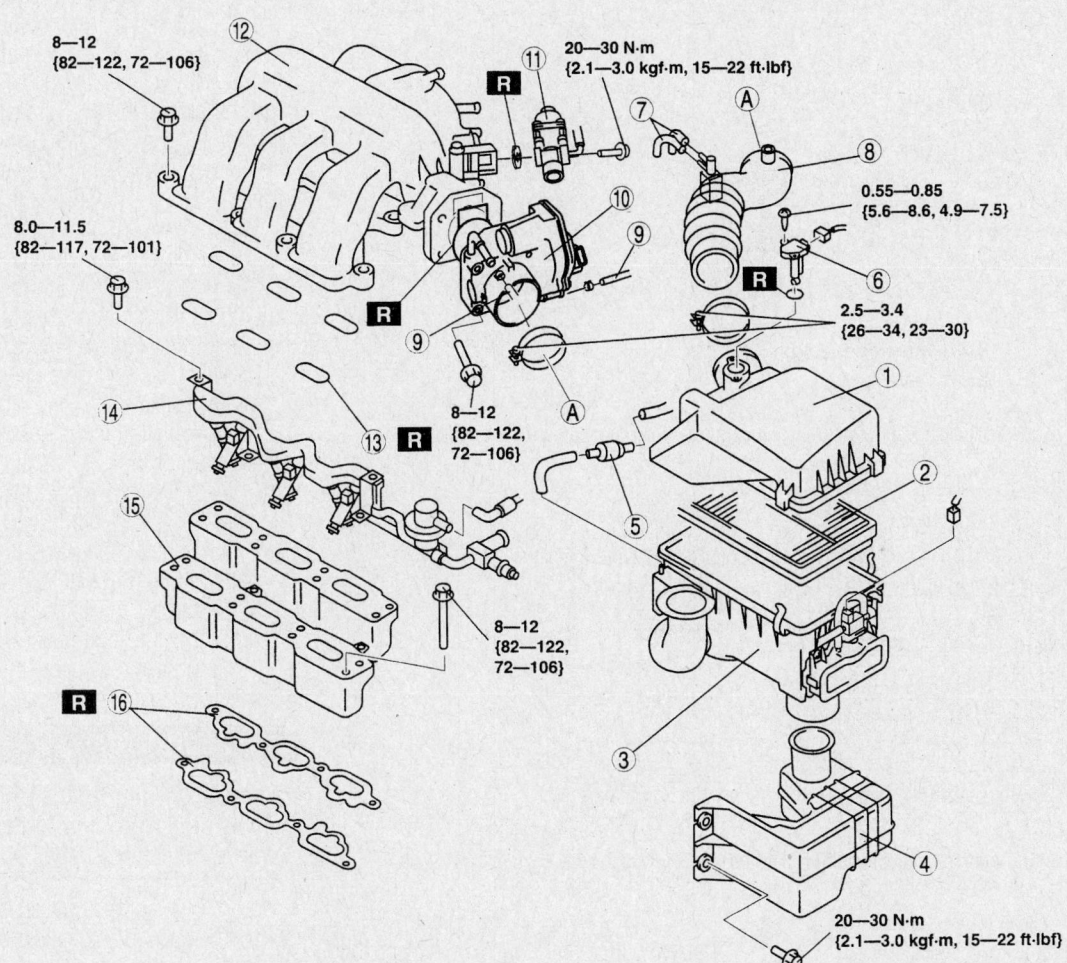

N·m {kgf·cm, in·lbf}

1	Air cleaner cover	9	Water hose
2	Air cleaner element	10	Throttle body
3	Air cleaner case	11	EGR valve
4	Resonance chamber	12	Dynamic chamber
5	VAD check valve (one-way)	13	Dynamic chamber gasket
6	MAF/IAT sensor	14	Fuel distributor
7	Vacuum hose (purge solenoid valve)	15	Intake manifold
8	Air hose	16	Intake manifold gasket

67162-MAZC-G152

Exploded view of the intake manifold assembly and related components—Mazda6 with the 3.0L (AJ) engine

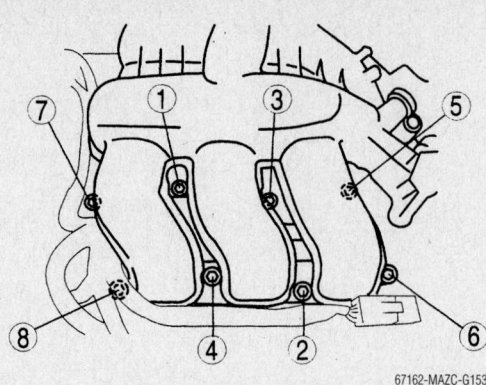

Tighten the dynamic chamber bolts in this sequence—Mazda6 with the 3.0L (AJ) engine

67162-MAZC-G153

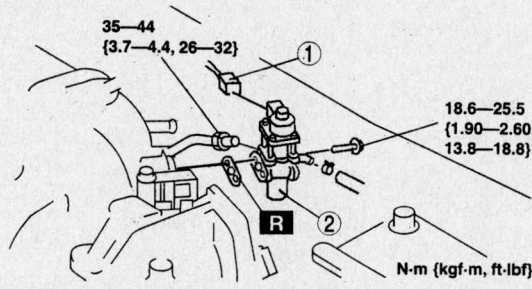

1   EGR valve connector
2   EGR valve

67162-MAZC-G154

Exploded view of the EGR assembly mounting and components—Mazda6 with the 3.0L (AJ) engine

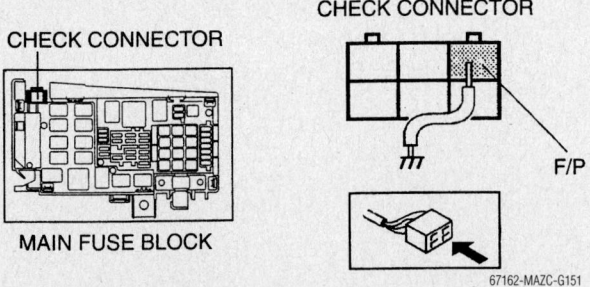

67162-MAZC-G151

Using a jumper wire, short the check connector to terminal F/P to a body ground—Mazda6 models

  d. Connect the EGR connector, and pipe.
14. Fill the cooling system.
15. Start the vehicle and check for leaks as follows:
  a. Using a jumper wire, short the check connector to terminal F/P to a body ground.
  b. Turn the ignition switch the ON position to activate the fuel pump.
  c. Check the hoses, clips and other fuel system components for leaks.
  d. If there are any leaks, replace the fuel hoses and clips. If the is damage to

the seal on the fuel pipe side, replace the pipe.
  e. The system must be leak free for five minutes with the terminal grounded. If any component is replaced because of a system, leak, turn the ignition key OFF; remove the jumper wire from the terminal. reapply the jumper wire, turn the ignition On and check for leaks.

Mazdaspeed6

1. Disconnect the negative battery cable.

2. Before servicing the vehicle, refer to the precautions section.
3. Remove the charge air cooler cover.
4. Remove the battery and tray.
5. Remove the air cleaner.
6. Remove the resonance chamber, remove the left hand front mudguard before removing the resonance chamber.
7. Disconnect the MAF/IAT sensor.
8. Remove the duct.
9. Remove the charge air cooler cover.
10. Remove the charge air cooler.
11. Remove the charge air cooler bracket.
12. Remove the air bypass valve.
13. Drain and recycle the engine coolant.
14. Remove the throttle body.
15. Disconnect the EGR valve connector.
16. Remove the air hose.
17. Remove the high pressure fuel pump bracket and the EGR pipe bracket.
18. Remove the fuel delivery pipe cover.
19. Remove the oil level gauge pipe.
20. Remove the drive belt.
21. Remove the power steering pump out of the way with the lines still attached.
22. Remove the vacuum hose.
23. Remove the intake manifold.
To install:
24. Installation is the reverse of removal, note the following:
  a. Tighten the intake manifold bolts to 13–16 ft. lbs. (17–23 Nm).
  b. Tighten the fuel delivery pipe cover bolts to 13–16 ft. lbs. (17–23 Nm).
  c. Tighten the charge air cooler bracket to 31–35 ft. lbs. (42–48 Nm).
  d. Tighten the charge air cooler to 14–19 ft. lbs. (42–48 Nm).

✳✳ CAUTION

When installing the charge air cooler cover, be careful not to damage the charge air cooler cover clips.

  e. Tighten the throttle body installation bolts to 71–101 inch lbs. (8–11.5 Nm).
  f. Perform the following when installing the air cleaner case:
  • Verify that the rubber mounts are set in the air cleaner bracket
  • Install the projections on the frame side.
  • Verify that the projections on the frame side are installed securely.
  • Install the projection on the engine side.
  • Verify that the projection on the engine side installed securely.

Exhaust Manifold

REMOVAL & INSTALLATION

MX-5 Miata

1. Before servicing the vehicle, refer to the precautions section.
2. Disconnect the front and oxygen sensor connectors.
3. Move the water pipe and heater hose slightly out of the way.
4. Disconnect the ventilation hose from the cylinder head
5. Remove the upper exhaust manifold insulator as follows:

   a. Remove the battery and battery tray.
   b. Remove the drive belt.
   c. Remove the under cover.
   d. Remove the suspension tower bar (joint) and (right side).
   e. Remove the alternator.
   f. Remove the upper insulator.
6. Remove the lower exhaust manifold insulator.
7. Remove the exhaust manifold bracket.
8. Remove the exhaust manifold.

To install:

9. Installation is the reverse of removal, please note the following:

   a. Install the manifold and tighten the retainers to 43–64 Nm (32–47 ft. lbs.)

   b. Tighten the exhaust manifold bracket to 38–52 Nm (29–38 ft. lbs.).
   c. Tighten the upper and lower insulators to 8–11 Nm (70–95 in. lbs.).
   d. If removed, tighten the front and oxygen sensors to 29–39 Nm (22–36 ft. lbs.)

Mazda3

2.0L (LF) ENGINE

1. Before servicing the vehicle, refer to the precautions section.
2. Remove the plug hole plate.
3. Remove the battery cover and duct.
4. Remove the under cover.

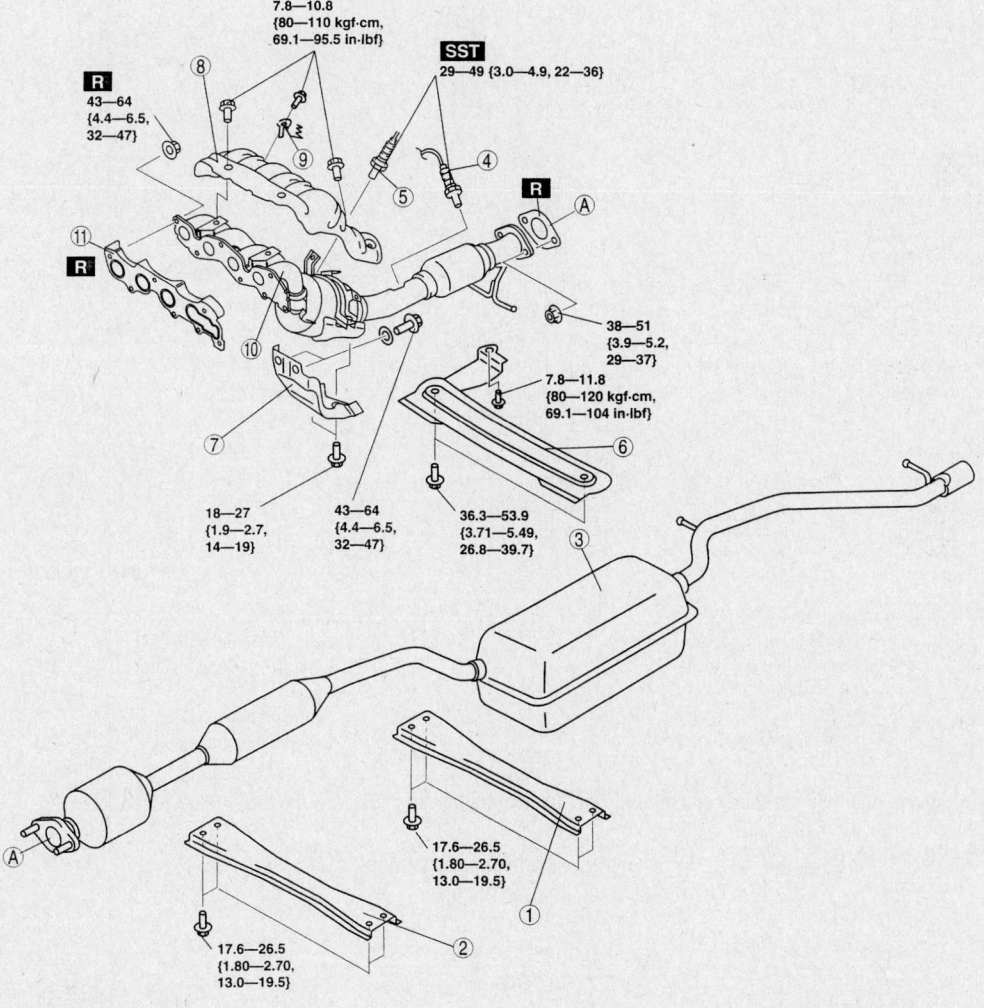

1	Rear tunnel member
2	Front tunnel member
3	Main silencer
4	Rear heated oxygen sensor
5	Front heated oxygen sensor
6	Member
7	Exhaust manifold bracket
8	Exhaust manifold insulator
9	Clip
10	Exhaust manifold
11	Exhaust manifold gasket

67162-MAZC-G23

Exploded view of the exhaust manifold assembly and related components–Non California emissions models–Mazda3 models

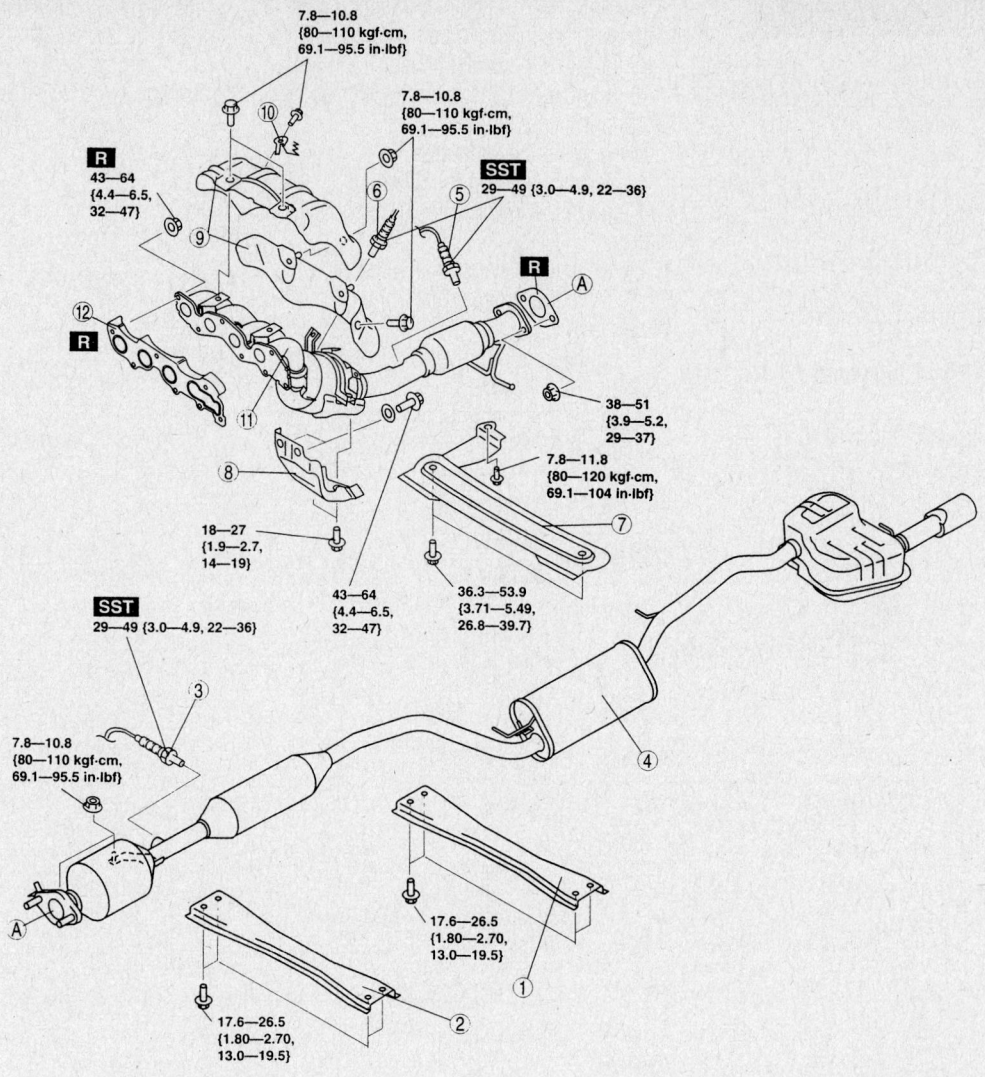

N·m {kgf·m, ft·lbf}

1	Rear tunnel member	7	Member
2	Front tunnel member	8	Exhaust manifold bracket
3	Rear heated oxygen sensor	9	Exhaust manifold insulator
4	Main silencer	10	Clip
5	Middle heated oxygen sensor	11	Exhaust manifold
6	Front heated oxygen sensor	12	Exhaust manifold gasket

67162-MAZC-G24

Exploded view of the exhaust manifold assembly and related components–California emissions models–Mazda3 models

5. Remove the rear and front tunnel members.

6. If necessary, remove the main silencer as follows:

   a. Disconnect the ABS sensor wiring harness connector.

   b. Disengage the brake pipe mount from the bracket and remove the lower shock absorber bolts.

   c. Loosen the rear crossmember bolts and lower the crossmember approximately 2.8 inches (70mm) and remove the main silencer.

7. Unplug both Oxygen Sensor (O2S) connectors.

8. Remove the exhaust manifold bracket, heat shield and clip.

9. Remove the front wheels.

10. Disconnect the steering shaft from the steering gear and linkage side.

11. Support the engine and remove the No. 1 engine mount.

12. Loosen the exhaust manifold bolts.

13. Remove the front stabilizer and stabilizer control link bolts.

14. Loosen the front crossmember bolts and lower the crossmember approximately 3.94 inches (100mm).

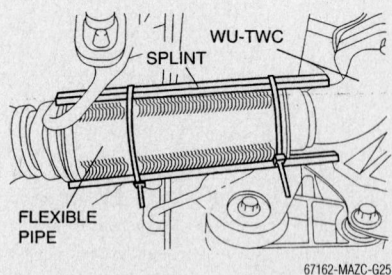

67162-MAZC-G25

Support the flexible pipe with a support wrap or splint–Mazda3 models

15. Support the flexible pipe with a support wrap or splint as illustrated.

16. Remove the exhaust manifold and gasket.

To install:

17. Installation a new exhaust manifold gasket and the manifold. Tighten the manifold retainers in the sequence illustrated to 32–47 ft. lbs. (43–64 Nm).

18. Install the remaining components in the reverse order of removal, using the exploded view of the exhaust system for component location and torque specifications.

Mazda6

2.3L (L3) ENGINE

1. Refer to the exploded view illustration for component location and torque specifications.

2. Before servicing the vehicle, refer to the precautions section.

3. Disconnect the negative battery cable.

4. Remove the main silencer and presilencer.

5. Remove the Three Way Converter (TWC).

6. Remove the manifold upper insulator.

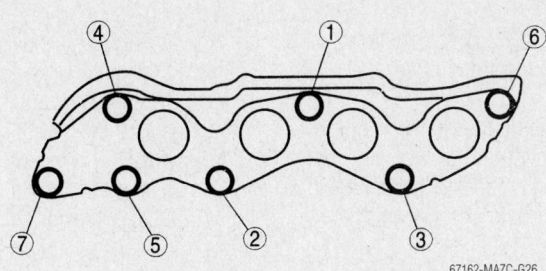

Tighten the manifold retainers in this sequence—Mazda3 models

67162-MAZC-G26

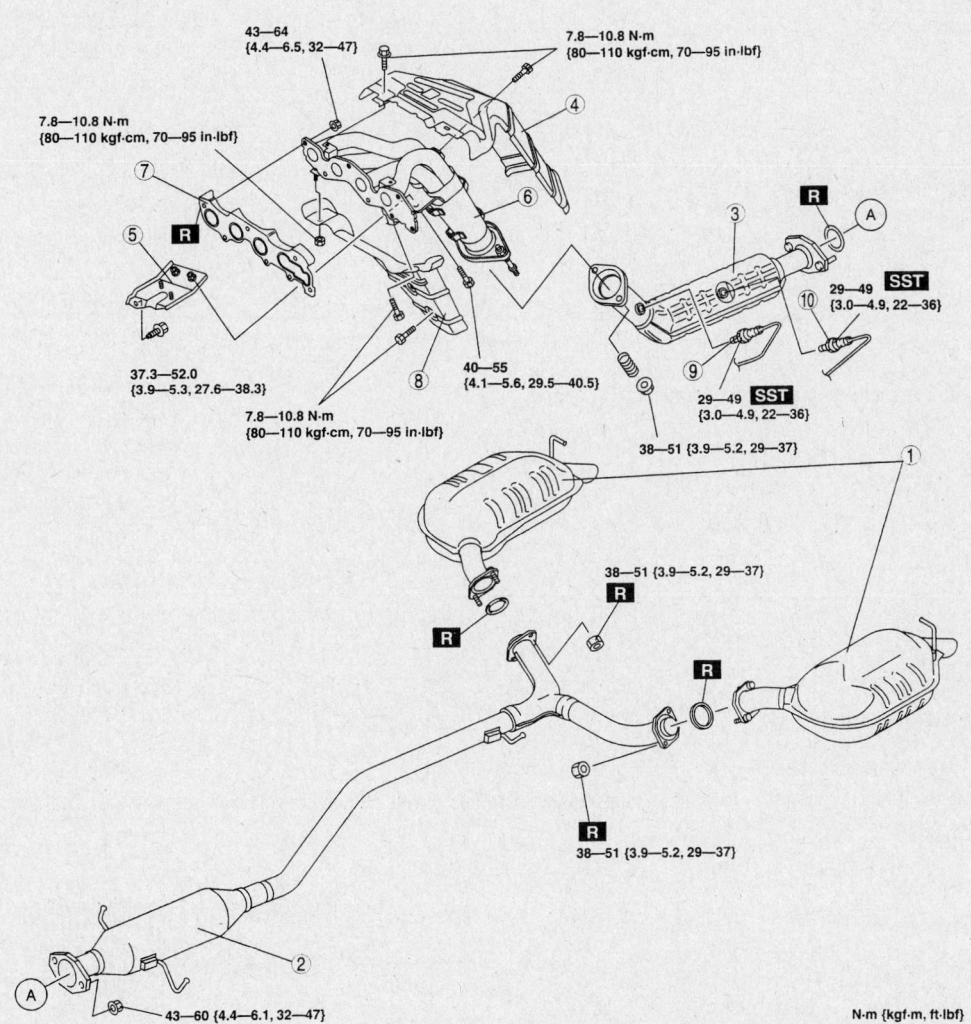

1	Main silencer	
2	Presilencer	
3	TWC	
4	Exhaust manifold insulator (upper)	
5	Bracket	
6	Exhaust manifold	
7	Exhaust manifold gasket	
8	Exhaust manifold insulator (lower)	
9	HO2S (front)	
10	HO2S (rear)	

67162-MAZC-G155

Exploded view of the exhaust manifold assembly mounting and related components—Mazda6 with the 2.3L (L3) engine

7. Remove the bracket.

8. Remove the exhaust manifold and gasket. Discard the gasket.

To install:

9. Clean the manifold mating surfaces.

10. Install a new gasket and the manifold.

11. Tighten the manifold retainers in the sequence illustrated to 32–47 ft. lbs. (43–64 Nm).

12. Install the bracket as follows:

a. Install the bracket and hand tighten the manifold side bolts.

b. Measure the gap between the manifold and bracket it should be 0.08–0.18 inch (2–4mm).

c. Tighten the cylinder block side bolt to 27–38 ft. lbs. (37–52 Nm).

d. Tighten the manifold side bolts to 29–40 ft. lbs. (40–55 Nm).

13. Tighten the manifold upper insulator bolts in the sequence illustrated to 70–95 inch lbs. (7–10 Nm).

14. Install the remaining components on the reverse order of removal, referring to the illustration for torque specifications.

3.0L (AJ) ENGINE

1. Refer to the exploded view illustration for component location and torque specifications.

2. Before servicing the vehicle, refer to the precautions section.

3. Disconnect the negative battery cable.

4. Remove the main silencer and presilencer.

5. Remove the Three Way Converter (TWC) from the left or right side depending on which manifold is being replaced.

6. Remove the stud bolt.

7. Remove the front pipe.

8. Unplug both Heated Oxygen Sensor (HO2S) connectors.

9. remove the Exhaust Gas Recirculation (EGR) pipe.

10. Remove the exhaust manifold and gasket, from the left or right side depending on which manifold is being replaced. If removing the right hand side manifold, remove the alternator bracket first.

11. Discard the manifold gasket.

To install:

12. Clean the manifold mating surfaces.

13. Install a new gasket and the manifold.

14. Tighten the manifold retainers in the sequence illustrated to 14–19 ft. lbs. (19–26 Nm).

15. Install the remaining components on the reverse order of removal, referring to the illustration for torque specifications.

Mazdaspeed6

1. Before servicing the vehicle, refer to the precautions section.

2. Disconnect the negative battery cable.

3. Remove the main silencer.

4. Remove the Three Way Converter (TWC).

5. Remove the rear oxygen sensor.

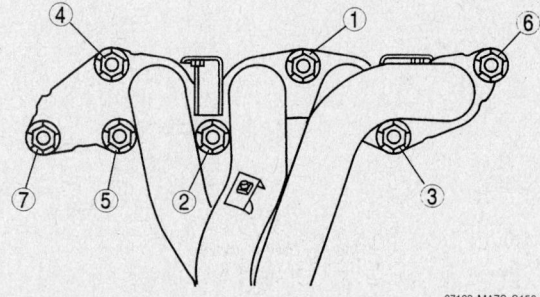

Exhaust manifold torque sequence—Mazda6 with the 2.3L (L3) engine

67162-MAZC-G156

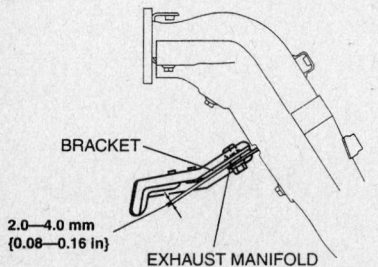

BRACKET

2.0–4.0 mm {0.08—0.16 in}

EXHAUST MANIFOLD

67162-MAZC-G157

Measure the gap between the manifold and bracket—Mazda6 with the 2.3L (L3) engine

67162-MAZC-G160

Left side exhaust manifold torque sequence—Mazda6 with the 3.0L (AJ) engine

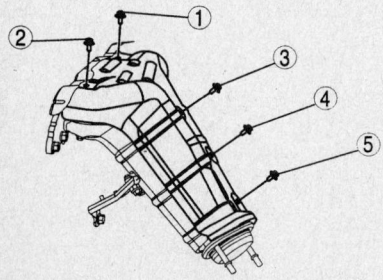

67162-MAZC-G158

Tighten the manifold upper insulator bolts in this sequence—Mazda6 with the 2.3L (L3) engine

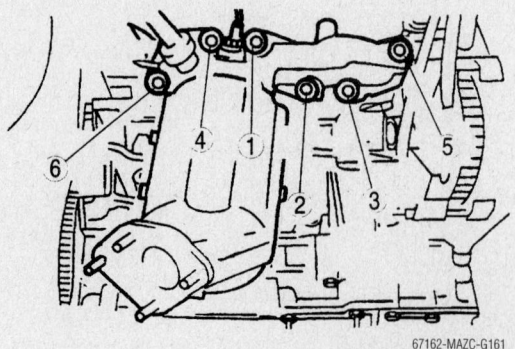

67162-MAZC-G161

Right side exhaust manifold torque sequence—Mazda6 with the 3.0L (AJ) engine

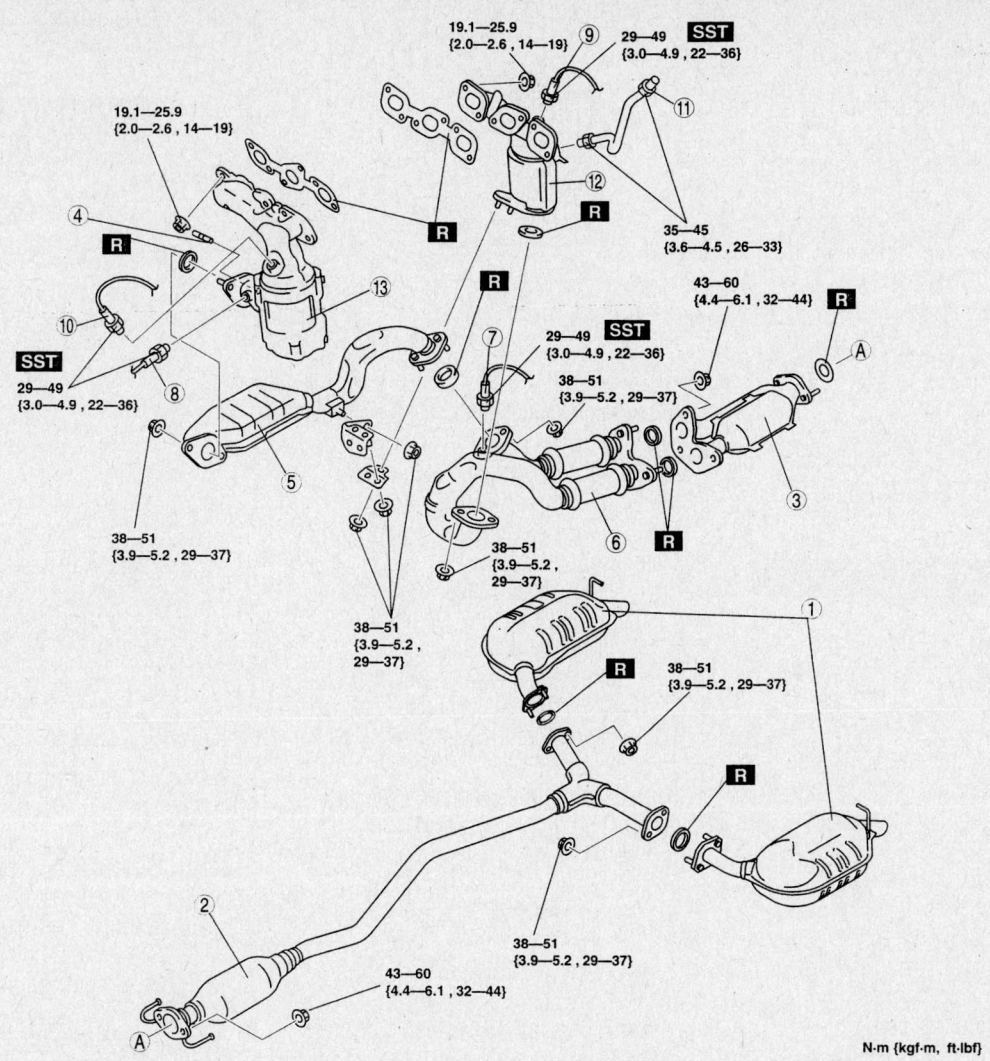

1   Main silencer
2   Presilencer
3   TWC (RH)
4   Stud bolt
5   TWC (LH)
6   Front pipe
7   HO2S (RR)
8   HO2S (LR)
9   HO2S (RF)
10 HO2S (LF)
11 EGR pipe
12 Exhaust manifold (RH)
13 Exhaust manifold (LH)

N·m {kgf·m, ft-lbf}

67162-MAZC-G159

Exploded view of the exhaust manifold assembly mounting and related components—Mazda6 with the 3.0L (AJ) engine

6. Remove the front pipe.

7. Remove the charge air cooler and the charge air cooler bracket.

8. Remove the cowl grille.

9. Remove the insulators

10. Remove the manifold upper and lower insulator.

11. Remove the front oxygen sensor.

12. Set the alternator out of the way.

13. Disconnect the vacuum hose on the master cylinder side.

14. Remove the warm up three way converter.

15. Remove the exhaust manifold and gasket. Discard the gasket.

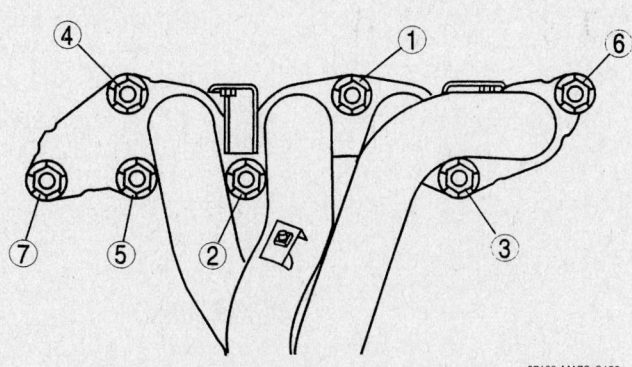

67162-MAZC-G156

Exhaust manifold torque sequence—Mazdaspeed6 with the 2.3L (L3) engine

To install:

16. Installation is the reverse of removal, note the following:

a. Clean the manifold mating surfaces.

b. Install a new gasket and the manifold.

c. Tighten the manifold retainers in the sequence illustrated to 32–47 ft. lbs. (43–64 Nm).

d. Tighten the installation nuts between the catalytic converter and middle pipe. Tighten to 32–47 ft. lbs. (43–64 Nm).

e. Tighten the installation nuts between main silencer and middle pipe in the order of left side, then right side. Tighten to 29–37 ft. lbs. (38–51 Nm).

Front Crankshaft Seal

REMOVAL & INSTALLATION

MX-5 Miata

1. Before servicing the vehicle, refer to the precautions section.

2. Remove the battery and battery tray.

3. Remove the air cleaner.

4. Remove the drive belt.

5. Remove the under cover.

6. Remove the front suspension tower bar.

7. Remove the ignition coil.

8. Remove the OCV connector.

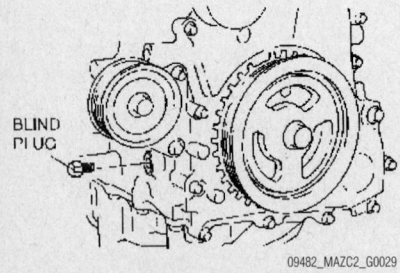

Remove the cylinder block lower blind plug—MX-5 Miata

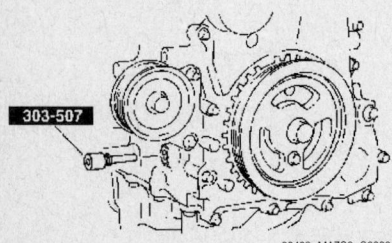

Install tool 303-507 in the lower blind plug opening—MX-5 Miata

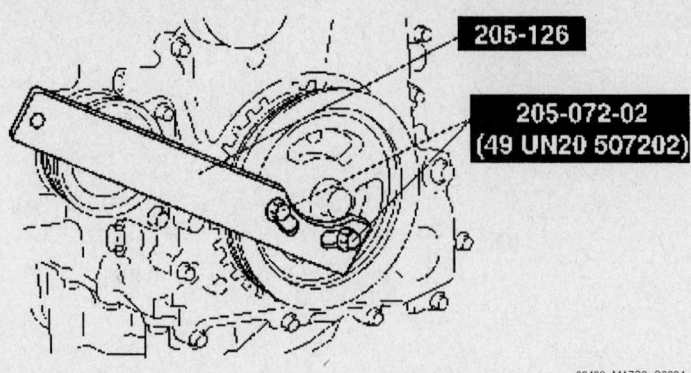

Hold the crankshaft pulley using the tools shown—MX-5 Miata

9. Remove the cylinder head cover.

10. Remove the CKP sensor.

11. Remove the crankshaft pulley lock bolt as follows:

a. Remove the cylinder block lower blind plug.

b. Install tool 303-507.

c. Turn the crankshaft clockwise until the crankshaft is in the No.1 cylinder TDC position (until the balance weight contacts tool 303-507).

d. Hold the crankshaft pulley using the tools shown in the illustration.

e. Remove the crankshaft pulley lock bolt.

12. Remove the front oil seal as follows:

a. Cut the oil seal lip using a razor knife.

b. Remove the oil seal using a screwdriver wrapped with a rag.

To install:

13. Install the oil seal as follows:

a. Apply clean engine oil to a new oil seal.

b. Push the front oil seal in the engine front cover by hand.

c. Tap the oil seal in evenly using tool shown and a hammer to 0–10 mm (0–0.039 in.).

14. Install the crankshaft pulley lock bolt as follows:

a. Install tool shown in the illustration on the camshaft.

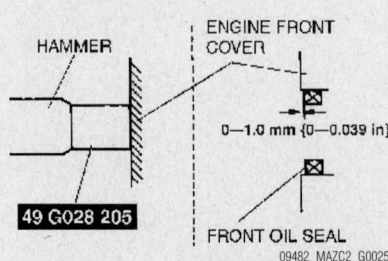

Tap the oil seal in evenly using tool shown and a hammer—MX-5 Miata

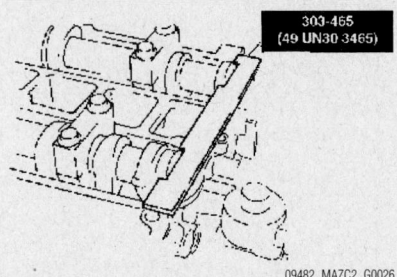

Install tool shown onto the camshaft—MX-5 Miata

b. Verify that cylinder No.1 is at TDC of the compression stroke. (Crankshaft balance weight contacts tool 303-507)

c. Position the crankshaft pulley by temporarily tightening it and, using a suitable length bolt 25–35mm (0.99–1.37 in.), fix the crankshaft pulley to the engine front cover.

d. Install the tools illustrated to the crankshaft pulley, lock the crankshaft against rotation.

e. Tighten the crankshaft pulley lock bolt to 96–104 Nm (71–77 ft. lbs.), plus 87–93 degrees.

f. Remove the bolt.

g. Remove the tool the camshaft.

h. Remove tool from the cylinder block lower blind plug.

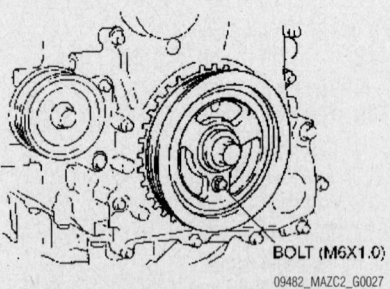

Position the crankshaft pulley by temporarily tightening it using a suitable length bolt—MX-5 Miata

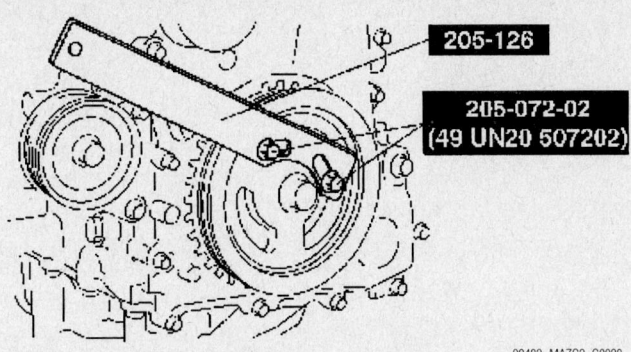

09482_MAZC2_G0028

Install the tools shown onto the crankshaft pulley to lock the crankshaft against rotation—MX-5 Miata

i. Remove the tool from the crankshaft pulley.

j. Rotate the crankshaft clockwise two turns until the TDC position.

k. If not aligned, loosen the crankshaft pulley lock bolt and repeat from the first step.

l. Install the cylinder block lower blind plug and tighten to 18–22 Nm (13–16 ft. lbs.)

15. Install the remaining components in the reverse order of removal.

Mazda3 and 6 Models

2.3L (L3) ENGINE

1. Before servicing the vehicle, refer to the precautions section.

2. Remove the plug hole plate and bracket.

3. Remove the battery cover.

4. Disconnect the negative battery cable.

5. Disconnect the wiring harness.

6. Remove the ignition coils, spark plugs, valve cover and accessory drive belt.

7. Remove the front wheels.

8. Remove the under cover and splash shield, if equipped.

9. On Mazda6 models, remove the right hand halfshaft from the joint shaft.

10. Remove the Crankshaft Position (CKP) sensor.

11. Remove the crankshaft pulley bolt and pulley.

12. Remove the cylinder block lower blind plug.

13. Install tool 49 JE01 061.

14. Turn the crankshaft clockwise until the crankshaft is in the number 1 cylinder Top Dead Center position (the balance weight should be attached to tool 49 JE01 061).

15. Hold the crankshaft pulley using the tools illustrated and remove the bolt.

16. Remove the crankshaft pulley.

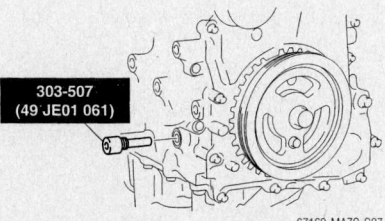

67162-MAZC-G27

Install tool 49 JE01 061—2.0L (LF) and 2.3L (L3) engines

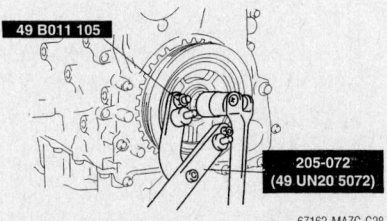

67162-MAZC-G28

Install tool 49 JE01 061 Hold the crankshaft pulley using the tools illustrated and remove the bolt—2.0L (LF) and 2.3L (L3) engines

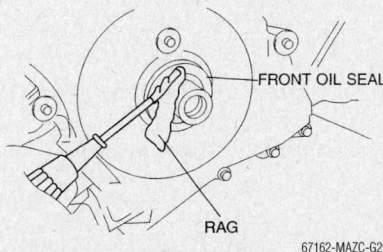

67162-MAZC-G29

Cut the seal lip using a suitable knife and using a suitable prytool wrapped in a rag, remove the seal—2.0L (LF) and 2.3L (L3) engines

17. Cut the seal lip using a suitable knife and using a suitable prytool wrapped in a rag, remove the seal.

To install:

18. Coat the new seal with clean engine oil.

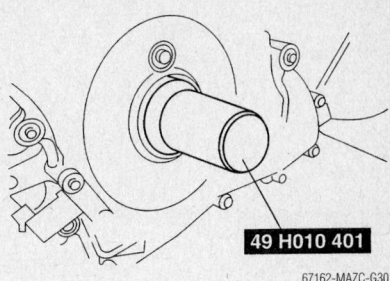

67162-MAZC-G30

Use seal installer tool 49 H010 401 and a hammer to install the seal—2.0L (LF) and 2.3L (L3) engines

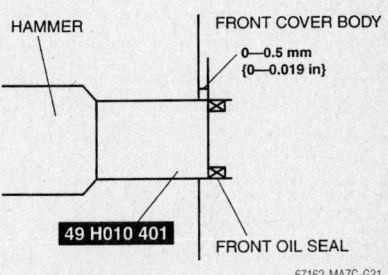

67162-MAZC-G31

The seal should be recessed 0–0.019 inch (0–0.5mm) as shown when properly installed—2.0L (LF) and 2.3L (L3) engines

19. Push the seal in by hand to get it started.

20. Use seal installer tool 49 H010 401 and a hammer to install the seal so that it is recessed 0–0.019 inch (0–0.5mm) as shown in the accompanying illustration.

21. Install the crankshaft pulley.

22. Install tool 49 UN30 3465 as illustrated on the camshaft.

23. Install and hand tighten the M6 x 1.0 bolt as illustrated.

24. Turn the crankshaft clockwise until the crankshaft is in number 1 TDC (the balance weight should be attached to the tool).

25. Hold the crankshaft pulley and tighten the lock bolt in two steps. First tighten to 71–77 ft. lbs. (96–104 Nm). Then final tighten an additional 87–93 degrees.

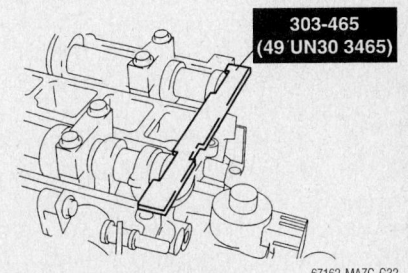

67162-MAZC-G32

Install tool 49 UN30 3465 on the camshaft—2.0L (LF) and 2.3L (L3) engines

67162-MAZC-G33

Install and hand tighten the M6 x 1.0 bolt—2.0L (LF) and 2.3L (L3) engines

26. Remove the M6 x 1.0 bolt.
27. Remove the tools from the camshaft and the cylinder block lower blind plug.
28. Rotate the crankshaft clockwise 2 turns until you reach TDC. If not aligned properly, loosen the crankshaft pulley lock bolt and reinstall the lock bolt again using the above procedure and special tools.
29. Install the cylinder block blind plug and tighten to 13–16 ft. lbs. (18–22 Nm).
30. When installing the CKP sensor perform the following:

   a. Remove the cylinder block lower blind plug.

   b. Install tool 49 JE01 061.

   c. Turn the crankshaft clockwise until the crankshaft is in the number 1 cylinder Top Dead Center position (the balance weight should be attached to tool 49 JE01 061).

   d. Using a ruler, mark the center line on the pulse wheel teeth on the crank pulley which is located at the 9th tooth counting counterclockwise from the empty space. Refer to the illustration for more detail.

✳✳ CAUTION

If you do not mark the center line correctly this will cause improper engine control for the ignition and fuel system, so be sure to mark the line carefully.

   e. Install the CKP sensor making sure the mark on the sensor is aligned with the mark on the pulse wheel. Tighten the sensor bolt to 49–66 inch lbs. (5.5–7.5 Nm) and attach the electrical connector.

   f. Remove the tool from the cylinder block lower blind plug.

   g. Rotate the crankshaft clockwise 2 turns until you reach TDC. If not aligned properly, loosen the crankshaft pulley lock bolt and reinstall the lock bolt.

   h. Install the cylinder block blind plug and tighten to 13–16 ft. lbs. (18–22 Nm).

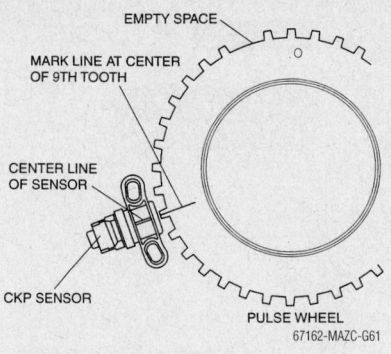

67162-MAZC-G61

Using a ruler, mark the center line on the pulse wheel teeth on the crank pulley which is located at the 9th tooth counting counterclockwise from the empty space— 2.0L (LF) and 2.3L (L3) engines

31. Install the remaining components in the reverse order of removal.
32. Start the vehicle and check for leaks, repair if necessary.

3.0L (AJ) ENGINE

1. Before servicing the vehicle, refer to the precautions section.
2. Disconnect the negative battery cable.
3. Remove the accessory drive belt.
4. Hold the crankshaft pulley using the tools illustrated and remove the pulley bolt.
5. Remove the A/C compressor and set aside with the lines still attached.
6. Remove the crankshaft pulley using

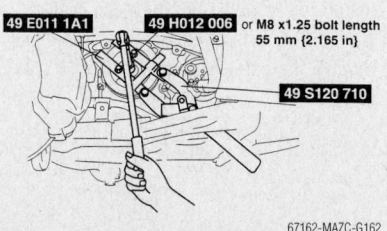

67162-MAZC-G162

Hold the crankshaft pulley using the tools illustrated and remove the pulley bolt— Mazda6 with the 3.0L (AJ) engine

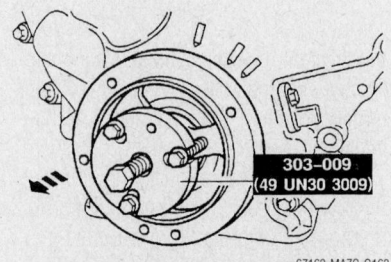

67162-MAZC-G163

Remove the crankshaft pulley using a suitable puller such as the one shown— Mazda6 with the 3.0L (AJ) engine

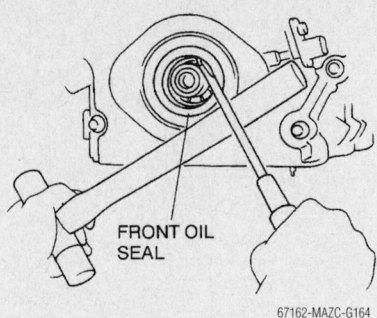

67162-MAZC-G164

Remove the front oil seal using a prytool as shown—Mazda6 with the 3.0L (AJ) engine

a suitable puller such as the one illustrated.

7. Remove the front oil seal using a prytool as illustrated.

   To install:

8. Assemble the seal using part (A) of tool 49 UN01 002 and tool 49 UN01 002 as illustrated.
9. Apply clean oil to the seal and push the seal in by hand.
10. Install the seal using the installation tools until the seal is recessed 0–0.039 inch (0–1mm).
11. Seal the crankshaft pulley using a silicone sealant.
12. Install the pulley using the tools illustrated.

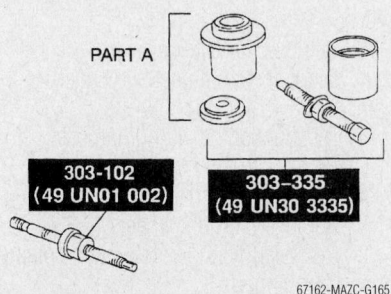

67162-MAZC-G165

Assemble the seal using part (A) of tool 49 UN01 002 and tool 49 UN01 002 as shown—Mazda6 with the 3.0L (AJ) engine

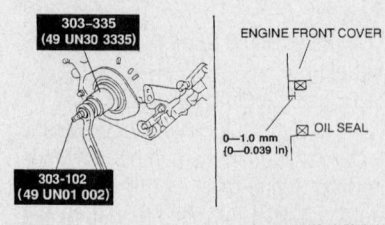

67162-MAZC-G166

Install the seal using the installation tools until the seal is recessed 0–0.039 inch (0–1mm)—Mazda6 with the 3.0L (AJ) engine

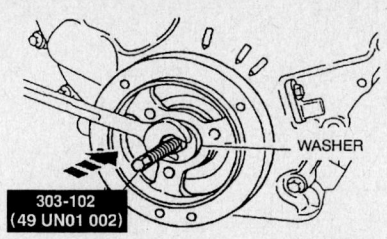

**303-102
(49 UN01 002)**

67162-MAZC-G167

**Install the pulley using the tools shown—
Mazda6 with the 3.0L (AJ) engine**

13. While holding the pulley, tighten the crankshaft pulley bolt as follows:

a. Tighten to 88 ft. lbs. (120 Nm).

b. Loosen one full turn.

c. Tighten to 35–39 ft. lbs. (47–53 Nm).

d. Tighten 85–95 degrees.

14. Install the A/C compressor.

15. Install the accessory drive belt.

16. Connect the negative battery cable.

17. Start the vehicle and check for leaks, repair if necessary.

Mazdaspeed6

1. Before servicing the vehicle, refer to the precautions section.

2. Disconnect the negative battery cable.

3. Remove the right side front tire.

4. Remove the under cover.

5. Remove the right side splash shield.

➡**If the high pressure fuel pump is removed, replace the O-ring with a new one.**

6. Remove the charge air cooler cover.

7. Disconnect the spill valve control solenoid valve connector.

8. Disconnect the quick release connector on the high pressure fuel pump.

9. Remove the battery and battery tray.

❄❄ WARNING

If the high pressure fuel pump joint nut is loosened, fuel leakage may occur resulting in death or serious injury, or damage to the equipment or the vehicle. Fuel can also irritate the skin and eyes. When removing the high pressure line pipe, always tighten the high pressure line pipe installation nut while fixing the high pressure fuel pump joint nut with a wrench. If the high pressure fuel pump joint nut has rotated, replace the high pressure fuel pump with a new one.

10. Disconnect the high pressure line pipe of the high pressure fuel pump.

11. Fix the joint nut with a wrench on the high pressure fuel pump side.

12. Loosen the high pressure line pipe installation nut.

13. Drain and recycle the engine coolant.

14. Loosen the water outlet case installation bolts securing the high pressure line pipe.

15. Remove the high pressure fuel pump.

16. Remove the high pressure fuel pump cover.

17. Remove the ignition coils, spark plugs, valve cover and accessory drive belt.

18. Remove the right hand halfshaft from the joint shaft.

19. Remove the Crankshaft Position (CKP) sensor.

20. Remove the crankshaft pulley bolt and pulley.

21. Remove the cylinder block lower blind plug.

22. Install tool 49 JE01 061.

23. Turn the crankshaft clockwise until the crankshaft is in the number 1 cylinder Top Dead Center position (the balance weight should be attached to tool 49 JE01 061).

24. Hold the crankshaft pulley using the tools illustrated and remove the bolt.

25. Remove the crankshaft pulley.

26. Cut the seal lip using a suitable knife and using a suitable prytool wrapped in a rag, remove the seal.

To install:

27. Coat the new seal with clean engine oil.

28. Push the seal in by hand to get it started.

29. Use seal installer tool 49 H010 401 and a hammer to install the seal so that it is recessed 0–0.019 inch (0–0.5mm) as shown in the accompanying illustration.

30. Install the crankshaft pulley.

31. Install tool 49 UN30 3465 as illustrated on the camshaft.

32. Install and hand tighten the M6 x 1.0 bolt as illustrated.

33. Turn the crankshaft clockwise until the crankshaft is in number 1 TDC (the balance weight should be attached to the tool).

34. Hold the crankshaft pulley and tighten the lock bolt in two steps. First tighten to 71–77 ft. lbs. (96–104 Nm). Then final tighten an additional 87–93 degrees.

35. Remove the M6 x 1.0 bolt.

36. Remove the tools from the camshaft and the cylinder block lower blind plug.

37. Rotate the crankshaft clockwise 2 turns until you reach TDC. If not aligned properly, loosen the crankshaft pulley lock bolt and reinstall the lock bolt again using the above procedure and special tools.

38. Install the cylinder block blind plug and tighten to 13–16 ft. lbs. (18–22 Nm).

39. When installing the CKP sensor perform the following:

a. Remove the cylinder block lower blind plug.

b. Install tool 49 JE01 061.

c. Turn the crankshaft clockwise until the crankshaft is in the number 1 cylinder Top Dead Center position (the balance weight should be attached to tool 49 JE01 061).

d. Using a ruler, mark the center line on the pulse wheel teeth on the crank pulley which is located at the 9th tooth counting counterclockwise from the empty space. Refer to the illustration for more detail.

❄❄ CAUTION

If you do not mark the center line correctly this will cause improper engine control for the ignition and fuel system, so be sure to mark the line carefully.

e. Install the CKP sensor making sure the mark on the sensor is aligned with the mark on the pulse wheel. Tighten the sensor bolt to 49–66 inch lbs. (5.5–7.5 Nm)and attach the electrical connector.

f. Remove the tool from the cylinder block lower blind plug.

g. Rotate the crankshaft clockwise 2 turns until you reach TDC. If not aligned properly, loosen the crankshaft pulley lock bolt and reinstall the lock bolt.

h. Install the cylinder block blind plug and tighten to 13–16 ft. lbs. (18–22 Nm).

40. Install the high pressure fuel pump cover and tighten to 13–16 ft. lbs. (18–22 Nm).

❄❄ CAUTION

If the high pressure fuel pump installation bolts are tightened with the high pressure fuel pump tilted, the high pressure fuel pump may not operate correctly. Tighten the high pressure fuel pump installation bolts in a few passes with equal torque. Tighten to 76–101 inch lbs. (8.5–11.5 Nm).

✳✳ WARNING

If the high pressure fuel pump joint nut is loosened, fuel leakage may occur resulting in death or serious injury, or damage to the equipment or the vehicle. Fuel can also irritate the skin and eyes. When installing the high pressure line pipe, always tighten the high pressure line pipe installation nut while fixing the high pressure fuel pump joint nut with a wrench. If the high pressure fuel pump joint nut has rotated, replace the high pressure fuel pump with a new one.

41. Assemble the high pressure line pipe.
42. Fix the joint nut with a wrench on the high pressure fuel pump side. Tighten the high pressure line pipe installation nut to 17–26 ft. lbs. (23–35 Nm).
43. Tighten the water outlet case installation bolts to 13–16 ft. lbs. (18–22 Nm).
44. Install the quick release connector.
45. Install the remaining components in the reverse order of removal.
46. Verify that the high pressure fuel pump is assembled securely.
47. Drive the vehicle starting from a standstill and brake suddenly five to six times at a low speed.
48. Stop the vehicle and verify from outside the vehicle that there is no fuel leakage around the high pressure fuel pump.

Camshaft

REMOVAL & INSTALLATION

MX-5 Miata

1. Before servicing the vehicle, refer to the precautions section.
2. Properly relieve the fuel system pressure.
3. Remove the battery and battery tray.
4. Drain the engine coolant.
5. Remove the front suspension tower bar.
6. Remove the air cleaner.
7. Remove the dynamic chamber.
8. Remove the ignition coil.
9. Remove the drive belt.
10. Remove the CKP sensor.
11. Remove the P/S oil pump with the oil hose still connected and position the P/S oil pump so that it is out of the way.
12. Remove the timing chain.
13. Remove the wiper arm.
14. Remove the cowl grille.

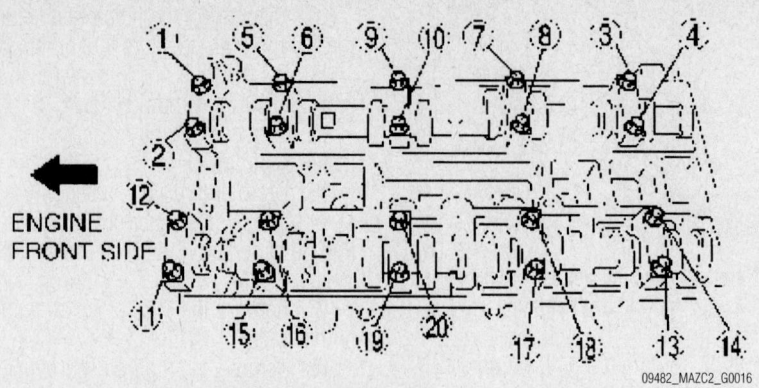

ENGINE FRONT SIDE

Camshaft cap bolt loosening sequence—MX-5 Miata

09482_MAZC2_G0016

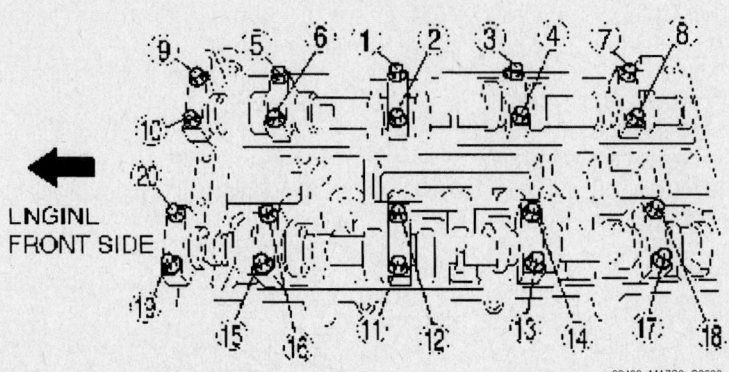

ENGINE FRONT SIDE

Camshaft cap bolt torque sequence—MX-5 Miata

09482_MAZC2_G0020

15. Remove the side cowl grille.
16. Remove the service hole cover.
17. Disconnect the alternator, but do not remove it from the vehicle.
18. Remove the exhaust manifold.
19. Remove the cylinder head cover.
20. Remove the OCV sensor, if equipped with variable valve timing.
21. Mark the camshaft cap locations so they may be reinstalled in their original positions.
22. Remove camshafts.
To install:
23. Set the cam position of the number 1 cylinder to top dead center, install the camshafts, making sure to install the caps in their original locations.
24. Tighten the camshaft cap bolts using two to three steps in the sequence illustrated to 5–9 Nm (44.3–79.6 in. lbs.) and then to 14–17 Nm (10.4–12.5 ft. lbs.)
25. Install the remaining components in the reverse order of removal
26. Start the engine and inspect for leaks.
27. Check the ignition timing, idle speed and idle mixture.

Mazda3

1. Before servicing the vehicle, refer to the precautions section.

2. Disconnect the negative battery cable.
3. Remove the timing chain.
4. Remove the intake manifold.
5. Disconnect the following components:
   - Warm Up–Three Way Converter (WU–TWC)
   - Upper radiator hose
   - Water and heater hose
   - Wiring harness
6. Support the engine using an engine jack and attachment as shown in the illustration.
7. Remove the camshafts.

➡ The cylinder head and camshaft caps are numbered and must be reassembled in their original locations. When removing a when removed, keep the caps with the cylinder head they were removed from, do not switch the caps.

8. Loosen the camshaft cap bolts using 2–3 passes in the sequence illustrated.
9. Once all the bolts have been removed, remove the caps, keeping them in the original positions and remove the camshafts.
To install:
10. Set the cam position of the number 1

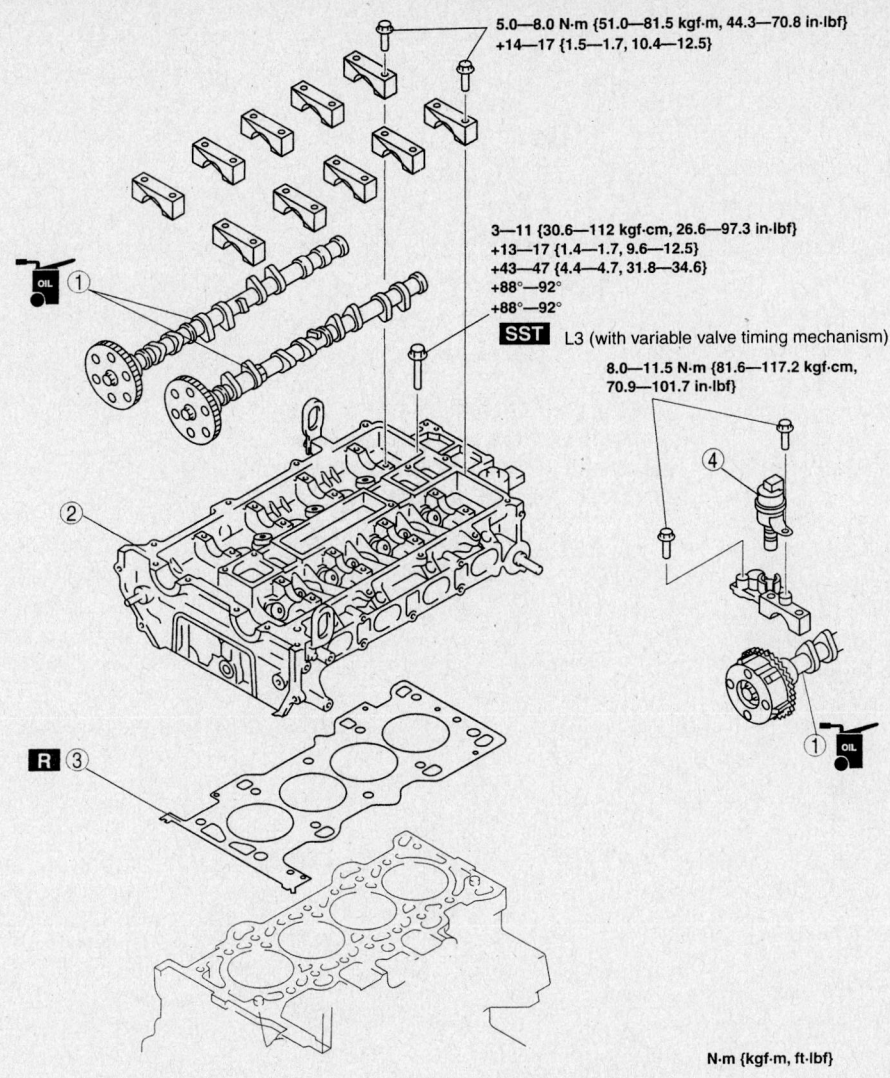

1   Camshaft
2   Cylinder head
3   Cylinder head gasket
4   Oil control valve (OCV) (L3 (with variable valve timing mechanism))

67162-MAZC-G34

Exploded view of the camshaft and cylinder head components—Mazda3 models

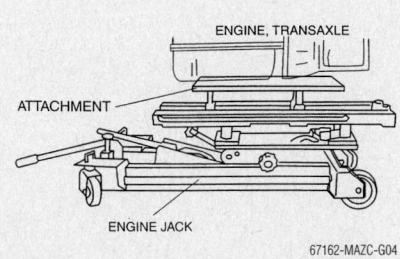

67162-MAZC-G04

Support the engine using an engine jack and attachment as shown—Mazda3 models

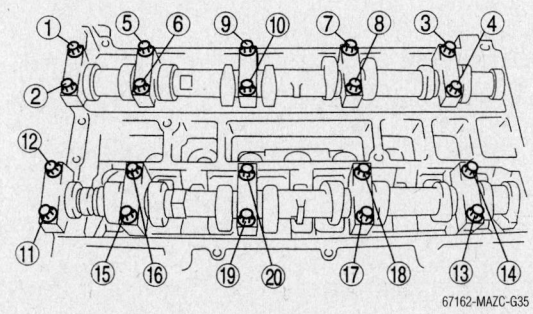

67162-MAZC-G35

Camshaft cap bolt removal sequence—Mazda3 models

cylinder at Top Dead center (TDC) and install the camshaft.

11. Temporarily install the camshaft caps in their original positions.

12. Tighten the camshaft cap bolt in two steps:

a. Step 1: 44–71 inch lbs. (5–8 Nm).

b. Step 2 : 10.4–12.5 ft. lbs. (14–17 Nm).

13. Remove the engine jack and attachment.

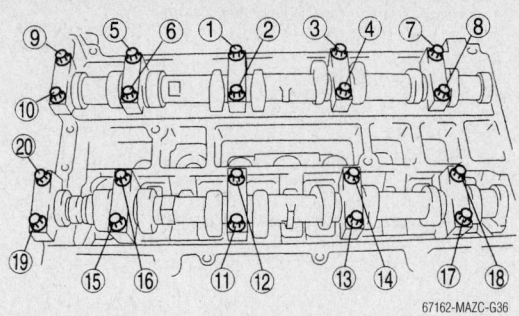

Camshaft cap bolt tightening sequence—Mazda3 models

14. Connect the following components:
- Wiring harness
- Water and heater hose
- Upper radiator hose
- WU–TWC converter

15. Install the intake manifold.
16. Install the timing chain.
17. Connect the negative battery cable.
18. Fill and bleed the cooling system.
19. Change the oil and filter.
20. Run the engine and check for proper operation.

Mazda6 and Mazdaspeed6

2.3L (L3) ENGINE

1. Before servicing the vehicle, refer to the precautions section.
2. Disconnect the negative battery cable.
3. Remove the timing chain.
4. Remove the ignition coil and spark plug wires.
5. Disconnect the alternator and move it aside.
6. Remove the front exhaust pipe.
7. Remove the intake manifold.
8. Remove the upper radiator hose, bypass and heater hoses.
9. Support the engine using an engine jack and attachment as shown in the illustration.
10. Remove the oil control valve.
11. Remove the camshafts.

➡ The cylinder head and camshaft caps are numbered and must be reassem-bled in their original locations. When removing a when removed, keep the caps with the cylinder head they were removed from, do not switch the caps.

12. Loosen the camshaft cap bolts using 2–3 passes in the sequence illustrated.

13. Once all the bolts have been removed, remove the caps, keeping them in the original positions and remove the camshafts.

To install:

14. Set the cam position of the number 1 cylinder at Top Dead center (TDC) and install the camshaft.
15. Temporarily install the camshaft caps in their original positions.
16. Tighten the camshaft cap bolt in two steps:
   a. Step 1: 44–71 inch lbs. (5–8 Nm).
   b. Step 2: 10.4–12.5 ft. lbs. (14–17 Nm).
17. Remove the engine jack and attachment.
18. Install the remaining components in the reverse order of removal.
19. Connect the negative battery cable.

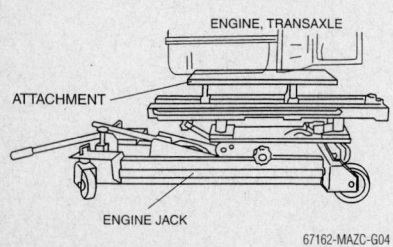

Support the engine using an engine jack and attachment as shown

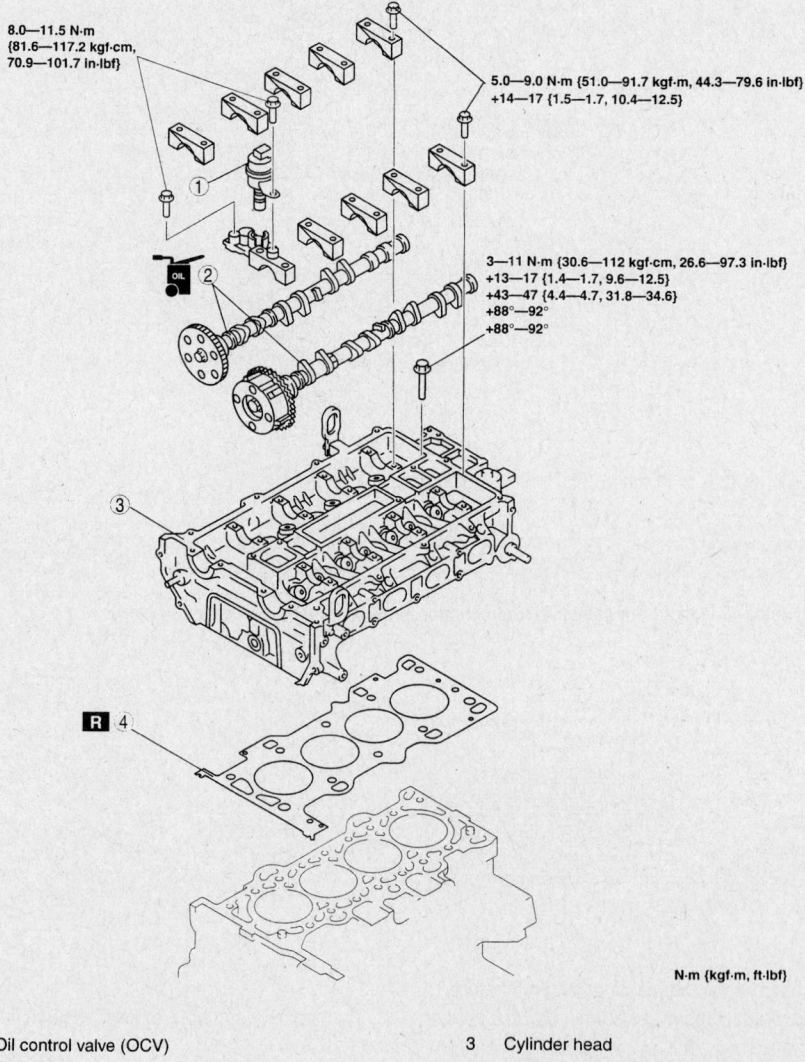

| 1 | Oil control valve (OCV) | 3 | Cylinder head |
| 2 | Camshaft | 4 | Cylinder head gasket |

Exploded view of the camshaft and cylinder head components—Mazda6 models with the 2.3L (L3) engine

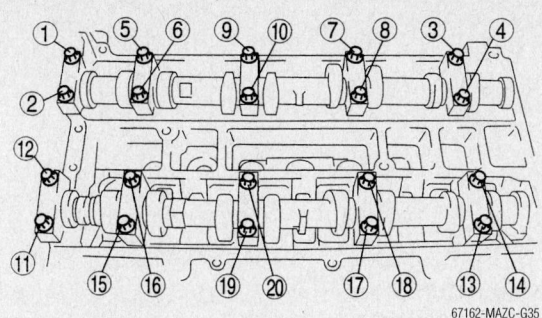

Camshaft cap bolt removal sequence—Mazda6 models with the 2.3L (L3) engine

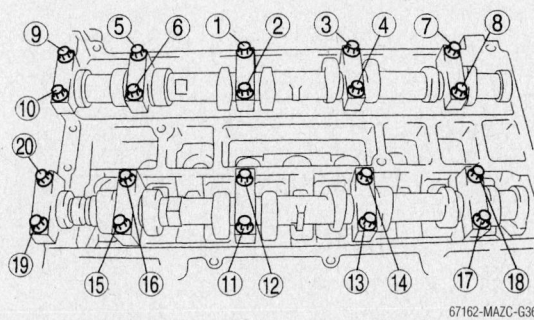

Camshaft cap bolt tightening sequence—Mazda6 models with the 2.3L (L3) engine

20. Fill and bleed the cooling system.
21. Change the oil and filter.
22. Run the engine and check for proper operation.

3.0L (AJ) ENGINE

1. Before servicing the vehicle, refer to the precautions section.
2. Properly relieve the fuel system pressure.
3. Drain the cooling system.
4. Disconnect the negative battery cable.
5. Remove the water pump.
6. Remove the timing chain.
7. Support the engine using an engine jack and attachment.
8. Remove the water bypass tube stud bolt and bolt and disconnect the tube from the cylinder head.
9. Remove the camshaft(s).
10. Remove the rocker arm(s).
11. Remove the camshafts.

➡If removing the right hand camshaft caps, remove thrust caps 1R and 5R first. Do not loosen any of the other cap bolts until the thrust caps are removed or you could damage the thrust caps. If removing the left hand camshaft caps, remove thrust caps 1L and 6L first. Do not loosen any of the other cap bolts until the thrust caps are removed or you could damage the thrust caps.

12. Loosen the camshaft caps using 7–8 passes in sequence after first removing the thrust caps to allow the camshafts to be slowly raised. Keep the caps in the order they were removed so they are re-installed in their original positions.

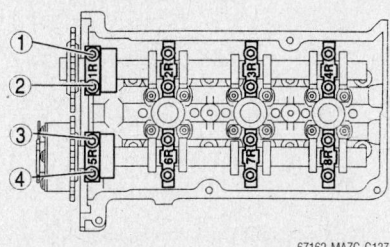

If removing the right hand camshaft caps, remove thrust caps 1R and 5R first— Mazda6 with the 3.0L (AJ) engine

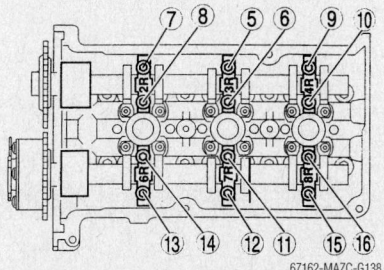

Loosen the right hand camshaft caps using 7–8 passes in this sequence after first removing the thrust caps—Mazda6 with the 3.0L (AJ) engine

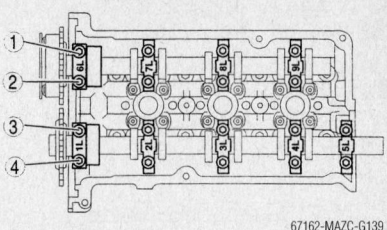

If removing the left hand camshaft caps, remove thrust caps 1L and 6L first— Mazda6 with the 3.0L (AJ) engine

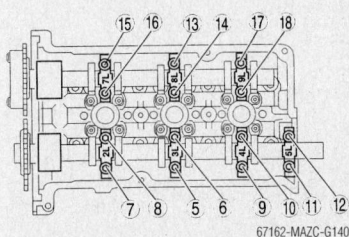

Loosen the left hand camshaft caps using 7–8 passes in this sequence after first removing the thrust caps—Mazda6 with the 3.0L (AJ) engine

To install:

13. Install the camshafts in their original positions.
14. Install the left hand camshaft as follows:

a. Place the crankshaft keyway at the 11 O'clock position by rotating the crankshaft clockwise.
b. Install the camshaft and the caps in their original positions and hand tighten the bolts.
c. Position the mark on the intake camshaft to the 9 O'clock position as illustrated.
d. Position the mark on the exhaust camshaft to the 12 O'clock position.
e. Install the timing chain.

➡Tighten the camshaft journal thrust caps last to avoid damaging the thrust caps.

f. Tighten the camshaft caps using several passes to 71–106 inch lbs. (8–12 Nm). After adjust the camshaft endplay using the thrust caps 1L and 6L, tighten the remaining caps.

15. Install the right hand camshaft as follows:

a. Install the camshaft and the caps in their original positions and hand tighten the bolts.
b. Place the crankshaft keyway at the 3 O'clock position.
c. Position the mark on the exhaust

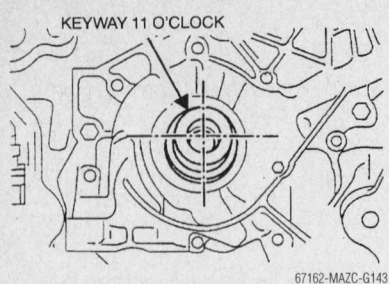

When installing the left hand camshaft place the crankshaft keyway at the 11 O'clock position—Mazda6 with the 3.0L (AJ) engine

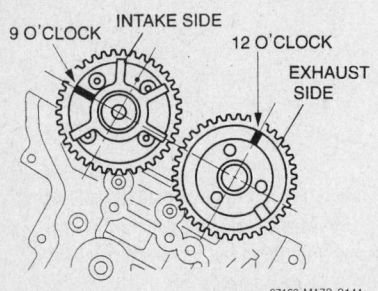

Install the left hand camshaft and position the mark on the intake camshaft to the 9 O'clock and the exhaust camshaft to the 12 O'clock position—Mazda6 with the 3.0L (AJ) engine

camshaft to the 12 O'clock position by rotating the crankshaft clockwise.

d. Position the mark on the intake camshaft to the 3 O'clock position as illustrated.

e. Install the timing chain.

➡Tighten the camshaft journal thrust caps last to avoid damaging the thrust caps.

f. Tighten the camshaft caps using several passes to 71–106 inch lbs. (8–12 Nm). After adjust the camshaft endplay using the thrust caps 1R and 5R, tighten the remaining caps.

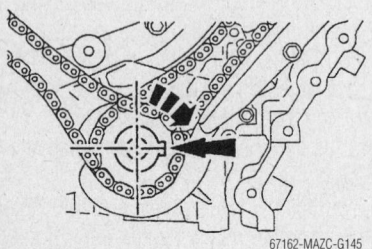

When installing the right hand camshaft place the crankshaft keyway at the 3 O'clock position—Mazda6 with the 3.0L (AJ) engine

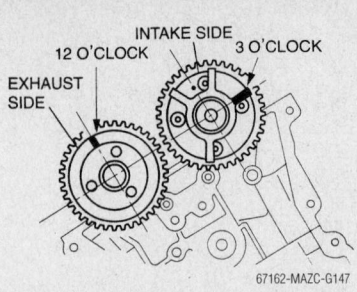

Install the right hand camshaft and position the mark on the intake camshaft to the 3 O'clock and the exhaust camshaft to the 12 O'clock position—Mazda6 with the 3.0L (AJ) engine

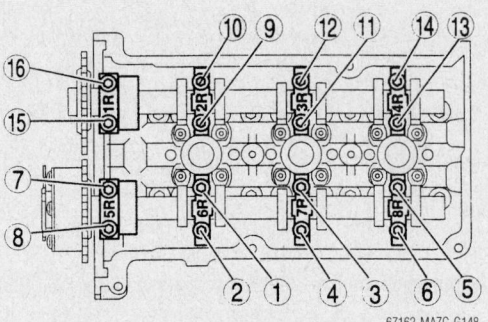

Right side camshaft cap torque sequence—Mazda6 with the 3.0L (AJ) engine

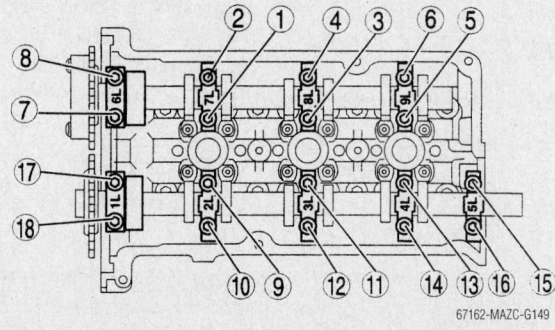

Left side camshaft cap torque sequence—Mazda6 with the 3.0L (AJ) engine

16. Install the water bypass tube.
17. Remove the engine jack and attachment.
18. Install the water pump.
19. Fill the cooling system.
20. Fill and bleed the cooling system.
21. Change the oil and filter.
22. Run the engine and check for proper operation.

Valve Lash

These engines use solid cam followers with a removable adjustment shim. The valve lash clearance is measured with the original shim installed and checked against the specification. If adjustment is necessary, the original shim is removed, and a thicker

or thinner shim is installed to obtain the proper clearance. Special tools are required in order to adjust the shim without removing the camshaft.

INSPECTION

MX-5 Miata

The valve clearance should be measured with the engine cold, the specifications are:

- Intake: 22–28mm (0.0087–0.0110 in.)
- Exhaust: 0.27–0.33 (0.0107–0.0129 in.)

1. Before servicing the vehicle, refer to the precautions section.
2. Remove the battery cover.
3. Disconnect the negative battery cable.
4. Remove the plug hole plate.
5. Disconnect the ventilation hose.
6. Remove the front suspension tower bar.
7. Remove the CMP sensor.
8. Disconnect the OCV connector.
9. Disconnect the P/S pressure switch connector.
10. Remove the ignition coils.
11. Remove the cylinder head cover.
12. Measure the valve clearance.
13. Turn the crankshaft clockwise so that

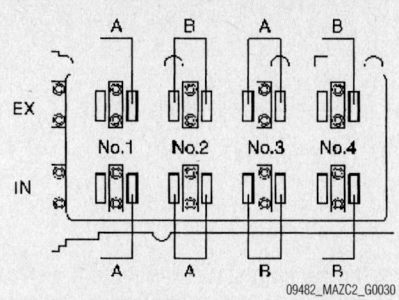

Measure the valve clearance at points A and B—MX-5 Miata

the No.1 piston is at TDC of the compression stroke.

14. Measure the valve clearance at point A in the illustration.

15. If the valve clearance is out of the specification, adjust it.

➡**Make sure to note down the measured values for choosing the suitable replacement tappets.**

16. Turn the crankshaft 360 degrees clockwise so that the No.4 piston is at TDC of the compression stroke.

17. Measure the valve clearance at point B in the illustration.

18. If the valve clearance is out of the specification, adjust it.

19. Install the cylinder head cover.

20. Install the ignition coils.

21. Connect the P/S pressure switch connector.

22. Connect the OCV connector.

23. Install the CMP sensor.

24. Install the front suspension tower bar.

25. Connect the ventilation hose.

26. Install the plug hole plate.

27. Connect the negative battery cable.

28. Install the battery cover.

REMOVAL & INSTALLATION

MX-5 Miata

The valve clearance should be measured with the engine cold, the specifications are:
- Intake: 22–28mm (0.0087–0.0110 in.)
- Exhaust: 0.27–0.33 (0.0107–0.0129 in.)

1. Before servicing the vehicle, refer to the precautions section.

2. Remove the battery cover.

3. Disconnect the negative battery cable.

4. Remove the plug hole plate.

5. Disconnect the ventilation hose.

6. Remove the front suspension tower bar.

7. Remove the CMP sensor.

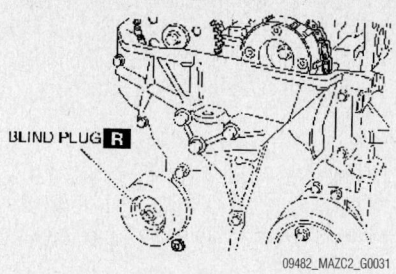

Remove the engine front cover lower blind plug—MX-5 Miata

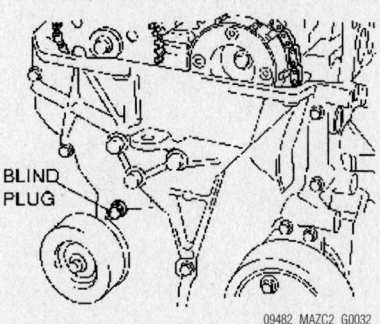

Remove the engine front cover upper blind plug—MX-5 Miata

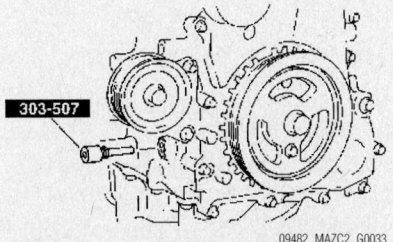

Remove the cylinder block upper blind plug and install tool 303-507—MX-5 Miata

8. Disconnect the OCV connector.

9. Disconnect the P/S pressure switch connector.

10. Remove the ignition coils.

11. Remove the cylinder head cover.

12. Remove the drive belt.

13. Remove the engine front cover lower blind plug.

14. Remove the engine front cover upper blind plug.

15. Remove the cylinder block lower blind plug.

16. Install tool 303-507 as shown.

17. Turn the crankshaft clockwise until the crankshaft is in the No.1 cylinder TDC position.

18. Loosen the timing chain.

19. Using a suitable screwdriver or equivalent tool, unlock the chain tensioner ratchet.

20. Turn the exhaust camshaft clockwise

using a suitable wrench on the cast hexagon and loosen the timing chain.

21. Place a M6 X 1 length bolt at the engine front cover upper blind plug, secure the chain guide at the position where the tension is released.

22. Hold the exhaust camshaft using a suitable wrench on the cast hexagon as illustrated.

23. Remove the exhaust camshaft sprocket.

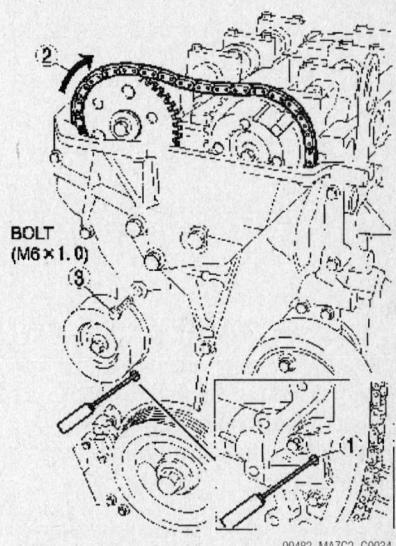

Place a M6 X 1 length bolt at the engine front cover upper blind plug, secure the chain guide at the position where the tension is released—MX-5 Miata

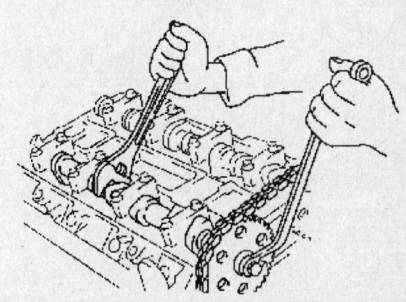

Hold the exhaust camshaft using a suitable wrench on the cast hexagon—MX-5 Miata

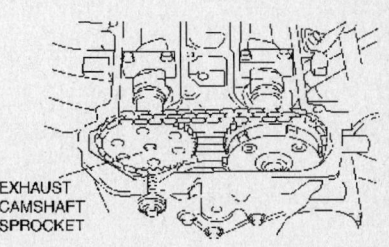

Remove the exhaust camshaft sprocket—MX-5 Miata

24. Loosen the camshaft cap bolts in several passes in the order shown.

➡ **The cylinder head and the camshaft caps are numbered to make sure they are reassembled in their original position. Do not mix the caps.**

25. Remove the camshafts.
26. Remove the tappet.
27. Select proper adjustment tappet using the following formula:

    a. New adjustment tappet = removed tappet thickness + Measured valve clearance − standard valve clearance. For the intake: 0.25 mm (0.0098 in.). For the exhaust: 0.30 mm (0.0118 in.).

28. Install the camshaft with No.1 cylinder camshaft lobes aligned with the TDC position.
29. Tighten the camshaft cap bolt using two steps, first to 5–9 Nm (44–80 in. lbs.), then to 14–17 Nm (10–12.5 ft. lbs.).
30. Install a new washer.
31. Install the exhaust camshaft sprocket.

➡ **Do not tighten the bolt for the camshaft sprocket during this step. First confirm the valve timing, then tighten the bolt.**

32. Install the tool on the camshaft as shown.
33. Remove the M6 X 1.0 bolt from the engine front cover to apply tension to the timing chain.
34. Turn the crankshaft clockwise until

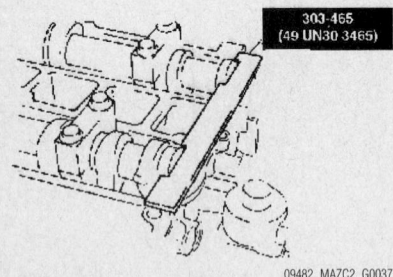

09482_MAZC2_G0037

Install the tool illustrated on the camshaft—MX-5 Miata

09482_MAZC2_G0038

Install a new washer—MX-5 Miata

the crankshaft is in the No.1 cylinder TDC position.

35. Hold the exhaust camshaft using a suitable wrench on the cast hexagon as shown.
36. Tighten the exhaust camshaft sprocket lock bolt to 69–75 Nm (51–55 ft. lbs.)
37. Remove the tool from the camshaft.
38. Remove the tool from the block lower blind plug.
39. Rotate the crankshaft clockwise two turns until the TDC position.
40. If not aligned, loosen the crankshaft pulley lock bolt and repeat the procedure.
41. Apply silicone sealant to the engine front cover upper blind plug.
42. Install the engine front cover upper blind plug and tighten to 8–11 Nm (71–101 in. lbs.)
43. Install the cylinder block lower blind plug and tighten to 18–2 Nm (13–16 ft. lbs.)
44. Install the new engine front cover lower blind plug and tighten to 10–14 Nm (89–123 ft. lbs.)
45. Install the drive belt.
46. Measure the valve clearance.
47. Turn the crankshaft clockwise so that the No.1 piston is at TDC of the compression stroke.
48. Measure the valve clearance at point A in the illustration.
49. If the valve clearance is out of the specification, adjust it.
50. Turn the crankshaft 360 degrees clockwise so that the No.4 piston is at TDC of the compression stroke.
51. Measure the valve clearance at B in the illustration.
52. If the valve clearance is out of specification, adjust it.
53. Install the cylinder head cover.
54. Install the ignition coils.
55. Connect the OCV connector.
56. Install the CMP sensor.
57. Install the front suspension tower bar.

58. Connect the ventilation hose.
59. Install the air cleaner.
60. Install the plug hole plate.
61. Install the battery and battery tray.

2.0L (LF) Engine

1. Before servicing the vehicle, refer to the precautions section.

➡ **With the engine cold, standard valve clearance is 0.087–0.0110 inch (0.22–0.28mm) on the intake side and 0.0107–0.0129 inch (0.27–0.33mm) on the exhaust side.**

2. Before servicing the vehicle, refer to the precautions section.
3. Remove the plug hole plate.
4. Remove the battery cover and disconnect the negative battery cable.
5. Disconnect the wiring harness.
6. Remove the right front wheel, engine under cover and splash shield.
7. Remove the ignition coils.
8. Remove the Positive Crankcase Ventilation (PCV) hose, if equipped.
9. Disconnect the oil control valve connector, if equipped.
10. Remove the cylinder head cover.
11. Measure the valve clearance by turning the crankshaft clockwise until the No. 1 piston is at Top Dead Center (TDC).
12. Measure the valve clearance at **A**. If the clearance exceeds specifications, replace the adjustment shim.
13. Turn the crankshaft clockwise 360 degrees until the No. 4 piston is at TDC. Measure the valve clearance at **B**. If the clearance exceeds specifications, replace the adjustment shim.
14. Repeat this procedure for all the camshafts.
15. Remove the right side halfshaft from the joint shaft on Mazda6 models.
16. Remove the engine front cover lower and upper blind plugs and the cylinder block lower blind plug.
17. Install tool 49 JE01 061.

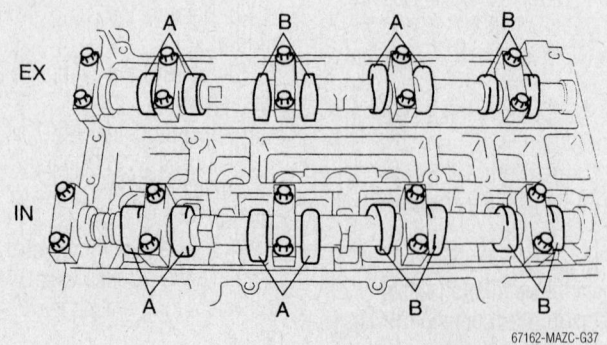

67162-MAZC-G37

Valve clearance checking positions—2.0L (LF) and 2.3L (L3) engines

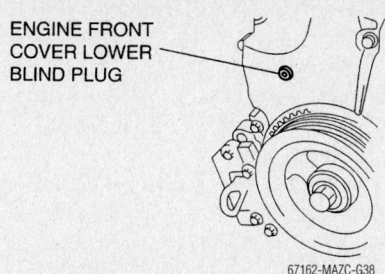

Remove the engine front cover lower blind plug—2.0L (LF) and 2.3L (L3) engines

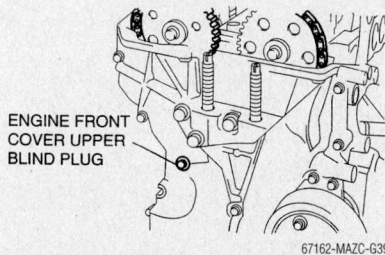

Remove the engine front cover upper blind plug—2.0L (LF) and 2.3L (L3) engines

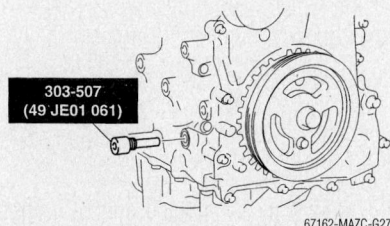

Install tool 49 JE01 061—2.0L (LF) and 2.3L (L3) engines

18. Turn the crankshaft clockwise until the crankshaft is in the number 1 cylinder Top Dead Center position (the balance weight should be attached to tool 49 JE01 061).

19. Loosen the timing chain as follows:

a. Unlock the chain tensioner ratchet using a screwdriver.

b. Turn the exhaust camshaft clockwise using a wrench on the cast hexagon portion of the camshaft and loosen the timing chain.

c. Place a M6 x 1.0 length bolt: 0.99–1.37 inch (25–35mm) at the engine front cover upper blind plug and secure the timing chain guide at the location where the chain tension is released.

20. Hold the exhaust camshaft using a wrench on the cast hexagon portion of the camshaft and remove the exhaust camshaft sprocket.

➡The cylinder head and camshaft caps are numbered and must be reassembled in their original locations. When

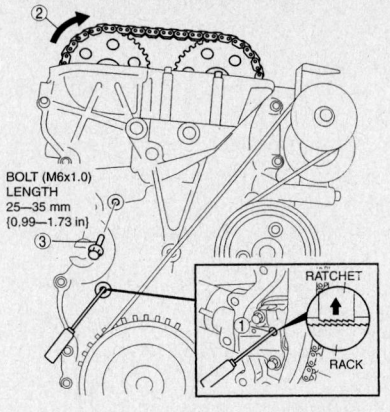

Loosen the timing chain tension, refer to the text for procedure—2.0L (LF) and 2.3L (L3) engines

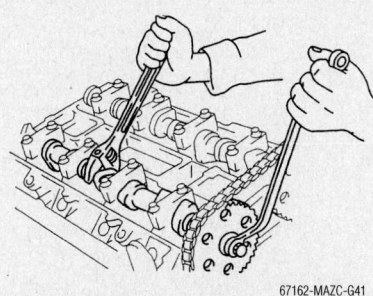

Hold the exhaust camshaft using a wrench on the cast hexagon portion of the camshaft and remove the exhaust camshaft sprocket—2.0L (LF) and 2.3L (L3) engines

removing a when removed, keep the caps with the cylinder head they were removed from, do not switch the caps.

21. Loosen the camshaft cap bolts using 2–3 passes in the sequence illustrated.

22. Once all the bolts have been removed, remove the caps, keeping them in the original positions and remove the camshafts.

23. Remove the tappets and select the proper adjustment shim. The formula for selecting a shim is as follows:

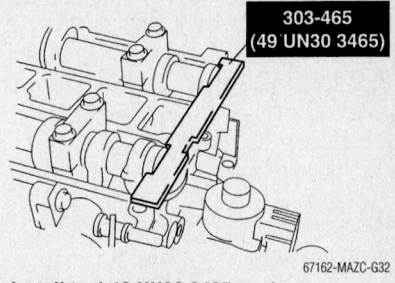

Install tool 49 UN30 3465 on the camshaft—2.0L (LF) and 2.3L (L3) engines

a. Removed shim thickness + measured valve clearance–standard valve clearance 0.0098 inch (0.25mm) for the intake and 0.0118 inch (0.30mm) for the exhaust.

24. Install the adjustment shim.

25. Set the cam position of the number 1 cylinder at Top Dead center (TDC) and install the camshaft.

26. Temporarily install the camshaft caps in their original positions.

27. Tighten the camshaft cap bolt in two steps:

a. Step 1: 44–71 inch lbs. (5–8 Nm).

b. Step 2: 10.4–12.5 ft. lbs. (14–17 Nm).

28. Install the exhaust camshaft sprocket and hand tighten the bolt.

29. Install tool 49 UN30 3465 as illustrated on the camshaft.

30. Remove the M6 x 1.0 bolt from the engine front cover to tension the timing chain.

31. Turn the crankshaft clockwise until the crankshaft is in the number 1 cylinder Top Dead Center position (the balance weight should be attached to tool 49 JE01 061).

32. Hold the exhaust camshaft using a wrench on the cast hexagon portion of the camshaft and tighten the exhaust camshaft sprocket bolt to 51–55 ft. lbs. (69–75 Nm).

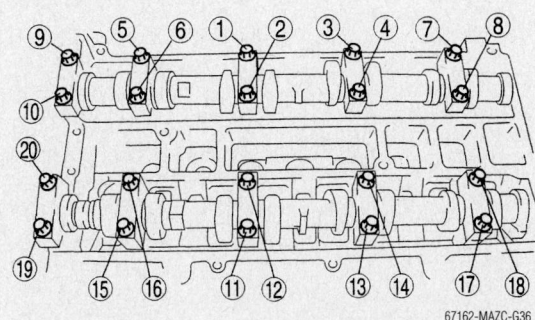

Camshaft cap bolt tightening sequence—2.0L (LF) and 2.3L (L3) engines

33. Remove the tools from the camshaft and from the block lower blind plug.

34. Rotate the engine two full turns clockwise to TDC. If not aligned, remove the crankshaft pulley lockbolt, remove the tappet and repeat the adjustment shim selection and replacement procedure.

35. Apply sealant to the engine front cover upper blind plug and tighten to 71–101 inch lbs. (8–11 Nm).

36. Install the cylinder block lower blind plug and tighten to 13–16 ft. lbs. (18–22 Nm).

37. Install a new engine front cover blind plug and tighten to 7.4–10 ft. lbs. (10–14 Nm).

38. Install the right side halfshaft from the joint shaft on Mazda6 models.

39. Install the cylinder head cover.

40. Install the PCV hose, if equipped.

41. Connect the oil control valve connector, if equipped.

42. Install the ignition coils.

43. Connect the wiring harness.

44. Connect the negative battery cable and install the battery cover.

45. Install the plug hole plate.

46. Install the splash shield, under cover and wheel.

2.3L (L3) Engine

1. Before servicing the vehicle, refer to the precautions section.

➡ **With the engine cold, standard valve clearance is 0.087–0.0110 inch (0.22–0.28mm) on the intake side and 0.0107–0.0129 inch (0.27–0.33mm) on the exhaust side.**

2. Before servicing the vehicle, refer to the precautions section.

3. Remove the plug hole plate.

4. Remove the battery cover and disconnect the negative battery cable.

5. Disconnect the wiring harness.

6. Remove the ignition coils.

7. Disconnect the Oil Control Valve (OCV) connector.

8. Remove the right front wheel, under cover and splash shield.

9. Remove the Positive Crankcase Ventilation (PCV) hose.

10. Remove the cylinder head cover.

11. Measure the valve clearance by turning the crankshaft clockwise until the No. 1 piston is at Top Dead Center (TDC).

12. Measure the valve clearance at **A**. If the clearance exceeds specifications, replace the adjustment shim.

13. Turn the crankshaft clockwise 360 degrees until the No. 4 piston is at TDC.

Measure the valve clearance at **B**. If the clearance exceeds specifications, replace the adjustment shim.

14. Repeat this procedure for all the camshafts.

15. Remove the engine front cover lower and upper blind plugs and the cylinder block lower blind plug.

16. Install tool 49 JE01 061.

17. Turn the crankshaft clockwise until the crankshaft is in the number 1 cylinder Top Dead Center position (the balance weight should be attached to tool 49 JE01 061).

18. Loosen the timing chain as follows:

a. Unlock the chain tensioner ratchet using a screwdriver.

b. Turn the exhaust camshaft clockwise using a wrench on the cast hexagon portion of the camshaft and loosen the timing chain.

c. Place a M6 x 1.0 length bolt: 0.99–1.37 inch (25–35mm) at the engine front cover upper blind plug and secure the timing chain guide at the location where the chain tension is released.

19. Hold the exhaust camshaft using a wrench on the cast hexagon portion of the camshaft and remove the exhaust camshaft sprocket.

➡ **The cylinder head and camshaft caps are numbered and must be reassembled in their original locations. When removing a when removed, keep the caps with the cylinder head they were removed from, do not switch the caps.**

20. Loosen the camshaft cap bolts using 2–3 passes in the sequence illustrated.

21. Once all the bolts have been removed, remove the caps, keeping them in the original positions and remove the camshafts.

22. Remove the tappets and select the proper adjustment shim. The formula for selecting a shim is as follows:

a. Removed shim thickness + measured valve clearance − standard valve clearance 0.0098 inch (0.25mm) for the intake and 0.0118 inch (0.30mm) for the exhaust.

23. Install the adjustment shim.

24. Set the cam position of the number 1 cylinder at Top Dead center (TDC) and install the camshaft.

25. Temporarily install the camshaft caps in their original positions.

26. Tighten the camshaft cap bolt in two steps:

a. Step 1: 44–71 inch lbs. (5–8 Nm).

b. Step 2: 10.4–12.5 ft. lbs. (14–17 Nm).

27. Install the exhaust camshaft sprocket and hand tighten the bolt.

28. Install tool 49 UN30 3465 as illustrated on the camshaft.

29. Remove the M6 x 1.0 bolt from the engine front cover to tension the timing chain.

30. Turn the crankshaft clockwise until the crankshaft is in the number 1 cylinder Top Dead Center position (the balance weight should be attached to tool 49 JE01 061).

31. Hold the exhaust camshaft using a wrench on the cast hexagon portion of the camshaft and tighten the exhaust camshaft sprocket bolt to 51–55 ft. lbs. (69–75 Nm).

32. Remove the tools from the camshaft and from the block lower blind plug.

33. Rotate the engine two full turns clockwise to TDC. If not aligned, remove the crankshaft pulley lockbolt, remove the tappet and repeat the adjustment shim selection and replacement procedure.

34. Apply sealant to the engine front cover upper blind plug and tighten to 71–101 inch lbs. (8–11 Nm).

35. Install the cylinder block lower blind plug and tighten to 13–16 ft. lbs. (18–22 Nm).

36. Install a new engine front cover blind plug and tighten to 7.4–10 ft. lbs. (10–14 Nm).

37. Install the cylinder head cover.

38. Install the PCV hose.

39. Connect the OCV connector.

40. Install the ignition coils.

41. Connect the wiring harness.

42. Connect the negative battery cable and install the battery cover.

43. Install the plug hole plate.

44. Install the under cover, splash shield and wheel.

3.0L (AJ) Engine

1. Before servicing the vehicle, refer to the precautions section.

2. Remove or disconnect the following:

- Negative battery cable
- Camshaft followers
- Hydraulic lash adjusters

➡ **Mark the position of the hydraulic lash adjusters to assure they are assembled in their original position**

3. Inspect the adjusters for scoring or uneven wear in the bore and replace them as required.

To install:

4. Install or connect the following:
- Hydraulic lash adjusters after lubricating them with clean engine oil
- Camshaft followers
- Negative battery cable

Mazdaspeed6

1. Before servicing the vehicle, refer to the precautions section.

➡ **With the engine cold, standard valve clearance is 0.087–0.0110 inch (0.22–0.28mm) on the intake side and 0.0107–0.0129 inch (0.27–0.33mm) on the exhaust side.**

2. Disconnect the negative battery cable.
3. Remove the right side front tire.
4. Remove the under cover.
5. Remove the right side splash shield.

➡ **If the high pressure fuel pump is removed, replace the O-ring with a new one.**

6. Remove the charge air cooler cover.
7. Disconnect the spill valve control solenoid valve connector.
8. Disconnect the quick release connector on the high pressure fuel pump.
9. Remove the battery and battery tray.

✳✳ WARNING

If the high pressure fuel pump joint nut is loosened, fuel leakage may occur resulting in death or serious injury, or damage to the equipment or the vehicle. Fuel can also irritate the skin and eyes. When removing the high pressure line pipe, always tighten the high pressure line pipe installation nut while fixing the high pressure fuel pump joint nut with a wrench. If the high pressure fuel pump joint nut has rotated, replace the high pressure fuel pump with a new one.

10. Disconnect the high pressure line pipe of the high pressure fuel pump.
11. Fix the joint nut with a wrench on the high pressure fuel pump side.
12. Loosen the high pressure line pipe installation nut.
13. Drain and recycle the engine coolant.
14. Loosen the water outlet case installation bolts securing the high pressure line pipe.
15. Remove the high pressure fuel pump.
16. Remove the high pressure fuel pump cover.
17. Remove the ignition coils, spark plugs, valve cover and accessory drive belt.
18. Measure the valve clearance by turning the crankshaft clockwise until the No. 1 piston is at Top Dead Center (TDC).
19. Measure the valve clearance at **A**. If the clearance exceeds specifications, replace the adjustment shim.
20. Turn the crankshaft clockwise 360 degrees until the No. 4 piston is at TDC. Measure the valve clearance at **B**. If the clearance exceeds specifications, replace the adjustment shim.
21. Repeat this procedure for all the camshafts.
22. Remove the engine front cover lower and upper blind plugs and the cylinder block lower blind plug.
23. Install tool 49 JE01 061.
24. Turn the crankshaft clockwise until the crankshaft is in the number 1 cylinder Top Dead Center position (the balance weight should be attached to tool 49 JE01 061).

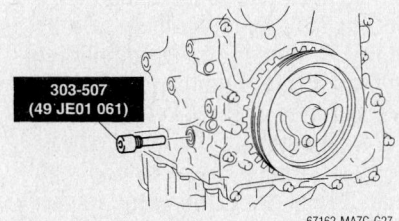

303-507 (49 JE01 061)

67162-MAZC-G27

Install tool 49 JE01 061—Mazdaspeed6

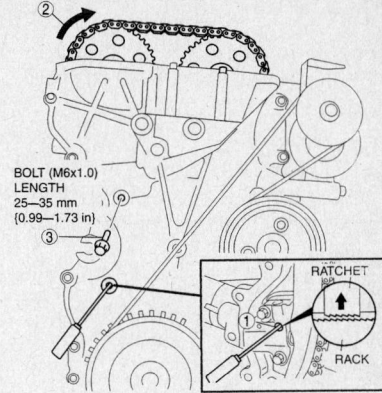

BOLT (M6x1.0) LENGTH 25—35 mm (0.99—1.73 in)

RATCHET

RACK

67162-MAZC-G40

Loosen the timing chain tension, refer to the text for procedure—Mazdaspeed6

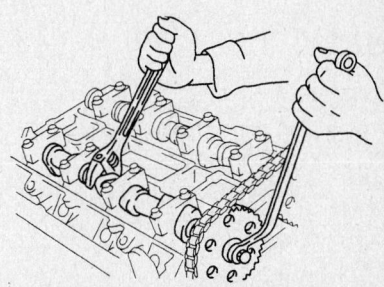

67162-MAZC-G41

Hold the exhaust camshaft using a wrench on the cast hexagon portion of the camshaft and remove the exhaust camshaft sprocket—Mazdaspeed6

25. Loosen the timing chain as follows:
   a. Unlock the chain tensioner ratchet using a screwdriver.
   b. Turn the exhaust camshaft clockwise using a wrench on the cast hexagon portion of the camshaft and loosen the timing chain.
   c. Place a M6 x 1.0 length bolt: 0.99–1.37 inch (25–35mm) at the engine front cover upper blind plug and secure the timing chain guide at the location where the chain tension is released.
26. Hold the exhaust camshaft using a wrench on the cast hexagon portion of the camshaft and remove the exhaust camshaft sprocket.

➡ **The cylinder head and camshaft caps are numbered and must be reassembled in their original locations. When removing a when removed, keep the caps with the cylinder head they were removed from, do not switch the caps.**

27. Loosen the camshaft cap bolts using 2–3 passes in the sequence illustrated.
28. Once all the bolts have been removed, remove the caps, keeping them in

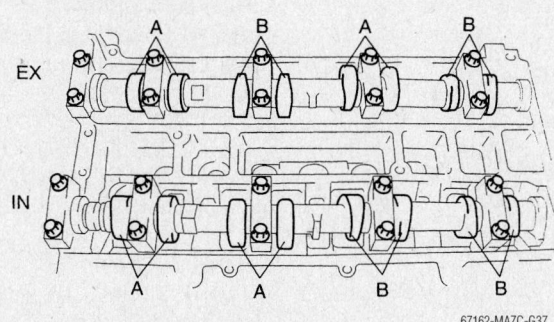

EX

IN

A B A B

A A B B

67162-MAZC-G37

Valve clearance checking positions—Mazdaspeed6

the original positions and remove the camshafts.

29. Remove the tappets and select the proper adjustment shim. The formula for selecting a shim is as follows:

a. Removed shim thickness + measured valve clearance – standard valve clearance 0.0098 inch (0.25mm) for the intake and 0.0118 inch (0.30mm) for the exhaust.

30. Install the adjustment shim.

31. Set the cam position of the number 1 cylinder at Top Dead center (TDC) and install the camshaft.

32. Temporarily install the camshaft caps in their original positions.

33. Tighten the camshaft cap bolt in two steps:

a. Step 1: 44–71 inch lbs. (5–8 Nm).

b. Step 2: 10.4–12.5 ft. lbs. (14–17 Nm).

34. Install the exhaust camshaft sprocket and hand tighten the bolt.

35. Install tool 49 UN30 3465 as illustrated on the camshaft.

36. Remove the M6 x 1.0 bolt from the engine front cover to tension the timing chain.

37. Turn the crankshaft clockwise until the crankshaft is in the number 1 cylinder Top Dead Center position (the balance weight should be attached to tool 49 JE01 061).

38. Hold the exhaust camshaft using a

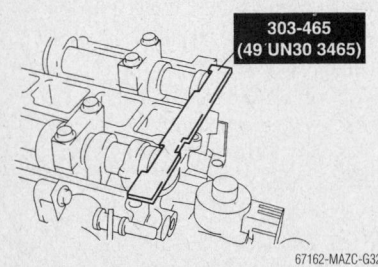

**303-465
(49 UN30 3465)**

67162-MAZC-G32

Install tool 49 UN30 3465 on the camshaft—Mazdaspeed6

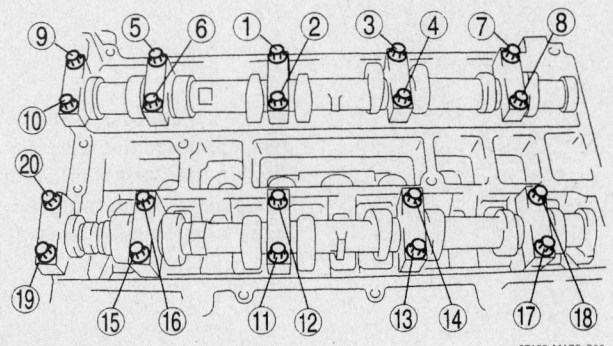

67162-MAZC-G36

Camshaft cap bolt tightening sequence—Mazdaspeed6

wrench on the cast hexagon portion of the camshaft and tighten the exhaust camshaft sprocket bolt to 51–55 ft. lbs. (69–75 Nm).

39. Remove the tools from the camshaft and from the block lower blind plug.

40. Rotate the engine two full turns clockwise to TDC. If not aligned, remove the crankshaft pulley lockbolt, remove the tappet and repeat the adjustment shim selection and replacement procedure.

41. Apply sealant to the engine front cover upper blind plug and tighten to 71–101 inch lbs. (8–11 Nm).

42. Install the cylinder block lower blind plug and tighten to 13–16 ft. lbs. (18–22 Nm).

43. Install a new engine front cover blind plug and tighten to 7.4–10 ft. lbs. (10–14 Nm).

44. Install the high pressure fuel pump cover and tighten to 13–16 ft. lbs. (18–22 Nm).

> ※※ **CAUTION**
>
> If the high pressure fuel pump installation bolts are tightened with the high pressure fuel pump tilted, the high pressure fuel pump may not operate correctly. Tighten the high pressure fuel pump installation bolts in a few passes with equal torque. Tighten to 76–101 inch lbs. (8.5–11.5 Nm).

> ※※ **WARNING**
>
> If the high pressure fuel pump joint nut is loosened, fuel leakage may occur resulting in death or serious injury, or damage to the equipment or the vehicle. Fuel can also irritate the skin and eyes. When installing the high pressure line pipe, always tighten the high pressure line pipe installation nut while fixing the high pressure fuel pump joint nut with a wrench. If the high pressure fuel pump joint nut has rotated, replace the high pressure fuel pump with a new one.

45. Assemble the high pressure line pipe.

46. Fix the joint nut with a wrench on the high pressure fuel pump side. Tighten the high pressure line pipe installation nut to 17–26 ft. lbs. (23–35 Nm).

47. Tighten the water outlet case installation bolts to 13–16 ft. lbs. (18–22 Nm).

48. Install the quick release connector.

49. Install the remaining components in the reverse order of removal.

50. Verify that the high pressure fuel pump is assembled securely.

51. Drive the vehicle starting from a standstill and brake suddenly five to six times at a low speed.

52. Stop the vehicle and verify from outside the vehicle that there is no fuel leakage around the high pressure fuel pump.

Starter Motor

REMOVAL & INSTALLATION

MX5-Miata

1. Before servicing the vehicle, refer to the precautions section.

> ※※ **CAUTION**
>
> Remove and install all parts when the engine is cold, otherwise they can cause severe burns or serious injury.

> ※※ **CAUTION**
>
> When the battery cables are connected, touching the vehicle body with starter terminal B will cause sparks. This can cause personal injury, fire, and damage to the electrical components. Make sure to always disconnect the negative battery cable before removing any electrical component.

2. Remove the battery cover.

3. Disconnect the negative battery cable.

4. Remove the left hand side cover.

5. Remove the under cover on models equipped with a manual transmission.

6. On models equipped with a manual transmission, remove the clutch release cylinder with the pipes still connected.

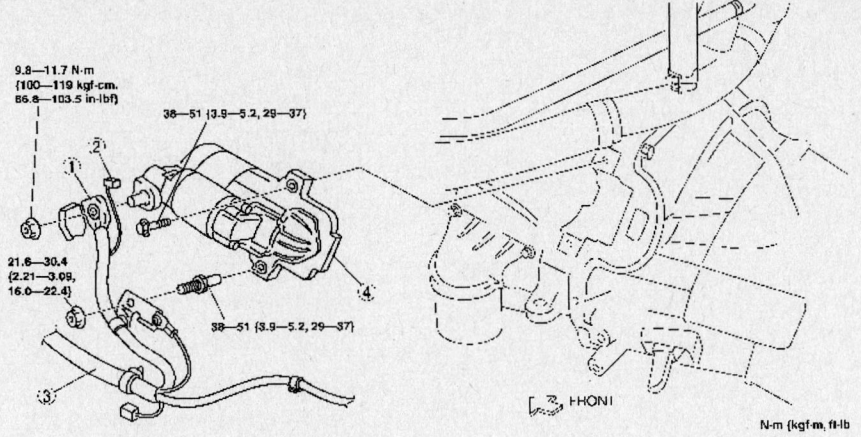

9.8—11.7 N·m
{100—119 kgf·cm,
86.8—103.5 in-lbf}

38—51 {3.9—5.2, 29—37}

21.6—30.4
{2.21—3.09,
16.0—22.4}

38—51 {3.9—5.2, 29—37}

FRONT

N·m {kgf·m, ft-lb

09482_MAZC2_G0039

Exploded view of the starter mounting—MX-5 Miata

Position the clutch release cylinder so that it is out of the way.

7. Remove the wiring from the starter, wiring harness bracket, starter bolts and the starter.

To install:

8. Installation is the reverse of removal, tighten the starter bolts to 38–51 Nm (29–37 ft. lbs.)

Mazda3

1. Before servicing the vehicle, refer to the precautions section.

2. Remove the battery cover and disconnect the negative battery cable.

3. Remove the under cover.

4. If equipped with a manual transaxle, remove the clutch release cylinder with the line still attached.

5. Remove the starter electrical connections and wiring harness bracket.

6. Remove the starter bolts and the starter.

To install:

7. Installation is the reverse of removal. Tighten the wiring harness bracket retainers to 70–104 inch lbs. (8–11 Nm) and the starter bolts to 28–38 ft. lbs. (38–51 Nm).

Mazda6

2.3L (L3) ENGINE

1. Before servicing the vehicle, refer to the precautions section.

2. Remove the battery cover and disconnect the negative battery cable.

3. Remove the under cover.

4. If equipped with a manual transaxle, remove the clutch release cylinder with the line still attached.

5. Remove the starter electrical connection.

6. Remove the starter bolts and the starter.

To install:

7. Installation is the reverse of removal. Tighten the starter bolts to 29–37 ft. lbs. (38–51 Nm).

3.0L (AJ) ENGINE

1. Before servicing the vehicle, refer to the precautions section.

2. Remove the battery and tray.

3. Disconnect the selector cable and bracket.

4. Remove the starter electrical connection.

5. Remove the starter bolts and the starter.

To install:

6. Installation is the reverse of removal. Tighten the starter bolts to 28–38 ft. lbs. (38–51 Nm).

Mazdaspeed6

1. Before servicing the vehicle, refer to the precautions section.

2. Remove the battery and tray.

3. Remove the air cleaner.

4. Remove the under cover.

5. Remove the oil filter and cooler component with the water hoses attached and set the assembly aside.

6. Remove the harness bracket.

7. Remove the heater pipe with the hoses attached and position aside.

8. Remove the starter electrical connection.

9. Remove the starter bolts and the starter.

To install:

10. Installation is the reverse of removal. Tighten the starter bolts to 29–37 ft. lbs. (38–51 Nm).

Oil Pan

REMOVAL & INSTALLATION

MX5-Miata

1. Before servicing the vehicle, refer to the precautions section.

✲✲ CAUTION

Remove and install all parts when the engine is cold, otherwise they can cause severe burns or serious injury.

2. Remove the battery and battery tray.

3. Remove the air cleaner.

4. Drain the engine oil.

5. Loosen the water pump pulley bolt and remove the drive belt.

6. Remove the front suspension tower bar from the joint, right side and left side.

7. Remove the plug hole plate.

8. Remove the ignition coils.

9. Remove the P/S oil pump with the hose and pipe sill connected. Position the P/S oil pump out of the way.

10. Remove the Crankshaft Position (CKP) sensor.

11. Remove the engine front cover.

12. Remove the transverse member.

13. Remove the member bracket if equipped with a manual transmission.

14. Remove the windshield wiper arm.

15. Remove the cowl grille.

16. Remove the side cowl grille.

17. Remove the engine compartment service hole cover.

18. Remove the front tires.

19. Support the engine using the an engine support tool such as 49 ED17 5AO.

20. Remove the engine mount rubber installation nuts

21. Lift up the engine approx. 25 mm (0.98 in.) to assure clearance for the oil pan removal, then remove the oil pan bolts.

22. Remove the oil pan using a separator tool to break the seal between the gasket and the pan/block.

To install:

➡ **Apply the silicon sealant in a single, unbroken line around the whole perimeter.**

➡ **Using bolts with the old seal adhering could cause cracks in the housing.**

23. Completely clean and remove any oil, dirt, sealant or other foreign material that may be adhering to the housing and oil pan.

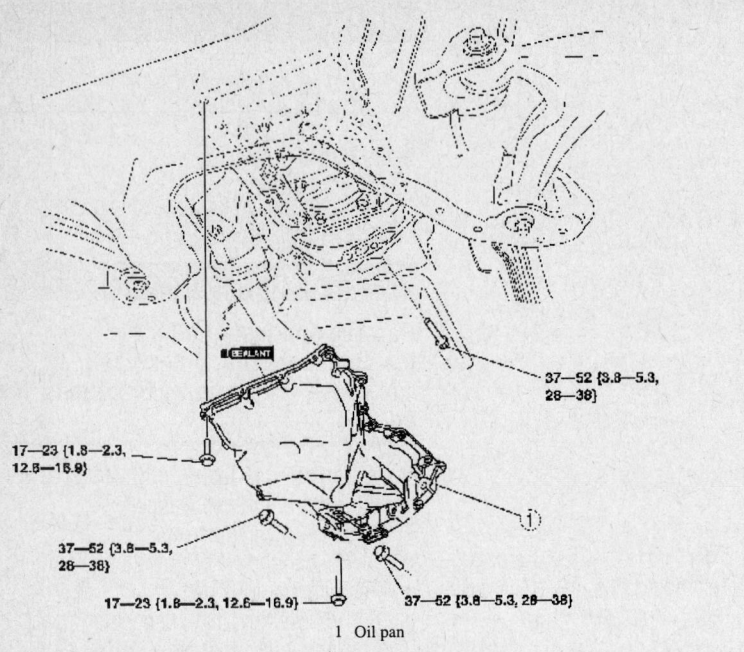

17—23 {1.8—2.3, 12.6—16.9}

37—52 {3.8—5.3, 28—38}

37—52 {3.8—5.3, 28—38}

17—23 {1.8—2.3, 12.6—16.9}

37—52 {3.8—5.3, 28—38}

N·m {kgf·m, ft·lb}

09482_MAZC2_G0040

1 Oil pan

Exploded view of the oil pan mounting—MX-5 Miata

24. When reusing the oil pan installation bolts, clean any old sealant from the bolts.

25. Use a square ruler to align the oil pan and the cylinder block junction side on the engine front cover side.

26. Apply a 2.2–3.2 mm (0.087–0.126 in.) bead of silicone sealant to the oil pan along the inside of the bolt holes as shown in the illustration.

27. Tighten the bolts in the sequence illustrated to 17–23 Nm (12.6–16.9 ft. lbs.).

28. Tighten the oil pan to transmission

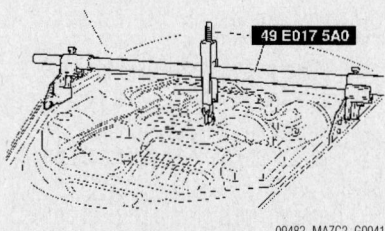

49 E017 5A0

09482_MAZC2_G0041

Support the engine using the an engine support tool such as 49 ED17 5A0—MX-5 Miata

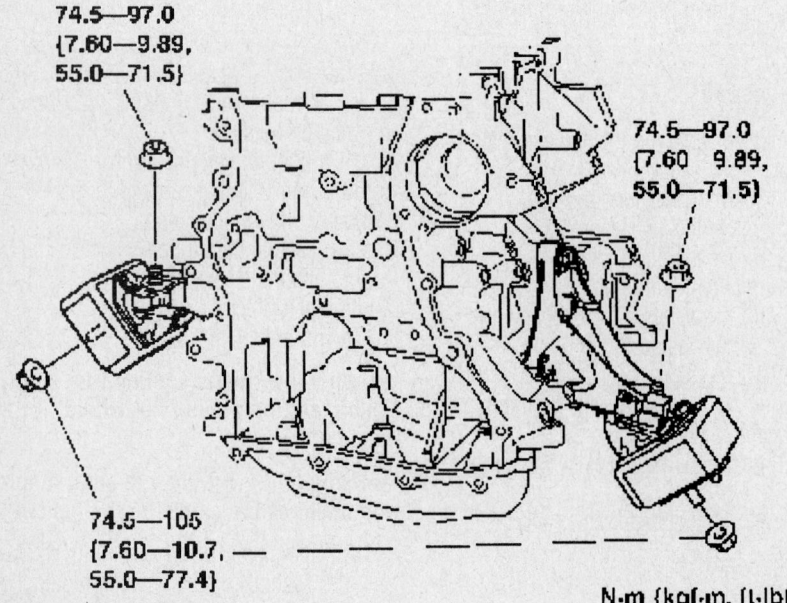

74.5—97.0 {7.60—9.89, 55.0—71.5}

74.5—97.0 {7.60—9.89, 55.0—71.5}

74.5—105 {7.60—10.7, 55.0—77.4}

N·m {kgf·m, ft·lbf}

09482_MAZC2_G0042

Remove the engine mount rubber installation nuts—MX-5 Miata

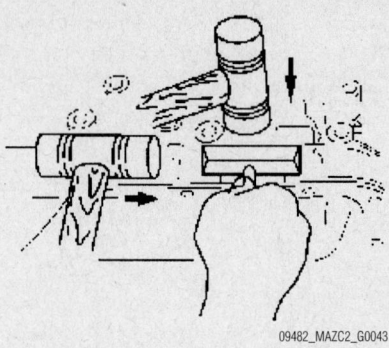

09482_MAZC2_G0043

Remove the oil pan using a separator tool to break the seal between the gasket and the pan/block—MX-5 Miata

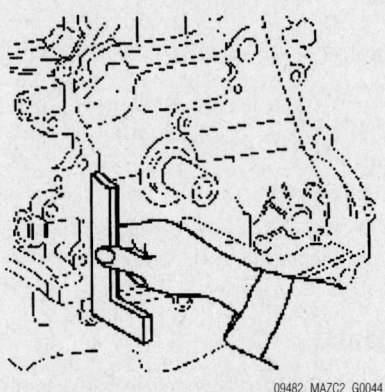

09482_MAZC2_G0044

Use a square ruler to align the oil pan and the cylinder block junction side on the engine front cover side—MX-5 Miata

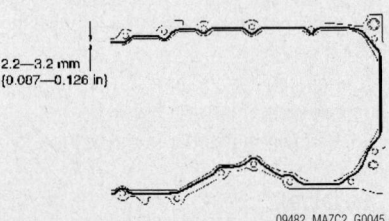

2.2–3.2 mm {0.087–0.126 in}

09482_MAZC2_G0045

Apply a 2.2–3.2 mm (0.087–0.126 in.) bead of silicone sealant to the oil pan along the inside of the bolt holes—MX-5 Miata

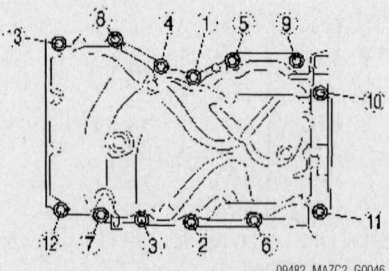

09482_MAZC2_G0046

Tighten the bolts in this sequence—MX-5 Miata

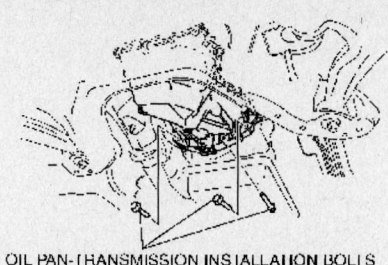

OIL PAN-TRANSMISSION INSTALLATION BOLTS
09482_MAZC2_G0047

Tighten the oil pan to transmission bolts —MX-5 Miata

installation bolts to 37–52 Nm (28–38 ft. lbs.).

29. Install the remaining components in the reverse order of removal, refer to the illustrations accompanying this procedure for component locations and related torque specifications.

30. Refill the engine oil, start the engine and check for leaks.

Mazda3

1. Before servicing the vehicle, refer to the precautions section.

2. Remove the battery cover and disconnect the negative battery cable.

3. Remove the engine under cover and splash shield as an assembly.

4. Remove the right front wheel.

5. Remove the plug hole plate.

6. Drain the engine oil.

7. Remove the drive belt.

8. Remove the coolant reservoir and position aside with the lines attached.

9. Remove the A/C compressor and position aside with the lines attached.

10. Remove the ignition coil and position the accelerator cable bracket aside.

11. Remove the Crankshaft Position (CKP) sensor.

12. Remove the engine front cover. Refer to the timing chain removal procedure in this section.

13. Remove the dipstick tube pipe and O–ring.

14. Remove the oil pan bolts and use a separator tool such as the one illustrated to separate the oil pan from the block.

15. Remove the oil pan.

To install:

16. Clean the oil pan mating surfaces.

17. Apply a 0.087–0.126 inch (2.2–3.2mm) bead of sealant around the perimeter of the oil pan as illustrated.

18. Use a square ruler to align the oil pan and the block junction side on the engine front cover side as illustrated.

19. Install the oil pan within 5 minutes of applying the sealant.

20. Install the lower oil pan bolts and

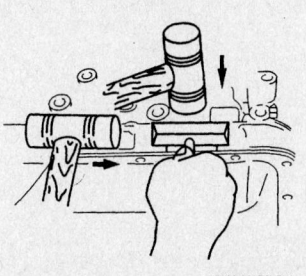

67162-MAZC-G43

Use a separator tool such as the one illustrated to separate the oil pan from the block—2.0L (LF) and 2.3L (L3) engines

tighten in sequence to 12.6–16.9 ft. lbs. (17–23 Nm) and the oil pan–to–transaxle bolts to 23–38 ft. lbs. (32–52 Nm).

21. When installing the CKP sensor perform the following:

a. Remove the cylinder block lower blind plug.

b. Install tool 49 JE01 061.

c. Turn the crankshaft clockwise until the crankshaft is in the number 1 cylinder Top Dead Center position (the balance weight should be attached to tool 49 JE01 061).

d. Using a ruler, mark the center line on the pulse wheel teeth on the crank pulley which is located at the 9th tooth counting counterclockwise from the empty space. Refer to the illustration for more detail.

☀☀ CAUTION

If you do not mark the center line correctly this will cause improper engine control for the ignition and fuel system, so be sure to mark the line carefully.

e. Install the CKP sensor making sure the mark on the sensor is aligned with the mark on the pulse wheel. Tighten the sensor bolt to 49–66 inch lbs. (5.5–7.5 Nm) and attach the electrical connector.

f. Remove the tool from the cylinder block lower blind plug.

g. Rotate the crankshaft clockwise 2 turns until you reach TDC. If not aligned properly, loosen the crankshaft pulley lock bolt and reinstall the lock bolt.

h. Install the cylinder block blind plug and tighten to 13–16 ft. lbs. (18–22 Nm).

22. Install the remaining components in the reverse order of removal.

23. Fill the engine with clean oil.

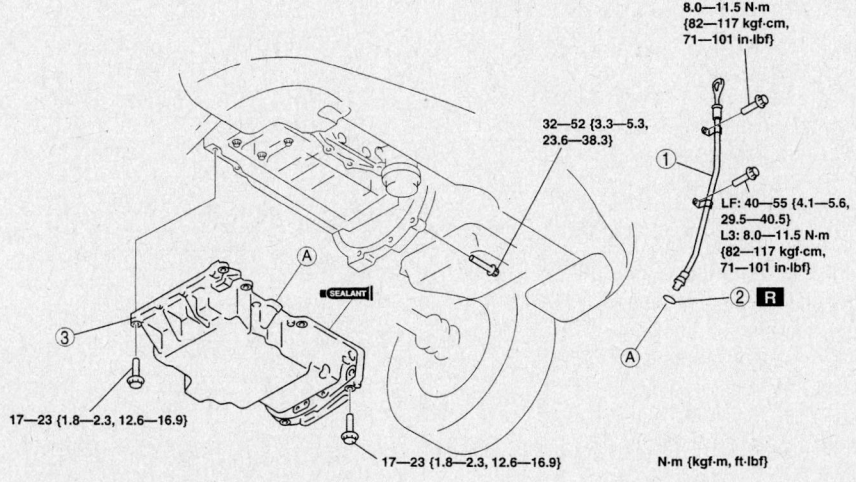

8.0—11.5 N·m
{82—117 kgf·cm, 71—101 in·lbf}

32—52 {3.3—5.3, 23.6—38.3}

LF: 40—55 {4.1—5.6, 29.5—40.5}
L3: 8.0—11.5 N·m {82—117 kgf·cm, 71—101 in·lbf}

17—23 {1.8—2.3, 12.6—16.9}

17—23 {1.8—2.3, 12.6—16.9}

N·m {kgf·m, ft·lbf}

| 1 | Oil level gauge pipe | 3 | Oil pan |
| 2 | O-ring | | |

67162-MAZC-G42

Exploded view of the oil pan and related components—Mazda3 with the 2.0L (LF) and 2.3L (L3) engines

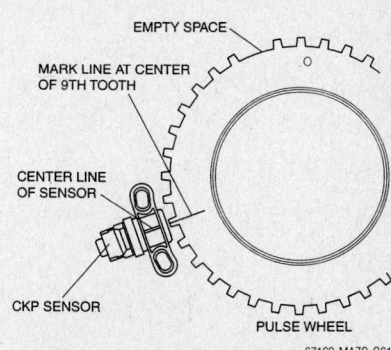

EMPTY SPACE

MARK LINE AT CENTER OF 9TH TOOTH

CENTER LINE OF SENSOR

CKP SENSOR

PULSE WHEEL

67162-MAZC-G61

Using a ruler, mark the center line on the pulse wheel teeth on the crank pulley which is located at the 9th tooth counting counterclockwise from the empty space— 2.0L (LF) and 2.3L (L3) engines

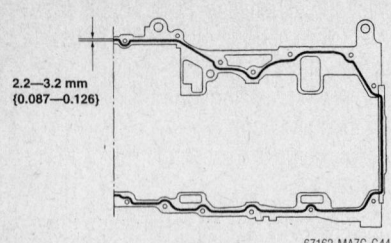

Apply a 0.087–0.126 inch (2.2–3.2mm) bead of sealant around the perimeter of the oil pan—Mazda3 with the 2.0L (LF) and 2.3L (L3) engines

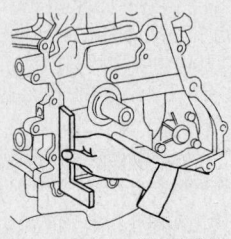

Use a square ruler to align the oil pan and the block junction side on the engine front cover side—2.0L (LF) and 2.3L (L3) engines

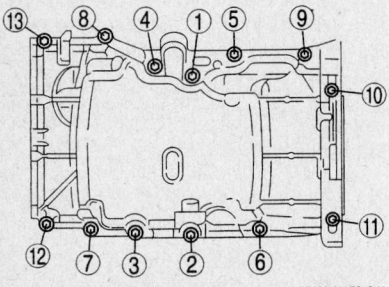

Oil pan bolt torque sequence—2.0L (LF) and 2.3L (L3) engines

24. Start the vehicle, check for leaks and repair if necessary.

Mazda6

2.3L (L3) ENGINE

1. Before servicing the vehicle, refer to the precautions section.
2. Disconnect the negative battery cable.
3. Remove the engine under cover.
4. Remove the right front wheel.
5. Drain the engine oil.
6. Remove the engine front cover. Refer to the timing chain removal procedure in this section.
7. Remove the dipstick tube pipe and O-ring.
8. Remove the oil pan bolts and use a separator tool to separate the oil pan from the block.

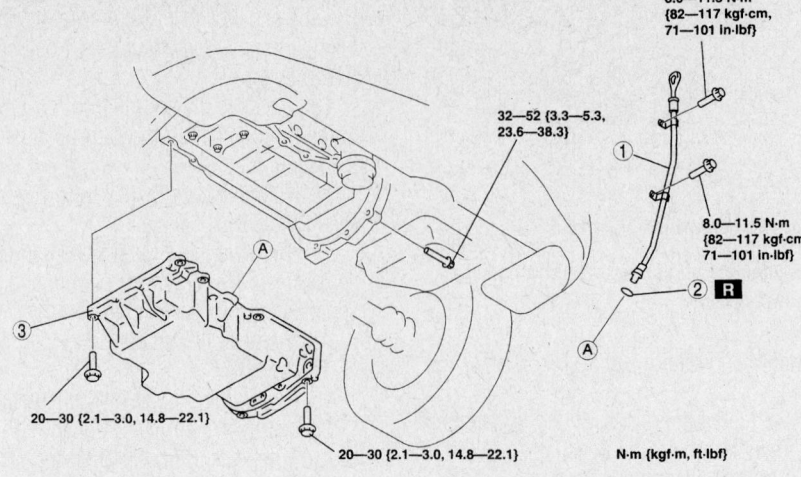

1 Oil level gauge pipe   2 Oil pan

Exploded view of the oil pan and related components—Mazda6 with the 2.3L (L3) engine

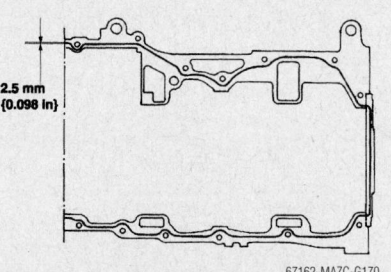

Apply a 0.098 inch (2.5mm) bead of sealant around the perimeter of the oil pan—Mazda6 with the 2.3L (L3) engine

9. Remove the oil pan.
To install:
10. Clean the oil pan mating surfaces.
11. Apply a 0.098 inch (2.5mm) bead of sealant around the perimeter of the oil pan as illustrated.
12. Use a square ruler to align the oil pan and the block junction side on the engine front cover side.
13. Install the oil pan within 5 minutes of applying the sealant.
14. Install the lower oil pan bolts and tighten in sequence to 15–22 ft. lbs. (20–30 Nm) and the oil pan–to–transaxle bolts to 23–38 ft. lbs. (32–52 Nm).
15. Install the remaining components in the reverse order of removal.
16. Fill the engine with clean oil.
17. Start the vehicle, check for leaks and repair if necessary.

3.0L (AJ) Engine

1. Before servicing the vehicle, refer to the precautions section.

2. Drain the engine oil.
3. Disconnect the negative battery cable.
4. Remove the right hand three way converter.
5. Remove the end plate cover.
6. Remove the oil pan–to–transaxle bolts.
7. Loosen the oil pan retainers using 2–3 passes in the order illustrated.
8. Thoroughly clean the gasket mating surfaces.
To install:
9. Apply a 0.39 inch (10mm) bead of silicone sealer to the oil pan at the locations illustrated.

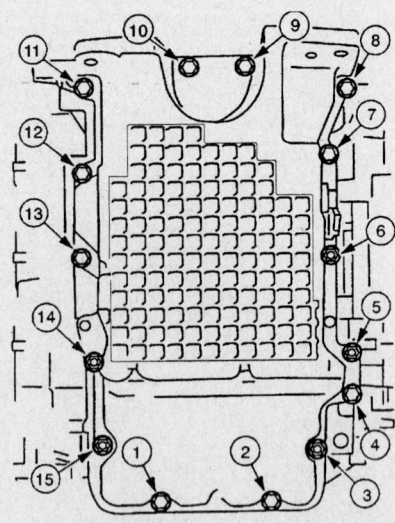

Loosen the oil pan retainers using 2–3 passes in the order shown—Mazda6 with the 3.0L (AJ) engine

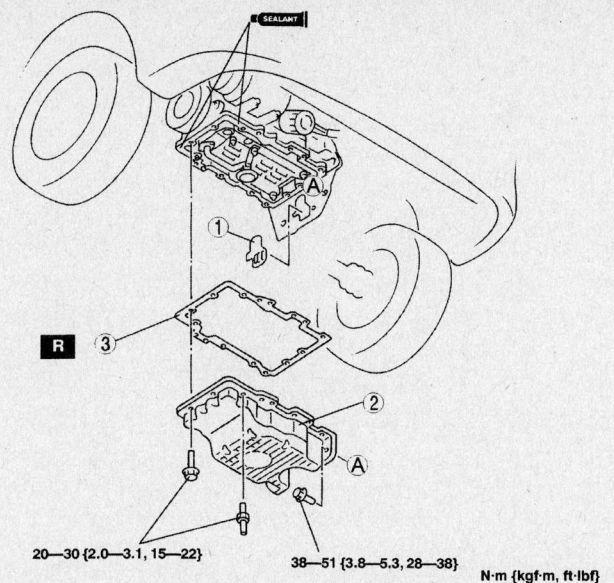

20—30 {2.0—3.1, 15—22}   38—51 {3.8—5.3, 28—38}   N·m {kgf·m, ft·lbf}

| 1 | End plate cover | 3 | Oil pan gasket |
| 2 | Oil Pan | | |

67162-MAZC-G173

Exploded view of the oil pan and related components—Mazda6 with the 3.0L (AJ) engine

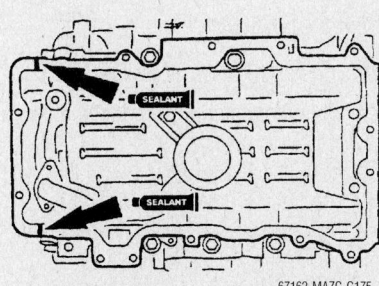

67162-MAZC-G175

Apply a 0.39 inch (10mm) bead of silicone sealer to the oil pan at the locations shown—Mazda6 with the 3.0L (AJ) engine

10. Install the oil pan retainers and hand tighten.

11. Install the oil pan–to–transaxle bolts.

12. Tighten the oil pan retainers in the sequence illustrated to 15–22 ft. lbs. (20–30 Nm).

13. Tighten the oil pan–to–transaxle bolts to 28–38 ft. lbs. (38–51 Nm)

14. Install the remaining components in the reverse order of removal.

15. Fill the engine with clean oil.

16. Start the vehicle, check for leaks and repair if necessary.

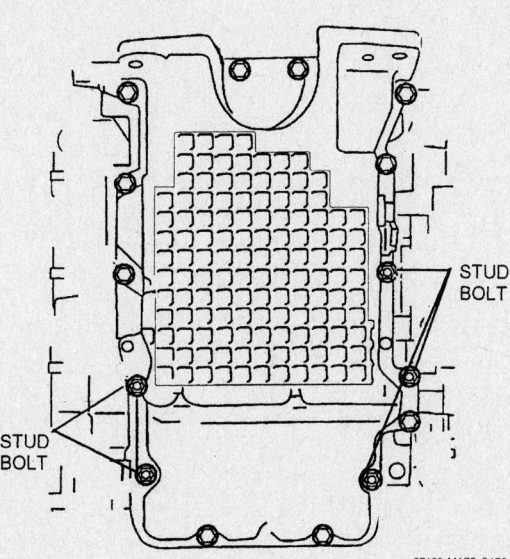

67162-MAZC-G176

Location of the oil pan stud bolts—Mazda6 with the 3.0L (AJ) engine

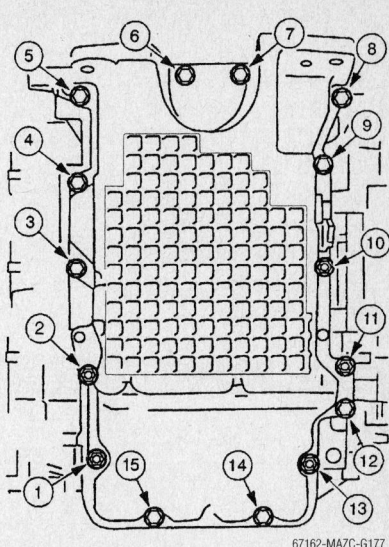

67162-MAZC-G177

Oil pan torque sequence—Mazda6 with the 3.0L (AJ) engine

Mazdaspeed6

1. Before servicing the vehicle, refer to the precautions section.

2. Disconnect the negative battery cable.

3. Remove the engine under cover.

4. Remove the right front wheel.

5. Drain the engine oil.

6. Remove the engine front cover. Refer to the timing chain removal procedure in this section.

7. Remove the dipstick tube pipe and O–ring.

8. Remove the oil pan bolts and use a separator tool to separate the oil pan from the block.

9. Remove the oil pan.

To install:

10. Clean the oil pan mating surfaces.

11. Apply a 0.087–126 inch (2.2–3.2mm) bead of sealant around the perimeter of the oil pan as illustrated.

12. Use a square ruler to align the oil pan and the block junction side on the engine front cover side.

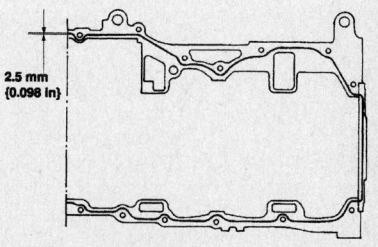

2.5 mm {0.098 in}

67162-MAZC-G170

Apply a 0.087–126 inch (2.2–3.2mm) bead of sealant around the perimeter of the oil pan—Mazdaspeed6

13. Install the oil pan within 5 minutes of applying the sealant.

14. Install the lower oil pan bolts and tighten in sequence to 12.6–17 ft. lbs. (17–23 Nm) and the oil pan–to–transaxle bolts to 28–38 ft. lbs. (37–52 Nm).

15. Install the remaining components in the reverse order of removal.

16. Fill the engine with clean oil.

17. Start the vehicle, check for leaks and repair if necessary.

Oil Pump

REMOVAL & INSTALLATION

MX5-Miata

1. Before servicing the vehicle, refer to the precautions section.

✳✳ CAUTION

Remove and install all parts when the engine is cold, otherwise they can cause severe burns or serious injury.

2. Remove the oil pan.

3. Remove the oil strainer.

4. Remove the oil pump chain guide.

5. Remove the oil pump chain tensioner.

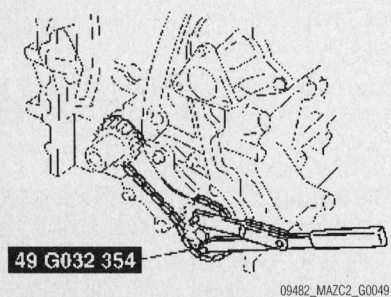

49 G032 354

09482_MAZC2_G0049

Install the tool 49 G032 354 onto the oil pump sprocket to stop the oil pump from rotating—MX-5 Miata

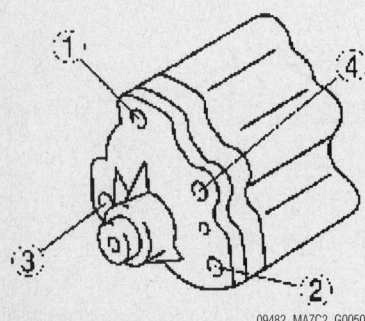

09482_MAZC2_G0050

Oil pump bolt torque sequence—MX-5 Miata

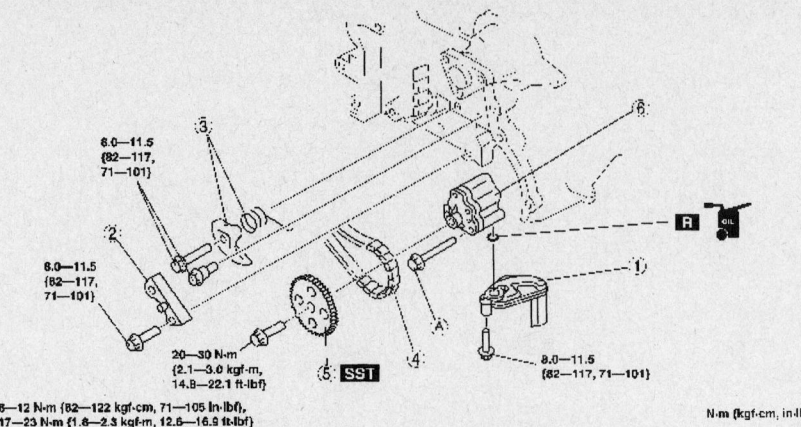

09482_MAZC2_G0048

1 Oil strainer	4 Oil pump chain
2 Oil pump chain guide	5 Oil pump sprocket
3 Oil pump chain tensioner	6 Oil pump

Exploded view of the oil pump and related components—MX-5 Miata

6. Remove the oil pump chain.

7. Install the tool 49 G032 354 onto the oil pump sprocket to stop the oil pump from rotating.

8. Remove the oil pump sprocket.

9. Remove the oil pump bolts and pump.

To install:

10. Install the oil pump, tighten the bolts in the sequence illustrated to 8–12 Nm (71–105 inch lbs.), then final tighten to 17–23 Nm (12.6–17 ft. lbs.).

11. Install the pump sprocket. Tighten the bolt to 20–30 Nm (14.8–22 ft. lbs.).

12. Install the pump chain and tensioner and chain guide. Tighten the tensioner and guide bolts to 8–11.5 Nm (71–101 inch lbs.).

13. Install the oil strainer. Tighten the bolts to 8–11.5 Nm (71–101 inch lbs.).

14. Install the oil pan.

Mazda3 and 6 Models

2.0L (LF) AND 2.3L (L3) ENGINES

1. Before servicing the vehicle, refer to the precautions section.

2. Remove the oil pan.

3. Remove the oil strainer.

4. Remove the oil pump chain guide, tensioner, spring and chain.

5. Remove the oil pump sprocket using

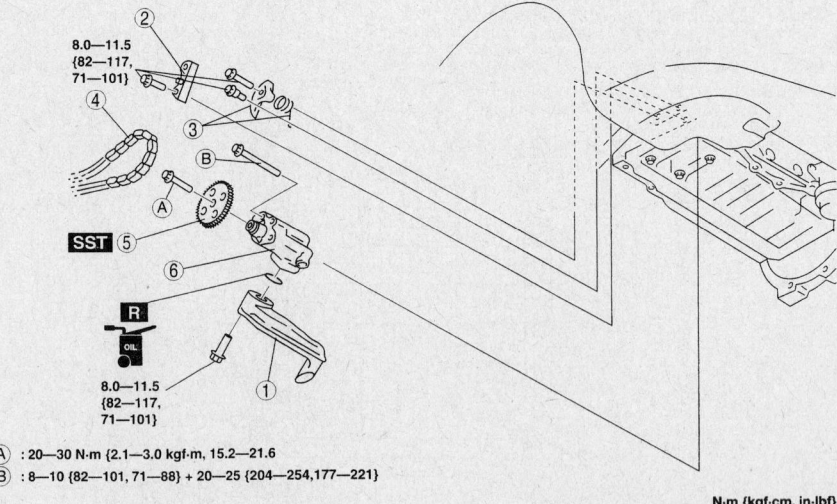

67162-MAZC-G47

1 Oil strainer	4 Oil pump chain
2 Oil pump chain guide	5 Oil pump sprocket
3 Oil pump chain tensioner and spring component	6 Oil pump

Exploded view of the oil pump and related components—2.0L (LF) and 2.3L (L3) engines

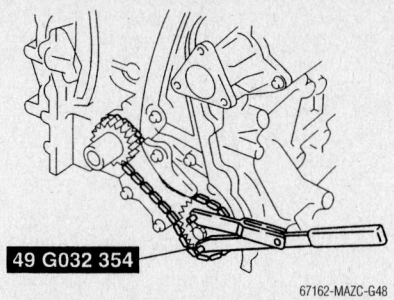

49 G032 354

67162-MAZC-G48

Remove and install the oil pump sprocket using tool 49 G032 354 to stop the pump rotating—2.0L (LF) and 2.3L (L3) engines

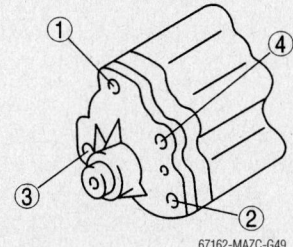

67162-MAZC-G49

Oil pump torque sequence—2.0L (LF) and 2.3L (L3) engines

tool 49 G032 354 to stop the pump rotating.

6. Remove the oil pump bolts and the pump.

To install:

7. Install the oil pump and tighten the bolts in two steps in the sequence illustrated. Tighten the bolts in sequence to 71–88 inch lbs. (8–10 Nm) and final tighten to 15.2–18.4 ft. lbs. (20–27 Nm).

8. Install the oil pump sprocket using tool 49 G032 354 to stop the pump rotating. Tighten the bolts to the specifications shown in the oil pump exploded view illustration.

9. Install the oil pump chain, spring, tensioner and guide. Tighten the bolts to the specifications shown in the oil pump exploded view illustration.

10. Install the oil strainer. Tighten the bolts to the specifications shown in the oil pump exploded view illustration.

11. Install the oil pan.

12. Fill the engine with clean oil.

13. Start the vehicle, check for leaks and repair if necessary.

3.0L (AJ) ENGINE

1. Before servicing the vehicle, refer to the precautions section.

2. Drain the engine oil.

3. Disconnect the negative battery cable.

4. Remove the oil pan.

5. Remove the timing chain.

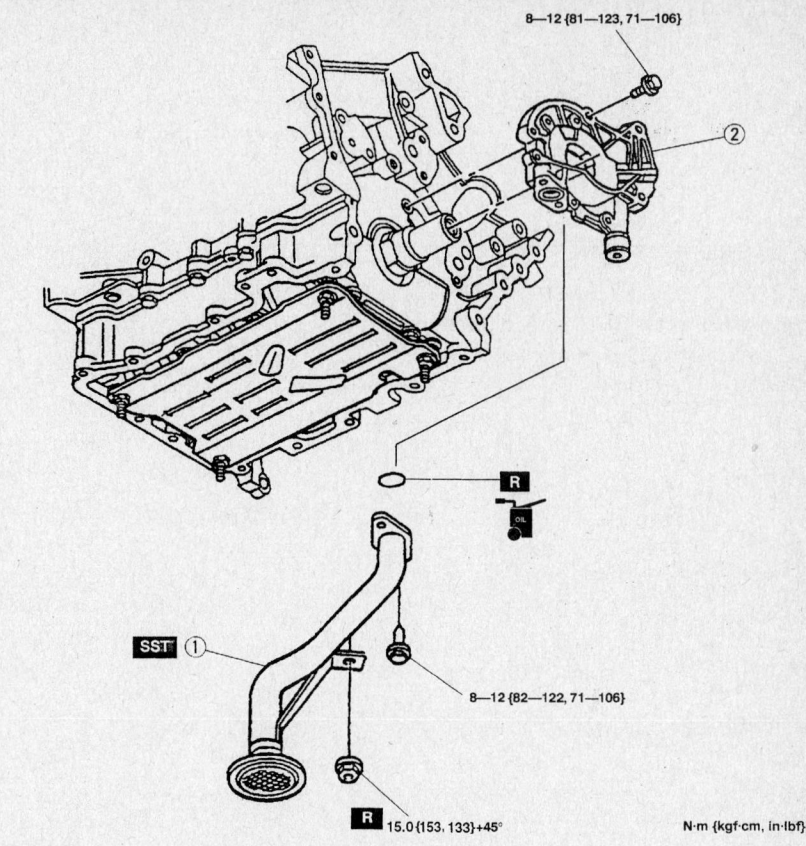

8—12 {81—123, 71—106}

8—12 {82—122, 71—106}

15.0 {153, 133}+45°

N·m {kgf·cm, in·lbf}

1 Oil strainer   2 Oil pump

67162-MAZC-G178

Exploded view of the oil pump and related components—Mazda6 with the 3.0L (AJ) engine

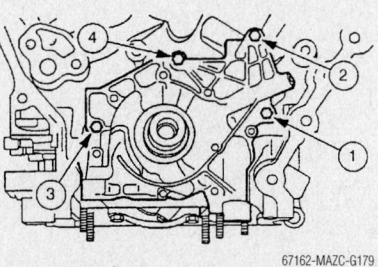

67162-MAZC-G179

Remove the oil pump bolts in the proper sequence—Mazda6 with the 3.0L (AJ) engine

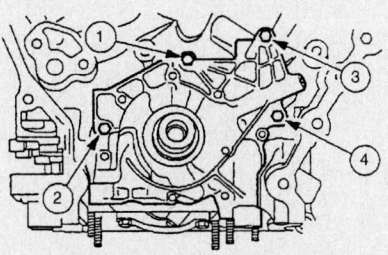

67162-MAZC-G180

Tighten the oil pump bolts in the proper sequence—Mazda6 with the 3.0L (AJ) engine

6. Remove the oil strainer.

7. Remove the oil pump bolts in the sequence illustrated.

8. Thoroughly clean the gasket mating surfaces.

To install:

9. Install the oil pump and tighten the bolts in the proper sequence. Torque the bolts to 71–106 inch lbs. (8–12 Nm).

10. Install the oil strainer. Tighten the strainer bolts to 71–106 inch lbs. (8–12 Nm) and the oil strainer stay nut to 133 inch lbs. (15 Nm) plus an additional 45 degree turn.

11. Install the remaining components in the reverse order of removal.

12. Refill the engine with clean oil.

13. Start the engine and check for leaks; repair if necessary.

Mazdaspeed6

1. Before servicing the vehicle, refer to the precautions section.

2. Remove the oil pan.

3. Remove the oil strainer.

4. Remove the oil pump chain guide, tensioner, spring and chain.

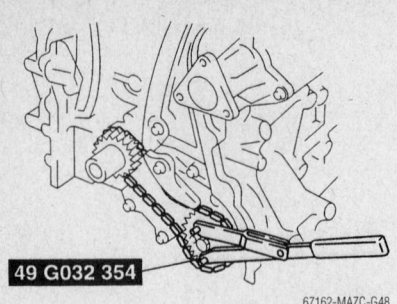

Remove and install the oil pump sprocket using tool 49 G032 354 to stop the pump rotating—Mazdaspeed6

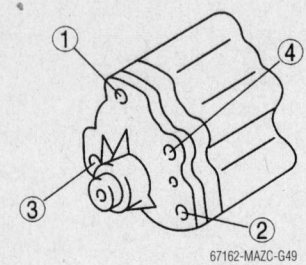

Oil pump torque sequence—Mazdaspeed6

5. Remove the oil pump sprocket using tool 49 G032 354 to stop the pump rotating.

6. Remove the oil pump bolts and the pump.

To install:

7. Install the oil pump and tighten the bolts in two steps in the sequence illustrated. Tighten the bolts in sequence to 71–105 inch lbs. (8–12 Nm) and final tighten to 12.6–17 ft. lbs. (17–23 Nm).

8. Install the oil pump sprocket using tool 49 G032 354 to stop the pump rotating. Tighten the bolts to the specifications shown in the oil pump exploded view illustration.

9. Install the oil pump chain, spring, tensioner and guide. Tighten the bolts to the specifications shown in the oil pump exploded view illustration.

10. Install the oil strainer. Tighten the bolts to the specifications shown in the oil pump exploded view illustration.

11. Install the oil pan.

12. Fill the engine with clean oil.

13. Start the vehicle, check for leaks and repair if necessary.

Rear Main Seal

REMOVAL & INSTALLATION

MX5-Miata

1. Before servicing the vehicle, refer to the precautions section.

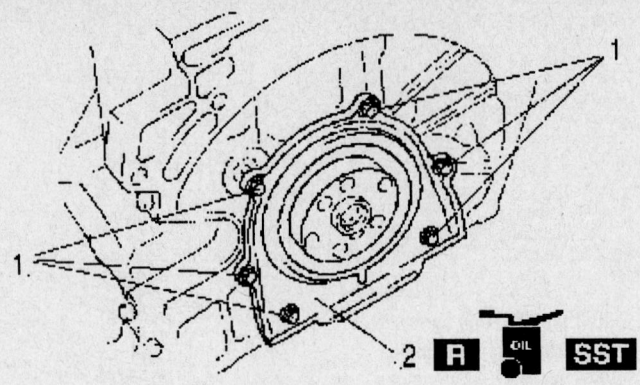

1 Bolt
2 Rear oil seal

Location of the rear main seal bolts—MX-5 Miata

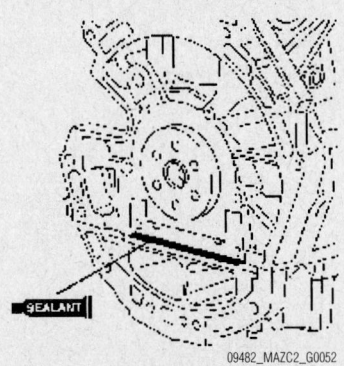

Apply a 4–6 mm bead of silicone sealant to the mating faces—MX-5 Miata

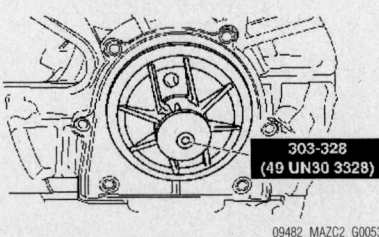

Install the rear oil seal using a tool such as 303-328—MX-5 Miata

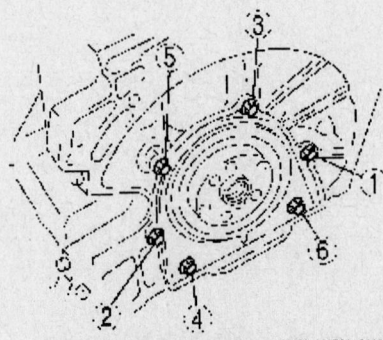

Tighten the rear oil seal bolts in this sequence—MX-5 Miata

2. Remove the flywheel on manual transmissions, or the drive plate on automatic transmissions.

3. Remove the bolt and the rear oil seal.

To install:

4. Apply a 4–6 mm bead of silicone sealant to the mating faces as illustrated.

5. Apply clean engine oil to the new oil seal lip.

6. Install the rear oil seal using a tool such as 303-328.

7. Tighten the rear oil seal bolts in the sequence illustrated to 8–11.5 Nm (71–101 inch lbs.).

8. Install the flywheel or drive plate.

Mazda3, Mazda6 and Mazdaspeed6 Models

2.0L (LF) AND 2.3L (L3) ENGINES

1. Before servicing the vehicle, refer to the precautions section.

2. Remove the transmission.

3. Remove the flywheel.

4. Remove the rear seal housing bolts.

5. Remove the seal.

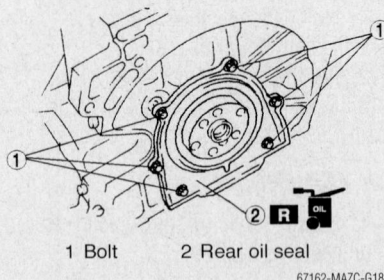

1 Bolt   2 Rear oil seal

Rear oil seal housing bolt removal sequence—2.0L (LF) and 2.3L (L3) engines

To install:

6. Apply a 0.16–0.23 inch (4–6mm) bead of silicone sealant to the seal mating surfaces as illustrated. Install the seal within 10 minutes of applying the sealant.

7. Apply a coat of clean engine oil to the new seal lip.

8. Using a suitable installer tool, install the seal as illustrated.

9. Tighten the seal housing bolts in sequence to 71–101 inch lbs. (8–11.5 Nm).

10. Install the flywheel, use tool 49 E011 1A0 to prevent the unit from turning. If reusing the old bolts coat them with a thread locking compound. If install new bolts no locking compound is needed.

11. On Mazda3 models, tighten bolts in sequence 80–85 ft. lbs. (108–116 Nm) on models equipped with a manual transaxle. On models equipped with an automatic transaxle, tighten the bolts to 80–86 ft. lbs. (96–103 Nm).

12. On Mazda6 models, if equipped with a G35M-R manual transaxle, tighten the bolts to 79–84 ft. lbs. (108–115 Nm). If equipped with a A65M-R manual transaxle, tighten the bolts to 54–64 ft. lbs. (73–87 Nm). On models equipped with an FN4A-EL automatic transaxle, tighten the bolts to 80–86 ft. lbs. (96–103 Nm). On models

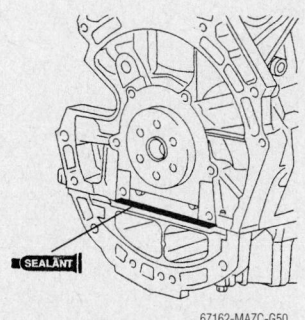

Apply a 0.16–0.23 inch (4–6mm) bead of silicone sealant to the seal mating surfaces—2.0L (LF) and 2.3L (L3) engines

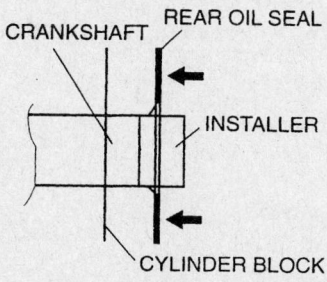

Using a suitable installer tool, install the seal as shown—2.0L (LF) and 2.3L (L3) engines

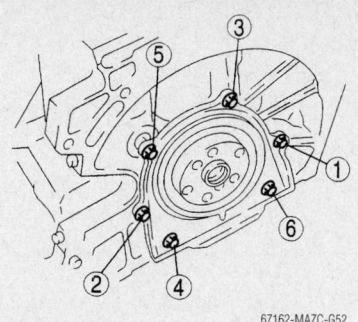

Rear oil seal housing bolt torque sequence—2.0L (LF) and 2.3L (L3) engines

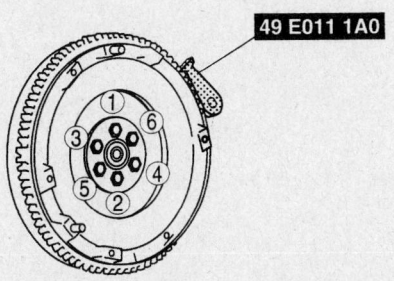

Flywheel bolt torque sequence—2.0L (LF) and 2.3L (L3) engines with a manual transaxle

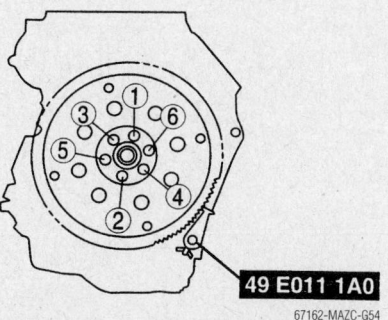

Flywheel bolt torque sequence—2.0L (LF) and 2.3L (L3) engines with an automatic transaxle

equipped with an JA5A-EL automatic transaxle, tighten the bolts to 54–64 ft. lbs. (73–87 Nm).

13. Install the transaxle.

3.0L (AJ) Engine

1. Before servicing the vehicle, refer to the precautions section.

2. Remove the transmission.

3. Remove the flywheel.

4. Cut the rear seal with a razor and remove the seal using a suitable prytool.

To install:

5. Assemble the new oil seal with part

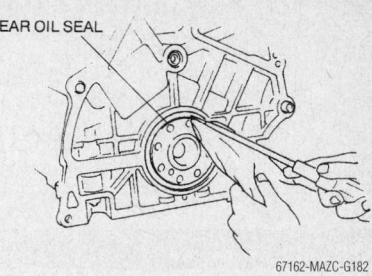

Remove the seal using a suitable pry-tool—Mazda6 with the 3.0L (AJ) engine

(A) of tool 49 UN01 070 and tool 49 UN30 3384.

6. Install the studs of tool 49 UN30 3384 as illustrated.

7. Coat the oil seal with clean engine oil.

8. Push the seal in by hand and install part (A) of tool 49 UN01 070 and compress the seal into the bore by tightening the nuts

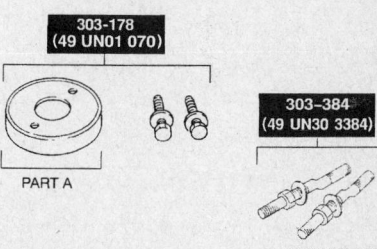

Assemble the new oil seal with part (A) of tool 49 UN01 070 and tool 49 UN30 3384—Mazda6 with the 3.0L (AJ) engine

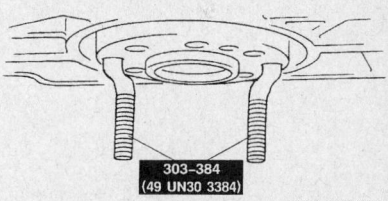

Install the studs of tool 49 UN30 3384—Mazda6 with the 3.0L (AJ) engine

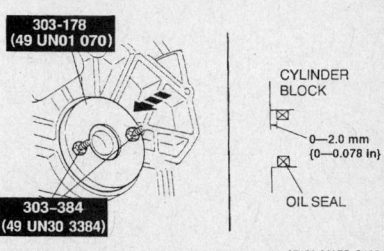

Compress the seal into the bore by tightening the nuts on tool 49 UN30 3384 until the seal is recessed 0–0.078 inch (0–2mm)—Mazda6 with the 3.0L (AJ) engine

on tool 49 UN30 3384 until the seal is recessed 0–0.078 inch (0–2mm).

9. Remove the tools.

10. Install the flywheel and tighten bolts in sequence 80–85 ft. lbs. (108–116 Nm).

11. Install the transaxle.

Timing Chain

REMOVAL & INSTALLATION

MX5-Miata

1. Before servicing the vehicle, refer to the precautions section.

2. Remove the battery and battery tray.

3. Remove the air cleaner.

4. Disconnect the ventilation hose.

5. Loosen the water pump pulley bolt and removal the drive belt.

6. Remove the front suspension tower bar.

7. Remove the Camshaft Position (CMP) sensor.

8. Disconnect the OCV connector.

9. Remove the ignition coils.

10. Remove the drive belt.

11. Remove the under cover.

12. Remove the Crankshaft Position (CKP) sensor.

13. Remove the P/S oil pump with the oil hose still connected and position the P/S oil pump so that it is out of the way.

14. Move the cooler pipe No.3 and heater pipe slightly out of the way.

15. Remove the dipstick.

16. Remove the cylinder head cover.

17. Remove the crankshaft pulley lock bolt as follows:

   a. Remove the cylinder block lower blind plug.

   b. Install tool 303-507.

   c. Turn the crankshaft clockwise until the crankshaft is in the No.1 cylinder TDC position (until the balance weight contacts tool 303-507).

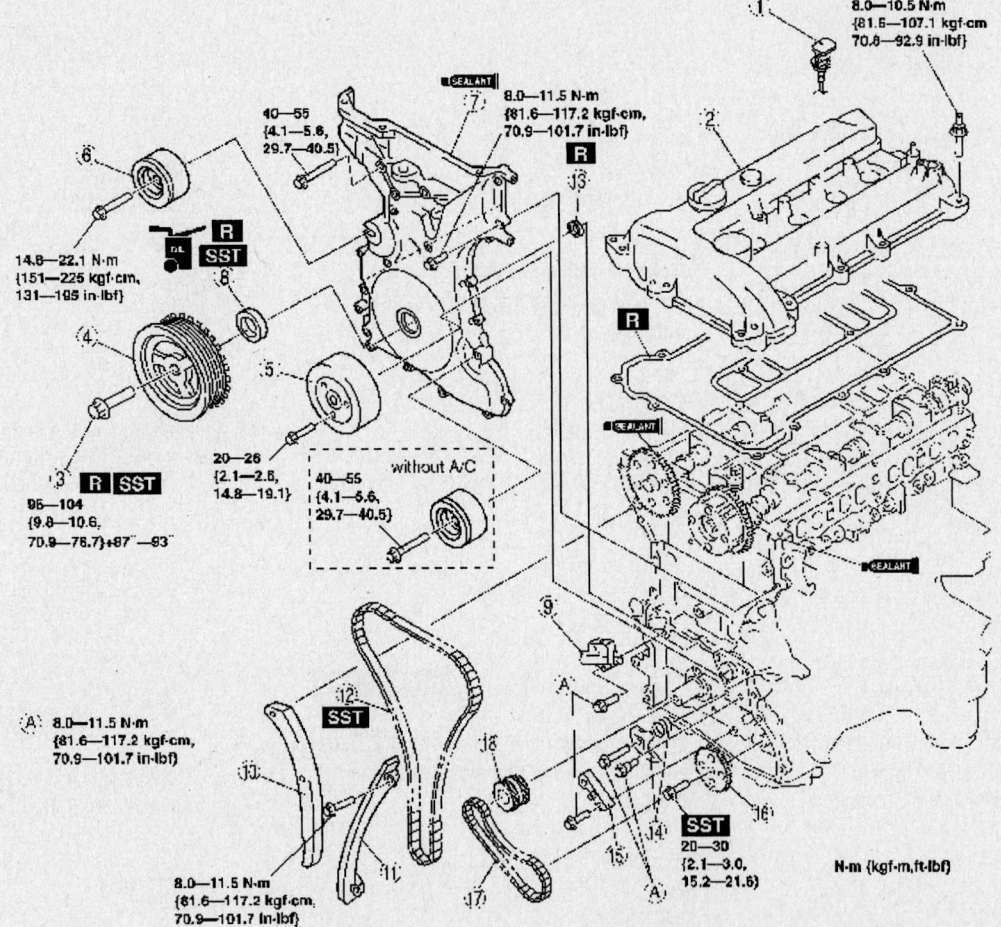

1	Dipstick	10	Tensioner arm
2	Cylinder head cover	11	Chain guide
3	Crankshaft pulley lock bolt	12	Timing chain
4	Crankshaft pulley	13	Seal
5	Water pump pulley	14	Oil pump chain tensioner
6	Drive belt idler pulley	15	Oil pump chain guide
7	Engine front cover	16	Oil pump sprocket
8	Front oil seal	17	Oil pump chain
9	Chain tensioner	18	Crankshaft sprocket

Exploded view of the timing chain and related components—MX-5 Miata

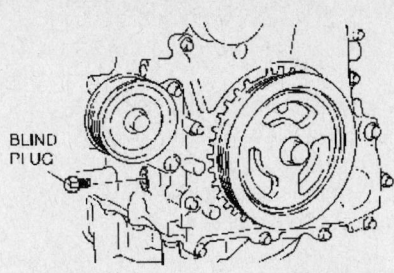

Remove the cylinder block lower blind plug—MX-5 Miata

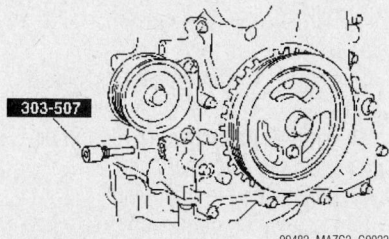

Install tool 303-507 in the lower blind plug opening—MX-5 Miata

d. Hold the crankshaft pulley using the tools shown in the illustration.

e. Remove the crankshaft pulley lock bolt.

18. Remove the crankshaft pulley.

19. Remove the front oil seal as follows:

a. Cut the oil seal lip using a razor knife.

b. Remove the oil seal using a screwdriver wrapped with a rag.

20. Remove the water pump pulley.

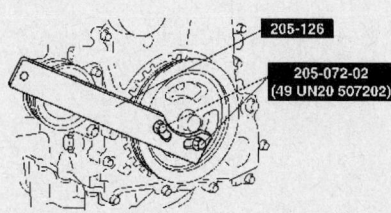

Hold the crankshaft pulley using the tools shown—MX-5 Miata

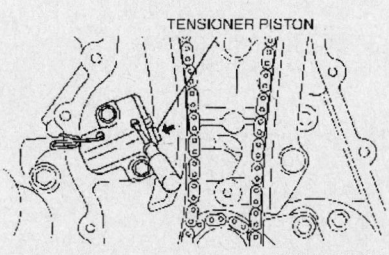

Hold the tensioner piston using a 1.5 mm (0.059 in.) wire or paper clip—MX-5 Miata

21. Remove the drive belt idler pulley.

22. Remove the engine front cover.

23. Remove the oil seal using a screwdriver.

24. Remove the chain tensioner as follows:

a. Using a thin screwdriver, hold the chain tensioner ratchet lock mechanism away from the ratchet stem.

b. Slowly compress the tensioner piston.

c. Hold the tensioner piston using a 1.5 mm (0.059 in.) wire or paper clip.

25. Remove the tensioner arm.

26. Remove the chain guide.

27. Remove the timing chain

To install:

28. Install tool 303-465 onto the camshaft as illustrated.

29. Install the timing chain.

30. Install the chain guide. Tighten the bolt to 8–11.5 Nm (71–101 inch lbs.).

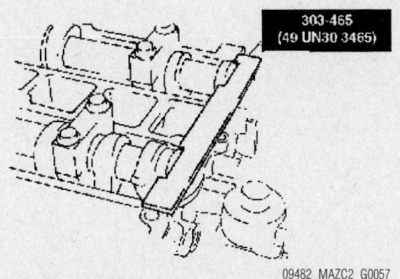

Install tool 303-465 onto the camshaft— MX-5 Miata

31. Install the tensioner arm. Tighten the bolt to 8–11.5 Nm (71–101 inch lbs.).

32. Remove the retaining wire or paper clip from the chain tensioner to apply tension to the timing chain.

33. Install the oil seal.

34. Install the engine front cover as follows:

a. Install the front oil seal before installing the cover.

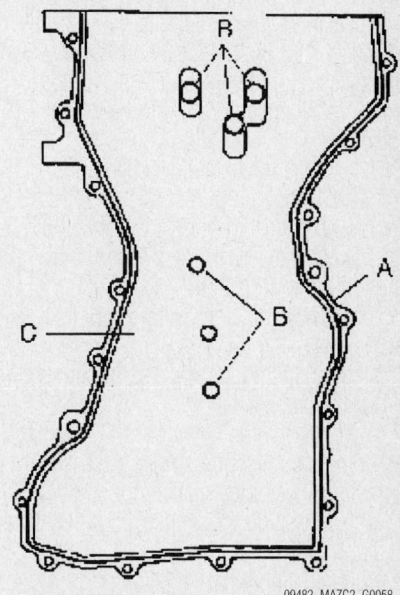

Apply sealant at the locations shown in the illustration. Refer to the illustration for sealant thickness—MX-5 Miata

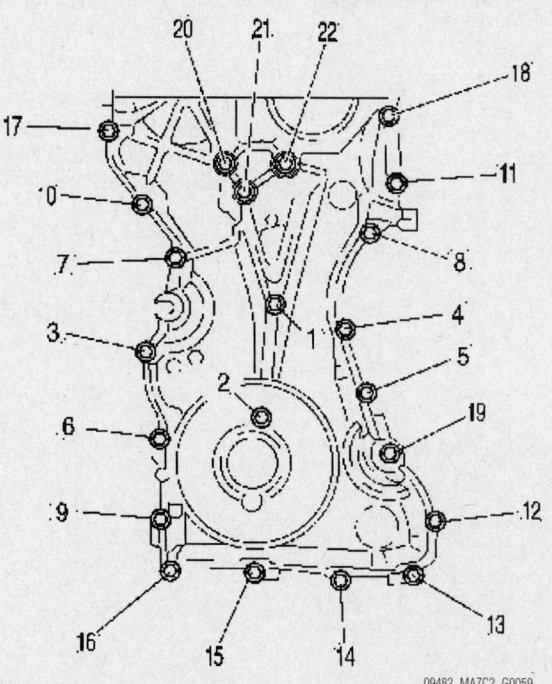

Front cover bolt locations—MX-5 Miata

b. Apply silicone sealant to the engine front cover as illustrated.

❋❋ CAUTION

Install the engine front cover within 10 minutes of applying the silicone sealant.

➡ Silicone sealant is not need in area C as shown.

35. Apply silicone sealant at location **A**. The sealant thickness should be 2–3mm (0.079–0.118 inch. At location **B** the sealant thickness should be 1.5–2.5mm (0.059–0.098 inch). Refer to the illustration for locations.

36. Install the front cover. On bolts 1 through 18 tighten to 8–11.5 Nm (71–101 in. lbs). Bolts 19 through 22, tighten to 20–55 Nm (30–40 ft. lbs.). Refer to the illustration for locations.

37. Install the crankshaft pulley lock bolt as follows:

a. Install tool shown in the illustration on the camshaft.

b. Verify that cylinder No.1 is at TDC of the compression stroke. (Crankshaft balance weight contacts tool 303-507)

c. Position the crankshaft pulley by temporarily tightening it and, using a suitable length bolt 25–35mm (0.99–1.37 in.), fix the crankshaft pulley to the engine front cover.

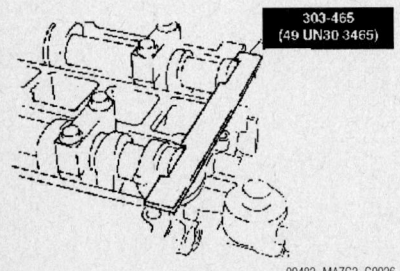

Install tool shown onto the camshaft—MX-5 Miata

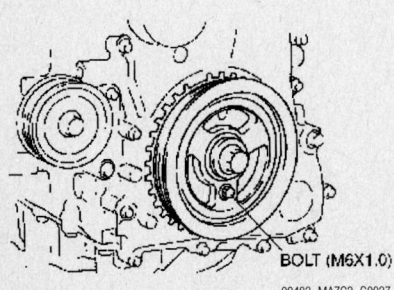

Position the crankshaft pulley by temporarily tightening it using a suitable length bolt—MX-5 Miata

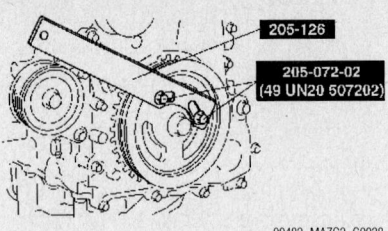

Install the tools shown onto the crankshaft pulley to lock the crankshaft against rotation—MX-5 Miata

d. Install the tools illustrated to the crankshaft pulley, lock the crankshaft against rotation.

e. Tighten the crankshaft pulley lock bolt to 96–104 Nm (71–77 ft. lbs.), plus 87–93 degrees.

f. Remove the bolt.

g. Remove the tool the camshaft.

h. Remove tool from the cylinder block lower blind plug.

i. Remove the tool from the crankshaft pulley.

j. Rotate the crankshaft clockwise two turns until the TDC position.

k. If not aligned, loosen the crankshaft pulley lock bolt and repeat from the first step.

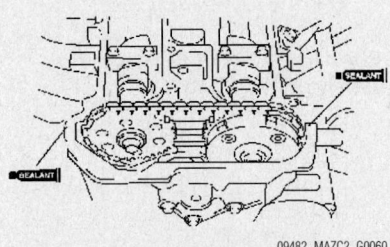

Prior to install the cylinder head cover, apply a 4–6mm (0.16–0.23 in) bead of sealant at these locations—MX-5 Miata

l. Install the cylinder block lower blind plug and tighten to 18–22 Nm (13–16 ft. lbs.).

38. Install the cylinder head cover bolts and tighten in the sequence illustrated to 8–10.5 Nm (70–93 n. lbs.).

39. Install the remaining components in the reverse order of removal.

Mazda3

2.0L (LF) AND 2.3L (L3) ENGINES

1. Before servicing the vehicle, refer to the precautions section.

2. Remove the plug hole plate and bracket.

3. Remove the accelerator cable and bracket.

4. Remove the battery cover and disconnect the negative battery cover.

5. Remove the ignition coils.

6. Remove the right hand front wheel.

7. Remove the engine under cover and splash shields.

8. Remove the Crankshaft Position (CKP) sensor.

9. Remove the accessory drive belt.

10. Remove the A/C compressor and set aside with the lines attached.

11. Remove the coolant reservoir tank and set aside with the lines still attached.

12. Remove the cylinder head cover.

13. Remove the cylinder block lower blind plug.

14. Install tool 49 JE01 061.

15. Turn the crankshaft clockwise until the crankshaft is in the number 1 cylinder Top Dead Center position (the balance weight should be attached to tool 49 JE01 061).

16. Hold the crankshaft pulley using the tools illustrated and remove the bolt.

17. Remove the crankshaft pulley.

18. Remove the water pump pulley.

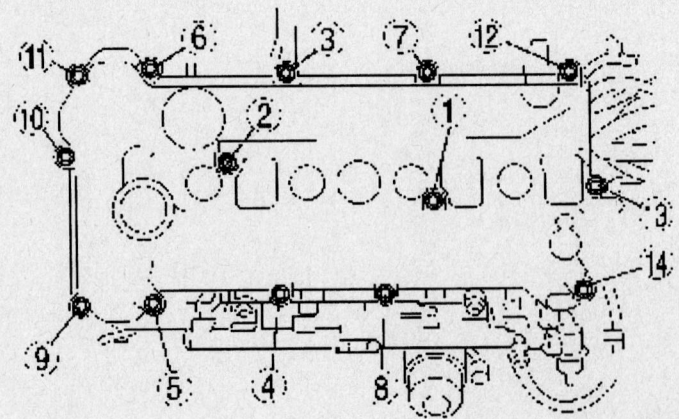

Cylinder head cover torque sequence—MX-5 Miata

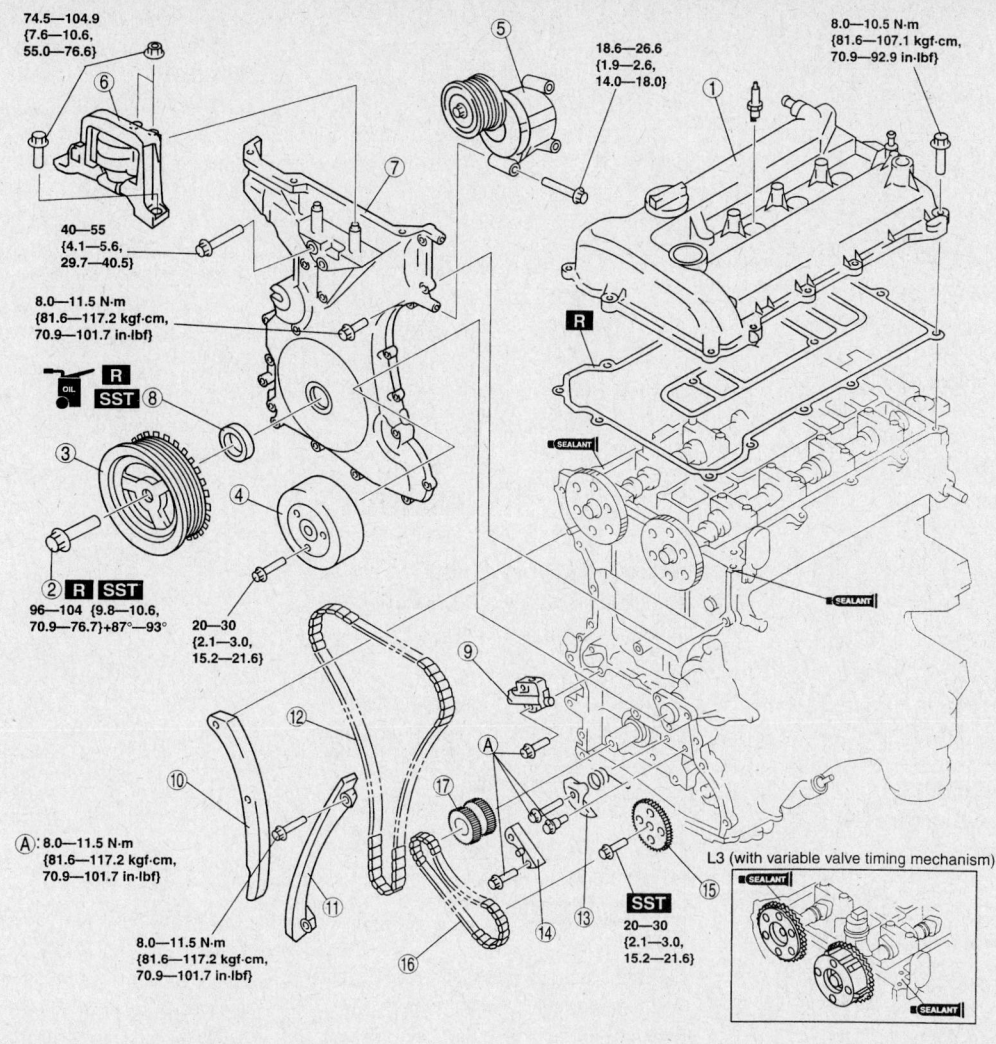

74.5—104.9
{7.6—10.6,
55.0—76.6}

40—55
{4.1—5.6,
29.7—40.5}

8.0—11.5 N·m
{81.6—117.2 kgf·cm,
70.9—101.7 in·lbf}

96—104 {9.8—10.6,
70.9—76.7}+87°—93°

20—30
{2.1—3.0,
15.2—21.6}

18.6—26.6
{1.9—2.6,
14.0—18.0}

8.0—10.5 N·m
{81.6—107.1 kgf·cm,
70.9—92.9 in·lbf}

Ⓐ 8.0—11.5 N·m
{81.6—117.2 kgf·cm,
70.9—101.7 in·lbf}

8.0—11.5 N·m
{81.6—117.2 kgf·cm,
70.9—101.7 in·lbf}

SST
20—30
{2.1—3.0,
15.2—21.6}

L3 (with variable valve timing mechanism)

N·m {kgf·m, ft·lbf}

1	Cylinder head cover	9	Chain tensioner
2	Crankshaft pulley lock bolt	10	Tensioner arm
3	Crankshaft pulley	11	Chain guide
4	Water pump pulley	12	Timing chain
5	Drive belt auto tensioner	13	Oil pump chain tensioner
6	No.3 engine mount rubber and No.3 engine joint bracket	14	Oil pump chain guide
7	Engine front cover	15	Oil pump sprocket
8	Front oil seal	16	Oil pump chain
		17	Crankshaft sprocket

67162-MAZC-G55

Exploded view of the timing chain assembly and related components—Mazda3 with the 2.0L (LF) and 2.3L (L3) engines

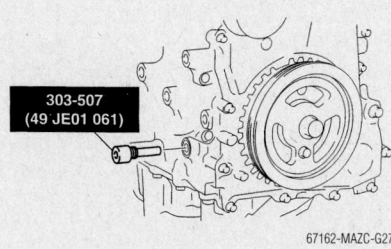

303-507
(49 JE01 061)

67162-MAZC-G27

Install tool 49 JE01 061—Mazda3 with the 2.0L (LF) and 2.3L (L3) engines

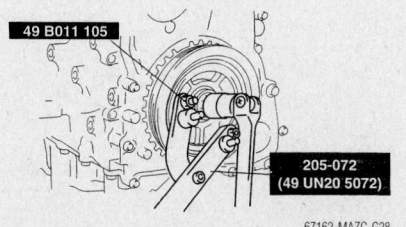

49 B011 105

205-072
(49 UN20 5072)

67162-MAZC-G28

Install tool 49 JE01 061 to hold the crankshaft pulley using the tools illustrated and remove the bolt—Mazda3 with the 2.0L (LF) and 2.3L (L3) engines

19. Remove the drive belt tensioner.
20. Remove the No. 3 engine mount and joint bracket as follows:

   a. Install two suitable pieces of wood between the front fender panel and upper apron reinforcement as illustrated. The wood size should be approximately 1.38 inch (35mm) on 4 door models or 2.36 inch (60mm) on 5 door models.

21. Install an engine support device such as 49 E017 5A0.
22. Remove the engine front cover and oil seal.

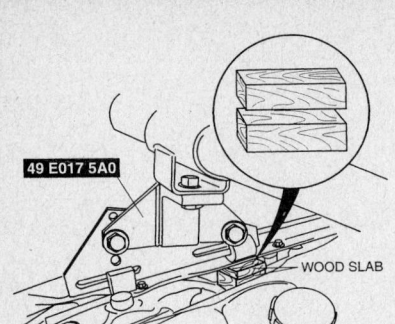

49 E017 5A0

WOOD SLAB

67162-MAZC-G56

Install two suitable pieces of wood between the front fender panel and upper apron reinforcement—Mazda3 with the 2.0L (LF) and 2.3L (L3) engines

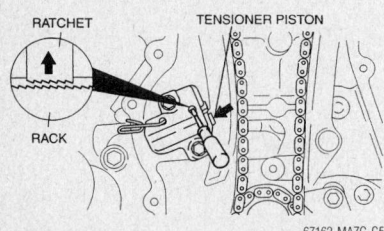

RATCHET   TENSIONER PISTON

RACK

67162-MAZC-G57

Compressing and retaining the chain tensioner piston—Mazda3 with the 2.0L (LF) and 2.3L (L3) engines

23. Unlock the chain tensioner using a suitable tool to slowly compress the tensioner piston. Insert a 0.059 inch (1.5mm) wire or a paper clip to hold the piston in its compressed position.

24. Remove the tensioner arm, chain guide and timing chain.

To install:

25. Install tool 49 UN30 3465 as illustrated on the camshaft.

26. Install the timing chain and remove the paper clip or wire retaining the tensioner piston to apply tension to the chain.

27. Install the timing chain guide and tighten the bolts to 71–101 inch lbs. (8–11.5 Nm).

28. Install the tensioner arm and tighten

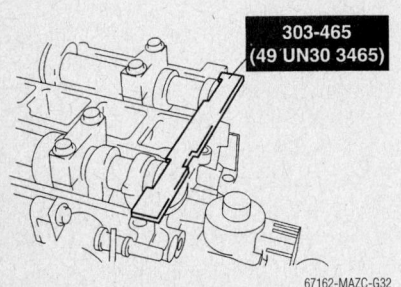

303-465
(49 UN30 3465)

67162-MAZC-G32

Install tool 49 UN30 3465 on the camshaft—Mazda3 with the 2.0L (LF) and 2.3L (L3) engines

the bolts to 71–101 inch lbs. (8–11.5 Nm).

29. Install a new oil seal in the front cover as follows:

a. Coat the new seal with clean engine oil.

b. Push the seal in by hand to get it started.

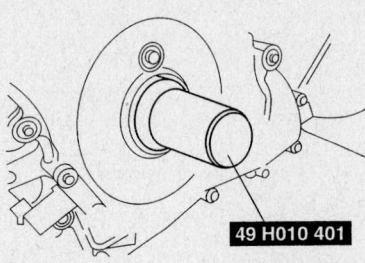

49 H010 401

67162-MAZC-G30

Use seal installer tool 49 H010 401 and a hammer to install the seal—Mazda3 with the 2.0L (LF) and 2.3L (L3) engines

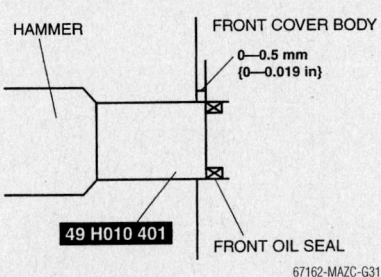

HAMMER   FRONT COVER BODY

0—0.5 mm
{0—0.019 in}

49 H010 401

FRONT OIL SEAL

67162-MAZC-G31

The seal should be recessed 0–0.019 inch (0–0.5mm) as shown when properly installed—Mazda3 with the 2.0L (LF) and 2.3L (L3) engines

c. Use seal installer tool 49 H010 401 and a hammer to install the seal so that it is recessed 0–0.019 inch (0–0.5mm) as shown in the accompanying illustration.

30. Apply sealant to the engine front cover. At point A the bead should be 0.087–0.125 inch (2.2–3.2mm) thick and at point B the bead should be 0.059–0.098 inch (1.5–2.5mm) thick. Refer to the accompanying illustration for the locations of points A and B. No sealant is needed at the points marked C on 2.3L (L3) engines with variable valve timing.

31. Install the cover within 10 minutes of apply the sealant. Tighten the cover bolts as follows:

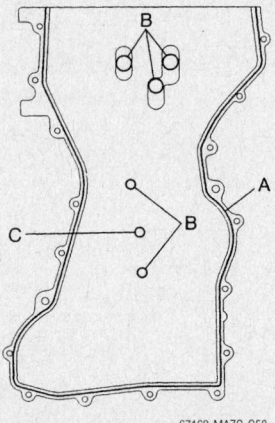

B

A

B

C

67162-MAZC-G58

Apply a bead of sealant to the engine front cover at the locations shown. Refer to the text for the bead thickness—Mazda3 with the 2.0L (LF) and 2.3L (L3) engines

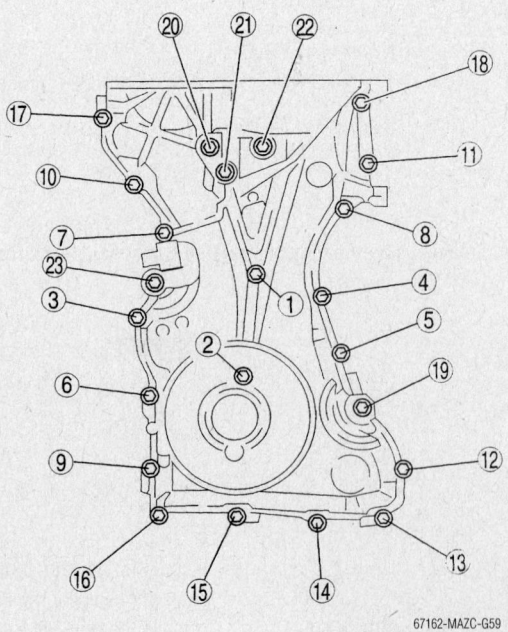

67162-MAZC-G59

Engine front cover bolts tightening sequence—Mazda3 with the 2.0L (LF) and 2.3L (L3) engines

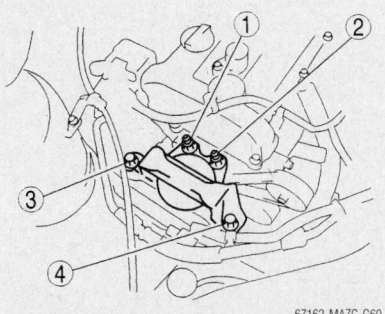

Location of the No. 3 engine mount bracket stud bolts—Mazda3 with the 2.0L (LF) and 2.3L (L3) engines

    a. Bolts 1 through 18: In sequence to 71–101 inch lbs. (8–11.5 Nm).

    b. Bolts 19 through 22: In sequence to 29.7–40.5 ft. lbs. (40–55 Nm).

    c. Bolt 23: 14.8–22 ft. lbs. (20–30 Nm).

32. Install the No. 3 engine mount and joint bracket as follows:

    a. Tighten the No. 3 mount bracket stud bolts to 62–115 inch lbs. (7–13 Nm). Tighten the stud bolt with the mount nut loosened.

    b. Hand tighten the mount rubber and bracket nuts and bolts, then tighten in the sequence illustrated to 55–77 ft. lbs. (74–105 Nm).

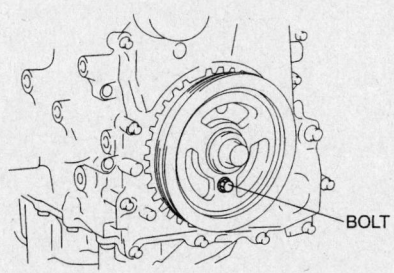

Tighten the mount rubber and bracket nuts and bolts in the sequence—Mazda3 with the 2.0L (LF) and 2.3L (L3) engines

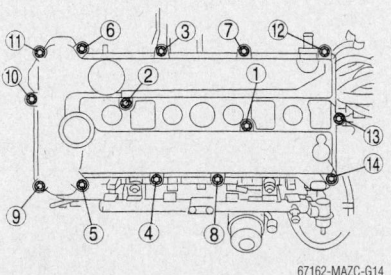

Install and hand tighten the M6 x 1.0 bolt—Mazda3 with the 2.0L (LF) and 2.3L (L3) engines

33. Install tool 49 UN30 3465 as illustrated on the camshaft.

34. Install and hand tighten the M6 x 1.0 bolt as illustrated.

35. Turn the crankshaft clockwise until the crankshaft is in number 1 TDC (the balance weight should be attached to the tool).

36. Hold the crankshaft pulley and tighten the lock bolt in two steps. First tighten to 71–77 ft. lbs. (96–104 Nm). Then final tighten an additional 87–93 degrees.

37. Remove the M6 x 1.0 bolt.

38. Remove the tools from the camshaft and the cylinder block lower blind plug.

39. Rotate the crankshaft clockwise 2 turns until you reach TDC. If not aligned properly, loosen the crankshaft pulley lock bolt and reinstall the lock bolt again using the above procedure and special tools.

40. Install the cylinder block blind plug and tighten to 13–16 ft. lbs. (18–22 Nm).

41. Apply a 0.16–0.24 inch (4–7mm) bead of silicone sealant to cylinder head at the areas illustrated. Make sure to install the cover within 10 minutes of applying the sealant.

42. Install the cylinder head cover with a new gasket. Torque the bolts in the sequence illustrated to 71–93 inch lbs. (8–11 Nm).

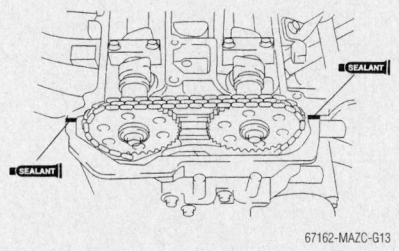

Apply silicone sealant to cylinder head at the areas illustrated—Mazda3 with the 2.0L (LF) and 2.3L (L3) engines

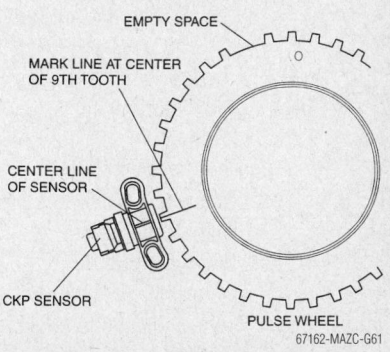

Tighten the cylinder head cover bolts in the sequence shown—Mazda3 with the 2.0L (LF) and 2.3L (L3) engines

43. Install the remaining components in the reverse order of removal.

44. When installing the CKP sensor perform the following:

    a. Remove the cylinder block lower blind plug.

    b. Install tool 49 JE01 061.

    c. Turn the crankshaft clockwise until the crankshaft is in the number 1 cylinder Top Dead Center position (the balance weight should be attached to tool 49 JE01 061).

    d. Using a ruler, mark the center line on the pulse wheel teeth on the crank pulley which is located at the 9th tooth counting counterclockwise from the empty space. Refer to the illustration for more detail.

✽✽ CAUTION

If you do not mark the center line correctly this will cause improper engine control for the ignition and fuel system, so be sure to mark the line carefully.

    e. Install the CKP sensor making sure the mark on the sensor is aligned with the mark on the pulse wheel. Tighten the sensor bolt to 49–66 inch lbs. (5.5–7.5 Nm)and attach the electrical connector.

    f. Remove the tool from the cylinder block lower blind plug.

    g. Rotate the crankshaft clockwise 2 turns until you reach TDC. If not aligned properly, loosen the crankshaft pulley lock bolt and reinstall the lock bolt.

    h. Install the cylinder block blind plug and tighten to 13–16 ft. lbs. (18–22 Nm).

45. Change the engine oil.

46. Start the engine.

47. Inspect for the following:

Using a ruler, mark the center line on the pulse wheel teeth on the crank pulley which is located at the 9th tooth counting counterclockwise from the empty space—Mazda3 with the 2.0L (LF) and 2.3L (L3) engines

- Pulley and belt for run–out and contact
- Any leaking fluids.
- Ignition timing, idle speed and exhaust emissions
- All remaining components for proper operation.

Mazda6

2.3L (L3) ENGINE

1. Before servicing the vehicle, refer to the precautions section.
2. Disconnect the negative battery cable.

3. Remove the cylinder head cover.
4. Remove the cylinder block lower blind plug.
5. Install tool 49 JE01 061.
6. Turn the crankshaft clockwise until the crankshaft is in the number 1 cylinder Top Dead Center position (the balance

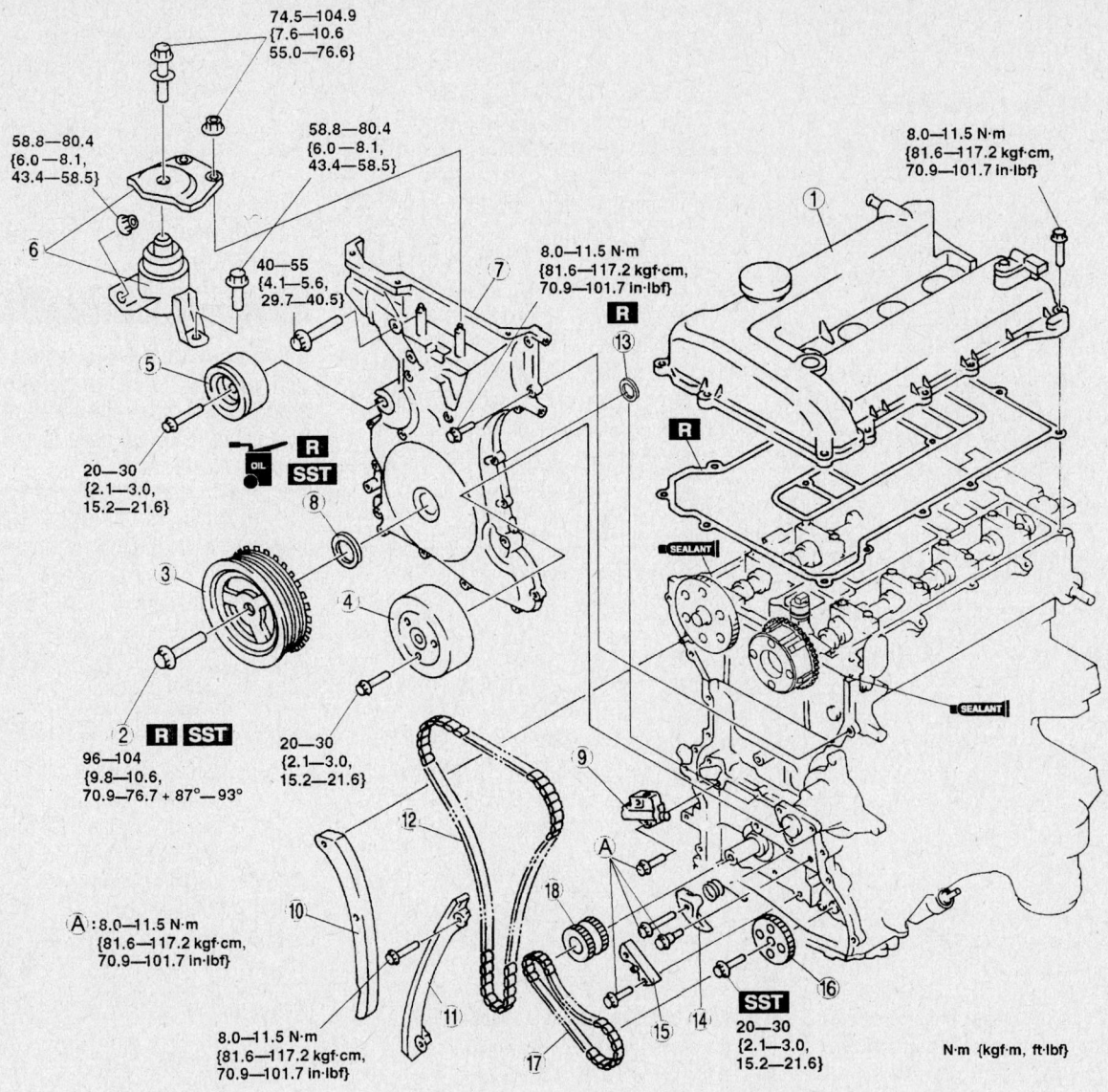

1	Cylinder head cover	10	Tensioner arm
2	Crankshaft pulley lock bolt	11	Chain guide
3	Crankshaft pulley	12	Timing chain
4	Water pump pulley	13	Seal
5	Drive belt idler pulley	14	Oil pump chain tensioner
6	No.3 engine mount rubber and No.3 engine joint bracket	15	Oil pump chain guide
7	Engine front cover	16	Oil pump sprocket
8	Front oil seal	17	Oil pump chain
9	Chain tensioner	18	Crankshaft sprocket

67162-MAZC-G186

Exploded view of the timing chain assembly and related components—Mazda6 with the 2.3L (L3) engine

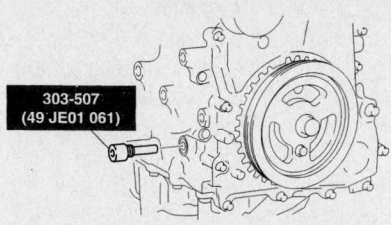

Install tool 49 JE01 061—Mazda6 with the 2.3L (L3) engine

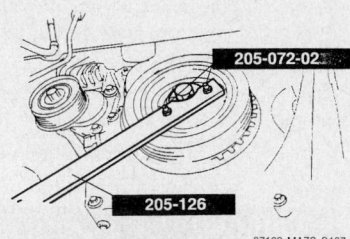

Install tools 205-072-02 and 205-126 to hold the crankshaft pulley using the tools illustrated and remove the bolt—Mazda6 with the 2.3L (L3) engine

weight should be attached to tool 49 JE01 061).

7. Hold the crankshaft pulley using the tools illustrated and remove the bolt.

8. Remove the crankshaft pulley.

9. Remove the water pump pulley.

10. Remove the drive belt idler pulley.

11. Install an engine support device such as 49 E017 5A0.

12. Remove the No. 3 engine mount and joint.

13. Remove the engine front cover and oil seal.

14. Unlock the chain tensioner using a suitable tool to slowly compress the tensioner piston. Insert a 0.059 inch (1.5mm) wire or a paper clip to hold the piston in its compressed position.

15. Remove the tensioner arm, chain guide and timing chain.

To install:

16. Install tool 49 UN30 3465 as illustrated on the camshaft.

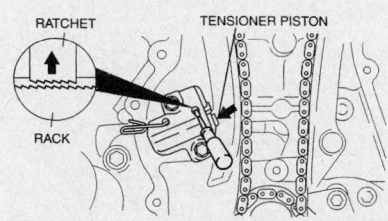

Compressing and retaining the chain tensioner piston—Mazda6 with the 2.3L (L3) engine

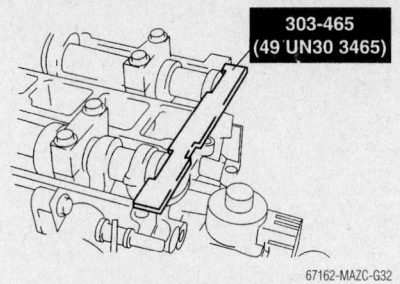

Install tool 49 UN30 3465 on the camshaft—Mazda6 with the 2.3L (L3) engine

17. Install the timing chain and remove the paper clip or wire retaining the tensioner piston to apply tension to the chain.

18. Install the timing chain guide and tighten the bolts to 71–101 inch lbs. (8–11.5 Nm).

19. Install the tensioner arm and tighten the bolts to 71–101 inch lbs. (8–11.5 Nm).

20. Install a new oil seal in the front cover as follows:

a. Coat the new seal with clean engine oil.

b. Push the seal in by hand to get it started.

c. Use seal installer tool 49 H010 401 and a hammer to install the seal so that it is recessed 0–0.019 inch (0–0.5mm) as shown in the accompanying illustration.

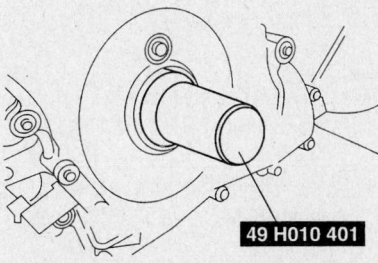

Use seal installer tool 49 H010 401 and a hammer to install the seal—Mazda6 with the 2.3L (L3) engine

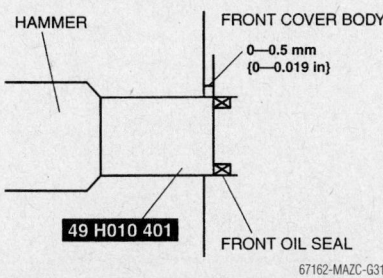

The seal should be recessed 0–0.019 inch (0–0.5mm) as shown when properly installed—Mazda6 with the 2.3L (L3) engine

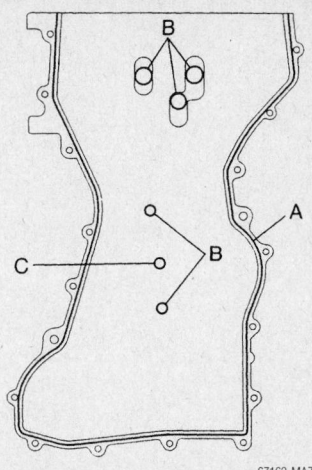

Apply a bead of sealant to the engine front cover at the locations shown. Refer to the text for the bead thickness—Mazda6 with the 2.3L (L3) engine

21. Apply sealant to the engine front cover. At point A the bead should be 0.087–0.125 inch (2.2–3.2mm) thick and at point B the bead should be 0.059–0.098 inch (1.5–2.5mm) thick. Refer to the accompanying illustration for the locations of points A and B.

22. Install the cover within 10 minutes of apply the sealant. Tighten the cover bolts as follows:

a. Bolts 1 through 18: In sequence to 71–101 inch lbs. (8–11.5 Nm).

b. Bolts 19 through 22: In sequence to 29.7–40.5 ft. lbs. (40–55 Nm).

c. Bolt 23: 14.8–22 ft. lbs. (20–30 Nm).

23. When installing the No. 3 engine joint bracket, tighten the mount bracket stud bolt to 62–115 inch lbs. (7–13 Nm) and the joint bracket bolt and nut in the order illustrated to 55–76 ft. lbs. (74–105 Nm).

24. Install the drive belt idler and water pump pulleys.

25. Install tool 49 UN30 3465 as illustrated on the camshaft.

26. Install and hand tighten the M6 x 1.0 bolt as illustrated.

27. Turn the crankshaft clockwise until the crankshaft is in number 1 TDC (the balance weight should be attached to the tool).

28. Hold the crankshaft pulley and tighten the lock bolt in two steps. First tighten to 71–77 ft. lbs. (96–104 Nm). Then final tighten an additional 87–93 degrees.

29. Remove the M6 x 1.0 bolt.

30. Remove the tools from the camshaft and the cylinder block lower blind plug.

31. Rotate the crankshaft clockwise 2 turns until you reach TDC. If not aligned properly, loosen the crankshaft pulley lock

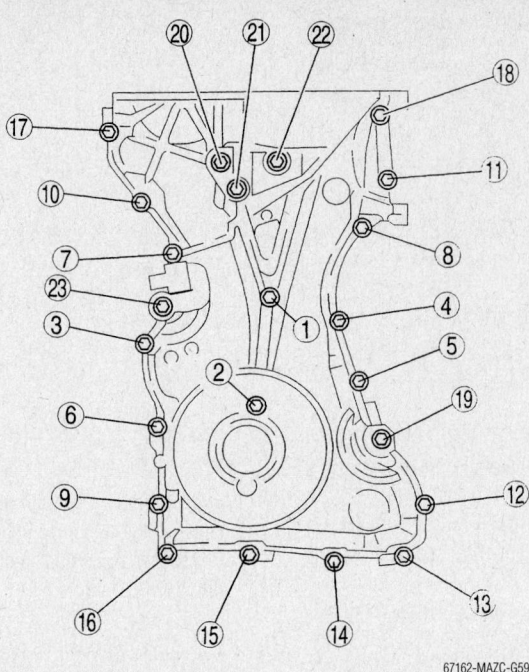

Engine front cover bolts tightening sequence—Mazda6 with the 2.3L (L3) engine

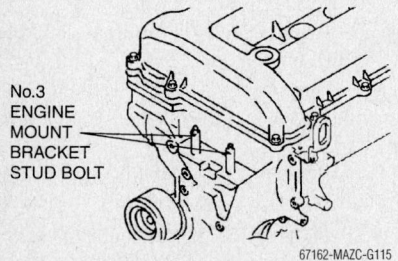

No. 3 engine mount bracket stud bolt locations—Mazda6 models with the 2.3L (L3) engine

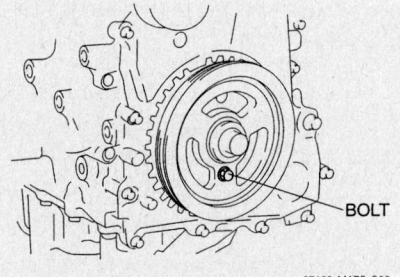

Install and hand tighten the M6 x 1.0 bolt—Mazda6 with the 2.3L (L3) engine

bolt and reinstall the lock bolt again using the above procedure and special tools.

32. Install the cylinder block blind plug and tighten to 13–16 ft. lbs. (18–22 Nm).

33. Install the cylinder head cover with a

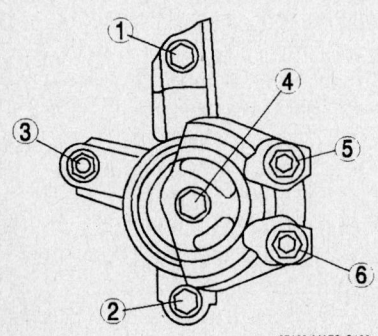

Tighten the No. 3 joint bracket bolts and nuts in the order shown—Mazda6 models with the 2.3L (L3) engine

new gasket. Torque the bolts in the sequence illustrated to 71–93 inch lbs. (8–11 Nm).

34. Install the remaining components in the reverse order of removal.

35. Change the engine oil.

36. Start the engine.

37. Inspect for the following:
- Pulley and belt for run–out and contact
- Any leaking fluids.
- Ignition timing, idle speed and exhaust emissions
- All remaining components for proper operation.

3.0L (AJ) ENGINE

1. Before servicing the vehicle, refer to the precautions section.

2. Disconnect the negative battery cable.

3. Remove the upper intake manifold (dynamic chamber).

4. Remove the ignition coils.

5. Remove the accessory drive belt.

6. Remove the power steering pump and bracket with the lines still attached and set aside.

7. Remove the A/C compressor with the lines still attached and set aside.

8. Unbolt the alternator with the connector still attached and wire it aside.

9. Remove the right hand halfshaft.

10. Remove the front crossmember as follows:

  a. Remove the stabilizer bar nut.

  b. Remove the front lower arm front and rear ball joints from the knuckle.

  c. Remove the front shock absorber lower bolt.

  d. Remove the No. 1 engine mount center bolt.

✳✳ WARNING

Support the crossmember with a suitable jack making sure it is securely damaged or it could fall causing injury or damage.

  e. Support the crossmember with a jack and remove the crossmember bracket.

  f. Remove the crossmember assembly retainers and lower the jack to remove the crossmember.

11. Remove the left hand three way converter.

12. Remove the oil pan.

13. Remove the plug hole plate.

14. Remove the water pump drive belt by rotating the belt tensioner counterclockwise.

15. Remove the water pump drive belt pulley as follows:

  a. Replace part of tool 49 UN30 3009 with tool 49 UN30 3457.

  b. Install the tools as illustrated and remove the water pump drive pulley.

16. Cut the camshaft oil seal using a razor knife and remove the seal using the tools illustrated.

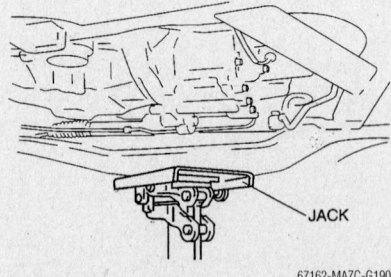

Support the crossmember with a jack and remove the crossmember bracket— Mazda6 with the 3.0L (AJ) engine

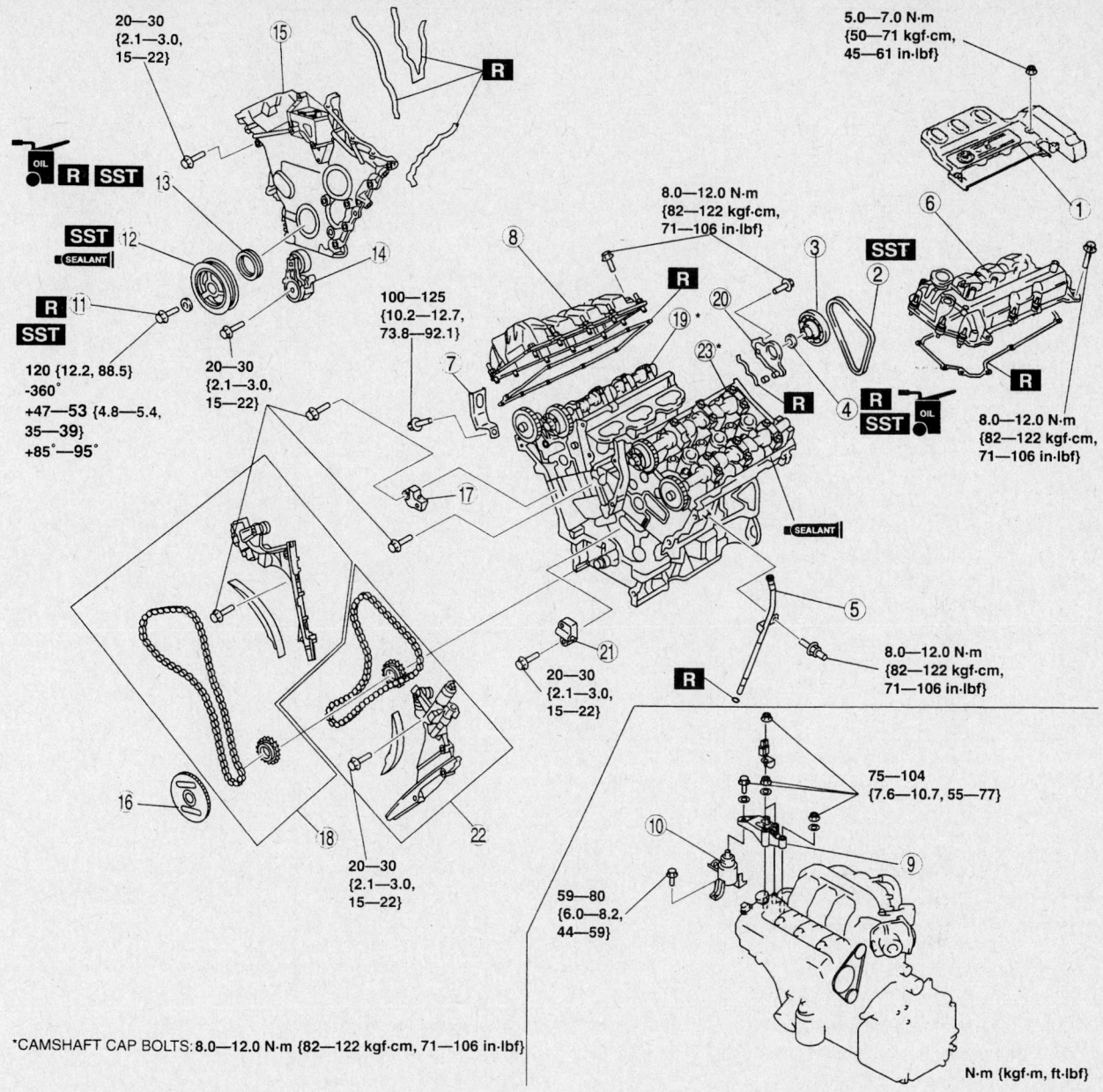

20—30
{2.1—3.0,
15—22}

5.0—7.0 N·m
{50—71 kgf·cm,
45—61 in·lbf}

8.0—12.0 N·m
{82—122 kgf·cm,
71—106 in·lbf}

SST
SEALANT

SST
OIL

100—125
{10.2—12.7,
73.8—92.1}

120 {12.2, 88.5}
-360°
+47—53 {4.8—5.4,
35—39}
+85°—95°

20—30
{2.1—3.0,
15—22}

8.0—12.0 N·m
{82—122 kgf·cm,
71—106 in·lbf}

SEALANT

8.0—12.0 N·m
{82—122 kgf·cm,
71—106 in·lbf}

75—104
{7.6—10.7, 55—77}

20—30
{2.1—3.0,
15—22}

20—30
{2.1—3.0,
15—22}

59—80
{6.0—8.2,
44—59}

N·m {kgf·m, ft·lbf}

*CAMSHAFT CAP BOLTS: 8.0—12.0 N·m {82—122 kgf·cm, 71—106 in·lbf}

1	Plug hole plate	13	Front oil seal
2	Water pump drive belt	14	Front drive belt auto tensioner
3	Water pump drive belt pulley	15	Engine front cover
4	Camshaft oil seal	16	CKP sensor pulse wheel
5	Oil level gauge pipe	17	Chain tensioner (RH)
6	Cylinder head cover (LH)	18	Timing chain component (RH)
7	Engine hanger (RH)	19	Camshaft cap (RH)
8	Cylinder head cover (RH)	20	Camshaft oil seal housing
9	No.3 engine joint bracket	21	Chain tensioner (LH)
10	No.3 engine mount rubber	22	Timing chain component (LH)
11	Crankshaft pulley lock bolt	23	Camshaft cap (LH)
12	Crankshaft pulley		

67162-MAZC-G189

Exploded view of the timing chain assembly and related components—Mazda6 with the 3.0L (AJ) engine

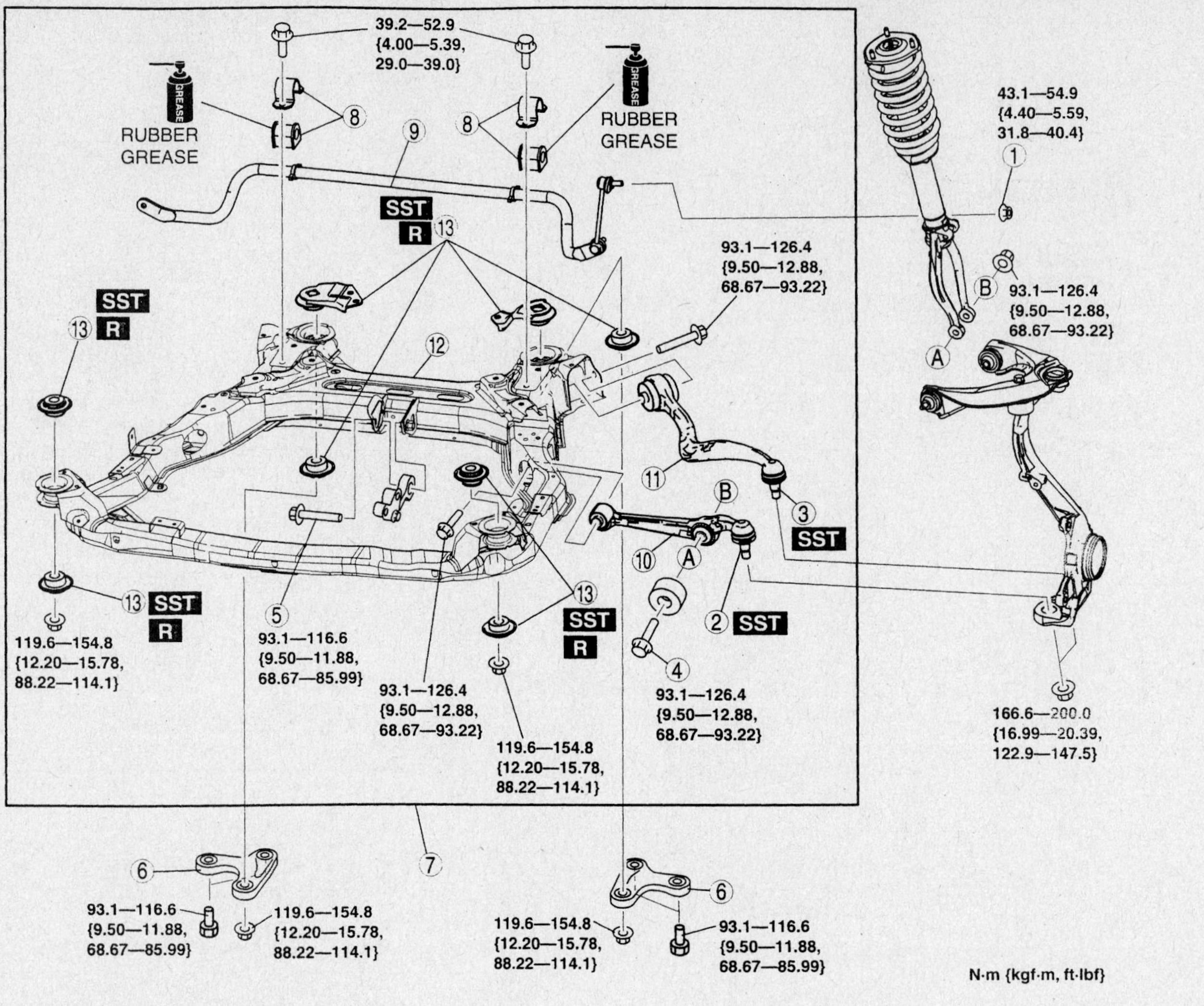

39.2—52.9
{4.00—5.39,
29.0—39.0}
RUBBER
GREASE
RUBBER
GREASE
GREASE
GREASE
43.1—54.9
{4.40—5.59,
31.8—40.4}
SST
R
SST
R
93.1—126.4
{9.50—12.88,
68.67—93.22}
93.1—126.4
{9.50—12.88,
68.67—93.22}
SST
SST
SST
R
SST
R
SST
119.6—154.8
{12.20—15.78,
88.22—114.1}
93.1—116.6
{9.50—11.88,
68.67—85.99}
93.1—126.4
{9.50—12.88,
68.67—93.22}
93.1—126.4
{9.50—12.88,
68.67—93.22}
119.6—154.8
{12.20—15.78,
88.22—114.1}
166.6—200.0
{16.99—20.39,
122.9—147.5}
93.1—116.6
{9.50—11.88,
68.67—85.99}
119.6—154.8
{12.20—15.78,
88.22—114.1}
119.6—154.8
{12.20—15.78,
88.22—114.1}
93.1—116.6
{9.50—11.88,
68.67—85.99}
N·m {kgf·m, ft·lbf}

1 Nut (stabilizer control link)
2 Front lower arm (front) ball joint
3 Front lower arm (rear) ball joint
4 Bolt (front shock absorber lower side)
5 No.1 engine mount center bolt
6 Crossmember bracket
7 Crossmember component
8 Stabilizer bracket and bushing
9 Front Stabilizer
10 Front lower arm (front)
11 Front lower arm (rear)
12 Front crossmember
13 Front crossmember bushing

67162-MAZC-G124

Exploded view of the crossmember assembly and related components—Mazda6 with the 3.0L (AJ) engine

17. Remove the oil dipstick tube.
18. Remove the left rocker arm cover.
19. Raise the vehicle and remove the engine hanger bolts and the hanger from below.
20. Remove the right rocker arm cover.
21. Remove the No. 3 joint bracket as follows:
    a. Install the right hand engine hanger.
    b. Use a M10 x 1.25 0.984 (25mm) length bolt to attach tool 49 UN30 3050 as illustrated. Tighten the bolt to 73–92 ft. lbs. (100–125 Nm).

303-009
(49 UN30 3009)
303-457
(49 UN30 3457)
PART A

67162-MAZC-G127

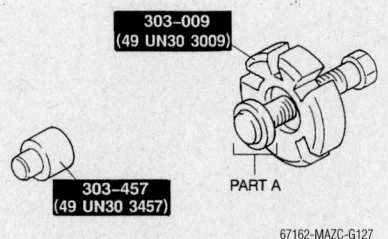

Replace part of tool 49 UN30 3009 with tool 49 UN30 3457—Mazda6 models with the 3.0L (AJ) engine

303-009
(49 UN30 3009)
303-457
(49 UN30 3457)
303-456
(49 UN30 3456)

67162-MAZC-G128

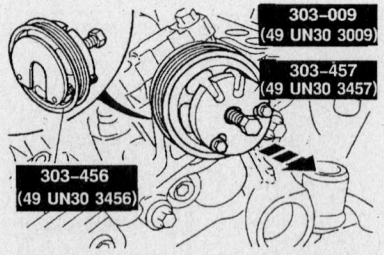

Install the tools as illustrated and remove the water pump drive pulley—Mazda6 models with the 3.0L (AJ) engine

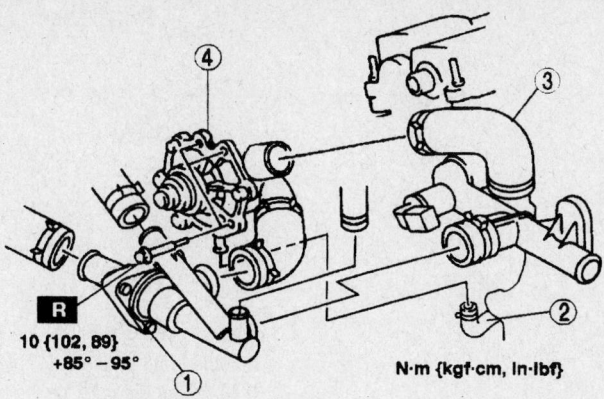

R

10 {102, 89}
+85° – 95°

N·m {kgf·cm, in·lbf}

1   Thermostat case
2   Heater hose
3   Water outlet pipe
4   Water pump

67162-MAZC-G126

Exploded view of the water pump and related components—Mazda6 models with the 3.0L (AJ) engine

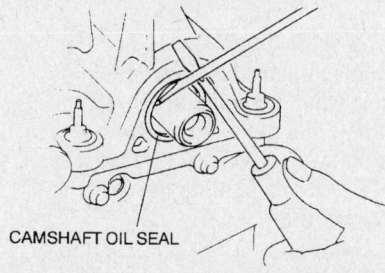

CAMSHAFT OIL SEAL

67162-MAZC-G192

Cut the camshaft oil seal using a razor knife and remove the seal as shown—Mazda6 with the 3.0L (AJ) engine

c. Suspend the engine using the engine support device and remove the joint bracket.

22. Remove the No. 3 mount rubber.

23. Remove the A/C compressor and set aside with the lines still attached.

24. Remove the crankshaft pulley using a suitable puller such as the one illustrated.

25. Remove the front oil seal using a prytool as illustrated.

26. Remove the drive belt tensioner.

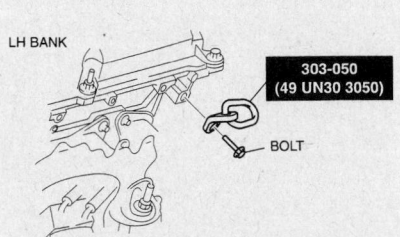

LH BANK

303-050
(49 UN30 3050)

BOLT

67162-MAZC-G193

Use a M10 x 1.25 0.984 (25mm) length bolt to attach tool 49 UN30 3050 as shown—Mazda6 with the 3.0L (AJ) engine

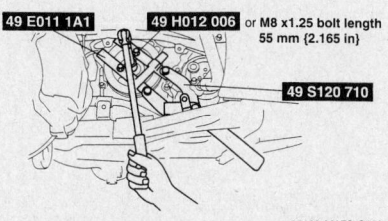

49 E011 1A1   49 H012 006 or M8 x1.25 bolt length 55 mm {2.165 in}

49 S120 710

67162-MAZC-G162

Hold the crankshaft pulley using the tools illustrated and remove the pulley bolt—Mazda6 with the 3.0L (AJ) engine

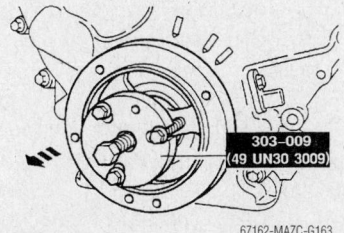

303-009
(49 UN30 3009)

67162-MAZC-G163

Remove the crankshaft pulley using a suitable puller such as the one shown—Mazda6 with the 3.0L (AJ) engine

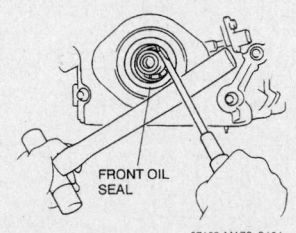

FRONT OIL SEAL

67162-MAZC-G164

Remove the front oil seal using a prytool as shown—Mazda6 with the 3.0L (AJ) engine

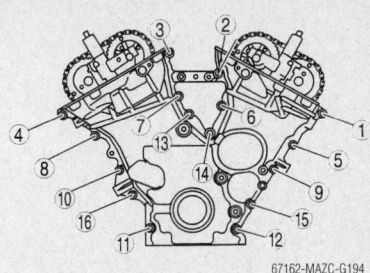

67162-MAZC-G194

Remove the engine front cover bolts in the sequence shown—Mazda6 with the 3.0L (AJ) engine

27. Remove the engine front cover bolts in the sequence illustrated.

28. Remove the Crankshaft position (CKP) sensor pulse wheel.

29. Remove the right side chain tensioner and chain assembly as follows:

a. Rotate the crankshaft clockwise so the crankshaft keyway is at the 3 O'clock position. The camshafts should be in the neutral position.

➤**Do not rotate the crankshaft counterclockwise or you may bind the chains causing engine damage.**

b. Remove the right side chain in this order:
- Chain tensioner
- Tensioner arm
- Timing chain
- Chain guide
- Timing chain crankshaft sprocket

30. Remove the right side camshaft caps. Refer to the camshaft removal procedure earlier in this section

31. Remove the left side chain tensioner and chain assembly as follows:

a. Rotate the crankshaft clockwise 1 ⅔ turns until the keyway is at the 11 O'clock position.

b. Remove the chain in this order:
- Chain tensioner
- Tensioner arm
- Timing chain
- Chain guide
- Timing chain crankshaft sprocket

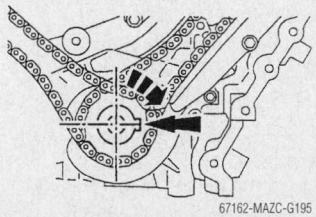

67162-MAZC-G195

Rotate the crankshaft clockwise so the crankshaft keyway is at the 3 O'clock position when removing the right side chain—Mazda6 with the 3.0L (AJ) engine

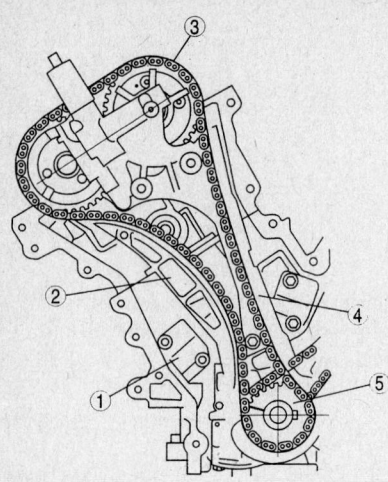

(1) Chain tensioner
(2) Tensioner arm
(3) Timing chain
(4) Chain guide
(5) Timing chain crankshaft sprocket

67162-MAZC-G196

Remove right side timing chain in numerical order—Mazda6 with the 3.0L (AJ) engine

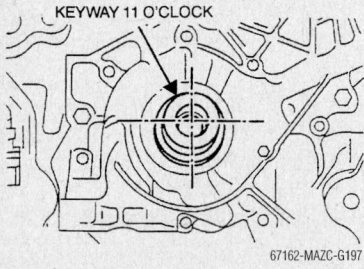

67162-MAZC-G197

Rotate the crankshaft clockwise 1 ⅔ turns until the keyway is at the 11 O'clock when removing the left side chain—Mazda6 with the 3.0L (AJ) engine

32. Remove the left side camshaft caps. Refer to the camshaft removal procedure earlier in this section

To install:

33. Install the left side chain assembly as follows:

  a. Make sure the crankshaft keyway is at the 11 O'clock position.

  b. Position the mark on the intake camshaft to 9 O'clock.

  c. Place the mark on the exhaust camshaft at the 12 O'clock position.

  d. Align the colored links on the timing chain with the marks on the timing sprockets.

  e. If the timing chain marks cannot be seen, use a marker or paint pen to mark the crankshaft and camshaft marks on the chain as follows:

  • Mark any link as the crankshaft link
  • Count 29 links from the crankshaft

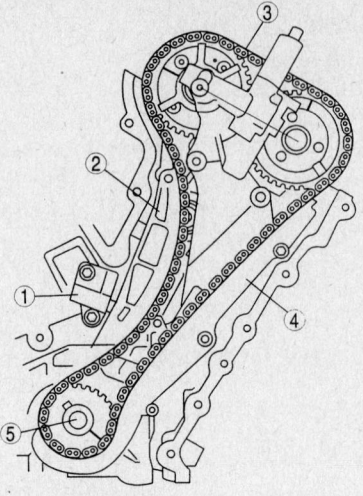

(1) Chain tensioner
(2) Tensioner arm
(3) Timing chain
(4) Chain guide
(5) Timing chain crankshaft sprocket

67162-MAZC-G198

Remove left side timing chain in numerical order—Mazda6 with the 3.0L (AJ) engine

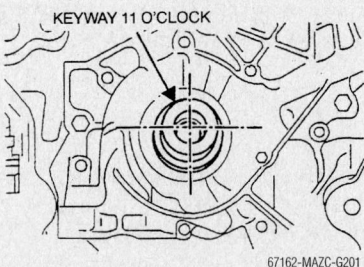

67162-MAZC-G201

When installing the left side chain, make sure the crankshaft keyway is at the 11 O'clock position—Mazda6 with the 3.0L (AJ) engine

mark and place a mark to be used as the exhaust camshaft sprocket mark

  • Continue counting to the 42 link mark and mark this link as the intake camshaft sprocket mark

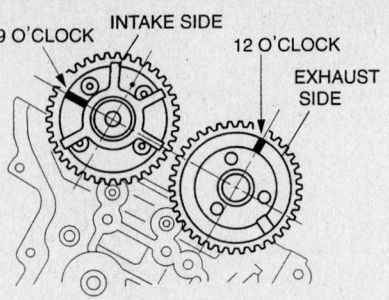

67162-MAZC-G202

When installing the left side chain, Position the mark on the intake camshaft to 9 O'clock and the mark on the exhaust camshaft at the 12 O'clock position—Mazda6 with the 3.0L (AJ) engine

34. Install the left hand chain in this order:

  • Timing chain crankshaft sprocket
  • Chain guide
  • Timing chain
  • Tensioner arm
  • Chain tensioner

➡**The chain guide should be installed to the actuator and allowed to hang freely when the bolts are installed. Do not hold the guide up when installing the bolts. The actuator causes wear to an O−ring and holding the guide will increase wear.**

  a. Install the left side camshaft caps. Refer to the camshaft removal procedure earlier in this section

  b. Install the left side chain tensioner as follows:

  • Place the tensioner in a vise with jaw protectors
  • Use a small screwdriver to hold the tensioner ratchet lock mechanism away from the ratchet stem

➡**Minimal force should be used to retract the piston, if binding occurs remove then reinstall the tensioner is the vise.**

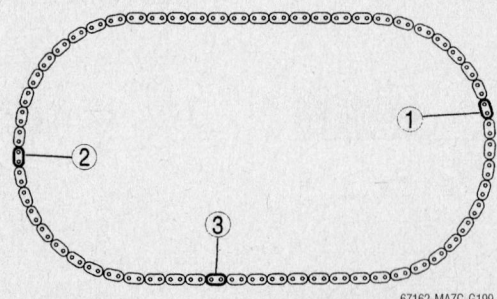

67162-MAZC-G199

If the timing chain marks cannot be seen, use a marker or paint pen to mark the crankshaft and camshaft marks on the chain. Refer to the text for marking steps.—Mazda6 with the 3.0L (AJ) engine

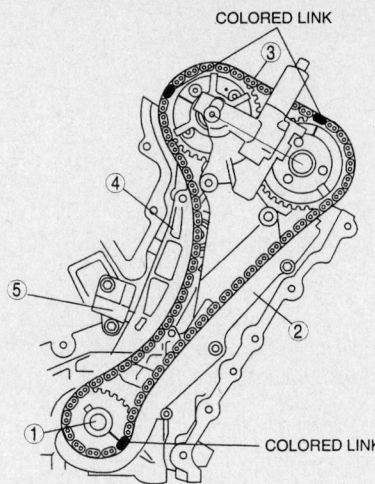

(1) Timing chain crankshaft sprocket
(2) Chain guide
(3) Timing chain
(4) Tensioner arm
(5) Chain tensioner

67162-MAZC-G200

Install the left side timing chain in numerical order and align the colored links with the timing marks on the sprockets—Mazda6 with the 3.0L (AJ) engine

• Slowly compress the tensioner piston and install a paper clip to hold the piston
• Install the tensioner, tighten the bolts to 15–22 ft. lbs. (20–30 Nm) and remove the paper clip

35. Install the right side chain assembly as follows:

   a. Make sure the crankshaft keyway is at the 3 O'clock position.

   b. Place the mark on the exhaust camshaft at the 12 O'clock position.

   c. Position the mark on the intake camshaft to 3 O'clock.

   d. Align the colored links on the timing chain with the marks on the timing sprockets.

   e. If the timing chain marks cannot be seen, use a marker or paint pen to mark

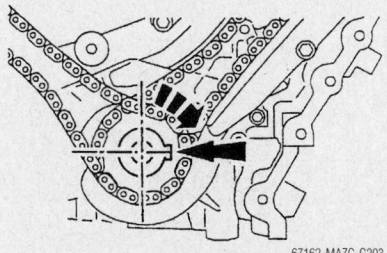

67162-MAZC-G203

When installing the right side chain, make sure the crankshaft keyway is at the 3 O'clock position—Mazda6 with the 3.0L (AJ) engine

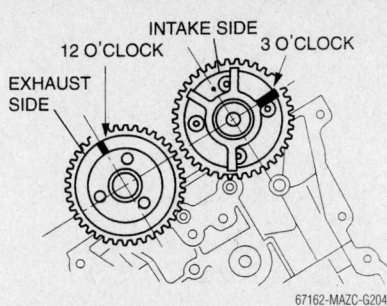

67162-MAZC-G204

When installing the right side chain, Position the mark on the exhaust camshaft to 12 O'clock and the mark on the intake camshaft at the 3 O'clock position—Mazda6 with the 3.0L (AJ) engine

the crankshaft and camshaft marks on the chain as follows:

• Mark any link as the crankshaft link
• Count 29 links from the crankshaft mark and place a mark to be used as the exhaust camshaft sprocket mark
• Continue counting to the 42 link mark and mark this link as the intake camshaft sprocket mark

36. Install the left hand chain in this order:

• Timing chain crankshaft sprocket
• Chain guide
• Timing chain
• Tensioner arm
• Chain tensioner

➡The chain guide should be installed to the actuator and allowed to hang freely when the bolts are installed. Do not hold the guide up when installing

the bolts. The actuator causes wear to an O–ring and holding the guide will increase wear.

   a. Install the right side camshaft caps. Refer to the camshaft removal procedure earlier in this section

   b. Install the right side chain tensioner as follows:

• Place the tensioner in a vise with jaw protectors
• Use a small screwdriver to hold the tensioner ratchet lock mechanism away from the ratchet stem

➡Minimal force should be used to retract the piston, if binding occurs remove then reinstall the tensioner is the vise.

• Slowly compress the tensioner piston and install a paper clip to hold the piston
• Install the tensioner, tighten the

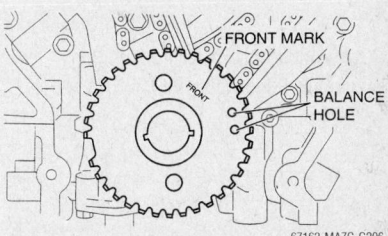

67162-MAZC-G206

Place the CKP sensor pulse wheel with theFrontmark facing towards you, using the keyway on the same side as the empty space as shown—Mazda6 with the 3.0L (AJ) engine

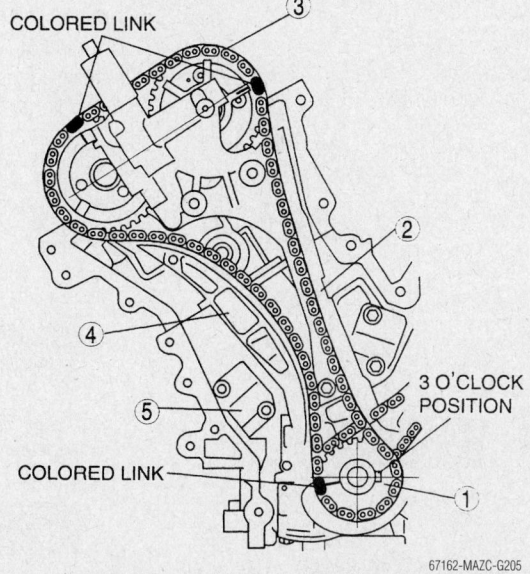

67162-MAZC-G205

Install the right side timing chain in numerical order and align the colored links with the timing marks on the sprockets—Mazda6 with the 3.0L (AJ) engine

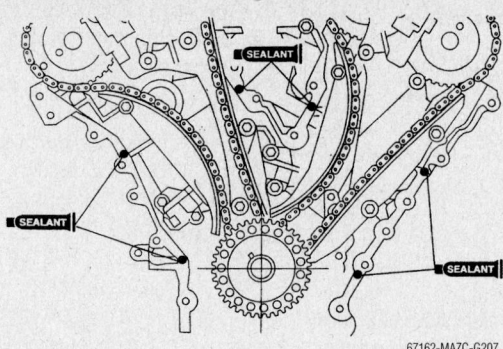

67162-MAZC-G207

Apply a 0.24 inch (6mm) bead of silicone sealant to the locations illustrated and install the front cover—Mazda6 with the 3.0L (AJ) engine

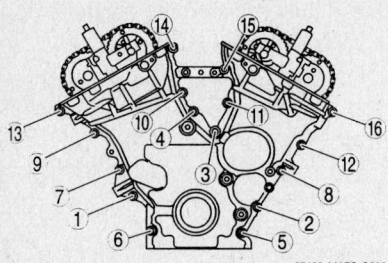

67162-MAZC-G208

Tighten the front cover bolts in this sequence—Mazda6 with the 3.0L (AJ) engine

bolts to 15–22 ft. lbs. (20–30 Nm) and remove the paper clip.

37. Place the CKP sensor pulse wheel with the **Front** mark facing towards you, using the keyway on the same side as the empty space as illustrated.

38. Apply a 0.24 inch (6mm) bead of silicone sealant to the locations illustrated and install the front cover.

39. Tighten the front cover bolts in the sequence illustrated and tighten the retainers to 15–20 ft. lbs. (20–30 Nm).

40. Assemble the seal using part (A) of tool 49 UN01 002 and tool 49 UN01 002 as illustrated.

41. Apply clean oil to the seal and push the seal in by hand.

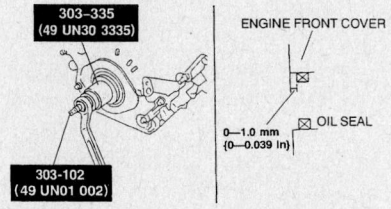

67162-MAZC-G166

Install the seal using the installation tools until the seal is recessed 0–0.039 inch (0–1mm)—Mazda6 with the 3.0L (AJ) engine

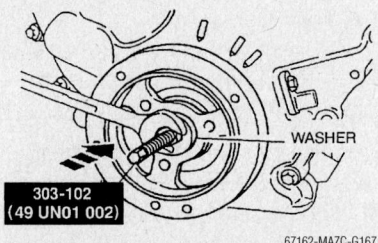

67162-MAZC-G167

Install the pulley using the tools shown—Mazda6 with the 3.0L (AJ) engine

42. Install the seal using the installation tools until the seal is recessed 0–0.039 inch (0–1mm).

43. Seal the crankshaft pulley using a silicone sealant.

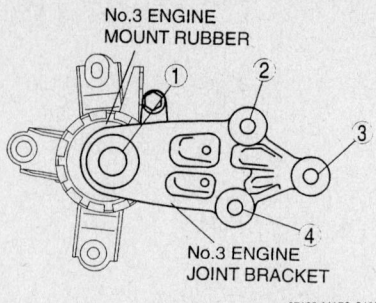

67162-MAZC-G123

Tighten the No. 3 engine joint bracket as shown—Mazda6 models with the 3.0L (AJ) engine

44. Install the pulley using the tools illustrated.

45. While holding the pulley, tighten the crankshaft pulley bolt as follows:
   a. Tighten to 88 ft. lbs. (120 Nm).
   b. Loosen one full turn.
   c. Tighten to 35–39 ft. lbs. (47–53 Nm).
   d. Tighten 85–95 degrees.

46. Install the no. 3 engine mount rubber.

47. When installing the No. 3 engine joint bracket, tighten the No.3 mount bracket bolts and nuts in the order illustrated as follows:
   a. Bolt 1: 55–77 ft. lbs. (74–105 Nm).
   b. Bolts 2, 3 and 4: 31–44 ft. lbs. (43–60 Nm).

48. Install the right hand valve cover.

49. Install the right hand engine hanger.

50. Install the left hand valve cover.

51. Install the dipstick tube.

52. Apply clean engine oil to the camshaft seal and install using the tools illustrated. The seal should be recessed 0.10–0.11 inch (2.5–3mm).

53. Install a new water pump drive belt pulley using tool 49 UN21 1185.

54. Install the water pump drive belt.

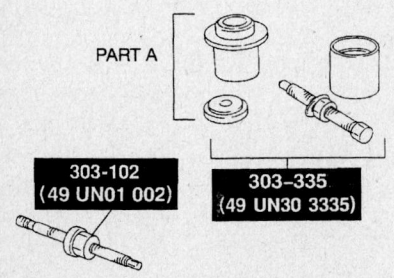

67162-MAZC-G165

Assemble the seal using part (A) of tool 49 UN01 002 and tool 49 UN01 002 as shown—Mazda6 with the 3.0L (AJ) engine

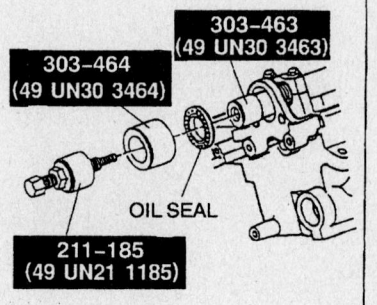
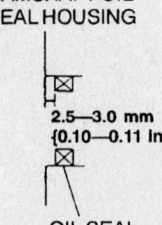

67162-MAZC-G209

Install the camshaft oil seal—Mazda6 with the 3.0L (AJ) engine

211–185
(49 UN21 1185)

67162-MAZC-G129

Install a new water pump drive belt pulley using tool 49 UN21 1185—Mazda6 models with the 3.0L (AJ) engine

55. Install the plug hole plate.
56. Install the remaining components in the reverse order of removal.
57. Chain the engine oil and filter.
58. Check all fluid levels and replenish as necessary.
59. road test the vehicle.

Mazdaspeed6

1. Before servicing the vehicle, refer to the precautions section.
2. Disconnect the negative battery cable.
3. Remove the right side front tire.
4. Remove the under cover.
5. Remove the right side splash shield.

➡**If the high pressure fuel pump is removed, replace the O-ring with a new one.**

6. Remove the charge air cooler cover.
7. Disconnect the spill valve control solenoid valve connector.
8. Disconnect the quick release connector on the high pressure fuel pump.
9. Remove the battery and battery tray.

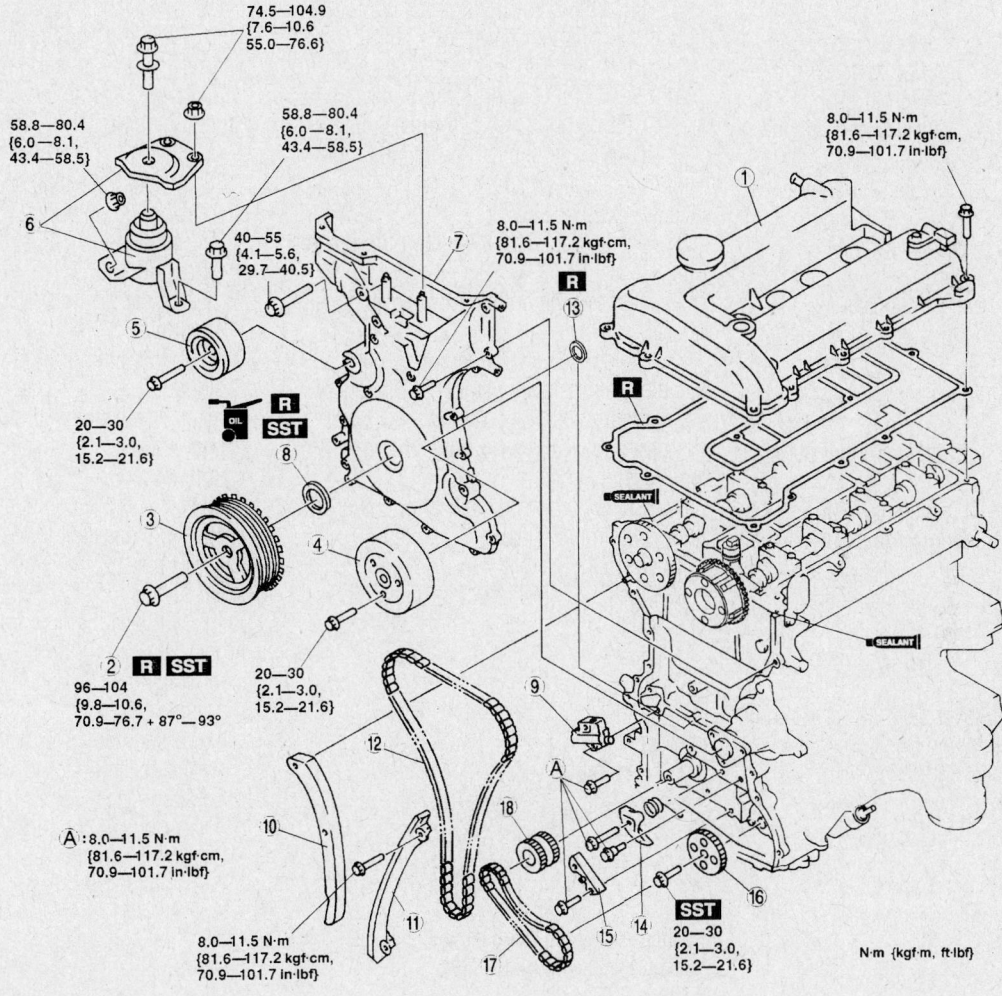

1	Cylinder head cover	10	Tensioner arm
2	Crankshaft pulley lock bolt	11	Chain guide
3	Crankshaft pulley	12	Timing chain
4	Water pump pulley	13	Seal
5	Drive belt idler pulley	14	Oil pump chain tensioner
6	No.3 engine mount rubber and No.3 engine joint bracket	15	Oil pump chain guide
7	Engine front cover	16	Oil pump sprocket
8	Front oil seal	17	Oil pump chain
9	Chain tensioner	18	Crankshaft sprocket

67162-MAZC-G186

Exploded view of the timing chain assembly and related components—Mazdaspeed6

✲✲ WARNING

If the high pressure fuel pump joint nut is loosened, fuel leakage may occur resulting in death or serious injury, or damage to the equipment or the vehicle. Fuel can also irritate the skin and eyes. When removing the high pressure line pipe, always tighten the high pressure line pipe installation nut while fixing the high pressure fuel pump joint nut with a wrench. If the high pressure fuel pump joint nut has rotated, replace the high pressure fuel pump with a new one.

10. Disconnect the high pressure line pipe of the high pressure fuel pump.

11. Fix the joint nut with a wrench on the high pressure fuel pump side.

12. Loosen the high pressure line pipe installation nut.

13. Drain and recycle the engine coolant.

14. Loosen the water outlet case installation bolts securing the high pressure line pipe.

15. Remove the high pressure fuel pump.

16. Remove the high pressure fuel pump cover.

17. Disconnect the Camshaft Position (CMP) and power steering oil pump connectors.

18. Remove the ignition coils, spark plugs, valve cover and accessory drive belt.

19. Remove the Crankshaft Position (CKP) sensor.

20. Remove the power steering oil pump with the lines attached and position aside.

21. Disconnect the right hand driveshaft from the joint shaft.

22. Remove the cylinder block lower blind plug.

23. Install tool 49 JE01 061.

24. Turn the crankshaft clockwise until the crankshaft is in the number 1 cylinder Top Dead Center position (the balance weight should be attached to tool 49 JE01 061).

67162-MAZC-G27

Install tool 49 JE01 061—Mazdaspeed6

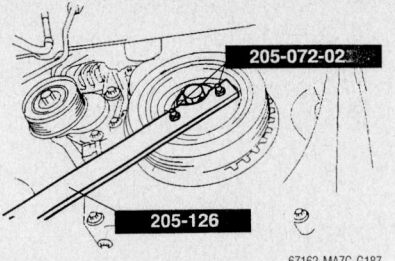

67162-MAZC-G187

Install tools 205-072-02 and 205-126 to hold the crankshaft pulley using the tools illustrated and remove the bolt—Mazdaspeed6

25. Hold the crankshaft pulley using the tools illustrated and remove the bolt.

26. Remove the crankshaft pulley.

27. Remove the water pump pulley.

28. Remove the drive belt idler pulley.

29. Disconnect the engine ground.

30. Install an engine support device such as 49 E017 5A0.

31. Remove the No. 3 engine mount and joint.

32. Remove the engine front cover and oil seal.

33. Unlock the chain tensioner using a suitable tool to slowly compress the tensioner piston. Insert a 0.059 inch (1.5mm) wire or a paper clip to hold the piston in its compressed position.

34. Remove the tensioner arm, chain guide and timing chain.

To install:

35. Install tool 49 UN30 3465 as illustrated on the camshaft.

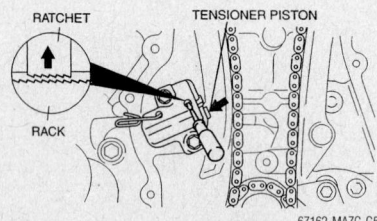

67162-MAZC-G57

Compressing and retaining the chain tensioner piston—Mazdaspeed6

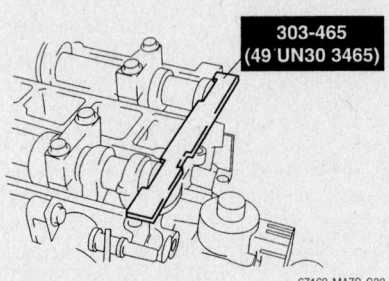

67162-MAZC-G32

Install tool 49 UN30 3465 on the camshaft—Mazdaspeed6

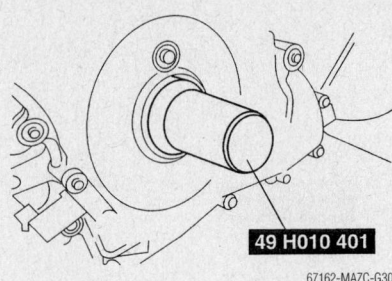

67162-MAZC-G30

Use seal installer tool 49 H010 401 and a hammer to install the seal—Mazdaspeed6

36. Install the timing chain and remove the paper clip or wire retaining the tensioner piston to apply tension to the chain.

37. Install the timing chain guide and tighten the bolts to 71–101 inch lbs. (8–11.5 Nm).

38. Install the tensioner arm and tighten the bolts to 71–101 inch lbs. (8–11.5 Nm).

39. Install a new oil seal in the front cover as follows:

    a. Coat the new seal with clean engine oil.

    b. Push the seal in by hand to get it started.

    c. Use seal installer tool 49 H010 401 and a hammer to install the seal so that it is recessed 0–0.019 inch (0–0.5mm) as shown in the accompanying illustration.

40. Apply sealant to the engine front cover. At point A the bead should be 0.086–0.126 inch (2.2–3.2mm) thick and at point B the bead should be 0.059–0.098 inch (1.5–2.5mm) thick. Refer to the accompanying illustration for the locations of points A and B.

41. Install the cover within 10 minutes of apply the sealant. Tighten the cover bolts as follows:

    a. Bolts 1 through 18: In sequence to 71–101 inch lbs. (8–11.5 Nm).

    b. Bolts 19 through 22: In sequence to 29.7–40.5 ft. lbs. (40–55 Nm).

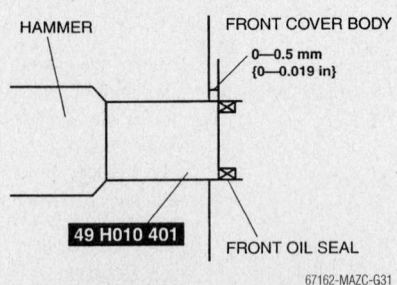

67162-MAZC-G31

The seal should be recessed 0–0.019 inch (0–0.5mm) as shown when properly installed—Mazdaspeed6

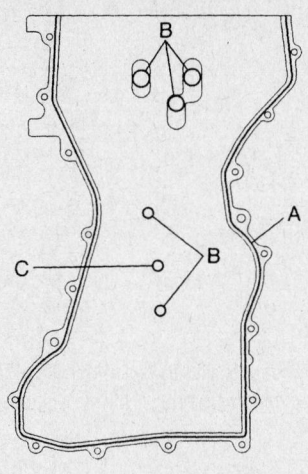

Apply a bead of sealant to the engine front cover at the locations shown. Refer to the text for the bead thickness—Mazdaspeed6

No. 3 engine mount bracket stud bolt locations—Mazdaspeed6

c. Bolt 23: 14.8–22 ft. lbs. (20–30 Nm).

42. When installing the No. 3 engine joint bracket, tighten the mount bracket stud bolt to 62–115 inch lbs. (7–13 Nm) and the

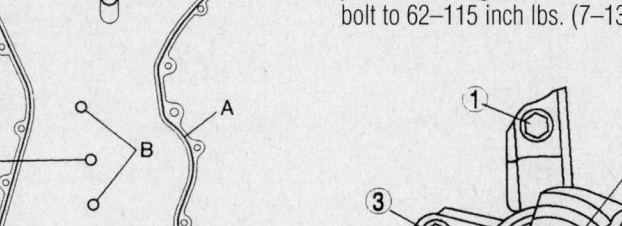

Tighten the No. 3 joint bracket bolts and nuts in the order shown—Mazdaspeed6

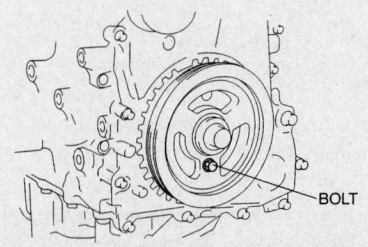

Install and hand tighten the M6 x 1.0 bolt—Mazdaspeed6

joint bracket bolt and nut in the order illustrated to 55–76 ft. lbs. (74–105 Nm).

43. Install the drive belt idler and water pump pulleys.

44. Install tool 49 UN30 3465 as illustrated on the camshaft.

45. Install and hand tighten the M6 x 1.0 bolt as illustrated.

46. Turn the crankshaft clockwise until the crankshaft is in number 1 TDC (the balance weight should be attached to the tool).

47. Hold the crankshaft pulley and tighten the lock bolt in two steps. First tighten to 71–77 ft. lbs. (96–104 Nm). Then final tighten an additional 87–93 degrees.

48. Remove the M6 x 1.0 bolt.

49. Remove the tools from the camshaft and the cylinder block lower blind plug.

50. Rotate the crankshaft clockwise 2 turns until you reach TDC. If not aligned properly, loosen the crankshaft pulley lock bolt and reinstall the lock bolt again using the above procedure and special tools.

51. Install the cylinder block blind plug and tighten to 13–16 ft. lbs. (18–22 Nm).

52. Install the cylinder head cover with a new gasket. Torque the bolts in the sequence illustrated to 71–93 inch lbs. (8–11 Nm).

53. Install the high pressure fuel pump cover and tighten to 13–16 ft. lbs. (18–22 Nm).

✳✳ CAUTION

If the high pressure fuel pump installation bolts are tightened with the high pressure fuel pump tilted, the high pressure fuel pump may not operate correctly. Tighten the high pressure fuel pump installation bolts in a few passes with equal torque. Tighten to 76–101 inch lbs. (8.5–11.5 Nm).

✳✳ WARNING

If the high pressure fuel pump joint nut is loosened, fuel leakage may occur resulting in death or serious injury, or damage to the equipment or the vehicle. Fuel can also irritate the skin and eyes. When installing the high pressure line pipe, always tighten the high pressure line pipe installation nut while fixing the high pressure fuel pump joint nut with a wrench. If the high pressure fuel pump joint nut has rotated, replace the high pressure fuel pump with a new one.

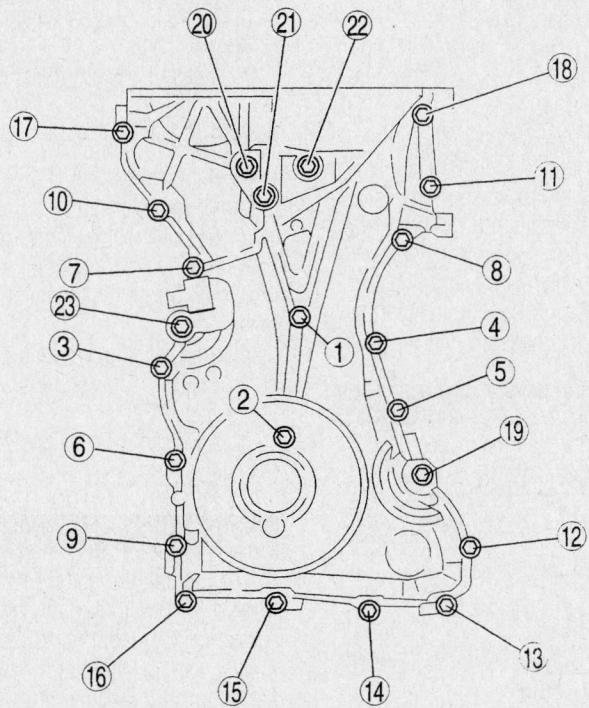

Engine front cover bolts tightening sequence—Mazdaspeed6

54. Assemble the high pressure line pipe.

55. Fix the joint nut with a wrench on the high pressure fuel pump side. Tighten the high pressure line pipe installation nut to 17–26 ft. lbs. (23–35 Nm).

56. Tighten the water outlet case installation bolts to 13–16 ft. lbs. (18–22 Nm).

57. Install the quick release connector.

58. Install the remaining components in the reverse order of removal.

59. Change the engine oil.

60. Start the engine.

61. Inspect for the following:
- Pulley and belt for run–out and contact
- Ignition timing, idle speed and exhaust emissions
- All remaining components for proper operation.

62. Verify that the high pressure fuel pump is assembled securely.

63. Drive the vehicle starting from a standstill and brake suddenly five to six times at a low speed.

64. Stop the vehicle and verify from outside the vehicle that there is no fuel leakage around the high pressure fuel pump.

Piston and Ring

POSITIONING

7923AG60

Before removing the caps from the connecting rods, be sure to matchmark them—Mazda engines

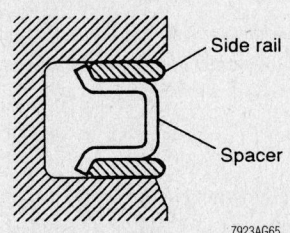

7923AG65

Compression ring identification and positioning—Mazda engines

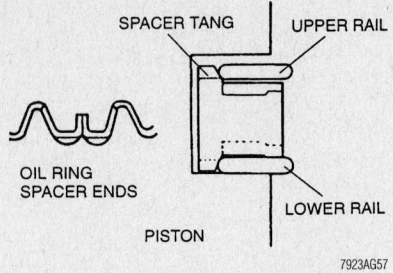

7923AG57

Upper, spacer and lower oil ring identification and positioning—Mazda engines

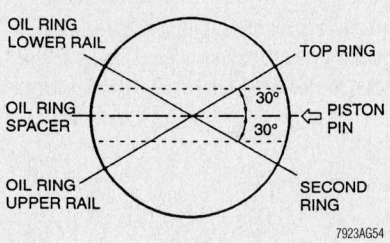

7923AG54

Piston ring end-gap spacing—Mazda engines

FUEL SYSTEM

Fuel System Service Precautions

Safety is the most important factor when performing not only fuel system maintenance but any type of maintenance. Failure to conduct maintenance and repairs in a safe manner may result in serious personal injury or death. Maintenance and testing of the vehicle's fuel system components can be accomplished safely and effectively by adhering to the following rules and guidelines.

1. To avoid the possibility of fire and personal injury, always disconnect the negative battery cable unless the repair or test procedure requires that battery voltage be applied.

2. Always relieve the fuel system pressure prior to disconnecting any fuel system component (injector, fuel rail, pressure regulator, etc.), fitting or fuel line connection. Exercise extreme caution whenever relieving fuel system pressure, to avoid exposing skin, face and eyes to fuel spray. Please be advised that fuel under pressure may penetrate the skin or any part of the body that it contacts.

3. Always place a shop towel or cloth around the fitting or connection prior to loosening to absorb any excess fuel due to spillage. Ensure that all fuel spillage (should it occur) is quickly removed from engine surfaces. Ensure that all fuel soaked cloths or towels are deposited into a suitable waste container.

4. Always keep a dry chemical (Class B) fire extinguisher near the work area.

5. Do not allow fuel spray or fuel vapors to come into contact with a spark or open flame.

6. Always use a back-up wrench when loosening and tightening fuel line connection fittings. This will prevent unnecessary stress and torsion to fuel line piping. Always follow the proper torque specifications.

7. Always replace worn fuel fitting O-rings with new. Do not substitute fuel hose where fuel pipe is installed.

Fuel System Pressure

RELIEVING

1. After servicing the vehicle, reinstall the relay.

MX-5 Miata

1. Before servicing the vehicle, refer to the precautions section.

2. Remove the fuel-filler cap to release the pressure inside the fuel tank.

3. Remove the fuel pump relay

4. Start the engine.

5. After the engine stalls, crank the engine several times.

6. Turn the ignition switch to the LOCK position.

7. Install the fuel pump relay.

Except MX-5 Miata

1. Before servicing the vehicle, refer to the precautions section.

2. Remove the filler cap.

3. Remove the fuel pump relay from the relay box, located in the engine compartment.

4. Start the engine.

5. After the engine stalls, turn the ignition switch OFF.

6. After servicing the vehicle, reinstall the relay.

Fuel Filter

The fuel filter on most Mazda cars can be located on a bracket in the left rear of the engine compartment, next to or beneath the brake master cylinder fluid reservoir or as part of the fuel pump assembly.

The filter on Mazda3 models is part of the fuel pump assembly.

REMOVAL & INSTALLATION

MX5-Miata

The fuel filter is part of the pump assembly. Once the fuel pump is removed, remove the filter retainer and filter. refer to the fuel pump procedure or removal and installation.

Fuel Pump

REMOVAL & INSTALLATION

MX-5 Miata

1. Before servicing the vehicle, refer to the precautions section.
2. Properly relieve the fuel system pressure.
3. Remove the battery cover.
4. Disconnect the negative battery cable.
5. Perform the following procedure to remove the service hole cover.

    a. Remove the console.
    b. Remove the quarter trim.
    c. Remove the scuff plate.
    d. Remove the tire house trim.
    e. Remove the aeroboard.
    f. Remove the front seat back bar garnish.
    g. Remove the remove the back trim.
    h. Remove the service hole cover.

6. Disconnect the quick release connector from the fuel pump unit.
7. Disconnect the fuel pump unit connector.
8. Remove plate, packing and pump assembly.

To install:

9. Installation is the reverse of removal.

Mazda3

1. Before servicing the vehicle, refer to the precautions section.
2. Relive the fuel system pressure.
3. Disconnect the quick release connector in the engine compartment.
4. Attach a long hose to the disconnect fuel pipe and drain the fuel into a suitable container as follows:

    a. Using a jumper wire, ground the Powertrain Control Module (PCM) terminals. If equipped with an immobilizer, ground terminal 1AR, if not equipped with an imobilizer, ground terminal 1AQ. Refer to the illustration for terminal location.
    b. Turn the ignition switch the ON position to activate the fuel pump.

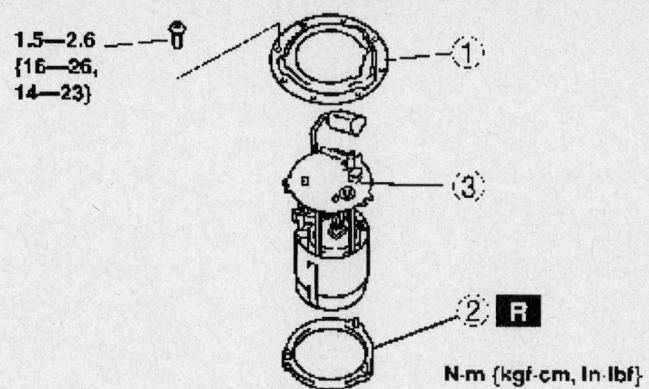

1.5—2.6
{16—26,
14—23}

N·m {kgf·cm, ln·lbf}

1 Plate
2 Packing
3 Fuel pump unit

09482_MAZC2_G0062

Exploded view of the fuel pump assembly—MX-5 Miata

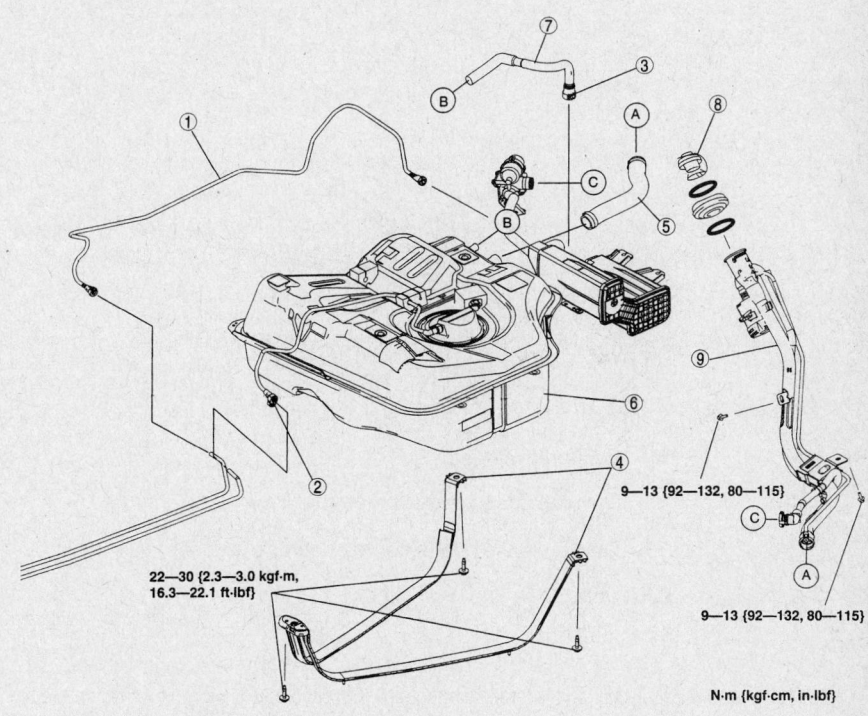

22—30 {2.3—3.0 kgf·m,
16.3—22.1 ft·lbf}

9—13 {92—132, 80—115}

9—13 {92—132, 80—115}

N·m {kgf·cm, in·lbf}

1	Evaporative hose	6	Fuel tank
2	Quick release connector (on fuel line)	7	Breather hose
3	Quick release connector (on charcoal canister)	8	Fuel-filler cap
4	Strap	9	Fuel-filler pipe
5	Joint hose		

67162-MAZC-G62

Exploded view of the fuel tank and related components—Mazda3 California emission models

✳✳ CAUTION

The fuel pump can be damaged if all fuel is removed from the tank, monitor the hose and stop when no fuel is be discharged from the tank.

    c. Once no more fuel is being discharged, turn the ignition OFF.
5. Disconnect the negative battery cable.
6. Remove the rear seat cushion and remove the pump service hole cover.
7. Disconnect the fuel pump connector and remove the charcoal canister.

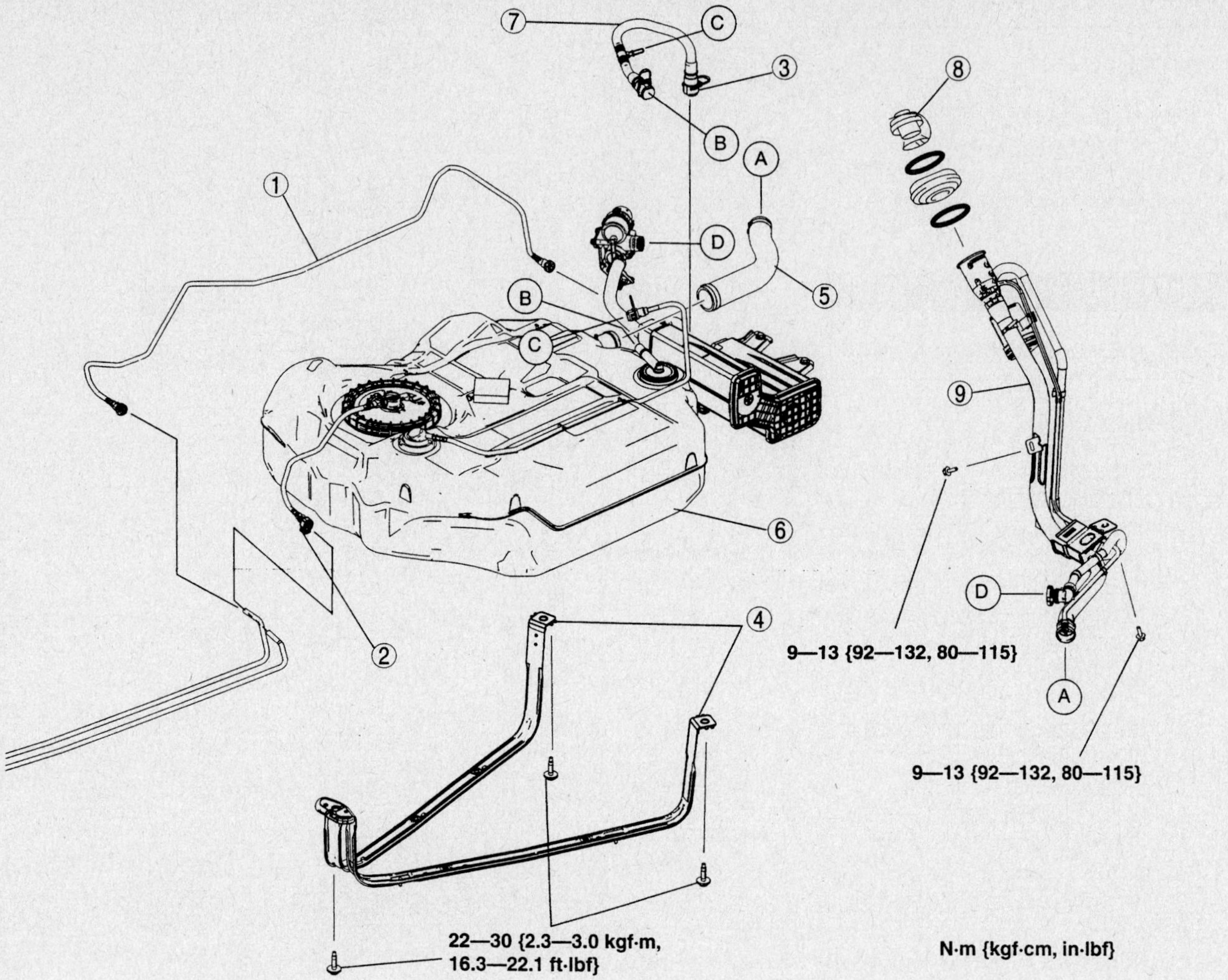

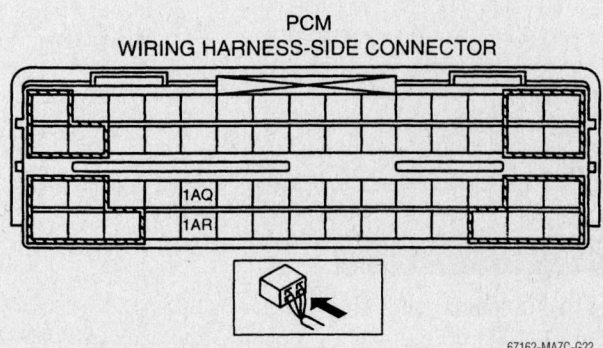

9—13 {92—132, 80—115}

9—13 {92—132, 80—115}

22—30 {2.3—3.0 kgf·m,
16.3—22.1 ft·lbf}

N·m {kgf·cm, in·lbf}

1	Evaporative hose	6	Fuel tank
2	Quick release connector (on fuel line)	7	Breather hose
3	Quick release connector (on charcoal canister)	8	Fuel-filler cap
4	Strap	9	Fuel-filler pipe
5	Joint hose		

67162-MAZC-G70

Exploded view of the fuel tank and related components—Mazda3 non–California emission models

PCM
WIRING HARNESS-SIDE CONNECTOR

1AQ
1AR

67162-MAZC-G22

If equipped with an immobilizer, ground PCM terminal 1AR, if not equipped with an imobilizer, ground PCM terminal 1AQ—Mazda3 models

8. Lower the exhaust main silencer so the insulator can be removed and remove the insulator.

9. Remove the rear left hand under cover.

10. Disconnect the evaporator hose, quick release fuel connector on the fuel line and charcoal canister.

11. Support the fuel tank and remove the strap.

12. Remove the filler pipe bolt, loosen the tie band and pull down on the filler pipe to disconnect the joint hose.

13. Lower and remove the fuel tank.

14. Disconnect the fuel release connector.

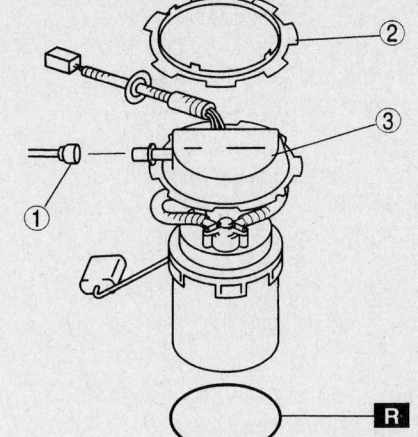

1   Quick release connector

2   Retainer ring

3   Fuel pump unit

67162-MAZC-G63

Exploded view of the pump assembly and related components—Mazda3 non–California emission models

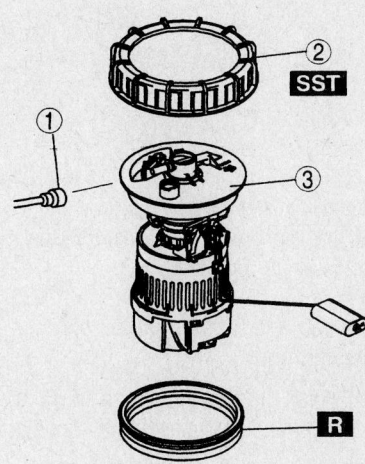

1   Quick release connector

2   Fuel pump cap

3   Fuel pump unit

67162-MAZC-G64

Exploded view of the pump assembly and related components—Mazda3 California emission models

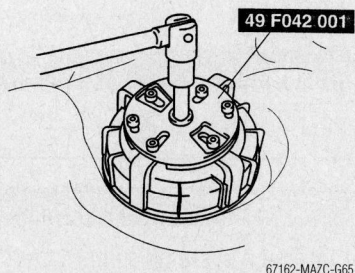

`49 F042 001`

67162-MAZC-G65

Use tool 49 F042 001 to remove the pump cap—Mazda3 non–California emission models

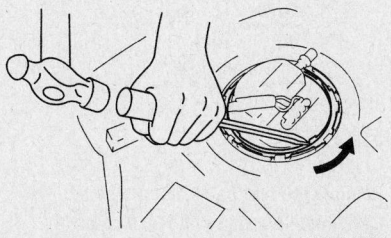

67162-MAZC-G66

Use a brass drift and a hammer to remove the fuel pump retaining ring—Mazda3 non–California emission models

15. Use tool 49 F042 001 to remove the pump cap on non–California emission models.

16. Use a brass drift and a hammer to remove the fuel pump retaining ring on California emission models.

17. Remove the fuel pump assembly.

To install:

18. Install the fuel pump assembly.

19. Clean all gasoline from the pump gasket to prevent it from turning during installation.

20. On non–California emission models, align the pump and tank assembly as illustrated.

21. Use tool 49 F042 001 to tighten the pump cap on non–California emission models. Tighten the cap to 59–66 ft. lbs. (80–90 Nm). If the torque cannot be reached, replace the pump cap and gasket. If the torque cannot still be reached, replace the fuel tank.

22. On California emission models, use a brass drift and a hammer to install the fuel pump retaining ring

23. Before installing the tank apply 1.7 inch Hg (5.9 kPa) of pressure to the tank to check for leakage around the pump.

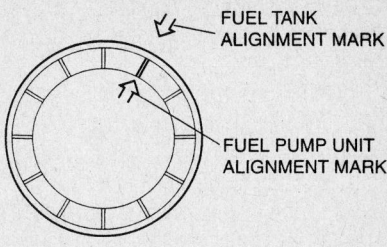

FUEL TANK ALIGNMENT MARK

FUEL PUMP UNIT ALIGNMENT MARK

67162-MAZC-G67

Align the pump and tank assembly as shown—Mazda3 non–California emission models

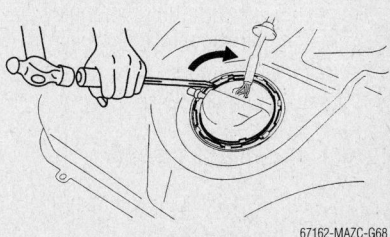

67162-MAZC-G68

Use a brass drift and a hammer install the fuel pump retaining ring—Mazda3 non–California emission models

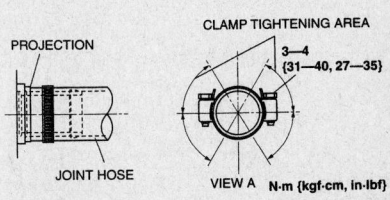

PROJECTION

CLAMP TIGHTENING AREA

3—4 {31—40, 27—35}

JOINT HOSE

VIEW A   N·m {kgf·cm, in·lbf}

67162-MAZC-G69

Install the joint hose and align the hose and clamp as shown—Mazda3

24. Connect the fuel release connector.

25. Raise the fuel tank into position.

26. Install the joint hose and align the hose and clamp as illustrated.

27. Install the strap. Tighten the bolts to 16–22 ft. lbs. (22–30 Nm).

28. Connect the evaporator hose, quick release fuel connector on the fuel line and charcoal canister.

29. Install the rear left hand under cover.

30. Install insulator and the exhaust main silencer.

31. Install the charcoal canister and connect the fuel pump connector.

32. Install the pump service hole cover and rear seat.

33. Disconnect the negative battery cable.

34. Start the vehicle and check for leaks as follows:

a. Using a jumper wire, ground the Powertrain Control Module (PCM) terminals. If equipped with an immobilizer, ground terminal 1AR, if not equipped with an imobilizer, ground terminal 1AQ. Refer to the illustration for terminal location.

b. Turn the ignition switch the ON position to activate the fuel pump.

c. Check the hoses, clips and other fuel system components for leaks.

d. If there are any leaks, replace the fuel hoses and clips. If the is damage to the seal on the fuel pipe side, replace the pipe.

e. The system must be leak free for five minutes with the terminal grounded. If any component is replaced because of a system, leak, turn the ignition key OFF; remove the jumper wire from the terminal. reapply the jumper wire, turn the ignition On and check for leaks.

Mazda6 and Mazdaspeed6

1. Before servicing the vehicle, refer to the precautions section.
2. Relieve the fuel system pressure.
3. Disconnect the negative battery cable.
4. Remove the rear seat cushion.
5. Remove the service hole cover.
6. Disconnect the fuel pump electrical connector.
7. Disconnect the hoses from the fuel tank.

➡ **The fuel pump cap can be damaged if there is a gap between the removal/installation tool 49 F042 001. Make sure the tool is attached securely to the cap and that no gap exists between them.**

8. Remove the fuel pump ring using tool 49 F042 001.
9. Remove the fuel pump and gaskets from the fuel tank.

To install:

10. Install the fuel pump using a new gasket.

➡ **The fuel pump cap can be damaged if there is a gap between the removal/installation tool 49 F042 001. Make sure the tool is attached securely**

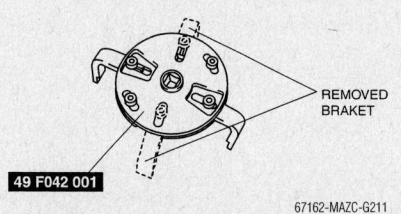

49 F042 001

REMOVED BRAKET

67162-MAZC-G211

Remove the fuel pump ring using tool 49 F042 001—Mazda6 and Mazdaspeed6 models

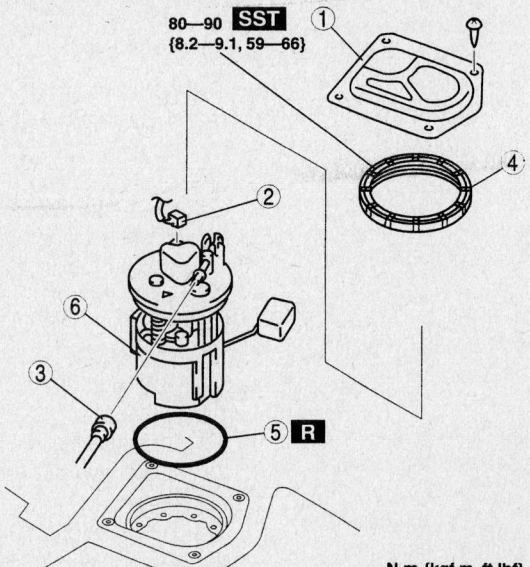

80—90 SST
{8.2—9.1, 59—66}

N·m {kgf·m, ft·lbf}

1	Service hole cover
2	Connector
3	Plastic fuel hose
4	Fuel pump cap
5	Packing
6	Fuel pump unit

67162-MAZC-G210

Exploded view of the fuel pump assembly—Mazda6 and Mazdaspeed6 models

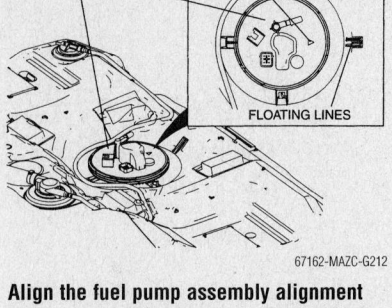

FUEL PUMP UNIT

ALIGNMENT MARK

FLOATING LINES

67162-MAZC-G212

Align the fuel pump assembly alignment marks and the floating lines—Mazda6 and Mazdaspeed6 models

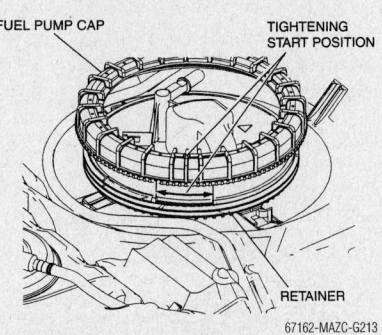

FUEL PUMP CAP

TIGHTENING START POSITION

RETAINER

67162-MAZC-G213

Align the tightening start positions of the cap and the retainer notch—Mazda6 and Mazdaspeed6 models

to the cap and that no gap exists between them.

11. Align the fuel pump assembly alignment marks and the floating lines as illustrated.
12. Align the tightening start positions of the cap and the retainer notch as illustrated and tighten them one full turn by hand. If the cap cannot be tightened by hand , remove the cap and make sure the cap is not damaged or misaligned on the retainer or cap and retighten.
13. Make sure the alignment mark and the floating lines are still aligned, tighten the cap to 59–99 ft. lbs. (80–135 Nm) or a rotation angle of 50–140 degrees with the total angle of step 2 and 3 being 410–500 degrees.

Fuel Injector

REMOVAL & INSTALLATION

✳✳ CAUTION

Fuel injection systems remain under pressure after the engine has been turned OFF. Properly relieve fuel pressure before disconnecting any

fuel lines. Failure to do so may result in fire or personal injury. Do not allow fuel spray or fuel vapors to come in contact with a spark or open flame. Keep a dry chemical fire extinguisher nearby. Never store fuel in an open container due to risk of fire or explosion.

MX5-Miata

1. Before servicing the vehicle, refer to the precautions section.
2. Properly relieve the fuel system pressure.
3. Remove the plug hole plate by lifting off and removing the plug hole plate from the installation areas (rubber and clips) as shown in the illustration.
4. Remove the battery cover.
5. Disconnect the negative battery cable.
6. Disconnect the fuel injector connector and move the harness slightly out of the way.
7. Disconnect the quick release connector.
8. Remove the fuel rail bolts and the injectors as an assembly.
9. Remove the injector clip.

⁂ CAUTION

Use of a deformed injector clip will cause the fuel injector to be connected incorrectly and could result in fuel leakage. It will also cause the injector to rotate. Therefore, always replace the clip when the injector is removed.

10. Insert a flathead screwdriver between the injector cup and clip finger.

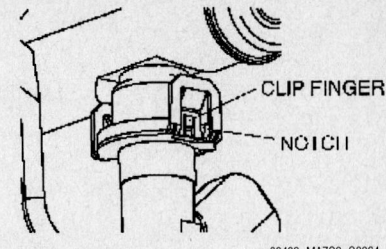

When pushing the clip finger outward, deform the finger until it is removed completely from the cup notch—MX-5 Miata

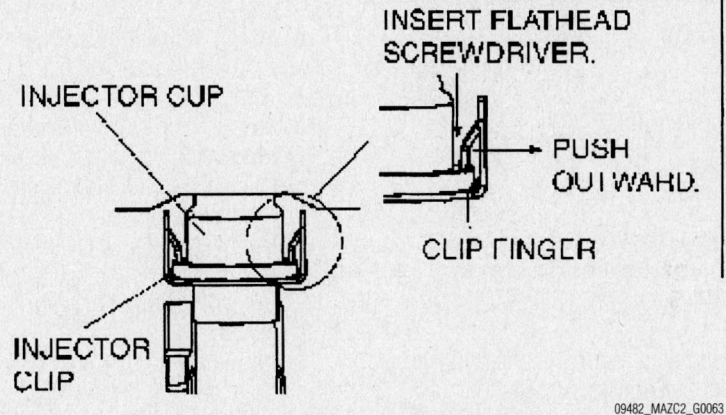

Insert a flathead screwdriver between the injector cup and clip finger—MX-5 Miata

➡**When pushing the clip finger outward, deform the finger until it is removed completely from the cup notch.**

11. Push the clip finger outward using a flathead screwdriver.
12. Remove the injector with the clip.
13. Hold the clip using pliers.
14. Pull the clip parallel to the injector groove and remove it from the injector. Discard the clip.

To install:

15. Apply a small amount of clean oil to the injector groove and the O-ring.
16. Temporarily attach a new clip to the injector groove.

➡**When the clip is attached correctly, the central area of the injector and the clip finger positions are aligned.**

17. Hold the injector firmly and push the clip into the injector until the clip stops sliding.
18. Verify that the injector connector position is correct.
19. Press the injector into the injector cup. Continue pressing until the clip contacts the lower surface of the injector cup.
20. Verify that the injector and clip are

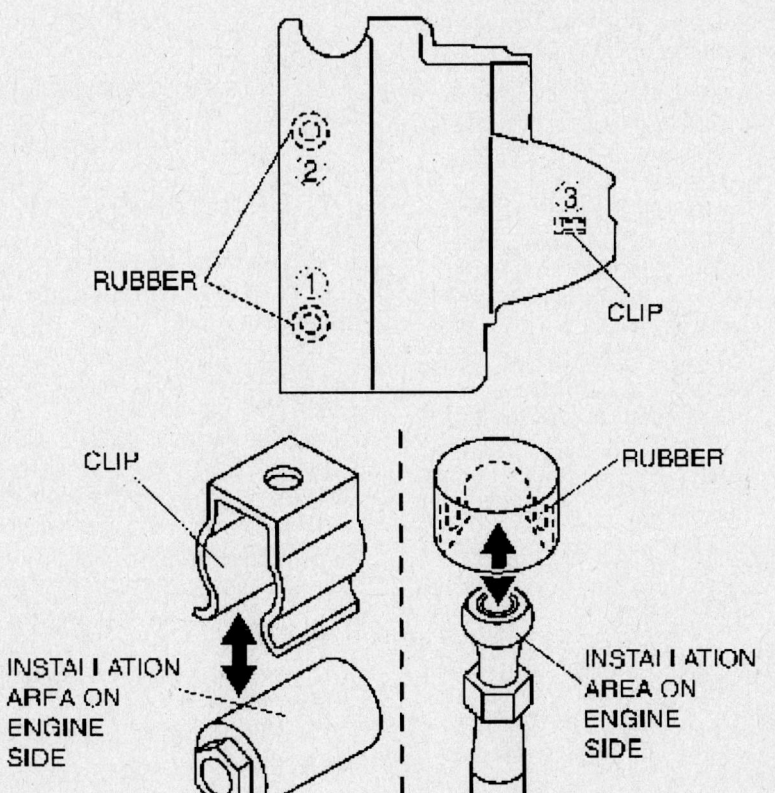

Remove the plug hole plate—MX-5 Miata

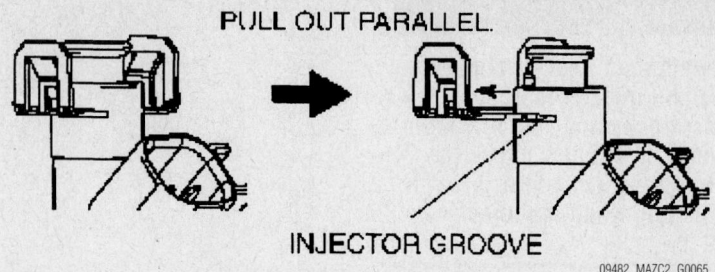

PULL OUT PARALLEL.

INJECTOR GROOVE

09482_MAZC2_G0065

Pull the clip parallel to the injector groove and remove it from the injector—MX-5 Miata

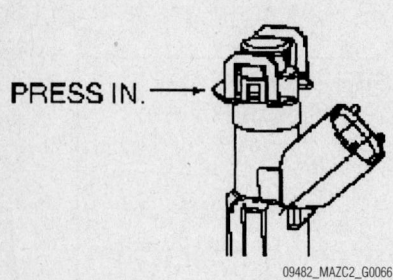

PRESS IN.

09482_MAZC2_G0066

Hold the injector firmly and push the clip into the injector until the clip stops sliding—MX-5 Miata

correctly installed with the clip locked onto the injector cup notch.

21. Install the fuel rail and tighten the bolts to 20–26 Nm (15–19 ft. lbs.).

22. Install the remaining components in the reverse order of removal, start the vehicle and check for leaks.

Mazda3

1. Before servicing the vehicle, refer to the precautions section.

2. Properly relieve the fuel system pressure.

3. Remove the plug hole plate.

4. Remove the battery cover and disconnect the negative battery cable.

5. Unplug the injector electrical connections.

6. Disconnect the fuel lines from the rail.

7. Remove the fuel rail bolts.

8. Remove the rail and injectors as an assembly.

9. Insert a flat head screwdriver between the injector cup and clip finger as illustrated.

10. Push the clip finger out using the screwdriver. Deform the finger until it is completely removed from the cup notch.

11. Remove the injector from the clip.

12. Use pliers to pull the clip parallel to the injector groove and remove the clip from the injector. Discard the clip.

To install:

13. Use new O–rings and coat them with clean engine oil.

14. Temporarily attach a new clip to the injector groove.

➡**When the clip is correctly installed, the central area of the injector and clip finger positions are aligned.**

15. While firmly holding the injector, push the clip into the injector until the clip stops. Make sure the injector connector is correctly positioned.

16. Press the injector into the cup until the cup contacts the lower surface of the cup. Make sure the injector and clip are installed properly and the clip is hooked into the injector cup notch.

17. Install the injectors and fuel rail as an assembly.

18. Install the fuel rail bolts and tighten to 15–19 ft. lbs. (20–26 Nm).

19. Attach the fuel lines to the rail.

20. Start the vehicle and check for leaks as follows:

a. Using a jumper wire, ground the Powertrain Control Module (PCM) terminals. If equipped with an immobilizer, ground terminal 1AR, if not equipped with an imobilizer, ground terminal 1AQ. Refer to illustration 67162-mazc-g22 for terminal location.

b. Turn the ignition switch the ON position to activate the fuel pump.

c. Check the hoses, clips and other fuel system components for leaks.

d. If there are any leaks, replace the fuel hoses and clips. If the is damage to the seal on the fuel pipe side, replace the pipe.

e. The system must be leak free for five minutes with the terminal grounded. If any component is replaced because of a system, leak, turn the ignition key OFF; remove the jumper wire from the terminal. reapply the jumper wire, turn the ignition On and check for leaks.

21. Install the remaining components.

Mazda6

2.3L (L3) ENGINE

1. Before servicing the vehicle, refer to the precautions section.

2. Properly relieve the fuel system pressure.

3. Disconnect the negative battery cable.

4. Unplug the injector electrical connections.

5. Disconnect the fuel lines from the rail.

6. Remove the fuel rail bolts.

7. Remove the rail and injectors as an assembly.

8. Insert a flat head screwdriver between the injector cup and clip finger as illustrated.

9. Push the clip finger out using the screwdriver. Deform the finger until it is completely removed from the cup notch.

10. Remove the injector from the clip.

11. Use pliers to pull the clip parallel to the injector groove and remove the clip from the injector. Discard the clip.

To install:

12. Use new O–rings and coat them with clean engine oil.

13. Temporarily attach a new clip to the injector groove.

➡**When the clip is correctly installed, the central area of the injector and clip finger positions are aligned.**

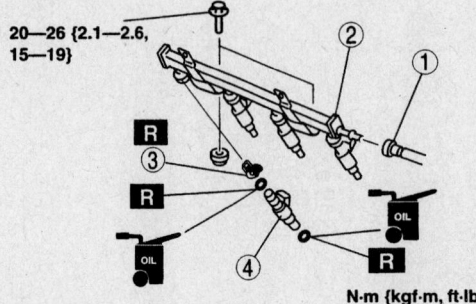

20—26 {2.1—2.6, 15—19}

1 Quick release connector
2 Fuel distributor
3 Injector clip
4 Fuel injector

N·m {kgf·m, ft·lbf}

67162-MAZC-G71

Exploded view of the fuel rail and injector assembly—Mazda3 models

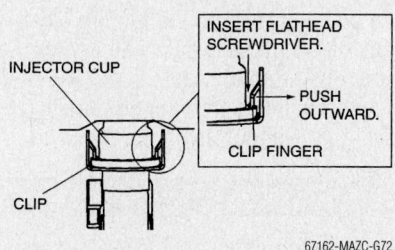

20—26
{2.1—2.6, 15—19}

1   Fuel injector connectors
2   Plastic fuel hose
3   Fuel distributor
4   Fuel injector
5   Snap ring
6   Pulsation damper

N·m {kgf·m, ft-lbf}

67162-MAZC-G214

Exploded view of the fuel rail and injector assembly—Mazda6 models with the 2.3L (L3) engine

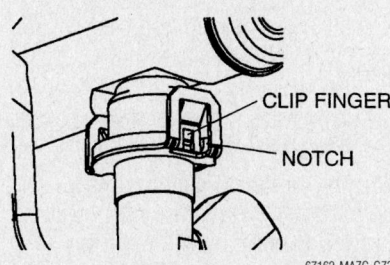

67162-MAZC-G72

Insert a flat head screwdriver between the injector cup and clip finger—Mazda3 and 6 models with the 2.3L (L3) engine

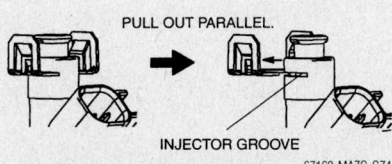

67162-MAZC-G73

Deform the finger until it is completely removed from the cup notch—Mazda3 and 6 models with the 2.3L (L3) engine

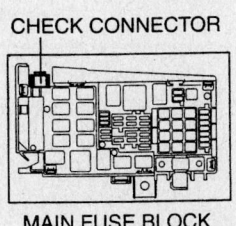

67162-MAZC-G74

Use pliers to pull the clip parallel to the injector groove and remove the clip from the injector—Mazda3 and 6 models with the 2.3L (L3) engine

14. While firmly holding the injector, push the clip into the injector until the clip stops. Make sure the injector connector is correctly positioned.

15. Press the injector into the cup until the cup contacts the lower surface of the cup. Make sure the injector and clip are installed properly and the clip is hooked into the injector cup notch.

16. Install the injectors and fuel rail as an assembly.

17. Install the fuel rail bolts and tighten to 15–19 ft. lbs. (20–26 Nm).

18. Attach the fuel lines to the rail.

19. Start the vehicle and check for leaks as follows:

   a. Using a jumper wire, short the check connector to terminal F/P to a body ground.

   b. Turn the ignition switch the ON position to activate the fuel pump.

   c. Check the hoses, clips and other fuel system components for leaks.

   d. If there are any leaks, replace the fuel hoses and clips. If the is damage to

CHECK CONNECTOR

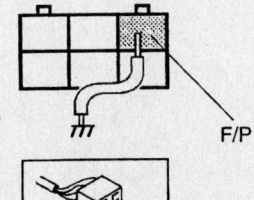

MAIN FUSE BLOCK

Using a jumper wire, short the check connector to terminal F/P to a body ground—Mazda6 models

the seal on the fuel pipe side, replace the pipe.

   e. The system must be leak free for five minutes with the terminal grounded. If any component is replaced because of a system, leak, turn the ignition key OFF; remove the jumper wire from the terminal. reapply the jumper wire, turn the ignition On and check for leaks.

20. Install the remaining components.

3.0L (AJ) ENGINE

1. Before servicing the vehicle, refer to the precautions section.

2. Release the fuel system pressure.

3. Disconnect the negative battery cable.

4. Remove the upper intake manifold (dynamic chamber).

5. Disconnect the fuel injector connectors.

6. Disconnect the fuel hoses from the rail.

7. Remove the rail and injectors as an assembly.

8. Gently twist the fuel injector out of the manifold

9. Check the O-rings and replace if damaged.

To install:

10. Install the fuel injector(s) using new O-rings lubricated with clean engine oil.

11. Attach the fuel injector into the supply manifold.

12. Install the fuel rail and injectors as an assembly. Tighten the rail bolts to 72–101 inch lbs. (8–11 Nm).

13. Connect the fuel hoses to the rail.

14. Install the injector connectors.

15. Install the upper intake manifold.

16. Connect the negative battery cable.

17. Start the vehicle and check for leaks as follows:

   a. Using a jumper wire, short the check connector to terminal F/P to a body ground.

   b. Turn the ignition switch the ON position to activate the fuel pump.

CHECK CONNECTOR

F/P

67162-MAZC-G151

8.0—11.5 {82—117, 72—101}

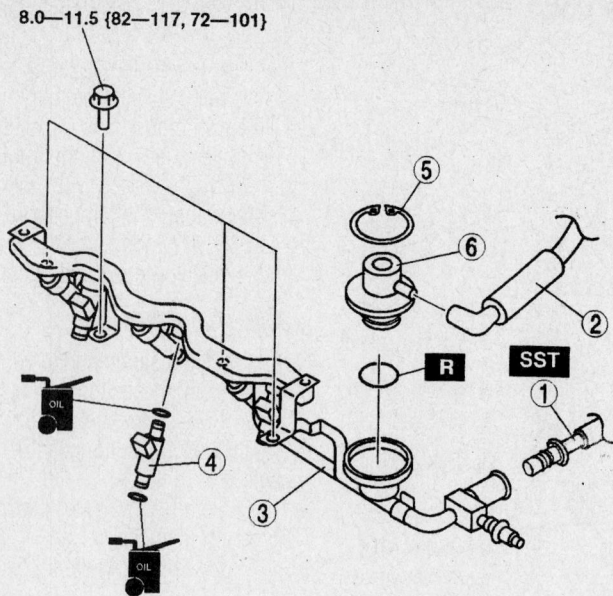

1   Plastic fuel hose
2   Hose
3   Fuel distributor
4   Fuel injector
5   Snap ring
6   Pulsation damper

67162-MAZC-G215

Exploded view of the fuel rail and injector assembly—Mazda6 models with the 3.0L (AJ) engine

c. Check the hoses, clips and other fuel system components for leaks.

d. If there are any leaks, replace the fuel hoses and clips. If the is damage to the seal on the fuel pipe side, replace the pipe.

e. The system must be leak free for five minutes with the terminal grounded. If any component is replaced because of a system, leak, turn the ignition key OFF; remove the jumper wire from the terminal. reapply the jumper wire, turn the ignition On and check for leaks.

Mazdaspeed6

1. Before servicing the vehicle, refer to the precautions section.

2. Properly relieve the fuel system pressure.

3. Disconnect the negative battery cable.

4. Remove the intake manifold.

5. Drain and recycle the engine coolant.

➡**If the high pressure fuel pump is removed, replace the O-ring with a new one.**

6. Remove the charge air cooler cover.

7. Disconnect the spill valve control solenoid valve connector.

8. Disconnect the quick release connector on the high pressure fuel pump.

9. Remove the battery and battery tray.

✳✳ WARNING

If the high pressure fuel pump joint nut is loosened, fuel leakage may occur resulting in death or serious injury, or damage to the equipment or the vehicle. Fuel can also irritate the skin and eyes. When removing the high pressure line pipe, always tighten the high pressure line pipe installation nut while fixing the high pressure fuel pump joint nut with a wrench. If the high pressure fuel pump joint nut has rotated, replace the high pressure fuel pump with a new one.

10. Disconnect the high pressure line pipe of the high pressure fuel pump.

11. Fix the joint nut with a wrench on the high pressure fuel pump side.

12. Loosen the high pressure line pipe installation nut.

13. Disconnect the fuel delivery pipe.

14. Unplug the injector electrical connections.

15. Remove the fuel injector bract.

16. Remove the injectors as follows:

a. Install the service tool onto the fuel injector.

✳✳ CAUTION

If the tool slips while ratcheting up the fuel injector, the fuel injector or surrounding parts could be damaged. Press fit the tool to the fuel injector firmly and operate carefully. When ratcheting up the fuel injector, the fuel injector connector may contact the cylinder head and damage the fuel injector. Ratchet up the fuel injector so that the fuel injector connector does not contact the cylinder head.

➡**If fuel injector No.3 contacts the oil separator, cut the tab on the oil separator. Carefully cut the tab so that the oil separator is not deformed or damaged, with no clearance on the mating surfaces of the oil separator and engine. Keep ratcheting the tool so that the fuel injector becomes free enough to ratchet up without using the tool.**

✳✳ CAUTION

Do not apply excessive force to the fuel injector connector because the fuel injector could be damaged. Pull out the fuel injector by ratcheting it upright.

b. Verify that there are no gasket in the cylinder heads after removing the fuel injectors.

✳✳ WARNING

If foreign material such as metal shavings penetrates the fuel injector installation hole on the cylinder block, the engine could be damaged. Remove all foreign material and cap the fuel injector installation hole after removing the fuel injector.

c. Clean the fuel injector and around the insertion hole using a vacuum cleaner.

To install:

17. Installation is the reverse of removal, note the following:

a. Use new O-rings and coat them with clean engine oil.

b. Tighten the fuel pipe delivery bolts to 13–16 ft. lbs. (17–23 Nm).

c. Assemble the high pressure line pipe.

d. Fix the joint nut with a wrench on the high pressure fuel pump side. Tighten the high pressure line pipe installation nut to 17–26 ft. lbs. (23–35 Nm).

e. Tighten the water outlet case installation bolts to 13–16 ft. lbs. (18–22 Nm).

f. Install the quick release connector.

18. Verify that the high pressure fuel pump is assembled securely.

19. Drive the vehicle starting from a standstill and brake suddenly five to six times at a low speed.

20. Stop the vehicle and verify from outside the vehicle that there is no fuel leakage around the high pressure fuel pump.

DRIVE TRAIN

Manual Transaxle/Transmission Assembly

REMOVAL & INSTALLATION

MX5-Miata

1. Before servicing the vehicle, refer to the precautions section.

➡**Refer the exploded view illustration for component locations and if applicable, their retainer torque specifications.**

2. Remove the battery cover.
3. Disconnect the negative battery cable.
4. Loosen the starter installation bolts only enough that the starter is loose, but not removed.
5. Remove the shift lever knob.
6. Remove the console.
7. Remove the outer, then inner shift insulator components.
8. Remove the shift lever component.
9. Remove the member bracket
10. Remove the tunnel member.
11. Remove the catalytic converter and middle pipe.
12. Remove the clutch release pipe and cylinder.
13. Remove the power plant frame as follows:

a. Support the transmission using a transmission jack.

b. Remove the power plant frame.

14. Remove the hanger bracket.
15. Remove the propeller shaft as follows:

➡**When replacing with a new propeller shaft, mark the companion flange to match the position of the tag on the propeller shaft.**

a. Before removing the propeller shaft, make alignment marks on the yoke and differential companion flange.

b. Remove the retainers and the shaft.

c. Install the tools illustrated onto the to the extension housing.

16. Disconnect the back-up light switch connector.

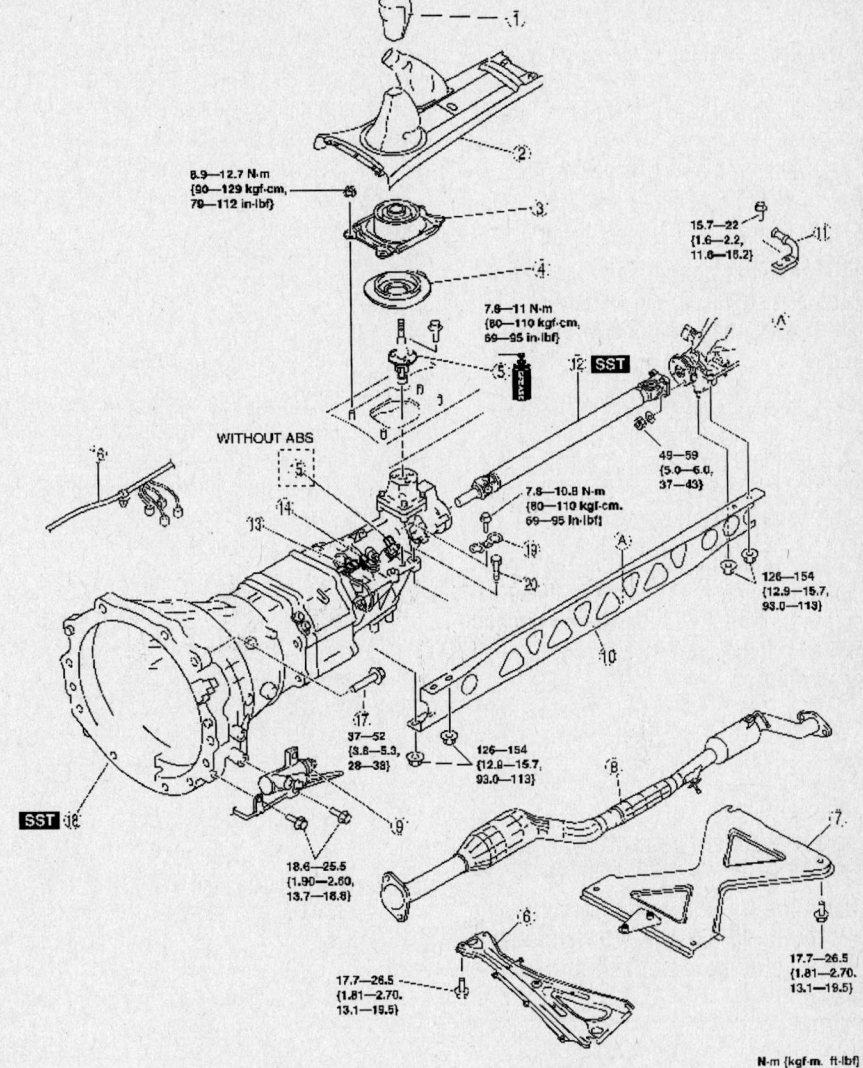

1	Shift lever knob	11	Hanger bracket
2	Console	12	Propeller shaft
3	Shift insulator component (outer)	13	Back-up light switch connector
4	Shift insulator component (inner)	14	Neutral switch connector
5	Shift lever component	15	Vehicle speed sensor connector
6	Member bracket	16	Wire
7	Tunnel member	17	Transmission installation bolt
8	Catalytic converter, middle pipe	18	Transmission
9	Clutch release cylinder	19	Stopper
10	Power plant frame	20	Bolt

09482_MAZC2_G0066A

Exploded view of the manual transmission assembly and related components—MX-5 Miata

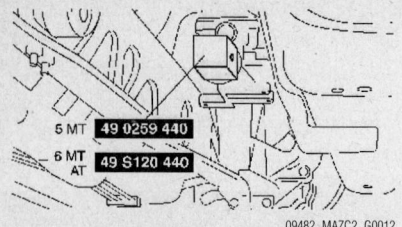

Install the tools shown depending on transmission type, onto the to the extension housing—MX-5 Miata

17. Disconnect the neutral switch connector.

18. Disconnect the vehicle speed sensor connector

19. Remove the wiring harness and any remaining electrical connections from the transmission.

➡**Remove the transmission carefully, holding it steady. If the transmission falls it could be damaged or cause injury.**

20. Support the transmission securely using a transmission jack.

21. Remove the transmission-to-engine bolts.

22. Remove the transmission.

➡**When removing/installing the transmission, be sure not to move the engine up and down more than necessary to prevent part interference with the engine.**

To install:

23. Shift to any gear position.

24. Install tool 49 0259-44 onto the main shaft.

✳✳ WARNING

Install the transmission carefully, holding it steady. If the transmission falls it could be damaged or cause injury.

25. Place the transmission on the transmission jack and raise it.

26. Slowly rotate the tool 49 0259-44 to

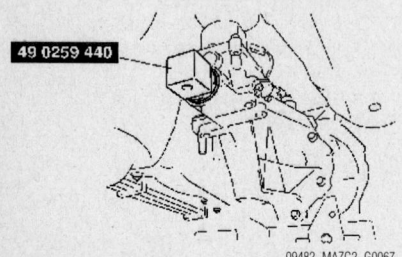

Install tool 49 0259-44 onto the main shaft—MX-5 Miata

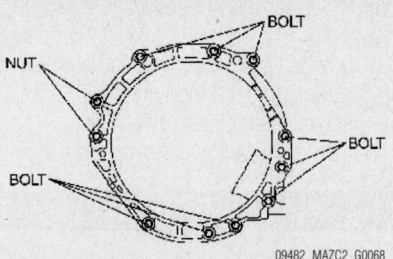

Location of the manual transmission bolts and nuts—MX-5 Miata

engage the clutch with the main drive gear spline, and install the transmission.

27. Tighten the transmission installation bolts and nuts to 37–52 Nm (28–38 Nm).

28. Installation of the remaining components is the reverse of removal, please note the following:

a. When installing the propeller shaft:

- Align the marks and install the propeller shaft.
- When installing a new propeller shaft, align the differential companion flange mark with the tag on the propeller shaft and assemble.

b. When installing the power plant frame:

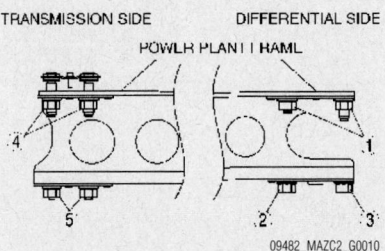

Temporarily tighten the nuts 1, 2, 3 in the order shown—MX-5 Miata

- Support the transmission and differential so that they are level using a transmission jack.
- Install the power plant frame.
- Temporarily tighten the nuts 1, 2, 3 in the order shown in the illustration.
- Tighten nut 2 until the power plant frame is seated in the rear differential.
- Temporarily tighten the nuts 4, 5 in order shown in the illustration.
- Install the middle pipe and tunnel member.
- Raise the front end of the power plant frame (transmission side) or the transmission with the transmission jack, and adjust dimension A to 26.7–34.7 mm (1.06–1.36 in) (lower surface of power plant frame-upper surface of the tunnel member) as shown in the illustration.
- Tighten the power plant frame installation nuts to 126–154 Nm (93–113.5 ft. lbs)
- Verify that dimension A is within the specification with the transmission jack and the adjustment bolt removed.

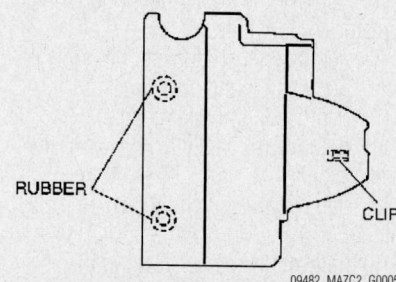

Install the plug hole plate by grasping rubber and pressing them in—MX-5 Miata

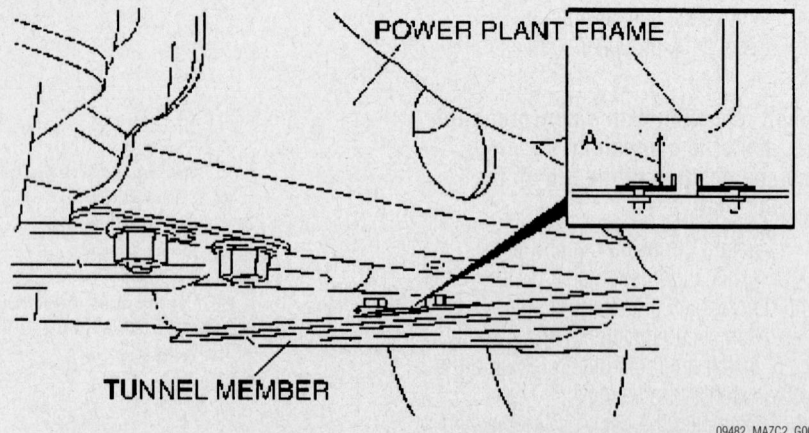

Raise the front end of the power plant frame (transmission side) or the transmission with the transmission jack, and adjust dimension A to 26.7–34.7 mm (1.06–1.36 in—MX-5 Miata

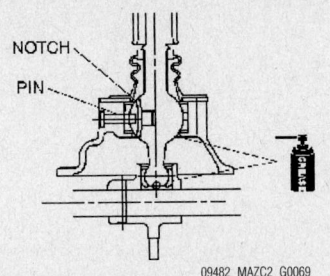

09482_MAZC2_G0069

Apply grease to the areas of the shift lever component indicated—MX-5 Miata

• If not within the specification, adjust dimension A again.

c. To position the plug hole plate, grasp rubber 1 and 2, as shown in the illustration, with your hands and press them in.

d. Apply grease to the areas of the shift lever component as shown in the illustration.

e. Align the shift lever component notch with the shift control case pin and install the shift lever component.

Mazda3

1. Before servicing the vehicle, refer to the precautions section.
2. Refer to the illustration for component location and torque specifications.
3. Drain the transaxle oil.
4. Remove or disconnect the following:
• Battery and battery box
• Air cleaner assembly
• Exhaust manifold insulator
• Wheels

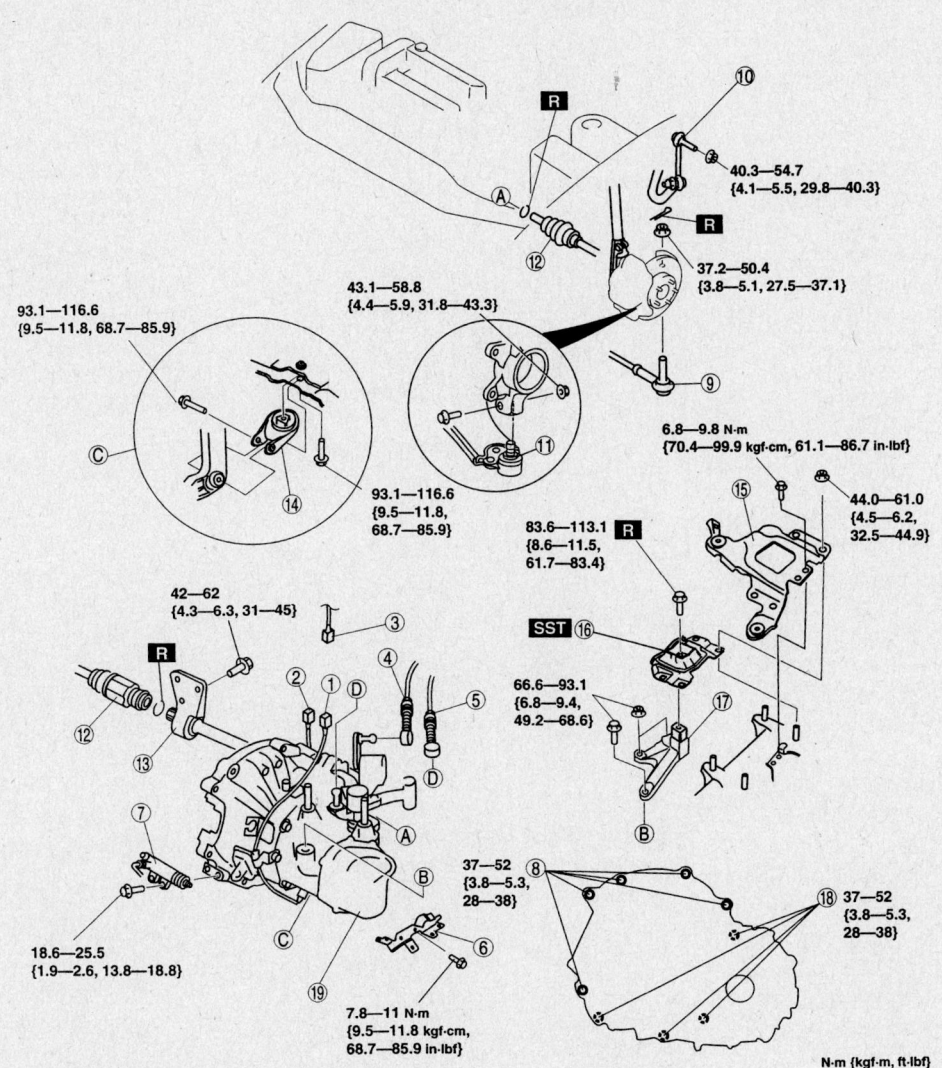

1	Back-up light switch connector	10	Stabilizer control link
2	Neutral switch connector	11	Lower arm ball joint
3	Vehicle speed sensor connector (without ABS)	12	Drive shaft
4	Select cable	13	Joint shaft
5	Shift cable	14	No.1 engine mount rubber
6	Harness bracket	15	Battery tray bracket
7	Clutch release cylinder	16	No.4 engine mount rubber
8	Transaxle mounting bolt (upper side)	17	No.4 engine mount bracket
9	Tie-rod end ball joint	18	Transaxle mounting bolt (lower side)
		19	Manual transaxle

67162-MAZC-G75

Exploded view of the manual transaxle assembly mounting—Mazda3 models

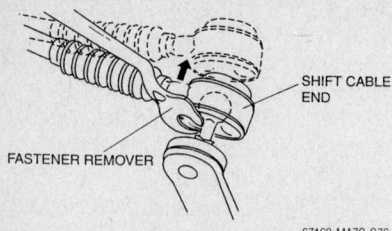

67162-MAZC-G76

Remove the shift cable and select cable using a fastener remover—Mazda3 models

- Splash shields
- Under cover
- Starter
- Back up light switch connector
- Neutral safety switch connector, if equipped
- Vehicle speed sensor
- Shift cable and select cable using a fastener remover as illustrated.
- Harness bracket
- Clutch release cylinder
- Transaxle upper mount bolts
- Tie rod ends from the knuckle
- Stabilizer bar link
- Lower control arm ball joint
- Halfshafts

5. Remove the joint shaft as follows:

a. Disconnect the right halfshaft from the joint shaft by tapping the transaxle side outer ring with a brass bar and a hammer. Disconnect the joint shaft bracket from the block and remove the joint shaft.

b. Install tool 49 G030 455 to hold the side gears after removal

6. Support the engine assembly with a engine support assembly such as 49 E017 5A0.

7. Remove or disconnect the following:
- Number 1 engine mount rubber
- Battery tray bracket

8. Remove the number 4 engine mount and joint bracket as follows:

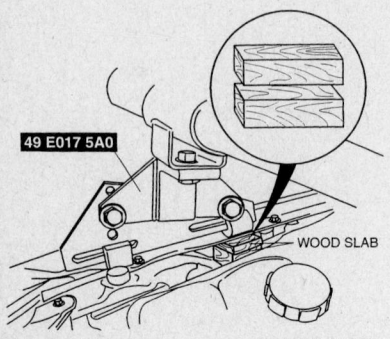

49 E017 5A0

WOOD SLAB

67162-MAZC-G56

Install two suitable pieces of wood between the front fender panel and upper apron reinforcement—Mazda3 with a manual transaxle

a. Install two suitable pieces of wood between the front fender panel and upper apron reinforcement as illustrated. The wood size should be approximately 1.38 inch (35mm) on 4 door models or 2.36 inch (60mm) on 5 door models.

9. Install an engine support device such as 49 E017 5A0.

10. Loosen the engine support assembly and lean the engine towards the transaxle.

11. Support the transaxle with a jack, remove the transaxle bolts and the transaxle.

To install:

12. Place the transaxle onto a jack and raise into position.

13. Install the transaxle bolts and tighten the upper bolts to 28–38 ft. lbs. (37–52 Nm), the lower bolts 28–38 ft. lbs. (37–52 Nm). Refer to the exploded view illustration for bolt locations.

14. Install or connect the following:
- Number 4 engine mount bracket to the transaxle case and tighten the bolts to 49–68 ft. lbs. (66–93 Nm)
- Number 1 engine mount rubber to the crossmember and hand tighten the bolts
- Number 4 engine mount rubber with the body stud passing through

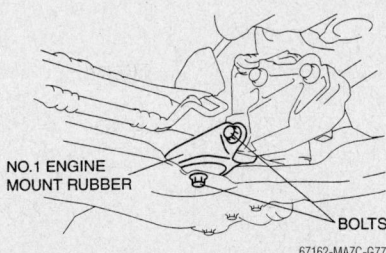

NO.1 ENGINE
MOUNT RUBBER

BOLTS

67162-MAZC-G77

Number 1 engine mount rubber bolt locations—Mazda3 models with a manual transaxle

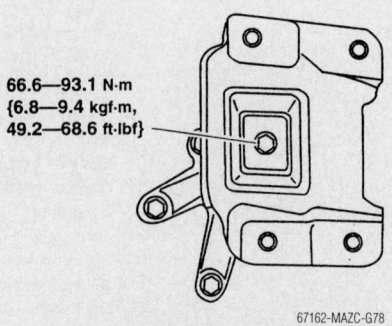

66.6—93.1 N·m
{6.8—9.4 kgf·m,
49.2—68.6 ft-lbf}

67162-MAZC-G78

Install the number 4 engine mount rubber with the body stud passing through the holes—Mazda3 models with a manual transaxle

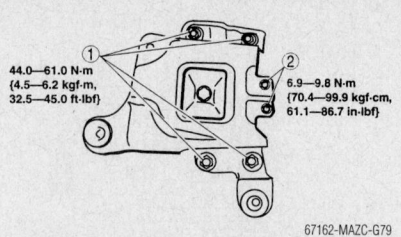

44.0—61.0 N·m
{4.5—6.2 kgf·m,
32.5—45.0 ft-lbf}

6.9—9.8 N·m
{70.4—99.9 kgf·cm,
61.1—86.7 in-lbf}

67162-MAZC-G79

Install the battery tray bracket on the number 4 engine mount rubber with the body stud bolts passing through the holes—Mazda3 models with a manual transaxle

the holes and tighten the bolts to 49–62 ft. lbs. (66–93 Nm)
- Battery tray bracket on the number 4 engine mount rubber with the body stud bolts passing through the holes. Tighten retainers identified in the illustration as 1 to 32–45 ft. lbs. (44–61 Nm) and the retainers identified as 2 in the same illustration to 61–86 inch lbs. (7–10 Nm).

15. Remove the engine support device and tighten the number 1 mount rubber bolts to 68–85 ft. lbs. (116 Nm).
- Joint shaft. Install a new circlip with the opening facing up. Hand tighten the bolts, then tighten the bolts to 31–45 ft. lbs. (42–62 Nm). Refer to the illustration for bolt location.
- Halfshafts
- Lower control arm ball joint. Tighten the retainers to 32–43 ft. lbs. (43–59 Nm).
- Stabilizer bar link. Tighten the retainers to 30–40 ft. lbs. (40–54 Nm).
- Tie rod ends to the knuckle. Tighten the retainers to 27–37 ft. lbs. (37–50 Nm).
- Clutch release cylinder. Tighten the retainers to 14–19 ft. lbs. (18–25 Nm).
- Harness bracket
- Shift and select cables
- Vehicle speed sensor
- Neutral safety switch connector
- Back up light switch connector
- Starter
- Under cover
- Splash shields
- Wheels
- Exhaust manifold insulator
- Air cleaner assembly
- Battery tray and battery

16. Fill the transaxle with fluid. Road test the vehicle and check for leaks. Top off all fluids as needed.

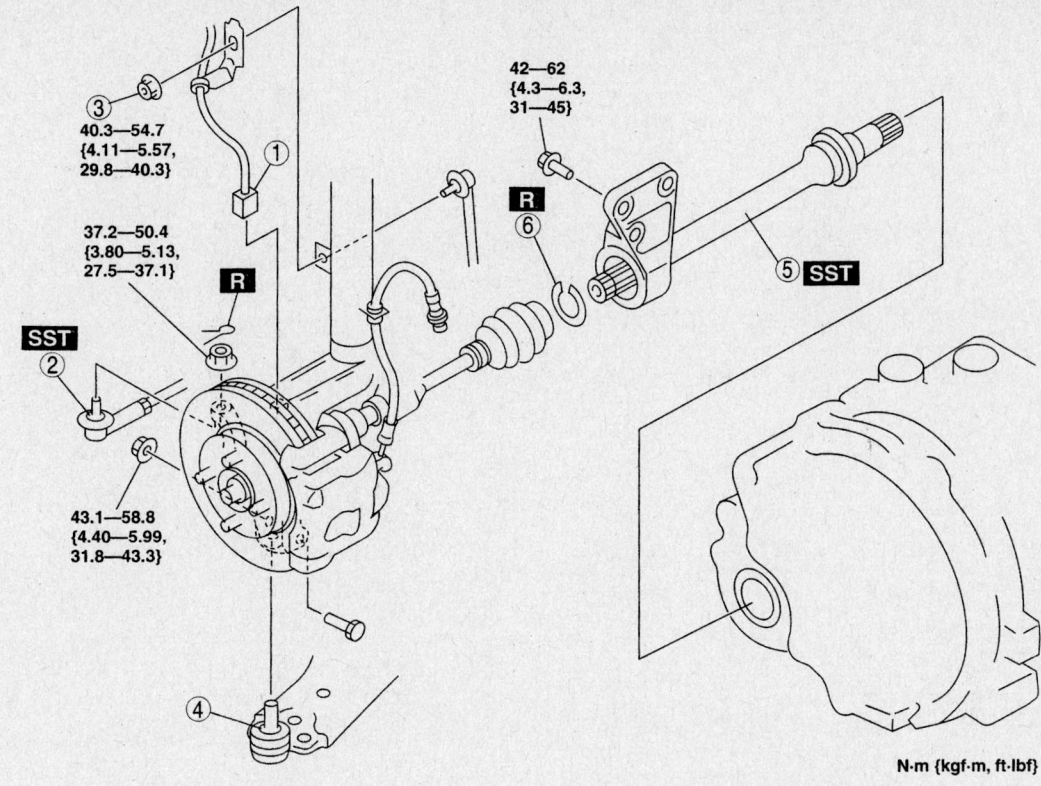

1 ABS wheel-speed sensor connector	4 Front lower arm ball joint
2 Tie-rod end ball joint	5 Joint shaft
3 Stabilizer control link upper nut	6 Clip

67162-MAZC-G80

Exploded view of the joint shaft assembly—Mazda3

Mazda6

G35M-R TRANSAXLE

1. Before servicing the vehicle, refer to the precautions section.

2. Refer to the illustration for component location and torque specifications.

3. Drain the transaxle oil.

4. Remove or disconnect the following:

- Battery and battery tray
- Air cleaner assembly
- Wheels
- Splash shields
- Under cover
- Steering gear, linkage and pipe assembly bolts from the crossmember and using mechanics wire, position the steering gear and linkage assembly aside.
- Heated Oxygen (HO_2S) connector
- Back up light switch connector
- Neutral safety switch connector
- Harness bracket
- Vehicle speed sensor
- Shift cable and select cable
- Selector cable bracket
- Clutch release cylinder
- Starter
- Endplate cover
- Transaxle upper mount bolts
- Lower control arm ball joints
- Damper fork
- Tie rod ends from the knuckle
- Stabilizer bar link
- Halfshafts.

5. Disconnect the right halfshaft from the joint shaft by tapping the transaxle side outer ring with a brass bar and a hammer. Disconnect the joint shaft bracket from the block and remove the joint shaft.

6. Install tool 49 G030 455 to hold the side gears after removal.

7. Support the engine assembly with a engine support assembly such as 49 E017 5A0.

8. Remove or disconnect the following:

- Number 1 engine mount bracket
- Lower front shock absorber bolt
- No. 1 engine mount center bolt

9. Support the crossmember with a jack and remove the nuts and the crossmember bracket.

- Crossmember

10. Remove the number 4 engine mount

- Dynamic damper
- Lower transaxle bolts

11. Loosen the engine support assembly and lean the engine towards the transaxle.

12. Support the transaxle with a jack, remove the transaxle bolts and the transaxle.

To install:

13. Place the transaxle onto a jack and raise into position.

14. Install the transaxle bolts and tighten the bolts to 28–38 ft. lbs. (37–52 Nm). Refer to the exploded view illustration for bolt locations.

15. Install the No. 1 engine mount and No. 4 engine mount bracket as follows:

a. Make sure the mount is installed as illustrated.

b. Align the bolt holes with the stud bolts, install the No. 4 mount bracket to the transaxle.

c. Align the holes with the stud bolts

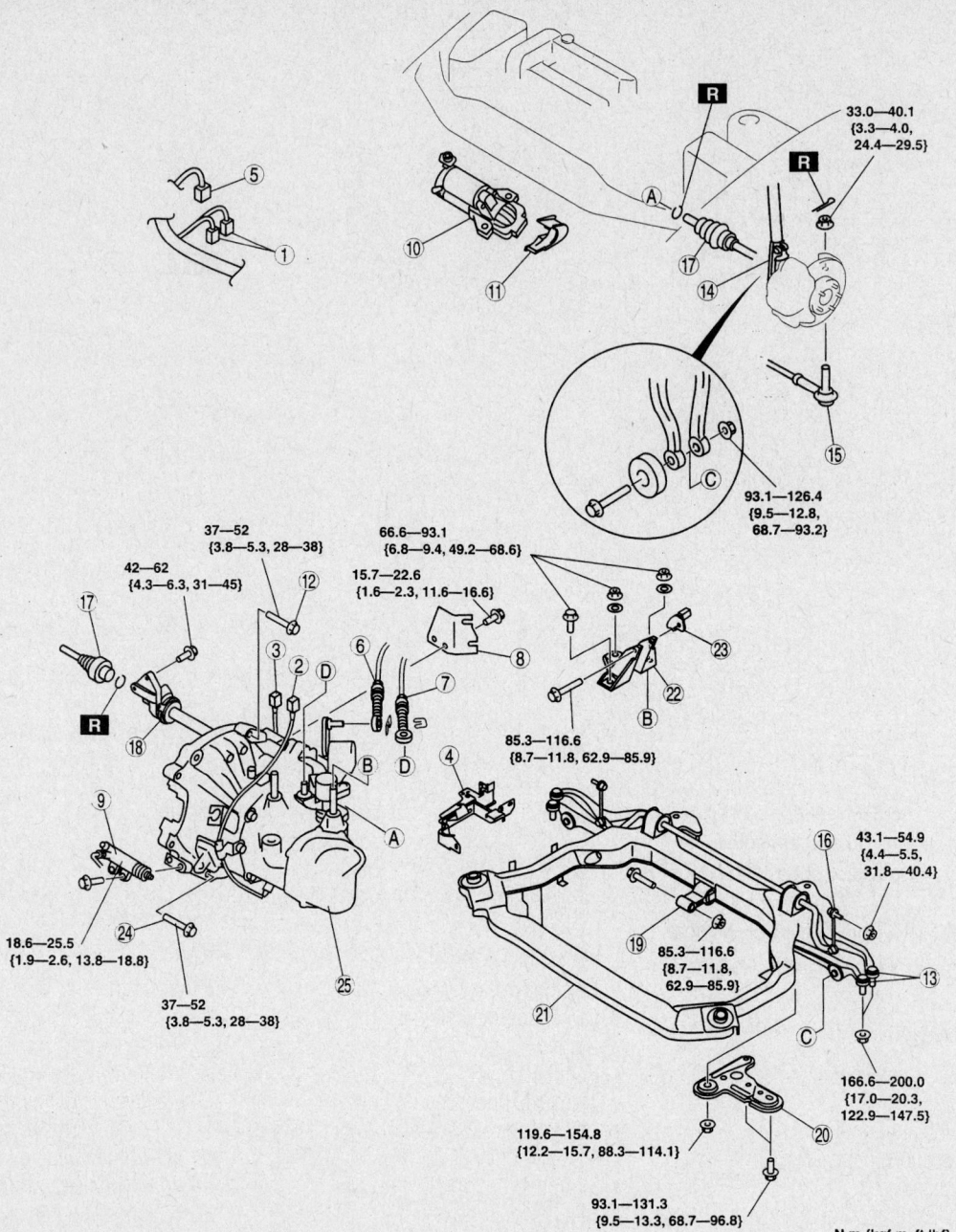

N·m {kgf·m, ft·lbf}

1	HO2S connector	14	Damper fork
2	Back-up light switch connector	15	Tie-rod end ball joint
3	Neutral switch connector	16	Stabilizer control link
4	Harness bracket	17	Drive shaft
5	Vehicle speedometer sensor connector (Without ABS)	18	Joint shaft
6	Selector cable	19	No.1 engine mount
7	Shift cable	20	Crossmember bracket
8	Selector cable bracket	21	Crossmember component
9	Clutch release cylinder	22	No.4 engine mount
10	Starter	23	Dynamic damper
11	Endplate cover	24	Transaxle mounting bolt (lower side)
12	Transaxle mounting bolt (upper side)	25	Manual transaxle
13	Lower arm (front, rear) ball joint		

67162-MAZC-G216

Exploded view of the G35M-R manual transaxle assembly mounting—Mazda6 models

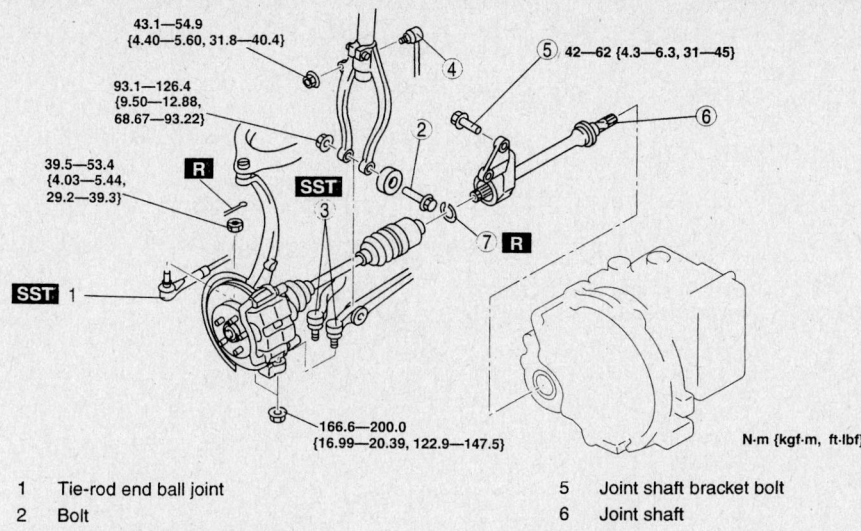

and install the No. 1 engine mount to the transaxle.

d. Align the hole on the No. 4 engine mount bracket with the No. 4 mount rubber on the car and hand tighten bolt D. Refer to the illustration for bolt locations.

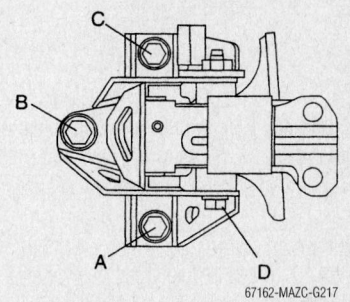

67162-MAZC-G217

Install the No. 1 mount and No. 4 mount bracket and tighten the bolts in the sequence and specifications outlined in the text—Mazda6 models with a G35M-R manual transaxle

1   Tie-rod end ball joint
2   Bolt
3   Lower arm (front, rear) ball joint
4   Stabilizer control link
5   Joint shaft bracket bolt
6   Joint shaft
7   Clip

67162-MAZC-G125

Exploded view of the joint shaft assembly—Mazda6 models with the 2.3L (L3) engine

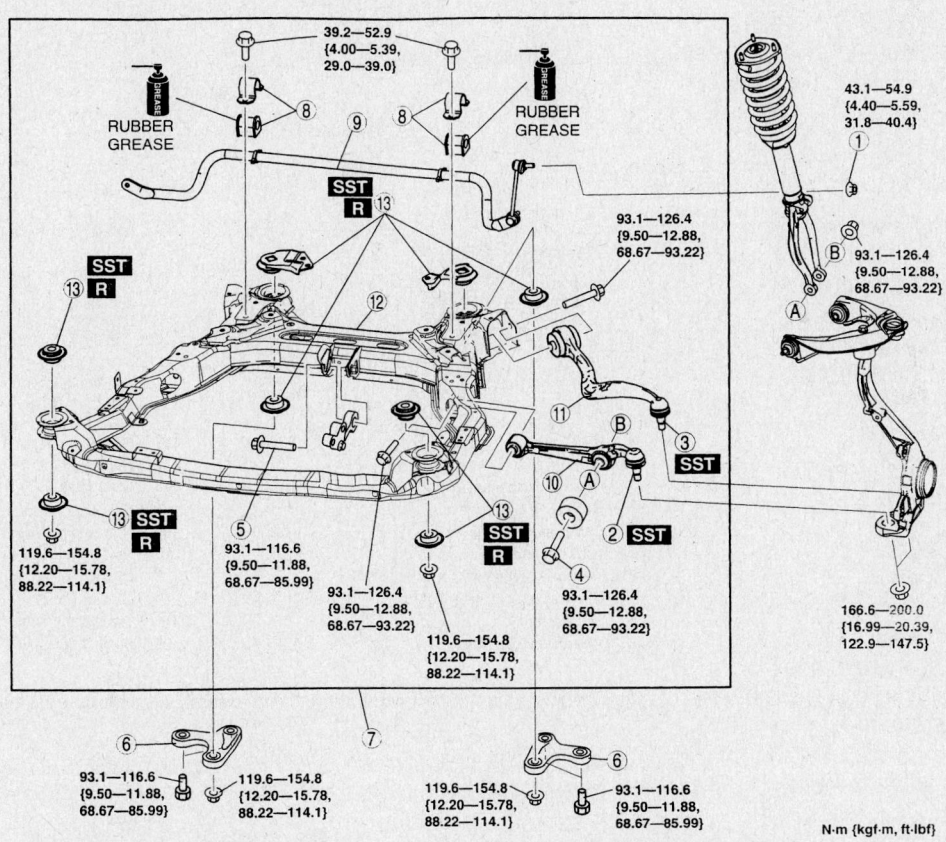

67162-MAZC-G124

1   Nut (stabilizer control link)
2   Front lower arm (front) ball joint
3   Front lower arm (rear) ball joint
4   Bolt (front shock absorber lower side)
5   No.1 engine mount center bolt
6   Crossmember bracket
7   Crossmember component
8   Stabilizer bracket and bushing
9   Front Stabilizer
10 Front lower arm (front)
11 Front lower arm (rear)
12 Front crossmember
13 Front crossmember bushing

Exploded view of the front crossmember, related components and their torque specifications—Mazda3 and 6 models

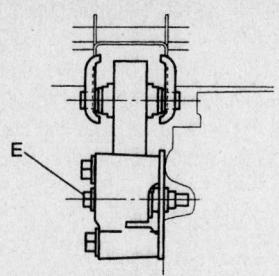

67162-MAZC-G218

Tighten bolt E on the No. 1 mount to the specification outlined in the text—Mazda6 models with a G35M-R manual transaxle

16. Hand tighten bolts B, C and A.

a. Tighten bolts B and C then bolt A to 49–68 ft. lbs. (66–93 Nm).

b. Tighten bolt D to 63–86 ft. lbs. (85–116 Nm).

c. Tighten bolt E on the No. 1 mount to 63–86 ft. lbs. (85–116 Nm).

17. Remove the engine support device.

18. Install the remaining components in the reverse order of removal.

19. Fill the transaxle with fluid. Road test the vehicle and check for leaks. Top off all fluids as needed.

A65M-R TRANSAXLE

1. Before servicing the vehicle, refer to the precautions section.

2. Refer to the illustration for component location and torque specifications.

3. Drain the transaxle oil.

4. Remove or disconnect the following:

- Battery and battery tray
- Air cleaner assembly
- Starter
- Wheels
- Splash shields
- Under cover
- Steering gear, linkage and pipe assembly bolts from the crossmember and using mechanics wire, position the steering gear and linkage assembly aside.
- Shift cable and select cable
- Back up light switch connector
- Neutral safety switch connector
- Clutch release cylinder
- Ground harness
- Transaxle upper mount bolts
- Lower control arm ball joints
- Damper fork
- Tie rod ends from the knuckle
- Stabilizer bar link
- Halfshafts

5. Disconnect the left halfshaft from the joint shaft by inserting a prybar between the

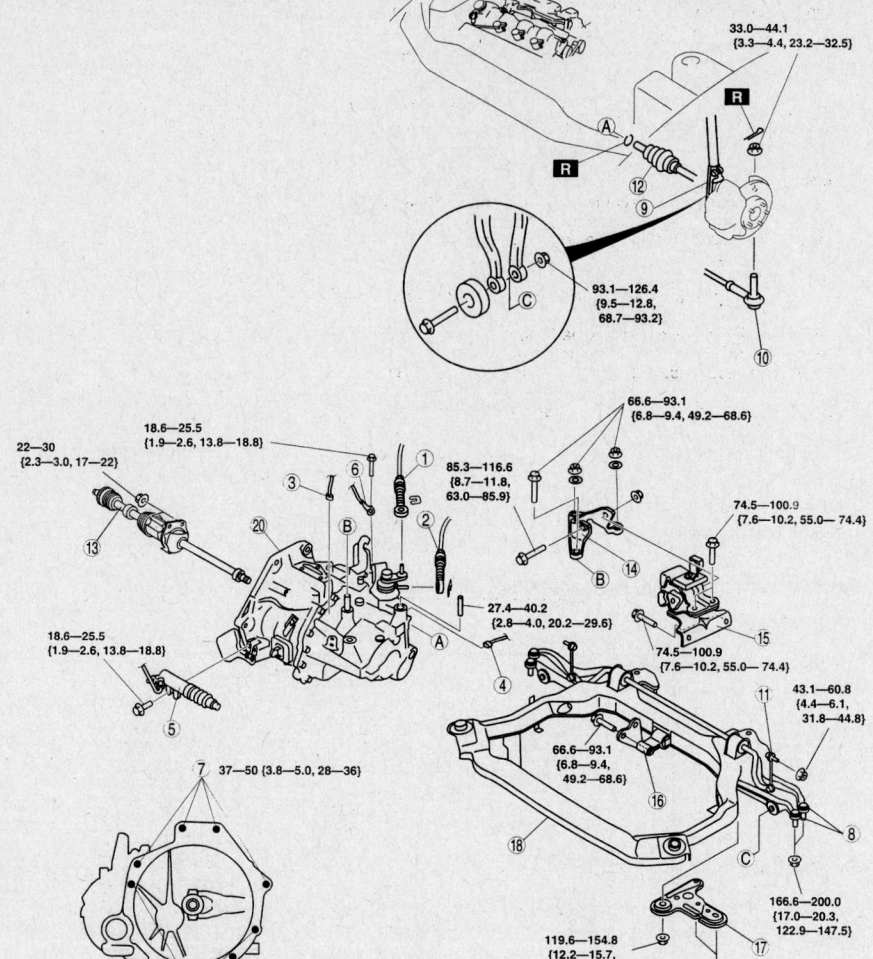

1	Shift cable
2	Select cable
3	Reverse switch connector
4	Neutral switch connector
5	Clutch release cylinder
6	GND harness
7	Transaxle mounting bolt (upper side)
8	Lower arm (front, rear) ball joint
9	Damper fork
10	Tie-rod end ball joint
11	Stabilizer control link
12	Drive shaft
13	Drive shaft, joint shaft
14	No.4 engine mount bracket
15	No.4 engine mount rubber
16	No.1 engine mount
17	Crossmember bracket
18	Crossmember
19	Transaxle mounting bolt (lower side)
20	Manual transaxle

67162-MAZC-G219

Exploded view of the A65M-R manual transaxle assembly mounting—Mazda6 models

transaxle and the halfshaft outer ring. Disconnect the joint shaft bracket from the block and remove the joint shaft.

6. Install tool 49 G030 455 to hold the side gears after removal

- No. 4 engine mount bracket and rubber

7. Support the engine assembly with a engine support assembly such as 49 E017 5A0.

8. Remove or disconnect the following:

- Number 1 engine mount bracket

- Lower front shock absorber bolt
- No. 1 engine mount. Remove the intake manifold and attach tool 49 UN30 3050 to the head as shown and remove the mount.

9. Support the crossmember with a jack and remove the nuts and the crossmember bracket.

- Crossmember
- Lower transaxle bolts

10. Loosen the engine support assembly and lean the engine towards the transaxle.

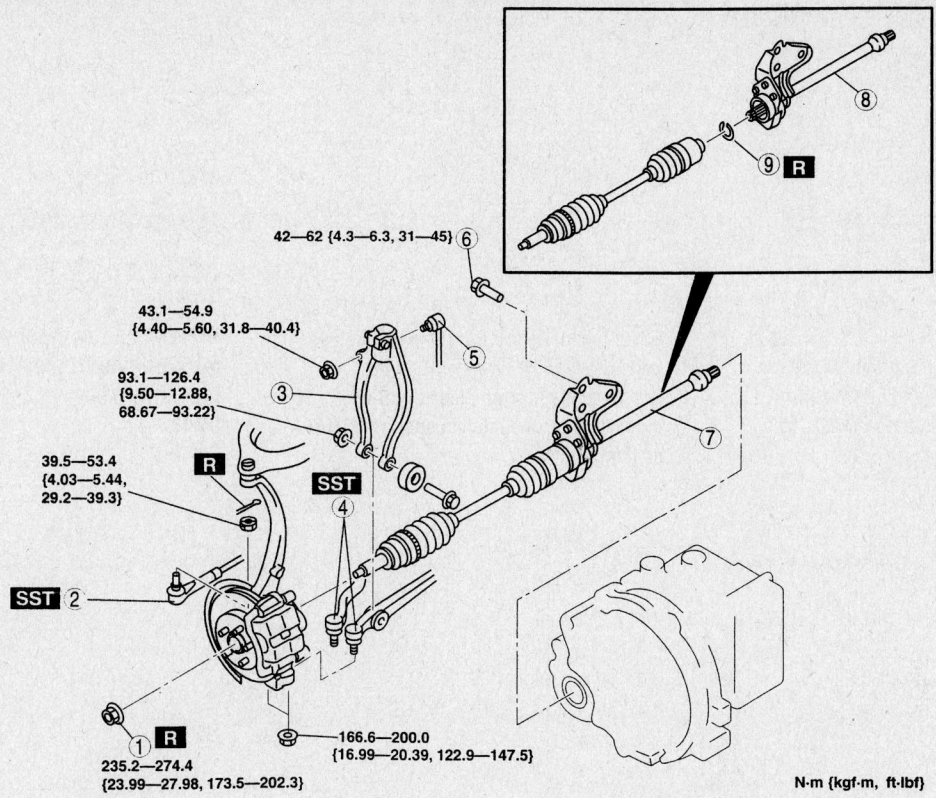

42—62 {4.3—6.3, 31—45} ⑥

43.1—54.9
{4.40—5.60, 31.8—40.4}

93.1—126.4
{9.50—12.88,
68.67—93.22}

39.5—53.4
{4.03—5.44,
29.2—39.3}

166.6—200.0
{16.99—20.39, 122.9—147.5}

235.2—274.4
{23.99—27.98, 173.5—202.3}

N·m {kgf·m, ft·lbf}

1	Locknut	5	Stabilizer control link
2	Tie-rod end ball joint	6	Joint shaft bracket bolt
3	Damper fork	7	Drive shaft and joint shaft
4	Lower arm (front, rear) ball joint	8	Joint shaft
		9	Clip

67162-MAZC-G121

Exploded view of the joint shaft assembly—Mazda6 models with the 3.0L (AJ) engine

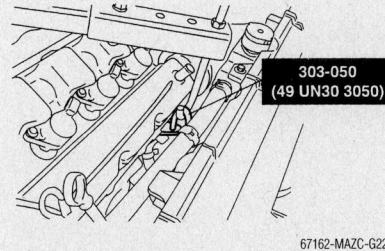

303-050
(49 UN30 3050)

67162-MAZC-G220

When removing the No. 1 engine mount, remove the intake manifold and attach tool 49 UN30 3050 to the head as shown—Mazda6 models with the A65M-R manual transaxle

11. Support the transaxle with a jack, remove the transaxle bolts and the transaxle.

To install:

12. Place the transaxle onto a jack and raise into position.

13. Install the transaxle bolts and tighten the bolts to 28–38 ft. lbs. (37–52 Nm). Refer to the exploded view illustration for bolt locations.

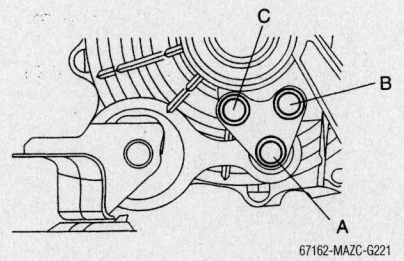

67162-MAZC-G221

Tighten the No. 1 mount in the sequence and specifications outlined in the text—Mazda6 models with a A65M-R manual transaxle

14. Install the crossmember in the reverse order of removal and refer to the exploded view illustration for component locations and torque specifications.

15. Align the hole of the No. 1 engine mount rubber with the bolt hole of the transaxle. Hand tighten bolt A and then tighten bolts B and C to 49–68 ft. lbs. (66–93 Nm) and then tighten bolt A to 49–68 ft. lbs. (66–93 Nm).

16. Install the No. 4 engine mount as follows:

a. Make sure the No. 4 mount is installed as illustrated.

b. Hand tighten bolts A and B.

c. Align the holes on the contact area of the front frame with bolt C hole.

d. Tighten bolt A and then B to 55–74 ft. lbs. (74–100 Nm).

e. Tighten bolt C and then D to 55–74 ft. lbs. (74–100 Nm).

f. Make sure the No. 4 mount is installed as illustrated.

g. Hand tighten bolt F.

h. Raise the transaxle with a floor jack and align the hole on the No. 4 engine mount bracket with the stud bolts on the transaxle.

i. Install bolts F, G and H and tighten in that sequence to 49–68 ft. lbs. (66–93 Nm).

j. Tighten bolt E to 63–86 ft. lbs. (85–116 Nm).

17. Remove the engine support device.

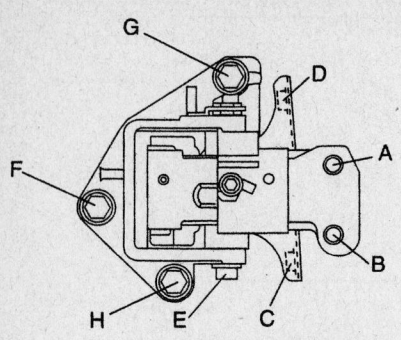

67162-MAZC-G222

Install the No. 4 mount and tighten bolts A thru D in the sequence and specifications outlined in the text—Mazda6 models with a A65M-R manual transaxle

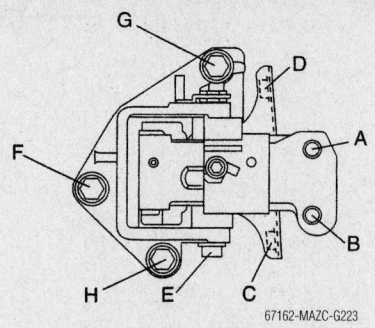

67162-MAZC-G223

When installing the No. 4 mount, tighten bolts F, G and H, then bolt E in the sequence and specifications outlined in the text—Mazda6 models with a A65M-R manual transaxle

18. Install the remaining components in the reverse order of removal.

19. Fill the transaxle with fluid. Road test the vehicle and check for leaks. Top off all fluids as needed.

Mazda6

A26MX-R TRANSAXLE

1. Before servicing the vehicle, refer to the precautions section.

→Refer the exploded view illustration for component locations and if applicable, their retainer torque specifications.

2. Disconnect the negative battery cable.

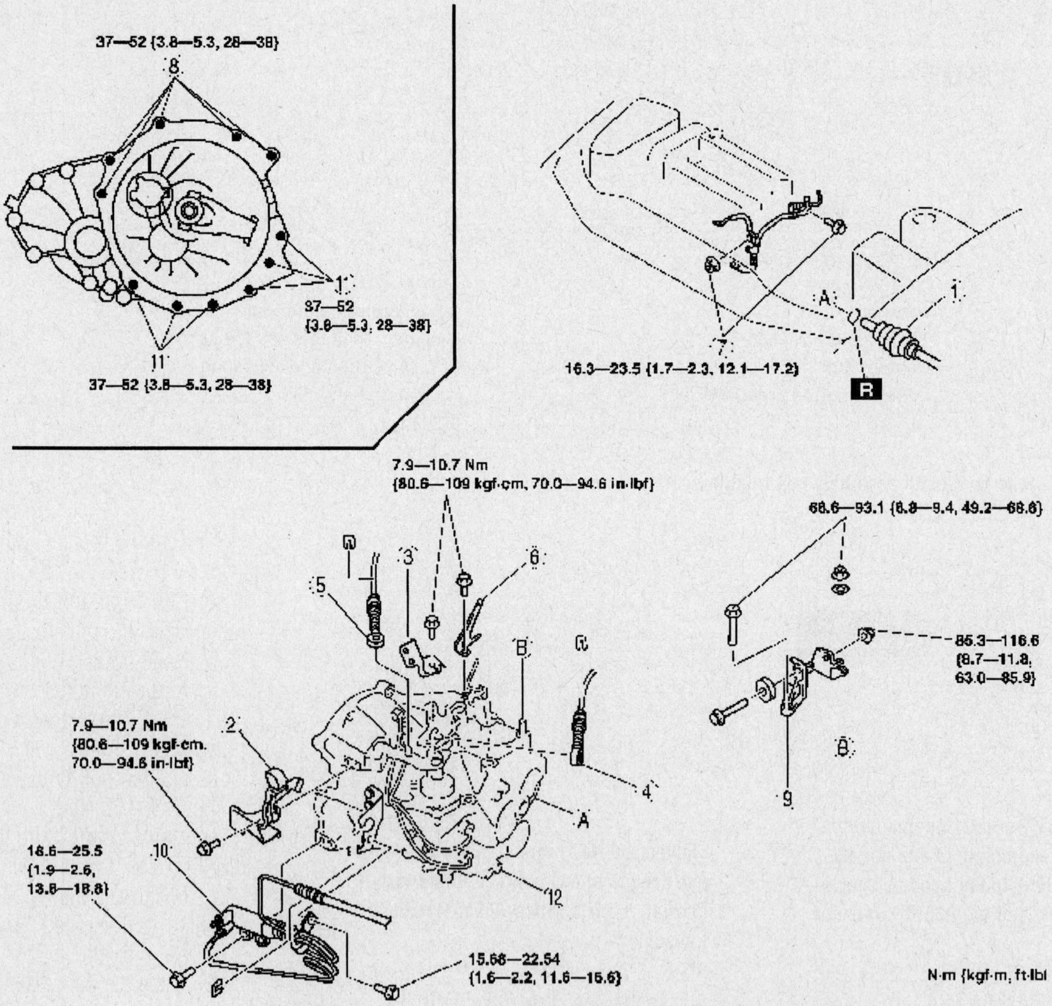

37—52 {3.8—5.3, 28—38}

37—52 {3.8—5.3, 28—38}

37—52 {3.8—5.3, 28—38}

16.3—23.5 {1.7—2.3, 12.1—17.2}

7.9—10.7 Nm {80.6—109 kgf·cm, 70.0—94.6 in·lbf}

7.9—10.7 Nm {80.6—109 kgf·cm, 70.0—94.6 in·lbf}

68.6—93.1 {6.9—9.4, 49.2—68.6}

85.3—116.6 {8.7—11.8, 63.0—85.9}

18.6—25.5 {1.9—2.6, 13.8—18.8}

15.68—22.54 {1.6—2.2, 11.6—16.6}

N·m {kgf·m, ft·lbf}

1	Drive shaft	7	Water pipe installation bolt and nut
2	Engine wiring harness bracket	8	Transaxle mounting bolt (upper side)
3	HO2S bracket	9	No.4 engine mount bracket
4	Selector cable	10	Clutch release cylinder
5	Shift cable	11	Transaxle mounting bolt (lower side)
6	Wiring harness bracket	12	Manual transaxle

Exploded view of the A26MX-R manual transaxle assembly mounting—Mazdaspeed6 models

09482_MAZC2_G0213

3. Remove the battery and battery tray.

4. Remove the wheels, tires and splash shields.

5. Remove the under cover.

6. Remove the steering gear and linkage, and pipe assembly installation bolts from the front crossmember, then suspend the steering gear and linkage with a cable.

7. Drain the transfer oil into a suitable container.

8. Remove the front pipe.

9. Disconnect the propeller shaft from the transfer side.

10. Remove the transfer oil cooler with the hose still connected.

11. Remove the lower arm (front, rear) ball joint.

12. Remove the damper fork.

13. Remove the tie-rod end ball joint.

14. Remove the stabilizer control link.

15. Remove the ABS sensor.

16. Remove the tie-rod end ball joint.

17. Remove the pinch bolt.

18. Separate the right side drive shaft from the joint shaft by tapping on a brass bar inserted between them.

19. Remove the joint shaft. Remove the bolts and pull the joint shaft straight out.

❊❊ CAUTION

The sharp edges of the joint shaft can slice or puncture the oil seal. Be careful when removing the joint shaft from the transaxle.

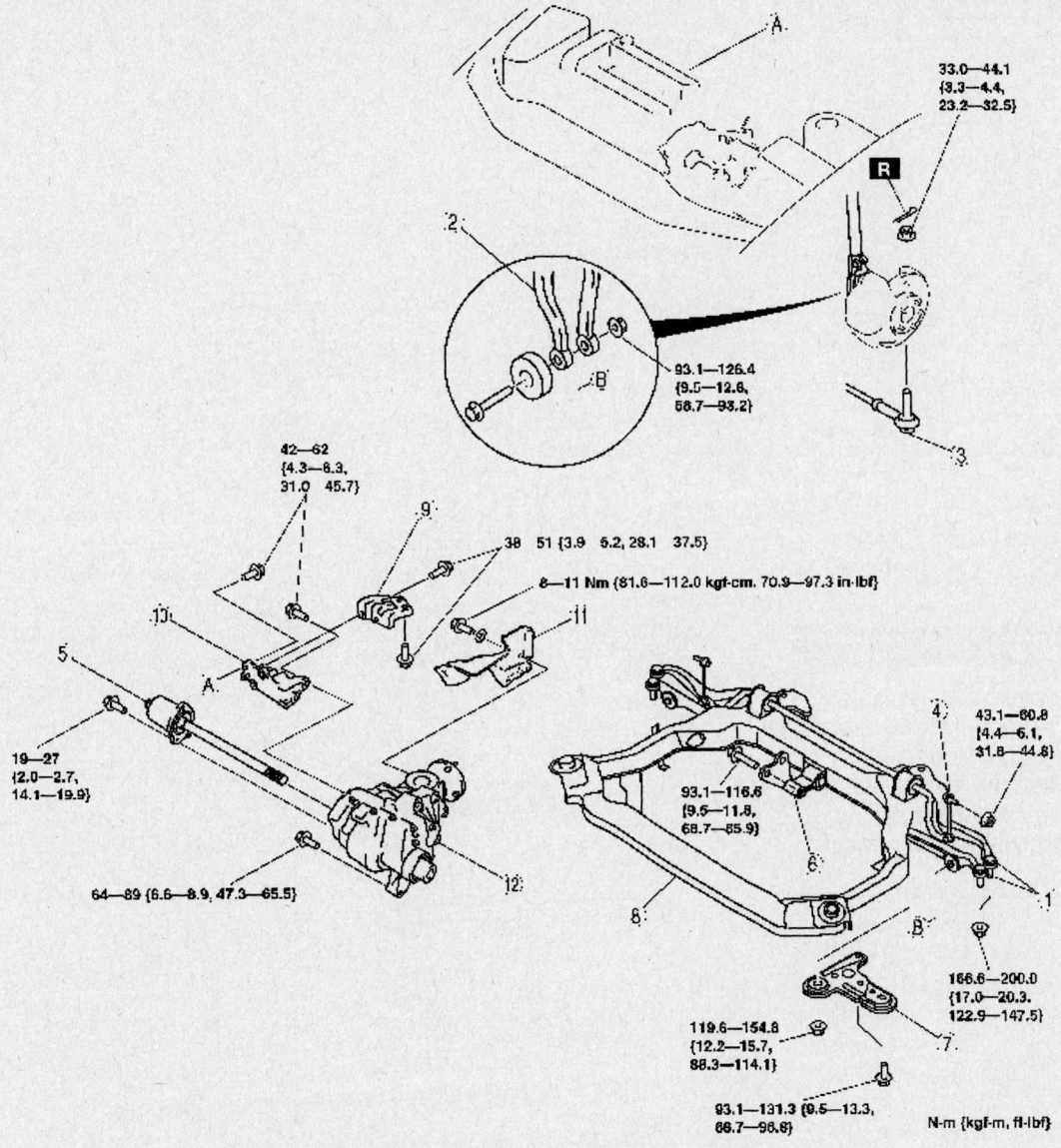

1	Lower arm (front, rear) ball joint	7	Crossmember bracket
2	Damper fork	8	Crossmember component
3	Tie-rod end ball joint	9	Catalytic converter bracket
4	Stabilizer control link	10	Transfer bracket
5	Joint shaft	11	Transfer heat insulator
6	No.1 engine mount	12	Transfer

09482_MAZC2_G0214

Exploded view of the transfer assembly mounting—Mazdaspeed6 models

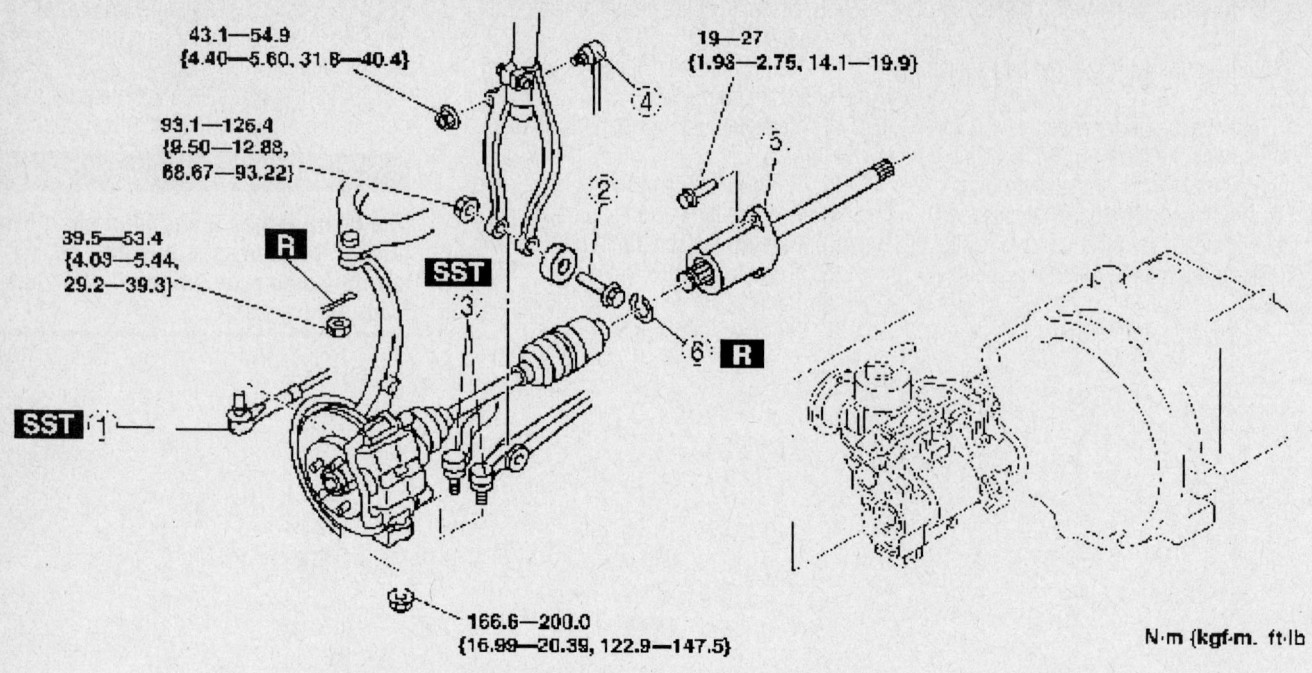

43.1—54.9
{4.40—5.60, 31.8—40.4}

93.1—126.4
{9.50—12.88,
68.67—93.22}

39.5—53.4
{4.09—5.44,
29.2—39.3}

19—27
{1.98—2.75, 14.1—19.9}

166.6—200.0
{16.99—20.39, 122.9—147.5}

N·m {kgf·m, ft·lb}

1 Tie-rod end ball joint	4 Stabilizer control link
2 Bolt	5 Joint shaft
3 Lower arm (front, rear) ball joint	6 Clip

09482_MAZC2_G0215

Exploded view of the joint shaft assembly mounting—Mazdaspeed6 models

20. Remove the clip.
21. Remove the No.1 engine mount.
22. Remove the crossmember bracket as follows:

✳✳ WARNING

Removing the crossmember is dangerous. The crossmember component could fall and cause serious injury or death. Verify that the jack securely supports the crossmember component before removing the crossmember bracket.

   a. Support the crossmember component with a jack and remove the nuts.
   b. Remove the crossmember bracket.
23. Remove the transverse member.
24. Remove the stabilizer control link nut.
25. Remove the front shock absorber lower side bolt.
26. Remove the crossmember component.
27. Remove the catalytic converter bracket.
28. Remove the transfer bracket, insulator and transfer.
29. Remove the air cleaner.
30. Remove the charge air cooler.
31. Disconnect the reverse and neutral switch connectors.

32. Drain the transaxle fluid.
33. Remove the drive shaft.
34. Remove the engine wiring harness bracket.
35. Remove the oxygen sensor bracket.
36. Remove the selector cable and shift cable.

   37. Remove the wiring harness bracket.
   38. Remove the water pipe installation bolt and nut.
   39. Remove the upper side transaxle mounting bolt.
   40. Support the engine with an engine

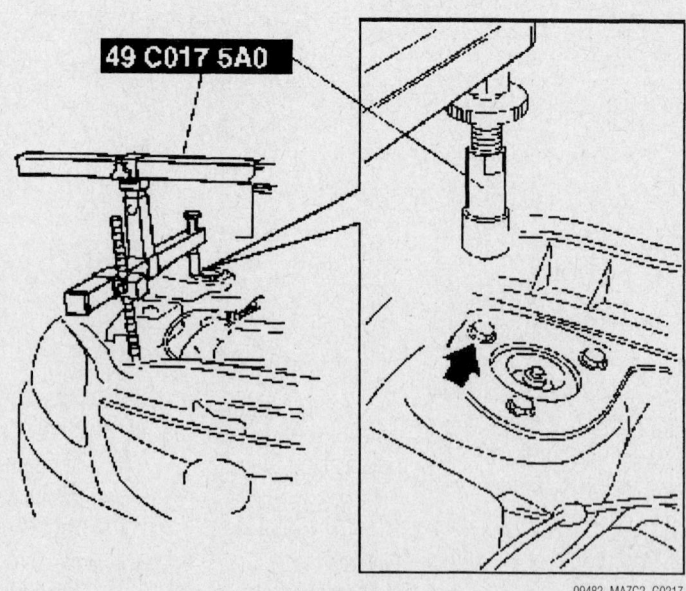

49 C017 5A0

09482_MAZC2_G0217

Install the left/right front shaft of the 49 C017 5AO with front foot No.2 to the bolt as shown—Mazdaspeed6 models

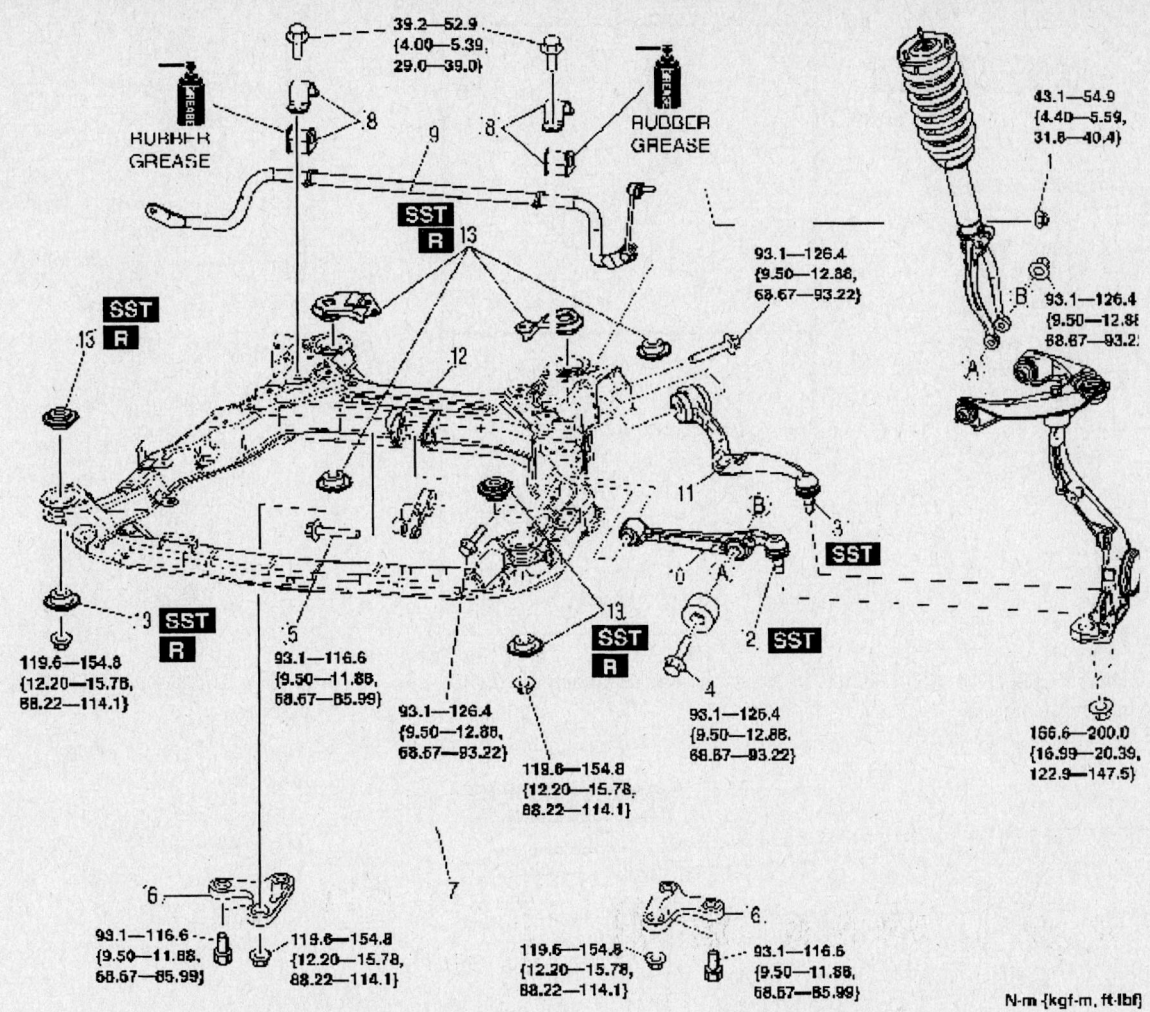

1   Nut (stabilizer control link)
2   Front lower arm (front) ball joint
3   Front lower arm (rear) ball joint
4   Bolt (front shock absorber lower side)
5   No.1 engine mount center bolt
6   Crossmember bracket
7   Crossmember component
8   Stabilizer bracket and bushing
9   Front Stabilizer
10 Front lower arm (front)
11 Front lower arm (rear)
12 Front crossmember
13 Front crossmember bushing

09482_MAZC2_G0216

Exploded view of the crossmember assembly mounting—Mazdaspeed6 models

support device such as 49 C017 5AO as illustrated.

✴✴ CAUTION

When setting tool 49 C017 5AO on the right side, make sure it doesn't interfere with the DSC HU/CM.

41. Remove the No.4 engine mount bracket.
42. Remove the clutch release cylinder.
43. Remove the lower side transaxle mounting bolt.
44. Remove the bolt shown in the figure and set the brake pipe out of the way.

45. Lean the engine toward the transaxle.
46. Support the transaxle on a jack.
47. Remove the transaxle mounting bolts.
48. Remove the transaxle.
To install:
49. Installation is the reverse of removal, please note the following:
    a. Install the transaxle mounting bolts and tighten to 28–38 ft. lbs. (37–52 Nm).
    b. Tighten the brake pipe bolt to 14–18 ft. lbs. (18–25 Nm).
    c. Install the No.4 engine mount bracket.
    d. Verify that the engine mount rubber is installed as shown.

    e. By aligning the holes with the stud bolts, install the No.4 engine mount bracket to the transaxle.
    f. Align the hole of the No.4 engine mount bracket with the No.4 engine mount rubber on vehicle, and temporarily tighten nut D.
    g. Lightly tighten nuts B, C and bolt A.
    h. Tighten nuts B, C in order of B through C, then bolt A.
    i. Tighten nut D.
    j. Bolts A, B and C torque to 49–67 ft. lbs. (66–93 Nm).
    k. Nut D torque to 63–86 ft. lbs. (85–116 Nm).

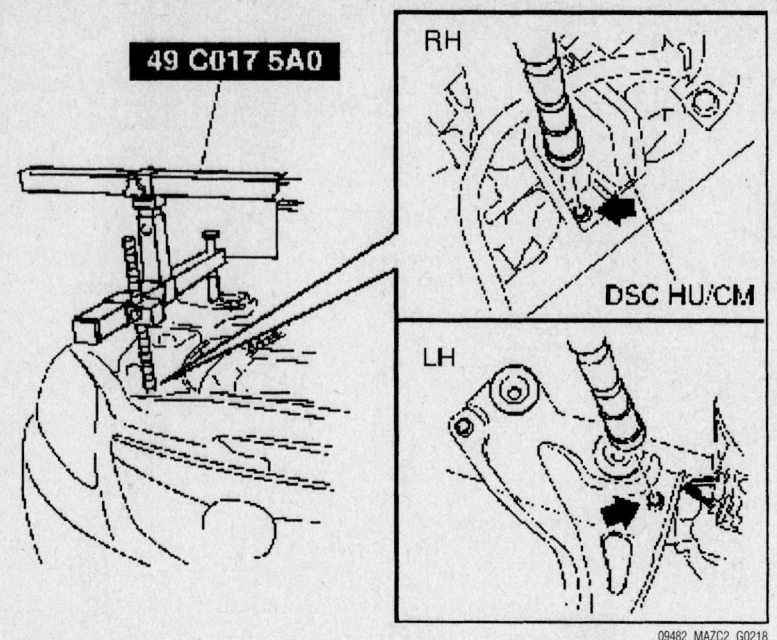

09482_MAZC2_G0218

Install the right rear shaft of the 49 C017 5AO to the bolt of the right shock absorber as shown—Mazdaspeed6 models

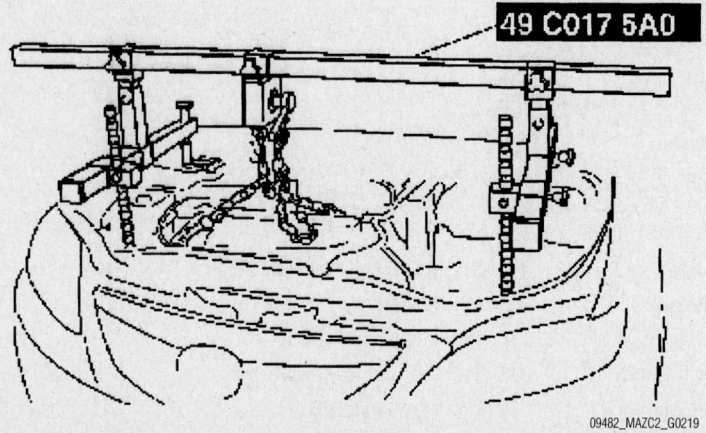

09482_MAZC2_G0219

Install the left rear shaft of the 49 C017 5AO to the bolt of the left shock absorber. (Identical position to the right side)—Mazdaspeed6 models

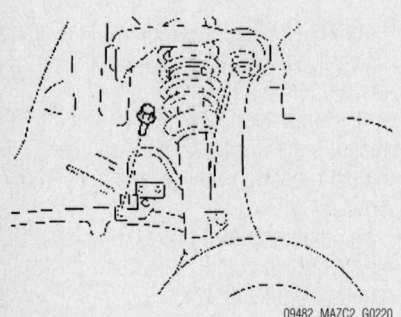

09482_MAZC2_G0220

Remove the bolt shown and set the brake pipe out of the way—Mazdaspeed6 models

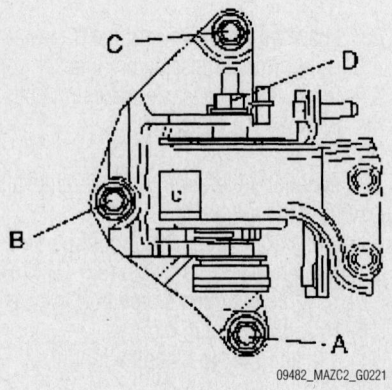

09482_MAZC2_G0221

No.4 engine mount bracket mounting—Mazdaspeed6 models

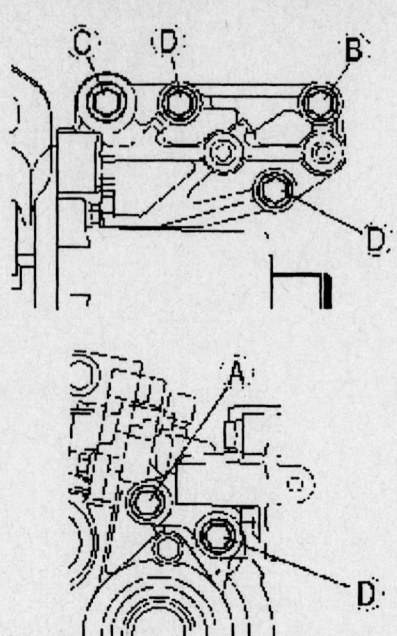

09482_MAZC2_G0222

Transfer bracket mounting—Mazdaspeed6 models

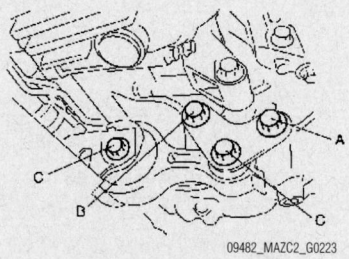

09482_MAZC2_G0223

No 1 engine mount assembly—Mazdaspeed6 models

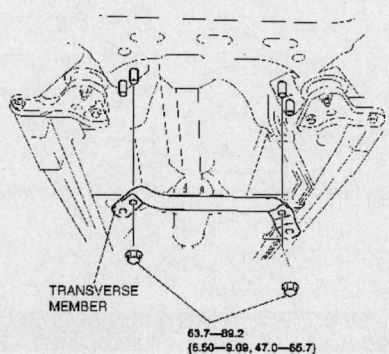

63.7–89.2
{6.50–9.09, 47.0–65.7}

N·m {kgf·m, ft·lbf}

09482_MAZC2_G0224

Transverse mount assembly—Mazdaspeed6 models

    l. Install the transfer bracket.
    m. Hand-tighten the bolt A and B.
    n. Tighten the bolts in order of C, A, B and D to 31–45 ft. lbs. (42–62 Nm).
    o. Install the No.1 engine mount.
    p. Align the hole of the No.1 engine

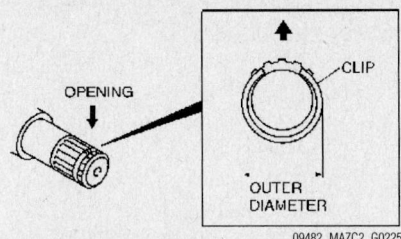

Install a new clip onto the joint shaft with the opening facing up and that the clip diameter does not exceed 1.34 inches (34mm)—Mazdaspeed6 models

mount rubber with the bolt hole of transaxle.

   q. Hand-tighten the bolt A, B, and C. Tighten the bolts to 68–86 ft. lbs. (93–116 Nm).

Automatic Transaxle/Transaxle Assembly

REMOVAL & INSTALLATION

MX5-Miata

   1. Before servicing the vehicle, refer to the precautions section.

➡**Refer the exploded view illustration for component locations and if applicable, their retainer torque specifications.**

   2. Remove the battery cover.
   3. Drain the transmission fluid.
   4. Disconnect the negative battery cable.
   5. Loosen the starter installation bolts only enough that the starter is loose, but not removed.
   6. Remove the tunnel member.
   7. Remove the catalytic converter and middle pipe.
   8. Mark the manual shaft lever component as shown in the illustration.
   9. Remove the manual shaft lever component installation nut.
   10. Remove the transverse member.
   11. Remove the under cover.
   12. Lock the drive plate using a flathead screwdriver as shown in the illustration.
   13. Remove the torque converter nuts.
   14. Remove the oil pipe and oil hose.
   15. Remove the insulator.
   16. Disconnect the TR switch, solenoid valve, VSS, turbine sensor, oil pressure (for oil filter) switch connectors.
   17. Remove the wiring harness.
   18. Remove the power plant frame as follows:

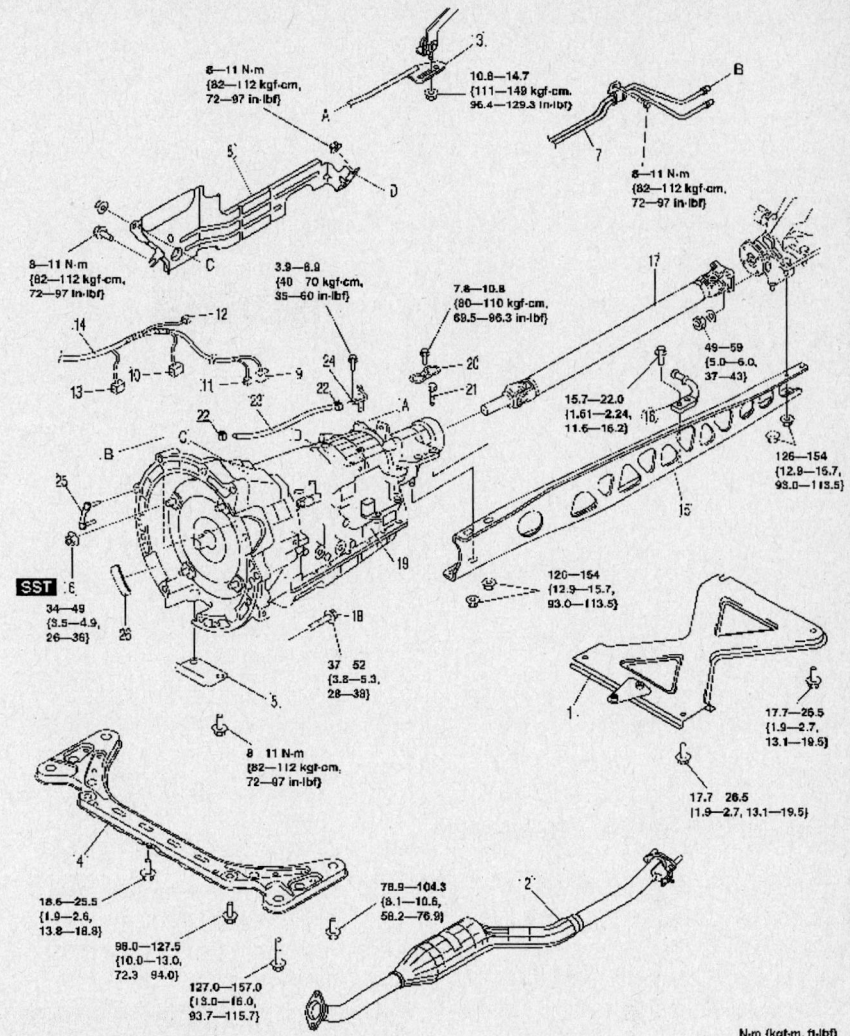

1	Tunnel member component	14	Wiring harness
2	Middle pipe	15	Power plant frame
3	Manual shaft lever component	16	Hanger bracket
4	Transverse member	17	Propeller shaft
5	Under cover	18	Transmission installation bolt and nut
6	Torque converter installation nuts	19	Transmission
7	Oil pipe, oil hose	20	Stopper
8	Insulator	21	Bolt
9	TR switch connector	22	Hose clamp
10	Solenoid valve connector	23	Breather hose
11	VSS connector	24	Breather tube
12	Turbine sensor connector	25	Stiffener
13	Oil pressure switch connector (for oil filter)	26	Side cover

Exploded view of the automatic transmission assembly and related components—MX-5 Miata

   a. Support the transmission using a transmission jack.
   b. Remove the power plant frame.
   19. Remove the hanger bracket.
   20. Remove the propeller shaft as follows:

➡**When replacing with a new propeller shaft, mark the companion flange to match the position of the tag on the propeller shaft.**

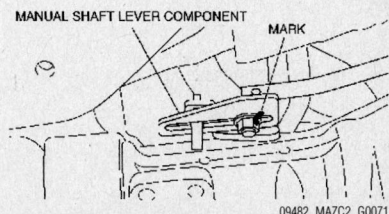

Mark the manual shaft lever component as shown—MX-5 Miata

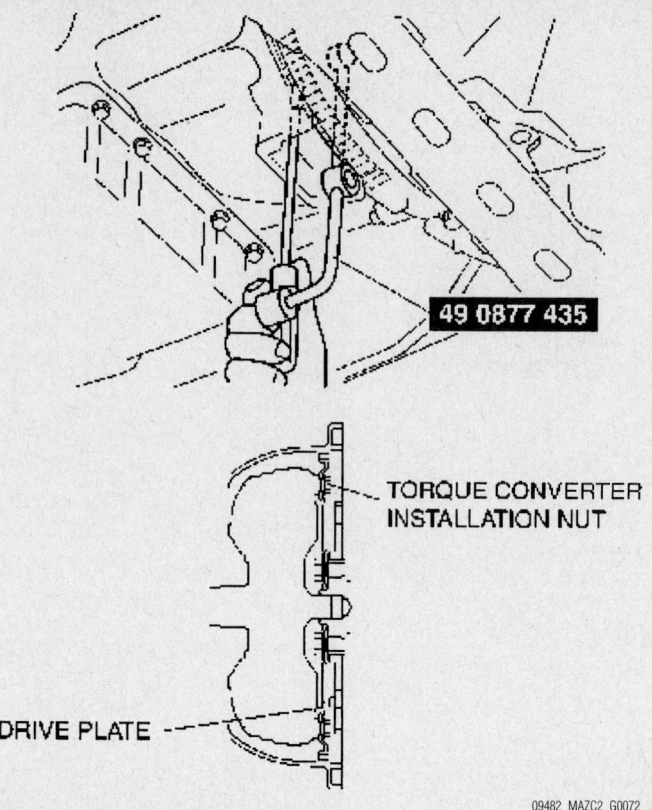

49 0877 435

TORQUE CONVERTER
INSTALLATION NUT

DRIVE PLATE

09482_MAZC2_G0072

Remove the torque converter nuts—MX-5 Miata

a. Before removing the propeller shaft, make alignment marks on the yoke and differential companion flange.

b. Remove the retainers and the shaft.

c. Install the tools illustrated onto the to the extension housing.

21. Verify that the transmission is securely supported by the jack. If the transmission falls, serious injury or death and damage to the vehicle could result. Before removing the transmission make sure that the jack is securely supporting the transmission.

✳✳ CAUTION

To prevent the torque converter and transmission from separating, remove the transmission without tilting it toward the torque converter.

5 MT 49 0259 440
6 MT 49 S120 440
AT

09482_MAZC2_G0012

Install the tools shown depending on transmission type, onto the to the extension housing—MX-5 Miata

22. Support the transmission securely using a transmission jack.

23. Remove the transmission bolts.

To install:

✳✳ WARNING

Verify that the transmission is securely supported by the jack. If the transmission falls, serious injury or death and damage to the vehicle could result. Before removing the transmission make sure that the jack is securely supporting the transmission.

24. Support the transmission securely using a transmission jack.

25. Tighten the transmission installation bolts and nuts to 37–52 Nm (28–38 Nm).

26. Installation of the remaining components is the reverse of removal, please note the following:

a. When installing the propeller shaft:
- Align the marks and install the propeller shaft.
- When installing a new propeller shaft, align the differential companion flange mark with the tag on the propeller shaft and assemble.

b. When installing the power plant frame:
- Support the transmission and differential so that they are level using a transmission jack.
- Install the power plant frame.
- Temporarily tighten the nuts 1, 2, 3 in the order shown in the illustration.
- Tighten nut 2 until the power plant frame is seated in the rear differential.
- Temporarily tighten the nuts 4, 5 in order shown in the illustration.
- Install the middle pipe and tunnel member.
- Raise the front end of the power plant frame (transmission side) or the transmission with the transmission jack, and adjust dimension A to 26.7–34.7 mm (1.06–1.36 in) (lower surface of power plant frame-upper surface of the tunnel member) as shown in the illustration.
- Tighten the power plant frame installation nuts to 126–154 Nm (93–113.5 ft. lbs)
- Verify that dimension A is within the specification with the transmission jack and the adjustment bolt removed.

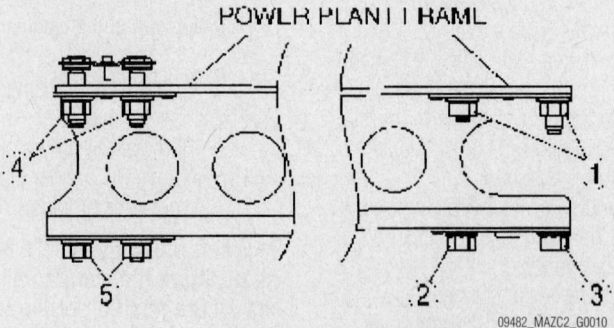

TRANSMISSION SIDE   DIFFERENTIAL SIDE

POWER PLANT FRAME

09482_MAZC2_G0010

Temporarily tighten the nuts 1, 2, 3 in the order shown—MX-5 Miata

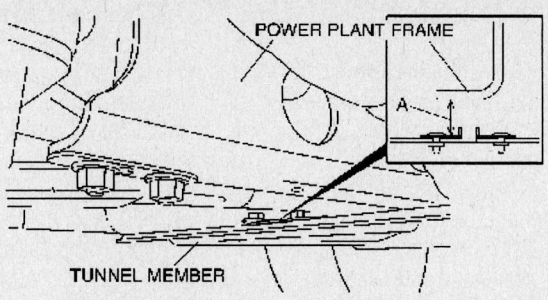

09482_MAZC2_G0011

Raise the front end of the power plant frame (transmission side) or the transmission with the transmission jack, and adjust dimension A to 26.7–34.7 mm (1.06–1.36 in—MX-5 Miata

- If not within the specification, adjust dimension A again.
- c. When securing the torque converter:
  - Align the holes by turning the torque converter.

27. Lock the drive plate using a flathead screwdriver.
  - Loosely and equally tighten the torque converter nuts, then further tighten them to 39–49 Nm (26–36 ft. lbs.).
  - a. When installing the manual shaft lever:
  - Align the mark of the manual shaft lever component as shown in the illustration
  - Install the manual shaft lever nut and tighten to 11–14 Nm (96–129 in. lbs.).

Mazda3

1. Before servicing the vehicle, refer to the precautions section.
2. Refer to the illustration for component location and torque specifications.
3. Drain the transaxle oil.
4. Remove or disconnect the following:
  - Battery duct and cover
  - Negative battery cable
  - Battery, tray and box
  - Air cleaner assembly
  - Exhaust manifold insulator
  - Wheels
  - Splash shields
  - Under cover
  - Input/turbine speed sensor connector
  - Vehicle speed sensor connector
  - Transaxle connector
  - Transaxle range switch connector
  - Ground wiring harness
  - Oil pressure switch connector
  - Harness bracket
  - Upper transaxle bolts
  - Stabilizer bar link

- Tie rod ends from the knuckle
- Lower control arm ball joint
- Halfshafts

5. Remove the joint shaft as follows:
  a. Disconnect the right halfshaft from the joint shaft by tapping the transaxle side outer ring with a brass bar and a hammer. Disconnect the joint shaft bracket from the block and remove the joint shaft.
  b. Install tool 49 G030 455 to hold the side gears after removal
6. Support the engine assembly with a

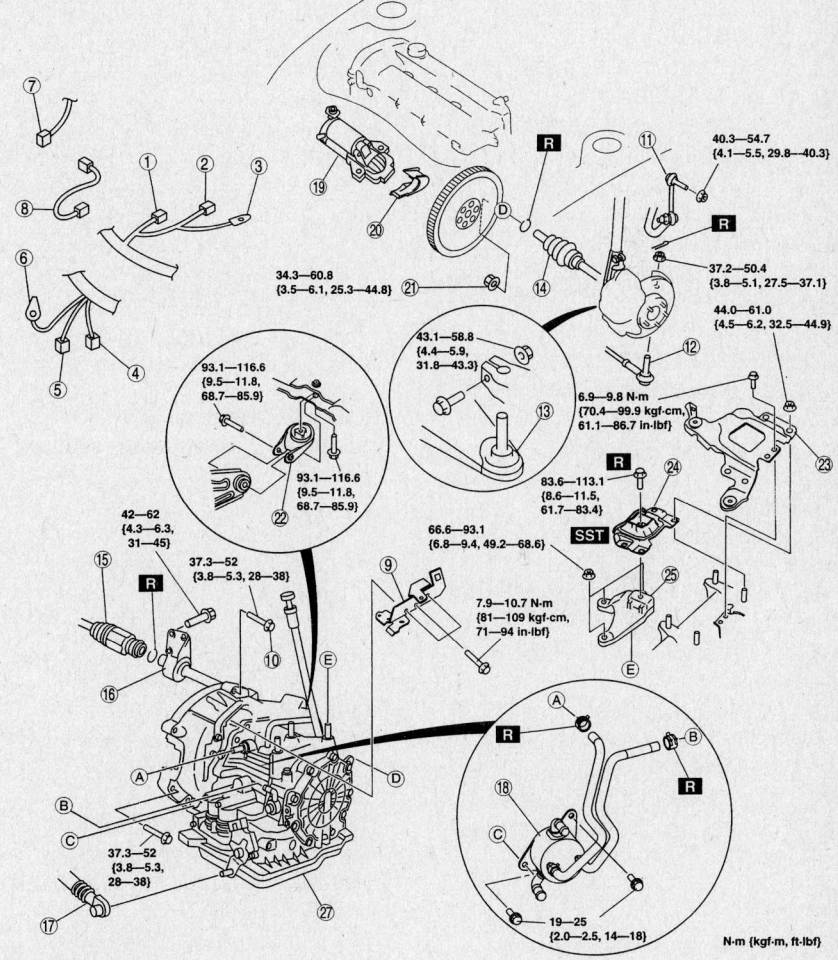

1	Input/turbine speed sensor connector	15	Drive shaft
2	VSS connector	16	Joint shaft
3	GND wiring harness	17	Selector cable
4	Transaxle connector	18	Oil cooler
5	TR switch connector	19	Starter
6	GND wiring harness	20	End plate cover
7	Oil pressure switch connector (for oil filter)	21	Torque converter installation nuts
8	Oil pressure switch connector (for L3 ATX)	22	No.1 engine mount rubber
9	Harness bracket	23	Battery tray bracket
10	Transaxle mounting bolt (upper side)	24	No.4 engine mount rubber
11	Stabilizer control link	25	No.4 engine mount bracket
12	Tie-rod end ball joint	26	Transaxle mounting bolt (lower side)
13	Lower arm ball joint	27	Transaxle
14	Drive shaft		

67162-MAZC-G81

Exploded view of the automatic transaxle assembly mounting—Mazda3 models

engine support assembly such as 49 E017 5A0.

7. Remove or disconnect the following:
- Selector cable

8. Remove the oil cooler as follows:
   a. Remove the water hose.
   b. Remove the oil hose.
   c. Remove the hose clamp.
   d. Remove the connector bolt, packing, oil pipe, packing and oil cooler.

9. Remove or disconnect the following:
- Starter
- End plate cover
- Torque converter bolts, through the starter opening

10. Install an engine support device such as 49 E017 5A0.
- Number 1 engine mount rubber
- Battery tray bracket

11. Remove the number 4 engine mount and joint bracket as follows:

   a. Install two suitable pieces of wood between the front fender panel and upper apron reinforcement as illustrated. The wood size should be approximately 1.38 inch (35mm) on 4 door models or 2.36 inch (60mm) on 5 door models.

12. Loosen the engine support assembly and lean the engine towards the transaxle.

13. Support the transaxle with a jack, remove the transaxle bolts and the transaxle.

To install:

14. Place the transaxle onto a jack and raise into position.

15. Install the transaxle bolts and tighten the bolts to 28–38 ft. lbs. (37–52 Nm). Refer to the exploded view illustration for bolt locations.

16. Install or connect the following:
- Number 4 engine mount bracket to the transaxle case and tighten the bolts to 49–68 ft. lbs. (66–93 Nm)
- Number 1 engine mount rubber to

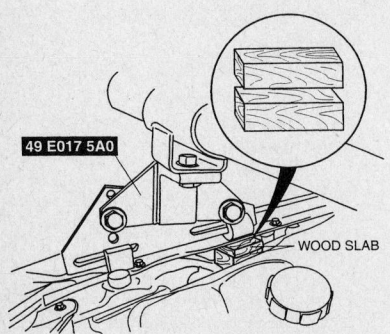

Install two suitable pieces of wood between the front fender panel and upper apron reinforcement—Mazda3 with an automatic transaxle

---

the crossmember and hand tighten the bolts
- Number 4 engine mount rubber with the body stud passing through the holes and tighten the bolts to 61–87 ft. lbs. (83–113 Nm)
- Battery tray bracket on the number 4 engine mount rubber with the body stud bolts passing through the holes. Tighten retainers identified in the illustration as 2 to 32–45 ft. lbs. (44–61 Nm) and the retainers identified as 1 in the same illustration to 61–86 inch lbs. (7–10 Nm).

17. Remove the engine support device and tighten the number 1 mount rubber bolts to 68–85 ft. lbs. (116 Nm).

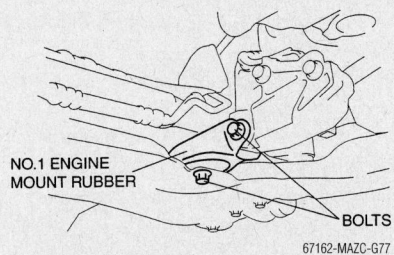

NO.1 ENGINE MOUNT RUBBER

BOLTS

67162-MAZC-G77

Number 1 engine mount rubber bolt locations—Mazda3 models with an automatic transaxle

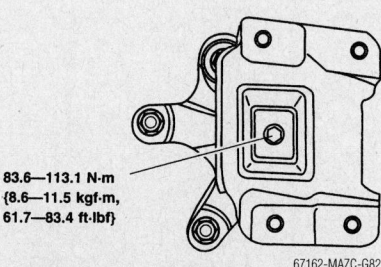

83.6—113.1 N·m {8.6—11.5 kgf·m, 61.7—83.4 ft·lbf}

67162-MAZC-G82

Install the number 4 engine mount rubber with the body stud passing through the holes—Mazda3 models with an automatic transaxle

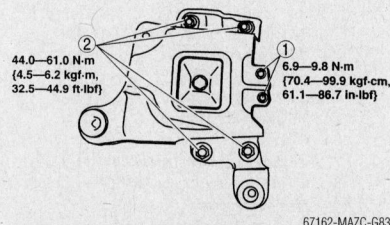

44.0—61.0 N·m {4.5—6.2 kgf·m, 32.5—44.9 ft·lbf}

6.9—9.8 N·m {70.4—99.9 kgf·cm, 61.1—86.7 in·lbf}

67162-MAZC-G83

Install the battery tray bracket on the number 4 engine mount rubber with the body stud bolts passing through the holes—Mazda3 models with an automatic transaxle

---

- Torque converter bolts and tighten to 25–44 ft. lbs. (34–60 Nm)
- End plate cover
- Starter

18. Install the oil cooler in the reverse order of removal keeping in mind the following steps and referring to the oil cooler exploded illustration for torque specifications:

   a. Apply compressed air to the cooler side opening to clear any debris. Apply the air for less than a minute only.

   b. If reusing the same oil hose, install a new clamp right on the mark left by the old clamp and apply force in the direction of the arrow to secure the clamp. Align the marks and slide the oil hose onto the pipe until seated as illustrated. Make sure the hose clamp does not interfere with any other component.

- Selector cable
- Joint shaft. Install a new circlip with the opening facing up. Hand tighten the bolts, then tighten the bolts to 31–45 ft. lbs. (42–62 Nm). Refer to the illustration for bolt location.
- Halfshafts
- Lower control arm ball joint. Tighten the retainers to 31–43 ft. lbs. (43–59 Nm).
- Tie rod ends to the knuckle. Tighten the retainers to 27–37 ft. lbs. (37–50 Nm).
- Stabilizer bar link. Tighten the retainers to 30–40 ft. lbs. (40–54 Nm).
- Harness bracket
- Oil pressure switch connector
- Ground wiring harness
- Transaxle range switch connector
- Transaxle connector
- Vehicle speed sensor
- Input/Turbine speed sensor connector
- Under cover
- Splash shields
- Wheels
- Exhaust manifold insulator
- Air cleaner assembly
- Battery tray and battery

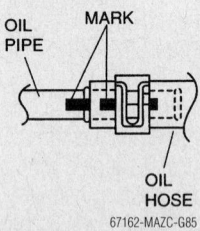

OIL PIPE

MARK

OIL HOSE

67162-MAZC-G85

Align the marks and slide the oil hose onto the pipe until seated—Mazda3 models with an automatic transaxle

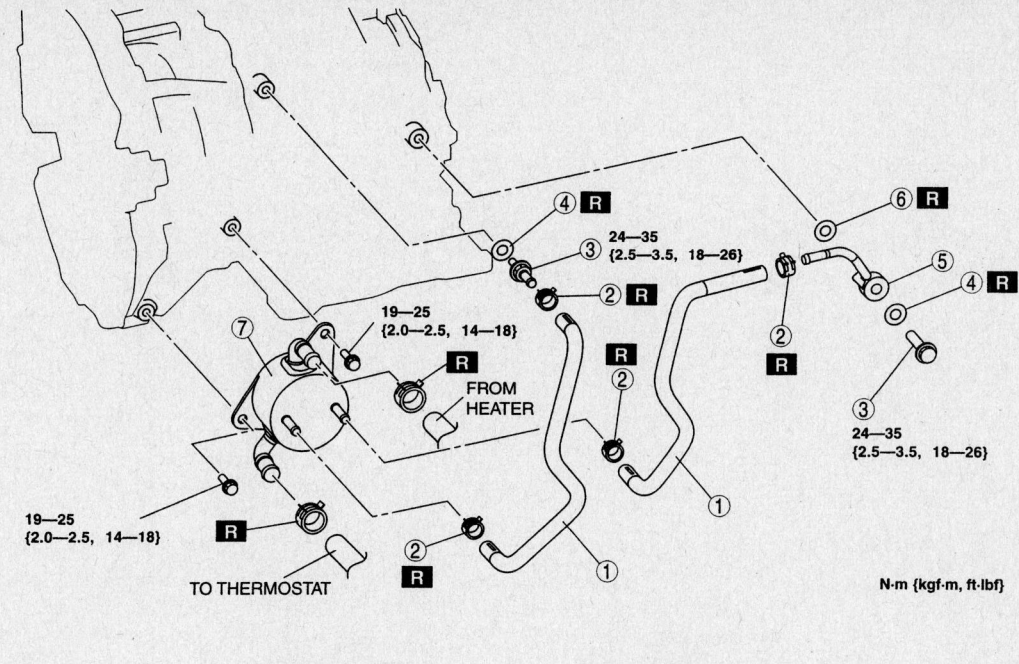

1 Oil hose	5 Oil pipe
2 Hose clamp	6 Packing
3 Connector bolt	7 Oil cooler
4 Packing	

67162-MAZC-G84

Exploded view of the oil cooler assembly and related components—Mazda3 models with an automatic transaxle

19. Fill the transaxle with fluid. Road test the vehicle and check for leaks. Top off all fluids as needed.

Mazda6

FN4A–EL TRANSAXLE

1. Before servicing the vehicle, refer to the precautions section.

2. Refer to the illustration for component location and torque specifications.

3. Drain the transaxle oil.

4. Remove or disconnect the following:
- Negative battery cable
- Battery and battery tray
- Air cleaner assembly
- Wheels
- Splash shields
- Under cover
- Steering gear, linkage and pipe assembly bolts from the crossmember and using mechanics wire, position the steering gear and linkage assembly aside.
- Heated Oxygen (HO$_2$S) connector
- Oil pressure switch connectors for the oil filter and transaxle
- Input/turbine speed connector
- Transmission range switch connector
- Transaxle connector
- Vehicle speed sensor connector

- Oil dipstick tube
- Harness bracket
- Oil hose
- Selector cable
- Transaxle upper mount bolts
- Starter
- Endplate cover
- Lower control arm ball joints
- Damper fork
- Tie rod ends from the knuckle
- Stabilizer bar link
- Halfshafts

5. Disconnect the right halfshaft from the joint shaft by tapping the transaxle side outer ring with a brass bar and a hammer. Disconnect the joint shaft bracket from the block and remove the joint shaft.

6. Install tool 49 G030 455 to hold the side gears after removal.

7. Support the engine assembly with a engine support assembly such as 49 E017 5A0.

8. Remove or disconnect the following:
- Number 1 engine mount
- Lower front shock absorber bolt
- No. 1 engine mount center bolt

9. Support the crossmember with a jack and remove the nuts and the crossmember bracket.
- Crossmember
- Torque converter nuts, hold the

crankshaft pulley to prevent the flywheel from turning while loosening the nuts and remove the nuts through the starter motor opening
- Number 4 engine mount
- Lower transaxle bolts

10. Loosen the engine support assembly and lean the engine towards the transaxle.

11. Support the transaxle with a jack, remove the transaxle bolts and the transaxle.

To install:

12. Place the transaxle onto a jack and raise into position.

13. Install the transaxle bolts and tighten the bolts to 28–38 ft. lbs. (37–52 Nm). Refer to the exploded view illustration for bolt locations.

14. Install the torque converter nuts and tighten to 25–32 ft. lbs. (34–44 Nm).

15. Install the No. 1 engine mount and No. 4 engine mount bracket as follows:

a. Make sure the mount is installed as illustrated.

b. Align the bolt holes with the stud bolts, install the No. 4 mount bracket to the transaxle.

c. Align the holes with the stud bolts and install the No. 1 engine mount to the transaxle.

d. Align the hole on the No. 4 engine

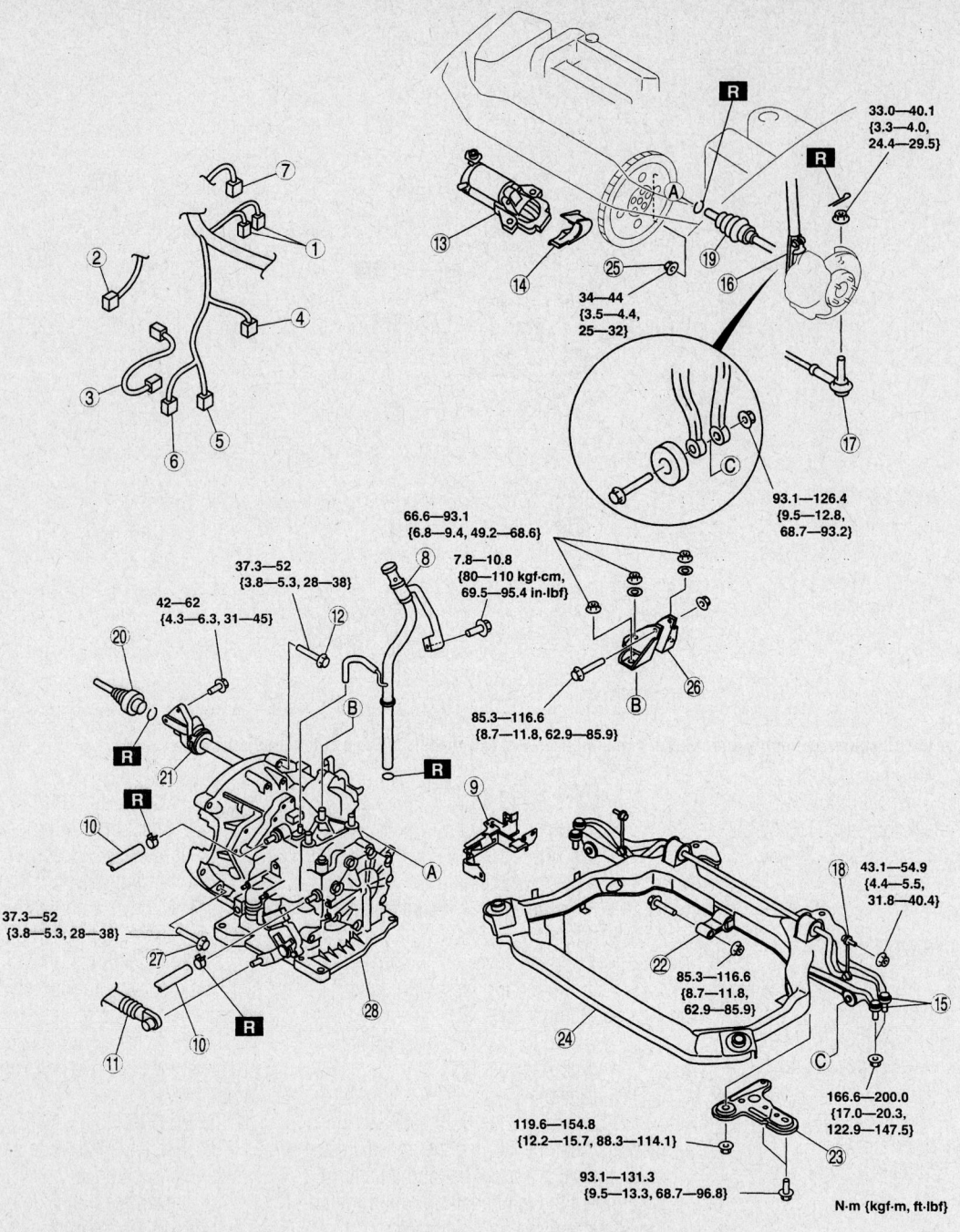

1	HO2S connector	15	Lower arm (front, rear) ball joint
2	Oil pressure switch connector (for oil filter)	16	Damper fork
3	Oil pressure switch connector (for ATX)	17	Tie-rod end ball joint
4	Input/turbine speed sensor connector	18	Stabilizer control link
5	TR switch connector	19	Drive shaft
6	Transaxle connector	20	Drive shaft
7	VSS connector (Without ABS)	21	Joint shaft
8	Oil dipstick and filler tube	22	No.1 engine mount
9	Harness bracket	23	Crossmember bracket
10	Oil hose	24	Crossmember
11	Selector cable	25	Torque converter installation nuts
12	Transaxle mounting bolt (Upper side)	26	No.4 engine mount
13	Starter	27	Transaxle mounting bolt (lower side)
14	Endplate cover	28	Transaxle

67162-MAZC-G224

Exploded view of the FN4A–EL automatic transaxle assembly mounting—Mazda6 models

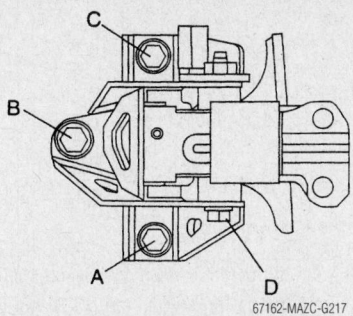

67162-MAZC-G217

Install the No. 1 mount and No. 4 mount bracket and tighten the bolts in the sequence and specifications outlined in the text—Mazda6 models with a FN4A-EL automatic transaxle

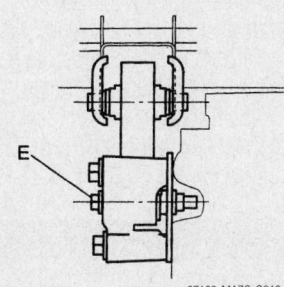

67162-MAZC-G218

Tighten bolt E on the No. 1 mount to the specification outlined in the text—Mazda6 models with a FN4A-EL automatic transaxle

mount bracket with the No. 4 mount rubber on the car and hand tighten bolt D. Refer to the illustration for bolt locations.

16. Hand tighten bolts B, C and A.

   a. Tighten bolts B and C then bolt A to 49–68 ft. lbs. (66–93 Nm).

   b. Tighten bolt D to 63–86 ft. lbs. (85–116 Nm).

   c. Tighten bolt E on the No. 1 mount to 63–86 ft. lbs. (85–116 Nm).

17. Remove the engine support device.

18. Install the remaining components in the reverse order of removal.

19. Fill the transaxle with fluid. Road test the vehicle and check for leaks. Top off all fluids as needed.

JA5A–EL TRANSAXLE

1. Before servicing the vehicle, refer to the precautions section.

2. Refer to the illustration for component location and torque specifications.

3. Drain the transaxle oil.

4. Remove or disconnect the following:
- Battery and battery tray
- Air cleaner assembly
- Starter
- Separate the heater pipe

- Wheels
- Splash shields
- Under cover
- Steering gear, linkage and pipe assembly bolts from the cross-member and using mechanics wire, position the steering gear and linkage assembly aside.
- Terminal No. 1, No. 2, Transmission range switch connectors and the ground harness
- Clip
- Selector cable
- Cable bracket
- Transaxle upper mount bolts
- Transaxle oil hoses (cooler lines)
- Lower control arm ball joints
- Damper fork

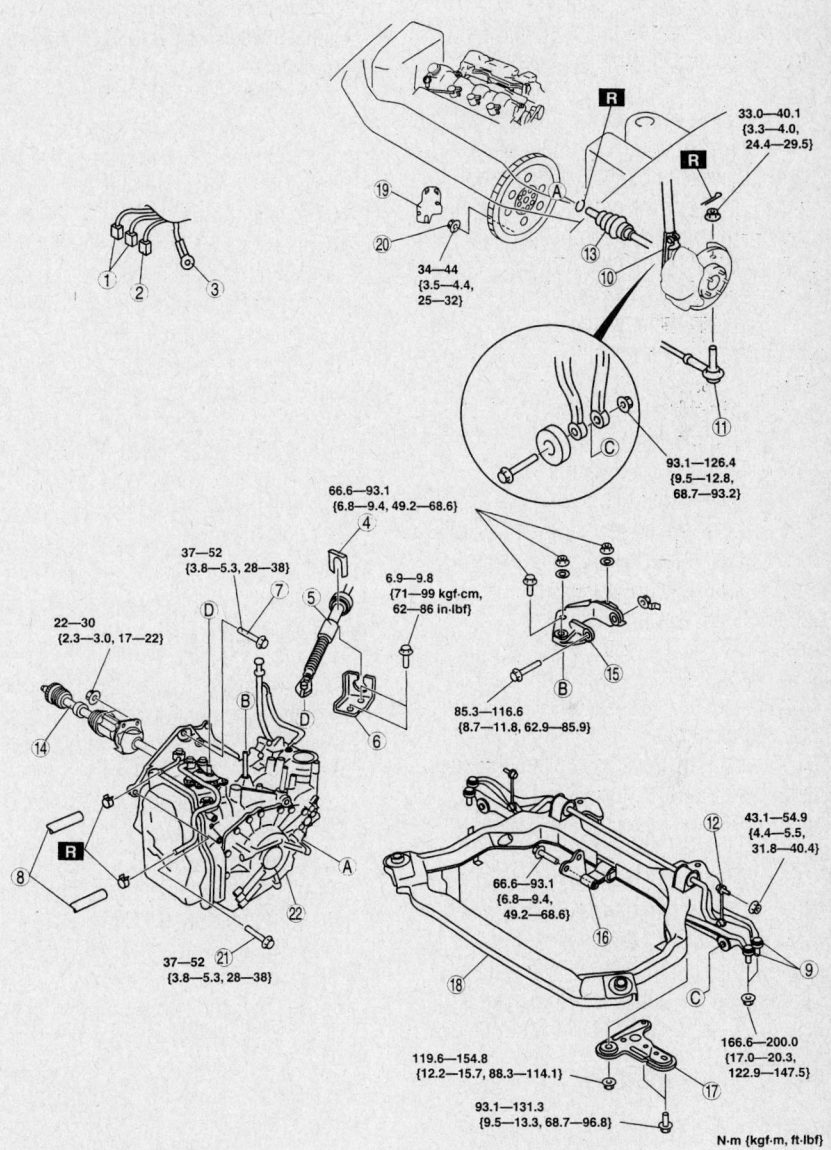

N·m {kgf·m, ft·lbf}

1	Terminal component No.1, No.2 connector	12	Stabilizer control link
2	TR switch connector	13	Drive shaft
3	GND harness	14	Drive shaft, joint shaft
4	Clip	15	No.4 engine mount
5	Selector cable	16	No.1 engine mount
6	Cable bracket	17	Crossmember bracket
7	Transaxle mounting bolt (Upper side)	18	Crossmember
8	Oil hose	19	Endplate cover
9	Lower arm (front, rear) ball joint	20	Torque converter installation nuts
10	Damper fork	21	Transaxle mounting bolt (lower side)
11	Tie-rod end ball joint	22	Transaxle

67162-MAZC-G225

Exploded view of the JA5A–EL automatic transaxle assembly mounting—Mazda6 models

- Tie rod ends from the knuckle
- Stabilizer bar link
- Halfshafts

5. Disconnect the left halfshaft from the joint shaft by inserting a prybar between the transaxle and the halfshaft outer ring. Disconnect the joint shaft bracket from the block and remove the joint shaft.

6. Install tool 49 G030 455 to hold the side gears after removal
- No. 4 engine mount

7. Support the engine assembly with a engine support assembly such as 49 E017 5A0.

8. Remove or disconnect the following:
- Number 1 engine mount bracket
- Lower front shock absorber bolt
- No. 1 engine mount. Remove the intake manifold and attach tool 49 UN30 3050 to the head as shown and remove the mount.

9. Support the crossmember with a jack and remove the nuts and the crossmember bracket.
- Crossmember
- Endplate cover
- Torque converter nuts
- Lower transaxle bolts

10. Loosen the engine support assembly and lean the engine towards the transaxle.

11. Support the transaxle with a jack, remove the transaxle bolts and the transaxle.

To install:

12. Place the transaxle onto a jack and raise into position.

13. Install the transaxle bolts and tighten the bolts to 28–38 ft. lbs. (37–52 Nm). Refer to the exploded view illustration for bolt locations.

14. Align the holes for the torque converter nuts by rotating the converter and then lock the drive plate by inserting a suitable tool through the starter motor opening. Hand tighten the nuts, then final tighten them to 25–32 ft. lbs. (34–44 Nm) using 2–3 passes.

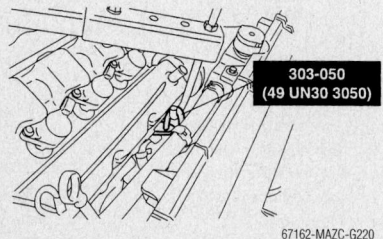

303-050
(49 UN30 3050)

67162-MAZC-G220

When removing the No. 1 engine mount, remove the intake manifold and attach tool 49 UN30 3050 to the head as shown—Mazda6 models with the JA5A-EL automatic transaxle

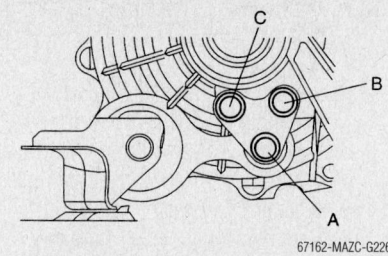

67162-MAZC-G226

Tighten the No. 1 mount in the sequence and specifications outlined in the text—Mazda6 models with a JA5A-EL automatic transaxle

15. Install the crossmember in the reverse order of removal and refer to the exploded view illustration for component locations and torque specifications. Refer to the exploded view of the crossmember assembly in the engine removal procedure for the Mazda6 models equipped the 3.0L (AJ) engine earlier in this section.

16. Align the hole of the No. 1 engine mount rubber with the bolt hole of the transaxle. Hand tighten bolt A and then tighten bolts B and C to 49–68 ft. lbs. (66–93 Nm) and then tighten bolt A to 49–68 ft. lbs. (66–93 Nm).

17. Install the No. 4 engine mount as follows:
  a. Make sure the No. 4 mount is installed as illustrated.
  b. Hand tighten bolt D.
  c. Raise the transaxle with a floor jack and align the hole on the No. 4 engine mount bracket with the stud bolts on the transaxle.
  d. Hand tighten bolt A and nuts B and C.
  e. Tighten bolts B, C to 49–68 ft. lbs. (66–93 Nm) and then tighten bolt A to 49–68 ft. lbs. (66–93 Nm).
  f. Tighten bolt D to 62–86 ft. lbs. (85–116 Nm).

18. Remove the engine support device.

19. Connect the selector lever to manual

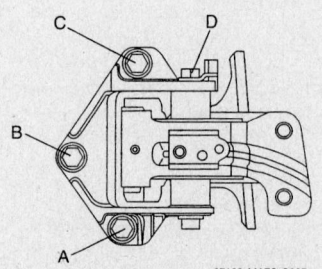

67162-MAZC-G227

Location of the No. 4 mount bolts, tighten in the sequence and specifications outlined in the text—Mazda6 models with a JA5A-EL automatic transaxle

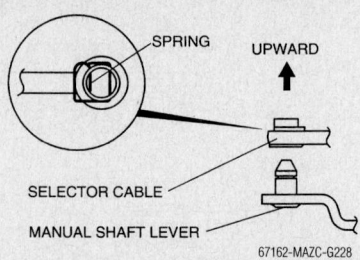

SPRING
UPWARD
SELECTOR CABLE
MANUAL SHAFT LEVER

67162-MAZC-G228

Connect the selector lever to manual shaft lever so the spring side of the cable end is facing upwards—Mazda6 models with a JA5A-EL automatic transaxle

shaft lever so the selector cable does not have a load on it and the spring side of the cable end is facing upwards as illustrated. Make sure the end of the manual shaft lever sticks out of the end of the selector cable.

20. Install the remaining components in the reverse order of removal.

21. Fill the transaxle with fluid. Road test the vehicle and check for leaks. Top off all fluids as needed.

Clutch

REMOVAL & INSTALLATION

MX5-Miata

1. Before servicing the vehicle, refer to the precautions section.

➡Refer the exploded view illustration for component locations and if applicable, their retainer torque specifications.

2. Clutch release cylinder.
3. Manual transmission.
4. Clutch release collar.
5. Clutch release fork.
6. Clutch cover bolts in a criss-cross pattern using several steps.
7. Clutch disc.

To install:

8. Installation is the reverse of removal. Tighten the clutch cover bolts to 25–33 Nm (18–24 ft. lbs.).

Mazda3, Mazda6 and Mazdaspeed6 Models

1. Before servicing the vehicle, refer to the precautions section.

2. Remove or disconnect the following:
- Negative battery cable
- Clutch release cylinder
- Transaxle
- Rubber boot
- Clutch release collar
- Clutch release fork

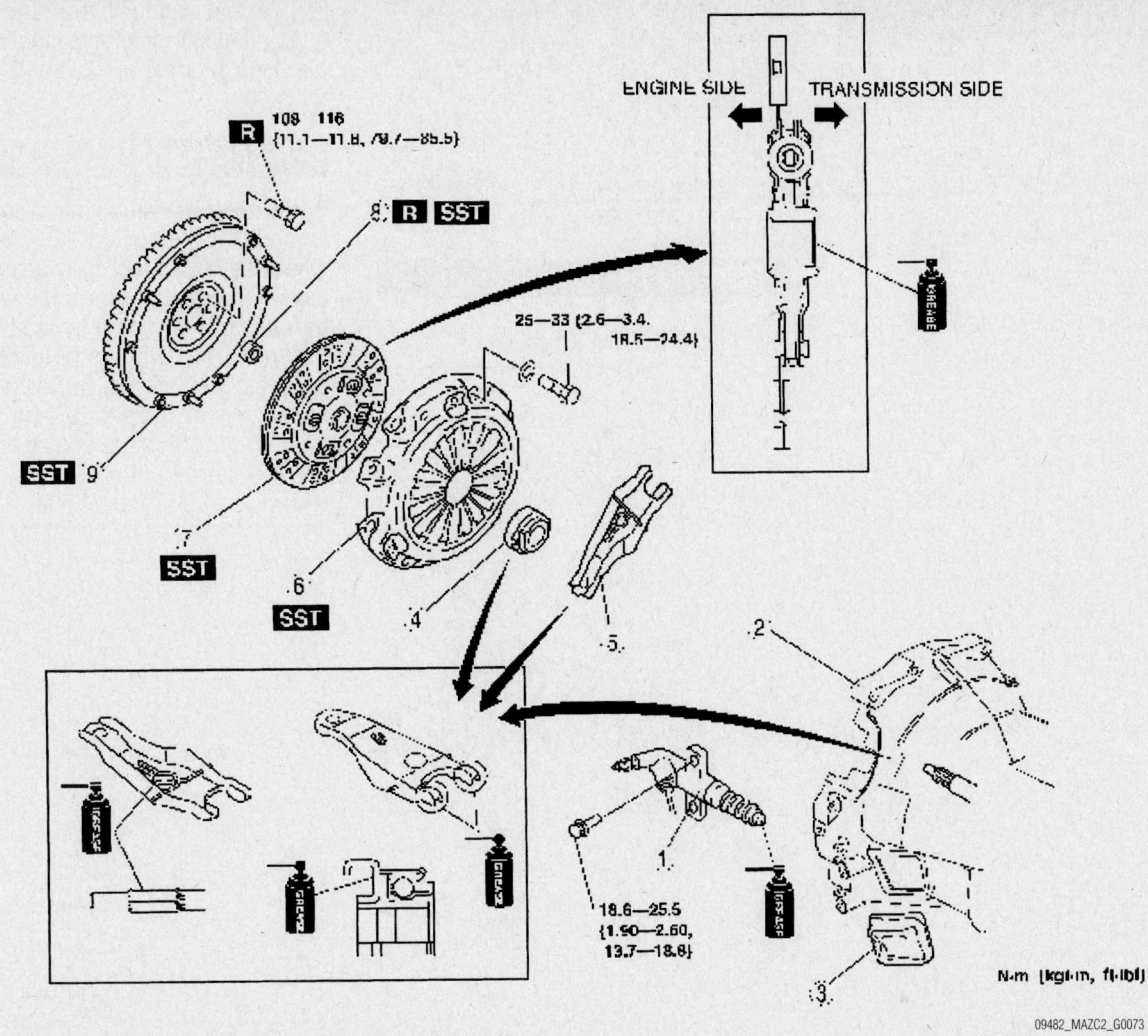

ENGINE SIDE   TRANSMISSION SIDE

108 116
{11.1—11.8, 79.7—85.5}

25—33 {2.6—3.4,
18.5—24.4}

18.6—25.5
{1.90—2.60,
13.7—18.8}

N·m {kgf·m, ft·lbf}

09482_MAZC2_G0073

Exploded view of the clutch assembly components—MX-5 Miata

- Pressure plate loosening the bolts one turn each in a criss–cross pattern
- Clutch disc

3. Inspect the pilot bearing. If it is worn or damaged and does not turn easily by hand, remove it using a puller/slide hammer.

4. Check the flywheel surface for scoring, cracks or burning and machine or replace, as necessary.

5. Install Holder tool 49 E011 1A0 to keep the flywheel from turning. Loosen the flywheel bolts evenly and gradually in a crisscross pattern. Remove the flywheel.

6. Inspect the clutch release bearing for wear. Replace it if it sticks or does not turn easily.

7. Inspect the release fork for wear or damage and replace as necessary.

To install:

8. Lubricate the release fork fingers and pivot with molybdenum grease and install in the release fork boot.

9. Install or connect the following:
- Clutch release bearing on the release fork
- New pilot bearing in the flywheel, if removed, using an installation tool

10. Be sure the flywheel mounting surface and the crankshaft or eccentric shaft mounting surfaces are clean. Remove any old sealant from the flywheel bolt hole threads and the flywheel bolts.
- Flywheel
- Sealant to the flywheel bolt threads and install them hand tight
- Flywheel holding tool. Tighten the bolts, in a crisscross pattern as follows:

a. Mazda3 models, tighten to 79–85 ft. lbs. (108–116 Nm).

b. Mazda6 models equipped with the G35M-R transaxle, tighten to 79–84 ft. lbs. (108–115 Nm).

c. Mazda6 models equipped with the A65M-R transaxle, tighten to 54–64 ft. lbs. (73–87 Nm).

d. Mazdaspeed6 models equipped with the A26MX-R transaxle, tighten to 111–118ft. lbs. (151–161 Nm).

11. Install or connect the following:
- Small amount of molybdenum grease to the clutch disc splines
- Clutch disc on the flywheel with the spring side toward the transaxle
- An alignment tool in the pilot bearing to position the clutch disc
- Clutch pressure plate by aligning the dowel holes with the flywheel dowels
- Pressure plate. Gradually, torque the bolts, in a crisscross pattern, to 19 ft. lbs. (26 Nm).

12. Remove the alignment tool.
- Clutch release fork
- Clutch release collar
- Rubber boot
- Transaxle
- Clutch release cylinder
- Negative battery cable

Hydraulic Clutch System

BLEEDING

1. Before servicing the vehicle, refer to the precautions section.
2. Remove the rubber cap from the bleeder screw on the release cylinder.
3. Place a bleeder tube over the end of the bleeder screw.
4. Submerge the other end of the tube in a jar half filled with hydraulic brake fluid.
5. Slowly pump the clutch pedal fully and allow it to return slowly, several times.
6. While pressing the clutch pedal to the floor, loosen the bleeder screw until the fluid starts to run out. Then, close the bleeder screw. Keep repeating this Step, while watching the hydraulic fluid in the jar. As soon as the air bubbles disappear, close the bleeder screw.
7. During the bleeding procedure the reservoir must be kept at least ¾ full.

Halfshafts

REMOVAL & INSTALLATION

MX5-Miata

1. Before servicing the vehicle, refer to the precautions section.

→Refer the exploded view illustration for component locations and if applicable, their retainer torque specifications.

✳✳ CAUTION

Performing the following procedures without first removing the ABS wheel-speed sensor may possibly cause an open circuit in the wiring harness if it is pulled by mistake. Before performing the following procedures, remove the ABS wheel-speed sensor (axle side) and fix it to an appropriate place where the sensor will not be pulled by mistake while servicing the vehicle.

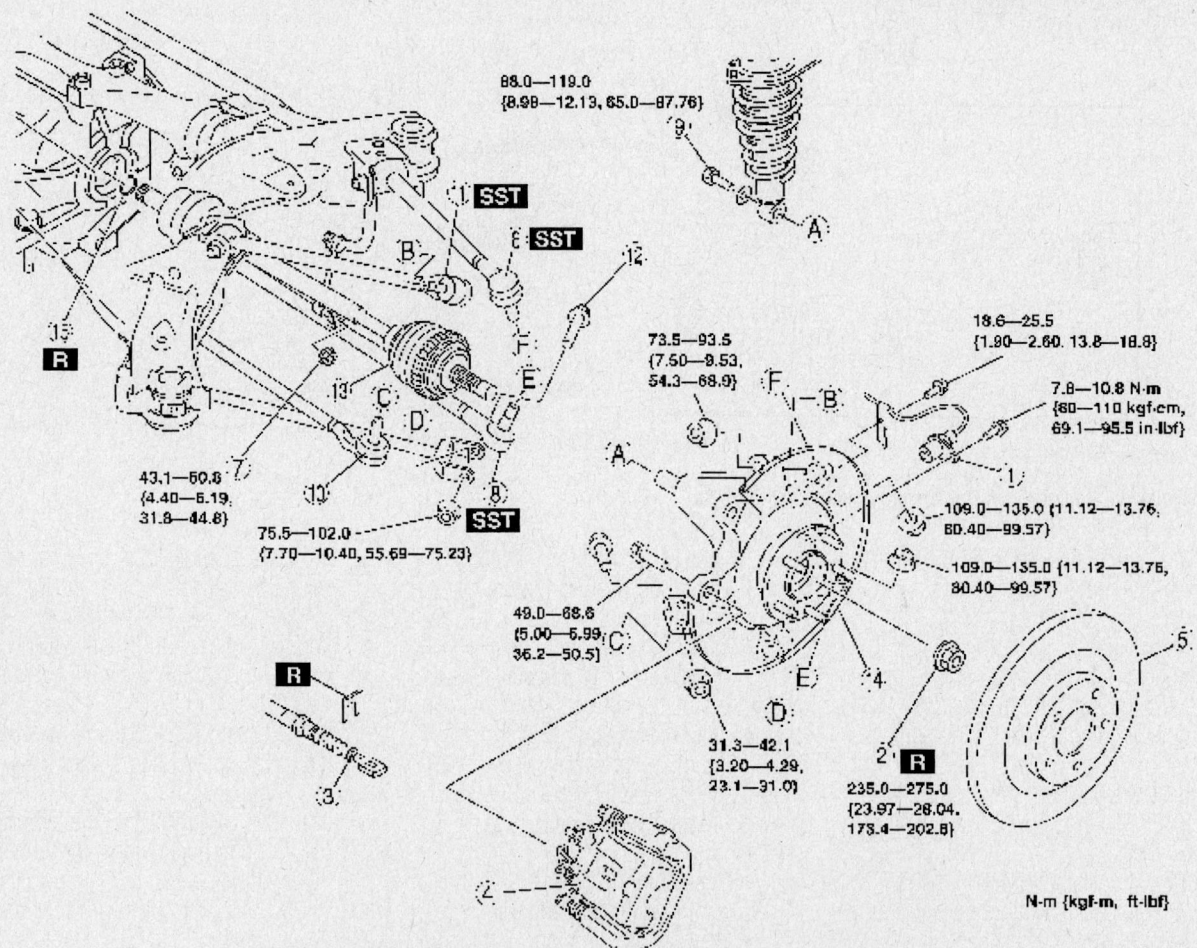

1	ABS wheel-speed sensor	9	Shock absorber bolt (lower)
2	Locknut	10	Toe control link ball joint
3	Parking brake cable	11	Rear trailing link (upper) ball joint
4	Brake caliper component	12	Rear trailing link (lower) bolt (outer side)
5	Disc plate	13	Rear drive shaft
6	Rear lateral link (upper) ball joint	14	Rear knuckle component
7	Stabilizer control link (lower)	15	Clip
8	Rear lateral link (lower) ball joint		

09482_MAZC2_G0074

Exploded view of the rear axle assembly—MX-5 Miata

2. Disconnect the ABS wheel-speed sensor.

3. Lock the disc plate by applying the brakes.

4. Knock the crimped portion of the locknut outward using a chisel and a hammer.

5. Remove the locknut.

6. Disconnect the parking brake cable.

7. Disconnect the brake caliper and position aside with the hose still attached. Secure the caliper so that no strain is put on the hose.

8. Remove the rotor.

9. Remove the rear upper lateral link ball joint.

10. Remove the lower stabilizer control link.

11. Remove the rear lower lateral link ball joint.

12. Remove the shock absorber lower bolt.

13. Remove the toe control link ball joint.

14. Remove the rear trailing link upper ball joint.

15. Remove the rear trailing link lower bolt on the outer side.

16. Temporarily install a spare nut to the end of the rear drive shaft.

17. Strike the nut with copper hammer lightly and remove the rear drive shaft from the wheel hub.

18. Separate the rear drive shaft from the wheel hub.

19. Insert a prybar between the rear differential and differential side outer ring, and then remove the rear drive shaft.

✳✳ CAUTION

The drive shaft edges are sharp and can damage the oil seal. Be careful not to damage the oil seal when removing the drive shaft from the differential.

20. Pull the rear drive shaft to the outer side of the vehicle and disconnect it from the rear differential.

21. Secure the rear knuckle component by installing the upper rear lateral link to the rear knuckle temporarily after disconnecting the rear drive shaft.

To install:

22. Apply differential oil to the differential oil seal lip.

✳✳ CAUTION

The drive shaft edges are sharp and can damage the oil seal. Be careful not to damage the oil seal when installing the rear drive shaft to the rear differential.

23. Insert the rear drive shaft into the rear differential with the clip opening facing upward. Point the opening of the new drive shaft clip upward, install it to the clip groove at the end of the rear drive shaft with the installation width within the specification shown in the illustration. After installing the clip, measure the outer diameter. If it exceeds the specification, reinstall the new clip.

24. After installation, verify that the rear drive shaft is securely held by the clip by pulling the outer ring on the differential side towards the axle.

25. Install the rear trailing link lower bolt.

26. Install the rear trailing link upper ball joint.

27. Install the toe control link ball joint.

28. Install the shock absorber lower bolt.

29. Install the rear lower lateral link ball joint.

30. Install the lower stabilizer control link.

31. Install the rear upper lateral link ball joint.

32. Install the rotor.

33. Install the brake caliper.

34. Connect the parking brake cable.

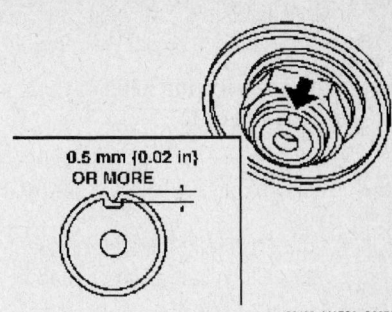

0.5 mm {0.02 in} OR MORE

09482_MAZC2_G0076

Crimp the locknut, using a chisel and hammer—MX-5 Miata

35. Install the locknut and tighten to 235–275 Nm (173–202 ft. lbs.).

36. Crimp the locknut, using a chisel and hammer.

37. Connect the parking brake cable.

38. Install the remaining components in the reverse order of removal.

Mazda3 Models

1. Before servicing the vehicle, refer to the precautions section.

2. Drain the transaxle oil.

3. Remove or disconnect the following:
   • Wheels
   • ABS sensor, if equipped
   • Splash shield, if equipped
   • Halfshaft lockbolt

4. Install a spare bolt onto the halfshaft, tap the bolt with a copper hammer and separate the halfshaft from the knuckle.

5. Remove or disconnect the following:
   • Stabilizer bar from the lower control arm
   • Cotter pin and nut from the tie rod end ball stud
   • Tie rod end from the knuckle
   • Lower ball joint pinch bolt and nut
   • Lower ball joint from the knuckle

6. Position a prybar between the inner CV-joint and transaxle case. Carefully pry the halfshaft from the transaxle being careful not to damage the oil seal.

7. Pull outward on the hub/knuckle assembly, push the outer CV-joint stub shaft through the hub and remove the halfshaft.

➡ **Install plug tool 49 G030 455 into the transaxle after removing the halfshaft, to keep the differential side gear in position. If the gear becomes positioned incorrectly, the differential may have to be removed to realign the gear.**

To install:

8. Install or connect the following:
   • New circlip on the end of the halfshaft, with the end gap facing upward

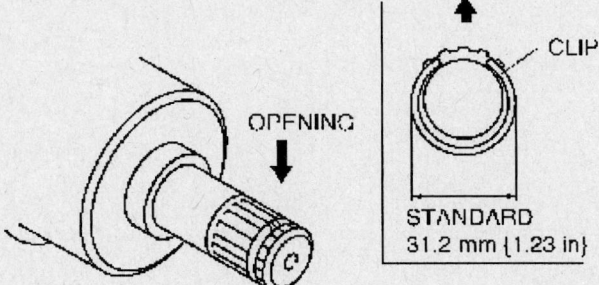

OPENING

CLIP

STANDARD 31.2 mm {1.23 in}

Standard
○ 31.2 mm {1.23 in}

09482_MAZC2_G0075

Drive shaft clip installation width—MX-5 Miata

- Halfshaft into the transaxle, being careful not to damage the oil seal

➡ **If equipped, push the halfshaft into the intermediate shaft.**

- Other end of the halfshaft through the hub. Loosely install a new lock-bolt
- Lower ball joint into the knuckle. Torque the pinch bolt to 31–43 ft. lbs. (43–58 Nm).
- Stabilizer bar to the lower control arm. Tighten to 29–40 ft. lbs. (40–54 Nm).
- Tie rod end to the steering knuckle. Torque the nut to 27–37 ft. lbs. (37–50 Nm.
- Tighten the nut, if necessary, to align the ball stud hole with the nut castellation.
- halfshaft lockbolt. Torque it to 23–28 ft. lbs. (31–38 Nm), then tighten an additional 85–95 degrees.

- ABS sensor
- Splash shield
- Wheels
9. Fill the transaxle.

Mazda6 and Mazdaspeed6 Models

2.3L (L3) ENGINES

1. Before servicing the vehicle, refer to the precautions section.
2. Drain the transaxle oil.
3. Raise the staked portion of the hub locknut with a hammer and chisel.
4. Lock the hub by applying the brakes and remove the nut.
5. Remove or disconnect the following:
- Wheels
- Tie rod end from the knuckle
- Stabilizer bar link from the damper fork
- Bolt attaching the fork to the knuckle
- Front and rear lower arm ball joints
6. Install a spare nut onto the drive

shaft so the nut is flush with the end of the drive shaft and tap the nut with a copper hammer to separate the halfshaft from the hub assembly.

➡ **The halfshaft edges can be quite sharp and can damage the oil seal. be careful when removing the halfshaft not to cause any damage.**

7. If removing the left side halfshaft, insert a prybar between the outer ring and the transaxle and pry the halfshaft from the transaxle.

8. If removing the right side halfshaft, separate it from the joint shaft by tapping on a bar positioned between the halfshaft and joint shaft.

➡ **Install plug tool 49 G030 455 into the transaxle after removing the halfshaft, to keep the differential side gear in position. If the gear becomes positioned incorrectly, the differential may have to be removed to realign the gear.**

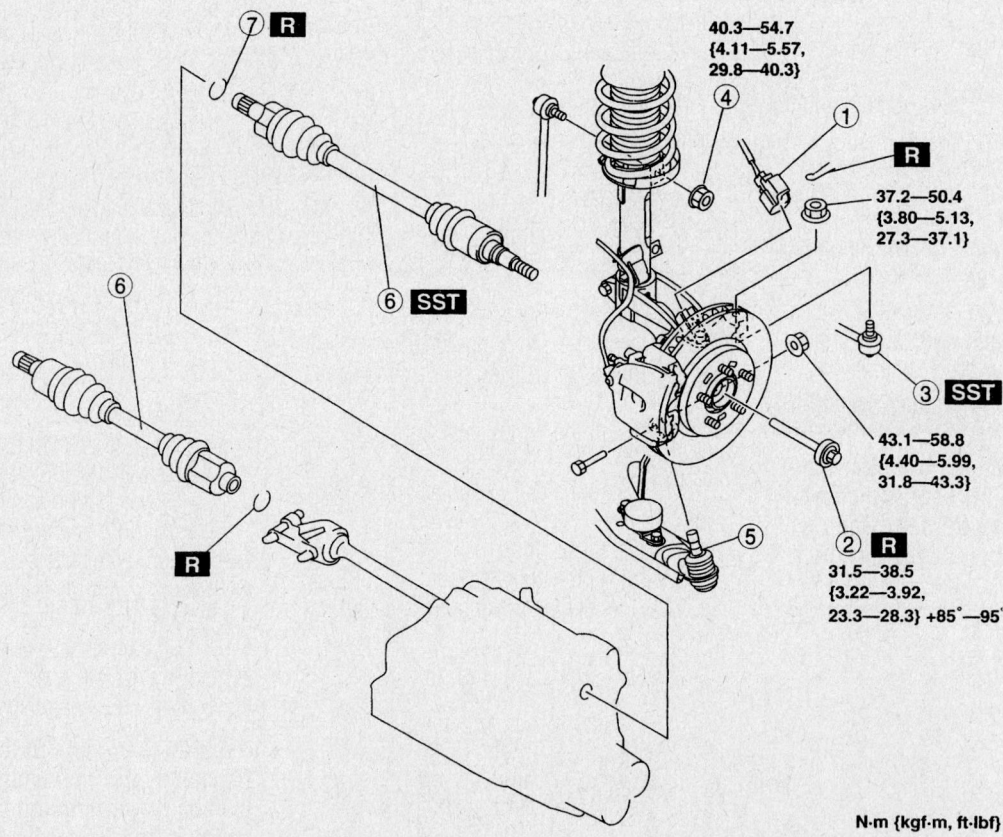

40.3—54.7
{4.11—5.57,
29.8—40.3}

37.2—50.4
{3.80—5.13,
27.3—37.1}

43.1—58.8
{4.40—5.99,
31.8—43.3}

31.5—38.5
{3.22—3.92,
23.3—28.3} +85°—95°

N·m {kgf·m, ft·lbf}

1	ABS wheel-speed sensor connector	4	Stabilizer control link upper nut
2	Lockbolt	5	Front lower arm ball joint
3	Tie-rod end ball joint	6	Drive shaft
		7	Clip

67162-MAZC-G86

Exploded view of a typical halfshaft mounting—Mazda3

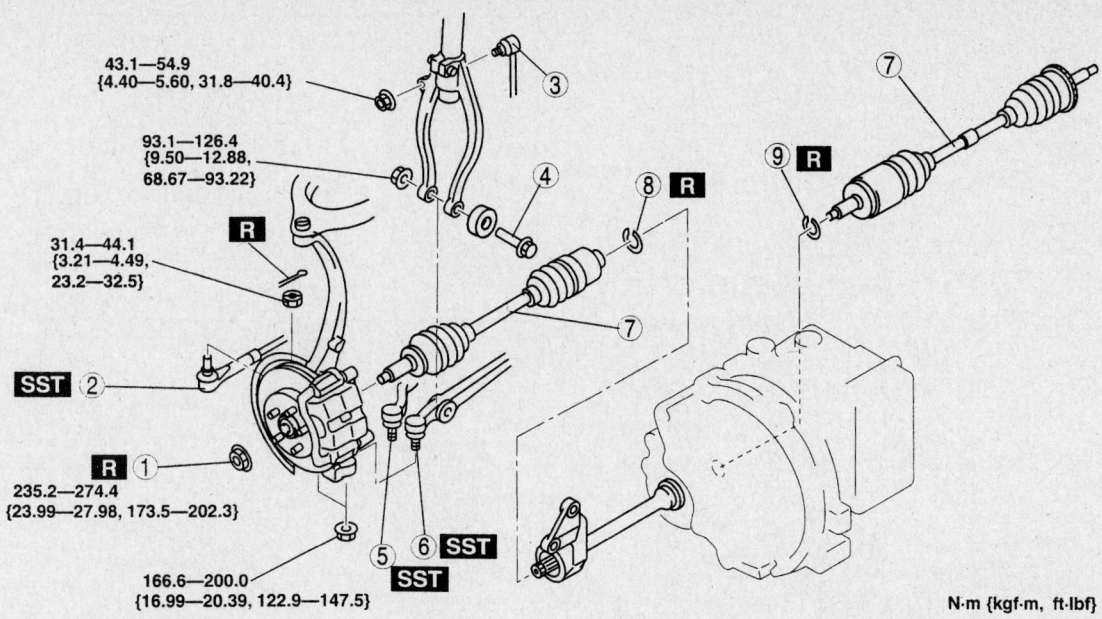

43.1—54.9
{4.40—5.60, 31.8—40.4}

93.1—126.4
{9.50—12.88, 68.67—93.22}

31.4—44.1
{3.21—4.49, 23.2—32.5}

235.2—274.4
{23.99—27.98, 173.5—202.3}

166.6—200.0
{16.99—20.39, 122.9—147.5}

N·m {kgf·m, ft·lbf}

1	Locknut	6	Front lower arm (front) ball joint
2	Tie-rod end ball joint	7	Drive shaft
3	Front stabilizer control link	8	Clip (right side)
4	Bolt	9	Clip (left side)
5	Front lower arm (rear) ball joint		

67162-MAZC-G229

Exploded view of the halfshaft assembly and related components—Mazda6 and Mazdaspeed6 models equipped with the 2.3L (L3) engine

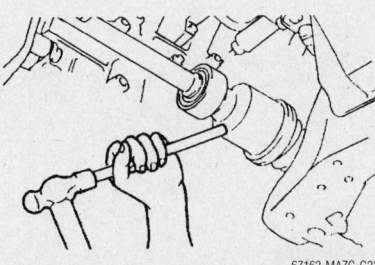

67162-MAZC-G230

Remove the right side halfshaft by tapping on a bar positioned between the halfshaft and joint shaft—Mazda6 and Mazdaspeed6 models equipped with the 2.3L (L3) engine

To install:

9. Install a new circlip on the end of the shaft, if removed, with the end gap facing upward. Measure the outer diameter of the clip, if it exceeds 1.19–1.23 inch (30–31mm) replace the clip.

10. Install the left side shaft as follows:
    a. Insert the shaft into the hub.
    b. Apply clean engine oil to the oil seal lip.
    c. Push the shaft into the transaxle and once the shaft clicks into place, pull on the outer ring to make sure the clip is engaged.

11. Install the right side shaft as follows:
    a. Insert the shaft into the hub.
    b. Insert the halfshaft into the joint shaft and once the shaft clicks into place, pull on the outer ring to make sure the clip is engaged.

12. Install the remaining components in the reverse order of removal. Refer to the exploded view of the halfshaft assembly and related components for torque specifications.

13. Install a new hub nut. Torque it to 173–202 ft. lbs. (235–274 Nm). After tightening, stake the locknut using a hammer and dull bladed chisel with an indent at least 0.02 inch (0.5mm).

14. Fill the transaxle.

3.0L (AJ) ENGINES

1. Before servicing the vehicle, refer to the precautions section.

2. Drain the transaxle oil.

3. Remove the knuckle and hub assembly.

4. Raise the staked portion of the hub locknut with a hammer and chisel.

5. Lock the hub by applying the brakes and remove the nut.

6. Remove or disconnect the following:
   • Wheels

   • Tie rod end from the knuckle
   • Damper fork
   • Front and rear lower arm ball joints
   • Stabilizer bar link from the damper fork

7. Install a spare nut onto the drive shaft so the nut is flush with the end of the drive shaft and tap the nut with a copper hammer to separate the halfshaft from the hub assembly.

➡The halfshaft edges can be quite sharp and can damage the oil seal. be careful when removing the halfshaft not to cause any damage.

8. If removing the left side halfshaft, insert a prybar between the outer ring and the transaxle and pry the halfshaft from the transaxle.

9. Remove the joint shaft bracket bolt and remove the right side halfshaft. Secure the shaft in a vise and insert a prybar between the halfshaft and joint shaft and separate them.

➡Install plug tool 49 G030 455 into the transaxle after removing the halfshaft, to keep the differential side gear in position. If the gear becomes positioned incorrectly, the differential may have to be removed to realign the gear.

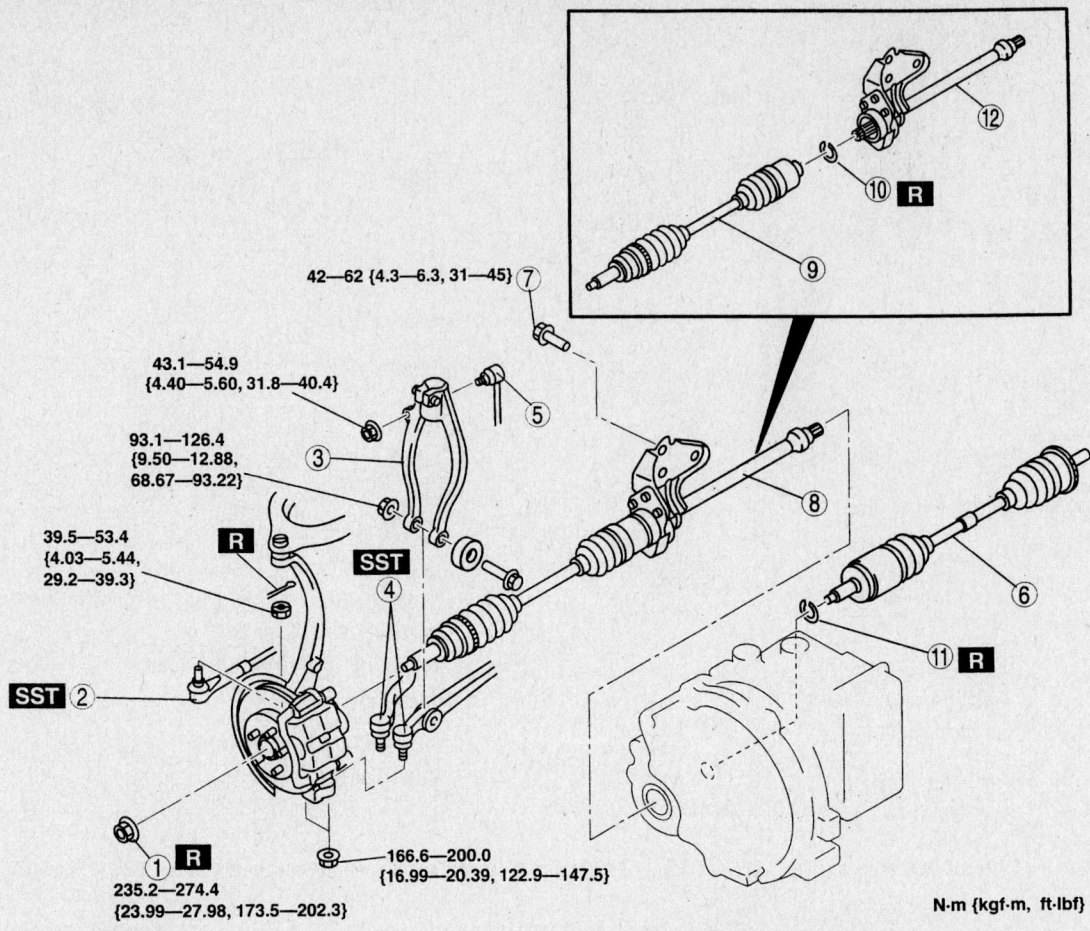

42—62 {4.3—6.3, 31—45} 7

43.1—54.9
{4.40—5.60, 31.8—40.4}

93.1—126.4
{9.50—12.88,
68.67—93.22}

39.5—53.4
{4.03—5.44,
29.2—39.3}

166.6—200.0
{16.99—20.39, 122.9—147.5}

235.2—274.4
{23.99—27.98, 173.5—202.3}

N·m {kgf·m, ft·lbf}

1	Locknut	7	Joint shaft bracket bolt
2	Tie-rod end	8	Drive shaft (right side) and joint shaft
3	Damper fork	9	Drive shaft (right side)
4	Lower arm (front, rear) ball joint	10	Clip (right side)
5	Stabilizer control link	11	Clip (left side)
6	Drive shaft (left side)	12	Joint shaft

67162-MAZC-G231

Exploded view of the halfshaft assembly and related components—Mazda6 models equipped with the 3.0L (AJ) engine

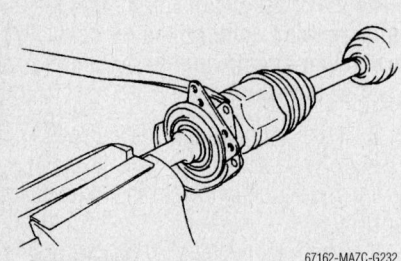

67162-MAZC-G232

Remove the right side halfshaft by securing the shaft in a vise, inserting a prybar between the halfshaft and joint shaft and separating them—Mazda6 models equipped with the 3.0L (AJ) engine

To install:

10. Install a new circlip on the end of the shaft, if removed, with the end gap facing upward. Measure the outer diameter of the clip, if it exceeds 1.26–1.30 inch (32–33mm) replace the clip.

11. Install the left side shaft as follows:

   a. Insert the shaft into the hub.

   b. Apply clean engine oil to the oil seal lip.

   c. Push the shaft into the transaxle and once the shaft clicks into place, pull on the outer ring to make sure the clip is engaged.

12. Install the right side shaft as follows:

   a. Place the halfshaft and joint shaft assembly in a vise.

   b. Insert the halfshaft into the joint shaft and use a plastic hammer to attach the joint shaft as illustrated.

   c. Tighten the joint shaft bracket bolt to 31–45 ft. lbs. (42–62 Nm).

13. Install the remaining components in the reverse order of removal. Refer to the exploded view of the halfshaft assembly and related components for torque specifications.

   • New hub nut. Torque it to 173–202 ft. lbs. (235–274 Nm). After tightening, stake the locknut using a hammer and dull bladed chisel with an indent at least 0.02 inch (0.5mm).

14. Fill the transaxle.

CV-Joints

OVERHAUL

MX5-Miata

1. Before servicing the vehicle, refer to the precautions section.

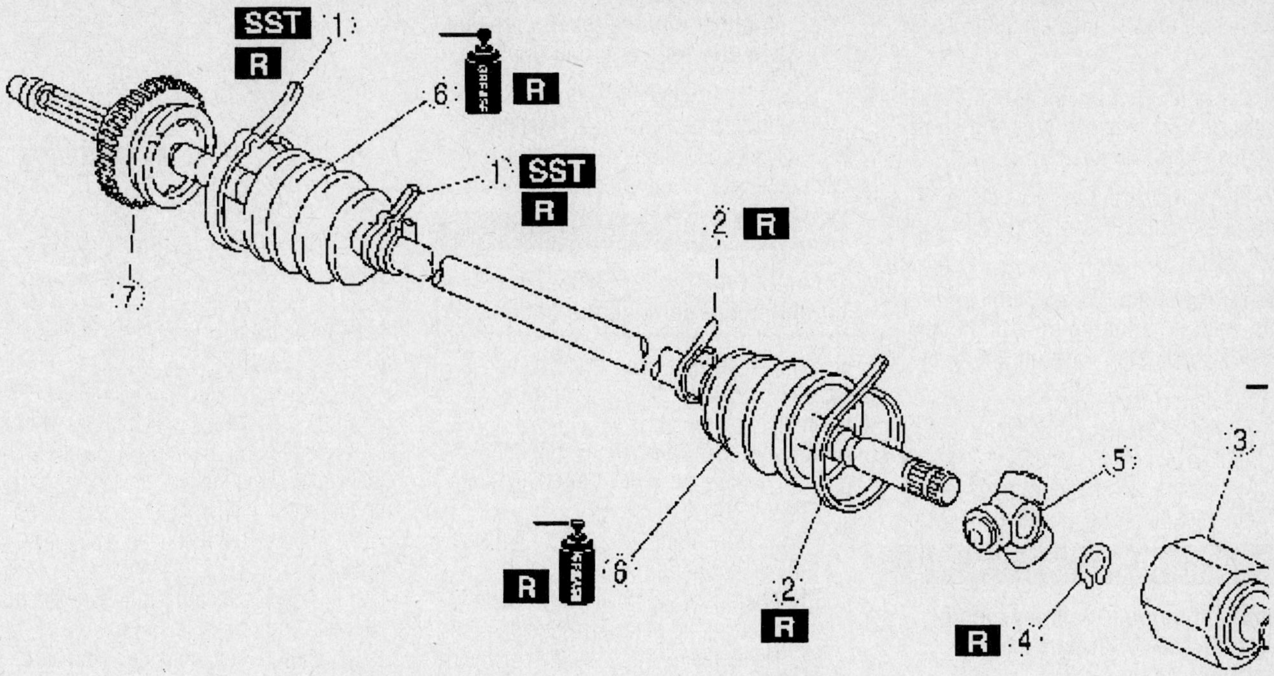

1 Boot band (axle side)
2 Boot band (differential side)
3 Tripod joint socket
4 Snap ring
5 Tripod joint
6 Boot
7 Shaft and ball joint component

09482_MAZC2_G0077

Exploded view of the rear driveshaft assembly—MX-5 Miata

➡Refer the exploded view illustration for component locations and if applicable, their retainer torque specifications.

2. Remove the boot band on the axle side using end clamp pliers.
3. Remove the boot band on the differential side using a flathead screwdriver.
4. Remove the tripod joint socket by placing an alignment mark on the drive shaft and the outer ring and then removing the outer ring.
5. Place an alignment mark on the shaft and tripod joint.
6. Remove the snap ring using snap ring pliers.
7. Remove the tripod joint from the shaft.

⁂ CAUTION

To prevent damage to the component, do not use a hammer when removing it.

8. Wrap the shaft spline with vinyl tape and remove the boot.
9. Remove the shaft and ball joint assembly.

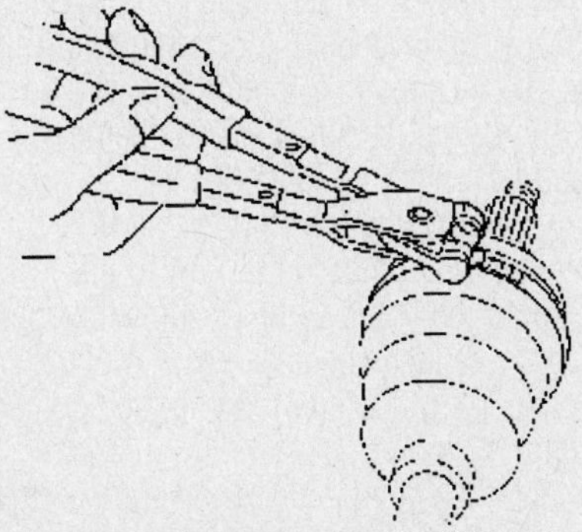

09482_MAZC2_G0078

Remove the boot band on the axle side using end clamp pliers—MX-5 Miata

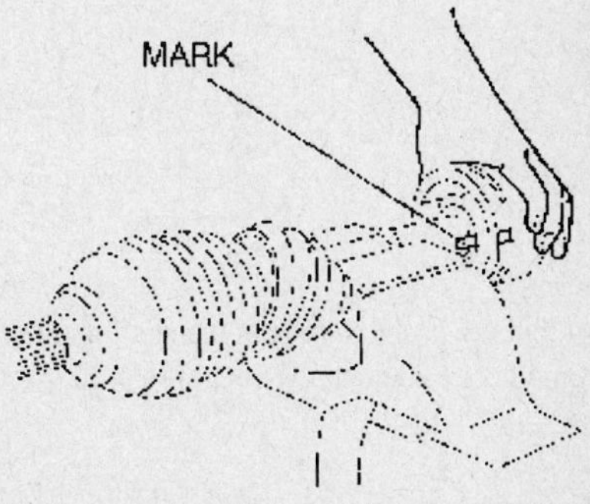

09482_MAZC2_G0080

Place an alignment mark on the shaft and tripod joint—MX-5 Miata

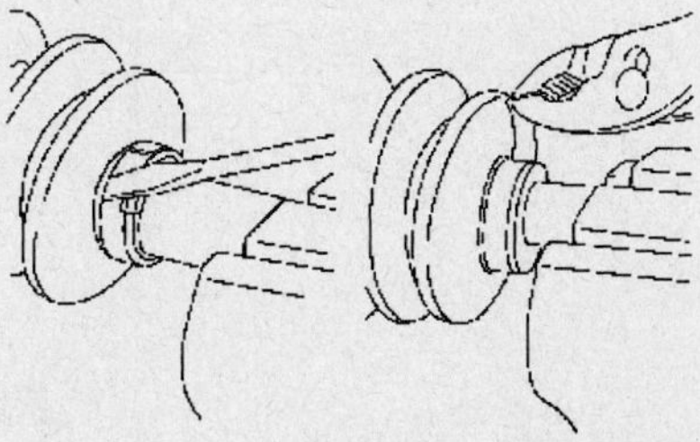

09482_MAZC2_G0079

Remove the boot band on the differential side using a flathead screwdriver—MX-5 Miata

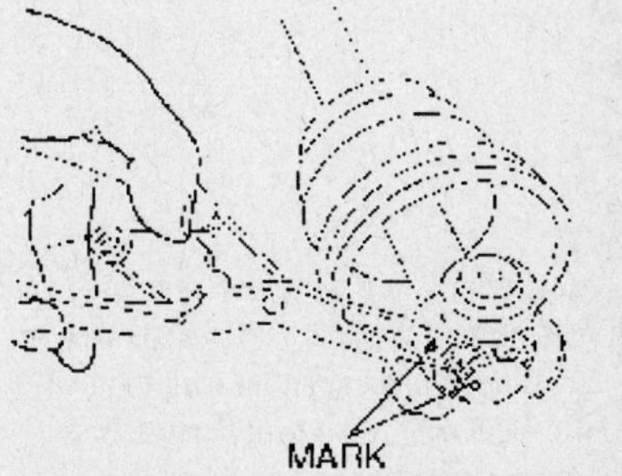

09482_MAZC2_G0081

Remove the snap ring using snap ring pliers—MX-5 Miata

To install:

10. Install the shaft and ball joint assembly.

➡**The boot shapes on the axle side and the differential side are different so do not miss install them.**

11. Fill the inside of the new dust boot on the wheel side with 90–110g (3.18–3.88 oz.) of grease.

➡**Do not touch the grease with your hand. Apply it from the tube to prevent foreign matter from entering the boot.**

12. Install the boot with the drive shaft spline still wrapped with vinyl tape.

13. Remove the vinyl tape.

14. Align the tripod joint with the shaft mark and insert it using a brass bar.

❊❊ CAUTION

To prevent damage to the assembly, do not tap the roller part when installing.

15. Install the new snap ring to the shaft installation slot securely using snap ring pliers.

16. Fill the outer ring and boot on the differential side with the 135–155g (4.77–5.46 oz.) of grease.

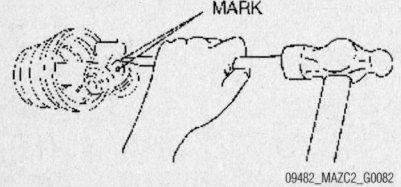

MARK

09482_MAZC2_G0082

Align the tripod joint with the shaft mark and insert it using a brass bar—MX-5 Miata

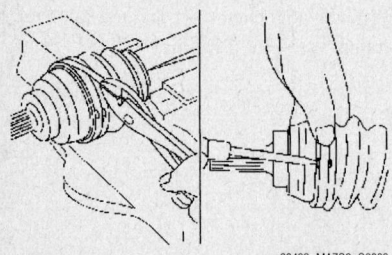

09482_MAZC2_G0083

Using pliers, pull the differential side boot band around the boot slot in opposite direction of drive shaft forward rotation. Insert the end of the boot band between the boot band clip and fold back the clip tabs using a flathead screwdriver to secure the boot band—MX-5 Miata

➡**Do not touch the grease with your hand. Apply it from the tube to prevent foreign matter from entering the boot.**

17. Assemble the outer ring.

18. Release any trapped air from the boots by carefully lifting up the small end of each boot with a cloth wrapped screwdriver.

❊❊ CAUTION

Do not let the grease leak and be careful not to damage the boot.

19. Set the drive shaft length to the specification when the inside of the boots is at ambient pressure.

20. The drive shaft length should be:
   a. On the left side: 778.5–788.5 mm (30.65–31.04 in.).
   b. On the right side: 818.5–828.5 mm (32.22–32.62 in.).

21. After installation, verify that there is no boot damage or grease leakage.

22. Using pliers, pull the differential side boot band around the boot slot in opposite direction of drive shaft forward rotation direction and tighten.

23. Insert the end of the boot band between the boot band clip and fold back the clip tabs using a flathead screwdriver to secure the boot band.

24. Verify that the boot band is installed to the boot slot securely.

25. Turn the adjusting bolt of tool 49 T025 001 and adjust the opening size **A** to the specification to 2.9 mm (0.11 in.).

26. Crimp the small side of the boot band using tool 49 T025 001.

27. Verify that the crimp value **B** is within 2.4–2.8 mm (0.095–0.110 in.).

28. If the crimp value exceeds the specification, reduce opening length **A** of the tool 49 T025 001 and recrimp the boot band.

29. If the crimp value **B** is less than the specification, increase opening length **A** of the tool 49 T025 001 and crimp the new boot band.

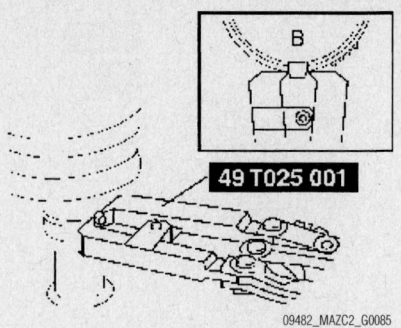

B

49 T025 001

09482_MAZC2_G0085

Crimping the boot bands—MX-5 Miata

30. Verify that the boot band does not protrude from the boot band installation area.

31. If the boot band protrudes from the installation area, replace it with a new band and repeat the band installation steps.

32. Fill the boot with the repair kit grease.

33. Adjust opening length **A** of the tool to the 3.2 mm (0.13 in.).

34. Crimp the large side boot band using the tool 49 T025 001.

35. Verify that the boot band crimp value **B** is within the 2.4–2.8 mm (0.095–0.110 in.) specification.

36. If crimp value **B** exceeds the specification, reduce opening length **A** of the tool 49 T025 001 and recrimp the boot band.

37. If the crimp value **B** is less than the specification, replace the boot band, increase the opening length of **A** of the tool 49 T025 001, and then recrimp the new boot band.

38. Verify that the boot band does not protrude from the boot band installation area.

39. If the boot band protrudes from the installation area, replace it with a new band.

Mazda3, Mazda6 and Mazdaspeed6 Models

1. Before servicing the vehicle, refer to the precautions section.

Two types of CV-joints are used. The inboard CV-joints are the tri-Pot type. All

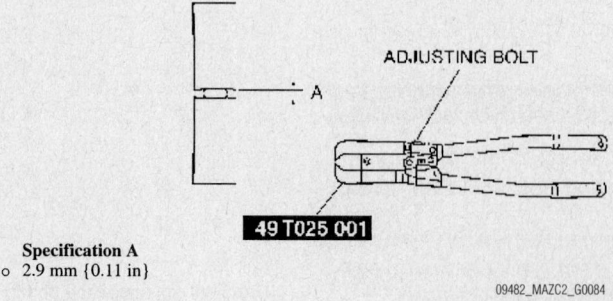

ADJUSTING BOLT

A

49 T025 001

Specification A
○ 2.9 mm {0.11 in}

09482_MAZC2_G0084

Turn the adjusting bolt of tool 49 T025 001 and adjust the opening size A to the specification to 2.9 mm (0.11 in.)—MX-5 Miata

outboard CV-joints are Birfield type. The Birfield CV-joint cannot be disassembled; if an outboard CV-joint boot needs replacement, the inboard CV-joint must be removed. If the outboard CV-joint needs to be replaced, replace the entire halfshaft as an assembly.

2. Remove the halfshaft from the vehicle and clamp it in a vise equipped with jaw caps, to prevent damage to the machined surfaces. Do not allow the vise to contact the boot or its clamps.

3. Remove the large boot clamp from the inboard CV-joint, using side cutters. After removing the clamp, roll the boot back over the shaft.

➡**Check the grease for contamination by rubbing it between 2 fingers. Any gritty feeling indicates a contaminated CV-joint, in which case the entire CV-joint must be disassembled, cleaned and inspected. If the grease is not contaminated and the CV-joint has been operating satisfactorily, continue with the boot replacement procedure and add the required lubricant.**

4. Paint alignment marks on the outer race and shaft for assembly reference. Remove the wire ring bearing retainer and remove the outer race.

5. Paint alignment marks on the tri-pot bearing and shaft for assembly reference. Remove the tri-pot bearing snapring and, using a brass drift and hammer, remove the tri-pot bearing from the shaft.

6. Remove the small clamp and remove the inner boot from the halfshaft. If the boot is to be reused, wrap the shaft splines with tape before removing.

7. If the outer CV-joint boot is to be replaced, remove the clamps and slide the boot off the shaft from the inboard side.

To install:

8. If the outboard boot was removed, slide the boot onto the shaft from the inboard side. Wrap tape on the splines before installing to protect the boot.

9. Install the inboard boot and remove the tape from the shaft.

10. Install the tri-pot assembly on the halfshaft. Tap the assembly onto the shaft using a hammer and brass drift. Install the tri-pot assembly retaining ring.

11. Fill the CV-joint outer race with high temperature CV-joint grease. Install the outer race over the tri-pot joint and install the wire ring bearing retainer.

12. Position the CV-joint boot(s). Make sure the boot is fully seated in the grooves in the shaft and outer race.

13. Insert a small prybar with rounded edges between the boot and the outer bearing race to allow trapped air to escape from the boot. Install new boot clamps.

14. Wrap the clamps around the boots in a clockwise direction, pull tight with pliers and bend the locking tabs to secure in position.

15. Work the CV-joint through its full range of travel at various angles. The joint should flex, extend and compress smoothly.

16. Install the halfshaft into the vehicle.

STEERING AND SUSPENSION

Air Bag

PRECAUTIONS

Several precautions must be observed when handling the inflator module to avoid accidental deployment and possible personal injury.

1. Never carry the inflator module by the wires or connector on the underside of the module.

2. When carrying a live inflator module, hold securely with both hands, and ensure that the bag and trim cover are pointed away.

3. Place the inflator module on a bench or other surface with the bag and trim cover facing up.

4. With the inflator module on the bench, never place anything on or close to the module which may be thrown in the event of an accidental deployment.

5. An air bag is an explosive device. Handle with extreme caution.

6. Always disconnect the battery and the air bag connector before removing the steering wheel or beginning work on the air bag system.

7. Air bag components must not be repaired or opened. Always use new parts, including the wiring harness.

8. Always place a removed air bag unit with the horn pad facing up. Put it in a safe place where it will not be disturbed.

9. The air bag unit must not be exposed to grease, fluids, or cleaning agents.

10. The air bag unit must not be exposed to temperatures above 194°F (90°C) at any time. Even the heat of a soldering iron can damage or ignite the charge.

11. Storage and transport of air bags is subject to rules governing explosive devices and should be done only in the original package.

12. Failure to follow proper safety precautions may result in personal injury through accidental firing of the air bag, or through failure of the air bag in an accident.

DISARMING

1. Before servicing the vehicle, refer to the precautions section.

2. If equipped, deactivate the audio anti-theft system.

3. Turn the ignition switch to LOCK.

4. Disconnect and isolate the negative battery cable and wait for more than 1 minute to allow the backup power supply to deplete its stored power.

ARMING

1. Before servicing the vehicle, refer to the precautions section.

2. Connect the negative battery cable, turn the ignition switch **ON** and verify the air bag warning light cones on for 6 seconds. If the light does not illuminate there are problems with the system.

3. If equipped, activate the audio anti-theft system.

Rack and Pinion Steering Gear

REMOVAL & INSTALLATION

Power

MX5-MIATA

1. Before servicing the vehicle, refer to the precautions section.

➡**Refer the exploded view illustration for component locations and if applicable, their retainer torque specifications.**

❊❊ CAUTION

Performing the following procedures without first removing the ABS wheel-speed sensor may possibly cause an open circuit in the harness if it is pulled by mistake. Before performing the following procedures, remove the ABS wheel-speed sensor (axle side) and fix it to an appropriate place where the sensor will not be pulled by mistake while servicing the vehicle.

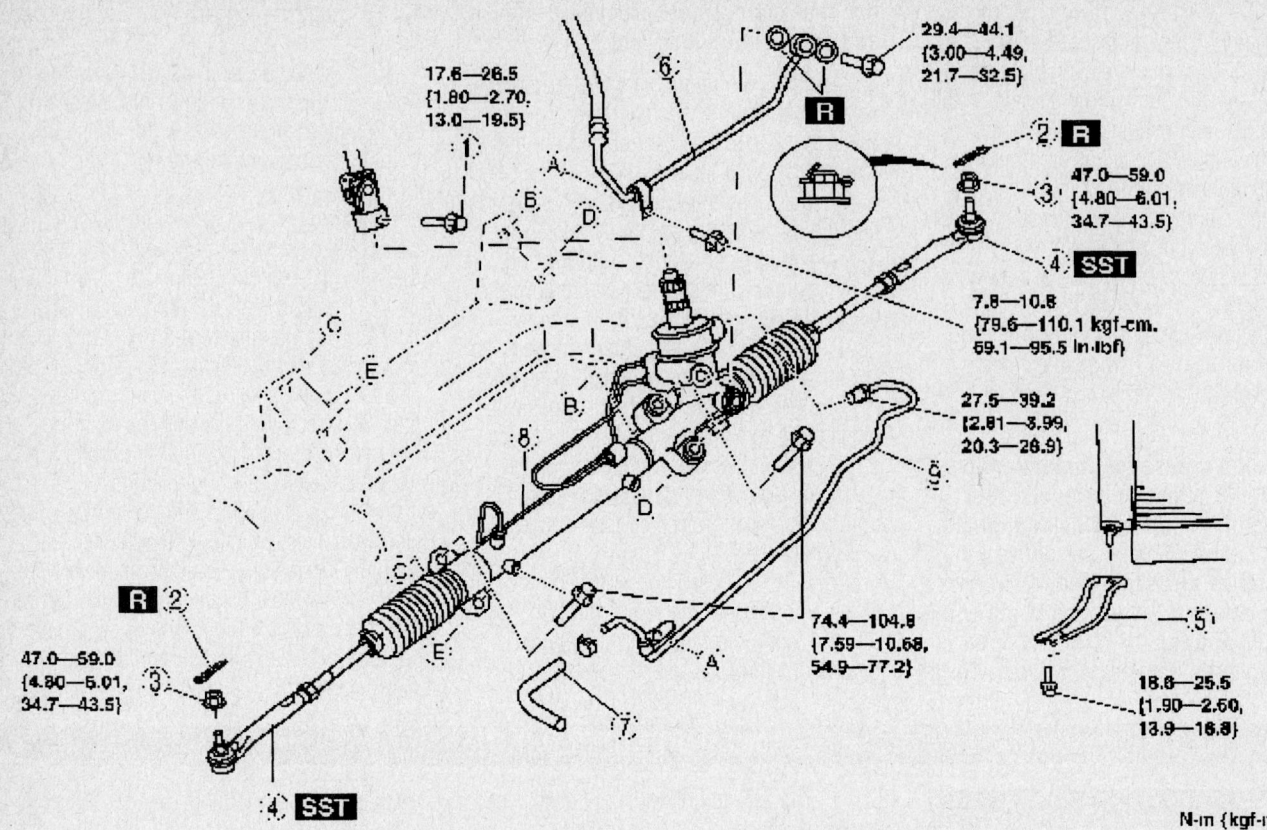

29.4—44.1
{3.00—4.49,
21.7—32.5}

17.6—26.5
{1.80—2.70,
13.0—19.5}

47.0—59.0
{4.80—6.01,
34.7—43.5}

7.8—10.8
{79.6—110.1 kgf-cm.
59.1—95.5 In·lbf}

27.5—39.2
{2.81—3.99,
20.3—26.9}

47.0—59.0
{4.80—5.01,
34.7—43.5}

74.4—104.8
{7.59—10.68,
54.9—77.2}

18.6—25.5
{1.90—2.60,
13.9—18.8}

N-m {kgf-m. ft·lb

1 Bolt (intermediate shaft)	6 Pressure pipe
2 Cotter pin	7 Return hose
3 Nuts (tie-rod end ball joint)	8 Steering gear and linkage
4 Tie-rod end ball joint	9 Return pipe
5 Lower mounting rubber bracket	

09482_MAZC2_G0086

Exploded view of the steering gear and linkage assembly—MX-5 Miata

2. Remove the ABS wheel-speed sensor.
3. Remove the radiator mount bracket and the front stabilizer control link and the stabilizer.
4. Mark the pinion shaft and gear housing prior to removal.
5. Remove the cotter pin.
6. Remove the tie-rod end and ball joint nuts.

7. Remove the tie-rod nut.
8. Separate the tie-rod end from the steering knuckle using the a suitable puller.
9. Remove the lower mounting rubber bracket.
10. Remove the pressure pipe and return hose
11. Remove the steering gear and linkage retainers.

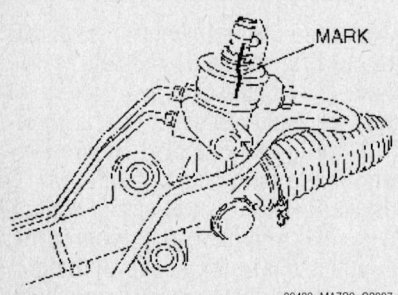

09482_MAZC2_G0087

Mark the pinion shaft and gear housing prior to removal—MX-5 Miata

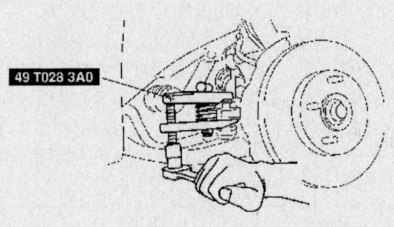

09482_MAZC2_G0088

Separate the tie-rod end from the steering knuckle using the a suitable puller—MX-5 Miata

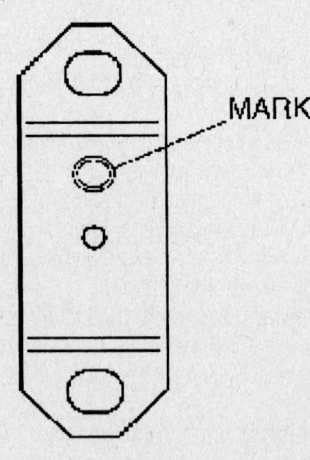

09482_MAZC2_G0089

Assemble the mounting bracket with the mark on the bracket facing the vehicle rear—MX-5 Miata

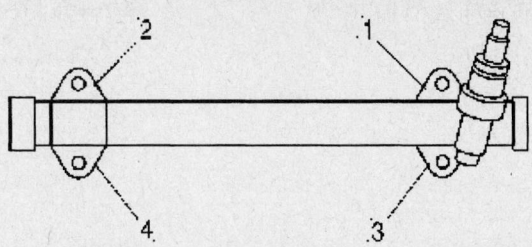

Tightening torque
o 74.4—104.8 N•m {7.587—10.68 kgf•m, 54.88—77.29 ft•lbf}

09482_MAZC2_G0090

Steering gear and linkage bracket bolt torque sequence—MX-5 Miata

12. Remove the steering gear and linkage by pulling it from the right side of the vehicle.

To install:

13. Install the gear and linkage assembly and loosely tighten the bolts.

14. Assemble the mounting bracket with the mark on the bracket facing the vehicle rear.

15. Tighten the mounting bracket bolts to the 74.4–104.8 Nm (54.88–77.29 ft. lbs.) in the order shown.

16. Install the remaining components in the reverse order of removal using the accompanying illustration for torque values.

17. When installing the intermediate shaft. Align the marks and install the intermediate shaft and bolt. Tighten to 17–26 Nm (13–19 ft. lbs.).

18. After installation, adjust alignment.

Mazda3

1. Before servicing the vehicle, refer to the precautions section.

2. Remove or disconnect the following:

- Negative battery cable
- Front wheels
- Intermediate shaft to steering gear pinion shaft bolt. Mark the shaft-to-gear location.
- Cotter pins and nuts from both steering tie rod ends and press the tie rod out from the knuckle arm
- Stabilizer bar link
- Lower arm ball joint from the knuckle
- Pressure line and return pipe from the steering gear
- Bolt and power steering angle sensor connector. Remove the coolant over flow tank and set aside with the lines still attached to access the sensor connector. Refer to the exploded view illustration for component location.

3. Support the engine assembly with an engine support assembly such as 49 E017 5A0.

- Crossmember bolts. Support with a jack before removing the bolts and nuts.

- Front crossmember, stabilizer, lower arm and steering gear as an assembly

4. Remove the stabilizer bar, the steering gear, insulator and linkage retaining bolts and remove from the crossmember.

To install:

5. Attach the steering gear and insulator to the crossmember and tighten the insulator bolts to 69–95 inch lbs. (8–10 Nm) and the steering gear bolts to 55–79 ft. lbs. (74–105 Nm).

6. Install the stabilizer bar to the crossmember and tighten the bolts to 30–40 ft. lbs. (40–55 Nm).

7. Install the lower arm to the crossmember, tighten the side mounting bolt to 96–110 ft. lbs. (130–150 Nm) and the inner arm bolt to 72–97 ft. lbs. (97–132 Nm).

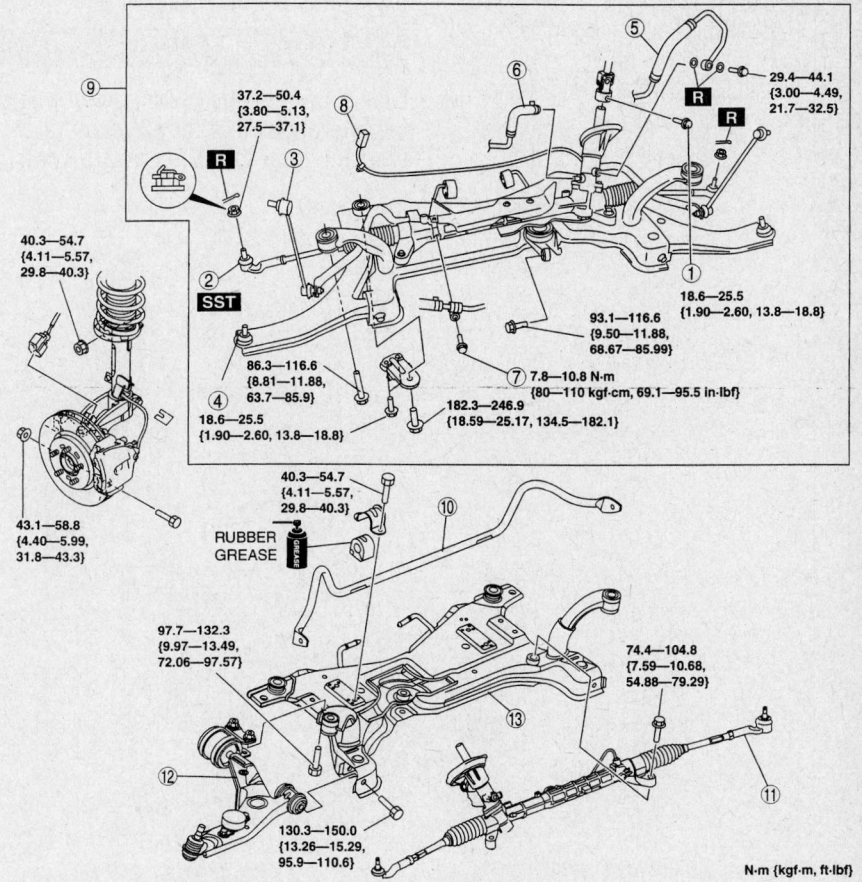

1 Bolt (intermediate shaft)
2 Tie-rod end ball joint
3 Stabilizer control link upper nut
4 Lower arm ball joint
5 Pressure pipe (gear side)
6 Return hose (gear side)
7 Bolt
8 P/S angle sensor connector

9 Front crossmember component, steering gear and linkage component
10 Front stabilizer
11 Steering gear and linkage
12 Front lower arm
13 Front crossmember

67162-MAZC-G87

Exploded view of the steering gear/front crossmember assembly and related components—Mazda3

8. Install the front crossmember, stabilizer, lower arm and steering gear as an assembly, using a transmission jack to raise the assembly into position. Refer to the exploded view illustration for the crossmember bolt locations and torque values.

9. Install the power steering angle sensor connector, the bolt and the coolant reservoir.

10. Install the power steering return hose and pressure pipe. Tighten the pipe retainer to 21–32 ft. lbs. (29–44 Nm).

11. Install the stabilizer shaft link nut upper link, and the tie rod end ball joint. Tighten the stabilizer bar link upper nut to 30–40 ft. lbs. (40–54 Nm) and the ball joint nut to 27–37 ft. lbs. (37–50 Nm).

12. Install the steering shaft to the steering gear pinion shaft, align the marks made during removal and tighten the bolt to 14–18 ft. lbs. (19–25.5 Nm).

13. Remove the engine support assembly.

14. Install the front wheels and connect the negative battery cable.

15. Check and/or adjust the front end alignment.

Mazda6 and Mazdaspeed6

2.3L (L3) ENGINE

Refer to the exploded view illustration for component location and torque specifications.

1. Before servicing the vehicle, refer to the precautions section.
2. Remove the ABS speed sensor.
3. Disconnect the negative battery cable.
4. Remove the front wheels.
5. Remove the intermediate shaft to steering gear pinion shaft bolt. Mark the shaft-to-gear location.
6. Remove the cotter pins and nuts from both steering tie rod ends
7. Press the tie rod out from the knuckle arm.
8. Disconnect the pressure line and return hose from the steering gear.

※※ CAUTION

Do not remove the crossmember nuts completely as this may cause the crossmember to fall. Leave the nuts threaded on the studs while loosening.

9. Support the crossmember with a jack.
10. Loosen the crossmember nuts but do not remove, lower the jack and crossmember assembly enough to access to steering gear retainers.
11. Remove the steering gear and linkage from the left side.

To install:
12. Install the steering gear and linkage.
13. Tighten the gear mounting bolts in the sequence illustrated to 55–77 ft. lbs. (74–104 Nm).

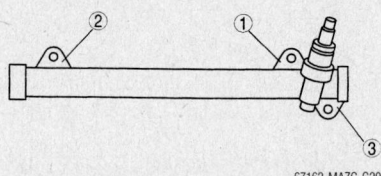

67162-MAZC-G299

Exploded view of the power steering gear assembly–Mazda6 and Mazdaspeed6 models

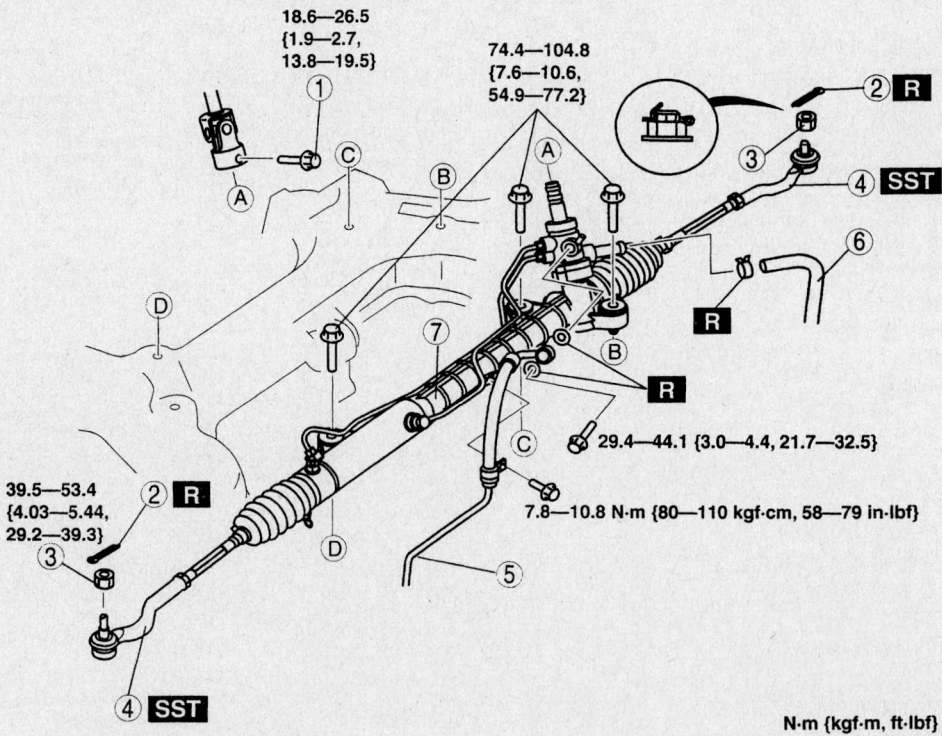

N·m {kgf·m, ft·lbf}

1	Bolt (intermediate shaft)	5	Pressure pipe
2	Cotter pin	6	Return hose
3	Nuts (tie-rod end ball joint)	7	Steering gear and linkage
4	Tie-rod end ball joint		

67162-MAZC-G233

Exploded view of the power steering gear assembly–Mazda6and Mazdaspeed6 with the 2.3L (L3) engine

14. Raise the crossmember assembly into position and tighten the nuts to 89–114 ft. lbs. (119–154 Nm) and the bolts to 68–86 ft. lbs. (93–116 Nm).

15. Install the return hose and pressure pipe. Tighten the pipe banjo bolt to 21–32 ft. lbs. (29–44 Nm).

16. Install the tie rod ends to the knuckle arm. Torque the nuts to 29–39 ft. lbs. (39–53 Nm). Install new cotter pins.

17. Align the intermediate shaft–to–gear marks made during removal and tighten the bolt to 13–19 ft. lbs. (18–26 Nm). Check the power steering fluid level.

18. Install the abs sensor and the wheels.

19. Connect the negative battery cable.

20. Check and/or adjust the front end alignment.

3.0L (AJ) ENGINE

Refer to the exploded view illustration for component location and torque specifications.

1. Before servicing the vehicle, refer to the precautions section.

2. Disconnect the negative battery cable.

3. Remove the ABS speed sensor.

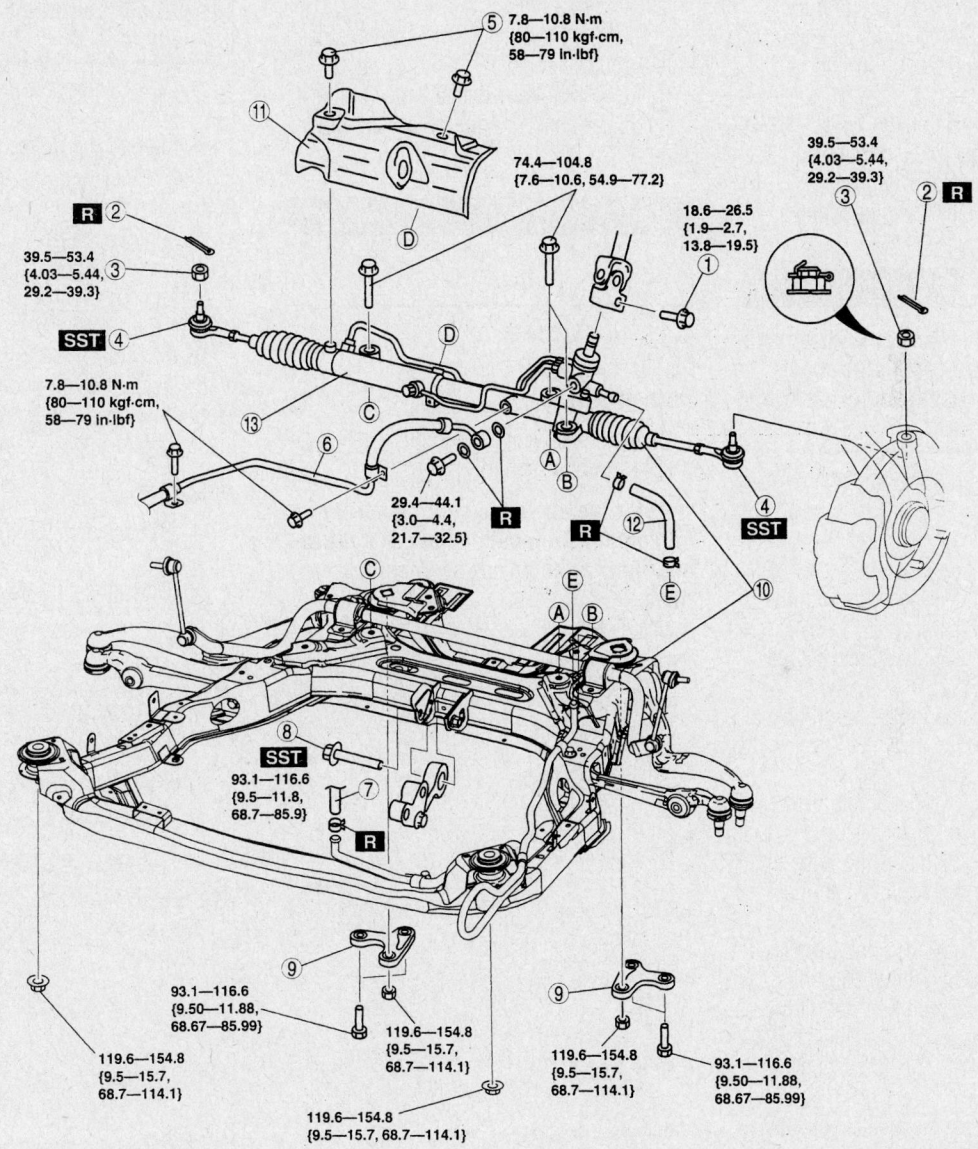

1   Bolt (intermediate shaft)
2   Cotter pin
3   Nuts (tie-rod end ball joint)
4   Tie-rod end ball joint
5   Insulator bolts
6   Pressure pipe
7   Return hose

8   No.1 engine mount center bolt
9   Crossmember bracket
10   Crossmember component, steering gear and linkage
11   Insulator
12   Return hose
13   Steering gear and linkage

67162-MAZC-G234

Exploded view of the power steering gear assembly–Mazda6 with the 3.0L (AJ) engine

4. Remove the front wheels, under cover and splash shield.

5. Separate the stabilizer bar link at the shock absorber side.

6. Separate the front lower arm front and rear ball joints from the knuckle.

7. Remove the lower side shock absorber bolt.

8. Remove the intermediate shaft to steering gear pinion shaft bolt. Mark the shaft-to-gear location.

9. Remove the cotter pins and nuts from both steering tie rod ends

10. Press the tie rod out from the knuckle arm.

11. Remove the heat shield bolts.

12. Disconnect the pressure line and return hose from the steering gear.

13. Remove the No. 1 engine mount center bolt.

✳✳ CAUTION

Support the crossmember with a jack and make sure the jack is attached securely to the crossmember before removing the bracket.

14. Support the crossmember with a jack.

15. Remove the crossmember bracket.

16. Remove the crossmember bolts and lower the crossmember assembly with the gear and linkage attached.

17. Remove the heat shield and return hose.

18. Remove the steering gear and linkage.

To install:

19. Install the steering gear and linkage to the crossmember assembly.

20. Tighten the gear mounting bolts in the sequence illustrated to 55–77 ft. lbs. (74–104 Nm).

21. Install the return hose and heat shield but do not tighten the shield bolts.

22. Raise the crossmember assembly into position and tighten the retainers to the specifications shown in the accompanying illustration.

23. Install the crossmember bracket and tighten the retainers to the specifications shown in the accompanying illustration.

24. Install the No. 1 engine mount center bolt and tighten to 68–86 ft. lbs. (93–116 Nm).

25. Install the return hose and pressure pipe. Tighten the pipe banjo bolt to 21–32 ft. lbs. (29–44 Nm).

26. Install the heat shield bolts and tighten to 79 inch lbs. (10 Nm).

27. Install the tie rod ends to the knuckle arm. Torque the nuts to 29–39 ft. lbs. (39–53 Nm). Install new cotter pins .

28. Align the intermediate shaft–to–gear marks made during removal and tighten the bolt to 13–19 ft. lbs. (18–26 Nm). Check the power steering fluid level.

29. Install the lower side shock absorber bolt.

30. Attach the front lower arm front and rear ball joints to the knuckle.

31. Attach the stabilizer bar link to the shock absorber side.

32. Install the abs sensor, wheels, splash shield and under cover.

33. Connect the negative battery cable.

34. Check and/or adjust the front end alignment.

Strut

REMOVAL & INSTALLATION

Front

MX5-MIATA

1. Before servicing the vehicle, refer to the precautions section.

➡**Refer the exploded view illustration for component locations and if applicable, their retainer torque specifications.**

✳✳ CAUTION

Performing the following procedures without first removing the ABS wheel-speed sensor may possibly cause an open circuit in the wiring harness if it is pulled by mistake. Before performing the following procedures, remove the ABS wheel-speed sensor (axle side) and fix it to an appropriate place where the sensor will not be pulled while servicing the vehicle.

2. Remove the front suspension tower bar.

3. Remove the brake hose bracket.

4. Remove the front upper arm ball joint from the knuckle.

5. Remove the front shock absorber, coil spring and front upper arm.

6. Remove the front shock absorber and coil spring.

To install:

7. Install the components in the reverse order of removal using the accompanying illustration for torque values.

MAZDA3

1. Before servicing the vehicle, refer to the precautions section.

2. Disconnect the ABS sensor connector.

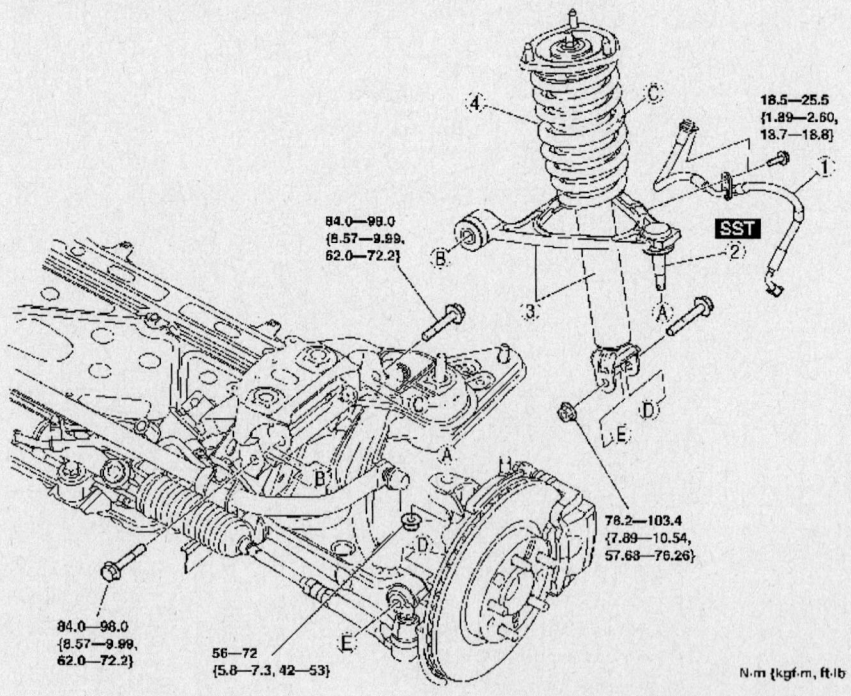

1 Brake hose bracket
2 Front upper arm ball joint
3 Front shock absorber, coil spring and front upper arm
4 Front shock absorber and coil spring

09482_MAZC2_G0091

Exploded view of the front strut assembly and related components—MX-5 Miata

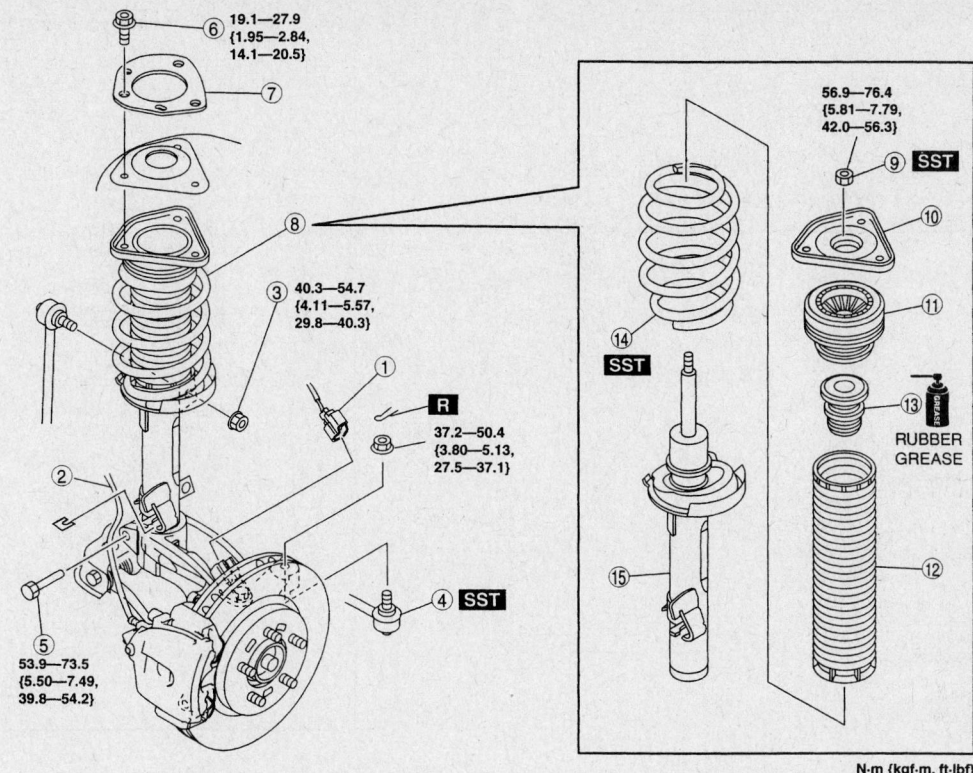

1 ABS wheel-speed sensor wiring harness connector
2 Brake hose
3 Stabilizer control link upper nut
4 Tie-rod end ball joint
5 Shock absorber lower bolt
6 Shock absorber upper bolt
7 Stiffener
8 Shock absorber and coil spring
9 Piston rod nut
10 Mounting rubber
11 Bearing
12 Dust boot
13 Bound stopper
14 Coil spring
15 Front shock absorber

N·m {kgf·m, ft-lbf}

67162-MAZC-G88

Exploded view of the front strut and spring assembly components—Mazda3

3. Disconnect the brake hose from the bracket on the strut assembly.

4. Remove the stabilizer bar link upper nut and disconnect the bar from the strut assembly.

5. Disconnect the tie rod from the knuckle.

6. Remove the shock absorber lower bolt.

7. Loosen the front lower arm inner bolt, the separate the shock from the hub assembly by tapping the knuckle with a hammer or mallet being careful not to damage any components.

8. Remove the shock absorber assembly.

To install:

9. Align the piston rod nut with the center part where the shock absorber is installed by placing the piston rod nut with lengths (A) all the same and tighten the shock absorber upper bolts. Refer to the accompanying illustration.

10. Use a jack to raise the lower control arm, attach the shock absorber and tighten the bolts

11. Tighten the upper shock nuts to 14–20 ft. lbs. (19–28 Nm) and the lower shock nut and bolt to 40–54 ft. lbs. (54–74 Nm).

12. Tighten the tie rod end nut to 27–37 ft. lbs. (37–50 Nm), install a new cotter pin. Tighten the nut, if necessary, to align the ball stud hole with the nut castellation.

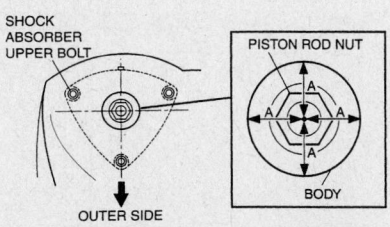

67162-MAZC-G89

Align the piston rod nut with the center part where the shock absorber is installed by placing the piston rod nut with lengths (A) all the same—Mazda3

13. Install the stabilizer bar link and tighten the upper nut to 30–40 ft. lbs. (40–54 Nm).

14. Install the brake hose and ABS sensor connector.

15. Check and/or adjust the front end alignment.

MAZDA6 AND MAZDASPEED6

1. Before servicing the vehicle, refer to the precautions section.

2. Disconnect the ABS sensor.

3. Disconnect the brake hose from the bracket on the strut assembly.

4. Remove the stabilizer bar link upper nut and disconnect the bar from the strut assembly.

5. Remove the shock absorber lower bolt, the dynamic damper and the damper fork.

6. Remove the upper shock absorber nuts.

7. Remove the shock absorber assembly.

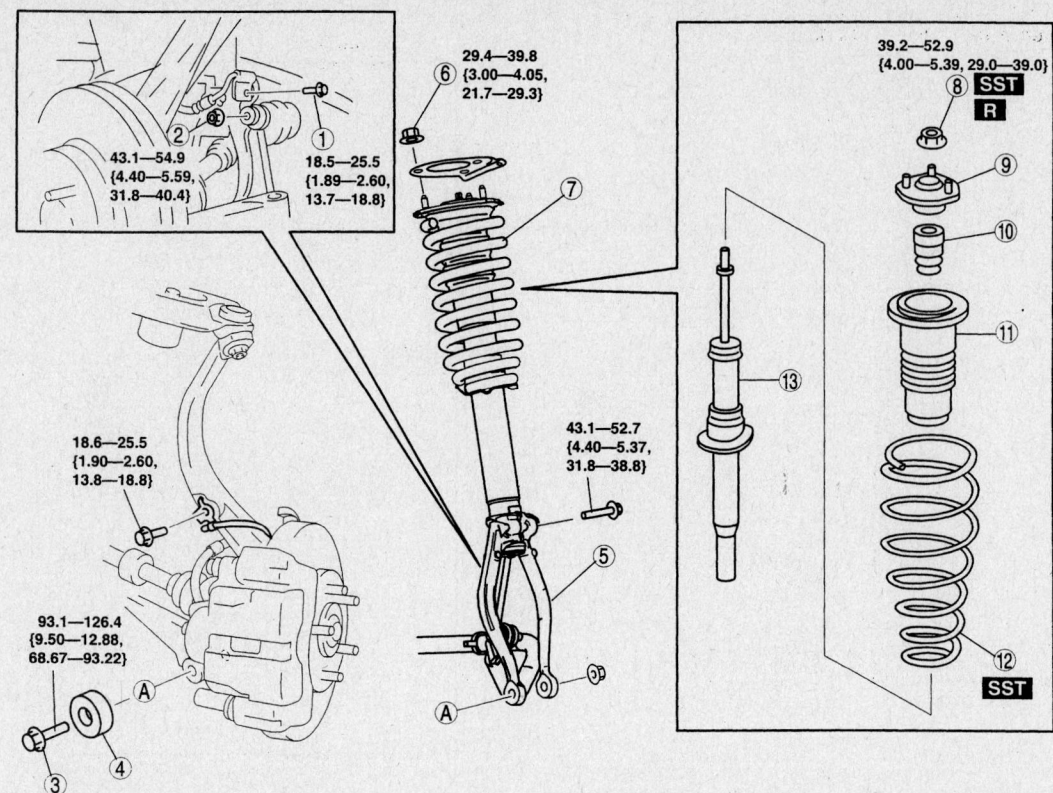

1   Bolt (brake hose bracket)
2   Nut (front stabilizer control link)
3   Bolt (front shock absorber lower side)
4   Dynamic damper
5   Damper fork
6   Nut (front shock absorber upper side)
7   Front shock absorber and coil spring
8   Piston rod nut
9   Mounting rubber
10 Bound stopper
11 Dust boot
12 Coil spring
13 Front shock absorber

N·m {kgf·m, ft·lbf}

67162-MAZC-G235

Exploded view of the front strut and spring assembly components—Mazda6 and Mazdaspeed6 models

To install:

8. Install the shock absorber assembly. Position the stud bolts at a 27–33 degree angle from where the stabilizer bar is installed (center line), towards the inner side of the vehicle. Refer to the accompanying illustration.

9. Tighten the upper nuts to 21–29 ft. lbs. (29–39 Nm).

10. Install the damper fork and align the gap of the fork with the projection of the damper as illustrated and install and tighten the bolt to 31–38 ft. lbs. (43–52 Nm)

11. Install the dynamic damper and the lower shock bolt and tighten to 68–93 ft. lbs. (93–126 Nm).

12. Install the stabilizer bar link and tighten the nut to 31–40 ft. lbs. (43–54 Nm).

13. Install the brake hose, tighten the bolt to 13–18 ft. lbs. (18–25 Nm).

14. Install the sensor.

15. Check and/or adjust the front end alignment.

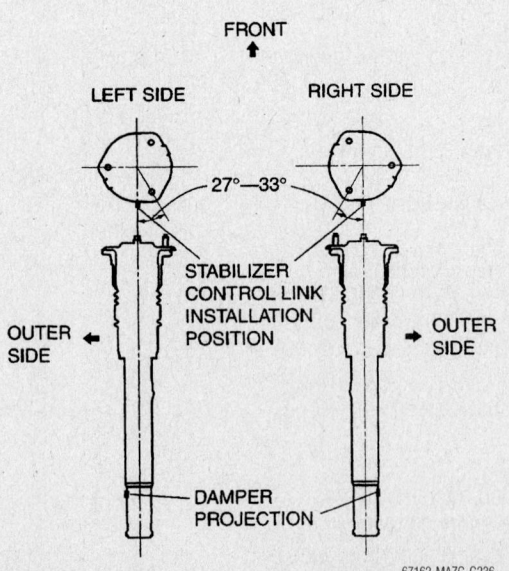

67162-MAZC-G236

Position the stud bolts at a 27–33 degree angle from where the stabilizer bar is installed (center line), towards the inner side of the vehicle when installing the front shock assembly—Mazda6 and Mazdaspeed6 models

Rear

MX5-MIATA

1. Before servicing the vehicle, refer to the precautions section.

➡**Refer the exploded view illustration for component locations and if applicable, their retainer torque specifications.**

✷✷ CAUTION

Performing the following procedures without first removing the ABS wheel-speed sensor may possibly cause an open circuit in the wiring harness if it is pulled by mistake. Before performing the following procedures, remove the ABS wheel-speed sensor (axle side) and fix it to an appropriate place where the sensor will not be pulled while servicing the vehicle.

2. If removing the left side, remove the fuel tank protector.
3. Remove the parking brake cable.
4. Remove the caliper without disconnecting the hose, and support the caliper with wire.
5. Remove the rear lateral link upper bolt.
6. Remove the stabilizer control link upper nut.
7. Remove the rear shock absorber lower bolt.
8. Remove the rear shock absorber and coil spring.

To install:

9. Install the components in the reverse order of removal using the accompanying illustration for torque values.

MAZDA3

1. Before servicing the vehicle, refer to the precautions section.

2. Support the rear axle assembly with a jack.
3. Remove or disconnect the following:
   • Rear wheel(s)
   • Top strut nuts

➡**The suspension will drop when the weight lifts off the wheels.**

   • Bottom strut mount bolt(s)
   • Strut assembly

To install:

4. Install or connect the following:
   • Strut assembly
   • Bottom strut mount bolt(s) and tighten to 56–74 ft. lbs. (76–101 Nm)
   • Upper mounting nuts and tighten to 15–21 ft. lbs. (21–28 Nm)
   • Rear wheel(s)

MAZDA6 AND MAZDASPEED6

1. Before servicing the vehicle, refer to the precautions section.
2. Support the rear axle assembly with a jack.
3. Remove or disconnect the following:
   • Rear wheel(s)
   • ABS sensor
   • Top strut and bracket retainers
   • Bracket

➡**The suspension will drop when the weight lifts off the wheels.**

   • Bottom strut mount bolt(s)
   • Strut assembly

To install:

4. Install or connect the following:
   • Strut assembly

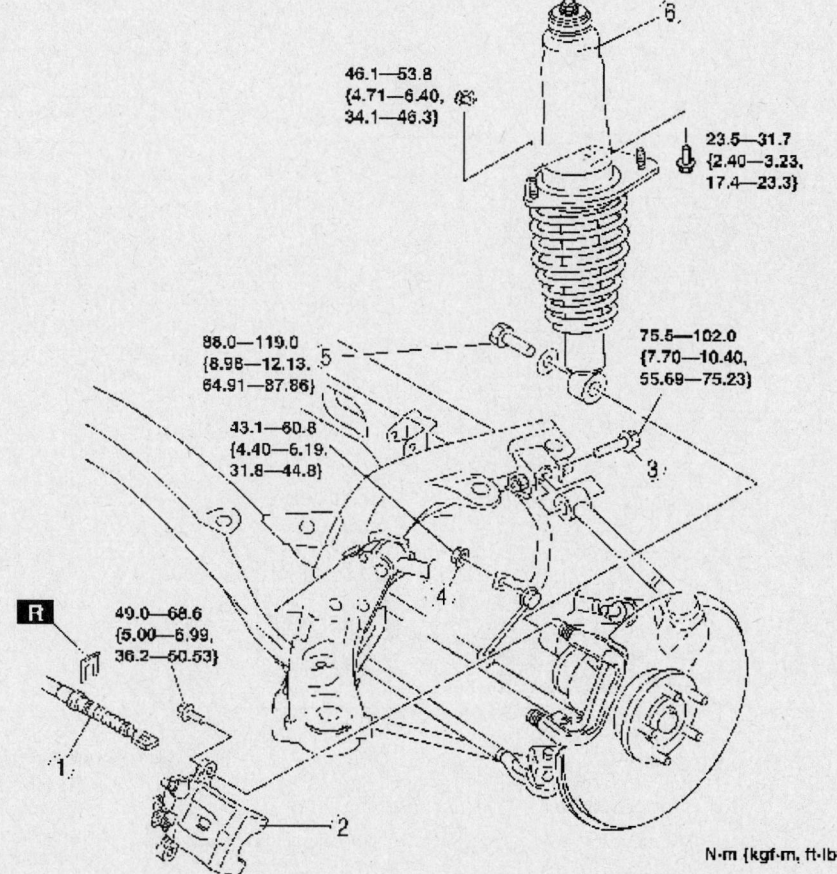

46.1—53.8
{4.71—6.40,
34.1—46.3}

23.5—31.7
{2.40—3.23,
17.4—23.3}

88.0—119.0
{8.98—12.13,
64.91—87.86}

75.5—102.0
{7.70—10.40,
55.69—75.23}

43.1—60.8
{4.40—6.19,
31.8—44.8}

R 49.0—68.6
{5.00—6.99,
36.2—50.53}

N·m {kgf·m, ft·lbf}

1	Parking brake cable	4 Stabilizer control link upper nut
2	Caliper	5 Rear shock absorber lower bolt
3	Rear lateral link (upper) bolt	6 Rear shock absorber and coil spring

09482_MAZC2_G0095A

Exploded view of the rear strut assembly and related components—MX-5 Miata

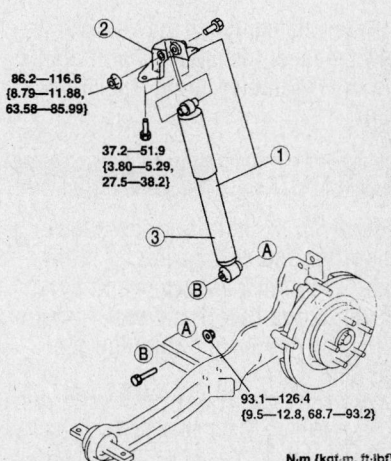

86.2—116.5
{8.79—11.88,
63.58—85.99}

37.2—51.9
{3.80—5.29,
27.5—38.2}

93.1—126.4
{9.5—12.8, 68.7—93.2}

N·m {kgf·m, ft·lbf}

1 Rear shock absorber and bracket
2 Bracket
3 Rear shock absorber

67162-MAZC-G237

Exploded view of the rear shock absorber assembly—Mazda6 and Mazdaspeed6 models

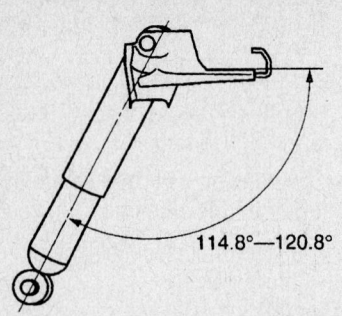

67162-MAZC-G238

Position the bracket and rear shock absorber assembly as shown when installing—Mazda6 and Mazdaspeed6 models

- Bracket and position the strut and bracket as illustrated. Tighten the bracket bolts to 27–38 ft. lbs. (37–51 Nm).
- Bottom strut mount bolt(s) and tighten to 68–93 ft. lbs. (93–126 Nm)
- Upper mounting retainers and tighten to 63–85 ft. lbs. (86–116 Nm)
- ABS sensor
- Rear wheel(s)

Coil Spring

REMOVAL & INSTALLATION

MX5-Miata

FRONT

1. Before servicing the vehicle, refer to the precautions section.

➡**Refer the exploded view illustration for component locations and if applicable, their retainer torque specifications.**

✳✳ CAUTION

Removing or installing the shock absorber and coil spring is dangerous. The shock absorber and coil spring could fly off and cause serious injury or death, and damage the vehicle.

2. Remove the front shock absorber and coil spring.

✳✳ WARNING

Before removing the piston rod nut, secure the shock absorber and spring in a suitable spring compressor. Otherwise, the shock absorber and

spring could fly off under tremendous pressure and cause serious injury or death, or damage to vehicle parts.

3. Protect the coil spring from scratches using a piece of cloth and install the spring compressor.
4. Compress the coil spring and remove the piston rod nut.
5. Remove the following:
   - Retainer
   - Bushing
   - Upper spring seat
   - Dust boot
   - Spacer
   - Bushing
   - Stopper casing and bound stopper
   - Bound stopper
   - Stopper casing
   - Coil spring
   - Shock absorber

To install:

6. Install the following:
   - Shock absorber
   - Coil spring
   - Stopper casing

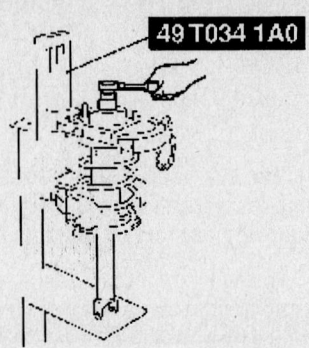

09482_MAZC2_G0093

Compress the front coil spring and remove the piston rod nut—MX-5 Miata

- Bound stopper
- Stopper casing and bound stopper
- Bushing
- Spacer
- Dust boot
- Upper spring seat
- Bushing
- Retainer

7. Protect the coil spring from scratches using a piece of cloth and install the spring compressor.

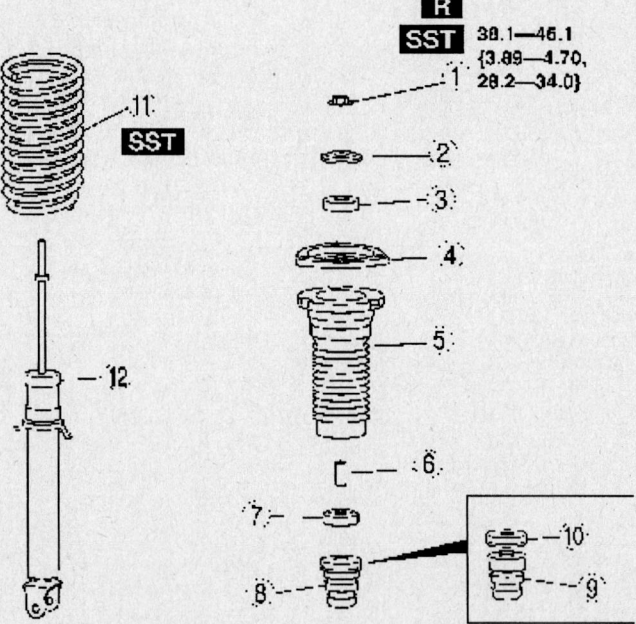

R
SST 39.1—46.1
{3.99—4.70,
28.2—34.0}

N·m {kgf·m, ft·lbf}

1	Piston rod nut	7	Bushing
2	Retainer	8	Stopper casing and bound stopper
3	Bushing	9	Bound stopper
4	Upper spring seat	10	Stopper casing
5	Dust boot	11	Coil spring
6	Spacer	12	Front shock absorber

09482_MAZC2_G0092

Exploded view of the front coil spring assembly—MX-5 Miata

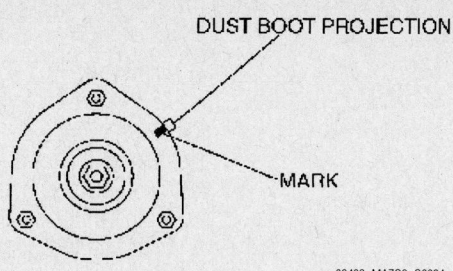

09482_MAZC2_G0094

Align the mark on the upper spring seat with the dust boot projection—MX-5 Miata

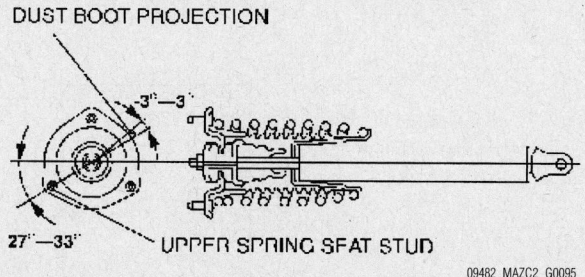

09482_MAZC2_G0095

Install the front strut upper spring seat so that the upper spring seat stud is at a 27–33 degree angle to the shock absorber installation shaft—MX-5 Miata

8. Compress the coil spring using the spring compressor.

9. Install the shock absorber so that the lower end of the coil spring is seated on the step of the lower spring seat.

10. When installing the upper spring seat:

a. Align the mark on the upper spring seat with the dust boot projection.

b. Install the upper spring seat so that the upper spring seat stud is at a 27–33 degree angle to the shock absorber installation shaft (lower side).

11. Install the piston rod nut and tighten to 38–46 Nm (28–34 ft. lbs.).

12. Install the strut assembly.

REAR

1. Before servicing the vehicle, refer to the precautions section.

➡**Refer the exploded view illustration for component locations and if applicable, their retainer torque specifications.**

⁑ CAUTION

Removing or installing the shock absorber and coil spring is dangerous. The shock absorber and coil spring could fly off and cause serious injury or death, and damage the vehicle.

2. Remove the rear shock absorber and coil spring.

⁑ WARNING

Before removing the piston rod nut, secure the shock absorber and spring in a suitable spring compressor. Otherwise, the shock absorber and spring could fly off under tremendous pressure and cause serious injury or death, or damage to vehicle parts.

3. Protect the coil spring from scratches using a piece of cloth and install the spring compressor.

4. Compress the coil spring and remove the piston rod nut.

5. Remove the following:
- Retainer
- Bushing
- Upper spring seat
- Spring seat rubber
- Spacer
- Bushing
- Stopper casing and bound stopper
- Bound stopper
- Collar

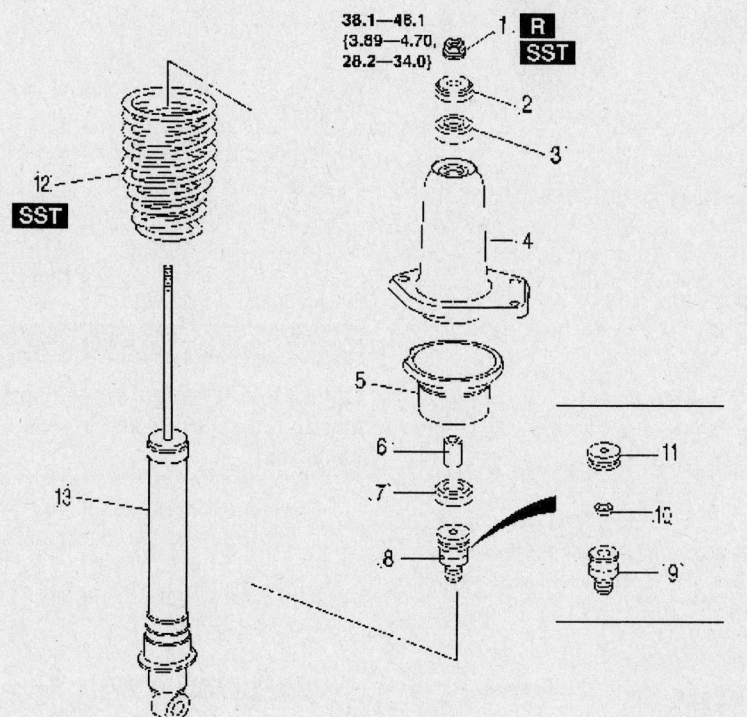

N•m {kgf•m, ft·lbf}

1	Piston rod nut	8	Bound stopper and stopper casing
2	Retainer	9	Bound stopper
3	Bushing	10	Collar
4	Upper spring seat	11	Stopper casing
5	Spring seat rubber	12	Coil spring
6	Spacer	13	Rear shock absorber
7	Bushing		

09482_MAZC2_G0096

Exploded view of the rear coil spring assembly—MX-5 Miata

- Stopper casing
- Coil spring
- Shock absorber

To install:

6. Install the following:

- Shock absorber so that the lower end of the coil spring is seated on the step of the lower spring seat
- Coil spring lead lower end of the lower spring seat facing the direction shown in the illustration
- Stopper casing
- Collar so that the tapered side is facing downward as shown in the illustration
- Bound stopper
- Stopper casing and bound stopper
- Bushing
- Spacer
- Spring seat rubber to the upper

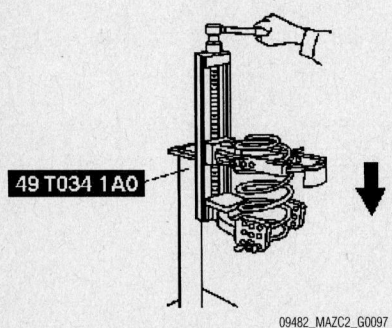

Installing the coil spring—MX-5 Miata

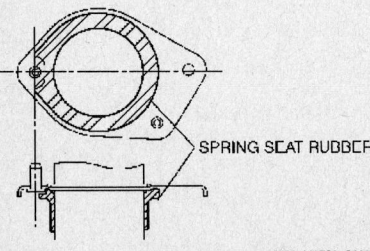

Install the collar on the rear strut so that the tapered side is facing downward—MX-5 Miata

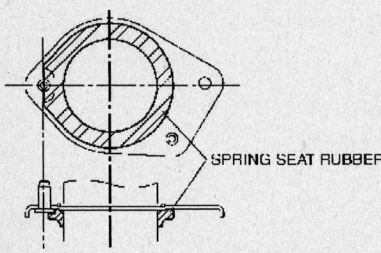

Install the spring seat rubber on the rear strut to the upper spring seat as shown—MX-5 Miata

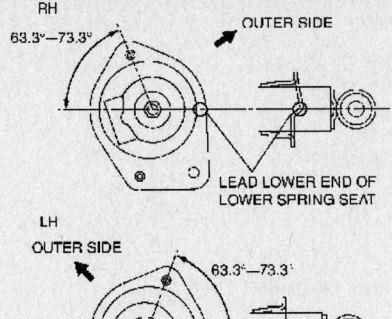

Install the coil spring lead lower end of the lower spring seat facing the direction shown on the rear strut —MX-5 Miata

spring seat as shown in the illustration

- Upper spring seat
- Bushing
- Retainer

7. Install the piston rod nut and tighten to 38–46 Nm (28–34 ft. lbs.).

8. Install the strut assembly.

Mazda3

1. Before servicing the vehicle, refer to the precautions section.

2. Remove the strut assembly from the vehicle.

3. Install a coil spring compressor and compress the coil spring.

✳✳ CAUTION

Failure to fully compress the spring and hold it securely can be extremely dangerous.

4. Remove the upper strut nut.

5. Slowly release the coil spring tension

6. Remove the mounting rubber, bear-

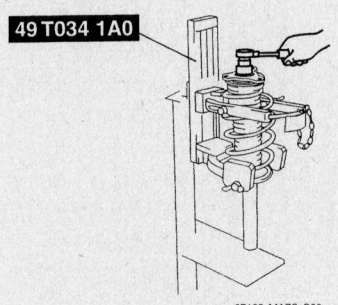

Install a coil spring compressor and compress the coil spring—Mazda3

ing, dust boot, bound stopper and coil spring.

To install:

7. Compress the coil spring in the compressor tool and install the shock absorber so the lower end of the spring is seated on the step of the lower spring seat.

8. Install the bound stopper.

9. Install the dust boot by hooking the bottom edge over the shock absorber lip as illustrated.

10. Install the bearing by hooking the upper end of the dust boot to the bearing lip

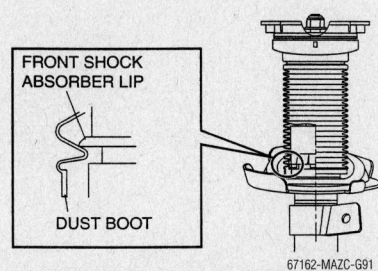

Install the dust boot by hooking the bottom edge over the shock absorber lip—Mazda3

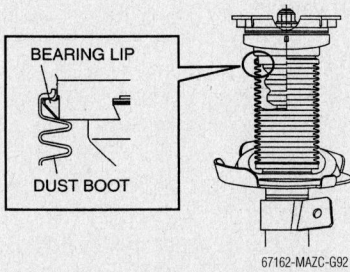

Install the bearing by hooking the upper end of the dust boot to the bearing . . . —Mazda3

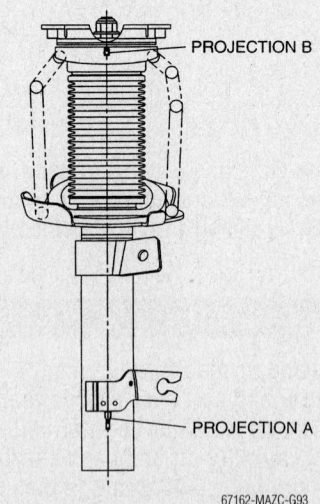

. . . and align the projection A on the lower part of the shock absorber with the bearing projection B—Mazda3

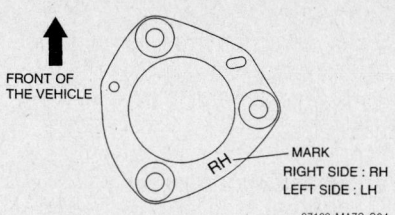

FRONT OF
THE VEHICLE

— MARK
RIGHT SIDE : RH
LEFT SIDE : LH

67162-MAZC-G94

The stiffener should be positioned as shown—Mazda3

and align the projection A on the lower part of the shock absorber with the bearing projection B, as illustrated.

11. Install the mounting rubber.

12. Install the piston rod nut and tighten to 42–56 ft. lbs. (57–76 Nm) and remove the spring compressor.

13. Install the stiffener so the mark LH or RH is facing upwards as shown.

14. Align the piston rod nut with the center part where the shock absorber is installed by placing the piston rod nut with lengths A all the same and tighten the shock absorber upper bolts.

15. Use a jack to raise the lower control arm, attach the shock absorber and tighten the bolts.

Mazda6 and Mazdaspeed6

FRONT

1. Before servicing the vehicle, refer to the precautions section.

2. Remove the strut assembly from the vehicle.

3. Install a coil spring compressor and compress the coil spring.

> ✲✲ **CAUTION**
>
> **Failure to fully compress the spring and hold it securely can be extremely dangerous.**

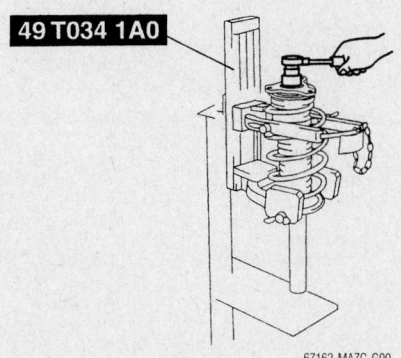

49 T034 1A0

67162-MAZC-G90

Install a coil spring compressor and compress the coil spring—Mazda6 and Mazdaspeed6 models

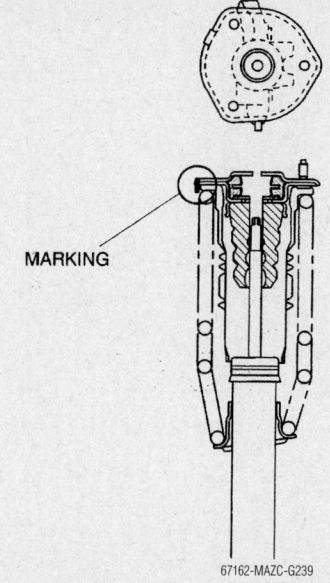

MARKING

67162-MAZC-G239

Mark the spring, dust boot and rubber for correct installation as shown when assembly the front shock assembly—Mazda6 and Mazdaspeed6 models

4. Remove the upper strut nut.

5. Slowly release the coil spring tension

6. Remove the mounting rubber, bound stopper, dust boot and coil spring.

To install:

7. Temporarily install the spring, dust boot and mounting so the lower end of the coil spring is seated on the step of the lower seat.

8. Mark the spring, dust boot and rubber for correct installation as shown in the accompanying illustration.

9. Align the mark of the coil spring and dust boot, use a piece of cloth to protect the spring and boot and install the compressor.

10. Compress the coil spring in the compressor tool and install the shock absorber so the lower end of the spring is seated on the step of the lower spring seat.

11. Ensure the marks on the shock absorber and dust boot are aligned to the marks made earlier, install the mounting rubber and the rod not and tighten the nut to 29–39 ft. lbs. (39–52 Nm)

12. Install the shock assembly.

REAR

1. Before servicing the vehicle, refer to the precautions section.

2. Support the lower control arm with a floor jack.

3. Remove the rear stabilizer bar bolt from the lower link side.

4. Loosen the inner bolt of the rear lower control arm.

5. Remove the outer bolts of the rear lower control arm.

6. Lower the control arm and remove the coil spring, upper spring seat rubber, lower spring seat, bound stopper from the spring side and the body side.

To install:

7. Install the bound stopper to the body and spring sides.

8. Install the lower spring seat and upper seat rubber.

9. Install the coil spring with the small outer diameter side faces downwards and raise the lower control arm using the jack.

10. Install the lower arm outer and inner bolts and tighten the outer bolt first then the inner bolt to 63–86 ft. lbs. (86–116 Nm).

11. Install the rear stabilizer link bolt and tighten to 31–44 ft. lbs. (43–60 Nm).

Lower Ball Joint

REMOVAL & INSTALLATION

The lower ball joint is an integral part of the lower control and cannot be replaced separately. If the lower ball joint is defective, the entire lower control arm must be replaced. Refer to the lower control arm procedure.

Upper Control Arm

REMOVAL & INSTALLATION

MX-5 Miata

1. Before servicing the vehicle, refer to the precautions section.

➡ **Refer the exploded view illustration for component locations and if applicable, their retainer torque specifications.**

> ✲✲ **CAUTION**
>
> **Performing the following procedures without first removing the ABS wheel-speed sensor may possibly cause an open circuit in the wiring harness if it is pulled by mistake. Before performing the following procedures, remove the ABS wheel-speed sensor (axle side) and fix it to an appropriate place where the sensor will not be pulled while servicing the vehicle.**

2. Remove the brake hose bracket.

3. Separate the front upper arm ball joint from the knuckle.

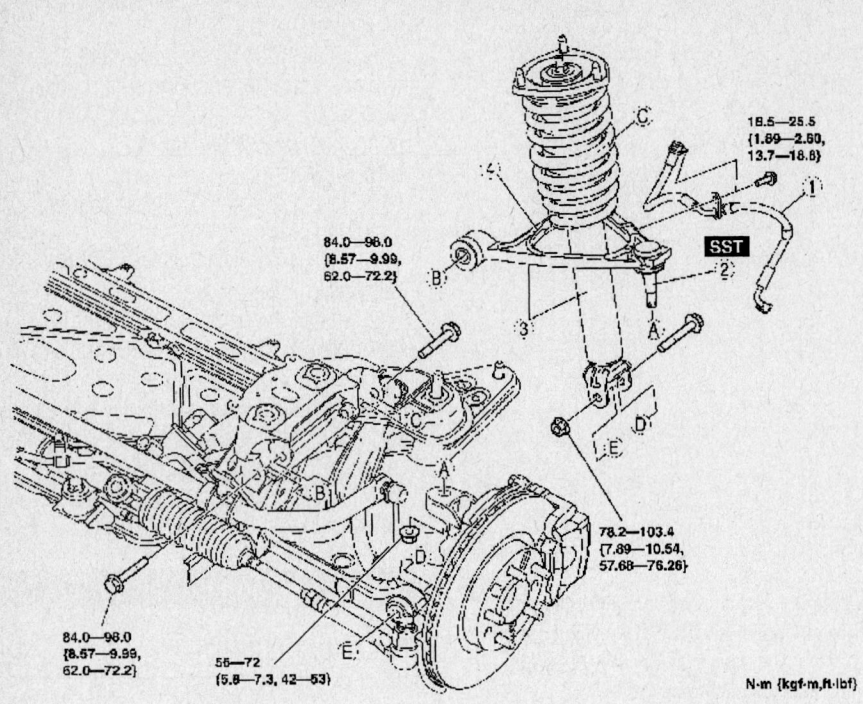

84.0—98.0
{8.57—9.99,
62.0—72.2}

18.5—25.5
{1.89—2.60,
13.7—18.8}

SST

84.0—98.0
{8.57—9.99,
62.0—72.2}

55—72
{5.6—7.3, 42—53}

78.2—103.4
{7.89—10.54,
57.68—76.26}

N·m {kgf·m, ft·lbf}

| 1 | Brake hose bracket | 3 | Front shock absorber, coil spring and front upper arm |
| 2 | Front upper arm ball joint | 4 | Front upper arm |

09482_MAZC2_G0101

Exploded view of the front upper control arm assembly—MX-5 Miata

4. Remove the front shock absorber, coil spring and front upper arm, by loosening the shock absorber upper nuts, then remove the front shock absorber lower bolt and nut.

5. Remove the front upper arm bolts.

6. Push down the front lower arm, and then remove the front upper arm from the gap between the shock absorber lower end and the front lower arm.

To install:

7. Install the components in the reverse order of removal using the accompanying illustration for torque values.

8. Inspect the front wheel alignment and adjust as necessary.

Mazda3

1. Before servicing the vehicle, refer to the precautions section.

2. Remove or disconnect the following:
   • Wheel

3. Support the lower control arm with a jack.

4. Remove or disconnect the following:
   • Coil spring
   • Upper control arm bolts and nuts
   • Upper control arm

To install:

5. Install the upper control arm.

6. Loosely tighten the bolt and nut.

   • Wheel

7. Torque upper control arm nuts and bolts to 55–75 ft. lbs. (75–101 Nm).

8. Check and/or adjust the front wheel alignment.

Mazda6 and Mazdaspeed6

FRONT

1. Before servicing the vehicle, refer to the precautions section.

2. Remove the wheel.

3. Remove the ABS sensor.

4. Remove the brake hose bracket bolt.

5. Remove the stabilizer bar link nut.

6. Support the lower control arm with a jack.

7. Separate the upper arm ball joint from the knuckle using a separator tool such as 49 T028 3A0.

8. Remove the upper arm rear bolts and the arm.

To install:

   • Install the upper arm and the bolts.

9. Attach the ball joint to the knuckle. Tighten the ball joint bolt to 29–39 ft. lbs. (39–53 Nm) and the rear upper arm bolts to 36–49 ft. lbs. (49–66 Nm).

10. Install the stabilizer bar link nut and tighten to 31–40 ft. lbs. (43–54 Nm).

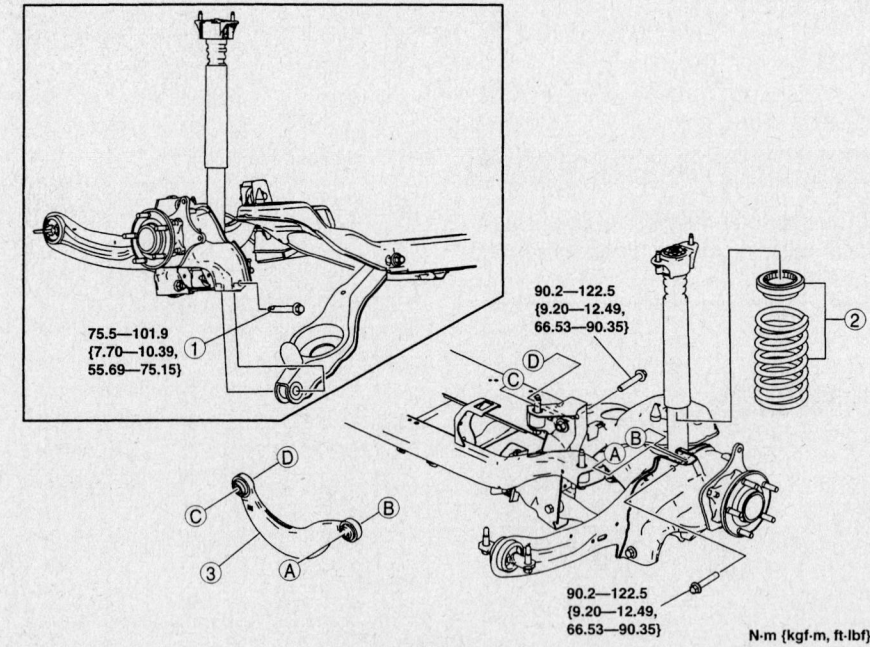

75.5—101.9
{7.70—10.39,
55.69—75.15}

90.2—122.5
{9.20—12.49,
66.53—90.35}

90.2—122.5
{9.20—12.49,
66.53—90.35}

N·m {kgf·m, ft·lbf}

1	Rear lower arm outer bolt
2	Rear coil spring component
3	Rear upper arm

67162-MAZC-G95

Exploded view of the rear upper control arm assembly and related components—Mazda3

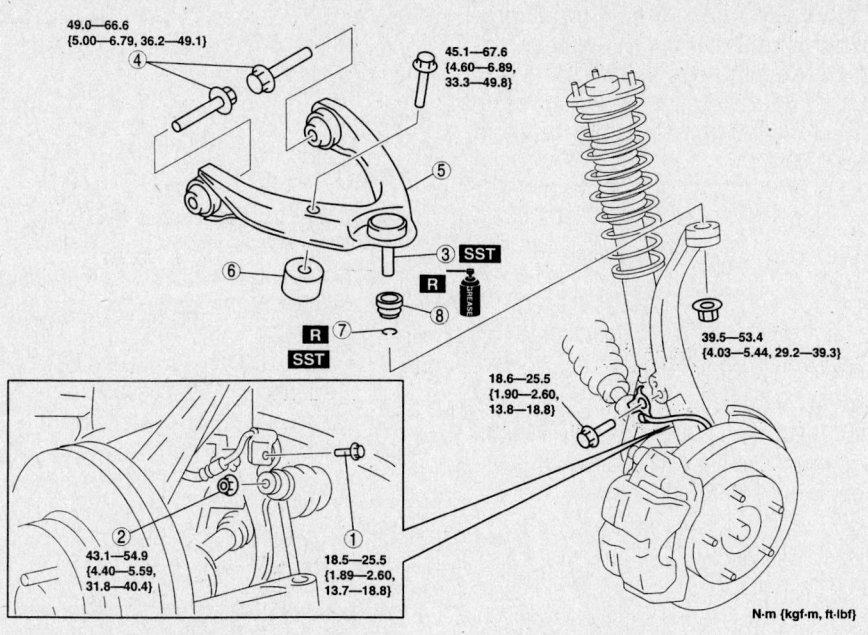

1 Bolt (brake hose bracket
2 Nut (stabilizer control link)
3 Front upper arm ball joint
4 Bolt (front upper arm)

5 Front upper arm
6 Dynamic Vamper
7 Clip
8 Dust boot

67162-MAZC-G240

Exploded view of the front upper control arm assembly—Mazda6 and Mazdaspeed6 models

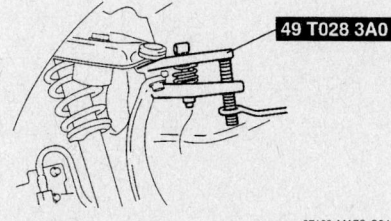

67162-MAZC-G241

Separate the upper arm ball joint from the knuckle using a separator tool—Mazda6 and Mazdaspeed6 models

11. Install the brake hose bracket bolt and tighten to 13–18 ft. lbs. (18–25 Nm).

12. Install the ABS sensor and in the wheel.

REAR

1. Before servicing the vehicle, refer to the precautions section.

2. Remove the wheel.

3. Remove the ABS sensor.

4. Disconnect the parking brake cable.

5. Support the rear trailing link with a jack.

6. Remove the rear shock absorber lower bolt.

7. Remove the front trailing link bolt.

8. Remove the rear upper arm bolt and the arm.

To install:

9. Install the arm and the rear bolt.

10. Attach the bolt to the trailing link front side.

11. Tighten the rear upper bolt to 68–93 ft. lbs. (93–126 Nm) and the trailing link bolt to 63–86 ft. lbs. (86–116 Nm).

12. Install the lower shock absorber bolt and tighten to 68–93 ft. lbs. (93–126 Nm).

13. Attach the parking brake cable.

14. Install the ABS sensor.

• Install the wheel.

CONTROL ARM BUSHING REPLACEMENT

All Mazda's use a pressed in control arm bushing, and the pressing can be done using two appropriately sized sockets (a press socket and a catch socket) and a large bench vise.

1. Position the control arm and the 2 sockets into a vise.

2. Position the press socket onto the control arm bushing.

3. Position the catch socket onto the control arm, opposite of the press socket.

4. Tighten the bench vise slowly and press the bushing into the catch socket.

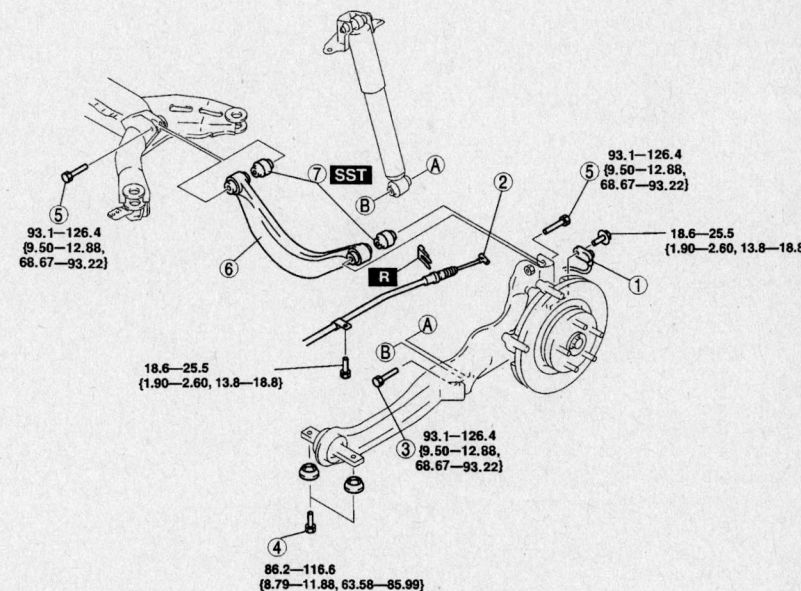

1 ABS wheel-speed sensor
2 Parking brake cable
3 Bolt (rear shock absorber lower side)
4 Bolt (trailing link front side)

5 Bolt (rear upper arm)
6 Rear upper arm
7 Rear upper arm bushing

67162-MAZC-G242

Exploded view of the rear upper control arm assembly—Mazda6 and Mazdaspeed6 models

To install:

5. Apply soapy water to the new control arm bushing.

6. Position the bushing against the control arm.

7. Using the same sockets, in the same positions, press the new bushing into the control arm.

Lower Control Arm

REMOVAL & INSTALLATION

MX-5 Miata

FRONT

1. Before servicing the vehicle, refer to the precautions section.

➡Refer the exploded view illustration for component locations and if applicable, their retainer torque specifications.

❋❋ CAUTION

Performing the following procedures without first removing the ABS wheel-speed sensor may possibly cause an open circuit in the wiring

harness if it is pulled by mistake. Before performing the following procedures, remove the ABS wheel-speed sensor (axle side) and fix it to an appropriate place where the sensor will not be pulled while servicing the vehicle.

2. Remove the caliper and mounting support from the steering knuckle and suspend it with a cable in a location out of the way.

3. Separate the front lower arm ball joint from the knuckle.

➡When removing the front lower arm ball joint, the steering knuckle bushing may also come off. If it comes off, replace the steering knuckle.

4. Disconnect the tie-rod end.

5. Separate the front upper arm ball joint from the knuckle

6. Remove the front hub and steering knuckle assembly.

7. Remove the stabilizer control link nut on the front lower arm side.

8. Remove the front lower arm

To install:

9. Install the front lower arm rear side bushing part horizontally.

10. Install the front lower arm front side bushing part.

11. Install all components in the reverse of removal.

Mazda3

FRONT

1. Before servicing the vehicle, refer to the precautions section.

2. Remove or disconnect the following:
   • Wheel
   • Cotter pin
   • Lower ball joint by loosening it

3. With the nut protecting the ball joint stud, separate the stud from the knuckle. Remove the nut.
   • Cotter pin and nut from the tie-rod end
   • Tie-rod end from the steering knuckle

➡If removing the right side lower arm, move the engine and transaxle slightly towards the front side of the vehicle so the engine does not interfere with the lower arm rear side bolt removal.

   • No. 1 engine mount center bolt.
   • Engine and transaxle assembly slightly towards the front off the vehicle if necessary to remove the right lower arm
   • Lower arm rear side bolt and the lower arm.
   • Dust boot, if necessary

To install:

4. Install or connect the following:
   • Dust boot, if removed using a press. Always fill the inside of the new boot with grease prior to installation.
   • Lower control arm by loosely tightening the bolts and nuts
   • No. 1 engine mount bolt and tighten to 68–86 ft. lbs. (93–116 Nm).

5. Tighten the lower control arm bolts to the specifications shown in the accompanying illustration.

6. Install the remaining components in the reverse order of removal, refer to the exploded view of the lower control arm assembly and related components illustration for component location and torque specifications.

7. Check and/or adjust the front wheel alignment.

REAR

1. Before servicing the vehicle, refer to the precautions section.

2. Remove the rear wheels.

18.5—25.5
{1.89—2.60,
13.7—18.8}

84.0—98.0
{8.57—9.99,
62.0—72.2}

78.2—103.4
{7.89—10.54,
57.68—76.26}

84.0—98.0
{8.57—9.99,
62.0—72.2}

56—72
{5.8—7.3, 42—53}

SST

N·m {kgf·m,ft·lbf}

| 1 | Brake hose bracket | 3 | Front shock absorber, coil spring and front upper arm |
| 2 | Front upper arm ball joint | 4 | Front upper arm |

09482_MAZC2_G0101

Exploded view of the front upper control arm assembly—MX-5 Miata

12. Install the rear stabilizer, tighten the bracket bolts in the sequence illustrated to 30–40 ft. lbs. (40–54 Nm). Tighten the link bolts to 30–40 ft. lbs. (40–54 Nm).

13. Install the crossmember bracket and wheels.

14. Check and/or adjust the wheel alignment.

Mazda6 and Mazdaspeed6

The Mazda6 models have two lower control arms in the front suspension assembly, they are identified as the front lower control arm (Front side) and front lower control arm (rear side). They also have one rear suspension lower control arm, which will be identified as rear lower control arm.

FRONT ARM (FRONT SIDE)

Refer to the accompanying illustration for component locations and torque specifications.

1. Before servicing the vehicle, refer to the precautions section.

2. Remove the three way converter on models with the 3.0L (AJ) engine.

3. Separate the lower control arm (front) ball joint from the knuckle.

4. Remove the front shock lower bolt and the dynamic damper.

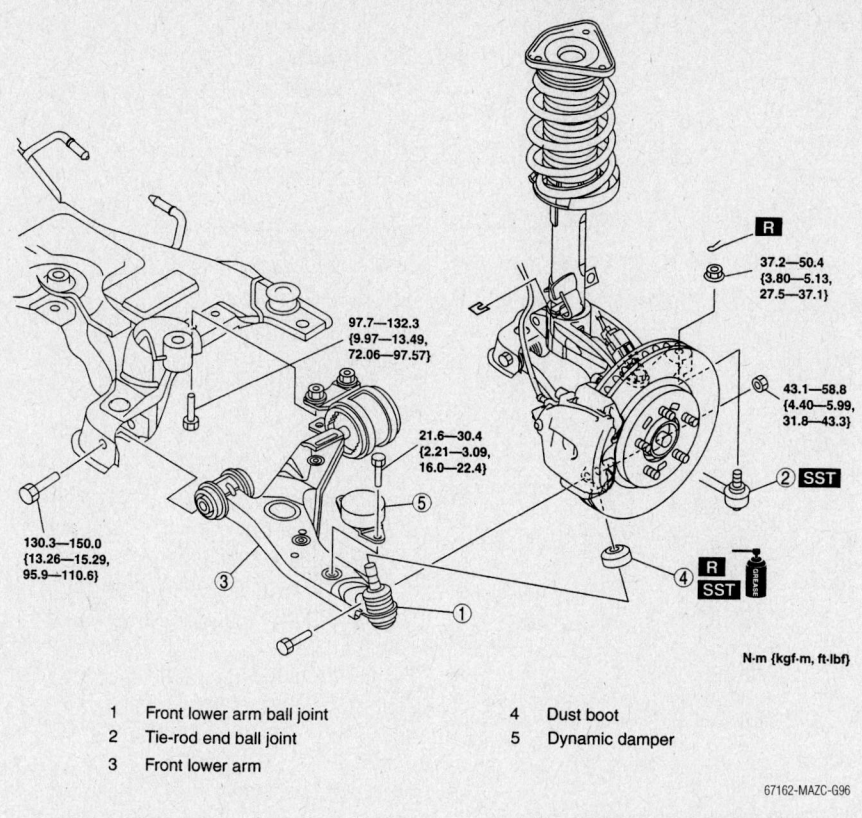

1 Front lower arm ball joint
2 Tie-rod end ball joint
3 Front lower arm
4 Dust boot
5 Dynamic damper

N·m {kgf·m, ft-lbf}

67162-MAZC-G96

Exploded view of the front lower control arm assembly and related components—Mazda3

3. Remove the rear crossmember bracket.

4. Remove the rear stabilizer.

5. Support the rear axle with a jack, loosen the rear lower control arm inner bolt and remove the outer bolt.

6. Remove the rear coil spring and upper spring seat rubber.

7. Remove the rear lower arm inner bolt and the arm.

To install:

8. Install the rear arm and the inner bolt and hand tighten.

9. Install the upper seat rubber and the coil spring.

10. Raise the axle into position with the jack.

11. Install the rear lower arm outer bolt and tighten the outer bolt to 55–75 ft. lbs. (75–102 Nm) and the inner bolt to 59–74 ft. lbs. (80–100 Nm).

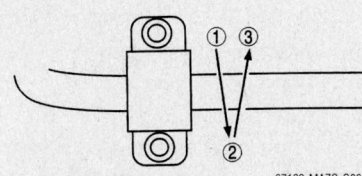

67162-MAZC-G98

Tighten the rear stabilizer bracket bolts in this sequence—Mazda3

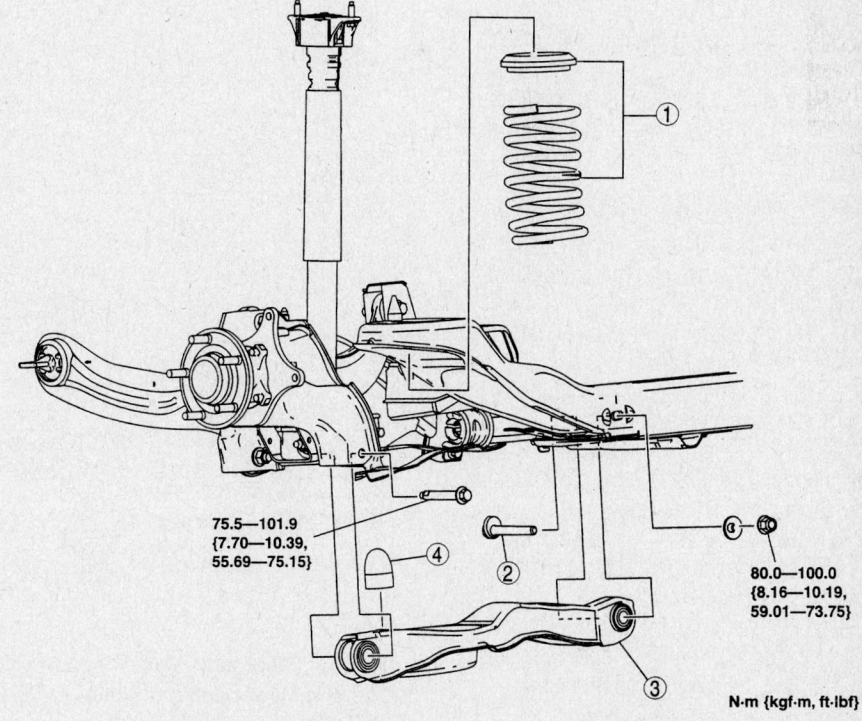

N·m {kgf·m, ft-lbf}

1 Rear coil spring component
2 Rear lower arm inner bolt
3 Rear lower arm
4 Bound stopper

67162-MAZC-G97

Exploded view of the rear lower control arm assembly and related components—Mazda3

REAR LOWER CONTROL ARM

1. Before servicing the vehicle, refer to the precautions section.

2. Support the lower control arm with a floor jack.

3. Remove the rear stabilizer bar bolt from the lower link side.

4. Loosen the inner bolt of the rear lower control arm.

5. Remove the outer bolts of the rear lower control arm.

6. Lower the control arm and remove the coil spring.

7. Remove the rear lower arm inner side bolt and the arm.

To install:

8. Install the rear lower arm and the inner side bolt. Tighten the bolt to 63–85 ft. lbs. (86–116 Nm).

9. Install the coil spring with the small outer diameter side faces downwards and raise the lower control arm using the jack.

10. Install the lower arm outer and inner bolts and tighten the outer bolt first then the inner bolt to 63–86 ft. lbs. (86–116 Nm).

11. Install the rear stabilizer link bolt and tighten to 31–44 ft. lbs. (43–60 Nm).

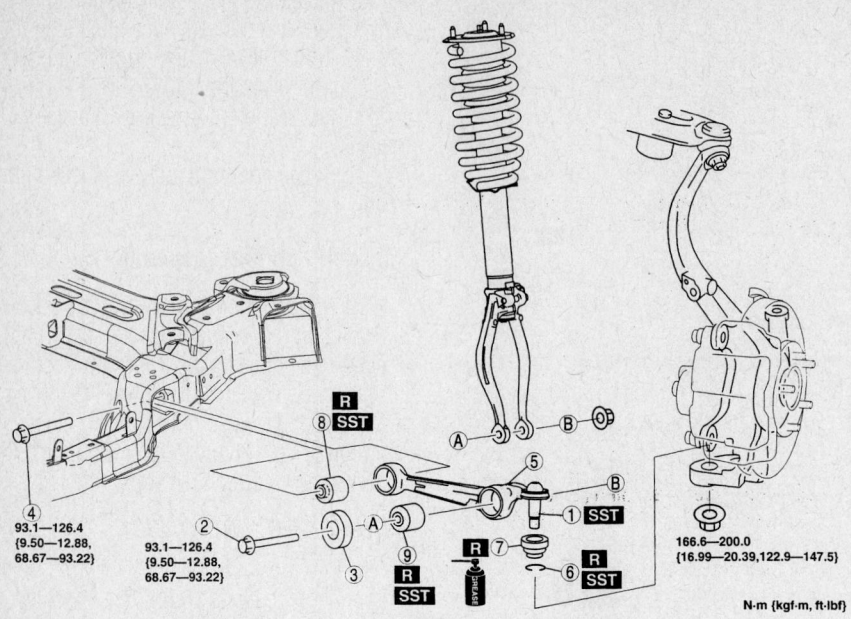

93.1—126.4
{9.50—12.88,
68.67—93.22}

93.1—126.4
{9.50—12.88,
68.67—93.22}

166.6—200.0
{16.99—20.39,122.9—147.5}

N·m {kgf·m, ft·lbf}

1	Front lower arm (front) ball joint	6	Clip
2	Bolt (front shock absorber lower side)	7	Dust boot
3	Dynamic damper	8	Front lower arm (front) bushing (inner side)
4	Bolt (front lower arm inner side)	9	Front lower arm (front) bushing (outer side)
5	Front lower arm (front) component		

67162-MAZC-G250

Exploded view of the front suspension lower control arm (front) assembly—Mazda6 and Mazdaspeed6 models

5. Remove the lower inner side arm bolt.

6. Remove the front lower arm (front).

To install:

7. Installation is the reverse of removal. refer to the illustration for bolt torque specifications.

FRONT ARM (REAR SIDE)

Refer to the accompanying illustration for component locations and torque specifications.

1. Before servicing the vehicle, refer to the precautions section.

2. Remove the wheels, engine under cover and splash shield.

3. Remove the steering gear and linkage bolts, pipe assembly from the crossmember and wire the assembly aside.

4. Remove the stabilizer bar link nut.

5. Separate the lower control arm (rear) ball joint from the knuckle.

6. Remove the No. 1 center bolt.

7. Support the crossmember with a jack and remove the crossmember bracket.

8. Remove the front lower arm (rear).

To install:

9. Installation is the reverse of removal. refer to the illustration for bolt torque specifications.

93.1—116.6
{9.50—11.88,
68.67—85.99}

166.6—200.0
{16.99—20.39,
122.9—147.5}

43.1—54.9
{4.40—5.59,
21.8—40.4}

93.1—126.4
{9.50—12.88,
68.67—93.22}

119.6—154.8
{12.20—15.78,
88.22—114.1}

93.1—116.6
{9.50—11.88,
68.67—85.99}

N·m {kgf·m, ft·lbf}

1	Nut (stabilizer control link lower side)	5	Front lower arm (rear)
2	Front lower arm (rear) ball joint	6	Clip
3	No.1 engine mount center bolt	7	Dust boot
4	Crossmember bracket	8	Front lower arm (rear) bushing

67162-MAZC-G251

Exploded view of the front suspension lower control arm (rear) assembly—Mazda6 and Mazdaspeed6 models

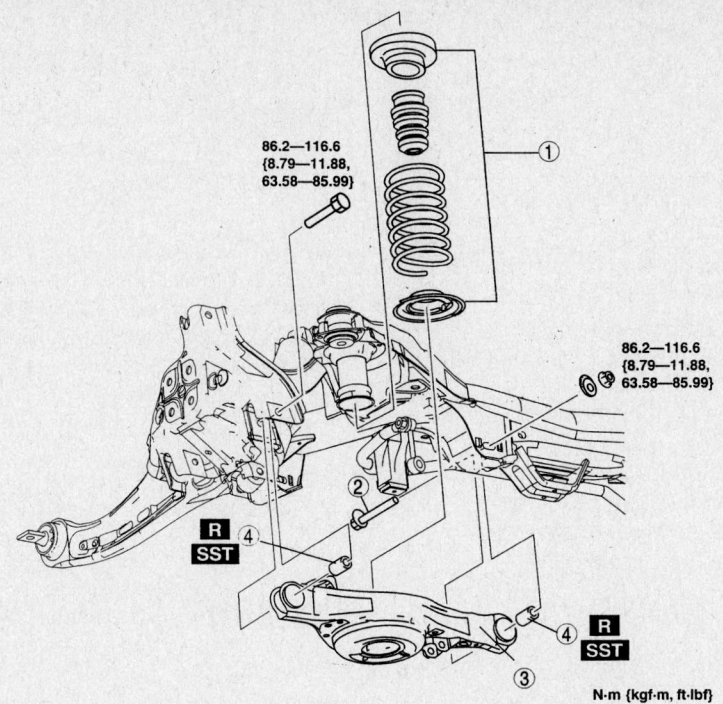

86.2—116.6
{8.79—11.88,
63.58—85.99}

86.2—116.6
{8.79—11.88,
63.58—85.99}

N·m {kgf·m, ft·lbf}

| 1 | Rear coil spring component | 3 | Rear lower arm |
| 2 | Bolt (rear lower arm inner side) | 4 | Rear lower arm bushing |

67162-MAZC-G249

Exploded view of the rear suspension lower control arm assembly—Mazda6 and Mazdaspeed6 models

CONTROL ARM BUSHING REPLACEMENT

MX-5 Miata Models

1. Before servicing the vehicle, refer to the precautions section.
2. Remove the lower control arm.
3. Remove the clip.
4. Remove the dust boot.

➡**Be careful not to damage the front lower arm. If it is damaged, replace it.**

5. Mark the front upper arm as shown in the illustration.
6. Remove the rear side bushing using the tools illustrated
7. Remove the front side bushing by cutting off the stopper plate rubber using a razor.

8. Cut off 5–6mm each side of the knob end of the front side bushing using a hacksaw.
9. Remove the bushing using the tools illustrated.

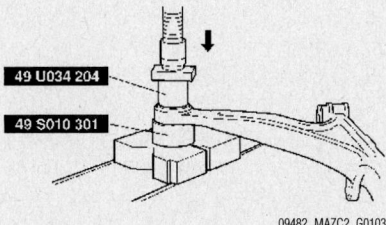

Remove the rear side bushing—MX-5 Miata

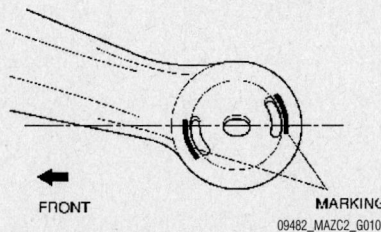

FRONT   MARKING
09482_MAZC2_G0102
Mark the front upper arm as shown—MX-5 Miata

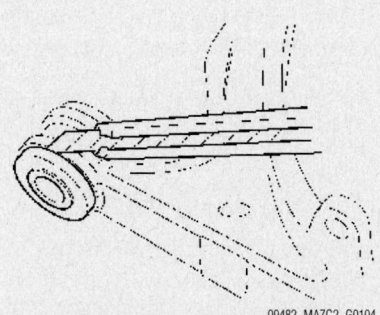

09482_MAZC2_G0104
Cut off the stopper plate rubber using a razor—MX-5 Miata

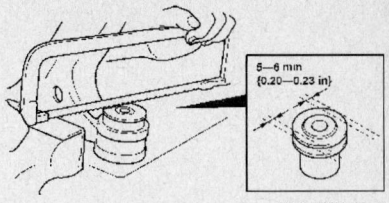

09482_MAZC2_G0105
Cut off 5–6mm each side of the knob end of the front side bushing—MX-5 Miata

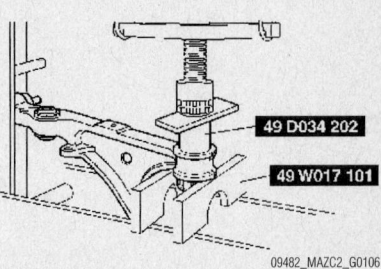

09482_MAZC2_G0106
Remove the front side bushing—MX-5 Miata

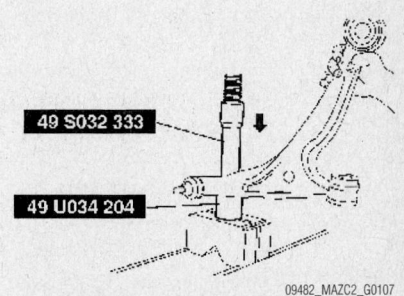

09482_MAZC2_G0107
Remove the shock absorber lower side bushing—MX-5 Miata

10. Remove the shock absorber lower side bushing using the tools illustrated.
To install:
11. Install the shock absorber lower side bushing by compressing the bushing using the tools illustrated.
12. Install the front side bushing as follows:

a. Press the bushing in using the tools illustrated.

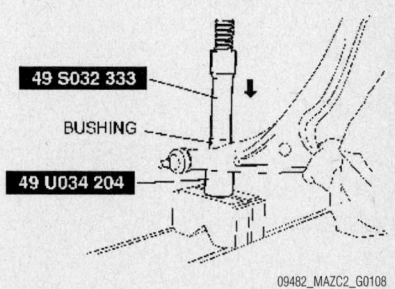

09482_MAZC2_G0108
Install the shock absorber lower side bushing by compressing the bushing as shown—MX-5 Miata

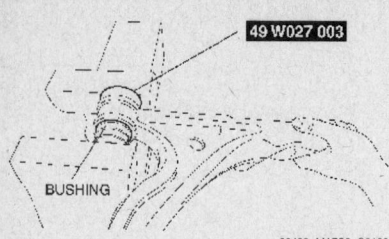

Press the front side bushing in—MX-5 Miata

09482_MAZC2_G0109

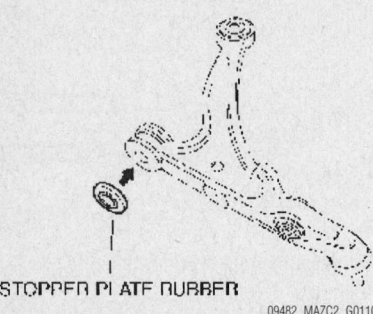

STOPPER PLATE RUBBER

09482_MAZC2_G0110

Insert the stopper plate rubber into the inner pipe of the bushing—MX-5 Miata

  b. Insert the stopper plate rubber into the inner pipe of the bushing (front side).
13. Install the rear side bushing as follows:
  a. Align the marks placed during removal of the bushing.
  b. Press the bushing in using the tools illustrated.
14. Install the clip as follows:
  a. Wipe the grease off the ball joint stud.
  b. Fill the inside of the new dust boot with grease.

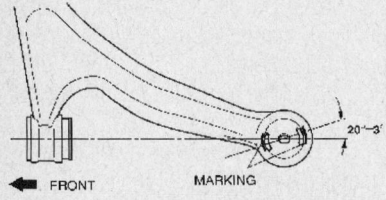

FRONT   MARKING   20°~3°

09482_MAZC2_G0111

Align the marks placed during removal of the rear side bushing—MX-5 Miata

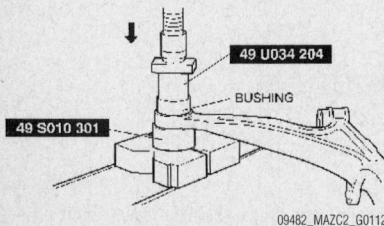

49 U034 204
BUSHING
49 S010 301

09482_MAZC2_G0112

Press in the rear side bushing—MX-5 Miata

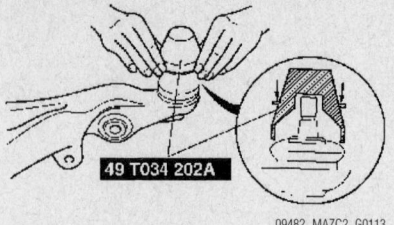

49 T034 202A

09482_MAZC2_G0113

Install the clip using the tools shown—MX-5 Miata

  c. Install the dust boot to the ball joint.
  d. Install the clip using the tools illustrated.
  e. Verify that the clip is installed securely to the groove.
  f. Wipe off any excess grease.
15. Install the control arm.

Mazda3 Models

1. Before servicing the vehicle, refer to the precautions section.
  All Mazda's use a pressed in control arm bushing, and the pressing can usually be done using 2 appropriately sized sockets (a press socket and a catch socket) and a large vise.
2. Position the control arm and the 2 sockets into a vise.
3. Position the press socket onto the control arm bushing.
4. Position the catch socket onto the control arm, opposite of the press socket.
5. Tighten the vise slowly and press the bushing into the catch socket.
  To install:
6. Apply soapy water to the new control arm bushing.
7. Position the bushing against the control arm.
8. Using the same sockets, in the same positions, press the new bushing into the control arm.

Mazda6

  The Mazda6 models have two lower control arms in the front suspension assembly, they are identified as the front lower control arm (Front side) and front lower control arm (rear side). They also have one rear suspension lower control arm, which will be identified as Rear lower control arm.

FRONT ARM (FRONT SIDE)

1. Before servicing the vehicle, refer to the precautions section.
2. Remove the front lower control arm (front).
3. Press the bushing from the front

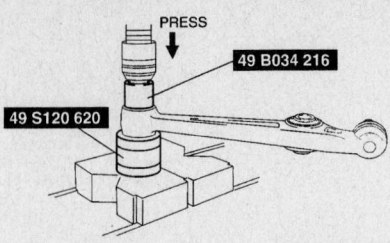

PRESS
49 B034 216
49 S120 620

67162-MAZC-G247

Press the bushing from/into the front lower control arm (front) inner side using the removal/installer tools shown—Mazda6 and Mazdaspeed6 models

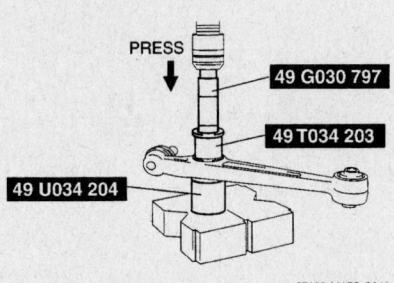

PRESS
49 G030 797
49 T034 203
49 U034 204

67162-MAZC-G248

Press the bushing from/into the front lower control arm (front) outer side using the removal/installer tools shown—Mazda6 and Mazdaspeed6 models

lower control arm (front) inner side using the tools illustrated.
4. Press the bushing from the front lower control arm (front) outer side using the tools illustrated.
  To install:
5. Mark the front bushing for the outer side on the front lower arm (front) as illustrated.
6. Press the bushing into the arm up to the marking made on the bushing using the tools and press used to remove the bushing. Make sure the clearance is 0.3445–0.4234 inch (8.75–10.75mm) as illustrated.
7. Mark the front bushing for the inner side on the front lower arm (front) as illustrated.
8. Press the bushing into the arm up to

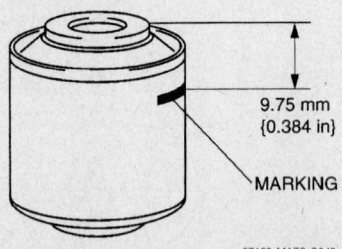

9.75 mm
{0.384 in}
MARKING

67162-MAZC-G245

Mark the front bushing for the outer side on the front lower arm (front) as shown—Mazda6 and Mazdaspeed6 models

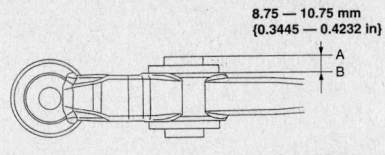

67162-MAZC-G254

Make sure the clearance is 0.3445–0.4234 inch (8.75–10.75mm) when installing the front bushing for the outer side on the front lower arm (front)—Mazda6 and Mazda-speed6 models

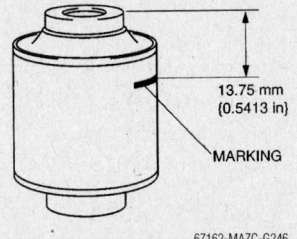

67162-MAZC-G246

Mark the front bushing for the inner side on the front lower arm (front) as shown—Mazda6 and Mazdaspeed6 models

the marking made on the bushing using the tools and press used to remove the outer side bushing.

FRONT ARM (REAR SIDE)

1. Before servicing the vehicle, refer to the precautions section.
2. Remove the front lower control arm (rear).
3. Press the bushing from the front lower control arm (rear) using the tools illustrated. remove the arm from the press and use a hammer to remove the bushing completely.

To install:
4. Mark the rear bushing on the front lower arm (rear) as illustrated.
5. Press the bushing into the arm up to the marking made on the bushing using the tools and press used to remove the bushing.
6. The clearance should be 0.936–1.013

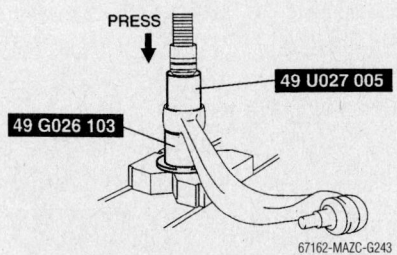

67162-MAZC-G243

Press the bushing from/into the front lower control arm (rear) using the removal/installer tools shown—Mazda6 and Mazdaspeed6 models

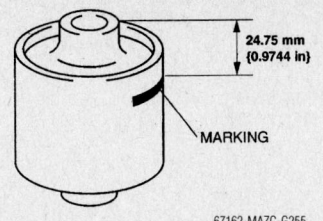

67162-MAZC-G255

Mark the rear bushing on the front lower arm (rear) as shown—Mazda6 and Mazda-speed6 models

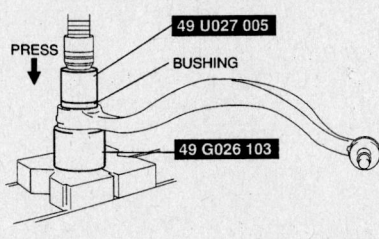

67162-MAZC-G256

Install the rear bushing on the front lower arm (rear) as shown—Mazda6 and Mazda-speed6 models

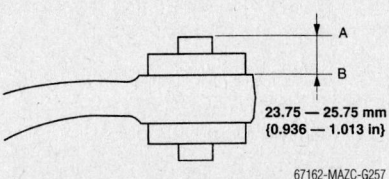

67162-MAZC-G257

The rear bushing on the front lower arm (rear) should be 0.936–1.013 inch (23–25mm) once the bushing is seated properly—Mazda6 and Mazdaspeed6 models

inch (23–25mm) as illustrated once the bushing is seated properly.

REAR LOWER CONTROL ARM

1. Before servicing the vehicle, refer to the precautions section.
2. Remove the lower control arm.
3. Press the bushing from the arm using the tools illustrated.

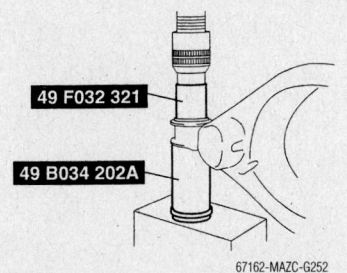

67162-MAZC-G252

Press the bushing from the rear suspension lower control arm using the tools shown—Mazda6 and Mazdaspeed6 models

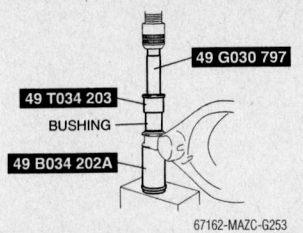

67162-MAZC-G253

Press the bushing into the rear suspension lower control arm using the tools shown—Mazda6 and Mazdaspeed6 models

To install:
4. Press the bushing into the arm using the tools illustrated.
5. Install the lower control arm.

Rear Trailing Links

REMOVAL & INSTALLATION

MX-5 Miata

1. Before servicing the vehicle, refer to the precautions section.

➡**Refer the exploded view illustration for component locations and if applicable, their retainer torque specifications.**

2. Separate the rear trailing link (upper) ball joint from the knuckle.

➡**When removing the rear trailing link (upper) ball joint, the rear knuckle bushing may also come off. If it comes off, replace the rear knuckle.**

3. Remove the rear upper trailing link.
4. Remove the dust boot.

To install:
5. Wipe the grease off the ball joint stud.
6. Fill the inside of the new dust boot with grease.
7. Using the tool shown, install the dust boot to the ball joint.
8. Wipe off the excess grease.

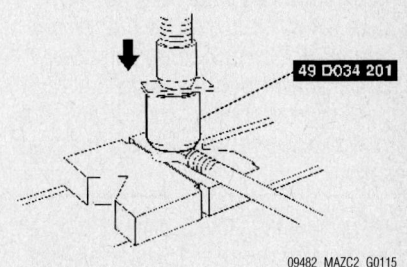

09482_MAZC2_G0115

Install the dust boot to the ball joint—MX-5 Miata

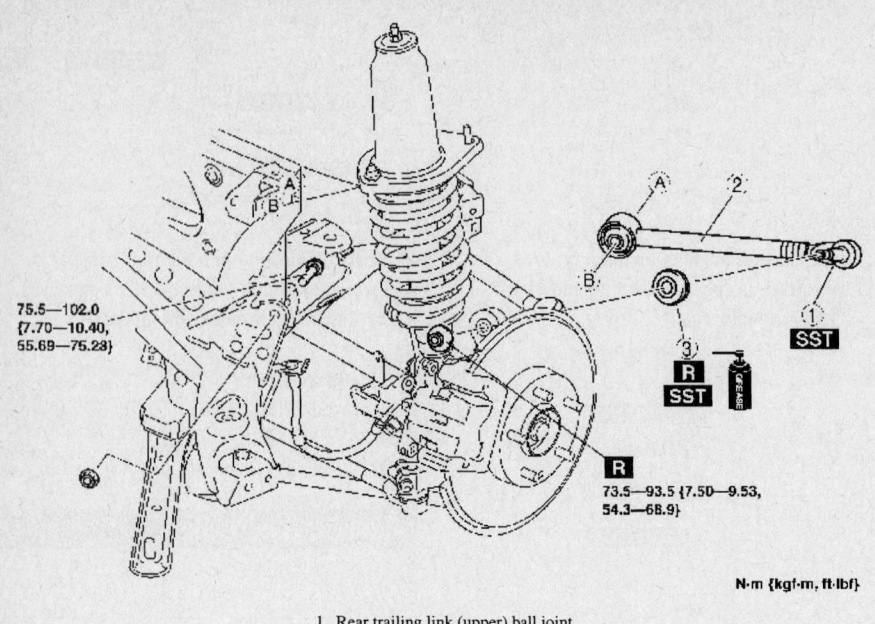

75.5—102.0
{7.70—10.40,
55.69—75.23}

73.5—93.5 {7.50—9.53,
54.3—68.9}

N·m {kgf·m, ft·lbf}

1 Rear trailing link (upper) ball joint
2 Rear trailing link (upper)
3 Dust boot

09482_MAZC2_G0114

Exploded view of rear trailing link assembly—MX-5 Miata

9. Install the remaining components in the reverse order of removal using the accompanying illustration for torque values.

10. Inspect the rear wheel alignment.

Rear Lateral Links

REMOVAL & INSTALLATION

MX-5 Miata

1. Before servicing the vehicle, refer to the precautions section.

➡**Refer the exploded view illustration for component locations and if applicable, their retainer torque specifications.**

✳✳ CAUTION

Performing the following procedures without first removing the ABS wheel-speed sensor may possibly cause an open circuit in the wiring harness if it is pulled by mistake. Before operations, remove the ABS wheel-speed sensor (axle side) and move the sensor away from the harnesses.

2. Using the tools illustrated, disconnect the rear lateral link (upper) ball joint.

➡**When removing the rear lateral link (upper) ball joint, the rear knuckle**

bushing may also come off. If it comes **off, replace the rear knuckle.**

3. Remove the rear lateral link (upper).

4. Remove the dust boot.

To install:

5. Wipe the grease off the ball joint stud.

6. Fill the inside of the new dust boot with grease.

7. Using the tool shown, install the dust boot to the ball joint.

8. Wipe off the excess grease.

9. Install the remaining components in the reverse order of removal using the accompanying illustration for torque values.

10. Inspect the rear wheel alignment.

Wheel Bearings

ADJUSTMENT

The front and rear wheel bearings are not adjustable. If the bearings become loose or make noise, they must be replaced.

REMOVAL & INSTALLATION

MX-5 Miata

FRONT

1. Before servicing the vehicle, refer to the precautions section.

➡**Refer the exploded view illustration for component locations and if applicable, their retainer torque specifications.**

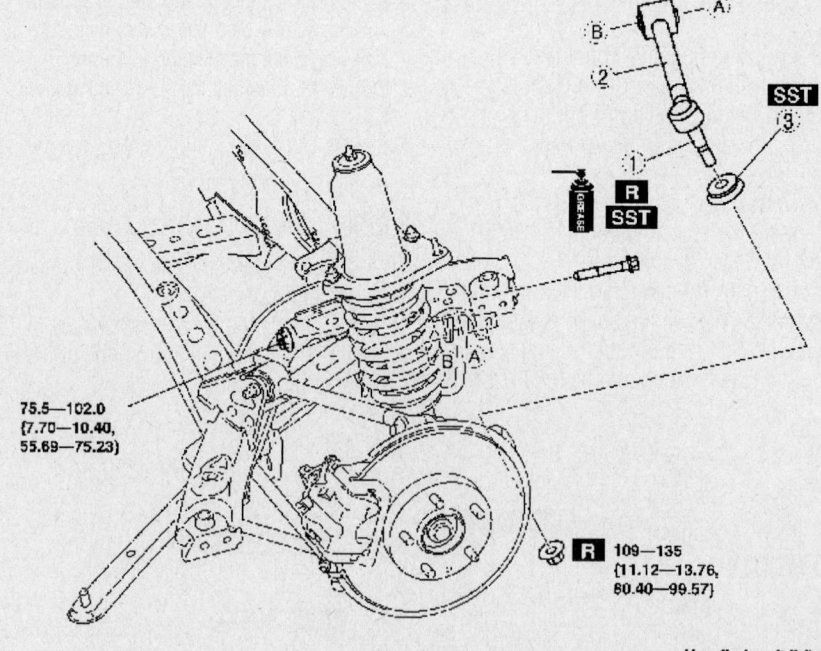

75.5—102.0
{7.70—10.40,
55.69—75.23}

109—135
{11.12—13.76,
80.40—99.57}

N·m {kgf·m, ft·lbf}

1 Rear lateral link (upper) ball joint
2 Rear lateral link (upper)
3 Dust boot

09482_MAZC2_G0116

Exploded view of rear lateral link assembly—MX-5 Miata

✳✳ CAUTION

Performing the following procedures without first removing the ABS wheel-speed sensor may possibly cause an open circuit in the wiring harness if it is pulled by mistake. Before operations, remove the ABS wheel-speed sensor (axle side) and move the sensor away from the harnesses.

---

2. Disconnect the ABS wheel-speed sensor.

3. Remove the caliper with the hose still attached and support the assembly with wire so as to not place strain on the hose.

4. Remove the rotor.

5. Disconnect the tie-rod end from the knuckle.

6. Disconnect the stabilizer control link nut (lower).

7. Disconnect the front upper arm ball joint from the knuckle.

8. Remove the front upper arm bolt.

9. Disconnect the front lower arm ball joint from the knuckle.

10. Remove the Steering knuckle assembly

11. Remove the dust cover.

12. Remove the wheel hub bolts from the wheel hub using a press.

13. Remove the hub assembly.

To install:

14. Install the hub assembly. tighten the bolts to 54–60 Nm (40–44 ft. lbs.).

15. Press in new wheel hub bolts into the wheel hub using a press.

16. Install the remaining components in the reverse order of removal using the accompanying illustration for torque values.

17. Inspect the wheel alignment.

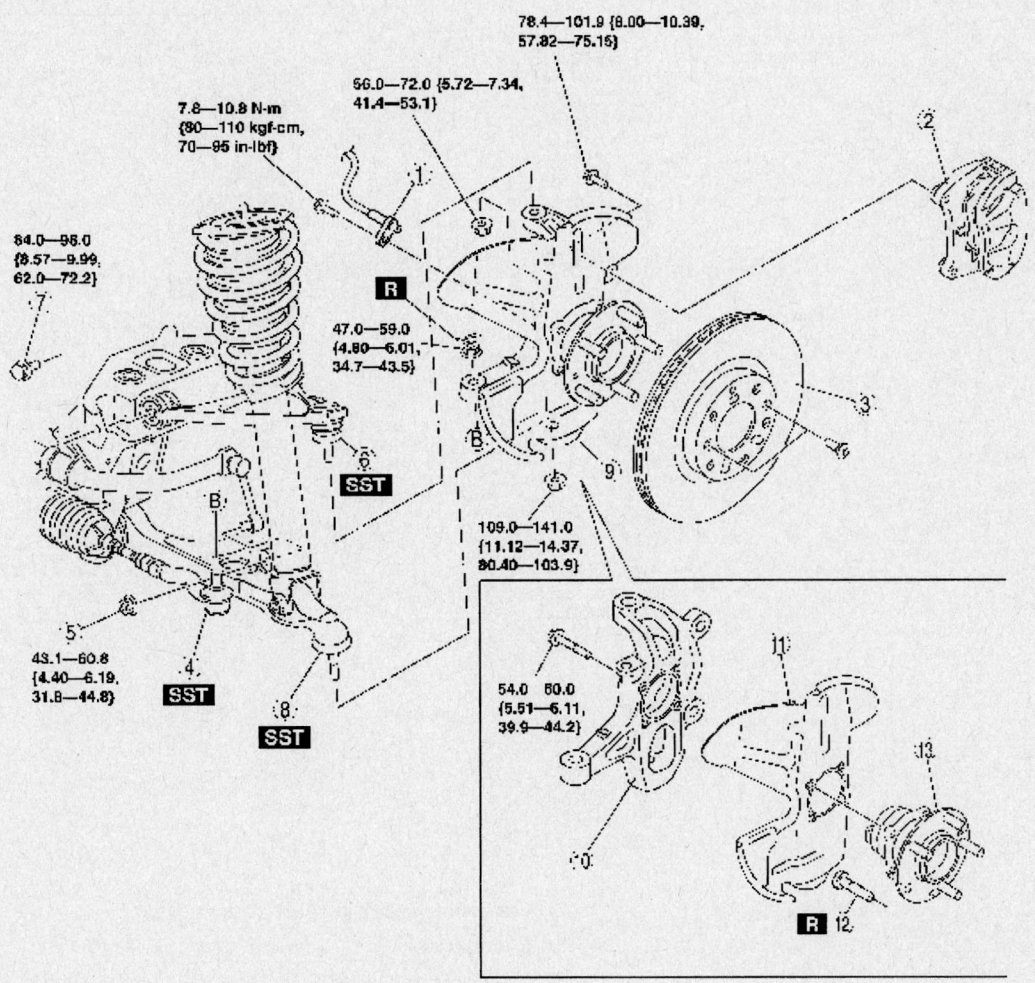

1	ABS wheel-speed sensor	8	Front lower arm ball joint
2	Brake caliper component	9	Steering knuckle component
3	Disc plate	10	Steering knuckle
4	Tie-rod end	11	Dust cover
5	Stabilizer control link nut (lower)	12	Wheel hub bolt
6	Front upper arm ball joint	13	Wheel hub component
7	Front upper arm bolt		

09482_MAZC2_G0117

Exploded view of front hub and knuckle assembly—MX-5 Miata

REAR

1. Before servicing the vehicle, refer to the precautions section.

→ Refer the exploded view illustration for component locations and if applicable, their retainer torque specifications.

✳✳ CAUTION

Performing the following procedures without first removing the ABS wheel-speed sensor may possibly cause an open circuit in the wiring

harness if it is pulled by mistake. Before operations, remove the ABS wheel-speed sensor (axle side) and move the sensor away from the harnesses.

2. Disconnect the ABS wheel-speed sensor.
3. Lock the disc plate by applying the brakes.
4. Knock the crimped portion of the locknut outward using a chisel and a hammer.
5. Remove the locknut.
6. Disconnect the parking brake cable.

7. Remove the caliper with the hose still attached and support the assembly with wire so as to not place strain on the hose.
8. Remove the rotor.
9. Disconnect the rear lateral link (upper) ball joint from the knuckle.
10. Remove the stabilizer control link lower nut.
11. Disconnect the rear lateral link lower ball joint from the knuckle.
12. Remove the lower shock absorber bolt.
13. Disconnect the toe control link ball joint.

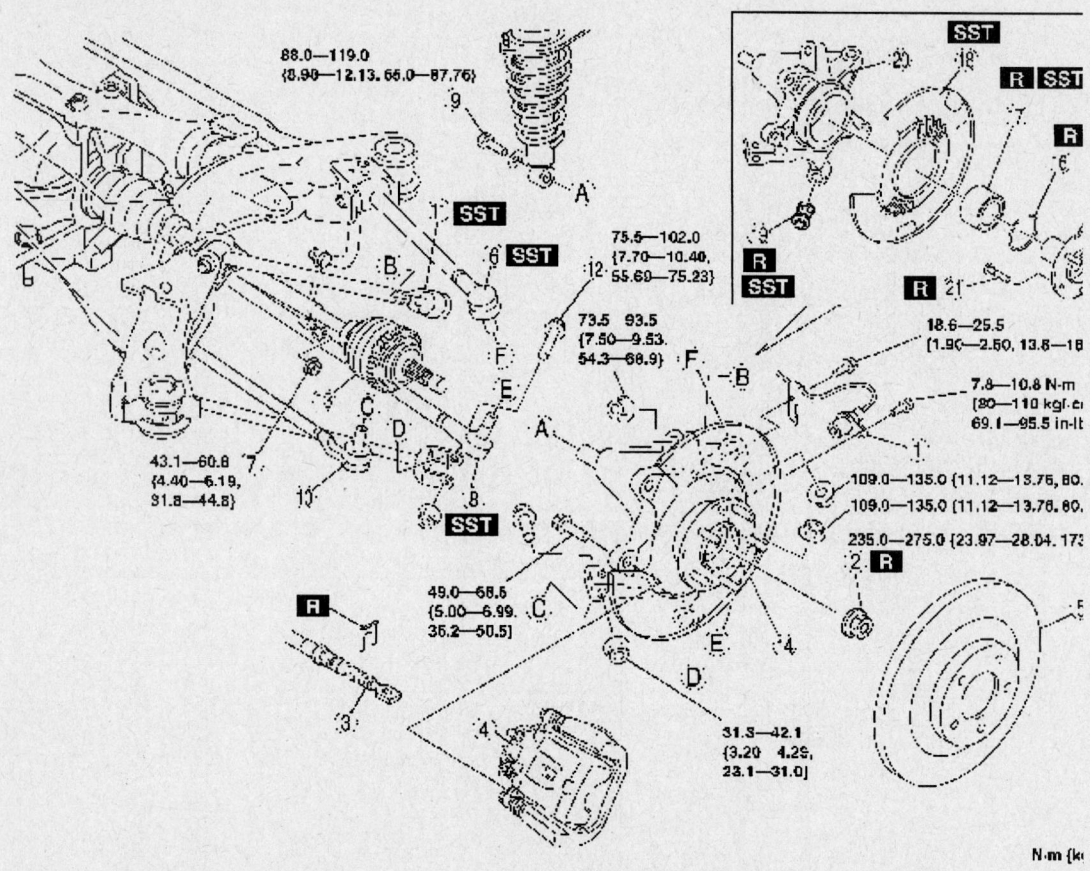

1	ABS wheel-speed sensor	12	Rear trailing link (lower) bolt (outer side)
2	Locknut	13	Rear drive shaft
3	Parking brake cable	14	Rear knuckle component
4	Brake caliper component	15	Wheel hub component
5	Disc plate	16	Retaining ring
6	Rear lateral link (upper) ball joint	17	Wheel bearing
7	Stabilizer control link nut (lower)	18	Dust cover
8	Rear lateral link (lower) ball joint	19	Bushing
9	Shock absorber bolt (lower)	20	Rear knuckle
10	Toe control link ball joint	21	Wheel hub bolt
11	Rear trailing link (upper) ball joint		

Exploded view of rear hub and knuckle assembly and related components—MX-5 Miata

09482_MAZC2_G0118

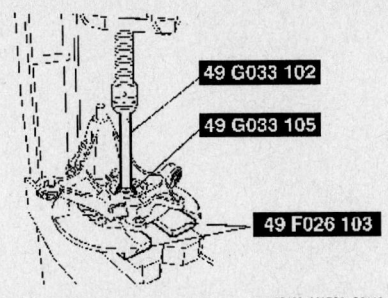

Remove the rear wheel hub component—MX-5 Miata

09482_MAZC2_G0119

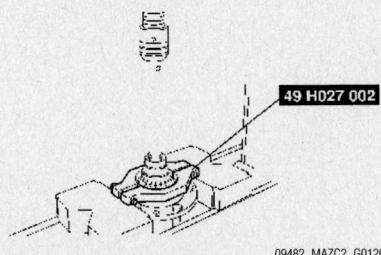

Remove the rear bearing inner race—MX-5 Miata

09482_MAZC2_G0120

14. Disconnect the rear trailing link (upper) ball joint.

15. Remove the rear lower trailing link bolt on the outer side.

16. Temporarily install a spare nut onto the end of the rear drive shaft.

17. Tap the nut with a copper hammer to loosen the drive shaft from the wheel hub.

18. Separate the rear drive shaft from the wheel hub.

19. Remove the rear knuckle component.

20. Remove the wheel Hub Component as follows:

   a. Wind the tool illustrated and backing plate contact area with packing tape two times.

   b. Remove the wheel hub component using the tools illustrated.

   c. If the bearing inner race remains on the wheel hub component, use a

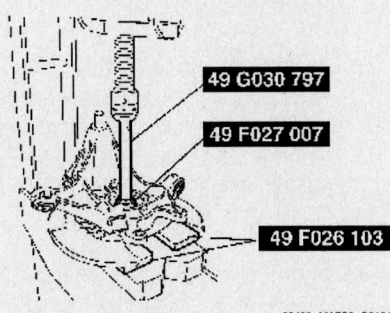

Remove the wheel bearing from the knuckle—MX-5 Miata

09482_MAZC2_G0121

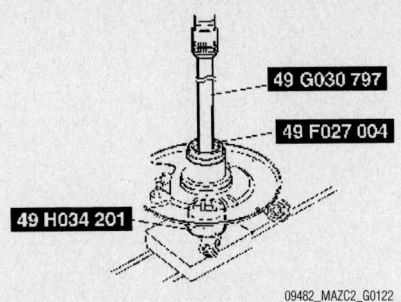

Install a new rear wheel bearing—MX-5 Miata

09482_MAZC2_G0122

chisel to secure a sufficient space for installing the service tool between wheel hub component and bearing inner race.

21. Remove the bearing inner race using the tool illustrated.

To install:

22. Install a new wheel bearing using the tool illustrated.

23. Install the wheel hub component using the tool illustrated.

24. Install the remaining components in the reverse order of removal using the accompanying illustration for torque values.

25. Inspect the wheel alignment.

Mazda3

FRONT

1. Before servicing the vehicle, refer to the precautions section.

2. Refer to the illustration for component location and torque specifications.

3. Remove or disconnect the following:

- Wheels
- ABS sensor connector and the sensor
- Halfshaft lockbolt
- Brake caliper and rotor
- Tie rod end from the knuckle

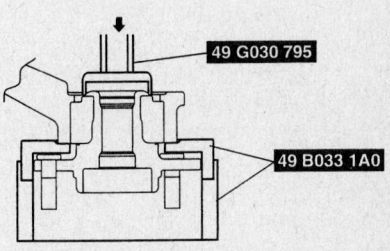

67162-MAZC-G100

Use a press and the tools illustrated to disassemble the hub/bearing assembly—Mazda3

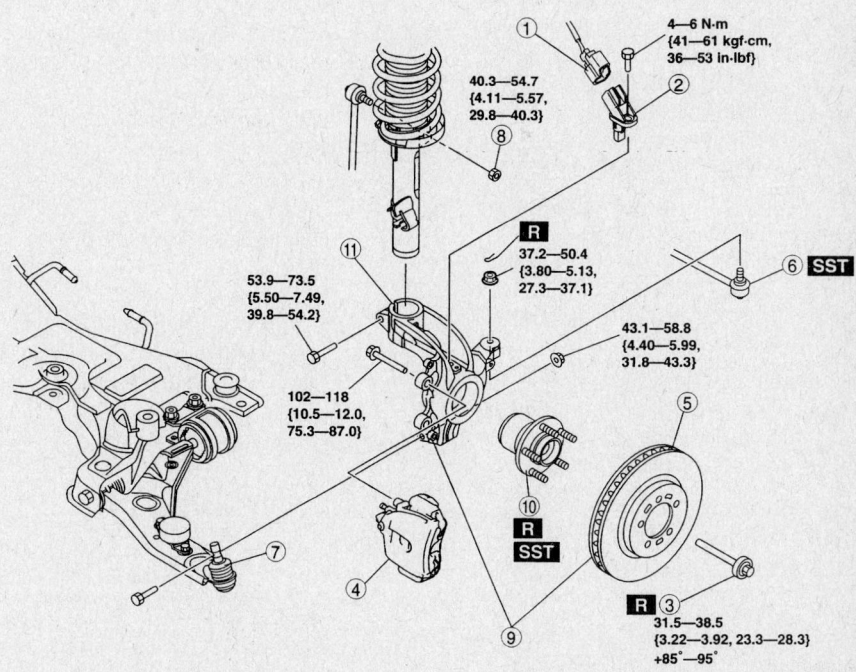

1	ABS wheel-speed sensor connector	
2	ABS wheel-speed sensor	
3	Lockbolt	
4	Brake caliper component	
5	Disc plate	
6	Tie-rod end ball joint	
7	Front lower arm ball joint	
8	Stabilizer control link upper nut	
9	Wheel hub, steering knuckle component	
10	Wheel hub component	
11	Steering knuckle	

Exploded view of the front wheel bearing and knuckle assembly—Mazda3

67162-MAZC-G99

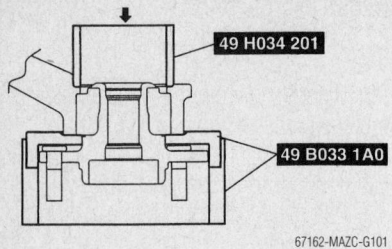

67162-MAZC-G101

Use a press and the tools illustrated to assemble the hub/bearing assembly—Mazda3

- Lower control arm ball joint from the knuckle
- Stabilizer bar link nut
- Knuckle assembly
- Hub bearing using a press and Mazda tools 49 G030 795 and 49 B033 1A0

4. Clean and inspect all parts but do not wash or clean the wheel bearing. The bearing must be replaced.

To install:

5. Using Mazda Press tools 49 H034 201 and 49 B033 1A0, press a new wheel bearing into the knuckle assembly. Make sure the installation tool engages to the bearing outer race properly to avoid damage.

6. Install or connect the following:

- Press the hub onto the knuckle
- Knuckle assembly, tighten the bolts to the specifications shown in the accompanying illustration
- Control arm ball joint to the knuckle tighten, the bolts to the specifications shown in the accompanying illustration
- Tie rod end to the knuckle, tighten the bolts to the specifications shown in the accompanying illustration
- Brake rotor and caliper
- New halfshaft lockbolt and tighten to 23–28 ft. lbs. (31–38 Nm), plus an additional 85–95 degrees
- ABS sensor and connector
- Wheels

REAR

1. Before servicing the vehicle, refer to the precautions section.

2. Refer to the illustration for component location and torque specifications.

➡ **The wheel bearings are not serviceable. If the bearings are bad, a new hub/bearing assembly must be installed.**

3. Remove or disconnect the following:

- Rear wheels
- Wheel speed sensor connector and sensor

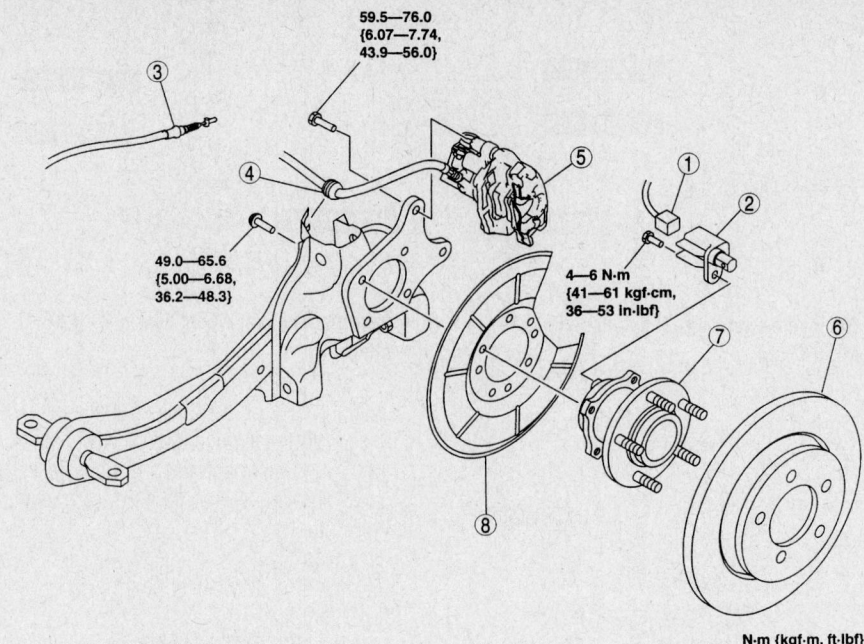

59.5—76.0
{6.07—7.74,
43.9—56.0}

49.0—65.6
{5.00—6.68,
36.2—48.3}

4—6 N·m
{41—61 kgf·cm,
36—53 in·lbf}

N·m {kgf·m, ft·lbf}

1	ABS wheel-speed sensor connector	5	Brake caliper component
2	ABS wheel-speed sensor	6	Disc plate
3	Rear parking brake cable	7	Wheel hub component
4	Brake hose	8	Dust cover

67162-MAZC-G102

Exploded view of the rear wheel bearing assembly and related components—Mazda3

- Rear parking brake cable
- Brake hose grommet and move the hose aside
- Brake caliper and rotor

To install:

4. Install or connect the following:

- Hub/bearing assembly. Torque the new nut to 36–48 ft. lbs. (49–65 Nm).
- Brake assembly
- Rear parking brake cable. Pass the cable inside the rear wheel speed sensor wiring harness as illustrated.
- Wheel speed sensor and connector
- Rear wheel

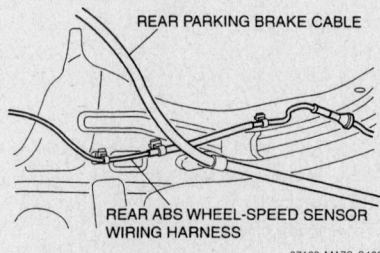

REAR PARKING BRAKE CABLE

REAR ABS WHEEL-SPEED SENSOR WIRING HARNESS

67162-MAZC-G103

Pass the cable inside the rear wheel speed sensor wiring harness when installing—Mazda3

Mazda6 and Mazdaspeed6

FRONT

1. Before servicing the vehicle, refer to the precautions section.

2. Refer to the illustration for component location and torque specifications.

3. Remove or disconnect the following:

- Wheels
- ABS sensor
- Halfshaft axle nut, unstake the nut prior to removal.
- Brake caliper and rotor
- Tie rod end from the knuckle
- Damper fork–to–control arm bolt
- Front lower control arm ball joints
- Front upper arm ball joint
- Wheel hub dust cover
- Hub bolts and the hub
- Hub using a press and Mazda tools 49 F026 10, 49 G033 102 and 49 G033 105. If the bearing inner race remains in the hub, grind a section of the bearing inner race until 0.02 inch (0.5mm) remains and use a chisel to remove it.
- Snap ring
- Bearing from the hub using a press and the tools illustrated
- Brake dust shield, if it is being replaced. Mark the cover and knuckle

for replacement purposes and use a chisel to remove the shield.

4. Clean and inspect all parts but do not wash or clean the wheel bearing. The bearing must be replaced.

To install:

5. Using the tools illustrated, install a new dust shield cover assembly to the knuckle, if removed.

6. Using the tools illustrated, press a new wheel bearing into the knuckle assembly and install the snap ring.

7. Install or connect the following:

- Wheel bearing retaining ring

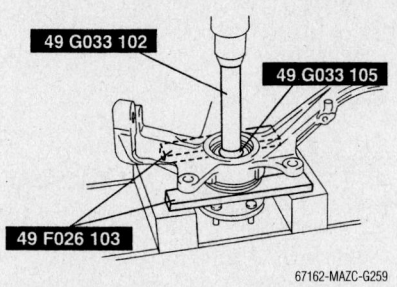

Use a press to remove the hub—Mazda6 and Mazdaspeed6 models

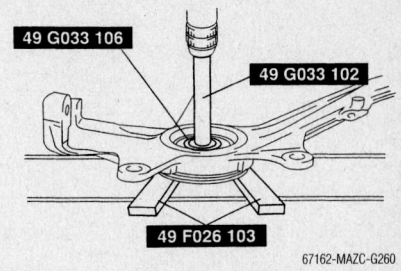

Use a press to remove the wheel bearing—Mazda6 and Mazdaspeed6 models

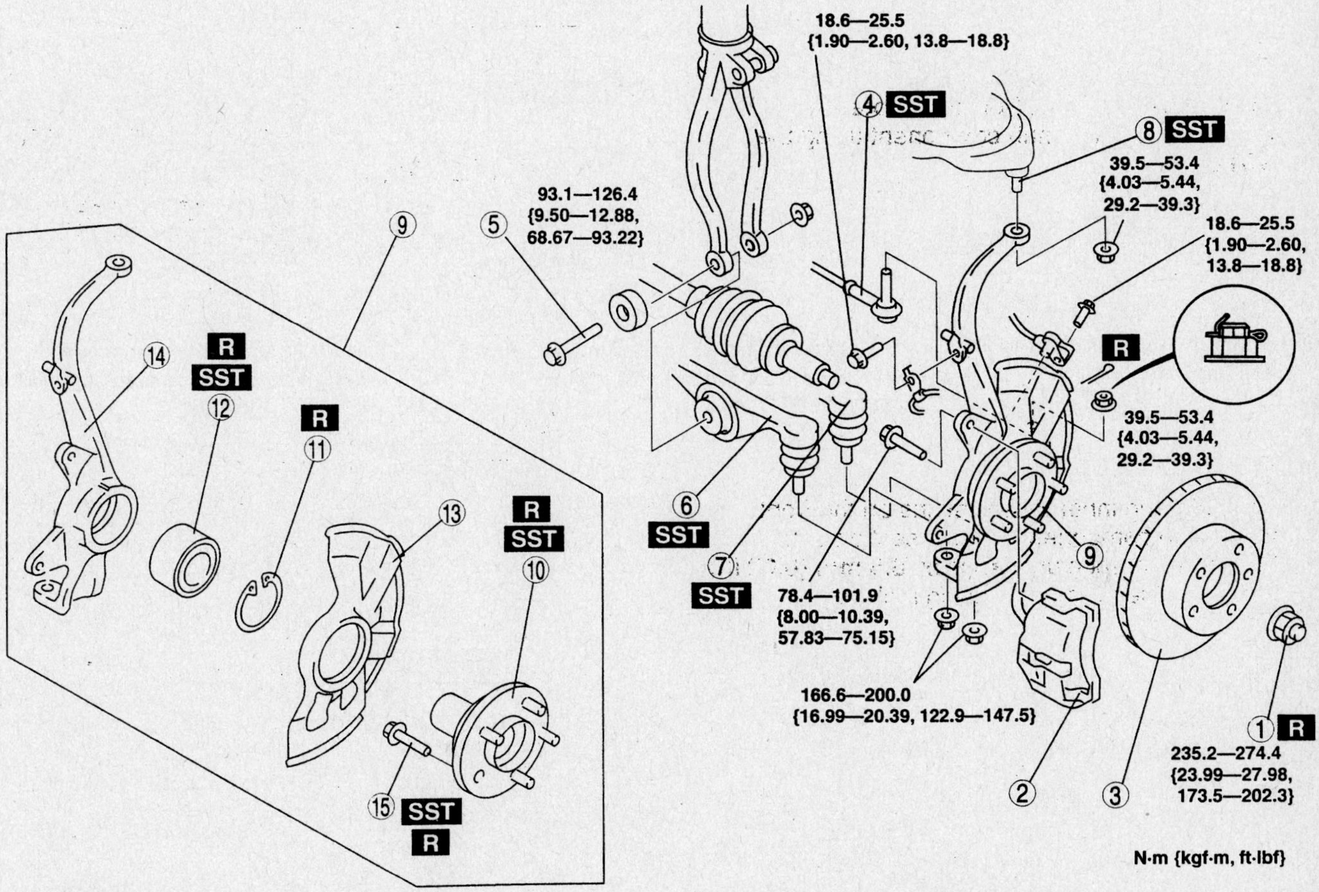

N·m {kgf·m, ft·lbf}

#	Part	#	Part
1	Locknut	9	Wheel hub, steering knuckle, dust cover
2	Brake caliper component	10	Wheel hub component
3	Disc plate	11	Retaining ring
4	Tie-rod end ball joint	12	Wheel bearing
5	Bolt	13	Dust cover
6	Front lower arm (front) ball joint	14	Steering knuckle
7	Front lower arm (rear) ball joint	15	Hub bolt
8	Front upper arm ball joint		

Exploded view of the front wheel bearing and knuckle assembly—Mazda6 and Mazdaspeed6 models

8. Using the tools illustrated, press in the hub assembly.
- Wheel hub dust cover
- Front upper arm ball joint and tighten the nut to 29–39 ft. lbs. (39–53 Nm)
- Front lower arm ball joints and tighten the nuts to 122–147 ft. lbs. (166–200 Nm)
- Damper fork bolt and tighten to 68–93 ft. lbs. (93–126 Nm)
- Tie rod end to the knuckle and tighten the nut to 29–39 ft. lbs. (39–53 Nm) and install a new cotter pin
- Brake caliper and rotor
- Halfshaft axle nut, tighten the nut to

173–202 ft. lbs. (235–274 Nm) and stake the nut
- Wheels

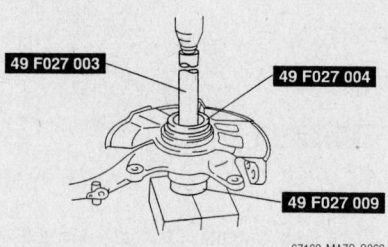

49 F027 003
49 F027 004
49 F027 009
67162-MAZC-G262

Install a new wheel bearing using a press—Mazda6 and Mazdaspeed6 models

REAR

1. Before servicing the vehicle, refer to the precautions section.
2. Refer to the illustration for component location and torque specifications.

➡ **The wheel bearings are not service-able. If the bearings are bad, a new hub/bearing assembly must be installed.**

3. Remove or disconnect the following:
- Hub cap
4. Raise the staked portion of the hub retaining nut with a hammer and chisel.
- Rear wheels
- ABS sensor

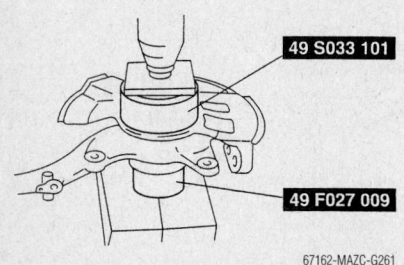

49 S033 101
49 F027 009
67162-MAZC-G261

Install the dust shield—Mazda6 and Mazdaspeed6 models

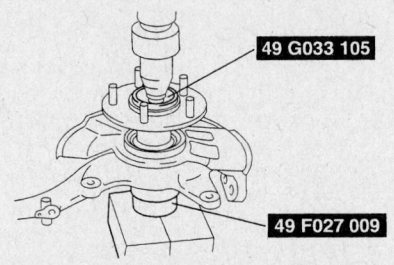

49 G033 105
49 F027 009
67162-MAZC-G263

Install a wheel hub using a press—Mazda6 and Mazdaspeed6 models

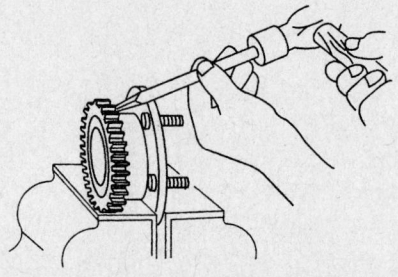

67162-MAZC-G265

Remove the ABS sensor rotor using a chisel—Mazda6 and Mazdaspeed6 models

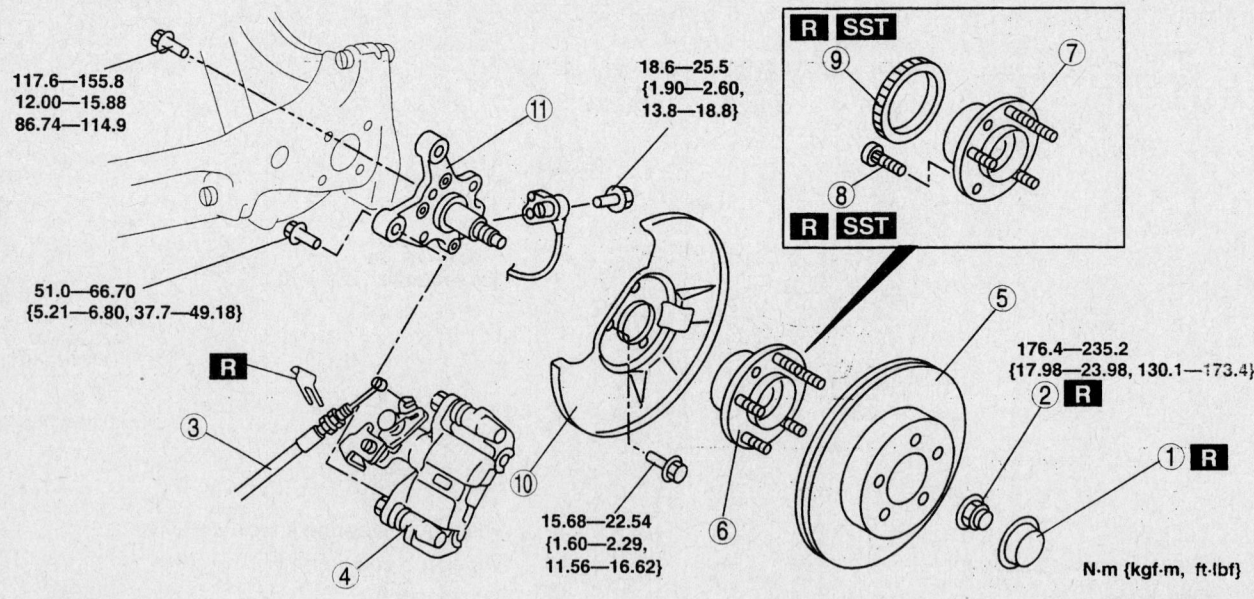

117.6—155.8
12.00—15.88
86.74—114.9

18.6—25.5
{1.90—2.60,
13.8—18.8}

51.0—66.70
{5.21—6.80, 37.7—49.18}

15.68—22.54
{1.60—2.29,
11.56—16.62}

176.4—235.2
{17.98—23.98, 130.1—173.4}

N·m {kgf·m, ft·lbf}

1	Hub cap	
2	Locknut	
3	Parking brake cable	
4	Brake caliper component	
5	Disc plate	
6	Wheel hub component	
7	Wheel hub	
8	Hub bolt	
9	ABS sensor rotor (with ABS)	
10	Dust cover	
11	Hub spindle	

Exploded view of the rear hub and spindle assembly—Mazda6 and Mazdaspeed6 models

67162-MAZC-G264

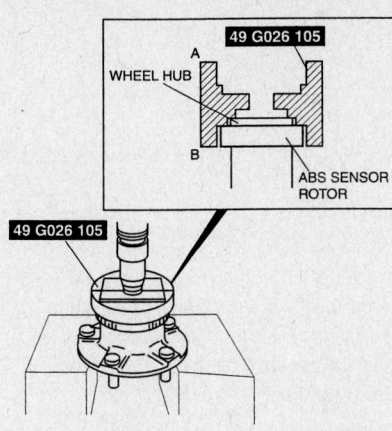

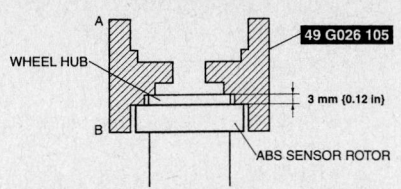

When installing the ABS sensor rotor, make sure there is a 0.12 inch (3mm) gap as shown B to the bottom—Mazda6 and Mazdaspeed6 models

- Parking brake cable
- Rear caliper and rotor assembly from the hub
- Hub retaining nut and discard it
- ABS sensor rotor using a chisel
- Hub bolts
- Hub and bearing assembly from the spindle

When using the installation tool for the ABS sensor rotor, face the carved side B to the bottom—Mazda6 and Mazdaspeed6 models

To install:
- Hub and bearing assembly on the spindle
- Hub bolts and tighten to the specifications shown in the illustration

❋❋ CAUTION

When using the installation tool for the ABS sensor rotor, face the carved side B to the rotor.

- ABS sensor rotor using a press and tools illustrated. Make sure there is a 0.12 inch (3mm) gap as illustrated
- Remaining components in the reverse order of removal, referring to the illustration for component location and torque specifications

BRAKES

Brake Caliper

REMOVAL & INSTALLATION

MX5-Miata

FRONT

1. Before servicing the vehicle, refer to the precautions section.

➡ **Refer the exploded view illustration for component locations and if applicable, their retainer torque specifications.**

2. Disconnect the brake hose.
3. Remove the caliper bolts and the caliper.
4. Remove the disc pads and shims.

To install:
5. Clean the exposed area of the piston.
6. Compress the piston using tool 49 0221 600C
7. Install the caliper. Tighten the bolts to 21–31 Nm (15–23 ft. lbs.).

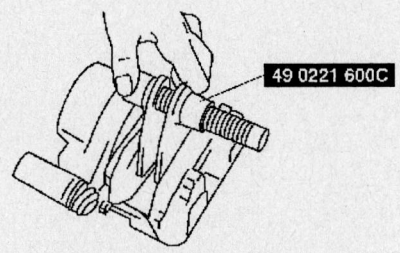

Compress the front caliper piston using tool 49 0221 600C—MX-5 Miata

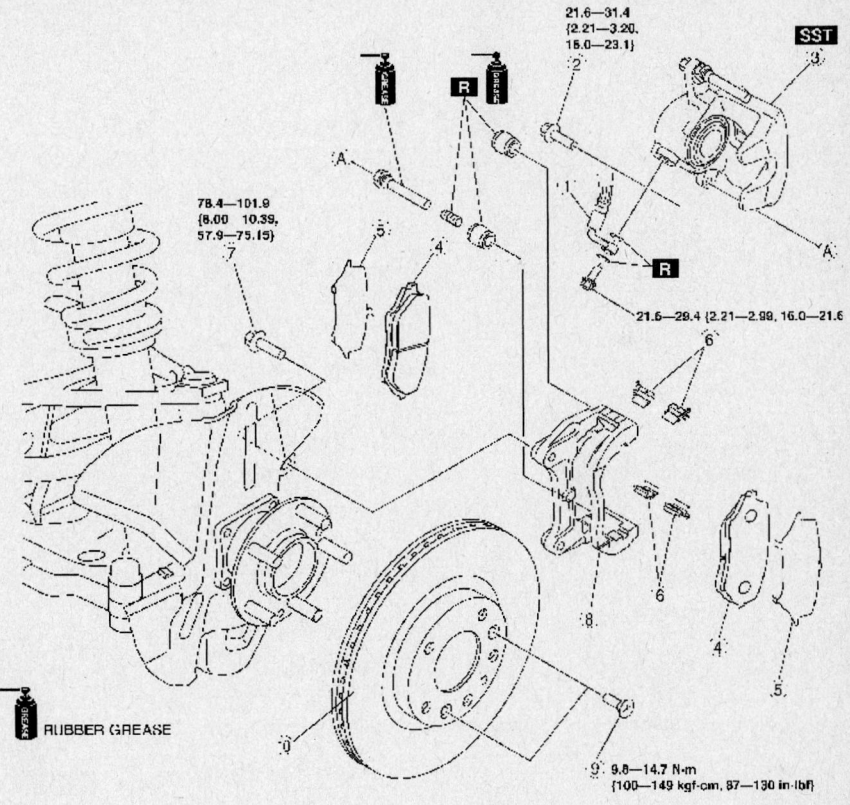

1	Brake hose	6	Guide plate
2	Bolt	7	Bolt
3	Caliper	8	Mounting support
4	Disc pad	9	Screw
5	Shim	10	Disc plate

Exploded view of the front disc brake assembly—MX-5 Miata

8. Connect the brake hose and tighten to 21–29 Nm (15–21 ft. lbs.).

9. Bleed the brake system.

10. Depress the brake pedal a few times and verify that the brakes do not drag

REAR

1. Before servicing the vehicle, refer to the precautions section.

➡**Refer the exploded view illustration for component locations and if applicable, their retainer torque specifications.**

2. Disconnect the parking brake cable.

3. Disconnect the brake hose.

4. Remove the caliper bolts and the caliper.

5. Remove the disc pads and shims.

To install:

6. Clean the exposed area of the piston.

7. Rotate the piston clockwise slowly using the tool shown to push the piston completely until the piston grooves are in the position shown in the figure.

8. Install the caliper. Tighten the bolts to 20–25 Nm (14–18 ft. lbs.).

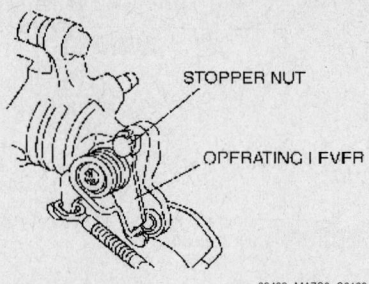

09482_MAZC2_G0128

After installing the parking brake cable, verify that the operating lever returns to the stopper nut with the parking brake lever released—MX-5 Miata

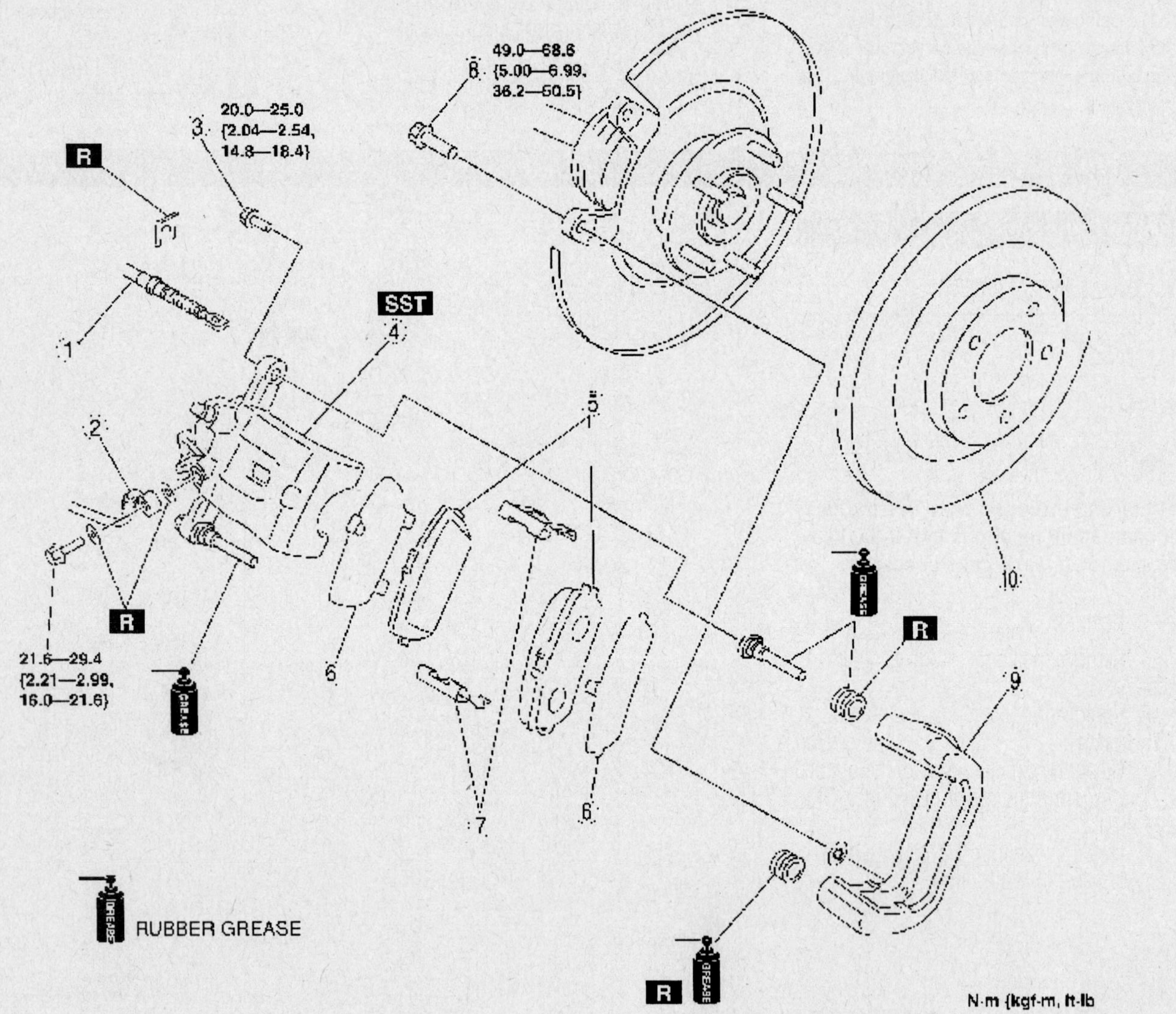

1	Parking brake cable	6	Shim
2	Brake hose	7	Guide plate
3	Bolt	8	Bolt
4	Caliper	9	Mounting support
5	Disc pad	10	Disc plate

N·m {kgf-m, ft-lb

09482_MAZC2_G0126

Exploded view of the rear disc brake assembly—MX-5 Miata

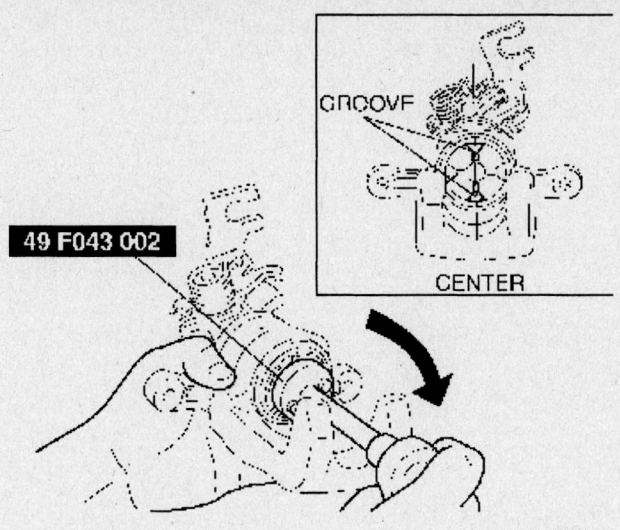

49 F043 002

GROOVE

CENTER

09482_MAZC2_G0127

Rotate the piston clockwise slowly using the tool shown—MX-5 Miata

9. Connect the brake hose and tighten to 21–29 Nm (15–21 ft. lbs.).
10. Bleed the brake system.
11. Install the parking brake cable. After installation the parking brake cable, verify that the operating lever returns to the stopper nut with the parking brake lever released.

12. Depress the brake pedal a few times and verify that the brakes do not drag

Mazda3

FRONT

1. Before servicing the vehicle, refer to the precautions section.

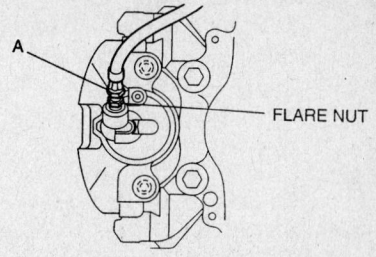

A

FLARE NUT

67162-MAZC-G105

Tighten the brake hose flare nut while holding the hose at point A—Mazda3

2. Remove or disconnect the following:
    • Wheels
    • Brake hose
    • Retaining clip
    • Cap from the caliper bolts
    • Caliper mounting bolts and the caliper

To install:

3. Install or connect the following:
    • Caliper. Torque the caliper mounting bolts to 19–22 ft. lbs. (25–30 Nm). Install the bolt caps.
    • Brake hose to the caliper. Tighten the flare nut while holding the hose at location A shown in the accompanying illustration with an open

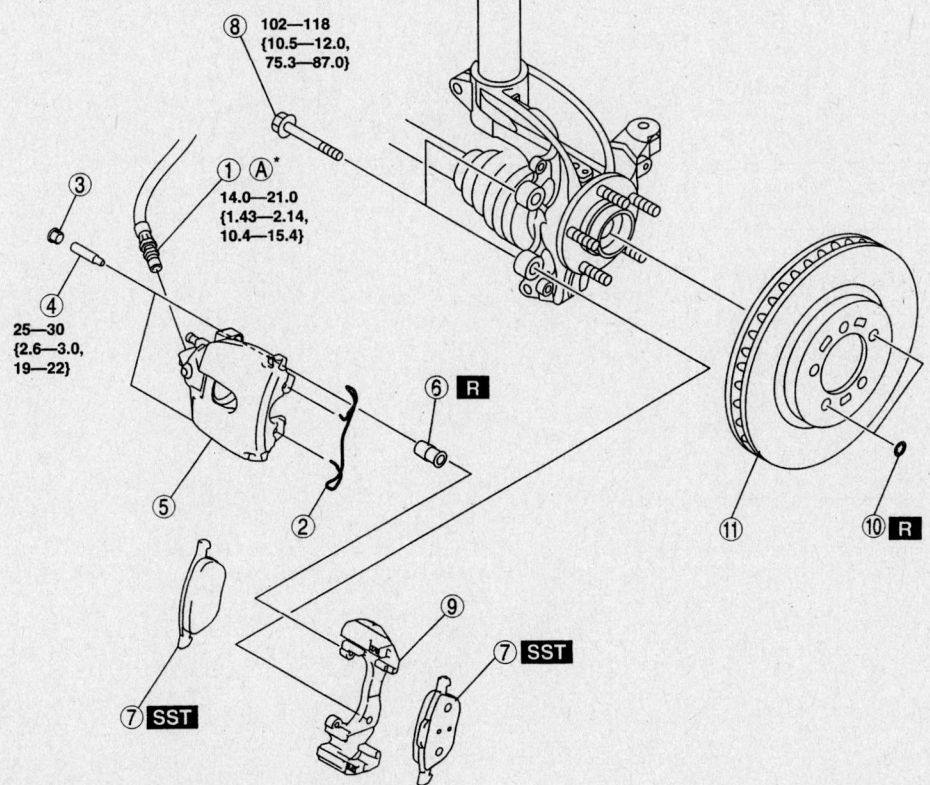

⑧ 102—118 {10.5—12.0, 75.3—87.0}

③

① Ⓐ* 14.0—21.0 {1.43—2.14, 10.4—15.4}

④

25—30 {2.6—3.0, 19—22}

⑥ **R**

⑤

②

⑦ **SST**

⑨

⑦ **SST**

⑪

⑩ **R**

1	Brake hose
2	Retaining clip
3	Cap
4	Bolt
5	Caliper
6	Boot
7	Disc pad
8	Bolt
9	Mounting support
10	Washer
11	Disc plate

Ⓐ* COMMERCIALLY AVAILABLE FLARE NUT WRENCH (FLARE NUT ACROSS FLAT 13 mm {0.51 in})

N·m {kgf·m, ft·lbf}

67162-MAZC-G104

Exploded view of the front disc brakes—Mazda3

end wrench. Tighten the nut to 10–15 ft. lbs. (14–21 Nm) and make sure the brake hose is not twisted, if it is unfasten the flare nut and retighten making sure the brake line remains straight. Fill the master cylinder with clean brake fluid and bleed the hydraulic system.
- Retaining clip
- Wheels

4. Pump the brake pedal several times to seat the pads.

REAR

1. Before servicing the vehicle, refer to the precautions section.
2. Remove or disconnect the following:
- Wheels
- Parking brake cable
- Brake pipe from the hose and the clip
- Pad retaining clip
- Caps from the caliper bolts
- Caliper brake hose
- Caliper mounting bolts and the caliper

To install:

3. Install or connect the following:
- Caliper. Torque the caliper mounting bolts to 19–22 ft. lbs. (25–30 Nm). Install the bolt caps.
- Brake hose and pipe and tighten to 14–16 ft. lbs. (19–23 Nm). Reinstall a new clip and fill the master cylinder with clean brake fluid and bleed the hydraulic system.
- Pad retaining clip and parking brake cable
- Wheels

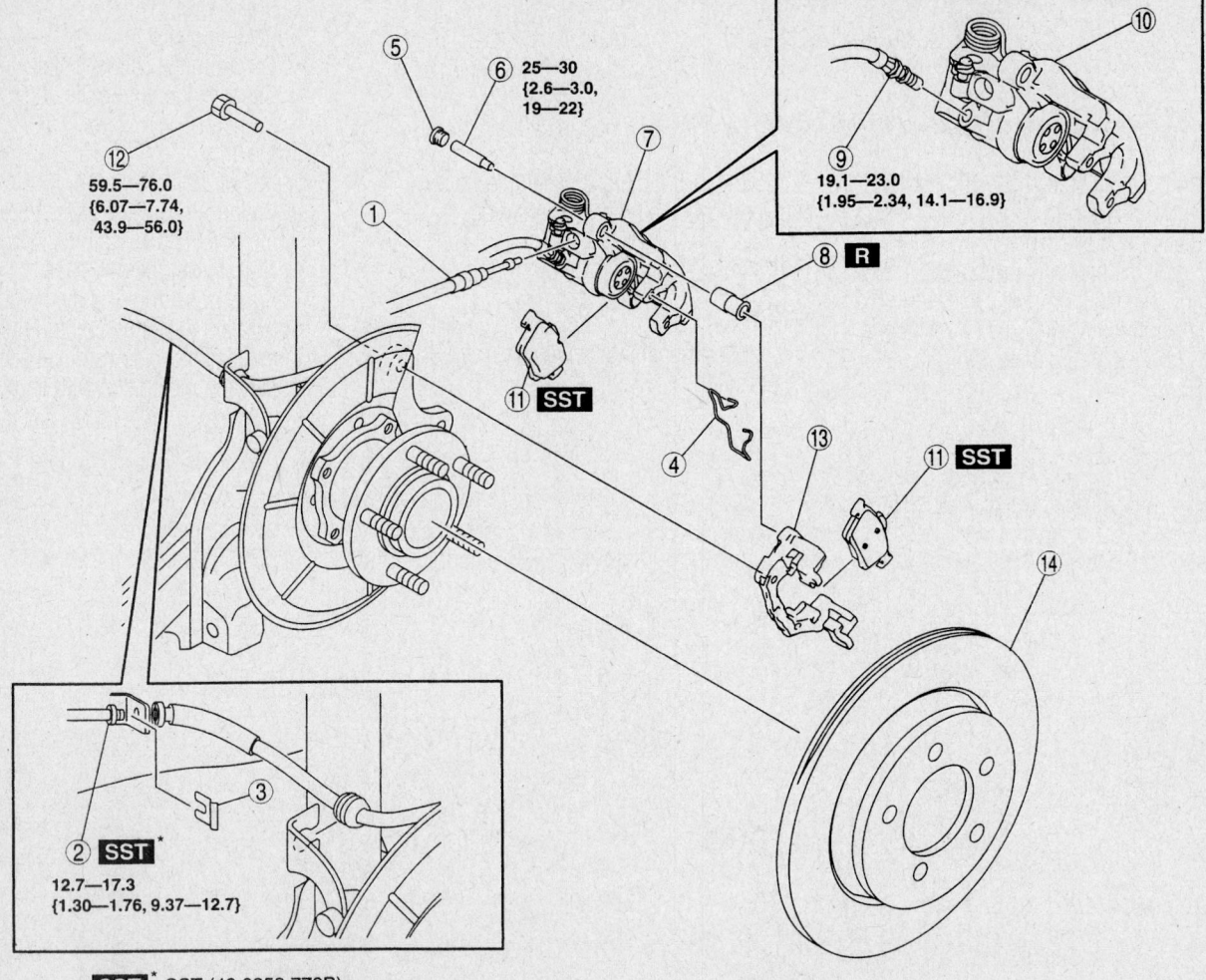

⑥ 25—30 {2.6—3.0, 19—22}

⑫ 59.5—76.0 {6.07—7.74, 43.9—56.0}

⑨ 19.1—23.0 {1.95—2.34, 14.1—16.9}

② SST* 12.7—17.3 {1.30—1.76, 9.37—12.7}

SST* SST (49 0259 770B) (FLARE NUT ACROSS FLAT 10 mm {0.39 in})

N·m {kgf·m, ft·lbf}

1	Parking brake cable	8	Boot
2	Brake pipe	9	Brake hose
3	Clip	10	Caliper
4	Retaining clip	11	Disc pad
5	Cap	12	Bolt
6	Bolt	13	Mounting support
7	Caliper, brake hose	14	Disc plate

67162-MAZC-G106

Exploded view of the rear disc brakes—Mazda3

4. Pump the brake pedal several times to seat the pads. Inspect the parking brake lever stroke and brake drag and adjust as necessary.

Mazda6 and Mazdaspeed6

FRONT

1. Before servicing the vehicle, refer to the precautions section.
2. Remove or disconnect the following:
   • Wheels

• Brake hose
• Cap from the caliper bolts
• Caliper mounting bolts and the caliper

To install:

3. Install or connect the following:
   • Caliper. Torque the caliper mounting bolts to 36-–39 ft. lbs. (49–54 Nm). Install the bolt caps.
   • Brake hose to the caliper. Tighten the bolt to 16–21 ft. lbs. (21–29 Nm) and make sure the brake hose is not twisted. Fill the master cylin-

der with clean brake fluid and bleed the hydraulic system.
   • Wheels
4. Pump the brake pedal several times to seat the pads.

REAR

1. Before servicing the vehicle, refer to the precautions section.
2. Remove or disconnect the following:
   • Wheels
   • Parking brake cable clip
   • Caliper brake hose bolt

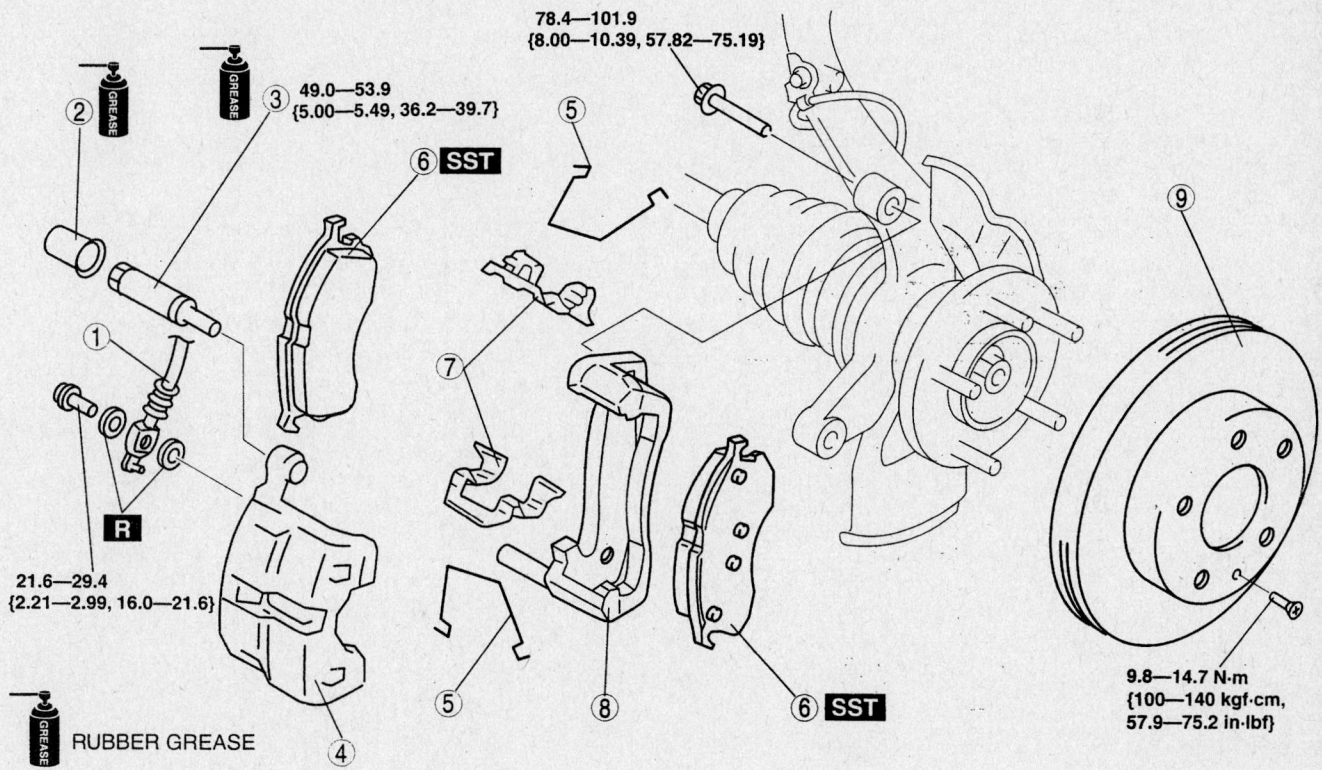

1	Flexible hose
2	Cap
3	Guide pin
4	Caliper
5	M-spring

6	Disc pad
7	Guide plate
8	Mounting support
9	Disc plate

67162-MAZC-G268

Exploded view of the front disc brakes—Mazda6 and Mazdaspeed6 models

- Caliper brake hose
- Caliper mounting bolts and the caliper

To install:

3. Install or connect the following:
   - Caliper. Torque the caliper mounting bolts to 27–36 ft. lbs. (37–49 Nm) on Mazda6models, or 16–23 ft. lbs. (21–31 Nm) on Mazda-speed6 models.
   - Brake hose and pipe and tighten to 16–21 ft. lbs. (21–29 Nm).
   - Parking brake cable clip and fill the master cylinder with clean brake fluid and bleed the hydraulic system.

- Wheels

4. Pump the brake pedal several times to seat the pads. Inspect the parking brake lever stroke and brake drag and adjust as necessary.

Disc Brake Pads

REMOVAL & INSTALLATION

MX5-Miata

FRONT

1. Before servicing the vehicle, refer to the precautions section.

➡**Refer the exploded view illustration for component locations and if applicable, their retainer torque specifications.**

2. Remove the lower caliper bolt and pivot the caliper up.

3. Remove the pads and shims.

4. Remove the guide plate.

To install:

5. Compress the piston into the bore.

6. Installation is the reverse of removal. Tighten the lower caliper bolt to 21–31 Nm (15–23 ft. lbs.).

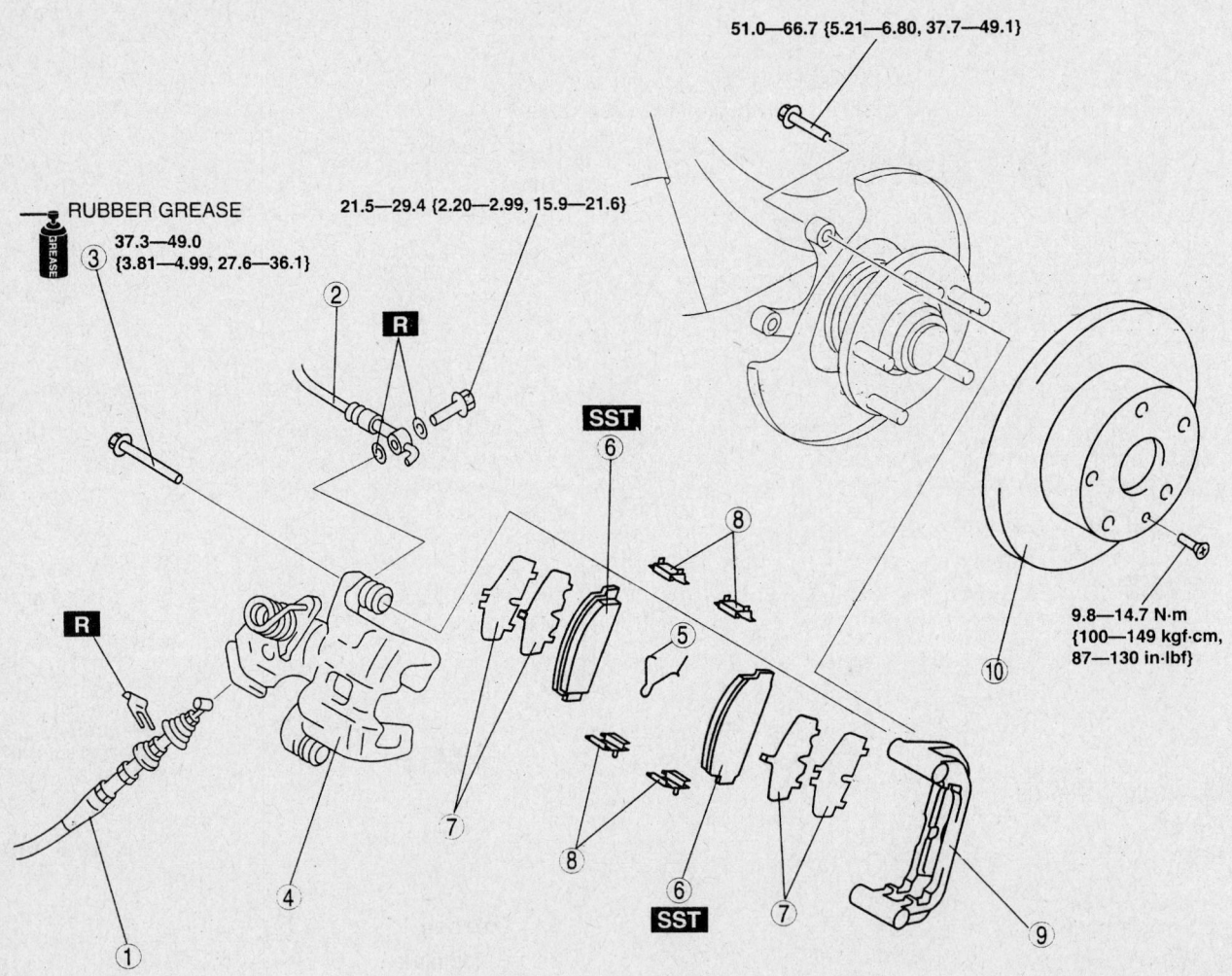

51.0—66.7 {5.21—6.80, 37.7—49.1}

RUBBER GREASE

37.3—49.0 {3.81—4.99, 27.6—36.1}

21.5—29.4 {2.20—2.99, 15.9—21.6}

9.8—14.7 N·m {100—149 kgf·cm, 87—130 in·lbf}

N·m {kgf·m, ft·lbf}

1	Parking brake cable, clip	6	Disc pad
2	Flexible hose	7	Shim
3	Bolt	8	Guide plate
4	Caliper	9	Mounting support
5	Spring	10	Disc plate

67162-MAZC-G269

Exploded view of the rear disc brakes—Mazda6, shown, Mazdaspeed6 models similar

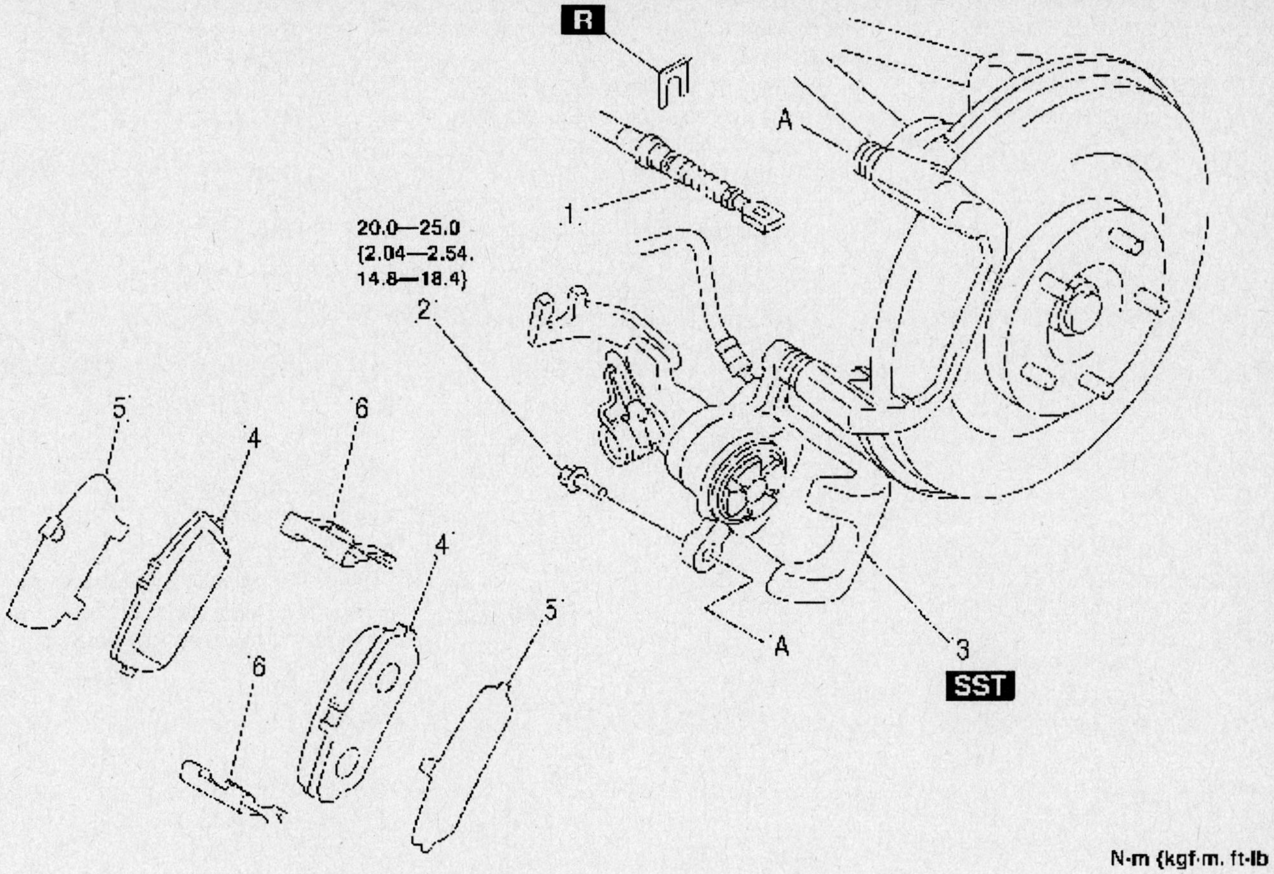

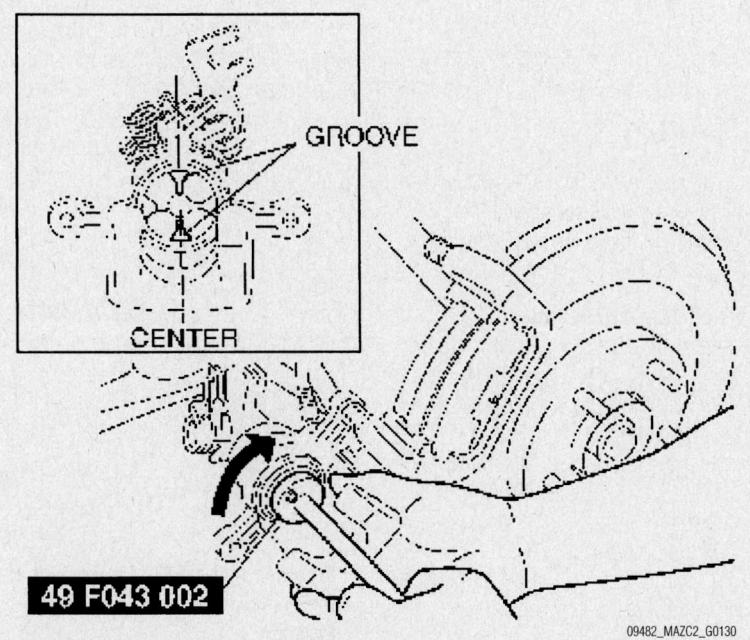

20.0—25.0
{2.04—2.54,
14.8—18.4}

N·m {kgf·m, ft·lb

1	Parking brake cable	4	Disc pad
2	Bolt	5	Shim
3	Caliper	6	Guide plate

09482_MAZC2_G0129

Front disc brake pad components—MX-5 Miata

GROOVE

CENTER

49 F043 002

09482_MAZC2_G0130

Compress the piston into the bore—MX-5 Miata

REAR

1. Before servicing the vehicle, refer to the precautions section.

➡**Refer the exploded view illustration for component locations and if applicable, their retainer torque specifications.**

2. Disconnect the parking brake cable.
3. Remove the upper caliper bolt and pivot the caliper downwards.
4. Remove the disc pads and shims.

To install:

5. Clean the exposed area of the piston.
6. Rotate the piston clockwise slowly using the tool shown to push the piston completely until the piston grooves are in the position shown in the figure.
7. Install the components in the reverse order of removal. Tighten the caliper bolts to 20–25 Nm (14–18 ft. lbs.).
8. Install the parking brake cable. After installation the parking brake cable, verify

that the operating lever returns to the stopper nut with the parking brake lever released.

9. Depress the brake pedal a few times and verify that the brakes do not drag

Mazda3

FRONT

1. Before servicing the vehicle, refer to the precautions section.
2. Remove or disconnect the following:
   - Wheels
   - Pad retaining clip
   - Cap from the bottom caliper bolt
   - Lower caliper mounting bolt and swing the caliper up
   - Brake pads

To install:
3. Press the caliper piston back into the cylinder using tool 49 0221 600C.
4. Install or connect the following:
   - Outer pad to the mounting support and inner pad to the caliper
   - Caliper. Torque the caliper lower mounting bolt to 19–22 ft. lbs. (25–30 Nm). Install the bolt cap.
   - Retaining clip
   - Wheels
5. Pump the brake pedal several times to seat the pads.

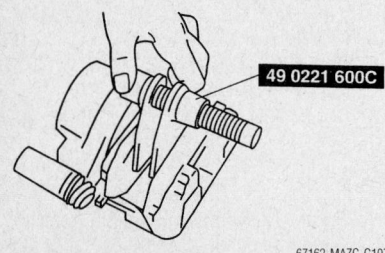

67162-MAZC-G107

Press the front caliper piston back into the cylinder—Mazda3

REAR

1. Before servicing the vehicle, refer to the precautions section.
2. Remove or disconnect the following:
   - Wheels
   - Parking brake cable
   - Pad retaining clip
   - Caps from the caliper bolts
   - Caliper mounting bolts and the caliper
   - Outer brake pad from the mount support and pull the inner pad from the caliper

To install:
3. Install the out pad to the mounting support and clean the piston area.

4. Using tool 49 F043 002, turn the piston clockwise slowly until the piston is fully seated in its bore.
5. Align the inner side pad spring with the piston groove and insert the pad. Refer to the illustration for spring location and inner pad installation arrow.
6. Install or connect the following:
   - Caliper. Torque the caliper mounting bolts to 19–22 ft. lbs. (25–30 Nm). Install the bolt caps.
   - Pad retaining clip and parking brake cable
   - Wheels
7. Pump the brake pedal several times to seat the pads. Inspect the parking brake lever stroke and brake drag and adjust as necessary.

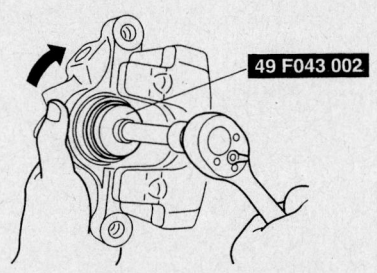

67162-MAZC-G108

Turn the rear caliper piston clockwise slowly until the piston is fully seated in its bore—Mazda3

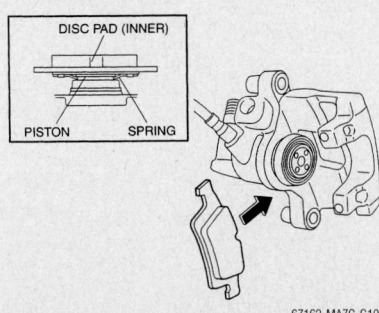

67162-MAZC-G109

Align the inner side pad spring with the piston groove and insert the pad—Mazda3

Mazda6 and Mazdaspeed6

FRONT

1. Before servicing the vehicle, refer to the precautions section.
2. Remove or disconnect the following:
   - Wheels
   - Brake pipe bolts
   - Top caliper bolt cap and bolt and swing the caliper downwards
   - M-springs the pads and shims

To install:
3. Install or connect the following:
4. Press the caliper pistons back into their cylinders using tool 49 0221 600C
   - Brake pads, shims and M-springs
   - Caliper. Torque the bolt to 36–39 ft. lbs. (49–52 Nm) and install the cap.
   - Wheels

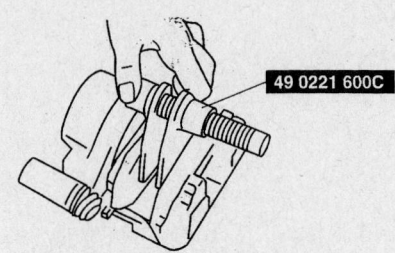

67162-MAZC-G270

Press the front caliper pistons back into their bores using tool 49 0221 600C— Mazda6 and Mazdaspeed6 models

REAR

1. Before servicing the vehicle, refer to the precautions section.
2. Remove or disconnect the following:
   - Wheels
   - Parking brake cable clip
   - Upper caliper bolt and rotate the caliper downwards
   - V-springs, pads and the shims from the pads

To install:
3. Press the rear caliper pistons back into their bores using tool 49 FA18 602.
4. Install or connect the following:
   - Pads, shims, and springs
   - Caliper and torque the bolt to 27–36 ft. lbs. (37–49 Nm) on Mazda6models, or 16–23 ft. lbs. (21–31 Nm) on Mazdaspeed6 models
   - Wheels

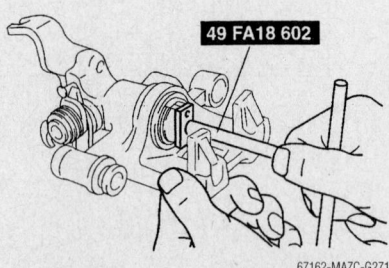

67162-MAZC-G271

Press the rear caliper pistons back into their bores using tool 49 FA18 602— Mazda6

MAZDA

B Series

SPECIFICATIONS AND MAINTENANCE CHARTS

ENGINE AND VEHICLE IDENTIFICATION

		Engine							Model Year	
Code ①	Liters (cc)	Cu. In.	Cyl.	Fuel Sys.	Type	Eng. Mfg.			Code ②	Year
D	2.3 (2261)	138	4	EFI	DOHC	Ford			2	2002
U	3.0 (2999)	183	6	EFI	OHV	Ford			3	2003
E	4.0 (4000)	244	6	EFI	SOHC	Ford			4	2004
EFI: Electronic fuel injection									5	2005
OHV: Overhead Valve									6	2006

DOHC: Dual Overhead Camshafts

SOHC: Single Overhead Camshaft

① 8th digit of the Vehicle Identification Number (VIN)

② 10th digit of the Vehicle Identification Number (VIN)

09482_MAZB_C0001

GENERAL ENGINE SPECIFICATIONS

Year	Model	Engine Displacement Liters	Engine (VIN)	Net Horsepower @ rpm	Net Torque @ rpm (ft. lbs.)	Bore x Stroke (in.)	Compression Ratio	Oil Pressure @ rpm
2002	B2300	2.3	D	143@5200	154@3750	3.44x3.70	9.7:1	29-39@2000
	B3000	3.0	U	147@5000	147@5000	3.50x3.14	9.3:1	40-60@2500
	B4000	4.0	E	160@4000	225@2500	3.94x3.31	9.0:1	40-60@2000
2003	B2300	2.3	D	143@5200	154@3750	3.44x3.70	9.7:1	29-39@2000
	B3000	3.0	U	147@5000	147@5000	3.50x3.14	9.3:1	40-60@2500
	B4000	4.0	E	160@4000	225@2500	3.94x3.31	9.0:1	40-60@2000
2004	B2300	2.3	D	143@5200	154@3750	3.44x3.70	9.7:1	29-39@2000
	B3000	3.0	U	147@5000	147@5000	3.50x3.14	9.3:1	40-60@2500
	B4000	4.0	E	160@4000	225@2500	3.94x3.31	.9.0:1	40-60@2000
2005	B2300	2.3	D	143@5200	154@3750	3.44x3.70	9.7:1	29-39@2000
	B3000	3.0	U	147@5000	147@5000	3.50x3.14	9.6:1	40-60@2500
	B4000	4.0	E	160@4000	225@2500	3.94x3.31	9.7:1	40-60@2000
2006	B2300	2.3	D	143@5250	154@3750	3.44x3.70	9.7:1	29-39@2000
	B3000	3.0	U	148@4900	180@3950	3.50x3.14	9.6:1	40-60@2500
	B4000	4.0	E	207@5250	238@3000	3.94x3.31	9.7:1	40-60@2000

09482_MAZB_C0002

GASOLINE ENGINE TUNE-UP SPECIFICATIONS

Year	Engine Displacement Liters	Engine VIN	Spark Plug Gap (in.)	Ignition Timing (deg.) ①		Fuel Pump (psi)	Idle Speed (rpm)		Valve Clearance	
				MT	AT		MT	AT	In.	Ex.
2002	2.3	D	0.041-0.045	10B	10B	56-72	①	①	0.008	0.010
	3.0	U	0.042-0.046	10B	10B	56-72	①	①	②	③
	4.0	E	0.061-0.068	10B	10B	56-72	①	①	②	③
2003	2.3	D	0.041-0.045	10B	10B	56-72	①	①	0.008	0.010
	3.0	U	0.042-0.046	10B	10B	56-72	①	①	②	③
	4.0	E	0.061-0.068	10B	10B	56-72	①	①	②	③
2004	2.3	D	0.041-0.045	10B	10B	56-72	①	①	0.008	0.010
	3.0	U	0.042-0.046	10B	10B	56-72	①	①	②	③
	4.0	E	0.061-0.068	10B	10B	56-72	①	①	②	③
2005	2.3	D	0.041-0.045	10B	10B	56-72	①	①	0.008	0.010
	3.0	U	0.042-0.046	10B	10B	56-72	①	①	②	③
	4.0	E	0.061-0.068	10B	10B	56-72	①	①	②	③
2006	2.3	D	0.041-0.045	10B	10B	56-72	①	①	0.008	0.010
	3.0	U	0.042-0.046	10B	10B	56-72	①	①	②	③
	4.0	E	0.061-0.068	10B	10B	56-72	①	①	②	③

NOTE: The Vehicle Emission Control Information label often reflects specification changes changes made during production. The label figures must be used
if they differ from those in this chart.

B: Before top dead center

HYD: Hydraulic

① Electronically controlled and cannot be adjusted

② 0.008-0.011

③ 0.010-0.013

09482_MAZB_C0003

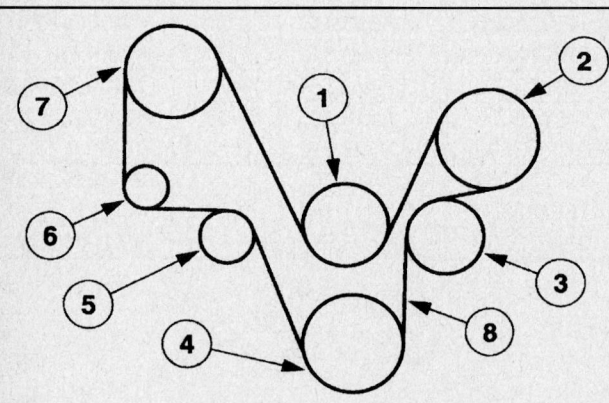

1 Fan pulley
2 Power steering pump pulley
3 Water pump pulley
4 Crankshaft pulley
5 Belt tensioner pulley
6 Generator pulley
7 A/C clutch pulley
8 Drive belt

67197-RANG-G97

Accessory drive belt routing—2.3L engine with A/C

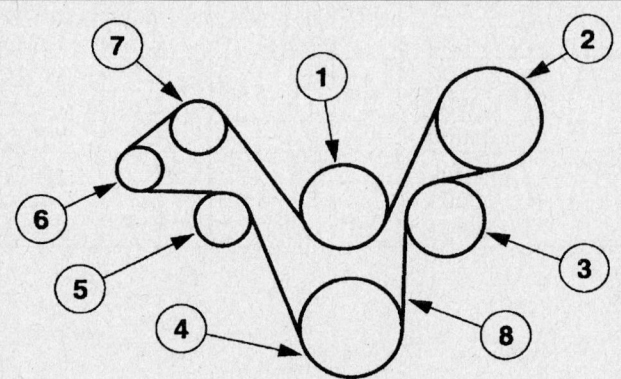

1 Fan pulley
2 Power steering pump pulley
3 Water pump pulley
4 Crankshaft pulley
5 Belt tensioner pulley
6 Generator pulley
7 Belt idler pulley
8 Drive belt

67197-RANG-G98

Accessory drive belt routing—2.3L engine without A/C

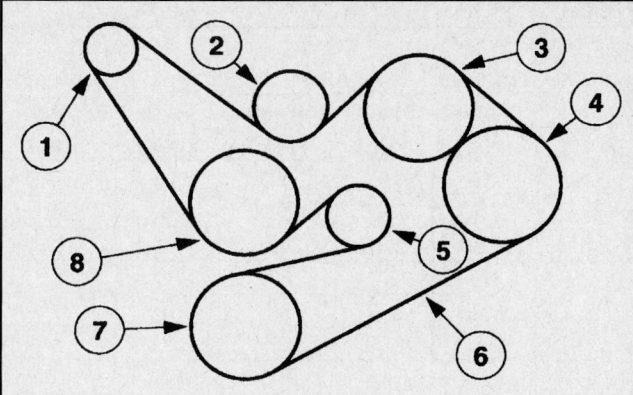

1 Generator pulley
2 Belt idler pulley
3 A/C clutch pulley
4 Power steering pump pulley
5 Belt tensioner
6 Drive belt
7 Crankshaft pulley
8 Water pump pulley

67197-RANG-G99

Accessory drive belt routing—3.0L engine with A/C

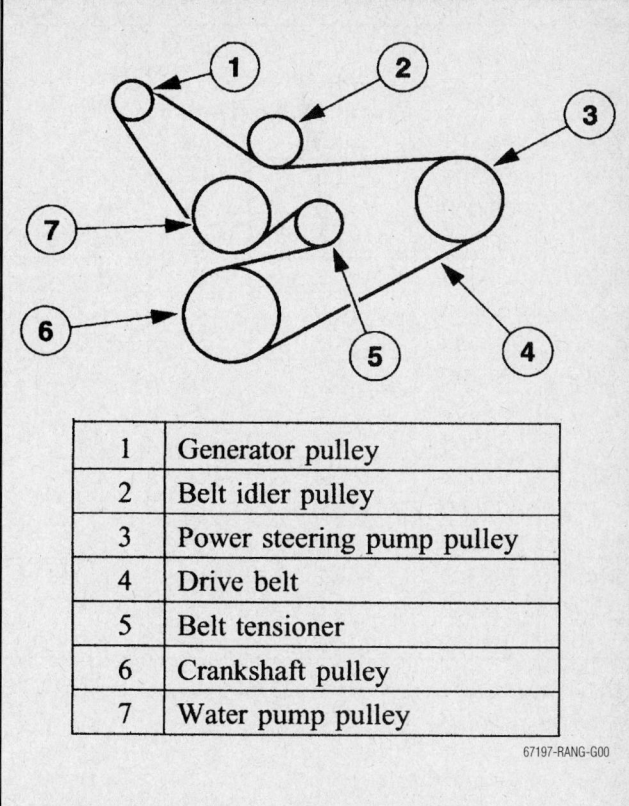

1	Generator pulley
2	Belt idler pulley
3	Power steering pump pulley
4	Drive belt
5	Belt tensioner
6	Crankshaft pulley
7	Water pump pulley

67197-RANG-G00

Accessory drive belt routing—3.0L engine without A/C

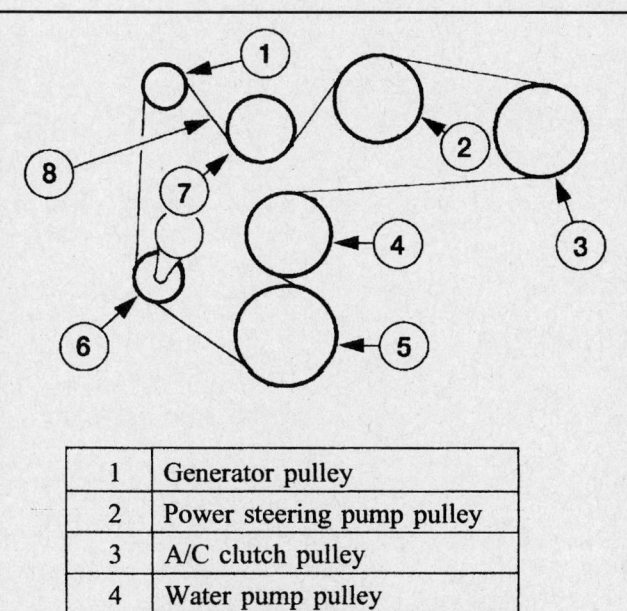

1	Generator pulley
2	Power steering pump pulley
3	A/C clutch pulley
4	Water pump pulley
5	Crankshaft pulley
6	Drive belt tensioner
7	Belt idler pulley
8	Drive belt

67197-RANG-G1A

Accessory drive belt routing—4.0L engine with A/C

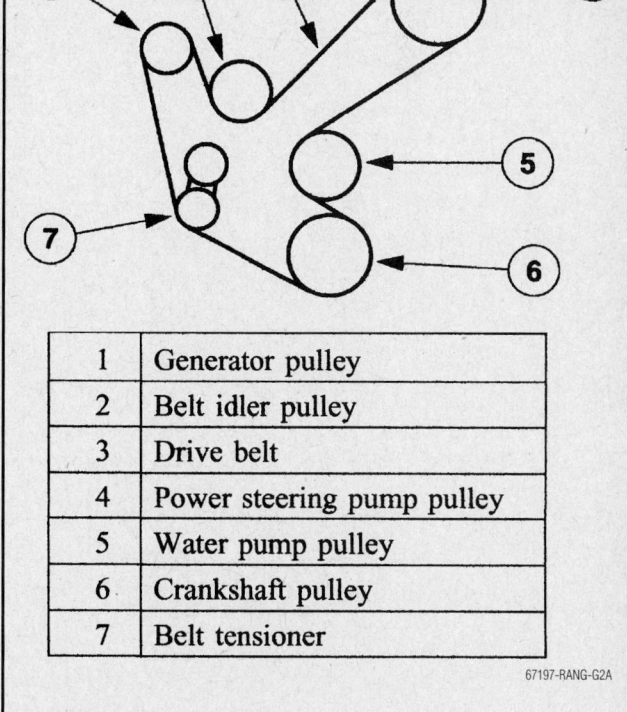

1	Generator pulley
2	Belt idler pulley
3	Drive belt
4	Power steering pump pulley
5	Water pump pulley
6	Crankshaft pulley
7	Belt tensioner

67197-RANG-G2A

Accessory drive belt routing—4.0L engine without A/C

CAPACITIES

Year	Model	Engine Displacement Liters	Engine VIN	Engine Oil with Filter (qts.)	Transmission (pts.) 5-Spd	Auto.	Transfer Case (pts.)	Drive Axle Front (pts.)	Rear (pts.)	Fuel Tank (gal.)	Cooling System (qts.)
2002	B2300	2.3	D	4.0	5.6	19.8	NA	NA	5.0	①	②
	B3000	3.0	U	4.5	5.6	③	2.50	2.70	5.0	①	④
	B4000	4.0	E	5.0	5.6	③	2.50	2.70	5.0	①	⑤
2003	B2300	2.3	D	4.0	5.6	19.8	NA	NA	5.0	①	②
	B3000	3.0	U	4.5	5.6	③	2.50	2.70	5.0	①	④
	B4000	4.0	E	5.0	5.6	③	2.50	2.70	5.0	①	⑤
2004	B2300	2.3	D	4.0	5.6	19.8	NA	NA	5.0	⑥	②
	B3000	3.0	U	4.5	5.6	③	2.50	3.25	5.0	⑥	④
	B4000	4.0	E	5.0	5.6	③	2.50	3.25	5.0	⑥	⑤
2005	B2300	2.3	D	4.0	5.6	19.8	NA	NA	5.0	⑥	②
	B3000	3.0	U	4.5	5.6	③	2.50	3.25	5.0	⑥	④
	B4000	4.0	E	5.0	5.6	③	2.50	3.25	5.0	⑥	⑤
2006	B2300	2.3	D	4.0	5.6	19.8	NA	NA	5.0	⑥	②
	B3000	3.0	U	4.5	5.6	③	2.50	3.25	5.0	⑥	④
	B4000	4.0	E	5.0	5.6	③	2.50	3.25	5.0	⑥	⑤

NOTE: All capacities are approximate. Add fluid gradually and check to be sure a proper fluid level is obtained.

NA: not available

① Std: 16.5
  Long Wheelbase: 20.0
  Super Cab: 19.5

② w/MT: 11.2
  w/AT: 10.9

③ 2wd: 20.0
  4wd: 20.6

④ w/MT: 15.2
  w/AT: 14.8

⑤ w/MT: 13.5
  w/AT: 13.2

⑥ Regular Cab short wheel base: 17 gal.
  Regular Cab long wheel base: 20.3 gal.
  Super Cab: 19.5 gal.

⑦ w/MT: 10.5
  w/AT: 10.2

09482_MAZB_C0004

VALVE SPECIFICATIONS

Year	Engine Displacement Liters	Engine VIN	Seat Angle (deg.)	Face Angle (deg.)	Spring Test Pressure (lbs. @ in.)	Spring Installed Height (in.)	Stem-to-Guide Clearance (in.)		Stem Diameter (in.)	
							Intake	Exhaust	Intake	Exhaust
2002	2.3	D	45	45	①	1.492	0.0009	0.0011	0.2153-0.2159	0.2151-0.2157
	3.0	U	44.5	45.62	NA	1.736-1.650	0.0010-0.0027	0.0015-0.0032	0.2744-0.2752	0.2739-0.2747
	4.0	E	45	45	202-225@ 1.413-1.445	1.569-1.601	0.0010-0.0020	0.0010-0.0030	0.2740-0.2750	0.2740
2003	2.3	D	45	45	①	1.492	0.0009	0.0011	0.2153-0.2159	0.2151-0.2157
	3.0	U	44.5	45.62	NA	1.736-1.650	0.0010-0.0027	0.0015-0.0032	0.2744-0.2752	0.2739-0.2747
	4.0	E	45	45	202-225@ 1.413-1.445	1.569-1.601	0.0010-0.0020	0.0010-0.0030	0.2740-0.2750	0.2740
2004	2.3	D	45	45	①	1.492	0.0009	0.0011	0.2153-0.2159	0.2151-0.2157
	3.0	U	44.5	45.62	NA	1.736-1.650	0.0010-0.0027	0.0015-0.0032	0.2744-0.2752	0.2739-0.2747
	4.0	E	45	45	202-225@ 1.413-1.445	1.569-1.601	0.0010-0.0020	0.0010-0.0030	0.2740-0.2750	0.2740
2005	2.3	D	45	45	①	1.492	0.0009	0.0011	0.2153-0.2159	0.2151-0.2157
	3.0	U	44.5	45.62	NA	1.736-1.650	0.0010-0.0027	0.0015-0.0032	0.2744-0.2752	0.2739-0.2747
	4.0	E	45	45	202-225@ 1.413-1.445	1.569-1.601	0.0010-0.0020	0.0010-0.0030	0.2740-0.2750	0.2740
2006	2.3	D	45	45	①	1.492	0.0009	0.0011	0.2153-0.2159	0.2151-0.2157
	3.0	U	44.5	45.62	NA	1.736-1.650	0.0010-0.0027	0.0015-0.0032	0.2744-0.2752	0.2739-0.2747
	4.0	E	45	45	202-225@ 1.413-1.445	1.569-1.608	0.0010-0.0020	0.0010-0.0030	0.2740-0.2750	0.2740

NA: not available

① Intake: 97.0@1.201
Exhaust: 93.3@1.201

09482_MAZB_C0005

CAMSHAFT SPECIFICATIONS CHART
All measurements are given in inches.

Year	Engine Displ. Liters	Engine VIN	Journal Dia.	Brg. Oil Clearance	Shaft End-play	Runout	Lobe Height Intake	Lobe Height Exhaust
2002	2.3	D	0.9820-0.9830	0.0010-0.0030	NA	①	0.3240	0.3070
	3.0	U	2.0074-2.0084	0.0010-0.0030	②	③	0.2600	0.2600
	4.0	E	1.0990-1.1010	0.0020-0.0040	0.0003-0.0070	0.0020	0.2590	0.2590
2003	2.3	D	0.9820-0.9830	0.0010-0.0030	NA	①	0.3240	0.3070
	3.0	U	2.0074-2.0084	0.0010-0.0030	②	③	0.2600	0.2600
	4.0	E	1.0990-1.1010	0.0020-0.0040	0.0003-0.0070	0.0020	0.2590	0.2590
2004	2.3	D	0.9820-0.9830	0.0010-0.0030	NA	①	0.3240	0.3070
	3.0	U	2.0074-2.0084	0.0010-0.0030	②	③	0.2600	0.2600
	4.0	E	1.0990-1.1010	0.0020-0.0040	0.0003-0.0070	0.0020	0.2590	0.2590
2005	2.3	D	0.9820-0.9830	0.0010-0.0030	NA	①	0.3240	0.3070
	3.0	U	2.0074-2.0084	0.0010-0.0030	②	③	0.2600	0.2600
	4.0	E	1.0990-1.1010	0.0020-0.0040	0.0003-0.0070	0.0020	0.2590	0.2590
2006	2.3	D	0.9820-0.9830	0.0010-0.0030	NA	①	0.3240	0.3070
	3.0	U	2.0074-2.0084	0.0010-0.0030	②	③	0.2600	0.2600
	4.0	E	1.0990-1.1010	0.0020-0.0040	0.0003-0.0070	0.0020	0.2590	0.2590

NA - Not available

① 0.001, no.3 journal supported by no.1 and no.5 journals

② 0.007 service limit

③ 0.002 limit. Runout no.2 or no.3 journal relative to no.1 and no.4 journals

09482_MAZB_C0006

CRANKSHAFT AND CONNECTING ROD SPECIFICATIONS

All measurements are given in inches.

Year	Engine Displacement Liters	Engine VIN	Crankshaft				Connecting Rod		
			Main Brg. Journal Dia.	Main Brg. Oil Clearance	Shaft End-play	Thrust on No.	Journal Diameter	Oil Clearance	Side Clearance
2002	2.3	D	2.0460-2.0470	0.0007-0.0013	0.0080-0.0160	3	1.9670-1.9680	0.0010-0.0020	0.0760-0.1200
	3.0	U	2.5190-2.5198	0.0010-0.0014	0.0040-0.0080	3	2.1253-2.1261	0.0010-0.0014	0.0060-0.0114
	4.0	E	2.2430-2.2440	0.0008-0.0015	0.0020-0.0126	3	2.1250-2.1260	0.0003-0.0024	0.0036-0.0106
2003	2.3	D	2.0460-2.0470	0.0007-0.0013	0.0080-0.0160	3	1.9670-1.9680	0.0010-0.0020	0.0760-0.1200
	3.0	U	2.5190-2.5198	0.0010-0.0014	0.0040-0.0080	3	2.1253-2.1261	0.0010-0.0014	0.0060-0.0114
	4.0	E	2.2430-2.2440	0.0008-0.0015	0.0020-0.0126	3	2.1250-2.1260	0.0003-0.0024	0.0036-0.0106
2004	2.3	D	2.0460-2.0470	0.0007-0.0013	0.0080-0.0160	3	1.9670-1.9680	0.0010-0.0020	0.0760-0.1200
	3.0	U	2.5190-2.5198	0.0010-0.0014	0.0040-0.0080	3	2.1253-2.1261	0.0010-0.0014	0.0060-0.0114
	4.0	E	2.2430-2.2440	0.0008-0.0015	0.0020-0.0126	3	2.1250-2.1260	0.0003-0.0024	0.0036-0.0106
2005	2.3	D	2.0460-2.0470	0.0007-0.0013	0.0080-0.0160	3	1.9670-1.9680	0.0010-0.0020	0.0760-0.1200
	3.0	U	2.5190-2.5198	0.0010-0.0014	0.0040-0.0080	3	2.1253-2.1261	0.0010-0.0014	0.0060-0.0114
	4.0	E	2.2430-2.2440	0.0008-0.0015	0.0020-0.0126	3	2.1250-2.1260	0.0003-0.0024	0.0036-0.0106
2006	2.3	D	2.0460-2.0470	0.0007-0.0013	0.0080-0.0160	3	1.9670-1.9680	0.0010-0.0020	0.0760-0.1200
	3.0	U	2.5190-2.5198	0.0010-0.0014	0.0040-0.0080	3	2.1253-2.1261	0.0010-0.0014	0.0060-0.0114
	4.0	E	2.2430-2.2440	0.0008-0.0015	0.0020-0.0126	3	2.1250-2.1260	0.0003-0.0024	0.0036-0.0106

09482_MAZB_C0007

PISTON AND RING SPECIFICATIONS

All measurements are given in inches.

Year	Engine Displacement Liters	Engine VIN	Piston Clearance	Ring Gap			Ring Side Clearance		
				Top Compression	Bottom Compression	Oil Control	Top Compression	Bottom Compression	Oil Control
2002	2.3	D	0.0009-0.0017	0.0060-0.0012	0.0120-0.0180	0.0070-0.0270	0.0008-0.0013	0.0004-0.0011	0.0025-0.0054
	3.0	U	0.0012-0.0022	0.010-0.020	0.010-0.020	0.010-0.049	0.0602-0.0612	0.0602-0.0612	0.1587-0.1596
	4.0	E	0.0008-0.0019	0.015-0.023	0.015-0.023	0.015-0.055	0.0010-0.0030	0.0010-0.0030	SNUG
2003	2.3	D	0.0009-0.0017	0.0060-0.0012	0.0120-0.0180	0.0070-0.0270	0.0008-0.0013	0.0004-0.0011	0.0025-0.0054
	3.0	U	0.0012-0.0022	0.010-0.020	0.010-0.020	0.010-0.049	0.0602-0.0612	0.0602-0.0612	0.1587-0.1596
	4.0	E	0.0008-0.0019	0.015-0.023	0.015-0.023	0.015-0.055	0.0010-0.0030	0.0010-0.0030	SNUG
2004	2.3	D	0.0009-0.0017	0.0060-0.0012	0.0120-0.0180	0.0070-0.0270	0.0008-0.0013	0.0004-0.0011	0.0025-0.0054
	3.0	U	0.0012-0.0022	0.010-0.020	0.010-0.020	0.010-0.049	0.0602-0.0612	0.0602-0.0612	0.1587-0.1596
	4.0	E	0.0008-0.0019	0.015-0.023	0.015-0.023	0.015-0.055	0.0010-0.0030	0.0010-0.0030	SNUG
2005	2.3	D	0.0009-0.0017	0.0060-0.0012	0.0120-0.0180	0.0070-0.0270	0.0008-0.0013	0.0004-0.0011	0.0025-0.0054
	3.0	U	0.0012-0.0022	0.010-0.020	0.010-0.020	0.010-0.049	0.0602-0.0612	0.0602-0.0612	0.1587-0.1596
	4.0	E	0.0008-0.0019	0.015-0.023	0.015-0.023	0.015-0.055	0.0010-0.0030	0.0010-0.0030	SNUG
2006	2.3	D	0.0009-0.0017	0.0060-0.0012	0.0120-0.0180	0.0070-0.0270	0.0008-0.0013	0.0004-0.0011	0.0025-0.0054
	3.0	U	0.0012-0.0022	0.010-0.020	0.010-0.020	0.010-0.049	0.0602-0.0612	0.0602-0.0612	0.1587-0.1596
	4.0	E	0.0008-0.0019	0.015-0.023	0.015-0.023	0.015-0.055	0.0010-0.0030	0.0010-0.0030	SNUG

09482_MAZB_C0008

TORQUE SPECIFICATIONS

All readings in ft. lbs.

	Engine Displacement Liters	Engine VIN	Cylinder Head Bolts	Main Bearing Bolts	Rod Bearing Bolts	Crankshaft Damper Bolts	Flywheel Bolts	Manifold		Spark Plugs	Oil Pan Drain Plug
								Intake *	Exhaust		
2002	2.3	D	①	NA	NA	②	③	13	40	9	21
	3.0	U	④	60	26	107	54-64	24	⑤	8-10	NA
	4.0	E	⑥	⑫	⑦	⑧	75-85	7	16	13	18
2003	2.3	D	①	NA	NA	②	③	13	40	9	21
	3.0	U	④	60	26	107	54-64	24	⑤	8-10	NA
	4.0	E	⑥	⑫	⑦	⑧	75-85	7	16	13	18
2004	2.3	D	①	NA	NA	②	③	13	40	9	21
	3.0	U	④	60	26	107	59	21	⑤	11	NA
	4.0	E	⑥	⑫	⑦	⑧	⑨	7	16	13	18
2005	2.3	D	①	NA	NA	②	③	13	40	9	21
	3.0	U	⑩	59	26	107	59	21	⑤	11	NA
	4.0	E	⑥	⑫	⑦	⑧	⑨	7	16	13	18
2006	2.3	D	①	NA	NA	②	③	13	40	9	21
	3.0	U	⑩	59	26	107	59	21	⑤	11	NA
	4.0	E	⑪	⑫	⑦	⑧	⑨	7	16	13	18

NA: Information not available

* NOTE: Applies to Lower Manifold only.

① Step 1: 44 inch lbs.
   Step 2: 11 ft. lbs.
   Step 3: 33 ft. lbs.
   Step 4: +90 degrees
   Step 5: +90 degrees

② Step 1: 74 ft. lbs.
   Step 2: +90 degrees

③ Step 1: 37 ft. lbs.
   Step 2: 59 ft. lbs.
   Step 3: 83 ft. lbs.

④ Step 1: 59 ft. lbs.
   Step 2: Back off 1 full turn
   Step 3: 34-40 ft. lbs.
   Step 4: 63-73 ft. lbs.

⑤ Step 1: 89 inch lbs.
   Step 2: 17 ft. lbs.

⑥ 8mm bolts: 24 ft. lbs.
   12mm bolts: 24 ft. lbs. +80 degrees, +80 degrees

⑦ Step 1: 15 ft. lbs.
   Step 2: +90 degrees

⑧ Step 1: 37 ft. lbs.
   Step 2: +90 degrees

⑨ Step 1: 10 ft.lbs.
   Step 2: 52 ft. lbs.

⑩ Step 1: 37 ft. lbs.
   Step 2: back off 1 full turn
   Step 3: 22 ft. lbs.
   Step 4: + 90 degrees
   Step 5: + 90 degrees

⑪ Step 1: 12mm bolts 9 ft. lbs.
   Step 2: 12mm bolts 18 ft. lbs.
   Step 3: 8mm bolts 24 ft. lbs.
   Step 4: 12 mm bolts +90 degrees
   Step 5: 12mm bolts +90 degrees

⑫ Step 1: 26 ft. lbs.
   Step 2: +57 degrees

09482_MAZB_C0009

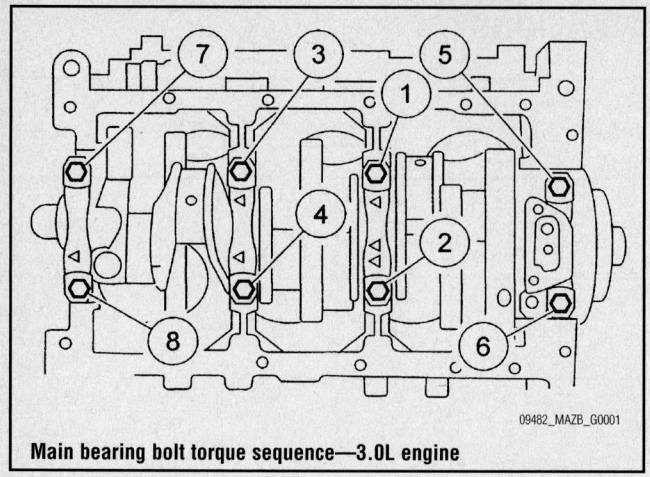

Main bearing bolt torque sequence—3.0L engine

09482_MAZB_G0001

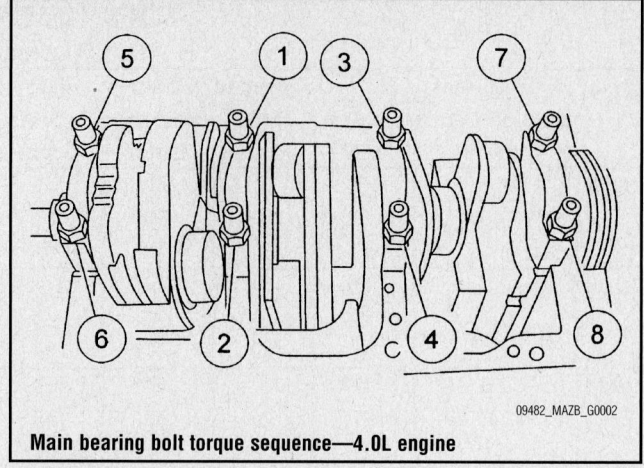

Main bearing bolt torque sequence—4.0L engine

09482_MAZB_G0002

WHEEL ALIGNMENT

| Year | Model | | Caster | | Camber | | Toe-in |
			Range (+/-Deg.)	Preferred Setting (Deg.)	Range (+/-Deg.)	Preferred Setting (Deg.)	(in.)
2002	B Series	2wd	1.0	①	0.70	-0.50	0.06+/-0.25
		4wd	1.0	②	0.70	-0.50	0.12+/-0.25
		Rear	—	—	0.75	0	0+/-0.30
2003	B Series	2wd	1.0	①	0.70	-0.50	0.06+/-0.25
		4wd	1.0	②	0.70	-0.50	0.12+/-0.25
		Rear	—	—	0.75	0	0+/-0.30
2004	B Series	2wd	1.0	③	0.70	④	0.06+/-0.25
		4wd	1.0	⑤	0.70	-0.50	0.12+/-0.25
		Rear	—	—	0.75	0	0+/-0.30
2005	B Series	2wd	1.0	③	0.70	-0.50	0.06+/-0.25
		4wd	1.0	⑤	0.70	-0.50	0.12+/-0.25
		Rear	—	—	0.75	0	0+/-0.30
2006	B Series	2wd	1.0	③	0.70	-0.50	0.06+/-0.25
		4wd	1.0	⑤	0.70	-0.50	0.12+/-0.25
		Rear	—	—	0.75	0	0+/-0.30

① Left: +4.0
  Right: +4.4
② Left: +3.9
  Right: +4.4
③ Left: +3.0
  Right: +3.4
④ Left: -0.5
  Right: -0.7
⑤ Left: +2.9
  Right: +3.6

09482_MAZB_C0010

TIRE, WHEEL AND BALL JOINT SPECIFICATIONS

Year	Model	OEM Tires Standard	OEM Tires Optional	Tire Pressures (psi) Front	Tire Pressures (psi) Rear	Wheel Size	Ball Joint Inspection	Lugnut Torque (ft. lbs.)
2002	B Series 2WD	P235/75R15	none	①	①	7-JJ	0.030 in. ②	100
	B Series 4WD	P245/75R16	none	①	①	7-JJ	0.030 in. ②	100
2003	B Series 2WD	P235/75R15	none	①	①	7-JJ	0.030 in. ②	100
	B Series 4WD	P245/75R16	none	①	①	7-JJ	0.030 in. ②	100
2004	B2300	P225/70R15	none	①	①	7-JJ	0.030 in. ②	100
	B3000	P225/70R15	P235/75R15SL	①	①	7-JJ	0.030 in. ②	100
	B4000	P235/75R15	P245/75R16	①	①	7-JJ	0.030 in. ②	100
2005	B2300	P225/70R15	none	①	①	7-JJ	0.030 in. ②	100
	B3000	P235/75R15	none	①	①	7-JJ	0.030 in. ②	100
	B4000	P235/75R15	P255/70R16	①	①	7-JJ	0.030 in. ②	100
2006	B2300	P225/70R15	none	①	①	7-JJ	0.030 in. ②	100
	B3000	P235/75R15	none	①	①	7-JJ	0.030 in. ②	100
	B4000	P235/75R15	P255/70R16	①	①	7-JJ	0.030 in. ②	100

OEM: Original Equipment Manufacturer

PSI: Pounds Per Square Inch

① See placard on door post

② Both upper and lower

09482_MAZB_C0011

BRAKE SPECIFICATIONS

All measurements in inches unless noted

Year	Model	Brake Disc Original Thickness	Brake Disc Minimum Thickness	Brake Disc Maximum Runout	Brake Drum Diameter Original Inside Diameter	Brake Drum Diameter Max. Wear Limit	Brake Drum Diameter Maximum Machine Diameter	Minimum Lining Thickness	Brake Caliper Bracket Bolts (ft. lbs.)	Brake Caliper Mounting Bolts (ft. lbs.)
2002	B-Series ①	0.850	0.810	0.0030	9.00	9.09	9.06	0.030	85	21-26
	②	0.850	0.810	0.0030	10.00	10.09	10.06	0.030	85	21-26
	③	0.850	0.810	0.0030	10.00	10.09	10.06	0.030	85	21-26
2003	B-Series ①	0.850	0.810	0.0030	9.00	9.09	9.06	0.030	85	21-26
	②	0.850	0.810	0.0030	10.00	10.09	10.06	0.030	85	21-26
	③	0.850	0.810	0.0030	10.00	10.09	10.06	0.030	85	21-26
2004	B-Series	NA	④	NA	NA	NA	⑤	⑥	85	21-26
2005	B-Series	NA	0.965	NA	NA	NA	⑤	⑥	85	21-26
2006	B-Series	NA	⑦	NA	NA	NA	⑤	⑥	85	27

NOTE: Due to changes made during production, refer to manufacturer's specifications if they differ from those in this chart

NA: Not Available

① With 9 inch brakes

② 4x2 with 10 inch brakes

③ 4x4 with 10 inch brakes

④ Molded into the disc

⑤ Molded into the drum

⑥ Front: 0.04 in.

  Rear: 0.03 in.

⑦ 0.960 2WD, 4WD discard thickness

09482_MAZB_C0012

SCHEDULED MAINTENANCE INTERVALS
2002 B SERIES

TO BE SERVICED	TYPE OF SERVICE	VEHICLE MILEAGE INTERVAL (x1000)												
		5	10	15	20	25	30	35	40	45	50	55	60	65
Engine oil & filter	R	✓	✓	✓	✓	✓	✓	✓	✓	✓	✓	✓	✓	✓
Tires	Rotate	✓	✓	✓	✓	✓	✓	✓	✓	✓	✓	✓	✓	✓
Auto trans. fluid	I			✓			✓			✓			✓	
Brake pads/shoes	I			✓			✓			✓			✓	
Coolant hoses	S/I			✓			✓			✓			✓	
Steering linkage	I			✓			✓			✓			✓	
Cabin air filter	R			✓			✓			✓			✓	
Ball joints (2wd)	L			✓			✓			✓			✓	
Exhaust system	I						✓						✓	
Engine air filter	R						✓						✓	
Fuel filter ①	R						✓						✓	
Auto trans fluid (4-speed)	R						✓						✓	
Green coolant ②	R									✓				
Wheel bearings (2wd)	L												✓	
Manual trans. fluid	R												✓	
Spark plugs	R	every 100,000 miles												
PCV valve	R	every 100,000 miles												
Orange coolant	R	every 150,000 miles												
Auto trans fluid (5-speed)	R	every 150,000 miles												
Differential fluid	R	every 150,000 miles												
Accessory drive belts	R	every 150,000 miles												
Transfer case fluid	R	every 150,000 miles												

R: Replace   S: Service   I: Inspect   L: Lubricate

① Recommended, but not required in Calif.

② Change every 30,000 miles or 36 months thereafter

FREQUENT OPERATION MAINTENANCE (SEVERE SERVICE)

If a vehicle is operated under any of the following conditions it is considered severe service:

- Towing a trailer or using a camper or car-top carrier.

- Repeated short trips of less than 5 miles in temperatures below freezing, or trips of less than 10 miles in any temperature.

- Extensive idling or low-speed driving for long distance as in heavy commercial use, such as delivery, taxi or police cars.

- Operating on rough, muddy or salt-covered roads.

- Operating on unpaved or dusty roads.

- Driving in extremely hot (over 90°) conditions.

Engine oil & filter: replace every 3000 miles.
Air cleaner filter: service or inspect every 6000 miles.
Exhaust system: check every 6000 miles.
Automatic transmission fluid & filter: change every 30,000 miles.
Transfer case fluid: change every 60,000 miles
Fule filter: change every 15,000 miles
Spark plugs: change every 60,000 miles
2wd front wheel bearings: lubricate every 30,000 miles

09482_MAZB_C0013

SCHEDULED MAINTENANCE INTERVALS
2003 B SERIES

TO BE SERVICED	TYPE OF SERVICE	VEHICLE MILEAGE INTERVAL (x1000)												
		5	10	15	20	25	30	35	40	45	50	55	60	65
Engine oil & filter	R	✓	✓	✓	✓	✓	✓	✓	✓	✓	✓	✓	✓	✓
Tires	Rotate	✓	✓	✓	✓	✓	✓	✓	✓	✓	✓	✓	✓	✓
Auto trans. fluid	I			✓			✓			✓			✓	
Brake pads/shoes	I			✓			✓			✓			✓	
Coolant hoses	S/I			✓			✓			✓			✓	
Steering linkage	I			✓			✓			✓			✓	
Suspension and driveshaft	I			✓			✓			✓			✓	
Cabin air filter	R			✓			✓			✓			✓	
Ball joints (2wd)	I			✓			✓			✓			✓	
Exhaust system	I						✓						✓	
Engine air filter	R						✓						✓	
Fuel filter ①	R						✓						✓	
Auto trans fluid ②	R						✓						✓	
Green coolant	R									✓				
Accessory drive belts	I	every 100,000 miles												
Spark plugs	R	every 100,000 miles												
PCV valve	R	every 100,000 miles												
Yellow coolant ③	R	first 100,000 miles or 5 years												
Auto trans fluid and filter ④	R	every 150,000 miles												
Rear axle fluid	R	every 150,000 miles												
Accessory drive belts ⑤	R	every 150,000 miles												

R: Replace   S: Service   I: Inspect   L: Lubricate

① Recommended, but not required in Calif.

② If equipped with AX4S, 4F50N, 4R100, 4F27E

③ Every 3 years or 50,000 miles thereafter

④ All except AX4S, 4F50N, 4R100, 4F27E

⑤ If not previously replaced

Special Operating Condition Requirements

When towing a trailer or using a camper or car-top carrier:
Change engine oil and install a new oil filter every 4,800 km (3,000 miles) or 3 months.
Change transfer case fluid every 96,000 km (60,000 miles).
Change manual transmission fluid as required.
Inspect and lubricate U-joints as required.

During extensive idling and/or low speed driving for long distances, as in heavy commercial use such as delivery, taxi, patrol car or livery:
Change engine oil and install a new oil filter, lube front lower control arm and steering linkage ball joints with
zerk fittings (if equipped) every 4,800 km (3,000 miles) or 3 months.
Inspect brake system and check battery electrolyte level (Patrol cars) every 8,000 km (5,000 miles).
Install a new fuel filter every 24,000 km (15,000 miles).
Change automatic transmission fluid, lubricate 4x2 wheel bearings,
 install new grease seals and adjust bearings every 48,000 km (30,000 miles).
Install new spark plugs and change transfer case fluid every 96,000 km (60,000 miles).
Install a new cabin air filter as required.

When operating in dusty conditions such as unpaved or dusty roads:
Change engine oil and install a new oil filter every 4,800 km (3,000 miles) or 3 months.
Install a new fuel filter every 24,000 km (15,000 miles).
Change automatic transmission fluid every 48,000 km (30,000 miles).
Change transfer case fluid every 96,000 km (60,000 miles).
Install a new engine air filter as required.
Install a new cabin air filter as required.

When operating in off-road conditions:
Change automatic transmission fluid every 48,000 km (30,000 miles).
Change transfer case fluid every 96,000 km (60,000 miles).
Install a new cabin air filter as required.
Inspect and lubricate U-joints.
Inspect and lubricate steering linkage ball joints with zerk fittings.

SCHEDULED MAINTENANCE INTERVALS
2004 B SERIES

TO BE SERVICED	TYPE OF SERVICE	VEHICLE MILEAGE INTERVAL (x1000)												
		5	10	15	20	25	30	35	40	45	50	55	60	65
Engine oil & filter	R	✓	✓	✓	✓	✓	✓	✓	✓	✓	✓	✓	✓	✓
Tires	Rotate	✓	✓	✓	✓	✓	✓	✓	✓	✓	✓	✓	✓	✓
Auto trans. fluid	I			✓			✓			✓			✓	
Brake pads/shoes	I			✓			✓			✓			✓	
Coolant hoses	S/I			✓			✓			✓			✓	
Steering linkage	I			✓			✓			✓			✓	
Suspension and driveshaft	I			✓			✓			✓			✓	
Cabin air filter	R			✓			✓			✓			✓	
Ball joints (2wd)	I/L			✓			✓			✓			✓	
Exhaust system	I						✓						✓	
Engine air filter	R						✓						✓	
Fuel filter	R						✓						✓	
Auto trans fluid ①	R						✓						✓	
Wheel bearings (2wd)	R	at 150,000 miles, if not previously replaced, including seals												
Coolant, exc. Premium Gold	R	every 105,000 miles												
Premium Gold coolant ③	R	at 5 years or 100,000 miles												
Spark plugs	R	every 100,000 miles												
PCV valve	R	every 100,000 miles												
Auto trans fluid ④	R	every 150,000 miles												
Fuel injectors	Clean	every 100,000 miles												
Rear axle fluid	R	every 150,000 miles												
Accessory drive belts	I	every 100,000 miles												
Accessory drive belts	R	every 150,000 miles, if not previously replaced												

R: Replace   S: Service   I: Inspect   L: Lubricate

① Vehicles equipped with AX4S, 4F50N, 4R100, 4F27E

② Change every 30,000 miles or 36 months thereafter

③ After the initial change, every 3 years or 50,000 miles

④ Except vehicles equipped with AX4S, 4F50N, 4R100, 4F27E

Special Operating Condition Requirements

When towing a trailer or using a camper or car-top carrier:

Change engine oil and install a new oil filter every 4,800 km (3,000 miles), 3 months or 200 hours of engine operation (whichever occurs first).

Change transfer case fluid every 96,000 km (60,000 miles).

Change manual transmission fluid as required.

Inspect and lubricate U-joints as required.

During extensive idling and/or low speed driving for long distances, as in heavy commercial use such as delivery, taxi, patrol car or livery:

Change engine oil and install a new oil filter every 4,800 km (3,000 miles), 3 months or 200 hours of engine operation (whichever occurs first).

Lube front lower control arm and steering linkage ball joints with zerk fittings (if equipped) every 4,800 km (3,000 miles) or 3 months.

Inspect brake system and check battery electrolyte level (Patrol cars) every 8,000 km (5,000 miles).

Install a new fuel filter every 24,000 km (15,000 miles).

Change automatic transmission fluid, lubricate 4x2 wheel bearings, install new grease seals and adjust bearings every 48,000 km (30,000 miles). If equipped, change the in-line service installed transmission fluid filter.

Install new spark plugs and change transfer case fluid every 96,000 km (60,000 miles).

Install a new cabin air filter as required.

09482_MAZB_C0015

SCHEDULED MAINTENANCE INTERVALS
2004 Mazda B-Series
Footnotes Continued

When operating in dusty conditions such as unpaved or dusty roads:

Change engine oil and install a new oil filter every 4,800 km (3,000 miles) or 3 months.

Install a new fuel filter every 24,000 km (15,000 miles).

Change automatic transmission fluid every 48,000 km (30,000 miles). If equipped, change the in-line service installed transmission fluid filter.

Change transfer case fluid every 96,000 km (60,000 miles).

Install a new engine air filter as required.

Install a new cabin air filter as required.

When operating in off-road conditions:

Change automatic transmission fluid every 48,000 km (30,000 miles). If equipped, change the in-line service installed transmission fluid filter.

Change transfer case fluid every 96,000 km (60,000 miles).

Install a new cabin air filter as required.

Inspect and lubricate U-joints.

Inspect and lubricate steering linkage ball joints with zerk fittings.

09482_MAZB_C0016

SCHEDULED MAINTENANCE INTERVALS
2005-2006 B SERIES

TO BE SERVICED	TYPE OF SERVICE	VEHICLE MILEAGE INTERVAL (x1000)												
		5	10	15	20	25	30	35	40	45	50	55	60	65
Engine oil & filter	R	✓	✓	✓	✓	✓	✓	✓	✓	✓	✓	✓	✓	✓
Tires	Rotate	✓	✓	✓	✓	✓	✓	✓	✓	✓	✓	✓	✓	✓
Auto trans. fluid	I									✓				
Brake pads/shoes	I			✓			✓			✓			✓	
Wheel ends ①	I			✓			✓			✓			✓	
Coolant hoses	S/I			✓			✓			✓			✓	
Steering linkage	I/L			✓			✓			✓			✓	
Suspension and driveshaft	I			✓			✓			✓			✓	
Cabin air filter	R			✓			✓			✓			✓	
Ball joints (2wd)	I/L			✓			✓			✓			✓	
Exhaust system	I						✓						✓	
Engine air filter	R						✓						✓	
Fuel filter	R						✓						✓	
Manual trans fluid	R												✓	
Wheel bearings (2wd)	R	at 150,000 miles, if not previously replaced, including seals												
Premium Gold coolant ②	R	at 5 years or 100,000 miles												
Spark plugs	R	every 100,000 miles												
PCV valve	R	every 100,000 miles												
Auto trans fluid	R	every 150,000 miles, if not previously replaced												
Rear axle fluid	R	every 150,000 miles												
Accessory drive belts	I	every 100,000 miles												
Accessory drive belts	R	every 150,000 miles, if not previously replaced												

R: Replace   S: Service   I: Inspect   L: Lubricate

① Check for play and noise

② After the initial change, every 3 years or 50,000 miles

Special Operating Condition Requirements

When towing a trailer or using a camper or car-top carrier:

Change engine oil and install a new oil filter every 4,800 km (3,000 miles), 3 months or 200 hours of engine operation (whichever occurs first).

Change transfer case fluid every 96,000 km (60,000 miles).

Change manual transmission fluid as required.

Inspect and lubricate U-joints as required.

During extensive idling and/or low speed driving for long distances, as in heavy commercial use such as delivery, taxi, patrol car or livery:

Change engine oil and install a new oil filter every 4,800 km (3,000 miles), 3 months or 200 hours of engine operation (whichever occurs first).

Lube front lower control arm and steering linkage ball joints with zerk fittings (if equipped) every 4,800 km (3,000 miles) or 3 months.

Inspect brake system and check battery electrolyte level (Patrol cars) every 8,000 km (5,000 miles).

Install a new fuel filter every 24,000 km (15,000 miles).

Change automatic transmission fluid, lubricate 4x2 wheel bearings, install new grease seals and adjust bearings every 48,000 km (30,000 miles). If equipped, change the in-line service installed transmission fluid filter.

Install new spark plugs and change transfer case fluid every 96,000 km (60,000 miles).

Install a new cabin air filter as required.

When operating in dusty conditions such as unpaved or dusty roads:

Change engine oil and install a new oil filter every 4,800 km (3,000 miles) or 3 months.

Install a new fuel filter every 24,000 km (15,000 miles).

Change automatic transmission fluid every 48,000 km (30,000 miles). If equipped, change the in-line service installed transmission fluid filter.

Change transfer case fluid every 96,000 km (60,000 miles).

Install a new engine air filter as required.

Install a new cabin air filter as required.

When operating in off-road conditions:

Change automatic transmission fluid every 48,000 km (30,000 miles). If equipped, change the in-line service installed transmission fluid filter.

Change transfer case fluid every 96,000 km (60,000 miles).

Install a new cabin air filter as required.

Inspect and lubricate U-joints.

Inspect and lubricate steering linkage ball joints with zerk fittings.

09482_MAZB_C0017A

ENGINE REPAIR

Alternator

REMOVAL

2.3L Engine

1. Before servicing the vehicle, refer to the Precautions Section.
2. Disconnect the negative battery cable.

➡**On some vehicles, when the battery cable is disconnected and reconnected, some abnormal drive symptoms may occur while the vehicle relearns its adaptive strategy. The vehicle may need to be driven to relearn its strategy.**

3. Remove or disconnect the following:
   • Air cleaner outlet tube
   • Accessory drive belt
   • Mounting bolts and alternator
   • Nut and the electrical connectors.
4. To install, reverse the removal procedure. Torque all mounting bolts to 18 ft. lbs. (25 Nm).

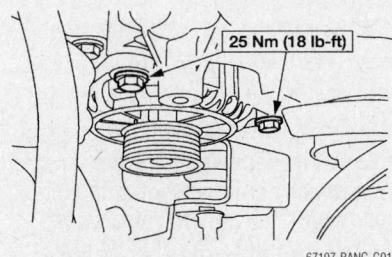

Alternator mounting bolts—2.3L engine

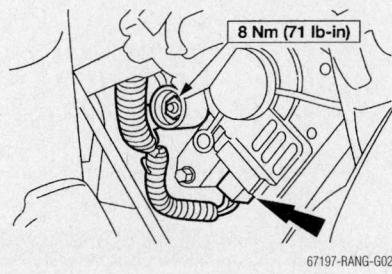

Alternator wiring connections—2.3L engine

3.0L Engine

1. Before servicing the vehicle, refer to the Precautions Section.
2. Disconnect the negative battery cable.

➡**On some vehicles, when the battery cable is disconnected and reconnected,**

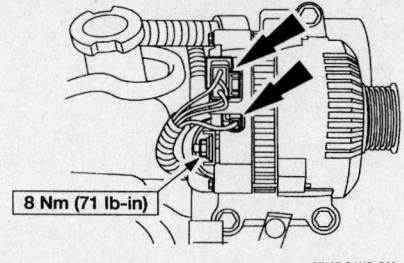

Alternator wiring connections—3.0L engine

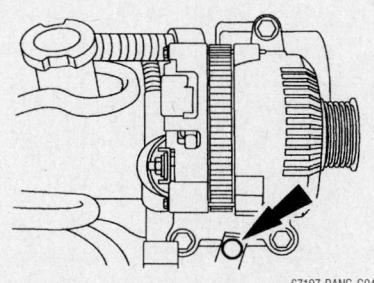

Wiring harness pin-type retainer—3.0L engine

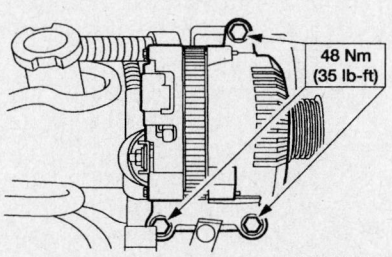

Alternator mounting bolts—3.0L engine

some abnormal drive symptoms may occur while the vehicle relearns its adaptive strategy. The vehicle may need to be driven to relearn its strategy.

3. Remove or disconnect the following:
   • Air cleaner outlet tube
   • Drive belt
   • Electrical connectors from the alternator
   • Wiring harness to alternator push pin
   • Alternator

To install:
4. Install or connect the following:
   • Alternator. Torque the bolts to 40 ft. lbs. (55 Nm) for 2002–2003; 35 ft. lbs (48 Nm) for 2004–2006.
   • Pushpin for the alternator wiring harness

• Electrical connectors to the alternator
• Drive belt
• Air cleaner outlet tube
• Negative battery cable

4.0L Engine

1. Before servicing the vehicle, refer to the Precautions Section.
2. Disconnect the negative battery cable.

➡**On some vehicles, when the battery cable is disconnected and reconnected, some abnormal drive symptoms may occur while the vehicle relearns its adaptive strategy. The vehicle may need to be driven to relearn its strategy.**

3. Remove or disconnect the following:
   • Air cleaner outlet tube

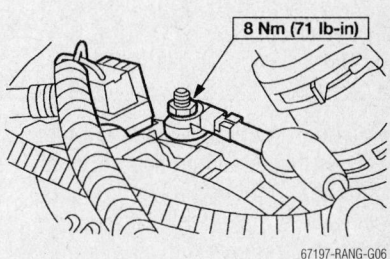

Alternator wiring connections—4.0L engine

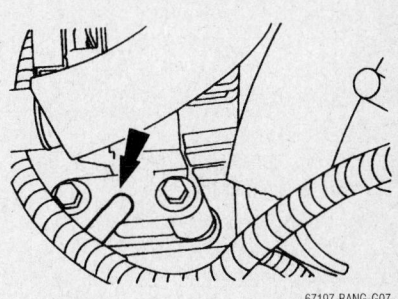

Alternator wiring pin-type retainer—4.0L engine

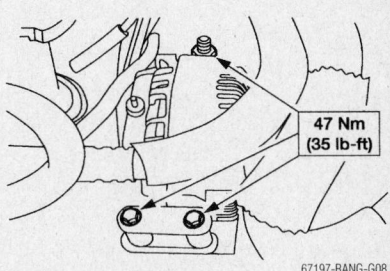

Alternator mounting bolts—4.0L engine

- Accessory drive belt
- Electrical connectors
- Wiring harness-to-alternator pin-type retainer
- Stud bolt, the bolts and the alternator

4. To install, reverse the removal procedure. Torque all mounting bolts to 35 ft. lbs. (47 Nm).

Ignition Timing

ADJUSTMENT

The ignition timing is preset and is not adjustable.

Engine Assembly

REMOVAL & INSTALLATION

2.3L Engine

1. Before servicing the vehicle, refer to the Precautions Section.
2. Relieve the fuel system pressure.
3. Disconnect the negative battery cable.

➡**On some vehicles, when the battery cable is disconnected and reconnected, some abnormal drive symptoms may occur while the vehicle relearns its adaptive strategy. The vehicle may need to be driven to relearn its strategy.**

4. Drain the cooling system.
5. Drain the engine oil.
6. Properly discharge the air conditioning system.
7. Remove or disconnect the following:
- Hood
- Accelerator control snow shield
- Air cleaner tube
- Upper radiator hose
- Lower radiator hose
- Fan and shroud
- PCM electrical connector. Remove the retaining nut on the harness clamp. Position the harness on the engine.
- Ground stud for the PCM
- Heater hoses
- All vacuum hoses
- Coolant reservoir hoses
- Air conditioning compressor clutch
- MAF electrical connector
- Air conditioning compressor manifold, plug the lines and the compressor ports
- Accelerator mazda speed control cables
- Power steering return hose

- PSP switch electrical connector
- High pressure power steering hose
- Fuel supply hose
- 42-pin electrical connector
- VMV vacuum regulator solenoid supply hose
- Evaporative purge hose
- Brake booster vacuum hose and the engine ground strap
- Solenoid control wire at the starter
- Starter wiring harness clamp bolt and position it out of the way.
- RH splash shield
- Alternator electrical connections
- Block heater electrical connector
- Front heated oxygen sensor electrical connector at the bell housing
- Oil pressure sensor electrical connector
- Engine wiring pushpins and position the engine wiring harnesses out of the way.
- Oil filter
- With automatic transmission, the bolt retaining the transmission cooling tubes to the engine. Remove the bracket.
- Transmission dust shield
- Starter motor
- Heated oxygen sensor electrical connector at the rear of the transmission
- Transmission wiring harness
- Vehicle speed sensor, transmission range sensor, backup light switch and the transmission electrical connectors. Disconnect the pushpins and position the harness forward to the engine.
- Oil filter adapter

➡**Leave two side bolts in until the engine is ready to be removed.**

- Nine of the transmission-to-engine bolts
- With automatic transmission, the transmission fluid indicator and tube assembly
- Starter dust shield

➡**Mark one stud and the flexplate for assembly reference.**

- With automatic transmission, the four torque converter nuts

8. Support the transmission with a floor jack.
9. Support the engine with a floor crane using a spreader bar.
10. Remove the two side transmission-to-engine bolts.
11. Remove the four engine support insulator.

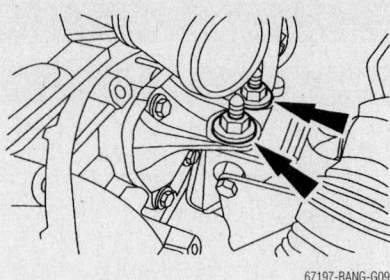

67197-RANG-G09

Engine support insulator nuts, left side shown, right side similar—2.3L engine

12. Remove the engine from the vehicle.
To install:
13. Installation is the reverse of removal. Observe the following torques:
- Torque converter bolts: 26 ft. lbs. (35 Nm)
- Nine transmission-to-engine bolts 35 ft. lbs. (48 Nm)
- Oil filter adapter: 18 ft. lbs. (25 Nm)
- Starter: 30 ft. lbs. (40 Nm)
- Engine support nuts: 75 ft. lbs. (102 Nm)

3.0L Engine

1. Before servicing the vehicle, refer to the Precautions Section.
2. Relieve the fuel system pressure.
3. Disconnect the negative battery cable.

➡**On some vehicles, when the battery cable is disconnected and reconnected, some abnormal drive symptoms may occur while the vehicle relearns its adaptive strategy. The vehicle may need to be driven to relearn its strategy.**

4. Drain the cooling system.
5. Drain the engine oil.
6. Properly discharge the air conditioning system.
7. Remove or disconnect the following:
- Hood
- Air cleaner outlet tube
- Upper and the lower radiator hoses

✳✳ WARNING

The fan clutch has left-hand threads.

- The fan clutch and blade as an assembly
- Drive belt
- Fan shroud
- Radiator
- Air conditioning manifold and tube. Remove the nut and position the line aside.
- Air conditioning compressor wiring

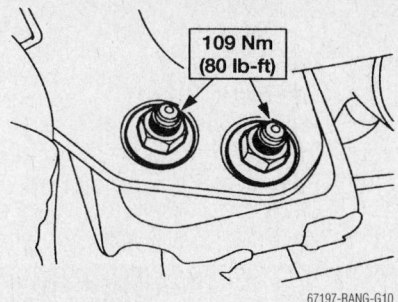

109 Nm
(80 lb-ft)

67197-RANG-G10

Engine support nuts, one side shown—
3.0L engine

- Air conditioning compressor and the air conditioning compressor mounting bracket
- Heater hoses
- Ground cable
- Fuel lines
- Snow shield
- Accelerator cable and the speed control actuator cable
- All vacuum lines
- 42-pin connector
- Powertrain control module connector
- Nut from the powertrain control module harness
- Stud bolt and the powertrain control module ground strap
- Alternator wiring and position aside
- Both heated oxygen sensors
- Transmission harness connectors
- MAF sensor
- LH heated oxygen sensor
- Dual converter Y pipe
- Starter motor and the starter grounding stud bolt
- Torque converter nuts
- 8 transmission-to-engine bolts

8. Install the lifting eyes.
9. Remove the four nuts.
10. Support the transmission.
11. Remove the engine from the vehicle.
To install:
12. Installation is the reverse of removal. Observe the following torques:
- Engine mount nuts: 80 ft. lbs. (109 Nm)
- Transmission-to-engine bolts: 33 ft. lbs. (45 Nm)
- Torque converter nuts: 2002–03 26 ft. lbs. (35 Nm); 2004–06 35 ft. lbs. (47 Nm).

4.0L Engine

1. Before servicing the vehicle, refer to the Precautions Section.
2. Relieve the fuel system pressure.
3. Disconnect the negative battery cable.

➥On some vehicles, when the battery cable is disconnected and reconnected, some abnormal drive symptoms may occur while the vehicle relearns its adaptive strategy. The vehicle may need to be driven to relearn its strategy.

✳✳ CAUTION

If the fuel supply manifold is used as a leverage device, damage may occur to the supply manifold. Care must be taken when working around the fuel supply manifold.

4. Remove or disconnect the following:
- Accelerator cable from engine
- Speed control cable from engine
- Radiator, the fan blade, and the fan shroud
- Accessory bracket bolts and position bracket aside
- Alternator wiring
- Wiring harness retainer and position alternator wiring away from engine
- Engine electrical connector
- PCM connector
- PCM ground wire
- Engine ground wire
- Brake booster vacuum hose
- Air conditioning high pressure switch electrical connector
- Bolt and position the air conditioning lines aside

➥Heater hose will be removed with engine.

- Heater hoses
- Fuel line
- Starter motor
- Engine oil
- Oil drain plug
- Transmission portion of wiring harness
- RH and LH heated oxygen sensor connectors
- Transmission control connector
- Output shaft speed sensor connector
- Digital transmission range sensor connector
- Catalyst monitor sensor electrical connector
- Transmission/transfer case portion of the wiring harness from any routing clips or pushpins. Route transmission/transfer case portion of the wiring harness to top of engine.
- Bolt, and position the transmission cooling line bracket aside
- Air conditioning line bracket nut and position it aside

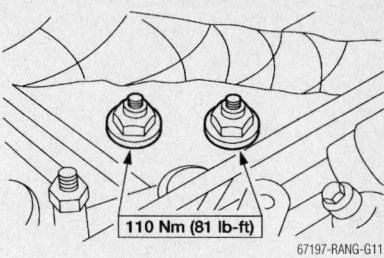

110 Nm (81 lb-ft)

67197-RANG-G11

Left side engine insulator nuts—4.0L engine

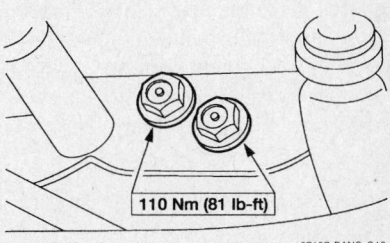

110 Nm (81 lb-ft)

67197-RANG-G12

Right side engine insulator nuts—4.0L engine

- Power steering return hose
- Power steering pressure hose
- Vapor management valve hose connector
- Eight bolts and the LH and the RH engine support insulator nuts

➥The lifting eyes should be installed on the exhaust manifold studs for number three and number four cylinders.

5. Install the lifting eyes.
6. Install the spreader bar to the lifting eyes.
7. Attach a floor crane to the spreader bar and remove the engine.
To install:
8. Installation is the reverse of removal. Observe the following torques:
- Left and right engine insulator nuts: 81 ft. lbs. (110 Nm)
- Engine mount nuts: 59 ft. lbs. (80 Nm)
- Transmission-to-engine bolts: 35 ft. lbs. (47 Nm)
- Torque converter nuts: 35 ft. lbs. (47 Nm)

Water Pump

REMOVAL & INSTALLATION

2.3L Engine

1. Before servicing the vehicle, refer to the Precautions Section.
2. Disconnect the negative battery cable.

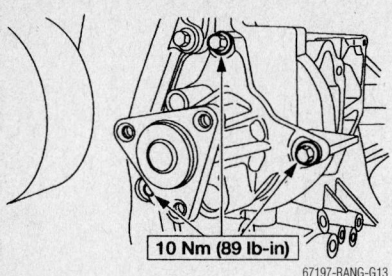

67197-RANG-G13

Water pump mounting bolts—2.3L engine

➡ On some vehicles, when the battery cable is disconnected and reconnected, some abnormal drive symptoms may occur while the vehicle relearns its adaptive strategy. The vehicle may need to be driven to relearn its strategy.

3. Drain the cooling system.
4. Remove the drive belt.
5. Remove the water pump pulley.
6. Remove the water pump.

To install:

7. Clean the mating surfaces where the water pump attaches to the engine.

➡ Lubricate the water pump O-ring, with MERPOL®, or equivalent.

8. To install, reverse the removal procedure. Torque the water pump mount bolts to 89 inch lbs. (10 Nm). Torque the pulley bolts to 18 ft. lbs. (25 Nm).

3.0L Engine

1. Before servicing the vehicle, refer to the Precautions Section.
2. Disconnect the negative battery cable.

➡ On some vehicles, when the battery cable is disconnected and reconnected, some abnormal drive symptoms may occur while the vehicle relearns its adaptive strategy. The vehicle may need to be driven to relearn its strategy.

3. Drain the cooling system.
4. Remove or disconnect the following:
   • Air cleaner outlet tube

67197-RANG-G14

Water pump mounting bolts—3.0L engine

• Fan and radiator shroud
• Water bypass tube
• Drive belt
• Heater hose
• Water pump pulley
• Lower radiator hose
• Air conditioning compressor and bracket assembly and move them aside
• Water pump

To install:

5. Clean the mating surfaces where the water pump attaches to the engine.
6. Install or connect the following:
   • Water pump. Torque the bolts to 89 inch lbs. (10 Nm).
   • Air conditioning compressor mounting bracket. Torque the bolts to 44 ft. lbs. (61 Nm).
   • Water pump pulley. Torque the bolts to 20 ft. lbs. (28 Nm) for 2002–04 models; 18 ft. lbs. (25 Nm) for 2006 models.
   • Drive belt
   • Heater hose
   • Lower radiator hose
   • Fan and shroud
   • Air cleaner outlet tube
   • Negative battery cable
7. Fill the cooling system.
8. Start the vehicle and check for leaks, repair if necessary.

4.0L Engine

1. Before servicing the vehicle, refer to the Precautions Section.
2. Disconnect the negative battery cable.

➡ On some vehicles, when the battery cable is disconnected and reconnected, some abnormal drive symptoms may occur while the vehicle relearns its adaptive strategy. The vehicle may need to be driven to relearn its strategy.

3. Drain the cooling system.
4. Remove or disconnect the following:
   • Fan shroud
   • Accessory drive belt

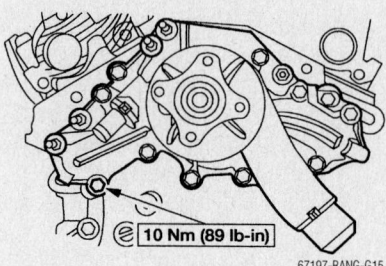

10 Nm (89 lb-in)

67197-RANG-G15

Water pump mounting bolts—4.0L engine

• Idler pulley
• Water bypass hose
• Heater hose
• Lower radiator hose
• Water pump pulley
• Water pump

To install:

5. Clean the mating surfaces where the water pump attaches to the engine.

✳✳ WARNING

Use care when scraping the water pump-to-engine block mating surfaces. Gouges in the aluminum could form leak paths.

6. Clean all the sealing surfaces.
7. To install, reverse the removal procedure. Torque the water pump bolts to 89 inch lbs. (10 Nm). Torque the pulley bolts to 18 ft. lbs. (25 Nm).

Heater Core

REMOVAL & INSTALLATION

2002–2003

1. Before servicing the vehicle, refer to the Precautions Section.
2. Disconnect the negative battery cable.

➡ On some vehicles, when the battery cable is disconnected and reconnected, some abnormal drive symptoms may occur while the vehicle relearns its adaptive strategy. The vehicle may need to be driven to relearn its strategy.

✳✳ CAUTION

After disconnecting the negative battery cable, wait for 1 minute for the SRS module to deplete its energy.

3. Drain the cooling system.
4. Remove the steering column by performing the following procedure:
   a. Position the front wheels in the straight-ahead direction.
   b. At the both sides of the steering wheel, remove the cover plugs, the steering wheel-to-air bag module screws, disconnect the air bag electrical connector and carefully remove the air bag module.

✳✳ CAUTION

Safely store the air bag module with the front side facing upward.

c. Remove the steering wheel-to-steering column nut.

d. Using a steering wheel puller, press the steering wheel from the steering column.

e. Remove the parking brake release handle screws and move the release handle aside.

f. Remove the hood release screws and move the hood release aside.

g. Remove the 2 instrument panel-to-steering column cover screws and the cover.

h. Remove the instrument panel steering column opening reinforcement bolts and the reinforcement.

i. Remove the ignition switch bolt and disconnect the ignition switch electrical connector.

j. At the base of the steering column, disconnect the electrical connectors.

k. If equipped with an automatic transmission, remove the transmission range indicator bolt and the cable.

l. If equipped with an automatic transmission, disconnect the shift cable from the steering column shift tube lever and the steering column bracket.

m. Disconnect the brake shift interlock solenoid electrical connector.

n. Remove the air bag sliding contact.

o. Remove the upper intermediate steering shaft-to-column shaft bolt and discard the bolt.

p. Remove the lower steering column-to-instrument panel nuts and the steering column.

5. Remove the instrument panel by performing the following procedure:

a. Remove the parking brake release handle screws and move the release handle aside.

b. Disconnect the Brake Pedal Position (BPP) switch electrical connector.

c. Remove both front door scuff plates.

d. Remove the push pins and remove both cowl side trim panels.

e. At the right side cowl panel, disconnect the electrical connectors and ground wires.

f. Remove both sides windshield garnish moldings.

g. Remove the instrument panel fuse door.

h. Disconnect the power distribution box from its bracket and move it aside.

i. In the engine compartment, loosen the bulkhead wiring harness bolts and disconnect the electrical connectors.

j. Pull the bulkhead electrical connector handle and disconnect the wiring harness.

k. Remove the passenger's side air bag module-to-instrument panel screws, disconnect the electrical connector and remove the air bag module.

✳✳ WARNING

Store the air bag module in a safe location with the front facing upward.

l. Disconnect the blend door actuator's electrical connector.

m. Disconnect the climate control vacuum harness connector.

n. Disconnect the radio's antenna connector.

o. Remove the glove compartment.

p. Remove the instrument panel defroster grille.

q. Remove the upper instrument panel bolts.

r. Under the steering column, remove the instrument panel brace bolt.

s. Remove both the right and left instrument panel-to-cowl bolts.

t. Pull the instrument panel away from the dash.

u. Loosen the instrument panel-to-body harness bolt and disconnect the harness.

v. Using an assistant, remove the instrument panel.

6. Remove the evaporator core by performing the following procedure:

a. Discharge and recover the air conditioning system refrigerant.

b. Remove the refrigerant lines from the evaporator core. Discard the O-rings.

c. If equipped, remove the air conditioning vacuum reservoir tank/bracket screws and reposition the tank.

d. If equipped, disconnect the speed

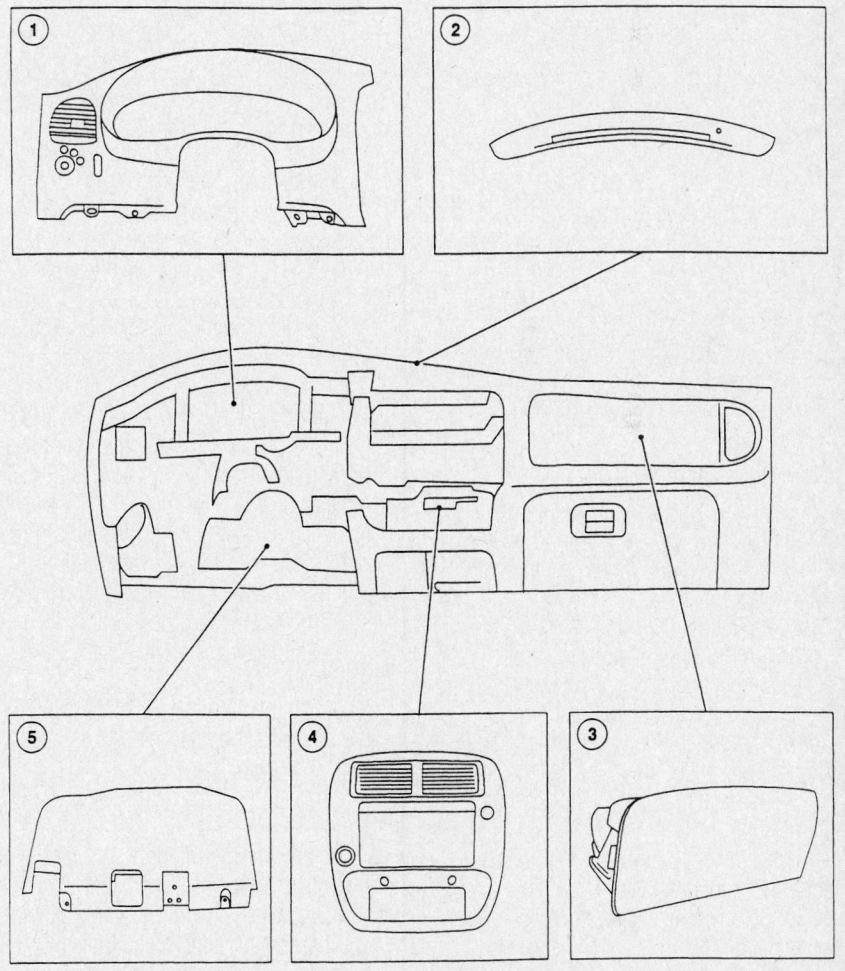

1 Instrument Panel Finish Panel
2 Instrument Panel Defroster
  Opening Grille Assembly
3 Passenger Side Air Bag Module
4 Instrument Panel Center Finish Panel
5 Instrument Panel Steering
  Column Cover

93113GL3

Instrument panel and related components—2002–2005

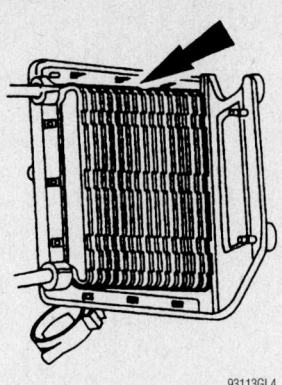

Evaporator core—2002–2003

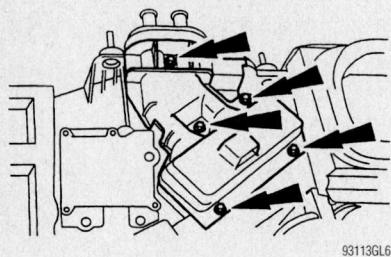

Heater core cover—2002–2003

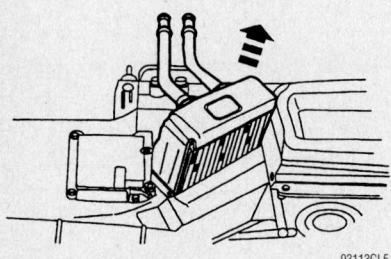

Heater core—2002–2003

control servo connector; then, remove the bolt and reposition the speed control servo.

e. If equipped with a 3.0L or 4.0L engine, remove the support bracket.

f. Disengage the windshield washer hose retainer and move it aside.

g. Disconnect the vacuum hose and the retainer; then, move the hose aside.

h. Remove the passenger's compartment nut.

i. At the back of the engine, remove the hose support bolts.

j. Remove the evaporator housing-to-chassis nuts.

k. Remove the air conditioning accumulator bracket screws.

l. Remove the evaporator housing cover screws, clips and the cover.

m. Remove the evaporator core from the housing.

7. Disconnect the heater hoses from the heater core.

8. Remove the heater housing plenum chamber nuts and the plenum chamber.

9. Remove the heater core-to-heater housing screws and the cover.

10. Remove the heater core.

To install:

11. Install the heater core.

12. Install the heater core-to-heater housing cover and the cover screws.

13. Install the heater housing plenum chamber and the plenum chamber nuts.

14. Connect the heater hoses to the heater core.

15. Install the evaporator core by performing the following procedure:

a. Install the evaporator core to the housing.

b. Install the evaporator housing cover, clips and the cover screws.

c. Install the air conditioning accumulator bracket screws.

d. Install the evaporator housing-to-chassis nuts.

e. At the back of the engine, install the hose support bolts.

f. Install the passenger's compartment nut.

g. Connect the vacuum hose and the retainer.

h. Engage the windshield washer hose retainer.

i. If equipped with a 3.0L or 4.0L engine, install the support bracket.

j. If equipped, install the speed control servo bolt and connect the connector.

k. If equipped, install the air conditioning vacuum reservoir tank and the bracket screws.

l. Using new O-rings, install the refrigerant lines to the evaporator core.

16. Install the instrument panel by performing the following procedure:

a. Using an assistant, install the instrument panel.

b. Connect the harness and tighten the instrument panel-to-body harness bolt.

c. Push the instrument panel toward the dash.

d. Install both the right and left instrument panel-to-cowl bolts.

e. Under the steering column, install the instrument panel brace bolt.

f. Install the upper instrument panel bolts.

g. Install the instrument panel defroster grille.

h. Install the glove compartment.

i. Connect the radio's antenna connector.

j. Connect the climate control vacuum harness connector.

k. Connect the blend door actuator's electrical connector.

l. Install the passenger's side air bag module, connect the electrical connector and torque the air bag module-to-instrument panel screws to 67–92 inch lbs. (7.6–10.4 Nm).

m. Connect the bulkhead electrical connector handle wiring harness.

n. In the engine compartment, connect the electrical connectors and tighten the bulkhead wiring harness bolts.

o. Connect the power distribution box to its bracket.

p. Install the instrument panel fuse door.

q. Install both sides windshield garnish moldings.

r. At the right side cowl panel, connect the electrical connectors and ground wires.

s. Install both cowl side trim panels and the push pins.

t. Install both front door scuff plates.

u. Connect the brake pedal position (BPP) switch electrical connector.

v. Install the parking brake release handle and the release handle aside screws.

17. Install the steering column by performing the following procedure:

a. Install the lower steering column and the steering column-to-instrument panel nuts; then, torque the nuts to 10–13 ft. lbs. (13–17 Nm).

b. Using a new bolt, install the upper intermediate steering shaft-to-column shaft bolt and torque to 19–25 ft. lbs. (26–34 Nm).

c. Install the air bag sliding contact.

d. Connect the brake shift interlock solenoid electrical connector.

e. If equipped with an automatic transmission, connect the shift cable from the steering column shift tube lever and the steering column bracket.

f. If equipped with an automatic transmission, install the transmission range indicator cable and bolt.

g. At the base of the steering column, connect the electrical connectors.

h. Connect the ignition switch electrical connector and install the ignition switch bolt.

i. Install the instrument panel steering column opening reinforcement and the reinforcement bolts.

j. Install the instrument panel-to-steering column cover and the 2 cover screws.

k. Install the hood release and the hood release screws.

l. Install the parking brake release handle and the release handle screws.

m. Install the steering wheel to the steering column.

n. Install the steering wheel-to-steering column nut and torque the nut to 25–34 ft. lbs. (34–46 Nm).

o. At the both sides of the steering wheel, install the air bag module, connect the air bag electrical connector, install the steering wheel-to-air bag module screws and the cover plugs.

18. Refill the cooling system.

19. Connect the negative battery cable.

20. Evacuate and charge the air conditioning system.

21. Run the engine to normal operating temperatures; then, check the climate control operation and check for leaks.

2004–2006

1. Before servicing the vehicle, refer to the Precautions Section.

2. Disconnect the negative battery cable.

➡**On some vehicles, when the battery cable is disconnected and reconnected, some abnormal drive symptoms may occur while the vehicle relearns its adaptive strategy. The vehicle may need to be driven to relearn its strategy.**

✳✳ CAUTION

After disconnecting the negative battery cable, wait for 1 minute for the SRS module to deplete its energy.

3. Depower the SRS system.

4. Discharge the air conditioning system.

5. Drain the cooling system.

6. Remove the suction accumulator.

7. If equipped with the 2.3L engine, remove the A/C compressor and the engine oil indicator and tube.

8. If equipped with the 3.0L or 4.0L engine, position the coolant reservoir and windshield washer reservoir aside.

9. Detach the speed control servo, on vehicles equipped with cruise control.

10. Disconnect the blower motor and blower motor resistor electrical connectors.

11. Disconnect the heater hoses.

12. Detach the pin-type retainer and position aside the windshield washer hose.

13. Disconnect and detach the heater control valve vacuum hose.

14. Disconnect the vacuum supply hose near the evaporator core housing.

15. Disconnect the condenser to evaporator line spring lock coupling from the evaporator core inlet. Discard the O-ring seals.

➡**This step is performed at the lower passenger side dash panel, inside the passenger compartment.**

16. Disconnect the vacuum hose connector and remove the nut.

17. If equipped with the 4.0L engine, remove the vehicle splash guard. Remove the splash shield. On the right splash shield, disconnect the vacuum lines from the vacuum storage reservoir.

➡**Remove the electrical connector locators from the splash shield prior to removal.**

18. Remove the nuts and the evaporator core housing.

19. Lock the steering column.

20. Remove the front seats.

21. Remove the screws and position the parking brake release handle aside.

22. Remove the left and right lower cowl kick panels.

23. Position the parking brake assembly aside.

24. Remove the screws and position the hood release handle aside.

25. Remove the instrument panel steering column cover.

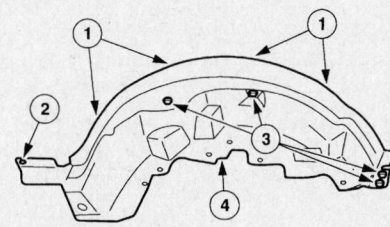

1 Remove the screws.
2 Remove the pushpin.
3 Remove the bolts.
4 Remove the splash shield.

09482_MAZB_G0003

Splash shield removal points

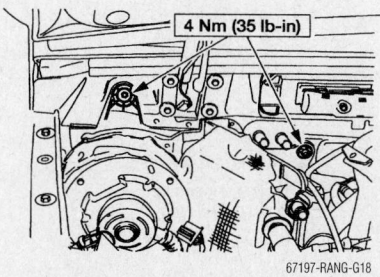

67197-RANG-G18

**Evaporator core housing bolts—
2004–2006**

26. Remove the instrument panel steering column opening cover reinforcement.

27. Disconnect the brake pedal position (BPP) switch electrical connector from the steering column shaft.

28. If equipped, disconnect the clutch pedal position (CPP) switch electrical connector.

29. If equipped, disconnect the shift cable from the steering column.

30. Remove the steering column pinch bolt and disconnect the steering column intermediate shaft.

➡**To avoid damage to the clockspring, do not allow the steering shaft to rotate while the intermediate shaft is disconnected.**

31. Remove the left and right side garnish moldings.

a. Remove the bolt covers and remove the bolts.

b. Remove the assist handle.

c. Remove the windshield side garnish molding.

32. Remove the door moldings.

33. On regular cab vehicles

a. Remove the screws.

b. Remove the scuff plate.

34. Disconnect the electrical connectors and the ground wire from the RH side lower kick panel.

➡**To avoid damaging the bulkhead electrical connectors, be sure the release tab is fully depressed before pulling release lever into the disconnect position.**

35. Disconnect the LH side bulkhead electrical connector. Press the release tab and pull the release lever.

36. Remove the audio unit. Insert the removal tool. Remove and support the audio unit.

37. Disconnect the audio unit electrical connector and antenna cable.

38. Lower the glove compartment. Press the release tabs inward while lowering the compartment.

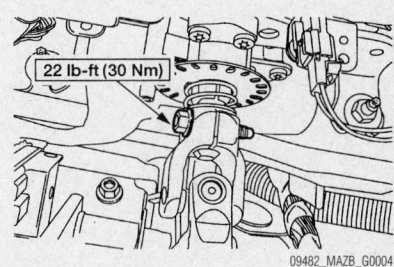

09482_MAZB_G0004

Steering column pinch bolt

39. Through the glove compartment opening, disconnect the blend door actuator electrical connector.

40. Through the glove compartment opening, disconnect the climate control vacuum harness connector.

41. Raise and secure the glove compartment. Press the release tabs inward while raising the glove compartment.

42. Remove the instrument panel defroster opening grille.

43. Remove the instrument panel cowl top bolts.

44. If equipped, remove the floor console.

45. If not equipped with the high-series floor console, remove the cup holders.

46. Release the clips and remove the restraints control module (RCM) cover.

47. If not equipped with high-series floor console, remove the consolette mat.

48. Remove the screws and remove the restraints control module (RCM) cover.

49. Remove the consolette base.

50. Remove the gearshift lever.

51. Remove the screws and remove the manual transmission consolette.

52. Disconnect the RCM electrical connector.

53. Pull the floor carpeting back.

54. On 2WD vehicles disconnect the instrument panel main harness.

55. On 4WD vehicles disconnect the instrument panel main harness. From underneath the vehicle, release the instrument panel main harness at the transfer case.

56. Remove the instrument panel side finish panel.

57. Disconnect the door harness electrical connector.

58. Remove the RH side instrument panel bolt. If necessary, transfer the components to the new instrument panel.

59. Remove the LH instrument panel cowl side bolts.

60. Remove the instrument panel.

61. Remove the powertrain control module (PCM).

62. Remove the PCM heat sink.

63. Remove the four nuts from the engine side of the dash panel. Position the plenum chamber on the vehicle floor.

64. Remove the heater core cover.

65. Remove the heater core.

To install:

66. To install, reverse the removal procedure.

67. During installation, be sure to install a new oval foam seal around the heater core inlet and outlet tubes.

68. Torque the steering column pinch bolt to 21 ft. lbs.

➡**To avoid damage to the clockspring, do not allow the steering shaft to rotate while the intermediate shaft is disconnected.**

69. Lubricate the refrigerant system with the correct amount of clean PAG oil or equivalent. Install new O-ring seals lubricated in clean mineral oil. Lubricate the coolant hoses with plain water only, if needed.

70. Evacuate, leak test, and charge the refrigerant system.

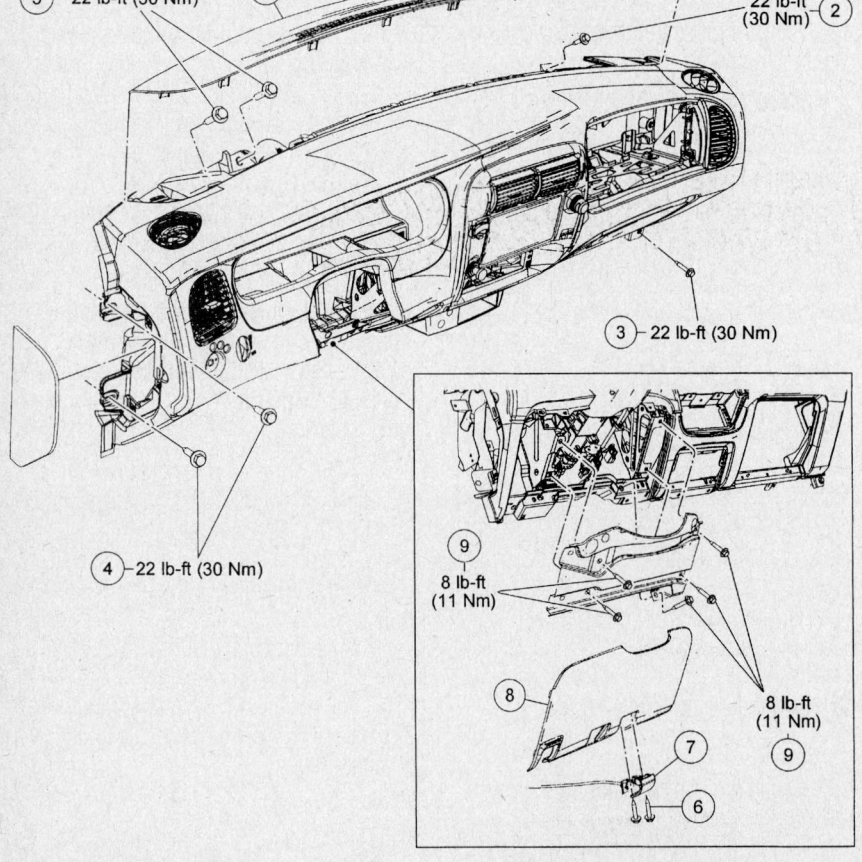

1	Instrument panel defroster opening grille	6	Hood release handle bolt (2 required)
2	Instrument panel cowl top bolt	7	Hood release handle
3	Instrument panel bolt	8	Steering column opening panel cover
4	Instrument panel bolts	9	Opening cover reinforcement panel bolts
5	Instrument panel cowl top bolts		

09482_MAZB_G0005

Instrument panel and related components—2006

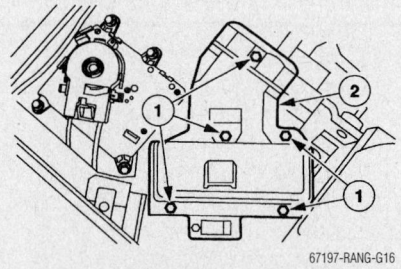

67197-RANG-G16

Heater core cover bolts—2004–2006

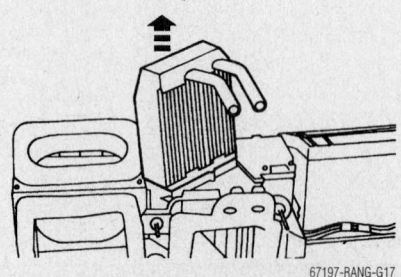

67197-RANG-G17

Heater core removal—2004–2006

Cylinder Head

REMOVAL & INSTALLATION

2.3L Engine

1. Before servicing the vehicle, refer to the Precautions Section.
2. Relieve the fuel system pressure.
3. Disconnect the negative battery cable.

➡**On some vehicles, when the battery cable is disconnected and reconnected, some abnormal drive symptoms may occur while the vehicle relearns its adaptive strategy. The vehicle may need to be driven to relearn its strategy.**

4. Drain the cooling system.
5. Properly discharge the air conditioning system.
6. Remove or disconnect the following:
   - Drive belt.
   - Engine oil level indicator assembly.
   - Engine oil level indicator.
   - Engine oil level indicator tube.
   - Water outlet tube.
   - Water outlet tube.
   - Air conditioning compressor.

➡**The alternator will be removed with the accessory bracket.**

   - Accessory bracket.
   - Right motor mount.
   - Coolant hose from the thermostat.
   - Coolant hose from the EGR valve.
   - Coolant tube assembly.
   - Exhaust manifold and gasket.
   - Block heater (if so equipped).
   - Water outlet.
   - EGR valve.
   - Power steering pump and reservoir as an assembly.
   - Idle air control (IAC) valve.
   - Throttle position (TP) sensor.
   - Manifold absolute pressure (MAP) sensor.
   - Swirl control valve monitor electrical connector.
   - CKP sensor and the wiring harness pin-type retainers.
   - Knock sensor (KS).
   - Electric thermostat.
   - Swirl control valve.
   - CMP sensor electrical connector and disconnect the PCV hose from the intake manifold.
   - Engine wiring harness pin-type retainers from the intake manifold.
   - Engine wiring harness connector

bracket. Position the engine wiring harness aside.
   - EGR tube.
   - Fuel supply line clip from the front of the intake manifold. Disconnect the vacuum hose from the intake manifold.
   - Intake manifold assembly.
   - Fuel injector electrical connectors. Detach the wiring harness pin-type retainers.
   - Ignition coil and the cylinder head temperature (CHT) sensor electrical connectors.
   - Engine wiring harness anchors from the valve cover studs. Remove the engine wiring harness.
   - Ignition coil.
   - Bypass hose.
   - Thermostat housing.
   - Knock sensor and the engine vent cover.
   - Left motor mount.
   - Fuel injector supply manifold with the injectors and the ground strap.
   - Water pump pulley.
   - Water pump.
   - CMP sensor.
   - CHT sensor.
   - Spark plugs.
   - Valve cover.
   - CKP sensor.
   - Crankshaft vibration damper

➡**There is one front cover bolt behind the cooling fan drive pulley. To remove this bolt, align one of the cooling fan drive pulley access holes with the bolt head to access the bolt.**

   - Front cover.
   - Timing chain tensioner.
   - Timing chain guides.
   - Timing chain assembly.

➡**Use a wrench on the flats between cylinders No. 1 and No. 2 to hold the camshaft in place.**

   - Camshaft drive sprockets.
   - Oil pump chain tensioner and guide.

➡**The oil pump chain sprocket must be held in place.**

   - Oil pump chain and sprockets.

➡**Note the position of the lobes on the No. 1 cylinder before removing the camshafts for assembly reference.**

7. Loosen the camshaft bearing cap bolts in sequence, one turn at a time. Repeat the first step until all tension is released from the camshaft bearing caps. Remove the camshaft bearing caps.

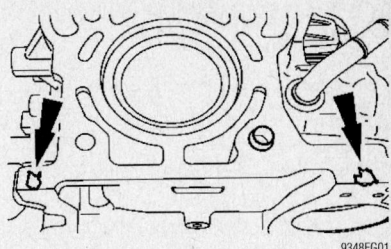

RTV sealer application—2.3L engine

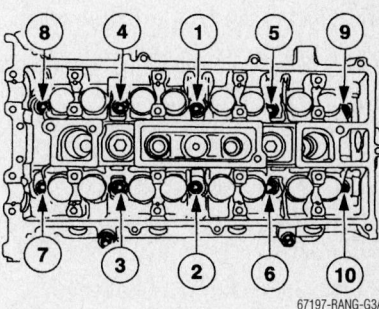

Cylinder head bolt torque sequence—2.3L engine

8. Remove or disconnect the following:
   - Camshafts.
   - Cylinder head bolts and the cylinder head.
   - Cylinder head gasket.
9. Installation is the reverse of removal. Apply RTV sealer to the places shown. The head must be installed within 4 minutes of application. Observe the following torques:
   a. Cylinder head:
   - Step 1: Tighten the bolts to 5 Nm (44 inch lbs.)
   - Step 2: Tighten the bolts to 15 Nm (11 ft. lbs.)
   - Step 3: tighten the bolts to 45 Nm (33 ft. lbs.)
   - Step 4: Tighten the bolts an additional 90 degrees (¼ turn)
   - Step 5: Tighten the bolts an additional 90 degrees (¼ turn)
   b. Camshafts:

➡**Install the camshafts with the alignment notches in the camshaft lined up so the camshaft alignment plate can be installed without rotating the camshafts. Make sure the lobes on the No. 1 cylinder are in the same position as noted in the disassembly procedure. Rotating the camshafts, or installing the camshafts 180 degrees out of position can cause severe damage to the valves and pistons. Lubricate the camshaft journals and bearing caps with clean engine oil. Install the camshafts and bearing caps. Tighten** the bolts in the sequence shown in three stages.

- Step 1: Tighten the camshaft bearing caps one turn at a time until tight.
- Step 2: Tighten the bolts to 7 Nm (62 inch lbs.)
- Step 3: Tighten the bolts to 16 Nm (12 ft. lbs.)
- c. Crankshaft vibration damper:

➡ **Do not reuse the crankshaft pulley bolt. Tighten the bolt in two stages.**

- Step 1: Tighten the bolt to 40 Nm (30 ft. lbs.)
- Step 2: Tighten the bolt and additional 90 degrees (¼ turn).

3.0L Engine

2002–2004

➡ **It may be easier to remove the engine from the vehicle. If removing the engine, refer to the engine removal procedure.**

1. Before servicing the vehicle, refer to the Precautions Section.
2. Disconnect the negative battery cable.

➡ **On some vehicles, when the battery cable is disconnected and reconnected, some abnormal drive symptoms may occur while the vehicle relearns its adaptive strategy. The vehicle may need to be driven to relearn its strategy.**

3. Evacuate the air conditioning system.
4. Drain the cooling system.
5. Drain the engine oil.
6. Remove or disconnect the following:
   - Lower intake manifold
   - Air conditioning compressor
   - Alternator
   - Power steering pump

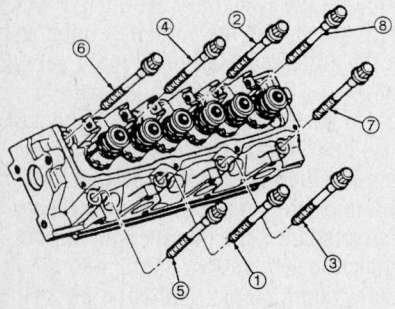

7924EG03

Cylinder head bolt torque sequence— 2002–2004 3.0L engine

- Alternator mounting bracket
- Air conditioning compressor mounting bracket
- Exhaust manifolds
- Cylinder head and discard the bolts and gasket

To install:

➡ **The "V" in the cylinder head gasket must face the front of the engine.**

7. Clean the mating surfaces where the head attaches to the engine.
8. Install a new cylinder head gasket and the cylinder head to the engine.
9. Torque the new cylinder head bolts in stages as follows:
   - a. Step 1: 59 ft. lbs. (80 Nm).
   - b. Step 2: Loosen the bolts one full turn.
   - c. Step 3: 40 ft. lbs. (55 Nm).
   - d. Step 4: 63 ft. lbs. (85 Nm).
10. Install or connect the following:
    - Lower intake manifold
    - Exhaust manifold
    - Air conditioning compressor mounting bracket. Torque the bolts to 44 ft. lbs. (66 Nm).
    - Alternator mounting bracket
    - Power steering pump
    - Alternator
    - Air conditioning compressor
    - Negative battery cable
11. Fill the engine with clean oil.
12. Fill the cooling system.
13. Recharge the air conditioning system
14. Start the vehicle and check for leaks, repair if necessary.

2005–2006

1. Before servicing the vehicle, refer to the Precautions Section.
2. With the vehicle in NEUTRAL, position it on a hoist.
3. Disconnect the negative battery cable.

➡ **On some vehicles, when the battery cable is disconnected and reconnected, some abnormal drive symptoms may occur while the vehicle relearns its adaptive strategy. The vehicle may need to be driven to relearn its strategy.**

4. Remove the lower intake manifold.

✳✳ WARNING

Refrigerant compressor oil (mineral oil) F73Z-19577-AA (Motorcraft YN-9-A) should be used to lubricate R-134a refrigerant system O-ring seals only and should not be added to the

R-134a refrigerant system as an A/C compressor lubricant. PAG refrigerant compressor oil F7AZ-19D589-DA (Motorcraft YN-12-C) or equivalent meeting Ford specification WSH-M1C231-B only should be used as an A/C compressor lubricant.

➡ **Installation of a new suction accumulator is not required when repairing the air conditioning system except when there is physical evidence of system contamination from a failed A/C compressor or damage to the suction accumulator.**

5. If flushing of the refrigerant system has not been carried out, recover the refrigerant.
6. Position the air cleaner outlet tube aside.
7. Remove the drive belt from the A/C compressor pulley.
8. Disconnect the electrical connector.
9. Remove the clamp nut on the compressor manifold and tube assembly.
10. Loosen the bolt and disconnect the compressor manifold and tube assembly from the A/C compressor.
11. Remove the bolts and the A/C compressor. Discard the O-ring seals.
12. Remove the alternator.
13. Remove the power steering pump pulley.
14. Disconnect the power steering pressure hose. Remove and discard the seal.
15. Disconnect the power steering return hose.
16. Disconnect the power steering return line hose.
17. Remove the bolts and the power steering pump.
18. Remove the alternator mounting bracket.
19. Remove the rocker arms and the push rods.
20. Remove the nuts and detach the fuel tube bracket and the wiring harness bracket.
21. Remove the two studbolts from the

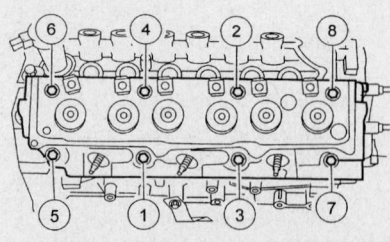

09482_MAZB_G0006

Cylinder head bolt torque sequence— 2005–2006 3.0L engine

rear of the A/C compressor mounting bracket.

22. Remove the bolts and remove the front of the A/C compressor mounting bracket.

23. Remove the exhaust manifolds.

24. Remove and discard the eight cylinder head bolts.

25. Remove the cylinder heads.

26. Remove and discard the cylinder head gaskets.

Installation

27. Clean all the sealing surfaces.

28. Check the cylinder head for flatness.

➡**The "V" notch in the head gasket faces the front of the engine.**

29. Install new cylinder head gaskets.

30. Install the cylinder heads.

31. Install the bolts. Tighten the bolts in five steps in the sequence shown.

- Step 1: Tighten to 50 Nm (37 ft. lbs.).
- Step 2: Loosen the bolts one full turn.
- Step 3: Tighten to 30 Nm (22 ft. lbs.).
- Step 4: Tighten the bolts 90 degrees.
- Step 5: Tighten the bolts an additional 90 degrees.

32. Install the lower intake manifold.

33. Install the exhaust manifolds.

34. Install the A/C compressor mounting bracket. Torque to 34 ft. lbs. (46 Nm).

35. Install the studbolts to the A/C compressor mounting bracket. Torque to 35 ft. lbs. (48 Nm).

36. Position the wiring harness bracket and the fuel tube bracket and install the nuts.

37. Install the push rods and the rocker arms. Torque to 24 ft. lbs. (32 Nm).

38. Install the alternator mounting bracket. Install the bolts. Torque to 34 ft. lbs. (46 Nm).

39. Install the power steering pump. Torque to 35 ft. lbs. (48 Nm).

40. Install the alternator.

41. Install the A/C compressor. Torque to 18 ft. lbs. (25 Nm).

42. Connect the battery ground cable.

43. Drain the engine oil.

44. Fill the engine with clean engine oil.

45. Fill and bleed the cooling system.

46. Fill the power steering system.

47. Charge the A/C system.

4.0L Engine

➡**If only one cylinder head is to be removed, only follow the procedures**

that apply. The following tools, or their equivalents are absolutely necessary to properly perform this procedure:

- Cam Chain Tensioner tool T97T-6K254-A
- Cam Gear Removal tool T97T-6256-F
- Cam Gear Torque adapter T97T-6256-G
- Camshaft Gear Positioning/Holding tool T97T-6256-B
- Camshaft Gear Positioning/Holding tool adapter T97T-6256-A
- Camshaft holding tool T97T-6256-C
- Crankshaft holding tool T97T-6303-A
- Camshaft holding tool adapter T97T-6256-D

1. Before servicing the vehicle, refer to the Precautions Section.

2. Properly relieve the fuel system pressure.

3. Disconnect the negative battery cable.

➡**On some vehicles, when the battery cable is disconnected and reconnected, some abnormal drive symptoms may occur while the vehicle relearns its adaptive strategy. The vehicle may need to be driven to relearn its strategy.**

4. Drain the cooling system.

5. Remove or disconnect the following:
- Lower intake manifold
- Fan blade and shroud
- Valve cover
- Roller followers, if equipped
- Drive belt
- Upper radiator hose and tube
- Alternator electrical connectors
- Alternator mounting bracket
- Engine accessory bracket and move it aside
- Camshaft Position (CMP) electrical connector
- Crankshaft Position (CKP) sensor electrical connector
- Engine Coolant Temperature (ECT) sensor electrical connector
- Coil pack electrical connector
- Exhaust Gas Recirculation (EGR) valve electrical connector
- EGR valve bracket and move it aside
- Heater hoses
- Fuel injector electrical connectors
- Water bypass hose
- Thermostat housing
- Spark plug wires

Cylinder head bolt loosening sequence–4.0L engine, early build head

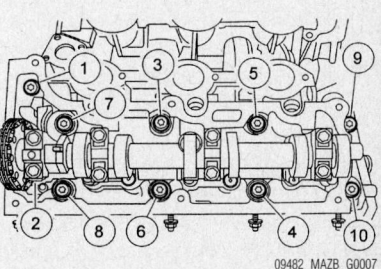

Cylinder head bolt loosening sequence—4.0L engine, late build head

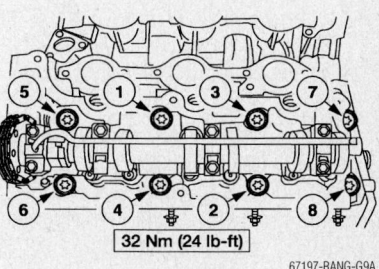

Cylinder head bolt torque sequence–4.0L engine, early build head

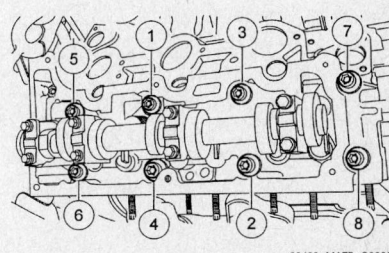

Cylinder head bolt torque sequence—4.0L engine, late build head

- Fuel injection supply manifold
- Fuel injectors
- Crankcase vent separator spring
- Oil dipstick housing
- Exhaust manifold
- Hydraulic chain tensioner
- Cassette retaining bolt
- Camshaft sprocket
- Cylinder head and discard the gasket. Discard the head bolts.

To install:

6. Thoroughly clean all gasket mating surfaces. Remove all traces of old gasket material, oil, grease or dirt.

7. Insure that the rubber band is holding the right-hand chain to the cassette.

8. Install a new head gasket and the cylinder head. Be sure to use new cylinder head bolts.

9. On 2002–2005 engines torque the new cylinder head bolts in sequence as follows:

  a. Install bolts 8 (12mm) and torque, in sequence to:
- Step 1: 24 ft. lbs. (32 Nm)
- Step 2: Plus 80 degrees
- Step 3: Plus an additional 80 degrees
- Install bolts 5 and 6 (8mm), and torque to 24 ft. lbs. (32 Nm)

10. On 2006 engines torque the new cylinder head bolts in sequence as follows:

  a. Install bolts 8 (12mm) and torque, in sequence to:
- Step 1: 9 ft. lbs. (12 Nm)
- Step 2: 18 ft. lbs. (25 Nm)
- Step 3: Install bolts 5 and 6 (8mm), and torque to 24 ft. lbs. (32 Nm)
- Step 4: 8 (12mm) bolts plus 90 degrees
- Step 5: 8 (12mm) bolts plus an additional 90 degrees

11. Install or connect the following:
- Camshaft sprocket in the cassette and make certain that the camshaft sprocket turns freely on the camshaft
- Cassette retaining bolt. Torque the bolt to 89 inch lbs. (10 Nm).
- Exhaust manifold
- Oil level indicator tube. Torque the bolt to 18 ft. lbs. (25 Nm).
- Crankcase vent separator and spring
- Thermostat housing. Torque the bolts to 8 ft. lbs. (11 Nm).
- Water bypass hose
- Heater hoses
- EGR bracket. Torque the bolt to 89 inch lbs. (10 Nm).
- EGR tube. Torque the nut to 30 ft. lbs. (40 Nm).
- ECT sensor electrical connector
- Electrical harness retainer. Torque the bolt to 89 inch lbs. (10 Nm).
- CKP and CMP electrical connectors
- Accessory bracket. Torque the bolts to 31 ft. lbs. (42 Nm).
- Alternator mounting bracket. Torque the bolts to 31 ft. lbs. (42 Nm).
- Alternator and electrical connectors

- Drive belt
- Fan shroud
- Roller followers
- Valve cover
- Lower intake manifold
- Negative battery cable

12. Change the engine oil and filter.

13. Refill the cooling system.

14. Start the engine and check for leaks, repair if necessary.

Rocker Arms/Shafts

REMOVAL & INSTALLATION

3.0L Engine

RIGHT SIDE

1. Before servicing the vehicle, refer to the Precautions Section.

2. Disconnect the negative battery cable.

➡**On some vehicles, when the battery cable is disconnected and reconnected, some abnormal drive symptoms may occur while the vehicle relearns its adaptive strategy. The vehicle may need to be driven to relearn its strategy.**

3. Disconnect the ignition coil electrical connector. Disconnect the spark plug wires.

4. Disconnect the heater hose support from the stud. Disconnect the engine control sensor wiring from the IAT sensor.

5. Disconnect the engine control sensor wiring from the MAF sensor. Position the engine control sensor wiring aside.

6. Disconnect the vacuum hose. Disconnect the engine control wire harness from the valve cover studs. Disconnect the PCV valve tube.

7. Remove the valve cover retaining bolts. Remove the valve cover from the engine. Discard the gasket.

➡**To prevent stretching the valve cover gasket, loosen silicone gasket and sealant from the cylinder head by sliding a flat knife between the valve cover gasket and the cylinder head.**

8. Remove the retaining bolt at each rocker arm.

9. The rocker arm and pushrod may then be removed from the engine. Keep all rocker arms and pushrods in order so they may be installed in their original locations.

To install:

10. Be sure to clean all old gasketing material from the sealing surfaces, using the proper tools. Do not use metal scrappers,

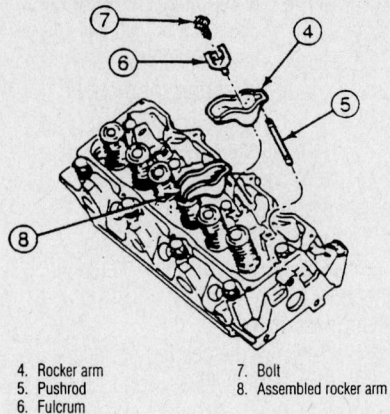

4. Rocker arm   7. Bolt
5. Pushrod   8. Assembled rocker arm
6. Fulcrum

7924EG27

Rocker arm and related components— 3.0L engine

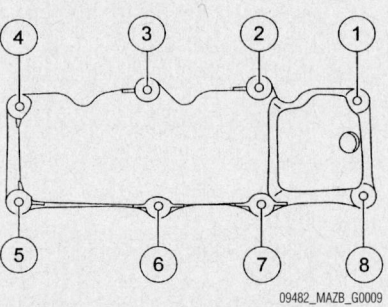

09482_MAZB_G0009

Valve cover bolt torque sequence—3.0L engine

wire brushes, power abrasive discs or other abrasive means to clean the sealing areas.

11. Installation is the reverse of the removal procedure.

12. Be sure to use a new valve cover gasket. Apply a bead of silicone gasket and sealant in two places where the cylinder head and lower intake manifold meet. The valve cover must be installed within 4 minutes of application.

13. Tighten the valve cover retaining bolts to 9 ft. lbs. in the proper sequence.

14. Lubricate the rocker arm assemblies with SAE 50W engine oil.

15. Ensure that the fulcrums are properly seated into the cylinder head. Torque the rocker arm fulcrum bolts to 19 ft. lbs. (26 Nm).

LEFT SIDE

1. Before servicing the vehicle, refer to the Precautions Section.

2. Disconnect the negative battery cable.

➡**On some vehicles, when the battery cable is disconnected and reconnected, some abnormal drive symptoms may occur while the vehicle relearns its** adaptive strategy. The vehicle may need to be driven to relearn its strategy.

3. Remove the upper intake manifold.
4. Disconnect the left side spark plug wire holders. Disconnect the engine control harness from the valve cover.
5. Remove the three bolts and the five stud bolts. Remove the valve cover from the engine. Discard the gasket.

➡To prevent stretching the valve cover gasket, loosen silicone gasket and sealant from the cylinder head by sliding a flat knife between the valve cover gasket and the cylinder head.

6. Remove the retaining bolt at each rocker arm.
7. The rocker arm and pushrod may then be removed from the engine. Keep all rocker arms and pushrods in order so they may be installed in their original locations.

To install:
8. Be sure to clean all old gasketing material from the sealing surfaces, using the proper tools. Do not use metal scrappers, wire brushes, power abrasive discs or other abrasive means to clean the sealing areas.
9. Installation is the reverse of the removal procedure.
10. Be sure to use a new valve cover gasket. Apply a bead of silicone gasket and sealant in two places where the cylinder head and lower intake manifold meet. The valve cover must be installed within 4 minutes of application.
11. Tighten the valve cover retaining bolts to 9 ft. lbs. in the proper sequence.
12. Lubricate the rocker arm assemblies with SAE 50W engine oil.
13. Ensure that the fulcrums are properly seated into the cylinder head. Torque the rocker arm fulcrum bolts to 19 ft. lbs. (26 Nm).

Intake Manifold

REMOVAL & INSTALLATION

2.3L Engine

2002–2003

1. Before servicing the vehicle, refer to the Precautions Section.
2. Relieve the fuel system pressure.
3. Disconnect the negative battery cable.

➡On some vehicles, when the battery cable is disconnected and reconnected, some abnormal drive symptoms may

occur while the vehicle relearns its adaptive strategy. The vehicle may need to be driven to relearn its strategy.

4. Remove the accelerator control snow shield. Remove the air cleaner outlet tube.
5. Disconnect the throttle cables from the intake manifold.
6. Disconnect the TP sensor. Disconnect the MAP sensor electrical connector.
7. Disconnect the vacuum hose and the idle air control (IAC) valve electrical connector.
8. Disconnect the engine vacuum harness and brake booster hose and the PCV breather hose.
9. Disconnect the engine wiring harness and position it to the side.
10. Remove the EGR (EEGR) tube bracket bolt. Disconnect the swirl control valve monitor electrical connector.
11. Disconnect the fuel line from the clip. Remove the bolts from the EEGR tube flange.
12. Detach the 42 pin electrical connector from the bracket and disconnect the intake vacuum hose.
13. Disconnect the KS sensor electrical connector and disconnect the connector from the intake manifold.
14. Remove the wiring harness push pin from the bottom of the intake manifold. Disconnect the fuel manifold vacuum hose.
15. Remove the intake manifold retaining bolts. Remove the intake manifold from the engine.

To install:
16. Be sure to clean all old gasketing material from the sealing surfaces, using proper tools.
17. Installation is the reverse of the removal procedure.
18. Be sure to use new gaskets.
19. Torque the bolts to 13 ft. lbs. (18 Nm).

20. Start the engine and check for fuel leaks, correct as required.

2004–2006

1. Before servicing the vehicle, refer to the Precautions Section.
2. Relieve the fuel system pressure.
3. Disconnect the negative battery cable.

➡On some vehicles, when the battery cable is disconnected and reconnected, some abnormal drive symptoms may occur while the vehicle relearns its adaptive strategy. The vehicle may need to be driven to relearn its strategy.

4. Remove the accelerator control splash shield. Remove the air cleaner outlet tube.
5. Disconnect the throttle cables from the intake manifold.
6. Disconnect the TP sensor. Disconnect the MAP sensor electrical connector.
7. Disconnect the vacuum hose and the idle air control (IAC) valve electrical connector.
8. Disconnect the engine vacuum harness and brake booster hose and the PCV breather hose.
9. Disconnect and plug the heater hoses.
10. Disconnect the two engine wiring harness pin type retainers from the intake manifold. Disconnect the pin type retainer from the rear of the intake manifold.
11. Remove the EGR (EEGR) tube bracket bolt. Disconnect the fuel hose from the clip. Remove the bolts from the EEGR tube flange.
12. Detach the 42 pin electrical connector from the bracket and disconnect the vapor purge hose.
13. Disconnect the KS sensor electrical connector and disconnect the connector from the intake manifold.

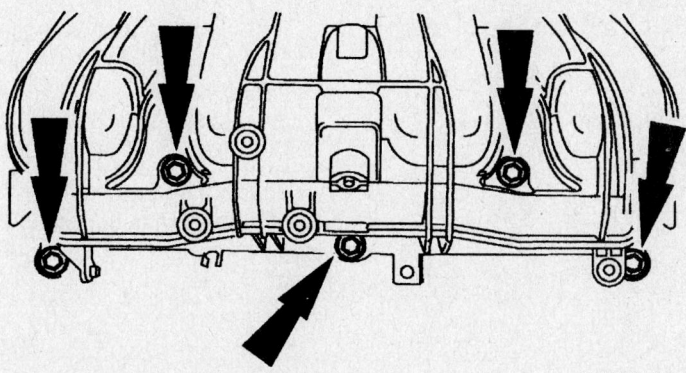

Intake manifold bolts—2.3L engine

67197-RANG-G19

14. Remove the wiring harness push pin from the bottom of the intake manifold. Remove the bracket. Disconnect the fuel supply line from the manifold.

15. Remove the intake manifold retaining bolts. Remove the intake manifold from the engine.

To install:

16. Be sure to clean all old gasketing material from the sealing surfaces, using the proper tools.

17. Installation is the reverse of the removal procedure.

18. Be sure to use new gaskets.

19. Torque the bolts to 13 ft. lbs. (18 Nm).

20. Be sure to fill the cooling system with the proper grade and type engine coolant.

21. Start the engine and check for leaks, correct as required.

3.0L Engine

UPPER MANIFOLD

1. Before servicing the vehicle, refer to the Precautions Section.

2. Disconnect the negative battery cable.

➡ On some vehicles, when the battery cable is disconnected and reconnected, some abnormal drive symptoms may occur while the vehicle relearns its adaptive strategy. The vehicle may need to be driven to relearn its strategy.

3. Remove the air cleaner outlet pipe.

4. Remove the bolt and the snow shield.

5. Disconnect the accelerator cable, speed control cable and the return spring.

6. Disconnect the accelerator cable and speed control cable from the accelerator bracket.

7. Disconnect the throttle position (TP) sensor and idle air control (IAC) valve electrical connectors.

8. Disconnect the two vacuum hoses.

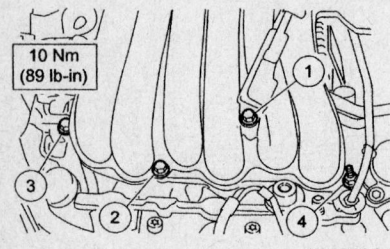

Upper intake manifold torque sequence—
3.0L engine

9. Detach the 42-pin wiring harness retainer from the stud bolt.

10. Remove the nut and the stud bolt. Position the 42-pin wiring harness and bracket aside.

11. Remove the oil indicator tube bracket nut and the upper intake manifold support bracket stud bolt.

12. Disconnect the brake booster vacuum hose and the positive crankcase ventilation (PCV) valve tube.

13. Remove the bolts, the stud bolt and the upper intake manifold. Remove and discard the gaskets.

To install:

14. Be sure to clean all old gasketing material from the sealing surfaces, using the proper tools.

15. Installation is the reverse of the removal procedure.

16. Be sure to use new gaskets.

17. Torque the bolts to 89 inch lbs. (10 Nm).

LOWER MANIFOLD

1. Before servicing the vehicle, refer to the Precautions Section.

2. Relieve the fuel system pressure.

3. Disconnect the negative battery cable.

➡ On some vehicles, when the battery cable is disconnected and reconnected, some abnormal drive symptoms may occur while the vehicle relearns its adaptive strategy. The vehicle may need to be driven to relearn its strategy.

4. Drain the engine cooling system.

5. Remove both the valve covers.

6. Disconnect the upper radiator hose, heater hose and coolant bypass hose.

7. Disconnect the engine coolant temperature (ECT) sensor and coolant temperature sender electrical connectors.

8. Disconnect the crankshaft position (CKP) sensor and detach the wiring from the stud bolt.

9. Disconnect the fuel injector electrical connectors and position the engine control wiring aside.

10. Disconnect the fuel tube.

11. Loosen the rocker arm bolts.

➡ Identify the location of each pushrod. Each pushrod is to be installed in the original location to prevent premature wear.

12. Remove all the pushrods.

13. Remove the bolts from the lower intake manifold.

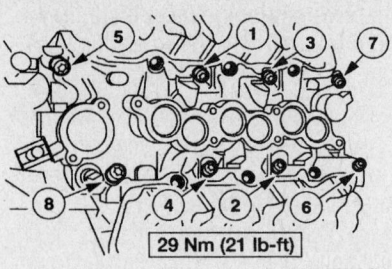

Lower intake manifold torque sequence—
3.0L engine

➡ Gently loosen the intake manifold to separate the silicone sealant from the cylinder block.

14. Remove the lower intake manifold.

15. Remove and discard the intake manifold gaskets and end seals.

To install:

➡ Clean and inspect all mating surfaces.

➡ If the lower intake manifold is not installed within four minutes, remove the sealant and reapply.

16. Apply a drop of silicone gasket and sealant at the four cylinder block-to-cylinder head seams.

17. Install the intake manifold gaskets and end seals.

18. Position the lower intake manifold.

19. Install the bolts and tighten in the sequence shown to 21 ft. lbs. (29 Nm).

20. Install the pushrods.

21. Tighten the rocker arms to 24 ft. lbs. (32 Nm).

22. Connect the fuel tube.

23. Connect the fuel injector electrical connectors.

24. Connect the crankshaft position (CKP) sensor and attach the wiring from the stud bolt.

25. Connect the engine coolant temperature (ECT) sensor and coolant temperature sender electrical connectors.

26. Connect the upper radiator hose, heater hose and coolant bypass hose.

27. Install both valve covers.

28. Fill and bleed the engine cooling system.

4.0L Engine

1. Before servicing the vehicle, refer to the Precautions Section.

2. Disconnect the negative battery cable.

➡ On some vehicles, when the battery cable is disconnected and reconnected, some abnormal drive symptoms may occur while the vehicle relearns its adaptive strategy. The vehicle may need to be driven to relearn its strategy.

3. Remove the bolts and the shield.

4. Remove the air cleaner outlet pipe.

5. Disconnect the idle air control (IAC) valve, throttle position (TP) sensor electrical connectors and the TP sensor wiring pin-type retainer.

6. Disconnect the MAF sensor wiring pin-type retainer.

7. Detach the accelerator and speed control cables from the throttle body.

8. Detach the accelerator and speed control cables from the bracket, and position the cables aside.

9. Disconnect the exhaust gas recirculation (EGR) valve vacuum hose and tube fitting.

10. Disconnect the EGR vacuum regulator solenoid valve electrical connector and vacuum hose.

11. Disconnect the hose.

12. Loosen the clamp and disconnect the brake booster vacuum hose.

❋❋ CAUTION

It is important to twist the spark plug wire boots while pulling upward to avoid possible damage to the spark plug wire.

➡**Mark the spark plug wire locations before removing them.**

13. Disconnect the RH spark plug wires from the coil. Remove the spark plug wire routing clip pin-type retainer and position the wires aside.

14. Remove the wiring harness bracket retainer, then position the wiring harness aside.

15. Remove the accelerator cable routing clip pin-type retainer and the wiring harness pin-type retainer.

16. Remove the bolts.

17. Remove the bolts and position the coil and bracket aside.

18. Disconnect the vacuum hoses.

19. Remove the nut.

20. Disconnect the powertrain control module (PCM) electrical connector.

21. Remove the retainer and position the ground wires aside.

22. Detach the electrical connector retainer.

23. Remove the intake manifold bolts and lift up the intake manifold.

24. Remove the heated positive crankcase ventilation (PCV) hose retainers and remove the heated PCV fitting.

25. Remove the intake manifold retaining bolts. Remove the intake manifold from the engine.

To install:

26. Be sure to clean all old gasketing material from the sealing surfaces, using the proper tools.

27. Installation is the reverse of the removal procedure.

28. Be sure to use new gaskets.

29. Torque the bolts to 89 inch lbs. (10 Nm).

Exhaust Manifold

REMOVAL & INSTALLATION

2.3L Engine

1. Before servicing the vehicle, refer to the Precautions Section.

2. Disconnect the negative battery cable.

➡**On some vehicles, when the battery cable is disconnected and reconnected, some abnormal drive symptoms may occur while the vehicle relearns its adaptive strategy. The vehicle may need to be driven to relearn its strategy.**

3. Remove or disconnect the following:
   • Exhaust flange nuts
   • Drive belt
   • Coolant
   • Upper radiator hose and the engine reservoir hose
   • Air conditioning compressor
   • Heater hose
   • Oil indicator and the upper bolt for the tube assembly
   • Lower bolt and remove the oil indicator tube assembly
   • Front radiator tube

4. Remove the pushpins and position the right inner fender splash shield out of the way.

5. Remove or disconnect the following:
   • Alternator electrical connectors
   • Lower front end accessory drive (FEAD) mounting bolts
   • Upper mounting bolt and the FEAD assembly
   • Two nuts and position the coolant tube out of the way
   • Exhaust manifold
   • Exhaust manifold gasket

To install:

6. Install or connect the following:

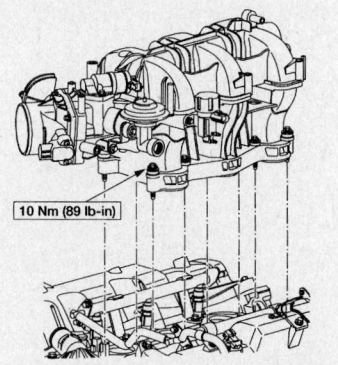

10 Nm (89 lb-in)

67197-RANG-G20

Intake manifold installation—4.0L engine

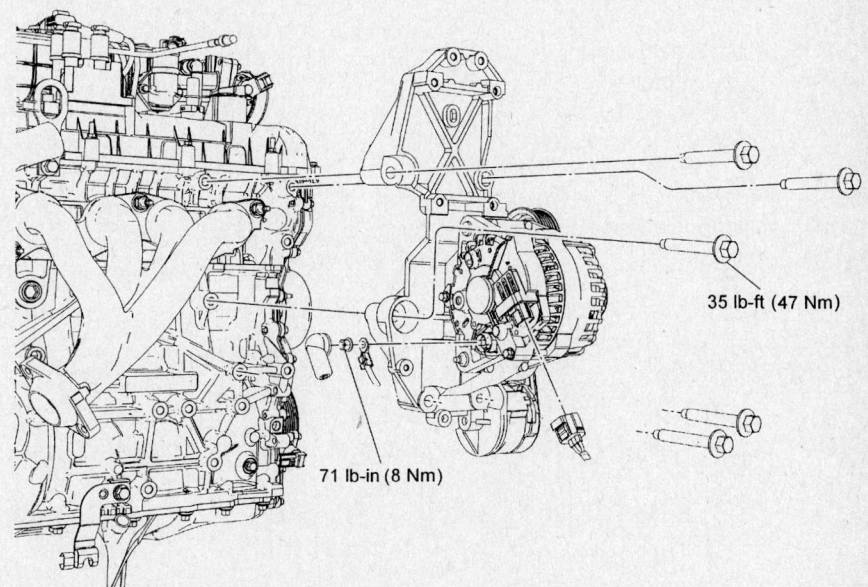

35 lb-ft (47 Nm)

71 lb-in (8 Nm)

09482_MAZB_G0010

Exhaust manifold and related components—2.3L engine

- Exhaust manifold gasket
- Exhaust manifold and the nuts
- Coolant tube and the nuts
- FEAD assembly and the upper mounting bolts, finger tight.
- Lower FEAD mounting bolts, finger tight. Then, torque all FEAD bolts, in the sequence shown, to 35 ft. lbs. (47 Nm).
- Alternator electrical connectors
- Right inner splash shield and pushpins
- Upper radiator tube and install the bolts
- Oil indicator tube assembly and the lower bolt
- Oil indicator tube upper bolt and the oil indicator
- Heater water hose
- Air conditioning compressor
- Upper radiator hose and the engine reservoir hose
7. Fill the cooling system.
8. Install the serpentine drive belt.
9. Install the exhaust flange nuts.
10. Connect the battery ground cable.

3.0L Engine
LEFT SIDE

1. Before servicing the vehicle, refer to the Precautions Section.
2. Disconnect the negative battery cable.

➡ **On some vehicles, when the battery cable is disconnected and reconnected, some abnormal drive symptoms may occur while the vehicle relearns its adaptive strategy. The vehicle may need to be driven to relearn its strategy.**

3. Install or connect the following:
- Exhaust flange nuts
- Exhaust Gas Recirculation (EGR) valve from the exhaust manifold tube
- Oil lever indicator and bracket
- Exhaust manifold and discard the gasket

To install:
4. Clean the mating surfaces for the exhaust manifold and cylinder head.
5. Install a new gasket and the exhaust manifold. Torque the bolts in sequence to:
   a. 89 inch lbs. (10 Nm).
   b. 15 ft. lbs. (20 Nm).
6. Install or connect the following:
- Oil lever indicator tube and bracket. Torque the bolt to 12 ft. lbs. (16 Nm).
- EGR valve to the exhaust manifold tube. Torque the fastener to 26 ft. lbs. (35 Nm).
- Exhaust flange. Torque the nuts to 25 ft. lbs. (34 Nm).
- Negative battery cable
7. Start the vehicle and check for leaks, repair if necessary.

RIGHT SIDE

1. Before servicing the vehicle, refer to the Precautions Section.
2. Disconnect the negative battery cable.

➡ **On some vehicles, when the battery cable is disconnected and reconnected, some abnormal drive symptoms may occur while the vehicle relearns its adaptive strategy. The vehicle may need to be driven to relearn its strategy.**

3. Remove or disconnect the following:
- Exhaust manifold flange
- Ignition coil support bracket
- Exhaust manifold and discard the gasket

To install:
4. Clean the mating surfaces for the exhaust manifold and cylinder head
5. Install a new gasket and the exhaust manifold. Torque the bolts is sequence to:
   a. 89 inch lbs. (10 Nm).
   b. 18 ft. lbs. (25 Nm).
6. Install or connect the following:
- Ignition coil support bracket. Torque the bolts to 15 ft. lbs. (20 Nm).
- Exhaust flange nuts. Torque the nuts to 33 ft. lbs. (46 Nm).
- Negative battery cable

7. Start the vehicle and check for leaks, repair if necessary.

4.0L Engine
LEFT SIDE

1. Before servicing the vehicle, refer to the Precautions Section.
2. Disconnect the negative battery cable.

➡ **On some vehicles, when the battery cable is disconnected and reconnected, some abnormal drive symptoms may occur while the vehicle relearns its adaptive strategy. The vehicle may need to be driven to relearn its strategy.**

3. Raise and support the vehicle safely. Remove the exhaust pipe flange retaining bolts.
4. Lower the vehicle. Disconnect the Differential Pressure Feedback EGR (DPFE) transducer hoses. Disconnect the EGR valve to exhaust manifold tube.
5. Disconnect the EGR valve to exhaust manifold tube from the EGR valve and remove the EGR valve to exhaust manifold tube.
6. Remove the exhaust manifold retaining nuts. Remove the exhaust manifold from the engine.
To install:
7. Clean the mating surfaces for the exhaust manifold and cylinder head
8. Install a new gasket and the exhaust manifold. Torque the bolts to specification.
9. Continue the installation in the reverse order of the removal procedure.

RIGHT SIDE

1. Before servicing the vehicle, refer to the Precautions Section.
2. Disconnect the negative battery cable.

➡ **On some vehicles, when the battery cable is disconnected and reconnected, some abnormal drive symptoms may occur while the vehicle relearns its adaptive strategy. The vehicle may**

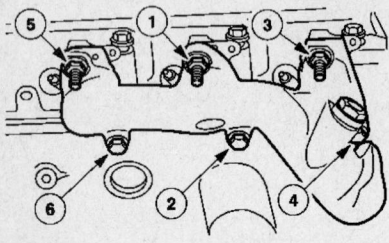

9308EG08

Left side exhaust manifold bolt torque sequence–3.0L engine

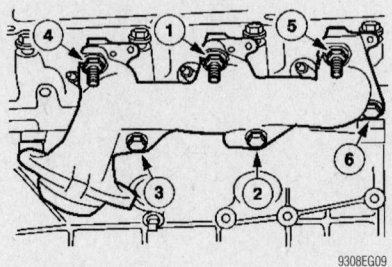

9308EG09

Right side exhaust manifold bolt torque sequence–3.0L engine

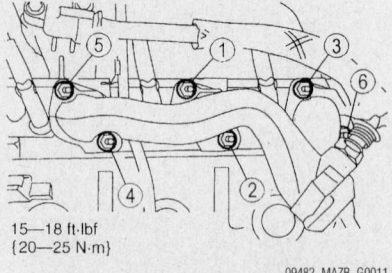

15—18 ft-lbf
{20—25 N·m}

09482_MAZB_G0011

Left side exhaust manifold bolt torque sequence—4.0L engine

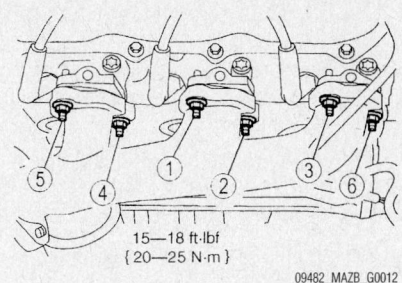

Right side exhaust manifold bolt torque sequence—4.0L engine

15—18 ft·lbf
{20—25 N·m}
09482_MAZB_G0012

need to be driven to relearn its strategy.

3. Raise and support the vehicle safely. Remove the exhaust pipe flange retaining bolts.
4. Lower the vehicle.
5. Remove the exhaust manifold retaining nuts. Remove the exhaust manifold from the engine.
To install:
6. Clean the mating surfaces for the exhaust manifold and cylinder head
7. Install a new gasket and the exhaust manifold. Torque the bolts to specification.
8. Continue the installation in the reverse order of the removal procedure.

Front Crankshaft Seal

REMOVAL & INSTALLATION

2.3L Engine

➡On some engines the crankshaft, crankshaft sprocket and the pulley are fitted together by friction, with diamond washers between the flange on each part. For that reason, the crankshaft sprocket is also unfastened if you loosen the pulley. Therefore the engine must be retimed each time the damper is removed. Otherwise severe damage can occur.

1. Before servicing the vehicle, refer to the Precautions Section.
2. Disconnect the negative battery cable.

➡On some vehicles, when the battery cable is disconnected and reconnected, some abnormal drive symptoms may occur while the vehicle relearns its adaptive strategy. The vehicle may need to be driven to relearn its strategy.

3. Remove the crankshaft pulley

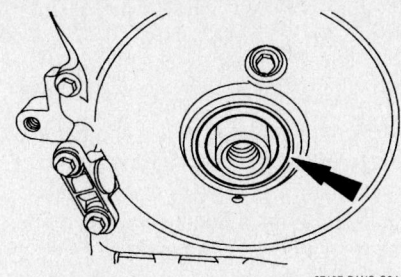

Front crankshaft seal (arrow) location—2.3L engine

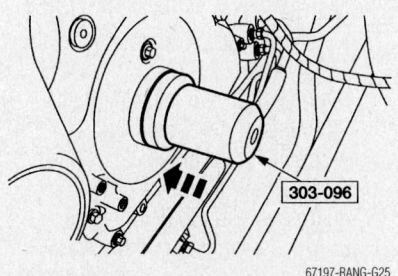
303-096
67197-RANG-G25

Front crankshaft seal installation using special tool—2.3L engine

❋❋ WARNING

Use care not to damage the engine front cover or the crankshaft when removing the seal.

4. Remove the crankshaft front oil seal by prying the seal out of the front cover
To install:
5. Using the special tool, install the crankshaft front oil seal.
6. Install the crankshaft pulley.
7. Tighten the crankshaft damper in two stages:
  • Step 1: Tighten to 74 ft. lbs. (100 Nm)
  • Step 2: Tighten an additional 90 degrees

3.0L Engine

1. Before servicing the vehicle, refer to the Precautions Section.
2. Disconnect the negative battery cable.

➡On some vehicles, when the battery cable is disconnected and reconnected, some abnormal drive symptoms may occur while the vehicle relearns its adaptive strategy. The vehicle may need to be driven to relearn its strategy.

3. Remove the air cleaner outlet tube.
4. Remove the cooling fan blade and fan shroud. Remove the drive belt.

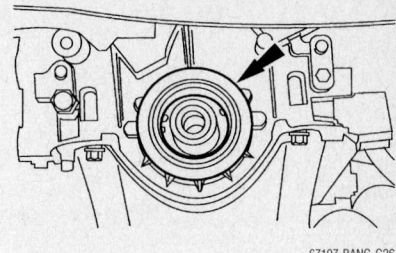
67197-RANG-G26

Front crankshaft seal (arrow) location—3.0L engine

5. Remove the crankshaft vibration damper bolt and washer. Remove the damper, using the proper removal tool.
6. If required, remove the crankshaft key.
7. Remove the crankshaft front seal.
To install:
8. Install the crankshaft front seal.
  a. Lubricate the seal lip with clean engine oil.
  b. Using a seal installer, install the crankshaft front seal.
9. Install the crankshaft damper. Torque to 107 ft. lbs. (145 Nm).
10. Continue the installation in the reverse order of the removal procedure.

4.0L Engine

1. Before servicing the vehicle, refer to the Precautions Section.
2. Disconnect the negative battery cable.

➡On some vehicles, when the battery cable is disconnected and reconnected, some abnormal drive symptoms may occur while the vehicle relearns its adaptive strategy. The vehicle may need to be driven to relearn its strategy.

3. Remove the fan shroud. Remove the accessory drive belt.
4. Using the proper tool remove the crankshaft pulley bolt.

➡This bolt must not be reused. It must be replaced.

5. Remove the crankshaft pulley
6. Using a seal removal tool, remove the crankshaft front oil seal.
To install:
7. Lubricate the seal lip with clean engine oil.
8. Using a seal driver, install the crankshaft front oil seal.
9. Apply silicone sealer to the Woodruff key slot on the crankshaft pulley.

➡If not secured within four minutes, the sealant must be removed and the sealing area cleaned.

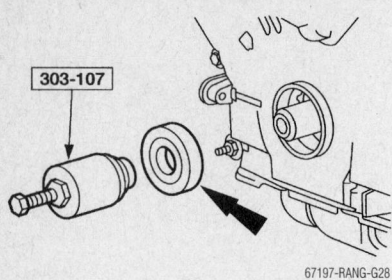

67197-RANG-G28

Front crankshaft seal removal using special tool—4.0L engine

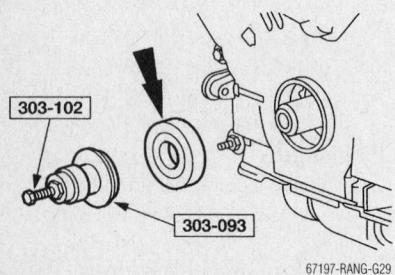

67197-RANG-G29

Front crankshaft seal installation using special tool—4.0L engine

10. Install the crankshaft pulley. Torque to 37 ft. lbs. (50 Nm), then 90 degrees.

11. Continue the installation in the reverse order of the removal procedure.

Camshaft and Valve Lifters

➡**Although Ford suggests that this component is removable while the engine is installed in the vehicle, depending on the particular options with which your truck is equipped, working clearance may be extremely tight and this procedure may be much easier to perform with the engine removed. Before commencing, read through this procedure and make certain enough clearance, or working room, exists with the engine in the vehicle; if there is not enough space, the engine should be removed.**

REMOVAL & INSTALLATION

2.3L Engine

➡**On some engines the crankshaft, crankshaft sprocket and the pulley are fitted together by friction, with diamond washers between the flange on each part. For that reason, the crankshaft sprocket is also unfastened if you loosen the pulley. Therefore the engine must be retimed each time the damper is removed. Otherwise severe damage can occur.**

1. Before servicing the vehicle, refer to the Precautions Section.
2. Relieve the fuel system pressure.
3. Disconnect the negative battery cable.

➡**On some vehicles, when the battery cable is disconnected and reconnected, some abnormal drive symptoms may occur while the vehicle relearns its adaptive strategy. The vehicle may need to be driven to relearn its strategy.**

4. Drain the cooling system.
5. Properly discharge the air conditioning system.
6. Remove or disconnect the following:
   - Drive belt.
   - Engine oil level indicator assembly.
   - Engine oil level indicator.
   - Engine oil level indicator tube.
   - Water outlet tube.
   - Water outlet tube.
   - Air conditioning compressor.

➡**The alternator will be removed with the accessory bracket.**

   - Accessory bracket
   - Right motor mount.
   - Coolant hose from the thermostat
   - Coolant hose from the EGR valve
   - Coolant tube assembly
   - Exhaust manifold and gasket
   - Block heater (if so equipped)
   - Water outlet
   - EGR valve
   - Power steering pump and reservoir as an assembly
   - Idle air control (IAC) valve
   - Throttle position (TP) sensor
   - Manifold absolute pressure (MAP) sensor
   - Swirl control valve monitor electrical connector
   - CKP sensor and the wiring harness pin-type retainers
   - Knock sensor (KS)
   - Electric thermostat
   - Swirl control valve
   - CMP sensor electrical connector and disconnect the PCV hose from the intake manifold
   - Engine wiring harness pin-type retainers from the intake manifold
   - Engine wiring harness connector bracket. Position the engine wiring harness aside
   - EGR tube
   - Fuel supply line clip from the front of the intake manifold. Disconnect the vacuum hose from the intake manifold
   - Intake manifold assembly

   - Fuel injector electrical connectors. Detach the wiring harness pin-type retainers
   - Ignition coil and the cylinder head temperature (CHT) sensor electrical connectors
   - Engine wiring harness anchors from the valve cover studs. Remove the engine wiring harness
   - Ignition coil
   - Bypass hose
   - Thermostat housing
   - Knock sensor and the engine vent cover
   - Left motor mount
   - Fuel injector supply manifold with the injectors and the ground strap
   - Water pump pulley
   - Water pump
   - CMP sensor
   - CHT sensor
   - Spark plugs
   - Valve cover
   - CKP sensor
   - Crankshaft vibration damper

➡**There is one front cover bolt behind the cooling fan drive pulley. To remove this bolt, align one of the cooling fan drive pulley access holes with the bolt head to access the bolt.**

   - Front cover
   - Timing chain tensioner
   - Timing chain guides
   - Timing chain assembly

➡**Use a wrench on the flats between cylinders No. 1 and No. 2 to hold the camshaft in place.**

   - Camshaft drive sprockets
   - Oil pump chain tensioner and guide

➡**The oil pump chain sprocket must be held in place**

   - Oil pump chain and sprockets

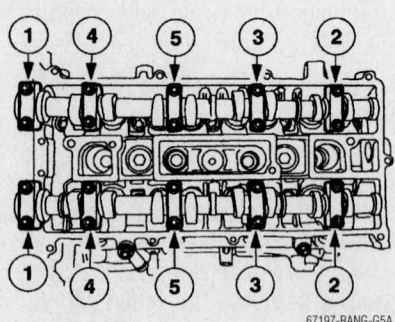

67197-RANG-G5A

Camshaft cap loosening sequence—2.3L engine

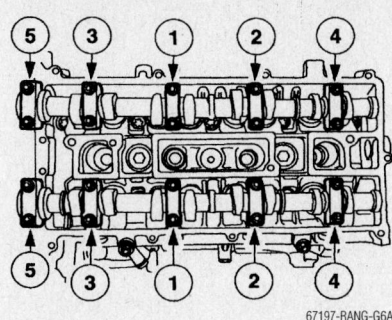

Camshaft cap torque sequence—2.3L engine

→**Note the position of the lobes on the No. 1 cylinder before removing the camshafts for assembly reference.**

7. Loosen the camshaft bearing cap bolts in sequence, one turn at a time. Repeat the first step until all tension is released from the camshaft bearing caps. Remove the camshaft bearing caps.

8. Remove the camshafts.

To install:

9. Installation is the reverse of removal.

→**Install the camshafts with the alignment notches in the camshaft lined up so the camshaft alignment plate can be installed without rotating the camshafts. Make sure the lobes on the No. 1 cylinder are in the same position as noted in the disassembly procedure. Rotating the camshafts, or installing the camshafts 180 degrees out of position can cause severe damage to the valves and pistons. Lubricate the camshaft journals and bearing caps with clean engine oil. Install the camshafts and bearing caps. Tighten the bolts in the sequence shown in three stages.**

- Step 1: Tighten the camshaft bearing caps one turn at a time until tight.
- Step 2: Tighten the bolts to 7 Nm (62 inch lbs.)
- Step 3: Tighten the bolts to 16 Nm (12 ft. lbs.)
  a. Crankshaft vibration damper:

→**Do not reuse the crankshaft pulley bolt. Tighten the bolt in two stages.**

- Step 1: Tighten the bolt to 40 Nm (30 ft. lbs.)
- Step 2: Tighten the bolt and additional 90 degrees (1/4 turn).

3.0L Engine

1. Before servicing the vehicle, refer to the Precautions Section.

2. Properly relieve the fuel system pressure.

3. Disconnect the negative battery cable.

→**On some vehicles, when the battery cable is disconnected and reconnected, some abnormal drive symptoms may occur while the vehicle relearns its adaptive strategy. The vehicle may need to be driven to relearn its strategy.**

4. Drain the cooling system.

5. Drain the engine oil.

6. Evacuate the air conditioning system.

7. Remove or disconnect the following:
- Air cleaner hoses
- Fan, spacer and shroud
- Radiator

8. Rotate the crankshaft so that No. 1 piston is at Top Dead Center (TDC) on the compression stroke.
- Air conditioning condenser
- Fuel lines from the fuel supply manifold
- Vacuum hoses
- Electrical wiring
- Engine front cover
- Water pump
- Alternator
- Power steering pump. Do not disconnect the hoses
- Air conditioning compressor. Do not disconnect the hoses
- Throttle body
- Fuel injection wire harness

9. Turn the engine by hand to TDC of the power stroke on No. 1 cylinder.
- Spark plug wires from the plugs
- Rocker arm covers
- Intake manifold
- Loosen the rocker arm bolts enough to pivot the rocker arms out of the way and remove the pushrods. Identify them for installation
- Lifters and identify them for installation
- Crankshaft pulley/damper
- Starter
- Oil pan
- Camshaft gear attaching bolt and washer, then slide the gear off the camshaft
- Camshaft thrust plate

10. Carefully slide the camshaft out of the engine block, using caution to avoid any damage to the camshaft bearings.

To install:

11. Oil the camshaft journals and cam lobes with heavy SJ engine oil (50W). Install the spacer ring with the chamfered

side toward the camshaft, then insert the camshaft key.

12. Install or connect the following:
- Camshaft using caution to avoid any damage to the camshaft bearings
- Thrust plate. Torque the screws to 84 inch lbs. (10 Nm).

13. Rotate the camshaft and crankshaft as necessary to align the timing marks. Install the camshaft gear and chain. Torque the bolt to 46 ft. lbs. (62 Nm).

14. Coat the tappets with 50W engine oil and place them in their original locations.

15. Apply 50W engine oil to both ends of the pushrods. Install the pushrods in their original locations.

16. Pivot the rocker arms into position. Torque the fulcrum bolts to 96 inch lbs. (11 Nm).

17. Rotate the engine until both timing marks are at the top of their sprockets and aligned. Torque the following fulcrum bolts to 18 ft. lbs. (24 Nm):
  a. No.1 intake.
  b. No.2 exhaust.
  c. No.4 intake.
  d. No.5 exhaust.

18. Rotate the engine until the camshaft timing mark is at the bottom of the sprocket and the crankshaft timing mark is at the top of the sprocket, and both are aligned. Torque the following fulcrum bolts to 18 ft. lbs. (24 Nm):
  a. No.1 exhaust.
  b. No.2 intake.
  c. No.3 intake and exhaust.
  d. No.4 exhaust.
  e. No.5 intake.
  f. No.6 intake and exhaust.

19. Torque all the bolts to 24 ft. lbs. (33 Nm).

20. Turn the engine by hand to 0 degrees Before Top Dead center (BTDC) of the power stroke on No. 1 cylinder.

21. Install or connect the following:
- Engine front cover and water pump assembly
- Oil pan
- Crankshaft damper/pulley and tighten the retaining bolt to 107 ft. lbs. (145 Nm).
- Intake manifold
- Starter
- Crankshaft pulley and damper
- Rocker arm covers
- Spark plug wires
- Fuel lines to the fuel supply manifold
- Fuel injection wire harness
- Throttle body
- Air conditioning compressor

- Power steering pump
- Alternator
- Water pump
- Engine front cover
- All electrical connectors and vacuum lines
- Air conditioning condenser
- Radiator
- Fan, spacer and shroud
- Air cleaner hoses
- Negative battery cable

22. Recharge the air conditioning system.

23. Refill the cooling system.

24. Replace the oil filter and refill the engine with the specified amount of engine oil.

25. Start the engine and check the ignition timing and idle speed. Adjust if necessary. Run the engine at fast idle and check for coolant, fuel, vacuum or oil leaks.

4.0L Engine

RIGHT SIDE

1. Before servicing the vehicle, refer to the Precautions Section.
2. Properly relieve the fuel system pressure.
3. Disconnect the negative battery cable.

➡**On some vehicles, when the battery cable is disconnected and reconnected, some abnormal drive symptoms may occur while the vehicle relearns its adaptive strategy. The vehicle may need to be driven to relearn its strategy.**

4. Remove the valve covers. Remove the fuel supply manifold.

➡**Mark each camshaft roller follower to ensure its original position during reassembly.**

5. Using tool 303-581, remove the roller followers.

6. On 2002–2005 vehicles remove the tire and wheel assembly. Remove the lower splash shield.

7. Remove the hydraulic chain tensioner.

8. Install camshaft sprocket holding tool, 303-575 or equivalent on the rear of the cylinder head.

➡**The camshaft sprocket is a left hand threaded bolt pattern.**

9. Loosen and remove the camshaft sprocket bolt. Position the camshaft sprocket aside.

➡**Mark the position of the camshaft bearing caps so that they can be reinstalled in their original positions.**

10. Remove the bolts in the proper sequence and remove the bearing caps.

11. If equipped, remove the oil supply tube.

12. Remove the camshaft.

To install:

13. Lubricate all of the moving parts with SAE 50W engine oil.

14. Install camshaft onto the cylinder head.

15. Position the oil rail, if equipped and install the bearing caps and bolts. Torque the bolts in 2 steps:

    a. Step 1—53.5 inch lbs. (6 Nm).

➡**The camshaft gear must turn freely on the camshaft. Do not tighten the bolt at this time.**

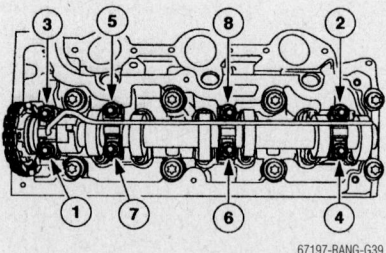

Camshaft bolt removal sequence—4.0L engine

16. Install the camshaft sprocket and loosely install the bolt.

➡**The camshafts must be retimed or engine damage may occur.**

17. Retime the camshafts.

18. Continue the installation in the reverse order of the removal procedure.

LEFT SIDE

1. Before servicing the vehicle, refer to the Precautions Section.
2. Properly relieve the fuel system pressure.
3. Disconnect the negative battery cable.

➡**On some vehicles, when the battery cable is disconnected and reconnected, some abnormal drive symptoms may occur while the vehicle relearns its adaptive strategy. The vehicle may need to be driven to relearn its strategy.**

4. Remove the valve covers. Remove the fuel supply manifold.

➡**Mark each camshaft roller follower to ensure its original position during reassembly.**

5. Using tool 303-581, remove the roller followers.

6. Remove the intake manifold. Remove the thermostat housing.

7. Remove the hydraulic chain tensioner.

8. On 2006 vehicles, install camshaft sprocket holding tool 303-564 and 303-578 on the front of the camshaft.

9. Loosen and remove the camshaft sprocket bolt. Position the camshaft sprocket aside.

➡**Mark the position of the camshaft bearing caps so that they can be reinstalled in their original positions.**

Right side hydraulic chain tensioner location—4.0L engine

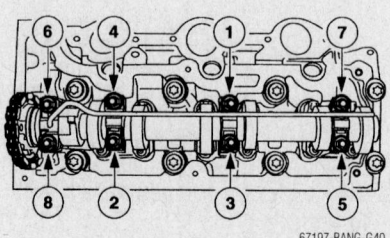

Camshaft bolt torque sequence—4.0L engine

Left side hydraulic chain tensioner location—4.0L engine

10. Remove the bolts in the proper sequence and remove the bearing caps.
11. If equipped, remove the oil supply tube.
12. Remove the camshaft.

To install:
13. Lubricate all of the moving parts with SAE 50W engine oil.
14. Install camshaft onto the cylinder head.
15. Position the oil rail, if equipped and install the bearing caps and bolts. Torque the bolts in 2 steps:
  a. Step 1—53.5 inch lbs. (6 Nm).

➡**The camshaft gear must turn freely on the camshaft. Do not tighten the bolt at this time.**

16. Install the camshaft sprocket and loosely install the bolt.

➡**The camshafts must be retimed or engine damage may occur.**

17. Retime the camshafts.
18. Continue the installation in the reverse order of the removal procedure.

Camshaft Timing

ADJUSTMENT

4.0L Engine

2002–2005

➡**You must retime the left and right camshafts when either camshaft is disturbed.**

1. Before servicing the vehicle, refer to the Precautions Section.
2. Properly relieve the fuel system pressure.
3. Disconnect the negative battery cable.

➡**On some vehicles, when the battery cable is disconnected and reconnected, some abnormal drive symptoms may occur while the vehicle relearns its adaptive strategy. The vehicle may need to be driven to relearn its strategy.**

4. Raise and support the vehicle safely.
5. Remove the intake manifold.
6. Remove the fuel supply manifold.
7. Remove the accessory drive belt.
8. Remove the thermostat housing.
9. Remove the valve covers.

➡**Mark each camshaft roller follower to ensure its original position during reassembly.**

Special tool 303-573 location and installation—4.0L engine

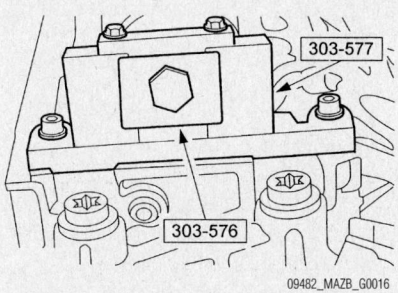
Special tool 303-577 and 303-576 location and installation—4.0L engine

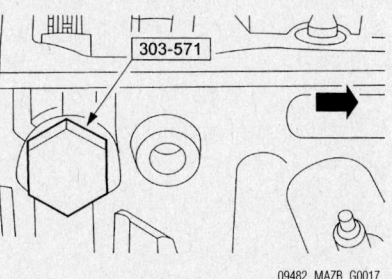
Special tool 303-571 location and installation—4.0L engine

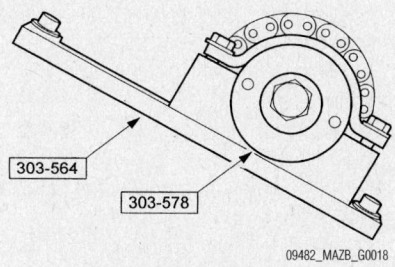
Special tool 303-564 and 303-578 location and installation—4.0L engine

10. Using tool 303-581, remove the roller followers.
11. Rotate the crankshaft clockwise to position the number one cylinder at TDC.
12. Install tool 303-573.

➡**Tool 303-573 must be installed on the damper and should contact the engine block to position the engine at TDC.**

13. Loosen the right side camshaft sprocket bolt, using tool 303-575.

➡**The camshaft sprocket is a left hand threaded bolt pattern.**

14. Position the camshaft timing slots below the centerline of the camshaft to correctly fit the special tools. Install special tools 303-576 and 303-577 on the front of the right cylinder head.

➡**The camshaft timing slots are off center.**

15. Remove the right side lower splash shield.
16. Remove the right side hydraulic camshaft tensioner.
17. Install special tools 303-564 and 303-578 on the rear of the right side cylinder head.

➡**Leave the top two special tool clamp bolts loose.**

18. Install special tool 303-571.
19. The camshaft sprocket is a left hand threaded bolt pattern. If necessary use camshaft gear torque adapter, tool 303-565 to tighten the camshaft sprocket bolt.
20. Tighten the special tool top two clamp bolts to 89 inch lbs. (10 Nm).
21. Tighten the camshaft bolt to 45 ft. lbs. Install the right side camshaft tensioner and torque to 32 ft. lbs (44 Nm).
22. Install the right side lower splash shield.
23. Remove the left side hydraulic camshaft tensioner.
24. Install special tools 303-564 and 303-578 on the front of the left side cylinder head. Torque the top two clamp bolts to 89 inch lbs.
25. Loosen the left side camshaft sprocket bolt. Loosen the top two clamp bolts on the special tool to allow the camshaft sprocket to rotate freely.
26. Position the camshaft timing slots below the centerline of the camshaft to correctly fit the special tools. Install special tools 303-576 and 303-577 on the rear of the left cylinder head.

➡**The camshaft timing slots are off center.**

27. Install special tool 303-571.
28. Tighten the special tool top two clamp bolts to 89 inch lbs. (10 Nm). Tighten the camshaft bolt to 63 ft. lbs.
29. Install the left side hydraulic camshaft tensioner and torque the bolt to 32 ft. lbs..
30. Install the roller followers.
31. Install the thermostat housing. Install the accessory drive belt.

32. Install the fuel supply rail. Install the intake manifold.

2006

➡ **You must retime the left and right camshafts when either camshaft is disturbed.**

1. Before servicing the vehicle, refer to the Precautions Section.
2. Properly relieve the fuel system pressure.
3. Disconnect the negative battery cable.

➡ **On some vehicles, when the battery cable is disconnected and reconnected, some abnormal drive symptoms may occur while the vehicle relearns its adaptive strategy. The vehicle may need to be driven to relearn its strategy.**

4. Raise and support the vehicle safely.
5. Remove the intake manifold.
6. Remove the fuel supply manifold.
7. Remove the accessory drive belt.
8. Remove the thermostat housing.
9. Remove the valve covers.

➡ **Mark each camshaft roller follower to ensure its original position during reassembly.**

10. Using tool 303-581, remove the roller followers.
11. Rotate the crankshaft clockwise to position the number one cylinder at TDC.
12. Remove the nut and position the air conditioning hose to the side. Install tool 303-573.

➡ **Tool 303-573 must be installed on the damper and should contact the engine block to position the engine at TDC.**

13. Install the camshaft sprocket holding tools 303-578 and 303-564 to the right side cylinder head and tighten the top two clamp bolts to 89 inch lbs.

➡ **The camshaft sprocket is a left hand threaded bolt pattern.**

14. Loosen the right side camshaft sprocket bolt. Loosen the top two special tool clamp bolts.
15. Position the camshaft timing slots below the centerline of the camshaft to correctly fit the special tools. Install special tools 303-576 and 303-577 on the front of the right cylinder head.

➡ **The camshaft timing slots are off center.**

16. Remove the right side lower splash shield.

17. Remove the right side hydraulic camshaft tensioner.
18. Install the timing chain tensioner.

➡ **The camshaft sprocket is a left hand threaded bolt pattern.**

19. Tighten the bolts. Tighten the special tool top two clamp bolts to 89 inch lbs. Using a torque wrench extension with the sprocket nut socket tool (303-565) tighten the camshaft bolt to 45 ft. lbs. Remove the timing chain tensioner tool.
20. Install the right side hydraulic camshaft tensioner and torque to 32 ft. lbs (44 Nm).
21. Install the right side lower splash shield.
22. Remove the special tools for the right side cylinder head.
23. Install special tools 303-564 and 303-578 on the front of the left side cylinder head. Torque the top two clamp bolts to 89 inch lbs.
24. Loosen the left side camshaft sprocket bolt. Loosen the top two clamp bolts on the special tool to allow the camshaft sprocket to rotate freely.
25. Position the camshaft timing slots below the centerline of the camshaft to correctly fit the special tools. Install special tools 303-576 and 303-577 on the rear of the left cylinder head.

➡ **The camshaft timing slots are off center.**

26. Remove the left side hydraulic camshaft tensioner.
27. Install special tool 303-571.
28. Tighten the special tool top two clamp bolts to 89 inch lbs. (10 Nm). Tighten the camshaft bolt to 63 ft. lbs. Remove the timing chain tensioner tool.
29. Install the left side hydraulic camshaft tensioner and torque the bolt to 32 ft. lbs.
30. Remove the special tools from the left cylinder head.
31. Remove special tool 303-573 from its mounting.
32. Position the air conditioning hose and install the bolt.
33. Install the roller followers.

INSPECTION

1. Before servicing the vehicle, refer to the Precautions Section.
2. Remove the camshaft from the engine.
3. Check the camshaft bearing journals for damage and binding.
4. If the journals are binding, check the cylinder head for damage.
5. Check the cylinder head for clogged oil holes.

6. Check the camshaft surface for abnormal wear and damage. Replace the camshaft, as required.
7. Measure the camshaft lobe surface and replace the camshaft if not within specification.
8. Measure the camshaft journal diameter and replace the camshaft if not within specification.
9. Measure the camshaft run out and replace the camshaft if not within specification.

Valve Lash

ADJUSTMENT

2.3L Engine

1. Before servicing the vehicle, refer to the Precautions Section.
2. Disconnect the negative battery cable.

➡ **On some vehicles, when the battery cable is disconnected and reconnected, some abnormal drive symptoms may occur while the vehicle relearns its adaptive strategy. The vehicle may need to be driven to relearn its strategy.**

3. Remove the valve cover.
4. Rotate the crankshaft in the clockwise direction and position the piston at TDC on the compression stroke.
5. Measure the valve clearance at the base circle with the lobe pointed away from the tappet, before removing the camshafts.

➡ **Failure to measure all clearances prior to removing the camshafts will necessitate repeated removal and installation and wasted labor time.**

6. Use a feeler gauge to measure each valves clearance and record the measurement.
7. A midrange clearance specification is most desirable. Intake specification is 0.008–0.011 inch and exhaust specification is 0.010–0.013 inch.
8. If not within specification select a new tappet.

➡ **Select tappets using the following formula: Tappet thickness equals measured clearance plus the base tappet thickness minus the most desirable thickness.**

9. Select the replacement tappets and mark the installation location.
10. Remove the camshafts and install the new tappets, as required.

Starter Motor

REMOVAL & INSTALLATION

1. Before servicing the vehicle, refer to the Precautions Section.
2. Disconnect the negative battery cable.

➡**On some vehicles, when the battery cable is disconnected and reconnected, some abnormal drive symptoms may occur while the vehicle relearns its adaptive strategy. The vehicle may need to be driven to relearn its strategy.**

3. Raise and support the vehicle safely.
4. On 2.3L engine, remove the left front inner fender splash shield.
5. Disconnect the electrical connections from the starter.
6. Disconnect the positive battery cable.
7. Remove the starter retaining bolts. Remove the starter from the engine.

To install:

8. Installation is the reverse of the removal procedure.
9. Torque the starter retaining bolts to 20 ft. lbs.

Oil Pan

REMOVAL & INSTALLATION

2.3L Engine

1. Before servicing the vehicle, refer to the Precautions Section.
2. Disconnect the negative battery cable.

➡**On some vehicles, when the battery cable is disconnected and reconnected, some abnormal drive symptoms may occur while the vehicle relearns its adaptive strategy. The vehicle may need to be driven to relearn its strategy.**

3. Drain the engine oil.
4. Remove or disconnect the following:

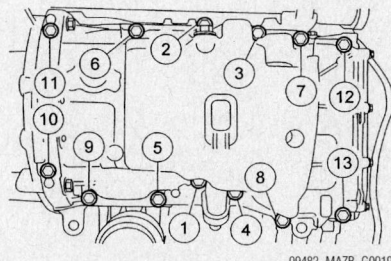

Oil pan torque sequence—2.3L engine

• Engine from the vehicle
• Engine oil level indicator assembly
• Engine oil pan bolts and oil pan

To install:

5. Clean and inspect all mating surfaces.

➡**The oil pan must be installed and the bolts tightened with four minutes of applying the silicone gasket and sealant.**

6. Apply a 2.5 mm bead of silicone gasket and sealant to the oil pan. Install the oil pan. Tighten the oil pan bolts to specification and in the proper sequence.
7. Lubricate the O-ring with clean engine oil and install the engine oil level indicator assembly.
8. Install the engine into the vehicle.

3.0L Engine

2002–2004 2WD

1. Before servicing the vehicle, refer to the Precautions Section.
2. Disconnect the negative battery cable.

➡**On some vehicles, when the battery cable is disconnected and reconnected, some abnormal drive symptoms may occur while the vehicle relearns its adaptive strategy. The vehicle may need to be driven to relearn its strategy.**

3. Drain the engine oil.
4. Remove or disconnect the following:

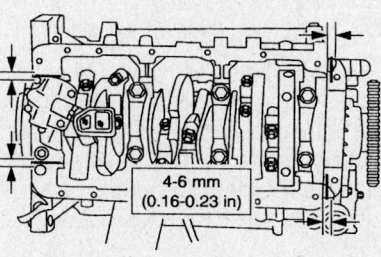

4-6 mm
(0.16-0.23 in)

67197-RANG-G41

Oil pan sealer application (arrows indicate location)—3.0L engine

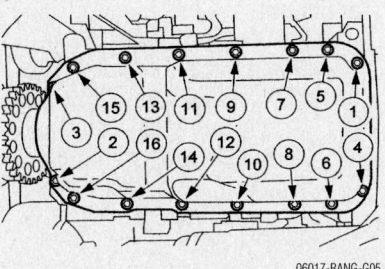

06017-RANG-G05

Oil pan bolt torque sequence—3.0L engine

• Oil level dipstick tube
• Fan shroud. Leave the fan shroud over the fan assembly
• Motor mount nuts from the frame
• Starter
• Transmission inspection cover
• Right hand axle I-beam. The brake caliper must be removed and secured out of the way.
• Oil pan attaching bolts, using a suitable lifting device, raise the engine about 2 in. (5cm)
• Oil pan and discard the gasket

➡**The oil pan fits tightly between the transmission spacer plate and oil pump pick-up tube. Use care when removing the oil pan from the engine.**

5. Clean all gasket surfaces on the engine and oil pan. Remove all traces of old gasket and/or sealer.

To install:

6. Apply a ⅛ (4mm) bead of RTV sealer to the junctions of the rear main bearing cap and block, and the front cover and block. The oil pan must be installed within 15 minutes, after applying the sealant.
7. Apply adhesive to the gasket surfaces and install the oil pan gasket.
8. Install or connect the following:

• Oil pan on the engine block. Torque the bolts to 9 ft. lbs. (12 Nm) and in the proper sequence.
• Right hand axle I-beam
• Brake caliper
• Transmission inspection cover
• Starter
• Fan shroud
• Motor mount retaining nuts
• Oil level dipstick tube
• Negative battery cable

9. Fill the engine with clean oil.
10. Start the vehicle and check for leaks, repair if necessary.

2002–2004 4WD

1. Before servicing the vehicle, refer to the Precautions Section.
2. Disconnect the negative battery cable.

➡**On some vehicles, when the battery cable is disconnected and reconnected, some abnormal drive symptoms may occur while the vehicle relearns its adaptive strategy. The vehicle may need to be driven to relearn its strategy.**

3. Drain the engine oil.
4. Remove or disconnect the following:
• Negative battery cable
• Engine from the vehicle and place it on a suitable engine stand

- Oil pan and discard the gasket

To install:

5. Apply a ⅛ (4mm) bead of RTV sealer to the junctions of the rear main bearing cap and block, and the front cover and block. The oil pan must be installed within 15 minutes, after applying the sealant.

6. Install or connect the following:
- New oil pan gasket
- Oil pan. Torque the bolts to 9 ft. lbs. (12 Nm) and in the proper sequence
- Engine
- Negative battery cable

7. Fill the engine with clean oil.

8. Start the vehicle and check for leaks, repair if necessary.

2005–2006

1. Before servicing the vehicle, refer to the Precautions Section.

2. Disconnect the negative battery cable.

➡ **On some vehicles, when the battery cable is disconnected and reconnected, some abnormal drive symptoms may occur while the vehicle relearns its adaptive strategy. The vehicle may need to be driven to relearn its strategy.**

3. Drain the engine oil.

4. Raise and support the vehicle safely.

5. Remove the transmission. Remove the flywheel or flexplate.

6. If equipped with coil springs, remove the air cleaner outlet tube. Remove the nuts from the engine mounts. Using an engine lifting device, raise the engine about 1.72 inches. Install blocks between the engine mounts and the support brackets.

7. Remove the oil pan retaining bolts. Remove the oil pan. Discard the oil pan gasket.

To install:

8. Apply a ⅛ (4mm) bead of RTV sealer to the junctions of the rear main bearing cap and block, and the front cover and block. The oil pan must be installed within 15 minutes, after applying the sealant.

9. Position a new pan gasket on the pan. Be sure to align the fastener holes and secure the gasket with gasket clips. Install the oil pan.

10. Torque the bolts to specification and in the proper sequence.

11. Continue the installation in the reverse order of the removal procedure.

4.0L Engine

➡ **The 2002 engine does not use an oil pan in the conventional sense. There is a separate access panel that unbolts from what would be considered the oil**

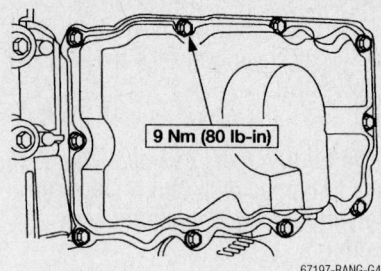

9 Nm (80 lb-in)

67197-RANG-G45

Oil pan installation—2002 4.0L engine

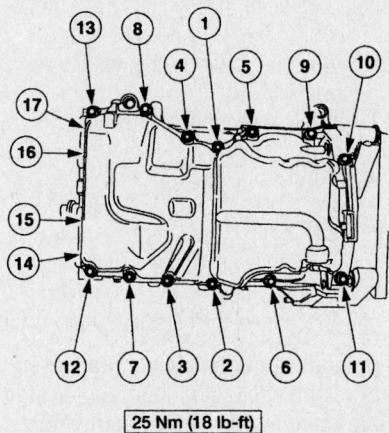

25 Nm (18 lb-ft)

67197-RANG-G7A

Oil pan bolt torque sequence—2003–2006 4.0L engine

pan (which is now known as the ladder frame).

1. Before servicing the vehicle, refer to the Precautions Section.

2. Disconnect the negative battery cable.

➡ **On some vehicles, when the battery cable is disconnected and reconnected, some abnormal drive symptoms may occur while the vehicle relearns its adaptive strategy. The vehicle may need to be driven to relearn its strategy.**

3. Raise and support the vehicle safely.

4. Drain the engine oil.

5. Remove the oil pan retaining bolts. Remove the oil pan and discard the gasket

To install:

6. Install or connect the following:
- New gasket and oil pan. On the 2002 vehicles, torque the bolts to 80 inch lbs. (9 Nm). On 2003–2006 vehicles torque the bolts to 18 ft. lbs. (25 Nm) and in the proper sequence.
- Negative battery cable

7. Fill the engine with clean oil.

8. Start the vehicle and check for leaks, repair if necessary.

Oil Pump

REMOVAL & INSTALLATION

2.3L Engine

➡ **The oil pump is located on the front of the engine and is turned by the timing belt.**

1. Before servicing the vehicle, refer to the Precautions Section.

2. Disconnect the negative battery cable.

➡ **On some vehicles, when the battery cable is disconnected and reconnected, some abnormal drive symptoms may occur while the vehicle relearns its adaptive strategy. The vehicle may need to be driven to relearn its strategy.**

3. Remove or disconnect the following:
- Timing chain
- Oil pump chain and sprockets
- Oil pan
- Oil pump pickup tube and gasket
- Oil pump assembly and gasket

To install:

4. Turn the crankshaft clockwise to position the No. 1 piston.

5. Remove the plug bolt.

6. Install the Engine Timing Peg 303-507.

➡ **Clean the gasket surface with metal surface cleaner.**

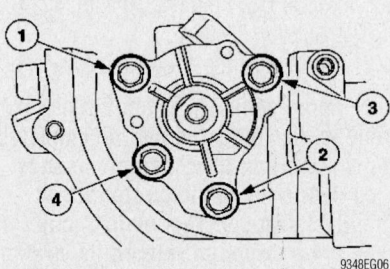

9348EG06

Oil pump bolt torque sequence—2.3L engine

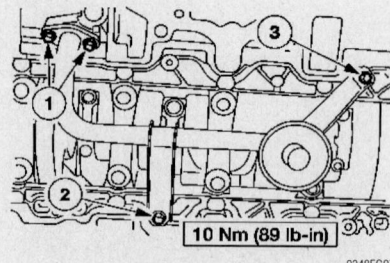

10 Nm (89 lb-in)

9348EG07

Oil pump pickup tube bolt torque sequence—2.3L engine

7. Install a new gasket and the oil pump assembly. Tighten the bolts in the sequence shown in two stages.
- Step 1: Tighten the bolts to 10 Nm (80 inch lbs.)
- Step 2: Tight the bolts to 23 Nm (17 ft. lbs.)

8. Install a new oil pump pickup tube gasket and the pickup tube. Tighten the bolts in the proper sequence to 10 Nm (89 inch lbs.)

9. Continue the installation in the reverse order of the removal procedure.

3.0 L Engine

1. Before servicing the vehicle, refer to the Precautions Section.
2. Disconnect the negative battery cable.

➡**On some vehicles, when the battery cable is disconnected and reconnected, some abnormal drive symptoms may occur while the vehicle relearns its adaptive strategy. The vehicle may need to be driven to relearn its strategy.**

3. Raise and support the vehicle safely.
4. Drain the engine oil.
5. Remove or disconnect the following:
- Oil pan
- Oil pick-up and tube assembly from the pump
- Oil pump retainer bolts and the oil pump

To install:
6. Prime the oil pump with clean engine oil by filling either the inlet or outlet port. Rotate the pump shaft to distribute the oil within the pump body.
7. Install the oil pump and tighten the mounting bolts to 30–40 ft. lbs. (41–54 Nm).
8. Install or connect the following:
- Oil pick-up and tube assembly
- Oil pan

9. Fill the engine with clean oil.
10. Start the vehicle and check for leaks, repair if necessary.

4.0L Engine

➡**The oil pump cannot be removed with the engine in the vehicle.**

1. Before servicing the vehicle, refer to the Precautions Section.
2. Disconnect the negative battery cable.

➡**On some vehicles, when the battery cable is disconnected and reconnected, some abnormal drive symptoms may occur while the vehicle relearns its adaptive strategy. The vehicle may need to be driven to relearn its strategy.**

3. Raise and support the vehicle safely.
4. Drain the engine oil.
5. Remove or disconnect the following:
- Engine from the vehicle
- Oil pan
- Unbolt the oil pick-up tube
- The 8 ladder frame bolts that were under the oil pan
- The 2 rear outer ladder frame bolts
- The 7 left-hand and the 8 right-hand ladder frame bolts
- The ladder frame from the engine
- The 2 oil pump attaching bolts and the pump.

To install:
6. Submerge the pump in clean engine oil to prime it.
7. Install or connect the following:
- The ladder frame on the engine
- The 8 right-hand and 7 left-hand ladder frame bolts
- The 2 rear outer and the 8 frame bolts under the pan
- The oil pump. Torque the bolts to 13–15 ft. lbs. (17–21 Nm).
- Oil pick-up tube
- Oil pan
- Engine to the vehicle
- Negative battery cable

8. Fill the engine with clean oil.
9. Start the vehicle and check for leaks, repair if necessary.

Rear Main Seal

REMOVAL & INSTALLATION

2.3L Engine

1. Before servicing the vehicle, refer to the Precautions Section.
2. Remove or disconnect the following:
- Flywheel or flexplate
- Oil pan
- Bolts and the crankshaft rear oil seal

To install
3. Install or connect the following:
- Rear oil seal on the Crankshaft Rear Main Oil Seal Installer
- Crankshaft Rear Main Oil Seal Installer and the crankshaft rear oil seal on the crankshaft

4. Tighten the bolts in the sequence shown to 10 Nm (89 inch lbs.)
5. Remove the Crankshaft Rear Main Oil Seal Installer.
6. Install the flywheel or flexplate.
7. Install the oil pan.

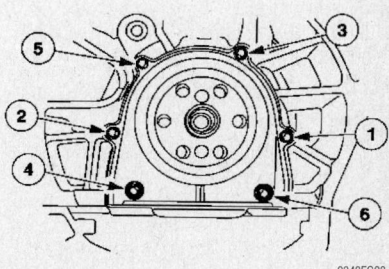

Rear main seal bolt torque sequence—2.3L engine

3.0L Engine

1. Before servicing the vehicle, refer to the Precautions Section.
2. Remove the flexplate or flywheel.

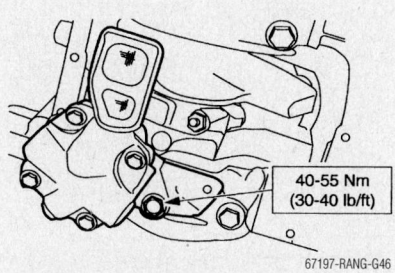

Oil pump installed—3.0L engine

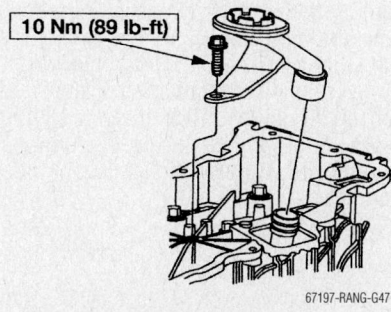

Oil pump installation—4.0L engine

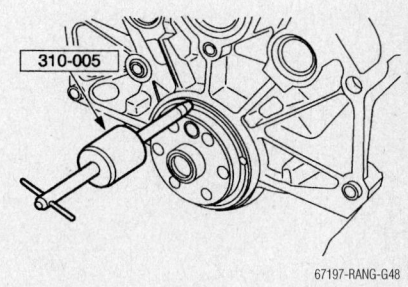

Rear main seal removal—3.0L engine

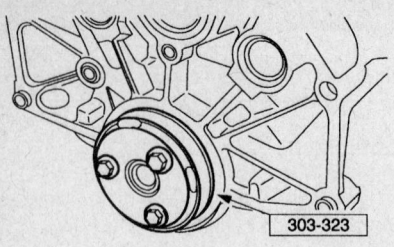

303-323

67197-RANG-G49

Rear main seal installation—3.0L engine

✳✳ WARNING

Use care to avoid scratching or damaging the oil seal surface or leakage may occur.

3. Using a sharp awl, punch one hole into the crankshaft rear oil seal metal surface between the seal lip and the cylinder block.

4. Screw the threaded end of the special tool into the oil seal. Use the special tool to remove the crankshaft rear oil seal.

To install:

5. Lubricate the outer lips and the inner seal on the crankshaft rear oil seal with clean engine oil.

6. Using the special tool, install the crankshaft rear oil seal. Alternate bolt tightening to correctly seat the crankshaft rear oil seal.

7. Install the flexplate or flywheel.

4.0L Engine

1. Before servicing the vehicle, refer to the Precautions Section.

2. Remove the flexplate or flywheel.

✳✳ WARNING

Avoid scratching or damaging the oil crankshaft seal running surface during removal of the crankshaft rear oil seal.

3. Using the special tool, remove the crankshaft rear oil seal.

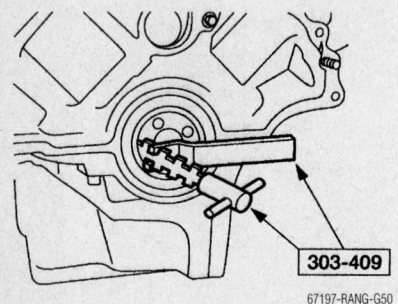

303-409

67197-RANG-G50

Rear main seal removal—4.0L engine

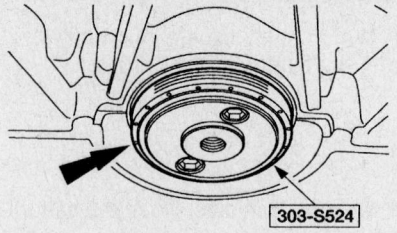

303-S524

67197-RANG-G51

Front part of rear main seal installation tool—4.0L engine

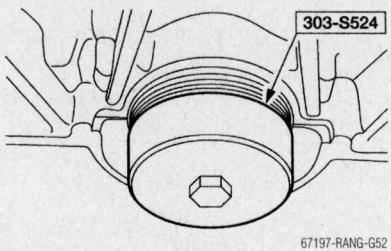

303-S524

67197-RANG-G52

Rear main seal installation—4.0L engine

To install:

➡ Be sure the crankshaft rear sealing surface is clean and free of any rust or corrosion. To clean the crankshaft rear sealing surface, use extra-fine emery cloth or extra-fine 0000 steel wool with metal surface cleaner.

4. Lubricate the crankshaft rear oil seal with clean engine oil and install on the special tool.

5. Using the special tool, install the crankshaft rear oil seal.

6. Install the flexplate or flywheel.

Timing Chain, Sprockets, Front Cover and Seal

REMOVAL & INSTALLATION

2.3L Engine

➡ On some engines the crankshaft, crankshaft sprocket and the pulley are fitted together by friction, with diamond washers between the flange on each part. For that reason, the crankshaft sprocket is also unfastened if you loosen the pulley. Therefore the engine must be retimed each time the damper is removed. Otherwise severe damage can occur.

1. Before servicing the vehicle, refer to the Precautions Section.

2. Disconnect the negative battery cable.

➡ On some vehicles, when the battery cable is disconnected and reconnected, some abnormal drive symptoms may occur while the vehicle relearns its adaptive strategy. The vehicle may need to be driven to relearn its strategy.

3. Remove or disconnect the following:
- Fan and shroud
- Drive belt
- Valve cover

4. Set No. 1 piston to TDC and install the Camshaft Alignment Plate 303-376, or equivalent.

5. Remove the plug for the crankshaft timing peg.

6. Install the Crankshaft Timing Peg 303-507, or equivalent.

7. Install an M6 bolt into the crankshaft pulley to verify the engine timing.

8. Remove or disconnect the following:
- Camshaft pulley
- Crankshaft position sensor
- Belt tensioner
- Water pump pulley
- Power steering high pressure hose. Remove the nylon O-ring.
- Power steering return hose
- Power steering pump

➡ This step is needed only if a new front cover is being installed.

9. Using a three-jaw puller, remove the fan drive pulley.

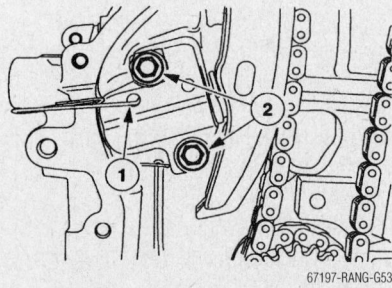

67197-RANG-G53

Timing chain tensioner removal; 1-paper clip, 2-bolts—2.3L engine

67197-RANG-G54

Right side timing chain guide—2.3L engine

➡There is one bolt behind the cooling fan drive pulley. This bolt can be accessed by lining up one of the holes in the pulley with the bolt.

10. Remove the bolts and the engine front cover.

11. Compress the timing chain tensioner and remove the tensioner.

12. Remove the right-hand timing chain guide.

13. Remove the timing chain.

14. Remove the bolts and the left-hand timing chain guide.

⁂ WARNING

Do not rely on the Camshaft Alignment Plate to prevent camshaft rotation. Damage to the tool or the camshaft can occur.

15. If necessary, remove the bolts and the camshaft sprockets. Use the flats on the camshaft to prevent camshaft rotation.

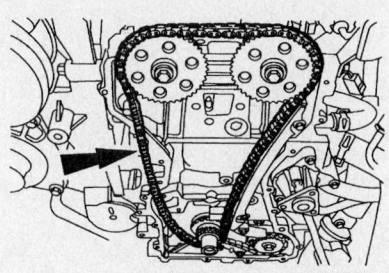

Timing chain removal—2.3L engine

67197-RANG-G55

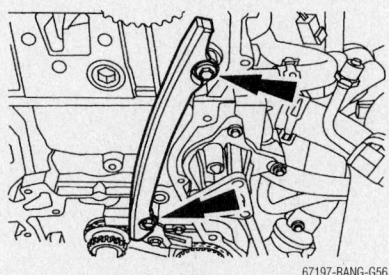

Left timing chain guide—2.3L engine

67197-RANG-G56

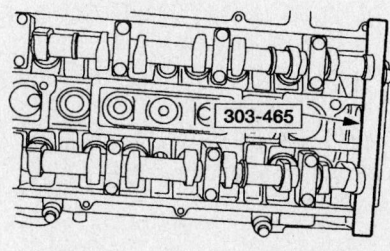

Camshaft alignment plate installed—2.3L engine

67197-RANG-G57

To install:

16. Remove the special tool.

⁂ WARNING

Do not rotate the camshafts. Damage to the valves and pistons can occur.

If the camshaft sprockets were not removed, use the flats on the camshafts to prevent camshaft rotation and loosen the sprocket bolts.

17. If removed, install the camshaft sprockets and the bolts. Do not tighten the bolts at this time.

18. Install or connect the following:
- Left-hand timing chain guide and bolts
- Timing chain
- Right-hand timing chain guide
- Timing chain tensioner and release the piston
- Timing chain tensioner and the bolts

19. Remove the drill rod to release the piston.

20. Install the special tool.

⁂ WARNING

Do not rely on the Camshaft Alignment Plate to prevent camshaft rotation. Damage to the tool or the camshafts can result. Using the flats on the camshafts to prevent camshaft rotation, tighten the bolts.

➡This step is needed only if a new front cover is being installed.

21. Install the fan drive pulley using a nut and bolt with flat washers.

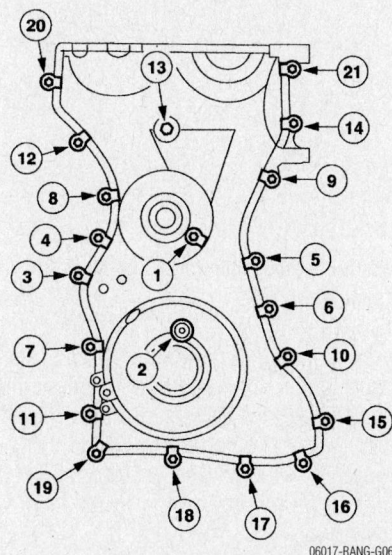

Front cover torque sequence—2.3L

06017-RANG-G06

22. Clean and inspect the mounting surfaces of the engine and the front cover.

➡The engine front cover must be installed and the bolts tightened within four minutes of applying the silicone gasket and sealant.

23. Apply a 2.5 mm bead of silicone gasket and sealant to the cylinder head and oil pan joint areas. Apply a 2.5 mm bead of silicone gasket and sealant to the front cover.

24. Install the front cover. Tighten the bolts in the sequence shown, to the following specifications:
- Step 1: 8 mm bolts to 10 Nm (89 inch lbs.)
- Step 2: 10 mm bolts to 25 Nm (18 ft. lbs.)
- Step 3: 13 mm bolts to 48 Nm (35 ft. lbs.)

25. Install or connect the following:
- Power steering pump and lower retaining bolt
- Power steering return hose
- New nylon O-ring and install the high pressure line.
- Water pump pulley
- Belt tensioner

➡Do not reuse the crankshaft damper bolt.

- Crankshaft pulley and hand-tighten the bolt

26. Install an M6 bolt in the crankshaft pulley. Tighten the crankshaft retaining bolt in two stages.
- Step 1: 40 Nm (30 ft. lbs.)
- Step 2: Rotate the bolt an additional 90 degrees.

27. Install the crankshaft position sensor, do not tighten the bolts at this time.

➡A new sensor must be installed whenever the old one is removed.

28. Adjust the crankshaft position sensor with the Alignment Tool, and tighten the mounting bolts.

29. Connect the crankshaft position sensor electrical connector.

30. Remove the M6 bolt from the crankshaft pulley.

31. Remove the Crankshaft Timing Peg 303-507.

32. Install the plug.

33. Remove the Camshaft Alignment Plate 303-376.

34. Install the valve cover.

35. Install the drive belt.

36. Install the fan and shroud.

37. Connect the battery ground cable.

3.0L Engine

1. Before servicing the vehicle, refer to the Precautions Section.

2. Disconnect the negative battery cable.

➡**On some vehicles, when the battery cable is disconnected and reconnected, some abnormal drive symptoms may occur while the vehicle relearns its adaptive strategy. The vehicle may need to be driven to relearn its strategy.**

3. Remove or disconnect the following:
   - Engine front cover
   - Rotate the crankshaft and align the timing marks
   - Sprocket bolt
   - Timing chain, camshaft sprocket and crankshaft sprocket as an assembly

To install:

4. Install or connect the following:
   - Timing chain, camshaft and crankshaft sprockets as an assembly

5. Align the timing marks.

6. Install or connect the following:
   - Sprocket bolt. Torque the bolt to 46 ft. lbs. (63 Nm).
   - Engine front cover
   - Negative battery cable

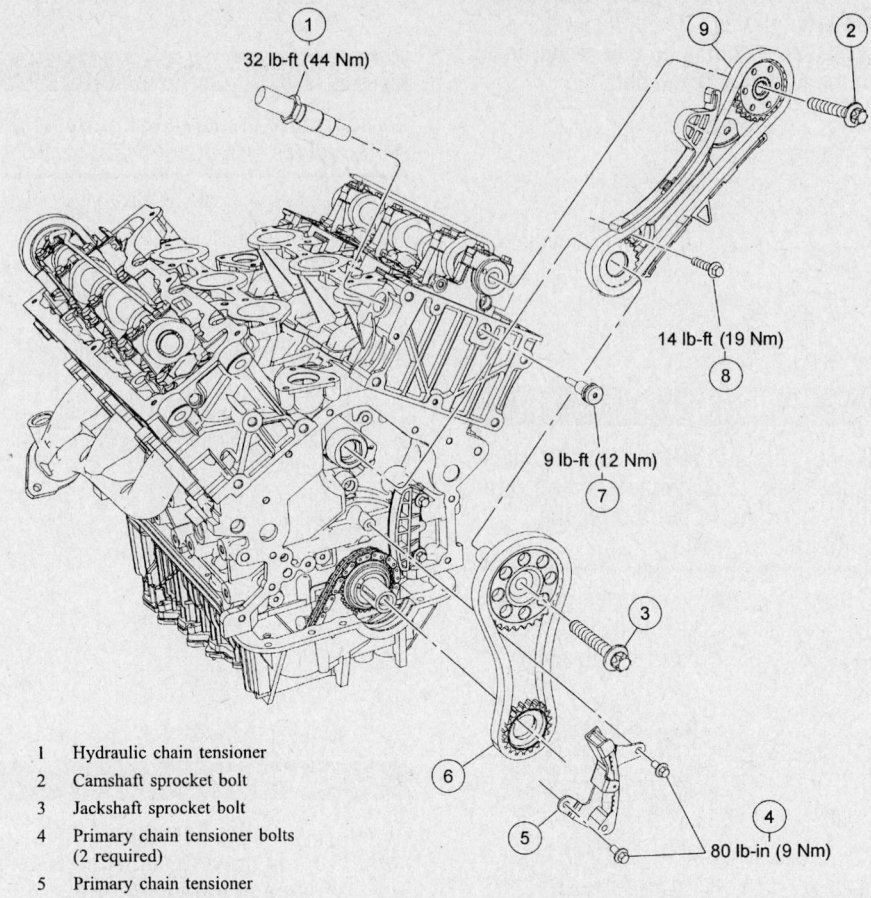

1	Hydraulic chain tensioner
2	Camshaft sprocket bolt
3	Jackshaft sprocket bolt
4	Primary chain tensioner bolts (2 required)
5	Primary chain tensioner
6	Primary chain and sprocket assembly
7	LH cassette upper bolt
8	LH cassette lower bolt
9	LH cassette

32 lb-ft (44 Nm)
14 lb-ft (19 Nm)
9 lb-ft (12 Nm)
80 lb-in (9 Nm)

09482_MAZB_G0020

Primary and left secondary timing chain assemblies—4.0L engine

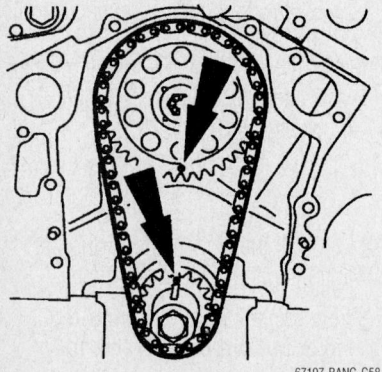

67197-RANG-G58

Timing mark alignment—3.0L engine

4.0L Engine

TIMING CHAINS

➡**This engine uses a primary chain and sprocket assembly and a right and left secondary chain and sprocket assembly. The procedure below outlines the primary chain assembly and the left secondary chain assembly (left lower cassette assembly).**

1. Before servicing the vehicle, refer to the Precautions Section.

2. Disconnect the negative battery cable.

09482_MAZB_G0021

Lower cassette upper bolt location—4.0L engine

➡**On some vehicles, when the battery cable is disconnected and reconnected, some abnormal drive symptoms may occur while the vehicle relearns its adaptive strategy. The vehicle may need to be driven to relearn its strategy.**

3. Remove the intake manifold.

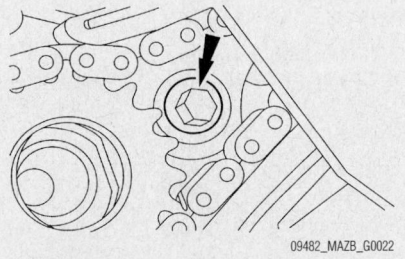

09482_MAZB_G0022

Lower cassette lower bolt location—4.0L engine

4. Remove the fuel supply manifold.

5. Remove the accessory drive belt.

6. Remove the thermostat housing.

7. Remove the valve covers.

➡**Mark each camshaft roller follower to ensure its original position during reassembly.**

8. Using tool 303-581, remove the roller followers.

9. Rotate the crankshaft clockwise to position the number one cylinder at TDC.

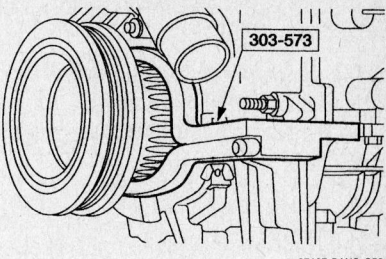

TDC positioning tool installed—4.0L engine

67197-RANG-G59

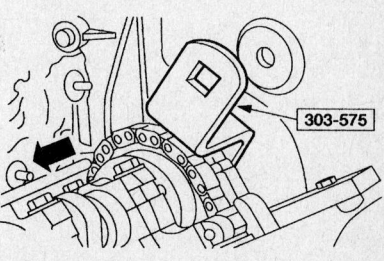

Loosening the right side camshaft sprocket bolt—4.0L engine

67197-RANG-G60

09482_MAZB_G0023

Lower cassette chain sprocket orientation—4.0L engine

Camshaft holding tool installed—4.0L engine

67197-RANG-G61

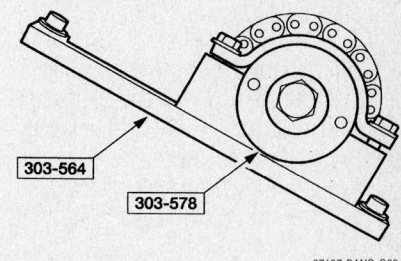

Camshaft gear holding tool and adapter—4.0L engine

67197-RANG-G62

10. Remove the left hydraulic chain tensioner. Install tool 303-578 and tighten the bolts to 89 inch lbs. Remove the left camshaft sprocket bolt.

11. Using tool 303-674, which prevents the crankshaft from turning, remove the jackshaft sprocket bolt. Remove the bolts and the primary chain tensioner. Remove these components as an assembly.

12. Remove the left cassette upper bolt. Remove the left cassette lower bolt and the left cassette.

To install:

➡ **The camshaft chain sprockets must be oriented correctly.**

13. Install the left lower cassette bolt. Torque to 14 ft. lbs.

14. Install the left upper cassette bolt. Torque to 9 ft. lbs.

15. Install the primary chain and sprockets, as an assembly.

16. Install the primary chain tensioner and the bolts. Torque to 80 inch lbs.

17. Using the special tool to prevent the crankshaft from turning, tighten the jackshaft sprocket bolt in two stages.

• Step 1: 33 ft. lbs. (45 Nm).

• Step 2: Rotate the bolt an additional 90 degrees.

18. Loosely install the camshaft sprocket bolt.

19. Retime the camshafts.

➡ **The camshafts must be retimed when either camshaft is disturbed.**

20. Continue the installation in the reverse order of the removal procedure.

TIMING DRIVE COMPONENTS (CAMSHAFT TIMING)

1. Before servicing the vehicle, refer to the Precautions Section.

2. Disconnect the negative battery cable.

➡ **On some vehicles, when the battery cable is disconnected and reconnected, some abnormal drive symptoms may occur while the vehicle relearns its adaptive strategy. The vehicle may need to be driven to relearn its strategy.**

3. With the vehicle in neutral, position it on a hoist.

4. Remove the intake manifold.

5. Remove the fuel supply manifold.

6. Remove the accessory drive belt.

7. Remove the thermostat housing.

8. Remove the roller followers.

➡ **You must retime the LH and RH camshafts when either camshaft is disturbed. Turn the crankshaft clockwise to** position the number one cylinder at top dead center (TDC).

➡ The special tool must be installed on the damper and should contact the engine block to position the engine at TDC.

9. Install the special tool.

➡ The right-hand camshaft sprocket bolt is a left-hand threaded bolt.

➡ If necessary, use camshaft gear torque adapter to loosen the camshaft sprocket bolt.

10. Using the special tool, loosen the RH camshaft sprocket bolt.

➡ The camshaft timing slots are off-center.

11. Position the camshaft timing slots below the centerline of the camshaft to correctly fit the special tools. Install the special tools on the front of the RH cylinder head.

12. Remove the RH lower splash shield.

13. Remove the RH camshaft tensioner.

➡ Leave the top two special tool clamp bolts loose.

14. Install the special tools on the rear of the RH cylinder head.

15. Install the special tool.

➡ The right-hand camshaft sprocket bolt is a left-hand threaded bolt.

➡ If necessary, use camshaft gear torque adapter to tighten the camshaft sprocket bolt.

16. Tighten the bolts.

17. Tighten the special tool top two clamp bolts to 10 Nm (89 inch lbs.).

18. Tighten the camshaft bolt.

19. Install the RH camshaft tensioner.

20. Install the RH lower splash shield.

21. Remove the LH camshaft tensioner.

22. Install the special tools on the front of the LH cylinder head and tighten the top two clamp bolts to 10 Nm (89 inch lbs.).

23. Loosen the LH camshaft sprocket bolt.

24. Loosen the top two clamp bolts on the special tool to allow the camshaft sprocket to rotate freely.

➡ The camshaft timing slots are off-center.

25. Position the camshaft timing slots below the centerline of the camshaft to correctly fit the special tools. Install the special tools on the rear of the LH cylinder head.

26. Install the special tool.

27. Tighten the bolts.

28. Tighten the special tool top two clamp bolts to 10 Nm (89 inch lbs.).

29. Tighten the camshaft bolt.

30. Install the LH camshaft tensioner.

31. Install the roller followers.

32. Install the thermostat housing.

33. Install the accessory drive belt.

34. Install the fuel supply manifold.

35. Install the intake manifold.

36. Install the RH valve cover.

37. Install the LH valve cover.

Piston and Ring

POSITIONING

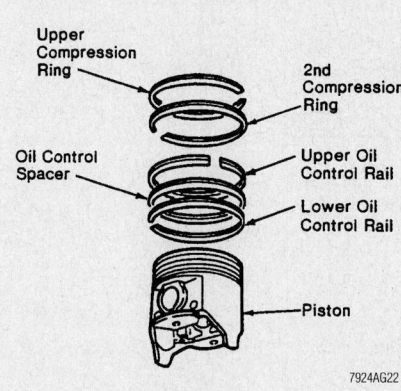

Piston ring positioning

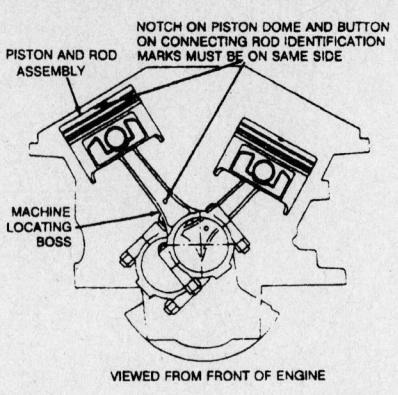

Piston and connecting rod positioning—3.0L engine

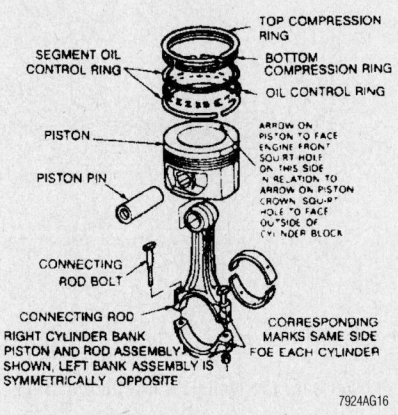

Piston and connecting rod positioning—4.0L engine

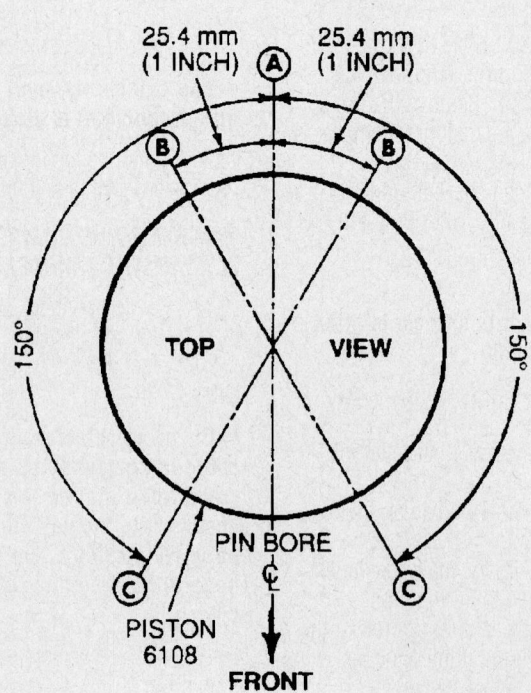

Piston ring end gap spacing

FUEL SYSTEM

Fuel System Service Precautions

Safety is the most important factor when performing not only fuel system maintenance, but any type of maintenance. Failure to conduct maintenance and repairs in a safe manner may result in serious personal injury or death. Work on a vehicle's fuel system components can be accomplished safely and effectively by adhering to the following rules and guidelines.

• To avoid the possibility of fire and personal injury, always disconnect the negative battery cable unless the repair or test procedure requires that battery voltage be applied.

• Always relieve the fuel system pressure prior to disconnecting any fuel system component (injector, fuel rail, pressure regulator, etc.) fitting or fuel line connection. Exercise extreme caution whenever relieving fuel system pressure, to avoid exposing your skin, face and eyes to fuel spray. Please be advised that fuel under pressure may penetrate the skin or any part of the body that it contacts.

• Always place a shop towel or cloth around the fitting or connection prior to loosening to absorb any excess fuel due to spillage. Ensure that all fuel spillage is quickly removed from engine surfaces. Ensure that all fuel-soaked cloths or towels are deposited into a flame-proof waste container with a lid.

• Always keep a dry chemical (Class B) fire extinguisher near the work area.

• Do not allow fuel spray or fuel vapors to come into contact with a light bulb, spark or open flame.

• Always use a second wrench when loosening or tightening fuel line connection fittings. This will prevent unnecessary stress and torsion to fuel piping. Always follow the proper torque specifications.

• Always replace worn fuel fitting O-rings with new ones. Do not substitute fuel hose where rigid pipe is installed.

Relieving Fuel System Pressure

All engines are equipped with a pressure relief valve located on the fuel supply manifold. Remove the fuel tank cap. Attach a fuel pressure gauge to the valve and release the fuel pressure. Be sure to drain the fuel into a suitable container and to avoid fuel spillage. Disconnect the negative battery cable.

Fuel Filter

REMOVAL & INSTALLATION

1. Before servicing the vehicle, refer to the Precautions Section.
2. Properly relieve the fuel system pressure.
3. Disconnect the negative battery cable.

➡**On some vehicles, when the battery cable is disconnected and reconnected, some abnormal drive symptoms may occur while the vehicle relearns its adaptive strategy. The vehicle may need to be driven to relearn its strategy.**

4. Remove or disconnect the following:
   • Push connect and R-clip fittings from the fuel filter
   • Fuel filter

To install:
5. Install or connect the following:
   • Fuel filter.
   • R-clip and push connect fittings
   • Negative battery cable
6. Start the vehicle, check for leaks and repair if necessary.

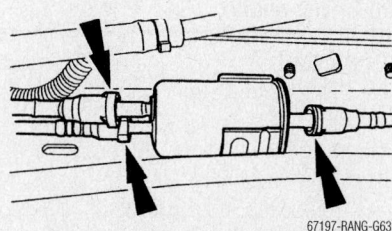

67197-RANG-G63

Fuel filter location and connections

Fuel Pump

REMOVAL & INSTALLATION

2002–2003

1. Before servicing the vehicle, refer to the Precautions Section.
2. Properly relieve the fuel system pressure.
3. Disconnect the negative battery cable.

➡**On some vehicles, when the battery cable is disconnected and reconnected, some abnormal drive symptoms may occur while the vehicle relearns its adaptive strategy. The vehicle may need to be driven to relearn its strategy.**

4. Remove or disconnect the following:
   • Fuel tank
   • Fuel tank pump locking retainer ring
   • Fuel pump mounting gasket and discard the gasket
   • Fuel pump

To install:
5. Install or connect the following:
   • Fuel pump and a new mounting gasket
   • Fuel tank pump locking retainer ring. Torque the ring to 66 ft. lbs. (90 Nm).
   • Fuel tank
   • Negative battery cable
6. Start the vehicle, check for leaks and repair if necessary.

2004–2006

1. Before servicing the vehicle, refer to the Precautions Section.
2. Properly relieve the fuel system pressure.
3. Disconnect the negative battery cable.

➡**On some vehicles, when the battery cable is disconnected and reconnected, some abnormal drive symptoms may occur while the vehicle relearns its adaptive strategy. The vehicle may need to be driven to relearn its strategy.**

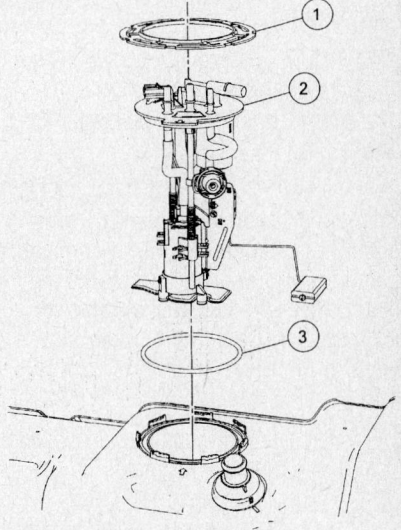

1   Fuel pump locking ring
2   Fuel pump
3   Fuel pump O-ring seal

09482_MAZB_G0024

Fuel pump and related components—2004–2006

4. Remove the fuel tank.

5. Clean the area around the fuel pump mounting flange.

6. Using the special tool, remove the fuel tank pump assembly locking retainer ring.

✳✳ WARNING

The fuel pump assembly must be removed and handled carefully to avoid damage to the float arm and filter.

7. Remove the fuel pump assembly.

8. Remove and discard the fuel pump mounting gasket.

To install:

9. Clean the fuel pump mounting flange and the fuel tank mounting surface.

10. Install a new fuel pump mounting gasket.

11. Install the fuel pump and sender assembly with the float toward the rear of the tank. Align the arrows molded into the tank and flange.

12. Install the locking ring while compressing the pump assembly into the tank.

13. Using the special tool, tighten the fuel pump assembly locking ring retainer ring until it locks in place.

14. Install the fuel tank.

Fuel Injectors

REMOVAL & INSTALLATION

2.3L Engine

1. Before servicing the vehicle, refer to the Precautions Section.

2. Properly relieve the fuel system pressure.

3. Disconnect the negative battery cable.

➡️ On some vehicles, when the battery cable is disconnected and reconnected, some abnormal drive symptoms may occur while the vehicle relearns its adaptive strategy. The vehicle may need to be driven to relearn its strategy.

4. Remove or disconnect the following:
- Upper intake manifold
- Fuel injector connectors
- Fuel injector harness from the fuel injector supply manifold
- Fuel line spring lock
- Fuel line
- Fuel injection supply manifold
- Fuel injector retaining clip
- Fuel injector

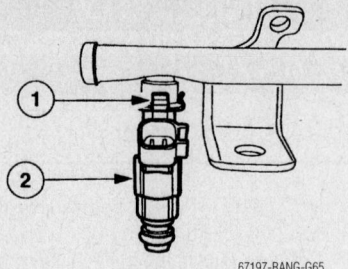

Fuel injector-to-fuel rail installation. 1-retaining clip; 2-injector—2.3L engine

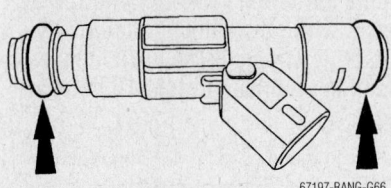

Fuel injector O-rings—2.3L engine

✳✳ WARNING

Use O-ring seals that are made of special fuel-resistant material. Use of ordinary O-ring seals can cause the fuel system to leak. Do not reuse the O-ring seals.

To install:

5. Installation is the reverse of removal.

6. Install new O-rings. Lubricate the O-rings with clean engine oil.

7. Torque the supply manifold bolts to 18 ft. lbs. (25 Nm).

3.0L and 4.0L Engines

1. Before servicing the vehicle, refer to the Precautions Section.

2. Disconnect the negative battery cable.

➡️ On some vehicles, when the battery cable is disconnected and reconnected, some abnormal drive symptoms may occur while the vehicle relearns its adaptive strategy. The vehicle may need to be driven to relearn its strategy.

3. Properly relieve the fuel system pressure.

4. Remove or disconnect the following:
- Upper intake manifold
- Engine control sensor wiring from the fuel injectors
- Fuel lines

Fuel injector wiring connectors—3.0L engine

Fuel rail bolts—3.0L engine

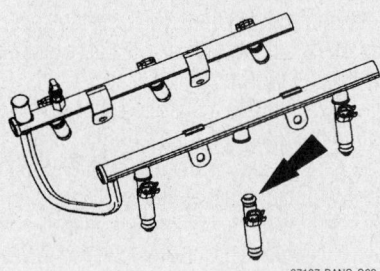

Fuel rail and injectors—3.0L engine

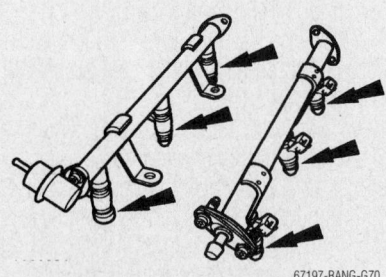

Fuel rail and injectors—4.0L engine

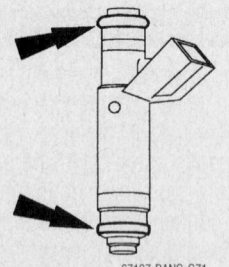

Fuel injector O-rings—4.0L engine

- Fuel injection supply manifold and injectors as an assembly
- Vacuum line
- Fuel injectors from the supply manifold
- Inspect the O-rings and replace them as needed

To install:

5. Install or connect the following:
   - Fuel injectors
   - Vacuum line
   - Fuel injection supply manifold. Torque the bolts to 89 inch lbs. (10 Nm).

- Fuel line
- Engine control sensor wiring to the fuel injectors
- Upper intake manifold
- Negative battery cable

6. Start the vehicle, check for leaks and repair if necessary.

DRIVE TRAIN

Manual Transmission

REMOVAL & INSTALLATION

2002–2003

1. Before servicing the vehicle, refer to the Precautions Section.
2. Disconnect the negative battery cable.

➡**On some vehicles, when the battery cable is disconnected and reconnected, some abnormal drive symptoms may occur while the vehicle relearns its adaptive strategy. The vehicle may need to be driven to relearn its strategy.**

3. Remove or disconnect the following:
   - Upper gearshift lever and the outer gearshift lever boot and console assembly as an assembly
4. If transmission disassembly is necessary, remove the drain plug, and drain the transmission fluid. Install the drain plug after draining all of the fluid.
5. Remove or disconnect the following:
   - Electrical connector from the reverse lamp switch
   - Electrical connector from the vehicle speed sensor (VSS)
   - Heated oxygen sensor (HO2S) electrical connector from the bracket
   - Wiring harness from the bracket
   - Electrical connectors from the heated oxygen sensors (HO2S)
   - Starter motor

➡**The driveshaft centering socket yoke fits tightly on the rear axle pinion**

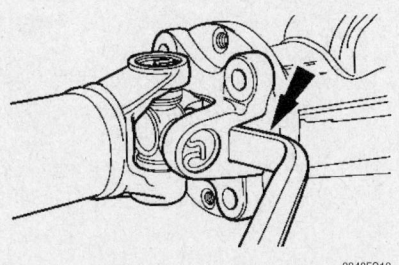

9348EG10

Pry here (arrow) for driveshaft removal

flange pilot. Never hammer on the driveshaft or any of its components to disconnect the yoke from the flange. Pry only in the area shown, with a suitable tool, to disconnect the yoke from the flange.

➡**If equipped, always disconnect the front driveshaft from the transfer case first. Otherwise, the weight of the driveshaft can cause the boot to tear.**

- Rear driveshaft, and the front driveshaft, if so equipped
- Bolts retaining the exhaust inlet crossover pipe to the exhaust manifold
- Bolts retaining the catalytic converter to the muffler. Discard the exhaust converter outlet gasket.
- Exhaust hanger from the insulator. Position the exhaust assembly aside.
- On 4WD vehicles, the transfer case
- Clutch hydraulic line from the clutch slave cylinder

※※ WARNING

Secure the transmission to the jack with a suitable safety strap. Failure to follow these instructions may result in personal injury.

6. Using a suitable transmission jack, support the transmission. Secure the transmission to the jack with a suitable safety strap.
7. Loosen, but do not remove the nuts retaining the transmission insulator to the crossmember.
8. Remove the six bolts retaining the crossmember to the frame.
9. Remove the nuts and the crossmember.

➡**Lower the transmission enough to gain access to the upper bolts retaining the transmission to the engine.**

10. Remove the nine bolts retaining the transmission to the engine.
11. Remove the transmission from the vehicle.

To install:

12. Installation is the reverse of removal. Observe the following torques:
    - Transmission-to-engine bolts: 44 ft. lbs. (60 Nm)
    - Crossmember-to-frame: 46 ft. lbs. (63 Nm)
    - Transmission insulator-to-crossmember: 72 ft. lbs. (98 Nm)

2004–2006

1. Before servicing the vehicle, refer to the Precautions Section.
2. Disconnect the negative battery cable.

➡**On some vehicles, when the battery cable is disconnected and reconnected, some abnormal drive symptoms may occur while the vehicle relearns its adaptive strategy. The vehicle may need to be driven to relearn its strategy.**

3. Remove the upper gearshift lever, the outer gearshift lever boot and the console as an assembly.
4. Raise and support the vehicle safely.
5. If transmission disassembly is required, remove the drain plug and drain the transmission fluid. Install the drain plug after draining all the fluid.
6. To maintain initial driveshaft balance, index-mark the driveshaft yoke to the axle flange, so they can be installed in their original positions. Remove the rear driveshaft.
7. If equipped with 4WD, remove the transfer case.
8. Using special tool 308-182, disconnect the clutch hydraulic line.
9. Remove the starter motor.
10. Place a suitable jack under the transmission. Secure the transmission to the jack with a safety strap.
11. Remove the transmission mount bolts and the transmission mount.
12. Remove the transmission bolt and reposition the exhaust bracket. Remove the exhaust inlet crossover pipe on 3.0L and 4.0L engines.
13. Lower the transmission enough to gain access to the upper transmission-to-

engine bolts. Remove the nine transmission-to-engine bolts.

14. Remove the transmission to engine bolts. Remove the transmission from the vehicle.

To install:

15. Installation is the reverse of the removal procedure.

16. Torque the transmission-to-engine bolts: 44 ft. lbs. (60 Nm)

17. Align the index marks when installing the front and rear driveshafts.

18. Check and, if necessary, fill the transmission with the specified type and quantity of fluid.

Automatic Transmission

REMOVAL & INSTALLATION

2002–2003

1. Before servicing the vehicle, refer to the Precautions Section.

2. Disconnect the negative battery cable.

➡**On some vehicles, when the battery cable is disconnected and reconnected, some abnormal drive symptoms may occur while the vehicle relearns its adaptive strategy. The vehicle may need to be driven to relearn its strategy.**

3. Place the selector lever in NEUTRAL position.

4. Remove or disconnect the following:
   • Fluid level indicator
   • The two bolts retaining the fan shroud to the radiator.

➡**If transmission disassembly is required, drain the transmission fluid.**

   • With 4wd, the transfer case

➡**Mark the driveshaft yoke and axle flange, so they may be installed in their original alignment.**

   • Rear driveshaft
   • Starter motor
   • Torque converter access cover

➡**Mark the torque converter and the flexplate for correct alignment at reinstallation.**

   • The four converter nuts
   • Shift cable
   • Transmission wiring harness
   • Three way catalytic converter
   • Left HO$_2$S sensor
   • Front exhaust crossover pipe
   • Transmission cooler lines

5. Position a High-Lift Jack under the transmission. Raise and support the transmission.

6. Remove or disconnect the following:
   • Crossmember.
   • Transmission mount
   • Transmission upper fill tube

➡**Lower the High-Lift Transmission Jack to gain access to screws.**

   • On 4x4 models, the vent tube assembly

※※ WARNING

Install the Torque Converter Holding Tool before lowering the transmission from the vehicle. Secure the transmission to the transmission jack with a safety chain. Failure to follow these instructions can result in personal injury.

7. Lower the transmission.

8. Installation is the reverse of removal. Observe the following torques:
   • Transmission-to-engine bolts: 41 ft. lbs. (55 Nm)
   • Exhaust bracket bolts: 81 ft. lbs. (110 Nm)
   • Crossmember-to-frame: 87 ft. lbs. (118 Nm)
   • Transmission mount-to-crossmember: 81 ft. lbs. (110 Nm)
   • Converter-to-flexplate: 30 ft. lbs. (40 Nm)
   • Rear driveshaft-to-flange bolts: 95 ft. lbs. (129 Nm)

2004–2006

1. Before servicing the vehicle, refer to the Precautions Section.

2. Disconnect the negative battery cable.

➡**On some vehicles, when the battery cable is disconnected and reconnected, some abnormal drive symptoms may occur while the vehicle relearns its adaptive strategy. The vehicle may need to be driven to relearn its strategy.**

➡**If the transmission is to be removed for a period of time, support the engine with a safety stand and a wood block.**

3. With the vehicle in NEUTRAL, position it on a hoist.

4. On 4.0L vehicles, remove the fluid level indicator tube bolt and remove the tube and indicator.

5. If transmission disassembly is required, drain the transmission fluid.

6. On 4x4 vehicles, remove the transfer case.

7. To maintain initial driveshaft balance, mark the driveshaft yoke and axle flange so they can be installed in their original alignment.

8. Remove the rear driveshaft.
   a. Remove the four bolts.
   b. Remove the driveshaft.

9. Remove the starter motor.

10. With a 2.3L engine:
   a. When removing the torque converter nuts, the crankshaft must be rotated only in the clockwise direction, otherwise engine damage can occur. The crankshaft, crankshaft sprocket and the pulley are fitted together by friction between the flange faces on each part. For that reason, the crankshaft sprocket can also be moved when the crankshaft pulley is turned in the counterclockwise direction.
   b. It may be necessary to gain access to the flexplate nuts through the wheel well.
   c. Mark the torque converter and the flexplate for correct alignment at reinstallation.
   d. Remove and discard the four torque converter nuts. Rotate the flexplate to access to all the nuts.

11. With 3.0L and 4.0L engines:
   a. Mark the torque converter and the flexplate for correct alignment at reinstallation. Remove the four nuts. Rotate the flexplate to access to all the nuts.
   b. Disconnect the shift cable.
   c. Disconnect the transmission wiring harness from the case.

12. Disconnect the transmission wiring harness. Remove the three way catalytic converter.

13. With a 2.3L engine, remove the rear engine cover plate.

➡**Care should be taken not to bend or damage the cooler lines.**

14. Hold the case fitting and remove the transmission cooler lines.

15. Remove the nuts.

16. Position a transmission jack under the transmission. Raise and support the transmission.

17. Remove the crossmember.

18. Remove the transmission mount.

19. Remove the transmission upper fill tube.

20. Lower the jack to gain access to screws. Remove the transmission-to-engine bolts.

21. Remove the lower screws.

22. Remove the HO$_2$S connector bracket from the transmission.

23. On 4x4 vehicles, remove the vent tube assembly.

❋❋ CAUTION

The torque converter is heavy and may result in injury if it falls out of the transmission. Secure the torque converter in the transmission. Failure to follow these instructions may result in personal injury. Install a converter locking tool before lowering the transmission from the vehicle.

❋❋ CAUTION

Secure the transmission to the transmission jack with a safety chain. Failure to follow these instructions may result in personal injury.

24. Lower the transmission.
25. If the transmission is being overhauled or if installing a new or remanufactured transmission, carry out transmission fluid cooler backflushing and cleaning.
To install:
26. On 4x4 vehicles, install the vent tube assembly.

❋❋ CAUTION

Secure the transmission to the transmission jack with a safety chain. Failure to follow these instructions may result in personal injury.

27. Raise and position the transmission.
28. Remove the holding tool.
29. On the 4.0L engine:
   a. Align the flexplate to the converter marks made at removal.
   b. Install the transmission-to-engine screws. Torque to 35 ft. lbs. (48 Nm).
30. With the 2.3L and 3.0L engines
   a. Align the flexplate to the converter marks made at removal.
   b. Install the upper fluid filler tube and bracket screw.
31. On the 2.3L engine:
   a. Install the HO2S connector bracket.
   b. Install the rear engine cover plate.
32. On 4x4 vehicles, install the transfer case.
33. Install the exhaust bracket. Torque the bolts to 73 ft. lbs. (99 Nm).
34. Install the crossmember. Tighten the bolts to 74 ft. lbs. (101 Nm).
35. Install the transmission mount into the crossmember and torque the nuts to 73 ft. lbs. (99 Nm).

➡️**Prior to installing the cooler lines to the case, inspect the O-rings. If damaged new O-rings will need to be installed.**

   a. Hold the case fitting and install the transmission cooler lines. Torque to 19 ft. lbs. (28 Nm).
   b. Install four new torque converter nuts. Rotate the crankshaft as needed to gain access to all the nuts. Torque to 26 ft. lbs. (35 Nm).

➡️**On 2.3L engine when installing the torque converter nuts, the crankshaft must be rotated only in the clockwise direction, otherwise engine damage can occur. The crankshaft, the crankshaft sprocket and the pulley are fitted together by friction between the flange faces on each part. For that reason, the crankshaft sprocket can also be moved when the crankshaft pulley is turned in the counterclockwise direction.**

36. Install the starter motor.
37. Install the catalytic converter assembly.
38. Position the transmission wiring harness in place.
39. Connect the transmission wiring harness.
40. Install the shift cable.
41. Align the driveshaft yoke and the axle shaft marks made at removal to maintain driveline balance. Install the rear driveshaft. Install the driveshaft bolts. Torque to 83 ft. lbs. (112 Nm).
42. Use the following guidelines for installing the in-line transmission fluid filter:
   a. If the transmission was overhauled and the vehicle was equipped with an in-line fluid filter, install a new in-line fluid filter.
   b. If the transmission was overhauled

and the vehicle was not equipped with an in-line fluid filter, install a new in-line fluid filter kit.
   c. If the transmission is being installed for a non-internal repair, do not install an in-line filter or filter kit.
   d. If installing a new or re-manufactured transmission, install the in-line transmission fluid filter that is supplied.
   e. Prior to lowering the vehicle, install a new in-line transmission filter or a filter kit.
43. Install the transmission fill tube and indicator as an assembly.

➡️**When the battery has been disconnected and reconnected, some abnormal drive symptoms can occur while the vehicle relearns its adaptive strategy.**

44. Connect the battery ground cable.
45. Fill the transmission with clean automatic transmission fluid to the specified level.
46. Check the transmission for correct operation.
47. Verify that the shift cable is correctly adjusted.

Clutch

REMOVAL & INSTALLATION

1. Before servicing the vehicle, refer to the Precautions Section.
2. Disconnect the negative battery cable.

➡️**On some vehicles, when the battery cable is disconnected and reconnected, some abnormal drive symptoms may occur while the vehicle relearns its adaptive strategy. The vehicle may need to be driven to relearn its strategy.**

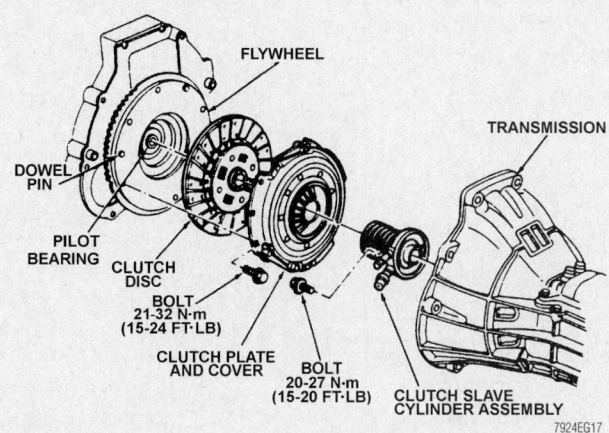

Clutch disc and related components

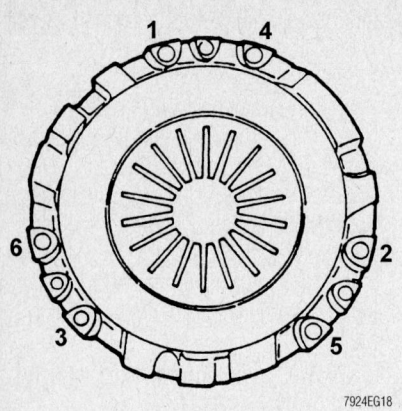

Clutch disc bolt torque sequence

7924EG18

3. Remove or disconnect the following:
- Transmission

➡**If the clutch disc and pressure plate are to be reinstalled, bolts must be removed evenly or permanent damage to the diaphragm spring will occur resulting in complete clutch release.**

- Bolts, clutch pressure plate and the clutch disc.

➡**If the parts are to be reused, index-mark the clutch pressure plate to the flywheel.**

To install:

4. Lubricate the transmission input shaft pilot bearing with front axle grease.
5. Using a suitable press, press downward on the pressure plate fingers until the adjusting ring moves freely.
6. Rotate the adjusting ring counter-clockwise to compress the tension springs. Hold the adjusting ring in this position.
7. Release the pressure on the fingers. The adjusting ring will stay in the reset position.
8. Position the clutch disc on the flywheel.

➡**If reusing the clutch pressure plate and flywheel, align the marks made during removal.**

9. Align the clutch disc and the clutch pressure plate. Install the bolts and tighten in a star pattern sequence to 24 ft. lbs. (35 Nm) for 2002 vehicles and 20 ft. lbs. (27 Nm) for 2003–2006 vehicles.
- Install the transmission.

ADJUSTMENT

1. Before servicing the vehicle, refer to the Precautions Section.
Because the clutch is hydraulically driven, there is no adjustment required.

In the event the clutch pedal develops a squeak or uneven feel when depressing, spray the pedal bushing assembly with penetrating oil and work the pedal back-and-forth.

Hydraulic Clutch System

BLEEDING

The following procedure is recommended for bleeding the clutch hydraulic system installed on the vehicle. It is recommended that the original clutch tube, with quick-connect fitting be replaced when servicing the hydraulic system, because air can be trapped in the quick-connect fitting and prevent complete bleeding of the system. The replacement tube does not include a quick-connect fitting.

1. Before servicing the vehicle, refer to the Precautions Section.
2. Clean the dirt and grease from the dust cap.
3. Remove the cap and diaphragm and fill the reservoir to the top with approved brake fluid C6AZ-19542-AA or BA, (ESA-M6C25-A).

➡**To keep brake fluid from entering the clutch housing, route a suitable rubber tube of appropriate inside diameter from the bleed screw to a container.**

4. Loosen the bleed screw, located in the slave cylinder body, next to the inlet connection. Fluid will now begin to move from the master cylinder down the tube to the slave cylinder.

➡**The reservoir must be kept full at all times during the bleeding operation, to ensure no additional air enters the system.**

5. Observe the bleed screw outlet. When the slave cylinder is full, a steady stream of fluid will flow from the outlet port. Tighten the bleed screw.
6. Depress the clutch pedal to the floor and hold for 1–2 seconds. Release the pedal as rapidly as possible. The pedal must be released completely. Pause for 1–2 seconds. Repeat 10 times.
7. Check the fluid level in the reservoir. The fluid should be level with the step when the diaphragm is removed.
8. Hold the pedal to the floor, slightly open the bleed screw to allow any additional air to escape. Close the bleed screw, and then release the pedal.
9. Check the fluid in the reservoir. The hydraulic system should now be fully bled, and should actuate the clutch.

10. Check the vehicle by starting, pushing the clutch pedal to the floor and selecting reverse gear. There should be no grating of gears. If there is, and the hydraulic system still contains air; repeat the bleeding procedure.

Transfer Case Assembly

REMOVAL & INSTALLATION

2002–2003

1. Before servicing the vehicle, refer to the Precautions Section.
2. Disconnect the negative battery cable.

➡**On some vehicles, when the battery cable is disconnected and reconnected, some abnormal drive symptoms may occur while the vehicle relearns its adaptive strategy. The vehicle may need to be driven to relearn its strategy.**

3. Place the transmission in neutral. Raise and support the vehicle safely.
4. Remove or disconnect the following:
- Skid plate
- Damper
- Transfer case harness connector and position it aside

5. If transfer case disassembly is necessary, remove the drain plug and drain the fluid.

➡**Index-mark the front output shaft assembly and the front driveshaft constant velocity (CV) joint. Always disconnect the front driveshaft from the transfer case first. Otherwise, the weight of the driveshaft can pinch the boot between the shaft and the boot can and cause the boot to tear.**

- Front driveshaft from the transfer case and position the driveshaft aside. Remove and discard the bolts and washers.

➡**Index-mark the front flange on the rear driveshaft and the flange on the transfer case.**

- Rear driveshaft

➡**Secure the transfer case to the jack with safety straps.**

6. Position a high lift jack under the transfer case.
7. Remove or disconnect the following:
- Five bolts retaining the transfer case to the extension housing

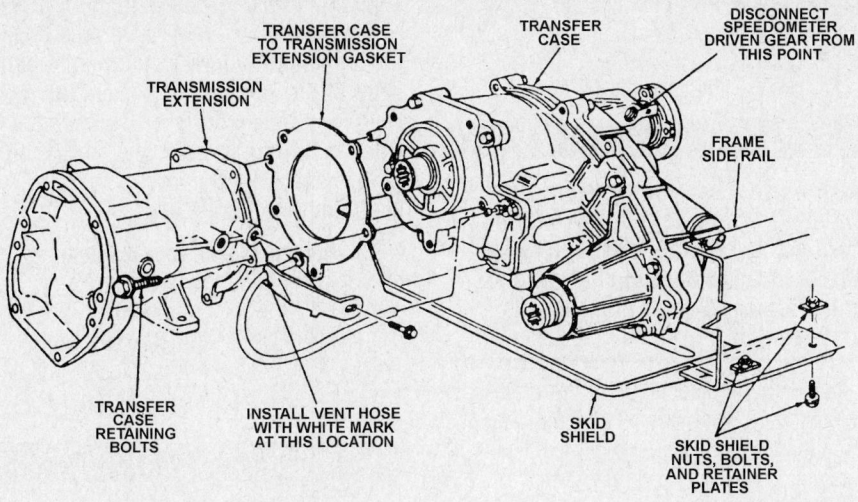

Transfer case and related components—2004–2006

- Transfer case rearward and off of the transmission output shaft

8. Remove and discard the front extension housing gasket and clean the mating surfaces.

To install:

9. Installation is the reverse of removal. Take note of the following:

- Install the transfer case with a new gasket.
- Tighten the bolts that retain the transfer case to the extension housing in a clockwise direction beginning with the upper LH bolt.
- Install the front and the rear driveshafts with new bolts. If new bolts are not available, coat the threads of the original bolts with Threadlock and Sealer E0AZ-19554-AA or equivalent meeting Ford specification WSK-M2G351-A5.
- When installing the front driveshaft, always connect it to the axle first and then connect it to the transfer case.
- Align the index marks when installing the front and rear driveshafts.

10. Observe the following torques:

- Nut retaining the flange to the rear output shaft: 262 ft. lbs. (355 Nm)
- Bolt retaining the rear driveshaft to the flange: 82 ft. lbs. (111 Nm)
- Bolt retaining the motor assembly and connector to the transfer case cover: 89 inch lbs. (10 Nm)
- Bolt retaining the skid plate to the frame: 18 ft. lbs. (24 Nm)
- Bolt retaining the damper to the transfer case: 30 ft. lbs. (40 Nm)
- Bolt retaining the driveshaft CV

joint to the front output shaft assembly: 22 ft. lbs. (30 Nm)
- Bolt retaining the transfer case to the extension housing: 40 ft. lbs. (54 Nm)
- Bolt retaining the front adapter to the transfer case: 30 ft. lbs. (40 Nm)
- Bolt retaining the transfer case to the transfer case cover: 27 ft. lbs. (36 Nm)
- Drain plug: 18 ft. lbs. (24 Nm)
- Fill plug: 18 ft. lbs. (24 Nm)

2004–2006

1. Before servicing the vehicle, refer to the Precautions Section.

2. Disconnect the negative battery cable.

➡On some vehicles, when the battery cable is disconnected and reconnected, some abnormal drive symptoms may occur while the vehicle relearns its adaptive strategy. The vehicle may need to be driven to relearn its strategy.

3. With the vehicle in NEUTRAL, raise and support the vehicle.

4. Remove the skid plate.

5. Remove the damper, if equipped.

6. Disconnect the transfer case harness connector and position it aside.

7. If transfer case disassembly is necessary, remove the drain plug and drain the fluid. Install the drain plug when all of the fluid has drained.

➡Index-mark the front output shaft assembly and the front driveshaft constant velocity (CV) joint.

Always disconnect the front driveshaft from the transfer case first. Otherwise, the weight of the driveshaft can pinch the boot between the shaft and the boot can and cause the boot to tear.

8. Index-mark the front output shaft assembly and the front driveshaft constant velocity (CV) joint.

9. Remove and discard the bolts and washers.

10. Disconnect the front driveshaft from the transfer case and position the driveshaft aside.

➡Index-mark the front flange on the rear driveshaft and the flange on the transfer case.

11. Remove the rear driveshaft.

Secure the transfer case to the jack with safety straps.

12. Position a high lift jack under the transfer case.

13. Remove the five bolts retaining the transfer case to the extension housing.

14. Slide the transfer case rearward and off of the transmission output shaft.

15. Remove and discard the front extension housing gasket, and clean the mating surfaces.

To install:

16. Install the transfer case with a new gasket.

17. Tighten the bolts that retain the transfer case to the extension housing in a clockwise direction beginning with the upper LH bolt. Torque to 40 ft. lbs. (54 Nm).

18. Install the front driveshaft with new bolts and washers and the rear driveshaft with new bolts. If new bolts are not available, coat the threads of the original bolts with Threadlock and Sealer E0AZ-19554-AA, or equivalent.

➡When installing the front driveshaft, always connect it to the axle first and then connect it to the transfer case.

➡Align the index marks when installing the front and rear driveshafts.

19. The remainder of installation is the reverse of the removal procedure.

20. Check and, if necessary, fill the transfer case with the specified type and quantity of fluid.

Halfshaft

REMOVAL & INSTALLATION

1. Before servicing the vehicle, refer to the Precautions Section.

2. With the vehicle in NEUTRAL, raise and support the vehicle.

3. Remove the front wheel and tire assembly.

➡**Do not reuse the torque prevailing design hub nut and washer assembly.**

4. Remove and discard the hub nut and washer assembly.

✳✳ WARNING

Do not allow the disc brake caliper to hang suspended from the brake hose. Provide a suitable support.

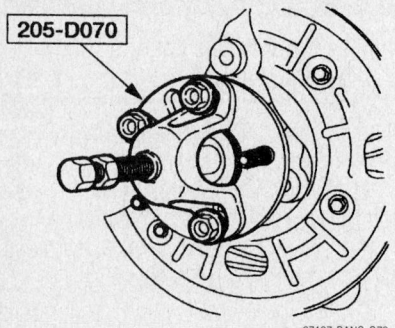

Halfshaft removal tool in position

67197-RANG-G72

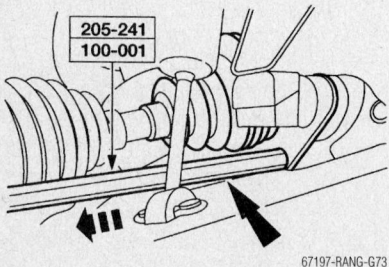

Separating the halfshaft from the axle housing using tools 205-241 and 100-001

67197-RANG-G73

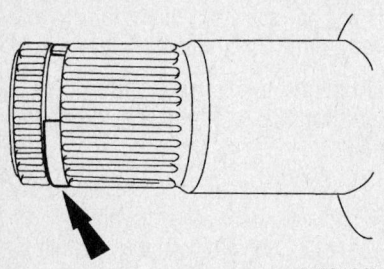

67197-RANG-G74

Circlip installed on halfshaft end

5. Remove the front disc brake caliper, anchor plate, and pads as an assembly, and position the assembly aside.

6. Remove the brake disc.

✳✳ WARNING

Do not use a hammer to separate the outboard front wheel halfshaft joint from the wheel hub. Damage to the outboard CV joint stub shaft threads and internal CV joint components may result.

7. Using the special tool, separate the outboard front wheel halfshaft joint from the wheel hub. Remove the special tool.

8. Support the front suspension lower arm.

9. Remove the nut and bolt retaining the upper ball joint to the front wheel knuckle.

10. Rotate the front wheel knuckle.

11. Compress the outboard front wheel halfshaft joint.

12. Remove the outboard front wheel halfshaft joint from the wheel hub.

13. Using the special tools, 205-241 and 100-001, or equivalent, separate the inboard front wheel halfshaft joint from the front axle housing.

14. Remove the halfshaft assembly from the vehicle with both hands. Do not damage the axle seal.

To install:

✳✳ WARNING

Install the halfshaft with a new hub nut and washer assembly. Do not use power or impact tools to tighten the hub nut and washer assembly.

➡**Install a new retainer circlip in the groove in the LH inboard CV joint housing stub shaft before installing the halfshaft in the vehicle. To prevent the new retainer circlip from over-expanding when installing it, start one end in the groove and work the circlip over the shaft and into the groove.**

15. Continue the installation in the reverse order of the removal procedure. Observe the following torques:

- Hub nut: 162 ft. lbs. (220 Nm)
- Upper ball joint-to-knuckle nut: 41 ft. lbs. (55 Nm)

Driveshaft

REMOVAL & INSTALLATION

Front

1. Before servicing the vehicle, refer to the Precautions Section.

2. With the vehicle in NEUTRAL, raise and support the vehicle.

3. Index the front axle pinion flange to the front driveshaft.

4. Index the transfer case pinion flange to the front driveshaft.

5. Remove the front driveshaft-to-transfer case bolts and retainers.

6. Remove the front driveshaft-to-front axle bolts.

7. Remove the front driveshaft.

To install:

8. To install, reverse the removal procedure.

9. Torque the bolts at the axle end to 15 ft. lbs. (20 Nm)

10. Torque the transfer case end bolts to 22 ft. lbs. (30 Nm).

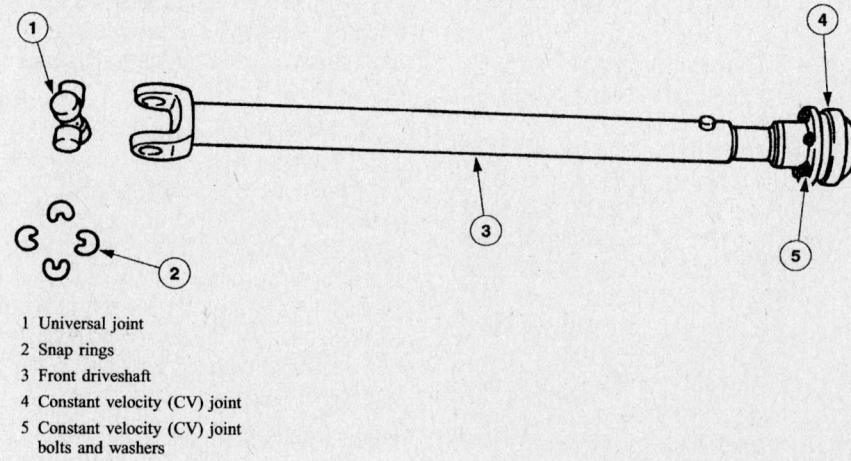

1 Universal joint
2 Snap rings
3 Front driveshaft
4 Constant velocity (CV) joint
5 Constant velocity (CV) joint bolts and washers

06017-RANG-G08

4WD front driveshaft and related components—2002–2005

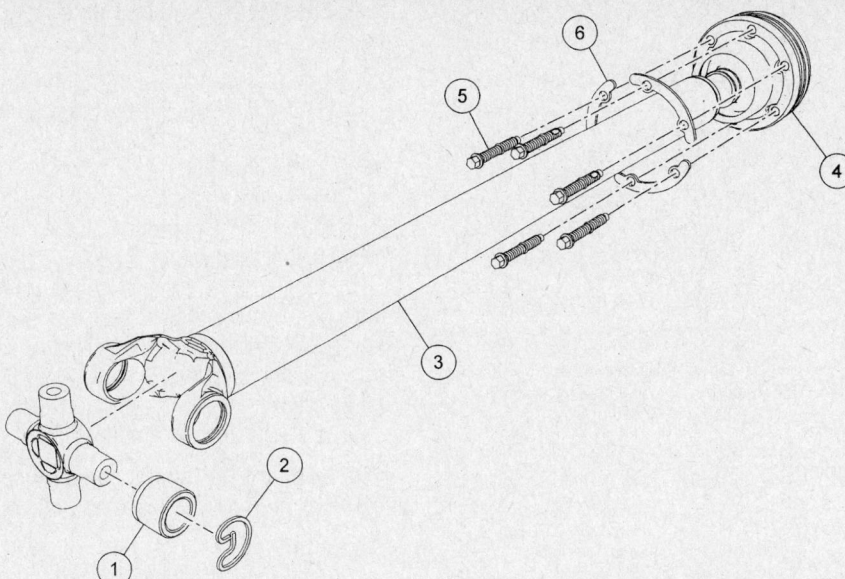

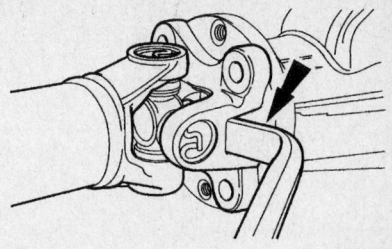

06017-RANG-G11

Pry only in the area shown, with a suitable tool, to disconnect the driveshaft flange from the pinion flange

1	Universal Joint	4	Constant Velocity (CV) Joint
2	Snap Rings	5	CV Joint Bolts and Washers
3	Propeller Shaft	6	CV Joint Retainers

09482_MAZB_G0025

4WD front driveshaft and related components—2006

Rear

2 WD

1. Before servicing the vehicle, refer to the Precautions Section.
2. With the vehicle in NEUTRAL, raise and support the vehicle.
3. Index-mark the driveshaft flange and rear axle pinion flange.

4. Index-mark the driveshaft and the extension housing.
5. Remove the four bolts.

✳✳ WARNING

The driveshaft flange fits tightly on the rear axle pinion flange pilot. Never hammer on the driveshaft or any of its components to disconnect

the driveshaft flange from the pinion flange. Pry only in the area shown, with a suitable tool, to disconnect the driveshaft flange from the pinion flange.

6. Using a suitable tool as shown, disconnect the driveshaft flange from the rear axle pinion flange.
7. Lower the driveshaft and slide it off the output shaft.
8. Plug the extension housing to prevent fluid loss.

To install:

✳✳ WARNING

If new bolts to retain the driveshaft to the axle are not available, coat the threads of the original bolts with Threadlock and Sealer EOAZ-19554-AA or equivalent meeting Ford specification WSK-M2G351-A5.

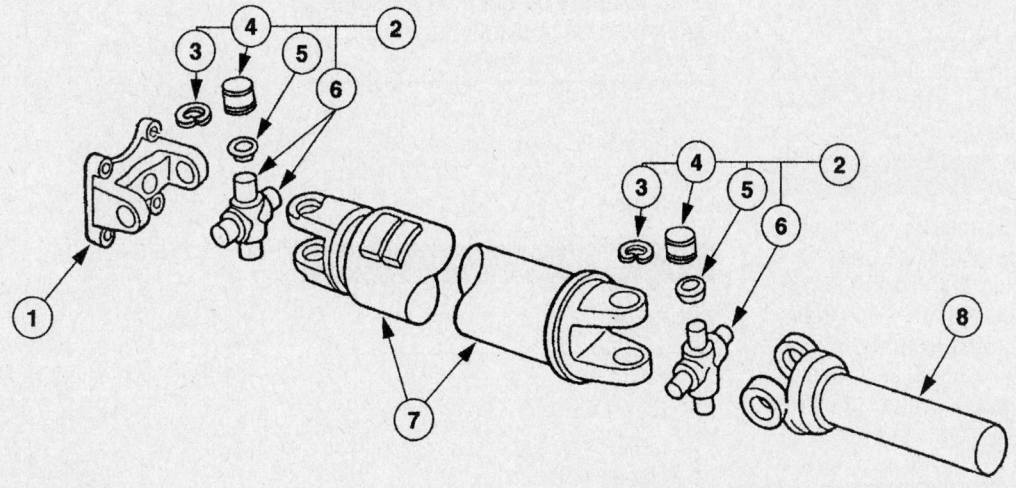

1 Flange	5 Seal
2 Universal joint	6 Spider
3 Snap ring	7 Driveshaft
4 Bearing	8 Driveshaft slip yoke

06017-RANG-G09

2WD rear driveshaft and related components

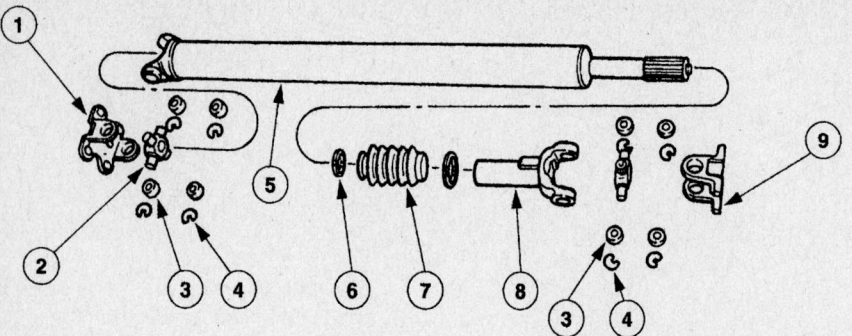

1 Rear flange
2 Spider
3 Bearing cup
4 Snap ring
5 Driveshaft

6 Driveshaft slip-yoke boot clamp
7 Universal joint slip-yoke boot
8 Driveshaft slip-yoke
9 Front flange

06017-RANG-G10

4WD rear driveshaft and related components

✳✳ WARNING

The driveshaft flange fits tightly on the rear axle pinion flange pilot. To make sure that the driveshaft flange seats squarely on the pinion flange, tighten the bolts evenly in a cross pattern.

9. Continue the installation in the reverse order of the removal procedure. Torque the bolts to 83 ft. lbs. (112 Nm).

4 WD

1. Before servicing the vehicle, refer to the Precautions Section.
2. With the vehicle in NEUTRAL, raise and support the vehicle.
3. Index-mark the driveshaft flange to the rear axle pinion flange.
4. Remove the four rear axle flange bolts.
5. Index-mark the driveshaft and the extension housing.

✳✳ WARNING

The driveshaft flange fits tightly on the rear axle pinion flange pilot. Never hammer on the driveshaft or any of its components to disconnect the driveshaft flange from the pinion flange. Pry only in the area shown, with a suitable tool, to disconnect the driveshaft flange from the pinion flange.

6. Using a suitable tool as shown, disconnect the driveshaft flange from the rear axle pinion flange.

➡**Make sure the index marks on the extension housing and driveshaft are aligned before separation.**

7. Lower the driveshaft and separate it from the transmission.
   To install:

✳✳ WARNING

If new bolts to retain the driveshaft to the axle are not available, coat the threads of the original bolts with Threadlock and Sealer E0AZ-19554-AA or equivalent meeting Ford specification WSK-M2G351-A5.

✳✳ WARNING

The driveshaft flange fits tightly on the rear axle pinion flange pilot. To make sure that the driveshaft flange seats squarely on the pinion flange, tighten the bolts evenly in a cross pattern.

8. Continue the installation in the reverse order of the removal procedure. Torque the bolts at each end to 83 ft. lbs. (112 Nm).

CV-Joints

OVERHAUL

2002–2003

1. Before servicing the vehicle, refer to the Precautions Section.
2. Disconnect the negative battery cable.

➡**On some vehicles, when the battery cable is disconnected and reconnected, some abnormal drive symptoms may occur while the vehicle relearns its adaptive strategy. The vehicle may need to be driven to relearn its strategy.**

3. Remove or disconnect the following:
   • Halfshaft and place it in a vice with the inboard joint lower than the outboard joint
4. Cut the inner boot clamps with side cutters and remove the clamp from the boot.
   • Larger boot end off the joint
   • Inboard CV-joint bolts and separate the spacer and grease cap
   • Snapring retaining the interconnecting shaft end to the CV-joint cage
   • CV-joint and discard the washer

➡**The outboard CV-joint is non-service-able other than to replace the boot.**

To install:

5. Install or connect the following:
   • Slide the boot over the shaft
6. Fill the CV-joint area with grease.
   • Assemble the outer boot to the outboard CV-joint and interconnecting shaft. Make certain that the boot is seated in the grooves on the outer race and on the shaft
   • New clamps to the boot
   • New inner boot to the shaft
   • New washer to the end of the shaft
   • Assemble the inboard CV-joint to the interconnecting shaft spline until it rests on the washer
   • Snapring
7. Fill the CV-joint area with grease.
   • Boot into position and make certain that it is seated in the grooves on the boot adapter and the shaft
   • New clamps and tighten the clamps with crimping pliers
   • Spacer to the CV-joint end pilot. Torque the bolts to 25 ft. lbs. (34 Nm).
   • Halfshaft
   • Negative battery cable

2004–2006

1. Before servicing the vehicle, refer to the Precautions Section.
2. Remove the front wheel halfshaft. Do not damage the halfshaft boot.
3. Remove the two inboard boot clamps.
4. Slide the inboard halfshaft boot off the inboard CV joint housing.
5. Separate the CV joint from the CV joint housing.
6. Matchmark the shaft and the inboard CV joint for correct alignment during assembly.
7. Remove the snapring.

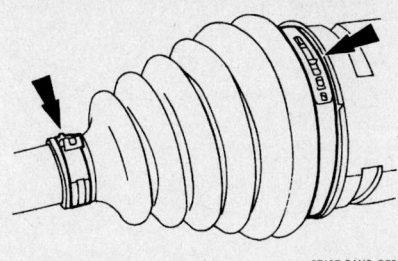

Inboard boot clamps

67197-RANG-G75

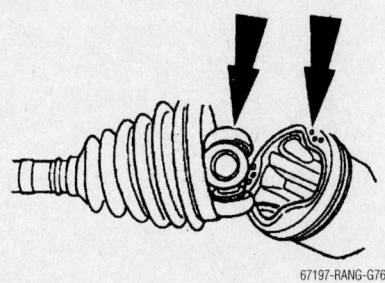

Separating the joint from the housing

67197-RANG-G76

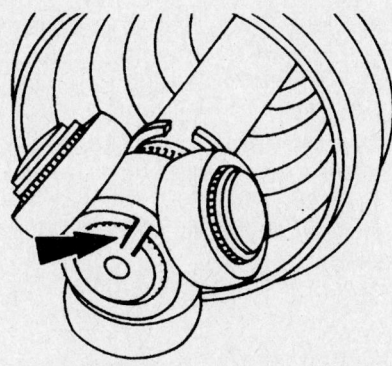

Make an alignment mark for reassembly

67197-RANG-G77

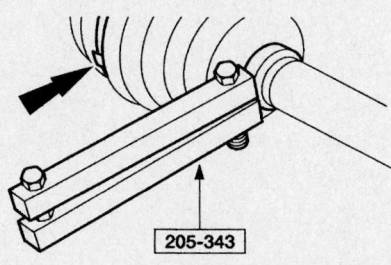

205-343

67197-RANG-G78

Boot clamp crimping tool

8. Remove the CV joint.

9. Remove the inboard halfshaft boot from the shaft assembly.

10. Remove the front wheel excluder seal, if necessary. Discard the seal. Tap uniformly around the seal to separate it from the joint.

11. Remove the two outboard boot clamps.

12. Remove the outboard halfshaft boot.

13. If the grease is contaminated, clean and inspect the joint for wear. Install a new outboard CV joint and shaft assembly if worn/damaged.

14. Inspect the assembly for contaminated grease.

To assemble

15. Pack the outboard CV joint with grease. Use Ford High Temp Constant Velocity Joint Grease E43Z-19590-A or equivalent meeting Ford specification ESP-M1C207-A. Spread any remaining grease from the service kit evenly inside the outboard halfshaft boot.

16. Clean the halfshaft boot mounting surfaces of excess grease before positioning the halfshaft boot into place.

17. Position the outboard halfshaft boot.

18. Position the boot clamps on the outboard halfshaft boot.

19. Tighten the through-bolt until the installer is in the closed position.

20. Install the outboard CV joint boot clamps. There are special tools made for this procedure.

21. Position the boot clamp on the halfshaft.

22. Position the inboard halfshaft boot.

23. Align the index marks on the halfshaft and the CV joint.

24. Install the CV joint on the halfshaft.

25. Install the snapring.

26. Lubricate the three CV joint needle bearings. Use Ford High Temp Constant Velocity Joint Grease E43Z-19590-A or equivalent meeting Ford specification ESP-M1C207-A.

27. Fill the inboard CV joint housing with 235 grams (8.3 oz.) of grease. Use Ford High Temp Constant Velocity Joint Grease E43Z-19590-A or equivalent meeting Ford specification ESP-M1C207-A.

28. Position the CV joint housing onto the CV joint.

29. Remove any excess grease from the inboard halfshaft boot mating surface before positioning it into place.

30. Position the inboard halfshaft boot into place.

31. Position the boot clamp.

32. Insert a dulled screwdriver blade to relieve built-up air pressure in the halfshaft boot.

33. Using the special tool, install the inboard boot clamps.

34. Using the special tool, install the new front wheel excluder seal, if removed. Seat the metal ring at the seal's inner diameter flat against the CV joint housing.

35. Install the front wheel halfshaft.

Front Axle Tube Bearing

REMOVAL & INSTALLATION

1. Before servicing the vehicle, refer to the Precautions Section.

2. Raise and support the vehicle safely. Remove the tire and wheel assembly.

3. Remove or disconnect the following:

- Right side halfshaft
- Right side axle shaft
- Axle seal, with a slide hammer
- Axle tube bearing, with a slide hammer

4. Clean the bearing and seal surfaces of any foreign debris.

To install:

5. Use an axle bearing replacer and the handle to replace the right side axle tube bearing.

6. Check the bearing depth as shown in the illustration.

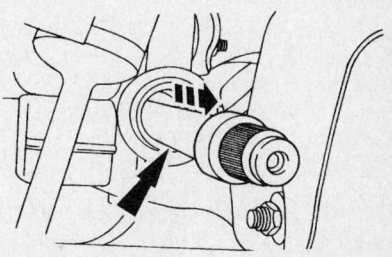

67197-RANG-G83

Right side axle shaft removal

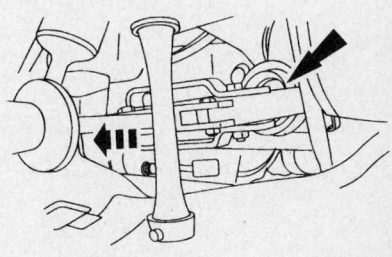

67197-RANG-G84

Axle seal removal

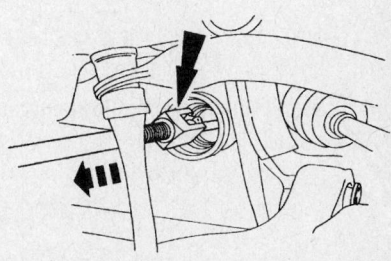

67197-RANG-G85

Axle tube bearing removal

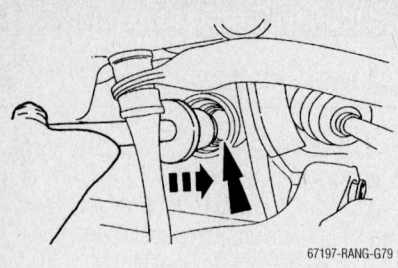

Axle tube bearing installation

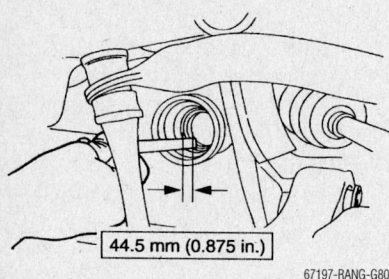

44.5 mm (0.875 in.)

67197-RANG-G80

Axle tube bearing depth

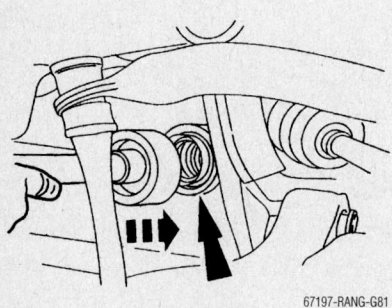

67197-RANG-G81

Axle tube seal installation

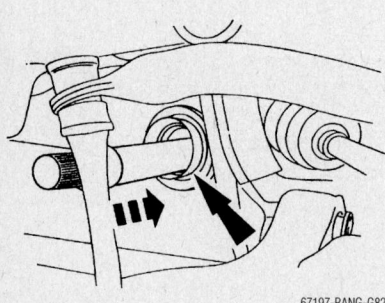

67197-RANG-G82

Axle shaft installation

7. Use an axle seal replacer and the handle to replace the axle tube seal.

➡**Care should be taken not to damage the axle seal surface.**

8. Install the axle shaft.
9. Refill the front drive axle to proper level using SAE 80W90.
10. Install the right halfshaft.

Rear Axle Shaft, Bearing and Seal

REMOVAL & INSTALLATION

2002–2003

1. Before servicing the vehicle, refer to the Precautions Section.
2. Disconnect the negative battery cable.

➡**On some vehicles, when the battery cable is disconnected and reconnected, some abnormal drive symptoms may occur while the vehicle relearns its adaptive strategy. The vehicle may need to be driven to relearn its strategy.**

3. Raise and support the vehicle safely. Remove the tire and wheel assembly.
4. Drain the axle housing fluid.
5. Remove or disconnect the following:
   - Rear wheel
   - Brake drum
   - Wheel speed sensor, if equipped
   - Axle housing cover
   - Bearing retainer nuts
   - Axle shaft and bearing
   - Axle shaft inner oil seal
6. If equipped with ABS, grind a flat spot on the wheel speed sensor tone ring, then split the ring with a chisel.
7. Press the wheel bearing off the axle shaft.
8. Remove the bearing retainer and the outer oil seal.

To install:
9. Install or connect the following:
   - Outer oil seal to the bearing retainer
   - Bearing retainer to the axle shaft
   - Bearing and retainer ring pressed onto the axle shaft
   - Wheel speed sensor tone ring pressed onto the axle shaft, if equipped
   - Axle shaft inner oil seal
   - Axle shaft and bearing
   - Bearing retainer nuts. Tighten them to 17 ft. lbs. (23 Nm).
   - Wheel speed sensor, if equipped
   - Brake drum
   - Rear wheel
   - Negative battery cable
10. Fill the rear differential to the correct level.

2004–2006

FORD 7 1/2 INCH RING GEAR

1. Before servicing the vehicle, refer to the Precautions Section.

2. Raise and support the vehicle safely.
3. Remove the wheel and tire assembly.
4. Remove the 10 differential housing cover bolts and drain the lubricant from the rear axle housing.
5. Remove the differential housing cover.
6. Remove the rear brake drums.
7. Loosen the differential pinion shaft lock bolt.

➡**The bolt will not be able to be removed at this time.**

8. Remove the differential pinion shaft.

➡**Do not damage the rubber O-rings in the axle shaft grooves.**

9. Push the axle shafts inboard.
10. Remove the U-washers.

➡**Do not damage the wheel bearing oil seal.**

11. Remove the two axle shafts.

➡**If only a new seal needs to be installed, use care to avoid damaging the seal bore.**

12. Using a suitable seal remover, remove the axle shaft oil seal. Discard the oil seal.
13. Inspect the rear wheel bearing and axle shaft for wear or damage.
14. If necessary, using a slidehammer, remove the rear wheel bearing.

To install:
15. Lubricate the new rear wheel bearing with lubricant.

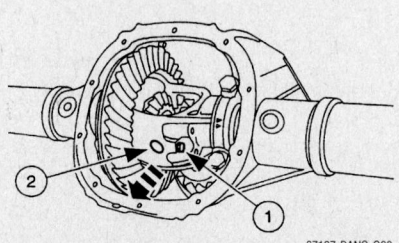

67197-RANG-G90

Differential pinion shaft removal. 1-lock bolt; 2-pinion shaft

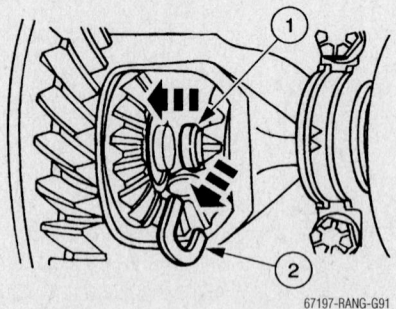

67197-RANG-G91

Removing the U-washers. 1-axle shaft; 2-U-washer

16. Using a driver, install the rear wheel bearing.

17. Lubricate the lip of the new wheel bearing oil seal with grease.

18. Using a driver, install the wheel bearing oil seal.

➡**Make sure the machined surfaces on both the rear axle housing and the differential housing cover are clean and free of oil before installing the new silicone sealant. The inside of the rear axle must be covered when cleaning the machined surface to prevent contamination.**

19. Clean the gasket mating surface of the rear axle and the differential housing cover.

20. Lubricate the lip of the wheel bearing oil seal with grease.

➡**Do not damage the wheel bearing oil seal.**

21. Install the axle shafts.

➡**Do not damage the rubber O-rings in the axle shaft grooves.**

22. Position the two U-washers on the button end of the axle shafts.

23. Pull the axle shafts outward.

➡**If a new pinion shaft lock bolt is unavailable, coat the threads with threadlock and sealer prior to installation.**

24. Install the differential pinion shaft.
   a. Align the hole in the differential pinion shaft with the case lock bolt hole.
   b. Install a new differential pinion shaft lock bolt. Torque to 20 ft. lbs. (33 Nm).

25. Install the rear brake drums.

26. Apply a new continuous bead of sealant of the specified thickness to the differential housing cover.

➡**The differential housing cover must be installed within 15 minutes of application of the silicone, or new sealant must be applied. If possible, allow one hour before filling with lubricant to make sure the silicone sealant has correctly cured.**

27. Install the different housing cover.

28. Install the 10 differential housing cover bolts. Torque to 33 ft. lbs. (45 Nm).

29. Fill the rear axle housing with the proper grade and type lubricant.

30. Install the wheels and tires.

31. Lower the vehicle.

FORD 8.8 INCH RING GEAR

1. Before servicing the vehicle, refer to the Precautions Section.

2. Raise and support the vehicle safely.

3. Remove the rear wheel and tire assembly.

4. Remove the brake drum.

5. Remove the differential housing cover and drain the lubricant.

6. Loosen the differential pinion shaft lock bolt.

➡**The bolt will not be able to be removed at this time.**

7. Remove the differential pinion shaft.

➡**Do not damage the rubber O-ring in the U-washer groove.**

8. Push the axle shaft inboard.

9. Remove the U-washer.

10. Do not damage the wheel bearing oil seal.

11. Remove the axle shaft.

➡**If only a new seal needs to be installed, use care to avoid damaging the seal bore. If the wheel bearing oil seal is leaking, the differential housing vent may be plugged with foreign material.**

12. Using a suitable seal remover, remove the axle shaft oil seal. Discard the oil seal.

13. Inspect the rear wheel bearing and axle shaft for wear or damage.

14. Using the special tools, remove the rear wheel bearing.

To install:

15. Lubricate the new rear wheel bearing with rear axle lubricant.

16. Using the special tools, install the rear wheel bearing.

17. Lubricate the lip of the new wheel bearing oil seal with grease.

18. Using the special tools, install the wheel bearing oil seal.

19. Install the axle shaft.

20. Lubricate the lip of the wheel bearing oil seal with grease.

➡**Do not damage the wheel bearing oil seal.**

21. Install the axle shaft.

22. Do not damage the rubber O-ring in the U-washer groove.

23. Position the U-washer on the button end of the axle shaft.

24. Pull the axle shaft outward.

➡**If a new bolt is unavailable, coat the bolt threads with threadlock prior to installation.**

25. Align the bolt hole in the differential pinion shaft with the bolt hole in the case.

26. Install the new bolt. Torque to 22 ft. lbs. (33 Nm).

27. Install the brake drum.

28. Install the differential housing cover and fill the differential housing with the specified lubricant.

29. Install the rear wheel and tire assembly.

30. Lower the vehicle.

Front Pinion Seal

REMOVAL & INSTALLATION

➡**This operation disturbs the differential pinion bearing preload. Carefully reset the preload during assembly.**

1. Before servicing the vehicle, refer to the Precautions Section.

2. Index-mark the front driveshaft and pinion flange.

3. Remove or disconnect the following:
   • Front driveshaft from the pinion flange, and position it aside

➡**Do not allow the driveshaft to hang unsupported.**

4. Using a torque wrench, measure the torque required to maintain pinion rotation. Record the measurement.

5. Matchmark the pinion flange and the pinion stem.

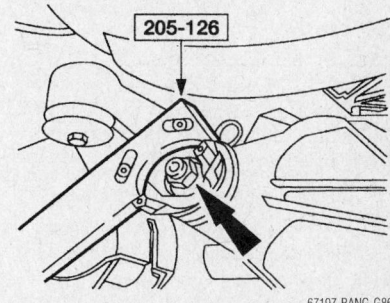

67197-RANG-G86

Holding the front axle pinion flange

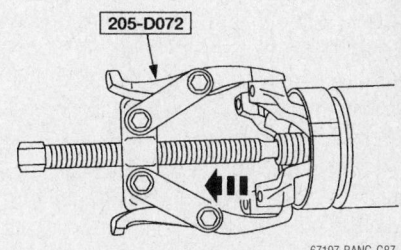

67197-RANG-G87

Removing the front axle pinion flange

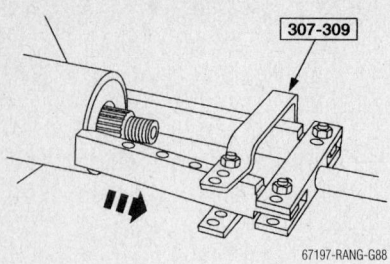

Removing the front axle pinion seal

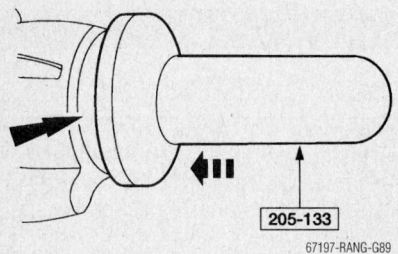

Installing the front axle pinion seal

6. Hold the pinion flange while removing the nut.

7. Place a drain pan under the differential housing.

8. Using a puller, remove the pinion flange.

9. Inspect the pinion flange for burrs and damage. Inspect the end of the pinion flange that contacts the bearing cone, the nut counterbore, and the seal surface for nicks. Discard the pinion flange as necessary.

10. Using a seal remover and impact slide hammer, remove the pinion seal.

11. Remove the front axle drive pinion shaft oil slinger and the differential pinion bearing.

12. Remove and discard the collapsible spacer.

To install:

13. Verify that the splines on the pinion stem are free of burrs. If burrs are evident, remove them with a fine crocus cloth. Work in a rotating motion to wipe the pinion clean.

14. Clean the pinion seal bore.

15. Install a new collapsible spacer.

16. Install the original differential pinion bearing and the front axle drive pinion shaft oil slinger.

17. Lubricate the pinion seal. Use Motorcraft SAE 80W90 Thermally Stable 4x4 Axle Lubricant meeting Ford specification WSP-M2C197-A.

18. Install the pinion seal.

19. Lubricate the pinion flange splines. Use Motorcraft SAE 80W90 Thermally Stable 4x4 Axle Lubricant meeting Ford specification WSP-M2C197-A.

➡**Never use a metal hammer on the pinion flange or install the flange with power tools. If necessary, use a plastic hammer to tap on a tight fitting flange.**

- Align the index marks and install the pinion flange.
- Install the new nut hand-tight.

➡**Do not loosen the nut to reduce preload. Install a new collapsible spacer and nut if preload reduction is necessary.**

20. Use the special tool to hold the pinion flange while tightening the nut to set the preload.

21. Tighten the nut, rotating the pinion occasionally to ensure the differential pinion bearings are seating correctly. Take frequent differential pinion bearing preload readings by rotating the pinion with torque wrench. The final reading must be 0.56 Nm (5 inch lbs.) more than the initial reading taken during removal.

22. Align the index marks and position the front driveshaft.

23. Install the universal joint spider retainers and bolts.

24. Check the fluid level and, if necessary, fill the axle to specification. Use Motorcraft SAE 80W90 Thermally Stable 4x4 Axle Lubricant meeting Ford specification WSP-M2C197-A.

25. Lower the vehicle.

STEERING

Air Bag

PRECAUTIONS

- Always wear safety glasses when servicing an air bag vehicle, and when handling an air bag.
- Never attempt to service the steering wheel or steering column on an air bag-equipped vehicle without first properly disarming the air bag system. The air bag system should be properly disarmed whenever ANY service procedure in this manual indicates that you should do so.
- When carrying a live air bag module, always make sure the bag and trim cover are pointed away from your body. In the unlikely event of an accidental deployment, the bag will then deploy with minimal chance of injury.
- When placing a live air bag on a bench or other surface, always face the bag and trim cover up, away from the surface. This will reduce the motion of the air bag if it is accidentally deployed.
- If you should come in contact with a deployed air bag, be advised that the air bag surface may contain deposits of sodium hydroxide, which is a product of the gas combustion and is irritating to the skin. Always wear gloves and safety glasses when handling a deployed air bag, and wash your hands with mild soap and water afterwards.

DEPOWERING

1. Before servicing the vehicle, refer to the Precautions Section.

2. Turn all vehicle accessories OFF.

3. Turn the ignition switch to OFF.

4. At the smart junction box (SJB), located below the RH side of the instrument panel, remove the RH lower cowl trim panel and remove the restraints control module (RCM) fuse from the SJB. For additional information, refer to the Owner's Manual.

5. Turn the ignition ON and visually monitor the air bag indicator for at least 30 seconds. The air bag indicator will remain lit continuously (no flashing) if the correct RCM fuse has been removed. If the air bag indicator does not remain lit continuously, remove the correct RCM fuse before proceeding.

6. Turn the ignition OFF.

✳✳ CAUTION

To avoid accidental deployment and possible personal injury, the backup power supply must be depleted before repairing or replacing any front or side air bag supplemental restraint system (SRS) components and before servicing, replacing, adjusting or striking components near the front or side air bag sensors or RCM, such as doors, instrument panel, console, door latches, strikers, seats and hood latches.

7. To deplete the backup power supply energy, disconnect the battery ground cable and wait at least one minute. Be sure to disconnect auxiliary batteries and power supplies (if equipped).

8. Disconnect the battery ground cable and wait at least one minute.

REPOWERING

1. Before servicing the vehicle, refer to the Precautions Section.

❊❊ CAUTION

The restraint system diagnostic tool is for restraint system service only. Remove from vehicle prior to road use. Failure to remove could result in injury and possible violation of vehicle safety standards. Make sure all restraint system diagnostic tool(s) that may have been installed during the repair have been removed from the vehicle and all SRS components are connected.

2. Turn the ignition switch from OFF to ON.
3. Install the RCM fuse to the SJB and install the RH lower cowl trim panel.

❊❊ CAUTION

Be sure that nobody is in the vehicle and that there is nothing blocking or set in front of any air bag module when the battery ground cable is connected.

4. Connect the battery ground cable.
5. Prove out the supplemental restraint system (SRS) as follows:
   a. Turn the ignition key from ON to OFF. Wait 10 seconds, and then turn the key back to ON and visually monitor the air bag indicator with the air bag modules installed. The air bag indicator will light continuously for approximately six seconds and then turn off. If an air bag supplemental restraint system (SRS) fault is present, the air bag indicator will either:
   - fail to light.
   - remain lit continuously.
   - flash.
   b. The flashing might not occur until approximately 30 seconds after the ignition switch has been turned from the OFF to the ON position. This is the time required for the restraints control module (RCM) to complete the testing of the SRS. If the air bag indicator is inoperative and a SRS fault exists, a chime will sound in a pattern of five sets of five beeps. If this occurs, the air bag indicator and any SRS fault discovered must be diagnosed and repaired.
   c. Clear all continuous DTCs from the restraints control module using a scan tool.

Power Steering Gear

REMOVAL & INSTALLATION

2WD

1. Before servicing the vehicle, refer to the Precautions Section.
2. Turn the wheel to the straight-ahead position and turn the ignition switch to the OFF position.
3. Raise and support the vehicle safely.
4. Remove the front wheel and tire assemblies.
5. Remove the fluid cooler.
6. Remove and discard the cotter pins and nuts.

➡ **Do not damage the tie-rod boot when installing the special tool.**

7. Using special tool, separate the tie-rod ends from the wheel knuckles.

➡ **Do not allow the intermediate shaft to rotate while it is disconnected from the steering gear or damage to the clockspring can result. If there is evidence that the intermediate shaft has rotated, the clockspring must be removed and recentered.**

8. Remove the pinch bolt and detach the intermediate shaft from the gear. Discard the bolt.
9. Remove the nut and disconnect the lines.
10. Plug or cap the power steering return hose, power steering pressure hose, and the steering gear ports to prevent the entry of dirt.

➡ **Hold the tops of the steering gear to crossmember stud bolts to avoid damaging the steering gear fluid transfer tubes.**

11. Remove the rack mounting nuts.
12. Remove the mounting stud, nut, washer and stop assemblies.
13. Remove the steering gear. Clean the mounting surfaces.
   To install:
14. To install, reverse the removal procedure.
   a. Install new seals on the power

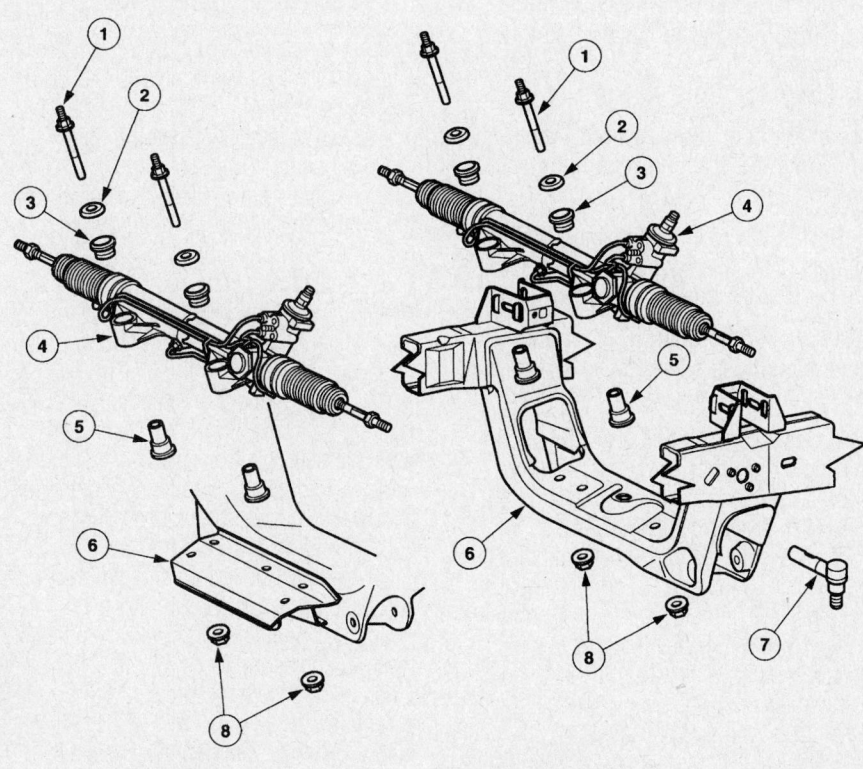

1	Stud	5	Insulator
2	Washer	6	Crossmember
3	Insulator	7	Tie-rod end — outer
4	Steering gear	8	Nut

67197-RANG-G91A

Steering gear mounting and related components—2WD (left) 4WD (right)

steering return hose and power steering pressure hose.

b. The dished side of the washer faces downward.

c. Install a new intermediate shaft pinch bolt.

15. Observe the following torques:
- Pinch bolt: 35 ft. lbs. (48 Nm), on 2002 vehicles and 40 ft. lbs. (55 Nm), on 2003–2006 vehicles
- Pressure line bracket nut: 18 ft. lbs. (25 Nm)
- Rack retaining nuts: 111 ft. lbs. (150 Nm)

16. Fill and leak check the power steering system.

17. Check and, if necessary, adjust the wheel alignment.

4WD

1. Before servicing the vehicle, refer to the Precautions Section.

2. Turn the wheel to the straight-ahead position and turn the ignition switch to the OFF position.

3. Raise and support the vehicle safely.

4. Remove the wheel and tire assemblies.

5. Remove the fluid cooler.

6. If equipped, remove the four air deflector retaining screws. Pull downward on the air deflector to disengage the retaining pins.

7. Loosen the LH tie-rod end jam nut.

8. Remove and discard the cotter pins and nuts.

➡**Do not damage the tie-rod boot when installing the special tool.**

9. Using a separator, separate the tie-rod ends from the wheel knuckles

10. Remove the LH tie-rod end. Count and record the number of turns required to remove the tie-rod end.

11. Remove the front stabilizer bar. Note or mark the driver side end of the sway bar for correct installation.

➡**Do not allow the intermediate shaft to rotate while it is disconnected from the steering gear or damage to the clockspring can result. If there is evidence that the intermediate shaft has rotated, the clockspring must be removed and recentered.**

12. Remove and discard the pinch bolt and detach the intermediate shaft from the gear.

13. Remove the nut and disconnect the lines.

14. Plug the ends of all fluid lines removed and ports in the steering gear to prevent entry of dirt.

➡**Hold the tops of the steering gear to crossmember stud bolts to avoid damaging the steering gear fluid transfer tubes.**

15. Remove the steering rack nuts.

16. Remove the stud bolts and washers.

17. Remove the steering gear to crossmember insulator bushings.

18. Rotate the steering gear control valve housing toward the front of the vehicle.

19. Turn the steering gear input shaft to the right until the stop is reached.

20. Move the steering gear as far to the RH side of the vehicle as possible.

21. Move the LH front wheel spindle tie-rod forward to clear the frame crossmember.

22. Remove the steering gear from the vehicle.

To install:

23. As required, using special tool 211-027, install new seals on the power steering return hose and power steering pressure hose.

➡**Make sure the steering gear input shaft is turned to the left until the stop is reached.**

➡**Handle the steering gear with caution to avoid damage to fluid transfer tubes and to avoid dimples in tie-rod boots.**

24. Turn the steering gear input shaft to the right until the stop is reached. Note the number of turns required.

➡**Make sure the steering gear control valve housing is turned toward the front of the vehicle.**

25. Install the steering gear into the RH opening of the crossmember.

26. Move the steering gear as far to the RH side of the vehicle as possible.

27. Move the LH front wheel spindle tie-rod into the opening in the crossmember and move the steering gear into position.

28. To place the steering gear in the straight ahead position, turn the steering gear input shaft to the left by half the number of turns recorded previously.

29. Rotate the steering gear control valve housing toward the rear of the vehicle.

30. Install the steering gear to crossmember insulator bushings.

a. The large end of the metal sleeve must be positioned downward.

b. Check that the mounting surfaces on the crossmember are clean and free of foreign material.

31. Install the steering gear to crossmember washers and stud bolts. The dished side of the washer faces downward.

➡**Hold the tops of the steering gear to crossmember stud bolts to avoid damaging the steering gear fluid transfer tubes.**

32. Install the rack mounting nuts. Torque to 111 ft. lbs. (150 Nm).

33. Install the lines and tighten the pressure line bracket nut to 18 ft. lbs. (25 Nm) on vehicles built through December 2005 and 26 ft. lbs. (35 Nm) on vehicles built after December 2005.

➡**Do not allow the intermediate shaft to rotate while it is disconnected from the steering gear or damage to the clockspring can result. If there is evidence that the intermediate shaft has rotated, the clockspring must be removed and recentered.**

34. Connect the intermediate shaft to the steering gear input shaft. Install a new lower steering column pinch bolt. Torque to 35 ft. lbs. (48 Nm), on 2002 vehicles and 40 ft. lbs. (55 Nm), on 2003–2006 vehicles.

35. Install the power steering fluid cooler.

36. Install the front stabilizer bar. Orient the front stabilizer bar as noted during removal.

37. Install the LH tie-rod end on the front wheel spindle tie-rod. Rotate the tie-rod end the number of turns recorded during removal.

38. Position the tie-rod ends on the steering knuckles. Install the castellated nuts and new cotter pins. Check that the brake dust shields are not bent and are not in contact with the outer tie-rod boot seals. Torque to 52 ft. lbs. (70 Nm).

39. Tighten the tie-rod end jam nut. Torque to 59 ft. lbs. (80 Nm).

40. Position the air deflector, and install the retaining screws.

41. Install the front wheel and tire assemblies.

42. Fill and leak check the system.

43. Check and, if necessary, adjust the wheel alignment.

FRONT SUSPENSION

Shock Absorber

REMOVAL & INSTALLATION

➡ Low pressure gas shocks are charged with nitrogen gas. Do not attempt to open, puncture or apply heat to them. Prior to installing a new shock absorber, hold it upright and extend it fully. Invert it and fully compress and extend it at least 3 times. This will bleed trapped air.

1. Before servicing the vehicle, refer to the Precautions Section.
2. Disconnect the negative battery cable.

➡ On some vehicles, when the battery cable is disconnected and reconnected, some abnormal drive symptoms may occur while the vehicle relearns its adaptive strategy. The vehicle may need to be driven to relearn its strategy.

3. Raise and support the vehicle safely.

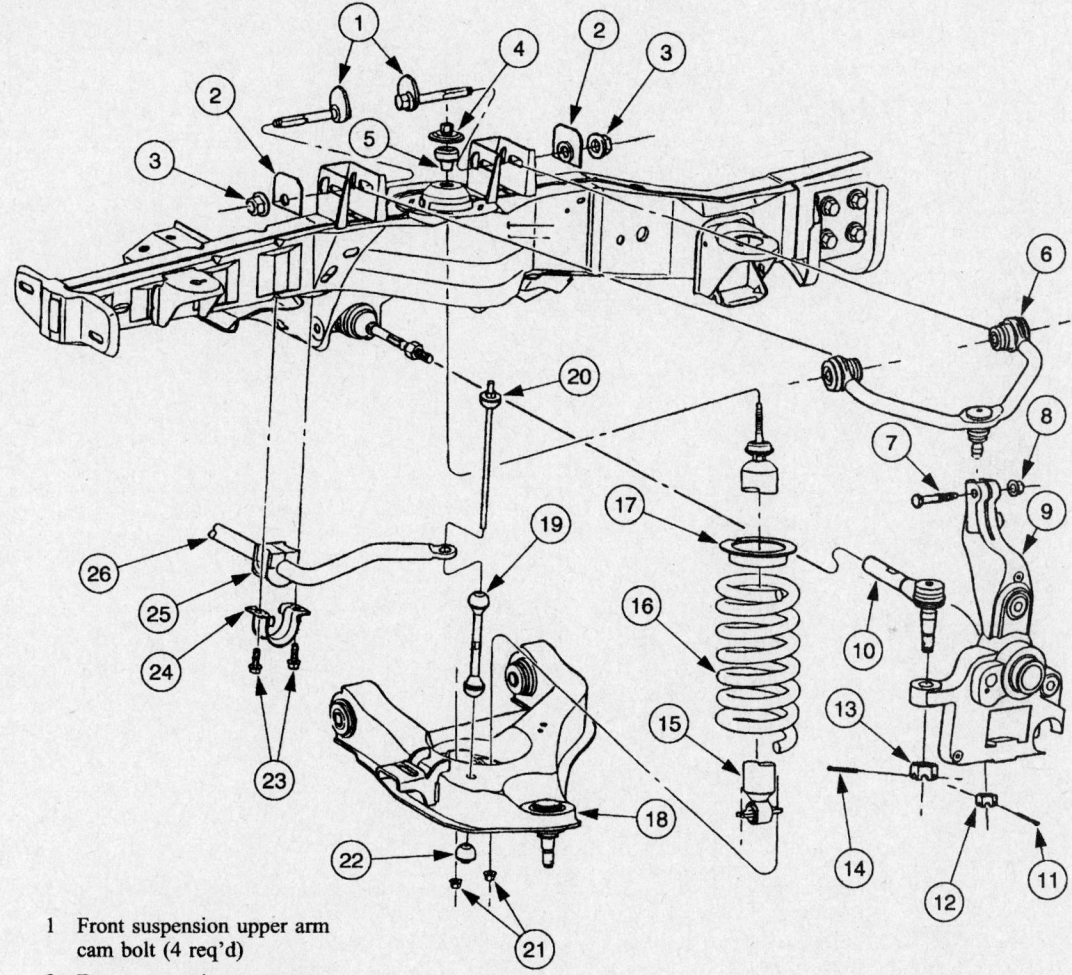

1	Front suspension upper arm cam bolt (4 req'd)
2	Front suspension upper arm cam assy (2 req'd)
3	Front suspension upper arm cam assy nut (2 req'd)
4	Front shock absorber upper nut/washer assy (2 req'd)
5	Front shock absorber upper bushing (2 req'd)
6	Front suspension upper arm
6	Front suspension upper arm
7	Front wheel spindle pinch bolt (2 req'd)
8	Front wheel spindle pinch bolt nut

9	Front wheel spindle
10	Tie-rod end
11	Cotter pin
12	Lower ball joint castellated nut (2 req'd)
13	Tie-rod end castellated nut (2 req'd)
14	Cotter pin
15	Front shock absorber
16	Front coil spring
17	Front spring insulator
18	Front suspension lower arm
18	Front suspension lower arm

19	Front stabilizer bar link
20	Front stabilizer bar stud and bushing assy (2 req'd)
21	Front shock absorber lower nut (4 req'd)
22	Front stabilizer bar nut and washer assy (2 req'd)
23	Front stabilizer bar mounting bolts (4 req'd)
24	Stabilizer bar bracket (2 req'd)
25	Front stabilizer bar bushing assy (2 req'd)
26	Front stabilizer bar

67197-RANG-G92

Front suspension exploded view—2WD

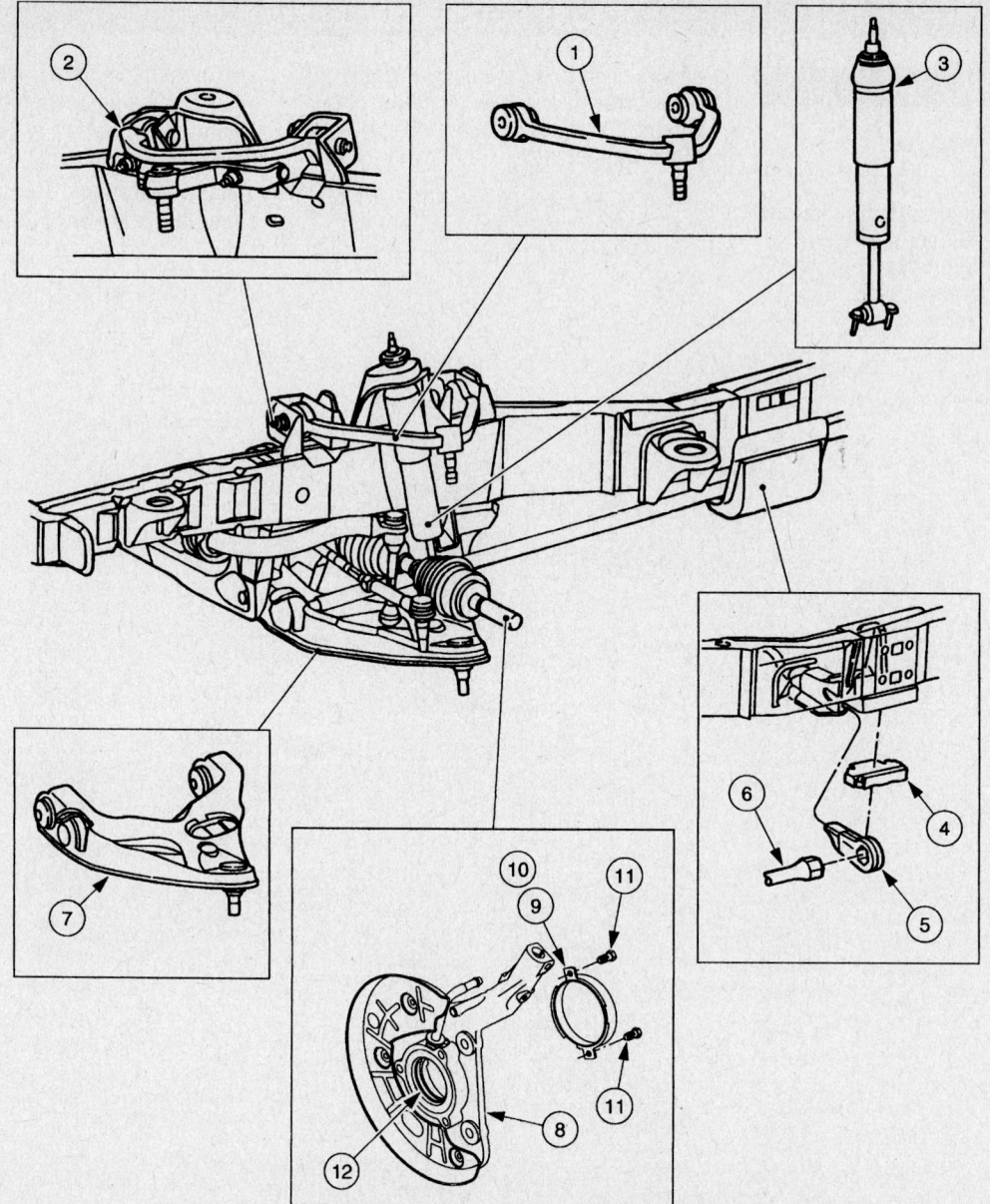

1 Upper arm, bushing and joint assembly (LH)
2 Upper arm, bushing and joint assembly (RH)
3 Shock assembly
4 Torsion bar adjuster plate
5 Torsion bar adjuster
6 Torsion bar (LH)
7 Lower arm, bushing and joint assembly (LH)
8 Knuckle assembly
9 Protection shield (RH)
10 Protection shield (LH)
11 Screw (self-tapping)
12 Oil seal

67197-RANG-G93

Front suspension exploded view—4WD

4. Remove or disconnect the following:
• Upper shock-to-frame attaching nut, washer and insulator assembly
• Lower shock-to-control arm attaching nuts
• Slightly compress the shock absorber by hand and remove it from the vehicle

To install:
5. Install or connect the following:
• Position the lower washer and insulator on the shock absorber rod and position the shock absorber to the upper frame bracket mount
• Position the upper insulator and washer on the shock absorber rod and install the attaching nut loosely.

• Position the lower shock absorber mounting studs into the control arm and install the attaching nuts loosely.
• Torque the lower shock attaching nuts to 15–21 ft. lbs. (21–29 Nm), and the upper shock attaching bolts to 30–40 ft. lbs. (40–55 Nm).
• Negative battery cable

Coil Spring

REMOVAL & INSTALLATION

2WD

1. Before servicing the vehicle, refer to the Precautions Section.
2. Raise and support the vehicle safely.
3. Remove or disconnect the following:
   - Wheel and tire assembly
   - Shock absorber
   - Front stabilizer bar link nut
4. Use a coil spring compressor to compress the coil spring.
5. Remove the cotter pin and castellated nut.
6. Using a suitable jack, support the front suspension lower control arm.
7. Separate the lower ball joint from the front wheel spindle, using the proper tool.
8. Carefully remove the jack from under the lower control arm.
9. Loosen the lower control arm bolts and allow the lower arm to swing down.
10. Position the front wheel spindle out of the way and remove the coil spring.

To install:

➡**The end of the coil spring must cover the first hole and should not be visible in the second hole.**

11. Install the coil spring in the lower arm.

❋❋ WARNING

Always install the cotter pin into the lower ball joint castellated nut from outboard to inboard. Failure to do so will result in damage to the wheel and tire assembly.

12. Install the lower ball joint.
13. Install the front stabilizer bar link nut.
14. Remove the coil spring compressor tool.
15. Install the front shock absorber and the two lower nuts.
16. Install the upper shock absorber bushing and nut/washer assembly.
17. Install the wheel and tire assembly.

Stabilizer Bar

REMOVAL & INSTALLATION

1. Before servicing the vehicle, refer to the Precautions Section.
2. Raise and support the vehicle safely.

3. Remove the wheel and tire assembly.
4. Remove the front stabilizer bar link nuts from the front suspension lower arms. Discard the nuts.
5. Remove the front stabilizer bar link studs and the front stabilizer bar links.
6. Remove the four bolts and two brackets. Discard the bolts.
7. Remove the front stabilizer bar.
8. Remove the stabilizer bar insulator.

To install:

9. Installation is the reverse of removal. Be sure to use new bolts and nuts. Observe the following torques:
   - End link nuts: 18 ft. lbs. (25 Nm)
   - Bracket bolts: 30 ft. lbs. (40 Nm)

➡**In the event the self-tapping bolts cannot be installed in the frame, there is a kit available with flag nuts.**

Torsion Bar

REMOVAL & INSTALLATION

4WD

1. Before servicing the vehicle, refer to the Precautions Section.
2. Raise and support the vehicle safely.
3. Remove the torsion bar cover plate.

➡**Before relieving the torsion bar tension, measure and record the measurement of the torsion bar adjustment bolt. This measurement will be used as the preset depth for the new torsion bar adjustment bolt during installation.**

4. Relieve the torsion bar tension.
   a. Position the torsion bar tool and adapters.
   b. Tighten the torsion bar tool until the torsion bar adjuster lifts off the adjustment bolt.

❋❋ CAUTION

The torsion bar adjustment bolt is coated with dry adhesive; and must be replaced if it is backed off or removed. Failure to do so can cause the adjustment bolt to loosen during operation and cause a loss of vehicle alignment.

   c. Remove the torsion bar adjustment bolt and nut.
   d. Loosen the torsion bar tool until the tension is removed from the torsion bar.
5. Mark the torsion bar and the adjuster for proper installation.
6. Remove the torsion bar insulator.

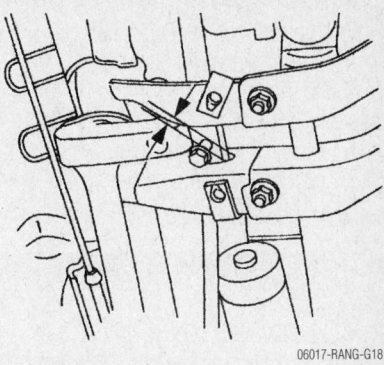

06017-RANG-G18

Measure and record the measurement of the torsion bar adjustment bolt

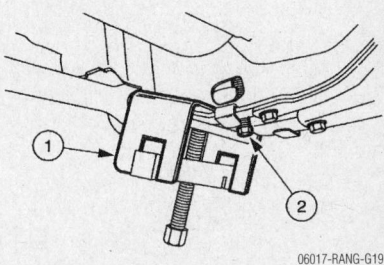

06017-RANG-G19

Relieving torsion bar tension (1) tool 204-185 (2) tool 204-204

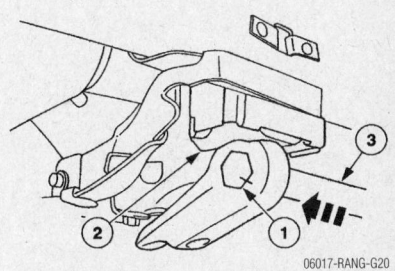

06017-RANG-G20

Torsion bar adjuster (1) locating point (2) mount (3) torsion bar

7. Grasp the torsion bar, and pull it free from the front suspension lower arm.

To install:

8. Position the torsion bar and the torsion bar adjuster.
9. Align the marks on the torsion bar and the torsion bar adjuster, then install the torsion bar adjuster.
10. Position the torsion bar insulator.
11. Install the torsion bar tool and the adapters.
12. Tighten the torsion bar tool until the new adjustment bolt and nut can be installed.
13. Turn the adjustment bolt until the preliminary adjustment measurement (recorded length of the old adjustment bolt) is reached.

14. Install the torsion bar cover plate. Torque the bolts to 46 ft. lbs. (63 Nm).

15. If equipped with air suspension, reactivate the system by turning on the air suspension switch.

16. Lower the vehicle.

17. Adjust the ride height.

18. Check the alignment.

Spindle

REMOVAL & INSTALLATION

2WD

2002–2005

1. Before servicing the vehicle, refer to the Precautions Section.

2. Raise and support the vehicle safely.

3. Remove the tire and wheel assembly.

4. Remove the front disc brake rotor shield.

5. Support the front suspension lower control arm, using a floor jack.

6. Remove the upper shock absorber retaining nut. Remove the lower shock absorber retaining bolts. Remove the shock absorber.

7. Remove the ball joint retaining nut and pinch bolt. Separate the ball joint from the front wheel spindle, using the proper tool.

8. Remove the tie rod castellated nut cotter pin and remove the nut.

9. Use a pitman arm puller tool to separate the tie rod end from the front wheel spindle.

10. Use a coil spring compressor tool and compress the coil spring.

11. Remove the lower ball joint castellated nut cotter pin and remove the nut.

12. Use a pitman arm puller tool to separate the lower ball joint from the front wheel spindle.

13. Remove the front spindle from the vehicle.

To install:

14. Installation is the reverse of the removal procedure. Be sure to use new bolts and nuts, as required.

❊❊ WARNING

Always install the cotter pin into the lower ball joint castellated nut from outboard to inboard. Failure to do so will result in damage to the wheel and tire assembly.

15. Check and adjust the front end alignment.

2006

1. Before servicing the vehicle, refer to the Precautions Section.

2. Raise and support the vehicle safely.

3. Remove the tire and wheel assembly.

4. Remove the brake disc shield. Remove and discard the wheel speed sensor harness bracket bolt.

5. Remove and discard the outer tie rod end nut.

➡**Do not use a hammer to separate the tie rod end from the wheel spindle or damage to the wheel spindle will result.**

6. Using the proper tool separate the outer tie rod end from the wheel spindle.

7. Using a floor jack support the front suspension under the lower control arm.

8. Remove and discard the upper ball joint nut and pinch bolt. Using the proper tool disconnect the upper ball joint from the wheel spindle.

9. Remove and discard the lower ball joint cotter pin and nut. Using the proper tool disconnect the lower ball joint from the wheel spindle.

➡**Do not use a hammer to separate the ball joint from the wheel spindle or damage to the wheel spindle will result.**

10. Remove the front spindle from the vehicle.

To install:

11. Installation is the reverse of the removal procedure.

12. Be sure to use new bolts and nuts, as required. Torque the upper ball joint to wheel spindle retaining nut to 46 ft. lbs. Torque the lower ball joint to wheel spindle retaining nut to 98 ft. lbs.

13. On vehicles built through December 2005, torque the outer tie rod nut to 52 ft. lbs. On vehicles built after December 2005, torque the nut to 59 ft. lbs.

❊❊ WARNING

Always install the cotter pin into the lower ball joint castellated nut from outboard to inboard. Failure to do so will result in damage to the wheel and tire assembly.

14. Check and adjust the front end alignment.

Knuckle

REMOVAL & INSTALLATION

4WD

1. Before servicing the vehicle, refer to the Precautions Section.

2. Raise and support the vehicle safely.

3. Remove the tire and wheel assembly.

4. Remove the front disc brake caliper and wire it to the side out of the way.

➡**Do not allow the caliper to hang by the brake line. Damage may result.**

5. Remove the rotor.

6. Remove the hub nut and washer.

➡**Do not use a hammer to separate the outer CV joint from the hub. Damage to the outboard CV threads and to internal components may result.**

7. Use special tool 205-D070, or equivalent and separate the outer CV joint from the hub.

8. Remove the brake dust shield. Remove the bolt and detach the wheel speed sensor from the wheel hub.

➡**Do not overextend the CV joint and boots when removing the hub and bearing assembly.**

9. Remove the three retaining bolts. Remove the wheel and hub.

10. Remove the torsion bar.

➡**Secure the front axle shaft to prevent it from overextending. Failure to do so can cause damage to the front axle shaft. Suspend the front axle shaft with wire.**

11. Remove the tie rod end cotter pin and nut.

12. Using the proper tool separate the tie rod end from the front wheel knuckle.

13. Remove the lower ball joint cotter pin and nut.

14. Using the proper tool separate the front wheel knuckle from the front lower suspension arm.

15. Remove the pinch bolt, nut and the front wheel knuckle.

To install:

➡**The knuckle main seal should only be removed from the knuckle if it is damaged. The main seal cannot be reused once it has been removed from the knuckle. Failure to lubricate the new seal prior to installation can cause premature wear of the seal, resulting** in damage to the axle shaft and or wheel knuckle.

16. Installation is the reverse of the removal procedure. Be sure to use new bolts and nuts, as required.

17. Torque the hub nut to 162 ft. lbs.

18. Check and adjust the front end alignment, as required.

Upper Ball Joint

REMOVAL & INSTALLATION

The ball joints are integral with the control arm. If the ball joint is defective, the entire control arm must be replaced.

Lower Ball Joint

REMOVAL & INSTALLATION

2002–2005

1. Before servicing the vehicle, refer to the Precautions Section.

2. Raise and support the vehicle safely.

3. Remove the tire and wheel assembly.

✲✲ WARNING

Do not allow the disc brake caliper to hang suspended from the brake hose. Provide a suitable support.

4. Remove the caliper support bracket bolts, then position the caliper and support bracket aside.

5. Disconnect the front anti-lock brake sensor (ABS) wire from the vehicle frame.

6. Using a suitable jack, support the front suspension lower arm.

7. Remove the tie-rod end castellated nut. Remove and discard the cotter pin and the castellated nut.

✲✲ WARNING

Do not use a hammer to separate the tie-rod from the wheel knuckle or damage to the wheel knuckle will result. Do not damage the tie-rod boot when installing the special tool.

8. Using the special tool, separate the tie-rod end from the front wheel knuckle.

9. Remove the lower ball joint castellated nut. Remove and discard the cotter pin and the castellated nut.

10. Separate the front wheel knuckle from the front suspension lower arm. Then, loosely install the lower ball joint castellated nut.

11. Remove the pinch bolt and nut.

12. Remove the hand-tightened lower ball joint castellated nut, and then remove the front wheel knuckle.

13. Remove the snapring from the ball joint. Discard the snapring.

14. Using a suitable ball joint remover tool, remove the ball joint.

To install:

✲✲ WARNING

Do not damage the ball joint boot when installing the special tool.

➡Clean and inspect the control arm ball joint bore for damage before installing a new ball joint.

➡Make sure the new ball joint snapring is fully seated.

15. Continue the installation in the reverse order of the removal procedure. Always install new castellated nuts and cotter pins. Observe the following torques:
- Pinch bolt and nut: 41 ft. lbs. (55 Nm)
- Lower ball stud nut: 98 ft. lbs. (133 Nm)
- Tie rod stud nut: 52 ft. lbs. (70 Nm)
- Caliper support bracket bolts: 83 ft. lbs. (112 Nm)

2006

1. Before servicing the vehicle, refer to the Precautions Section.

2. Raise and support the vehicle safely.

3. Remove the tire and wheel assembly.

4. On 2WD vehicles, remove the front spindle. On 4WD vehicles remove the front knuckle.

5. Remove and discard the lower ball joint snapring.

6. Using a ball joint removal too, remove the ball joint from its mounting.

To install:

✲✲ WARNING

Do not damage the ball joint boot when installing the special tool.

➡Clean and inspect the control arm ball joint bore for damage before installing a new ball joint.

➡Make sure the new ball joint snapring is fully seated.

7. Continue the installation in the reverse order of the removal procedure. Always use new bolts and nuts, as required.

8. Check and adjust the front end, as required.

Upper Control Arm

REMOVAL & INSTALLATION

2002–2005

2WD

1. Before servicing the vehicle, refer to the Precautions Section.

2. Raise and support the vehicle safely.

3. Remove or disconnect the following:
- Wheel and tire assembly
- Brake disc shield

4. Use a jack to support the front suspension lower arm.

5. Mark the position of the front suspension upper arm adjustment cams.

6. Remove the upper ball joint retaining nut and pinch bolt.

7. Separate the ball joint from the front wheel spindle.

8. Remove the front suspension upper arm.

9. Installation is the reverse of removal. Align the marks made during removal on the front suspension upper arm adjustment cam. The forward front suspension upper arm nut must be tightened first while the arm is held at the curb position ride height.
Observe the following torques:
- Control arm attaching nuts: 98 ft. lbs. (133 Nm)
- Pinch bolt: 46 ft. lbs. (63 Nm)

4WD

1. Before servicing the vehicle, refer to the Precautions Section.

2. Raise and support the vehicle safely.

3. Raise the vehicle on a hoist.

4. Remove the wheel and tire assembly.

5. Use a suitable jack stand to support the front suspension lower arm.

6. Remove the pinch bolt.

✲✲ WARNING

Before separating the front suspension upper arm from the front wheel knuckle, secure the front wheel knuckle to prevent it from tilting outward. Failure to do so can cause damage to the front axle shaft.

7. Separate the front suspension upper arm from the front wheel knuckle.

8. Remove the front suspension upper arm:

a. Remove the two nuts and alignment plates.

b. Remove the two bolts and the cams.

c. Remove the front suspension upper arm.

To install:

➡**When installing the front suspension upper arm, replace the alignment plates with new alignment cams.**

9. Install the front suspension upper arm.

a. Position the front suspension arm bushing joint.

b. Install the two bolts, four cams and two nuts. Torque to 83–112 ft. lbs. (113–153 Nm).

10. Install the pinch bolt and nut.

a. Position the upper arm into the front wheel knuckle.

b. Install the pinch bolt and nut. Torque to 41 ft. lbs. (55 Nm).

11. Remove the jack stand from under the front suspension lower arm.

12. Install the wheel and tire assembly.

13. Lower the vehicle.

14. Check the wheel alignment.

2006

1. Before servicing the vehicle, refer to the Precautions Section.

2. Raise and support the vehicle safely.

3. Remove the tire and wheel assembly.

4. Using a suitable jack, support the lower control arm assembly.

➡**To avoid possible damage to the wheel spindle (2WD) or knuckle (4WD) secure the assembly to keep it from tilting before removing the pinch bolt and nut.**

5. Remove and discard the upper ball joint nut and pinch bolt. Separate the ball joint from the wheel spindle (2WD) or the knuckle (4WD).

6. Remove and discard the two upper control arm nuts.

7. Remove the upper control arm from the vehicle.

To install:

8. Installation is the reverse of the removal procedure. Be sure to use new nuts and bolts, as required.

9. The upper control arm retaining nuts must be fully tightened at curb height. Torque them to 98 ft. lbs.

10. Make sure that the shims are installed in their original locations.

11. Torque the upper ball joint nut and pinch bolt to 46 ft. lbs.

12. Check and adjust the front end alignment, as required.

UPPER CONTROL ARM BUSHING REPLACEMENT

The control arm bushings are not serviceable. If they require service, arm must be replaced.

Lower Control Arm

REMOVAL & INSTALLATION

2002–2005

2WD

1. Before servicing the vehicle, refer to the Precautions Section.

2. Raise and support the vehicle safely.

3. Remove or disconnect the following:

- Front wheel
- Brake rotor shield
- Shock absorber
- Stabilizer bar link hardware

4. Using a spring compressor tool, compress the coil spring.

- Lower ball joint from the spindle
- Lower control arm bolts
- Lower control arm and coil spring

To install:

5. Install or connect the following:

- Coil spring to the lower control arm

➡**The end of the coil spring must cover the first hole and should not be visible in the second hole.**

- Lower arm and front coil spring
- The two front suspension lower arm bolts and nuts. Do not tighten the nuts at this time.

➡**On the RH front suspension lower arm, install the rear bolt adjustment cam, and nut in the center of the frame slot.**

✳✳ CAUTION

Always install the cotter pin into the lower ball joint castellated nut from outboard to inboard, with the fingers bent together at a right angle. Failure to do so will result in damage to the wheel and tire assembly.

- Lower ball joint. Torque the nut to 113 ft. lbs. (153 Nm).

6. Remove the Coil Spring Compressor.

7. Install or connect the following:

- Front stabilizer bar link nut. Torque the nut to 21 ft. lbs. (29 Nm).
- Shock absorber and the two lower nuts

- Upper shock absorber bushing and nut/washer assembly

8. Support the lower control arm with a jackstand. Torque the bolts to 129 ft. lbs. (175 Nm).

- Brake disc shield
- Wheel and tire assembly

9. Inspect and adjust the front end alignment.

4WD

1. Before servicing the vehicle, refer to the Precautions Section.

2. Raise the vehicle on a hoist.

3. Remove the wheel and tire assembly.

4. Remove the stabilizer link nut, washer and bushing.

5. Remove the front shock absorber-to-front suspension lower arm nuts.

6. Remove the torsion bar.

7. Remove the lower ball joint castellated nut.

➡**Do not use a hammer to separate the ball joint from the wheel knuckle or damage to the wheel knuckle will result. Do not damage the ball joint boot while installing the special tool.**

8. Using the special tool, separate the front suspension lower arm from the front wheel knuckle/spindle.

9. Remove the front suspension lower arm bolts and nuts.

10. Remove the front suspension lower arm.

To install:

➡**Tighten the front suspension lower arm pivot bolts and nuts until snug. Do not tighten to specification until the installation procedure is complete.**

11. Position the front suspension lower arm to the front suspension crossmember.

12. Install the pivot bolts and nuts and tighten until snug.

✳✳ WARNING

Install the cotter pin into the lower ball joint from outboard to inboard with the fingers bent together at a right angle. Failure to do so will cause damage to the wheel and tire assembly.

13. Position the lower ball joint into the front wheel knuckle/spindle.

14. Install the new castellated nut. Torque to 83–112 ft. lbs. (113–153 Nm).

15. Install a new cotter pin.

16. Install the front shock absorber-to-front suspension lower arm nuts. Torque to 15–21 ft. lbs. (21–29 Nm).

17. Install the stabilizer link bushing, washer, and nut. Torque to 15–21 ft. lbs. (21–29 Nm).

➡**Whenever the torsion bar or torsion bar adjuster is removed, the vehicle ride height must be checked.**

18. Install the torsion bar.
19. Install the tire and wheel assembly.
20. Lower the vehicle.
21. Tighten the front suspension lower arm nuts. Torque to 111–148 ft. lbs. (150–200 Nm).
22. Inspect and adjust the front end alignment.

2006

2WD

1. Before servicing the vehicle, refer to the Precautions Section.
2. Raise and support the vehicle safely.
3. Remove the tire and wheel assembly.
4. Remove the coil spring.
5. Remove the lower control arm retaining bolts. Remove the lower control arm from the vehicle.
To install:
6. Installation is the reverse of the removal procedure.
7. Be sure to use new bolts and nuts, as required.
8. Torque the left side lower control arm rearward bolt to 148 ft. lbs.
9. Torque the right side lower control arm rearward bolt to 129 ft. lbs.
10. Torque the lower control arm forward bolts to 148 ft. lbs.
11. Check and adjust the front end alignment, as required.

4WD

1. Before servicing the vehicle, refer to the Precautions Section.
2. Raise and support the vehicle safely.
3. Remove the tire and wheel assembly.
4. Remove the torsion bar.
5. Remove and discard the stabilizer bar link nut and grommet.
6. Remove the stabilizer link stud and grommet and the stabilizer link assembly.
7. Remove and discard the two lower shock absorber retaining nuts.
8. Remove and discard the lower ball joint cotter pin and nut.

➡**Do not use a hammer to separate the ball joint from the knuckle or damage to the knuckle will result.**

9. Using the proper tool, separate the ball joint from the knuckle.
10. Remove the lower control arm retain-

ing bolts. Remove the lower control arm from the vehicle.
To install:
11. Installation is the reverse of the removal procedure.
12. Be sure to use new bolts and nuts, as required.

✳✳ WARNING

Always install the cotter pin into the lower ball joint castellated nut from outboard to inboard. Failure to do so will result in damage to the wheel and tire assembly.

13. Torque the lower control arm bolts to 148 ft. lbs.
14. Check and adjust the front end alignment, as required.

LOWER CONTROL ARM BUSHING REPLACEMENT

The control arm bushings are not serviceable. If they require service, arm must be replaced.

Wheel Bearings

ADJUSTMENT

2WD

➡**If the wheel and tire assembly is loose on the spindle or does not rotate freely, adjust the wheel bearings.**

1. Before servicing the vehicle, refer to the Precautions Section.
2. Raise and support the vehicle safely.
3. Remove the grease cap from the hub and wipe the excess grease from the end of the spindle. Remove the cotter pin and retainer. Discard the cotter pin.
4. Loosen the adjusting nut 3 turns.

✳✳ WARNING

Obtain running clearance between the disc brake rotor surface and shoe linings by rocking the entire wheel assembly in and out several times in order to push the caliper and brake pads away from the rotor. An alternate method to obtain proper running clearance is to tap lightly on the caliper housing. Be sure not to tap on any other area that may damage the disc brake rotor or the brake lining surfaces. Do not pry on the phenolic caliper piston. The running clearance must be maintained throughout the adjustment procedure. If proper

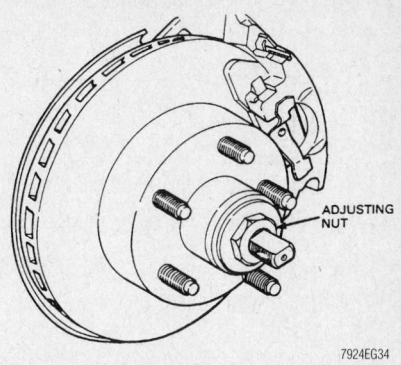

7924EG34

Loosen the adjusting nut 3 turns, then rock the entire wheel assembly in-and-out to spread the brake pads before attempting to adjust the bearing—2WD

clearance cannot be maintained, the caliper must be removed from its mounting.

5. Tighten the adjusting nut while turning the rotor in the opposite direction to seat the bearings.
6. Loosen the adjusting nut ½ half turn. Retighten the nut to 18–20 inch lbs. (2.0–2.2 Nm).
7. Place the retainer on the adjusting nut. The castellations on the retainer must be in alignment with the cotter pin holes in the spindle. Once this is accomplished install a new cotter pin and bend the ends to insure its being locked in place.
8. Check for proper wheel rotation. End play should be 0.0002–0.005 inch or a breakaway torque of 25 inch lbs. If correct, install the grease cap.
9. Lower the vehicle and tighten the lug nuts to 100 ft. lbs., (136 Nm) if the wheel was removed. Before driving the vehicle, pump the brake pedal several times to restore normal brake pedal travel.

✳✳ CAUTION

If the wheel was removed, retighten the wheel lug nuts to specification after about 500 miles (804km) of driving. Failure to do this could result in the wheel coming off while the vehicle is in motion causing loss of vehicle control or collision.

REMOVAL & INSTALLATION

2WD

1. Before servicing the vehicle, refer to the Precautions Section.
2. Raise and support the vehicle safely. Remove the tire and wheel assembly.

3. Remove or disconnect the following:
- Disc brake caliper anchor plate
- Hub grease cap
- Cotter pin
- Nut retainer
- Spindle nut
- Wheel outer bearing retainer washer
- Outer front wheel bearing
- Brake disc and hub
- Hub grease seal
- Inner wheel bearing

To install:

4. Thoroughly clean and inspect the front wheel bearings and the brake disc and hub.

5. Lubricate the front wheel bearings.

6. Install the inner front wheel bearing.

7. Install a new wheel hub grease seal.

8. Position the brake disc and hub.

9. Assemble all parts and adjust the bearings.

4WD

1. Before servicing the vehicle, refer to the Precautions Section.

2. Raise and support the vehicle safely. Remove the tire and wheel assembly.

3. Remove the brake disc.

4. Remove and discard the nut and washer assembly.

> ✳✳ **WARNING**
>
> **Do not use a hammer to separate the outboard CV joint from the hub. Damage to the outboard CV threads and to internal components may result.**

5. Using the special tool, separate the outboard CV joint from the hub.

6. Remove the dust shield.

7. Remove the bolt and detach the anti-lock sensor from the wheel hub.

> ✳✳ **WARNING**
>
> **Do not overextend the CV joint and boots when removing the hub and bearing assembly.**

8. Remove the bolts and the wheel hub.

9. To install, reverse the removal procedure. Observe the following torques:
- Hub/bearing bolts: 85 ft. lbs. (115 Nm)
- Dust shield: 9 ft. lbs. (12 Nm)
- Halfshaft nut: 162 ft. lbs. (220 Nm)

REAR SUSPENSION

Shock Absorber

REMOVAL & INSTALLATION

➡**Low pressure gas shocks are charged with nitrogen gas. Do not attempt to open, puncture or apply heat to them. Prior to installing a new shock absorber, hold it upright and extend it fully. Invert it and fully compress and extend it at least 3 times. This will bleed trapped air.**

1. Before servicing the vehicle, refer to the Precautions Section.

2. Raise and support the vehicle safely. Remove the tire and wheel assembly.

3. Remove or disconnect the following:
- Upper shock-to-frame attaching nut
- Lower shock nut
- Slightly compress the shock absorber by hand and remove it from the vehicle

To install:

4. Install or connect the following:
- Shock absorber upper end and nut
- Shock absorber lower end and nut
- Torque the upper shock attaching nuts to 46 ft. lbs. (63 Nm)
- Torque the lower shock attaching nuts to 59 ft. lbs. (80 Nm)

Leaf Springs

REMOVAL & INSTALLATION

1. Before servicing the vehicle, refer to the Precautions Section.

2. Raise and support the vehicle safely. Remove the tire and wheel assembly.

3. Properly support the rear axle assembly.

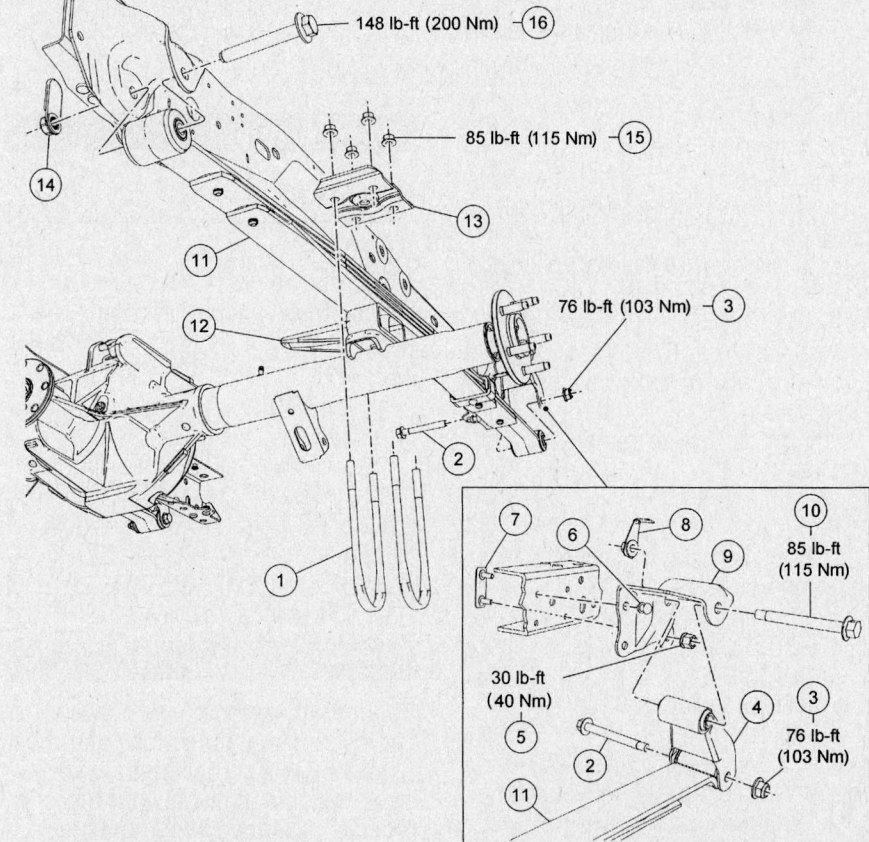

1	U-bolt (2 required)
2	Spring shackle-to-spring bolt
3	Spring shackle-to-spring nut
4	Spring shackle
5	Frame bracket nut (2 required)
6	Frame bracket rivet (2 required)
7	Frame bracket bolt and retainer assembly
8	Spring shackle-to-frame bracket flag nut
9	Frame bracket

09482_MAZB_G0026

Rear leaf spring and related components

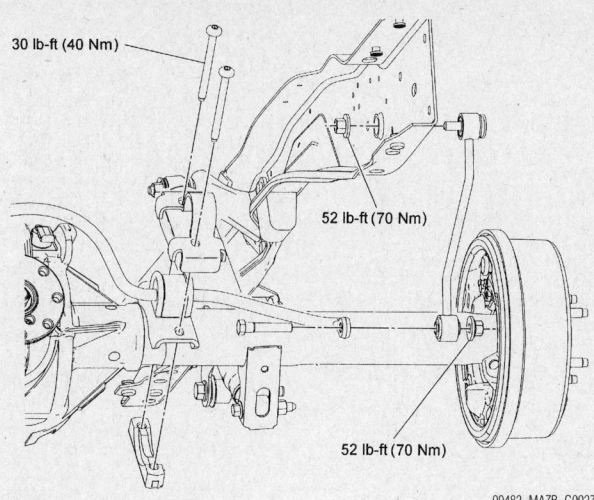

30 lb-ft (40 Nm)

52 lb-ft (70 Nm)

52 lb-ft (70 Nm)

09482_MAZB_G0027

Rear stabilizer bar and related components

4. Remove or disconnect the following:
- U-bolts from the rear spring plate. Discard the bolts.
- Lower shock absorber bolt and nut. Discard the bolt and nut.
- Hardware from the spring to bracket at the front of the rear spring
- Upper and lower shackle bolts at the rear of the spring. Discard the bolts.

- Spring and shackle from the bracket

To install:
5. Installation is the reverse of the removal procedure.
6. Be sure to use new bolts and nuts.
7. Install or connect the following:
- Spring and shackle to the bracket
- Upper and lower shackle bolts at

the rear of the spring. Torque the nuts to 85 ft. lbs. (115 Nm).
- U-bolts to the spring plate. Torque the nuts 76 ft. lbs. (103 Nm).
- Rear wheels

Stabilizer Bar

REMOVAL & INSTALLATION

1. Before servicing the vehicle, refer to the Precautions Section.
2. Raise the vehicle and install safety stands.
3. Remove the wheel and tire assembly.
4. Remove the rear stabilizer bar link nuts and bolt. Discard the bolt.
5. Remove the rear stabilizer bar link.
6. Remove the rear stabilizer bar mounting bracket bolts and remove the brackets. Discard the bolts.
7. Remove the rear stabilizer bar.

To install:
8. Installation is the reverse of the removal procedure.
9. Observe the following torques:
- End link nuts: 52 ft. lbs. (80 Nm)
- Bracket bolts: 30 ft. lbs. (46 Nm)

BRAKES

Brake Caliper

REMOVAL & INSTALLATION

1. Before servicing the vehicle, refer to the Precautions Section.
2. Loosen the wheel lug nuts.
3. Raise and safely support the front of the vehicle. Remove the wheel.
4. Remove the two caliper slide pin bolts and lift the caliper from the anchor plate.

➡**Use care to retain as much of the original caliper slide pin grease as possible.**

5. Position the caliper on a frame member or suspend it with some wire. Do not allow the caliper to hang by the brake hose.
6. Disconnect and plug the brake hose at the caliper. Remove the caliper from the rotor.
To install:
7. Position the caliper over the brake pads and align the slide pin mounting holes.
8. Install the slide pin bolts and torque to specification.
9. Install the caliper brake hose using

new washers. Tighten the bolt to 25 ft. lbs. (34 Nm)
10. Install the wheel and snug the lug nuts.
11. Lower the vehicle and tighten the lug nuts to 100 ft. lbs. (135 Nm).

➡**The first couple of times you apply the brakes, the pedal may go to the floor. Continue to pump the brake pedal until it feels firm.**

12. Start the engine and apply the brakes several times to readjust the caliper pistons. Ensure that the pedal feels firm before operating the vehicle.
13. Check and adjust the brake fluid level, as required.

Disc Brake Pads

REMOVAL & INSTALLATION

1. Before servicing the vehicle, refer to the Precautions Section.
2. Raise and safely support the front of the vehicle. Remove the wheel.
3. Remove the two caliper slide pin bolts and lift the caliper from the anchor plate.

➡**Use care to retain as much of the original caliper slide pin grease as possible.**

4. Position the caliper on a frame member or suspend it with some wire. Do not allow the caliper to hang by the brake hose.
5. Remove the brake pads and, if necessary, the anti-rattle clips from the anchor plate.
6. Remove the shims, if any, from the brake pads for re-use.
To install:
7. If removed, install the anti-rattle clips.
8. Install the brake pads to the anchor plate.
9. Position the caliper over the brake pads and align the slide pin mounting holes.
10. Install the slide pin bolts and tighten them to specification.
11. Install the wheel and snug the lug nuts.
12. Lower the vehicle and tighten the lug nuts to 100 ft. lbs. (135 Nm).

➡**The first couple of times you apply the brakes, the pedal may go to the floor. Continue to pump the brake pedal until it feels firm.**

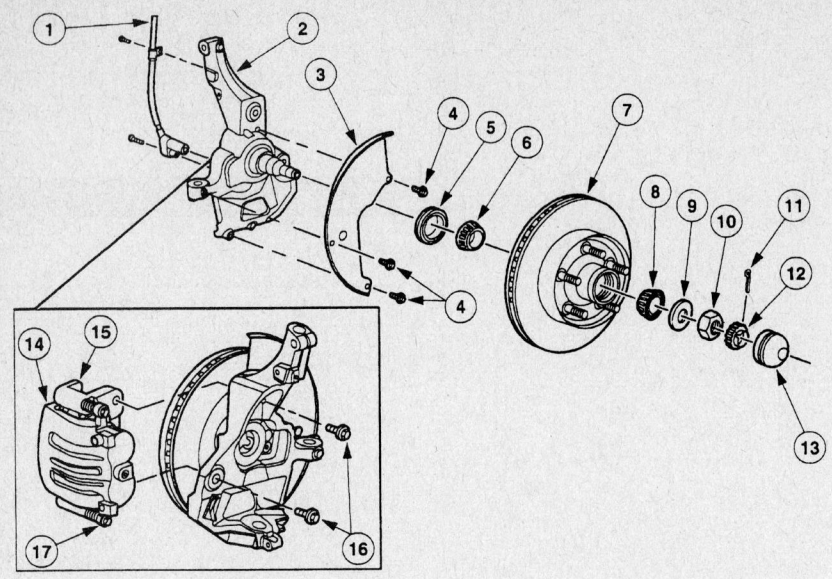

1. Front Brake Anti-Lock Sensor
2. Front Wheel Spindle
3. Front Disc Brake Rotor Shield
4. Rotor Shield Bolt
5. Grease Seal
6. Front Wheel Bearing
7. Front Disc Brake Hub and Rotor
8. Front Wheel Bearing
9. Front Wheel Outer Bearing Retainer Washer
10. Hub Spindle Nut
11. Cotter Pin
12. Nut Retainer
13. Hub Grease Cap
14. Disc Brake Caliper
15. Front Disc Brake Caliper Anchor Plate
16. Caliper Anchor Plate Bolts
17. Disc Brake Caliper Bolt

93026G22

Front disc brake and related components—2WD

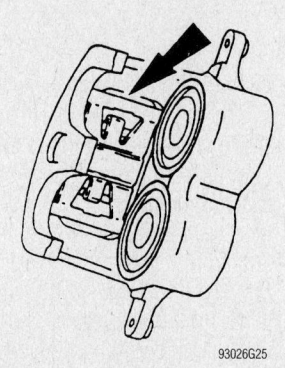

1 stainless slippers
2 pads

93026G24

Position of the front disc brake components

View of the front disc brake anti-rattle spring

93026G25

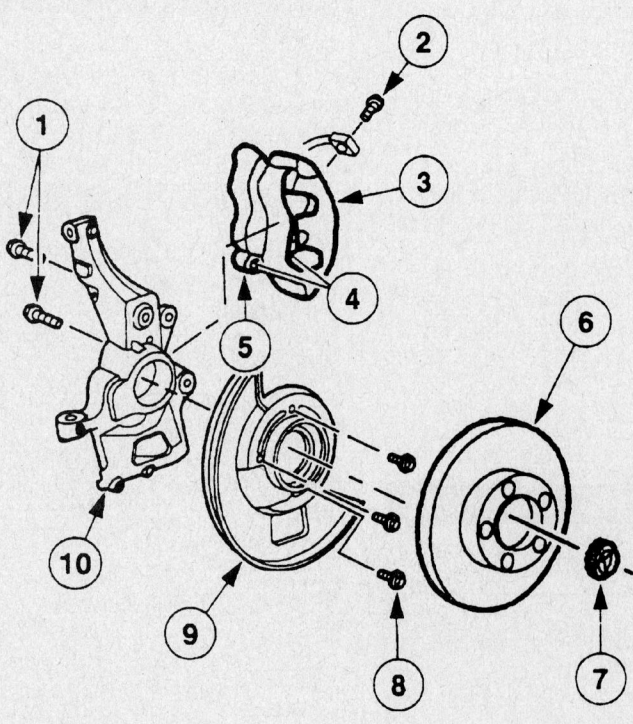

1 Front Disc Brake Caliper Anchor Plate Bolt (2 Req'd)

2 Front Brake Hose Bolt

3 Disc Brake Caliper

4 Pads

5 Front Disc Brake Caliper Anchor Plate

6 Front Disc Brake Rotor

7 Front Axle Wheel Hub Retainer

8 Front Disc Brake Rotor Shield Bolt (3 Req'd)

9 Front Disc Brake Rotor Shield

10 Front Wheel Knuckle

93026G23

Front disc brake and related components—4WD

13. Start the engine and apply the brakes several times to readjust the caliper pistons. Ensure that the pedal feels firm before operating the vehicle.

14. Check and adjust the brake fluid level, as required.

Brake Drums

REMOVAL & INSTALLATION

1. Before servicing the vehicle, refer to the Precautions Section.

2. Raise and safely support the vehicle. Remove the wheel and tire assembly.

3. Remove the retaining nuts, if equipped, and remove the brake drum.

4. Inspect the brake drum surface for wear, scoring and runout. Machine or replace, as necessary.

To install:

5. Install the brake drum and secure in place with the retainer nuts, if equipped.

6. Adjust the rear brakes.

7. Install the wheel. Lower the vehicle.

Brake Shoes

REMOVAL & INSTALLATION

1. Before servicing the vehicle, refer to the Precautions Section.

2. Raise and safely support the vehicle. Remove the wheel and tire assembly and the brake drum.

3. Pull backward on the adjusting lever cable to disengage the adjusting lever from the adjusting screw. Move the outboard side of the adjusting screw upward and back off the pivot nut as far as it will go.

4. Pull the adjusting lever, cable and automatic adjuster spring down and toward the rear to unhook the pivot hook from the large hole in the secondary shoe web. Do not pry the pivot hook from the hole.

5. Remove the automatic adjuster spring and adjusting lever.

6. Remove the secondary shoe-to-anchor spring using a suitable brake spring removal/installation tool. Using the tool, remove the primary shoe-to-anchor spring and unhook the cable anchor. Remove the anchor pin plate, if equipped.

7. Remove the cable guide from the secondary shoe.

8. Remove the shoe hold-down springs, shoes, adjusting screw, pivot nut and socket. Note the color and position of each hold-down spring so they can be reassembled in the same position.

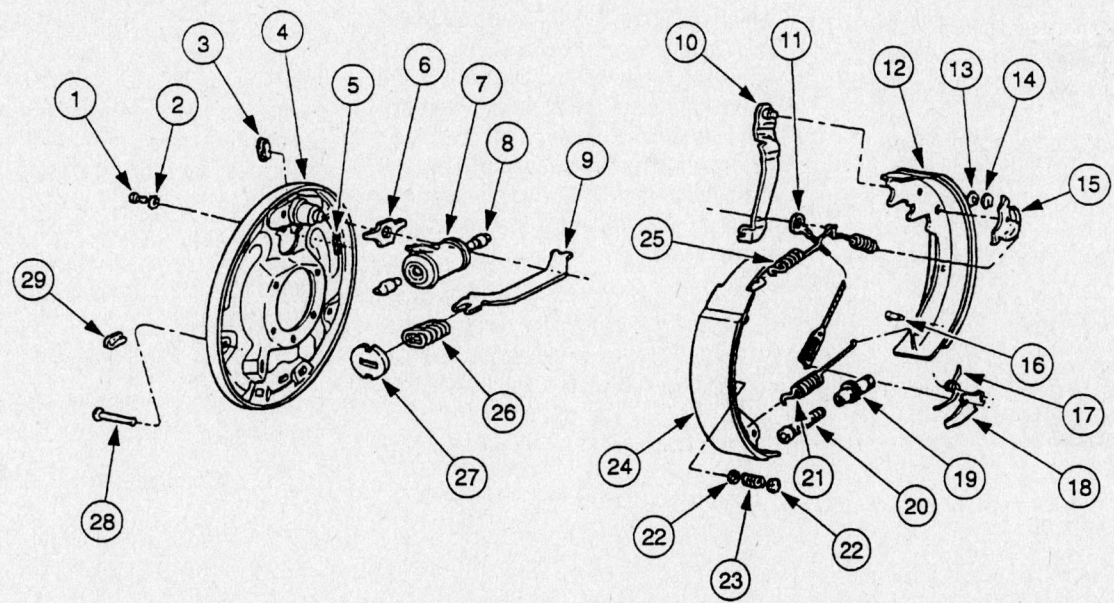

1	Wheel Cylinder-to-Backing Plate Bolt (2 Req'd)	12	Rear Brake Shoe and Lining, Secondary
2	Washer	13	Washer
3	Inspection Hole Cover	14	Parking Brake Lever Pin Retainer
4	Brake Backing Plate	15	Cable Guide
5	Lining Inspection Hole	16	Adjusting Lever Pin
6	Anchor Pin Guide Plate	17	Adjusting Lever Return Spring
7	Rear Wheel Cylinder	18	Brake Shoe Adjusting Lever
8	Wheel Cylinder Brake Shoe Link	19	Brake Shoe Adjusting Screw Nut
9	Parking Brake Strut	20	Brake Adjuster Screw
10	Parking Brake Lever	21	Brake Shoe Adjusting Screw Spring
11	Brake Shoe Adjusting Lever Cable		

22	Brake Shoe Hold-Down Spring Cup
23	Brake Shoe Hold-Down Spring
24	Rear Brake Shoe and Lining, Primary
25	Brake Shoe Retracting Spring, Short
26	Parking Brake Link Spring
27	Parking Brake Spring Retainer
28	Brake Shoe Hold-Down Spring Pin
29	Brake Adjusting Hole Cover

93026G21

Rear brake assembly and related components

9. Remove the parking brake link and spring. Disconnect the parking brake cable from the parking brake lever.

10. Remove the secondary brake shoe. On 9 in. (22.8cm) rear brakes, remove the parking brake lever from the shoe. On 10 in. (25.4cm) rear brakes, remove the retainer clip and spring washer and remove the parking brake lever.

To install:

11. Clean the backing plate ledge pads and sand lightly. Apply a light coating of high temperature lithium grease to the points where the brake shoes touch the backing plate. Lubricate the adjusting cable eye and the anchor pin area.

12. Install the parking brake lever on the secondary shoe. On 10 in. (25.4cm) brakes, secure with the spring washer and retaining clip.

13. Position the brake shoes on the backing plate and install the hold-down spring pins, springs and cups. Install the parking brake link, spring and washer. Connect the parking brake cable to the parking brake lever.

14. Install the anchor pin plate, if equipped, and place the cable anchor over the anchor pin with the crimped side toward the backing plate.

15. Install the primary shoe-to-anchor spring using the brake spring removal/installation tool.

16. Install the cable guide on the secondary shoe with the flanged hole fitted into the hole in the secondary shoe. Thread the cable around the cable guide groove.

➡**Make sure the cable is positioned in the groove and not between the guide and shoe web.**

17. Install the secondary shoe-to-anchor (long) spring.

➡**Make sure the cable end is not cocked or binding on the anchor pin when installed. All parts should be flat on the anchor pin.**

18. Apply high temperature lithium grease to the threads and the socket end of the adjusting screw. Turn the adjusting screw into the adjusting pivot nut to the end of the threads and then loosen, ½ turn.

19. Place the adjusting socket on the screw and install the assembly between the shoe ends with the adjusting screw nearest the secondary shoe.

➡**Be sure to install the adjusting screw on the same side of the vehicle from which it was removed. To prevent incorrect installation, the socket end of each adjusting screw is stamped with R or L, to indicate installation on the right or left side of the vehicle. The adjusting pivot nuts have lines machined around the body of the nut, 2 lines indicating the right side nut and 1 line indicating the left side nut.**

20. Hook the cable hook into the hole in the adjusting lever from the outboard plate side. The adjusting levers are also stamped with an **R** or **L** to indicate right or left side installation.

21. Place the hooked end of the adjuster spring in the large hole in the primary shoe web and connect the loop end of the spring to the adjuster lever hole.

22. Pull the adjuster lever, cable and automatic adjuster spring down toward the rear to engage the pivot hook in the large hole in the secondary shoe web.

23. After installation, check the action of the adjuster by pulling the section of the cable between the cable guide and the adjusting lever toward the secondary shoe web far enough to lift the lever past a tooth on the adjusting screw wheel. The lever should snap into position behind the next tooth and releasing the cable should cause the adjuster spring to return the lever to its original position. This return action will turn the adjusting screw 1 tooth.

24. If pulling the cable does not produce the action described previously, or if lever action is sluggish instead of positive and sharp, check the position of the lever on the adjusting screw toothed wheel. With the brake in a vertical position, anchor at the top, the lever should contact the adjusting wheel 1 tooth above the centerline of the adjusting screw. If the contact point is below the centerline, the lever will not lock on the adjusting screw wheel teeth and the screw will not turn, since the lever is actuated by the cable.

25. Adjust the brake shoes using either a brake adjustment gauge or manually with the drums installed.

26. Install the wheels, and lower the vehicle.

27. Check and adjust the brake fluid level, as required.

MAZDA

MPV

SPECIFICATIONS AND MAINTENANCE CHARTS

ENGINE AND VEHICLE IDENTIFICATION

Code ①	Liters (cc)	Cu. In.	Cyl.	Fuel Sys.	Engine Type	Eng. Mfg.
		Engine				
AJ	3.0 (NA)	NA	6	MFI	DOHC	Mazda

Code ②	Year
	Model Year
2	2002
3	2003
4	2004
5	2005
6	2006

MFI: Multi-port Fuel Injection

NA: Not available

DOHC: Double Overhead Camshaft

① 8th digit of the Vehicle Identification Number (VIN)

② 10th digit of the Vehicle Identification Number (VIN)

09482-MMPV-C0001

GENERAL ENGINE SPECIFICATIONS

Year	Model	Engine Displacement Liters	Engine ID	Net Horsepower @ rpm	Net Torque @ rpm (ft. lbs.)	Bore x Stroke (in.)	Com-pression Ratio	Oil Pressure @ rpm
2002	MPV	3.0	AJ	200@6200	200@3000	NA	NA	20-45@1500
2003	MPV	3.0	AJ	200@6200	200@3000	NA	NA	20-45@1500
2004	MPV	3.0	AJ	200@6200	200@3000	NA	NA	20-45@1500
2005	MPV	3.0	AJ	200@6200	200@3000	NA	NA	20-45@1500
2006	MPV	3.0	AJ	200@6200	200@3000	NA	NA	20-45@1500

NA: Not available

09482-MMPV-C0002

ENGINE TUNE-UP SPECIFICATIONS

Year	Engine Displacement Liters	Engine ID	Spark Plug Gap (in.)	Ignition Timing (deg.) MT	AT	Fuel Pump (psi)	Idle Speed (rpm) MT	AT	Valve Clearance Intake	Exhaust
2002	3.0	AJ	0.051-0.055	—	10B	86-113	—	650-750	HYD	HYD
2003	3.0	AJ	0.051-0.055	—	10B	86-113	—	650-750	HYD	HYD
2004	3.0	AJ	0.051-0.055	—	10B	86-116	—	650-750	HYD	HYD
2005	3.0	AJ	0.051-0.055	—	10B	86-116	—	650-750	HYD	HYD
2006	3.0	AJ	0.051-0.055	—	10B	86-116	—	650-750	HYD	HYD

NOTE: The Vehicle Emission Control Information label often reflects specification changes made during production. The label figures must be used if they differ from those in this chart.

B: Before top dead center

HYD: Hydraulic

NA: Information not available

09482-MMPV-C0003

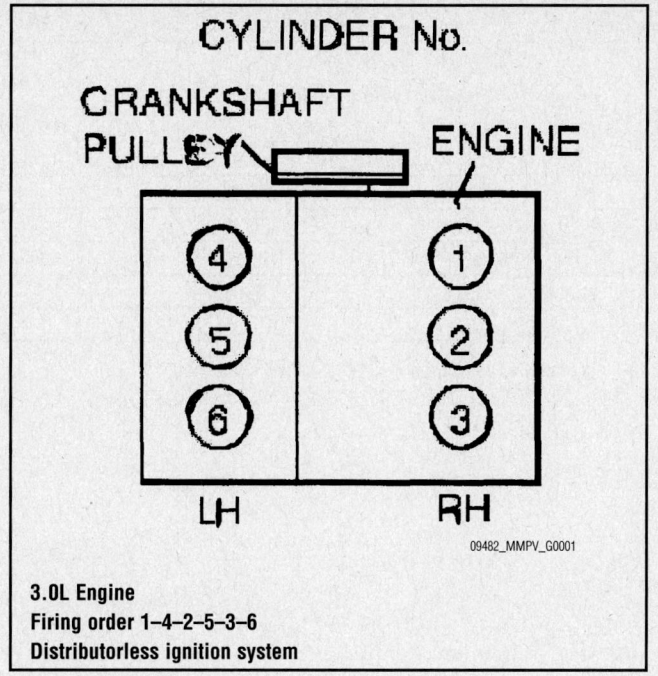

3.0L Engine
Firing order 1–4–2–5–3–6
Distributorless ignition system

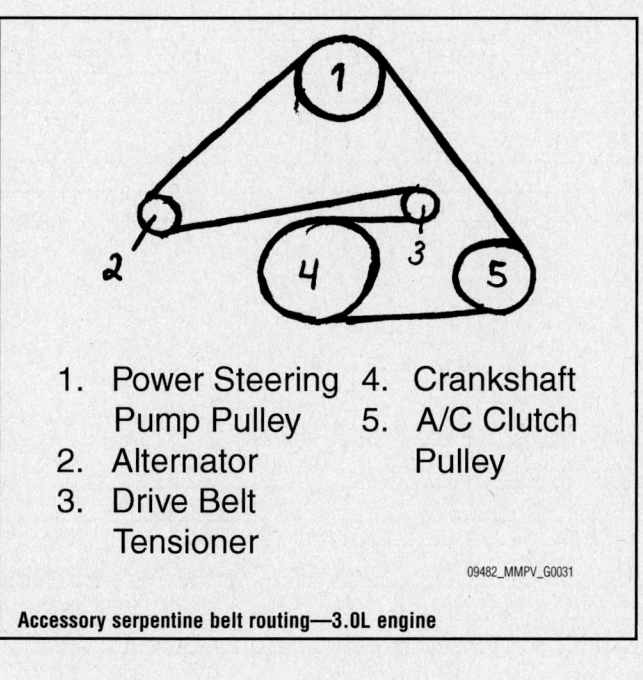

1. Power Steering Pump Pulley
2. Alternator
3. Drive Belt Tensioner
4. Crankshaft
5. A/C Clutch Pulley

Accessory serpentine belt routing—3.0L engine

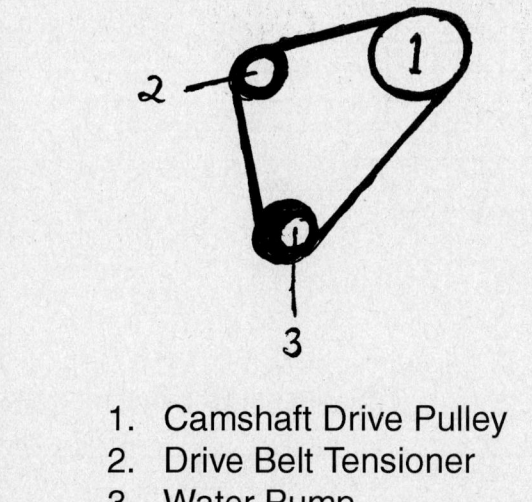

1. Camshaft Drive Pulley
2. Drive Belt Tensioner
3. Water Pump

Water pump belt routing—3.0L engine

CAPACITIES

Year	Model	Engine Displacement Liters	Engine ID	Engine Oil with Filter (qts.)	Transmission (pts.) Manual	Auto.	Transfer Case (pts.)	Drive Axle Front (pts.)	Rear (pts.)	Fuel Tank (gal.)	Cooling System (qts.)
2002	MPV	3.0	AJ	5.5	—	20.6	—	—	—	19.8	①
2003	MPV	3.0	AJ	5.5	—	20.6	—	—	—	19.8	①
2004	MPV	3.0	AJ	6.0	—	20.6	—	—	—	19.8	②
2005	MPV	3.0	AJ	6.0	—	20.6	—	—	—	19.8	②
2006	MPV	3.0	AJ	6.0	—	20.6	—	—	—	19.8	②

NOTE: All capacities are approximate. Add fluid gradually and ensure a proper fluid level is obtained.

① With rear heater: 13.1
Without rear heater: 11.2

② With rear heater: 12.3
Without rear heater: 10.4

09482-MMPV-C0004

VALVE SPECIFICATIONS

Year	Engine Displacement Liters	Engine ID	Seat Angle (deg.)	Face Angle (deg.)	Spring Test Pressure (lbs. @ in.)	Spring Installed Height (in.)	Stem-to-Guide Clearance (in.) Intake	Exhaust	Stem Diameter (in.) Intake	Exhaust
2002	3.0	AJ	NA	NA	NA	NA	NA	NA	NA	NA
2003	3.0	AJ	NA	NA	NA	NA	NA	NA	NA	NA
2004	3.0	AJ	NA	NA	NA	NA	NA	NA	NA	NA
2005	3.0	AJ	NA	NA	NA	NA	NA	NA	NA	NA
2006	3.0	AJ	NA	NA	NA	NA	NA	NA	NA	NA

NA: Information not available

09482-MMPV-C0005

CRANKSHAFT AND CONNECTING ROD SPECIFICATIONS
All measurements are given in inches.

Year	Engine Displacement Liters	Engine ID	Crankshaft Main Brg. Journal Dia.	Main Brg. Oil Clearance	Shaft End-play	Thrust on No.	Connecting Rod Journal Diameter	Oil Clearance	Side Clearance
2002	3.0	AJ	NA	NA	NA	NA	NA	NA	NA
2003	3.0	AJ	NA	NA	NA	NA	NA	NA	NA
2004	3.0	AJ	NA	NA	NA	NA	NA	NA	NA
2005	3.0	AJ	NA	NA	NA	NA	NA	NA	NA
2006	3.0	AJ	NA	NA	NA	NA	NA	NA	NA

NA: Information not available

09482-MMPV-C0006

PISTON AND RING SPECIFICATIONS

All measurements are given in inches.

| Year | Engine Displacement Liters | Engine ID | Piston Clearance | Ring Gap | | | Ring Side Clearance | | |
				Top Compression	Bottom Compression	Oil Control	Top Compression	Bottom Compression	Oil Control
2002	3.0	AJ	NA	NA	NA	NA	NA	NA	NA
2003	3.0	AJ	NA	NA	NA	NA	NA	NA	NA
2004	3.0	AJ	NA	NA	NA	NA	NA	NA	NA
2005	3.0	AJ	NA	NA	NA	NA	NA	NA	NA
2006	3.0	AJ	NA	NA	NA	NA	NA	NA	NA

NA: Information not available

09482-MMPV-C0007

TORQUE SPECIFICATIONS

All readings in ft. lbs.

| Year | Engine Displacement Liters | Engine ID | Cylinder Head Bolts | Main Bearing Bolts | Rod Bearing Bolts | Crankshaft Damper Bolts | Flywheel Bolts | Manifold | | Spark Plugs | Oil Pan Drain Plug |
								Intake	Exhaust		
2002	3.0	AJ	①	NA	NA	②	NA	7.5	15-18	8-14	16-22
2003	3.0	AJ	①	NA	NA	②	NA	7.5	15-18	8-14	16-22
2004	3.0	AJ	①	NA	NA	②	NA	7.5	15-18	8-14	16-22
2005	3.0	AJ	①	NA	NA	②	NA	7.5	15-18	8-14	16-22
2006	3.0	AJ	①	NA	NA	②	NA	7.5	15-18	8-14	16-22

NA: Information not available

① Step 1: 24-28 ft. lbs.
   Step 2: plus 90 degrees
   Step 3: back off one full turn
   Step 4: 24-28 ft. lbs.
   Step 5: plus 90 degrees
   Step 6: plus an additional 90 degrees

② Step 1: 88 ft. lbs.
   Step 2: back off one full turn
   Step 3: 35-39 ft. lbs.
   Step 4: plus 85-90 degrees

09482-MMPV-C0008

WHEEL ALIGNMENT

| Year | Model | | Caster ① | | Camber ① | | Toe-in (in.) | Kingpin Angle (Deg.) |
			Range (+/-Deg.)	Preferred Setting (Deg.)	Range (+/-Deg.)	Preferred Setting (Deg.)		
2002	MPV	F	1.0	+2.03	1.0	-0.23	0.08+/-0.16	11.18
		R	—	—	1.0	-1.00	0.12+/-0.16	—
2003	MPV	F	1.0	+1.98	1.0	-0.13	0.08+/-0.16	11.10
		R	—	—	1.0	-1.00	0.17+/-0.22	—
2004	MPV	F	1.0	+1.98	1.0	-0.13	0.08+/-0.16	11.10
		R	—	—	1.0	-1.00	0.17+/-0.22	—
2005	MPV	F	1.0	+1.98	1.0	-0.13	0.08+/-0.16	11.10
		R	—	—	1.0	-1.00	0.17+/-0.22	—
2006	MPV	F	1.0	+1.98	1.0	-0.13	0.08+/-0.16	11.10
		R	—	—	1.0	-1.00	0.17+/-0.22	—

① Empty vehicle

09482-MMPV-C0009

TIRE, WHEEL AND BALL JOINT SPECIFICATIONS

Year	Model	OEM Tires Standard	OEM Tires Optional	Tire Pressures (psi) Front	Tire Pressures (psi) Rear	Wheel Size	Ball Joint Inspection	Lugnut Torque (ft. lbs.)
2002	MPV	205/65R15	215/60R16 P215/60R17	①	①	std: 6-JJ opt: 6.5-JJ, 7-JJ	NA	108
2003	MPV	205/65R15	215/60R16 P215/60R17	①	①	std: 6-JJ opt: 6.5-JJ, 7-JJ	NA	108
2004	MPV	205/65R15	215/60R16 P215/60R17	①	①	std: 6-JJ opt: 6.5-JJ, 7-JJ	NA	108
2005	MPV	205/65R15	215/60R16 P215/60R17	①	①	std: 6-JJ opt: 6.5-JJ, 7-JJ	NA	108
2006	MPV	205/65R15	215/60R16 P215/60R17	①	①	std: 6-JJ opt: 6.5-JJ, 7-JJ	NA	108

NA: Information not available

OEM: Original Equipment Manufacturer

PSI: Pounds Per Square Inch

① See placard on vehicle

09482-MMPV-C0010

BRAKE SPECIFICATIONS
All measurements in inches unless noted

Year	Model		Brake Disc Original Thickness	Brake Disc Minimum Thickness	Brake Disc Maximum Runout	Brake Drum Diameter Original Inside Diameter	Brake Drum Diameter Max. Wear Limit	Brake Drum Diameter Maximum Machine Diameter	Minimum Lining Thickness Front	Minimum Lining Thickness Rear	Brake Caliper Bracket Bolts (ft. lbs.)	Brake Caliper Mounting Bolts (ft. lbs.)
2002	MPV		NA	1.030	0.002	①	①	10.060	0.080	NA	65-79	62-68
2003	MPV		NA	1.030	0.002	①	①	10.060	0.080	NA	65-79	62-68
2004	MPV	F	NA	1.030	0.002	—	—	—	0.080	—	65-79	61-69
		R	NA	0.630	0.002	①	①	10.060	—	②	36-51	16-23
2005	MPV	F	NA	1.030	0.002	—	—	—	0.080	—	65-79	61-69
		R	NA	0.630	0.002	①	①	10.060	—	②	36-51	16-23
2006	MPV	F	NA	1.030	0.002	—	—	—	0.080	—	65-79	61-69
		R	NA	0.630	0.002	①	①	10.060	—	②	36-51	16-23

NA: Information not available

① Stamped on drum

② Disc brakes: 0.08
   Drum brakes: 0.04

09482-MMPV-C0011

SCHEDULED MAINTENANCE INTERVALS
MAZDA—2002-03 MPV

TO BE SERVICED	TYPE OF SERVICE	VEHICLE MILEAGE INTERVAL (x1000)												
		7.5	15	22.5	30	37.5	45	52.5	60	67.5	75	82.5	90	97.5
Engine oil & filter	R	✓	✓	✓	✓	✓	✓	✓	✓	✓	✓	✓	✓	✓
Drive belt(s)	S/I				✓				✓				✓	
PCV valve	I								✓					
Spark plugs(platinum tip)	R	every 100,000 miles												
Air cleaner filter	R				✓				✓				✓	
Brake lines and hoses	I				✓				✓				✓	
Brake fluid	R				✓				✓				✓	
Brake pads/shoes	I				✓				✓				✓	
Bolts & nuts on chassis & body	S/I				✓				✓				✓	
Cooling system hoses	S/I				✓				✓				✓	
Driveshaft dust boots	S/I				✓				✓				✓	
Exhaust system heat shields	S/I				✓				✓				✓	
Front suspension ball joints	S/I				✓				✓				✓	
Fuel lines & hoses	S/I				✓				✓				✓	
Steering operation & linkages	S/I				✓				✓				✓	
Engine coolant	R						✓				✓			

R: Replace   S/I: Service or Inspect

FREQUENT OPERATION MAINTENANCE (SEVERE SERVICE)

If a vehicle is operated under any of the following conditions it is considered severe service:

- Extremely dusty areas.

- 50% or more of the vehicle operation is in 32°C (90°F) or higher temperatures, or constant operation in temperatures below 0°C (32°F).

- Prolonged idling (vehicle operation in stop and go traffic).

- Frequent short running periods (engine does not warm to normal operating temperatures).

- Police, taxi, delivery usage or trailer towing usage.

Air cleaner filter: service or inspect every 15,000 miles

Engine oil & filter: replace every 5000 miles.

Ball joints & dust covers: service or inspect every 7500 miles.

Bolts & nuts on chassis & body: tighten every 15,000 miles.

Automatic transmission fluid & filter: replace every 30,000 miles.

09482-MMPV-C0012

SCHEDULED MAINTENANCE INTERVALS
MAZDA—2004-06 MPV

TO BE SERVICED	TYPE OF SERVICE	VEHICLE MILEAGE INTERVAL (x1000)												
		7.5	15	22.5	30	37.5	45	52.5	60	67.5	75	82.5	90	97.5
Engine oil & filter	R	✓	✓	✓	✓	✓	✓	✓	✓	✓	✓	✓	✓	✓
Drive belt(s)	S/I				✓				✓				✓	
PCV valve	I								✓					
Spark plugs(platinum tip)	R	every 100,000 miles												
Air cleaner filter	R					✓					✓			
Brake lines and hoses	I			✓					✓				✓	
Cabin air filter	R	Every 25,000 miles												
Disc brake pads	I		✓		✓		✓		✓		✓		✓	
Drum brake shoes	I				✓				✓				✓	
Tires	Rotate	✓	✓	✓	✓	✓	✓	✓	✓	✓	✓	✓	✓	✓
All locks and hinges	L	✓	✓	✓	✓	✓	✓	✓	✓	✓	✓	✓	✓	✓
Cooling system hoses	S/I				✓				✓				✓	
Driveshaft dust boots	S/I				✓				✓				✓	
Exhaust system heat shields	S/I				✓				✓				✓	
Front suspension ball joints	S/I				✓				✓				✓	
Fuel lines & hoses	S/I				✓				✓				✓	
Engine valve clearance	I ①													
Steering operation & linkages	S/I				✓				✓				✓	
Engine coolant	R ②								✓					

R: Replace   S/I: Service or Inspect   L: Lubricate

① Inspect audibly, and, if noisy, adjust

② Then, every 2 years afterward

FREQUENT OPERATION MAINTENANCE (SEVERE SERVICE)

If a vehicle is operated under any of the following conditions it is considered severe service:

- Extremely dusty areas.

- 50% or more of the vehicle operation is in 32°C (90°F) or higher temperatures, or constant operation in temperatures below 0°C (32°F).

- Prolonged idling (vehicle operation in stop and go traffic).

- Frequent short running periods (engine does not warm to normal operating temperatures).

- Police, taxi, delivery usage or trailer towing usage.

Air cleaner filter: service or inspect every 15,000 miles

Engine oil & filter: replace every 5000 miles.

Ball joints & dust covers: service or inspect every 7500 miles.

Bolts & nuts on chassis & body: tighten every 15,000 miles.

Automatic transmission fluid & filter: replace every 30,000 miles.

09482-MMPV-C0013

ENGINE REPAIR

➡**Disconnecting the negative battery cable on some vehicles may interfere with the functions of the on board computer system. The computer may undergo a relearning process once the negative battery cable is reconnected.**

Distributor

This engine is equipped with a Distributorless Ignition System (DIS).

Alternator

REMOVAL & INSTALLATION

1. Before servicing the vehicle, refer to the Precautions Section.
2. Remove or disconnect the following:
   - Negative battery cable
   - Accessory drive belt
   - Exhaust front pipe
   - Right axle halfshaft and center shaft assembly
   - Alternator harness connectors
   - Center shaft support bracket
   - Alternator

To install:
3. Install or connect the following:
   - Alternator. Tighten the bolts to 29–41 ft. lbs. (40–50 Nm).
   - Center shaft support bracket. Tighten the bolts to 32–45 ft. lbs. (43–61 Nm).
   - Alternator harness connectors. Tighten the battery terminal nut to 87–130 inch lbs. (10–15 Nm).
   - Right axle halfshaft and center shaft assembly
   - Exhaust front pipe
   - Accessory drive belt
   - Negative battery cable

Engine

REMOVAL & INSTALLATION

1. Before servicing the vehicle, refer to the Precautions Section.
2. Drain the cooling system.
3. Drain the engine oil.
4. Drain the transaxle.
5. Relieve the fuel system pressure.
6. Disconnect the negative batter cable.
7. Install a support fixture to the engine lifting eyes.
8. Remove or disconnect the following:
   - Timing chain plug hole plate
   - Power steering hoses

- Splash shield
- Axle halfshafts
- Battery and tray
- Air intake assembly
- Accelerator cable and bracket
- Gear select cable
- Transaxle dipstick tube
- Cruise control actuator
- Radiator
- Fuel line
- Powertrain Control Module (PCM) connector. Pull the harness through the firewall into the engine compartment.
- Exhaust front pipe
- Accessory drive belt
- A/C compressor with lines still attached
- Alternator and bracket
- Engine mounts

To install:
9. Installation is the reverse of removal. Observe the following torques:
   - Subframe center section. Tighten the bolts to 48–65 ft. lbs. (64–89 Nm) and the nut to 50–67 ft. lbs. (67–93 Nm).
   - Left and rear engine mount through bolts to 63–86 ft. lbs. (85–116 Nm).
   - Front engine mount nuts to 50–67 ft. lbs. (67–93 Nm).
   - Right engine mount bracket nuts to 56–76 ft. lbs. (75–104 Nm).

Water Pump

REMOVAL & INSTALLATION

1. Before servicing the vehicle, refer to the Precautions Section.
2. Drain the cooling system.
3. Remove or disconnect the following:
   - Battery and tray
   - Water pump drive belt
   - Water pump drive pulley
   - Thermostat housing
   - Water pump belt tensioner
   - Oil cooler hose
   - Water outlet pipe
   - Water pump

To install:
4. Install or connect the following:
   - Water pump. Tighten the bolts to 89 inch lbs. (10 Nm) plus 90 degrees.
   - Water outlet pipe
   - Oil cooler hose
   - Water pump belt tensioner

- Thermostat housing
- Water pump drive pulley
- Water pump drive belt
- Battery and tray
5. Fill the cooling system.
6. Start the engine and check for leaks.

Heater Core

REMOVAL & INSTALLATION

Front System

1. Before servicing the vehicle, refer to the Precautions Section.
2. Disconnect the negative battery cable.

✴✴ CAUTION

After disconnecting the battery, wait for more than 1 minute for the air bag system to deplete its stored energy.

3. Drain the cooling system into a clean container for reuse.
4. Discharge and recover the air conditioning system refrigerant.

✴✴ WARNING

If moisture or foreign material enters the refrigeration cycle, cooling ability will be lowered and abnormal noise will occur. Always immediately plug open fittings after removing any refrigeration cycle parts to keep moisture or foreign material out of the cycle.

5. Remove the dashboard as follows:
   a. Bend the glove compartment door stoppers inward, then remove.
   b. Pull the glove compartment toward you while pushing it downward, then remove the clip.
   c. Remove by sliding the glove compartment toward the driver's side door.
   d. Turn the ignition switch to LOCK position.
   e. Disengage the wiring connector from the passenger-side air bag module (if equipped with the Standard Deployment Control System), or wiring connector A from the air bag module (if equipped with the Two-Stage Deployment Control System).
   f. Remove the air bag module mounting bolts from under the air bag module.
   g. If equipped with the Two-Stage Deployment Control System, lift up the

air bag module and disengage wiring connector B.

h. Remove the passenger-side air bag module from the vehicle.

※※ CAUTION

Place air bag modules in a safe place with the module always facing upward.

i. Pull down the adjusting lever of the tilt steering wheel and push the steering wheel down.

j. Remove the meter hood.

k. Pull down the selector lever to the D range position.

l. Remove the instrument cluster mounting screws and disconnect the electrical connectors, then remove the cluster from the vehicle.

※※ WARNING

When removing the instrument cluster, in order to prevent damage to the lens, cover the steering shaft with a cloth.

m. Remove the console lower cover and remove the dashboard lower panel.

n. Place the steering wheel in the straight-ahead position and turn the ignition switch to the LOCK position.

o. Remove the steering wheel-to-air bag module bolts.

p. Disconnect the clock spring connector(s).

q. Carefully, lift the air bag module from the steering wheel.

r. Remove the steering wheel-to-column nut and, using a steering wheel puller, press the steering wheel from the steering column.

s. Remove the steering column cover.

t. Remove the clock spring assembly.

u. Remove the combination switch.

v. Remove the clip and disconnect the selector cable from the selector lever.

w. Remove the selector lever.

x. Remove the column lower panel.

y. Remove the steering column-to-instrument panel bolts and lower the steering column.

z. Turn the seaming welt over at the right and left A-pillar trim.

aa. Disengage the clips in the top portion of the A-pillar trim.

bb. Pull the A-pillar trim upward, then disengage the bottom and lower hooks, then remove the right and left A-pillar trim.

cc. Remove each side panel.

dd. Remove the front scuff plates.

ee. Turn the seaming welt over.

ff. Pull the front side trim toward you, then disengage clip and pin from the body, and remove the right and left front side trim.

gg. Disconnect the antenna plug.

hh. For vehicles equipped with the wire-type climate control unit, disconnect the wires for the front A/C unit and disconnect the air mix and airflow mode wires from each wire clamp and link.

ii. After removing the screws, pull the front climate control unit out and disconnect the connectors, then remove the front climate control unit.

jj. Disconnect any remaining electrical connectors to the dashboard and/or front A/C unit.

kk. Remove the dashboard mounting bolts.

※※ WARNING

Removing the dashboard without supporting it can be dangerous. The dashboard may fall and injure you. Always perform these procedure together with at least another person.

ll. Disconnect the thermosensor electrical connector from the front A/C unit.

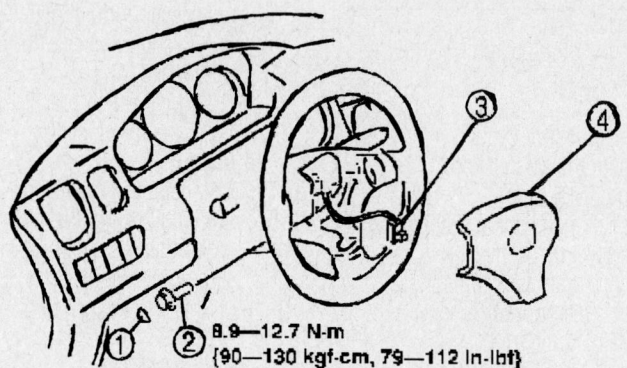

8.9—12.7 N·m
{90—130 kgf·cm, 79—112 in-lbf}

1. Cover
2. Bolt
3. Connector
4. Driver-side Air Bag Module

09482_MMPV_G0007

Exploded view of the steering wheel and air bag module—standard deployment control system shown

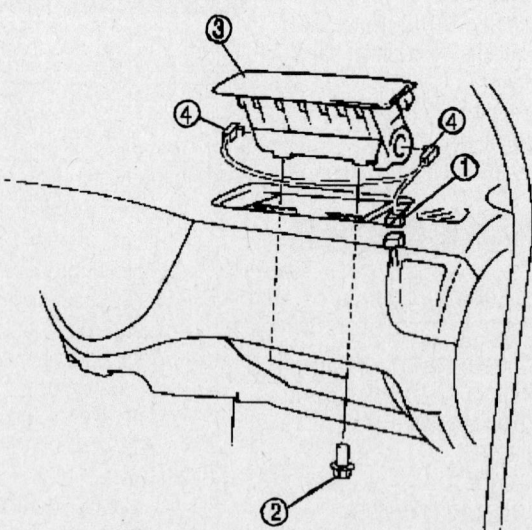

1. Connector A
2. Bolt
3. Passenger-side Air Bag Module
4. Connector B

09482_MMPV_G0008

Exploded view of the passenger-side air bag module—Two-step deployment control system shown

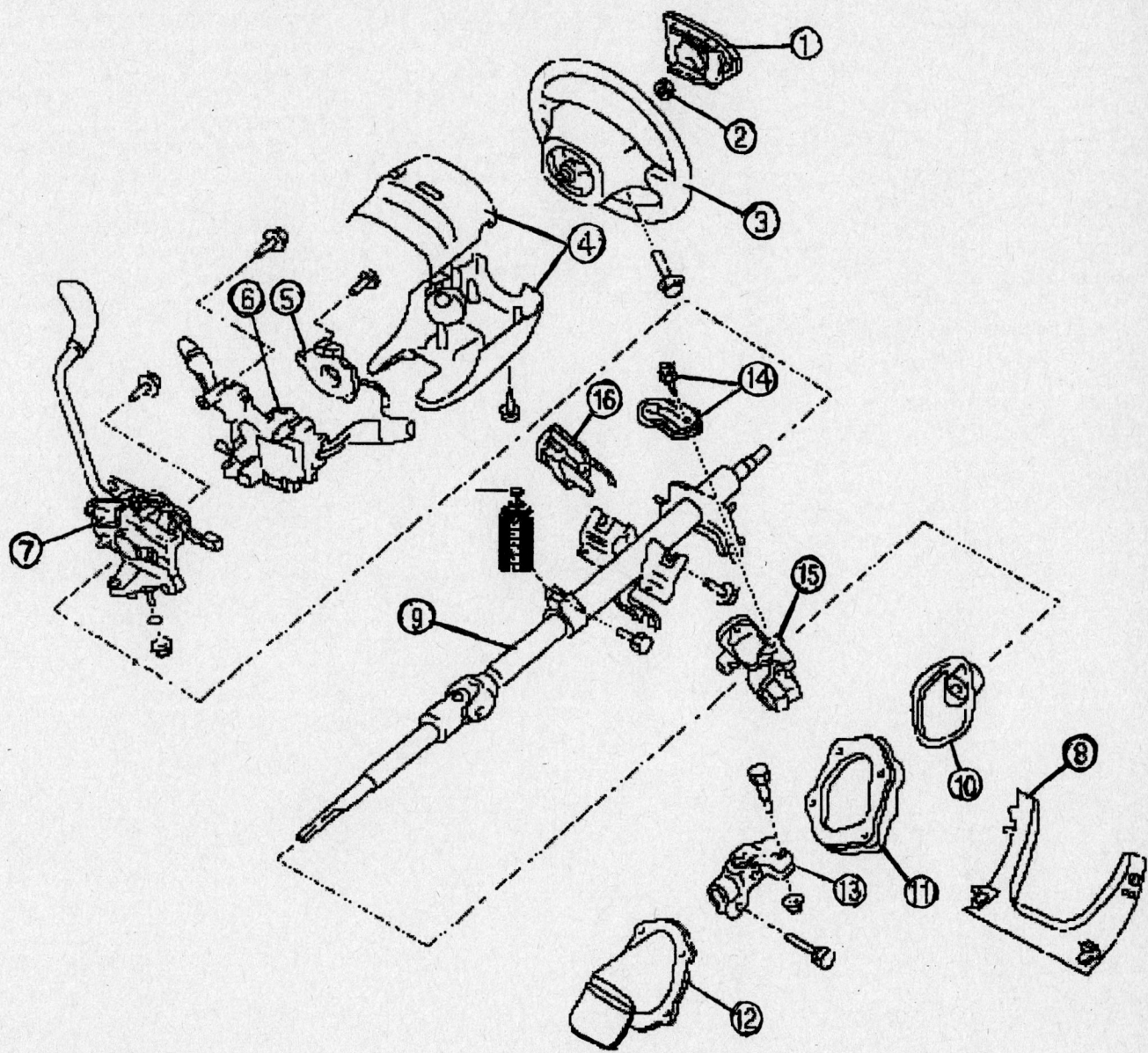

1. Air Bag Module
2. Locknut
3. Steering Wheel
4. Column Cover
5. Clock Spring
6. Combination Switch
7. Selector Lever
8. Lower Panel

9. Steering Shaft
10. Shaft Seal
11. Set Plate
12. Dust Cover
13. Universal Joint
14. Steering Lock Mounting Bolts and Bracket
15. Steering Lock Component
16. Cyclinder Outer Component

09482_MMPV_G0009

Exploded view of the steering column and related components

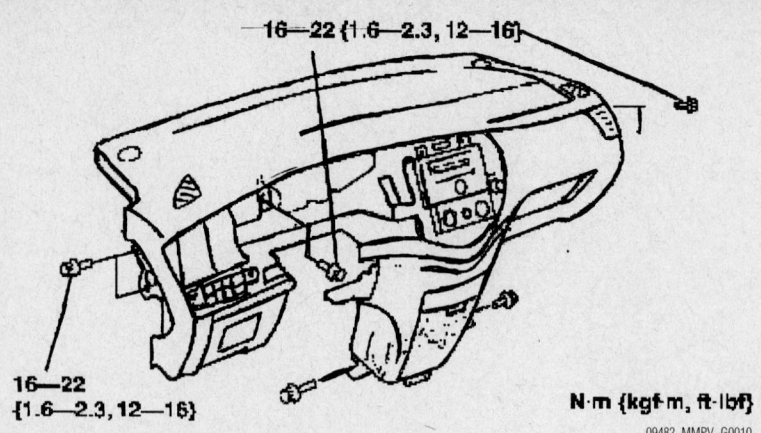

16—22 {1.6—2.3, 12—16}

16—22
{1.6—2.3, 12—16}

N·m {kgf·m, ft·lbf}

09482_MMPV_G0010

View of the instrument panel and related components

mm. Incline the dashboard slowly forward, then lift it while it is inclined forward and remove from vehicle through the passenger-side front door.

6. Disconnect the heater hoses from the heater core.

7. Disconnect the cooler pipe and air duct from the front A/C unit.

8. Remove the mounting fasteners and remove the front A/C unit housing from the vehicle.

9. Disassemble the heater housing and remove the heater core.

To install:

10. Install the heater core and assemble the heater housing.

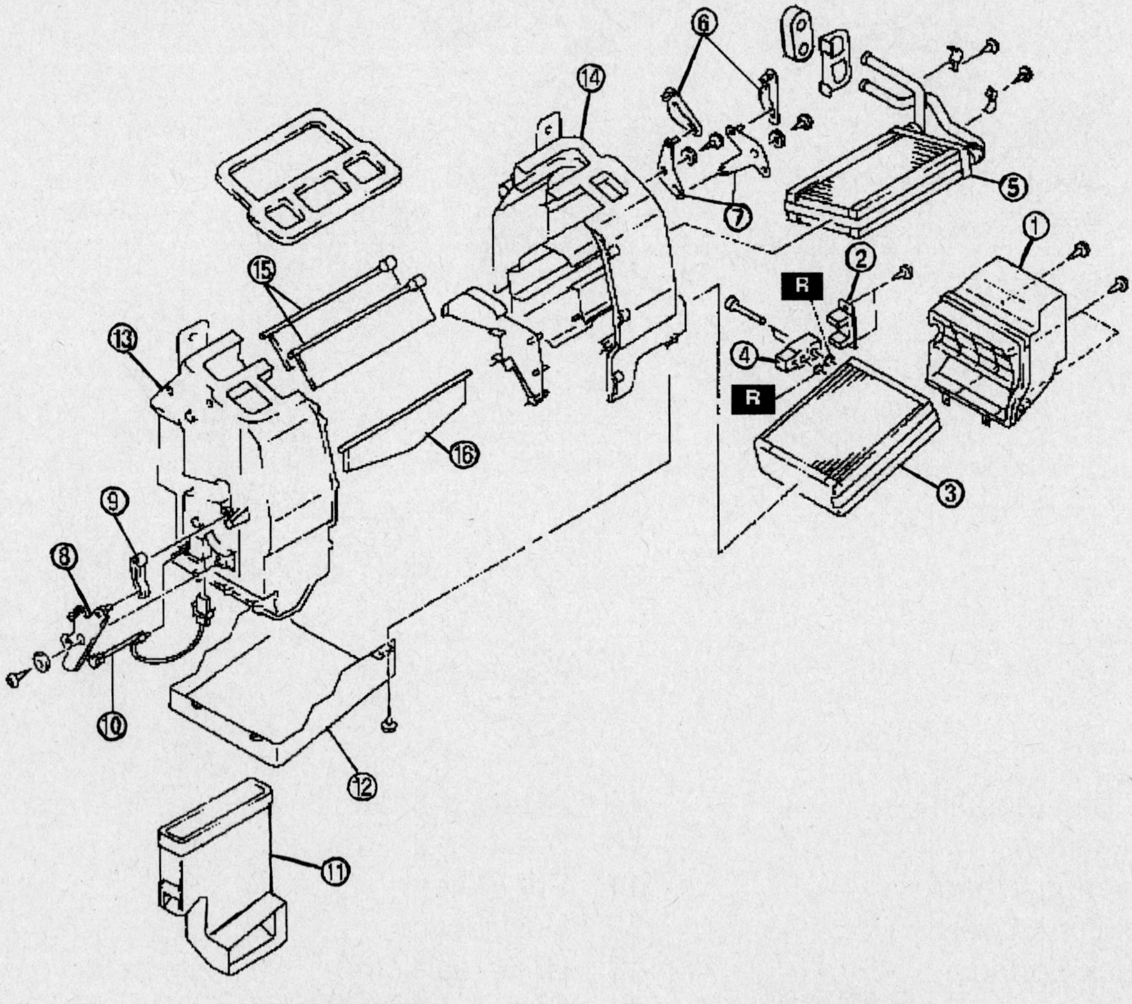

1.	Heater Case (5)	9.	Air Mix Crank
2.	Bracket	10.	Thermosensor
3.	Front Evaporator	11.	Heater Case (4)
4.	Expansion Valve	12.	Heater Case (3)
5.	Front Heater Core	13.	Heater Case (2)
6.	Airflow Mode Crank	14.	Heater Case (1)
7.	Airflow Mode Link	15.	Airflow Mode Door
8.	Air Mix Link	16.	Air Mix Door

09482_MMPV_G0011

Exploded view of the heater core, front A/C unit housing and related components

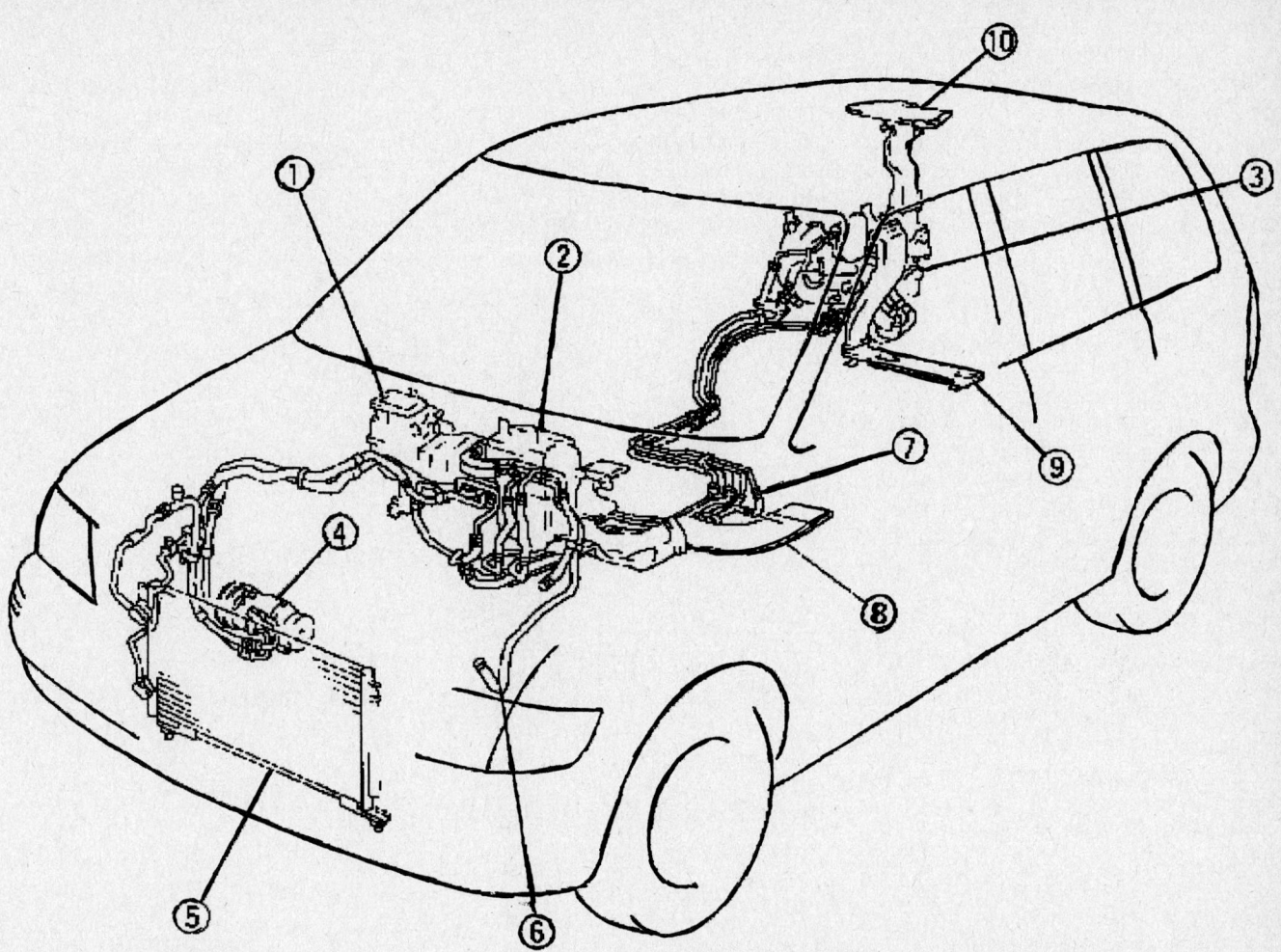

1. Blower Unit
2. Front A/C Unit
3. Rear A/C Unit
4. A/C Compressor
5. Condenser
6. Heater Hose and Pipe Component
7. Refrigerator Lines
8. Rear Heat Duct Component (Second Row Seat)
9. Rear Heat Duct Component (Third Row Seat)
10. Rear Cooler Duct

09482_MMPV_G0012

Exploded view of the front and rear HVAC system

11. Install the heater housing and tighten the mounting fasteners.

12. Connect the cooler pipe and air duct to the front A/C unit.

13. Connect the heater hoses to the heater core.

14. Install the instrument panel by performing the following procedure:

a. Using an assistant, carefully, install the instrument panel.

b. Connect the thermosensor electrical connector to the front A/C unit.

c. Connect any remaining electrical connectors to the dashboard and/or front A/C unit.

d. Install the instrument panel mounting bolts.

e. Install the front climate control unit into position and engage the connectors.

f. For vehicles equipped with the wire-type climate control unit, connect the wires to the front A/C unit and connect the air mix and airflow mode wires to each wire clamp and link.

g. Connect the antenna plug.

h. Install the right and left front side trim.

i. Install the right and left side panels.

j. Install the right and left A-pillar trim.

k. Install the steering column and the steering column-to-instrument panel bolts. Torque the bolts to 12–16 ft. lbs. (16–22 Nm).

l. Install the column lower panel.

m. Install the selector lever.

n. Install the combination switch.

o. Install the clock spring assembly.

p. Install the steering column cover.

q. Install the steering wheel and the steering wheel-to-column nut. Torque the steering wheel nut to 29–36 ft. lbs. (40–49 Nm).

r. Carefully, install the air bag module to the steering wheel.

s. Connect the clock spring connector(s).

t. Install the steering wheel-to-air bag module bolts. Torque the bolts to 79–112 inch lbs. (8.9–12.7 Nm).

u. Install the console lower cover and dashboard lower panel.

v. Install the instrument cluster and the meter hood.

w. Carefully, install the passenger-side air bag module and connect the electrical connector(s).

x. Install the passenger-side air bag module mounting bolts. Torque the bolts to 61–85 inch lbs. (7–9 Nm).

y. Install the glove compartment by sliding it to the right; then, push it downward to install the clip and bend the door stoppers inward to engage.

15. Connect the heater hoses to the heater core.

16. Refill the cooling system.

17. Connect the negative battery cable.

18. Recharge the air conditioning system.

19. Run the engine to normal operating temperatures; then, check the climate control operation and check for leaks.

Rear Auxiliary System

1. Disconnect the negative battery cable.

2. Drain the cooling system into a clean container for reuse.

3. Discharge and recover the air conditioning system refrigerant.

4. Remove the right rear side trim as follows:

a. Turn the weatherstrip over (liftgate side and sliding door side).

b. Remove the mat set end plate.

c. Remove the rear scuff plate.

d. Pull the cup holder upward, then disengage clips A and pin B from the right rear side trim, and remove the cup holder.

e. Remove the bolts and fastener.

f. Pull the rear side trim toward you, then disengage clips C, pin D, tabs E and F from the body.

g. Disconnect the rear climate control unit connectors and rear tweeter connectors (if equipped).

h. Remove the right rear side trim.

⁂ WARNING

If moisture or foreign material enters the refrigeration cycle, cooling ability will be lowered and abnormal noise will occur. Always immediately plug open fittings after removing any refrigeration cycle parts to keep moisture or foreign material out of the cycle.

5. Disconnect the rear heater hose No. 1.
6. Disconnect the rear heater hose No. 2.

7. Disconnect the rear cooler pipe No. 2.

8. Disconnect the rear A/C unit wire connectors.

9. Remove the rear A/C unit attaching bolts and remove the assembly.

10. Separate the rear A/C unit attaching bolts and remove the heater core.

To install:

11. Install the heater core and assemble the rear A/C unit.

12. Install the assembly into the vehicle.

13. Connect the rear A/C unit wire connectors.

14. Connect the heater hoses and rear cooler pipe to the assembly.

15. Install the right rear side trim.

16. Refill the cooling system.

17. Recharge the air conditioning system.

18. Connect the negative battery cable.

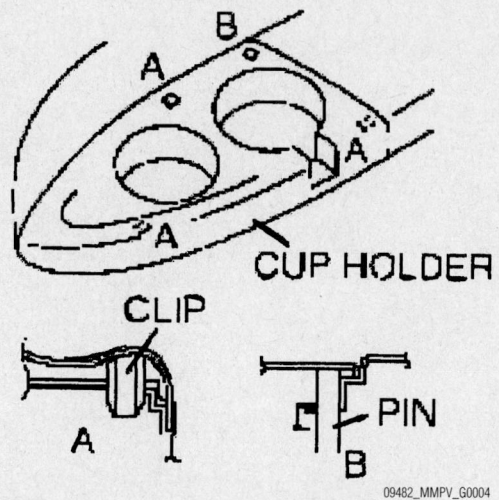

09482_MMPV_G0004

Removal of right rear side trim cup holder assembly—Mazda MPV

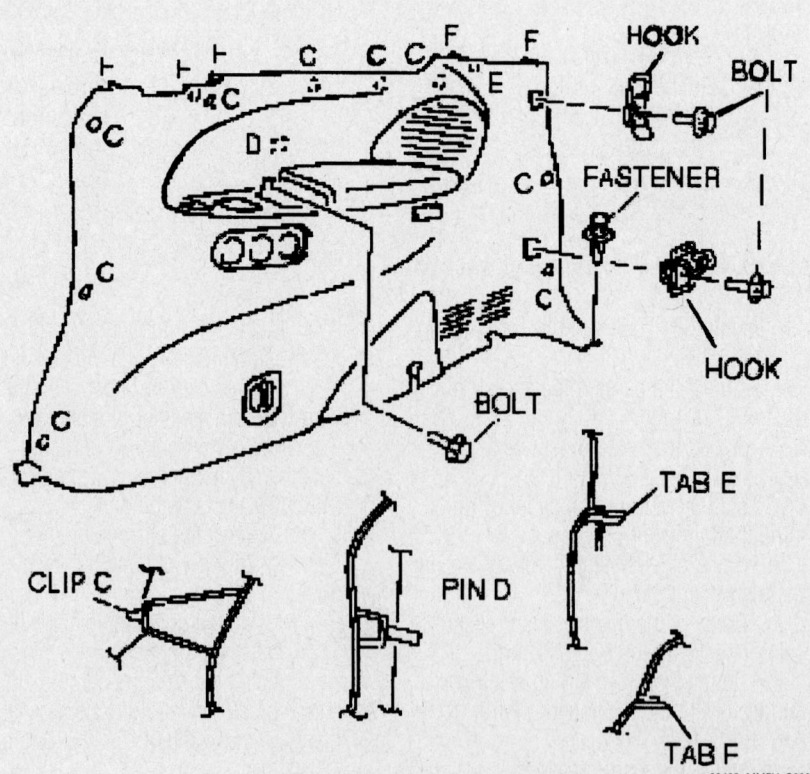

09482_MMPV_G0005

Exploded view of the right rear side trim—Mazda MPV

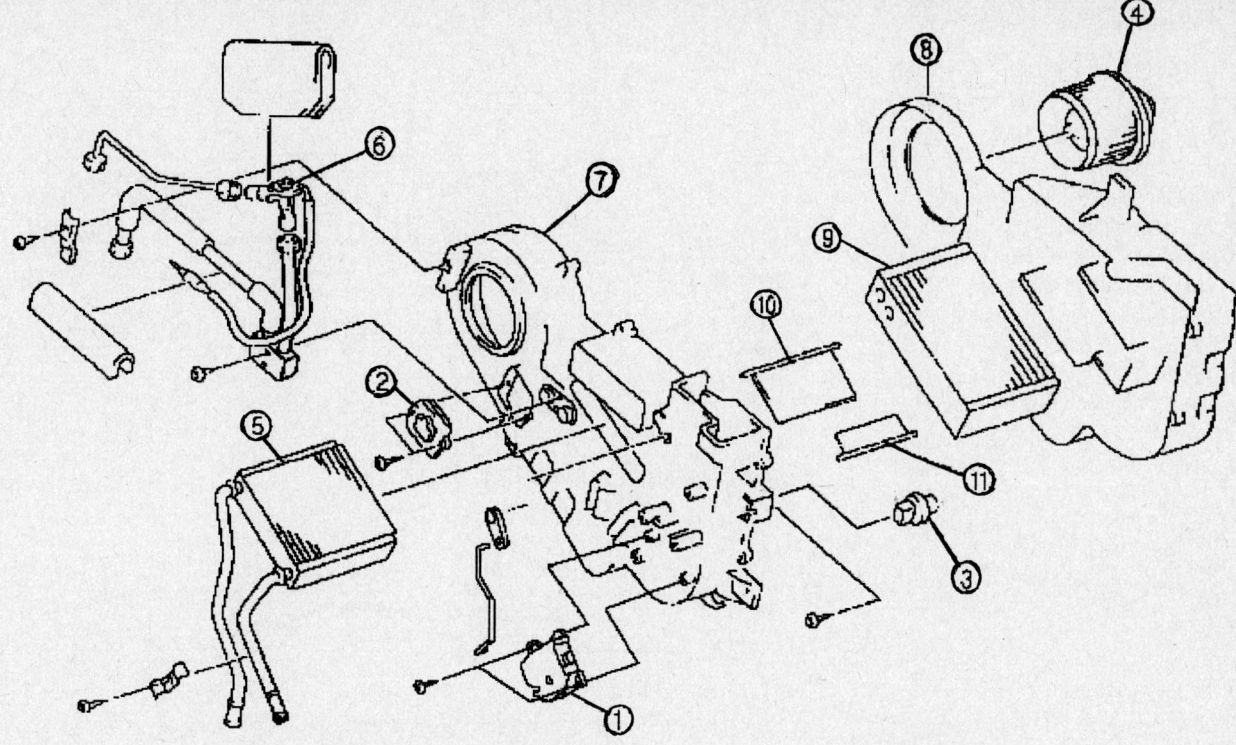

1. Rear Airflow Mode and Air Mix Actuator
2. Rear Resistor
3. Rear Blower Relay
4. Rear Blower Motor
5. Rear Heater Core
6. Rear Expansion Valve
7. Case (Left)
8. Case (Right)
9. Rear Evaporator
10. Airflow Mode Door
11. Air Mix Door

09482_MMPV_G0006

Exploded view of the rear auxiliary A/C unit—Mazda MPV

19. Run the engine to normal operating temperatures; then, check the rear climate control operation and check for leaks.

Cylinder Head

REMOVAL & INSTALLATION

1. Before servicing the vehicle, refer to the Precautions Section.
2. Drain the cooling system.
3. Drain the engine oil.
4. Relieve the fuel system pressure.
5. Remove or disconnect the following:
   - Accessory drive belt
   - Water pump and drive pulley
   - Timing chains
   - No.3 engine mount rubber and joint bracket
   - Ventilation pipe
   - Water bypass tube
   - Camshafts

➡ **Remove the Nos. 1 and 5 caps first. Don't loosen any other cap bolts until these caps are removed.**

- Rocker arms
- Cylinder heads. Loosen the bolts in several passes and in the sequence shown.

To install:

6. Installation is the reverse of removal. Observe the following torques:

➡ **The cylinder head bolts are a torque-to-yield design and must be replaced.**

7. Install the cylinder heads with new gaskets. Tighten the bolts in sequence as follows:
   a. Step 1: 24–28 ft. lbs. (32–38 Nm).
   b. Step 2: Plus 90 degrees.
   c. Step 3: Loosen one full turn.
   d. Step 4: 24–28 ft. lbs. (32–38 Nm).
   e. Step 5: Plus 90 degrees.
   f. Step 6: Plus 90 degrees.

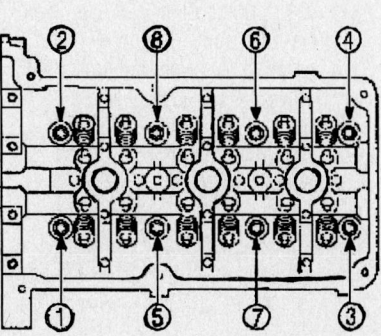

09482_MMPV_G0002

Cylinder head loosening sequence—3.0L engine

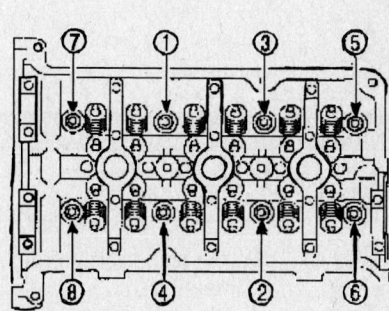

09482_MMPV_G0003

Cylinder head torque sequence—3.0L engine

Rocker Arms/Shafts

REMOVAL & INSTALLATION

1. Before servicing the vehicle, refer to the Precautions Section.
2. Relieve the fuel system pressure.
3. Drain the engine oil.
4. Remove or disconnect the following:
   - Negative battery cable
   - Intake manifold
   - Accessory drive belt
   - Power steering pump
   - Intake Manifold Runner Control (IMRC) actuator
   - Spark plug wires
   - Ignition coil
   - Water pump drive belt and pulley
   - Camshaft seal housing
   - Wiring harness connector bracket
   - Valve covers
   - Oil pan
   - Front cover
   - Timing chains
   - Camshafts
   - Rocker arms

➡ **Keep all valvetrain components in order for assembly.**

To install:

5. Install or connect the following:
   - Rocker arms
   - Camshafts
   - Timing chains
   - Front cover
   - Oil pan
   - Valve covers
   - Wiring harness connector bracket
   - Camshaft seal housing
   - Water pump drive belt and pulley
   - Ignition coil
   - Spark plug wires
   - Intake Manifold Runner Control (IMRC) actuator
   - Power steering pump
   - Accessory drive belt
   - Intake manifold
   - Negative battery cable
6. Fill the crankcase to the correct level.
7. Start the engine and check for leaks.

Intake Manifold

REMOVAL & INSTALLATION

1. Before servicing the vehicle, refer to the Precautions Section.
2. Relieve the fuel system pressure.
3. Remove or disconnect the following:
   - Negative battery cable

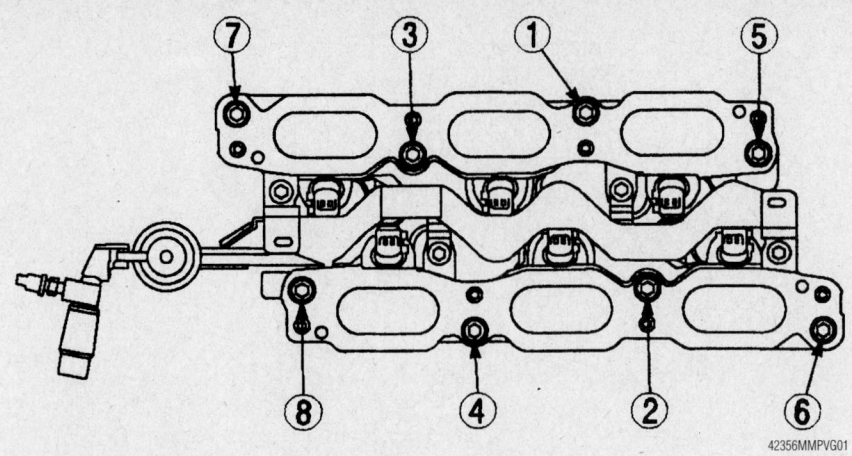

Intake manifold torque sequence—3.0L engine

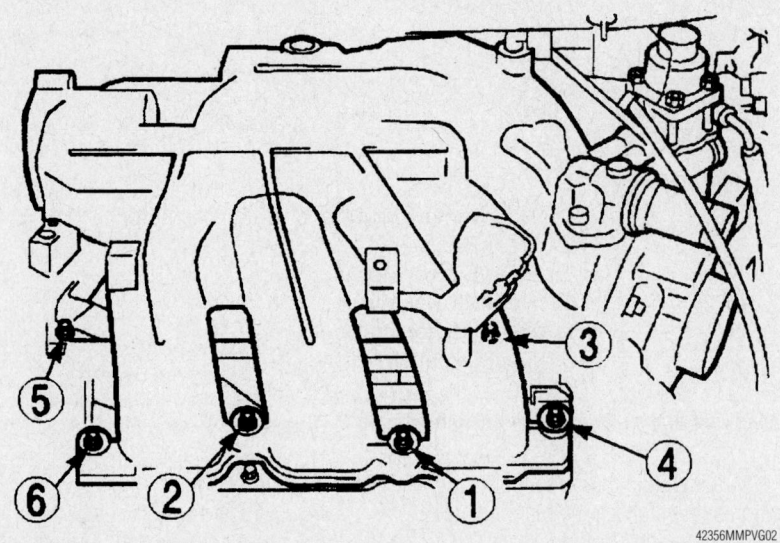

Dynamic chamber torque sequence—3.0L engine

   - Air cleaner housing and fresh air duct
   - Resonance chamber
   - Mass Air Flow (MAF) sensor
   - Throttle body intake hose
   - Accelerator cable and bracket
   - IMCC actuator
   - Throttle body
   - Exhaust Gas Recirculation (EGR) valve
   - Idle Air Control (IAC) valve
   - Dynamic chamber
   - Intake manifold

To install:

4. Install or connect the following:
   - Intake manifold. Tighten the bolts in sequence to 72–105 inch lbs. (8–12 Nm).
   - Dynamic chamber. Tighten the bolts in sequence to 72–105 inch lbs. (8–12 Nm).

5. The remainder if installation is the reverse of removal.
6. Start the engine and check for leaks.

Exhaust Manifold

REMOVAL & INSTALLATION

1. Before servicing the vehicle, refer to the Precautions Section.
2. Remove or disconnect the following:

   - Negative battery cable
   - Subframe transverse section
   - Heated Oxygen (HO$_2$S) sensor connectors
   - Exhaust front pipe
   - Exhaust Gas Recirculation (EGR) pipe
   - Exhaust manifolds

Right exhaust manifold torque sequence–3.0L engine

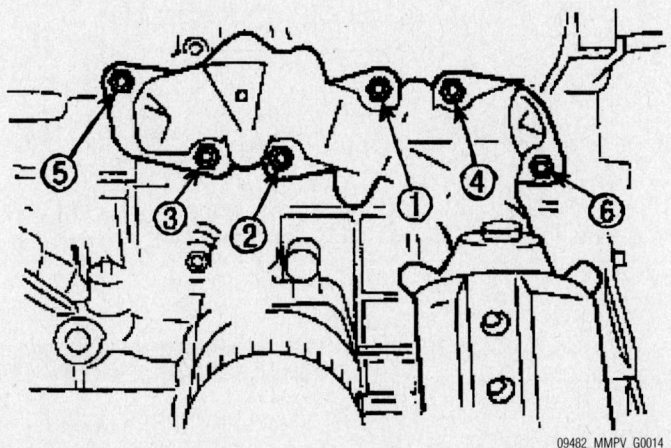

Left exhaust manifold torque sequence–3.0L engine

To install:

3. Install or connect the following:
- Exhaust manifolds. Tighten the nuts in sequence to 15–18 ft. lbs. (20–25 Nm).
- EGR pipe
- Exhaust front pipe
- HO₂S sensor connectors
- Subframe transverse section. Tighten the bolts to 69–96 ft. lbs. (93–131 Nm).
- Negative battery cable

4. Start the engine and check for leaks.

Front Crankshaft Seal

REMOVAL & INSTALLATION

Refer to the Timing Chain, Sprockets, Front Cover and Seal procedure in this section.

Camshaft and Valve Lifters

REMOVAL & INSTALLATION

1. Before servicing the vehicle, refer to the Precautions Section.
2. Drain the cooling system.
3. Drain the engine oil.
4. Relieve the fuel system pressure.
5. Remove or disconnect the following:
- Accessory drive belt
- Water pump and drive pulley
- Timing chains
- No.3 engine mount rubber and joint bracket
- Ventilation pipe
- Water bypass tube
- Camshafts

➡**Remove the Nos. 1 and 5 caps first. Don't loosen any other cap bolts until these caps are removed.**

- Rocker arms

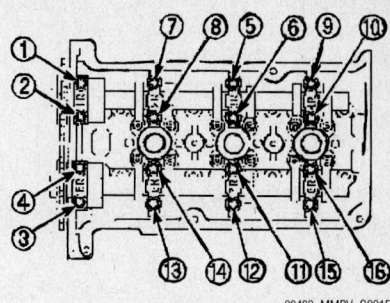

Right bank camshaft bearing and thrust cap loosening sequence—3.0L engine

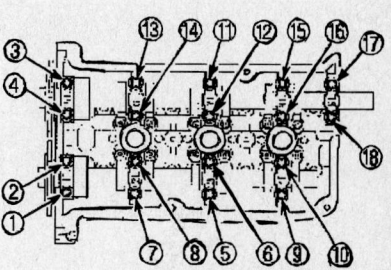

Left bank camshaft bearing and thrust cap loosening sequence—3.0L engine

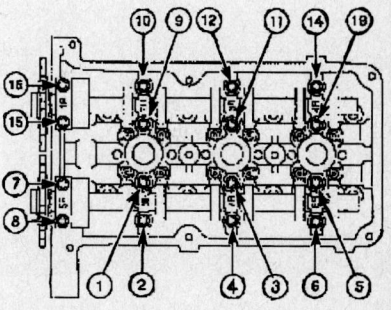

Right bank camshaft bearing and thrust cap torque sequence—3.0L engine

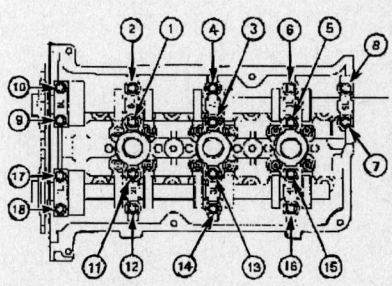

Left bank camshaft bearing and thrust cap torque sequence—3.0L engine

To install:

6. Installation is the reverse of removal. Observe the following torques:

7. Install the camshaft caps. Tighten the bolts in sequence to 71–106 inch lbs. (8–12 Nm)

Valve Lash

ADJUSTMENT

The engine covered in this section are equipped with hydraulic lash adjusters. Valve clearance adjustments are not possible.

Starter Motor

REMOVAL & INSTALLATION

1. Before servicing the vehicle, refer to the Precautions Section.
2. Remove or disconnect the following:
   - Battery and tray
   - Air intake assembly
   - Gear select cable
   - Starter harness connectors
   - Starter motor

To install:

3. Install or connect the following:
   - Starter motor. Tighten the bolts to 28–38 ft. lbs. (38–51 Nm).
   - Starter harness connectors. Tighten the battery cable nut to 87–104 inch lbs. (10–12 Nm).
   - Gear select cable
   - Air intake assembly
   - Battery and tray

Oil Pan

REMOVAL & INSTALLATION

1. Before servicing the vehicle, refer to the Precautions Section.
2. Drain the engine oil.
3. Remove or disconnect the following:
   - Negative battery cable
   - Exhaust front pipe
   - Flywheel access panel
   - Transaxle housing bolts
   - Oil pan bolts. Loosen the bolts in the sequence shown.
   - Oil pan

To install:

4. Apply a bead of silicone sealer to the gasket area where the pan meets the parting lines of the lower cylinder block and the front engine cover.

5. Install or connect the following:

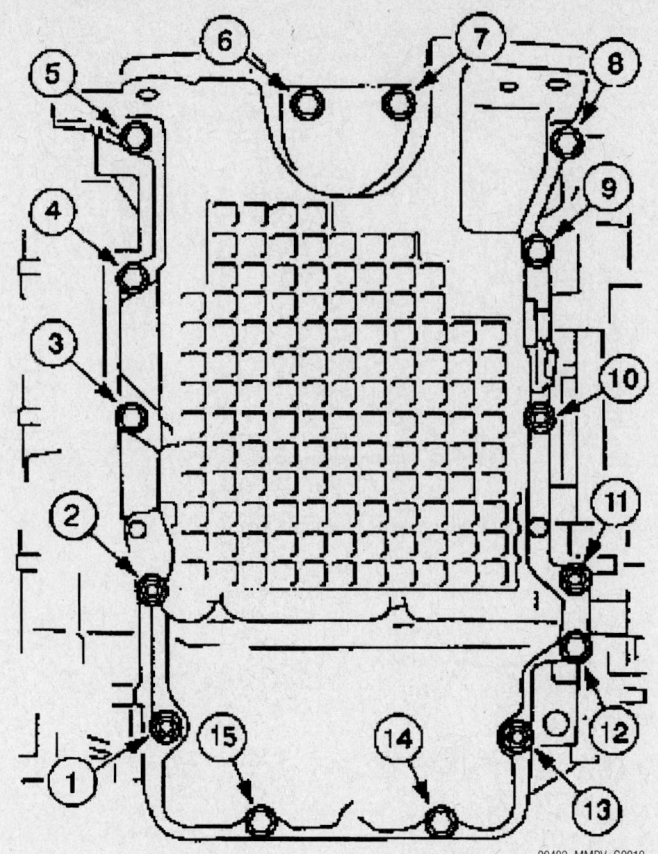

Oil pan torque sequence—3.0L engine

- Oil pan. Use a new gasket, tighten the pan bolts in several passes to 15–22 ft. lbs. (20–30 Nm), then tighten the transaxle case bolts to 28–38 ft. lbs. (38–51 Nm).
- Flywheel access panel
- Exhaust front pipe
- Negative battery cable
6. Fill the crankcase to the correct level.
7. Start the engine and check for leaks.

Oil Pump

REMOVAL & INSTALLATION

1. Before servicing the vehicle, refer to the Precautions Section.
2. Drain the engine oil.
3. Remove or disconnect the following:
   - Negative battery cable
   - Oil pan

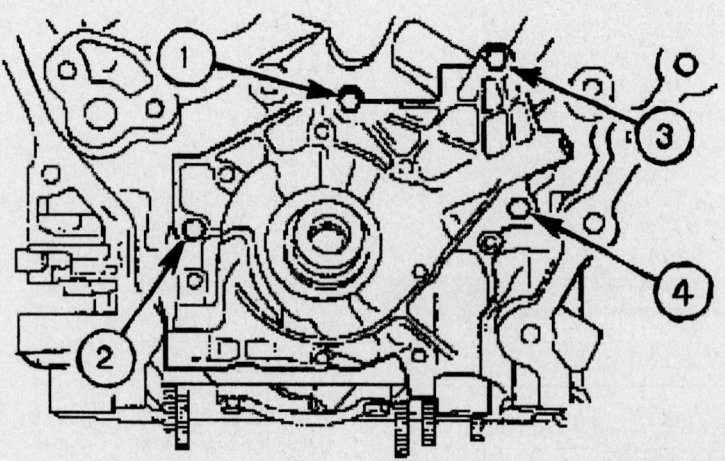

Oil pump torque sequence—3.0L engine

- Timing chains
- Oil pump pick up tube
- Oil pump. Loosen the bolts in reverse of the tightening sequence.

To install:

4. Install or connect the following:
- Oil pump. Tighten the bolts in sequence to 71–106 inch lbs. (8–12 Nm).
- Oil pump pick up tube. Tighten the bolts to 71–106 inch lbs. (8–12 Nm) and the nut to 44 inch lbs. (5 Nm) plus 45 degrees.
- Timing chains
- Oil pan
- Negative battery cable

5. Fill the crankcase to the correct level.
6. Start the engine and check for leaks.

Rear Main Seal

REMOVAL & INSTALLATION

1. Before servicing the vehicle, refer to the Precautions Section.
2. Remove or disconnect the following:
- Negative battery cable
- Transaxle
- Flywheel
- Oil seal

To install:

3. Install or connect the following:
- Oil seal. Press the seal in evenly with Special Service Tools 49 UN01 070 and 303-384 as shown.
- Flywheel. Tighten the bolts to 54–64 ft. lbs. (73–87 Nm).
- Transaxle
- Negative battery cable

4. Start the engine and check for leaks.

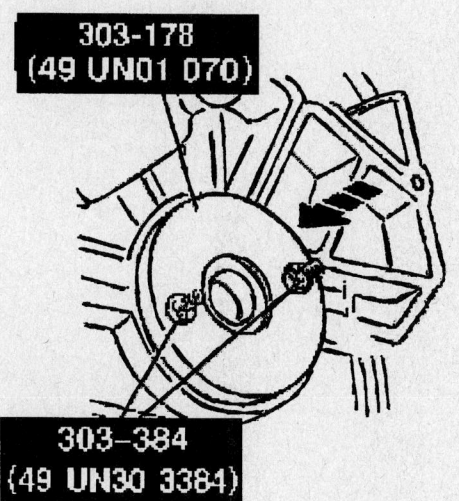

Rear main seal installation—3.0L engine

Timing Chain, Sprockets, Front Cover and Seal

REMOVAL & INSTALLATION

1. Before servicing the vehicle, refer to the Precautions Section.
2. Relieve the fuel system pressure.
3. Drain the engine oil.
4. Remove or disconnect the following:
- Negative battery cable
- Intake manifold
- Accessory drive belt
- Intake Manifold Runner Control (IMRC) actuator
- Spark plug wires
- Ignition coil
- Exhaust front pipe
- Oil pan
- Alternator and bracket
- A/C compressor
- Wiring harness connector bracket
- Water pump drive belt and pulley
- Camshaft oil seal housing
- Valve covers
- Right motor mount and bracket
- Crankshaft pulley
- Front crankshaft seal
- Front cover. Loosen the bolts in the sequence shown.
- Crankshaft Position (CKP) sensor pulse wheel

5. Rotate the crankshaft so that the keyway is at the 11 o'clock position to locate the crankshaft at TDC for No. 1 cylinder.

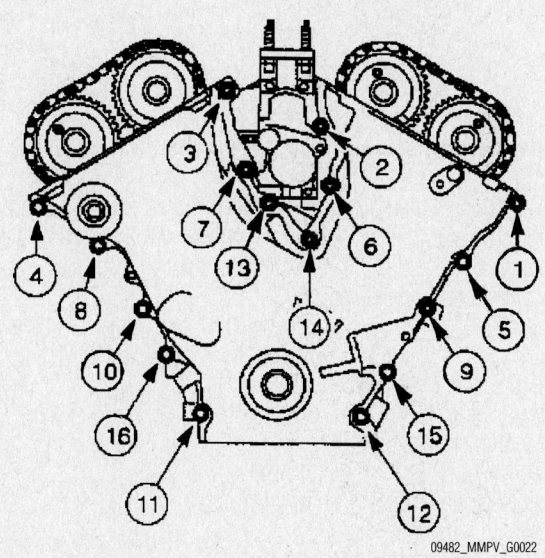

Front cover bolt removal sequence—3.0L engine

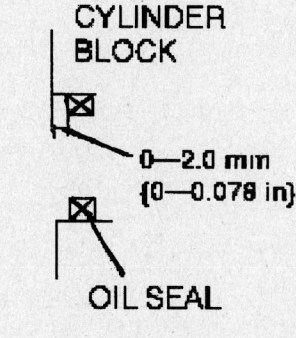

6. Verify that the alignment arrows on the camshafts are aligned. If not, rotate the crankshaft 1 complete revolution and recheck.

7. Rotate the crankshaft so that the keyway is at the 3 o'clock position. This positions the right cylinder head camshafts to the neutral position.

➡**Keep all valvetrain components in order for assembly.**

8. Remove or disconnect the following:
- Right timing chain tensioner
- Right timing chain tensioner arm
- Right timing chain and crankshaft timing sprocket
- Right bank camshafts

9. Rotate the crankshaft 1 and ⅔ turns and set the crankshaft keyway at the 11 o'clock position. This places the left bank camshafts in the neutral position.

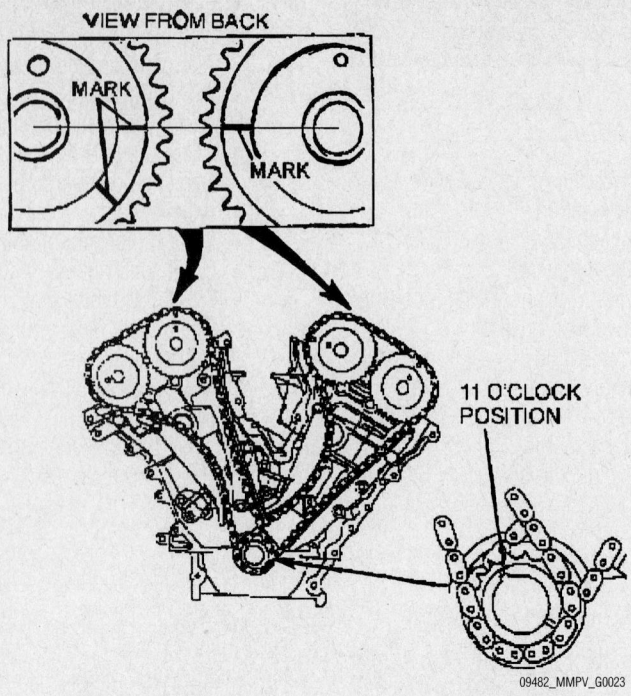

Camshaft alignment with the crankshaft in the 11 o'clock position—3.0L engine

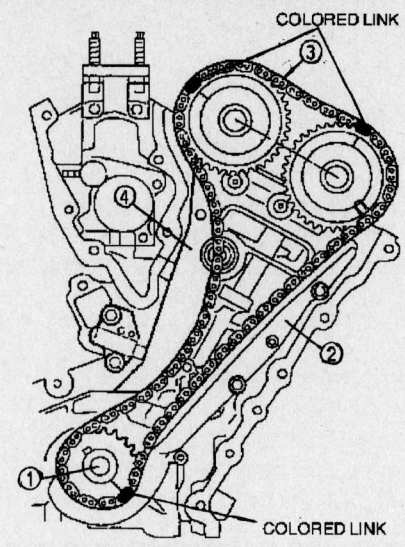

1. Timing Chain
   Crankshaft Sprocket
2. Chain Guide
3. Timing Chain
4. Tensioner Arm

09482_MMPV_G0025

Left bank timing chain alignment—3.0L engine

10. Remove or disconnect the following:
- Left timing chain tensioner
- Left timing chain tensioner arm
- Left timing chain and crankshaft timing sprocket

To install:

11. Prepare the timing chain tensioners for installation as follows:

a. Place the left chain tensioner in a vise.

b. Using a small prytool, release and hold the timing chain tensioner ratchet/pawl mechanism through the access hole in the timing chain tensioner.

c. Slowly compress the tensioner.

d. Lock the piston with a 1.5mm wire or paperclip.

e. Repeat for the right chain tensioner.

➡**Be sure that the crankshaft keyway is still at the 11 o'clock position.**

12. Install or connect the following:
- Left timing chain and crankshaft sprocket. Align the colored links with the index marks on the camshaft and crankshaft sprockets.
- Left timing chain tensioner arm
- Left timing chain tensioner. Tighten the retaining bolts to 15–22 ft. lbs. (20–30 Nm).

13. Remove the left timing chain tensioner retaining wire.

14. Rotate the crankshaft so that the keyway is at the 3 o'clock position.

15. Install the right bank camshafts with the exhaust camshaft index mark at 12 o'clock and the intake camshaft index mark at 3 o'clock as shown.

16. Install or connect the following:
- Right timing chain and crankshaft sprocket. Align the colored links with the index marks on the camshaft and crankshaft sprockets.
- Right timing chain tensioner arm
- Right timing chain tensioner. Tighten the retaining bolts to 15–22 ft. lbs. (20–30 Nm).

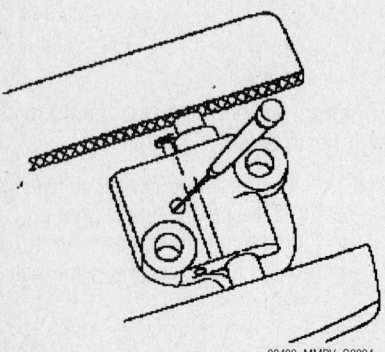

Using a thin prytool, release and hold the timing chain tensioner ratchet/pawl mechanism—3.0L engine

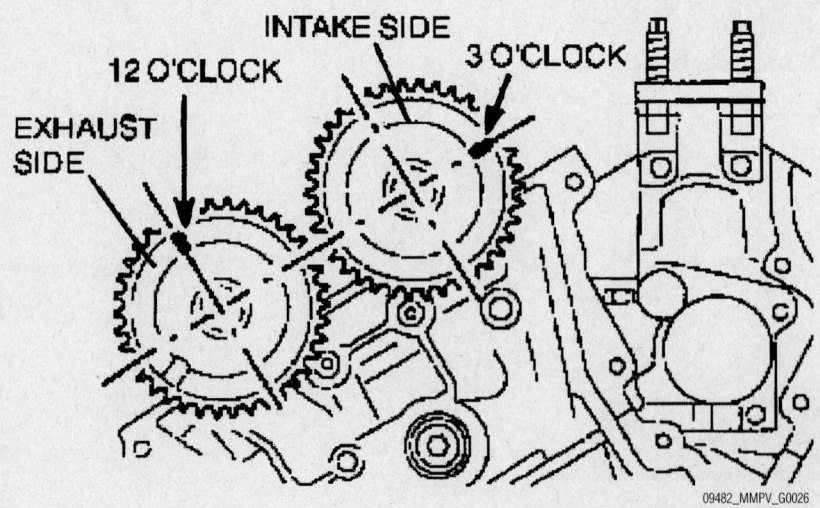

Right bank camshaft positioning—3.0L engine

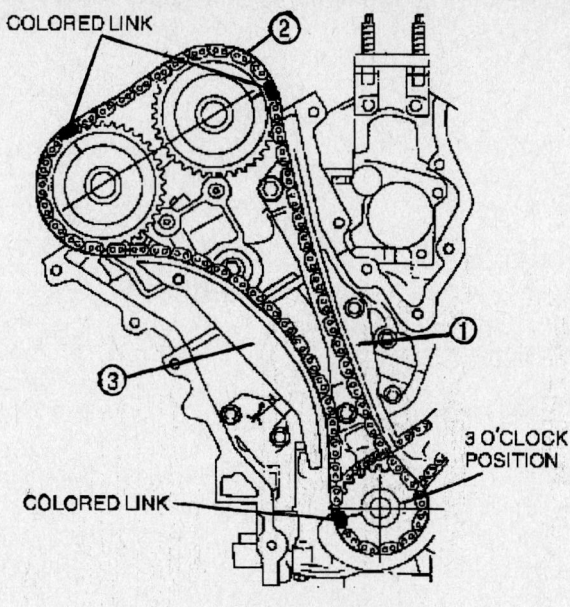

1. Chain Guide
2. Timing Chain
3. Tensioner Arm

09482_MMPV_G0027

Right bank timing chain alignment—3.0L engine

17. Remove the right timing chain tensioner retaining wire.
18. Install or connect the following:
  • CKP sensor pulse wheel
  • Front cover. Tighten the bolts in the reverse of the loosening sequence to 15–22 ft. lbs. (20–30 Nm).
  • Front crankshaft seal
  • Crankshaft pulley
19. Tighten the crankshaft pulley bolt as follows:
  a. Step 1: 88 ft. lbs. (120 Nm).
  b. Step 2: Loosen the bolt one turn.
  c. Step 3: 35–39 ft. lbs. (47–53 Nm).

  d. Step 4: Plus 85–95 degrees.
20. Install or connect the following:
  • Right motor mount and bracket
  • Valve covers
  • Camshaft oil seal housing
  • Water pump drive belt and pulley
  • Wiring harness connector bracket
  • A/C compressor
  • Alternator and bracket
  • Oil pan
  • Exhaust front pipe
  • Ignition coil
  • Spark plug wires
  • IMRC actuator

  • Accessory drive belt
  • Intake manifold
  • Negative battery cable
21. Fill the crankcase to the correct level.
22. Start the engine and check for leaks.

Piston and Ring

POSITIONING

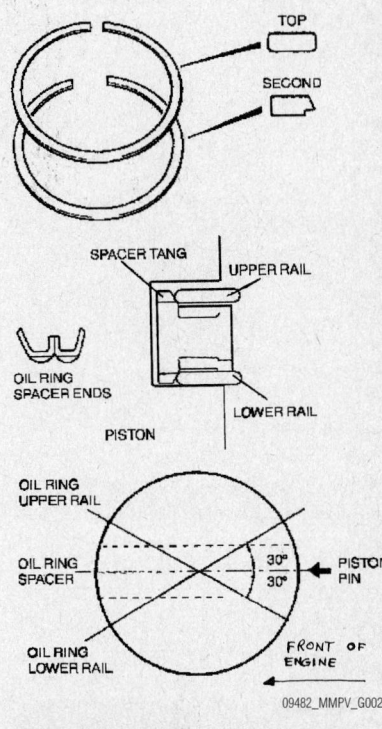

09482_MMPV_G0028

Piston ring positioning, end-gap spacing, and piston positioning. The small directional arrow must face the front of the engine—3.0L engine

FUEL SYSTEM

Fuel System Service Precautions

Safety is the most important factor when performing not only fuel system maintenance but any type of maintenance. Failure to conduct maintenance and repairs in a safe manner may result in serious personal injury or death. Maintenance and testing of the vehicle's fuel system components can be accomplished safely and effectively by adhering to the following rules and guidelines.

  • To avoid the possibility of fire and personal injury, always disconnect the negative battery cable unless the repair or test procedure requires that battery voltage be applied.

  • Always relieve the fuel system pressure prior to disconnecting any fuel system component (injector, fuel rail, pressure regulator, etc.), fitting or fuel line connection. Exercise extreme caution whenever relieving fuel system pressure, to avoid exposing skin, face and eyes to fuel spray. Please be advised that fuel under pressure may penetrate the skin or any part of the body that it contacts.

  • Always place a shop towel or cloth around the fitting or connection prior to loosening to absorb any excess fuel due to spillage. Ensure that all fuel spillage (should it occur) is quickly removed from engine surfaces. Ensure that all fuel soaked cloths or towels are deposited into a suitable waste container.

  • Always keep a dry chemical (Class B) fire extinguisher near the work area.

  • Do not allow fuel spray or fuel vapors to come into contact with a spark or open flame.

  • Always use a back-up wrench when loosening and tightening fuel line connection fittings. This will prevent unnecessary stress and torsion to fuel line piping. Always follow the proper tighten specifications.

  • Always replace worn fuel fitting O-rings with new. Do not substitute fuel hose or equivalent, where fuel pipe is installed.

Fuel System Pressure

RELIEVING

1. Before servicing the vehicle, refer to the Precautions Section.
2. Disconnect the fuel pump relay, located at the ECM.
3. Start the engine.
4. After the engine stalls, crank the engine several times.
5. Turn the ignition switch **OFF**.
6. When repairs are complete, connect the fuel pump relay.

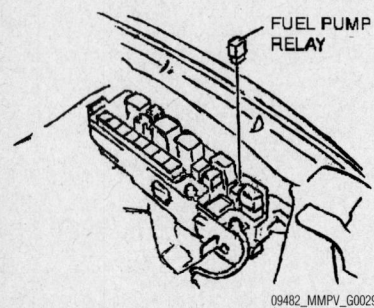

Fuel pump relay location

Fuel Filter

REMOVAL & INSTALLATION

The fuel filter is located in the fuel tank as part of the fuel pump module.

Fuel Pump

REMOVAL & INSTALLATION

1. Before servicing the vehicle, refer to the Precautions Section.
2. Relieve the fuel system pressure.
3. Remove or disconnect the following:
   - Negative battery cable
   - Front seats

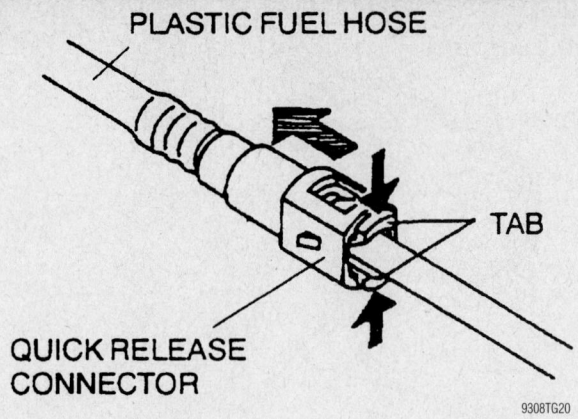

Fuel hose quick release connector

- Center console
- Door sill plates
- Parking brake lever
- Carpet
- Access panel
- Fuel lines
- Fuel pump module harness connector
- Fuel pump module

To install:

4. Install or connect the following:
   - Fuel pump module
   - Fuel pump module harness connector
   - Fuel lines
   - Access panel
   - Carpet
   - Parking brake lever
   - Door sill plates
   - Center console
   - Front seats
   - Negative battery cable
5. Start the engine and check for leaks.

Fuel Injector

REMOVAL & INSTALLATION

1. Before servicing the vehicle, refer to the Precautions Section.

2. Relieve the fuel system pressure.
3. Remove or disconnect the following:
   - Negative battery cable
   - Air cleaner housing and fresh air duct
   - Mass Air Flow (MAF) sensor
   - Intake manifold
   - Fuel injector harness connectors
   - Fuel lines
   - Pressure regulator vacuum hose
   - Fuel supply manifold with injectors attached
   - Fuel injectors

To install:

4. Install or connect the following:
   - Fuel injectors with new O-ring seals
   - Fuel supply manifold with injectors attached. Tighten the bolts to 72–101 inch lbs. (8–11 Nm).
   - Pressure regulator vacuum hose
   - Fuel lines
   - Fuel injector harness connectors
   - Intake manifold
   - MAF sensor
   - Air cleaner housing and fresh air duct
   - Negative battery cable
5. Start the engine and check for leaks.

DRIVE TRAIN

Automatic Transaxle Assembly

REMOVAL & INSTALLATION

1. Before servicing the vehicle, refer to the Precautions Section.
2. Drain the transaxle fluid.
3. Attach a support fixture to the engine lifting eyes.
4. Remove or disconnect the following:
   - Battery and tray
   - Air cleaner assembly
   - Mass Air Flow (MAF) sensor
   - Front wheels
   - Inner fender liner
   - Starter motor
   - Transaxle solenoid valve connector
   - Range switch connector
   - Wiring harness bracket
   - Turbine speed sensor connector
   - Vehicle Speed (VSS) sensor connector
   - Shift cable
   - Transaxle oil cooler hoses
   - Subframe transverse section
   - Axle halfshafts
   - Flywheel access panel
   - Torque converter
   - Left engine mount bracket
   - Subframe center section
   - Rear engine mount
   - Transaxle flange bolts. Support the transaxle.
5. Lower the transaxle from the vehicle.

To install:

6. Install or connect the following:
   - Transaxle. Tighten the flange bolts to 28–38 ft. lbs. (38–51 Nm).
   - Rear engine mount. Tighten the bracket bolts to 50–68 ft. lbs. (67–93 Nm) and the through bolt to 63–86 ft. lbs. (86–116 Nm).
   - Subframe center section. Tighten the bolts to 48–65 ft. lbs. (64–89 Nm) and the nuts to 50–67 ft. lbs. (67–93 Nm).
   - Left engine mount bracket. Tighten the bracket fasteners to 50–68 ft. lbs. (67–93 Nm) and the through bolt to 63–86 ft. lbs. (86–116 Nm).
   - Torque converter. Tighten the nuts to 26–36 ft. lbs. (35–49 Nm).
   - Flywheel access panel
   - Axle halfshafts
   - Subframe transverse section. Tighten the bolts to 69–97 ft. lbs. (94–131 Nm).
   - Transaxle oil cooler hoses
   - Shift cable

- VSS sensor connector
- Turbine speed sensor connector
- Wiring harness bracket
- Range switch connector
- Transaxle solenoid valve connector
- Starter motor
- Inner fender liner
- Front wheels
- MAF sensor
- Air cleaner assembly
- Battery and tray

7. Fill the transaxle to the correct level.
8. Start the engine and check for leaks.

Halfshaft

REMOVAL & INSTALLATION

Left

1. Before servicing the vehicle, refer to the Precautions Section.
2. Drain the transaxle fluid.
3. Remove or disconnect the following:
   - Front wheel
   - Wheel speed sensor
   - Hub locknut
   - Outer tie rod end
   - Lower ball joint
   - Stabilizer bar link
4. Separate the stub shaft from the hub and pry the inner joint from the transaxle.

To install:

→**Use a new circlip, split pin and locknut for assembly.**

5. Insert the stub shaft into the wheel hub.
6. Lubricate the oil seal with transaxle fluid, then push the axle halfshaft into the transaxle. Pull on the inner joint to confirm that the circlip is seated.
7. Install or connect the following:
   - Stabilizer bar link
   - Lower ball joint. Tighten the pinch bolt to 32–43 ft. lbs. (44–58 Nm).
   - Outer tie rod end. Tighten the nut to 24–32 ft. lbs. (32–44 Nm).
   - Hub locknut. Tighten the nut to 174–235 ft. lbs. (235–318 Nm) for 2002–03; 174–202 ft. lbs. (235–274 Nm) for 2004–06.
   - Wheel speed sensor
   - Front wheel

Right

✳✳ WARNING

Attempting to remove the right axle halfshaft while the center shaft sup-

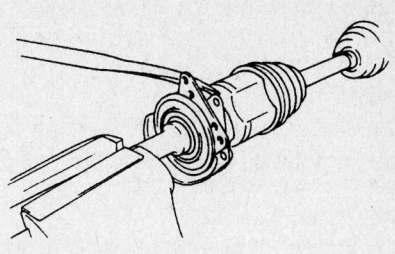

9308TG27

Separating the axle halfshaft from the center shaft

port bracket is installed may result in damage to the center shaft support bracket.

1. Before servicing the vehicle, refer to the Precautions Section.
2. Remove or disconnect the following:
   - Front wheel
   - Wheel speed sensor
   - Hub locknut
   - Brake caliper and rotor
   - Outer tie rod end
   - Lower ball joint
   - Strut bracket bolts
   - Steering knuckle. Separate the stub shaft from the wheel hub.
   - Center shaft support bracket
   - Axle halfshaft and center shaft assembly
3. Separate the axle halfshaft and the center shaft as follows:
   a. Step 1: Place the center shaft in a vise.
   b. Step 2: Insert a pry tool between the center shaft and the axle halfshaft.
   c. Step 3: Tap on the pry tool to separate the axle halfshaft from the center shaft.

To install:

→**Use a new split pin, locknut, and new circlips for assembly.**

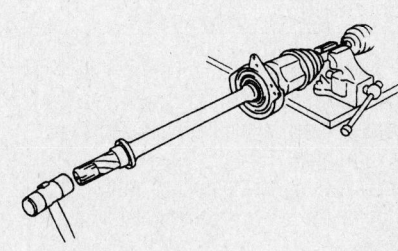

9308TG28

Installing the center shaft

4. Place the axle halfshaft in a vise and install the center shaft by tapping it with a plastic hammer as shown.

5. Lubricate the oil seal with transaxle fluid, then push the center shaft into the transaxle. Pull on the inner joint to confirm that the circlip is seated.

6. Install or connect the following:
- Center shaft support bracket. Tighten the nuts to 16–22 ft. lbs. (22–30 Nm).
- Steering knuckle. Guide the stub shaft into the wheel hub.
- Strut bracket bolts. Tighten the bolts to 76–90 ft. lbs. (103–122 Nm).
- Lower ball joint. Tighten the pinch bolt to 32–43 ft. lbs. (44–58 Nm).
- Outer tie rod end. Tighten the nut to 24–32 ft. lbs. (32–44 Nm).
- Hub locknut. Tighten the nut to 174–235 ft. lbs. (235–318 Nm) for 2002–03; 174–202 ft. lbs. (235–274 Nm) for 2004–06.
- Brake caliper and rotor. Tighten the caliper bracket bolts to 66–79 ft. lbs. (89–107 Nm).
- Wheel speed sensor
- Front wheel

7. Check the wheel alignment and adjust as necessary.

CV-Joint

OVERHAUL

Outer CV-Joint

The outer CV-joint is serviced with the axle halfshaft as an assembly. The outer CV-joint boot may be serviced by removing the inner joint.

Inner CV-Joint

1. Before servicing the vehicle, refer to the Precautions Section.
2. Remove or disconnect the following:
- Axle halfshaft from the vehicle
- Inner CV-joint boot clamps
- Housing retainer clip
- CV-joint housing
- CV-joint balls and cage
- Snapring
- CV-joint inner race
- CV-joint boot

To install:

➡**Use new snaprings, clips, and boot clamps for assembly.**

3. Install or connect the following:
- CV-joint boot
- CV-joint inner race
- Snapring
- CV-joint balls and cage
- CV-joint housing
- Housing retainer clip

4. Fill the CV-joint housing and boot with CV-joint grease and tighten the boot clamps.
5. Install the axle halfshaft.

Inner Tri-Pot Joint

1. Before servicing the vehicle, refer to the Precautions Section.
2. Remove or disconnect the following:
- Axle halfshaft from the vehicle
- Inner tri-pot joint boot clamps
- Tri-pot joint housing
- Snapring
- Tri-pot joint

To install:

➡**Use new snaprings, clips, and boot clamps for assembly.**

3. Install or connect the following:
- Tri-pot joint
- Snapring
- Tri-pot joint housing

4. Fill the tri-pot joint housing and boot with grease and tighten the boot clamps.
5. Install the axle halfshaft.

STEERING

Air Bag

✳✳ CAUTION

These vehicles are equipped with an air bag system. The system must be disarmed before performing service on, or around, system components, the steering column, instrument panel components, wiring and sensors. Failure to follow the safety precautions and the disarming procedure could result in accidental air bag deployment, possible injury and unnecessary system repairs.

PRECAUTIONS

Several precautions must be observed when handling the inflator module to avoid accidental deployment and possible personal injury.
- Never carry the inflator module by the wires or connector on the underside of the module.
- When carrying a live inflator module, hold securely with both hands, and ensure

that the bag and trim cover are pointed away.
- Place the inflator module on a bench or other surface with the bag and trim cover facing up.
- With the inflator module on the bench, never place anything on or close to the module which may be thrown in the event of an accidental deployment.

DISARMING

1. Turn the ignition switch to the **LOCK** position.
2. Disconnect the negative battery cable and wait at least 1 minute to allow the back-up power supply to deplete its stored power.
3. When repairs are complete, connect the negative battery cable.

Power Rack and Pinion Steering Gear

REMOVAL & INSTALLATION

1. Before servicing the vehicle, refer to the Precautions Section.

2. Attach a support fixture to the engine lifting eyes.
3. Remove or disconnect the following:
- Front wheels
- Wheel speed sensors
- Steering shaft pinch bolt
- Outer tie rod ends
- Subframe transverse section
- Subframe center section
- Power steering pressure and return lines
- Steering gear

To install:

4. Install or connect the following:
- Steering gear. Tighten the fasteners in sequence to 55–77 ft. lbs. (75–104 Nm).
- Power steering pressure and return lines
- Subframe center section. Tighten the bolts to 48–65 ft. lbs. (64–89 Nm) and the nuts to 50–68 ft. lbs. (67–93 Nm).
- Subframe transverse section. Tighten the bolts to 69–96 ft. lbs. (94–131 Nm).

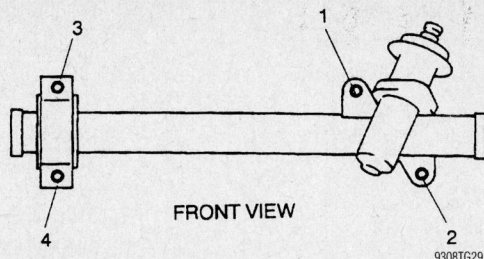

FRONT VIEW

9308TG29

Steering gear torque sequence

- Outer tie rod ends. Tighten the nuts to 24–32 ft. lbs. (32–44 Nm).
- Steering shaft pinch bolt. Tighten the bolt to 14–19 ft. lbs. (19–26 Nm).
- Wheel speed sensors
- Front wheels

5. Fill the power steering reservoir.
6. Check the wheel alignment and adjust as necessary.

FRONT SUSPENSION

Strut

REMOVAL & INSTALLATION

1. Before servicing the vehicle, refer to the Precautions Section.
2. Remove or disconnect the following:
   - Front wheel
   - Brake hose clip
   - Stabilizer bar link
   - Steering knuckle bolts
   - Upper strut mount nuts
   - Strut assembly

To install:

3. Install or connect the following:
   - Strut assembly. Tighten the upper strut mount nuts to 34–46 ft. lbs. (47–62 Nm) and the steering knuckle bolts to 76–90 ft. lbs. (103–122 Nm).
   - Stabilizer bar link. Tighten the nut to 32–44 ft. lbs. (44–60 Nm).
   - Brake hose clip
   - Front wheel

4. Check the wheel alignment and adjust as necessary.

Coil Spring

REMOVAL & INSTALLATION

1. Before servicing the vehicle, refer to the Precautions Section.
2. Remove the strut assembly from the vehicle.
3. Compress the coil spring and remove the piston rod nut.
4. Remove or disconnect the following:
   - Upper strut mount
   - Strut mount bearing
   - Spring upper seat
   - Coil spring

To install:

5. Install or connect the following:
   - Coil spring
   - Spring upper seat
   - Strut mount bearing

- Upper strut mount. Tighten the piston rod nut to 66–94 ft. lbs. (90–127 Nm).

6. Remove the spring compressor and install the strut assembly to the vehicle.
7. Check the wheel alignment and adjust as necessary.

Lower Ball Joint

REMOVAL & INSTALLATION

The lower ball joint is serviced with the lower control arm as an assembly.

Lower Control Arm

REMOVAL & INSTALLATION

1. Before servicing the vehicle, refer to the Precautions Section.
2. Support the arm.
3. Remove or disconnect the following:
   - Front wheel
   - Pivot bolt
   - Dynamic damper
   - Ball joint bolt
   - Ball joint bracket
   - Nut
   - Lower arm

4. Installation is the reverse of removal. Observe the following torques:
   - Nut: 76–97 ft. lbs. (103–131 Nm)
   - Bracket bolt: 32–43 ft. lbs. (43–59 Nm)
   - Pivot bolt: 74–101 ft. lbs. (101–137 Nm)

CONTROL ARM BUSHING REPLACEMENT

1. Before servicing the vehicle, refer to the Precautions Section.
2. Remove the control arm from the vehicle.
3. Mark the control arm to indicate the alignment of the rear bushing as shown.
4. Remove the control arm bushings with a hydraulic press.

To install:

5. Lubricate the control arm bushings with liquid soap.
6. If replacing the rear bushing, align the direction marks as shown.
7. Press the bushings into the control arm until the bushing flange contacts the housing edge of the control arm.
8. Install the control arm to the vehicle.
9. Check the wheel alignment and adjust as necessary.

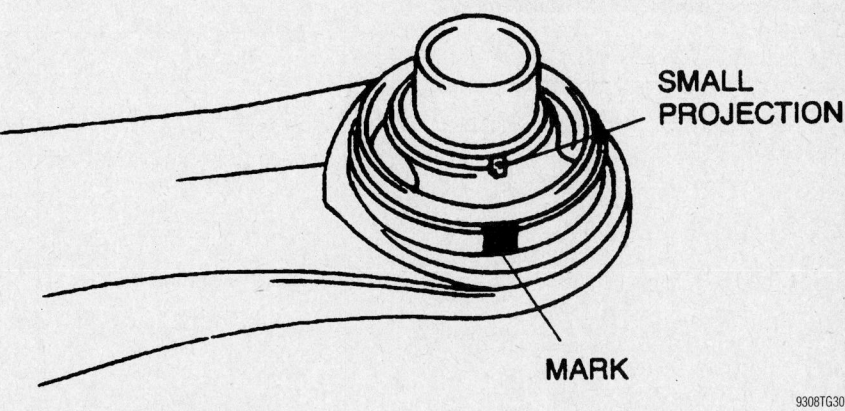

SMALL PROJECTION

MARK

9308TG30

Rear bushing alignment marks

Wheel Bearing

ADJUSTMENT

1. Before servicing the vehicle, refer to the Precautions Section.
2. Remove or disconnect the following:
   • Front wheel
   • Brake caliper and rotor
3. Position a dial indicator gauge against the wheel hub. Push and pull the wheel hub in and out and measure the end-play of the wheel bearing.
4. End-play should not exceed 0.002 in. (0.05mm).
5. If end-play is excessive, replace the hub retainer locknut and tighten it to specification. Recheck the end-play.
6. If end-play is not within specification, replace the wheel bearing assembly.

REMOVAL & INSTALLATION

1. Before servicing the vehicle, refer to the Precautions Section.
2. Remove or disconnect the following:
   • Front wheel
   • Brake caliper and rotor
   • Wheel speed sensor
   • Outer tie rod end
   • Lower ball joint
   • Hub retainer locknut
   • Strut bracket bolts
   • Steering knuckle
   • Inner oil seal
   • Hub
   • Snapring
   • Wheel bearing cartridge

To install:

➡ **Use new locknuts, split pins and oil seals for assembly.**

3. Install or connect the following:
   • Wheel bearing cartridge
   • Snapring
   • Hub
   • Inner oil seal
   • Steering knuckle. Tighten the strut bracket bolts to 76–90 ft. lbs. (103–122 Nm).
   • Hub retainer locknut. Tighten the nut to 174–235 ft. lbs. (236–318 Nm) for 2002; 174–202 ft. lbs. (235–274 Nm) for 2003–06.
   • Lower ball joint. Tighten the pinch bolt to 32–43 ft. lbs. (44–58 Nm).
   • Outer tie rod end. Tighten the nut to 24–32 ft. lbs. (32–44 Nm).
   • Wheel speed sensor. Tighten the bolt to 14–18 ft. lbs. (19–25 Nm).
   • Brake caliper and rotor
   • Front wheel

REAR SUSPENSION

Shock Absorber

REMOVAL & INSTALLATION

1. Before servicing the vehicle, refer to the Precautions Section.
2. Support the rear axle with a jack or stands.
3. Remove or disconnect the following:
   • Rear wheel
   • Shock absorber

To install:

4. Install or connect the following:
   • Shock absorber. Tighten the upper nut to 56–76 ft. lbs. (76–102 Nm). Tighten the lower bolt to 71–94 ft. lbs. (97–127 Nm) for 2002; 76–102 ft. lbs. (103–139 Nm) for 2003–06.
   • Rear wheel

Coil Spring

REMOVAL & INSTALLATION

1. Before servicing the vehicle, refer to the Precautions Section.

2. Support the vehicle at the frame and support the axle with a jack.
3. Remove or disconnect the following:
   • Rear wheels
   • Lateral rod
   • Shock absorber
4. Lower the rear axle and remove the coil springs.

To install:

5. Place the coil springs on the spring seats and raise the axle into position.
6. Install or connect the following:
   • Shock absorber. Tighten the upper nut to 56–76 ft. lbs. (76–102 Nm) and the lower bolt to 71–94 ft. lbs. (97–127 Nm).
   • Lateral rod. Tighten the fastener to 76–101 ft. lbs. (102–137 Nm).
   • Rear wheels

Wheel Bearing

ADJUSTMENT

1. Before servicing the vehicle, refer to the Precautions Section.
2. Remove the rear wheel.
3. Position a dial indicator gauge against the wheel hub. Push and pull the wheel hub in and out and measure the end-play of the wheel bearing.
4. End-play should not exceed 0.002 in. (0.05mm).
5. If end-play is excessive, replace the hub retainer locknut and tighten it to specification. Recheck the end-play.
6. If end-play is not within specification, replace the wheel bearing assembly.

REMOVAL & INSTALLATION

1. Before servicing the vehicle, refer to the Precautions Section.
2. Remove or disconnect the following:
   • Rear wheel
   • Brake drum
   • Dust cap
   • Hub retaining lock nut
   • Wheel bearing and hub assembly

To install:

3. Install or connect the following:
   • Wheel bearing and hub assembly. Tighten the locknut to 131–173 ft. lbs. (177–235 Nm).
   • Dust cap
   • Brake drum
   • Rear wheel

FRONT BRAKES

Disc Brake Caliper

REMOVAL AND INSTALLATION

1. Before servicing the vehicle, refer to the Precautions Section.

2. Remove or disconnect the following:
- Wheel assembly
- Banjo bolt and disconnect the brake hose from the caliper. Plug the hose to prevent fluid leakage.
- Caliper mounting bolts

3. Installation is the reverse of the removal procedure. Lubricate the caliper mounting bolts or bolt and pin prior to installation.

4. Tighten the caliper mounting bolts to 61–69 ft. lbs. (83–93 Nm).

5. Bleed the brake system.

Disc Brake Pads

REMOVAL AND INSTALLATION

1. Before servicing the vehicle, refer to the Precautions Section.

2. Remove or disconnect the following:
- Wheel assembly
- Lower lock pin bolt from the caliper

3. Rotate the caliper upward and remove the brake pads, shims, guide plates and if equipped, the springs.

To install:

4. Remove the master cylinder reservoir cap and remove about ½ of the fluid from the reservoir.

5. Using a large C-clamp and piece of wood, depress the caliper piston(s) until they bottom in their bores.

6. Install the shims, guide plates, new pads and if removed, the springs.

7. Reposition the caliper and install the caliper mounting bolt. Torque the mounting bolt to 61–69 ft. lbs. (83–93 Nm).

8. Install the wheels, lower the vehicle, refill the master cylinder and depress the brake pedal a few times to restore pressure. Bleed the system if required.

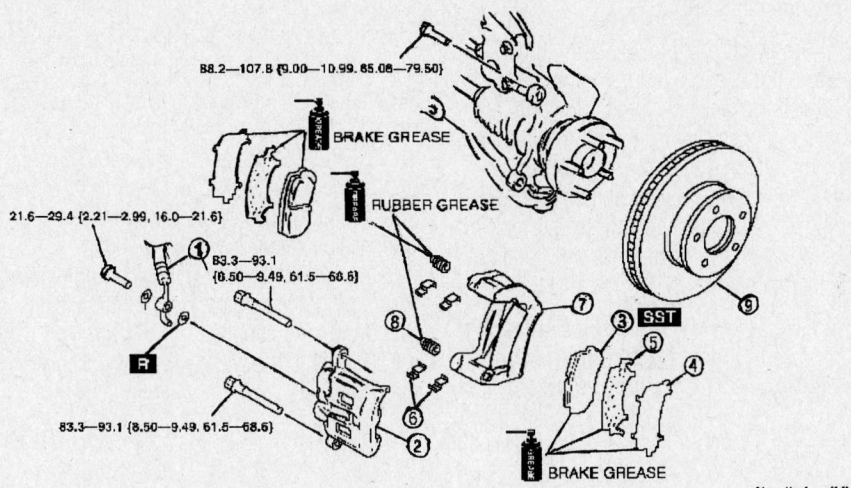

1. Flexible Hose
2. Caliper
3. Disc Pad
4. Outer Shim
5. Inner Shim
6. Guide Plate
7. Mounting Support
8. Dust Boot
9. Disc Plate

09482_MMPV_G0030

Front disc brake assembly

REAR BRAKES

Disc Brake Caliper

REMOVAL AND INSTALLATION

1. Before servicing the vehicle, refer to the Precautions Section.

2. Raise and support the rear of the vehicle.

3. Remove the wheel(s).

4. If the caliper is being replaced or repaired, disconnect and plug the brake hose at the caliper.

5. Remove the caliper pins and lift off the caliper.

6. Installation is the reverse of removal. Torque the caliper bolts to 16–23 ft. lbs. (21–31 Nm).

Disc Brake Pads

REMOVAL AND INSTALLATION

1. Before servicing the vehicle, refer to the Precautions Section.

2. Raise and support the rear of the vehicle.

3. Remove the wheel(s).

4. Remove the caliper.

5. Remove the brake pads, shims and guide plates.

6. Installation is the reverse of removal. Torque the caliper bolts to 16–23 ft. lbs. (21–31 Nm).

Brake Drum

REMOVAL AND INSTALLATION

1. Raise and safely support the vehicle. Remove the wheel and tire assembly.

2. Remove the screws, if equipped, and remove the brake drum.

3. Inspect the brake drum surface for wear, scoring and runout. Machine or replace, as necessary.

To install:

4. Install the brake drum and secure in place with the screws, if equipped. Torque the screws to 10 ft. lbs. (14 Nm).

5. Adjust the rear brakes.

6. Install the wheel. Lower the vehicle.

Brake Shoes

REMOVAL AND INSTALLATION

1. Raise and safely support the vehicle. Remove the wheel and tire assembly and the brake drum.

2. Pull backward on the adjusting lever cable to disengage the adjusting lever from the adjusting screw. Move the outboard side of the adjusting screw upward and back off the pivot nut as far as it will go.

3. Pull the adjusting lever, cable and automatic adjuster spring down and toward the rear to unhook the pivot hook from the large hole in the secondary shoe web. Do not pry the pivot hook from the hole.

4. Remove the automatic adjuster spring and adjusting lever.

5. Remove the secondary shoe-to-anchor spring using a suitable brake spring removal/installation tool. Using the tool, remove the primary shoe-to-anchor spring and unhook the cable anchor. Remove the anchor pin plate, if equipped.

6. Remove the cable guide from the secondary shoe.

7. Remove the shoe hold-down springs, shoes, adjusting screw, pivot nut and socket. Note the color and position of each hold-down spring so they can be reassembled in the same position.

8. Remove the parking brake link and spring. Disconnect the parking brake cable from the parking brake lever.

9. Remove the secondary brake shoe. Remove the retainer clip and spring washer and remove the parking brake lever.

To install:

10. Clean the backing plate ledge pads and sand lightly. Apply a light coating of high temperature lithium grease to the points where the brake shoes touch the backing plate. Lubricate the adjusting cable eye and the anchor pin area.

11. Install the parking brake lever on the secondary shoe.

12. Position the brake shoes on the backing plate and install the hold-down spring pins, springs and cups. Install the parking brake link, spring and washer. Connect the parking brake cable to the parking brake lever.

13. Install the anchor pin plate, if equipped, and place the cable anchor over the anchor pin with the crimped side toward the backing plate.

14. Install the primary shoe-to-anchor spring using the brake spring removal/installation tool.

15. Install the cable guide on the sec-

ondary shoe with the flanged hole fitted into the hole in the secondary shoe. Thread the cable around the cable guide groove.

➡**Make sure the cable is positioned in the groove and not between the guide and shoe web.**

16. Install the secondary shoe-to-anchor (long) spring.

➡**Make sure the cable end is not cocked or binding on the anchor pin when installed. All parts should be flat on the anchor pin.**

17. Apply high temperature lithium grease to the threads and the socket end of the adjusting screw. Turn the adjusting screw into the adjusting pivot nut to the end of the threads and then loosen, ½ turn.

18. Place the adjusting socket on the screw and install the assembly between the shoe ends with the adjusting screw nearest the secondary shoe.

➡**Be sure to install the adjusting screw on the same side of the vehicle from which it came. To prevent incorrect**

installation, the socket end of each adjusting screw is stamped with R or L, to indicate installation on the right or left side of the vehicle. The adjusting pivot nuts have lines machined around the body of the nut, 2 lines indicating the right side nut and 1 line indicating the left side nut.

19. Hook the cable hook into the hole in the adjusting lever from the outboard plate side. The adjusting levers are also stamped with an **R** or **L** to indicate right or left side installation.

20. Place the hooked end of the adjuster spring in the large hole in the primary shoe web and connect the loop end of the spring to the adjuster lever hole.

21. Pull the adjuster lever, cable and automatic adjuster spring down toward the rear to engage the pivot hook in the large hole in the secondary shoe web.

22. Adjust the brake shoes using either a brake adjustment gauge or manually with the drums installed.

23. Install the wheels, and lower the vehicle.

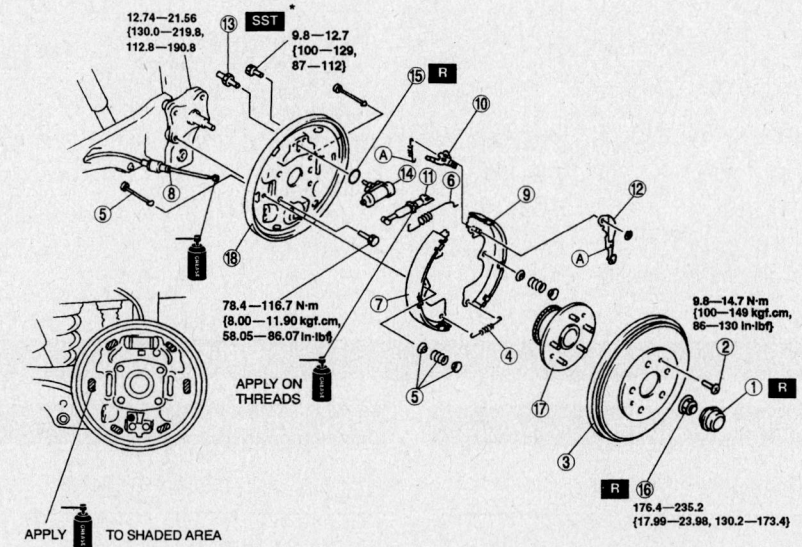

1	Hub cap
2	Screw
3	Brake drum
4	Return spring
5	Hold pin and hold spring
6	Anti-rattle spring
7	Leading shoe
8	Parking brake cable
9	Trailing shoe

10	Ajuster lever
11	Adjuster component
12	Operating lever
13	Brake pipe
14	Wheel cylinder
15	O-ring
16	Locknut
17	Wheel hub
18	Backing plate

42356MMPVG05

Exploded view of the rear brake shoes and components

MAZDA

RX-8

9

SPECIFICATIONS AND MAINTENANCE CHARTS

VEHICLE AND ENGINE IDENTIFICATION CHART

Code ①	Liters	Cu. In.	Cyl.	Fuel Sys.	Engine Type	Eng. Mfg.
N	1.3	79	③	MPI	Rotary	Mazda
3	1.3	79	③	MPI	Rotary	Mazda

Engine spans Code① through Eng. Mfg.

Code ②	Year
4	2004
5	2005
6	2006

Model Year spans Code② and Year.

MPI: Multi point fuel injection

① 8th position of VIN

② 10th position of VIN

③ Twin rotary

09482_RX8_C0001

GENERAL ENGINE SPECIFICATIONS

Year	Engine Displacement Liters	Engine VIN	Net Horsepower @ rpm	Net Torque @ rpm (ft. lbs.)	Bore x Stroke (in.)	Compression Ratio	Oil Pressure @ rpm
2004	1.3	N	197@7200	164@5000	①	10.0:1	51@3000
	1.3	3	238@8500	159@5500	①	10.0:1	51@3000
2005	1.3	N	197@7200	164@5000	①	10.0:1	51@3000
	1.3	3	238@8500	159@5500	①	10.0:1	51@3000
2006	1.3	N	212@7500	159@5500	①	10.0:1	51@3000
	1.3	3	232@8500	159@5500	①	10.0:1	51@3000

① Rotary chamber not measured.

09482_RX8_C0002

ENGINE TUNE-UP SPECIFICATIONS

Year	Engine Displacement Liters	Engine VIN	Spark Plugs Gap (in.)	Ignition Timing (deg.) MT	Ignition Timing (deg.) AT	Fuel Pump (psi)	Idle Speed (rpm) MT	Idle Speed (rpm) AT	Valve Clearance In.	Valve Clearance Ex.
2004	1.3	①	0.046-0.049	②	②	③	750-850	760-860	NA	NA
2005	1.3	①	0.046-0.049	②	②	③	750-850	760-860	NA	NA
2006	1.3	①	0.046-0.049	②	②	③	750-850	760-860	NA	NA

NOTE: The Vehicle Emission Control Information label often reflects specification changes changes made during production.

The label figures must be used if they differ from those in this chart.

NA: Not applicable.

① N & 3 Engines

② Controlled by the Powertrain Control Module and cannot be adjusted.

③ Fuel line hold pressure 55-65 psi

09482_RX8_C0003

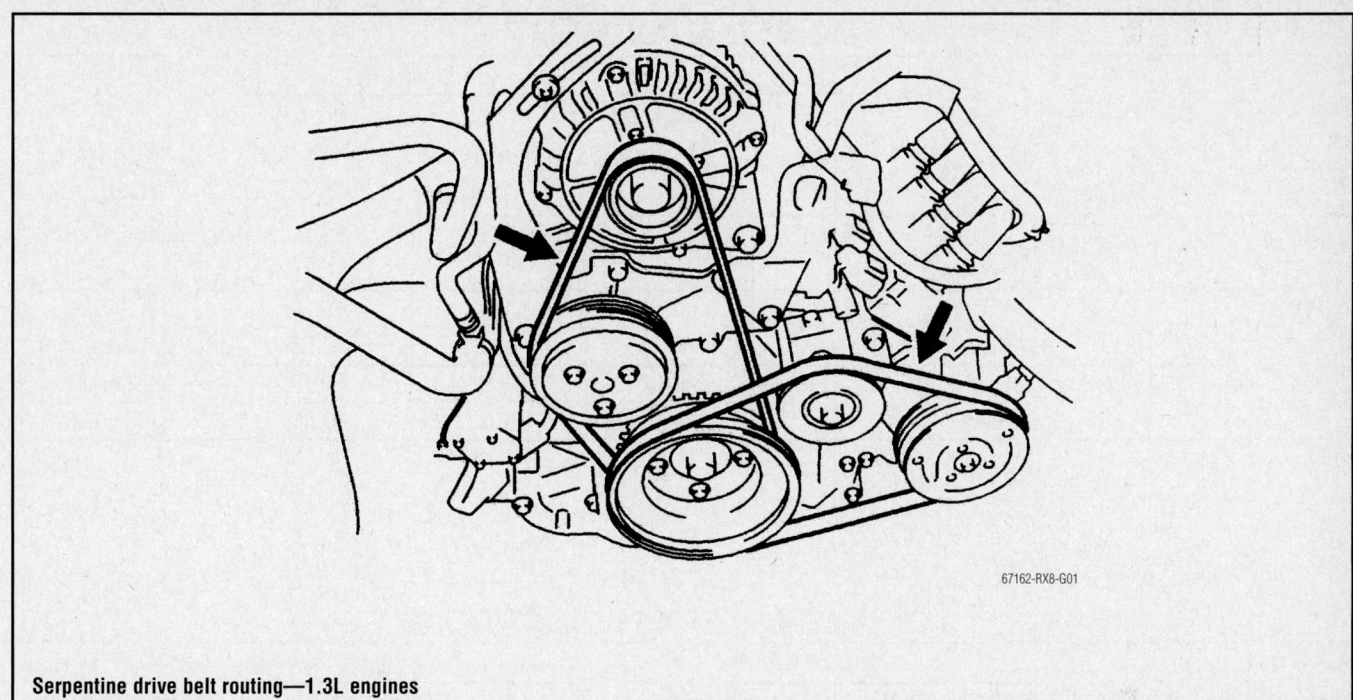

67162-RX8-G01

Serpentine drive belt routing—1.3L engines

CAPACITIES

Year	Model	Engine Displacement Liters	Engine ID/VIN	Engine Oil with Filter (qts.)	Transmission (pts.) 6-Spd	Transmission (pts.) Auto.	Drive Axle Front (pts.)	Drive Axle Rear (pts.)	Fuel Tank (gal.)	Cooling System (qts.)
2004	RX-8	1.3	①	3.7	3.7	18.4	—	2.7	15.9	8.7
2005	RX-8	1.3	①	3.7	3.7	18.4	—	2.7	15.9	8.7
2006	RX-8	1.3	①	3.7	3.7	18.4	—	2.7	15.9	8.7

NOTE: All capacities are approximate. Add fluid gradually and check to be sure a proper fluid level is obtained.

① N and 3 Engines

09482_RX8_C0004

TORQUE SPECIFICATIONS
All readings in ft. lbs.

Year	Engine VIN	Engine Displacement Liters	Cylinder Head Bolts	Main Bearing Bolts	Rod Bearing Bolts	Crankshaft Damper Bolts	Flywheel Locknut	Manifold Intake	Manifold Exhaust	Spark Plugs	Oil Pan Drain Plug
2004	①	1.3	NA	NA	NA	NA	30-42	14-19	31-44	9-13	22-29
2005	①	1.3	NA	NA	NA	NA	30-42	14-19	31-44	9-13	22-29
2006	①	1.3	NA	NA	NA	NA	30-42	14-19	31-44	9-13	22-29

NA: Not applicable

① N and 3 Engines

09482_RX8_C0005

WHEEL ALIGNMENT

Year	Model		Caster Range (+/-Deg.)	Caster Preferred Setting (Deg.)	Camber Range (+/-Deg.)	Camber Preferred Setting (Deg.)	Toe-in (Deg.)
2004	RX-8	Front	1.0	+6.06	1.0	+0.04	0.11+/-21
		Rear	—	—	1.0	-0.56	0.16+/-20
2005	RX-8	Front	1.0	+6.06	1.0	+0.04	0.11+/-21
		Rear	—	—	1.0	-0.56	0.16+/-20
2006	RX-8	Front	1.0	+6.06	1.0	+0.04	0.11+/-21
		Rear	—	—	1.0	-0.56	0.16+/-20

09482_RX8_C0006

TIRE AND WHEEL SPECIFICATIONS

Year	Model	OEM Tires Standard	OEM Tires Optional	Tire Pressures (psi) Front	Tire Pressures (psi) Rear	Wheel Size	Wheel Lug Nut Torque (Ft. Lbs.)
2004	RX-8	P225/55R16	P225/45R18	32	32	①	65-87
2005	RX-8	P225/55R16	P225/45R18	32	32	①	65-87
2006	RX-8	P225/55R16	P225/45R18	32	32	①	65-87

OEM: Original Equipment Manufacturer

PSI: Pounds Per Square Inch

① Not available

09482_RX8_C0007

BRAKE SPECIFICATIONS

All measurements in inches unless noted

Year	Model		Brake Disc Original Thickness	Brake Disc Minimum Thickness	Brake Disc Maximum Runout	Minimum Lining Thickness Front	Minimum Lining Thickness Rear	Brake Caliper Bracket Bolts (ft. lbs.)	Brake Caliper Mounting Bolts (ft. lbs.)
2004	RX-8	F	①	0.870	0.002	0.790	—	58-75	16-23
		R	①	0.630	0.002	—	0.079	36-51	16-23
2005	RX-8	F	①	0.870	0.002	0.790	—	58-75	16-23
		R	①	0.630	0.002	—	0.079	36-51	16-23
2006	RX-8	F	①	0.870	0.002	0.790	—	58-75	16-23
		R	①	0.630	0.002	—	0.079	36-51	16-2

① Not available

09482_RX8_C0008

SCHEDULED MAINTENANCE INTERVALS
MAZDA RX-8

TO BE SERVICED	TYPE OF SERVICE	7.5	15	22.5	30	37.5	45	52.5	60	67.5	75	82.5	90	97.5	105	113	120
VEHICLE MILEAGE INTERVAL (x1000)																	
Engine oil & filter	R	✓	✓	✓	✓	✓	✓	✓	✓	✓	✓	✓	✓	✓	✓	✓	✓
Cabin air filter	R				✓				✓				✓				✓
Engine coolant strength hoses & clamps	S/I				✓				✓				✓				✓
Air cleaner filter	R					✓							✓				✓
Brake fluid	R			✓					✓				✓				✓
Engine coolant ①	R								✓								✓
Spark plugs	R					✓					✓						
Drive belts	S/I			✓					✓				✓				✓
Exhaust system & heat shields	S/I			✓					✓				✓				✓
Manual transmission oil	R								✓								✓
Rear differential oil									✓								✓
Front & rear brakes	S/I		✓			✓				✓			✓				✓
Fuel filter	R													✓			✓

R: Replace   S/I: Service or Inspect

① Engine coolant: change initially at 60,000 miles and every 24 months thereafter.

FREQUENT OPERATION MAINTENANCE (SEVERE SERVICE)

If a vehicle is operated under any of the following conditions it is considered severe service:

- Extremely dusty areas.
- 50% or more of the vehicle operation is in 32°C (90°F) or higher temperatures, or constant operation in temperatures below 0°C (32°F).
- Prolonged idling (vehicle operation in stop and go traffic.
- Frequent short running periods (engine does not warm to normal operating temperatures).
- Police, taxi, delivery usage or trailer towing usage.

Engine oil & filter: replace every 5000 miles.

Air cleaner filter: change every 35,000 miles.

Engine coolant: replace every 60,000 miles.

Exhaust system: check every 30,000 miles.

09482_RX8_C0009

ENGINE REPAIR

Distributor

REMOVAL

1.3L Engine

1. The 1.3L engine is equipped with a distributorless ignition system.

Alternator

REMOVAL

1. Before servicing the vehicle, refer to the Precautions Section.
2. Remove or disconnect the following:
   - Negative battery cable
   - Engine cover
   - Rear engine cross brace
   - Intake air duct
   - Accessory drive belt
   - Electrical connectors from the alternator
   - Alternator bolts
   - Alternator

INSTALLATION

1. Installation is the reverse of the removal procedure, noting the following:
   a. Tighten the left side engine cross

brace nut to 40 ft. lbs. (Nm) and the right side nut to 16 ft. lbs.

Ignition Timing

ADJUSTMENT

➡ **The ignition timing cannot be adjusted. To check the timing, connect a scan tool to the Data Link Connector (DLC). Connect a timing light to the front rotor housing on the leading side. Place the scan tool in the test mode. Start the engine and verify that the white alignment marks on the front cover and eccentric shaft plate are aligned.**

Engine Assembly

REMOVAL & INSTALLATION

➡ **The procedure for pulling the engine requires removing the transaxle and front crossmember along with it. A suitable support fixture that will support the entire assembly must be used for removal.**

1. Before servicing the vehicle, refer to the Precautions Section.

2. Properly relieve the fuel system pressure.
3. Drain the engine oil.
4. Drain the transaxle fluid.
5. Drain the cooling system.
6. Place the front wheels in the straight ahead position.
7. Raise and support the vehicle.
8. Remove or disconnect the following:
   - Front wheels
   - Engine cover
   - Rear engine cross brace
   - Battery cover, battery, box and tray
   - Air cleaner, intake duct and insulator
   - Powertrain Control Module (PCM)
   - Secondary Air Injection (AIR) pump
   - Brake vacuum hose
   - Charcoal canister connector
   - Fuel lines
   - Ignition coil
   - Accessory drive belts
   - A/C compressor and wire aside
   - Engine wiring harness from main fuse block
   - Engine splash shield
   - ABS speed sensor connector
   - Radiator, heater and coolant tank hoses
   - Selector link on automatic transmission
   - Manual transmission shift lever
   - Clutch slave cylinder and wire aside
   - Steering shaft pinch bolt. Refer to the accompanying illustration for location.
   - Oil lines. Refer to the accompanying illustration for location.
   - Transmission cooler lines. Refer to the accompanying illustration for location.
   - Brake caliper. Refer to the accompanying illustration for location.
   - Lower strut bolt. Refer to the accompanying illustration for location.
   - Rear crossmembers. Refer to the accompanying illustration for location.
   - Catalytic converter and exhaust system. Refer to the accompanying illustration for location.
   - Heat insulator. Refer to the accompanying illustration for location.
   - Driveshaft. Refer to the accompanying illustration for location.
   - Front crossmember. Refer to the accompanying illustration for location.

9.8—14.7 {100—149, 86.8—130.1}

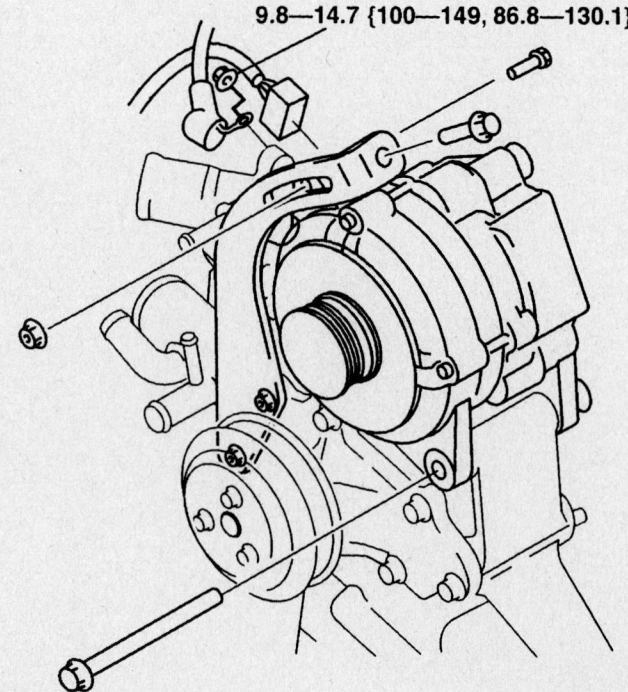

67162-RX8-G02

Alternator mounting—1.3L engine

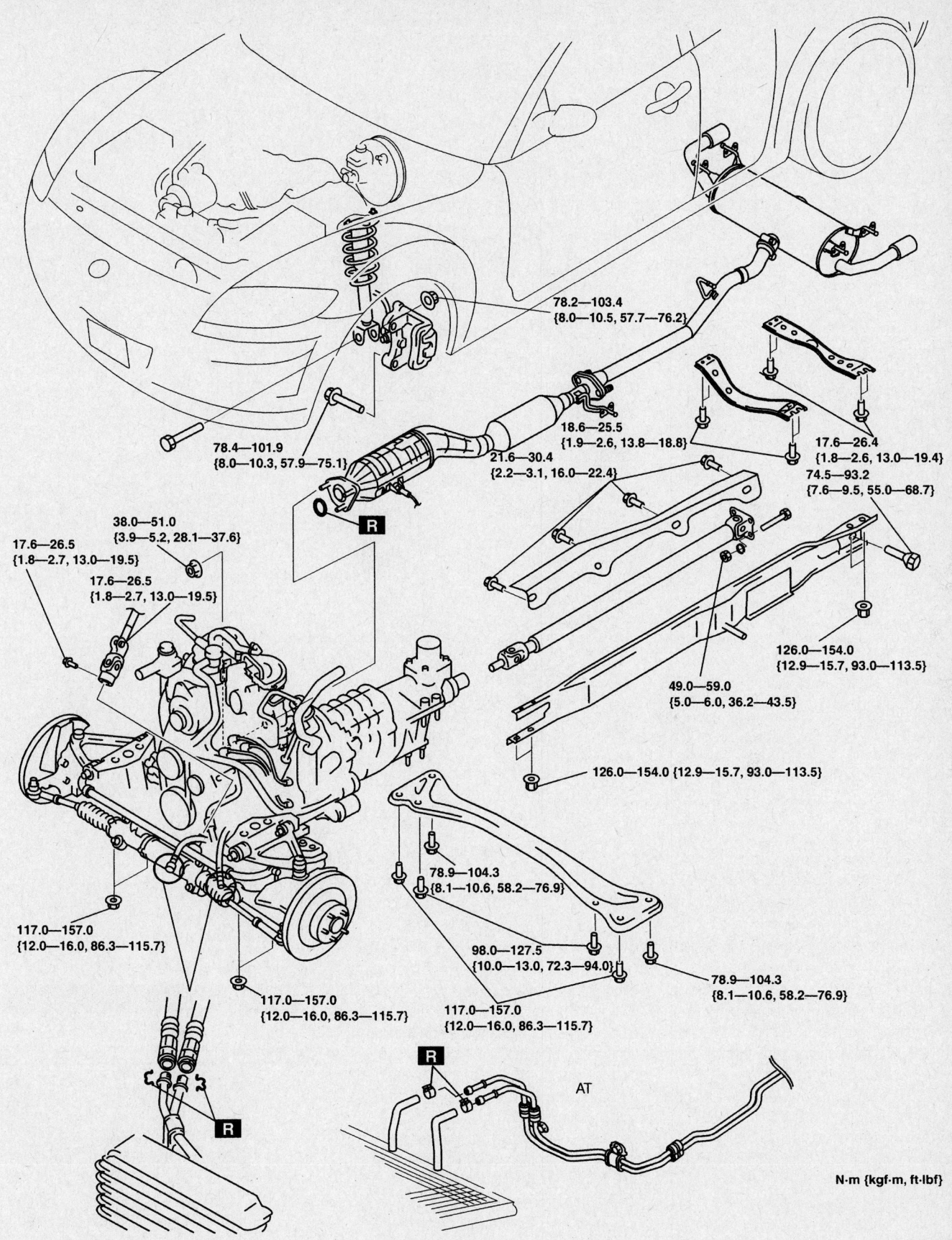

78.2—103.4
{8.0—10.5, 57.7—76.2}

78.4—101.9
{8.0—10.3, 57.9—75.1}

18.6—25.5
{1.9—2.6, 13.8—18.8}

21.6—30.4
{2.2—3.1, 16.0—22.4}

17.6—26.4
{1.8—2.6, 13.0—19.4}

74.5—93.2
{7.6—9.5, 55.0—68.7}

38.0—51.0
{3.9—5.2, 28.1—37.6}

17.6—26.5
{1.8—2.7, 13.0—19.5}

17.6—26.5
{1.8—2.7, 13.0—19.5}

126.0—154.0
{12.9—15.7, 93.0—113.5}

49.0—59.0
{5.0—6.0, 36.2—43.5}

126.0—154.0 {12.9—15.7, 93.0—113.5}

117.0—157.0
{12.0—16.0, 86.3—115.7}

78.9—104.3
{8.1—10.6, 58.2—76.9}

117.0—157.0
{12.0—16.0, 86.3—115.7}

98.0—127.5
{10.0—13.0, 72.3—94.0}

78.9—104.3
{8.1—10.6, 58.2—76.9}

117.0—157.0
{12.0—16.0, 86.3—115.7}

AT

N·m {kgf·m, ft·lbf}

67162-RX8-G03

Location of the engine mounting components and their torque specifications—RX-8 1.3L engine

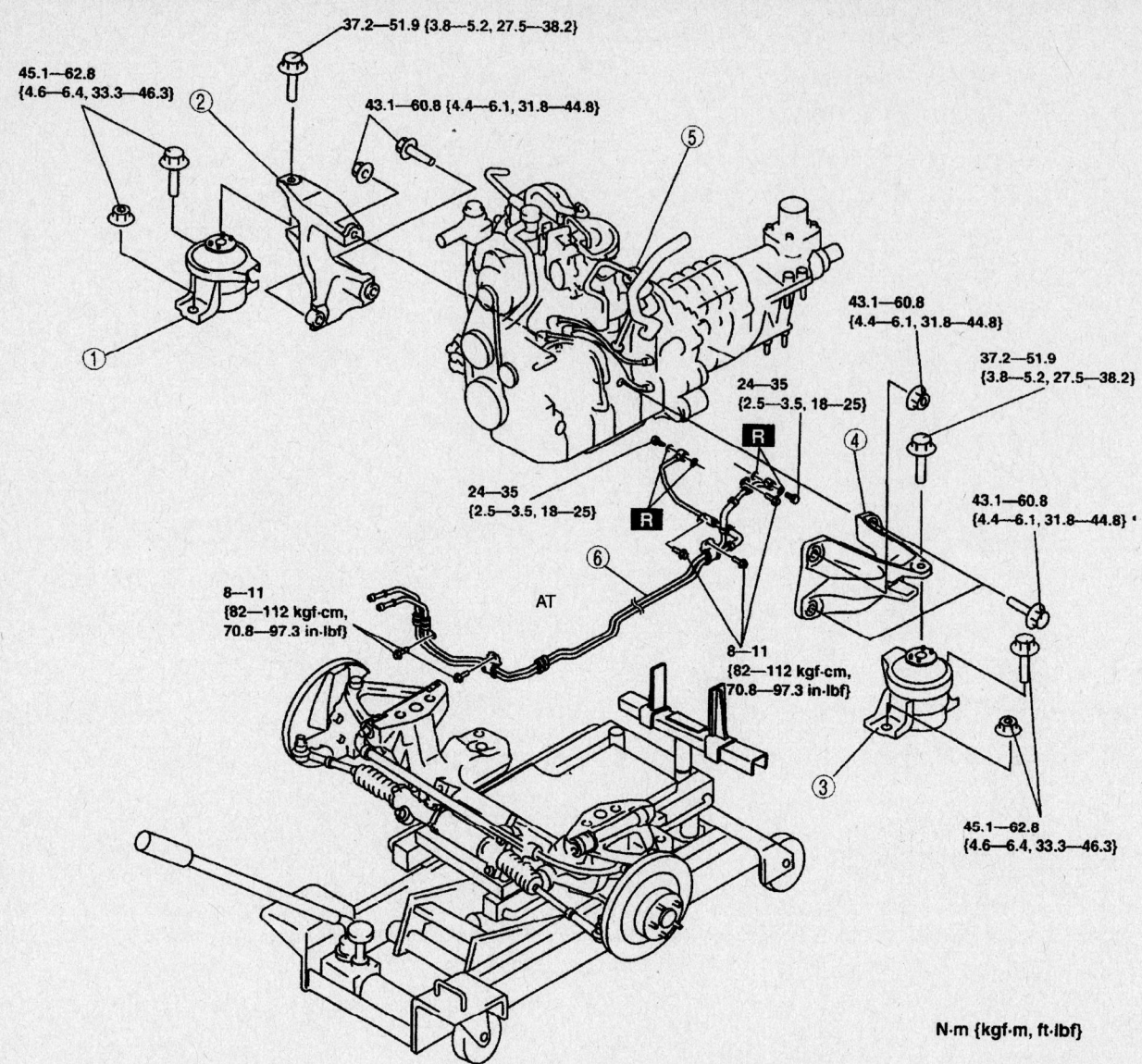

37.2—51.9 {3.8—5.2, 27.5—38.2}

45.1—62.8 {4.6—6.4, 33.3—46.3}

43.1—60.8 {4.4—6.1, 31.8—44.8}

43.1—60.8 {4.4—6.1, 31.8—44.8}

37.2—51.9 {3.8—5.2, 27.5—38.2}

24—35 {2.5—3.5, 18—25}

43.1—60.8 {4.4—6.1, 31.8—44.8}

24—35 {2.5—3.5, 18—25}

8—11 {82—112 kgf·cm, 70.8—97.3 in·lbf}

AT

8—11 {82—112 kgf·cm, 70.8—97.3 in·lbf}

45.1—62.8 {4.6—6.4, 33.3—46.3}

N·m {kgf·m, ft·lbf}

1 Engine mount rubber (RH)
2 Engine mount bracket (RH)
3 Engine mount rubber (LH)
4 Engine mount bracket (LH)
5 Engine, transaxle
6 AT oil cooler pipe

67162-RX8-G04

Separating the engine/transmission assembly from the front crossmember— RX-8 1.3L engine

- Power plant frame. Refer to the accompanying illustration for location.
- Engine, transmission and front suspension frame assembly

9. Use an engine hoist and separate the engine/transmission assembly from the front suspension crossmember using the steps shown in the accompanying illustration.

To install:

10. Installation is the reverse of removal. Tighten the fasteners to the specifications shown in the accompanying illustrations.

11. When possible, leave the engine mounting nuts/bolts loose (hand tight) until all mounts are aligned and bolted. This may help in aligning the engine and transmission assembly in the vehicle.

12. Fill the engine and the transaxle with the proper type and amount of fluids. Fill the cooling system.

13. Connect the battery cables.

14. On models with dynamic suspension, perform the steering angle sensor initialization procedure.

15. Check the ignition timing and the idle speed.

16. Check the front wheel alignment.

17. Check all fluid levels.

Water Pump

REMOVAL & INSTALLATION

1. Before servicing the vehicle, refer to the Precautions Section.

2. Drain the cooling system.

3. Remove or disconnect the following:

- Battery cables, battery box and tray
- Loosen the water pump pulley bolt
- Drive belt
- Water pump pulley
- Front engine hangar

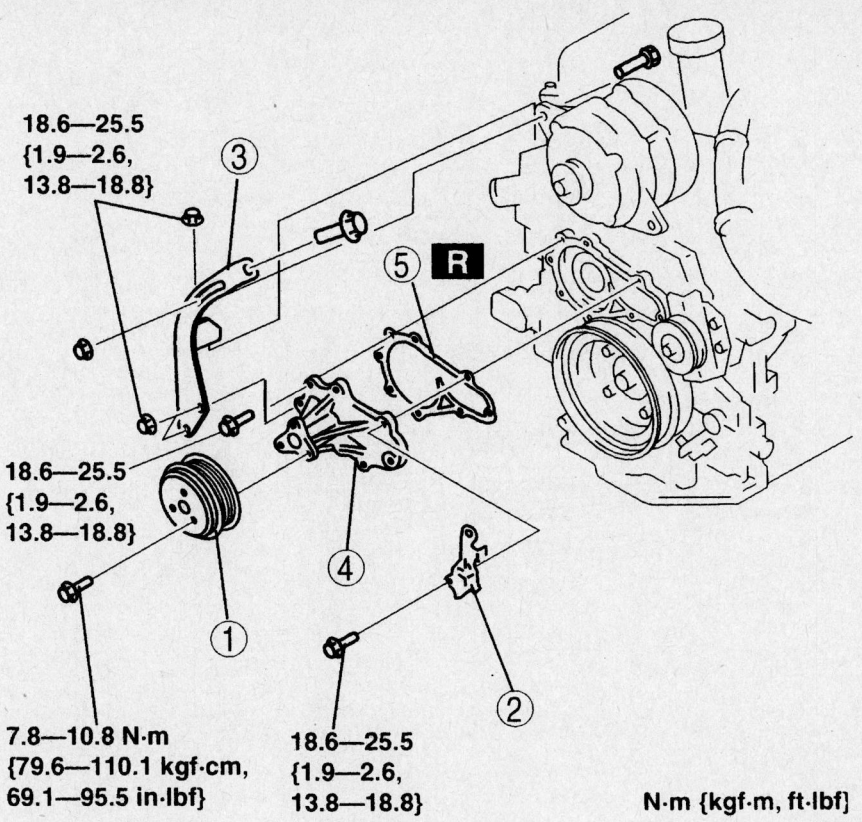

18.6—25.5
{1.9—2.6,
13.8—18.8}

18.6—25.5
{1.9—2.6,
13.8—18.8}

7.8—10.8 N·m
{79.6—110.1 kgf·cm,
69.1—95.5 in·lbf}

18.6—25.5
{1.9—2.6,
13.8—18.8}

N·m {kgf·m, ft·lbf}

67162-RX8-G05

Exploded view of the water pump assembly—1.3L engine

- Alternator strap
- Water pump and gasket

To install:

4. Installation is the reverse of removal. Tighten the fasteners to the specifications shown in the accompanying illustration.

5. Fill and bleed the cooling system.

6. On models with dynamic suspension, perform the steering angle sensor initialization procedure.

7. Start the engine, check for leaks and repair if necessary.

Heater Core

REMOVAL AND INSTALLATION

1. Before servicing the vehicle, refer to the Precautions Section.

2. Place the ignition switch in the **LOCK** position.

3. Disconnect the negative battery cable.

✳✳ CAUTION

After disconnecting the battery, wait for more than 1 minute for the SAS to deplete its stored energy.

4. Drain the cooling system into a clean container for reuse.

5. Disconnect the heater hoses from the heater core.

6. Discharge and recover the air conditioning system refrigerant.

7. Remove or disconnect the following:
- Center console upper panel
- Ash tray panel
- Cigar lighter connector
- Ash tray light
- Storage compartment
- Front and rear center consoles
- Console under cover
- Glove box
- Lower scuff plate
- Lower door side trim plate
- Lower dashboard side panel
- Lower dashboard front panel
- Steering column upper cover
- Ignition key light
- Steering column lower cover

8. At the driver's side, remove the SAS module and the steering wheel by removing or disconnecting the following:
- Place the wheel in the straight-ahead position and turn the ignition switch to LOCK
- Cover clips at both sides of the steering wheel
- Steering wheel-to-SAS module bolts
- SAS module from the steering wheel and disconnect the electrical connector
- Steering wheel-to-column nut
- Steering wheel from the steering column using a suitable puller
- Steering column mounting bolts and lower the column
- Both A pillar trims
- Instrument panel-to-chassis fasteners in the order shown
- Instrument panel with the help of an assistant

9. Remove or disconnect the following:
- A/C unit

10. Separate the heater core from the A/C unit.

To install:

11. Install the heater core to the A/C unit.

12. Install or connect the following:
- A/C unit
- Instrument panel and tighten the fasteners as shown
- Both A pillar trims
- Raise the steering column and tighten the bolts to 14 ft. lbs. (19 Nm)
- Steering wheel to steering column

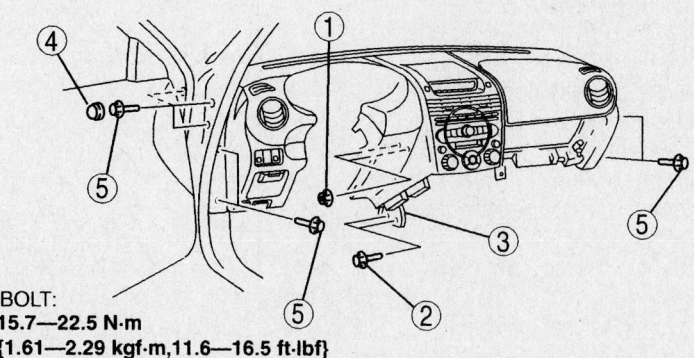

BOLT:
15.7—22.5 N·m
{1.61—2.29 kgf·m, 11.6—16.5 ft·lbf}

67162-RX8-G06

Instrument panel fastener removal sequence—RX-8

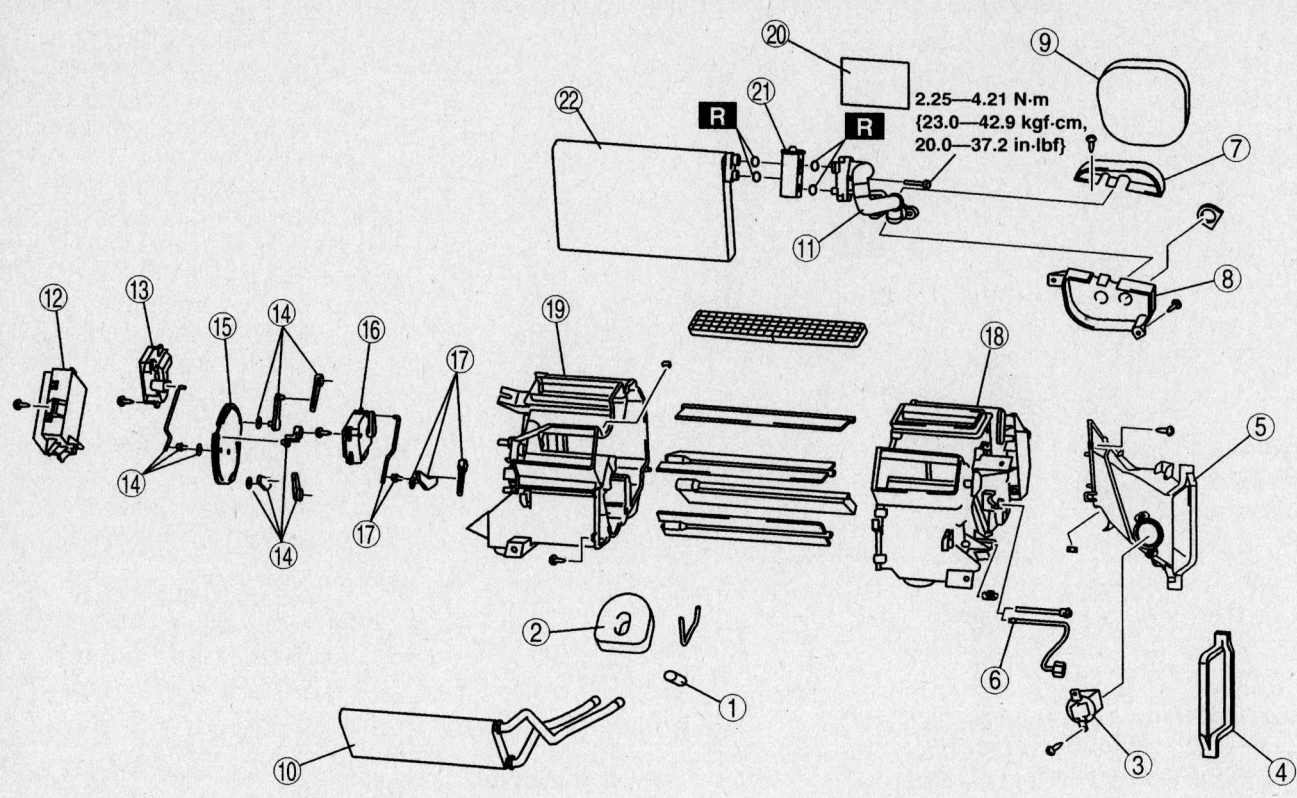

1	Drain hose
2	Polyurethane foam (1)
3	Resistor
4	Adhesive polyurethane (1)
5	Air duct
6	Evaporator temperature sensor
7	Polyurethane foam (2)
8	Bracket (1)
9	Bracket (2)
10	Heater core
11	Evaporator pipe
12	A/C amplifier
13	Airflow mode actuator
14	Airflow mode link set
15	Airflow mode main link
16	Air mix actuator
17	Air mix link set
18	A/C case (1)
19	A/C case (2)
20	Adhesive polyurethane (2)
21	Expansion valve
22	Evaporator

67162-RX8-G07

Exploded view of the A/C unit with heater core—RX8

- Steering wheel-to-column nut and tighten to 33 ft. lbs. (45 Nm)
- SAS module to the steering wheel and connect the electrical connector
- Steering wheel-to-SAS module bolts and tighten to 70–103 inch lbs. (8–12 Nm)
- Cover clips at both sides of the steering wheel
- Steering column lower cover
- Ignition key light
- Steering column upper cover
- Lower dashboard front panel
- Lower dashboard side panel
- Lower door side trim plate

- Lower scuff plate
- Glove box
- Console under cover
- Front and rear center consoles
- Storage compartment
- Ash tray light
- Cigar lighter connector
- Ash tray panel
- Center console upper panel

13. Connect the heater hoses to the heater core.
14. Refill the cooling system.
15. Connect the negative battery cable.
16. Evacuate, charge and leak test the air conditioning system refrigerant.

17. On models with dynamic suspension, perform the steering angle sensor initialization procedure.
18. Operate the engine to normal operating temperatures; then, check the climate control operation and check for leaks.

Intake Manifold

REMOVAL & INSTALLATION

1. Before servicing the vehicle, refer to the Precautions Section.
2. Relieve the fuel system pressure.

3. Drain the cooling system.
4. Remove the engine and transmission assembly.
5. Remove or disconnect the following:
- Air hose
- Air cleaner cover
- Variable Fresh Air Duct (VFAD) solenoid valve on high output engines
- Vacuum chamber on high output engines
- Air cleaner case
- Throttle body
- Upper extension manifold

- Lower extension manifold on high output engines
- Oil filler pipe
- Air Injection Reactor (AIR) solenoid valve
- Secondary Shutter Valve (SSV) solenoid
- Variable Dynamic Effect Intake (VDI) solenoid valve.
- Air cleaner insulator
- Auxiliary Port Valve (APV) bracket and motor on high output engines
- Fresh air intake duct

- Fuel distributors
- Intake manifold and discard the gasket

To install:
6. Clean all gasket mating surfaces.
7. Apply clean oil to the APV valves as shown
8. Install or connect the following:
- Intake manifold using a new gasket in the sequence shown. Torque the bolts to 14–19 ft. lbs. (19–26 Nm). Retighten bolt no. 1.
- Fuel distributors and torque the bolts to 14–19 ft. lbs. (19–26 Nm)

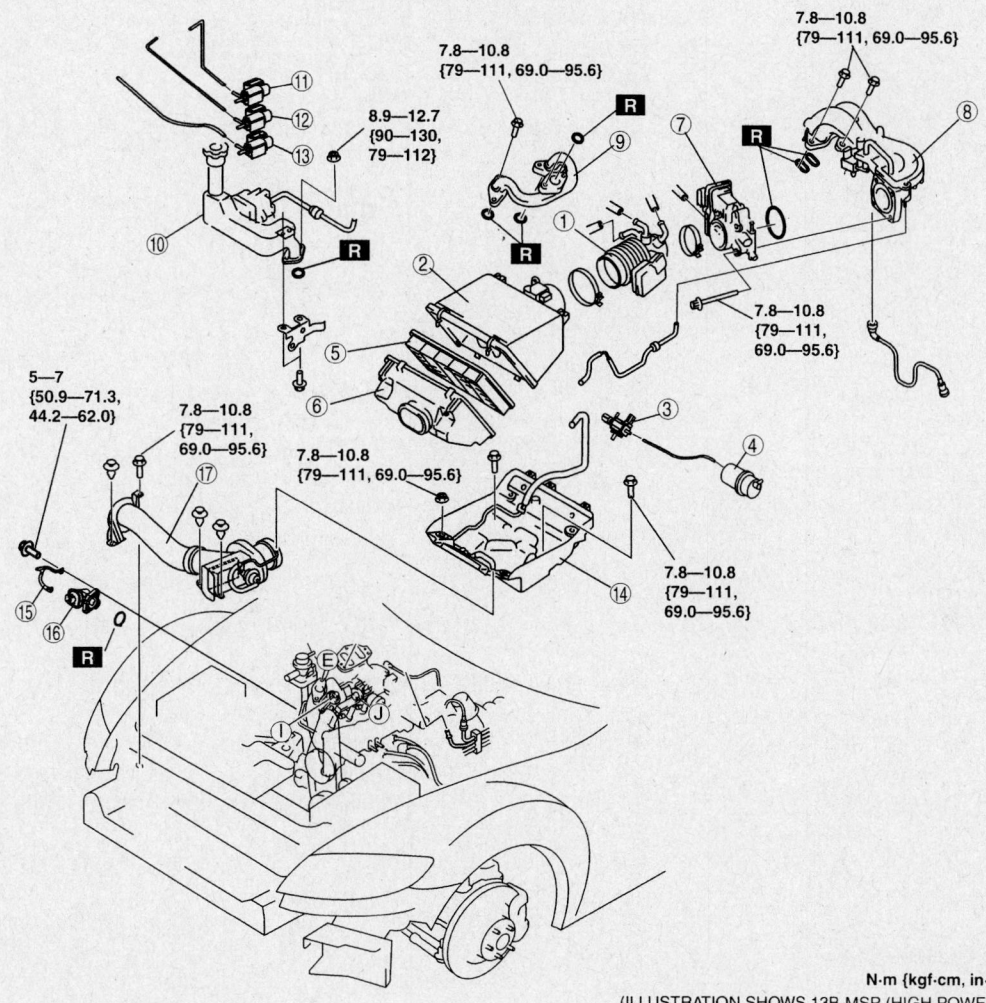

N·m {kgf·cm, in·lbf}
(ILLUSTRATION SHOWS 13B-MSP (HIGH POWER))

1	Air hose	10	Oil filler pipe
2	Air cleaner cover	11	AIR solenoid valve
3	VFAD solenoid valve (13B-MSP (High power))	12	SSV solenoid valve
4	Vacuum chamber (13B-MSP (High power))	13	VDI solenoid valve
5	Air cleaner element	14	Air cleaner insulator
6	Air cleaner case	15	Bracket (13B-MSP (High power))
7	Throttle body	16	APV motor (13B-MSP (High power))
8	Extension manifold (upper)	17	Fresh-air duct
9	Extension manifold (lower) (13B-MSP (High power))		

67162-RX8-G08

Exploded view of the air intake system—RX-8

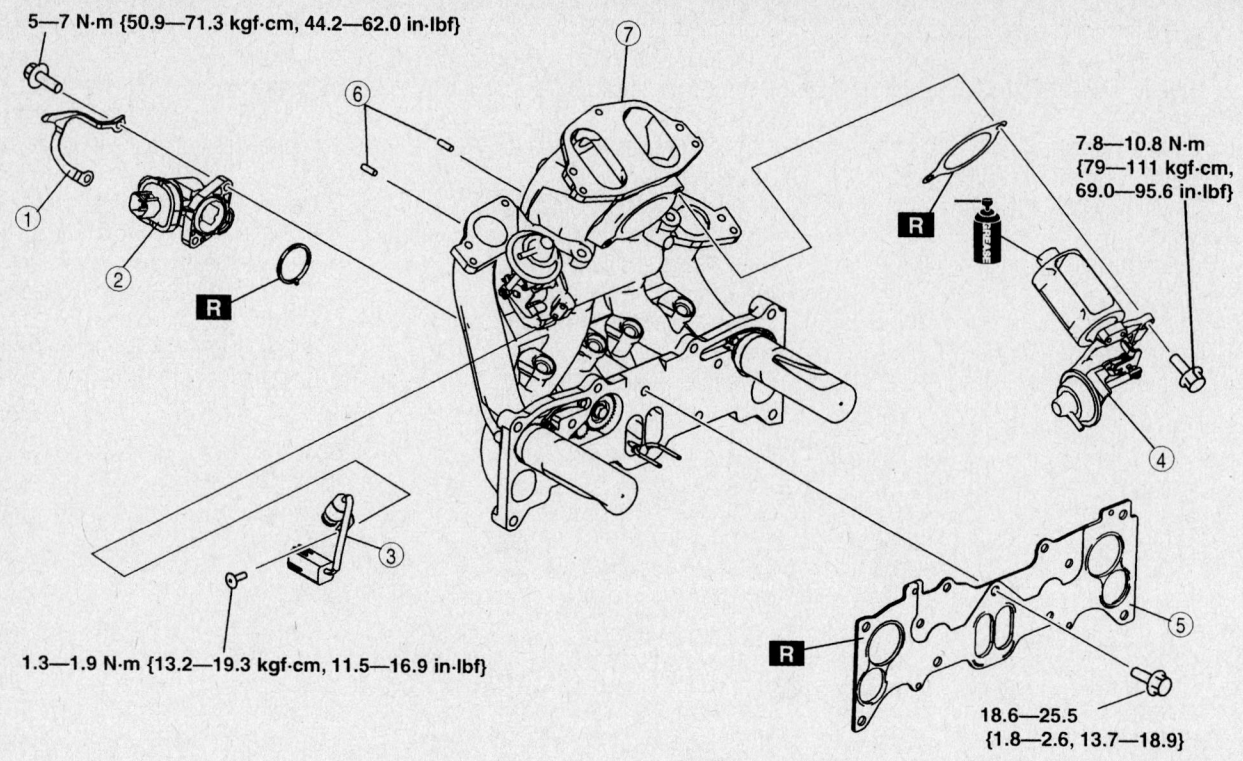

5—7 N·m {50.9—71.3 kgf·cm, 44.2—62.0 in·lbf}

7.8—10.8 N·m {79—111 kgf·cm, 69.0—95.6 in·lbf}

1.3—1.9 N·m {13.2—19.3 kgf·cm, 11.5—16.9 in·lbf}

18.6—25.5 {1.8—2.6, 13.7—18.9}

N·m {kgf·m, ft·lbf}

(ILLUSTRATION SHOWS 13B-MSP (HIGH POWER))

1	Bracket (13B-MSP (HIGH POWER))		5	Gasket
2	APV motor (13B-MSP (HIGH POWER))		6	Blind cap
3	SSV switch		7	Intake manifold
4	VDI valve			

67162-RX8-G09

Exploded view of the intake manifold—RX-8

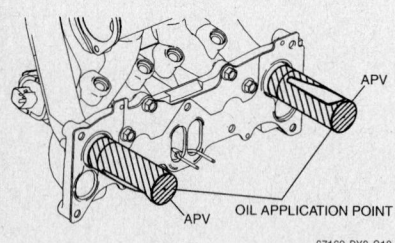

67162-RX8-G10

Applying oil to the APV valves—RX-8

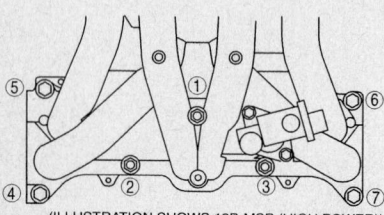

(ILLUSTRATION SHOWS 13B-MSP (HIGH POWER))

67162-RX8-G11

Intake manifold tightening sequence—RX-8

- Fresh air intake duct and torque the bolts to 69–96 inch lbs. (8–11 Nm)
- Auxiliary Port Valve (APV) bracket and motor on high output engines
- Air cleaner insulator and torque the bolts to 69–96 inch lbs. (8–11 Nm)
- Variable Dynamic Effect Intake (VDI) solenoid valve.
- Secondary Shutter Valve (SSV) solenoid
- Air Injection Reactor (AIR) solenoid valve
- Oil filler pipe and torque the bolts to 79–112 inch lbs. (9–13 Nm)
- Lower extension manifold on high output engines and torque the bolts to 79–112 inch lbs. (9–13 Nm)
- Upper extension manifold and torque the bolts to 69–96 inch lbs. (8–11 Nm)
- Throttle body and torque the bolts to 69–96 inch lbs. (8–11 Nm)
- Air cleaner case

- Vacuum chamber on high output engines
- Variable Fresh Air Duct (VFAD) solenoid valve on high output engines
- Air cleaner cover
- Air hose
- Engine and transmission assembly
- Negative battery cable

9. Fill the cooling system.

10. On models with dynamic suspension, perform the steering angle sensor initialization procedure.

11. Run the engine and check for leaks.

Exhaust Manifold

REMOVAL & INSTALLATION

1. Before servicing the vehicle, refer to the Precautions Section.

2. Remove or disconnect the following:
- Negative battery cable
- Front and rear tunnel crossmembers

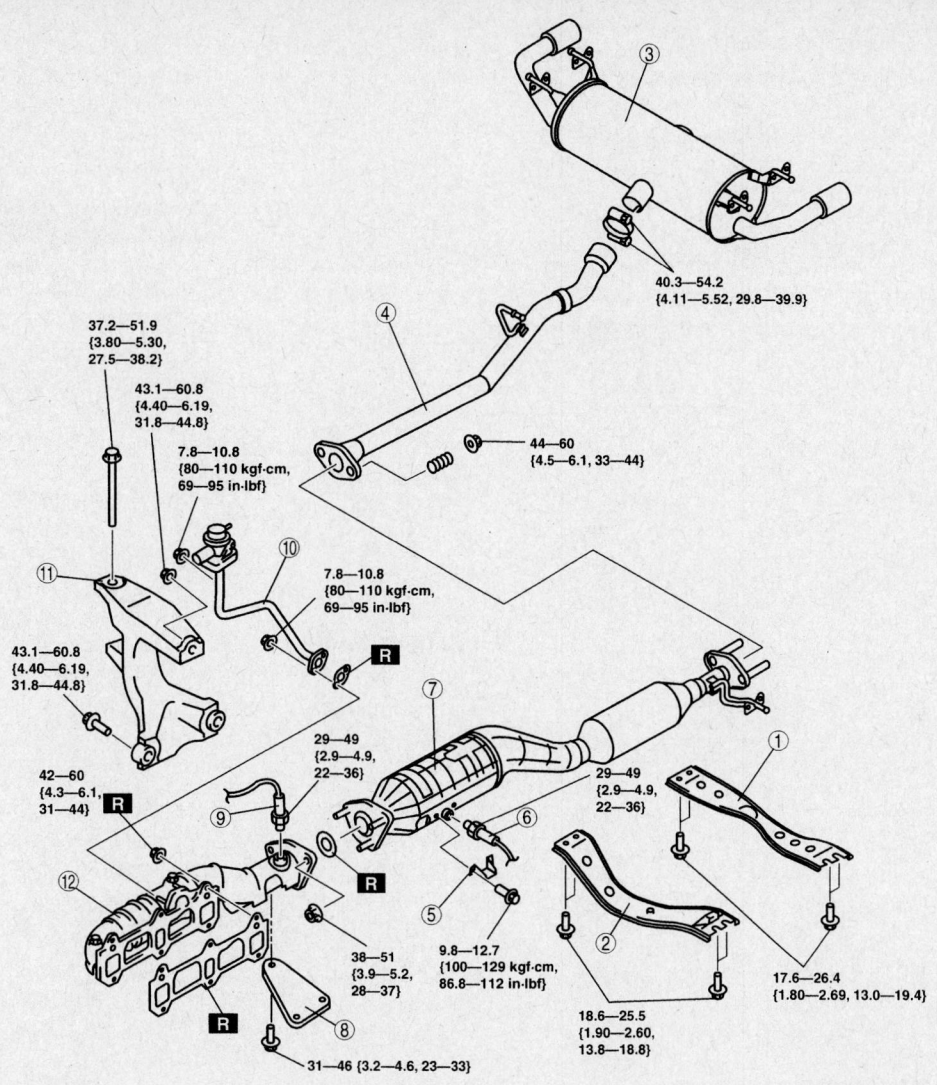

40.3—54.2
{4.11—5.52, 29.8—39.9}

37.2—51.9
{3.80—5.30,
27.5—38.2}

43.1—60.8
{4.40—6.19,
31.8—44.8}

7.8—10.8
{80—110 kgf·cm,
69—95 in·lbf}

44—60
{4.5—6.1, 33—44}

7.8—10.8
{80—110 kgf·cm,
69—95 in·lbf}

43.1—60.8
{4.40—6.19,
31.8—44.8}

42—60
{4.3—6.1,
31—44}

29—49
{2.9—4.9,
22—36}

29—49
{2.9—4.9,
22—36}

17.6—26.4
{1.80—2.69, 13.0—19.4}

38—51
{3.9—5.2,
28—37}

9.8—12.7
{100—129 kgf·cm,
86.8—112 in·lbf}

18.6—25.5
{1.90—2.60,
13.8—18.8}

31—46 {3.2—4.6, 23—33}

N·m {kgf·m, ft·lbf}

1	Rear tunnel member	7	Catalytic converter
2	Front tunnel member	8	Bracket
3	Main silencer	9	Front heated oxygen sensor
4	Middle pipe	10	AIR pipe
5	Protector	11	Engine mount bracket (RH)
6	Rear heated oxygen sensor	12	Exhaust manifold

67162-RX8-G12

Exploded view of the exhaust system—RX-8

- Main silencer
- Middle exhaust pipe
- Protector
- Rear oxygen sensor
- Catalytic converter
- Bracket
- Front oxygen sensor
- Air Injection reactor (AIR pipe)

3. Use a suitable overhead engine lift and support the engine.

4. Remove the right side engine mounting rubber and bracket.

5. Remove the exhaust manifold.

To install:

6. Clean all gasket mating surfaces.

➡**Use new self-locking nuts. The exhaust manifold gasket has crimps attached to it. Ensure that all the crimps are in place when installing the gasket, or the gasket will leak.**

7. Install or connect the following:
- Exhaust manifold. Torque the nuts to 31–44 ft. lbs. (43–61 Nm).
- Right side engine mounting bracket. Torque the bolts as shown.

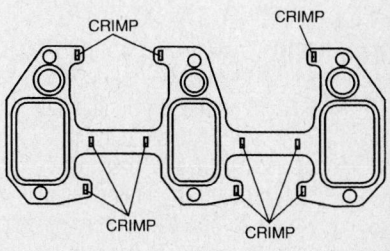

67162-RX8-G13

Exhaust manifold gasket identification— RX-8

37.2—51.9 {3.80—5.30, 27.5—38.2}

43.1—60.8 {4.40—6.19, 31.8—44.8}

37.2—51.9 {3.80—5.30, 27.5—38.2}

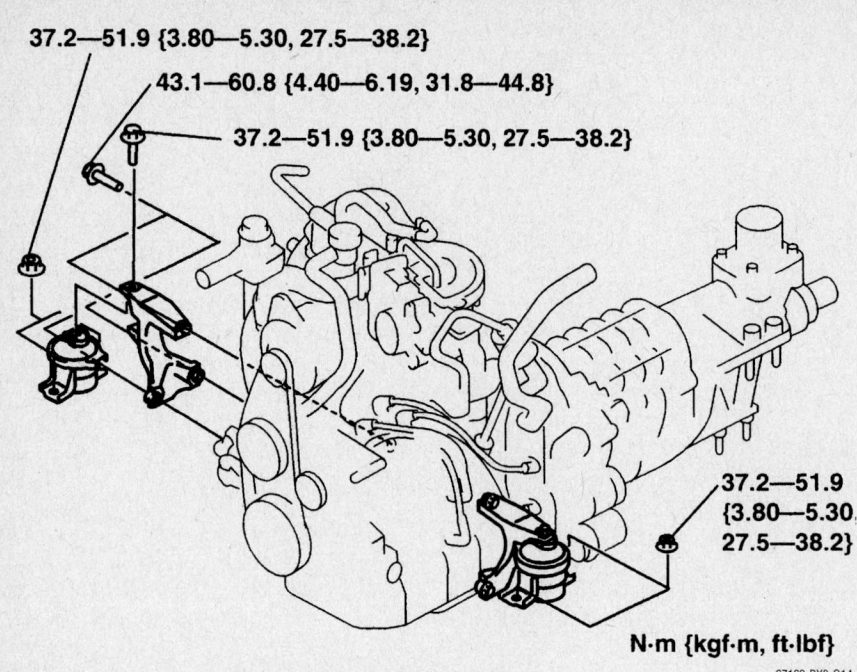

37.2—51.9
{3.80—5.30,
27.5—38.2}

N·m {kgf·m, ft·lbf}

67162-RX8-G14

Removing and installing the engine mount bracket—RX-8

- Air Injection reactor (AIR pipe)
- Bracket
- Catalytic converter
- Rear oxygen sensor
- Protector
- Middle exhaust pipe
- Main silencer
- Front and rear tunnel crossmembers and torque the bolts to 14–19 ft. lbs. (19–26 Nm)
- Negative battery cable

8. On models with dynamic suspension, perform the steering angle sensor initialization procedure.

Starter Motor

REMOVAL & INSTALLATION

1. Remove or disconnect the following:
   - Engine cover
   - Negative battery cable
   - Air cleaner
   - Starter electrical connectors
   - Starter

To install:

2. Install or connect the following:
   - Starter and loosely tighten the lower starter mounting bolt
   - Starter electrical connectors
   - Starter bolts. Torque the bolts 14–18 ft. lbs. (19–25 Nm) on automatic transmission models, or 29–37 ft. lbs. (38–51 Nm) on manual transmission models.

- Air cleaner
- Negative battery cable
- Engine cover

3. On models with dynamic suspension, perform the steering angle sensor initialization procedure.

Oil Pan

REMOVAL & INSTALLATION

1. Before servicing the vehicle, refer to the Precautions Section.
2. Drain the engine oil.
3. Remove or disconnect the following:

- Engine cover
- Battery cover
- Negative battery cable
- Electrical connector
- Oil pan bolts and the oil pan using a seal cutter, then insert a flat pry tool into the locations illustrated.

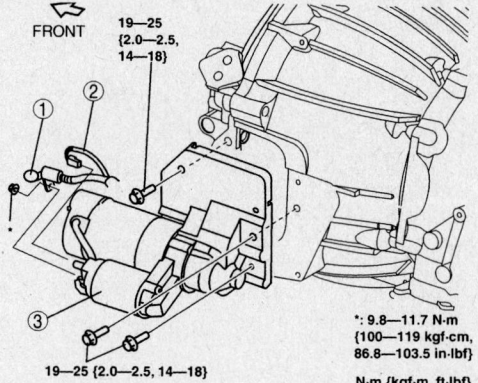

FRONT

19—25 {2.0—2.5, 14—18}

① ②

③

19—25 {2.0—2.5, 14—18}

*: 9.8—11.7 N·m
{100—119 kgf·cm,
86.8—103.5 in·lbf}

N·m {kgf·m, ft·lbf}

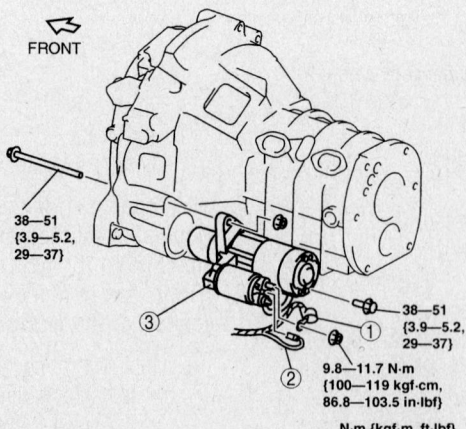

FRONT

38—51
{3.9—5.2,
29—37}

③

①

38—51
{3.9—5.2,
29—37}

②

9.8—11.7 N·m
{100—119 kgf·cm,
86.8—103.5 in·lbf}

N·m {kgf·m, ft·lbf}

67162-RX8-G15

Starter mounting—A/T top and M/T bottom

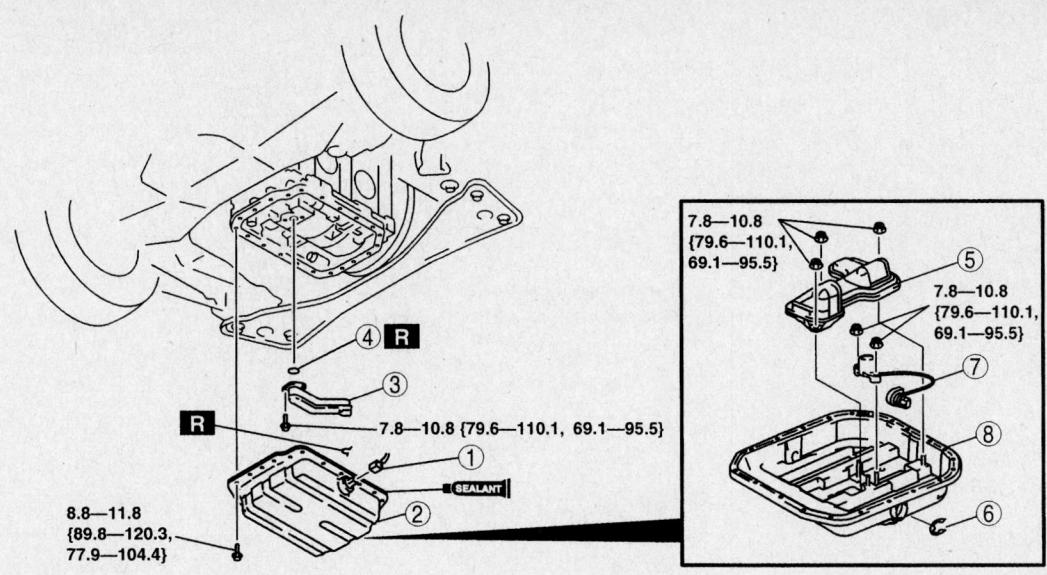

1 Connector
2 Oil pan component
3 Oil strainer
4 O-ring
5 Oil baffle plate
6 Clip
7 Oil-level switch
8 Oil pan

N·m {kgf·cm, in·lbf}

67162-RX8-G16

Exploded view of the oil pan and related components—RX-8

To install:

4. Clean the oil pan. Clean all dirt, oil and old sealant from the oil pan and cylinder block contact surfaces.

5. Apply a continuous bead of silicone sealant around the perimeter of the oil pan.

6. Install the oil pan and tighten the bolts to 78–104 inch lbs. (9–12 Nm).
• Electrical connector
• Negative battery cable
• Battery cover
• Engine cover

7. Fill the engine with clean oil.

8. Start the vehicle, check for leaks and repair if necessary.

9. On models with dynamic suspension, perform the steering angle sensor initialization procedure.

Oil Pump

REMOVAL & INSTALLATION

1. Before servicing the vehicle, refer to the Precautions Section.
2. Drain the engine oil.
3. Remove or disconnect the following:
• Engine cover
• Battery cables
• Battery cover, battery box and tray

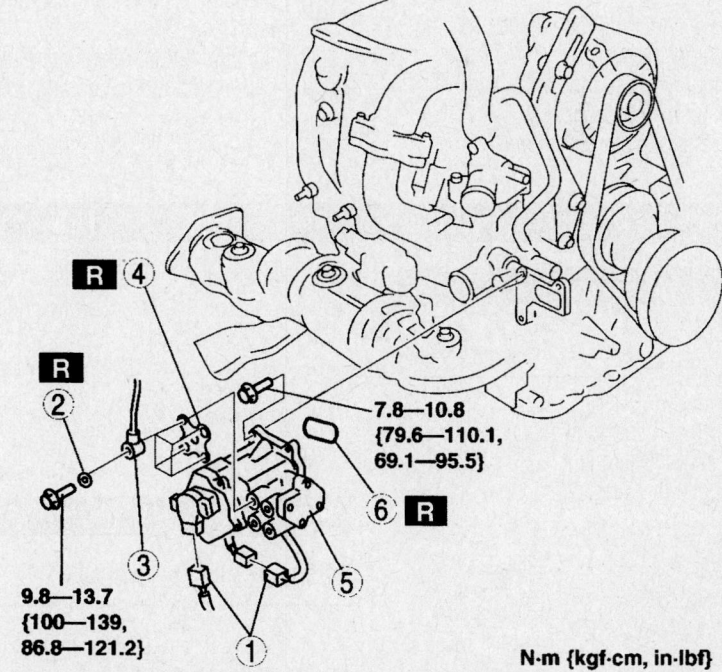

N·m {kgf·cm, in·lbf}

1) CONNECTORS
2) WASHER
3) OIL PIPE
4) GASKET
5) METERING OIL PUMP
6) O-RING

09482_RX-8_G0001

Exploded view of the oil pump mounting—RX-8

- Upper and lower extension manifolds
- Electrical connectors
- Oil pipe
- Gasket
- Oil pump and o-ring

To install:

4. Clean the oil, dirt and old sealant from all contact surfaces.

5. Install or connect the following:
   - New O-rings on the oil pump
   - Oil pump. Torque the bolts to 87–122 inch lbs. (10–14 Nm).
   - Gasket
   - Oil pipe
   - Electrical connectors
   - Upper and lower extension manifolds
   - Battery cover, battery box and tray
   - Battery cables
   - Engine cover

6. Fill the engine with clean oil.

7. Start the vehicle, check for leaks and repair if necessary.

8. On models with dynamic suspension, perform the steering angle sensor initialization procedure.

Rear Main Seal

REMOVAL & INSTALLATION

1. Before servicing the vehicle, refer to the Precautions Section.

2. Remove or disconnect the following:
   - Negative battery cable
   - Transmission assembly

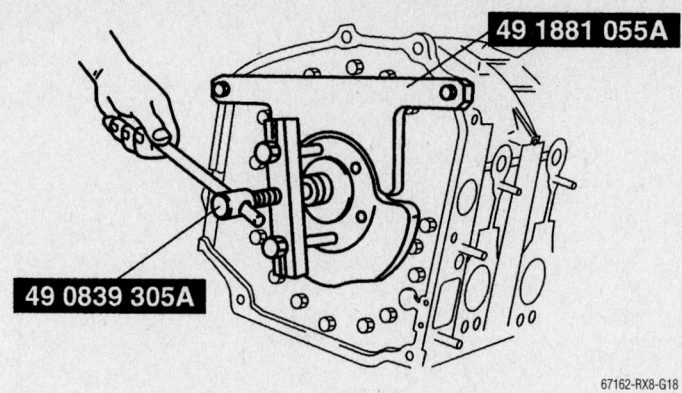

49 1881 055A

49 0839 305A

67162-RX8-G18

Removing the counterweight using special tools—RX-8

- Clutch/flywheel assembly, if equipped with a manual transaxle/transmission
- Flexplate/shim plates, if equipped with an automatic transaxle/transmission
- Using special tools 49 1881-055A and 49-0820-035, remove the counterweight locknut.
- Using special tools 49 1881-055A and 49-0839-305A, remove the counterweight
- Place a rag over the eccentric shaft and using a prytool, carefully pry the oil seal from the oil seal housing.

3. Clean the gasket mounting surfaces.

To install:

4. Clean the oil seal housing. Coat the lip of the oil seal and the housing with clean engine oil.

5. Install or connect the following:

- New oil seal into the housing by tapping it evenly into place with a hammer and a seal installer until it is flush with the edge of the rear cover
- Install the key into the eccentric shaft and install the counterweight
- Apply sealant to the seating face then install the locknut and loosely tighten.
- Lock the counterweight using the special tools, then tighten the locknut to 290–361 ft. lbs. (392–490 Nm).
- Clutch/flywheel assembly or the flexplate, as applicable
- Transaxle/transmission
- Negative battery cable

6. On models with dynamic suspension, perform the steering angle sensor initialization procedure.

FUEL SYSTEM

Fuel System Pressure

RELIEVING

1. Before servicing the vehicle, refer to the Precautions Section.

2. Remove the filler cap.

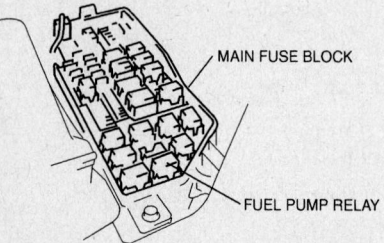

MAIN FUSE BLOCK

FUEL PUMP RELAY

67162-RX8-G19

Fuel pump relay location—RX-8

3. Remove the fuel pump relay from the relay box, located in the main fuse block.

4. Start the engine.

5. After the engine stalls, turn the ignition switch **OFF**.

6. After servicing the vehicle, reinstall the relay.

Fuel Pump

REMOVAL & INSTALLATION

1. Before servicing the vehicle, refer to the Precautions Section.

2. Relieve the fuel system pressure.

3. Remove or disconnect the following:
   - Negative battery cable
   - Rear seat cushion

4. Drain the fuel from the tank.
   - Service hole cover

- Fuel pump electrical connector
- All fuel hoses from the fuel pump unit
- Fuel pump ring using tool 49 T042 001
- Fuel pump and gaskets from the fuel tank

To install:

5. Align the fuel pump alignment mark with the notch in the retainer and install the fuel pump using a new gasket.

6. Align the cap with the retainer and tighten one full turn by hand.

7. Tighten the cap with the special tool to 75 ft. lbs. (102 Nm).

8. Install or connect the following:
   - Fuel hoses to the fuel pump
   - Fuel pump electrical connector
   - Service hole cover
   - Rear seat cushion
   - Negative battery cable

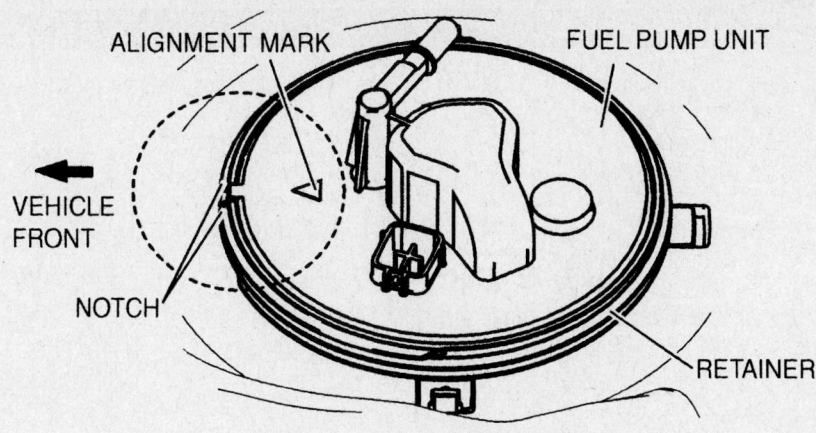

Installing the fuel the fuel pump ring—RX-8

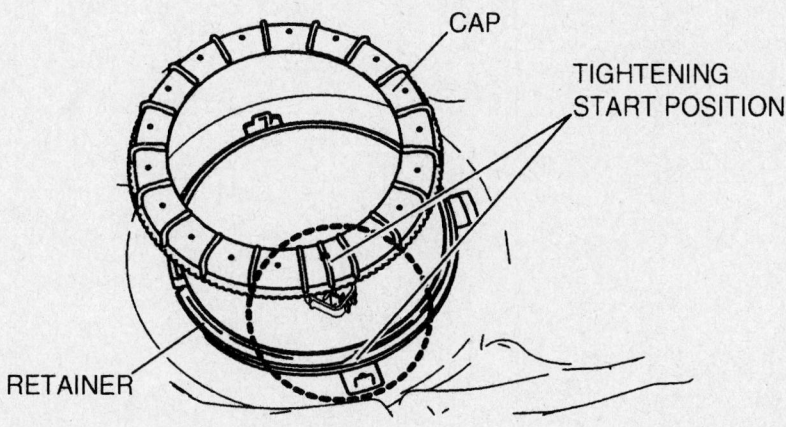

9. Add a minimum of 10 gallons of fuel to the tank and check for leaks.

10. On models with dynamic suspension, perform the steering angle sensor initialization procedure.

Fuel Injector

REMOVAL & INSTALLATION

✳✳ CAUTION

Fuel injection systems remain under pressure after the engine has been turned OFF. Properly relieve fuel pressure before disconnecting any fuel lines. Failure to do so may result in fire or personal injury. Do not allow fuel spray or fuel vapors to come in contact with a spark or open flame. Keep a dry chemical fire extinguisher nearby. Never store fuel in an open container due to risk of fire or explosion.

1. Before servicing the vehicle, refer to the Precautions Section.

2. Relieve the fuel system pressure.

3. Remove or disconnect the following:
- Negative battery cable
- Upper and lower extension manifolds
- Variable Dynamic Effect (VDI) actuator and position out of the way
- Fuel injector wiring harness
- Fuel lines at the fuel rail
- Fuel distributor from intake manifold and housing sides
- Fuel rail with the injectors attached
- Fuel injectors, grommets and O-rings from the fuel rail
- O-rings from the fuel injectors

To install:

4. Install or connect the following:
- New O-rings and grommets lubricated with engine oil on the fuel injectors.
- Insulators and injectors on the intake manifold
- Grommets and the fuel rail onto the injectors. Torque the bolts to 14–18 ft. lbs. (19–25 Nm).
- Fuel lines to the fuel rail
- Fuel distributors
- Fuel injector wiring harness
- VDI actuator
- Negative battery cable

5. Turn the ignition switch **ON** to pressurize the fuel system.

6. Check for leaks and correct as necessary, before starting the engine.

7. On models with dynamic suspension, perform the steering angle sensor initialization procedure.

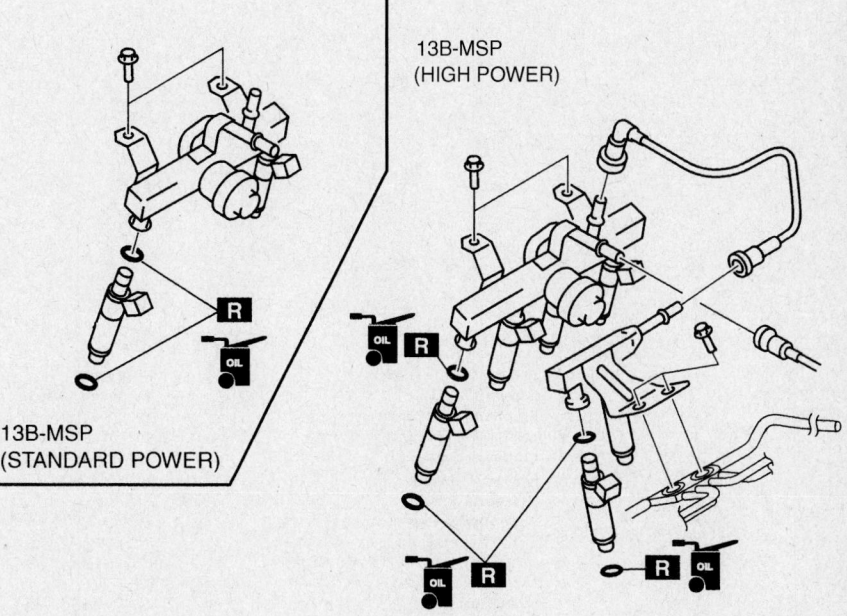

Exploded view of the fuel rail and injector assembly—RX-8

DRIVE TRAIN

Manual Transmission Assembly

REMOVAL & INSTALLATION

1. Before servicing the vehicle, refer to the Precautions Section.

2. Refer to the illustration for component location and torque specifications.

3. Drain the transmission oil.

4. Remove or disconnect the following:
- Engine cover
- Negative battery cable
- Shifter knob
- Shifter panel
- Upper and lower shift insulators
- Shift lever
- Front and rear tunnel supports

- Oxygen sensor connector and bracket
- Catalytic converter
- Middle exhaust pipe
- Rear silencer
- Exhaust manifold stay
- Heat insulator
- Starter
- Clutch slave cylinder

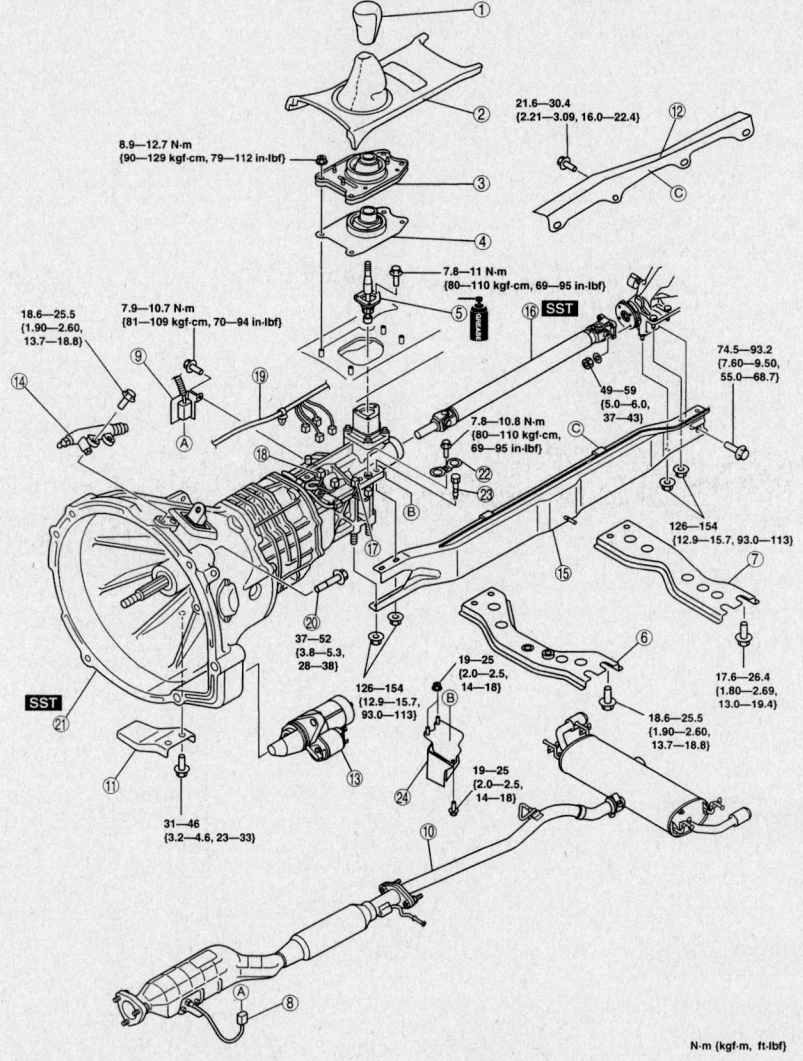

1	Shift lever knob	13	Starter
2	Upper panel	14	Clutch release cylinder
3	Shift insulator component (outer)	15	Power plant frame
4	Shift insulator component (inner)	16	Propeller shaft
5	Shift lever component	17	Back-up light switch connector
6	Front tunnel member	18	Neutral switch connector
7	Rear tunnel member	19	Wire
8	Heated oxygen sensor connector	20	Transmission installation bolt
9	Heated oxygen sensor connector bracket	21	Transmission
10	Catalytic converter, middle pipe, main silencer	22	Stopper
11	Exhaust manifold stay	23	Bolt
12	Heat insulator	24	Dynamic damper

67162-RX8-G22

Exploded view of the manual transmission mounting—RX-8

5. Support the transmission with a suitable jack

6. Support the rear differential with a block of wood.

7. Place match marks on the driveshaft, then remove the driveshaft.

8. Remove or disconnect the following:

- Power plant frame
- Back up light switch connector
- Neutral safety switch connector
- Transmission bolts and the transmission

To install:

9. Install the transmission and tighten the mounting bolts to 28–38 ft. lbs. (37–52 Nm).

10. Support the transmission with a jack and install the power plant frame. Temporarily tighten the bolts in the sequence shown.

11. Raise the front end of the frame and adjust dimension A to 1.91–2.22 inches (48.4–56.4mm).

12. Tighten the power plant frame bolts in the sequence shown to 93–113 ft. lbs.

(126–154 Nm) for bolts 1 and 2 and 55–69 ft. lbs. (75–93 Nm) for bolt 3.

13. Install or connect the following:

- Neutral safety switch connector
- Back up light switch connector
- Clutch release cylinder
- Drive shaft
- Starter
- Heat insulator
- Exhaust manifold stay
- Rear silencer
- Middle exhaust pipe
- Catalytic converter
- Oxygen sensor connector and bracket
- Front and rear tunnel supports
- Shift lever

14. Apply grease to the shift lever components as illustrated.

15. Install or connect the following:

- Upper and lower shift insulators
- Shifter panel
- Shifter knob
- Negative battery cable
- Engine cover

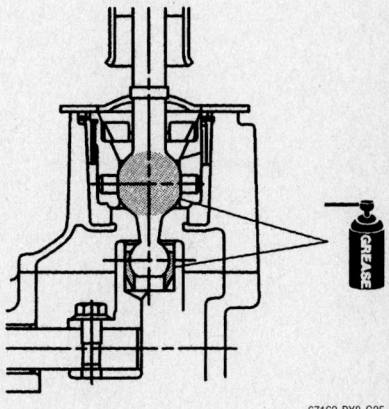

67162-RX8-G24

Measure the distance A which should be 1.91–2.22 inch (48.4–56.4mm)—RX-8

67162-RX8-G25

Apply grease to the shift lever components—RX-8

16. Fill the transmission with fluid. Road test the vehicle and check for leaks. Top off all fluids as needed.

17. On models with dynamic suspension, perform the steering angle sensor initialization procedure.

Automatic Transmission Assembly

REMOVAL & INSTALLATION

1. Before servicing the vehicle, refer to the Precautions Section.

2. Refer to the illustration for component location and torque specifications.

3. Drain the transaxle oil.

4. Remove or disconnect the following:

- Engine cover
- Front and rear tunnel supports
- Oxygen sensor connector
- Catalytic converter
- Middle exhaust pipe
- Rear silencer
- Exhaust manifold stay
- Manual shaft lever after match marking its position
- Heat insulator
- Transverse member

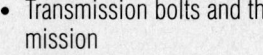

REAR DIFFERENTIAL SIDE

POWER PLANT FRAME

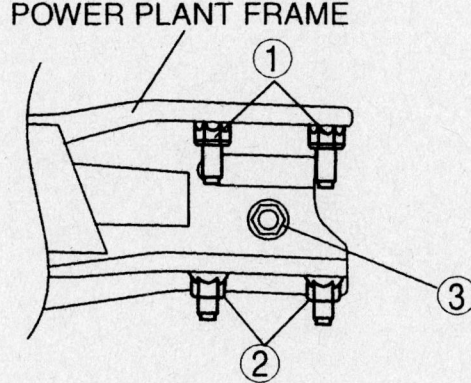

TRANSMISSION SIDE

POWER PLANT FRAME

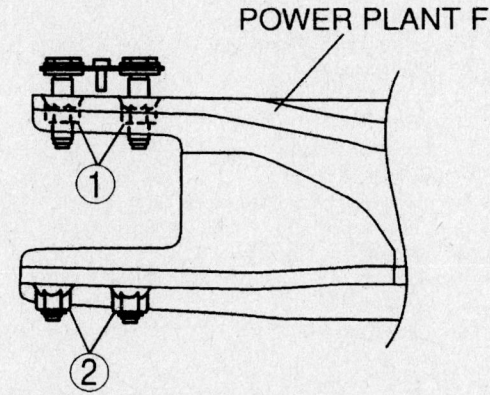

67162-RX8-G23

Tighten the power plant frame bolts using this sequence—RX-8

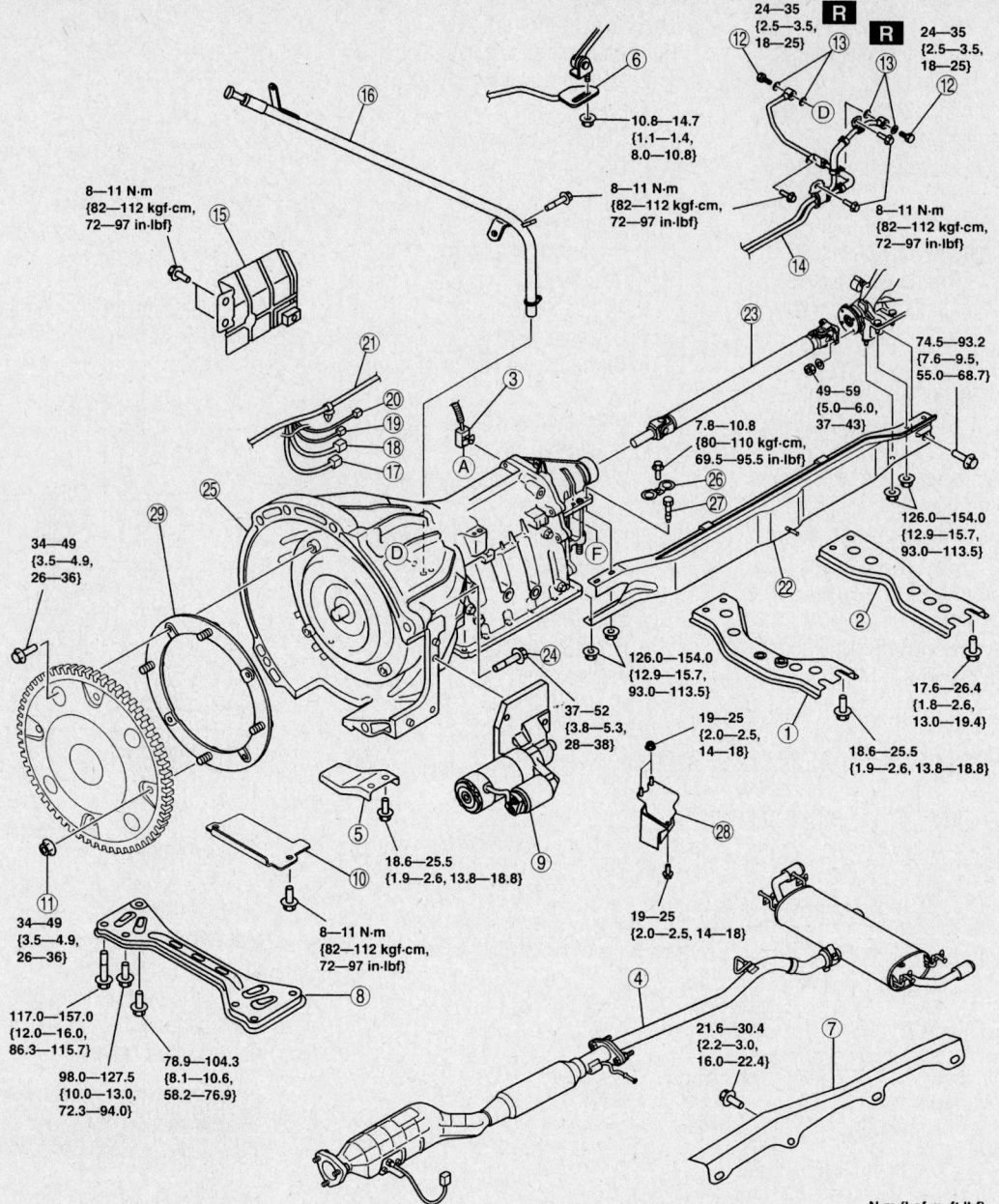

24—35
{2.5—3.5,
18—25}

R **R**

24—35
{2.5—3.5,
18—25}

10.8—14.7
{1.1—1.4,
8.0—10.8}

8—11 N·m
{82—112 kgf·cm,
72—97 in·lbf}

8—11 N·m
{82—112 kgf·cm,
72—97 in·lbf}

8—11 N·m
{82—112 kgf·cm,
72—97 in·lbf}

74.5—93.2
{7.6—9.5,
55.0—68.7}

49—59
{5.0—6.0,
37—43}

7.8—10.8
{80—110 kgf·cm,
69.5—95.5 in·lbf}

126.0—154.0
{12.9—15.7,
93.0—113.5}

34—49
{3.5—4.9,
26—36}

126.0—154.0
{12.9—15.7,
93.0—113.5}

37—52
{3.8—5.3,
28—38}

19—25
{2.0—2.5,
14—18}

17.6—26.4
{1.8—2.6,
13.0—19.4}

18.6—25.5
{1.9—2.6, 13.8—18.8}

18.6—25.5
{1.9—2.6, 13.8—18.8}

8—11 N·m
{82—112 kgf·cm,
72—97 in·lbf}

19—25
{2.0—2.5, 14—18}

34—49
{3.5—4.9,
26—36}

117.0—157.0
{12.0—16.0,
86.3—115.7}

98.0—127.5
{10.0—13.0,
72.3—94.0}

78.9—104.3
{8.1—10.6,
58.2—76.9}

21.6—30.4
{2.2—3.0,
16.0—22.4}

N·m {kgf·m, ft·lbf}

1	Front tunnel member	16	Oil filter tube, Dipstick
2	Rear tunnel member	17	TR switch connector
3	Heated oxygen sensor connector	18	Solenoid valve connector
4	Catalytic converter, middle pipe, main silencer	19	VSS connector
5	Exhaust manifold stay	20	Turbine sensor connector
6	Manual shaft lever component	21	Wire
7	Heat insulator	22	Power plant frame
8	Transverse member	23	Propeller shaft
9	Starter	24	Transmission installation bolt
10	Under cover	25	Transmission
11	Torque converter installation nuts	26	Stopper
12	Connector bolt	27	Bolt
13	Washer	28	Dynamic dumper
14	Oil pipe, oil hose	29	Driven plate
15	Insulator		

67162-RX8-G26

Exploded view of the automatic transaxle assembly mounting—RX-8

- Starter
- Under cover
- Torque converter nuts
- Drive plate bolts
- Connector bolt
- Washer
- Oil pipe and hose
- Insulator
- Oil filler tube
- Transmission range switch connector
- Solenoid valve connector
- VSS connector
- Turbine sensor connector

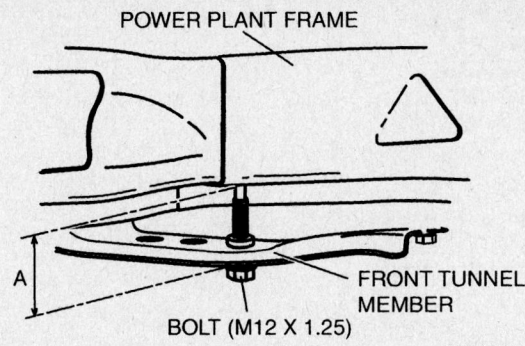

Measure the distance A which should be 1.91–2.22 inch (48.4–56.4mm)—RX-8

REAR DIFFERENTIAL SIDE

POWER PLANT FRAME

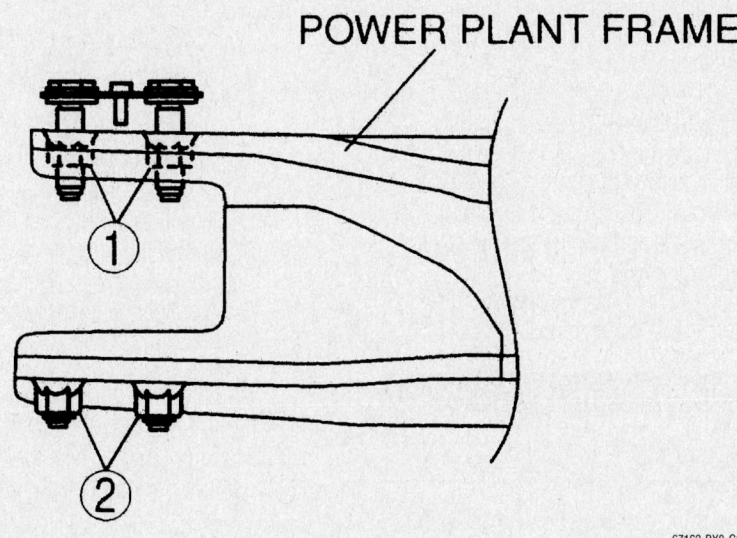

TRANSMISSION SIDE

POWER PLANT FRAME

Tighten the power plant frame bolts using this sequence—RX-8

- Connector wire loom
- Power plant frame

5. Support the transmission with a suitable jack

6. Support the rear differential with a block of wood.

7. Place match marks on the driveshaft, then remove the driveshaft.

8. Remove the transmission bolts and the transmission.

To install:

9. Install the transmission and tighten the mounting bolts to 28–38 ft. lbs. (37–52 Nm).

10. Support the transmission with a jack and install the power plant frame. Temporarily tighten the bolts in the sequence shown.

11. Raise the front end of the frame and adjust dimension A to 1.91–2.22 inches (48.4–56.4mm).

12. Tighten the power plant frame bolts in the sequence shown to 93–113 ft. lbs. (126–154 Nm) for bolts 1 and 2 and 55–69 ft. lbs. (75–93 Nm) for bolt 3.

13. Install the torque converter nuts and drive plate bolts and tighten them equally and evenly to 26–36 ft. lbs. (34–49 Nm).

14. Install or connect the following:
- Driveshaft
- Connector wire loom
- Turbine sensor connector
- VSS connector
- Solenoid valve connector
- Transmission range switch connector
- Oil filler tube
- Insulator
- Oil pipe and hose
- Washer
- Connector bolt
- Under cover
- Starter
- Transverse member
- Heat insulator
- Manual shaft lever

- Exhaust manifold stay
- Rear silencer
- Middle exhaust pipe
- Catalytic converter
- Oxygen sensor connector
- Front and rear tunnel supports
- Engine cover

15. Fill the transmission with fluid. Road test the vehicle and check for leaks. Top off all fluids as needed.

16. On models with dynamic suspension, perform the steering angle sensor initialization procedure.

Clutch

REMOVAL & INSTALLATION

1. Before servicing the vehicle, refer to the Precautions Section.

2. Remove or disconnect the following:
- Negative battery cable
- Clutch release cylinder
- Transmission
- Rubber boot
- Clutch release collar
- Clutch release fork
- Clutch cover
- Pressure plate loosening the bolts one turn each in a criss–cross pattern
- Clutch disc

3. Inspect the pilot bearing. If it is worn or damaged and does not turn easily by hand, remove it using a puller/slide hammer.

4. Check the flywheel surface for scoring, cracks or burning and machine or replace, as necessary.

5. Install Holder tool 49 F011 101 to keep the flywheel from turning, remove the flywheel lock bolt and remove the flywheel.

6. Inspect the clutch release bearing for wear. Replace it if it sticks or does not turn easily.

7. Inspect the release fork for wear or damage and replace as necessary.

To install:

8. Lubricate the release fork fingers and pivot with molybdenum grease and install in the release fork boot.

9. Install or connect the following:
- Clutch release bearing on the release fork
- New pilot bearing in the flywheel, if removed, using a installation tool

10. Be sure the flywheel mounting surface and the crankshaft or eccentric shaft mounting surfaces are clean. Remove any old sealant from the flywheel lock bolt and threads.

- Flywheel
- Sealant to the flywheel lock bolt threads and install it tight
- Flywheel holding tool. Tighten the lock bolt to 289–361 ft. lbs. (392–490 Nm)..
- Small amount of molybdenum grease to the clutch disc splines
- Clutch disc on the flywheel with the spring side toward the transaxle
- An alignment tool in the pilot bearing to position the clutch disc
- Clutch pressure plate by aligning the dowel holes with the flywheel dowels
- Pressure plate. Gradually, torque the bolts, in a crisscross pattern, to 13–19 ft. lbs. (18–27 Nm).

11. Remove the alignment tool.
- Clutch release fork
- Clutch release collar
- Rubber boot
- Transaxle
- Clutch release cylinder
- Negative battery cable

12. On models with dynamic suspension, perform the steering angle sensor initialization procedure.

Hydraulic Clutch System

BLEEDING

1. Before servicing the vehicle, refer to the Precautions Section.

2. Remove the rubber cap from the bleeder screw on the release cylinder.

3. Place a bleeder tube over the end of the bleeder screw.

4. Submerge the other end of the tube in a jar half filled with hydraulic brake fluid.

5. Slowly pump the clutch pedal fully and allow it to return slowly, several times.

6. While pressing the clutch pedal to the floor, loosen the bleeder screw until the fluid starts to run out. Then, close the bleeder screw. Keep repeating this Step, while watching the hydraulic fluid in the jar. As soon as the air bubbles disappear, close the bleeder screw.

7. During the bleeding procedure the reservoir must be kept at least ¾ full.

Halfshafts

REMOVAL & INSTALLATION

1. Before servicing the vehicle, refer to the Precautions Section.

2. Drain the differential oil.

3. Remove or disconnect the following:
- Rear wheels

4. Raise the staked portion of the hub locknut with a hammer and chisel.

5. Lock the hub by applying the brakes and remove the nut.

6. Remove or disconnect the following:
- Parking brake cable
- Brake caliper and wire aside
- Later link upper ball joint
- Stabilizer link
- Lower ball joint
- Lower shock absorber bolt
- Trailing link upper ball joint
- Outer toe control link

7. Position a prybar between the inner CV-joint and differential case. Carefully pry the halfshaft from the transaxle being careful not to damage the oil seal.

8. Pull outward on the hub/knuckle assembly, push the outer CV-joint stub shaft through the hub and remove the halfshaft. If the halfshaft is stuck in the hub, install the old hub nut to protect the stub shaft threads. Tap on the nut, using only a soft mallet, to remove the halfshaft.

➡ **Reinstall the rear lateral upper link to the knuckle to temporarily hold it in position.**

To install:

9. Install or connect the following:
- New circlip on the end of the halfshaft, if removed, with the end gap facing upward.
- Differential oil to the oil seal lip
- Halfshaft into the diferential, being careful not to damage the oil seal
- Other end of the halfshaft through the hub. Loosely install a new locknut
- Outer toe control link
- Trailing link upper ball joint
- Lower shock absorber bolt
- Lower ball joint
- Stabilizer link
- Later link upper ball joint
- Brake caliper and wire aside
- Parking brake cable
- Wheels
- New hub nut. Torque it to 174–203 ft. lbs. (235–275 Nm). After tightening, stake the locknut using a hammer and dull bladed chisel.

10. Fill the differential.

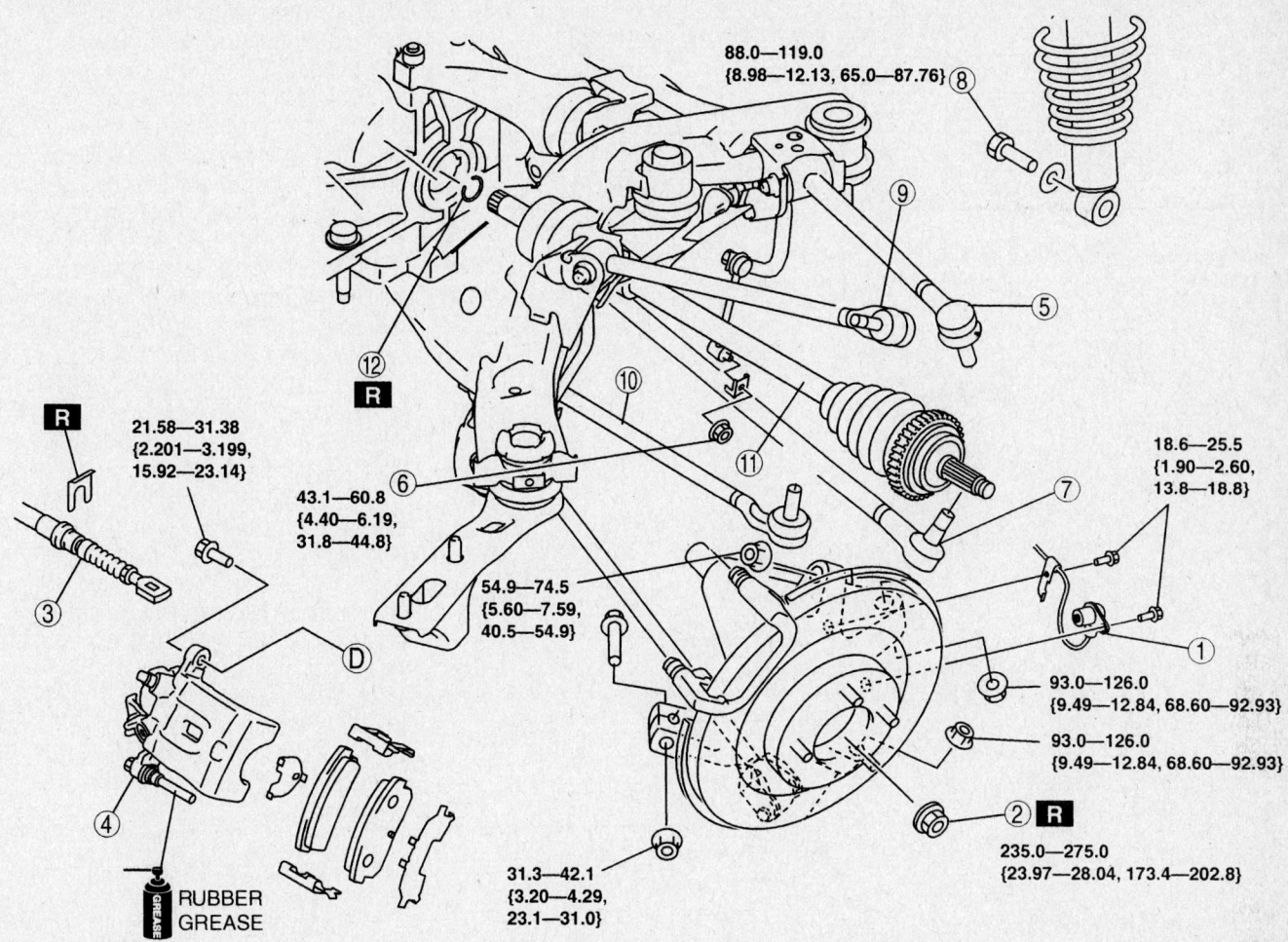

88.0—119.0
{8.98—12.13, 65.0—87.76}   8

9

5

21.58—31.38
{2.201—3.199,
15.92—23.14}

43.1—60.8
{4.40—6.19,
31.8—44.8}

18.6—25.5
{1.90—2.60,
13.8—18.8}

12

R

R

6

10

7

11

3

D

54.9—74.5
{5.60—7.59,
40.5—54.9}

1

93.0—126.0
{9.49—12.84, 68.60—92.93}

4

93.0—126.0
{9.49—12.84, 68.60—92.93}

RUBBER
GREASE

2   R

31.3—42.1
{3.20—4.29,
23.1—31.0}

235.0—275.0
{23.97—28.04, 173.4—202.8}

N·m {kgf·m, ft·lbf}

1	ABS wheel-speed sensor	7	Rear lateral link (lower) ball joint
2	Locknut	8	Shock absorber bolt (lower)
3	Parking brake cable	9	Rear trailing link (upper) ball joint
4	Brake caliper component	10	Toe control link (outer)
5	Rear lateral link (upper) ball joint	11	Rear drive shaft
6	Stabilizer control link (lower)	12	Clip

67162-RX8-G27

Exploded view of rear halfshaft mounting—RX-8

CV-Joints

OVERHAUL

1. Before servicing the vehicle, refer to the Precautions Section.

2. Remove the halfshaft from the vehicle and clamp it in a vise equipped with jaw caps, to prevent damage to the machined surfaces. Do not allow the vise to contact the boot or its clamps.

3. Remove the large boot clamp from the inboard CV-joint, using side cutters.

After removing the clamp, roll the boot back over the shaft.

➡**Check the grease for contamination by rubbing it between 2 fingers. Any gritty feeling indicates a contaminated CV-joint, in which case the entire CV-joint must be disassembled, cleaned and inspected. If the grease is not contaminated and the CV-joint has been operating satisfactorily, continue with the boot replacement procedure and add the required lubricant.**

4. Paint alignment marks on the outer race and shaft for assembly reference. Remove the wire ring bearing retainer and remove the outer race.

5. Paint alignment marks on the tri-pot bearing and shaft for assembly reference. Remove the tri-pot bearing snaring and, using a brass drift and hammer, remove the tri-pot bearing from the shaft.

6. Remove the small clamp and remove the inner boot from the halfshaft. If the boot is to be reused, wrap the shaft splines with tape before removing.

Standard

		Drive shaft length (mm {in})
MT	Left side	792.6—802.6 {31.21—31.59}
	Right side	832.6—842.6 {32.78—33.17}
AT	Left side	791.1—801.1 {31.15—31.53}
	Right side	831.1—841.1 {32.71—33.11}

67162-RX8-G28

Halfshaft length specifications—RX-8

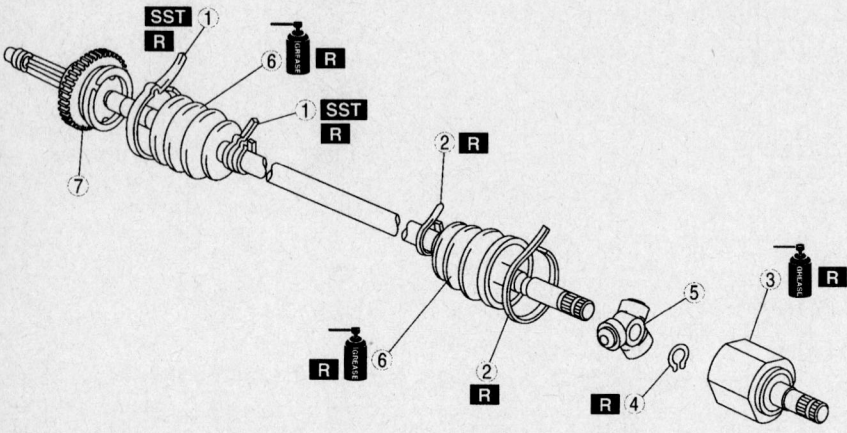

1) BOOT BANDS–AXLE SIDE
2) BOOT BANDS–DIFFERENTIAL SIDE
3) OUTER RING
4) SNAP RING

5) TRIPOD JOINT
6) BOOT
7) SHAFT AND BALL JOINT COMPONENT

09482_RX-8_G0002

Exploded view of the halfshaft—RX-8

7. If the outer CV-joint boot is to be replaced, remove the clamps and slide the boot off the shaft from the inboard side.

To install:

8. If the outboard boot was removed, slide the boot onto the shaft from the inboard side. Wrap tape on the splines before installing to protect the boot.

9. Install the inboard boot and remove the tape from the shaft.

10. Install the tri-pot assembly on the halfshaft. Tap the assembly onto the shaft using a hammer and brass drift. Install the tri-pot assembly retaining ring.

11. Fill the CV-joint outer race with high temperature CV-joint grease. Install the outer race over the tri-pot joint and install the wire ring bearing retainer.

12. Position the CV-joint boot(s). Make sure the boot is fully seated in the grooves in the shaft and outer race.

13. Insert a small prybar with rounded edges between the boot and the outer bearing race to allow trapped air to escape from the boot. Install new boot clamps.

14. Wrap the clamps around the boots in a clockwise direction, pull tight with pliers and bend the locking tabs to secure in position.

15. Set the halfshaft length to the specification as shown.

16. Work the CV-joint through its full range of travel at various angles. The joint should flex, extend and compress smoothly.

17. Install the halfshaft into the vehicle.

STEERING AND SUSPENSION

Air Bag

DISARMING

1. Before servicing the vehicle, refer to the Precautions Section.

2. If equipped, deactivate the audio anti-theft system.

3. Turn the ignition switch to LOCK.

4. Disconnect and isolate the negative battery cable and wait for more than 1 minute to allow the backup power supply to deplete its stored power.

ARMING

1. Before servicing the vehicle, refer to the Precautions Section.

2. Connect the negative battery cable, turn the ignition switch **ON** and verify the air bag warning light cones on for 6 seconds. If the light does not illuminate there are problems with the system.

3. If equipped, activate the audio anti-theft system.

Steering Angle Sensor

INITIALIZATION

1. Connect the negative battery cable.

2. Turn the ignition switch to the **ON** position.

3. Confirm that the Dynamic Stability Control (DSC) indicator light comes on and the DSC OFF light flashes.

4. Turn the steering wheel to the right full lock position and then all the way to the left full lock position.

5. Confirm that the DSC OFF light goes out.

6. Turn the ignition off.

7. Turn the ignition on again and confirm that the DSC indicator light is out. If the light is not out, repeat the procedure.

8. Drive the vehicle for about 10 min-

utes and confirm that the ABS and DSC indicator lights are out.

Rack and Pinion Steering Gear

REMOVAL & INSTALLATION

1. Before servicing the vehicle, refer to the Precautions Section.

2. Remove or disconnect the following:
   - Negative battery cable
   - Front wheels
   - Splash shield
   - Radiator bracket
   - Cotter pins and nuts from both steering tie rod ends

3. Press the tie rod out from the knuckle arm.
   - Intermediate shaft to steering gear pinion shaft bolt. Mark the shaft-to-gear location.
   - Shaft from the steering gear
   - Torque sensor connector

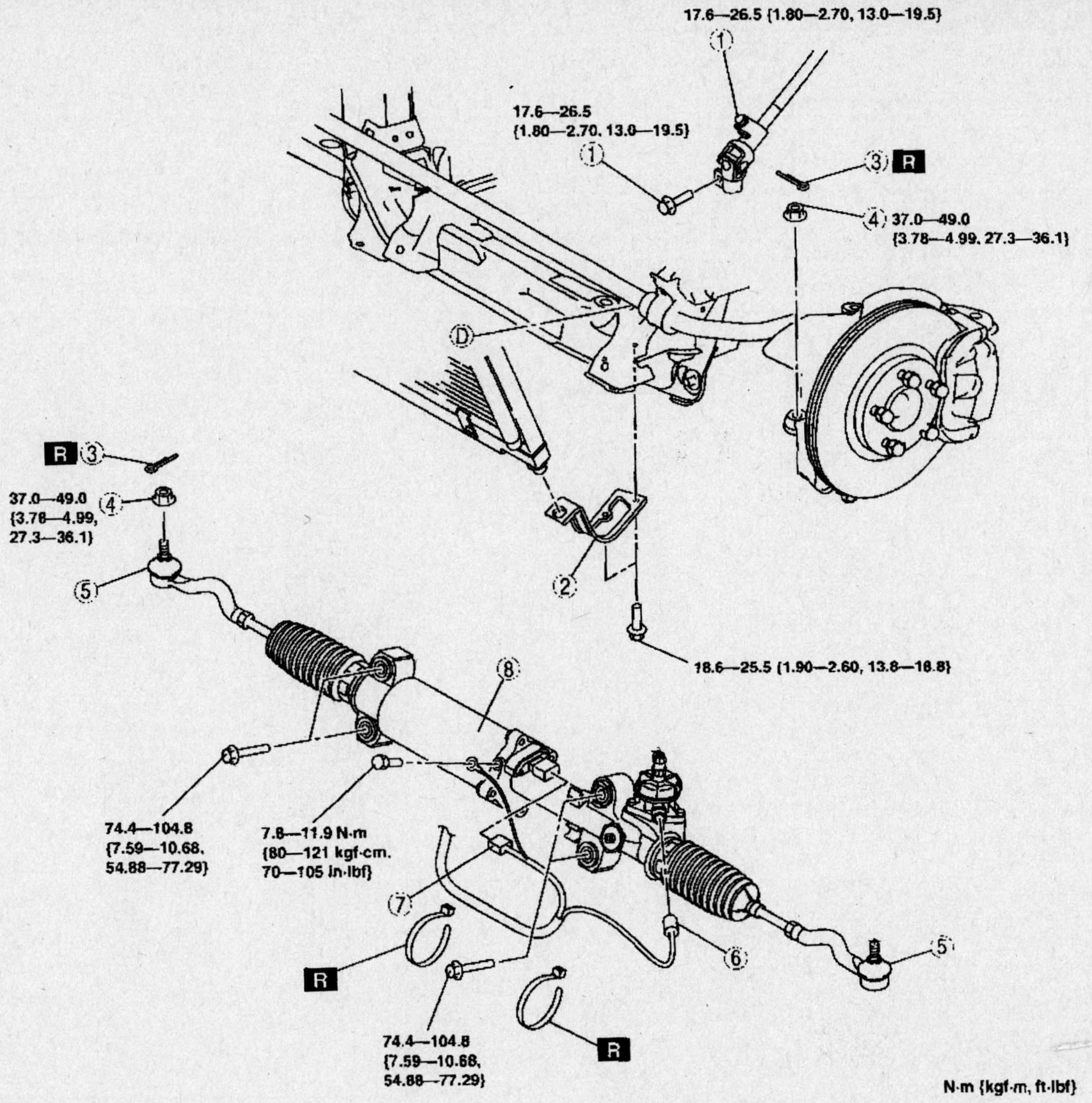

17.6—26.5 {1.80—2.70, 13.0—19.5}
①

17.6—26.5
{1.80—2.70, 13.0—19.5}
①

③ R

④ 37.0—49.0
{3.78—4.99, 27.3—36.1}

Ⓓ

R ③

37.0—49.0
{3.78—4.99,
27.3—36.1}
④

⑤

②

18.6—25.5 {1.90—2.60, 13.8—18.8}

⑧

74.4—104.8
{7.59—10.68,
54.88—77.29}

7.8—11.9 N·m
{80—121 kgf·cm,
70—105 in·lbf}

⑦

R

⑥

⑤

74.4—104.8
{7.59—10.68,
54.88—77.29}

R

N·m {kgf·m, ft·lbf}

1	Bolt (intermediate shaft)	5	Tie-rod end
2	Radiator bracket	6	Torque sensor connector
3	Cotter pin	7	EPS motor connector
4	Locknut (tie-rod end)	8	Steering gear and linkage

.67162-RX8-G29

Exploded view of steering gear mounting—RX-8

- Electronic Power Steering (ESP) motor connector
- Steering gear mounting nuts
- Steering gear and linkage from the vehicle

To install:

4. Install or connect the following:
- Steering gear and linkage to the vehicle. Torque the bolts in sequence to 55–77 ft. lbs. (75–104 Nm).
- Steering shaft to the steering gear pinion shaft, align the marks made during removal and tighten the bolt to 19 ft. lbs. (26 Nm)
- Tie rod ends to the knuckle arm.

Torque the nuts to 27–36 ft. lbs. (37–49 Nm).
- Electronic Power Steering (ESP) motor connector
- Torque sensor connector
- New cotter pins
- Radiator bracket
- Splash shield
- Wheels
- Negative battery cable

5. Check and/or adjust the front end alignment.

6. On models with dynamic suspension, perform the steering angle sensor initialization procedure.

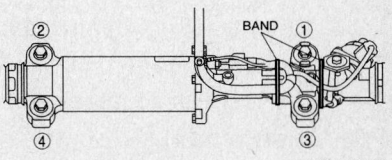

67162-RX8-G30

Steering gear bolt tightening sequence—RX-8

Shock Absorber

REMOVAL & INSTALLATION

Front

1. Before servicing the vehicle, refer to the Precautions Section.
2. Remove the strut cross brace in the engine compartment and the upper shock mounting nuts.
3. Remove or disconnect the following:

- Front wheel
- Brake hose bracket
- Stabilizer bar nut
- Upper arm ball joint

4. Remove the upper and lower shock absorber mounting nuts and bolts and remove the shock absorber/coil spring assembly.

To install:

5. Installation is the reverse of removal. Tighten the upper shock nuts to 34–46 ft. lbs. (46–62 Nm) and the lower shock nut and bolt to 58–76 ft. lbs. (78–103 Nm)
6. Check and/or adjust the front end alignment.

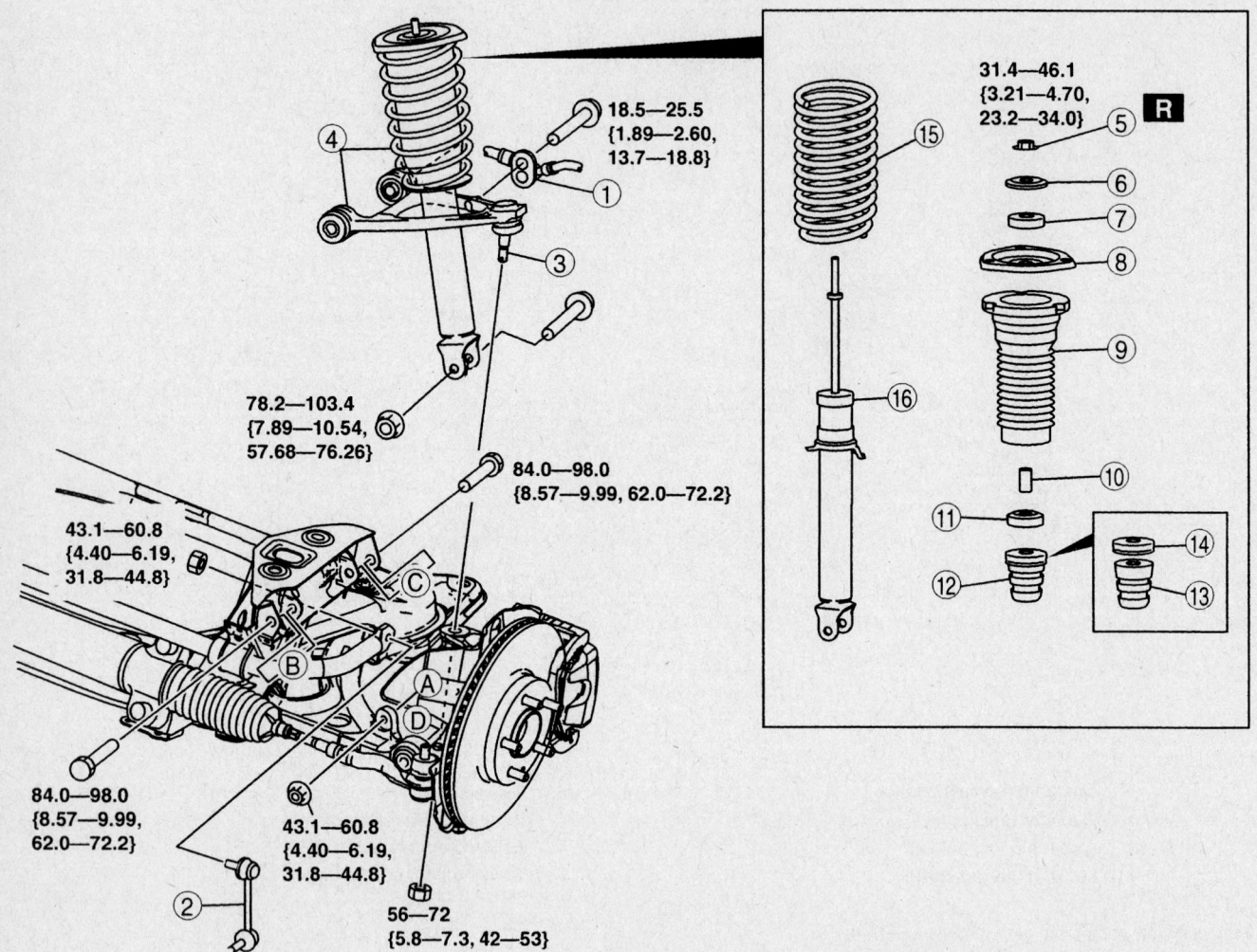

18.5—25.5 {1.89—2.60, 13.7—18.8}

78.2—103.4 {7.89—10.54, 57.68—76.26}

43.1—60.8 {4.40—6.19, 31.8—44.8}

84.0—98.0 {8.57—9.99, 62.0—72.2}

84.0—98.0 {8.57—9.99, 62.0—72.2}

43.1—60.8 {4.40—6.19, 31.8—44.8}

56—72 {5.8—7.3, 42—53}

31.4—46.1 {3.21—4.70, 23.2—34.0} R

N·m {kgf·m, ft·lbf}

1	Brake hose bracket	9	Dust boot
2	Front stabilizer control link	10	Spacer
3	Front upper arm ball joint	11	Bushing
4	Front shock absorber and coil spring	12	Stopper casing and bound stopper
5	Piston rod nut	13	Bound stopper
6	Retainer	14	Stopper casing
7	Bushing	15	Coil spring
8	Upper spring seat	16	Front shock absorber

Exploded view of front shock absorber mounting—RX-8

67162-RX8-G31

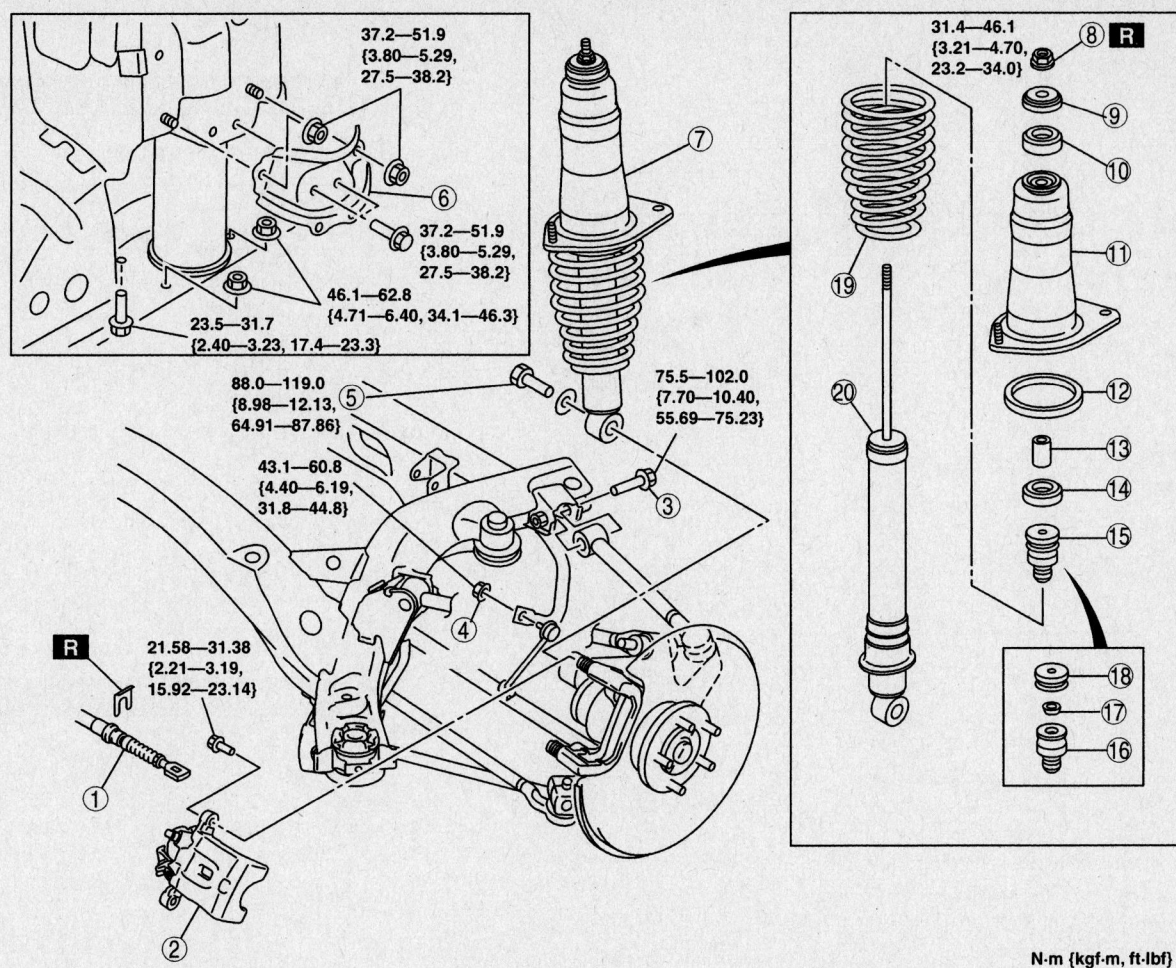

N·m {kgf·m, ft·lbf}

1	Parking brake cable	11	Upper spring seat
2	Caliper	12	Spring seat rubber
3	Rear lateral link (upper) inner bolt	13	Bushing
4	Stabilizer control link upper nut	14	Spacer
5	Rear shock absorber lower bolt	15	Bound stopper and stopper casing
6	Rear shock absorber bracket	16	Bound stopper
7	Rear shock absorber and coil spring	17	Collar
8	Piston rod nut	18	Stopper casing
9	Retainer	19	Coil spring
10	Bushing	20	Rear shock absorber

67162-RX8-G32

Exploded view of rear shock absorber mounting—RX-8

Rear

1. Before servicing the vehicle, refer to the Precautions Section.
2. Remove or disconnect the following:

- Rear wheel
- Brake caliper and wire aside
- Rear lateral link upper inner bolt
- Stabilizer bar upper nut
- Shock absorber lower bolt
- Inside trunk end and side trim panels

- Shock absorber upper bracket and mounting nuts
- Shock absorber/coil spring assembly

To install:

3. Installation is the reverse of removal. Tighten the upper shock nuts to 34–46 ft. lbs. (46–62 Nm) and the lower shock nut and bolt to 65–88 ft. lbs. (88–119 Nm). Tighten the shock bracket bolts and nuts to 28–38 ft. lbs. (37–52 Nm).

4. Check and/or adjust the front end alignment.

Coil Spring

REMOVAL & INSTALLATION

1. Before servicing the vehicle, refer to the Precautions Section.
2. Remove or disconnect the following:

- Shock absorber/coil spring
- Place the shock/coil spring in a spring compressor and compress the spring
- Piston rod nut

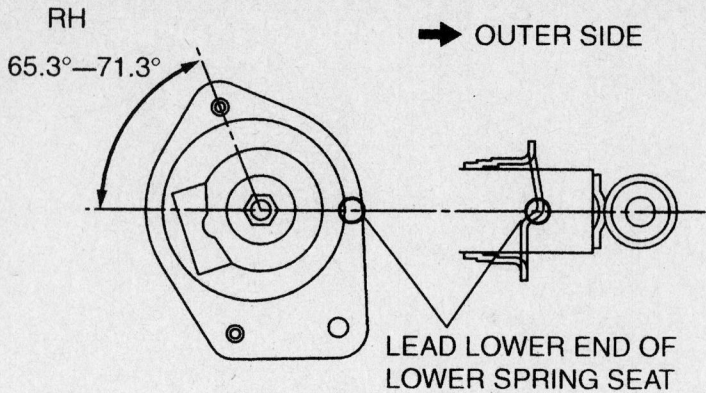

RH

65.3°—71.3°

→ OUTER SIDE

LEAD LOWER END OF
LOWER SPRING SEAT

LH

OUTER SIDE ←

65.3°—71.3°

LEAD LOWER END OF
LOWER SPRING SEAT

67162-RX8-G33

Mark the lower seat, shock and spring assembly as illustrated for reassembly–RX-8

- Upper retainer, bushing spring seat and rubber insulator
- Bushing, spacer, bound stopper and casing
- Coil spring

3. While pushing on the piston rod, be sure that the pull stroke is even and that there is no unusual noise or resistance. Also inspect for any oil leakage around the piston rod.

4. Push the piston rod in, then release it. Be sure that the return rate is constant.

5. If the shock absorber does not operate as described, replace it.

To install:

6. Install or connect the following:
- Strut assembly into a vise
- Bound stopper and casing onto the piston rod
- Temporarily install the lower spring seat, seat rubber and spring. Mark the seat, shock and spring assembly as illustrated for reassembly. Align the marks of the upper seat and coil spring. Protect the assembly with cloth and install the spring compressor.
- Coil spring

7. Compress the coil spring with the spring compressor

8. Install or connect the following:
- Bushing, spacer, bound stopper and casing
- Upper retainer, bushing spring seat and rubber insulator
- Piston rod upper nut

9. Be sure that the spring upper seat notched portion is facing inward and tighten the piston rod upper nut to 23–34 ft. lbs. (32–46 Nm).

10. Be sure that the spring is well seated in the upper seats.

11. Install the shock to the vehicle.

Lower Ball Joint

REMOVAL & INSTALLATION

1. Before servicing the vehicle, refer to the Precautions Section.

2. Remove or disconnect the following:
- Wheel
- Ball joint clip
- Ball joint using a ball joint remover

To install:

3. Install or connect the following:
- Ball joint to lower control arm using a ball joint installer
- Ball joint clip
- Wheel

4. Check and/or adjust the front wheel alignment.

Upper Ball Joint

REMOVAL & INSTALLATION

1. Before servicing the vehicle, refer to the Precautions Section.

2. Remove or disconnect the following:
- Wheel
- Lower nut on the vehicle side
- Ball joint using a ball joint remover

To install:

3. Install or connect the following:
- Ball joint to lower control arm using a ball joint installer
- Tighten the nut to 42–53 ft. lbs. (56–72 Nm).
- Wheel

4. Check and/or adjust the front wheel alignment.

Upper Control Arm

REMOVAL & INSTALLATION

1. Before servicing the vehicle, refer to the Precautions Section.

2. Remove or disconnect the following:
- Front wheel
- Brake caliper and wire aside
- Brake rotor
- Stabilizer bar nut
- Lower shock mounting bolt and nut
- Upper arm ball joint

3. Remove the front upper arm bolts, push down on the lower arm and remove the upper arm through the lower shock and lower arm gap.

To install:

4. Install the upper control arm.

5. Loosely tighten the bolt and nut.

6. Loosely install the lower strut mounting bolt.
- Upper arm ball joint. Torque the nut 42–53 ft. lbs. (56–72 Nm).
- Wheel

7. Torque upper control arm bolt to 62–72 ft. lbs. (84–98 Nm) and the lower strut mounting bolt to 58–76 ft. lbs. (78–103 Nm).

8. Check and/or adjust the front wheel alignment.

CONTROL ARM BUSHING REPLACEMENT

1. Before servicing the vehicle, refer to the Precautions Section.

2. Position the control arm into a vise. Cut away the projecting rubber from the

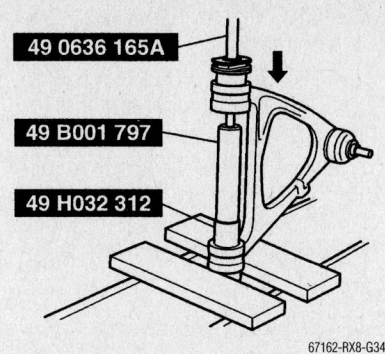

Removing upper control arm bushing – RX-8

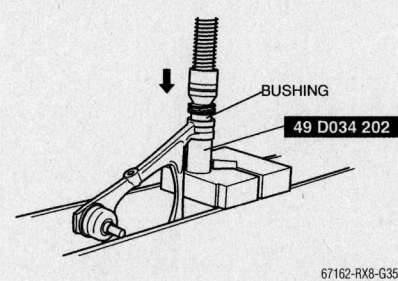

Installing upper control arm bushing –RX-8

bushing. Use tools 49 B0636 165A, 49 B0001 797 and 49 H032 312 to remove the bushing.

To install:

3. Apply soapy water to the new control arm bushing.

4. Install the rear bushing using and tool 49 D034 202.

Lower Control Arm

REMOVAL & INSTALLATION

1. Before servicing the vehicle, refer to the Precautions Section.

2. Remove the strut cross brace in the engine compartment.

3. Remove or disconnect the following:
- Front wheel
- Brake caliper and wire aside
- Brake rotor
- Tie rod end
- Lower ball joint
- Upper ball joint
- Stabilizer link from the lower control arm
- Lower control arm bolts and nuts
- Lower control arm

To install:

4. Install or connect the following:
- Lower control arm
- Lower control arm bolt and nut and

tighten to 62–72 ft. lbs. (84–98 Nm)
- Stabilizer link to the lower control arm and tighten to 32–45 ft. lbs. (43–61 Nm)
- Upper ball joint
- Lower ball joint
- Tie rod end
- Brake rotor
- Brake caliper and wire aside
- Front wheel

5. Check and/or adjust the front wheel alignment.

CONTROL ARM BUSHING REPLACEMENT

1. Before servicing the vehicle, refer to the Precautions Section.

2. Position the control arm into a vise. Cut away the stopper rubber and the projecting rubber from the front bushing and press out the bushing

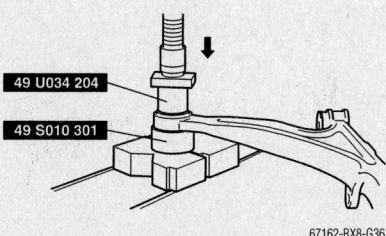

Removing and installing the lower control arm front bushing—RX-8

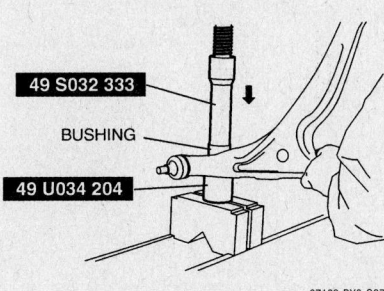

Removing and installing the lower control arm shock absorber bushing—RX-8

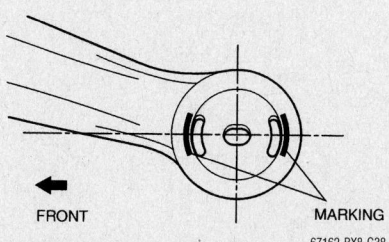

Align the mark of the lower arm rear bushing when installing the bushing—RX-8

3. Remove the rear bushing using a press and tools 49 U034 204 and 49 S010 301.

4. Remove the shock absorber bushing using a press and tools 49 U034 204 and 49 S032 333.

To install:

5. When installing the rear bushing on the lower arm, align the mark of the lower arm and the holes of the bushing as illustrated, set the arm onto the press and press the bushings into position.

6. Apply soapy water to the new control arm bushing.

7. Install the rear bushing using a press and tools 49 U034 204 and 49 S010 301.

8. Install the front bushing using a vise.

9. Install the shock absorber bushing using a press and tools 49 U034 204 and 49 S032 333.

Wheel Bearings

ADJUSTMENT

The front and rear wheel bearings are not adjustable. If the bearings become loose or make noise, they must be replaced.

REMOVAL & INSTALLATION

Front

1. Before servicing the vehicle, refer to the Precautions Section.

2. Refer to the illustration for component location and torque specifications.

3. Remove or disconnect the following:
- Wheels
- ABS wheel speed sensor
- Brake caliper and rotor
- Tie rod end from the knuckle
- Stabilizer bar link
- Upper control arm ball joint from the knuckle
- Upper arm bolt
- Lower ball joint
- Wheel hub and knuckle assembly
- Separate the knuckle from the wheel hub and remove the backing plate.

4. Clean and inspect all parts but do not wash or clean the wheel bearing. The wheel hub/bearing must be replaced.

To install:
- Install the backing plate to the hub, then install the knuckle to the hub and tighten the bolts to 40–44 ft. lbs. (54–60 Nm).
- Lower ball joint
- Upper arm bolt and tighten to 62–72 ft. lbs. (84–96 Nm).

- Upper control arm ball joint
- Stabilizer bar link and tighten to 32–45 ft. lbs. (43–61 Nm).
- Tie rod end
- ABS wheel speed sensor
- Brake caliper and rotor
- Wheels

Rear

1. Before servicing the vehicle, refer to the Precautions Section.
2. Refer to the illustration for component location and torque specifications.

➡ **The wheel bearings are not service-able. If the bearings are bad, a new hub/bearing assembly must be installed.**

3. Remove or disconnect the following:
- Rear wheels
- ABS wheel speed sensor
4. Raise the staked portion of the axle retaining nut with a hammer and chisel.
5. Remove or disconnect the following:
- Axle nut
- Parking brake cable

- Rear caliper and rotor assembly from the hub
- Rear lateral link upper ball joint
- Stabilizer bar lower link
- Rear lateral link lower ball joint
- Lower shock absorber bolt
- Rear trailing link lower outside bolt
- Rear trailing link upper ball joint
- Toe control link outside bolt
- Axle shaft
- Wheel hub/knuckle assembly
6. Press the wheel hub from the knuckle.

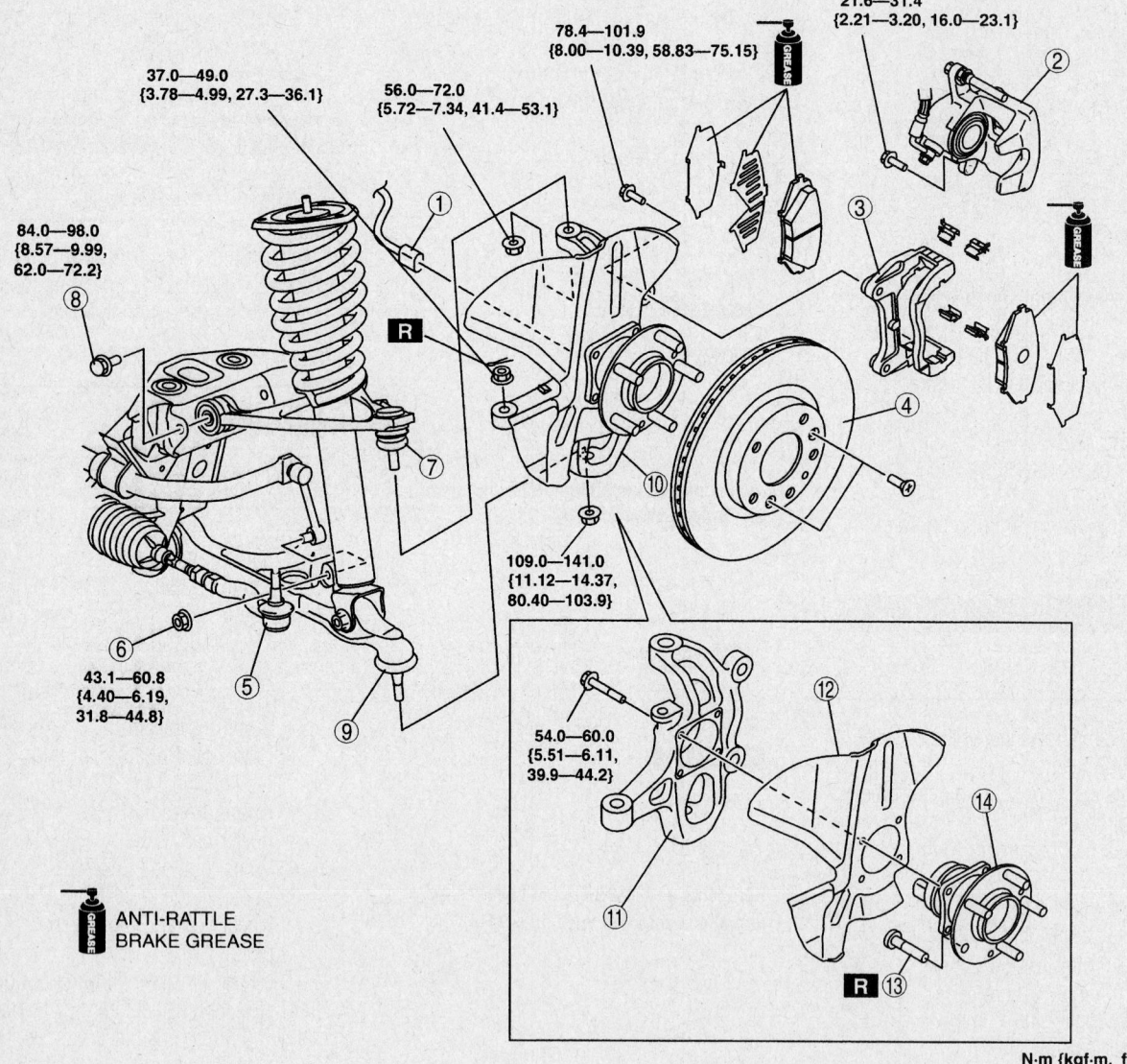

ANTI-RATTLE
BRAKE GREASE

N·m {kgf·m, ft·lbf}

1	ABS wheel-speed sensor connector	5	Tie-rod end
2	Brake caliper component	6	Stabilizer control link (lower)
3	Mounting support	7	Front upper arm ball joint
4	Disc plate	8	Front upper arm bolt

Exploded view of the front wheel bearing and knuckle assembly–RX-8

67162-RX8-G39

7. Press the bearing inner race from the wheel hub.

8. Press the wheel bearing from the rear knuckle.

To install:

9. Press the wheel bearing into the rear knuckle.

10. Press the bearing inner race into the wheel hub.

11. Press the wheel hub into the knuckle.

12. Install or connect the following:
- Wheel hub/knuckle assembly. Tighten the bolts to 23–31 ft. lbs. (31–42 Nm).
- Axle shaft
- Toe control link outside bolt
- Rear trailing link upper ball joint
- Rear trailing link lower outside bolt and tighten to 56–75 ft. lbs. (76–102 Nm).
- Lower shock absorber bolt and

tighten to 65–88 ft. lbs. (86–119 Nm).
- Rear lateral link lower ball joint
- Stabilizer bar lower link and tighten to 32–45 ft. lbs. (43–61 Nm).
- Rear lateral link upper ball joint
- Rear caliper and rotor assembly
- Parking brake cable
- Axle nut and tighten to 174–203 ft. lbs. (235–275 Nm).

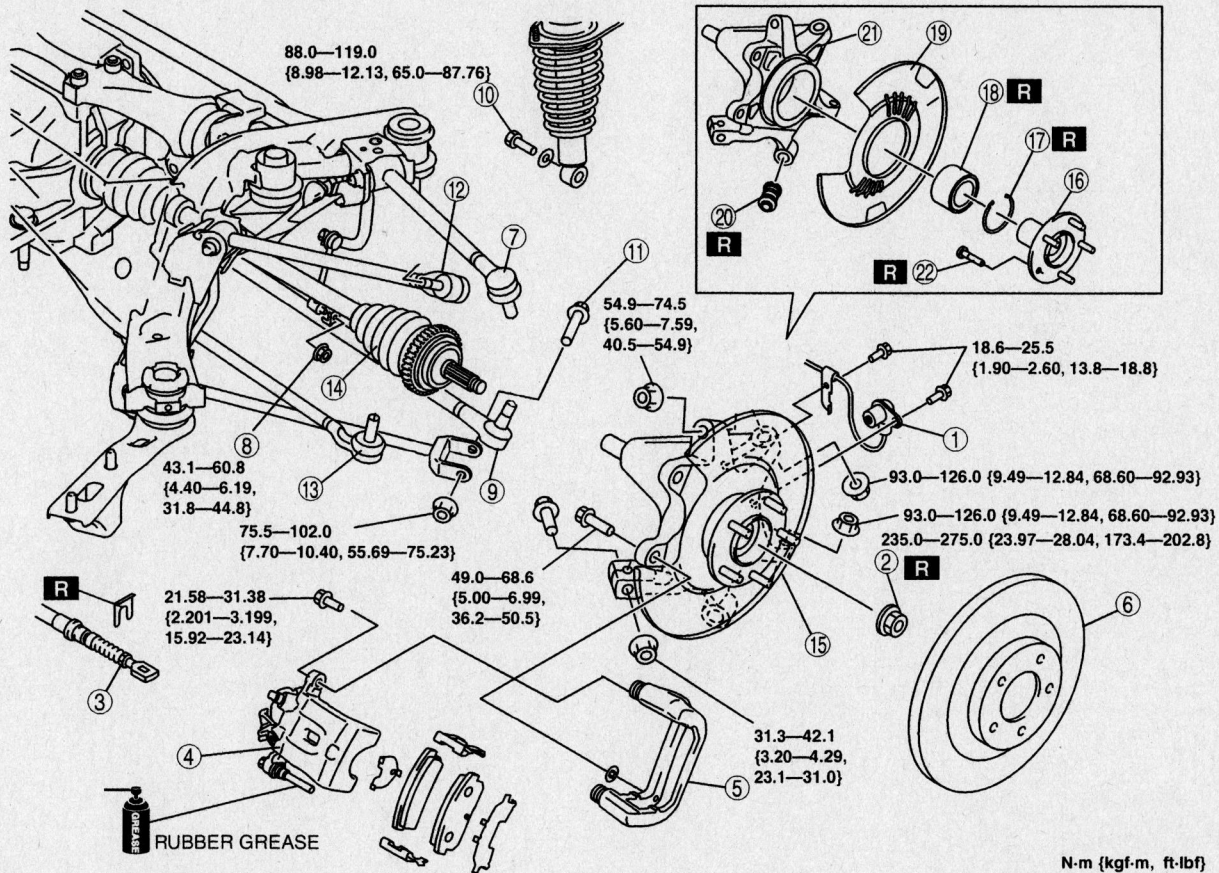

N·m {kgf·m, ft·lbf}

1	ABS wheel-speed sensor	12	Rear trailing link (upper) ball joint
2	Locknut	13	Toe control link outside bolt
3	Parking brake cable	14	Rear drive shaft
4	Brake caliper component	15	Rear knuckle component
5	Mounting support	16	Wheel hub component
6	Disc plate	17	Retaining ring
7	Rear lateral link (upper) ball joint	18	Wheel bearing
8	Stabilizer control link (lower)	19	Dust cover
9	Rear lateral link (lower) ball joint	20	Bushing
10	Shock absorber bolt (lower)	21	Rear knuckle
11	Rear trailing link (lower) outside bolt	22	Wheel hub bolt

Exploded view of the rear wheel hub and bearing–RX-8

BRAKES

Brake Caliper

REMOVAL & INSTALLATION

Front

1. Before servicing the vehicle, refer to the Precautions Section.

2. Remove or disconnect the following:
   - Wheels
   - Flexible brake hose from the caliper
   - Caliper bolt
   - Caliper

To install:

3. Install or connect the following:
   - Caliper on the brake disc

- Caliper mounting bolts and tighten the bolts 16–23 ft. lbs. (22–31 Nm)
- Brake hose to the caliper and tighten the hose nut to 16–21 ft. lbs. (22–29 Nm)

4. Bleed the brake system.
5. Install the wheels.

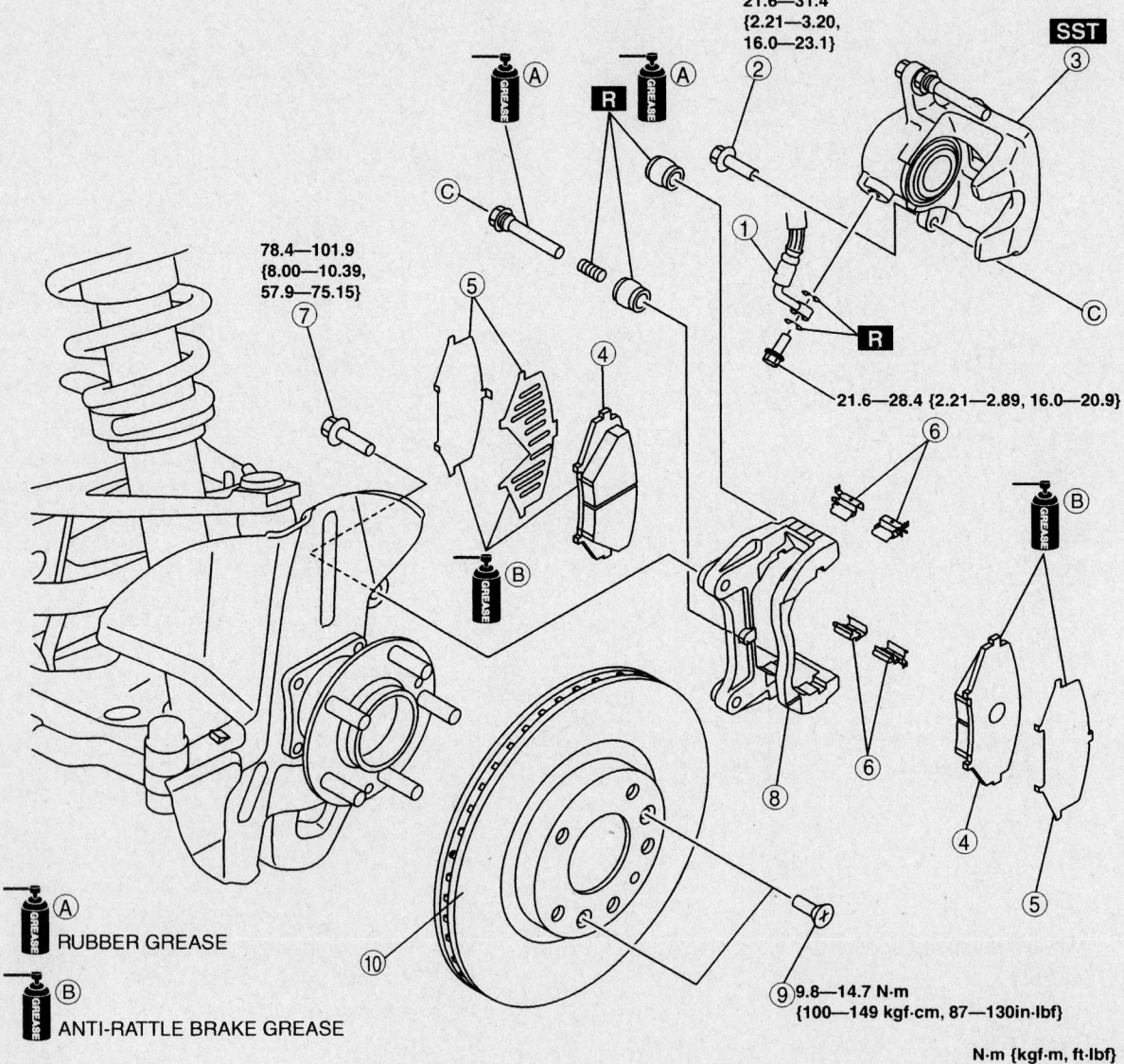

@ RUBBER GREASE

@ ANTI-RATTLE BRAKE GREASE

21.6—31.4 {2.21—3.20, 16.0—23.1}

78.4—101.9 {8.00—10.39, 57.9—75.15}

21.6—28.4 {2.21—2.89, 16.0—20.9}

9.8—14.7 N·m {100—149 kgf·cm, 87—130in·lbf}

N·m {kgf·m, ft·lbf}

1	Brake hose	6	Guide plate
2	Bolt	7	Bolt
3	Caliper	8	Mounting support
4	Disc pad	9	Screw
5	Shim	10	Disc plate

67162-RX8-G41

Exploded view of the front brake caliper–RX-8

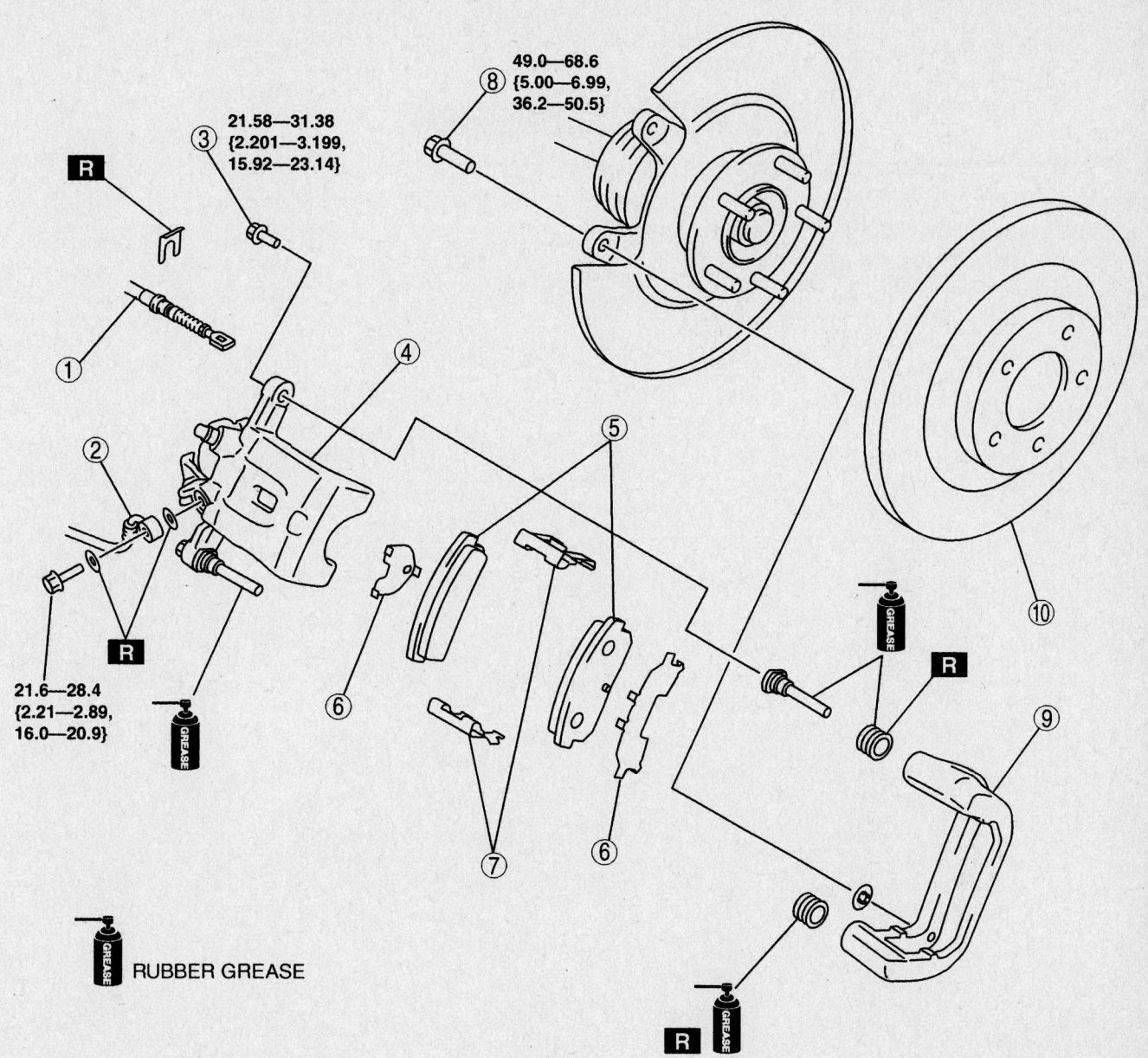

49.0—68.6
{5.00—6.99,
36.2—50.5}

21.58—31.38
{2.201—3.199,
15.92—23.14}

21.6—28.4
{2.21—2.89,
16.0—20.9}

GREASE RUBBER GREASE

N·m {kgf·m, ft·lbf}

1	Parking brake cable	6	Shim
2	Brake hose	7	Guide plate
3	Bolt	8	Bolt
4	Caliper	9	Mounting support
5	Disc pad	10	Disc plate

67162-RX8-G42

Exploded view of rear brake system–RX-8

Rear

1. Before servicing the vehicle, refer to the Precautions Section.
2. Remove or disconnect the following:
   - Wheels
   - Flexible brake line from the caliper assembly
   - Caliper mounting bolts
   - Caliper

To install:

3. Install or connect the following:
   - Caliper. Torque the caliper mount bolts to 16–23 ft. lbs. (22–32 Nm).

- Brake hose. Torque the line bolt to 16–22 ft. lbs. (22–30 Nm).
- Parking brake cable

4. Install the wheels and bleed the brake system.

Disc Brake Pads

REMOVAL & INSTALLATION

Front and Rear

1. Before servicing the vehicle, refer to the Precautions Section.

2. Remove or disconnect the following:
   - Wheels
   - Caliper
   - Brake pads
   - Shim
   - Guide plate

To install:

3. Install or connect the following:
   - Guide plate
   - Shim
   - Brake pads
   - Caliper and tighten the bolt to 16–23 ft. lbs. (22–32 Nm).

4. Install the wheels.

SPECIFICATIONS AND MAINTENANCE CHARTS

ENGINE AND VEHICLE IDENTIFICATION

Engine							Model Year	
Code ①	Liters	Cu. In.	Cyl.	Fuel Sys.	Engine Type	Eng. Mfg.	Code ②	Year
B	2.0	121	4	SFI	DOHC	Ford	2	2002
B	2.3	137	4	SFI	DOHC	Ford	3	2003
1	3.0	182	6	SFI	DOHC	Ford	4	2004
							5	2005
							6	2006

SFI: Multi-port Fuel Injection

DOHC: Double Overhead Camshafts

① 8th digit of VIN

② 10th digit of VIN

09482_TRIB_C0001

GENERAL ENGINE SPECIFICATIONS

Year	Model	Engine Displacement Liters	Engine VIN	Net Horsepower @ rpm	Net Torque @ rpm (ft. lbs.)	Bore x Stroke (in.)	Compression Ratio	Oil Pressure @ rpm
2002	Tribute	2.0	B	127@5500	135@4500	3.34x3.46	9.6:1	54-80 ①
	Tribute	3.0	1	200@5500	200@4500	3.50x3.13	10.0:1	45 ①
2003	Tribute	2.0	B	127@5500	135@4500	3.34x3.46	9.6:1	54-80 ①
	Tribute	3.0	1	200@5500	200@4500	3.50x3.13	10.0:1	45 ①
2004	Tribute	2.0	B	127@5500	135@4500	3.34x3.46	9.6:1	54-80 ①
	Tribute	3.0	1	200@5500	200@4500	3.50x3.13	10.0:1	45 ①
2005	Tribute	2.3	B	153@5800	152@4250	3.44x3.70	9.7:1	29-39@2000
	Tribute	3.0	1	200@6000	193@4850	3.50x3.13	10.0:1	11@1500②
2006	Tribute	2.3	B	153@5800	152@4250	3.44x3.70	9.7:1	29-39@2000
	Tribute	3.0	1	200@6000	193@4850	3.50x3.13	10.0:1	11@1500②

① The manufacturer does not provide an engine speed specification for oil pump pressure.

② Minimum hot

09482_TRIB_C0002

ENGINE TUNE-UP SPECIFICATIONS

Year	Engine Displacement Liters	Engine VIN	Spark Plug Gap (in.)	Ignition Timing (deg.)		Fuel Pump (psi)	Idle Speed (rpm)		Valve Clearance	
				MT	AT		MT	AT	Intake	Exhaust
2002	2.0	B	0.051	10 BTDC	—	65②	①	—	HYD.	HYD.
	3.0	1	0.052-0.056	10 BTDC	10 BTDC	65②	①	①	HYD.	HYD.
2003	2.0	B	0.051	10 BTDC	—	65②	①	—	HYD.	HYD.
	3.0	1	0.052-0.056	10 BTDC	10 BTDC	65②	①	①	HYD.	HYD.
2004	2.0	B	0.051	10 BTDC	—	65②	①	—	HYD.	HYD.
	3.0	1	0.052-0.056	10 BTDC	10 BTDC	65②	①	①	HYD.	HYD.
2005	2.3	B	0.049-0.053	①	①	35-55	①	①	0.008-0.011	0.010-0.013
	3.0	1	0.052-0.056	①	①	35-55	①	①	HYD.	HYD.
2006	2.3	B	0.049-0.053	①	①	35-55	①	①	0.008-0.011	0.010-0.013
	3.0	1	0.052-0.056	①	①	35-55	①	①	HYD.	HYD.

BTDC: Before Top Dead Center

HYD: Hydraulic lash adjusters

① Refer to Vehicle Emission Control Information Label

② Key on; engine off

09482_TRIB_C0003

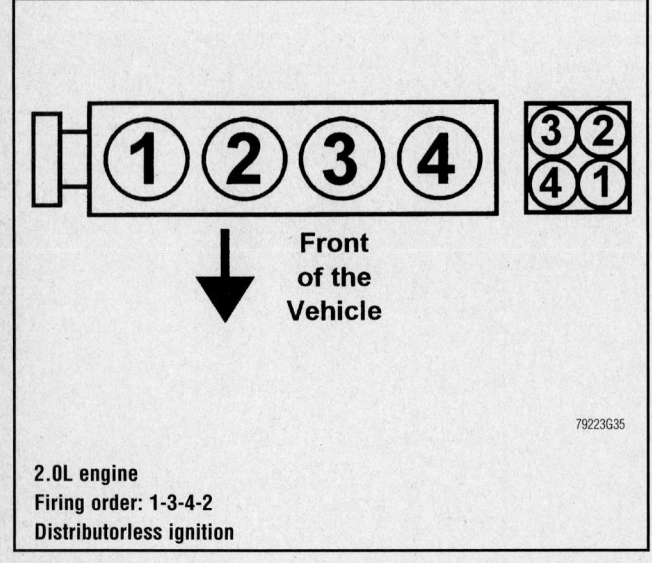

2.0L engine
Firing order: 1-3-4-2
Distributorless ignition

79223G35

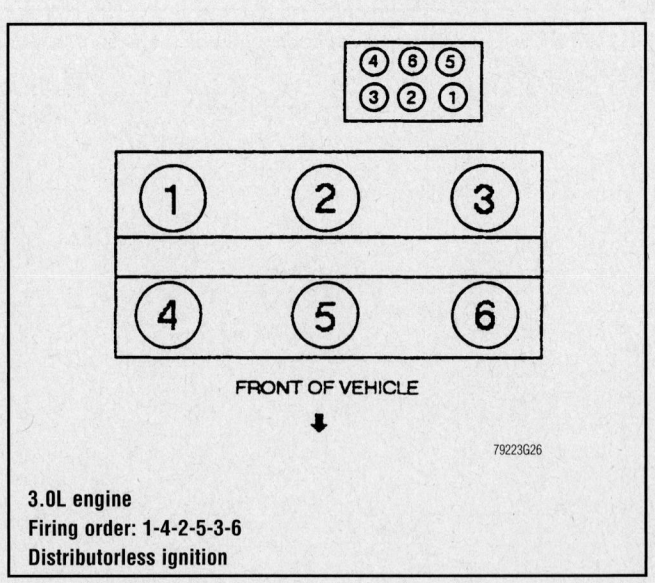

FRONT OF VEHICLE

3.0L engine
Firing order: 1-4-2-5-3-6
Distributorless ignition

79223G26

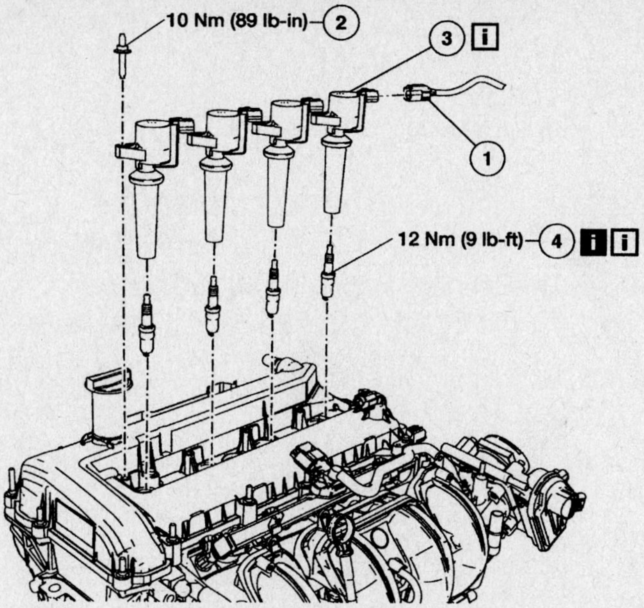

10 Nm (89 lb-in)—②

③ i

①

12 Nm (9 lb-ft)—④ i i

1 Ignition coil-on-plug electrical connectors
2 Ignition coil-to-valve cover bolts
3 Ignition coils
4 Spark plugs

67197-ESCA-G61

Coil and spark plug arrangement—2.3L engine

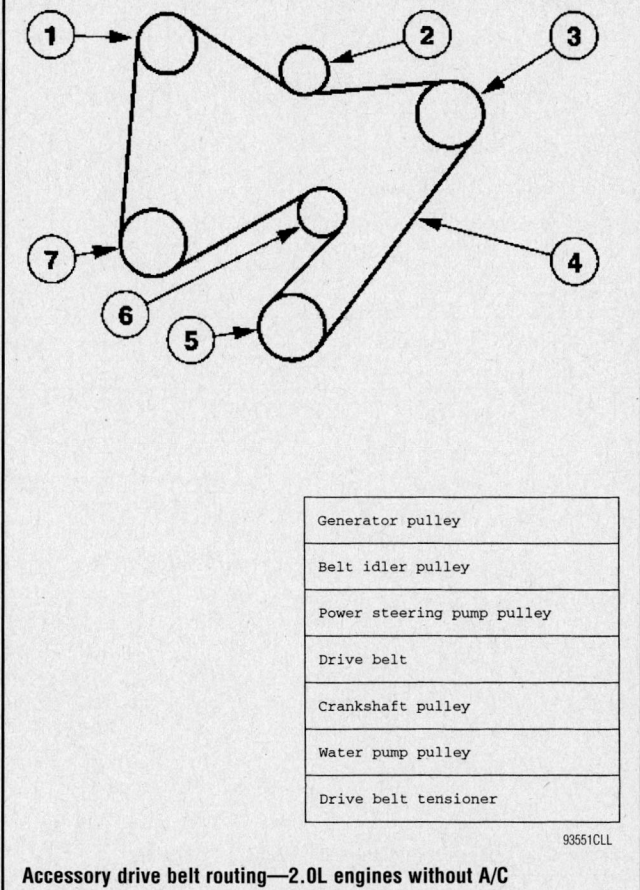

Generator pulley
Belt idler pulley
Power steering pump pulley
Drive belt
Crankshaft pulley
Water pump pulley
Drive belt tensioner

93551CLL

Accessory drive belt routing—2.0L engines without A/C

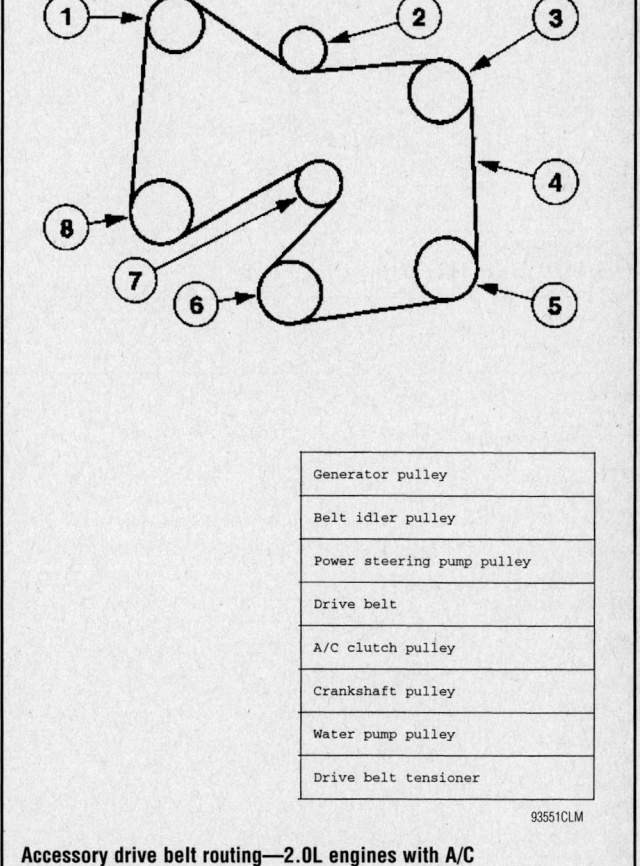

Generator pulley
Belt idler pulley
Power steering pump pulley
Drive belt
A/C clutch pulley
Crankshaft pulley
Water pump pulley
Drive belt tensioner

93551CLM

Accessory drive belt routing—2.0L engines with A/C

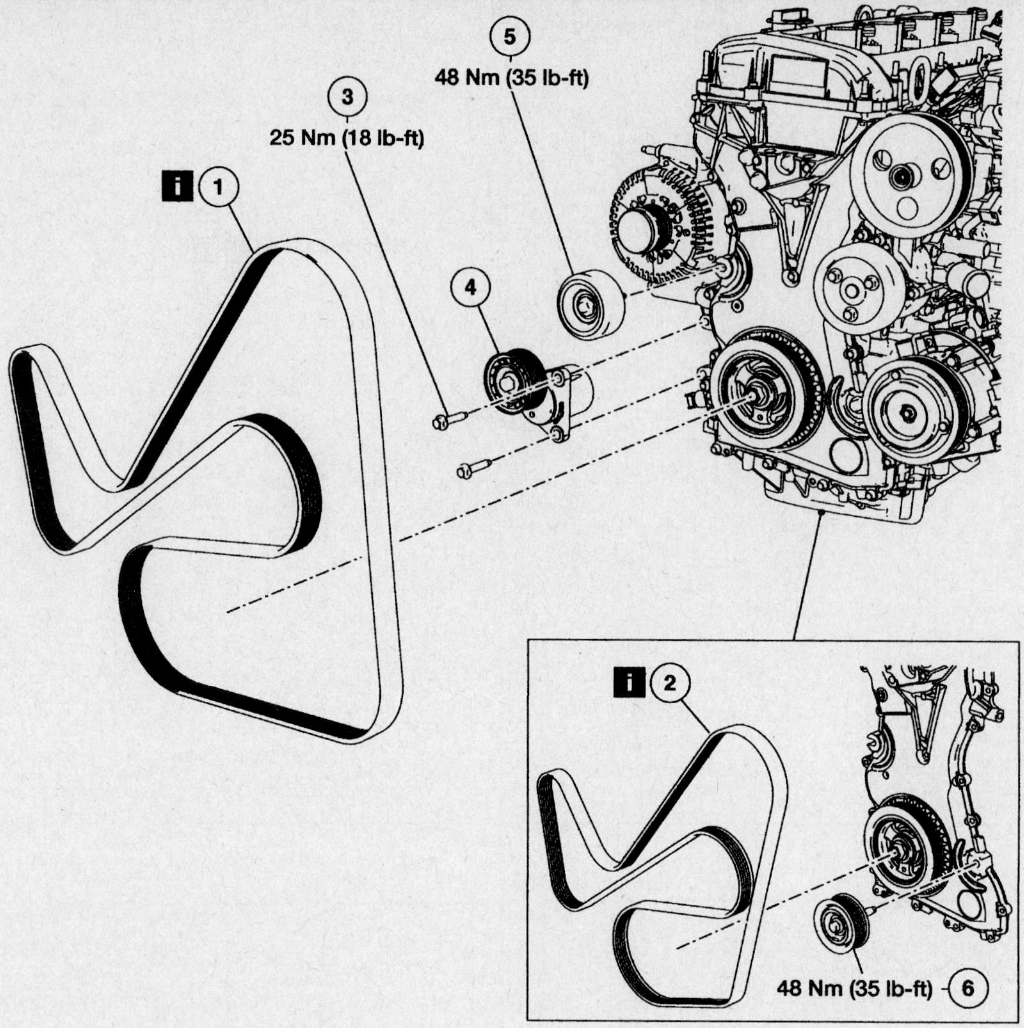

⑤
48 Nm (35 lb-ft)

③
25 Nm (18 lb-ft)

ℹ①

④

⑤

ℹ②

48 Nm (35 lb-ft) ⑥

1 Accessory drive belt (with A/C)
2 Accessory drive belt (without A/C)
3 Accessory drive belt tensioner bolts
4 Accessory drive belt tensioner
5 Accessory drive belt idler pulley
6 Accessory drive belt idler pulley (without A/C only)

67197-ESCA-G62

Accessory drive belt routings—2.3L engine

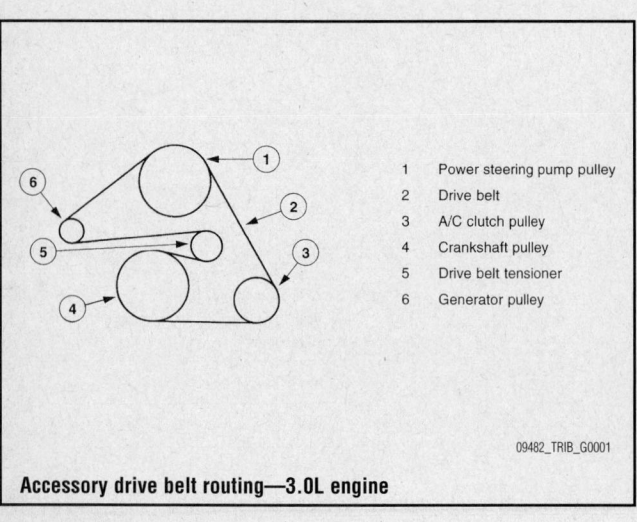

1 Power steering pump pulley
2 Drive belt
3 A/C clutch pulley
4 Crankshaft pulley
5 Drive belt tensioner
6 Generator pulley

09482_TRIB_G0001

Accessory drive belt routing—3.0L engine

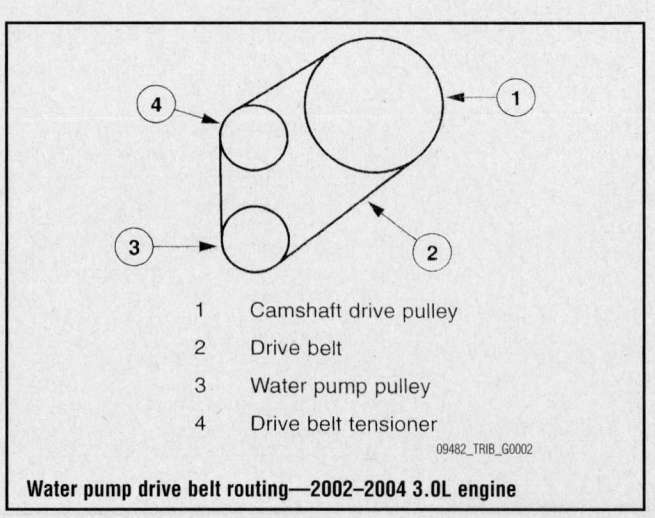

1 Camshaft drive pulley
2 Drive belt
3 Water pump pulley
4 Drive belt tensioner

09482_TRIB_G0002

Water pump drive belt routing—2002–2004 3.0L engine

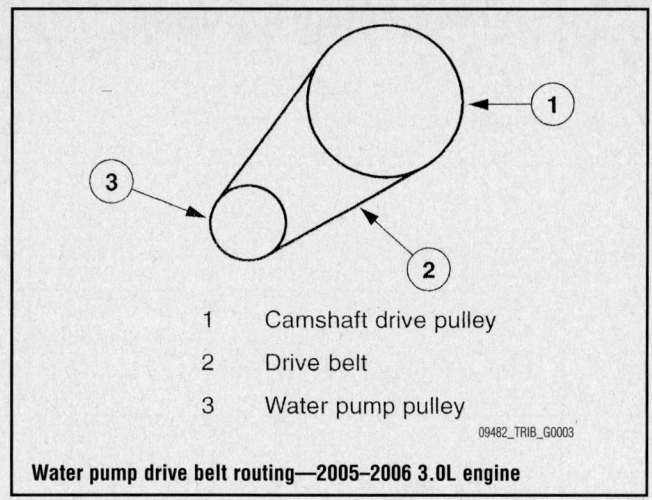

1   Camshaft drive pulley

2   Drive belt

3   Water pump pulley

09482_TRIB_G0003

Water pump drive belt routing—2005–2006 3.0L engine

CAPACITIES

Year	Model	Engine Displacement Liters	Engine VIN	Engine Oil with Filter (qts.)	Transmission (pts.) Manual	Transmission (pts.) Auto.	Transfer Case (pts.)	Drive Axle Front (pts.)	Drive Axle Rear (pts.)	Fuel Tank (gal.)	Cooling System (qts.)
2002	Tribute	2.0	B	4.5	5.7	—	①	2.6	3.0	15.0	7.0
	Tribute	3.0	1	5.5	5.7	20.0	①	2.6	3.0	16.0	10.5
2003	Tribute	2.0	B	4.5	5.7	—	—	—	3.0	15.0	7.0
	Tribute	3.0	1	5.5	5.7	20.0	①	2.6	3.0	16.0	10.5
2004	Tribute	2.0	B	4.5	5.7	—	—	—	3.0	15.0	7.0
	Tribute	3.0	1	5.5	5.7	20.0	①	2.6	3.0	16.0	10.5
2005	Tribute	2.3	B	4.5	5.0	20.0	2.95	—	3.0	16.5	②
	Tribute	3.0	1	5.5	NA	20.0	2.95	—	3.0	16.5	10.5
2006	Tribute	2.3	B	4.5	5.0	20.0	2.95	—	3.0	16.5	②
	Tribute	3.0	1	5.5	NA	20.0	2.95	—	3.0	16.5	10.5

NA: Information not available

NOTE: All capacities are approximate. Add fluid gradually and check to be sure a proper fluid level is obtained.

① The transfer case is lubricated for life and is not to be checked unless a leak is suspected or a repair is necessary.

② Manual transaxle: 5.3
   Automatic transaxle: 6.3

09482_TRIB_C0004

VALVE SPECIFICATIONS

Year	Engine Displacement Liters	Engine VIN	Seat Angle (deg.)	Face Angle (deg.)	Spring Test Pressure (lbs. @ in.)	Spring Installed Height (in.)	Stem-to-Guide Clearance (in.)		Stem Diameter (in.)	
							Intake	Exhaust	Intake	Exhaust
2002	2.0	B	45	45	①	1.346	0.0007-0.0025	0.0007-0.0025	0.2374	0.2374
	3.0	1	44.75	45.5	153@ 1.18	1.570	0.0008-0.0027	0.0018-0.0037	0.2352-0.2360	0.2343-0.2350
2003	2.0	B	45	45	①	1.346	0.0007-0.0025	0.0007-0.0025	0.2374	0.2374
	3.0	1	44.75	45.5	153@ 1.18	1.570	0.0008-0.0027	0.0018-0.0037	0.2352-0.2360	0.2343-0.2350
2004	2.0	B	45	45	①	1.346	0.0007-0.0025	0.0007-0.0025	0.2374	0.2374
	3.0	1	44.75	45.5	153@ 1.18	1.570	0.0008-0.0027	0.0018-0.0037	0.2352-0.2360	0.2343-0.2350
2005	2.3	B	45	NA	38.6@1.49	1.496	0.0009	0.0011	0.2153-0.2159	0.2151-0.2157
	3.0	1	44.75	45.5	156@ 1.18	1.570	0.0007-0.0027	0.0017-0.0037	0.2350-0.2358	0.2343-0.2350
2006	2.3	B	45	NA	38.6@1.49	1.496	0.0009	0.0011	0.2153-0.2159	0.2151-0.2157
	3.0	1	44.75	45.5	156@ 1.18	1.570	0.0007-0.0027	0.0017-0.0037	0.2350-0.2358	0.2343-0.2350

NA: Information not available

① Intake: 82.1@ 0.988

　Exhaust: 95@ 1.0275

09482_TRIB_C0005

CAMSHAFT SPECIFICATIONS CHART

All measurements are given in inches.

Year	Engine Displ. Liters	Engine VIN	Journal Dia.	Brg. Oil Clearance	Shaft End-play	Runout	Lobe Height	
							Intake	Exhaust
2002	2.0	B	1.0221-1.0227	NA	0.00315-0.00866	NA	0.0137	0.0134
	3.0	1	1.0600-1.0610	NA	0.00100-0.00740	NA	0.1880	0.1880
2003	2.0	B	1.0221-1.0227	NA	0.00315-0.00866	NA	0.0137	0.0134
	3.0	1	1.0600-1.0610	NA	0.00100-0.00740	NA	0.1880	0.1880
2003	2.0	B	1.0221-1.0227	NA	0.00315-0.00866	NA	0.0137	0.0134
	3.0	1	1.0600-1.0610	NA	0.00100-0.00740	NA	0.1880	0.1880
2005	2.3	B	0.9820-0.9830	NA	NA	0.0010	0.3240	0.3070
	3.0	1	1.0600-1.0610	NA	0.00748 ①	NA	0.1890	0.1890
2006	2.3	B	0.9820-0.9830	NA	NA	0.0010	0.3240	0.3070
	3.0	1	1.0600-1.0610	NA	0.00748 ①	NA	0.1890	0.1890

NA: Information not available

① service limit

09482_TRIB_C0006

CRANKSHAFT AND CONNECTING ROD SPECIFICATIONS

All measurements are given in inches.

Year	Engine Displacement Liters)	Engine VIN	Crankshaft				Connecting Rod		
			Main Brg. Journal Dia.	Main Brg. Oil Clearance	Shaft End-play	Thrust on No.	Journal Diameter	Oil Clearance	Side Clearance
2002	2.0	B	2.2827-2.2835	①	0.0035-0.0102	3	1.8460-1.8468	0.0006-0.0028	0.0040-0.0110
	3.0	1	2.4670-2.4790	0.0009-0.0019	0.0040-0.0090	3	1.9670-1.9680	0.0010-0.0025	0.0039-0.0118
2003	2.0	B	2.2827-2.2835	①	0.0035-0.0102	3	1.8460-1.8468	0.0006-0.0028	0.0040-0.0110
	3.0	1	2.4670-2.4790	0.0009-0.0019	0.0040-0.0090	3	1.9670-1.9680	0.0010-0.0025	0.0039-0.0118
2004	2.0	B	2.2827-2.2835	①	0.0035-0.0102	3	1.8460-1.8468	0.0006-0.0028	0.0040-0.0110
	3.0	1	2.4670-2.4790	0.0009-0.0019	0.0040-0.0090	3	1.9670-1.9680	0.0010-0.0025	0.0039-0.0118
2005	2.3	B	2.0460-2.0470	0.0007-0.0013	0.0080-0.0160	NA	1.9670-1.9680	0.0010-0.0020	0.0760-0.1200
	3.0	1	2.4670-2.4790	0.0009-0.0019	0.0050-0.0100	3	1.9670-1.9680	0.0010-0.0025	0.0039-0.0118
2006	2.3	B	2.0460-2.0470	0.0007-0.0013	0.0080-0.0160	NA	1.9670-1.9680	0.0010-0.0020	0.0760-0.1200
	3.0	1	2.4670-2.4790	0.0009-0.0019	0.0050-0.0100	3	1.9670-1.9680	0.0010-0.0025	0.0039-0.0118

NA: Information not available

① Journals 1, 2 and 4: 0.0010 - 0.0017 in.
   Journal 3: 0.0012 - 0.0019 in.

09482_TRIB_C0007

PISTON AND RING SPECIFICATIONS
All measurements are given in inches.

Year	Engine Displacement Liters	Engine VIN	Piston Clearance	Ring Gap			Ring Side Clearance		
				Top Compression	Bottom Compression	Oil Control	Top Compression	Bottom Compression	Oil Control
2002	2.0	B	0.0004-0.0012	0.0100-0.0300	0.0100-0.0300	0.0160-0.0660	0.0015-0.0032	0.0015-0.0035	snug
	3.0	1	0.0005-0.0009	0.0039-0.0098	0.0106-0.0165	0.0059-0.0256	0.0016-0.0030	0.0016-0.0033	snug
2003	2.0	B	0.0004-0.0012	0.0100-0.0300	0.0100-0.0300	0.0160-0.0660	0.0015-0.0032	0.0015-0.0035	snug
	3.0	1	0.0005-0.0009	0.0039-0.0098	0.0106-0.0165	0.0059-0.0256	0.0016-0.0030	0.0016-0.0033	snug
2004	2.0	B	0.0004-0.0012	0.0100-0.0300	0.0100-0.0300	0.0160-0.0660	0.0015-0.0032	0.0015-0.0035	snug
	3.0	1	0.0005-0.0009	0.0039-0.0098	0.0106-0.0165	0.0059-0.0256	0.0016-0.0030	0.0016-0.0033	snug
2005	2.3	B	0.0009-0.0017	0.0473-0.0474	0.0460-0.0468	0.0984-0.0985	NA	NA	NA
	3.0	1	0.0005-0.0009	0.0039-0.0098	0.0106-0.0165	0.0059-0.0256	0.0016-0.0030	0.0016-0.0033	snug
2006	2.3	B	0.0009-0.0017	0.0473-0.0474	0.0460-0.0468	0.0984-0.0985	NA	NA	NA
	3.0	1	0.0005-0.0009	0.0039-0.0098	0.0106-0.0165	0.0059-0.0256	0.0016-0.0030	0.0016-0.0033	snug

NA: Information not available

09482_TRIB_C0008

TORQUE SPECIFICATIONS
All readings in ft. lbs.

Year	Engine Displacement Liters	Engine VIN	Cylinder Head Bolts	Main Bearing Bolts	Rod Bearing Bolts	Crankshaft Damper Bolts	Flywheel Bolts	Manifold Intake	Manifold Exhaust	Spark Plugs	Oil Pan Drain Plug
2002	2.0	B	①	②	③	80-87	83	13	12	11	18
	3.0	1	④	⑤	⑥	⑦	59	⑧	15	11	NA
2003	2.0	B	①	②	③	80-87	83	13	12	11	18
	3.0	1	④	⑤	⑥	⑦	59	⑧	15	11	NA
2004	2.0	B	①	②	③	85	83	13	12	11	18
	3.0	1	④	⑤	⑥	⑦	59	⑧	15	11	NA
2005	2.3	B	⑨	NA	NA	⑩	⑪	13	35	11	21
	3.0	1	④	⑤	⑥	⑦	59	⑧	15	11	NA
2006	2.3	B	⑨	NA	NA	⑩	⑪	13	35	11	21
	3.0	1	④	⑤	⑥	⑦	59	⑧	15	11	NA

NA: Information not available

① Step 1: 15 ft. lbs. (20 Nm).
Step 2: 30 ft. lbs. (40 Nm).
Step 3: Plus an additional 90 degrees.

② Step 1: 18 ft. lbs.
Step 2: +60 degrees

③ Step 1: 26 ft. lbs.
Step 2: +90 degrees

④ Step 1: 30 ft. lbs. (40 Nm).
Step 2: Tighten the bolts 90 degrees.
Step 3: Loosen the bolts one full turn.
Step 4: 30 ft. lbs. (40 Nm).
Step 5: Tighten the bolts 90 degrees.
Step 6: Tighten the bolts 90 degrees.

⑤ Step 1: Fasteners 1-8: 18 ft. lbs.
Step 2: Fasteners 9-19: 30 ft. lbs.
Step 3: Fasteners 1-16: +90 degrees
Step 4: fasteners 17-22: 18 ft. lbs.

⑥ Step 1: 17 ft. lbs.
Step 2: 32 ft. lbs.

⑦ Step 1: 89 ft. lbs.
Step 2: Loosen 1 full turn
Step 3: 37 ft. lbs.
Step 4: 66 ft. lbs.

⑧ 89 inch lbs.

⑨ Step 1: 44 inch lbs.
Step 2: 11 ft. lbs.
Step 3: 33 ft. lbs.
Step 4: +90 degrees
Step 5: + 90 degrees

⑩ Step2: 74 ft. lbs.
Step 2: plus 90 degrees

⑪ Step 1: 37 ft. lbs.
Step 2: 59 ft. lbs
Step 3: 83 ft. lbs.

09482_TRIB_C0009

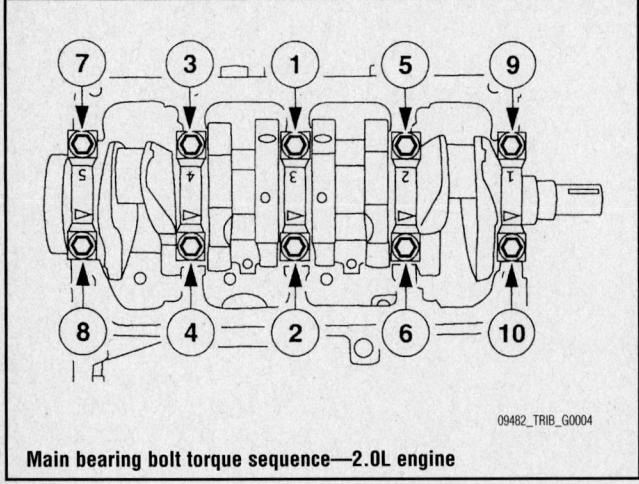

Main bearing bolt torque sequence—2.0L engine

09482_TRIB_G0004

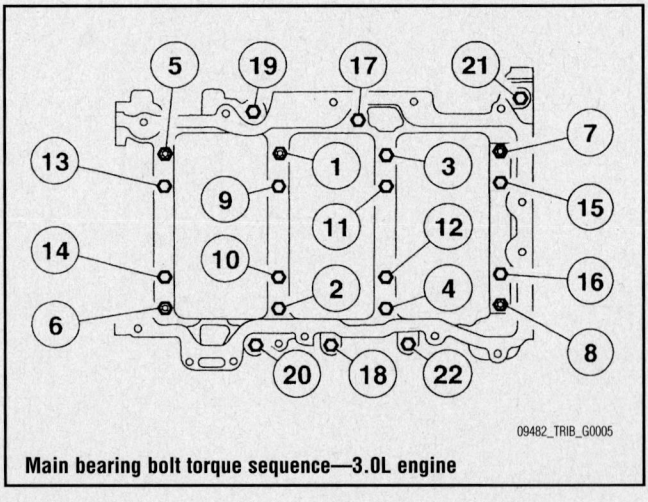

Main bearing bolt torque sequence—3.0L engine

09482_TRIB_G0005

WHEEL ALIGNMENT

Year	Model		Caster Range (+/-Deg.)	Caster Preferred Setting (Deg.)	Camber Range (+/-Deg.)	Camber Preferred Setting (Deg.)	Toe-in (in.)
2002	Tribute	F	NA	+1.93	NA	-0.84	0.12+/-0.12
		R	NA	NA	NA	-0.04	0.09+/-0.11
2003	Tribute	F	NA	+1.93	NA	-0.84	0.12+/-0.12
		R	NA	NA	NA	-0.04	0.09+/-0.11
2004①	4-cyl.	F	1.00	+1.72	1.00	-0.48	-0.08+/-0.32
		R	NA	NA	1.00	+0.13	0.04+/-0.17
	6-cyl.	F	1.00	+1.72	1.00	-0.84	0.23+/-0.32
		R	NA	NA	1.00	+0.13	0.10+/-0.17
2005	Tribute	F	1.00	+1.60	1.00	-1.00	-0.23+/-0.23
		R	NA	NA	0.75	+0.10	0.14+/-0.20
2006	Tribute	F	1.00	+1.60	1.00	-1.00	0.23+/-0.32
		R	NA	NA	0.75	+0.10	0.14+/-0.20

NA: Information not available

① Assumes 8 gallons of gas

09482_TRIB_C0010

TIRE, WHEEL AND BALL JOINT SPECIFICATIONS

Year	Model	OEM Tires Standard	OEM Tires Optional	Tire Pressures (psi) Front	Tire Pressures (psi) Rear	Wheel Size	Ball Joint Inspection	Lug Nuts (ft. lbs.)
2002	Tribute	P215/70R16	P225/70SR15 P235/70R16	①	①	NA	0.030 in.	98
2003	Tribute	P215/70R16	P225/70SR15 P235/70R16	①	①	NA	0.030 in.	98
2004	Tribute	P225/70SR15	P235/70R16	①	①	6.5	0.030 in.	98
2005	Tribute	P225/70SR15	P235/70R16	①	①	7.5	0.030 in.	98
2006	Tribute	P235/70R16		①	①	7.5	0.030 in.	98

OEM: Original Equipment Manufacturer

PSI: Pounds Per Square Inch

NA: Not Available

① See certification label located on the inside driver's side front door jamb.

09482_TRIB_C0011

BRAKE SPECIFICATIONS
All measurements in inches unless noted

Year	Model		Brake Disc			Brake Drum		Minimum Lining Thickness	Brake Caliper	
			Original Thickness	Minimum Thickness	Maximum Run-out	Original Inside Diameter	Maximum Machine Diameter		Bracket Bolts (ft. lbs.)	Mounting Bolts (ft. lbs.)
2002	Tribute		0.940	0.860	0.004	9.00	9.06	0.039	111	26
2003	Tribute		0.940	0.860	0.002	9.00	9.06	0.039	111	26
2004	Tribute		0.940	0.860	0.002	9.00	9.06	0.039	111	26
2005	Tribute	F	NA	①	0.004	NA	9.05	0.118	129	②
		R	NA	0.430	0.004	—	—	③	NA	26
2006	Tribute	F	NA	①	0.004	NA	9.05	0.118	129	②
		R	NA	0.430	0.004	—	—	③	NA	26

NA: Information not available

① With disc/drum: 0.880 in.

   With 4-wheel disc: 0.950 in.

② With disc/drum: 26 ft. lbs.

   With 4-wheel disc: 33 ft. lbs.

③ Drum brake shoe: 0.030

   Disk brake pad: 0.118

09482_TRIB_C0012

SCHEDULED MAINTENANCE INTERVALS
2002-03 Mazda Tribute

TO BE SERVICED	TYPE OF SERVICE	VEHICLE MILEAGE INTERVAL (x1000)												
		5	10	15	20	25	30	35	40	45	50	55	60	65
Air cleaner filter	R						✓						✓	
Accessory drive belt	S/I	Every 100,000 miles												
Accessory drive belt	R	At 120,000 miles if not previously done so												
Auto. Trans. Fluid level	I			✓			✓			✓			✓	
Auto. Trans. Fluid	②						✓						✓	
Auto. Trans. Fluid	③	Every 150,000 miles												
Brake system ①	S/I			✓			✓			✓			✓	
Cabin air filter	R			✓			✓			✓			✓	
Camshaft belt (2.0L)	R	Every 120,000 miles												
Cooling system hoses and clamps	S/I			✓			✓			✓			✓	
Engine coolant (green coolant)	R	Every 75,000 miles and every 30,000 miles thereafter												
Engine coolant (yellow coolant)	R	At 5 years or 100,000 miles												
Engine oil & filter	R	✓	✓	✓	✓	✓	✓	✓	✓	✓	✓	✓	✓	✓
PCV valve	R	Every 100,000 miles												
Exhaust system & heat shields	S/I						✓						✓	
Fuel filter	R						✓						✓	
Rear axle lubricant	R	Every 150,000 miles												
Rotate tires	S/I	✓	✓	✓	✓	✓	✓	✓	✓	✓	✓	✓	✓	✓
Steering linkage	S/I			✓			✓			✓			✓	
Spark plugs	R	Change at 100,000 miles												
Suspension components	S/I			✓			✓			✓			✓	
Wheels for play and noise	I			✓			✓			✓			✓	

R: Replace   S/I: Inspect and service, if necessary   L: Lubricate   A: Adjust   C: Clean

① Inspect the reservoir fluid level, rotor and or drum, brake lines, hoses, calipers and or wheel cylinders

② Change automatic transmission/transaxle fluid on all vehicles equipped with AX4S, 4F50N, 4R100, 4F27E

Special Operating Condition Requirements

When towing a trailer or using a camper or car-top carrier:

Change engine oil and install a new oil filter every 4,800 km (3,000 miles) or 3 months.

Change transfer case fluid every 96,000 km (60,000 miles).

Change manual transmission fluid as required.

Inspect and lubricate U-joints as required.

During extensive idling and/or low speed driving for long distances, as in heavy commercial use such as delivery, taxi, patrol car or livery:

Change engine oil and install a new oil filter, lube front lower control arm and steering linkage ball joints with

zerk fittings (if equipped) every 4,800 km (3,000 miles) or 3 months.

Inspect brake system and check battery electrolyte level (Patrol cars) every 8,000 km (5,000 miles).

Install a new fuel filter every 24,000 km (15,000 miles).

Change automatic transmission fluid, lubricate 4x2 wheel bearings,

 install new grease seals and adjust bearings every 48,000 km (30,000 miles).

Install new spark plugs and change transfer case fluid every 96,000 km (60,000 miles).

Install a new cabin air filter as required.

When operating in dusty conditions such as unpaved or dusty roads:

Change engine oil and install a new oil filter every 4,800 km (3,000 miles) or 3 months.

Install a new fuel filter every 24,000 km (15,000 miles).

Change automatic transmission fluid every 48,000 km (30,000 miles).

Change transfer case fluid every 96,000 km (60,000 miles).

Install a new engine air filter as required.

Install a new cabin air filter as required.

SCHEDULED MAINTENANCE INTERVALS
2002-03 Mazda Tribute
Footnotes Continued

When operating in off-road conditions:

Change automatic transmission fluid every 48,000 km (30,000 miles).

Change transfer case fluid every 96,000 km (60,000 miles).

Install a new cabin air filter as required.

Inspect and lubricate U-joints.

Inspect and lubricate steering linkage ball joints with zerk fittings.

09482_TRIB_C0014

SCHEDULED MAINTENANCE INTERVALS
2004 Mazda Tribute

TO BE SERVICED	TYPE OF SERVICE	VEHICLE MILEAGE INTERVAL (x1000)												
		5	10	15	20	25	30	35	40	45	50	55	60	65
Air cleaner filter	R						✓						✓	
Accessory drive belt	S/I	Every 100,000 miles												
Accessory drive belt	R	At 150,000 miles if not previously done so												
Auto. Trans. Fluid level	I			✓			✓			✓			✓	
Auto. Trans. Fluid	②						✓						✓	
Auto. Trans. Fluid	③	Every 150,000 miles												
Ball joints (2wd)	L			✓			✓			✓			✓	
Brake system ①	S/I			✓			✓			✓			✓	
Cabin air filter	R			✓			✓			✓			✓	
Camshaft belt (2.0L)	R	Every 120,000 miles												
Cooling system hoses and clamps	S/I			✓			✓			✓			✓	
Engine coolant (exc. Premium gold)	R	Every 105,000 miles, then every 3 years or 50,000 miles												
Engine coolant (Premium gold)	R	At 5 years or 100,000 miles												
Engine oil & filter	R	✓	✓	✓	✓	✓	✓	✓	✓	✓	✓	✓	✓	✓
PCV valve (external)	R	Every 100,000 miles												
Exhaust system & heat shields	S/I						✓						✓	
Front wheel bearings and seals	R	Every 150,000 miles, if not previously done so												
Fuel filter	R						✓						✓	
Rear axle lubricant	R	Every 150,000 miles												
Rotate tires	S/I	✓	✓	✓	✓	✓	✓	✓	✓	✓	✓	✓	✓	✓
Steering linkage	S/I			✓			✓			✓			✓	
Spark plugs	R	Change at 100,000 miles												
Suspension components	S/I			✓			✓			✓			✓	
Wheels for play and noise	I			✓			✓			✓			✓	

R: Replace   S/I: Inspect and service, if necessary   L: Lubricate   A: Adjust   C: Clean

① Inspect the reservoir fluid level, rotor and or drum, brake lines, hoses, calipers and or wheel cylinders

② Change automatic transmission/transaxle fluid and filter on all vehicles equipped with AX4S, 4F50N, 4R100, 4F27E.

③ All transaxles

Special Operating Condition Requirements

When towing a trailer or using a camper or car-top carrier:

Change engine oil and install a new oil filter every 4,800 km (3,000 miles) or 3 months.

Change transfer case fluid every 96,000 km (60,000 miles).

Change manual transmission fluid as required.

Inspect and lubricate U-joints as required.

During extensive idling and/or low speed driving for long distances, as in heavy commercial use such as delivery, taxi, patrol car or livery:

Change engine oil and install a new oil filter, lube front lower control arm and steering linkage ball joints with

zerk fittings (if equipped) every 4,800 km (3,000 miles) or 3 months.

Inspect brake system and check battery electrolyte level (Patrol cars) every 8,000 km (5,000 miles).

Install a new fuel filter every 24,000 km (15,000 miles).

Change automatic transmission fluid, lubricate 4x2 wheel bearings,

 install new grease seals and adjust bearings every 48,000 km (30,000 miles).

Install new spark plugs and change transfer case fluid every 96,000 km (60,000 miles).

Install a new cabin air filter as required.

When operating in dusty conditions such as unpaved or dusty roads:

Change engine oil and install a new oil filter every 4,800 km (3,000 miles) or 3 months.

Install a new fuel filter every 24,000 km (15,000 miles).

Change automatic transmission fluid every 48,000 km (30,000 miles).

Change transfer case fluid every 96,000 km (60,000 miles).

SCHEDULED MAINTENANCE INTERVALS
2004 Mazda Tribute
Footnotes Continued

Install a new engine air filter as required.

Install a new cabin air filter as required.

When operating in off-road conditions:

Change automatic transmission fluid every 48,000 km (30,000 miles).

Change transfer case fluid every 96,000 km (60,000 miles).

Install a new cabin air filter as required.

Inspect and lubricate U-joints.

Inspect and lubricate steering linkage ball joints with zerk fittings.

09482_TRIB_C0016

SCHEDULED MAINTENANCE INTERVALS
2005-06 Mazda Tribute

TO BE SERVICED	TYPE OF SERVICE	5	10	15	20	25	30	35	40	45	50	55	60	65
		\multicolumn{13}{VEHICLE MILEAGE INTERVAL (x1000)}
Air cleaner filter	R						✓						✓	
Accessory drive belt	I ⑤						✓						✓	
Auto. Trans. fluid level	I		✓				✓			✓			✓	
Auto. Trans. Fluid	③ ④						✓						✓	
Ball joints (2wd)	L			✓			✓			✓			✓	
Brake system ①	S/I			✓			✓			✓			✓	
Cabin air filter	R			✓			✓			✓			✓	
Cooling system hoses and clamps	S/I			✓			✓			✓			✓	
Driveshafts & halfshafts	S/I			✓			✓			✓			✓	
Engine coolant (Premium Gold)	R	Five years or 100,000 miles, then every 3 years or 50,000 miles												
Engine coolant (exc. Premium Gold)	R	Every 105,000 miles												
Engine oil & filter	R	✓	✓	✓	✓	✓	✓	✓	✓	✓	✓	✓	✓	✓
Front wheel bearings and seals (2wd)	R	Every 150,000 miles, if not previously done												
Fuel filter	R						✓						✓	
Man. Trans. Fluid	R	Every 90,000 miles												
PCV valve, 3.0L engine	S/I	Every 100,000 miles												
Exhaust system & heat shields	S/I						✓						✓	
Rear axle lubricant (4wd)	R	Every 150,000 miles												
Rotate tires	S/I	✓	✓	✓	✓	✓	✓	✓	✓	✓	✓	✓	✓	✓
Steering linkage	S/I			✓			✓			✓			✓	
Spark plugs	R	Change at 75,000 miles												
Suspension components	S/I			✓			✓			✓			✓	
Transfer case	R	Every 150,000 miles												
Valve adjustment 2.3L engine	S/I						✓						✓	
Wheels ②	I			✓			✓			✓			✓	

R: Replace   S/I: Inspect and service, if necessary   L: Lubricate   A: Adjust   C: Clean

① Inspect the reservoir fluid level, rotor and or drum, brake lines, hoses, calipers and or wheel cylinders

② Inspect for end play and noise

③ Change automatic transmission/transaxle fluid and filter on all vehicles equipped with 4F50N, 4R100 and 4F27E.

④ Change every 150,000 miles for all transaxles

⑤ Replace at 100,000 miles, if not previously done

Special Operating Condition Requirements

When towing a trailer or using a camper or car-top carrier:

Change engine oil and install a new oil filter every 4,800 km (3,000 miles) or 3 months.

Change transfer case fluid every 96,000 km (60,000 miles).

Change manual transmission fluid as required.

Inspect and lubricate U-joints as required.

During extensive idling and/or low speed driving for long distances, as in heavy commercial use such as delivery, taxi, patrol car or livery:

Change engine oil and install a new oil filter, lube front lower control arm and steering linkage ball joints with zerk fittings (if equipped) every 4,800 km (3,000 miles) or 3 months.

Inspect brake system and check battery electrolyte level (Patrol cars) every 8,000 km (5,000 miles).

Install a new fuel filter every 24,000 km (15,000 miles).

Change automatic transmission fluid, lubricate 4x2 wheel bearings, install new grease seals and adjust bearings every 48,000 km (30,000 miles).

Install new spark plugs and change transfer case fluid every 96,000 km (60,000 miles).

Install a new cabin air filter as required.

SCHEDULED MAINTENANCE INTERVALS
2005-06 Mazda Tribute
Footnotes Continued

When operating in dusty conditions such as unpaved or dusty roads:

Change engine oil and install a new oil filter every 4,800 km (3,000 miles) or 3 months.

Install a new fuel filter every 24,000 km (15,000 miles).

Change automatic transmission fluid every 48,000 km (30,000 miles).

Change transfer case fluid every 96,000 km (60,000 miles).

Install a new engine air filter as required.

Install a new cabin air filter as required.

When operating in off-road conditions:

Change automatic transmission fluid every 48,000 km (30,000 miles).

Change transfer case fluid every 96,000 km (60,000 miles).

Install a new cabin air filter as required.

Inspect and lubricate U-joints.

Inspect and lubricate steering linkage ball joints with zerk fittings.

09482_TRIB_C0018

ENGINE REPAIR

Distributor

This vehicle uses a Direct Ignition System (DIS). No distributor is used.

Alternator

REMOVAL & INSTALLATION

2.0L Engine

1. Remove or disconnect the following:
   - Negative battery cable
   - Drive belt
   - Alternator electrical connectors and loosen the upper alternator bolt while moving the alternator to the rear of the engine
   - Alternator. Torque the lower bolt to 35 ft. lbs. (48Nm). Torque the upper bolt to 18 ft. lbs. (25 Nm).

To install:

2. Install or connect the following:
   - Alternator with the upper bolt in the alternator before installation. Torque the bolts to 18 ft. lbs. (25 Nm).
   - Alternator electrical connectors
   - Drive belt
   - Negative battery cable

2.3L Engine

1. Disconnect the battery.
2. Remove the front end accessory drive belt tensioner. Rotate the front end accessory drive belt tensioner counterclockwise to loosen tension on the front end accessory drive belt.
3. Remove the front end accessory drive belt.
4. Remove the alternator B+ terminal.
5. Remove the alternator electrical connector.
6. Remove the alternator lower air duct bolt.
7. Remove the alternator lower air duct. Press the locking tab to release the lower air duct from the upper air duct.
8. Remove the pin-type retainer.
9. Remove the alternator shield .
10. Remove the alternator stud nut.
11. Remove the alternator stud.
12. Remove the alternator bolts.
13. Remove the alternator.
14. To install, reverse the removal procedure. Observe the following torques:
   - Alternator mounting bolts: 35 ft. lbs. (47 Nm)
   - Alternator stud: 18 ft. lbs. (24 Nm)
   - Stud nut: 35 ft. lbs. (47 Nm)

11 - 47 Nm (35 lb-ft)
24 Nm (18 lb-ft) - 12
20 Nm (15 lb-ft)
47 Nm (35 lb-ft) - 13
6 - 4 Nm (35 lb-in)

1 Front end accessory drive belt tensioner
2 Front end accessory drive belt
3 Generator B+ terminal nut
4 Generator B+ terminal
5 Generator electrical connector
6 Generator lower air duct bolt
7 Generator lower air duct
8 Pin-type retainer
9 Generator shield nut
10 Generator shield
11 Generator stud nut
12 Generator stud
13 Generator bolts
14 Generator

67197-ESCA-G01

Alternator and related components—2.3L Engine

- Shield nut: 15 ft. lbs. (20 Nm)
- Lower air duct bolt: 35 inch lbs. (4 Nm)

3.0L Engine

2002–2004

1. Remove or disconnect the following:
   - Negative battery cable
   - Right side intermediate axle shaft
   - Right side splash shield and retainers
   - Drive belt
   - Alternator electrical connectors
   - Alternator. Torque the mounting and adjusting bolts to 35 ft. lbs. (48Nm).

To install:

2. Install or connect the following:
   - Alternator. Torque the bolts to 35 ft. lbs. (48 Nm).
   - Alternator electrical connectors
   - Drive belt
   - Negative battery cable

2005–2006

1. Before servicing the vehicle, refer to the Precautions Section.
2. Disconnect the negative battery cable.

➡ **To deplete the backup power supply energy, disconnect the negative battery cable and wait at least one minute before working on the vehicle. Be sure to disconnect auxiliary batteries and power supplies, if equipped.**

3. Remove the right side intermediate axle shaft.
4. Remove the lower splash shield retaining bolts and pin type retainers. Remove the lower splash shields.

➡ **The left lower splash shield must be removed first.**

5. Remove the nut and position the bracket and wiring harness to the side.
6. Rotate the front end accessory drive tensioner counterclockwise and position the accessory drive belt to the side.
7. Position the alternator B+ protective cover aside and remove the alternator B+ terminal.
8. Disconnect the alternator electrical connector.

➡ **Discard the removed clip. A new harness tie wrap with clip, must be used for installation. Part number 1F20-61-147.**

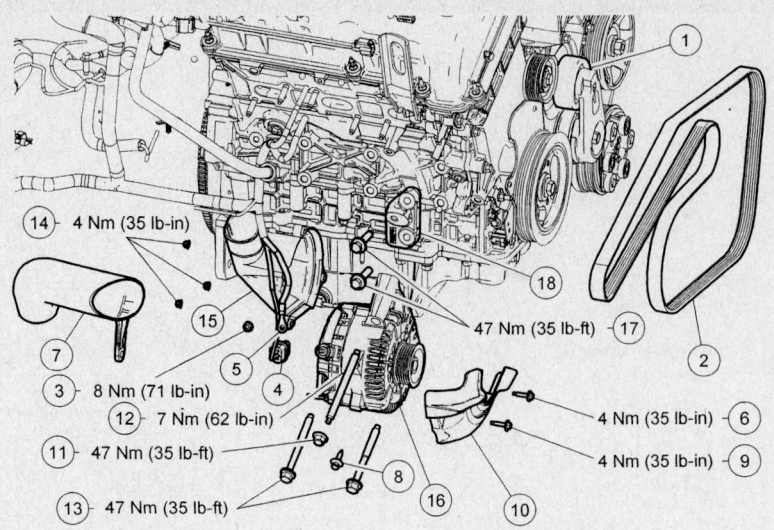

14 · 4 Nm (35 lb-in)

15

7

3 · 8 Nm (71 lb-in)
5
4
12 · 7 Nm (62 lb-in)
11 · 47 Nm (35 lb-ft)
8
13 · 47 Nm (35 lb-ft)
16
18
47 Nm (35 lb-ft) · 17
2
4 Nm (35 lb-in) · 6
4 Nm (35 lb-in) · 9
10

1	Front end accessory drive belt tensioner	10	Generator shield
2	Front end accessory drive belt	11	Generator stud nut
3	Generator B+ terminal nut	12	Generator stud
4	Generator electrical connector	13	Generator bolts (2 required)
5	Generator B+ cable	14	Generator lower air duct nuts (3 required)
6	Generator upper air duct bolt	15	Generator lower air duct
7	Generator upper air duct	16	Generator
8	Pin-type retainer	17	Generator bracket bolts (2 required)
9	Generator shield bolt	18	Generator bracket

09482_TRIB_G0006

Alternator and related components—2005–2006 3.0L engine

9. Remove the clip and cable from the alternator. Remove the alternator upper air duct bolt.

10. Loosen the two lower alternator bolts and remove the upper air duct.

11. Remove the pin type retainer from the alternator shield. Remove the bolt and alternator shield.

12. Remove the alternator stud nut. Remove the alternator stud. Remove the two alternator bolts.

13. Position the alternator to the side and remove the three alternator lower duct nuts.

14. Rotate the alternator to gain enough clearance and remove it through the opening in the fender well.

To install:

15. Installation is the reverse of the removal procedure.

Ignition Timing

ADJUSTMENT

Ignition timing is controlled by the Powertrain Control Module (PCM). No adjustment is necessary or possible.

Engine Assembly

REMOVAL & INSTALLATION

2.0L Engine

MANUAL TRANSAXLE

1. Before servicing the vehicle, refer to the Precautions Section.

2. Properly recover the air conditioning system refrigerant.

3. Properly relieve the fuel system pressure.

4. Drain the cooling system.

5. Drain the engine oil.

6. Remove or disconnect the following:

- Hood
- Battery and battery tray
- Air cleaner housing
- Fuel lines
- Throttle cable and speed control cable, if equipped
- Exhaust Gas Recirculation (EGR) vacuum valve regulator
- EGR electrical connectors and vacuum hoses
- Brake booster vacuum hose
- Powertrain Control Module (PCM) wire harness and ground
- Wire harness connector
- Power distribution board electrical connectors
- Evaporative emissions (EVAP) canister vacuum lines
- Upper radiator hose
- Power steering line bracket
- Upper power steering pump bolts
- Coolant hose
- Heater hoses
- Speed control unit, if equipped
- Catalytic converter
- A/C compressor
- Both halfshafts
- Shifter linkages
- Block heater electrical connector, if equipped
- Front transaxle through bolt
- Engine-to-transaxle bolts
- Lower radiator hose
- Power steering pump
- Clutch slave cylinder line from the bracket and move it aside
- Rear transaxle mount
- Left side transaxle mount
- Lower ground cable
- Engine mount upper bracket
- Engine and transaxle as an assembly by using a proper lifting device
- Alternator electrical connectors
- Knock Sensor (KS) electrical connector
- Oil pressure sender electrical connector
- Starter electrical connector
- Vehicle Speed Sensor (VSS) electrical connector
- Park Neutral Position (PNP) electrical connector
- Fuel charging wire harness electrical connector
- PCM wire harness from the bracket
- PCM ground wire
- Back up lamp switch electrical connector
- Wire harness
- Differential Pressure Feedback (DPFEE) EGR sensor

7. Separate the engine from the transaxle.

8. Lock the flywheel to the engine.

9. Clutch pressure plate and disc.

10. Flywheel and rear cover plates.

To install:

11. Install or connect the following:

- Flywheel. Torque the bolts to 83 ft. lbs. (112 Nm).
- Clutch disc to the flywheel.
- Pressure plate to the flywheel.

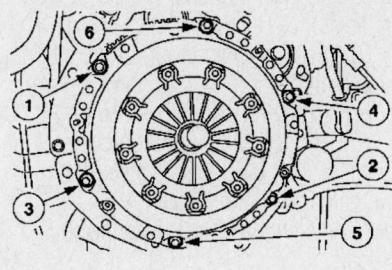

9308TG07

Pressure plate bolt torque sequence–2.0L engine

Torque the bolts in the proper sequence to 18 ft. lbs. (25 Nm).

- Transaxle to the engine. Torque the bolts to 33 ft. lbs. (45 Nm).
- Starter. Torque the bolts to 18 ft. lbs. (25 Nm).
- Wire harness and attach it to the powertrain assembly
- DPFEE sensor electrical connector
- Reverse lamp switch electrical connector
- Ground wire. Torque the bolt to 80 inch lbs. (9 Nm).
- PCM wire harness to the bracket
- Fuel charging wire harness electrical connector
- PNP switch electrical connector
- VSS electrical connector
- KS, Oil pressure sender and starter electrical connector. Torque the fasteners to 9 ft. lbs. (12 Nm).
- Alternator electrical connectors. Torque the fasteners to 71 inch lbs. (8 Nm).
- Powertrain assembly in the vehicle
- Left side transaxle mount. Torque the side bolts to 41 ft. lbs. (55 Nm) and the center bolt to 66 ft. lbs. (90 Nm).
- Engine mount upper bracket. Torque the side bolts to 72 ft. lbs. (98 Nm) and the center bolt to 57 ft. lbs. (77 Nm).
- Ground wire. Torque the bolt to 25 ft. lbs. (34 Nm).
- Rear transaxle mount. Torque the bolts to 41 ft. lbs. (55 Nm).
- Speed control unit, if equipped. Torque the bolts to 89 inch lbs. (10 Nm).
- Power steering pump and hand tighten the bolts
- Lower power steering line bracket. Torque the bolt to 89 inch lbs. (10 Nm).
- Upper power steering line bolt. Torque the bolt to 15 ft. lbs. (20 Nm).

- Slave cylinder line and clip. Torque the bolts to 16 ft. lbs. (22 Nm).
- Power steering lines. Torque the retaining bolts to 89 inch lbs. (10 Nm). Torque the power steering pump bolts to 18 ft. lbs. (25 Nm).
- Lower radiator hose
- Engine-to-transaxle bolts. Torque the bolts to 33 ft. lbs. (45 Nm).
- Front transaxle through bolt. Torque the bolt to 66 ft. lbs. (90 Nm).
- Block heater electrical connector
- Shifter linkages. Torque the upper bolt to 33 ft. lbs. (45 Nm) and the lower bolt to 15 ft. lbs. (20 Nm).
- Coolant hose
- Catalytic converter
- Heater hoses
- Upper radiator hose
- EVAP canister vacuum lines
- Power distribution box electrical connector. Torque the fastener to 9 ft. lbs. (12 Nm).
- Wire harness electrical connector
- Ground wires
- PCM wire harness and ground
- Brake booster vacuum supply hose to the intake manifold
- EGR vacuum regulator valve hoses and electrical connector
- Throttle cable and speed control cable, if equipped
- Fuel lines
- Battery tray and battery
- Air cleaner
- Hood

12. Fill the engine with clean oil.
13. Fill and bleed the cooling system.
14. Recharge the A/C system.
15. Start the engine, check for leaks and repair if necessary.

2.3L Engine

MANUAL TRANSAXLE

All vehicles

1. With the vehicle in NEUTRAL, position it on a hoist.
2. Release the fuel system pressure.
3. Remove the engine air cleaner and air cleaner outlet pipe.
4. Remove the battery tray.
5. Drain the engine oil.
6. Drain the cooling system.
7. Remove the starter.
8. Remove the catalytic converter.
9. Remove the accessory drive belt.
10. Remove the bolts and the lateral support crossmember.
11. Remove the LH front drive halfshaft.
12. Remove the front drive intermediate halfshaft

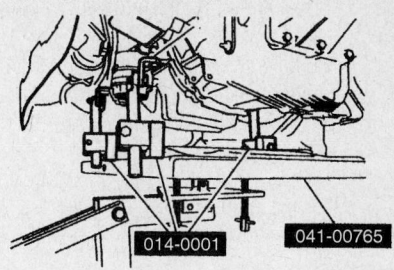

67197-ESCA-G02

Engine secured to lift table—2.3L engine w/manual transaxle

4x4 vehicles

13. Remove the six bolts holding the driveshaft to the transfer case.
14. Position the driveshaft aside.

All vehicles

15. If equipped, remove the bolt and ground eyelet.
16. Remove the power distribution box cover.
17. Remove the nuts and disconnect the cables.
18. Disconnect the electrical connector from the power distribution box.
19. Remove the bolt and disconnect the ground strap. Loosen the bolt and disconnect the 42-pin electrical connector.
20. Detach the wiring harness retainers from the battery tray bracket and position the wiring harness out of the way.
21. Disconnect the clutch hydraulic tube fitting. Detach the tube from the spring clip and position aside.
22. Remove the retaining clips and disconnect the transaxle control cable.
23. Remove the retaining clips and disconnect the transaxle control cable.
24. Disconnect the vehicle speed sensor electrical connector and pin-type retainer.
25. Disconnect the reversing lamp indicator switch and detach the wiring harness retainers.
26. If equipped, disconnect the block heater electrical connector. Detach all the block heater wiring harness retainers and position the wiring harness aside.
27. Disconnect the upper radiator and coolant vent hoses.
28. Remove the nuts and the coolant vent hose brackets. Position the coolant vent hose aside.
29. Detach the heater hose support strap from the stud.
30. Disconnect the heater hoses from the heater core.
31. Remove the retainers and the accelerator cable snow shield.
32. Disconnect the accelerator cable and speed control cable (if equipped).

33. Remove the nut from the accelerator control cable bracket.

34. Remove the nut from the accelerator control cable bracket and position the accelerator control cable and bracket assembly aside.

35. Remove the nut and position the power steering tube and bracket aside.

36. Disconnect the vacuum supply tube and position aside.

37. Disconnect the fuel vapor return tube and position aside.

38. Disconnect the vacuum reservoir tube and position aside.

39. Disconnect the fuel supply tube and retainer and position aside.

40. Detach the electrical connector retainers.

41. Disconnect the powertrain control module (PCM) electrical connectors. Remove the nut and position the harness aside.

42. Remove the bolt and detach the ground wire.

43. Remove the two power steering pump bolts.

44. Disconnect the lower radiator hose from the radiator.

45. Disconnect the A/C compressor electrical connector and remove the four bolts. Position the A/C compressor aside and support the compressor with a length of mechanics wire.

46. Disconnect the power steering pressure (PSP) sensor electrical connector.

➡The bolt under the power steering pressure tube will remain with the power steering pump.

47. Remove the bolts and position the power steering pump aside.

48. Remove the front roll restrictor bolt and the two bolts for the engine support crossmember.

49. Remove the rear nut and the engine support crossmember.

➡The transaxle-to-engine bolts differ in length. Mark the bolts for correct installation.

50. Remove the two transaxle-to-engine bolts.

➡The transaxle-to-engine bolts differ in length. Mark the bolts for correct installation.

51. Remove the two transaxle-to-engine bolts.

52. Using the special tools, secure the engine to the lift table.

53. Remove the engine mount bracket bolt.

54. Remove the nuts and the engine mount bracket.

55. Remove the bolt from the transaxle rear mount.

56. Remove bolt from the LH transaxle mount.

57. Lower the engine and transaxle from the vehicle.

58. Using the engine crane and spreader bar, remove the engine and transaxle from the lift table.

➡The transaxle-to-engine bolts differ in length. Mark the bolts for correct installation.

59. Remove the remaining six engine-to-transaxle bolts and separate the engine and transaxle.

To install:
All vehicles

60. Using the engine crane and spreader bar, position the engine and transaxle together. Install the six upper transaxle-to-engine bolts. Torque to 35 ft. lbs. (48 Nm).

61. Using the engine crane and spreader bar, position the engine and transaxle onto the lift table.

62. Using the special tools, secure the engine to the lift table.

63. Raise the engine and transaxle into the vehicle.

64. Install the bolt in the LH transaxle mount. Torque to 76 ft. lbs. (103 Nm).

65. Install the bolt in the rear transaxle mount. Torque to 85 ft. lbs. (115 Nm).

66. Install the engine mount bracket. Torque to 66 ft. lbs. (90 Nm).

67. Install the engine mount bracket bolt. Torque to 66 ft. lbs. (90 Nm).

68. Install the 4 lower transaxle-to-engine bolts. Torque to 35 ft. lbs. (48 Nm).

69. Install the engine support crossmember and nut. Torque to 66 ft. lbs. (90 Nm).

70. Install the two bolts for the engine support crossmember. Torque to 66 ft. lbs. (90 Nm).

71. Install the front roll restrictor bolt. Torque to 85 ft. lbs. (115 Nm).

➡The bolt under the power steering pressure tube will remain with the power steering pump.

72. Position the power steering pump and install the bolts. Torque to 18 ft. lbs. (25 Nm).

73. Connect the power steering pressure (PSP) sensor electrical connector.

74. Install the A/C compressor and connect the A/C compressor electrical connector. Torque to 18 ft. lbs. (25 Nm).

75. Connect the lower radiator hose to the radiator.

76. Install the two lower power steering pump bolts. Torque to 18 ft. lbs. (25 Nm).

77. Install the ground wire and bolt.

78. Connect the powertrain control module (PCM) electrical connectors. Position the harness and install the nut.

79. Attach the electrical connector retainers.

80. Connect the fuel supply tube.

81. Connect the vacuum reservoir tube.

82. Connect the fuel vapor return tube and retainer.

83. Connect the vacuum supply tube.

84. Install the power steering tube and bracket.

85. Position the accelerator control cable and bracket and install the nut.

86. Install the accelerator control cable and bracket and nut.

87. Install the accelerator cable and speed control cable (if equipped).

88. Install the accelerator cable snow shield and retainers.

89. Connect the heater hoses to the heater core.

90. Attach the heater hose support strap to the stud.

91. Position the coolant vent hose and install the coolant vent hose brackets and nuts.

92. Connect the upper radiator and coolant vent hoses.

93. If equipped, route the block heater wiring harness and attach all retainers. Connect the block heater electrical connector.

94. Connect the reversing lamp indicator switch and attach the wiring harness retainers.

95. Connect the vehicle speed sensor (VSS) electrical connector and pin-type retainer.

96. Connect the transaxle control cable and install the retaining clips.

97. Connect the transaxle control cable and install the retaining clips.

98. Connect the clutch hydraulic tube fitting. Attach the tube to the spring clip.

99. Attach the wiring harness retainers to the battery tray bracket.

100. Connect the 42-pin electrical connector and tighten the bolt. Install the ground strap and bolt.

101. Connect the electrical connector to the power distribution box.

102. Connect the cables and install the nuts.

103. Install the power distribution box cover.

104. If equipped, install the ground eyelet and bolt.

4x4 vehicles

105. Install the driveshaft. Torque to 15 ft. lbs. (20 Nm).

All vehicles

106. Install the front drive intermediate halfshaft.

107. Install the LH front drive halfshaft.

108. Install the lateral support crossmember. Torque to 85 ft. lbs. (115 Nm).

109. Install the accessory drive belt.

110. Install the catalytic converter.

111. Install the starter.

112. Install the battery tray and battery.

113. Install the engine air cleaner and air cleaner outlet pipe.

114. Fill the engine with clean engine oil.

115. Fill and bleed the cooling system.

116. Bleed the clutch system.

AUTOMATIC TRANSAXLE

All vehicles

1. With the vehicle in NEUTRAL, position it on a hoist.

2. Release the fuel system pressure.

3. Remove the engine air cleaner and air cleaner outlet pipe

4. Remove the battery tray.

5. Drain the engine oil.

6. Drain the cooling system.

7. Remove the starter.

8. Remove the catalytic converter.

9. Remove the accessory drive belt.

10. Remove the left front drive halfshafts.

4wd vehicles

11. Remove the transfer case.

2wd vehicles

12. Remove the bolts and the lateral support crossmember.

13. Remove the front drive intermediate halfshaft.

All vehicles

14. If equipped, remove the bolt and ground eyelet.

15. Remove the power distribution box cover.

16. Remove the nuts and disconnect the cables.

17. Disconnect the electrical connector from the power distribution box.

18. Remove the bolt and disconnect the ground strap. Loosen the bolt and disconnect the 42-pin electrical connector.

19. Detach the wiring harness retainers from the battery tray bracket and position the wiring harness out of the way.

20. Disconnect the transaxle electrical connector.

21. Disconnect the shift cable from the transaxle manual lever.

22. Position the transaxle control cable and bracket aside.

23. Disconnect the transaxle range (TR) sensor electrical connector.

24. Detach the transaxle control harness from the brackets.

25. Disconnect the fluid cooler tube.

26. Disconnect the output shaft speed (OSS) sensor electrical connector (black).

27. Disconnect the turbine shaft speed (TSS) sensor electrical connector (white connector).

28. Remove the transaxle fluid cooler retaining bracket bolt.

29. Position the fluid cooler tube aside.

30. Remove the bolt and the OSS sensor.

31. Detach the transaxle control harness from the retaining clip.

32. If equipped, disconnect the block heater electrical connector. Detach all the block heater wiring harness retainers and position the wiring harness aside.

33. Disconnect the upper radiator and coolant vent hoses.

34. Remove the nuts and the coolant vent hose brackets. Position the coolant vent hose aside.

35. Detach the heater hose support strap from the stud.

36. Disconnect the heater hoses from the heater core.

37. Remove the retainers and the accelerator cable snow shield.

38. Disconnect the accelerator cable and speed control cable (if equipped).

39. Remove the nut from the accelerator control cable bracket.

40. Remove the nut from the accelerator control cable bracket and position the accelerator control cable and bracket assembly aside.

41. Remove the nut and position the power steering tube and bracket aside.

42. Disconnect the vacuum supply tube and position aside.

43. Disconnect the fuel vapor return tube and retainer and position aside.

44. Disconnect the vacuum reservoir tube and position aside.

45. Disconnect the fuel supply tube and position aside.

46. Detach the electrical connector retainers.

47. Disconnect the powertrain control module (PCM) electrical connectors. Remove the nut and position the harness aside.

48. Remove the bolt and detach the ground wire.

49. Remove the two power steering pump bolts.

50. Disconnect the lower radiator hose from the radiator.

51. Disconnect the A/C compressor electrical connector and remove the four

bolts. Position the A/C compressor aside and support the compressor with a length of mechanics wire.

52. Disconnect the power steering pressure (PSP) sensor electrical connector.

➡**The bolt under the power steering pressure tube will remain with the power steering pump.**

53. Remove the bolts and position the power steering pump aside.

54. Remove the front roll restrictor bolt and the two bolts for the engine support crossmember.

55. Remove the rear nut and the engine support crossmember.

➡**The transaxle-to-engine bolts differ in length. Mark the bolts for correct installation.**

56. Remove the two transaxle-to-engine bolts

➡**The transaxle-to-engine bolts differ in length. Mark the bolts for correct installation.**

57. Remove the two transaxle-to-engine bolts

58. Using the special tools, secure the engine to the lift table.

59. Remove the engine mount bracket bolt.

60. Remove the nuts and the engine mount bracket.

61. Remove the bolt from the transaxle rear mount.

62. Remove the bolt from the left transaxle mount.

63. Lower the engine and transaxle from the vehicle.

64. Using the engine crane and spreader bar remove the engine and transaxle from the lift table.

65. Remove the starter motor isolator.

66. Remove and discard the four torque converter nuts.

➡**The transaxle-to-engine bolts differ in length. Mark the bolts for correct installation.**

67. Remove the remaining six engine-to-transaxle bolts and separate the engine and transaxle.

To install:
All vehicles

68. Using the engine crane and spreader bar, position the engine and transaxle together. Install the six upper transaxle-to-engine bolts. Torque to 35 ft. lbs. (48 Nm).

69. Install new torque converter nuts. Torque to 26 ft. lbs. (35 Nm).

70. Install the starter motor isolator.

71. Using the engine crane and spreader bar, position the engine and transaxle onto the lift table.

72. Using the special tools, secure the engine to the lift table.

73. Raise the engine and transaxle into the vehicle.

74. Install the bolt in the left transaxle mount. Torque to 76 ft. lbs. (103 Nm).

75. Install the bolt in the rear transaxle mount. Torque to 85 ft. lbs. (115 Nm).

76. Install the engine mount bracket. Torque to 66 ft. lbs. (90 Nm).

77. Install the engine mount bracket bolt. Torque to 66 ft. lbs. (90 Nm).

78. Install the 4 transaxle-to-engine bolts. Torque to 35 ft. lbs. (48 Nm).

79. Install the engine support crossmember and nut. Torque to 66 ft. lbs. (90 Nm).

80. Install the two bolts for the engine support crossmember. Torque to 66 ft. lbs. (90 Nm).

81. Install the front roll restrictor bolt. Torque to 85 ft. lbs. (115 Nm).

82. Position the power steering pump and install the upper bolts. Torque to 18 ft. lbs. (25 Nm).

83. Connect the power steering pressure (PSP) sensor electrical connector.

84. Install the A/C compressor and connect the A/C compressor electrical connector. Torque to 18 ft. lbs. (25 Nm).

85. Connect the lower radiator hose to the radiator.

86. Install the two lower power steering pump bolts. Torque to 18 ft. lbs. (25 Nm).

87. Install the ground wire and bolt.

88. Connect the powertrain control module (PCM) electrical connectors. Position the harness and install the nut.

89. Attach the electrical connector retainers.

90. Connect the fuel supply tube.

91. Connect the vacuum reservoir tube.

92. Connect the fuel vapor return tube and retainer.

93. Connect the vacuum supply tube.

94. Position the power steering tube and bracket and install the nut.

95. Position the accelerator control cable and bracket and install the nut.

96. Install the accelerator control cable and bracket and nut.

97. Install the accelerator cable and speed control cable (if equipped).

98. Install the accelerator cable snow shield and the retainers.

99. Connect the heater hoses to the heater core.

100. Attach the heater hose support strap to the stud.

101. Position the coolant vent hose and install the coolant vent hose brackets and nuts.

102. Connect the upper radiator and coolant vent hoses.

103. If equipped, route the block heater wiring harness and attach all retainers. Connect the block heater electrical connector.

104. Attach the transaxle control harness to the retaining clip.

105. Install the output shaft speed (OSS) sensor and bolt.

106. Install the fluid cooler tube.

107. Connect the transaxle fluid cooler tube.

108. Attach the transaxle control harness to the brackets.

109. Connect the transaxle range (TR) sensor electrical connector.

110. Install the transaxle control cable and bracket.

111. Connect the shift cable to the transaxle manual lever.

112. Connect the transaxle electrical connector.

113. Attach the wiring harness retainers to the battery tray bracket.

114. Connect the 42-pin electrical connector and tighten the bolt. Install the ground strap and bolt.

115. Connect the electrical connector to the power distribution box.

116. Connect the cables and install the nuts.

117. Install the power distribution box cover.

118. If equipped, install the ground eyelet and bolt.

2wd vehicles

119. Install the front drive intermediate halfshaft.

120. Install the lateral support crossmember. Torque to 85 ft. lbs. (115 Nm).

4wd vehicles

121. Install the transfer case.

All vehicles

122. Install the left front drive halfshaft.

123. Install the accessory drive belt.

124. Install the catalytic converter.

125. Install the starter.

126. Install the battery tray and battery.

127. Install the engine air cleaner and air cleaner outlet pipe.

128. Fill the engine with clean engine oil.

129. Fill and bleed the cooling system.

3.0L Engine

1. Before servicing the vehicle, refer to the Precautions Section.

2. Properly recover the air conditioning system refrigerant.

3. Properly relieve the fuel system pressure.

4. Drain the cooling system.

5. Drain the engine oil.

6. Remove or disconnect the following:
- Hood
- Battery and battery tray
- Air cleaner outlet tube and housing
- Lower radiator air deflectors
- Fuel lines
- Water pump drive belt
- Accelerator cable and speed control cable, if equipped
- Vapor Management Valve (VMV)
- Powertrain Control Module (PCM)
- PCM ground wire
- Thermostat housing and hose assembly and move them aside
- Power distribution box electrical connector
- Power distribution box cover
- Nuts and cables from inside the power distribution box
- Transaxle linkage
- Brake booster vacuum hose
- Heater hoses
- Power steering return line
- Power Steering Pressure (PSP) switch electrical connector
- Power steering supply line
- Oil level indicator
- Catalytic converter
- A/C compressor
- Both front wheels
- Intermediate drive shaft, if equipped

7. Separate both side ball joints.

8. Separate both side tie rod ends from the steering knuckles.

9. Separate both sway bar links from the strut mounts.

10. Separate the struts from the steering knuckles.

11. Remove or disconnect the following:
- Both wheel speed sensors, if equipped
- Brake calipers from the steering knuckles and properly support the struts
- Steering shaft from the rack
- Transaxle line bracket bolt
- Transaxle cooler lines
- Torque converter inspection cover
- Torque converter nuts
- Block heater wiring, if equipped

12. Install a powertrain lifting devise and raise the vehicle.
- Engine support bracket
- Transaxle support
- 2 rear subframe bolts
- 2 subframe side bolts

- Motor mount support bolts
- Engine and transaxle as an assembly
- Heated Oxygen (HO2S) sensor
- Transaxle Range (TR) sensor
- Transaxle harness electronic control switch
- Transaxle control harness from the bracket
- Starter and wire harness
- Knock Sensor (KS) electrical connector
- Output Shaft Speed (OSS) sensor electrical connector
- HO2S sensor and Exhaust Gas Recirculation (EGR) tube from the exhaust manifold
- Alternator and electrical connectors
- Right side exhaust manifold and gasket
- Halfshaft support bracket and move it aside

13. Separate the engine from the transaxle assembly

To install:

14. Install or connect the following:
- Powertrain assembly on the subframe
- Transaxle-to-engine bolts. Torque the bolts to 30 ft. lbs. (40 Nm).
- Halfshaft bracket. Torque the bolts to 18 ft. lbs. (25 Nm).
- Right side exhaust manifold and new gasket. Torque the bolts to 15 ft. lbs. (25 Nm).
- Alternator. Torque the larger bolts to 18 ft. lbs. (25 Nm) and smaller bolt to 89 inch lbs. (10 Nm).
- EGR tube and HO2S sensor electrical connectors
- OSS sensor electrical connector
- KS jumper electrical connector
- Starter. Torque the bolts to 18 ft. lbs. (25 Nm).
- Transaxle control harness to the bracket. Torque the bolt to 18 ft. lbs. (25 Nm).
- Transaxle harness
- Transaxle range sensor
- Powertrain assembly
- Motor mount support. Torque the bolts to 66 ft. lbs. (90 Nm).
- Subframe side nuts. Torque the nuts to 76 ft. lbs. (103 Nm). Raise the vehicle and support the powertrain assembly with a lifting device.
- Transaxle mount. Torque the bolts to side bolts to 66 ft. lbs. (90 Nm) and the other bolts to 76 ft. lbs. (103 Nm).
- Motor mount. Torque the bolts to side bolts to 66 ft. lbs. (90 Nm)

and the other bolts to 76 ft. lbs. (103 Nm). Remove the powertrain lift.
- Block heater electrical connector, if equipped
- Torque converter. Torque the nuts to 27 ft. lbs. (37 Nm).
- Transaxle cover plate and plug
- Transaxle cooler lines
- Transaxle cooler line bracket. Torque the bolt to 15 ft. lbs. (20 Nm).
- Steering shaft to the rack. Torque the bolt to 18 ft. lbs. (25 Nm).
- Struts to the steering knuckles. Torque the bolts to 75 ft. lbs. (102 Nm).
- Brake calipers to the steering knuckles
- Wheel speed sensors, if equipped. Torque the bolts to 89 inch lbs. (10 Nm).
- Sway bar links to the strut mount. Torque the bolts to 41 ft. lbs. (55 Nm).
- Tie rods to the steering knuckles. Torque the bolts to 41 ft. lbs. (55 Nm).
- Ball joints. Torque the bolts to 52 ft. lbs. (70 Nm).
- Intermediate drive shaft, if equipped
- Both front wheels
- A/C compressor
- Lower radiator air deflectors
- Catalytic converter
- Oil level indicator dipstick tube
- Power steering line and bracket. Torque the bolt to 13 ft. lbs. (17 Nm).
- PSP switch electrical connector
- Power steering return line
- Heater hoses
- Vacuum lines
- Transaxle linkage
- Wire harness cables and nuts to the power distribution box. Torque the nuts to 89 inch lbs. (10 Nm).
- Power distribution box wire harness
- Thermostat housing and connect the hoses
- Ground wire. Torque the bolt to 89 inch lbs. (10 Nm).
- PCM electrical connector
- VMV electrical connector
- Accelerator cable and speed control cable, if equipped
- Air cleaner assembly
- Water pump drive belt
- Battery and tray

15. Fill and bleed the cooling system.
16. Fill the engine with clean oil.

17. Recharge the A/C system.
18. Inspect and top off the power steering fluid.
19. Start the engine, check for leaks and repair if necessary.

Water Pump

REMOVAL & INSTALLATION

2.0L Engine

1. Before servicing the vehicle, refer to the Precautions Section.
2. Drain the cooling system.
3. Remove or disconnect the following:
- Negative battery cable
- Right front wheel
- Splash shield
- Drive belt
- Water pump pulley
- Water pump

To install:
4. Install or connect the following:
- Water pump. Torque the bolts to 89 inch lbs. (10 Nm).
- Water pump pulley. Torque the bolts to 89 inch lbs. (10 Nm).
- Drive belt
- Splash shield
- Right front wheel
- Negative battery cable
5. Refill the cooling system.
6. Start the engine and check for leaks, repair if necessary.

2.3L Engine

1. Before servicing the vehicle, refer to the Precautions Section.
2. Disconnect the negative battery cable.

➡**To deplete the backup power supply energy, disconnect the negative battery cable and wait at least one minute. Be sure to disconnect auxiliary batteries and power supplies, if equipped.**

3. Drain the cooling system.
4. Raise and support the vehicle safely.
5. Drain the cooling system.
6. Remove the accessory drive belt.
7. Remove the water pump pulley bolts.
8. Remove the water pump pulley.
9. Remove the water pump bolts.
10. Remove the water pump.
11. Remove the water pump O-ring seal.
12. To install, reverse the removal procedure. Torque the water pump bolts to 89 inch lbs. (10 Nm). Torque the pulley bolts to 18 ft. lbs. (25 Nm).
13. Fill and bleed the cooling system.

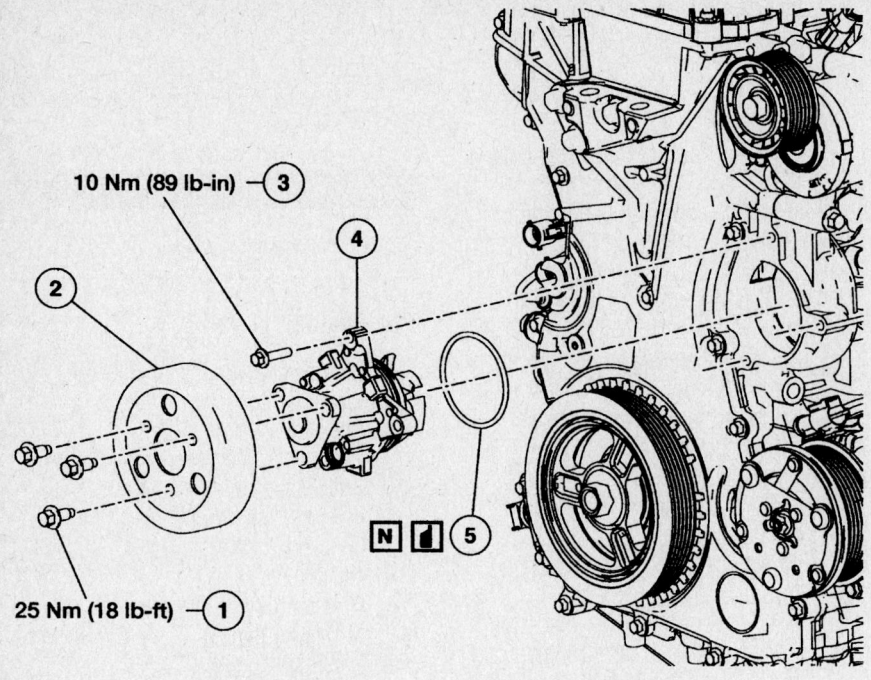

10 Nm (89 lb-in) — ③

④

②

25 Nm (18 lb-ft) — ①

Ⓝ ⒭ ⑤

1 Coolant pump pulley bolts
2 Coolant pump pulley
3 Coolant pump bolts
4 Coolant pump
5 Coolant pump O-ring seal

67197-ESCA-G03

Water pump and related components—2.3L engine

2006 LATE BUILD ENGINES

1. Before servicing the vehicle, refer to the Precautions Section.

2. Disconnect the negative battery cable.

➡ To deplete the backup power supply energy, disconnect the negative battery cable and wait at least one minute. Be sure to disconnect auxiliary batteries and power supplies, if equipped.

3. Raise and support the vehicle safely.

4. Drain the cooling system.

5. Disconnect the crankcase ventilation tube and position it to the side.

6. Remove the water pump drive belt.

7. Using a water pump pulley removal tool, remove the water pump drive pulley.

8. Disconnect the heater hose from the water pump. Disconnect the water pump to engine hose and position it to the side.

9. Remove the three bolts from the water pump. Reposition the water pump to thermostat housing clamp. Remove the water pump and hose, as an assembly.

To install:

10. Position the water and install the retaining bolts. Torque the bolts to 89 inch lbs. (10 Nm) and than an additional 90 degrees.

3.0L Engine

EXCEPT 2006 LATE BUILD ENGINES

1. Before servicing the vehicle, refer to the Precautions Section.

2. Drain the cooling system.

3. Disconnect the negative battery cable.

➡ To deplete the backup power supply energy, disconnect the negative battery cable and wait at least one minute. Be sure to disconnect auxiliary batteries and power supplies, if equipped.

4. Drain the cooling system.

5. Remove or disconnect the following:
- Air cleaner outlet tube
- Water pump belt tensioner
- Coolant hoses
- Water pump
- Water pump from the housing

To install:
- Water pump. Torque the bolts to 89 inch lbs. (10 Nm).
- Coolant hoses
- Water pump belt tensioner
- Air cleaner outlet tube
- Negative battery cable

6. Refill the cooling system. Bleed the cooling system.

7. Start the engine and check for leaks, repair if necessary.

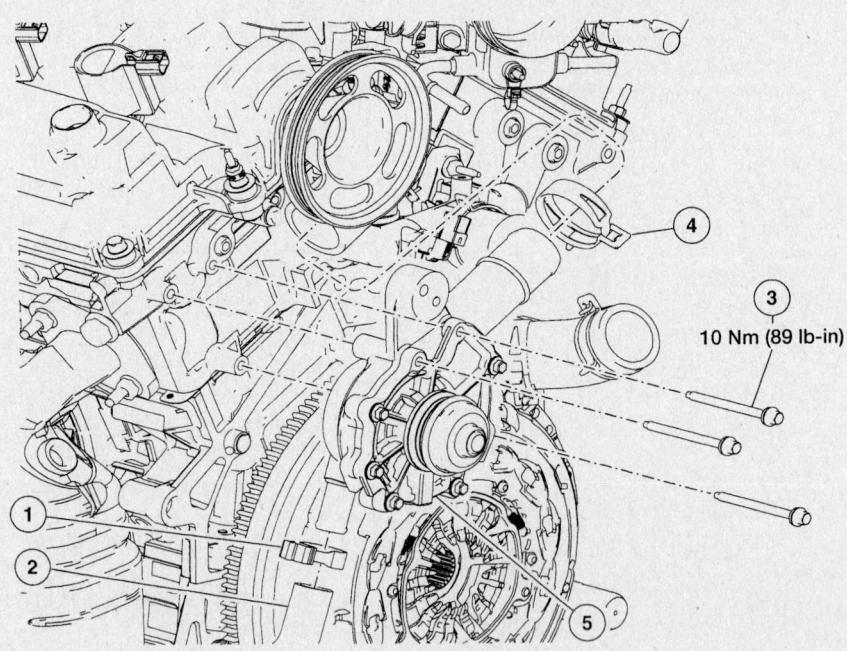

④

③

10 Nm (89 lb-in)

①

②

⑤

1 Clamp

2 Coolant pump-to-oil cooler hose

3 Coolant pump assembly bolts (3 required)

4 Hose clamp

5 Coolant pump assembly

09482_TRIB_G0007

Water pump and related components—except late build 3.0L engine

1   Heater hose clamp

2   Heater hose

3   Coolant pump assembly bolts (3 required)

4   Coolant pump-to-engine hose clamp

5   Coolant pump-to-engine hose

6   Coolant pump-to-thermostat housing hose clamp

7   Coolant pump assembly

8   Coolant pump drive pulley

09482_TRIB_G0008

Water pump and related components—late build 3.0L engine

11. Connect the hoses.

12. Install the water pump drive pulley flush with the end of the camshaft.

13. Continue the installation in the reverse order of the removal procedure.

14. Fill the cooling system with the proper grade and type coolant. Bleed the cooling system.

15. Start the engine and check for leaks, repair if necessary.

Heater Core

REMOVAL & INSTALLATION

2002–2004

1. Before servicing the vehicle, refer to the Precautions Section.

2. Disconnect the negative battery cable.

➡**To deplete the backup power supply energy, disconnect the negative battery cable and wait at least one minute. Be sure to disconnect auxiliary batteries and power supplies, if equipped.**

3. Drain the engine coolant.

4. Disconnect the heater hoses from the heater core.

5. Remove the driver air bag module.

6. Remove the two front door scuff plates.

7. Remove the four pin-type retainers.

8. Remove the two front door scuff plates.

9. Remove the two A-pillar lower trim panels.

10. Remove the two pin-type retainers.

11. Remove the two A-pillar lower trim panels.

12. Disconnect the electrical connectors located by the LH cowl.

13. Position the hood latch release handle aside.

14. Remove the bolts.

15. Position the hood latch release handle aside.

16. Remove the utility compartment.

17. Remove the four pin-type retainers.

18. Remove the utility compartment.

19. Disconnect the electrical connector.

20. Remove the instrument panel steering column cover.

21. Release the upper clips and rotate the cover outward to release the lower pivot retainers.

22. Remove the steering column lower cover.

23. Remove the screws.

24. Remove the steering column lower cover.

25. If equipped, disconnect the shift cable.

26. Disconnect the shift cable.

27. Disconnect the shift cable from the retaining bracket.

28. Remove the steering column coupler access cover.

29. Disconnect the steering column coupler.

30. Remove the steering column coupler bolt and nut.

31. Disconnect the steering column coupler.

32. Remove the cover panel.

33. Remove the pin-type retainer.

34. Release the retaining clip.

35. Disconnect the electrical connectors.

36. Disconnect the climate control vacuum harness connector.

37. Disconnect the in-line electrical connector.

38. Remove the four instrument panel center brace bolts.

39. Remove the passenger air bag module.

40. Disconnect the vacuum harness connector.

41. Disconnect the temperature control cable.

42. Position the locator pin.

43. Release the locking tab.

44. Disconnect the temperature control cable from the blend door shaft.

45. Close the glove compartment.

46. Press the release tabs inward while raising the glove compartment.

47. Disconnect the electrical connectors at the blower motor.

48. Disconnect the antenna cable in-line connector.

49. Open the four A-pillar passenger assist handle covers.

50. Remove the two A-pillar passenger assist handles.

51. Remove the four bolts.
52. Remove the two A-pillar passenger assist handles.
53. Remove the two windshield side garnish moldings.
54. Remove the instrument panel cowl top cover.
55. Remove the instrument panel cowl top bolt.
56. Loosen the tilt lever (if equipped) and lower steering column.
57. Position the transaxle range selector lever (if equipped) down to provide access to the instrument cluster finish panel and instrument cluster.
58. Remove the screws and the instrument cluster finish panel.
59. Remove the screws.
60. Disconnect the electrical connectors and remove the instrument cluster.
61. Through the instrument cluster opening, remove the instrument panel nut.
62. Remove the two instrument panel finish end panels.
63. Remove the four instrument panel cowl side bolts.

➡**This step requires an assistant.**

64. Remove the instrument panel.
65. Remove the heater blending door levers.
66. Remove the screw for heater blending door.
67. Remove the levers for the blending door.
68. Remove the heater core.
69. Remove the three screws.
70. Remove the cover for the heater core and pull the heater core out of the housing.

➡**Before installing the temperature control cable, make sure the blend door, cable and temperature switch are correctly positioned.**

71. To install, reverse the removal procedure.

✳✳ CAUTION

Electronic modules are sensitive to static electrical charges. If exposed to these charges, damage may result.

✳✳ CAUTION

Once the new module is installed, it is necessary to download the module configuration information from the scan tool into the new instrument cluster.

2005–2006

1. Before servicing the vehicle, refer to the Precautions Section.
2. De-power the supplemental restraint system.
3. Disconnect the negative battery cable.

➡**To deplete the backup power supply energy, disconnect the negative battery cable and wait at least one minute. Be sure to disconnect auxiliary batteries and power supplies, if equipped.**

4. Drain the engine coolant.
5. Release the 2 clamps and disconnect the 2 heater hoses from the heater core.
6. Position the seats forward and remove the 2 rear bolts.
7. Position the seats rearward and disconnect the battery.
8. Position the parking brake handle to the full-up position.
9. Remove the transaxle selector lever bezel.
10. Release the parking brake handle boot from the floor console finish panel.
11. If equipped, remove the floor console top panel.

➡**If removing the floor console storage bin, squeeze the front and rear of the storage bin to release the retaining tabs from the floor console finish panel.**

12. Remove the floor console finish panel. Disconnect the electrical connectors.
13. Remove the 6 bolts and remove the floor console.
14. Remove the 4 pin-type retainers and the 2 front door scuff plates.
15. Remove the 2 pin-type retainers and the 2 A-pillar lower trim panels.
16. Remove the instrument panel steering column opening cover.
17. Remove the left and right instrument panel end trim panels.
18. Lower the tilt steering column to the lowest position.
19. Remove the 2 screws and the instrument cluster finish panel.
20. Remove the 4 instrument cluster screws and the instrument cluster. Disconnect the electrical connector.
21. Disconnect the main electrical connector.
22. Remove the ground wire bolt and position the ground wire aside.
23. Remove the 2 bolts and position the hood release handle aside.
24. Remove the steering column coupler access cover.

25. Remove the steering column pinch bolt and disconnect the intermediate shaft.
26. Remove the cover panel pin-type retainer and the cover panel. Release the retaining clip from the instrument panel center brace.
27. Disconnect the climate control vacuum harness connector.
28. Disconnect the restraint control module (RCM) electrical connector.
29. Disconnect the temperature control cable from the blend door shaft.
   a. Align the locator holes.
   b. Release the locking tab.
   c. Disconnect the temperature control cable.
30. Remove the 4 transaxle selector lever bolts and position the transaxle selector lever aside. Disconnect the electrical connectors.
31. Remove the bolts and position the parking brake control aside Disconnect the electrical connectors.
32. Open the glove compartment. Press the release tabs inward while lowering the glove compartment.
33. Disconnect the blower motor electrical connectors.
34. Disconnect the antenna cable in-line connector.
35. Remove the cover and the instrument panel cowl top bolt.
36. Remove the 4 instrument panel center brace bolts.
37. Remove the instrument panel cluster opening nut through the instrument cluster opening.
38. Remove the 4 instrument panel cowl side bolts.

➡**This step requires an assistant.**

39. Remove the instrument panel.
40. Remove the screw and the temperature blend door lever.
41. Remove the 3 screws and the heater core cover.
42. Remove the heater core from the housing.

➡**Make sure the temperature blend door actuator and switch are in the correct position when installing the temperature blend door lever.**

➡**Lubricate the coolant hoses with plain water only, if needed.**

43. To install, reverse the removal procedure. Observe the following torques:
   - the 4 instrument panel cowl side bolts: 9 Nm (80 lb-in)
   - the instrument panel cluster opening nut: 9 Nm (80 lb-in)

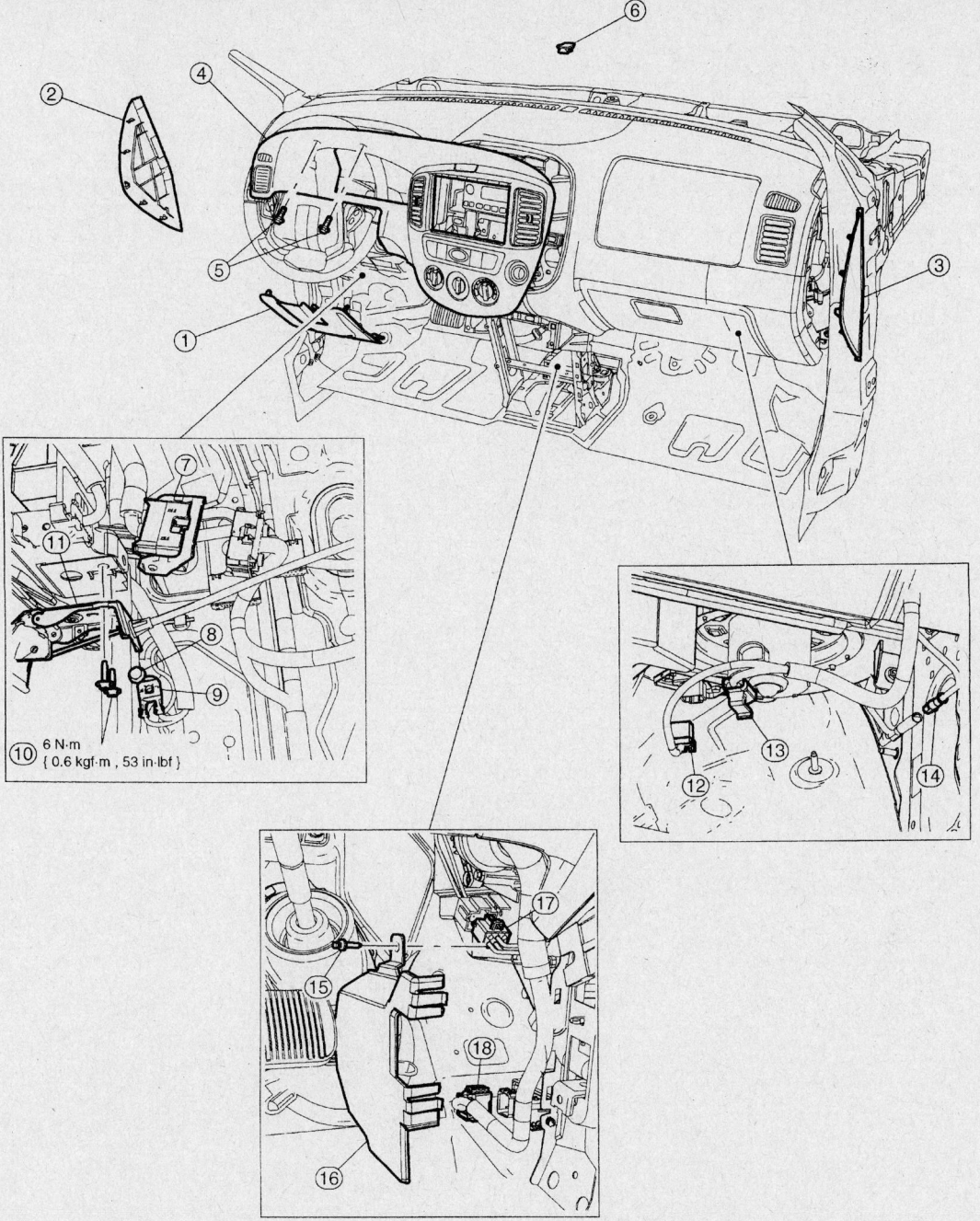

1	Instrument panel steering column opening cover	10	Hood release handle bolts
2	LH instrument panel end trim panel	11	Hood release handle
3	RH instrument panel end trim panel	12	Blower motor resistor connector
4	Instrument panel trim panel	13	Blower motor connector
5	Cluster finish panel screws	14	Audio antenna connector
6	Cowl top cover	15	Cover panel pin-type retainer
7	Main electrical connector	16	Cover panel
8	Ground wire bolt	17	Vacuum connector
9	Ground wire	18	Restraints control module connector

09482_TRIB_G0009

Instrument panel and related components—2005–2006 vehicles, view one

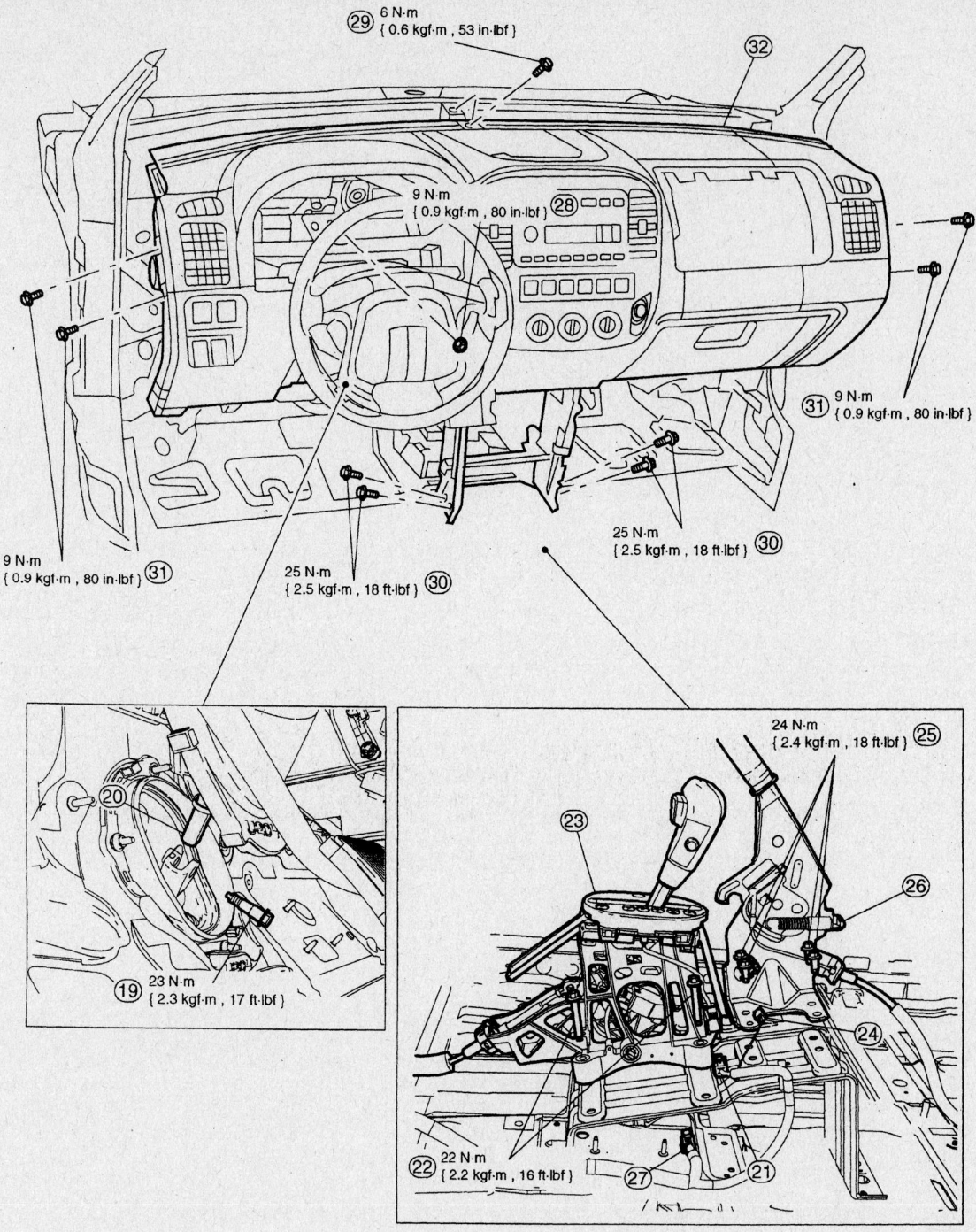

6 N·m
{ 0.6 kgf·m , 53 in·lbf }

9 N·m
{ 0.9 kgf·m , 80 in·lbf }

9 N·m
{ 0.9 kgf·m , 80 in·lbf }

9 N·m
{ 0.9 kgf·m , 80 in·lbf }

25 N·m
{ 2.5 kgf·m , 18 ft·lbf }

25 N·m
{ 2.5 kgf·m , 18 ft·lbf }

23 N·m
{ 2.3 kgf·m , 17 ft·lbf }

24 N·m
{ 2.4 kgf·m , 18 ft·lbf }

22 N·m
{ 2.2 kgf·m , 16 ft·lbf }

19	Pinch bolt	25	Parking brake handle bolts
20	Intermediate shaft	26	Parking brake handle
21	Transmission selector lever electrical connector	27	Electrical connector
22	Transmission selector lever bolts	28	Instrument cluster opening panel nut
23	Transmission selector lever (automatic transaxle shown/manual transaxle similar)	29	Cowl bolt
		30	Center brace bolts
24	Parking brake electrical connector	31	Instrument panel cowl side bolts
		32	Instrument panel

09482_TRIB_G0010

Instrument panel and related components—2005–2006 vehicles, view two

- The 4 instrument panel center brace bolts: 25 Nm (18 lb-ft)
- The instrument panel cowl top bolt: 6 Nm (53 lb-in)
- The parking brake control: 24 Nm (18 lb-ft)
- The 4 transaxle selector lever bolts: 22 Nm (16 lb-ft) for automatic transaxle; 20 Nm (15 lb-ft) for manual transaxle
- The steering column pinch bolt: 23 Nm (17 lb-ft)
- The hood release handle: 6 Nm (53 lb-in)

44. Fill and bleed the engine cooling system.

45. If equipped with automatic transaxle, adjust the linkage.

46. Repower the supplemental restraint system.

47. Following installation of the new instrument cluster, download the module configuration information from the diagnostic tool into the new module as follows:

Using the Vehicle Communication Module (VCM) When the Original Body Chassis Electrical Module is Not Available

48. Install the new module.

49. Using the VCM and the latest version of the service function card, SELECT: Programmable Module Installation.

50. Select the module being installed.

51. Follow the on-screen instructions.

52. SELECT: Retrieve Module Configuration—Old ECU and press trigger.

53. Follow the on-screen instructions.

54. The VCM attempts to retrieve the module data from the powertrain control module (PCM). If the module data is available, go to Step A. If the VCM displays: Call As-Built Data Center, go to Step B.

Step A

55. SELECT: Restore Configuration—New ECU. Press trigger.

56. The VCM completes loading the retrieved data and displays Module Download Successful.

57. Test the module for correct operation.

Step B

58. Press the trigger.

59. If the VCM asks for vehicle data, enter the vehicle data, then press store.

60. The VCM asks for module data line 1. Enter the data and press store.

61. The VCM then asks if there is an additional line of data available for that address. Select YES or NO depending on the information in the As Built Data Sheet.

62. Repeat Steps 3 and 4 until the answer is NO for Step 4.

63. The VCM should show a screen stating that the module data was stored. Press the trigger.

64. Follow the on-screen instructions.

65. SELECT: Restore Configuration—New ECU. Press the trigger.

66. The VCM completes loading the retrieved data and displays Module Download Successful.

67. Test the module for correct operation.

Using the Vehicle Communication Module (VCM) When the Original Body Chassis Electrical Module is Available

68. With the original module still installed, using the VCM and the latest version of the service function card, SELECT: Programmable Module Installation.

69. Select the module being installed and press the trigger.

70. Follow the on-screen instructions.

71. SELECT: Retrieve Module Configuration—Old ECU. Press the trigger.

72. Follow the on-screen instructions.

73. INSTALL new module, SELECT: Restore Configuration—New ECU. Press the trigger.

74. The VCM completes loading the retrieved data and displays Module Download Successful.

75. Test the module for correct operation.

Using the Worldwide Diagnostic System (WDS) When the Original Body Chassis Electrical Module is Not Available

76. Install the new module.

77. Connect the WDS and ID the vehicle as normal.

78. From the Toolbox icon, select and highlight Module Programming. Then highlight the module that was installed and press the check mark.

79. Select and highlight Programmable Module Installation. Then highlight the module that was installed and press the check mark.

80. Follow the on-screen instructions, turn the ignition key to the OFF position and press the check mark.

81. The WDS retrieves the module data from the PCM, automatically downloads the data into the new module, and displays Module Configuration Complete.

82. If the data is not available in the PCM, the WDS displays a screen stating to contact the As-Built Data Center. Retrieve the data from WWW.FMCDEALER.COM at this time and press the check mark.

83. Enter the module data (the module address and line are displayed to the left of

the 3 entry boxes) and press the check mark.

84. The WDS downloads the data into the new module and displays Operation Successful—Programming Complete.

85. Test the module for correct operation.

Using the Worldwide Diagnostic System (WDS) When the Original Body Chassis Electrical Module is Available

86. Connect the WDS and ID the vehicle as normal.

87. From the Toolbox icon, select and highlight Module Programming and press the check mark.

88. Select and highlight Programmable Module Installation.

89. Follow the on-screen instructions, turn the ignition key to the OFF position, and press the check mark.

90. Install the new module and press the check mark.

91. Follow the on-screen instructions, turn the ignition key to the ON position, and press the check mark.

92. The module configuration is complete.

93. Test the module for correct operation.

Cylinder Head

REMOVAL & INSTALLATION

2.0L Engine

1. Before servicing the vehicle, refer to the Precautions Section.

2. Properly relieve the fuel system pressure.

3. Drain the engine oil.

4. Remove or disconnect the following:
- Negative battery cable
- Ignition coil bracket
- Thermostat housing
- Positive Crankcase Ventilation (PCV) tube
- Intake manifold
- Exhaust manifold
- Power steering bracket and move it aside
- Valve tappets
- Engine mount lower bracket
- Engine mount upper bracket
- Cylinder head bolts in the proper sequence and discard the gasket

To install:

5. Install a new head gasket and the cylinder head.

6. Lubricate the cylinder head bolt threads.

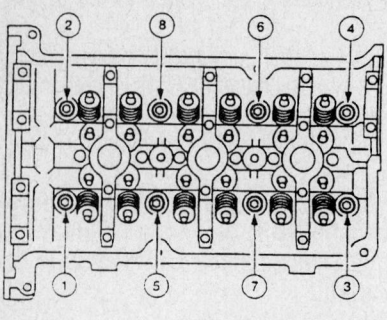

Cylinder head bolt torque sequence—2.0L engine

9308TG04

7. Torque the cylinder head bolts in the proper sequence as follows:

    a. Step 1: 15 ft. lbs. (20 Nm).

    b. Step 2: 30 ft. lbs. (40 Nm).

    c. Step 3: Plus an additional 90 degrees.

8. Install or connect the following:

- Engine mount upper bracket. Torque the 2 upper bolts to 72 ft. lbs. (98 Nm) and the center bolt to 57 ft. lbs. (77 Nm).
- Engine mount lower bracket. Torque the bolts to 37 ft. lbs. (50 Nm).
- Valve tappets
- Power steering pump bracket. Torque the bolts to 20 ft. lbs. (28 Nm).
- Exhaust manifold
- Intake manifold
- PCV tube
- Thermostat housing
- Ignition coil bracket
- Negative battery cable

9. Fill the engine with clean oil and replace the filter.

10. Start the engine and check for leaks, repair if necessary.

2.3L Engine

✳✳ WARNING

During engine repair procedures, cleanliness is extremely important. Any foreign material, including any material created while cleaning gasket surfaces, that enters the oil passages, coolant passages or the oil pan can cause engine failure.

1. Disconnect the negative battery cable.

➡ To deplete the backup power supply energy, disconnect the negative battery cable and wait at least one minute. Be

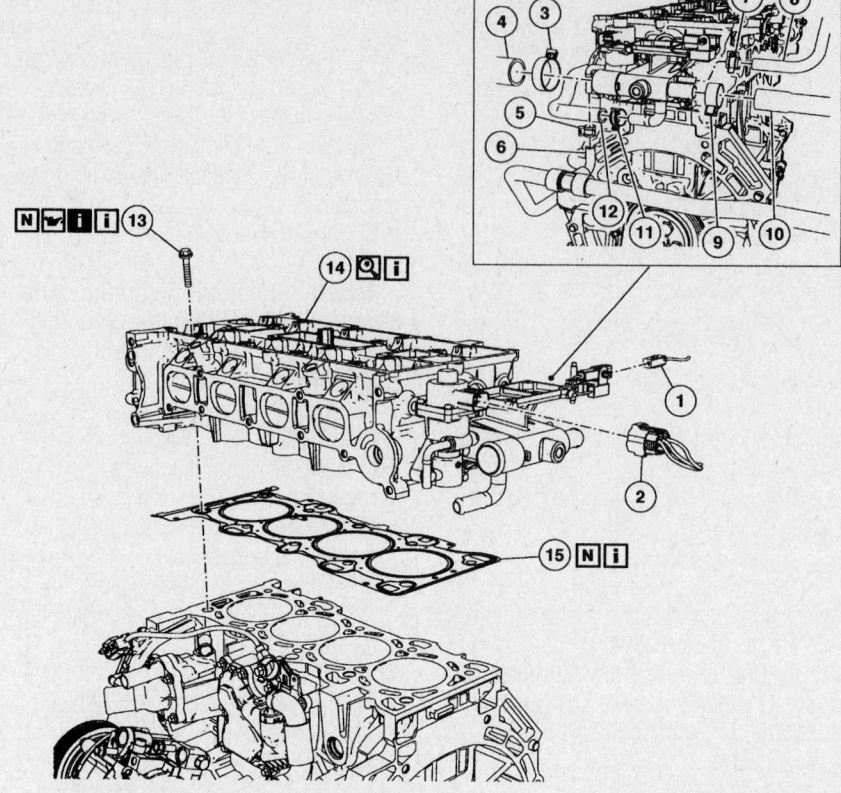

1 Radio ignition interference capacitor electrical connector	9 Heater hose clamp
2 Exhaust gas recirculation (EGR) valve electrical connector	10 Heater hose (position aside)
3 Upper radiator hose clamp	11 Bypass hose clamp
4 Upper radiator hose (position aside)	12 Bypass hose (position aside)
5 EGR coolant tube clamp	13 Cylinder head bolt
6 EGR coolant hose (part of heater hose) (position aside)	14 Cylinder head
7 Engine coolant vent hose clamp	15 Cylinder head gasket
8 Engine coolant vent hose (position aside)	

67197-ESCA-G04

Cylinder head removal—2.3L engine

sure to disconnect auxiliary batteries and power supplies, if equipped.

2. Raise and support the vehicle safely.

3. Drain the engine coolant.

4. Relieve the fuel system pressure.

5. Remove the alternator.

6. Remove the camshafts.

7. Remove the intake manifold.

8. Remove the catalytic converter.

9. Disconnect the radio ignition interference capacitor electrical connector

10. Disconnect the exhaust gas recirculation (EGR) valve electrical connector

11. Remove the upper radiator hose.

12. Remove the EGR coolant tube clamp.

13. Remove the EGR coolant hose.

14. Remove the engine coolant vent hose.

15. Remove the heater hose.

16. Remove the bypass hose.

17. Remove and discard the cylinder head bolts.

18. Remove the cylinder head.

19. Remove the cylinder head gasket.

20. Inspect the cylinder head for distortion.

✳✳ WARNING

Do not use metal scrapers, wire brushes, power abrasive discs or other abrasive means to clean the sealing surfaces. These tools cause scratches and gouges that make leak paths. Use a plastic scraping tool to remove all traces of the head gasket.

✳✳ WARNING

Observe all warnings or cautions and follow all application directions contained on the packaging of the silicone gasket remover and the metal surface prep.

➡**If there is no residual gasket material present, metal surface prep can be used to clean and prepare the surfaces.**

21. Clean the cylinder head-to-cylinder block mating surface of both the cylinder head and the cylinder block.

22. Remove any large deposits of silicone or gasket material with a plastic scraper.

23. Apply silicone gasket remover, following package directions, and allow to set for several minutes.

24. Remove the silicone gasket remover with a plastic scraper. A second application of silicone gasket remover may be required if residual traces of silicone or gasket material remain.

25. Apply metal surface prep, following package directions, to remove any traces of oil or coolant, and to prepare the surfaces to bond with the new gasket. Do not attempt to make the metal shiny. Some staining of the metal surfaces is normal.

26. Apply silicone gasket and sealant to the locations shown (Arrow indicates sealant).

27. Install a new head gasket.

➡**The cylinder head bolts are torque-to-yield and must not be reused. New cylinder head bolts must be installed.**

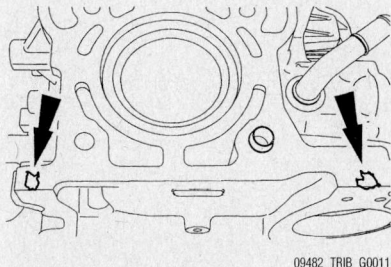

Sealant application location (arrow indicates sealant)—2.3L engine

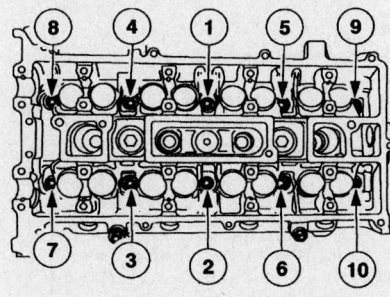

Cylinder head bolt torque sequence—2.3L engine

➡**Lubricate the bolts with clean engine oil prior to installation.**

28. Install new cylinder head bolts. Tighten the bolts in the sequence shown in five stages.

   a. Tighten the bolts to 5 Nm (44 inch lbs.).

   b. Tighten the bolts to 15 Nm (11 ft. lbs.).

   c. Tighten the bolts to 45 Nm (33 ft. lbs.).

   d. Turn the bolts 90 degrees.

   e. Turn the bolts an additional 90 degrees.

29. To install, reverse the removal procedure.

3.0L Engine

The procedure for the left side cylinder head and right side are similar. Changes in the procedure will be noted for either side cylinder head.

1. Before servicing the vehicle, refer to the Precautions Section.

2. Properly relieve the fuel system pressure.

3. Disconnect the negative battery cable.

➡**To deplete the backup power supply energy, disconnect the negative battery cable and wait at least one minute. Be sure to disconnect auxiliary batteries and power supplies, if equipped.**

4. Drain the cooling system.

5. Remove or disconnect the following:

- Camshafts
- Exhaust Gas Recirculation (EGR) tube, right side only
- Exhaust manifold
- Camshaft followers
- Hydraulic lash adjusters and matchmark them for proper installation
- Cylinder head bolts in sequence and discard them
- Cylinder head and discard the gasket

To install:

6. Install a new head gasket and the cylinder head.

7. Be sure to use new cylinder head bolts. Lubricate the cylinder head bolt threads.

8. Torque the cylinder head bolts in the proper sequence as follows:

   a. Step 1: 30 ft. lbs. (40 Nm).

   b. Step 2: Additional 90 degrees.

   c. Step 3: Loosen the bolts one full turn.

   d. Step 4: 30 ft. lbs. (40 Nm).

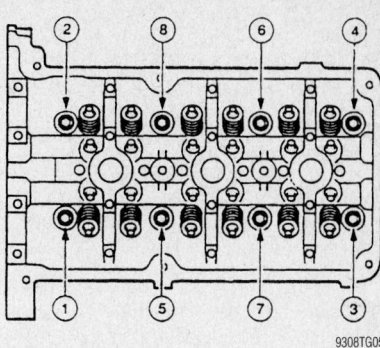

Left side cylinder head bolt torque sequence—3.0L engine

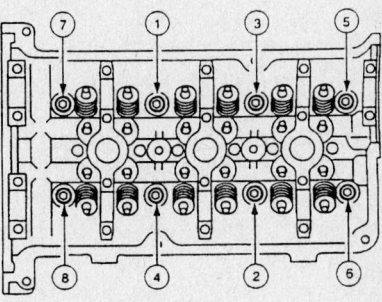

Right side cylinder head bolt torque sequence—3.0L engine

   e. Step 5: Plus an additional 90 degrees.

   f. Step 6: Plus an additional 90 degrees.

9. Install or connect the following:

- Hydraulic lash adjusters. They must be installed in their original positions.
- Camshaft followers
- Camshaft
- Exhaust manifold. Torque the bolts in sequence to 15 ft. lbs. (20 Nm), right side only
- EGR tube, right side only
- Coolant bypass tube
- Negative battery cable

10. Fill the coolant to the proper level.

11. Start the engine and check for leaks, repair if necessary.

Intake Manifold

REMOVAL & INSTALLATION

2.0L Engine

1. Before servicing the vehicle, refer to the Precautions Section.

2. Properly relieve the fuel system pressure.

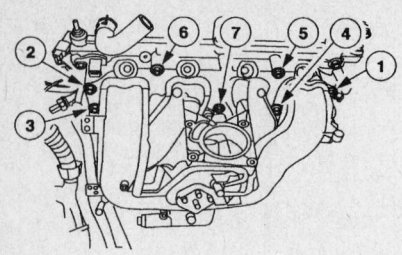

67197-ESCA-G64

Intake manifold bolt loosening sequence—2.0L engine

3. Remove or disconnect the following:
   - Negative battery cable
   - Fuel injection supply manifold
   - Throttle Position (TP) sensor electrical connector
   - Idle Air Control (IAC) electrical connector and unclip the harness from the bracket
   - Main engine control sensor wiring
   - Connector from the bracket
   - Powertrain Control Module (PCM) wire harness from the bracket
   - Brake booster vacuum hose
   - 4 additional vacuum lines
   - Positive Crankcase Ventilation (PCV) hose from the intake manifold
   - Knock Sensor (KS) electrical connector
   - Alternator
   - Intake manifold and discard the gasket
4. Clean the mating surfaces.

To install:
5. Install or connect the following:
   - New gasket
   - Intake manifold. Torque the bolts, in sequence, to 13 ft. lbs. (18 Nm).
   - Alternator
   - KS electrical connector
   - PCV vacuum line
   - 4 vacuum lines
   - Brake booster vacuum supply hose
   - PCM wire harness to the bracket
   - Main engine control sensor wiring

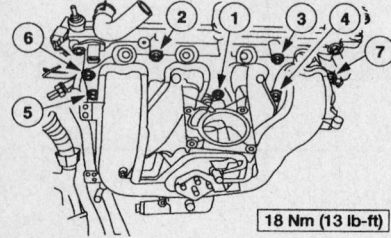

18 Nm (13 lb-ft)

67197-ESCA-G63

Intake manifold bolt torque sequence— 2.0L engine

- IAC valve electrical connector and attach the harness to the bracket
- TP sensor electrical connector
- Fuel injection supply manifold
- Negative battery cable
6. Start the engine and check for leaks, repair if necessary.

2.3L Engine

1. Before servicing the vehicle, refer to the Precautions Section.
2. Disconnect the negative battery cable.

➡ **To deplete the backup power supply energy, disconnect the negative battery cable and wait at least one minute. Be sure to disconnect auxiliary batteries and power supplies, if equipped.**

3. Raise and support the vehicle safely.
4. Relieve the fuel system pressure.
5. Remove the throttle body.
6. Remove the fuel rail.
7. Remove the oil level indicator tube.
8. Remove the vacuum tube.

9. Remove the vacuum supply hose.
10. Remove the fuel vapor return hose.
11. Remove the idle air control (IAC) motor electrical connector.
12. Remove the swirl control valve electrical connector.
13. Remove the knock sensor (KS) electrical connector.
14. Remove the temperature manifold absolute pressure (TMAP) sensor electrical connector.
15. Remove the oil pressure sender electrical connector.
16. Remove the engine control wiring harness.
17. Remove the intake manifold bolts.

➡ **There are three different size bolts used. Mark the location of the bolts to make sure they are installed in the correct location.**

18. Remove the bolts and position the intake manifold aside to access the crankcase vent hose clamp and the EGR tube.

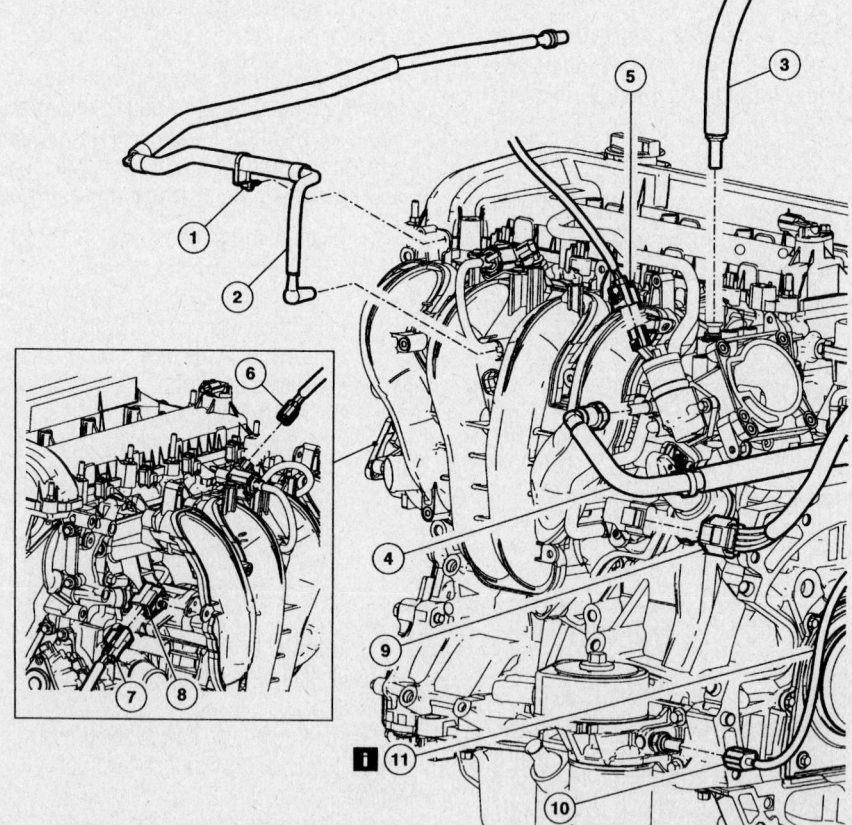

1 Vacuum tube retainer	5 Idle air control (IAC) motor electrical connector	9 Temperature manifold absolute pressure (TMAP) sensor electrical connector
2 Vacuum tube	6 Swirl control valve electrical connector	10 Oil pressure sender electrical connector
3 Vacuum supply hose	7 Knock sensor (KS) electrical connector	11 Engine control wiring harness
4 Fuel vapor return hose	8 Pin-type retainer	

67197-ESCA-G06

Intake manifold mounting location—2.3L engine

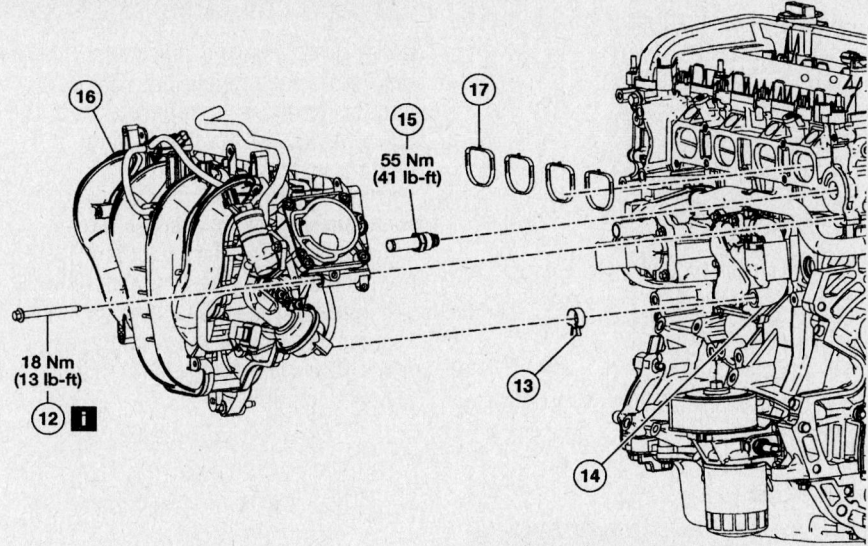

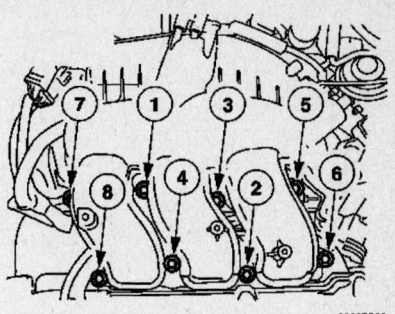

9308TG02

Upper intake manifold bolt torque sequence—3.0L engine

12 Intake manifold bolts
13 Crankcase vent hose clamp
14 Crankcase vent hose (position aside)
15 Exhaust gas recirculation (EGR) tube
16 Intake manifold
17 Intake manifold gasket

67197-ESCA-G07

Intake manifold and related components—2.3L engine

19. Remove the crankcase vent hose.
20. Remove the exhaust gas recirculation (EGR) tube.
21. Remove the intake manifold.
22. Remove the intake manifold gasket.

To install:

23. Installation is the reverse the removal procedure.
24. Be sure to use new intake manifold gaskets.
25. Torque the intake manifold bolts to 13 ft. lbs. (18 Nm).

3.0L Engine

UPPER

1. Before servicing the vehicle, refer to the Precautions Section.
2. Properly relieve the fuel system pressure.
3. Disconnect the negative battery cable.

➡ **To deplete the backup power supply energy, disconnect the negative battery cable and wait at least one minute. Be sure to disconnect auxiliary batteries and power supplies, if equipped.**

4. Drain the coolant system.
5. Remove or disconnect the following:
- Air cleaner outlet tube
- Engine appearance cover
- Throttle cable
- Speed control cable, if equipped
- Throttle cable bracket
- Throttle Position (TP) sensor electrical connector

- Idle Air Control (IAC) valve electrical connector
- Exhaust Gas Recirculation (EGR) valve vacuum hose and tube
- EGR vacuum regulator valve electrical connector and hose
- Chassis vacuum hose
- Engine vacuum hose
- Positive Crankcase Ventilation (PCV) hose
- Vapor Management Valve (VMV) vacuum hose
- Brake booster vacuum hose and clamp, as necessary
- Electrical connectors from the left side of the upper intake manifold
- Power Steering Pressure (PSP) sensor electrical connector, if equipped
- On 2005 and 2006 vehicles, remove the two throttle body coolant hose clamps and hoses
- Upper intake manifold and discard the gasket

6. Clean the mating surfaces.

To install:

7. Install or connect the following:
- New gasket
- Intake manifold. Torque the bolts, in sequence, to 89 inch lbs. (10 Nm).
- PSP electrical connector, as required
- Brake booster vacuum hose and clamp, as necessary
- Two throttle body coolant hose clamps and hoses, as necessary

- Electrical connectors on the left side of the upper intake manifold
- VMV vacuum hose
- Chassis, engine and PCV hoses
- EGR valve vacuum regulator
- EGR valve vacuum hose and tube. Torque the nut to 30 ft. lbs. (40 Nm).
- TP sensor electrical connector
- IAC valve electrical connector
- Throttle cable and speed control cable, if equipped. Torque the bracket bolts to 89 inch lbs. (10 Nm).
- Air cleaner outlet tube
- Engine appearance cover. Torque the bolts to 53 inch lbs. (6 Nm).
- Negative battery cable

8. Fill the coolant system to the proper level.
9. Start the engine and check for leaks, repair if necessary.

LOWER

1. Before servicing the vehicle, refer to the Precautions Section.
2. Properly relieve the fuel system pressure.
3. Disconnect the negative battery cable.

➡ **To deplete the backup power supply energy, disconnect the negative battery cable and wait at least one minute. Be sure to disconnect auxiliary batteries and power supplies, if equipped.**

4. Remove or disconnect the following:
- Fuel line spring lock coupling
- Upper intake manifold
- Fuel rail
- Fuel injector electrical connectors
- Fuel pressure damper vacuum line
- Lower intake manifold retaining bolts. Lower intake manifold.
- Lower intake manifold from the fuel rail
- Fuel injectors from the manifold and discard the gasket

5. Clean the mating surfaces.

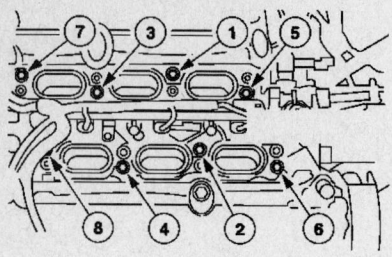

Lower intake manifold bolt torque sequence—3.0L engine

9308TG03

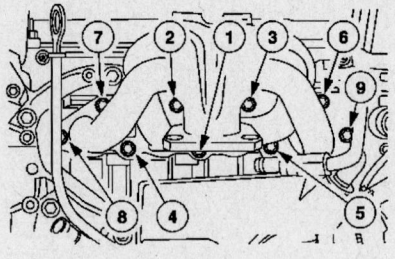

Exhaust manifold bolt torque sequence–2.0L engine

9308TG10

To install:

6. Inspect the fuel injector O-rings and replace if necessary.

7. Install or connect the following:
- Fuel injectors into the lower intake manifold
- Fuel rail. Torque the bolts to 89 inch lbs. (10 Nm).
- New gasket
- Intake manifold. Torque the bolts, in sequence, to 89 inch lbs. (10 Nm).
- Fuel rail electrical connectors
- Fuel injector electrical connectors
- Fuel pressure damper vacuum line
- Upper intake manifold
- Fuel line spring lock coupling
- Negative battery cable

8. Start the engine and check for leaks, repair if necessary.

Exhaust Manifold

REMOVAL & INSTALLATION

2.0L Engine

1. Before servicing the vehicle, refer to the Precautions Section.

2. Remove or disconnect the following:
- Negative battery cable
- Catalytic converter
- Oil level indicator tube and bracket
- Exhaust manifold and discard the gasket

To install:

3. Clean the sealing surfaces of any old gasket material.

4. Install or connect the following:
- Exhaust manifold and new gasket. Torque the bolts to 12 ft. lbs. (16 Nm).
- Oil level indicator tube and bracket. Torque the bolt to 89 inch lbs. (10 Nm).
- Catalytic converter
- Negative battery cable

5. Start the engine and check for leaks, repair if necessary.

2.3L Engine

✷✷ WARNING

Do not use oil or grease-based lubri-cants on the insulators. They may cause deterioration of the rubber.

✷✷ WARNING

Oil or grease-based lubricants on the insulators may cause the exhaust hanger insulator to separate from the exhaust hanger bracket during vehicle operation.

➡**Exhaust fasteners are of a torque prevailing design. Use only new fasten-ers with the same part number as the original. Torque values must be used as specified during reassembly to make sure of correct retention of exhaust components.**

1. Remove the flex pipe nuts.
2. Remove the flex pipe gasket.
3. Remove the manifold bracket bolts.
4. Remove the heat shield.
5. Remove the exhaust manifold nuts.
6. Remove the catalyst monitor sensor.
7. Remove the heated oxygen sensor.
8. Remove the exhaust manifold.

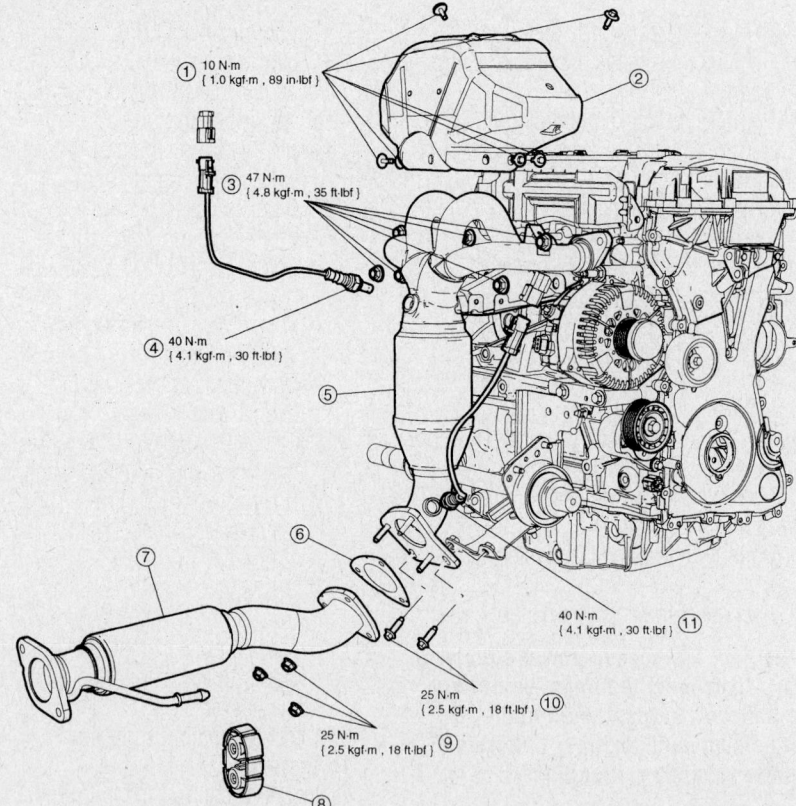

1	Heat shield bolts (6 required)	7	Flex pipe
2	Heat shield	8	Exhaust hanger
3	Exhaust manifold nuts (7 required)	9	Flex pipe nuts (3 required)
4	Heated oxygen sensor	10	Manifold bracket bolts (2 required)
5	Exhaust manifold	11	Catalyst monitor sensor
6	Flex pipe gasket		

09482_TRIB_G0012

Exhaust manifold and related components—2.3L engine

9. To install, reverse the removal procedure. Make sure to apply anti-seize lubricant to the threads of the sensors before installation. Failure to tighten the exhaust manifold nuts to specification before installing the manifold bracket bolts will cause the manifold to develop an exhaust gas leak.

10. Observe the following torques:
- Exhaust manifold-to-head: 35 ft. lbs. (47 Nm)
- Flex pipe-to-manifold: 18 ft. lbs. (25 Nm)
- Heated oxygen sensor: 35 ft. lbs. (47 Nm)
- Catalyst monitor sensor: 30 ft. lbs. (40 Nm)

11. Check the exhaust system for proper alignment.

3.0L Engine

LEFT SIDE

1. Before servicing the vehicle, refer to the Precautions Section.
2. Disconnect the negative battery cable.

➡ **To deplete the backup power supply energy, disconnect the negative battery cable and wait at least one minute. Be sure to disconnect auxiliary batteries and power supplies, if equipped.**

3. Remove or disconnect the following:
- Heated Oxygen (HO$_2$S) sensor and catalyst monitor
- Splash shield
- Exhaust crossover pipe
- Drive belt
- A/C compressor and move it aside
- Exhaust manifold and discard the gasket

To install:

4. Clean the sealing surfaces of any old gasket material.
5. Install or connect the following:

- Exhaust manifold and new gasket. Torque the bolts to 15 ft. lbs. (20 Nm).
- A/C compressor. Torque the bolts to 18 ft. lbs. (20 Nm).
- Drive belt
- Exhaust crossover pipe. Torque the bolts to 30 ft. lbs. (40 Nm).
- Splash shield. Torque the bolts to 80 inch lbs. (9 Nm).
- Left side HO$_2$S sensor and catalyst monitor
- Negative battery cable

6. Start the engine and check for leaks, repair if necessary.

RIGHT SIDE

1. Before servicing the vehicle, refer to the Precautions Section.
2. Disconnect the negative battery cable.

➡ **To deplete the backup power supply energy, disconnect the negative battery cable and wait at least one minute. Be sure to disconnect auxiliary batteries and power supplies, if equipped.**

3. Remove or disconnect the following:
- Exhaust Gas Recirculation (EGR) tube
- Alternator
- Right side Heated Oxygen (HO$_2$S) sensor
- Right side exhaust manifold and discard the gasket

To install:

4. Clean the sealing surfaces of any old gasket material.
5. Install or connect the following:
- Exhaust manifold and new gasket. Torque the bolts to 15 ft. lbs. (20 Nm).
- Right side HO$_2$S sensor
- Alternator

- EGR tube
- Negative battery cable

6. Start the engine and check for leaks, repair if necessary.

Front Crankshaft Seal

REMOVAL & INSTALLATION

2.0L Engine

1. Before servicing the vehicle, refer to the Precautions Section.
2. Remove or disconnect the following:

- Negative battery cable
- Timing belt
- Crankshaft sprocket and timing belt guide
- Crankshaft oil seal

➡ **Be careful not to damage the seal surface of the cover.**

To install:

3. Install or connect the following:
- New front crankshaft oil seal
- Timing belt guide and crankshaft sprocket
- Timing belt
- Negative battery cable

4. Start the engine and check for leaks, repair if necessary.

2.3L Engine

❋❋ **WARNING**

During engine repair procedures, cleanliness is extremely important. Any foreign material, including any material created while cleaning gasket surfaces, that enters the oil passages, coolant passages or the oil pan can cause engine failure.

9308TG11

Left side exhaust manifold bolt torque sequence–3.0L engine

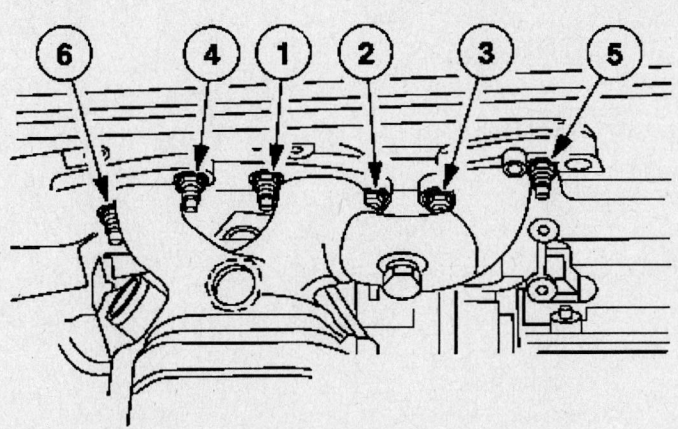

9308TG12

Right side exhaust manifold bolt torque sequence–3.0L engine

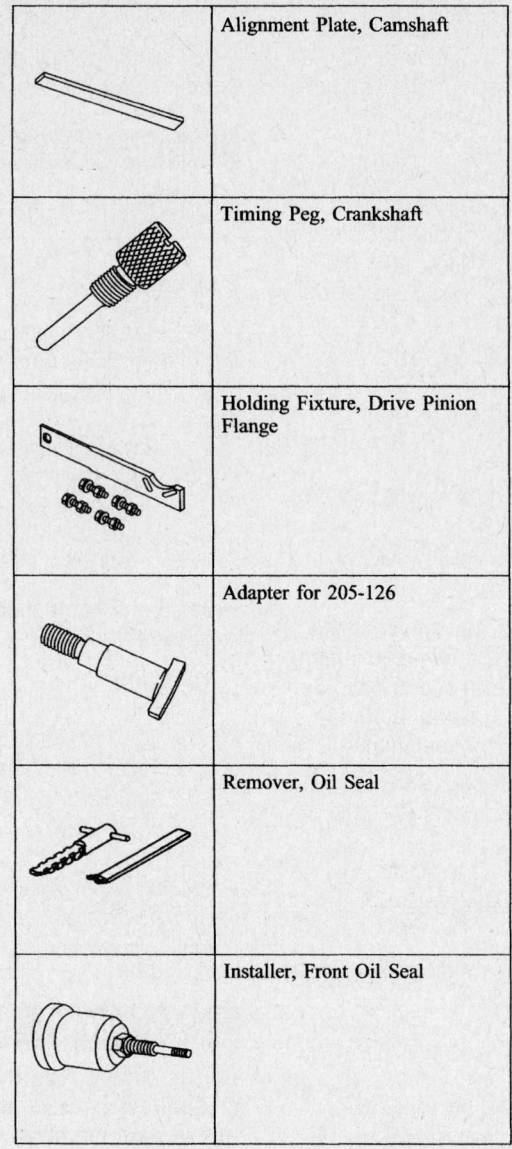

	Alignment Plate, Camshaft
	Timing Peg, Crankshaft
	Holding Fixture, Drive Pinion Flange
	Adapter for 205-126
	Remover, Oil Seal
	Installer, Front Oil Seal

67197-ESCA-G08

Tools necessary for this job—2.3L engine

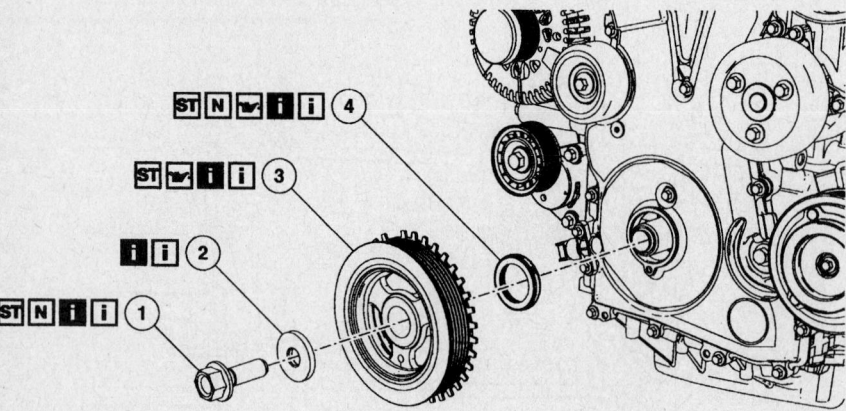

1 Crankshaft pulley bolt   3 Crankshaft pulley
2 Crankshaft pulley washer   4 Crankshaft front seal

67197-ESCA-G09

Crankshaft pulley and related components—2.3L engine

✳✳ WARNING

The crankshaft, the crankshaft sprocket and the pulley are fitted together by friction, using diamond washers between the flange faces on each part. For that reason, the crankshaft sprocket is also unfastened if you loosen the pulley. Therefore, the engine must be retimed each time the damper is removed. Otherwise severe engine damage can occur.

1. Raise and support the vehicle safely.
2. Remove the accessory drive belt.
3. Remove the valve cover.

✳✳ WARNING

Failure to position the No. 1 piston at top dead center (TDC) can result in damage to the engine. Turn the engine in the normal direction of rotation only.

4. Using the crankshaft pulley bolt, turn the crankshaft clockwise to position the No. 1 piston at top dead center (TDC).

✳✳ WARNING

The special tool 303-465 is for camshaft alignment only. Using this tool to prevent engine rotation can result in engine damage.

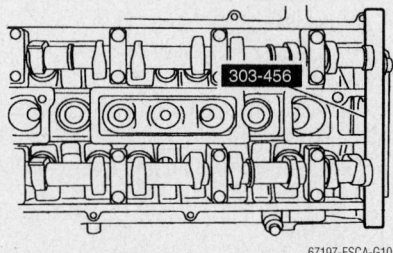

67197-ESCA-G10

Camshaft holding tool—2.3L engine

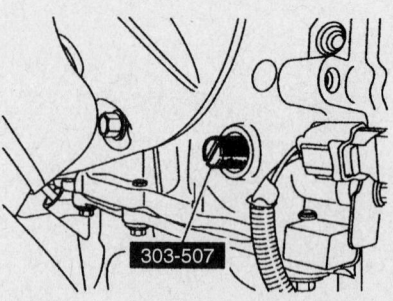

67197-ESCA-G11

Special tool 303-507 installation—2.3L engine

➡The camshaft timing slots are offset. If the special tool cannot be installed, rotate the crankshaft one complete revolution clockwise to correctly position the camshafts.

5. Install the special tool in the slots on the rear of both camshafts.

➡Installing the special tool in this step will prevent the engine from being rotated in the clockwise direction.

6. Install special tool 303-507.
7. Crankshaft pulley bolt and washer.
8. Remove the engine plug bolt.
9. Install the crankshaft holding tools.

✳✳ WARNING

Failure to hold the crankshaft pulley in place while loosening the bolt can result in damage to the engine.

10. Remove the crankshaft pulley. Remove the crankshaft pulley bolt and washer.
11. Remove the crankshaft pulley.

✳✳ WARNING

Use care not to damage the engine front cover or the crankshaft when removing the seal.

12. Using the special tool, remove the crankshaft front oil seal.

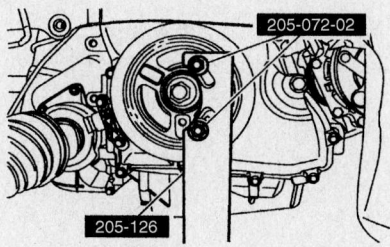

Crankshaft holding tool installation—2.3L engine

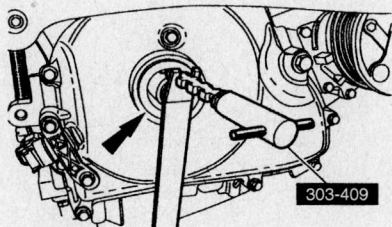

Crankshaft oil seal installation using special tool 303-409—2.3L engine

➡Remove the through-bolt from the special tool.

➡Lubricate the oil seal with clean engine oil.

13. Using the special tool, install the crankshaft front oil seal.

➡Do not reuse the crankshaft damper bolt.

➡Apply clean engine oil on the seal area before installing.

14. Install the crankshaft pulley and hand-tighten the bolt.

✳✳ WARNING

Only hand-tighten the bolt or damage to the front cover can occur.

➡This step will correctly align the crankshaft pulley to the crankshaft.

15. Install a standard 6 mm x 18 mm bolt through the crankshaft pulley and thread it into the front cover. Rotate the pulley as necessary to align the bolt holes.

✳✳ WARNING

Failure to hold the crankshaft pulley in place while tightening the bolt can cause damage to the engine front cover.

16. Using the special tools to hold the crankshaft pulley in place, tighten the crankshaft pulley bolt in two stages:
   a. Stage 1: Tighten to 100 Nm (74 ft. lbs.).
   b. Stage 2: Tighten an additional 90 degrees (¼ turn).
17. Remove the 6 mm x 18 mm bolt.
18. Remove the special tools.

➡Only turn the engine in the normal direction of rotation.

19. Turn the engine two complete revolutions.
20. Turn the crankshaft until the No. 1 piston is at TDC.
21. Install special tool 303-507.

✳✳ WARNING

Only hand-tighten the bolt or damage to the front cover can occur.

22. Using the 6 mm x 18 mm bolt, check the position of the crankshaft pulley. If it is not possible to install the bolt, correct the engine timing.
23. Using special tool 303-465, check the position of the camshafts. If it is not

possible to install the special tool, correct the engine timing.
24. Remove the 6 mm x 18 mm bolt.
25. Install the engine plug bolt.

3.0L Engine

1. Before servicing the vehicle, refer to the Precautions Section.
2. Disconnect the negative battery cable.

➡To deplete the backup power supply energy, disconnect the negative battery cable and wait at least one minute. Be sure to disconnect auxiliary batteries and power supplies, if equipped.

3. Remove or disconnect the following:
   • Crankshaft pulley
   • Front oil seal

To install:
4. Install or connect the following:
   • New front crankshaft oil seal
   • Crankshaft pulley
   • Negative battery cable
5. Start the engine and check for leaks, repair if necessary.

Camshaft and Valve Lifters

REMOVAL & INSTALLATION

2.0L Engine

1. Before servicing the vehicle, refer to the Precautions Section.
2. Remove or disconnect the following:
   • Negative battery cable
   • Camshaft timing sprocket and verify the valve clearance
   • Camshaft journal cap bolts by loosening them in several passes in the proper sequence
   • Camshafts

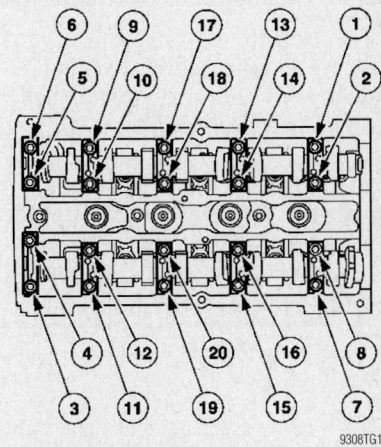

Camshaft bearing cap removal sequence–2.0L engine

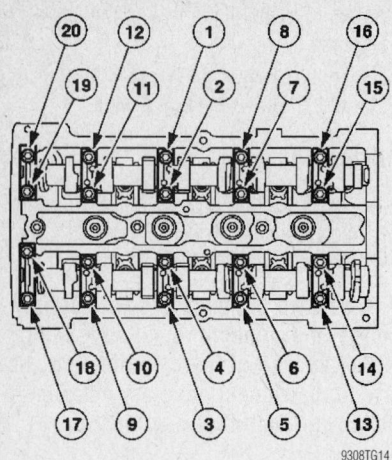

Camshaft bearing cap torque
sequence–2.0L engine

9308TG14

3. Inspect the camshaft for wear and discard the oil seals

To install:

4. Install or connect the following:
   - Camshaft cam followers, lubricate the bearing journals thoroughly. Torque the caps to 14 ft. lbs. (19 Nm).
   - Exhaust camshaft oil seal
   - Camshaft timing sprocket
   - Negative battery cable

2.3L Engine

✳✳ WARNING

During engine repair procedures, cleanliness is extremely important. Any foreign material, including any material created while cleaning gasket surfaces, that enters the oil passages, coolant passages or the oil pan can cause engine failure.

✳✳ WARNING

The crankshaft, the crankshaft sprocket and the pulley are fitted together by friction, using diamond washers between the flange faces on each part. For that reason, the crankshaft sprocket is also unfastened if you loosen the pulley. Therefore, the engine must be retimed each time the damper is removed. Otherwise severe engine damage can occur.

1. With the vehicle in NEUTRAL, position it on a hoist.

➡Valve tappets are select fit and the valve clearance must be checked before removing the tappets.

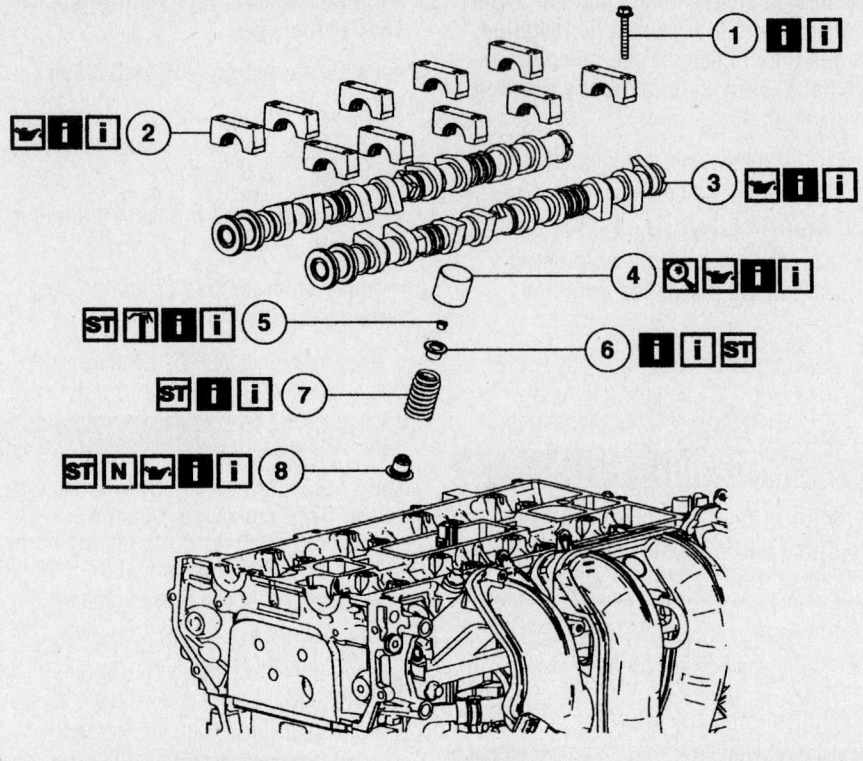

1 Camshaft bearing cap bolt	4 Valve tappet	7 Valve spring
2 Camshaft bearing cap	5 Valve collet	8 Valve seal
3 Camshaft	6 Valve spring retainer	

67197-ESCA-G16

Camshafts and related components—2.3L engine

✳✳ WARNING

Turn the engine clockwise only, and only use the crankshaft bolt.

➡Before removing the camshafts, measure the clearance of each valve at base circle, with the lobe pointed away from the tappet. Failure to measure all clearances prior to removing the camshafts will necessitate repeated removal and installation and wasted labor time.

2. Use a feeler gauge to measure the clearance of each valve and record its location.

➡The number on the valve tappet only reflects the digits that follow the decimal. For example, a tappet with the number 0.650 has the thickness of 3.650 mm.

➡A midrange clearance is the most desirable:

- Intake: 0.22–0.28 mm (0.008–0.011 inch)
- Exhaust: 0.27–0.33 mm (0.010–0.013 inch)

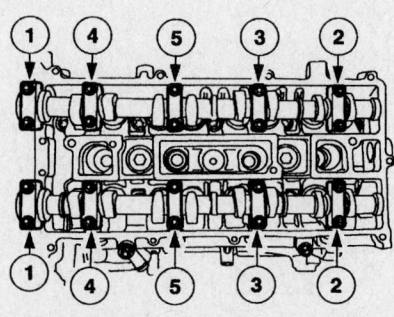

67197-ESCA-G17

Camshaft bearing cap removal sequence—2.3L engine

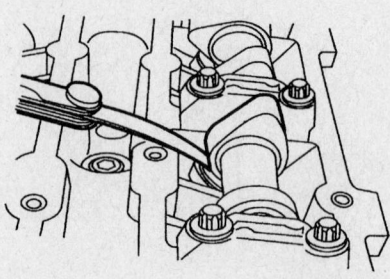

67197-ESCA-G14

Valve clearance check—2.3L engine

3. Select tappets using this formula: tappet thickness = measured clearance + the base tappet thickness–most desirable thickness.

4. Select the tappets and mark the installation location.

5. If any tappets do not measure within specifications, install new tappets in these locations.

6. Remove the timing chain and sprockets.

7. Mark the position of the camshaft lobes on the No. 1 cylinder for assembly reference.

✳✳ WARNING

Failure to follow the camshaft loosening procedure can result in damage to the camshafts.

8. Loosen the camshaft bearing bolts in the sequence shown, one turn at a time. Repeat until all the tension is released.

9. Remove the camshaft bearing caps.

✳✳ WARNING

If the camshafts and valve tappets are to be reused, mark the location of the valve tappets to make sure they are assembled in their original positions.

➡ **The number on the valve tappets only reflects the digits that follow the decimal. For example, a tappet with the number 0.650 has the thickness of 3.650 mm.**

10. Remove the camshafts.
11. Remove the valve tappets.
To install:

12. Installation is the, reverse the removal procedure. Coat the valve tappets with clean engine oil and insert them.

✳✳ WARNING

Install the camshafts with the alignment slots in the camshafts lined up so the Camshaft Alignment Plate can be installed without rotating the camshafts. Make sure the lobes on the No. 1 cylinder are in the same position as noted in the removal procedure. Rotating the camshafts when the timing chain is removed, or installing the camshafts 180 degrees out of position can cause severe damage to the valves and pistons.

➡ **Lubricate the camshaft journals and bearing caps with clean engine oil.**

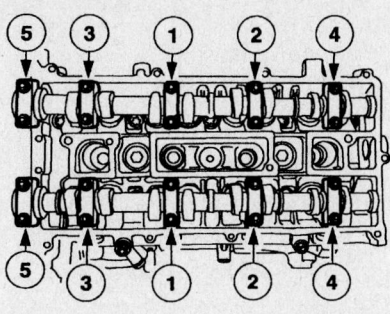

Camshaft bearing cap torque sequence— 2.3L engine

13. Install the camshafts and bearing caps. Tighten the bolts in the sequence shown in three stages.
   a. Stage 1: Tighten the camshaft bearing bolt caps one turn at a time until tight.
   b. Stage 2: Tighten the bolts to 7 Nm (62 inch lbs.).
   c. Stage 3: Tighten the bolts to 16 Nm (12 ft. lbs.).

3.0L Engine

LEFT SIDE

✳✳ WARNING

During engine repair procedures, cleanliness is extremely important. Any foreign material, including any material created while cleaning gasket surfaces, that enters the oil passages, coolant passages or the oil pan can cause engine failure.

➡ **Early build vehicles are equipped with the water pump drive pulley on the left intake camshaft. Late build vehicles are equipped with the water pump pulley on the left exhaust manifold.**

1. Before servicing the vehicle, refer to the Precautions Section.
2. Disconnect the negative battery cable.

➡ **To deplete the backup power supply energy, disconnect the negative battery cable and wait at least one minute. Be sure to disconnect auxiliary batteries and power supplies, if equipped.**

3. Remove or disconnect the following:
   • Water pump belt
   • Rocker cover
   • Timing drive components
   • Using the proper tools remove the water pump pulley, as necessary.
   • Camshaft oil seal
   • Camshaft oil seal retainer. Discard the gasket.

• Camshaft cap bolts by loosening them in sequence

➡ **After loosening all of the camshaft bearing cap bolts, remove the camshaft bearing thrust caps first, or damage to the thrust caps may occur. The thrust caps are marked 1L and 5L on early build engines and 1L and 6L on late build engines.**

4. Remove the camshaft caps. Remove the camshafts.
To install:
5. Install or connect the following:
   • Camshaft bearing caps in their original position

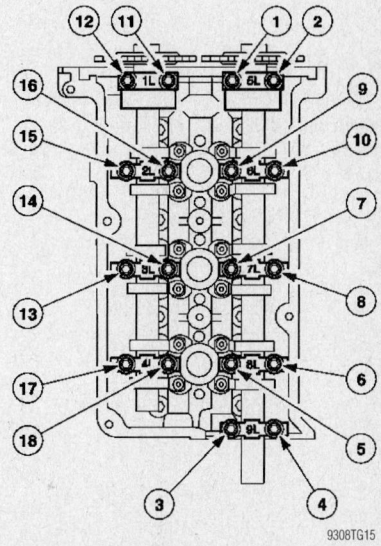

Left side camshaft bearing cap removal sequence–early build 3.0L engine

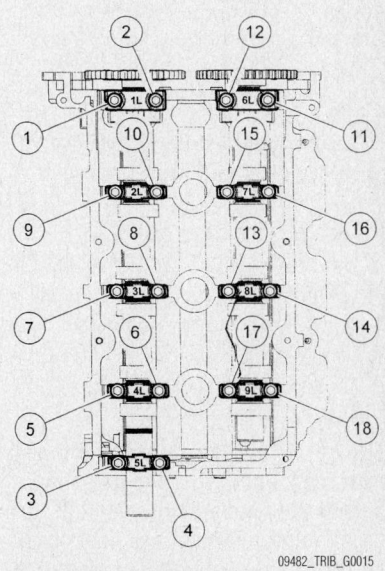

Left side camshaft bearing cap removal sequence–late build 3.0L engine

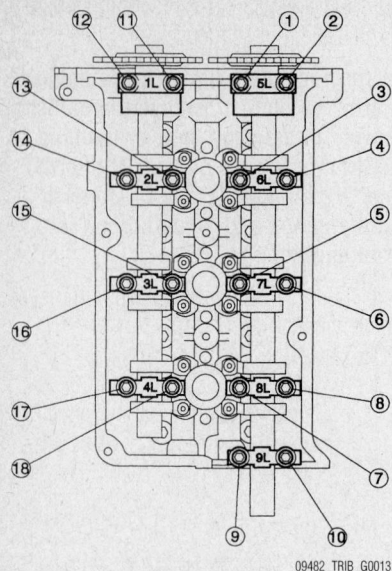

Left side camshaft bearing cap torque sequence–2002–2005 early build 3.0L engine

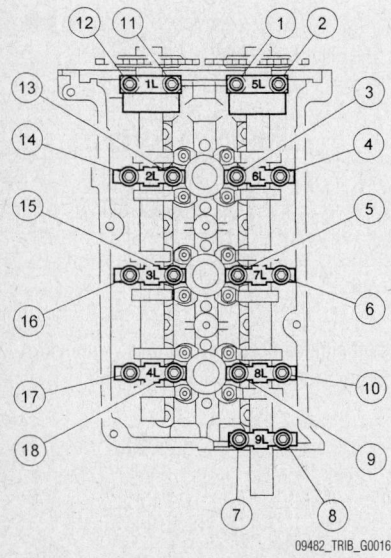

Left side camshaft bearing cap torque sequence–2006 early build 3.0L engine

- Align the camshafts
- Bearing thrust caps and hand tighten the bolts. When aligned properly, torque the bolts to 89 inch lbs. (10 Nm) on 2002–2005 vehicles and 71 inch lbs. (8 Nm) on 2006 vehicles.

➡Do not install the camshaft journal thrust caps until all of the camshaft bearing caps have been installed, or damage to the thrust caps may occur. The thrust caps are marked 1L and 5L on early build engines and 1L and 6L on late build engines.

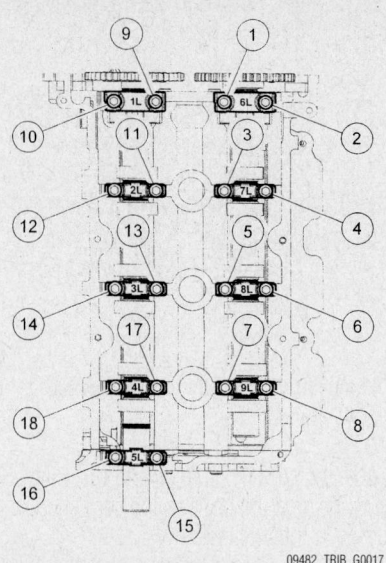

Left side camshaft bearing cap torque sequence–2006 late build 3.0L engine

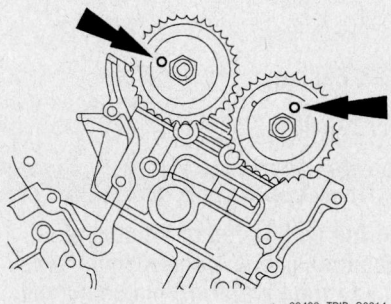

Left side camshaft alignment—3.0L engine

- Timing drive components
- Camshaft oil seal retainer, using a new gasket
- Crankshaft oil seal
- Water pump drive pulley

➡On early build engines, the original pulley is pressed on 0.18 inch past the flush end of the camshaft. On late build engines, the original pulley is pressed on flush to the end of the camshaft. Service pulleys on all engines are pressed on flush to the end of the camshaft.

- Water pump belt
- Negative battery cable

RIGHT SIDE

❋❋ WARNING

During engine repair procedures, cleanliness is extremely important. Any foreign material, including any material created while cleaning gas-

ket surfaces, that enters the oil passages, coolant passages or the oil pan can cause engine failure.

1. Before servicing the vehicle, refer to the Precautions Section.
2. Disconnect the negative battery cable.

➡To deplete the backup power supply energy, disconnect the negative battery cable and wait at least one minute. Be sure to disconnect auxiliary batteries and power supplies, if equipped.

3. Remove or disconnect the following:
- Timing drive components
- Rocker cover
- Camshaft cap bolts by loosening them in sequence

➡After loosening all of the camshaft bearing cap bolts, remove the camshaft bearing thrust caps first, or damage to the thrust caps may occur. The thrust caps are marked 5R and 1R.

4. Remove the camshaft caps. Remove the camshafts.

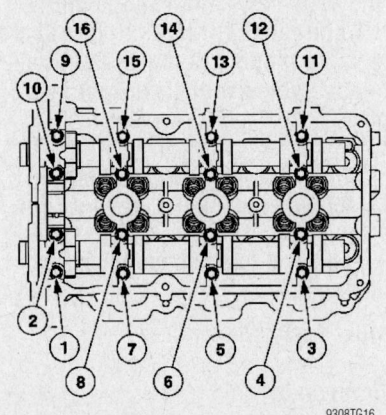

Right side camshaft bearing cap removal sequence–3.0L engine

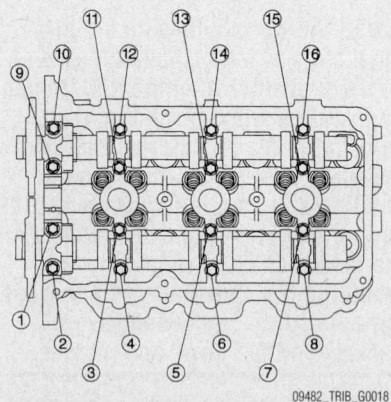

Right side camshaft bearing cap torque sequence–3.0L engine

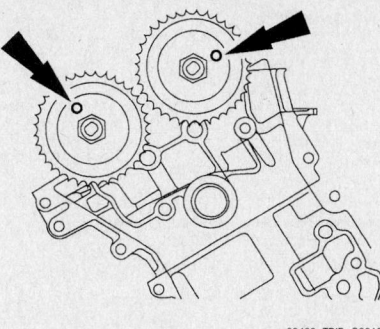

09482_TRIB_G0019

Right side camshaft alignment—3.0L engine

To install:
5. Install or connect the following:
- Camshaft bearing caps in their original position
- Align the camshafts
- Bearing caps and hand tighten the bolts
- Bearing thrust caps and hand tighten the bolts. When aligned properly, torque the bolts to 89 inch lbs. (10 Nm).

➡**Do not install the camshaft journal thrust caps until all of the camshaft bearing caps have been installed, or damage to the thrust caps may occur. The thrust caps are marked 5R and 1R.**

- Timing drive components
- Negative battery cable

INSPECTION

1. Before servicing the vehicle, refer to the Precautions Section.
2. Remove the camshaft from the engine.
3. Check the camshaft bearing journals for damage and binding.
4. If the journals are binding, check the cylinder head for damage.
5. Check the cylinder head for clogged oil holes.
6. Check the camshaft surface for abnormal wear and damage. Replace the camshaft, as required.
7. Measure the camshaft lobe surface and replace the camshaft if not within specification.
8. Measure the camshaft journal diameter and replace the camshaft if not within specification.
9. Measure the camshaft run out and replace the camshaft if not within specification.

Valve Lash

ADJUSTMENT

2.0L Engine

1. Before servicing the vehicle, refer to the Precautions Section.
2. Remove or disconnect the following:
- Negative battery cable
- Timing belt
3. Measure each valve's clearance at the base circle with the lobe facing away from the tappet.
4. Use a feeler gauge to measure and record each valve's clearance
5. Remove or disconnect the following:
- Camshafts
- Valve tappets from the cylinder head
6. A mid range clearance is recommended as follows:
  a. Intake: 0.006 inch (0.15mm).
  b. Exhaust: 0.012 inch (0.3mm).
To install:
7. Install or connect the following:
- Valve tappets after lubricating them with clean engine oil
- Camshafts and verify each valve's clearance at the base circle with the lobe facing away from the tappet
- Timing belt
- Negative battery cable

2.3L Engine

➡**Before removing the camshafts, measure the clearance of each valve at base circle, with the lobe pointed away from the tappet. Failure to measure all clearances prior to removing the camshafts will necessitate repeated removal and installation and wasted labor time.**

1. Before servicing the vehicle, refer to the Precautions Section.
2. Disconnect the negative battery cable.

➡**To deplete the backup power supply energy, disconnect the negative battery cable and wait at least one minute. Be sure to disconnect auxiliary batteries and power supplies, if equipped.**

3. Remove the rocker cover.
4. Use a feeler gauge to measure the clearance of each valve and record its location.

➡**The number on the valve tappet only reflects the digits that follow the decimal. For example, a tappet with the number 0.650 has the thickness of 3.650 mm.**

➡**A midrange clearance is the most desirable:**

- Intake: 0.22–0.28 mm (0.008–0.011 inch)
- Exhaust: 0.27–0.33 mm (0.010–0.013 inch)
5. Select tappets using this formula: tappet thickness = measured clearance + the base tappet thickness–most desirable thickness.
6. Select the tappets and mark the installation location.
7. If any tappets do not measure within specifications, install new tappets in these locations.

3.0L Engine

1. Before servicing the vehicle, refer to the Precautions Section.
2. Disconnect the negative battery cable.

➡**To deplete the backup power supply energy, disconnect the negative battery cable and wait at least one minute. Be sure to disconnect auxiliary batteries and power supplies, if equipped.**

3. Remove or disconnect the following:
- Camshaft followers
- Hydraulic lash adjusters

➡**Mark the position of the hydraulic lash adjusters to assure they are assembled in their original position**

4. Inspect the adjusters for scoring or uneven wear in the bore and replace them as required.
To install:
5. Install or connect the following:
- Hydraulic lash adjusters after lubricating them with clean engine oil
- Camshaft followers
- Negative battery cable

Starter Motor

REMOVAL & INSTALLATION

2.0L Engine

1. Before servicing the vehicle, refer to the Precautions Section.
2. Remove or disconnect the following:

- Negative battery cable
- Starter bolts
- Exhaust system, AWD vehicles only
- Halfshaft support bracket bolts
- Starter electrical connectors
- Starter

To install:

3. Install or connect the following:
- Starter. Torque bolts to 20 ft. lbs. (27 Nm).
- Starter electrical connectors
- Halfshaft support bracket. Torque the bolts to 11 ft. lbs. (15 Nm).
- Exhaust system on AWD vehicles. Torque the bolts to 18 ft. lbs. (25 Nm).
- Negative battery cable

2.3L Engine

> ✳✳ **WARNING**
>
> When performing maintenance on the starting system, be aware that heavy gauge leads are connected directly to the battery. Make sure protective caps are in place when maintenance is completed.

1. Before servicing the vehicle, refer to the Precautions Section.
2. Disconnect the negative battery cable.

➡ To deplete the backup power supply energy, disconnect the negative battery cable and wait at least one minute. Be sure to disconnect auxiliary batteries and power supplies, if equipped.

3. Raise and support the vehicle safely.
4. Remove the halfshaft support bracket bolts and nuts.
5. Remove the starter motor solenoid terminal cover, solenoid wire and the solenoid battery cable.
6. Disconnect the wiring harness retainer and ground strap.
7. Remove the starter motor stud bolts, starter motor bracket bolt and starter motor.

To install:

8. To install, reverse the removal procedure.
9. Torque the starter and bracket bolts to 20 ft. lbs. (27 Nm).

3.0L Engine

> ✳✳ **WARNING**
>
> When performing maintenance on the starting system, be aware that heavy gauge leads are connected directly to the battery. Make sure protective caps are in place when maintenance is completed.

1. Before servicing the vehicle, refer to the Precautions Section.
2. Disconnect the negative battery cable.

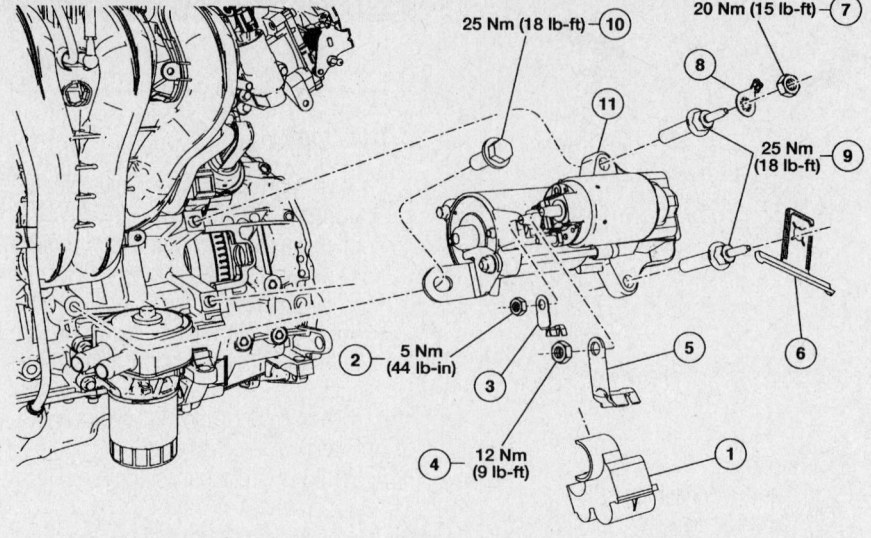

1	Starter motor solenoid terminal cover	5	Starter solenoid battery cable	
2	Starter solenoid wire nut	6	Wiring harness retainer	
3	Starter solenoid wire	7	Ground strap nut	
4	Starter solenoid battery cable nut	8	Ground strap	

9	Starter motor stud bolts
10	Starter motor bracket bolt
11	Starter motor

67197-ESCA-G19

Starter and related components—2.3L engine

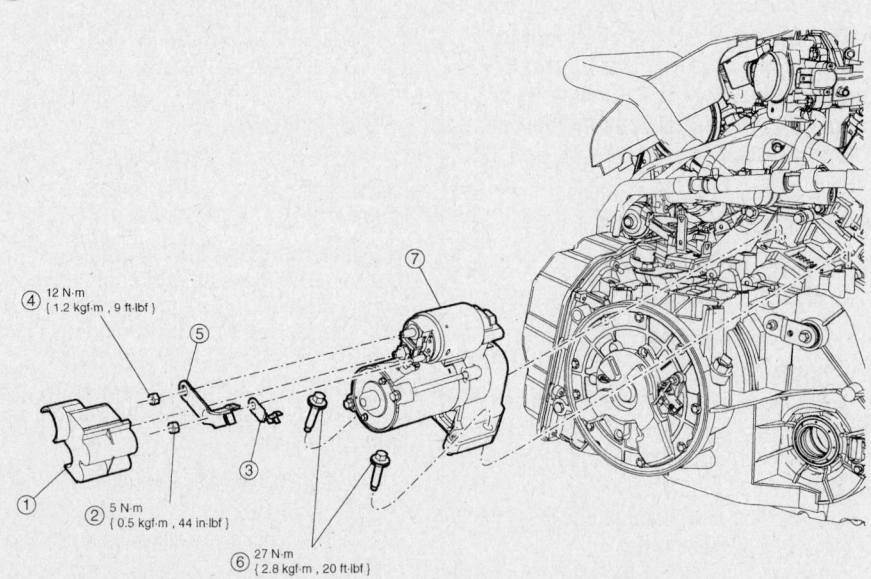

1	Starter motor solenoid terminal cover	5	Starter motor solenoid battery cable nut
2	Starter motor solenoid wire nut	6	Starter motor bolts
3	Starter motor solenoid wire	7	Starter motor
4	Starter motor solenoid battery cable nut		

09482_TRIB_G0020

Starter and related components—3.0L engine

➡ To deplete the backup power supply energy, disconnect the negative battery cable and wait at least one minute. Be sure to disconnect auxiliary batteries and power supplies, if equipped.

3. Drain the cooling system.
4. Remove the air cleaner assembly.

5. Disconnect the hoses and position the thermostat to the side.
6. Disconnect the starter motor electrical connections, and position them to the side.
7. Remove the starter mounting bolts. Remove the starter from the vehicle.

To install:

8. To install, reverse the removal procedure.

9. Torque the starter and bracket bolts to 20 ft. lbs. (27 Nm).

10. Fill the cooling system, using the proper grade and type coolant.

11. Start the engine and check for leaks.

Oil Pan

REMOVAL & INSTALLATION

2.0L Engine

1. Before servicing the vehicle, refer to the Precautions Section.
2. Drain the engine oil.
3. Support the powertrain assembly.
4. Remove or disconnect the following:
   - Negative battery cable
   - Catalytic converter
   - Oil pan and gasket
5. Thoroughly clean the gasket mating surfaces.

To install:

6. Apply silicone sealer to the oil pan.
7. Install a new gasket on the oil pan.
8. Oil pan. Torque the bolts in sequence to:
   a. Step 1: 53 inch lbs. (6 Nm).
   b. Step 2: 106 in lbs. (12 Nm).
9. Install or connect the following:
   - Catalytic converter
   - Negative battery cable
10. Fill the engine with clean oil.
11. Start the engine and check for leaks, repair if necessary.

2.3L Engine

1. Before servicing the vehicle, refer to the Precautions Section.
2. Disconnect the negative battery cable.

➡To deplete the backup power supply energy, disconnect the negative battery cable and wait at least one minute. Be sure to disconnect auxiliary batteries and power supplies, if equipped.

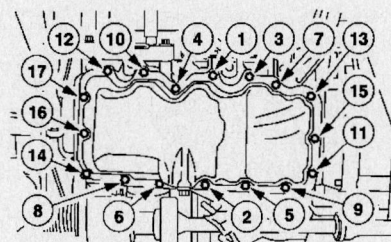

Oil pan bolt torque sequence–2.0L engine

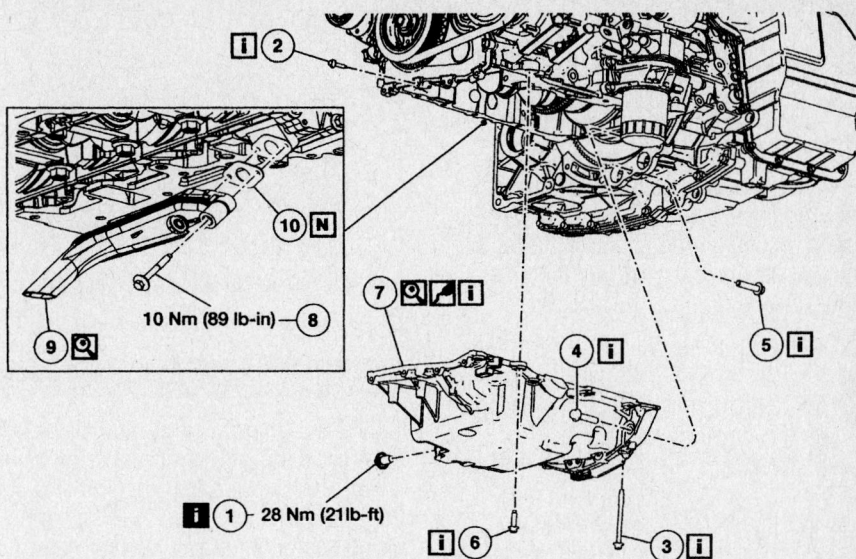

10 Nm (89 lb-in)

28 Nm (21lb-ft)

1 Drain plug
2 Engine front cover bolt
3 Oil pan bolt
4 Oil pan-to-bell housing bolt
5 Oil pan-to-bell housing bolt
6 Oil pan bolt
7 Oil pan
8 Oil pump screen and pickup tube bolt
9 Oil pump screen and pickup tube
10 Oil pump screen and pickup tube gasket

67197-ESCA-G20

Oil pan and related components—2.3L engine

3. Raise and support the vehicle safely.
4. Remove the oil level indicator and tube.
5. Drain the oil.
6. Remove the 4 front cover-to-oil pan bolts.
7. Remove the 2 rear oil pan bolts.
8. Remove the oil pan-to-bell housing bolts.
9. Remove the oil pan-to-block bolts.

To install:

※※ WARNING

CAUTION: Do not use metal scrapers, wire brushes, power abrasive discs or other abrasive means to clean the sealing surfaces. These tools cause scratches and gouges, which make leak paths. Use a plastic scraping tool to remove traces of sealant. Clean and inspect all mating surfaces.

➡If the oil pan is not secured within four minutes of sealant application the sealant must be removed and the sealing area cleaned with metal surface cleaner. Allow to dry until there is no sign of wetness, or four minutes, whichever is longer. Failure to follow this procedure can cause future oil leakage.

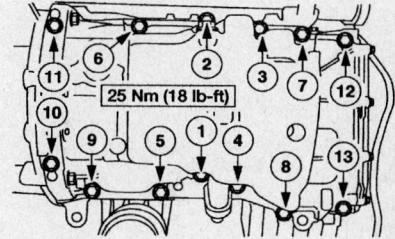

25 Nm (18 lb-ft)

67197-ESCA-G21

Oil pan bolt torque sequence—2.3L engine

10. Apply a 2.5 mm bead of silicone gasket and sealant to the oil pan. Install the oil pan.

➡The oil pan must be installed and the bolts tightened within four minutes of applying the silicone gasket and sealant.

11. Install the front cover bolts. Torque to 89 inch lbs. (10 Nm).
12. Install the oil pan-to-bell housing bolts. Torque to 35 ft. lbs. (48 Nm).
13. Install the rear oil pan bolts. Torque to 35 ft. lbs. (48 Nm).
14. Install and tighten the oil pan bolts in the sequence shown to 18 ft. lbs. (25 Nm).
15. Fill the engine with clean engine oil.

3.0L Engine

1. Before servicing the vehicle, refer to the Precautions Section.

2. Disconnect the negative battery cable.

➡ **To deplete the backup power supply energy, disconnect the negative battery cable and wait at least one minute. Be sure to disconnect auxiliary batteries and power supplies, if equipped.**

3. Raise and support the vehicle safely.

4. Remove the flexible exhaust pipe.

5. Drain the engine oil. Remove and discard the oil filter.

6. If equipped, remove the downstream catalyst monitor sensor.

7. Remove the oil pan retaining bolts. Remove the three oil pan to transaxle bolts.

8. Remove the oil pan from the vehicle. Discard the gasket.

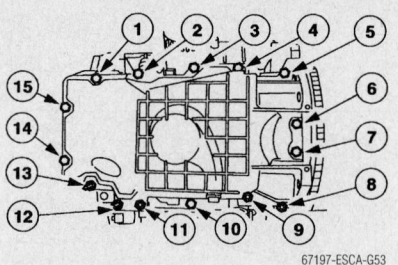

67197-ESCA-G53

Oil pan torque sequence—3.0L engine

To install:

✳✳ WARNING

CAUTION: Do not use metal scrapers, wire brushes, power abrasive discs or other abrasive means to clean the sealing surfaces. These tools cause scratches and gouges, which make leak paths. Use a plastic scraping tool to remove traces of sealant. Clean and inspect all mating surfaces.

➡ **If the oil pan is not secured within four minutes of sealant application the sealant must be removed and the sealing area cleaned with metal surface cleaner. Allow to dry until there is no sign of wetness, or four minutes, whichever is longer. Failure to follow this procedure can cause future oil leakage.**

9. Apply a 0.40 inch (10 mm) dot of silicone gasket and sealer to the oil pan bolt hole locations.

10. Install the oil pan. Torque the retaining bolts to specification.

11. Continue the installation in the reverse order of the removal procedure.

12. Install a new oil filter. Fill the engine with clean engine oil.

13. Start the engine and check for leaks. Correct as required.

Oil Pump

REMOVAL & INSTALLATION

2.0L Engine

1. Before servicing the vehicle, refer to the Precautions Section.

2. Drain the engine oil.

3. Remove or disconnect the following:
   - Negative battery cable
   - Oil pan
   - Oil pump screen cover and tube
   - Oil pump and discard the gasket

4. Thoroughly clean the gasket mating surfaces.

To install:

5. Install or connect the following:
   - Oil pump screen cover and tube with a new gasket. Torque the bolts to 89 inch lbs. (10 Nm).
   - Oil pump to the oil pan
   - Oil pan
   - Negative battery cable

6. Refill the engine with clean oil.

7. Start the engine and check for leaks; repair if necessary.

2.3L Engine

1. Before servicing the vehicle, refer to the Precautions Section.

2. Remove the engine from the vehicle and mount it on an engine stand.

3. Remove the oil pan.

4. Remove the oil pump pickup tube and screen.

5. Remove the front cover and the timing chain.

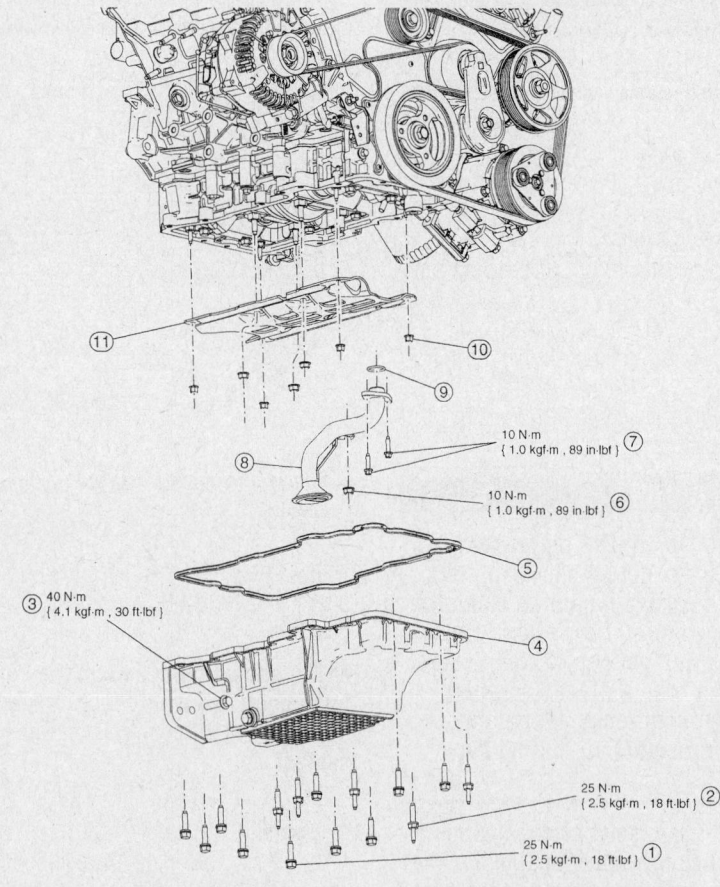

10 N·m
{ 1.0 kgf·m , 89 in·lbf } ⑦

10 N·m
{ 1.0 kgf·m , 89 in·lbf } ⑥

③ 40 N·m
{ 4.1 kgf·m , 30 ft·lbf }

25 N·m
{ 2.5 kgf·m , 18 ft·lbf } ②

25 N·m
{ 2.5 kgf·m , 18 ft·lbf } ①

1	Oil pan-to-engine bolts (10 required)	7	Oil pump screen and pickup tube bolts (2 required)
2	Oil pan-to-engine stud bolts (5 required)		
3	Oil pan-to-transaxle bolts (2 required)	8	Oil pump screen and pickup tube
4	Oil pan	9	Oil pump screen and pickup tube O-ring
5	Oil pan gasket	10	Oil pan baffle nuts (7 required)
6	Oil pump screen and pickup tube nut	11	Oil pan baffle

09482_TRIB_G0021

Oil pan and related components—3.0L engine

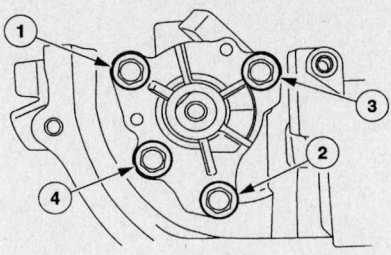

67197-ESCA-G65

Oil pump bolt torque sequence—2.3L engine

6. Release the tension on the tensioner spring.

7. Remove the tensioner and the shoulder bolt.

8. Remove the guide.

➡**The oil pump chain sprocket must be held in place.**

9. Remove the oil pump chain and sprockets.

10. Remove the oil pump assembly and gasket.

To install:

11. Install the oil pump with a new gasket. Tighten the bolts in sequence as follows:

    a. Step 1: 89 inch lbs. (10 Nm).

    b. Step 2: 17 ft. lbs. (23 Nm).

12. Install the pump chain and sprockets. Tighten the pump sprocket bolt to 18 ft. lbs. (25 Nm).

13. Install the chain guide, tensioner, and shoulder bolt. Tighten the bolts to 89 inch lbs. (10 Nm).

14. Hook the tensioner spring around the shoulder bolt.

15. Install the oil pump pickup tube and screen with a new gasket. Tighten the bolts to 89 ft. lbs. (10 Nm).

16. Install the oil pan.

17. Install the timing chain and front cover.

18. Install the engine into the vehicle.

3.0L Engine

✳✳ WARNING

During engine repair procedures, cleanliness is extremely important. Any foreign material, including any material created while cleaning gasket surfaces, that enters the oil passages, coolant passages or the oil pan can cause engine failure.

1. Before servicing the vehicle, refer to the Precautions Section.

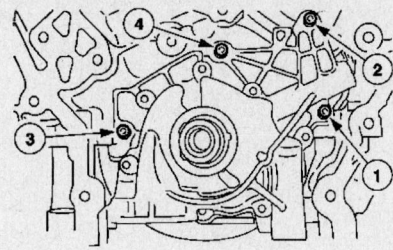

9308TG19

Oil pump bolt removal sequence—3.0L engine

9308TG20

Oil pump bolt torque sequence—3.0L engine

2. Disconnect the negative battery cable.

➡**To deplete the backup power supply energy, disconnect the negative battery cable and wait at least one minute. Be sure to disconnect auxiliary batteries and power supplies, if equipped.**

3. Raise and support the vehicle safely.

4. Drain the engine oil. Remove and discard the oil filter.

5. Remove the oil pan.

6. Remove the oil pump screen cover and tube. Remove the oil pan baffle retaining nuts. Remove the oil pan baffle.

7. Remove the timing drive components.

8. Remove the oil pump retaining bolts in the proper sequence.

9. Remove the oil pump from the engine.

To install:

10. Thoroughly clean the gasket mating surfaces.

11. Install the oil pump. Torque the bolts to 89 inch lbs. (10 Nm).

12. Continue the installation in the reverse order of the removal procedure.

13. Use a new oil filter. Refill the engine with clean oil.

14. Start the engine and check for leaks; repair if necessary.

Rear Main Seal

REMOVAL & INSTALLATION

2.0L Engine

1. Before servicing the vehicle, refer to the Precautions Section.

2. Remove or disconnect the following:

- Negative battery cable
- Flywheel
- Rear main seal

To install:

3. Coat the oil seal with clean engine oil.

4. Install or connect the following:

- Crankshaft rear oil seal
- Flywheel
- Negative battery cable

2.3L Engine

1. With the vehicle in NEUTRAL, position it on a hoist.

2. If equipped, remove the automatic transaxle.

3. If equipped, remove the manual transaxle and clutch.

4. Remove the flexplate or flywheel.

5. Remove the engine front cover bolts.

6. Remove the oil pan.

7. Remove the crankshaft rear oil seal with retainer plate

To install:

8. Using a seal installer, position the crankshaft rear oil seal with retainer plate onto the crankshaft.

9. Install the crankshaft rear oil seal with retainer plate. Tighten the bolts in the sequence shown to 10 Nm (89 inch lbs.).

10. Install the oil pan.

➡**Special bolts are used for flywheel (manual transaxle) or flexplate (automatic transaxle) installation. Do not use standard bolts.**

11. Install the flywheel (manual transaxle) or flexplate (automatic transaxle). Tighten the bolts in the sequence shown in three stages.

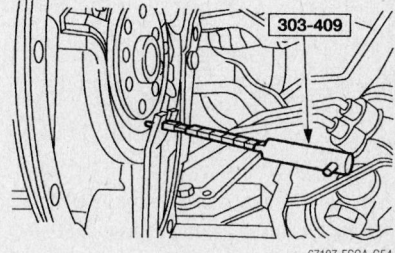

67197-ESCA-G54

Rear main seal removal—2.0L engine

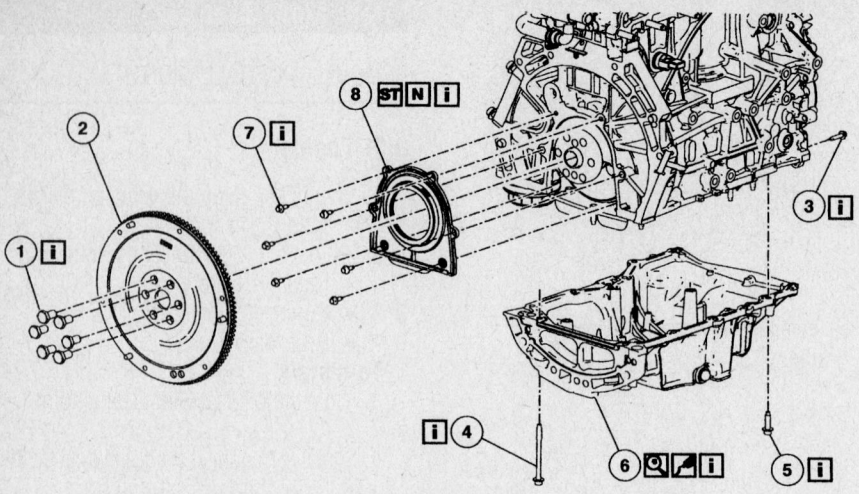

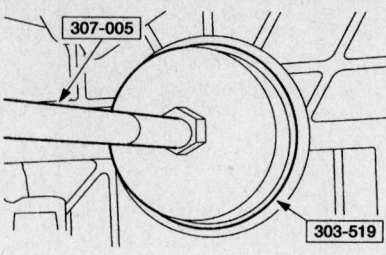

Rear main seal removal—3.0L engine

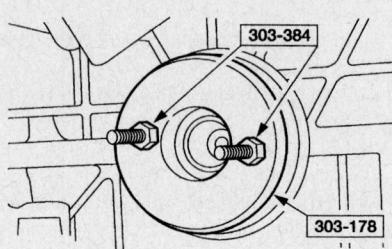

Rear main seal installation—3.0L engine

1 Flexplate or flywheel bolt
2 Flexplate or flywheel
3 Engine front cover bolt
4 Oil pan bolt
5 Oil pan bolt
6 Oil pan
7 Crankshaft rear oil seal with retainer plate bolt
8 Crankshaft rear oil seal with retainer plate

67197-ESCA-G22

Rear main seal and related components—2.3L engine

   a. Stage 1: Tighten to 50 Nm (37 ft. lbs.).
   b. Stage 2: Tighten to 80 Nm (59 ft. lbs.).
   c. Stage 3: Tighten to 112 Nm (83 ft. lbs.).

3.0L Engine

   1. Before servicing the vehicle, refer to the Precautions Section.
   2. Raise and support the vehicle safely.
   3. Remove the transaxle.
   4. Remove the flywheel (manual transaxle) or flexplate (automatic transaxle).
   5. Using the rear main seal removal tool, remove the rear main seal from its mounting.

To install:

   6. Coat the oil seal with clean engine oil.
   7. Using the rear main seal installation tool, install the new rear main seal to its mounting.
   8. Continue the installation in the reverse order of the removal procedure.

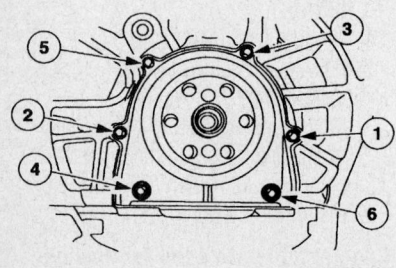

67197-ESCA-G23

Retainer plate torque sequence—2.3L engine

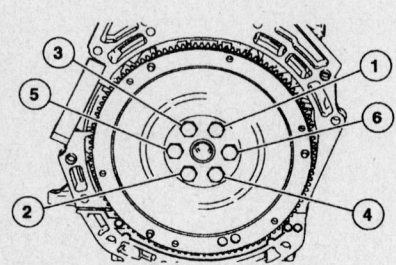

67197-ESCA-G24

Flywheel (manual transaxle) or flexplate (automatic transaxle) torque sequence—2.3L engine

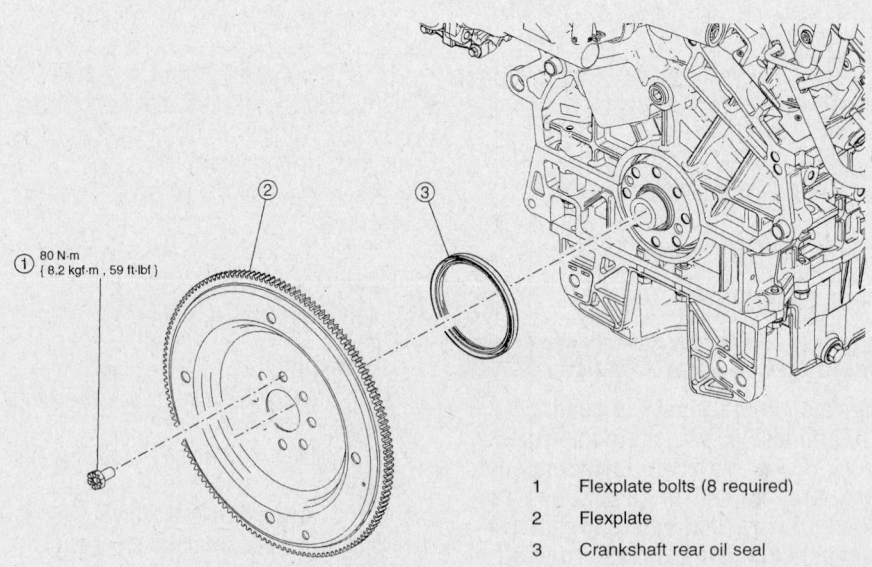

1 Flexplate bolts (8 required)
2 Flexplate
3 Crankshaft rear oil seal

09482_TRIB_G0022

Rear main seal and related components—3.0L engine

Timing Belt, Cover and Crankshaft Seal

REMOVAL & INSTALLATION

2.0L Engine

1. Before servicing the vehicle, refer to the Precautions Section.
2. With the vehicle in NEUTRAL, position it on a hoist.
3. Disconnect the battery ground cable.
4. Remove the air cleaner outlet pipe.
5. Remove the spark plugs.
6. Disconnect the throttle cables. If equipped, disconnect the speed control actuator cable. Disconnect the accelerator cable.
7. Disconnect the catalyst monitor and heated oxygen sensor electrical connectors.
8. Detach the electrical connectors from the bracket.
9. Remove the wiring harness anchors from the valve cover studs and position the wiring harness aside.
10. Remove the valve cover.

✳✳ WARNING

The valve cover sealing surfaces are soft materials. Do not use abrasive grinding discs to remove gasket material. Use only a plastic scraping tool. Do not scratch or gouge sealing surfaces or oil leaks can occur.

11. Clean and inspect the sealing surfaces of the valve cover and cylinder head. Both surfaces must be clean and flat.
12. Remove the heated oxygen sensor (HO$_2$S).
13. Disconnect the exhaust gas recirculation (EGR) tube.
   - Remove the two bolts.
   - Loosen the EGR tube nut.
14. Remove the catalytic converter nuts. Discard the nuts.
15. Remove the catalyst monitor sensor.
16. Remove the bolts and the catalytic converter bracket.
17. Remove the flagnuts and bolts. Discard the flagnuts, springs and bolts.
18. Remove the catalytic converter bracket nuts.
19. Remove the catalytic converter and rear support bracket. Discard the ring seal and gasket.
20. Remove the bolt and nut and position the coolant tube aside.
21. Remove the wheel and tire assembly.
22. Remove the right lower splash shield.

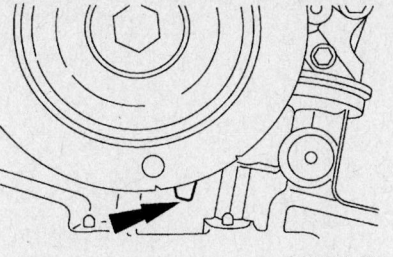

Rotate the crankshaft to just before top dead center (TDC) (No. 1 cylinder)—2.0L engine

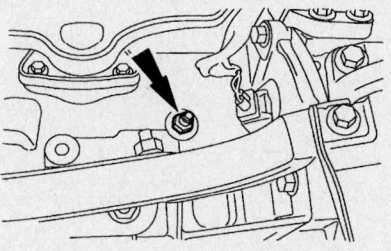

Stud removal location—2.0L engine

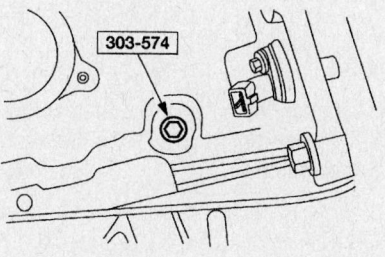

Install special tool 303-574 in the stud hole—2.0L engine

23. Rotate the crankshaft to just before top dead center (TDC) (No. 1 cylinder). Remove the stud. Install special tool 303-574.

➥**Make sure the correct (second) notch in the pulley is indexed to the lower cylinder block.**

24. Rotate the crankshaft clockwise against the peg to bring it to TDC (No. 1 cylinder).
25. Loosen the coolant pump pulley bolts.
26. Loosen the crankshaft pulley bolt.
27. Rotate the tensioner and remove the accessory drive belt.
28. Remove the bolt and the crankshaft pulley.

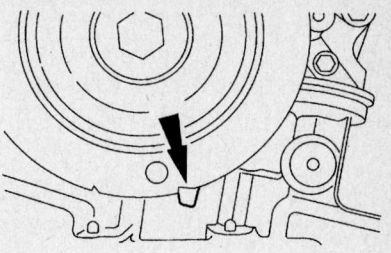

Rotate the crankshaft clockwise against the peg to bring it to TDC (No. 1 cylinder)—2.0L engine

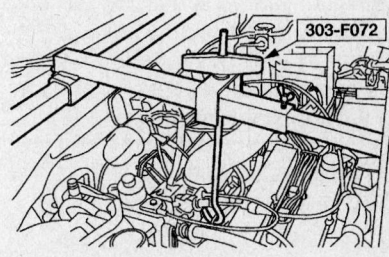

Install special tool 303-F072 to support the engine—2.0L engine

Release the tension on the timing belt by disconnecting the tensioner tab from the timing cover back plate—2.0L engine

29. Remove the bolts and the lower timing belt cover.
30. Install special tool 303-F072 to support the engine.
31. Detach the ground strap.
32. Remove the nuts, the bolt and the engine mount upper bracket.
33. Remove the studs.
34. Detach the knock sensor electrical connector from the upper timing cover.
35. Remove the bolts and the upper timing cover.
36. Remove the coolant pump pulley.
37. Remove the accessory drive belt idler pulley.
38. Remove the bolts and the engine mount lower bracket.

39. Relieve the tension on the timing belt tensioner pulley.
- Loosen the tensioner pulley bolt.
- Release the tension on the timing belt by disconnecting the tensioner tab from the timing cover back plate.

40. Remove the timing belt. Slide the timing belt off the camshaft sprockets and the crankshaft sprocket.

41. Inspect the timing belt for wear. Install a new timing belt if required.

To install:

✻✻ WARNING

The camshaft must be held stationary at the hexagons with locking pliers. Do not use the alignment tool to hold the camshaft in position or damage to the camshaft may occur.

➥**To loosen the camshaft pulleys, hold the camshafts by the hexagon.**

42. Remove the blanking plug from the exhaust camshaft pulley.

43. Loosen the exhaust camshaft pulley bolt.

44. Loosen the intake camshaft pulley bolt.

➥**Rotate the camshafts clockwise as necessary.**

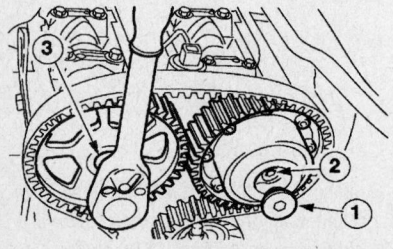

06017-ESCA-G31

(1) blanking plug, (2) exhaust camshaft pulley bolt, (3) intake camshaft pulley bolt—2.0L engine

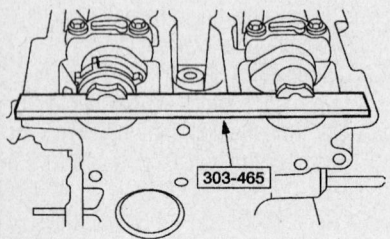

303-465

06017-ESCA-G32

Install the camshaft alignment timing tool on the back of the camshafts—2.0L engine

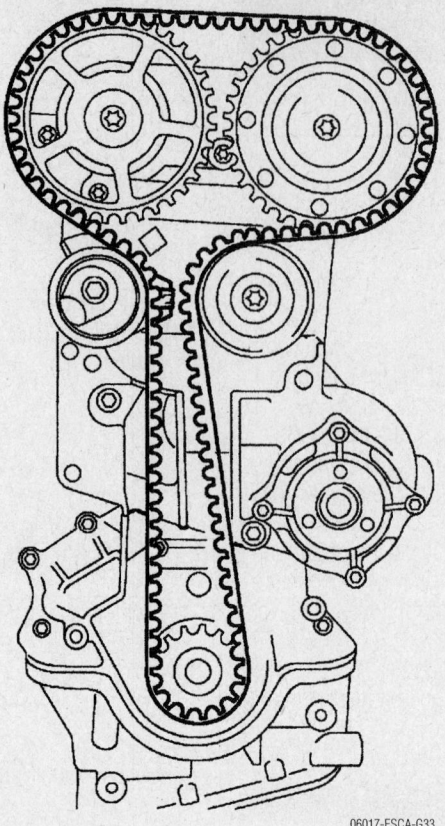

06017-ESCA-G33

Position a new timing belt in place—2.0L engine

45. Install the camshaft alignment timing tool on the back of the camshafts.

➥**Cylinder No. 1 is at top dead center (TDC) when the keyway is in the 12 o'clock position.**

46. Confirm the crankshaft position is at TDC (No. 1 cylinder) by rotating it clockwise against the alignment peg.

✻✻ WARNING

Do not rotate the crankshaft; as necessary check that it is still resting against the timing pin.

➥**The lug of the belt tensioner should not be hooked in the sheet metal cover during timing belt installation.**

47. Position a new timing belt in place. Starting from the crankshaft timing belt pulley and working counterclockwise, position the timing belt in place while keeping it under tension.

➥**Incorrect timing belt tension will cause incorrect valve timing.**

48. Pre-tension the timing belt.
    a. Rotate the tensioner locating tab counterclockwise and insert the locating tab into the slot in the rear timing cover.

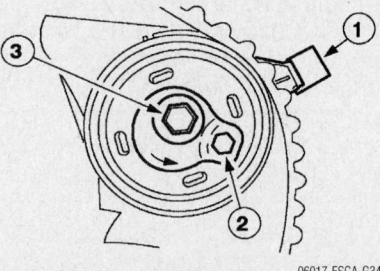

06017-ESCA-G34

Tensioner locating tab (1), tensioner hex slot (2), attaching bolt (3)—2.0L engine

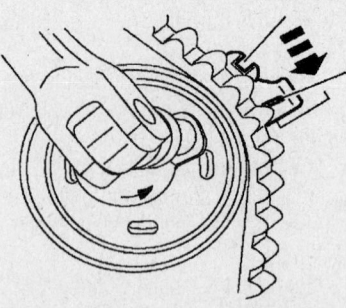

06017-ESCA-G35

Using the hex key, rotate the adjusting washer counterclockwise until the notch in the pointer is centered over the index line on the locating tab (the pointer will move clockwise during adjustment)—2.0L engine

b. Position the hex key slot in the tensioner adjusting washer to the 4 o'clock position.

c. Tighten the attaching bolt enough to seat the tensioner firmly against the rear timing cover, but still allow the tensioner adjusting washer to be rotated using a 6 mm hex key.

✳✳ WARNING

Tension the timing belt, working counterclockwise.

49. Using the hex key, rotate the adjusting washer counterclockwise until the notch in the pointer is centered over the index line on the locating tab (the pointer will move clockwise during adjustment).

50. While holding the adjusting washer in position, tighten the bolt to 25 Nm (18 ft. lbs.).

✳✳ WARNING

The camshaft must be held stationary at the hexagons with locking pliers. Do not use the alignment tool to hold the camshaft in position or damage to the camshaft may occur.

51. Tighten the bolt on the intake camshaft sprocket to 68 Nm (50 ft. lbs.).

✳✳ WARNING

The camshaft must be held stationary at the hexagons with locking pliers. Do not use the alignment tool to hold the camshaft in position or damage to the camshaft may occur.

52. Tighten the bolt on the exhaust camshaft sprocket in three stages.
- Stage 1: Tighten the bolt to 50 Nm (36 ft. lbs.).
- Stage 2: Remove the TDC peg and the camshaft alignment timing tool.
- Stage 3: Tighten the bolt to 115–125 Nm (85–92 ft. lbs.).

➡**Install a new oil plug seal.**

53. Screw in the new oil plug on the variable camshaft timing assembly. Torque to 37 Nm (27 ft. lbs.).

➡**Turn the engine two turns in the normal direction of rotation by the crankshaft.**

54. Check the valve timing by inserting the special tools and correct the alignment as necessary.

a. Screw in special tool 303-574 and make sure that the crankshaft is resting against the special tool.

b. Insert special tool 303-465 into the camshafts. If necessary loosen the timing pulleys and correct the camshaft alignment.

c. Remove the special tools.

55. Install the front engine mount lower bracket. Torque to 50 Nm (37 ft. lbs.).

56. Install the accessory drive belt idler pulley. Torque to 45 Nm (35 ft. lbs.).

57. Install the coolant pump pulley. Hand-tighten the bolts.

58. Position the upper timing belt cover and install the bolts. Torque to 8 Nm (71 inch lbs.).

59. Install the studs. Torque to 34 Nm (25 ft. lbs.).

60. Connect the knock sensor electrical connector to the upper timing belt cover.

61. Install the engine mount upper bracket, the nuts and the bolt. Torque the upper nuts to 98 Nm (72 ft. lbs.); the lower bolt to 77 Nm (57 ft. lbs.).

62. Attach the ground strap and install the nut. Torque the nut to 34 Nm (25 ft. lbs.).

63. Position the lower timing belt cover and install the bolts. Torque to 7 Nm (62 inch lbs.).

64. Position the crankshaft pulley and install the bolt. Hand-tighten the bolt.

65. Rotate the tensioner and install the accessory drive belt.

66. Tighten the crankshaft pulley bolt. Torque to 115 Nm (85 ft. lbs.).

67. Tighten the coolant pump pulley bolts to 23 Nm (17 ft. lbs.).

68. Position the right lower splash shield and install the bolts.

69. Install the wheel and tire assembly.

70. Install the stud. Torque to 24 Nm (18 ft. lbs.).

71. Install the coolant tube. Torque to 20 Nm (15 ft. lbs.).

72. Position the catalytic converter into place and loosely install the two nuts. Install a new gasket and ring seal. Install new nuts.

73. Connect the converter outlet to the muffler inlet pipe. Position the rear support bracket and install the washers and nuts. Loosely install the new bolts, springs and flagnuts.

74. Lower the vehicle.

75. Install the third catalytic converter nut and tighten all the nuts. Install a new nut. Torque to 47 Nm (35 ft. lbs.).

76. Connect the EGR tube and install the bracket bolts. Torque the lower end to 63 Nm (46 ft. lbs.); the upper end to 9 Nm (80 inch lbs.).

77. Raise the vehicle.

78. Tighten the converter outlet to the muffler inlet bolts to 40 Nm (30 ft. lbs.).

79. Tighten the catalytic converter rear support bracket nuts to 47 Nm (35 ft. lbs.).

80. Install the catalytic converter bracket and bolts. Torque to 25 Nm (18 ft. lbs.).

81. Install the catalyst monitor sensor. Torque to 40 Nm (30 ft. lbs.).

82. Install the HO2S. Torque to 40 Nm (30 ft. lbs.).

83. Inspect the valve cover gasket. Install a new gasket, if necessary.

84. Position the valve cover. Install the studbolts. Install the bolts. Torque to 9 Nm) 80 inch lbs.).

85. Connect the wiring harness anchors to the valve cover studs.

86. Attach the catalyst monitor and heated oxygen sensor electrical connectors to the bracket.

87. Connect the electrical connectors.

88. Connect the accelerator cable. If equipped, connect the speed control actuator cable.

89. Install the spark plugs.

90. Install the air cleaner outlet pipe.

91. Connect the battery ground cable.

Front Cover

REMOVAL & INSTALLATION

3.0L Engine

2002–2005

✳✳ CAUTION

During engine repair procedures, cleanliness is extremely important. Any foreign material, including any material created while cleaning gasket surfaces, that enters the oil passages, coolant passages or the oil pan can cause engine failure.

1. Before servicing the vehicle, refer to the Precautions Section.

2. Disconnect the negative battery cable.

➡**To deplete the backup power supply energy, disconnect the negative battery cable and wait at least one minute. Be sure to disconnect auxiliary batteries and power supplies, if equipped.**

3. Remove the valve covers. Remove the oil pan. Remove the alternator.

4. Remove the bolt and the camshaft position sensor.

5. Remove the power steering pump and position it to the side.

➡ On some vehicles it may be necessary to remove the power steering pump pulley. The bolt positioned behind the power steering pressure tube will remain in the pump housing.

6. Install engine support tools 3030F072 and 303-050, to support the engine assembly.

7. Remove the engine support insulator.

8. Remove the bolt and the belt tensioner.

9. On 2005 vehicles, remove the accessory drive belt idler pulley bolt and the accessory drive belt idler pulley.

10. Remove the crankshaft pulley. Remove the CKP sensor.

➡> Remove the front cover with the vehicle raised in the air.

11. Raise and support the vehicle safely.

12. Remove the front cover retaining bolts and studs. Remove the front cover from the engine.

To install:

✳✳ CAUTION

Do not use metal scrapers, wire brushes, power abrasive disks or other abrasive means to clean sealing surfaces. These tools cause scratches and gouges which make leak paths.

13. Clean and inspect the mounting surfaces of the engine and the front cover.

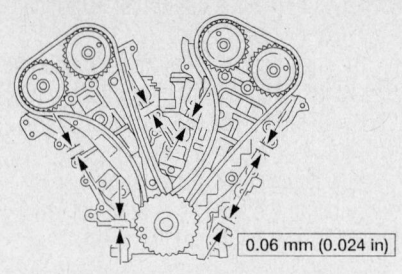

`0.06 mm (0.024 in)`

09482_TRIB_G0024

Sealant application point location—3.0L engine

14. Install three new gaskets in the front cover.

➡ The engine front cover must be installed and the bolts tightened within four minutes of applying the silicone gasket and sealant.

15. Apply 0.24 inch diameter dot of silicone gasket and sealer to the cylinder block, lower cylinder block and cylinder head mating surfaces.

16. Position the cover on the engine and install the bolts and studs. Torque the bolts and studs to 18 ft. lbs (25 Nm).

➡ On 2002–2004 engines, fasteners numbered 1, 3, 4, 8, 10, 11, 14, 15 and 16 are studs. On 2005 engines, fasteners numbered 1, 8, 13 and 16 are stud bolts.

17. Continue the installation in the reverse order of the removal procedure.

2006

✳✳ CAUTION

During engine repair procedures, cleanliness is extremely important. Any foreign material, including any material created while cleaning gasket surfaces, that enters the oil passages, coolant passages or the oil pan can cause engine failure.

1. Before servicing the vehicle, refer to the Precautions Section.

2. Properly relieve the fuel system pressure.

3. Disconnect the negative battery cable.

➡ To deplete the backup power supply energy, disconnect the negative battery cable and wait at least one minute. Be sure to disconnect auxiliary batteries and power supplies, if equipped.

4. Raise and support the vehicle safely. Remove the crankshaft front oil seal.

5. Remove the five bolts and the pin type retainer and the lower right splash shield.

6. Remove the alternator upper air duct bolt. Remove the air duct.

7. Remove the alternator shield pin type retainer, bolt and alternator shield.

8. Remove the alternator bolts. Remove the alternator stud nut and position the alternator to the side.

9. Remove the valve covers. Remove the engine support insulator.

10. Remove the bolt and the accessory drive belt tensioner. Remove the bolt and the accessory drive belt idler pulley.

11. Remove the nut and position the heated oxygen sensor electrical connector aside. Disconnect the CPK sensor electrical connector. Disconnect the CMP sensor electrical connector.

12. Remove the power steering pump and position it to the side.

➡ The bolt positioned behind the power steering pressure tube will remain in the pump housing.

13. If equipped, remove the three nuts and position the speed control actuator aside.

14. Remove the front cover retaining bolts. Remove the two oil pan to front cover retaining bolts.

15. Remove the bolts and studs and remove the cover from the engine. Discard the gaskets.

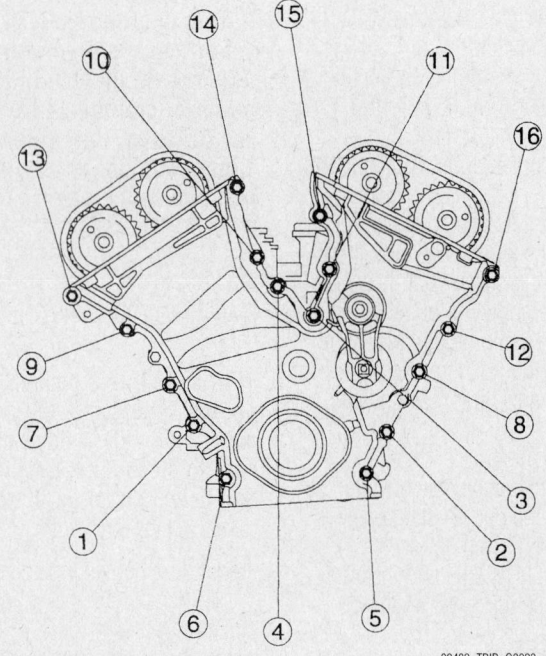

09482_TRIB_G0023

Front cover bolt torque sequence—3.0L engine

To install:

✳✳ CAUTION

Do not use metal scrapers, wire brushes, power abrasive disks or other abrasive means to clean sealing surfaces. These tools cause scratches and gouges which make leak paths.

16. Clean and inspect the mounting surfaces of the engine and the front cover.

17. Install three new gaskets in the front cover.

➡**The engine front cover must be installed and the bolts tightened within four minutes of applying the silicone gasket and sealant.**

18. Apply 0.24 inch diameter dot of silicone gasket and sealer to the cylinder block, lower cylinder block and cylinder head mating surfaces.

19. Position the cover on the engine and install the bolts and studs. Torque the bolts and studs to 18 ft. lbs (25 Nm).

➡**Fasteners numbered 1, 8, 13 and 16 are stud bolts.**

20. Continue the installation in the reverse order of the removal procedure.

Timing Chain, Sprockets, Front Cover and Seal

REMOVAL & INSTALLATION

2.3L Engine

✳✳ CAUTION

During engine repair procedures, cleanliness is extremely important. Any foreign material, including any material created while cleaning gasket surfaces, that enters the oil passages, coolant passages or the oil pan can cause engine failure.

✳✳ CAUTION

The crankshaft, the crankshaft sprocket and the pulley are fitted together by friction, using diamond washers between the flange faces on each part. For that reason, the crankshaft sprocket is also unfastened if you loosen the pulley. Therefore, the engine must be retimed each time the damper is removed. Otherwise severe engine damage can occur.

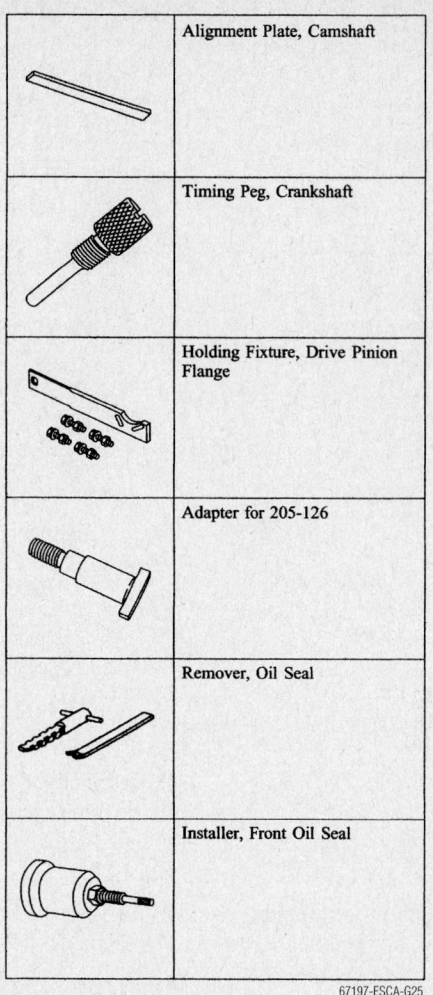

	Alignment Plate, Camshaft
	Timing Peg, Crankshaft
	Holding Fixture, Drive Pinion Flange
	Adapter for 205-126
	Remover, Oil Seal
	Installer, Front Oil Seal

67197-ESCA-G25

Tools needed for timing chain and gears replacement—2.3L engine

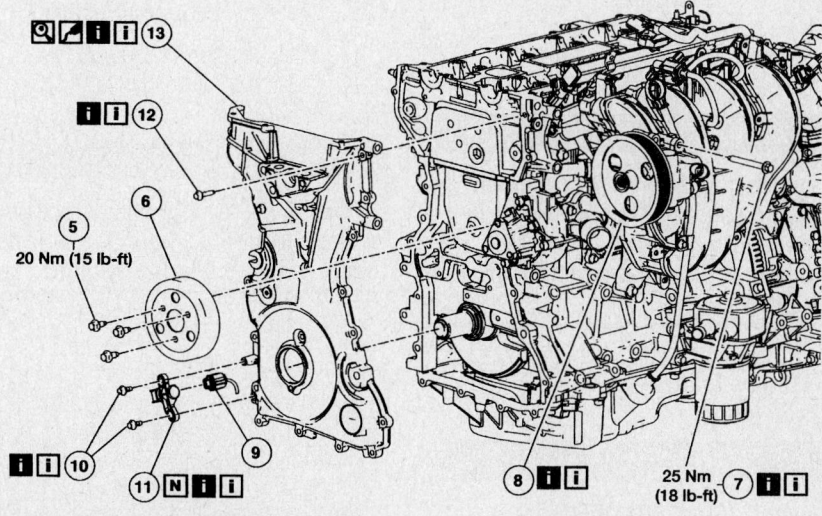

5 Coolant pump pulley bolt
6 Coolant pump pulley
7 Power steering pump bolt
8 Power steering pump (position aside)
9 Crankshaft position (CKP) sensor electrical connector
10 CKP sensor bolts
11 CKP sensor
12 Engine front cover bolt
13 Engine front cover

67197-ESCA-G26

Front cover and related components—2.3L engine

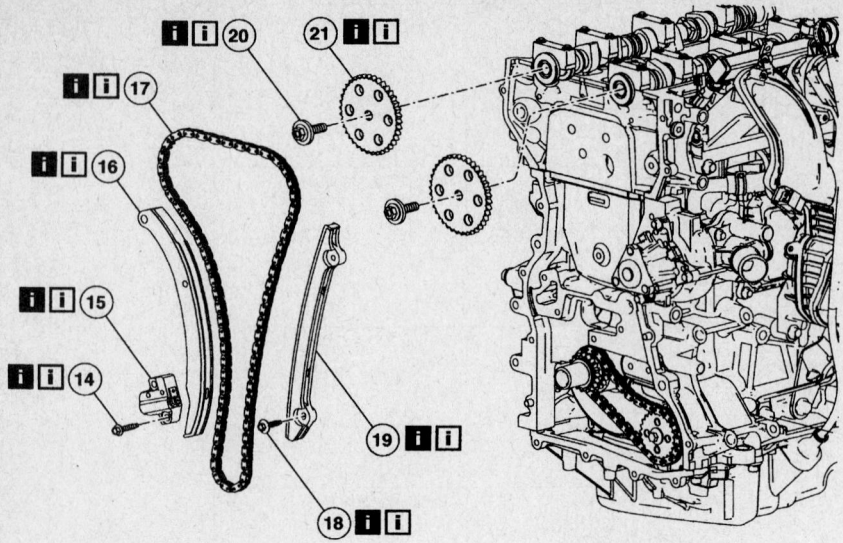

14 Timing chain tensioner bolt
15 Timing chain tensioner
16 RH timing chain guide
17 Timing chain

18 LH timing chain guide bolt
19 LH timing chain guide
20 Camshaft sprocket bolt
21 Camshaft sprocket

67197-ESCA-G27

Timing chain and related components—2.3L engine

1. Before servicing the vehicle, refer to the Precautions Section.
2. Disconnect the negative battery cable.

➡ **To deplete the backup power supply energy, disconnect the negative battery cable and wait at least one minute. Be sure to disconnect auxiliary batteries and power supplies, if equipped.**

3. Raise and support the vehicle safely.
4. Remove the accessory drive belt and idler pulleys.
5. Remove the engine mount.
6. Remove the valve cover.

✳✳ CAUTION

Failure to position the No. 1 piston at top dead center (TDC) can result in damage to the engine. Turn the engine in the normal direction of rotation only.

7. Using the crankshaft pulley bolt, turn the crankshaft clockwise to position the No. 1 piston at TDC.

✳✳ CAUTION

The special tool 303-465 is for camshaft alignment only. Using this tool to prevent engine rotation can result in engine damage.

➡ **The camshaft timing slots are offset. If the special tool cannot be installed,**

rotate the crankshaft one complete revolution clockwise to correctly position the camshafts.

8. Install special tool 303-465 in the slots on the rear of both camshafts.
9. Remove the engine plug bolt.

➡ **Only turn the engine in the normal direction of rotation.**

➡ **Installing the special tool in this step will prevent the engine from being rotated in the clockwise direction.**

10. Install special tool 303-507.
11. Install the special tools 205-126 and 205-072-02.

✳✳ CAUTION

Failure to hold the crankshaft pulley in place while loosening the bolt can result in damage to the engine.

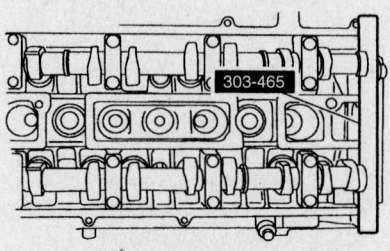

Install special tool 303-465 in the slots on the rear of both camshafts—2.3L engine

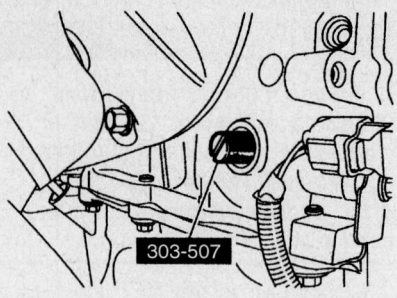

Install special tool 303-507—2.3L engine

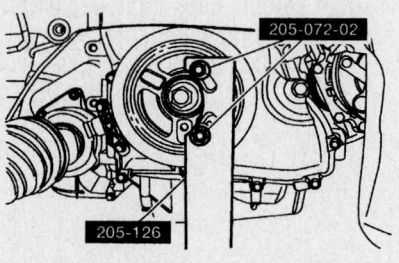

Install the special tools 205-126 and 205-072-02—2.3L engine

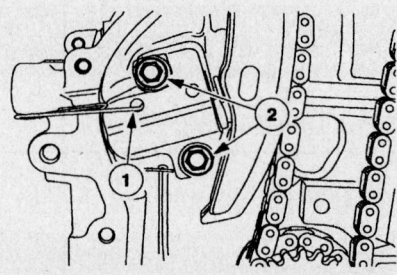

Compress the timing chain tensioner, and insert a paper clip into the hole to retain the tensioner—2.3L engine

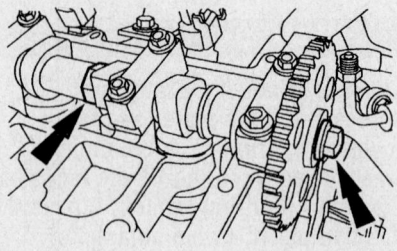

Use the flats on the camshaft to prevent camshaft rotation—2.3L engine

12. Remove the crankshaft pulley bolt and washer.

13. Remove the crankshaft pulley.

14. Remove the crankshaft front seal.

15. Remove the coolant pump pulley.

16. Remove the power steering pump and position it aside.

➡**The bolt under the power steering pressure tube will remain with the power steering pump.**

17. Remove the CKP sensor.

➡**Whenever the crankshaft position (CKP) sensor is removed, a new one must be installed, using the alignment jig supplied with the new part.**

18. Remove the engine front cover bolts (there are 22).

19. Remove the engine front cover.

20. Remove the timing chain tensioner. Compress the timing chain tensioner, and insert a paper clip into the hole to retain the tensioner.

21. Remove the RH timing chain guide.

22. Remove the timing chain.

23. Remove the LH timing chain guide.

24. Remove the camshaft sprocket bolts.

25. Remove the camshaft sprockets.

✳✳ CAUTION

Do not rely on the Camshaft Alignment Plate to prevent camshaft rotation. Damage to the tool or the camshaft can occur. Use the flats on the camshaft to prevent camshaft rotation.

To install:

26. Installation is the reverse of removal. Note the following:

✳✳ CAUTION

Do not use metal scrapers, wire brushes, power abrasive disks or other abrasive means to clean sealing surfaces. These tools cause scratches and gouges which make leak paths.

27. Clean and inspect the mounting surfaces of the engine and the front cover.

➡**The engine front cover must be installed and the bolts tightened within four minutes of applying the silicone gasket and sealant.**

28. Apply a 2.5 mm bead of silicone gasket and sealant to the cylinder head and oil pan joint areas. Apply a 2.5 mm bead of

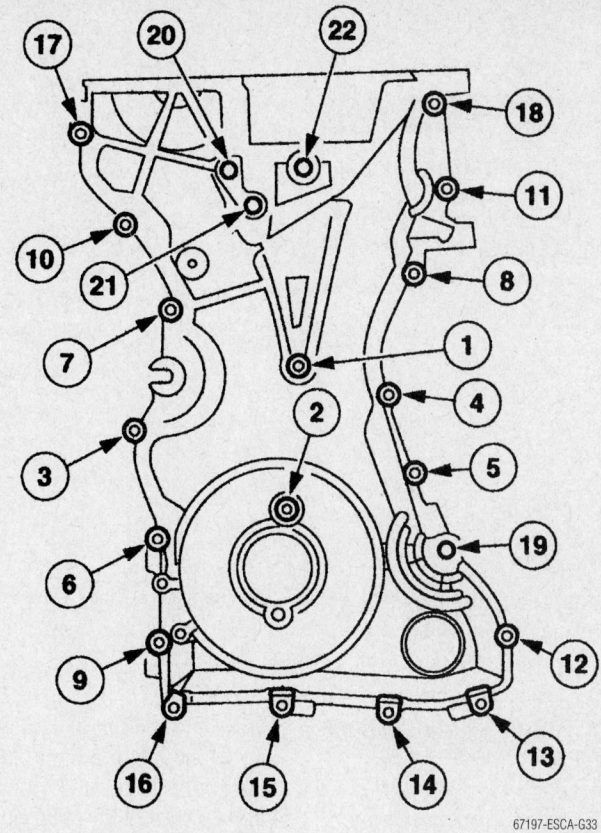

67197-ESCA-G33

Front cover bolt torque sequence—2.3L engine

silicone gasket and sealant to the front cover.

29. Install the engine front cover. Tighten the bolts in the sequence shown, to the following specifications:

  a. Tighten the 8 mm bolts to 10 Nm (89 inch lbs.).

  b. Tighten the 13 mm bolts to 48 Nm (35 ft. lbs.).

30. Position the power steering pump and install the bolts.

➡**Remove the through-bolt from the special tool.**

➡**Lubricate the oil seal with clean engine oil.**

31. Using a seal driver, install the crankshaft front oil seal.

➡**Do not reuse the crankshaft damper bolt.**

➡**Apply clean engine oil on the seal area before installing.**

32. Install the crankshaft pulley and hand-tighten the bolt.

✳✳ CAUTION

Only hand-tighten the bolt or damage to the front cover can occur.

➡**This step will correctly align the crankshaft pulley to the crankshaft.**

33. Install a standard 6 mm x 18 mm bolt through the crankshaft pulley and thread it into the front cover. Rotate the pulley as necessary to align the bolt holes.

✳✳ CAUTION

Failure to hold the crankshaft pulley in place while tightening the bolt can cause damage to the engine front cover.

34. Using the special tools to hold the crankshaft pulley in place, tighten the crankshaft pulley bolt in two stages:

  a. Stage 1: Tighten to 100 Nm (74 ft. lbs.).

  b. Stage 2: Tighten an additional 90 degrees (¼ turn).

35. Remove the 6 mm x 18 mm bolt.

36. Remove special tool 303-507.

37. Remove special tool 303-465.

➡**Only turn the engine in the normal direction of rotation.**

38. Turn the engine two complete revolutions.

➡**Only turn the engine in the normal direction of rotation.**

39. Turn the crankshaft until the No. 1 piston is at TDC.

40. Install special tool 303-507.

⚙ CAUTION

Only hand-tighten the bolt or damage to the front cover can occur.

41. Using the 6 mm x 18 mm bolt, check the position of the crankshaft pulley. If it is not possible to install the bolt, correct the engine timing.

42. Using special tool 303-465, check the position of the camshafts. If it is not possible to install the special tool, correct the engine timing.

43. Install the CKP sensor. Do not tighten the bolts at this time.

44. Adjust the CKP sensor alignment jig and tighten the bolts.

45. Remove the 6 mm x 18 mm bolt.

46. Install the engine plug bolt.

3.0L Engine

⚙ CAUTION

During engine repair procedures, cleanliness is extremely important. Any foreign material, including any material created while cleaning gasket surfaces, that enters the oil passages, coolant passages or the oil pan can cause engine failure.

➡**Failure to verify correct timing drive component alignment will result in severe engine damage.**

1. Before servicing the vehicle, refer to the Precautions Section.

2. Raise and support the vehicle safely.

3. Properly relieve the fuel system pressure.

4. Disconnect the negative battery cable.

➡**To deplete the backup power supply energy, disconnect the negative battery cable and wait at least one minute. Be sure to disconnect auxiliary batteries and power supplies, if equipped.**

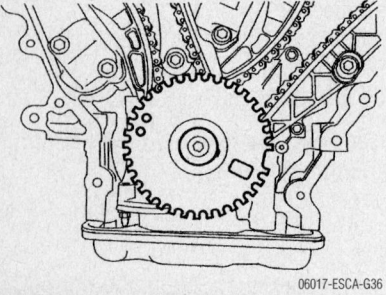

Ignition pulse wheel—3.0L engine
06017-ESCA-G36

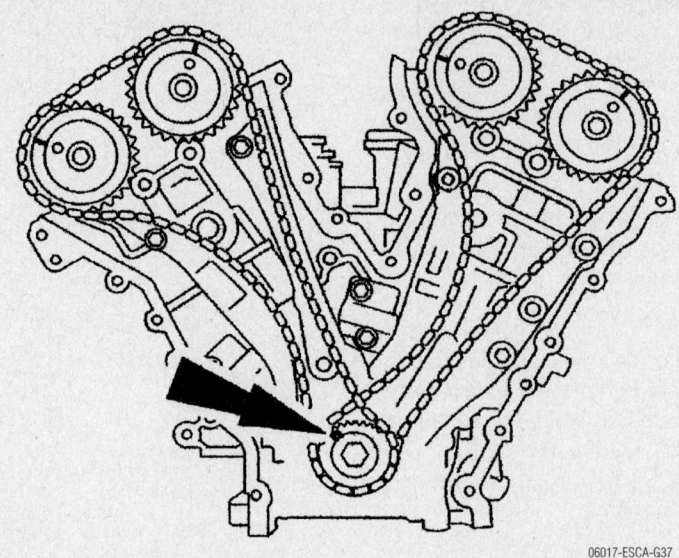

06017-ESCA-G37

Rotate the crankshaft clockwise to position the keyway at the 11 o'clock position—3.0L engine

5. Remove the crankshaft front oil seal.

6. Remove the front cover.

➡**Install the pulse wheel with the keyway in the slot stamped 20–25–34Y–30M (Color Blue) on 2002–2004 engines. On 2005–2006 engines, install the pulse wheel with the keyway in the slot stamped "30" or "30RFF" (orange in color).**

7. Remove the ignition pulse wheel. Install the damper bolt. Remove the spark plugs.

8. Rotate the crankshaft clockwise to position the keyway at the 11 o'clock position and the camshafts in the correct positions. The No. 1 cylinder will be at Top Dead Center (TDC).

➡**Verify that the camshafts are correctly located, if not rotate the crankshaft one additional turn and recheck.**

9. Rotate the crankshaft clockwise 120 degrees to the 3 o'clock position to locate the right side camshafts in the neutral posi-

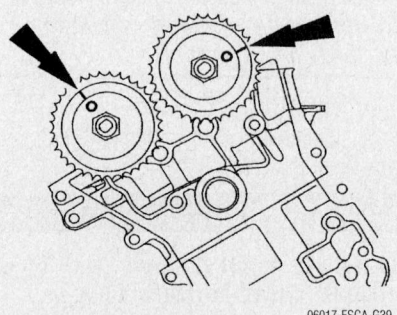

06017-ESCA-G38

Rotate the crankshaft clockwise 120 degrees to the 3 o'clock position to locate the right side camshafts in the neutral position—3.0L engine

tion. Verify that the right camshafts are in the neutral position.

10. Remove or disconnect the following:
- Right side timing chain and tensioner
- Tensioner arm and timing chain guide

11. Rotate the crankshaft clockwise 1⅔ times to position the keyway at the 11 o'clock position. This will position the left side camshafts in the neutral position.

12. Verify that the left side camshafts are

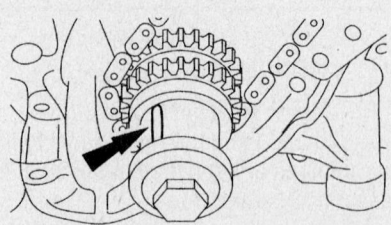

06017-ESCA-G39

Right side camshaft alignment in the neutral position—3.0L engine

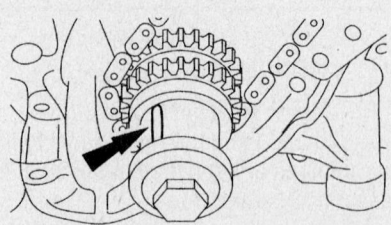

06017-ESCA-G40

Rotate the crankshaft clockwise 1⅔ times to position the keyway at the 11 o'clock position—3.0L engine

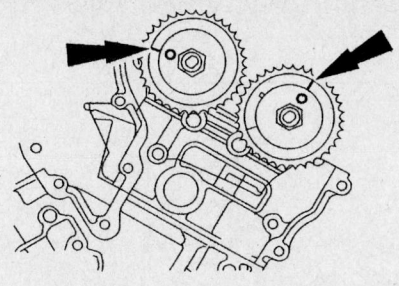

06017-ESCA-G41

Left side camshaft alignment in the neutral position—3.0L engine

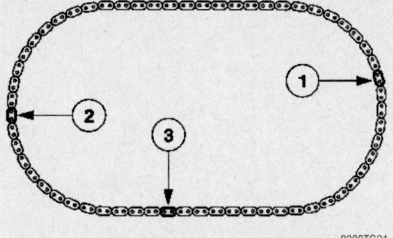

9308TG21

Mark the timing chain in the proper sequence (1) mark any link (2) count 29 links (3) count 42 links—3.0L engine

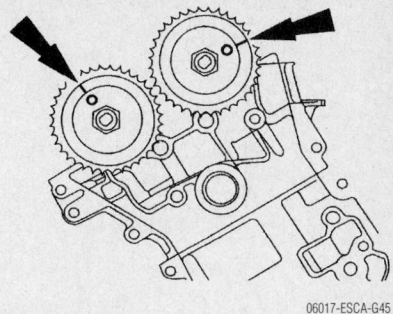

06017-ESCA-G45

Right camshafts are correctly positioned—3.0L engine

in the neutral position and mark the link position on the crankshaft sprocket.

13. Remove or disconnect the following:
- Left side timing chain and tensioner
- Tensioner arm and timing chain guide
- Damper bolt and crankshaft sprockets

To install:

14. Install the crankshaft sprockets.

15. Position the timing chain tensioner in a soft jaw vise. Hold the ratchet lock mechanism away from the ratchet stem and slowly compress the timing chain tensioner. Retain the piston with a 1.5mm wire or paper clip.

16. If the timing marks on the chain are

not visible, use a permanent marker to mark the left and right side timing chains. Mark the timing chains in the following sequence:

a. Mark any link to use as the crankshaft timing mark.

b. Count 29 links (including the starting link) in the counterclockwise direction from the crankshaft timing mark and mark the link as the left hand chain exhaust/right hand chain intake cam sprocket timing mark.

c. Continue counting to 42 and mark the link as the left hand chain exhaust/right hand chain intake sprocket timing mark.

17. Install the guide. Torque the bolts to 18 ft. lbs. (25 Nm).

18. Install the left side timing chain and align the chain in the following sequence:

a. Mark any link to use as the crankshaft timing mark.

b. Count 29 links from the crankshaft timing mark and mark the link as the exhaust cam sprocket timing mark.

c. Continue counting to 42 and mark the link as the intake sprocket timing mark

19. Install or connect the following:
- Left side timing chain and tensioner arm. Torque the bolts to 18 ft. lbs. (25 Nm).
- Crankshaft damper bolt and rotate the keyway to the 3 o'clock position.

20. Verify that the right side camshafts are properly positioned and install the right side timing chain and guide. Torque the bolts to 18 ft. lbs. (25 Nm).

21. Make certain that the timing chain aligns with the marks on the camshaft and crankshaft sprockets

22. Install or connect the following:
- Right side timing chain tensioner and arm. Torque the bolts to 18 ft. lbs. (25 Nm) and remove the damper bolt
- LH and RH chain tensioner position retaining wires

23. Remove the crankshaft damper bolt.

24. Verify valve timing, there should be 12 chain links between the front camshaft timing marks. There should be 27 chain links between the camshaft and crankshaft timing marks. There should be 30 chain

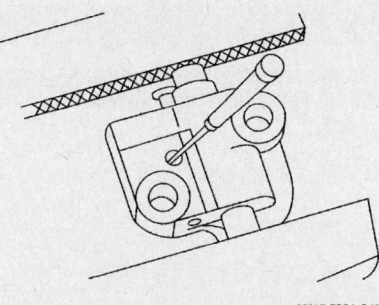

06017-ESCA-G42

Hold the ratchet lock mechanism away from the ratchet stem—3.0L engine

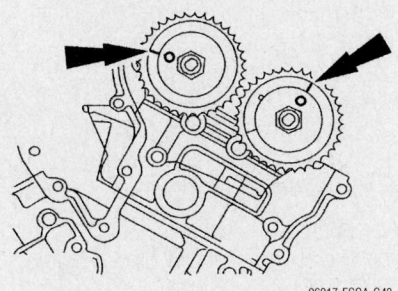

06017-ESCA-G43

Left camshafts are correctly positioned—3.0L engine

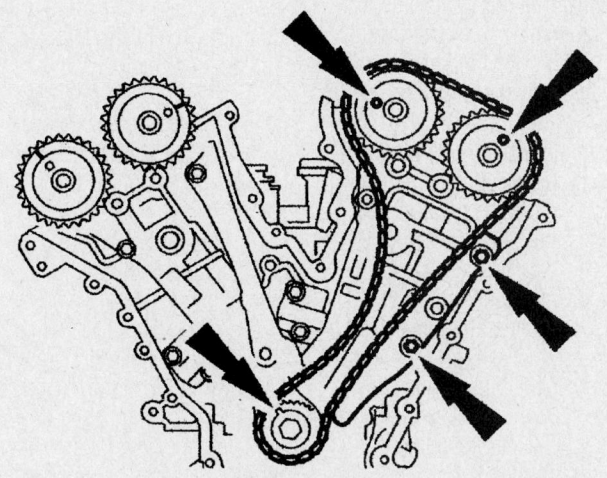

06017-ESCA-G44

Left side timing chain installed—3.0L engine

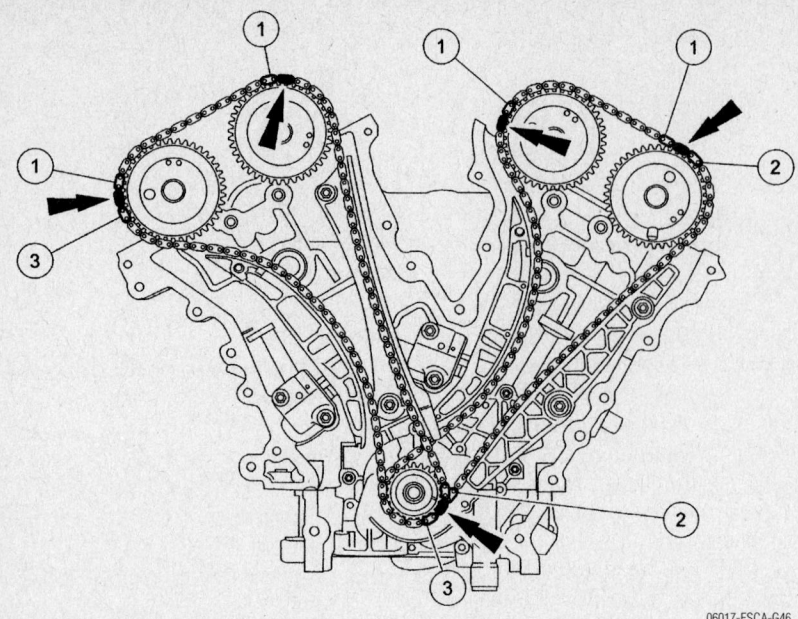

Timing chains alignment (1)12 chain links (2) 27 chain links (3) 30 chain links—3.0L engine

06017-ESCA-G46

Piston and Ring

POSITIONING

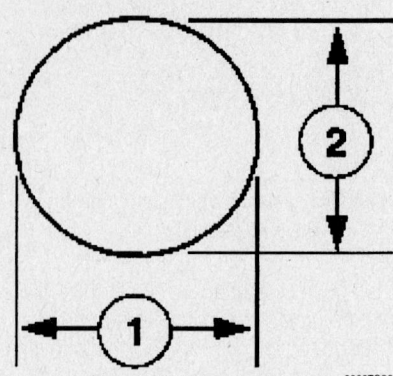

9308TG26

2.0L and 2.3L engine —piston ring end-gap spacing

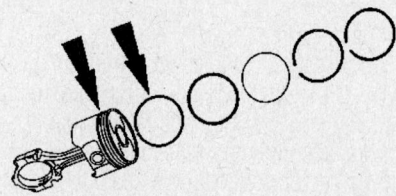

9308TG25

3.0L (VIN 1) engine—piston ring end-gap spacing

links between the camshaft and crankshaft timing marks.

➡**Failure to verify correct timing drive component alignment will result in severe engine damage.**

25. Install the ignition pulse wheel.

➡**Install the pulse wheel with the key-way in the slot stamped 20–25–34Y–**
30M (Color Blue) on 2002–2004 engines. On 2005–2006 engines, install the pulse wheel with the keyway in the slot stamped "30" or "30RFF" (orange in color).

- Spark plugs
- Engine front cover
- Negative battery cable

FUEL SYSTEM

Fuel System Service Precautions

Safety is the most important factor when performing not only fuel system maintenance but also any type of maintenance. Failure to conduct maintenance and repairs in a safe manner may result in serious personal injury or death. Maintenance and testing of the vehicle's fuel system components can be accomplished safely and effectively by adhering to the following rules and guidelines.

1. To avoid the possibility of fire and personal injury, always disconnect the negative battery cable unless the repair or test procedure requires that battery voltage be applied.

2. Always relieve the fuel system pressure prior to disconnecting any fuel system component (injector, fuel rail, pressure regulator, etc.), fitting or fuel line connection. Exercise extreme caution whenever relieving fuel system pressure, to avoid exposing skin, face and eyes to fuel spray. Please be advised that fuel under pressure may penetrate the skin or any part of the body that it contacts.

3. Always place a shop towel or cloth around the fitting or connection prior to loosening to absorb any excess fuel due to spillage. Ensure that all fuel spillage (should it occur) is quickly removed from engine surfaces. Ensure that all fuel soaked cloths or towels are deposited into a suitable waste container.

4. Always keep a dry chemical (Class B) fire extinguisher near the work area.

5. Do not allow fuel spray or fuel vapors to come into contact with a spark or open flame.

6. Always use a backup wrench when loosening and tightening fuel line connection fittings. This will prevent unnecessary stress and torsion to fuel line piping.

7. Always replace worn fuel fitting O-rings with new. Do not substitute fuel hose or equivalent, where fuel pipe is installed.

Before servicing the vehicle, make sure to refer to the precautions in the beginning of this section as well.

Fuel System Pressure

RELIEVING

2.0L Engine

1. Before servicing the vehicle, refer to the Precautions Section.

2. Remove or disconnect the following:

3. Remove the fuel pump relay and start the engine.

4. After the engine stalls, crank the engine 2 more times to be certain that all fuel pressure has been relieved.

5. Turn the ignition switch to the **OFF** position.

6. Install the fuel pump relay.

2.3L Engine

1. Remove the fuel pump relay.

2. Start the engine and allow it to idle until it stalls.

3. After the engine stalls, crank the engine for approximately 5 seconds to make

sure the fuel injection supply manifold pressure has been released.

4. Turn the ignition switch to the **OFF** position.

5. When fuel system service is complete, install the fuel pump relay.

➡**It may take more than one key cycle to pressurize the fuel system.**

6. Cycle the ignition key and wait three seconds to pressurize the fuel system. Check for leaks before starting the engine.

7. Start the engine and check the fuel system for leaks.

3.0L Engine

2002–2004

1. Before servicing the vehicle, refer to the Precautions Section.

2. Remove or disconnect the following:

3. Remove the schrader valve cap at the end of the fuel injection supply manifold and attach a fuel pressure gauge.

4. Open the manual valve slowly and drain the fuel into a suitable container.

5. Continue draining the fuel system to relieve fuel pressure.

2005–2006

1. Remove the fuel pump relay.

2. Start the engine and allow it to idle until it stalls.

3. After the engine stalls, crank the engine for approximately 5 seconds to make sure the fuel injection supply manifold pressure has been released.

4. Turn the ignition switch to the **OFF** position.

5. When fuel system service is complete, install the fuel pump relay.

➡**It may take more than one key cycle to pressurize the fuel system.**

6. Cycle the ignition key and wait three seconds to pressurize the fuel system. Check for leaks before starting the engine.

7. Start the engine and check the fuel system for leaks.

Fuel Filter

REMOVAL & INSTALLATION

1. Before servicing the vehicle, refer to the Precautions Section.

2. Properly relieve the fuel system pressure.

3. Disconnect the negative battery cable.

➡**To deplete the backup power supply energy, disconnect the negative battery**

cable and wait at least one minute. Be sure to disconnect auxiliary batteries and power supplies, if equipped.

4. Disconnect the fuel line at the fuel filter.

5. Loosen the clamp and remove the filter.

To install:

6. Install or connect the following:
   - New clips to the fuel lines
   - Fuel filter and tighten the clamp
   - Fuel lines to the fuel filter
   - Negative battery cable

7. Start the engine and check for leaks, repair if necessary.

Fuel Pump

REMOVAL & INSTALLATION

2002–2004

1. Before servicing the vehicle, refer to the Precautions Section.

2. Properly relieve the fuel system pressure.

3. Disconnect the negative battery cable.

➡**To deplete the backup power supply energy, disconnect the negative battery cable and wait at least one minute. Be sure to disconnect auxiliary batteries and power supplies, if equipped.**

4. Remove or disconnect the following:
   - Gas cap to relieve any additional fuel pressure
   - Left rear seat cushion and lift the access cover on the scuff plate
   - Pin type retainers and move the carpet aside
   - Screws from the fuel pump module access cover
   - Fuel pump module electrical connectors
   - Fuel and vapor lines from the fuel tank
   - Fuel pump module and discard the gasket

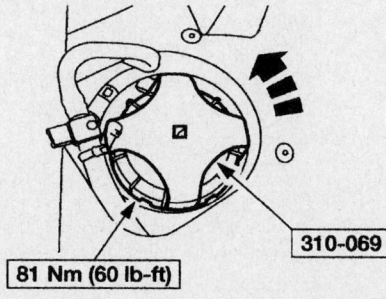

81 Nm (60 lb-ft)   310-069

67197-ESCA-G57
Fuel pump location—2002–2004 vehicles

To install:

5. Install or connect the following:
   - New fuel pump module gasket
   - Fuel pump module. Torque the module to 60 ft. lbs. (81 Nm).
   - Fuel and vapor lines to the fuel tank
   - Fuel pump module electrical connectors
   - Fuel pump module access cover and tighten the screws securely
   - Pin type retainers and reposition the carpet
   - Left rear seat cushion
   - Gas cap
   - Negative battery cable

6. Start the engine and check for leaks, repair if necessary.

2005–2006

1. Before servicing the vehicle, refer to the Precautions Section.

2. Release the fuel system pressure.

3. Disconnect the negative battery cable.

➡**To deplete the backup power supply energy, disconnect the negative battery cable and wait at least one minute. Be sure to disconnect auxiliary batteries and power supplies, if equipped.**

4. Lift the left side rear seat cushion, position the carpet aside and remove the screws and the fuel pump module access cover.

5. Disconnect the fuel pump module electrical connector.

6. Fuel vapor control tube assembly valve electrical connector. Remove the fuel cap.

7. Using a suitable fuel pump lock ring remover, rotate the lock ring counterclockwise and remove.

✳✳ WARNING

The fuel pump module must be handled carefully to avoid damage to the float arm and filter.

✳✳ WARNING

Some fuel will remain in the fuel pump module after draining the fuel tank. Carefully drain the fuel pump module into a suitable container.

8. Prior to completely removing the fuel pump, position it aside and using a suitable fuel recovery system, drain the fuel tank.

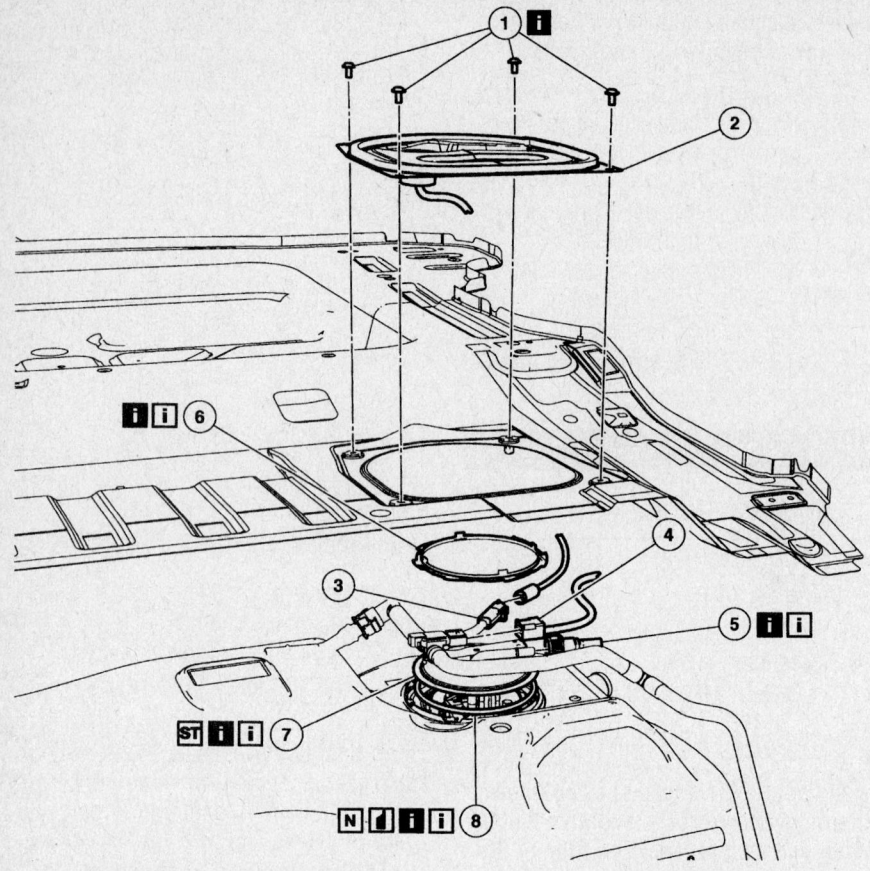

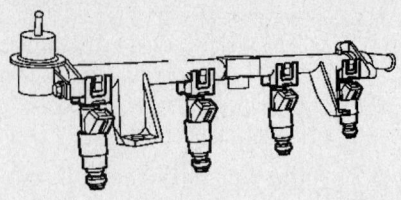

9308TG22

Fuel rail and injectors–2.0L engine

1 Fuel pump module access cover screws
2 Fuel pump module access cover
3 Fuel pump module electrical connector
4 Fuel vapor control tube assembly valve electrical connector
5 Fuel supply tube quick connect coupling
6 Fuel pump module lock ring
7 Fuel pump module
8 Fuel pump module O-ring seal

67197-ESCA-G34

Fuel pump module and related components—2005–2006 vehicles

➡ **To release the bottom-mounted fuel pump module, reach into the fuel pump module opening and squeeze the retainer tabs on the pump module housing and pull upward.**

9. Remove the fuel pump module. Discard the O-ring seal.

To install:
10. Installation is the reverse of the removal procedure.
11. Be sure to use a new fuel pump mounting gasket.
12. When installing the pump, be sure to align the locator tabs on the fuel tank mounting flange.

➡ **When installing the fuel pump module locking ring, hold the O-ring while turning the locking ring until it stops against the retainer tabs.**

13. Start the engine and visually inspect the fuel system for leaks. Correct, as required.

Fuel Injectors

REMOVAL & INSTALLATION

2.0L Engine

1. Before servicing the vehicle, refer to the Precautions Section.
2. Properly release the fuel system pressure.
3. Remove or disconnect the following:
   - Negative battery cable
   - Fuel injection supply manifold
   - Retaining clips and gently twist the fuel injector out of the manifold
4. Check the O-rings and replace if damaged.

To install:
5. Install or connect the following:
   - Fuel injector(s) using new O-rings lubricated with clean engine oil. Never use silicone grease. Silicone grease may clog the fuel injectors.

- Fuel injector into the supply manifold
- Retaining clips when the fuel injectors are seated properly
- Fuel injection supply manifold
- Negative battery cable

6. Start the engine and check for leaks, repair if necessary.

2.3L Engine

※ WARNING

Do not smoke or carry lighted tobacco or open flame of any type when working on or near fuel-related components. Highly flammable vapors are always present and can ignite. Failure to follow these instructions can result in personal injury.

※ WARNING

This procedure involves fuel handling. Be prepared for fuel spillage at all times and always observe fuel handling precautions. Failure to follow these instructions can result in personal injury.

1. Before servicing the vehicle, refer to the Precautions Section.
2. Properly release the fuel pressure.
3. Disconnect the negative battery cable.

➡ **To deplete the backup power supply energy, disconnect the negative battery cable and wait at least one minute. Be sure to disconnect auxiliary batteries and power supplies, if equipped.**

4. Properly disconnect the fuel line.
5. Disconnect the fuel injector electrical connectors.
6. Disconnect the fuel pressure regulator electrical connector and the vacuum hose.
7. Disconnect the fuel injector harness retaining clips from the fuel injection supply manifold.

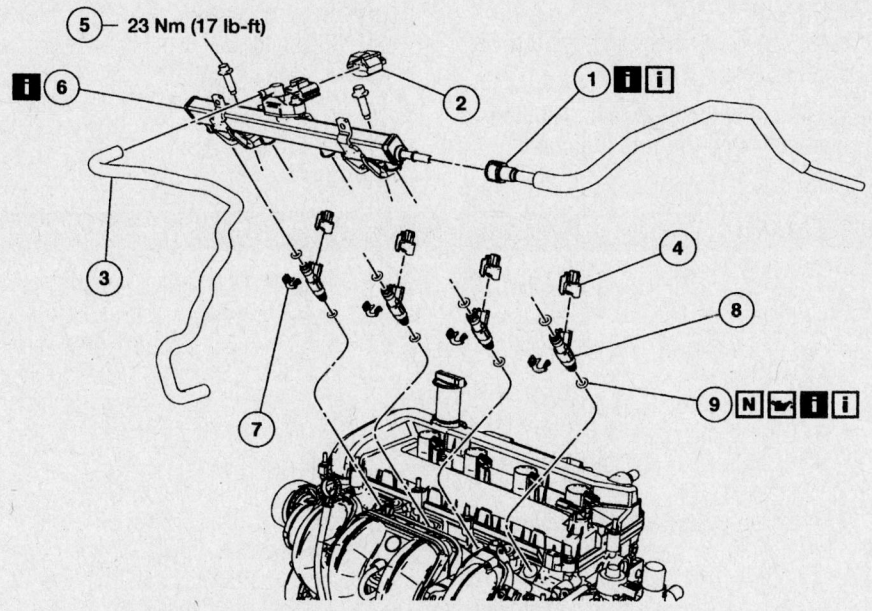

1 Fuel tube quick release coupling (position aside)
2 Fuel rail pressure and temperature sensor electrical connector
3 Fuel rail pressure and temperature sensor vacuum tube (position aside)
4 Fuel injector electrical connectors
5 Fuel rail bolts
6 Fuel rail
7 Fuel injector clips
8 Fuel injectors
9 Fuel injector O-ring seals

67197-ESCA-G35

Fuel rail and related components—2.3L engine

8. Disconnect the fuel tube.

9. Remove the bolts and the fuel injection supply manifold and injectors, as an assembly.

10. If equipped, remove the fuel rail spacers.

➡**Remove and discard the fuel injector O-rings.**

11. Remove the fuel injectors from the fuel rail assembly.

To install:

12. Install new O-rings and lubricate them with clean engine oil.

➡**Never use silicone grease. Silicone grease may clog the fuel injectors.**

13. Install the injectors to the fuel rail.

14. Continue the installation in the reverse order of the removal procedure.

15. Tighten the fuel supply manifold bolts to 18 ft. lbs. (25 Nm).

16. Start the engine and check for leaks. Correct, as required.

3.0L Engine

⁂ **WARNING**

Do not smoke or carry lighted tobacco or open flame of any type when working on or near fuel-related components. Highly flammable vapors are always present and can ignite. Failure to follow these instructions can result in personal injury.

⁂ **WARNING**

This procedure involves fuel handling. Be prepared for fuel spillage at all times and always observe fuel handling precautions. Failure to follow these instructions can result in personal injury.

1. Before servicing the vehicle, refer to the Precautions Section.

2. Properly release the fuel pressure.

3. Disconnect the negative battery cable.

➡**To deplete the backup power supply energy, disconnect the negative battery cable and wait at least one minute. Be sure to disconnect auxiliary batteries and power supplies, if equipped.**

4. Remove the upper intake manifold.

5. Disconnect the fuel tube quick release coupling.

6. Disconnect the fuel rail pressure and temperature sensor vacuum tube

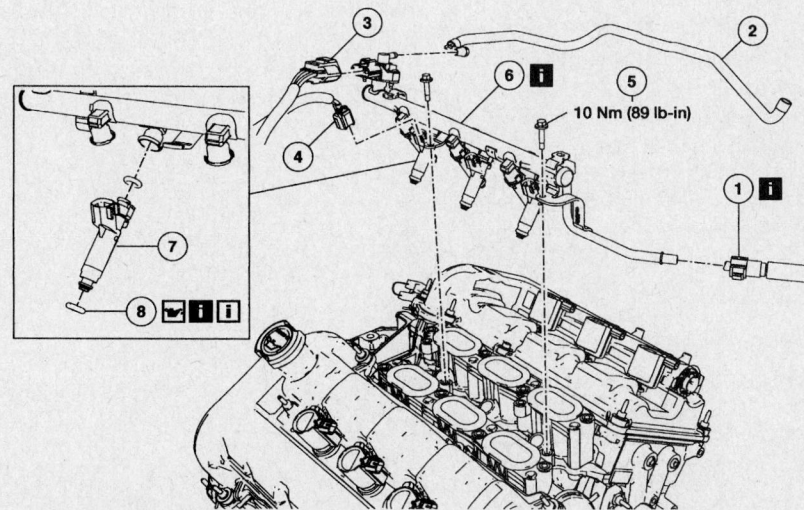

1 Fuel tube quick release coupling (position aside)
2 Fuel rail pressure and temperature sensor vacuum tube (position aside)
3 Fuel rail pressure and temperature sensor electrical connector
4 Fuel injector electrical connector
5 Fuel rail bolt
6 Fuel rail
7 Fuel injector
8 Fuel injector O-ring seals

67197-ESCA-G36

Fuel rail and related components—3.0L engine

7. Disconnect the fuel rail pressure and temperature sensor electrical connector.

8. Disconnect the fuel injector electrical connectors.

9. Remove the fuel rail bolts.

10. Remove the fuel rail and fuel injectors, as an assembly.

11. Remove the fuel injector.

12. Remove the fuel injector O-ring seals.

To install:

13. Install new O-rings and lubricate them with clean engine oil.

➡**Never use silicone grease. Silicone grease may clog the fuel injectors.**

14. Install the injectors to the fuel rail.

15. Continue the installation in the reverse order of the removal procedure.

16. Tighten the fuel supply manifold bolts to 89 inch lbs. (10 Nm).

17. Start the engine and check for leaks. Correct, as required.

DRIVE TRAIN

Transaxle

REMOVAL & INSTALLATION

Manual

1. Before servicing the vehicle, refer to the Precautions Section.

2. Disconnect the negative battery cable.

➡**To deplete the backup power supply energy, disconnect the negative battery cable and wait at least one minute. Be sure to disconnect auxiliary batteries and power supplies, if equipped.**

3. Raise and support the vehicle safely.

4. Drain the transaxle fluid.

5. Remove or disconnect the following:
- Battery and tray
- Mass Air Flow (MAF) sensor electrical connector
- Accelerator cable from the air cleaner outlet tube
- Emission management tube and hose
- Crankcase ventilation hose
- Air cleaner outlet tube
- Air cleaner housing
- Back-up lamp switch electrical connector
- Front wire harness bracket and move it aside
- Front wire harness bracket spacer
- Wire harness from the rear harness bracket
- Park Neutral Position (PNP) electrical connector
- Rear wire harness bracket and move it aside
- Vehicle Speed Sensor (VSS) electrical connector
- Clutch slave cylinder line from the bracket and move it aside while properly supporting the engine
- Left side transaxle support insulator and bracket
- Rear transaxle support insulator
- Front transaxle support insulator and bracket
- Starter and move it aside

- Top transaxle flywheel housing bolts
- Front transaxle flywheel housing bolts
- Transfer case, if equipped
- Left side halfshaft
- Rear transaxle support insulator bracket
- Shifter linkage and stabilizer bar
- Transverse crossmember
- Front to aft crossmember
- Left side splash shield and properly support the transaxle
- Remaining transaxle flywheel housing bolts
- Transaxle and separate the right side halfshaft from the transaxle

To install:

6. Align the right side halfshaft to the transaxle and position the transaxle to the engine.

7. Install or connect the following:
- Transaxle flywheel housing bolts. Torque the bolts to 33 ft. lbs. (45 Nm) and remove the transaxle support
- Left side splash shield
- Front-to-aft crossmember. Torque the bolts to 66 ft. lbs. (90 Nm).
- Transverse crossmember. Torque the bolts to 85 ft. lbs. (115 Nm).
- Shifter linkage. Torque the bolt to 15 ft. lbs. (20 Nm).
- Stabilizer bar. Torque the bolt to 30 ft. lbs. (40 Nm).
- Rear transaxle support bracket. Torque the bolts to 66 ft. lbs. (90 Nm).
- Left side halfshaft
- Transfer case, if equipped
- Front transaxle flywheel housing bolts. Torque the bolts to 33 ft. lbs. (45 Nm).
- Top transaxle flywheel housing bolts. Torque the bolts to 33 ft. lbs. (45 Nm).
- Starter. Torque the bolts to 33 ft. lbs. (45 Nm).
- Front transaxle support insulator and bracket. Torque the lower bolt

to 66 ft. lbs. (90 Nm) and the 3 upper bolts to 41 ft. lbs. (55 Nm).
- Rear transaxle support insulator bolt. Torque the bolt to 66 ft. lbs. (90 Nm).
- Left side transaxle support insulator bracket. Torque the bolts to 66 ft. lbs. (90 Nm).
- Left side transaxle support insulator. Torque the large bolt to 66 ft. lbs. (90 Nm) and the 3 remaining bolts to 41 ft. lbs. (55 Nm).
- Clutch slave cylinder. Torque the bolt to 15 ft. lbs. (20 Nm).
- Clutch slave cylinder line to the bracket and install the retaining clip
- VSS electrical connector
- Rear wire harness bracket. Torque the bolts to 80 inch lbs. (9 Nm).
- PNP switch electrical connector
- Wire harness to the rear bracket
- Front wire harness bracket spacer and bracket. Torque the bolt to 9 ft. lbs. (12 Nm).
- Back-up lamp switch electrical connector
- Air cleaner housing
- MAF sensor electrical connector
- Air cleaner outlet tube
- Crankcase ventilation hose
- Emission management tube and hose
- Accelerator cable to the air cleaner outlet tube
- Battery and tray
- Both battery cables

8. Fill the transaxle to the proper level.

9. Start the engine and check for leaks, repair if necessary.

Automatic

2002–2004

1. Before servicing the vehicle, refer to the Precautions Section.

2. Disconnect the negative battery cable.

➡**To deplete the backup power supply energy, disconnect the negative battery cable and wait at least one minute. Be sure to disconnect auxiliary batteries and power supplies, if equipped.**

3. Raise and support the vehicle safely.
4. Drain the transaxle fluid.
5. Remove or disconnect the following:
- Battery and tray
- Breather tube
- Mass Air Flow (MAF) sensor
- Intake tube and air cleaner cover
- Air cleaner assembly
- Transaxle Range (TR) sensor
- Heated Oxygen (HO_2S) sensors
- Transaxle harness connector and bracket
- Wire harness bracket spacer and move the bracket aside
- Shift cable
- Shift cable bracket and move the bracket aside
- Starter electrical connectors
- Starter
- Electrical connectors from the valve cover and install an engine support bar
- Upper transaxle retaining bolts
- Left side upper transaxle mounting plate
- Rear transaxle mount
- Right side engine mount bolt and slightly raise the engine
- Both front wheels and splash shields
- Right side halfshaft and intermediate shaft assembly after matchmarking them
- Cross brace
- Center exhaust pipe and rubber hanger
- Front exhaust pipe and flange
- Rear exhaust pipe flange
- Driveshaft
- PTU vent tube
- Lower transaxle bracket
- Access cover
- Flexplate nuts
- Output Shaft Speed (OSS) sensor
- Turbine Shaft Speed (TSS) sensor
- Fluid cooler tube and move it aside
- Fluid cooler line and install a transaxle jack
- Bolts from the PTU unit
- Transaxle with the PTU unit attached

To install:
6. Install or connect the following:
- Transaxle with the PTU unit. Torque the engine-to-transaxle mounting bolts to 30 ft. lbs. (40 Nm).
- Fluid cooler line. Torque the fastener to 17 ft. lbs. (23 Nm) and remove the transaxle jack.
- Fluid cooler tube. Torque the bolt to 17 ft. lbs. (23 Nm).
- OSS sensor
- TSS sensor

- Flexplate nuts. Torque the nuts to 27 ft. lbs. (36 Nm).
- Access cover
- Cross brace. Torque the bolts to 96 ft. lbs. (130 Nm).
- Transaxle bracket. Torque the bolts to 30 ft. lbs. (40 Nm).
- PTU vent tube
- Driveshaft. Torque the bolts to 15 ft. lbs. (20 Nm).
- Exhaust pipe and flange. Torque the bolts to 21 ft. lbs. (29 Nm).
- Exhaust pipe and rubber hanger. Torque the bolts to 21 ft. lbs. (29 Nm).
- Left side halfshaft assembly
- Right side halfshaft and intermediate shaft assembly by aligning the matchmarks
- Splash shields
- Both front wheels and lower the engine on to the right side engine mount
- Right side engine mount bolt. Torque the bolt to 89 ft. lbs. (120 Nm).
- Rear transaxle mount. Torque the upper bolt to 89 ft. lbs. (120 Nm) and the lower bolts to 35 ft. lbs. (45 Nm).
- Transaxle mount assemble. Torque the bolts to 30 ft. lbs. (40 Nm) and remove the engine support bar.
- Electrical connectors to the valve cover
- Starter. Torque the bolts to 20 ft. lbs. (27 Nm).
- Starter electrical connectors
- Shifter cable and bracket. Torque the bolt to 14 ft. lbs. (19 Nm) and connect the shifter cable.
- Wire harness and install the harness bracket spacer
- Wire harness bracket. Torque the bolt to 89 inch lbs. (10 Nm).
- HO_2S sensor
- TR sensor and make certain it is properly aligned
- Air cleaner assembly
- Intake tube and air cleaner cover
- Breather tube
- MAF sensor
- Battery tray
- Battery and cables

7. Fill the transaxle with clean fluid to the proper level.
8. Start the engine and check for leaks, repair if necessary

2005–2006 WITH 2.3L ENGINE

1. Before servicing the vehicle, refer to the Precautions Section.

2. Disconnect the negative battery cable.

➡**To deplete the backup power supply energy, disconnect the negative battery cable and wait at least one minute. Be sure to disconnect auxiliary batteries and power supplies, if equipped.**

3. Raise and support the vehicle safely.
4. Drain the transaxle fluid.
5. Remove the air cleaner assembly. Remove the battery and the battery tray.
6. If equipped, disconnect the PTO vent hose from the clip located on the fill tube.
7. Remove the three bolts holding the shift cable bracket and ground wires and unplug the transaxle electrical connector.
8. Remove the two bolts from the shift cable bracket and disconnect the shift cable from the transaxle manual lever.
9. Disconnect the transaxle range (TR) sensor electrical connector.
10. Remove the three upper transaxle retaining bolts. Install engine support tools 303-290A, 303-290A-01, 303-290A-03A, 303-D089 and 303-050.
11. Loosen but do not remove the four retaining nuts holding the bracket to the transaxle case. Remove the left side upper transaxle mount bolt.
12. Remove the six bolts holding the halfshaft to the PTO. Position the halfshaft to the side.
13. Remove the four bolts and remove the cross brace. Remove the seven retainers and the left side splash shields.
14. Remove the seven retainers and the right side splash shield. Remove the bolt for the mount and the two bolts for the cross brace. Remove the rear nut and the cross brace.
15. Remove and discard the left side front axle wheel hub nut. Remove the frame bolt from the left side and right side control arms.
16. Using special tool 205-D070, separate the left side halfshaft from the front wheel knuckle. Using tools 100-001 and 205-241 to remove the left and right side halfshaft.
17. Remove the two intermediate shaft retaining nuts. Remove the intermediate shaft.
18. Remove the two nuts which hold the exhaust bracket to the rear engine mount.
19. Remove the two nuts and bolts on the outer end of the exhaust to rear engine mount bracket and remove the bracket.
20. Remove the three nuts and separate the flexpipe from the exhaust manifold.
21. For vehicles so equipped, remove the six bolts holding the engine bracket to

the PTO. Remove the bracket. Remove the three bolts and the PTO assembly.

22. For vehicles not equipped with a PTO, remove the dampener.

23. Remove the three bolts holding the transaxle front mount plate. Remove the fluid cooler line. Remove the fluid cooler tube. Remove the output shaft speed (OSS) sensor electrical connector and remove the sensor.

24. Remove the starter. Remove the starter insulator.

25. Remove and discard the four torque converter retaining nuts. Push the torque converter back from the flexplate.

26. Use a suitable transaxle jack to support the transaxle and remove the bolt from the lower front case.

27. Remove the four remaining transaxle to engine mounting bolts. Remove the transaxle rear mount.

➡ Move the transaxle back far enough to install tool 307-346, which locks the torque converter in position prior to lower the transaxle from the vehicle.

28. Lower the transaxle from the engine compartment.

To install:

29. Installation is the reverse of the removal procedure.

30. Be sure to use new bolts and nuts, as required.

31. Fill the transaxle with the proper grade and type automatic transaxle fluid.

32. Start the engine and check for transaxle fluid leaks, correct as required.

33. Check and correct the fluid level as required.

2005–2006 WITH 3.0L ENGINE

1. Before servicing the vehicle, refer to the Precautions Section.

2. Disconnect the negative battery cable.

➡ To deplete the backup power supply energy, disconnect the negative battery cable and wait at least one minute. Be sure to disconnect auxiliary batteries and power supplies, if equipped.

3. Raise and support the vehicle safely.

4. Drain the transaxle fluid.

5. Remove the air cleaner assembly. Remove the battery and the battery tray.

6. Disconnect the transaxle range (TR) sensor electrical connector. Disconnect the transaxle harness connector. Remove the wire harness bracket nut and position the harness aside.

7. Remove the main control cover vent tube. Disconnect the EVAP and PTO vent

hoses from the transaxle filler tube. Disconnect the wire harness from the battery tray hold down bracket.

8. Disconnect the shift cable from the manual lever. Remove the two retaining bolts from the shift cable bracket. Position the cable and bracket to the side. Remove the starter.

9. Install engine support tools 303-290A, 303-290A-01 and 303-290A-03A, 303-D089 and 303-050.

➡ Remove and disconnect both electrical connectors from the upper intake to gain access to the engine for installing the lifting bracket

10. Remove the upper transaxle retaining bolts. Loosen but do not remove the four retaining nuts holding the bracket to the transaxle case. Remove the left side upper transaxle mount bolt.

11. Remove the right side upper engine mount bolt.

12. Using the engine lifting tool, raise the front of the engine. It is necessary to raise the engine a couple of inches in order to remove the transaxle.

13. Remove the seven retainers and the left side splash shield. Rekmove the seven retainers and the right side splash shield.

14. Disconnect the sway bar link. Remove the tie rod end cotter pin, lower knuckle bolt and tie rod retaining nut.

15. Use special tool 211-105 and disconnect the left and right tie rod end from the steering knuckle.

16. Carefully pry down on the left and right lower control arms and disconnect the steering knuckle from the lower ball joint and position the steering knuckle aside.

17. Using tools 100-001 and 205-241, remove the right side outer halfshaft. Using a prybar between the transaxle case and the left side halfshaft carefully disconnect the halfshaft from the transaxle case.

18. Remove the alternator shield retaining screw and remove the shield. Disconnect the heated oxygen sensor connector and unclip the two clips from the oil pan bolt studs.

19. Remove the two intermediate shaft retaining nuts. Remove the intermediate shaft.

20. Remove the cross brace. Disconnect the exhaust pipe. Disconnect the front exhaust pipe flange. Disconnect the rear exhaust pipe flange and remove the exhaust pipe.

21. If equipped with a PTO, remove the rear driveshaft and position it to the side. Remove the bolts and the PTO bracket.

➡ On four wheel drive vehicles, remove the right exhaust manifold to gain access to the PTO bracket.

22. Remove the bolt for the mount and the two bolts for the cross brace. Remove the rear nut and the cross brace.

23. Disconnect the electrical connectors from the lower mount bracket.

24. For vehicles not equipped with a PTO, remover the dampener assembly.

25. If equipped, remove the transfer case retaining bolts. Remove the transfer case.

26. Remove the center bolt from the rear transaxle mount.

27. Lower the transaxle enough to clear the frame.

➡ It is necessary to lower the transaxle in order to clear the subframe in order to remove the transaxle.

28. Remove the torque converter access cover. Remove and discard the four torque converter retaining nuts. Push the torque converter back from the flexplate.

29. Remove the fluid cooler line. Remove the output shaft speed (OSS) sensor electrical connector and remove the sensor. Remove the fluid cooler tube.

30. Use a suitable transaxle jack to support the transaxle and remove the transaxle to engine retaining bolts. Remove the transaxle mount bracket.

➡ Move the transaxle back far enough to install tool 307-346, which locks the torque converter in position prior to lower the transaxle from the vehicle.

31. Lower the transaxle from the engine compartment.

To install:

32. Installation is the reverse of the removal procedure.

33. Be sure to use new bolts and nuts, as required.

34. Fill the transaxle with the proper grade and type automatic transaxle fluid.

35. Start the engine and check for transaxle fluid leaks, correct as required.

36. Check and correct the fluid level as required.

Clutch

REMOVAL & INSTALLATION

1. Before servicing the vehicle, refer to the Precautions Section.

2. Disconnect the negative battery cable.

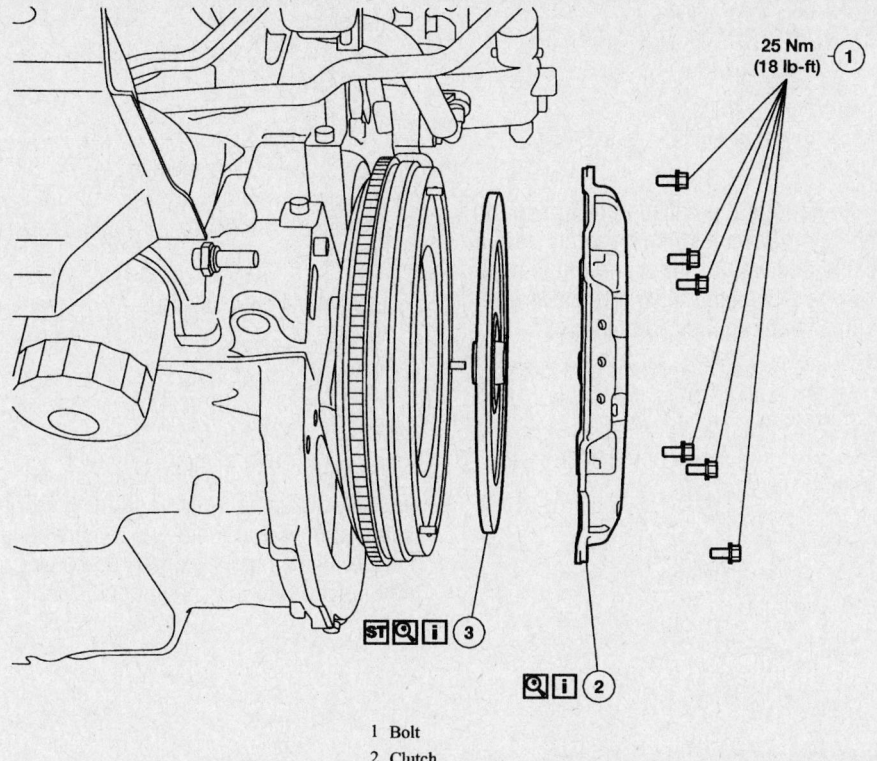

1 Bolt
2 Clutch
3 Clutch disc

67197-ESCA-G58

Clutch and related components

➡ **To deplete the backup power supply energy, disconnect the negative battery cable and wait at least one minute. Be sure to disconnect auxiliary batteries and power supplies, if equipped.**

3. Raise and support the vehicle safely.
4. Remove the transaxle from the vehicle.
5. Remove the pressure plate bolts by loosening them evenly.
6. Remove the clutch pressure plate and disc.
7. Clean the pressure plate and inspect it for burn marks, scores, flatness or ridges, replace if damaged.
8. Inspect the pressure plate diaphragm finger for wear, replace if damaged.
9. Measure the depth of the rivet heads. Minimum depth is 0.012 inch (0.3mm).
10. Inspect the clutch disc for signs of wear and replace if needed.
11. Check the clutch disc runout. Replace the disc if not with specification: 0.027 inch (0.7mm).

To install:
12. Install or connect the following:
- Clutch disc to the flywheel
- Pressure plate to the flywheel. Torque the bolts in sequence to 21 ft. lbs. (29 Nm).

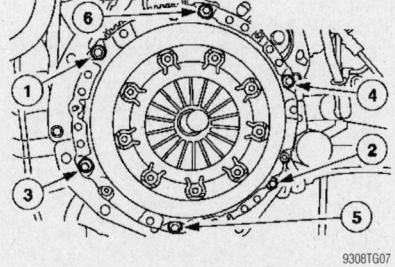

9308TG07

Pressure plate bolt torque sequence

- Transaxle
- Negative battery cable
13. Check the transaxle fluid level and top off if necessary.

ADJUSTMENTS

The clutch is hydraulically driven and therefore no adjustment is required.

Hydraulic Clutch System

BLEEDING

The following procedure is recommended for bleeding the clutch hydraulic system installed on the vehicle. It is recommended

that the original clutch tube, with quick-connect fitting be replaced when servicing the hydraulic system, because air can be trapped in the quick-connect fitting and prevent complete bleeding of the system.

1. Before servicing the vehicle, refer to the Precautions Section.
2. Clean the dirt and grease from the dust cap.
3. Remove the cap and diaphragm and fill the reservoir ¾ of the way with approved fluid.
4. Loosen the bleeder screw cover from the slave cylinder and attach a hose to the screw.
5. Place the hose in a container and slowly pump the clutch pedal several times.
6. With the clutch pedal depressed, loosen the bleeder screw to release the fluid and air.
7. Remove the hose and tighten the bleeder screw.
8. Repeat this procedure until all the air is removed from the hydraulic system.

Transfer Case

REMOVAL & INSTALLATION

2002–2004

WITH 2.0L ENGINE

1. Before servicing the vehicle, refer to the Precautions Section.
2. Raise and support the vehicle safely.
3. Remove the driveshaft.
4. Remove the four bolts and remove the crossmember brace.
5. Remove the two bolts and the two nuts.
6. Remove the transfer case.
7. Remove and discard the O-ring seal.
To install:
8. Install a new O-ring seal.
9. Install the transfer case.
10. Install the two bolts and the two nuts. Torque to 33 ft. lbs. (45 Nm).
11. Install the crossmember brace and the four bolts. Torque to 30 ft. lbs. (40 Nm).
12. Install the driveshaft.

EXCEPT 2.0L ENGINE

1. Before servicing the vehicle, refer to the Precautions Section.
2. Raise and support the vehicle safely.
3. Disconnect the battery.
4. Drain the transfer case.
5. Remove the driveshaft.
6. Remove the 4 bolts and the crossmember brace.
7. Remove the generator.

8. Disconnect the RH catalyst monitor.

9. Remove and discard the 2 nuts and separate the flexible Y-pipe from the manifold.

10. Remove and discard the nuts. Discard the gasket.

11. Remove and discard the nuts. Discard the gasket.

12. Remove the flexible pipe. Disconnect the hanger.

13. Disconnect the EGR valve to exhaust manifold tube at the manifold.

14. Disconnect the RH heated oxygen sensor.

15. Remove the RH exhaust manifold. Discard the gasket.

16. Remove the support bracket.

17. Remove the 3 transfer case bolts.

➥**The transfer case driven gear seal must be replaced whenever the link shaft or transfer case is removed from the vehicle.**

➥**If necessary, replace the RH differential fluid seal.**

18. Remove the bolt and the transfer case.
To install:

➥**The transfer case driven gear seal must be replaced whenever the link shaft or transfer case is removed from the vehicle. If necessary, replace the RH differential fluid seal.**

19. Install the transfer case. Install the bolts. Torque to 33 ft. lbs. (45 Nm).

20. Install the side mount bolt. Torque to 33 ft. lbs. (45 Nm).

21. Install the bracket and the 6 bolts. Torque to 30 ft. lbs. (40 Nm).

22. Install the RH exhaust manifold.

23. Connect the RH heated oxygen sensor.

24. Connect the EGR to exhaust manifold tube. Torque to 30 ft. lbs. (40 Nm).

25. Install the flex pipe and connect the hanger.

26. Install the gasket and the nuts. Torque to 37 ft. lbs. (50 Nm).

27. Install the flex pipe gasket and the nuts. Torque to 37 ft. lbs. (50 Nm).

28. Connect the flexible Y-pipe to the exhaust manifold and install the two nuts. Torque to 30 ft. lbs. (40 Nm).

29. Connect the RH catalyst monitor sensor.

30. Install the generator.

31. Install the crossmember brace and the four bolts. Torque to 30 ft. lbs. (40 Nm).

32. Check the transfer case fluid level.

33. Install the driveshaft.

34. Connect the battery ground cable.

35. Check the transaxle fluid level.

2005–2006

MANUAL TRANSAXLE

1. Before servicing the vehicle, refer to the Precautions Section.

2. Disconnect the negative battery cable.

➥**To deplete the backup power supply energy, disconnect the negative battery cable and wait at least one minute. Be sure to disconnect auxiliary batteries and power supplies, if equipped.**

3. Raise and support the vehicle safely.

4. Drain the transfer case.

5. Remove the driveshaft.

6. Remove the 4 bolts and the crossmember brace.

7. Remove the transfer case-to-transaxle bolts.

8. Remove the transfer case-to-transaxle nut.

9. Remove the transfer case.
To install:

10. To install, reverse the removal procedure.

✴✴ WARNING

The O-ring must be properly installed before mating the transfer case to the manual transaxle. Failure to properly install the O-ring may cause the O-ring to be damaged resulting in transaxle oil leak.

11. Install a new O-ring seal. Observe the following torques:

- Crossmember bolts: 30 ft. lbs. (40 Nm).
- Transfer case mounting nuts: 33 ft. lbs. (45 Nm).
- Transfer case mounting bolts: 33 ft. lbs. (45 Nm).

12. Fill the transfer case with the proper grade and type fluid.

AUTOMATIC TRANSAXLE

1. Before servicing the vehicle, refer to the Precautions Section.

2. Disconnect the negative battery cable.

➥**To deplete the backup power supply energy, disconnect the negative battery cable and wait at least one minute. Be sure to disconnect auxiliary batteries and power supplies, if equipped.**

3. Raise and support the vehicle safely.

4. Drain the transfer case.

5. Remove the front right side intermediate shaft.

6. Remove the driveshaft.

7. Remove the 4 bolts and the crossmember brace.

8. Remove the alternator.

9. Remove the exhaust as required.

10. Remove the heat shield.

11. Remove the transfer case-to-transaxle bolts.

12. Remove the transfer case.

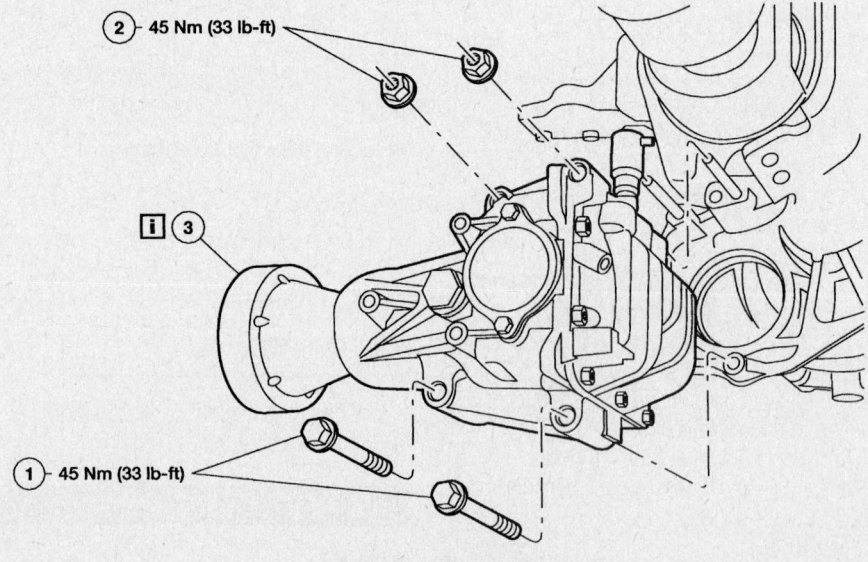

2 — 45 Nm (33 lb-ft)

1 — 45 Nm (33 lb-ft)

1 Transfer case-to-transaxle bolts
2 Transfer case-to-transaxle nuts
3 Transfer case

67197-ESCA-G37

Manual transaxle transfer case mounting—2005–2006 vehicles

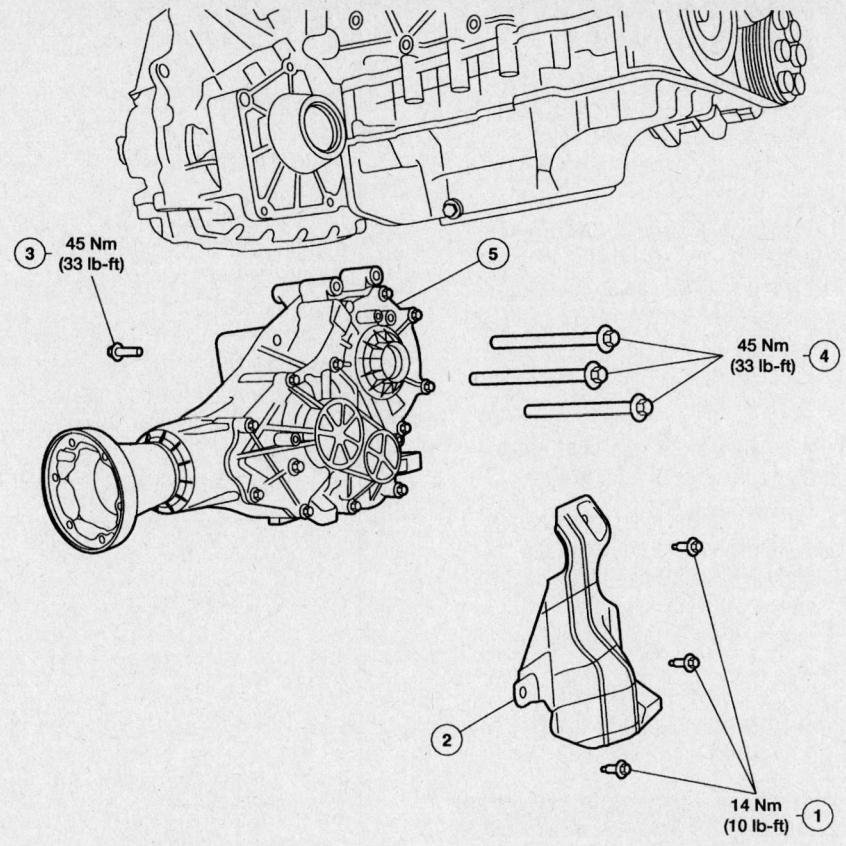

1 Heat shield bolt
2 Heat shield
3 Transfer case-to-transaxle bolts

4 Transfer case-to-transaxle bolts
5 Transfer case

67197-ESCA-G38

Automatic transaxle transfer case mounting—2005–2006 vehicles

To install:

> ❋❋ **WARNING**

A new transfer case driven gear seal must be installed whenever the intermediate shaft or transfer case is removed from the vehicle.

➡ **If necessary, replace the right side differential fluid seal.**

13. To install, reverse the removal procedure. Observe the following torques:
- Crossmember bolts: 30 ft. lbs. (40 Nm).
- Transfer case-to-transaxle bolts: 33 ft. lbs. (45 Nm).

14. Fill the transfer case with the proper grade and type fluid.

Front Halfshaft

REMOVAL & INSTALLATION

1. Before servicing the vehicle, refer to the Precautions Section.
2. Disconnect the negative battery cable.

➡ To deplete the backup power supply energy, disconnect the negative battery cable and wait at least one minute. Be sure to disconnect auxiliary batteries and power supplies, if equipped.

3. Raise and support the vehicle safely.
4. Remove or disconnect the following:
- Front wheel and tire assembly
- Front brake disc
- Front axle wheel hub nut and discard the nut
- Tie rod end and separate the lower ball from the steering knuckle
- Halfshaft from the steering knuckle
- Halfshaft

To install:

5. When seated properly, the halfshaft bearing retainer circlip will snap into the differential side gear groove.
6. Position the halfshaft and joint so that the splines align with differential side gear splines. Push the halfshaft into side gear.
7. Install or connect the following:
- Halfshaft into the steering knuckle
- Lower ball joint to steering knuckle. Torque the pinch bolt to 52 ft. lbs. (70 Nm).
- Tie rod end. Torque the nut to 41 ft. lbs. (55 Nm).
- New front axle wheel hub nut. Torque the nut to 214 ft. lbs. (290 Nm) on 2002–2005 vehicles and to 184 ft. lbs. (250 Nm) on 2006 vehicles.

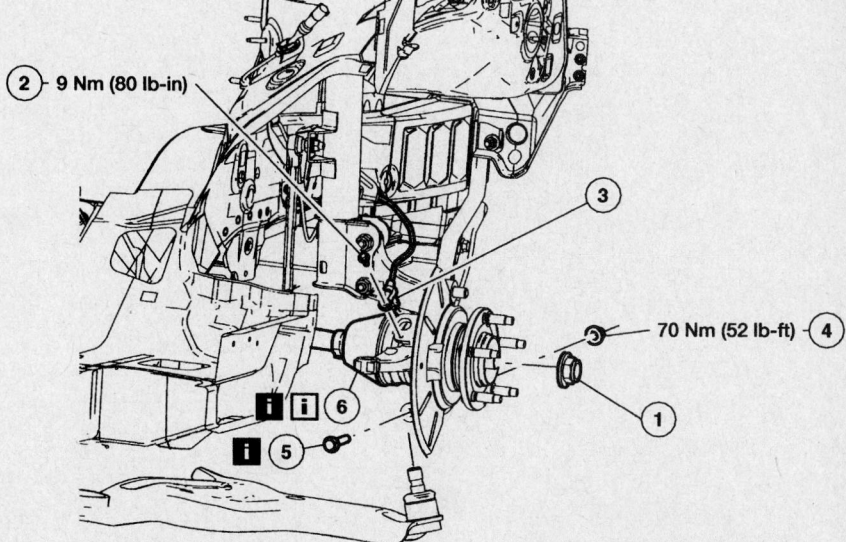

1 Front axle wheel hub nut
2 Wheel speed sensor bolt
3 Wheel speed sensor

4 Lower control arm pinch nut
5 Lower control arm pinch bolt
6 Front halfshaft

67197-ESCA-G39

Front halfshaft and related components

- Front brake disc
- Front wheel
- Negative battery cable

8. Check the fluid level and adjust as needed.

Rear Halfshaft

REMOVAL & INSTALLATION

1. Before servicing the vehicle, refer to the Precautions Section.
2. Disconnect the negative battery cable.
3. Raise and support the vehicle safely.

✳✳ WARNING

Do not loosen the rear axle wheel hub retainer until after the wheel and tire assembly are removed from the vehicle. Wheel bearing damage will occur if the wheel bearing is unloaded with the weight of the vehicle applied.

4. Remove the tire and wheel assembly.
5. Remove the rear brake drum or brake disc.
6. Remove the rear coil spring.
7. Remove the rear axle wheel hub nut.
8. Using tool 205-D070 separate the halfshaft from the knuckle.

➡Do not use a hammer to separate the rear axle halfshaft assembly from the hub. Damage to the threads and the internal CV joint can result.

9. Properly support the rear knuckle, with the proper support equipment. Remove the nut and separate the lower ball joints.
10. Using tool 205-241, remove the halfshaft.
11. Using tools 307-309 and 100-001 remove the stub and shaft seal. Using tools 308-125 and 100-001 remove the stub shaft pilot bearing seal.

To install:
12. Installation is the reverse of the removal procedure.

➡**Lubricate the new stub shaft pilot bearing housing seal with grease.**

13. Using the special tools, install the stub shaft pilot bearing housing seal.
14. Install a new circlip on the inboard CV joint.
15. Install the halfshaft end into the hub assembly.
16. Observe the following torques:
- Lower ball joint: 85 ft. lbs. (115 Nm).
- Halfshaft nut: 214 ft. lbs. (290 Nm).

CV-Joints

OVERHAUL

Front Halfshaft

1. Before servicing the vehicle, refer to the Precautions Section.

2. Remove the front halfshaft.
3. Position the halfshaft in a soft-jawed vise.
4. Remove the inboard halfshaft boot clamp.
5. Remove the boot from the inboard CV-joint housing.
6. Remove or disconnect the following:
- Tripod joint from the CV-joint housing and matchmark the tripod joint to the halfshaft
- Snapring and boot from the halfshaft
- Outboard halfshaft boot clamps
- Outboard boot back to expose the CV-joint and matchmark the joint to the halfshaft
- Outboard CV-joint from the halfshaft
- Halfshaft retainer circlip and discard it
- Boot from the halfshaft

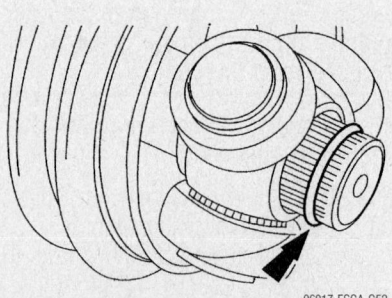
06017-ESCA-G51
Matchmark the tripod joint to the halfshaft—front halfshaft

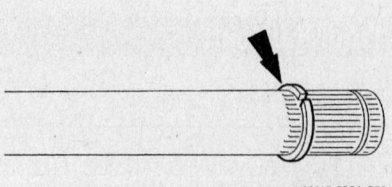
06017-ESCA-G52
Remove the snapring from the from the shaft end—front halfshaft

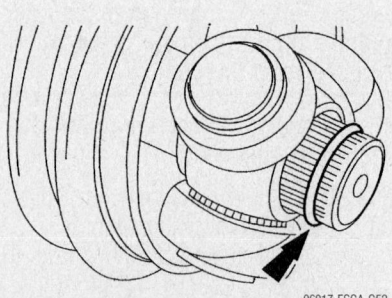
Rear halfshaft and related components
67197-ESCA-G40

1 Rear axle wheel hub retainer
2 Rear axle hub assembly
3 Lower ball joint pinch bolt nut
4 Rear axle halfshaft assembly
5 Stub shaft seal and bearing

N 1 — 290 Nm (214 lb-ft)
3 — 115 Nm (85 lb-ft)

Halfshaft retainer circlip—front halfshaft
06017-ESCA-G53

To install:

7. Lubricate the outer CV-joint with grease.

8. Install or connect the following:
- Outboard CV-joint and boot
- New halfshaft bearing circlip
- Inboard CV-joint to the halfshaft
- Outboard halfshaft boot forward on to the outboard CV-joint
- New outboard halfshaft boot clamps
- Inboard halfshaft boot
- Tripod joint on the halfshaft by aligning the matchmarks
- New snapring to the tripod joint and lubricate the needle bearings while filling the housing with CV-joint grease, E43Z–19590–A

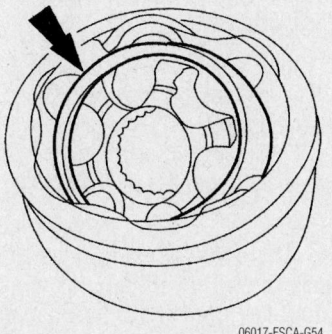

06017-ESCA-G54

Lubricate the outer CV-joint with grease— front halfshaft

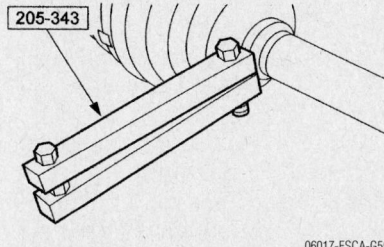

205-343

06017-ESCA-G55

Installing new outer CV-joint boot clamps—front halfshaft

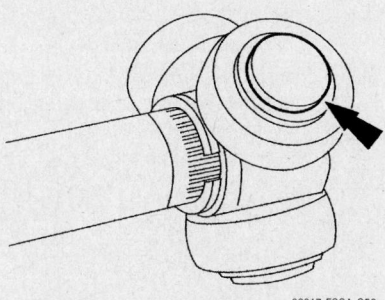

06017-ESCA-G56

Lubricate the needle bearings while filling the housing with CV-joint grease—front halfshaft

- Inboard halfshaft boot with new clamps
- Halfshaft

Rear Halfshaft

1. Before servicing the vehicle, refer to the Precautions Section.

2. Remove the halfshaft.

3. Secure the halfshaft in a vise using protective jaw covers.

4. Remove the 2 inner halfshaft boot clamps.

5. Slide the inner halfshaft boot off the inner CV-joint housing.

6. If reinstalling the original inner joint, mark the inner CV-joint and the halfshaft to be sure of correct installation.

7. Using a soft face hammer, separate the halfshaft from the inner joint housing.

8. Remove and discard the bearing retainer circlip.

9. Remove and discard the snap ring.

10. Remove the inner halfshaft boot from the halfshaft.

11. Remove the 2 outer halfshaft boot clamps.

12. Slide the outer halfshaft joint boot back out of the way exposing the outer CV-joint.

13. If reinstalling the original inner CV-joint, mark the outer CV-joint and the halfshaft to be sure of correct installation.

14. Using a soft-face hammer, separate the outer CV-joint by gently tapping it off the halfshaft.

15. Remove and discard the bearing retainer circlip.

16. Remove the snap ring from the halfshaft.

17. Slide the outer halfshaft boot off the halfshaft.

To install:

18. Lubricate the inner and outer CV-joints with joint grease.

19. Install the outer halfshaft boot.

20. Install the snap ring on the halfshaft.

21. Install a new halfshaft bearing retainer circlip.

22. Using a soft-face hammer, install the inner CV-joint by gently tapping it onto the halfshaft.

23. Remove any excess grease on the mating surfaces and slide the outer halfshaft joint boot forward onto the outer CV-joint.

24. Remove any excess air trapped in the outer halfshaft boot using a cloth-covered screwdriver after adjusting the outer halfshaft boot spacing.

25. Using a crimping tool, install 2 new outer halfshaft boot clamps.

26. Position the inner halfshaft boot.

27. Install the snap ring.

28. Install the bearing retainer circlip.

29. Using a soft face hammer, install the halfshaft on the inner CV-joint.

30. Position the inner halfshaft boot.

31. Remove any excess air trapped in the inner halfshaft boot using a cloth-covered screwdriver after adjusting the inner halfshaft boot spacing.

32. Using a crimping tool, install 2 new inner CV-joint boot clamps.

33. Install the halfshaft.

Rear Differential Mass Damper

REMOVAL & INSTALLATION

1. Before servicing the vehicle, refer to the Precautions Section.

2. Remove the rear differential mass damper.

3. Remove the bolts.

4. Remove the rear differential mass damper.

To install:

✳✳ WARNING

The mass damper bolts must be installed in the sequence shown or damage to the vehicle may occur.

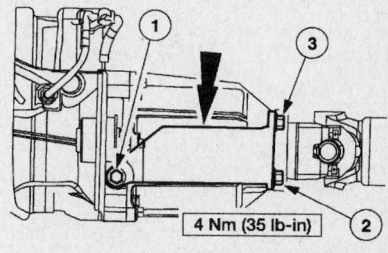

4 Nm (35 lb-in)

06017-ESCA-G59

Tighten the rear differential mass damper bolts to 35 inch lbs. in the sequence shown—2002–2004 vehicles

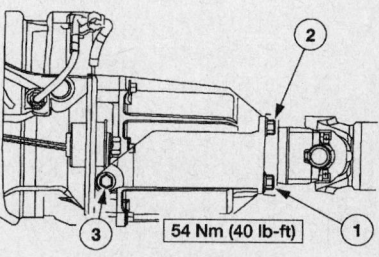

54 Nm (40 lb-ft)

06017-ESCA-G60

Tighten the rear differential mass damper bolts to 40 ft. lbs. in the sequence shown—2002–2004 vehicles

➡️**Install the mass damper bolts by hand until finger tight.**

5. Install the rear differential mass damper.

6. Position the rear differential mass damper and install the bolts in the sequence shown to 35 inch lbs. (4 Nm).

7. Final torque the mass damper bolts in the sequence shown to 40 ft. lbs. (54 Nm).

Rear Drive Axle Housing

REMOVAL & INSTALLATION

2002–2004

1. Before servicing the vehicle, refer to the Precautions Section.

2. Remove the negative battery cable

3. Remove the halfshaft.

4. Disconnect the electrical connector.

➡️**Drain the differential fluid into a suitable drain pan.**

5. Remove the bolts and the rear differential cover.

6. Drain the differential fluid from the housing.

7. Remove the six bolts and rotary blade coupling assembly.

8. Remove or disconnect:
- Rear halfshafts
- Axle assembly-to-front bracket bolts

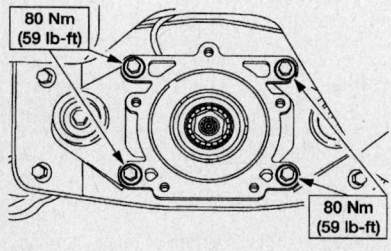

Differential housing-to-front insulator bracket bolts—2002–2004 vehicles

06017-ESCA-G61

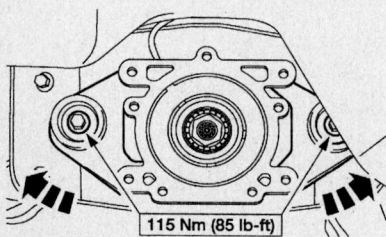

Front insulator bracket-to-subframe bolts—2002–2004 vehicles

06017-ESCA-G62

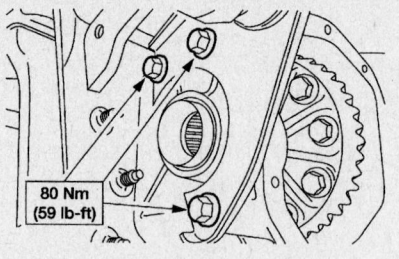

06017-ESCA-G63

Differential housing-to-side insulator bracket bolts—2002–2004 vehicles

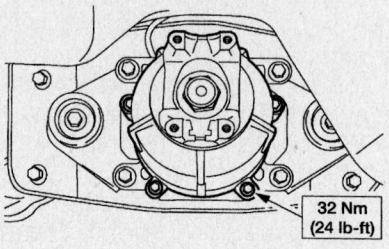

06017-ESCA-G57

Rotary blade coupling—2002–2004 vehicles

- Rear axle-to-side bracket bolts
- Axle assembly

To install:

9. Install or connect the following:
- Axle assembly
- Rear axle-to-side-bearing bolts. Torque the bolts to 59 ft. lbs. (80 Nm).
- Axle assembly-to-front bracket bolts. Torque the bolts to 59 ft. lbs. (80 Nm).

✳✳ WARNING

Make sure the machined surfaces on the rear axle housing, the differential housing cover and the rotary blade coupling housing are clean and free of oil before installing the new silicone sealant. The inside of the rear axle must be covered when cleaning the machined surface to prevent contamination.

10. Clean the gasket mating surfaces of the differential housing, the differential housing cover and the rotary blade coupling housing.

➡️**Make sure the grommet is seated all the way in the slot.**

➡️**The rotary blade coupling assembly must be installed within 15 minutes of application of the silicone, or new sealant must be applied. If possible,**

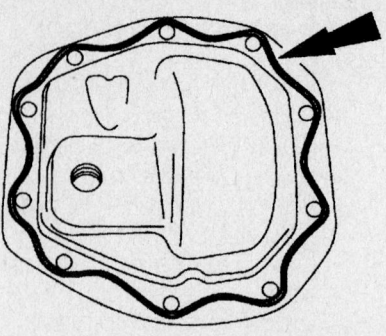

06017-ESCA-G58

Apply a new continuous bead of clear silicone rubber to the differential housing cover as shown—2002–2004 vehicles

allow one hour before filling with lubricant to make sure the silicone sealant has properly cured.

11. Apply a new continuous bead of clear silicone rubber to the rotary blade coupling housing.

12. Install the rotary blade coupling assembly and six bolts. Torque to 24 ft. lbs. (32 Nm).

➡️**The differential housing cover must be installed within 15 minutes of application of the silicone, or new sealant must be applied. If possible, allow one hour before filling with lubricant to make sure the silicone sealant has properly cured.**

13. Apply a new continuous bead of clear silicone rubber to the differential housing cover as shown in the illustration.

14. Install the differential housing cover and bolts. Torque to 17 ft. lbs. (23 Nm).

15. Connect the electrical connector.

16. Install the driveshaft.

17. Fill the rear axle with 1.4 liters (2.95 pints) of rear axle lubricant. Tighten the filler plug. Torque to 20 ft. lbs. (27 Nm).

18. Connect the negative battery cable.

2005–2006

1. Before servicing the vehicle, refer to the Precautions Section.

2. Remove the spare tire.

3. Remove the rear driveshaft assembly.

4. Remove the rear halfshafts.

5. Position a suitable transaxle hydraulic jack to the axle housing. Securely strap the jack to the housing.

6. Remove the electrical connector at the axle.

7. Remove the rear axle differential housing-to-front insulator bracket bolts.

8. Remove the front insulator-to-bracket subframe bolts.

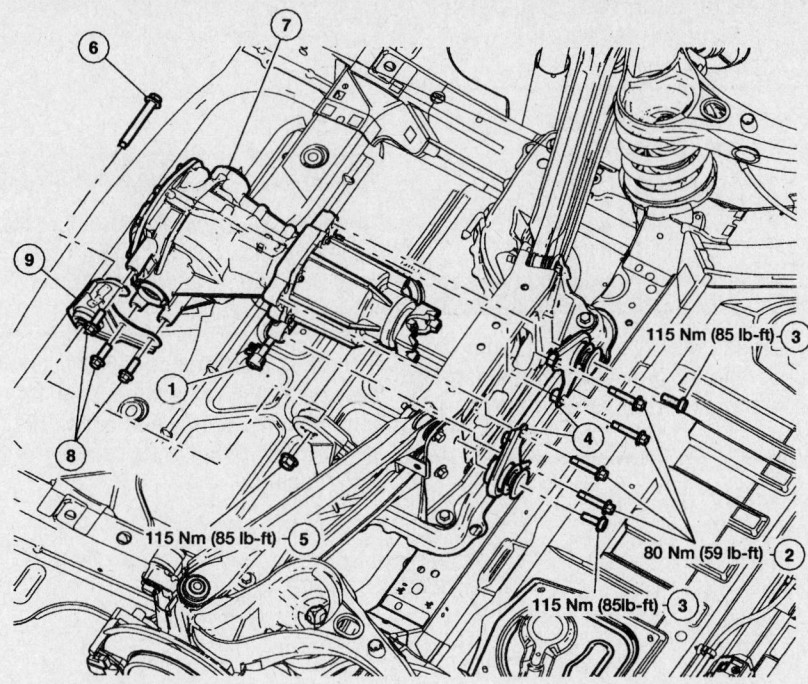

1 Electrical connector
2 Rear axle differential housing-to-front insulator bracket bolts
3 Front insulator-to-bracket subframe bolts
4 Front insulator brackets
5 Side insulator bracket-to-subframe nut
6 Side insulator bracket-to-subframe bolt
7 Rear axle assembly
8 Side insulator bracket-to-rear axle differential bolts
9 Side insulator bracket

67197-ESCA-G41

Rear drive axle removal—2005–2006 vehicles

9. Remove the front insulator brackets.
10. Remove the side insulator bracket-to-subframe nut.
11. Remove the side insulator bracket-to-subframe bolt.
12. Remove the rear axle assembly.
13. Remove the side insulator bracket-to-rear axle differential bolts.
14. Remove the side insulator bracket.
15. To install, reverse the removal procedure. Observe the following torques:
  • Axle housing-to-front insulator bracket: 59 ft. lbs. (80 Nm)
  • Front insulator-to-bracket subframe bolts: 85 ft. lbs. (115 Nm)
  • Side insulator bracket-to-subframe nut: 85 ft. lbs. (115 Nm)
  • Side insulator bracket-to-differential bolts: 59 ft. lbs. (80 Nm) on 2005 vehicles and 66 ft. lbs. (90 Nm) for 2006 vehicles

Differential Pinion Seal

REMOVAL & INSTALLATION

Rear Differential

1. Before servicing the vehicle, refer to the Precautions Section.

2. With the vehicle in NEUTRAL, position it on a hoist.
3. Index-mark the pinion and pinion flange to the rear of the driveshaft.
4. Remove the 4 bolts and the 2 cap straps. Disconnect and support the driveshaft.

➡Discard the nut after removing it. Install a new nut during installation.

5. Using the special tool, hold the pinion flange while removing the nut. Remove the nut.
6. Index-mark the location of the pinion to the yoke.
7. Using a puller, remove the pinion flange.
8. Using the special tool, remove the seal.

To install:

➡Make sure that the mating surface is clean before installing the new seal.

9. Using a seal driver, install the seal.

➡Lubricate the pinion flange with premium long-life grease.

10. Line up the index marks and position the pinion flange.
11. Using the special tool, install the

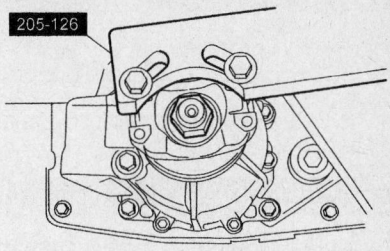
Flange holding tool installation

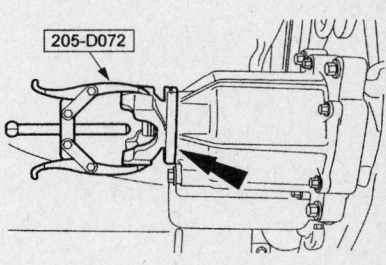
Removing the pinion flange

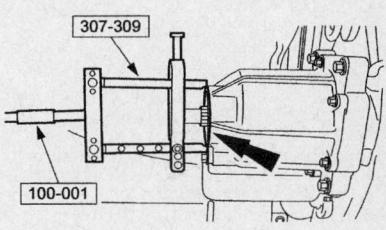
Removing the pinion seal

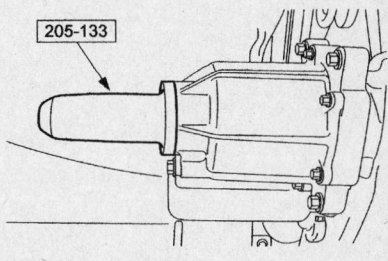
Installing the pinion seal

pinion nut. Tighten to 244 Nm (180 ft. lbs.).
12. Line up the index marks and position the rear driveshaft.
13. Install the 2 cap straps and the 4 bolts. Tighten to 23 Nm (17 ft. lbs.).

Rear Driveshaft

REMOVAL & INSTALLATION

1. Before servicing the vehicle, refer to the Precautions Section.

✳✳ CAUTION

The normal operating temperature of the exhaust system is very high. Never attempt to remove any part of the system until it has cooled. Be especially careful when working around the catalytic converters. The temperature of the converter rises to a high level after only a few minutes of engine operation. Failure to follow these instructions may result in personal injury.

➡Do not swap driveshaft assembles from different vehicles

2. With the vehicle in NEUTRAL, position it on a hoist.
3. Remove the ground strap bolt.

✳✳ WARNING

Do not reuse the CV-joint bolts and washers. Install new bolts and washers or damage to the vehicle may occur.

4. Remove and discard the 6 front driveshaft-to-transfer case bolts and washers.
Index-mark the front driveshaft to the center bearing.

✳✳ WARNING

Do not reuse the bolts and cap straps for the center U-joint. Install new bolts and cap straps or damage to the vehicle may occur.

➡There is a difference in the length of the head of the replacement cap strap bolts from the production bolts. The longer head pinion bolts can be used in either location.

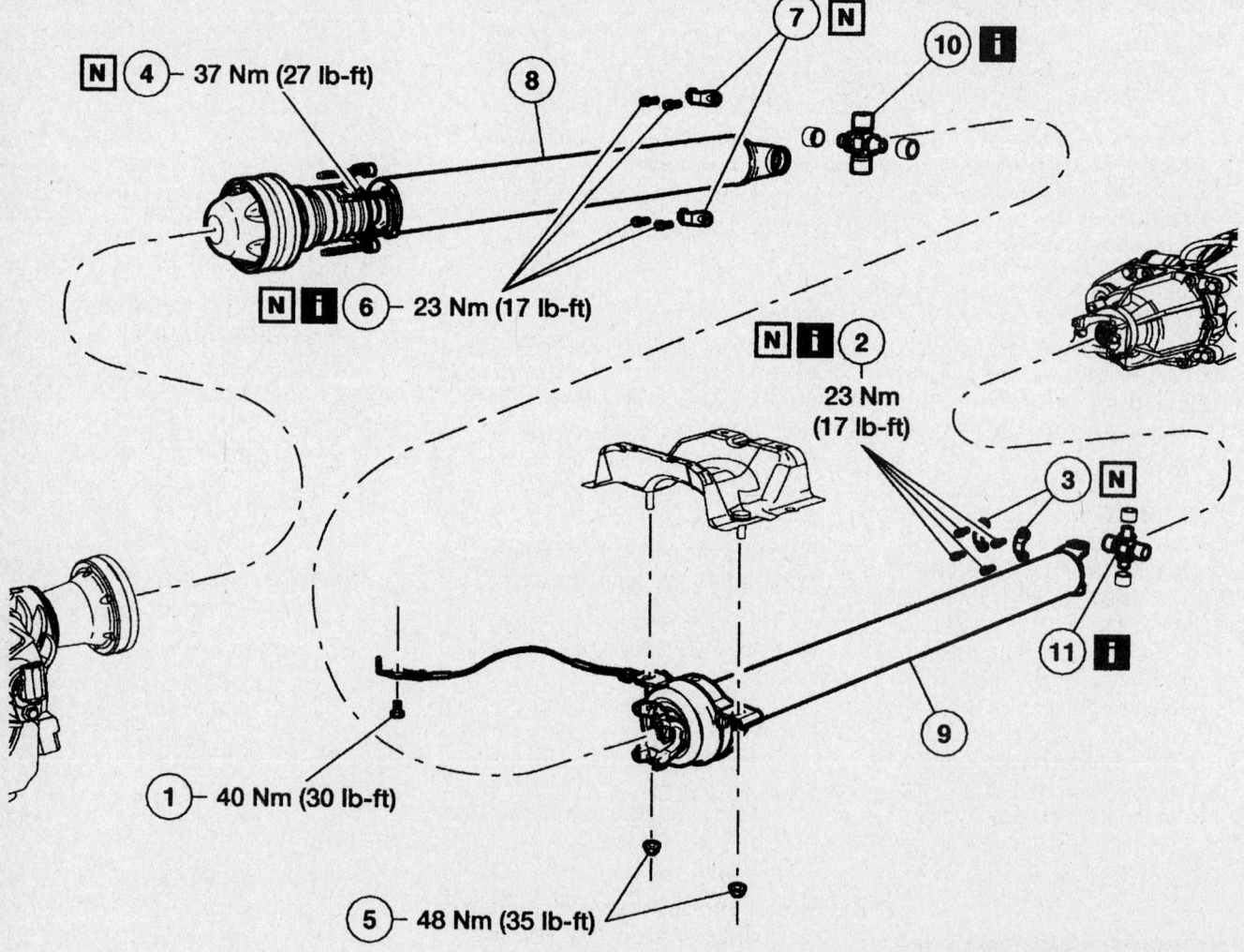

N ④ – 37 Nm (27 lb-ft)

⑧

⑦ N

⑩ i

N i ⑥ – 23 Nm (17 lb-ft)

N i ② 23 Nm (17 lb-ft)

③ N

⑪ i

⑨

① – 40 Nm (30 lb-ft)

⑤ – 48 Nm (35 lb-ft)

1 Ground strap bolt	5 Center bearing support nuts	9 Rear driveshaft
2 Universal joint cap bolts	6 Universal joint cap bolts	10 Front driveshaft U-joint
3 Universal joint cap straps	7 Universal joint cap straps	11 Rear driveshaft U-joint
4 Front driveshaft-to-transfer case bolts	8 Front driveshaft	

Rear driveshaft and related components

67197-ESCA-G42

5. Remove and discard the 4 universal joint cap strap bolts and 2 cap straps and remove the front driveshaft.

6. Index-mark the pinion and yoke to the driveshaft.

※※ WARNING

Do not reuse the bolts and cap straps for the rear U-joint. Install new bolts and cap straps.

➡ **There is a difference in the length of the head of the replacement strap bolts from the production bolts. The longer head pinion bolts can be used in either location.**

7. Remove and discard the 4 universal joint cap bolts and 2 cap straps from the rear driveshaft universal joint.

8. With the help of an assistant, remove the center bearing support nuts and the driveshaft.

9. To install, reverse the removal procedure. Observe the following torques:

- Center bearing support nuts: 48 Nm (35 ft. lbs.)
- Rear universal joint cap bolts: 23 Nm (17 ft. lbs.)
- Front universal joint cap strap bolts: 23 Nm (17 ft. lbs.)
- The 6 front driveshaft-to-transfer case bolts: 37 Nm (27 ft. lbs.)
- Ground strap bolt: 40 Nm (30 ft. lbs.)

10. If a driveshaft is installed and driveshaft vibration is encountered after installation, index the driveshaft.

a. With the vehicle in NEUTRAL, position it on a hoist.

※※ WARNING

Do not reuse the CV-joint bolts and washers. Install new bolts and washers or damage to the vehicle may occur.

11. Remove and discard the 6 front driveshaft-to-transfer case bolts and washers.

12. Rotate the flange 60 degrees.

13. Connect the front driveshaft and install the 6 new bolts and washers. Tighten to 37 Nm (27 ft. lbs.).

※※ WARNING

Do not reuse the bolts and cap straps for the pinion yoke. Install new bolts and cap straps or damage to the vehicle may occur.

14. Disconnect the rear driveshaft universal joint. Discard the 4 bolts and the 2 cap straps.

15. Rotate the rear pinion 180 degrees.

16. Connect the rear driveshaft and install 4 new bolts and 2 new cap straps. Tighten to 23 Nm (17 ft. lbs.).

17. Lower the vehicle and test drive.

18. Repeat the procedure if necessary.

Intermediate Shaft

REMOVAL & INSTALLATION

1. Before servicing the vehicle, refer to the Precautions Section.

➡ **If removing the intermediate shaft in order to repair a separate component, it should only be removed as an assembly with the right front halfshaft.**

2. Remove the right front halfshaft.

3. Remove the inner halfshaft bearing retainer nuts

4. Remove the intermediate shaft

5. To install, reverse the removal procedure. Apply a thin coat of grease to the splines of the intermediate shaft.

6. Verify the front axle lubricant level is to specifications.

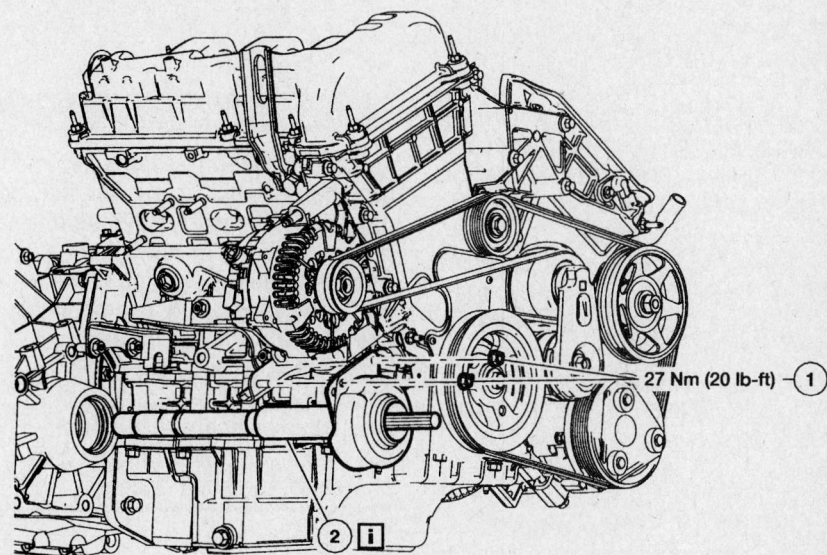

1 Inner halfshaft bearing retainer nuts
2 Intermediate shaft

67197-ESCA-G43

Intermediate shaft and related components

STEERING

Air Bag

✳✳ CAUTION

All vehicles are equipped with an air bag system. The system MUST BE disabled before performing service on or around system components, steering column, instrument panel components, wiring and sensors. Failure to follow safety and disabling procedures could result in accidental air bag deployment, possible personal injury and unnecessary system repairs.

PRECAUTIONS

- Always wear safety glasses when repairing an air bag supplemental restraint system (SRS) vehicle and when handling an air bag module. This will reduce the risk of injury in the event of an accidental deployment.
- After deployment, the air bag surface can contain deposits of sodium hydroxide, a product of the gas generate combustion that is irritating to the skin. Wash your hands with soap and water afterwards.
- Never probe the connectors on the air bag module. Doing so can result in air bag deployment, which can result in personal injury.
- Never probe the connectors on the safety canopy module. Doing so can result in safety canopy deployment, which can result in personal injury.
- The safety belt pretensioner is a pyrotechnic device. Always wear safety glasses when repairing an air bag equipped vehicle and when handling a safety belt buckle pretensioner or safety belt retractor pretensioner. Never probe a pretensioner electrical connector. Doing so could result in pretensioner or air bag deployment and could result in personal injury.
- To reduce the risk of personal injury, do not use any memory saver devices.
- Never carry the inflator module by the wires or connector on the underside of the module.
- When carrying a live inflator module, hold securely with both hands and ensure that the bag and trim cover are pointed away.
- Place the inflator module on a bench or other surface with the bag and trim cover facing up.
- With the inflator module on the bench,

never place anything on or close to the module, which may be thrown in the event of an accidental deployment.

DEPOWERING

1. Before servicing the vehicle, refer to the Precautions Section.
2. Turn all vehicle accessories OFF.
3. Turn the ignition switch to OFF.
4. At the central junction box (SJB), located below the left side of the instrument panel, remove the cover and the restraints control module (RCM) fuse(s) from the SJB. See the Owner's Manual.
5. Turn the ignition ON and visually monitor the air bag indicator for at least 30 seconds. The air bag indicator will remain lit continuously (no flashing) if the correct RCM fuse has been removed. If the air bag indicator does not remain lit continuously, remove the correct RCM fuse before proceeding.
6. Turn the ignition OFF.

✳✳ CAUTION

To avoid accidental deployment and possible personal injury, the backup power supply must be depleted before repairing or replacing any front or side air bag supplemental restraint system (SRS) components and before servicing, replacing, adjusting or striking components near the front or side air bag sensors, such as doors, instrument panel, console, door latches, strikers, seats and hood latches.

The front impact severity sensor is located on the radiator support bracket.
The first row side impact sensors (if equipped) are located at or near the base of the B-pillars.
The second row side impact sensors (if equipped) are located on the C-pillar.

➡ **To deplete the backup power supply energy, disconnect the battery ground cable and wait at least one minute. Be sure to disconnect auxiliary batteries and power supplies (if equipped).**

7. Disconnect the battery ground cable and wait at least one minute.

REPOWERING

1. Before servicing the vehicle, refer to the Precautions Section.

✳✳ CAUTION

The restraint system diagnostic tool is for restraint system service only. Remove from vehicle prior to road use. Failure to remove could result in injury and possible violation of vehicle safety standards.

2. Make sure all restraint system diagnostic tool(s) that may have been installed during the repair have been removed from the vehicle and all SRS components are connected.
3. Turn the ignition switch from OFF to ON.
4. Install RCM fuse(s) to the SJB and close the cover.

✳✳ CAUTION

Be sure that nobody is in the vehicle and that there is nothing blocking or set in front of any air bag module when the battery ground cable is connected.

5. Connect the battery ground cable.
6. Prove out the supplemental restraint system (SRS) as follows:
   a. Turn the ignition key from ON to OFF. Wait 10 seconds, and then turn the key back to ON and visually monitor the air bag indicator with the air bag modules installed. The air bag indicator will light continuously for approximately six seconds and then turn off. If an air bag supplemental restraint system (SRS) fault is present, the air bag indicator will either:
   - fail to light.
   - remain lit continuously.
   - flash.
   b. The flashing might not occur until approximately 30 seconds after the ignition switch has been turned from the OFF to the ON position. This is the time required for the restraints control module (RCM) to complete the testing of the SRS. If the air bag indicator is inoperative and a SRS fault exists, a chime will sound in a pattern of five sets of five beeps. If this occurs, the air bag indicator and any SRS fault discovered must be diagnosed and repaired.
7. Clear all continuous DTCs from the restraints control module using a scan tool.

DISARMING

The Supplemental Inflatable Restraint (SIR) system must be disarmed before performing service around SIR system components or SIR system wiring. Failure to do so may cause accidental deployment of the air bag, resulting in unnecessary SIR system repairs and/or personal injury.

The positive battery cable must be disconnected for a minimum of 1 minute before beginning any air bag work to de-energize the back-up power supply. It is a good idea to disengage both the positive and negative battery cables to ensure that the Air Bag system is definitely discharged.

ARMING THE SYSTEM

If the air bag simulators have been used, the air bag simulators must be removed and the air bags reconnected when the system is reactivated to avoid non-deployment in a collision resulting in possible personal injury.

1. Disconnect the positive battery cable.
2. Wait 1 minute, this is required for the back-up power supply in the air bag diagnostic monitor to deplete its stored energy.
3. Remove the air bag simulator from the air bag sliding contact connector at the top of the steering column. Reconnect the driver's side air bag module assembly. Position the driver's air bag module on the steering wheel and secure with the 2 bolts and washers. Tighten the bolt and washer assembly to 8–10 ft. lbs. (10–14 Nm).
4. Connect the positive battery cable.
5. Turn the ignition switch from the **OFF** to **RUN** and visually monitor the air bag warning indicator. The light will illuminate continuously for approximately 6 seconds and then turn off. If a fault occurs, the air bag indicator will either fail to light, remain lighted continuously or flash. The flashing may not occur until approximately 30 seconds after the ignition switch has been turned from **OFF** to **RUN**. This is the time needed for the air bag diagnostic monitor to complete testing the system. If the air bag indicator is inoperative, an air bag system fault exists, a tone will sound in a pattern of 5 sets of 5 beeps. If this occurs, the air bag

indicator will need to be serviced before further diagnostics can be done.

Power Steering Gear

REMOVAL & INSTALLATION

2002–2004

1. Before servicing the vehicle, refer to the Precautions Section.
2. Place the steering wheel in the straight-ahead position. Lock the steering wheel in place, using a steering wheel holder.

➡**Locking the steering wheel keeps the clockspring in alignment position.**

3. Drain the power steering fluid.
4. Remove or disconnect the following:

- Negative battery cable
- Rear transaxle insulator
- Rear transaxle insulator bracket, if equipped with an automatic transaxle
- Both front wheels
- Rear transaxle insulator bracket, if equipped with a manual transaxle
- Tie rod end cotter pin and nut
- Tie rod end from the steering knuckle and record the number of turns required to remove the tie rod end
- Steering gear coupling pinch bolt
- Power steering pressure and return lines and bracket
- Steering gear mounting bolts
- Steering gear and separate the steering coupling from the steering gear shaft
- Steering gear

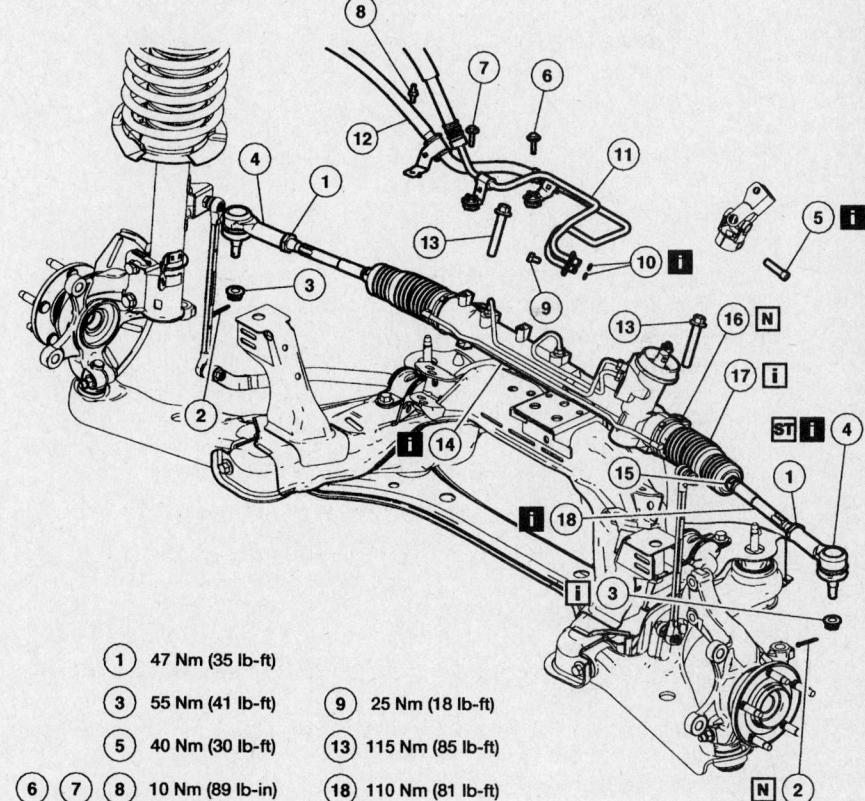

①	47 Nm (35 lb-ft)		
③	55 Nm (41 lb-ft)	⑨	25 Nm (18 lb-ft)
⑤	40 Nm (30 lb-ft)	⑬	115 Nm (85 lb-ft)
⑥ ⑦ ⑧	10 Nm (89 lb-in)	⑱	110 Nm (81 lb-ft)

1 Tie-rod jam nuts (loosen)
2 Cotter pins
3 Tie-rod end nuts
4 Tie-rod end, outer
5 Steering column coupling-to-steering gear pinch bolt
6 Power steering return line bracket-to-subframe bolt
7 Power steering pressure line bracket-to-steering gear bolt
8 Power steering return line bracket-to-steering gear stud
9 Power steering line clamp plate bolt
10 O-rings
11 Power steering pressure line
12 Power steering return line
13 Steering gear mounting bolts
14 Steering gear
15 Tie-rod boot clamp, outer
16 Tie-rod boot clamp, inner
17 Tie-rod boot
18 Tie-rod, inner

67197-ESCA-G44

Steering gear and related components—2002–2004 vehicles

To install:
- Slide the steering gear rearward to connect the steering coupling to the steering gear shaft

5. Install or connect the following:
- Steering gear mounting bolts. Torque the bolts to 93 ft. lbs. (126 Nm).
- Pressure and return lines and bracket. Torque the bracket bolts to 89 inch lbs. (10 Nm).
- Power steering pressure and return lines to the steering gear. Torque the bolt to 18 ft. lbs. (25 Nm).
- Steering gear pinch bolt and reposition the boot. Torque the bolt to 18 ft. lbs. (25 Nm).
- Tie rod end to the tie rod using the number of turns required to remove the tie rod end
- Jam nuts. Torque the nuts to 35 ft. lbs. (47 Nm).
- Tie rod end to the steering knuckle. Torque the nut to 41 ft. lbs. (57 Nm) and install a new cotter pin
- Rear transaxle insulator bracket. Torque the bolts to 66 ft. lbs. (90 Nm).
- Both front wheels
- Rear transaxle insulator bracket. Torque the bolts to 66 ft. lbs. (90 Nm).
- Rear transaxle insulator. Torque the bolts to 66 ft. lbs. (90 Nm).
- Negative battery cable

6. Fill and bleed the power steering system.

7. Start the vehicle and check for leaks, repair if necessary.

8. Check and adjust the front end alignment.

2005–2006

1. Before servicing the vehicle, refer to the Precautions Section.

2. With the vehicle in NEUTRAL, position it on a hoist.

3. Vehicles equipped with 3.0L engine, remove the EGR valve.

   a. Disconnect the exhaust manifold-to-exhaust gas recirculation (EGR) tube fitting from the EGR valve.

   b. Disconnect the vacuum tube fitting from the EGR valve.

➡**The EGR valve sealing surfaces are soft metals. Do not reuse the EGR valve gasket.**

   c. Remove the 2 bolts and the EGR valve. Remove and discard the EGR valve gasket.

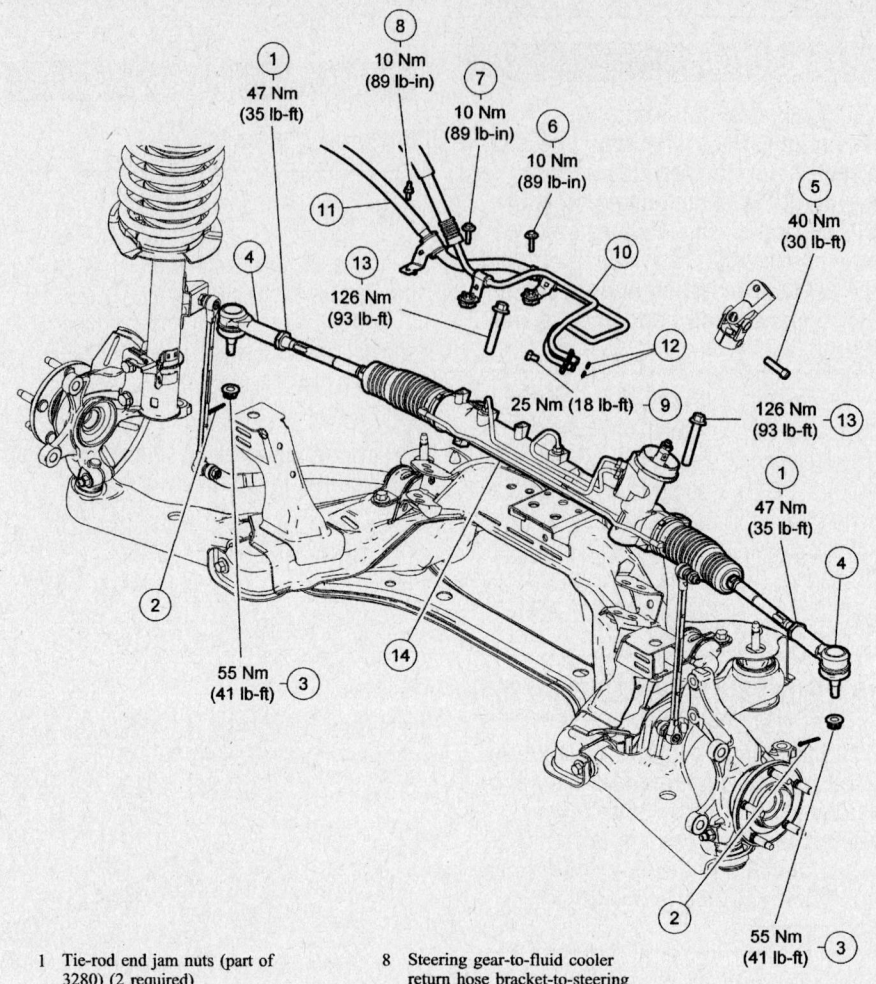

1	Tie-rod end jam nuts (part of 3280) (2 required)	8	Steering gear-to-fluid cooler return hose bracket-to-steering gear stud
2	Cotter pins (2 required)	9	Power steering line clamp plate bolt
3	Outer tie-rod end nuts (2 required)	10	Power steering pressure line
4	Outer tie-rod end	11	Steering gear-to-fluid cooler return hose
5	Steering column coupling-to-steering gear pinch bolt	12	O-rings (1 each required)
6	Steering gear-to-fluid cooler return hose bracket-to-subframe bolt	13	Steering gear mounting bolts (2 required)
7	Power steering pressure line bracket-to-steering gear bolt	14	Steering gear

06017-ESCA-G68

Steering gear and related components—2005–2006 vehicles

4. Carefully clean the EGR valve sealing surfaces.

5. Remove the rear transaxle insulator bracket.

6. Remove the 3 bolts and the rear transaxle mounting plate.

➡**Do not loosen the tie-rod end nut to align the slot in the nut with the tie-rod end stud through-hole.**

7. Remove the 2 outer tie-rod end cotter pins and the 2 tie-rod end nuts.

8. Separate the tie-rod end from the knuckle.

✳✳ WARNING

Do not allow the steering wheel to rotate while the intermediate shaft is disconnected or damage to the clockspring can result. If there is evidence that the shaft has rotated, the clockspring must be removed and recentered.

9. Hold the steering wheel in a straight-ahead position using a suitable tool.

10. Remove the nuts and the steering column boot.

11. Remove the steering column coupling-to-steering gear pinch bolt.

12. Remove the steering gear-to-fluid cooler return hose bracket-to-subframe bolt.

13. Remove the power steering pressure line bracket-to-steering gear bolt.

14. Remove the steering gear-to-fluid return hose bracket-to-steering gear stud.

15. Remove the power steering line clamp plate bolt.

➡**Install a new high pressure hose O-ring seal and a new return hose O-ring seal.**

16. Remove the power steering pressure and return lines.

17. Remove the 2 steering gear mounting bolts.

➡**Remove the steering gear from the right side of the vehicle.**

18. Remove the steering gear.

19. To install, reverse the removal procedure. Observe the following torques:
- Steering gear mounting bolts: 126 Nm (93 ft. lbs.)
- Power steering line clamp plate bolt: 25 Nm (18 ft. lbs.)
- Steering gear-to-fluid return hose bracket-to-steering gear stud: 10 Nm (89 inch lbs.)
- Power steering pressure line bracket-to-steering gear bolt: 10 Nm (89 inch lbs.)
- Steering gear-to-fluid cooler return hose bracket-to-subframe bolt: 10 Nm (89 inch lbs.)
- Steering column coupling-to-steering gear pinch bolt: 40 Nm (30 ft. lbs.)
- Outer tie-rod end cotter pins and the 2 tie-rod end nuts: 55 Nm (41 ft. lbs.). If necessary, continue to tighten the tie-rod end nut until the slot in the nut aligns with the tie-rod end stud through-hole.
- Rear transaxle mounting plate: 90 Nm (66 ft. lbs.)
- Rear transaxle insulator-to-bracket through-bolt: 90 Nm (66 ft. lbs.)
- Rear transaxle insulator bracket nuts: 90 Nm (66 ft. lbs.)
- Rear transaxle insulator bracket bolt: 90 Nm (66 ft. lbs.)
- EGR valve: 25 Nm (18 ft. lbs.)
- EGR tube fitting: 40 Nm (30 ft. lbs.)

20. Fill the power steering system.

21. Check and, if necessary, align the front end.

Clockspring

REMOVAL

1. Before servicing the vehicle, refer to the Precautions Section.

2. Disconnect the negative battery cable.

➡**To deplete the backup power supply energy, disconnect the negative battery cable and wait at least one minute. Be sure to disconnect auxiliary batteries and power supplies, if equipped.**

3. Depower the air bag system.

4. Make sure the road wheels are in the straight-ahead position.

5. Position the steering wheel in the straight-ahead position and remove the ignition key. Rotate the steering wheel until the steering column locks into position.

6. Open the cover on the underside of the steering wheel.

7. Remove the steering wheel pinion bolt.

8. Remove the steering wheel. Position the steering wheel rearward. Disconnect the 2 electrical connectors.

9. If equipped with tilt steering, position the steering column completely downward and lock in place.

10. Push in where indicated, releasing the retaining tabs, and remove the upper steering column shroud.

11. Release the tilt column locking lever, allowing the steering column to move upward. Do not lock the tilt column locking lever back in place.

12. Remove the 3 screws and position the lower steering column shroud aside.

13. If installing the same clockspring, apply 2 strips of masking tape across the clockspring to prevent accidental rotation when the clockspring is removed.

14. Remove the 3 clockspring screws.

➡**If the clockspring is to be reinstalled, do not allow the clockspring to turn from its removal position.**

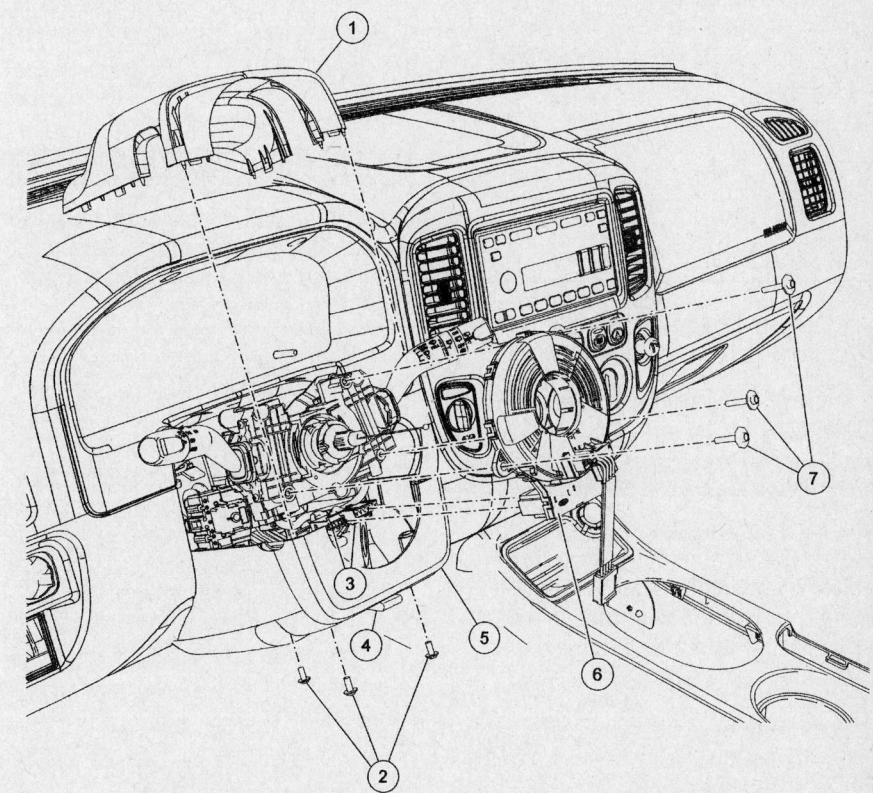

1 Upper steering column shroud
2 Lower steering column shroud screws
3 Clockspring electrical connectors
4 Tilt column locking lever (if equipped)
5 Lower steering column shroud
6 Clockspring
7 Clockspring screws

06017-ESCA-G69

Clockspring location and related components

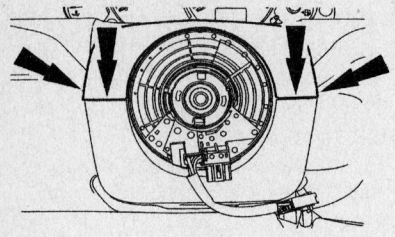

Push in where indicated, releasing the retaining tabs, and remove the upper steering column shroud

06017-ESCA-G70

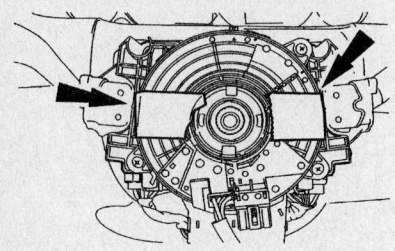

06017-ESCA-G71

If installing the same clockspring, apply 2 strips of masking tape across the clock-spring to prevent accidental rotation when the clockspring is removed

15. Partially remove the clockspring, then disconnect the 2 electrical connectors and remove the clockspring.

INSTALLATION

Vehicles Needing Clockspring Recentering

1. Before servicing the vehicle, refer to the Precautions Section.

�֍ CAUTION

Incorrect centralization may result in premature component failure. If in doubt when centralizing the clock-spring, repeat the centralizing procedure. Failure to follow this instruction may result in personal injury.

➡ **Make sure the road wheels are in the straight-ahead position.**

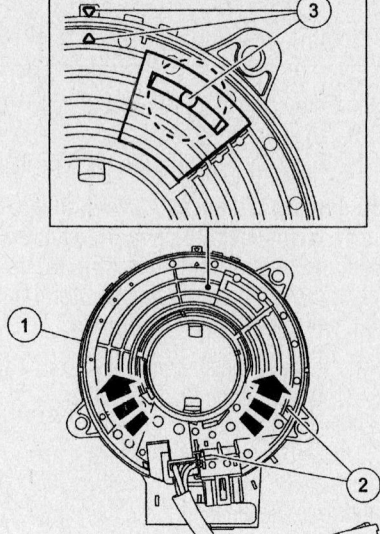

06017-ESCA-G72

Outer housing (1), rotor (2), clockspring aligning

2. If the vehicle's clockspring has rotated out of center, follow these steps to center the clockspring.
  a. Hold the clockspring outer housing stationary.

✷✷ WARNING

Overturning will destroy the clock-spring. The internal ribbon wire acts as the stop and can be broken from its internal connection.

  b. While turning the rotor clockwise, carefully feel for the ribbon wire to run out of length and for a slight resistance. Stop turning at this point.
  c. Turn the clockspring counterclock-wise until the yellow indicator shows anywhere in the window (window will be near the 1 o'clock position) and the arrow on the rotor lines up with the arrow on the top of the housing. The clockspring is now centered. Do not allow the rotor to turn from this position.

All Vehicles

1. Before servicing the vehicle, refer to the Precautions Section.

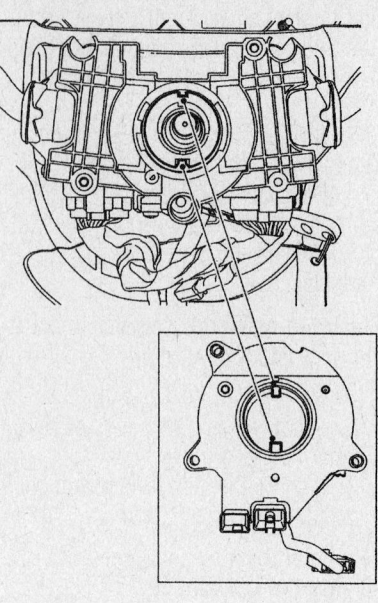

06017-ESCA-G73

Clockspring installation

2. Connect the 2 clockspring electrical connectors to the clockspring.

➡ **Slight turning of the clockspring rotor is allowable for alignment purposes to the steering column.**

3. Align the clockspring for installation.
  a. Align the large slot to the large tab in the clockspring.
  b. Align the small slot to the small tab in the clockspring.
4. Install the 3 clockspring screws.
5. For vehicles reusing a clockspring that was removed, remove the tape. For vehicles installing a new clockspring, remove the retaining pin.
6. Install the lower steering column shroud and the 3 screws.
7. Position the steering column completely downward and lock in place.
8. Install the upper steering column shroud and engage the retaining tabs.
9. Install the steering wheel. Torque the steering wheel pinion bolt to 12 Nm (9 ft. lbs.).
10. Repower the system.

FRONT SUSPENSION

Strut

REMOVAL & INSTALLATION

2002–2004

1. Before servicing the vehicle, refer to the Precautions Section.
2. Install or connect the following:
   - Negative battery cable
   - Front wheel
   - Brake hose grommet from the bracket
   - Antilock Brake System (ABS) harness from the strut assembly and move the brake hose bracket aside
   - Stabilizer bar link nut and move the bar aside
   - Strut to steering knuckle bolts and support the strut assembly
   - Upper strut nuts
   - Strut and coil spring assembly

To install:
3. Install or connect the following:
   - Strut and spring assembly. Torque the upper nuts to 59 ft. lbs. (80 Nm).
   - Lower strut assembly to the steering knuckle. Torque the lower bolts to 85 ft. lbs. (115 Nm).
   - Stabilizer bar into position. Torque the bolts to 35 ft. lbs. (48 Nm).
   - Brake hose bracket. Torque the bolts to 14 ft. lbs. (18 Nm).
   - ABS harness to the strut assembly, if equipped
   - Brake hose grommet to the bracket
   - Front wheel
   - Negative battery cable

2005–2006

➡ **Make sure the steering wheel is in the unlocked position.**

➡ **Use the hex holding feature to prevent the ball studs from turning while removing or installing the stabilizer bar link nuts.**

1. Raise and support the vehicle safely.
2. Remove the brake jounce hose clip.
3. Remove the brake jounce hose. Pull the brake jounce hose downward slightly to remove the hose from the bracket.
4. Remove the ABS sensor harness bolt.
5. Remove the stabilizer bar link nut.
6. Remove the strut-to-knuckle nuts.
7. Remove the strut-to-knuckle bolts.

1 Brake jounce hose clip	5 Strut-to-knuckle nuts
2 Brake jounce hose (LH/RH)	6 Strut-to-knuckle bolts
3 ABS sensor harness bolt	7 Strut upper bushing nuts
4 Stabilizer bar link nut	8 Strut and spring assembly

67197-ESCA-G45

Front strut and related components—2005–2006

8. Remove the strut upper bushing nuts. Reference mark the strut mounting plate nuts.
9. Remove the strut and spring assembly.

✳✳ WARNING

Do not allow the axle shaft to move outboard. Over-extension of the tripod CV joint can result in separation of internal parts, causing failure of the axle shaft.

To install:
10. To install, reverse the removal procedure.
11. Align the strut mounting plate nuts to the reference marks.
12. Check the front end alignment and adjust as necessary.

DISASSEMBLY & ASSEMBLY

1. Before servicing the vehicle, refer to the Precautions Section.
2. Install or connect the following:
   - Negative battery cable
   - Front wheel
   - Strut and spring assembly and

mount the strut assembly in a holding fixture and compress the coil spring using a suitable tool
   - Strut piston rod nut
3. Coil spring by disassembling the strut in the following sequence:
   a. Step 1: Metal sheet plate.
   b. Step 2: Upper strut mount.
   c. Step 3: Thrust bearing plate.
   d. Step 4: Thrust bearing.
   e. Step 5: Upper spring seat.
   f. Step 6: Upper spring seat isolator.
   g. Step 7: Coil spring.
   h. Step 8: Dust boot.
   i. Step 9: Rubber bump stopper.
   j. Step 10: Lower spring seat.

To install:
   - Assemble the strut assembly in the reverse order of the removal procedure
4. Install or connect the following:
   - Strut piston rod nut. Torque the nut to 76 ft. lbs. (103 Nm) and remove the assembly from the holding fixture
   - Strut and spring assembly
   - Front wheel
   - Negative battery cable

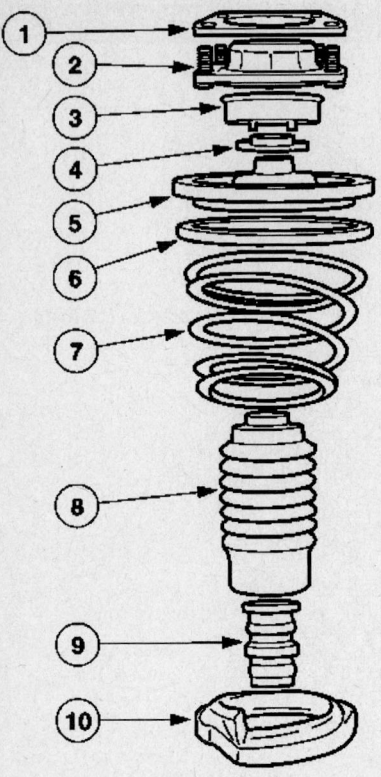

1   Metal sheet plate
2   Upper strut mount
3   Thrust bearing plate
4   Thrust bearing
5   Upper spring seat
6   Upper spring seat isolator
7   Spring
8   Dust boot
9   Rubber bump stopper
10 Lower spring seat

9308TG23

Front strut components

Stabilizer Bar

REMOVAL & INSTALLATION

1. Before servicing the vehicle, refer to the Precautions Section.
2. Raise and support the vehicle safely.
3. Remove the tire and wheel assembly.
4. Remove the stabilizer bar bushing bracket bolts.

➡**Use the hex holding feature to prevent the ball stud from turning while removing or installing the stabilizer link nut.**

5. Remove the 2 lower stabilizer bar link nuts.

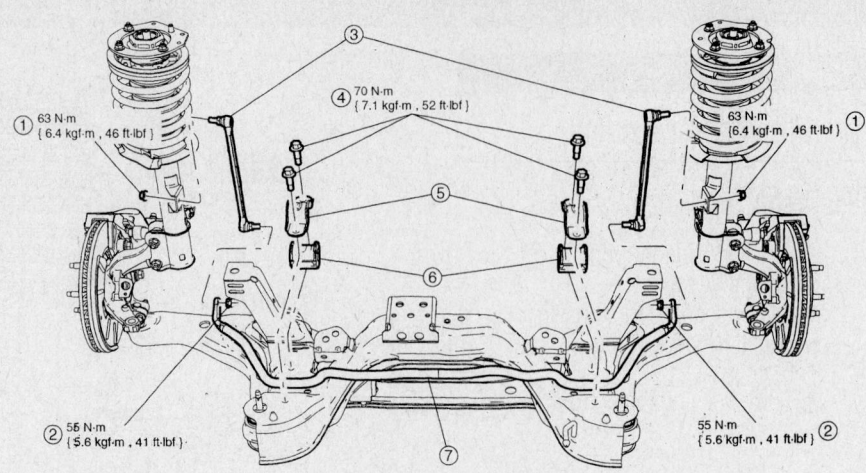

1   Stabilizer bar link nuts (upper)
2   Stabilizer bar link nuts (lower)
3   Stabilizer bar links
4   Stabilizer bar bushing bracket bolts
5   Stabilizer bar bushing bracket
6   Stabilizer bar bushing
7   Stabilizer bar

09482_TRIB_G0025

Front stabilizer bar and related components

➡**Access the stabilizer bar through the left wheel opening.**

6. Remove the stabilizer bar.

To install:

7. To install, reverse the removal procedure.

➡**When installing the stabilizer bar bushings, make sure the bushings are correctly oriented with the bushing flanges in the up position and the bushing split pointing to the rear of the vehicle.**

8. Observe the following torques:
- 2002–2004 link nuts: 35 ft. lbs. (48 Nm)
- 2005–2006 link nuts: 41 ft. lbs. (55 Nm)
- Bushing bracket bolts: 52 ft. lbs. (70 Nm)

Lower Control Arm

REMOVAL & INSTALLATION

2002–2003

1. Before servicing the vehicle, refer to the Precautions Section.
2. Raise and support the vehicle safely.
3. Remove or disconnect the following:
- Negative battery cable

- Front wheel
- Lower ball joint from the knuckle and support the subframe
- Lower control arm

To install:

4. Install or connect the following:
- Lower control arm bolts and hand tighten them
- Pinch bolt to the wheel knuckle. Torque the nut to 52 ft. lbs. (70 Nm) and remove the subframe support
- Front wheel and jounce the vehicle

5. Torque the inner lower control arm bolt to 148 ft. lbs. (200 Nm) and outer bolt 85 ft. lbs. (115 Nm).

2004–2006

1. Before servicing the vehicle, refer to the Precautions Section.
2. Raise and support the vehicle safely.
3. Remove the tire and wheel assembly.
4. Lift the lower arm with a floor jack until the vehicle starts to lift.
5. Record the ride height. It's measure from the center of the halfshaft to the fender lip.
6. Remove the floor jack.
7. Disconnect the ball joint from the knuckle.
8. Support the sub-frame and remove the lower arm.

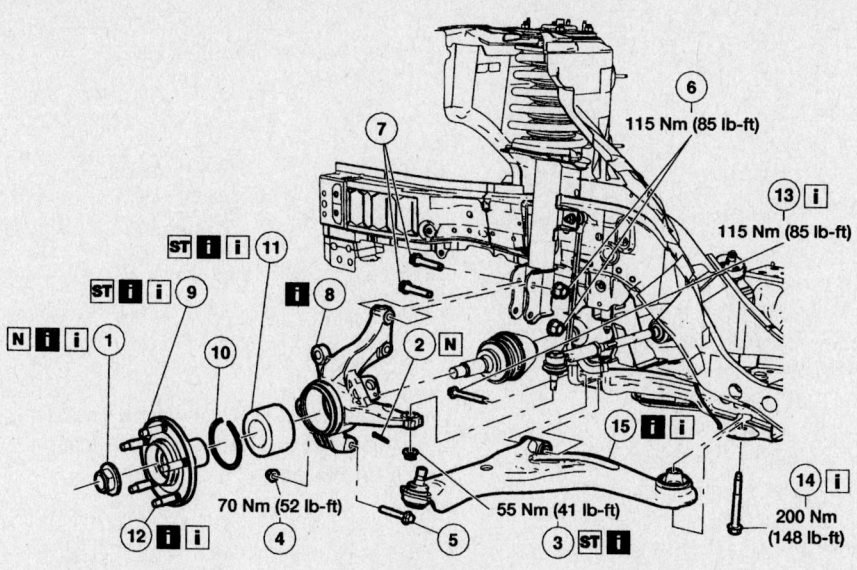

1 Wheel hub nut
2 Cotter pin
3 Tie rod end-to-knuckle nut
4 Lower ball joint pinch bolt
5 Lower ball joint pinch bolt
6 Strut-to-knuckle nuts
7 Strut-to-knuckle bolts
8 Wheel knuckle (LH/RH)
9 Wheel hub
10 Snap ring
11 Bearing
12 Wheel stud
13 Lower arm bolt (front)
14 Lower arm bolt (rear)
15 Lower arm

67197-ESCA-G48

Front lower control arm and related components—2004–2006

To install:

9. Install the lower arm, with the bolts loose.

10. Connect the ball joint. Torque the bolt to 52 ft. lbs. (70 Nm).

11. Remove the support.

12. Position the jack under the ball joint and raise the arm to the previously recorded ride height.

13. Tighten the lower arm bolts. Horizontal 85 ft. lbs. (115 Nm); vertical 148 ft. lbs. (200 Nm).

14. Tighten the wheel hub nut to 214 ft. lbs. (290 Nm) on 2004–2005 vehicles and 221 ft. lbs. (300 Nm) on 2006 vehicles.

Knuckle

REMOVAL & INSTALLATION

1. Before servicing the vehicle, refer to the Precautions Section.

2. Raise and support the vehicle safely.

3. Remove the tire and wheel assembly.

4. Remove the brake disc.

5. Remove and discard the wheel hub nut.

6. Separate the outer CV-joint spindle from the wheel hub.

7. Remove the cotter pin and the tie-rod end-to-knuckle nut.

✳✳ WARNING

Do not use a hammer to separate the tie-rod end from the wheel knuckle or damage to the wheel knuckle can result. Do not damage the tie-rod end boot while installing the special tool.

8. Separate the tie-rod from the wheel knuckle.

9. Remove the lower ball joint pinch bolt nut and the pinch bolt.

10. Remove the anti-lock brake system (ABS) wheel speed sensor bolt and position the sensor aside.

Separate the lower ball joint from the wheel knuckle.

11. Remove the 2 strut-to-knuckle nuts, bolts and the wheel knuckle.

To install:

12. Position the wheel knuckle and install the 2 strut-to-knuckle bolts and nuts. Tighten to 115 Nm (85 ft. lbs.).

13. Position and align the ball joint stud into the wheel knuckle.

14. Install the lower ball joint pinch bolt and nut. Tighten to 70 Nm (52 ft. lbs.).

15. Install the ABS wheel speed sensor and the bolt. Tighten to 9 Nm (80 inch lbs.).

16. Position the tie rod-end into the wheel knuckle and install the tie-rod end-to-knuckle nut and a new cotter pin. Tighten to 55 Nm (41 ft. lbs.).

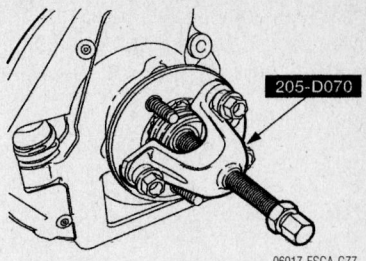

Separate the outer CV-joint spindle from the front wheel hub

06017-ESCA-G77

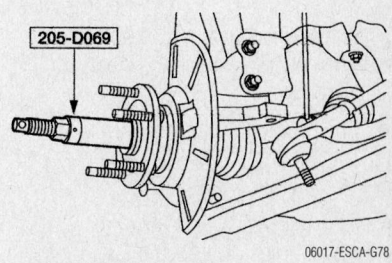

Insert the halfshaft into the front wheel hub using tool 205-D069

06017-ESCA-G78

17. Insert the halfshaft into the wheel hub.

18. Install the wheel hub nut. Tighten to 290 Nm (214 ft. lbs.).

19. Install the brake disc.

20. Check and, if necessary, align the front end.

Wheel Bearings

REMOVAL & INSTALLATION

1. Before servicing the vehicle, refer to the Precautions Section.

2. Raise and support the vehicle safely.

3. Remove the tire and wheel assembly.

➡**If removing the wheel hub, the wheel bearing must be replaced.**

4. Remove the wheel knuckle.

5. Using the special tool, press the wheel hub from the wheel bearing.

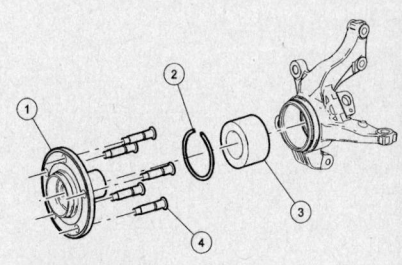

1 Wheel hub
2 Snap ring
3 Wheel bearing
4 Wheel studs (5 required)

06017-ESCA-G86

Front hub and related components

➡This step may not be necessary if the inner wheel bearing race remains in the wheel knuckle after removing the wheel hub.

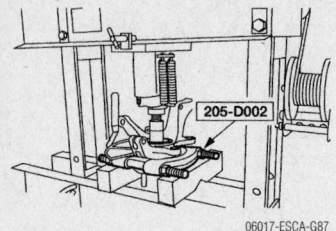

06017-ESCA-G87

Using the special tool, press the wheel hub from the wheel bearing—front hub/bearing

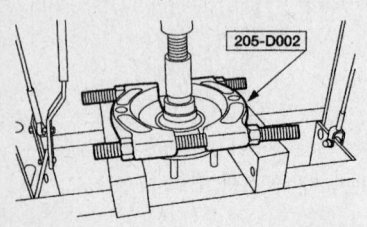

06017-ESCA-G88

Using the special tool, press the inner wheel bearing race from the wheel hub—front hub/bearing

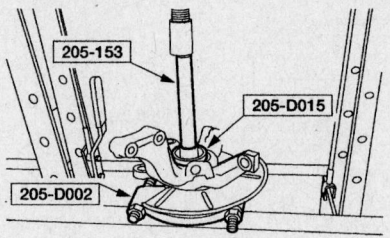

06017-ESCA-G89

Using the special tools, press the outer wheel bearing race from the wheel knuckle—front hub/bearing

6. Using the special tool, press the inner wheel bearing race from the wheel hub.

7. Remove the snapring.

8. Using the special tools, press the outer wheel bearing race from the wheel knuckle.

To install:

9. Position the wheel knuckle in a vise.

➡**Special Tool 205-278 is not seen in place. It is located behind the wheel knuckle.**

10. Using the special tools, install the wheel bearing into the wheel knuckle.

11. Install the snapring.

12. Using the special tool, press the wheel hub into the wheel bearing.

13. Install the knuckle.

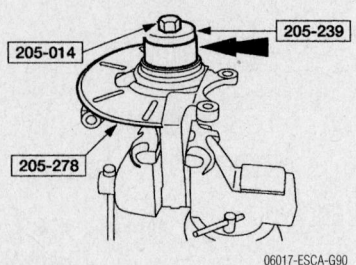

06017-ESCA-G90

Using the special tools, install the wheel bearing into the wheel knuckle—front hub/bearing

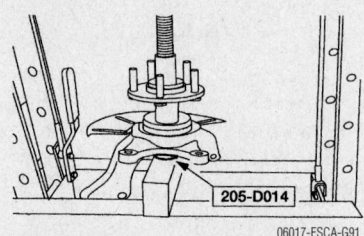

06017-ESCA-G91

Using the special tool, press the wheel hub into the wheel bearing—front hub/bearing

REAR SUSPENSION

Shock Absorber

REMOVAL & INSTALLATION

2002–2004

1. Before servicing the vehicle, refer to the Precautions Section.

2. Remove or disconnect the following:
   - Negative battery cable
   - Rear quarter trim panel
   - Upper shock absorber nut and raise the vehicle enough to relax the suspension
   - Lower shock absorber nut
   - Shock absorber

To install:

3. Install or connect the following:
   - Shock absorber. Torque the lower nut to 85 ft. lbs. (115 Nm).
   - Upper shock absorber nut. Torque the nut to 13 ft. lbs. (18 Nm).
   - Rear quarter trim panel
   - Negative battery cable

2005–2006

1. Before servicing the vehicle, refer to the Precautions Section.

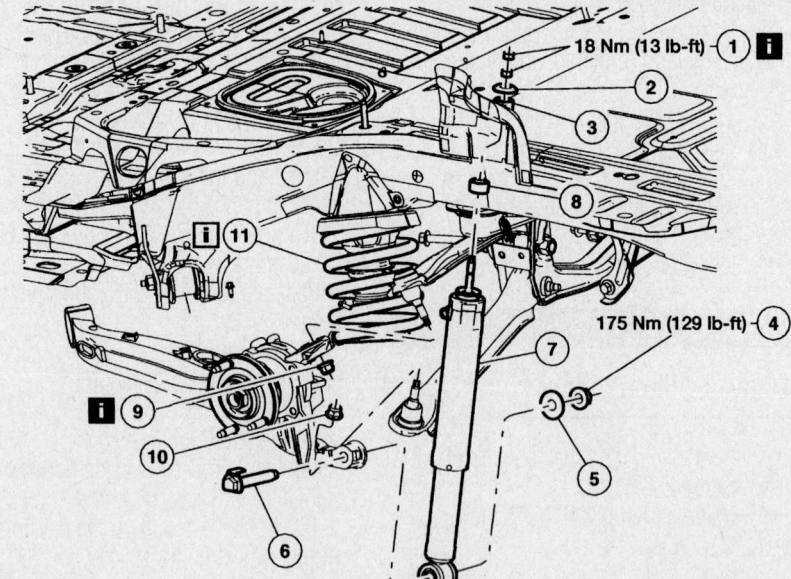

1	Upper shock absorber nuts	5	Washer	9	Upper ball joint nut
2	Washer	6	Lower shock absorber bolt	10	Lower ball joint nut
3	Bushing	7	Shock absorber	11	Coil spring
4	Lower shock absorber nut	8	Bushing		

67197-ESCA-G46

Rear shock absorber and related components—2005–2006

2. Raise and support the vehicle safely.

3. Remove the tire and wheel assembly.

4. Remove the rear quarter trim panel. Remove the upper shock absorber nut, bushing and washer.

5. Remove the lower shock absorber nut, bolt and washer.

6. Remove the shock absorber and bushing.

7. To install, reverse the removal procedure. Torque the upper nut to 13 ft. lbs.; the lower nut to 129 ft. lbs. (175 Nm).

Coil Spring

REMOVAL & INSTALLATION

2002–2004

1. Before servicing the vehicle, refer to the Precautions Section.

2. Raise and support the vehicle safely.

3. Remove the tire and wheel assembly.

4. Remove or disconnect the following:

- Wheel and install 1 lug nut to retain the brake drum
- Brake line from the wheel cylinder
- Brake line bracket
- Bolts from the Antilock Braking System (ABS) sensor bracket and move the sensor aside, if equipped
- Rear knuckle and loosen the inside upper and lower arm bolts
- Shock absorber lower nut
- Spring

To install:

5. Install or connect the following:

- Spring to the shock absorber
- Lower shock absorber nut. Torque the nut to 85 ft. lbs. (115 Nm).
- Inside upper and lower arm bolts. Torque the bolts to 85 ft. lbs. (115 Nm).
- ABS sensor bracket, if equipped. Torque the bolts to 80 inch lbs. (9 Nm).
- Brake line bracket. Torque the bolt to 15 ft. lbs. (20 Nm).
- Brake line to the wheel cylinder. Torque the fastener to 11 ft. lbs. (15 Nm) and remove the lug nut
- Wheel
- Negative battery cable

2005–2006

1. Before servicing the vehicle, refer to the Precautions Section.

2. Raise and support the vehicle safely.

3. Remove the tire and wheel assembly.

4. Remove the lower shock absorber nut and washer

5. Remove the lower shock absorber bolt

6. Support the wheel knuckle.

7. Remove the upper ball joint nut.

8. Remove the upper arm inner bolt.

9. Loosen the lower arm inner bolt.

➡**Note the position of the coil spring insulator and coil spring for installation.**

10. Carefully lower the wheel knuckle support.

11. Remove the coil spring.

To install:

12. To install, reverse the removal procedure.

13. Align the coil spring and coil spring insulator to the previously noted position.

Lower Control Arm

REMOVAL & INSTALLATION

1. Before servicing the vehicle, refer to the Precautions Section.

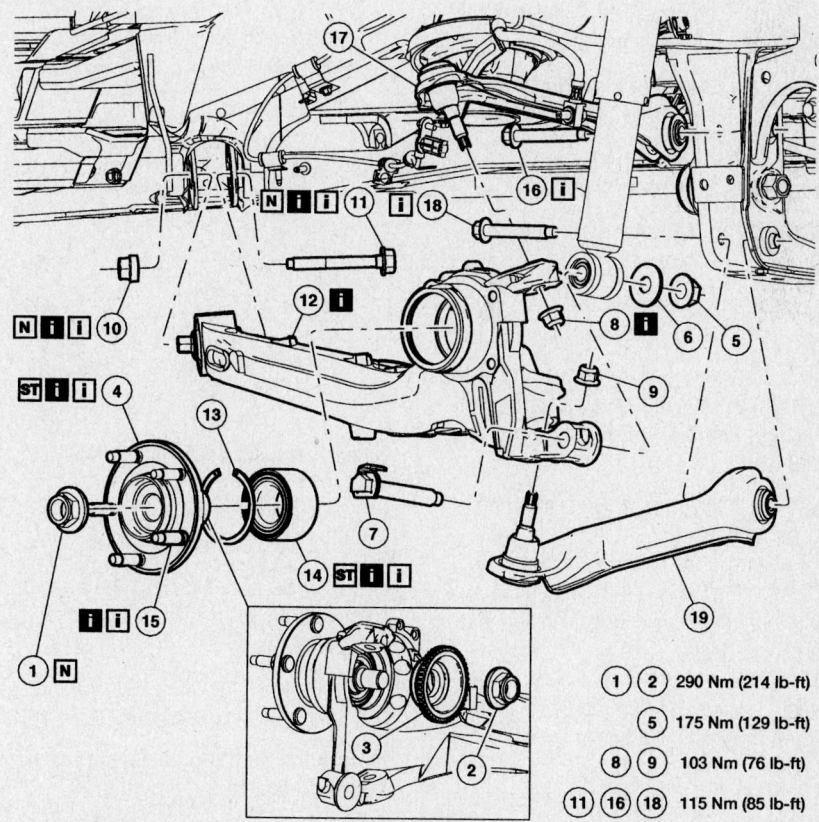

2. Raise and support the vehicle safely.

3. Remove the tire and wheel assembly.

4. Remove or disconnect the following:

- Lower ball joint from the knuckle while holding the ball joint stud from moving
- Lower ball joint nut
- Lower control arm
- Lower control arm inner bolt

To install:

5. Install or connect the following:

- Lower control arm inner bolt
- Lower control arm.
- Lower ball joint nut
- Lower ball joint the knuckle.
- Rear wheel and tire assembly

Upper Control Arm

REMOVAL & INSTALLATION

1. Before servicing the vehicle, refer to the Precautions Section.

2. Raise and support the vehicle safely.

1	2	290 Nm (214 lb-ft)
5		175 Nm (129 lb-ft)
8	9	103 Nm (76 lb-ft)
11	16 18	115 Nm (85 lb-ft)

1 Wheel hub nut (4WD)	8 Upper ball joint nut
2 Wheel hub nut (2WD)	9 Lower ball joint nut
3 ABS sensor ring (2WD)	10 Cam nut
4 Wheel hub	11 Wheel knuckle bolt
5 Lower shock absorber nut	12 Knuckle assembly (LH/RH)
6 Washer	13 Wheel bearing snap ring
7 Lower shock absorber bolt	14 Wheel bearing

15 Wheel stud
16 Upper arm inner bolt
17 Upper arm
18 Lower arm inner bolt
19 Lower arm

Rear lower control arm and related components

67197-ESCA-G47

3. Remove the tire and wheel assembly.

➡ **It may be necessary to hold the ball joint stud to keep it from turning while removing the nut.**

4. Separate the upper arm from the wheel knuckle. Remove the upper ball joint nut.
5. Remove the upper arm inner bolt.
6. Remove the upper arm.
7. To install, reverse the removal procedure. Observe the following torques:
   - Ball joint nut: 76 ft. lbs. (103 Nm)
   - Lower arm bolts: 85 ft. lbs. (115 Nm)

Rear Knuckle

REMOVAL & INSTALLATION

2002–2004

2WD

1. Before servicing the vehicle, refer to the Precautions Section.
2. Raise and support the vehicle safely.
3. Remove the tire and wheel assembly.

➡ **When a new wheel knuckle is installed a new wheel bearing must also be installed.**

4. Remove the brake shoes.

➡ **If equipped, remove ABS sensor ring.**

5. Remove the wheel hub nut.
6. Remove the wheel hub.

➡ **This step may not be necessary if the inner wheel bearing race remains in the wheel knuckle, after removing the wheel hub.**

7. Press the inner wheel bearing race from the wheel hub.
8. Remove the snapring.
9. Vehicles with ABS: Remove the ABS sensor bolt and bracket bolts. Position the ABS sensor aside.
10. Disconnect the parking brake cable from the brake backing plate. Remove the parking brake cable bolt from the wheel knuckle.
11. Disconnect the brake line from the wheel cylinder. Remove the brake line bracket bolt.
12. Support the wheel knuckle.
13. Remove the lower shock absorber nut.

➡ **It may be necessary to hold the ball joint stud to keep it from turning while removing the nut.**

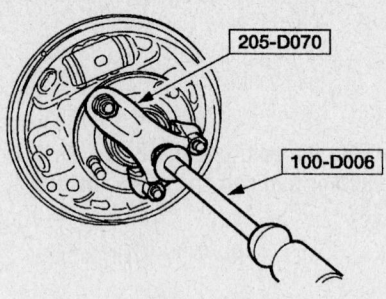

06017-ESCA-G79

Remove the rear wheel hub using special tools

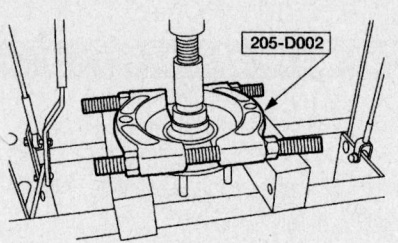

06017-ESCA-G80

Press the inner wheel bearing race from the rear wheel hub

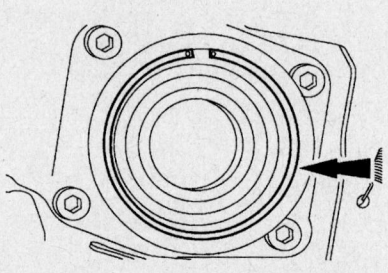

06017-ESCA-G81

Rear wheel hub Snapring location

14. Remove the nut. Separate the lower ball joint from the wheel knuckle.

➡ **It may be necessary to hold the ball joint stud to keep it from turning while removing the nut.**

15. Remove the nut. Separate the upper ball joint from the wheel knuckle.

➡ **Note the position of the spring insulator and spring for installation.**

16. Remove the spring. Lower the support to the wheel knuckle.
17. Mark the position of the adjustment cam notch.
18. Remove the wheel knuckle cam bolt and nut.
19. Remove the wheel knuckle cam.
20. Remove the wheel knuckle.

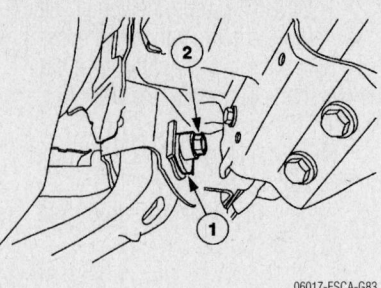

06017-ESCA-G83

Cam notch adjustment (1) cam notch (2) wheel knuckle cam bolt and nut

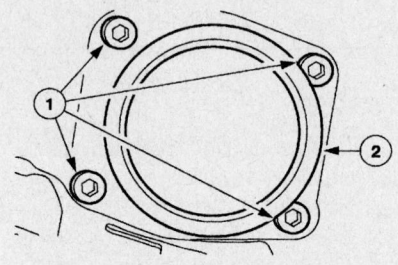

06017-ESCA-G84

Remove the brake backing plate from the wheel knuckle

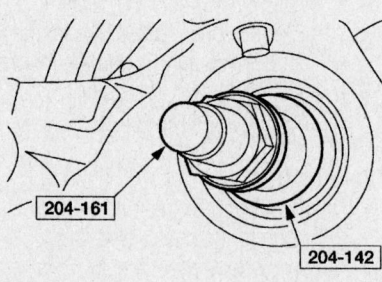

06017-ESCA-G85

Install the rear wheel hub into the wheel bearing using special tools

21. Remove the brake backing plate from the rear wheel knuckle.
   To install:
22. Install a new wheel bearing into the wheel knuckle.
23. Install the snapring into the wheel knuckle.
24. Position the brake backing plate to the wheel knuckle and install the bolts. Torque to 49 ft. lbs. (66 Nm).
25. Using the wheel hub washer and the special tool, install the wheel hub into the wheel bearing.

➡ **If equipped, install the ABS sensor ring.**

26. Install the wheel hub nut. Torque to 214 ft. lbs. (290Nm).

27. Position the wheel knuckle and install the wheel knuckle cam.

28. Install the wheel knuckle cam bolt and nut loosely.

29. Support the wheel knuckle and install the spring.

30. Install the lower shock absorber nut. Torque to 85 ft. lbs. (115 Nm).

31. Position the upper ball joint and install the nut. Torque to 76 ft. lbs. (103 Nm).

32. Position the lower ball joint and install the nut. Torque to 76 ft. lbs. (103 Nm).

33. Tighten the wheel knuckle cam bolt and nut.

    a. Align the wheel knuckle cam.

    b. Tighten the wheel knuckle cam bolt and nut to 85 ft. lbs. (115 Nm).

34. Connect the brake line to the wheel cylinder. Install the brake line bracket bolt. Torque the brake line-to-wheel cylinder to 13 ft. lbs. (17 Nm). Torque the 16 ft. lbs. (22 Nm).

35. Install the parking brake cable through the brake backing plate. Torque the bolt to 16 ft. lbs. (22 Nm).

36. Vehicles with ABS: Position the ABS sensor and bracket bolts. Install the bolts. Torque to 80 inch lbs. (9 Nm).

37. Install the brake shoes.

38. Bleed the brake system.

39. Check and adjust the wheel alignment as necessary.

4WD

1. Before servicing the vehicle, refer to the Precautions Section.

2. Raise and support the vehicle safely.

3. Remove the tire and wheel assembly.

➡**When a new wheel knuckle is installed a new wheel bearing must also be installed.**

⁕⁕ WARNING

Use of a brake drum puller or a torch is not recommended. Brake drum distortion can occur.

➡**If the brake drum is rusted to the axle shaft pilot diameter, tap the center of the brake drum between the wheel studs.**

4. Remove the brake drum.

5. If the brake drum will not come off, follow these steps.

    a. Move the brake shoe adjusting lever off the brake adjuster screw.

    b. Loosen the brake shoe adjuster screw nut by adjusting the nut upward.

6. Remove the rear halfshaft nut.

7. Loosen the halfshaft from the wheel hub.

8. Remove the wheel hub.

➡**This step may not be necessary if the inner wheel bearing race remains in the wheel knuckle, after removing the wheel hub.**

9. Press the inner wheel bearing race from the wheel hub.

10. Remove the snapring.

11. Vehicles with ABS: Remove the ABS sensor bolt and bracket bolts. Position the ABS sensor aside.

12. Disconnect the parking brake cable. Remove the parking brake cable bolt from the wheel knuckle.

13. Remove brake line bracket bolt.

14. Support the wheel knuckle.

15. Remove the lower shock absorber nut.

➡**It may be necessary to hold the ball joint stud to keep it from turning while removing the nut.**

16. Disconnect the lower ball joint.

➡**It may be necessary to hold the ball joint stud to keep it from turning while removing the nut.**

17. Disconnect the upper ball joint.

➡**Note the position of the spring insulator and spring for installation.**

18. Remove the spring. Lower the support to the wheel knuckle.

19. Mark the position of the adjustment cam notch.

20. Remove the wheel knuckle cam bolt and nut.

21. Remove the wheel knuckle cam.

22. Remove the wheel knuckle.

23. Remove the brake backing plate from the wheel knuckle.

To install:

24. Using the special tools, install a new wheel bearing into the wheel knuckle.

25. Install the snapring into the wheel knuckle.

26. Position the wheel knuckle and install the wheel knuckle cam.

27. Install the wheel knuckle cam bolt loosely.

➡**Brake shoes shown removed for clarity.**

28. Install the brake backing plate into the wheel knuckle. Install the bolts. Torque to 49 ft. lbs. (66 Nm).

29. Install the wheel hub.

30. Position the halfshaft into the wheel hub.

31. Support the wheel knuckle and install the spring.

32. Install the lower shock absorber nut. Torque to 85 ft. lbs. (115 Nm).

33. Remove the support from the wheel knuckle.

34. Install the upper ball joint. Install the nut. Torque to 76 ft. lbs. (103 Nm).

35. Install the lower ball joint. Install the nut. Torque to 76 ft. lbs. (103 Nm).

36. Align the wheel knuckle cam.

37. Tighten the wheel knuckle cam bolt and nut to 85 ft. lbs. (115 Nm).

38. Install brake line bracket bolt. Torque to 16 ft. lbs. (22 Nm).

39. Install the parking brake cable bolt. Torque to 16 ft. lbs. (22 Nm).

40. Vehicles with ABS: Position the ABS brake sensor and bracket bolts. Install the bolts. Torque to 80 inch lbs. (9 Nm).

41. Install the halfshaft end into the hub assembly.

42. Install the halfshaft nut. Torque to 214 ft. lbs. (290 Nm).

43. Install the brake drum.

44. Install the wheel and tire assembly.

45. Check and adjust the wheel alignment as necessary.

2005–2006

1. Before servicing the vehicle, refer to the Precautions Section.

2. Raise and support the vehicle safely.

3. Remove the tire and wheel assembly.

4. Drum brake vehicles:

    a. Remove the brake shoes.

    b. Disconnect the parking brake cable from the brake backing plate. Remove the parking brake cable from the brake backing plate.

    c. Disconnect the brake line from the wheel cylinder and remove the brake line bracket bolt.

5. Disc brake vehicles, remove the parking brake shoes.

6. Remove and discard the wheel hub nut.

⁕⁕ WARNING

Do not use a hammer to separate the outer constant velocity (CV) joint from the hub.

7. Damage to the threads and internal CV joint components can result.

8. With 4wd, separate the outer CV-joint from the wheel hub.

9. Remove the anti-lock brake system (ABS) wheel speed sensor bolt and the 2 ABS wheel speed sensor wire bolts.

10. Remove and position the wheel speed sensor and harness aside.
11. Remove the coil spring.
12. Remove the lower ball joint nut.

➡**The joint surfaces must be clean. Clean the general area of the joint to prevent debris from entering the joint. Clean using only mild liquids.**

13. Reference mark the notch on the cam nut adjustment cam.
14. Remove and discard the wheel knuckle bolt.
15. Remove and discard the cam nut.
16. Remove the wheel knuckle.

To install:

❊❊ WARNING

The joint area must be free of debris to ensure correct clamping.

➡**The joint surfaces and the bushing sleeve serrations must be clean before assembly.**

17. Clean the joint surfaces and the bushing sleeve serrations with a wire brush.

➡**Align the notch on the cam nut with the reference marks.**

18. Position the wheel knuckle and install a new wheel knuckle bolt and cam nut.
19. Using a suitable tool, hold the cam nut stationary while tightening the wheel knuckle bolt. Tighten to 150 Nm (111 ft. lbs.).
20. Position the ABS wheel speed sensor harness and the sensor.
21. Position the lower ball joint into the wheel knuckle and install the lower ball joint nut. Tighten to 103 Nm (76 ft. lbs.).
22. Install the coil spring.
23. Install the ABS wheel speed sensor bolt and the 2 ABS wheel speed sensor wire bolts. Tighten to 9 Nm (80 inch lbs.).
24. With 4wd, install the outer CV joint into the wheel hub.
25. Install a new wheel hub nut. Tighten to 290 Nm (214 ft. lbs.).
26. Install the brake shoes.
27. Connect the brake line to the wheel cylinder. Tighten to 17 Nm (13 ft. lbs.).
28. Install the brake line bracket bolt. Tighten to 22 Nm (16 ft. lbs.).
29. Connect the parking brake cable to the brake backing plate and install the parking brake cable bracket bolt. Tighten to 23 Nm (17 ft. lbs.).
30. Install the parking brake shoes.
31. Check and adjust the wheel alignment as necessary.

Wheel Bearings

REMOVAL & INSTALLATION

2WD

1. Before servicing the vehicle, refer to the Precautions Section.

2. Raise and support the vehicle safely.
3. Remove the tire and wheel assembly.
4. Remove or disconnect the following:
   • Rear brake drum
   • Wheel hub nut
   • Wheel hub
   • Inner wheel bearing race from the hub

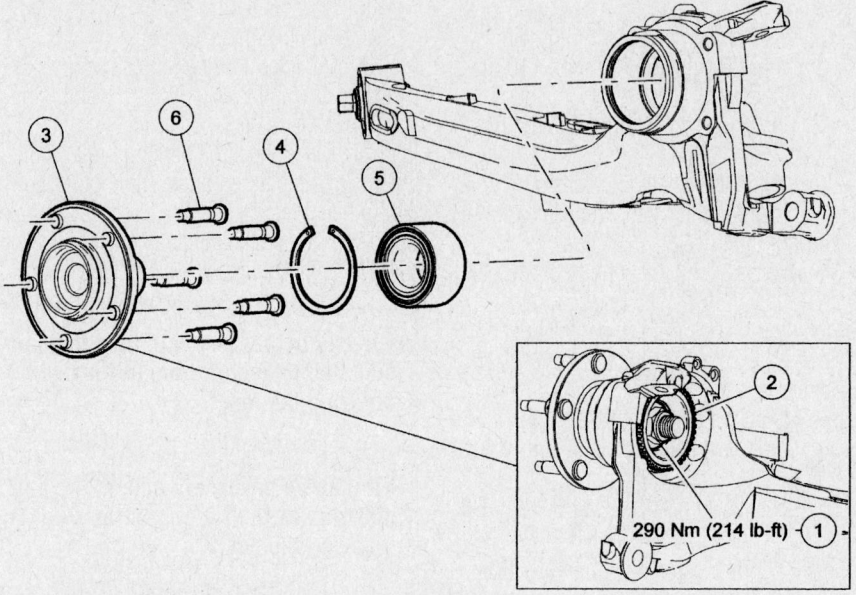

1 Wheel hub nut (FWD)	3 Wheel hub
2 Anti-lock brake system (ABS) wheel speed sensor ring (FWD)	4 Wheel bearing snap ring
	5 Wheel bearing
	6 Wheel studs

06017-ESCA-G92

Rear hub and related components

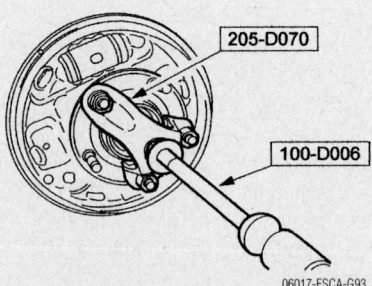

06017-ESCA-G93

Rear hub removal using special tools

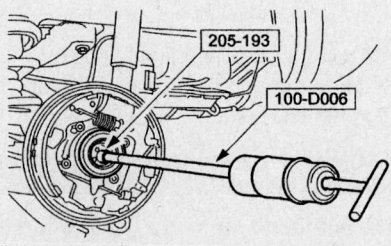

06017-ESCA-G95

Rear wheel bearing removal—2WD

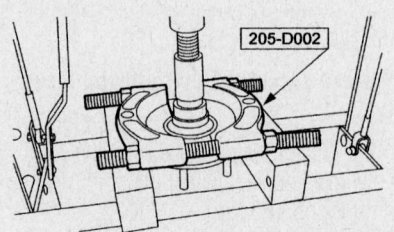

06017-ESCA-G94

Inner wheel bearing removal—rear hub/bearing

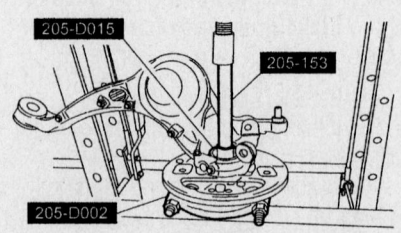

06017-ESCA-G96

Rear wheel bearing removal—4WD

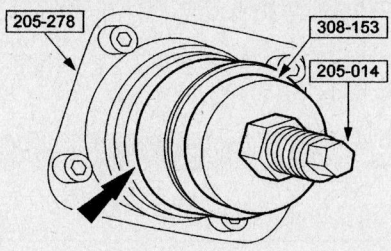

Rear wheel bearing installation using special tools

06017-ESCA-G97

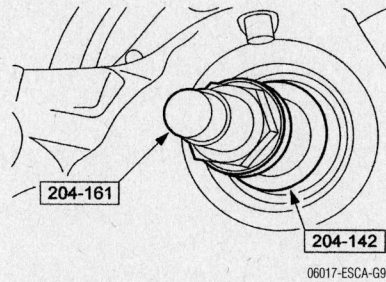

Rear hub installation using special tools

06017-ESCA-G98

- Snapring
- Wheel bearing outer race from the knuckle

To install:

5. Install or connect the following:
- Wheel bearing in to the knuckle
- Snapring
- Wheel hub into the wheel bearing

- Wheel hub nut. Torque the nut to 214 ft. lbs. (290 Nm).
- Brake drum
- Rear wheel

4WD

1. Before servicing the vehicle, refer to the Precautions Section.
2. Raise and support the vehicle safely.
3. Remove the tire and wheel assembly.
4. Remove or disconnect the following:
- Rear brake shoes
- Rear halfshaft nut and loosen the halfshaft from the hub
- Wheel hub and place it in a vise
- Inner wheel bearing race from the hub
- Antilock Brake System (ABS) sensor bracket and move the sensor aside, if equipped
- Parking brake cable from the steering knuckle
- Brake line from the wheel cylinder and support the knuckle
- Lower shock absorber nut
- Lower ball joint by holding the ball joint stud
- Upper ball joint
- Coil spring while noting the location of the insulator
- Steering knuckle cam
- Steering knuckle
- Snapring and press out the outer wheel bearing race from the knuckle

To install:

5. Install or connect the following:
- New wheel bearing into the steering knuckle
- Snapring to the knuckle
- Wheel hub
- Steering knuckle cam and hand tighten the bolt
- Coil spring
- Shock absorber lower nut. Torque the nut to 85 ft. lbs. (115 Nm) for 2002–04 models; 129 ft. lbs. (175 Nm).
- Upper ball joint. Torque the nut to 76 ft. lbs. (103 Nm).
- Lower ball joint. Torque the nut to 76 ft. lbs. (103 Nm). Align the steering knuckle cam and torque the bolt to 85 ft. lbs. (115 Nm).
- Brake line to the wheel cylinder. Torque the brake line bracket bolt to 15 ft. lbs. (20 Nm) and the brake line fastener to 11 ft. lbs. (15 Nm).
- Parking brake cable to the backing plate. Torque the bolt to 16 ft. lbs. (22 Nm).
- ABS sensor bracket. Torque the bolt to 80 inch lbs. (9 Nm), if equipped
- Halfshaft nut. Torque the nut to 214 ft. lbs. (290 Nm).
- Brake shoes
- Rear wheel

6. Fill and bleed the brake system.
7. Check and adjust the wheel alignment as needed.

BRAKES

Brake Caliper

REMOVAL & INSTALLATION

Front

2002–2004

1. Before servicing the vehicle, refer to the Precautions Section.
2. Raise and support the vehicle safely.
3. Remove the wheel and tire assembly.
4. Remove the brake caliper clip.
5. Remove the brake caliper bolt caps and bolts.
6. Position the caliper aside.
7. Disconnect and cap the brake line from the caliper and remove the caliper.

To install:

8. Installation is the reverse of the removal procedure. Torque the mounting bolts to 26 ft. lbs. (35Nm). Torque the brake line to 15 ft. lbs. (20Nm).
9. Bleed the brake system.

2005–2006

1. Before servicing the vehicle, refer to the Precautions Section.
2. Raise and support the vehicle safely.
3. Remove the wheel and tire assembly.
4. Remove the brake caliper clip.
5. Remove the brake caliper dust boot caps.
6. Remove the brake caliper guide bolts.
7. Remove the brake caliper.

✳✳ CAUTION

Do not allow the brake caliper to hang by the flexible brake hose.

8. Remove the disc brake pads.
9. Remove the brake caliper jounce hose. Loosen the jounce hose fitting prior to removing the brake caliper.

To install:

10. Installation is the reverse of the removal procedure.

11. Torque the caliper pin bolts to 26 ft. lbs. (35 Nm) on disc/drum systems; 33 ft. lbs. (45 Nm) on 4-wheel disc systems.
12. If the hydraulic system was opened, bleed the brake system.

➡**Thread the brake caliper jounce hose onto the brake caliper before installing the brake caliper.**

➡**Make sure that the brake caliper jounce hose is not twisted.**

13. Position the brake caliper to the anchor plate and tighten the brake caliper jounce hose.

Rear

1. Before servicing the vehicle, refer to the Precautions Section.
2. Raise and support the vehicle safely.
3. Remove the wheel and tire assembly.
4. Remove the brake caliper guide bolts.
5. Remove the caliper.

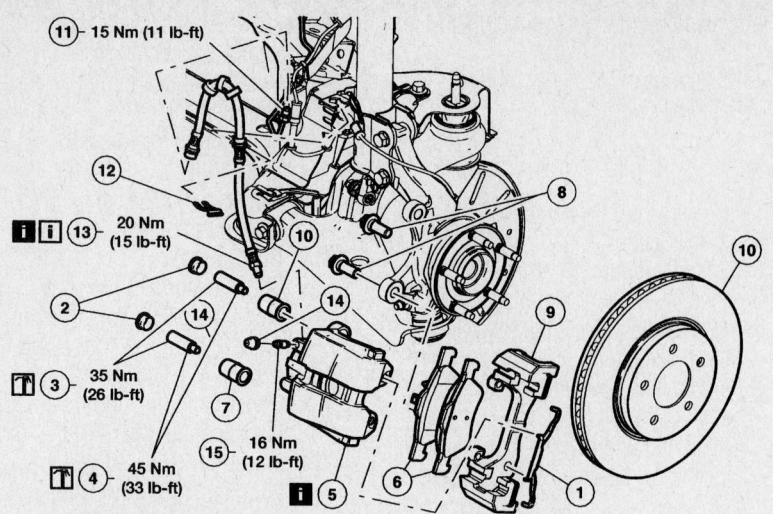

Front caliper and related components—2005–2006

1 Brake caliper clip	9 Brake caliper anchor plate
2 Brake caliper dust boot caps	10 Brake disc
3 Brake caliper guide bolts (disc-drum system)	11 Brake line fitting nut
4 Brake caliper guide bolts (4-wheel disc brake system)	12 Brake caliper jounce hose retaining clip
5 Brake caliper (RH/LH)	13 Brake caliper jounce hose
6 Disc brake pads (kit)	14 Bleeder screw cap
7 Brake caliper dust boots	15 Bleeder screw
8 Brake caliper anchor plate bolts	

67197-ESCA-G49

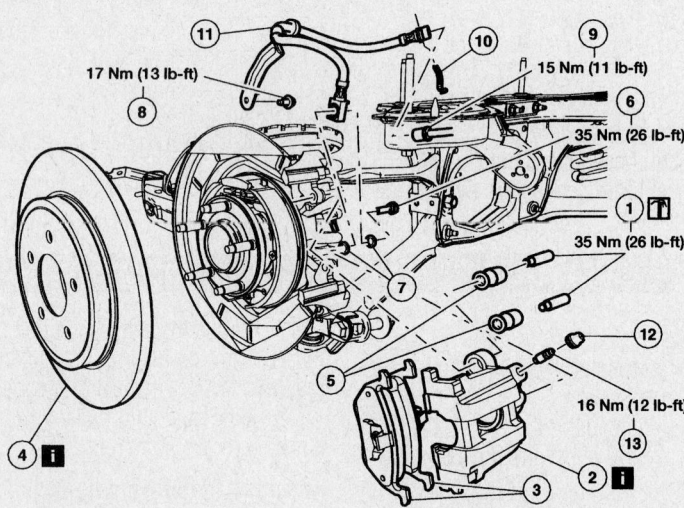

Rear caliper and related components—2005–2006

1 Brake caliper guide bolts	8 Brake caliper jounce hose bracket bolt
2 Caliper (RH/LH)	9 Brake line fitting nut
3 Brake disc pads	10 Brake caliper jounce hose retaining clip
4 Brake disc	11 Jounce hose (RH/LH)
5 Brake caliper guide bolt	12 Bleeder screw cap
6 Brake caliper hose flow bolt	13 Bleeder screw
7 Copper washers	

67197-ESCA-G50

✳✳ CAUTION

Do not allow the brake caliper to hang by the flexible brake hose.

6. Remove the brake disc pads.
7. Remove the brake caliper hose flow bolt.

8. Remove and discard the copper washers.
To install:
9. Installation is the reverse of the removal procedure. Use new copper washers.
10. Torque the caliper pin bolts to 26 ft.

lbs. (35 Nm); torque the flow bolt to 26 ft. lbs. (35 Nm).
11. If the hydraulic system was opened, bleed the brake system.

Brake Pads

REMOVAL & INSTALLATION

Front

1. Before servicing the vehicle, refer to the Precautions Section.
2. Raise and support the vehicle safely.
3. Remove the wheel and tire assembly.
4. Remove the brake caliper clip.
5. Position the caliper aside. Do not allow the caliper to hang with being supported.
6. Remove brake caliper bolt caps and the bolts.
7. Position the caliper aside and support.
8. Remove the brake pads.
9. Remove the outer brake pad from the anchor.
10. Remove the inner brake pad from the caliper piston.
To install:
11. Installation is the reverse of the removal procedure.
12. If the hydraulic system was opened, bleed the brake system.

Rear

1. Before servicing the vehicle, refer to the Precautions Section.
2. Raise and support the vehicle safely.
3. Remove the wheel and tire assembly.
4. Remove the brake caliper guide bolts.
5. Remove the caliper.

✳✳ CAUTION

Do not allow the brake caliper to hang by the flexible brake hose.

6. Remove the brake disc pads.
To install:
7. Installation is the reverse of the removal procedure.
8. If the hydraulic system was opened, bleed the brake system.

Brake Rotor

REMOVAL & INSTALLATION

Front

1. Before servicing the vehicle, refer to the Precautions Section.

2. Raise and support the vehicle safely.
3. Remove the wheel and tire assembly.
4. Remove the brake caliper anchor plate.
5. Remove the brake disc retaining clips (if equipped) and the brake disc.
6. To install, reverse the removal procedure. Torque the anchor plate bolts to 111 ft. lbs. (150Nm).

Rear

1. Before servicing the vehicle, refer to the Precautions Section.
2. Raise and support the vehicle safely.
3. Remove the wheel and tire assembly.
4. Remove the caliper.
5. Remove the rotor.
6. Installation is the reverse of removal.

Brake Drum

REMOVAL & INSTALLATION

1. Before servicing the vehicle, refer to the Precautions Section.
2. Raise and support the vehicle safely.
3. Remove the wheel and tire assembly.

✳✳ CAUTION

Use of a brake drum puller or a torch is not recommended. Brake drum distortion can result.

➡If the brake drum is rusted to the axle shaft pilot diameter, tap the center of the brake drum between the wheel studs.

4. Remove the brake drum.
5. If equipped, remove the brake drum retaining clips.
6. If the brake drums will not come off, follow these steps.
7. Move the brake shoe adjusting lever off the brake adjuster screw.
8. Loosen the brake shoe adjuster screw nut by adjusting the nut upward.
9. Using the special tool, 134-R0191, measure the brake drum inside diameter.
10. Install a new brake drum if the maximum inside diameter exceeds specification.
To install:

✳✳ WARNING

Whenever a wheel is installed, always remove any corrosion, dirt or foreign material present on the mounting surfaces of the wheel or the surface of the wheel hub, brake drum or brake disc that contacts the wheel. Installing wheels without cor-

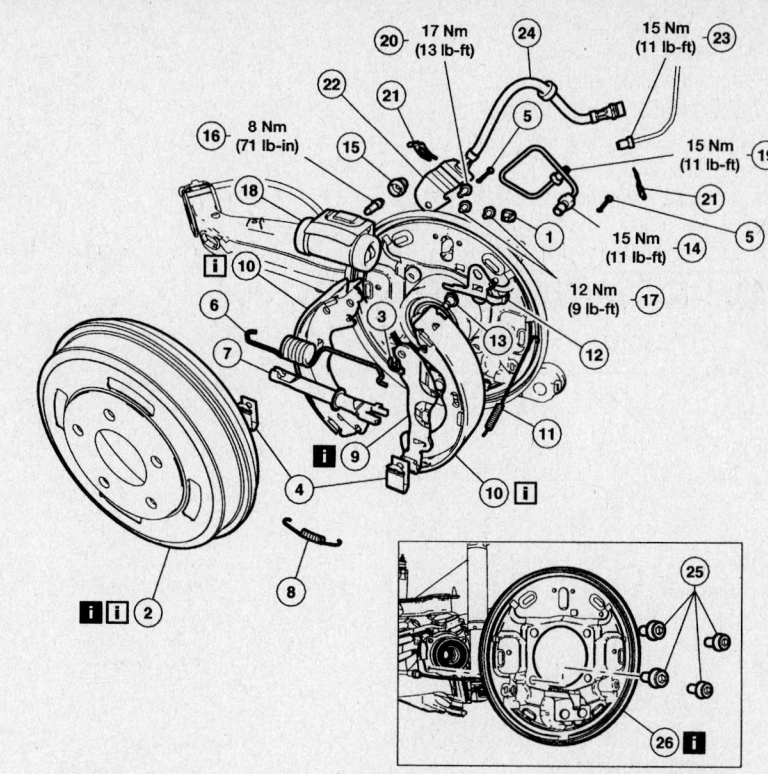

1 Plug	10 Brake shoe (kit)
2 Brake drum	11 Adjuster lever spring
3 Parking brake lever clip	12 Adjuster lever (LH/RH)
4 Brake shoe retaining clips	13 Pivot pin (part of 2200)
5 Brake shoe retaining pins	14 Brake line fitting nut
6 Upper return spring	15 Bleeder screw cap
7 Adjuster assembly (LH/RH)	16 Bleeder screw
8 Lower return spring	17 Wheel cylinder bolts
9 Parking brake actuator lever (LH/RH)	18 Wheel cylinder
	19 Brake line fitting nut
	20 Jounce hose bracket bolt
	21 Jounce hose retaining clips
	22 Jounce hose bracket
	23 Brake line fitting nut
	24 Jounce hose (LH/RH)
	25 Backing plate bolts
	26 Backing plate

67197-ESCA-G51

Drum brake and related components

rect metal-to-metal contact at the wheel mounting surfaces can cause the wheel nuts to loosen and the wheel to come off while the vehicle is in motion, causing loss of control. Failure to follow these instructions may result in personal injury.

11. Clean the wheel hub mounting surface and wheel pilot.
12. Install the tire and wheel assembly.

Brake Shoes

REMOVAL & INSTALLATION

1. Before servicing the vehicle, refer to the Precautions Section.
2. Raise and support the vehicle safely.
3. Remove the wheel and tire assembly.
4. Remove the brake drum.
5. Use the Brake/Clutch/Service Vacuum to remove brake dust and dirt from the brake assemblies.

➡If new rear brake shoes and linings are being installed, resurface the brake drums to remove glazing and to ensure an equal friction surface from side-to-side. Resurfacing will also correct out-of-round and bell conditions.

6. Using the special tool, measure the braking surface diameter. If the inside diameter measures more than the maximum specification shown on the outside of the brake drum, install a new brake drum.
7. Remove the parking brake cable from the parking brake cable lever.
8. Remove the hold-down clips and pins.
9. Remove the lower spring.
10. Remove the rear brake shoes.
11. Pull the bottom of the brake shoe forward.
12. Release the upper return spring.
13. Remove both brake shoes together.
14. Remove the self adjuster lever.
15. Remove the self adjuster and spring assembly.
16. Return the self adjuster to the fully seated position.

17. Remove the parking brake lever.
18. Remove the horseshoe clip.
19. Remove the parking brake lever.
20. Inspect the rear brake shoes for minimum thickness above the backing plate, and install new as necessary.
21. To install, reverse the removal procedure.

Wheel Cylinder

REMOVAL & INSTALLATION

1. Before servicing the vehicle, refer to the Precautions Section.
2. Raise and support the vehicle safely.
3. Remove the wheel and tire assembly.
4. Remove the brakes shoes.
5. Disconnect the brake line at the wheel cylinder.
6. Remove the wheel cylinder bolts.

To install:

7. Installation is the reverse of the removal procedure.
8. Torque the bolts to 9 ft. lbs. (12 Nm).
9. Bleed the brake system.
10. Correct and adjust the brake fluid level.

Parking Brake Shoes

REMOVAL & INSTALLATION

1. Before servicing the vehicle, refer to the Precautions Section.
2. Raise and support the vehicle safely.
3. Remove the wheel and tire assembly.
4. Remove the rear brake disc.
5. Remove the parking brake shoe upper return spring.
6. Remove the 2 parking brake shoe retaining pins.
7. Remove the 2 parking brake shoe retaining springs.
8. Remove the parking brake shoe lower return spring.
9. Remove the parking brake shoe adjuster.
10. Remove the parking brake shoes.

To install:

11. Installation is the reverse of the removal procedure.
- Using anti-seize lubricant, lubricate the parking brake shoe contact points before installation of the rear parking brake shoes.
- Lubricate the adjust screw threads with anti-seize lubricant.
- Adjust the parking brake shoe and lining.
- Check the parking brake for normal operation.

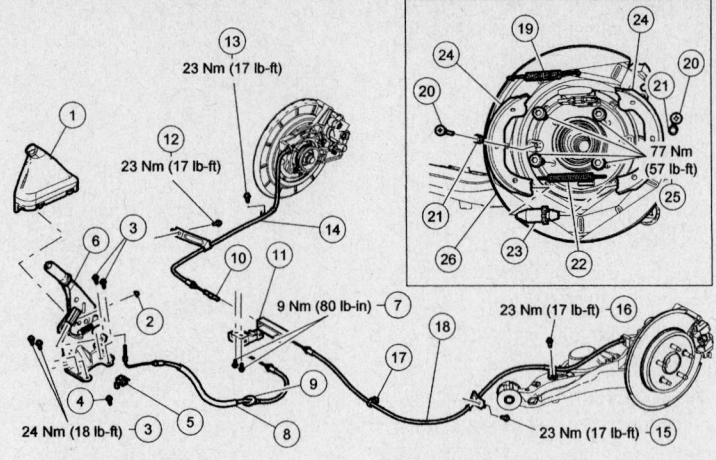

1	Parking brake control boot
2	Front cable adjuster nut
3	Parking brake control bolts (4 required)
4	Warning indicator switch screw
5	Warning indicator switch
6	Parking brake control
7	Parking brake equalizer bracket bolts (2 required)
8	Front parking brake cable
9	Grommet
10	Cable connector
11	Parking brake equalizer and bracket
12	Cable bracket-to-body bolt
13	Cable bracket-to-trailing arm bolt
14	Rear parking brake cable (RH)
15	Cable bracket-to-body bolt
16	Cable bracket-to-trailing arm bolt
17	Cable-to-fuel tank strap clip
18	Rear parking brake cable (LH)
19	Parking brake shoe upper return spring
20	Parking brake shoe retaining pins (2 required)
21	Parking brake shoe retaining springs (2 required)
22	Parking brake shoe lower return spring
23	Parking brake shoe adjuster
24	Parking brake shoe (LH/RH)
25	Support plate bolts (4 required)
26	Support plate (LH/RH)

06017-ESCA-G99

Parking brake assembly and related components—rear disc brakes

ADJUSTMENT

1. Before servicing the vehicle, refer to the Precautions Section.
2. Raise and support the vehicle safely.
3. Remove the wheel and tire assembly.

➡**Make sure the parking brake is fully released.**

4. Using the release handle, release the parking brake control.
5. Remove the rear brake disc.
6. Using the special tool, measure the

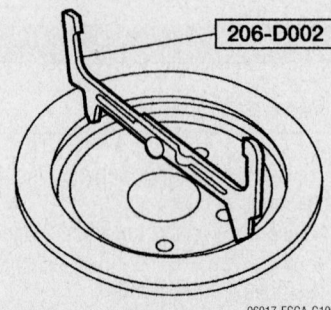

06017-ESCA-G100

Using the special tool, measure the inside diameter of the drum portion of the rear brake disc and set the locking screw

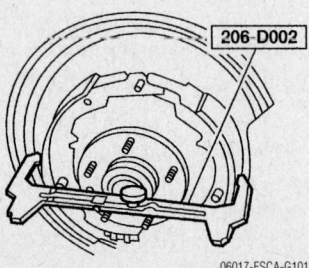

06017-ESCA-G101

Place the special tool over the widest diameter of the rear disc parking brake shoes

inside diameter of the drum portion of the rear brake disc and set the locking screw. Record the measurement.

7. Place the special tool over the widest diameter of the parking brake shoes.
8. Adjust the parking brake shoe clearance to 1.07 mm (0.04 in) less than the inside diameter of the drum portion of the rear brake disc. Rotate the parking brake shoe adjuster to achieve the correct parking brake shoe-to-brake disc clearance.
9. Install the rear brake disc.
10. Test the parking brake for normal operation.

MITSUBISHI

Diamante • Eclipse • Evolution • Galant • Lancer • Mirage

SPECIFICATIONS AND MAINTENANCE CHARTS

ENGINE AND VEHICLE IDENTIFICATION

Code ①	Liters (cc)	Cu. In.	Cyl.	Fuel Sys.	Type	Eng. Mfg.
4G15/A	1.5 (1468)	87	4	MFI	SOHC	Mitsubishi
4G93/C	1.8 (1834)	112	4	MFI	SOHC	Mitsubishi
4G63/D	2.0 (1997)	122	4	MFI	DOHC	Mitsubishi
4G63/F	2.0 (1997)	122	4	MFI	DOHC	Mitsubishi
4G63/G	2.0 (1997)	122	4	MFI	DOHC	Mitsubishi
4G94/E	2.0 (1999)	122	4	MFI	SOHC	Mitsubishi
4G69/F	2.4 (2378)	145	4	MFI	SOHC	Mitsubishi
4G64/G	2.4 (2351)	143	4	MFI	SOHC	Mitsubishi
6G72/H	3.0 (2972)	181	6	MFI	SOHC	Mitsubishi
6G72/L	3.0 (2972)	181	6	MFI	SOHC	Mitsubishi
6G74/P	3.5 (3497)	213	6	MFI	SOHC	Mitsubishi
6G75S	3.8 (3828)	234	6	MFI	SOHC	Mitsubishi
6G75/T	3.8 (3828)	234	6	MFI	SOHC	Mitsubishi

Code ②	Year
2	2002
3	2003
4	2004
5	2005
6	2006

MFI: Multiport fuel injection

SOHC: Single overhead cam

DOHC: Double overhead camshafts

① Engine ID / 8th digit of the VIN

② 10th digit of the VIN

09482_GALA_C0001

GENERAL ENGINE SPECIFICATIONS

Year	Model	Engine Displacement Liters	Engine ID/VIN	Net Horsepower @ rpm	Net Torque @ rpm (ft. lbs.)	Bore x Stroke (in.)	Compression Ratio	Oil Pressure @ rpm
2002	Mirage	1.5	4G15/A	92@6000	93@3000	2.97x3.23	9.2:1	54@2000
	Mirage	1.8	4G93/C	113@6000	116@4500	3.19x3.50	9.5:1	41@2000
	Lancer	2.0	4G94/E	120@5500	130@4250	3.21x3.77	9.5:1	43-100@3500
	Eclipse/Spyder	2.4	4G64/G	①	148@3000	3.41x3.94	9.5:1	41@2000
	Eclipse/Spyder	3.0	6G72/L	175@5500	185@3000	3.59x2.99	8.9:1	30-80@2000
	Galant	2.4	4G64/G	①	148@3000	3.41x3.94	9.5:1	41@2000
	Galant	3.0	6G72/L	175@5500	185@3000	3.59x2.99	8.9:1	30-80@2000
	Diamante	3.5	6G74/P	214@5000	228@3000	3.66x3.38	9.5:1	30-80@2000
2003	Lancer	2.0	4G94/E	120@5500	130@4250	3.21x3.77	9.5:1	43-100@3500
	Eclipse/Spyder	2.4	4G64/G	①	148@3000	3.41x3.94	9.5:1	41@2000
	Eclipse/Spyder	3.0	6G72/H	175@5500	185@3000	3.59x2.99	8.9:1	30-80@2000
	Galant	2.4	4G64/G	①	148@3000	3.41x3.94	9.5:1	41@2000
	Galant	3.0	6G72/H	175@5500	185@3000	3.59x2.99	8.9:1	30-80@2000
	Diamante	3.5	6G74/P	214@5000	228@3000	3.66x3.38	9.5:1	30-80@2000
2004	Lancer	2.0	4G94/E	120@5500	130@4250	3.21x3.77	9.5:1	43-100@3500
	Lancer	2.4	4G69/F	162@5750	162@4000	3.43x3.94	9.5:1	43-100@3500
	Evolution	2.0	4G63/D	271@6500	273@3500	3.35x3.46	8.8:1	43-100@3500
	Sportback	2.4	4G69/F	②	③	3.43x3.94	9.5:1	43-100@3500
	Eclipse/Spyder	2.4	4G64/G	④	⑤	3.41x3.94	9.0:1	43-100@3500
	Eclipse/Spyder	3.0	6G72/H	⑥	⑦	3.59x2.99	⑧	43-100@3500
	Galant	2.4	4G69/F	160@5500	157@4000	3.43x3.90	9.5:1	43-100@3500
	Galant	3.8	6G75/S	230@5250	250@4000	3.74x3.54	10:01	43-100@3500
	Diamante	3.5	6G74/P	214@5000	228@3000	3.66x3.38	9.5:1	30-80@2000
2005	Lancer	2.0	4G94/E	120@5500	130@4250	3.21x3.77	9.5:1	43-100@3500
	Lancer	2.4	4G69/F	162@5750	162@4000	3.43x3.94	9.5:1	43-100@3500
	Evolution	2.0	4G63/F	276@6500	285@3500	3.35x3.46	8.8:1	43-100@3500
	Eclipse/Spyder	2.4	4G64/G	④	⑤	3.41x3.94	9.0:1	43-100@3500
	Eclipse/Spyder	3.0	6G72/H	⑥	⑦	3.59x2.99	⑧	43-100@3500
	Galant	2.4	4G69/F	160@5500	157@4000	3.43x3.90	9.5:1	43-100@3500
	Galant	3.8	6G75/S	230@5250	250@4000	3.74x3.54	10:01	43-100@3500
2006	Lancer	2.0	4G94/E	120@5500	130@4250	3.21x3.77	9.5:1	43-100@3500
	Lancer	2.4	4G69/F	162@5750	162@4000	3.43x3.94	9.5:1	43-100@3500
	Evolution	2.0	4G63/C	276@6500	285@3500	3.35x3.46	8.8:1	43-100@3500
	Eclipse	2.4	4G69/F	162@6000	162@4000	3.43x3.94	9.5:1	43-100@3500
	Eclipse	3.8	6G75/T	263@5750	260@4500	3.74x3.54	10.5:1	43-100@3500
	Galant	2.4	4G69/F	160@5500	157@4000	3.43x3.90	9.5:1	43-100@3500
	Galant	3.8	6G75/S	230@5250	250@4000	3.74x3.54	10:01	43-100@3500

① California: 138@5500
Except California: 141@5500

② LS: 160@5750
Ralliart: 162@5750

③ LS: 161@4000
Ralliart: 162@4000

④ With M/T: 147@5500
With A/T: 142@5500

⑤ With M/T: 158@4000
With A/T: 155@4000

⑥ Eclipse GT: 200@5500
Eclipse GTS and Spyder: 210@5750

⑦ Eclipse GT: 205@4000
Eclipse GTS and Spyder: 205@3750

⑧ With IMT: 10.0:1
Without IMT: 9.0:1

09482_GALA_C0002

ENGINE TUNE-UP SPECIFICATIONS

Year	Engine Displacement Liters	Engine ID/VIN	Spark Plugs Gap (in.)	Ignition Timing (deg.) MT	AT	Fuel Pump (psi)	Idle Speed (rpm) MT	AT	Valve Clearance In.	Ex.
2002	1.5	4G15/A	0.039-0.043	2-8B	2-8B	38	600-800	600-800	HYD	HYD
	1.8	4G93/C	0.039-0.043	2-8B	2-8B	38	600-800	600-800	HYD	HYD
	2.0	4G94/E	0.039-0.043	2-8B	2-8B	38	600-800	600-800	HYD	HYD
	2.4	4G64/G	0.039-0.043	2-8B	2-8B	38	650-850	650-850	HYD	HYD
	3.0	6G72/L	0.039-0.043	5B	5B	38	600-800	600-800	HYD	HYD
	3.5	6G74/P	0.039-0.043	—	2-8B	38	—	600-800	HYD	HYD
2003	2.0	4G94/E	0.039-0.043	2-8B	2-8B	38	600-800	600-800	HYD	HYD
	2.4	4G64/G	0.039-0.043	2-8B	2-8B	38	600-800	650-850	HYD	HYD
	3.0	6G72/H	0.028-0.031	2-8B	2-8B	38	600-800	600-800	HYD	HYD
	3.5	6G74/P	0.039-0.043	—	2-8B	38	—	600-800	HYD	HYD
2004	2.0	4G94/E	0.039-0.043	2-8B	2-8B	38	600-800	600-800	HYD	HYD
	2.0	4G63/D	0.024-0.027	2-8B	—	33	800-900	—	HYD	HYD
	2.4	4G69/F	0.028-0.031	2-8B	2-8B	38	600-800	600-800	HYD	HYD
	2.4	4G64/G	0.039-0.043	2-8B	2-8B	38	600-800	650-850	HYD	HYD
	3.0	6G72/H	0.028-0.031	2-8B	2-8B	38	600-800	600-800	HYD	HYD
	3.5	6G74/P	0.039-0.043	—	2-8B	38	—	600-800	HYD	HYD
	3.8	6G75/S	0.028-0.031	—	2-8B	38	—	550-750	HYD	HYD
2005	2.0	4G94/E	0.039-0.043	2-8B	2-8B	47	600-800	600-800	HYD	HYD
	2.0	4G63/F	0.024-0.027	2-8B	—	33	800-900	—	HYD	HYD
	2.4	4G69/F	0.020-0.024	2-8B	2-8B	47	600-800	600-800	①	①
	2.4	4G64/G	0.039-0.043	2-8B	2-8B	38	600-800	650-850	HYD	HYD
	3.0	6G72/H	0.028-0.031	2-8B	2-8B	38	600-800	600-800	HYD	HYD
	3.8	6G75/S	0.028-0.031	—	2-8B	38	—	550-750	HYD	HYD
2006	2.0	4G94/E	0.039-0.043	2-8B	2-8B	47	600-800	600-800	HYD	HYD
	2.0	4G63/C	0.024-0.027	2-8B	—	33	800-900	—	HYD	HYD
	2.4	4G69/F	0.020-0.024	2-8B	2-8B	47	600-800	600-800	①	①
	3.8	6G75/T	0.028-0.031	2-8B	2-8B	47	580-780	600-800	HYD	HYD
	3.8	6G75/S	0.028-0.031	—	2-8B	38	—	580-780	HYD	HYD

NOTE: The Vehicle Emission Control Information label often reflects specification changes made during production.

   The label figures must be used if they differ from those in this chart.

B: Before top dead center

HYD: Hydraulic

① Intake: 0.008
   Exhaust: 0.012

09482_GALA_C0003

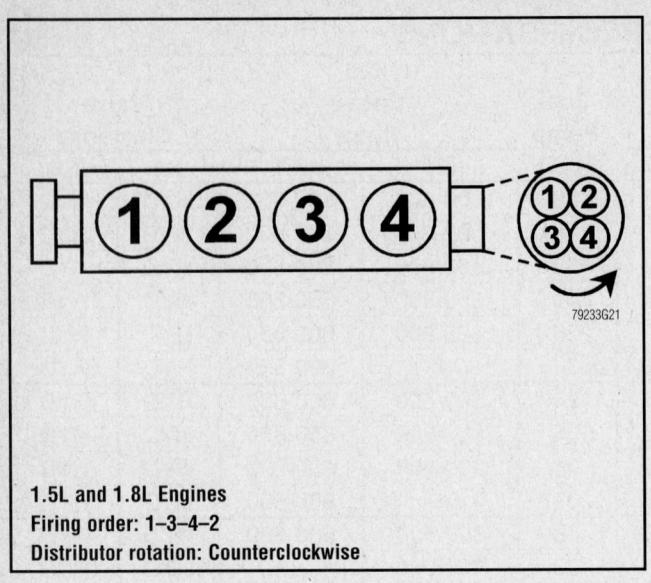

1.5L and 1.8L Engines
Firing order: 1–3–4–2
Distributor rotation: Counterclockwise

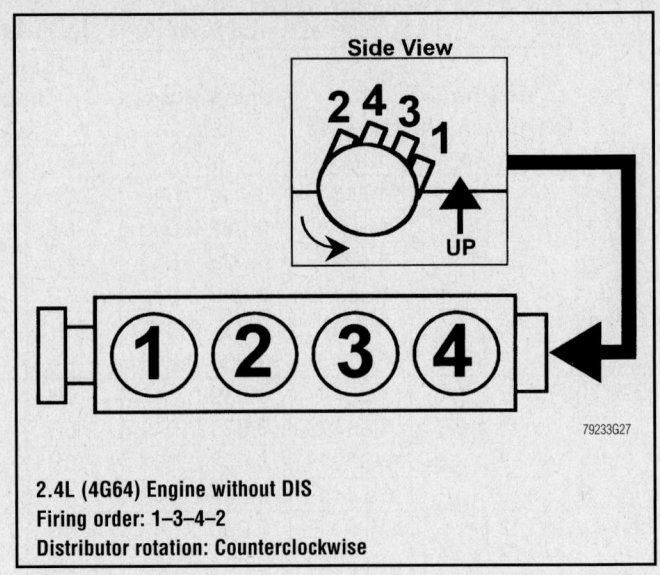

2.4L (4G64) Engine without DIS
Firing order: 1–3–4–2
Distributor rotation: Counterclockwise

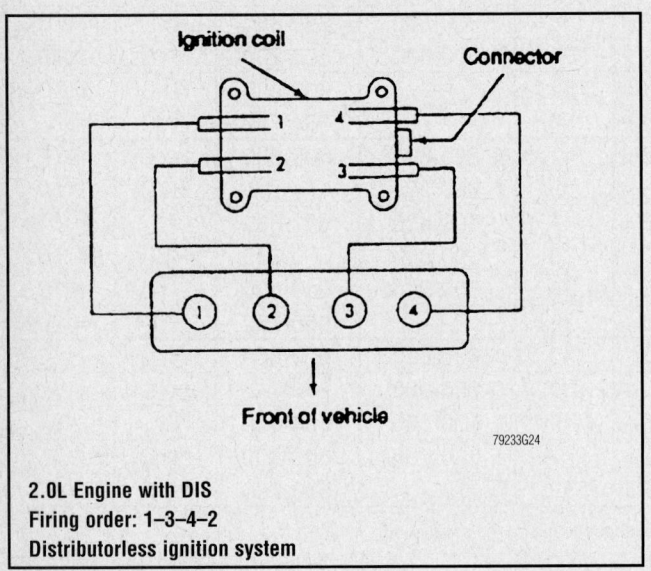

2.0L Engine with DIS
Firing order: 1–3–4–2
Distributorless ignition system

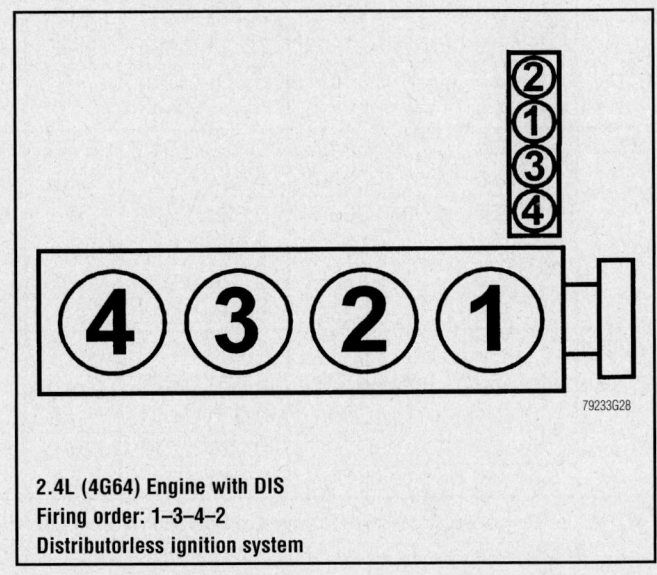

2.4L (4G64) Engine with DIS
Firing order: 1–3–4–2
Distributorless ignition system

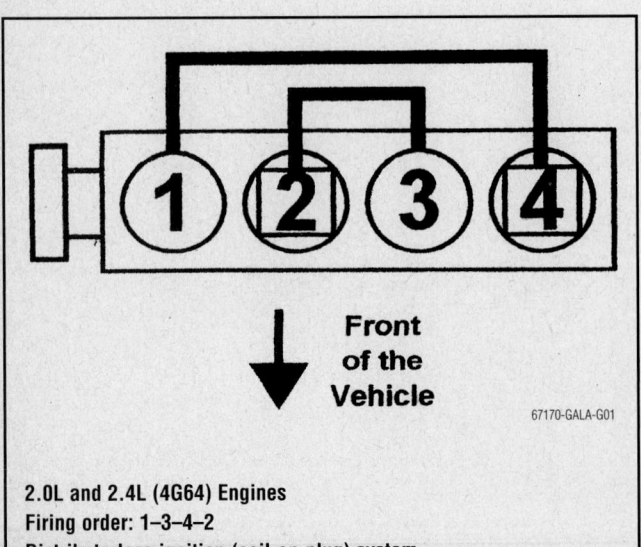

2.0L and 2.4L (4G64) Engines
Firing order: 1–3–4–2
Distributorless ignition (coil-on-plug) system

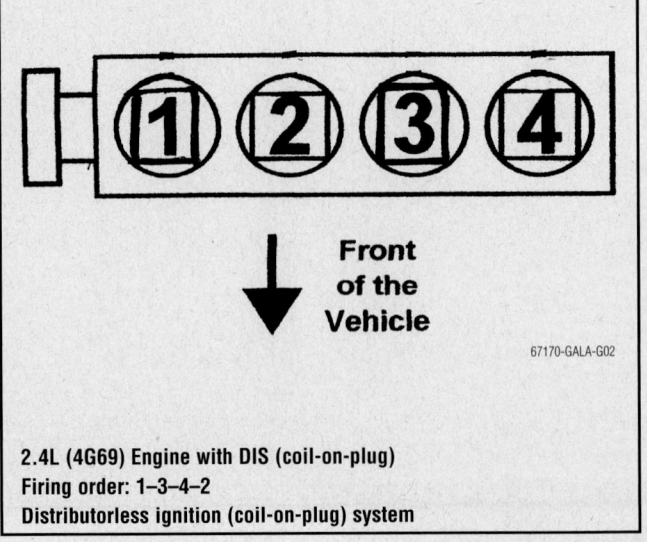

2.4L (4G69) Engine with DIS (coil-on-plug)
Firing order: 1–3–4–2
Distributorless ignition (coil-on-plug) system

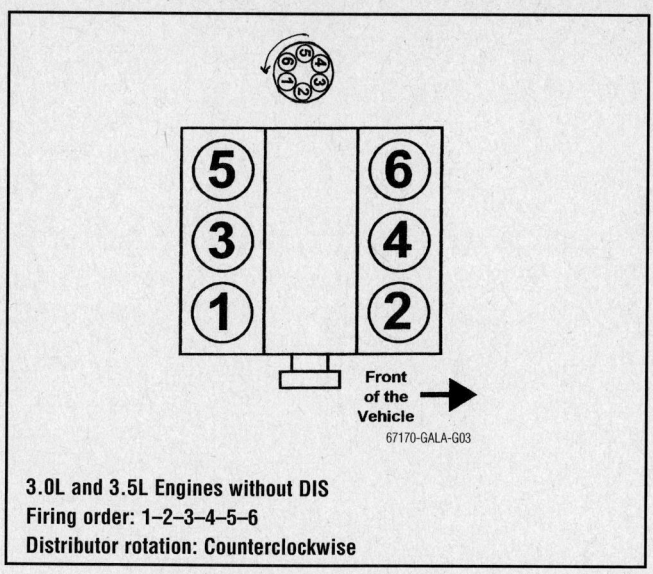

3.0L and 3.5L Engines without DIS
Firing order: 1–2–3–4–5–6
Distributor rotation: Counterclockwise

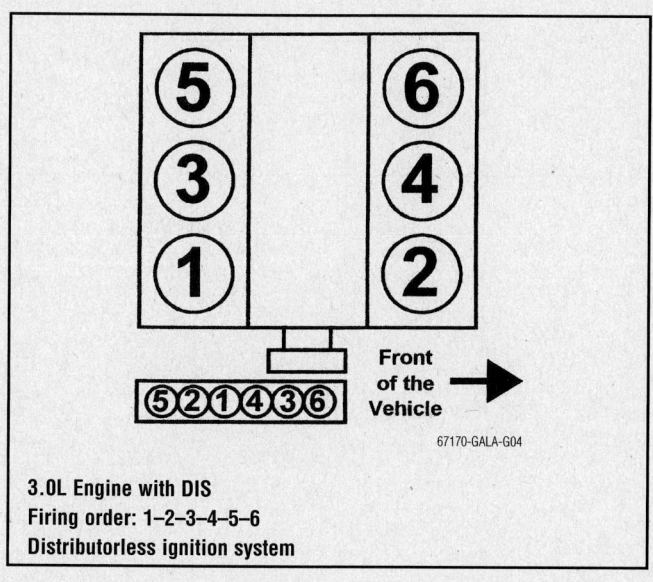

3.0L Engine with DIS
Firing order: 1–2–3–4–5–6
Distributorless ignition system

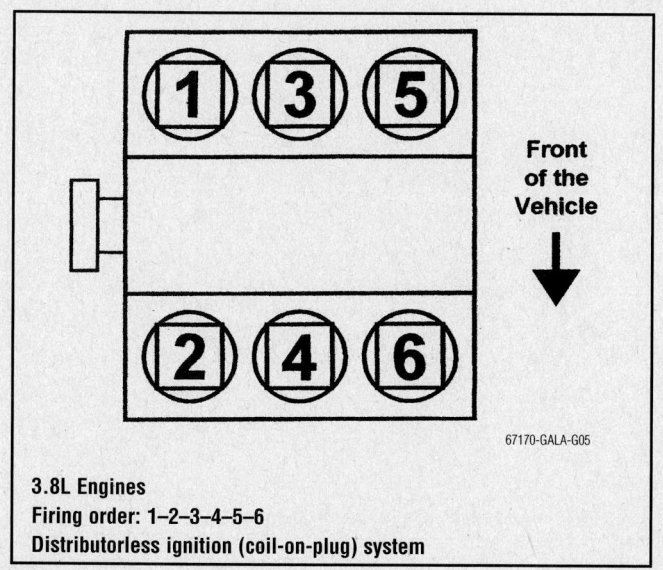

3.8L Engines
Firing order: 1–2–3–4–5–6
Distributorless ignition (coil-on-plug) system

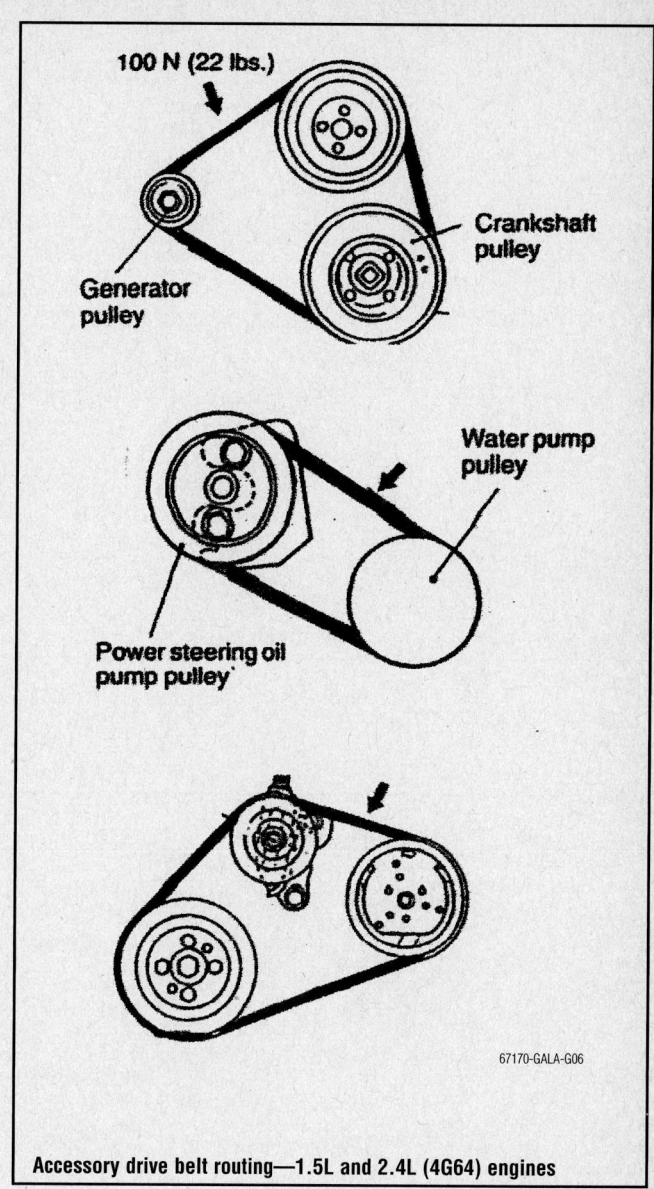

Accessory drive belt routing—1.5L and 2.4L (4G64) engines

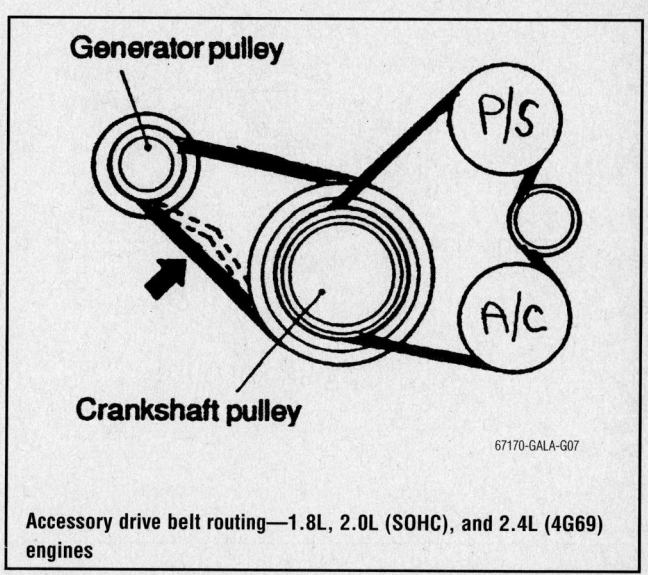

Accessory drive belt routing—1.8L, 2.0L (SOHC), and 2.4L (4G69) engines

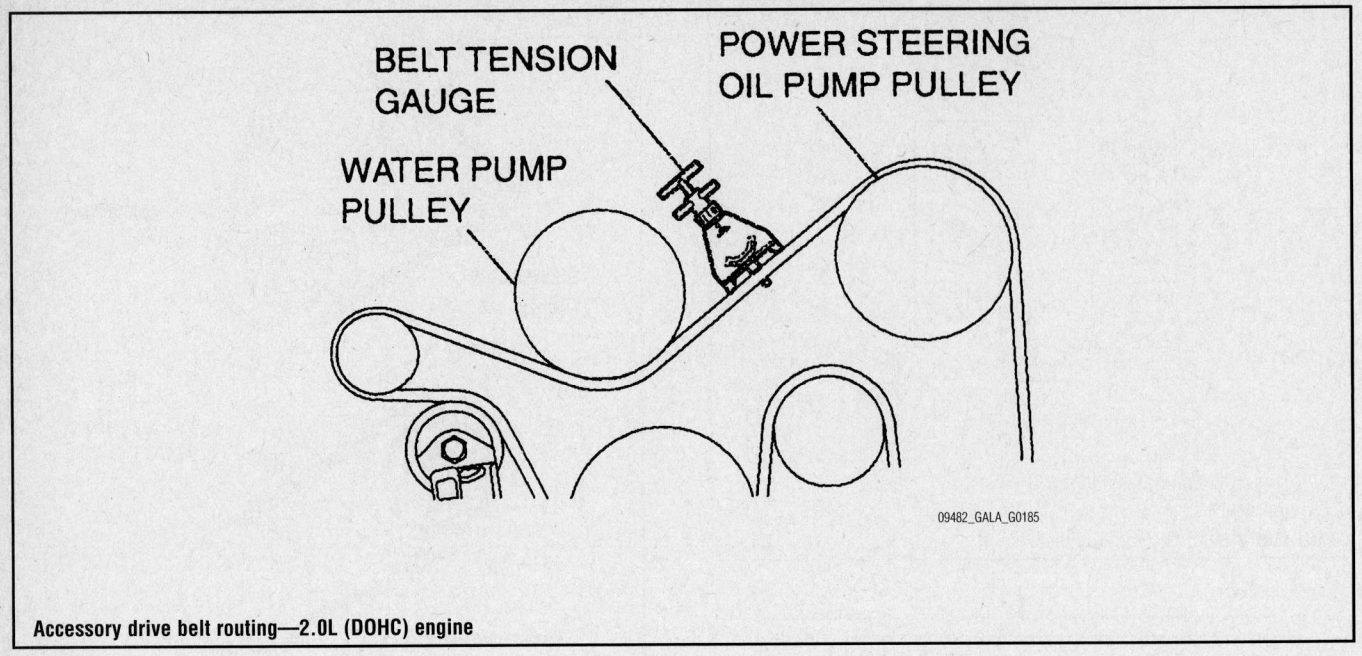

Accessory drive belt routing—2.0L (DOHC) engine

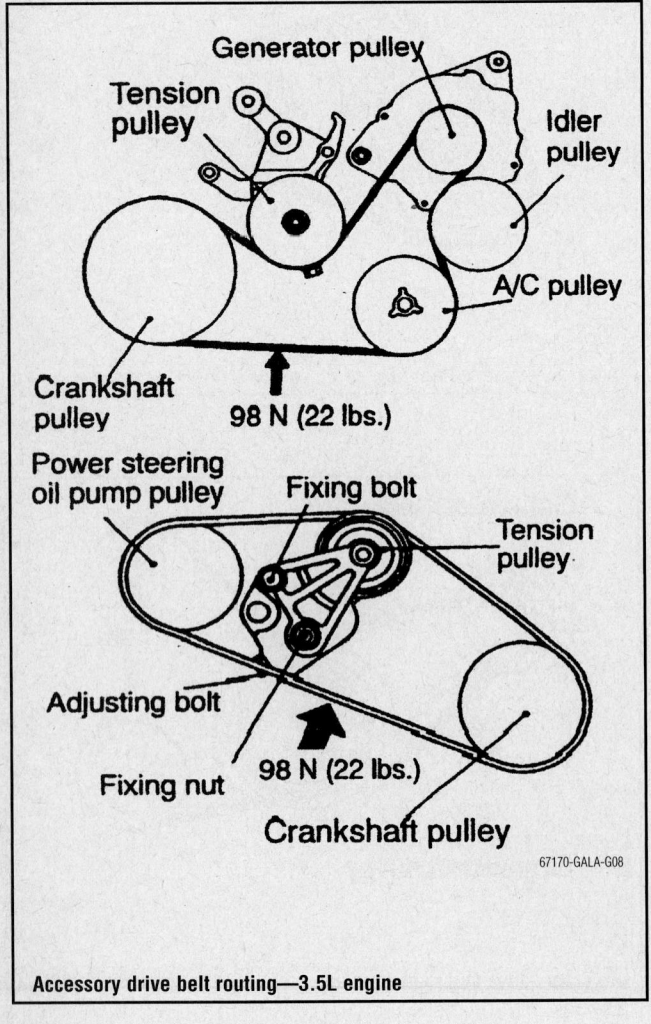

Accessory drive belt routing—3.5L engine

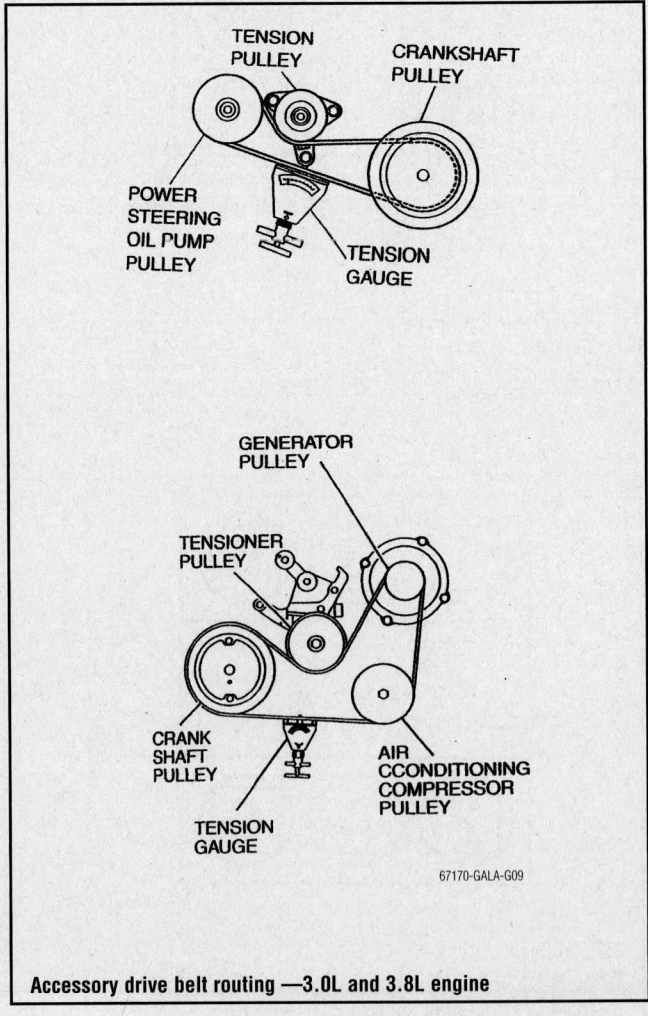

Accessory drive belt routing —3.0L and 3.8L engine

CAPACITIES

Year	Model	Engine Displacement Liters	Engine ID/VIN	Engine Oil with Filter (qts.)	Transaxle (pts.) Manual	Transaxle (pts.) Auto.	Transfer Case (pts.)	Drive Axle Front (pts.)	Drive Axle Rear (pts.)	Fuel Tank (gal.)	Cooling System (qts.)
2002	Mirage	1.5	4G15/A	3.3	4.4	16.2	—	—	—	12.4	5.3
	Mirage	1.8	4G93/C	3.9	4.6	16.2	—	—	—	12.4	6.3
	Lancer	2.0	4G94/E	3.8	4.6	16.2	—	—	—	13.2	5.3
	Eclipse/Spyder	2.4	4G64/G	4.5	4.6	16.4	—	—	—	16.4	7.4
	Eclipse/Spyder	3.0	6G72/L	4.5	6.0	17.8	—	—	—	16.4	8.5
	Galant	2.4	4G64/G	4.5	—	16.2	—	—	—	16.3	14.8
	Galant	3.0	6G72/L	4.5	—	17.8	—	—	—	16.3	8.5
	Diamante	3.5	6G74/P	4.5	—	17.8	—	—	—	18.7	10.0
2003	Lancer	2.0	4G94/E	3.8	4.6	16.2	—	—	—	13.2	5.3
	Eclipse/Spyder	2.4	4G64/G	4.5	4.6	16.2	—	—	—	16.4	①
	Eclipse/Spyder	3.0	6G72/H	4.5	6.0	17.6	—	—	—	16.4	9.6
	Galant	2.4	4G64/G	4.5	—	16.2	—	—	—	16.3	14.8
	Galant	3.0	6G72/L	4.5	—	17.8	—	—	—	16.3	8.5
	Diamante	3.5	6G74/P	4.5	—	17.8	—	—	—	18.7	10.0
2004	Lancer	2.0	4G94/E	4.0	4.6	16.2	—	—	—	13.2	6.3
	Lancer	2.4	4G69/F	4.5	4.6	16.2	—	—	—	13.2	7.4
	Evolution	2.0	4G63/D	5.4	6.0	—	1.16	—	1.16	14.0	6.3
	Eclipse/Spyder	2.4	4G64/G	4.5	4.6	16.2	—	—	—	16.4	①
	Eclipse/Spyder	3.0	6G72/H	4.5	6.0	17.6	—	—	—	16.4	9.6
	Galant	2.4	4G69/F	4.5	—	16.2	—	—	—	17.7	8.5
	Galant	3.8	6G75/S	4.8	—	17.8	—	—	—	17.7	9.2
	Diamante	3.5	6G74/P	4.5	—	17.8	—	—	—	18.7	10.0
2005	Lancer	2.0	4G94/E	4.0	4.6	16.2	—	—	—	13.2	6.3
	Lancer	2.4	4G69/F	4.5	4.6	16.2	—	—	—	13.2	7.4
	Evolution	2.0	4G63/F	5.4	②	—	1.16	—	1.16	14.0	6.3
	Eclipse/Spyder	2.4	4G64/G	4.5	4.6	16.2	—	—	—	16.4	7.4
	Eclipse/Spyder	3.0	6G72/H	4.5	6.0	17.6	—	—	—	16.4	8.5
	Galant	2.4	4G69/F	4.5	—	16.2	—	—	—	17.7	8.5
	Galant	3.8	6G75/S	4.8	—	17.8	—	—	—	17.7	9.2
2006	Lancer	2.0	4G94/E	4.0	4.6	16.2	—	—	—	13.2	6.3
	Lancer	2.4	4G69/F	4.5	4.6	16.2	—	—	—	13.2	7.4
	Evolution	2.0	4G63/C	5.4	6.0	—	1.16	—	1.16	14.0	6.3
	Eclipse	2.4	4G69/F	4.5	4.6	18.0	—	—	—	17.7	③
	Eclipse	3.8	6G75/T	4.5	4.6	18.0	—	—	—	17.7	③
	Galant	2.4	4G69/F	4.5	—	16.2	—	—	—	17.7	8.5
	Galant	3.8	6G75/S	4.8	—	17.8	—	—	—	17.7	9.2

NOTE: All capacities are approximate. Add fluid gradually and ensure a proper fluid level is obtained.

① Manual transaxle: 8.5 qts.
   Automatic transaxle: 8.3 qts.

② 5 speed:6.0M/T: 8.5 qts.
   6 speed: 5.5

③ Manual transaxle: 8.5M/T: 8.5 qts.
   Automatic transaxle: 8.6

VALVE SPECIFICATIONS

Year	Engine Displacement Liters	Engine ID/VIN	Seat Angle (deg.)	Face Angle (deg.)	Spring Test Pressure (lbs. @ in.)	Spring Installed Height (in.)	Stem-to-Guide Clearance (in.)		Stem Diameter (in.)	
							Intake	Exhaust	Intake	Exhaust
2002	1.5	4G15/A	44.5-45	45-45.5	①	1.570	0.0008-0.0020	0.0020-0.0035	0.260	0.260
	1.8	4G93/C	44.5-45	45-45.5	59@1.740	1.740	0.0008-0.0020	0.0020 0.0035	0.234	0.234
	2.0	4G94/E	44.5-45	45-45.5	44.2@1.740	1.740	0.0008-0.0020	0.0016-0.0024	0.236	0.236
	2.4	4G64/G	44.5-45	45-45.5	60@1.740	1.740	0.0008-0.0020	0.0008-0.0028	0.236	0.232
	3.0	6G72/L	44.5-45	45-45.5	40.4@1.591	1.591	0.0012-0.0024	0.0020-0.0035	0.315	0.311
	3.5	6G74/P	44-44.5	45-45.5	60@1.740	1.740	0.0008-0.0020	0.0016-0.0028	0.236	0.236
2003	2.0	4G94/E	44.5-45	45-45.5	44.2@1.740	1.740	0.0008-0.0020	0.0016-0.0024	0.236	0.236
	2.4	4G64/G	44.5-45	45-45.5	60@1.740	1.740	0.0008-0.0020	0.0008-0.0028	0.236	0.232
	3.0	6G72/H	44.5-45	45-45.5	40.4@1.591	1.591	0.0012-0.0024	0.0020-0.0035	0.315	0.311
	3.5	6G74/P	44-44.5	45-45.5	60@1.740	1.740	0.0008-0.0020	0.0016-0.0028	0.236	0.236
2004	2.0	4G94/E	44.5-45	45-45.5	44.2@1.740	1.740	0.0008-0.0020	0.0016-0.0024	0.236	0.236
	2.0	4G63/D	44.5-45	45-45.5	63@1.570	1.570	0.0008-0.0020	0.0020 0.0035	0.236	0.236
	2.4	4G69/F	43-43.5	43.5-44	60@1.740	1.740	0.0008-0.0030	0.0016-0.0050	0.240	0.240
	2.4	4G64/G	44.5-45	45-45.5	60@1.740	1.740	0.0008-0.0020	0.0008-0.0028	0.236	0.232
	3.0	6G72/H	44.5-45	45-45.5	40.4@1.591	1.591	0.0012-0.0024	0.0020-0.0035	0.315	0.311
	3.5	6G74/P	44-44.5	45-45.5	60@1.740	1.740	0.0008-0.0020	0.0016-0.0028	0.236	0.236
	3.8	6G75/S	43-43.5	43.5-44	60@1.740	1.740	0.0008-0.0030	0.0016-0.0050	0.240	0.240
2005	2.0	4G94/E	44.5-45	45-45.5	44.2@1.740	1.740	0.0008-0.0020	0.0016-0.0024	0.236	0.236
	2.0	4G63/F	44.5-45	45-45.5	63@1.570	1.570	0.0008-0.0020	0.0020 0.0035	0.236	0.236
	2.4	4G69/F	43-43.5	43.5-44	②	1.740	0.0008-0.0030	0.0016-0.0024	0.240	0.240
	2.4	4G64/G	44.5-45	45-45.5	60@1.740	1.740	0.0008-0.0020	0.0008-0.0028	0.236	0.232
	3.0	6G72/H	44.5-45	45-45.5	40.4@1.591	1.591	0.0012-0.0024	0.0020-0.0035	0.315	0.311
	3.8	6G75/S	43-43.5	43.5-44	60@1.740	1.740	0.0008-0.0019	0.0014-0.0024	0.240	0.240
2006	2.0	4G94/E	44.5-45	45-45.5	44.2@1.740	1.740	0.0008-0.0020	0.0016-0.0024	0.236	0.236
	2.0	4G63/C	44.5-45	45-45.5	63@1.570	1.570	0.0008-0.0020	0.0020 0.0035	0.236	0.236

09482_GALA_C0005

VALVE SPECIFICATIONS

Year	Engine Displacement Liters	Engine ID/VIN	Seat Angle (deg.)	Face Angle (deg.)	Spring Test Pressure (lbs. @ in.)	Spring Installed Height (in.)	Stem-to-Guide Clearance (in.)		Stem Diameter (in.)	
							Intake	Exhaust	Intake	Exhaust
2006 Cont.	2.4	4G69/F	43-43.5	43.5-44	60@1.740	1.740	0.0008-0.0019	0.0014-0.0024	0.240	0.240
	3.8	6G75/T	43.5-43.5	43.5-44	③	1.740	0.0008-0.0019	0.0016-0.0023	0.240	0.240
	3.8	6G75/S	43-43.5	43.5-44	60@1.740	1.740	0.0008-0.0019	0.0014-0.0024	0.240	0.240

① Intake: 51@1.57
   Exhaust: 64@1.57

② Intake: 53@1.74
   Exhaust: 51@1.74

③ Intake: 59@1.74
   Exhaust: 53@1.74

09482_GALA_C0006

CRANKSHAFT AND CONNECTING ROD SPECIFICATIONS
All measurements are given in inches.

Year	Engine Displacement Liters	Engine ID/VIN	Crankshaft				Connecting Rod		
			Main Brg. Journal Dia.	Main Brg. Oil Clearance	Shaft End-play	Thrust on No.	Journal Diameter	Oil Clearance	Side Clearance
2002	1.5	4G15/A	1.8900	0.0008-0.0040	0.0020-0.0120	3	1.6500	0.0008-0.0040	0.0039-0.0160
	1.8	4G93/C	1.9678-1.9685	0.0008-0.0040	0.0020-0.0098	3	1.7709-1.7717	0.0008-0.0040	0.0039-0.0160
	2.0	4G94/E	1.9700	0.0008-0.0012	0.0020-0.0098	3	NA	0.0008-0.0016	0.0039-0.0098
	2.4	4G64/G	2.2436-2.2441	0.0008-0.0040	0.0020-0.0098	3	1.7709-1.7717	0.0008-0.0040	0.0039-0.0160
	3.0	6G72/L	2.3614-2.3622	0.0008-0.0040	0.0020-0.0120	3	2.1646-2.1654	0.0008-0.0040	0.0039-0.0160
	3.5	6G74/P	2.3614-2.3622	0.0008-0.0040	0.0020-0.0120	3	1.9700	0.0008-0.0040	0.0039-0.0160
2003	2.0	4G94/E	1.9700	0.0008-0.0012	0.0020-0.0098	3	NA	0.0008-0.0016	0.0039-0.0098
	2.4	4G64/G	2.2436-2.2441	0.0008-0.0040	0.0020-0.0098	3	1.7709-1.7717	0.0008-0.0040	0.0039-0.0160
	3.0	6G72/H	2.3614-2.3622	0.0008-0.0040	0.0020-0.0120	3	2.1646-2.1654	0.0008-0.0040	0.0039-0.0160
	3.5	6G74/P	2.3614-2.3622	0.0008-0.0040	0.0020-0.0120	3	1.9700	0.0008-0.0040	0.0039-0.0160
2004	2.0	4G94/E	1.9700	0.0008-0.0012	0.0020-0.0098	3	NA	0.0008-0.0016	0.0039-0.0098
	2.0	4G63/D	2.2400	0.0008-0.0040	0.0020-0.0098	3	1.770	0.0008-0.0016	0.0039-0.0098
	2.4	4G69/F	2.2400	0.0008-0.0030	0.0020-0.0090	3	1.770	0.0008-0.0030	0.0040-0.0150
	2.4	4G64/G	2.2436-2.2441	0.0008-0.0040	0.0020-0.0098	3	1.7709-1.7717	0.0008-0.0040	0.0039-0.0160
	3.0	6G72/H	2.3614-2.3622	0.0008-0.0040	0.0020-0.0120	3	2.1646-2.1654	0.0008-0.0040	0.0039-0.0160
	3.5	6G74/P	2.3614-2.3622	0.0008-0.0040	0.0020-0.0120	3	1.9700	0.0008-0.0040	0.0039-0.0160
	3.8	6G75/S	2.5200	①	0.0020-0.0100	3	2.1650	0.0008-0.0019	0.0030-0.0090
2005	2.0	4G94/E	1.9700	0.0008-0.0012	0.0020-0.0098	3	NA	0.0008-0.0016	0.0039-0.0098
	2.0	4G63/F	2.2400	0.0008-0.0016	0.0020-0.0098	3	1.770	0.0008-0.0016	0.0039-0.0098
	2.4	4G69/F	2.2400	0.0008-0.0015	0.0020-0.0090	3	1.770	0.0008-0.0019	0.0040-0.0090
	2.4	4G64/G	2.2436-2.2441	0.0008-0.0040	0.0020-0.0098	3	1.7709-1.7717	0.0008-0.0040	0.0039-0.0160
	3.0	6G72/H	2.3614-2.3622	0.0008-0.0040	0.0020-0.0120	3	2.1646-2.1654	0.0008-0.0040	0.0039-0.0160
	3.8	6G75/S	2.5200	①	0.0020-0.0100	3	2.1650	0.0008-0.0019	0.0030-0.0090
2006	2.0	4G94/E	1.9700	②	0.0020-0.0098	3	NA	0.0008-0.0016	0.0039-0.0098

CRANKSHAFT AND CONNECTING ROD SPECIFICATIONS

All measurements are given in inches.

Year	Engine Displacement Liters	Engine ID/VIN	Crankshaft				Connecting Rod		
			Main Brg. Journal Dia.	Main Brg. Oil Clearance	Shaft End-play	Thrust on No.	Journal Diameter	Oil Clearance	Side Clearance
2006 Cont.	2.0	4G63/C	2.2400	0.0008-0.0016	0.0020-0.0098	3	1.770	0.0008-0.0016	0.0039-0.0098
	2.4	4G69/F	2.2400	0.0008-0.0015	0.0020-0.0090	3	1.770	0.0008-0.0030	0.0040-0.0150
	3.8	6G75/T	2.5200	①	0.0020-0.0090	3	2.1650	0.0008-0.0019	0.0030-0.0900
	3.8	6G75/S	2.5200	③	0.0020-0.0100	3	2.1650	0.0008-0.0019	0.0030-0.0090

NA - Not Available
① Number 1 & 4 mains: 0.0008-0.0012
　 Number 2 & 3 mains: 0.0012-0.0016
② Number 1, 2, 4 & 5 mains: 0.0008-0.0015
　 Number 3 main: 0.0012-0.0015
③ Number 1 & 4 mains: 0.0007-0.0014
　 Number 2 & 3 mains: 0.0009-0.0017

09482_GALA_C0008

PISTON AND RING SPECIFICATIONS
All measurements are given in inches.

Year	Engine Displacement Liters	Engine ID/VIN	Piston Clearance	Ring Gap			Ring Side Clearance		
				Top Compression	Bottom Compression	Oil Control	Top Compression	Bottom Compression	Oil Control
2002	1.5	4G15/A	0.0008-0.0016	0.0079-0.0310	0.0079-0.0310	0.0079-0.0390	0.0012-0.0040	0.0008-0.0040	NA
	1.8	4G93/C	0.0008-0.0016	0.0098-0.0310	0.0157-0.0310	0.0078-0.0390	0.0012-0.0039	0.0008-0.0039	NA
	2.0	4G94/E	0.0008-0.0016	0.0059-0.0118	0.0157-0.0217	0.0039-0.0138	0.0012-0.0028	0.0008-0.0024	NA
	2.4	4G64/G	0.0008-0.0016	0.0098-0.0310	0.0157-0.0310	0.0039-0.0390	0.0012-0.0040	0.0012-0.0040	NA
	3.0	6G72/L	0.0008-0.0020	0.0118-0.0310	0.0177-0.0310	0.0079-0.0390	0.0012-0.0040	0.0008-0.0040	NA
	3.5	6G74/P	0.0008-0.0020	0.0118-0.0310	0.0177-0.0310	0.0079-0.0390	0.0012-0.0040	0.0008-0.0040	NA
2003	2.0	4G94/E	0.0008-0.0016	0.0059-0.0118	0.0157-0.0217	0.0039-0.0138	0.0012-0.0028	0.0008-0.0024	NA
	2.4	4G64/G	0.0008-0.0016	0.0098-0.0310	0.0157-0.0310	0.0039-0.0390	0.0012-0.0040	0.0012-0.0040	NA
	3.0	6G72/H	0.0008-0.0020	0.0118-0.0310	0.0177-0.0310	0.0079-0.0390	0.0012-0.0040	0.0008-0.0040	NA
	3.5	6G74/P	0.0008-0.0020	0.0118-0.0310	0.0177-0.0310	0.0079-0.0390	0.0012-0.0040	0.0008-0.0040	NA
2004	2.0	4G94/E	0.0008-0.0016	0.0059-0.0118	0.0157-0.0217	0.0039-0.0138	0.0012-0.0028	0.0008-0.0024	NA
	2.0	4G63/D	0.0008-0.0016	0.0079-0.0118	0.0128-0.0197	0.0039-0.0157	0.0012-0.0028	0.0008-0.0024	NA
	2.4	4G69/F	0.0008-0.0015	0.0060-0.0300	0.0110-0.0300	0.0040-0.0300	0.0012-0.0030	0.0008-0.0030	NA
	2.4	4G64/G	0.0008-0.0016	0.0098-0.0310	0.0157-0.0310	0.0039-0.0390	0.0012-0.0040	0.0012-0.0040	NA
	3.0	6G72/H	0.0008-0.0020	0.0118-0.0310	0.0177-0.0310	0.0079-0.0390	0.0012-0.0040	0.0008-0.0040	NA
	3.5	6G74/P	0.0008-0.0020	0.0118-0.0310	0.0177-0.0310	0.0079-0.0390	0.0012-0.0040	0.0008-0.0040	NA
	3.8	6G75/S	0.0008-0.0016	0.0100-0.0300	0.0140-0.0300	0.0030-0.0300	0.0012-0.0030	0.0008-0.0030	NA
2005	2.0	4G94/E	0.0008-0.0016	0.0059-0.0118	0.0157-0.0217	0.0039-0.0138	0.0012-0.0028	0.0008-0.0024	NA
	2.0	4G63/F	0.0008-0.0016	0.0079-0.0118	0.0128-0.0197	0.0039-0.0157	0.0012-0.0028	0.0008-0.0024	NA
	2.4	4G69/F	0.0008-0.0015	0.0060-0.0120	0.0110-0.0170	0.0040-0.0160	0.0012-0.0028	0.0008-0.0024	NA
	2.4	4G64/G	0.0008-0.0016	0.0098-0.0310	0.0157-0.0310	0.0039-0.0390	0.0012-0.0040	0.0012-0.0040	NA
	3.0	6G72/H	0.0008-0.0020	0.0118-0.0310	0.0177-0.0310	0.0079-0.0390	0.0012-0.0040	0.0008-0.0040	NA
	3.8	6G75/S	0.0008-0.0016	0.0100-0.0160	0.0140-0.0190	0.0030-0.0140	0.0120-0.0270	0.0008-0.0023	NA
2006	2.0	4G94/E	0.0008-0.0016	0.0059-0.0118	0.0157-0.0217	0.0039-0.0138	0.0016-0.0032	0.0008-0.0024	NA

PISTON AND RING SPECIFICATIONS

All measurements are given in inches.

Year	Engine Displacement Liters	Engine ID/VIN	Piston Clearance	Ring Gap			Ring Side Clearance		
				Top Compression	Bottom Compression	Oil Control	Top Compression	Bottom Compression	Oil Control
2006 Cont.	2.0	4G63/C	0.0008-0.0016	0.0079-0.0118	0.0128-0.0197	0.0039-0.0157	0.0012-0.0028	0.0008-0.0024	NA
	2.4	4G69/F	0.0008-0.0015	0.0060-0.0300	0.0110-0.0300	0.0040-0.0300	0.0012-0.0030	0.0008-0.0030	NA
	3.8	6G75/T	0.0008-0.0016	0.0100-0.0160	0.0140-0.0190	0.0030 0.0140	0.0120 0.0270	0.0008-0.0023	NA
	3.8	6G75/S	0.0008-0.0016	0.0100-0.0160	0.0140-0.0190	0.0030 0.0140	0.0120 0.0270	0.0008-0.0023	NA

NA - Not Available

09482_GALA_C0010

TORQUE SPECIFICATIONS
All readings in ft. lbs.

Year	Engine Displacement Liters	Engine ID/VIN	Cylinder Head Bolts	Main Bearing Bolts	Rod Bearing Bolts	Crankshaft Damper Bolts	Flywheel Bolts	Manifold Intake	Manifold Exhaust	Spark Plugs	Oil Pan Drain Plug
2002	1.5	4G15/A	①	37	12 ②	93	95	12	12	18	65-80
	1.8	4G93/C	③	18 ②	15 ②	131	71	15	④	18	65-80
	2.0	4G94/E	③	18 ②	15 ②	134	73	14	④	19	66-80
	2.4	4G64/G	⑤	14.5 ②	14.5 ②	87	98	13	⑥	18	25-33
	3.0	6G72/L	⑦	57	38	136	54	13	14	18	25-33
	3.5	6G74/P	80	67	38	134	54	16	36	18	29
2003	2.0	4G94/E	③	18 ②	15 ②	134	73	14	④	19	66-80
	2.4	4G64/G	⑤	14.5 ②	14.5 ②	87	98	13	⑥	18	25-33
	3.0	6G72/H	⑦	57	38	136	54	13	14	18	25-33
	3.5	6G74/P	80	67	38	134	54	16	36	18	29
2004	2.0	4G94/E	③	18 ②	15 ②	134	73	14	④	19	66-80
	2.0	4G63/D	⑩	18 ②	15 ②	17-21	95-101	⑪	⑫	16-22	26-32
	2.4	4G69/F	⑤	18 ②	15 ②	17-21	95-101	⑬	33-39	16-22	26-32
	2.4	4G64/G	⑤	14.5 ②	14.5 ②	87	98	13	⑥	18	25-33
	3.0	6G72/H	⑦	57	38	136	54	13	14	18	25-33
	3.5	6G74/P	80	67	38	134	54	16	36	18	29
	3.8	6G75/S	⑧	51-57	20 ②	134-140	54	15-17	⑨	14-22	25-33
2005	2.0	4G94/E	③	18 ②	15 ②	134	73	14	④	19	29
	2.0	4G63/F	⑩	18 ②	15 ②	17-21	95-101	⑪	⑫	16-22	26-32
	2.4	4G69/F	⑤	18 ②	15 ②	17-21	95-101	⑬	33-39	16-22	26-32
	2.4	4G64/G	⑤	14.5 ②	14.5 ②	87	98	13	⑥	18	25-33
	3.0	6G72/H	⑦	57	38	136	54	13	14	18	25-33
	3.8	6G75/S	⑧	51-57	20 ②	134-140	54	15-17	⑨	14-22	25-33
2006	2.0	4G94/E	③	18 ②	15 ②	134	73	14	④	19	29
	2.0	4G63/C	⑩	18 ②	15 ②	17-21	95-101	⑪	⑫	16-22	26-32
	2.4	4G69/F	⑤	18 ②	15 ②	17-21	95-101	⑬	33-39	16-22	26-32
	3.8	6G75/T	⑧	51-57	20 ②	134-140	54	15-17	⑨	14-22	25-33
	3.8	6G75/S	⑧	51-57	20 ②	134-140	54	15-17	⑨	14-22	25-33

① Step 1: Tighten all bolts to 35 ft. lbs.
Step 2: Loosen all bolts to 0 ft. lbs.
Step 3: Tighten all bolts to 15 ft. lbs.
Step 4: Tighten all bolts 90 degrees.
Step 5: Tighten all bolts an additional 90 degrees.

② Torque to specification plus an additional 90 degrees.

③ Step 1: Tighten all bolts to 54 ft. lbs.
Step 2: Loosen all bolts to 0 ft. lbs.
Step 3: Tighten all bolts to 15 ft. lbs.
Step 4: Tighten all bolts 90 degrees.
Step 5: Tighten all bolts an additional 90 degrees.

④ 8mm: 13 ft. lbs.
10mm: 21 ft. lbs

⑤ Step 1: Tighten all bolts to 58 ft. lbs.
Step 2: Loosen all bolts to 0 inch lbs.
Step 3: Tighten all bolts to 15 ft. lbs.
Step 4: Tighten all bolts 90 degrees.
Step 5: Tighten all bolts an additional 90 degrees.

⑥ 8mm: 20 ft. lbs.
10mm: 21 ft. lbs.

⑦ Step 1: Tighten all bolts in 3 steps to 80 ft. lbs.
Step 2: Loosen all bolts to 0 ft. lbs.
Step 3: Tighten all bolts in 3 steps to 80 ft. lbs.

⑧ Step 1: Tighten all bolts to 76-84 ft. lbs.
Step 2: Loosen all bolts to 0 inch lbs.
Step 3: Tighten all bolts to 76-84 ft. lbs.

⑨ 8mm: 12-16 ft. lbs.
10mm: 28-38 ft. lbs.
12mm bolt: 49-63 ft. lbs.
Mounting nut: 29-37 ft. lbs.

⑩ Step 1: Tighten all bolts to 58 ft. lbs.
Step 2: Loosen all bolts to 0 ft. lbs.
Step 3: Tighten all bolts to 15 ft. lbs.
Step 4: Tighten all bolts 90 degrees.
Step 5: Tighten all bolts an additional 90 degrees.

⑪ 8mm bolt: 15 ft. lbs.
10mm bolt: 22-30 ft. lbs.
Stay bolt: 21-25 ft. lbs.

⑫ 8mm: 20-28 ft. lbs.
10mm: 35-47 ft. lbs

⑬ Bolts: 18 ft. lbs.
Nuts: 15 ft. lbs

09482_GALA_C0011

WHEEL ALIGNMENT

Year	Model		Caster Range (+/-Deg.)	Caster Preferred Setting (Deg.)	Camber Range (+/-Deg.)	Camber Preferred Setting (Deg.)	Toe-in (in.)
2002	Diamante	F	0.50	+3.00	0.50	0	0 +/- 0.13
		R	—	—	0.50	-0.81	0.13 +/- 0.13
	Eclipse ①	F	1.50	+4.69	0.50	-0.09	0 +/- 0.13
		R	—	—	0.50	-1.31	0.13 +/- 0.13
	Eclipse ②	F	1.50	+4.69	0.50	-0.31	0 +/- 0.13
		R	—	—	0.50	-1.69	0.13 +/- 0.13
	Lancer	F	0.30	+2.50	0.30	0	0.09 +/- 0.13
		R	—	—	0.30	-0.40	0.12 +/- 0.08
	Galant	F	0.50	+3.00	0.50	0	0 +/- 0.13
		R	—	—	0.50	-1.00	0 +/- 0.13
	Mirage	F	0.50	+2.84	0.50	0	0 +/- 0.13
		R	—	—	0.50	-0.69	0.13 +/- 0.10
2003	Diamante	F	0.50	+3.00	0.50	0	0 +/- 0.13
		R	—	—	0.50	-0.81	0.13 +/- 0.13
	Eclipse ①	F	+/- 30	+3.00	+/- 30	0	0 +/- 0.12
		R	—	—	+/- 30	④	0.12 +/- 0.12
	Eclipse ②	F	+/- 30	+3.00	+/- 30	0	0 +/- 0.12
		R	—	—	+/- 30	④	0.13 +/- 0.13
	Lancer	F	+/- 30	+2.50	0.30	0	0.09 +/- 0.13
		R	—	—	0.30	-0.40	0.12 +/- 0.08
	Galant	F	0.50	+3.00	0.50	0	0 +/- 0.13
		R	—	—	0.50	-1.00	0 +/- 0.13
2004	Diamante	F	0.50	+3.00	0.50	0	0 +/- 0.13
		R	—	—	0.50	-0.81	0.13 +/- 0.13
	Eclipse	F	+/- 30	+3.00	+/- 30	0	0 +/- 0.12
		R	—	—	+/- 30	-0.50	0.12 +/- 0.12
	Lancer ①	F	+/- 30	+2.55	+/- 30	0	0.04 +/- 0.07
		R	—	—	+/- 30	-0.40	0.12 +/- 0.08
	Lancer ②	F	+/- 30	+2.55	+/- 30	-0.05	0.04 +/- 0.07
		R	—	—	+/- 30	-0.40	0.12 +/- 0.08
	Lancer ③	F	+/- 30	+2.45	+/- 30	0.05	0.04 +/- 0.07
		R	—	—	+/- 30	-0.40	0.12 +/- 0.08
	Evolution	F	+/- 30	+3.55	+/- 30	⑤	0 +/- 0.08
		R	—	—	+/- 30	-1.00	0.12 +/- 0.07
	Sportback ③	F	+/- 30	+2.35	+/- 30	0.05	0.04 +/- 0.07
		R	—	—	+/- 30	-0.40	0.12 +/- 0.08
	Sportback ②	F	+/- 30	+2.45	+/- 30	-0.05	0.04 +/- 0.07
		R	—	—	+/- 30	-0.40	0.12 +/- 0.08
	Galant	F	+/- 30	+3.00	+/- 30	0	0 +/- 0.12
		R	—	—	+/- 30	-0.50	0.12 +/- 0.12
2005	Eclipse	F	+/- 30	+3.00	+/- 30	0	0 +/- 0.12
		R	—	—	+/- 30	-0.50	0.12 +/- 0.12
	Lancer ①	F	+/- 30	+2.55	+/- 30	0	0.04 +/- 0.07
		R	—	—	+/- 30	-0.40	0.12 +/- 0.08
	Lancer ②	F	+/- 30	+2.55	+/- 30	-0.05	0.04 +/- 0.07
		R	—	—	+/- 30	-0.40	0.12 +/- 0.08
	Lancer ③	F	+/- 30	+2.45	+/- 30	0.05	0.04 +/- 0.07
		R	—	—	+/- 30	-0.40	0.12 +/- 0.08
	Evolution	F	+/- 30	+3.55	+/- 30	⑤	0 +/- 0.08
		R	—	—	+/- 30	-1.00	0.12 +/- 0.07

WHEEL ALIGNMENT

Year	Model		Caster		Camber		Toe-in (in.)
			Range (+/-Deg.)	Preferred Setting (Deg.)	Range (+/-Deg.)	Preferred Setting (Deg.)	
2005 Cont.	Galant	F	+/- 30	+3.00	+/- 30	0	0 +/- 0.12
		R	—	—	+/- 30	-0.50	0.12 +/- 0.12
2006	Eclipse	F	+/- 30	+3.00	+/- 30	0	0 +/- 0.12
		R	—	—	+/- 30	-0.50	0.12 +/- 0.12
	Lancer ①	F	+/- 30	+2.55	+/- 30	0	0.04 +/- 0.07
		R	—	—	+/- 30	-0.40	0.12 +/- 0.08
	Lancer ②	F	+/- 30	+2.55	+/- 30	-0.05	0.04 +/- 0.07
		R	—	—	+/- 30	-0.40	0.12 +/- 0.08
	Lancer ③	F	+/- 30	+2.45	+/- 30	0.05	0.04 +/- 0.07
		R	—	—	+/- 30	-0.40	0.12 +/- 0.08
	Evolution	F	+/- 30	+3.55	+/- 30	⑤	0 +/- 0.08
		R	—	—	+/- 30	-1.00	0.12 +/- 0.07
	Galant	F	+/- 30	+3.00	+/- 30	0	0 +/- 0.12
		R	—	—	+/- 30	-0.50	0.12 +/- 0.12

① With 14 in. wheels

② With 16 in. wheels

③ With 15 in. wheels

④ Eclipse: -1.20

   Eclipse Spyder: -1.10

⑤ With 17 in. wheels

⑤ -1.00

   Optional: -2.00

09482_GALA_C0013

TIRE, WHEEL AND BALL JOINT SPECIFICATIONS

Year	Model	OEM Tires Standard	OEM Tires Optional	Tire Pressures (psi) Front	Tire Pressures (psi) Rear	Wheel Size	Ball Joint Inspection	Wheel Lug Torque (ft. lbs.)
2002	Diamante	215/60VR16	None	32	30	6-JJ	87-109 in. ①	65-80
	Galant LS/GT-Z/ES V6	205/55R16	None	32	29	6-JJ	3-13 in. ①	66-80
	Galant DE/ES 4-cyl	185/70HR14	None	29	26	5.5-JJ	3-13 in. ①	66-80
	Mirage DE	P175/70R13	None	31	31	5-J	9-56 in. ①	66-80
	Mirage LS	P185/65R14	None	31	31	5.5-JJ	9-56 in. ①	66-80
	Mirage GS Spyder	P215/50VR17	None	32	30	6.5-JJ	9-56 in. ①	66-80
	Lancer ES	P185/65R14	None	32	29	5-JJ	0-35 in. ①	66-80
	Lancer LS/OZ/Rally	P195/60R15	None	32	29	6-JJ	0-35 in. ①	66-80
	Eclipse RS, GS	P195/70HR14	None	32	30	5.5-JJ	3-13 in. ①	66-80
	Eclipse GS-T	P205/55HR16	P215/50VR17	32	30	6-JJ	3-13 in. ①	66-80
2003	Diamante	215/60VR16	None	32	30	6-JJ	87-109 in. ①	65-80
	Galant LS/GT-Z/ES V6	205/55R16	None	32	29	6-JJ	3-13 in. ①	66-80
	Galant DE/ES 4-cyl	185/70HR14	None	29	26	5.5-JJ	3-13 in. ①	66-80
	Lancer ES	P185/65R14	None	32	29	5-JJ	0-35 in. ①	66-80
	Lancer LS/OZ/Rally	P195/60R15	None	32	29	6-JJ	0-35 in. ①	66-80
	Eclipse RS, GS	P195/70HR14	None	32	29	5.5-JJ	3-13 in. ①	66-80
	Eclipse GS-T	P205/55HR16	P215/50VR17	32	29	6-JJ	3-13 in. ①	66-80
2004	Diamante	215/60VR16	None	32	30	6-JJ	87-109 in. ①	65-80
	Galant	P215/60R16	②	32	32	③	31-61 ①	66-80
	Lancer ES	P185/65R14	None	29	26	5.5-JJ	0-35 in. ①	66-80
	Lancer LS/OZ Rally	P195/60R15	None	29	26	6-JJ	0-35 in. ①	66-80
	Lancer Ralliart	P205/50R16	None	32	32	6-JJ	0-35 in. ①	66-80
	Evolution	P235/45R17	None	32	29	8-JJ	4.4-30 in. ①	66-80
	Lancer Sportback LS	P195/60R15	None	29	26	6-JJ	0-35 in. ①	66-80
	Lancer Sportback Ralliart	P205/50R16	None	32	32	6-JJ	0-35 in. ①	66-80
	Eclipse	P195/65HR15	④	32	29	⑤	3-13 in. ①	66-80
2005	Galant	P215/60R16	②	32	32	③	31-61 ①	66-80
	Lancer ES, DE	P185/65R14	None	29	26	5.5-JJ	0-35 in. ①	66-80
	Lancer LS/OZ Rally	P195/60R15	None	29	26	6-JJ	0-35 in. ①	66-80
	Lancer Ralliart	P205/50R16	None	32	32	6-JJ	0-35 in. ①	66-80
	Evolution	P235/45R17	None	32	29	8-JJ	4.4-30 in. ①	66-80
	Eclipse	P195/65HR15	④	32	29	⑤	3-13 in. ①	66-80
2006	Galant	P215/60R16	②	32	32	③	31-61 ①	66-80
	Lancer ES, DE	P185/65R14	None	29	26	5.5-JJ	0-35 in. ①	66-80
	Lancer ES/OZ Rally	P195/60R15	None	29	26	6-JJ	0-35 in. ①	66-80
	Lancer Ralliart	P205/50R16	None	32	32	6-JJ	0-35 in. ①	66-80
	Evolution	P235/45R17	None	32	29	8-JJ	4.4-30 in. ①	66-80
	Eclipse	P225/50R17	⑦	32	29	⑥	3-13 in. ①	66-80

NOTE: If the PSI specification differs from the one on the vehicle door, use the specification on the door.

OEM: Original Equipment Manufacturer

PSI: Pounds Per Square Inch

① Torque required in inch lbs. to rotate ball joint when removed from the knuckle.

② 3.8L engine: P215/55R17

③ STD: 6.5-JJ 16" wheels
   OPT: 7-JJ 17" wheels

④ P205/55R16 or P215/50R17

⑤ STD: 6-JJ 15" and 16" wheels
   OPT: 6.5-JJ 17" wheels

⑥ STD: 7.5-JJ 17" wheels
   OPT: 8-JJ 18" wheels

⑦ 3.8L engine: P225/45R18

09482_GALA_C0014

BRAKE SPECIFICATIONS
All measurements in inches unless noted

Year	Model		Brake Disc — Original Thickness	Brake Disc — Minimum Thickness	Brake Disc — Maximum Runout	Brake Drum Diameter — Original Inside Diameter	Brake Drum Diameter — Max. Wear Limit	Brake Drum Diameter — Maximum Machine Diameter	Minimum Lining Thickness — Front	Minimum Lining Thickness — Rear	Brake Caliper — Bracket Bolts (ft. lbs.)	Brake Caliper — Mounting Bolts (ft. lbs.)
2002	Diamante	F	0.940	0.880	0.002	—	—	—	0.080	—	54	65
		R	0.410	0.330	0.0023	—	—	—	—	0.039	24	36-43
	Galant	F	①	②	0.002	—	—	—	0.080	—	③	65
		R	0.390	0.331	0.002	9.000	—	9.100	—	0.040	36-43	54
	Mirage	F	0.940	0.882	0.0012	—	—	—	0.080	—	36	67-81
		R	—	—	—	8.00	8.10	8.10	—	0.039	—	—
	Lancer	F	0.900	0.880	0.002	—	—	—	0.080	—	37	62
		R	—	—	—	7.99	8.07	8.07	—	0.040	—	—
	Eclipse	F	0.940	0.882	0.002	—	—	—	0.080	—	58-72	46-62
		R	—	—	—	9.000	—	9.100	—	0.039	—	—
	Eclipse w/rear disc	F	0.940	0.882	0.003	—	—	—	0.080	—	③	46-62
		R	0.390	0.331	0.003	—	—	—	—	0.080	36-43	54
2003	Diamante	F	0.940	0.880	0.002	—	—	—	0.080	—	54	65
		R	0.410	0.330	0.0023	—	—	—	—	0.039	24	36-43
	Galant	F	①	②	0.002	—	—	—	0.080	—	③	65
		R	0.390	0.331	0.002	9.000	—	9.100	—	0.040	36-43	54
	Lancer	F	0.900	0.880	0.002	—	—	—	0.080	—	37	62
		R	—	—	—	7.99	8.07	8.07	—	0.040	—	—
	Eclipse	F	0.940	0.882	0.002	—	—	—	0.080	—	58-72	46-62
		R	—	—	—	9.000	—	9.100	—	0.039	—	—
	Eclipse w/rear disc	F	0.940	0.882	0.003	—	—	—	0.080	—	③	46-62
		R	0.390	0.331	0.003	—	—	—	—	0.080	36-43	54
2004	Diamante	F	0.940	0.880	0.002	—	—	—	0.080	—	54	65
		R	0.410	0.330	0.0023	—	—	—	—	0.039	24	36-43
	Galant	F	1.020	0.960	0.0039	—	—	—	0.080	—	25-31	77-81
		R	0.390	0.330	0.0016	—	—	—	—	0.080	28-36	42-48
	Lancer	F	1.020	0.960	0.0006	—	—	—	0.080	—	④	67-81
		R	—	—	—	7.99	8.07	8.07	—	0.040	—	—
	Lancer w/rear disc	F	1.020	0.960	0.0006	—	—	—	0.080	—	④	67-81
		R	0.390	0.330	0.0015	—	—	—	—	0.080	28-36	42-48
	Evolution	F	1.260	1.170	0.0012	—	—	—	0.080	—	—	73-87
		R	0.870	0.80	0.0012	—	—	—	—	0.080	—	36-44
	Sportback	F	1.020	0.960	0.0015	—	—	—	0.080	—	25-31	67-81
		R	0.390	0.330	0.0015	—	—	—	—	0.080	28-36	42-48
	Eclipse	F	0.940	0.882	0.002	—	—	—	0.080	—	⑤	67-81
		R	—	—	—	9.000	—	9.100	—	0.039	—	—
	Eclipse w/rear disc	F	0.940	0.882	0.003	—	—	—	0.080	—	⑤	67-81
		R	0.390	0.331	0.003	—	—	—	—	0.080	36-43	41-44
2005	Galant	F	1.020	0.960	0.0039	—	—	—	0.039	—	25-31	67-81
		R	0.390	0.330	0.0016	—	—	—	—	0.039	28-36	42-48
	Lancer	F	1.020	0.960	0.0006	—	—	—	0.080	—	④	67-81
		R	—	—	—	7.99	8.07	8.07	—	0.040	—	—
	Lancer w/rear disc	F	1.020	0.960	0.0006	—	—	—	0.080	—	④	67-81
		R	0.390	0.330	0.0015	—	—	—	—	0.080	28-36	42-48
	Evolution	F	1.260	1.170	0.0012	—	—	—	0.080	—	—	73-87
		R	0.870	0.80	0.0012	—	—	—	—	0.080	—	36-44
	Eclipse	F	1.020	0.960	0.002	—	—	—	0.080	—	25-31	67-81
		R	—	—	—	9.000	—	9.100	—	0.080	—	—

BRAKE SPECIFICATIONS

All measurements in inches unless noted

Year	Model		Brake Disc Original Thickness	Brake Disc Minimum Thickness	Brake Disc Maximum Runout	Brake Drum Diameter Original Inside Diameter	Brake Drum Diameter Max. Wear Limit	Brake Drum Diameter Maximum Machine Diameter	Minimum Lining Thickness Front	Minimum Lining Thickness Rear	Brake Caliper Bracket Bolts (ft. lbs.)	Brake Caliper Mounting Bolts (ft. lbs.)
2005 Cont.	Eclipse w/rear disc	F	1.020	0.960	0.002	—	—	—	0.080	—	25-31	67-81
		R	0.400	0.331	0.003	—	—	—	—	0.080	28-36	⑥
2006	Galant	F	1.020	0.960	0.0039	—	—	—	0.039	—	25-31	67-81
		R	0.390	0.330	0.0016	—	—	—	—	0.039	28-36	42-48
	Lancer	F	1.020	0.960	0.0006	—	—	—	0.080	—	④	67-81
		R	—	—	—	7.99	8.07	8.07	—	0.040	—	—
	Lancer w/rear disc	F	1.020	0.960	0.0006	—	—	—	0.080	—	④	67-81
		R	0.390	0.330	0.0015	—	—	—	—	0.080	28-36	42-48
	Evolution	F	1.260	1.170	0.0012	—	—	—	0.080	—	—	73-87
		R	0.870	0.80	0.0012	—	—	—	—	0.080	—	36-44
	Eclipse	F	1.020	0.960	0.0039	—	—	—	0.080	—	25-31	67-81
		R	0.390	0.330	0.0016	—	—	—	—	0.080	28-36	42-48

F: Front

R: Rear

① 2.4L: 0.940
  3.0L: 1.020

② 2.4L: 0.88
  3.0L: 0.96

③ Lock pin (2.4L): 55 ft. lbs.
  Lock bolt (3.0L): 28 ft. lbs.

④ Main slide pin (2.0L): 59-65 ft. lbs.
  Sub slide pin (2.0L): 34-40 ft. lbs.
  Lock pin bolt (2.4L): 25-31 ft. lbs.

⑤ Lock pin (2.4L): 46-62 ft. lbs.
  Lock bolt (3.0L): 25-31 ft. lbs.

⑤ Lock pin (2.4L): 46-62 ft. lbs.
  Lock bolt (3.0L): 25-31 ft. lbs.

⑥ Bolt and washer: 38-44ft. lbs.
  Flange bolt: 41-47 ft. lbs.

09482_GALA_C0016

SCHEDULED MAINTENANCE INTERVALS
Mitsubishi—Diamante, Eclipse, Galant, Mirage, Lancer & Evolution

TO BE SERVICED	TYPE OF SERVICE	VEHICLE MILEAGE INTERVAL (x1000)													
		7.5	15	22.5	30	37.5	45	52.5	60	67.5	75	82.5	90	97.5	105
Engine oil & filter	R	✔	✔	✔	✔	✔	✔	✔	✔	✔	✔	✔	✔	✔	✔
Automatic transaxle fluid & filter	S/I		✔		✔		✔		✔		✔		✔		
Brake hoses	S/I		✔		✔		✔		✔		✔		✔		
Disc brake pads	S/I		✔		✔		✔		✔		✔		✔		
Driveshaft boots	S/I		✔		✔		✔		✔		✔		✔		
Valve clearance	S/I				✔				✔				✔		
Air cleaner element	R				✔				✔				✔		
Engine coolant (except Evolution)	R								✔				✔		
Engine coolant (Evolution)	R					✔			✔				✔		
Spark plugs (standard type)	R					✔			✔				✔		
Spark plugs (iridium type) (except Evolution)	R														✔
Spark plugs (iridium type) (Evolution)	R								✔						
Spark plugs (platinum type)	R								✔		✔				
Ball joints & steering linkage seals	S/I				✔				✔				✔		
Drive belt(s)	S/I				✔				✔				✔		
Exhaust system	S/I				✔				✔				✔		
Fuel hoses	S/I				✔				✔				✔		
Manual transaxle fluid	S/I				✔				✔				✔		
Manual transaxle fluid (including transfer case Evolution)	S/I				✔				✔				✔		
Rear axle oil (Evolution AWD)	S/I				✔				✔				✔		
Rear drum brake linings & rear wheel cylinders	S/I				✔				✔				✔		
Ignition cables	R								✔						
Timing belt(s)	R								✔						
Distributor cap & rotor	S/I								✔						
Rotate tires	S/I	✔	✔	✔	✔	✔	✔	✔	✔	✔	✔	✔	✔	✔	✔
EVAP system (except canister)	S/I								✔						
Fuel system (tank, pipe line, connection & fuel tank filler tube cap)	S/I								✔						

R: Replace   S/I: Service or Inspect

09482_GALA_C0017

FREQUENT OPERATION MAINTENANCE (SEVERE SERVICE)

If a vehicle is operated under any of the following conditions it is considered severe service:
- Extremely dusty or sandy areas.
- Extensive use of brakes while driving.
- 50% or more of the vehicle operation is in 32°C (90°F) or higher temperatures,
  or constant operation in temperatures below 0°C (32°F).
- Prolonged idling (vehicle operation in stop and go traffic).
- Frequent short running periods (engine does not warm to normal operating temperatures).
- Police, taxi, delivery usage or trailer towing usage.

Oil & oil filter: change every 3000 miles.

Disc brake pads: service or inspect ever 6000 miles.

Air filter element: service or inspect every 15,000 miles.

Automatic transaxle fluid & filter: replace every 15,000 miles.

Spark plugs: replace every 15,000 miles.

Manual transaxle oil (including transfer case Evolution): replace every 30,000 miles.

09482_GALA_C0018

ENGINE REPAIR

➡Disconnecting the negative battery cable on some vehicles may interfere with the functions of the on board computer systems and may require the computer to undergo a relearning process, once the negative battery cable is reconnected.

Distributor

REMOVAL

Before removing the distributor, position No. 1 cylinder at TDC on the compression stroke and align the timing marks.

1. Before servicing the vehicle, refer to the Precautions Section.
2. Remove or disconnect the following:
   - Negative battery cable
   - Ignition wire cover, if equipped
   - Distributor harness connector
   - Distributor cap with all ignition wires still connected
   - Coil wire, if necessary
3. Matchmark the rotor to the distributor housing, and the distributor housing to the engine.
4. Remove or disconnect the following:
   - Hold-down nut
   - Distributor from the engine

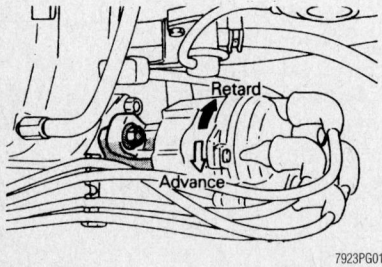

Adjusting the distributor—1.5L engine

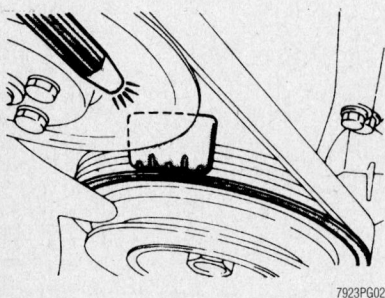

Checking the ignition timing—1.5L engine

INSTALLATION

Timing Not Disturbed

1. Install or connect the following:
   - New distributor housing O-ring and lubricate with clean oil
   - Distributor in the engine, match-marks aligned
   - Hold-down nut
   - Distributor harness connectors
   - Distributor cap
   - Coil wire, if removed
   - Negative battery cable
2. Adjust the ignition timing and tighten the hold-down nut to 96 inch lbs. (11 Nm).

Timing Disturbed

1. Install a new distributor housing O-ring and lubricate with clean oil.
2. Position the engine so the No. 1 piston is at TDC of its compression stroke and the mark on the vibration damper is aligned with **0** on the timing indicator.
3. Align the distributor housing and gear mating marks. Install the distributor in the engine so the slot or groove of the distributor's installation flange aligns with the distributor installation stud in the engine block. Be sure the distributor is fully seated. Inspect alignment of the distributor rotor making sure the rotor is aligned with the position of the No. 1 ignition wire in the distributor cap.
4. Install or connect the following:
   - Hold-down nut
   - Distributor harness connectors
   - Distributor cap
   - Negative battery cable
5. Adjust the ignition timing and tighten the hold-down nut to 96 inch lbs. (11 Nm).

Alternator

REMOVAL & INSTALLATION

1. Before servicing the vehicle, refer to the Precautions Section.
2. Disconnect the negative battery cable.
3. Remove the necessary components in order to gain access to the alternator assembly.
4. Disconnect the alternator electrical connections.
5. Remove the alternator mounting bolts.
6. Remove the alternator from the vehicle.

➡On some vehicles it may be easier to remove the alternator from underneath of the vehicle.

To install:
7. Position the alternator assembly to its mounting.
8. Install the retaining bolts.
9. Connect the electrical connectors.
10. Continue the installation in the reverse order of the removal procedure.

Ignition Timing

ADJUSTMENT

The ignition timing is controlled by the Electronic Control Module (ECM) and is not adjustable. However it can be inspected using a scan tool.

Engine Assembly

REMOVAL & INSTALLATION

Diamante

1. Before servicing the vehicle, refer to the Precautions Section.
2. Remove the hood assembly.
3. Relieve fuel system pressure.
4. Remove or disconnect the following:
   - Negative, then the positive battery cable
   - Battery
   - Air cleaner assembly and all adjoining air intake duct work
5. Drain the engine coolant and remove the radiator assembly and coolant reservoir (and bracket).
6. Remove or disconnect the following:
   - Engine undercover, if equipped
   - Front exhaust pipe
   - Transaxle assembly
   - Accelerator cable from the throttle body
   - Vacuum hoses from the intake manifold, label for installation
   - High pressure fuel line and the fuel return line
   - Vacuum hoses from the solenoid valves
   - Vacuum hoses from the purge canister
   - Heater hose connections from the engine
   - Harness for the Exhaust Gas Recirculation (EGR) temperature sensor connection, if equipped

- Engine drive belts
- Power steering pump oil pressure switch connection from the pump
- Power steering pump and secure away from the work area
- Air conditioning compressor. Wire the compressor aside. Do not discharge or disconnect the air conditioning lines.
- Wiring to the alternator
- Harness plugs for the Barometer (BARO) sensor, Idle Speed Control (ISC) motor, Throttle Position (TP) sensor, fuel injectors and Knock (KS) sensor
- Harness plugs for the Engine Coolant Temperature (ECT) switch, sensor and gauge
- Harness plugs for the ignition coil, condenser and ignition power transistor
- Harness plugs for the variable induction control motor and the Manifold Absolute Pressure (MAP) sensor
- Harness plugs for the Crankshaft Position (CKP) and Camshaft Position (CMP) sensors
- Radiator overflow tank and remove the mounting bracket
- Ground cable connections

7. Attach a hoist to the engine and take up the engine weight. Remove the engine mount bracket. Remove any torque control brackets (roll stoppers).

8. Lift the engine slowly and remove from the engine compartment.

To install:

9. Install or connect the following:
- Engine and secure all control brackets
- Transaxle assembly
- Engine ground cable connections
- Harness plugs for the CKP and CMP sensors
- Harness plugs for the variable induction control motor and the MAP sensor

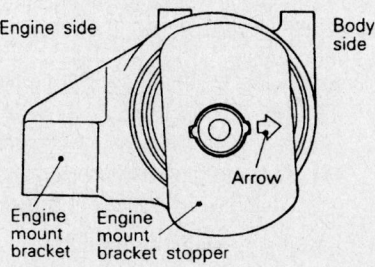

Alignment of the engine mount stopper bracket—3.5L engine

- Harness plugs for the ignition coil, condenser and ignition power transistor
- Harness plugs for the ECT switch, sensor and gauge
- Harness plugs for the BARO sensor, ISC motor, TP sensor, fuel injectors and KS sensor
- Wiring to the alternator
- Air conditioning compressor assembly
- Power steering pump assembly
- Power steering pump oil pressure switch harness plug to the pump
- Engine drive belts, adjust
- Harness for the EGR temperature sensor
- Heater hose connections to the engine, using new hose clamps
- Vacuum hoses to the purge canister
- Vacuum hoses to the solenoid valves
- High pressure fuel line and the fuel return line, using new clamps or O-rings
- Vacuum hoses
- Accelerator cable to the throttle body
- Air cleaner assembly and all adjoining air intake duct work
- Radiator and coolant reservoir assembly
- Transaxle assembly
- Exhaust system to the engine, using new gaskets
- Battery to the vehicle
- Positive, then the negative battery cables
- Engine undercover, if equipped

10. Fill the engine with the proper amount of engine oil and coolant.
11. Install the hood.
12. Start the engine and check for leaks.

Eclipse

2002–2005 2.4L ENGINE

1. Before servicing the vehicle, refer to the Precautions Section.
2. Relieve the fuel system pressure.
3. Remove or disconnect the following:
- Negative battery cable
- Hood
- Intake air duct
4. Drain the engine coolant.
5. Remove or disconnect the following:
- Hoses and remove the radiator
- Engine undercover
6. Attach an engine lifting fixture to the engine and remove the transaxle assembly.
7. Disconnect the following connectors:
- Air conditioning compressor

- Power steering pressure switch
- Heated Oxygen (HO$_2$S) sensor
- Engine Coolant Temperature (ECT) gauge sender
- ECT sensor
- Manifold Absolute Pressure (MAP) sensor
- Intake Air Temperature (IAT) sensor
8. Remove or disconnect the following:
- Power steering pump from the bracket and position the pump out of the way
- Air conditioning compressor from the bracket and position it out of the way. Do not disconnect the hoses.
- Accelerator cable from the throttle body and mounting bracket
9. Disconnect the following connectors:
- Idle Air Control (IAC) motor
- Knock (KS) sensor
- Ignition module (power transistor)
- Exhaust Gas Recirculation (EGR) solenoid
- Oil pressure switch
- Throttle Position (TP) sensor
- Condenser
- Injectors
- Ignition coil
- Camshaft Position (CMP) sensor
- Crankshaft Position (CKP) sensor
- Engine control wiring harness
10. Remove or disconnect the following:
- Heater hoses from the engine
- Fuel lines from the fuel supply rail
- Purge air hose and the brake booster vacuum hose
- Front exhaust pipe from the manifold

11. Place a floor jack against the oil pan with a piece of wood in between to protect the oil pan.
12. Raise the engine with the jack and remove the engine support fixture.
13. Install a chain hoist to the top of the engine.
14. Remove the engine mount bracket.
15. Lift the engine up slowly out of the engine compartment.

To install:

16. Slowly lower the engine assembly into the vehicle.
17. Position the floor jack under the oil pan with a piece of wood in between. Use the floor jack to adjust the height of the engine while installing the engine mount bracket.
18. Remove the chain hoist and install the engine support fixture.
19. Install or connect the following:
- Front exhaust pipe to the manifold
- Brake booster vacuum hose

- New O-ring on the high pressure fuel line. Apply a small amount of clean engine oil to the O-ring and connect the fuel lines to the fuel supply rail.
20. Connect the following connectors:
  - IAC motor
  - KS sensor
  - Ignition module (power transistor)
  - EGR solenoid
  - Oil pressure switch
  - TP sensor
  - Condenser
  - Injectors
  - Ignition coil
  - CMP sensor
  - CKP sensor
  - Engine control wiring harness
21. Install or connect the following:
  - Accelerator cable, adjust
  - Air conditioning compressor and the power steering pump in their brackets
  - IAT sensor, MAP sensor, ECT sensor and gauge sender, HO_2S sensor, power steering pressure switch and the air conditioning compressor harness connectors
  - Radiator and hoses
  - Transaxle and remove the engine support fixture
  - Engine undercovers
  - Intake air duct
  - Negative battery cable
  - Hood
22. Refill the engine with the proper amount of coolant.

2006 2.4L (4G69) ENGINE

1. Before servicing the vehicle, refer to the Precautions Section.
2. Disconnect the negative battery cable. Relieve the fuel system pressure.
3. Remove the engine under cover. Drain the engine oil. Drain the engine coolant. Drain the transaxle fluid.
4. Remove the hood. Remove the battery and battery tray.
5. Remove the engine air cleaner. Remove the ECM/PCM.
6. Remove the upper and lower radiator hoses. Remove the radiator. Remove the number one and two exhaust pipes.
7. Disconnect the engine electrical wiring harness. Disconnect the EVAP hose connection.
8. Disconnect the brake booster vacuum hose connection. Remove the drive belt.
9. Remove the power steering pump and bracket assembly. Lay the power steering pump to the side. Do not disconnect the fluid lines from the pump.

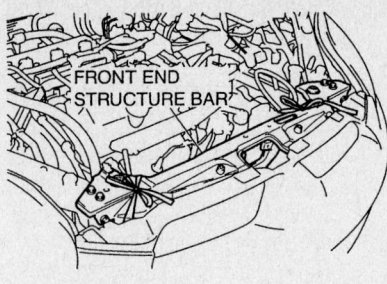

Front end structure bar—2006 Eclipse 2.4L and 3.0L engines

10. Remove the air conditioning compressor. Lay the air conditioning compressor to the side. Do not disconnect the fluid lines from the pump.
11. With the hose installed, remove the automatic transaxle fluid cooler and bracket assembly from the transaxle case front stopper bracket.
12. Remove the heater hoses. Remove the fuel high pressure hose connection.
13. Secure the engine front end structure bar tool in a location that will not interfere with the engine removal.
14. Remove the transaxle assembly from the vehicle.
15. Remove the grounding cable connection and power steering fluid reservoir.
16. Support the engine with a garage jack. Remove tool MB991895, MB991454 and MB991527.

➡ **These special tools were needed and installed to remove the transaxle from the vehicle.**

17. Hold the engine assembly in place with the engine lifting device. Properly position a garage jack against the engine oil pan with a piece of wood in between so that the weight of the engine assembly is no longer being applied to the front mounting bracket.
18. Loosen the engine front mounting bracket mounting nuts and blots. Remove the front engine mounting bracket.

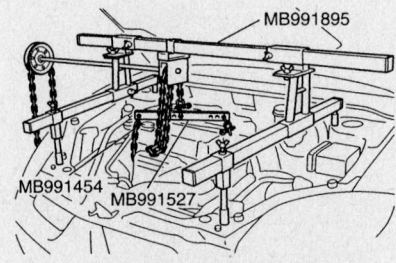

These special tools are needed to support the engine while the transaxle is removed—2006 Eclipse 2.4L engine

19. After checking that all cables, hoses and wiring are disconnected from the engine, lift the engine hoist slowly and remove the engine from the vehicle.

To install:

20. Slowly lower the engine assembly into the vehicle.
21. The installation is the reverse of the removal procedure.
22. Be sure to fill the engine with the proper grade and type fluids.
23. Be sure to fill the transaxle with the proper grade and type fluid.
24. Start the engine and check for leaks, correct as required.
25. Complete the vehicle initialization procedure.

➡ **To complete the initialization procedure the following tools are needed. MB991958 scan tool, MB991824 VCI, MB991827 MUT III USB cable, MB991910 MUT III Main harness "A"**

26. Connect the scan tool to the data link connector. To prevent damage to the scan tool be sure that the ignition switch is in the LOCK position before connecting the scan tool.
27. Turn the ignition switch to the ON position.
28. Select "check mode" from the menu screen.
29. Select "erase memory" from the menu screen.
30. Initialize the learning value.
31. After initialization complete the idle learning procedure.

➡ **This procedure must be performed when the PCM is replaced, or when the learning value is initialized, as the idling is not stabilized because the learning value in the MFI engine is not completed.**

32. Start the engine. Allow the coolant temperature to reach 176 degrees or more.
33. Stop the engine and place the ignition switch in the LOCK position.
34. After ten seconds, restart the engine.
35. For ten minutes carry out the idling procedure below to confirm that the engine has normal idling.
  - Position the transaxle selector lever in the "P" range
  - The engine fan is not to be operated
  - The engine coolant temperature should be 176 degrees or more

➡ **If the engine stalls during idling, check the throttle valve of the throttle body for dirt. Correct and perform the procedure again.**

36. Road test the vehicle.

37. Check the front end alignment and correct as required.

38. Reprogram the radio stations and security codes, as necessary.

2002–2005 3.0L ENGINE

1. Before servicing the vehicle, refer to the Precautions Section.

2. Disconnect the negative battery cable.

3. Drain the engine coolant.

4. Drain the engine oil and the transmission oil.

5. Relieve the fuel system pressure.

6. Remove or disconnect the following:
- All wires, cables and hoses connected to the engine
- Hood
- Air intake and breather hoses
- Radiator hoses and remove the radiator
- Front exhaust pipe
- Power steering pump and position it aside
- Air conditioning compressor drive belt
- Compressor from its mount and hang it out of the way. Do not disconnect the hoses and do not allow the compressor to hang by the hoses.

7. Install engine hoist equipment and make certain the attaching points on the engine are secure.

8. Raise the hoist enough to support the engine.

9. Remove or disconnect the following:
- Front and rear engine roll stoppers
- Left engine mount and support bracket

10. Slowly lift the engine and remove it from the vehicle.

3.8L ENGINE

1. Before servicing the vehicle, refer to the Precautions Section.

2. Disconnect the negative battery cable. Relieve the fuel system pressure.

3. Remove the engine under cover. Drain the engine oil. Drain the engine coolant. Drain the transaxle fluid.

4. Remove the hood. Remove the battery and battery tray.

5. Remove the engine air cleaner. Remove the ECM/PCM.

6. Remove the front exhaust pipe.

7. Remove the top engine cover. Disconnect the control wiring harness connection. Disconnect the EVAP hose connection.

8. Disconnect the high pressure fuel hose connection. Disconnect and remove the top and bottom radiator hoses. Remove the radiator.

9. Remove the driveshaft.

10. Remove the exhaust manifold right bank. Remove the intake manifold plenum.

11. Remove the drive belt. Remove the power steering pump belt. Disconnect the power steering pressure switch connector.

12. Remove the power steering pump from its mounting. Position it to the side where it will not interfere with the engine removal.

13. Remove the starter. Remove the intake manifold plenum stay, throttle body stay, engine oil dipstick assembly and engine hanger, and install special tools MB992012 and MB992013 to the engine.

14. Remove the transaxle from the vehicle.

15. Remove the air conditioning compressor and bracket. Position the unit to the side where it will not interfere with the engine removal.

16. Support the engine with a garage jack. Remove tool MB991895, MB991454, MB992012 and MB992013.

➡**These special tools were needed and installed to remove the transaxle from the vehicle.**

17. Hold the engine assembly in place with the engine lifting device. Properly position a garage jack against the engine oil pan with a piece of wood in between so that the weight of the engine assembly is no longer being applied to the front mounting bracket.

18. Loosen the engine front mounting bracket mounting nuts and blots. Remove the front engine mounting bracket.

19. After checking that all cables, hoses and wiring are disconnected from the engine, lift the engine hoist slowly and remove the engine from the vehicle.

To install:

20. Slowly lower the engine assembly into the vehicle.

21. The installation is the reverse of the removal procedure.

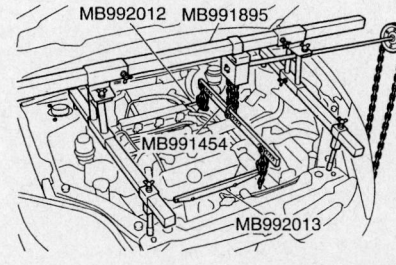

09482_GALA_G0003

These special tools are needed to support the engine while the transaxle is removed—2006 Eclipse 3.8L engine

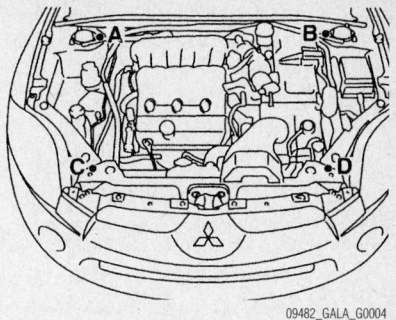

09482_GALA_G0004

Special tool installation–strut mounting nuts A and B. Front end structure bar assembling bolts (C and D)—2006 Eclipse 3.8L engine

22. Place a garage jack against the engine oil pan, with a piece of wood in between. Install the engine mount while adjusting the position of the engine.

23. Support the engine with a garage jack, Remove the chain block (engine lifting device).

24. Install engine hanger tool MB991895 to the strut mounting nuts and the front end structure bar assembling bolts.

25. Continue the installation in the reverse order of the removal procedure.

26. Be sure to fill the engine with the proper grade and type fluids.

27. Be sure to fill the transaxle with the proper grade and type fluid.

28. Start the engine and check for leaks, correct as required.

29. Complete the vehicle initialization procedure.

➡**To complete the initialization procedure the following tools are needed. MB991958 scan tool, MB991824 VCI, MB991827 MUT III USB cable, MB991910 MUT III Main harness "A"**

30. Connect the scan tool to the data link connector. To prevent damage to the scan tool be sure that the ignition switch is in the LOCK position before connecting the scan tool.

31. Turn the ignition switch to the ON position.

32. Select "check mode" from the menu screen.

33. Select "erase memory" from the menu screen.

34. Initialize the learning value.

35. After initialization complete the idle learning procedure.

➡**This procedure must be performed when the PCM is replaced, or when the learning value is initialized, as the idling is not stabilized because the learning value in the MFI engine is not completed.**

36. Start the engine. Allow the coolant temperature to reach 176 degrees or more.

37. Stop the engine and place the ignition switch in the LOCK position.

38. After ten seconds, restart the engine.

39. For ten minutes carry out the idling procedure below to confirm that the engine has normal idling.

- Position the transaxle selector lever in the "P" range
- The engine fan is not to be operated
- The engine coolant temperature should be 176 degrees or more

➡**If the engine stalls during idling, check the throttle valve of the throttle body for dirt. Correct and perform the procedure again.**

40. Road test the vehicle.

41. Check the front end alignment and correct as required.

42. Reprogram the radio stations and security codes, as necessary.

Galant

2002–2003 2.4L AND 3.0L ENGINES

1. Before servicing the vehicle, refer to the Precautions Section.

2. Disconnect the negative battery cable.

3. Drain the engine coolant.

4. Drain the engine oil and the transmission oil.

5. Relieve the fuel system pressure.

6. Remove or disconnect the following:

- Hood
- Transaxle assembly
- Radiator hoses and remove the radiator
- Accelerator cable and remove the bracket
- Air intake and breather hoses
- Heater hoses
- Brake booster vacuum hose at the engine
- Vacuum hoses at the throttle body, label
- Fuel feed and return hoses

7. Disconnect the following:

- Power steering pressure switch
- Alternator
- Oil pressure switch
- Air conditioning compressor
- Each injector
- Power transistor
- Ignition coil
- Throttle Position (TP) sensor
- Idle Air Control (IAC) motor
- Engine Coolant Temperature (ECT) switch
- ECT sensor
- Exhaust Gas Recirculation (EGR) temperature sensor
- Engine control wiring harness
- Heated Oxygen (HO2S) sensor
- Crankshaft Position (CKP) sensor
- Camshaft Position (CMP) sensor
- Refrigerant temperature switch
- Condenser connection

8. Remove or disconnect the following:

- Power steering pump and position it aside
- Air conditioning compressor drive belt
- Compressor from its mount and hang it out of the way. Do not disconnect the hoses and do not allow the compressor to hang by the hoses.
- Front exhaust pipe

9. Install engine hoist equipment and make certain the attaching points on the engine are secure.

10. Raise the hoist enough to support the engine.

11. Remove or disconnect the following:

- Front and rear engine roll stoppers
- Left engine mount and support bracket

12. Slowly lift the engine and remove it from the vehicle.

To install:

13. Lower the engine into the vehicle.

14. Install the front and rear roll stoppers and the left engine mount. Do not torque the through-bolts at this time.

15. Remove the lifting apparatus from the engine.

16. Connect the exhaust system to the manifold, using a new gasket and new locking nuts. Tighten the nuts and the small bolt to 33 ft. lbs. (44 Nm)

17. Tighten the engine mount nuts and bolts. Correct torque values are:

   a. Upper mount to engine nuts: 42 ft. lbs. (57 Nm).

   b. Upper mount to engine bolt: 108 inch lbs. (12 Nm).

   c. Upper mount through-bolt: 72–87 ft. lbs. (98–118 Nm).

   d. Rear roll stopper through-bolt: 32 ft. lbs. (44 Nm).

   e. Front roll stopper through-bolt: 41 ft. lbs. (57 Nm).

18. Install or connect the following:

- Air conditioning compressor, tightening the mounting bolts to 18 ft. lbs. (25 Nm)
- Power steering pump, tightening the front bolts to 21 ft. lbs. (28 Nm) and the rear bolt to 16 ft. lbs. (22 Nm)

- Accessory drive belts

19. Connect the following:

- Power steering pressure switch
- Alternator
- Oil pressure switch
- Air conditioning compressor
- Each injector
- Power transistor
- Ignition coil
- TP sensor
- IAC motor
- ECT switch
- ECT sensor
- EGR temperature sensor
- Engine control wiring harness
- HO2S sensor
- CKP sensor
- CMP sensor
- Refrigerant temperature switch
- Condenser connection

20. Install or connect the following:

- Fuel return hose and secure with the retaining clamp
- New O-ring, connect the high pressure fuel line and tighten the bolts to 48 inch lbs. (6 Nm)
- Vacuum lines running to the throttle body
- Heater hoses
- Accelerator cable bracket, tightening the bolts to 48 inch lbs. (6 Nm), and connect the accelerator cable
- Radiator and connect the hoses
- Transaxle

21. Fill the coolant system.

22. Connect the negative battery cable.

23. Start the engine and check for leaks.

24. Install the hood.

2004–2006 2.4L (4G69) ENGINE

1. Before servicing the vehicle, refer to the Precautions Section.

2. Disconnect the negative battery cable. Relieve the fuel system pressure.

3. Remove the engine under cover. Drain the engine oil. Drain the engine coolant. Drain the transaxle fluid.

4. Remove the hood. Remove the battery and battery tray.

5. Remove the engine air cleaner. Remove the ECM/PCM.

6. Remove the upper and lower radiator hoses. Remove the radiator. Remove the number one and two exhaust pipes.

7. Disconnect the engine electrical wiring harness. Disconnect the EVAP hose connection.

8. Disconnect the brake booster vacuum hose connection. Remove the drive belt.

9. Remove the power steering pump

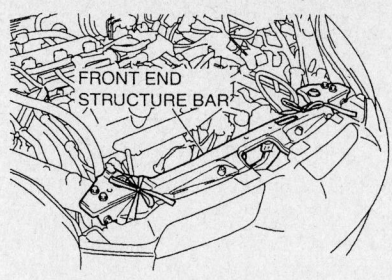

09482_GALA_G0005

Front end structure bar—2004–2006 Galant 2.4L engine

and bracket assembly. Lay the power steering pump to the side. Do not disconnect the fluid lines from the pump.

10. Remove the air conditioning compressor. Lay the air conditioning compressor to the side. Do not disconnect the fluid lines from the pump.

11. With the hose installed, remove the automatic transaxle fluid cooler and bracket assembly from the transaxle case front stopper bracket.

12. Remove the heater hoses. Remove the fuel high pressure hose connection.

13. Secure the engine front end structure bar tool in a location that will not interfere with the engine removal.

14. Remove the transaxle assembly from the vehicle.

15. Remove the grounding cable connection and power steering fluid reservoir.

16. Support the engine with a garage jack. Remove tool MB991895, MB991454 and MB991527.

➡ **These special tools were needed and installed to remove the transaxle from the vehicle.**

17. Hold the engine assembly in place with the engine lifting device. Properly position a garage jack against the engine oil pan with a piece of wood in between so that the weight of the engine assembly is no longer being applied to the front mounting bracket.

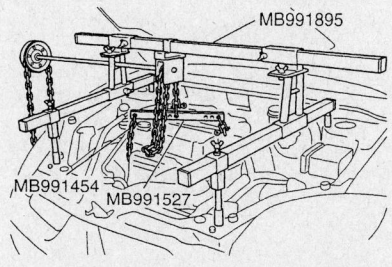

09482_GALA_G0006

These special tools are needed to support the engine while the transaxle is removed—2004–2006 Galant 2.4L engine

18. Loosen the engine front mounting bracket mounting nuts and blots. Remove the front engine mounting bracket.

19. After checking that all cables, hoses and wiring are disconnected from the engine, lift the engine hoist slowly and remove the engine from the vehicle.

To install:

20. Slowly lower the engine assembly into the vehicle.

21. The installation is the reverse of the removal procedure.

22. Be sure to fill the engine with the proper grade and type fluids.

23. Be sure to fill the transaxle with the proper grade and type fluid.

24. Start the engine and check for leaks, correct as required.

25. Complete the vehicle initialization procedure.

➡ **To complete the initialization procedure the following tools are needed. MB991958 scan tool, MB991824 VCI, MB991827 MUT III USB cable, MB991910 MUT III Main harness "A"**

26. Connect the scan tool to the data link connector. To prevent damage to the scan tool be sure that the ignition switch is in the LOCK position before connecting the scan tool.

27. Turn the ignition switch to the ON position.

28. Select "check mode" from the menu screen.

29. Select "erase memory" from the menu screen.

30. Initialize the learning value.

31. After initialization complete the idle learning procedure.

➡ **This procedure must be performed when the PCM is replaced, or when the learning value is initialized, as the idling is not stabilized because the learning value in the MFI engine is not completed.**

32. Start the engine. Allow the coolant temperature to reach 176 degrees or more.

33. Stop the engine and place the ignition switch in the LOCK position.

34. After ten seconds, restart the engine.

35. For ten minutes carry out the idling procedure below to confirm that the engine has normal idling.

- Position the transaxle selector lever in the "P" range
- The engine fan is not to be operated
- The engine coolant temperature should be 176 degrees or more

➡ **If the engine stalls during idling, check the throttle valve of the throttle body for dirt. Correct and perform the procedure again.**

36. Road test the vehicle.

37. Check the front end alignment and correct as required.

38. Reprogram the radio stations and security codes, as necessary.

2004–2006 3.8L ENGINE

1. Before servicing the vehicle, refer to the Precautions Section.

2. Disconnect the negative battery cable. Relieve the fuel system pressure.

3. Remove the engine under cover. Drain the engine oil. Drain the engine coolant. Drain the transaxle fluid.

4. Remove the hood. Remove the battery and battery tray.

5. Remove the engine air cleaner. Remove the ECM/PCM.

6. Remove the front exhaust pipe. Remove the strut tower bar. Remove the radiator grille.

7. Remove the top engine cover. Disconnect the control wiring harness connection. Disconnect the EVAP hose connection.

8. Disconnect the high pressure fuel hose connection. Disconnect and remove the top and bottom radiator hoses. Remove the radiator.

9. Remove the driveshaft.

10. Remove the exhaust manifold right bank. Remove the intake manifold plenum.

11. Remove the drive belt. Remove the power steering pump belt. Disconnect the power steering pressure switch connector.

12. Remove the power steering pump from its mounting. Position it to the side where it will not interfere with the engine removal.

13. Remove the starter. Remove the intake manifold plenum stay, throttle body stay, engine oil dipstick assembly and engine hanger, and install special tools MB992012 and MB992013 to the engine.

14. Remove the transaxle from the vehicle.

15. Remove the air conditioning compressor and bracket. Position the unit to the side where it will not interfere with the engine removal.

16. Support the engine with a garage jack. Remove tool MB991895, MB991454, MB992012 and MB992013.

➡ **These special tools were needed and installed to remove the transaxle from the vehicle.**

17. Hold the engine assembly in place with the engine lifting device. Properly

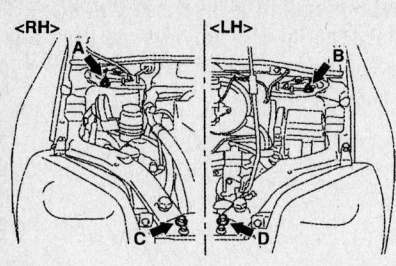

09482_GALA_G0007

Special tool installation–strut mounting nuts A and B. Front end structure bar assembling bolts (C and D)—2004–2006 Galant 3.8L engine

position a garage jack against the engine oil pan with a piece of wood in between so that the weight of the engine assembly is no longer being applied to the front mounting bracket.

18. Loosen the engine front mounting bracket mounting nuts and blots. Remove the front engine mounting bracket.

19. After checking that all cables, hoses and wiring are disconnected from the engine, lift the engine hoist slowly and remove the engine from the vehicle.

To install:

20. Slowly lower the engine assembly into the vehicle.

21. The installation is the reverse of the removal procedure.

22. Place a garage jack against the engine oil pan, with a piece of wood in between. Install the engine mount while adjusting the position of the engine.

23. Support the engine with a garage jack, Remove the chain block (engine lifting device).

24. Install engine hanger tool MB991895 to the strut mounting nuts and the front end structure bar assembling bolts.

25. Continue the installation in the reverse order of the removal procedure.

26. Be sure to fill the engine with the proper grade and type fluids.

27. Be sure to fill the transaxle with the proper grade and type fluid.

28. Start the engine and check for leaks, correct as required.

29. Complete the vehicle initialization procedure.

➡**To complete the initialization procedure the following tools are needed. MB991958 scan tool, MB991824 VCI, MB991827 MUT III USB cable, MB991910 MUT III Main harness "A"**

30. Connect the scan tool to the data link connector. To prevent damage to the scan tool be sure that the ignition switch is

in the LOCK position before connecting the scan tool.

31. Turn the ignition switch to the ON position.

32. Select "check mode" from the menu screen.

33. Select "erase memory" from the menu screen.

34. Initialize the learning value.

35. After initialization complete the idle learning procedure.

➡**This procedure must be performed when the PCM is replaced, or when the learning value is initialized, as the idling is not stabilized because the learning value in the MFI engine is not completed.**

36. Start the engine. Allow the coolant temperature to reach 176 degrees or more.

37. Stop the engine and place the ignition switch in the LOCK position.

38. After ten seconds, restart the engine.

39. For ten minutes carry out the idling procedure below to confirm that the engine has normal idling.

   • Position the transaxle selector lever in the "P" range
   • The engine fan is not to be operated
   • The engine coolant temperature should be 176 degrees or more

➡**If the engine stalls during idling, check the throttle valve of the throttle body for dirt. Correct and perform the procedure again.**

40. Road test the vehicle.

41. Check the front end alignment and correct as required.

42. Reprogram the radio stations and security codes, as necessary.

Lancer

1. Before servicing the vehicle, refer to the Precautions Section.

2. Relieve fuel system pressure.

3. Remove or disconnect the following:
   • Negative battery cable
   • Undercover, if equipped
   • Hood assembly
   • Air cleaner assembly and all adjoining air intake duct work

4. Drain the engine coolant and engine oil.

5. Remove or disconnect the following:
   • Radiator
   • Front exhaust pipe
   • Battery and battery tray
   • Accelerator cable

6. Detach the electrical connectors from the following components:
   • Air Conditioning (A/C) compressor
   • Power steering oil pressure switch
   • Crank angle sensor
   • Manifold differential pressure sensor
   • Evaporative emission (EVAP) purge solenoid
   • Exhaust Gas Recirculation (EGR) solenoid valve
   • Ignition coil
   • Fuel injectors
   • Throttle Position (TP) sensor
   • Idle Air Control (IAC) motor
   • Engine Coolant Temperature (ECT) sensor
   • Camshaft Position (CMP) sensor
   • Knock Sensor (KS)
   • ECT gauge unit
   • Heated Oxygen (HO$_2$S) sensor
   • Starter and alternator
   • Oil pressure switch

7. Remove or disconnect the following:
   • Brake booster vacuum hose
   • Power steering pump and A/C compressor drive belt
   • Power steering pump and brace. Position the assembly aside, but do NOT disconnect the fluid line.
   • A/C compressor, but do NOT disconnect the lines

➡**Matchmark the installed position of the radiator hoses before disconnecting them.**

   • Upper and lower radiator hoses
   • Heater and purge hoses
   • Fuel lines. Discard the O-rings
   • Transaxle assembly. Do NOT remove the flywheel bolt indicated by the arrow in the accompanying illustration. Removal of this bolt will cause the flywheel to be out of balance.

8. Remove the engine mount insulator and bracket as follows:

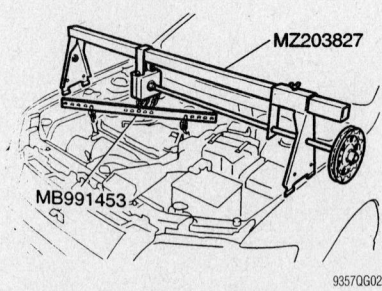

MZ203827

MB991453

9357QG02

These special tools are needed to support the engine while the transaxle is removed—Lancer 2.0L engine

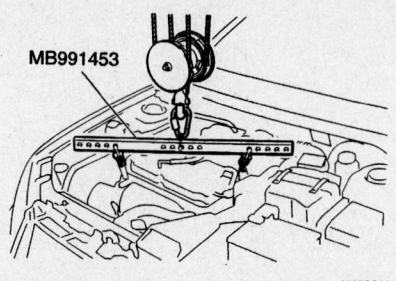

View of the special tool needed to support the engine during mount removal—Lancer 2.0L engine

a. Support the engine with a suitable floor jack.

b. Remove the special tools that were installed for transaxle removal.

c. Support the engine with special tool MB991453 attached to a chain block or engine hoist.

d. Place a jack under the oil pan with a block of wood in between to protect the pan. Jack up the engine to take the weight off the engine mount insulator and bracket, then remove the insulator and bracket.

9. Make sure that all cables, hoses and harnesses are disconnected from the engine, then use the engine hose to slowly lift the engine up and out of the engine compartment

To install:

10. Installation is the reverse of the removal procedure.

11. Fill the coolant system and engine crankcase.

12. Connect the negative battery cable.

13. On 2006 vehicles, complete the vehicle initialization procedure.

➡**To complete the initialization procedure the following tools are needed. MB991958 scan tool, MB991824 VCI, MB991827 MUT III USB cable, MB991910 MUT III Main harness "A"**

14. Connect the scan tool to the data link connector. To prevent damage to the scan tool be sure that the ignition switch is in the LOCK position before connecting the scan tool.

15. Turn the ignition switch to the ON position.

16. Select "check mode" from the menu screen.

17. Select "erase memory" from the menu screen.

18. Initialize the learning value.

19. After initialization, on 2006 vehicles, complete the idle learning procedure.

➡**This procedure must be performed when the PCM is replaced, or when the learning value is initialized, as the idling is not stabilized because the learning value in the MFI engine is not completed.**

20. Start the engine. Allow the coolant temperature to reach 176 degrees or more.

21. Stop the engine and place the ignition switch in the LOCK position.

22. After ten seconds, restart the engine.

23. For ten minutes carry out the idling procedure below to confirm that the engine has normal idling.

• Position the transaxle selector lever in the "P" range
• The engine fan is not to be operated
• The engine coolant temperature should be 176 degrees or more

➡**If the engine stalls during idling, check the throttle valve of the throttle body for dirt. Correct and perform the procedure again.**

Lancer Evolution

1. Before servicing the vehicle, refer to the Precautions Section.

2. Relieve fuel system pressure.

3. Disconnect the negative battery cable. Remove the battery and battery tray.

4. Remove the engine under cover. Remove the hood. Remove the side cover.

5. Drain the engine oil. Drain the coolant. Drain the transaxle. Drain the transfer case.

6. Remove the strut tower bar. Remove the air cleaner assembly.

7. Remove the air cooler air pipes. Remove the accelerator cable.

8. Remove the rocker cover center cover. Remove the upper and lower radiator hoses. Remove the radiator.

9. Remove the front axle crossmember bar. Remove the front exhaust pipe. Remove the turbocharger air outlet fitting.

10. Disconnect the ignition coil connectors.

11. Disconnect the Heated Oxygen (HO$_2$S) sensor connector.

12. Disconnect the Crankshaft Position (CKP) sensor connector.

13. Disconnect the manifold differential pressure sensor connector.

14. Disconnect the fuel pressure solenoid connector.

15. Disconnect the Knock Sensor (KS) connector.

16. Disconnect the evaporative emission purge solenoid connector.

17. Disconnect the Throttle Position (TP) sensor connector.

18. Disconnect the Idle Air Control (IAC) motor connector.

19. Disconnect the fuel injector connectors.

20. Disconnect the Camshaft Position (CMP) sensor connector.

21. Disconnect the Engine Coolant Temperature (ECT) gauge unit connector.

22. Disconnect the ECT sensor connector.

23. Remove the control wiring harness and transaxle wiring harness combination.

24. Disconnect the ground cable connection.

25. Disconnect the alternator wiring connectors.

26. Disconnect the Exhaust Gas Recirculation (EGR) vacuum regulator solenoid valve connector.

27. Disconnect the engine oil pressure switch connector.

28. Remove the turbocharger wastegate actuator bolts.

29. Remove the accessory drive belt.

30. Disconnect the brake booster vacuum hose.

31. Disconnect the purge hose connection.

32. Disconnect the power steering oil pressure switch connector.

33. Remove the power steering oil pump heat protector.

34. Remove the power steering oil pump, mounting bracket and reservoir assembly.

35. Disconnect the A/C compressor connector.

36. Remove the A/C compressor and clutch assembly.

37. Disconnect the engine oil cooler feed and return hose connections.

38. Disconnect the heater water hose connections.

39. Disconnect the fuel return line and high-pressure hose connections. Discard the O-rings.

40. Remove the upper and lower radiator hoses.

41. Remove the transaxle and transfer case assemblies from the engine.

42. Remove the engine front roll stopper bracket and engine front mounting bracket through-bolts as follows:

a. Support the engine with a suitable floor jack.

b. Remove the special tools that were installed for transaxle removal.

c. Support the engine with special tool MB991453 attached to a chain block or engine hoist.

d. Place a jack under the oil pan with a block of wood in between to protect the pan. Jack up the engine to take the weight off the engine mount insulator and bracket, then remove the front roll stopper bracket and engine front mounting bracket through-bolts.

43. Make sure that all cables, hoses and harnesses are disconnected from the engine, then use the engine hose to slowly lift the engine up and out of the engine compartment.

To install:

44. Install the engine and secure in position. The engine mount through-bolts should not be tightened until the full weight of the engine is on the mounts. Tighten the engine front roll stopper bracket through-bolt to 34–44 ft. lbs. (45–59 Nm). Tighten the engine front mounting bracket bolt to 66–80 ft. lbs. (88–108 Nm).

45. Continue installation in the reverse of the removal procedure.

46. Be sure to fill the engine with the proper grade and type fluids.

47. Be sure to fill the transaxle with the proper grade and type fluid.

48. Start the engine and check for leaks, correct as required.

49. Complete the vehicle initialization procedure.

➡**To complete the initialization procedure the following tools are needed. MB991958 scan tool, MB991824 VCI, MB991827 MUT III USB cable, MB991910 MUT III Main harness "A"**

50. Connect the scan tool to the data link connector. To prevent damage to the scan tool be sure that the ignition switch is in the LOCK position before connecting the scan tool.

51. Turn the ignition switch to the ON position.

52. Select "check mode" from the menu screen.

53. Select "erase memory" from the menu screen.

54. Initialize the learning value.

55. After initialization complete the idle learning procedure.

➡**This procedure must be performed when the PCM is replaced, or when the learning value is initialized, as the idling is not stabilized because the learning value in the MFI engine is not completed.**

56. Start the engine. Allow the coolant temperature to reach 176 degrees or more.

57. Stop the engine and place the ignition switch in the LOCK position.

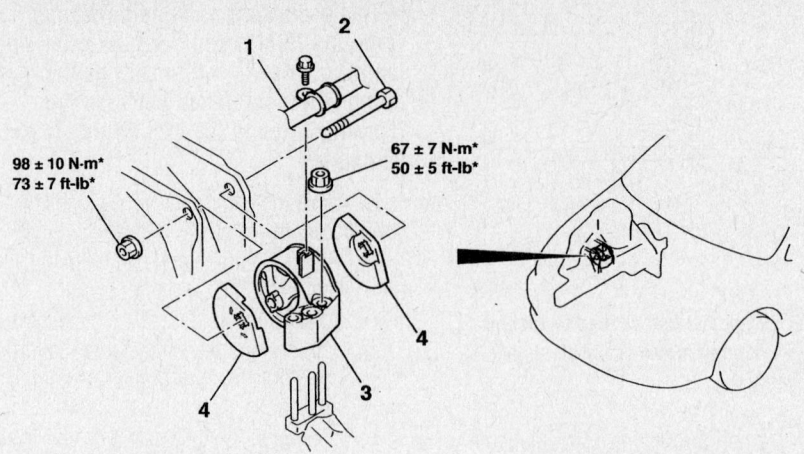

98 ± 10 N·m*
73 ± 7 ft-lb*

67 ± 7 N·m*
50 ± 5 ft-lb*

1. POWER STEERING OIL PRESSURE HOSE
2. ENGINE MOUNTING BOLT
3. ENGINE FRONT MOUNTING BRACKET
4. ENGINE FRONT MOUNTING CUSHION STOPPERS

67170-GALA-G11

Front mounting bracket and related components—Lancer Evolution 2.0L engine

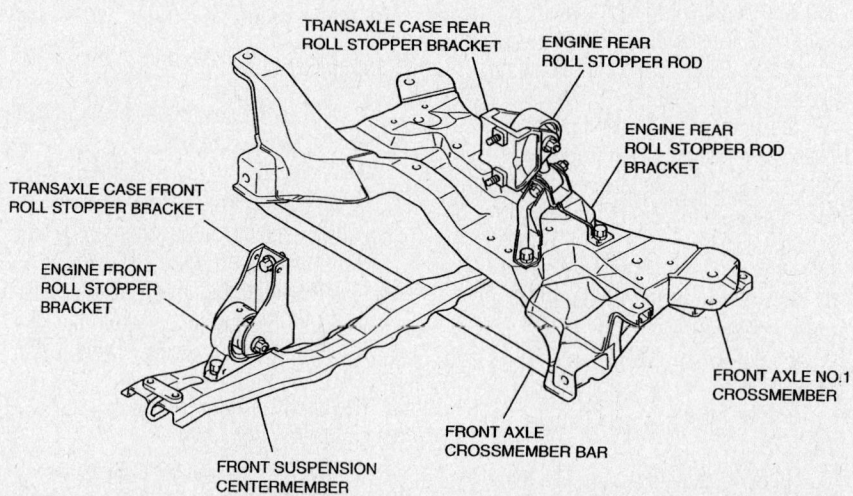

TRANSAXLE CASE REAR ROLL STOPPER BRACKET
ENGINE REAR ROLL STOPPER ROD
ENGINE REAR ROLL STOPPER ROD BRACKET
TRANSAXLE CASE FRONT ROLL STOPPER BRACKET
ENGINE FRONT ROLL STOPPER BRACKET
FRONT AXLE NO.1 CROSSMEMBER
FRONT AXLE CROSSMEMBER BAR
FRONT SUSPENSION CENTERMEMBER

67170-GALA-G12

Engine front roll stopper bracket and related components— Lancer Evolution 2.0L engine

58. After ten seconds, restart the engine.

59. For ten minutes carry out the idling procedure below to confirm that the engine has normal idling.

- Position the transaxle selector lever in the "P" range
- The engine fan is not to be operated
- The engine coolant temperature should be 176 degrees or more

➡**If the engine stalls during idling, check the throttle valve of the throttle body for dirt. Correct and perform the procedure again.**

60. Road test the vehicle.

61. Check the front end alignment and correct as required.

62. Reprogram the radio stations and security codes, as necessary.

Mirage

1. Before servicing the vehicle, refer to the Precautions Section.

2. Relieve fuel system pressure.

3. Remove or disconnect the following:
- Negative battery cable
- Undercover, if equipped
- Hood assembly
- Air cleaner assembly and all adjoining air intake duct work

4. Drain the engine coolant.

5. Remove or disconnect the following:
- Radiator assembly and coolant reservoir

- Transaxle assembly
- Ground cable, accelerator cable, breather hose and heater hose connections from the engine

6. Note locations and remove vacuum hoses from engine.

7. Remove or disconnect the following:
- Fuel feed and return hoses
- Crankshaft Position (CKP) and Camshaft Position (CMP) sensor wiring
- Heated Oxygen (HO2S sensor), Engine Coolant (ECT) gauge and ECT sensor connections
- Oil pressure switch
- Thermo switch, with automatic transmissions
- Harness connections for the Idle Speed Control (ISC) motor and Throttle Position (TP) sensor
- Intake Air Temperature (IAT) sensor
- Exhaust Gas Recirculation (EGR) temperature sensor (California)
- Injector harness plugs
- Power transistor and the ignition coil connections
- Alternator and power steering switch wiring
- Air conditioning compressor and hang it out of the way. Do NOT allow the compressor to hang by the hoses.
- Power steering pump and hang it out of the way–Do not allow the pump to hang by the hoses.
- Starter and alternator harness clamp, for 1.8L engines
- Exhaust manifold to head pipe nuts

8. Attach a hoist to the engine and support the engine weight. Remove the engine mount bracket. Remove any torque control brackets (roll stoppers).

9. Remove the engine assembly from the vehicle.

To install:

10. Install the engine and secure in position. The front lower mount through-bolt nut should not be tightened until the full weight of the engine is on the mount. Tighten through-bolt to 72 ft. lbs. (100 Nm) and bracket mounting bolts to 42 ft. lbs. (58 Nm). Tighten bracket mounting nut to 38 ft. lbs. (53 Nm).

11. Using a new gasket, position exhaust pipe onto the manifold and tighten the flange nuts to 36 ft. lbs. (50 Nm).

12. Install or connect the following:
- Power steering pump, alternator and air conditioning compressor
- Accessory drive belts
- Alternator and power steering wiring

- Alternator and starter harness clamp for 1.8L engines
- Ignition coil and power transistor connections
- Fuel injector harness connections
- EGR temperature sensor plug–California models
- IAT sensor
- IAC and TPS connectors
- Thermo switch, automatic transmission
- Oil pressure switch wiring
- HO2S sensor, ECT gauge and ECT sensor
- CKP and CMP sensors
- Fuel feed hose and tighten bolts to 44 inch lbs. (5 Nm), using new O-rings
- Fuel return hose, using a new hose clamp
- Vacuum hoses and the brake booster vacuum supply
- Breather hose, heater hoses, accelerator cable and ground cables. Inspect accelerator cable for proper adjustment.
- Transaxle assembly
- Radiator assembly and refill the cooling system
- Air cleaner and hood assembly
- Negative battery cable

Water Pump

REMOVAL & INSTALLATION

Diamante

1. Before servicing the vehicle, refer to the Precautions Section.
2. Drain the cooling system.
3. Disconnect the negative battery cable.
4. Remove the timing belt.
5. Remove the coolant hoses from the pump, if equipped.
6. Remove the alternator brace.

➡**The water pump bolts are different in size. Note their locations for installation.**

7. Remove the water pump, gasket and O-ring where the water inlet pipe joins the pump.

To install:

8. Thoroughly clean both gasket surfaces of the water pump and block.
9. Install a new O-ring into the groove on the front end of the water inlet pipe. Do not apply oils or grease to the O-ring. Wet with water only.
10. Install the water pump assembly to

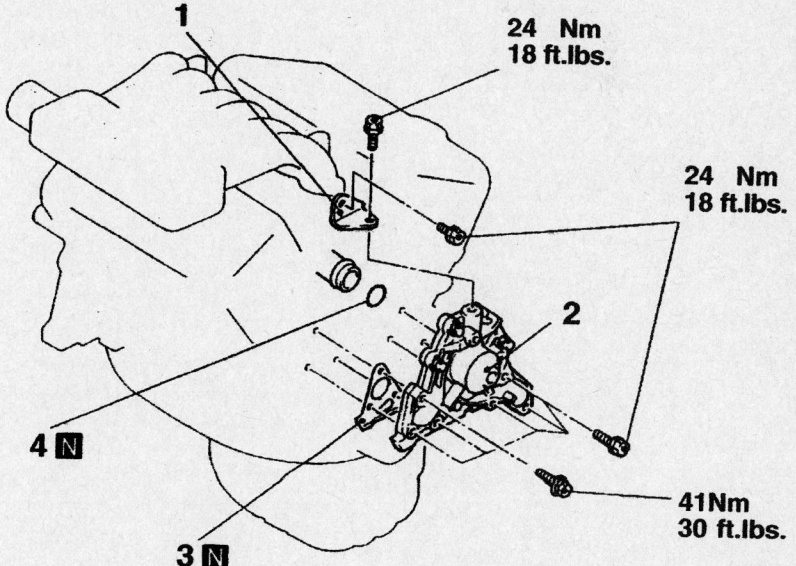

1. GENERATOR BRACE
2. WATER PUMP
3. WATER PUMP GASKET
4. O-RING

Water pump and related components—Diamante

67170-GALA-G13

the engine block, with new gasket. Torque the 8mm mounting bolts to 18 ft. lbs. (24 Nm), and the 10mm mounting bolt to 30 ft. lbs. (41 Nm).

11. Connect the hoses to the pump
12. Install the timing belt.
13. Install the engine drive belts.
14. Fill the system with coolant.
15. Connect the negative battery cable, run the vehicle until the thermostat opens and fill the radiator completely.
16. Once the vehicle has cooled, recheck the coolant level.

Eclipse

2002–2005 2.4L AND 3.0L ENGINES

1. Before servicing the vehicle, refer to the Precautions Section.
2. Disconnect the negative battery cable.
3. Drain the engine coolant.
4. Remove or disconnect the following:

- Timing belt tensioner pulley on 2.4L engine
- Timing belt on 3.0L engine
- Alternator brace from the water pump
- On 3.0L engine, remove the thermostat
- Timing belt rear cover

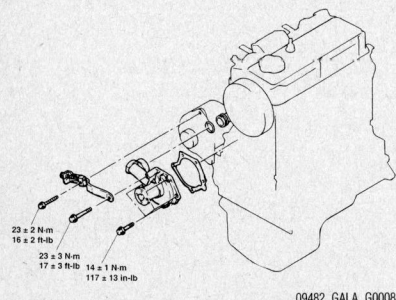

Water pump and related components— 2002–2004 Eclipse 2.4L engine

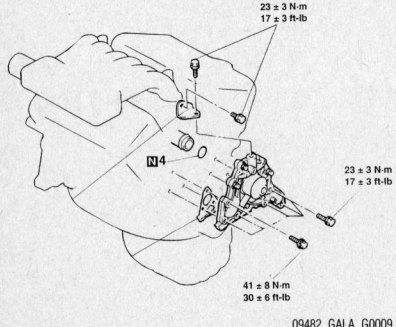

09482_GALA_G0009

Water pump and related components— 2002–2004 Eclipse 3.0L engine

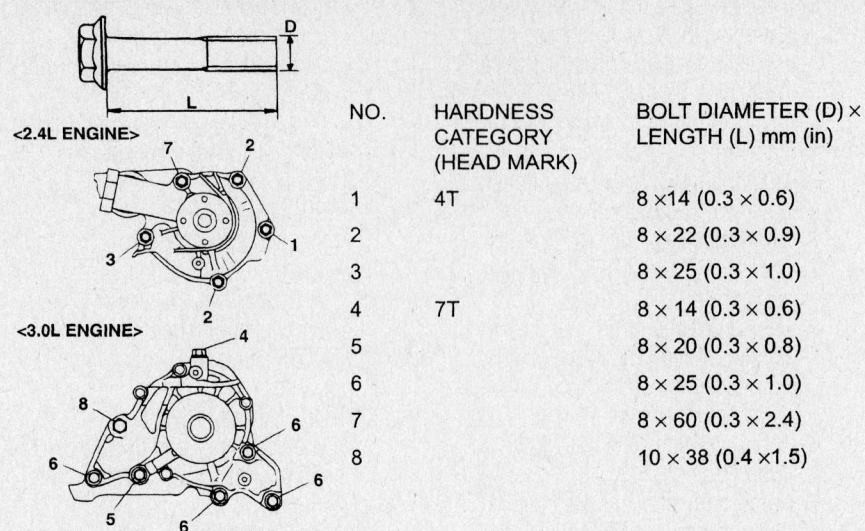

NO.	HARDNESS CATEGORY (HEAD MARK)	BOLT DIAMETER (D) × LENGTH (L) mm (in)
1	4T	8 × 14 (0.3 × 0.6)
2		8 × 22 (0.3 × 0.9)
3		8 × 25 (0.3 × 1.0)
4	7T	8 × 14 (0.3 × 0.6)
5		8 × 20 (0.3 × 0.8)
6		8 × 25 (0.3 × 1.0)
7		8 × 60 (0.3 × 2.4)
8		10 × 38 (0.4 × 1.5)

09482_GALA_G0010

Water pump bolt location and identification—2002–2004 Eclipse with 2.4L and 3.0L engines

- Water pump mounting bolts
- Water pump, gasket and O-ring

To install:

5. Install or connect the following:
- New O-ring on the water inlet pipe. Coat the O-ring with water or coolant. Do not allow oil or other grease to contact the O-ring.
- Water pump to the engine block, with new gasket. Torque the mounting bolts to 10 ft. lbs. (13 Nm)
- Alternator brace on the water pump. Torque the brace pivot bolt to 17 ft. lbs. (24 Nm).
- Timing belt rear cover
- Timing belt

- Remaining components
6. Refill the engine with coolant.
7. Connect the negative battery cable, start the engine and check for leaks.

2006 2.4L (4G69) ENGINE

1. Before servicing the vehicle, refer to the Precautions Section.
2. Disconnect the negative battery cable.
3. Drain the engine coolant.
4. Remove the timing belt.
5. Remove the water pump retaining bolts.
6. Remove the water pump from the engine. Discard the water pump gasket and O-ring.

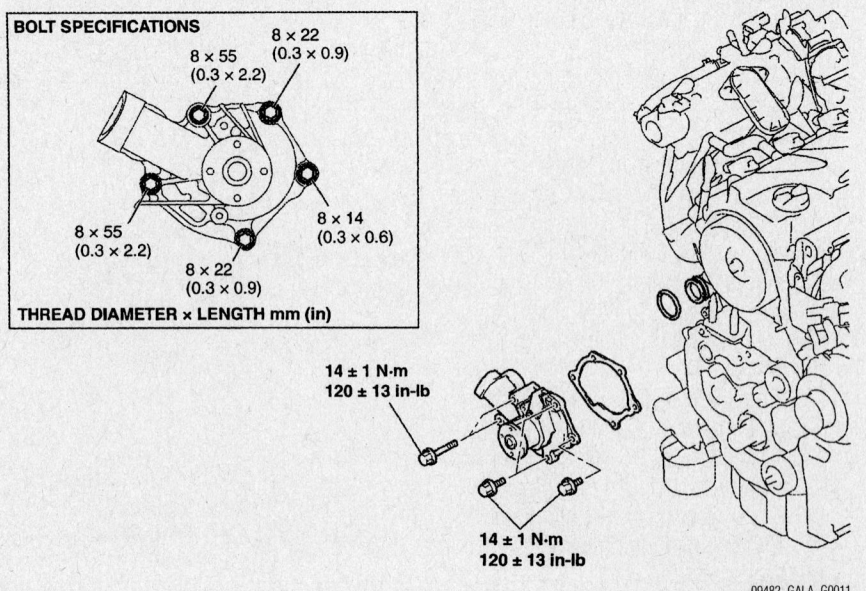

09482_GALA_G0011

Water pump and related components—2006 Eclipse 2.4L engine and 2004–2006 Galant 2.4L engine

To install:

7. Install or connect the following:
- New O-ring on the water inlet pipe. Coat the O-ring with water or coolant. Do not allow oil or other grease to contact the O-ring.
- Water pump to the engine block, with new gasket. Tighten the mounting bolts.

8. Continue the installation in the reverse order of the removal procedure.

9. Fill the engine with the proper grade and type engine coolant. Start the engine and check for leaks.

3.8L ENGINE

1. Before servicing the vehicle, refer to the Precautions Section.
2. Disconnect the negative battery cable.
3. Drain the engine coolant.
4. Remove the timing belt.
5. Remove the crankshaft position sensor connector clip. Remove the crankshaft position sensor mounting bolt. Remove the sensor.
6. Remove the water pump retaining bolts.
7. Remove the water pump from the engine. Discard the water pump gasket and O-ring.

To install:

8. Install or connect the following:
- New O-ring on the water inlet pipe. Coat the O-ring with water or coolant. Do not allow oil or other grease to contact the O-ring.
- Water pump to the engine block, with new gasket. Tighten the mounting bolts.

9. Continue the installation in the reverse order of the removal procedure.

10. Fill the engine with the proper grade and type engine coolant. Start the engine and check for leaks.

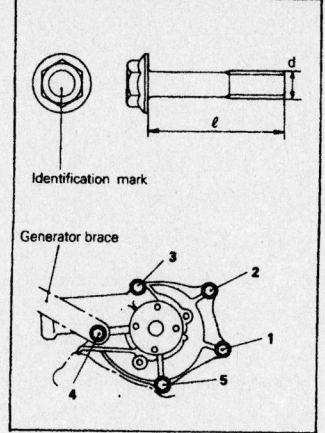

No.	Identification mark	Bolt diameter (d) x length (ℓ) mm (in.)	Torque Nm (ft.lbs.)
1	4	8 x 14 (.31 x .55)	
2	4	8 x 22 (.31 x .87)	12–15 (9–10)
3	4	8 x 30 (.31 x 1.18)	
4	7	8 x 65 (.31 x 2.56)	20–27 (15–19)
5	4	8 x 28 (.31 x 1.10)	12–15 (9–10)

7923PG11

Water pump bolt location and identification—2002–2003 Galant 2.4L engine

Galant

2002–2003 2.4L AND 3.0L ENGINES

1. Before servicing the vehicle, refer to the Precautions Section.
2. Disconnect the negative battery cable.
3. Drain the cooling system.
4. Remove or disconnect the following:
- Engine undercover
- Clamp bolt from the power steering hose

5. Support the engine with the appropriate equipment and remove the engine mount bracket.
6. Remove or disconnect the following:
- Engine drive belts and the air conditioning tensioner bracket
- Timing belt covers from the front of the engine
- Camshaft and silent shaft timing belts
- Alternator brace
- Water pump, gasket and O-ring where the water inlet pipe(s) joins the pump

To install:

7. Thoroughly clean both gasket surfaces of the water pump and block.
8. Install a new O-ring into the groove on the front end of the water inlet pipe and wet with clean antifreeze only. Do not apply oils or grease to the O-ring.
9. Using a new gasket, install the water pump assembly. Tighten bolts with the head mark **4** to 10 ft. lbs. (14 Nm) and bolts with the head mark **7** to 18 ft. lbs. (24 Nm).
10. Install or connect the following:
- Timing belts
- Engine drive belts
- Engine mount bracket
- Engine undercover

11. Fill the system with coolant.
12. Connect the negative battery cable, run the vehicle until the thermostat opens and fill the radiator completely.
13. Once the vehicle has cooled, recheck the coolant level.

2004–2006 2.4L (4G69) ENGINE

1. Before servicing the vehicle, refer to the Precautions Section.
2. Disconnect the negative battery cable.
3. Drain the engine coolant.
4. Remove the timing belt.
5. Remove the water pump retaining bolts.
6. Remove the water pump from the engine. Discard the water pump gasket and O-ring.

To install:

7. Install or connect the following:
- New O-ring on the water inlet pipe. Coat the O-ring with water or coolant. Do not allow oil or other grease to contact the O-ring.
- Water pump to the engine block, with new gasket. Tighten the mounting bolts.

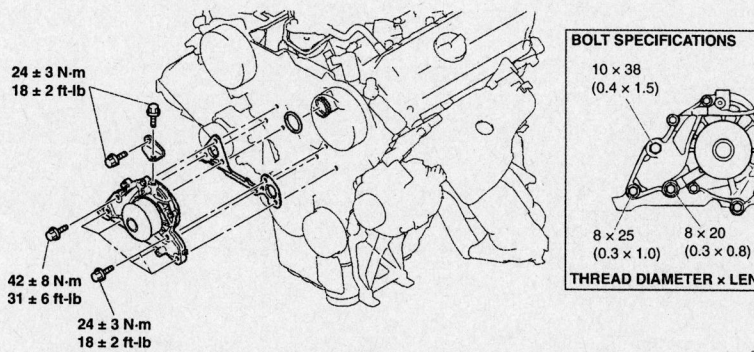

24 ± 3 N·m
18 ± 2 ft-lb

42 ± 8 N·m
31 ± 6 ft-lb

24 ± 3 N·m
18 ± 2 ft-lb

BOLT SPECIFICATIONS

10 x 38 (0.4 x 1.5)

8 x 25 (0.3 x 1.0)

8 x 25 (0.3 x 1.0)

8 x 20 (0.3 x 0.8)

THREAD DIAMETER x LENGTH mm (in)

09482_GALA_G0012

Water pump and related components—2006 Eclipse 3.8L engine and 2004–2006 Galant 3.8L engine

8. Continue the installation in the reverse order of the removal procedure.

9. Fill the engine with the proper grade and type engine coolant. Start the engine and check for leaks.

2004–2006 3.8L ENGINE

1. Before servicing the vehicle, refer to the Precautions Section.

2. Disconnect the negative battery cable.

3. Drain the engine coolant.

4. Remove the timing belt.

5. Remove the crankshaft position sensor connector clip. Remove the crankshaft position sensor mounting bolt. Remove the sensor.

6. Remove the water pump retaining bolts.

7. Remove the water pump from the engine. Discard the water pump gasket and O-ring.

To install:

8. Install or connect the following:

- New O-ring on the water inlet pipe. Coat the O-ring with water or coolant. Do not allow oil or other grease to contact the O-ring.
- Water pump to the engine block, with new gasket. Tighten the mounting bolts.

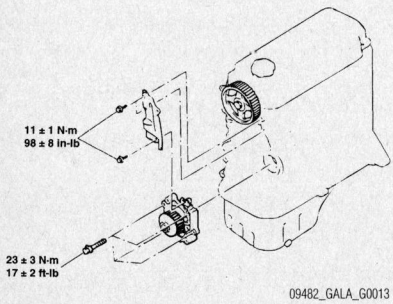

09482_GALA_G0013

Exploded view of the water pump—Lancer 2.0L engine

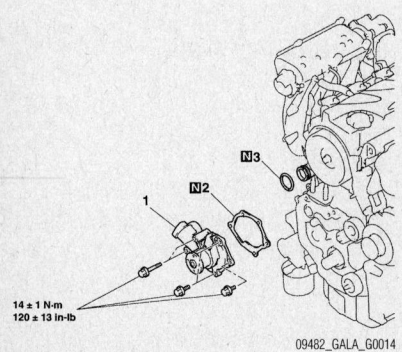

09482_GALA_G0014

Exploded view of the water pump—Lancer 2.4L Engine

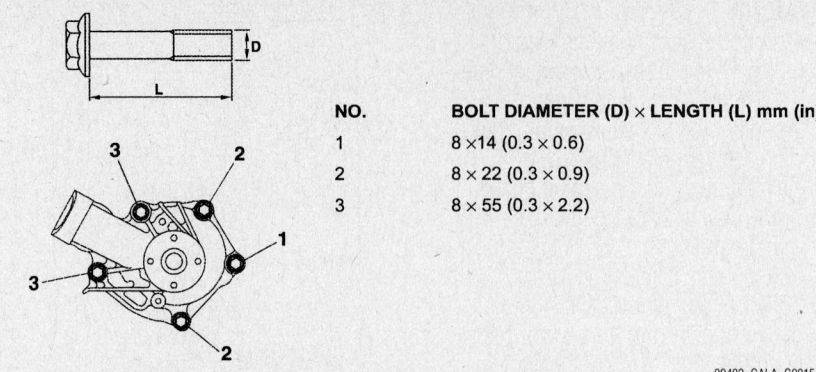

NO.	BOLT DIAMETER (D) × LENGTH (L) mm (in)
1	8 ×14 (0.3 × 0.6)
2	8 × 22 (0.3 × 0.9)
3	8 × 55 (0.3 × 2.2)

09482_GALA_G0015

Water pump bolt location and identification—Lancer 2.4L engine

9. Continue the installation in the reverse order of the removal procedure.

10. Fill the engine with the proper grade and type engine coolant. Start the engine and check for leaks.

Lancer

1. Before servicing the vehicle, refer to the Precautions Section.

2. Disconnect the negative battery cable.

3. Drain the cooling system.

4. Remove the timing belt.

5. Remove the water pump bolts(s) and pump.

6. Remove the water pump gasket and O-ring.

To install:

7. Install the water pump along with a new gasket and O-ring. Tighten the mounting bolt(s) to 15–19 ft. lbs. (20–26 Nm).

8. Install the timing belt.

9. Refill the cooling system and connect the negative battery cable.

Evolution

1. Before servicing the vehicle, refer to the Precautions Section.

2. Disconnect the negative battery cable.

3. Remove the engine under cover.

4. Drain the engine coolant.

5. Remove the timing belt tension adjuster.

6. Remove the alternator brace.

7. Remove the water pump retaining bolts.

8. Remove the water pump from the engine. Discard the water pump gasket and O-ring.

To install:

9. Install or connect the following:

- New O-ring on the water inlet pipe. Coat the O-ring with water or coolant. Do not allow oil or other grease to contact the O-ring.
- Water pump to the engine block, with new gasket. Tighten the mounting bolts.

10. Continue the installation in the reverse order of the removal procedure.

11. Fill the engine with the proper grade and type engine coolant. Start the engine and check for leaks.

Mirage

1. Before servicing the vehicle, refer to the Precautions Section.

2. Disconnect the negative battery cable.

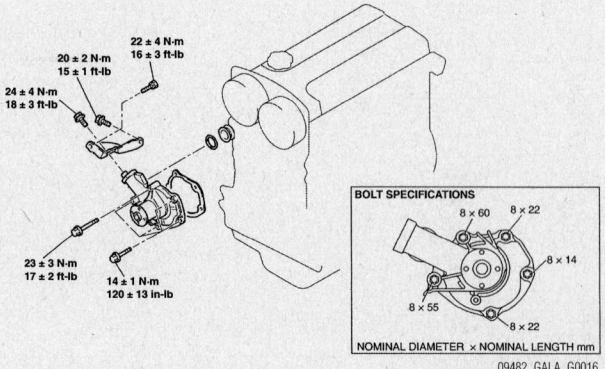

09482_GALA_G0016

Exploded view of the water pump—Lancer Evolution

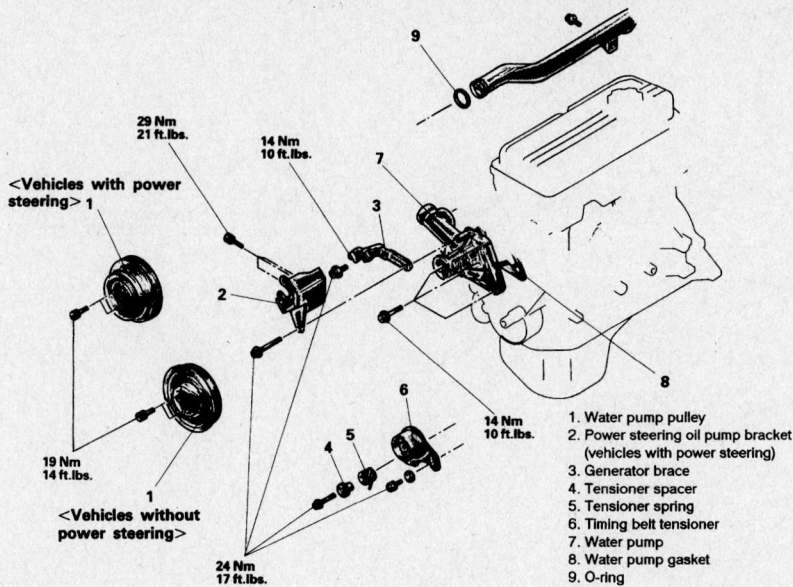

29 Nm
21 ft.lbs.

14 Nm
10 ft.lbs.

<Vehicles with power steering> 1

19 Nm
14 ft.lbs.

<Vehicles without power steering>

24 Nm
17 ft.lbs.

14 Nm
10 ft.lbs.

1. Water pump pulley
2. Power steering oil pump bracket (vehicles with power steering)
3. Generator brace
4. Tensioner spacer
5. Tensioner spring
6. Timing belt tensioner
7. Water pump
8. Water pump gasket
9. O-ring

7923PG07

Water pump and related components—Mirage 1.5L engine

10 Nm
7 ft.lbs.

24 Nm
18 ft.lbs.

1. Timing belt rear cover
2. Water pump

7923PG08

Water pump and related components—Mirage 1.8L engine

3. Drain the cooling system.

4. Remove or disconnect the following:
• Engine undercover
• Clamp bolt from the power steering hose
• Engine drive belts

5. Support the engine with the appropriate equipment and remove the engine mount bracket.

6. Remove or disconnect the following:
• Timing belt
• Power steering pump bracket
• Alternator brace

➡The water pump mounting bolts are different in length, note their positioning for reassembly.

7. Remove the water pump, gasket and O-ring where the water inlet pipe(s) joins the pump.

To install:

8. Thoroughly clean both gasket surfaces of the water pump and block.

9. For 1.5L engines, install a new O-ring into the groove on the front end of the water inlet pipe. Do not apply oils or grease to the O-ring. Wet the O-ring with water only.

10. For 1.8L engines, apply a 0.09–0.12 in. (2.5–3.0mm) continuous bead of sealant to water pump and install the pump assembly. Install the water pump within 15 minutes of the application of the sealant. Wait 1 hour after installation of the water pump to refill the cooling system or starting the engine.

11. Install or connect the following:
• Gasket and pump assembly and tighten the bolts to 17 ft. lbs. (24 Nm)
• Remaining components in the reverse order of removal

12. Fill the system with coolant.

13. Connect the negative battery cable, run the vehicle until the thermostat opens and fill the radiator completely.

14. Once the vehicle has cooled, recheck the coolant level.

Heater Core

REMOVAL & INSTALLATION

Diamante

1. Before servicing the vehicle, refer to the Precautions Section.

2. Disconnect the negative battery cable.

3. Drain the cooling system into a clean container for reuse.

4. Discharge and recover the air conditioning system refrigerant.

5. Remove or disconnect the following:
• Heater hoses from the heater core
• Refrigerant lines from the evaporator core and discard the O-rings

❊❊ CAUTION

After disconnecting the negative battery cable, wait at least 60 seconds before working on the SRS module or instrument panel.

6. Remove the passenger's side air bag by removing or disconnecting the following:
• Dash undercover
• Glove box assembly
• Glove box case
• Air bag-to-dash bolts and the air bag; then, disconnect the electrical connector

7. Remove or disconnect the following:
• Floor console
• Front pillar trim at both sides

8. Remove the instrument panel by removing or disconnecting the following:
• Steering column covers
• Hood lock release handle

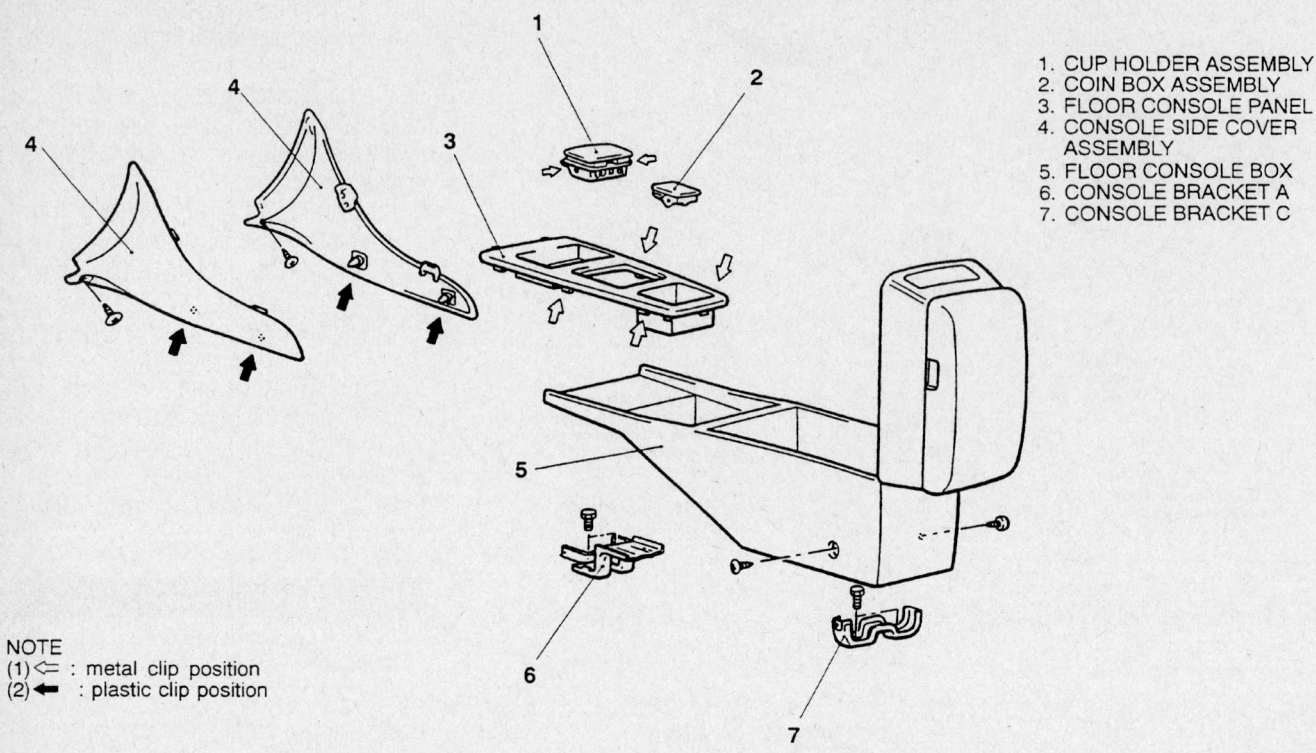

1. CUP HOLDER ASSEMBLY
2. COIN BOX ASSEMBLY
3. FLOOR CONSOLE PANEL
4. CONSOLE SIDE COVER ASSEMBLY
5. FLOOR CONSOLE BOX
6. CONSOLE BRACKET A
7. CONSOLE BRACKET C

NOTE
(1) ⇐ : metal clip position
(2) ◄ : plastic clip position

93112GF0

Floor console and related components—Diamante

- Parking brake release handle
- Lower left side instrument panel cover
- Ignition key cylinder panel
- Instrument panel Electronic Control Unit (ECU) and remove the ECU
- Instrument panel meter bezel and the combination meter
- Center air outlet assembly
- Ashtray
- Air conditioning control panel assembly and the audio unit
- Console side cover assembly
- Floor carpet rear reinforcement
- Electrical harness connector and plug
- Steering column mounting bolts and lower the steering column assembly
- Instrument panel with the help of an assistant

9. Remove or disconnect the following:
- ECU bracket
- Center stay assembly
- Heater hose connection and the center duct assembly
- Foot distribution duct and the breather hose
- Refrigerant lines from the evaporator and discard the O-rings

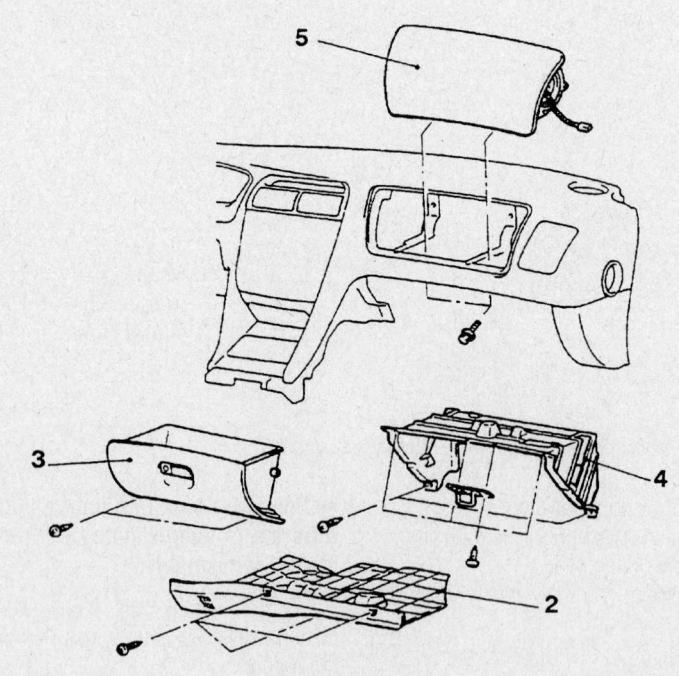

2. UNDERCOVER
3. GLOVE BOX ASSEMBLY
4. GLOVE BOX CASE
5. AIR BAG MODULE

93112GG1

Passenger's air bag module and related components—Diamante

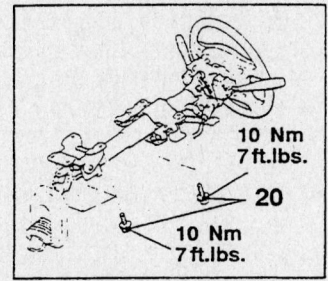

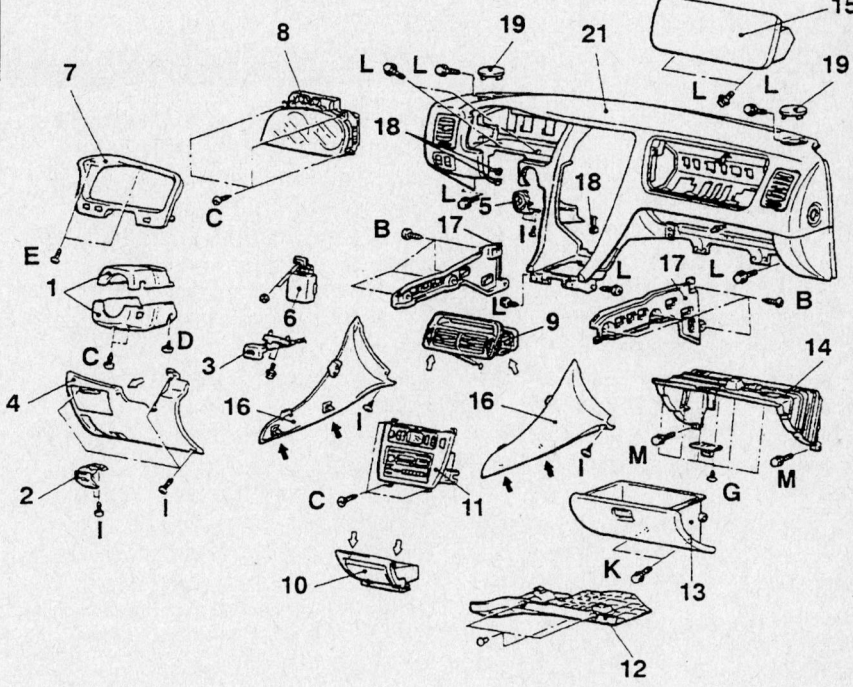

1. COLUMN COVER
2. HOOD LOCK RELEASE HANDLE
3. PARKING BRAKE RELEASE HANDLE
4. INSTRUMENT PANEL LOWER COVER ASSEMBLY (LH)
5. KEY CYLINDER PANEL
6. INSTRUMENT PANEL ECU
7. METER BEZEL
8. COMBINATION METER
9. CENTER AIR OUTLET ASSEMBLY
10. ASHTRAY
11. AIR CONTROL PANEL ASSEMBLY & AUDIO UNIT
12. UNDERCOVER ASSEMBLY
13. GLOVEBOX ASSEMBLY
14. GLOVEBOX OUTER CASE
15. PASSENGER SIDE AIRBAG MODULE
16. CONSOLE SIDE COVER ASSEMBLY
17. FLOOR CARPET REAR REINFORCEMENT
18. HARNESS CONNECTOR
19. PLUG
20. STEERING COLUMN MOUNTIN BOLT
21. INSTRUMENT PANEL

NOTE
(1) ⇦ : metal clip position
(2) ◄ : plastic clip position

93112GG2

Instrument panel and steering column assembly—Diamante

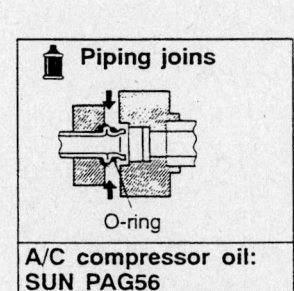

Piping joins

O-ring

A/C compressor oil:
SUN PAG56

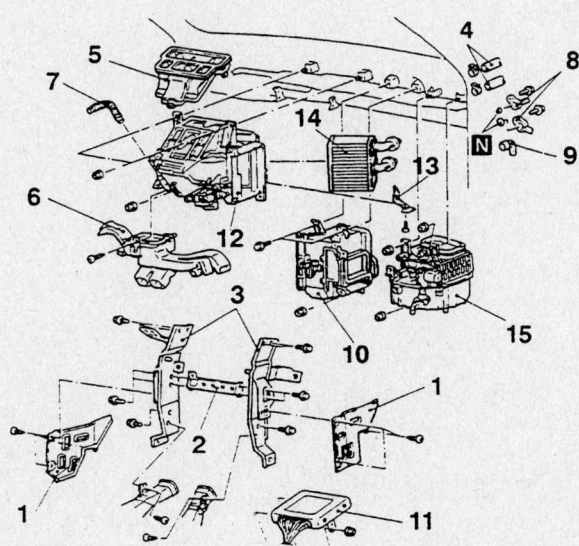

1. FLOOR CARPET FRONT REINFORCEMENT
3. ECU BRACKET
4. CENTER STAY ASSEMBLY
5. HEATER HOSE CONNECTION
6. CENTER DUCT ASSEMBLY
7. FOOT DISTRIBUTION DUCT
8. BREATHER HOSE
9. SUCTION PIPE, LIQUID PIPE B AND COOLING UNIT CONNECTION
10. DRAIN HOSE
11. EVAPORATOR
12. ENGINE CONTROL MODULE
13. HEATER UNIT
14. HEATER CORE SUPPORT
15. HEATER CORE

93112GG3

Heater core and related components—Diamante

- Air conditioning housing drain hose and remove the evaporator housing
- Heater housing unit
- Heater core support and the heater core

To install:

10. Install or connect the following:
  - Heater core support and the heater core
  - Heater housing unit
  - Air conditioning housing drain hose and Install the evaporator housing
  - Refrigerant lines to the evaporator using new O-rings
  - Foot distribution duct and the breather hose
  - Heater hose connection and the center duct assembly
  - Center stay assembly
  - ECU bracket
11. Install the instrument panel by installing or connecting the following:
  - Instrument panel with the help of an assistant
  - Steering column assembly and the steering column mounting bolts. Torque the bolts to 84 inch lbs. (10 Nm).
  - Electrical harness connector and plug
  - Floor carpet rear reinforcement
  - Console side cover assembly
  - Air conditioning control panel assembly and the audio unit
  - Ashtray
  - Center air outlet assembly
  - Instrument panel meter bezel and the combination meter
  - Instrument panel ECU and connect the ECU electrical connector
  - Ignition key cylinder panel
  - Lower left side instrument panel cover
  - Parking brake release handle
  - Hood lock release handle
  - Steering column covers
12. Front pillar trim at both sides
13. Floor console
14. Install the passenger's side air bag by installing or connecting the following:
  - Air bag-to-dash bolts and the air bag; then, connect the electrical connector
  - Glove box case
  - Glove box assembly
  - Dash undercover
  - Refrigerant lines to the evaporator core using new O-rings
  - Heater hoses to the heater core
15. Refill the cooling system.

16. Connect the negative battery cable.
17. Evacuate, charge and leak test the air conditioning system.
18. Operate the engine to normal operating temperatures; then, check the climate control operation and check for leaks.

Eclipse

2002–2003

❊❊ CAUTION

Wait for 1 minute after disconnecting the negative battery cable before working inside the vehicle. The air bag system is set to deploy for a short period of time after the battery is disconnected.

1. Before servicing the vehicle, refer to the Precautions Section.
2. Disconnect the negative battery cable.
3. Drain the cooling system into a clean container for reuse.
4. Disconnect the heater hoses from the heater core tubes at the firewall. Do not allow the coolant to damage the vehicle speed sensor located below the heater hoses on the manual transmission vehicles.

❊❊ WARNING

To prevent damage to the air bag control unit during removal or installation of the floor console, avoid shocks or impact. Do not drop.

5. Remove the floor console by removing or disconnecting the following:
  - Center console trim panel
  - Ashtray and cup holder assembly
  - Shift lever knob on manual transmission
  - Retaining screws
  - Floor console assembly
6. Locate the rectangular plugs in the knee protector on either side of the steering column. Pry these plugs out and remove the screws.
7. Remove or disconnect the following:
  - Driver's side air bag assembly, the steering wheel and the passenger's side air bag assembly
  - Lap cooler duct and steering column covers
  - Instrument cluster bezel and then the instrument cluster
  - Radio
  - Glove box
  - Center air outlet assembly

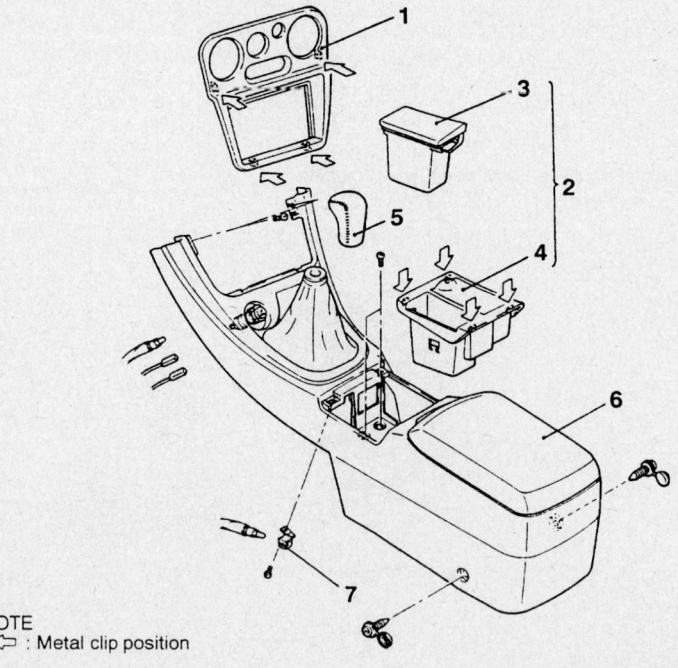

NOTE
⬅ : Metal clip position

1. Center console panel
2. Ashtray and cupholder assembly
3. Ashtray
4. Cup holder
5. Shift lever knob <M/T>
6. Floor console assembly
7. Ashtray illumination light bracket

93112G56

Floor console and related components—2002–2003 Eclipse

- Hood release handle and the lower cover
- Heater control assembly
- Front speakers and the instrument panel switch
- Steering shaft support bolts and lower the steering column

- Instrument panel mounting hardware and remove the instrument panel from the vehicle
- Stamped steel center reinforcement
- Lower ductwork from the heater box
- Evaporator case mounting bolt and

nut to allow clearance for the heater unit removal
- Heater unit
- Heater core from the heater unit

To install:

8. Install or connect the following:
  - Heater core to the heater unit

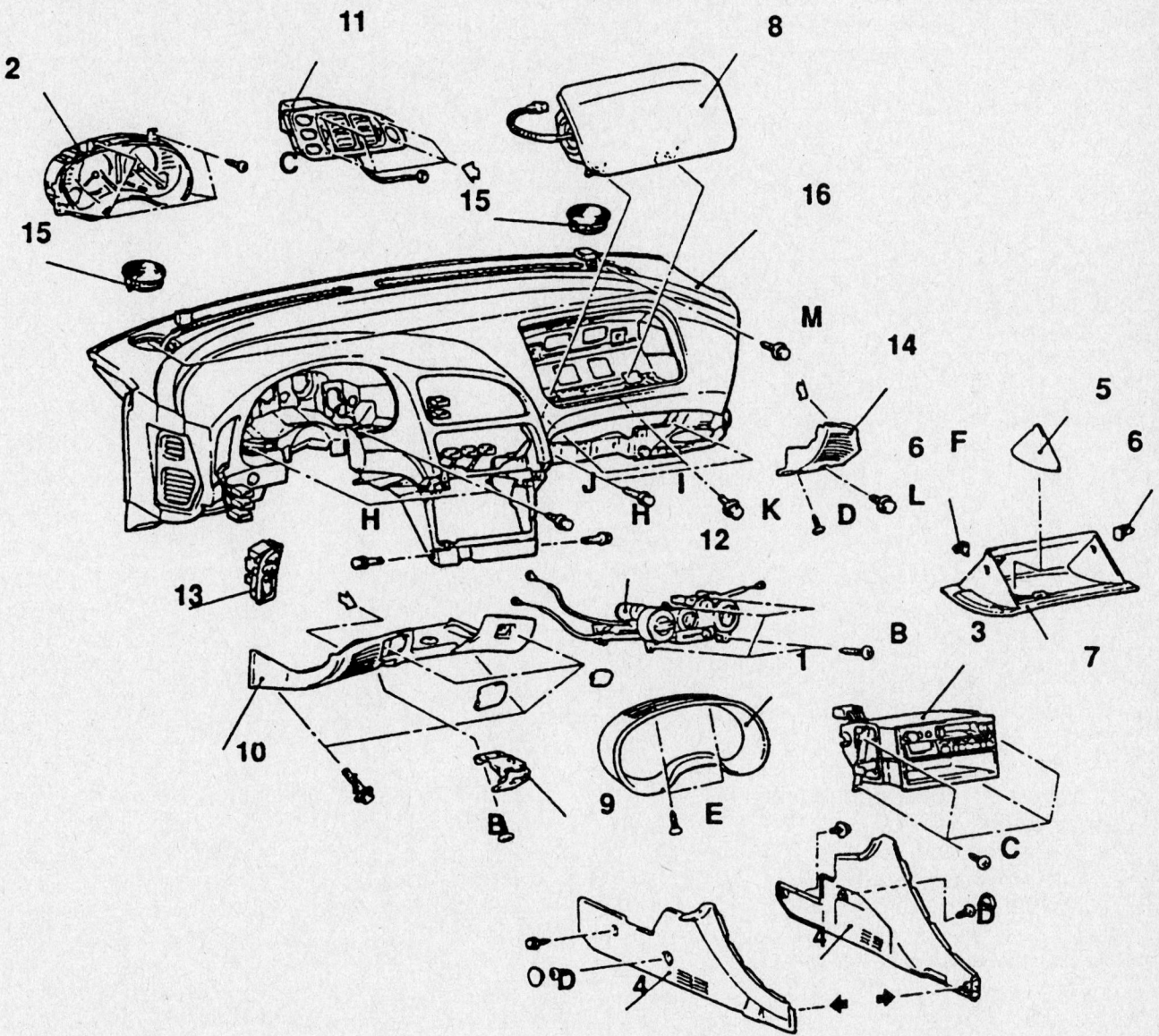

1. Meter bezel
2. Combination meter
3. Radio and tape player, and box
4. Console side cover
5. Sunglasses holder
6. Stopper
7. Glove box
8. Passenger's side air bag module assembly
9. Hood lock release handle
10. Instrument under cover L.H.
11. Center air outlet assembly
12. Heater control assembly
13. Instrument panel switch
14. Instrument under cover R.H.
15. Front speaker
16. Instrument panel assembly

93112G74

Instrument panel and related components—2002–2003 Eclipse

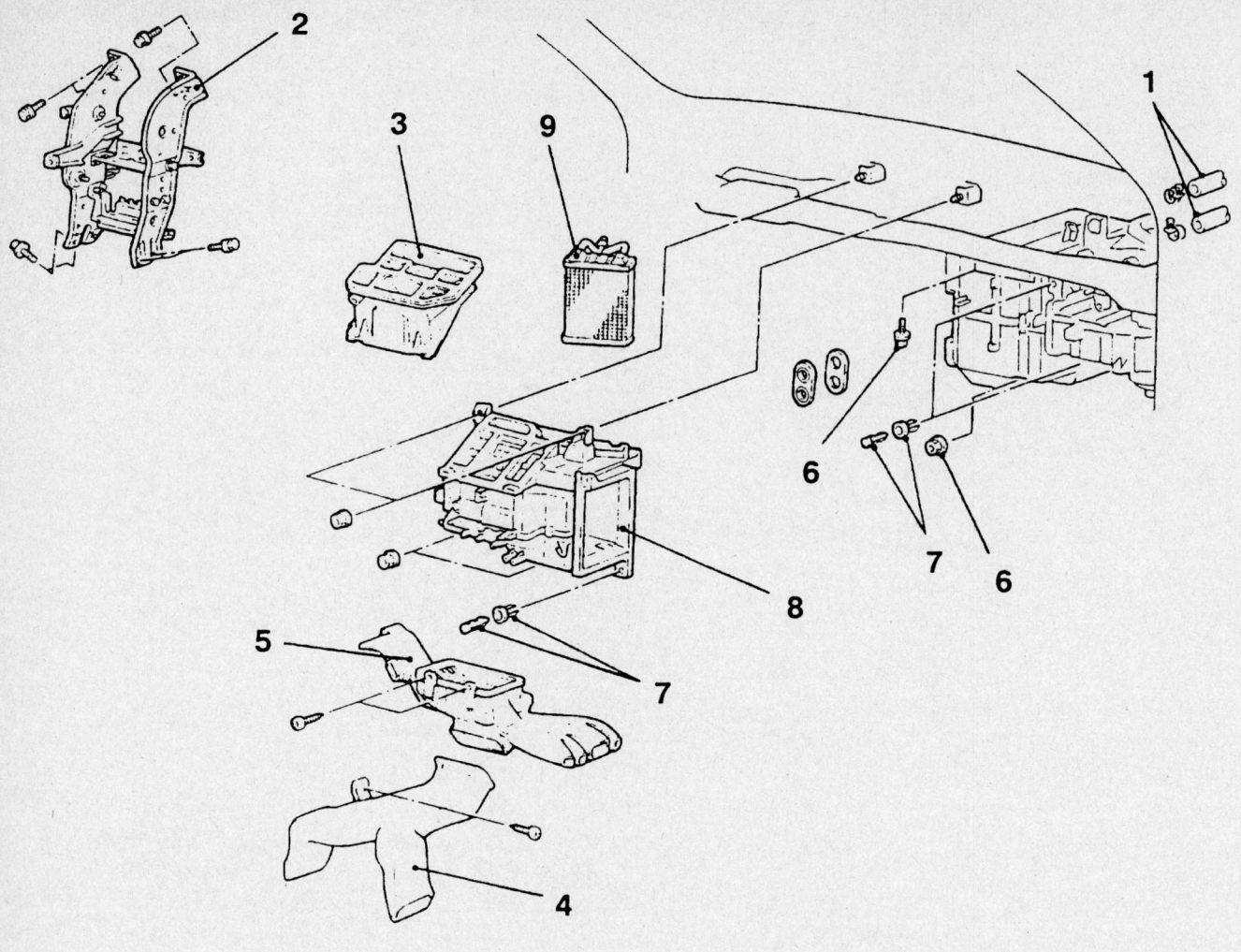

1. Heater hose connection
2. Center stay
3. Center duct
4. Semi rear heater duct
5. Foot distribution duct
6. Cooling unit installation bolt and nut
7. Clip
8. Heater unit
9. Heater core

93112GE0

Heater core and related components—2002–2003 Eclipse

- Heater unit
- Evaporator case mounting bolt and nut
- Lower ductwork to the heater box
- Stamped steel center reinforcement
- Instrument panel and the instrument panel mounting hardware
- Steering column and the steering shaft support bolts
- Front speakers and the instrument panel switch
- Heater control assembly
- Hood release handle and the lower cover

- Center air outlet assembly
- Glove box
- Radio
- Instrument cluster and the instrument cluster bezel
- Steering column covers and the lap cooler duct
- Steering wheel, the driver's side air bag assembly and the passenger's side air bag assembly
- Screws and the rectangular plugs in the knee protector on either side of the steering column

9. Install the floor console by installing or connecting the following:

- Floor console assembly
- Retaining screws
- Shift lever knob on manual transmission
- Ashtray and cup holder assembly
- Center console trim panel
- Center console trim panel

10. Connect the heater hoses to the heater core tubes at the firewall.

11. Refill the cooling system.

12. Connect the negative battery cable.

13. Operate the engine to normal operating temperatures; then, check the climate control operation and check for leaks.

2004–2005

1. Before servicing the vehicle, refer to the Precautions Section.

2. Disconnect the negative battery cable. Drain the cooling system. Discharge the air conditioning system. Be sure the front tires are in the straight ahead position.

3. Remove the center console box assembly, lid assembly (except Eclipse RS), door mirror control switch and harness, accessory socket harness, ashtray, top switch and harness (Spyder), shift lever panel assembly and boot (manual transaxle) and garnish (automatic transaxle)

4. Remove the front and rear floor console retaining screws. Remove the front and rear consoles.

5. Remove the driver's air bag module. Remove the steering wheel nut. Remove the steering wheel. Remove the cover.

6. Remove the auto-cruise control switch. Remove the instrument panel under cover switch. Remove the lower column cover. Remove the upper column cover.

7. Remove the clock spring and column switch assembly.

8. If equipped with automatic transaxle remove the cover and key interlock cable.

9. Remove the steering shaft assembly. Remove the steering cover assembly.

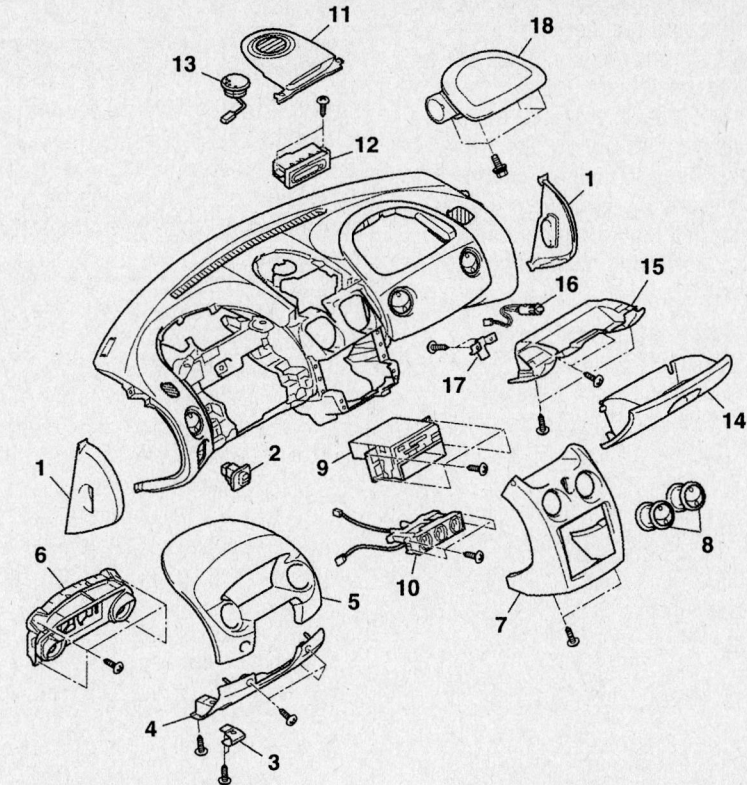

1. INSTRUMENT PANEL SIDE COVER
2. FOG LAMP SWITCH ASSEMBLY
3. HOOD LOCK RELEASE HANDLE
4. INSTRUMENT PANEL UNDER COVER
5. METER BEZEL ASSEMBLY
6. COMBINATION METER
7. CENTER PANEL ASSEMBLY
8. AIR OUTLET (CENTER SIDE)
9. RADIO AND TAPE PLAYER ASSEMBLY
10. HEATER CONTROL ASSEMBLY
11. CENTER HOOD
12. MULTI CENTER DISPLAY
13. CENTER SPEAKER
14. GLOVE BOX, OUTER
15. GLOVE BOX, INNER
16. GLOVE BOX LAMP SWITCH ASSEMBLY

09482_GALA_G0017

Heater core and related components—2004–2005 Eclipse

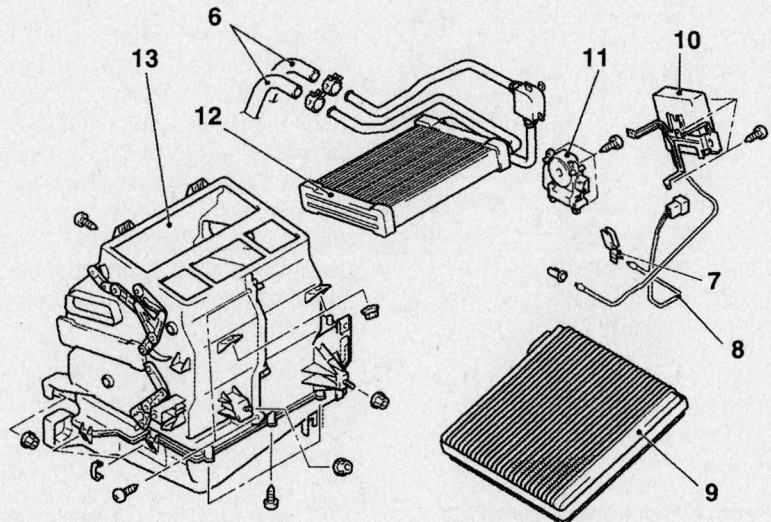

1. STRUT TOWER BAR ASSEMBLY
2. AIR INTAKE HOSE
3. A/C PIPE CONNECTION
4. EXPANSION VALVE
5. O-RING
6. HEATER HOSE
7. THERMISTOR SENSOR CLIP
8. THERMISTOR SENSOR
9. EVAPORATOR
10. AUTOMATIC COMPRESSOR
11. WATER SHUT MOTOR
12. HEATER CORE
13. HEATER/COOLER UNIT

09482_GALA_G0018

Instrument panel and related components—2004–2005 Eclipse

➡The tilt lever should be held in the lock position until the steering column is reinstalled in the vehicle. If the column is removed with the lever released, or the lever released after the column is removed from the vehicle, the steering column cannot be reinstalled correctly. If the steering column is installed incorrectly, the collision energy absorbing mechanism may be damaged.

10. Remove the front pillar trim panel. Remove the instrument panel side cover. Remove the fog lamp switch assembly.

11. Remove the hood release handle. Remove the meter bezel assembly. Remove the center panel assembly.

12. Remove the center air outlet. Remove the radio assembly. Remove the heater control assembly.

13. Remove the center hood. Remove the multi center display. Remove the center speaker. Remove the inner and outer glove box and light switch. Remove the glove box striker.

14. Remove the passenger's side air bag module. Remove the instrument panel assembly.

15. Remove the instrument panel reinforcement and both knee absorbers.

16. Remove the auto-cruise control ECU and the immobilizer ECU.

17. Remove the instrument panel center reinforcement. Remove the joint duct.

18. Remove the strut tower bar assembly.

19. Remove the air intake hose. Remove the air conditioning pipe connection, expansion valve, heater hose, thermistor sensor.

20. Remove the evaporator. Remove the automatic compressor controller. Remove the water shut motor.

21. Remove the heater core.

To install:

22. Installation is the reverse of the removal procedure.

➡The tilt lever should be held in the lock position until the steering column is reinstalled in the vehicle. If the column is removed with the lever released, or the lever released after the column is removed from the vehicle, the steering column cannot be reinstalled correctly. If the steering column is installed incorrectly, the collision energy absorbing mechanism may be damaged.

23. Be sure to fill the cooling system with the proper grade and type coolant.

24. Recharge the air conditioning system.

2006

✳✳ CAUTION

Wait for 1 minute after disconnecting the negative battery cable before working inside the vehicle. The air bag system is set to deploy for a short period of time after the battery is disconnected.

1. Before servicing the vehicle, refer to the Precautions Section.

2. Disconnect the negative battery cable. Drain the cooling system. Discharge the air conditioning system. Be sure the front tires are in the straight ahead position.

3. Remove the front seat.

4. Remove the center panel assembly. Remove the floor console center panel assembly. Remove the parking brake boot panel, cup holder insert, heated switch controls and accessory socket.

5. Remove the console retaining screws. Remove the console from the vehicle.

6. Remove the glove box, glove box lock, and glove box lid lock cylinder. Remove the glove box damper.

7. Remove the instrument panel parcel box. Remove the front pillar trim. Remove the front speaker grilles and remove the speakers.

8. Remove the passenger's side air bag module.

9. Remove the instrument panel center cover. Remove the multi function center display assembly. Remove the radio, CD player and/or changer assembly.

10. Remove the hood lock release handle. Remove the switch panel assembly. Remove the instrument lower panel assembly.

11. Remove the driver's side air bag module and clock spring. Remove the steering wheel.

12. Remove the steering column upper cover. Remove the steering column lower cover.

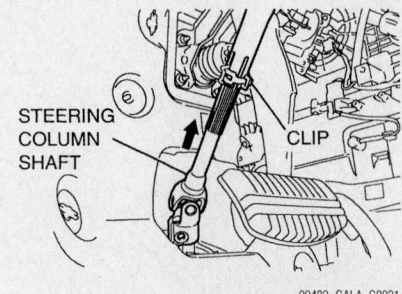

STEERING COLUMN SHAFT   CLIP

09482_GALA_G0021

Steering column shaft retention—2006 Eclipse, 2004–2006 Galant and 2004–2006 Evolution

13. Disconnect the electrical connections for the ignition switch. Remove the key interlock cable.

14. Remove the steering shaft pad. Remove the steering column retaining bolts. Remove the steering column assembly. Pinch the steering column shaft clip with a pliers, and pull up the shaft to disengage the steering column assembly.

➡The tilt lever should be held in the lock position until the steering column is reinstalled in the vehicle. If the column is removed with the lever released, or the lever released after the column is removed from the vehicle, the steering column cannot be reinstalled correctly. If the steering column is installed incorrectly, the collision energy absorbing mechanism may be damaged.

15. Remove the combination meter assembly. Remove the cowl side trim. Remove the passenger's side air bag module.

16. Remove the instrument panel assembly from the vehicle.

17. If the vehicle is equipped with the 3.8L engine, remove the strut tower bar retaining nuts. Remove the strut tower bar.

18. Remove the heater hose connection. Remove the suction pipe connection. Remove the liquid pipe connection. Remove the front deck stay. Disconnect all the electrical connectors from the necessary components in order to remove the heater unit and front deck crossmember. Remove the heater unit and front deck crossmember assembly.

To install:

19. Installation is the reverse of the removal procedure.

➡The tilt lever should be held in the lock position until the steering column is reinstalled in the vehicle. If the column is removed with the lever released, or the lever released after the column is removed from the vehicle, the steering column cannot be reinstalled correctly. If the steering column is installed incorrectly, the collision energy absorbing mechanism may be damaged.

20. If the steering column shaft was removed accidentally, remove the steering column assembly and be sure to insert the steering column shaft using the alignment illustration.

21. Be sure to fill the cooling system with the proper grade and type coolant.

22. Recharge the air conditioning system.

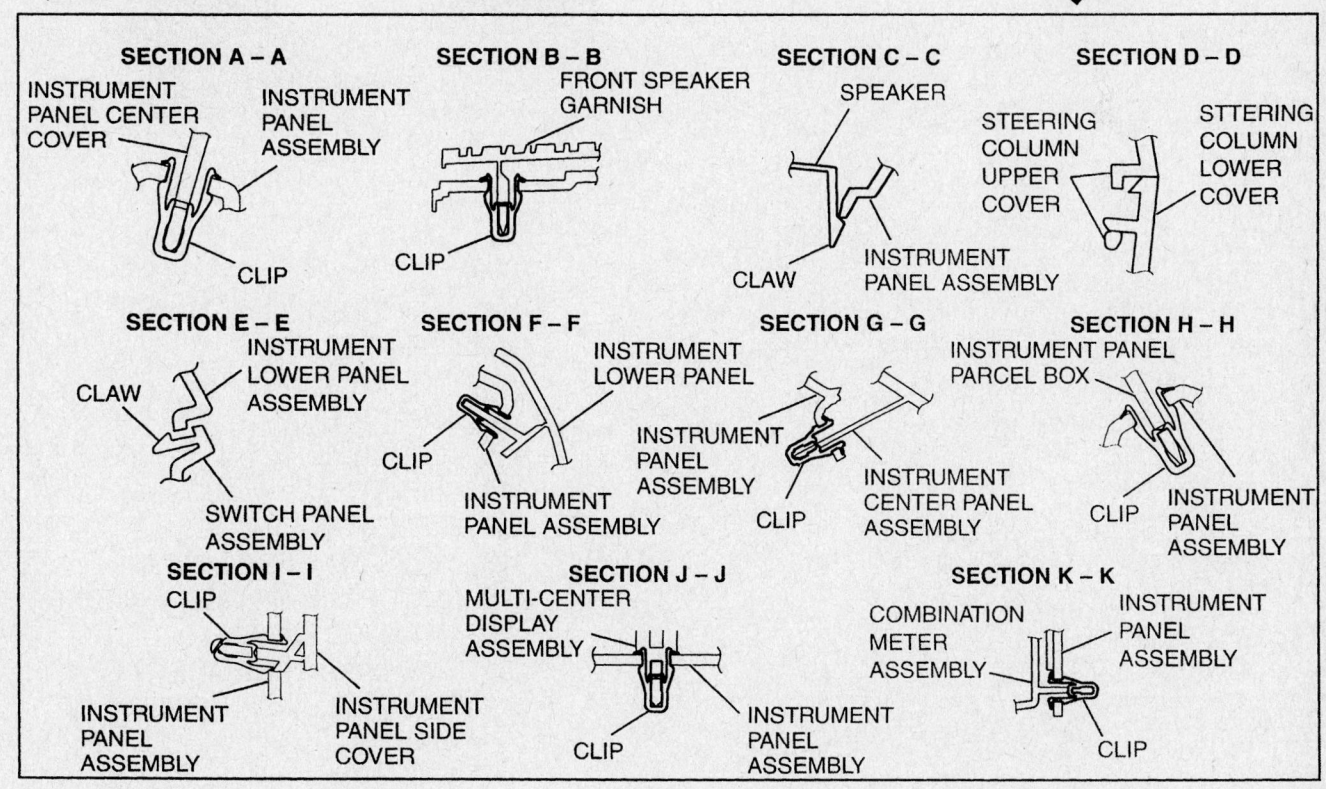

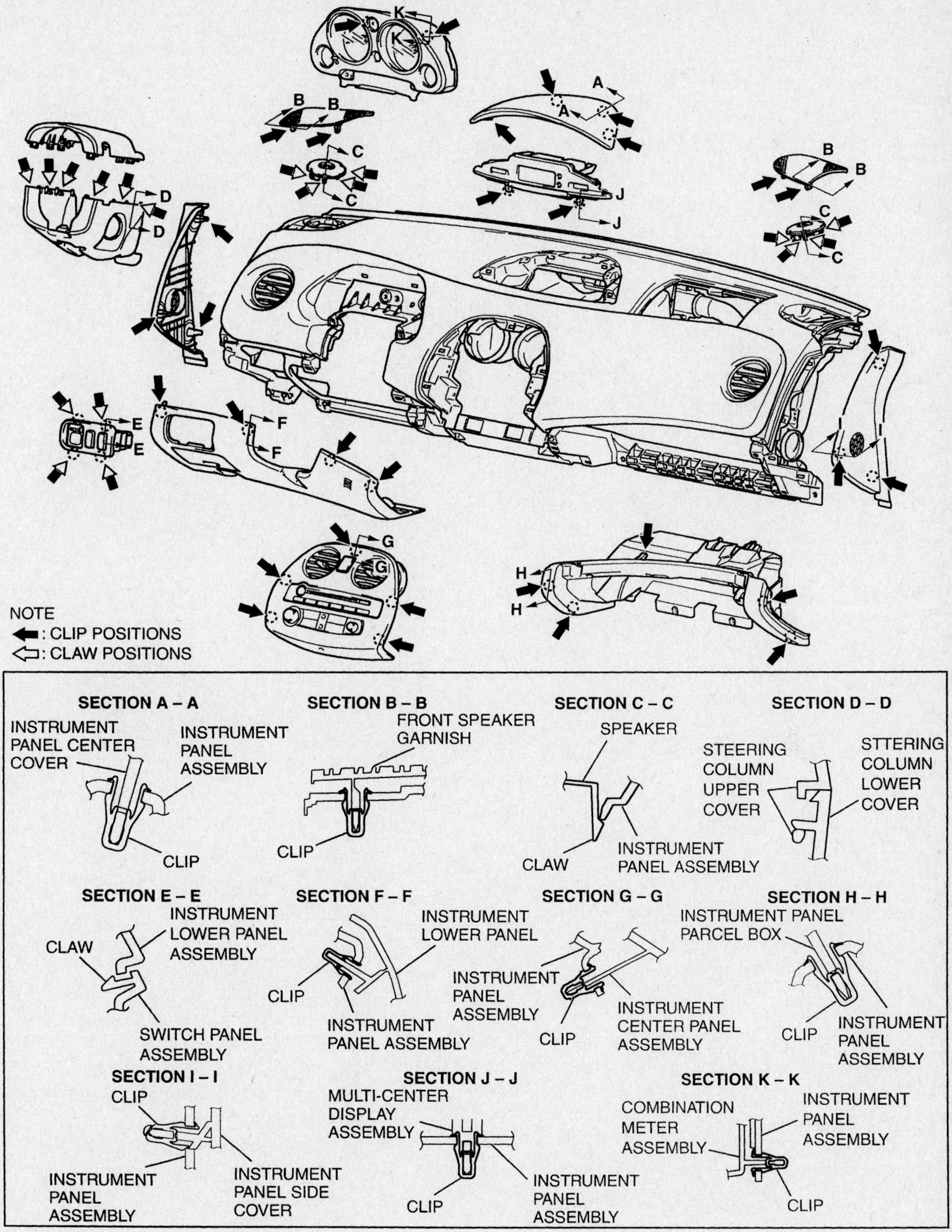

NOTE
← : CLIP POSITIONS
⇐ : CLAW POSITIONS

SECTION A – A

INSTRUMENT PANEL CENTER COVER

INSTRUMENT PANEL ASSEMBLY

CLIP

SECTION B – B

FRONT SPEAKER GARNISH

CLIP

SECTION C – C

SPEAKER

CLAW

INSTRUMENT PANEL ASSEMBLY

SECTION D – D

STEERING COLUMN UPPER COVER

STTERING COLUMN LOWER COVER

SECTION E – E

CLAW

INSTRUMENT LOWER PANEL ASSEMBLY

SWITCH PANEL ASSEMBLY

SECTION F – F

INSTRUMENT LOWER PANEL

CLIP

INSTRUMENT PANEL ASSEMBLY

SECTION G – G

INSTRUMENT PANEL ASSEMBLY

INSTRUMENT CENTER PANEL ASSEMBLY

CLIP

SECTION H – H

INSTRUMENT PANEL PARCEL BOX

CLIP

INSTRUMENT PANEL ASSEMBLY

SECTION I – I

CLIP

INSTRUMENT PANEL ASSEMBLY

INSTRUMENT PANEL SIDE COVER

SECTION J – J

MULTI-CENTER DISPLAY ASSEMBLY

CLIP

INSTRUMENT PANEL ASSEMBLY

SECTION K – K

COMBINATION METER ASSEMBLY

INSTRUMENT PANEL ASSEMBLY

CLIP

09482_GALA_G0019

Instrument panel and related components—2006 Eclipse

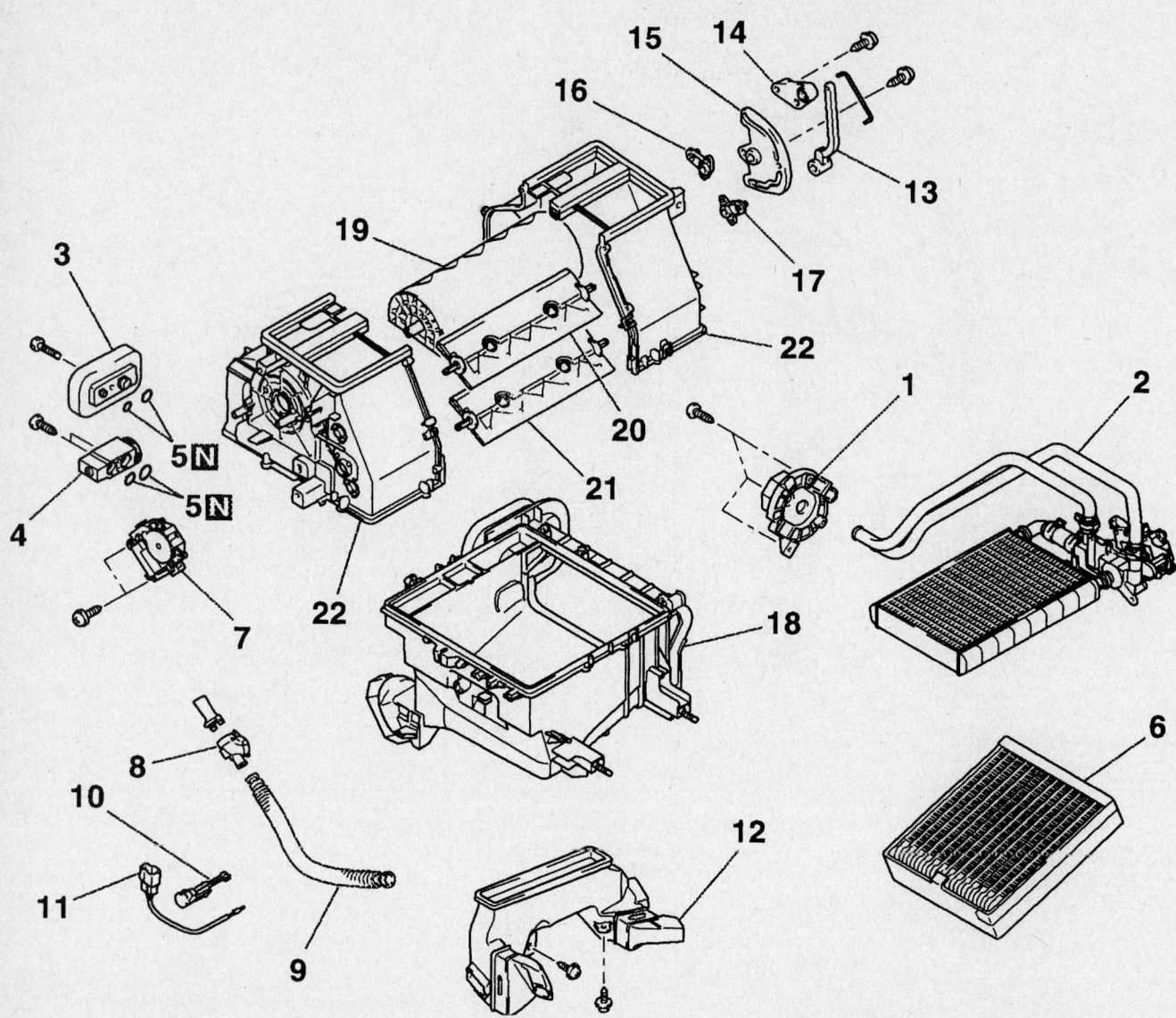

1. PACKING
2. HEATER CORE ASSEMBLY
3. EXPANSION VALVE JOINT
4. EXPANSION VALVE
5. O-RING
6. EVAPORATOR
7. MODE SELECTION DAMPER CONTROL MOTOR AND POTENTIOMETER
8. ASPIRATOR
9. ASPIRATOR HOSE
10. AIR THERMO SENSOR CLIP

11. AIR THERMO SENSOR
12. FOOT DUCT
13. LEVER A
14. LEVER B
15. LEVER C
16. LEVER D
17. LEVER E
18. HEATER CASE LOWER
19. MODE SELECTION DAMPER
20. MAX A/C DAMPER
21. AIR MIXING DAMPER
22. HEATER CASE UPPER

09482_GALA_G0020

Heater core and related components—2006 Eclipse

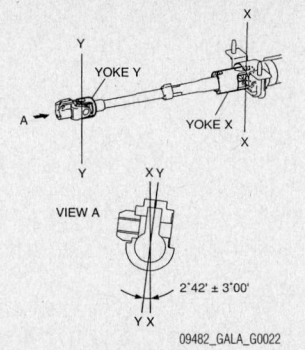

09482_GALA_G0022

**Steering column shaft alignment—2006
Eclipse and 2004–2006 Galant**

Galant

2002–2003

1. Before servicing the vehicle, refer to the Precautions Section.
2. Disconnect the negative battery cable.

✳✳ CAUTION

After disconnecting the negative battery cable, wait at least 60 seconds before working on the SRS module or instrument panel.

3. Drain the cooling system into a clean container for reuse.
4. Disconnect the heater hoses from the heater core at the firewall.

✳✳ WARNING

To prevent damage to the air bag control unit during removal or installation of the floor console, avoid shocks or impact. Do not drop.

5. Remove the floor console by removing or disconnecting the following:
   • Shift lever knob on manual trans-

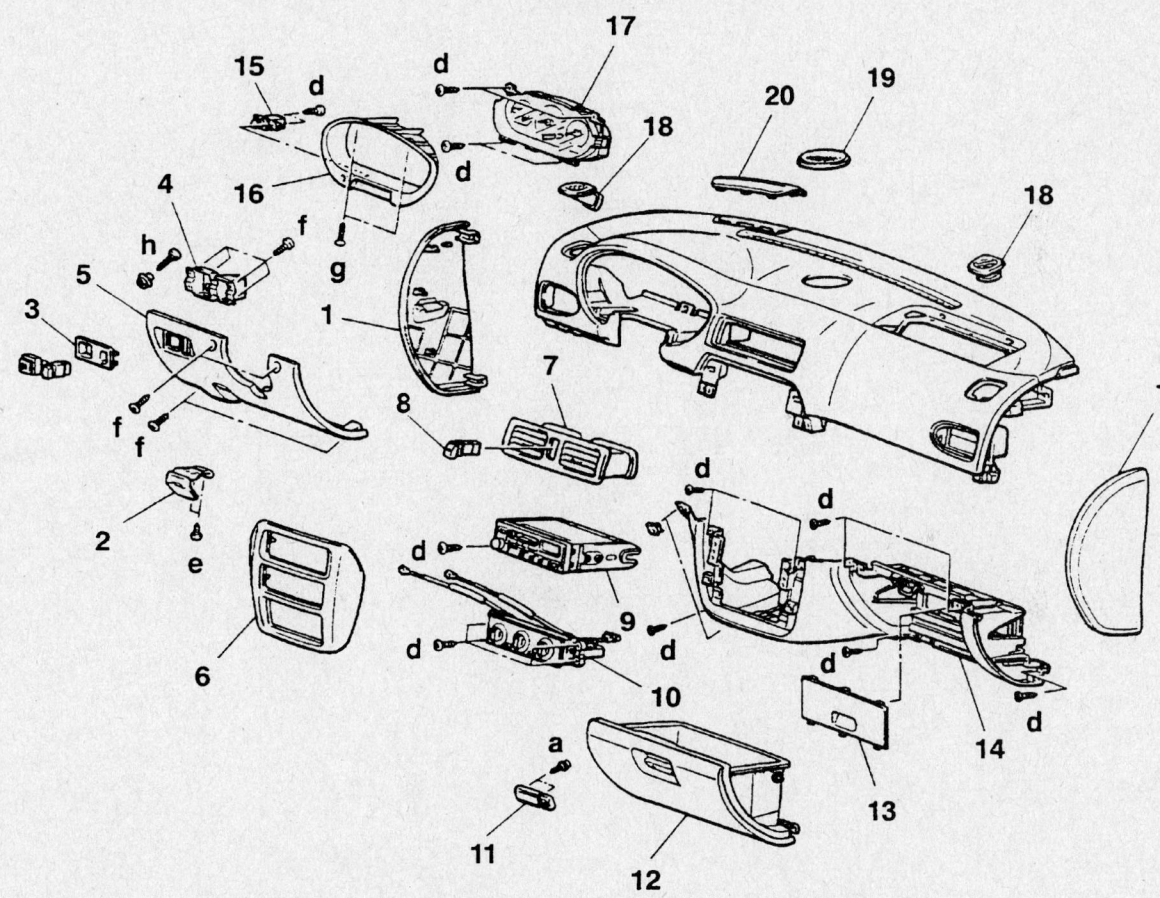

1. INSTRUMENT PANEL SIDE COVER
2. HOOD LOCK RELEASE HANDLE
3. SWITCH PANEL ASSEMBLY
4. CONNECTOR HOLDER
5. FRONT DRIVER'S SIDE UNDER COVER
6. CENTER PANEL ASSEMBLY
7. CENTER AIR OUTLET ASSEMBLY
8. HAZARD WARNING LIGHT SWITCH
9. RADIO AND TAPE PLAYER
10. HEATER CONTROL ASSEMBLY
11. GLOVE BOX STRIKER
12. GLOVE BOX
13. FRONT PASSENGER'S UNDER COVER PLUG
14. FRONT PASSENGER'S SIDE UNDER COVER
15. RHEOSTAT
16. METER BEZEL
17. COMBINATION METER
18. SIDE DEFROSTER GRILLE
19. SPEAKER GRILLE
20. INSTRUMENT PANEL UPPER PLUG

93112GF1

Instrument panel and related components—2002–2003 Galant

mission vehicles or the shift indicator plate on automatic transmissions
- Coin holder behind the shifter, then the center console trim cover in front of the shifter
- Center console retaining bolt cover plugs, then remove the bolts
- Console assembly, then the brackets
6. Remove or disconnect the following:
- Steering column covers
- Instrument cluster bezel and then the instrument cluster
- Instrument panel switch, hood lock

release handle and the lower duct work
- Driver's knee protector and the left side air outlet cover
- Center panel assembly
- Glove box undercover, then the glove box and the right side panel cover
- Radio and cup holder
- Cables from the heater assembly and the blower, then pull out the heater control panel assembly, noting the location of the boss in the center reinforcement
- Cool air bypass damper lever cable connection

- Passenger's side air bag module and disconnect the harness connector, if equipped
- Steering column bolts and lower the column
- Harness connector at the lower left side of the instrument panel
- Instrument panel mounting hardware and remove the instrument panel from the vehicle
- Joint duct between the heater case and the blower assembly (on models without air conditioning)
- Both stamped steel center reinforcement piece

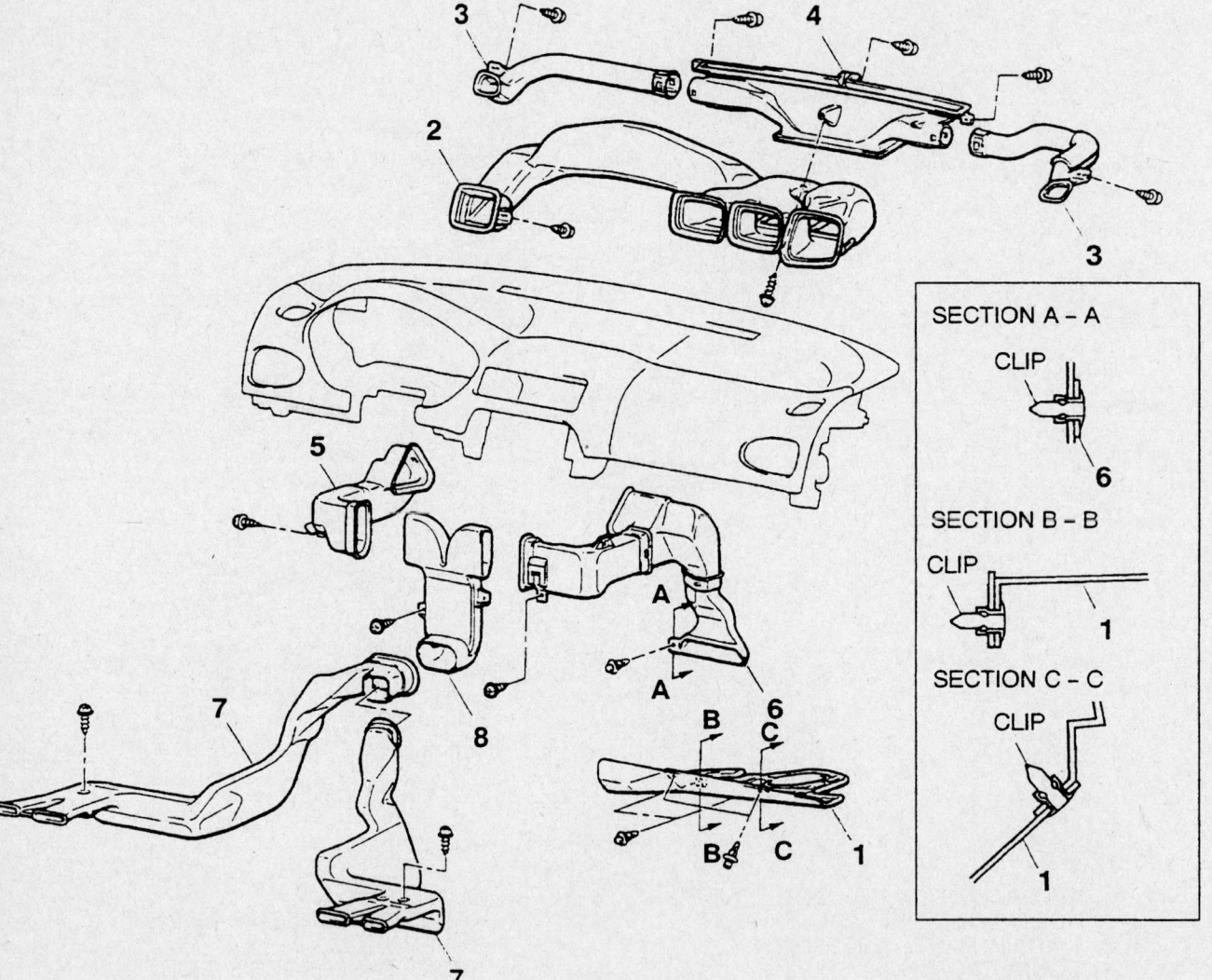

1. UNDER COVER
2. DISTRIBUTION DUCT
3. SIDE DEFROSTER DUCT
4. DEFROSTER NOZZLE ASSEMBLY
5. FOOT DUCT (LH)
6. FOOT DUCT (RH)
7. REAR HEATER DUCT
8. FOOT CENTER DUCT

93112GF2

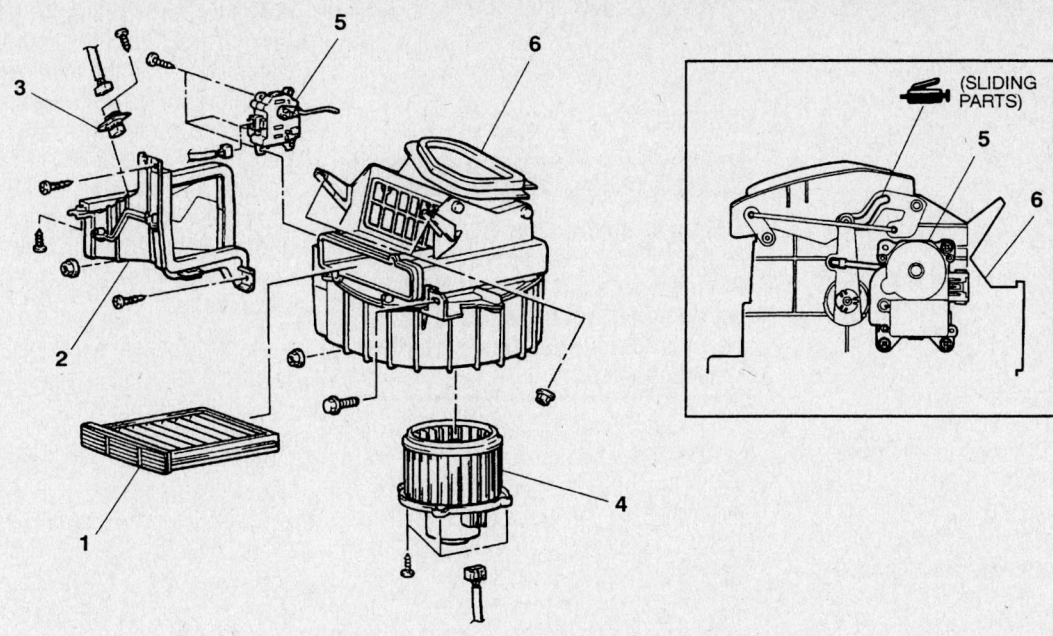

1. AIR PURIFIER ASSEMBLY
2. JOINT DUCT
3. RESISTOR
4. BLOWER FAN AND MOTOR
5. INSIDE/OUTSIDE AIR
   CHANGEOVER DAMPER MOTOR
6. BLOWER ASSEMBLY

93112GF3

Blower motor assembly and related components—2002–2003 Galant

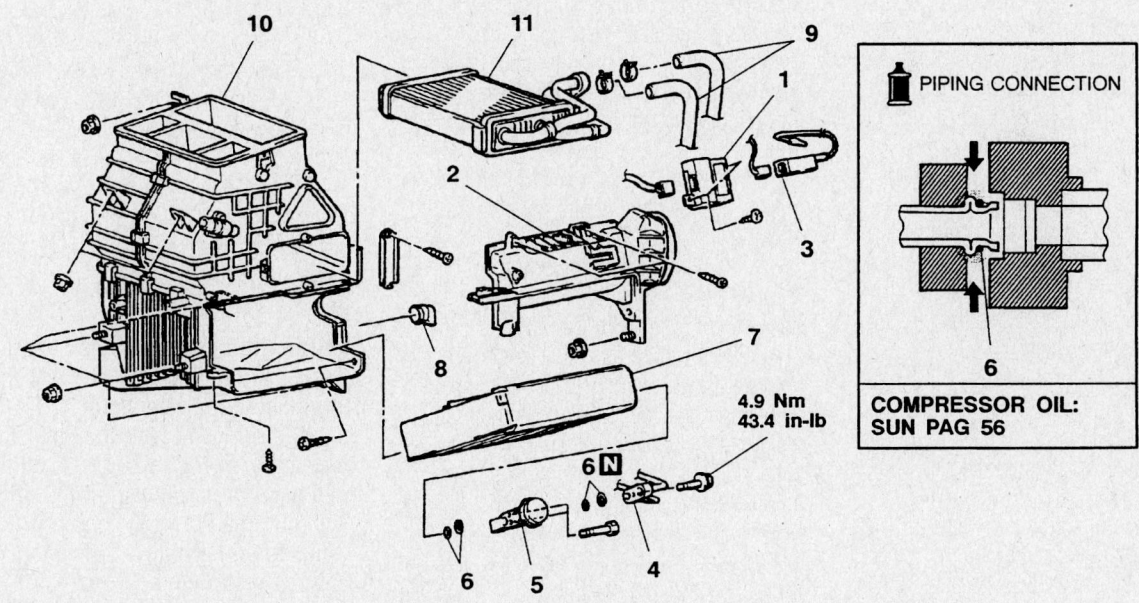

1. BELT LOCK CONTROLLER
2. COVER
3. AUTOMATIC COMPRESSOR
   CONTROLLER
4. A/C PIPE
5. EXPANSION VALVE
6. O-RING
7. EVAPORATOR
8. DRAIN HOSE
9. HEATER HOSE
10. HEATER/COOLER UNIT
11. HEATER CORE

93112GF4

Heater core and related components—2002–2003 Galant

- Electronic Control Module (ECM) bracket
- Evaporator retaining nut and remove the heater case assembly, if equipped with air conditioning
- Heater core from the case

To install:

7. Install or connect the following:
- Heater core to the case
- Evaporator retaining nut and Install the heater case assembly, if equipped with air conditioning
- ECM bracket
- Both stamped steel center reinforcement pieces
- Joint duct between the heater case and the blower assembly (on models without air conditioning)
- Instrument panel and install the instrument panel mounting hardware
- Harness connector at the lower left side of the instrument panel
- Steering column bolts
- Passenger's side air bag module and connect the harness connector, if equipped
- Cool air bypass damper lever cable connection
- Cables to the heater assembly and the blower and install the heater control panel assembly
- Radio and cup holder
- Glove box undercover, then the glove box and the right side panel cover
- Center panel assembly
- Left side air outlet cover and the driver's knee protector
- Lower duct work, the instrument panel switch and hood lock release handle
- Instrument cluster and the instrument cluster bezel
- Steering column covers

8. Install the floor console installing or connecting the following:
- Console assembly brackets and the console
- Center console retaining bolts, then the bolt cover plugs
- Coin holder behind the shifter, then the center console trim cover in front of the shifter
- Shift lever knob on manual transmission vehicles or the shift indicator plate on automatic transmissions
- Heater hoses to the heater core at the firewall

9. Refill the cooling system.
10. Connect the negative battery cable.

11. Operate the engine to normal operating temperatures; then, check the climate control operation and check for leaks.

2004–2006

✳✳ CAUTION

Wait for 1 minute after disconnecting the negative battery cable before working inside the vehicle. The air bag system is set to deploy for a short period of time after the battery is disconnected.

1. Before servicing the vehicle, refer to the Precautions Section.
2. Disconnect the negative battery cable. Drain the cooling system. Discharge the air conditioning system. Be sure the front tires are in the straight ahead position.
3. Remove the front seat. Remove the glove box assembly.
4. Remove the instrument panel parcel box and glove box lock. Remove the hood release handle.
5. Remove the instrument panel under passenger's side cover. Remove the instrument lower panel. Remove the instrument center panel assembly.
6. Remove the radio and CD player. Remove the CD changer, if equipped. Remove the accessory box.
7. Remove the console meter hood. Remove the passenger's side air bag indicator light and passenger's seat belt warning light.
8. If equipped, remove the multi center display. Remove the center console assembly.
9. Remove the instrument panel cover. Remove the combination meter assembly. Remove the front pillar trim.
10. Remove the front speaker covers. Remove the speakers. Remove the instrument panel side cover. Remove the instrument panel side air outlet.
11. Remove the driver's side air bag module. Remove the steering wheel. Remove the steering column cover. Remove the clock spring connector and clock spring and column switch assembly.
12. Disconnect the electrical connections for the ignition switch. Remove the key interlock cable.
13. Remove the steering shaft pad. Remove the steering column retaining bolts. Remove the steering column assembly. Pinch the steering column shaft clip with a pliers, and pull up the shaft to disengage the steering column assembly.

➡ The tilt lever should be held in the lock position until the steering column is reinstalled in the vehicle. If the column is removed with the lever released, or the lever released after the column is removed from the vehicle, the steering column cannot be reinstalled correctly. If the steering column is installed incorrectly, the collision energy absorbing mechanism may be damaged.

14. If equipped with automatic temperature, remove the interior temperature sensor and photo sensor. Remove the instrument panel front end garnish.
15. Remove the gearshift lever handle, heated seat switch, accessory socket, socket cover, front plate and front box.
16. Remove the lid lock lever, hinge, lid, liner box, accessory socket and cover.
17. Remove the floor console retaining screws. Remove the floor console. Remove the accessory socket harness and rear bracket.
18. Remove the cowl side trim. Remove the passenger's side air bag module.
19. Remove the instrument panel assembly from the vehicle.
20. If equipped remove the strut tower retaining bolts and remove the strut tower bar. Remove the battery. Remove the air cleaner body.
21. Remove the heater hoses, and air condition lines from the evaporator.
22. Remove the rear heat duct. Remove the heater unit and deck crossmember assembly.
23. Remove the heater core from the heater case.

To install:

24. Installation is the reverse of the removal procedure.

➡ The tilt lever should be held in the lock position until the steering column is reinstalled in the vehicle. If the column is removed with the lever released, or the lever released after the column is removed from the vehicle, the steering column cannot be reinstalled correctly. If the steering column is installed incorrectly, the collision energy absorbing mechanism may be damaged.

25. If the steering column shaft was removed accidentally, remove the steering column assembly and be sure to insert the steering column shaft using the alignment illustration.
26. Be sure to fill the cooling system with the proper grade and type coolant.

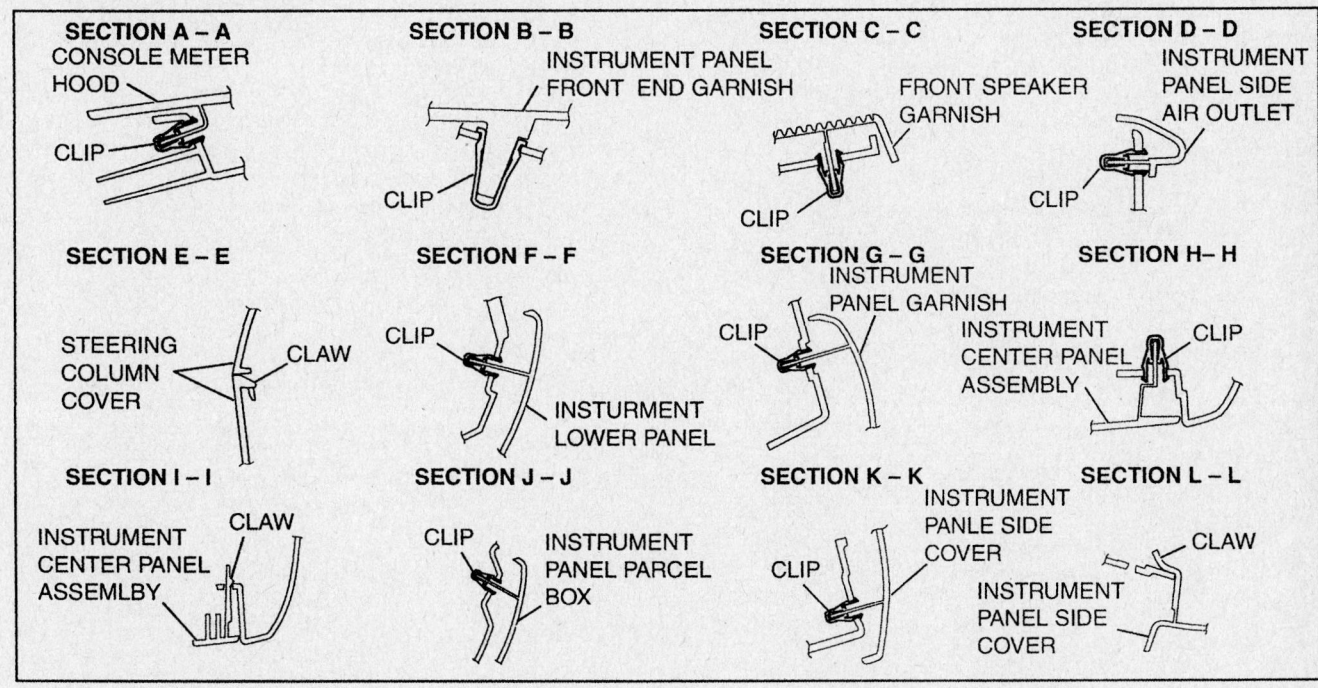

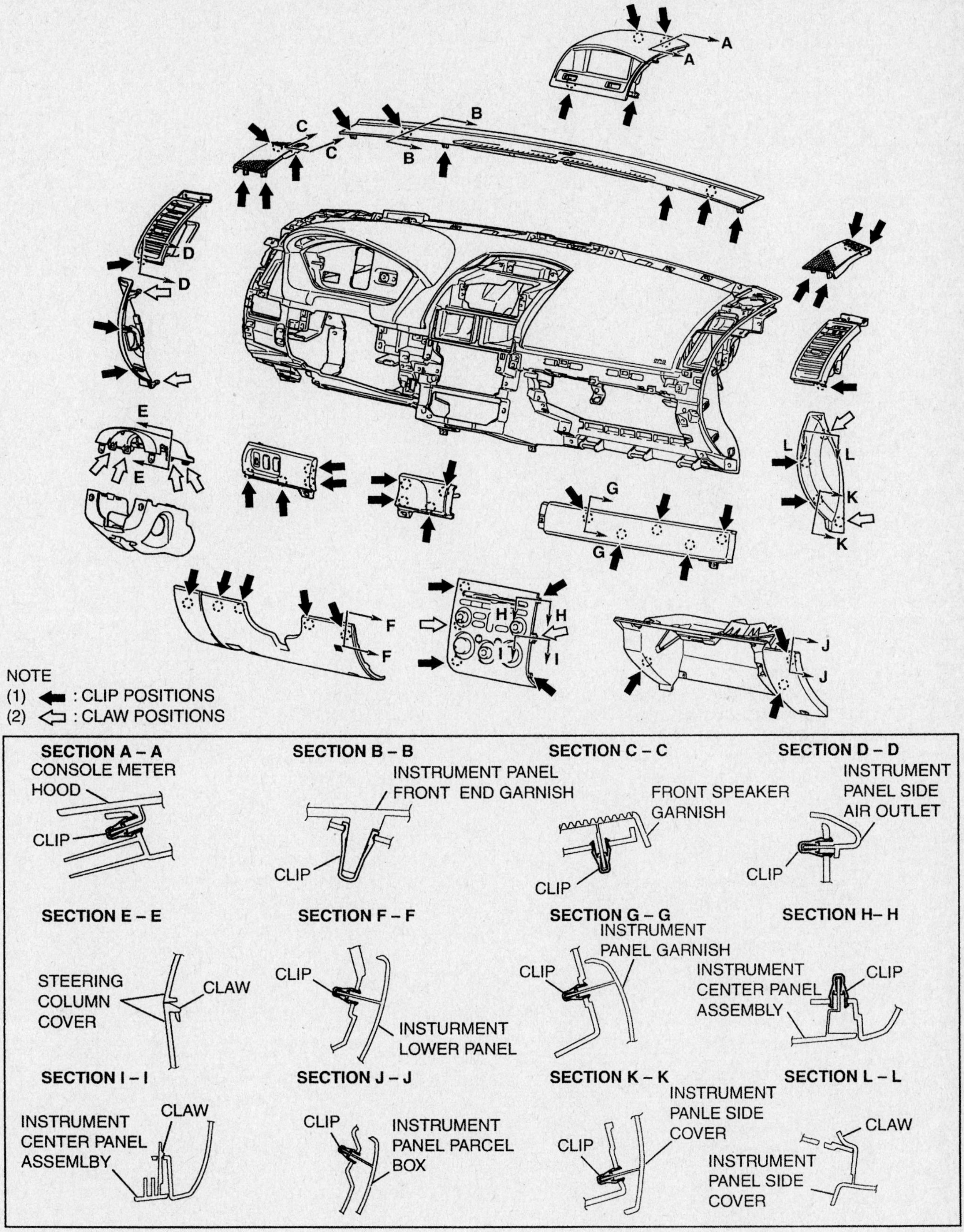

NOTE
(1) ◄— : CLIP POSITIONS
(2) ⇦ : CLAW POSITIONS

SECTION A – A
CONSOLE METER HOOD
CLIP

SECTION B – B
INSTRUMENT PANEL FRONT END GARNISH
CLIP

SECTION C – C
FRONT SPEAKER GARNISH
CLIP

SECTION D – D
INSTRUMENT PANEL SIDE AIR OUTLET
CLIP

SECTION E – E
STEERING COLUMN COVER
CLAW

SECTION F – F
CLIP
INSTURMENT LOWER PANEL

SECTION G – G
INSTRUMENT PANEL GARNISH
CLIP

SECTION H– H
INSTRUMENT CENTER PANEL ASSEMBLY
CLIP

SECTION I – I
INSTRUMENT CENTER PANEL ASSEMLBY
CLAW

SECTION J – J
CLIP
INSTRUMENT PANEL PARCEL BOX

SECTION K – K
INSTRUMENT PANLE SIDE COVER
CLIP

SECTION L – L
CLAW
INSTRUMENT PANEL SIDE COVER

09482_GALA_G0023

Instrument panel and related components—2004–2006 Galant

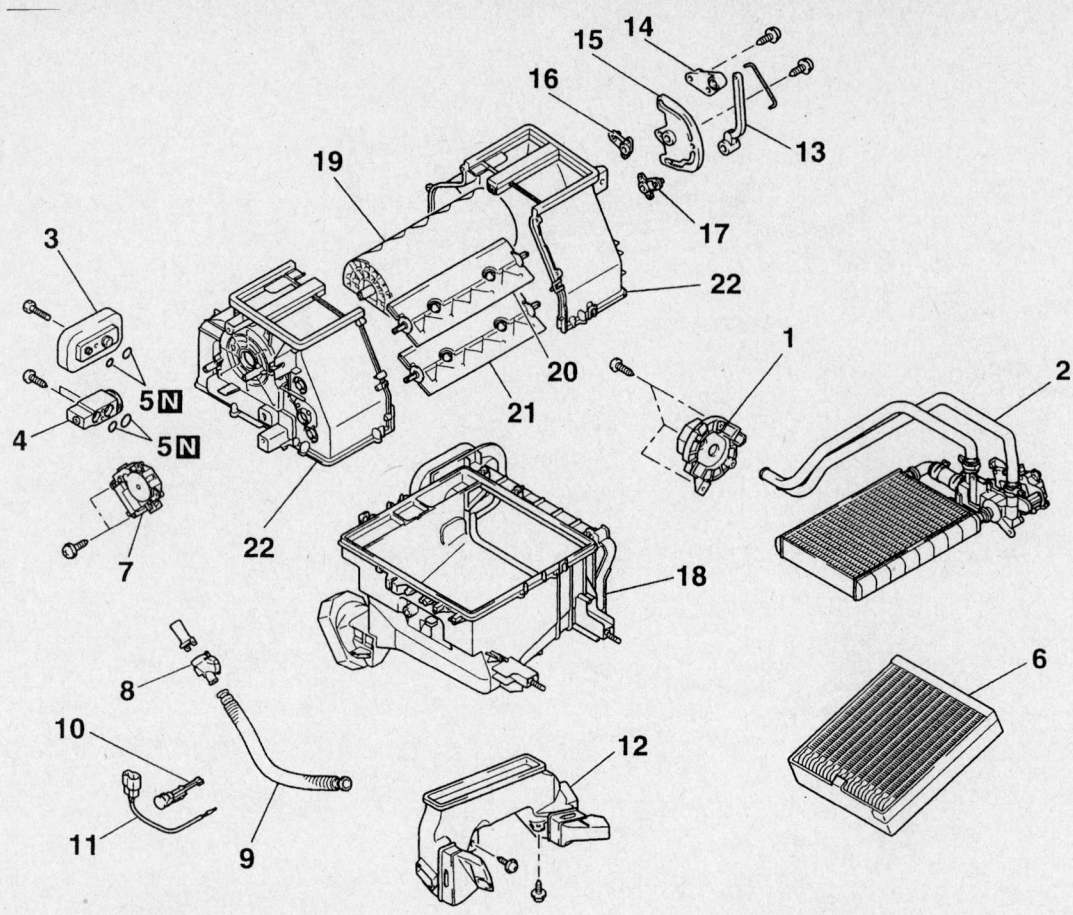

1. PACKING
2. HEATER CORE ASSEMBLY
3. EXPANSION VALVE JOINT
4. EXPANSION VALVE
5. O-RING
6. EVAPORATOR
7. MODE SELECTION DAMPER CONTROL MOTOR AND POTENTIOMETER
8. ASPIRATOR
9. ASPIRATOR HOSE
10. AIR THERMO SENSOR CLIP
11. AIR THERMO SENSOR
12. FOOT DUCT
13. LEVER A
14. LEVER B
15. LEVER C
16. LEVER D
17. LEVER E
18. HEATER CASE LOWER
19. MODE SELECTION DAMPER
20. MAX A/C DAMPER
21. AIR MIXING DAMPER
22. HEATER CASE UPPER

09482_GALA_G0024

Heater core and related components—2004–2006 Galant

27. Recharge the air conditioning system.

Lancer

1. Before servicing the vehicle, refer to the Precautions Section.

2. Disconnect the negative battery cable.

3. Drain the cooling system into a clean container for reuse.

4. Discharge and recover the air conditioning system refrigerant.

5. Remove or disconnect the following:
• Instrument panel
• Front seat assembly
• Front console assembly
• Front floor carpet
• Steering shaft attachment bolt
• Front deck crossmember
• Heater hoses
• Flexible suction hose
• Liquid pipe B connection
• Center duct
• Heater unit
• Intake duct
• Blower assembly

6. Disassemble the heater unit as necessary for access to components,

To install:

7. Assemble the heater unit as necessary.

8. Install or connect the following:
• Blower assembly
• Intake duct
• Heater unit
• Center duct
• Liquid pipe B connection
• Flexible suction hose
• Heater hoses
• Front deck crossmember
• Steering shaft attachment bolt
• Front floor carpet

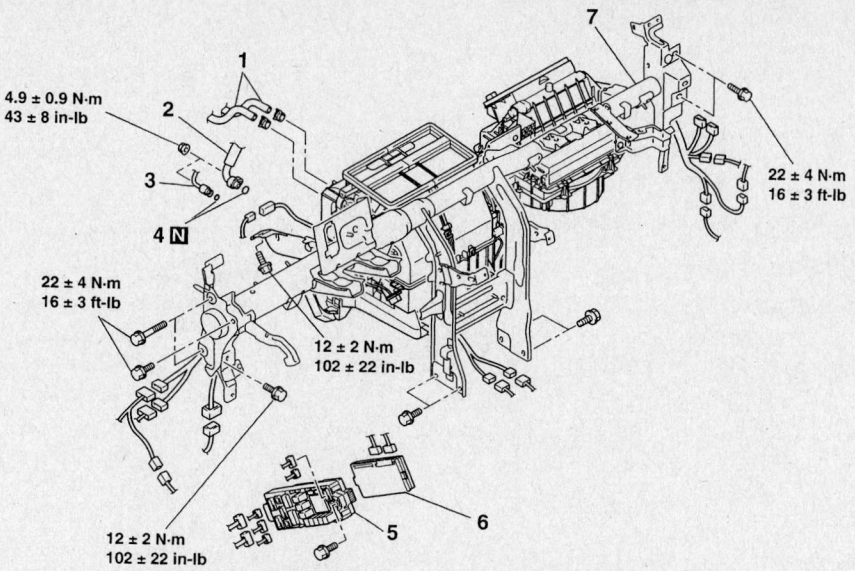

1. HEATER HOSE CONNECTION
2. SUCTION PIPE CONNECTION
3. LIQUID PIPE CONNECTION
4. O-RING

5. JUNCTION BLOCK
6. ETACS-ECU
7. HEATER UNIT AND DECK
   CROSSMEMBER ASSEMBLY

09482_GALA_G0025

Typical heater unit crossmember assembly—all models

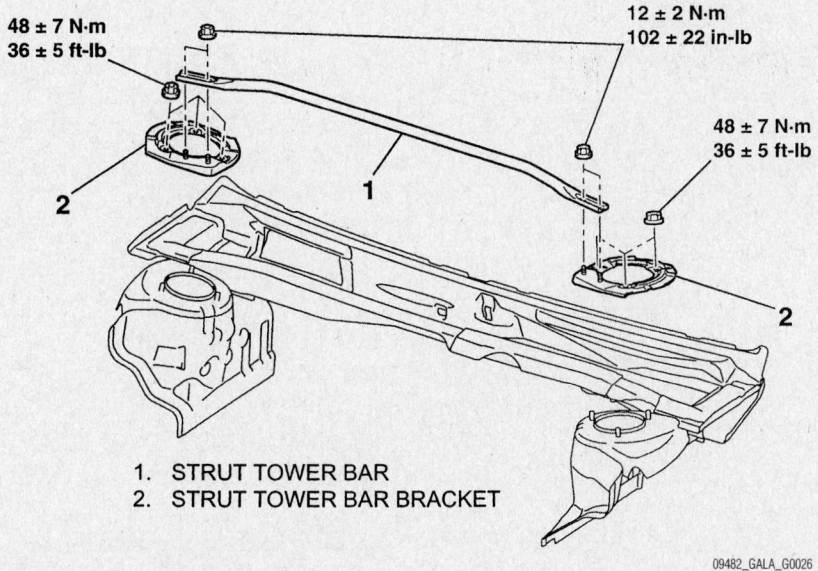

1. STRUT TOWER BAR
2. STRUT TOWER BAR BRACKET

09482_GALA_G0026

Typical strut tower bar assembly—all models

- Front console assembly
- Front seat assembly
- Instrument panel
9. Refill the cooling system.
10. Connect the negative battery cable.
11. Evacuate, charge and leak test the air conditioning system.
12. Operate the engine to normal operating temperatures; then, check the climate control operation and check for leaks.

Evolution

❋❋ CAUTION

Wait for 1 minute after disconnecting the negative battery cable before working inside the vehicle. The air bag system is set to deploy for a short period of time after the battery is disconnected.

1. Before servicing the vehicle, refer to the Precautions Section.
2. Disconnect the negative battery cable. Drain the cooling system. Discharge the air conditioning system. Be sure the front tires are in the straight ahead position.
3. Remove the front seat.
4. Remove the column cover. Remove the meter bezel. Remove the combination meter.
5. Remove the instrument panel ornament. Remove the under cover, switch panel, switch holder and lower frame.
6. Remove the heater control knob, center panel and heater control assembly.
7. Remove the radio and CD player.
8. Remove the center air outlet panel and the center lower case, if equipped.
9. If equipped with a CD changer, remove the center lower panel.
10. Remove the stopper, glove box and harness cover. Remove the instrument panel side cover.
11. Remove the passenger's side air bag module mounting bolt and air bag module.
12. Remove the steering column cover. Remove the driver's side air bag module. Remove the steering wheel.
13. Remove the steering column clock spring and column switch assembly. Remove the shaft cover.
14. Remove the steering column shaft assembly. Remove the cover assembly.

➡**The tilt lever should be held in the lock position until the steering column is reinstalled in the vehicle. If the column is removed with the lever released, or the lever released after the column is removed from the vehicle, the steering column cannot be reinstalled correctly. If the steering column is installed incorrectly, the collision energy absorbing mechanism may be damaged.**

15. Remove the instrument panel assembly.
16. Remove the console side cover, shift lever knob and front floor console assembly.
17. Remove the front floor console panel. If equipped with manual transaxle remove the shift lever cover.
18. Remove the shift lever bezel, lighter and ashtray.
19. Remove the console retaining screws. Remove the console from the vehicle. Remove the electrical harness and bracket assembly, as required.
20. Remove the front deck crossmember. Remove the heater hoses. Remove the air conditioning lines. Remove the center duct.

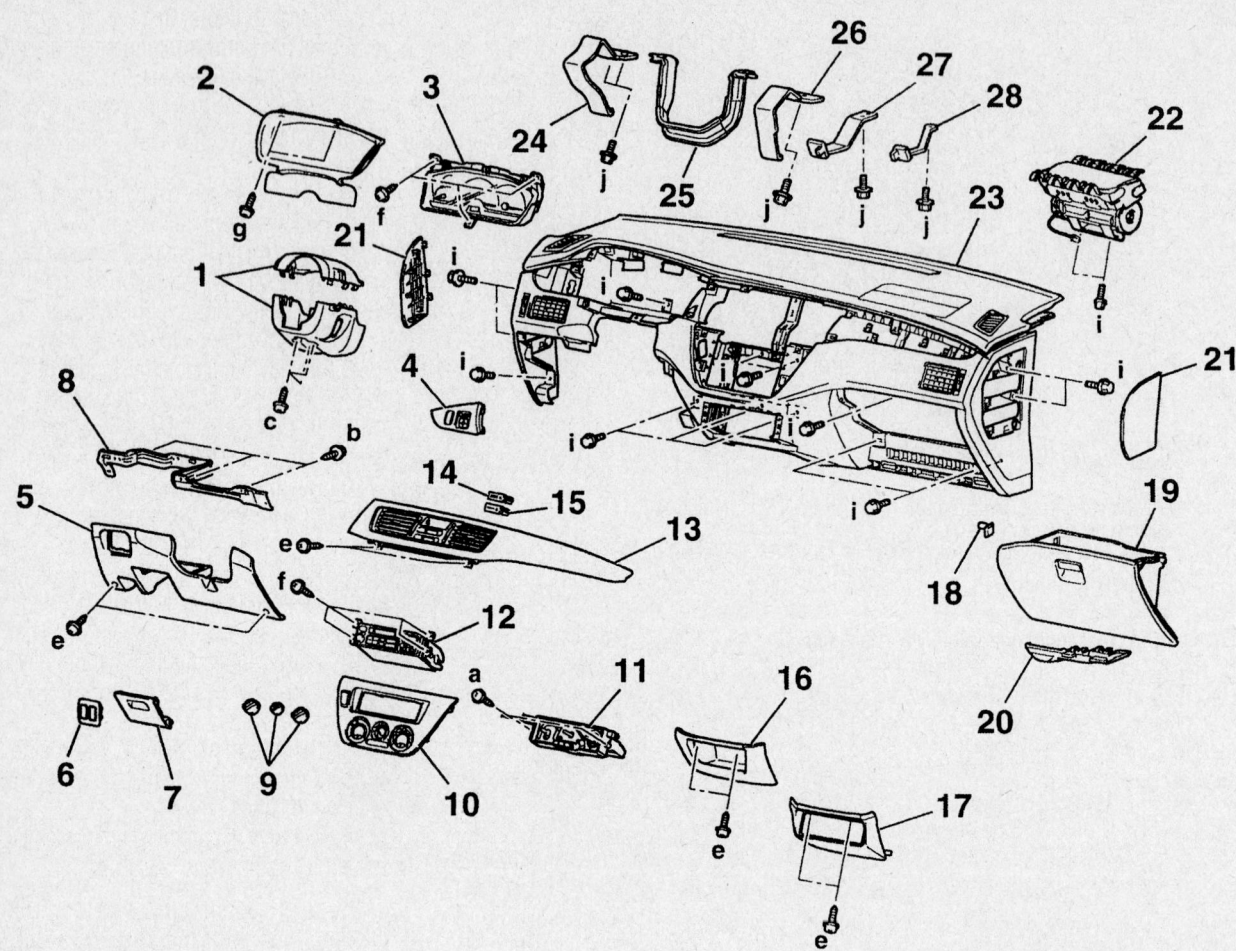

1. COLUMN COVER
2. METER BEZEL
3. COMBINATION METER
4. INSTRUMENT PANEL ORNAMENT
5. UNDER COVER
6. SWITCH PANEL
7. FUSE LID
8. LOWER FRAME
9. HEATER CONTROL KNOB
10. CENTER PANEL
11. HEATER CONTROL ASSEMBLY
12. RADIO AND TAPE PLAYER
13. CENTER AIR OUTLET PANEL
14. AIR BAG OFF INDECATOR LIGHT
15. SEAT BELT WARNING LIGHT
16. CENTER LOWER CASE

17. CENTER LOWER PANEL <VEHICLE WITH CD AUTOMATIC CHANGER>
18. STOPPER
19. GLOVE BOX
20. HARNESS COVER
21. INSTRUMENT PANEL SIDE COVER
22. SRS FRONT PASSENGER'S AIR BAG MODULE
23. INSTRUMENT PANEL ASSEMBLY
24. KNEE ABSORBER DRIVER'S SIDE LH
25. STEERING UNDER BRACKET
26. KNEE ABSORBER DRIVER'S SIDE RH
27. KNEE ABSORBER PASSENGER'S LH
28. KNEE ABSORBER PASSENGER'S SIDE RH

09482_GALA_G0027

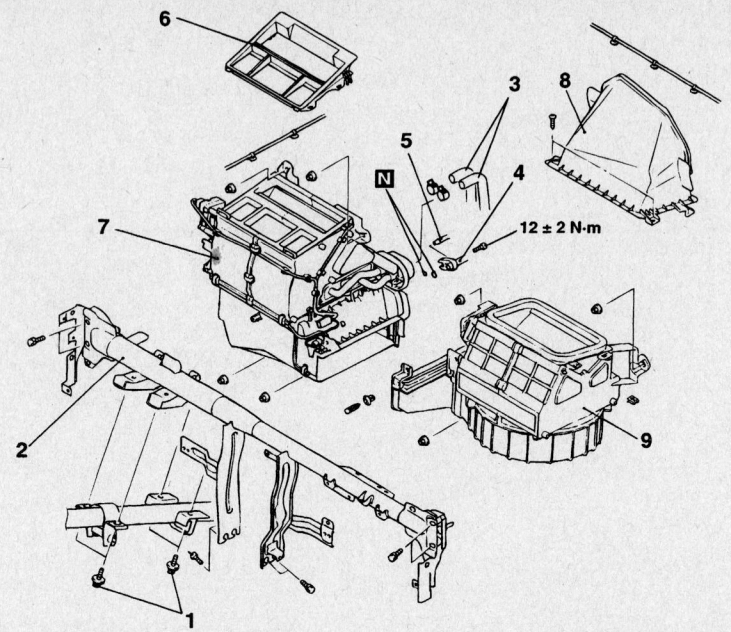

1. STEERING SHAFT ATTACHMENT BOLT
2. FRONT DECK CROSSMEMBER
3. HEATER HOSE CONNECTION
4. FLEXIBLE SUCTION HOSE CONNECTION
5. LIQUID PIPE B CONNECTION
6. CENTER DUCT
7. HEATER UNIT
8. INTAKE DUCT
9. BLOWER ASSEMBLY

42356-GALA-G01

Heater unit and related components—Lancer

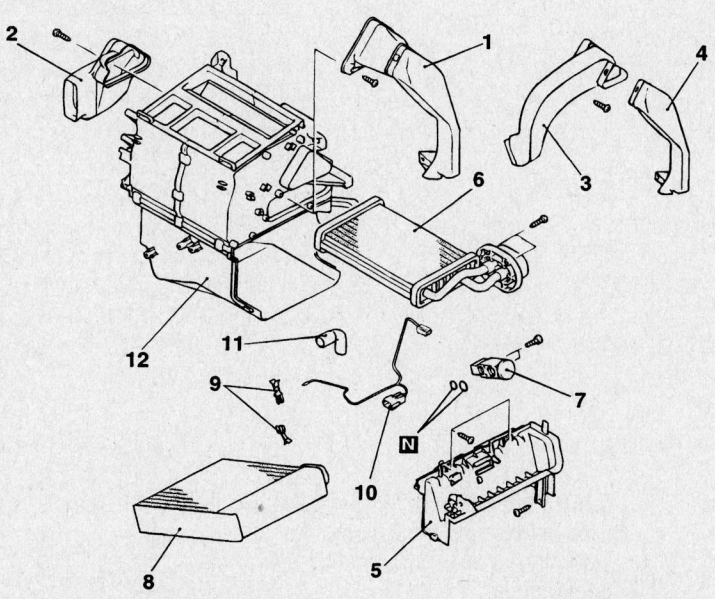

1. RIGHT-HAND FOOT DUCT
2. LEFT-HAND FOOT DUCT
3. LEFT-HAND FOOT DUCT <VEHICLE WITH REAR HEATER DUCT>
4. LEFT-HAND UPPER REAR HEATER DUCT A <VEHICLE WITH REAR HEATER DUCT>
5. EVAPORATOR COVER
6. HEATER CORE
7. EXPANSION VALVE
8. EVAPORATOR
9. AIR THERMO SENSOR CLIP
10. AIR THERMO SENSOR
11. DRAIN PLUG
12. HEATER CASE

42356-GALA-G02

Heater core and related components—Lancer

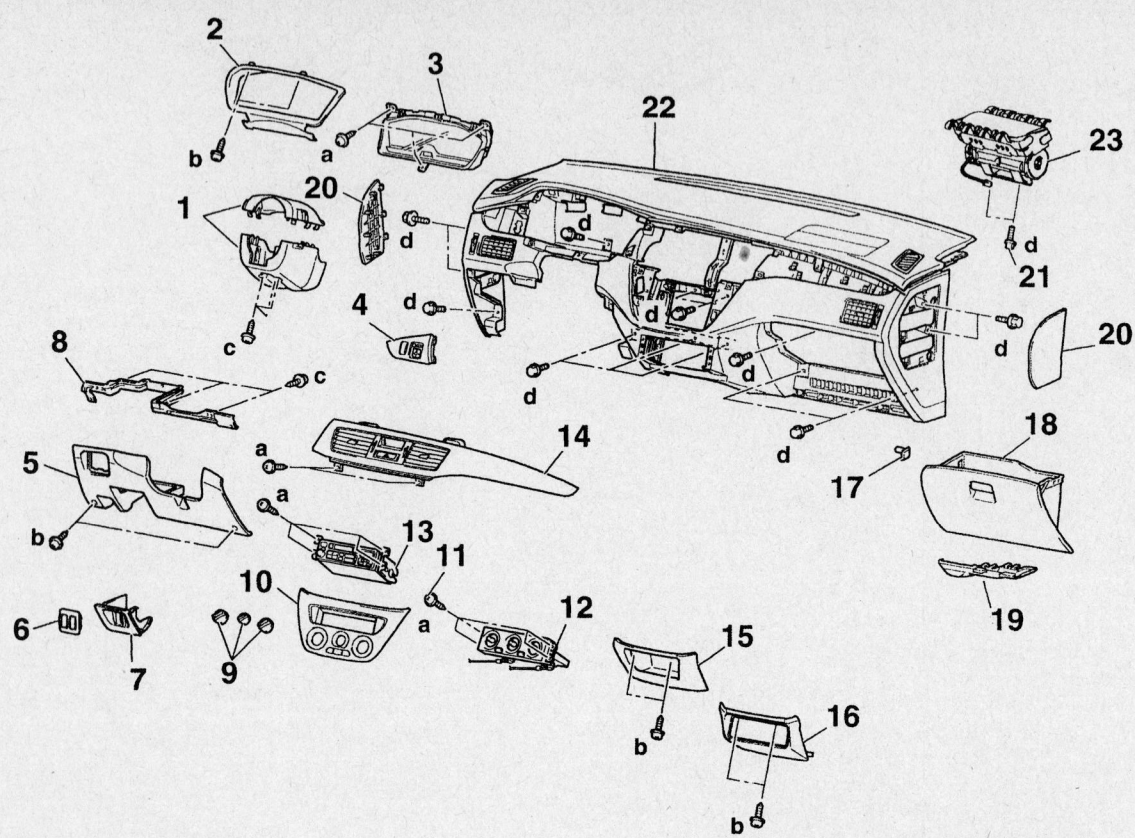

1. COLUMN COVER
2. METER BEZEL
3. COMBINATION METER
4. INSTRUMENT PANEL ORNAMENT
5. UNDER COVER
6. SWITCH PANEL
7. SWITCH HOLDER
8. LOWER FRAME
9. HEATER CONTROL KNOB
10. CENTER PANEL
11. HEATER CONTROL ASSEMBLY
    MOUNTING SCREW
12. HEATER CONTROL ASSEMBLY
13. RADIO AND TAPE PLAYER

14. CENTER AIR OUTLET PANEL
15. CENTER LOWER CASE (VEHICLES
    WITHOUT CD AUTO CHANGER)
16. CENTER LOWER PANEL (VEHICLE
    WITH CD AUTO CHANGER)
17. STOPPER
18. GLOVE BOX
19. HARNESS COVER
20. INSTRUMENT PANEL SIDE COVER
21. PASSENGER'S (FRONT) AIR BAG
    MODULE MOUNTING BOLT
22. INSTRUMENT PANEL ASSEMBLY
23. PASSENGER'S (FRONT) AIR BAG
    MODULE

09482_GALA_G0028

Instrument panel and related components—Evolution

Remove the heater unit, intake duct and blower assembly from the vehicle.

21. Remove the heater core.

To install:

22. Installation is the reverse of the removal procedure.

➡**The tilt lever should be held in the lock position until the steering column is reinstalled in the vehicle. If the column is removed with the lever released, or the lever released after the column is removed from the vehicle, the steering column cannot be reinstalled correctly. If the steering column is installed incorrectly, the collision energy absorbing mechanism may be damaged.**

23. If the steering column shaft was removed accidentally, remove the steering column assembly and be sure to insert the steering column shaft using the alignment illustration.

24. Be sure to fill the cooling system with the proper grade and type coolant.

25. Recharge the air conditioning system.

Mirage

1. Before servicing the vehicle, refer to the Precautions Section.

2. Disconnect the negative battery cable.

3. Drain the cooling system into a clean container for reuse.

4. Remove the air cleaner cover and the air intake hose.

5. Disconnect the heater hoses from the heater core.

6. If equipped with air conditioning, discharge and recover the air conditioning system refrigerant.

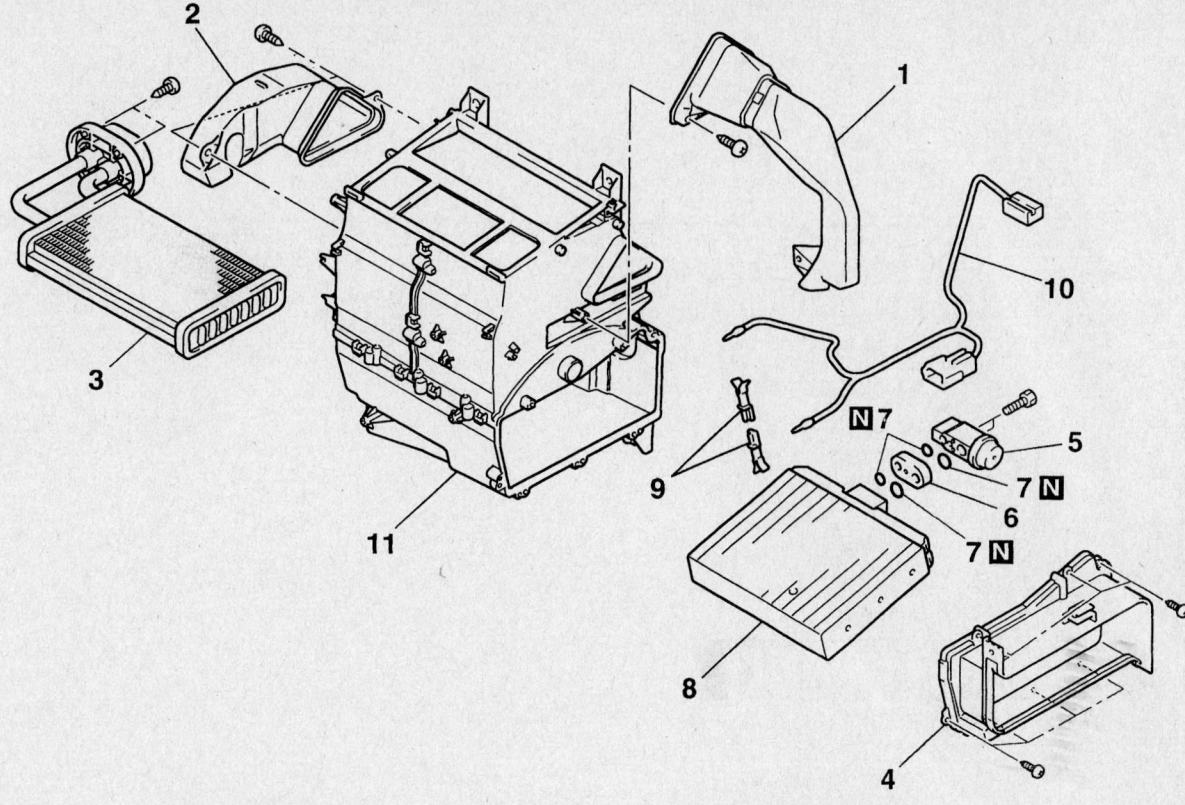

1. FOOT DUCT (RH)
2. FOOT DUCT (LH)
3. HEATER CORE
4. EVAPORATOR COVER
5. EXPANSION VALVE
6. JOINT
7. O-RING
8. EVAPORATOR
9. AIR THERMO SENSOR CLIP

09482_GALA_G0029

Heater core and related components—Evolution

7. If equipped with air conditioning, disconnect the refrigerant lines from the evaporator core and discard the O-rings.

❄❄ CAUTION

After disconnecting the negative battery cable, wait at least 60 seconds before working on the SRS module or instrument panel.

8. Remove the passenger's side air bag by removing or disconnecting the following:
- Glove box assembly
- Air bag-to-dash bolts and the air bag; then, disconnect the electrical connector

9. Remove the floor console.

10. Remove the instrument panel by removing or disconnecting the following:
- Rheostat
- Hood release handle
- Knee protector plug and the knee protector assembly
- Steering column cover

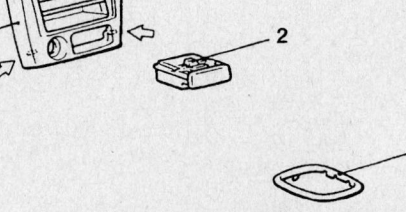
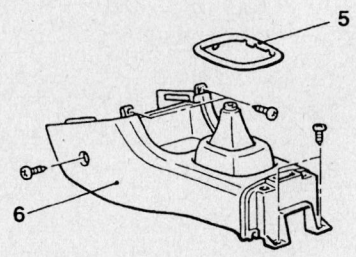

NOTE
⇦ : metal clip position

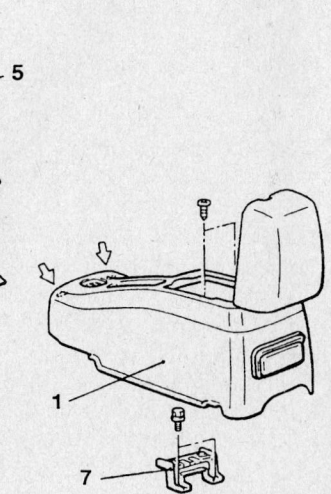

1. Rear floor console assembly
2. Ashtray
3. Audio panel
4. Box
● Shift lever knob
5. A/T panel
6. Front floor console assembly
7. Rear console bracket

Floor console and related components—Mirage

93112GF5

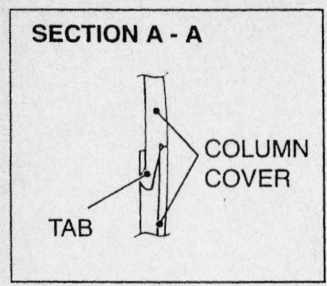

SECTION A - A

COLUMN COVER

TAB

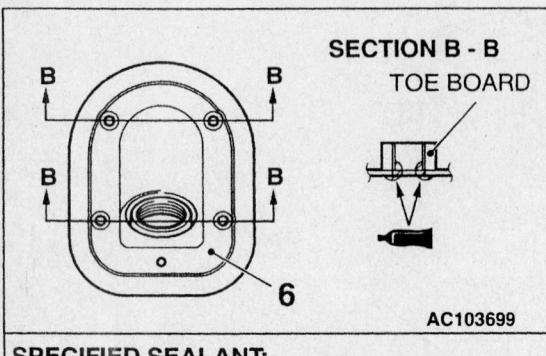

SECTION B - B

TOE BOARD

B B
B B
6

AC103699

SPECIFIED SEALANT:
3M™ AAD PART NO.8633 WINDO-WELD
RESEALANT OR EQUIVALENT

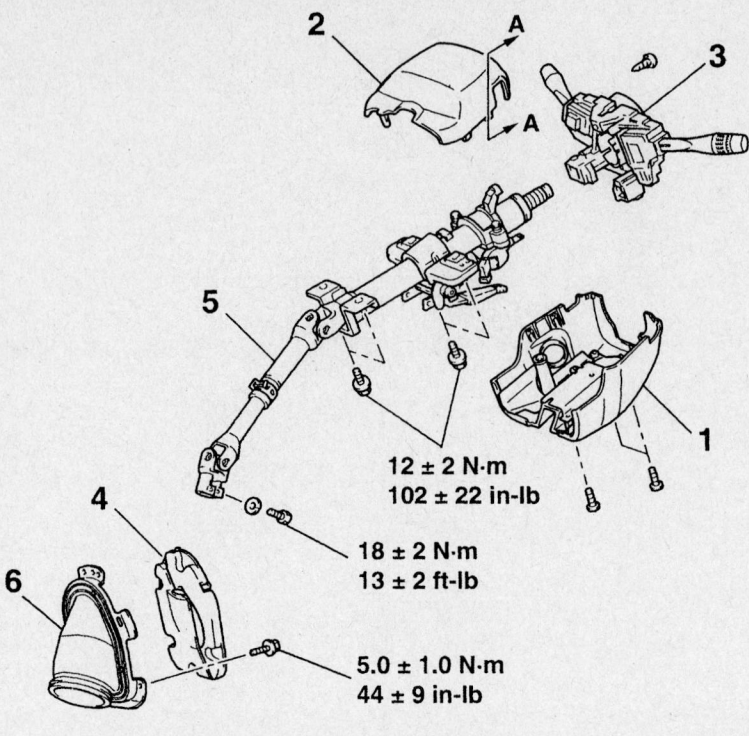

12 ± 2 N·m
102 ± 22 in-lb

18 ± 2 N·m
13 ± 2 ft-lb

5.0 ± 1.0 N·m
44 ± 9 in-lb

1. LOWER COLUMN COVER
2. UPPER COLUMN COVER
3. CLOCK SPRING AND COLUMN
   SWITCH ASSEMBLY
4. SHAFT COVER

5. STEERING COLUMN SHAFT
   ASSEMBLY
6. COVER ASSEMBLY

09482_GALA_G0030

Steering column installation—Evolution

- Meter bezel and the combination meter
- Mirror control switch or plug, if equipped
- Auto-cruise main switch, fog light switch or plug, if equipped
- Side air outlet assembly
- Radio and tape player
- Cup holder
- Heater control panel
- Heater control assembly
- Steering column bolts and lower the steering column
- Instrument panel assembly with the help of an assistant

11. If not equipped with air conditioning, remove the blower motor-to-heater housing joint duct.

12. If equipped with air conditioning, remove the evaporator housing-to-heater housing fasteners and remove the evaporator housing.

13. Remove or disconnect the following:
- Center reinforcement

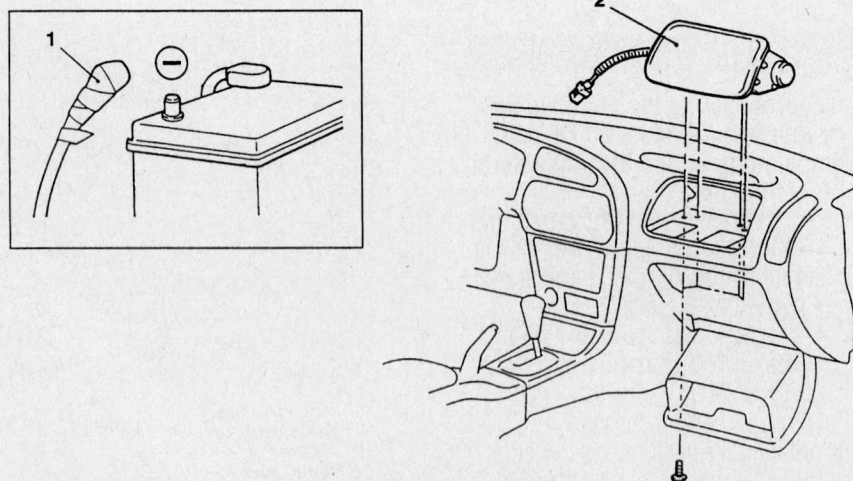

1. Negative (–) battery cable connection
2. Air bag module

93112GF6

Passenger's air bag module—Mirage

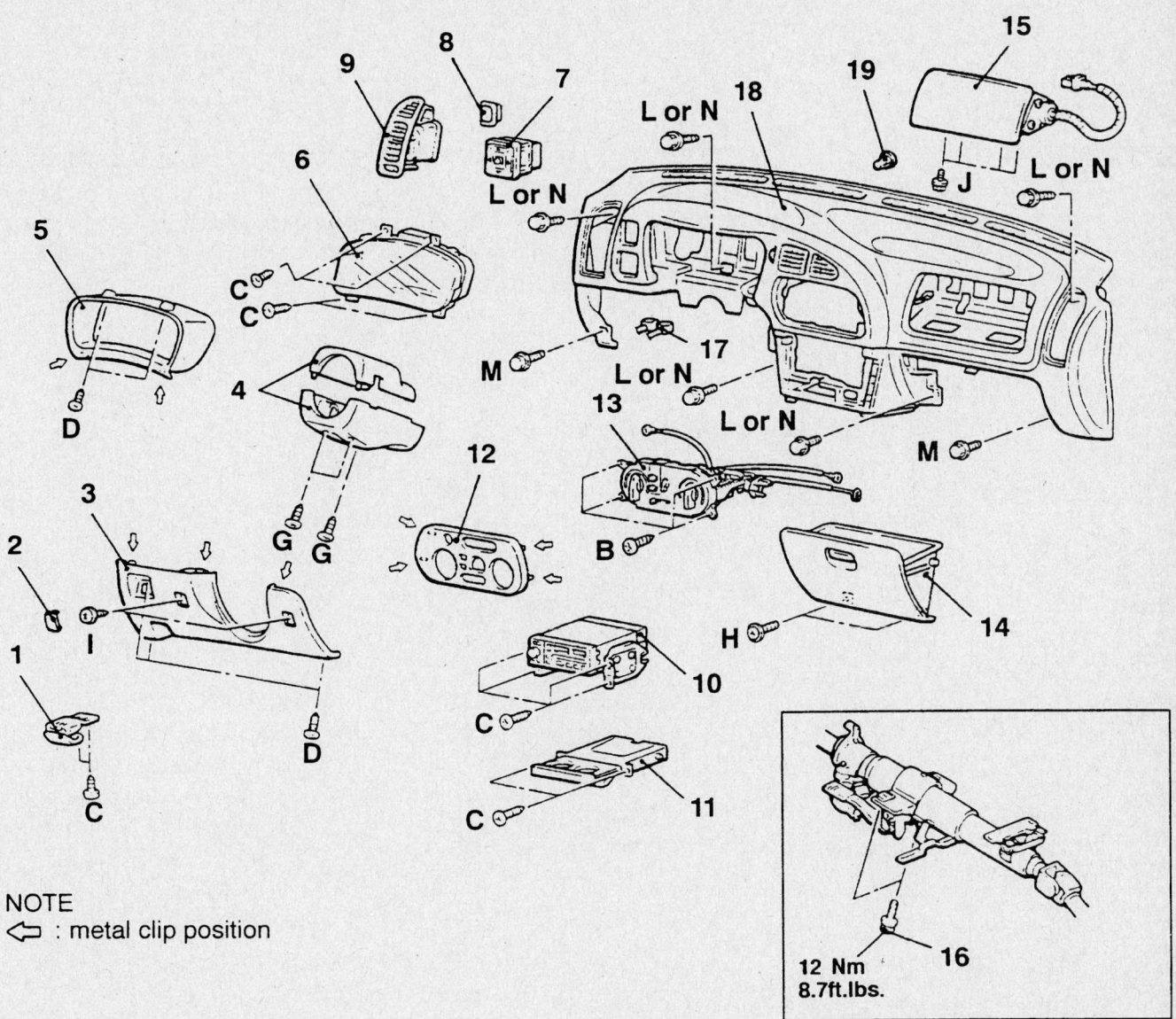

NOTE
⇦ : metal clip position

12 Nm
8.7ft.lbs.

1. Hood lock release handle
2. Knee protector plug
3. Knee protector assembly
4. Column cover
5. Meter bezel
6. Combination meter
7. Door mirror control switch or plug
8. Auto-cruise control main switch, fog light switch or plug
9. Side air outlet assembly
10. Radio and tape player
11. Cup holder
12. Heater control panel
13. Heater control assembly
14. Glove box
15. Front passenger's air bag module assembly
16. Steering column assembly installation bolt
17. Harness connector
18. Instrument panel assembly
19. Grommet

93112GF7

Instrument panel and steering column assembly—Mirage

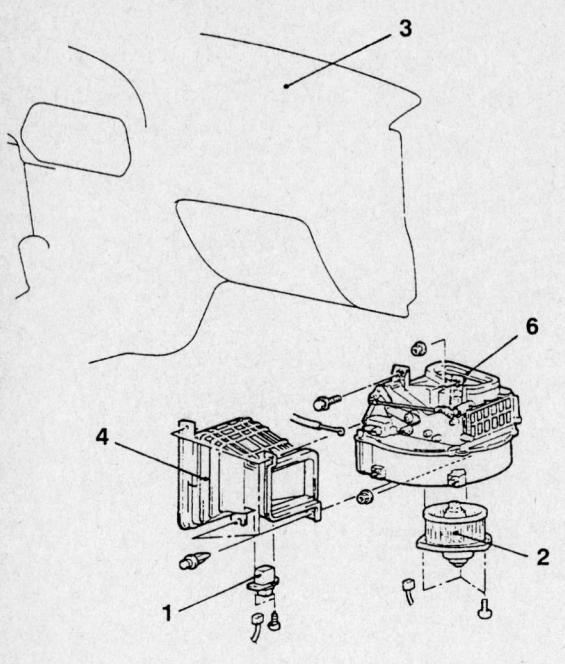

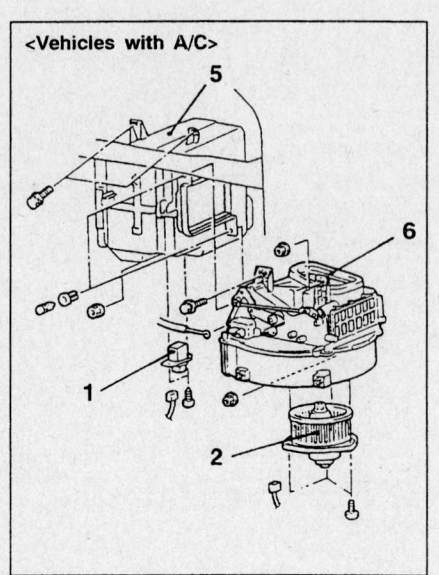

<Vehicles with A/C>

1. Resistor
2. Blower fan and motor
3. Instrument panel
4. Joint duct
5. Evaporator
6. Blower unit assembly

93112GF9

Blower motor assembly—Mirage

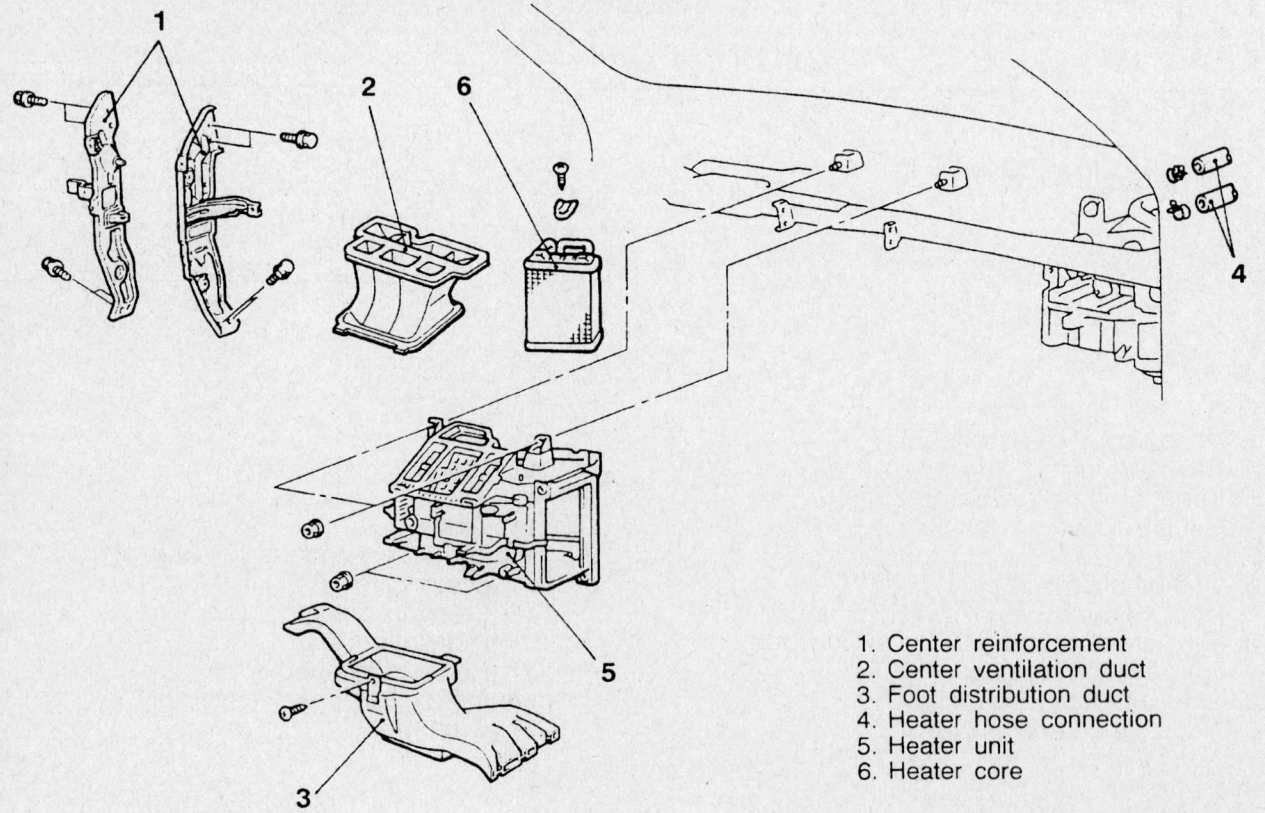

1. Center reinforcement
2. Center ventilation duct
3. Foot distribution duct
4. Heater hose connection
5. Heater unit
6. Heater core

93112GF8

Heater core and related components—Mirage

- Center ventilation duct
- Foot distribution duct
- Heater housing
- Heater core from the heater housing

To install:

14. Install or connect the following:
- Heater core to the heater housing
- Heater housing
- Foot distribution duct
- Center ventilation duct
- Center reinforcement

15. If equipped with air conditioning, install the evaporator housing and the evaporator housing-to-heater housing fasteners.

16. If not equipped with air conditioning, install the blower motor-to-heater housing joint duct.

17. Install the instrument panel by installing or connecting the following:
- Instrument panel assembly with the help of an assistant
- Steering column and install the steering column bolts
- Heater control assembly
- Heater control panel
- Cup holder
- Radio and tape player
- Side air outlet assembly
- Auto-cruise main switch, fog light switch or plug, if equipped
- Mirror control switch or plug, if equipped
- Combination meter and the meter bezel
- Steering column cover
- Knee protector assembly and the knee protector plug
- Hood release handle
- Rheostat
- Floor console

18. Install the passenger's side air bag by installing or connecting the following:
- Electrical connector; then, install the air bag and the air bag-to-dash bolts
- Glove box assembly

19. If equipped with air conditioning, use new O-rings and connect the refrigerant lines to the evaporator core.

20. Install or connect the following:
- Heater hoses to the heater core
- Air cleaner cover and the air intake hose

21. Refill the cooling system.

22. Connect the negative battery cable.

23. If equipped with air conditioning, evacuate, charge and leak test the air conditioning system refrigerant.

24. Operate the engine to normal operating temperatures; then, check the climate control operation and check for leaks.

Cylinder Head

REMOVAL & INSTALLATION

Diamante

1. Before servicing the vehicle, refer to the Precautions Section.
2. Disconnect the negative battery cable.
3. Drain the engine coolant
4. Remove or disconnect the following:
- Timing belt
- Intake and exhaust manifolds
- Spark plug wires
- Cylinder head covers
- Timing belt rear center cover

5. Loosen the cylinder head bolts gradually in 3 stages, in the opposite of the installation sequence.

6. Remove the cylinder head.

To install:

7. Clean the cylinder head and mounting surface on the engine block.

8. Install the cylinder head using a new gasket.

9. Tighten the bolts in sequence using 3 stages to 80 ft. lbs. (105 Nm).

10. Install or connect the following:
- Timing belt rear center cover
- Cylinder head covers using new gaskets. Tighten the bolts to 24–36 inch lbs. (3–4 Nm).
- Spark plug wires
- Intake and exhaust manifolds
- Timing belt
- Remaining components

11. Refill the cooling system.

12. Connect the negative battery cable.

Eclipse

2002–2003 2.4L ENGINE

1. Before servicing the vehicle, refer to the Precautions Section.
2. Relieve the fuel system pressure.
3. Disconnect the negative battery cable.
4. Remove the air cleaner with all air intake hoses.
5. Drain the cooling system.
6. Remove or disconnect the following:
- Accelerator cable
- Cable mounting brackets and position the cable aside
- Breather hose
- Vacuum lines at the throttle body, label for identification
- High pressure fuel line, plug
- Fuel return hose, plug

7. Disconnect the following connectors:
- Air conditioning compressor
- Power steering pressure switch
- Heated Oxygen (HO2S) sensor
- Engine Coolant Temperature (ECT) gauge sender
- ECT sensor
- Manifold Absolute Pressure (MAP) sensor
- Intake Air Temperature (IAT) sensor
- Throttle Position (TP) sensor
- Idle Air Control (IAC) motor
- Injector harness
- Ignition coil
- Camshaft Position (CMP) sensor
- Exhaust Gas Recirculation (EGR) solenoid valve

8. Remove or disconnect the following:
- Spark plug wire cover and wires

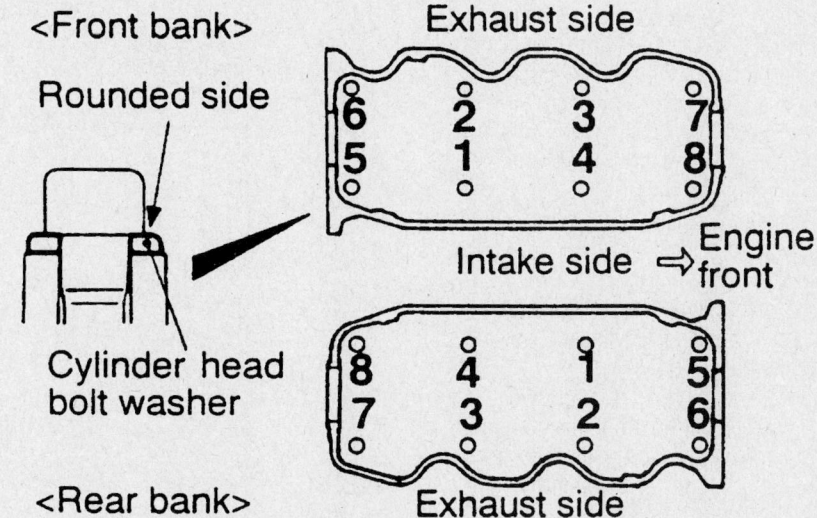

Cylinder head bolt torque sequence—3.5L engine

7923PGD2

Intake side

Front of engine ⇨

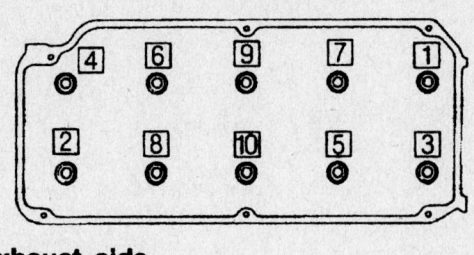

Exhaust side

67170-GALA-G15

Cylinder head bolt removal sequence—2002–2003 Eclipse and Galant 2.4L engine

- Coolant hoses and unbolt the thermostat case from the engine, at the thermostat case assembly
- Upper timing belt cover
9. Align all timing marks.
10. Secure the timing belt to the camshaft sprocket with cord or a wire tie.
11. Remove or disconnect the following:
  - Camshaft sprocket
  - Valve cover and the half-round seal
  - Intake manifold stay bracket from the intake manifold
  - Exhaust pipe self-locking nuts and separate the exhaust pipe from the exhaust manifold. Discard the gasket.
12. Loosen the cylinder head mounting bolts in 3 steps, starting from the outside and working inward. Lift off the cylinder head assembly and remove the head gasket.

To install:

13. Thoroughly clean the mating surfaces of the head and block.
14. Place a new head gasket on the cylinder block with the identification marks at the front top (upward) position. Do not use sealer on the gasket.
15. Inspect the cylinder head bolt length prior to installation. If the length exceeds 3.91 in. (99.4mm), the bolt must be replaced. Install the washer onto the bolt so the chamfer on the washer faces towards the head of the bolt.
16. Carefully install the cylinder head on the block and tighten the cylinder head bolts as follows:
  a. Following the proper tightening sequence, tighten the cylinder head bolts to 58 ft. lbs. (78 Nm).
  b. Loosen all bolts completely.
  c. Torque bolts to 15 ft. lbs. (20 Nm).
  d. Tighten bolts an additional ¼ turn.
  e. Tighten bolts an additional ¼ turn.
17. Install or connect the following:
  - New exhaust pipe gasket and connect the exhaust pipe to the mani-

fold. Tighten the bolts to 33 ft. lbs. (44 Nm).
  - Thermostat case and tighten the mounting bolts to 18 ft. lbs. (24 Nm)
  - Coolant hoses to the thermostat case
18. Apply sealer to the perimeter of the half-round seal and to the lower edges of the half-round portions of the belt-side of the new gasket. Install the valve cover.
19. Install or connect the following:
  - Camshaft sprocket with the timing belt attached. Remove the cord or wire tie.

Intake side

Front of engine ⇨

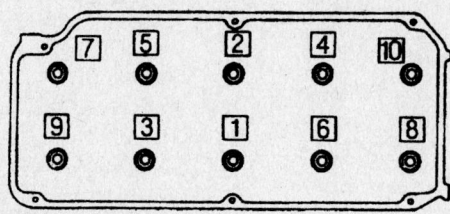

Exhaust side

67170-GALA-G16

Cylinder head bolt torque sequence—2002–2003 Eclipse and Galant 2.4L engine

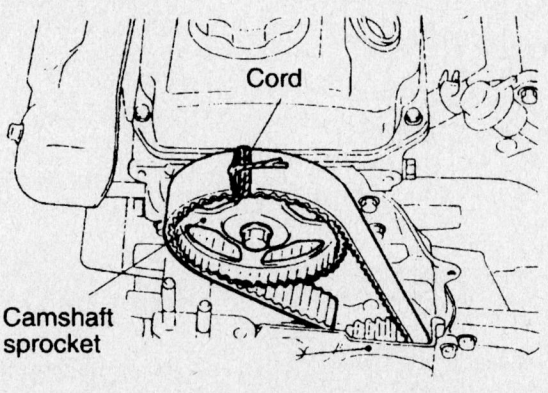

7923PG23

Secure the timing belt to the camshaft sprocket and remove the sprocket—2002–2003 Eclipse and Galant 2.4L engine

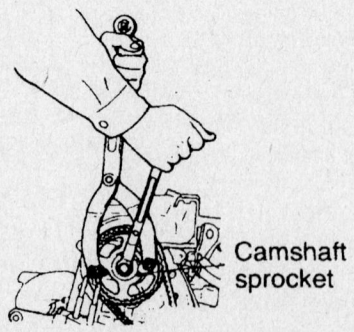

- Upper timing belt cover
- Intake manifold stay and tighten the mounting bolts to 22 ft. lbs. (30 Nm)
20. Connect the following connectors:
- Air conditioning compressor
- Power steering pressure switch
- HO2S sensor
- ECT gauge sender
- ECT sensor
- MAP sensor
- IAT sensor
- TP sensor
- IAC motor
- Injector harness
- Ignition coil
- CMP sensor
- EGR solenoid valve
21. Remove or disconnect the following:
- Spark plug wires and cover
- Fuel lines using new O-rings
- Air cleaner and intake hose
- Breather hose
22. Fill the cooling system.
23. Connect the negative battery cable

2004–2005 2.4L ENGINE

1. Before servicing the vehicle, refer to the Precautions Section.
2. Relieve the fuel system pressure. Disconnect the negative battery cable. Drain the cooling system.
3. Remove the air cleaner assembly. Drain the engine oil. If equipped, remove the strut tower bar.
4. Remove the thermostat case assembly. Remove the front exhaust pipe.
5. Disconnect the accelerator cable connection, purge hose connection and brake booster vacuum hose.
6. Disconnect the ignition coil connector, injector connector and manifold absolute pressure sensor connector.
7. Disconnect the throttle position sensor connector, heated front oxygen sensor connector and the capacitor connector.
8. Disconnect the engine coolant temperature sensor, camshaft position sensor connector and the knock sensor connector.
9. Disconnect the engine coolant temperature gauge unit connector, the idle air control motor connector and the EVAP solenoid valve connector.
10. Disconnect the EGR vacuum regulator solenoid valve connector, the fuel high pressure hose connection, fuel return hose connection and the pressure hose connection.
11. Remove the engine oil dipstick and dipstick guide. Remove the spark plug cables and ignition coil.
12. Remove the upper radiator hose.

Remove the PCV valve hose connection and breather hose connection.
13. Remove the rocker arm cover. Remove the spark plug guide oil seal. Remove the water hose connection.
14. Remove the timing belt.
15. Remove the power steering pressure switch connector. Remove the power steering pump and bracket assembly, position the unit to the side as not to interfere with removal of the cylinder head.
16. Remove the exhaust manifold bracket.

➡**Be careful not to damage or deform the plug guides when removing the cylinder head bolts. Plug guides cannot be replaced separately.**

17. Loosen all cylinder head bolts in two or three steps and in the proper bolt removal sequence. Remove the cylinder head from the engine. Discard the cylinder head gasket.

To install:
18. Thoroughly clean the mating surfaces of the head and block.
19. Place a new head gasket on the cylinder block.
20. Inspect the cylinder head bolt length prior to installation. If the length exceeds 3.91 in. (99.4mm), the bolt must be replaced. Apply a small amount of engine oil to the thread section and washer of the bolt.

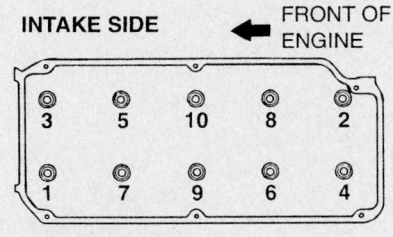

Cylinder head bolt removal sequence— 2004–2005 Eclipse 2.4L engine

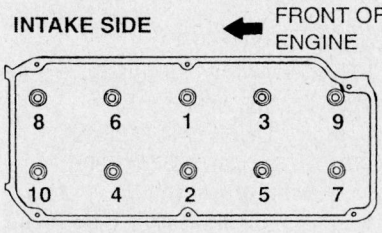

Cylinder head bolt torque sequence— 2004–2005 Eclipse 2.4L engine

21. Position the cylinder head on the engine. Install and torque the retaining bolts to specification and in the proper sequence.
22. Continue the installation in the reverse order of the removal procedure.
23. Fill the engine with the proper grade and type engine oil.
24. Fill the cooling system with the proper grade and type engine coolant.
25. Start the engine and check for leaks.
26. Road test the vehicle.

2006 2.4L (4G69) ENGINE

1. Before servicing the vehicle, refer to the Precautions Section.
2. Relieve the fuel system pressure. Disconnect the negative battery cable. Drain the cooling system.
3. Remove the ECM, if the vehicle is equipped with manual transaxle. Remove the PCM if the vehicle is equipped with automatic transaxle.
4. Remove the air cleaner assembly. Remove the battery and tray.
5. Disconnect the control wiring harness, radiator hose lower clamp and the water hose clamp.
6. Disconnect the EVAP hose connection. Disconnect the brake booster vacuum hose, pressure hose clamp and knock sensor connector.
7. Remove the engine oil dipstick and dipstick guide. Remove the intake manifold stay.
8. Remove the exhaust manifold. Remove the timing belt upper cover. Remove the engine front mounting bracket.
9. Turn the crankshaft clockwise, align the timing marks on the camshaft sprocket to position number one piston to TDC of its compression stroke. Never rotate the crankshaft in the counterclockwise direction.
10. Remove the timing belt under cover rubber plug, using tool MD998738. Screw the tool in until it contacts the timing belt tensioner arm.

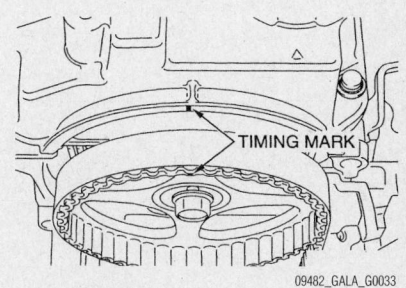

Timing mark alignment—2.0L (4G94) engine and 2.4L (4G69) engine

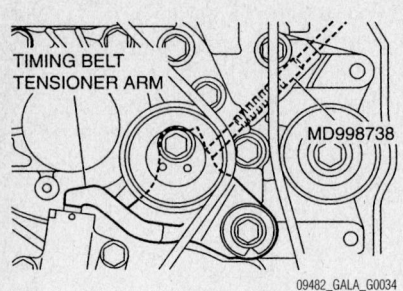

09482_GALA_G0034

Tool MD998738 installation—2.0L (4G94) engine and 2.4L (4G69) engine

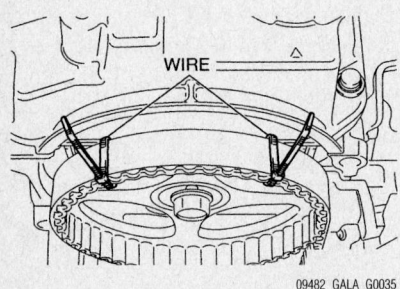

09482_GALA_G0035

Securing the camshaft sprocket—2.0L (4G94) engine and 2.4L (4G69) engine

11. Secure the camshaft and valve timing belt, with wire to prevent slippage between the camshaft sprocket and valve timing belt. Remove the camshaft sprocket. Do not turn the crankshaft after the camshaft sprocket is removed.

12. Remove the upper radiator hose connection, water cooler hose connection, if equipped with automatic transaxle and water hose connection. Disconnect the high pressure hose connection.

13. Remove the valve cover.

14. Loosen all cylinder head bolts in two or three steps and in the proper bolt removal sequence. Remove the cylinder head from the engine. Discard the cylinder head gasket.

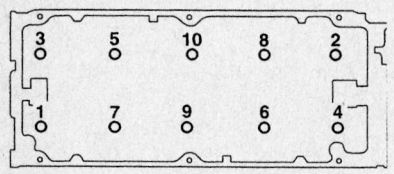

← ENGINE FRONT

09482_GALA_G0036

Cylinder head bolt removal sequence—2006 Eclipse and 2004–2006 Galant 2.4L engine

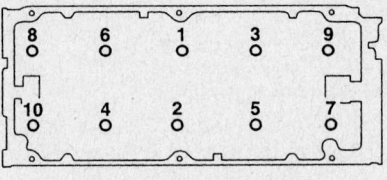

← ENGINE FRONT

09482_GALA_G0037

Cylinder head bolt torque sequence—2006 Eclipse and 2004–2006 Galant 2.4L engine

➡If the cylinder head bolts cannot be pulled out due to the washer being trapped in the valve spring, raise the bolt slightly, then remove it while holding it with a magnet.

To install:

15. Thoroughly clean the mating surfaces of the head and block.

16. Place a new head gasket on the cylinder block. Position the gasket on the block so that the identification mark "381" is at the top surface and on the exhaust side.

17. Inspect the cylinder head bolt length prior to installation. If the length exceeds 3.91 in. (99.4mm), the bolt must be replaced. Apply a small amount of engine oil to the thread section and washer of the bolt.

18. Position the cylinder head on the engine. Install and torque the retaining bolts to specification and in the proper sequence.

19. Torque the camshaft bolt to 58—72 ft. lbs. (80—98 Nm)

20. Continue the installation in the reverse order of the removal procedure.

21. Fill the engine with the proper grade and type engine oil.

22. Fill the cooling system with the proper grade and type engine coolant.

23. Initialize the ECM, if equipped with manual transaxle or the PCM if equipped with automatic transaxle by turning the ignition switch ON then OFF. Keep it off for at least ten seconds.

24. Start the engine and check for leaks.

25. Road test the vehicle.

2002–2005 3.0L ENGINE

1. Before servicing the vehicle, refer to the Precautions Section.

2. Relieve the fuel system pressure. Disconnect the negative battery cable.

3. Drain the cooling system.

4. Remove or disconnect the following:
   • Air intake hose
   • Exhaust manifold
   • Air intake plenum and intake manifold

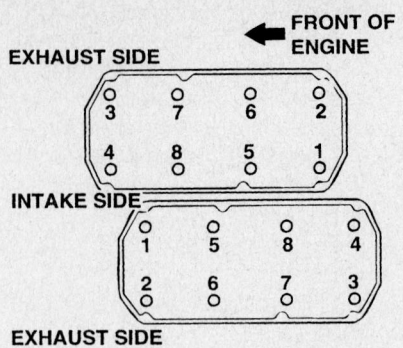

09482_GALA_G0038

Cylinder head bolt removal sequence—2002–2005 Eclipse and 2002–2003 Galant 3.0L engine

   • Timing belt
   • Camshaft sprockets and the rear timing belt cover
   • Power steering pump bracket. If removing the rear head, remove the alternator brace.
   • Water inlet pipe
   • Purge pipe assembly
   • Valve cover

5. Loosen all cylinder head bolts in two

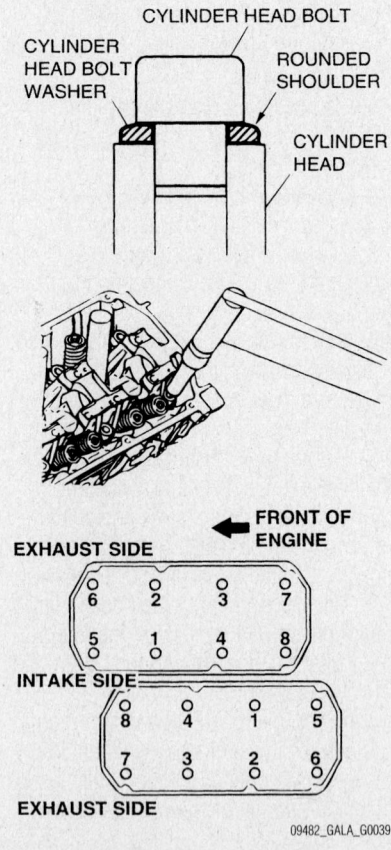

09482_GALA_G0039

Cylinder head bolt torque sequence—2002–2005 Eclipse and 2002–2003 Galant 3.0L engine

or three steps and in the proper bolt removal sequence. Remove the cylinder head from the engine. Discard the cylinder head gasket.

To install:

6. Thoroughly clean the sealing surfaces of the head and block.

7. Place a new head gasket on the cylinder block making sure the identification mark on the cylinder head gasket is in the front top (upward) location. Do not use sealer on the gasket.

8. Carefully install the cylinder head on the block. Be sure the head bolt washers are installed with the chamfered edge upward. Torque the cylinder head bolts to specification.

9. Apply sealer to the lower edges of the half-round portions and install the valve cover. Tighten valve cover bolts to 84 inch lbs. (9 Nm).

10. Install or connect the following:
- Purge pipe assembly
- Water inlet pipe
- Power steering pump bracket and alternator brace
- Rear timing belt cover and camshaft sprockets
- Timing belt
- Intake manifold, air intake plenum and exhaust manifold, using new gaskets
- Air intake hose

11. Fill the system with coolant.

12. Connect the negative battery cable.

13. Start the engine.

14. Once the vehicle has cooled, recheck the coolant level.

3.8L ENGINE

1. Before servicing the vehicle, refer to the Precautions Section.

2. Relieve the fuel system pressure. Disconnect the negative battery cable. Drain the cooling system.

3. Remove the intake manifold. Remove the exhaust manifolds. Remove the timing belt.

4. Remove the thermostat housing. Disconnect the PCV hose connection. Remove the PCV valve.

5. Disconnect the ignition coil connector and remove the ignition coil.

6. Disconnect the engine control wiring harness clamp. Remove the harness bracket.

7. Remove the rocker arm covers and gaskets. Remove the camshaft position sensor connector.

8. Remove the engine oil control valve connector. Remove the grounding bolt. Remove the engine oil dipstick assembly.

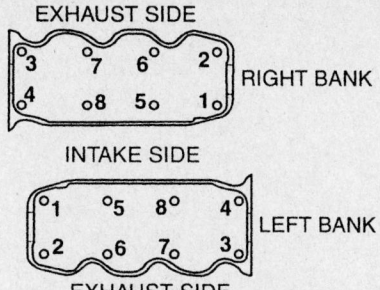

Cylinder head bolt removal sequence—2006 Eclipse and 2004–2006 Galant 3.8L engine

9. Remove the camshaft sprockets, using tool MB990767 or equivalent. Remove the timing belt rear center cover.

10. Remove the power steering oil pump assembly and position it to the side. Remove the power steering oil pump bracket bolt. Remove the engine oil control valve connector.

11. Loosen all cylinder head bolts in two or three steps and in the proper bolt removal sequence. Remove the cylinder head from the engine. Discard the cylinder head gasket.

To install:

12. Thoroughly clean the sealing surfaces of the head and block.

13. Place a new head gasket on the cylinder block making sure the identification mark on the cylinder head gasket is in the front top (upward) location. Do not use sealer on the gasket.

14. Carefully install the cylinder head on

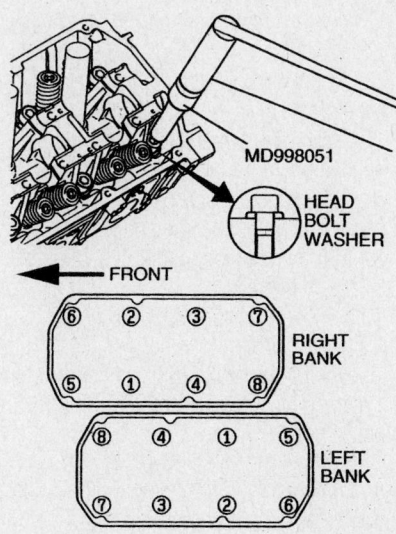

Cylinder head bolt torque sequence—2006 Eclipse and 2004–2006 Galant 3.8L engine

the block. Be sure the head bolt washers are installed with the beveled side facing upward. Torque the cylinder head bolts to specification.

Galant

2002–2003 2.4L ENGINE

1. Before servicing the vehicle, refer to the Precautions Section.

2. Relieve the fuel system pressure.

3. Disconnect the negative battery cable.

4. Remove the air cleaner with all air intake hoses.

5. Drain the cooling system.

6. Remove or disconnect the following:
- Accelerator cable
- Cable mounting brackets and position the cable aside
- Breather hose
- Vacuum lines at the throttle body, label for identification
- High pressure fuel line and plug the line to avoid contamination
- Fuel return hose and plug the hose to avoid contamination

7. Disconnect the following connectors:
- Air conditioning compressor
- Power steering pressure switch
- Heated Oxygen (HO2S) sensor
- Engine Coolant Temperature (ECT) gauge sender
- ECT sensor
- Manifold Absolute Pressure (MAP) sensor
- Intake Air Temperature (IAT) sensor
- Throttle Position (TP) sensor
- Idle Air Control (IAC) motor
- Injector harness
- Ignition coil
- Camshaft Position (CMP) sensor
- Exhaust Gas Recirculation (EGR) solenoid valve

8. Remove or disconnect the following:
- Spark plug wire cover and wires
- Coolant hoses and unbolt the thermostat case from the engine, at the thermostat case assembly
- Upper timing belt cover

9. Align all timing marks.

10. Secure the timing belt to the camshaft sprocket with cord or a wire tie.

11. Remove or disconnect the following:
- Camshaft sprocket
- Valve cover and the half-round seal
- Intake manifold stay bracket from the intake manifold
- Exhaust pipe self-locking nuts and separate the exhaust pipe from the exhaust manifold. Discard the gasket.

12. Loosen the cylinder head mounting bolts in 3 steps, starting from the outside and working inward. Lift off the cylinder head assembly and remove the head gasket.

To install:

13. Thoroughly clean the mating surfaces of the head and block.

14. Place a new head gasket on the cylinder block with the identification marks at the front top (upward) position. Do not use sealer on the gasket.

15. Inspect the cylinder head bolt length prior to installation. If the length exceeds 3.91 in. (99.4mm), the bolt must be replaced. Install the washer onto the bolt so the chamfer on the washer faces towards the head of the bolt.

16. Carefully install the cylinder head on the block and tighten the cylinder head bolts as follows:

   a. Following the proper tightening sequence, tighten the cylinder head bolts to 58 ft. lbs. (78 Nm).

   b. Loosen all bolts completely.

   c. Torque bolts to 15 ft. lbs. (20 Nm).

   d. Tighten bolts an additional ¼ turn.

   e. Tighten bolts an additional ¼ turn.

17. Install or connect the following:

- New exhaust pipe gasket and connect the exhaust pipe to the manifold. Tighten the bolts to 33 ft. lbs. (44 Nm).
- Thermostat case and tighten the mounting bolts to 18 ft. lbs. (24 Nm)
- Coolant hoses to the thermostat case

18. Apply sealer to the perimeter of the half-round seal and to the lower edges of the half-round portions of the belt-side of the new gasket. Install the valve cover.

19. Install or connect the following:

- Camshaft sprocket with the timing belt attached. Remove the cord or wire tie.
- Upper timing belt cover
- Intake manifold stay and tighten the mounting bolts to 22 ft. lbs. (30 Nm)

20. Connect the following connectors:

- Air conditioning compressor
- Power steering pressure switch
- HO$_2$S sensor
- ECT gauge sender
- ECT sensor
- MAP sensor
- IAT sensor
- TP sensor
- IAC motor
- Injector harness
- Ignition coil

- CMP sensor
- EGR solenoid valve

21. Remove or disconnect the following:

- Spark plug wires and cover
- Fuel lines using new O-rings
- Air cleaner and intake hose
- Breather hose

22. Fill the cooling system.

23. Connect the negative battery cable

2004–2006 2.4L (4G69) ENGINE

1. Before servicing the vehicle, refer to the Precautions Section.

2. Relieve the fuel system pressure. Disconnect the negative battery cable. Drain the cooling system.

3. Remove the ECM, if the vehicle is equipped with manual transaxle. Remove the PCM if the vehicle is equipped with automatic transaxle.

4. Remove the air cleaner assembly. Remove the battery and tray.

5. Disconnect the control wiring harness, radiator hose lower clamp and the water hose clamp.

6. Disconnect the EVAP hose connection. Disconnect the brake booster vacuum hose, pressure hose clamp and knock sensor connector.

7. Remove the engine oil dipstick and dipstick guide. Remove the intake manifold stay.

8. Remove the exhaust manifold. Remove the timing belt upper cover. Remove the engine front mounting bracket.

9. Turn the crankshaft clockwise, align the timing marks on the camshaft sprocket to position number one piston to TDC of its compression stroke. Never rotate the crankshaft in the counterclockwise direction.

10. Remove the timing belt under cover rubber plug, using tool MD998738. Screw the tool in until it contacts the timing belt tensioner arm.

11. Secure the camshaft and valve timing belt, with wire to prevent slippage between the camshaft sprocket and valve timing belt. Remove the camshaft sprocket. Do not turn the crankshaft after the camshaft sprocket is removed.

12. Remove the upper radiator hose connection, water cooler hose connection, if equipped with automatic transaxle and water hose connection. Disconnect the high pressure hose connection.

13. Remove the valve cover.

14. Loosen all cylinder head bolts in two or three steps and in the proper bolt removal sequence. Remove the cylinder head from the engine. Discard the cylinder head gasket.

➡**If the cylinder head bolts cannot be pulled out due to the washer being trapped in the valve spring, raise the bolt slightly, then remove it while holding it with a magnet.**

To install:

15. Thoroughly clean the mating surfaces of the head and block.

16. Place a new head gasket on the cylinder block. Position the gasket on the block so that the identification mark "381" is at the top surface and on the exhaust side.

17. Inspect the cylinder head bolt length prior to installation. If the length exceeds 3.91 in. (99.4mm), the bolt must be replaced. Apply a small amount of engine oil to the thread section and washer of the bolt.

18. Position the cylinder head on the engine. Install and torque the retaining bolts to specification and in the proper sequence.

19. Torque the camshaft bolt to 58—72 ft. lbs. (80—98 Nm)

20. Continue the installation in the reverse order of the removal procedure.

21. Fill the engine with the proper grade and type engine oil.

22. Fill the cooling system with the proper grade and type engine coolant.

23. Initialize the ECM, if equipped with manual transaxle or the PCM if equipped with automatic transaxle by turning the ignition switch ON then OFF. Keep it off for at least ten seconds.

24. Start the engine and check for leaks.

25. Road test the vehicle.

2002–2003 3.0L ENGINE

1. Before servicing the vehicle, refer to the Precautions Section.

2. Relieve the fuel system pressure. Disconnect the negative battery cable.

3. Drain the cooling system.

4. Remove or disconnect the following:

- Air intake hose
- Exhaust manifold
- Air intake plenum and intake manifold
- Timing belt
- Camshaft sprockets and the rear timing belt cover
- Power steering pump bracket. If removing the rear head, remove the alternator brace.
- Water inlet pipe
- Purge pipe assembly
- Valve cover

5. Loosen all cylinder head bolts in two or three steps and in the proper bolt removal sequence. Remove the cylinder

head from the engine. Discard the cylinder head gasket.

To install:

6. Thoroughly clean the sealing surfaces of the head and block.

7. Place a new head gasket on the cylinder block making sure the identification mark on the cylinder head gasket is in the front top (upward) location. Do not use sealer on the gasket.

8. Carefully install the cylinder head on the block. Be sure the head bolt washers are installed with the chamfered edge upward. Torque the cylinder head bolts to specification.

9. Apply sealer to the lower edges of the half-round portions and install the valve cover. Tighten valve cover bolts to 84 inch lbs. (9 Nm).

10. Install or connect the following:
- Purge pipe assembly
- Water inlet pipe
- Power steering pump bracket and alternator brace
- Rear timing belt cover and camshaft sprockets. Torque the retaining bolt to 65 ft. lbs. (90 Nm).
- Timing belt
- Intake manifold, air intake plenum and exhaust manifold, using new gaskets
- Air intake hose

11. Fill the system with coolant.

12. Connect the negative battery cable.

13. Start the engine.

14. Check and adjust the idle speed and ignition timing.

15. Once the vehicle has cooled, recheck the coolant level.

2004–2006 3.8L ENGINE

1. Before servicing the vehicle, refer to the Precautions Section.

2. Relieve the fuel system pressure. Disconnect the negative battery cable. Drain the cooling system.

3. Remove the intake manifold. Remove the exhaust manifolds. Remove the timing belt.

4. Remove the thermostat housing. Disconnect the PCV hose connection. Remove the PCV valve.

5. Disconnect the ignition coil connector and remove the ignition coil.

6. Disconnect the engine control wiring harness clamp. Remove the harness bracket.

7. Remove the rocker arm covers and gaskets. Remove the camshaft position sensor connector.

8. Remove the engine oil control valve connector. Remove the grounding bolt. Remove the engine oil dipstick assembly.

9. Remove the camshaft sprockets, using tool MB990767 or equivalent. Remove the timing belt rear center cover.

10. Remove the power steering oil pump assembly and position it to the side. Remove the power steering oil pump bracket bolt. Remove the engine oil control valve connector.

11. Loosen all cylinder head bolts in two or three steps and in the proper bolt removal sequence. Remove the cylinder head from the engine. Discard the cylinder head gasket.

To install:

12. Thoroughly clean the sealing surfaces of the head and block.

13. Place a new head gasket on the cylinder block making sure the identification mark on the cylinder head gasket is in the front top (upward) location. Do not use sealer on the gasket.

14. Carefully install the cylinder head on the block. Be sure the head bolt washers are installed with the beveled side facing upward. Torque the cylinder head bolts to specification.

Lancer

2002–2006 2.0L ENGINE

1. Before servicing the vehicle, refer to the Precautions Section.

2. Relieve the fuel system pressure. Disconnect the negative battery cable. Drain the cooling system.

3. Remove the engine under cover. Remove the air cleaner assembly.

4. Remove the exhaust manifold. Remove the water hose and pipe. Disconnect the accelerator cable connection.

5. Disconnect the air conditioning compressor connector, power steering oil pressure switch connector and the crankshaft position sensor connector.

6. Disconnect the manifold absolute pressure sensor connector, EVAP solenoid connector and EGR solenoid valve connector.

7. Disconnect the ignition coil connector, injector connector and throttle position sensor connector.

8. Disconnect the idle air control motor connector, engine coolant temperature sensor connector and the camshaft position sensor connector.

9. Disconnect the knock sensor connector, engine coolant temperature gauge unit connector and the heated oxygen sensor connector.

10. Disconnect the brake booster vacuum hose. Disconnect the high pressure hose connection. Remove the timing belt upper cover.

11. Turn the crankshaft clockwise, align the timing marks on the camshaft sprocket to position number one piston to TDC of its compression stroke. Never rotate the crankshaft in the counterclockwise direction.

12. Remove the timing belt under cover rubber plug, using tool MD998738. Screw the tool in until it contacts the timing belt tensioner arm.

13. Secure the camshaft and valve timing belt, with wire to prevent slippage between the camshaft sprocket and valve timing belt. Remove the camshaft sprocket. Do not turn the crankshaft after the camshaft sprocket is removed.

14. Remove the intake manifold stay connecting bolts.

➡**Be careful not to damage or deform the plug guides when removing the cylinder head bolts. Plug guides cannot be replaced separately.**

15. Loosen all cylinder head bolts in two or three steps and in the proper bolt removal sequence. Remove the cylinder head from the engine. Discard the cylinder head gasket.

To install:

16. Thoroughly clean the mating surfaces of the head and block.

17. Place a new head gasket on the cylinder block.

18. Inspect the cylinder head bolt length prior to installation. If the length exceeds 3.91 in. (99.4mm), the bolt must be replaced. Apply a small amount of engine oil to the thread section and washer of the bolt.

19. Position the cylinder head on the

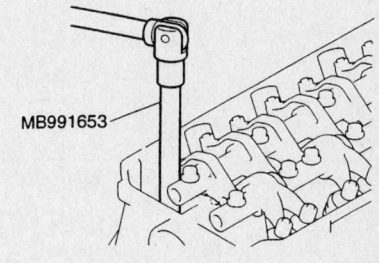

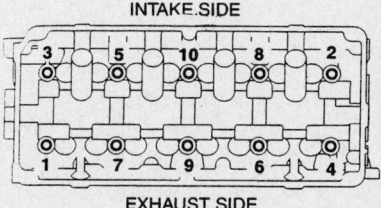

9357QG08

Cylinder head bolt removal sequence— Lancer 2.0L and 2.4L engines

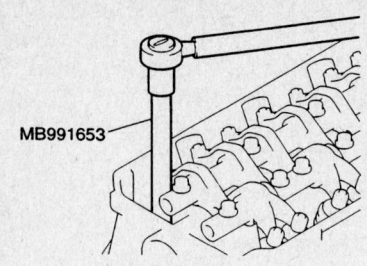

MB991653

INTAKE SIDE

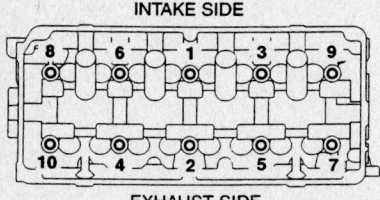

EXHAUST SIDE

93570G10

Cylinder head bolt torque sequence—Lancer 2.0L and 2.4L engines

engine. Install and torque the retaining bolts to specification and in the proper sequence.

20. Continue the installation in the reverse order of the removal procedure.

21. Fill the engine with the proper grade and type engine oil.

22. Fill the cooling system with the proper grade and type engine coolant.

23. Start the engine and check for leaks.

24. Road test the vehicle.

2004–2006 2.4L ENGINE

1. Before servicing the vehicle, refer to the Precautions Section.

2. Relieve the fuel system pressure. Disconnect the negative battery cable. Drain the cooling system.

3. Remove the air cleaner assembly. Remove the battery and tray.

4. Disconnect the control wiring harness, radiator hose lower clamp and the water hose clamp.

5. Disconnect the EVAP hose connection. Disconnect the brake booster vacuum hose, pressure hose clamp and knock sensor connector.

6. Remove the engine oil dipstick and dipstick guide. Remove the intake manifold stay.

7. Remove the exhaust manifold. Remove the timing belt upper cover. Remove the engine front mounting bracket.

8. Turn the crankshaft clockwise, align the timing marks on the camshaft sprocket to position number one piston to TDC of its compression stroke. Never rotate the crankshaft in the counterclockwise direction.

9. Remove the timing belt under cover rubber plug, using tool MD998738. Screw

the tool in until it contacts the timing belt tensioner arm.

10. Secure the camshaft and valve timing belt, with wire to prevent slippage between the camshaft sprocket and valve timing belt. Remove the camshaft sprocket. Do not turn the crankshaft after the camshaft sprocket is removed.

11. Remove the upper radiator hose connection, water cooler hose connection, if equipped with automatic transaxle and water hose connection. Disconnect the high pressure hose connection.

12. Remove the valve cover.

13. Loosen all cylinder head bolts in two or three steps and in the proper bolt removal sequence. Remove the cylinder head from the engine. Discard the cylinder head gasket.

➡ **If the cylinder head bolts cannot be pulled out due to the washer being trapped in the valve spring, raise the bolt slightly, then remove it while holding it with a magnet.**

To install:

14. Thoroughly clean the mating surfaces of the head and block.

15. Place a new head gasket on the cylinder block. Position the gasket on the block so that the identification mark "381" is at the top surface and on the exhaust side.

16. Inspect the cylinder head bolt length prior to installation. If the length exceeds 3.91 in. (99.4mm), the bolt must be replaced. Apply a small amount of engine oil to the thread section and washer of the bolt.

17. Position the cylinder head on the engine. Install and torque the retaining bolts to specification and in the proper sequence.

18. Torque the camshaft bolt to 58—72 ft. lbs. (80—98 Nm)

19. Continue the installation in the reverse order of the removal procedure.

20. Fill the engine with the proper grade and type engine oil.

21. Fill the cooling system with the proper grade and type engine coolant.

22. Start the engine and check for leaks.

23. Road test the vehicle.

Evolution

1. Before servicing the vehicle, refer to the Precautions Section.

2. Relieve the fuel system pressure. Disconnect the negative battery cable. Drain the cooling system.

3. Remove the engine under cover. Remove the side cover.

4. Drain the engine oil. Remove the strut tower bar assembly. Remove the air cleaner assembly.

5. Remove the turbocharger bypass valve assembly, air bypass hose, air hose, air pipe and secondary air hose.

6. Remove the battery. Disconnect the accelerator cable. Remove the rocker cover center cover.

7. Remove the radiator. Remove the front axle crossmember bar. Remove the front exhaust pipe. Remove the starter.

8. Remove the timing belt.

9. Disconnect the control wiring harness and ground cable connector.

10. Disconnect the EGR vacuum regulator solenoid valve connector, EGR vacuum regulator solenoid valve and bracket assembly and the brake booster vacuum hose.

11. Remove the engine oil dipstick and dipstick guide. Remove the purge hose connection. Remove the alternator brace bolt. Remove the battery wiring harness connection.

12. Remove the intake manifold stay. Remove the eye bolt, gasket, check valve assembly, oil feeder control valve pipe and filter.

13. Remove the oil pipe joint, gasket, oil return tube, oil return tube gasket and turbocharger bracket.

14. Remove the water outlet fitting and thermostat case assembly.

15. Remove the rocker arm cover, water return hose connector, heater hose connection and fuel line return hose connection.

16. Disconnect the fuel high pressure hose connection and remove the O-ring.

17. Loosen all cylinder head bolts in two or three steps and in the proper bolt removal sequence. Remove the cylinder head from the engine. Discard the cylinder head gasket.

➡ **If the cylinder head bolts cannot be pulled out due to the washer being trapped in the valve spring, raise the bolt slightly, then remove it while holding it with a magnet.**

⬅ ENGINE FRONT

```
   3     5    10    8     2
   O     O    O     O     O

   1     7     9    6     4
   O     O     O    O     O
```

09482_GALA_G0041

Cylinder head bolt removal sequence—Evolution

← ENGINE FRONT

```
8    6    1    3    9
O    O    O    O    O

10   4    2    5    7
O    O    O    O    O
```

09482_GALA_G0042

Cylinder head bolt torque sequence—Evolution

To install:

18. Thoroughly clean the mating surfaces of the head and block.

19. Place a new head gasket on the cylinder block. Position the gasket on the block so that the identification mark "63" is at the top surface and on the exhaust side.

20. Inspect the cylinder head bolt length prior to installation. If the length exceeds 3.91 in. (99.4mm), the bolt must be replaced. Apply a small amount of engine oil to the thread section and washer of the bolt.

21. Position the cylinder head on the engine. Install and torque the retaining bolts to specification and in the proper sequence.

Mirage

1.5L ENGINE

1. Before servicing the vehicle, refer to the Precautions Section.

2. Relieve the fuel system pressure. Disconnect the negative battery cable.

3. Drain the cooling system.

4. Remove or disconnect the following:
- Air intake hose and the air cleaner assembly
- Ground cable connection and the accelerator cable
- Positive Crankcase Ventilation (PCV) and the breather hose connection
- Vacuum hoses from the intake and throttle body, label for reference
- Vacuum line for the brake booster
- Upper radiator hose, throttle body hoses, bypass hose and heater hose connections
- Fuel feed and return lines
- Spark plug wires

5. Disconnect the electrical harness plugs from the following:
- Crankshaft Position (CKP) and Camshaft Position (CMP) sensors
- Heated Oxygen (HO2S) sensor
- Engine Coolant Temperature (ECT) sensor and gauge sender

- Idle Speed Control (ISC) motor
- Throttle Position (TP) sensor
- Intake Air Temperature (IAT) sensor
- Exhaust Gas Recirculation (EGR) temperature sensor

6. Remove or disconnect the following:
- Electrical harness plugs from the ignition distributor, fuel injectors, power transistor and ground cable
- Engine control wiring harness
- Clamp that holds the power steering pressure hose to the engine mounting bracket

7. Place a jack and wood block under the oil pan and carefully lift just enough to take the weight off the engine mounting bracket and remove the bracket.
- Valve cover
- Timing belt upper cover

8. Rotate the crankshaft clockwise and align the timing marks.

9. Attach the timing belt to the camshaft sprocket with cord or a wire tie.

10. Secure the camshaft from turning and remove the camshaft sprocket with the timing belt attached.

11. Remove the timing belt rear upper cover.

12. Remove the exhaust pipe from the exhaust manifold.

13. Loosen the cylinder head mounting bolts in sequence using 3 steps. Remove the cylinder head.

To install:

14. Thoroughly clean the mating surfaces of the head and block.

15. Place a new head gasket on the cylinder block with the identification marks facing upward. Do not use sealer on the gasket.

16. Carefully install the cylinder head on the block. Tighten the cylinder head bolts as follows:

   a. 36 ft. lbs. (49 Nm) in the correct sequence.

   b. Loosen the bolts completely in the reverse order.

   c. Tighten the bolts in sequence to 14 ft. lbs. (20 Nm).

   d. Tighten each bolt in sequence 90 degrees.

   e. Tighten each bolt in sequence an additional 90 degrees.

17. Install or connect the following:
- New exhaust pipe gasket and connect the exhaust pipe to the manifold
- Upper rear timing cover

18. Align the timing marks and install the cam sprocket. Torque the retaining bolt

Front of engine

Intake side ⇨

```
 4   6    9    7    1

 2   8   10    5    3
```

Exhaust side

7923PG13

Cylinder head bolt removal sequence—1.5L engine

Intake side ⇦ Front of engine

```
 8    6    1    3    9

10    4    2    5    7
```

Exhaust side

7923PG14

Cylinder head bolt torque sequence—1.5L engine

to 51 ft. lbs. (70 Nm). Check the belt tension and adjust, if necessary. Install the outer timing cover.

19. Install or connect the following:
   • Valve cover and torque the retaining bolts to 16 inch lbs. (1.8 Nm)
   • Engine mount bracket and remove the support jack
   • Clamp that holds the power steering pressure hose to the engine mounting bracket
20. Connect the following:
   • CKP and CMP sensors
   • HO2S sensor
   • ECT sensor and gauge sender
   • ISC motor
   • TP sensor
   • IAT
   • EGR temperature sensor
21. Install or connect the following:
   • Ignition distributor, fuel injectors, power transistor and ground cable
   • Engine control wiring harness
   • Fuel lines with new O-rings
   • Air cleaner assembly
   • Breather hose
22. Fill the system with coolant.
23. Connect the negative battery cable.

1.8L ENGINE

1. Before servicing the vehicle, refer to the Precautions Section.
2. Relieve fuel system pressure. Disconnect the negative battery cable.
3. Remove the air cleaner assembly.
4. Drain the cooling system.
5. Disconnect the brake booster vacuum hose and PVC valve connection.
6. Note the locations and disconnect the vacuum hoses from the intake and throttle body.
7. Remove or disconnect the following:
   • Upper radiator hose, overflow tube and the water hose from the thermostat to the throttle body
   • Fuel feed and return lines
   • Accelerator cable connection from the throttle body
   • Oil pressure switch
8. Disconnect the following:
   • Heated Oxygen (HO2S) sensor
   • Engine Coolant Temperature (ECT) sensor and gauge sender
   • IAC motor
   • Exhaust Gas Recirculation (EGR) temperature sensor
   • Throttle Position (TP) sensor
   • Knock (KS) sensor
   • Fuel injectors
   • Spark plug wires
   • Control harness assembly and position aside

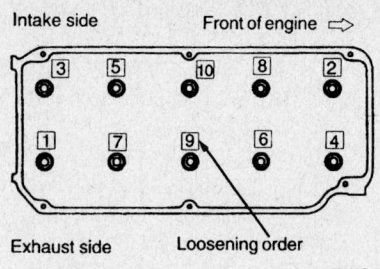

Intake side   Front of engine ⇨

Exhaust side   Loosening order

7923PG15

Cylinder head bolt removal sequence—1.8L engine

   • Thermostat housing, thermostat and the thermostat case with O-ring from the engine
   • Rocker cover
   • Timing belt upper cover
9. Rotate the crankshaft clockwise and align the timing marks.
10. Attach the timing belt to the camshaft sprocket with cord or a wire tie.
11. Secure the camshaft from turning and remove the camshaft sprocket with the timing belt attached.
12. Remove the timing belt rear upper cover.
13. Loosen the cylinder head bolts in 2 or 3 steps in the proper sequence.
14. Remove the cylinder head from the engine.

❊❊ CAUTION

When removing the cylinder head, take care not to bend or damage the plug guide. The plug guide can not be replaced.

To install:
15. Thoroughly clean the mating surfaces of the head and block.
16. Place a new head gasket on the cylinder block with the identification marks facing upward. Do not use sealer on the gasket.
17. Carefully install the cylinder head on the block.

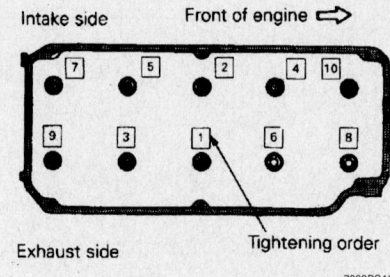

Intake side   Front of engine ⇨

Exhaust side   Tightening order

7923PG16

Cylinder head bolt torque sequence—1.8L engine

18. Measure the cylinder head bolts prior to installation. Replace any that exceed 3.795 in. (96.4mm).
19. Apply a small amount of engine oil to the thread section of the bolt and install so the chamfer of the washer faces upward.
20. Tighten the cylinder head bolts as follows:
   a. In the proper tightening sequence, torque bolts to 54 ft. lbs. (75 Nm).
   b. In the reverse order of the tightening sequence, fully loosen all bolts.
   c. In the proper tightening sequence, torque bolts to 14 ft. lbs. (20 Nm).
   d. In the proper tightening sequence, tighten bolts ¼ turn (90 degrees).
   e. In the proper tightening sequence, tighten bolts an additional ¼ turn (90 degrees).
21. Install the camshaft sprocket and tighten the bolt to 65 ft. lbs. (90 Nm), while holding the sprocket in place using the appropriate wrench. Confirm proper timing mark alignment.
22. Install the upper timing belt cover and rocker cover. Torque the rocker cover bolts to 29 inch lbs. (3.3 Nm).
23. Loosen the water pipe mounting bolt for ease of thermostat housing installation.
24. Apply a thin bead of RTV sealant to the water tube connection on the thermostat case.
25. Apply a small amount of water to the O-ring of the water inlet pipe and press the thermostat case assembly onto the water inlet pipe. Install the thermostat case assembly mounting bolt tightening to 16 ft. lbs. (22 Nm).
26. Tighten the water pipe mounting bolt.
27. Install the thermostat into the housing so the jiggle valve is located at the top. Tighten the housing bolts to 10 ft. lbs. (14 Nm).
28. Connect the following:
   • HO2S sensor
   • ECT sensor and gauge sender
   • IAC motor
   • EGR temperature sensor
   • TP sensor
   • KS sensor
   • Fuel injectors
29. Install or connect the following:
   • Upper radiator hose to the thermostat housing
   • Accelerator cable connection to the throttle body
   • Oil pressure switch
   • Spark plug wires
   • Control harness assembly
30. Replace the O-ring for the high pressure hose and install a new clamp on the return hose and reconnect the fuel lines.

- Air intake hose
- Breather hose and air cleaner case cover
- Brake booster and the PCV vacuum hoses

31. Fill the system with coolant.
32. Connect the negative battery cable

Rocker Arms/Shafts

REMOVAL & INSTALLATION

Diamante

1. Before servicing the vehicle, refer to the Precautions Section.
2. Disconnect the negative battery cable.
3. Remove the rocker arm cover.
4. Install the lash adjuster clips on the rocker arms, then loosen the bearing cap bolts. Do not remove the bolts from the bearing caps.
5. Remove the rocker arms, shafts and bearing caps as an assembly.

To install:

6. Install the bearing caps/rocker arm assemblies. Tighten the bolts to 23 ft. lbs. (31 Nm).
7. Remove the lash adjuster clips.
8. Install the rocker arm cover using a new gasket.
9. Connect the negative battery cable.

Eclipse

2002–2005 2.4L ENGINE

1. Before servicing the vehicle, refer to the Precautions Section.
2. Remove or disconnect the following:
- Negative battery cable
- Accelerator cable, remove the cable clamp mounting screws and position the accelerator cable out of the way.
- Air intake hose
- Breather hose and the Positive Crankcase Ventilation (PCV) hose
- Spark plug cables from the spark plugs
- Rocker cover and gasket
3. Install the lash adjuster retainer tools to the rocker arm.
4. Remove the rocker shaft hold-down bolts gradually and evenly and remove the rocker shaft/arm assemblies.
5. Disassemble the rockers and the rocker shaft springs from the rocker shafts. If they are to be reused, note the location and positioning of all rocker arm shaft components. It is recommended that all lash adjusters and rockers be replaced as a complete set.

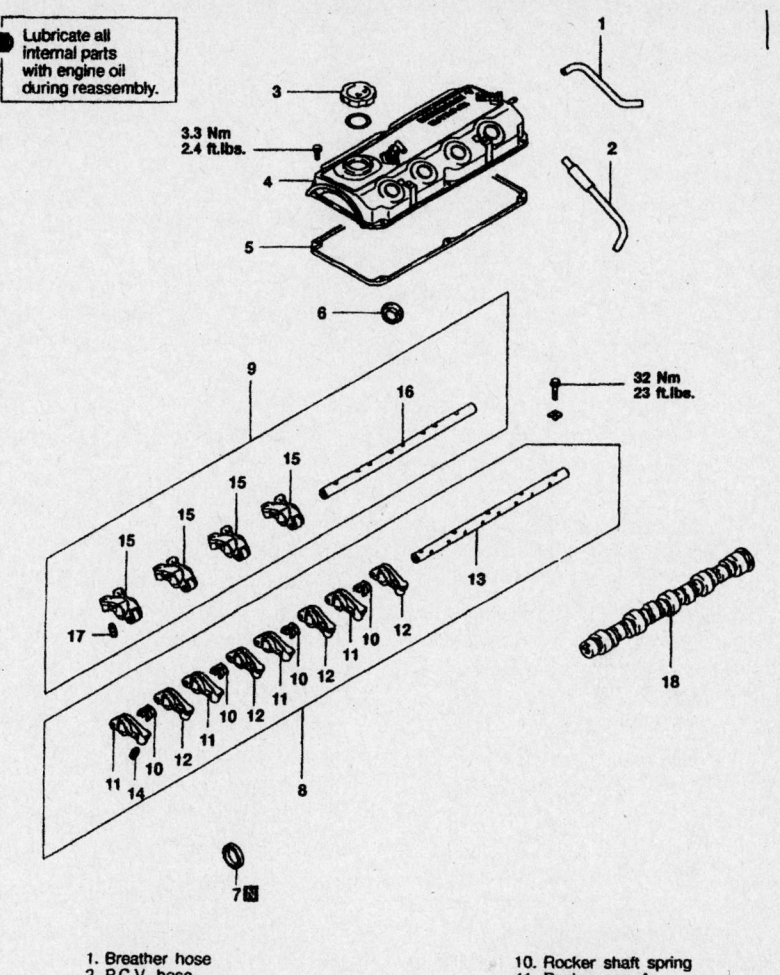

1. Breather hose
2. P.C.V. hose
3. Oil filler cap
4. Rocker cover
5. Rocker cover gasket
6. Oil seal
7. Oil seal
8. Rocker arms and rocker arm shaft
9. Rocker arms and rocker arm shaft
10. Rocker shaft spring
11. Rocker arm A
12. Rocker arm B
13. Rocker arm shaft (Intake side)
14. Lash adjuster
15. Rocker arm C
16. Rocker arm shaft (Exhaust side)
17. Lash adjuster
18. Camshaft

67170-GALA-G19

Rocker arm shafts and related components—2002–2005 Eclipse 2.4L engine

To install:

6. Immerse the lash adjusters in clean diesel fuel, and using a small wire, move the plunger up and down 4 or 5 times. While pushing down lightly on the check ball in order to bleed the air from the adjuster.
7. Install the lash adjusters to the rocker arms and attach the special holding tool.
8. Lubricate the rocker shaft with clean engine oil and install the rocker arms.
9. Temporarily tighten the rocker shaft assembly with the mounting bolts so that all rocker arms on the inlet valve side do not push on the valves.
10. Fit the rocker shaft springs from above and position them so that they are at right angles to the plug side. Install the rocker springs before installing the exhaust side rocker shaft and rocker arm assembly.

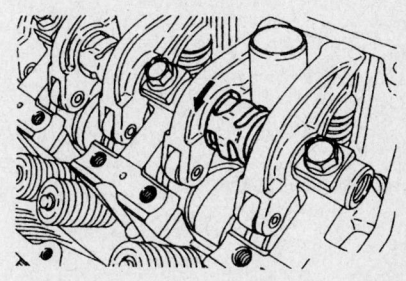

67170-GALA-G20

Installing the rocker shaft springs— 2002–2005 Eclipse 2.4L engine

11. Install the exhaust side rocker shaft assembly in the engine. Tighten the rocker shaft mounting bolts gradually and evenly to 23 ft. lbs. (32 Nm).
12. Remove the lash adjuster retaining tools.

13. Install or connect the following:
- Rocker cover and tighten the mounting bolts to 29 inch lbs. (3.3 Nm)
- Spark plug wires to the spark plugs
- PCV and breather hoses
- Air intake hose
- Accelerator cable brackets and reconnect the accelerator cable
- Negative battery cable

2006 2.4L (4G69) ENGINE

1. Before servicing the vehicle, refer to the Precautions Section.

2. Disconnect the negative battery cable.

3. If equipped with manual transaxle, remove the EMC. If equipped with automatic transaxle remove the PCM.

4. Remove the air cleaner assembly. Remove the battery. Remove the ignition coils.

5. Remove the timing belt upper cover.

6. Disconnect the control wiring harness connection. Remove the rocker cover PCV connection.

7. Remove the rocker cover breather hose connection. Remove the engine oil control valve and O-ring. Remove the oil pressure switch.

8. Remove the rocker cover retaining bolts. Remove the rocker cover. Discard the gasket.

9. Remove the spark plug guide oil seals. Remove the accumulator assembly.

10. Remove the exhaust rocker arm shaft caps. Remove the exhaust rocker arm shaft assembly.

11. Remove the intake rocker arm shaft caps. Remove the intake rocker arm shaft assembly.

➡ **Never disassemble the exhaust or intake rocker arm assembly.**

To install:

12. Position the intake rocker arm assembly so that its 0.22 inch hole faces

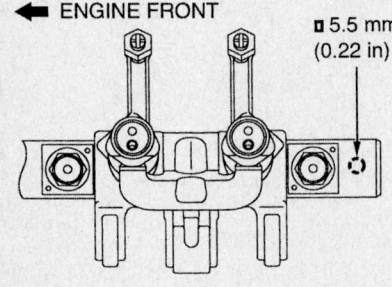

09482_GALA_G0043

Intake rocker arm shaft installation locating point—2006 Eclipse and 2004–2006 Galant 2.4L engine

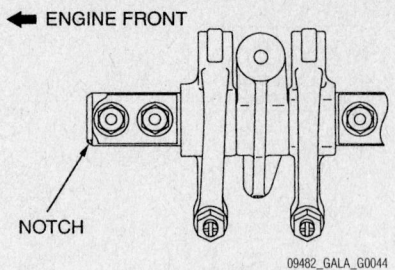

⬅ ENGINE FRONT

NOTCH

09482_GALA_G0044

Exhaust rocker arm shaft installation locating point—2006 Eclipse and 2004–2006 Galant 2.4L engine

toward the cylinder head. Install the rocker shaft mounting bolts. Torque to 21—25 ft. lbs. 29—34 Nm).

13. Position the exhaust rocker arm assembly so that its notch is positioned as shown in the illustration. Install the rocker shaft mounting bolts. Torque to 106—124 inch lbs. 12—14 Nm).

14. Continue the installation in the reverse order of the removal procedure.

15. Adjust the valves.

16. To initialize the ECM, manual transaxle equipped vehicles, or the PCM, automatic transaxle equipped vehicles, turn the ignition switch ON then OFF and keep it in the OFF position for at least ten seconds.

2002–2003 3.0L ENGINE

The hydraulic lash adjusters are built into the rocker arms.

1. Before servicing the vehicle, refer to the Precautions Section.

2. Disconnect the negative battery cable.

3. Remove the valve cover. Install lash adjuster retainer tools to prevent the auto-lash adjuster from falling out of the rocker arm.

4. Loosen rocker arm and shaft assembly evenly in several steps. Remove the rocker arm and shaft assembly as a complete unit.

5. Remove the rear camshaft bearing cap and slide the rocker arms, springs and washers from the shaft. If they are to be reused, note the location and positioning of all rocker shaft components. It is recommended that all lash adjusters and rockers be replaced as a complete set.

To install:

6. Immerse the lash adjusters in clean diesel fuel. Using a small wire, move the plunger of the lash adjuster up and down 4 or 5 times while pushing down lightly on the check ball in order to bleed out the air. Install the lash adjusters in the rocker arms.

7. Using a light coat of engine oil,

assemble the rocker arms to the shaft. Install the rear camshaft bearing cap.

8. Lubricate the camshaft and rocker shaft with clean engine oil and position on the cylinder head.

9. Apply a drop of sealant to the rear edges of the end caps.

10. Install or connect the following:
- Assembly making sure the notches in the rocker shafts are facing up
- Cap bolts and tighten evenly and gradually to 14 ft. lbs. (20 Nm). Remove the lash adjuster retainers.
- Valve cover
- Negative battery cable

2004–2005 3.0L ENGINE

1. Before servicing the vehicle, refer to the Precautions Section.

2. Disconnect the negative battery cable.

3. On the left side, remove the thermostat housing assembly. Remove the PCV valve hose connection, spark plug cable and battery cable connection.

4. On the right side, remove the intake manifold plenum, breather hose connection and spark plug cable.

5. Remove the rocker arm cover retaining bolts. Remove the rocker arm cover. Discard the gasket.

6. Install tool MD998443 on the rocker arms, so that the lash adjusters will not fall out.

7. Loosen the rocker arm and shaft assembly mounting bolts. Remove the rocker arm and shaft assembly with the bolts still attached.

➡ **Never disassemble the rocker arm and shaft assembly.**

To install:

8. Rotate the camshaft until the dowel pin on its front end is located as shown in the illustration.

➡ **Placing the camshaft in the illustrated position minimizes the amount of cam lift, making it easier to install the rocker arm shaft assembly.**

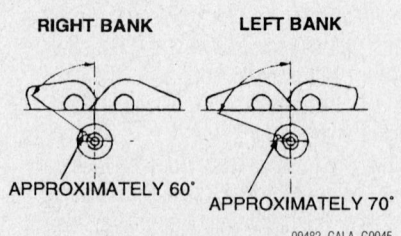

09482_GALA_G0045

Rocker arm and shaft positioning—2004–2005 Eclipse 3.0L engine

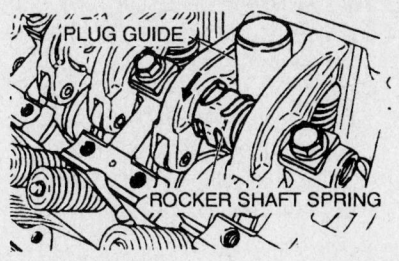

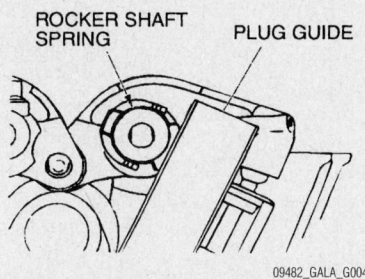

Rocker shaft spring alignment—
2004–2005 Eclipse 3.0L engine

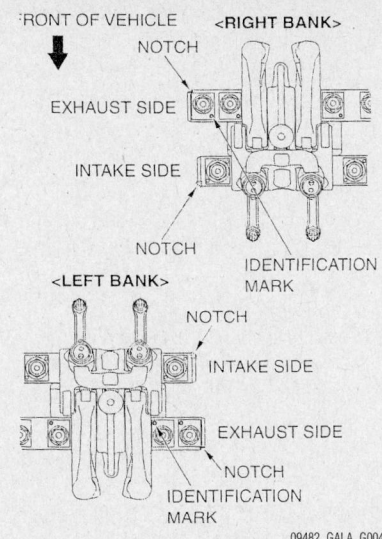

Rocker arm and shaft positioning—2006
Eclipse and 2004–2006 Galant 3.8L engine

9. Temporarily tighten the rocker shaft with the bolts so that all rocker arms on the inlet valve side do not push the valves.

10. Push the rocker shaft spring so that it takes a right angle against the plug guide.

➡Set the rocker shaft spring before installing the rocker arm and shaft assembly on the exhaust side.

11. Tighten the rocker arm and shaft assembly mounting bolts to 21—25 ft. lbs. 28—34 Nm). Remove tool MD998443.

12. Continue the installation in the reverse order of the removal procedure. Be sure to use new valve cover gaskets.

3.8L ENGINE

1. Before servicing the vehicle, refer to the Precautions Section.

2. Disconnect the negative battery cable. Drain the cooling system.

3. On the left side, remove the thermostat housing assembly. Remove the PCV valve hose connection, ignition coil connector and ignition coil. Disconnect the engine control wiring harness clamp and harness bracket.

4. On the right side, remove the intake manifold plenum, breather hose connection, heater hose and water hose connection. Remove the ignition coil connector and coil.

5. Remove the rocker arm cover retaining bolts. Remove the rocker arm cover. Discard the gasket.

6. Install tool MD998443 on the rocker arms, so that the lash adjusters will not fall out.

7. Loosen the rocker arm and shaft

assembly mounting bolts. Remove the rocker arm and shaft assembly with the bolts still attached.

➡Never disassemble the rocker arm and shaft assembly.

To install:

8. Install the rocker arm, shaft and lash adjuster assembly.

9. Check that the notches in the rocker shaft are facing the direction, shown in the illustration. Install the rocker shaft cap with its identification mark as shown in the illustration.

10. Tighten the intake side rocker arm and shaft assembly mounting bolts to 21—25 ft. lbs. 28—34 Nm). Tighten the exhaust side rocker arm and shaft assembly mounting bolts to 106—124 inch. lbs. 12—14 Nm). Remove tool MD998443.

11. Continue the installation in the reverse order of the removal procedure. Be sure to use new valve cover gaskets.

Galant

2002–2003 2.4L ENGINE

1. Before servicing the vehicle, refer to the Precautions Section.

2. Disconnect the negative battery cable.

3. Disconnect the accelerator cable, remove the cable clamp mounting screws and position the accelerator cable out of the way.

4. Remove the air intake hose.

5. Disconnect the breather hose and the Positive Crankcase Ventilation (PCV) hose.

6. Disconnect the spark plug cables from the spark plugs.

7. Remove the rocker cover and gasket.

8. Install lash adjuster retainer tools to the rocker arm.

9. Remove the rocker shaft hold-down bolts gradually and evenly and remove the rocker shaft/arm assemblies.

10. Disassemble the rockers and the rocker shaft springs from the rocker shafts. If they are to be reused, note the location and positioning of all rocker shaft components. It is recommended that all lash adjusters and rockers be replaced as a complete set.

To install:

11. Immerse the lash adjusters in clean diesel fuel, and using a small wire, move the plunger up and down 4 or 5 times. While pushing down lightly on the check ball in order to bleed the air from the adjuster.

12. Install the lash adjusters to the rocker arms and attach the special holding tool.

13. Lubricate the rocker shaft with clean engine oil and install the rocker arms.

14. Temporarily tighten the rocker shaft assembly with the mounting bolts so that all rocker arms on the inlet valve side do not push on the valves.

15. Fit the rocker shaft springs from above and position them so that they are at right angles to the plug side. Install the rocker springs before installing the exhaust side rocker shaft and rocker arm assembly.

16. Install the exhaust side rocker shaft assembly in the engine. Tighten the rocker shaft mounting bolts gradually and evenly to 23 ft. lbs. (32 Nm).

17. Remove the lash adjuster retaining tools.

18. Install the rocker cover and tighten the mounting bolts to 29 inch lbs. (3.3 Nm).

19. Connect the spark plug wires to the spark plugs.

20. Connect the PCV and breather hoses.

21. Install the air intake hose.

22. Install the accelerator cable brackets and reconnect the accelerator cable.

23. Connect the negative battery cable.

2004–2006 2.4L (4G69) ENGINE

1. Before servicing the vehicle, refer to the Precautions Section.

2. Disconnect the negative battery cable.

3. If equipped with manual transaxle, remove the EMC. If equipped with automatic transaxle remove the PCM.

4. Remove the air cleaner assembly. Remove the battery. Remove the ignition coils.

5. Remove the timing belt upper cover.

6. Disconnect the control wiring harness connection. Remove the rocker cover PCV connection.

7. Remove the rocker cover breather hose connection. Remove the engine oil control valve and O-ring. Remove the oil pressure switch.

8. Remove the rocker cover retaining bolts. Remove the rocker cover. Discard the gasket.

9. Remove the spark plug guide oil seals. Remove the accumulator assembly.

10. Remove the exhaust rocker arm shaft caps. Remove the exhaust rocker arm shaft assembly.

11. Remove the intake rocker arm shaft caps. Remove the intake rocker arm shaft assembly.

➡ **Never disassemble the exhaust or intake rocker arm assembly.**

To install:

12. Position the intake rocker arm assembly so that its 0.22 inch hole faces toward the cylinder head. Install the rocker shaft mounting bolts. Torque to 21—25 ft. lbs. 29—34 Nm).

13. Position the exhaust rocker arm assembly so that its notch is positioned as shown in the illustration. Install the rocker shaft mounting bolts. Torque to 106—124 inch lbs. 12—14 Nm).

14. Continue the installation in the reverse order of the removal procedure.

15. Adjust the valves.

16. To initialize the ECM, manual transaxle equipped vehicles, or the PCM, automatic transaxle equipped vehicles, turn the ignition switch ON then OFF and keep it in the OFF position for at least ten seconds.

2002–2003 3.0L ENGINE

The hydraulic lash adjusters are built into the rocker arms.

1. Before servicing the vehicle, refer to the Precautions Section.

2. Disconnect the negative battery cable.

3. Remove the valve cover. Install lash adjuster retainer tools to prevent the auto-lash adjuster from falling out of the rocker arm.

4. Loosen rocker arm and shaft assembly evenly in several steps. Remove the rocker arm and shaft assembly as a complete unit.

5. Remove the rear camshaft bearing cap and slide the rocker arms, springs and washers from the shaft. If they are to be reused, note the location and positioning of all rocker shaft components. It is recom-

mended that all lash adjusters and rockers be replaced as a complete set.

To install:

6. Immerse the lash adjusters in clean diesel fuel. Using a small wire, move the plunger of the lash adjuster up and down 4 or 5 times while pushing down lightly on the check ball in order to bleed out the air. Install the lash adjusters in the rocker arms.

7. Using a light coat of engine oil, assemble the rocker arms to the shaft. Install the rear camshaft bearing cap.

8. Lubricate the camshaft and rocker shaft with clean engine oil and position on the cylinder head.

9. Apply a drop of sealant to the rear edges of the end caps.

10. Install the assembly making sure the notches in the rocker shafts are facing up.

11. Install the cap bolts and tighten evenly and gradually to 14 ft. lbs. (20 Nm). Remove the lash adjuster retainers.

12. Install the valve cover.

13. Connect the negative battery cable.

2004–2006 3.8L ENGINE

1. Before servicing the vehicle, refer to the Precautions Section.

2. Disconnect the negative battery cable. Drain the cooling system.

3. On the left side, remove the thermostat housing assembly. Remove the PCV valve hose connection, ignition coil connector and ignition coil. Disconnect the engine control wiring harness clamp and harness bracket.

4. On the right side, remove the intake manifold plenum, breather hose connection, heater hose and water hose connection. Remove the ignition coil connector and coil.

5. Remove the rocker arm cover retaining bolts. Remove the rocker arm cover. Discard the gasket.

6. Install tool MD998443 on the rocker arms, so that the lash adjusters will not fall out.

7. Loosen the rocker arm and shaft assembly mounting bolts. Remove the rocker arm and shaft assembly with the bolts still attached.

➡ **Never disassemble the rocker arm and shaft assembly.**

To install:

8. Install the rocker arm, shaft and lash adjuster assembly.

9. Check that the notches in the rocker shaft are facing the direction, shown in the illustration. Install the rocker shaft cap with its identification mark as shown in the illustration.

10. Tighten the intake side rocker arm

and shaft assembly mounting bolts to 21—25 ft. lbs. 28—34 Nm). Tighten the exhaust side rocker arm and shaft assembly mounting bolts to 106—124 inch. lbs. 12—14 Nm). Remove tool MD998443.

11. Continue the installation in the reverse order of the removal procedure. Be sure to use new valve cover gaskets.

Lancer
2.0L ENGINE

1. Before servicing the vehicle, refer to the Precautions Section.

2. Disconnect the negative battery cable. Remove the air cleaner assembly. Remove the ignition coil.

3. Disconnect the breather hose connection, PCV valve connection and accelerator cable clamp.

4. Remove the rocker arm cover retaining bolts. Remove the rocker arm cover. Discard the gasket.

5. Loosen the rocker arm and shaft assembly mounting bolts. Remove the rocker arm and shaft assembly with the bolts still attached.

➡ **Never disassemble the rocker arm and shaft assembly.**

To install:

6. Installation is the reverse of the removal procedure.

2.4L ENGINE

1. Before servicing the vehicle, refer to the Precautions Section.

2. Disconnect the negative battery cable.

3. Remove the air cleaner assembly. Remove the battery. Remove the ignition coils.

4. Remove the timing belt upper cover.

5. Disconnect the control wiring harness connection. Remove the rocker cover PCV connection.

6. Remove the rocker cover breather hose connection. Remove the engine oil control valve and O-ring. Remove the oil pressure switch.

7. Remove the rocker cover retaining bolts. Remove the rocker cover. Discard the gasket.

8. Remove the spark plug guide oil seals. Remove the accumulator assembly.

9. Remove the exhaust rocker arm shaft caps. Remove the exhaust rocker arm shaft assembly.

10. Remove the intake rocker arm shaft caps. Remove the intake rocker arm shaft assembly.

➡**Never disassemble the exhaust or intake rocker arm assembly.**

To install:

11. Position the intake rocker arm assembly so that its 0.22 inch hole faces toward the cylinder head. Install the rocker shaft mounting bolts. Torque to 21—25 ft. lbs. 29—34 Nm).

12. Position the exhaust rocker arm assembly so that its notch is positioned as shown in the illustration. Install the rocker shaft mounting bolts. Torque to 106—124 inch lbs. 12—14 Nm).

13. Continue the installation in the reverse order of the removal procedure.

14. Adjust the valves.

Evolution

1. Before servicing the vehicle, refer to the Precautions Section.

2. Remove the engine under cover. Remove the side cover.

3. Disconnect the battery cable. Drain the engine coolant.

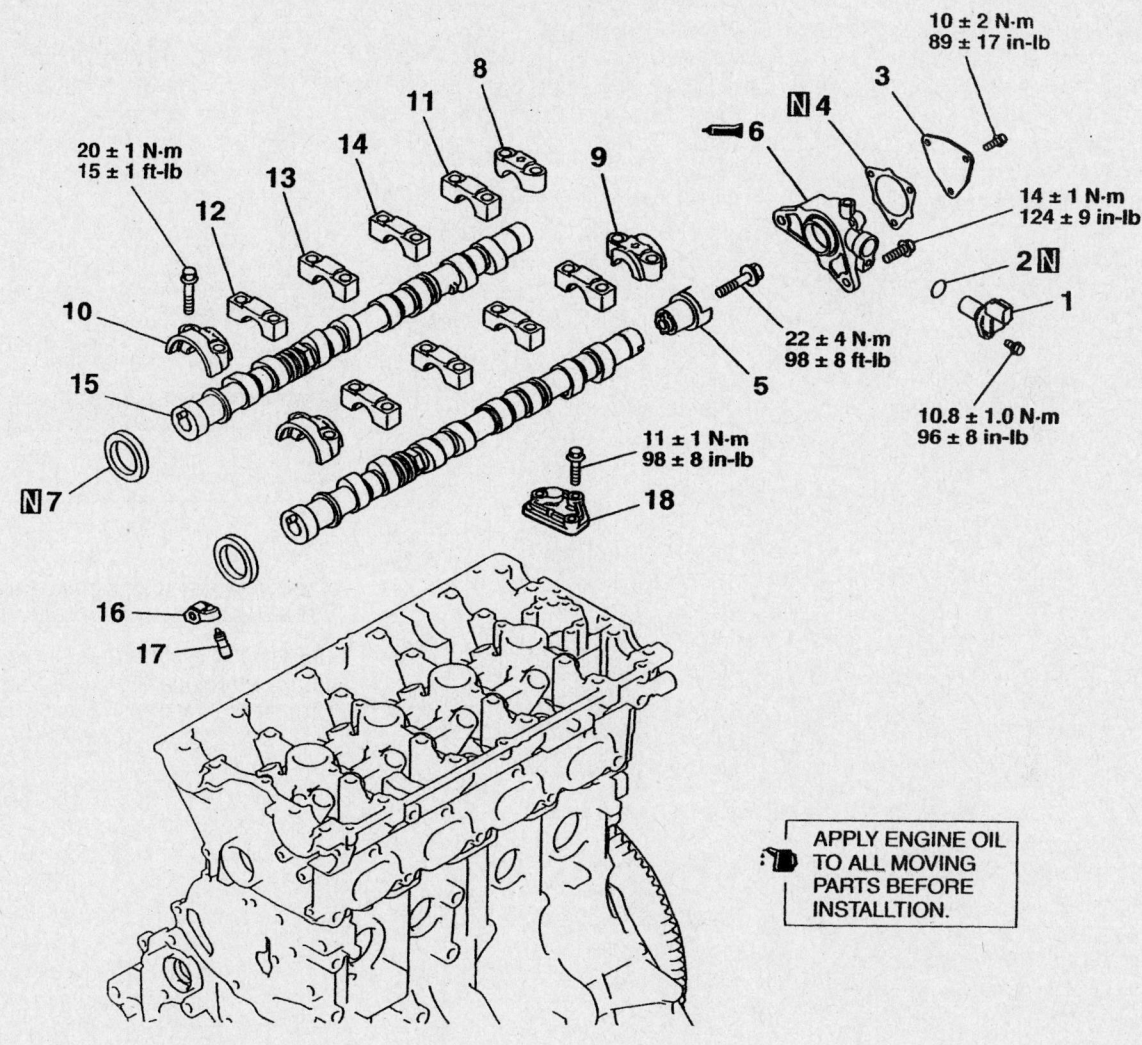

1. CAMSHAFT POSITION SENSOR
2. O-RING
3. COVER
4. GASKET
5. CAMSHAFT POSITION SENSING CYLINDER
6. CAMSHAFT POSITION SENSOR SUPPORT
7. CAMSHAFT OIL SEAL
8. BEARING CAP, REAR RIGHT

9. BEARING CAP, REAR LEFT
10. BEARING CAP, FRONT
11. BEARING CAP NO.5
12. BEARING CAP NO.2
13. BEARING CAP NO.3
14. BEARING CAP NO.4
15. CAMSHAFT
16. ROCKER ARM
17. LASH ADJUSTER
18. OIL DELIVERY BODY

67170-GALA-G23

Rocker arm assemblies and related components—Evolution

4. Remove the air duct assembly. Remove the air pipe "C" assembly.

5. Remove the accelerator cable. Remove the timing belt.

6. Disconnect the oil feed control valve connector and remove the oil feed control valve and O-ring.

7. Remove the rocker cover center cover. Remove the spark plug cables and ignition coils.

8. Disconnect the front heated oxygen sensor connector, crankshaft position sensor connector, control wiring harness connection and rocker cover PCV hose.

9. Remove the rocker cover breather hose, upper radiator hose and camshaft position sensor connector (exhaust and inlet side).

10. Disconnect the ground terminal, the control wiring harness connection and vacuum hose and pipe assembly.

11. Remove the valve cover retaining bolts. Remove the valve cover. Remove the spark plug hole gaskets. Remove and discard the valve cover gasket.

12. Remove the camshaft position sensor support cover, and gasket. Remove the camshaft position sensing cylinder (exhaust side).

13. Remove the camshaft position sensor support, cover and gasket.

14. Remove the camshaft sensing cylinder (intake side). Remove the camshaft position sensor support. Remove the exhaust camshaft sprocket and oil seal. Remove the intake camshaft sprocket and oil seal.

15. Remove the front camshaft bearing cap. Remove the rear camshaft bearing cap.

16. Remove bearing cap number 2, 5, 3 and 4. Remove the exhaust camshaft sprocket and washer

17. Remove the intake camshaft sprocket and oil seal. Remove the camshaft bearing front cap. Remove bearing cap number 2, 5, 3 and 4. Remove the intake camshaft.

18. Remove the rocker arms.

To install:

19. Clean the lash adjusters if they are being reused.

20. Install the oil delivery body and tighten the bolt to 90–106 inch lbs. (10–12 Nm).

21. Install the lash adjusters into the rocker arms.

22. Install the rocker arms.

23. Install the camshaft(s). The exhaust camshaft has a 0.16 inch (4mm) wide slit at the rear end, so not to confuse it with the intake camshaft. Make sure that the dowel pin on the end of each camshaft is in the top position.

24. Install the camshaft bearing caps. Intake and exhaust bearing caps for intake Nos. 2 through 5 are identical, except for an identification mark on each of them. (Example: E2=exhaust side No. 2; I3=Intake side No. 3.) Be sure to install the correct cap in the correct location.

25. Install the front and rear camshaft bearing caps. Apply fresh sealant to the mating surfaces between the caps and cylinder head.

26. Once the camshaft bearing caps are installed, tighten each bolt in 2 or 3 passes. Finally tighten the bolts to 15 ft. lbs. (20 Nm). Completely wipe off any squeezed out sealant.

27. Install the camshaft oil seals.

28. Apply a 0.12 inch (3mm) bead of sealant around the circumference of the sensor support that mates to the bearing cap/cylinder head side. Install the CMP sensor support.

29. Install the CMP sensing cylinder and tighten the bolt to 90–106 ft. lbs. (18–26 Nm).

30. Install the CMP sensor support cover and gasket.

31. Install CMP sensor.

32. Install the rocker arm cover gasket and cover. Tighten the bolts to 27–35 inch lbs. (3.0–4.0 Nm).

33. Install or connect the oil filler cap.

34. Install or connect the PCV hose.

35. Install or connect the breather hose.

36. Fill the crankcase with oil and connect the negative battery cable.

Mirage

1. Before servicing the vehicle, refer to the Precautions Section.

2. Remove or disconnect the following:
   - Negative battery cable
   - Spark plug cables—for 1.8L engine
   - Accelerator cable, breather hose and Positive Crankcase Ventilation (PCV) hose connections
   - Rocker cover

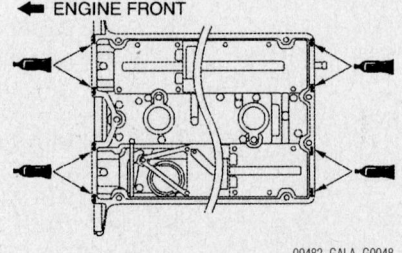

← ENGINE FRONT

09482_GALA_G0048

Rocker arm cover sealant application—Evolution

3. Loosen both rocker arm shaft assemblies gradually and evenly and remove the rocket shafts from the vehicle.

4. If disassembly is required, keep all parts in the exact order of removal.

To install:

5. Lubricate the rocker shaft with clean engine oil and install the rockers and springs.

6. Install the rocker arm and shaft assemblies. Tighten the rocker arm shaft retainer bolts to 23 ft. lbs. (32 Nm).

7. Check valve adjustment and install the valve cover. Tighten the valve cover bolts to 16 inch lbs. (1.8 Nm) for the 1.5L engine or to 29 inch lbs. (3.3 Nm) for the 1.8L engine.

8. Install or connect the following:
   - Spark plug cables, if detached
   - Accelerator cable, breather hose and PCV hose
   - Negative battery cable

Turbocharger

REMOVAL & INSTALLATION

Evolution

❋❋ CAUTION

The air bag system must be disarmed before removing the turbocharger.

1. Before servicing the vehicle, refer to the Precautions Section.

2. Disconnect the negative battery cable.

3. Drain the engine coolant.

4. Disengage the retaining clips and remove the front under cover, located under the front of the vehicle behind the front bumper face.

5. Remove the radiator assembly.

6. Remove the air intake hose from the air cleaner assembly.

7. Remove the air pipes & hoses from the charge air cooler assembly.

8. Remove the front axle crossmember bar from underneath the vehicle.

9. Remove the front exhaust pipe from the exhaust manifold.

10. Remove the exhaust manifold cover.

11. Remove the front Heated Oxygen (HO2S) sensor.

12. Remove the turbocharger heat protector.

13. Disconnect the turbocharger water feed pipe and return hose connections and remove gaskets.

14. Disconnect the turbocharger oil feed and oil return pipes and remove gaskets.

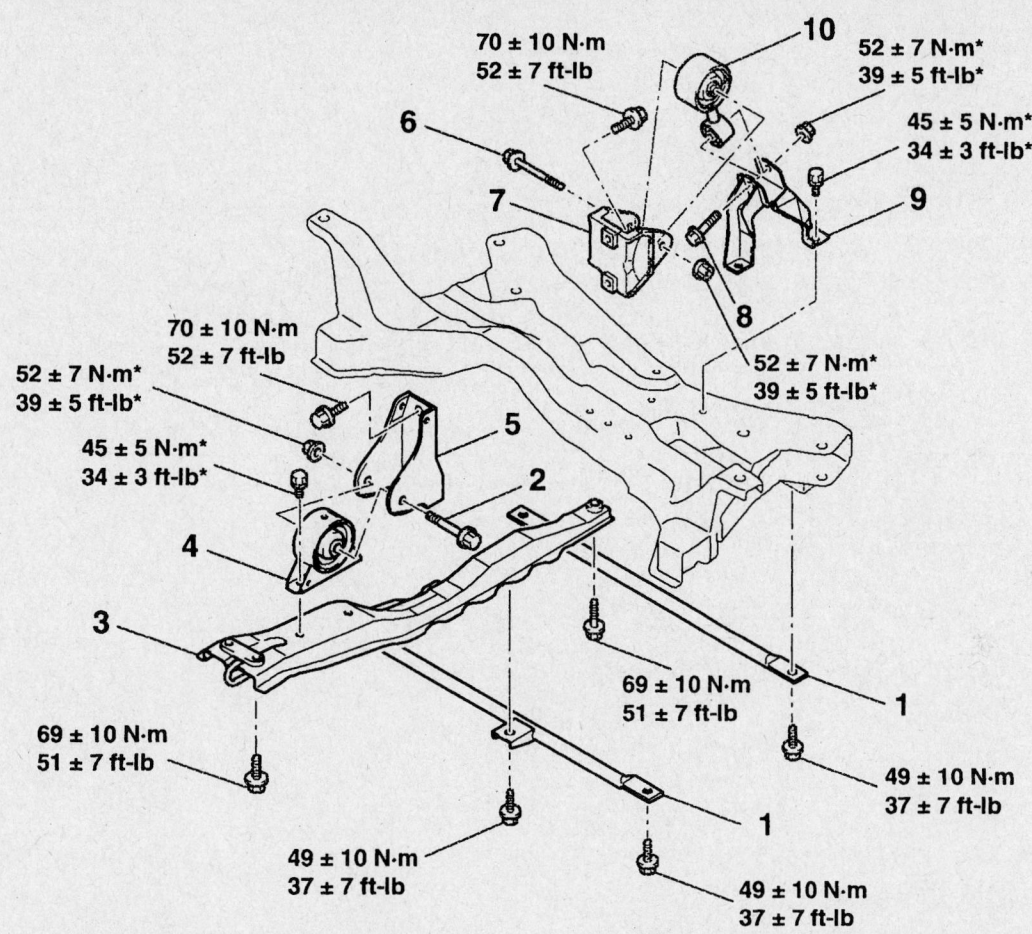

70 ± 10 N·m
52 ± 7 ft-lb

52 ± 7 N·m*
39 ± 5 ft-lb*

45 ± 5 N·m*
34 ± 3 ft-lb*

70 ± 10 N·m
52 ± 7 ft-lb

52 ± 7 N·m*
39 ± 5 ft-lb*

52 ± 7 N·m*
39 ± 5 ft-lb*

45 ± 5 N·m*
34 ± 3 ft-lb*

69 ± 10 N·m
51 ± 7 ft-lb

69 ± 10 N·m
51 ± 7 ft-lb

49 ± 10 N·m
37 ± 7 ft-lb

49 ± 10 N·m
37 ± 7 ft-lb

49 ± 10 N·m
37 ± 7 ft-lb

1. FRONT AXLE CROSSMEMBER BARS
2. ENGINE FRONT ROLL STOPPER BRACKET BOLT
3. FRONT SUSPENSION CENTERMEMBER
4. ENGINE FRONT ROLL STOPPER BRACKET
5. TRANSAXLE CASE FRONT ROLL STOPPER BRACKET
6. ENGINE REAR ROLL STOPPER ROD BOLT
7. TRANSAXLE CASE REAR ROLL STOPPER BRACKET
8. ENGINE REAR ROLL STOPPER ROD BOLT
9. ENGINE REAR ROLL STOPPER ROD BRACKET
10. ENGINE REAR ROLL STOPPER ROD

09482_GALA_G0049

Front axle crossmember assembly—Evolution

15. Remove the starter motor.

16. Disconnect the vacuum hose connections.

17. Remove the air outlet fitting and gasket.

18. Remove the exhaust fitting bracket.

19. Remove the turbocharger and exhaust fitting assembly. Remove the gasket.

20. Remove the exhaust manifold and gasket.

To install:

21. Clean all gasket material from the mating surfaces.

22. Install a new gasket and exhaust manifold. Tighten the retainers to 32–40 ft. lbs. (44–54 Nm).

23. Install the turbocharger assembly. Be sure to clean the oil and water pipe fittings, inside of eye bolts and individual pipes of any clogs. Clean or use compressed air to remove any carbon matter stuck in the oil passage of the turbocharger. Refill new engine oil at the oil feed pipe fitting hole.

24. Install the exhaust fitting assembly, along with a new gasket.

25. Install the exhaust fitting bracket.

26. Install the air outlet fitting and gasket.

27. Connect the vacuum hose connections.

28. Install the starter motor.

29. Connect the turbocharger oil feed and oil return pipes, along with new gaskets. Tighten the oil feed pipe fastener to 10–14 ft. lbs. (15–19 Nm). Tighten the oil return pipe fasteners to 71–89 inch lbs. (8–10 Nm).

30. Connect the turbocharger water feed pipe and return hoses, along with new gaskets. Tighten the pipe fasteners to 26–36 ft. lbs. (35–49 Nm).

31. Install the turbocharger heat protector.

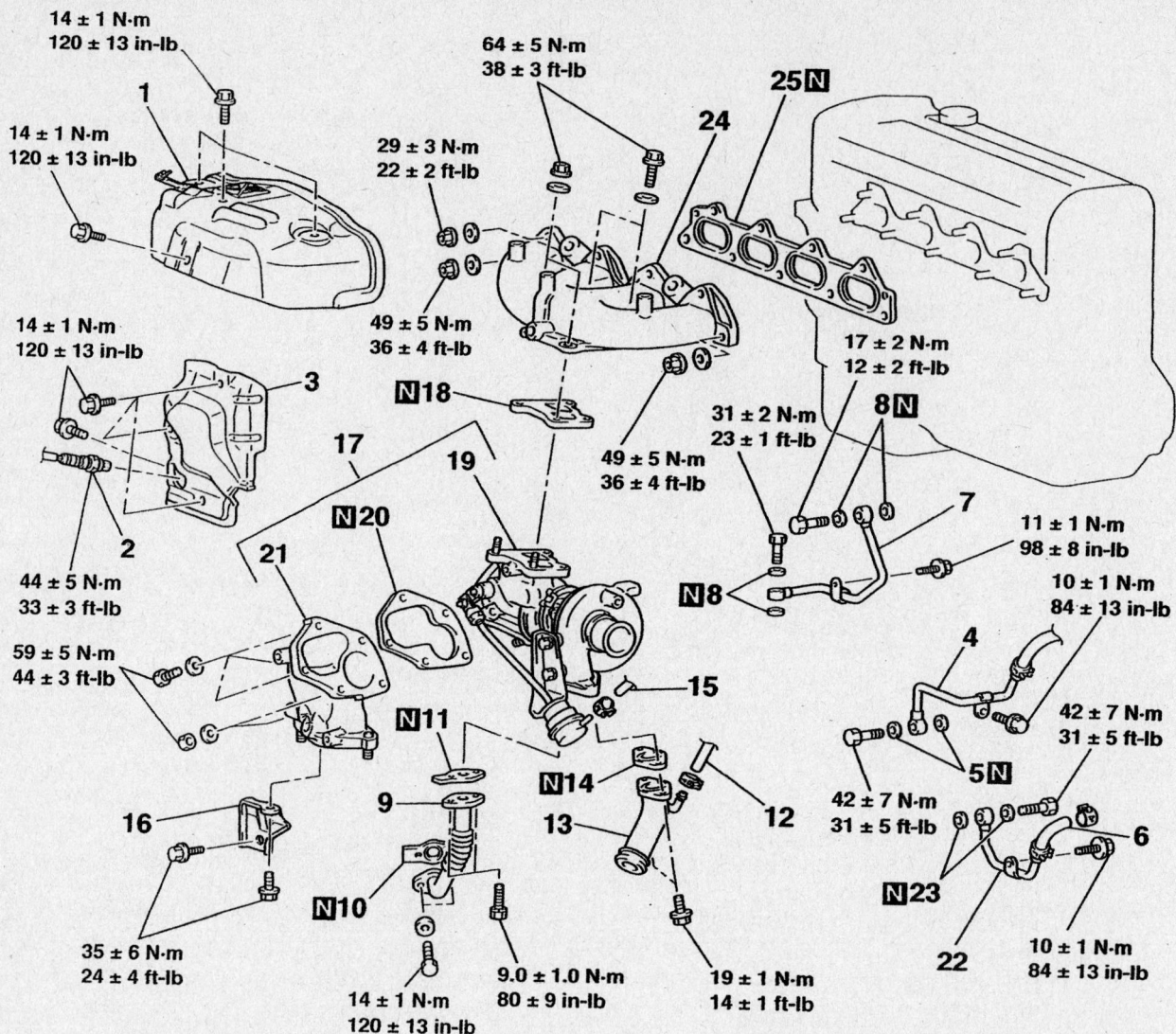

14 ± 1 N·m
120 ± 13 in-lb

14 ± 1 N·m
120 ± 13 in-lb

14 ± 1 N·m
120 ± 13 in-lb

44 ± 5 N·m
33 ± 3 ft-lb

59 ± 5 N·m
44 ± 3 ft-lb

35 ± 6 N·m
24 ± 4 ft-lb

29 ± 3 N·m
22 ± 2 ft-lb

49 ± 5 N·m
36 ± 4 ft-lb

64 ± 5 N·m
38 ± 3 ft-lb

49 ± 5 N·m
36 ± 4 ft-lb

31 ± 2 N·m
23 ± 1 ft-lb

14 ± 1 N·m
120 ± 13 in-lb

9.0 ± 1.0 N·m
80 ± 9 in-lb

19 ± 1 N·m
14 ± 1 ft-lb

42 ± 7 N·m
31 ± 5 ft-lb

17 ± 2 N·m
12 ± 2 ft-lb

11 ± 1 N·m
98 ± 8 in-lb

10 ± 1 N·m
84 ± 13 in-lb

42 ± 7 N·m
31 ± 5 ft-lb

10 ± 1 N·m
84 ± 13 in-lb

1. EXHAUST MANIFOLD COVER
2. HEATED OXYGEN SENSOR
    (FRONT)
3. TURBOCHARGER HEAT
    PROTECTOR
4. TURBOCHARGER WATER FEED
    PIPE CONNECTION
5. GASKET
6. TURBOCHARGER WATER RETURN
    HOSE CONNECTION
7. TURBOCHARGER OIL FEED PIPE
8. GASKET
•   STARTER MOTOR

9. TURBOCHARGER OIL RETURN
    PIPE

10. TURBOCHARGER OIL RETURN
    PIPE GASKET
11. TURBOCHARGER OIL RETURN
    PIPE GASKET
12. VACUUM HOSE CONNECTION
13. AIR OUTLET FITTING
14. AIR OUTLET FITTING GASKET
15. VACUUM HOSE CONNECTION
16. EXHAUST FITTING BRACKET
17. TURBOCHARGER AND EXHAUST
    FITTING ASSEMBLY
18. TURBOCHARGER GASKET
19. TURBOCHARGER ASSEMBLY
20. EXHAUST FITTING GASKET
21. EXHAUST FITTING ASSEMBLY
22. TURBOCHARGER WATER RETURN
    PIPE AND HOSE ASSEMBLY
23. GASKET
24. EXHAUST MANIFOLD
25. EXHAUST MANIFOLD GASKET

67170-GALA-G25

Turbocharger and exhaust manifold assembly—Evolution

32. Install the front Heated Oxygen (HO2S) sensor.

33. Install the exhaust manifold cover.

34. Install the front exhaust pipe to the exhaust manifold.

35. Install the front axle crossmember bar from underneath the vehicle.

36. Install the air pipes & hoses to the charge air cooler assembly.

37. Install the air intake hose to the air cleaner assembly.

38. Install the radiator assembly.

39. Install the front under cover and secure with the retaining clips.

40. Refill the cooling system with the correct amount and type of coolant.

41. Connect the negative battery cable.

Intake Manifold Plenum

REMOVAL & INSTALLATION

3.0L Engine

1. Before servicing the vehicle, refer to the Precautions Section.

2. Relieve the fuel system pressure. Drain the cooling system. Remove the engine strut tower.

3. Remove the air cleaner assembly. Remove the throttle body.

4. Disconnect the MAP sensor connector, the control wiring harness and power steering wiring harness connector.

5. Disconnect the EGR vacuum regulator solenoid valve connector, the EVAP solenoid valve connector and the knock sensor connector.

6. Disconnect the CKP sensor connector, right bank heated oxygen sensor and injector connector.

7. Disconnect the distributor connector, vacuum hose connection, vacuum pipe and EGR vacuum solenoid valve.

8. Disconnect the brake booster vacuum hose connection. Remove the EGR valve and gasket. Remove the EGR pipe connection

9. Remove the power steering pump drive belt. Remove the power steering oil pump stay bracket.

10. Remove the intake manifold plenum stay, front. Remove the intake manifold plenum stay, rear. Remove the engine mount stay.

11. Remove the intake manifold plenum retaining bolts. Remove the intake manifold plenum from the engine.

12. Remove and discard the gasket.

To install:

13. Clean all gasket material from the mating surfaces.

14. Install a new gasket and the intake manifold plenum.

15. Continue the installation in the reverse order of the removal procedure.

3.8L Engine

1. Before servicing the vehicle, refer to the Precautions Section.

2. Disconnect the negative battery cable. Drain the cooling system.

3. Remove the air cleaner assembly. Remove the strut tower bar assembly.

4. Disconnect the throttle position sensor connector, water feed hose connection and the water return feed connection.

5. Remove the throttle body assembly retaining bolts. Remove the throttle body and gasket.

6. Disconnect the manifold absolute pressure sensor connector, control wiring harness and injector wiring harness combination connector.

7. Disconnect the CKP sensor connector, knock sensor connector, power steering pressure switch connector and EVAP solenoid connector.

8. Disconnect the EGR valve connector, brake booster vacuum hose connection and EVAP purge hose connection.

9. Disconnect the vacuum hose connection, control wiring harness and engine cover bracket assembly.

10. Disconnect the PCV valve hose connection. Remove the intake manifold plenum stay (rear) connecting bolt, EGR pipe and gasket. Remove the intake manifold plenum stay (front) connecting bolt.

11. Remove the throttle body stay connecting bolt.

12. Remove the intake manifold plenum retaining bolts. Remove the intake manifold plenum and gasket from the engine.

To install:

13. Clean all gasket material from the mating surfaces.

14. Install a new gasket and the intake manifold plenum.

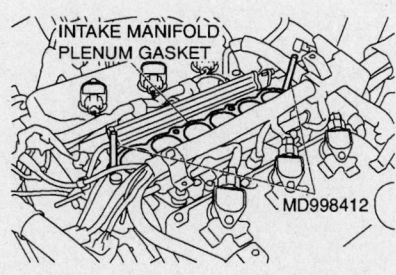

09482_GALA_G0050

Intake manifold plenum installation tool location—3.8L engine

➡ **Be sure to use tool MD998412 to install the intake manifold plenum.**

15. Continue the installation in the reverse order of the removal procedure.

16. Check and adjust all fluid levels, as required.

Intake Manifold

REMOVAL & INSTALLATION

Diamante

1. Before servicing the vehicle, refer to the Precautions Section.

2. Relieve the fuel system pressure.

3. Remove or disconnect the following:

- Negative cable and drain the cooling system
- Air intake hose(s)
- Accelerator control cables from the throttle body
- Vacuum hoses including the brake booster hose
- Wiring harness connectors
- High pressure and return fuel hoses
- Exhaust Gas Recirculation (EGR) pipe and remove the EGR valve and EGR temperature sensor from the intake plenum assembly
- Manifold Absolute Pressure (MAP) sensor, if equipped
- Plenum retaining bracket
- Plenum retaining nuts and bolts and remove the air intake plenum from the intake manifold
- Upper timing belt covers
- Water pump stay bracket

➡ **It is not necessary to remove the fuel injectors from the intake unless the manifold assembly is being replaced.**

- Fuel rail with the injectors attached
- Coolant hoses from the intake manifold
- Intake manifold mounting nuts and remove the intake manifold

4. Clean the gasket mounting surfaces.

To install:

5. Thoroughly clean the mating surfaces of the heads, intake manifold and air intake plenum.

6. Install new intake manifold gaskets to the cylinder heads with the adhesive side facing up.

7. Place the manifold on the cylinder heads.

8. Lubricate the studs lightly with oil and install the nuts.

9. Tighten the mounting nuts as follows:

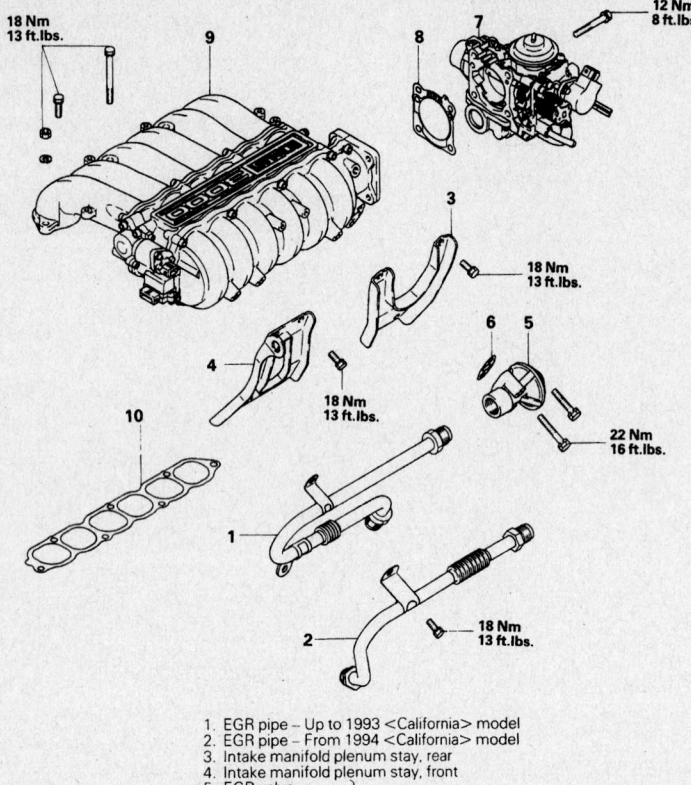

1. EGR pipe – Up to 1993 <California> model
2. EGR pipe – From 1994 <California> model
3. Intake manifold plenum stay, rear
4. Intake manifold plenum stay, front
5. EGR valve
6. EGR valve gasket } <For California>
7. Throttle body
8. Throttle body gasket
9. Intake manifold plenum
10. Intake manifold plenum gasket

7923PG42

Intake manifold plenum assembly—3.5L engine

a. Front bank nuts: 48–72 inch lbs. (5–8 Nm).
b. Rear bank nuts: 14–17 ft. lbs. (20–23 Nm).
c. Front bank nuts: 14–17 ft. lbs. (20–23 Nm).
10. Connect the coolant hoses to the intake manifold.
11. Using new O-rings, install the fuel rail assembly, if removed. Tighten the mounting bolts to 84–108 inch lbs. (10–13 Nm).
12. Install or connect the following:
- Plenum, with new gasket. Tighten the retaining nuts and bolts evenly and gradually to 13 ft. lbs. (18 Nm).
- Retaining bracket and tighten the retaining bolts to 13 ft. lbs. (18 Nm)
- MAP sensor, if removed
- EGR valve, using a new gasket. Tighten the bolts to 16 ft. lbs. (22 Nm).
- EGR temperature sensor and tighten the fitting to 84–108 inch lbs. (10–12 Nm)
- EGR pipe and tighten the fittings to 43 ft. lbs. (60 Nm)

1. Connection for high-pressure fuel hose
2. O-ring
3. Connection for fuel return hose
4. Connection for vacuum hoses
5. Wiring harness connector
6. Oxygen sensor <For California from 1994 models>
7. Fuel rail (with injectors)
8. Insulators
9. Timing belt upper cover
10. Water pump stay mounting bolt
11. Intake manifold mounting nut
12. Intake manifold mounting nut
13. Cone disc spring
14. Intake manifold
15. Intake manifold gasket

- High pressure fuel hose, use a new O-ring. Tighten the retaining bolts to 48 inch lbs. (5 Nm).
- Fuel return hose, using a new clamp
- Water pump stay bracket
- Upper timing belt covers
- Harness connectors and vacuum hoses
- Accelerator cables, adjust
- Air intake hose(s)
13. Fill the system with coolant.
14. Connect the negative battery cable.

Eclipse

2002–2003 2.4L ENGINE

1. Before servicing the vehicle, refer to the Precautions Section.
2. Relieve the fuel system pressure.
3. Remove the battery.
4. Drain the engine coolant.
5. Remove or disconnect the following:
- Accelerator cable
- Air intake hose
- Ignition coil and the module wiring connectors
- Manifold Absolute Pressure (MAP) sensor
- Condenser
- Throttle Position (TP) sensor and the Idle Air Control (IAC) motor connectors

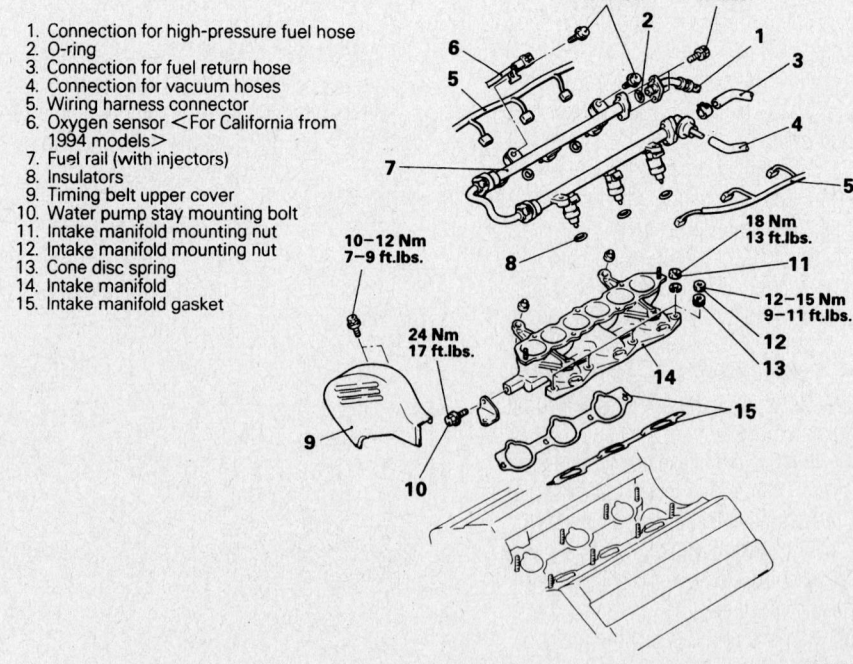

7923PG43

Intake manifold and related components—3.5L engine

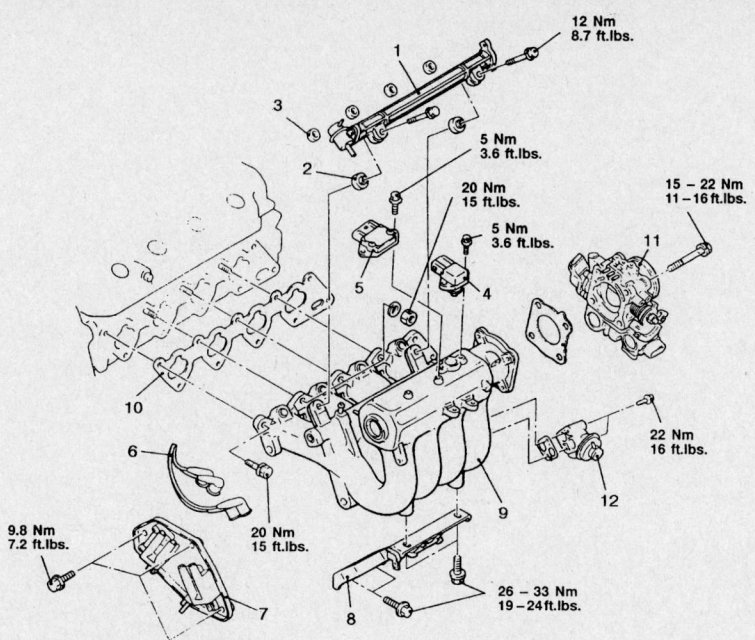

1. Fuel rail, fuel injector and pressure regulator assembly
2. Insulator
3. Insulator
4. Manifold differential pressure sensor
5. Ignition power transistor
6. Spark plug cable connection
7. Ignition coil
8. Intake manifold stay
9. Intake manifold
10. Intake manifold gasket
11. Throttle body
12. EGR valve assembly

7923PG41

Intake manifold and related components—2002–2003 Eclipse and Galant 2.4L engine

- Heated Oxygen (HO2S) sensor connector
- Crankshaft Position (CKP) sensor connector
- Air conditioning compressor connector
- Engine control wiring harness bracket and position the harness out of the way
- Vacuum hoses, label for reference
- Spark plug wires from the ignition coil
- Fuel lines from the fuel rail
- Heater hoses
- Fuel rail assembly
- Ignition coil and module
- Exhaust Gas Recirculation (EGR) valve assembly
- Intake manifold stay and the engine hanger
- Intake manifold

To install:

6. Install or connect the following:
- Intake manifold. Torque the intake manifold bolts to 15 ft. lbs. (20 Nm).
- Intake manifold stay and the engine hanger. Torque the mounting bolts to 19–24 ft. lbs. (26–33 Nm).
- EGR assembly
- Ignition coil and module

- Fuel rail and insulators and reconnect the high-pressure fuel hose
- Heater hoses and fuel lines
- Spark plug wires to the coil towers
- Vacuum hoses
- Engine harness in the proper position
- MAP sensor
- TP sensor and the IAC motor connectors
- HO2S
- Ignition condenser
- Accelerator cable, adjust
- Battery

7. Refill the engine with coolant.

2004–2005 2.4L ENGINE

1. Before servicing the vehicle, refer to the Precautions Section.

2. Relieve the fuel system pressure. Disconnect the negative battery cable.

3. Remove the air cleaner assembly. Drain the cooling system.

4. Disconnect the accelerator cable connection, throttle position sensor connector and idle air control motor connector. Remove the thermostat case assembly.

5. Disconnect the vacuum hose connection. Disconnect the water hose connection.

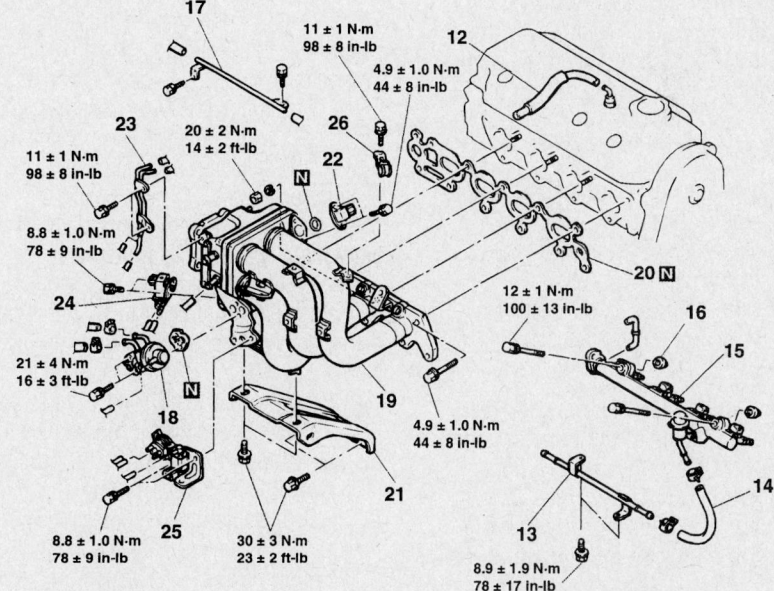

12. PCV HOSE
13. FUEL RETURN PIPE
14. FUEL HOSE
15. FUEL RAIL, INJECTOR AND FUEL PRESSURE REGULATOR
16. INSULATOR
17. VACUUM PIPE
18. EGR VALVE
19. INTAKE MANIFOLD
20. INTAKE MANIFOLD GASKET
21. INTAKE MANIFOLD STAY
22. MANIFOLD ABSOLUTE PRESSURE SENSOR
23. VACUUM PIPE
24. EVAPORATIVE EMISSION PURGE SOLENOID VALVE
25. EGR VACUUM REGULATOR SOLENOID VALVE AND VACUUM CONTROL VALVE
26. ACCELERATOR CABLE CLAMP

09482_GALA_G0051

Intake manifold and related components—2004–2005 Eclipse 2.4L engine

6. Remove the throttle body bracket. Remove the throttle body retaining bolts. Remove the throttle body. Discard the gasket.

7. Disconnect the hose connection, brake booster vacuum hose connection, ignition coil connector and injector connector.

8. Disconnect the manifold absolute pressure sensor connector, EVAP solenoid valve connector, EGR vacuum regulator solenoid valve connector and the high pressure hose connection.

9. Disconnect the fuel return hose connection. Remove the oil dipstick and tube assembly. Disconnect the pressure hose connection. Remove the PCV hose.

10. Remove the fuel return pipe, fuel hose, fuel injector rail, injector and pressure regulator.

11. Remove the vacuum pipe and EGR valve.

12. Remove the intake manifold retaining bolts. Remove the intake manifold from the engine. Remove and discard the gasket.

To install:

13. Be sure to use a new intake manifold gasket. Position the gasket on the engine.

14. Install the intake manifold. Torque the retaining bolts to specification.

15. Continue the installation in the reverse order of the removal procedure.

➡**Be sure to install the throttle body gasket in the proper direction. If this gasket is not installed correctly, poor engine idling may result.**

2006 2.4L (4G69) ENGINE

1. Before servicing the vehicle, refer to the Precautions Section.

2. Relieve the fuel system pressure. Disconnect the negative battery cable.

3. Remove the air cleaner assembly. Drain the cooling system. Remove the battery.

➡**Do not loosen the retaining screws for the resin cover of the throttle body assembly. If these screws are loosened, the sensor incorporated in the resin cover becomes misaligned and the throttle body can not function properly.**

4. Disconnect the accelerator cable connection, throttle position sensor connector and water return hose connection.

5. Remove the throttle body bracket. Remove the throttle body retaining bolts. Remove the throttle body. Discard the gasket.

6. Remove the delivery pipe and injector assembly.

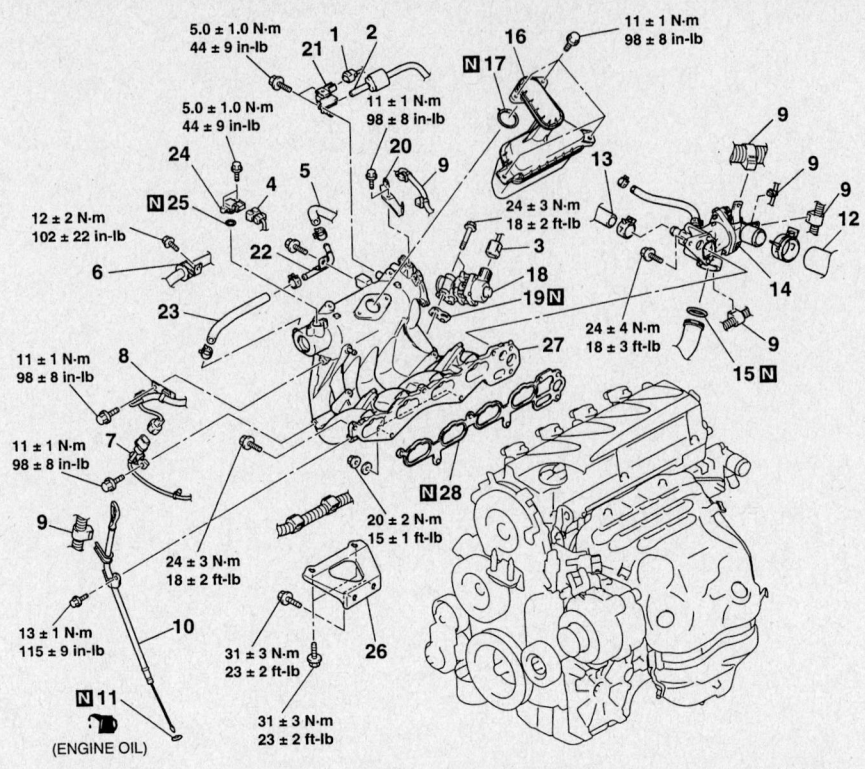

Intake manifold and related components—2006 Eclipse, 2004–2006 Galant and 2004–2006 Lancer 2.4L engine

7. Disconnect the EVAP purge solenoid valve connector, the evaporative emission purge hose connection and the EGR valve connector.

8. Disconnect the MAP sensor connector, brake booster vacuum hose connection, pressure hose clamp and knock sensor connector bracket.

9. Disconnect the harness clamp bracket and harness clamp. Remove the engine oil dipstick and tube assembly. Disconnect the lower radiator hose. Remove the thermostat case assembly, resonator and gasket.

10. Remove the EGR valve and gasket. Remove the harness clamp bracket, EVAP purge solenoid valve and brake booster vacuum pipe.

11. Remove the brake booster vacuum hose and MAP sensor. Remove the intake manifold bracket.

12. Remove the intake manifold retaining bolts. Remove the intake manifold from the engine. Discard the gasket.

To install:

13. Be sure to use a new intake manifold gasket. Position the gasket on the engine.

14. Install the intake manifold. Torque the retaining bolts to specification.

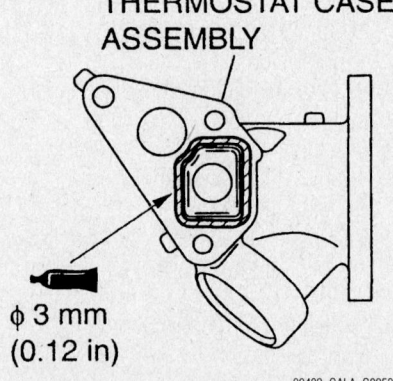

Thermostat case assembly sealant application—2006 Eclipse, 2004–2006 Galant and 2004–2006 Lancer 2.4L engine

15. When installing the thermostat case assembly be sure to apply a bead of sealant, part number 8672 or equivalent, to the cylinder head mating surface of the case assembly.

➡**Be sure to assemble the components with 15 minutes of applying the sealant to the surfaces.**

16. Continue the installation in the reverse order of the removal procedure.

➡**Be sure to install the throttle body gasket in the proper direction. If this gasket is not installed correctly, poor engine idling may result.**

17. Complete the vehicle initialization procedure.

➡**To complete the initialization procedure the following tools are needed. MB991958 scan tool, MB991824 VCI, MB991827 MUT III USB cable, MB991910 MUT III Main harness "A"**

18. Connect the scan tool to the data link connector. To prevent damage to the scan tool be sure that the ignition switch is in the LOCK position before connecting the scan tool.

19. Turn the ignition switch to the ON position.

20. Select "check mode" from the menu screen.

21. Select "erase memory" from the menu screen.

22. Initialize the learning value.

23. After initialization complete the idle learning procedure.

➡**This procedure must be performed when the PCM is replaced, or when the learning value is initialized, as the idling is not stabilized because the learning value in the MFI engine is not completed.**

24. Start the engine. Allow the coolant temperature to reach 176 degrees or more.

25. Stop the engine and place the ignition switch in the LOCK position.

26. After ten seconds, restart the engine.

27. For ten minutes carry out the idling procedure below to confirm that the engine has normal idling.
- Position the transaxle selector lever in the "P" range
- The engine fan is not to be operated
- The engine coolant temperature should be 176 degrees or more

➡**If the engine stalls during idling, check the throttle valve of the throttle body for dirt. Correct and perform the procedure again.**

2002–2003 3.0L ENGINE

1. Before servicing the vehicle, refer to the Precautions Section.

2. Relieve the fuel system pressure.

3. Disconnect the negative battery cable and drain the cooling system.

4. Remove both the front and rear intake manifold plenum stay brackets.

5. Remove the EGR valve and gasket.

6. Remove the EGR pipe and gasket.

7. Remove the Manifold Absolute Pressure (MAP) sensor.

8. Remove the throttle body and gasket.

9. If equipped with an induction control valve, remove the following:
- Induction control valve assembly
- Intake manifold upper
- Intake manifold plenum gasket

10. If NOT equipped with an induction control valve, remove the following:
- Intake manifold plenum
- Intake manifold plenum gasket

11. Disconnect the fuel injector harness.

12. Remove the fuel rail and injectors assembly, along with the insulators.

13. Remove the intake manifold mounting nuts and remove the intake manifold and gaskets.

14. Clean the gasket mounting surfaces.

To install:

15. Thoroughly clean the mating surfaces of the heads, intake manifold and air intake plenum.

16. Install new intake manifold gaskets to the cylinder heads with the adhesive side facing up.

17. Place the manifold on the cylinder heads.

18. Lubricate the studs lightly with oil and install the nuts.

19. Tighten the mounting nuts as follows:
  a. Right bank nuts: 44–70 inch lbs. (5–8 Nm).
  b. Left bank nuts: 21–23 ft. lbs. (15–17 Nm).
  c. Right bank nuts: 21–23 ft. lbs. (15–17 Nm).
  d. Left bank nuts: 21–23 ft. lbs. (15–17 Nm).
  e. Right bank nuts: 21–23 ft. lbs. (15–17 Nm).

20. Using new O-rings, install the fuel rail and injectors assembly.

21. Connect the fuel injector wiring harness.

22. If equipped with an induction control valve, install the following:
- Intake manifold plenum gasket
- Intake manifold upper. Tighten retaining nuts to 12–14 ft. lbs. (16–20 Nm).
- Install the induction control valve gasket. Apply sealant to the gasket surface and install while the sealant is still wet (within 15 minutes).
- Induction control valve assembly. Tighten fasteners to 71–89 inch lbs. (8–10 Nm).

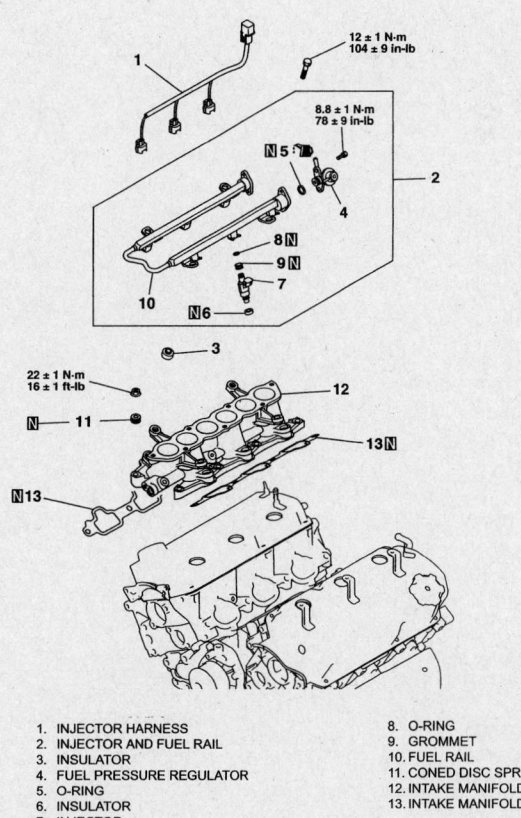

1. INJECTOR HARNESS	8. O-RING
2. INJECTOR AND FUEL RAIL	9. GROMMET
3. INSULATOR	10. FUEL RAIL
4. FUEL PRESSURE REGULATOR	11. CONED DISC SPRING
5. O-RING	12. INTAKE MANIFOLD
6. INSULATOR	13. INTAKE MANIFOLD GASKET
7. INJECTOR	

67170-GALA-G27

Intake manifold and related components—2002–2003 Eclipse and Galant 3.0L engine

23. If NOT equipped with an induction control valve, install the following:
- Intake manifold plenum gasket
- Intake manifold plenum. Tighten retaining nuts to 12–14 ft. lbs. (16–20 Nm).

24. Install a new throttle body gasket and install the throttle body assembly.

25. Install the MAP sensor.

26. Install the EGR pipe and tighten the fittings to 12–14 ft. lbs. (16–20 Nm).

27. Install the EGR valve, using a new gasket. Tighten the bolts to 16 ft. lbs. (22 Nm).

28. Install the front and rear intake manifold plenum stay brackets.

29. Fill the system with coolant.

30. Connect the negative battery cable.

2004–2005 3.0L ENGINE

1. Before servicing the vehicle, refer to the Precautions Section.

2. Relieve the fuel system pressure. Disconnect the negative battery cable.

3. Remove the air cleaner assembly. Drain the cooling system.

4. Remove the intake manifold plenum assembly.

5. Disconnect the injector connector wiring assembly, high pressure hose connection, O-ring and fuel return hose.

6. Disconnect the vacuum hose connection. Remove the fuel rail, injector and fuel pressure regulator.

7. Remove the PCV hose connection. Remove the timing belt front upper cover and right bracket.

8. Remove the intake manifold retaining bolts. Remove the intake manifold from the engine. Discard the gaskets.

To install:

9. Be sure to use a new intake manifold gasket. Position the gasket on the engine.

10. Install the intake manifold. Coat the retaining bolts in clean engine oil. Torque the retaining bolts to specification.

11. Continue the installation in the reverse order of the removal procedure.

3.8L ENGINE

1. Before servicing the vehicle, refer to the Precautions Section.

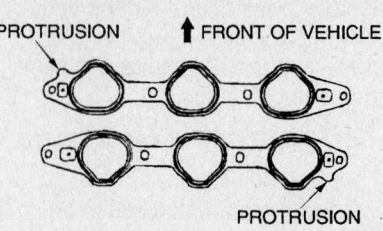

Intake manifold gasket positioning–2004—2005 Eclipse 3.0L engine and 2006 Eclipse 3.8L engine

2. Relieve the fuel system pressure. Disconnect the negative battery cable.

3. Remove the air cleaner assembly. Drain the cooling system.

4. Remove the intake manifold plenum assembly.

5. Disconnect the injector connector wiring assembly, high pressure hose connection, O-ring and fuel return hose.

6. Disconnect the vacuum hose connection. Remove the fuel rail, injector and fuel pressure regulator. Remove the harness bracket.

7. Remove the PCV hose connection. Remove the left timing belt front upper cover. Remove the right timing belt front upper cover. Remove the water pump bracket.

8. Remove the intake manifold retaining bolts. Remove the intake manifold from the engine. Discard the gaskets.

To install:

9. Be sure to use a new intake manifold gasket. Position the gasket on the engine.

10. Install the intake manifold. Coat the retaining bolts in clean engine oil. Torque the retaining bolts to specification.

11. Continue the installation in the reverse order of the removal procedure.

Galant

2002–2003 2.4L ENGINE

1. Before servicing the vehicle, refer to the Precautions Section.

2. Relieve the fuel system pressure.

3. Remove the battery.

4. Drain the engine coolant.

5. Remove or disconnect the following:
- Accelerator cable
- Air intake hose
- Ignition coil and the module wiring connectors
- Manifold Absolute Pressure (MAP) sensor
- Condenser
- Throttle Position (TP) sensor and the Idle Air Control (IAC) motor connectors

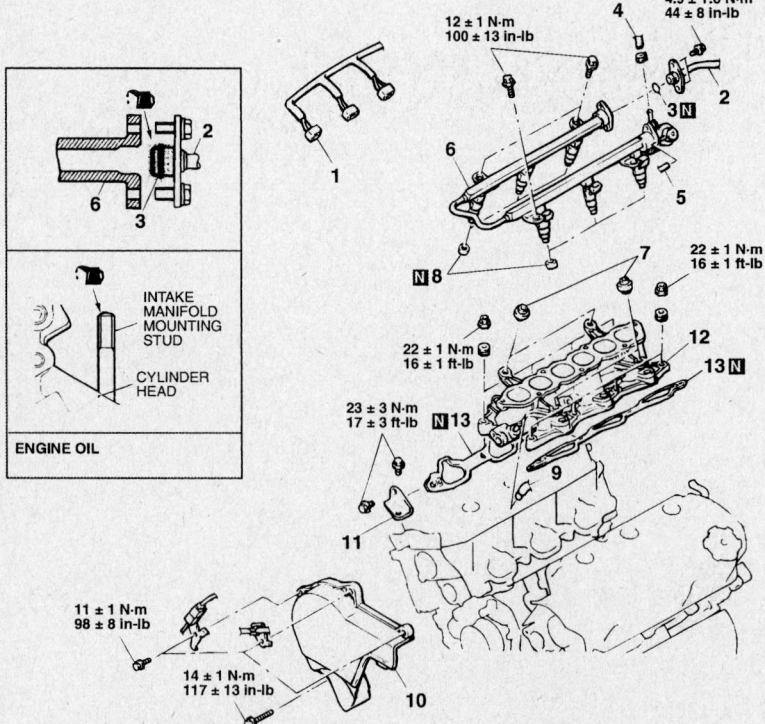

1. INJECTOR CONNECTOR
2. FUEL HIGH-PRESSURE HOSE CONNECTION
3. O-RING
4. FUEL RETURN HOSE CONNECTION
5. VACUUM HOSE CONNECTION
6. FUEL RAIL, INJECTOR AND FUEL PRESSURE REGULATOR
7. INSULATORS
8. INSULATORS
9. PCV HOSE CONNECTION
10. TIMING BELT FRONT UPPER COVER, RIGHT
11. BRACKET
12. INTAKE MANIFOLD
13. INTAKE MANIFOLD GASKET

Intake manifold and related components–2004—2005 Eclipse 3.0L engine

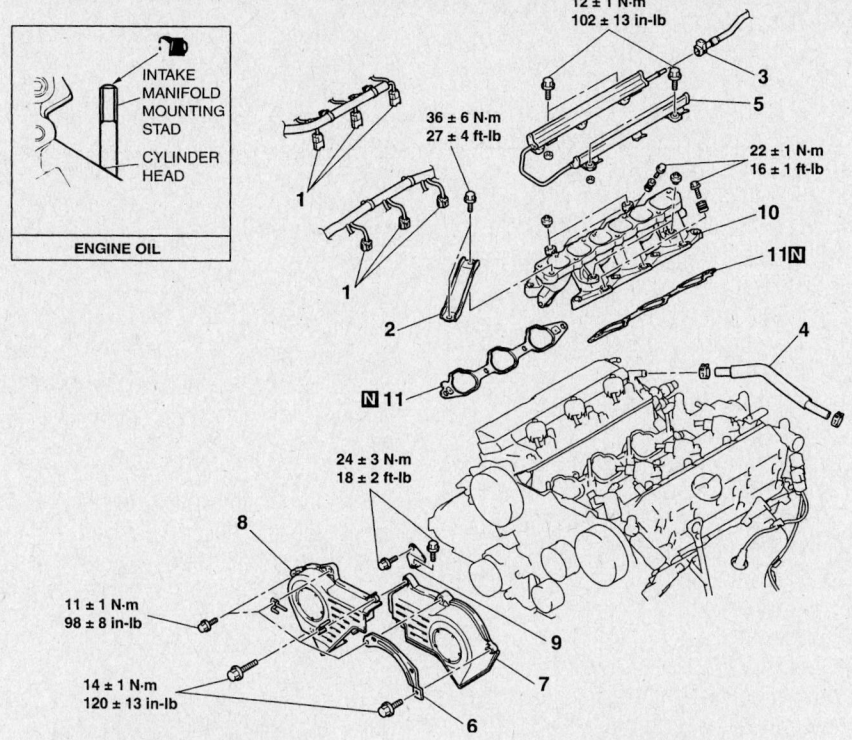

INTAKE MANIFOLD MOUNTING STAD

CYLINDER HEAD

ENGINE OIL

12 ± 1 N·m
102 ± 13 in-lb

36 ± 6 N·m
27 ± 4 ft-lb

22 ± 1 N·m
16 ± 1 ft-lb

24 ± 3 N·m
18 ± 2 ft-lb

11 ± 1 N·m
98 ± 8 in-lb

14 ± 1 N·m
120 ± 13 in-lb

1. INJECTOR CONNECTOR
2. ENGINE MOUNTING STAY
3. FUEL HIGH-PRESSURE HOSE CONNECTION
4. PCV HOSE CONNECTION
5. FUEL RAIL AND INJECTOR
6. HARNESS BRACKET
7. TIMING BELT FRONT UPPER COVER, LEFT
8. TIMING BELT FRONT UPPER COVER, RIGHT
9. WATER PUMP BRACKET
10. INTAKE MANIFOLD
11. INTAKE MANIFOLD GASKET

09482_GALA_G0056

Intake manifold and related components—2006 Eclipse and 2004—2006 Galant 3.8L engine

- Heated Oxygen (HO2S) sensor connector
- Crankshaft Position (CKP) sensor connector
- Air conditioning compressor connector
- Engine control wiring harness bracket and position the harness out of the way
- Vacuum hoses, label for reference
- Spark plug wires from the ignition coil
- Fuel lines from the fuel rail
- Heater hoses
- Fuel rail assembly
- Ignition coil and module
- Exhaust Gas Recirculation (EGR) valve assembly
- Intake manifold stay and the engine hanger
- Intake manifold

To install:

6. Install or connect the following:
- Intake manifold. Torque the intake manifold bolts to 15 ft. lbs. (20 Nm).
- Intake manifold stay and the engine hanger. Torque the mounting bolts to 19–24 ft. lbs. (26–33 Nm).
- EGR assembly
- Ignition coil and module
- Fuel rail and insulators and reconnect the high-pressure fuel hose
- Heater hoses and fuel lines
- Spark plug wires to the coil towers
- Vacuum hoses
- Engine harness in the proper position
- MAP sensor
- TP sensor and the IAC motor connectors
- HO2S
- Ignition condenser
- Accelerator cable, adjust
- Battery
7. Refill the engine with coolant.

2004–2006 2.4L (4G69) ENGINE

1. Before servicing the vehicle, refer to the Precautions Section.
2. Relieve the fuel system pressure. Disconnect the negative battery cable.
3. Remove the air cleaner assembly. Drain the cooling system. Remove the battery.

➡Do not loosen the retaining screws for the resin cover of the throttle body assembly. If these screws are loosened, the sensor incorporated in the resin cover becomes misaligned and the throttle body can not function properly.

4. Disconnect the accelerator cable connection, throttle position sensor connector and water return hose connection
5. Remove the throttle body bracket. Remove the throttle body retaining bolts. Remove the throttle body. Discard the gasket.
6. Remove the delivery pipe and injector assembly.
7. Disconnect the EVAP purge solenoid valve connector, the evaporative emission purge hose connection and the EGR valve connector.
8. Disconnect the MAP sensor connector, brake booster vacuum hose connection, pressure hose clamp and knock sensor connector bracket.
9. Disconnect the harness clamp bracket and harness clamp. Remove the engine oil dipstick and tube assembly. Disconnect the lower radiator hose. Remove the thermostat case assembly, resonator and gasket.
10. Remove the EGR valve and gasket. Remove the harness clamp bracket, EVAP purge solenoid valve and brake booster vacuum pipe.
11. Remove the brake booster vacuum hose and MAP sensor. Remove the intake manifold bracket.
12. Remove the intake manifold retaining bolts. Remove the intake manifold from the engine. Discard the gasket.

To install:

13. Be sure to use a new intake manifold gasket. Position the gasket on the engine.
14. Install the intake manifold. Torque the retaining bolts to specification.
15. When installing the thermostat case assembly be sure to apply a bead of sealant, part number 8672 or equivalent, to the cylinder head mating surface of the case assembly.

➡Be sure to assemble the components with 15 minutes of applying the sealant to the surfaces.

16. Continue the installation in the reverse order of the removal procedure.

➡Be sure to install the throttle body gasket in the proper direction. If this gasket is not installed correctly, poor engine idling may result.

17. Complete the vehicle initialization procedure.

➡ **To complete the initialization procedure the following tools are needed. MB991958 scan tool, MB991824 VCI, MB991827 MUT III USB cable, MB991910 MUT III Main harness "A"**

18. Connect the scan tool to the data link connector. To prevent damage to the scan tool be sure that the ignition switch is in the LOCK position before connecting the scan tool.

19. Turn the ignition switch to the ON position.

20. Select "check mode" from the menu screen.

21. Select "erase memory" from the menu screen.

22. Initialize the learning value.

23. After initialization complete the idle learning procedure.

➡ **This procedure must be performed when the PCM is replaced, or when the learning value is initialized, as the idling is not stabilized because the learning value in the MFI engine is not completed.**

24. Start the engine. Allow the coolant temperature to reach 176 degrees or more.

25. Stop the engine and place the ignition switch in the LOCK position.

26. After ten seconds, restart the engine.

27. For ten minutes carry out the idling procedure below to confirm that the engine has normal idling.

- Position the transaxle selector lever in the "P" range
- The engine fan is not to be operated
- The engine coolant temperature should be 176 degrees or more

➡ **If the engine stalls during idling, check the throttle valve of the throttle body for dirt. Correct and perform the procedure again.**

2002–2003 3.0L ENGINE

1. Before servicing the vehicle, refer to the Precautions Section.

2. Relieve the fuel system pressure.

3. Disconnect battery negative cable.

4. Drain the cooling system.

5. Remove or disconnect the following:
- Accelerator cable
- Air intake hose
- Coolant hose from the throttle housing
- Vacuum lines, label for reference
- High pressure fuel line and fuel return hose

- Throttle control cable brackets

6. Disconnect the following:
- Engine Coolant Temperature (ECT) sensor and gauge sender
- Idle Air Control (IAC) motor
- Exhaust Gas Recirculation (EGR) temperature sensor
- Ignition coil
- Knock (KS) sensor
- Heated Oxygen (HO2S) sensor
- Throttle Position (TP) sensor
- Distributor (if equipped)
- Air conditioning temperature sensor
- Ignition power transistor
- Fuel injectors

7. Remove or disconnect the following:
- Spark plug wires
- Intake manifold stay bracket
- Intake manifold mounting bolts and remove the intake manifold assembly

To install:

8. Clean all gasket material from the cylinder head intake mounting surface and intake manifold assembly.

9. Install or connect the following:
- Intake manifold, using a new gasket. Torque the manifold in a criss-cross pattern, starting from the inside and working outwards to 15 ft. lbs. (20 Nm).
- Fuel injectors, fuel rail and pressure regulator to the engine. Torque the retaining bolts to 48 inch lbs. (6 Nm).
- Intake manifold brace bracket and tighten bolts to 21 ft. lbs. (29 Nm)
- Spark plug wires

10. Connect the following:
- ECT sensor and gauge sender
- IAC motor
- EGR temperature sensor
- Ignition coil
- KS sensor
- HO2S sensor
- TP sensor
- Distributor (if equipped)
- Air conditioning temperature sensor
- Ignition power transistor
- Fuel injectors

11. Install or connect the following:
- Throttle control cable brackets
- High pressure fuel line and fuel return hose
- Vacuum lines
- Coolant hoses
- Accelerator cable
- Air intake hose

12. Fill the system with coolant.

13. Connect the negative battery cable.

2004–2006 3.8L ENGINE

1. Before servicing the vehicle, refer to the Precautions Section.

2. Relieve the fuel system pressure. Disconnect the negative battery cable.

3. Remove the air cleaner assembly. Drain the cooling system.

4. Remove the intake manifold plenum assembly.

5. Disconnect the injector connector wiring assembly, high pressure hose connection, O-ring and fuel return hose.

6. Disconnect the vacuum hose connection. Remove the fuel rail, injector and fuel pressure regulator. Remove the harness bracket.

7. Remove the PCV hose connection. Remove the left timing belt front upper cover. Remove the right timing belt front upper cover. Remove the water pump bracket.

8. Remove the intake manifold retaining bolts. Remove the intake manifold from the engine. Discard the gaskets.

To install:

9. Be sure to use a new intake manifold gasket. Position the gasket on the engine.

10. Install the intake manifold. Coat the retaining bolts in clean engine oil. Torque the retaining bolts to specification.

11. Continue the installation in the reverse order of the removal procedure.

Lancer

2.0L ENGINE

1. Before servicing the vehicle, refer to the Precautions Section.

2. Relieve the fuel system pressure.

3. Disconnect the negative battery cable and drain the cooling system.

4. Remove the air cleaner assembly.

5. Remove the throttle body.

➡ **If the throttle body uses a resin cover, do not loosen the retaining screws for the resin cover of the throttle body assembly. If these screws are loosened, the sensor incorporated in the resin cover becomes misaligned and the throttle body can not function properly.**

6. Remove the Exhaust Gas Recirculation (EGR) valve and gasket.

7. Remove the fuel rail and injector assembly.

8. Remove the bracket and the engine hanger.

9. Remove the Manifold Differential Pressure (MDP) sensor and O-ring.

10. Remove the intake manifold stay bracket.

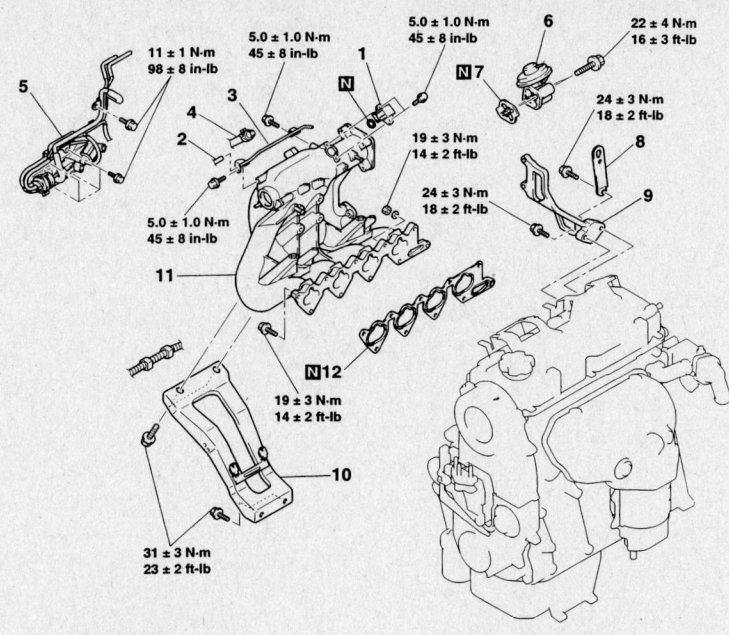

REMOVAL STEPS
1. MANIFOLD DIFFERENTIAL PRESSURE SENSOR
2. AUTO CRUISE VACUUM HOSE CONNECTION
3. VACUUM PIPE
4. BRAKE BOOSTER VACUUM HOSE CONNECTION
5. VACUUM HOSE AND PIPE ASSEMBLY

REMOVAL STEPS (Continued)
6. EGR VALVE
7. EGR VALVE GASKET
8. ENGINE HANGER
9. THROTTLE BODY STAY
10. INTAKE MANIFOLD STAY
11. INTAKE MANIFOLD
12. INTAKE MANIFOLD GASKET

9357QG13

Intake manifold and related components—2.0L engine

11. Remove the intake manifold bolt and manifold.

12. Remove the intake manifold gasket and discard.

To install:

13. Clean all gasket material from the cylinder head intake mounting surface and intake manifold assembly.

14. Install the intake manifold, using a new gasket. Torque the manifold in a criss-cross pattern, starting from the inside and working outwards to 12–16 ft. lbs. (16–22 Nm).

15. Install the intake manifold stay.

16. Install the throttle body stay.

17. Install the engine hanger.

18. Install the EGR valve and gasket.

19. Install the vacuum hose and pipe assembly.

20. Connect the brake booster vacuum hose connection.

21. Connect the vacuum pipe.

22. Connect the auto cruise vacuum hose connection.

23. Install the MDP sensor.

24. Install the fuel rail assembly.

25. Install the throttle body.

26. Install the air cleaner assembly.

27. Fill the system with coolant.

28. Connect the negative battery cable.

29. On 2006 vehicles, complete the vehicle initialization procedure.

➡**To complete the initialization procedure the following tools are needed. MB991958 scan tool, MB991824 VCI, MB991827 MUT III USB cable, MB991910 MUT III Main harness "A"**

30. Connect the scan tool to the data link connector. To prevent damage to the scan tool be sure that the ignition switch is in the LOCK position before connecting the scan tool.

31. Turn the ignition switch to the ON position.

32. Select "check mode" from the menu screen.

33. Select "erase memory" from the menu screen.

34. Initialize the learning value.

35. After initialization complete the idle learning procedure.

➡**This procedure must be performed when the PCM is replaced, or when the learning value is initialized, as the idling is not stabilized because the learning value in the MFI engine is not completed.**

36. Start the engine. Allow the coolant temperature to reach 176 degrees or more.

37. Stop the engine and place the ignition switch in the LOCK position.

38. After ten seconds, restart the engine.

39. For ten minutes carry out the idling procedure below to confirm that the engine has normal idling.

- Position the transaxle selector lever in the "P" range
- The engine fan is not to be operated
- The engine coolant temperature should be 176 degrees or more

➡**If the engine stalls during idling, check the throttle valve of the throttle body for dirt. Correct and perform the procedure again.**

2.4L ENGINE

1. Before servicing the vehicle, refer to the Precautions Section.

2. Relieve the fuel system pressure. Disconnect the negative battery cable.

3. Remove the air cleaner assembly. Drain the cooling system. Remove the battery.

➡**Do not loosen the retaining screws for the resin cover of the throttle body assembly. If these screws are loosened, the sensor incorporated in the resin cover becomes misaligned and the throttle body can not function properly.**

4. Disconnect the accelerator cable connection, throttle position sensor connector and water return hose connection

5. Remove the throttle body bracket. Remove the throttle body retaining bolts. Remove the throttle body. Discard the gasket.

6. Remove the delivery pipe and injector assembly.

7. Disconnect the EVAP purge solenoid valve connector, the evaporative emission purge hose connection and the EGR valve connector.

8. Disconnect the MAP sensor connector, brake booster vacuum hose connection, pressure hose clamp and knock sensor connector bracket.

9. Disconnect the harness clamp bracket and harness clamp. Remove the engine oil dipstick and tube assembly. Disconnect the lower radiator hose. Remove the thermostat case assembly, resonator and gasket.

10. Remove the EGR valve and gasket. Remove the harness clamp bracket, EVAP purge solenoid valve and brake booster vacuum pipe.

11. Remove the brake booster vacuum hose and MAP sensor. Remove the intake manifold bracket.

12. Remove the intake manifold retaining bolts. Remove the intake manifold from the engine. Discard the gasket.

To install:

13. Be sure to use a new intake manifold gasket. Position the gasket on the engine.

14. Install the intake manifold. Torque the retaining bolts to specification.

15. When installing the thermostat case assembly be sure to apply a bead of sealant, part number 8672 or equivalent, to the cylinder head mating surface of the case assembly.

➡ **Be sure to assemble the components with 15 minutes of applying the sealant to the surfaces.**

16. Continue the installation in the reverse order of the removal procedure.

➡ **Be sure to install the throttle body gasket in the proper direction. If this gasket is not installed correctly, poor engine idling may result.**

17. Complete the vehicle initialization procedure.

➡ **To complete the initialization procedure the following tools are needed. MB991958 scan tool, MB991824 VCI, MB991827 MUT III USB cable, MB991910 MUT III Main harness "A"**

18. Connect the scan tool to the data link connector. To prevent damage to the scan tool be sure that the ignition switch is in the LOCK position before connecting the scan tool.

19. Turn the ignition switch to the ON position.

20. Select "check mode" from the menu screen.

21. Select "erase memory" from the menu screen.

22. Initialize the learning value.

23. After initialization complete the idle learning procedure.

➡ **This procedure must be performed when the PCM is replaced, or when the learning value is initialized, as the idling is not stabilized because the learning value in the MFI engine is not completed.**

24. Start the engine. Allow the coolant temperature to reach 176 degrees or more.

25. Stop the engine and place the ignition switch in the LOCK position.

26. After ten seconds, restart the engine.

27. For ten minutes carry out the idling

procedure below to confirm that the engine has normal idling.

- Position the transaxle selector lever in the "P" range
- The engine fan is not to be operated
- The engine coolant temperature should be 176 degrees or more

➡ **If the engine stalls during idling, check the throttle valve of the throttle body for dirt. Correct and perform the procedure again.**

Evolution

1. Before servicing the vehicle, refer to the Precautions Section.

2. Relieve the fuel system pressure. Remove the engine under cover.

3. Disconnect the negative battery cable. Drain the cooling system.

4. Remove the air cleaner assembly. Remove the engine strut tower bar assembly.

5. Disconnect the accelerator cable connection, throttle position sensor connector, idle air control motor connector, vacuum hose connection and water hose connection.

6. Remove the throttle body retaining bolts. Remove the throttle body from its mounting. Discard the gasket.

7. Remove the front axle crossmember

bars. Remove the engine front roll stopper bracket bolt. Remove the front suspension crossmember.

8. Remove the front roll stopper bracket. Remove the transaxle case front roll stopper bracket.

9. Remove the rear roll stopper rod bolt. Remove the transaxle case rear roll stopper bracket.

10. Remove the engine rear roll stopper rod bolt. Remove the engine rear roll stopper rod bracket. Remove the rear roll stopper rod.

11. Remove the front exhaust pipe. Remove the engine center cover.

12. Disconnect the ignition coil connector, heated oxygen sensor (front) connector, CPS connector and knock sensor connector.

13. Disconnect the fuel injector connector, EVAP purge solenoid connector and fuel pressure solenoid connector.

14. Disconnect the MAP pressure sensor connector, vacuum hoses, fuel return hose connection and the high pressure fuel hose connection.

15. Remove the fuel rail, fuel injector, and fuel pressure regulator assembly. Remove the engine oil and dipstick holder assembly.

16. Disconnect the brake booster vacuum hose connection and the EVAP purge hose connection and the knock sensor connector.

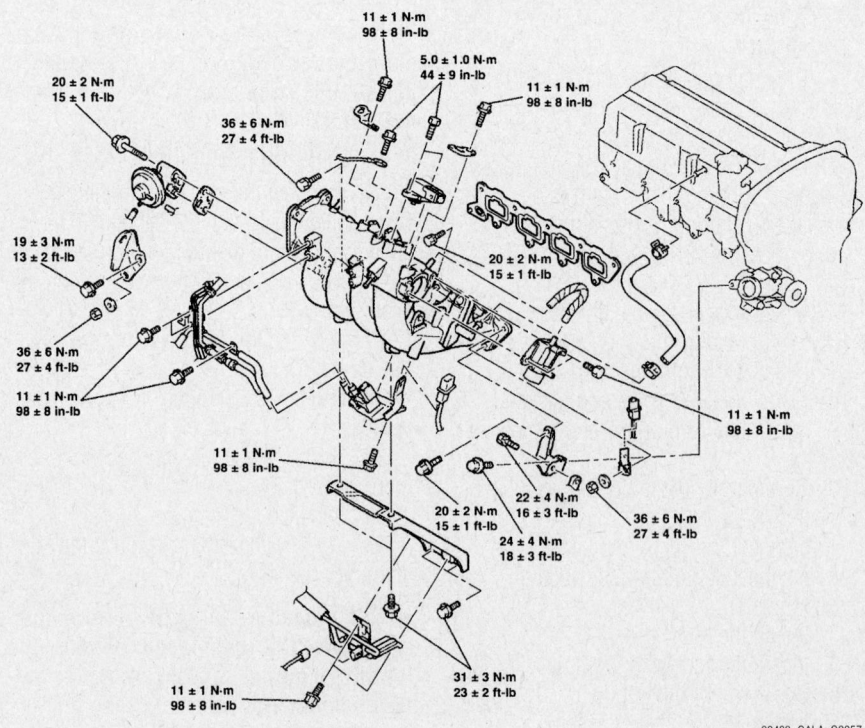

09482_GALA_G0057

Intake manifold and related components—Evolution

17. Remove the EVAP purge solenoid. Disconnect the vacuum hose, fuel pressure solenoid and PCV hose.

18. Remove the alternator. Remove the EGR vacuum regulator solenoid connector.

19. Remove the EGR vacuum regulator solenoid and vacuum pipe and hose assembly. Remove the intake manifold bracket. Remove the ERG valve and gasket. Disconnect the ground cable connection.

20. Remove the MAP pressure sensor, CPK sensor and alternator brace.

21. Remove the engine hanger and harness bracket.

22. Remove the intake manifold retaining bolts. Remove the intake manifold from the engine.

To install:

23. Be sure to use a new intake manifold gasket. Position the gasket on the engine.

24. Install the intake manifold. Coat the retaining bolts in clean engine oil. Torque the retaining bolts to specification.

25. Continue the installation in the reverse order of the removal procedure.

Mirage

1.5L ENGINE

1. Before servicing the vehicle, refer to the Precautions Section.
2. Relieve the fuel system pressure.
3. Remove or disconnect the following:
   - Battery negative cable and drain the cooling system
   - Upper radiator hose, heater hose and water bypass hose
   - Thermostat housing from intake manifold
   - Accelerator cable, breather hose and air intake hose
   - Vacuum hoses, label for reference
   - Throttle body assembly
   - High pressure fuel line and the fuel return hose
4. Disconnect the following:
   - Heated Oxygen (HO2S) sensor
   - Engine Coolant Temperature (ECT) sensor
   - Idle Air Control (IAC) motor
   - Intake Air Temperature (IAT) sensor
   - Distributor (if equipped)
   - Exhaust Gas Recirculation (EGR) temperature sensor
5. Remove or disconnect the following:
   - Spark plug wires
   - Fuel rail, fuel injectors, pressure regulator and insulators
   - EGR valve from the intake manifold
   - Intake manifold support bracket and remove the engine mount support bracket

   - Intake manifold mounting bolts and remove the intake manifold assembly

To install:

6. Clean all gasket material from the cylinder head intake mounting surface and intake manifold assembly.

7. Install or connect the following:
   - Intake manifold gasket, using a new gasket. Torque the manifold in a crisscross pattern, starting from the inside and working outwards to 13 ft. lbs. (18 Nm).
   - Intake manifold support bracket and tighten the mounting bolts to 16 ft. lbs. (22 Nm)
   - Engine mount support bracket and tighten the mounting bolts to 26 ft. lbs. (36 Nm)
   - EGR valve and tighten the mounting bolts to 15 ft. lbs. (21 Nm)
   - Install the fuel rail, fuel injectors and pressure regulator to the engine, using new insulators and O-rings. Torque the retaining bolts to 84–108 inch lbs. (10–13 Nm).
   - Spark plug wires
8. Connect the following:

   - HO2S sensor
   - ECT sensor
   - IAC motor
   - IAT sensor
   - Distributor (if equipped)
   - EGR temperature sensor
9. Install or connect the following:
   - Fuel feed and return lines
   - Throttle body assembly
   - Vacuum hoses and pipes as necessary, including the brake booster vacuum line
   - Accelerator cable
   - Breather and air intake hose
   - Thermostat housing to the intake manifold and tighten the mounting bolts to 13 ft. lbs. (18 Nm)
   - Upper radiator hose, heater hose and water bypass hose
10. Fill the system with coolant.
11. Connect the negative battery cable.

1.8L ENGINE

1. Before servicing the vehicle, refer to the Precautions Section.
2. Relieve the fuel system pressure.
3. Remove or disconnect the following:

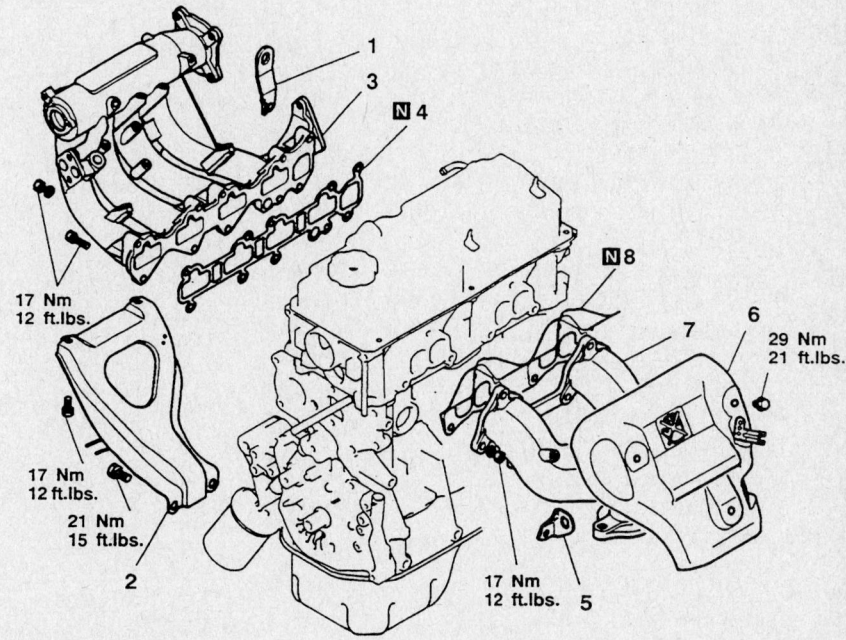

17 Nm
12 ft.lbs.

17 Nm
12 ft.lbs.

21 Nm
15 ft.lbs.

29 Nm
21 ft.lbs.

17 Nm
12 ft.lbs.

1. Engine hanger
2. Intake manifold stay
3. Intake manifold
4. Intake manifold gasket

5. Engine hanger
6. Exhaust manifold cover
7. Exhaust manifold
8. Exhaust manifold gasket

7923PG38

Intake and exhaust manifold mounting—1.5L engine

- Battery negative cable and drain the cooling system
- Accelerator cable and the air intake hose

4. Disconnect the following:
- Heated Oxygen (HO₂S) sensor
- Engine Coolant Temperature (ECT) sensor
- Idle Air Control (IAC) motor
- Exhaust Gas Recirculation (EGR) temperature sensor
- Throttle Position (TP) sensor
- Oil pressure switch
- Distributor (if equipped)
- Fuel injectors

5. Label and remove all vacuum hoses.
6. Remove or disconnect the following:
- Upper radiator hose, heater hose and water bypass hose
- High pressure fuel line and the fuel return hose
- Fuel rail, fuel injectors, pressure regulator and insulators
- Intake manifold support bracket
- Thermostat housing, if necessary for clearance
- Intake manifold mounting bolts/nuts and remove the intake manifold assembly

To install:

7. Clean all gasket material from the cylinder head intake mounting surface and intake manifold assembly.
8. Install or connect the following:
- Intake manifold, using a new gasket. Torque the manifold in a criss-cross pattern, starting from the inside and working outwards to 14 ft. lbs. (20 Nm).
- Thermostat housing
- Intake manifold brace bracket
- Fuel rail, fuel injectors and pressure regulator to the engine. Torque the retaining bolts to 108 inch lbs. (12 Nm).
- Fuel feed and return lines
- Upper radiator hose, heater hose and water bypass hoses
- Vacuum hoses

9. Connect the following:
- HO₂S sensor
- ECT sensor
- IAC motor
- EGR temperature sensor
- TP sensor
- Oil pressure switch
- Distributor (if equipped)
- Fuel injectors

10. Connect and adjust the accelerator cable and install the air intake hose.
11. Fill the system with coolant.
12. Connect the negative battery cable.

Exhaust Manifold

REMOVAL & INSTALLATION

Diamante

1. Before servicing the vehicle, refer to the Precautions Section.
2. Remove or disconnect the following:
- Battery negative cable
- Exhaust pipe from the exhaust manifold
- Condenser electric cooling fan assembly

3. If removing the front manifold, remove the oil dipstick and tube from the engine.
4. If removing the rear manifold, disconnect the Exhaust Gas Recirculation (EGR) tube.
5. If removing the rear manifold, remove the intake plenum stay and the roll stopper bracket.
6. Remove or disconnect the following:
- Electrical connector and remove the Heated Oxygen (HO₂S) sensor
- Exhaust manifold mounting bolts the manifold

To install:

7. Clean all gasket material from the mating surfaces.
8. Install or connect the following:
- New gasket and install the manifold. Tighten the nuts in a criss-cross pattern to 21 ft. lbs. (30 Nm) for the J- engine or to 14 ft. lbs. (19 Nm) for the H- engine.
- Heat shields
- EGR tube and intake plenum stay and roll stopper bracket, if removed
- HO₂S sensor
- Electric cooling fan assembly, air conditioning compressor, dipstick tube and alternator, as required
- New flange gasket and connect the exhaust pipe or converter assembly
- Negative battery cable and check for exhaust leaks

Eclipse

2.4L ENGINE

1. Before servicing the vehicle, refer to the Precautions Section.
2. Disconnect the negative battery cable.
3. Disconnect the heated oxygen sensor electrical connection. Remove the heated oxygen sensor.

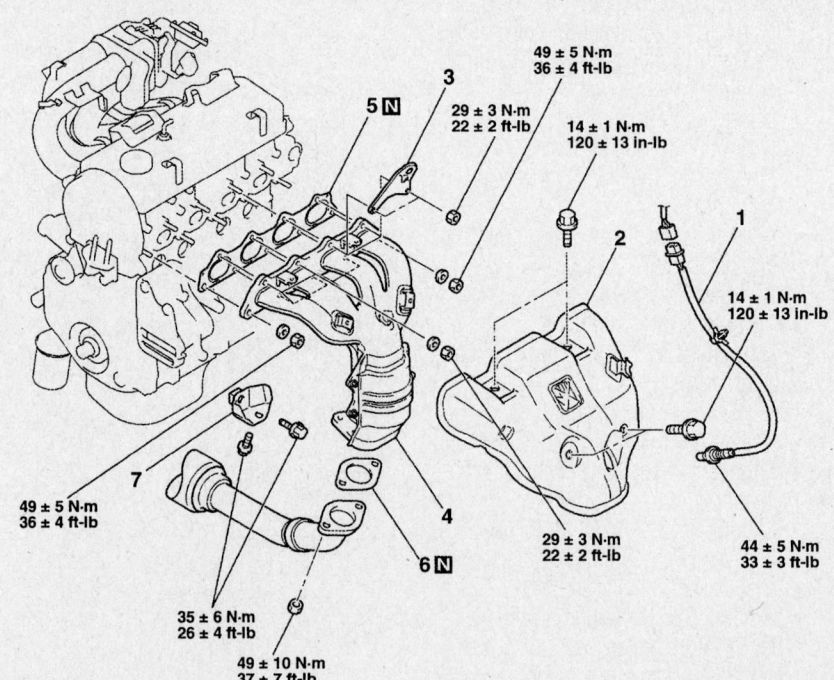

1. HEATED OXYGEN SENSOR (FRONT)
2. HEAT PROTECTOR
3. ENGINE HANGER
4. EXHAUST MANIFOLD
5. EXHAUST MANIFOLD GASKET
6. GASKET
7. EXHAUST MANIFOLD BRACKET

09482_GALA_G0058

Exhaust manifold and related components—2002–2005 Eclipse 2.4L engine

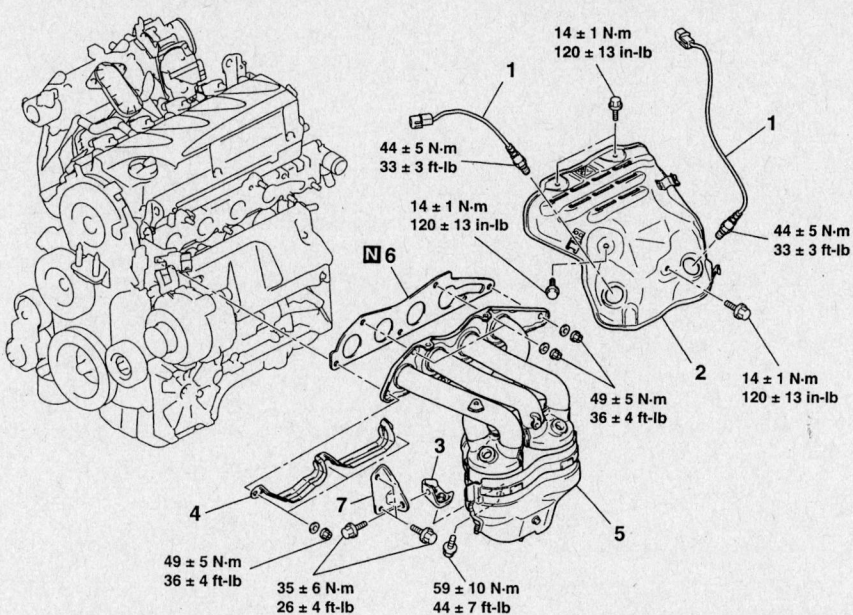

1. HEATED OXYGEN SENSOR
2. EXHAUST MANIFOLD COVER
3. EXHAUST MANIFOLD BRACKET B
4. LOWER HEAT PROTECTOR <A/T>
5. EXHAUST MANIFOLD
6. EXHAUST MANIFOLD GASKET
7. EXHAUST MANIFOLD BRACKET A

09482_GALA_G0059

Exhaust manifold and related components—2006 Eclipse 2.4L engine

4. Remove the exhaust manifold cover retaining screws. Remove the cover.

5. Remove the exhaust manifold bracket. Remove the lower heat shield.

6. Remove the exhaust manifold retaining bolts. Remove the exhaust manifold from the engine.

7. Discard the old gasket.

To install:

8. Position a new exhaust manifold gasket to the mating surface of the cylinder.

9. Install the exhaust manifold to the engine. Torque the retaining bolts to specification.

10. Continue the installation in the reverse order of the removal procedure.

2002–2003 3.0L ENGINE

1. Before servicing the vehicle, refer to the Precautions Section.

2. Disconnect the negative battery cable.

3. Remove the air cleaner assembly.

4. Remove the exhaust manifold heat protector shields.

5. Remove the Oxygen (O2S) sensor.

6. Remove the exhaust manifold mounting nuts and exhaust manifold(s).

7. Remove the exhaust manifold gasket(s), and discard.

8. Remove the engine hanger.

To install:

9. Install the engine hanger.

10. Clean all gasket material from the mating surfaces.

11. Install a new gasket and the exhaust manifold. Torque the manifold nuts in a crisscross pattern, starting from the inside and working outwards to 29–37 ft. lbs. (39–49 Nm).

12. Install the O2S sensor. Tighten to 29–37 ft. lbs. (39–49 Nm).

13. Install the exhaust manifold heat protector shields.

14. Install the air cleaner assembly.

15. Connect the negative battery cable.

2004–2005 3.0L ENGINE

1. Before servicing the vehicle, refer to the Precautions Section.

2. Disconnect the negative battery cable. Remove the battery and battery tray.

3. Disconnect the front exhaust pipe. Remove the strut tower bar assembly.

4. Remove the air cleaner assembly. Remove the engine oil dipstick and dipstick tube assembly.

5. Remove the upper heat shield protector.

6. On the left side remove the front

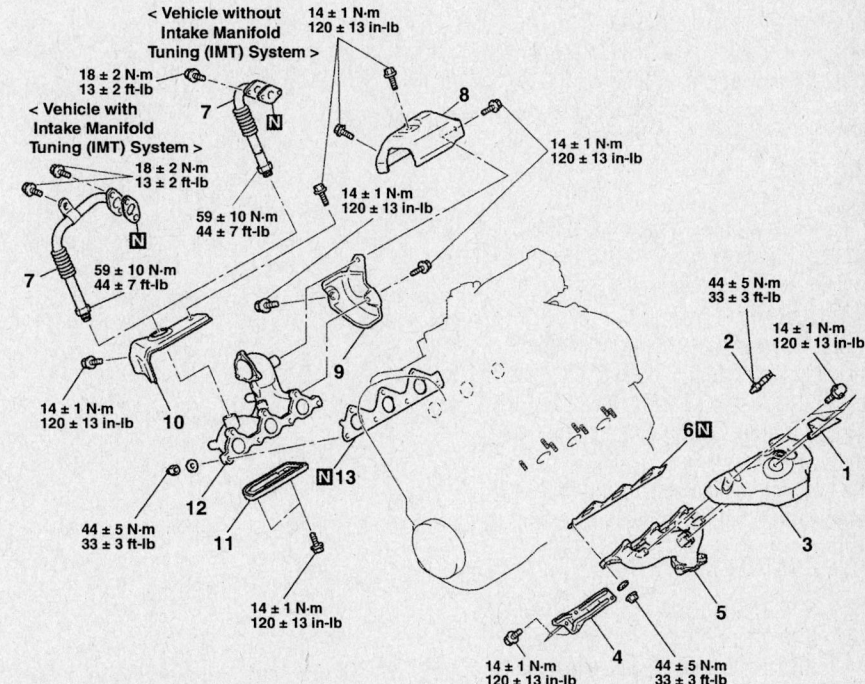

1. HEAT UPPER PROTECTOR <LEFT BANK>
2. LEFT BANK HEATED OXYGEN SENSOR (FRONT)
3. HEAT PROTECTOR <LEFT BANK>
4. HEAT LOWER PROTECTOR <LEFT BANK>
5. EXHAUST MANIFOLD <LEFT BANK>

09482_GALA_G0060

Exhaust manifold and related components—2004–2005 Eclipse 3.0L engine

heated oxygen sensor, heat protector shield, and lower heat protector shield.

7. Remove the left exhaust manifold retaining bolts. Remove the manifold from the engine. Discard the gasket.

8. Remove the intake manifold plenum, if equipped with IMT system. Remove the EGR pipe.

9. On the right side remove the upper heat protector shield, front heat protector shield, and lower heat protector shield.

10. Remove the right exhaust manifold retaining bolts. Remove the manifold from the engine. Discard the gasket.

To install:

11. Position a new exhaust manifold gasket to the mating surface of the cylinder.

12. Install the exhaust manifold to the engine. Torque the retaining bolts to specification.

13. Continue the installation in the reverse order of the removal procedure.

3.8L ENGINE (LEFT SIDE)

1. Before servicing the vehicle, refer to the Precautions Section.

2. Disconnect the negative battery cable.

3. Remove the engine under cover. Remove the air cleaner intake duct assembly.

4. Remove the front heated oxygen sensor electrical connector. Remove the sensor.

5. Remove the rear heated oxygen sensor electrical connector. Remove the sensor.

6. Remove the heat protector shield.

7. Remove the front exhaust pipe connecting bolts. Remove and discard the gasket. Remove the exhaust manifold bracket.

8. Remove the exhaust manifold retaining bolts. Remove the manifold from the engine. Discard the gasket.

To install:

9. Position a new exhaust manifold gasket to the mating surface of the cylinder.

10. Install the exhaust manifold to the engine. Torque the retaining bolts to specification.

11. Continue the installation in the reverse order of the removal procedure.

3.8L ENGINE (RIGHT SIDE)

1. Before servicing the vehicle, refer to the Precautions Section.

2. Disconnect the negative battery cable. Remove the battery and battery tray.

3. Remove the engine under cover. Remove the air cleaner intake duct assembly.

4. Remove the front exhaust pipe. Remove the center exhaust pipe.

5. Remove the strut tower bar assembly. Drain the engine coolant.

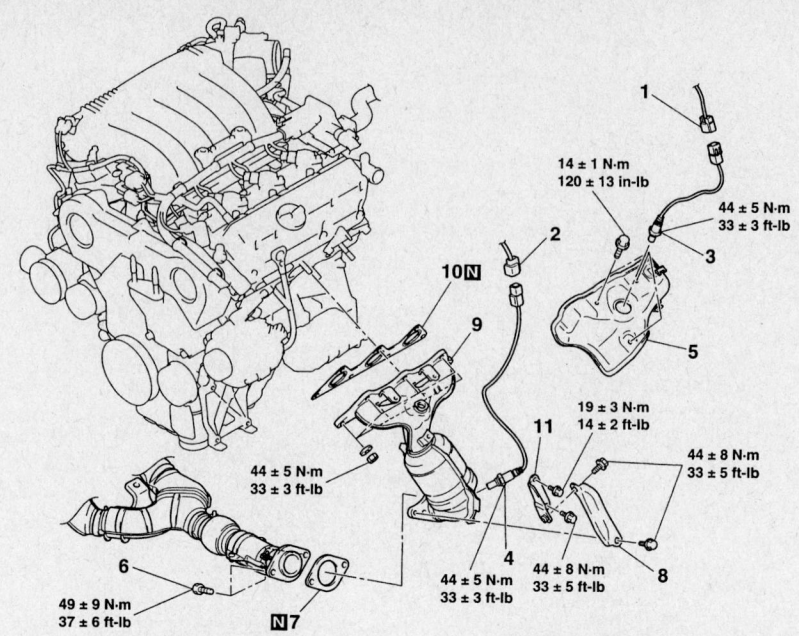

1. LEFT BANK HEATED OXYGEN SENSOR (FRONT) CONNECTOR	6. FRONT EXHAUST PIPE CONNECTING BOLTS
2. LEFT BANK HEATED OXYGEN SENSOR (REAR) CONNECTOR	7. FRONT EXHAUST PIPE GASKET
3. LEFT BANK HEATED OXYGEN SENSOR (FRONT)	8. EXHAUST MANIFOLD STAY, LEFT B
4. LEFT BANK HEATED OXYGEN SENSOR (REAR)	9. EXHAUST MANIFOLD
5. HEAT PROTECTOR	10. EXHAUST MANIFOLD GASKET
	11. EXHAUST MANIFOLD STAY, LEFT A

09482_GALA_G0061

Exhaust manifold and related components—2006 Eclipse and 2004–2006 Galant 3.8L engine (left side)

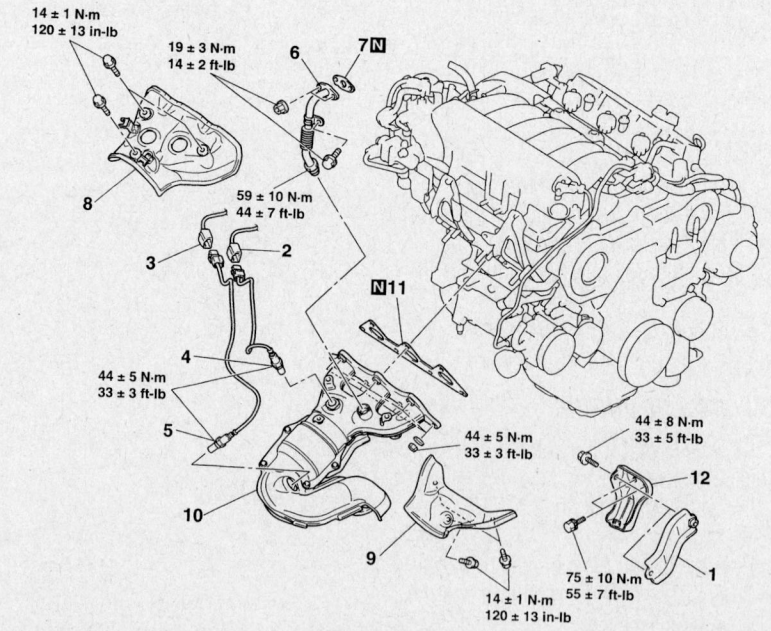

1. EXHAUST MANIFOLD STAY, RIGHT B	4. RIGHT BANK HEATED OXYGEN SENSOR (FRONT)	9. LOWER HEAT PROTECTOR
2. RIGHT BANK HEATED OXYGEN SENSOR (FRONT) CONNECTOR	5. RIGHT BANK HEATED OXYGEN SENSOR (REAR)	10. EXHAUST MANIFOLD
3. RIGHT BANK HEATED OXYGEN SENSOR (REAR) CONNECTOR	6. EGR PIPE	11. EXHAUST MANIFOLD GASKET
	7. EGR PIPE GASKET	12. EXHAUST MANIFOLD STAY, RIGHT A
	8. UPPER HEAT PROTECTOR	

09482_GALA_G0062

Exhaust manifold and related components—2006 Eclipse 3.8L engine (right side)

6. Remove the steering gear and linkage protector.

7. Remove the center under floor heat protector shield. Remove the exhaust manifold bracket.

8. Remove the front heated oxygen sensor electrical connector. Remove the sensor.

9. Remove the rear heated oxygen sensor electrical connector. Remove the sensor.

10. Remove the EGR pipe and gasket. Remove the upper and lower heat shield protector.

11. Remove the exhaust manifold retaining bolts. Remove the manifold from the engine. Discard the gasket.

To install:

12. Position a new exhaust manifold gasket to the mating surface of the cylinder.

13. Install the exhaust manifold to the engine. Torque the retaining bolts to specification.

14. Continue the installation in the reverse order of the removal procedure.

Galant

2.4L ENGINE

1. Before servicing the vehicle, refer to the Precautions Section.

2. Disconnect the negative battery cable.

3. Disconnect the heated oxygen sensor electrical connection. Remove the heated oxygen sensor.

4. Remove the exhaust manifold cover retaining screws. Remove the cover.

5. Remove the exhaust manifold bracket. Remove the lower heat shield.

6. Remove the exhaust manifold retaining bolts. Remove the exhaust manifold from the engine.

7. Discard the old gasket.

To install:

8. Position a new exhaust manifold gasket to the mating surface of the cylinder.

9. Install the exhaust manifold to the engine. Torque the retaining bolts to specification.

10. Continue the installation in the reverse order of the removal procedure.

2002–2003 3.0L ENGINE

1. Before servicing the vehicle, refer to the Precautions Section.

2. Disconnect the negative battery cable.

3. Remove the air cleaner assembly.

4. Remove the exhaust manifold heat protector shields.

5. Remove the Oxygen sensor.

6. Remove the exhaust manifold mounting nuts and exhaust manifold(s).

7. Remove the exhaust manifold gasket(s), and discard.

8. Remove the engine hanger.

To install:

9. Install the engine hanger.

10. Clean all gasket material from the mating surfaces.

11. Install a new gasket and the exhaust manifold. Torque the manifold nuts in a crisscross pattern, starting from the inside and working outwards to 29–37 ft. lbs. (39–49 Nm).

12. Install the oxygen sensor. Tighten to 29–37 ft. lbs. (39–49 Nm).

13. Install the exhaust manifold heat protector shields.

14. Install the air cleaner assembly.

15. Connect the negative battery cable.

2004–2006 3.8L ENGINE (LEFT SIDE)

1. Before servicing the vehicle, refer to the Precautions Section.

2. Disconnect the negative battery cable.

3. Remove the engine under cover. Remove the air cleaner intake duct assembly.

4. Remove the front heated oxygen sensor electrical connector. Remove the sensor.

5. Remove the rear heated oxygen sensor electrical connector. Remove the sensor.

6. Remove the heat protector shield.

7. Remove the front exhaust pipe connecting bolts. Remove and discard the gasket. Remove the exhaust manifold bracket.

8. Remove the exhaust manifold retaining bolts. Remove the manifold from the engine. Discard the gasket.

To install:

9. Position a new exhaust manifold gasket to the mating surface of the cylinder.

10. Install the exhaust manifold to the engine. Torque the retaining bolts to specification.

11. Continue the installation in the reverse order of the removal procedure.

2004–2006 3.8L ENGINE (RIGHT SIDE)

1. Before servicing the vehicle, refer to the Precautions Section.

2. Disconnect the negative battery cable. Remove the battery and battery tray.

3. Remove the engine under cover. Remove the air cleaner intake duct assembly.

4. Remove the front exhaust pipe. Remove the center exhaust pipe.

5. Remove the strut tower bar assembly. Drain the engine coolant.

6. Remove the steering gear and linkage protector.

7. Remove the center under floor heat protector shield. Remove the exhaust manifold bracket.

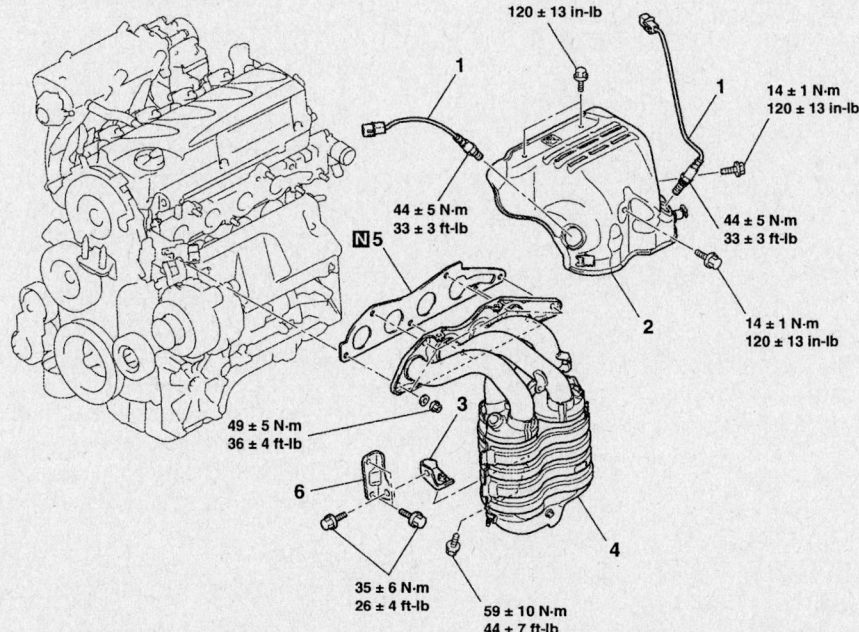

14 ± 1 N·m
120 ± 13 in-lb

14 ± 1 N·m
120 ± 13 in-lb

44 ± 5 N·m
33 ± 3 ft-lb

44 ± 5 N·m
33 ± 3 ft-lb

14 ± 1 N·m
120 ± 13 in-lb

49 ± 5 N·m
36 ± 4 ft-lb

35 ± 6 N·m
26 ± 4 ft-lb

59 ± 10 N·m
44 ± 7 ft-lb

1. HEATED OXYGEN SENSOR
2. EXHAUST MANIFOLD COVER
3. EXHAUST MANIFOLD BRACKET B
4. EXHAUST MANIFOLD
5. EXHAUST MANIFOLD GASKET
6. EXHAUST MANIFOLD BRACKET A

09482_GALA_G0063

Exhaust manifold and related components—2004–2006 Galant 2.4L engine

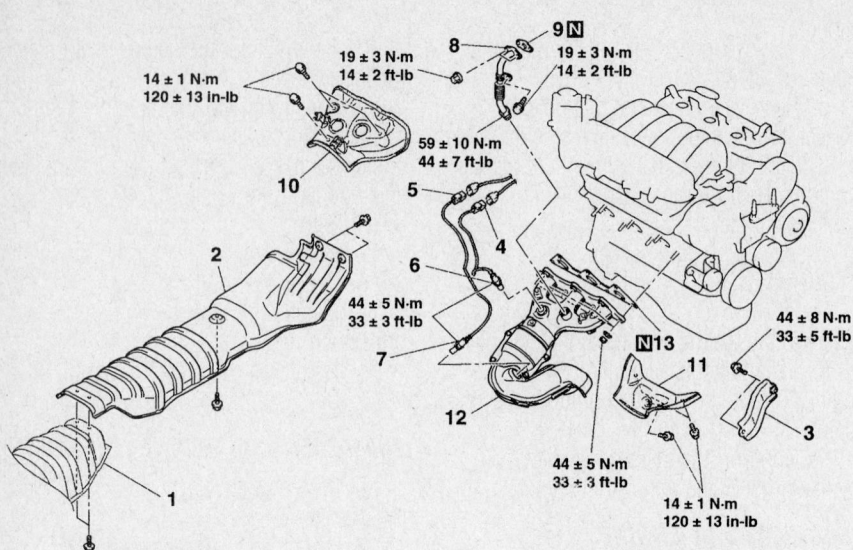

14 ± 1 N·m
120 ± 13 in-lb

19 ± 3 N·m
14 ± 2 ft-lb

19 ± 3 N·m
14 ± 2 ft-lb

59 ± 10 N·m
44 ± 7 ft-lb

44 ± 5 N·m
33 ± 3 ft-lb

44 ± 8 N·m
33 ± 5 ft-lb

44 ± 5 N·m
33 ± 3 ft-lb

14 ± 1 N·m
120 ± 13 in-lb

1. CENTER UNDER FLOOR HEAT PROTECTOR
2. FRONT UNDER FLOOR HEAT PROTECTOR
3. EXHAUST MANIFOLD STAY, RIGHT B
4. RIGHT BANK HEATED OXYGEN SENSOR (FRONT) CONNECTOR
5. RIGHT BANK HEATED OXYGEN SENSOR (REAR) CONNECTOR
6. RIGHT BANK HEATED OXYGEN SENSOR (FRONT)
7. RIGHT BANK HEATED OXYGEN SENSOR (REAR)
8. EGR PIPE
9. EGR PIPE GASKET
10. UPPER HEAT PROTECTOR
11. LOWER HEAT PROTECTOR
12. EXHAUST MANIFOLD
13. EXHAUST MANIFOLD GASKET

09482_GALA_G0064

Exhaust manifold and related components—2004–2006 Galant 3.8L engine (right side)

8. Remove the front heated oxygen sensor electrical connector. Remove the sensor.

9. Remove the rear heated oxygen sensor electrical connector. Remove the sensor.

10. Remove the EGR pipe and gasket. Remove the upper and lower heat shield protector.

11. Remove the exhaust manifold retaining bolts. Remove the manifold from the engine. Discard the gasket.

To install:

12. Position a new exhaust manifold gasket to the mating surface of the cylinder.

13. Install the exhaust manifold to the engine. Torque the retaining bolts to specification.

14. Continue the installation in the reverse order of the removal procedure.

Lancer

2.0L ENGINE

1. Before servicing the vehicle, refer to the Precautions Section.

2. Remove or disconnect the following:
- Negative battery cable
- Engine undercover
- Pressure hose clamp bolt
- Power steering pump bracket stay bolt
- Drive belt
- Power steering pump and bracket assembly and position it aside. Do NOT disconnect the fluid lines.
- Front exhaust pipe and gasket from the manifold. Discard the gasket.
- Exhaust manifold bracket "B" (see illustration)
- Heated Oxygen (HO2S) sensor
- Heat shield
- Exhaust manifold retainers and manifold
- Exhaust manifold gasket, and discard
- Exhaust manifold bracket "A" (see illustration)

To install:

3. Clean all gasket material from the mating surfaces.

4. Install or connect the following:
- Exhaust manifold bracket "A"
- New gasket and exhaust manifold. Tighten the upper retainers to 20–24 ft. lbs. (27–33 Nm) and the lower retainers to 10–14 ft. lbs. (15–19 Nm).
- Heat shield
- HO2S
- Exhaust manifold bracket "B"
- Front exhaust pipe to the exhaust manifold, using a new gasket

- Power steering pump assembly
- Drive belt
- Power steering pump bracket stay bolt
- Pressure hose clamp bolt
- Engine undercover
- Negative battery cable

2.4L ENGINE

1. Before servicing the vehicle, refer to the Precautions Section.

2. Disconnect the negative battery cable.

3. Remove the air cleaner assembly.

4. Remove the engine hanger.

5. Remove the exhaust manifold cover.

6. Remove the exhaust manifold mounting nuts and exhaust manifold.

7. Remove the exhaust manifold gasket, and discard.

To install:

8. Clean all gasket material from the mating surfaces.

9. Install a new gasket and the exhaust manifold. Torque the manifold nuts to specification.

10. Install the exhaust manifold cover.

11. Install the engine hanger.

12. Install the air cleaner assembly.

13. Connect the negative battery cable.

Evolution

The removal and installation procedure for the exhaust manifold is a part of the Turbocharger procedure, located earlier in this section.

Mirage

1. Before servicing the vehicle, refer to the Precautions Section.

2. Remove or disconnect the following:
- Battery negative cable
- Exhaust pipe from the exhaust manifold
- Electric cooling fan assembly
- Heated Oxygen (HO2S) sensor
- Exhaust Gas Recirculation (EGR) pipe
- Outer exhaust manifold heat shield and engine hanger
- Exhaust manifold mounting bolts, the inner heat shield and the exhaust manifold

To install:

3. Clean all gasket material from the mating surfaces.

4. Using a new gasket and install the manifold. For 1.5L engines, tighten the nuts on a crisscross patter to 13 ft. lbs. (18 Nm). For 1.8L engines, tighten the inner nuts to in a crisscross pattern to 13 ft. lbs. (18 Nm)

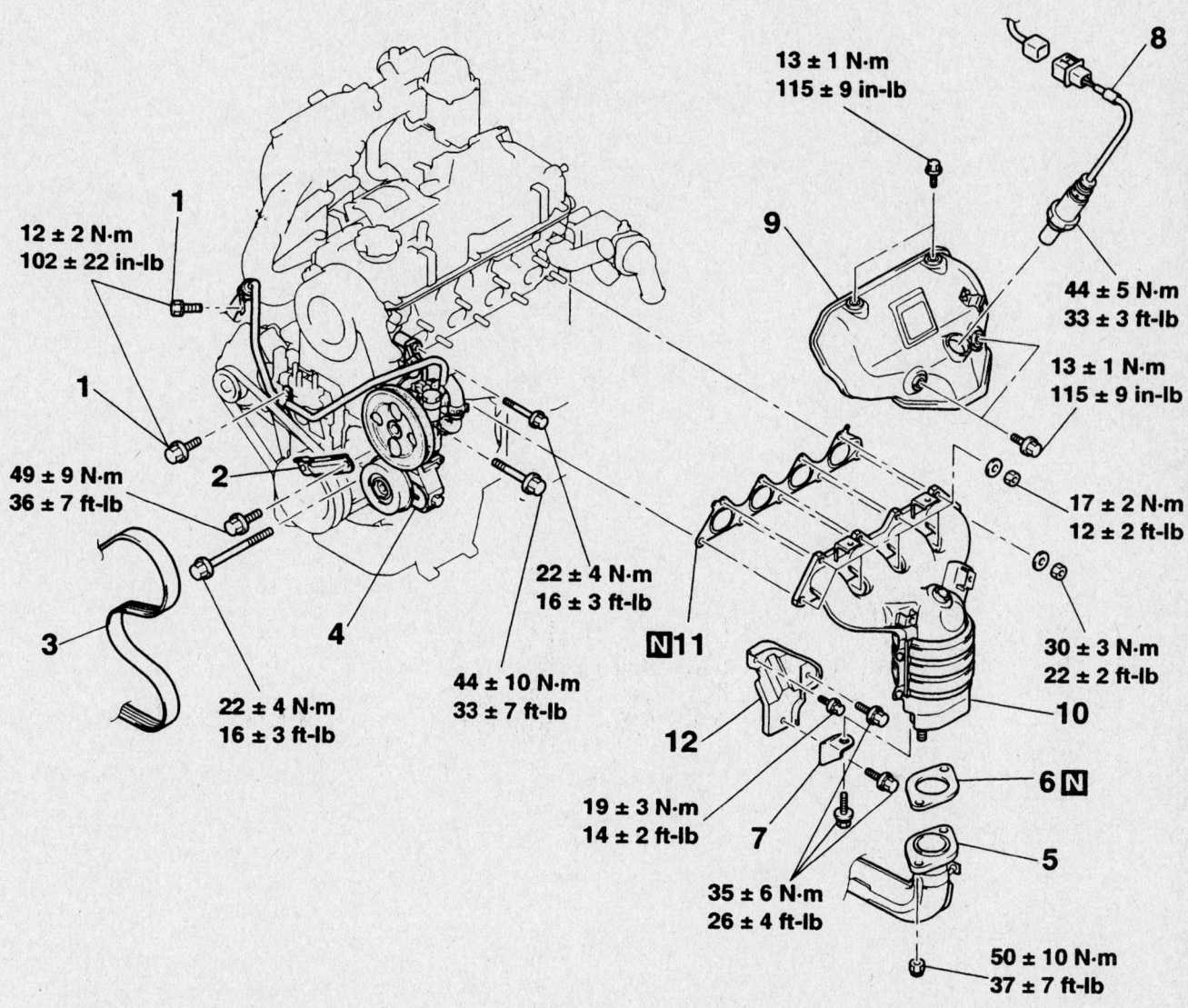

1. PRESSURE HOSE CLAMP BOLT
2. POWER STEERING PUMP
   BRACKET STAY BOLT
3. POWER STEERING PUMP AND A/C
   COMPRESSOR DRIVE BELT
4. POWER STEERING OIL PUMP AND
   BRACKET ASSEMBLY
5. FRONT EXHAUST PIPE
   CONNECTION

6. FRONT EXHAUST PIPE GASKET
7. EXHAUST MANIFOLD BRACKET B
8. HEATED OXYGEN SENSOR
9. HEAT PROTECTOR
10. EXHAUST MANIFOLD
11. EXHAUST MANIFOLD GASKET
12. EXHAUST MANIFOLD BRACKET A

9357QG14

Exhaust manifold and related components—2.0L engine

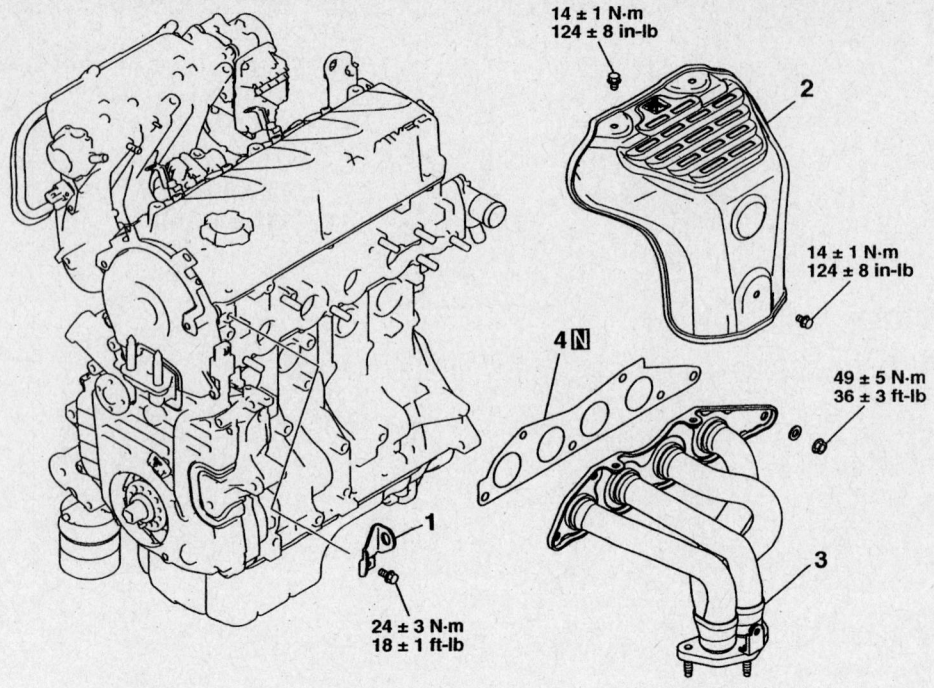

14 ± 1 N·m
124 ± 8 in-lb

14 ± 1 N·m
124 ± 8 in-lb

49 ± 5 N·m
36 ± 3 ft-lb

24 ± 3 N·m
18 ± 1 ft-lb

1. ENGINE HANGER
2. EXHAUST MANIFOLD COVER

3. EXHAUST MANIFOLD
4. EXHAUST MANIFOLD GASKET

67170-GALA-G31

Exhaust manifold and related components—2004–2006 Lancer 2.4L engine (except California)

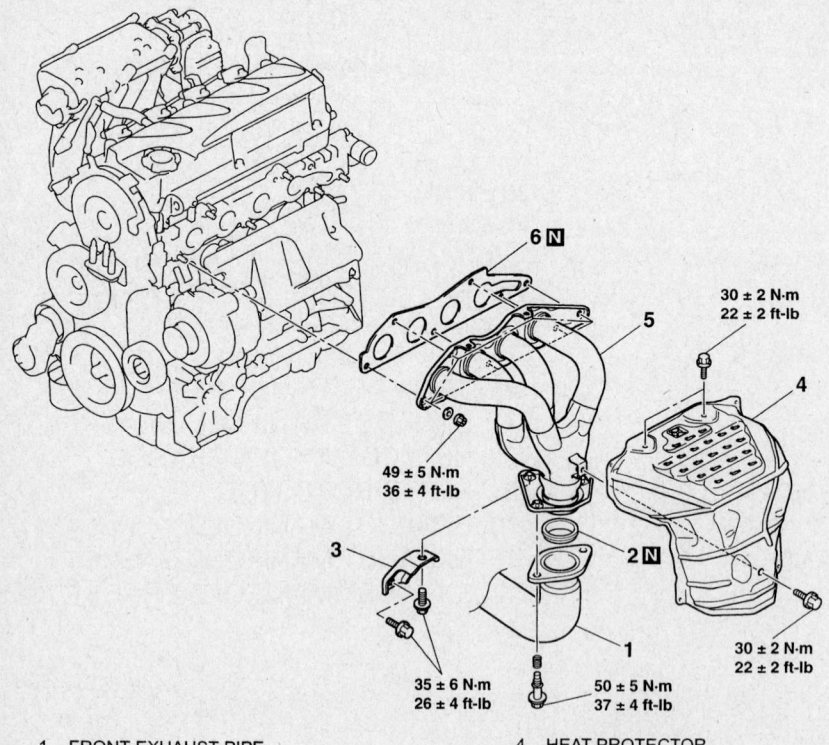

30 ± 2 N·m
22 ± 2 ft-lb

49 ± 5 N·m
36 ± 4 ft-lb

35 ± 6 N·m
26 ± 4 ft-lb

50 ± 5 N·m
37 ± 4 ft-lb

30 ± 2 N·m
22 ± 2 ft-lb

1. FRONT EXHAUST PIPE
   CONNECTION
2. SEAL RING
3. EXHAUST MANIFOLD BRACKET

4. HEAT PROTECTOR
5. EXHAUST MANIFOLD
6. EXHAUST MANIFOLD GASKET

09482_GALA_G0065

Exhaust manifold and related components—2005–2006 Lancer 2.4L engine (California)

and tighten the 2 outer (larger) nuts to 22 ft. lbs. (30 Nm).

5. Install or connect the following:
   - Heat shields
   - EGR pipe
   - HO$_2$S sensor
   - Electric cooling fan assembly
   - New flange gasket and connect the exhaust pipe
   - Negative battery cable and check for exhaust leaks

Front Crankshaft Seal

REMOVAL & INSTALLATION

1.5L and 1.8L Engines

1. Before servicing the vehicle, refer to the Precautions Section.
2. Remove or disconnect the following:
   - Negative battery cable
   - Crankshaft pulley retainer bolts and remove the pulley
   - Vibration damper retainer bolt and washer and remove damper
   - Timing belt
   - Crankshaft sprocket
3. Pry out the oil seal from front of engine.

To install:

4. Using proper size driver, install a new front seal.

5. Lubricate the lips of the new seal with clean engine oil.

6. Install or connect the following:
- Timing belt, timing covers, valve cover and remaining components
- Crankshaft sprocket and vibration damper
- Engine undercover and connect the negative battery cable

2.0L SOHC Engine

1. Before servicing the vehicle, refer to the Precautions Section.

2. Remove or disconnect the following:
- Negative battery cable
- Timing belt
- Crank angle sensor
- Crankshaft sprocket
- Spring pin
- Crankshaft sending blade
- Crankshaft spacer
- Crankshaft front oil seal

To install:

3. Install or connect the following:
- Engine oil to the oil seal lip
- Crankshaft front using seal into the front case, using special driver tool MD998717

4. Remove all oil or other lubricants from the mounting surfaces on the crankshaft, spacer, sensing blade and crankshaft sprocket.

5. Assemble the spring pin, sensing blade, and crankshaft spacer together.

6. Install or connect the following:
- Crankshaft sprocket assembly onto the crankshaft
- Crank angle sensor
- Timing belt
- Negative battery cable

To install:

2.0L DOHC Engine

1. Before servicing the vehicle, refer to the Precautions Section.

2. Disconnect the negative battery cable.

3. Remove the timing belt.

4. Remove the crankshaft balancer shaft drive sprocket.

5. Remove the crankshaft (woodruff) key.

6. Carefully pry the oil seal out of the front case. Be careful not to damage the oil seal bore or the crankshaft sealing surface.

To install:

7. Apply clean engine oil to the oil seal lip. Using a seal driver, install the oil seal.

8. Install the woodruff key.

9. Install the crankshaft balancer shaft drive sprocket.

10. Install the timing belt.

11. Connect the negative battery cable.

2.4L Engine

1. Before servicing the vehicle, refer to the Precautions Section.

2. Remove or disconnect the following:
- Negative battery cable
- Timing belt
- Crankshaft sprocket

3. Carefully pry the oil seal out of the front case. Be careful not to damage the oil seal bore or the crankshaft sealing surface.

To install:

4. Apply clean engine oil to the oil seal lip. Using a seal driver, install the oil seal.

5. Install or connect the following:
- Crankshaft sprocket. If equipped, tighten the crankshaft bolt to 87 ft. lbs. (118 Nm).
- Timing belt
- Negative battery cable

3.0L and 3.8L Engines

1. Before servicing the vehicle, refer to the Precautions Section.

2. Disconnect the negative battery cable.

3. Remove the timing belt.

4. Remove the crankshaft sprocket.

5. Remove the Crankshaft Position (CKP) sensor.

6. Remove the crankshaft sensing blade.

7. Remove the crankshaft spacer and woodruff key.

8. Carefully pry the oil seal out of the front case. Be careful not to damage the oil seal bore or the crankshaft sealing surface.

To install:

9. Apply clean engine oil to the oil seal lip. Using a seal driver, install the oil seal.

10. Install the woodruff key and crankshaft spacer.

11. Install the crankshaft sensing blade.

12. Install the CKP sensor.

13. Install the crankshaft sprocket.

14. Install the timing belt.

15. Connect the negative battery cable.

3.5L Engine

1. Before servicing the vehicle, refer to the Precautions Section.

2. Remove or disconnect the following:
- Negative battery cable
- Drive belts
- Crankshaft pulley

- Timing belt covers and the timing belt
- Crankshaft Position (CKP) sensor
- Crankshaft sprocket, the sensing blade, spacer and Woodruff® key

3. Pry the seal from the bore, using a suitable tool.

To install:

4. Using a seal driver, install the new crankshaft seal. Lubricate the lips of the seal with clean engine oil.

5. Install or connect the following:
- Woodruff® key, spacer, sensing blade and the crankshaft sprocket
- CKP sensor and tighten the retaining bolts to 84 inch lbs. (9 Nm)
- Timing belt and the timing belt covers
- Crankshaft pulley and retaining bolt. Torque the retaining bolt to 130–137 ft. lbs. (180–190 Nm) for the DOHC engine or to 108–116 ft. lbs. (150–160 Nm) for the SOHC engine.
- Drive belts, adjust
- Negative battery cable

Camshaft and Valve Lifters

REMOVAL & INSTALLATION

Diamante

3.5L ENGINE

1. Before servicing the vehicle, refer to the Precautions Section.

2. Remove or disconnect the following:
- Negative battery cable
- Timing belt
- Rocker arm cover
- Lash adjuster clips on the rocker arms, then loosen the bearing cap bolts. Do not remove the bolts from the bearing caps.
- Rocker arms, shafts and bearing caps as an assembly
- Camshafts

To install:

3. Lubricate the camshafts with engine oil and position them on the cylinder heads.

4. Position the dowel pins as shown in the drawing.

5. Install or connect the following:
- Bearing caps/rocker arm assemblies. Tighten the bolts to 23 ft. lbs. (31 Nm).
- Rocker arm cover using a new gasket
- Timing belt and remaining components
- Negative battery cable

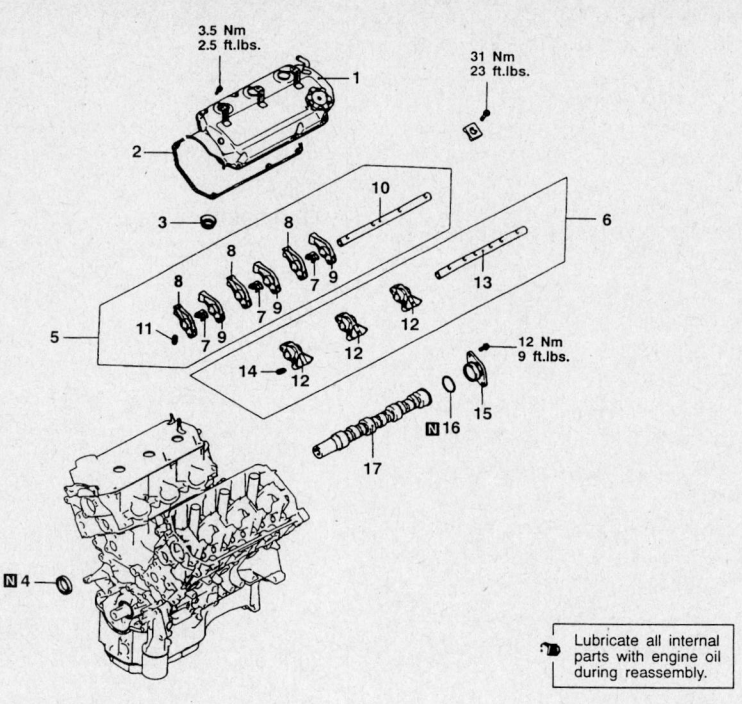

Removal steps
1. Rocker cover
2. Rocker cover gasket
3. Oil seal
4. Camshaft oil seal
5. Rocker arm, rocker arm shaft
6. Rocker arm, rocker arm shaft
7. Rocker shaft spring
8. Rocker arm A
9. Rocker arm B
10. Rocker arm shaft
11. Lash adjuster
12. Rocker arm C
13. Rocker arm shaft
14. Lash adjuster
15. Thrust case
16. O ring
17. Camshaft

7923PGD3

Camshaft and related components—3.5L engine

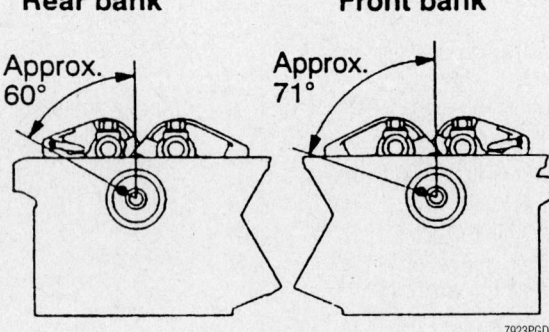

7923PGD4

Camshaft dowel position during installation—3.5L engine

Eclipse

2002–2003 2.4L ENGINE

1. Before servicing the vehicle, refer to the Precautions Section.
2. Remove or disconnect the following:
   - Remove the battery
   - Accelerator cable bracket and position the cable aside
   - Air intake hose
   - Breather hose and disconnect the Positive Crankcase Ventilation (PCV) hose
   - Spark plug cables
   - Rocker cover
3. Install lash adjuster retainer tools to the rocker arm.
   - Timing belt covers and the timing belt
   - Camshaft sprocket retainer bolt

and remove the sprocket from the shaft
   - Camshaft oil seal
   - Both rocker arm shaft assemblies from the head
   - Camshaft from the cylinder head

To install:

4. Lubricate the camshaft journals and camshaft with clean engine oil and install the camshaft in the cylinder head.
5. Install the rocker arm and shaft assemblies. Tighten the rocker arm shaft retainer bolts to 21–25 ft. lbs. (29–35 Nm).
6. Apply a coating of engine oil to the oil seal. Using the proper size driver, press-fit the seal into the cylinder head.
7. Install or connect the following:
   - Camshaft sprocket and retainer bolt to 65 ft. lbs. (90 Nm)
   - Timing belt and belt covers
8. Remove the lash adjuster retaining tools.
9. Install or connect the following:
   - Rocker cover using new gasket material on mating surfaces
   - Spark plug cables
   - Air intake hose
   - Breather hose and connect the PCV hose
   - Battery
10. Run the engine at idle until normal operating temperature is reached. Check idle speed and ignition timing; adjust as required.

2004–2005 2.4L ENGINE

1. Before servicing the vehicle, refer to the Precautions Section.
2. Disconnect the negative battery cable. Remove the air cleaner assembly.
3. Remove the timing belt.
4. Disconnect the ignition coil connector, spark plug cable, PCV hose and breather hose.
5. Remove the ignition coil. Disconnect the control harness connection.
6. Remove the rocker arm cover retaining bolts. Remove the rocker arm cover. Discard the gasket.
7. Remove the camshaft position sensor support and camshaft position sensing cylinder.
8. Use tool MB990767 to hold the camshaft sprocket in place. Remove the camshaft sprocket retaining bolt. Remove the camshaft sprocket. Remove the spark plug guide oil seal.
9. Remove the intake rocker arm assembly. Remove the exhaust rocker arm assembly.
10. Remove the camshaft from its mounting on the engine.

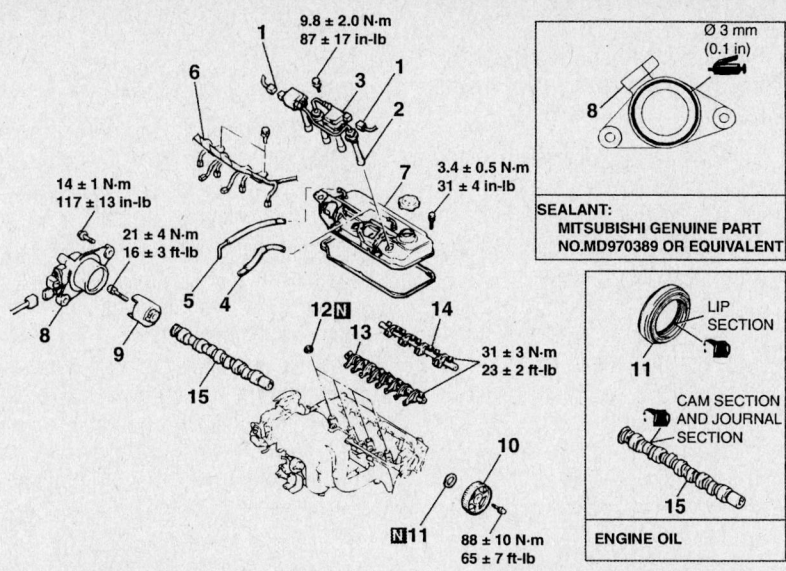

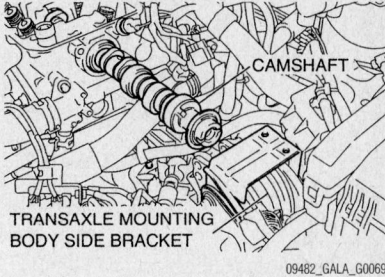

SEALANT:
MITSUBISHI GENUINE PART
NO.MD970389 OR EQUIVALENT

ENGINE OIL

1. IGNITION COIL CONNECTOR
2. SPARK PLUG CABLE
3. IGNITION COIL
4. PCV HOSE
5. BREATHER HOSE
6. CONTROL HARNESS CONNECTION
7. ROCKER COVER
8. CAMSHAFT POSITION SENSOR SUPPORT
9. CAMSHAFT POSITION SENSING CYLINDER

10. CAMSHAFT SPROCKET
12. SPARK PLUG GUIDE OIL SEAL
13. ROCKER ARM AND SHAFT ASSEMBLY (INTAKE SIDE)
14. ROCKER ARM AND SHAFT ASSEMBLY (EXHAUST SIDE)
15. CAMSHAFT
10. CAMSHAFT SPROCKET
11. CAMSHAFT OIL SEAL

09482_GALA_G0066

Camshaft and related components—2004–2005 Eclipse 2.4L engine

To install:

11. Lubricate the camshaft with clean engine oil. Install the camshaft on its mounting.

12. Install the rocker arm assemblies.

13. Continue the installation in the reverse order of the removal procedure.

2006 2.4L (4G69) ENGINE

1. Before servicing the vehicle, refer to the Precautions Section.

2. Disconnect the negative battery cable.

3. If equipped with manual transaxle, remove the EMC. If equipped with automatic transaxle remove the PCM.

4. Remove the air cleaner assembly. Remove the battery. Remove the ignition coils.

5. Remove the timing belt.

6. Disconnect the control wiring harness connection. Remove the rocker cover PCV connection.

7. Remove the rocker cover breather hose connection. Remove the engine oil control valve and O-ring. Remove the oil pressure switch.

8. Remove the spark plug guide oil seals. Remove the accumulator assembly.

9. Remove the connector bracket.

CAMSHAFT

TRANSAXLE MOUNTING
BODY SIDE BRACKET

09482_GALA_G0069

Camshaft and transaxle mounting side bracket location—2006 Eclipse, 2004–2006 Galant and 2004–2006 Lancer 2.4L engine

Remove the camshaft position sensor support and camshaft position sensing cylinder.

10. Use tool MB998719 to hold the camshaft sprocket in place. Remove the camshaft sprocket retaining bolt. Remove the camshaft sprocket. Remove the oil seal.

11. Remove the rocker cover retaining bolts. Remove the rocker cover. Discard the gasket.

12. Remove the exhaust rocker arm shaft caps. Remove the exhaust rocker arm shaft assembly.

13. Remove the intake rocker arm shaft caps. Remove the intake rocker arm shaft assembly.

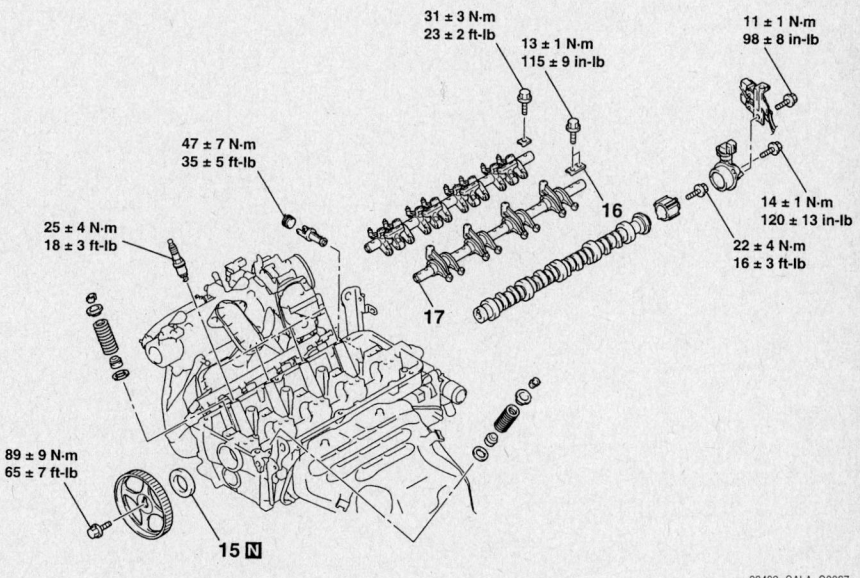

09482_GALA_G0067

Camshaft and related components—2006 Eclipse, 2004–2006 Galant and 2004–2006 Lancer 2.4L engine

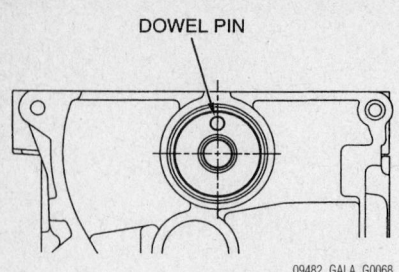

DOWEL PIN

09482_GALA_G0068

Camshaft installation alignment—2006 Eclipse, 2004–2006 Galant and 2004–2006 Lancer 2.4L engine

14. Raise the transaxle until the camshaft and transaxle mounting side bracket does not touch it.

15. Remove the camshaft from its mounting on the engine.

To install:

16. Lubricate the camshaft with clean engine oil. Install the camshaft. Set the dowel pin of the camshaft in the position as shown in the illustration.

17. Install the rocker arm assemblies.

18. Continue the installation in the reverse order of the removal procedure.

19. Adjust the valves.

20. To initialize the ECM, manual transaxle equipped vehicles, or the PCM, automatic transaxle equipped vehicles, turn the ignition switch ON then OFF and keep it in the OFF position for at least ten seconds.

2002–2003 3.0L ENGINE

1. Before servicing the vehicle, refer to the Precautions Section.

2. Remove or disconnect the following:
- Negative battery cable
- Intake manifold plenum stay bracket
- Camshaft Position (CMP) sensor
- Valve covers and the timing belt

3. Using a camshaft sprocket holding tool, hold the sprocket and loosen the bolt.

4. Remove the bolt and note the positioning of the knock pin at the end of the camshaft and remove the sprocket.

5. Install auto lash adjuster retainer tools on the rocker arms.

➡Be sure to note the position of the rocker arms, rocker shafts and bearing caps for reinstallation purposes.

6. Remove or disconnect the following:
- Camshaft bearing caps but do not remove the bolts from the caps
- Rocker arms, rocker shafts and bearing caps, as an assembly
- Camshaft from the cylinder head

7. Inspect the bearing journals on the camshaft, cylinder head, and bearing caps.

To install:

➡The right bank camshaft is identified by a 4mm slit at the rear end of the camshaft.

8. Lubricate the camshaft journals and camshaft with clean engine oil and install the camshaft in the cylinder head. Be sure to properly position the knock pin of the camshaft as noted during removal.

9. Apply sealer at the ends of the bearing caps and install the rocker arms, rocker shafts and bearing caps as an assembly. Properly position the arrows on the bearing caps.

10. Torque the bearing cap bolts in the following sequence: No. 3, No. 2, No. 1 and No. 4 to 85 inch lbs. (10 Nm).

11. Repeat the sequence increasing the torque to 14 ft. lbs. (20 Nm).

12. Remove the auto lash adjuster retainer tools from the rocker arms.

13. Install the camshaft sprocket and bolt.

14. Using a camshaft sprocket holding tool, hold the sprocket and tighten the bolt to 65 ft. lbs. (90 Nm).

15. Install or connect the following:
- Timing belt and valve covers
- CMP sensor
- Intake manifold plenum stay bracket

- Negative battery cable and check for leaks

2004–2005 3.0L ENGINE

1. Before servicing the vehicle, refer to the Precautions Section.

2. Disconnect the negative battery cable. Remove the timing belt.

3. On the left side, remove the thermostat housing assembly. Remove the PCV valve hose connection, spark plug cable and battery cable connection.

4. On the right side, remove the intake manifold plenum, distributor, breather hose connection and spark plug cable.

5. Remove the rocker arm cover retaining bolts. Remove the rocker arm cover. Discard the gasket.

6. Install tool MD998443 on the rocker arms, so that the lash adjusters will not fall out.

7. Loosen the rocker arm and shaft assembly mounting bolts. Remove the rocker arm and shaft assembly with the bolts still attached.

8. Use tool MB990767 to hold the camshaft sprocket in place. Remove the camshaft sprocket retaining bolt. Remove the camshaft sprocket. Remove the oil seal.

9. Remove the camshaft from the cylinder head.

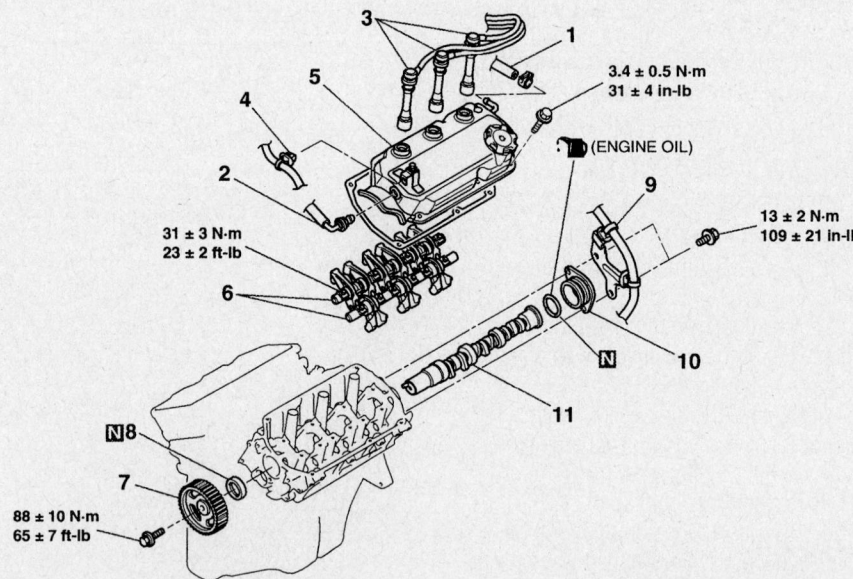

3.4 ± 0.5 N·m
31 ± 4 in-lb

(ENGINE OIL)

13 ± 2 N·m
109 ± 21 in-lb

31 ± 3 N·m
23 ± 2 ft-lb

88 ± 10 N·m
65 ± 7 ft-lb

1.	BLOW-BY HOSE CONNECTION	7.	CAMSHAFT SPROCKET
2.	PCV HOSE CONNECTION	8.	CAMSHAFT OIL SEAL
3.	SPARK PLUG CABLE	9.	BATTERY CABLE AND
4.	BATTERY CABLE CONNECTION		HARNESS BRACKET
5.	ROCKER COVER	10.	THRUST CASE
6.	ROCKER ARM AND SHAFT	11.	CAMSHAFT
	ASSEMBLY		

09482_GALA_G0070

Camshaft and related components—2004–2005 Eclipse 3.0L engine (left side)

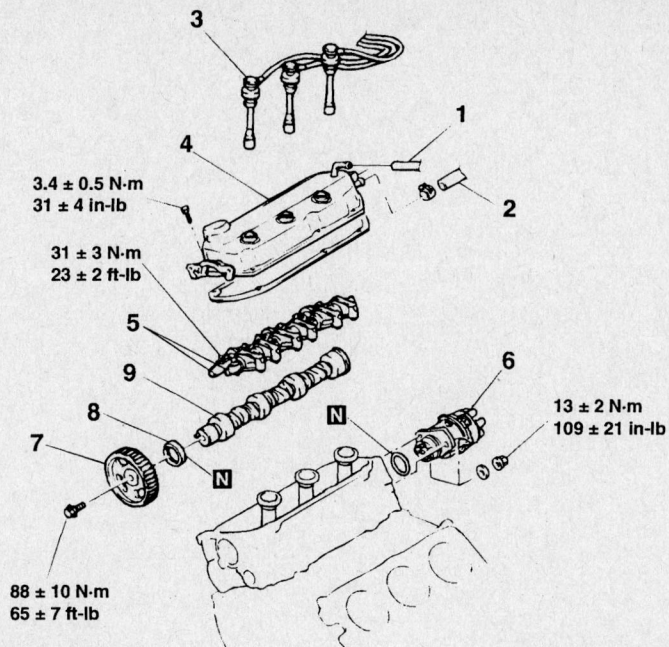

3.4 ± 0.5 N·m
31 ± 4 in-lb

31 ± 3 N·m
23 ± 2 ft-lb

88 ± 10 N·m
65 ± 7 ft-lb

13 ± 2 N·m
109 ± 21 in-lb

1. BREATHER HOSE
   CONNECTION
2. BLOW-BY HOSE CONNECTION
3. SPARK PLUG CABLE
4. ROCKER COVER

5. ROCKER ARM AND SHAFT
   ASSEMBLY
6. DISTRIBUTOR
7. CAMSHAFT SPROCKET
8. CAMSHAFT OIL SEAL
9. CAMSHAFT

09482_GALA_G0071

Camshaft and related components—2004–2005 Eclipse 3.0L engine (right side)

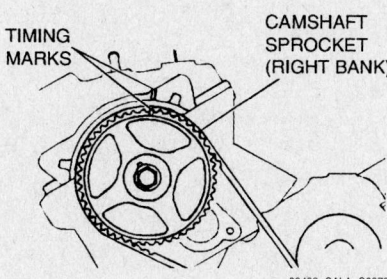

LEFT BANK   RIGHT BANK

MD998713   MB991559   MD998713

09482_GALA_G0072

Camshaft oil seal installation—2004–2005
Eclipse 3.0L engine

To install:

10. Coat the camshaft oil seal lip with clean engine oil. Use special tool MB991559 and MD998713 to press fit the camshaft oil seal. Lubricate the camshaft with clean engine oil. Install the camshaft.

11. Install the camshaft sprocket.

12. Align the timing mark of the camshaft sprocket (right side) with that of the cylinder head. Align the mating marks on the distributor housing and coupling. Install the distributor.

13. Continue the installation in the reverse order of the removal procedure.

3.8L ENGINE

1. Before servicing the vehicle, refer to the Precautions Section.

2. Disconnect the negative battery cable. Drain the cooling system.

3. Remove the timing belt.

4. On the left side, remove the thermostat housing assembly. Remove the PCV valve hose connection, ignition coil connector and ignition coil. Disconnect the engine control wiring harness clamp and harness bracket.

5. On the right side, remove the intake manifold plenum, breather hose connection, heater hose and water hose connection. Remove the ignition coil connector and coil.

6. Remove the rocker arm cover retaining bolts. Remove the rocker arm cover. Discard the gasket.

7. Install tool MD998443 on the rocker arms, so that the lash adjusters will not fall out.

8. Loosen the rocker arm and shaft assembly mounting bolts. Remove the rocker arm and shaft assembly with the bolts still attached.

➡**Never disassemble the rocker arm and shaft assembly.**

9. On the left side, disconnect the camshaft position sensor and remove the sensor support. Remove the camshaft position sensing cylinder. Remove the camshaft sprocket. Remove the camshaft and camshaft oil seal.

10. On the right side, disconnect the engine oil control valve connector and engine oil pressure switch connector. Remove the oil feed control valve right housing assembly. Remove the camshaft sprocket. Remove the camshaft and camshaft oil seal.

To install:

11. Coat the camshaft oil seal lip with clean engine oil.

12. Lubricate the camshaft with clean engine oil. Install the camshaft.

13. Continue the installation in the reverse order of the removal procedure.

Galant

2002–2003 2.4L ENGINE

1. Before servicing the vehicle, refer to the Precautions Section.

2. Relieve the fuel system pressure.

3. Remove or disconnect the following:
- Negative battery cable
- Accelerator cable, Positive Crankcase Ventilation (PCV) hoses, breather hoses, spark plug cables

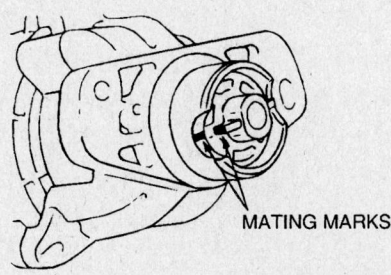

TIMING
MARKS

CAMSHAFT
SPROCKET
(RIGHT BANK)

MATING MARKS

09482_GALA_G0073

09482_GALA_G0074

Timing mark alignment—2004–2005
Eclipse 3.0L engine

Distributor housing alignment—
2004–2005 Eclipse 3.0L engine

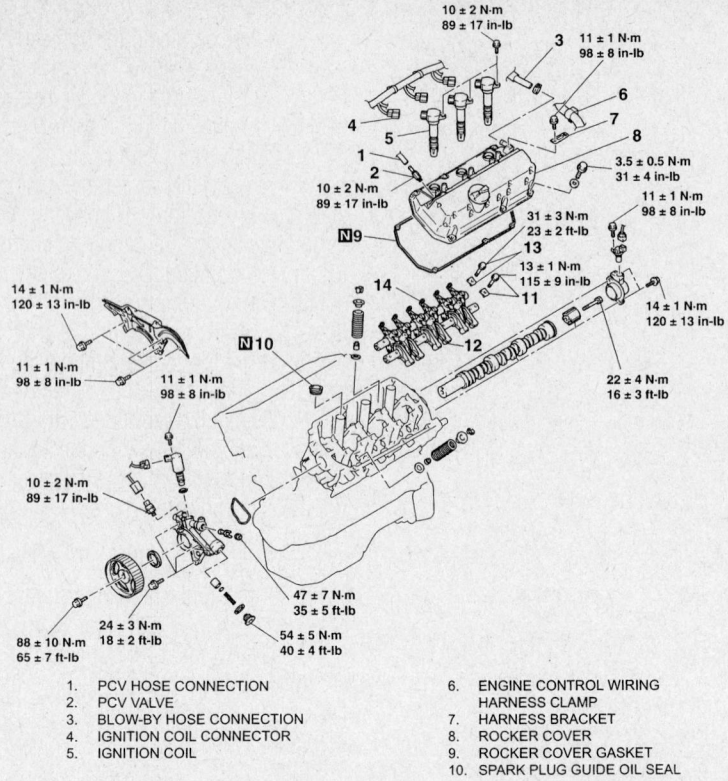

1. PCV HOSE CONNECTION
2. PCV VALVE
3. BLOW-BY HOSE CONNECTION
4. IGNITION COIL CONNECTOR
5. IGNITION COIL
6. ENGINE CONTROL WIRING HARNESS CLAMP
7. HARNESS BRACKET
8. ROCKER COVER
9. ROCKER COVER GASKET
10. SPARK PLUG GUIDE OIL SEAL

09482_GALA_G0075

Camshaft and related components—2006 Eclipse and 2004–2006 Galant 3.8L engine (left side)

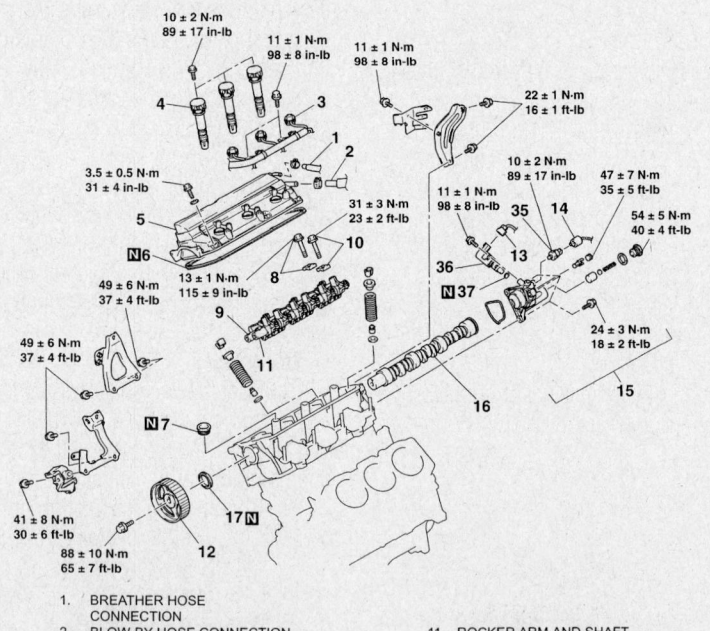

1. BREATHER HOSE CONNECTION
2. BLOW-BY HOSE CONNECTION
3. IGNITION COIL CONNECTOR
4. IGNITION COIL
5. ROCKER COVER
6. ROCKER COVER GASKET
7. SPARK PLUG GUIDE OIL SEAL
8. ROCKER SHAFT BOLTS AND CAP (EXHAUST SIDE)
9. ROCKER ARM AND SHAFT ASSEMBLY (EXHAUST SIDE)
10. ROCKER SHAFT BOLTS AND CAP (INTAKE SIDE)
11. ROCKER ARM AND SHAFT ASSEMBLY (INTAKE SIDE)
12. CAMSHAFT SPROCKET
13. ENGINE OIL CONTROL VALVE CONNECTOR
14. ENGINE OIL PRESSURE SWITCH CONNECTOR
15. OIL FEEDER CONTROL VALVE RIGHT HOUSING ASSEMBLY
16. CAMSHAFT
17. CAMSHAFT OIL SEAL

09482_GALA_G0076

Camshaft and related components—2004–2006 Eclipse and 2004–2006 Galant 3.8L engine (right side)

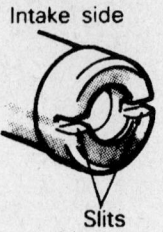

Intake side   Exhaust side

Slits

7923PG57

Camshaft identification—2002–2003 Eclipse and Galant 2.4L engine

- Valve cover
- Timing belt upper and lower covers
- Timing belt
- Camshaft sprockets

4. Loosen the bearing cap bolts in 2–3 steps. Label and remove all camshaft bearing caps.

- Intake and exhaust camshafts
- Rocker arms and lash adjusters

To install:

5. Install the lash adjusters and rocker arms into the cylinder head. Lubricate lightly with clean oil prior to installation.

6. Lubricate the camshafts with clean engine oil and position the camshafts on the cylinder head.

7. Be sure the dowel pin on both camshaft sprocket ends are located on the top.

8. Install the bearing caps. Tighten the caps in sequence and in 2 or 3 steps. No. 2 and 5 caps are of the same shape. Check the markings on the caps to identify the cap number and intake/exhaust symbol. Only **L** (intake) or **R** (exhaust) is stamped on No. 1 bearing cap. Also, be sure the rocker arm is correctly mounted on the lash adjuster and the valve stem end. Torque the retaining bolts to 15 ft. lbs. (20 Nm).

9. Apply a coating of engine oil to the oil seal. Using the proper size driver, press-fit the seal into the cylinder head.

10. Install or connect the following:

- Camshaft sprockets and tighten the sprocket bolts to 58–72 ft. lbs. (80–100 Nm)
- Timing belt, covers and related components
- Valve cover and reconnect all related components
- Negative battery cable

2004–2006 2.4L ENGINE

1. Before servicing the vehicle, refer to the Precautions Section.

2. Disconnect the negative battery cable.

3. If equipped with manual transaxle,

remove the EMC. If equipped with automatic transaxle remove the PCM.

4. Remove the air cleaner assembly. Remove the battery. Remove the ignition coils.

5. Remove the timing belt.

6. Disconnect the control wiring harness connection. Remove the rocker cover PCV connection.

7. Remove the rocker cover breather hose connection. Remove the engine oil control valve and O-ring. Remove the oil pressure switch.

8. Remove the spark plug guide oil seals. Remove the accumulator assembly.

9. Remove the connector bracket. Remove the camshaft position sensor support and camshaft position sensing cylinder.

10. Use tool MB998719 to hold the camshaft sprocket in place. Remove the camshaft sprocket retaining bolt. Remove the camshaft sprocket. Remove the oil seal.

11. Remove the rocker cover retaining bolts. Remove the rocker cover. Discard the gasket.

12. Remove the exhaust rocker arm shaft caps. Remove the exhaust rocker arm shaft assembly.

13. Remove the intake rocker arm shaft caps. Remove the intake rocker arm shaft assembly.

14. Raise the transaxle until the camshaft and transaxle mounting side bracket does not touch it.

15. Remove the camshaft from its mounting on the engine.

To install:

16. Lubricate the camshaft with clean engine oil. Install the camshaft. Set the dowel pin of the camshaft in the position as shown in the illustration.

17. Install the rocker arm assemblies.

18. Continue the installation in the reverse order of the removal procedure.

19. Adjust the valves.

20. To initialize the ECM, manual transaxle equipped vehicles, or the PCM, automatic transaxle equipped vehicles, turn the ignition switch ON then OFF and keep it in the OFF position for at least ten seconds.

2002–2003 3.0L ENGINE

The hydraulic lash adjusters are built into the rocker arms.

1. Before servicing the vehicle, refer to the Precautions Section.

2. Remove or disconnect the following:
- Negative battery cable
- Intake manifold plenum stay bracket
- Camshaft Position (CMP) sensor
- Valve covers and the timing belt

3. Using a camshaft sprocket holding tool, hold the sprocket and loosen the bolt.

4. Remove the bolt and note the positioning of the knock pin at the end of the camshaft and remove the sprocket.

5. Install auto lash adjuster retainer tools on the rocker arms.

➡ **Be sure to note the position of the rocker arms, rocker shafts and bearing caps for reinstallation purposes.**

6. Remove or disconnect the following:
- Camshaft bearing caps but do not remove the bolts from the caps
- Rocker arms, rocker shafts and bearing caps, as an assembly
- Camshaft from the cylinder head

7. Inspect the bearing journals on the camshaft, cylinder head, and bearing caps.

To install:

➡ **The right bank camshaft is identified by a 4mm slit at the rear end of the camshaft.**

8. Lubricate the camshaft journals and camshaft with clean engine oil and install the camshaft in the cylinder head. Be sure to properly position the knock pin of the camshaft as noted during removal.

9. Apply sealer at the ends of the bearing caps and install the rocker arms, rocker shafts and bearing caps as an assembly. Properly position the arrows on the bearing caps.

10. Torque the bearing cap bolts in the following sequence: No. 3, No. 2, No. 1 and No. 4 to 85 inch lbs. (10 Nm).

11. Repeat the sequence increasing the torque to 14 ft. lbs. (20 Nm).

12. Remove the auto lash adjuster retainer tools from the rocker arms.

13. Install the camshaft sprocket and bolt.

14. Using a camshaft sprocket holding tool, hold the sprocket and tighten the bolt to 65 ft. lbs. (90 Nm).

15. Install or connect the following:
- Timing belt and valve covers
- CMP sensor
- Intake manifold plenum stay bracket
- Negative battery cable and check for leaks

2004–2006 3.8L ENGINE

1. Before servicing the vehicle, refer to the Precautions Section.

2. Disconnect the negative battery cable. Drain the cooling system.

3. Remove the timing belt.

4. On the left side, remove the thermostat housing assembly. Remove the PCV

valve hose connection, ignition coil connector and ignition coil. Disconnect the engine control wiring harness clamp and harness bracket.

5. On the right side, remove the intake manifold plenum, breather hose connection, heater hose and water hose connection. Remove the ignition coil connector and coil.

6. Remove the rocker arm cover retaining bolts. Remove the rocker arm cover. Discard the gasket.

7. Install tool MD998443 on the rocker arms, so that the lash adjusters will not fall out.

8. Loosen the rocker arm and shaft assembly mounting bolts. Remove the rocker arm and shaft assembly with the bolts still attached.

➡ **Never disassemble the rocker arm and shaft assembly.**

9. On the left side, disconnect the camshaft position sensor and remove the sensor support. Remove the camshaft position sensing cylinder. Remove the camshaft sprocket. Remove the camshaft and camshaft oil seal.

10. On the right side, disconnect the engine oil control valve connector and engine oil pressure switch connector. Remove the oil feed control valve right housing assembly. Remove the camshaft sprocket. Remove the camshaft and camshaft oil seal.

To install:

11. Coat the camshaft oil seal lip with clean engine oil.

12. Lubricate the camshaft with clean engine oil. Install the camshaft.

13. Continue the installation in the reverse order of the removal procedure.

Lancer

2.0L ENGINE

1. Before servicing the vehicle, refer to the Precautions Section.

2. Remove or disconnect the following:
- Negative battery cable
- Air cleaner assembly
- Ignition coil
- Breather hose
- Positive Crankcase Ventilation (PCV) hose
- Rocker arm (valve) cover and gasket
- Spark plug guide
- Timing belt front upper cover

3. Remove the camshaft sprocket, as follows:

a. Secure the cam sprocket and timing belt with wire ties to prevent them from slipping out of place.

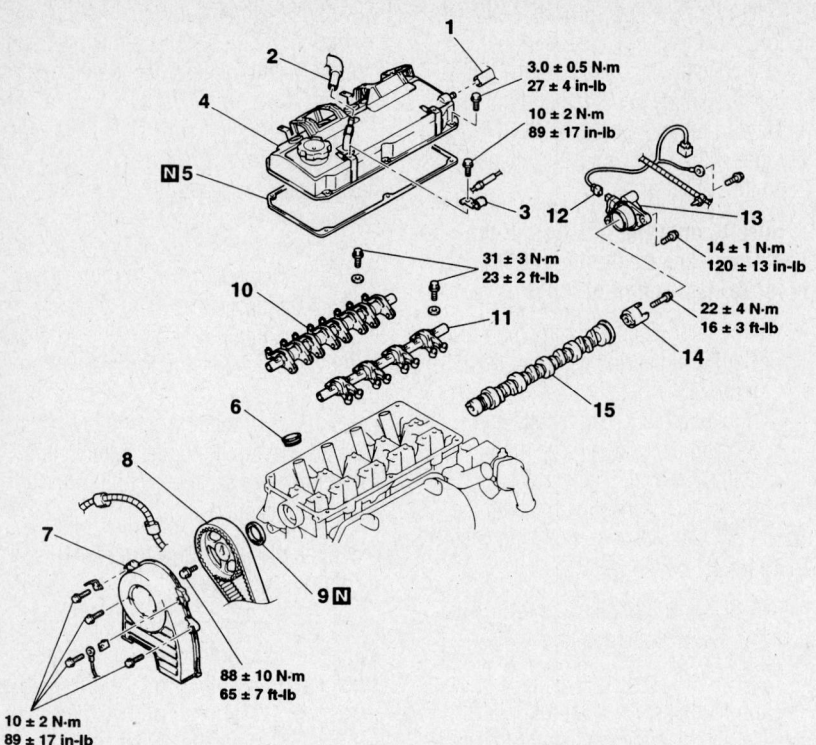

3.0 ± 0.5 N·m
27 ± 4 in-lb

10 ± 2 N·m
89 ± 17 in-lb

14 ± 1 N·m
120 ± 13 in-lb

31 ± 3 N·m
23 ± 2 ft-lb

22 ± 4 N·m
16 ± 3 ft-lb

88 ± 10 N·m
65 ± 7 ft-lb

10 ± 2 N·m
89 ± 17 in-lb

1. BREATHER HOSE
   CONNECTION
2. PCV HOSE CONNECTION
3. ACCELERATOR CABLE CLAMP
4. ROCKER COVER
5. ROCKER COVER GASKET
6. SPARK PLUG GUIDE
7. TIMING BELT FRONT UPPER
   COVER
8. CAMSHAFT SPROCKET

9. CAMSHAFT OIL SEAL
10. INTAKE ROCKER ARM AND
    SHAFT ASSEMBLY
11. EXHAUST ROCKER ARM AND
    SHAFT ASSEMBLY
12. CAMSHAFT POSITION SENSOR
    CONNECTOR
13. CAMSHAFT POSITION SENSOR
    SUPPORT

93570G17

Camshaft and related components—2.0L engine

b. While holding the sprocket from turning with special tools MB990767 and MD998719, remove the camshaft sprocket bolt and sprocket.

4. Remove or disconnect the following:
- Intake and exhaust rocker arm and shaft assemblies. Loosen both rocker arm assemblies gradually and evenly and remove the rocker shafts from the vehicle. Do NOT disassemble!
- Camshaft Position (CMP) sensor connector
- CMP sensor support
- CMP sensor sensing cylinder
- Camshaft

To install:
5. Install or connect the following:
- Camshaft
- CMP sensor sensing cylinder, support and connector
- Rocker arm and shaft assemblies
- Camshaft oil seal

- Camshaft sprocket. Tighten the bolt to 58–72 ft. lbs. (78–98 Nm).
- Timing belt front upper cover
- Spark plug guide
- Rocker arm cover, with a new gasket. Tighten the bolts to 23–31 inch lbs. (2.5–3.5 Nm).
- Accelerator cable clamp
- PCV and breather hoses
- Negative battery cable

2.4L ENGINE

1. Before servicing the vehicle, refer to the Precautions Section.
2. Disconnect the negative battery cable.
3. Remove the air cleaner assembly. Remove the battery. Remove the ignition coils.
4. Remove the timing belt.
5. Disconnect the control wiring harness connection. Remove the rocker cover PCV connection.
6. Remove the rocker cover breather

hose connection. Remove the engine oil control valve and O-ring. Remove the oil pressure switch.
7. Remove the spark plug guide oil seals. Remove the accumulator assembly.
8. Remove the connector bracket. Remove the camshaft position sensor support and camshaft position sensing cylinder.
9. Use tool MB998719 to hold the camshaft sprocket in place. Remove the camshaft sprocket retaining bolt. Remove the camshaft sprocket. Remove the oil seal.
10. Remove the rocker cover retaining bolts. Remove the rocker cover. Discard the gasket.
11. Remove the exhaust rocker arm shaft caps. Remove the exhaust rocker arm shaft assembly.
12. Remove the intake rocker arm shaft caps. Remove the intake rocker arm shaft assembly.
13. Raise the transaxle until the camshaft and transaxle mounting side bracket does not touch it.
14. Remove the camshaft from its mounting on the engine.

To install:
15. Lubricate the camshaft with clean engine oil. Install the camshaft. Set the dowel pin of the camshaft in the position as shown in the illustration.
16. Install the rocker arm assemblies.
17. Continue the installation in the reverse order of the removal procedure.
18. Adjust the valves.

Evolution

2.0L ENGINE

1. Before servicing the vehicle, refer to the Precautions Section.
2. Remove the engine under cover. Remove the side cover.
3. Disconnect the battery cable. Drain the engine coolant.
4. Remove the air duct assembly. Remove the air pipe "C" assembly.
5. Remove the accelerator cable. Remove the timing belt.
6. Disconnect the oil feed control valve connector and remove the oil feed control valve and O-ring.
7. Remove the rocker cover center cover. Remove the spark plug cables and ignition coils.
8. Disconnect the front heated oxygen sensor connector, crankshaft position sensor connector, control wiring harness connection and rocker cover PCV hose.
9. Remove the rocker cover breather hose, upper radiator hose and camshaft

position sensor connector (exhaust and inlet side).

10. Disconnect the ground terminal, the control wiring harness connection and vacuum hose and pipe assembly.

11. Remove the valve cover retaining bolts. Remove the valve cover. Remove the spark plug hole gaskets. Remove and discard the valve cover gasket.

12. Remove the camshaft position sensor support cover, and gasket. Remove the camshaft position sensing cylinder (exhaust side).

13. Remove the camshaft position sensor support, cover and gasket.

14. Remove the camshaft sensing cylinder (intake side). Remove the camshaft position sensor support. Remove the exhaust camshaft sprocket and oil seal. Remove the intake camshaft sprocket and oil seal.

15. Remove the front camshaft bearing cap. Remove the rear camshaft bearing cap.

16. Remove bearing cap number 2, 5, 3 and 4. Remove the exhaust camshaft sprocket and washer

17. Remove the intake camshaft sprocket and oil seal. Remove the camshaft bearing front cap. Remove bearing cap number 2, 5, 3 and 4. Remove the intake camshaft.

To install:

18. Coat the camshafts with clean engine oil. Position the camshafts on the cylinder head.

➡**Do not install the wrong camshaft on the wrong side. The exhaust camshaft has a slit at the rear surface.**

19. Set the dowel pin of the camshaft to the position shown in the illustration.

20. Since the shape of the bearing caps numbers 2 thru 5 are identical, check the identification marks so that the bearing cap number (intake or exhaust) is installed according to the direction in the illustration.

21. Apply sealant, part number 8679/8678 or equivalent, to the positions of the upper side of the cylinder head as shown in the illustration.

22. Tighten the bearing caps, increasing the pressure in two to three passes and finally tightening the bolts to 20 ft. lbs.

23. Continue the installation in the reverse order of the removal procedure.

Mirage

1.5L ENGINE

1. Before servicing the vehicle, refer to the Precautions Section.

2. Remove or disconnect the following:
   • Negative battery cable

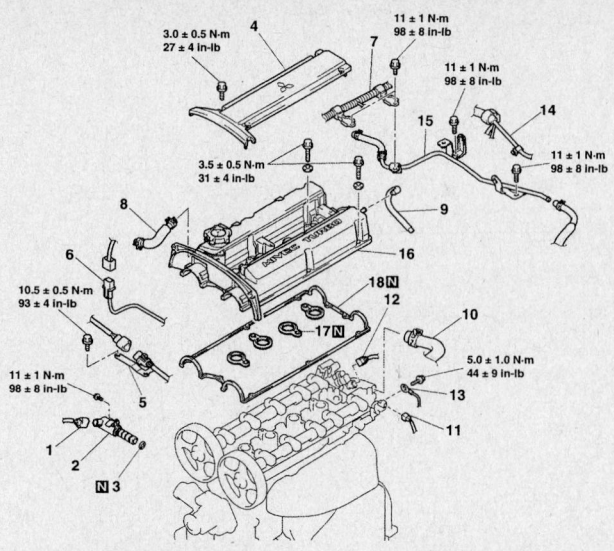

1. OIL FEEDER CONTROL VALVE CONNECTOR
2. OIL FEEDER CONTROL VALVE
3. O-RING
4. ROCKER COVER CENTER COVER
• SPARK PLUG CABLES AND IGNITION COILS (REFER TO GROUP 16, IGNITION COIL P.16-39.)
5. HEATED OXYGEN SENSOR (FRONT) CONNECTOR
6. CRANKSHAFT POSITION SENSOR CONNECTOR
7. CONTROL WIRING HARNESS CONNECTION
8. ROCKER COVER PCV HOSE
9. ROCKER COVER BREATHER HOSE
10. RADIATOR UPPER HOSE CONNECTION
11. CAMSHAFT POSITION SENSOR CONNECTOR (EXHAUST SIDE)
12. CAMSHAFT POSITION SENSOR CONNECTOR (INLET SIDE)
13. GROUND TERMINAL
14. CONTROL WIRING HARNESS CONNECTION
15. VACUUM HOSE AND PIPE ASSEMBLY
16. ROCKER COVER
17. SPARK PLUG HOLE GASKETS
18. ROCKER COVER GASKET

09482_GALA_G0077

Camshafts and related components—Evolution (view one)

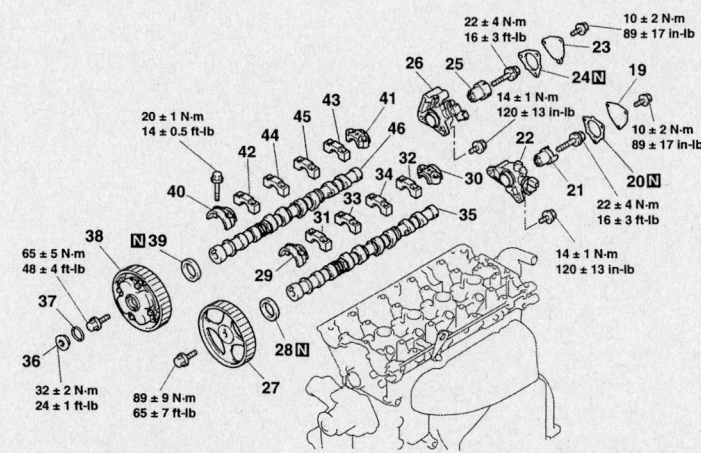

19. CAMSHAFT POSITION SENSOR SUPPORT COVER
20. CAMSHAFT POSITION SENSOR SUPPORT COVER GASKET
21. CAMSHAFT POSITION SENSING CYLINDER (EXHAUST SIDE)
22. CAMSHAFT POSITION SENSOR SUPPORT
23. CAMSHAFT POSITION SENSOR SUPPORT COVER
24. CAMSHAFT POSITION SENSOR SUPPORT COVER GASKET
25. CAMSHAFT POSITION SENSING CYLINDER (INLET SIDE)
26. CAMSHAFT POSITION SENSOR SUPPORT
27. CAMSHAFT SPROCKET (EXHAUST SIDE)
28. CAMSHAFT OIL SEAL
29. CAMSHAFT BEARING CAP, FRONT
30. CAMSHAFT BEARING CAP, REAR
31. CAMSHAFT BEARING CAP, No.2
32. CAMSHAFT BEARING CAP, No.5
33. CAMSHAFT BEARING CAP, No.3
34. CAMSHAFT BEARING CAP, No.4
35. EXHAUST CAMSHAFT
36. CAMSHAFT SPROCKET CAP
37. WASHER
38. CAMSHAFT SPROCKET (INLET SIDE)
39. CAMSHAFT OIL SEAL
40. CAMSHAFT BEARING CAP, FRONT
41. CAMSHAFT BEARING CAP, REAR
42. CAMSHAFT BEARING CAP, No.2
43. CAMSHAFT BEARING CAP, No.5
44. CAMSHAFT BEARING CAP, No.3
45. CAMSHAFT BEARING CAP, No.4
46. INLET CAMSHAFT

09482_GALA_G0078

Camshafts and related components—Evolution (view two)

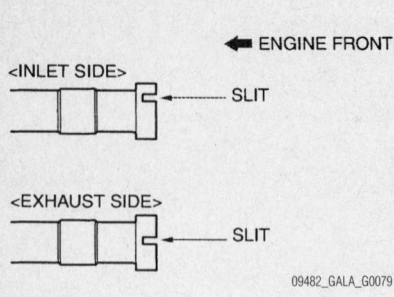

Camshaft identification—Evolution

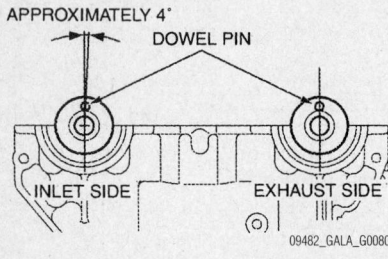

Camshaft alignment—Evolution

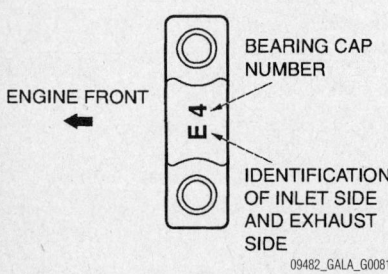

Camshaft bearing cap installation identification—Evolution

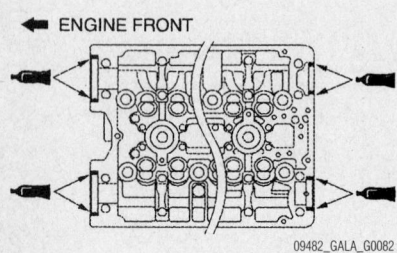

Sealant application—Evolution

- Accelerator cable, breather hose and Positive Crankcase Ventilation (PCV) hose connections
- Distributor, if equipped
- Valve cover and discard the gasket

3. Loosen both rocker arm assemblies gradually and evenly and remove the rocker shafts from the vehicle.

4. Remove or disconnect the following:
- Timing belt covers
- Timing belt
- Camshaft sprocket from the camshaft. Note the positioning of the dowel pin at the end of the camshaft.
- Camshaft oil seal from the front of the cylinder head
- Camshaft from the head

To install:

5. Lubricate the camshaft with clean engine oil and slide it into the head. Be sure to position the dowel pin at the 12 o'clock position.

6. Install or connect the following:
- New camshaft oil seal. Be sure to lubricate the lips of the seal with clean engine oil.
- Camshaft sprocket and install the mounting bolt. Tighten the bolt to 51 ft. lbs. (70 Nm)
- Timing belt
- Timing belt covers
- Rocker shaft assemblies. Torque the bolts gradually and evenly to 23 ft. lbs. (32 Nm).

7. Install the valve cover with a new gasket. Tighten the valve cover bolt to 16 inch lbs. (1.8 Nm).

8. Install or connect the following:
- Distributor, if equipped
- Accelerator cable, breather hose and PCV hose
- Negative battery cable and check the ignition timing

1.8L ENGINE

1. Before servicing the vehicle, refer to the Precautions Section.

2. Remove or disconnect the following:
- Negative battery cable
- Spark plug cables
- Manifold Absolute Pressure (MAF) sensor connector and remove the air cleaner case cover
- Accelerator cable, breather hose and Positive Crankcase Ventilation (PCV) hose connections
- Rocker cover and discard the gasket

3. Loosen both rocker arm shaft assemblies gradually and evenly and remove the rocket shafts from the vehicle.

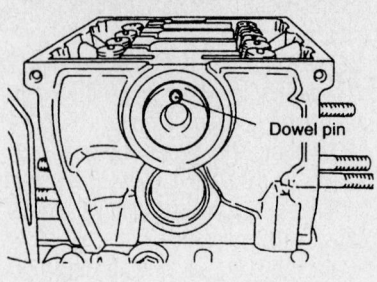

Positioning of the camshaft dowel pin—1.5L engine

4. Remove or disconnect the following:
- Timing belt covers
- Timing belt
- Camshaft sprocket from the camshaft. Note the positioning of the dowel pin at the end of the camshaft.
- Camshaft oil seal from the front of the cylinder head
- Camshaft from the head

To install:

5. Lubricate the camshaft journals and camshaft with clean engine oil and install the camshaft in the cylinder head. Be sure to position the dowel pin at the end of the camshaft as noted during the removal procedure.

6. Install or connect the following:
- New camshaft oil seal.
- Camshaft sprocket and tighten the retainer bolt to 65 ft. lbs. (90 Nm)
- Timing belt
- Timing belt covers
- Rocker arm and shaft assemblies. Tighten the rocker arm shaft retainer bolts to 23 ft. lbs. (32 Nm).
- Valve cover with a new gasket. Tighten the valve cover bolts to 29 inch lbs. (3.3 Nm).
- Spark plug cables
- Accelerator cable, breather hose and PCV hose
- MAP sensor connector and install the air cleaner case cover
- Negative battery cable

Valve Lash

ADJUSTMENT

2.4L (4G69) Engine

➡**Before adjusting the valves, check that the engine coolant temperature is between 176—203 degrees and all lights and accessories are off.**

INTAKE VALVE SIDE

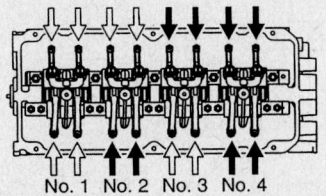

EXHAUST VALVE SIDE

09482_GALA_G0083

Valve adjustment—2.4L (4G69) Engine

1. Remove the ignition coils. Remove the valve cover.

2. Turn the crankshaft clockwise until the notch on the pulley is lined up with the "T" mark on the timing indicator.

3. Move the rocker arms on the number one and number four cylinders up and down, by hand to determine which cylinder has its piston at the TDC position on the compression stroke.

➡**If both intake and exhaust valve rocker arms have a valve lash, the piston in the cylinder corresponding to these rocker arms is at TDC on the compression stroke.**

4. Valve clearance can be checked and adjusted on the rocker arms indicated by the white arrow, in the illustration, when the number one cylinder piston is at TDC on the compression stroke and on the rocker arms indicated by the black arrow, in the illustration, when the number four piston is at TDC on the compression stroke.

5. Check the valve clearance.

6. If adjustment is required, loosen the rocker arm lock nut and adjust the clearance to specification, while turning the adjusting screw.

7. After adjustment, hold the adjusting screw with a screwdriver (to prevent it from turning) and tighten the locknut to 8 inch lbs.

8. Rotate the crankshaft 360 degrees to line up the notch on the crankshaft pulley with the "T" mark on the timing indicator.

9. Check the valve clearance.

10. If adjustment is required, loosen the rocker arm lock nut and adjust the clearance to specification, while turning the adjusting screw.

11. After adjustment, hold the adjusting screw with a screwdriver (to prevent it from turning) and tighten the locknut to 8 inch lbs.

12. Install the valve cover. Install the ignition coils.

Oil Pan

REMOVAL & INSTALLATION

Diamante

1. Before servicing the vehicle, refer to the Precautions Section.

2. Disconnect the negative battery cable.

3. Drain the engine oil.

4. Remove the mounting bolts from the lower oil pan.

5. Place a block of wood against the side of the pan and tap the block with a hammer to break the seal and remove the lower pan.

6. Remove or disconnect the following:
- Starter
- Dipstick tube
- Upper oil pan

❋❋ WARNING

Do not pry or use seal breaker tool to remove the oil pan. Damage to the aluminum surface can result.

7. Screw a bolt into the threaded hole to force the oil pan from the engine block and remove the pan.

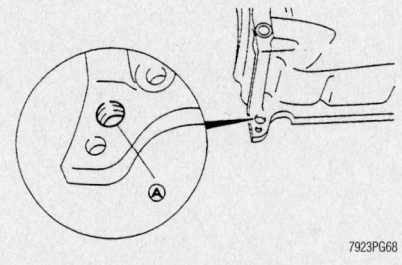

7923PG68

Install a bolt in the threaded hole to force the oil pan from the engine block—3.5L engine

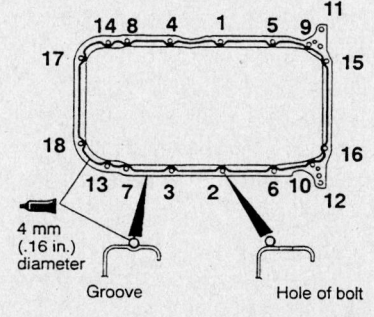

7923PG69

Apply sealant and tighten the bolts in the order shown—3.5L engine, upper oil pan

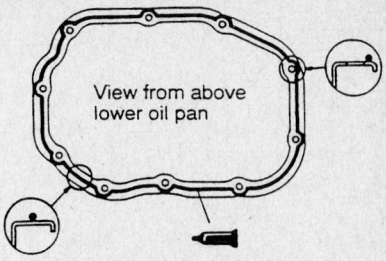

View from above lower oil pan

Flange bolt tightening sequence

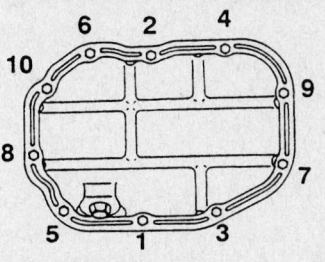

7923PG70

Apply sealant and tighten the bolts in the order shown—3.5L engine, lower oil pan

8. Remove the bolt used to remove the pan.

To install:

9. Clean and degrease the sealing surfaces of the upper oil pan and engine block.

10. Apply a bead of silicone sealant along the mounting surface of the upper oil pan.

11. Install or connect the following:
- Upper oil pan. Tighten the bolts in sequence to 48 inch lbs. (6 Nm).
- Dipstick tube using a new O-ring
- Starter assembly

12. Clean and degrease the sealing surface of the lower oil pan.

13. Place a bead of sealant on the mounting surface of the lower oil pan. Install the lower pan. Tighten the bolts in sequence to 84–108 inch lbs. (10–12 Nm).

14. Install the drain plug using a new washer. Tighten the drain plug to 29 ft. lbs. (39 Nm).

15. Lower the vehicle and fill the crankcase to the correct level.

16. Connect the negative battery cable.

17. Start the engine and check for leaks.

Eclipse

2002–2005 2.4L ENGINE

1. Before servicing the vehicle, refer to the Precautions Section.

2. Remove the negative battery cable.

3. Drain the engine oil.

4. Remove or disconnect the following:
- Engine dipstick and tube assembly
- Front exhaust pipe
- Bell housing inspection cover

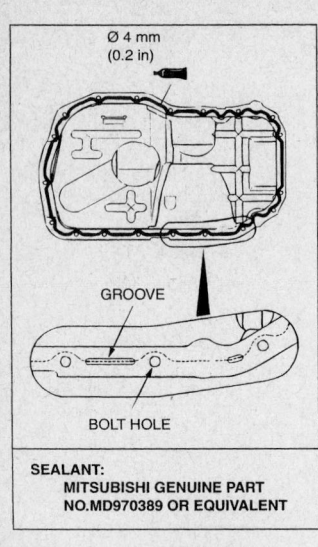

SEALANT:
MITSUBISHI GENUINE PART
NO.MD970389 OR EQUIVALENT

1. DRAIN PLUG
2. DRAIN PLUG GASKET

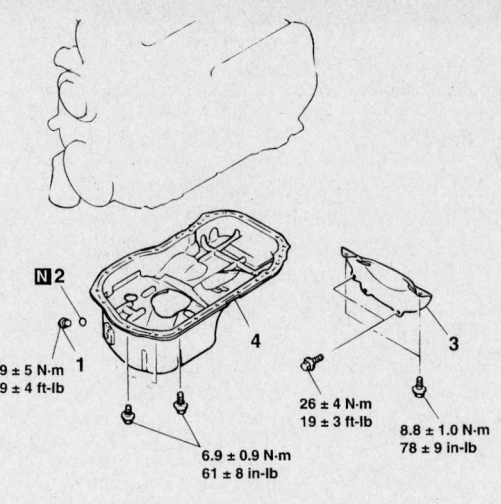

3. BELL HOUSING COVER
4. OIL PAN

09482_GALA_G0084

Oil pan and related components—2002–2005 Eclipse 2.4L Engine

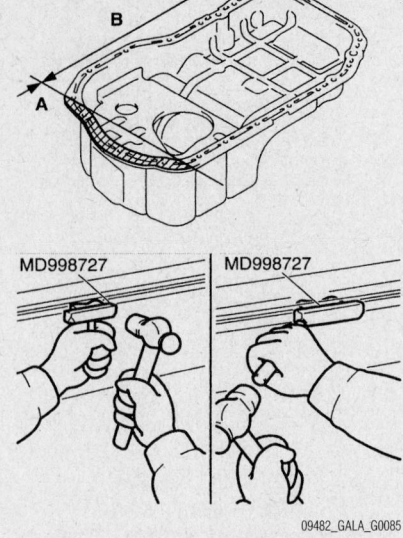

09482_GALA_G0085

Tool installation and positioning—2006 Eclipse, 2004–2006 Galant and 2004–2006 Lancer 2.4L Engine

- Bolts attaching the oil pan to the cylinder block
5. Remove the oil pan assembly.

To install:

6. Clean the sealing surface on the oil pan and engine block. Apply a continuous bead of sealant to the oil pan.

7. Install or connect the following:
- Oil pan to the cylinder block and tighten the bolts to 60 inch lbs. (7 Nm)
- Bell housing inspection cover. Torque the bolts to 84 inch lbs. (9 Nm).
- Front exhaust pipe
- Engine dipstick and tube assembly using a new O-ring

8. Refill the engine with oil. Connect the negative battery cable. Start the engine and check for leaks.

2006 2.4L ENGINE

1. Before servicing the vehicle, refer to the Precautions Section.
2. Remove the negative battery cable.
3. Drain the engine oil.
4. Remove the engine under cover. Remove the front exhaust pipe.
5. Remove the torque converter housing front lower cover.
6. Remove the oil pan retaining bolts. Remove the oil pan. Discard the gasket.

➡ **Use tool MD998727 to remove the oil pan. Tap the tool into the range (B) between the cylinder block and the engine oil pan, and then slide the tool sideways. Do not position the tool in**

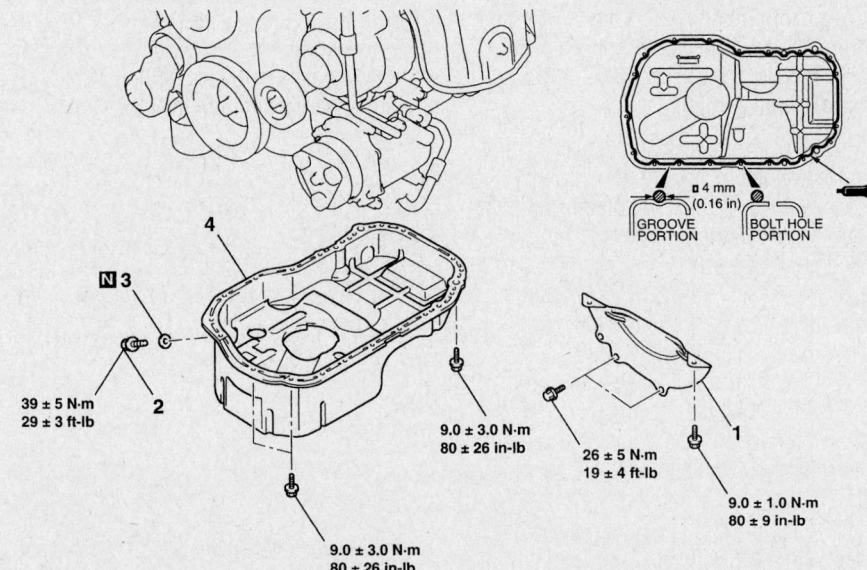

1. TORQUE CONVERTER HOUSING FRONT LOWER COVER
2. ENGINE OIL PAN DRAIN PLUG
3. ENGINE OIL PAN DRAIN PLUG GASKET
4. ENGINE OIL PAN

09482_GALA_G0086

Oil pan and related components—2006 Eclipse, 2004–2006 Galant and 2004–2006 Lancer 2.4L Engine

area (A) of the engine oil pan as this may cause deformation of the front case because the front case is made of aluminum.

To install:

7. Clean the sealing surface on the oil pan and engine block. Apply a continuous bead of sealant to the oil pan.

8. Install the oil pan. Torque the bolts to specification.

9. Continue the installation in the reverse order of the removal procedure.

2002–2003 3.0L ENGINE

1. Before servicing the vehicle, refer to the Precautions Section.
2. Remove or disconnect the following:
- Negative battery cable
- Oil pan drain plug and drain the engine oil

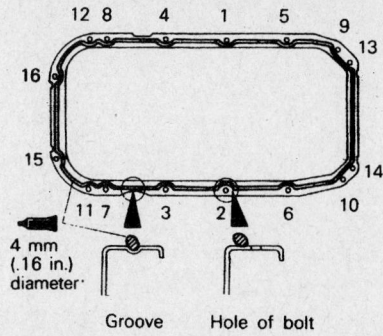

4 mm (.16 in.) diameter

Groove   Hole of bolt

67170-GALA-G32

Oil pan bolt tightening sequence and application of sealant to the pan—2002–2003 Eclipse and Galant 3.0L engine

- Left side crossmember.
- Starter motor
- Roll stopper stay bracket, from the rear transaxle stay bracket
- Transaxle stay brackets
- Bell housing lower cover
- Oil pan mounting bolts
- The engine oil pan

To install:

3. Apply a 0.16 in. (4mm) continuous bead of sealer around the surface of the oil pan.

➡**Assemble the oil pan to the cylinder block within 15 minutes after applying the sealant.**

4. Install the oil pan mounting bolts. Following proper sequence, tighten mounting bolts to 48 inch lbs. (6 Nm).

5. Install or connect the following:
- Lower bell housing cover and the starter motor
- Transaxle stay brackets and connect the roll stopper bracket
- Crossmember(s) and tighten the mounting bolts to 43–51 ft. lbs. (60–70 Nm)

6. Fill the engine with the proper amount of oil.

7. Connect the negative battery cable and check for leaks.

09482_GALA_G0087

Transaxle to engine bolt location and installation—2004–2005 Eclipse 3.04L Engine

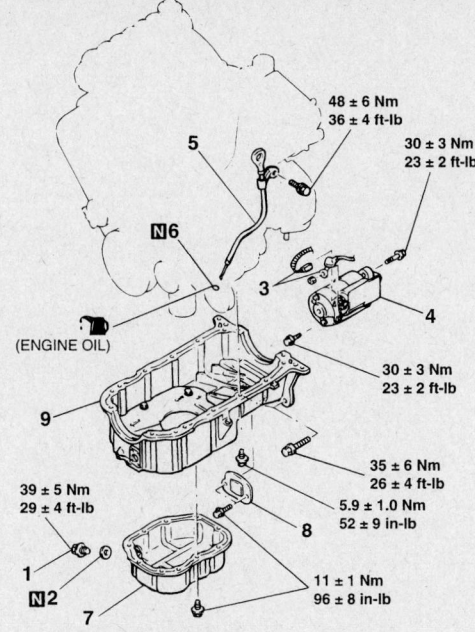

48 ± 6 Nm / 36 ± 4 ft-lb
30 ± 3 Nm / 23 ± 2 ft-lb
30 ± 3 Nm / 23 ± 2 ft-lb
35 ± 6 Nm / 26 ± 4 ft-lb
5.9 ± 1.0 Nm / 52 ± 9 in-lb
39 ± 5 Nm / 29 ± 4 ft-lb
11 ± 1 Nm / 96 ± 8 in-lb

1. DRAIN PLUG
2. DRAIN PLUG GASKET
7. LOWER OIL PAN
UPPER OIL PAN REMOVAL STEPS
1. DRAIN PLUG
2. DRAIN PLUG GASKET
3. STARTER CONNECTOR
4. STARTER
5. OIL DIPSTICK AND DIPSTICK GUIDE
6. O-RING
7. LOWER OIL PAN
8. OIL PAN BOLT HOLE COVER
9. UPPER OIL PAN

Oil pan and related components—2004–2005 Eclipse 3.04L Engine

2004–2005 3.0L ENGINE

1. Before servicing the vehicle, refer to the Precautions Section.
2. Remove the negative battery cable.
3. Drain the engine oil. Remove the front exhaust pipe.
4. To remove the lower oil pan, remove the lower oil pan retaining bolts. Remove the lower oil pan. Discard the gasket.
5. To remove the upper oil pan, remove the engine oil dipstick and guide assembly. Remove the starter.
6. Remove the upper oil pan retaining bolts. Remove the upper oil pan from the engine. Discard the gasket.

➡**After removing the upper oil pan retaining bolts, screw two M10 bolts securing the upper oil pan to the transaxle assembly, then remove the upper oil pan.**

To install:

7. Clean the sealing surface on the oil pan and engine block. Apply a continuous bead of sealant to the oil pan.
8. Install the upper oil pan. Torque the bolts to specification.

➡**The bolt holes for bolt 13 and 14 are cut away on the transaxle side. Be careful not to insert these bolts at an angle.**

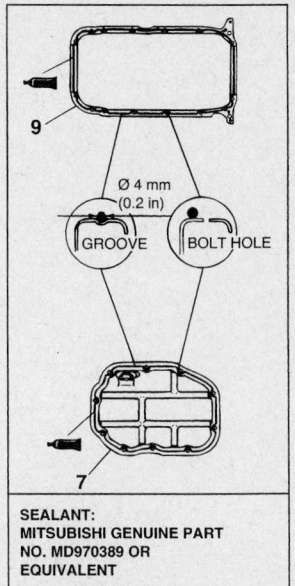

SEALANT: MITSUBISHI GENUINE PART NO. MD970389 OR EQUIVALENT

09482_GALA_G0088

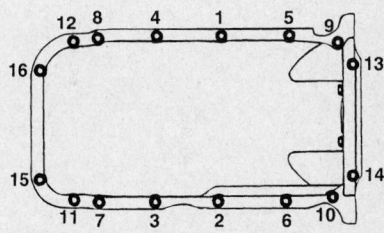

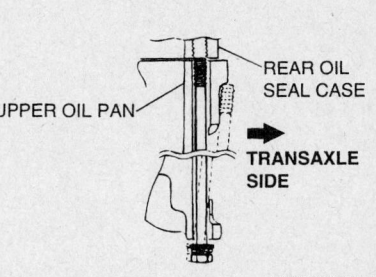

09482_GALA_G0089

Upper oil pan torque sequence—2004–2005 Eclipse 3.0L Engine

9. Install the lower oil pan. Torque the bolts to specification.
10. Continue the installation in the reverse order of the removal procedure.

3.8L ENGINE

1. Before servicing the vehicle, refer to the Precautions Section.

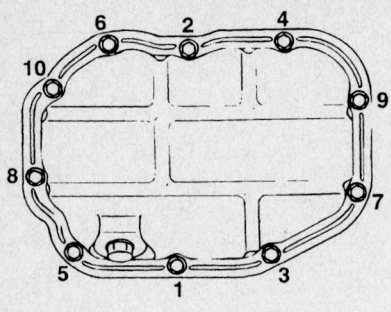

Lower oil pan torque sequence—2004–2005 Eclipse 3.0L Engine

09482_GALA_G0090

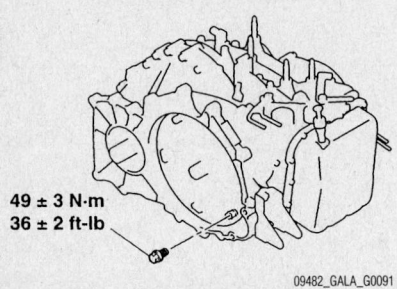

49 ± 3 N·m
36 ± 2 ft-lb

09482_GALA_G0091

Automatic transaxle torque converter connecting bolt location—2006 Eclipse and 2004–2006 Galant 3.8L Engine

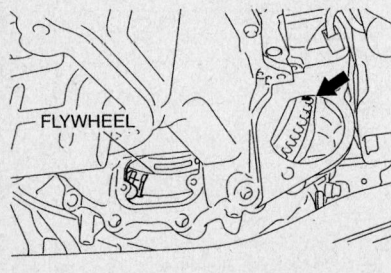

FLYWHEEL

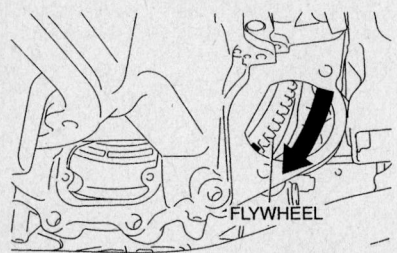

FLYWHEEL

09482_GALA_G0092

Manual transaxle alignment location—2006 Eclipse and 2004–2006 Galant 3.8L Engine

2. Remove the negative battery cable.

3. Remove the engine under cover. Remove the starter. Remove the engine oil dipstick and guide.

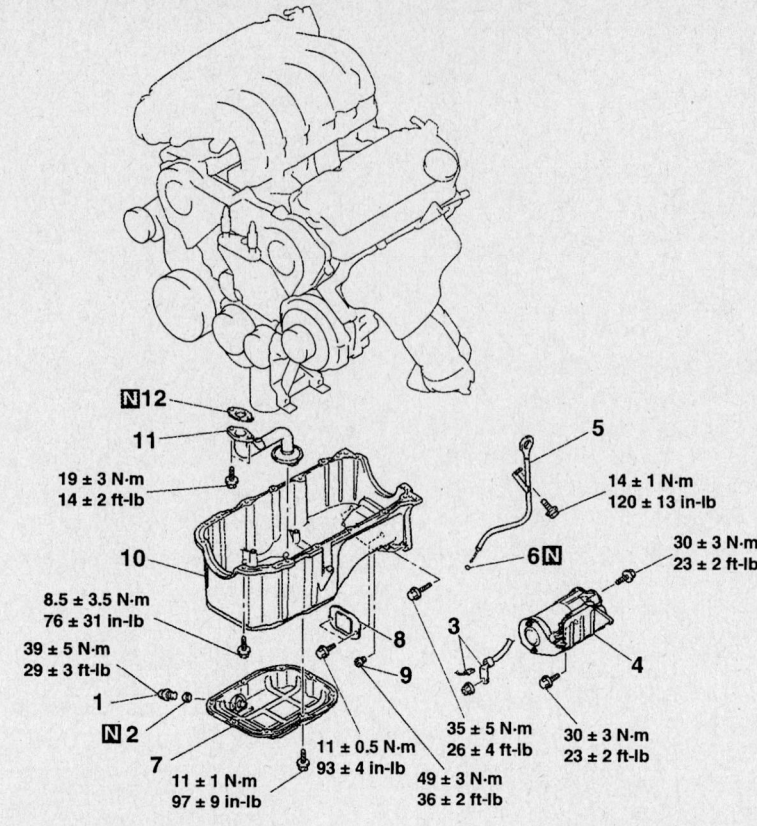

N 12

11

19 ± 3 N·m
14 ± 2 ft-lb

10

8.5 ± 3.5 N·m
76 ± 31 in-lb

39 ± 5 N·m
29 ± 3 ft-lb

1

N 2

7

11 ± 1 N·m
97 ± 9 in-lb

8

9

11 ± 0.5 N·m
93 ± 4 in-lb

3

35 ± 5 N·m
26 ± 4 ft-lb

49 ± 3 N·m
36 ± 2 ft-lb

5

14 ± 1 N·m
120 ± 13 in-lb

6 N

30 ± 3 N·m
23 ± 2 ft-lb

4

30 ± 3 N·m
23 ± 2 ft-lb

1. ENGINE OIL PAN DRAIN PLUG
2. ENGINE OIL PAN DRAIN PLUG GASKET
3. STARTER CONNECTOR
4. STARTER ASSEMBLY
5. ENGINE OIL DIPSTICK ASSEMBLY
6. O-RING
7. ENGINE LOWER OIL PAN

• FRONT NO.1 EXHAUST PIPE
8. COVER
9. TORQUE CONVERTER CONNECTING BOLT
10. ENGINE UPPER OIL PAN
11. OIL SCREEN
12. GASKET

09482_GALA_G0093

Oil pan and related components—2006 Eclipse and 2004–2006 Galant 3.8L Engine

4. If removing the upper oil pan, remove the front exhaust pipe. If equipped with automatic transaxle, remove the cover and connecting bolt.

5. Remove the lower engine oil pan retaining bolts. Remove the engine lower oil pan. Discard the gasket.

6. Remove the upper oil pan retaining bolts. Remove the upper oil pan. Discard the gasket.

➡If the vehicle is equipped with manual transaxle, align the recessed area in the flywheel with the location shown in the illustration. Mark the flywheel. Turn the crankshaft so that the align mark is positioned as shown in the illustration.

To install:

7. Clean the sealing surface on the oil pan and engine block. Apply a continuous bead of sealant to the oil pan.

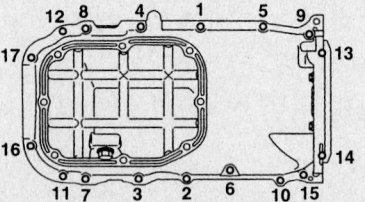

12 8 4 1 5 9
17   13
16   14
11 7 3 2 6 10 15

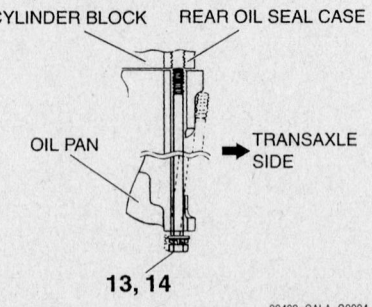

CYLINDER BLOCK   REAR OIL SEAL CASE

OIL PAN   TRANSAXLE SIDE

13, 14

09482_GALA_G0094

Upper oil pan torque sequence—2006 Eclipse and 2004–2006 Galant 3.8L Engine

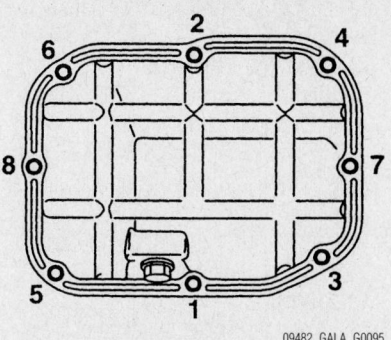

Lower oil pan torque sequence—2006
Eclipse and 2004–2006 Galant 3.8L Engine

8. Install the upper oil pan. Torque the bolts to specification.

➡**The bolt holes for bolt 13 and 14 are cut away on the transaxle side. Be careful not to insert these bolts at an angle.**

9. Install the lower oil pan. Torque the bolts to specification.

10. Continue the installation in the reverse order of the removal procedure.

Galant

2002–2003 2.4L ENGINE

1. Before servicing the vehicle, refer to the Precautions Section.
2. Remove or disconnect the following:
   - Negative battery cable
   - Oil pan drain plug and drain the engine oil
   - Oil dipstick and tube assembly
   - Heated Oxygen (HO$_2$S) sensor connector
   - Front exhaust pipe from the vehicle
   - Bell housing cover
   - Oil pan retainer bolts
3. Tap in between the engine block and the oil pan.

➡**Do not use a pry tool when removing the oil pan. Damage to engine components may occur.**

To install:
4. Apply sealant around the gasket surfaces of the oil pan.
5. Install or connect the following:
   - Oil pan onto the cylinder block within 15 minutes after applying sealant. Tighten to 72 inch lbs. (8 Nm).
   - Oil drain plug and tighten to 29 ft. lbs. (39 Nm)
   - Bell housing cover. Tighten the mounting bolts to 84 inch lbs. (9 Nm).
   - Front exhaust pipe and tighten the

bolts at the catalytic converter to 36 ft. lbs. (49 Nm). Tighten the nuts at the exhaust manifold to 32 ft. lbs. (44 Nm).
   - HO$_2$S sensor connector
6. Fill the crankcase to the proper level.
7. Connect the negative battery cable. Start the engine and check for leaks.

2004–2006 2.4L ENGINE

1. Before servicing the vehicle, refer to the Precautions Section.
2. Remove the negative battery cable.
3. Drain the engine oil.
4. Remove the engine under cover. Remove the front exhaust pipe.
5. Remove the torque converter housing front lower cover.
6. Remove the oil pan retaining bolts. Remove the oil pan. Discard the gasket.

➡**Use tool MD998727 to remove the oil pan. Tap the tool into the range (B) between the cylinder block and the engine oil pan, and then slide the tool sideways. Do not position the tool in area (A) of the engine oil pan as this may cause deformation of the front case because the front case is made of aluminum.**

To install:
7. Clean the sealing surface on the oil pan and engine block. Apply a continuous bead of sealant to the oil pan.
8. Install the oil pan. Torque the bolts to specification.
9. Continue the installation in the reverse order of the removal procedure.

2002–2003 3.0L ENGINE

1. Before servicing the vehicle, refer to the Precautions Section.
2. Remove or disconnect the following:
   - Negative battery cable
   - Oil pan drain plug and drain the engine oil
   - Left side crossmember.
   - Starter motor
   - Roll stopper stay bracket, from the rear transaxle stay bracket
   - Transaxle stay brackets
   - Bell housing lower cover
   - Oil pan mounting bolts
   - The engine oil pan

To install:
3. Apply a 0.16 in. (4mm) continuous bead of sealer around the surface of the oil pan.

➡**Assemble the oil pan to the cylinder block within 15 minutes after applying the sealant.**

4. Install the oil pan mounting bolts. Following proper sequence, tighten mounting bolts to 48 inch lbs. (6 Nm).
5. Install or connect the following:
   - Lower bell housing cover and the starter motor
   - Transaxle stay brackets and connect the roll stopper bracket
   - Crossmember(s) and tighten the mounting bolts to 43–51 ft. lbs. (60–70 Nm)
6. Fill the engine with the proper amount of oil.
7. Connect the negative battery cable and check for leaks.

2004–2006 3.8L ENGINE

1. Before servicing the vehicle, refer to the Precautions Section.
2. Remove the negative battery cable.
3. Remove the engine under cover. Remove the starter. Remove the engine oil dipstick and guide.
4. If removing the upper oil pan, remove the front exhaust pipe. If equipped with automatic transaxle, remove the cover and connecting bolt.
5. Remove the lower engine oil pan retaining bolts. Remove the engine lower oil pan. Discard the gasket.
6. Remove the upper oil pan retaining bolts. Remove the upper oil pan. Discard the gasket.

➡**If the vehicle is equipped with manual transaxle, align the recessed area in the flywheel with the location shown in the illustration. Mark the flywheel. Turn the crankshaft so that the align mark is positioned as shown in the illustration.**

To install:
7. Clean the sealing surface on the oil pan and engine block. Apply a continuous bead of sealant to the oil pan.
8. Install the upper oil pan. Torque the bolts to specification.

➡**The bolt holes for bolt 13 and 14 are cut away on the transaxle side. Be careful not to insert these bolts at an angle.**

9. Install the lower oil pan. Torque the bolts to specification.
10. Continue the installation in the reverse order of the removal procedure.

Lancer

2.0L ENGINE

1. Before servicing the vehicle, refer to the Precautions Section.

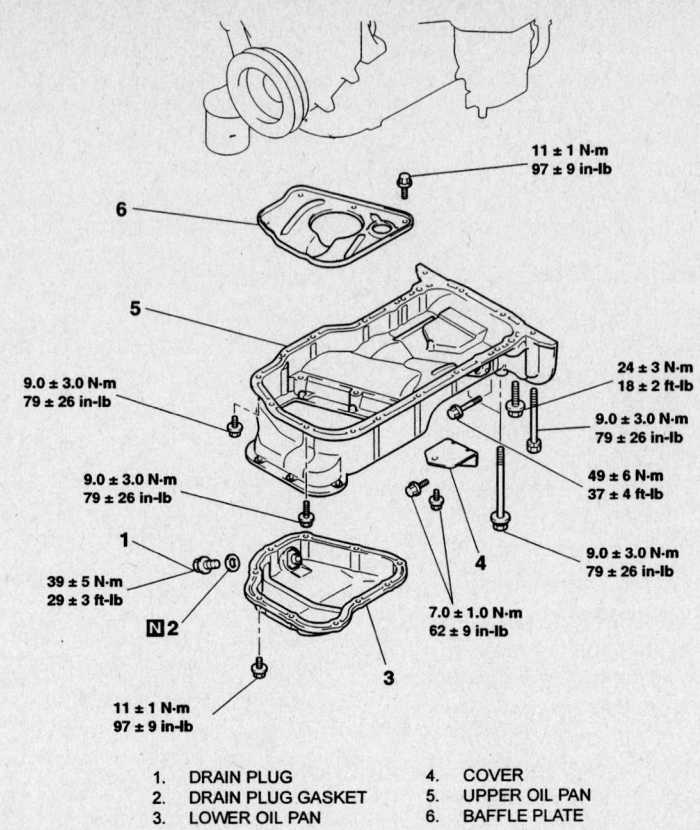

11 ± 1 N·m
97 ± 9 in-lb

24 ± 3 N·m
18 ± 2 ft-lb

9.0 ± 3.0 N·m
79 ± 26 in-lb

49 ± 6 N·m
37 ± 4 ft-lb

9.0 ± 3.0 N·m
79 ± 26 in-lb

9.0 ± 3.0 N·m
79 ± 26 in-lb

9.0 ± 3.0 N·m
79 ± 26 in-lb

39 ± 5 N·m
29 ± 3 ft-lb

7.0 ± 1.0 N·m
62 ± 9 in-lb

11 ± 1 N·m
97 ± 9 in-lb

1. DRAIN PLUG
2. DRAIN PLUG GASKET
3. LOWER OIL PAN
4. COVER
5. UPPER OIL PAN
6. BAFFLE PLATE

93570G33

Oil pan and related components—2.0L engine

2. Disconnect the negative battery cable.
3. Drain the engine oil.
4. Remove or disconnect the following:
- Engine undercover
- Front exhaust pipe
- Lower oil pan bolts and lower pan
- Cover
- Upper oil pan bolt and upper pan
- Baffle plate

To install:

5. Clean all gasket surfaces of the cylinder block and the upper and lower oil pan.
6. Install or connect the following:
- Baffle plate
7. Apply a 0.16 in. (4mm) bead of sealant to the gasket surfaces of the upper oil pan.
- Upper oil pan onto the cylinder block within 15 minutes after applying sealant. Tighten the bolts as shown in the accompanying figure.
8. Apply 0.16 in. (4mm) bead of sealant to the gasket surfaces of the lower oil pan.
- Lower oil pan and tighten the bolts, in the sequence shown, to 88–106 inch lbs. (10–12 Nm)
- Front exhaust pipe

Designation	Symbol	Qty	Diameter × length mm (in)	Tightening torque
Flange Bolt	A	2	6 × 10 (0.2 × 0.4)	7.0 ± 1.0 N·m (62 ± 9 in-lb)
	B	10	6 × 18 (0.2 × 0.7)	9.0 ± 3.0 N·m (79 ± 26 in-lb)
	C	2	6 × 22 (0.2 × 0.9)	
	D	2	8 × 40 (0.3 × 1.6)	24 ± 3 N·m (18 ± 2 ft-lb)
	E	2	10 × 40 (0.4 × 1.6)	49 ± 6 N·m (37 ± 4 ft-lb)
Bolts with Washers	F	2	6 × 50 (0.2 × 2.0)	9.0 ± 3.0 N·m (79 ± 26 in-lb)
	G	2	6 × 127 (0.2 × 5.0)	

93570G18

Upper oil pan bolt location and torque sequence—2.0L engine

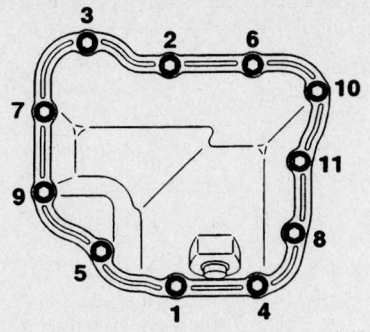

Lower oil pan bolt tightening sequence—2.0L engine

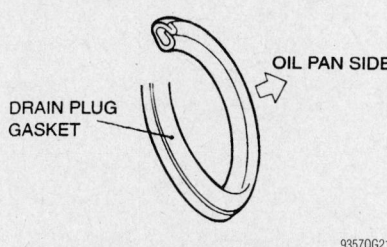

Make sure to the install the new drain plug gasket as shown, or leaks will occur

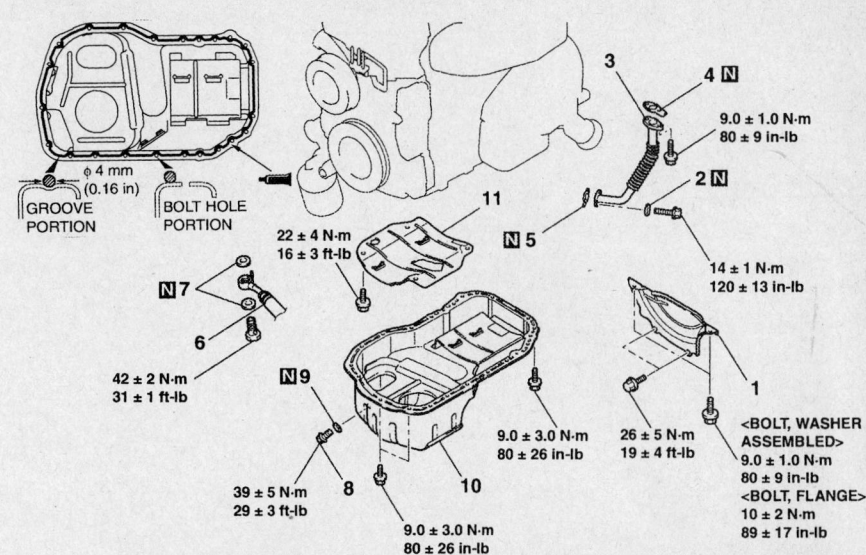

1. FLYWHEEL HOUSING FRONT LOWER COVER
2. GASKET
3. OIL RETURN TUBE
4. OIL RETURN TUBE GASKET
5. OIL RETURN TUBE GASKET
6. ENGINE OIL COOLER RETURN HOSE
7. ENGINE OIL COOLER TUBE GASKETS
8. ENGINE OIL PAN DRAIN PLUG
9. ENGINE OIL PAN DRAIN PLUG GASKET
10. ENGINE OIL PAN
11. CYLINDER BLOCK BAFFLE PLATE

Oil pan and related components—Evolution

• Engine undercover
• Oil drain plug with a new gasket and tighten to 29 ft. lbs. (40 Nm)
9. Lower the vehicle and fill the crankcase to the proper level with clean engine oil.
10. Connect the negative battery cable. Start the engine and check for leaks.

2.4L ENGINE

1. Before servicing the vehicle, refer to the Precautions Section.
2. Remove the negative battery cable.
3. Drain the engine oil.
4. Remove the engine under cover. Remove the front exhaust pipe.
5. Remove the torque converter housing front lower cover.
6. Remove the oil pan retaining bolts. Remove the oil pan. Discard the gasket.

➡Use tool MD998727 to remove the oil pan. Tap the tool into the range (B) between the cylinder block and the engine oil pan, and then slide the tool sideways. Do not position the tool in area (A) of the engine oil pan as this may cause deformation of the front case because the front case is made of aluminum.

To install:
7. Clean the sealing surface on the oil pan and engine block. Apply a continuous bead of sealant to the oil pan.

8. Install the oil pan. Torque the bolts to specification.
9. Continue the installation in the reverse order of the removal procedure.

Evolution

1. Before servicing the vehicle, refer to the Precautions Section.
2. Disconnect the negative battery cable. Remove the engine under cover.
3. Drain the engine oil. Remove the front axle crossmember bar. Remove the front exhaust pipe.
4. Remove the starter. Remove the flywheel housing front lower cover.
5. Remove the oil dipstick and tube assembly.
6. Remove the engine oil cooler return hose and gaskets.
7. Remove the engine oil pan retaining bolts. Remove the oil pan. Discard the gasket.

To install:
8. Clean the sealing surface on the oil pan and engine block. Apply a continuous bead of sealant to the oil pan. Install the oil pan within 15 minutes of applying the sealant.
9. Install the oil pan. Torque the bolts to specification.

10. Continue the installation in the reverse order of the removal procedure.

Mirage

1.5L ENGINE

1. Before servicing the vehicle, refer to the Precautions Section.
2. Disconnect the negative battery cable.
3. Drain the engine oil.
4. Remove or disconnect the following:
• Bell housing lower cover
• Oil pan retainer bolts

➡Do not use a pry tool when removing the oil pan.

To install:
5. Clean all gasket surfaces of the cylinder block and the oil pan.
6. Apply sealant to the gasket surfaces of the oil pan.
7. Install or connect the following:
• Oil pan onto the cylinder block within 15 minutes after applying sealant. Tighten to 60 inch lbs. (7 Nm).
• Bell housing cover
• Oil drain plug with a new seal and tighten to 29 ft. lbs. (40 Nm)
8. Lower the vehicle and fill the

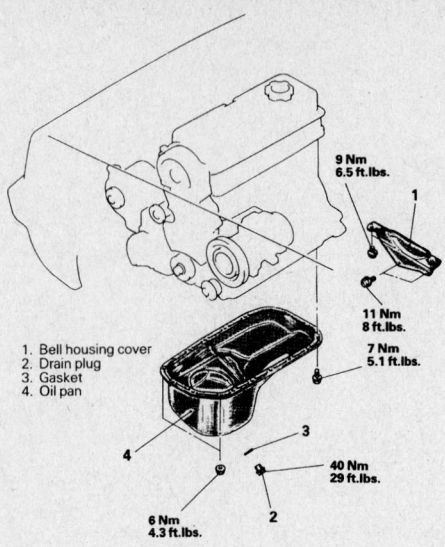

1. Bell housing cover
2. Drain plug
3. Gasket
4. Oil pan

9 Nm
6.5 ft.lbs.

11 Nm
8 ft.lbs.

7 Nm
5.1 ft.lbs.

40 Nm
29 ft.lbs.

6 Nm
4.3 ft.lbs.

7923PG65

Oil pan and related components—1.5L engine

crankcase to the proper level with clean engine oil.

9. Connect the negative battery cable. Start the engine and check for leaks.

1.8L ENGINE

1. Before servicing the vehicle, refer to the Precautions Section.
2. Disconnect the negative battery cable.
3. Raise the vehicle and support safely.
4. Remove or disconnect the following:
   - Oil pan drain plug and drain the engine oil
   - Exhaust pipe from the engine manifold
   - Bell housing lower cover
   - Oil pan retainer bolts and remove the oil pan

➡**Do not use a pry tool when removing the oil pan.**

To install:

5. Clean all gasket surfaces of the cylinder block and the oil pan.
6. Apply sealant around the gasket surfaces of the oil pan.
7. Install or connect the following:
   - Oil pan onto the cylinder block within 15 minutes after applying sealant. Tighten to 60 inch lbs. (5 Nm).
   - Bell housing cover
   - Exhaust pipe to the engine manifold with new gasket in place. Tighten the exhaust pipe to manifold flange nuts to 33 ft. lbs. (45 Nm). Install and tighten the support bolt to 18 ft. lbs. (25 Nm).
   - Oil drain plug and tighten to 29 ft. lbs. (40 Nm)

8. Fill the crankcase to the proper level.
9. Connect the negative battery cable. Start the engine and check for leaks.

Oil Pump

REMOVAL & INSTALLATION

1.5L and 1.8L Engines

➡**Whenever the oil pump is disassembled or the cover removed, the gear cavity must be filled with petroleum jelly to seal the pump and act as a prime. Do not use grease.**

1. Before servicing the vehicle, refer to the Precautions Section.
2. Remove or disconnect the following:
   - Negative battery cable
   - Front engine mount bracket and accessory drive belts
   - Timing belt upper and lower covers
   - Timing belt and crankshaft sprocket
   - Oil pan and remove the oil screen
   - Front cover mounting bolts. Note the lengths of the mounting bolts as they are removed for proper installation.
   - Front case assembly and oil pump assembly
   - Oil pump cover
   - Inner and outer gears from the front case

To install

3. Remove all gasket material from the mating surfaces and clean all parts.
4. Thoroughly coat both oil pump gears with clean engine oil and install them in the correct direction of rotation.

5. Install the pump cover and tighten the bolts to 84 inch lbs. (10 Nm).
6. Coat the relief valve and spring with clean engine oil. Install them and tighten the plug to 33 ft. lbs. (45 Nm).
7. Install or connect the following:
   - New front crankshaft seal and coat the lips of the seal with clean engine oil
   - Front case and oil pump assembly to the engine block using a new gasket. Tighten the bolts to 10 ft. lbs. (14 Nm).
   - Oil screen with new gasket. Torque the screen bolts to 14 ft. lbs. (19 Nm).
   - Oil pan
   - Crankshaft sprocket and timing belt
8. Fill the crankcase to the proper level.
9. Connect the negative battery cable.

2.0L SOHC Engine

1. Before servicing the vehicle, refer to the Precautions Section.
2. Drain the engine oil.
3. Remove or disconnect the following:
   - Negative battery cable
   - Oil pressure switch
   - Oil filter
   - Drain plug and gasket. Discard the gasket.
   - Cover
   - Upper oil pan. Remove the 5 in. (127mm) bolt, which is closest to the flywheel/flexplate first, then, the other bolts.
   - Baffle plate
   - Lower oil pan. Remove the 5 in. (127mm) bolt, which is closest to the flywheel/flexplate first, then, the other bolts.
   - Oil screen and gasket
   - Relief plug and spring
   - Oil seal
   - Front case
   - O-ring
   - Oil pump case cover

➡**Matchmark the installed position of the pump rotors before removing them.**

   - Outer and inner oil pump rotors

To install:

4. Install or connect the following:
   - Inner and outer rotors, making sure the alignment marks are matched up
   - Oil pump case cover
   - O-ring

➡**After installation or the front case, wait at least one hour before filling the crankcase with oil or starting the engine.**

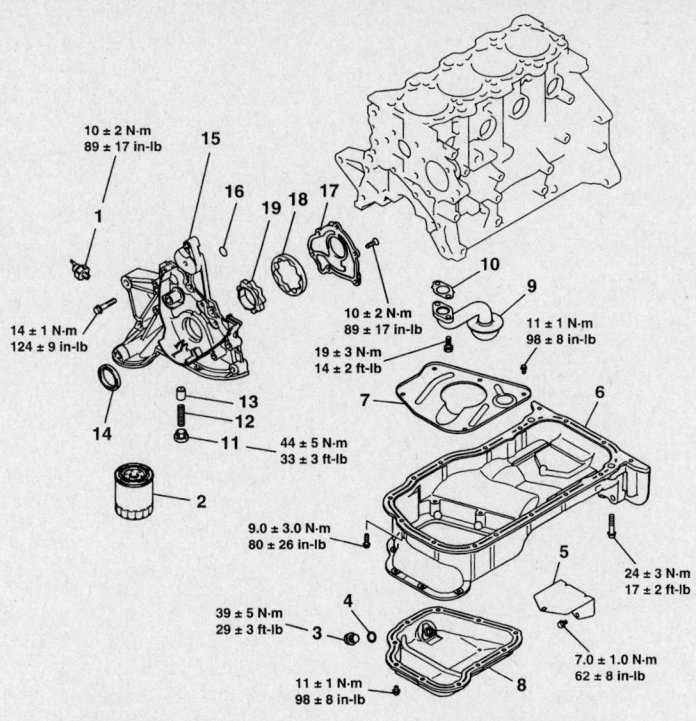

1. ENGINE OIL PRESSURE SWITCH
2. OIL FILTER
3. DRAIN PLUG
4. GASKET
5. COVER
6. UPPER OIL PAN
7. BAFFLE PLATER
8. LOWER OIL PAN
9. OIL SCREEN
10. OIL SCREEN GASKET
11. RELIEF PLUG
12. RELIEF SPRING
13. RELIEF PLUNGER
14. OIL SEAL
15. OIL PUMP CASE
16. O-RING
17. OIL PUMP CASE COVER
18. OUTER ROTOR
19. INNER ROTOR

09482_GALA_G0098

Oil pump and related components—2002–2006 Lancer 2.0L Engine

- Front case. Apply a 0.12 inch (3mm) bead of sealant, then tighten the case bolts to 124 inch lbs. (14 Nm).
- Oil seal
- Relief plunger, spring and plug
- Oil screen gasket and screen
- Lower oil pan. Refer to the oil pan procedure for sealant application and torque specifications.
- Baffle plate
- Upper oil pan. Refer to the oil pan procedure for sealant application and torque specifications.
- Cover
- Drain plug with a new gasket. Tighten the plug to 29 ft. lbs. (35 Nm).
- Oil filter
- Oil pressure switch

5. Fill the engine with the correct amount of oil.
6. Connect the negative battery cable.
7. Start the engine and check for leaks.

2.0L DOHC and 2.4L Engines

➡Whenever the oil pump is disassembled or the cover removed, the gear cavity must be filled with petroleum jelly to seal the pump and act as a prime. Do not use grease.

1. Before servicing the vehicle, refer to the Precautions Section.
2. Disconnect the negative battery cable. Rotate the engine so No. 1 cylinder is on Top Dead Center (TDC) of its compression stroke.
3. Drain the engine oil.
4. Using the proper equipment, support the weight of the engine. Remove the front engine mount bracket and accessory drive belts.
5. Remove or disconnect the following:
   - Timing belt upper and lower covers
   - Timing belt and crankshaft sprocket
   - Electrical connector from the oil pressure sending unit
   - Oil pressure sensor
   - Oil filter and the oil filter bracket
   - Oil pan, oil screen and gasket
6. Using special tool MD998162, remove the plug cap in the engine front cover.
7. Remove or disconnect the following:
   - Plug on the side of the engine block. Insert a steel rod with a shank diameter of 0.32 in. (8mm) into the plug hole. This will hold the silent shaft.
   - Driven gear bolt that secures the oil pump driven gear to the silent shaft
   - Front cover mounting bolts. Note the lengths of the mounting bolts as they are removed for proper installation.
   - Front case cover and oil pump assembly. If necessary, the silent shaft can come out with the cover assembly.
   - Oil pump cover, located on the back of the engine front cover. Remove the oil pump drive and driven gears.

8. After disassembling the oil pump, clean all components and remove gasket material from mating surfaces.
9. Assemble the oil pump gears into the front case and rotate it to ensure smooth rotation and no looseness. Be sure there is no ridge wear on the contact surface between the front case and the gear surface of the oil pump front cover.

To install

10. Align the timing mark on the oil pump drive gear with that on the driven gear and install them into the engine front case. Apply engine oil to the gears.
11. Install the oil pump cover and tighten the retainer bolts to 13 ft. lbs. (18 Nm) on Eclipse models and 17 ft. lbs. (24 Nm) on Galant models.
12. Using the appropriate driver, install a new crankshaft seal into the front case.
13. Position new front case gasket in place. Set seal guide tool MD998285 on the front end of the crankshaft to protect the seal from damage. Apply a thin coat of oil to the outer circumference of the seal pilot tool.
14. Install the front case assembly through a new front case gasket and temporarily tighten the flange bolts.
15. Mount the oil filter on the bracket with new oil filter bracket gasket in place. Install the bolts with washers and tighten to 14 ft. lbs. (19 Nm).
16. Insert a Phillips screwdriver into the hole in the left side of the engine block to lock the silent shaft in place.
17. Install or connect the following:
   - Oil pump drive gear onto the left silent shaft. Tighten the driven gear bolt to 27 ft. lbs. (37 Nm).
   - New O-ring to the groove in the front case and install the plug cap. Tighten the cap to 17 ft. lbs. (24 Nm).

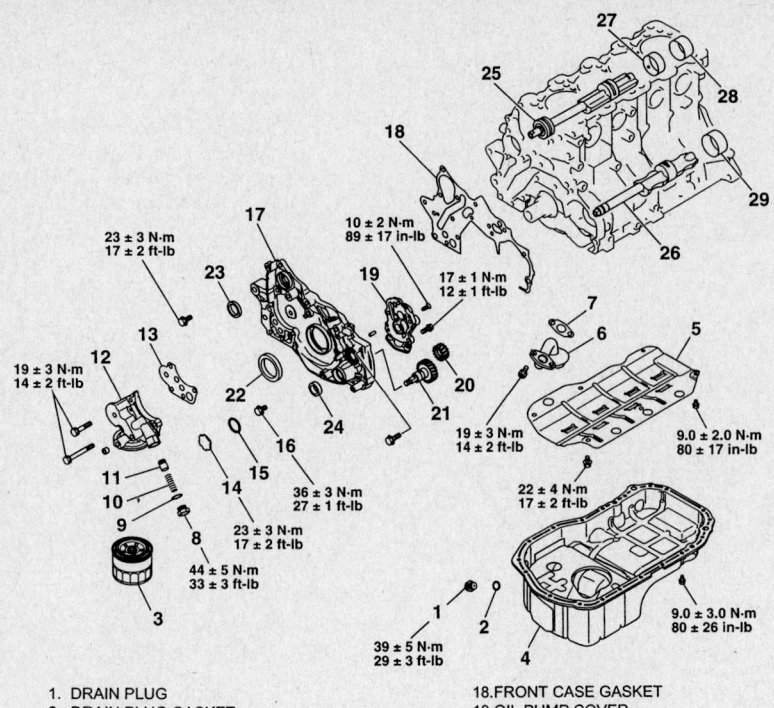

1. DRAIN PLUG
2. DRAIN PLUG GASKET
3. OIL FILTER
4. OIL PAN
5. BAFFLE PLATE
6. OIL SCREEN
7. OIL SCREEN GASKET
8. RELIEF PLUG
9. GASKET
10. RELIEF SPRING
11. RELIEF PLUNGER
12. OIL FILTER BRACKET
13. OIL FILTER BRACKET GASKET
14. PLUG
15. O-RING
16. FLANGE BOLT
17. FRONT CASE

18. FRONT CASE GASKET
19. OIL PUMP COVER
20. OIL PUMP DRIVEN GEAR
21. OIL PUMP DRIVE GEAR
22. CRANKSHAFT FRONT OIL SEAL
23. OIL PUMP OIL SEAL
24. COUNTERBALANCE SHAFT OIL SEAL
25. COUNTERBALANCE SHAFT, LEFT
26. COUNTERBALANCE SHAFT, RIGHT
27. COUNTERBALANCE SHAFT, FRONT BEARING
28. COUNTERBALANCE SHAFT, REAR BEARING, RIGHT
29. COUNTERBALANCE SHAFT, REAR BEARING, LEFT

09482_GALA_G0097

Oil pump and related components—2006 Eclipse, 2004–2006 Galant and 2004–2006 Lancer 2.4L Engine

- Oil screen in position with new gasket in place

18. Clean both mating surfaces of the oil pan and the cylinder block. Apply sealant in the groove in the oil pan flange.

➡**After applying sealant to the oil pan, do not exceed 15 minutes before installing the oil pan.**

19. Install or connect the following:
- Oil pan to the engine and secure with the retainers. Tighten bolts to 60 inch lbs. (7 Nm).
- Oil pressure gauge unit and the oil pressure switch
- Electrical harness connector
- Oil cooler. Oil cooler bolt to 31 ft. lbs. (43 Nm).

20. Refill the crankcase. Install new oil filter.

21. Connect the negative battery cable and start the engine. Verify correct oil pressure. Inspect for leaks.

3.0L Engine

➡**Whenever the oil pump is disassembled or the cover removed, the gear cavity must be filled with petroleum jelly to seal the pump and act as a prime. Do not use grease.**

1. Before servicing the vehicle, refer to the Precautions Section.
2. Disconnect the negative battery cable.
3. Drain the engine oil.
4. Remove or disconnect the following:

- Front engine mount bracket and accessory drive belts
- Timing belt upper and lower covers
- Timing belt and crankshaft sprocket
- Oil pan
- Oil screen and gasket
- Front cover mounting bolts. Note the lengths of the mounting bolts as they are removed for proper installation.

- Front case cover and oil pump assembly

To install:

5. Thoroughly clean all gasket material from all mounting surfaces.
6. Apply engine oil to the entire surface of the gears or rotors.
7. Assemble the front case cover and oil pump assembly to the engine block.
8. Install or connect the following:
- Oil screen with new gasket
- Oil pan
- Crankshaft sprocket and timing belt
- Timing belt covers
- Drive belts and the front engine mount bracket
- Negative battery cable, refill the crankcase and check for adequate oil pressure

3.5L and 3.8L Engines

1. Before servicing the vehicle, refer to the Precautions Section.
2. Remove or disconnect the following:
- Negative battery cable
- Timing belt
3. Drain the engine oil.
4. Remove or disconnect the following:

- Splash shield from the wheel well, as needed
- Oil filter adapter
- Lower and upper oil pans
- Lower baffle, oil pump pick-up and upper baffle
- Oil pump case mounting bolts and the oil pump case
- Oil pump gear cover

5. Make matchmarks on the oil pump rotors before removing them.
- Crankshaft seal from the oil pump case

To install:

6. Install or connect the following:
- New crankshaft seal in the oil pump cover

7. Apply engine oil to the rotors, then align the matchmarks and install the rotors in the oil pump case.
- Rotor cover and tighten the bolts to 84 inch lbs. (10 Nm)

8. Apply a 0.113 in. (3mm) bead of sealant to the back of the oil pump case. Install the case on the engine and tighten the bolts to 10 ft. lbs. (13 Nm).
- Upper baffle plate and oil pump pick-up using a new gasket—Tighten the baffle bolts to 84 inch lbs. (10 Nm) and the pick-up bolts to 13 ft. lbs. (18 Nm).
- Lower baffle in the upper oil pan.

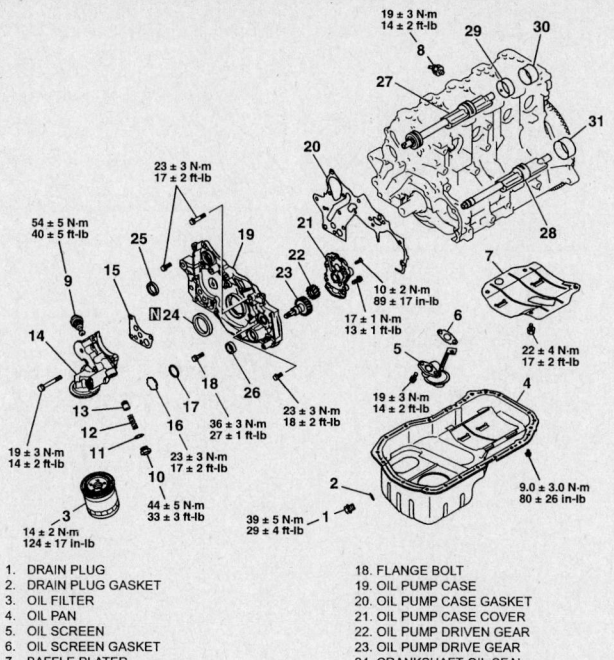

1. DRAIN PLUG
2. DRAIN PLUG GASKET
3. OIL FILTER
4. OIL PAN
5. OIL SCREEN
6. OIL SCREEN GASKET
7. BAFFLE PLATER
8. OIL PRESSURE SWITCH
9. OIL COOLER BY-PASS VALVE
10. RELIEF PLUG
11. GASKET
12. RELIEF SPRING
13. RELIEF PLUNGER
14. OIL FILTER BRACKET
15. OIL FILTER BRACKET GASKET
16. PLUG CAP
17. O-RING
18. FLANGE BOLT
19. OIL PUMP CASE
20. OIL PUMP CASE GASKET
21. OIL PUMP CASE COVER
22. OIL PUMP DRIVEN GEAR
23. OIL PUMP DRIVE GEAR
24. CRANKSHAFT OIL SEAL
25. COUNTERBALANCE SHAFT OIL SEAL
26. OIL PUMP OIL SEAL
27. COUNTERBALANCE SHAFT, RIGHT
28. COUNTERBALANCE SHAFT, LEFT
29. COUNTERBALANCE SHAFT FRONT BEARING
30. COUNTERBALANCE SHAFT REAR BEARING, RIGHT
31. COUNTERBALANCE SHAFT REAR BEARING, LEFT

09482_GALA_G0099

Oil pump and related components—Evolution

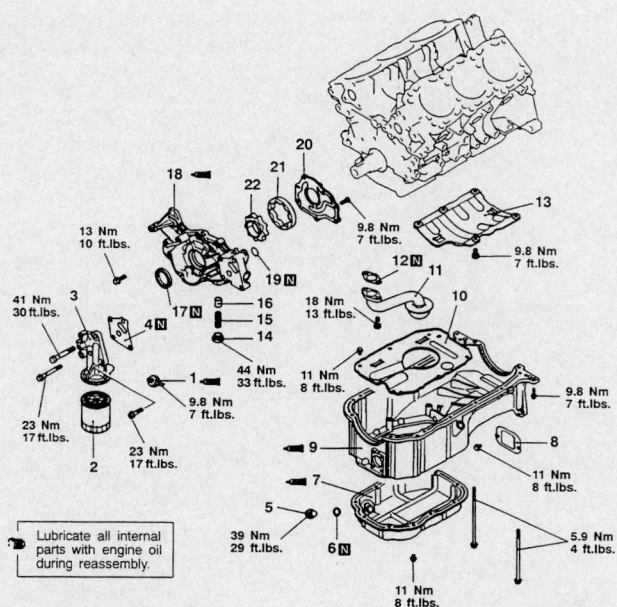

Lubricate all internal parts with engine oil during reassembly.

Removal steps

1. Oil pressure gauge unit
2. Oil filter
3. Oil filter bracket
4. Oil filter bracket gasket
5. Drain plug
6. Drain plug gasket
7. Oil pan, lower
8. Cover
9. Oil pan, upper
10. Baffle plate
11. Oil screen
12. Oil screen gasket
13. Baffle plate
14. Plug
15. Relief spring
16. Relief plunger
17. Crankshaft oil seal
18. Oil pump case
19. O-ring
20. Oil pump cover
21. Oil pump outer rotor
22. Oil pump inner rotor

7923PG77

Oil pump and related components—3.0L, 3.5L and 3.8L engines

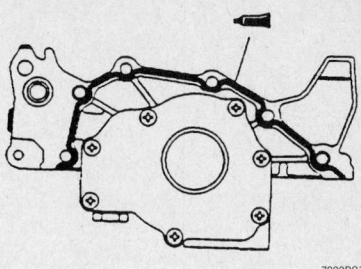

7923PG78

Rear oil pump case sealant application—3.5L engine

Tighten the bolts to 96 inch lbs. (11 Nm).

- Oil pans
- Oil filter adapter using a new gasket. Tighten the larger bolt to 30 ft. lbs. (41 Nm) and the smaller bolt to 17 ft. lbs. (23 Nm).
- Timing belt and remaining components

9. Fill the engine with the correct amount of oil.
10. Connect the negative battery cable.
11. Start the engine and check for leaks.

1.5L and 1.8L Engines

➡**Whenever the oil pump is disassembled or the cover removed, the gear cavity must be filled with petroleum jelly to seal the pump and act as a prime. Do not use grease.**

1. Before servicing the vehicle, refer to the Precautions Section.
2. Remove or disconnect the following:

- Negative battery cable
- Front engine mount bracket and accessory drive belts
- Timing belt upper and lower covers
- Timing belt and crankshaft sprocket
- Oil pan and remove the oil screen
- Front cover mounting bolts. Note the lengths of the mounting bolts as they are removed for proper installation.
- Front case assembly and oil pump assembly
- Oil pump cover
- Inner and outer gears from the front case

To install

3. Remove all gasket material from the mating surfaces and clean all parts.
4. Thoroughly coat both oil pump gears with clean engine oil and install them in the correct direction of rotation.
5. Install the pump cover and tighten the bolts to 84 inch lbs. (10 Nm).
6. Coat the relief valve and spring with

clean engine oil. Install them and tighten the plug to 33 ft. lbs. (45 Nm).

7. Install or connect the following:
- New front crankshaft seal and coat the lips of the seal with clean engine oil
- Front case and oil pump assembly to the engine block using a new gasket. Tighten the bolts to 10 ft. lbs. (14 Nm).
- Oil screen with new gasket. Torque the screen bolts to 14 ft. lbs. (19 Nm).
- Oil pan
- Crankshaft sprocket and timing belt

8. Fill the crankcase to the proper level.
9. Connect the negative battery cable.

Rear Main Seal

REMOVAL & INSTALLATION

1. Before servicing the vehicle, refer to the Precautions Section.
2. Remove or disconnect the following:
- Transaxle
- Clutch cover and disc
- Flywheel or flexplate from the crankshaft

3. Carefully pry the seal out of the oil seal case without damaging the sealing surface of the crankshaft.

To install:

4. Apply engine oil to the lip of the new seal and install the seal in the case using the proper size seal driver.
5. Install or connect the following:

- Flywheel or flexplate
- Clutch cover and disc
- Transaxle

Timing Belt, Cover and Crankshaft Seal

REMOVAL & INSTALLATION

Diamante

1. Before servicing the vehicle, refer to the Precautions Section.
2. Position the engine so the No. 1 cylinder is at Top Dead Center (TDC) of its compression stroke.
3. Remove all necessary components for access to the timing belt covers, then remove the covers from the engine.

✳✳ CAUTION

Be sure to disconnect the negative battery cable. Wait at least 90 seconds after the negative battery cable is disconnected to prevent possible deployment of the air bag.

4. If the same timing belt will be reused, mark the direction of the timing belt's rotation for installation in the same direction. Be sure the engine is positioned so the No. 1 cylinder is at the TDC of its compression stroke and the timing marks are aligned with the engine's timing mark indicators on the rear timing covers.
5. Remove the timing belt.

✳✳ WARNING

Turning the camshaft sprocket when the timing belt is removed could cause the valves to contact with the pistons, resulting in severe engine damage.

6. Remove the bolts that secure the auto-tensioner to the engine block and remove the tensioner.
To install:
➡**The auto-tensioner assembly must be reset to correctly adjust belt tension.**

7. Loosen the center bolt of tensioner pulley to provide timing belt slack. Remove the timing belt tensioner assembly.
8. Position the auto-tensioner into a vise with soft jaws. The plug at the rear of tensioner protrudes, be sure to use a washer as a spacer to protect the plug from contacting vise jaws.
9. Slowly push the rod into the tensioner until the set hole in rod is aligned with set hole in the auto-tensioner.
10. Insert a 0.055 in. (1.4mm) wire into the aligned set holes. Unclamp the tensioner from the vise and install it on the engine. Tighten tensioner mounting bolts to 17 ft. lbs. (24 Nm).

✳✳ WARNING

DO NOT rotate or turn the camshafts when removing the sprockets or severe engine damage will result from internal component interference.

11. Align the mark on the crankshaft sprocket with the mark on the front case. Then, move the crankshaft sprocket 3 teeth counterclockwise.
12. Align the timing marks of the camshafts with the marks on the rear covers.
13. Realign the crankshaft pulley with timing mark on the housing.

➡**Be sure camshafts-to-cylinder heads and crankshaft-to-front cover timing marks are aligned.**

14. Install the timing belt around the pulleys in the following order:
    a. Crankshaft pulley.
    b. Idler pulley.
    c. Left camshaft sprocket.
    d. Water pump pulley.
    e. Right camshaft sprocket.
    f. Tensioner pulley.

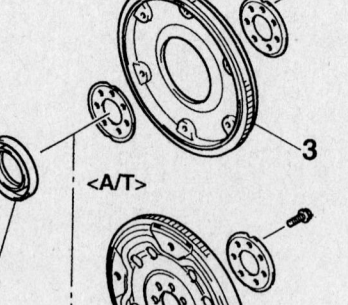

<M/T>

<A/T>

09482_GALA_G0100

Exploded view of a typical rear main seal

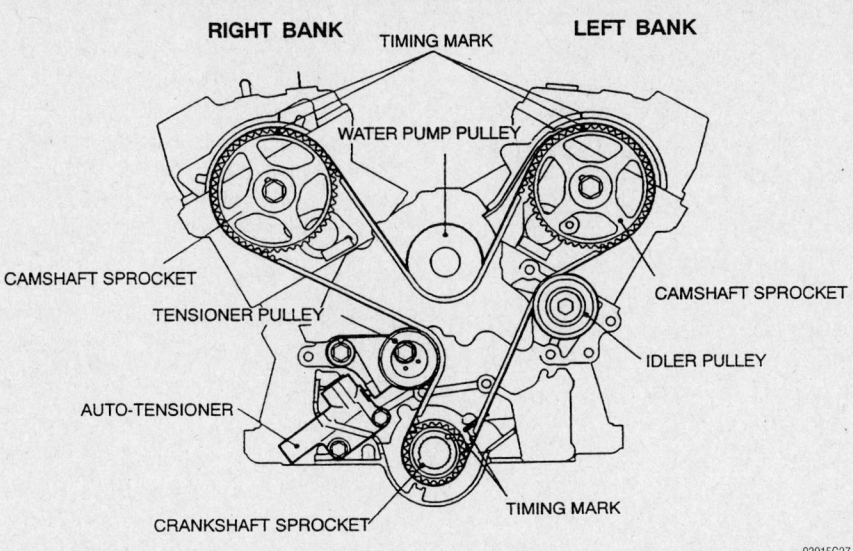

Timing belt sprocket alignment —3.5L engine

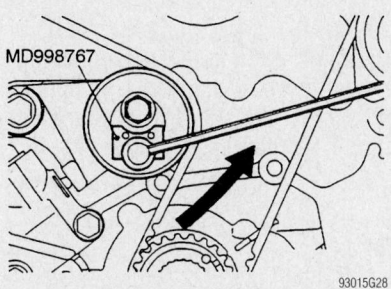

Special tool used for tightening timing belt—3.5L engine

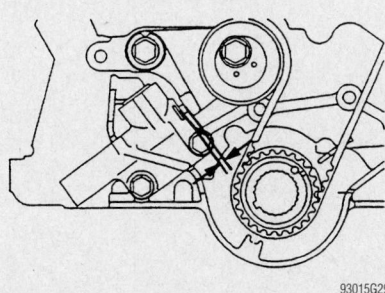

Measuring the standard value of the timing belt tensioner—3.5L engine

➡ **Since the camshaft sprockets turn easily because of spring action, be careful not to get your fingers caught.**

15. Align all timing mark on the crankshaft and raise tensioner pulley against belt to remove slack, snug tensioner bolt.

16. Check the alignment of all the timing.

17. Using special tool MD998769, rotate the crankshaft ¼ turn counterclockwise,

then rotate the crankshaft clockwise to align the timing marks. Check that all the timing marks are in alignment.

18. Loosen the center bolt on the tensioner pulley. Using tool MD998767 and a torque wrench, apply 3.3 ft. lbs. (4.4 Nm) to the tool on the tensioner. Tighten the ten-

sioner bolt to 33 ft. lbs. (44 Nm) and be sure the tensioner does not rotate with the bolt.

19. Rotate the crankshaft two complete turns clockwise and let it sit for approximately five minutes. Then, check that the set pin can easily be inserted and removed from the hole in the auto-tensioner.

20. Remove the set wire attached to the auto-tensioner.

21. Measure the auto-tensioner protrusion (the distance between the tensioner arm and auto-tensioner body) to ensure that it is within 0.150–0.196 in. (3.8–5.0mm). If out of specification, repeat adjustment procedure until the specified value is obtained.

22. Check again that the timing marks on all sprockets are in proper alignment.

23. Install the timing belt covers and all other applicable components.

Eclipse

2002–2005 2.4L ENGINE

1. Before servicing the vehicle, refer to the Precautions Section.

2. Position the engine so that the No. 1 piston is at Top Dead Center (TDC).

3. Remove the timing belt covers.

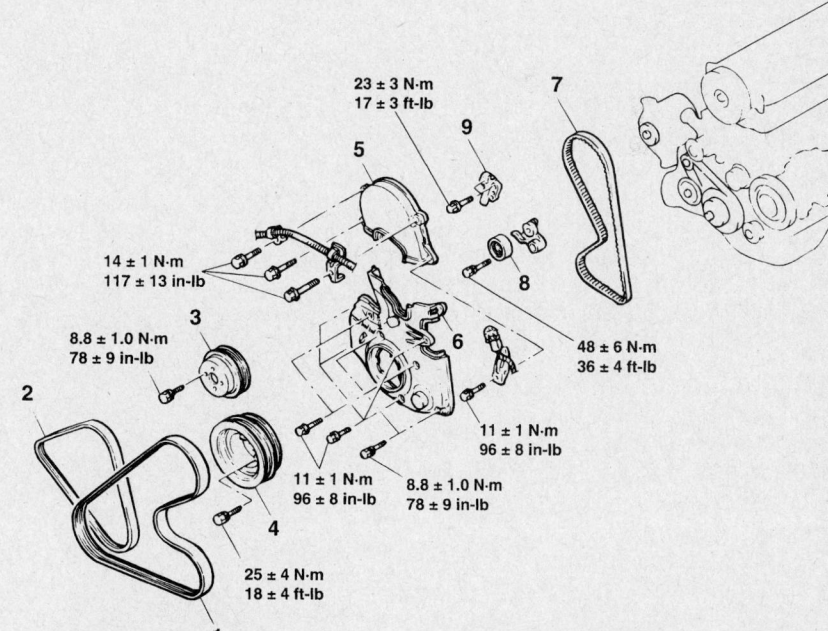

1. DRIVE BELT (POWER STEERING OIL PUMP AND A/C COMPRESSOR)
2. DRIVE BELT (GENERATOR)
3. WATER PUMP PULLEY
4. CRANKSHAFT PULLEY
5. TIMING BELT UPPER COVER ASSEMBLY
6. TIMING BELT LOWER COVER ASSEMBLY
• TIMING BELT TENSION ADJUSTMENT
7. TIMING BELT
8. TENSIONER PULLEY
9. AUTO-TENSIONER

Timing belt and related components—2002–2005 Eclipse and 2002–2003 Galant 2.4L engine

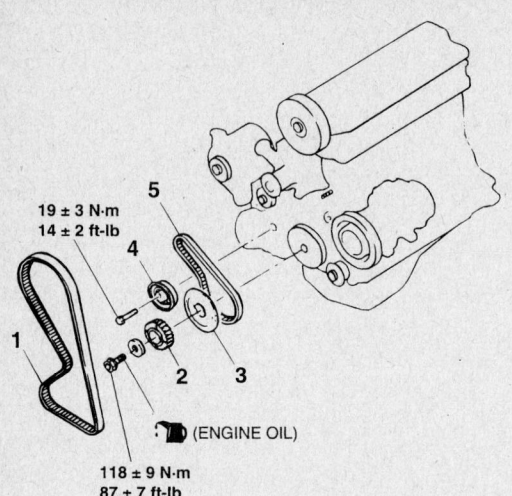

1. TIMING BELT
2. CRANKSHAFT SPROCKET
3. CRANKSHAFT SENSING BLADE
4. TIMING BELT B TENSIONER
5. TIMING BELT B

19 ± 3 N·m
14 ± 2 ft-lb

118 ± 9 N·m
87 ± 7 ft-lb

(ENGINE OIL)

09482_GALA_G0104

Timing belt "B" (balancer belt) and related components—2002–2005 Eclipse and 2002–2003 Galant 2.4L engine

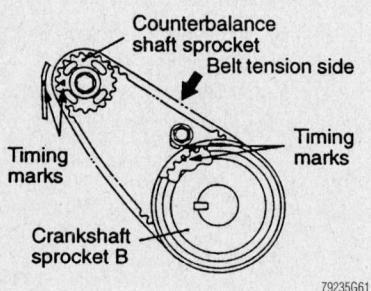

Sprocket alignment timing belt "B" (balancer belt)—2002–2005 Eclipse and 2002–2003 Galant 2.4L engine

➡ If the timing belts are going to be reused, mark the direction of rotation on the belt. This will ensure the belt is reinstalled in same direction, extending belt life.

4. To loosen the timing (outer) belt tensioner, install Mitsubishi Special tool MD998738 to the slot and screw inward to move the tensioner toward the water pump. Once the tension has been relieved, remove the outer timing belt.

5. If tensioner replacement is required, align the pin hole in the tensioner rod to the hole in the tensioner cylinder. Insert a 0.055 in. (1.4mm) wire in the hole and remove the special tool from the slot. With the cylinder tension relieved, remove the auto-tensioner cylinder assembly two mounting bolts.

6. Remove the outer crankshaft sprocket and flange.

7. Loosen the silent shaft (inner) belt tensioner and remove the belt.

To install:

<mark style="background: black; color: white;">✳✳ WARNING</mark>

Do not spray or immerse the sprockets or tensioners in cleaning solvent. The sprocket may absorb the solvent and transfer it to the belt. The tensioners are internally lubricated and the solvent will dilute or dissolve the lubricant.

8. Align the timing marks of the silent shaft sprockets and the crankshaft sprocket with the timing marks on the front case. Route the timing belt around the sprockets so there is no slack in the upper span of the belt and the timing marks are still aligned.

9. Install the tensioner pulley and move the pulley by hand so the long side of the belt deflects approximately ¼ in. (6mm).

10. Hold the pulley tightly so the pulley cannot rotate when the bolt is tightened. Tighten the bolt to 14 ft. lbs. (19 Nm) and recheck the deflection.

11. Align the timing marks of the camshaft, crankshaft and oil pump sprockets with their corresponding marks on the front case or rear cover.

➡ There is a possibility to align all timing marks and have the oil pump sprocket and silent shaft out of time, causing an engine vibration during operation. If the following step is not followed exactly, there is a 50 percent chance that the silent shaft alignment will be 180 degrees (½ turn) off.

12. Before installing the timing belt, ensure that the left side (rear) silent shaft

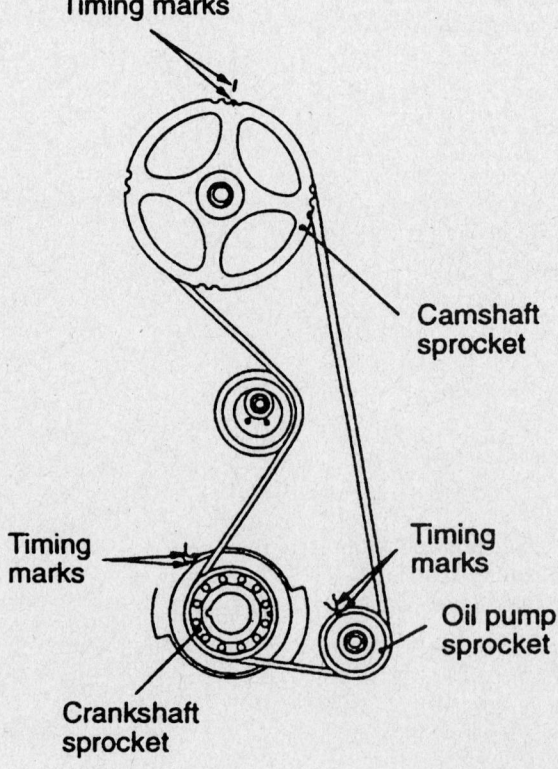

Timing belt sprocket alignment—2002–2002–2005 Eclipse and 2002–2003 Galant 2.4L engine

(oil pump sprocket) is in the correct position as follows:

a. Remove the plug from the rear side of the block and insert a tool with shaft diameter of 0.31 in. (8mm) into the hole.

b. With the timing marks still aligned, the shaft of the tool must be able to go in at least 2 ½ in. (63.5mm). If the tool can only go in approximately 1 in. (25mm), the shaft is not in the correct orientation and will cause a vibration during engine operation. Remove the tool from the hole and turn the oil pump sprocket one complete revolution. Realign the timing marks and insert the tool. The shaft of the tool must go in at least 2 ¼ in. (63.5mm).

c. Recheck and realign the timing marks.

d. Leave the tool in place to hold the silent shaft while continuing.

13. If the camshaft belt tensioner was removed, use a vise to carefully push the auto-tensioner rod in until the set hole in the rod is aligned with the hole in the cylinder. Place a wire into the hole to retain the rod. Mount the tensioner to the engine block and tighten the mounting bolt to 17 ft. lbs. (23 Nm).

14. Install the belt to the crankshaft sprocket, oil pump sprocket, then camshaft sprocket, in that order. While doing so, be sure there is no slack between the sprocket except where the tensioner is installed.

15. To adjust the timing (outer) belt perform the following steps:

a. Turn the crankshaft ¼ turn counterclockwise, then turn it clockwise to move No. 1 cylinder to TDC.

b. Loosen the center bolt. Using tool MD998752 and a torque wrench, apply a torque of 2.6 ft. lbs. (3.6 Nm) to the tensioner. Tighten the center bolt.

c. Screw the special tool into the engine left support bracket until its end makes contact with the tensioner arm. At this point, screw the special tool in some more and remove the set wire attached to the auto-tensioner, if the wire was not previously removed. Then, remove the special tool.

d. Rotate the crankshaft two complete turns clockwise and let it sit for approximately 15 minutes. Then, measure the auto-tensioner protrusion (the distance between the tensioner arm and auto-tensioner body) to ensure that it is within 0.15–0.18 in. (3.8–4.5mm). If out of specification, repeat sub steps **a** through **d** until the specified value is obtained.

➡**Do not manually overtighten the belt or it will howl.**

16. Install the upper and lower timing belt covers.

2006 2.4L ENGINE

1. Before servicing the vehicle, refer to the Precautions Section.

2. Disconnect the negative battery cable. Remove the engine under cover.

3. Remove the crankshaft damper pulley.

4. Disconnect the control wiring harness connector, battery wiring harness connector and connector bracket to engine mounting insulator.

5. Remove the harness bracket. Remove the upper timing belt cover. Remove the water pump pulley. Remove the idler pulley.

6. Remove the auto tensioner. Remove the lower timing belt cover.

7. Turn the crankshaft clockwise and align each timing mark to set the number one piston to TDC of its compression stroke.

8. Remove the timing belt under cover

rubber plug and then install special tool MD998738. Screw the special tool until it contacts the timing belt tensioner arm.

➡**The special tool must be screwed in gradually at the rate of a 30 degree turn per second. If it is screwed in all at once, the timing belt tensioner adjuster rod will not easily retract and the tool may bend.**

9. Gradually screw in the special tool and then align the timing belt tensioner adjuster rod set hole "A" with the timing belt tensioner adjuster cylinder set hole "B".

10. Insert a wire or pin in the set holes to lock the assembly in place. After removing the special tool, loosen the timing belt tensioner pulley mounting bolts and remove the timing belt. If the belt is being reused be sure to mark the direction of rotation (clockwise) on the belt.

11. Remove the timing belt tensioner pulley, tensioner arm and adjuster.

12. Remove the timing belt idler pulley.

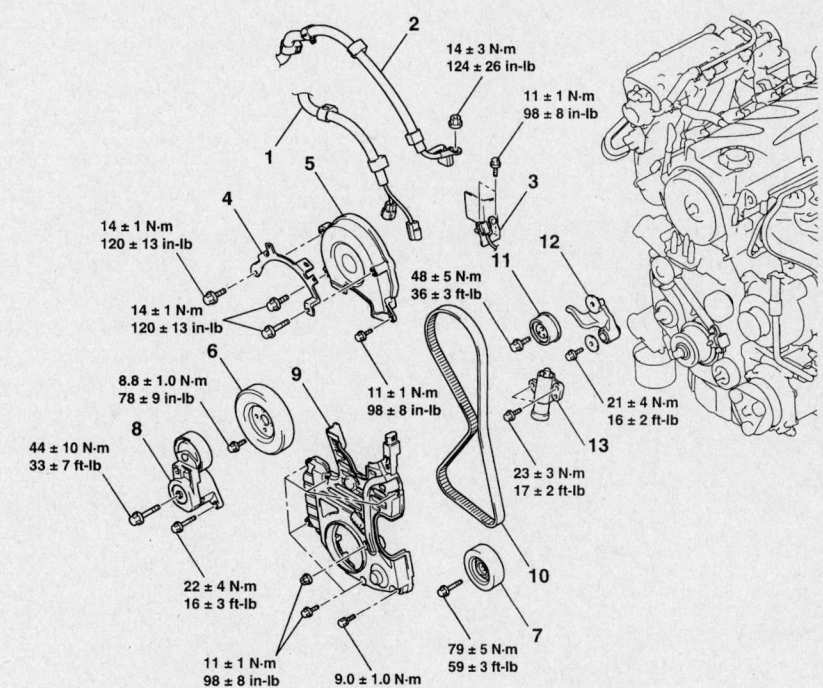

1. CONTROL WIRING HARNESS CONNECTION
2. BATTERY WIRING HARNESS CONNECTION
3. CONNECTOR BRACKET
4. HARNESS BRACKET
5. TIMING BELT UPPER COVER
6. WATER PUMP PULLEY
7. IDLER PULLEY
8. AUTO-TENSIONER
9. TIMING BELT LOWER COVER
• VALVE TIMING BELT TENSION ADJUSTMENT (INSTALLATION ONLY)
10. VALVE TIMING BELT
11. TIMING BELT TENSIONER PULLEY
12. TIMING BELT TENSIONER ARM
13. TIMING BELT TENSIONER ADJUSTER

09482_GALA_G0105

Timing belt and related components—2006 Eclipse, 2004–2006 Galant and 2004–2006 Lancer 2.4L engine

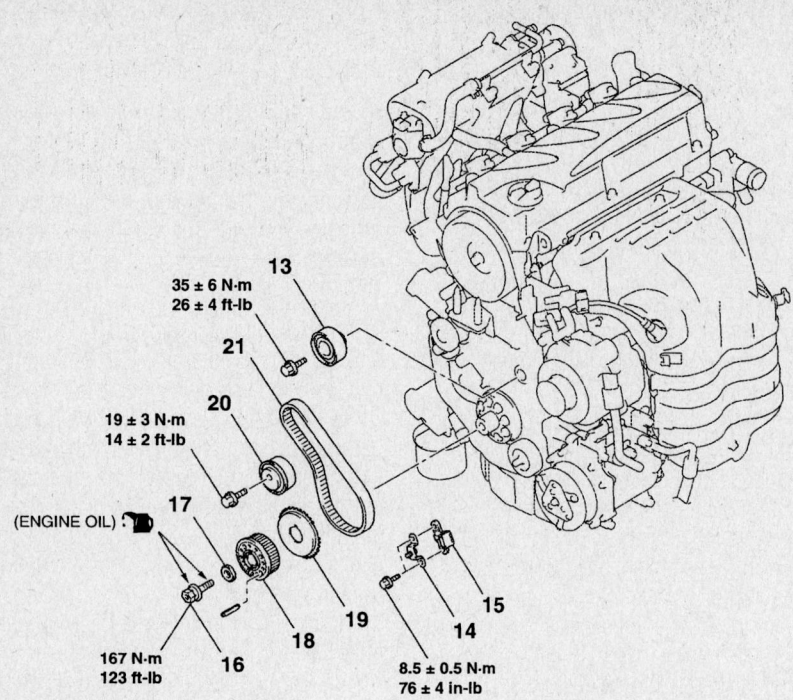

35 ± 6 N·m
26 ± 4 ft-lb

19 ± 3 N·m
14 ± 2 ft-lb

(ENGINE OIL)

167 N·m
123 ft-lb

8.5 ± 0.5 N·m
76 ± 4 in-lb

13. TIMING BELT IDLER PULLEY
14. TIMING BELT LOWER COVER
    BRACKET
15. CRANKSHAFT POSITION
    SENSOR
16. CRANKSHAFT PULLEY CENTER
    BOLT
17. CRANKSHAFT PULLEY WASHER
18. CRANKSHAFT CAMSHAFT
    DRIVE SPROCKET

19. CRANKSHAFT ANGLE SENSING
    BLADE
 • BALANCER TIMING BELT
    TENSION ADJUSTMENT
    (INSTALLATION ONLY)
20. BALANCER TIMING BELT
    TENSIONER
21. BALANCER TIMING BELT

09482_GALA_G0106

Timing belt "B" (balancer belt) and related components—2006 Eclipse, 2004–2006 Galant and 2004–2006 Lancer 2.4L engine

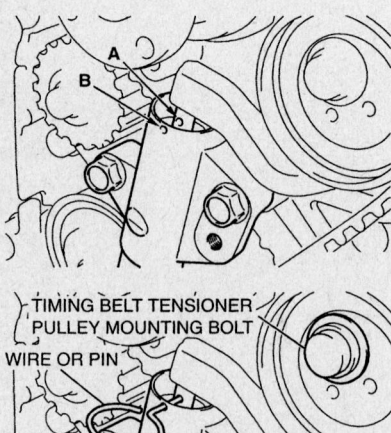

09482_GALA_G0107

Timing belt tensioner adjuster rod set hole "A" and cylinder set hole "B"—2006 Eclipse, 2004–2006 Galant and 2004–2006 Lancer 2.4L engine

Remove the timing belt lower cover bracket. Remove the crankshaft position sensor.

13. Remove the crankshaft pulley center bolt and washer. Remove the drive sprocket. Remove the crankshaft angle sensing blade.

14. Remove the balancer timing belt tensioner. Remove the timing belt "B" (balancer belt) from its mounting.

To install:

15. Be sure that the crankshaft balancer shaft drive sprocket timing marks and balancer shaft sprocket timing marks are aligned.

16. Install the timing belt "B" (balancer belt) on the crankshaft balancer drive sprocket and balancer shaft sprocket. There should be no slack on the tension side.

17. Assemble an temporarily fix the center of the pulley of the balancer timing belt tensioner so that it is at the top left from the center of the assembling bolt, and the pulley flange is at the front side of the engine.

➡**When tightening the mounting bolts, ensure that the tensioner does not rotate with the bolts. Allowing it to**

rotate can cause excessive tension of the belt.

18. Lift your fingers and the balancer timing belt tensioner in the counterclockwise position. Apply minimal torque to the balancer timing belt so the belt is tense without looseness. Tighten the assembling bolt to 12—16 ft. lbs. (16—22 Nm).

19. Turn the crankshaft clockwise two turns to set the number one piston to TDC on the compression stroke. Check that the sprocket timing marks are aligned.

20. Apply slight pressure at the center of the belt between both sprockets. Inspect whether the belt deflection is within specification (0.20—0.27 inch.). Correct as required.

21. Install the crankshaft angle sensing blade, crankshaft drive sprocket, pulley washer and pulley center bolt. Apply clean engine oil to the retaining bolt prior to installation. Tighten the assembling bolt to 123 ft. lbs. (167 Nm).

22. Slowly compress the timing belt tensioner adjuster rod using a press or vise. Align the set hole "A" of the rod with set hole "B" of the timing belt tensioner adjuster cylinder.

➡**Do not compress the assembly too fast as damage to the rod may occur.**

23. Insert a wire or pin in the aligned set holes to lock the assembly in place.

24. Install the timing belt tensioner adjuster to the engine and tighten the mounting bolt to 15—19 ft. lbs. (20—26 Nm).

25. Temporarily tighten the timing belt tensioner pulley. Align the timing marks on the camshaft sprocket, crankshaft drive sprocket and engine oil pump sprocket.

26. Adjust the timing mark of the engine oil pump sprocket, by removing the cylinder block plug. Insert a bolt (M6, section width 10MM, nominal length 45MM) from the plug hole.

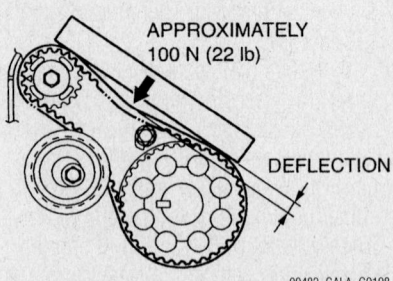

APPROXIMATELY
100 N (22 lb)

DEFLECTION

09482_GALA_G0108

Belt tension checking—2006 Eclipse, 2004–2006 Galant and 2004–2006 Lancer 2.4L engine

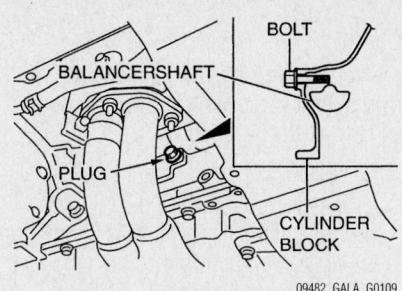

Cylinder block plug location—2006
Eclipse, 2004–2006 Galant and 2004–2006
Lancer 2.4L engine

➡If the bolt comes in contact with the balancer shaft, turn the engine oil sprocket one rotation. Re-adjust the timing mark and check to see that the bolt fits. Do not remove the bolt until the valve timing belt is assembled.

27. Position the timing belt on the timing belt tensioner pulley and crankshaft driver sprocket, support it with your hand so that it does not slide.

28. Position the belt on the engine oil pump sprocket while pulling it with your other hand.

29. Position the timing belt on the timing belt idler pulley.

➡Incorporate the timing belt. Then apply reverse rotation (counterclockwise) pressure to the camshaft sprocket. Recheck to see that each timing mark is aligned while the tension side of the belt is right.

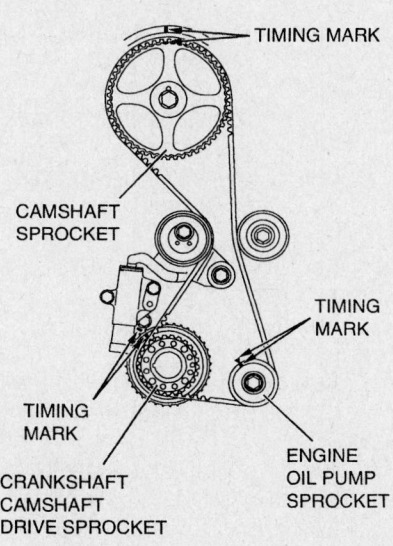

Timing belt sprocket alignment—
2002–2006 Eclipse, 2004–2006 Galant and
2004–2006 Lancer 2.4L engine

30. Position the timing mark on the camshaft sprocket.

31. Turn the timing belt tensioner pulley upward using tool MD998767 to apply tension to the belt. Temporarily tighten and fix the belt tensioner pulley mounting bolt.

32. Check that the timing marks are aligned.

33. Remove the bolt that was inserted in the cylinder block plug hole. Replace the cylinder block plug hole bolt and torque to 21—25 ft. lbs. (27—33 Nm).

➡Install tool MD998738 with your hands. Do not use any other tools to install the special tool as damage to the wire or pin inserted in the timing belt tensioner may occur.

34. Gradually screw the tool into position so that the wire or pin inserted in the timing belt tensioner adjuster moves slightly.

35. Turn the crankshaft in the clockwise direction; align each timing mark to set number one piston to TDC on the compression stroke.

36. Loosen the timing belt tensioner pulley mounting bolt. With tool MD998767 and a torque wrench, apply tension torque 31 inch lbs. to the timing belt. Tighten the timing belt tensioner pulley mounting bolt to 33—39 ft. lbs. (43—53 Nm).

37. Remove the wire or pin inserted in the timing belt tensioner adjuster. Remove tool MD998738. Install the rubber plug of the timing belt under cover.

38. Rotate the crankshaft clockwise two revolutions and leave it for about 15 minutes.

39. Insert the wire or pin, previously removed, and ensure that it can be pulled out with light load. When the wire or pin can be pulled out appropriate tension is applied on the timing belt.

40. If the wire or pin can not be pulled out easily repeat the above until it can.

➡Always check the tightening torque of the crankshaft pulley center bolt when turning the crankshaft pulley center bolt counterclockwise. Retighten if it is loose.

41. Check again that the timing marks on the sprockets are aligned.

42. Continue the installation is the reverse order of the removal procedure.

2002–2005 3.0L ENGINE

1. Before servicing the vehicle, refer to the Precautions Section.

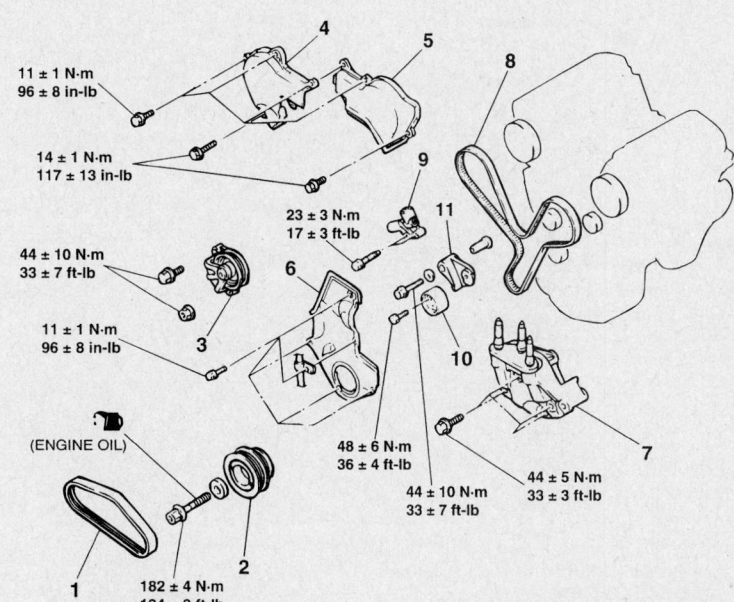

1. DRIVE BELT (POWER STEERING OIL PUMP)
2. CRANKSHAFT PULLEY
3. TENSIONER PULLEY ASSEMBLY (POWER STEERING OIL PUMP)
4. TIMING BELT FRONT UPPER COVER, RIGHT
5. TIMING BELT FRONT UPPER COVER, LEFT
6. TIMING BELT FRONT LOWER COVER
7. ENGINE SUPPORT BRACKET, RIGHT
8. TIMING BELT
9. AUTO-TENSIONER
10. TENSIONER PULLEY
11. TENSIONER ARM

Timing belt and related components—2002–2005 Eclipse and 2002–2003 Galant 3.0L engine

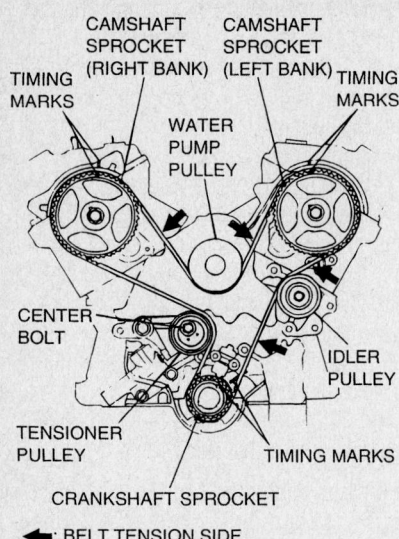

Timing belt sprocket alignment—
2002–3.0L and 3.8L engines

2. Disconnect the negative battery cable.

3. Remove the alternator. Remove the engine mount bracket. Remove the power steering fluid pump drive belt.

4. Remove the crankshaft pulley. Remove the tensioner pulley assembly.

5. Remove the front right upper timing belt cover. Remove the front left upper timing belt cover.

6. Remove the front lower timing belt cover. Remove the right engine support bracket.

7. Turn the crankshaft clockwise to align each timing mark and set the number one cylinder to TDC on the compression stroke. If the belt is going to be reused, mark the direction of rotation.
Remove the timing belt. Remove the auto tensioner, tensioner pulley and tensioner arm

To install:

8. Align the timing marks on the camshaft sprockets with those on the rocker cover and the timing mark on the crankshaft sprocket with that on the engine block.

➡The right side camshaft sprocket can turn easily due to the spring force applied, so be careful not to get your fingers caught.

9. Install the timing belt first on the crankshaft sprocket, than on the idler pulley, then on the left camshaft sprocket, then on the water pump pulley, then on the right camshaft sprocket and finally on the tensioner pulley.

10. Turn the right camshaft sprocket

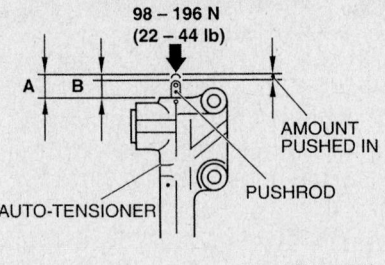

Auto tensioner pushrod test and specification—3.0L and 3.8L engines

counterclockwise until the tension side of the timing belt is firmly stretched. Check all timing marks, again.

11. Use special tool MD998767 to push the tensioner pulley into the timing belt, and then temporarily tighten the center bolt.

12. Use special tool MD998769 to turn the crankshaft ¨ turn counterclockwise, then turn it again clockwise until the timing marks are aligned.

13. Holding the auto tensioner with your hand press the end of the pushrod against a metal surface with a force of about 32 lbs. Measure how far the push rod is pushed in. Specification should be 0.04 inch. If not within specification replace the auto tensioner assembly.

14. Position the auto tensioner perpendicular in a vise. If the tensioner has a plug at the base, be sure to use a washer to protect the plug.

15. Slowly compress the pushrod of the auto tensioner until the pin hole "A" is aligned the pin hole "B" in the cylinder. Insert a pin into both holes once they are

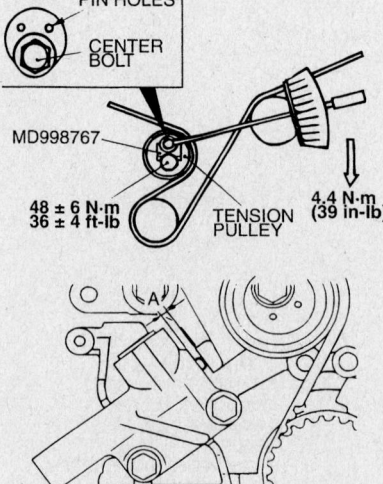

Auto tensioner center bolt and pushrod check—3.0L and 3.8L engines

aligned. Install the auto tensioner to the engine.

➡When tightening the center bolt, be careful that the tensioner pulley does not turn with the bolt.

16. Loosen the center bolt of the tensioner pulley. Use tool MD998767 and a torque wrench and apply tension torque to the timing belt in a downward motion. Tighten the center bolt to 32–40 ft. lbs. (42–54 Nm).

17. Remove the tensioner set pin. Turn the crankshaft clockwise twice to align the timing marks. Wait at least 5 minutes, then check that the auto tensioner pushrod extends within the standard value range "A" of 0.15–0.20 inch. If not repeat the operation again.

18. Check again that the timing marks of the sprockets are aligned.

19. Continue the installation in the reverse order of the removal procedure.

3.8L ENGINE

1. Before servicing the vehicle, refer to the Precautions Section.

2. Disconnect the negative battery cable. Remove the engine under cover. Remove the engine cover. Remove the engine side cover.

3. Remove the alternator. Remove the power steering fluid pump drive belt.

4. Remove the crankshaft pulley. Disconnect the control wiring harness and injector wiring connector.

5. Disconnect the knock sensor connector and crankshaft position sensor connector.

6. Remove the connector bracket. Disconnect the oil pressure switch connector and engine oil control valve connector.

7. Remove the engine mounting stay and the connector bracket.

8. Remove the front right upper timing belt cover. Remove the front left upper timing belt cover.

9. Remove the tensioner pulley and tensioner bracket.

10. Remove the crankshaft position sensor harness clamp.

11. Remove the front lower timing belt cover. Remove the engine support bracket.

12. Turn the crankshaft clockwise to align each timing mark and set the number one cylinder to TDC on the compression stroke. If the belt is going to be reused, mark the direction of rotation.
Remove the timing belt. Remove the auto tensioner, tensioner pulley and tensioner arm

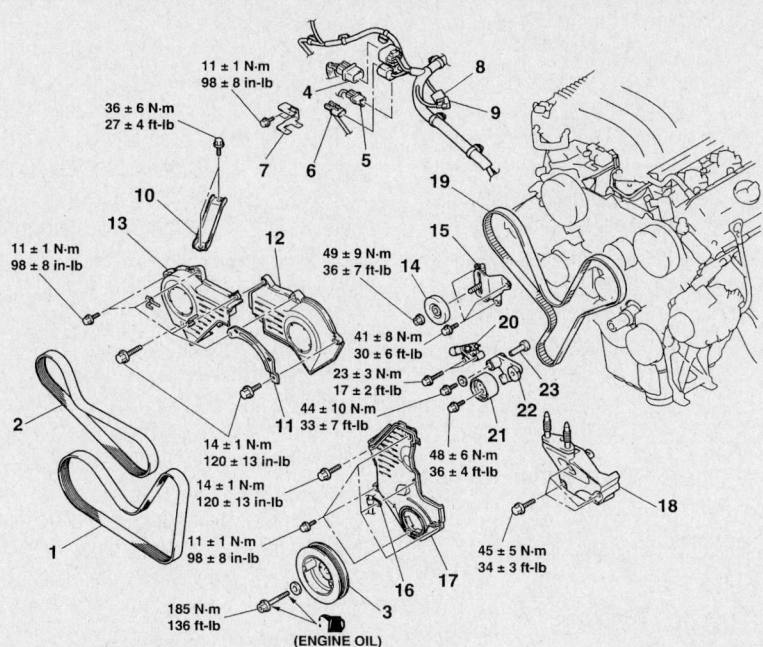

1. GENERATOR DRIVE BELT
2. POWER STEERING OIL PUMP DRIVE BELT
3. CRANKSHAFT PULLEY
4. CONTROL WIRING HARNESS AND INJECTOR WIRING HARNESS COMBINATION CONNECTOR
5. KNOCK SENSOR CONNECTOR
6. CRANKSHAFT POSITION SENSOR CONNECTOR
7. CONNECTOR BRACKET
8. ENGINE OIL PRESSURE SWITCH CONNECTOR
9. ENGINE OIL CONTROL VALVE CONNECTOR
10. ENGINE MOUNTING STAY
11. CONNECTOR BRACKET
12. TIMING BELT FRONT UPPER COVER, LEFT
13. TIMING BELT FRONT UPPER COVER, RIGHT
14. TENSIONER PULLEY
15. TENSIONER BRACKET
16. CRANKSHAFT POSITION SENSOR HARNESS CLAMP
17. TIMING BELT LOWER COVER
18. ENGINE SUPPORT BRACKET
19. TIMING BELT
20. AUTO-TENSIONER
21. TENSIONER PULLEY
22. TENSIONER ARM
23. SHAFT

09482_GALA_G0115

Timing belt and related components—2006 Eclipse and 2004–2006 Galant 3.8L engine

To install:

13. Align the timing marks on the camshaft sprockets with those on the rocker cover and the timing mark on the crankshaft sprocket with that on the engine block.

➡**The right side camshaft sprocket can turn easily due to the spring force applied, so be careful not to get your fingers caught.**

14. Install the timing belt first on the crankshaft sprocket, than on the idler pulley, then on the left camshaft sprocket, then on the water pump pulley, then on the right camshaft sprocket and finally on the tensioner pulley.

15. Turn the right camshaft sprocket counterclockwise until the tension side of the timing belt is firmly stretched. Check all timing marks, again.

16. Use special tool MD998767 to push the tensioner pulley into the timing belt, and then temporarily tighten the center bolt.

17. Use special tool MD998769 to turn the crankshaft " turn counterclockwise, then turn it again clockwise until the timing marks are aligned.

18. Holding the auto tensioner with your hand press the end of the pushrod against a metal surface with a force of about 32 lbs. Measure how far the push rod is pushed in. Specification should be 0.04 inch. If not within specification replace the auto tensioner assembly.

19. Position the auto tensioner perpendicular in a vise. If the tensioner has a plug at the base, be sure to use a washer to protect the plug.

20. Slowly compress the pushrod of the auto tensioner until the pin hole "A" is aligned the pin hole "B" in the cylinder. Insert a pin into both holes once they are aligned. Install the auto tensioner to the engine.

➡**When tightening the center bolt, be careful that the tensioner pulley does not turn with the bolt.**

21. Loosen the center bolt of the tensioner pulley. Use tool MD998767 and a torque wrench and apply tension torque to the timing belt in a downward motion. Tighten the center bolt to 32–40 ft. lbs. (42–54 Nm).

22. Remove the tensioner set pin. Turn the crankshaft clockwise twice to align the timing marks. Wait at least 5 minutes, then check that the auto tensioner pushrod extends within the standard value range "A" of 0.19–0.24 inch. If not repeat the operation again.

23. Check again that the timing marks of the sprockets are aligned.

24. Continue the installation in the reverse order of the removal procedure.

Galant

2002–2003 2.4L ENGINE

1. Before servicing the vehicle, refer to the Precautions Section.

2. Position the engine so that the No. 1 piston is at Top Dead Center (TDC).

3. Remove the timing belt covers.

➡**If the timing belts are going to be reused, mark the direction of rotation on the belt. This will ensure the belt is reinstalled in same direction, extending belt life.**

4. To loosen the timing (outer) belt tensioner, install Mitsubishi Special tool MD998738 to the slot and screw inward to move the tensioner toward the water pump. Once the tension has been relieved, remove the outer timing belt.

5. If tensioner replacement is required, align the pin hole in the tensioner rod to the hole in the tensioner cylinder. Insert a 0.055 in. (1.4mm) wire in the hole and remove the special tool from the slot. With the cylinder tension relieved, remove the auto-tensioner cylinder assembly two mounting bolts.

6. Remove the outer crankshaft sprocket and flange.

7. Loosen the silent shaft (inner) belt tensioner and remove the belt.

To install:

✳✳ WARNING

Do not spray or immerse the sprockets or tensioners in cleaning solvent. The sprocket may absorb the solvent and transfer it to the belt. The tensioners are internally lubricated and the solvent will dilute or dissolve the lubricant.

8. Align the timing marks of the silent shaft sprockets and the crankshaft sprocket with the timing marks on the front case. Route the timing belt around the sprockets so there is no slack in the upper span of the belt and the timing marks are still aligned.

9. Install the tensioner pulley and move the pulley by hand so the long side of the belt deflects approximately ¼ in. (6mm).

10. Hold the pulley tightly so the pulley

cannot rotate when the bolt is tightened. Tighten the bolt to 14 ft. lbs. (19 Nm) and recheck the deflection.

11. Align the timing marks of the camshaft, crankshaft and oil pump sprockets with their corresponding marks on the front case or rear cover.

➡**There is a possibility to align all timing marks and have the oil pump sprocket and silent shaft out of time, causing an engine vibration during operation. If the following step is not followed exactly, there is a 50 percent chance that the silent shaft alignment will be 180 degrees (½ turn) off.**

12. Before installing the timing belt, ensure that the left side (rear) silent shaft (oil pump sprocket) is in the correct position as follows:

a. Remove the plug from the rear side of the block and insert a tool with shaft diameter of 0.31 in. (8mm) into the hole.

b. With the timing marks still aligned, the shaft of the tool must be able to go in at least 2 ½ in. (63.5mm). If the tool can only go in approximately 1 in. (25mm), the shaft is not in the correct orientation and will cause a vibration during engine operation. Remove the tool from the hole and turn the oil pump sprocket one complete revolution. Realign the timing marks and insert the tool. The shaft of the tool must go in at least 2 ¼ in. (63.5mm).

c. Recheck and realign the timing marks.

d. Leave the tool in place to hold the silent shaft while continuing.

13. If the camshaft belt tensioner was removed, use a vise to carefully push the auto-tensioner rod in until the set hole in the rod is aligned with the hole in the cylinder. Place a wire into the hole to retain the rod. Mount the tensioner to the engine block and tighten the mounting bolt to 17 ft. lbs. (23 Nm).

14. Install the belt to the crankshaft sprocket, oil pump sprocket, then camshaft sprocket, in that order. While doing so, be sure there is no slack between the sprocket except where the tensioner is installed.

15. To adjust the timing (outer) belt perform the following steps:

a. Turn the crankshaft ¼ turn counterclockwise, then turn it clockwise to move No. 1 cylinder to TDC.

b. Loosen the center bolt. Using tool MD998752 and a torque wrench, apply a torque of 2.6 ft. lbs. (3.6 Nm) to the tensioner. Tighten the center bolt.

c. Screw the special tool into the engine left support bracket until its end makes contact with the tensioner arm. At this point, screw the special tool in some more and remove the set wire attached to the auto-tensioner, if the wire was not previously removed. Then, remove the special tool.

d. Rotate the crankshaft two complete turns clockwise and let it sit for approximately 15 minutes. Then, measure the auto-tensioner protrusion (the distance between the tensioner arm and auto-tensioner body) to ensure that it is within 0.15–0.18 in. (3.8–4.5mm). If out of specification, repeat sub steps **a** through **d** until the specified value is obtained.

➡**Do not manually overtighten the belt or it will howl.**

16. Install the upper and lower timing belt covers.

2004–2006 2.4L ENGINE

1. Before servicing the vehicle, refer to the Precautions Section.

2. Disconnect the negative battery cable. Remove the engine under cover.

3. Remove the crankshaft damper pulley.

4. Disconnect the control wiring harness connector, battery wiring harness connector and connector bracket to engine mounting insulator.

5. Remove the harness bracket. Remove the upper timing belt cover. Remove the water pump pulley. Remove the idler pulley.

6. Remove the auto tensioner. Remove the lower timing belt cover.

7. Turn the crankshaft clockwise and align each timing mark to set the number one piston to TDC of its compression stroke.

8. Remove the timing belt under cover rubber plug and then install special tool MD998738. Screw the special tool until it contacts the timing belt tensioner arm.

➡**The special tool must be screwed in gradually at the rate of a 30 degree turn per second. If it is screwed in all at once, the timing belt tensioner adjuster rod will not easily retract and the tool may bend.**

9. Gradually screw in the special tool and then align the timing belt tensioner adjuster rod set hole "A" with the timing belt tensioner adjuster cylinder set hole "B".

10. Insert a wire or pin in the set holes to lock the assembly in place. After removing the special tool, loosen the timing belt tensioner pulley mounting bolts and remove the timing belt. If the belt is being reused be sure to mark the direction of rotation (clockwise) on the belt.

11. Remove the timing belt tensioner pulley, tensioner arm and adjuster.

12. Remove the timing belt idler pulley. Remove the timing belt lower cover bracket. Remove the crankshaft position sensor.

13. Remove the crankshaft pulley center bolt and washer. Remove the drive sprocket. Remove the crankshaft angle sensing blade.

14. Remove the balancer timing belt tensioner. Remove the timing belt "B" (balancer belt) from its mounting.

To install:

15. Be sure that the crankshaft balancer shaft drive sprocket timing marks and balancer shaft sprocket timing marks are aligned.

16. Install the timing belt "B" (balancer belt) on the crankshaft balancer drive sprocket and balancer shaft sprocket. There should be no slack on the tension side.

17. Assemble an temporarily fix the center of the pulley of the balancer timing belt tensioner so that it is at the top left from the center of the assembling bolt, and the pulley flange is at the front side of the engine.

➡**When tightening the mounting bolts, ensure that the tensioner does not rotate with the bolts. Allowing it to rotate can cause excessive tension of the belt.**

18. Lift your fingers and the balancer timing belt tensioner in the counterclockwise position. Apply minimal torque to the balancer timing belt so the belt is tense without looseness. Tighten the assembling bolt to 12–16 ft. lbs. (16–22 Nm).

19. Turn the crankshaft clockwise two turns to set the number one piston to TDC on the compression stroke. Check that the sprocket timing marks are aligned.

20. Apply slight pressure at the center of the belt between both sprockets. Inspect whether the belt deflection is within specification (0.20–0.27 inch.). Correct as required.

21. Install the crankshaft angle sensing blade, crankshaft drive sprocket, pulley washer and pulley center bolt. Apply clean engine oil to the retaining bolt prior to installation. Tighten the assembling bolt to 123 ft. lbs. (167 Nm).

22. Slowly compress the timing belt tensioner adjuster rod using a press or vise. Align the set hole "A" of the rod with set hole "B" of the timing belt tensioner adjuster cylinder.

➡**Do not compress the assembly too fast as damage to the rod may occur.**

23. Insert a wire or pin in the aligned set holes to lock the assembly in place.

24. Install the timing belt tensioner adjuster to the engine and tighten the mounting bolt to 15–19 ft. lbs. (20–26 Nm).

25. Temporarily tighten the timing belt tensioner pulley. Align the timing marks on the camshaft sprocket, crankshaft drive sprocket and engine oil pump sprocket.

26. Adjust the timing mark of the engine oil pump sprocket, by removing the cylinder block plug. Insert a bolt (M6, section width 10MM, nominal length 45MM) from the plug hole.

➡️**If the bolt comes in contact with the balancer shaft, turn the engine oil sprocket one rotation. Re-adjust the timing mark and check to see that the bolt fits. Do not remove the bolt until the valve timing belt is assembled.**

27. Position the timing belt on the timing belt tensioner pulley and crankshaft driver sprocket, support it with your hand so that it does not slide.

28. Position the belt on the engine oil pump sprocket while pulling it with your other hand.

29. Position the timing belt on the timing belt idler pulley.

➡️**Incorporate the timing belt. Then apply reverse rotation (counterclockwise) pressure to the camshaft sprocket. Recheck to see that each timing mark is aligned while the tension side of the belt is right.**

30. Position the timing mark on the camshaft sprocket.

31. Turn the timing belt tensioner pulley upward using tool MD998767 to apply tension to the belt. Temporarily tighten and fix the belt tensioner pulley mounting bolt.

32. Check that the timing marks are aligned.

33. Remove the bolt that was inserted in the cylinder block plug hole. Replace the cylinder block plug hole bolt and torque to 21–25 ft. lbs. (27–33 Nm).

➡️**Install tool MD998738 with your hands. Do not use any other tools to install the special tool as damage to the wire or pin inserted in the timing belt tensioner may occur.**

34. Gradually screw the tool into position so that the wire or pin inserted in the timing belt tensioner adjuster moves slightly.

35. Turn the crankshaft in the clockwise direction; align each timing mark to set

number one piston to TDC on the compression stroke.

36. Loosen the timing belt tensioner pulley mounting bolt. With tool MD998767 and a torque wrench, apply tension torque 31 inch lbs. to the timing belt. Tighten the timing belt tensioner pulley mounting bolt to 33–39 ft. lbs. (43–53 Nm).

37. Remove the wire or pin inserted in the timing belt tensioner adjuster. Remove tool MD998738. Install the rubber plug of the timing belt under cover.

38. Rotate the crankshaft clockwise two revolutions and leave it for about 15 minutes.

39. Insert the wire or pin, previously removed, and ensure that it can be pulled out with light load. When the wire or pin can be pulled out appropriate tension is applied on the timing belt.

40. If the wire or pin can not be pulled out easily repeat the above until it can.

➡️**Always check the tightening torque of the crankshaft pulley center bolt when turning the crankshaft pulley center bolt counterclockwise. Retighten if it is loose.**

41. Check again that the timing marks on the sprockets are aligned.

42. Continue the installation is the reverse order of the removal procedure.

2002–2003 3.0L ENGINE

1. Before servicing the vehicle, refer to the Precautions Section.

2. Disconnect the negative battery cable.

3. Remove the alternator. Remove the engine mount bracket. Remove the power steering fluid pump drive belt.

4. Remove the crankshaft pulley. Remove the tensioner pulley assembly.

5. Remove the front right upper timing belt cover. Remove the front left upper timing belt cover.

6. Remove the front lower timing belt cover. Remove the right engine support bracket.

7. Turn the crankshaft clockwise to align each timing mark and set the number one cylinder to TDC on the compression stroke. If the belt is going to be reused, mark the direction of rotation.

Remove the timing belt. Remove the auto tensioner, tensioner pulley and tensioner arm

To install:

8. Align the timing marks on the camshaft sprockets with those on the rocker cover and the timing mark on the crankshaft sprocket with that on the engine block.

➡️**The right side camshaft sprocket can turn easily due to the spring force applied, so be careful not to get your fingers caught.**

9. Install the timing belt first on the crankshaft sprocket, than on the idler pulley, then on the left camshaft sprocket, then on the water pump pulley, then on the right camshaft sprocket and finally on the tensioner pulley.

10. Turn the right camshaft sprocket counterclockwise until the tension side of the timing belt is firmly stretched. Check all timing marks, again.

11. Use special tool MD998767 to push the tensioner pulley into the timing belt, and then temporarily tighten the center bolt.

12. Use special tool MD998769 to turn the crankshaft ¼ turn counterclockwise, then turn it again clockwise until the timing marks are aligned.

13. Holding the auto tensioner with your hand press the end of the pushrod against a metal surface with a force of about 32 lbs. Measure how far the push rod is pushed in. Specification should be 0.04 inch. If not within specification replace the auto tensioner assembly.

14. Position the auto tensioner perpendicular in a vise. If the tensioner has a plug at the base, be sure to use a washer to protect the plug.

15. Slowly compress the pushrod of the auto tensioner until the pin hole "A" is aligned the pin hole "B" in the cylinder. Insert a pin into both holes once they are aligned. Install the auto tensioner to the engine.

➡️**When tightening the center bolt, be careful that the tensioner pulley does not turn with the bolt.**

16. Loosen the center bolt of the tensioner pulley. Use tool MD998767 and a torque wrench and apply tension torque to the timing belt in a downward motion. Tighten the center bolt to 32–40 ft. lbs. (42–54 Nm).

17. Remove the tensioner set pin. Turn the crankshaft clockwise twice to align the timing marks. Wait at least 5 minutes, then check that the auto tensioner pushrod extends within the standard value range "A" of 0.15–0.20 inch. If not repeat the operation again.

18. Check again that the timing marks of the sprockets are aligned.

19. Continue the installation in the reverse order of the removal procedure.

3.8L ENGINE

1. Before servicing the vehicle, refer to the Precautions Section.

2. Disconnect the negative battery cable. Remove the engine under cover. Remove the engine cover. Remove the engine side cover.

3. Remove the alternator. Remove the power steering fluid pump drive belt.

4. Remove the crankshaft pulley. Disconnect the control wiring harness and injector wiring connector.

5. Disconnect the knock sensor connector and crankshaft position sensor connector.

6. Remove the connector bracket. Disconnect the oil pressure switch connector and engine oil control valve connector.

7. Remove the engine mounting stay and the connector bracket.

8. Remove the front right upper timing belt cover. Remove the front left upper timing belt cover.

9. Remove the tensioner pulley and tensioner bracket.

10. Remove the crankshaft position sensor harness clamp.

11. Remove the front lower timing belt cover. Remove the engine support bracket.

12. Turn the crankshaft clockwise to align each timing mark and set the number one cylinder to TDC on the compression stroke. If the belt is going to be reused, mark the direction of rotation.

Remove the timing belt. Remove the auto tensioner, tensioner pulley and tensioner arm

To install:

13. Align the timing marks on the camshaft sprockets with those on the rocker cover and the timing mark on the crankshaft sprocket with that on the engine block.

➡ **The right side camshaft sprocket can turn easily due to the spring force applied, so be careful not to get your fingers caught.**

14. Install the timing belt first on the crankshaft sprocket, than on the idler pulley, then on the left camshaft sprocket, then on the water pump pulley, then on the right camshaft sprocket and finally on the tensioner pulley.

15. Turn the right camshaft sprocket counterclockwise until the tension side of the timing belt is firmly stretched. Check all timing marks, again.

16. Use special tool MD998767 to push the tensioner pulley into the timing belt, and then temporarily tighten the center bolt.

17. Use special tool MD998769 to turn the crankshaft ¼ turn counterclockwise, then turn it again clockwise until the timing marks are aligned.

18. Holding the auto tensioner with your hand press the end of the pushrod against a metal surface with a force of about 32 lbs. Measure how far the push rod is pushed in. Specification should be 0.04 inch. If not within specification replace the auto tensioner assembly.

19. Position the auto tensioner perpendicular in a vise. If the tensioner has a plug at the base, be sure to use a washer to protect the plug.

20. Slowly compress the pushrod of the auto tensioner until the pin hole "A" is aligned the pin hole "B" in the cylinder. Insert a pin into both holes once they are aligned. Install the auto tensioner to the engine.

➡ **When tightening the center bolt, be careful that the tensioner pulley does not turn with the bolt.**

21. Loosen the center bolt of the tensioner pulley. Use tool MD998767 and a torque wrench and apply tension torque to the timing belt in a downward motion. Tighten the center bolt to 32–40 ft. lbs. (42–54 Nm).

22. Remove the tensioner set pin. Turn the crankshaft clockwise twice to align the timing marks. Wait at least 5 minutes, then check that the auto tensioner pushrod extends within the standard value range "A" of 0.19–0.24 inch. If not repeat the operation again.

23. Check again that the timing marks of the sprockets are aligned.

24. Continue the installation in the reverse order of the removal procedure.

Lancer

2.0L ENGINE

1. Before servicing the vehicle, refer to the Precautions Section.

2. Remove the left side engine under cover. Disconnect the negative battery cable. Remove the engine mounting insulator.

3. Remove the crankshaft pulley.

4. Disconnect the crankshaft position sensor connection. Remove the timing belt front upper cover.

5. Disconnect the pressure hose clamp. Remove the power steering oil line bracket.

6. Remove the alternator adjusting brace. Remove the timing belt lower cover.

7. Remove the power steering pump bracket stay.

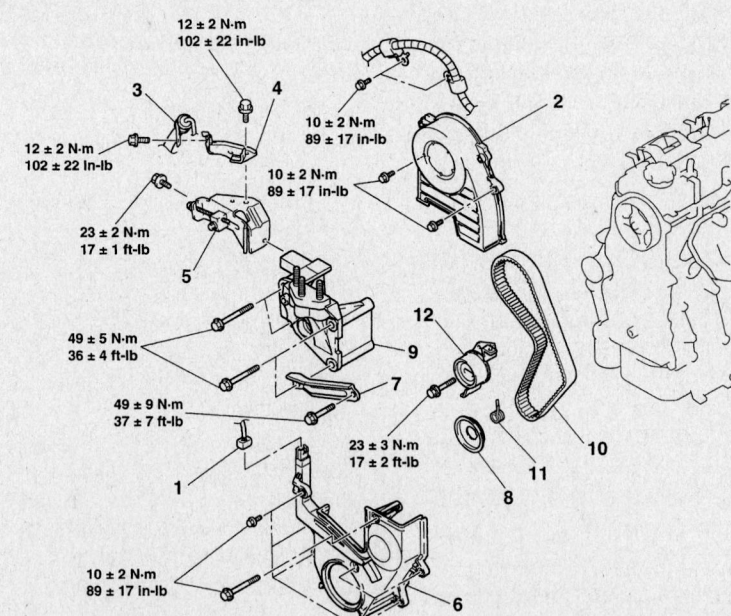

12 ± 2 N·m
102 ± 22 in-lb

10 ± 2 N·m
89 ± 17 in-lb

12 ± 2 N·m
102 ± 22 in-lb

10 ± 2 N·m
89 ± 17 in-lb

23 ± 2 N·m
17 ± 1 ft-lb

49 ± 5 N·m
36 ± 4 ft-lb

49 ± 9 N·m
37 ± 7 ft-lb

23 ± 3 N·m
17 ± 2 ft-lb

10 ± 2 N·m
89 ± 17 in-lb

1.	CRANKSHAFT POSITION SENSOR CONNECTION	7.	POWER STEERING OIL PUMP BRACKET STAY
2.	TIMING BELT FRONT UPPER COVER	8.	FLANGE
3.	PRESSURE HOSE CLAMP	9.	ENGINE SUPPORT BRACKET
4.	POWER STEERING OIL LINE BRACKET	10.	TIMING BELT
5.	GENERATOR ADJUSTING BRACE	11.	TENSIONER SPRING
6.	TIMING BELT LOWER COVER	12.	TIMING BELT TENSIONER

09482_GALA_G0116

Timing belt and related components—Lancer 2.0L engine

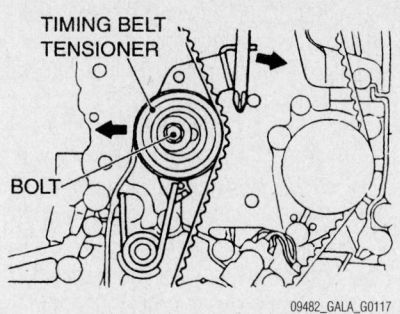

09482_GALA_G0117

Timing belt tensioner pulley and belt removal—Lancer 2.0L engine

8. Remove the flange. Remove the engine support bracket.

9. Turn the crankshaft clockwise to align each timing mark and set the number one cylinder to TDC on the compression stroke. If the belt is going to be reused, mark the direction of rotation.

10. Loosen the tension pulley retaining bolt. Position a suitable tool against the tensioner pulley and pry it fully back in the direction of the arrow in the illustration. Temporarily tighten the tensioner pulley bolt.

11. Remove the timing belt. Remove the auto tensioner, tensioner pulley and tensioner arm

To install:

12. With the timing belt tensioner pulley bolt loosened, use a suitable tool and pry the tensioner pulley as close to the engine mount as possible. Temporarily tighten the tensioner bolt.

13. Align each of the crankshaft and camshaft sprocket timing marks.

14. Install the timing belt in the following order. Crankshaft sprocket, then water

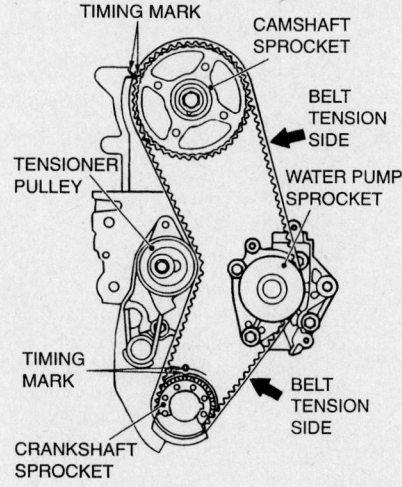

09482_GALA_G0118

Timing belt sprocket alignment—Lancer 2.0L engine

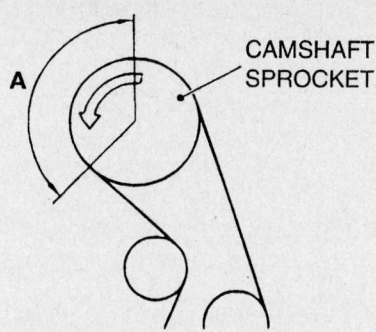

09482_GALA_G0119

Timing belt teeth checking location—Lancer 2.0L engine

pump sprocket, then camshaft sprocket and finally tensioner pulley.

➡ **After installing the belt, try to rotate the camshaft sprocket in the reverse direction. Recheck to be sure that the belt is fully tensioned and that each timing mark is in the proper direction.**

15. Initially loosen the fixing bolt of the tensioner pulley fixed to the engine mount side by ¼ to ½ turn. Use the force of the tensioner spring to apply tension to the belt.

➡ **As the purpose of this procedure is to apply the proper amount of tension to the tension side of the timing belt by using the cam driving torque, turn the crankshaft only by the amount given below. Do not turn the crankshaft in the opposite direction (counterclockwise).**

16. Turn the crankshaft in the clockwise direction for two rotations, and recheck to be sure that the timing marks on each sprocket are aligned.

17. After checking to be sure that no belt teeth in the section marked "A" in the illustration are lifted up and that the teeth in each sprocket are engaged, secure the tensioner pulley.

18. Install the flange.

19. Continue the installation in the reverse order of the removal procedure.

2.4L ENGINE

1. Before servicing the vehicle, refer to the Precautions Section.

2. Disconnect the negative battery cable. Remove the engine under cover.

3. Remove the crankshaft damper pulley.

4. Disconnect the control wiring harness connector, battery wiring harness connector and connector bracket to engine mounting insulator.

5. Remove the harness bracket. Remove the upper timing belt cover. Remove the water pump pulley. Remove the idler pulley.

6. Remove the auto tensioner. Remove the lower timing belt cover.

7. Turn the crankshaft clockwise and align each timing mark to set the number one piston to TDC of its compression stroke.

8. Remove the timing belt under cover rubber plug and then install special tool MD998738. Screw the special tool until it contacts the timing belt tensioner arm.

➡ **The special tool must be screwed in gradually at the rate of a 30 degree turn per second. If it is screwed in all at once, the timing belt tensioner adjuster rod will not easily retract and the tool may bend.**

9. Gradually screw in the special tool and then align the timing belt tensioner adjuster rod set hole "A" with the timing belt tensioner adjuster cylinder set hole "B".

10. Insert a wire or pin in the set holes to lock the assembly in place. After removing the special tool, loosen the timing belt tensioner pulley mounting bolts and remove the timing belt. If the belt is being reused be sure to mark the direction of rotation (clockwise) on the belt.

11. Remove the timing belt tensioner pulley, tensioner arm and adjuster.

12. Remove the timing belt idler pulley. Remove the timing belt lower cover bracket. Remove the crankshaft position sensor.

13. Remove the crankshaft pulley center bolt and washer. Remove the drive sprocket. Remove the crankshaft angle sensing blade.

14. Remove the balancer timing belt tensioner. Remove the timing belt "B" (balancer belt) from its mounting.

To install:

15. Be sure that the crankshaft balancer shaft drive sprocket timing marks and balancer shaft sprocket timing marks are aligned.

16. Install the timing belt "B" (balancer belt) on the crankshaft balancer drive sprocket and balancer shaft sprocket. There should be no slack on the tension side.

17. Assemble an temporarily fix the center of the pulley of the balancer timing belt tensioner so that it is at the top left from the center of the assembling bolt, and the pulley flange is at the front side of the engine.

➡ **When tightening the mounting bolts, ensure that the tensioner does not rotate with the bolts. Allowing it to rotate can cause excessive tension of the belt.**

18. Lift your fingers and the balancer timing belt tensioner in the counterclockwise position. Apply minimal torque to the

balancer timing belt so the belt is tense without looseness. Tighten the assembling bolt to 12–16 ft. lbs. (16–22 Nm).

19. Turn the crankshaft clockwise two turns to set the number one piston to TDC on the compression stroke. Check that the sprocket timing marks are aligned.

20. Apply slight pressure at the center of the belt between both sprockets. Inspect whether the belt deflection is within specification (0.20–0.27 inch.). Correct as required.

21. Install the crankshaft angle sensing blade, crankshaft drive sprocket, pulley washer and pulley center bolt. Apply clean engine oil to the retaining bolt prior to installation. Tighten the assembling bolt to 123 ft. lbs. (167 Nm).

22. Slowly compress the timing belt tensioner adjuster rod using a press or vise. Align the set hole "A" of the rod with set hole "B" of the timing belt tensioner adjuster cylinder.

➡**Do not compress the assembly too fast as damage to the rod may occur.**

23. Insert a wire or pin in the aligned set holes to lock the assembly in place.

24. Install the timing belt tensioner adjuster to the engine and tighten the mounting bolt to 15–19 ft. lbs. (20–26 Nm).

25. Temporarily tighten the timing belt tensioner pulley. Align the timing marks on the camshaft sprocket, crankshaft drive sprocket and engine oil pump sprocket.

26. Adjust the timing mark of the engine oil pump sprocket, by removing the cylinder block plug. Insert a bolt (M6, section width 10MM, nominal length 45MM) from the plug hole.

➡**If the bolt comes in contact with the balancer shaft, turn the engine oil sprocket one rotation. Re-adjust the timing mark and check to see that the bolt fits. Do not remove the bolt until the valve timing belt is assembled.**

27. Position the timing belt on the timing belt tensioner pulley and crankshaft driver sprocket, support it with your hand so that it does not slide.

28. Position the belt on the engine oil pump sprocket while pulling it with your other hand.

29. Position the timing belt on the timing belt idler pulley.

➡**Incorporate the timing belt. Then apply reverse rotation (counterclockwise) pressure to the camshaft sprocket. Recheck to see that each tim-** ing mark is aligned while the tension side of the belt is right.

30. Position the timing mark on the camshaft sprocket.

31. Turn the timing belt tensioner pulley upward using tool MD998767 to apply tension to the belt. Temporarily tighten and fix the belt tensioner pulley mounting bolt.

32. Check that the timing marks are aligned.

33. Remove the bolt that was inserted in the cylinder block plug hole. Replace the cylinder block plug hole bolt and torque to 21–25 ft. lbs. (27–33 Nm).

➡**Install tool MD998738 with your hands. Do not use any other tools to install the special tool as damage to the wire or pin inserted in the timing belt tensioner may occur.**

34. Gradually screw the tool into position so that the wire or pin inserted in the timing belt tensioner adjuster moves slightly.

35. Turn the crankshaft in the clockwise direction; align each timing mark to set number one piston to TDC on the compression stroke.

36. Loosen the timing belt tensioner pul- ley mounting bolt. With tool MD998767 and a torque wrench, apply tension torque 31 inch lbs. to the timing belt. Tighten the timing belt tensioner pulley mounting bolt to 33–39 ft. lbs. (43–53 Nm).

37. Remove the wire or pin inserted in the timing belt tensioner adjuster. Remove tool MD998738. Install the rubber plug of the timing belt under cover.

38. Rotate the crankshaft clockwise two revolutions and leave it for about 15 minutes.

39. Insert the wire or pin, previously removed, and ensure that it can be pulled out with light load. When the wire or pin can be pulled out appropriate tension is applied on the timing belt.

40. If the wire or pin can not be pulled out easily repeat the above until it can.

➡**Always check the tightening torque of the crankshaft pulley center bolt when turning the crankshaft pulley center bolt counterclockwise. Retighten if it is loose.**

41. Check again that the timing marks on the sprockets are aligned.

42. Continue the installation is the reverse order of the removal procedure.

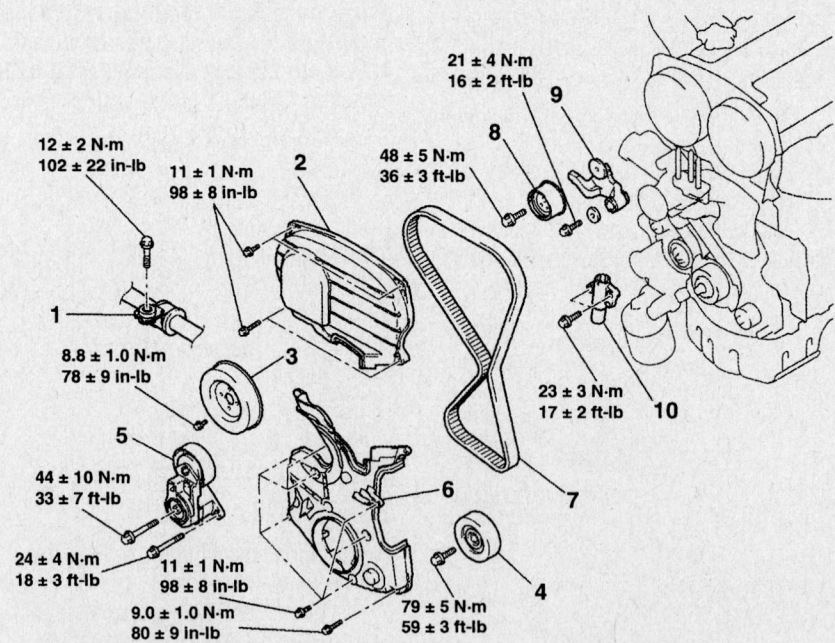

1. POWER STEERING PRESSURE HOSE CONNECTION
2. TIMING BELT UPPER COVER
3. WATER PUMP PULLEY
4. IDLER PULLEY
5. AUTO-TENSIONER
6. TIMING BELT LOWER COVER
7. VALVE TIMING BELT
8. TIMING BELT TENSIONER PULLEY
9. TIMING BELT TENSIONER ARM
10. TIMING BELT TENSIONER ADJUSTER

Timing belt and related components—Evolution

09482_GALA_G0120

Evolution

1. Before servicing the vehicle, refer to the Precautions Section.

2. Disconnect the negative battery cable. Remove the engine under cover.

3. Remove the side cover. Remove the crankshaft damper pulley.

4. Remove the front axle crossmember bar. Remove the front exhaust pipe.

5. Disconnect the power steering pressure hose connection. Remove the timing belt upper cover. Remove the water pump pulley.

6. Remove the idler pulley and auto tensioner.

7. Remove the timing belt lower cover. Remove the engine mounting insulator.

8. Turn the crankshaft clockwise and align each timing mark to set the number one piston to TDC of its compression stroke.

9. Remove the timing belt under cover rubber plug and then install special tool MD998738. Screw the special tool until it contacts the timing belt tensioner arm.

➡**The special tool must be screwed in gradually at the rate of a 30 degree turn per second. If it is screwed in all at once, the timing belt tensioner adjuster rod will not easily retract and the tool may bend.**

10. Gradually screw in the special tool and then align the timing belt tensioner adjuster rod set hole "A" with the timing belt tensioner adjuster cylinder set hole "B".

11. Insert a wire or pin in the set holes to lock the assembly in place. After removing the special tool, loosen the timing belt tensioner pulley mounting bolts and remove the timing belt. If the belt is being reused be sure to mark the direction of rotation (clockwise) on the belt.

12. Remove the timing belt tensioner pulley, tensioner arm and adjuster.

13. Remove the power steering pressure switch connector, fluid pump heat protector, fluid pump, bracket and reservoir assembly. Remove the power steering fluid pump bracket.

14. Remove the timing belt idler pulley. Remove the crankshaft position sensor.

15. Remove the crankshaft pulley center bolt and washer. Remove the drive sprocket. Remove the crankshaft angle sensing blade.

16. Remove the balancer timing belt tensioner. Remove the timing belt "B" (balancer belt) from its mounting.

To install:

17. Be sure that the crankshaft balancer shaft drive sprocket timing marks and bal-

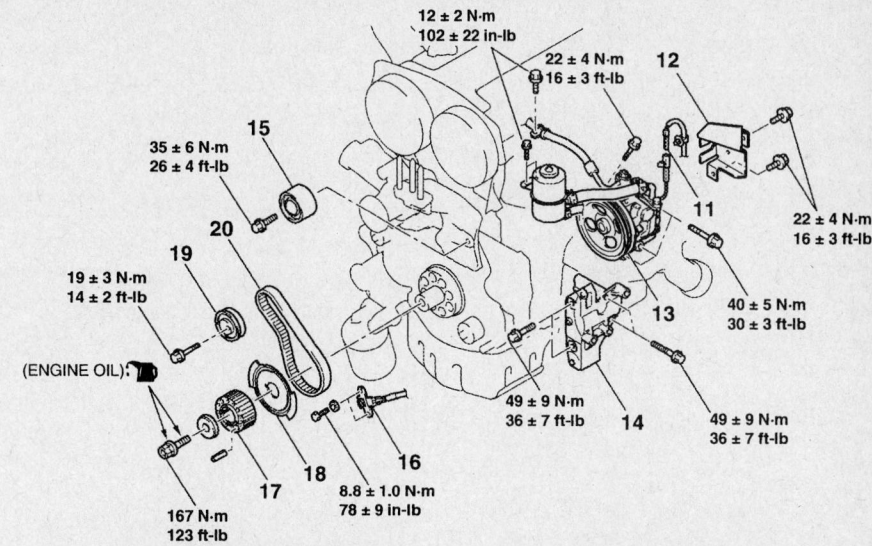

11. POWER STEERING PRESSURE SWITCH CONNECTOR
12. POWER STEERING OIL PUMP HEAT PROTECTOR
13. POWER STEERING OIL PUMP, BRACKET AND RESERVOIR ASSEMBLY
14. POWER STEERING OIL PUMP BRACKET
15. TIMING BELT IDLER PULLEY
16. CRANKSHAFT POSITION SENSOR
17. CRANKSHAFT CAMSHAFT DRIVE SPROCKET
18. CRANKSHAFT ANGLE SENSING BLADE
19. BALANCER TIMING BELT TENSIONER
20. BALANCER TIMING BELT

09482_GALA_G0121

Timing belt "B" (balancer belt) and related components—Evolution

ancer shaft sprocket timing marks are aligned.

18. Install the timing belt "B" (balancer belt) on the crankshaft balancer drive sprocket and balancer shaft sprocket. There should be no slack on the tension side.

19. Assemble an temporarily fix the center of the pulley of the balancer timing belt tensioner so that it is at the top left from the center of the assembling bolt, and the pulley flange is at the front side of the engine.

➡**When tightening the mounting bolts, ensure that the tensioner does not rotate with the bolts. Allowing it to rotate can cause excessive tension of the belt.**

20. Lift your fingers and the balancer timing belt tensioner in the counterclockwise position. Apply minimal torque to the balancer timing belt so the belt is tense without looseness. Tighten the assembling bolt to 12–16 ft. lbs. (16–22 Nm).

21. Turn the crankshaft clockwise two turns to set the number one piston to TDC on the compression stroke. Check that the sprocket timing marks are aligned.

22. Apply slight pressure at the center of the belt between both sprockets. Inspect

whether the belt deflection is within specification (0.20–0.27 inch.). Correct as required.

23. Install the crankshaft angle sensing blade, crankshaft drive sprocket, pulley washer and pulley center bolt. Apply clean engine oil to the retaining bolt prior to installation. Tighten the assembling bolt to 123 ft. lbs. (167 Nm).

24. Slowly compress the timing belt tensioner adjuster rod using a press or vise. Align the set hole "A" of the rod with set hole "B" of the timing belt tensioner adjuster cylinder.

➡**Do not compress the assembly too fast as damage to the rod may occur.**

25. Insert a wire or pin in the aligned set holes to lock the assembly in place.

26. Install the timing belt tensioner adjuster to the engine and tighten the mounting bolt to 15–19 ft. lbs. (20–26 Nm).

27. Temporarily tighten the timing belt tensioner pulley. Align the timing marks on the camshaft sprocket, crankshaft drive sprocket and engine oil pump sprocket.

28. Adjust the timing mark of the engine oil pump sprocket, by removing the cylinder block plug. Insert a bolt or the shaft of a screwdriver 2.36 inches or more.

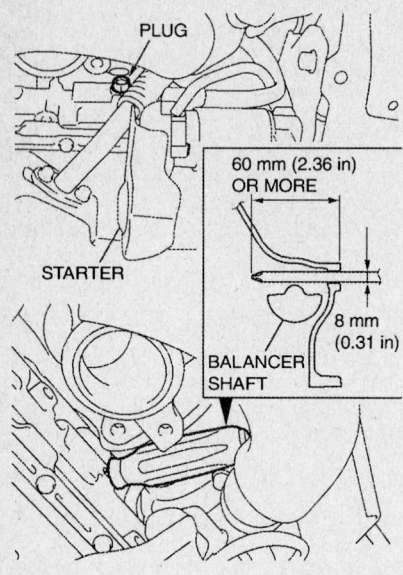

Tool contact against balancer shaft—Evolution

→If the bolt or screwdriver shaft comes in contact with the balancer shaft and can only be inserted 0.79–0.98 inch, turn the engine oil sprocket one rotation. Re-adjust the timing mark and check to see that the bolt or screwdriver shaft fits. Do not remove the bolt or screwdriver until the valve timing belt is assembled.

29. To install the timing belt, pass the belt around the crankshaft camshaft drive sprocket, the engine oil pump sprocket and the timing belt idler pulley, in that order.

30. Pass the timing belt around the camshaft sprocket (exhaust side) and hold it

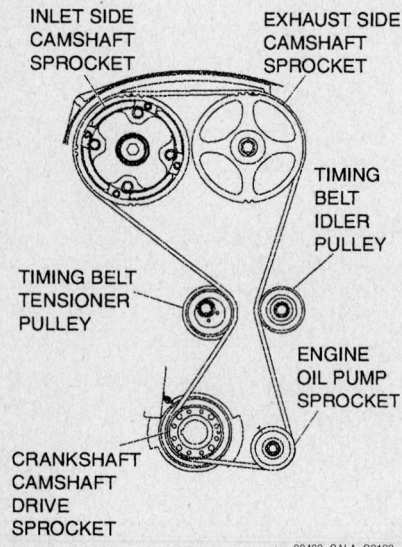

Timing belt sprocket alignment—Evolution

Paper clamp installation and locating point—Evolution

with paper clamps in the location shown in the illustration

31. Use a wrench to align the timing mark on the rocker cover with that on the camshaft sprocket. Pass the timing belt around the tensioner pulley.

→Install the timing belt. Then apply reverse rotation (counterclockwise) pressure to the camshaft sprocket. Recheck to see that each timing mark is aligned while the tension side of the belt is right.

32. Remove the two paper clamps.

33. Turn the timing belt tensioner pulley upward using tool MD998767 to apply tension to the belt. Temporarily tighten and fix the belt tensioner pulley mounting bolt.

34. Check that the timing marks are aligned.

35. Remove the bolt or screwdriver shaft that was inserted in the cylinder block plug hole. Replace the cylinder block plug hole bolt and torque to 21–25 ft. lbs. (27–33 Nm).

→Install tool MD998738 with your hands. Do not use any other tools to install the special tool as damage to the wire or pin inserted in the timing belt tensioner may occur.

36. Gradually screw the tool into position so that the wire or pin inserted in the timing belt tensioner adjuster moves slightly.

37. Turn the crankshaft in the clockwise direction; align each timing mark to set number one piston to TDC on the compression stroke.

38. Loosen the timing belt tensioner pulley mounting bolt. With tool MD998767 and a torque wrench, apply tension torque 31 inch lbs. to the timing belt. Tighten the timing belt tensioner pulley mounting bolt to 33–39 ft. lbs. (43–53 Nm).

39. Remove the wire or pin inserted in the timing belt tensioner adjuster. Remove tool MD998738. Install the rubber plug of the timing belt under cover.

40. Rotate the crankshaft clockwise two revolutions and leave it for about 15 minutes.

41. Insert the wire or pin, previously removed, and ensure that it can be pulled out with light load. When the wire or pin can be pulled out appropriate tension is applied on the timing belt.

42. If the wire or pin can not be pulled out easily repeat the above until it can.

→Always check the tightening torque of the crankshaft pulley center bolt when turning the crankshaft pulley center bolt counterclockwise. Retighten if it is loose.

43. Check again that the timing marks on the sprockets are aligned.

44. Continue the installation is the reverse order of the removal procedure.

Mirage

1. Before servicing the vehicle, refer to the Precautions Section.

2. Remove or disconnect the following:
   • Negative battery cable
   • Electrical connectors, tag before disconnecting
   • Timing belt upper and lower covers

3. Make a mark on the back of the timing belt indicating the direction of rotation so it may be reassembled in the same direction if it is to be reused. Loosen the timing belt tensioner and move the tensioner to provide slack to the timing belt. Tighten the tensioner in this position.
   • Timing belt

✱✱ WARNING

Coolant and engine oil will damage the rubber in the timing belt, drastically reducing its life. Do not allow engine oil or coolant to contact the timing belt, the sprockets or tensioner assembly.

4. If defective, replace the tensioner spacer, tensioner spring and tensioner assembly.
   1.8L (4G93) engine

To install:

5. Position the tensioner, tensioner spring and tensioner spacer on engine block.

6. Align the timing marks on the camshaft sprocket and crankshaft sprocket. This will position No. 1 piston on Top Dead Center (TDC) on the compression stroke.

7. Position the timing belt on the crankshaft sprocket and keeping the tension side of the belt tight, set it on the camshaft sprocket, then the tensioner.

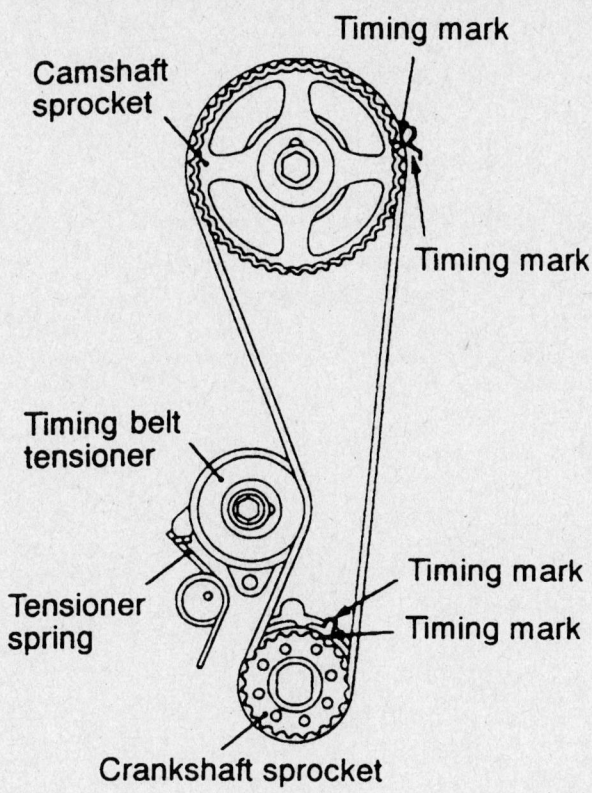

Timing mark sprocket alignment—1.5L engine

79235G56

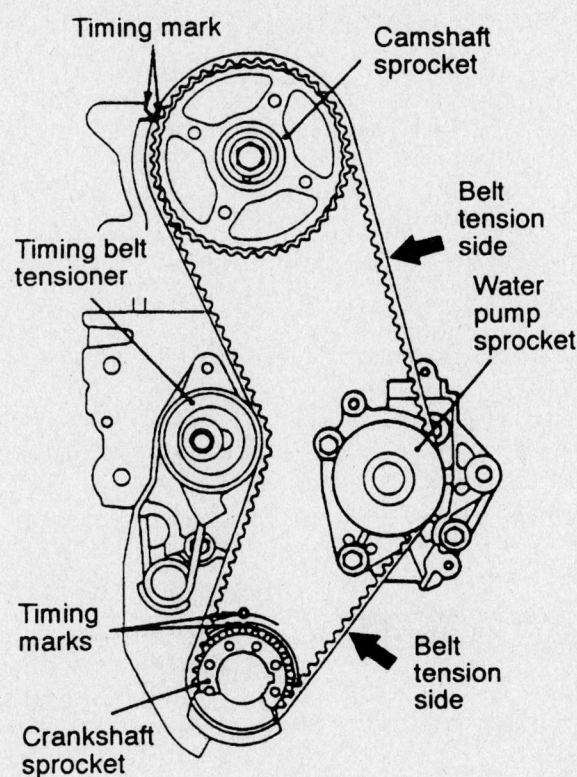

Timing mark sprocket alignment—1.8L engine

79235G57

8. Apply slight counterclockwise force to the camshaft sprocket to give tension to the belt and be sure all timing marks are aligned.

9. Loosen the pivot side tensioner bolt and the slot side bolt. Allow the spring to remove the slack.

10. Tighten the slot side tensioner bolt, then the pivot side bolt. If the pivot side bolt is tightened first, the tensioner could turn with bolt, causing over tension.

11. For 1.5L engines, turn the crankshaft clockwise. Loosen the pivot side tensioner bolt, then the slot side bolt to allow the spring to take up any remaining slack. Tighten the slot bolt, then the pivot side bolt to 17 ft. lbs. (24 Nm).

12. For 1.8L engines, turn the crankshaft clockwise two rotations and tighten the adjuster bolt to 18 ft. lbs. (24 Nm) and the pivot (spring) bolt to 35 ft. lbs. (45 Nm).

13. Install the timing belt covers and tighten the cover bolts to 84–96 inch lbs. (10–11 Nm).

14. Install the remaining components in the reverse order of removal.

Piston and Ring

POSITIONING

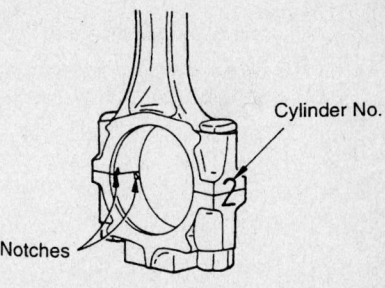

7923AG69

Before removing the caps from the connecting rods, be sure to matchmark them as shown

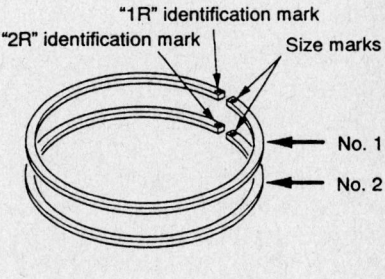

7923AG63

Compression ring identification mark— 1.5L , 2.0L DOHC and 2.4L (4G64)

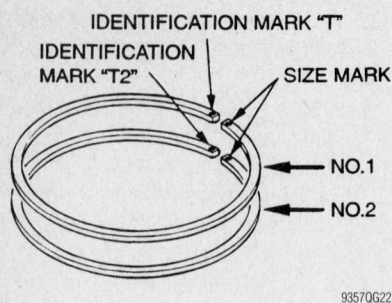

Compression ring identification mark—1.8L and 2.0L SOHC engines

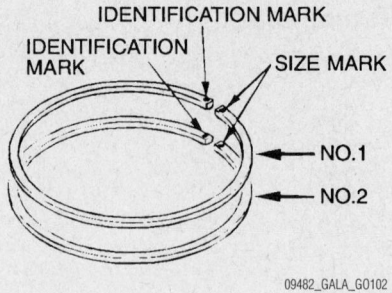

Compression ring identification mark—3.5L and 3.8L engines

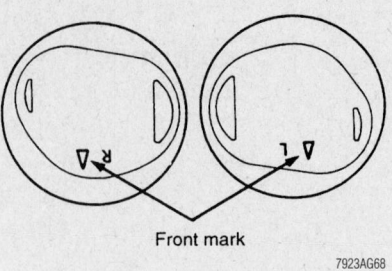

3.0L engine—Piston-to-engine block mark locations—3.0L engine

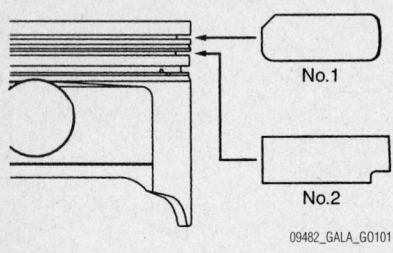

Compression ring identification mark—2.4L (4G69) engine

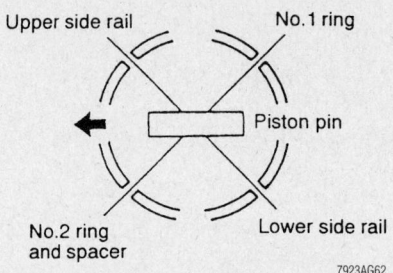

Piston ring end-gap spacing

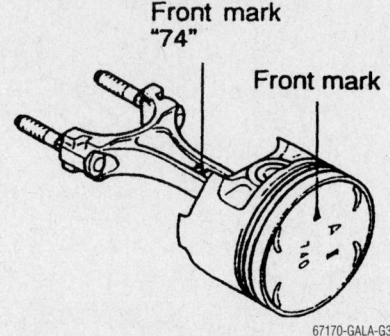

Piston and connecting rod assembly positioning—3.0L, 3.5L and 3.8L engines

FUEL SYSTEM

Fuel System Service Precautions

Safety is the most important factor when performing not only fuel system maintenance but any type of maintenance. Failure to conduct maintenance and repairs in a safe manner may result in serious personal injury or death. Maintenance and testing of the vehicle's fuel system components can be accomplished safely and effectively by adhering to the following rules and guidelines.

• To avoid the possibility of fire and personal injury, always disconnect the negative battery cable unless the repair or test procedure requires that battery voltage be applied.

• Always relieve the fuel system pressure prior to disconnecting any fuel system component (injector, fuel rail, pressure regulator, etc.), fitting or fuel line connection. Exercise extreme caution whenever relieving fuel system pressure, to avoid exposing skin, face and eyes to fuel spray. Please be advised that fuel under pressure may penetrate the skin or any part of the body that it contacts.

• Always place a shop towel or cloth around the fitting or connection prior to loosening to absorb any excess fuel due to spillage. Ensure that all fuel spillage (should it occur) is quickly removed from engine surfaces. Ensure that all fuel soaked cloths or towels are deposited into a suitable waste container.

• Always keep a dry chemical (Class B) fire extinguisher near the work area.

• Do not allow fuel spray or fuel vapors to come into contact with a spark or open flame.

• Always use a back-up wrench when loosening and tightening fuel line connection fittings. This will prevent unnecessary stress and torsion to fuel line piping. Always follow the proper torque specifications.

• Always replace worn fuel fitting O-rings with new. Do not substitute fuel hose or equivalent, where fuel pipe is installed.

Fuel System Pressure

RELIEVING

2002–2003 Eclipse

1. Before servicing the vehicle, refer to the Precautions Section.

2. Turn the ignition to the OFF position.
3. Loosen the fuel filler cap to release fuel tank pressure.
4. Remove the driver side instrument panel side cover and disconnect the fuel pump relay.
5. Start the vehicle and allow it to run until it stalls from lack of fuel. Turn the key to the OFF position.
6. Disconnect the negative battery cable, then reconnect the fuel pump relay. Install the side cover.
7. Wrap shop towels around the fitting that is being disconnected to absorb residual fuel in the lines.
8. Place shop towels into proper safety container.

Except 2002–2003 Eclipse

1. Before servicing the vehicle, refer to the Precautions Section.
2. Turn the ignition to the OFF position.
3. Loosen the fuel filler cap to release fuel tank pressure.
4. Remove the rear seat cushion, then remove the service cover and disconnect the fuel pump harness connector.

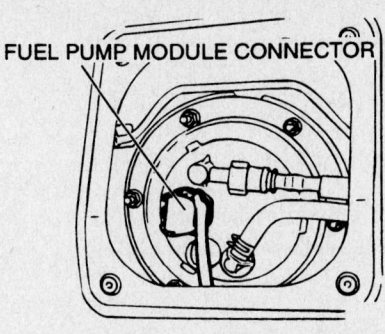

FUEL PUMP MODULE CONNECTOR

9357QG23

Location of the fuel pump connector—Lancer shown, other models similar

5. Start the vehicle and allow it to run until it stalls from lack of fuel. Turn the key to the **LOCK (OFF)** position.

6. On 2005–2006 Eclipse, crank the engine for two seconds or more. If the engine does not start, turn the ignition switch to the **LOCK (OFF)** position. If the engine starts, stop it naturally and turn the ignition switch to the **LOCK (OFF)** position.

7. Disconnect the negative battery cable, then reconnect the fuel pump connector. Install the access cover and rear seat cushion.

8. Wrap shop towels around the fitting that is being disconnected to absorb residual fuel in the lines.

9. Place shop towels into proper safety container.

Fuel Filter

REMOVAL & INSTALLATION

Diamante

1. Before servicing the vehicle, refer to the Precautions Section.

2. Properly relieve the fuel pressure.

3. Disconnect the negative battery cable.

➡**The filter is located in the engine compartment, mounted on the inner fender panel.**

4. Remove or disconnect the following:
- Air cleaner assembly and intake hoses
- Battery and battery tray
- Fuel lines from the filter
- Mounting bolts and remove the fuel filter from the vehicle

To install:

➡**Install new gaskets or O-rings whenever fuel connections have been disassembled.**

5. Install or connect the following:
- Filter to its bracket finger-tight
- New gaskets and connect the high pressure hose and eye bolt, then the main pipe and eye bolt. Tighten the eye bolts to 22 ft. lbs. (30 Nm). Tighten the flare nut to 25 ft. lbs. (35 Nm).

6. Tighten the mounting bolts fully.

7. Install or connect the following:
- Air cleaner assembly
- Battery and battery tray
- Negative battery cable, install the fuel filler cap, turn the key to the **ON** position to pressurize the fuel system and check for leaks.

Lancer, Evolution, Mirage and Galant

2002–2003

➡**The fuel filter is located in the engine compartment.**

1. Before servicing the vehicle, refer to the Precautions Section.

2. Properly relieve the fuel system pressure.

3. Remove or disconnect the following:
- Negative battery cable
- Air intake hose and the battery
- Fuel lines from the filter
- Mounting bolts and the fuel filter from the vehicle

To install:

4. If equipped with flare fitting, tighten the fitting by hand before installing the filter to the vehicle.

5. Install or connect the following:
- Filter to its bracket finger-tight
- New gaskets and connect the high pressure hose and eye bolt, then the main pipe. Tighten the eye bolts to 22 ft. lbs. (30 Nm). Tighten the flare nut to 27 ft. lbs. (37 Nm).

6. Tighten the filter mounting bolts fully.

7. Install or connect the following:
- Air intake hose
- Battery
- Negative battery cable, install the fuel filler cap, turn the key to the **ON** position to pressurize the fuel system and check for leaks.

Eclipse

2002–2003

1. Before servicing the vehicle, refer to the Precautions Section.

2. Properly relieve the fuel system pressure.

3. Disconnect the negative battery cable.

4. On models equipped with the 2.4L engine, remove the battery and the air intake hose.

5. Remove the fuel lines from the filter.

6. Remove clamp and the hose from the fuel pressure regulator.

7. Remove or disconnect the following:
- Fuel filter mounting bracket bolts and remove the fuel filter
- Bracket screw and remove the fuel filter from the mounting bracket

To install:

8. Install or connect the following:
- Fuel filter to the mounting bracket with the screw
- Fuel filter to the vehicle with the bracket mounting bolts
- Main fuel pipe to the fuel filter connector or the filter itself. Torque the flare nut to 27 ft. lbs. (36 Nm).

9. Reconnect the high pressure fuel hose to the fuel filter. Torque the eye bolt to 22 ft. lbs. (29 Nm).

10. On 2.4L engine models, install the battery and the air intake hose.

11. Reconnect the negative battery cable, start the engine and check for fuel leaks.

Fuel Pump

REMOVAL & INSTALLATION

Diamante

1. Before servicing the vehicle, refer to the Precautions Section.

2. Properly relieve the fuel system pressure.

3. Disconnect the negative battery cable.

4. Remove the left rear wheel well liner, if equipped.

5. Disconnect the center exhaust system from the main muffler. Disconnect the rear exhaust hangers, lower the exhaust and secure aside.

6. Remove the tank drain plug and drain the fuel into an approved container.

7. Remove or disconnect the following:
- Fuel return hose, high pressure hose and vent hose from the sending unit
- Electrical connector
- Filler and vent hoses. Place a support under the tank and remove the retaining nuts.

8. Lower the tank from the vehicle.

9. Remove the fuel pump retaining nuts and remove the assembly from the tank.

To install:

10. Install or connect the following:
- Pump assembly to the tank and

tighten the retaining nuts to 24 inch lbs. (3 Nm)

- Fuel tank and connect the filler and vent hoses. Tighten the tank retaining nuts and bolts to 19 ft. lbs. (26 Nm).
- Return hose, high pressure hose and all other hoses and connectors connected to the pump/sending unit
- Power cylinder unit and tighten the mounting bolts to 31 ft. lbs. (43 Nm), if equipped with 4WS
- Exhaust pipe and secure the rear hangers
- Left rear wheel well liner, if removed

11. Lower the vehicle and return fuel to the gas tank.

12. Connect the negative battery cable and check the entire system for proper operation and leaks.

Except Diamante and Mirage

1. Before servicing the vehicle, refer to the Precautions Section.

2. Relieve the fuel system pressure.

3. Remove or disconnect the following:

- Negative battery cable

- Rear seat cushion
- Inspection cover
- Harness connector and the fuel lines
- Fuel pump assembly from the tank.

To install:

4. Install or connect the following:

- Fuel pump in the tank
- Hoses and the harness connector
- Inspection cover
- Rear seat
- Negative battery cable

Mirage

1. Before servicing the vehicle, refer to the Precautions Section.

2. Properly relieve the fuel system pressure.

3. Disconnect the negative battery cable.

4. Raise and safely support the vehicle.

5. Drain the fuel from the fuel tank into an approved container.

6. Disconnect the filler and vent hoses.

7. Support the tank with a transmission jack. Disconnect the retainer straps and lower the tank to gain access to the fitting on top of the tank.

8. Remove or disconnect the following:

- Return hose, high pressure hose and vapor hoses from the pump/sending unit
- Electrical connectors at the pump/sending unit
- Fuel tank from the vehicle
- Access plate to the fuel tank and remove the pump assembly

To install:

9. Install fuel pump into fuel tank, with new packing gasket, and tighten mounting nuts.

10. Raise the tank in position under the vehicle.

11. Attach all connections to the top of the tank.

12. Raise the tank completely and position the retainer straps around the fuel tank. Install new fuel tank self-locking nuts and tighten to 22 ft. lbs. (31 Nm).

13. Connect the return hose and high pressure hoses.

14. Install the vapor hose and the filler hose. Install the filler hose retainer screws to the fender, if removed.

15. Lower the vehicle and pour the drained fuel into the gas tank.

16. Connect the negative battery cable. Check the fuel pump for proper pressure and inspect the entire system for leaks.

Fuel Injector

REMOVAL & INSTALLATION

Diamante

1. Relieve the fuel system pressure.

2. Disconnect the negative battery cable.

❊❊ CAUTION

Work MUST NOT be started until at least 90 seconds after the ignition switch is turned to the LOCK position and the negative battery cable is disconnected from the battery. This will allow time for the air bag system backup power supply to deplete its stored energy preventing accidental air bag deployment which could result in unnecessary air bag system repairs and/or personal injury.

3. Drain the cooling system.

4. Disconnect all components from the air intake plenum and remove the plenum from the intake manifold.

5. Wrap the connection with a shop towel and disconnect the high pressure fuel line at the fuel rail.

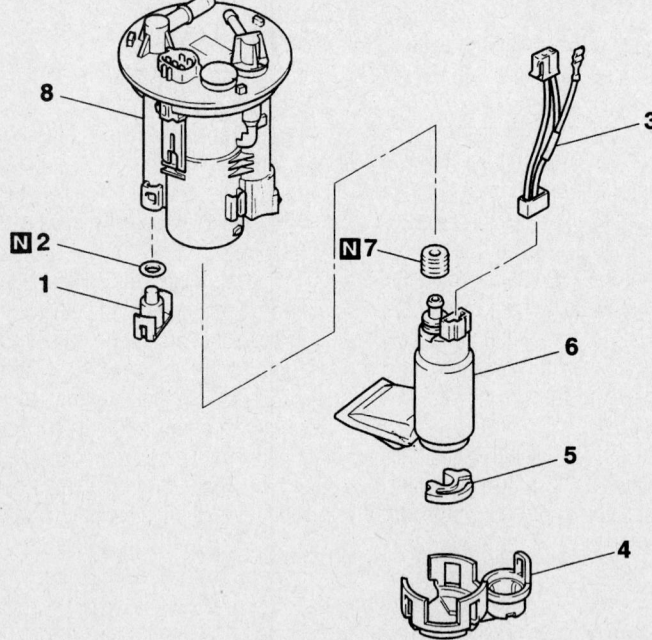

1. CAP	5. FUEL PUMP CUSHION
2. O-RING	6. FUEL PUMP
3. PUMP HARNESS	7. GROMMET
4. FUEL PUMP BRACKET	8. FUEL FILTER ASSEMBLY

9357QG24

Exploded view of the fuel pump module—Lancer

✱✱ CAUTION

Observe all applicable safety precautions when working around fuel. Whenever servicing the fuel system, always work in a well ventilated area. Do not allow fuel spray or vapors to come in contact with a spark or open flame. Keep a dry chemical fire extinguisher near the work area. Always keep fuel in a container specifically designed for fuel storage; also, always properly seal fuel containers to avoid the possibility of fire or explosion.

6. Remove or disconnect the following:
   • Fuel return hose and remove the O-ring
   • Vacuum hose from the fuel pressure regulator
   • Electrical connectors from each injector
   • Fuel pipe connecting the fuel rails
   • Injector rail retaining bolts. Make sure the rubber mounting bushings do not get lost.
7. Lift the rail assemblies up and away from the engine.
8. Remove the injectors from the rail by pulling gently. Discard the lower insulator.

To install:

➡Some of the vehicles may have a clip that secures the injector to the fuel rail. Be sure to remove or install the injector clip where necessary.

9. Install or connect the following:
   • New grommet and O-ring to the injector. Coat the O-ring with light oil.
   • Injector to the fuel rail
10. Replace the seats in the intake manifold. Install the fuel rails and injectors to the manifold. Make sure the rubber bushings are in place before tightening the mounting bolts.
11. Tighten the retaining bolts to 84–108 inch lbs. (10–13 Nm). Install the fuel pipe with new gasket.
12. Install or connect the following:
   • Electrical connectors to the injectors
   • Fuel return hose
   • O-ring, lightly lubricate it and connect the high pressure fuel line
   • Intake plenum and all related items, using new gaskets
13. Fill the cooling system.
14. Connect the negative battery cable and check the entire system for proper operation and leaks.

Eclipse

2002–2003 2.4L ENGINE

1. Relieve the fuel system pressure.
2. Label and disconnect the spark plug wires. Position the wires aside.
3. Remove or disconnect the following:
   • Positive Crankcase Ventilation (PCV) hose from the valve cover
   • High pressure fuel line to the fuel rail and disconnect the line. Be prepared to contain fuel spillage; plug the line to keep out dirt and debris.

✱✱ CAUTION

Observe all applicable safety precautions when working around fuel. Whenever servicing the fuel system, always work in a well ventilated area. Do not allow fuel spray or vapors to come in contact with a spark or open flame. Keep a dry chemical fire extinguisher near the work area. Always keep fuel in a container specifically designed for fuel storage; also, always properly seal fuel containers to avoid the possibility of fire or explosion.

4. Remove or disconnect the following:
   • Vacuum hose from the fuel pressure regulator
   • Fuel return hose from the pressure regulator
   • Electrical connector from each injector, label for reference
   • Bolt(s) holding the fuel rail to the manifold. Carefully lift the rail up and remove it with the injectors attached. Take great care not to drop an injector. Place the rail and injectors in a safe location on the workbench; protect the tips of the injectors from dirt and/or impact.
   • Injector insulators from the intake manifold, discard. The insulators are not reusable.
   • Injectors from the fuel rail by pulling gently in a straight outward motion. Make certain the grommet and O-ring come off with the injector.

To install:
5. Install a new insulator in each injector port in the manifold.
6. Remove the old grommet and O-ring from each injector. Install a new grommet and O-ring; coat the O-ring lightly with clean, thin oil.
7. If the fuel pressure regulator was removed, replace the O-ring with a new one and coat it lightly with clean, thin oil. Insert

the regulator straight into the rail, then check that it can be rotated freely. If it does not rotate smoothly, remove it and inspect the O-ring for deformation or jamming. When properly installed, align the mounting holes and tighten the retaining bolts to 84 inch lbs. (9 Nm). This procedure must be followed even if the fuel rail was not removed.
8. Install or connect the following:
   • Injector into the fuel rail, constantly turning the injector left and right during installation. When fully installed, the injector should still turn freely in the rail. If it does not, remove the injector and inspect the O-ring for deformation or damage.
   • Delivery pipe and injectors to the engine. Make certain that each injector fits correctly into its port and that the rubber insulators for the fuel rail mounts are in position.
   • Fuel rail retaining bolts and tighten them to 108 inch lbs. (12 Nm)
   • Wiring harnesses to the appropriate injector
   • Fuel return hose to the pressure regulator, then connect the vacuum hose
   • O-ring on the high pressure fuel line, coat the O-ring lightly with clean, thin oil and install the line to the fuel rail. Tighten the mounting bolts to 48 inch lbs. (6 Nm).
   • PCV hose and spark plug wires
   • Negative battery cable
9. Pressurize the fuel system and inspect all connections for leaks.

2004–2005 2.4L ENGINE

1. Relieve the fuel system pressure.
2. Disconnect the negative battery cable. Drain the engine coolant. Remove the air cleaner assembly.
3. Disconnect the accelerator cable connection. Disconnect the throttle position sensor connector. Disconnect the idle air control motor connector. Disconnect the vacuum hose connection.
4. Disconnect the water hose connection. Remove the throttle body retaining bolts. Remove the throttle body from its mounting. Discard the gasket.
5. Disconnect the PCV hose connection, ignition coil connector and injector connector.
6. Disconnect the ignition failure sensor connector. Disconnect the high pressure fuel hose connection. Disconnect the hose connection.
7. Disconnect the vacuum hose connec-

tion. Remove the fuel pressure regulator. Remove the fuel rail, insulators and injectors. Discard the O-rings.

To install:

8. Installation is the reverse of the removal procedure. Be sure to use new O-rings and gaskets as required.

9. Start the engine and check for leaks, correct as required.

2006 2.4L ENGINE

1. Relieve the fuel system pressure.
2. Disconnect the negative battery cable. Remove the air cleaner cover and air intake hose assembly.
3. Disconnect the PCV valve hose connection, ignition coil connectors, EGR valve connector and fuel injector connectors.
4. Disconnect the throttle position sensor connector, manifold absolute pressure sensor connector, EVAP emission purge solenoid connector and knock sensor connector.
5. Disconnect the power steering pressure switch connector. Remove the rocker cover bracket bolts.
6. Disconnect the high pressure hose connection. Remove the fuel rail, insulators, grommets and injectors. Discard the O-rings.

To install:

7. Installation is the reverse of the removal procedure. Be sure to use new O-rings and gaskets as required.

8. Complete the vehicle initialization procedure.

➡**To complete the initialization procedure the following tools are needed. MB991958 scan tool, MB991824 VCI, MB991827 MUT III USB cable, MB991910 MUT III Main harness "A"**

9. Connect the scan tool to the data link connector. To prevent damage to the scan tool be sure that the ignition switch is in the LOCK position before connecting the scan tool.
10. Turn the ignition switch to the ON position.
11. Select "check mode" from the menu screen.
12. Select "erase memory" from the menu screen.
13. Initialize the learning value.
14. Start the engine and check for leaks, correct as required.

3.0L ENGINE

1. Relieve the fuel system pressure.
2. Disconnect the negative battery cable.

✳✳ CAUTION

Work MUST NOT be started until at least 90 seconds after the ignition switch is turned to the LOCK position and the negative battery cable is disconnected from the battery. This will allow time for the air bag system backup power supply to deplete its stored energy preventing accidental air bag deployment which could result in unnecessary air bag system repairs and/or personal injury.

3. Drain the cooling system.
4. Disconnect all components from the air intake plenum and remove the plenum from the intake manifold.
5. Wrap the connection with a shop towel and disconnect the high pressure fuel line at the fuel rail.

✳✳ CAUTION

Observe all applicable safety precautions when working around fuel. Whenever servicing the fuel system, always work in a well ventilated area. Do not allow fuel spray or vapors to come in contact with a spark or open flame. Keep a dry chemical fire extinguisher near the work area. Always keep fuel in a container specifically designed for fuel storage; also, always properly seal fuel containers to avoid the possibility of fire or explosion.

6. Remove or disconnect the following:
   • Fuel return hose and remove the O-ring
   • Vacuum hose from the fuel pressure regulator
   • Electrical connectors from each injector
   • Fuel pipe connecting the fuel rails
   • Injector rail retaining bolts. Make sure the rubber mounting bushings do not get lost.

7. Lift the rail assemblies up and away from the engine.

8. Remove the injectors from the rail by pulling gently. Discard the lower insulator.

To install:

➡**Some of the vehicles may have a clip that secures the injector to the fuel rail. Be sure to remove or install the injector clip where necessary.**

9. Install or connect the following:
   • New grommet and O-ring to the injector. Coat the O-ring with light oil.

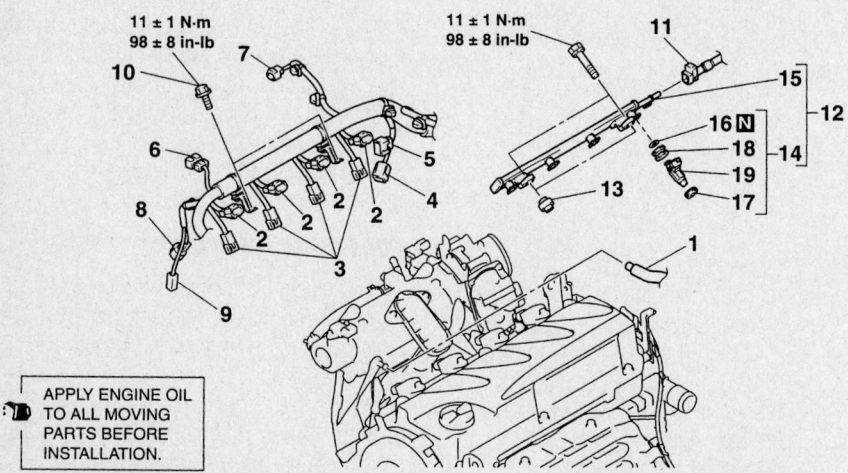

APPLY ENGINE OIL TO ALL MOVING PARTS BEFORE INSTALLATION.

1. PCV HOSE CONNECTION
2. IGNITION COIL CONNECTORS
3. FUEL INJECTOR CONNECTORS
4. EGR VALVE CONNECTOR
5. THROTTLE POSITION SENSOR CONNECTOR
6. MANIFOLD ABSOLUTE PRESSURE SENSOR CONNECTOR
7. EVAPORATIVE EMISSION PURGE SOLENOID CONNECTOR
8. KNOCK SENSOR CONNECTOR
9. POWER STEERING PRESSURE SWITCH CONNECTOR

10. ROCKER COVER BRACKET INSTALLATION BOLTS
11. FUEL HIGH-PRESSURE HOSE CONNECTION
12. FUEL RAIL AND FUEL INJECTOR ASSEMBLY
13. INSULATORS
14. FUEL INJECTOR ASSEMBLY
15. FUEL RAIL
16. O-RING
17. INSULATORS
18. GROMMETS
19. FUEL INJECTORS

09482_GALA_G0125

Fuel injectors and related components—2.4L engine—2006 shown

• Injector to the fuel rail

10. Replace the seats in the intake manifold. Install the fuel rails and injectors to the manifold. Make sure the rubber bushings are in place before tightening the mounting bolts.

11. Tighten the retaining bolts to 84–108 inch lbs. (10–13 Nm). Install the fuel pipe with new gasket.

12. Install or connect the following:
• Electrical connectors to the injectors
• Fuel return hose
• O-ring, lightly lubricate it and connect the high pressure fuel line
• Intake plenum and all related items, using new gaskets

13. Fill the cooling system.

14. Connect the negative battery cable and check the entire system for proper operation and leaks.

3.8L ENGINE

1. Relieve the fuel system pressure.

2. Disconnect the negative battery cable. Remove the intake manifold plenum assembly.

3. Disconnect the fuel injector connectors. Remove the control wiring harness bracket mounting bolts. Remove the engine mounting stay.

4. Disconnect the high pressure fuel hose connection. Remove the fuel rail and fuel injector assembly

5. Remove the insulators, O-rings and

fuel injectors. Discard the gaskets and O-rings.

To install:

6. Installation is the reverse of the removal procedure. Be sure to use new O-rings and gaskets as required.

7. Complete the vehicle initialization procedure.

➡ **To complete the initialization procedure the following tools are needed. MB991958 scan tool, MB991824 VCI, MB991827 MUT III USB cable, MB991910 MUT III Main harness "A"**

8. Connect the scan tool to the data link connector. To prevent damage to the scan tool be sure that the ignition switch is in the LOCK position before connecting the scan tool.

9. Turn the ignition switch to the ON position.

10. Select "check mode" from the menu screen.

11. Select "erase memory" from the menu screen.

12. Initialize the learning value.

13. Start the engine and check for leaks, correct as required.

Galant

2002–2003 2.4L ENGINE

1. Relieve the fuel system pressure.

2. Label and disconnect the spark plug wires. Position the wires aside.

3. Remove or disconnect the following:
• Positive Crankcase Ventilation (PCV) hose from the valve cover
• High pressure fuel line to the fuel rail and disconnect the line. Be prepared to contain fuel spillage; plug the line to keep out dirt and debris.

❊❊ CAUTION

Observe all applicable safety precautions when working around fuel. Whenever servicing the fuel system, always work in a well ventilated area. Do not allow fuel spray or vapors to come in contact with a spark or open flame. Keep a dry chemical fire extinguisher near the work area. Always keep fuel in a container specifically designed for fuel storage; also, always properly seal fuel containers to avoid the possibility of fire or explosion.

4. Remove or disconnect the following:
• Vacuum hose from the fuel pressure regulator
• Fuel return hose from the pressure regulator
• Electrical connector from each injector, label for reference
• Bolt(s) holding the fuel rail to the manifold. Carefully lift the rail up and remove it with the injectors attached. Take great care not to drop an injector. Place the rail and injectors in a safe location on the workbench; protect the tips of the injectors from dirt and/or impact.
• Injector insulators from the intake manifold, discard. The insulators are not reusable.
• Injectors from the fuel rail by pulling gently in a straight outward motion. Make certain the grommet and O-ring come off with the injector.

To install:

5. Install a new insulator in each injector port in the manifold.

6. Remove the old grommet and O-ring from each injector. Install a new grommet and O-ring; coat the O-ring lightly with clean, thin oil.

7. If the fuel pressure regulator was removed, replace the O-ring with a new one and coat it lightly with clean, thin oil. Insert the regulator straight into the rail, then check that it can be rotated freely. If it does not rotate smoothly, remove it and inspect the O-ring for deformation or jamming. When properly installed, align the mounting

APPLY ENGINE OIL TO ALL MOVING PARTS BEFORE INSTALLATION.

11 ± 1 N·m
98 ± 8 ft-lb

36 ± 6 N·m
27 ± 4 ft-lb

12 ± 1 N·m
102 ± 13 in-lb

12 ± 1 N·m
102 ± 13 in-lb

1. FUEL INJECTOR CONNECTORS
2. CONTROL WIRING HARNESS BRACKET MOUNTING BOLTS
3. ENGINE MOUNTING STAY
4. FUEL HIGH-PRESSURE HOSE CONNECTION (FUEL RAIL SIDE)
5. FUEL RAIL AND FUEL INJECTOR ASSEMBLY
6. INSULATORS
7. O-RINGS
8. FUEL INJECTORS
9. FUEL RAIL
10. INSULATORS

09482_GALA_G0126

Fuel injectors and related components—3.8L engine

holes and tighten the retaining bolts to 84 inch lbs. (9 Nm). This procedure must be followed even if the fuel rail was not removed.

8. Install or connect the following:
- Injector into the fuel rail, constantly turning the injector left and right during installation. When fully installed, the injector should still turn freely in the rail. If it does not, remove the injector and inspect the O-ring for deformation or damage.
- Delivery pipe and injectors to the engine. Make certain that each injector fits correctly into its port and that the rubber insulators for the fuel rail mounts are in position.
- Fuel rail retaining bolts and tighten them to 108 inch lbs. (12 Nm)
- Wiring harnesses to the appropriate injector
- Fuel return hose to the pressure regulator, then connect the vacuum hose
- O-ring on the high pressure fuel line, coat the O-ring lightly with clean, thin oil and install the line to the fuel rail. Tighten the mounting bolts to 48 inch lbs. (6 Nm).
- PCV hose and spark plug wires
- Negative battery cable

9. Pressurize the fuel system and inspect all connections for leaks.

2004–2006 2.4L ENGINE

1. Relieve the fuel system pressure.
2. Disconnect the negative battery cable. Remove the air cleaner cover and air intake hose assembly.
3. Disconnect the PCV valve hose connection, ignition coil connectors, EGR valve connector and fuel injector connectors.
4. Disconnect the throttle position sensor connector, manifold absolute pressure sensor connector, EVAP emission purge solenoid connector and knock sensor connector.
5. Disconnect the power steering pressure switch connector. Remove the rocker cover bracket bolts.
6. Disconnect the high pressure hose connection. Remove the fuel rail, insulators, grommets and injectors. Discard the O-rings.

To install:

7. Installation is the reverse of the removal procedure. Be sure to use new O-rings and gaskets as required.
8. On 2006 vehicles, complete the vehicle initialization procedure.

➡**To complete the initialization procedure the following tools are needed. MB991958 scan tool, MB991824 VCI, MB991827 MUT III USB cable, MB991910 MUT III Main harness "A"**

9. Connect the scan tool to the data link connector. To prevent damage to the scan tool be sure that the ignition switch is in the LOCK position before connecting the scan tool.
10. Turn the ignition switch to the ON position.
11. Select "check mode" from the menu screen.
12. Select "erase memory" from the menu screen.
13. Initialize the learning value.
14. Start the engine and check for leaks, correct as required.

3.0L ENGINE

1. Relieve the fuel system pressure.
2. Disconnect the negative battery cable.

✳✳ CAUTION

Work MUST NOT be started until at least 90 seconds after the ignition switch is turned to the LOCK position and the negative battery cable is disconnected from the battery. This will allow time for the air bag system backup power supply to deplete its stored energy preventing accidental air bag deployment which could result in unnecessary air bag system repairs and/or personal injury.

3. Drain the cooling system.
4. Disconnect all components from the air intake plenum and remove the plenum from the intake manifold.
5. Wrap the connection with a shop towel and disconnect the high pressure fuel line at the fuel rail.

✳✳ CAUTION

Observe all applicable safety precautions when working around fuel. Whenever servicing the fuel system, always work in a well ventilated area. Do not allow fuel spray or vapors to come in contact with a spark or open flame. Keep a dry chemical fire extinguisher near the work area. Always keep fuel in a container specifically designed for fuel storage; also, always properly seal fuel containers to avoid the possibility of fire or explosion.

6. Remove or disconnect the following:
- Fuel return hose and remove the O-ring
- Vacuum hose from the fuel pressure regulator
- Electrical connectors from each injector
- Fuel pipe connecting the fuel rails
- Injector rail retaining bolts. Make sure the rubber mounting bushings do not get lost.

7. Lift the rail assemblies up and away from the engine.
8. Remove the injectors from the rail by pulling gently. Discard the lower insulator.

To install:

➡**Some of the vehicles may have a clip that secures the injector to the fuel rail. Be sure to remove or install the injector clip where necessary.**

9. Install or connect the following:
- New grommet and O-ring to the injector. Coat the O-ring with light oil.
- Injector to the fuel rail

10. Replace the seats in the intake manifold. Install the fuel rails and injectors to the manifold. Make sure the rubber bushings are in place before tightening the mounting bolts.
11. Tighten the retaining bolts to 84–108 inch lbs. (10–13 Nm). Install the fuel pipe with new gasket.
12. Install or connect the following:
- Electrical connectors to the injectors
- Fuel return hose
- O-ring, lightly lubricate it and connect the high pressure fuel line
- Intake plenum and all related items, using new gaskets

13. Fill the cooling system.
14. Connect the negative battery cable and check the entire system for proper operation and leaks.

3.8L ENGINE

1. Relieve the fuel system pressure.
2. Disconnect the negative battery cable. Remove the intake manifold plenum assembly.
3. Disconnect the fuel injector connectors. Remove the control wiring harness bracket mounting bolts. Remove the engine mounting stay.
4. Disconnect the high pressure fuel hose connection. Remove the fuel rail and fuel injector assembly
5. Remove the insulators, O-rings and fuel injectors. Discard the gaskets and O-rings.

To install:

6. Installation is the reverse of the removal procedure. Be sure to use new O-rings and gaskets as required.

7. On 2006 vehicles, complete the vehicle initialization procedure.

➡**To complete the initialization procedure the following tools are needed.**
MB991958 scan tool, MB991824 VCI,
MB991827 MUT III USB cable,
MB991910 MUT III Main harness "A"

8. Connect the scan tool to the data link connector. To prevent damage to the scan tool be sure that the ignition switch is in the LOCK position before connecting the scan tool.

9. Turn the ignition switch to the ON position.

10. Select "check mode" from the menu screen.

11. Select "erase memory" from the menu screen.

12. Initialize the learning value.

13. Start the engine and check for leaks, correct as required.

Lancer

2.0L ENGINE

1. Relieve the fuel system pressure as described in this section.

2. Disconnect the negative battery cable.

3. Wrap the connection with a shop towel and disconnect the high pressure fuel line at the fuel rail.

✳✳ CAUTION

Observe all applicable safety precautions when working around fuel. Whenever servicing the fuel system, always work in a well ventilated area. Do not allow fuel spray or vapors to come in contact with a spark or open flame. Keep a dry chemical fire extinguisher near the work area. Always keep fuel in a container specifically designed for fuel storage; also, always properly seal fuel containers to avoid the possibility of fire or explosion.

4. Remove or disconnect the following:
   - Positive Crankcase Ventilation (PCV) hose
   - Exhaust Gas Recirculation (EGR) solenoid valve connector
   - Manifold Differential Pressure (MDP) sensor connector
   - Purge control solenoid valve connector
   - Throttle Position (TP) sensor connector

   - Idle Air Control (IAC) motor connector
   - Electrical connector from each injector, label for reference
   - High pressure fuel hose
   - Fuel return hose
   - Vacuum hose(s)
   - Fuel pressure regulator
   - Fuel hose
   - Fuel return pipe
   - Bolt(s) holding the fuel rail to the manifold. Carefully lift the rail up and remove it with the injectors attached. Take great care not to drop an injector. Place the rail and injectors in a safe location on the workbench; protect the tips of the injectors from dirt and/or impact.
   - Injector insulators from the intake manifold, discard. The insulators are not reusable.
   - Injectors from the fuel rail by pulling gently in a straight outward motion. Make certain the grommet and O-ring come off with the injector.

To install:

5. Install a new insulator in each injector port in the manifold.

6. Remove the old grommet and O-ring from each injector. Install a new grommet and O-ring; coat the O-ring lightly with clean, thin oil.

7. If the fuel pressure regulator was removed, replace the O-ring with a new one and coat it lightly with clean, thin oil. Insert the regulator straight into the rail, then check that it can be rotated freely. If it does not rotate smoothly, remove it and inspect the O-ring for deformation or jamming. When properly installed, align the mounting holes and tighten the retaining bolts to 84 inch lbs. (9 Nm). This procedure must be followed even if the fuel rail was not removed.

8. Install or connect the following:
   - Injector into the fuel rail, constantly turning the injector left and right during installation. When fully installed, the injector should still turn freely in the rail. If it does not, remove the injector and inspect the O-ring for deformation or damage.
   - Fuel rail and injectors to the engine. Make certain that each injector fits correctly into its port and that the rubber insulators for the fuel rail mounts are in position.
   - Fuel return pipe and fuel hose
   - Fuel return hose to the fuel rail
   - High pressure fuel line to the fuel rail
   - Fuel injector connectors
   - IAC motor connector
   - TP sensor connector
   - Purge control solenoid connector

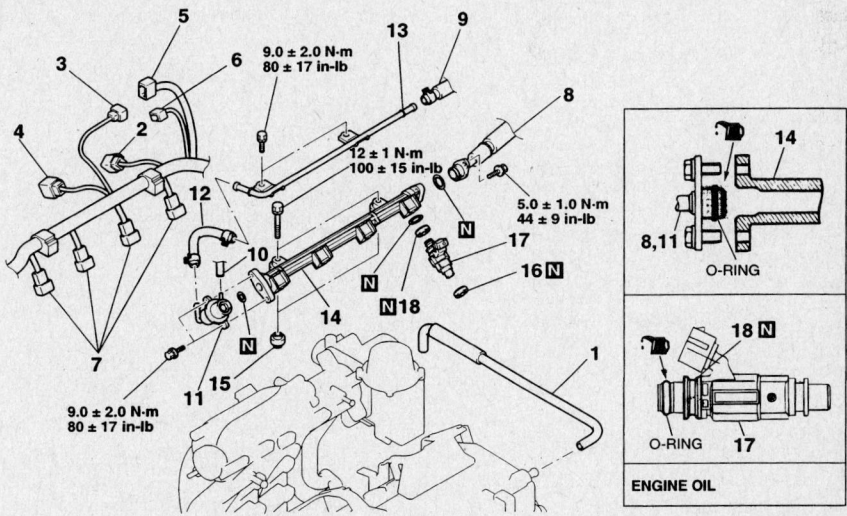

1. PCV HOSE CONNECTION
2. EGR SOLENOID VALVE CONNECTOR
3. MANIFOLD DIFFERENTIAL PRESSURE SENSOR CONNECTOR
4. PURGE CONTROL SOLENOID VALVE CONNECTOR
5. THROTTLE POSITION SENSOR CONNECTOR
6. IDLE AIR CONTROL MOTOR CONNECTOR
7. INJECTOR CONNECTOR

93570G25

Fuel injectors and related components—Lancer 2.0L engine

- MDP sensor connector
- EGR solenoid valve connector
- PCV hose connector
- Negative battery cable

9. Pressurize the fuel system and inspect all connections for leaks.

2.4L ENGINE

1. Relieve the fuel system pressure.
2. Disconnect the negative battery cable. Remove the air cleaner cover and air intake hose assembly.
3. Disconnect the PCV valve hose connection, capacitor connector, ignition coil connectors, EGR valve connector and fuel injector connectors.
4. Remove the rocker cover bracket bolts. Disconnect the high pressure hose connection.
5. Remove the fuel rail, insulators, grommets and injectors. Discard the O-rings.

To install:

6. Installation is the reverse of the removal procedure. Be sure to use new O-rings and gaskets as required.
7. On 2006 vehicles, complete the vehicle initialization procedure.

➡To complete the initialization procedure the following tools are needed. MB991958 scan tool, MB991824 VCI, MB991827 MUT III USB cable, MB991910 MUT III Main harness "A"

8. Connect the scan tool to the data link connector. To prevent damage to the scan tool be sure that the ignition switch is in the LOCK position before connecting the scan tool.
9. Turn the ignition switch to the ON position.
10. Select "check mode" from the menu screen.
11. Select "erase memory" from the menu screen.
12. Initialize the learning value.
13. Start the engine and check for leaks, correct as required.

Evolution

1. Relieve the fuel system pressure.
2. Disconnect the negative battery cable.
3. Remove the strut tower bar. Remove the air hose, air by pass hose and air pipe.
4. Disconnect the control harness connector, accelerator cable assembly connection and the injector connector.
5. Disconnect the high pressure fuel hose. Disconnect the fuel return hose. Disconnect the vacuum hose connection.
6. Remove the fuel pressure regulator, discard the O-ring. Remove the fuel return pipe.

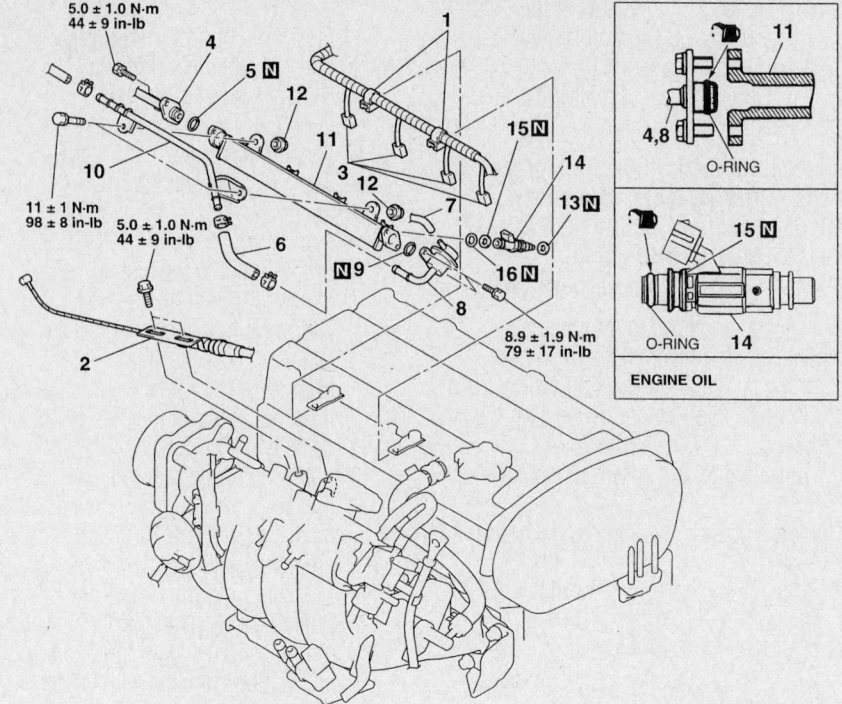

1. CONTROL HARNESS CONNECTOR
2. ACCELERATOR CABLE ASSEMBLY CONNECTION (THROTTLE BODY SIDE)
3. INJECTOR CONNECTOR
4. FUEL HIGH-PRESSURE HOSE CONNECTION
5. O-RING
6. FUEL RETURN HOSE CONNECTION
7. VACUUM HOSE CONNECTION
8. FUEL PRESSURE REGULATOR
9. O-RING
10. FUEL RETURN PIPE
11. FUEL RAIL
12. INSULATORS
13. INSULATORS
14. INJECTORS

09482_GALA_G0127

Fuel injectors and related components—Evolution

7. Remove the fuel rail, insulators and injectors.

To install:

8. Installation is the reverse of the removal procedure. Be sure to use new O-rings and gaskets as required.

Mirage

1.5L AND 1.8L ENGINES

1. Relieve the fuel system pressure as described in this section.
2. Remove or disconnect the following:
- Positive Crankcase Ventilation (PCV) hose from the valve cover
- Breather hose at the opposite end of the valve cover
- High pressure fuel line

✳✳ CAUTION

Observe all applicable safety precautions when working around fuel. Whenever servicing the fuel system, always work in a well ventilated area. Do not allow fuel spray or vapors to come in contact with a spark or open flame. Keep a dry chemical fire extinguisher near the work area. Always keep fuel in a container specifically designed for fuel storage; also, always properly seal fuel containers to avoid the possibility of fire or explosion.

3. Remove or disconnect the following:
- Vacuum hose from the fuel pressure regulator
- Fuel return hose from the pressure regulator
- Electrical connector from each injector. Label for reference
- Bolt(s) holding the fuel rail to the manifold. Carefully lift the rail up and remove it with the injectors attached. Take great care not to drop an injector. Place the rail and injectors in a safe location on the workbench; protect the tips of the injectors from dirt and/or impact.
- Injector insulators from the intake

manifold, discard. The insulators are not reusable.

- Injectors from the fuel rail by pulling gently in a straight outward motion. Make certain the grommet and O-ring come off with the injector.

To install:

4. Install a new insulator in each injector port in the manifold.

5. Remove the old grommet and O-ring from each injector. Install a new grommet and O-ring; coat the O-ring lightly with clean, thin oil.

6. If the fuel pressure regulator was removed, replace the O-ring with a new one and coat it lightly with clean, thin oil. Insert the regulator straight into the rail, then

check that it can be rotated freely. If it does not rotate smoothly, remove it and inspect the O-ring for deformation or jamming. When properly installed, align the mounting holes and tighten the retaining bolts to 84 inch lbs. (9 Nm). This procedure must be followed even if the fuel rail was not removed.

7. Install or connect the following:

- Injector into the fuel rail, constantly turning the injector left and right during installation. When fully installed, the injector should still turn freely in the rail. If it does not, remove the injector and inspect the O-ring for deformation or damage.
- Delivery pipe and injectors to the engine. Make certain that each

injector fits correctly into its port and that the rubber insulators for the fuel rail mounts are in position.

- Fuel rail retaining bolts and tighten them to 108 inch lbs. (12 Nm)
- Wiring harnesses to the appropriate injector
- Fuel return hose to the pressure regulator, then connect the vacuum hose

8. Replace the O-ring on the high pressure fuel line, coat the O-ring lightly with clean, thin oil and install the line to the fuel rail. Tighten the mounting bolts.

- PCV hose and the breather hose
- Negative battery cable

9. Pressurize the fuel system and inspect all connections for leaks.

DRIVE TRAIN

Transaxle Assembly

REMOVAL & INSTALLATION

Manual

2002–2005 ECLIPSE

1. Before servicing the vehicle, refer to the Precautions Section.

2. Remove or disconnect the following:

- Battery and the air intake hoses
- Battery tray and support
- Auto-cruise actuator and bracket, if equipped with cruise control
- Charcoal canister and bracket
- Shift and select cables from the transaxle
- Back-up light switch and the Vehicle Speed Sensor (VSS) connectors
- Starter assembly
- Engine support fixture to the engine and remove the transaxle mounting bolts

3. Remove or disconnect the following:

- Rear roll stopper bracket mounting bolts
- Transaxle mounting bracket mounting nuts
- Engine undercovers
- Axle shafts
- Slave cylinder from the bell housing without disconnecting the fluid line. Position it out of the way.
- Bell housing cover and the right-hand center member stay (support)
- Center member

4. Place a transmission jack under the transaxle and remove the transaxle mounting bolt.

5. Remove the transaxle mounting and lower the transaxle.

To install:

6. Raise the transaxle into position and install the transaxle mounting. Torque the through-bolt to 50 ft. lbs. (69 Nm).

7. Install or connect the following:

- Transaxle assembly mounting bolt. Torque the bolt to 22–25 ft. lbs. (30–34 Nm).
- Center member assembly and the right-hand stay
- Bell housing cover and the slave cylinder
- Axle shafts. Be sure to install the washer in the proper direction.
- Engine undercovers and lower the vehicle
- Transaxle mounting bracket mounting nuts
- Rear roll stopper bracket mounting bolts
- Transaxle assembly mounting bolts. Torque the mounting bolts to 35 ft. lbs. (48 Nm).

8. Remove the engine support fixture.

9. Install or connect the following:

- Starter assembly
- VSS and the back-up light connectors
- Cruise control actuator if removed
- Battery tray support and the tray
- Charcoal canister bracket and the canister
- Air duct and the air cleaner assembly

2006 ECLIPSE (FIVE SPEED)

1. Before servicing the vehicle, refer to the Precautions Section.

2. Disconnect the negative battery cable. Drain the transaxle fluid.

3. Remove the front number one exhaust pipe. Remove the driveshaft.

4. Remove the battery and the battery tray. Remove the air cleaner assembly.

5. Remove the ECM. Remove the starter.

6. Disconnect the backup light electrical connector, shift cable and select cable connection.

7. Remove the clutch release cylinder, without disconnecting the fluid lines. Position the unit to the side.

8. Disconnect the vehicle speed sensor connector. Remove the upper transaxle retaining bolts. Remove the bell housing cover.

9. Remove the center member assembly. Remove the rear roll stopper bracket. Jack up the vehicle and remove the transaxle mounting body side bracket. Remove the transaxle mounting stopper.

10. Install engine support tool MB991895. Set the tool to the front fender

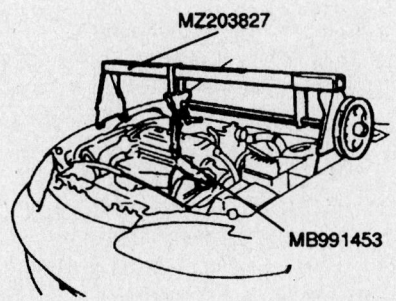

MZ203827

MB991453

7923PG82

Typical engine fixture

rear mounting bolts and the upper radiator support insulator mounting bolts, which are located in the engine compartment.

11. Use tools MB991527 (engine lifting fixture) and tool MB991454 (chain) to hold the engine and transaxle assembly in place.

12. Remove the lower transaxle retaining bolts. Remove the transaxle assembly from the vehicle.

To install:

13. Raise the transaxle into position and install the lower transaxle retaining bolts.

14. Continue the installation in the reverse order of the removal procedure.

15. Check and adjust fluid levels, as required.

16. Check and adjust the front end wheel alignment, as necessary.

17. To initialize the ECM, turn the ignition switch ON then OFF and keep it in the OFF position for at least ten seconds.

2006 ECLIPSE (SIX SPEED)

1. Before servicing the vehicle, refer to the Precautions Section.

2. Disconnect the negative battery cable.

3. Drain the transaxle fluid. Drain the clutch fluid.

4. Remove the number one exhaust pipe and the number two exhaust pipe.

5. Remove the driveshaft, and the driveshaft and inner shaft assembly.

6. Remove the engine strut tower assembly. Remove the battery and the battery tray. Remove the air cleaner assembly and air cleaner resonator.

7. Remove the ECM. Remove the starter. Drain the engine coolant. Remove the upper radiator hose.

8. Remove the intake manifold plenum.

9. Remove the left side heated oxygen sensor (front rear) connector and the right side heated oxygen sensor (front rear) connector. Remove the engine oil dipstick and dipstick tube.

10. Disconnect the backup light switch electrical connector, shift cable and cable connection, clutch release concentric cylinder and VSS connector.

11. Remove the transaxle upper retaining bolts. Remove the heat protector.

12. Remove the center member assembly. Remove the rear roll stopper bracket. Jack up the vehicle and remove the transaxle mounting body side bracket. Remove the transaxle mounting stopper.

13. Remove the engine hanger. Remove the intake manifold rear plenum stay and engine hanger. Install special tool MB992012 and MB992013 to each cylinder head.

14. Install engine support tool MB991895, to hold the engine and transaxle in position. Set the tool to the front fender rear mounting bolts and the upper radiator support insulator mounting bolts, which are located in the engine compartment.

15. Remove the lower transaxle retaining bolts. Remove the transaxle assembly from the vehicle.

To install:

16. Raise the transaxle into position and install the lower transaxle retaining bolts.

17. Continue the installation in the reverse order of the removal procedure.

18. Check and adjust fluid levels, as required.

19. Check and adjust the front end wheel alignment, as necessary.

20. To initialize the ECM, turn the ignition switch ON then OFF and keep it in the OFF position for at least ten seconds.

LANCER

1. Before servicing the vehicle, refer to the Precautions Section.

2. Drain the transaxle fluid.

3. Remove or disconnect the following:

- Negative battery cable
- Engine undercover
- Evaporative canister
- Positive battery cable, battery and battery tray
- Shifter cables
- Back-up light switch and Vehicle Speed Sensor (VSS) connector
- Starter motor
- Clutch hose
- Upper engine-to-transaxle bolts
- Transaxle mount
- Transaxle mount stopper

4. Install a suitable engine support assembly, then raise and safely support the vehicle.

- Stabilizer bar
- Wheel Speed Sensor (WSS) connector, if equipped with Anti-lock Brakes (ABS)
- Brake hose clamp
- Tie rod end
- Lower control arm
- Centermember
- Halfshafts by inserting a prybar between the transaxle case and the driveshaft and prying the shaft from the transaxle. Do not pull on the driveshaft.
- Bell housing lower cover
- Transaxle to engine bolts and lower the transaxle from the vehicle

To install:

5. Install or connect the following:

- Transaxle to the engine and install the lower mounting bolts
- Bell housing cover

➡ **When installing the halfshafts, use new circlips on the axle ends.**

6. Install or connect the following:

- Halfshafts into the transaxle
- Centermember
- Lower control arm
- Tie rod end
- Brake hose clamp
- WSS connector, if equipped
- Stabilizer bar

7. Lower the vehicle, then remove the engine support.

- Transaxle mount bracket. Tighten the nuts to 35 ft. lbs. (47 Nm).
- Transaxle mount stopper. Tighten the nuts to 61 ft. lbs. (82 Nm).
- Transaxle mount
- Upper transaxle-to-engine bolts and torque to 36 ft. lbs. (48 Nm)
- Clutch line
- Starter motor
- VSS connector
- Back-up light switch connector
- Shifter cables, adjust
- Evaporative canister
- Battery and battery tray
- Engine undercover
- Positive and negative battery cables
- Transaxle with fluid

8. Bleed the clutch, check and adjust the front wheel alignment, then check the transaxle for proper operation.

EVOLUTION (FIVE SPEED)

1. Before servicing the vehicle, refer to the Precautions Section.

2. Disconnect the negative battery cable.

3. Remove the transfer case. Remove the starter. Remove the air cleaner bracket.

4. Remove the rear roll rod assembly.

5. Disconnect the transaxle harness clamp, backup light switch connector and the VSS connector.

6. Remove the clutch release cylinder and clutch oil pipe. Remove the snap pin. Remove the cable bracket and cable assembly (transaxle side). Remove the rear roll mount bracket.

7. While supporting the engine and transaxle with a floor jack, remove the transaxle mounting insulator assembly.

8. Remove the upper transaxle assembly retaining bolts. Remove the upper transaxle mounting insulator assembly. Remove the transaxle mounting insulator

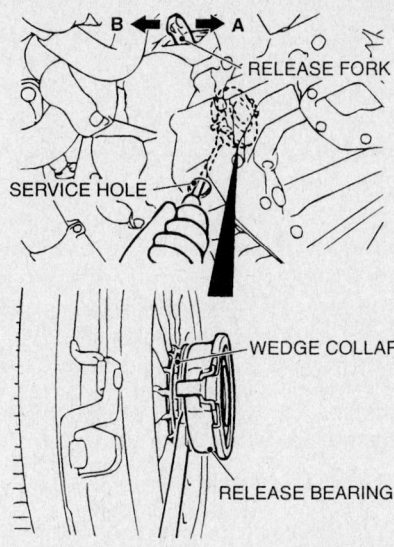

Clutch release bearing separation location points—Evolution

stopper. Remove the transaxle mounting insulator.

9. Install engine support tool MB991895. Set the tool to the front fender rear mounting bolts and the upper radiator support insulator mounting bolts, which are located in the engine compartment.

10. Use tool MB991454 to hold the engine and transaxle assembly in place.

11. Remove the cover from the clutch release service hole in the clutch housing.

➡**If it is hard to turn the suitable tool, to pry off the release bearing, remove the tool and repeat the procedure below, after pushing the release fork fully in direction "A" two or three times. Forcibly prying can cause the release bearing to be damaged.**

12. While pushing the release fork by hand in direction "A" (in the illustration), insert a suitable tool between the release bearing and the wedge collar.

➡**Be sure to push the release fork in direction "A" before inserting the suitable tool.**

13. Separate the release bearing from the wedge collar by prying the suitable tool at a 90 degree angle.

➡**The release fork is forced to move fully in direction "B" (in the illustration) by the return spring as soon as it is separated from the wedge collar.**

14. Remove the transaxle assembly lower retaining bolts. Remove the transaxle from the vehicle.

To install:

15. Raise the transaxle into position and install the lower transaxle retaining bolts.

16. Continue the installation in the reverse order of the removal procedure.

17. Check and adjust fluid levels, as required.

EVOLUTION (SIX SPEED)

1. Before servicing the vehicle, refer to the Precautions Section.

2. Disconnect the negative battery cable. Remove the battery. Remove the battery tray.

3. Remove the transfer case. Remove the starter. Remove the air cleaner bracket.

4. Remove the strut tower bar assembly. Remove the air duct, air cleaner assembly and air intake hose.

5. Remove the air bypass hose, air hoses and air pipe. Remove the starter.

6. Disconnect the main harness clamp connection. Disconnect the backup light switch connector, VSS sensor electrical connector and the snap pin.

7. Disconnect the shift cable connection and the select cable connection. Disconnect the cable control cable assembly and bracket bolt (transaxle side). Remove the release cylinder and clutch oil pipe. Remove the transaxle upper retaining bolts.

8. Remove the harness clamp and clamp mounting bolt.

9. Using a floor jack support the engine and transaxle assembly. Remove the transaxle mounting insulator assembly.

10. Remove the transaxle mounting insulator stopper. Remove the transaxle mounting insulator.

11. Install engine support tool MB991895. Set the tool to the front fender rear mounting bolts and the upper radiator support insulator mounting bolts, which are located in the engine compartment.

12. Use tool MB991454 to hold the engine and transaxle assembly in place.

13. Remove the cover from the clutch release service hole in the clutch housing.

➡**If it is hard to turn the suitable tool, to pry off the release bearing, remove the tool and repeat the procedure below, after pushing the release fork fully in direction "A" two or three times. Forcibly prying can cause the release bearing to be damaged.**

14. While pushing the release fork by hand in direction "A" (in the illustration), insert a suitable tool between the release bearing and the wedge collar.

➡**Be sure to push the release fork in direction "A" before inserting the suitable tool.**

15. Separate the release bearing from the wedge collar by prying the suitable tool at a 90 degree angle.

➡**The release fork is forced to move fully in direction "B" (in the illustration) by the return spring as soon as it is separated from the wedge collar.**

16. Remove the transaxle assembly lower retaining bolts. Remove the transaxle from the vehicle.

To install:

17. Raise the transaxle into position and install the lower transaxle retaining bolts.

18. Continue the installation in the reverse order of the removal procedure.

19. Check and adjust fluid levels, as required.

MIRAGE

1. Before servicing the vehicle, refer to the Precautions Section.

2. Remove or disconnect the following:

- Negative battery cable
- Front wheels and the inner wheel panels
- Air cleaner assembly and vacuum hoses

3. Note the locations and disconnect the shifter cables.

- Back-up lamp switch connector
- Speedometer cable and remove the starter motor
- Upper transaxle-to-engine mounting bolts

4. Remove the undercover and splash pan.

5. Drain the transaxle oil.

6. Support the engine and remove the crossmember.

7. Remove or disconnect the following:

- Upper transaxle mounting bolt and bracket
- Stabilizer bar, tie rod ends and the lower ball joint connections
- Clutch release cylinder and clutch oil line bracket. Disconnect the clutch cable, if equipped with cable controlled clutch system.
- Halfshafts by inserting a prybar between the transaxle case and the driveshaft and prying the shaft from the transaxle. Do not pull on the driveshaft.
- Bell housing lower cover
- Transaxle to engine bolts and lower the transaxle from the vehicle

To install:

8. Install or connect the following:
- Transaxle to the engine and install the mounting bolts
- Bell housing cover

➡ **When installing the halfshafts, use new circlips on the axle ends.**

9. Install or connect the following:
- Halfshafts into the transaxle
- Slave cylinder or connect the clutch cable
- Ball joints, tie rod ends and stabilizer bar connections
- Upper transaxle mounting bracket and bolt
- Crossmember
- Undercover
- Upper transaxle-to-engine mounting bolts
- Starter motor
- Back-up light switch connector and speedometer cable
- Shifter cables, adjust
- Air cleaner assembly
- Front wheels
- Negative battery cable and check the transaxle for proper operation

Automatic

DIAMANTE

1. Before servicing the vehicle, refer to the Precautions Section.
2. Properly disarm the Supplemental Restraint System (SRS) system.
3. Raise and safely support the vehicle.
4. Remove or disconnect the following:
- Front wheels
- Engine side cover and undercovers
5. Drain the transaxle assembly.
6. If equipped, remove the front catalytic converter.
7. Remove or disconnect the following:
- Exhaust pipe, main muffler and catalytic converter
- Tie rod end and ball joint from the steering knuckle
- Support bearing for the left side halfshaft
- Halfshafts by inserting a prybar between the transaxle case and the driveshaft and prying the shaft from the transaxle
- Air cleaner assembly and adjoining ductwork
- Engine harness connection
- Compressor assembly, if the vehicle is equipped with Active Electronic Controlled Suspension (Active-ECS)—suspend with

wire—Do not allow the compressor to hang from the air hose.
- Roll stopper stay bracket, if equipped
- Speedometer cable from the transaxle
- The clip that secures the shifter
- Shifter control cable from the transaxle
- Plug the oil cooler hoses from the transaxle
8. Disconnect the following:
- Park/neutral switch electrical harness
- Kickdown servo switch
- Pulse generator
- Oil temperature sensor electrical harness
- Shift control solenoid valve harness.
9. Support the transaxle and remove the transaxle mounting bracket.
10. Remove the 3 upper transaxle-to-engine mounting bolts.
11. For vehicles equipped with Active-ECS, disconnect the height sensor rod from the lower control arm.
12. Remove or disconnect the following:
- Bolt that secures the Heated Oxygen (HO2S) sensor harness to the right side crossmember
- Starter assembly
- Mounting brackets for access to the bell housing cover
- Bell housing/oil pan covers assembly
- Bolts holding the flexplate to the torque converter
- Lower transaxle to engine bolts and remove the transaxle assembly

To install:

13. Install or connect the following:
- Transaxle assembly to the engine block and install the mounting bolts
- Bolts that secure the torque converter to the driveplate. Tighten the bolts to 34–38 ft. lbs. (46–53 Nm).
- Bell housing/oil pan covers
- Transaxle stay brackets that were removed for access to the bell housing cover
- Starter assembly and connect the wiring
- Bolt that secures the HO2S sensor harness to the right side crossmember and tighten the bolt to 84–108 inch lbs. (10–12 Nm)
14. For vehicles equipped with Active-ECS, connect the height sensor rod from the lower control arm. Check the height sensor rod for a length (A) of 10.59–10.63 in. (269–270mm).

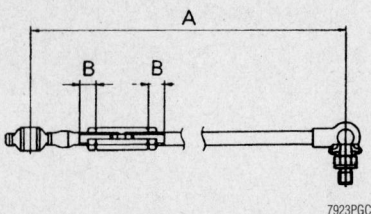

Height sensor rod adjustment—Diamante

15. Install the 3 upper transaxle-to-engine mounting bolts. Tighten the mounting bolts to 54 ft. lbs. (75 Nm).

➡ **One of the upper bolts has a grounding strap to secure under the bolt.**

16. Install or connect the following:
- Transaxle mounting bracket. Tighten the mounting nut and bolts to 51 ft. lbs. (70 Nm).
- Shift control solenoid valve harness
- Kickdown servo switch, pulse generator and oil temperature sensor electrical harness
- Park/neutral switch electrical harness
- Oil cooler hoses to the transaxle, using new hose clamps
- Shifter control cable to the transaxle and secure the cable with clip
- Speedometer cable to the transaxle
- Roll stopper stay bracket and tighten the one through nut and bolt to 36–43 ft. lbs. (50–60 Nm), if removed. Tighten the 2 mounting bolts to 16 ft. lbs. (22 Nm).
- Active-ECS compressor assembly, if removed. Tighten the mounting bolts to 48 inch lbs. (5 Nm) and connect the electrical harness.
- Engine harness connection
- Air cleaner assembly and adjoining ductwork
- Halfshafts and seat halfshafts into the transaxle, using new circlips
- Bolt that secure the left side support bearing and tighten the bolts to 33 ft. lbs. (45 Nm)
- Ball joint and tie rod end to the steering knuckle. Using new nuts, tighten the ball joint castle nut to 43–52 ft. lbs. (60–72 Nm) and tighten the tie rod castle nut to 22 ft. lbs. (30 Nm). Install new cotter pins.
- Exhaust system, using new gaskets
- Front catalytic converter, if removed
- Engine undercovers
- Negative battery cable
17. Fill the transaxle to the correct level.
18. Start the engine and check for leaks.

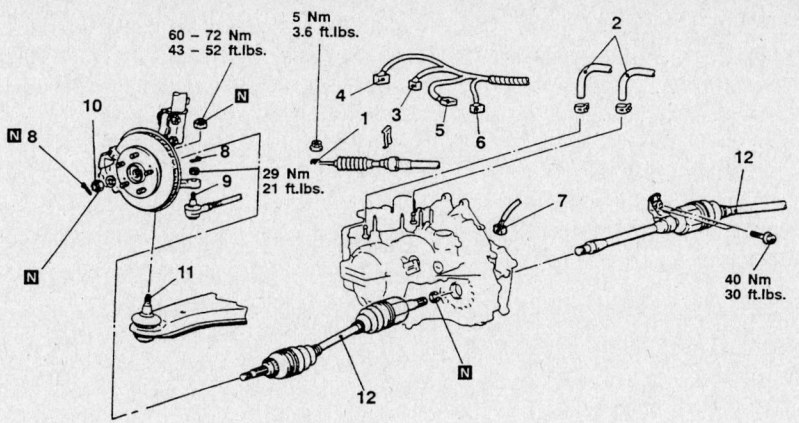

60 – 72 Nm
43 – 52 ft.lbs.

5 Nm
3.6 ft.lbs.

29 Nm
21 ft.lbs.

40 Nm
30 ft.lbs.

Removal steps
1. Transaxle control cable connection
2. Transaxle oil cooler hoses connection
3. PNP switch connector
4. A/T control solenoid valve connector
5. Input shaft speed sensor connector
6. Output shaft speed sensor connector
7. Vehicle speed sensor connector
8. Split pin
9. Connection of the tie rod end
10. Drive shaft nut
11. Connection for the lower arm ball joint
12. Drive shaft and inner shaft assembly (RH) and the drive shaft (LH)

Caution
Mounting locations marked by * should be provisionally tightened, and then fully tightened when the body is supporting the full weight of the engine.

7923PG84

Automatic transaxle and related components—Diamante (view one)

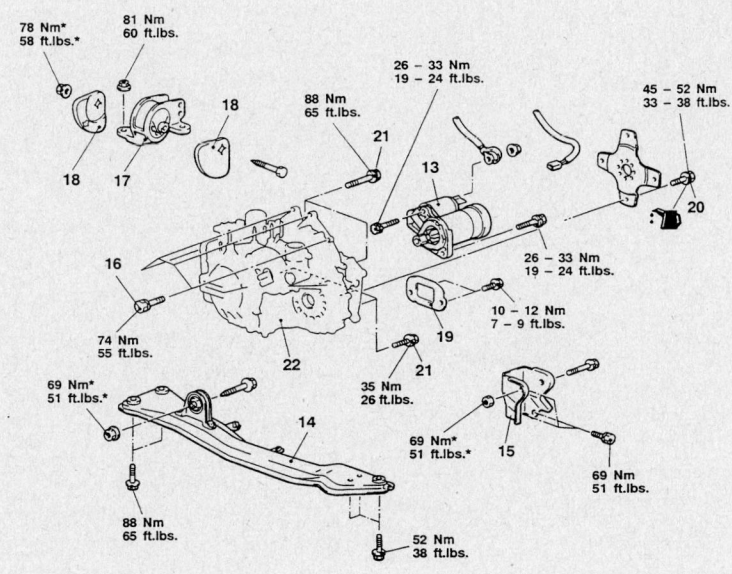

78 Nm*
58 ft.lbs.*

81 Nm
60 ft.lbs.

26 – 33 Nm
19 – 24 ft.lbs.

45 – 52 Nm
33 – 38 ft.lbs.

88 Nm
65 ft.lbs.

26 – 33 Nm
19 – 24 ft.lbs.

10 – 12 Nm
7 – 9 ft.lbs.

74 Nm
55 ft.lbs.

69 Nm*
51 ft.lbs.*

35 Nm
26 ft.lbs.

69 Nm*
51 ft.lbs.*

69 Nm
51 ft.lbs.

88 Nm
65 ft.lbs.

52 Nm
38 ft.lbs.

Lifting up of the vehicle
13. Starter motor
14. Center member assembly
15. Rear roll stopper bracket
16. Transaxle upper portion fixing bolt
17. Transaxle mounting bracket
18. Transaxle mount stopper
• Support the engine and transaxle assembly
19. Bell housing cover
20. Drive plate attaching bolt
21. Transaxle lower portion fixing bolt
22. Transaxle assembly

Caution
Mounting locations marked by * should be provisionally tightened, and then fully tightened when the body is supporting the full weight of the engine.

7923PG85

Automatic transaxle and related components—Diamante (view two)

2002–2005 ECLIPSE

1. Before servicing the vehicle, refer to the Precautions Section.
2. Remove or disconnect the following:
   • Battery and the air intake hoses
   • Battery tray and support
   • Auto-cruise actuator and bracket, if equipped with cruise control
   • Charcoal canister and bracket
   • Shift and select cables from the transaxle
   • Back-up light switch and the vehicle speed sensor connectors
   • Dipstick and tube assembly
   • Starter assembly
   • Park/neutral switch
   • Oil temperature sensor
   • Kick down servo switch
   • Solenoid valve
   • Pulse generator
   • Speedometer connections
3. Attach an engine support fixture to the engine and remove the transaxle mounting bolts.
4. Remove or disconnect the following:
   • Rear roll stopper bracket mounting bolts
   • Transaxle mounting bracket mounting nuts
5. Raise the vehicle and remove the engine undercovers.
6. Remove or disconnect the following:
   • Front exhaust pipe
   • Axle shafts
   • Bell housing cover and the right-hand center member stay (support)
   • Center member
   • Drive plate connecting bolts
7. Place a transmission jack under the transaxle and remove the transaxle mounting bolt.
8. Lower the transaxle.

To install:
9. Raise the transaxle into position and install the transaxle mounting bracket. Torque the through-bolt to 51 ft. lbs. (69 Nm).
10. Install or connect the following:
   • Transaxle assembly mounting bolt. Torque the bolt to 22–25 ft. lbs. (29–34 Nm).
   • Drive plate connecting bolts. Torque the bolts to 33–38 ft. lbs. (45–52 Nm).
   • Center member assembly and the right-hand stay
   • Bell housing cover and the slave cylinder
   • Axle shafts.
   • Front exhaust pipe
   • Engine undercovers and lower the vehicle

- Transaxle mounting bracket mounting nuts
- Rear roll stopper bracket mounting bolts
- Transaxle assembly mounting bolts. Torque the bolts to 35 ft. lbs. (48 Nm).

11. Remove the engine support fixture.

- Park/neutral switch
- Oil temperature sensor
- Kick down servo switch
- Solenoid valve
- Pulse generator
- Speedometer connections
- Starter assembly
- Dipstick and tube assembly
- Vehicle speed sensor and the back-up light connectors
- Cruise control actuator if removed
- Battery tray support and the tray
- Charcoal canister bracket and the canister
- Air duct and the air cleaner assembly

12. Refill the transaxle and the transfer case, if equipped, with the proper fluid.

2006 ECLIPSE (FOUR SPEED)

1. Before servicing the vehicle, refer to the Precautions Section.

2. Disconnect the negative battery cable. Drain the engine coolant. Drain the transaxle fluid.

3. Remove the left side under cover. Remove the air cleaner assembly.

4. Remove the PCM. Remove the battery and the battery tray.

5. Remove the transaxle control cable adjusting nut. Remove the transaxle control cable.

6. Disconnect the transaxle range switch connection. Disconnect the A/T control solenoid valve assembly connector. Disconnect the input shaft speed sensor connector.

7. Disconnect the output shaft speed sensor connector. Disconnect the output shaft speed sensor connector.

8. Remove the front tire and wheel assembly. Remove the cotter pin. Remove the locknut. Remove the wheel speed sensor, sensor bracket and brake hose clamp.

9. Remove the stabilizer link connection (strut side) Remove the tie rod end self locking nut. Separate the tie rod end from the steering knuckle.

10. Remove the self locking nut for the lower ball joint connection.

11. Using tools MB990242, MB990244, MB991354 and MB990767, remove the left side driveshaft, circlip, right side driveshaft and circlip.

12. Disconnect the transaxle cooler lines. Remove the starter.

13. Remove the transaxle upper retaining bolts. Remove the bell housing cover. Remove the torque converter and drive plate retaining bolts.

14. Remove the front exhaust pipe. Remove the center member assembly.

15. Remove the rear roll stopper bracket. Remove the transaxle mounting bracket assembly.

16. Remove the transaxle mounting stopper. Remove the transaxle mounting body side bracket.

17. Install engine support tool MB991895. Set the tool to the front fender rear mounting bolts and the upper radiator support insulator mounting bolts, which are located in the engine compartment.

18. Use tools MB991527 (engine lifting fixture) and tool MB991454 (chain) to hold the engine and transaxle assembly in place.

19. Remove the lower transaxle retaining bolts. Remove the transaxle assembly from the vehicle.

To install:

20. Engage the torque converter into the transaxle, securely. Raise the transaxle into position and install the lower transaxle retaining bolts. Install the transaxle mounting stopper.

21. Install the driveshaft. Be sure to properly install the driveshaft washer.

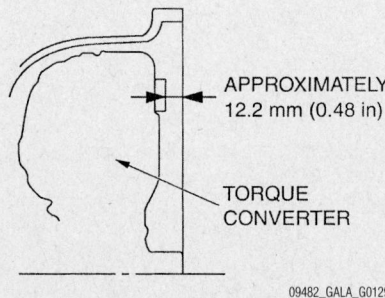

APPROXIMATELY 12.2 mm (0.48 in)

TORQUE CONVERTER

09482_GALA_G0129

Torque converter positioning —Eclipse, Galant and Lancer (four speed)

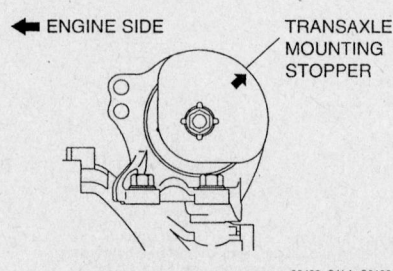

← ENGINE SIDE

TRANSAXLE MOUNTING STOPPER

09482_GALA_G0130

Automatic transaxle mount stopper alignment—Eclipse, Galant and Lancer

→Before securely tightening the driveshaft nuts, make sure there is no load on the wheel bearings. Otherwise the wheel bearing will be damaged. Using tool MB990767, to hold the assembly in place, torque the retaining bolt to 146–188 ft. lbs. (197–155 Nm).

22. Place the selector lever and the manual control lever in the neutral "N" position. Position the cable stud into the manual control lever slot and loosely install the nut. Gently push the transaxle control cable into the manual control lever slot until the cable is taut. Tighten the nut to 90–124 inch lbs. (10–14 Nm).

23. Continue the installation in the reverse order of the removal procedure.

24. Check and adjust fluid levels, as required.

25. Check and adjust the front alignment.

26. To initialize the PCM, turn the ignition switch ON then OFF and keep it in the OFF position for at least ten seconds.

2006 ECLIPSE (FIVE SPEED)

1. Before servicing the vehicle, refer to the Precautions Section.

2. Disconnect the negative battery cable. Drain the engine coolant. Drain the transaxle fluid.

3. Remove the left side under cover. Remove the air cleaner assembly.

4. Remove the engine strut tower bar. Remove the engine cover assembly.

5. Remove the PCM. Remove the battery and the battery tray.

6. Disconnect the front exhaust pipe. Remove the starter. Remove the upper radiator hose.

7. Remove the intake manifold plenum assembly. Remove the engine oil dipstick and tube assembly.

8. Disconnect the left side heated oxygen sensor (front, rear) connector. Disconnect the right side heated oxygen sensor (front, rear) connector.

9. Remove the transaxle control cable adjusting nut. Remove the transaxle control cable. Inhibitor switch sensor connector.

10. Disconnect the A/T control solenoid valve assembly connector. Disconnect the input shaft speed sensor connector. Disconnect the output shaft speed sensor connector.

11. Remove the front tire and wheel assembly. Remove the cotter pin. Remove the locknut. Remove the wheel speed sensor, sensor bracket and brake hose clamp.

12. Remove the stabilizer link connection (strut side) Remove the tie rod end self

locking nut. Separate the tie rod end from the steering knuckle.

13. Remove the self locking nut for the lower ball joint connection. Remove the lower arm ball joint connection.

14. Using tools MB990242, MB990244, MB991354 and MB990767, remove the left side driveshaft, circlip, right side driveshaft and circlip.

15. Disconnect the transaxle cooler lines. Remove the starter.

16. Remove the transaxle upper retaining bolts. Remove the bell housing cover. Remove the torque converter and drive plate retaining bolts. Remove the center member assembly.

17. Remove the engine oil pan to transaxle retaining bolts.

18. Remove the rear roll stopper bracket. Remove the transaxle mounting bracket assembly.

19. Remove the transaxle mounting stopper. Remove the transaxle mounting body side bracket.

20. Remove the engine hanger. Remove the intake manifold rear plenum stay and engine hanger. Install special tool MB992012 and MB992013 to each cylinder head.

21. Install engine support tool MB991895, to hold the engine and transaxle in position. Set the tool to the front fender rear mounting bolts and the upper radiator support insulator mounting bolts, which are located in the engine compartment.

22. Use tool MB991454 to hold the engine and transaxle assembly in place.

23. Remove the lower transaxle retaining bolts. Remove the transaxle assembly from the vehicle.

To install:

24. Engage the torque converter into the transaxle, securely. Raise the transaxle into position and install the lower transaxle retaining bolts. Install the transaxle mounting stopper.

25. Install the driveshaft. Be sure to properly install the driveshaft washer.

➡**Before securely tightening the driveshaft nuts, make sure there is no load on the wheel bearings. Otherwise the wheel bearing will be damaged. Using tool MB990767, to hold the assembly in place, torque the retaining bolt to 146–188 ft. lbs. (197–155 Nm).**

26. Place the selector lever and the manual control lever in the neutral "N" position. Position the cable stud into the manual control lever slot and loosely install the nut. Gently push the transaxle control cable into

the manual control lever slot until the cable is taut. Tighten the nut to 90–124 inch lbs. (10–14 Nm).

27. Continue the installation in the reverse order of the removal procedure.

28. Check and adjust fluid levels, as required.

29. Check and adjust the front alignment.

30. To initialize the PCM, turn the ignition switch ON then OFF and keep it in the OFF position for at least ten seconds.

2002–2003 GALANT

1. Before servicing the vehicle, refer to the Precautions Section.

2. Remove or disconnect the following:
   - Negative battery cable
   - Air cleaner and intake hoses

3. Drain the transaxle into a suitable waste container.
   - Nut securing the shifter lever to the transaxle
   - Cable retaining clip and remove the cable from the transaxle
   - Shifter cable mounting bracket
   - Electrical connectors for the speedometer, neutral safety switch (inhibitor switch), the pulse generator, kickdown servo switch, and the oil temperature sensor
   - Oil cooler lines, at the transaxle
   - Dipstick and tube from the transaxle
   - Starter motor and position it aside

4. Support the engine assembly.
   - Rear roll stopper mounting bracket
   - Transaxle mount bracket
   - Upper transaxle mounting bolts

5. Raise and safely support the vehicle.

6. Remove or disconnect the following:
   - Front wheel assemblies
   - Right-hand undercover
   - Tie rod end from the steering knuckle
   - Stabilizer bar link from the damper fork
   - Damper fork from the lateral lower control arm
   - Later lower arm and the compression arm lower ball joints, from the steering knuckle
   - Halfshafts from the transaxle and secure aside
   - Cover from the transaxle bell housing
   - Engine front roll stopper through-bolt
   - Crossmember and the triangular right-hand stay
   - Bolts holding the flexplate to the torque converter

7. Support the transaxle using a transmission jack, and remove the transaxle lower coupling bolt.

➡**The coupling bolt threads from the engine side into the transaxle and is located just above the halfshaft opening.**

8. Slide the transaxle rearward and carefully lower it from the vehicle.

To install:

9. After the torque converter has been mounted on the transaxle, install the transaxle assembly to the engine. Install the mounting bolts and tighten to 35 ft. lbs. (48 Nm). Install the transaxle lower coupling bolt and tighten to 21–25 ft. lbs. (29–34 Nm).

10. Install or connect the following:
    - Torque converter to the flexplate and tighten the bolts to 33–38 ft. lbs. (45–52 Nm)
    - Cover to the transaxle bell housing and tighten the mounting bolts to 84 inch lbs. (9 Nm)
    - Crossmember and tighten the front mounting bolts to 65 ft. lbs. (88 Nm) and the rear bolt to 54 ft. lbs. (73 Nm)
    - Front engine roll stopper through-bolt and lightly tighten. Once the full weight of the engine is on the mounts, tighten the bolt to 42 ft. lbs. (57 Nm).
    - Triangular stay bracket and tighten the mounting bolts to 65 ft. lbs. (88 Nm)
    - Halfshafts, using new circlips on the axle ends
    - Tie rod and ball joints to the steering knuckle. Tighten the ball joint self-locking nuts to 48 ft. lbs. (65 Nm). Tighten the tie rod end nut to 21 ft. lbs. (28 Nm) and secure with a new cotter pin.
    - Damper fork to the lower control arm and tighten the through-bolt to 65 ft. lbs. (88 Nm)
    - Stabilizer link to the damper fork, and tighten the self-locking nut to 29 ft. lbs. (39 Nm)
    - Underpan
    - Wheels and lower the vehicle
    - Transaxle mount bracket to the transaxle, and tighten the mounting nuts to 32 ft. lbs. (43 Nm)
    - Rear roll stopper mounting bracket
    - Engine support. Tighten the transaxle mount through-bolt to 51 ft. lbs. (69 Nm) and tighten the front engine roll stopper through-bolt.

- Upper transaxle mounting bolts and tighten to 35 ft. lbs. (48 Nm)
- Starter motor
- Dipstick tube and the dipstick
- Shifter cable mounting bracket
- Shifter lever and tighten the retaining nut to 14 ft. lbs. (19 Nm)
- Oil cooler lines and secure with clamps
- Electrical connectors for the speedometer, neutral safety switch (inhibitor switch), pulse generator, kickdown servo switch and oil temperature sensor
- Air cleaner and the air intake hose
- Negative battery cable

11. Fill the transaxle to the correct level.

2002–2006 GALANT (FOUR SPEED)

1. Before servicing the vehicle, refer to the Precautions Section.

2. Disconnect the negative battery cable. Drain the engine coolant. Drain the transaxle fluid.

3. Remove the left side under cover. Remove the air cleaner assembly. Remove the front exhaust pipe.

4. Remove the PCM. Remove the battery and the battery tray.

5. Remove the transaxle control cable adjusting nut. Remove the transaxle control cable.

6. Disconnect the transaxle range switch connection. Disconnect the A/T control solenoid valve assembly connector. Disconnect the input shaft speed sensor connector.

7. Disconnect the output shaft speed sensor connector. Disconnect the output shaft speed sensor connector.

8. Remove the front tire and wheel assembly. Remove the cotter pin. Remove the locknut. Remove the wheel speed sensor, sensor bracket and brake hose clamp.

9. Remove the stabilizer link connection (strut side) Remove the tie rod end self locking nut. Separate the tie rod end from the steering knuckle.

10. Remove the self locking nut for the lower ball joint connection.

11. Using tools MB990242, MB990244, MB991354 and MB990767, remove the left side driveshaft, circlip, right side driveshaft and circlip.

12. Disconnect the transaxle cooler lines. Remove the starter.

13. Remove the transaxle upper retaining bolts. Remove the bell housing cover. Remove the torque converter and drive plate retaining bolts. Remove the center member assembly.

14. Remove the rear roll stopper bracket.

Remove the transaxle mounting bracket assembly.

15. Remove the transaxle mounting stopper. Remove the transaxle mounting body side bracket.

16. Install engine support tool MB991895. Set the tool to the front fender rear mounting bolts and the upper radiator support insulator mounting bolts, which are located in the engine compartment.

17. Use tools MB991527 (engine lifting fixture) and tool MB991454 (chain) to hold the engine and transaxle assembly in place.

18. Remove the lower transaxle retaining bolts. Remove the transaxle assembly from the vehicle.

To install:

19. Engage the torque converter into the transaxle, securely. Raise the transaxle into position and install the lower transaxle retaining bolts. Install the transaxle mounting stopper.

20. Install the driveshaft. Be sure to properly install the driveshaft washer.

➡**Before securely tightening the driveshaft nuts, make sure there is no load on the wheel bearings. Otherwise the wheel bearing will be damaged. Using tool MB990767, to hold the assembly in place, torque the retaining bolt to 146–188 ft. lbs. (197–155 Nm).**

21. Place the selector lever and the manual control lever in the neutral "N" position. Position the cable stud into the manual control lever slot and loosely install the nut. Gently push the transaxle control cable into the manual control lever slot until the cable is taut. Tighten the nut to 90–124 inch lbs. (10–14 Nm).

22. Continue the installation in the reverse order of the removal procedure.

23. Check and adjust fluid levels, as required.

24. Check and adjust the front alignment.

25. To initialize the PCM, turn the ignition switch ON then OFF and keep it in the OFF position for at least ten seconds.

2004–2005 GALANT (FIVE SPEED)

1. Before servicing the vehicle, refer to the Precautions Section.

2. Disconnect the negative battery cable. Drain the transaxle fluid.

3. Remove the side under cover. Remove the air cleaner assembly. Remove the engine cover assembly.

4. Remove the PCM. Remove the battery and the battery tray.

5. Disconnect the front exhaust pipe. Remove the starter.

6. Remove the transaxle control cable adjusting nut. Remove the transaxle control cable. Inhibitor switch sensor connector.

7. Disconnect the A/T control solenoid valve assembly connector. Disconnect the input shaft speed sensor connector. Disconnect the output shaft speed sensor connector.

8. Remove the front tire and wheel assembly. Remove the cotter pin. Remove the locknut. Remove the wheel speed sensor, sensor bracket and brake hose clamp.

9. Remove the stabilizer link connection (strut side) Remove the tie rod end self locking nut. Separate the tie rod end from the steering knuckle.

10. Remove the self locking nut for the lower ball joint connection. Remove the lower arm ball joint connection.

11. Using tools MB990242, MB990244, MB991354 and MB990767, remove the left side driveshaft, circlip, right side driveshaft and circlip.

12. Disconnect the transaxle cooler lines. Remove the starter.

13. Remove the transaxle upper retaining bolts. Remove the bell housing cover. Remove the torque converter and drive plate retaining bolts. Remove the center member assembly.

14. Remove the engine oil pan to transaxle retaining bolts.

15. Remove the rear roll stopper bracket. Remove the transaxle mounting bracket assembly.

16. Remove the transaxle mounting stopper. Remove the transaxle mounting body side bracket.

17. Remove the engine hanger. Remove the intake manifold rear plenum stay and engine hanger. Install special tool MB992012 and MB992013 to each cylinder head.

18. Install engine support tool MB991895, to hold the engine and transaxle in position. Set the tool to the front fender rear mounting bolts and the upper radiator support insulator mounting bolts, which are located in the engine compartment.

19. Use tool MB991454 to hold the engine and transaxle assembly in place.

20. Remove the lower transaxle retaining bolts. Remove the transaxle assembly from the vehicle.

To install:

21. Engage the torque converter into the transaxle, securely. Raise the transaxle into position and install the lower transaxle retaining bolts. Install the transaxle mounting stopper.

22. Install the driveshaft. Be sure to properly install the driveshaft washer.

➡**Before securely tightening the drive-shaft nuts, make sure there is no load on the wheel bearings. Otherwise the wheel bearing will be damaged. Using tool MB990767, to hold the assembly in place, torque the retaining bolt to 146–188 ft. lbs. (197–155 Nm).**

23. Place the selector lever and the manual control lever in the neutral "N" position. Position the cable stud into the manual control lever slot and loosely install the nut. Gently push the transaxle control cable into the manual control lever slot until the cable is taut. Tighten the nut to 90–124 inch lbs. (10–14 Nm).

24. Continue the installation in the reverse order of the removal procedure.

25. Check and adjust fluid levels, as required.

26. Check and adjust the front alignment.

27. To initialize the PCM, turn the ignition switch ON then OFF and keep it in the OFF position for at least ten seconds.

2006 GALANT (FIVE SPEED)

1. Before servicing the vehicle, refer to the Precautions Section.

2. Disconnect the negative battery cable. Drain the engine coolant. Drain the transaxle fluid.

3. Remove the side under cover. Front under cover. Remove the air cleaner assembly.

4. Remove the PCM. Remove the battery and the battery tray.

5. Disconnect the front exhaust pipe. Remove the starter. Remove the upper radiator hose.

6. Remove the intake manifold plenum assembly. Remove the engine oil dipstick and tube assembly.

7. Remove the transaxle control cable adjusting nut. Remove the transaxle control cable. Inhibitor switch sensor connector.

8. Disconnect the A/T control solenoid valve assembly connector. Disconnect the input shaft speed sensor connector. Disconnect the output shaft speed sensor connector.

9. Remove the front tire and wheel assembly. Remove the cotter pin. Remove the locknut. Remove the wheel speed sensor, sensor bracket and brake hose clamp.

10. Remove the stabilizer link connection (strut side) Remove the tie rod end self locking nut. Separate the tie rod end from the steering knuckle.

11. Remove the self locking nut for the lower ball joint connection. Remove the lower arm ball joint connection.

12. Using tools MB990242, MB990244, MB991354 and MB990767, remove the left side driveshaft, circlip, right side driveshaft and circlip.

13. Disconnect the transaxle cooler lines. Remove the starter.

14. Remove the transaxle upper retaining bolts. Remove the bell housing cover. Remove the torque converter and drive plate retaining bolts. Remove the center member assembly.

15. Remove the engine oil pan to transaxle retaining bolts.

16. Remove the rear roll stopper bracket. Remove the transaxle mounting bracket assembly.

17. Remove the transaxle mounting stopper. Remove the transaxle mounting body side bracket.

18. Remove the engine hanger. Remove the intake manifold rear plenum stay and engine hanger. Install special tool MB992012 and MB992013 to each cylinder head.

19. Install engine support tool MB991895, to hold the engine and transaxle in position. Set the tool to the front fender rear mounting bolts and the upper radiator support insulator mounting bolts, which are located in the engine compartment.

20. Use tool MB991454 to hold the engine and transaxle assembly in place.

21. Remove the lower transaxle retaining bolts. Remove the transaxle assembly from the vehicle.

To install:

22. Engage the torque converter into the transaxle, securely. Raise the transaxle into position and install the lower transaxle retaining bolts. Install the transaxle mounting stopper.

23. Install the driveshaft. Be sure to properly install the driveshaft washer.

➡**Before securely tightening the drive-shaft nuts, make sure there is no load on the wheel bearings. Otherwise the wheel bearing will be damaged. Using tool MB990767, to hold the assembly in place, torque the retaining bolt to 146–188 ft. lbs. (197–155 Nm).**

24. Place the selector lever and the manual control lever in the neutral "N" position. Position the cable stud into the manual control lever slot and loosely install the nut. Gently push the transaxle control cable into the manual control lever slot until the cable is taut. Tighten the nut to 90–124 inch lbs. (10–14 Nm).

25. Continue the installation in the reverse order of the removal procedure.

26. Check and adjust fluid levels, as required.

27. Check and adjust the front alignment.

28. To initialize the PCM, turn the ignition switch ON then OFF and keep it in the OFF position for at least ten seconds.

2002–2003 LANCER

1. Before servicing the vehicle, refer to the Precautions Section.

2. Install or connect the following:
   - Negative battery cable
   - Engine undercover

3. Drain the transaxle oil and engine coolant.
   - Front exhaust pipe
   - Battery and battery tray
   - Air cleaner assembly
   - Transaxle control cable
   - Vehicle Speed Sensor (VSS) connector
   - Input and output shaft speed sensor connectors
   - Inhibitor switch sensor connector
   - A/C control solenoid valve connector
   - Starter
   - Transaxle oil cooler line
   - Upper engine-to-transaxle bolts
   - Transaxle mount
   - Transaxle mount stopper

4. Install a suitable engine support assembly, then raise and safely support the vehicle.
   - Wheel Speed Sensor (WSS)
   - Brake hose clamp
   - Stabilizer bar
   - Lower control arm
   - Tie rod end
   - Lower control arm
   - Halfshafts by inserting a prybar between the transaxle case and the driveshaft and prying the shaft from the transaxle. Do not pull on the driveshaft.
   - Centermember
   - Bell housing lower cover
   - Transaxle to engine bolts and lower the transaxle from the vehicle
   - Front roll stopper installation bolt
   - Bell housing cover

5. Support the transaxle with a suitable jack.
   - Driveplate bolts
   - Lower transaxle-to-engine mounting bolts
   - Transaxle from the vehicle

To install:

6. Installation is the reverse of the removal procedure, noting the following:
   - Driveplate bolts: 36 ft. lbs. (49 Nm)

- Centermember bolts: 51 ft. lbs. (69 Nm)
- Hub nut: 181 ft. lbs. (245 Nm)
- Transaxle mount bracket nuts: 36 ft. lbs. (49 Nm)
- Transaxle mount stopper nuts: 60 ft. lbs. (81 Nm)
- Transaxle-to-engine upper mounting bolts: 36 ft. lbs. (49 Nm)

7. Fill the transaxle and the engine cooling system to the correct level.

8. Check and adjust the front wheel alignment.

9. Check the speedometer and gear selector for proper operation.

10. Start the engine and check for leaks.

2004–2006 LANCER

1. Before servicing the vehicle, refer to the Precautions Section.

2. Disconnect the negative battery cable. Drain the engine coolant. Drain the transaxle fluid.

3. Remove the left side under cover. Remove the air cleaner assembly. Remove the front exhaust pipe.

4. Remove the battery and the battery tray.

5. Remove the transaxle control cable adjusting nut. Remove the transaxle control cable.

6. Disconnect the transaxle range switch connection. Disconnect the A/T control solenoid valve assembly connector. Disconnect the input shaft speed sensor connector.

7. Disconnect the output shaft speed sensor connector. Disconnect the output shaft speed sensor connector.

8. Remove the front tire and wheel assembly. Remove the cotter pin. Remove the locknut. Remove the wheel speed sensor, sensor bracket and brake hose clamp.

9. Remove the stabilizer link connection (strut side) Remove the tie rod end self locking nut. Separate the tie rod end from the steering knuckle.

10. Remove the self locking nut for the lower ball joint connection.

11. Using tools MB990242, MB990244, MB991354 and MB990767, remove the left side driveshaft, circlip, right side driveshaft and circlip.

12. Disconnect the transaxle cooler lines. Remove the starter.

13. Remove the transaxle upper retaining bolts. Remove the bell housing cover. Remove the torque converter and drive plate retaining bolts. Remove the center member assembly.

14. Remove the rear roll stopper bracket. Remove the transaxle mounting bracket assembly.

15. Remove the transaxle mounting stopper. Remove the transaxle mounting body side bracket.

16. Install engine support tool MB991895. Set the tool to the front fender rear mounting bolts and the upper radiator support insulator mounting bolts, which are located in the engine compartment.

17. Use tools MB991527 (engine lifting fixture) and tool MB991454 (chain) to hold the engine and transaxle assembly in place.

18. Remove the lower transaxle retaining bolts. Remove the transaxle assembly from the vehicle.

To install:

19. Engage the torque converter into the transaxle, securely. Raise the transaxle into position and install the lower transaxle retaining bolts. Install the transaxle mounting stopper.

20. Install the driveshaft. Be sure to properly install the driveshaft washer.

➡ **Before securely tightening the driveshaft nuts, make sure there is no load on the wheel bearings. Otherwise the wheel bearing will be damaged. Using tool MB990767, to hold the assembly in place, torque the retaining bolt to 146–188 ft. lbs. (197–155 Nm).**

21. Place the selector lever and the manual control lever in the neutral "N" position. Position the cable stud into the manual control lever slot and loosely install the nut. Gently push the transaxle control cable into the manual control lever slot until the cable is taut. Tighten the nut to 90–124 inch lbs. (10–14 Nm).

22. Continue the installation in the reverse order of the removal procedure.

23. Check and adjust fluid levels, as required.

24. Check and adjust the front alignment.

MIRAGE

1. Before servicing the vehicle, refer to the Precautions Section.

2. Remove or disconnect the following:
- Negative battery cable
- Battery and battery tray
- Air hose and air cleaner assembly
- Under guard pan

3. Drain the transaxle oil.
- Control cable and cooler lines
- Shift control solenoid valve connector
- Inhibitor switch, kickdown servo switch, the pulse generator and oil temperature sensor, if equipped
- Speedometer cable and remove the starter

- Transaxle mounting bolts and bracket
- Stabilizer bar from the lower control arm
- Steering tie rod end and the ball joint from the steering arm
- Halfshafts at the inboard side from the transaxle. Tie the joint assembly aside.

4. Support the engine and remove the center member.
- Bell housing cover and remove the driveplate bolts
- Transaxle assembly lower connecting bolt, located just over the halfshaft opening
- Transaxle

To install:

5. Install or connect the following:
- Transaxle assembly on the engine. Tighten the driveplate bolts to 33–38 ft. lbs. (46–53 Nm).
- Bell housing cover
- Center member
- Halfshafts to the transaxle, using new circlips
- Tie rods, ball joints and stabilizer links to the steering arm
- Transaxle mounting bracket and bolts
- Starter
- Speedometer cable
- Inhibitor switch, kickdown servo switch, the pulse generator and oil temperature sensor, if disconnected
- Shift control solenoid valve connector
- Control cables and oil cooler lines
- Air cleaner assembly
- Battery tray and battery. Connect the positive, then the negative terminal.

6. Fill the transaxle to the correct level.

7. Start the engine and check for leaks.

Clutch

REMOVAL & INSTALLATION

Eclipse and Mirage

1. Before servicing the vehicle, refer to the Precautions Section.

2. Remove or disconnect the following:
- Negative battery cable
- Transaxle assembly from the vehicle
- Pressure plate attaching bolts, pressure plate and clutch disc. If the pressure plate is to be reused, loosen the bolts in a diagonal pat-

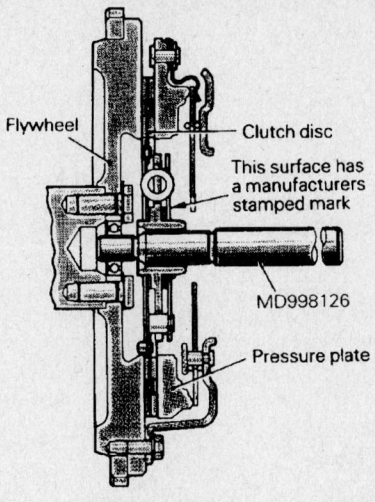

Use the alignment dowel to center the disc on the flywheel—Mirage

tern, 1 or 2 turns at a time. This will prevent warping the clutch cover assembly.

• Return clip and the pressure plate release bearing. Do not use solvent to clean the bearing.

3. Inspect the clutch release fork and fulcrum for damage or wear. If necessary, remove the release fork and unthread the fulcrum from the transaxle.

4. Carefully inspect the condition of the clutch components and replace any worn or damaged parts.

To install:

5. Inspect the flywheel for heat damage or cracks. Resurface or replace the flywheel as required.

6. Install the fulcrum and tighten to 25 ft. lbs. (35 Nm). Install the release fork. Apply a coating of multi-purpose grease to the point of contact with the fulcrum and the point of contact with the release bearing. Apply a coating of multi-purpose grease to the end of the release cylinder pushrod and the pushrod hole in the release fork.

7. Apply multi-purpose grease to the clutch release bearing. Pack the bearing inner surface and the groove with grease. Do not apply grease to the resin portion of the bearing. Place the bearing in position and install the return clip.

8. Using the proper alignment tool, install the clutch disc to the flywheel. Install the pressure plate assembly. Install the retainer bolts and tighten a little at a time, in a diagonal sequence. Tighten them to a final torque of 16 ft. lbs. (22 Nm). Remove the aligning tool.

9. Install the transaxle assembly.

10. Check for proper clutch operation.

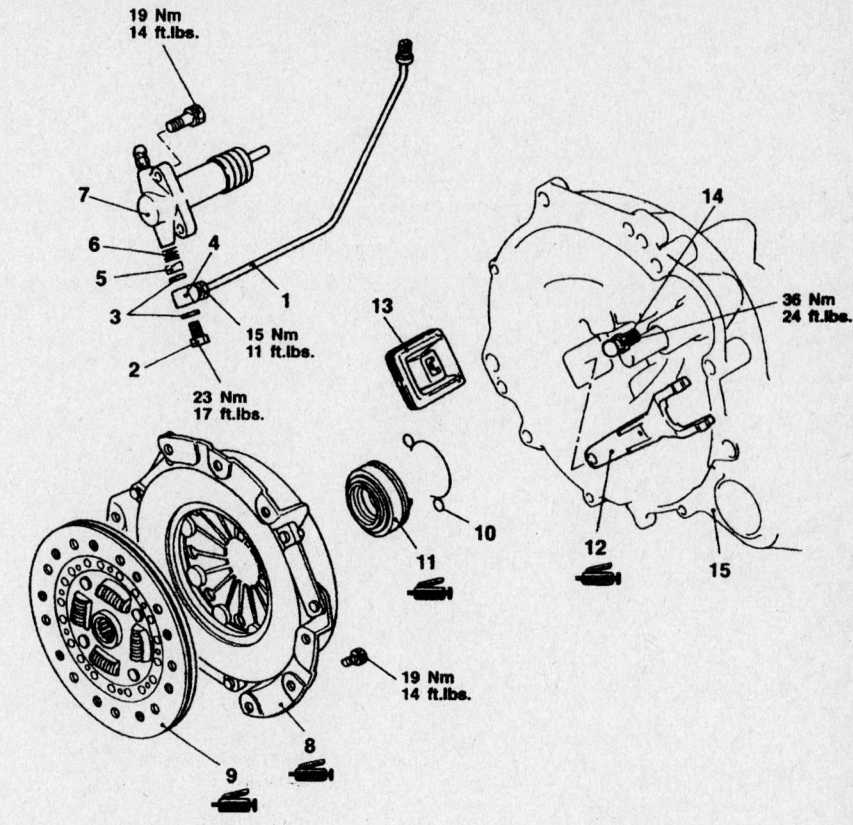

1. Clutch oil tube
2. Union bolt
3. Gasket
4. Union
5. Valve plate
6. Valve plate spring
7. Clutch release cylinder
8. Clutch cover
9. Clutch disc
10. Return clip
11. Clutch release bearing
12. Release fork
13. Release fork boot
14. Fulcrum
15. Transaxle

Clutch assembly and related components—Eclipse shown

Lancer and Evolution

1. Before servicing the vehicle, refer to the Precautions Section.

2. Remove or disconnect the following:

• Negative battery cable
• Transaxle assembly from the vehicle
• Clutch fluid line bracket, insulator and washer
• Clutch fluid line
• Clutch slave (release) cylinder
• Boot
• Clutch cover (pressure plate) attaching bolts, cover plate and clutch disc. If the pressure plate is to be reused, loosen the bolts in a diagonal pattern, 1 or 2 turns at a time. This will prevent warping the clutch cover assembly.

3. Carefully inspect the condition of the clutch components and replace any worn or damaged parts.

To install:

4. Inspect the flywheel for heat damage or cracks. Resurface or replace the flywheel as required.

5. Apply multi-purpose grease to the clutch release bearing. Pack the bearing inner surface and the groove with grease. Do not apply grease to the resin portion of the bearing. Place the bearing in position and install the return clip.

6. Using the proper alignment tool, install the clutch disc to the flywheel. Install the clutch cover (pressure plate) assembly. Install the retainer bolts and tighten a little at a time, in a diagonal sequence. Tighten them to a final torque of 14 ft. lbs. (19 Nm). Remove the aligning tool.

7. Install or connect the following:

• Boot
• Clutch line, washer, insulator and bracket
• Transaxle assembly

8. Check for proper clutch operation.

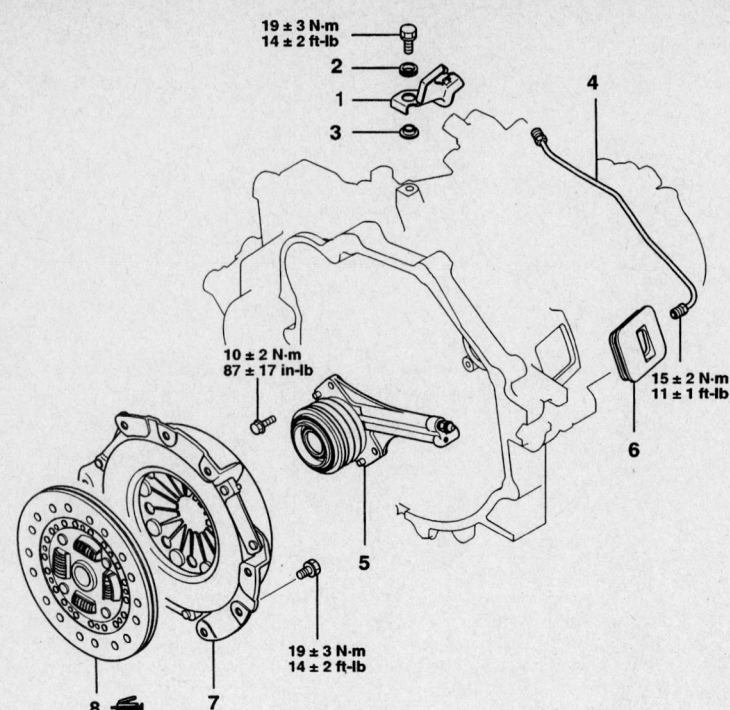

1. CLUTCH FLUID LINE BRACKET
2. INSULATOR
3. WASHER
4. CLUTCH TUBE
5. CLUTCH RELEASE CONCENTRIC CYLINDER
6. BOOT
7. CLUTCH COVER
8. CLUTCH DISC

9357QG26

Clutch assembly and related components—Lancer shown

ADJUSTMENT

Eclipse and Lancer

1. Before servicing the vehicle, refer to the Precautions Section.
2. Turn back the carpet under the clutch pedal.
3. On Eclipse, measure the clutch pedal height, specification should be 6.67 inches for 5 speed transaxle and 7.02 inches for 6 speed transaxle. Measure the clutch pedal

clevis pin play, specification should be 0.04—0.12 inch.
4. On Lancer, measure the clutch pedal height, specification should be 8.11—8.26. Measure the clutch pedal clevis pin play, specification should be 0.04—0.12 inch.
5. If the clutch pedal height is not with

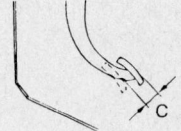

CLUTCH PEDAL FREE PLAY

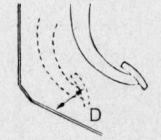

DISTANCE BETWEEN THE CLUTCH PEDAL AND THE FLOORBOARD WHEN THE CLUTCH IS DISENGAGED

09482_GALA_G0162

Clutch adjustment measurement locations for measurement "C" and "D"—Eclipse and Lancer

specification, loosen the locknut and adjust the pedal height to specification using the adjusting bolt or pushrod.

➡**Do not push the clutch master cylinder push rod at this time.**

6. If the clutch pedal play is not within specification, loosen the locking nut and move the push rod to adjust it to specification.
7. After adjustment, confirm that the clutch pedal free play and the distance between the clutch pedal and the floorboard when the clutch is disengaged is within specification ("C" 0.2—0.5 inch, "D" 2.83 inches or more for Eclipse and "C" 0.16—0.51 inch, "D" 4.1 inches or more for Lancer)

➡**If the measured free play and distance do not agree with specification, it is probably due to air in the system, faulty clutch master cylinder or clutch. Correct as required.**

8. Reinstall the carpet.

Evolution

1. Before servicing the vehicle, refer to the Precautions Section.

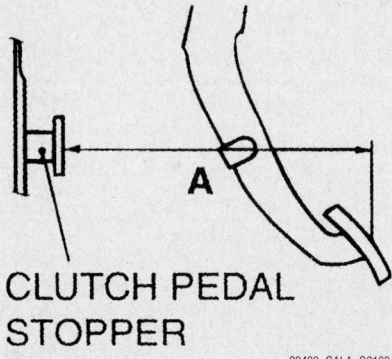

CLUTCH PEDAL STOPPER

09482_GALA_G0163

Clutch pedal height (A) measurement— Evolution

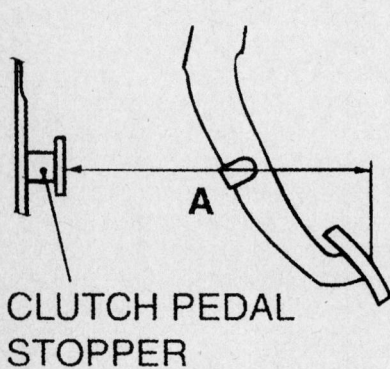

CLUTCH PEDAL STOPPER

09482_GALA_G0163

Clutch pedal height (A) measurement— Lancer

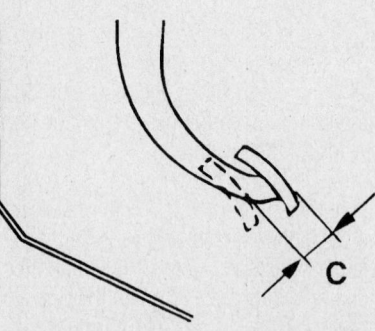

09482_GALA_G0164

Clutch adjustment measurement locations for measurement "C"—Evolution

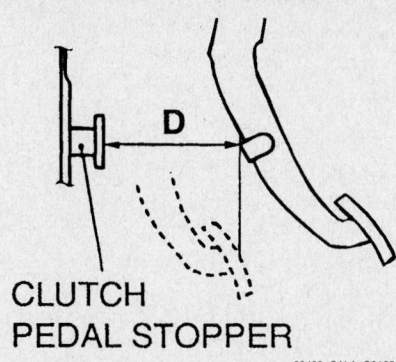

CLUTCH PEDAL STOPPER

09482_GALA_G0165

Clutch adjustment measurement locations for measurement "D"—Evolution

2. Turn back the carpet under the clutch pedal.

3. Measure the clutch pedal height, specification should be 8.0—8.1 inches. Measure the clutch pedal clevis pin play, specification should be 0.04—0.12 inch.

4. If the clutch pedal height is not with specification, loosen the locknut and adjust the pedal height to specification using the adjusting bolt or pushrod.

➡ **Do not push the clutch master cylinder push rod at this time.**

5. If the clutch pedal play is not within specification, loosen the locking nut and move the push rod to adjust it to specification.

6. After adjustment, confirm that the clutch pedal free play and the distance between the clutch pedal and the floorboard when the clutch is disengaged is within specification ("C" 0.16—0.51 inch, "D" 4.5).

➡ **If the measured free play and distance do not agree with specification, it is probably due to air in the system,**

faulty clutch master cylinder or clutch. Correct as required.

7. Reinstall the carpet.

Mirage

1. Before servicing the vehicle, refer to the Precautions Section.

2. Measure the clutch pedal height (measurement A). The specification is 6.38–6.50 in. (162–165mm).

➡ **The clutch pedal height is not adjustable. If not within specifications, part replacement is required.**

3. Depress clutch pedal several times and check the pedal free-play (measurement B).

4. If measurement is not 0.67–0.87 in. (17–22mm), adjustment is required.

5. To adjust turn the outer cable adjusting nut, located at the firewall, until free-play is within range.

6. Depress clutch pedal several times and recheck measurement.

Hydraulic Clutch System

BLEEDING

1. Before servicing the vehicle, refer to the Precautions Section.

2. Fill the reservoir with clean brake fluid meeting DOT 3 specifications.

3. Press the clutch pedal to the floor, then open the bleeder screw on the slave cylinder.

4. Tighten the bleed screw and release the clutch pedal.

5. Repeat the procedure until the fluid is free of air bubbles.

6. Be sure to fill the clutch master cylinder to the proper level when the procedure is complete.

Transfer Case Assembly

REMOVAL & INSTALLATION

Evolution (5 speed Transaxle)

1. Before servicing the vehicle, refer to the Precautions Section.

2. Disconnect the negative battery cable. Remove the under cover assembly. Drain the transaxle fluid.

3. Drain the transfer case fluid. Drain the engine coolant.

4. Remove the front axle crossmember assembly.

5. Remove the front exhaust pipe. Remove the battery and the battery tray.

6. Remove the air cleaner and air intake hose assembly.

7. Remove the strut tower bar assembly.

8. Remove the air hose, air by-pass hose and air by-pass valve.

9. Remove the radiator.

10. Remove the output shaft. Remove the driveshaft.

11. Remove the rear roll stopper connection bolt. Remove the crossmember assembly.

12. Disconnect the pressure hose connection and discard the gasket.

13. Remove the dust seal guard.

14. Remove the transfer case retaining bolts. Remove the transfer case from the vehicle. Discard the O-ring.

To install:

15. Position the transfer case on a suitable holding fixture. Install the transfer case to its mounting in the vehicle.

16. Continue the installation in the reverse order of the removal procedure.

17. Be sure to check and adjust all fluid levels, as necessary.

Evolution (6 speed Transaxle)

➡ **Prior to removal of the steering rack, center the front wheels and remove the ignition key. Failure to do so may damage the SRS clock spring and render SRS system inoperative.**

1. Before servicing the vehicle, refer to the Precautions Section.

2. Disconnect the negative battery cable. Remove the driver's side air bag module.

3. Remove the steering wheel. Remove the column lower cover. Remove the clock spring.

4. Remove the under cover. Remove the steering shaft cover. Disconnect the steering gear and joint connection.

Clutch pedal height

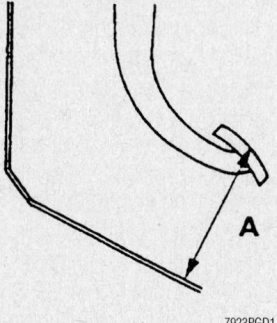

7923PGD1

Clutch pedal height (A) measurement—Mirage and Eclipse

7923PG91

Bleeding a typical clutch hydraulic system

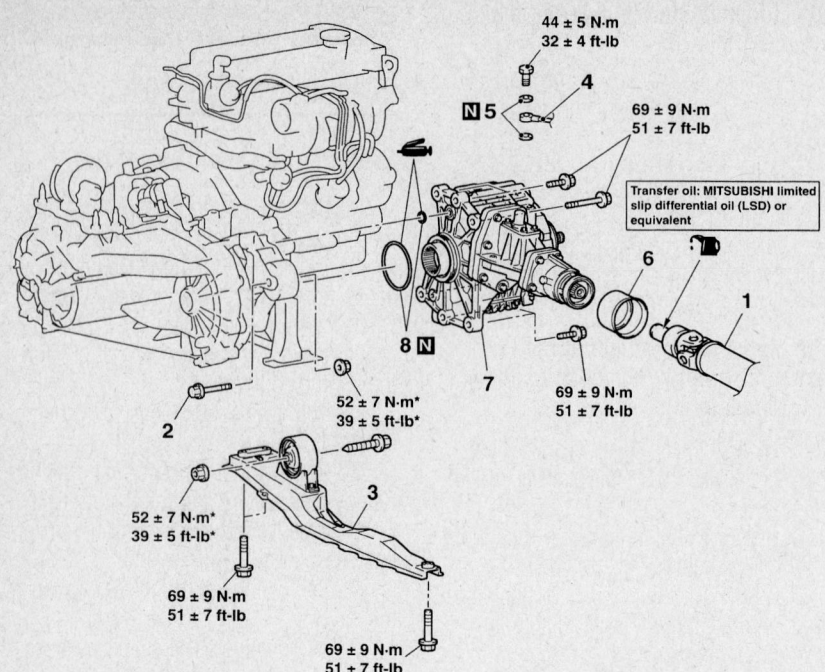

1. PROPELLER SHAFT
2. REAR ROLL STOPPER CONNECTION BOLT
3. CENTERMEMBER ASSEMBLY
4. PRESSURE HOSE CONNECTION
5. GASKET
6. DUST SEAL GUARD
7. TRANSFER ASSEMBLY
8. O-RING

09482_GALA_G0166

Transfer case and related components—Evolution with five speed transaxle

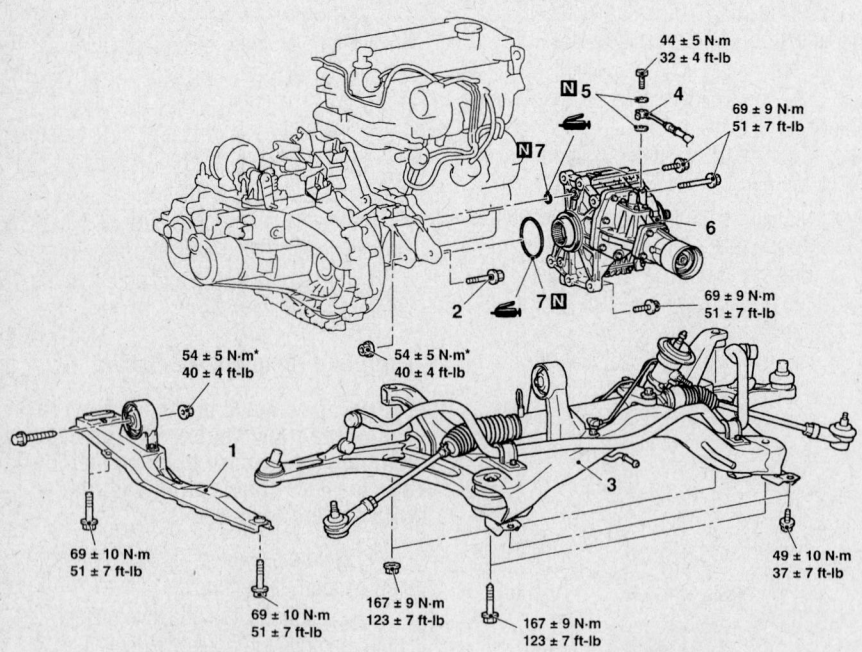

1. CENTER MEMBER ASSEMBLY
2. REAR ROLL STOPPER CONNECTING BOLT
3. FRONT AXLE NO.1 CROSSMEMBER ASSEMBLY
4. PRESSURE HOSE CONNECTION
5. GASKET
6. TRANSFER ASSEMBLY
7. O-RING

09482_GALA_G0167

Transfer case and related components—Evolution with six speed transaxle

5. Remove the front axle crossmember bar. Disconnect the front exhaust pipe.

6. Remove the driveshaft. Remove the output shaft.

7. Drain the power steering fluid. Drain the transaxle. Drain the transfer case.

8. Remove the center member assembly. Disconnect and plug the power steering hoses.

9. Remove the rear roll stopper connecting bolt.

10. Remove the front axle number one crossmember assembly. Disconnect the pressure hose connection and discard the gasket.

11. Remove the transfer case retaining bolts. Remove the transfer case from the vehicle. Discard the O-ring.

To install:

12. Position the transfer case on a suitable holding fixture. Install the transfer case to its mounting in the vehicle.

13. Continue the installation in the reverse order of the removal procedure.

14. Be sure to check and adjust all fluid levels, as necessary.

Halfshaft

REMOVAL & INSTALLATION

Diamante

1. Before servicing the vehicle, refer to the Precautions Section.

2. Raise the vehicle and support it safely.

3. Remove the cotter pin, halfshaft nut and washer.

4. If equipped with Anti-Lock Brake (ABS), remove the front wheel speed sensor.

5. If equipped with Active Electronic Control Suspension (Active-ECS) perform the following:

a. Loosen the nut that secures the air line to the to the top of the strut and discard the O-ring.

b. Remove the bolts that secure the actuator to the top of the strut and remove the component. Disconnect the wiring harness.

6. Disconnect the lower ball joint and the tie rod end from the steering knuckle.

7. If removing the left side axle with an inner shaft, remove the center support bearing bracket bolts and washers. Then, remove the halfshaft by setting up a puller on the outside wheel hub and pushing the halfshaft from the front hub. Tap the shaft union at the joint case with a plastic hammer to remove the halfshaft and inner shaft from the transaxle.

8. If removing right side axle shafts without an inner shaft, remove the halfshaft by setting up a puller on the outside wheel hub and pushing the halfshaft from the front hub. After pressing the outer shaft, insert a prybar between the transaxle case and the halfshaft and pry the shaft from the transaxle.

➡**Do not pull on the shaft; doing so damages the inboard joint.**

To install:

9. Replace the circlips on the ends of the halfshafts.

10. Insert the halfshaft into the transaxle. Be sure it is fully seated.

11. Pull the strut assembly out and install the other end to the hub.

12. Install the center bearing bracket bolts and tighten to 33 ft. lbs. (45 Nm).

13. Install the washer so the chamfered edge faces outward. Install the nut and tighten to 145–188 ft. lbs. (200–260 Nm) Secure with a new cotter pin.

14. Connect the ball joint to the steering knuckle. Torque the new retaining nut to 43–52 ft. lbs. (60–72 Nm) and secure with a new cotter pin.

15. Connect the tie rod end to the steering knuckle. Torque the retaining nut to 21 ft. lbs. (29 Nm) and secure with a new cotter pin.

16. If equipped with ABS, install the front wheel speed sensor.

17. If equipped with Active-ECS, perform the following:

   a. Install the air line with a new O-ring.

   b. Install the actuator to the top of the strut. Connect the wiring harness.

18. Install the wheel and lower the vehicle to the floor.

Eclipse

2002–2003

1. Before servicing the vehicle, refer to the Precautions Section.

2. Raise and safely support the vehicle.

3. Remove or disconnect the following:
   - Front wheel
   - Halfshaft nut and washer
   - Tie rod end from the knuckle
   - Stabilizer link from the damper fork
   - Compression and lateral arm ball joint studs from the knuckle

4. Mount a puller on the wheel studs and push the halfshaft through the hub assembly.

5. Detach the inner halfshaft from the transaxle by carefully prying the CV-joint housing out.

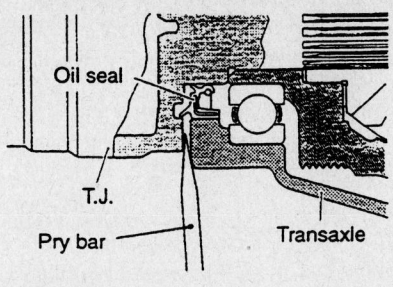

Oil seal

T.J.

Pry bar   Transaxle

7923PG94

Proper method for removing the inner halfshaft from the transaxle or differential

6. Pull the knuckle assembly outward and remove the halfshaft.

To install:

7. Place a new circlip on the inner halfshaft and install the halfshaft in the transaxle.

8. Push out on the knuckle assembly and install the halfshaft through the hub.

9. Using new nuts, install the lateral and compression arm ball joint studs in the knuckle. Tighten the nuts to 43–52 ft. lbs. (59–71 Nm). Install new cotter pins.

10. Install the damper fork on the knuckle. Do nut tighten the nut at this time.

11. Attach the stabilizer link to the damper fork. Tighten the nut to 29 ft. lbs. (39 Nm).

12. Install the washer and nut on the halfshaft. Prevent the hub from turning and tighten the nut to 145–188 ft. lbs. (196–255 Nm).

13. Install the wheel and lower the vehicle to the floor. Tighten the damper fork nut to 65 ft. lbs. (88 Nm).

2004–2005

1. Before servicing the vehicle, refer to the Precautions Section.

2. Raise and support the vehicle safely.

3. If equipped with ABS, disconnect the speed sensor connection. Remove the brake hose clip.

4. Remove the cotter pin. Install tool MB990767 to the hub and remove the halfshaft nut. Remove the washer.

5. Remove the lower ball joint cotter pin.

➡**Do not remove the nut from the ball joint. Loosen it and use special tool MB991897 to avoid possible damage to the ball joint threads. Hang the special tool in place with wire or string to prevent it from falling.**

6. Install the special tool. Turn the bolt and knob as necessary to make the jaws of the tool parallel. Tighten the bolt by hand and confirm that the jaws are still parallel.

➡**When adjusting the jaws in parallel, make sure the knob is in the vertical (upward) position.**

7. Tighten the bolt with a wrench to disconnect the lower arm ball joint connection.

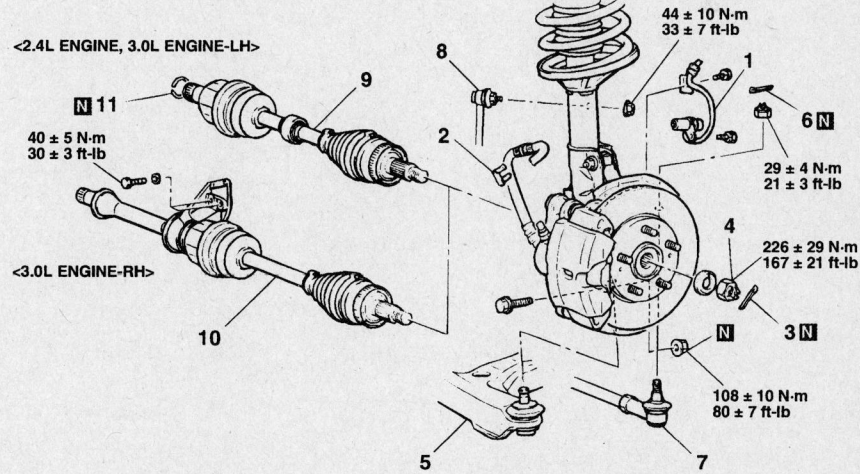

<2.4L ENGINE, 3.0L ENGINE-LH>

44 ± 10 N·m
33 ± 7 ft-lb

40 ± 5 N·m
30 ± 3 ft-lb

<3.0L ENGINE-RH>

29 ± 4 N·m
21 ± 3 ft-lb

226 ± 29 N·m
167 ± 21 ft-lb

108 ± 10 N·m
80 ± 7 ft-lb

1.	SPEED SENSOR CABLE CONNECTION <VEHICLES WITH ABS>	6. COTTER PIN
2.	BRAKE HOSE CLIP	7. TIE ROD END CONNECTION
3.	COTTER PIN	8. STABILIZER LINK CONNECTION
4.	DRIVESHAFT NUT	9. DRIVESHAFT
5.	LOWER ARM BALL JOINT CONNECTION	10. DRIVESHAFT AND INNER SHAFT
		11. CIRCLIP

09482_GALA_G0168

Halfshaft and related components—2004–2005 Eclipse

8. Remove the tie rod end cotter pin. Install tool MB990767 to the hub and remove the halfshaft nut. Remove the washer.

➡**Do not remove the nut from the tie rod end. Loosen it and use special tool MB991897 to avoid possible damage to the ball joint threads. Hang the special tool in place with wire or string to prevent it from falling.**

9. Install the special tool. Turn the bolt and knob as necessary to make the jaws of the tool parallel. Tighten the bolt by hand and confirm that the jaws are still parallel.

➡**When adjusting the jaws in parallel, make sure the knob is in the vertical (upward) position.**

10. Tighten the bolt with a wrench to disconnect the tie rod end.
11. Disconnect the stabilizer link connection.

➡**Do not damage the ABS rotor attached to the BJ outer race, on vehicle equipped with ABS.**

12. Use special tools MB991354, MB990242 and MB990767 to push the halfshaft out from the hub.

➡**Do not pull on the halfshaft, doing so will damage the TJ. Be sure to use a prybar. Do not insert the prybar so deep as to damage the oil seal.**

13. Insert a prybar between the transaxle case and the halfshaft to remove the halfshaft on vehicles equipped with 2.4L engine and vehicles equipped with 3.0L engine (left side shaft).

14. If the inner shaft and transaxle are tightly joined, tap the center bearing bracket with a plastic hammer to remove the halfshaft and inner shaft from the transaxle on vehicles with 3.0L engine (right side).

To install:

15. Installation is the reverse of the removal procedure.
16. Check and adjust the front end alignment, as necessary.

2006

1. Before servicing the vehicle, refer to the Precautions Section.
2. Raise and support the vehicle safely.
3. Remove the front under cover. Remove the side undercover. Drain the transaxle fluid.
4. On vehicles equipped with the 3.8L engine, disconnect the front exhaust pipe if working on the right side halfshaft.
5. Disconnect the speed sensor con-

nection. Remove the wheel speed sensor. Remove the brake hose clip.

6. Remove the cotter pin. Install tool MB990767 to the hub and remove the halfshaft nut. Remove the washer.
7. Remove the lower ball joint cotter pin.

➡**Do not remove the nut from the ball joint. Loosen it and use special tool MB991897 to avoid possible damage to the ball joint threads. Hang the special tool in place with wire or string to prevent it from falling.**

8. Install the special tool. Turn the bolt and knob as necessary to make the jaws of the tool parallel. Tighten the bolt by hand and confirm that the jaws are still parallel.

➡**When adjusting the jaws in parallel, make sure the knob is in the vertical (upward) position.**

9. Tighten the bolt with a wrench to disconnect the lower arm ball joint connection.
10. Remove the tie rod end cotter pin. Install tool MB990767 to the hub and remove the halfshaft nut. Remove the washer.

➡**Do not remove the nut from the tie rod end. Loosen it and use special tool MB991897 to avoid possible damage to the ball joint threads. Hang the special tool in place with wire or string to prevent it from falling.**

11. Install the special tool. Turn the bolt and knob as necessary to make the jaws of the tool parallel. Tighten the bolt by hand and confirm that the jaws are still parallel.

➡**When adjusting the jaws in parallel, make sure the knob is in the vertical (upward) position.**

12. Tighten the bolt with a wrench to disconnect the tie rod end.
13. Disconnect the stabilizer link connection.

➡**Do not strike the ABS rotor attached to the BJ or EBJ outer race, of the halfshaft against other parts when removing the halfshaft as damage to the rotors will result.**

14. Use special tools MB991354, MB990242 and MB990767 to push the halfshaft out from the hub.

➡**Do not pull on the halfshaft, doing so will damage the TJ or PTJ. Be sure to use a prybar. Do not insert the prybar so deep as to damage the oil seal.**

15. Remove the halfshaft from the hub by pulling the bottom of the brake disc towards you.
16. Insert a prybar between the transaxle case and the halfshaft, and than pry and remove the halfshaft from the transaxle.

➡**Insert a prybar, taking care not to damage the protrusion of the transaxle case when removing the halfshaft (left side).**

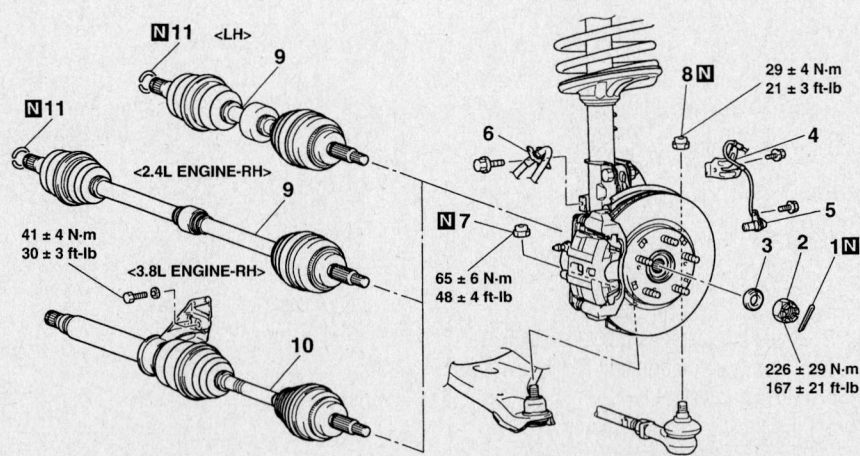

1. SPLIT PIN
2. DRIVE SHAFT NUT
3. WASHER
4. FRONT WHEEL SPEED SENSOR BRACKET
5. FRONT WHEEL SPEED SENSOR
6. BRAKE HOSE BRACKET
7. SELF LOCKING NUT (LOWER ARM BALL JOINT CONNECTION)
8. SELF LOCKING NUT (TIE ROD END CONNECTION)
9. DRIVE SHAFT
10. DRIVE SHAFT AND INNER SHAFT ASSEMBLY<3.8L ENGINE-RH>
11. CIRCLIP

09482_GALA_G0169

Halfshaft and related components—2006 Eclipse and 2004–2006 Galant

17. If the inner shaft and transaxle are tightly joined, tap the center bearing bracket with a plastic hammer to remove the half-shaft and inner shaft from the transaxle on vehicles with 3.0L engine (right side).

To install:

18. Installation is the reverse of the removal procedure.

19. Check and adjust the front end alignment, as necessary.

Galant

2002–2003

1. Before servicing the vehicle, refer to the Precautions Section.

2. Raise the vehicle and support it safely.

3. Remove the cotter pin, halfshaft nut and washer.

4. If equipped with Anti-Lock Brake (ABS), remove the front wheel speed sensor.

5. If equipped with Active Electronic Control Suspension (Active-ECS) perform the following:

    a. Loosen the nut that secures the air line to the to the top of the strut and discard the O-ring.

    b. Remove the bolts that secure the actuator to the top of the strut and remove the component. Disconnect the wiring harness.

6. Disconnect the lower ball joint and the tie rod end from the steering knuckle.

7. If removing the left side axle with an inner shaft, remove the center support bearing bracket bolts and washers. Then, remove the halfshaft by setting up a puller on the outside wheel hub and pushing the halfshaft from the front hub. Tap the shaft union at the joint case with a plastic hammer to remove the halfshaft and inner shaft from the transaxle.

8. If removing right side axle shafts without an inner shaft, remove the halfshaft by setting up a puller on the outside wheel hub and pushing the halfshaft from the front hub. After pressing the outer shaft, insert a prybar between the transaxle case and the halfshaft and pry the shaft from the transaxle.

➡ **Do not pull on the shaft; doing so damages the inboard joint.**

To install:

9. Replace the circlips on the ends of the halfshafts.

10. Insert the halfshaft into the transaxle. Be sure it is fully seated.

11. Pull the strut assembly out and install the other end to the hub.

12. Install the center bearing bracket bolts and tighten to 33 ft. lbs. (45 Nm).

13. Install the washer so the chamfered edge faces outward. Install the nut and tighten to 145–188 ft. lbs. (200–260 Nm). Secure with a new cotter pin.

14. Connect the ball joint to the steering knuckle. Torque the new retaining nut to 43–52 ft. lbs. (60–72 Nm) and secure with a new cotter pin.

15. Connect the tie rod end to the steering knuckle. Torque the retaining nut to 21 ft. lbs. (29 Nm) and secure with a new cotter pin.

16. If equipped with ABS, install the front wheel speed sensor.

17. If equipped with Active-ECS, perform the following:

    a. Install the air line with a new O-ring.

    b. Install the actuator to the top of the strut. Connect the wiring harness.

18. Install the wheel and lower the vehicle to the floor.

2004–2006

1. Before servicing the vehicle, refer to the Precautions Section.

2. Raise and support the vehicle safely.

3. Remove the front under cover. Remove the side undercover. Drain the transaxle fluid.

4. On vehicles equipped with the 3.8L engine, disconnect the front exhaust pipe if working on the right side halfshaft.

5. Disconnect the speed sensor connection. Remove the wheel speed sensor. Remove the brake hose clip.

6. Remove the cotter pin. Install tool MB990767 to the hub and remove the halfshaft nut. Remove the washer.

7. Remove the lower ball joint cotter pin.

➡ **Do not remove the nut from the ball joint. Loosen it and use special tool MB991897 to avoid possible damage to the ball joint threads. Hang the special tool in place with wire or string to prevent it from falling.**

8. Install the special tool. Turn the bolt and knob as necessary to make the jaws of the tool parallel. Tighten the bolt by hand and confirm that the jaws are still parallel.

➡ **When adjusting the jaws in parallel, make sure the knob is in the vertical (upward) position.**

9. Tighten the bolt with a wrench to disconnect the lower arm ball joint connection.

10. Remove the tie rod end cotter pin. Install tool MB990767 to the hub and remove the halfshaft nut. Remove the washer.

➡ **Do not remove the nut from the tie rod end. Loosen it and use special tool MB991897 to avoid possible damage to the ball joint threads. Hang the special tool in place with wire or string to prevent it from falling.**

11. Install the special tool. Turn the bolt and knob as necessary to make the jaws of the tool parallel. Tighten the bolt by hand and confirm that the jaws are still parallel.

➡ **When adjusting the jaws in parallel, make sure the knob is in the vertical (upward) position.**

12. Tighten the bolt with a wrench to disconnect the tie rod end.

13. Disconnect the stabilizer link connection.

➡ **Do not strike the ABS rotor attached to the BJ or EBJ outer race, of the halfshaft against other parts when removing the halfshaft as damage to the rotors will result.**

14. Use special tools MB991354, MB990242 and MB990767 to push the halfshaft out from the hub.

➡ **Do not pull on the halfshaft, doing so will damage the TJ or PTJ. Be sure to use a prybar. Do not insert the prybar so deep as to damage the oil seal.**

15. Remove the halfshaft from the hub by pulling the bottom of the brake disc towards you.

16. Insert a prybar between the transaxle case and the halfshaft, and than pry and remove the halfshaft from the transaxle.

➡ **Insert a prybar, taking care not to damage the protrusion of the transaxle case when removing the halfshaft (left side).**

17. If the inner shaft and transaxle are tightly joined, tap the center bearing bracket with a plastic hammer to remove the half-shaft and inner shaft from the transaxle on vehicles with 3.0L engine (right side).

To install:

18. Installation is the reverse of the removal procedure.

19. Check and adjust the front end alignment, as necessary.

Lancer

2002–2003

1. Before servicing the vehicle, refer to the Precautions Section.

2. Raise the vehicle and support it safely.

3. Remove the cotter pin, halfshaft nut and washer.

4. If equipped with Anti-Lock Brake (ABS), remove the front wheel speed sensor.

5. If equipped with Active Electronic Control Suspension (Active-ECS) perform the following:

a. Loosen the nut that secures the air line to the to the top of the strut and discard the O-ring.

b. Remove the bolts that secure the actuator to the top of the strut and remove the component. Disconnect the wiring harness.

6. Disconnect the lower ball joint and the tie rod end from the steering knuckle.

7. If removing the left side axle with an inner shaft, remove the center support bearing bracket bolts and washers. Then, remove the halfshaft by setting up a puller on the outside wheel hub and pushing the halfshaft from the front hub. Tap the shaft union at the joint case with a plastic hammer to remove the halfshaft and inner shaft from the transaxle.

8. If removing right side axle shafts without an inner shaft, remove the halfshaft by setting up a puller on the outside wheel hub and pushing the halfshaft from the front hub. After pressing the outer shaft, insert a prybar between the transaxle case and the halfshaft and pry the shaft from the transaxle.

➡ **Do not pull on the shaft; doing so damages the inboard joint.**

To install:

9. Replace the circlips on the ends of the halfshafts.

10. Insert the halfshaft into the transaxle. Be sure it is fully seated.

11. Pull the strut assembly out and install the other end to the hub.

12. Install the center bearing bracket bolts and tighten to 33 ft. lbs. (45 Nm).

13. Install the washer so the chamfered edge faces outward. Install the nut and tighten to 181 ft. lbs. (245 Nm). Secure with a new cotter pin.

14. Connect the ball joint to the steering knuckle. Torque the new retaining nut to 43–52 ft. lbs. (60–72 Nm) and secure with a new cotter pin.

15. Connect the tie rod end to the steering knuckle. Torque the retaining nut to 21 ft. lbs. (29 Nm) and secure with a new cotter pin.

16. If equipped with ABS, install the front wheel speed sensor.

17. If equipped with Active-ECS, perform the following:

a. Install the air line with a new O-ring.

b. Install the actuator to the top of the strut. Connect the wiring harness.

18. Install the wheel and lower the vehicle to the floor.

2004–2006

1. Before servicing the vehicle, refer to the Precautions Section.

2. Raise and support the vehicle safely.

3. If equipped with ABS, disconnect the speed sensor connection. Remove the ABS sensor and bracket. Remove the brake hose clip.

4. Remove the stabilizer bar locknut, rubber insulator and collar.

5. Remove the cotter pin. Install tool MB990767 to the hub and remove the halfshaft nut. Remove the washer.

6. Remove the lower ball joint cotter pin.

➡ **Do not remove the nut from the ball joint. Loosen it and use special tool MB991897 to avoid possible damage to the ball joint threads. Hang the special tool in place with wire or string to prevent it from falling.**

7. Install the special tool. Turn the bolt and knob as necessary to make the jaws of the tool parallel. Tighten the bolt by hand and confirm that the jaws are still parallel.

➡ **When adjusting the jaws in parallel, make sure the knob is in the vertical (upward) position.**

8. Tighten the bolt with a wrench to disconnect the lower arm ball joint connection.

9. Remove the tie rod end cotter pin. Install tool MB990767 to the hub and remove the halfshaft nut. Remove the washer.

➡ **Do not remove the nut from the tie rod end. Loosen it and use special tool MB991897 to avoid possible damage to the ball joint threads. Hang the special tool in place with wire or string to prevent it from falling.**

10. Install the special tool. Turn the bolt and knob as necessary to make the jaws of the tool parallel. Tighten the bolt by hand and confirm that the jaws are still parallel.

➡ **When adjusting the jaws in parallel, make sure the knob is in the vertical (upward) position.**

11. Tighten the bolt with a wrench to disconnect the tie rod end.

➡ **Do not damage the ABS rotor attached to the BJ outer race, on vehicle equipped with ABS.**

12. Use special tools MB990241 and MB990767 (vehicles without center bearing)

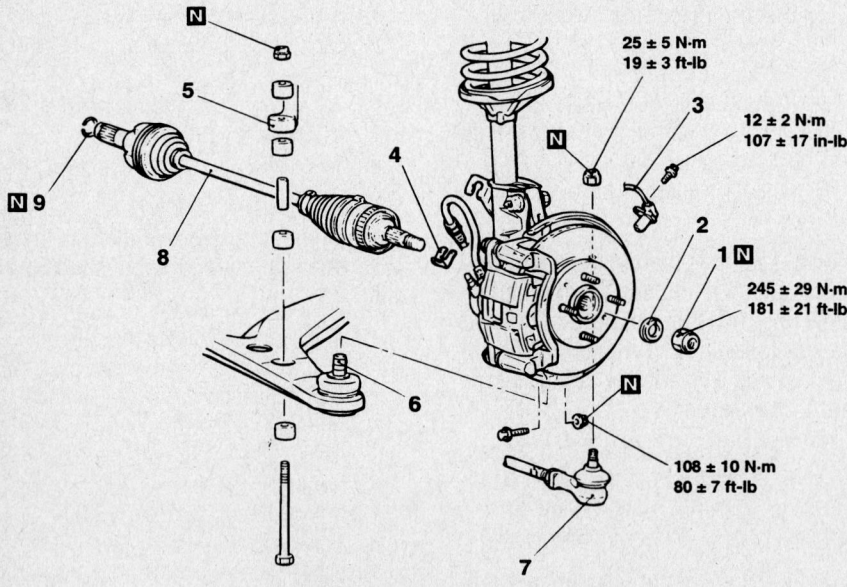

25 ± 5 N·m
19 ± 3 ft-lb

12 ± 2 N·m
107 ± 17 in-lb

245 ± 29 N·m
181 ± 21 ft-lb

108 ± 10 N·m
80 ± 7 ft-lb

1. DRIVESHAFT NUT
2. WASHER
3. FRONT SPEED SENSOR <VEHICLES WITH ABS>
4. BRAKE HOSE CRAMP
5. STABILIZER BAR CONNECTION
6. LOWER ARM BALL JOINT CONNECTION
7. TIE ROD END CONNECTION

93570G27

Halfshaft and related components—Lancer

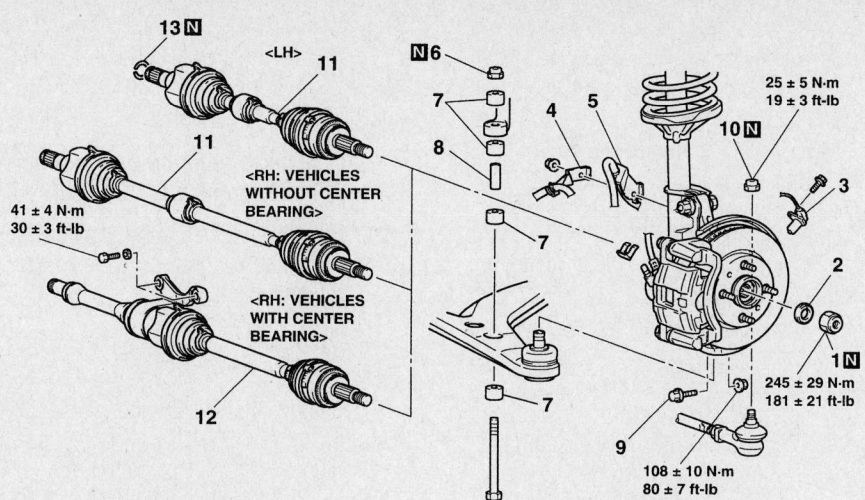

1. DRIVESHAFT NUT
2. WASHER
3. FRONT ABS SENSOR
   <VEHICLES WITH ABS>
4. FRONT ABS SENSOR BRACKET
   <VEHICLES WITH ABS>
5. BRAKE HOSE BRACKET
6. JAM NUT (STABILIZER BAR CONNECTION)
7. STABILIZER RUBBER
8. COLLAR
9. LOWER ARM CONNECTING BOLT
10. JAM NUT (TIE ROD END CONNECTION)
11. DRIVESHAFT

09482_GALA_G0170

Halfshaft and related components—2004–2006 Lancer

4. Remove the cotter pin. Install tool MB990767 to the hub and remove the driveshaft nut. Remove the washer.

5. Disconnect the speed sensor connection. Remove the ABS sensor and bracket. Remove the brake hose clip.

6. Remove the stabilizer bar locknut.

7. Remove the lower ball joint cotter pin.

➡ **Do not remove the nut from the ball joint. Loosen it and use special tool MB991897 to avoid possible damage to the ball joint threads. Hang the special tool in place with wire or string to prevent it from falling.**

8. Install the special tool. Turn the bolt and knob as necessary to make the jaws of the tool parallel. Tighten the bolt by hand and confirm that the jaws are still parallel.

➡ **When adjusting the jaws in parallel, make sure the knob is in the vertical (upward) position.**

9. Tighten the bolt with a wrench to disconnect the lower arm ball joint connection.

or tools MB991354, MB990242 and MB990244 (vehicles equipped with center bearing) to push the halfshaft out from the hub and knuckle.

13. Remove the halfshaft from the hub by pulling the bottom of the brake disc toward you, and than remove the hub retaining bolts.

➡ **Do not pull on the halfshaft, doing so will damage the TJ or ETJ. Be sure to use a prybar. Do not insert the prybar so deep as to damage the oil seal.**

14. Insert a prybar between the transaxle case and the halfshaft, and then pry the halfshaft from the transaxle.

15. If the inner shaft and transaxle are tightly joined, tap the center bearing bracket with a plastic hammer to remove the halfshaft and inner shaft from the transaxle.

To install:

16. Installation is the reverse of the removal procedure.

17. Check and adjust the front end alignment, as necessary.

Evolution

1. Before servicing the vehicle, refer to the Precautions Section.

2. Raise and support the vehicle safely.

3. Remove the under cover. Remove the side cover. Drain the transaxle fluid. Drain the transfer case.

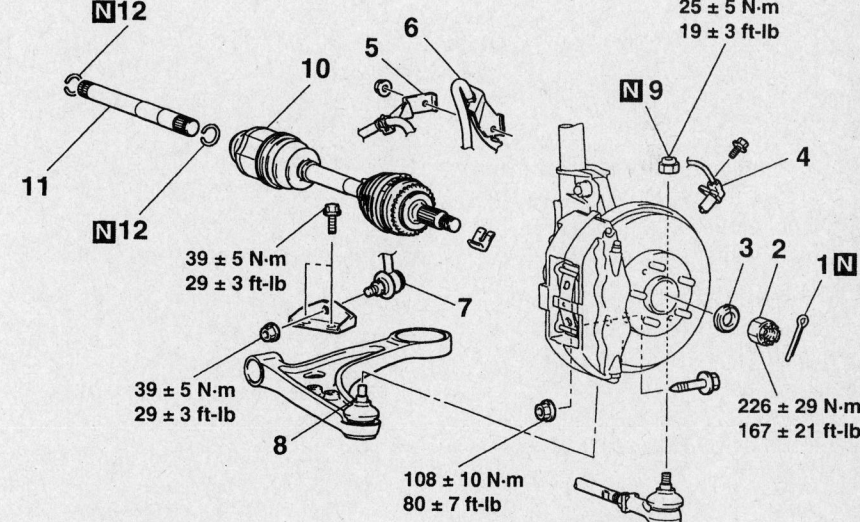

1. COTTER PIN
2. DRIVESHAFT NUT
3. WASHER
4. FRONT ABS SENSOR
5. FRONT ABS SENSOR HARNESS BRACKET
6. BRAKE HOSE BRACKET
7. STABILIZER BAR LINK CONNECTION
8. LOWER ARM BALL JOINT CONNECTION
9. SELF LOCKING NUT (TIE ROD END CONNECTION)
10. DRIVESHAFT
11. OUTPUT SHAFT
12. CIRCLIP

09482_GALA_G0171

Halfshaft and related components—Evolution

10. Remove the tie rod end cotter pin. Install tool MB990767 to the hub and remove the halfshaft nut. Remove the washer.

➡**Do not remove the nut from the tie rod end. Loosen it and use special tool MB991897 to avoid possible damage to the ball joint threads. Hang the special tool in place with wire or string to prevent it from falling.**

11. Install the special tool. Turn the bolt and knob as necessary to make the jaws of the tool parallel. Tighten the bolt by hand and confirm that the jaws are still parallel.

➡**When adjusting the jaws in parallel, make sure the knob is in the vertical (upward) position.**

12. Tighten the bolt with a wrench to disconnect the tie rod end.

➡**Do not strike the ABS rotor attached to the EBJ outer race, of the halfshaft against other parts when removing the halfshaft as damage to the rotors will result.**

13. Use special tools MB990241 (MB990242 and MB990244), MB991354 and MB990767 to push the halfshaft out from the hub.

14. Remove the halfshaft from the hub by pulling the bottom of the brake disc toward you, and than remove the hub retaining bolts.

➡**Do not pull on the halfshaft, doing so will damage the TJ. Be sure to use a prybar. Do not insert the prybar so deep as to damage the oil seal.**

15. Insert a prybar between the transaxle case and the halfshaft, and then pry and remove the driveshaft from the transaxle.

To install:

16. Installation is the reverse of the removal procedure.

17. Check and adjust the front end alignment, as necessary.

Mirage

1. Before servicing the vehicle, refer to the Precautions Section.

2. Raise the vehicle and support it safely.

3. Remove the cotter pin, halfshaft nut and washer.

4. If equipped with Anti-Lock Brake (ABS), remove the front wheel speed sensor.

5. If equipped with Active Electronic Control Suspension (Active-ECS) perform the following:

a. Loosen the nut that secures the air line to the to the top of the strut and discard the O-ring.

b. Remove the bolts that secure the actuator to the top of the strut and remove the component. Disconnect the wiring harness.

6. Disconnect the lower ball joint and the tie rod end from the steering knuckle.

7. If removing the left side axle with an inner shaft, remove the center support bearing bracket bolts and washers. Then, remove the halfshaft by setting up a puller on the outside wheel hub and pushing the halfshaft from the front hub. Tap the shaft union at the joint case with a plastic hammer to remove the halfshaft and inner shaft from the transaxle.

8. If removing right side axle shafts without an inner shaft, remove the halfshaft by setting up a puller on the outside wheel hub and pushing the halfshaft from the front hub. After pressing the outer shaft, insert a prybar between the transaxle case and the halfshaft and pry the shaft from the transaxle.

➡**Do not pull on the shaft; doing so damages the inboard joint.**

To install:

9. Replace the circlips on the ends of the halfshafts.

10. Insert the halfshaft into the transaxle. Be sure it is fully seated.

11. Pull the strut assembly out and install the other end to the hub.

12. Install the center bearing bracket bolts and tighten to 33 ft. lbs. (45 Nm).

13. Install the washer so the chamfered edge faces outward. Install the nut and tighten to 145–188 ft. lbs. (200–260 Nm). Secure with a new cotter pin.

14. Connect the ball joint to the steering knuckle. Torque the new retaining nut to 43–52 ft. lbs. (60–72 Nm) and secure with a new cotter pin.

15. Connect the tie rod end to the steering knuckle. Torque the retaining nut to 21 ft. lbs. (29 Nm) and secure with a new cotter pin.

16. If equipped with ABS, install the front wheel speed sensor.

17. If equipped with Active-ECS, perform the following:

a. Install the air line with a new O-ring.

b. Install the actuator to the top of the strut. Connect the wiring harness.

18. Install the wheel and lower the vehicle to the floor.

OVERHAUL

Eclipse

2004–2005

➡**Never disassemble the BJ assembly, except when replacing the BJ boot.**

1. Before servicing the vehicle, refer to the Precautions Section.

2. Remove the halfshaft from the vehicle.

3. Remove the large and small TJ boot bands.

4. Remove the TJ case and inner shaft assembly.

5. Remove the TJ case. Remove the seal plate. Remove the inner shaft using tool MB991248. Remove the dust cover.

6. Remove the bracket assembly. Remove the inner and outer dust seals.

7. Remove the center bearing bracket (3.0L engine, right side). Use tools MB990930 and MB990938 to press the center bearing out of the center bearing bracket.

8. Remove the circlip. Remove the snapring. Remove the spider assembly. Wrap tape around the spline part of the BJ assembly so that the TJ boot is not damaged when it is removed. Remove the TJ boot. Remove the BJ assembly.

9. Remove the damper band (2.4L engine and 3.0L engine, left side). Remove the dynamic damper (2.4L engine and 3.0L engine, left side).

10. Remove the large and small BJ boot clamp. Remove the BJ boot.

To install:

11. Installation is the reverse of the removal procedure.

12. Install the dynamic damper as shown in the illustration. Measurement "A" should be 9.5 +/- 0.12 for the left side and 10.0 +/- 0.12 for the right side.

➡**There should be no grease stuck to the rubber part of the dynamic damper. The damper band and TJ boot band (small) are different in shape. Be sure not to assemble a wrong band in the wrong place.**

13. The damper band is Blue. The TJ boot band is Pink for the 3.0L engine without ABS and Brown for the 3.0L engine with ABS. Secure the damper bands.

14. Wrap plastic tape around the shaft spline, then install the TJ boot band (small) and the TJ boot.

15. To install the spider assembly, apply grease to the spider axles and rollers of the spider assembly. Do not mix old and new grease or different types of grease.

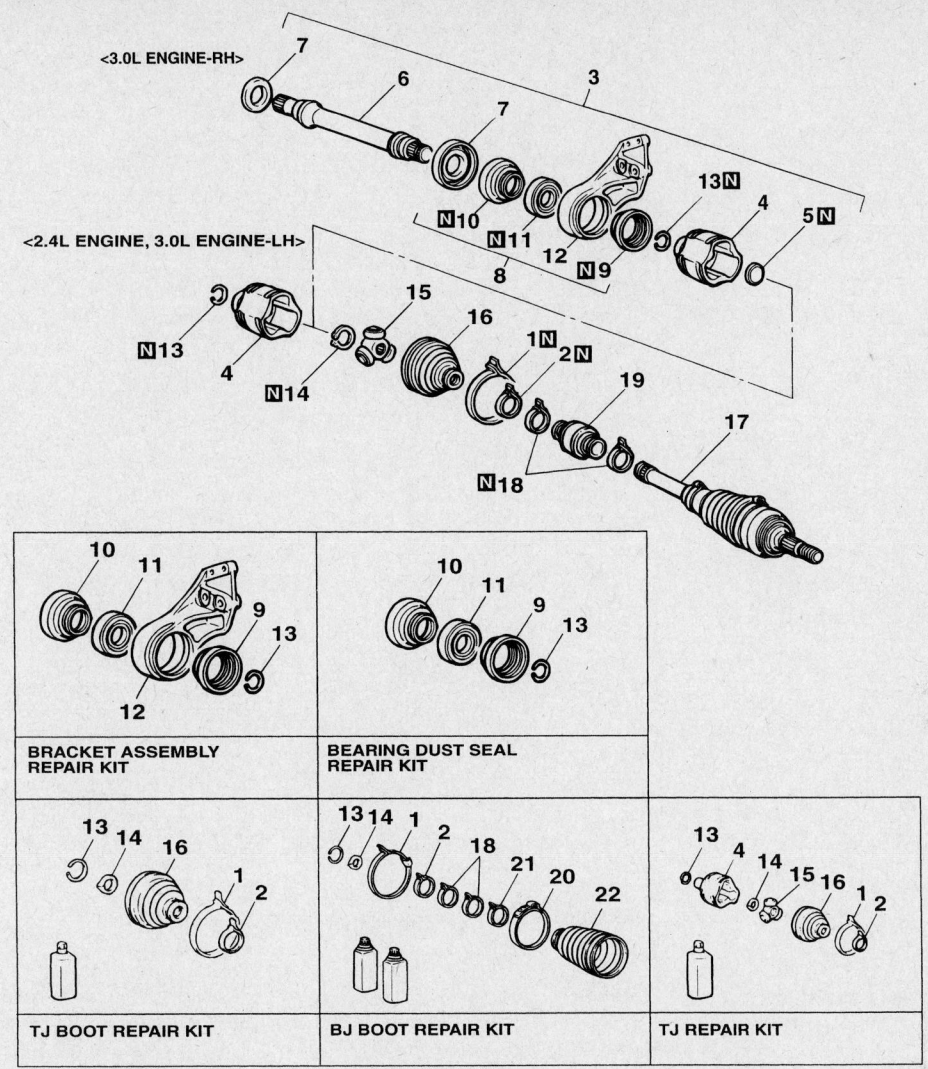

1. TJ BOOT BAND (LARGE)
2. TJ BOOT BAND (SMALL)
3. TJ CASE AND INNER SHAFT ASSEMBLY
4. TJ CASE
5. SEAL PLATE
6. INNER SHAFT
7. DUST COVER
8. BRACKET ASSEMBLY
9. DUST SEAL OUTER
10. DUST SEAL INNER
11. CENTER BEARING
12. CENTER BEARING BRACKET
13. CIRCLIP
14. SNAP RING
15. SPIDER ASSEMBLY
16. TJ BOOT
17. BJ ASSEMBLY
18. DAMPER BAND <2.4L ENGINE, 3.0L ENGINE-LH>
19. DYNAMIC DAMPER <2.4L ENGINE, 3.0L ENGINE-LH>
20. BJ BOOT BAND (LARGE)
21. BJ BOOT BAND (SMALL)
22. BJ BOOT

BRACKET ASSEMBLY REPAIR KIT

BEARING DUST SEAL REPAIR KIT

TJ BOOT REPAIR KIT

BJ BOOT REPAIR KIT

TJ REPAIR KIT

09482_GALA_G0172

Disassembled halfshaft assembly—2004–2005 Eclipse

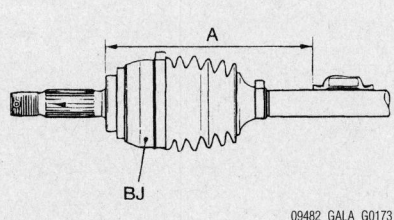

09482_GALA_G0173

Dynamic damper installation measurement—2004–2005 Eclipse

16. Face the chamfered portion of the spider assembly's spline toward the halfshaft, and then install the spider assembly to the halfshaft.

17. On 3.0L engine, right side, use tools MB990930 and MB990938 to press the center bearing into the center bearing bracket. Pack grease in the inner and outer seal and use the special tools to press the oil seal into the center bearing bracket. The dust inner seal should take about 0.5–0.7 ounces and the outer seal should take about 0.3–0.4 ounces of grease. Apply grease to the lip of the seal, not to the outside of the lip. Use tool MB991172 to hold the center

bearing inner race, and then press in the inner shaft.

18. Apply grease to the inner shaft serration and then press the inner shaft assembly to the TJ case.

19. Fill the TJ case with grease and insert the halfshaft. Refill the TJ case with grease. Do not mix old and new grease or different types of grease. The 2.4L engine uses 3.4 +/- 0.4 ounces of grease and the 3.0L engine uses 6.2 +/- 0.4 ounces of grease.

20. Position the TJ outer race so that the distance between the boot bands is 3.3 +/- 0.12 inch for the 2.4L engine and 3.5 +/- inch for the 3.0L engine.

DUST SEAL INNER

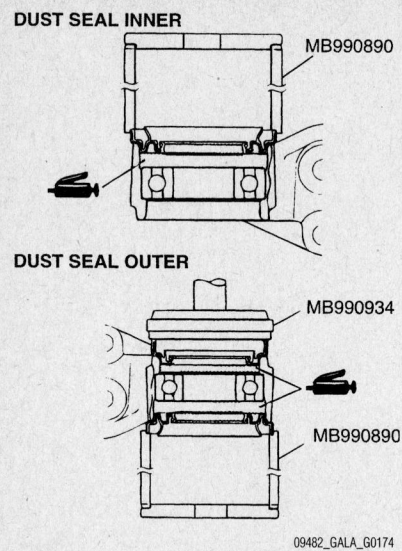

MB990890

DUST SEAL OUTER

MB990934

MB990890

09482_GALA_G0174

Dust seal grease application location—2004–2006 Eclipse with center bearing halfshaft

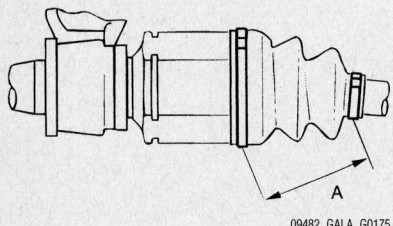

09482_GALA_G0175

TJ boot band measurement—2004–2005 Eclipse

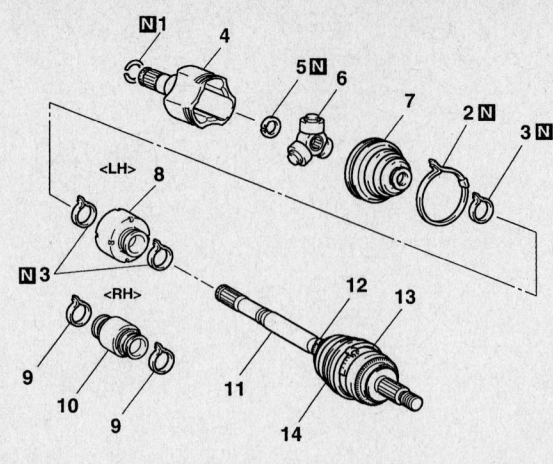

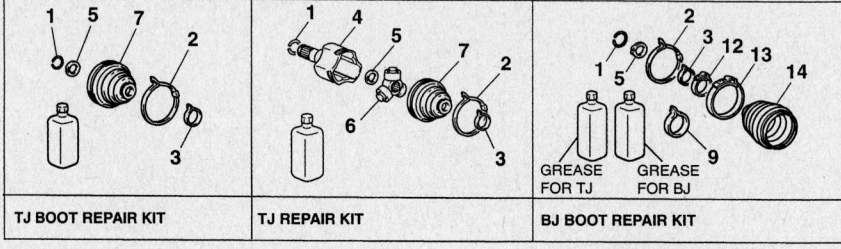

GREASE FOR TJ   GREASE FOR BJ

TJ BOOT REPAIR KIT   **TJ REPAIR KIT**   **BJ BOOT REPAIR KIT**

1. CIRCLIP
2. TJ BOOT BAND (LARGE)
3. TJ BOOT BAND (SMALL)
4. TJ CASE
5. SNAP RING
6. SPIDER ASSEMBLY
7. TJ BOOT

8. DYNAMIC DAMPER <LH>
9. DAMPER BAND <RH>
10. DYNAMIC DAMPER <RH>
11. BJ ASSEMBLY
12. BJ BOOT BAND (SMALL)
13. BJ BOOT BAND (LARGE)
14. BJ BOOT

09482_GALA_G0176

Disassembled halfshaft assembly—2006 Eclipse with 2.4L engine

21. Remove part of the TJ outer race to release the air pressure inside the boot.

2006 (2.4L ENGINE)

➡ **Be careful not to damage the ABS rotor, which is attached to the outer race during disassembly and reassembly. Never disassemble the BJ assembly, except when replacing the BJ boot.**

1. Before servicing the vehicle, refer to the Precautions Section.
2. Remove the halfshaft from the vehicle.
3. Remove the large and small TJ boot bands.
4. Remove the TJ case. Remove the snaping.
5. Remove the spider assembly. Remove the TJ boot. On left side remove the dynamic damper. On right side remove the damper band and dynamic damper.
6. Remove the BJ assembly. Remove the small and large BJ boot bands. Remove the BJ boot.

To install:

7. Installation is the reverse of the removal procedure.
8. Install the dynamic damper as shown in the illustration. Measurement "L" should be 10.2 +/- 0.12 inch for the left side and 16.0 +/- 0.12 inch for the right side.

➡ **There should be no grease stuck to the rubber part of the dynamic damper.**

9. Secure the damper bands. Wrap plastic tape around the shaft spline, then install the TJ boot band (small) and the TJ boot.
10. To install the spider assembly, apply grease to the spider axles and rollers of the spider assembly. Do not mix old and new grease or different types of grease.
11. Install the spider assembly to the shaft from the direction of the spline chamfered side.
12. Apply grease to the TJ case, then insert the halfshaft and apply more grease. The left side uses 4.9 +/- 0.3 ounces of grease and the right side uses 4.6 +/- 0.3 ounces of grease.
13. Position the TJ boot bands at 3.35 +/- 0.12 inch in order to adjust the amount of air inside the TJ boot and then tighten the TJ boot band (small) then the TJ boot band (large) securely.

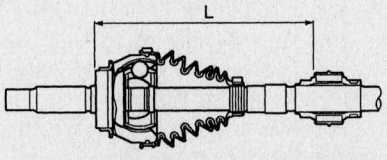

L

09482_GALA_G0177

Dynamic damper installation measurement—2006 Eclipse, 2004–2006 Galant and 2004–2006 Lancer

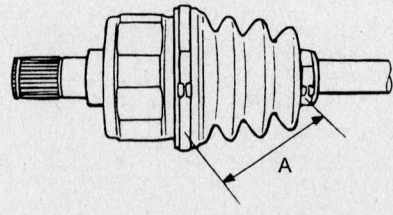

A

09482_GALA_G0178

TJ boot band measurement—2006 Eclipse, 2004–2006 Galant, 2004–2006 Lancer and Evolution

2006 (3.8L ENGINE)

➡ **Be careful not to damage the ABS rotor, which is attached to the outer race during disassembly and reassembly. Never disassemble the EBJ assembly, except when replacing the EBJ boot.**

1. Before servicing the vehicle, refer to the Precautions Section.

2. Remove the halfshaft from the vehicle.

3. Remove the large and small EBJ boot bands. Use a slotted screwdriver to make a hole in the seal plate inside the EBJ case and remove it.

4. Use special tool MB991248 to remove the inner shaft assembly from the EBJ case.

5. Remove the seal plate. Remove the EBJ case. Remove the inner shaft. Remove the dust cover. Remove the bracket assembly.

6. Remove the inner and outer dust seal. Remove the center bearing from the center bearing bracket using tools MB990938 and MB990930. Remove the center bearing bracket.

7. Remove the circlip. Remove the snapring. Remove the spider assembly.

8. Remove the damper band. Remove the dynamic damper.

9. Remove the EBJ assembly. Remove the EBJ small and large boot bands. Remove the EBJ boot.

To install:

10. Installation is the reverse of the removal procedure.

11. Install the dynamic damper as shown in the illustration. Measurement "L" should be 10.3 +/- 0.12 inch for automatic transaxle and 8.9.0 +/- 0.12 inch for manual transaxle.

➡ **There should be no grease stuck to the rubber part of the dynamic damper.**

12. Secure the damper bands. Wrap plastic tape around the shaft spline, then install the EBJ boot band (small) and the EBJ boot.

13. To install the spider assembly, apply grease to the spider axles and rollers of the spider assembly. Do not mix old and new grease or different types of grease.

14. Install the spider assembly to the shaft from the direction of the spline chamfered side.

15. Use tools MB990930 and MB990938 to press the center bearing into the center bearing bracket. Pack grease in the inner and outer seal and use the special tools to press the oil seal into the center bearing bracket. The dust inner seal should take about 0.5–0.7 ounces and the

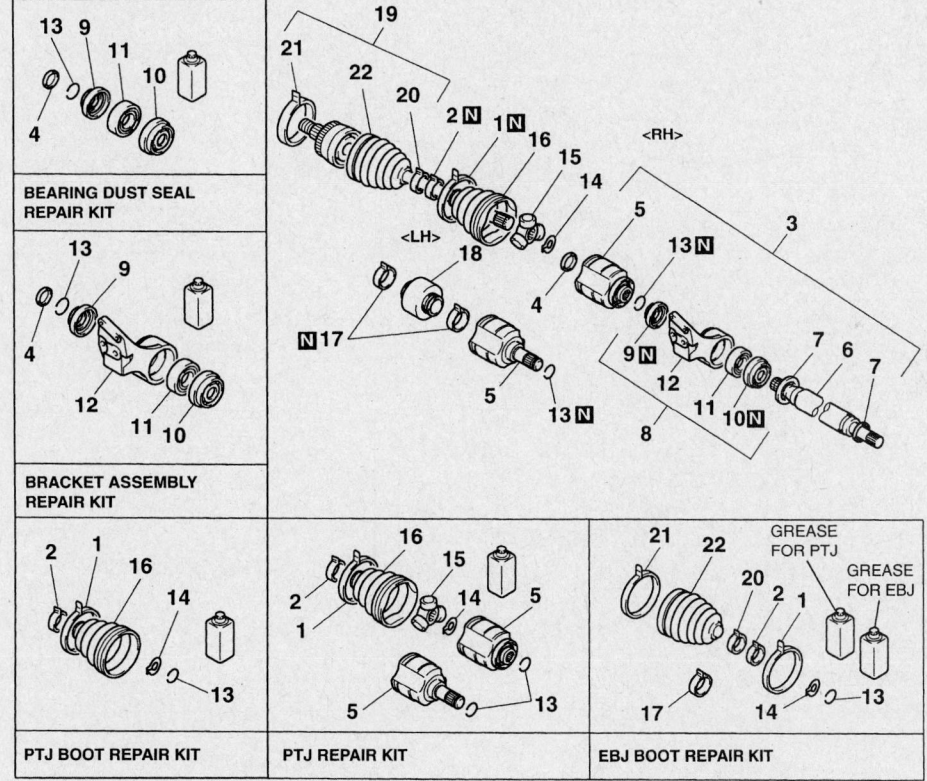

BEARING DUST SEAL REPAIR KIT

BRACKET ASSEMBLY REPAIR KIT

PTJ BOOT REPAIR KIT

PTJ REPAIR KIT

EBJ BOOT REPAIR KIT

1.	PTJ BOOT BAND (LARGE)	11.	CENTER BEARING
2.	PTJ BOOT BAND (SMALL)	12.	CENTER BEARING BRACKET
3.	PTJ CASE AND INNER SHAFT ASSEMBLY	13.	CIRCLIP
4.	SEAL PLATE	14.	SNAP RING
5.	PTJ CASE	15.	SPIDER ASSEMBLY
6.	INNER SHAFT	16.	PTJ BOOT
7.	DUST COVER	17.	DAMPER BAND
8.	BRACKET ASSEMBLY	18.	DYNAMIC DAMPER
9.	DUST SEAL OUTER	19.	EBJ ASSEMBLY
10.	DUST SEAL INNER	20.	EBJ BOOT BAND (SMALL)
		21.	EBJ BOOT BAND (LARGE)
		22.	EBJ BOOT

09482_GALA_G0179

Disassembled halfshaft assembly—2006 Eclipse with 3.8L engine

outer seal should take about 0.3–0.4 ounces of grease. Apply grease to the lip of the seal, not to the outside of the lip. Use tool MB991172 to hold the center bearing inner race, and then press in the inner shaft.

16. Apply grease to the inner shaft spline and then press the inner shaft assembly to the EBJ case.

➡ **When press fitting the inner shaft into the EBJ case, apply a thin coat of repair grease to the dust seal outer lip and to the outside edge of the EBJ axial component.**

17. Use special tool MB998369 and press in the seal plate.

18. Fill the EBJ case with grease and insert the halfshaft. Refill the EBJ case with grease. Do not mix old and new grease or different types of grease. Manual transaxle left side uses 9.2 +/- 0.3 ounces of grease and automatic transaxle (both sides) and manual transaxle right side uses 8.6 +/- 0.3 ounces of grease.

19. Position the EBJ boot bands at 3.35 +/- 0.12 inch in order to adjust the amount of air inside the EBJ boot. Tighten the EBJ boot band (small) and the EBJ boot band (large) securely.

Galant

2004–2006

➡ **Be careful not to damage the ABS rotor, which is attached to the outer race during disassembly and reassembly. Never disassemble the BJ assembly, except when replacing the BJ boot.**

1. Before servicing the vehicle, refer to the Precautions Section.
2. Remove the halfshaft from the vehicle.
3. Remove both TJ boot bands, if equipped with the 2.4L engine. Remove

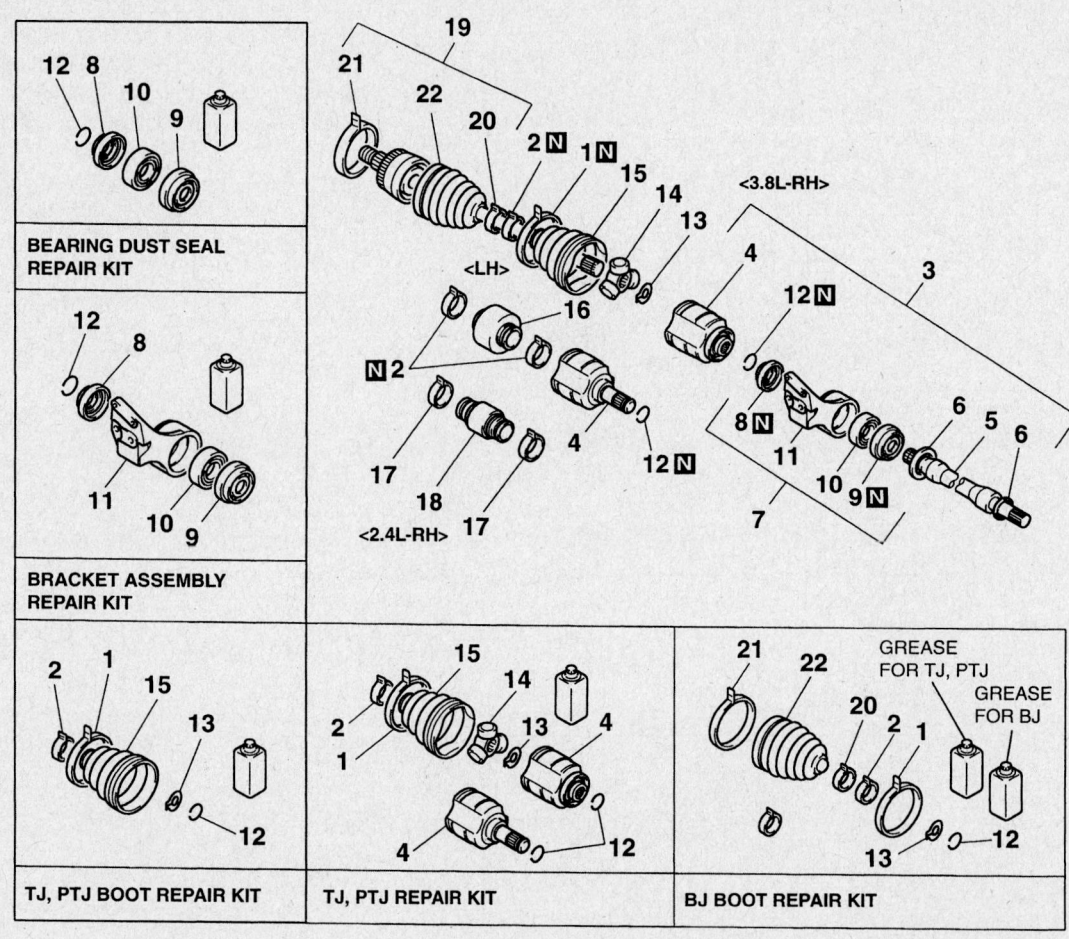

Disassembled halfshaft assembly—Galant

1. TJ BOOT BAND (LARGE) <2.4L>, PTJ BOOT BAND (LARGE) <3.8L>
2. TJ BOOT BAND (SMALL) <2.4L>, PTJ BOOT BAND (SMALL) <3.8L>
3. PTJ CASE AND INNER SHAFT ASSEMBLY <3.8L-RH>
4. TJ CASE <2.4L>, PTJ CASE <3.8L>
5. INNER SHAFT <3.8L-RH>
6. DUST COVER <3.8L-RH>
7. BRACKET ASSEMBLY <3.8L-RH>
8. DUST SEAL OUTER <3.8L-RH>

9. DUST SEAL INNER <3.8L-RH>
10. CENTER BEARING <3.8L-RH>
11. CENTER BEARING BRACKET <3.8L-RH>
12. CIRCLIP
13. SNAP RING
14. SPIDER ASSEMBLY
15. TJ BOOT <2.4L>, PTJ BOOT <3.8L>
16. DYNAMIC DAMPER <2.4L-LH, 3.8L-LH>
17. DAMPER BAND <2.4L-RH>

18. DYNAMIC DAMPER <2.4L-RH>
19. BJ ASSEMBLY
20. BJ BOOT BAND (SMALL)
21. BJ BOOT BAND (LARGE)
22. BJ BOOT

09482_GALA_G0180

both PTJ boot bands if, equipped with the 3.8L engine.

4. Using tool MB991248 remove the PTJ case inner shaft assembly, if equipped with the 3.8L engine.

5. Remove the center bearing bracket from the inner shaft using tool MB990810, if equipped with the 3.8L engine.

6. Remove the TJ case, if equipped with the 2.4L engine. Remove the PTJ case, if equipped with the 3.8L engine.

7. If equipped with the 3.8L engine (right side), remove the inner shaft, dust cover, bracket assembly, dust outer seal, dust inner seal, center bearing and center bearing bracket.

8. Use tools MB990938 and MB990930 to remove the center bearing from the center bearing bracket, if equipped with the 3.8L engine.

9. Remove the snapring. Remove the spider assembly.

10. Remove the TJ boot, if equipped with the 2.4L engine. Remove the PTJ boot if equipped with the 3.8L engine.

11. Remove the dynamic damper (left side). Remove the damper band and damper if equipped with the 2.4L engine (right side).

12. Remove the BJ assembly. Remove the small and large BJ boot bands. Remove the BJ boot.

To install:

13. Installation is the reverse of the removal procedure.

14. Install the dynamic damper as shown in the illustration. Measurement "L" should be 10.2 +/- 0.12 inch for the 2.4L engine left side and 16.0 +/- 0.12 inch for the 2.4L engine right side and 10.3 +/-0.12 inch for the 3.8L engine.

➡There should be no grease stuck to the rubber part of the dynamic damper.

15. Secure the damper bands. Wrap plastic tape around the shaft spline, then install the TJ boot band (small) and the TJ boot or the PTJ boot band (small) and the PTJ boot.

16. To install the spider assembly, apply grease to the spider axles and rollers of the spider assembly. Do not mix old and new grease or different types of grease.

17. Install the spider assembly to the shaft from the direction of the spline chamfered side.

18. Use tools MB990930 and MB990938 to press the center bearing into the center bearing bracket. Pack grease in the inner and outer seal and use the special tools to press the oil seal into the center bearing bracket. The dust inner seal should

take about 0.5–0.7 ounces and the outer seal should take about 0.3–0.4 ounces of grease. Apply grease to the lip of the seal, not to the outside of the lip. Use tool MB991172 to hold the center bearing inner race, and then press in the inner shaft.

19. Apply grease to the inner shaft spline and then press the inner shaft assembly to the PTJ case.

➡When press fitting the inner shaft into the PTJ case, apply a thin coat of repair grease to the dust seal outer lip and to the outside edge of the PTJ axial component.

20. Use special tool MB991172 to hold the center bearing inner race, then press in the inner shaft.

21. Fill the TJ or PTJ case with grease and insert the halfshaft. Refill the TJ or PTJ case with grease. Do not mix old and new grease or different types of grease. If equipped with the 2.4L engine the left side uses 4.9 +/- 0.3 ounces of grease and the right side uses 4.6 +/- 0.3 ounces of grease. The 3.8L engine uses 7.8 +/- 0.3 ounces of grease.

22. Position the TJ or PTJ boot bands at 3.35 +/- 0.12 inch in order to adjust the amount of air inside the TJ or PTJ boot.

Tighten the TJ or PTJ boot band (small) and the TJ or PTJ boot band (large) securely.

Lancer

2004–2006 (WITHOUT CENTER BEARING)

➡Be careful not to damage the ABS rotor, which is attached to the outer race during disassembly and reassembly. Never disassemble the BJ assembly, except when replacing the BJ boot.

1. Before servicing the vehicle, refer to the Precautions Section.

2. Remove the halfshaft from the vehicle.

3. Remove the large and small TJ boot bands.

4. Remove the TJ case. Remove the snapring.

5. Remove the spider assembly. Remove the TJ boot. Remove the dynamic damper. Remove the damper band.

6. Remove the BJ assembly. Remove the small and large BJ boot bands. Remove the BJ boot.

To install:

7. Installation is the reverse of the removal procedure.

8. Install the dynamic damper as shown

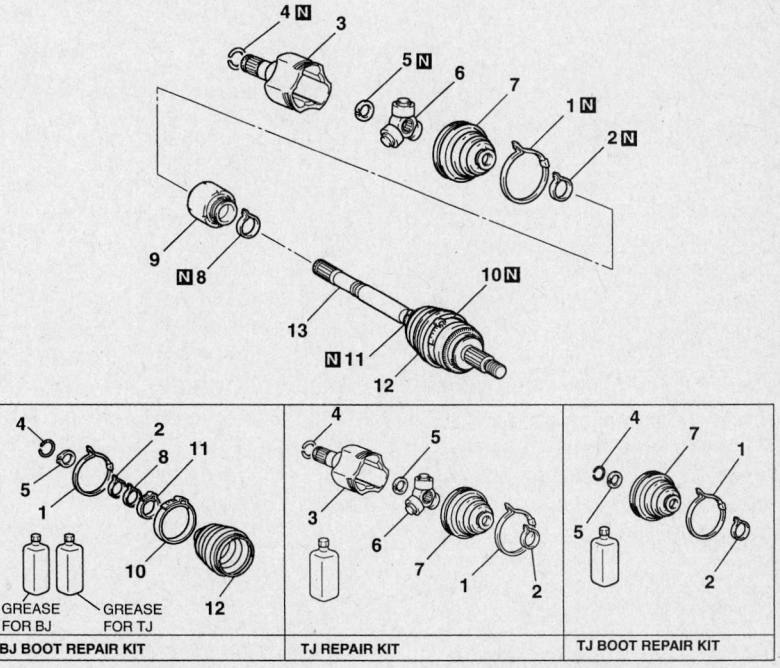

1.	TJ BOOT BAND (LARGE)		8.	DAMPER BAND
2.	TJ BOOT BAND (SMALL)		9.	DYNAMIC DAMPER
3.	TJ CASE		10.	BJ BOOT BAND (LARGE)
4.	CIRCLIP		11.	BJ BOOT BAND (SMALL)
5.	SNAP RING		12.	BJ BOOT
6.	SPIDER ASSEMBLY		13.	BJ ASSEMBLY
7.	TJ BOOT			

09482_GALA_G0181

Disassembled halfshaft assembly—Lancer without center bearing

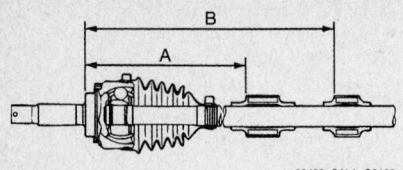

09482_GALA_G0182

Dynamic damper installation measurement—2004–2006 Lancer without center bearing

in the illustration. Measurement "A" should be 9.1 +/- 0.12 inch for the left side and measurement "B" should be 16.3 +/- 0.12 inch for the right side.

➡ There should be no grease stuck to the rubber part of the dynamic damper. The damper band and TJ boot band (small) are different in shape. Locate the identification numbers stamped on the band levers.

9. The damper band ID number is 8382. The TJ boot band ID number is E687. Secure the damper bands.

10. Wrap plastic tape around the shaft spline, then install the TJ boot band (small) and the TJ boot.

11. To install the spider assembly, apply grease to the spider axles and rollers of the spider assembly. Do not mix old and new grease or different types of grease.

12. Install the spider assembly to the shaft from the direction of the spline beveled section.

13. Apply grease to the TJ case, then insert the halfshaft and apply more grease. The assembly uses 4.2 +/- 0.3 ounces of grease.

14. Position the TJ boot bands at 3.54 +/- 0.12 inch in order to adjust the amount of air inside the TJ boot and then tighten the TJ boot band (small) then the TJ boot band (large) securely.

2004–2006 (WITH CENTER BEARING)

➡ Be careful not to damage the ABS rotor, which is attached to the outer race during disassembly and reassembly. Never disassemble the EBJ assembly, except when replacing the EBJ boot.

1. Before servicing the vehicle, refer to the Precautions Section.

2. Remove the halfshaft from the vehicle.

3. Remove the large and small ETJ boot bands. Remove the ETJ case and inner shaft assembly.

4. Use special tool MB991248 to remove the inner shaft assembly from the ETJ case.

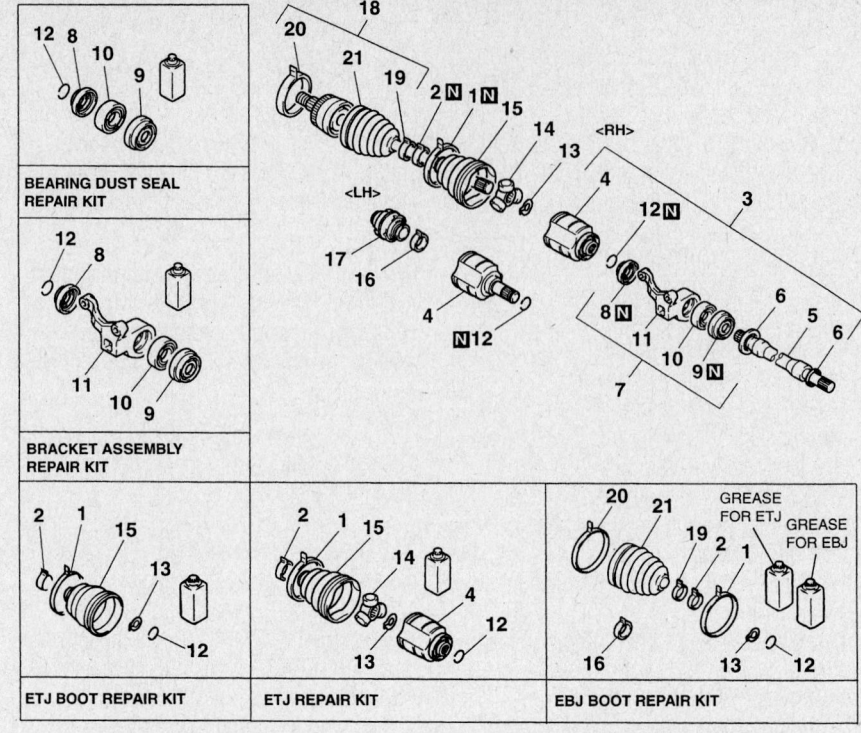

09482_GALA_G0183

Disassembled halfshaft assembly—Lancer with center bearing

1. ETJ BOOT BAND (LARGE)
2. ETJ BOOT BAND (SMALL)
3. ETJ CASE AND INNER SHAFT ASSEMBLY
4. ETJ CASE
5. INNER SHAFT
6. DUST COVER
7. BRACKET ASSEMBLY
8. DUST SEAL OUTER
9. DUST SEAL INNER
10. CENTER BEARING
11. CENTER BEARING BRACKET
12. CIRCLIP
13. SNAP RING
14. SPIDER ASSEMBLY
15. ETJ BOOT
16. DAMPER BAND
17. DYNAMIC DAMPER
18. EBJ ASSEMBLY
19. EBJ BOOT BAND (SMALL)
20. EBJ BOOT BAND (LARGE)
21. EBJ BOOT

5. Use tool MB990810 to remove the center bearing bracket from the inner shaft.

6. Remove the dust cover. Remove the bracket assembly.

7. Remove the inner and outer dust seal. Remove the center bearing from the center bearing bracket using tools MB990938 and MB990930. Remove the center bearing bracket.

8. Remove the circlip. Remove the snapring. Remove the spider assembly.

9. Remove the damper band. Remove the dynamic damper.

10. Remove the EBJ assembly. Remove the EBJ small and large boot bands. Remove the EBJ boot.

To install:

11. Installation is the reverse of the removal procedure.

12. Install the dynamic damper as shown in the illustration. Measurement "L" should be 9.1 +/- 0.12 inch.

➡ There should be no grease stuck to the rubber part of the dynamic damper.

13. Secure the damper bands. Wrap plastic tape around the shaft spline, then install the EBJ boot band (small) and the EBJ boot.

14. To install the spider assembly, apply grease to the spider axles and rollers of the spider assembly. Do not mix old and new grease or different types of grease.

15. Install the spider assembly to the shaft from the direction of the spline chamfered side.

16. Use tools MB990930 and MB990938 to press the center bearing into the center bearing bracket. Pack grease in the inner and outer seal and use the special tools to press the oil seal into the center bearing bracket. The dust inner seal should take about 0.5–0.7 ounces and the outer seal should take about 0.3–0.4 ounces of grease. Apply grease to the lip of the seal, not to the outside of the lip.

17. Use special tools MB990890, MB990938 and MB990934 to press the dust seals into the center bearing bracket

until they are flush with each other. Apply grease to the lip of each dust seal.

18. Use tool MB991172 to hold the center bearing inner race, and then press in the inner shaft.

19. Apply grease to the inner shaft spline and then press the inner shaft assembly to the EBJ case.

➡ **When press fitting the inner shaft into the EBJ case, apply a thin coat of repair grease to the dust seal outer lip and to the outside edge of the EBJ axial component.**

20. Use special tool MB998369 and press in the seal plate.

21. Fill the EBJ case with grease and insert the halfshaft. Refill the EBJ case with grease. Do not mix old and new grease or different types of grease. Left side uses 4.6 +/- 0.3 ounces of grease and right side uses 4.2 +/- 0.3 ounces of grease.

22. Position the EBJ boot bands at 3.15 +/- 0.12 inch in order to adjust the amount of air inside the EBJ boot. Tighten the EBJ boot band (small) and the EBJ boot band (large) securely.

Evolution

➡ **Be careful not to damage the ABS rotor, which is attached to the outer race during disassembly and reassembly. Never disassemble the EBJ assembly, except when replacing the EBJ boot.**

1. Before servicing the vehicle, refer to the Precautions Section.
2. Remove the halfshaft from the vehicle.
3. Remove the large and small TJ boot bands.
4. Remove the TJ case. Remove the snapring.
5. Remove the spider assembly. Remove the TJ boot.
6. Remove the EBJ assembly. Remove

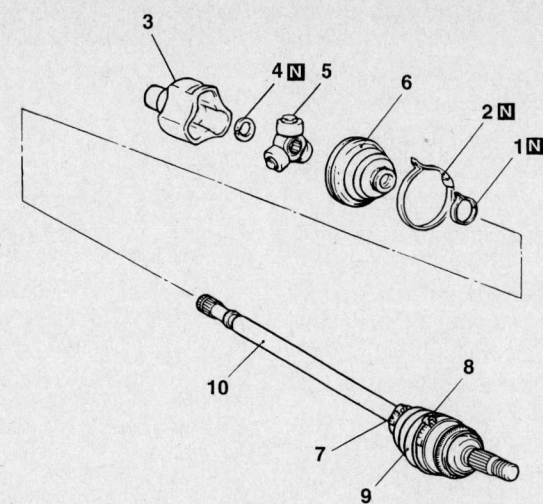

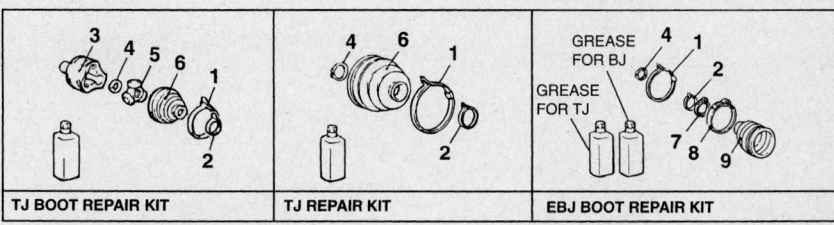

1.	TJ BOOT BAND (LARGE)	6.	TJ BOOT
2.	TJ BOOT BAND (SMALL)	7.	EBJ BOOT BAND (SMALL)
3.	TJ CASE	8.	EBJ BOOT BAND (LARGE)
4.	SNAP RING	9.	EBJ BOOT
5.	SPIDER ASSEMBLY	10.	EBJ ASSEMBLY

09482_GALA_G0184

Disassembled halfshaft assembly—Evolution

the small and large EBJ boot bands. Remove the EBJ boot.

To install:

7. Installation is the reverse of the removal procedure.

8. Wrap plastic tape around the shaft spline, then install the EBJ boot band (small) and the EBJ boot.

9. To install the spider assembly, apply grease to the spider axles and rollers of the spider assembly. Do not mix old and new grease or different types of grease.

10. Install the spider assembly to the shaft from the direction of the spline chamfered side.

11. Fill the TJ case with grease and insert the halfshaft. Refill the TJ case with grease. Do not mix old and new grease or different types of grease. Uses 5.1 +/- 0.3 ounces of grease.

12. Position the TJ boot bands at 3.35 +/- 0.12 inch in order to adjust the amount of air inside the TJ boot. Tighten the TJ boot band (small) and the TJ boot band (large) securely.

STEERING

Air Bag

✳✳ CAUTION

All vehicles are equipped with an air bag system. The system must be disabled before performing service on or around system components, steering column, instrument panel components, wiring and sensors. Failure to follow safety and disabling procedures could result in accidental air bag deployment, possible personal injury and unnecessary system repairs.

PRECAUTIONS

Several precautions must be observed when handling the inflator module to avoid accidental deployment and possible personal injury.

• Never carry the inflator module by the wires or connector on the underside of the module.

• When carrying a live inflator module, hold securely with both hands, and ensure that the bag and trim cover are pointed away.

• Place the inflator module on a bench or other surface with the bag and trim cover facing up.

• With the inflator module on the bench, never place anything on or close to the module which may be thrown in the event of an accidental deployment.

DISARMING

1. Before servicing the vehicle, refer to the Precautions Section.

2. Position the front wheels in the straight-ahead position and place the key in the **LOCK** position. Remove the key from the ignition lock cylinder.

3. Disconnect the negative battery cable and insulate the cable end with high-quality electrical tape or similar non-conductive wrapping.

4. Wait at least 1 minute before working on the vehicle. The air bag system is designed to retain enough voltage to deploy the air bag for a short period of time after the battery has been disconnected.

REARMING

1. Connect the negative battery cable, turn the ignition switch to the **ON** position

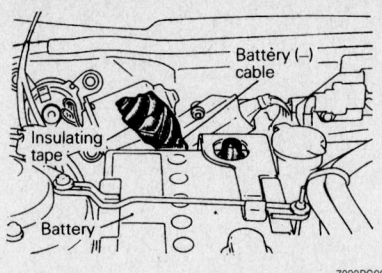

Insulate the negative battery cable to prevent accidental deployment of the air bag.

and check the Supplemental Restraint (SRS) warning light for proper operation.

Power Steering Rack

REMOVAL & INSTALLATION

Diamante

➡**Prior to removal of the steering rack, center the front wheels and remove the ignition key. Failure to do so may dam-** age the Supplemental Restraint System (SRS) clock spring and render SRS system inoperative.

1. Before servicing the vehicle, refer to the Precautions Section.

2. Remove or disconnect the following:
• Negative battery cable.
• Front exhaust pipe
• Transfer case assembly, if equipped with All Wheel Drive (AWD)
• Bolt holding the lower steering column joint to the rack and pinion input shaft
• Tie rod ends
• Left and right frame members
• Stabilizer bar bracket
• Lines going to the rear pump, if equipped with Four Wheel Steering (4WS)
• Rack and pinion steering assembly and its rubber mounts. Move the rack to the right to remove it from the crossmember.

To install:

3. Install or connect the following:
• Rack and install the mounting

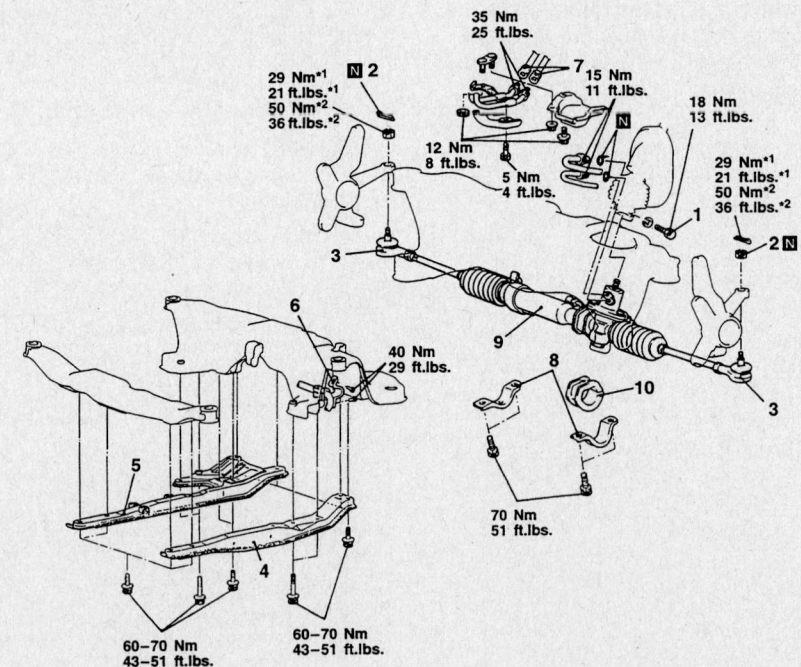

1. Joint assembly and gear box connecting bolt
2. Cotter pin
3. Tie-rod end and knuckle connecting nut
4. Left member
5. Right member
6. Stabilizer bar bracket
7. Connection of steering gear box with 4WS oil line
8. Clamp
9. Gear box assembly
10. Mounting rubber

NOTE
*1: FWD
*2: AWD

Power steering rack and related components—Diamante

7923PGA5

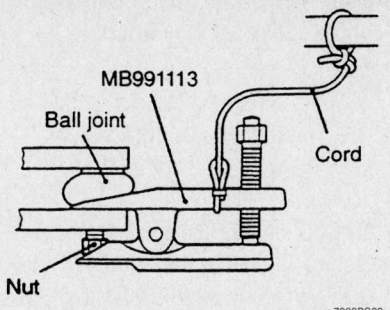

Tie rod tool installation

bolts, tightening bolts to 51 ft. lbs. (70 Nm). When installing the rubber rack mounts, align the projection of the mounting rubber with the indentation in the crossmember. Install the pinch bolt.

- Pressure and return lines to the rack and to the rear pump, if equipped
- Frame members and tighten the bolts to 43–51 ft. lbs. (60–70 Nm)
- Tie rods and install new cotter pins
- Transfer case and front exhaust pipe

4. Refill the reservoir and bleed the system.

5. Perform a front end alignment.

Eclipse

2002–2003

➡️**Prior to removal of the steering rack, center the front wheels and remove the ignition key. Failure to do so may damage the SRS clock spring and render SRS system inoperative.**

1. Before servicing the vehicle, refer to the Precautions Section.
2. Install or connect the following:
   - Negative battery cable
3. Drain the power steering fluid.
   - Stabilizer bar
   - Windshield washer reservoir
   - Pinch bolt from the joint assembly
   - Fluid lines from the steering rack
   - Tie rod ends from the steering knuckles
   - Left and right stays (supports)
4. Support the engine and remove the center member.
   - Clamp and the mounting bolts
   - Left lower compression arm from the body side of the vehicle and support it with wire or string
   - Steering rack from the joint assembly and remove the rack from the left side of the vehicle

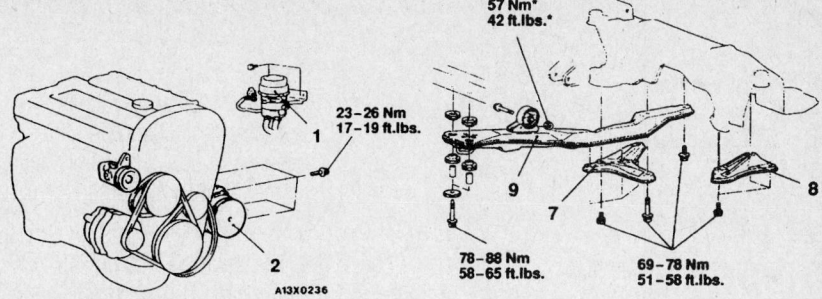

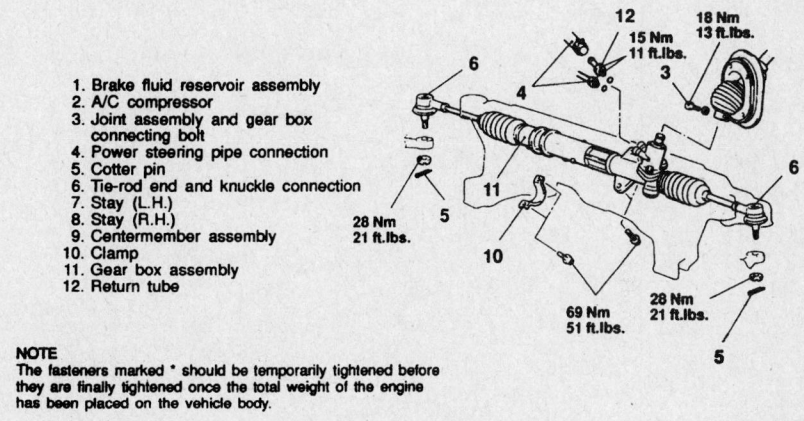

1. Brake fluid reservoir assembly
2. A/C compressor
3. Joint assembly and gear box connecting bolt
4. Power steering pipe connection
5. Cotter pin
6. Tie-rod end and knuckle connection
7. Stay (L.H.)
8. Stay (R.H.)
9. Centermember assembly
10. Clamp
11. Gear box assembly
12. Return tube

NOTE
The fasteners marked * should be temporarily tightened before they are finally tightened once the total weight of the engine has been placed on the vehicle body.

Power steering rack and related components—2002–2003 Eclipse

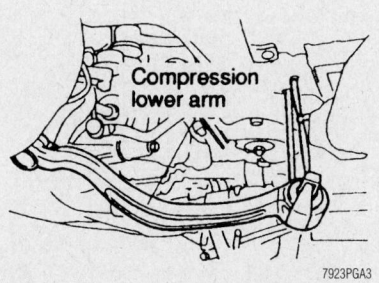

Disconnect the lower compression arm from the body—2002–2003 Eclipse

To install:

5. Position the steering rack in the vehicle and install the clamp and the mounting bolts. Be sure the rack is centered before connecting it to the joint assembly.

6. Install or connect the following:
   - Left lower compression arm to the body
   - Center member
   - Left and right stays and remove the engine support fixture or jack
   - Tie rods to the steering knuckles
   - Fluid lines to the steering rack
   - Pinch bolt in the joint assembly
   - Stabilizer bar and the windshield washer reservoir

7. Safely lower the vehicle.
8. Connect the negative battery cable.
9. Refill and bleed the power steering system.
10. Perform a front end alignment.

2004–2005

➡️**Prior to removal of the steering rack, center the front wheels and remove the ignition key. Failure to do so may damage the SRS clock spring and render SRS system inoperative.**

1. Before servicing the vehicle, refer to the Precautions Section.
2. Disconnect the negative battery cable. Remove the driver's side air bag module.
3. Remove the steering wheel. Remove the column lower cover. Remove the clock spring.
4. Drain the power steering fluid.
5. Remove the front exhaust pipe. Remove the stabilizer bar. Remove the roll stopper bracket.
6. Remove the steering rack cover assembly. Remove the steering shaft assembly and rack retaining bolt.
7. On the 2.4L engine, remove the stay.
8. Disconnect the tie rod end and knuckle connection. Disconnect and plug the pressure tube connection.

9. Remove the power steering rack retaining clamp. Remove the assembly from the vehicle.

To install:

10. Position the steering rack in the vehicle and install the clamp and the mounting bolts. Be sure the rack is centered before connecting it to the joint assembly.

➡**When installing the clock spring be sure that the springs mating marks are properly aligned. If not the steering wheel may not rotate completely during a turn, or the flat cable in the clock spring could be damaged. This would** prevent normal SRS operation and possibly cause serious injury to the driver.

11. After aligning the mating surfaces of the clock spring, turn the front wheels to the straight ahead position. Install the clock spring to the column switch.

➡**Turn the clock spring clockwise fully. Then turn it back approximately 3 turns counterclockwise to align the mating marks.**

12. Continue the installation in the reverse order of the removal procedure.

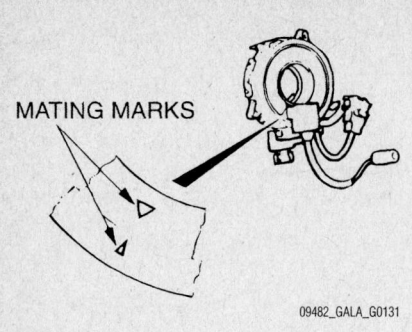

MATING MARKS

09482_GALA_G0131

Clock spring mating marks—2004–2006 Eclipse

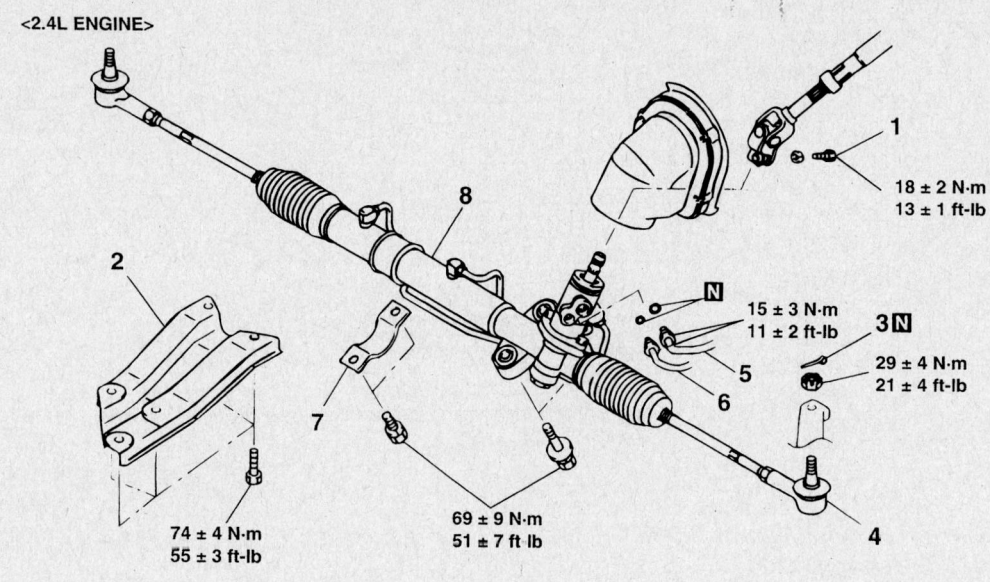

<2.4L ENGINE>

1 — 18 ± 2 N·m / 13 ± 1 ft-lb

2

8

7

5 — 15 ± 3 N·m / 11 ± 2 ft-lb

3 N — 29 ± 4 N·m / 21 ± 4 ft-lb

6

4

74 ± 4 N·m / 55 ± 3 ft-lb

69 ± 9 N·m / 51 ± 7 ft-lb

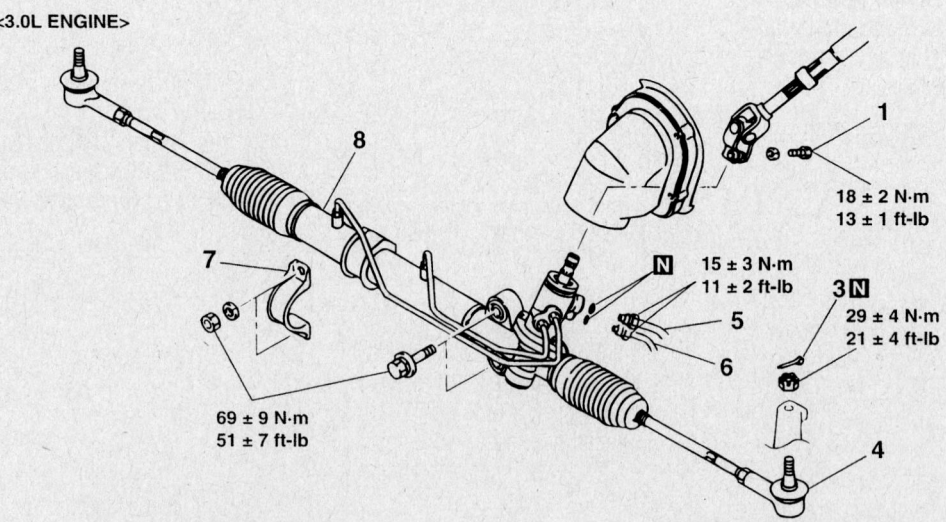

<3.0L ENGINE>

8

1 — 18 ± 2 N·m / 13 ± 1 ft-lb

7

5 — 15 ± 3 N·m / 11 ± 2 ft-lb

3 N — 29 ± 4 N·m / 21 ± 4 ft-lb

6

4

69 ± 9 N·m / 51 ± 7 ft-lb

1. STEERING SHAFT ASSEMBLY AND GEAR BOX CONNECTING BOLT
2. STAY <2.4L ENGINE>
3. COTTER PIN
4. TIE ROD END AND KNUCKLE CONNECTION
5. RETURN HOSE CONNECTION
6. PRESSURE TUBE CONNECTION
7. CYLINDER CLAMP
8. GEAR BOX ASSEMBLY

09482_GALA_G0132

Power steering rack and related components—2004–2005 Eclipse

13. Check and adjust the front end alignment, as required.

2006

➥**Prior to removal of the steering rack, center the front wheels and remove the ignition key. Failure to do so may damage the SRS clock spring and render SRS system inoperative.**

1. Before servicing the vehicle, refer to the Precautions Section.

2. Disconnect the negative battery cable. Remove the front under cover.

3. Remove the center member assembly. Remove the lower arm assembly.

4. Disconnect the stabilizer link and stabilizer bar. Remove the rear roll stopper.

5. Remove the driver's side air bag module. Remove the steering wheel. Remove the column lower cover. Remove the clock spring.

6. Drain the power steering fluid. Remove the steering shaft pad.

7. Disconnect the steering column assembly and steering rack connection. Remove the rear roll stopper connecting bolt.

8. Remove the front axle crossmember bracket. Remove the return tube clamp bolt and nut. Remove the pressure tube clamp bolt and nut.

9. Disconnect and plug the return tube connection. Disconnect and plug the pressure hose connection. Discard the O-ring.

10. Disconnect the pinch clamp. Pinch the steering column shaft clip with a pliers, and pull upward on the shaft to disengage the steering column assembly.

➥**If the steering column shaft is removed accidentally, remove the steering column assembly and be sure to insert the steering column shaft into the steering column.**

11. From inside the vehicle loosen the three clips from the body panel, near the steering column to floor hole.

12. Use a transmission jack to hold the crossmember, and then remove the crossmember mounting nuts and bolts. Lower the crossmember with the rear roll stopper, then the stabilizer bar and steering rack.

13. Remove the steering column dash panel cover. Remove the steering rack linkage protector, on the 3.8L engine.

14. Remove the power steering rack bracket. Remove the rack retaining bolts. Remove the assembly from the vehicle.

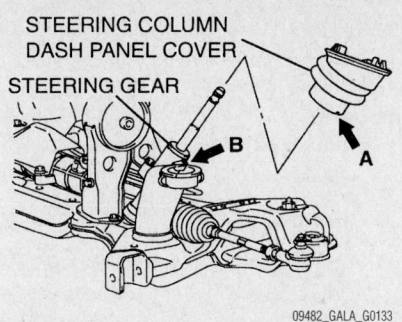

09482_GALA_G0133

Steering gear pinch cover installation—2006 Eclipse and 2004–2006 Galant

To install:

15. Position the steering rack in the vehicle and install the clamp and the mounting bolts. Be sure the rack is centered before connecting it to the joint assembly.

16. When installing the steering column dash panel cover pinch clamp, align the steering column dash panel cover notch "A" with the steering gear lug (arrow "B" and then install the steering column dash panel cover to the steering gear.

➥**When installing the clock spring be sure that the springs mating marks are properly aligned. If not the steering wheel may not rotate completely during a turn, or the flat cable in the clock spring could be damaged. This would prevent normal SRS operation and possibly cause serious injury to the driver.**

17. After aligning the mating surfaces of the clock spring, turn the front wheels to the straight ahead position. Install the clock spring to the column switch.

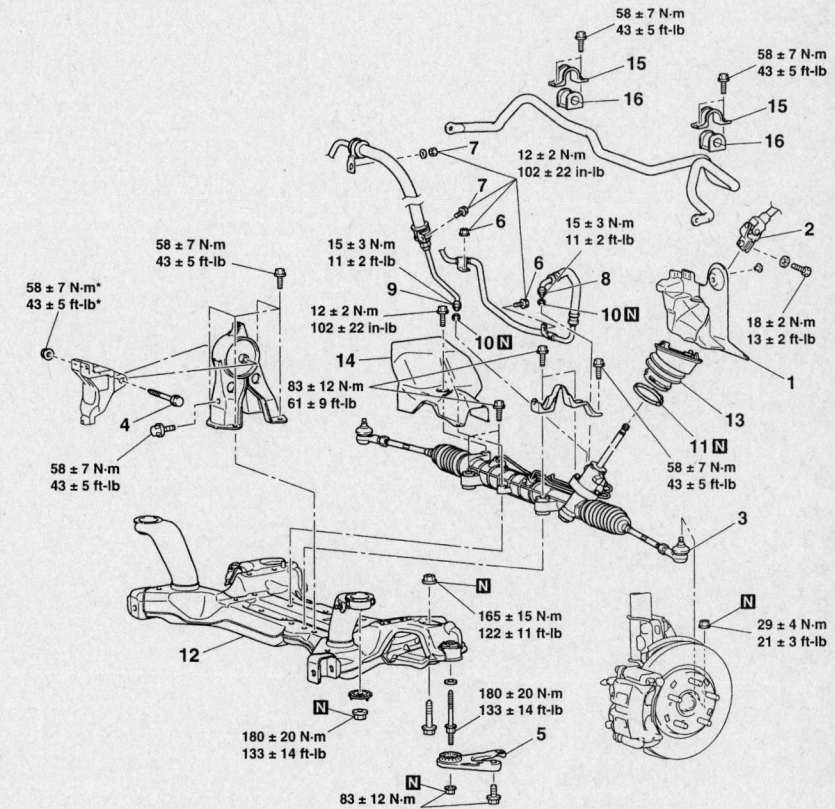

1. STEERING SHAFT PAD
2. STEERING COLUMN ASSEMBLY AND STEERING GEAR CONNECTION
3. TIE ROD END AND KNUCKLE CONNECTION
4. REAR ROLL STOPPER CONNECTING BOLT
5. FRONT AXLE CROSSMEMBER STAY
6. RETURN TUBE CLAMP BOLT AND NUT
7. PRESSURE TUBE CLAMP BOLT AND NUT
8. RETURN TUBE CONNECTION
9. PRESSURE HOSE CONNECTION
10. O-RING
11. PINCH CLAMP
12. CROSSMEMBER ASSEMBLY
13. STEERING COLUMN DASH PANEL COVER
14. STEERING GEAR AND LINKAGE PROTECTOR <3.8L ENGINE>
15. STABILIZER BRACKET
16. STABILIZER BUSHING

09482_GALA_G0134

Power steering rack and related components—2006 Eclipse

➡Turn the clock spring clockwise fully. Then turn it back approximately ¾ turns counterclockwise to align the mating marks.

18. Continue the installation in the reverse order of the removal procedure.

19. Check and adjust the front end alignment, as required.

Galant

2002–2003

✴✴ WARNING

Prior to removal of the steering rack, center the front wheels and remove the ignition key. Failure to do so may damage the Supplemental Restraint System (SRS) clock spring and render SRS system inoperative.

1. Before servicing the vehicle, refer to the Precautions Section.
2. Install or connect the following:
   • Negative battery cable
   • Both front wheel assemblies
   • Bolt holding lower steering column joint to the rack and pinion input shaft
   • Stabilizer bar
   • Cotter pins and the tie rod ends from the steering knuckle
3. On vehicles equipped with Electronic Control Power steering (EPS), disconnect the wiring harness from the solenoid connector.
4. Locate the 2 triangular braces near the crossmember and remove both.
5. Support the center crossmember. Remove the through-bolt from the front round roll stopper and remove the bolts securing the center crossmember.
6. Remove the center crossmember.
7. Properly support the engine and remove the rear roll stopper through-bolt.
8. Remove or disconnect the following:
   • Power steering fluid pressure pipe and return hose from the rack fittings. Plug the fittings to prevent excessive fluid leakage.
   • Clamp bolts and the 2 bolts securing the rack assembly to the chassis
   • Rack and pinion steering assembly and its rubber mounts

➡When removing the rack and pinion assembly, tilt the assembly to the vehicle side of the compression lower arm and remove from the left side of the vehicle.

To install:

9. Center the rack assembly and insert the pinion into the steering column shaft.
10. Install or connect the following:
    • Rack and with the mounting bolts. Torque the mounting bolts to 51 ft. lbs. (69 Nm).
    • Pinch bolt and tighten the bolt to 13 ft. lbs. (18 Nm)
    • Power steering fluid lines to the rack and tighten the pressure hose fitting to 11 ft. lbs. (15 Nm). Secure the return hose with the clamp.
11. Raise the engine into position. Install the rear roll stopper through-bolt and tighten to 32 ft. lbs. (43 Nm).
12. Raise the crossmember into position. Install the center member mounting bolts and tighten the front bolts to 58–65 ft. lbs. (78–88 Nm) and the rear bolt to 51–58 ft. lbs. (69–78 Nm).
13. Install or connect the following:
    • Front roll stopper bolt and tighten the nut to 32 ft. lbs. (43 Nm)
    • 2 triangular braces and tighten the mounting bolts to 50–56 ft. lbs. (69–78 Nm)

• Stabilizer bar
• Tie rod ends and tighten nuts to 20 ft. lbs. (27 Nm)
14. On vehicles equipped with EPS, connect the wiring harness to the solenoid connector.
15. Install the wheel assemblies and lower the vehicle.
16. Refill the reservoir with power steering fluid and bleed the system.
17. Perform a front end alignment.

2004–2006

➡Prior to removal of the steering rack, center the front wheels and remove the ignition key. Failure to do so may damage the SRS clock spring and render SRS system inoperative.

1. Before servicing the vehicle, refer to the Precautions Section.
2. Disconnect the negative battery cable. Drain the power steering fluid. Remove the front under cover.
3. Remove the center crossmember. Remove the lower arm assembly.
4. Remove the driver's side air bag module. Remove the steering wheel.

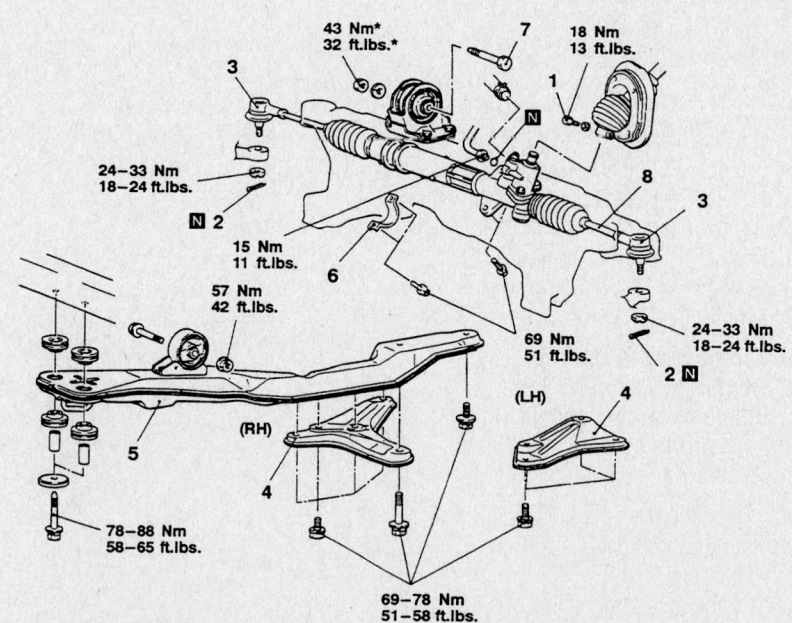

1. Joint assembly and gear box connecting bolt
2. Cotter pin
3. Connection for tie rod end and knuckle
4. Stay
5. Center member assembly
6. Clamp
7. Bolt
8. Gear box assembly

Caution
The fasteners marked * should be temporarily tightened before they are finally tightened once the total weight of the engine has been placed on the vehicle body.

7923PGA4

Power steering rack and related components—2002–2003 Galant

Remove the column lower cover. Remove the clock spring.

5. Remove the floor console assembly. Remove the front scuff plate and cowl side trim. Remove the trunk lid opener cover.

6. Remove the accelerator pedal stopper. Remove the front floor carpet.

7. Disconnect the stabilizer link and stabilizer bar. Remove the steering shaft pad.

8. Disconnect the steering column assembly and steering rack connection. Remove the rear roll stopper connecting bolt.

9. Remove the front axle crossmember bracket. Remove the return tube clamp bolt and nut. Remove the pressure tube clamp bolt and nut.

10. Disconnect and plug the return tube connection. Disconnect and plug the pressure hose connection. Discard the O-ring.

11. Disconnect the pinch clamp. Pinch the steering column shaft clip with a pliers, and pull upward on the shaft to disengage the steering column assembly.

➡**If the steering column shaft is removed accidentally, remove the steering column assembly and be sure to insert the steering column shaft into the steering column.**

12. From inside the vehicle loosen the three clips from the body panel, near the steering column to floor hole.

13. Use a transmission jack to hold the crossmember, and then remove the crossmember mounting nuts and bolts. Lower the crossmember with the rear roll stopper, then the stabilizer bar and steering rack.

14. Remove the steering column dash panel cover. Remove the steering rack linkage protector, on the 3.8L engine.

15. Remove the power steering rack bracket. Remove the rack retaining bolts. Remove the assembly from the vehicle.

To install:

16. Position the steering rack in the vehicle and install the clamp and the mounting bolts. Be sure the rack is centered before connecting it to the joint assembly.

17. When installing the steering column dash panel cover pinch clamp, align the steering column dash panel cover notch "A" with the steering gear lug (arrow "B" and then install the steering column dash panel cover to the steering gear.

➡**When installing the clock spring be sure that the springs mating marks are properly aligned. If not the steering wheel may not rotate completely during a turn, or the flat cable in the clock spring could be damaged. This would**

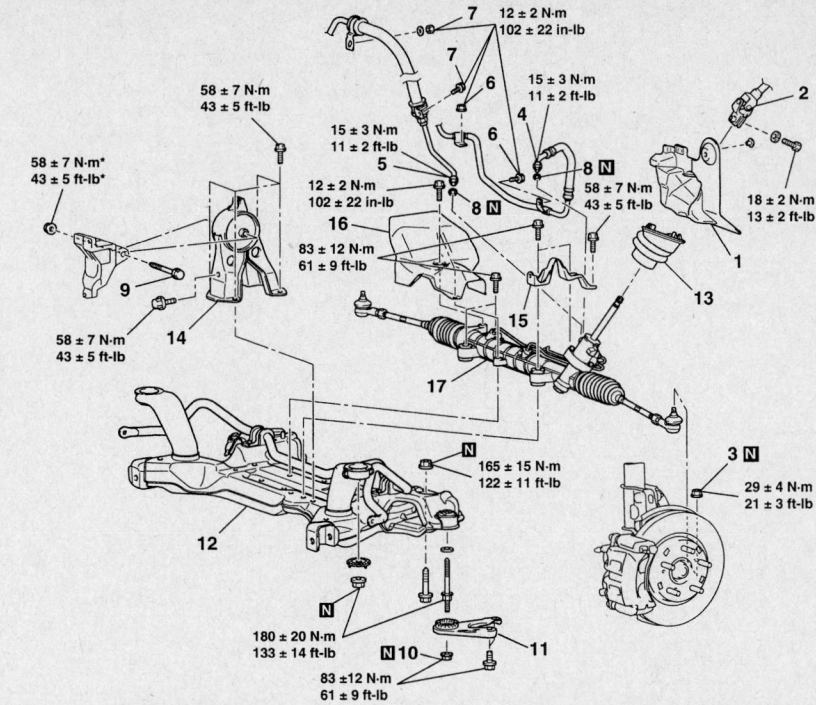

1. STEERING SHAFT PAD
2. STEERING COLUMN SHAFT ASSEMBLY AND STEERING GEAR CONNECTION
3. JAM NUT (TIE ROD END AND KNUCKLE CONNECTION)
4. RETURN TUBE CONNECTION
5. PRESSURE HOSE CONNECTION
6. RETURN TUBE CLAMP
7. PRESSURE HOSE CLAMP
8. O-RING
9. REAR ROLL STOPPER CONNECTING BOLT
10. JAM NUT
11. FRONT AXLE CROSSMEMBER STAY
12. CROSSMEMBER ASSEMBLY
13. STEERING COLUMN DASH PANEL COVER
14. REAR ROLL STOPPER
15. POWER STEERING GEAR BRACKET
16. STEERING GEAR AND LINKAGE PROTECTOR <3.8L ENGINE>
17. POWER STEERING GEAR AND LINKAGE

09482_GALA_G0135

Power steering rack and related components—2004–2006 Galant

prevent normal SRS operation and possibly cause serious injury to the driver.

18. After aligning the mating surfaces of the clock spring, turn the front wheels to the straight ahead position. Install the clock spring to the column switch.

➡**Turn the clock spring clockwise fully. Then turn it back approximately ¾ turns counterclockwise to align the mating marks.**

19. Continue the installation in the reverse order of the removal procedure.

20. Check and adjust the front end alignment, as required.

Lancer

➡**Prior to removal of the steering rack, center the front wheels and remove the ignition key. Failure to do so may damage the SRS clock spring and render SRS system inoperative.**

1. Before servicing the vehicle, refer to the Precautions Section.

2. Disconnect the negative battery cable. Drain the power steering fluid.

3. Remove the driver's side air bag module. Remove the steering wheel. Remove the column lower cover. Remove the clock spring.

4. Disconnect the front exhaust pipe.

5. Separate the steering rack ends from their mounting.

6. Remove the crossmember. Remove the rear roll stopper.

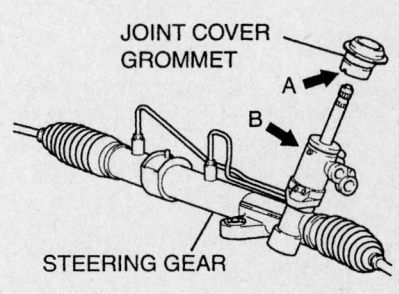

JOINT COVER GROMMET

A

B

STEERING GEAR

09482_GALA_G0136

Joint cover installation—Lancer and Evolution

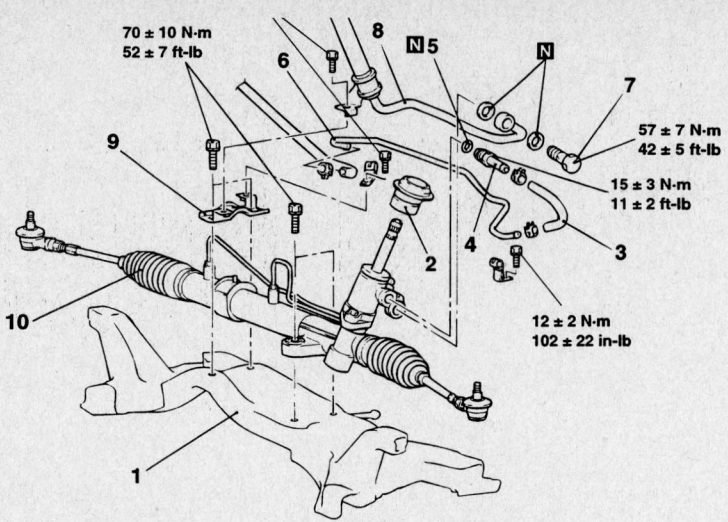

70 ± 10 N·m
52 ± 7 ft-lb

57 ± 7 N·m
42 ± 5 ft-lb

15 ± 3 N·m
11 ± 2 ft-lb

12 ± 2 N·m
102 ± 22 in-lb

1. CROSSMEMBER
2. JOINT COVER GROMMET
3. RETURN HOSE
4. RETURN TUBE
5. O-RING
6. RETURN TUBE
7. EYE BOLT
8. PRESSURE HOSE ASSEMBLY
9. CLAMP
10. STEERING GEAR AND LINKAGE

9357QG28

Power steering rack and related components—Lancer

7. Remove the joint cover grommet. Disconnect and plug the fluid return hose.

8. Disconnect the return tube and remove it. Discard the O-ring.

9. Remove the eye bolt. Remove the pressure hose assembly clamp.

10. Remove the steering rack retaining bolts. Remove the rack from the vehicle.

To install:

11. Position the steering rack in the vehicle and install the clamp and the mounting bolts. Be sure the rack is centered before connecting it to the joint assembly.

12. To install the joint cover grommet, align the joint cover grommet notch (arrow "A") with the steering gear lug (arrow "B"). Install the steering joint cover to the steering gear.

13. Continue the installation in the reverse order of the removal procedure.

14. Check and adjust the front end alignment, as required.

Evolution

➡**Prior to removal of the steering rack, center the front wheels and remove the ignition key. Failure to do so may damage the SRS clock spring and render SRS system inoperative.**

1. Before servicing the vehicle, refer to the Precautions Section.

2. Disconnect the negative battery cable. Drain the power steering fluid.

3. Remove the driver's side air bag

module. Remove the steering wheel. Remove the column lower cover. Remove the clock spring.

4. Remove the front floor carpet. Remove the center bar.

5. Disconnect the lower arm ball joint connection. Disconnect the stabilizer link. Disconnect the lower arm assembly. Remove the shaft cover.

6. Remove the steering shaft assembly and steering gear connecting bolt.

7. Remove the tie rod end and knuckle connection self locking nut.

8. Disconnect the return hose and the return line. Discard the O-ring.

9. Disconnect the high pressure hose connection. Discard the gasket.

10. Remove the rear roll stopper rod connecting bolt.

11. Position a floor jack under the crossmember. Remove the crossmember retaining bolts and remove the crossmember.

12. Remove the engine rear roll stopper rod and engine rear roll stopper rod bracket.

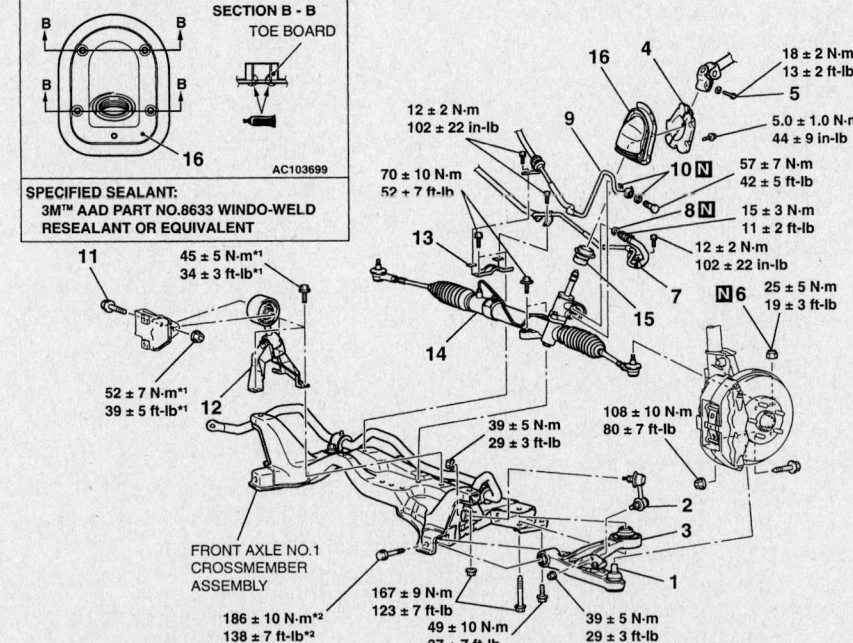

SECTION B - B
TOE BOARD

AC103699

SPECIFIED SEALANT:
3M™ AAD PART NO.8633 WINDO-WELD
RESEALANT OR EQUIVALENT

45 ± 5 N·m*1
34 ± 3 ft-lb*1

52 ± 7 N·m*1
39 ± 5 ft-lb*1

12 ± 2 N·m
102 ± 22 in-lb

70 ± 10 N·m
52 ± 7 ft-lb

18 ± 2 N·m
13 ± 2 ft-lb

5.0 ± 1.0 N·m
44 ± 9 in-lb

57 ± 7 N·m
42 ± 5 ft-lb

15 ± 3 N·m
11 ± 2 ft-lb

12 ± 2 N·m
102 ± 22 in-lb

25 ± 5 N·m
19 ± 3 ft-lb

108 ± 10 N·m
80 ± 7 ft-lb

39 ± 5 N·m
29 ± 3 ft-lb

FRONT AXLE NO.1
CROSSMEMBER
ASSEMBLY

167 ± 9 N·m
123 ± 7 ft-lb

186 ± 10 N·m*2
138 ± 7 ft-lb*2

49 ± 10 N·m
37 ± 7 ft-lb

39 ± 5 N·m
29 ± 3 ft-lb

1. LOWER ARM BALL JOINT CONNECTION
2. STABILIZER LINK
3. LOWER ARM ASSEMBLY
4. SHAFT COVER
5. STEERING COLUMN SHAFT ASSEMBLY AND STEERING GEAR CONNECTING BOLT
6. SELF LOCKING NUT (TIE ROD END AND KNUCKLE CONNECTION)
7. RETURN HOSE AND RETURN TUBE
8. O-RING
9. PRESSURE HOSE CONNECTION
10. GASKET
11. ENGINE REAR ROLL STOPPER ROD CONNECTING BOLT
12. ENGINE REAR ROLL STOPPER ROD AND ENGINE REAR ROLL STOPPER ROD BRACKET
13. CLAMP
14. STEERING GEAR
15. JOINT COVER GROMMET
16. COVER ASSEMBLY

09482_GALA_G0137

Power steering rack and related components—Evolution

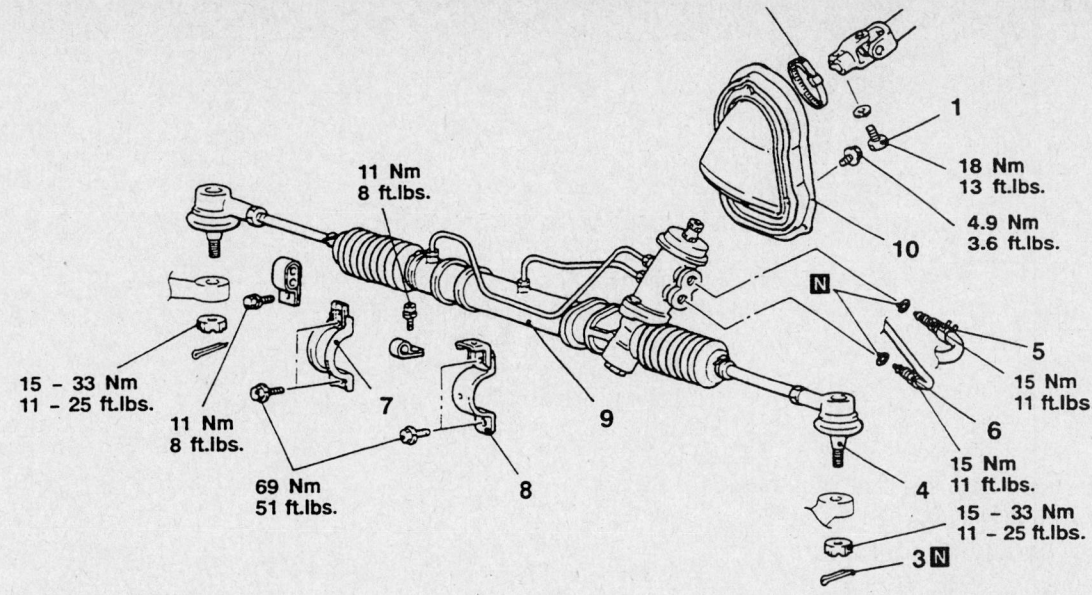

11 Nm
8 ft.lbs.

18 Nm
13 ft.lbs.

4.9 Nm
3.6 ft.lbs.

10

1

15 – 33 Nm
11 – 25 ft.lbs.

11 Nm
8 ft.lbs.

7

N

5

15 Nm
11 ft.lbs.

6

69 Nm
51 ft.lbs.

8

9

15 Nm
11 ft.lbs.

4

15 – 33 Nm
11 – 25 ft.lbs.

3 N

1. Steering shaft assembly and gear box connecting bolt
2. Band
3. Cotter pin
4. Tie-rod end and knuckle connection
5. Return tube connection

6. Pressure tube connection
7. Cylinder clamp
8. Gear housing clamp
9. Gear box assembly
10. Steering cover assembly

7923PGA1

Power steering rack and related components —Mirage

13. Remove the joint cover grommet. Remove the cover assembly.

14. Remove the steering rack mounting bolts. Remove the steering rack from the vehicle.

15. Remove the steering rack retaining bolts. Remove the rack from the vehicle.

To install:

16. Position the steering rack in the vehicle and install the clamp and the mounting bolts. Be sure the rack is centered before connecting it to the joint assembly.

17. To install the joint cover grommet, align the joint cover grommet notch (arrow "A") with the steering gear lug (arrow "B"). Install the steering joint cover to the steering gear.

18. Continue the installation in the reverse order of the removal procedure.

19. Check and adjust the front end alignment, as required.

Mirage

➡**Prior to removal of the steering rack, center the front wheels and remove the ignition key. Failure to do so may damage the Supplemental Restraint System**

(SRS) clockspring and render SRS system inoperative.

1. Before servicing the vehicle, refer to the Precautions Section.

2. Drain the power steering system.

3. Remove or disconnect the following:
   • Battery negative cable. Raise the vehicle and support safely
   • Heated Oxygen (HO$_2$S) sensor and remove the front exhaust pipe, if necessary

4. Properly support the engine.
   • Both roll stopper mounting bolts and the 4 center member installation bolts. Remove the center member.
   • Center member

➡**Matchmark the pinion input shaft of the rack to the lower steering column joint for installation purposes.**

5. Remove or disconnect the following:
   • Pinch bolt holding the lower steering column joint to the rack and pinion input shaft
   • Cotter pins and disconnect the tie rod ends from the steering knuckle
   • Power steering fluid pressure pipe

and return hose from the rack fittings
   • Rack and pinion steering assembly and its rubber mounts from the right side of the vehicle

To install:

6. Align the matchmarks of the input shaft and install the rack to the vehicle.

7. Secure the rack using the retainer clamps and bolts. Torque the bolts to 52 ft. lbs. (70 Nm).

8. Torque the steering column pinch bolt to 13 ft. lbs. (18 Nm).

9. Using new O-rings, connect the power steering fluid lines to the rack fittings.

10. Install or connect the following:
    • Center member
    • Front exhaust pipe
    • HO$_2$S sensor
    • Tie rod ends to the steering knuckles and tighten the castle nuts to 25 ft. lbs. (34 Nm). Install new cotter pins.
    • Wheels and connect the negative battery cable

11. Refill the reservoir and bleed the system.

12. Perform a front end alignment.

FRONT SUSPENSION

Strut

REMOVAL & INSTALLATION

Diamante

1. Before servicing the vehicle, refer to the Precautions Section.
2. Disconnect the negative battery cable.
3. Raise and safely support the vehicle.
4. Remove the brake hose and the tube bracket.

➡**Do not pry the brake hose and tube clamp away when removing it.**

5. If equipped with Anti-Lock Brake (ABS), disconnect the front speed sensor mounting clamp from the strut.
6. Support the lower arm and remove the strut to knuckle bolts. Use a piece of wire to suspend the knuckle to keep the weight off the brake hose.
7. If equipped with Active Electronic Control Suspension (Active-ECS) perform the following:
   a. Loosen the nut that secures the air line to the to the top of the strut and discard the O-ring.
   b. Remove the bolts that secure the actuator to the top of the strut and remove the component. Disconnect the wiring harness.

➡**Before removing the top bolts, make matchmarks on the body and the strut insulator for proper reassembly.**

8. Remove the strut upper nuts and remove the strut assembly from the vehicle.
9. Compress the coil spring using a spring compressor until the spring just comes away from one of the seats.
10. Remove or disconnect the following:
    - Center nut from the strut and remove the upper mounting bracket and bushings
    - Coil spring

To install:

11. Install or connect the following:
    - Compressed spring on the strut assembly
    - Upper bushings and the mounting bracket
    - Nut and tighten it to 43 ft. lbs. (59 Nm)
    - Strut to the vehicle and tighten the upper mounting nuts to 33 ft. lbs. (45 Nm)

12. Align the strut to the knuckle and connect with the mounting bolts. Torque the mounting bolts to 70–76 ft. lbs. (90–105 Nm).

13. If equipped with Active-ECS, perform the following:
    a. Install the air line with a new O-ring.
    b. Install the actuator to the top of the strut. Connect the wiring harness.

14. Install or connect the following:
    - Brake hose bracket and the ABS clamp, if equipped
    - Wheel and tire assembly

15. Perform a front end alignment.

Eclipse

2002–2003

1. Before servicing the vehicle, refer to the Precautions Section.
2. Raise and safely support the vehicle.
3. Remove or disconnect the following:
   - Front wheel
   - 3 upper shock absorber mounting nuts. Do not remove the larger nut in the center of the strut at this time.
   - Stabilizer link from the damper fork
   - Damper fork mounting bolt
   - Shock absorber assembly from the vehicle

4. Use a coil spring compressor and compress the coil spring.
5. While holding the piston rod, remove the self-locking nut.
6. Remove or disconnect the following:
   - Upper bracket assembly and spring pad
   - Collar, upper bushing, cup assembly, bump rubber and dust cover
   - Coil spring from the shock absorber

To install:

7. Align the end of the coil spring with the stepped part of the spring seat and

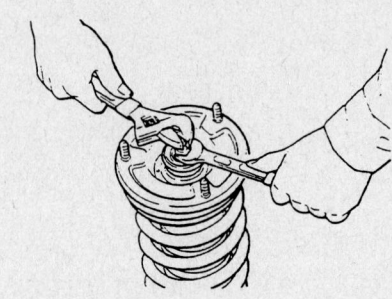

Removing the self-locking nut— 2002–2003 Eclipse

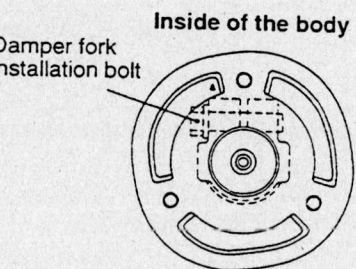

Upper bracket assembly alignment— 2002–2003 Eclipse

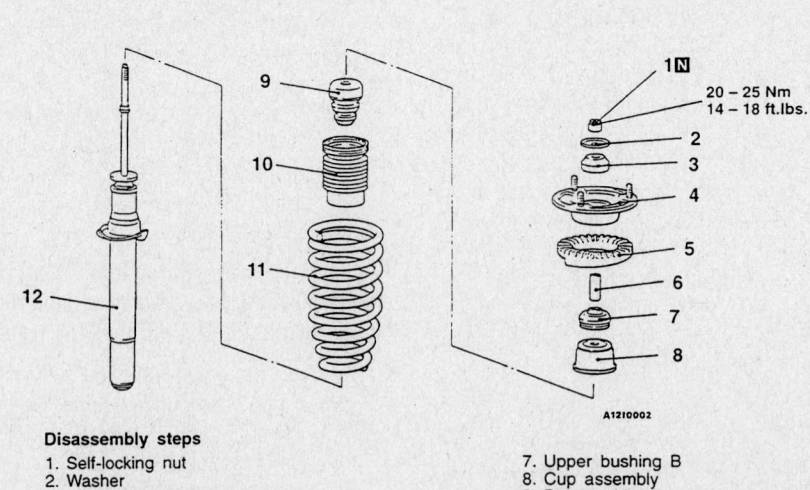

Disassembly steps
1. Self-locking nut
2. Washer
3. Upper bushing A
4. Upper bracket assembly
5. Upper spring pad
6. Collar
7. Upper bushing B
8. Cup assembly
9. Bump rubber
10. Dust cover
11. Coil spring
12. Shock absorber assembly

Coil spring removal procedure—2002–2003 Eclipse

install the compressed coil spring on the shock.

8. Install the dust cover, bump rubber, cup assembly, upper bushing, collar, upper spring pad and bracket assembly on the strut.

9. Install or connect the following:
   • Upper bushing and washer on the piston rod
   • New self-locking nut on the piston rod. Temporarily tighten the nut.

10. Carefully remove the spring compressor from the spring. Torque the self-locking nut to 16 ft. lbs. (25 Nm).

11. Position the shock absorber assembly in the damper fork and install the mounting bolt.

12. Pass the studs in the upper bracket assembly through the holes in the inner fender and install the 3 mounting nuts.

13. Connect the stabilizer link to the damper fork.

14. Install the wheel assembly.

15. Safely lower the vehicle to the floor.

16. Check the front wheel alignment and adjust if necessary.

2004–2006

1. Before servicing the vehicle, refer to the Precautions Section.

2. Raise and safely support the vehicle.

3. On 2004–2005 Spyder, remove the strut tower bar.

4. Disconnect the stabilizer link. Remove the brake hose bracket.

5. On vehicles equipped with ABS, remove the wheel speed sensor clamp.

6. Remove the lower strut bolts and nuts. Remove the upper strut nuts.

7. Remove the strut assembly from the vehicle.

To install:

8. Installation is the reverse of the removal procedure.

9. Be sure to check and adjust the front end alignment, as required.

Galant

2002–2003

1. Before servicing the vehicle, refer to the Precautions Section.

2. Disconnect the negative battery cable.

3. Raise and safely support vehicle.

4. Remove or disconnect the following:

   • Appropriate wheel assembly
   • Sway bar link from the damper fork
   • Damper fork lower through-bolt and upper pinch bolt
   • Damper fork assembly
   • Shock absorber upper nuts and remove the strut assembly from the vehicle

5. Compress the coil spring with a special compression tool.

   • Self-locking nut and washer
   • Upper bushing, upper bracket assembly, the upper spring pad, and the collar
   • Other upper bushing, cup assembly, bump rubber, dust cover, and the coil spring. Carefully remove the coil spring compression tool

To install:

6. Install or connect the following:

   • Compressed coil spring to the shock absorber assembly. Be sure to align the edge of coil spring to the stepped part of the spring seat. Install the dust cover, bump rubber, cup assembly, upper bushing, collar, and upper spring pad.
   • Upper bracket assembly and position it so that the 3 bolts are in the correct position
   • Upper bushing, washer, and locknut. Torque the locknut to 18 ft. lbs. (24 Nm).
   • Shock absorber and tighten the upper mounting nuts to 32 ft. lbs. (44 Nm)

7. Align the shock to the damper fork and install the damper fork. Tighten the lower through-bolt/nut to 65 ft. lbs. (88 Nm) and the upper pinch bolt to 76 ft. lbs. (103 Nm).

8. Connect the sway bar link to the damper fork and tighten the link nut to 29 ft. lbs. (39 Nm).

9. Install the wheel and tire assembly.

10. Perform a front end alignment.

2004–2006

1. Before servicing the vehicle, refer to the Precautions Section.

2. Raise and safely support the vehicle.

3. On GTS remove the strut tower bar bracket.

4. Disconnect the stabilizer link. Remove the brake hose bracket.

5. On vehicles equipped with ABS, remove the wheel speed sensor clamp.

6. Remove the lower strut bolts and nuts. Remove the upper strut nuts.

7. Remove the strut assembly from the vehicle.

To install:

8. Installation is the reverse of the removal procedure.

9. Be sure to check and adjust the front end alignment, as required.

Lancer

2002–2003

1. Before servicing the vehicle, refer to the Precautions Section.

2. Disconnect the negative battery cable.

3. Raise and safely support vehicle.

4. Remove the brake hose and tube bracket retainer bolt and bracket from the front strut. Do not pry the brake hose and tube clamp away when removing.

5. If equipped with ABS, disconnect the

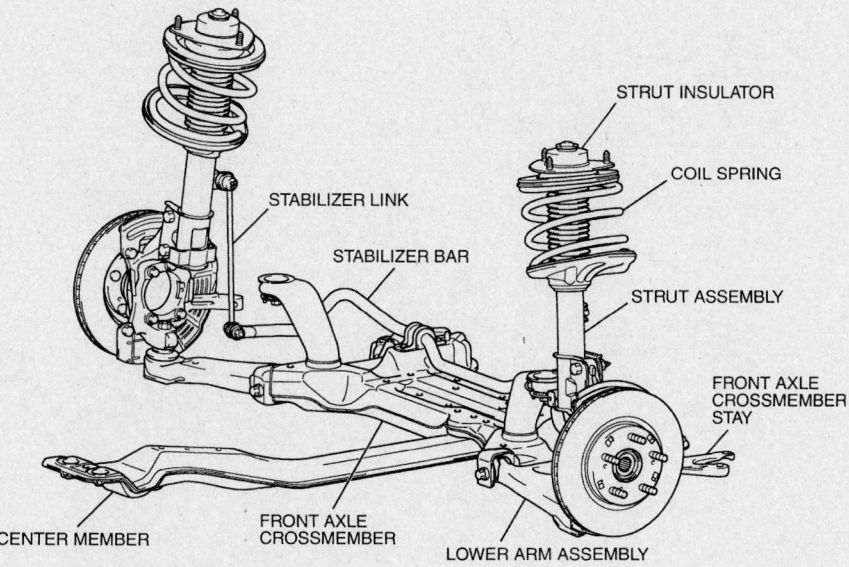

Front suspension and related components—2004–2006 Eclipse and Galant

STABILIZER LINK / STABILIZER BAR / CENTER MEMBER / FRONT AXLE CROSSMEMBER / LOWER ARM ASSEMBLY / STRUT INSULATOR / COIL SPRING / STRUT ASSEMBLY / FRONT AXLE CROSSMEMBER STAY

09482_GALA_G0138

front speed sensor mounting clamp from the strut.

6. Support the lower arm using floor jack. Remove the lower strut to knuckle bolts.

➡**Before removing the top bolts, make matchmarks on the body and the strut insulator for proper reassembly.**

7. Remove the strut upper mounting bolts. Remove the strut assembly from the vehicle.

8. Compress the coil spring using a spring compressor until the spring just comes away from one of the seats.

9. Remove or disconnect the following:
   - Center nut from the strut and remove the upper mounting bracket and bushings
   - Coil spring

To install:

10. Install or connect the following:
   - Compressed spring on the strut assembly
   - Upper bushings and the mounting bracket
   - Nut and tighten it to 43 ft. lbs. (59 Nm)
   - Strut to the vehicle and install the top mounting bolts. Tighten the mounting bolts to 29 ft. lbs. (40 Nm) for Mirage and 33 ft. lbs. (44 Nm) for Lancer.

11. Position the strut on the knuckle and install the mounting bolts. While holding the head of the lower mounting bolt, tighten the nuts to 123 ft. lbs. (167 Nm).

12. Install or connect the following:
   - Brake hose bracket and the ABS clamp, if equipped
   - Wheel and tire assembly

13. Perform a front end alignment.

2004–2006

1. Before servicing the vehicle, refer to the Precautions Section.

2. Raise and safely support the vehicle.

3. Remove the brake hose bracket.

4. On vehicles equipped with ABS, remove the wheel speed sensor clamp.

5. Remove the lower strut bolts and nuts from the knuckle connection. Remove the upper strut nuts.

6. Remove the strut assembly from the vehicle.

To install:

7. Installation is the reverse of the removal procedure.

8. Be sure to check and adjust the front end alignment, as required.

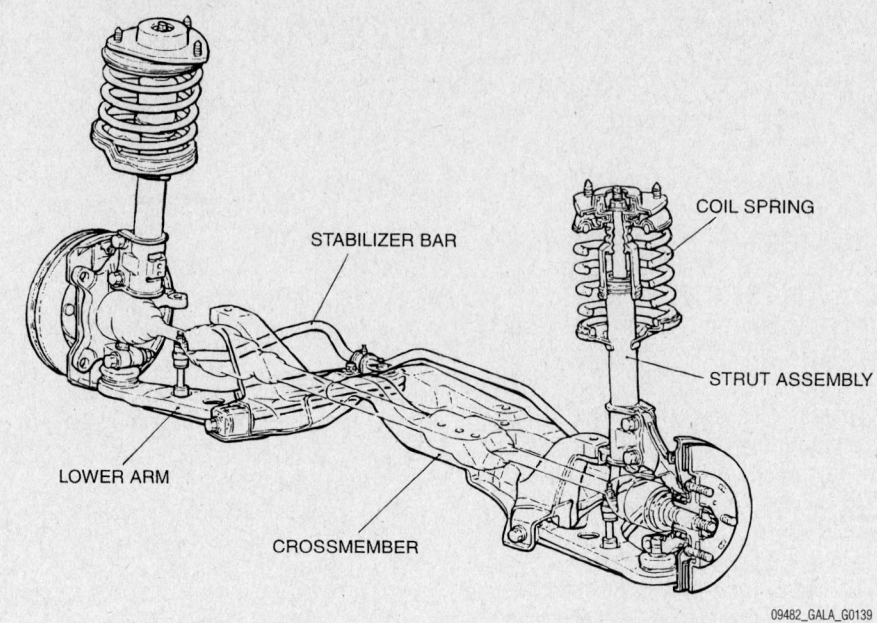

Front suspension and related components—2004–2006 Lancer

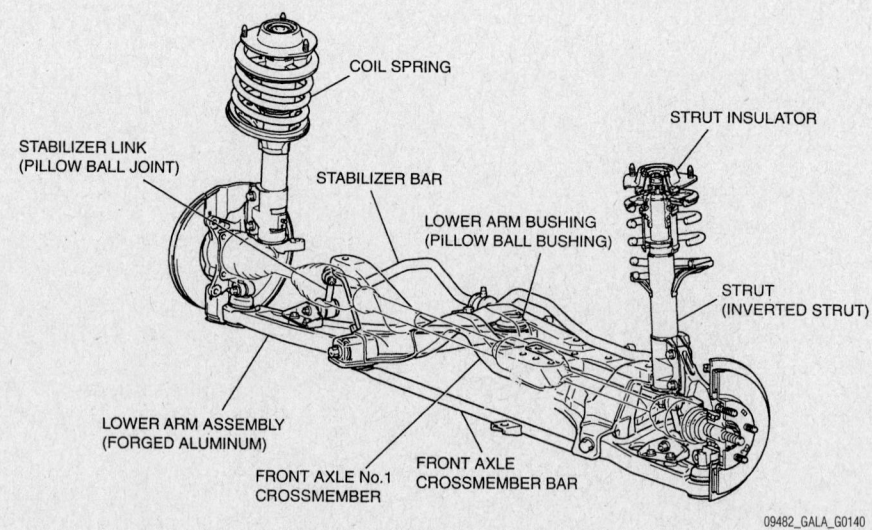

Front suspension and related components—Evolution

Evolution

1. Before servicing the vehicle, refer to the Precautions Section.
2. Raise and safely support the vehicle.
3. Remove the brake hose bracket.
4. Remove the wheel speed sensor harness bracket.
5. Remove the lower strut bolts and nuts from the knuckle connection. Remove the upper strut nuts.
6. Remove the strut assembly from the vehicle.

To install:

7. Installation is the reverse of the removal procedure.
8. Be sure to check and adjust the front end alignment, as required.

Mirage

1. Before servicing the vehicle, refer to the Precautions Section.
2. Disconnect the negative battery cable.
3. Raise and safely support vehicle.
4. Remove the brake hose and tube bracket retainer bolt and bracket from the front strut. Do not pry the brake hose and tube clamp away when removing.
5. If equipped with ABS, disconnect the front speed sensor mounting clamp from the strut.
6. Support the lower arm using floor jack. Remove the lower strut to knuckle bolts.

➡**Before removing the top bolts, make matchmarks on the body and the strut insulator for proper reassembly.**

7. Remove the strut upper mounting bolts. Remove the strut assembly from the vehicle.
8. Compress the coil spring using a spring compressor until the spring just comes away from one of the seats.
9. Remove or disconnect the following:
   • Center nut from the strut and remove the upper mounting bracket and bushings
   • Coil spring

To install:

10. Install or connect the following:
    • Compressed spring on the strut assembly
    • Upper bushings and the mounting bracket
    • Nut and tighten it to 43 ft. lbs. (59 Nm)
    • Strut to the vehicle and install the top mounting bolts. Tighten the mounting bolts to 29 ft. lbs. (40 Nm) for Mirage and 33 ft. lbs. (44 Nm) for Lancer.

11. Position the strut on the knuckle and install the mounting bolts. While holding the head of the lower mounting bolt, tighten the nuts to 80–94 ft. lbs. (110–130 Nm).
12. Install or connect the following:
    • Brake hose bracket and the ABS clamp, if equipped
    • Wheel and tire assembly
13. Perform a front end alignment.

DISASSEMBLY AND REASSEMBLY

Eclipse and Galant—2004–2006

1. Before servicing the vehicle, refer to the Precautions Section.
2. Remove the strut from the vehicle.
3. Position the strut assembly using fixture tools MB991793, MB991794, MB991795 and MB9910830 or equivalent.

➡**Position the strut assembly in the disassembly fixture using the bolt and nut that were removed from the vehicle. When installing the bolt and nut, lightly tighten them by hand.**

4. Compress the coil spring approximately 0.20 inch, using the spring compressor.

➡**Do not use an impact wrench to tighten the strut nut, otherwise the strut nut will be damaged. Vibration of the impact wrench will cause the valve inside the strut to drop out.**

5. If equipped with the 2.4L engine, use special tool MB991682 to secure the strut, and then remove the strut self locking nut using tool MB991681.
6. If equipped with the 3.8L engine, use a hexagon wrench and a pipe to secure the strut, and then remove the strut self locking nut using tool MB991681.
7. Remove the strut insulator, strut bearing, spring upper seat, spring upper pad, damper, coil spring, and spring lower pad.
8. To assemble, engage the three lugs of the spring lower pad into the holes on the strut.
9. If the upper plate and lower plate of the strut have been disassembled, reassemble them.
10. Position the assembly in the holding fixture. Be sure that the bearing is seated correctly.

➡**When the strut piston rod is positioned to the hole of the strut insulator with compressing the coil spring, be careful that your hand is not jammed by the coil spring.**

11. Compress the coil spring slowly using tools MB991793, MB991794, MB991795 and MB9910830 or equivalent.
12. While the coil spring is being compressed, by the special tools, temporarily tighten the strut nut (self locking nut).

➡**Do not use an impact wrench to tighten the strut nut, otherwise the strut nut will be damaged. Vibration of the impact wrench will cause the valve inside the strut to drop out.**

13. If equipped with the 2.4L engine, use special tools MB991681 and MB991682 to tighten the strut nut (self locking nut) to 60–70 ft. lbs. (44–52 Nm).
14. If equipped with the 3.8L engine, use special tool MB991681, a hexagon wrench and a pipe to tighten the strut (self locking nut) to 60–70 ft. lbs. (44–52 Nm).
15. Install the strut in the vehicle.

Lancer and Evolution—2004–2006

1. Before servicing the vehicle, refer to the Precautions Section.
2. Remove the strut from the vehicle.
3. Position the strut assembly using fixture tools MB991238 and MB9911237 or equivalent.

➡**Do not tighten the special tool bolt too tight or the tool will break. Install the tools evenly and so that the maximum length will be attained within the installation range. Do not use an impact wrench to tighten the bolt of special tool MB991237, otherwise the tool will be damaged.**

4. Using the special tools, compress the spring.

➡**To prevent the piston rod jam nut inside the strut from loosening, do not use an impact wrench when the jam nut is loosened.**

5. Use special tools MB991681 and MB991682 to secure the strut, and then remove the jam nut.
6. Remove the strut insulator assembly, bearing, upper bearing seat, bump rubber or dust cover (depending on the vehicle), coil spring and lower strut pad.
7. If the upper spring seat peeled off the pad adhere the seat and the pad with double sided tape.
8. Position the strut assembly in the holding fixture. Ensure that the bearing is seated correctly.

➡**Do not use an impact wrench to tighten the bolt of the special tool or the tool will break.**

9. While the coil spring is being compressed by the special tools, temporarily tighten the jam nut.

10. Align the hole in the strut assembly lower spring seat with the hole in the upper spring seat. You can use a rod to facilitate the alignment process.

11. Align both ends of the coil spring with the grooves in the spring seat, and then loosen the special tools.

➡ **Do not use an impact wrench to tighten the jam nut, otherwise the jam nut will not be tightened securely.**

12. Use special tools MB991681 and MB991682 and tighten the jam nut to 38–52 ft. lbs. (50–70 Nm).

13. Install the strut in the vehicle.

Stabilizer Bar

REMOVAL & INSTALLATION

Eclipse

2004–2005

1. Before servicing the vehicle, refer to the Precautions Section.

2. Raise and safely support the vehicle.

3. Remove the front exhaust pipe. Remove the center member.

4. On 2.4L engine, remove the stay pan.

5. Remove the lower control arm.

6. Disconnect the stabilizer link. Disconnect the stabilizer bracket and bushing.

7. On the 3.0L engine, turn the steering wheel to the right to remove the left end of the stabilizer bar.

8. Remove the stabilizer bar from the vehicle.

To install:

9. Installation is the reverse of the removal procedure.

10. Check and adjust the front end alignment, as required.

2006

➡ **Prior to removal of the stabilizer bar, center the front wheels and remove the ignition key. Failure to do so may damage the SRS clock spring and render SRS system inoperative.**

1. Before servicing the vehicle, refer to the Precautions Section.

2. Disconnect the negative battery cable. Drain the power steering fluid.

3. Raise and safely support the vehicle. Remove the center member. Remove the lower control arm.

4. Remove the driver's side air bag module. Remove the steering wheel. Remove the column lower cover. Remove the clock spring. Remove the steering damper

5. Disconnect the upper stabilizer link. Remove the rear roll stopper. Remove the steering shaft pad.

6. Disconnect the steering column assembly and steering rack connection. Remove the rear roll stopper connecting bolt. Disconnect the lower stabilizer link.

7. Remove the front axle crossmember bracket. Remove the return tube clamp bolt and nut. Remove the pressure tube clamp bolt and nut.

8. Disconnect and plug the return tube connection. Disconnect and plug the pressure hose connection. Discard the O-ring.

9. Disconnect the pinch clamp. Pinch the steering column shaft clip with a pliers, and pull upward on the shaft to disengage the steering column assembly.

➡ **If the steering column shaft is removed accidentally, remove the steering column assembly and be sure to insert the steering column shaft into the steering column.**

10. From inside the vehicle loosen the three clips from the body panel, near the steering column to floor hole.

11. Use a transmission jack to hold the crossmember, and then remove the crossmember mounting nuts and bolts. Lower the crossmember with the rear roll stopper, then the stabilizer bar and steering rack.

12. Remove the steering column dash panel cover. Remove the steering rack linkage protector, on the 3.8L engine.

13. Remove the power steering rack bracket. Remove the rack retaining bolts. Remove the assembly from the vehicle.

14. Remove the stabilizer bracket, bushing and bar.

To install:

15. Align the stabilizer bar identification mark with the right end of the bushing. Install the stabilizer bar.

16. Position the steering rack in the vehicle and install the clamp and the mounting bolts. Be sure the rack is centered before connecting it to the joint assembly.

17. When installing the steering column dash panel cover pinch clamp, align the steering column dash panel cover notch "A" with the steering gear lug (arrow "B" and then install the steering column dash panel cover to the steering gear.

➡ **When installing the clock spring be sure that the springs mating marks are properly aligned. If not the steering wheel may not rotate completely during a turn, or the flat cable in the clock spring could be damaged. This would prevent normal SRS operation and possibly cause serious injury to the driver.**

18. After aligning the mating surfaces of the clock spring, turn the front wheels to the straight ahead position. Install the clock spring to the column switch.

➡ **Turn the clock spring clockwise fully. Then turn it back approximately ¾ turns counterclockwise to align the mating marks.**

19. Continue the installation in the reverse order of the removal procedure.

20. Check and adjust the front end alignment, as required.

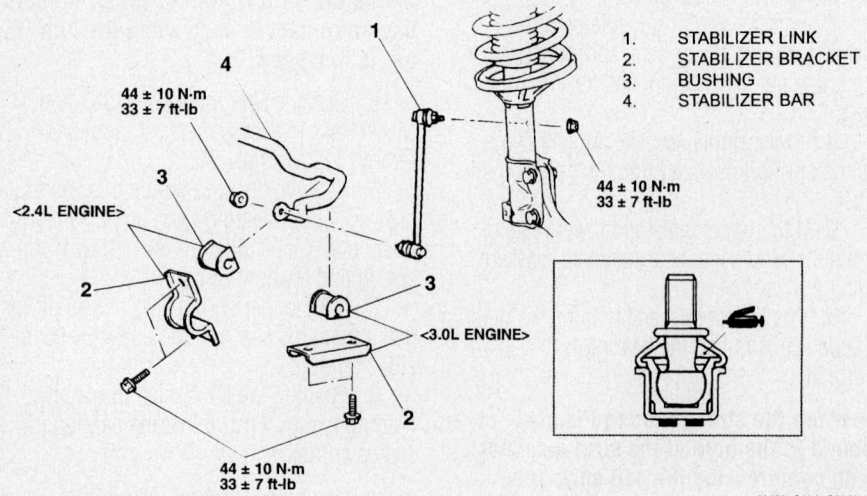

44 ± 10 N·m
33 ± 7 ft-lb

4

3
<2.4L ENGINE>

2

1

3
<3.0L ENGINE>

2

44 ± 10 N·m
33 ± 7 ft-lb

1. STABILIZER LINK
2. STABILIZER BRACKET
3. BUSHING
4. STABILIZER BAR

44 ± 10 N·m
33 ± 7 ft-lb

09482_GALA_G0141

Front stabilizer bar and related components—2004–2005 Eclipse

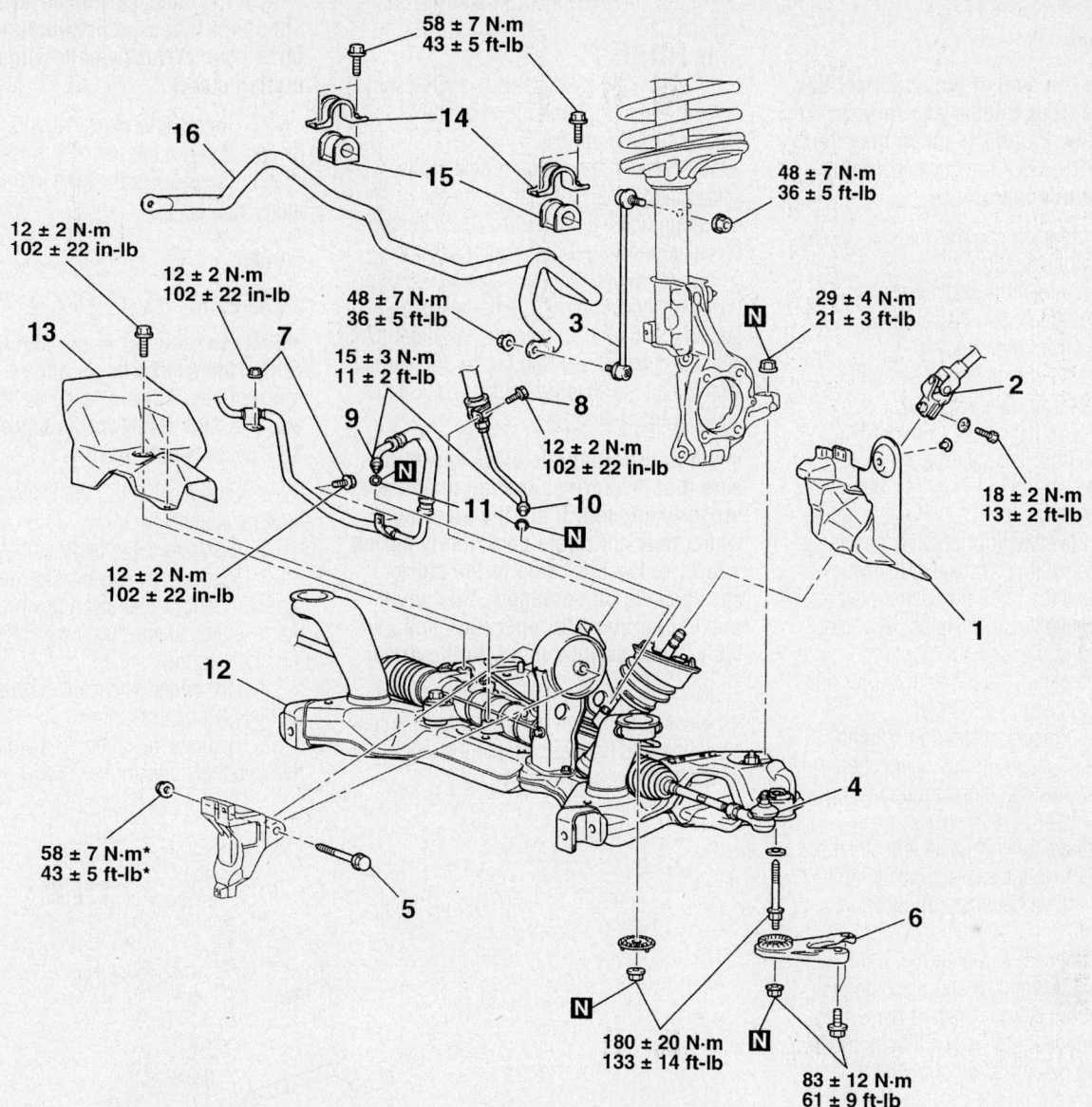

58 ± 7 N·m
43 ± 5 ft-lb

16

14

15

12 ± 2 N·m
102 ± 22 in-lb

13

12 ± 2 N·m
102 ± 22 in-lb

7

48 ± 7 N·m
36 ± 5 ft-lb

15 ± 3 N·m
11 ± 2 ft-lb

9

3

8

11

10

12 ± 2 N·m
102 ± 22 in-lb

48 ± 7 N·m
36 ± 5 ft-lb

29 ± 4 N·m
21 ± 3 ft-lb

2

18 ± 2 N·m
13 ± 2 ft-lb

1

12 ± 2 N·m
102 ± 22 in-lb

12

58 ± 7 N·m*
43 ± 5 ft-lb*

5

4

6

180 ± 20 N·m
133 ± 14 ft-lb

83 ± 12 N·m
61 ± 9 ft-lb

1. STEERING SHAFT PAD
2. STEERING GEAR AND STEERING COLUMN ASSEMBLY CONNECTION
3. STABILIZER LINK
4. TIE ROD END AND KNUCKLE CONNECTION
5. REAR ROLL STOPPER CONNECTION BOLT
6. FRONT AXLE CROSSMEMBER STAY
7. RETURN TUBE CLAMP NUT AND BOLT
8. PRESSURE TUBE CLAMP BOLT
9. STEERING GEAR AND RETURN TUBE CONNECTION

10. STEERING GEAR AND PRESSURE TUBE CONNECTION
11. O-RING
12. FRONT AXLE CROSSMEMBER, REAR ROLL STOPPER AND STEERING GEAR ASSEMBLY
13. STEERING GEAR AND LINKAGE PROTECTOR
14. STABILIZER BRACKET
15. STABILIZER BUSHING
16. STABILIZER BAR

09482_GALA_G0142

Front stabilizer bar and related components—2006 Eclipse and 2004–2006 Galant

Galant

2004–2006

➡ **Prior to removal of the stabilizer bar, center the front wheels and remove the ignition key. Failure to do so may damage the SRS clock spring and render SRS system inoperative.**

1. Before servicing the vehicle, refer to the Precautions Section.

2. Disconnect the negative battery cable. Drain the power steering fluid. Remove the front under cover.

3. Remove the center crossmember. Remove the lower arm assembly.

4. Remove the driver's side air bag module. Remove the steering wheel. Remove the column lower cover. Remove the clock spring.

5. Remove the floor console assembly. Remove the front scuff plate and cowl side trim. Remove the trunk lid opener cover.

6. Remove the accelerator pedal stopper. Remove the front floor carpet.

7. Disconnect the upper stabilizer link. Remove the steering shaft pad.

8. Disconnect the steering column assembly and steering rack connection. Remove the rear roll stopper connecting bolt.

9. Remove the front axle crossmember bracket. Remove the return tube clamp bolt and nut. Remove the pressure tube clamp bolt and nut.

10. Disconnect and plug the return tube connection. Disconnect and plug the pressure hose connection. Discard the O-ring.

11. Disconnect the pinch clamp. Pinch the steering column shaft clip with a pliers, and pull upward on the shaft to disengage the steering column assembly.

➡ **If the steering column shaft is removed accidentally, remove the steering column assembly and be sure to insert the steering column shaft into the steering column.**

12. From inside the vehicle loosen the three clips from the body panel, near the steering column to floor hole.

13. Use a transmission jack to hold the crossmember, and then remove the cross-member mounting nuts and bolts. Lower the crossmember with the rear roll stopper, then the stabilizer bar and steering rack.

14. Remove the steering column dash panel cover. Remove the steering rack linkage protector, on the 3.8L engine.

15. Remove the power steering rack bracket. Remove the rack retaining bolts. Remove the assembly from the vehicle.

16. Remove the stabilizer bracket, bushing and bar.

To install:

17. Align the stabilizer bar identification mark with the right end of the bushing. Install the stabilizer bar.

18. Position the steering rack in the vehicle and install the clamp and the mounting bolts. Be sure the rack is centered before connecting it to the joint assembly.

19. When installing the steering column dash panel cover pinch clamp, align the steering column dash panel cover notch "A" with the steering gear lug (arrow "B" and then install the steering column dash panel cover to the steering gear.

➡ **When installing the clock spring be sure that the springs mating marks are properly aligned. If not the steering wheel may not rotate completely during a turn, or the flat cable in the clock spring could be damaged. This would prevent normal SRS operation and possibly cause serious injury to the driver.**

20. After aligning the mating surfaces of the clock spring, turn the front wheels to the straight ahead position. Install the clock spring to the column switch.

➡ **Turn the clock spring clockwise fully. Then turn it back approximately 3 ¾ turns counterclockwise to align the mating marks.**

21. Continue the installation in the reverse order of the removal procedure.

22. Check and adjust the front end alignment, as required.

Lancer

2004–2006

➡ **Prior to removal of the stabilizer bar, center the front wheels and remove the ignition key. Failure to do so may damage the SRS clock spring and render SRS system inoperative.**

1. Before servicing the vehicle, refer to the Precautions Section.

2. Disconnect the negative battery cable.

3. Remove the driver's side air bag module. Remove the steering wheel. Remove the column lower cover. Remove the clock spring.

4. Disconnect the front exhaust pipe. Remove the center member.

5. Remove the stabilizer bar locknut. Remove the stabilizer bar rubber and collar.

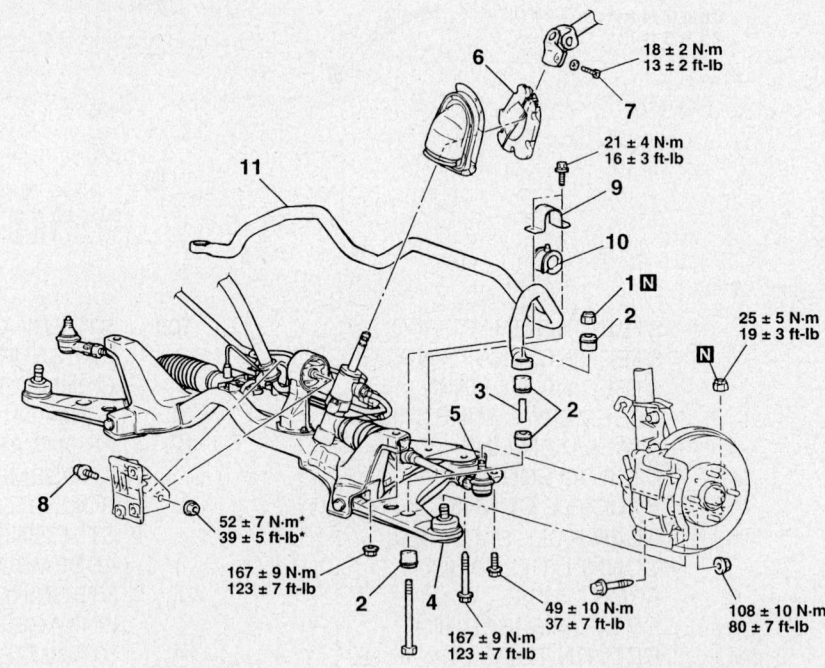

| 18 ± 2 N·m / 13 ± 2 ft-lb |
| 21 ± 4 N·m / 16 ± 3 ft-lb |
| 25 ± 5 N·m / 19 ± 3 ft-lb |
| 52 ± 7 N·m* / 39 ± 5 ft-lb* |
| 167 ± 9 N·m / 123 ± 7 ft-lb |
| 167 ± 9 N·m / 123 ± 7 ft-lb |
| 49 ± 10 N·m / 37 ± 7 ft-lb |
| 108 ± 10 N·m / 80 ± 7 ft-lb |

1.	JAM NUT
2.	STABILIZER RUBBER
3.	COLLAR
4.	LOWER ARM AND KNUCKLE CONNECTION
5.	TIE ROD END AND KNUCKLE CONNECTION
6.	STEERING SHAFT COVER
7.	STEERING GEAR AND JOINT CONNECTING BOLT
8.	REAR ROLL STOPPER CONNECTING BOLT
9.	FIXTURE
10.	BUSHING
11.	STABILIZER BAR

09482_GALA_G0143

Front stabilizer bar and related components—2004–2006 Lancer

6. Separate the lower arm and knuckle connection

7. Separate the tie rod end and knuckle connection.

8. Remove the steering shaft cover. Remove the steering gear joint connecting bolt.

9. Remove the rear stopper connecting bolt.

10. Use a transaxle jack to hold the crossmember in place. Remove the crossmember mounting bolts and nuts.

➡**Be careful not to lower the crossmember too much, otherwise the power steering return hose bracket may deform.**

11. Lower the crossmember until the fixture, bushing and stabilizer can be removed.

12. Remove the stabilizer retaining bolts. Remove the stabilizer from the vehicle.

To install:

13. Align the stabilizer bar identification mark with the right end of the bushing. Install the stabilizer bar.

14. Continue the installation in the reverse order of the removal procedure.

15. Check and adjust the front end alignment, as required.

Evolution

➡**Prior to removal of the stabilizer bar, center the front wheels and remove the ignition key. Failure to do so may damage the SRS clock spring and render SRS system inoperative.**

1. Before servicing the vehicle, refer to the Precautions Section.

2. Disconnect the negative battery cable.

3. Remove the driver's side air bag module. Remove the steering wheel. Remove the column lower cover. Remove the clock spring.

4. Disconnect the front exhaust pipe. Remove the center member.

5. Remove the stabilizer bar locknut. Remove the stabilizer bar rubber and collar.

6. Separate the lower arm and knuckle connection

7. Separate the tie rod end and knuckle connection.

8. Remove the steering shaft cover. Remove the steering gear joint connecting bolt.

9. Remove the rear stopper connecting bolt.

10. Use a transaxle jack to hold the crossmember in place. Remove the crossmember mounting bolts and nuts.

➡**Be careful not to lower the crossmember too much, otherwise the**

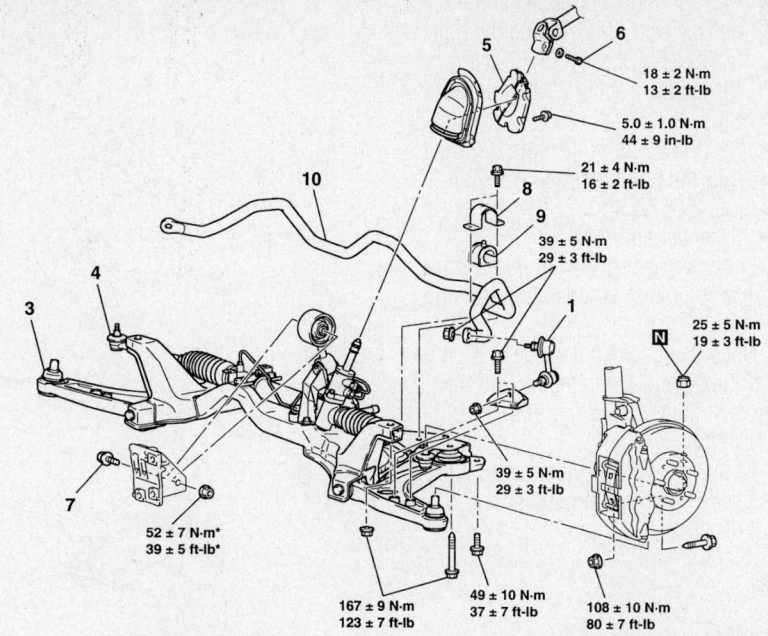

1.	STABILIZER LINK
2.	STABILIZER BAR BRACKET
3.	LOWER ARM AND KNUCKLE CONNECTION
4.	TIE ROD END KNUCKLE CONNECTION
5.	STEERING SHAFT COVER
6.	STEERING GEAR AND JOINT CONNECTION BOLT
7.	REAR ROLL STOPPER CONNECTING BOLT
8.	STABILIZER BRACKET
9.	STABILIZER BUSHING
10.	STABILIZER BAR

09482_GALA_G0144

Front stabilizer bar and related components—Evolution

power steering return hose bracket may deform.

11. Lower the crossmember until the fixture, bushing and stabilizer can be removed.

12. Remove the stabilizer retaining bolts. Remove the stabilizer from the vehicle.

To install:

13. Align the stabilizer bar identification mark with the right end of the bushing. Install the stabilizer bar.

14. Continue the installation in the reverse order of the removal procedure.

15. Check and adjust the front end alignment, as required.

Upper Control Arm

These vehicles use a strut type front suspension. No upper control arm is used.

Lower Ball Joint

REMOVAL & INSTALLATION

The lower ball joint is an integral part of the lower control arm assembly, and can not be serviced separately. A worn or damaged ball joint requires replacement of lower control arm assembly.

Lower Control Arm

REMOVAL & INSTALLATION

Diamante

1. Before servicing the vehicle, refer to the Precautions Section.

2. Disconnect the negative battery cable.

3. Raise the vehicle and support safely allowing wheels and suspension to hang freely.

4. Remove or disconnect the following:
   • Sway bar links from the lower control arm
   • Ball joint stud from the steering knuckle
   • Inner mounting frame through-bolt and nut
   • Rear mount bolts. Remove the clamp if equipped
   • Rear rod bushing if servicing

To install:

5. Assemble the control arm and bushing.

6. Install or connect the following:
   • Control arm to the vehicle and install the through-bolt. Replace the nut and snug temporarily.

- Rear mount clamp, bolts and replacement nuts. Torque the bolts to 72–87 ft. lbs. (100–120 Nm). Torque the nuts to 29 ft. lbs. (40 Nm).
- Ball joint stud to the knuckle
- New nut and tighten to 43–52 ft. lbs. (60–72 Nm)
- Sway bar and links

7. Lower the vehicle to the floor for the final tightening of the frame mount through-bolt.

8. Once the full weight of the vehicle is on the floor, tighten the frame mount through-bolt nuts to 75–90 ft. lbs. (102–122 Nm).

9. Connect the negative battery cable.

10. Check the wheel alignment and adjust if necessary.

Eclipse

2002–2003

The lower lateral arm ball joint and the compression arm ball joint are integral components of the lateral arm and the compression arm respectively. If the ball joints are to be serviced, the arms must be replaced.

1. Before servicing the vehicle, refer to the Precautions Section.

2. Raise and support the vehicle safely.

3. Disconnect both ball joint studs from the steering knuckle.

4. To remove the lower lateral arm, remove the crossmember brackets.

5. Remove or disconnect the following:
- Inner lateral arm mounting bolts and nut
- Arm from the vehicle
- 2 bolts holding the compression arm
- Compression arm

To install:

6. Assemble the control arms and bushings.

7. Install or connect the following:
- Lateral control arm to the vehicle and install the inner mounting bolts. Install a new nut and snug temporarily.
- Compression arm to the vehicle
- Ball joint studs to the knuckle
- New nuts and tighten to 43–51 ft. lbs. (59–71 Nm)

8. Lower the vehicle to the floor for the final tightening.

9. Once the full weight of the vehicle is on the suspension, tighten the lateral arm rear bolt to 71–85 ft. lbs. (98–118 Nm) and the front bolt to the damper fork to 64 ft. lbs. (88 Nm).

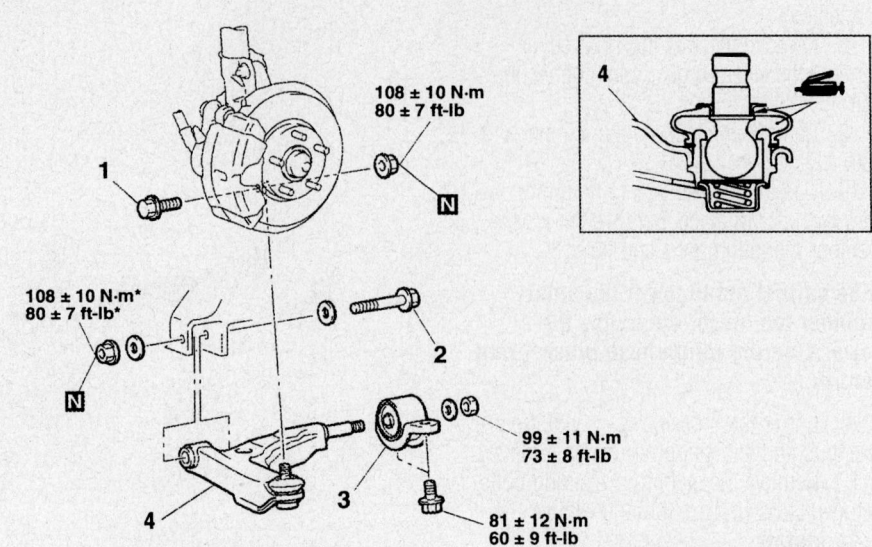

1. LOWER ARM AND KNUCKLE CONNECTION
2. LOWER ARM MOUNTING BOLT
3. LOWER ARM CLAMP
4. LOWER ARM

09482_GALA_G0145

Front lower control arm assembly—2004–2005 Eclipse

1. LOWER ARM AND KNUCKLE CONNECTION
2. LOWER ARM ASSEMBLY

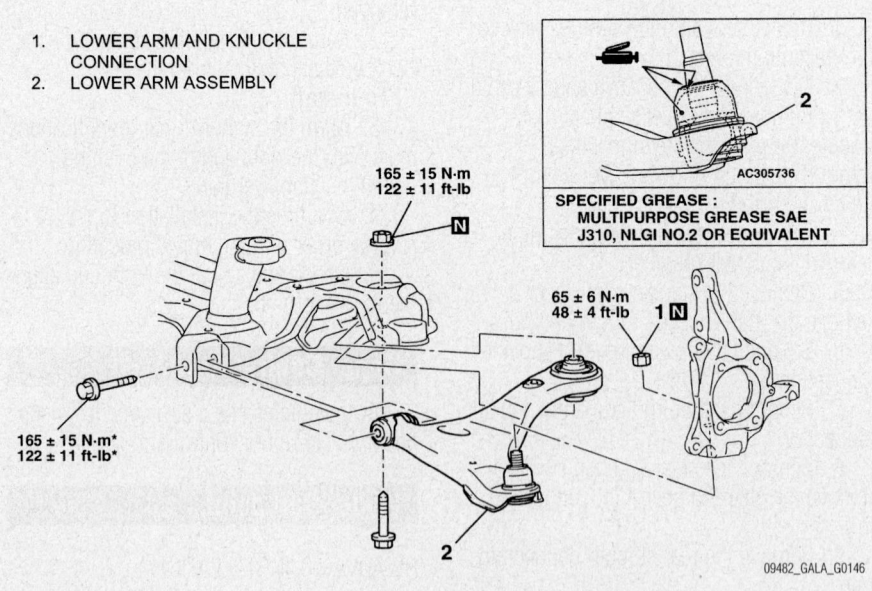

09482_GALA_G0146

Front lower control arm assembly—2006 Eclipse and 2004–2006 Galant

10. Torque the bolts for the compression arm to 60 ft. lbs. (83 Nm).

11. Reinstall the crossmember brackets with their mounting bolts. Torque the mounting bolts to 51–58 ft. lbs. (69–78 Nm).

12. Perform an alignment on the vehicle.

2004–2006

1. Before servicing the vehicle, refer to the Precautions Section.

2. Raise and support the vehicle safely.

➡**Do not remove the nut from the ball joint. Loosen it and use special tool MB991897 to avoid possible damage to the ball joint threads. Hang the special tool in place with wire or string to prevent it from falling.**

3. To disconnect the lower arm from the knuckle, replace the self locking nut for the lower arm ball joint with a regular nut. This is done because the original one is too large to install the special tool. Install special tool MB991897.

4. Turn the bolt and knob as necessary to make the jaws of the tool parallel. Tighten the bolt by hand and confirm that the jaws are still parallel.

➡**When adjusting the jaws in parallel, make sure the knob is in the vertical (upward) position.**

5. Tighten the bolt with a wrench to disconnect the ball joint.

6. Remove the lower control arm retaining bolts. Remove the lower control arm from the vehicle.

To install:

7. Installation is the reverse of the removal procedure.

8. Check and adjust the front end alignment, as required.

Galant

2002–2003

The lower lateral arm ball joint and the compression arm ball joint are integral components of the lateral arm and the compression arm respectively. If the ball joints are to be serviced, the arms must be replaced.

1. Before servicing the vehicle, refer to the Precautions Section.

2. Raise and support the vehicle safely.

3. Disconnect both ball joint studs from the steering knuckle.

4. To remove the lower lateral arm, remove the crossmember brackets.

5. Remove or disconnect the following:
- Inner lateral arm mounting bolts and nut

- Arm from the vehicle
- 2 bolts holding the compression arm
- Compression arm

To install:

6. Assemble the control arms and bushings.

7. Install or connect the following:
- Lateral control arm to the vehicle and install the inner mounting bolts. Install a new nut and snug temporarily.
- Compression arm to the vehicle
- Ball joint studs to the knuckle
- New nuts and tighten to 43–51 ft. lbs. (59–71 Nm)

8. Lower the vehicle to the floor for the final tightening.

9. Once the full weight of the vehicle is on the suspension, tighten the lateral arm rear bolt to 71–85 ft. lbs. (98–118 Nm) and the front bolt to the damper fork to 64 ft. lbs. (88 Nm).

10. Torque the bolts for the compression arm to 60 ft. lbs. (83 Nm).

11. Reinstall the crossmember brackets with their mounting bolts. Torque the mounting bolts to 51–58 ft. lbs. (69–78 Nm).

12. Perform an alignment on the vehicle.

2004–2006

1. Before servicing the vehicle, refer to the Precautions Section.

2. Raise and support the vehicle safely.

➡**Do not remove the nut from the ball joint. Loosen it and use special tool MB991897 to avoid possible damage to the ball joint threads. Hang the special tool in place with wire or string to prevent it from falling.**

3. To disconnect the lower arm from the knuckle, replace the self locking nut for the lower arm ball joint with a regular nut. This is done because the original one is too large to install the special tool. Install special tool MB991897.

4. Turn the bolt and knob as necessary to make the jaws of the tool parallel. Tighten the bolt by hand and confirm that the jaws are still parallel.

➡**When adjusting the jaws in parallel, make sure the knob is in the vertical (upward) position.**

5. Tighten the bolt with a wrench to disconnect the ball joint.

6. Remove the lower control arm retaining bolts. Remove the lower control arm from the vehicle.

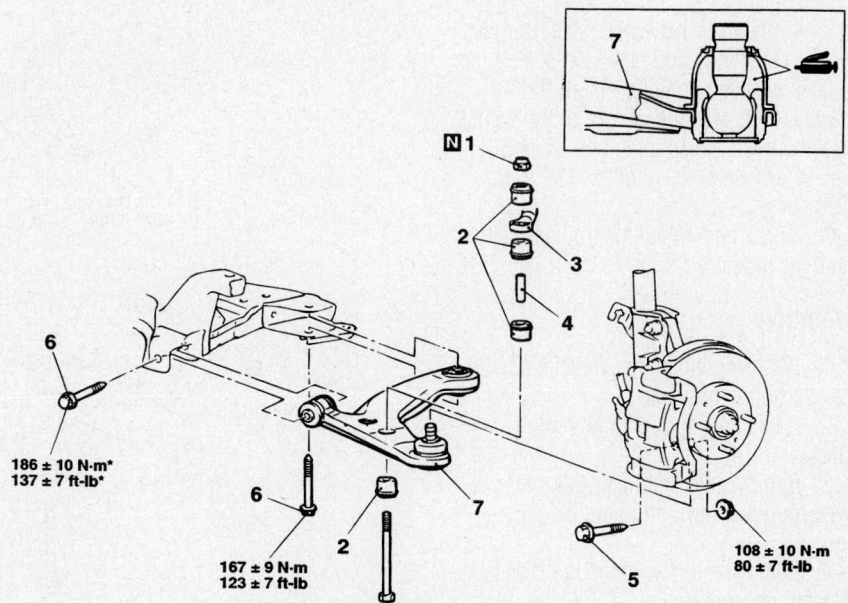

186 ± 10 N·m*
137 ± 7 ft-lb*

167 ± 9 N·m
123 ± 7 ft-lb

108 ± 10 N·m
80 ± 7 ft-lb

1. SELF-LOCKING NUT
2. STABILIZER RUBBER
3. STABILIZER BAR
4. COLLAR
5. LOWER ARM AND KNUCCKLE CONNECTION
6. LOWER ARM AND CROSSMEMBER CONNECTION
7. LOWER ARM ASSEMBLY

Front lower control arm assembly —Lancer

9357QG30

To install:

7. Installation is the reverse of the removal procedure.

8. Check and adjust the front end alignment, as required.

Lancer

1. Before servicing the vehicle, refer to the Precautions Section.

2. Raise the vehicle and support safely.

3. Remove or disconnect the following:
   - Wheel and tire assembly
   - Stabilizer bar self-locking nut, rubber bushings, stabilizer bar and collar. Discard the nut.
   - Lower control arm-to-knuckle bolt and nut

4. Lift the transaxle with a jack, then remove the front control arm-to-crossmember bolt.
   - Lower control arm

To install:

5. Install or connect the following:
   - Lower control arm into the vehicle
   - Lower control arm-to-crossmember bolts. Torque the bottom bolt to 123 ft. lbs. (167 Nm) and the side bolt snug until the vehicle is lowered.
   - Lower control arm-to-steering knuckle bolt. Torque to 80 ft. lbs. (108 Nm).
   - Stabilizer bar collar, stabilizer bar, bushings and new self-locking nut.

6. Lower the vehicle, install the wheels, then with the weight of the vehicle on the wheels, torque the side control arm-to-crossmember bolt to 137 ft. lbs. (186 Nm).

7. Check and adjust the front end alignment, as required.

Evolution

1. Before servicing the vehicle, refer to the Precautions Section.

2. Raise and support the vehicle safely.

3. Remove the stabilizer link nut. Separate the lower arm and knuckle connection.

4. Remove the lower arm and crossmember connection.

5. Remove the crossmember bar bracket.

6. Remove the lower control arm assembly.

To install:

7. Installation is the reverse of the removal procedure.

8. Check and adjust the front end alignment, as required.

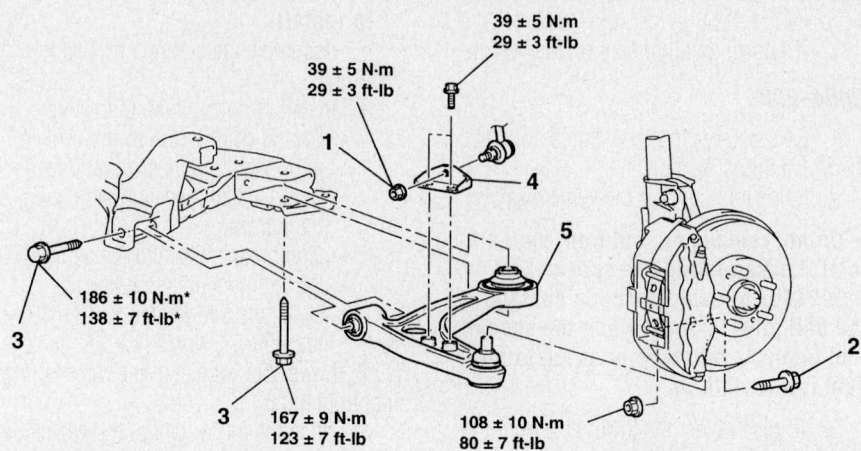

39 ± 5 N·m
29 ± 3 ft-lb

39 ± 5 N·m
29 ± 3 ft-lb

186 ± 10 N·m*
138 ± 7 ft-lb*

167 ± 9 N·m
123 ± 7 ft-lb

108 ± 10 N·m
80 ± 7 ft-lb

1. STABILIZER LINK NUT
2. LOWER ARM AND KNUCKLE CONNECTION
3. LOWER ARM AND CROSSMEMBER CONNECTION
4. STABILIZER BAR BRACKET
5. LOWER ARM ASSEMBLY

09482_GALA_G0147

Front lower control arm assembly—Evolution

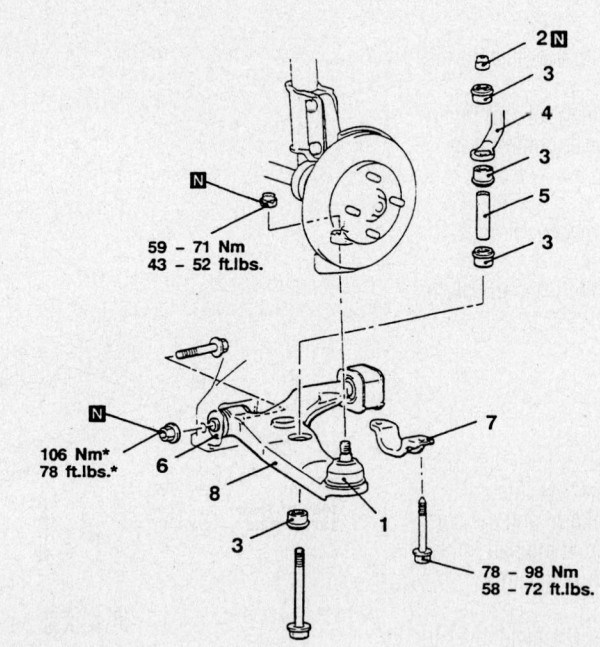

59 – 71 Nm
43 – 52 ft.lbs.

106 Nm*
78 ft.lbs.*

78 – 98 Nm
58 – 72 ft.lbs.

Removal steps

1. Lower arm ball joint connection
2. Self-locking nut
3. Stabilizer rubber
4. Stabilizer bar
5. Collar
6. Lower arm front bushing connection
7. Support bracket
8. Lower arm assembly

Caution

*: Indicates parts which should be temporarily tightened, and then fully tightened with the vehicle on the ground in the unladen condition.

7923PGB0

Front lower control arm assembly—Mirage

Mirage

➡**The suspension components should not be tightened until the vehicle's weight is resting on its wheels.**

1. Before servicing the vehicle, refer to the Precautions Section.
2. Raise the vehicle and support safely.
3. Remove or disconnect the following:
   - Wheel and tire assembly
   - Stabilizer bar links or mounting nuts and bolts from lower control arm. Remove the joint cups and bushings.
   - Ball joint stud from the steering knuckle
   - Inner lower arm mounting bolt and nut
   - Rear mount bolts from the retaining clamp. Remove the rear retainer clamp if equipped.
   - Arm from the vehicle

To install:

4. Install or connect the following:
   - Control arm to the vehicle and install the inner mounting bolt. Install new nut and tighten to 78 ft. lbs. (108 Nm).
   - Rear mount clamp and bolts. Torque the clamp mounting bolts to 65 ft. lbs. (90 Nm).
   - Ball joint stud to the knuckle. Install a new nut and tighten to 43–52 ft. lbs. (60–72 Nm).
   - Sway bar and links
5. Lower the vehicle to the floor for the final tightening of the inner frame mount bolt.
6. Install the wheel and tire assembly.

Wheel Bearings

ADJUSTMENT

The front wheel bearings are not adjustable. If the bearings are noisy or become loose, they must be replaced.

REMOVAL & INSTALLATION

Diamante and Mirage

1. Before servicing the vehicle, refer to the Precautions Section.
2. Disconnect the negative battery cable.
3. Raise the vehicle and support safely. Remove the halfshaft nut.
4. If equipped with Anti-Lock Brake (ABS), remove the front wheel speed sensor.
5. If equipped with Active Electronic Control Suspension (Active-ECS), discon-

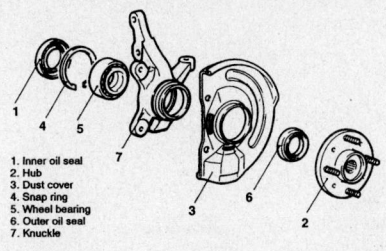

1. Inner oil seal
2. Hub
3. Dust cover
4. Snap ring
5. Wheel bearing
6. Outer oil seal
7. Knuckle

7923PGC1

Front wheel bearing assembly—Diamante and Mirage

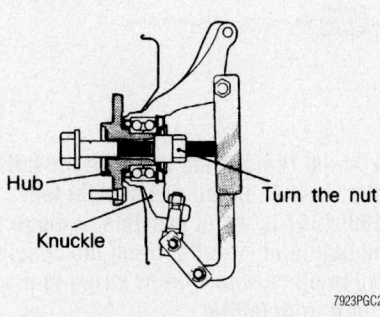

Hub
Knuckle
Turn the nut

7923PGC2

Use of press tool for hub removal— Diamante and Mirage

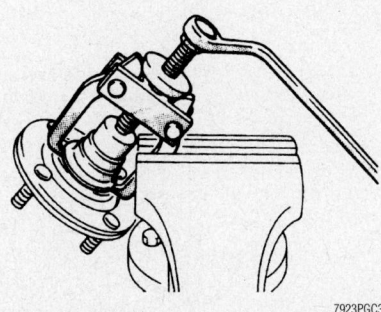

7923PGC3

Removing inner race from hub— Diamante and Mirage

nect the height sensor from the lower control arm.

6. Remove the caliper assembly and brake pads. Suspend the caliper with a wire.
7. Ball joint and tie rod end from the steering knuckle.
8. Remove the halfshaft from the hub.
9. Unbolt the lower end of the strut and remove the hub and steering knuckle assembly from the vehicle.
10. Press the hub from the bearing and remove the bearing races from the knuckle.

To install:

11. Press the wheel bearing into the knuckle. Once the bearing is installed, install the inner race.

12. Install the grease seal.
13. Using a pressing, mount the front hub assembly into the knuckle. Tighten the nut of the pressing tool to 144–188 ft. lbs. (200–260 Nm). Rotate the hub to seat the bearing.
14. Install the hub and knuckle assembly onto the vehicle. Install the lower ball joint stud into the steering knuckle and install a new nut. Tighten to 52 ft. lbs. (72 Nm).
15. Install the halfshaft into the hub/knuckle assembly.
16. Install 2 front strut lower mounting bolts and tighten to 80–94 ft. lbs. (110–130 Nm) on Mirage or 65–76 ft. lbs. (90–105 Nm) on Diamante models.
17. Install the tie rod end and tighten the nut to 25 ft. lbs. (34 Nm) for Mirage and 21 ft. lbs. (29 Nm) on Diamante models.
18. Install the brake disc and caliper assembly.
19. If equipped with Active-ECS, connect the height sensor and tighten the mounting bolt to 15 ft. lbs. (20 Nm).
20. Install the front speed sensor, if removed.
21. Install the washer and new locknut to the end of the halfshaft. Tighten the locknut snugly to 144–188 ft. lbs. (200–260 Nm).
22. Install the tire and wheel assembly onto the vehicle. Lower the vehicle to the ground.

Eclipse

2002–2003

1. Before servicing the vehicle, refer to the Precautions Section.
2. Remove or disconnect the following:
   - Front wheel
   - Axle nut
   - Wheel speed sensor, vehicles with Anti-Lock Brake (ABS)
   - Caliper and suspend it out of the way with wire or string
   - Brake rotor
   - Steering knuckle from the upper arm
3. Pull the knuckle away from the vehicle to access the hub mounting bolts on the inboard side of the hub. Be careful not to damage the ball joint boot or the ABS rotor if equipped.
4. Remove the mounting bolts and the front hub assembly.

➡**Do not disassemble the hub assembly. If binding or damaged, it must be replaced as a unit.**

To install:

5. Install or connect the following:
   - Hub to the knuckle. Torque the

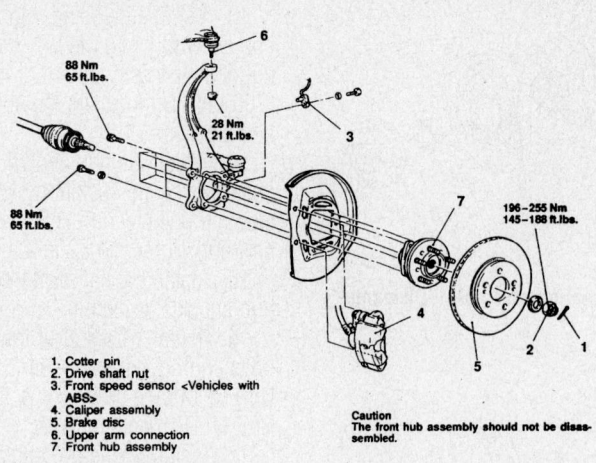

1. Cotter pin
2. Drive shaft nut
3. Front speed sensor <Vehicles with ABS>
4. Caliper assembly
5. Brake disc
6. Upper arm connection
7. Front hub assembly

88 Nm
65 ft.lbs.

28 Nm
21 ft.lbs.

88 Nm
65 ft.lbs.

196–255 Nm
145–188 ft.lbs.

Caution
The front hub assembly should not be disassembled.

7923PGC4

Front hub assembly—2002–2003 Eclipse

mounting bolts to 65 ft. lbs. (88 Nm).

- Knuckle to the upper arm
- Brake rotor and the caliper
- Wheel speed sensor if removed
- Axle nut and tighten to 145–188 ft. lbs. (196–255 Nm)
- Wheel and lower the vehicle to the floor

2004–2006

1. Before servicing the vehicle, refer to the Precautions Section.

2. Raise and support the vehicle safely. Remove the tire and wheel assembly.

3. Remove the dust cover. Remove the cotter pin. Remove the driveshaft locknut. Remove the washer.

4. Remove the front wheel speed sensor bracket. Remove the front wheel speed sensor.

5. Remove the brake hose bracket. Remove the brake caliper and wire it to the side. Do not disconnect the brake hose from the caliper.

6. Remove the rotor.

7. Remove the front wheel hub mounting bolts while pushing out the driveshaft by hand.

8. If it is difficult to push out the driveshaft by hand, use special tools MB990242, MB990244, MB991354 and MB990767 to push out the driveshaft from the hub and knuckle.

9. If the front wheel hub is seized, remove the knuckle together with the front wheel hub.

10. Support the driveshaft on the vehicle body with rope.

11. To remove the knuckle proceed as follows:

➡ **Do not remove the nut from the ball joint. Loosen it and use special tool MB991897 to avoid possible damage to the ball joint threads. Hang the special tool in place with wire or string to prevent it from falling.**

12. To disconnect the lower arm from the knuckle, replace the self locking nut for the lower arm ball joint with a regular nut. This is done because the original one is too large to install the special tool. Install special tool MB991897.

13. Turn the bolt and knob as necessary to make the jaws of the tool parallel. Tighten the bolt by hand and confirm that the jaws are still parallel.

➡ **When adjusting the jaws in parallel, make sure the knob is in the vertical (upward) position.**

14. Tighten the bolt with a wrench to disconnect the ball joint.

15. Remove the steering knuckle dust cover.

16. Remove the strut to knuckle retaining bolts. Remove the steering knuckle from the vehicle.

To install:

17. Installation is the reverse of the removal procedure.

18. Be sure to check and adjust the front end alignment, as required.

Galant

2002–2003

1. Before servicing the vehicle, refer to the Precautions Section.

2. Raise the vehicle and support safely.

3. Remove or disconnect the following:

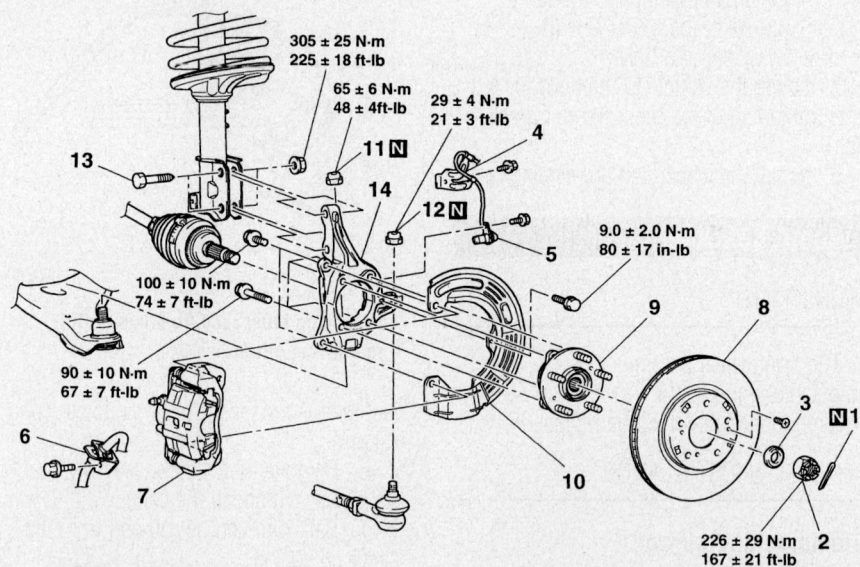

305 ± 25 N·m
225 ± 18 ft-lb

65 ± 6 N·m
48 ± 4ft-lb

29 ± 4 N·m
21 ± 3 ft-lb

9.0 ± 2.0 N·m
80 ± 17 in-lb

100 ± 10 N·m
74 ± 7 ft-lb

90 ± 10 N·m
67 ± 7 ft-lb

226 ± 29 N·m
167 ± 21 ft-lb

1. SPLIT PIN
2. DRIVE SHAFT NUT
3. WASHER
4. FRONT WHEEL SPEED SENSOR BRACKET
5. FRONT WHEEL SPEED SENSOR
6. BRAKE HOSE BRACKET
7. CALIPER ASSEMBLY
8. BRAKE DISC
9. FRONT WHEEL HUB ASSEMBLY
10. DUST COVER
11. SELF LOCKING NUT (CONNECTION FOR LOWER ARM BALL JOINT)
12. SELF LOCKING NUT (CONNECTION FOR TIE ROD END)
13. FRONT STRUT TO KNUCKLE MOUNTING BOLT AND NUT
14. KNUCKLE

09482_GALA_G0148

Front hub assembly—2004–2006 Eclipse and Galant

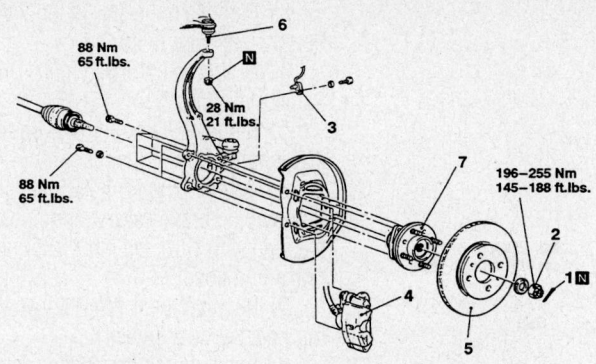

88 Nm
65 ft.lbs.

88 Nm
65 ft.lbs.

28 Nm
21 ft.lbs.

196–255 Nm
145–188 ft.lbs.

Removal steps
1. Cotter pin
2. Drive shaft nut
3. Front speed sensor <Vehicles with ABS>
4. Caliper assembly

5. Brake disc
6. Connection for upper arm
7. Front hub assembly

Caution
The front hub assembly should not be disassembled.

7923PGC5

Front hub assembly—2003–2003 Galant

- Appropriate wheel assembly
- Cotter pin, halfshaft nut and washer
- Vehicle Speed Sensor (VSS(), if equipped with Anti-Lock Brake (ABS)
- Caliper and brake pads. Support the caliper out of the way using wire.
- Brake rotor from the hub assembly
- Upper ball joint from the steering knuckle and pull the knuckle outward

4. From the back of the knuckle, remove the 4 bolts securing the hub to the knuckle.

5. Remove the hub and bearing assembly from the knuckle.

➡**The hub assembly is not serviceable and should not be disassembled.**

To install

6. Install or connect the following:
- Hub to the steering knuckle and tighten the mounting bolts to 65 ft. lbs. (88 Nm)
- Upper ball joint to the steering knuckle and tighten the self-locking nut to 21 ft. lbs. (28 Nm)
- Axle washer and nut. Tighten the nut to 145–188 ft. lbs. (200–260 Nm).

7. Position the rotor on the hub.

8. Install the caliper holder and the brake caliper.

9. If equipped with ABS, install the VSS.

10. Install the wheel assembly and lower the vehicle.

2004–2006

1. Before servicing the vehicle, refer to the Precautions Section.

2. Raise and support the vehicle safely. Remove the tire and wheel assembly.

3. Remove the dust cover. Remove the cotter pin. Remove the driveshaft locknut. Remove the washer.

4. Remove the front wheel speed sensor bracket. Remove the front wheel speed sensor.

5. Remove the brake hose bracket. Remove the brake caliper and wire it to the side. Do not disconnect the brake hose from the caliper.

6. Remove the rotor.

7. Remove the front wheel hub mounting bolts while pushing out the driveshaft by hand.

8. If it is difficult to push out the driveshaft by hand, use special tools MB990242, MB990244, MB991354 and MB990767 to push out the driveshaft from the hub and knuckle.

9. If the front wheel hub is seized, remove the knuckle together with the front wheel hub.

10. Support the driveshaft on the vehicle body with rope.

11. To remove the knuckle proceed as follows.

➡**Do not remove the nut from the ball joint. Loosen it and use special tool MB991897 to avoid possible damage to the ball joint threads. Hang the special tool in place with wire or string to prevent it from falling.**

12. To disconnect the lower arm from the knuckle, replace the self locking nut for the lower arm ball joint with a regular nut. This is done because the original one is too large to install the special tool. Install special tool MB991897.

13. Turn the bolt and knob as necessary to make the jaws of the tool parallel. Tighten the bolt by hand and confirm that the jaws are still parallel.

➡**When adjusting the jaws in parallel, make sure the knob is in the vertical (upward) position.**

14. Tighten the bolt with a wrench to disconnect the ball joint.

15. Remove the steering knuckle dust cover.

16. Remove the strut to knuckle retaining bolts. Remove the steering knuckle from the vehicle.

To install:

17. Installation is the reverse of the removal procedure.

18. Be sure to check and adjust the front end alignment, as required.

Lancer

1. Before servicing the vehicle, refer to the Precautions Section.

2. Raise and safely support the vehicle.

3. Remove or disconnect the following:
- Front wheel
- Wheel Speed Sensor (WSS), vehicles with Anti-Lock Brake (ABS)
- Caliper and suspend it out of the way with wire or string
- Brake rotor
- Axle nut, using special tool MB990767 to hold the hub secure while removing the nut. Discard the nut.
- Stabilizer bar from the lower control arm
- Lower ball joint from the steering knuckle
- Tie rod end from the steering knuckle. Do not remove the nut from the tie rod end. Loosen the nut and use special tool MB991113 or MB990635 to avoid damaging the threads.
- Halfshaft from the hub and knuckle using the proper puller
- Front strut-to-hub and knuckle mounting bolt and nut
- Hub and knuckle from the vehicle

➡**Do not disassemble the hub assembly. If binding or damaged, it must be replaced as a unit.**

To install:

4. Install or connect the following:
- Hub and knuckle assembly
- Front strut-to-hub bolt and nut. Torque to 123 ft. lbs. (167 Nm).
- Halfshaft

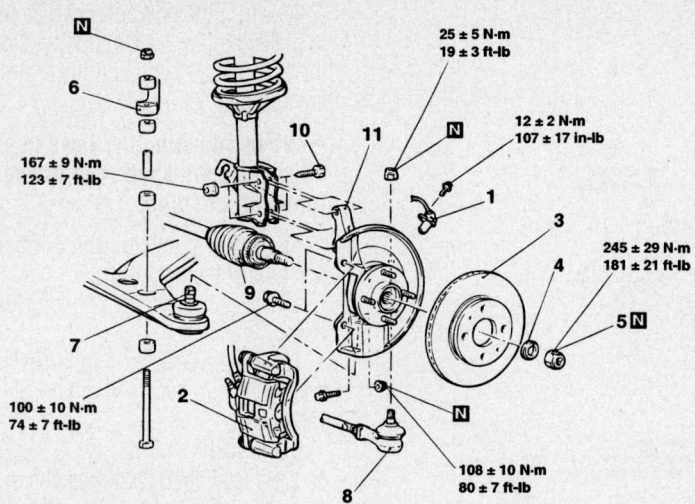

25 ± 5 N·m
19 ± 3 ft-lb

12 ± 2 N·m
107 ± 17 in-lb

167 ± 9 N·m
123 ± 7 ft-lb

245 ± 29 N·m
181 ± 21 ft-lb

100 ± 10 N·m
74 ± 7 ft-lb

108 ± 10 N·m
80 ± 7 ft-lb

9357QG31

1. FRONT ABS SPEED SENSOR
   <VEHICLES WITH ABS>
2. CALIPER ASSEMBLY
3. BRAKE DISC
4. WASHER
5. DRIVESHAFT NUT
6. CONNECTION FOR STABILIZER
   BAR
7. CONNECTION FOR LOWER ARM
   BALL JOINT
8. CONNECTION FOR TIE ROD END
9. DRIVESHAFT
10. FRONT STRUT TO HUB AND
    KNUCKLE MOUNTING BOLT AND
    NUT
11. HUB AND KNUCKLE

Front hub assembly—Lancer

- Tie rod end. Torque the nut to 19 ft. lbs. (25 Nm).
- Lower control arm ball joint. Tighten the nut to 19 ft. lbs. (25 Nm).
- Stabilizer bar
- New axle nut and washer and tighten to 181 ft. lbs. (245 Nm)
- Brake rotor and caliper. Torque the caliper bolts to 74 ft. lbs. (100 Nm).

- Front WSS, if equipped
- Front wheel
5. Lower the vehicle

Evolution

1. Before servicing the vehicle, refer to the Precautions Section.
2. Raise and safely support the vehicle. Remove the front tire and wheel assembly.

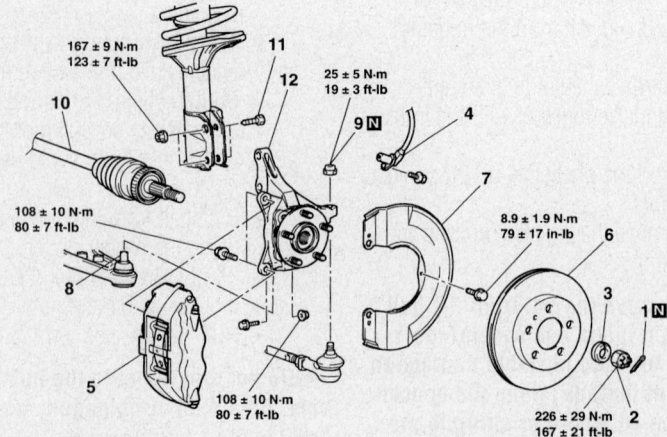

167 ± 9 N·m
123 ± 7 ft-lb

25 ± 5 N·m
19 ± 3 ft-lb

108 ± 10 N·m
80 ± 7 ft-lb

8.9 ± 1.9 N·m
79 ± 17 in-lb

108 ± 10 N·m
80 ± 7 ft-lb

226 ± 29 N·m
167 ± 21 ft-lb

1. COTTER PIN
2. DRIVESHAFT NUT
3. WASHER
4. FRONT ABS SENSOR
5. CALIPER ASSEMBLY
6. BRAKE DISC
7. DUST COVER
8. CONNECTION FOR LOWER ARM
   BALL JOINT
9. SELF LOCKING NUT (CONNECTION
   FOR TIE ROD END)
10. DRIVESHAFT
11. FRONT STRUT TO HUB AND
    KNUCKLE MOUNTING BOLT
12. HUB AND KNUCKLE ASSEMBLY

09482_GALA_G0149

Front hub assembly—Evolution

3. Remove the front under cover. Remove the side under cover.
4. Drain the transaxle fluid. Drain the transfer case fluid.
5. Remove the cotter pin. Remove the driveshaft nut and washer.
6. Remove the front ABS sensor. Remove the brake caliper and position it to the side with wire. Do not disconnect the brake line hose.
7. Remove the rotor. Remove the steering knuckle dust cover.
8. Separate the connection for the lower arm ball joint.
9. Remove the locknut for the tie rod end connection.

➡**Do not remove the nut from the ball joint. Loosen it and use special tool MB991897 to avoid possible damage to the ball joint threads. Hang the special tool in place with wire or string to prevent it from falling.**

10. To disconnect the lower arm from the knuckle, replace the self locking nut for the lower arm ball joint with a regular nut. This is done because the original one is too large to install the special tool. Install special tool MB991897.
11. Turn the bolt and knob as necessary to make the jaws of the tool parallel. Tighten the bolt by hand and confirm that the jaws are still parallel.

➡**When adjusting the jaws in parallel, make sure the knob is in the vertical (upward) position.**

12. Tighten the bolt with a wrench to disconnect the ball joint.
13. Using special tool MB990241 (MB990242, MB990244), MB991354 and MB990767 to push out the driveshaft from the hub and knuckle.
14. Withdraw the driveshaft from the hub by pulling the bottom of the hub and knuckle toward you. Support the driveshaft on the vehicle body with a rope.
15. Remove the front strut to hub and knuckle mounting bolt.
16. Remove the hub and knuckle assembly from the vehicle.
17. Separate the knuckle from the hub and bearing assembly.

To install:

18. Installation is the reverse of the removal procedure.
19. Be sure to refill the transfer case and the transaxle, with the proper grade and type fluids.
20. Be sure to check and adjust the front end alignment, as required.

REAR SUSPENSION

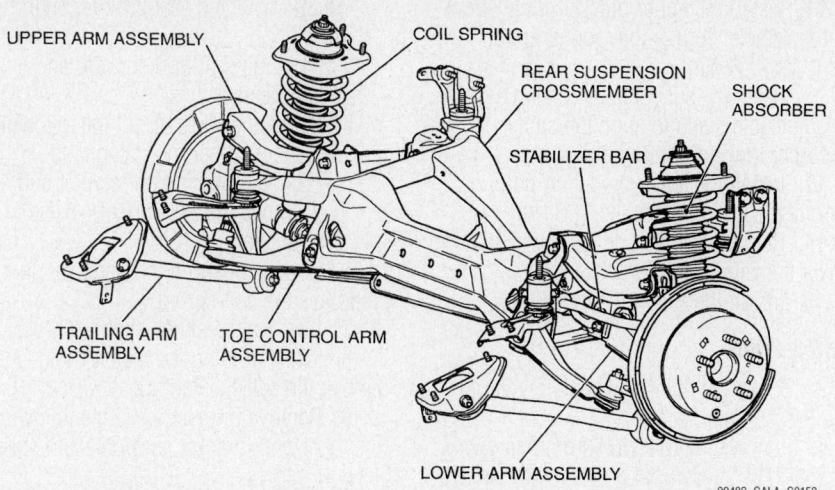

Rear suspension and related components—2006 Eclipse and 2004–2006 Galant

09482_GALA_G0150

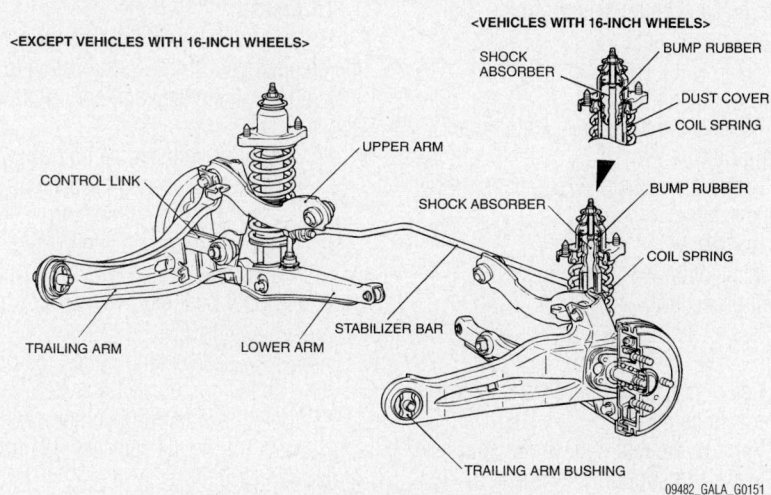

Rear suspension and related components—Lancer

09482_GALA_G0151

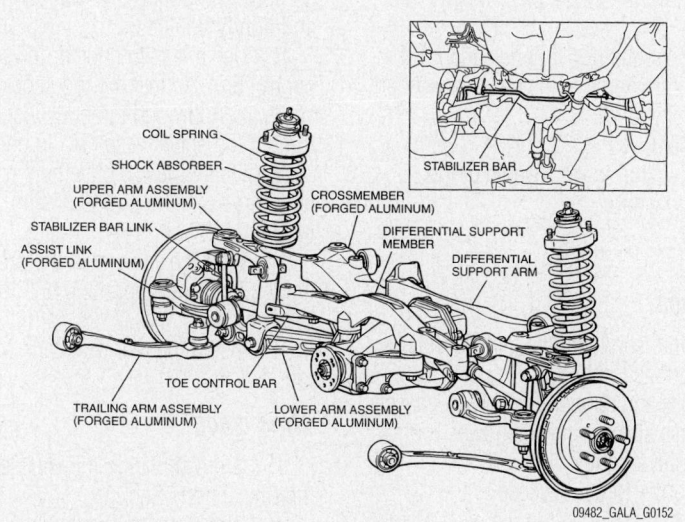

Rear suspension and related components—Evolution

09482_GALA_G0152

Shock Absorber

REMOVAL & INSTALLATION

Diamante

1. Before servicing the vehicle, refer to the Precautions Section.

2. Disconnect the negative battery cable.

3. Raise and properly support vehicle. Remove both rear wheels.

4. Support the lower control arm with a jack.

5. Matchmark the positioning of the upper spring plate to the vehicle for reinstallation purposes.

6. If equipped with Active Electronic Control Suspension (Active-ECS), perform the following:

    a. Loosen the nut that secures the air line to the to the top of the strut and discard the O-ring.

    b. Remove the bolts that secure the actuator to the top of the strut and remove the component. Disconnect the wiring harness.

7. Remove the shock absorber lower mounting bolt and remove the 2 nuts that secure the shock upper plate to the vehicle.

8. Lower the support jack and remove the shock from the vehicle.

To install

9. Position the upper spring plate and install the strut. Use the support jack to assist with installation.

10. Tighten the upper strut mounting nuts to 33 ft. lbs. (45 Nm).

11. Tighten the lower strut mounting bolt to 71 ft. lbs. (98 Nm).

12. If equipped with Active-ECS perform the following:

    a. Using a new O-ring, tighten the nut that secures the air line to the to the top of the strut to 84 inch lbs. (9 Nm).

    b. Install the actuator to the top of the shock absorber and secure with mounting bolts. Connect the wiring harness.

13. Remove the support jack, install wheels and lower vehicle.

14. Connect the negative battery cable.

Eclipse

2002–2003

1. Before servicing the vehicle, refer to the Precautions Section.

2. Remove or disconnect the following:

- Service lid in the luggage compartment
- Cap and flange nuts securing the upper mounting bracket to the body of the vehicle

3. Raise and safely support the vehicle.

4. Remove the bolt attaching the lower end of the shock to the knuckle and remove the shock absorber from the vehicle.

5. Use a coil spring compressor and compress the coil spring.

6. While holding the piston rod, remove the self-locking nut.

7. Remove or disconnect the following:
- Upper bracket assembly and spring pad
- Collar, upper bushing, cup assembly, bump rubber and dust cover
- Coil spring from the shock absorber

To install:

8. Align the end of the coil spring with the stepped part of the spring seat and install the compressed coil spring on the shock absorber.

9. Install or connect the following:
- Dust cover, bump rubber, cup assembly, upper bushing, collar, upper spring pad and bracket assembly on the shock absorber
- Upper bushing and washer on the piston rod
- New self-locking nut on the piston rod. Temporarily tighten the nut.

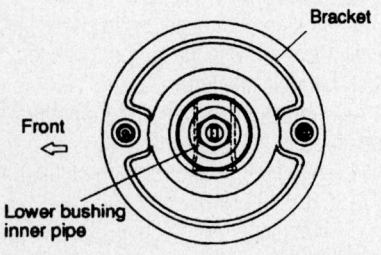

Rear upper bracket, installed position—2003–2003 Eclipse

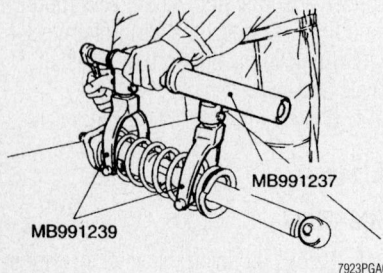

Correct method for compressing the coil spring

10. Remove the spring compressor from the spring. Torque the self-locking nut to 16 ft. lbs. (25 Nm).

11. Install the upper bracket of the shock to the vehicle. Torque the mounting nuts to 32 ft. lbs. (44 Nm).

12. Raise the suspension up with a jack or adjustable stand to align the shock absorber lower mounting holes.

13. Install the lower mounting bolt. Torque the bolt to 71 ft. lbs. (96 Nm).

14. Remove the jack or stand and safely lower the vehicle to the floor.

15. Install the cap and service lid.

2004–2005

1. Before servicing the vehicle, refer to the Precautions Section.

2. Remove the rear trunk trim panel, on Eclipse. On Spyder make sure the top is in the fully closed position.

3. Raise and support the vehicle safely.

4. Remove the drain turf mounting hose, so that the cap can be seen from the passenger compartment.

5. Remove the shock absorber cap. Remove the mounting nuts.

6. Remove the lower shock absorber retaining bolt.

7. Remove the shock absorber assembly from the vehicle.

To install:

8. Installation is the reverse of the removal procedure.

2006

1. Before servicing the vehicle, refer to the Precautions Section.

2. Remove the rear sub woofer speaker.

3. Raise and support the vehicle safely.

4. Remove the shock absorber assembly and knuckle retaining bolt.

5. Remove the shock absorber service cover.

6. Remove the upper shock absorber retaining nuts. Remove the shock absorber from the vehicle.

To install:

7. Installation is the reverse of the removal procedure.

Galant

2002–2003

1. Before servicing the vehicle, refer to the Precautions Section.

2. Raise and support the vehicle chassis.

3. Raise and support the lower control arm assembly slightly.

4. In order to gain access to the top mounting nuts, remove the rear seat as follows:

a. While pulling the rear seat stopper outward, lift the lower cushion upward. Remove the lower cushion.

b. Remove the seat back mounting bolts.

c. Lift the seat back upward and remove the seat.

5. Remove or disconnect the following:
- Shock upper mounting nuts
- Shock lower mounting bolt and remove the assembly from the vehicle

6. Use a coil spring compressor and compress the coil spring.

7. Remove the shock cap.

8. While holding the piston rod, remove the self-locking nut.

9. Remove or disconnect the following:
- Upper bracket assembly and spring pad
- Collar, upper bushing, cup assembly, bump rubber and dust cover
- Coil spring from the shock

To install

10. Align the end of the coil spring with the stepped part of the spring seat and install the compressed coil spring on the shock.

11. Install or connect the following:
- Dust cover, bump rubber, cup assembly, upper bushing, collar, upper spring pad and bracket assembly on the shock
- Upper bushing and washer on the piston rod
- New self-locking nut on the piston rod. Temporarily tighten the nut.

12. Remove the spring compressor from the spring. Torque the self-locking nut to 16 ft. lbs. (25 Nm).

13. Install the shock cap.

14. Position the shock assembly so that the lower mounting bolt can be installed and lightly tightened.

15. Use a jack to raise or lower the lower control arm, so that the top shock plate studs align through the body. Raise the jack to hold the shock assembly in position.

16. Install the top plate nuts on the studs and tighten the mounting nuts to 32 ft. lbs. (44 Nm).

17. With the vehicle on the ground, tighten the lower mounting bolt to 71 ft. lbs. (98 Nm).

18. Install the rear seat back and cushion.

2004–2006

1. Before servicing the vehicle, refer to the Precautions Section.

2. Remove the rear sub woofer speaker.

3. Raise and support the vehicle safely.

4. Remove the shock absorber assembly and knuckle retaining bolt.

5. Remove the shock absorber service cover.

6. Remove the upper shock absorber retaining nuts. Remove the shock absorber from the vehicle.

To install:

7. Installation is the reverse of the removal procedure.

Lancer

1. Before servicing the vehicle, refer to the Precautions Section.

2. Remove or disconnect the following:
  • Stabilizer link connection

3. Support the lower control arm with a jack.
  • Lower control arm and trailing arm bolt
  • Upper shock absorber mounting nut
  • Shock absorber-to-lower control arm attaching bolt
  • Shock absorber assembly

To install:

4. Position the shock absorber into the vehicle. Install the spring seat stepped section so that it points toward the rear side of the vehicle.

5. Install or connect the following:
  • Shock absorber-to-lower control arm bolt and nut. Tighten the nut to 70 ft. lbs. (95 Nm).
  • Upper shock mounting nut and tighten to 32 ft. lbs. (44 Nm).
  • Lower control arm and trailing arm
  • Stabilizer link. Tighten the self-locking nuts so that the end of the stabilizer line bolt protrudes 0.24–0.31 in. (6–8mm).

Evolution

1. Before servicing the vehicle, refer to the Precautions Section.

2. Raise and support the vehicle safely.

3. Remove the upper shock absorber retaining nuts.

4. Remove the lower shock absorber retaining nuts.

5. Remove the shock absorber from the vehicle.

To install:

6. Installation is the reverse of the removal procedure.

Mirage

1. Before servicing the vehicle, refer to the Precautions Section.

2. Remove or disconnect the following:

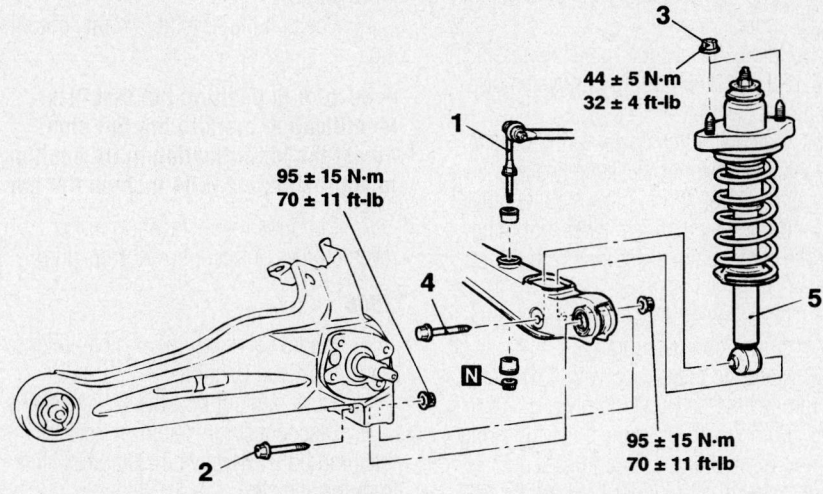

1. STABILIZER LINK CONNECTION
2. LOWER ARM AND TRAILING ARM CONNECTION
3. SHOCK ABSORBER MOUNTING NUT
4. SHOCK ABSORBER AND LOWER ARM CONNECTING BOLT
5. SHOCK ABSORBER ASSEMBLY

9357QG29

Rear shock absorber and related components—Lancer

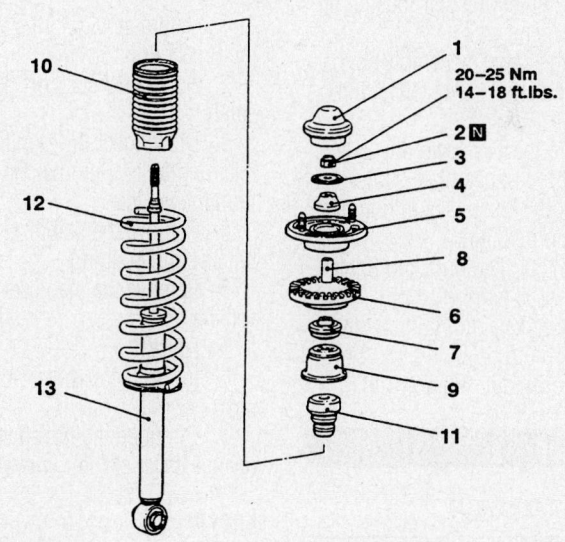

1. Cap
2. Self-locking nut
3. Washer
4. Upper bushing A
5. Bracket
6. Spring pad
7. Upper bushing B
8. Collar
9. Cup
10. Dust cover
11. Bump rubber
12. Coil spring
13. Shock absorber assembly

7923PGB4

Rear shock absorber and related components—Mirage

- Trunk interior trim to gain access to the top mounting nuts
- Top cap and upper shock mounting nuts

3. Raise and support vehicle chassis.

4. Support the trailing arm assembly with a jack.

5. Matchmark the upper spring plate to the vehicle chassis for reassembly and remove the upper spring plate mounting nuts.

6. Remove the shock lower mounting bolt and remove the assembly from the vehicle.

7. Compress the coil spring using the proper spring compressor.

8. Hold the piston rod with a wrench and remove the self-locking nut.

9. Remove or disconnect the following:
- Washer, upper bushing A, bracket, spring pad, upper bushing B, collar, cup, dust cover and bump rubber
- Coil spring

To install

10. Install or connect the following:
- Coil spring on the shock
- Bump rubber, dust cover, cup, collar, upper bushing A, spring pad, bracket, upper bushing B and the washer

11. Temporarily install a new self-locking nut, carefully release the spring from the compressor and tighten the self-locking nut to specifications.

12. Position the shock assembly so that lower mounting bolt can be installed and lightly tightened.

13. Use jack to raise or lower the axle assembly so that top shock plate studs aligns through body. Raise jack to hold the shock assembly in position.

14. Install the top plate nuts and tighten them to 20 ft. lbs. (28 Nm).

15. Lower the vehicle and tighten the lower mounting bolt to 65 ft. lbs. (90 Nm).

16. Install top cap and interior trim.

Stabilizer Bar

REMOVAL & INSTALLATION

Eclipse

2004–2005

1. Before servicing the vehicle, refer to the Precautions Section.

2. Raise and support the vehicle safely.

3. Remove the stabilizer link mounting nuts. Remove the stabilizer link.

4. Remove the stabilizer bar retaining bracket bolts and bushing.

5. Remove the stabilizer from the vehicle.

To install

6. Position the assembly on its mounting.

➡**Be sure to position the stabilizer identification mark to the left side. Adjust the identification mark position to approximately 0.04 inch on the bar.**

7. Continue the installation in the reverse order of the removal procedure.

2006

1. Before servicing the vehicle, refer to the Precautions Section.

2. Raise and support the vehicle safely.

3. Disconnect the stabilizer bar link retaining bolts. Remove the stabilizer link from the vehicle.

4. Remove the stabilizer bar retaining bracket and bushing.

5. Remove the stabilizer bar from the vehicle.

To install

6. Install the stabilizer bar to its mounting.

7. Continue the installation in the reverse order of the removal procedure.

Galant

2004–2006

1. Before servicing the vehicle, refer to the Precautions Section.

2. Raise and support the vehicle safely.

3. Disconnect the stabilizer bar link retaining bolts. Remove the stabilizer link from the vehicle.

4. Remove the stabilizer bar retaining bracket and bushing.

5. Remove the stabilizer bar from the vehicle.

To install

6. Install the stabilizer bar to its mounting.

7. Continue the installation in the reverse order of the removal procedure.

Lancer

1. Before servicing the vehicle, refer to the Precautions Section.

2. Raise and support the vehicle safely.

3. Remove the locknut. Remove the stabilizer rubber bushing. Remove the stabilizer link.

4. Remove the stabilizer retaining bracket bolts, and bushing.

5. Remove the stabilizer bar from the vehicle.

To install

6. Install the stabilizer bar to its mounting.

7. Align the identification color on the left side of the stabilizer bar with the right end of the bushing.

8. Continue the installation in the reverse order of the removal procedure.

Evolution

1. Before servicing the vehicle, refer to the Precautions Section.

2. Raise and support the vehicle safely.

3. Remove the stabilizer nut. Remove the stabilizer bar link.

4. Remove the stabilizer bar bracket retaining bolts. Remove the brackets and bushings.

5. Remove the stabilizer bar from the vehicle.

To install

6. Install the stabilizer bar to its mounting.

7. Align the stabilizer bar until the identification color of the stabilizer bar is out about 3.9 inches from the stabilizer bushing to the vehicle center.

8. Continue the installation in the reverse order of the removal procedure.

Lower Control Arm

REMOVAL & INSTALLATION

Eclipse

2004–2005

1. Before servicing the vehicle, refer to the Precautions Section.

2. Raise and support the vehicle safely.

3. Disconnect the stabilizer link connection. Disconnect the wheel speed sensor mounting bolts.

4. Properly support the lower control arm assembly.

5. Remove the lower arm assembly and knuckle connecting bolt. Remove the lower arm mounting bolt.

6. Remove the lower control arm assembly from the vehicle.

To install

7. Install the lower control arm to its mounting.

8. Continue the installation in the reverse order of the removal procedure.

9. Check and adjust the rear alignment, as necessary.

2006

1. Before servicing the vehicle, refer to the Precautions Section.

2. Raise and support the vehicle safely.

3. Remove the shock absorber assembly and knuckle connection retaining bolt.

4. Properly support the lower control arm assembly.

5. Remove the lower control arm assembly and stabilizer link assembly connection.

6. Remove the lower control arm assembly and knuckle connection bolt.

7. Remove the lower arm bolt and arm plate.

8. Remove the lower control arm from the vehicle.

To install

9. Install the lower control arm to its mounting.

10. Continue the installation in the reverse order of the removal procedure.

11. Check and adjust the rear alignment, as necessary.

Galant

2004–2006

1. Before servicing the vehicle, refer to the Precautions Section.

2. Raise and support the vehicle safely.

3. Remove the shock absorber assembly and knuckle connection retaining bolt.

4. Properly support the lower control arm assembly.

5. Remove the lower control arm assembly and stabilizer link assembly connection.

6. Remove the lower control arm assembly and knuckle connection bolt.

7. Remove the lower arm bolt and arm plate.

8. Remove the lower control arm from the vehicle.

To install

9. Install the lower control arm to its mounting.

10. Continue the installation in the reverse order of the removal procedure.

11. Check and adjust the rear alignment, as necessary.

Lancer

1. Before servicing the vehicle, refer to the Precautions Section.

2. Raise and support the vehicle safely.

3. Support the lower arm assembly, using a floor jack.

4. After making a mating mark on the toe-in or camber adjusting bolt, remove the control link or lower arm.

5. Remove the lower arm and trailing arm bolt and nut.

6. Remove the lower shock absorber retaining bolt and nut.

7. Remove the lower control arm from the vehicle.

To install

8. Install the lower control arm to its mounting.

9. Continue the installation in the reverse order of the removal procedure.

10. Check and adjust the rear alignment, as necessary.

Evolution

1. Before servicing the vehicle, refer to the Precautions Section.

2. Raise and support the vehicle safely.

3. After making a mating mark on the toe-in or camber adjusting bolt, remove the control link or lower arm.

4. Properly support the lower control arm assembly.

5. Remove the lower control arm assembly mounting bolt.

6. Remove the lower control arm assembly.

To install

7. Install the lower control arm to its mounting.

8. Continue the installation in the reverse order of the removal procedure.

9. Check and adjust the rear alignment, as necessary.

Upper Control Arm

REMOVAL & INSTALLATION

Eclipse

2004–2005

1. Before servicing the vehicle, refer to the Precautions Section.

2. Raise and support the vehicle safely.

3. Position a floor jack to properly support the knuckle before removing the upper control arm.

4. Remove the upper control arm and knuckle connecting bolt.

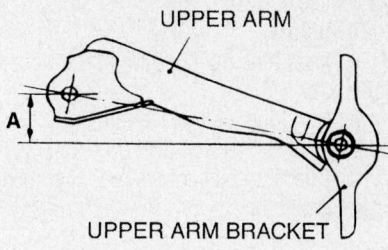

09482_GALA_G0153

Rear upper control arm bracket alignment—2004–2005 Eclipse

5. Remove the upper control arm assembly mounting bolts.

6. Remove the upper control arm bracket assembly.

7. Remove the upper control arm bracket. Remove the upper control arm from the vehicle.

To install

8. Install the upper control arm to its mounting.

9. Position the upper control arm so that value "A" is 1.38–1.54 inches, as shown in the illustration.

10. Check the installation angle "B" and "C" as shown in the illustration.

11. Continue the installation in the reverse order of the removal procedure.

12. Check and adjust the rear alignment, as necessary.

2006

1. Before servicing the vehicle, refer to the Precautions Section.

2. Raise and support the vehicle safely.

3. If equipped with ABS, remove the wheel speed sensor retaining bolts.

4. Position a floor jack to properly support the knuckle before removing the upper control arm.

5. Remove the upper arm assembly and knuckle retaining bolt and nut.

6. Remove the upper arm assembly retaining bolts and nuts.

B: 220.1 mm (8.67 in)
C: 274.4 mm (10.80 in)

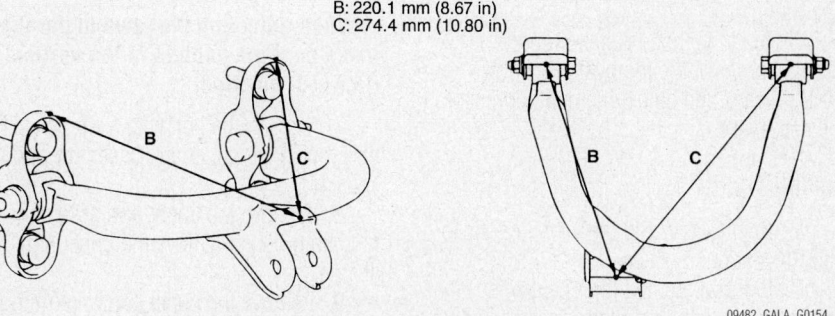

09482_GALA_G0154

Rear upper control arm installation angle—2004–2005 Eclipse

7. Remove the upper control arm from the vehicle.

To install

8. Install the upper control arm to its mounting.

9. Continue the installation in the reverse order of the removal procedure.

10. Check and adjust the rear alignment, as necessary.

Galant

2004–2006

1. Before servicing the vehicle, refer to the Precautions Section.

2. Raise and support the vehicle safely.

3. If equipped with ABS, remove the wheel speed sensor retaining bolts.

4. Position a floor jack to properly support the knuckle before removing the upper control arm.

5. Remove the upper arm assembly and knuckle retaining bolt and nut.

6. Remove the upper arm assembly retaining bolts and nuts.

7. Remove the upper control arm from the vehicle.

To install

8. Install the upper control arm to its mounting.

9. Continue the installation in the reverse order of the removal procedure.

10. Check and adjust the rear alignment, as necessary.

Lancer

1. Before servicing the vehicle, refer to the Precautions Section.

2. Raise and support the vehicle safely.

3. Support the lower arm assembly, using a floor jack.

4. After making a mating mark on the toe-in or camber adjusting bolt, remove the control link or lower arm.

5. Remove the upper control arm retaining bolts. Remove the upper control arm from the vehicle.

To install

6. Install the upper control arm to its mounting.

7. Continue the installation in the reverse order of the removal procedure.

8. Check and adjust the rear alignment, as necessary.

Evolution

1. Before servicing the vehicle, refer to the Precautions Section.

2. Remove the fuel filler cap, when removing the left side upper arm assembly.

3. Raise and support the vehicle safely.

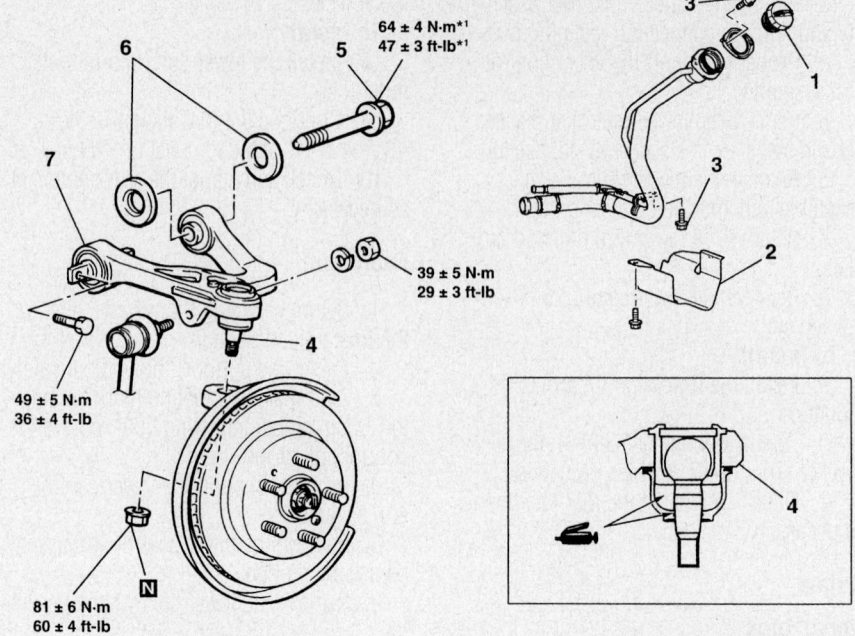

1. FUEL FILLER CAP
2. PROTECTOR
3. BOLT
4. UPPER ARM ASSEMBLY AND KNUCKLE CONNECTION

5. UPPER ARM ASSEMBLY MOUNTING BOLT
6. UPPER ARM STOPPER
7. UPPER ARM ASSEMBLY

09482_GALA_G0155

Rear upper control arm and related components—Evolution

4. Remove the fuel protector and bolt, when removing the left side upper arm assembly.

5. Position a floor jack to properly support the rotor and hub assembly before removing the upper control

➡**Do not remove the nut from the ball joint. Loosen it and use special tool MB991897 to avoid possible damage to the ball joint threads. Hang the special tool in place with wire or string to prevent it from falling.**

6. Install the special tool. Turn the bolt and knob as necessary to make the jaws of the tool parallel. Tighten the bolt by hand and confirm that the jaws are still parallel.

➡**When adjusting the jaws in parallel, make sure the knob is in the vertical (upward) position.**

7. Tighten the bolt with a wrench to disconnect the upper arm assembly and the knuckle.

8. Remove the upper arm assembly mounting bolts. Remove the upper arm stopper.

9. Remove the upper control arm from the vehicle.

To install

10. Install the upper control arm to its mounting.

11. Continue the installation in the reverse order of the removal procedure.

12. Check and adjust the rear alignment, as necessary.

Wheel Bearings

ADJUSTMENT

The rear wheel bearings are not adjustable. If the bearings are noisy or become loose, they must be replaced.

REMOVAL & INSTALLATION

Diamante

1. Before servicing the vehicle, refer to the Precautions Section.

2. Raise the vehicle and support safely.

3. Remove the appropriate wheel assembly.

4. If equipped with Anti-Lock Brake (ABS), remove the Vehicle Speed Sensor (VSS).

5. Remove the brake drum from the hub assembly.

6. From the back of the knuckle, remove the 4 bolts securing the hub to the knuckle.

7. Remove the hub and bearing assembly from the knuckle.

➥**The hub assembly is not serviceable and should not be disassembled.**

8. If replacing the hub, use special socket MB991248 and a press, to remove the wheel sensor rotor from the hub.

To install

9. Press the wheel sensor rotor onto the hub.

10. Install or connect the following:
- Hub to the knuckle and tighten the mounting bolts to 54–65 ft. lbs. (74–88 Nm)
- Brake drum on the hub
- VSS, if equipped with ABS
- Wheel assembly and lower the vehicle

Eclipse

➥**The hub and bearing assembly is serviced as a unit.**

1. Before servicing the vehicle, refer to the Precautions Section.

2. Disconnect the negative battery cable.

3. Raise and safely support the vehicle.

4. Remove or disconnect the following:
- Wheel and tire assembly
- Rear wheel speed sensor if equipped with Anti-Lock Brake (ABS)
- Brake drum. Or, if equipped with disc brakes, remove the caliper assembly and rotor. Suspend the caliper out of the way with wire.

5. On vehicles with rear disc brakes, remove the parking brake shoes.

6. Remove the hub mounting bolts from behind the backing plate and remove the hub.

➥**The rotor for the ABS must be removed and installed using a press.**

To install:

7. Press the rotor (ABS) to the hub.

8. Install or connect the following:
- Hub and tighten the mounting bolts to 54–65 ft. lbs. (74–88 Nm)
- Parking brake shoes if equipped
- Rotor and caliper or drum
- Speed sensor if equipped
- Wheel and tire assembly

9. Lower the vehicle to the floor.

10. Connect the negative battery cable.

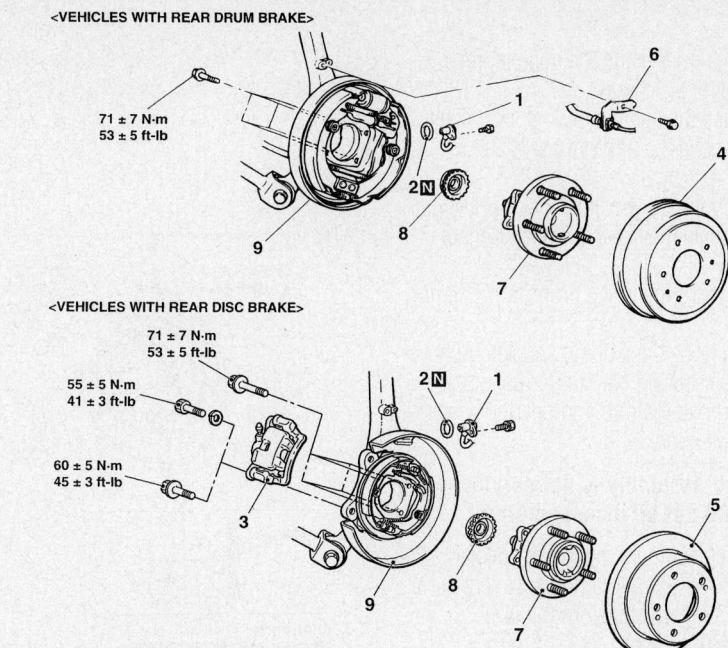

1. REAR WHEEL SPEED SENSOR <VEHICLES WITH ABS> (REFER TO GROUP 35B, WHEEL SPEED SENSOR P.35B-56.)
2. O-RING
3. CALIPER ASSEMBLY
4. BRAKE DRUM
5. BRAKE DISC
6. BRAKE HOSE INSTALLATION BRACKET
7. REAR HUB ASSEMBLY
8. ABS ROTOR <VEHICLES WITH ABS>
9. BACKING PLATE

09482_GALA_G0156

Rear axle hub and related components—2002–2005 Eclipse

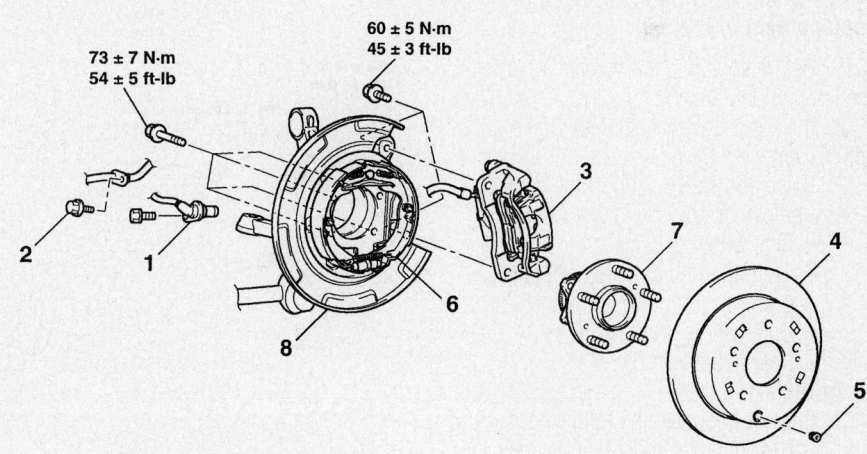

1. REAR WHEEL SPEED SENSOR
2. BRAKE HOSE CLAMP BOLT
3. CALIPER ASSEMBLY
4. BRAKE DISC
5. PLUG
6. REAR BRAKE ASSEMBLY
7. REAR HUB ASSEMBLY
8. BACKING PLATE

09482_GALA_G0157

Rear axle hub and related components—2006 Eclipse and 2004–2006 Galant

Galant

1. Before servicing the vehicle, refer to the Precautions Section.

2. Raise the vehicle and support safely.

3. Remove the appropriate wheel assembly.

4. If equipped with Anti-Lock Brake (ABS), remove the Vehicle Speed Sensor (VSS).

5. Remove the brake drum from the hub assembly.

6. From the back of the knuckle, remove the 4 bolts securing the hub to the knuckle.

7. Remove the hub and bearing assembly from the knuckle.

➡ **The hub assembly is not serviceable and should not be disassembled.**

8. If replacing the hub, use special socket MB991248 and a press, to remove the wheel sensor rotor from the hub.

To install

9. Press the wheel sensor rotor onto the hub.

10. Install or connect the following:
- Hub to the knuckle and tighten the mounting bolts to 54–65 ft. lbs. (74–88 Nm)
- Brake drum on the hub
- VSS, if equipped with ABS
- Wheel assembly and lower the vehicle

Lancer

➡ **The wheel bearing is serviced by replacement of the hub.**

1. Before servicing the vehicle, refer to the Precautions Section.

2. If equipped with Anti-Lock Brake (ABS), remove the wheel speed sensor.

3. Raise and safely support the vehicle.

4. Remove or disconnect the following:
- Rear wheel
- Caliper and brake disc or brake drum
- Dust cap and flange nut
- Rear hub assembly

To install:

5. Install or connect the following:
- Rear hub assembly using a new flange nut. Torque the flange nut to 130 ft. lbs. (180 Nm).
- Dust cap
- Wheel speed sensor if removed. The air gap should be 0.012–0.035 in. (0.3–0.9mm).
- Brake disc and caliper, or brake drum
- Rear wheel assembly and lower the vehicle to the floor

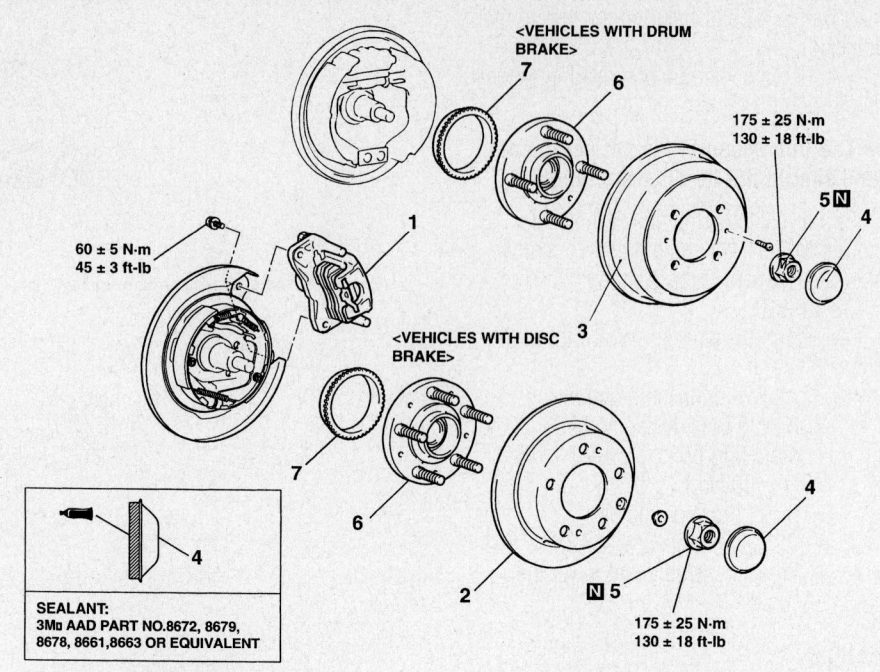

SEALANT:
3Mᵇ AAD PART NO.8672, 8679, 8678, 8661,8663 OR EQUIVALENT

1. CALIPER ASSEMBLY
2. BRAKE DISC
3. REAR DRUM
4. HUB CAP
5. JAM NUT
6. REAR HUB ASSEMBLY
7. ABS ROTOR <VEHICLES WITH ABS>

09482_GALA_G0158

Rear axle hub and related components—Lancer

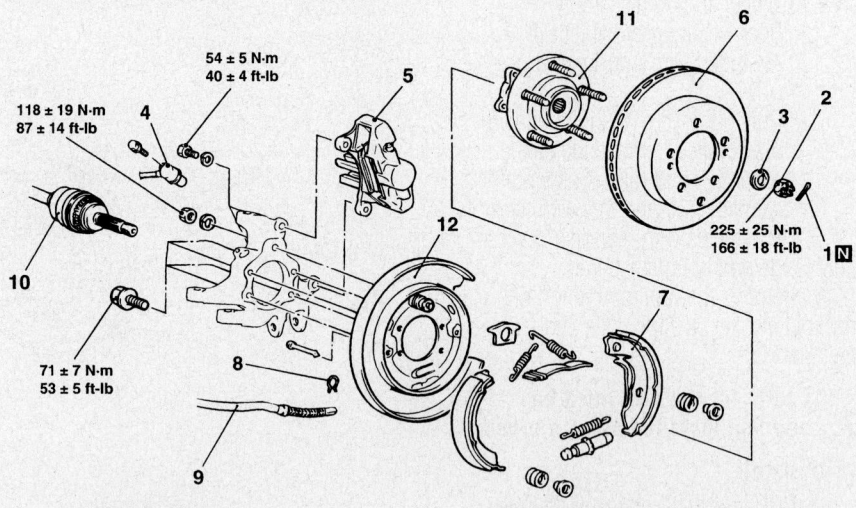

1. COTTER PIN
2. DRIVESHAFT NUT
3. WASHER
4. REAR ABS SENSOR
5. CALIPER ASSEMBLY
6. BRAKE DISC
7. PARKING BRAKE SHOE AND LINING ASSEMBLY
8. CLIP
9. PARKING BRAKE CABLE CONNECTION
10. REAR DRIVESHAFT CONNECTION
11. REAR WHEEL HUB ASSEMBLY
12. BACKING PLATE

09482_GALA_G0159

Rear axle hub and related components—Evolution

Evolution

➡ **The wheel bearing is serviced by replacement of the hub.**

1. Before servicing the vehicle, refer to the Precautions Section.

2. Raise and support the vehicle safely. Remove the tire and wheel assembly.

3. Remove the cotter pin. Remove the driveshaft nut and washer.

4. Remove the rear ABS sensor. Remove the caliper. Position it to the side with wire. Do not disconnect the brake hose.

5. Remove the parking brake shoe and lining assembly.

6. Remove the clip. Remove the parking brake cable connection.

7. Disconnect the rear driveshaft con-nection. Remove the wheel hub assembly retaining bolts.

8. Remove the rear hub assembly from the vehicle.

To install:

9. Installation is the reverse of the removal procedure.

10. Adjust the parking brake.

Mirage

➡ **The wheel bearing is serviced by replacement of the hub.**

1. Before servicing the vehicle, refer to the Precautions Section.

2. If equipped with Anti-Lock Brake (ABS), remove the wheel speed sensor.

3. Raise and safely support the vehicle.

4. Remove or disconnect the following:
   • Rear wheel
   • Caliper and brake disc or brake drum
   • Dust cap and flange nut
   • Rear hub assembly

To install:

5. Install or connect the following:
   • Rear hub assembly using a new flange nut. Torque the flange nut to 130 ft. lbs. (180 Nm).
   • Dust cap
   • Wheel speed sensor if removed. The air gap should be 0.012–0.035 in. (0.3–0.9mm).
   • Brake disc and caliper, or brake drum
   • Rear wheel assembly and lower the vehicle to the floor

BRAKES

Brake Caliper

REMOVAL & INSTALLATION

Front

DIAMANTE

1. Before servicing the vehicle, refer to the Precautions Section.

2. As required, partially drain the master cylinder.

3. Remove or disconnect the follow-ing:
   • Wheels
   • Brake hose from the caliper
   • Caliper guide and lock pins and lift the caliper assembly from the caliper support

To install

4. Install or connect the following:
   • Caliper onto the caliper support
   • Guide pin and lock pin and tighten to specification

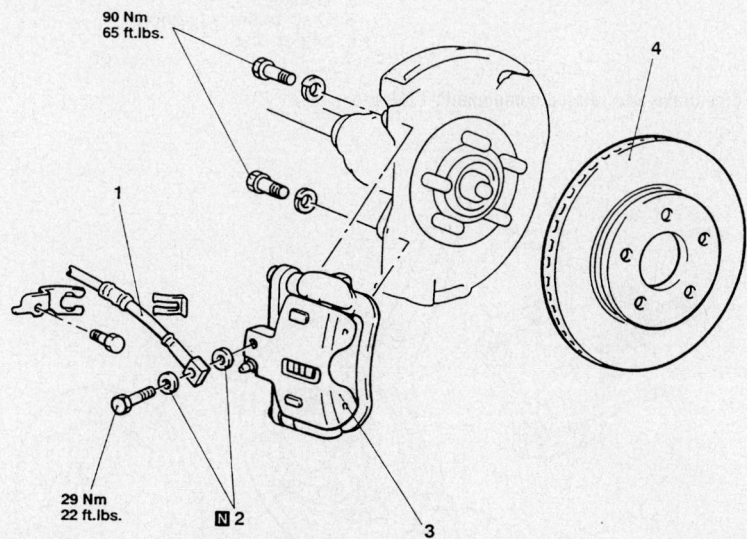

90 Nm
65 ft.lbs.

29 Nm
22 ft.lbs.   N 2

1. CONNECTION FOR THE BRAKE HOSE
2. GASKET
3. FRONT BRAKE ASSEMBLY
4. BRAKE DISC

93016G39

Front disc brake and related components—Diamante

   • Brake hose or banjo bolt with new washers

5. Bleed the brake system.

6. Install the wheels.

ECLIPSE

1. Before servicing the vehicle, refer to the Precautions Section.

2. As required, partially drain the master cylinder.

3. Remove or disconnect the following:
   • Wheels
   • Brake hose
   • Caliper guide and lock pins
   • Caliper assembly from the caliper support

To install

4. Install or connect the following:
   • Brake caliper into position on the caliper support
   • Guide and lock pins
   • Brake hose. Bleed the brake system.
   • Wheels

GALANT

1. Before servicing the vehicle, refer to the Precautions Section.

2. As required, partially drain the master cylinder.

3. Remove or disconnect the following:
   • Wheels
   • Brake hose from the caliper
   • Caliper guide and lock pins and lift the caliper assembly from the caliper support

To install

4. Install or connect the following:
   • Caliper onto the caliper support
   • Guide pin and lock pin and tighten to specification

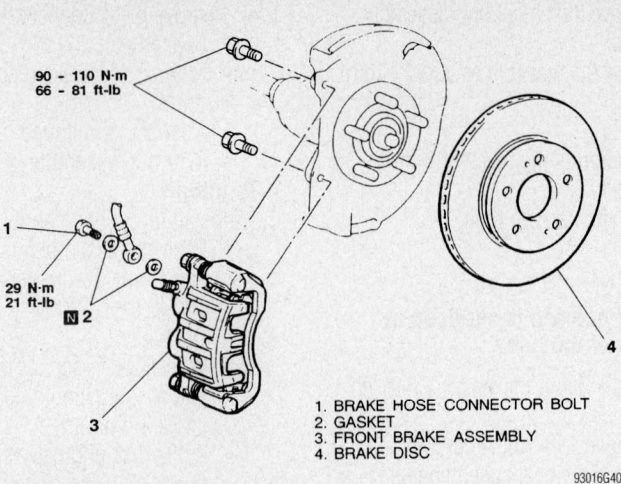

Front disc brake and related components—Galant

1. BRAKE HOSE CONNECTOR BOLT
2. GASKET
3. FRONT BRAKE ASSEMBLY
4. BRAKE DISC

93016G40

- Brake hose or banjo bolt with new washers
5. Bleed the brake system.
6. Install the wheels.

LANCER

1. Before servicing the vehicle, refer to the Precautions Section.
2. As required, partially drain the master cylinder.
3. Remove or disconnect the following:
- Wheels
- Brake hose
- Caliper guide and lock pins
- Caliper assembly from the caliper support

To install

4. Install or connect the following:
- Brake caliper into position on the caliper support
- Guide and lock pins
- Brake hose. Bleed the brake system.
- Wheels

EVOLUTION

1. Before servicing the vehicle, refer to the Precautions Section.
2. As required, partially drain the master cylinder.
3. Raise and support the vehicle safely.
4. Remove the tire and wheel assembly.
5. Disconnect and plug the brake hose connection. Discard the gasket.
6. Remove the caliper retaining bolts. Remove the caliper from the vehicle.

To install

7. Position the caliper to its mounting on the vehicle.
8. Install the caliper retaining bolts.
9. Continue the installation in the reverse order of the removal procedure.
10. Be sure to check and adjust the brake fluid level, as necessary.

11. Bleed the brake system, as necessary.

MIRAGE

1. Before servicing the vehicle, refer to the Precautions Section.
2. As required, partially drain the master cylinder.
3. Remove or disconnect the following:
- Wheels
- Brake hose
- Caliper guide and lock pins
- Caliper assembly from the caliper support

To install

4. Install or connect the following:
- Brake caliper into position on the caliper support

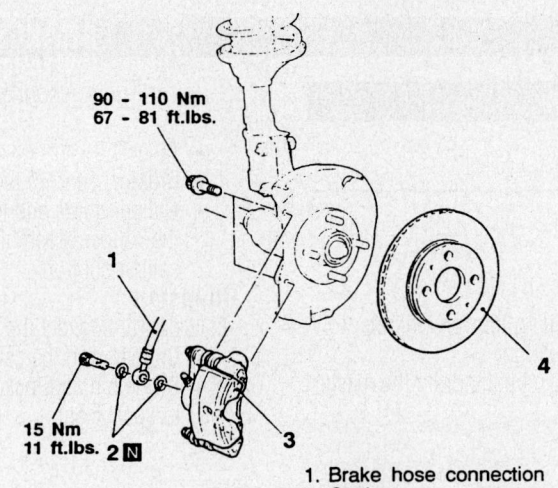

1. Brake hose connection
2. Gasket
3. Disc brake assembly
4. Brake disc

93016G36

Front disc brake and related components—Mirage

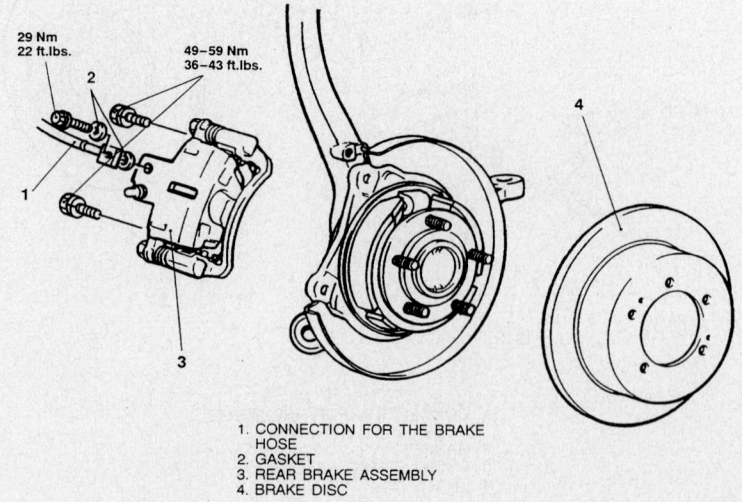

1. CONNECTION FOR THE BRAKE HOSE
2. GASKET
3. REAR BRAKE ASSEMBLY
4. BRAKE DISC

93016G41

Rear disc brake and related components—Diamante

- Guide and lock pins
- Brake hose. Bleed the brake system.
- Wheels

Rear

DIAMANTE

1. Before servicing the vehicle, refer to the Precautions Section.
2. As required, partially drain the master cylinder.
3. Remove or disconnect the following:
- Wheels
- Brake hose from the caliper
- Caliper guide and lock pins and lift the caliper assembly from the caliper support

To install

4. Install or connect the following:
- Caliper onto the caliper support
- Guide pin and lock pin and tighten to specification
- Brake hose or banjo bolt with new washers
5. Bleed the brake system.
6. Install the wheels.

ECLIPSE

1. Before servicing the vehicle, refer to the Precautions Section.
2. Loosen the parking brake cable adjustment from inside the vehicle.
3. As required, partially drain the master cylinder.
4. Remove or disconnect the following:
- Wheels
- Brake hose
- Caliper lock pin. Pivot the caliper upward, and slide the caliper assembly from the caliper support.

To install:

5. Install or connect the following:
- Caliper over the brake pads
- Lock pin, after lubricating it, and tighten to 23 ft. lbs. (32 Nm)
- Brake hose to the caliper
6. Bleed the brake system.
- Wheels

GALANT

1. Before servicing the vehicle, refer to the Precautions Section.
2. As required, partially drain the master cylinder.
3. Remove or disconnect the following:
- Wheels
- Brake hose from the caliper
- Caliper guide and lock pins and lift the caliper assembly from the caliper support

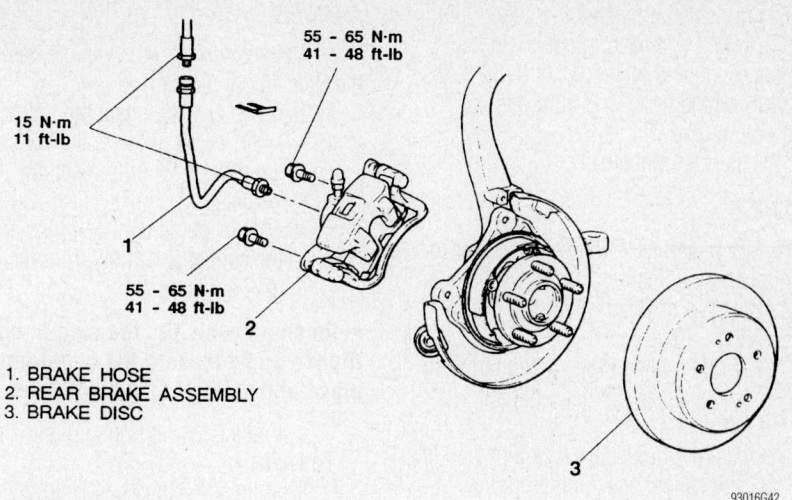

15 N·m
11 ft-lb

55 - 65 N·m
41 - 48 ft-lb

1

55 - 65 N·m
41 - 48 ft-lb

2

1. BRAKE HOSE
2. REAR BRAKE ASSEMBLY
3. BRAKE DISC

3

93016G42

Rear disc brakes—Galant

To install

4. Install or connect the following:
- Caliper onto the caliper support
- Guide pin and lock pin and tighten to specification
- Brake hose or banjo bolt with new washers
5. Bleed the brake system.
6. Install the wheels.

LANCER

1. Before servicing the vehicle, refer to the Precautions Section.
2. As required, partially drain the master cylinder.
3. Raise and support the vehicle safely.
4. Remove the tire and wheel assembly.
5. Disconnect and plug the brake hose connection. Discard the gasket.
6. Remove the caliper retaining bolts. Remove the caliper from the vehicle.

To install

7. Position the caliper to its mounting on the vehicle.
8. Install the caliper retaining bolts.
9. Continue the installation in the reverse order of the removal procedure.
10. Be sure to check and adjust the brake fluid level, as necessary.
11. Bleed the brake system, as necessary.

EVOLUTION

1. Before servicing the vehicle, refer to the Precautions Section.
2. As required, partially drain the master cylinder.
3. Raise and support the vehicle safely.
4. Remove the tire and wheel assembly.
5. Disconnect and plug the brake hose connection. Discard the gasket.

6. Remove the caliper retaining bolts. Remove the caliper from the vehicle.

To install

7. Position the caliper to its mounting on the vehicle.
8. Install the caliper retaining bolts.
9. Continue the installation in the reverse order of the removal procedure.
10. Be sure to check and adjust the brake fluid level, as necessary.
11. Bleed the brake system, as necessary.

Disc Brake Pads

REMOVAL & INSTALLATION

Front

DIAMANTE

1. Before servicing the vehicle, refer to the Precautions Section.
2. Remove or disconnect the following:
- Drain some of the brake fluid from the master cylinder reservoir
- Wheels
- Caliper guide and lock pins and lift the caliper assembly from the caliper support.

➡**On some vehicles, the caliper can be flipped up by leaving the upper pin in place and using it as a pivot point.**

- Brake pads, spring clip and shims

To install:

3. Compress the pistons back into the caliper bore.
4. Lubricate slide points and install the brake pads, shims and spring clip onto the caliper support. Install the caliper over the brake pads.

5. Lubricate and install the caliper guide and lock pins in their original positions. Tighten guide and locking pins to 54 ft. lbs. (75 Nm) on the front, and 20 ft. lbs. (27 Nm) on the rear.

6. Install the wheels.

ECLIPSE

1. Before servicing the vehicle, refer to the Precautions Section.

2. Remove or disconnect the following:
- Drain some of the brake fluid from the master cylinder reservoir
- Wheels
- Caliper guide and lock pins and lift the caliper assembly from the caliper support.

➡️On some vehicles, the caliper can be flipped up by leaving the upper pin in place and using it as a pivot point.

- Brake pads, spring clip and shims

To install:

3. Compress pistons back into the caliper bore.

4. Lubricate slide points and install the brake pads, shims and spring clips onto the caliper support. Install the caliper over the brake pads.

5. Lubricate and install the caliper guide and lock pins in their original positions.

6. Install the wheels.

GALANT

1. Before servicing the vehicle, refer to the Precautions Section.

2. Remove or disconnect the following:
- Drain some of the brake fluid from the master cylinder reservoir
- Wheels
- Caliper guide and lock pins and lift the caliper assembly from the caliper support.

➡️On some vehicles, the caliper can be flipped up by leaving the upper pin in place and using it as a pivot point.

- Brake pads, spring clip and shims

To install:

3. Compress the pistons back into the caliper bore.

4. Lubricate slide points and install the brake pads, shims and spring clip onto the caliper support. Install the caliper over the brake pads.

5. Lubricate and install the caliper guide and lock pins in their original positions. Tighten guide and locking pins to specification.

6. Install the wheels.

LANCER

1. Before servicing the vehicle, refer to the Precautions Section.

2. Remove or disconnect the following:
- Drain some of the brake fluid from the master cylinder reservoir
- Wheels
- Caliper guide and lock pins and lift the caliper assembly from the caliper support.

➡️On some vehicles, the caliper can be flipped up by leaving the upper pin in place and using it as a pivot point.

- Brake pads, spring clip and shims

To install:

3. Compress pistons back into the caliper bore.

4. Lubricate slide points and install the brake pads, shims and spring clips onto the caliper support. Install the caliper over the brake pads.

5. Lubricate and install the caliper guide and lock pins in their original positions.

6. Install the wheels.

EVOLUTION

1. Before servicing the vehicle, refer to the Precautions Section.

2. As required, partially drain the master cylinder.

3. Raise and support the vehicle safely.

4. Remove the tire and wheel assembly.

5. Remove the caliper retaining bolts. Remove the caliper from the vehicle.

6. Remove the brake pads and shims from the caliper.

To install

7. Position the brake pads and shims in the caliper.

8. Position the caliper to its mounting on the vehicle.

9. Install the caliper retaining bolts.

10. Continue the installation in the reverse order of the removal procedure.

11. Be sure to check and adjust the brake fluid level, as necessary.

12. Bleed the brake system, as necessary.

MIRAGE

1. Before servicing the vehicle, refer to the Precautions Section.

2. Remove or disconnect the following:
- Drain some of the brake fluid from the master cylinder reservoir
- Wheels
- Caliper guide and lock pins and lift the caliper assembly from the caliper support.

➡️On some vehicles, the caliper can be flipped up by leaving the upper pin in place and using it as a pivot point.

- Brake pads, spring clip and shims

To install:

3. Compress pistons back into the caliper bore.

4. Lubricate slide points and install the brake pads, shims and spring clips onto the caliper support. Install the caliper over the brake pads.

5. Lubricate and install the caliper guide and lock pins in their original positions.

6. Install the wheels.

Rear

DIAMANTE

1. Before servicing the vehicle, refer to the Precautions Section.

2. Remove or disconnect the following:
- Drain some of the brake fluid from the master cylinder reservoir
- Wheels
- Caliper guide and lock pins and lift the caliper assembly from the caliper support.

➡️On some vehicles, the caliper can be flipped up by leaving the upper pin in place and using it as a pivot point.

- Brake pads, spring clip and shims

To install:

3. Compress the pistons back into the caliper bore.

4. Lubricate slide points and install the brake pads, shims and spring clip onto the caliper support. Install the caliper over the brake pads.

5. Lubricate and install the caliper guide and lock pins in their original positions. Tighten guide and locking pins to 54 ft. lbs. (75 Nm) on the front, and 20 ft. lbs. (27 Nm) on the rear.

6. Install the wheels.

ECLIPSE

1. Before servicing the vehicle, refer to the Precautions Section.

2. Remove or disconnect the following:
- Drain some of the brake fluid from the master cylinder reservoir

3. Loosen the parking brake cable adjustment from inside the vehicle.
- Wheels
- Parking brake cable
- Caliper lower pin and swing the caliper assembly upwards
- Outer shim, brake pads and spring clips from the caliper support.

4. Compress the piston into the caliper bore.

To install:

5. Lubricate all sliding and pivot points. Install the brake pads, shims and spring clip to the caliper support. Install the caliper over the brake pads.

6. Lubricate, install and tighten the lower pin.

7. Install the wheels.

GALANT

1. Before servicing the vehicle, refer to the Precautions Section.

2. Remove or disconnect the following:
- Drain some of the brake fluid from the master cylinder reservoir
- Wheels
- Caliper guide and lock pins and lift the caliper assembly from the caliper support.

➡**On some vehicles, the caliper can be flipped up by leaving the upper pin in place and using it as a pivot point.**

- Brake pads, spring clip and shims

To install:

3. Compress the pistons back into the caliper bore.

4. Lubricate slide points and install the brake pads, shims and spring clip onto the caliper support. Install the caliper over the brake pads.

5. Lubricate and install the caliper guide and lock pins in their original positions. Tighten guide and locking pins to specification.

6. Install the wheels.

LANCER

1. Before servicing the vehicle, refer to the Precautions Section.

2. Remove or disconnect the following:
- Drain some of the brake fluid from the master cylinder reservoir

3. Loosen the parking brake cable adjustment from inside the vehicle.
- Wheels
- Parking brake cable
- Caliper lower pin and swing the caliper assembly upwards
- Outer shim, brake pads and spring clips from the caliper support.

To install:

4. Lubricate all sliding and pivot points. Install the brake pads, shims and spring clip to the caliper support. Install the caliper over the brake pads.

5. Lubricate, install and tighten the lower pin.

6. Install the wheels.

EVOLUTION

1. Before servicing the vehicle, refer to the Precautions Section.

2. As required, partially drain the master cylinder.

3. Raise and support the vehicle safely.

4. Remove the tire and wheel assembly.

5. Remove the caliper retaining bolts. Remove the caliper from the vehicle.

6. Remove the brake pads and shims from the caliper.

To install

7. Position the brake pads and shims in the caliper.

8. Position the caliper to its mounting on the vehicle.

9. Install the caliper retaining bolts.

10. Continue the installation in the reverse order of the removal procedure.

11. Be sure to check and adjust the brake fluid level, as necessary.

12. Bleed the brake system, as necessary.

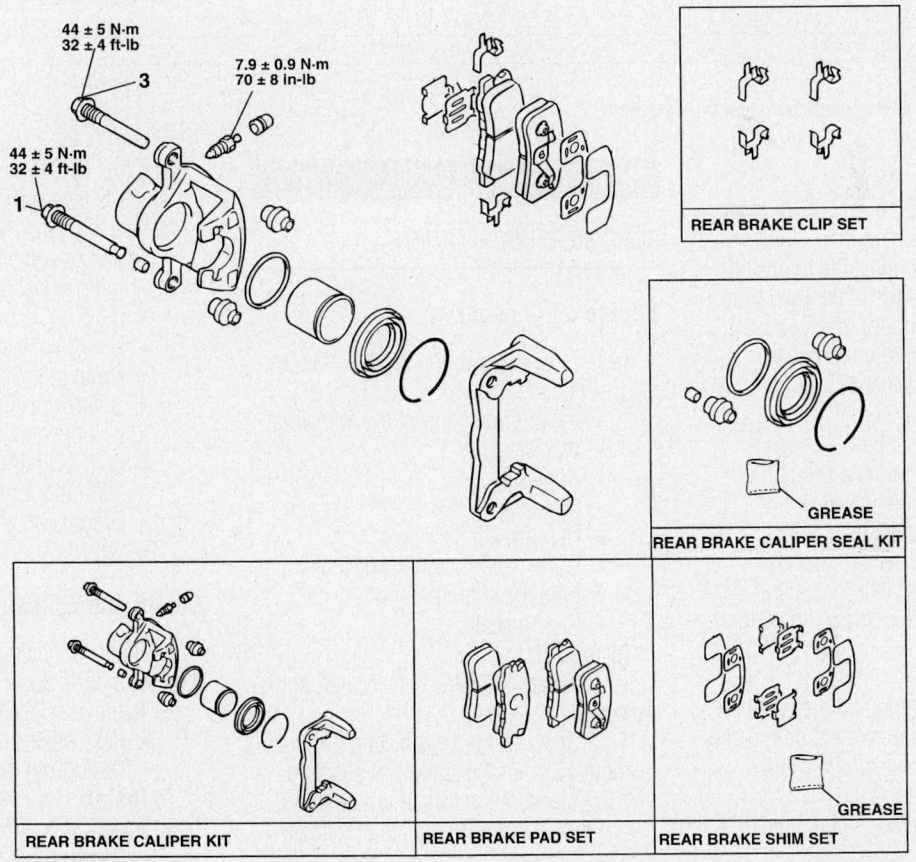

44 ± 5 N·m
32 ± 4 ft-lb

7.9 ± 0.9 N·m
70 ± 8 in-lb

44 ± 5 N·m
32 ± 4 ft-lb

REAR BRAKE CLIP SET

GREASE

REAR BRAKE CALIPER SEAL KIT

REAR BRAKE CALIPER KIT

REAR BRAKE PAD SET

REAR BRAKE SHIM SET

GREASE

09482_GALA_G0160

Rear disc brake pads and related components—Lancer

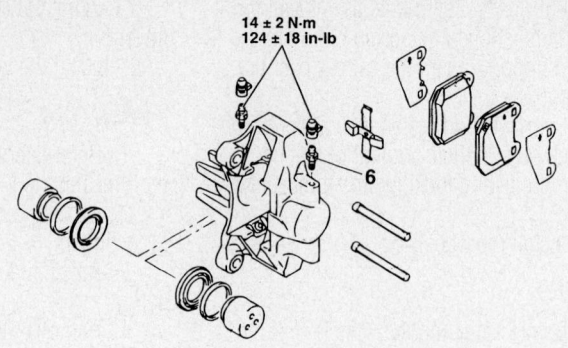

14 ± 2 N·m
124 ± 18 in-lb

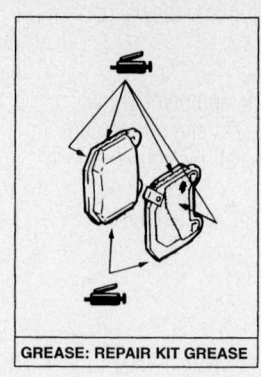

GREASE: REPAIR KIT GREASE

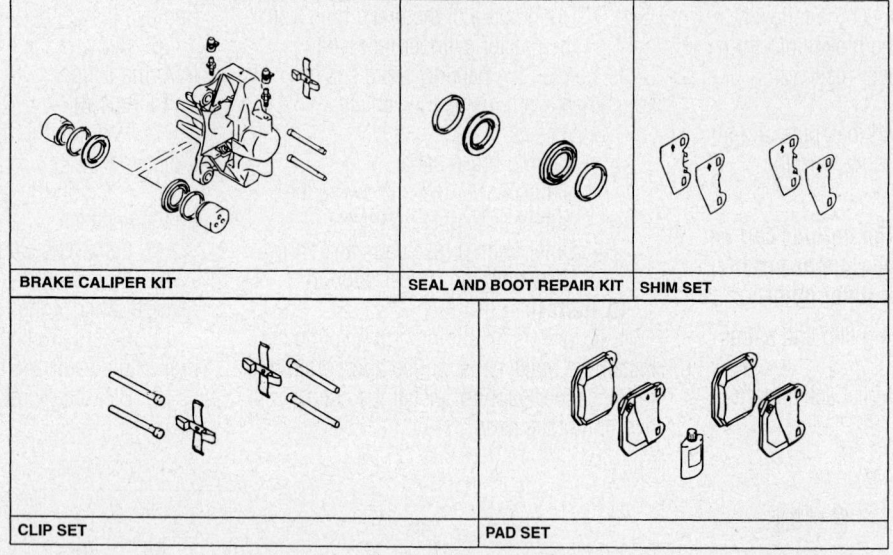

BRAKE CALIPER KIT

SEAL AND BOOT REPAIR KIT

SHIM SET

CLIP SET

PAD SET

09482_GALA_G0161

Rear disc brake pads and related components—Evolution

MIRAGE

1. Before servicing the vehicle, refer to the Precautions Section.
2. Remove or disconnect the following:
   - Drain some of the brake fluid from the master cylinder reservoir
3. Loosen the parking brake cable adjustment from inside the vehicle.
   - Wheels
   - Parking brake cable
   - Caliper lower pin and swing the caliper assembly upwards
   - Outer shim, brake pads and spring clips from the caliper support.
4. Thread the piston into the caliper bore clockwise using disc brake driver tool MB990652.

To install:

5. Lubricate all sliding and pivot points. Install the brake pads, shims and spring clip to the caliper support. Install the caliper over the brake pads.
6. Lubricate, install and tighten the lower pin.
7. Install the wheels.

Brake Drums

REMOVAL & INSTALLATION

Mirage and Lancer

1. Before servicing the vehicle, refer to the Precautions Section.
2. Remove or disconnect the following:
   - Wheels
   - Dust cap
   - Self-locking nut
   - Outer wheel bearing
   - Drum with the inner wheel bearing from the spindle
   - Grease seal

To install:

3. To determine if the self-locking nut is reusable:
   a. Screw in the self-locking nut until about ⅛ in. of the spindle is showing.
   b. Measure the torque required to turn the self-locking nut counterclockwise.
   c. The lowest allowable torque is 48 inch lbs. (5.5 Nm). If the measured torque is less than the specification, replace the nut.
4. Install or connect the following:
   - Inner wheel bearing, after lubricating it
   - New grease seal
   - Drum to the spindle
   - Outer wheel bearing, making sure to lubricate it first
   - Self-locking nut. Torque the nut to 108–145 ft. lbs. (150–200 Nm).
   - Grease cap
   - Wheels

Galant and Eclipse

1. Before servicing the vehicle, refer to the Precautions Section.
2. Remove or disconnect the following:
   - Rear wheels
   - Drum from the rear hub assembly

To install:

3. Install or connect the following:
   - Drum on the rear hub assembly
   - Wheels

Brake Shoes

REMOVAL & INSTALLATION

Mirage and Lancer

1. Before servicing the vehicle, refer to the Precautions Section.
2. Remove or disconnect the following:
   - Wheels
   - Brake drum
   - Shoe-to-shoe spring
   - Shoe-to-lever spring and adjuster assembly
   - Shoe hold-down clips and the brake shoes
   - Parking brake cable from the rear shoes by spreading the horseshoe clip apart.

   To install
3. Lubricate the backing plate bosses, anchor pin, and parking brake actuating mechanism with lithium-based grease.
4. Install or connect the following:
   - Parking brake arm to the appropriate brake shoe
   - Brake shoes and the shoe hold-down clips
   - Adjuster assembly and the shoe-to-lever spring
   - Shoe-to-shoe spring
5. Pre-adjust the shoes so the drum slides on with a light drag and install brake drum.
   - New wheel bearing self-locking nut and torque to 130 ft. lbs. (180 Nm)
   - Wheel bearing dust cap and adjust the rear brake shoes

Galant and Eclipse

1. Before servicing the vehicle, refer to the Precautions Section.
2. Remove or disconnect the following:
   - Rear wheels and drums
   - Lever return spring
   - Shoe-to-lever spring
   - Adjuster lever
   - Auto-adjuster assembly
   - Retainer spring
   - Brake shoe hold-down springs and spring cups
   - Shoe-to-shoe spring
   - Brake shoes
   - Parking brake cable from the lever on the rear shoe

   To install:
3. Remove the parking brake lever from the used shoe and install it on the new brake shoe. Make sure the wave washer is installed in the proper direction.
4. Clean the backing plate and lightly

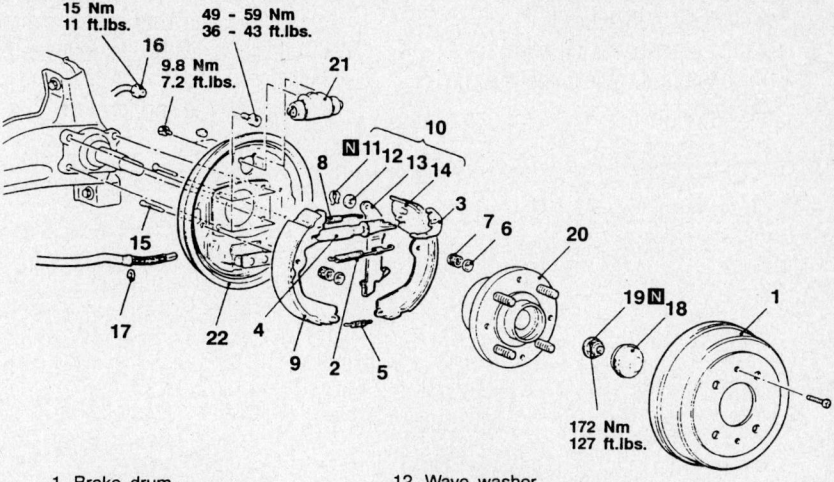

1. Brake drum
2. Shoe-to-lever spring
3. Adjuster lever
4. Auto adjuster assembly
5. Retainer spring
6. Shoe hold-down cup
7. Shoe hold-down spring
8. Shoe-to-shoe spring
9. Shoe and lining assembly
10. Shoe, lining and lever assembly
11. Retainer
12. Wave washer
13. Parking lever
14. Shoe and lining assembly
15. Shoe hold-down pin
16. Brake pipe connection
17. Snap ring
18. Hub cap
19. Flange nut
20. Rear hub assembly
21. Wheel cylinder
22. Backing plate

93016G46

Rear drum brake assembly—Mirage

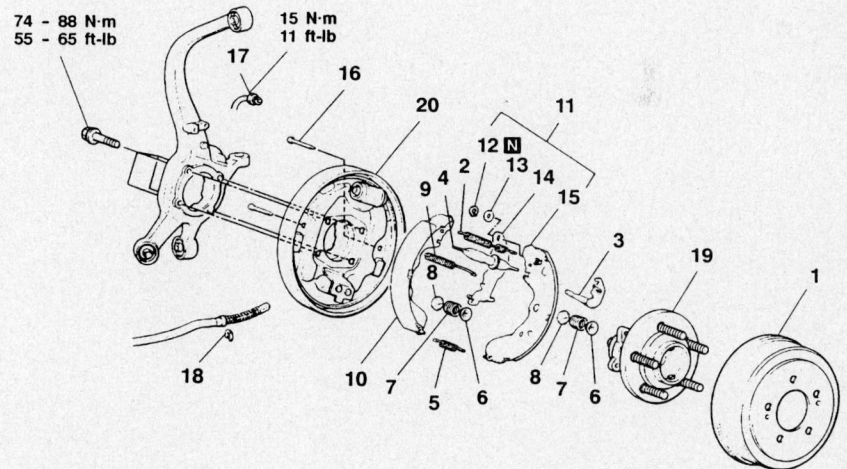

1. BRAKE DRUM
2. SHOE-TO-LEVER SPRING
3. ADJUSTER LEVER
4. AUTO ADJUSTER ASSEMBLY
5. RETAINER SPRING
6. SHOE HOLD-DOWN CUP
7. SHOE HOLD-DOWN SPRING
8. SHOE HOLD-DOWN CUP
9. SHOE-TO-SHOE SPRING
10. SHOE AND LINING ASSEMBLY
11. SHOE AND LEVER ASSEMBLY
12. RETAINER
13. WAVE WASHER
14. PARKING LEVER
15. SHOE AND LINING ASSEMBLY
16. SHOE HOLD-DOWN PIN
17. BRAKE TUBE CONNECTION
18. SNAP RING
19. REAR HUB ASSEMBLY
20. BACKING PLATE

93016G47

Rear drum brake assembly—2002–2003 Galant and 2002–2005 Eclipse

apply brake grease to the 6 shoe support pads.

5. Clean the adjuster assembly and apply brake grease to the threads.

6. Install or connect the following:
   - Parking brake cable to the lever on the rear shoe
   - Rear shoe on the backing plate and the hold-down spring and pin
   - Front shoe on the backing plate and the hold-down spring and pin
   - Adjuster assembly between the 2 shoes
   - Shoe-to-shoe spring
   - Retainer spring
   - Adjuster lever
   - Shoe-to-lever spring
   - Lever return spring

7. Adjust the brake shoes
   - Drum
   - Wheels

SPECIFICATIONS AND MAINTENANCE CHARTS

ENGINE AND VEHICLE IDENTIFICATION CHART

Engine							Model Year	
Code	Liters (cc)	Cu. In.	Cyl.	Fuel Sys.	Engine Type	Eng. Mfg.	Code	Year
6G75/S	3.8 (3828)	233.6	6	MFI	SOHC	Mitsubishi	4	2004
							5	2005
							6	2006

MFI: Multi-port Fuel Injection

09482_ENDE_C0001

GENERAL ENGINE SPECIFICATIONS

Year	Engine Displacement Liters	Engine ID/VIN	Net Horsepower @ rpm	Net Torque @ rpm (ft. lbs.)	Bore x Stroke (in.)	Compression Ratio	Oil Pressure @ rpm
2004	3.8	6G75/S	225@5000	255@3750	3.74x3.54	10.0:1	①
2005	3.8	6G75/S	225@5000	255@3750	3.74x3.54	10.0:1	①
2006	3.8	6G75/S	225@5000	255@3750	3.74x3.54	10.0:1	①

① 4.2 or more psi @ idle

09482_ENDE_C0002

ENGINE TUNE-UP SPECIFICATIONS

Year	Engine Displacement Liters	Engine ID/VIN	Spark Plugs Gap (in.)	Ignition Timing (deg.) MT	Ignition Timing (deg.) AT	Fuel Pump (psi)	Idle Speed (rpm) MT	Idle Speed (rpm) AT	Valve Clearance In.	Valve Clearance Ex.
2004	3.8	6G75/S	0.028-0.031	—	①	47	—	②	HYD	HYD
2005	3.8	6G75/S	0.028-0.031	—	①	47	—	②	HYD	HYD
2006	3.8	6G75/S	0.028-0.031	—	①	47	—	②	HYD	HYD

HYD: Hydraulic

① Base ignition timing: 2-8 degrees BTDC
   Actual ignition timing: 10 degrees BTDC

② 580-780 rpm's

09482_ENDE_C0003

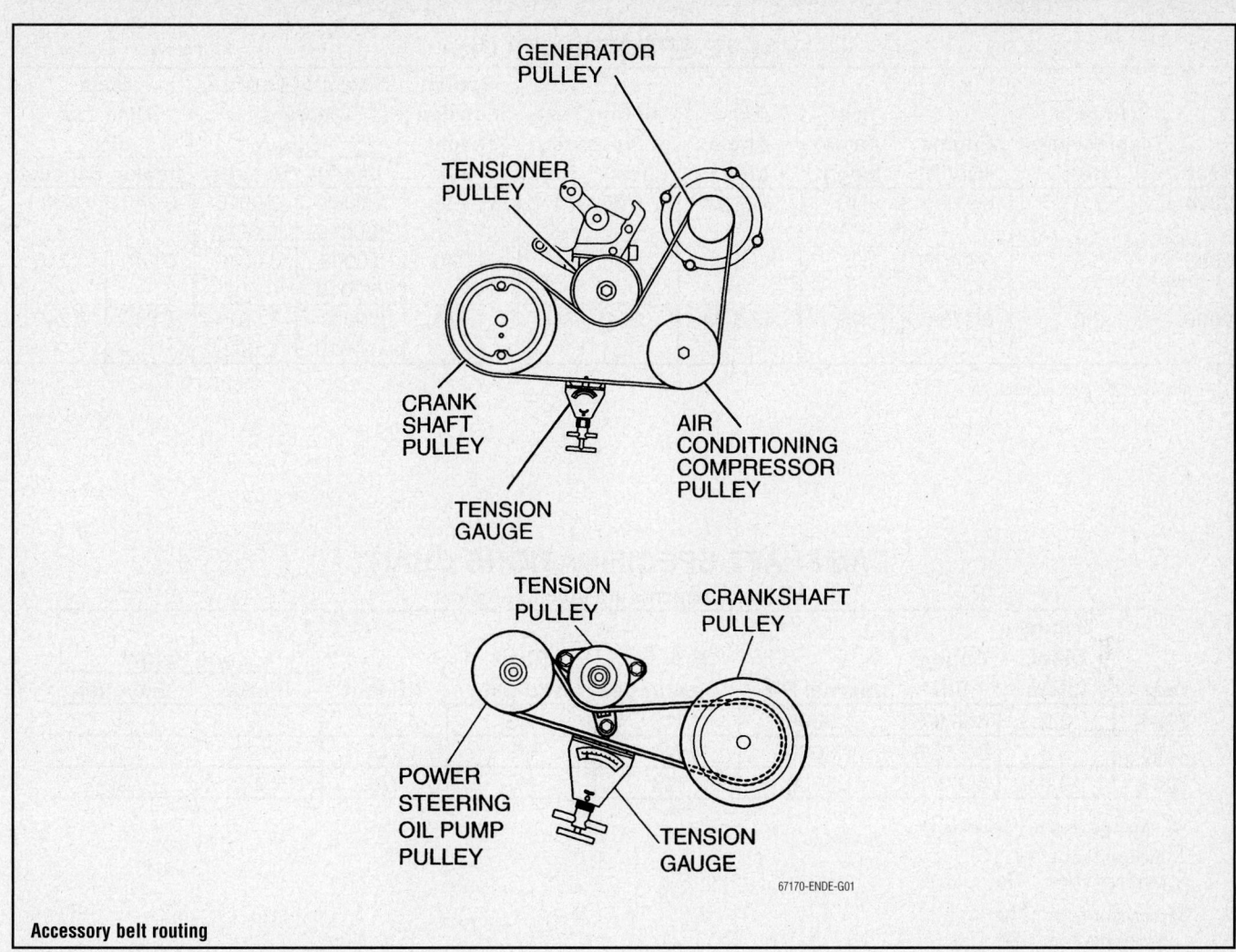

Accessory belt routing

67170-ENDE-G01

CAPACITIES

Year	Model	Engine Displacement Liters	Engine ID/VIN	Engine Oil with Filter (qts.)	Transmission (pts.) 5-Spd	Auto.	Transfer Case (pts.)	Drive Axle Front (pts.)	Rear (pts.)	Fuel Tank (gal.)	Cooling System (qts.)
2004	Endeavor	3.8	6G75/S	4.5	—	①	1.12	—	2.4	21.4	②
2005	Endeavor	3.8	6G75/S	4.5	—	①	1.12	—	2.4	21.4	②
2006	Endeavor	3.8	6G75/S	4.5	—	①	1.12	—	2.4	21.4	②

① FWD: 17.8 pts.
  AWD: 18.6 pts.

② FWD & AWD without tow kit: 9.5 qts.
  AWD with tow kit: 10.1 qts.

09482_ENDE_C0004

VALVE SPECIFICATIONS

Year	Engine Displacement Liters	Engine ID/VIN	Seat Angle (deg.)	Face Angle (deg.)	Spring Test Pressure (lbs. @ in.)	Spring Installed Height (in.)	Stem-to-Guide Clearance (in.)		Stem Diameter (in.)	
							Intake	Exhaust	Intake	Exhaust
2004	3.8	6G75/S	NS	43.5-44	60@1.74	1.740	0.0008-0.0019	0.0016-0.0023	0.240	0.240
2005	3.8	6G75/S	NS	43.5-44	60@1.74	1.740	0.0008-0.0019	0.0016-0.0023	0.240	0.240
2006	3.8	6G75/S	NS	43.5-44	60@1.74	1.740	0.0008-0.0019	0.0014-0.0024	0.240	0.240

NS - Not specified by manufacturer

09482_ENDE_C0005

CAMSHAFT SPECIFICATIONS CHART
All measurements are given in inches.

Year	Engine Displ. Liters	Engine VIN	Journal Dia.	Brg. Oil Clearance	Shaft End-play	Runout	Lobe Height	
							Intake	Exhaust
2004	3.8	6G75/S	1.8000	NS	NS	NS	①	②
2005	3.8	6G75/S	1.8000	NS	NS	NS	①	②
2006	3.8	6G75/S	1.8000	NS	NS	NS	①	②

NS - Not specified by manufacturer

① Standard value: 1.472
  Minimum value: 1.452

② Standard value: 1.462
  Minimum value: 1.443

09482_ENDE_C0006

CRANKSHAFT AND CONNECTING ROD SPECIFICATIONS
All measurements are given in inches.

Year	Engine Displacement Liters	Engine ID/VIN	Crankshaft				Connecting Rod		
			Main Brg. Journal Dia.	Main Brg. Oil Clearance	Shaft End-play	Thrust on No.	Journal Diameter	Oil Clearance	Side Clearance
2004	3.8	6G75/S	2.2500	①	0.0020-0.0090	NS	1.9700	0.0008-0.0015	0.0030-0.0090
2005	3.8	6G75/S	2.2500	①	0.0020-0.0090	NS	1.9700	0.0008-0.0015	0.0030-0.0090
2006	3.8	6G75/S	2.2500	②	0.0020-0.0090	NS	1.9700	0.0008-0.0015	0.0030-0.0090

NS - Not specified by manufacturer

① Nos. 1 & 4: 0.0008-0.0012 inch
  Nos. 2 & 3: 0.0012-0.0016 inch

② Nos. 1 & 4: 0.0007-0.0017 inch
  Nos. 2 & 3: 0.0009-0.0017 inch

09482_ENDE_C0007

PISTON AND RING SPECIFICATIONS

All measurements are given in inches.

Year	Engine Displacement Liters	Engine ID/VIN	Piston Clearance	Ring Gap			Ring Side Clearance		
				Top Compression	Bottom Compression	Oil Control	Top Compression	Bottom Compression	Oil Control
2004	3.8	6G75/S	0.0008-0.0015	0.0100-0.0170	0.0140-0.0190	0.0030-0.0140	0.0012-0.0027	0.0008-0.0023	Snug
2005	3.8	6G75/S	0.0008-0.0015	0.0100-0.0170	0.0140-0.0190	0.0030-0.0140	0.0012-0.0027	0.0008-0.0023	Snug
2006	3.8	6G75/S	0.0008-0.0015	0.0100-0.0160	0.0140-0.0200	0.0030-0.0140	0.0012-0.0027	0.0008-0.0023	Snug

09482_ENDE_C0008

TORQUE SPECIFICATIONS

All readings in ft. lbs.

Year	Engine Displacement Liters	Engine ID/VIN	Cylinder Head Bolts	Main Bearing Bolts	Rod Bearing Bolts	Crankshaft Damper Bolts	Flywheel Bolts	Manifold		Spark Plugs	Oil Pan Drain Plug
								Intake	Exhaust		
2004	3.8	6G75/S	①	④	②	136	55	⑦	③	⑥	⑤
2005	3.8	6G75/S	①	④	②	136	55	⑦	③	⑥	⑤
2006	3.8	6G75/S	①	④	②	136	55	⑦	③	⑥	⑤

① Step 1: 76-84 ft. lbs.
   Step 2: Loosen completely
   Step 3: 76-84 ft. lbs.

② 20 ft. lbs., plus an additional 90-94 degrees

③ Exhaust manifold nut: 29-37 ft. lbs.
   Exhaust manifold stay bolt (M8): 12-16 ft. lbs.
   Exhaust manifold stay bolt (M10): 28-38 ft. lbs.
   Exhaust manifold stay bolt (M12): 49-63 ft. lbs.

④ 51-57 ft. lbs.
⑤ 25-33 ft. lbs.
⑥ 14-22 ft. lbs.
⑦ Step 1: 45-71 inch lbs.
   Step 2: 15-17 ft. lbs.

09482_ENDE_C0009

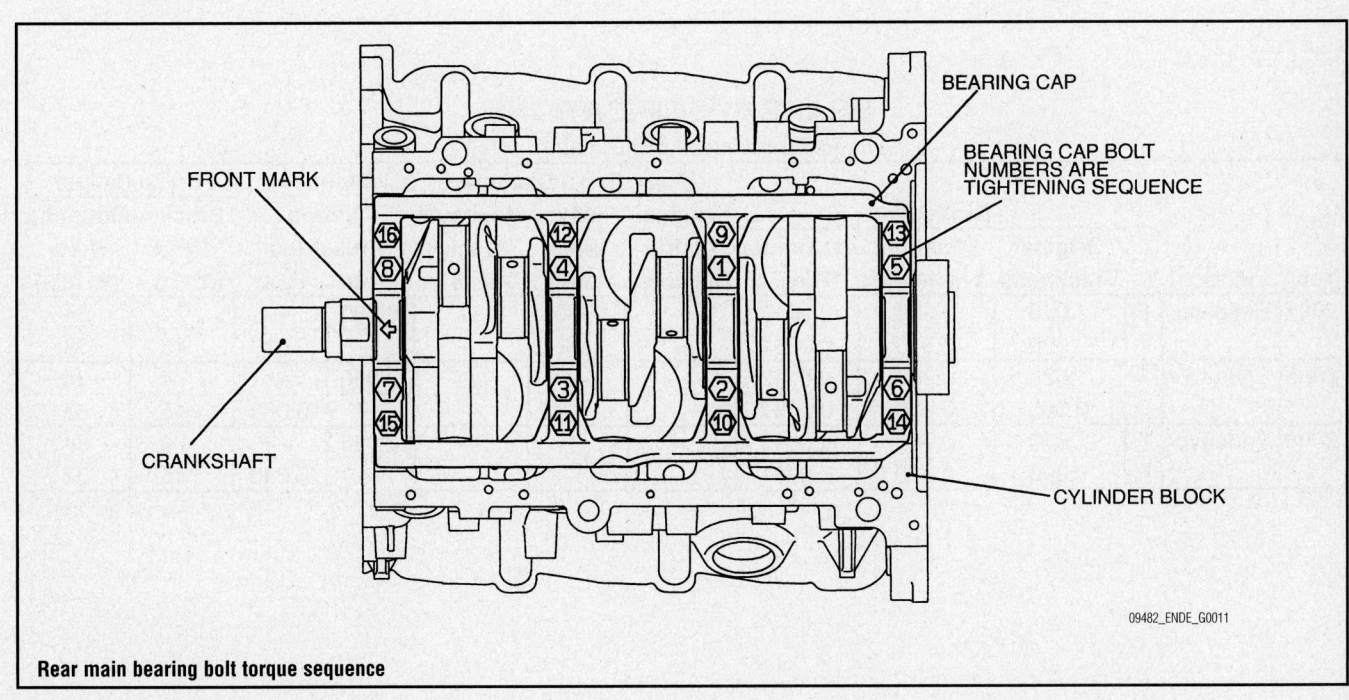

09482_ENDE_G0011

Rear main bearing bolt torque sequence

WHEEL ALIGNMENT

Year	Model		Caster Range (+/-Deg.)	Caster Preferred Setting (Deg.)	Camber Range (+/-Deg.)	Camber Preferred Setting (Deg.)	Toe-in (in.)
2004	Endeavor	F	0.30	+3.00	0.30	0	0.0+/-0.12
		R	—	—	0.30	-0.50	0.12+/-0.12
2005	Endeavor	F	0.30	+3.00	0.30	0	0.0+/-0.12
		R	—	—	0.30	-0.50	0.12+/-0.12
2006	Endeavor	F	0.30	+3.00	0.30	0	0.0+/-0.12
		R	—	—	0.30	-0.50	0.12+/-0.12

09482_ENDE_C0010

TIRE, WHEEL AND BALL JOINT SPECIFICATIONS

Year	Model	OEM Tires Standard	OEM Tires Optional	Tire Pressures (psi) Front	Tire Pressures (psi) Rear	Wheel Size	Ball Joint Inspection	Lug Nut (ft. lbs.)
2004	Endeavor	P235/65R17	None	29	29	7-JJ	U: 7-30 in. ① L: 0.010 in.	②
2005	Endeavor	P235/65R17	None	29	29	7-JJ	U: 7-30 in. ① L: 0.010 in.	②
2006	Endeavor	P235/65R17	None	29	29	7-JJ	U: 7-30 in. ① L: 0.010 in.	②

OEM: Original Equipment Manufacturer

PSI: Pounds Per Square Inch

STD: Standard

OPT: Optional

① Torque required in inch lbs. to rotate ball joint when removed from the knuckle

② 66-80 ft. lbs.

09482_ENDE_C0011

BRAKE SPECIFICATIONS
All measurements in inches unless noted

Year	Model		Brake Disc Original Thickness	Brake Disc Minimum Thickness	Brake Disc Maximum Runout	Brake Drum Diameter Original Inside Diameter	Brake Drum Diameter Max. Wear Limit	Brake Drum Diameter Maximum Machine Diameter	Minimum Lining Thickness Front	Minimum Lining Thickness Rear	Brake Caliper Bracket Bolts (ft. lbs.)	Brake Caliper Mounting Bolts (ft. lbs.)
2004	Endeavor	F	1.020	0.960	0.0012	—	—	—	0.080	—	74	34
		R	0.390	0.330	0.0031	—	—	—	—	0.080	45	32
2005	Endeavor	F	1.020	0.960	0.0012	—	—	—	0.080	—	74	34
		R	0.390	0.330	0.0031	—	—	—	—	0.080	45	32
2006	Endeavor	F	1.020	0.960	0.0012	—	—	—	0.080	—	74	34
		R	0.390	0.330	0.0031	—	—	—	—	0.080	45	32

09482_ENDE_C0012

SCHEDULED MAINTENANCE INTERVALS
Mitsubishi—Endeavor

TO BE SERVICED	TYPE OF SERVICE	VEHICLE MILEAGE INTERVAL (x1000)												
		7.5	15	22.5	30	37.5	45	52.5	60	67.5	75	82.5	90	97.5
Engine oil & filter	R	✓	✓	✓	✓	✓	✓	✓	✓	✓	✓	✓	✓	✓
Automatic transaxle & transfer case oil (AWD)	S/I		✓		✓		✓		✓		✓		✓	
Automatic transaxle & transfer case oil (AWD)	R				✓				✓				✓	
Brake hoses	S/I		✓		✓		✓		✓		✓		✓	
Disc brake pads & rotors	S/I		✓		✓		✓		✓		✓		✓	
Drive shaft boots	S/I		✓		✓		✓		✓		✓		✓	
Air cleaner filter	R				✓				✓				✓	
Engine coolant (2004-2005)	R				✓				✓				✓	
Ball joints & steering linkage seals	S/I				✓				✓				✓	
Drive belt(s)	S/I				✓				✓				✓	
Exhaust system	S/I				✓				✓				✓	
Front & rear axle	S/I				✓				✓				✓	
Fuel hoses	S/I				✓				✓				✓	
Propeller shaft joint	S/I				✓				✓				✓	
Ignition cables	R								✓					
Timing belt (2004-2005)	R								✓					
Timing belt (2006)	R													✓
Distributor cap & rotor	S/I								✓					
EVAP system (except EVAP canister)	S/I								✓					
EGR valve ①	S/I				✓				✓				✓	
EVAP canister ①	S/I				✓				✓				✓	
PCV system	S/I	✓	✓	✓	✓	✓	✓	✓	✓	✓	✓	✓	✓	✓
Spark plugs ②	R													
Tires (rotate)	R	✓	✓	✓	✓	✓	✓	✓	✓	✓	✓	✓	✓	✓

R: Replace   S/I: Service or Inspect
① Replace at 100,000 miles.
② Iron tips: 30,000 miles
   Platinum tips: 60,000 miles
   Irridium tips: 100,000 miles

FREQUENT OPERATION MAINTENANCE (SEVERE SERVICE)
If a vehicle is operated under any of the following conditions it is considered severe service:
- Extremely dusty areas.
- 50% or more of the vehicle operation is in 32°C (90°F) or higher temperatures, or constant operation in temperatures below 0°C (32°F).
- Prolonged idling (vehicle operation in stop and go traffic).
- Frequent short running periods (engine does not warm to normal operating temperatures).
- Police, taxi, delivery usage or trailer towing usage.

Oil & oil filter: replace every 3000 miles.

Front disc brake pads (dusty or salty conditions): service or inspect every 6000 miles.

Front disc brake pads: service or inspect every 7500 miles.

Air cleaner filter: service or inspect every 15,000 miles.

Spark plugs (iron tip): replace every 15,000 miles.

PCV system: service or inspect every 60,000 miles.

09482_ENDE_C0013

ENGINE REPAIR

→Disconnecting the negative battery cable on some vehicles may interfere with the functions of the on board computer systems and may require the computer to undergo a relearning process, once the negative battery cable is reconnected.

Distributor

This engine is distributorless.

Alternator

REMOVAL & INSTALLATION

1. Before servicing the vehicle, refer to the Precautions Section.
2. Remove or disconnect the following:
   • Negative battery cable
   • Under cover
   • Side under cover
   • Drive belt
   • Alternator connector
   • A/C compressor connector
   • A/C compressor and position aside. Do NOT disconnect the refrigerant lines.

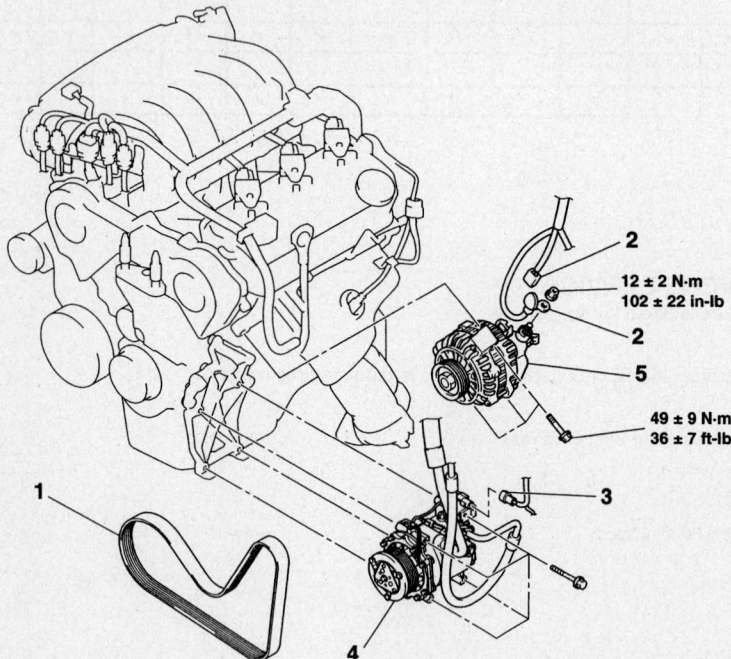

1. GENERATOR DRIVE BELT
2. GENERATOR CONNECTOR
3. A/C COMPRESSOR ASSEMBLY CONNECTOR

4. A/C COMPRESSOR ASSEMBLY
5. GENERATOR

67170-ENDE-G02

Alternator mounting and related components

• Alternator

To install:
3. Install or connect the following:
   • Alternator. Torque the bolts to 36 ft. lbs. (49 Nm).
   • A/C compressor
   • A/C compressor connector
   • Alternator connector. Torque the nut to 102 inch lbs. (12 Nm).
   • Drive belt
   • Side under cover
   • Under cover
   • Negative battery cable

Ignition Timing

ADJUSTMENT

The ignition timing is controlled by the ECM and is not adjustable. The ECM determines the timing based on input from the crankshaft position sensor.

TIMING CHECK

1. Before servicing the vehicle, refer to the Precautions Section.

Before attempting to adjust the ignition timing, be sure of the following:
   • The engine should be at normal operating temperature.
   • The lights and all accessories should be OFF.
   • The transaxle should be in **P** or **N**.
2. Connect scan tool MB991958 to the data link connector
3. Set up the timing light.
4. Start the engine and run at idle.
5. Verify that the idle speed is about 680 rpm.
6. Select scan tool MB991958 actuator test "item number 17".
7. Check that basic timing is with in standard, it should be 2–8˚ BTDC.
8. If the base timing is out of specification:
   a. Check to see if the distributor is aligned properly
   b. Check to see if the timing belt cover and Crankshaft Position (CKP) sensor installation is conditions.
   c. Crankshaft sensing blade conditions.
9. Press the clear key on the scan tool, select forced drive stop mode and cancel the actuator test.

❄❄ **CAUTION**

If the actuator test is not canceled, the forced drive will continue for 27 minutes. Driving in this state could lead to engine failure.

10. Check that the actual ignition timing is approximately 10˚ BTDC.

→**Keep in mind that the ignition timing may fluctuate by as much as +/-7˚ BTDC even under normal operation conditions. It is also further advanced by 5–10˚ BTDC at higher altitudes.**

Engine Assembly

REMOVAL & INSTALLATION

1. Before servicing the vehicle, refer to the Precautions Section.
2. Relieve the fuel system pressure.

❄❄ **CAUTION**

The fuel injection system remains under pressure after the engine has been OFF. Properly relieve fuel pressure before disconnecting any fuel lines. Failure to do so may result in fire or personal injury.

3. Disconnect the negative battery cable. Remove the engine under cover.

4. Drain the engine oil. Drain the cooling system.

5. Matchmark and remove the engine hood. Remove the air cleaner.

6. Remove the Powertrain Control Module (PCM). Remove the battery. Remove the battery tray.

7. Remove the radiator grille. Remove the radiator assembly.

8. Remove the right side exhaust manifold.

9. Remove the transaxle.

10. Disconnect the EVAP solenoid connector. Disconnect the EVAP hose. Remove the EVAP solenoid.

11. Disconnect the vacuum hose connection. Disconnect the purge hose connection. Remove the front wiring harness and control wiring harness.

12. Remove the ground wire retaining screw. Remove the ground wire. Disconnect the fuel pipe retainer, and main pipe connector.

13. Disconnect the heater hose connection. Remove the reservoir assembly.

14. Remove the alternator drive belt. Remove the power steering pump belt.

15. Disconnect the air conditioning compressor electrical connector. Remove the air conditioning compressor from its mounting and position it to the side. Do not disconnect the refrigerant lines.

16. Disconnect the power steering pressure switch connector.

17. Remove the power steering pump and position it to the side. Do not disconnect the fluid lines.

18. Remove the power steering pressure hose clamp. Remove the power steering pressure hose clamp bracket.

19. Remove the engine mount stay. Remove the ground cable. Remove the locknuts.

20. Support the engine with a floor jack. Install the engine lifting fixture. Raise the engine slightly to take the tension of the engine mount.

21. Loosen the engine mount retaining nuts and bolts. Remove the engine mount.

22. After checking that all cables, hoses and wiring harness connectors are disconnected from the engine, remove the engine from the vehicle.

To install:

23. Lower the engine into position and install the engine mount nuts and bolts. Tighten the nuts to 18–20 ft. lbs. (25–27 Nm), and the bolts to tighten to 33 ft. lbs. (44 Nm).

24. Continue the installation in the reverse order of the removal procedure.

25. Be sure to check and adjust all fluid levels, as required.

26. Check fuel system for leaks.

27. On 2005–2006 vehicles, complete the vehicle initialization procedure.

➡**To complete the initialization procedure the following tools are needed. MB991958 scan tool, MB991824 VCI, MB991827 MUT III USB cable, MB991910 MUT III Main harness "A".**

28. Connect the scan tool to the data link connector. To prevent damage to the scan tool be sure that the ignition switch is in the **LOCK** position before connecting the scan tool.

29. Turn the ignition switch to the **ON** position.

30. Select "check mode" from the menu screen.

31. Select "erase memory" from the menu screen.

32. Initialize the learning value.

33. After initialization complete the idle learning procedure.

➡**This procedure must be performed when the PCM is replaced, or when the learning value is initialized, as the idling is not stabilized because the learning value in the MFI engine is not completed.**

34. Start the engine. Allow the coolant temperature to reach 176 degrees or more.

35. Stop the engine and place the ignition switch in the **LOCK** position.

36. After ten seconds, restart the engine.

37. For ten minutes carry out the idling procedure below to confirm that the engine has normal idling.
  • Position the transaxle selector lever in the **P** range

  • The engine fan is not to be operated
  • The engine coolant temperature should be 176 degrees or more

➡**If the engine stalls during idling, check the throttle valve of the throttle body for dirt. Correct and perform the procedure again.**

38. Connect the negative battery cable. Turn the ignition **ON** and then **OFF**, and keep it off for at least ten seconds.

Water Pump

REMOVAL & INSTALLATION

1. Before servicing the vehicle, refer to the Precautions Section.

2. If necessary, properly release the fuel pressure.

3. Drain the cooling system.

4. Remove or disconnect the following:
  • Negative battery cable

✳✳ CAUTION

Wait at least 90 seconds after the negative battery cable is disconnected to prevent possible deployment of the air bag.

  • Timing belt
  • Crankshaft Position (CKP) sensor
  • Water pump bracket
  • Retaining bolts, water pump assembly, gasket and O-ring

To install:

5. Clean and dry the mating surfaces of the block and water pump

6. Install or connect the following:
  • Water pump assembly, using a new

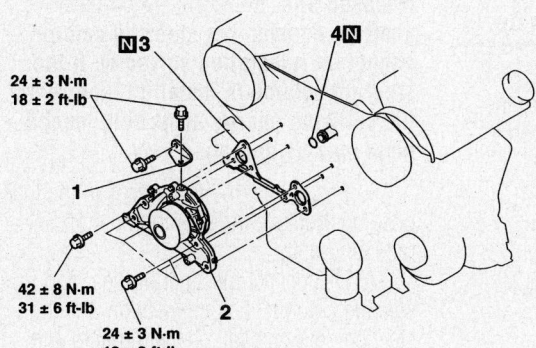
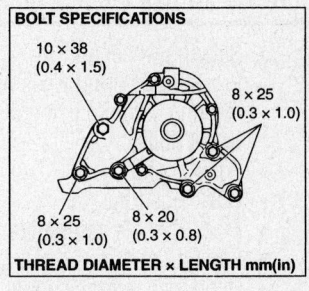

1. WATER PUMP BRACKET
2. WATER PUMP
3. WATER PUMP GASKET
4. O-RING

Water pump and bolt torque specifications

67170-ENDE-G03

gasket and O-ring. Torque the water pump bolts to the specifications shown in the accompanying figure.
- Water outlet fitting bracket
- CKP sensor
- Timing belt

7. Refill the radiator with coolant. This cooling system has a self-bleeding thermostat, so system bleeding is not required.

8. Run the vehicle until the thermostat opens and fill the overflow tank. Check for leaks.

9. Once the vehicle has cooled, recheck the coolant level.

Heater Core

REMOVAL & INSTALLATION

1. Before servicing the vehicle, refer to the Precautions Section.
2. Place the wheels in the straight-ahead position.
3. Drain the cooling system into a clean container for reuse.
4. Discharge and recover the air conditioning system refrigerant.
5. Remove or disconnect the following:
- Negative battery cable

❊❊ CAUTION

Wait at least 60 seconds after disconnecting the battery cable before performing any work on the air bag or instrument panel.

- Steering column-to-instrument panel cover screws and the cover
- Air bag module, carefully, from the steering wheel
- Electrical connectors from the air bag module

❊❊ CAUTION

Store the air bag module facing up.

- Steering wheel nut
- Press the steering wheel from the steering column, using a steering wheel puller
- Front scuff plate
- Cowl side trim
- Floor console assembly
- Front pillar trim
- Hood lock release handle
- Side cover
- Lower panel
- Plug
- Fog light switch or switch plug
- Rheostat
- Steering column lower cover

- Steering column cover protector
- Steering column upper cover
- Combination meter assembly
- Center panel assembly
- Air outlet
- Radio panel
- Heater/air conditioning control assembly
- Console box
- Heated seat switch, if equipped
- Hazard warning lamp switch
- Console meter hood
- Multi-center display unit
- Radio and CD player assembly
- Air bag caution label
- Glove box assembly, inner cover, lock assembly and damper
- Speaker panel
- Tweeter
- Clock spring connector
- Column switch connector
- Clock spring and column switch assembly
- Photo sensor and sensor harness clamp, if equipped
- Radio harness clamp
- Interior temperature sensor, if equipped
- Stay
- Passenger's air bag module nut and the air bag module
- Electrical connector from the air bag module
- Instrument panel assembly, carefully, with the help of an assistant
- Floor console side cover
- Front floor carpet

➥Before the steering column is removed from the vehicle the tilt lever should be in the lock position, and remain there, until the steering column shaft is reinstalled in the vehicle. If the steering column is removed with the tilt lever released, or the tilt lever is released after the steering column shaft is removed the steering column cannot be reinstalled correctly. If the steering column is installed incorrectly, the collision energy absorbing mechanism may be damaged.

6. Ensure that the tilt lever is in the lock position. Remove the steering column mounting bolts.

7. Disconnect the steering gear and steering column shaft connection. Pinch the steering column shaft clip with pliers and pull the shaft upwards to disengage the column shaft from the steering gear.

➥If the steering column shaft is removed accidentally, remove the

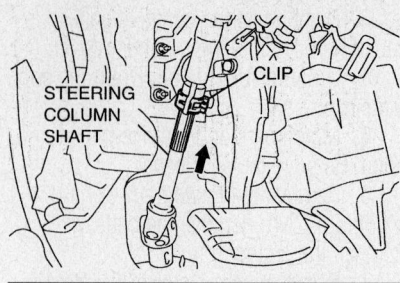

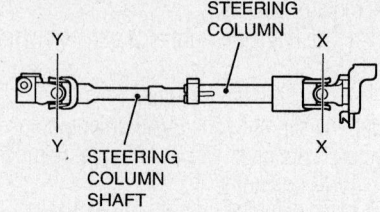

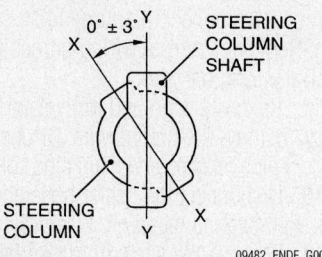

09482_ENDE_G0023

Steering gear to steering column disconnection

steering column assembly and be sure to insert the column shaft into the steering column as shown in the illustration.

8. Remove the steering column shaft assembly.
- Blower motor assembly
- Both foot ducts
- Joint duct from the air conditioning evaporator housing assembly
- Foot distribution duct
- Center reinforcement
- Center ventilation duct
- Drain hose from the air conditioning evaporator housing assembly
- Heater hoses from the heater housing assembly
- Refrigerant lines from the air conditioning evaporator housing assembly and discard the O-rings
- Heater housing assembly
- Center duct assembly
- Heater core from the heater housing

To install:

➥When installing the steering column assembly, be sure the tilt lever is in the lock position and that it has not been released since the column was removed from the vehicle. Do not release the tilt lever until the column has been installed in the vehicle.

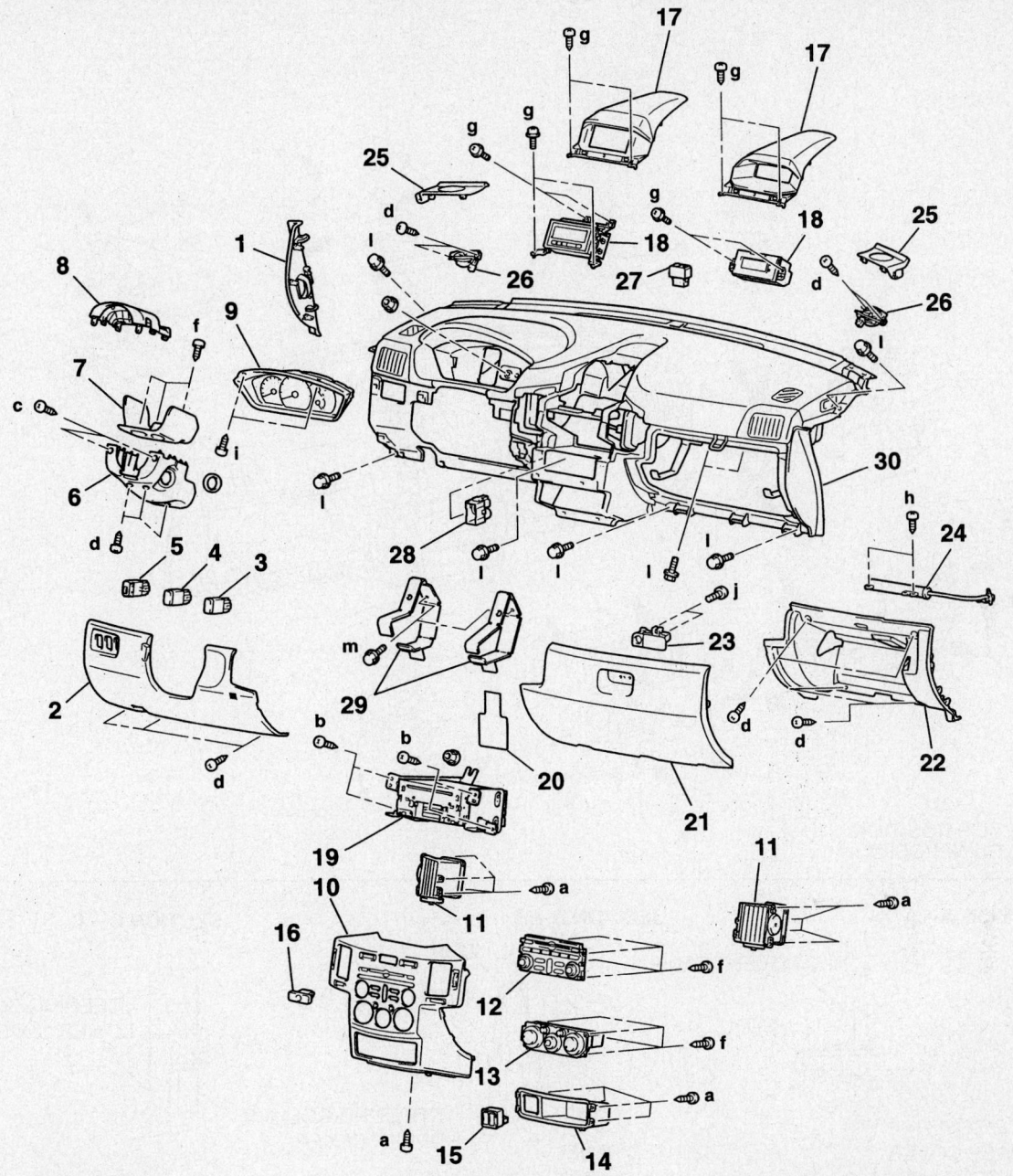

1. SIDE COVER
2. LOWER PANEL
3. PLUG
4. FOG LIGHT SWITCH OR FOG LIGHT SWITCH PLUG
5. RHEOSTAT
6. STEERING COLUMN LOWER COVER
7. STEERING COLUMN COVER PROTECTOR
8. STEERING COLUMN UPPER COVER
9. COMBINATION METER ASSEMBLY
10. CENTER PANEL ASSEMBLY
11. AIR OUTLET

12. RADIO PANEL
13. HEATER CONTROL (A/C-ECU)
14. CONSOLE BOX
15. HEATED SEAT SWITCH <VEHICLES WITH HEATED SEAT>
16. HAZARD WARNING LAMP SWITCH
17. CONSOLE METER HOOD
18. MULTI-CENTER DISPLAY UNIT
19. RADIO AND CD PLAYER OR RADIO, CD PLAYER AND CD CHANGER
20. AIR BAG CAUTION LABEL
21. GLOVE BOX ASSEMBLY
22. GLOVE BOX INNER COVER

23. GLOVE BOX LOCK ASSEMBLY
24. GLOVE BOX DAMPER
25. SPEAKER PANEL
26. TWEETER
27. PHOTO SENSOR <VEHICLES WITH AUTO A/C>
28. INTERIOR TEMPERATURE SENSOR <VEHICLES WITH AUTO A/C>
29. STAY
30. INSTRUMENT PANEL ASSEMB

Instrument panel and related components

67170-ENDE-G04

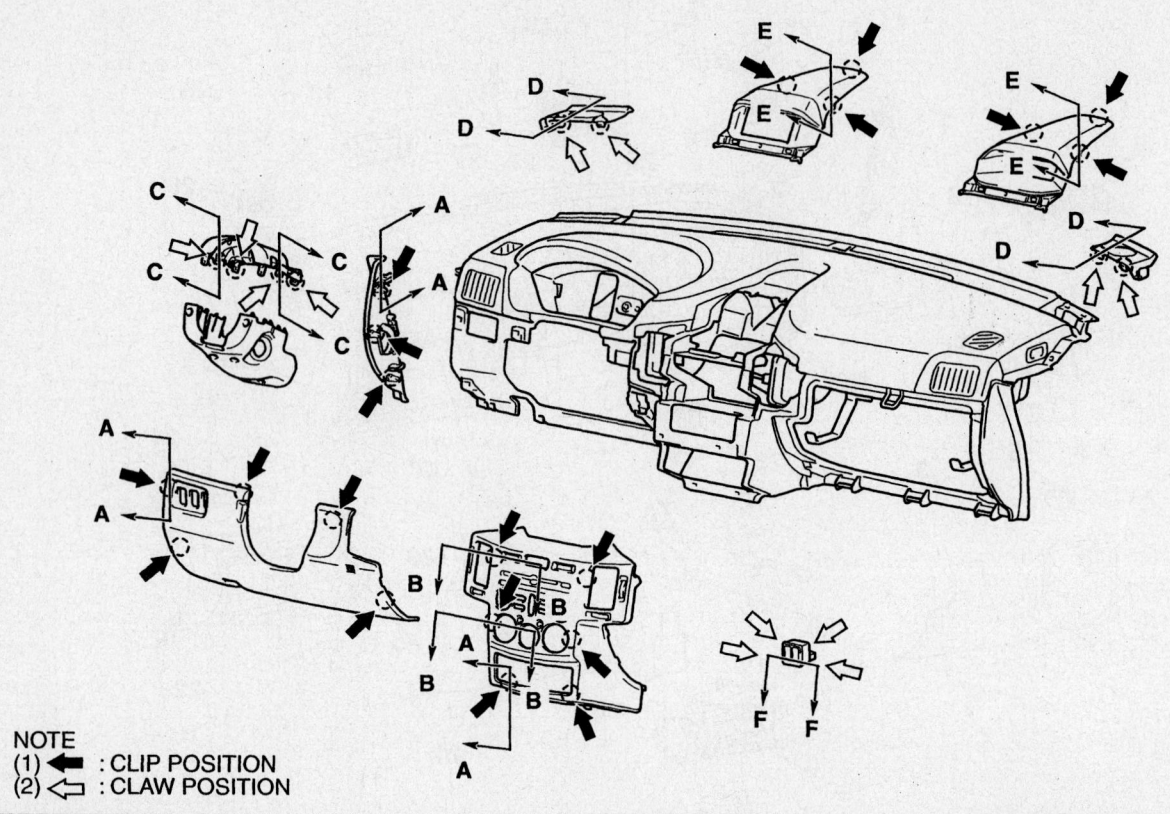

NOTE
(1) ◀ : CLIP POSITION
(2) ◁ : CLAW POSITION

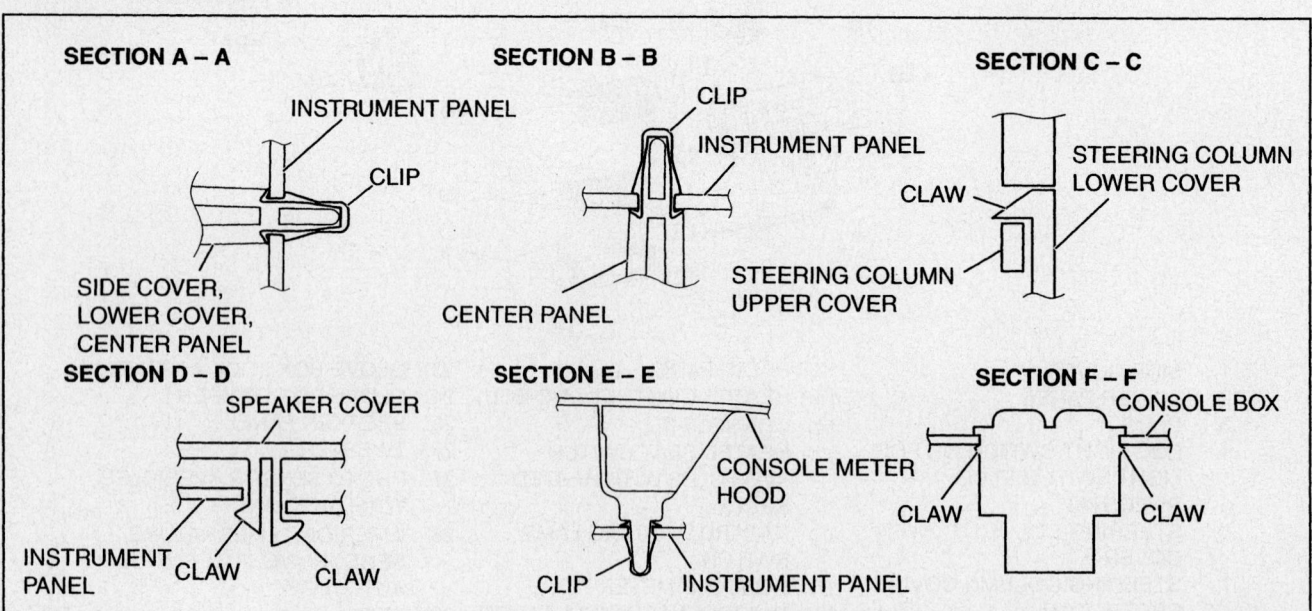

SECTION A – A

INSTRUMENT PANEL

CLIP

SIDE COVER,
LOWER COVER,
CENTER PANEL

SECTION B – B

CLIP

INSTRUMENT PANEL

CENTER PANEL

SECTION C – C

STEERING COLUMN
LOWER COVER

CLAW

STEERING COLUMN
UPPER COVER

SECTION D – D

SPEAKER COVER

INSTRUMENT
PANEL   CLAW   CLAW

SECTION E – E

CONSOLE METER
HOOD

CLIP   INSTRUMENT PANEL

SECTION F – F

CONSOLE BOX

CLAW   CLAW

67170-ENDE-G05

Location of the instrument panel retaining clips and claws

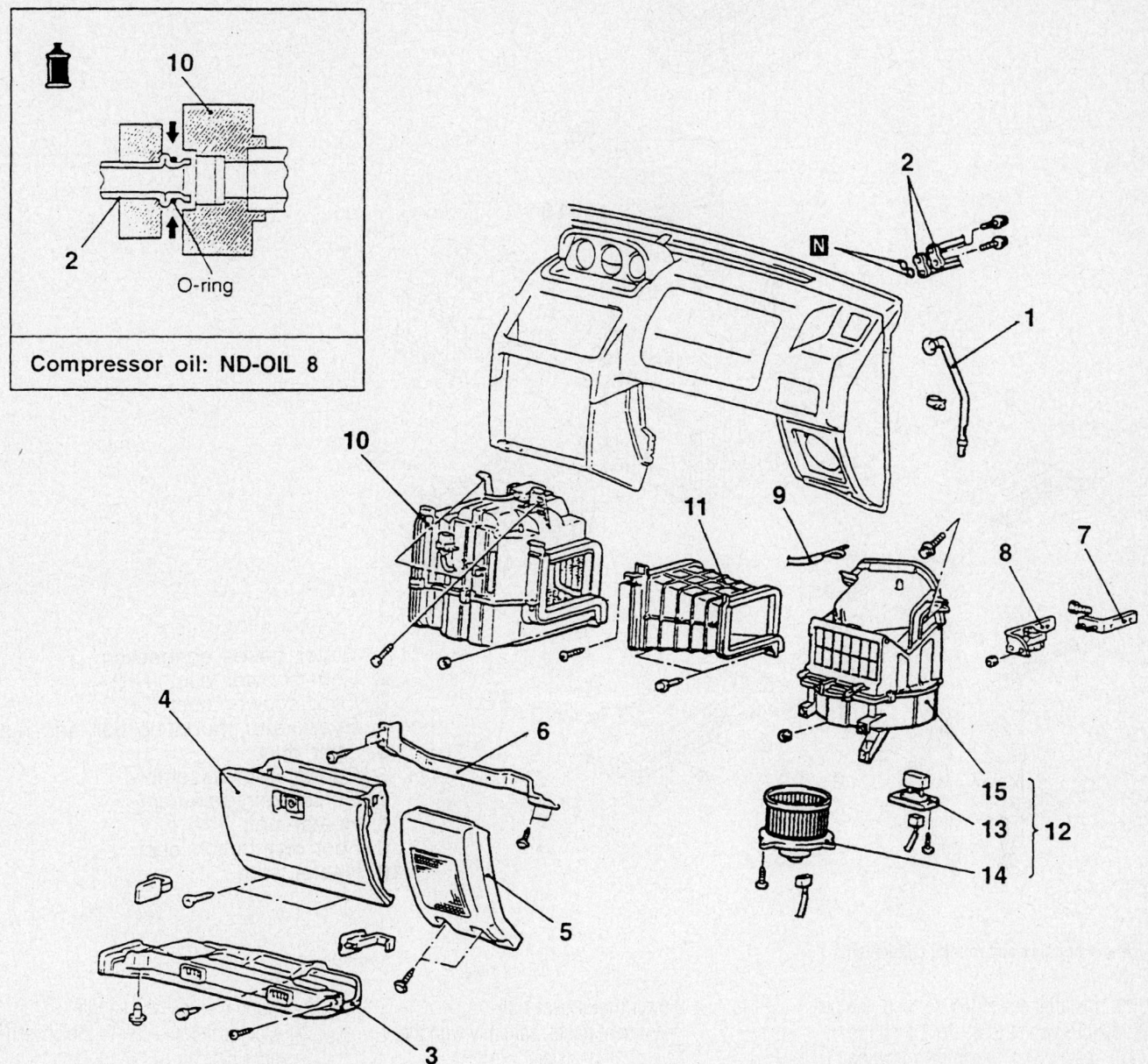

Compressor oil: ND-OIL 8

O-ring

1. Drain hose
2. Liquid pipe and suction hose connection
3. Foot shower duct (R.H.)
4. Glove box
5. Corner cover
6. Lower frame
7. Engine control relay assembly
8. Bracket
9. Air selection control wire connection
10. Evaporator
11. Duct joint
12. Blower assembly
13. Resistor
14. Blower motor assembly
15. Blower case assembly

93113GE6

Air conditioning evaporator housing and related components

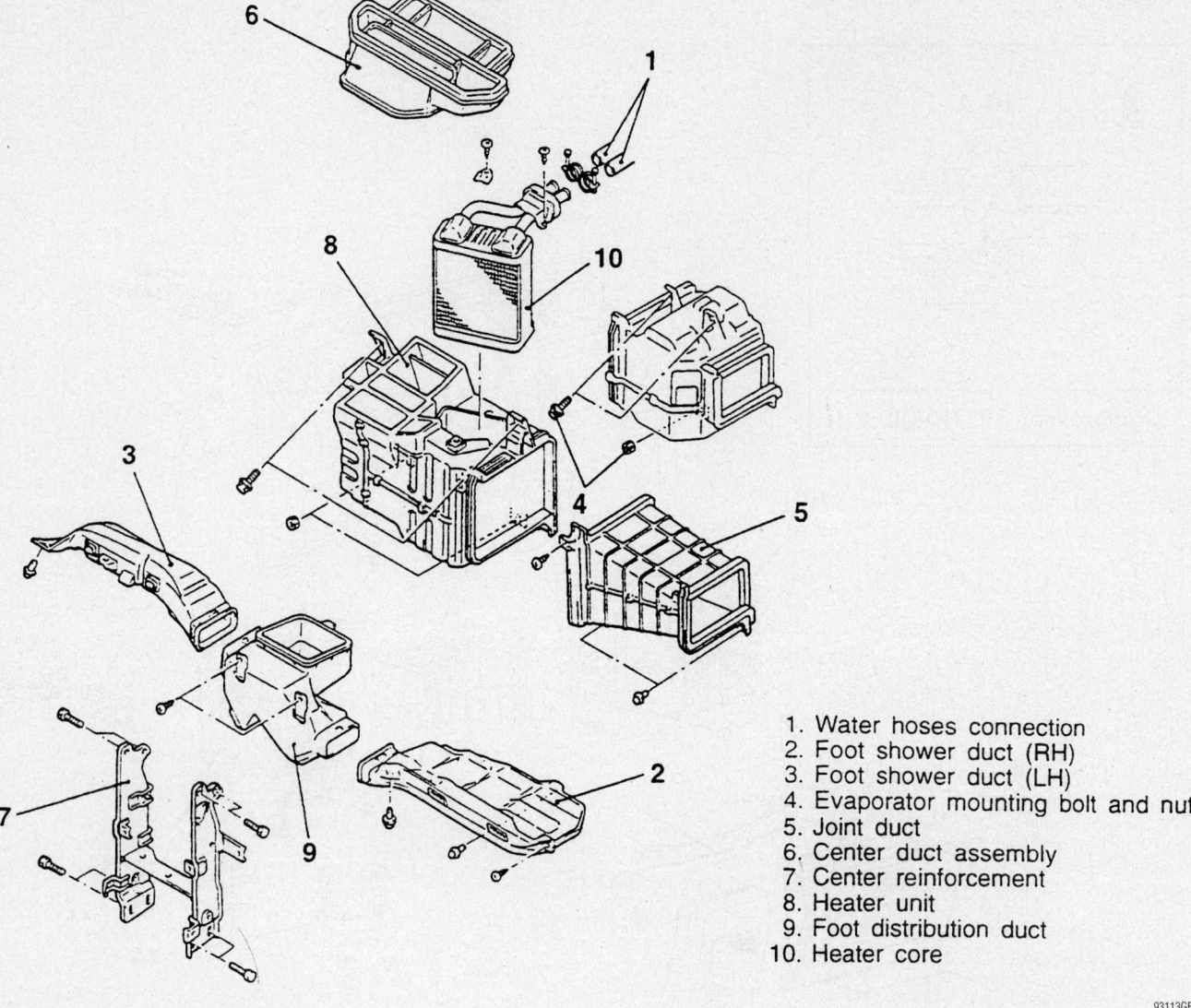

1. Water hoses connection
2. Foot shower duct (RH)
3. Foot shower duct (LH)
4. Evaporator mounting bolt and nut
5. Joint duct
6. Center duct assembly
7. Center reinforcement
8. Heater unit
9. Foot distribution duct
10. Heater core

93113GE7

Heater housing and related components

9. Install the steering column, torque the retaining bolts to 80–124 inch lbs.
10. Install or connect the following:
- Heater core to the heater housing
- Center duct assembly
- Heater housing assembly
- Refrigerant lines to the air conditioning evaporator housing assembly, using new O-rings
- Heater hoses to the heater housing assembly
- Drain hose to the air conditioning evaporator housing assembly
- Center ventilation duct
- Center reinforcement
- Foot distribution duct
- Joint duct to the air conditioning evaporator housing assembly
- Both foot shower ducts

- Blower motor assembly
- Instrument panel, carefully with the help of an assistant
- Side defroster grille
- Multi-meter assembly and the multi-meter panel
- Upper glove box frame and the glove box striker
- Speaker and the combination meter
- Heater/air conditioning control assembly
- Radio and tape/CD player
- Center under cover assembly
- Glove box assembly
- Under cover, the corner cover and the stopper
- Meter bezel assembly
- Knee protector assembly and bracket
- Filler door lock release handle

- Hood lock release handle
- Electrical connector to the passenger's side air bag module
- Passenger's air bag module and the air bag module nut
- Glove box and the glove box stoppers
- Foot shower duct

➡**When installing the clock spring, ensure that the mating marks are properly aligned. If not the steering wheel may not rotate completely during a turn, or the flat cable in the clock spring could be damaged. This would prevent normal SRS operation and possibly cause serious injury to the driver.**

11. To align the mating marks of the clock spring, turn the clock spring counterclockwise fully. Then turn it back approxi-

mately 3¾ turns counterclockwise to align the mating marks.

12. Turn the front wheels to the straight ahead position. Then install the clock spring to the column switch.

➡**If the vehicle is equipped with Active Skid Control (ASC), ensure that the steering wheel sensor's mating marks are properly aligned. If not, the steering wheel sensor could be damaged.**

13. To align the mating marks, if the vehicle is equipped with Active Skid Control (ASC), turn the steering wheel sensor clockwise fully. Then turn it back approximately 2¾ turns counterclockwise to align the mating marks. Align the mating marks on the clock spring and the steering wheel sensor, and install the steering wheel sensor to the column switch assembly. Connect the steering wheel sensor connector.

14. Install or connect the following:
- Steering column-to-instrument panel cover and the cover screws
- Floor console assembly

15. Refill the cooling system.

16. Connect the negative battery cable. On 2005–2006 vehicles, turn the ignition **ON** and then **OFF**, and keep it off for at least ten seconds.

➡**If the vehicle is equipped with Active Skid Control (ASC), the steering wheel sensor must be recalibrated after the steering shaft has been installed. This is necessary because the TCL/ASC-ECU should update the steering neutral point.**

➡**To complete the steering sensor calibration procedure the following tools are needed. MB991958 scan tool, MB991824 VCI, MB991827 MUT III USB cable, MB991910 MUT III Main harness "A".**

17. Connect the scan tool to the data link connector. To prevent damage to the scan tool be sure that the ignition switch is in the **LOCK (OFF)** position before connecting the scan tool.

18. Park the vehicle on a level surface.

19. Turn the ignition switch to the **ON** position.

20. Select "interactive diagnosis" from the menu screen.

21. Select "special function" from the menu screen.

22. Select "steering angle sensor" from the menu screen.

23. Select "calibration" from the menu screen.

24. Turn the ignition switch to the **LOCK (OFF)** position.

25. Disconnect the scan tool.

26. Evacuate and charge the air conditioning system refrigerant.

27. Run the engine to normal operating temperatures; then, check the climate control operation and check for leaks.

Cylinder Head

REMOVAL & INSTALLATION

1. Before servicing the vehicle, refer to the Precautions Section.

❊❊ CAUTION

The fuel injection system remains under pressure after the engine has been OFF. Properly relieve fuel pressure before disconnecting any fuel lines. Failure to do so may result in fire or personal injury.

2. Relieve fuel system pressure.

3. Drain the cooling system.

4. Remove or disconnect the following:
- Negative battery cable

❊❊ CAUTION

Work must be started after 90 seconds from the time the ignition switch is turned to theLOCK position and the negative battery cable is disconnected.

- Intake manifold
- Exhaust manifold
- Timing belt
- Thermostat housing
- Alternator
- Blow-by hose from the left and right rocker arm covers
- Positive Crankcase Ventilation (PCV) hose from the left and right rocker arm covers
- Spark plug wires; tag before disconnecting
- Ignition coil connectors and ignition coils
- Engine control wiring harness clamp
- Rocker covers and gaskets
- Camshaft Position (CMP) sensor connector
- Grounding
- Timing belt rear center cover

5. Loosen the left cylinder head mounting bolts in 3 steps, in the sequence shown.

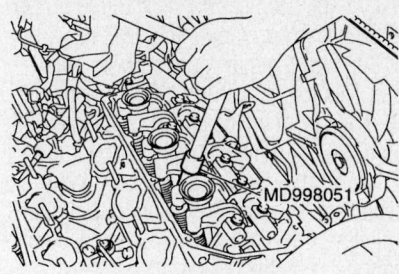

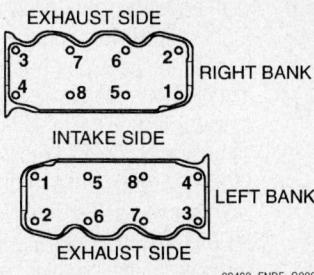

EXHAUST SIDE

RIGHT BANK

INTAKE SIDE

LEFT BANK

EXHAUST SIDE

09482_ENDE_G0001

Cylinder head bolt removal sequence

Lift off the left cylinder head assembly and remove the head gasket.
- Power steering oil pump bracket
- Exhaust Gas Recirculation (EGR) pipe B and gasket

6. Loosen the right cylinder head mounting bolts in 3 steps, in the sequence shown. Lift off the right cylinder head assembly and remove the head gasket.

To install:

7. Thoroughly clean and dry the mating surfaces of the head and block. Check the cylinder head for cracks, damage or engine coolant leakage. Remove scale, sealing compound and carbon. Clean oil passages thoroughly. Check the head for flatness. End

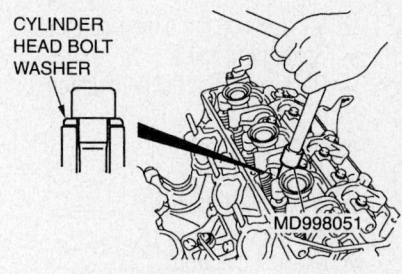

CYLINDER HEAD BOLT WASHER

MD998051

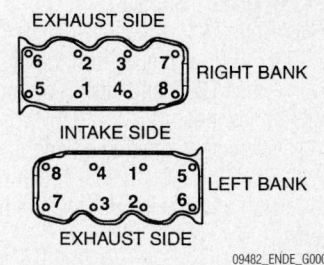

EXHAUST SIDE

RIGHT BANK

INTAKE SIDE

LEFT BANK

EXHAUST SIDE

09482_ENDE_G0002

Cylinder head bolt torque sequence

to end, the head should be within 0.0012 in. (0.030mm), normally with 0.008 in. (0.203mm) the maximum allowed out of true. The total thickness allowed to be removed from the head and block is 0.008 in. (0.203mm) maximum.

8. Place a new head gasket on the cylinder block with the identification marks in the front top (upward) position. Do not use sealer on the gasket.

9. Install or connect the following:

- Right cylinder head on the block. Be sure the head bolt washers are installed with the beveled edge upward. Torque the head bolts in sequence, to 77–83 ft. lbs. (105–113 Nm) with a torque wrench and Special Tool No. MD998501, then loosen the bolts completely and retighten in sequence to 77–83 ft. lbs. (105–113 Nm).
- Exhaust Gas Recirculation (EGR) pipe B and gasket
- Power steering oil pump bracket
- Left cylinder head on the block. Be sure the head bolt washers are installed with the beveled edge upward. Torque the head bolts in sequence, to 77–83 ft. lbs. (105–113 Nm) with a torque wrench and Special Tool No. MD998501, then loosen the bolts completely and retighten in sequence to 77–83 ft. lbs. (105–113 Nm).
- Timing belt rear center cover
- Grounding
- CMP sensor connector
- Rocker covers with new gaskets
- Engine control wiring harness clamp
- Ignition coils and their connectors
- Spark plug wires
- Positive Crankcase Ventilation (PCV) hoses
- Blow-by hoses
- Alternator
- Thermostat housing
- Timing belt
- Exhaust manifold
- Intake manifold

10. Connect the negative battery cable, On 2005–2006 vehicles, turn the ignition **ON** and then **OFF**, and keep it off for at least ten seconds.

11. Change the engine oil and oil filter.

12. Refill the system with coolant.

13. Run the vehicle until the thermostat opens.

14. Once the vehicle has cooled, recheck the coolant level.

Rocker Arms/Shafts

REMOVAL & INSTALLATION

Left Bank

1. Before servicing the vehicle, refer to the Precautions Section.

2. Disconnect the negative battery cable. Drain the cooling system.

3. Remove the upper timing belt covers.

4. Remove the thermostat housing. Remove the blow-by hose connection.

5. Remove the PCV valve connection. Remove the PCV valve.

6. Remove the ignition coil connector. Remove the ignition coil. Remove the control wiring harness clamp.

7. Remove the rocker arm cover retaining bolts. Remove the rocker arm cover.

8. Remove the spark plug guide oil seal.

➡**Install auto lash adjuster retainers SST MD998443 on the rocker arms.**

9. Remove the intake and exhaust rocker arms and shafts by loosening the mounting bolts and removing the rocker arm and shaft assembly with the bolt still attached.

To install:

➡**Lubricate the valve train components with clean engine oil.**

10. Bleed and install the lash adjusters in their original bores in the cylinder head.

11. Install the rocker arms shafts and lash adjuster assembly. Torque the bolts to 21–25 ft. lbs. (28–34 Nm). Remove the lash adjuster retaining tool.

12. Install the valve cover oil seals and gasket. Torque the bolts to 27–35 inch lbs. (3.0–4.0 Nm).

13. Install the remaining components in the reverse of the removal procedure.

14. Connect the negative battery cable. On 2005–2006 vehicles, turn the ignition **ON** and then **OFF**, and keep it off for at least ten seconds.

15. Run the engine and check for leaks, correct as required.

Right Bank

1. Before servicing the vehicle, refer to the Precautions Section.

2. Disconnect the negative battery cable. Drain the cooling system.

3. Remove the intake manifold plenum assembly.

4. Remove the upper timing belt covers. Remove the thermostat housing.

5. Remove the breather hose connection. Remove the blow-by hose connection.

6. Remove the ignition coil connector. Remove the ignition coil. Remove the control wiring harness clamp.

7. Remove the rocker arm cover retaining bolts. Remove the rocker arm cover.

8. Remove the spark plug guide oil seal.

➡**Install auto lash adjuster retainers SST MD998443 on the rocker arms.**

9. Remove the intake and exhaust rocker arms and shafts by loosening the mounting bolts and removing the rocker arm and shaft assembly with the bolt still attached.

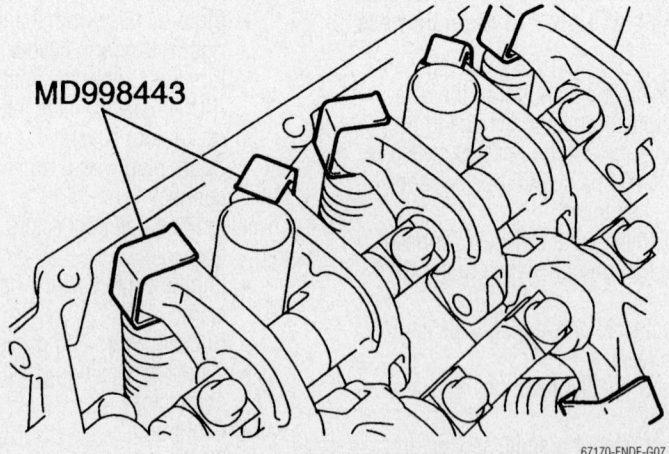

MD998443

67170-ENDE-G07

Install the special tool to prevent the lash adjusters from falling to the floor during rocker arm removal

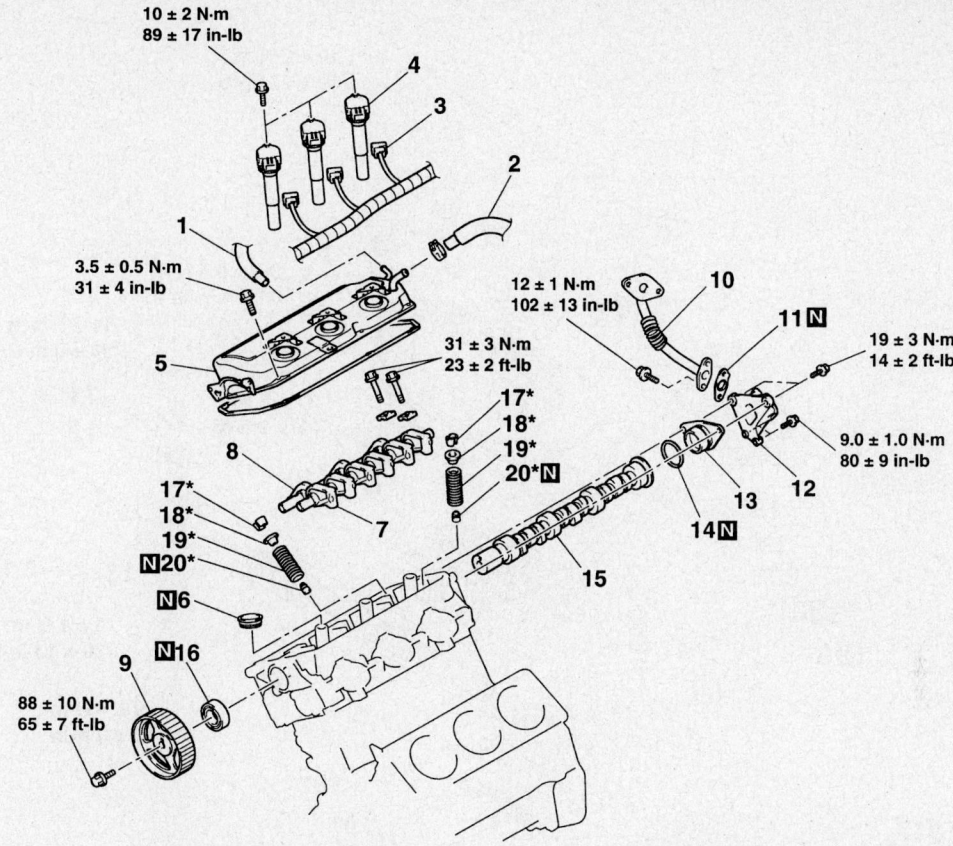

10 ± 2 N·m
89 ± 17 in-lb

3.5 ± 0.5 N·m
31 ± 4 in-lb

31 ± 3 N·m
23 ± 2 ft-lb

12 ± 1 N·m
102 ± 13 in-lb

19 ± 3 N·m
14 ± 2 ft-lb

9.0 ± 1.0 N·m
80 ± 9 in-lb

88 ± 10 N·m
65 ± 7 ft-lb

1. BREATHER HOSE CONNECTION
2. BLOW-BY HOSE CONNECTION
3. IGNITION COIL CONNECTOR
4. IGNITION COIL
5. ROCKER COVER
6. SPARK PLUG GUIDE OIL SEAL
7. ROCKER ARM, SHAFT AND LASH ADJUSTER ASSEMBLY (INTAKE SIDE)
8. ROCKER ARM, SHAFT AND

LASH ADJUSTER ASSEMBLY (EXHAUST SIDE)
9. CAMSHAFT SPROCKET
10. EGR PIPE B
11. GASKET
12. EGR VALVE SUPPORT
13. THRUST CASE
14. O-RING
15. CAMSHAFT
16. CAMSHAFT OIL SEAL

09482_ENDE_G0004

Rocker arms and related components—right bank

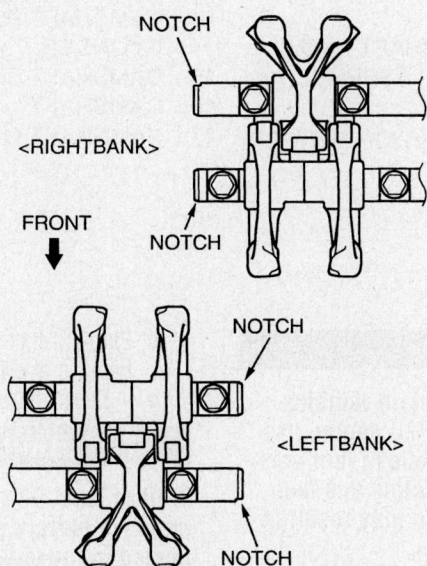

NOTCH

<RIGHTBANK>

FRONT

NOTCH

NOTCH

<LEFTBANK>

NOTCH

09482_ENDE_G0003

Rocker arm installation identification points

To install:

➡Lubricate the valve train components with clean engine oil.

10. Bleed and install the lash adjusters in their original bores in the cylinder head.

11. Install the rocker arms shafts and lash adjuster assembly. Torque the bolts to 21–25 ft. lbs. (28–34 Nm). Remove the lash adjuster retaining tool.

12. Install the valve cover oil seals and gasket. Torque the bolts to 27–35 inch lbs. (3.0–4.0 Nm).

13. Install the remaining components in the reverse of the removal procedure.

14. Connect the negative battery cable. On 2005–2006 vehicles, turn the ignition **ON** and then **OFF**, and keep it off for at least ten seconds.

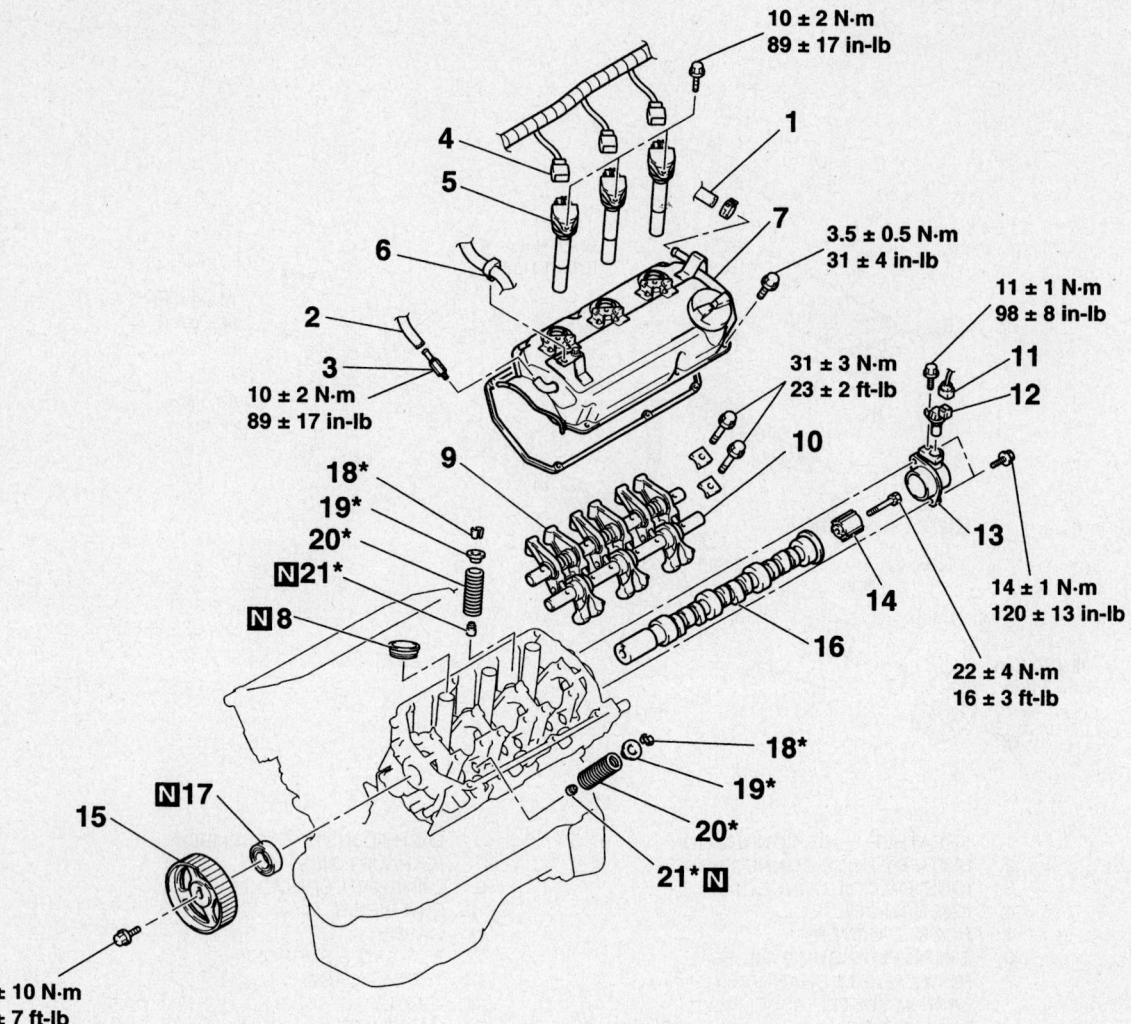

10 ± 2 N·m
89 ± 17 in-lb

3.5 ± 0.5 N·m
31 ± 4 in-lb

11 ± 1 N·m
98 ± 8 in-lb

31 ± 3 N·m
23 ± 2 ft-lb

14 ± 1 N·m
120 ± 13 in-lb

22 ± 4 N·m
16 ± 3 ft-lb

10 ± 2 N·m
89 ± 17 in-lb

88 ± 10 N·m
65 ± 7 ft-lb

1. BLOW-BY HOSE CONNECTION
2. PCV HOSE CONNECTION
3. PCV VALVE
4. IGNITION COIL CONNECTOR
5. IGNITION COIL
6. ENGINE CONTROL WIRING HARNESS CLAMP
7. ROCKER COVER

8. SPARK PLUG GUIDE OIL SEAL
9. ROCKER ARM, SHAFT AND LASH ADJUSTER ASSEMBLY (INTAKE SIDE)
10. ROCKER ARM, SHAFT AND LASH ADJUSTER ASSEMBLY (EXHAUST SIDE)
11. CAMSHAFT POSITION SENSOR CONNECTOR

12. CAMSHAFT POSITION SENSOR
13. CAMSHAFT POSITION SENSOR SUPPORT
14. CAMSHAFT POSITION SENSING CYLINDER
15. CAMSHAFT SPROCKET
16. CAMSHAFT
17. CAMSHAFT OIL SEAL

67170-ENDE-G08

Rocker arms and related components—left bank

15. Run the engine and check for leaks, correct as required.

Intake Manifold

REMOVAL & INSTALLATION

1. Before servicing the vehicle, refer to the Precautions Section.

✳✳ **CAUTION**

The fuel injection system remains under pressure after the engine has been OFF. Properly relieve fuel pressure before disconnecting any fuel lines. Failure to do so may result in fire or personal injury.

2. Relieve the fuel pressure.

3. Partially drain the cooling system.
4. Remove or disconnect the following:
   • Negative battery cable

✳✳ **CAUTION**

Wait at least 90 seconds after the negative battery cable is disconnected to prevent possible deployment of the air bag.

- Air cleaner assembly
- Throttle body
- Manifold Differential Pressure (MDP) sensor connector
- Knock Sensor (KS) connector
- Crankshaft Position (CKP) sensor connector
- Control wiring harness and injector wiring harness combination connector

- Right bank Heated Oxygen Sensor (HO$_2$S) connector connections and clamps
- Connector bracket
- Evaporative emission (EVAP) purge solenoid connector
- Purge hose
- Purge hose connection
- EVAP purge solenoid
- Front intake manifold plenum stay

- Exhaust Gas Recirculation (EGR) pipe A clamp
- Power steering pressure hose clamp
- Power steering pressure hose clamp bracket
- Rear intake manifold plenum stay
- EGR pipe B connection
- Retaining bolts and intake manifold plenum

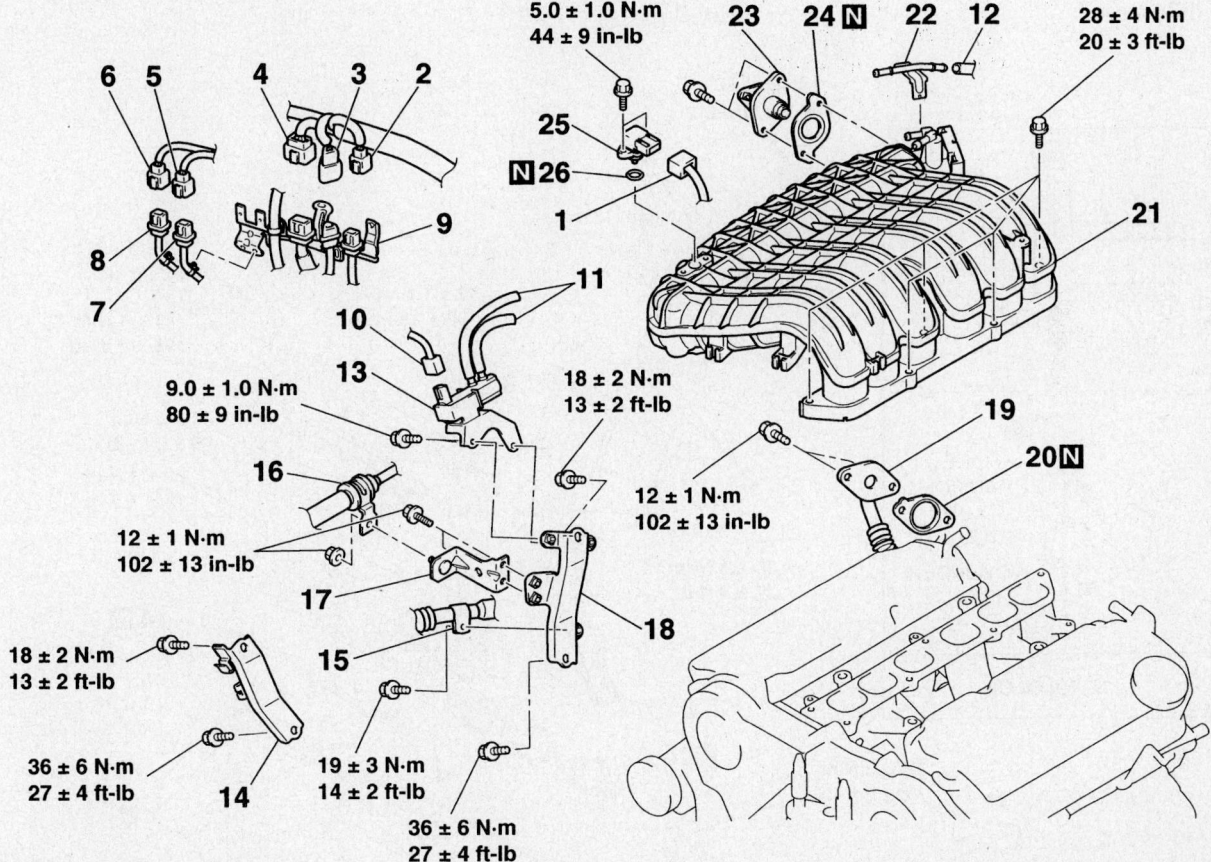

1. MANIFOLD DIFFERENTIAL PRESSURE SENSOR CONNECTOR
2. KNOCK SENSOR CONNECTOR
3. CRANKSHAFT POSITION SENSOR CONNECTOR
4. CONTROL WIRING HARNESS AND INJECTOR WIRING HARNESS COMBINATION CONNECTOR
5. RIGHT BANK HEATED OXYGEN SENSOR (FRONT) CONNECTOR
6. RIGHT BANK HEATED OXYGEN SENSOR (REAR) CONNECTOR
7. RIGHT BANK HEATED OXYGEN SENSOR (FRONT) CONNECTOR CLAMP

8. RIGHT BANK HEATED OXYGEN SENSOR (REAR) CONNECTOR CLAMP
9. CONNECTOR BRACKET
10. EVAPORATIVE EMISSION PURGE SOLENOID CONNECTOR
11. PURGE HOSE
12. PURGE HOSE CONNECTION
13. EVAPORATIVE EMISSION PURGE SOLENOID
14. INTAKE MANIFOLD PLENUM STAY, FRONT
15. EGR PIPE A CLAMP
16. POWER STEERING PRESSURE HOSE CLAMP

17. POWER STEERING PRESSURE HOSE CLAMP BRACKET
18. INTAKE MANIFOLD PLENUM STAY, REAR
19. EGR PIPE B CONNECTION
20. GASKET
21. INTAKE MANIFOLD PLENUM
22. VACUUM PIPE
23. EGR ADAPTER
24. EGR ADAPTER GASKET
25. MANIFOLD DIFFERENTIAL PRESSURE SENSOR
26. O-RING

Intake manifold plenum and related components

- Intake manifold plenum gasket
- EGR adapter and gasket
- MDP sensor and O-ring
- Ignition coil connector and coil
- Injector connector
- Engine mount stay
- High pressure fuel hose and O-ring
- Blow-by hose
- Fuel rail and injectors. Also remove the damper assembly, if equipped.
- Positive Crankcase Ventilation (PCV) hose

- Right and left front upper timing belt cover
- Water pump bracket
- Intake manifold retainers, manifold and gasket. Thoroughly clean and dry the mating surfaces of the manifold and heads.

To install:

5. Install or connect the following:
- New intake manifold gasket. Make sure the gaskets are installed with the protrusions as shown in the illustration.

- Intake manifold

6. Coat the intake manifold retaining studs with clean engine oil. Install the intake manifold bolts and tighten as follows:

a. 1st step: Right bank nuts to 45–71 inch lbs. (5–8 Nm).

b. 2nd step: Left bank nuts to 15–17 ft. lbs. (21–23 Nm).

c. 3rd step: Right bank nuts to 15–17 ft. lbs. (21–23 Nm).

d. 4th step: Left bank nuts to 15–17 ft. lbs. (21–23 Nm).

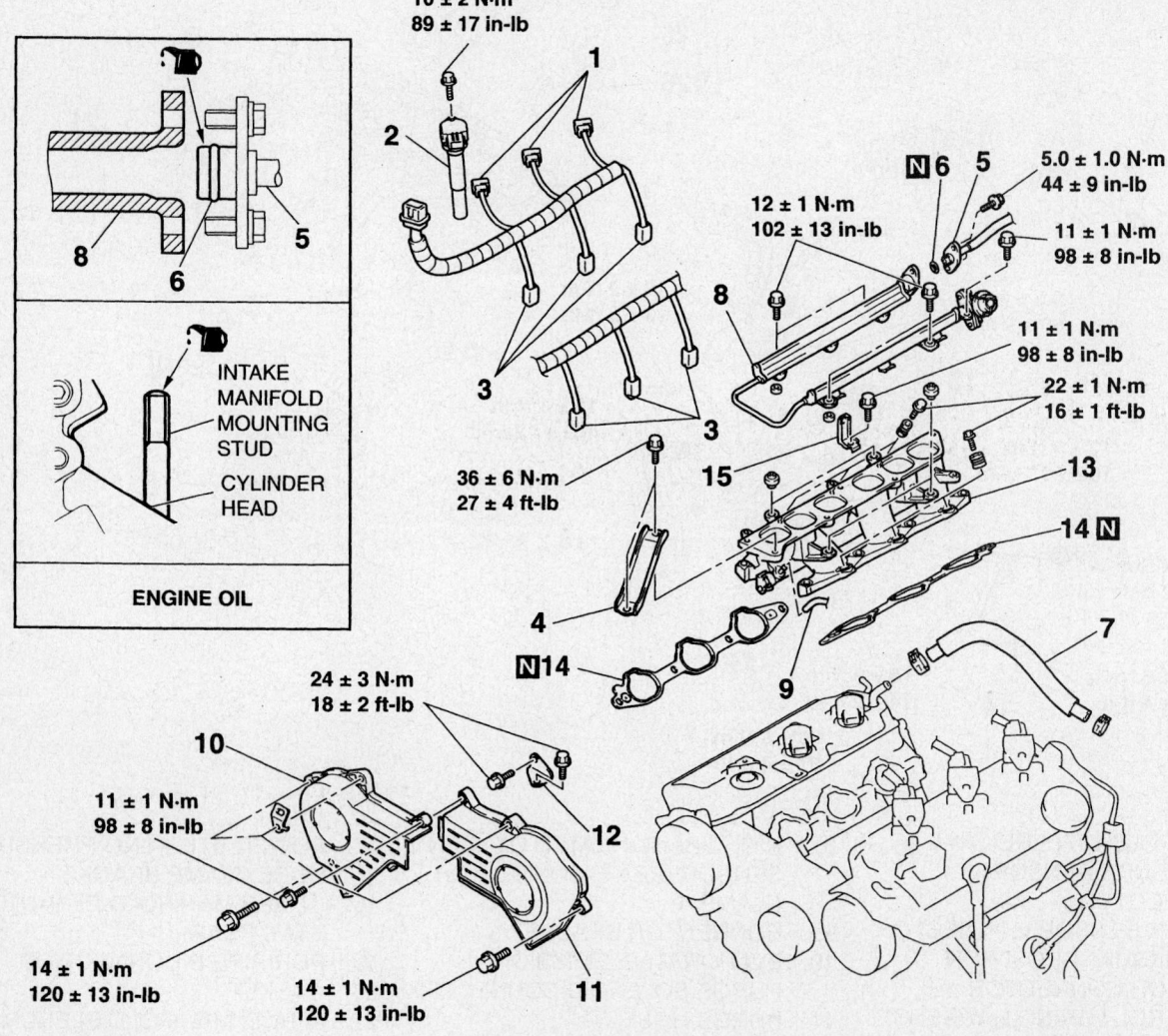

1. IGNITION COIL CONNECTOR
2. IGNITION COIL
3. INJECTOR CONNECTOR
4. ENGINE MOUNT STAY
5. FUEL HIGH-PRESSURE HOSE CONNECTION
6. O-RING
7. BLOW-BY HOSE
8. FUEL RAIL, INJECTOR AND FUEL DAMPER <UP TO DECEMBER 2003>, FUEL RAIL AND INJECTOR <FROM JANUARY 2004>
9. PCV HOSE CONNECTION
10. TIMING BELT FRONT UPPER COVER, RIGHT
11. TIMING BELT FRONT UPPER COVER, LEFT
12. WATER PUMP BRACKET
13. INTAKE MANIFOLD
14. INTAKE MANIFOLD GASKET
15. CONTROL HARNESS CLAMP

Intake manifold and related components

67170-ENDE-G10

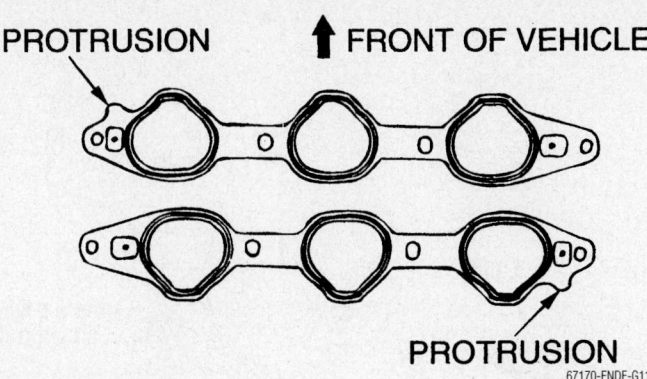

PROTRUSION   ↑ FRONT OF VEHICLE

PROTRUSION

67170-ENDE-G11

Intake manifold gasket positioning

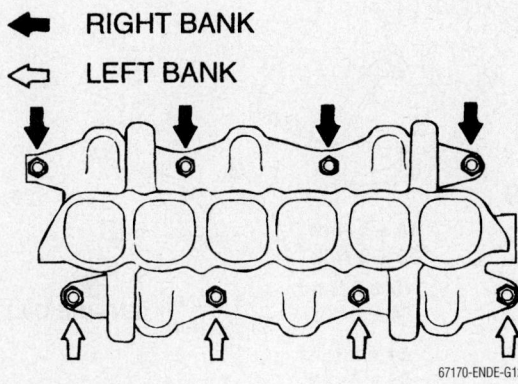

← RIGHT BANK
⇐ LEFT BANK

67170-ENDE-G12

Intake manifold retaining nut location

e. 5th step: Right bank nuts to 15–17 ft. lbs. (21–23 Nm).
- Water pump bracket
- Right and left front upper timing belt cover
- PCV hose
- Fuel rail and injectors. Also remove the damper assembly, if equipped.
- Blow-by hose
- High pressure fuel hose with a new O-ring
- Engine mount stay
- Injector connectors
- Ignition coil connectors and coils
- MDP sensor and O-ring
- EGR adapter and gasket
- New intake manifold plenum gasket
- Intake manifold plenum. Tighten the plenum mounting bolts to 20 ft. lbs. (28 Nm).
- EGR pipe B connection
- Rear intake manifold plenum stay
- Power steering pressure hose clamp bracket
- Power steering pressure hose clamp
- EGR pipe A clamp
- Front intake manifold plenum stay
- EVAP purge solenoid, hose and connector

- Connector bracket
- Right bank HO2S clamps and connectors
- Control wiring harness and injector wiring harness combination connector
- CKP sensor connector
- KS connector
- MDP sensor connector
- Throttle body
- Air cleaner assembly
7. Refill the radiator with coolant.
8. Connect the negative battery cable. On 2005–2006 vehicles, turn the ignition **ON** and then **OFF**, and keep it off for at least ten seconds.
9. Run the engine and check for fuel leaks, correct as required. Correct and adjust all fluid levels, as required.

Exhaust Manifold

REMOVAL & INSTALLATION

Left Bank

1. Before servicing the vehicle, refer to the Precautions Section.
2. Remove or disconnect the following:

- Negative battery cable
- Engine under cover
- Air duct
- Left Heated Oxygen Sensor (HO2S) connectors and sensors
- Engine oil dipstick, guide and O-ring
- Heat shield
- Front exhaust pipe from the manifold. Remove and discard the gasket.
- Exhaust manifold stay (left B)
- Left side exhaust manifold and gasket
- Exhaust manifold stay (left A)
3. Clean the gasket mounting surfaces. Inspect the manifolds for cracks, flatness and/or damage.
To install:
4. Install or connect the following:
- Exhaust manifold stay (left A)
- New gasket and left side exhaust manifold. Torque the nuts to specification.
- Exhaust manifold stay (left B)
- Exhaust pipe to the manifold with a new gasket. Torque the bolts to 31–43 ft. lbs. (40–58 Nm).
- Left side heat shield
- Oil dipstick, guide and new O-ring
- Left HO2S and connectors
- Air duct
- Under cover
5. Connect the negative battery cable. On 2005–2006 vehicles, turn the ignition **ON** and then **OFF**, and keep it off for at least ten seconds.
6. Start the engine and check for exhaust leaks.

Right Bank

1. Before servicing the vehicle, refer to the Precautions Section.
2. Drain the transfer case fluid and the engine coolant.
3. Remove or disconnect the following:
- Negative battery cable
- Air cleaner assembly
- Battery and battery tray
- Under cover
- Front exhaust pipe from the right exhaust manifold
- Propeller shaft, if AWD
- Manifold Differential Pressure (MDP) sensor connector
- Knock Sensor (KS) connector
- Crankshaft Position (CKP) sensor connector
- Control wiring harness and injector wiring harness combination connector

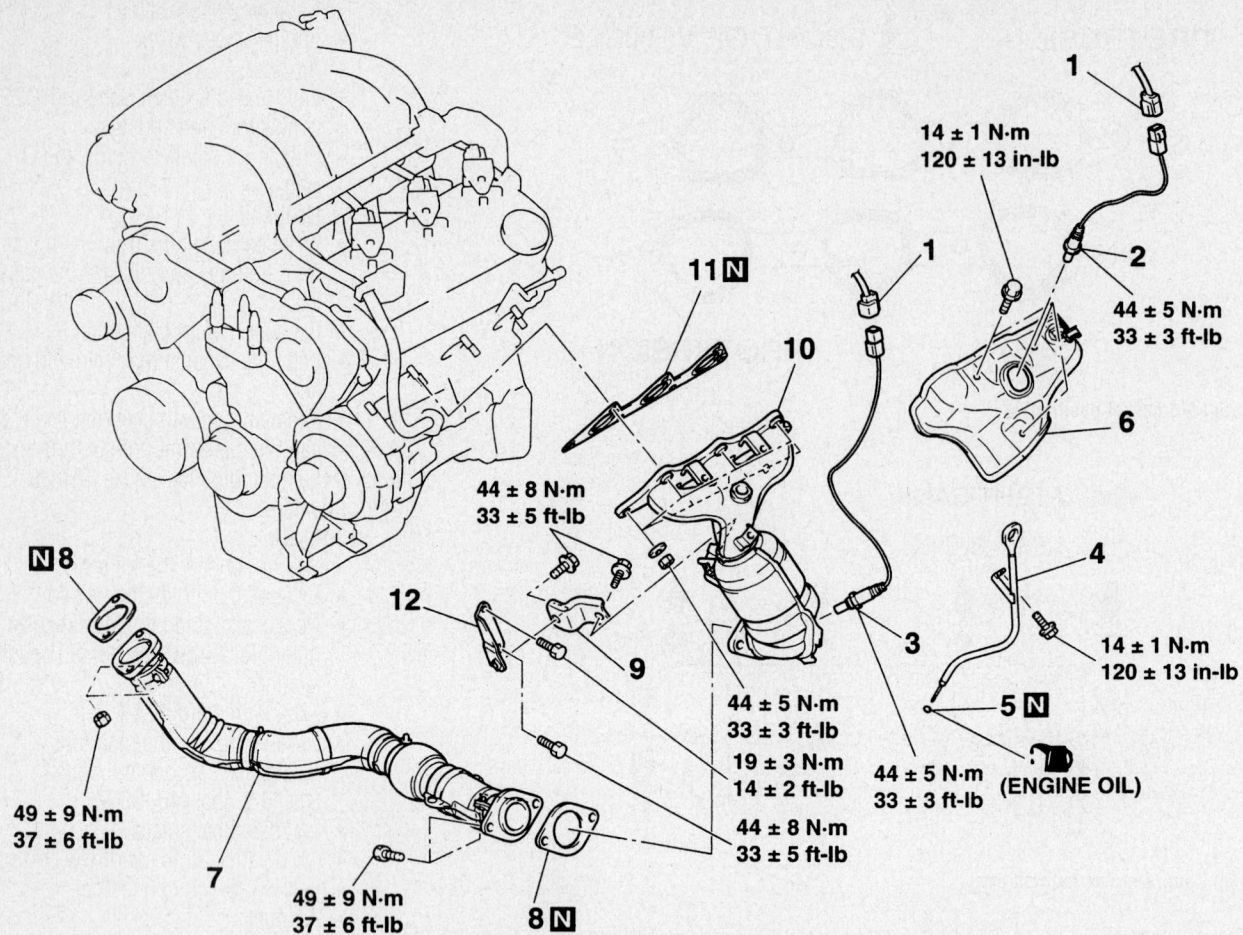

14 ± 1 N·m
120 ± 13 in-lb

44 ± 5 N·m
33 ± 3 ft-lb

44 ± 8 N·m
33 ± 5 ft-lb

44 ± 5 N·m
33 ± 3 ft-lb

19 ± 3 N·m
14 ± 2 ft-lb

44 ± 8 N·m
33 ± 5 ft-lb

44 ± 5 N·m
33 ± 3 ft-lb

14 ± 1 N·m
120 ± 13 in-lb

(ENGINE OIL)

49 ± 9 N·m
37 ± 6 ft-lb

49 ± 9 N·m
37 ± 6 ft-lb

1. LEFT HEATED OXYGEN SENSOR CONNECTOR
2. LEFT BANK HEATED OXYGEN SENSOR (FRONT)
3. LEFT BANK HEATED OXYGEN SENSOR (REAR)
4. ENGINE OIL DIPSTICK
5. O-RING
6. HEAT PROTECTOR
7. FRONT EXHAUST PIPE
8. FRONT EXHAUST PIPE GASKET
9. EXHAUST MANIFOLD STAY, LEFT B
10. EXHAUST MANIFOLD
11. EXHAUST MANIFOLD GASKET
12. EXHAUST MANIFOLD STAY, LEFT A

67170-ENDE-G13

Left bank exhaust manifold and related components

- Right Heated Oxygen Sensor (HO2S) connectors
- Connector bracket
- Right HO2S sensor connector clamps and sensors
- Exhaust Gas Recirculation (EGR) pipe A connection
- Water hoses
- EGR pipe A and gasket
- Exhaust manifold stay (right B)
- Steering gear and linkage protector
- Front floor backbone brace
- Front heat protector panel
- Transfer extension housing and O-ring, if AWD

- Lower and upper heat shields
- Power steering return pipe clamp connecting bolt and nut, if AWD
- Right side exhaust manifold and gasket

4. Clean the gasket mounting surfaces. Inspect the manifolds for cracks, flatness and/or damage.

To install:

5. Install or connect the following:
- New gasket and right side exhaust manifold. Torque the nuts to specification.
- Power steering return pipe clamp connecting bolt and nut, if AWD

- Lower and upper heat shields
- Transfer extension housing and O-ring, if AWD
- Front heat protector panel
- Front floor backbone brace
- Steering gear and linkage protector
- Exhaust manifold stay (right B)
- EGR pipe A and gasket
- Water hoses
- EGR pipe A connection
- Right HO2S sensors and connector clamps
- Connector bracket
- Right HO2S connectors
- Control wiring harness and injector

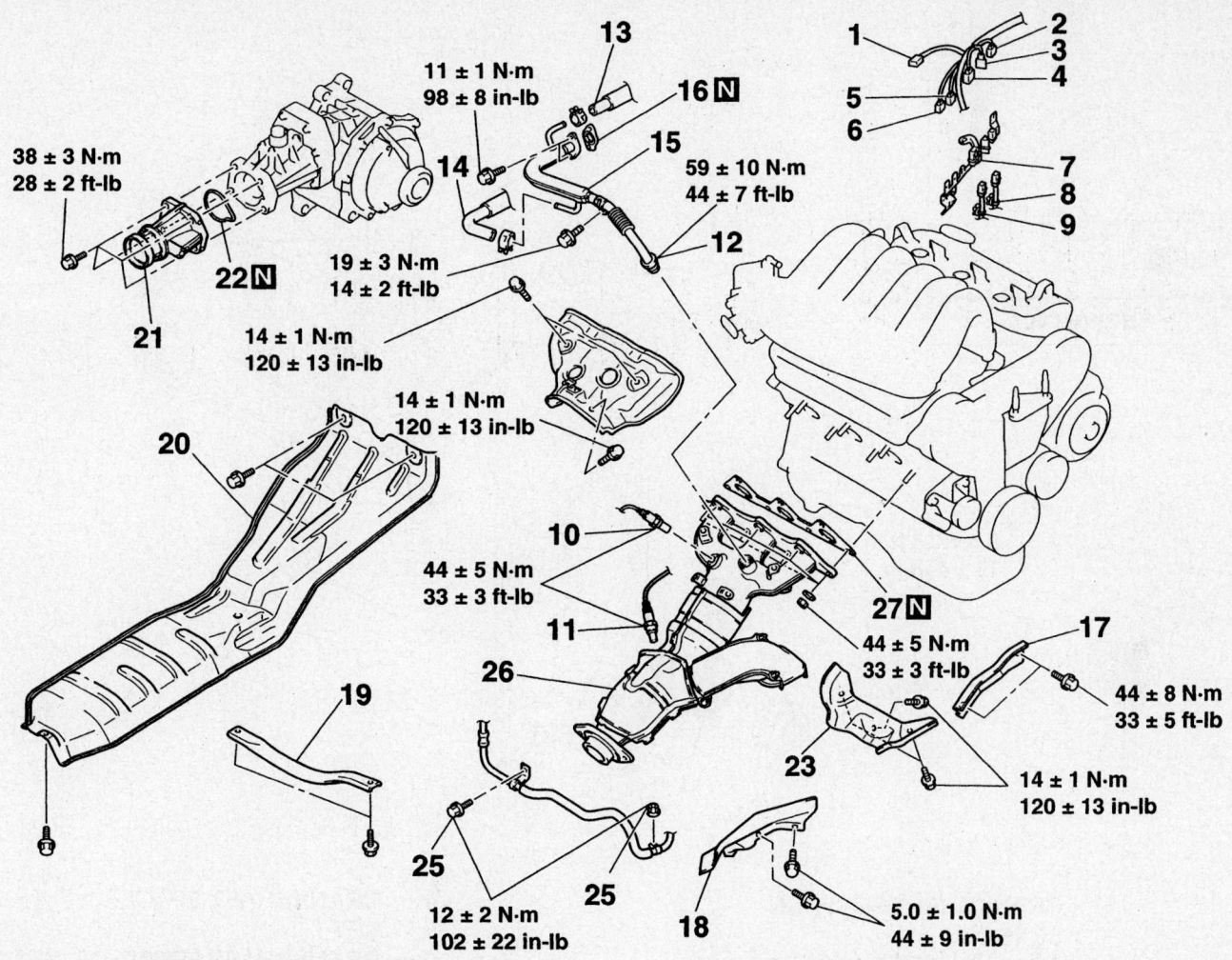

13

11 ± 1 N·m
98 ± 8 in-lb

16

15

59 ± 10 N·m
44 ± 7 ft-lb

14

12

1
2
3
4
5
6

7
8
9

38 ± 3 N·m
28 ± 2 ft-lb

19 ± 3 N·m
14 ± 2 ft-lb

22

21

14 ± 1 N·m
120 ± 13 in-lb

14 ± 1 N·m
120 ± 13 in-lb

20

10

44 ± 5 N·m
33 ± 3 ft-lb

11

26

27

44 ± 5 N·m
33 ± 3 ft-lb

17

44 ± 8 N·m
33 ± 5 ft-lb

14 ± 1 N·m
120 ± 13 in-lb

23

19

25

25

12 ± 2 N·m
102 ± 22 in-lb

18

5.0 ± 1.0 N·m
44 ± 9 in-lb

1. MANIFOLD DIFFERENTIAL
   PRESSURE SENSOR
   CONNECTOR
2. KNOCK SENSOR CONNECTOR
3. CRANKSHAFT POSITION
   SENSOR CONNECTOR

4. CONTROL WIRING HARNESS
   AND INJECTOR WIRING
   HARNESS COMBINATION
   CONNECTOR
5. RIGHT BANK HEATED OXYGEN
   SENSOR (FRONT) CONNECTOR
6. RIGHT BANK HEATED OXYGEN
   SENSOR (REAR) CONNECTOR

67170-ENDE-G14

Right bank exhaust manifold and related components

wiring harness combination con-
nector
- CKP sensor connector
- Knock Sensor (KS) connector
- Manifold Differential Pressure
  (MDP) sensor connector
- Propeller shaft, if AWD
- Front exhaust pipe to the exhaust
  manifold
- Under cover
- Battery and battery tray
- Air cleaner assembly
6. Connect the negative battery cable.
On 2005–2006 vehicles, turn the ignition

ON and then **OFF**, and keep it off for at
least ten seconds.
   7. Fill the transfer case with fluid. Fill
the engine cooling system.
   8. Start the engine and check for
exhaust leaks.

Front Crankshaft Seal

REMOVAL & INSTALLATION

   1. Before servicing the vehicle, refer to
the Precautions Section.

   2. Drain the engine oil. Drain the engine
coolant.
   3. Remove or disconnect the following:
- Negative battery cable

※※ CAUTION

**Wait at least 90 seconds after the
negative battery cable is disconnected to prevent possible deployment of the air bag.**

- Timing belt
- Crankshaft Position (CKP) sensor
- Crankshaft sprocket

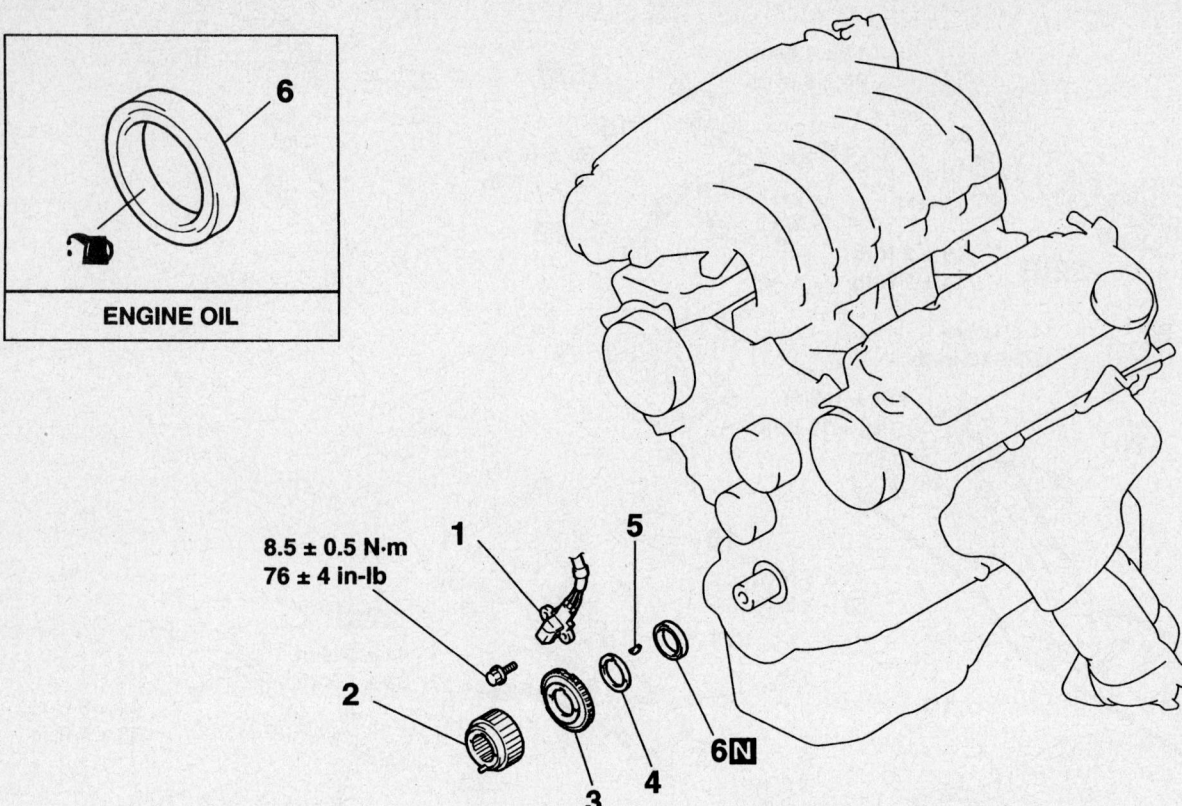

ENGINE OIL

8.5 ± 0.5 N·m
76 ± 4 in-lb

1. CRANKSHAFT POSITION
   SENSOR
2. CRANKSHAFT SPROCKET
3. CRANKSHAFT SENSING BLADE
4. CRANKSHAFT SPACER
5. KEY
6. CRANKSHAFT FRONT OIL SEAL

67170-ENDE-G15

Front oil seal and related components

- Crankshaft sensing blade
- Crankshaft spacer and key
- Front oil seal

To install:

- Front oil seal. Apply oil to the seal, and install using Crankshaft Front Oil Seal Installer tool no. MD998717.
- Crankshaft key and spacer.
- Crankshaft sensing blade
- CKP sensor

➡**To be sure the crankshaft pulley bolt does not loosen, make sure the clean the mating areas of the crankshaft, spacer, sensing blade and sprocket.**

- Crankshaft sprocket
- Timing belt

4. Connect the negative battery cable. On 2005–2006 vehicles, turn the ignition **ON** and then **OFF**, and keep it off for at least ten seconds.

Camshaft and Valve Lifters

REMOVAL & INSTALLATION

Left Bank

1. Before servicing the vehicle, refer to the Precautions Section.
2. Disconnect the negative battery cable. Drain the cooling system.
3. Remove the timing belt.
4. Remove the thermostat housing. Remove the blow-by hose connection.
5. Remove the PCV valve connection. Remove the PCV valve.
6. Remove the ignition coil connector. Remove the ignition coil. Remove the control wiring harness clamp.
7. Remove the rocker arm cover retaining bolts. Remove the rocker arm cover.
8. Remove the spark plug guide oil seal.

➡**Install auto lash adjuster retainers SST MD998443 on the rocker arms.**

9. Remove the intake and exhaust rocker arms and shafts by loosening the mounting bolts and removing the rocker arm and shaft assembly with the bolt still attached.
10. Disconnect the camshaft position sensor connector. Remove the camshaft position sensor. Remove the camshaft position sensor support.
11. Remove the camshaft position sensing cylinder.
12. Using a wrench and special tool MD998715, remove the camshaft sprocket.
13. Remove the camshaft. Remove the camshaft oil seal.

To install:

14. Installation is the reverse of the removal procedure.
15. Apply grease to the camshaft oil seal, prior to installation. Use special tools

MD998713 and MB991559 to press fit the camshaft oil seal into position.

16. Connect the negative battery cable. On 2005–2006 vehicles, turn the ignition **ON** and then **OFF**, and keep it off for at least ten seconds.

17. Run the engine and check for leaks, correct as required.

Right Bank

1. Before servicing the vehicle, refer to the Precautions Section.

2. Disconnect the negative battery cable. Drain the cooling system.

3. Remove the intake manifold plenum assembly.

4. Remove the timing belt. Remove the thermostat housing.

5. Remove the breather hose connection. Remove the blow-by hose connection.

6. Remove the ignition coil connector. Remove the ignition coil. Remove the control wiring harness clamp.

7. Remove the rocker arm cover retaining bolts. Remove the rocker arm cover.

8. Remove the spark plug guide oil seal.

➡**Install auto lash adjuster retainers SST MD998443 on the rocker arms.**

9. Remove the intake and exhaust rocker arms and shafts by loosening the mounting bolts and removing the rocker arm and shaft assembly with the bolt still attached.

10. Using a wrench and special tool MD998715, remove the camshaft sprocket.

11. Remove the EGR pipe "B", gasket and valve support.

12. Remove the thrust case and O-ring.

13. Remove the camshaft. Remove the camshaft oil seal.

To install:

14. Installation is the reverse of the removal procedure.

15. Apply grease to the camshaft oil seal, prior to installation. Use special tools MD998713 and MB991559 to press fit the camshaft oil seal into position.

16. Connect the negative battery cable. On 2005–2006 vehicles, turn the ignition **ON** and then **OFF**, and keep it off for at least ten seconds.

17. Run the engine and check for leaks, correct as required.

INSPECTION

1. Before servicing the vehicle, refer to the Precautions Section.

2. Remove the camshaft from the engine.

3. Check the camshaft bearing journals for damage and binding.

4. If the journals are binding, check the cylinder head for damage.

5. Check the cylinder head for clogged oil holes.

6. As required, check the tooth surface of the distributor drive gear teeth of the camshaft. Replace the camshaft if wear is evident.

7. Check the camshaft surface for abnormal wear and damage. Replace the camshaft, as required.

8. Measure the cam height and replace the camshaft if not within specification.

9. Intake camshaft standard specification is 1.472 inches. Minimum limit is 1.452 inches.

10. Exhaust camshaft standard specification is 1.452 inches. Minimum limit is 1.443 inches.

Valve Lash

ADJUSTMENT

1. Before servicing the vehicle, refer to the Precautions Section.

2. Check the engine oil. Change or adjust level as required.

3. Idle the engine for one to three minutes, to allow it to warm up.

4. Repeat the operation as indicated in the illustration. At a no load condition check for abnormal noise.

➡**Usually the abnormal noise is eliminated after repetition of the operation ten to thirty times. If no change is observed, suspect that the abnormal noise is due to other factors.**

5. After elimination of the abnormal noise, repeat the operation (in the illustration) five more times.

6. Run the engine at idle for one to three minutes to be sure the abnormal noise has been eliminated.

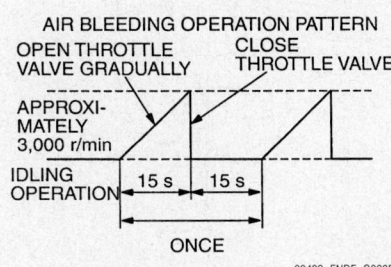

09482_ENDE_G0005

Air bleeding operation pattern

Oil Pan

REMOVAL & INSTALLATION

1. Before servicing the vehicle, refer to the Precautions Section.

2. Drain the engine oil.

3. Remove or disconnect the following:
- Negative battery cable
- Engine undercover
- Front exhaust pipe
- Oil pan drain plug and gasket
- Starter motor
- Engine oil dipstick and O-ring
- Lower oil pan. If necessary, use a block of wood and hammer to carefully dislodge the lower oil pan.
- Cover
- 2 lower torque converter connecting bolts
- Upper oil pan. Screw the M10 bolts holding the oil pan to the transaxle assembly into the bolt hole shown in the illustration to remove the pan.
- Oil screen
- Oil pan gasket

To install:

4. Before installing, thoroughly clean the oil pan and cylinder block mating surfaces.

5. Apply liquid gasket around the surface of the oil pan.

➡**Assemble the oil pan to the cylinder block within 30 minutes after applying the liquid gasket.**

6. Install or connect the following:
- Oil screen. Torque the bolts to 12–16 ft. lbs. (16–22 Nm).
- Upper oil pan. Torque the bolts to specification and, in the proper sequence. The bolt holes for bolts 13 and 14 are cut away on the transaxle side. Be sure you do not insert the bolts at an angle.
- Two lower torque converter connecting bolts. Tighten the bolts to 36 ft. lbs. (49 Nm).
- Cover
- Lower oil pan. Torque the bolts to specification and in the proper sequence.
- Engine oil dipstick and O-ring
- Starter motor
- Oil pan drain plug with a new gasket. Tighten to specification.
- Front exhaust pipe
- Engine undercover

7. Connect the negative battery cable. On 2005–2006 vehicles, turn the ignition

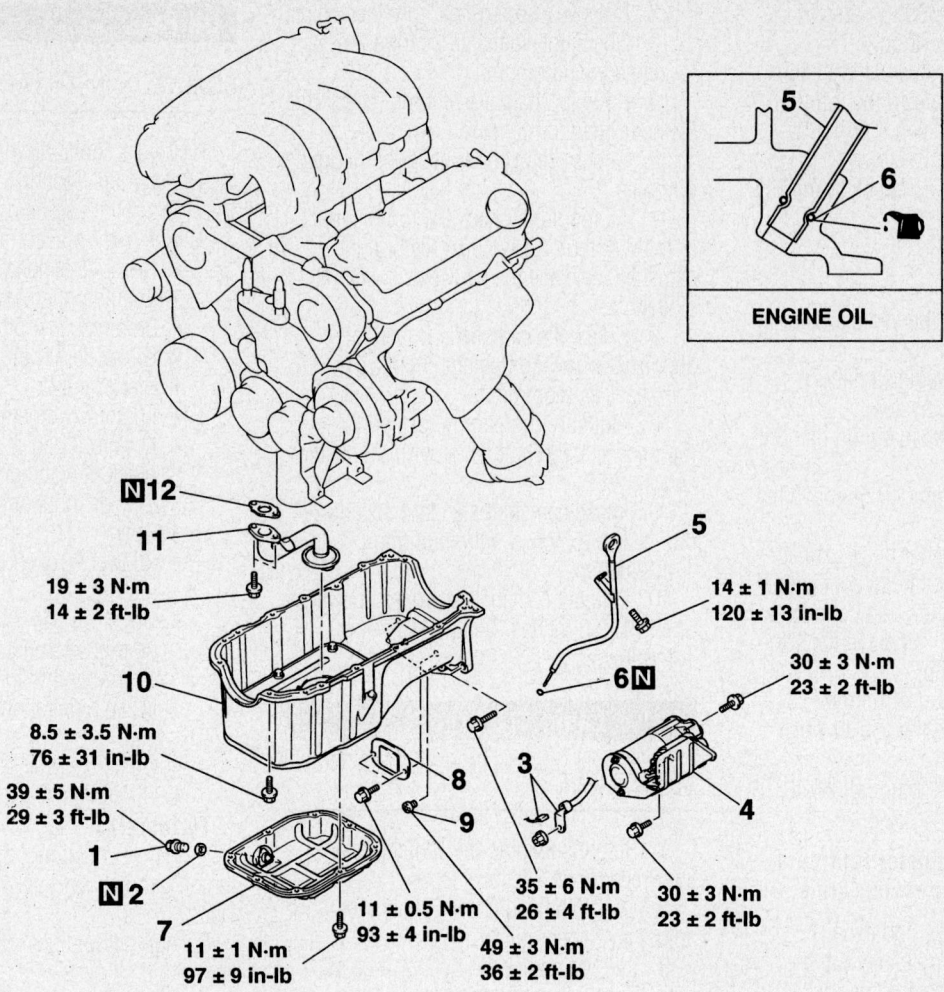

ENGINE OIL

N12

11

19 ± 3 N·m
14 ± 2 ft-lb

10

8.5 ± 3.5 N·m
76 ± 31 in-lb

39 ± 5 N·m
29 ± 3 ft-lb

1

N2

7

11 ± 1 N·m
97 ± 9 in-lb

11 ± 0.5 N·m
93 ± 4 in-lb

8

3

9

49 ± 3 N·m
36 ± 2 ft-lb

35 ± 6 N·m
26 ± 4 ft-lb

30 ± 3 N·m
23 ± 2 ft-lb

5

14 ± 1 N·m
120 ± 13 in-lb

6N

4

30 ± 3 N·m
23 ± 2 ft-lb

1. ENGINE OIL PAN DRAIN PLUG
2. ENGINE OIL PAN DRAIN PLUG
   GASKET
3. STARTER CONNECTOR
4. STARTER ASSEMBLY
5. ENGINE OIL DIPSTICK
   ASSEMBLY
6. O-RING
7. ENGINE LOWER OIL PAN
8. COVER
9. TORQUE CONVERTER
   CONNECTING BOLTS
10. ENGINE UPPER OIL PAN
11. OIL SCREEN
12. GASKET

67170-ENDE-G18

Oil pan and related components

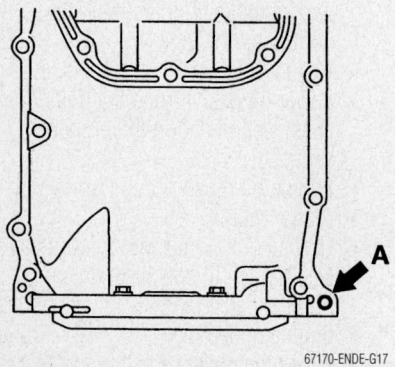

67170-ENDE-G17

Oil pan removal bolt installation location

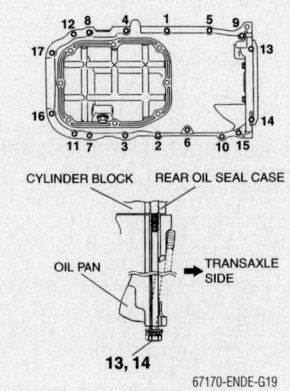

CYLINDER BLOCK   REAR OIL SEAL CASE

OIL PAN

TRANSAXLE
SIDE

13, 14

67170-ENDE-G19

Upper oil pan bolt torque sequence

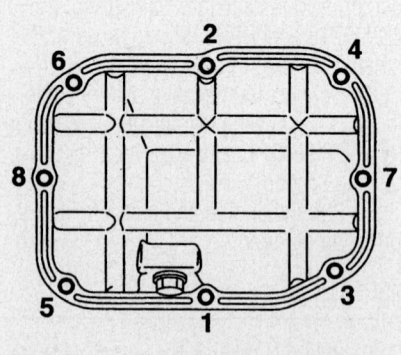

67170-ENDE-G20

Lower oil pan bolt torque sequence

ON and then OFF, and keep it off for at least ten seconds.

8. Fill the crankcase with oil. Start the engine and check for leaks.

Oil Pump

REMOVAL & INSTALLATION

1. Before servicing the vehicle, refer to the Precautions Section.
2. Drain the engine oil.
3. Remove or disconnect the following:
- Negative battery cable
- Timing belt

- Oil pressure switch
- Oil dipstick
- Oil pans from the engine
- Oil baffle and screen
- Oil pump mounting bolts and the pump from the front of the engine

➡Note the position of each oil pump case retaining bolts to facilitate installation. The bolts are of different length.

To install:

4. Clean the gasket mounting surfaces of the pump and engine block.
5. Prime the pump by pouring fresh oil into the inlet and turning the rotors or by

packing pump with petroleum jelly. Using a new gasket, install the oil pump on the engine and tighten all bolts to 10 ft. lbs. (14 Nm).

6. Clean the gasket mounting surfaces of the pump and engine block.
7. Apply a 0.1 inch diameter bead of sealant (part number MD970389 or equivalent) to the oil pump case.

➡Apply sealant as indicated by the broken line in the illustration. The grooves must be traced and the bolt holes must be surrounded with a bead of sealant.

8. Install the oil pump case assembly to the front of the cylinder block.

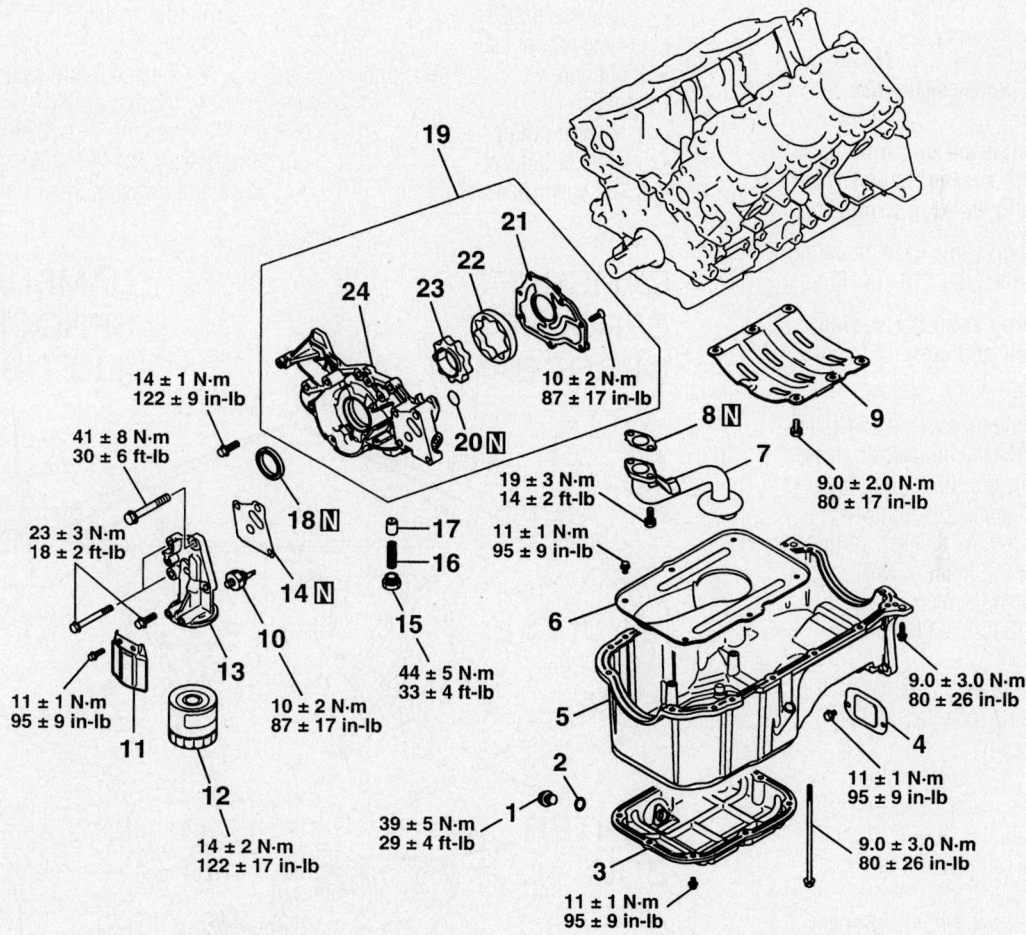

1. DRAIN PLUG
2. DRAIN PLUG GASKET
3. OIL PAN, LOWER
4. COVER
5. OIL PAN, UPPER
6. BAFFLE PLATE
7. OIL SCREEN
8. OIL SCREEN GASKET
9. BAFFLE PLATE
10. ENGINE OIL PRESSURE SWITCH
11. OIL FILTER COVER
12. OIL FILTER
13. OIL FILTER BRACKET
14. OIL FILTER BRACKET GASKET
15. RELIEF PLUG
16. RELIEF SPRING
17. RELIEF PLUNGER
18. CRANKSHAFT FRONT OIL SEAL
19. OIL PUMP CASE ASSEMBLY
20. O-RING
21. OIL PUMP COVER
22. OIL PUMP OUTER ROTOR
23. OIL PUMP INNER ROTOR
24. OIL PUMP CASE

09482_ENDE_G0007

Oil pump and related components

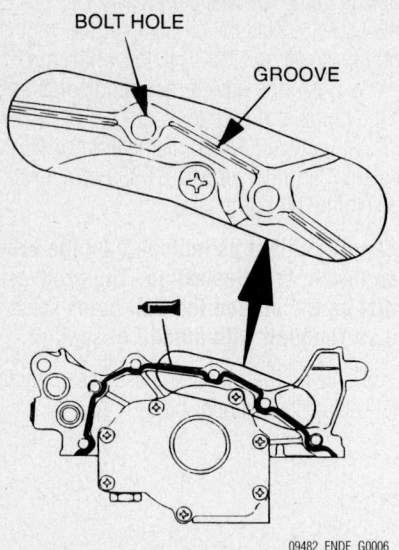

BOLT HOLE

GROOVE

09482_ENDE_G0006

Oil pump to case sealant application

➡ **Be sure to install the oil pump case quickly, while the sealant is wet. There is a fifteen minute working window.**

9. Torque the oil pump case mounting bolts to 113–131 inch lbs. (13–15 Nm).

➡ **After installation keep the sealed area free from oil and coolant for at least one hour.**

10. Clean out the oil pick-up or replace as required. Replace the oil pick-up gasket ring and install the pick-up to the pump.

11. Install or connect the following:
- Oil filter and the bracket. Torque the bolts to 17 ft. lbs. (23 Nm).
- Oil baffle and screen. Torque the bolts to 13 ft. lbs. (18 Nm).
- Oil pans
- Oil pressure switch. Torque the switch to 87 inch lbs. (9.8 Nm).
- Timing belt
- Dipstick

12. Connect the negative battery cable. On 2005–2006 vehicles, turn the ignition **ON** and then **OFF**, and keep it off for at least ten seconds.

13. Refill the engine with the proper amount of oil.

14. Start the engine and check for proper oil pressure. Check for leaks.

Timing Belt, Cover and Crankshaft Seal

REMOVAL & INSTALLATION

1. Before servicing the vehicle, refer to the Precautions Section.
2. Drain the engine coolant.

3. Remove or disconnect the following:
- Negative battery cable
- Engine under cover and side under cover
- Drive belts
- Crankshaft pulley, using special tools MB991800 and MB991802
- Manifold Differential Pressure (MDP) sensor connector
- Knock Sensor (KS) connector
- Crankshaft Position (CKP) sensor connector
- Control wiring harness and injector wiring harness combination connector
- Right bank Heated Oxygen Sensor (HO2S) connectors
- Connector bracket
- Engine mount stay
- Right and left timing belt upper covers
- Tensioner pulley
- Tensioner bracket
- CKP sensor harness clamp
- Lower timing belt cover
- Engine mount
- Engine support bracket

4. Turn the crankshaft clockwise to align the timing marks and set the No. 1 cylinder at Top Dead Center (TDC). If you are reusing the timing belt, mark the flat side of the belt with an arrow showing the clockwise direction.

5. Loosen the center bolt of the tension pulley, and then remove the timing belt.
- Auto tensioner
- Tensioner pulley
- Tensioner arm
- Shaft

To install:

6. Install or connect the following:
- Idler pulley
- Shaft
- Tensioner arm assembly
- Tension pulley

7. Press the end of the auto-tensioner inward with 72–145 ft. lbs. (98–196 Nm) of force and measure the distance that the

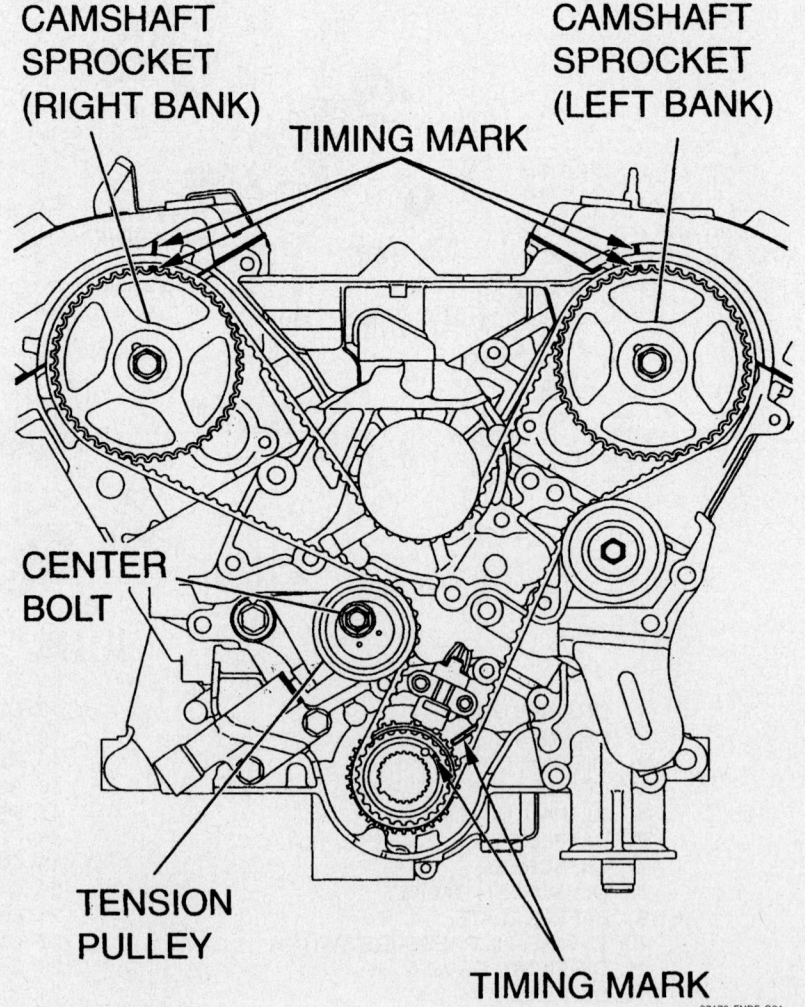

CAMSHAFT SPROCKET (RIGHT BANK)

CAMSHAFT SPROCKET (LEFT BANK)

TIMING MARK

CENTER BOLT

TENSION PULLEY

TIMING MARK

67170-ENDE-G21

Timing belt alignment

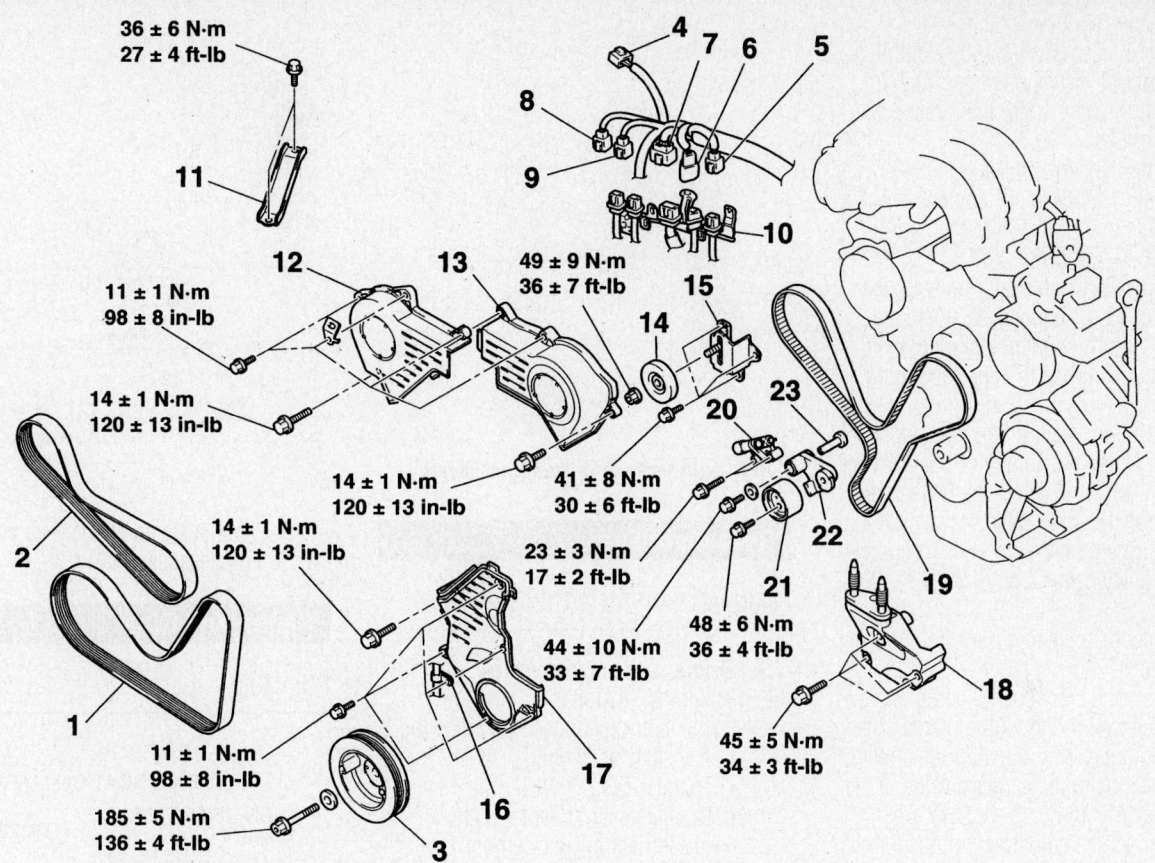

36 ± 6 N·m
27 ± 4 ft-lb

11 ± 1 N·m
98 ± 8 in-lb

14 ± 1 N·m
120 ± 13 in-lb

14 ± 1 N·m
120 ± 13 in-lb

49 ± 9 N·m
36 ± 7 ft-lb

14 ± 1 N·m
120 ± 13 in-lb

41 ± 8 N·m
30 ± 6 ft-lb

23 ± 3 N·m
17 ± 2 ft-lb

44 ± 10 N·m
33 ± 7 ft-lb

48 ± 6 N·m
36 ± 4 ft-lb

11 ± 1 N·m
98 ± 8 in-lb

185 ± 5 N·m
136 ± 4 ft-lb

45 ± 5 N·m
34 ± 3 ft-lb

67170-ENDE-G22

1. GENERATOR DRIVE BELT
2. POWER STEERING OIL PUMP DRIVE BELT
3. CRANKSHAFT PULLEY
4. MANIFOLD DIFFERENTIAL PRESSURE SENSOR CONNECTOR
5. KNOCK SENSOR CONNECTOR
6. CRANKSHAFT POSITION SENSOR CONNECTOR
7. CONTROL WIRING HARNESS AND INJECTOR WIRING HARNESS COMBINATION CONNECTOR
8. RIGHT BANK HEATED OXYGEN SENSOR (REAR) CONNECTOR
9. RIGHT BANK HEATED OXYGEN SENSOR (FRONT) CONNECTOR
10. CONNECTOR BRACKET
11. ENGINE MOUNT STAY
12. TIMING BELT FRONT UPPER COVER, RIGHT
13. TIMING BELT FRONT UPPER COVER, LEFT
14. TENSIONER PULLEY
15. TENSIONER BRACKET
16. CRANKSHAFT POSITION SENSOR HARNESS CLAMP
17. TIMING BELT LOWER COVER

Timing belt and related components

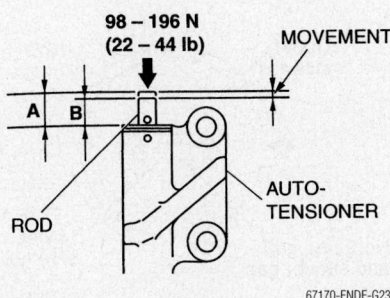

98 – 196 N
(22 – 44 lb)

MOVEMENT

A   B

ROD

AUTO-TENSIONER

67170-ENDE-G23

Auto-tensioner inspection

pushrod is pushed in. If the standard distance is not 0.04 in. (1mm), replace the auto-tensioner.

8. Position the auto-tensioner in a soft-jawed vise and SLOWLY compress the pushrod until the pushrod and housing holes align; then, install a setting pin to secure the auto-tensioner in the retracted position.

➡ **If you are installing a new auto-tensioner, the pin will already be inserted into the pin holes of the new tensioner.**

9. Install the auto-tensioner.
10. Align the camshaft and crankshaft TDC timing marks.
11. Install the timing belt (noting its rotational direction) so that there is no deflection between the sprockets and pulleys in the following manner:
   • Crankshaft sprocket
   • Idler pulley
   • Left camshaft sprocket
   • Water pump pulley
   • Right camshaft sprocket

- Tension pulley

12. Turn the camshaft sprocket counterclockwise until the tension side of the timing belt is firmly stretched, then, recheck the timing marks.

13. Using the Tension Pulley Socket Wrench tool MD998767, or equivalent, push the tensioner pulley into the timing belt and secure the center bolt.

14. Using the Crankshaft Pulley Spacer tool MD998769, or equivalent, rotate the crankshaft ¼ turn counterclockwise, then, turn it again clockwise to align the timing marks.

15. Loosen the timing belt tensioner center bolt. Using the Tension Pulley Socket Wrench tool MD998767, or equivalent, and a torque wrench, apply 39 inch lbs. (4.4 Nm) pressure on the timing belt. Torque the tensioner pulley center bolt to 35 ft. lbs. (48 Nm).

16. Remove the setting pin from the auto-tensioner.

17. Rotate the crankshaft 2 complete revolutions and realign the timing marks. Then, wait for 5 minutes until the auto-tensioner pushrod extends to its standard value. If the standard value is not 0.19–0.24 in. (4.8–6.0mm), repeat the adjustment procedure. If the standard value is still not achieved, replace the auto-tensioner.

- Install or connect the following:
- Engine support bracket
- Engine mount
- Lower timing belt cover
- CKP sensor harness clamp
- Tensioner bracket
- Tensioner pulley
- Left and right upper timing belt covers
- Engine mount stay
- Connector bracket
- HO2S connectors
- Wiring harness and injector connector
- CKP sensor connector
- KS connector
- MDP sensor connector

18. Install the crankshaft pulley. Using Pulley Holder MB991800 and the 2 Crankshaft Pulley Holder Pin tools MD991802, or equivalent to hold the crankshaft pulley, and a socket torque wrench, torque the crankshaft pulley bolt to 136 ft. lbs. (185 Nm).

- Drive belts
- Side under cover
- Under cover

19. Connect the negative battery cable. On 2005–2006 vehicles, turn the ignition **ON** and then **OFF**, and keep it off for at least ten seconds.

20. Refill the cooling system.

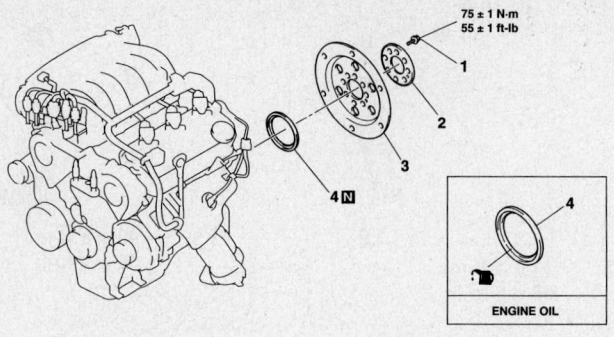

1. DRIVE PLATE BOLTS
2. ADAPTOR PLATE
3. DRIVE PLATE
4. CRANKSHAFT REAR OIL SEAL

67170-ENDE-G24

Rear main seal and related components

Rear Main Seal

REMOVAL & INSTALLATION

1. Before servicing the vehicle, refer to the Precautions Section.
2. Remove or disconnect the following:
   - Transaxle assembly
   - Transfer case, if AWD
   - Driveplate and adapter plate, matchmark for reassembly. Use Mitsubishi tool (MD998781) to hold the driveplate in position to remove the bolts.
3. Remove the rear oil seal as follows:
   a. Cut out a portion in the crankshaft oil seal lip.
   b. Cover the tip of a small prytool with a cloth and apply it to the cutout in the oil seal to pry the oil seal out.

✳✳ CAUTION

Take care not to damage the crankshaft and oil seal case.

To install:

4. Inspect the sealing surface at the rear of the crankshaft. If a deep groove is worn into the surface, the crankshaft will have to be replaced. Coat the sealing lip of the seal with fresh, clean engine oil. Press the new seal into the case with a seal installing tool. The seal must be pressed in squarely until it bottoms in the case. It is necessary to use the proper tool (MD998718-01) to fit the seal into place.

5. Install or connect the following:
   - Drive plate and adapter. Use Mitsubishi tool (MD998781) to hold the driveplate in position while tightening the bolts to 55 ft. lbs. (75 Nm).
   - Transfer case, if AWD

- Transaxle and related components as necessary

Piston and Ring

POSITION

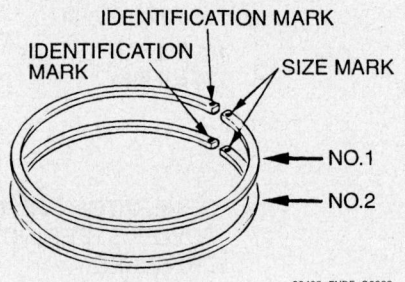

09482_ENDE_G0009

Piston ring identification

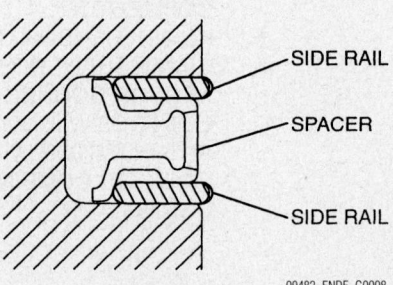

09482_ENDE_G0008

Oil ring identification

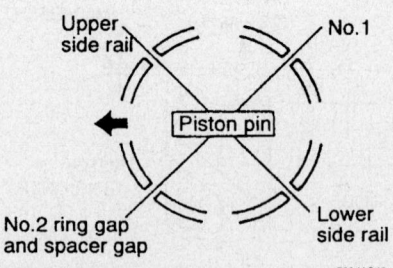

7924AG49

Piston ring end-gap spacing

FUEL SYSTEM

Fuel System Service Precautions

Safety is the most important factor when performing not only fuel system maintenance but any type of maintenance. Failure to conduct maintenance and repairs in a safe manner may result in serious personal injury or death. Maintenance and testing of the vehicle's fuel system components can be accomplished safely and effectively by adhering to the following rules and guidelines.

• To avoid the possibility of fire and personal injury, always disconnect the negative battery cable unless the repair or test procedure requires that battery voltage be applied.

• Always relieve the fuel system pressure prior to disconnecting any fuel system component (injector, fuel rail, pressure regulator, etc.), fitting or fuel line connection. Exercise extreme caution when relieving fuel system pressure, to avoid exposing your skin, face and eyes to fuel spray. Please be advised that fuel under pressure may penetrate the skin or any part of the body that it contacts.

• Always place a shop towel or cloth around the fitting or connection prior to loosening to absorb any excess fuel due to spillage. Ensure that all fuel spillage (should it occur) is quickly removed from engine surfaces. Ensure that all fuel soaked cloths or towels are deposited into a suitable waste container.

• Always keep a dry chemical (Class B) fire extinguisher near the work area.

• Do not allow fuel spray or fuel vapors to come into contact with a spark or open flame.

• Always use a back-up wrench when loosening and tightening fuel line connection fittings. This will prevent unnecessary stress and torsion to fuel line piping. Always follow the proper torque specifications.

• Always replace worn fuel fitting O-rings with new. Do not substitute fuel hose where fuel pipe is installed.

Fuel System Pressure

RELIEVING

✳✳ CAUTION

The fuel system is under constant pressure, even with the engine off. This pressure must be relieved

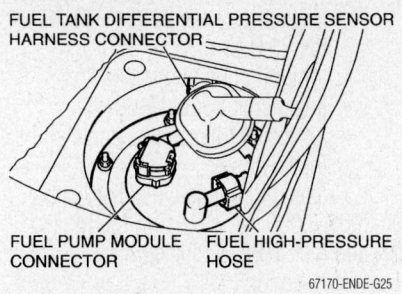

FUEL TANK DIFFERENTIAL PRESSURE SENSOR HARNESS CONNECTOR

FUEL PUMP MODULE CONNECTOR   FUEL HIGH-PRESSURE HOSE

67170-ENDE-G25

Fuel pump connector location

before disconnecting any fuel system component, fitting or fuel line connection. Failure to do so may result in personal injury.

1. Turn the ignition switch to **LOCK**.
2. Remove the left rear seat cushion mounting bolts, and lift the seat cushion to tie it to the head restraint.
3. Push on the floor mat to find the notches, then cut the carpet along the notches to access the service hole cover. Remove the service hole cover.
4. Detach the fuel pump module connector.
5. Start the engine and let it run out of fuel.
6. Attach the fuel pump module connector.
7. Install the service hole cover, floor mat and left rear seat cushion.

Fuel Filter

REMOVAL & INSTALLATION

The manufacturer does not provide a procedure or maintenance interval for replacing the fuel filter.

Fuel Pump

REMOVAL & INSTALLATION

1. Before servicing the vehicle, refer to the Precautions Section.
2. Relieve the fuel system pressure.
3. Remove or disconnect the following:
   • Negative battery cable

✳✳ CAUTION

The fuel injection system remains under pressure even after the engine has been turned OFF. Properly relieve fuel pressure before discon-

necting any fuel lines. Failure to do so may result in fire or personal injury. Do not allow fuel spray or fuel vapors to come in contact with a spark or open flame. Keep a dry chemical fire extinguisher nearby. Never store fuel in an open container due to risk of fire or explosion.

• Left rear seat cushion and tie out of the way to the head restraint.
4. Push on the floor mat to find the notches, then cut the carpet along the notches to access the service hole cover. Remove the service hole cover.
   • Fuel pump module connector
   • Fuel tank differential pressure sensor connector
   • Fuel high-pressure hose
   • Mounting nuts and plate and fuel pump module assembly

✳✳ WARNING

When removing the fuel pump module from the tank, be careful not to damage the module unit and float.

To install:
5. Install or connect the following:
   • Fuel pump assembly into the fuel tank. Torque the nuts to 24 inch lbs. (2.5 Nm).
   • Fuel lines
   • Fuel tank differential pressure sensor connector
   • Fuel pump module connector
   • Fuel pump cover. Torque the bolts to 14 inch lbs. (1.5 Nm).
   • Rear floor carpeting.
6. Connect the negative battery cable. On 2005–2006 vehicles, turn the ignition **ON** and then **OFF**, and keep it off for at least ten seconds.
7. Start the vehicle, check for leaks and proper operation.

Fuel Rail

REMOVAL & INSTALLATION

1. Before servicing the vehicle, refer to the Precautions Section.
2. Properly relieve the fuel system pressure. Disconnect the negative battery cable.
3. Remove or disconnect the following:
   • Air cleaner assembly, ducts and air intake hose
   • Intake manifold plenum

- Ignition coil connectors and coils
- Injector connectors
- High pressure fuel hose and O-ring
- Blow-by hose connection
- Fuel rail, injector and damper assembly, as necessary. Only earlier production vehicles will have a fuel damper.
- Insulators from the fuel rail

To install:

4. Install or connect the following:
- Fuel rail and insulators. Torque the bolts to 102 inch lbs. (12 Nm).

- Injector connectors
- Blow-by hose connection
- High pressure fuel hose and O-ring. Torque the bolts to 44 inch lbs. (5.0 Nm).
- Ignition coils and connectors
- Intake manifold plenum
- Air cleaner assembly, ducts and air intake hose

5. Connect the negative battery cable. On 2005–2006 vehicles, turn the ignition **ON** and then **OFF**, and keep it off for at least ten seconds.

Fuel Injectors

REMOVAL & INSTALLATION

1. Before servicing the vehicle, refer to the Precautions Section.
2. Properly relieve the fuel system pressure. Disconnect the negative battery cable.
3. Remove the fuel rail.
4. Remove the insulator from the fuel injector. Remove the fuel injector from the fuel rail. Discard the O-ring.

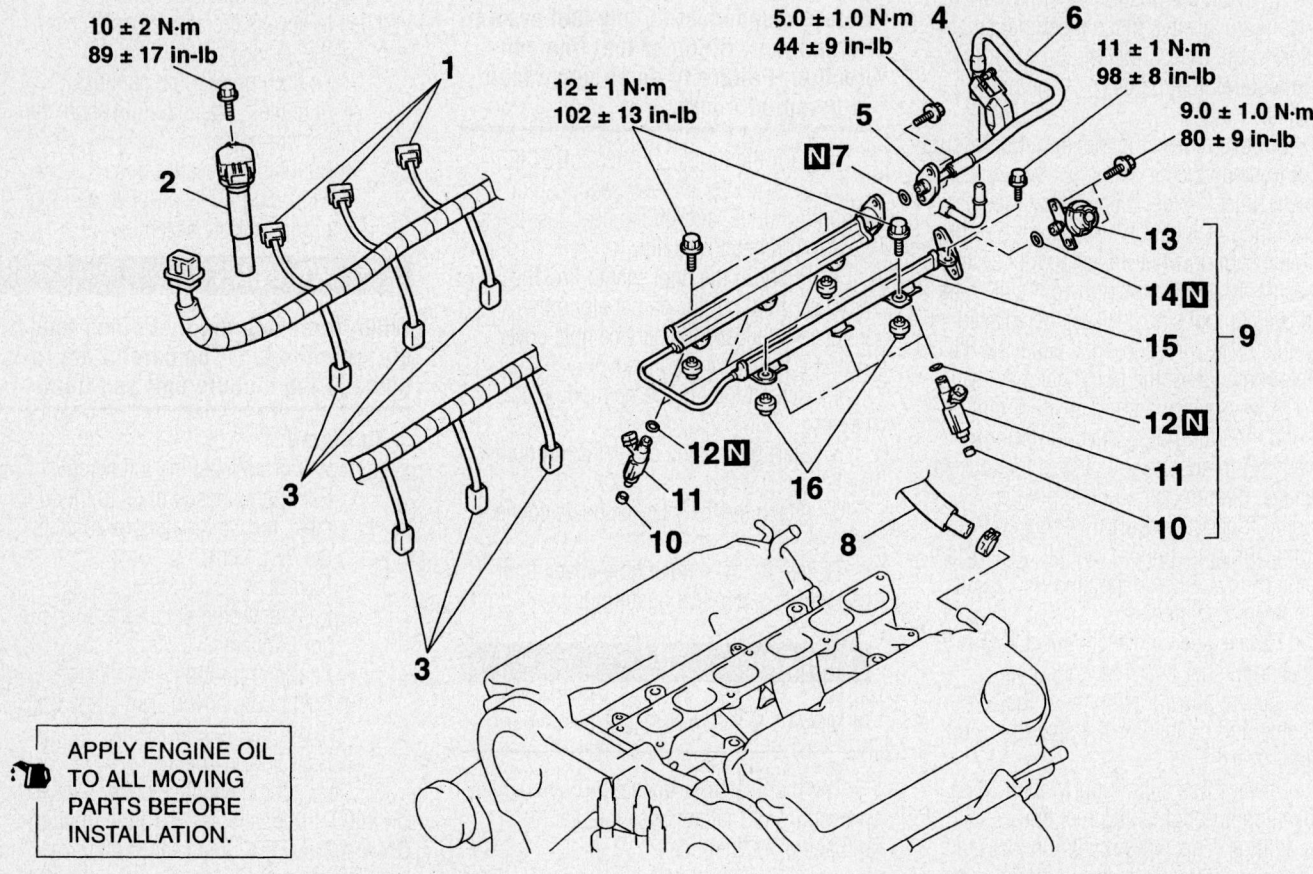

APPLY ENGINE OIL TO ALL MOVING PARTS BEFORE INSTALLATION.

1. IGNITION COIL CONNECTOR
2. IGNITION COIL
3. INJECTOR CONNECTOR
4. FUEL HIGH-PRESSURE HOSE CONNECTION (FUEL LINE PIPE SIDE)
5. FUEL HIGH-PRESSURE HOSE CONNECTION (FUEL RAIL SIDE)
6. FUEL HIGH-PRESSURE HOSE
7. O-RING
8. BLOW-BY HOSE CONNECTION
9. FUEL RAIL, INJECTOR AND FUEL DAMPER <UP TO DECEMBER 2003>, FUEL RAIL AND INJECTOR <FROM JANUARY 2004>
10. INSULATORS
11. FUEL INJECTORS
12. O-RING
13. FUEL DAMPER <UP TO DECEMBER 2003>
14. O-RING
15. FUEL RAIL
16. INSULATORS

Fuel rail and related components

67170-ENDE-G26

To install:

5. Installation is the reverse of the removal procedure.

6. Be sure to use new gaskets and O-rings, as required.

7. Connect the negative battery cable. On 2005–2006 vehicles, turn the ignition **ON** and then **OFF**, and keep it off for at least ten seconds.

8. Run the engine and check for leaks, correct as required.

9. Adjust all fluid levels, as required.

Throttle Body

REMOVAL & INSTALLATION

➡**On some 2005 vehicles and all 2006 vehicles, when replacing the throttle body, you must initialize the learning value in a MFI engine.**

1. Before servicing the vehicle, refer to the Precautions Section.

2. Properly relieve the fuel system pressure. Disconnect the negative battery cable.

3. Drain the engine coolant. Remove the air intake hose. Remove the battery.

4. Disconnect the throttle position sensor connector. Disconnect the water hose connection.

5. Remove the throttle body retaining bolts. Remove the throttle body from the engine. Discard the gasket.

To install:

➡**Do not loosen the retaining screws for the resin cover of the throttle body assembly. If the screws are loosened, the sensor incorporated in the resin cover becomes misaligned and the throttle body can not work properly.**

6. Align the recess on the intake manifold plenum with the projection of the throttle body gasket. Install the gasket.

➡**Poor idling etc. may result if the throttle body gasket is not installed properly.**

7. Install the throttle body to the engine. Torque the retaining bolts to 18–24 ft. lbs. (24–32 Nm).

8. Continue the installation in the reverse order of the removal procedure.

9. Connect the negative battery cable. On 2005–2006 vehicles, turn the ignition **ON** and then **OFF**, and keep it off for at least ten seconds.

10. On 2005–2006 vehicles, complete the vehicle initialization procedure.

➡**To complete the initialization procedure the following tools are needed. MB991958 scan tool, MB991824 VCI, MB991827 MUT III USB cable, MB991910 MUT III Main harness "A".**

11. Connect the scan tool to the data link connector. To prevent damage to the scan tool be sure that the ignition switch is in the **LOCK** position before connecting the scan tool.

12. Turn the ignition switch to the **ON** position.

13. Select "check mode" from the menu screen.

14. Select "erase memory" from the menu screen.

15. Initialize the learning value.

16. After initialization complete the idle learning procedure.

➡**This procedure must be performed when the PCM is replaced, or when the learning value is initialized, as the idling is not stabilized because the learning value in the MFI engine is not completed.**

17. Start the engine. Allow the coolant temperature to reach 176 degrees or more.

18. Stop the engine and place the ignition switch in the **LOCK** position.

19. After ten seconds, restart the engine.

20. For ten minutes carry out the idling procedure below to confirm that the engine has normal idling.

- Position the transaxle selector lever in the **P** range
- The engine fan is not to be operated
- The engine coolant temperature should be 176 degrees or more

➡**If the engine stalls during idling, check the throttle valve of the throttle body for dirt. Correct and perform the procedure again.**

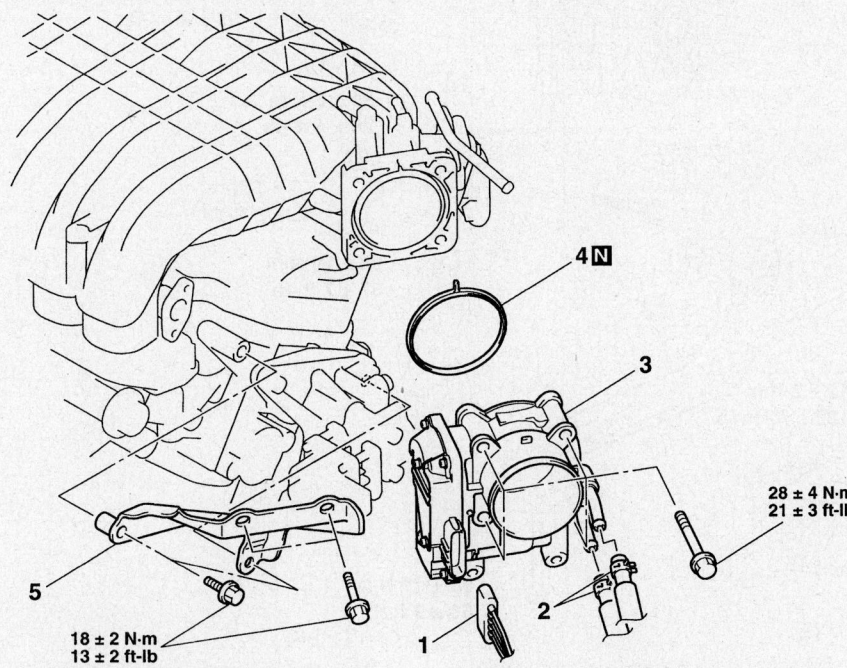

28 ± 4 N·m
21 ± 3 ft-lb

18 ± 2 N·m
13 ± 2 ft-lb

1.	THROTTLE POSITION SENSOR CONNECTOR	3.	THROTTLE BODY
2.	WATER HOSE CONNECTION	4.	THROTTLE BODY GASKET
		5.	THROTTLE BODY STAY

09482_ENDE_G0010

Throttle body and related components

DRIVE TRAIN

Transaxle

REMOVAL & INSTALLATION

1. Before servicing the vehicle, refer to the Precautions Section.
2. Remove the engine under cover.
3. Drain the transaxle fluid.
4. On AWD vehicles, drain the engine coolant.
5. Remove or disconnect the following:
   - Hood
   - Negative battery cable

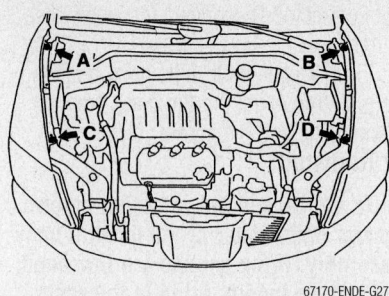

67170-ENDE-G27

Front fender assembling bolts for special tool installation

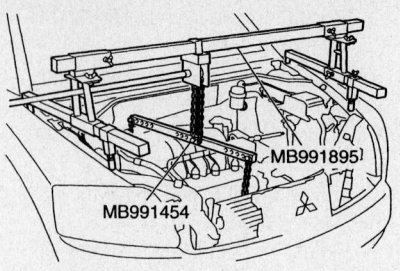

MB991895
MB991454

67170-ENDE-G29

View of the engine hangers installed on the engine

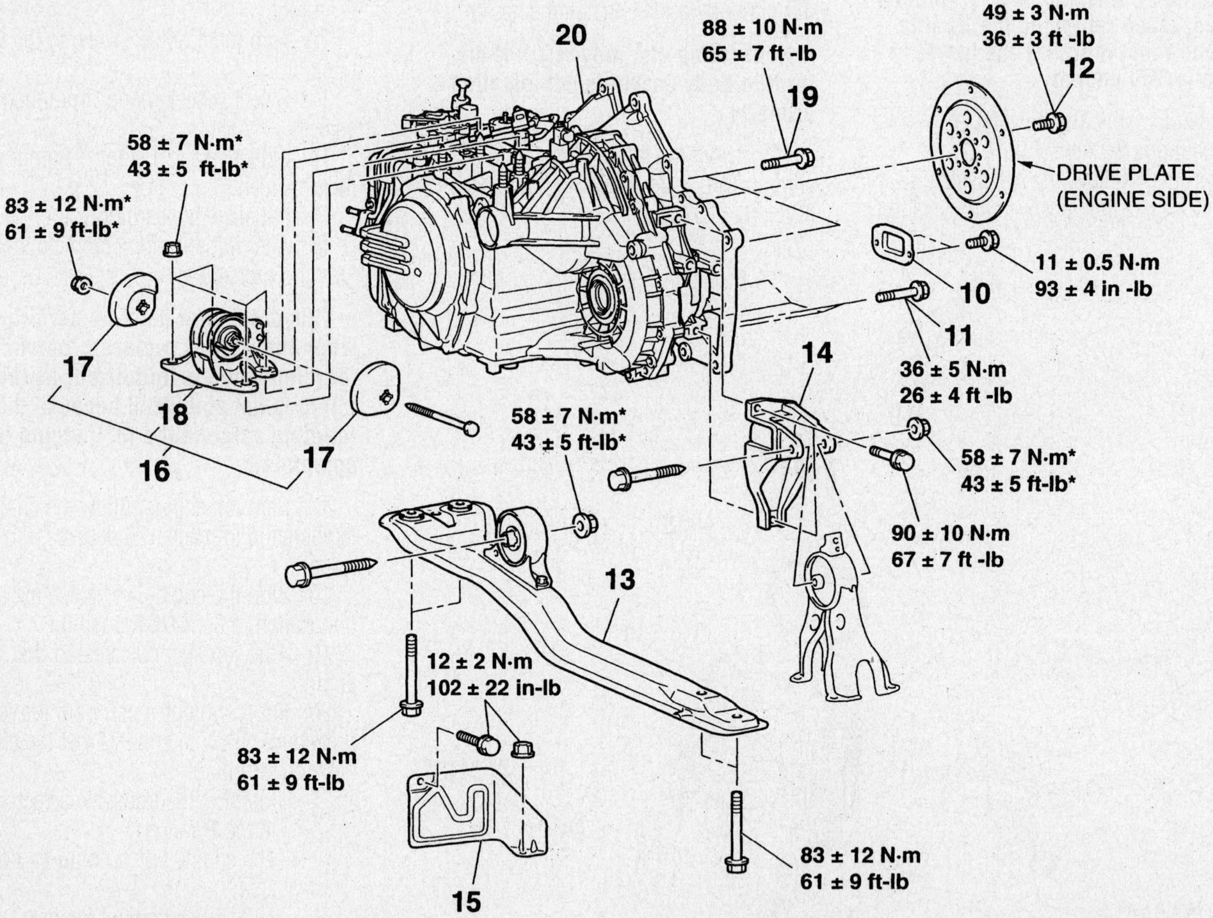

20

88 ± 10 N·m
65 ± 7 ft -lb
19

49 ± 3 N·m
36 ± 3 ft -lb
12

DRIVE PLATE
(ENGINE SIDE)

58 ± 7 N·m*
43 ± 5 ft-lb*

83 ± 12 N·m*
61 ± 9 ft-lb*

11 ± 0.5 N·m
93 ± 4 in -lb
10

36 ± 5 N·m
26 ± 4 ft -lb
11

17

18

16

17

58 ± 7 N·m*
43 ± 5 ft-lb*

14

58 ± 7 N·m*
43 ± 5 ft-lb*

90 ± 10 N·m
67 ± 7 ft -lb

13

12 ± 2 N·m
102 ± 22 in-lb

83 ± 12 N·m
61 ± 9 ft-lb

15

83 ± 12 N·m
61 ± 9 ft-lb

10. COVER
11. ENGINE OIL PAN AND TRANSAXLE ASSEMBLY COUPLING BOLTS
12. DRIVE PLATE BOLTS
13. CENTERMEMBER ASSEMBLY
14. REAR ROLL STOPPER BRACKET
15. CROSSMEMBER PLATE <LH SIDE>

16. TRANSAXLE MOUNT BRACKET ASSEMBLY
17. TRANSAXLE MOUNT STOPPER
18. TRANSAXLE MOUNT BRACKET
19. TRANSAXLE ASSEMBLY LOWER PART COUPLING BOLTS
20. TRANSAXLE ASSEMBLY

67170-ENDE-G28

Automatic transaxle and related components

- Air cleaner assembly
- Powertrain Control Module (PCM)
- Battery and battery tray

6. On 2006 vehicles, remove the intake manifold rear plenum stay, EGR pipe and intake manifold plenum. Remove the upper radiator hose, engine oil dipstick and engine hanger.

- Front exhaust pipe from the 2 exhaust manifolds, then disconnect it from the intermediate pipe/catalytic converter (make certain to retain the bolts and nuts for reassembly).
- Adjusting nut
- Transaxle control cable connection
- Inhibitor switch connector
- A/T control solenoid valve connector
- Input and output shaft speed sensor connectors
- Driveshaft (2WD) or driveshaft and output shaft assembly (AWD)
- Transfer case assembly

- Starter
- Transaxle oil cooler hose
- Transaxle assembly upper part coupling bolts
- Cover
- Engine oil pan and transaxle assembly coupling bolts
- Drive plate bolts. Use Mitsubishi tool MD998781, or equivalent, to hold the driveplate in position to remove the bolts.
- Centermember assembly
- Rear roll stopper bracket
- Crossmember plate, left side
- Air cleaner bracket
- Transaxle mount bracket assembly
- Transaxle mount stopper
- Manifold Differential Pressure (MDP) sensor

7. Install Engine Hanger MB991895 to the front fender bolts, as shown in the illustration.

8. Install Engine Hanger Balancer MB991454, or equivalent, to the engine/transaxle assembly. Place a thick towel on the end of the tool to prevent damaging the firewall.

9. Remove the transaxle from the vehicle.

To install:

10. Installation is the reverse of the removal procedure. Note the torque specifications in the illustration.

11. Refill the transaxle and transfer case with oil.

12. Start the vehicle and check for any leaks.

Transfer Case Assembly

REMOVAL & INSTALLATION

1. Before servicing the vehicle, refer to the Precautions Section.

2. Drain the transaxle and transfer case fluid. Drain the engine coolant.

3. Remove or disconnect the following:

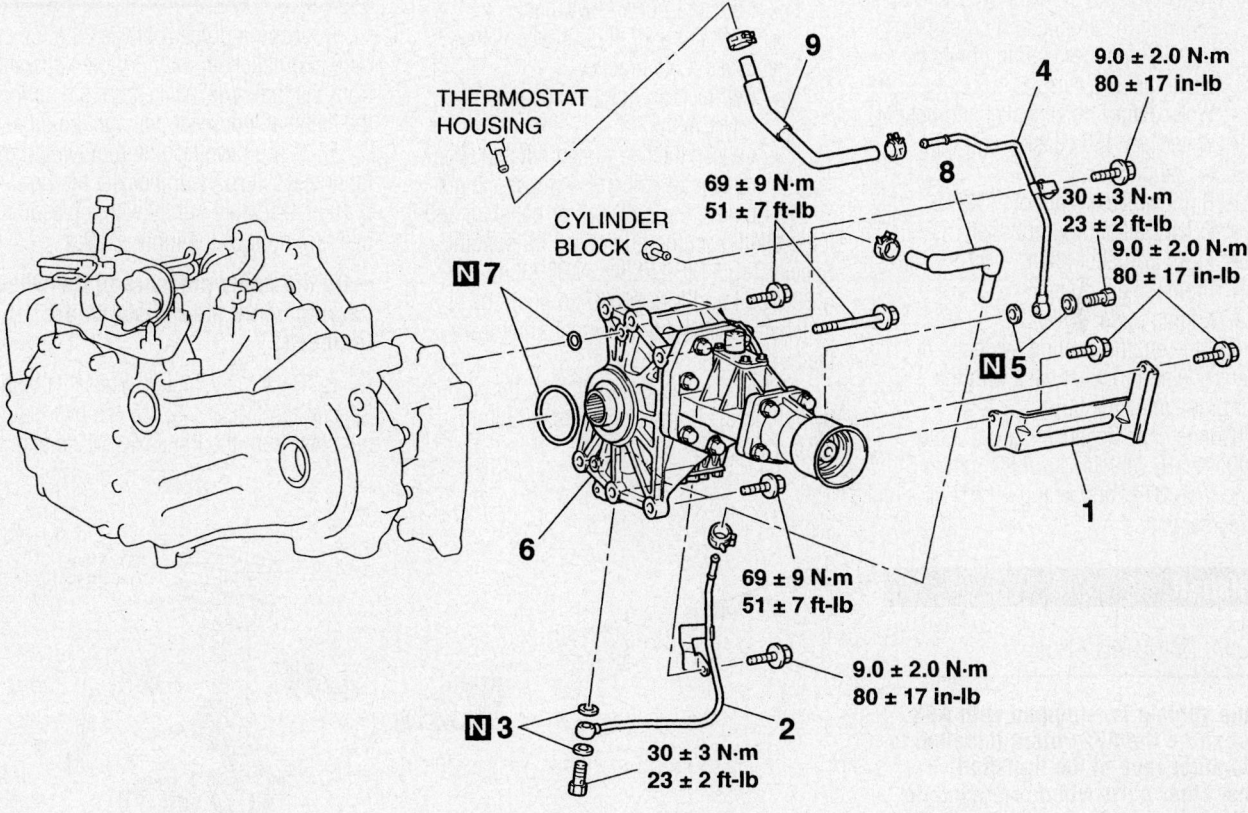

1. AIR GUIDE
2. WATER FEED TUBE ASSEMBLY
3. GASKET
4. WATER RETURN HOSE
5. GASKET
6. TRANSFER
7. O-RING
8. WATER FEED HOSE
9. WATER RETURN HOSE

Transfer case and related components

67170-ENDE-G30

- Engine under cover
- Front exhaust pipe
- Front propeller shaft
- Driveshaft and output shaft
- Right exhaust manifold stay "B"
- Air guide
- Water feed tube assembly and gasket
- Water return hose and gasket
- Retaining bolts and detach the transfer case assembly from the transaxle. Lower the transfer case between the engine and crossmember. Remove and discard the O-rings.

4. As necessary, remove the water feed hose, air cleaner assembly, water return hose, battery and battery tray.

To install:

5. Install or connect the following:
- Water return hose
- Battery tray and battery
- Air cleaner assembly
- Water feed hose
- New O-rings onto the transfer case
- Transfer case, by maneuvering it up between the engine and crossmember. Tighten the retaining bolts to 51 ft. lbs. (69 Nm).
- Water return hose gasket and hose
- Water feed tube gasket and tube
- Air guide
- Right exhaust manifold stay "B"
- Driveshaft and output shaft
- Front propeller shaft
- Front exhaust pipe
- Engine under cover

6. Fill the engine cooling system. Fill the transaxle and transfer case with the proper grade and type fluid.

7. Connect the negative battery cable. On 2005–2006 vehicles, turn the ignition **ON** and then **OFF**, and keep it off for at least ten seconds.

Halfshaft

REMOVAL & INSTALLATION

➡**If the vehicle is equipped with ABS, do not strike the ABS rotors installed to the BJ outer race of the halfshaft against other parts when removing or installing the halfshaft. Otherwise the ABS rotors will be damaged.**

1. Before servicing the vehicle, refer to the Precautions Section.

2. Drain the transaxle and transfer case fluid, as necessary.

3. Remove or disconnect the following:
- Negative battery cable

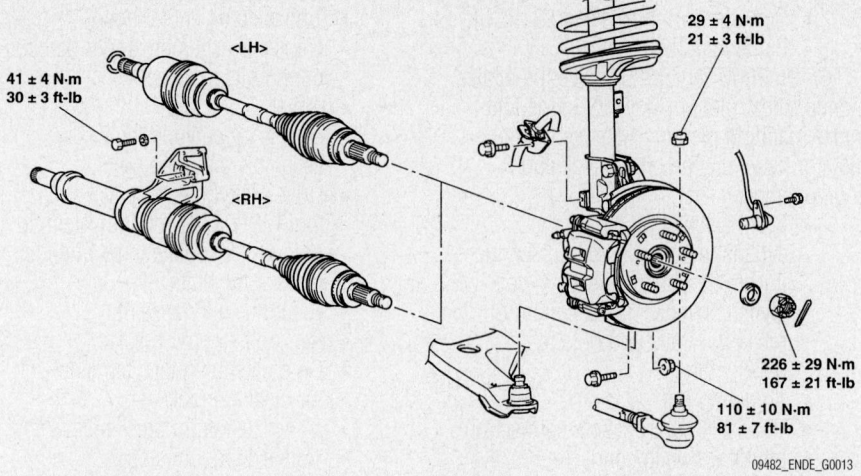

<LH>
41 ± 4 N·m
30 ± 3 ft-lb

<RH>

29 ± 4 N·m
21 ± 3 ft-lb

226 ± 29 N·m
167 ± 21 ft-lb

110 ± 10 N·m
81 ± 7 ft-lb

09482_ENDE_G0013

Front halfshaft and related components—FWD vehicles

- Front and side under covers
- Front exhaust pipe
- Transfer case heat shield, if equipped
- Wheels
- Cotter pin
- Halfshaft nut and washer
- Speed sensor, if equipped with ABS
- Brake hose bracket
- Self locking nut and separate the lower ball joint
- Loosen the tie rod end nut and separate the tie rod end. Remove the nut.

4. Remove the halfshaft or halfshaft and inner shaft from the hub using Mitsubishi Special Tools MB990242, MB990244, MB991354 and MB990767 to push the halfshaft or halfshaft and inner shaft from the hub.

5. Remove the halfshaft from the hub by pulling the bottom of the brake rotor toward you.

❊❊ WARNING

Never pull on the halfshaft or you risk damaging the joints. Only use a pry-bar to remove the halfshaft from the transaxle.

6. Insert a prybar between the transaxle case and halfshaft, then pry the halfshaft from the transaxle. Make sure the spline of the halfshaft does not damage the oil seal.

7. If you have trouble removing the inner shaft from the transaxle, hit the bracket assembly lightly with a plastic hammer and remove the inner shaft.

➡**Do not apply pressure to the wheel bearing while the halfshafts are removed.**

8. On AWD vehicles, use Mitsubishi Special tool MB991721 to remove the output shaft from the transaxle. Make sure the

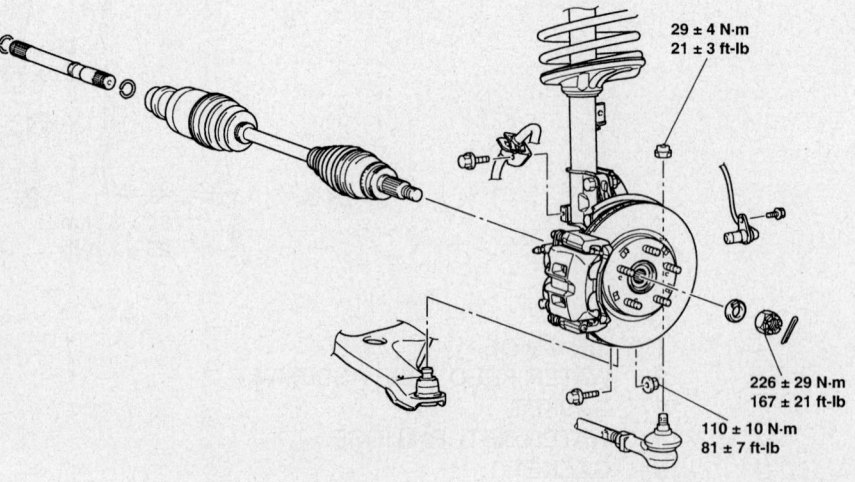

29 ± 4 N·m
21 ± 3 ft-lb

226 ± 29 N·m
167 ± 21 ft-lb

110 ± 10 N·m
81 ± 7 ft-lb

09482_ENDE_G0012

Front halfshaft and related components—AWD vehicles

splined part of the output shaft does not damage the oil seal.

- Circlips from the ends of the shafts

To install:

9. Install or connect the following:
- New circlips
- Output shaft, AWD vehicles
- Halfshaft or halfshaft and inner shaft assembly, as necessary
- Tie rod end and tighten the nut to 21 ft. lbs. (29 Nm)
- Lower ball joint and tighten the nuts to 81 ft. lbs. (110 Nm).
- Brake hose bracket
- Speed sensor, if equipped with ABS
- New halfshaft washer, with the beveled edge facing out

- Halfshaft nut. With no load on the wheel bearings, use Mitsubishi special tool MB990767 to tighten the nut to 174 ft. lbs. (236 Nm).
- New cotter pin
- Wheels
- Transfer case heat shield, if equipped
- Front exhaust pipe
- Front and side under covers

10. Connect the negative battery cable. On 2005–2006 vehicles, turn the ignition **ON** and then **OFF**, and keep it off for at least ten seconds.

11. Fill the transaxle and transfer case (if equipped) with the proper grade and type fluid

CV-Joints

OVERHAUL

FWD Vehicles

➡ **If the vehicle is equipped with ABS, do not strike the ABS rotors installed to the BJ outer race of the halfshaft against other parts when removing or installing the halfshaft. Otherwise the ABS rotors will be damaged.**

1. Before servicing the vehicle, refer to the Precautions Section.
2. Remove the halfshaft from the vehicle.

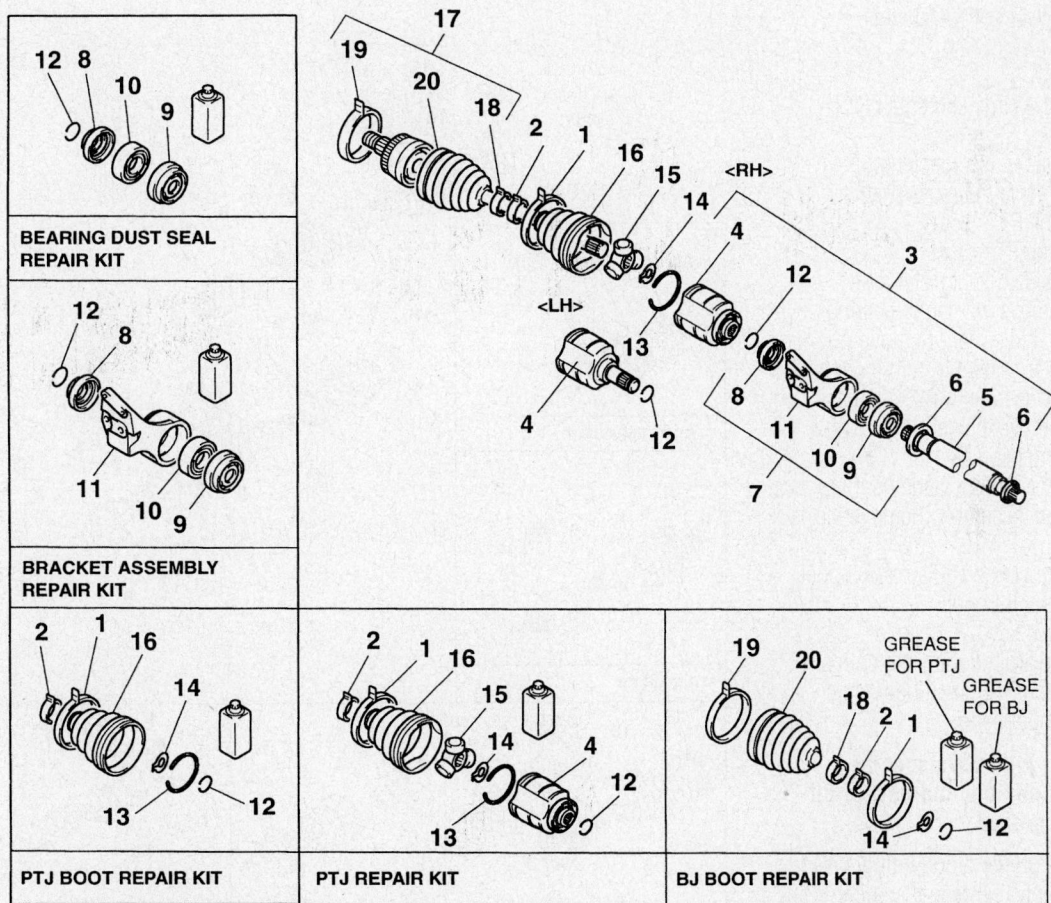

1. PTJ BOOT BAND (LARGE)
2. PTJ BOOT BAND (SMALL)
3. PTJ CASE AND INNER SHAFT ASSEMBLY <RH>
4. PTJ CASE
5. INNER SHAFT <RH>
6. DUST COVER <RH>
7. BRACKET ASSEMBLY <RH>
8. DUST SEAL OUTER <RH>
9. DUST SEAL INNER <RH>
10. CENTER BEARING <RH>
11. CENTER BEARING BRACKET <RH>
12. CIRCLIP
13. CIRCLIP
14. SNAP RING
15. SPIDER ASSEMBLY
16. PTJ BOOT
17. BJ ASSEMBLY
18. BJ BOOT BAND (SMALL)
19. BJ BOOT BAND (LARGE)
20. BJ BOOT

09482_ENDE_G0014

Halfshaft and related components—FWD vehicles

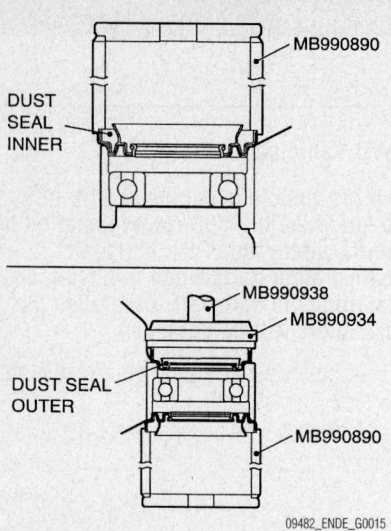

DUST SEAL INNER — MB990890

MB990938
MB990934
DUST SEAL OUTER
MB990890

09482_ENDE_G0015

Inner and outer dust seal installation—FWD vehicles

3. Remove the large and small PTJ boot band.

4. On right side, use special tool MB991248 to remove the inner shaft assembly from the PTJ case.

5. Remove the PTJ case.

6. On right side, use special tool MB990810 to remove the center bearing bracket from the inner shaft. Remove the dust cover, bracket assembly, and inner and outer dust seal. Use tools MB990938 and MB990930 to remove the center bearing and center bearing bracket.

7. Remove the circlips and snapring. Remove the spider assembly. Remove the PTJ boot.

8. Remove the large and small BJ boot bands. Remove the BJ boot.

To install:

9. Wrap tape around the shaft spline, and then install the PTJ small boot band. Install the large boot band.

➡ **The halfshaft joint uses special grease. Do not mix old and new or different types of grease.**

10. Apply the proper amount of grease to the spider assembly, between the spider axle and the roller. Install the spider assembly to the shaft from the direction of the spline chamfered side.

11. On the right side, use special tools MB990938 and MB990932 to press fit the center bearing into the center bearing bracket.

12. On the right side, apply 0.5–0.7 ounce of grease to the rear surface of the inner dust seal. Apply 0.3–0.4 ounce of grease to the rear surface of the outer dust seal.

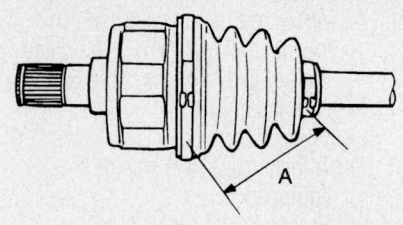

09482_ENDE_G0016

Boot band measurement

13. Use special tools MB990890, MB990938 and MB990934 to press fit the dust seals into the center bearing bracket until they are flush with each other. Apply grease to the lip of each dust seal.

14. Use special tool MB991172 to hold the center bearing inner race and press in the inner shaft. Apply grease to the inner shaft spline, and then press it into the PTJ case.

➡ **When press fitting the inner shaft into the PTJ case, apply a thin coat of grease to the dust seal outer lip par and the outside edge of the PTJ axial part.**

15. After applying 7.5–8.1 ounces of grease to the PTJ case, insert the halfshaft and apply more grease.

➡ **The total amount of grease should be divided in half for use, respectively, at the joint and inside the boot.**

16. Set the PTJ boot bands at the required distance "A" (3.23–3.47 inches) in

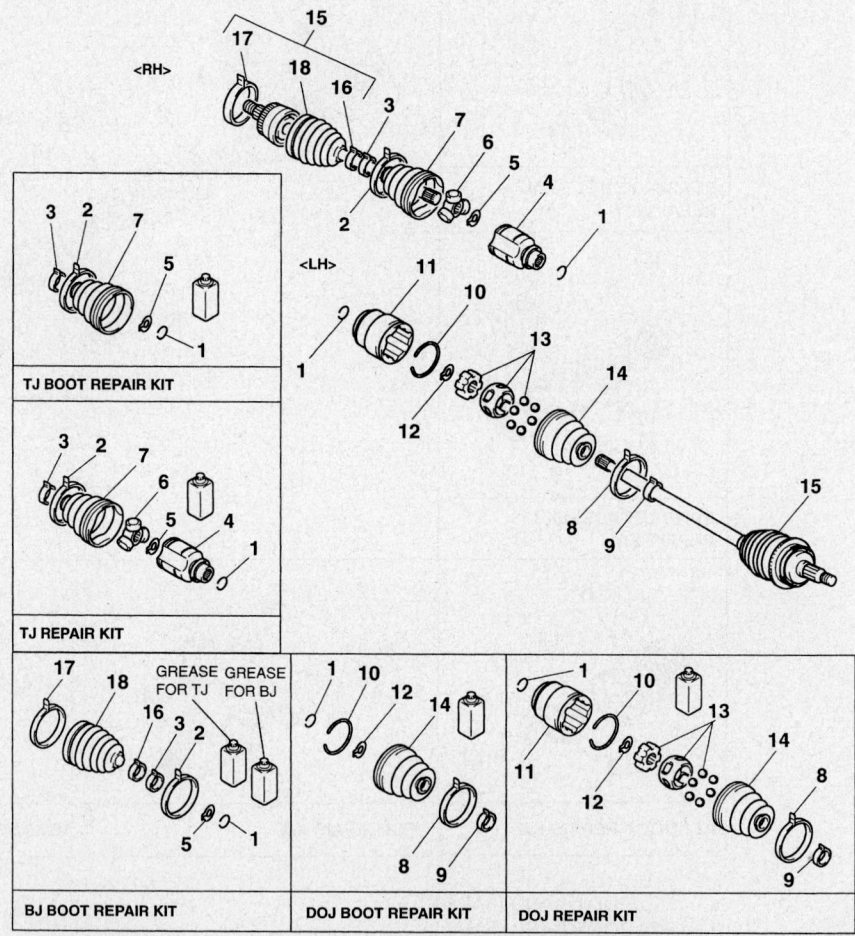

| TJ BOOT REPAIR KIT | TJ REPAIR KIT |
| BJ BOOT REPAIR KIT | DOJ BOOT REPAIR KIT | DOJ REPAIR KIT |

GREASE FOR TJ   GREASE FOR BJ

1. CIRCLIP
2. TJ BOOT BAND (LARGE) <RH>
3. TJ BOOT BAND (SMALL) <RH>
4. TJ CASE<RH>
5. SNAP RING
6. SPIDER ASSEMBLY
7. TJ BOOT<RH>
8. DOJ BOOT BAND (LARGE) <LH>
9. DOJ BOOT BAND (SMALL) <LH>
10. CIRCLIP<LH>

11. DOJ OUTER RACE<LH>
12. SNAP RING<LH>
13. INNER RACE, CAGE AND BALL ASSEMBLY<LH>
14. DOJ BOOT<LH>
15. BJ ASSEMBLY
16. BJ BOOT BAND (SMALL)
17. BJ BOOT BAND (LARGE)
18. BJ BOOT

09482_ENDE_G0017

Halfshaft and related components—AWD vehicles

order to adjust the amount of air inside the PTJ boot. Tighten the small and large boot bands securely.

AWD Vehicles

➡ **If the vehicle is equipped with ABS, do not strike the ABS rotors installed to the BJ outer race of the halfshaft against other parts when removing or installing the halfshaft. Otherwise the ABS rotors will be damaged.**

1. Before servicing the vehicle, refer to the Precautions Section.
2. Remove the halfshaft from the vehicle.
3. Remove the large and small TJ boot band.
4. Remove the TJ case.
5. Remove the circlips and snapring. Remove the spider assembly.
6. On the right side, remove the TJ boot.
7. On the left side, remove the DOJ

large and small boot band, circlip, outer race, snapring, inner race, cage and ball assembly and DOJ boot.

8. On the right side, remove the large and small BJ boot bands. Remove the BJ boot.

To install:

9. Wrap tape around the shaft spline, and then install the DOJ small boot band. Install the large boot band.
10. Set the DOJ boot bands at the required distance "A" (3.23–3.47 inches) in order to adjust the amount of air inside the DOJ boot. Tighten the small and large boot bands securely.

➡ **The halfshaft joint uses special grease. Do not mix old and new or different types of grease.**

11. Apply the proper amount of grease to the spider assembly, between the spider axle and the roller. Install the spider assembly to the shaft from the direction of the spline chamfered side.
12. Apply 4.1–4.7 ounces of grease to

the TJ case. Insert the halfshaft and apply more grease.

➡ **The total amount of grease should be divided in half for use, respectively, at the joint and inside the boot.**

13. Set the TJ boot bands at the required distance "A" (3.42–3.66 inches) in order to adjust the amount of air inside the TJ boot. Tighten the small and large boot bands securely.

Differential Carrier Assembly

REMOVAL & INSTALLATION

1. Before servicing the vehicle, refer to the Precautions Section.
2. Drain the differential gear oil. Raise and support the vehicle safely.
3. Remove or disconnect the following:
   • Center exhaust pipe
   • Rear driveshaft
   • Stabilizer bar bushing

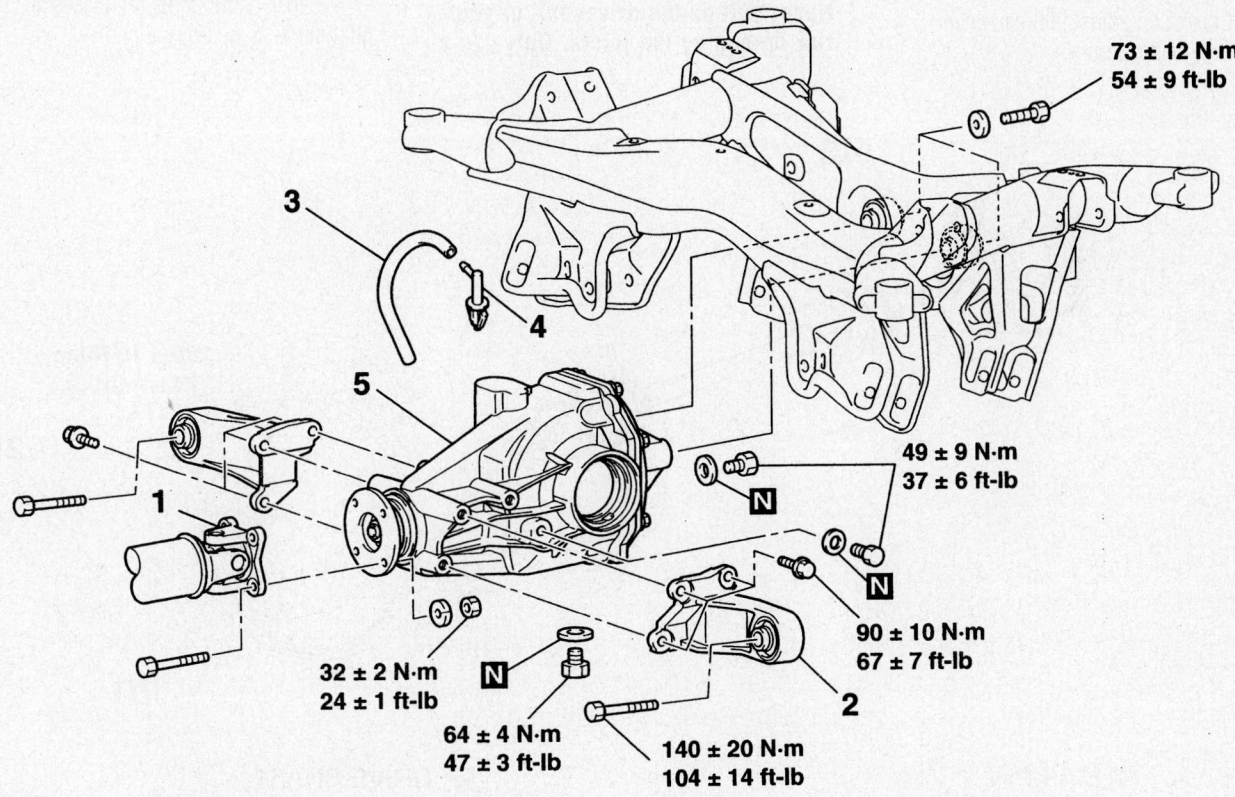

1. PROPELLER SHAFT CONNECTION
2. DIFFERENTIAL MOUNT BRACKET
3. HOSE
4. NIPPLE
5. DIFFERENTIAL CARRIER ASSEMBLY

67170-ENDE-G33

Rear differential carrier and related components

➡Matchmark the installed relation of the propeller shaft to the differential carrier before removal.

- Propeller shaft from the differential carrier. Suspend the propeller shaft from the body to avoid damaging or bending the shaft.
- Differential mount bracket
- Hose and nipple
- Retainers and differential carrier. Discard the washers.

To install:

4. Install or connect the following:
- Differential carrier. Tighten the retainers, with new washers, to the specifications shown in the illustration.
- Nipple and hose
- Differential mount bracket. Tighten the retainers, with new washers, to the specifications shown in the illustration.
- Propeller shaft, aligning the matchmarks made during removal.

5. Fill the differential with the proper grade and type gear oil.

6. Check and adjust the rear wheel alignment, as necessary.

Driveshaft

REMOVAL & INSTALLATION

1. Before servicing the vehicle, refer to the Precautions Section.
2. Remove or disconnect the following:
- Undercover
- Wheels
- Clevis pin
- Driveshaft nut, using and end yoke holder to hold the hub
- Washer
- Rear wheel speed sensor, if equipped with ABS
- Lower control arm, shock absorber, trailing arm and toe control arm connection
- Driveshaft. Use Mitsubishi Special Tools MB990242, MB990244, MB991354 and MB990767 to push the driveshaft from the hub.

✳✳ WARNING

Never pull on the driveshaft or you risk damaging the joints. Only use a prybar to remove the driveshaft from the transaxle.

3. Use a slide hammer to remove the driveshaft from the differential carrier. Make sure the spline of the driveshaft does not damage the oil seal.
- Circlip

To install:

4. Install or connect the following:
- New circlip on the driveshaft
- Driveshaft into the differential carrier and hub
- Lower control arm, shock absorber, trailing arm and toe control arm connection
- Rear wheel speed sensor, if equipped with ABS
- New driveshaft washer, with the beveled edge facing out
- Driveshaft nut. With no load on the wheel bearings, use Mitsubishi special tool MB990767 to tighten the nut to 174 ft. lbs. (236 Nm).
- New cotter pin
- Wheels

5. Check and adjust the rear wheel alignment, as necessary.

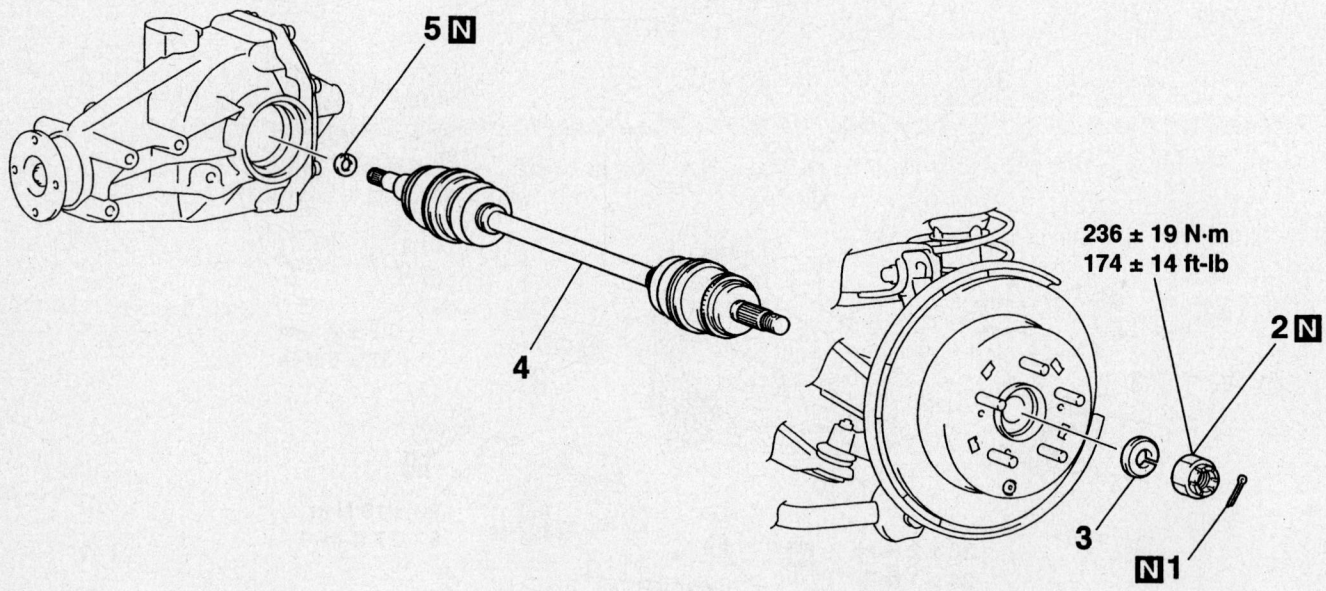

236 ± 19 N·m
174 ± 14 ft-lb

1. CLEVIS PIN
2. DRIVE SHAFT NUT
3. WASHER
4. DRIVE SHAFT
5. CIRCLIP

67170-ENDE-G32

Rear driveshaft and related components—AWD vehicles

STEERING

Air Bag

❊❊ CAUTION

The air bag system must be disabled before performing service on or around system components, steering column, instrument panel components, wiring and sensors. Failure to follow safety and disabling procedures could result in accidental air bag deployment, possible personal injury and unnecessary system repairs.

PRECAUTIONS

Several precautions must be observed when handling the inflator module to avoid accidental deployment and possible personal injury.

• Never carry the inflator module by the wires or connector on the underside of the module.

• When carrying a live inflator module, hold securely with both hands, and ensure that the bag and trim cover are pointed away.

• Place the inflator module on a bench or other surface with the bag and trim cover facing up.

• With the inflator module on the bench, never place anything on or close to the module that may be thrown in the event of an accidental deployment.

DISARMING THE SYSTEM

To avoid personal injury when working on vehicles equipped with an air bag, the negative battery cable must be disconnected and at least 60 seconds must elapse before working on the system. Failure to do so may result in deployment of the air bag. You should also wrap or isolate the negative battery cable with electrical or other non-conductive tape.

Power Steering Gear

REMOVAL & INSTALLATION

1. Before servicing the vehicle, refer to the Precautions Section.

2. Position the front wheel in the straight ahead position and remove the ignition key to prevent the steering wheel from turning.

➡**Prior to removal of the steering gear box, center the front wheels and remove the ignition key. Failure to do so may damage the Supplemental Restraint System (SRS) clockspring and render SRS system inoperative.**

3. Drain the power steering system.

4. Disconnect the negative battery cable. Insulate the cable end with high-quality electrical tape or similar non-conductive wrapping Wait at least 1 minute before proceeding.

5. Remove the steering wheel, as follows:

a. Pry the cover from the steering wheel.

b. Detach the connector, and retainers, as necessary then remove the air bag module assembly

c. Remove the steering wheel nut, and then use a puller to remove the steering wheel.

d. Detach the sub-harness.

6. Remove the clock spring and put aside in a safe place.

7. Remove or disconnect the following:

• Front scuff plate and cowl side trim
• Front floor left side console cover
• Front floor carpet
• Stabilizer link and lower control arm assembly
• Center member

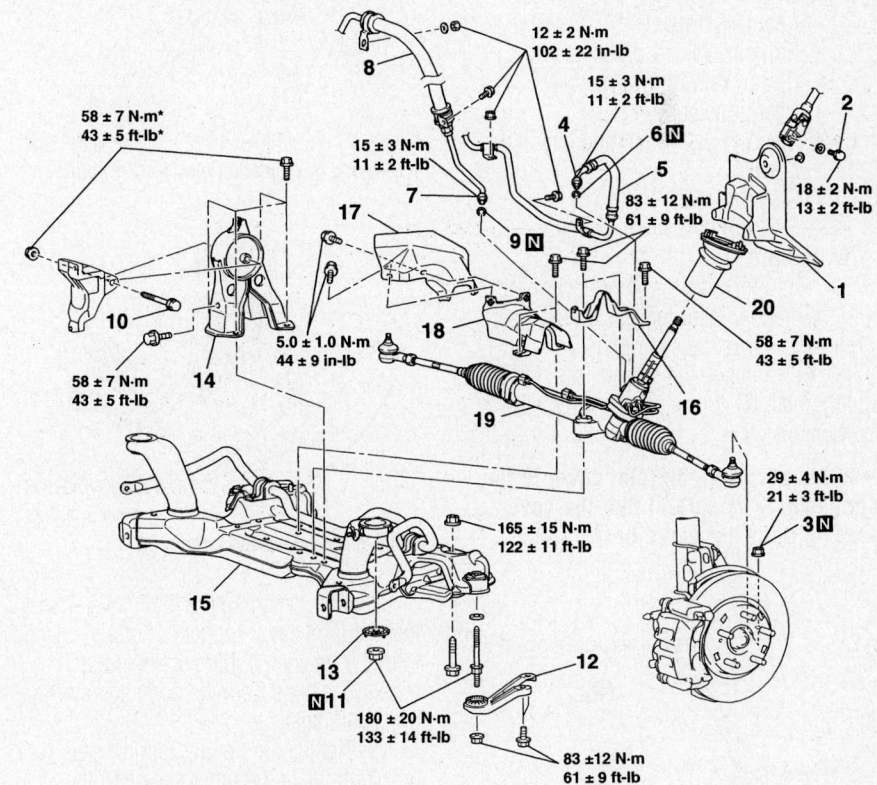

1. STEERING SHAFT PAD
2. STEERING COLUMN SHAFT ASSEMBLY AND STEERING GEAR CONNECTING BOLT
3. SELF LOCKING NUT (TIE ROD END AND KNUCKLE CONNECTION)
4. RETURN TUBE CONNECTION
5. RETURN TUBE
6. O-RING
7. PRESSURE TUBE CONNECTION
8. PRESSURE TUBE/PRESSURE HOSE
9. O-RING
10. REAR ROLL STOPPER CONNECTING BOLT
11. SELF LOCKING NUT
12. FRONT AXLE CROSSMEMBER STAY
13. PLATE STOPPER
14. REAR ROLL STOPPER
15. FRONT AXLE NO.1 CROSSMEMBER
16. POWER STEERING GEAR BRACKET
17. HEAT PROTECTOR
18. STEERING GEAR MOUNTING GEAR SIDE BRACKET
19. STEERING GEAR
20. STEERING JOINT COVER

Power steering gear and related components

67170-ENDE-G34

- Shaft pad
- Steering column shaft assembly and steering gear connecting bolt
- Tie rod end from the steering knuckle, using a suitable puller
- Fluid return hose and return tube. Plug the hoses. Remove and discard the O-ring.
- Pressure hose connection and hose gasket
- Rear roll stopper connecting bolt
- Self-locking nut
- Front axle crossmember stay
- Plate stopper

8. Use a transaxle jack to support the crossmember, then remove the crossmember mounting nuts and bolts. Lower the crossmember with the rear roll stopper, stabilizer bar, return tube and steering gear.

9. Remove the following from the crossmember.

- Rear roll stopper
- Front axle No. 1 crossmember
- Power steering gear bracket
- Heat protector
- Steering gear mounting gear side bracket
- Steering gear
- Steering gear joint cover

To install:

10. Align the steering joint cover notch with the steering gear lug. Install the steering joint cover to the steering gear.

11. From inside the vehicle, pull tab "A" and then tab "B" to secure the three clips to the body panel.

➡**When securing the joint cover to the body panel, be careful that the cover seal lip does not move backwards.**

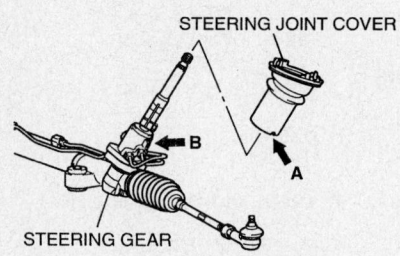

STEERING JOINT COVER

STEERING GEAR

09482_ENDE_G0018

Steering joint cover notch alignment

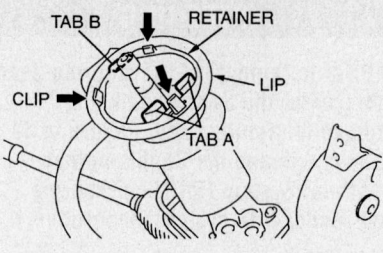

TAB B   RETAINER

CLIP   LIP

TAB A

09482_ENDE_G0019

Steering joint cover tab "A" and tab "B" location

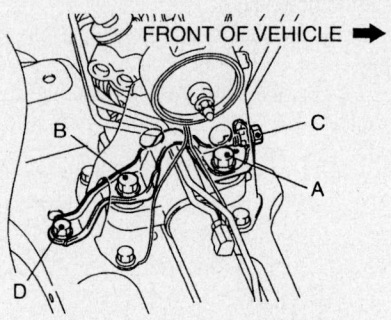

FRONT OF VEHICLE ➡

09482_ENDE_G0020

Steering gear bracket installation points

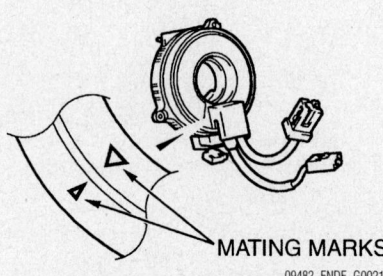

MATING MARKS

09482_ENDE_G0021

Air bag clock spring alignment

12. After installing the joint cover, check that the joint cover rubber is not disengaged from the retainer. If there is any doubt, release the clips from the body and start the procedure again.

13. Tighten the power steering gear bracket bolts in the order shown in the illustration.

14. Continue the installation in the reverse order of the removal procedure.

➡**When installing the clock spring, ensure that the mating marks are prop-**

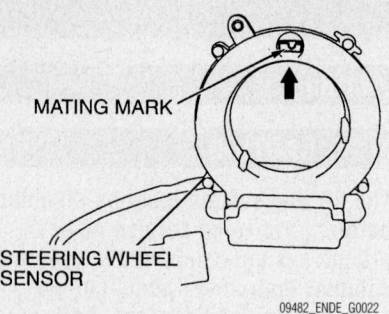

MATING MARK

STEERING WHEEL SENSOR

09482_ENDE_G0022

Active Skid Control (ASC) sensor alignment

erly aligned. If not the steering wheel may not rotate completely during a turn, or the flat cable in the clock spring could be damaged. This would prevent normal SRS operation and possibly cause serious injury to the driver.

15. To align the mating marks of the clock spring, turn the clock spring counterclockwise fully. Then turn it back approximately 3¾ turns counterclockwise to align the mating marks.

16. Turn the front wheels to the straight ahead position. Then install the clock spring to the column switch.

➡**If the vehicle is equipped with Active Skid Control (ASC), ensure that the steering wheel sensor's mating marks are properly aligned. If not, the steering wheel sensor could be damaged.**

17. To align the mating marks, if the vehicle is equipped with Active Skid Control (ASC), turn the steering wheel sensor clockwise fully. Then turn it back approximately 2¾ turns counterclockwise to align the mating marks. Align the mating marks on the clock spring and the steering wheel sensor, and install the steering wheel sensor to the column switch assembly. Connect the steering wheel sensor connector.

18. Connect the negative battery cable. On 2005–2006 vehicles, turn the ignition **ON** and then **OFF**, and keep it off for at least ten seconds.

19. Refill the reservoir and bleed the system.

20. Check and adjust the front end alignment, as required.

FRONT SUSPENSION

Strut

REMOVAL & INSTALLATION

1. Before servicing the vehicle, refer to the Precautions Section.
2. Remove the windshield wiper arm assemblies and front deck garnish.
3. Raise and support the vehicle safely.
4. Remove or disconnect the following:

- Tire and wheel
- Stabilizer link
- Brake hose bracket
- Lower strut mounting bolt and nut
- Front ABS sensor clamp, if equipped
- Upper strut mounting nuts
- Strut from the vehicle

To install:

5. Install or connect the following
- Strut into the vehicle
- Upper strut mounting nuts and tighten to 35 ft. lbs. (47 Nm)
- Front ABS sensor clamp, if equipped
- Lower strut mounting bolt and nut.

Torque the nut to 225 ft. lbs. (305 Nm).
- Brake hose bracket
- Stabilizer link
- Tire and wheel
- Front deck garnish and wiper arms

DISASSEMBLY & ASSEMBLY

1. Before servicing the vehicle, refer to the Precautions Section.
2. Remove or disconnect the following:
- Strut
- Strut cover
3. Compress the coil spring until there is a clearance on both ends
- Strut center nut
- Insulator
- Bearing
- Spring upper seat
- Spring upper pad
- Strut cover
- Strut damper
- Coil spring
- Spring lower pad
- Strut
4. Assembly is the reverse of disassembly. Torque the center nut to 48 ft. lbs. (65 Nm).

Stabilizer Bar

REMOVAL & INSTALLATION

➡**To avoid personal injury when working on vehicles equipped with an air bag, the negative battery cable must be disconnected and at least 60 seconds must elapse before working on the system. Failure to do so may result in deployment of the air bag. You should also wrap or isolate the negative battery cable with electrical or other non-conductive tape.**

1. Before servicing the vehicle, refer to the Precautions Section.
2. Disarm the SRS system. Disconnect the negative battery cable. Raise and support the vehicle safely.
3. Drain the power steering fluid. Remove the front under cover.
4. Remove the centermember.
5. Remove the steering wheel under-cover access plate. Remove the air bag module. Remove the steering damper. Remove the steering wheel.
6. To remove the clock spring, remove the steering column lower cover. Disconnect the clock spring electrical connectors. Remove the clock spring.
7. Remove the floor console side cover. Remove the front scruff plate and cowl side trim. Pull back the floor carpeting.
8. Remove the stabilizer link retaining nuts. Remove the stabilizer link.
9. Remove the steering gear and linkage protector. Disconnect the stabilizer bar retaining brackets and bushings.
10. Separate the tie rod end from the knuckle connection, using tool MB991897, or equivalent.
11. Separate the lower arm assembly from the knuckle connection, using the proper tool.
12. Disconnect the steering gear and steering column shaft connection. Pinch the steering column shaft clip with pliers and pull the shaft upwards to disengage the column shaft from the steering gear.

➡**If the steering column shaft is removed accidentally, remove the steering column assembly and be sure to insert the column shaft into the steering column as shown in the illustration.**

13. Disconnect the steering gear return tube connection. Remove the tube clamp. Disconnect the pressure hose clamp. Dis-

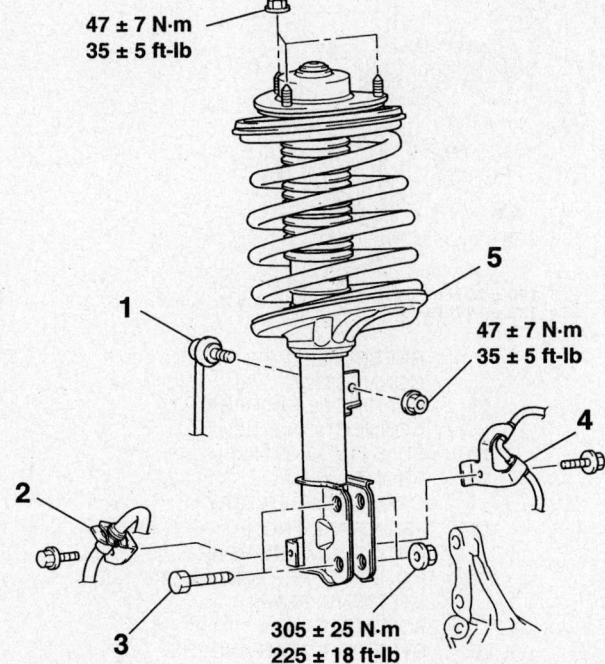

47 ± 7 N·m
35 ± 5 ft-lb

5

47 ± 7 N·m
35 ± 5 ft-lb

1

4

2

3

305 ± 25 N·m
225 ± 18 ft-lb

1. STABILIZER LINK
2. BRAKE HOSE BRACKET
3. STRUT BOLT
4. FRONT ABS SENSOR CLAMP <VEHICLES WITH ABS>
5. STRUT ASSEMBLY

67170-ENDE-G35

Front strut and related components

connect the steering gear pressure tube connection.

14. Remove the rear stopper connection bolt.

15. Remove the front axle crossmember stay locknut. Remove the front axle crossmember stay.

16. Remove the front axle number one crossmember, rear roll stopper and steering gear assembly.

17. Remove the steering gear and linkage protector. Remove the stabilizer bracket, bushing and stabilizer bar from its mounting.

To install:

18. Installation is the reverse of the removal procedure.

19. Be sure to align the stabilizer bar identification mark with the right end of the stabilizer bushing.

20. Align the steering joint cover notch with the steering gear lug. Install the steering joint cover to the steering gear.

➡**When installing the clock spring, ensure that the mating marks are properly aligned. If not the steering wheel may not rotate completely during a turn, or the flat cable in the clock spring could be damaged. This would prevent normal SRS operation and possibly cause serious injury to the driver.**

21. To align the mating marks of the clock spring, turn the clock spring counterclockwise fully. Then turn it back approxi-

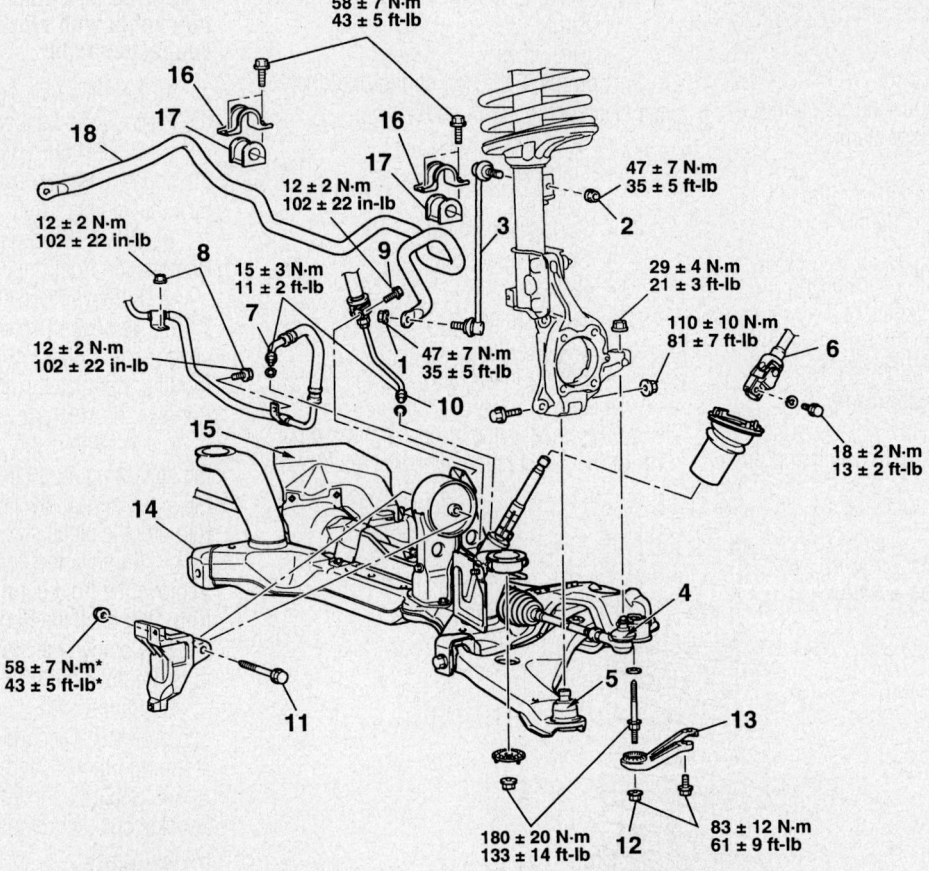

1. STABILIZER NUT
2. STABILIZER NUT
3. STABILIZER LINK
4. TIE ROD END AND KNUCKLE CONNECTION
5. LOWER ARM ASSEMBLY AND KNUCKLE CONNECTION
6. STEERING GEAR AND STEERING COLUMN SHAFT CONNECTION
7. STEERING GEAR AND RETURN TUBE CONNECTION
8. RETURN TUBE CLAMP
9. PRESSURE HOSE CLAMP
10. STEERING GEAR AND PRESSURE TUBE CONNECTION
11. REAR ROLL STOPPER CONNECTION BOLT
12. SELF-LOCKING NUT
13. FRONT AXLE CROSSMEMBER STAY
14. FRONT AXLE NO.1 CROSSMEMBER, REAR ROLL STOPPER AND STEERING GEAR ASSEMBLY
15. STEERING GEAR AND LINKAGE PROTECTOR
16. STABILIZER BRACKET
17. STABILIZER BUSHING
18. STABILIZER BAR

Front stabilizer bar and related components

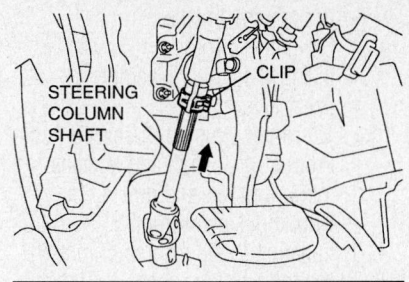

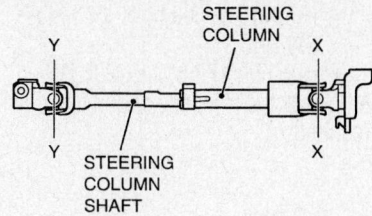

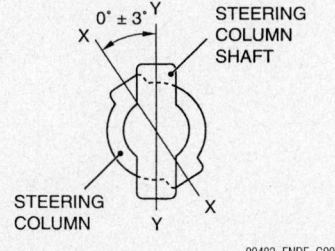

09482_ENDE_G0023

Steering gear to steering column disconnection

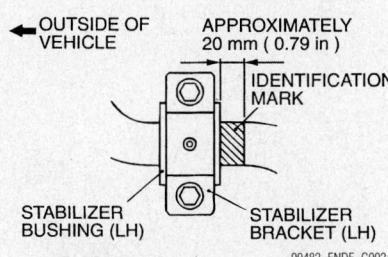

09482_ENDE_G0024

Front stabilizer bar bushing alignment

mately 3¾ turns counterclockwise to align the mating marks.

22. Turn the front wheels to the straight ahead position. Then install the clock spring to the column switch.

➥**If the vehicle is equipped with Active Skid Control (ASC), ensure that the steering wheel sensor's mating marks are properly aligned. If not, the steering wheel sensor could be damaged.**

23. To align the mating marks, if the vehicle is equipped with Active Skid Control (ASC), turn the steering wheel sensor clockwise fully. Then turn it back approximately 2¾ turns counterclockwise to align the mating marks. Align the mating marks on

the clock spring and the steering wheel sensor, and install the steering wheel sensor to the column switch assembly. Connect the steering wheel sensor connector.

24. Connect the negative battery cable. On 2005–2006 vehicles, turn the ignition **ON** and then **OFF**, and keep it off for at least ten seconds.

25. Check and adjust the power steering fluid level,

26. Check and adjust the front end alignment, as required.

Lower Control Arm

1. Before servicing the vehicle, refer to the Precautions Section.
2. Raise and support the vehicle safely.
3. Remove or disconnect the following:
   - Wheel
   - Lower control arm mounting bolts and nuts
   - Lower ball joint from knuckle
   - Lower control arm

To install:
4. Install or connect the following:
   - Lower control arm
   - Lower control arm retainers. Tighten the knuckle nut and bolt to 81 ft. lbs. (110 Nm) and the other retainers hand-tight only at this time.
   - Wheel
5. Once the weight of the vehicle is rest-

ing on the suspension, torque the lower control arm mounting bolts and nuts 122 ft. lbs. (165 Nm).

6. Check and adjust the front wheel alignment.

CONTROL ARM BUSHING REPLACEMENT

1. Before servicing the vehicle, refer to the Precautions Section.
2. Raise and support the vehicle safely.
3. Remove the wheel.
4. Remove the lower control arm and place in a vise.
5. Using tools MB991963 and MB990890 remove the bushing

To install:
6. Position the bushing with the larger end facing the front of the vehicle.

➥**Coat the bushing with a soap solution and take care not to twist.**

7. Using tools MB991963, MB990889 and MB990890, press the bushing into it's mounting on the lower arm.

➥**When pressing in the lower control arm bushing, take care not to damage the lower control arm.**

8. Install the lower control arm.
9. Install the wheel.
10. Check and adjust the front end alignment.

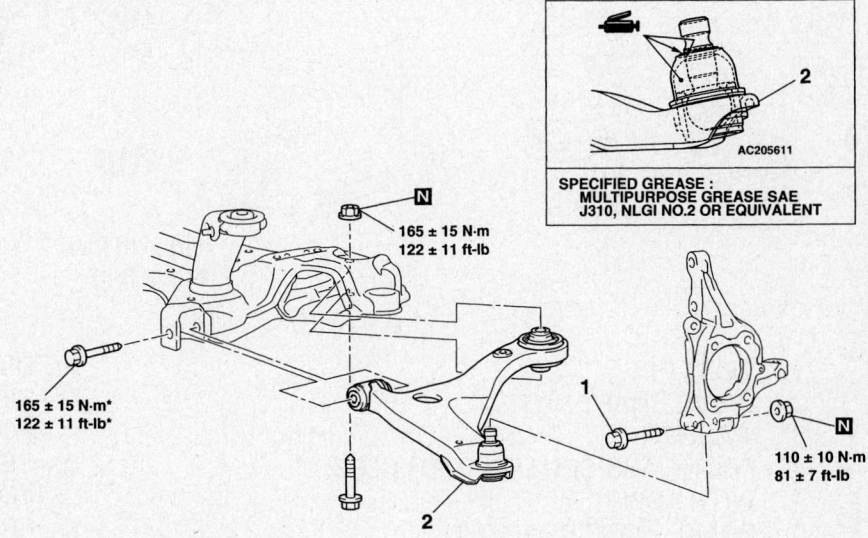

SPECIFIED GREASE :
MULTIPURPOSE GREASE SAE J310, NLGI NO.2 OR EQUIVALENT

AC205611

165 ± 15 N·m
122 ± 11 ft-lb

165 ± 15 N·m*
122 ± 11 ft-lb*

110 ± 10 N·m
81 ± 7 ft-lb

1. LOWER ARM BOLT
2. LOWER ARM ASSEMBLY

67170-ENDE-G40

Front lower control arm and related components

Wheel Bearings

ADJUSTMENT

The wheel bearings are not adjustable. If the bearings are noisy or become loose, they must be replaced.

REMOVAL & INSTALLATION

The wheel bearings are not replaceable. If defective, the hub/bearing assembly must be replaced.

➡**If the vehicle is equipped with ABS, do not strike the ABS rotors installed to** the BJ outer race of the halfshaft against other parts when removing or installing the halfshaft. Otherwise the ABS rotors will be damaged. Be careful not to strike the pole piece at the tip of the front ABS sensor, as damage may result.

1. Before servicing the vehicle, refer to the Precautions Section.
2. Raise and support the vehicle safely.
3. Remove or disconnect the following:
   - Tire and wheel assembly
   - Cotter pin
   - Driveshaft nut
   - Washer
   - Speed sensor, if equipped
   - Brake hose bracket
   - Caliper and rotor
   - Front wheel hub assembly
   - Dust cover
   - Tie rod end, using the proper tools
   - Front strut-to-hub and knuckle mounting bolt and nut
   - Hub/knuckle assembly
4. Installation is the reverse of removal. Observe the following torques:
   - Hub-to-knuckle bolts: 67 ft. lbs. (90 Nm)
   - Hub nut: 174 ft. lbs. (236 Nm)
5. Check the front end alignment, adjust as required.

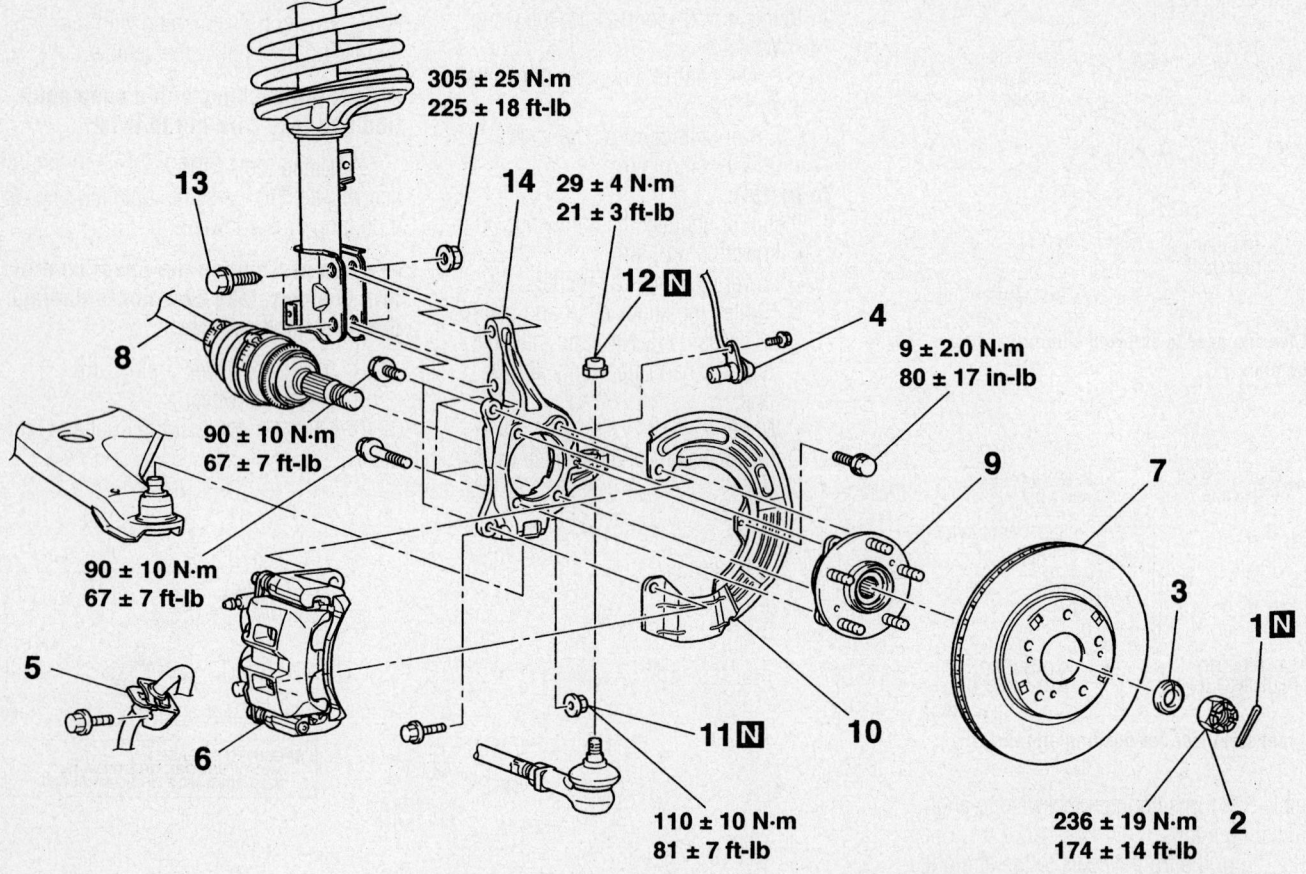

1. SPLIT PIN
2. DRIVE SHAFT NUT
3. WASHER
4. FRONT ABS SENSOR <VEHICLES WITH ABS>
5. BRAKE HOSE BRACKET
6. CALIPER ASSEMBLY
7. BRAKE DISC
8. DRIVE SHAFT
9. FRONT WHEEL HUB ASSEMBLY
10. DUST COVER
11. SELF LOCKING NUT (CONNECTION FOR LOWER ARM BALL JOINT)
12. SELF LOCKING NUT (CONNECTION FOR TIE ROD END)
13. FRONT STRUT TO HUB AND KNUCKLE MOUNTING BOLT AND NUT
14. KNUCKLE

Front hub and related components

67170-ENDE-G41

REAR SUSPENSION

Shock Absorber

REMOVAL & INSTALLATION

1. Before servicing the vehicle, refer to the Precautions Section.
2. Remove or disconnect the following:
   - Luggage floor rear board
   - Tonneau cover
   - Luggage floor front board
   - Parcel strap hook
   - Luggage floor side board
   - Rear end trim
   - Luggage floor carpet bracket
3. Raise and support the vehicle safely.
   - Wheel and tire
   - Lower shock absorber bolt and washer, and separate the shock from the knuckle

- Upper shock mounting nuts
- Shock absorber from the vehicle

To install:
4. Install or connect the following:
   - Shock absorber. Tighten the upper nuts to 34 ft. lbs. (45 Nm) and the lower shock-to-knuckle bolt hand-tight.
   - Wheel and tire
   - Luggage floor carpet bracket
   - Rear end trim
   - Luggage floor side board
   - Parcel strap hook
   - Luggage floor front board
   - Tonneau cover
   - Luggage floor rear board
5. With the weight of the vehicle resting on the suspension, tighten the lower shock-to-knuckle bolt to 74 ft. lbs. (100 Nm).

Stabilizer Bar

REMOVAL & INSTALLATION

1. Before servicing the vehicle, refer to the Precautions Section.
2. Raise and support the vehicle safely.
3. Remove the tire and wheel assembly.
4. Remove the stabilizer link retaining bolts. Remove the stabilizer link assembly.
5. Remove the stabilizer bar bracket retaining bolts. Remove the bushing.
6. Remove the stabilizer from the vehicle.

To install:
7. Position the stabilizer bar to it's mounting.
8. Be sure that the identification tape or

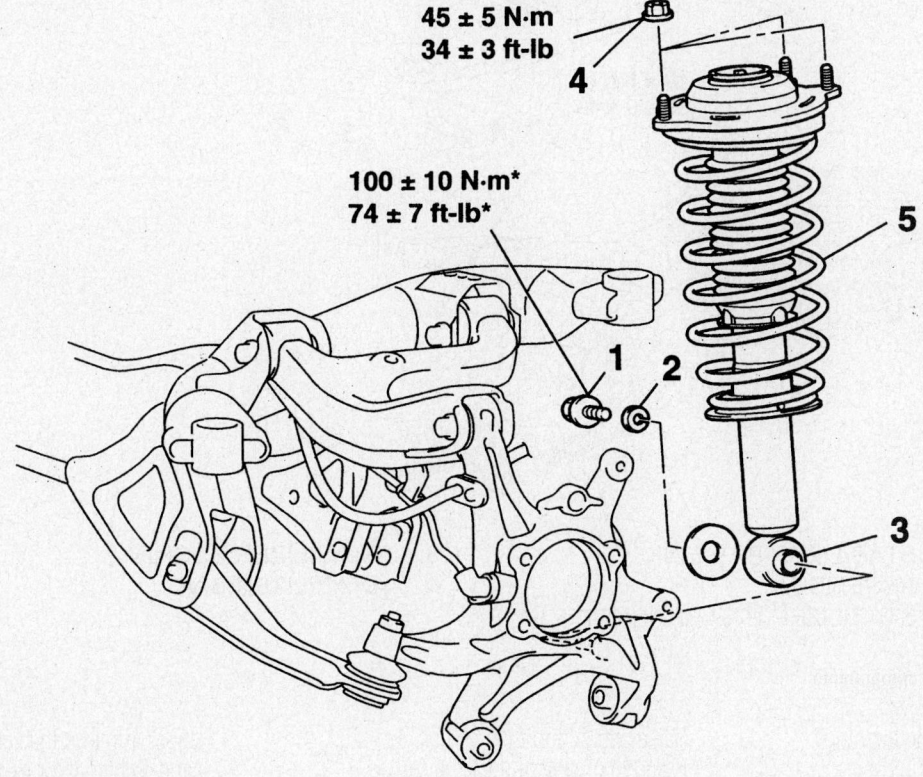

1. COIL SPRING BOLT
2. COIL SPRING WASHER
3. SHOCK ABSORBER ASSEMBLY AND KNUCKLE CONNECTION
4. COIL SPRING NUT
5. SHOCK ABSORBER ASSEMBLY

67170-ENDE-G37

Rear shock absorber and related components

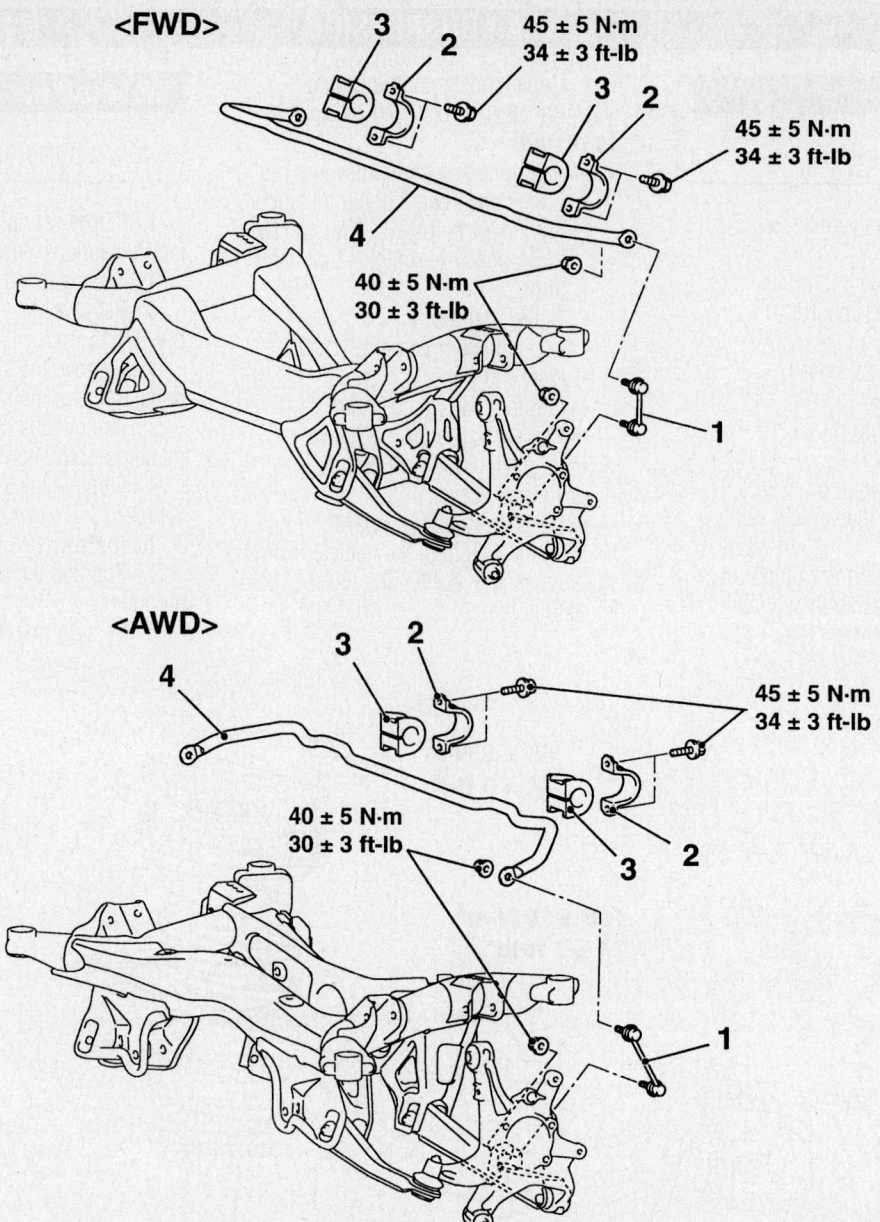

<FWD>

45 ± 5 N·m
34 ± 3 ft-lb

45 ± 5 N·m
34 ± 3 ft-lb

40 ± 5 N·m
30 ± 3 ft-lb

<AWD>

45 ± 5 N·m
34 ± 3 ft-lb

40 ± 5 N·m
30 ± 3 ft-lb

1. STABILIZER BAR LINK ASSEMBLY
2. STABILIZER BAR BRACKET
3. STABILIZER BUSHING
4. STABILIZER BAR

09482_ENDE_G0026

Rear stabilizer bar and related components

paint mark is at the center of the bushings. Tighten the retaining bolts.

9. Continue the installation in the reverse order of the removal procedure.

Upper Control Arm

REMOVAL & INSTALLATION

1. Before servicing the vehicle, refer to the Precautions Section.

2. Raise and support the vehicle safely.
3. Remove or disconnect the following:
   - Wheel assembly
   - Upper control arm from the knuckle
   - ABS sensor clamp bolts, if equipped
   - Upper control arm
   - Upper arm stopper

To install:

4. Install or connect the following:
   - Upper arm stoppers
   - Upper arm into the vehicle and hand-tighten the retainers
   - Upper control arm-to-knuckle bolt and nut and tighten hand-tight
   - ABS sensor clamp bolts
   - Wheel assembly

5. Once the weight of the vehicle is on the suspension, tighten the control arm-to-knuckle nut and the upper control arm mounting nuts to 83 ft. lbs. (113 Nm).

6. Check and adjust the rear wheel alignment.

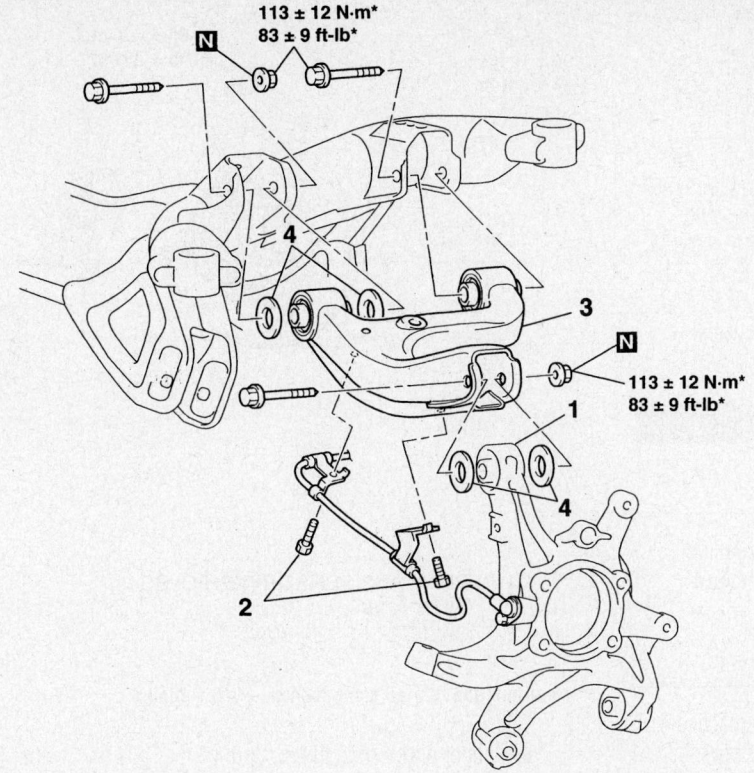

113 ± 12 N·m*
83 ± 9 ft-lb*

113 ± 12 N·m*
83 ± 9 ft-lb*

1. UPPER ARM ASSEMBLY AND
   KNUCKLE CONNECTION
2. ABS EQUIPMENT BOLT
3. UPPER ARM ASSEMBLY
4. UPPER ARM STOPPER

67170-ENDE-G38

Rear upper control arm and related components

Lower Control Arm

REMOVAL & INSTALLATION

1. Before servicing the vehicle, refer to the Precautions Section.
2. Raise and support the vehicle safely.
3. Remove or disconnect the following:

- Rear wheel
- Lower control arm from the knuckle
- Retaining nut, then separate the lower control arm from the stabilizer bar link

➡**Matchmark the crossmember and the plate before removing the lower control arm bolt.**

4. Remove the lower control arm bolt, plate and lower control arm assembly.

   To install:

5. Install or connect the following:

- Lower control arm
- Lower control arm plate
- Lower control arm bolt, hand-tight only at this time
- Lower control arm to the stabilizer

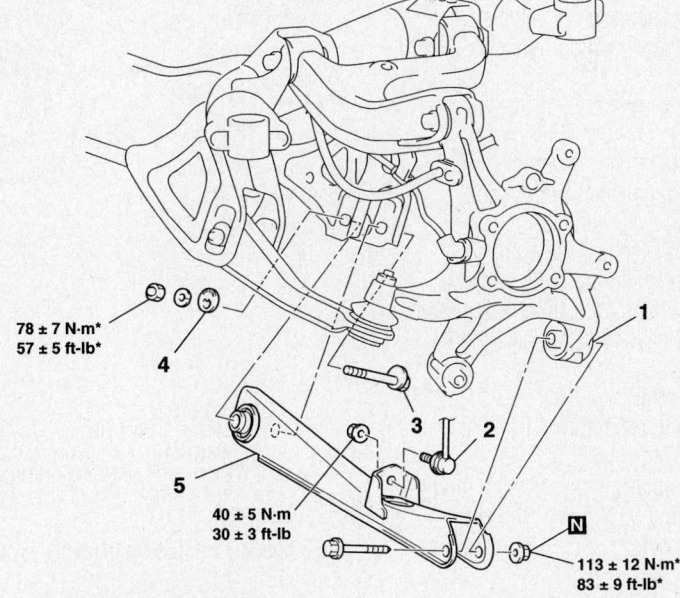

78 ± 7 N·m*
57 ± 5 ft-lb*

40 ± 5 N·m
30 ± 3 ft-lb

113 ± 12 N·m*
83 ± 9 ft-lb*

1. LOWER ARM ASSEMBLY AND
   KNUCKLE CONNECTION
2. LOWER ARM ASSEMBLY AND
   STABILIZER BAR LINK ASSEMBLY
   CONNECTION
3. LOWER ARM BOLT
4. LOWER ARM PLATE
5. LOWER ARM ASSEMBLY

67170-ENDE-G39

Rear lower control arm and related components

bar link. Tighten the retaining to 30 ft. lbs. (40 Nm).
- Lower control arm to the knuckle. Use a new nut and secure hand-tight.
- Rear wheel

6. Once the weight of the vehicle is resting on the suspension, torque the lower control arm mounting bolt to 57 ft. lbs. (78 Nm), and the lower control arm to knuckle nut to 83 ft. lbs. (113 Nm).

7. Check and adjust the rear wheel alignment.

Wheel Bearings

ADJUSTMENT

The wheel bearings are not adjustable. If the bearings are noisy or become loose, they must be replaced.

REMOVAL & INSTALLATION

The wheel bearings are not replaceable. If defective, the hub/bearing assembly must be replaced.

➡ **If the vehicle is equipped with ABS, do not to strike the pole piece at the tip of the rear ABS sensor, as damage may result. The hub assembly should not be disassembled.**

1. Before servicing the vehicle, refer to the Precautions Section.

2. Raise and support the vehicle safely.

3. Loosen the wheel lug nuts only ½ a turn.

4. Remove or disconnect the following:

- Wheel(s) from the vehicle
- Cotter pin, driveshaft nut and washer, AWD vehicles
- Rear ABS sensor, if equipped
- Caliper mounting bolts and use wire to hang the caliper aside. You do not have to disconnect the fluid line.
- Brake rotor
- Rear hub assembly

To install:
- Rear hub assembly. Tighten mounting bolts to 67 ft. lbs. (90 Nm).
- Brake rotor

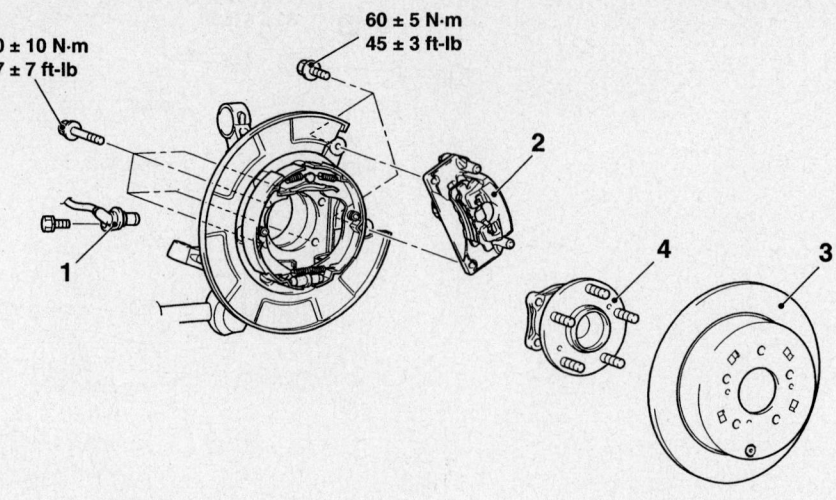

1. REAR ABS SENSOR<VEHICLES WITH ABS>
2. CALIPER ASSEMBLY
3. BRAKE DISC
4. REAR HUB ASSEMBLY

67170-ENDE-G42

Rear hub and related components—FWD vehicles

- Caliper and torque the mounting bolts to 45 ft. lbs. (60 Nm)
- Rear ABS sensor, if equipped
- Washer and driveshaft nut, AWD vehicles. Torque the nut to 174 ft. lbs. (236 Nm). Install a new cotter pin.
- Wheels

5. Road test the vehicle and check for leaks.

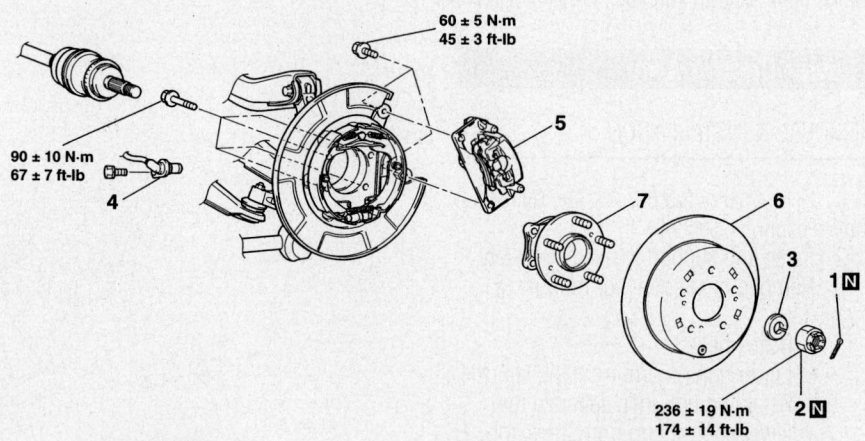

1. SPLIT PIN
2. DRIVE SHAFT NUT
3. WASHER
4. REAR ABS SENSOR<VEHICLES WITH ABS>
5. CALIPER ASSEMBLY
6. BRAKE DISC
7. REAR WHEEL HUB ASSEMBLY

67170-ENDE-G43

Rear hub and related components—AWD vehicles

BRAKES

Brake Caliper

REMOVAL & INSTALLATION

Front and Rear

1. Before servicing the vehicle, refer to the Precautions Section.
2. Raise and safely support the vehicle.
3. Remove or disconnect the following:
   - Wheel and tire assembly
   - Brake hose connection, if you are replacing the caliper
   - Caliper mounting bolts
   - Caliper

To install:

4. Position the caliper over the rotor so the caliper engages the adapter correctly
5. Install or connect the following:
   - Caliper. Tighten the front caliper mounting bolts to 74 ft. lbs. (100

Nm) or the rear mounting bolts to 45 ft. lbs. (60 Nm).
   - Brake hose
   - Wheel and tire assembly
6. Bleed the brake system.

Disc Brake Pads

REMOVAL & INSTALLATION

Front and Rear

1. Before servicing the vehicle, refer to the Precautions Section.
2. Remove ½ of the brake fluid from the master cylinder.
3. Raise and safely support the vehicle.
4. Remove or disconnect the following:

   - Wheel and tire assembly
   - Lower caliper guide pin bolt

   - Caliper from the caliper support
   - Disc brake pads, shims, and the clips from the caliper support

To install:

5. Clean the exposed portion of the caliper piston, then press the piston back into the caliper bore using the old inner brake pad and a C-clamp.
6. Install or connect the following:
   - Disc brake pads, shims, and the clips. Make sure the shims and clips are properly positioned.
   - Caliper over the rotor so the caliper engages the adapter correctly
   - Mounting pin(s)
   - Wheel and tire assembly and lower the vehicle
7. Apply the brake pedal several times until a firm pedal is obtained. Check the fluid level in the master cylinder and add fluid, as necessary.

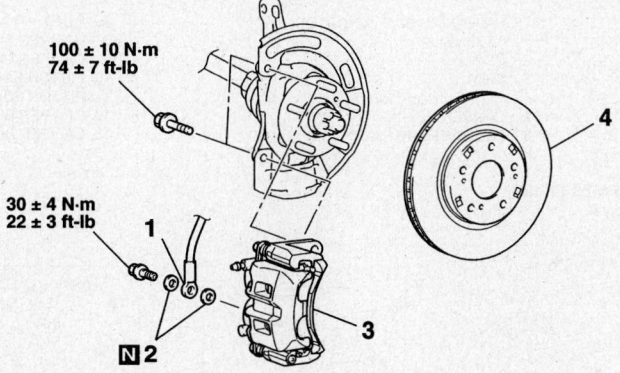

100 ± 10 N·m
74 ± 7 ft-lb

30 ± 4 N·m
22 ± 3 ft-lb

<REAR>

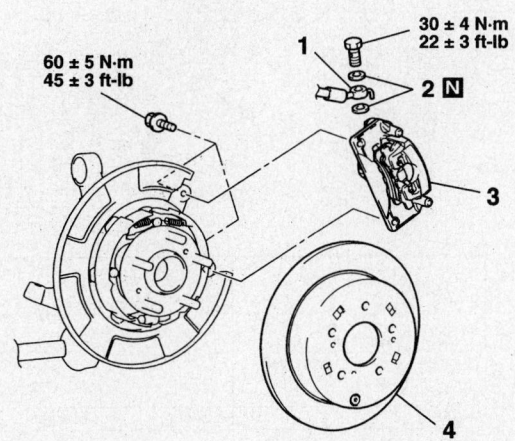

60 ± 5 N·m
45 ± 3 ft-lb

30 ± 4 N·m
22 ± 3 ft-lb

1. BRAKE HOSE CONNECTION
2. GASKET
3. BRAKE CALIPER ASSEMBLY
4. BRAKE DISC

67170-ENDE-G44

Front and rear calipers and related components

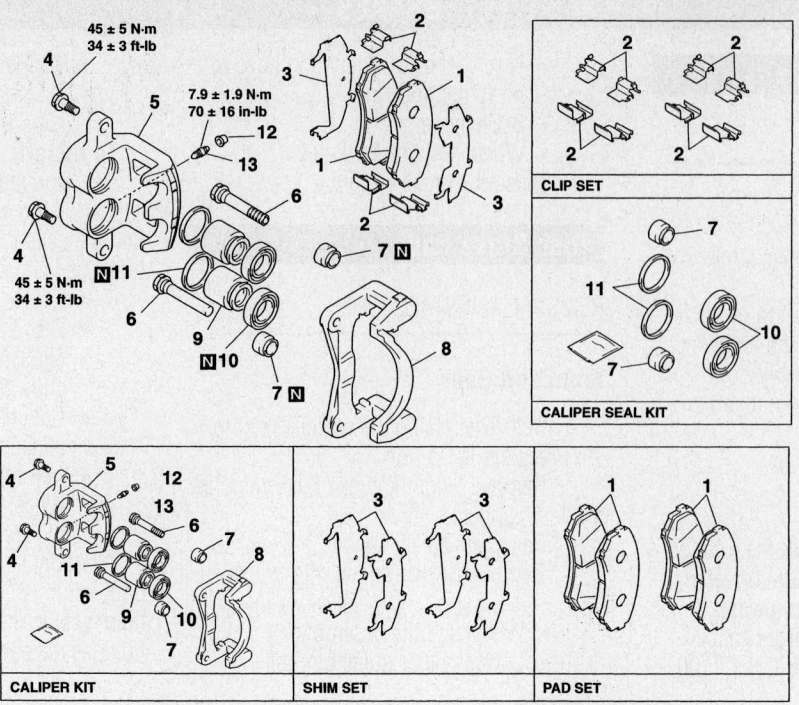

1. PAD (AND WEAR INDICATOR) ASSEMBLY
2. CLIP
3. SHIM
4. FRONT BRAKE BOLT
5. CALIPER BODY
6. FRONT BRAKE PIN
7. BOOT
8. CALIPER SUPPORT
9. CALIPER PISTON
10. PISTON BOOT
11. PISTON SEAL
12. CALIPER BLEEDER CAP
13. CALIPER BLEEDER

67170-ENDE-G45

Front brake pads and related components

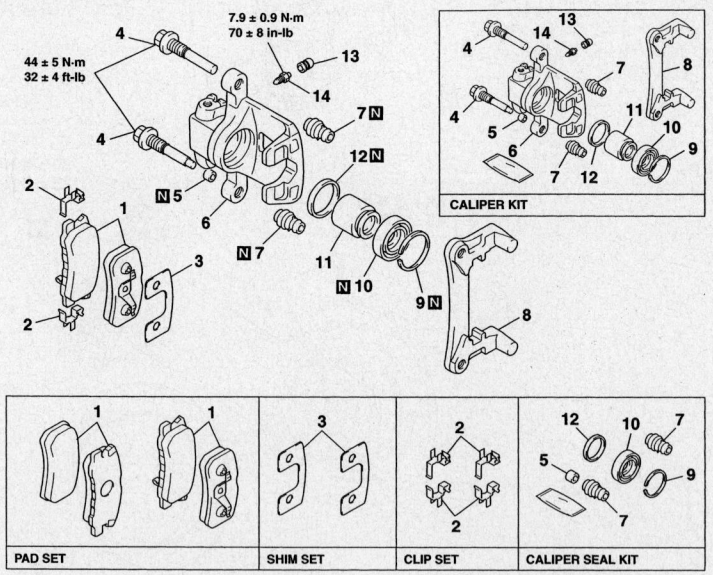

1. PAD (AND WEAR INDICATOR) ASSEMBLY
2. CLIP
3. SHIM
4. REAR BRAKE PIN
5. REAR BRAKE BUSHING
6. CALIPER BODY
7. PIN BOOT
8. CALIPER SUPPORT
9. BOOT RING
10. PISTON BOOT
11. CALIPER PISTON
12. PISTON SEAL
13. REAR BRAKE CAP
14. CALIPER BLEEDER

67170-ENDE-G46

Rear brake pads and related components

SPECIFICATIONS AND MAINTENANCE CHARTS

ENGINE AND VEHICLE IDENTIFICATION CHART

Engine							Model Year	
Code	Liters (cc)	Cu. In.	Cyl.	Fuel Sys.	Engine Type	Eng. Mfg.	Code	Year
P	3.0 (2972)	181.4	6	MFI	SOHC	Mitsubishi	2	2002
H	3.0 (2972)	181.4	6	MFI	SOHC	Mitsubishi	3	2003
M	3.5 (3479)	213.4	6	MFI	SOHC	Mitsubishi	4	2004
R	3.5 (3479)	213.4	6	MFI	SOHC	Mitsubishi	5	2005
S	3.8 (3828)	233.6	6	MFI	SOHC	Mitsubishi	6	2006

MFI: Multi-port Fuel Injection

09482_MONT_C0001

GENERAL ENGINE SPECIFICATIONS

Year	Engine Displacement Liters	Engine VIN	Net Horsepower @ rpm	Net Torque @ rpm (ft. lbs.)	Bore x Stroke (in.)	Compression Ratio	Oil Pressure @ rpm
2002	3.0	P	173@5500	188@4500	3.59x2.99	9.0:1	30-80@2000
	3.5	M	200@5000	228@3500	3.66x3.38	9.0:1	30-80@2000
2003	3.0	H	173@5500	188@4500	3.59x2.99	9.0:1	30-80@2000
	3.5	R	200@5000	228@3500	3.66x3.38	9.0:1	30-80@2000
	3.8	S	215@5500	248@3250	3.74x3.54	10.0:1	①
2004	3.5	R	200@5000	228@3500	3.66x3.38	9.0:1	30-80@2000
	3.8	S	215@5500	248@3250	3.74x3.54	10.0:1	①
2005	3.8	S	215@5500	248@3250	3.74x3.54	10.0:1	①
2006	3.8	S	215@5500	248@3250	3.74x3.54	10.0:1	①

① 11.6 psi @ idle

09482_MONT_C0002

ENGINE TUNE-UP SPECIFICATIONS

Year	Engine Displacement Liters	Engine VIN	Spark Plugs Gap (in.)	Ignition Timing (deg.) MT	AT	Fuel Pump (psi)	Idle Speed (rpm) MT	AT	Valve Clearance In.	Ex.
2002	3.0	P	0.039-0.043	—	5B	38 ①	—	750	HYD	HYD
	3.5	M	0.039-0.043	—	5B	38 ①	—	700	HYD	HYD
2003	3.0	H	0.039-0.043	—	5B	38 ①	—	750	HYD	HYD
	3.5	R	0.039-0.043	—	5B	38 ①	—	700	HYD	HYD
	3.8	S	0.028-0.031	—	②	③	—	⑤	HYD	HYD
2004	3.5	R	0.039-0.043	—	②	④	—	⑤	HYD	HYD
	3.8	S	0.028-0.031	—	②	③	—	⑤	HYD	HYD
2005	3.8	S	0.028-0.031	—	②	③	—	⑤	HYD	HYD
2006	3.8	S	0.028-0.031	—	②	③	—	⑤	HYD	HYD

B: Before top dead center

HYD: Hydraulic

① With vacuum hose connected

② 2-8 BTDC

③ 47-50 at idle. Vacuum hose disconnected

④ 97-103 at idle. Vacuum hose disconnected

⑤ 650-750

09482_MONT_C0003

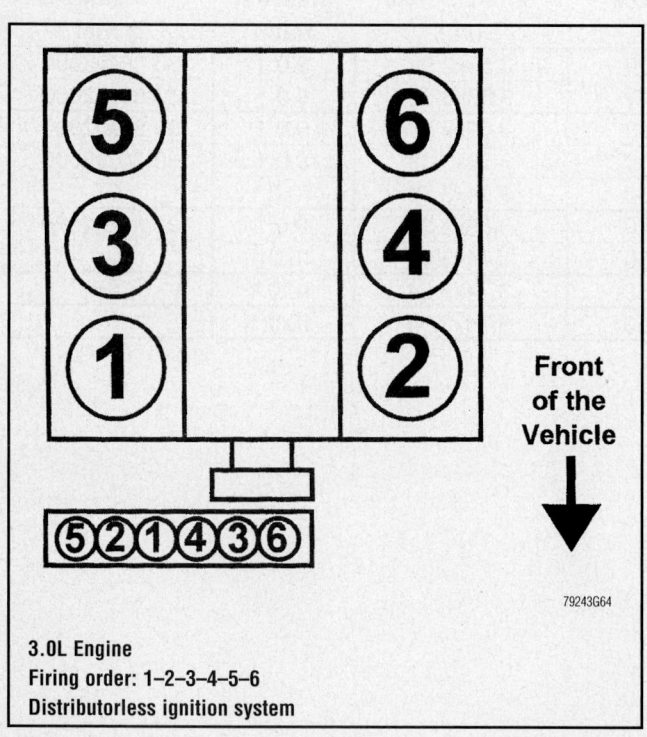

3.0L Engine
Firing order: 1-2-3-4-5-6
Distributorless ignition system

79243G64

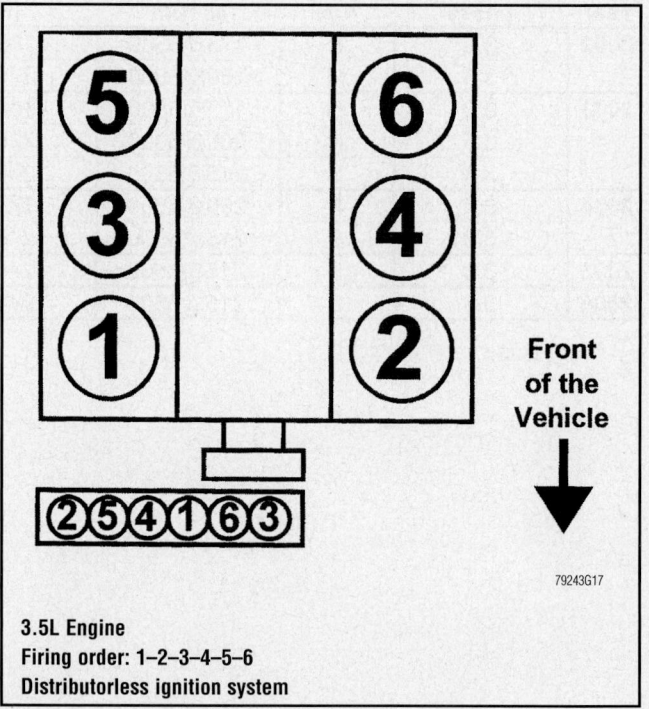

3.5L Engine
Firing order: 1-2-3-4-5-6
Distributorless ignition system

79243G17

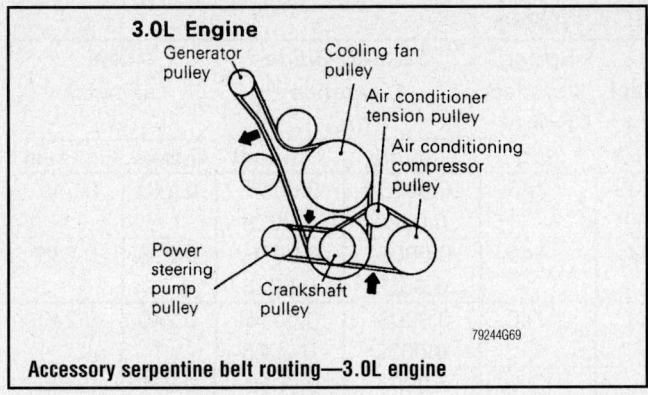

Accessory serpentine belt routing—3.0L engine

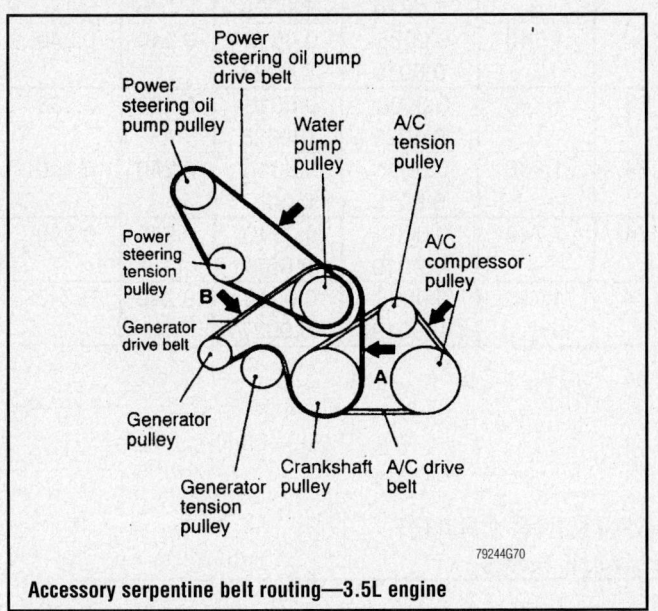

Accessory serpentine belt routing—3.5L engine

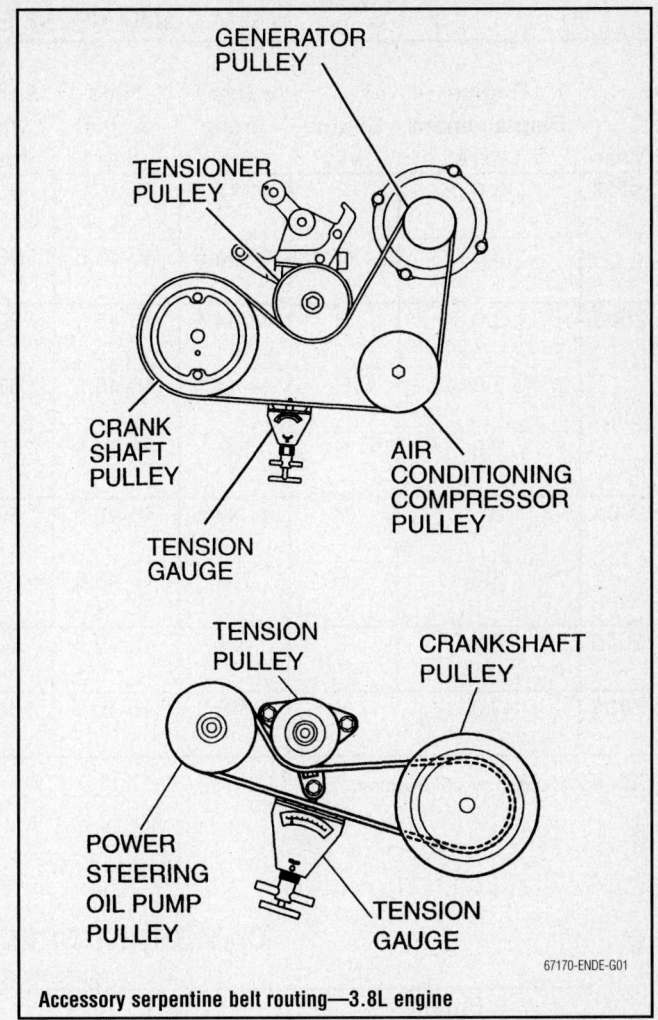

Accessory serpentine belt routing—3.8L engine

CAPACITIES

Year	Model	Engine Displacement Liters	Engine VIN	Engine Oil with Filter (qts.)	Transmission (pts.) 5-Spd	Transmission (pts.) Auto.	Transfer Case (pts.)	Drive Axle Front (pts.)	Drive Axle Rear (pts.)	Fuel Tank (gal.)	Cooling System (qts.)
2002	Montero	3.5	M	5.1	—	19.6	6.0	2.6	3.4	23.8	9.5
	Montero Sport	3.0	P	4.6	—	20.6	5.2	1.9	6.8	19.5	①
		3.5	M	4.7	—	20.6	5.2	1.9	5.6	19.5	①
2003	Montero	3.8	S	4.8	—	19.6	6.0	2.6	3.4	23.8	9.5
	Montero Sport	3.0	H	4.6	—	20.6	5.2	1.9	6.8	19.5	①
		3.5	R	4.7	—	20.6	5.2	1.9	5.6	19.5	①
2004	Montero	3.8	S	4.8	—	19.6	6.0	2.6	3.4	23.8	9.5
	Montero Sport	3.5	R	4.7	—	20.6	5.2	1.9	5.6	19.5	①
2005	Montero	3.8	S	4.8	—	19.6	6.0	2.6	3.4	23.8	9.5
2006	Montero	3.8	S	4.8	—	19.6	6.0	2.6	3.4	23.8	9.5

① without rear heater: 9.5
with rear heater: 10.6

VALVE SPECIFICATIONS

Year	Engine Displacement Liters	Engine VIN	Seat Angle (deg.)	Face Angle (deg.)	Spring Test Pressure (lbs. @ in.)	Spring Installed Height (in.)	Stem-to-Guide Clearance (in.)		Stem Diameter (in.)	
							Intake	Exhaust	Intake	Exhaust
2002	3.0	P	44-44.5	45-45.5	60@1.74	1.740	0.0008-0.0020	0.0016-0.0028	0.240	0.240
	3.5	M	44-44.5	45-45.5	60@1.74	1.740	0.0008-0.0020	0.0016-0.0028	0.236	0.236
2003	3.0	H	44-44.5	45-45.5	60@1.74	1.740	0.0008-0.0020	0.0016-0.0028	0.240	0.240
	3.5	R	44-44.5	45-45.5	60@1.74	1.740	0.0008-0.0020	0.0016-0.0028	0.236	0.236
	3.8	S	NS	45-45.5	60@1.74	1.740	0.0008-0.0019	0.0016-0.0027	0.240	0.240
2004	3.5	M	44-44.5	45-45.5	60@1.74	1.740	0.0008-0.0020	0.0016-0.0028	0.236	0.236
	3.8	S	NS	45-45.5	60@1.74	1.740	0.0008-0.0019	0.0016-0.0027	0.240	0.240
2005	3.8	S	NS	45-45.5	60@1.74	1.740	0.0008-0.0019	0.0016-0.0027	0.240	0.240
2006	3.8	S	NS	45-45.5	60@1.74	1.740	0.0008-0.0019	0.0016-0.0027	0.240	0.240

NS - Not specified by manufacturer

09482_MONT_C0005

CAMSHAFT SPECIFICATIONS CHART
All measurements are given in inches.

Year	Engine Displ. Liters	Engine VIN	Journal Dia.	Brg. Oil Clearance	Shaft End-play	Runout	Lobe Height	
							Intake	Exhaust
2002	3.0	P	1.8000	NS	NS	NS	①	②
	3.5	M	1.8000	NS	NS	NS	①	②
2003	3.0	H	1.8000	NS	NS	NS	①	②
	3.5	R	1.8000	NS	NS	NS	①	②
	3.8	S	1.8000	NS	NS	NS	①	③
2004	3.5	M	1.8000	NS	NS	NS	①	②
	3.8	S	1.8000	NS	NS	NS	①	③
2005	3.8	S	1.8000	NS	NS	NS	①	③
2006	3.8	S	1.8000	NS	NS	NS	①	③

NS - Not specified by manufacturer

① Standard value: 1.485
   Minimum value: 1.465

② Standard value: 1.462
   Minimum value: 1.443

③ Standard value: 1.472
   Minimum value: 1.452

09482_MONT_C0006

CRANKSHAFT AND CONNECTING ROD SPECIFICATIONS

All measurements are given in inches.

Year	Engine Displacement Liters	Engine VIN	Crankshaft				Connecting Rod		
			Main Brg. Journal Dia.	Main Brg. Oil Clearance	Shaft End-play	Thrust on No.	Journal Diameter	Oil Clearance	Side Clearance
2002	3.0	P	2.3614-2.3622	0.0008-0.0040	0.0020-0.0120	3	2.1646-2.1654	0.0008-0.0040	0.0039-0.0160
	3.5	M	2.3614-2.3622	0.0008-0.0040	0.0020-0.0120	3	1.9700	0.0008-0.0040	0.0039-0.0160
2003	3.0	H	2.3614-2.3622	0.0008-0.0040	0.0020-0.0120	3	2.1646-2.1654	0.0008-0.0040	0.0039-0.0160
	3.5	R	2.3614-2.3622	0.0008-0.0040	0.0020-0.0120	3	1.9700	0.0008-0.0040	0.0039-0.0160
	3.8	S	①	0.0008-0.0015	0.0020-0.0090	NS	②	0.0008-0.0019	0.0030-0.0090
2004	3.5	R	2.3614-2.3622	0.0008-0.0040	0.0020-0.0120	3	1.9700	0.0008-0.0040	0.0039-0.0160
	3.8	S	①	0.0008-0.0015	0.0020-0.0090	NS	②	0.0008-0.0019	0.0030-0.0090
2005	3.8	S	①	0.0008-0.0015	0.0020-0.0090	NS	②	0.0008-0.0019	0.0030-0.0090
2006	3.8	S	①	0.0008-0.0015	0.0020-0.0090	NS	②	0.0008-0.0019	0.0030-0.0090

NS - Not specified by manufacturer

① 2.520 outside diameter

② 2.165 pin outside diameter

09482_MONT_C0007

PISTON AND RING SPECIFICATIONS

All measurements are given in inches.

Year	Engine Displacement Liters	Engine VIN	Piston Clearance	Ring Gap			Ring Side Clearance		
				Top Compression	Bottom Compression	Oil Control	Top Compression	Bottom Compression	Oil Control
2002	3.0	P	0.0008-0.0020	0.0118-0.0310	0.0177-0.0310	0.0079-0.0390	0.0012-0.0040	0.0008-0.0040	Snug
	3.5	M	0.0008-0.0020	0.0118-0.0310	0.0177-0.0310	0.0079-0.0390	0.0012-0.0040	0.0008-0.0040	Snug
2003	3.0	H	0.0008-0.0020	0.0118-0.0310	0.0177-0.0310	0.0079-0.0390	0.0012-0.0040	0.0008-0.0040	Snug
	3.5	R	0.0008-0.0020	0.0118-0.0310	0.0177-0.0310	0.0079-0.0390	0.0012-0.0040	0.0008-0.0040	Snug
	3.8	S	0.0008-0.0015	0.0012-0.0170	0.0180-0.0230	0.0080-0.0230	0.0012-0.0027	0.0008-0.0023	Snug
2004	3.5	R	0.0008-0.0020	0.0118-0.0310	0.0177-0.0310	0.0079-0.0390	0.0012-0.0040	0.0008-0.0040	Snug
	3.8	S	0.0008-0.0015	0.0012-0.0170	0.0180-0.0230	0.0080-0.0230	0.0012-0.0027	0.0008-0.0023	Snug
2005	3.8	S	0.0008-0.0015	0.0012-0.0170	0.0180-0.0230	0.0080-0.0230	0.0012-0.0027	0.0008-0.0023	Snug
2006	3.8	S	0.0008-0.0015	0.0012-0.0170	0.0180-0.0230	0.0080-0.0230	0.0012-0.0027	0.0008-0.0023	Snug

09482_MONT_C0008

TORQUE SPECIFICATIONS
All readings in ft. lbs.

Year	Engine Displacement Liters	Engine ID/VIN	Cylinder Head Bolts	Main Bearing Bolts	Rod Bearing Bolts	Crankshaft Damper Bolts	Flywheel Bolts	Manifold Intake	Manifold Exhaust	Spark Plugs	Oil Pan Drain Plug
2002	3.0	P	80	69	37	134	55	16	33	18	29
	3.5	M	80	54	38	134	54	16	22	18	29
2003	3.0	H	80	69	37	134	55	16	33	18	29
	3.5	R	80	54	38	134	54	16	22	18	29
	3.8	S	③	74	①	137	54	②	33	18	29
2004	3.5	R	80	54	38	134	54	16	22	18	29
	3.8	S	③	④	①	137	54	⑤	⑥	18	⑦
2005	3.8	S	③	④	①	137	54	⑤	⑥	18	⑦
2006	3.8	S	③	④	①	137	54	⑤	⑥	18	⑦

① 20 ft. lbs., plus an additional 90 degrees

② 1st step (right bank nuts): 58 inch lbs.
2nd step (left bank nuts): 16 ft. lbs.
3rd step (right bank nuts): 16 ft. lbs.
4th step (left bank nuts): 16 ft. lbs.
5th step (right bank nuts): 16 ft. lbs.

③ Step 1: 76-84 ft. lbs.
Step 2: Loosen completely
Step 3: 76-84 ft. lbs.

④ 51-57 ft. lbs.

⑤ 15-17 ft. lbs.

⑥ 29-37 ft. lbs.

⑦ 25-33 ft. lbs.

09482_MONT_C0009

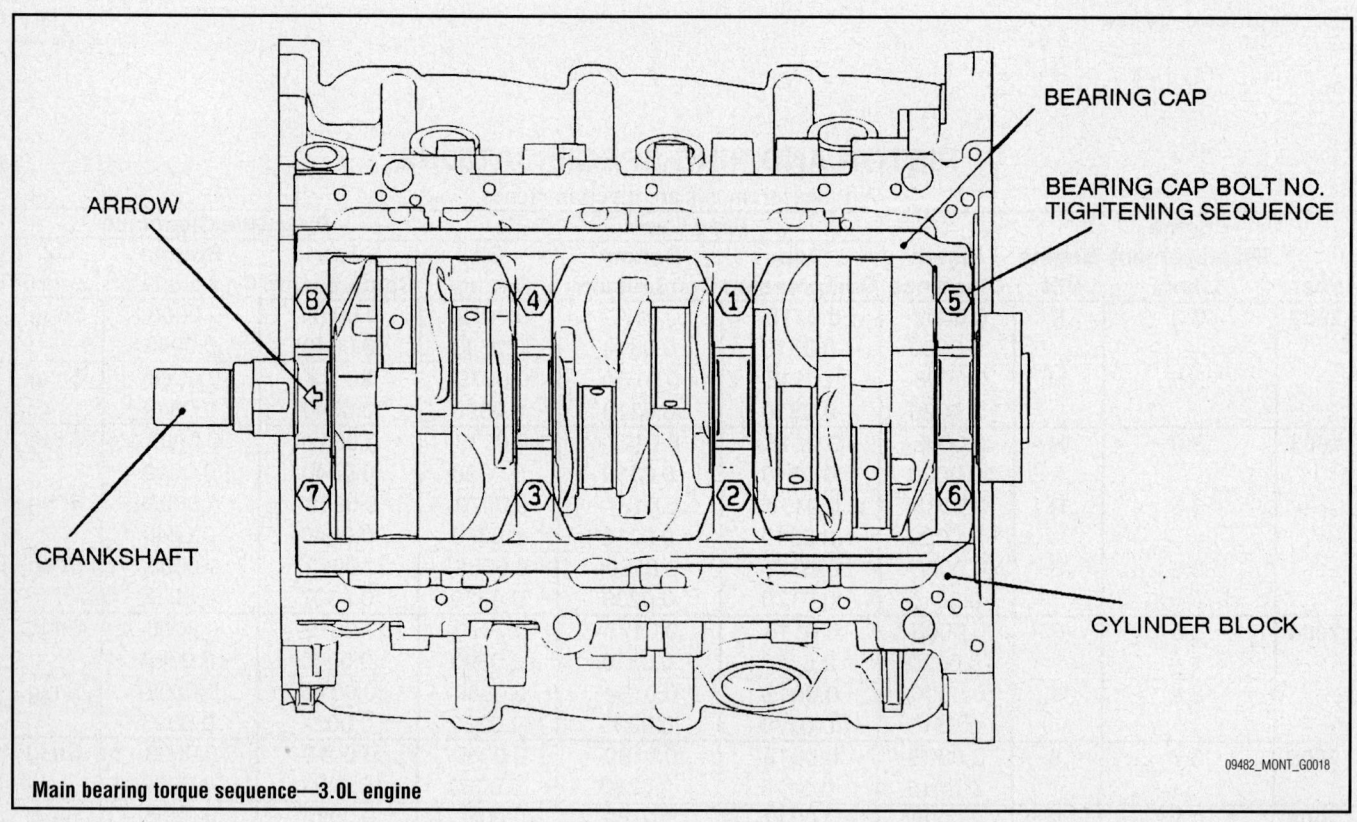

Main bearing torque sequence—3.0L engine

09482_MONT_G0018

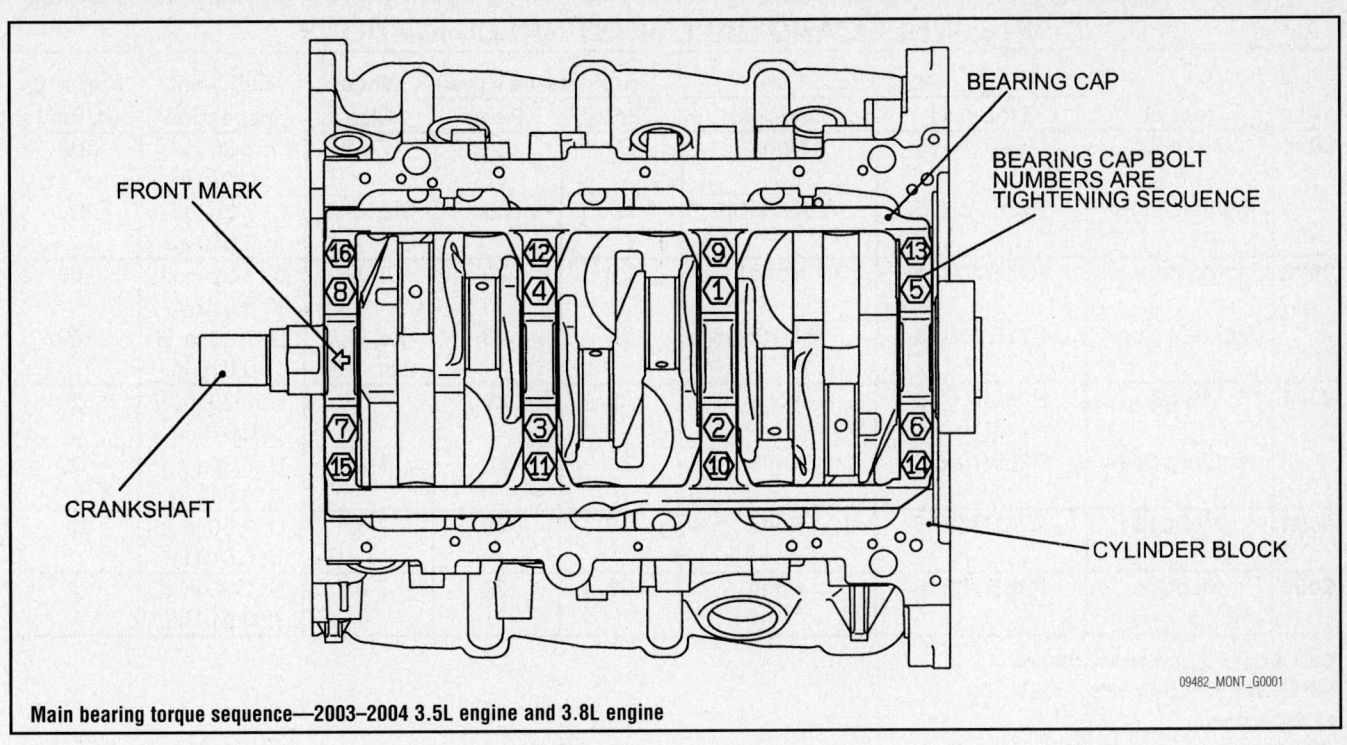

FRONT MARK

CRANKSHAFT

BEARING CAP

BEARING CAP BOLT
NUMBERS ARE
TIGHTENING SEQUENCE

CYLINDER BLOCK

09482_MONT_G0001

Main bearing torque sequence—2003–2004 3.5L engine and 3.8L engine

WHEEL ALIGNMENT

Year	Model		Caster Range (+/-Deg.)	Caster Preferred Setting (Deg.)	Camber Range (+/-Deg.)	Camber Preferred Setting (Deg.)	Toe-in (in.)
2002	Montero	F	0.50	+3.50	0.50	0	0.10+/-0.10
		R	—	—	0.50	0	0.12+/-0.12
	Montero Sport	F	1.00	+2.66	0.50	①	0.14+/-0.14
		R	—	—	—	0	0
2003	Montero	F	0.30	+3.50	0.30	0	0.10+/-0.10
		R	—	—	0.30	0	0.12+/-0.12
	Montero Sport	F	1.00	+2.66	0.50	①	0.14+/-0.14
		R	—	—	—	0	0
2004	Montero	F	0.30	+3.50	0.30	0	0.10+/-0.10
		R	—	—	0.30	0	0.12+/-0.12
	Montero Sport	F	1.00	+2.66	0.50	①	0.14+/-0.14
		R	—	—	—	0	0
2005	Montero	F	0.30	+3.50	0.30	0	0.10+/-0.10
		R	—	—	0.30	0	0.12+/-0.12
2006	Montero	F	0.30	+3.50	0.30	0	0.10+/-0.10
		R	—	—	0.30	0	0.12+/-0.12

① Right: 0.42
   Left: 0.92

09482_MONT_C0010

TIRE, WHEEL AND BALL JOINT SPECIFICATIONS

Year	Model	OEM Tires Standard	OEM Tires Optional	Tire Pressures (psi) Front	Tire Pressures (psi) Rear	Wheel Size	Ball Joint Inspection	Lug Nut (ft. lbs.)
2002	Montero	P265/70HR16	None	29	29	7-JJ	U: 7-30 in. ① / L: 0.010 in.	100
	Montero Sport	P235/75R15	P255/70R16	26	26	Std: 6-JJ / Opt: 7-JJ	U: 7-30 in. ① / L: 0.010 in.	100
2003	Montero	P265/70R16	None	29	29	7-JJ	U: 7-30 in. ① / L: 0.010 in.	100
	Montero Sport	P235/75R15	P255/70R16	26	26	Std: 6-JJ / Opt: 7-JJ	U: 7-30 in. ① / L: 0.010 in.	100
2004	Montero	P265/70R16	None	29	29	7-JJ	U: 7-30 in. ① / L: 0.010 in.	②
	Montero Sport	P255/70R16	None	26	26	7-JJ	U: 7-30 in. ① / L: 0.010 in.	②
2005	Montero	P265/70R16	None	29	29	7-JJ	U: 7-30 in. ① / L: 0.010 in.	②
2006	Montero	P265/70R16	None	29	29	7-JJ	U: 7-30 in. ① / L: 0.010 in.	②

OEM: Original Equipment Manufacturer

PSI: Pounds Per Square Inch

STD: Standard

OPT: Optional

① Torque required in inch lbs. to rotate ball joint when removed from the knuckle

② 72-87 ft. lbs.

BRAKE SPECIFICATIONS

All measurements in inches unless noted

Year	Model		Brake Disc Original Thickness	Brake Disc Minimum Thickness	Brake Disc Maximum Runout	Brake Drum Diameter Original Inside Diameter	Brake Drum Diameter Max. Wear Limit	Brake Drum Diameter Maximum Machine Diameter	Minimum Lining Thickness Front	Minimum Lining Thickness Rear	Brake Caliper Bracket Bolts (ft. lbs.)	Brake Caliper Mounting Bolts (ft. lbs.)
2002	Montero	F	1.023	0.960	0.002	—	—	—	0.079	—	65	54
		R	0.866	0.803	0.003	—	—	—	—	0.079	65	32
	Montero Sport	F	0.940	0.880	0.001	—	—	—	0.079	—	65	55
		R	0.710	0.650	0.003	10.63	—	10.71	—	①	94	32
2003	Montero	F	1.023	0.960	0.002	—	—	—	0.080	—	83	66
		R	0.870	0.800	0.002	—	—	—	—	0.080	74	33
	Montero Sport	F	0.940	0.880	0.001	—	—	—	0.079	—	65	55
		R	0.710	0.650	0.003	10.63	—	10.71	—	①	94	32
2004	Montero	F	1.020	0.960	0.002	—	—	—	0.080	—	②	③
		R	0.870	0.800	0.002	—	—	—	—	0.080	④	⑤
	Montero Sport	F	0.940	0.880	0.001	—	—	—	0.079	—	65	55
		R	0.710	0.650	0.003	10.63	—	10.71	—	①	94	32
2005	Montero	F	1.020	0.960	0.002	—	—	—	0.080	—	②	③
		R	0.870	0.800	0.002	—	—	—	—	0.080	④	⑤
2006	Montero	F	1.020	0.960	0.002	—	—	—	0.080	—	②	③
		R	0.870	0.800	0.002	—	—	—	—	0.080	④	⑤

① disc pad: 0.79

   brake shoe: 0.04

② 76-90

③ 63-69

④ 67-81

⑤ 30-36

SCHEDULED MAINTENANCE INTERVALS
Mitsubishi—Montero, Montero Sport

TO BE SERVICED	TYPE OF SERVICE	VEHICLE MILEAGE INTERVAL (x1000)												
		7.5	15	22.5	30	37.5	45	52.5	60	67.5	75	82.5	90	97.5
Engine oil & filter	R	✓	✓	✓	✓	✓	✓	✓	✓	✓	✓	✓	✓	✓
Automatic transmission & transfer oil	S/I		✓		✓		✓		✓		✓		✓	
Brake hoses	S/I		✓		✓		✓		✓		✓		✓	
Disc brake pads & rotors	S/I		✓		✓		✓		✓		✓		✓	
Drive shaft boots	S/I		✓		✓		✓		✓		✓		✓	
Air cleaner filter	R				✓				✓				✓	
Engine coolant	R								✓				✓	
Ball joints & steering linkage seals	S/I				✓				✓				✓	
Drive belt(s)	S/I				✓				✓				✓	
Drum brake linings & wheel cylinders	S/I				✓				✓				✓	
Exhaust system	S/I				✓				✓				✓	
Front & rear axle	S/I				✓				✓				✓	
Fuel hoses	S/I				✓				✓				✓	
Manual transmission & transfer oil (4WD)	S/I				✓				✓				✓	
Propeller shaft joint	S/I				✓				✓				✓	
Ignition cables	R								✓					
Timing belt	R								✓					
EVAP system (except EVAP canister)	S/I								✓					
EGR valve ①	S/I													
EVAP canister ①	S/I													
PCV system ②	S/I													
Spark plugs ③	R													
Tires (rotate)	S/I	✓	✓	✓	✓	✓	✓	✓	✓	✓	✓	✓	✓	✓

R: Replace   S/I: Service or Inspect

① Replace at 100,000 miles.

② PCV system (except EVAP canister): service or inspect at 100,000 miles.

③ Iron tips: 30,000 miles
    Platinum tips: 60,000 miles
    Irridium tips: 100,000 miles

FREQUENT OPERATION MAINTENANCE (SEVERE SERVICE)

If a vehicle is operated under any of the following conditions it is considered severe service:

- Extremely dusty areas.

- 50% or more of the vehicle operation is in 32°C (90°F) or higher temperatures, or constant operation in temperatures below 0°C (32°F).

- Prolonged idling (vehicle operation in stop and go traffic).

- Frequent short running periods (engine does not warm to normal operating temperatures).

- Police, taxi, delivery usage or trailer towing usage.

Oil & oil filter: replace every 3000 miles.

Front disc brake pads (dusty or salty conditions): service or inspect every 6000 miles.

Front disc brake pads: service or inspect every 7500 miles.

Air cleaner filter: service or inspect every 15,000 miles.

Rear drum brake linings & rear wheel cylinders: service or inspect every 15,000 miles.

Spark plugs (iron tip): replace every 15,000 miles.

PCV system: service or inspect every 60,000 miles.

09482_MONT_C0013

ENGINE REPAIR

➥Disconnecting the negative battery cable on some vehicles may interfere with the functions of the on board computer systems and may require the computer to undergo a relearning process, once the negative battery cable is reconnected.

Distributor

All of the engines covered in this section are distributorless.

Alternator

REMOVAL

1. Before servicing the vehicle, refer to the Precautions Section.
2. Remove or disconnect the following:
   • Negative battery cable
   • Under cover
   • Air cleaner assembly, ducts and air intake hose
   • Drive belt(s)
   • Wires
   • Mounting bracket, if equipped
   • Alternator

To install:
3. Install or connect the following:
   • Alternator. Torque the through-bolt to 32–40 ft. lbs. (44–54 Nm). Torque the other r retaining bolt to 14–18 ft. lbs. (20–25 Nm).
   • Mounting bracket, if equipped. Torque the bolt to 14–18 ft. lbs. (20–25 Nm).

   • Wires. Torque the nut to 98–150 inch lbs. (11–17 Nm).
   • Drive belt(s)
   • Air cleaner assembly, ducts and air intake hose
   • Under cover
   • Negative battery cable

Ignition Timing

ADJUSTMENT

The ignition timing is controlled by the ECM and is not adjustable. The ECM determines the timing based on input from the crankshaft position sensor.

TIMING CHECK

1. Before servicing the vehicle, refer to the Precautions Section.
Before attempting to adjust the ignition timing, be sure of the following:
   • The engine should be at normal operating temperature.
   • The lights and all accessories should be OFF.
   • If equipped with an automatic transmission, the transmission should be in **P** or **N**.
   • Connect scan tool MB991958 to the data link connector
   • Set up the timing light.
   • Start the engine and run at idle.
   • Verify that the idle speed is 600–800 rpm.

   • Select scan tool MB991958 actuator test "item number 17".
   • Check that basic timing is within specification, it should be 2–8° BTDC.
   • Press the clear key on the scan tool, select forced drive stop mode and cancel the actuator test.

✳✳ CAUTION

If the actuator test is not canceled, the forced drive will continue for 27 minutes. Driving in this state could lead to engine failure.

2. If the base timing is out of specification:
3. Check to see if the timing belt cover and Crankshaft Position (CKP) sensor installation is correct.
4. Check the crankshaft sensing blade condition.

Engine Assembly

REMOVAL & INSTALLATION

3.0L and 3.5L Engines
2002–2003

1. Before servicing the vehicle, refer to the Precautions Section.
2. Relieve the fuel system pressure.

✳✳ CAUTION

The fuel injection system remains under pressure after the engine has beenOFF. Properly relieve fuel pressure before disconnecting any fuel lines. Failure to do so may result in fire or personal injury.

3. Drain the engine oil.
4. Drain the cooling system.
5. Remove or disconnect the following:
   • Battery
   • Hood, matchmark for reassembly
   • Oil dipstick
   • Engine undercover
   • Starter
   • Exhaust pipe from the exhaust manifolds
   • Transfer case, if equipped with 4WD
   • Transmission, if equipped with a manual transmission
6. If equipped with an automatic transmission and 2WD:
   a. Remove the inspection plate.

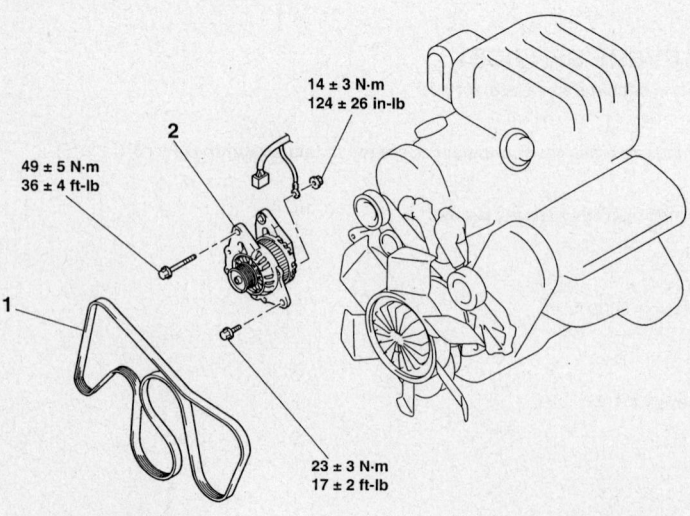

14 ± 3 N·m
124 ± 26 in-lb

49 ± 5 N·m
36 ± 4 ft-lb

23 ± 3 N·m
17 ± 2 ft-lb

1. DRIVE BELT
2. ALTERNATOR

09482_MONT_G0045

Alternator and related components

b. Matchmark the flexplate to the converter; remove the torque converter bolts and move the torque converter back as far as it will go.

c. Remove the lower bell housing bolts.

7. Remove or disconnect the following:

- Air cleaner assembly, ducts and air intake hose
- Linkages and cables from the throttle body
- Fuel lines and plug the lines
- Air conditioning compressor, if equipped and position it aside. It is not necessary to remove the lines from the compressor.
- Radiator, shroud
- Cooling fan
- Heater hoses
- Accessory belts
- Power steering pump and wires from its brackets and position it to the side. Do not remove the hoses from the pump.
- Alternator and wires
- Ignition coil and power transistor assembly, if equipped
- MDP sensor connector
- EGR connector
- TP sensor connector
- IAC motor connector
- Magnetic clutch and refrigerant temperature switch connector
- EVAP Purge Solenoid
- ECT sensor and gauge connectors
- Front and injector wiring harness
- CMP sensor
- CKP sensor
- Distributor signal Generator
- Compactor connector
- Left and right heated O_2 sensor
- Oil pressure switch connector
- Vacuum hoses

8. Attach an engine removal device to the engine support eyes on the engine.

9. If equipped with an automatic transmission, support the transmission with a floor jack. Remove the remaining bell housing bolts.

10. Remove the engine mount nuts and remove the engine from the vehicle.

To install:

11. Lower the engine into position and install the engine mount nuts and bolts. Tighten the nuts to 18–20 ft. lbs. (25–27 Nm), and the bolts to tighten to 33 ft. lbs. (44 Nm).

12. Install or connect the following:

- Bell housing bolts

13. Remove the engine removal device and the transmission support.

14. Install or connect the following:

- Transfer case, if equipped
- Manual transmission, if equipped
- Automatic transmission, if equipped align the torque converter and flexplate and the bolts.
- Inspection plate
- Starter motor
- Exhaust pipe to the exhaust manifolds using new gaskets
- Lower radiator hose
- Heater hoses
- Alternator and wires
- Power steering pump and brackets
- Air conditioning compressor
- Linkages and cables to the carburetor or throttle body
- Ignition coil and power transistor assembly, if equipped
- MDP sensor connector
- EGR connector
- TP sensor connector
- IAC motor connector
- Magnetic clutch and refrigerant temperature switch connector
- EVAP Purge Solenoid
- ECT sensor and gauge connectors
- Front and injector wiring harness
- CMP sensor
- CKP sensor
- Distributor signal Generator
- Compactor connector
- Left and right heated O_2 sensor
- Oil pressure switch connector
- Vacuum hoses
- Air cleaner assembly, ducts and air intake hose
- Accessory belts
- Radiator, shroud and upper hose
- Cooling fan
- Battery
- Oil dipstick
- Hood

15. Refill the engine with the specified amount of oil.

16. Refill the radiator with coolant.

17. Check fuel system for leaks.

18. Check the automatic transmission fluid level, if equipped.

19. Recheck all engine adjustments.

3.5L Engine

2004

1. Before servicing the vehicle, refer to the Precautions Section.

2. Relieve the fuel system pressure.

❄❄ CAUTION

The fuel injection system remains under pressure after the engine has been OFF. Properly relieve fuel pressure before disconnecting any fuel lines. Failure to do so may result in fire or personal injury.

3. Disconnect the negative battery cable. Matchmark and remove the hood.

4. Remove the air cleaner assembly. Remove the battery. Drain the engine oil.

5. Drain the cooling system. Remove the radiator. Disconnect the exhaust pipes at the catalytic converters.

6. Remove the transmission assembly.

7. Disconnect all the ignition coil electrical connectors. Disconnect the camshaft position electrical connector.

8. Disconnect the crankshaft position sensor electrical connector, ignition power transistor connector, capacitor connector and engine coolant temperature sensor connector.

9. Disconnect the engine coolant temperature gauge unit connector, front wiring harness and injection wiring harness combination connector.

10. Disconnect the MAP sensor electrical connector, EGR vacuum regulator solenoid valve connector, EVAP solenoid connector and the front right and front left heated oxygen sensor electrical connectors.

11. Disconnect the TPS sensor electrical connector, idle air control motor connector, alternator connector and the power steering pressure switch connector.

12. Disconnect the magnetic clutch and refrigerant temperature switch connector. Disconnect the oil pressure switch connector.

13. Disconnect the throttle cable connection, all vacuum hose connections, fuel return hose connection and the fuel high pressure hose connection.

14. Disconnect the heater hoses. Disconnect the rear heater hose, if the vehicle is equipped with rear heater.

15. Remove the power steering drive belt cover and the power steering belt. Remove the power steering pump retaining bolts. Remove the power steering pump. Position the assembly to the side; it is not necessary to disconnect the fluid lines.

16. Remove the air conditioning belt. Remove the air conditioning compressor retaining bolts. Remove the air conditioning compressor. Position the assembly to the side; it is not necessary to disconnect the refrigerant lines.

17. Remove the front insulator stopper. Remove the engine mount insulator.

18. Install engine lifting tool MB991683 or equivalent.

19. Check that all cables, hoses, harness

connectors, etc are disconnected from the engine.

20. Carefully remove the engine assembly from the vehicle.

To install:

21. Installation is the reverse of the removal procedure.

22. Be sure to check and adjust all fluid levels.

23. Start the engine and check for leaks. Correct, as required.

3.8L Engine

1. Before servicing the vehicle, refer to the Precautions Section.

2. Relieve the fuel system pressure.

✳✳ CAUTION

The fuel injection system remains under pressure after the engine has been OFF. Properly relieve fuel pressure before disconnecting any fuel lines. Failure to do so may result in fire or personal injury.

3. Disconnect the negative battery cable. Remove the battery. Matchmark and remove the hood.

4. Remove the air cleaner assembly. Remove the battery. Drain the engine oil.

5. Drain the cooling system. Remove the radiator. Remove the cooling fan and clutch assembly.

6. Remove the front exhaust pipe.

7. Remove the transmission assembly.

8. Disconnect the TPS sensor electrical connector, EGR connector, right and left front heated oxygen sensor connector and the MAP sensor connector.

9. Disconnect the noise condenser connector, control wiring harness and camshaft position sensor wiring harness connector.

10. Disconnect the knock sensor connector, ignition coil connector, control wiring harness and injection wiring harness combination connector.

11. Disconnect the power steering pump switch connector, air conditioning compressor assembly connector and the intake manifold tuning solenoid connector.

12. Disconnect the engine coolant temperature gauge unit connector, engine coolant temperature gauge sensor connector and all engine grounding cables.

13. Disconnect the purge hose connection. Disconnect the heater hoses. Disconnect the fuel return line and the high pressure line. Discard the O-rings.

14. Remove the power steering pump retaining bolts. Remove the power steering pump. Position the assembly to the side; it

is not necessary to disconnect the fluid lines.

15. Remove the air conditioning compressor retaining bolts. Remove the air conditioning compressor. Position the assembly to the side; it is not necessary to disconnect the refrigerant lines. Remove the eye bolts and gaskets.

16. Remove the heat protector. Remove the engine front mount insulator.

17. Install engine lifting tool MB991683 or equivalent.

18. Check that all cables, hoses, harness connectors, etc are disconnected from the engine.

19. Carefully remove the engine assembly from the vehicle.

To install:

20. Installation is the reverse of the removal procedure.

21. Be sure to check and adjust all fluid levels.

22. Start the engine and check for leaks. Correct, as required.

23. On 2006 vehicles, complete the vehicle initialization procedure.

➡**To complete the initialization procedure the following tools are needed. MB991958 scan tool, MB991824 VCI, MB991827 MUT III USB cable, MB991910 MUT III Main harness "A"**

24. Connect the scan tool to the data link connector. To prevent damage to the scan tool be sure that the ignition switch is in the **LOCK** position before connecting the scan tool.

25. Turn the ignition switch to the **ON** position.

26. Select "check mode" from the menu screen.

27. Select "erase memory" from the menu screen.

28. Initialize the learning value.

29. After initialization complete the idle learning procedure.

➡**This procedure must be performed when the PCM is replaced, or when the learning value is initialized, as the idling is not stabilized because the learning value in the MFI engine is not completed.**

30. Start the engine. Allow the coolant temperature to reach 176 degrees or more.

31. Stop the engine and place the ignition switch in the **LOCK** position.

32. After ten seconds, restart the engine.

33. For ten minutes carry out the idling procedure below to confirm that the engine has normal idling.

• Position the transaxle selector lever in the **P** range
• The engine fan is not to be operated
• The engine coolant temperature should be 176 degrees or more

➡**If the engine stalls during idling, check the throttle valve of the throttle body for dirt. Correct and perform the procedure again.**

Water Pump

REMOVAL & INSTALLATION

3.0L and 3.5L Engines

1. Before servicing the vehicle, refer to the Precautions Section.

2. If necessary, properly release the fuel pressure.

3. Drain the cooling system.

4. Remove or disconnect the following:
   • Negative battery cable

✳✳ CAUTION

Wait at least 90 seconds after the negative battery cable is disconnected to prevent possible deployment of the air bag.

• Upper radiator shroud
• Accessory belts
• Air conditioning compressor tensioner pulley, if equipped
• Cooling fan and clutch assembly and the water pump pulley
• Thermostat and housing
• Water outlet, gasket and houses
• Radiator hoses from the water pump
• Crankshaft pulley(s)
• Timing belt covers. If the same timing belt will be reused, mark the direction of the timing belt's rotation, for installation in the same direction. Be sure the engine is positioned so the No. 1 cylinder is at the TDC of its compression stroke and the sprockets timing marks are aligned with the engine's timing mark indicators.
• Timing belt
• Water pump bolts are different lengths, note their positions before removing.
• Water pump from the block
• Water pipe connection and O-ring

To install:

5. Clean and dry the mating surfaces of the block and water pump

6. Install or connect the following:
- New O-ring on the water pipe connection, wet the new O-ring with water to aid in installation
- Water pump, with a new gasket, Torque the bolts to 17 ft. lbs. (23 Nm) on 3.0L and 3.5L engines
- Alternator bracket bolt to 17 ft. lbs. (23 Nm)
- Timing belt(s) and covers
- Crankshaft pulley(s)
- Thermostat and housing on 3.0L, 3.5L engines. Torque the bolts to 12–14 ft. lbs. (17–20 Nm).
- Radiator hose to the water pump
- Water outlet, new gasket and houses. Torque the bolts to 12–14 ft. lbs. (17–20 Nm).
- Water pump pulley
- Cooling fan and clutch assembly
- Air conditioning compressor tensioner pulley, if equipped
- Accessory belts
- Upper radiator shroud
- Thermostat and housing
- Negative battery cable

7. Refill the radiator with coolant. This cooling system has a self-bleeding thermostat, so system bleeding is not required.
8. Run the vehicle until the thermostat

opens and fill the overflow tank. Check for leaks.
9. Once the vehicle has cooled, recheck the coolant level.

3.8L Engine

1. Before servicing the vehicle, refer to the Precautions Section.
2. If necessary, properly release the fuel pressure.
3. Drain the cooling system.
4. Remove or disconnect the following:
- Negative battery cable

❊❊ CAUTION

Wait at least 90 seconds after the negative battery cable is disconnected to prevent possible deployment of the air bag.

- Timing belt
- Camshaft sprocket
- Engine Coolant Temperature (ECT) sensor connector

- ECT gauge unit connector
- Spark plug cable support
- Upper radiator hose
- Water hoses
- Water outlet fitting bracket
- Water outlet fitting, O-ring and gasket
- Water pump assembly, gasket and O-ring
- Fitting, gaskets and thermostat case

To install:

5. Clean and dry the mating surfaces of the block and water pump
6. Install or connect the following:
- Thermostat case with new gaskets
- Fitting
- Water pump assembly, using a new gasket and O-ring. Torque the water pump bolts to 16–20 ft. lbs. (21–27 Nm).
- Water outlet fitting, O-ring and gasket
- Water outlet fitting bracket
- Water hoses

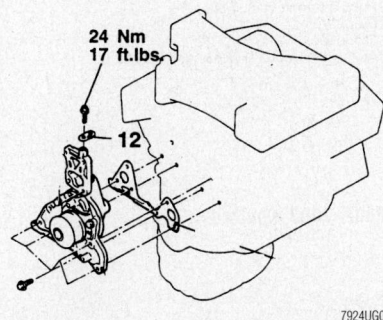

Water pump and related components— 3.0L engine

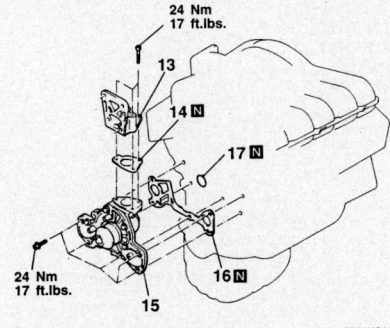

Water pump and related components— 3.5L engine

1. ENGINE COOLANT TEMPERATURE SENSOR CONNECTOR
2. ENGINE COOLANT TEMPERATURE GAUGE UNIT CONNECTOR
3. SPARK PLUG CABLE SUPPORT
4. RADIATOR UPPER HOSE CONNECTION
5. WATER HOSE
6. WATER HOSE
7. WATER OUTLET FITTING BRACKET
8. WATER OUTLET FITTING
9. O-RING
10. GASKET
11. WATER PUMP ASSEMBLY
12. GASKET
13. O-RING
14. FITTING
15. GASKET
16. GASKET
17. THERMOSTAT CASE

Water pump and related components—3.8L engine

- Upper radiator hose
- Spark plug cable support
- ECT gauge unit connector
- ECT sensor connector
- Camshaft sprocket
- Timing belt

7. Refill the radiator with coolant. This cooling system has a self-bleeding thermostat, so system bleeding is not required.

8. Run the vehicle until the thermostat opens and fill the overflow tank. Check for leaks.

9. Once the vehicle has cooled, recheck the coolant level.

Heater Core

REMOVAL & INSTALLATION

Montero

2002–2003

1. Before servicing the vehicle, refer to the Precautions Section.

2. Place the wheels in the straight-ahead position.

3. Drain the cooling system.

4. Discharge and recover the air conditioning system refrigerant.

5. Remove or disconnect the following:
- Negative battery cable

❈ CAUTION

Wait at least 90 seconds after disconnecting the battery cable before performing any work on the air bag or instrument panel.

- Floor console assembly
- Steering column-to-instrument panel cover screws and the cover
- Air bag module, carefully, from the steering wheel
- Electrical connectors from the air bag module

❈❈ CAUTION

Store the air bag module facing up.

- Steering wheel nut
- Press the steering wheel from the steering column, using a steering wheel puller
- Passenger's side foot shower duct
- Glove box stoppers and the glove box
- Passenger's air bag module nut and the air bag module
- Electrical connector from the air bag module

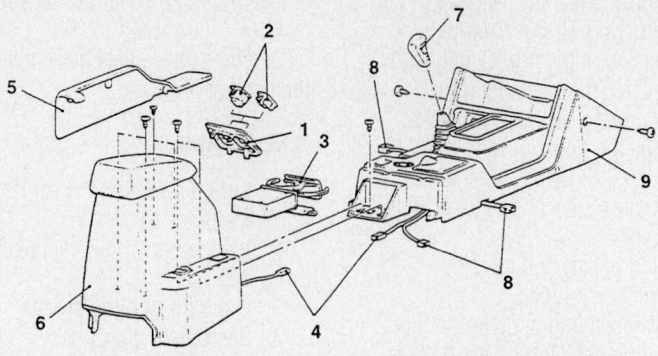

1. Switch panel
2. Suspension control switch or hole cover
3. Cup holder assembly
4. Rear console harness connector
5. Side panel A
6. Rear console assembly
7. Transfer shift lever knob
8. Floor console harness connector
9. Front console assembly

93113GE2

Floor console and related components—2002–2003 Montero

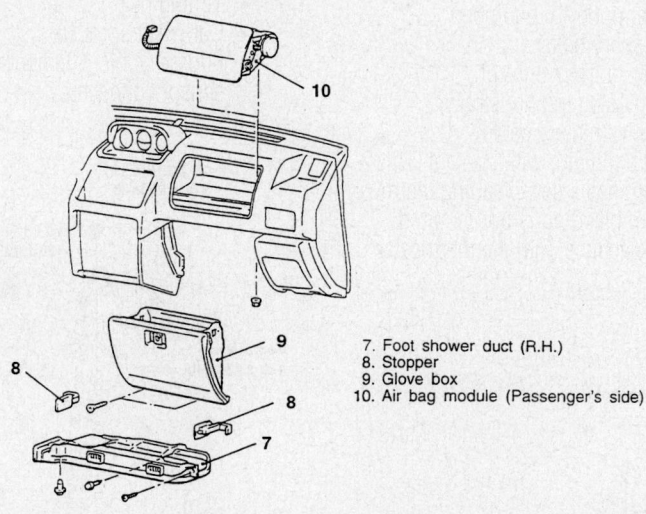

7. Foot shower duct (R.H.)
8. Stopper
9. Glove box
10. Air bag module (Passenger's side)

93113GE3

Passenger air bag module and related components—2002–2003 Montero

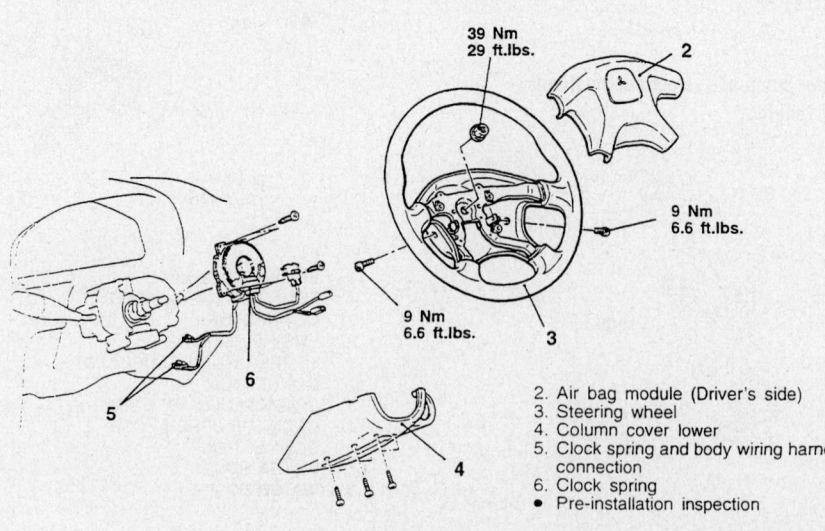

2. Air bag module (Driver's side)
3. Steering wheel
4. Column cover lower
5. Clock spring and body wiring harness connection
6. Clock spring
- Pre-installation inspection

93113GE4

Driver air bag module and related components—2002–2003 Montero

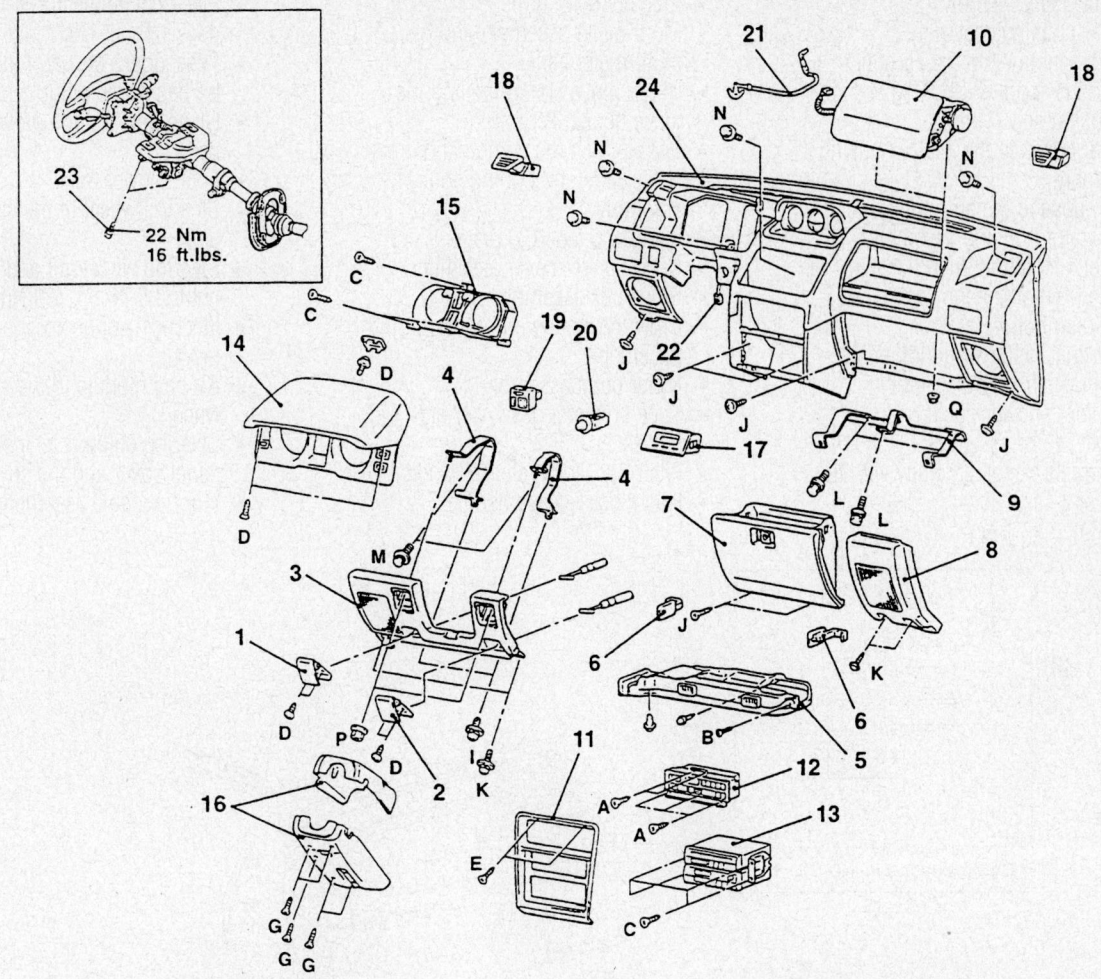

1. Hood lock release handle
2. Fuel filler door lock release handle
3. Knee protector
4. Stay
5. Foot shower duct (R.H.)
6. Glove box stopper
7. Glove box assembly
8. Corner cover
9. Stay
10. Passenger-side air bag module assembly
11. Center panel
12. Heater control assembly
13. Radio and tape player
14. Meter bezel assembly
15. Combination meter
16. Column cover
17. Clock
18. Side defroster garnish
19. Door mirror control switch
20. Rheostat
21. Ventilation control wire
22. Harness connector
23. Steering column installation bolts
24. Instrument panel assembly

93113GE5

Instrument panel and related components—2002–2003 Montero

- Hood lock release handle
- Fuel filler door lock release handle
- Knee protector assembly and bracket
- Meter bezel assembly
- Under cover, the corner cover and the stopper
- Glove box assembly
- Center under cover assembly
- Radio and tape/CD player
- Heater/air conditioning control assembly
- Combination meter and the speaker

- Glove box striker and the upper glove box frame
- Multi-meter panel and the multi-meter assembly
- Side defroster grille
- Instrument panel assembly, carefully, with the help of an assistant
- Blower motor assembly
- Both foot ducts
- Joint duct from the air conditioning evaporator housing assembly
- Foot distribution duct
- Center reinforcement
- Center ventilation duct

- Drain hose from the air conditioning evaporator housing assembly
- Heater hoses from the heater housing assembly
- Refrigerant lines from the air conditioning evaporator housing assembly and discard the O-rings
- Heater housing assembly
- Center duct assembly
- Heater core from the heater housing

To install:
6. Install or connect the following:
- Heater core to the heater housing

- Center duct assembly
- Heater housing assembly
- Refrigerant lines to the air conditioning evaporator housing assembly, using new O-rings
- Heater hoses to the heater housing assembly
- Drain hose to the air conditioning evaporator housing assembly
- Center ventilation duct
- Center reinforcement
- Foot distribution duct
- Joint duct to the air conditioning evaporator housing assembly
- Both foot shower ducts
- Blower motor assembly
- Instrument panel, carefully with the help of an assistant

- Side defroster grille
- Multi-meter assembly and the multi-meter panel
- Upper glove box frame and the glove box striker
- Speaker and the combination meter
- Heater/air conditioning control assembly
- Radio and tape/CD player
- Center under cover assembly
- Glove box assembly
- Under cover, the corner cover and the stopper
- Meter bezel assembly
- Knee protector assembly and bracket
- Filler door lock release handle
- Hood lock release handle

- Electrical connector to the passenger's side air bag module
- Passenger's air bag module and the air bag module nut
- Glove box and the glove box stoppers
- Foot shower duct
- Steering wheel to the steering column
- Steering wheel nut and torque the nut to 29 ft. lbs. (39 Nm)
- Electrical connectors to the air bag module
- Air bag module to the steering wheel
- Steering column-to-instrument panel cover and the cover screws
- Floor console assembly

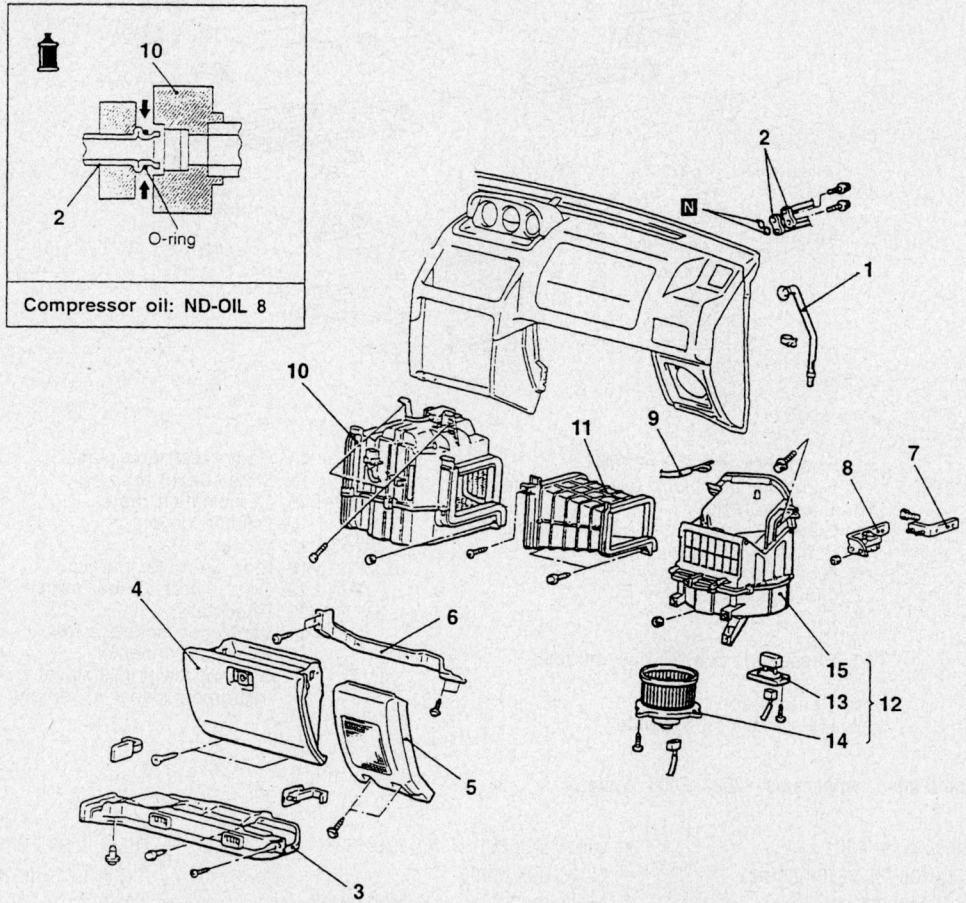

Compressor oil: ND-OIL 8

1. Drain hose
2. Liquid pipe and suction hose connection
3. Foot shower duct (R.H.)
4. Glove box
5. Corner cover
6. Lower frame
7. Engine control relay assembly
8. Bracket
9. Air selection control wire connection
10. Evaporator
11. Duct joint
12. Blower assembly
13. Resistor
14. Blower motor assembly
15. Blower case assembly

93113GE6

Heater/evaporator housing and related components—2002–2003 Montero

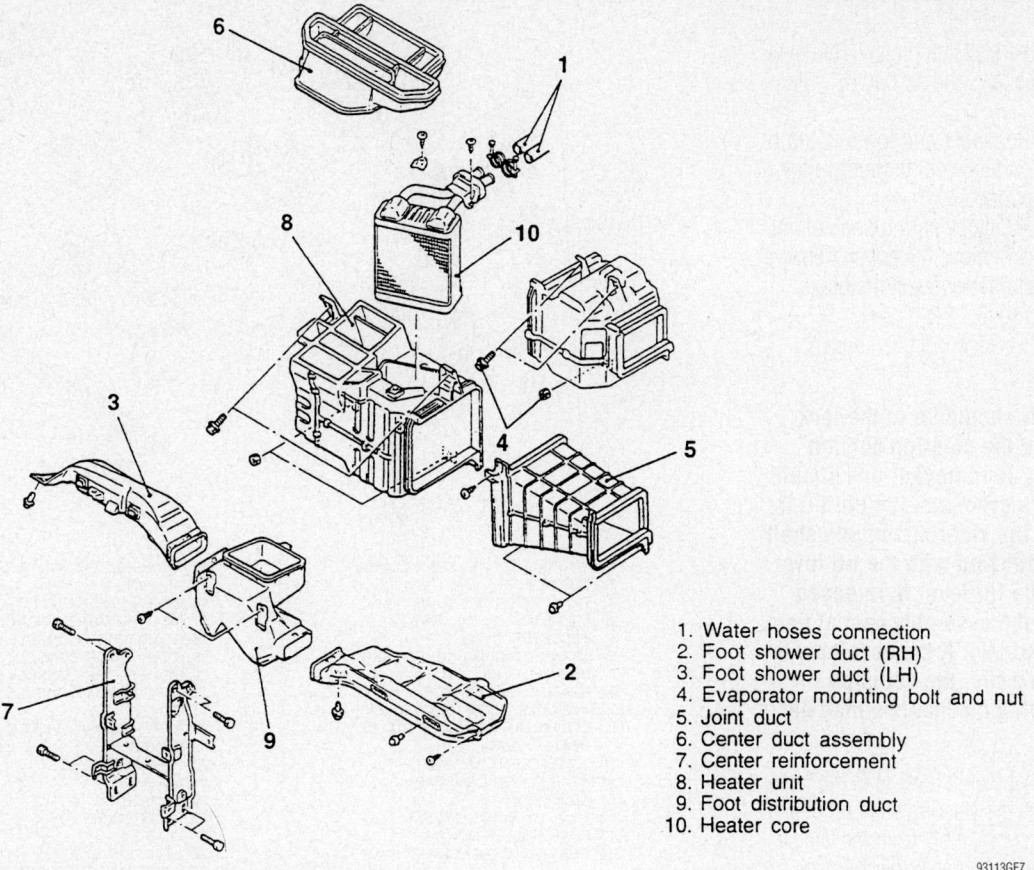

1. Water hoses connection
2. Foot shower duct (RH)
3. Foot shower duct (LH)
4. Evaporator mounting bolt and nut
5. Joint duct
6. Center duct assembly
7. Center reinforcement
8. Heater unit
9. Foot distribution duct
10. Heater core

93113GE7

Heater core and related components—2002–2003 Montero

7. Refill the cooling system.

8. Connect the negative battery.

9. Evacuate and charge the air conditioning system refrigerant.

10. Run the engine to normal operating temperatures; then, check the climate control operation and check for leaks.

2004–2006

1. Before servicing the vehicle, refer to the Precautions Section.

2. Place the wheels in the straight-ahead position.

3. Disconnect the negative battery cable.

✳✳ CAUTION

Wait at least 90 seconds after disconnecting the battery cable before performing any work on the air bag or instrument panel.

4. Drain the cooling system. Discharge the air conditioning system.

5. Remove the air cleaner, air intake hose and air intake duct.

6. Remove the parking brake lever cover, the cup holder, the transfer shift

knob, if equipped, and the selector lever knob.

7. Remove the indicator panel. Remove the heated seat switch, if equipped. Remove the coin plug, if equipped.

8. Remove the lower center panel. Remove the rear floor console assembly.

9. Remove the parking brake lever mounting bolt. Remove the front floor console assembly.

10. Remove the rear heater duct "A". Remove the rear heater duct "B". Remove the heater duct nozzle.

11. Remove the center panel assembly. Remove the clock or RV meter. Remove the radio and tape player and/or box.

12. Remove the heater control assembly. Remove the glove box stopper, glove box, lock, light switch connector, light switch, and light.

13. Remove the upper glove box striker, upper glove box, and lid lock.

14. Disconnect the front passenger's side air bag module harness connector.

15. Remove the instrument panel side cover. Remove the air outlet assembly. Remove the instrument cluster trim bezel. Remove the instrument cluster assembly.

16. Remove the lower panel. Remove the

switch panel. Remove the column cover. Remove the "B" stay. Remove the corner lower panel.

17. Remove the steering wheel cover, located under the steering wheel.

18. By sliding section "A" of the clock spring connector in the direction of the

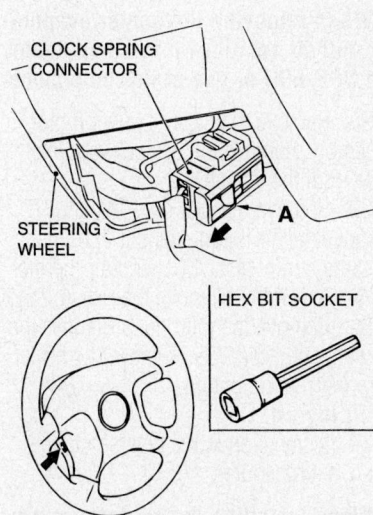

09482_MONT_G0002

Driver air bag module and related components—2004–2006 Montero

arrow, in the illustration, disconnect the connector.

19. Loosen the bolt completely. Remove the steering wheel and the air bag module assembly.

20. Remove the upper and lower column covers, if not previously done. Remove the wiring harness protector.

21. Remove the clock spring and column switch assembly. Remove the cover. Remove the interlock cable. Disconnect the brake pedal return spring.

22. Remove the steering shaft cover installation bolt.

➡ The tilt lever should be in the lock position before the steering column shaft assembly is removed, and should remain in the locked position until it is reinstalled. If the steering column shaft assembly is removed with the tilt lever released, or the tilt lever is released after removal the assembly cannot be reinstalled correctly. If the assembly is installed incorrectly, the collision energy absorbing mechanism may be damaged.

23. Ensure that the tilt lever is in the lock position. Remove the steering shaft mounting bolts. Remove the assembly from the vehicle.

24. Remove the photo sensor harness connector.

25. Remove the instrument panel assembly from the vehicle.

26. Disconnect the heater hoses. Disconnect and plug the air conditioning refrigerant lines.

27. Disconnect the harness connectors. Remove foot duct "B" and duct "D". Remove the front deck crossmember bar. Remove the flange bracket.

➡ When removing the heater/evaporator unit do not allow it to bump against the SRS-ECU or any of its components.

28. Remove the heater blower motor assembly. Remove the heater/evaporator case from the vehicle.

29. Remove the foot duct "A" and "C". Remove the air thermo sensor clip. Remove the auto compressor/ECU and air thermo sensor assembly. Remove the resistor.

30. Remove the joint duct. Remove the air duct sub assembly. Remove the heater core from its mounting.

To install:

31. Installation is the reverse of the removal procedure.

➡ When installing the heater/evaporator unit do not allow it to bump against the SRS-ECU or any of its components.

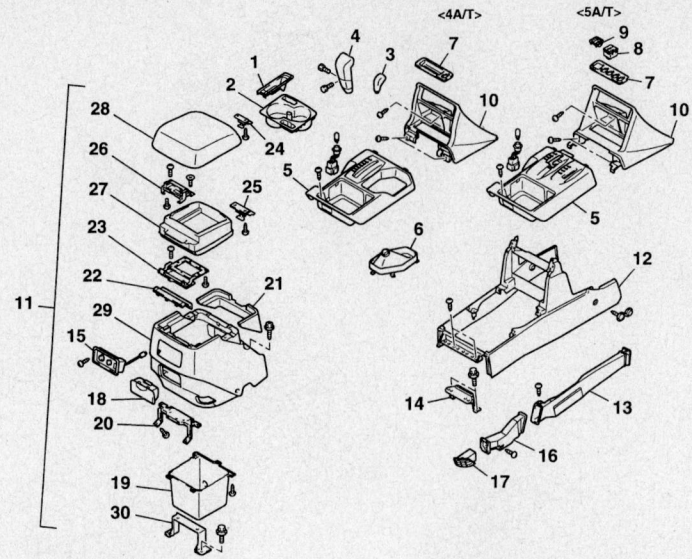

1. PARKING BRAKE LEVER COVER
2. CUP HOLDER
3. TRANSFER SHIFT KNOB
4. SELECTOR LEVER KNOB
5. INDICATOR PANEL
6. TRANSFER SHIFT COVER <4A/T>
7. SWITCH PANEL
8. HEATED SEAT SWITCH <VEHICLES WITH HEATED SEAT>
9. COIN PLUG <VEHICLES WITHOUT HEATED SEAT>
10. LOWER CENTER PANEL
11. REAR FLOOR CONSOLE ASSEMBLY
12. FRONT FLOOR CONSOLE
13. REAR HEATER DUCT A
14. CONSOLE BRACKET A
15. REAR HEATER CONTROL ASSEMBLY
16. REAR HEATER DUCT B
17. REAR HEATER DUCT NOZZLE
18. REAR FLOOR CONSOLE ASHTRAY
19. REAR FLOOR CONSOLE BOX
20. REAR FLOOR CONSOLE BRACKET
21. REAR FLOOR CONSOLE UPPER COVER
22. REAR FLOOR CONSOLE HINGE COVER
23. HINGE BRACKET
24. LOCK A
25. LOCK B
26. LID BRACKET
27. LID BOX
28. LID ASSEMBLY
29. REAR FLOOR CONSOLE
30. CONSOLE BRACKET B

09482_MONT_G0003

Floor console and related components—2004–2006 Montero

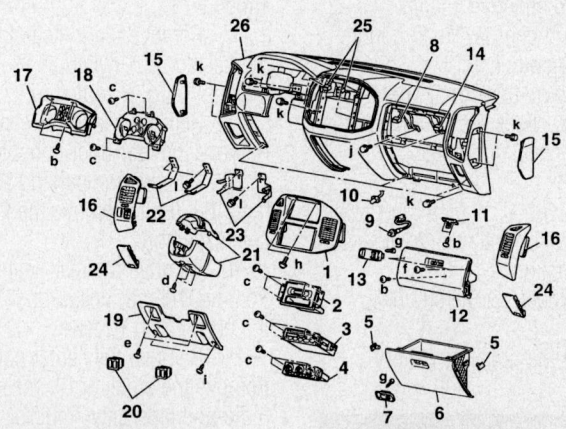

1. CENTER PANEL ASSEMBLY
2. CLOCK OR RV METER
3. RADIO AND TAPE PLAYER OR BOX
4. HEATER CONTROL ASSEMBLY
5. GLOVE BOX STOPPER
6. GLOVE BOX ASSEMBLY
7. GLOVE BOX LOCK
8. GLOVE BOX LIGHT SWITCH HARNESS CONNECTOR
9. GLOVE BOX LIGHT SWITCH
10. GLOVE BOX LIGHT
11. UPPER GLOVE BOX STRIKER
12. UPPER GLOVE BOX ASSEMBLY
13. UPPER GLOVE BOX LID LOCK
14. FRONT PASSENGER'S SIDE AIR BAG MODULE HARNESS CONNECTOR
15. INSTRUMENT PANEL SIDE COVER
16. AIR OUTLET ASSEMBLY
17. METER BEZEL
18. COMBINATION METER
19. LOWER PANEL
20. SWITCH PANEL
21. COLUMN COVER
22. STAY A
23. STAY B
24. CORNER LOWER PANEL
25. PHOTO SENSOR HARNESS CONNECTOR
26. INSTRUMENT PANEL ASSEMBLY

09482_MONT_G0004

Instrument panel and related components—2004–2006 Montero

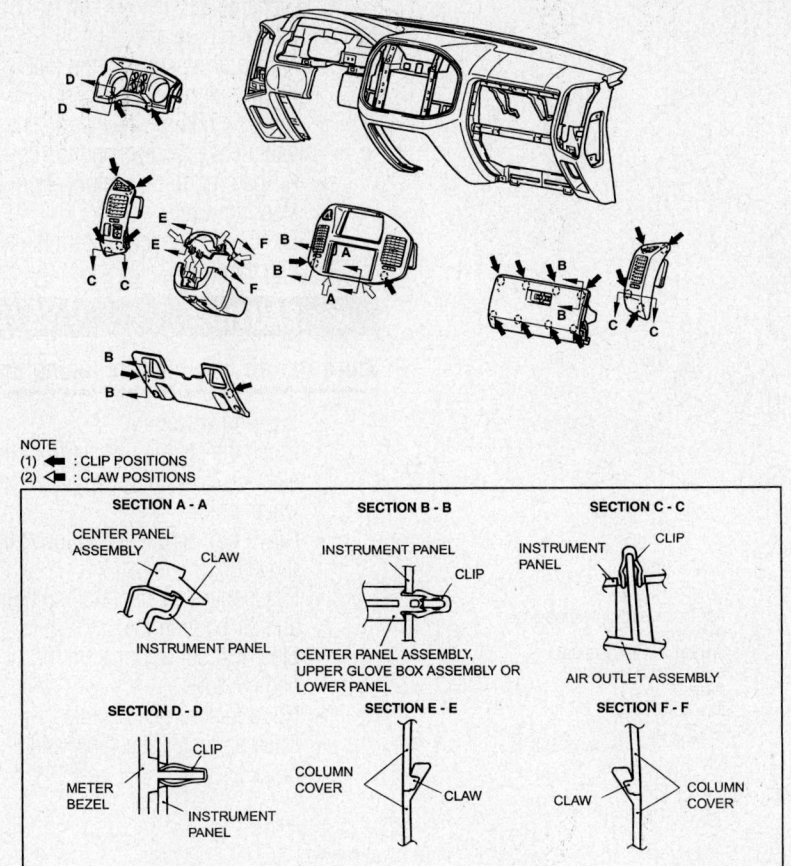

NOTE
(1) ◀ : CLIP POSITIONS
(2) ◁ : CLAW POSITIONS

SECTION A - A
CENTER PANEL ASSEMBLY
CLAW
INSTRUMENT PANEL

SECTION B - B
INSTRUMENT PANEL
CLIP
CENTER PANEL ASSEMBLY, UPPER GLOVE BOX ASSEMBLY OR LOWER PANEL

SECTION C - C
INSTRUMENT PANEL
CLIP
AIR OUTLET ASSEMBLY

SECTION D - D
METER BEZEL
CLIP
INSTRUMENT PANEL

SECTION E - E
COLUMN COVER
CLAW

SECTION F - F
CLAW
COLUMN COVER

09482_MONT_G0005

Instrument panel retention clip locations—2004–2006 Montero

32. Ensure that the tilt lever is in the lock position. Install the steering column shaft.

➡**Do not release the tilt lever until the steering column shaft has been installed. Torque the retaining bolts 13–19 ft. lbs.**

33. Before installing the steering shaft cover installation bolt, coat the mounting hole on the toe board with sealant, part number 8633 or equivalent.

➡**When installing the clock spring, ensure that the mating marks are properly aligned. If not the steering wheel may not rotate completely during a turn, or the flat cable in the clock spring could be damaged. This would prevent normal SRS operation and possibly cause serious injury to the driver.**

34. Turn the front wheels to the straight ahead position. Then install the clock spring to the column switch.

35. To align the mating marks of the clock spring, turn the clock spring counterclockwise fully. Then turn it back approximately 3-3/4 turns counterclockwise to align the mating marks.

➡**When installing the steering wheel, be sure that the harness of the clock spring does not become caught or tangled.**

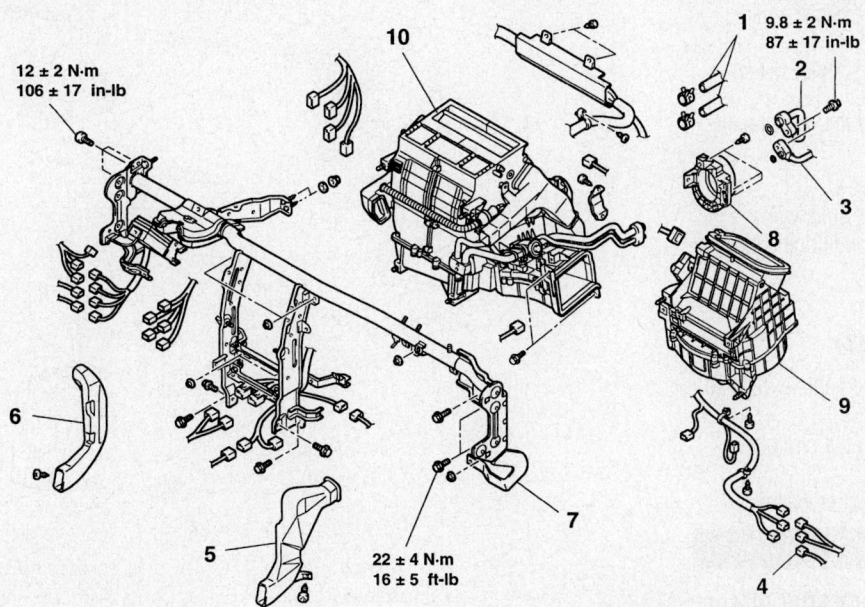

12 ± 2 N·m
106 ± 17 in-lb

9.8 ± 2 N·m
87 ± 17 in-lb

22 ± 4 N·m
16 ± 5 ft-lb

1. HEATER HOSE
2. SUCTION HOSE
3. LIQUID PIPE
4. HARNESS CONNECTORS
5. FOOT DUCT B

6. FOOT DUCT D
7. FRONT DECK CROSSMEMBER
8. FLANGE BRACKET
9. BLOWER ASSEMBLY
10. HEATER UNIT

09482_MONT_G0006

Heater/evaporator housing and related components—2004–2006 Montero

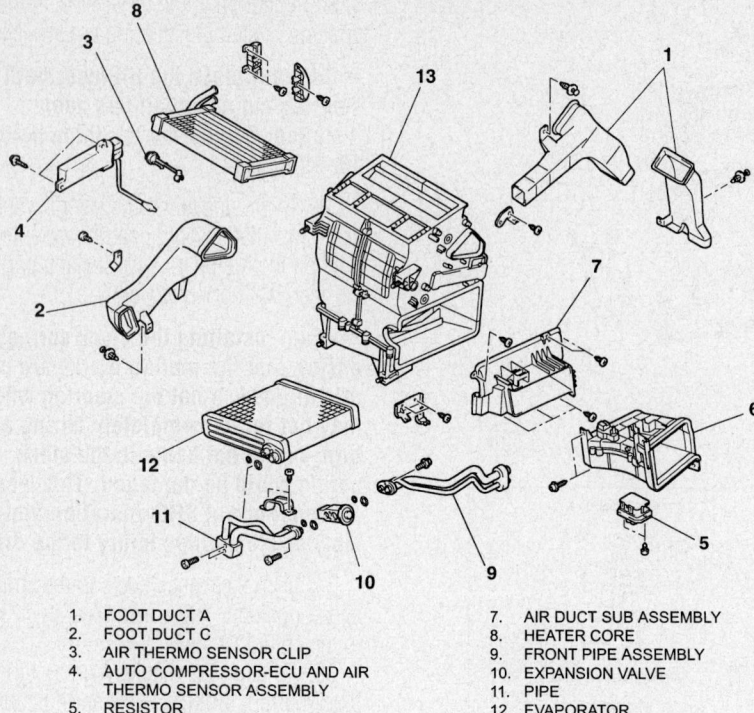

1. FOOT DUCT A
2. FOOT DUCT C
3. AIR THERMO SENSOR CLIP
4. AUTO COMPRESSOR-ECU AND AIR THERMO SENSOR ASSEMBLY
5. RESISTOR
6. JOINT DUCT
7. AIR DUCT SUB ASSEMBLY
8. HEATER CORE
9. FRONT PIPE ASSEMBLY
10. EXPANSION VALVE
11. PIPE
12. EVAPORATOR
13. CASE

09482_MONT_G0007

Heater core and related components—2004–2006 Montero

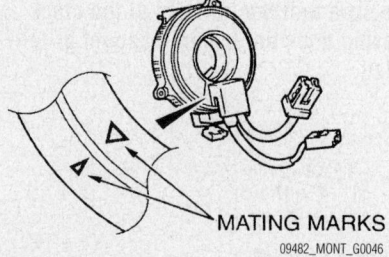

MATING MARKS

09482_MONT_G0046

Clock spring alignment—2004–2006 Montero

36. Check and adjust all fluid levels. Properly recharge the air conditioning system.

Montero Sport

FRONT HEATER SYSTEM

1. Before servicing the vehicle, refer to the Precautions Section.
2. Place the wheels in the straight-ahead position.
3. Disconnect the negative battery.

✳✳ CAUTION

Wait at least 90 seconds after disconnecting the battery cable before performing any work on the air bag or instrument panel.

4. Drain the cooling system into a clean container for reuse.

5. Discharge and recover the air conditioning system refrigerant.
6. Remove or disconnect the following:
- Floor console assembly
- Steering column-to-instrument panel cover screws and the cover
- Air bag module, carefully, from the steering wheel
- Electrical connectors from the air bag module

✳✳ CAUTION

Store the air bag module facing up.

- Steering wheel nut
- Press the steering wheel from the steering column, using a steering wheel puller
- Glove box stoppers and lower the glove box
- Passenger's air bag module bolts and air bag module
- Electrical connector from the air bag module
- Hood lock release handle
- Knee protector assembly and bracket

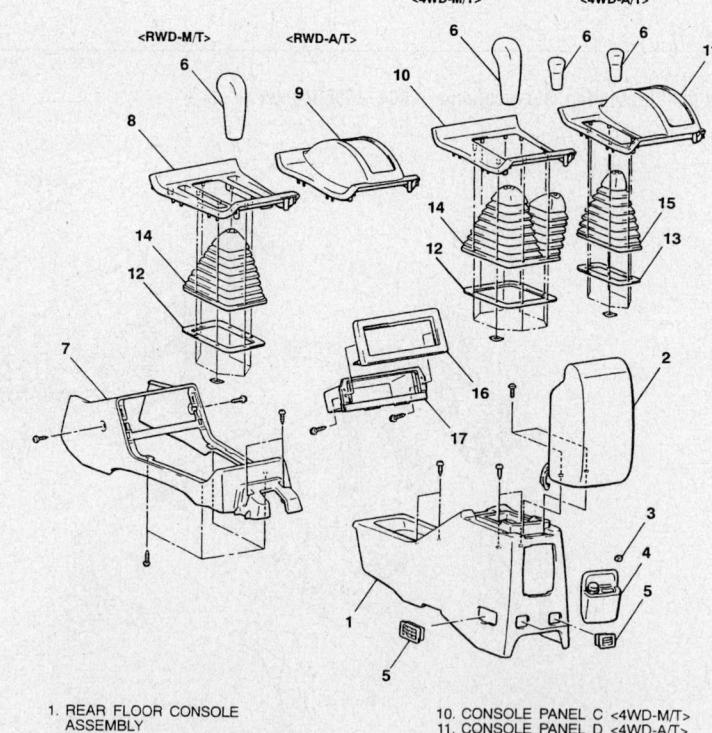

1. REAR FLOOR CONSOLE ASSEMBLY
2. CONSOLE LID ASSEMBLY
3. KNOB
4. REAR HEATER CONTROL PANEL ASSEMBLY
5. FOOT GRILL
6. SHIFT LEVER KNOB
7. FRONT FLOOR CONSOLE ASSEMBLY
8. CONSOLE PANEL A <RWD-M/T>
9. CONSOLE PANEL B <RWD-A/T>
10. CONSOLE PANEL C <4WD-M/T>
11. CONSOLE PANEL D <4WD-A/T>
12. SHIFT LEVER BOOT REINFORCEMENT <M/T>
13. TRANSFER LEVER BOOT REINFORCEMENT <4WD-A/T>
14. SHIFT LEVER BOOT <M/T>
15. TRANSFER LEVER BOOT <4WD-A/T>
16. CONSOLE PANEL
17. BOX

93113GD6

Floor console and related components—Montero Sport

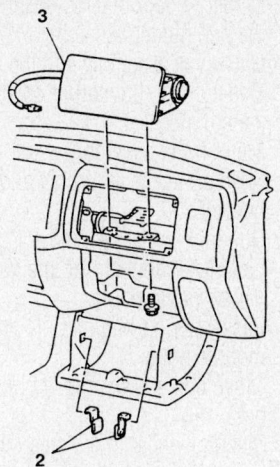

1. NEGATIVE (-) BATTERY CABLE CONNECTION
2. STOPPER
3. AIR BAG MODULE
• PRE-INSTALLATION INSPECTION

93113GD7

Passenger air bag module—Montero Sport

2. AIR BAG MODULE
3. STEERING WHEEL
4. COLUMN COVER LOWER
5. CLOCK SPRING

93113GD8

Driver air bag module—Montero Sport

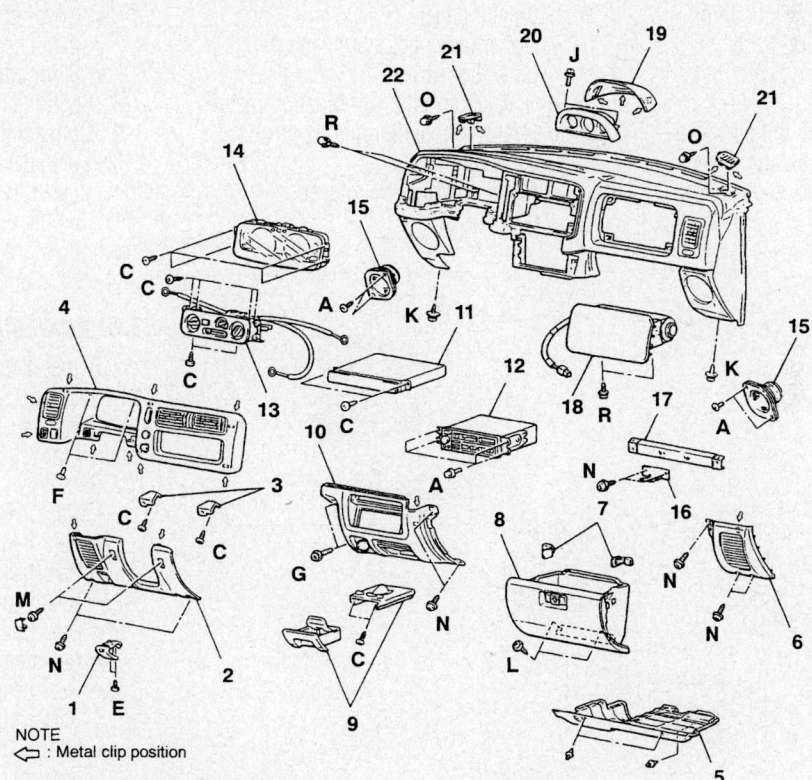

NOTE
⇦ : Metal clip position

1. HOOD LOCK RELEASE HANDLE
2. KNEE PROTECTOR ASSEMBLY
3. KNEE PROTECTOR BRACKET
4. METER BEZEL ASSEMBLY
5. UNDER COVER
6. CORNER COVER
7. STOPPER
8. GLOVE BOX ASSEMBLY
9. ASHTRAY ASSEMBLY
10. CENTER UNDER COVER ASSEMBLY
11. CUP HOLDER ASSEMBLY
12. RADIO AND TAPE PLAYER
13. HEATER CONTROL ASSEMBLY

14. COMBINATION METER
15. SPEAKER
16. GLOVE BOX STRIKER
17. GLOVE BOX UPPER FRAME
18. FRONT PASSENGER'S SIDE AIR BAG MODULE
19. MULTI-METER PANEL
20. MULTI-METER ASSEMBLY
21. SIDE DEFROSTER GRILL
22. INSTRUMENT PANEL ASSEMBLY

93113GD9

Instrument panel and related components—Montero Sport

- Meter bezel assembly
- Under cover, the corner cover and the stopper
- Glove box assembly and the ashtray
- Center under cover assembly and the cup holder assembly
- Radio and tape/CD player
- Heater/air conditioning control assembly
- Combination meter and the speaker
- Glove box striker and the upper glove box frame
- Multi-meter panel and the multimeter assembly
- Side defroster grille
- Instrument panel assembly, carefully, with the help of an assistant
- Blower motor assembly
- Joint duct from the air conditioning evaporator housing assembly
- Center reinforcement
- Center ventilation duct
- Drain hose from the air conditioning evaporator housing assembly
- Heater hoses from the heater housing assembly
- Refrigerant lines from the air conditioning evaporator housing assembly and discard the O-rings

- Heater housing assembly
- Heater core from the heater housing

To install:

7. Install or connect the following:
- Heater core to the heater housing
- Heater housing assembly
- Refrigerant lines to the air conditioning evaporator housing assembly, using new O-rings
- Heater hoses to the heater housing assembly
- Drain hose to the air conditioning evaporator housing assembly
- Center ventilation duct
- Center reinforcement
- Joint duct to the air conditioning evaporator housing assembly
- Blower motor assembly
- Instrument panel assembly, carefully, with the help of an assistant
- Side defroster grille
- Multi-meter assembly and the multimeter panel
- Upper glove box frame and the glove box striker
- Speaker and the combination meter
- Heater/air conditioning control assembly
- Radio and tape/CD player

- Center under cover assembly and the cup holder assembly
- Glove box assembly and the ashtray
- Under cover, the corner cover and the stopper
- Meter bezel assembly
- Knee protector assembly and bracket
- Hood lock release handle
- Electrical connector to the passenger air bag module
- Passenger air bag module and the module bolts
- Glove box and the glove box stoppers
- Steering wheel to the steering column
- Steering wheel nut and torque the nut to 29 ft. lbs. (39 Nm)
- Electrical connectors to the air bag module
- Air bag module to the steering wheel
- Steering column-to-instrument panel cover and the cover screws
- Floor console assembly

8. Refill the cooling system.
9. Connect the negative battery.
10. Evacuate and charge the air conditioning system refrigerant.
11. Run the engine to normal operating temperatures; then, check the climate control operation and check for leaks.

REAR AUXILIARY SYSTEM

1. Before servicing the vehicle, refer to the Precautions Section.
2. Drain the cooling system into a clean container for reuse.
3. Remove or disconnect the following:
- Negative battery cable
- Rear heater unit switch knob
- Rear heater control panel assembly
- Rear heater switch
- Rear floor console
- Resistor
- Rear heater hoses from the rear heater core
- Rear heater core from the rear heater housing

To install:

4. Install or connect the following:
- Rear heater core to the rear heater housing
- Rear heater hoses to the rear heater core
- Resistor
- Rear floor console
- Rear heater switch
- Rear heater control panel assembly
- Rear heater unit switch knob
- Negative battery cable

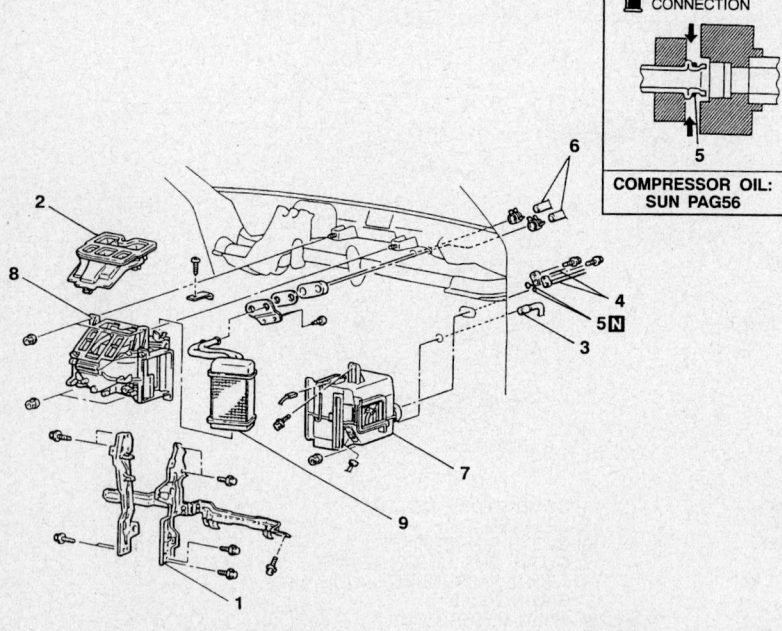

PIPING CONNECTION

COMPRESSOR OIL: SUN PAG56

1. CENTER REINFORCEMENT
2. CENTER VENTILATION DUCT
3. DRAIN HOSE <VEHICLES WITH A/C>
4. SUCTION PIPE OR HOSE AND DISCHARGE PIPE CONNECTION <VEHICLES WITH A/C>
5. O-RING
6. HEATER HOSE CONNECTION
7. EVAPORATOR <VEHICLES WITH A/C>
8. HEATER UNIT
9. HEATER CORE

93113GD0

Heater/evaporator housing and related components—Montero Sport

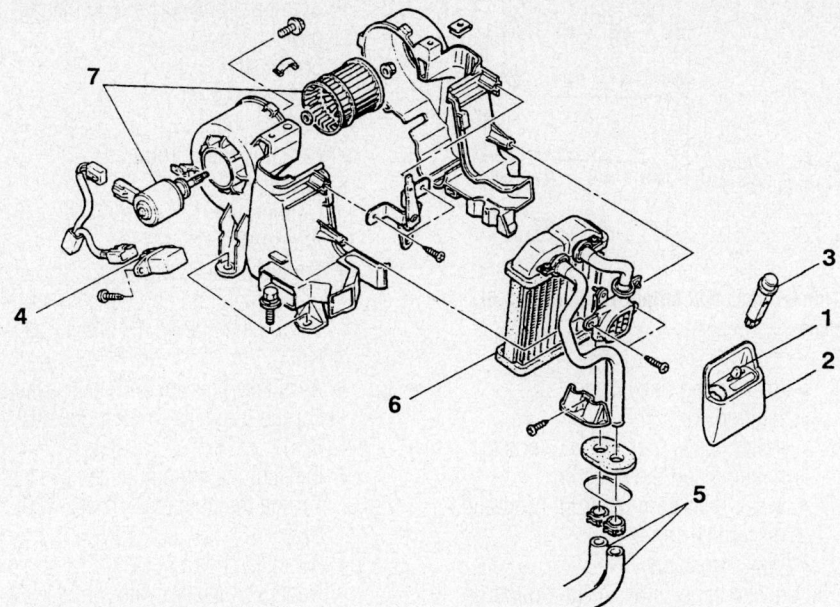

1. KNOB
2. REAR HEATER CONTROL PANEL
   ASSEMBLY
3. REAR HEATER SWITCH

4. RESISTOR
• DRAINING AND SUPPLYING OF
   COOLANT
5. REAR HEATER HOSE
   CONNECTION
6. REAR HEATER CORE ASSEMBLY
7. REAR BLOWER MOTOR
   ASSEMBLY

93113GE1A

Rear heater core and related components—Montero Sport

5. Refill the cooling system.

6. Run the engine to normal operating temperatures; then, check the climate control operation and check for leaks.

Cylinder Head

REMOVAL & INSTALLATION

3.0L Engine

1. Before servicing the vehicle, refer to the Precautions Section.

2. Relieve the fuel system pressure.

3. Drain the cooling system.

4. Remove or disconnect the following:

• Negative battery cable
• Air cleaner assembly, ducts and air intake hose
• Upper radiator hose
• Accessory drive belts
• Cooling fan and pulleys
• Air conditioning compressor, if equipped
• Power steering pump and mounting brackets and position them to the side, without disconnecting the lines.
• Timing belt covers

5. Remove the timing belt as follows:

a. Rotate the crankshaft and bring the No. 1 piston to Top Dead Center (TDC) on the compression stroke. Align the camshaft and crankshaft sprocket timing marks.

b. Mark the timing belt in the direction of rotation for reinstallation purposes.

c. Loosen the timing belt tensioner bolt and turn the tensioner counterclockwise.

✳✳ WARNING

Do not rotate the crankshaft or camshaft sprockets after the timing belt has been removed.

6. Remove or disconnect the following:

• Timing belt
• Fuel lines and plug
• Wiring connectors, vacuum lines and hoses from the air intake plenum, intake manifold and cylinder head.
• Air intake plenum
• Intake manifold
• Exhaust manifold
• Camshaft sprocket bolt and camshaft sprocket, if necessary

• Alternator bracket and/or timing belt rear cover
• Oil dipstick, on left side only
• Crankshaft position (CKP) sensor on left side only
• Spark plug wires from the spark plugs
• Valve cover
• Cylinder head bolts starting from the outside and working inward
• Cylinder head from the engine

To install:

7. Clean the gasket mounting surfaces.

8. Install or connect the following:

• New cylinder head gasket
• Cylinder head on the engine. Torque the cylinder head bolts in sequence using 3 even steps, to 80 ft. lbs. (108 Nm).
• Exhaust manifold
• Intake manifold and air intake plenum
• Fuel lines
• Wiring connectors, vacuum lines and hoses to the air intake plenum, intake manifold and cylinder head
• Valve cover
• Spark plug wires
• CKP sensor, if removed
• Oil dipstick, if removed
• Alternator bracket and/or timing belt rear cover
• Camshaft sprocket bolt and camshaft sprocket, if necessary

9. Be sure the camshaft and crankshaft sprocket timing marks are aligned.

10. Turn the timing belt tensioner to the extreme counter-clockwise position and temporarily tighten the bolt.

11. Install the timing belt in the original rotation direction. Loosen the timing belt tensioner bolt and allow the spring force of the tensioner to tension the belt.

12. Turn the crankshaft 2 turns in the normal direction of rotation and check the timing mark alignment.

13. If the timing is correct, tighten the tensioner bolt to 21 ft. lbs. (30 Nm). If the

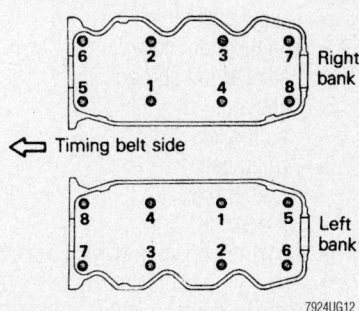

7924UG12

Cylinder head bolt torque sequence—3.0L engine

timing is incorrect, repeat the belt installation procedure.

14. Install or connect the following:
- Timing belt covers
- Alternator, alternator cover and alternator stay, if removed
- Air conditioning compressor
- Power steering pump with the brackets
- Pulleys. Torque the crankshaft pulley bolt to 134 ft. lbs. (181 Nm).
- Cooling fan
- Accessory drive belts
- Air cleaner assembly, ducts and air intake hose
- Upper radiator hose
- Negative battery

15. Refill the cooling system.
16. Start the engine and check for leaks. Check the ignition timing.

3.5L Engine

1. Before servicing the vehicle, refer to the Precautions Section.

2. Relieve fuel system pressure.
3. Drain the cooling system.
4. Remove or disconnect the following:
- Negative battery cable

- Air intake hoses
- Air intake plenum and intake manifold
- Exhaust manifold
- Engine under cover
- Radiator and shroud
- Alternator
- Cooling fan
- Timing belt
- Breather hose
- Oil dipstick
- Camshaft Position (CMP) sensor
- Spark plug cable center cover and remove the spark plug cables
- Valve cover
- Intake camshaft sprocket

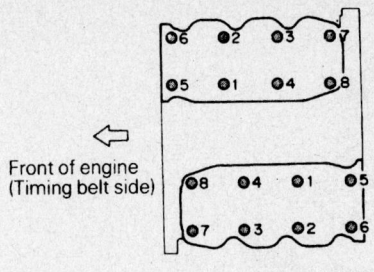

Front of engine
(Timing belt side)

7924UG13

Cylinder head bolt torque sequence—3.5L engine

- Rear timing belt cover
- Ignition coil
- Water hoses from the thermostat housing and the housing
- Water inlet from the front head and discard O-ring
- Water passage

5. Loosen the cylinder head mounting bolts in 3 steps, starting from the outside and working inward. Lift off the cylinder head assembly and remove the head gasket.

To install:

6. Thoroughly clean and dry the mating surfaces of the head and block. Check the cylinder head for cracks, damage or engine coolant leakage. Remove scale, sealing compound and carbon. Clean oil passages thoroughly. Check the head for flatness. End to end, the head should be within 0.0012 in. (0.030mm), normally with 0.008 in. (0.203mm) the maximum allowed out of true. The total thickness allowed to be removed from the head and block is 0.008 in. (0.203mm) maximum.

7. Place a new head gasket on the cylinder block with the identification marks in the front top (upward) position. Do not use sealer on the gasket.

8. Install or connect the following:
- Cylinder head on the block. Be sure the head bolt washers are installed with the chamfered edge upward. Using 3 even steps, torque the head bolts in sequence, to 76–83 ft. lbs. (105–115 Nm).
- New O-ring and the water inlet to the front head
- New gaskets, thermostat housing and connect the hoses
- Water passage and new gaskets
- Ignition coil and center rear timing belt cover
- Intake camshaft sprocket. Use hex flange on camshaft to secure and tighten the retaining bolt to 65 ft. lbs. (90 Nm).
- New gasket and the valve cover. Torque the bolts to 84 inch lbs. (10 Nm).

- Spark plug cables and the center cover
- Oil dipstick
- CMP sensor
- Breather hose
- Radiator and shroud
- Timing belt
- Cooling fan
- Alternator
- Engine under cover
- Intake manifold and new gasket. Torque the nuts to 16 ft. lbs. (21 Nm).
- Air intake plenum and new gaskets. Torque the bolts to 13 ft. lbs. (18 Nm).
- Exhaust manifold and new gaskets. Torque the nuts to 22 ft. lbs. (29 Nm).
- Air intake hoses
- Negative battery cable

9. Change the engine oil and oil filter.
10. Refill the system with coolant.
11. Run the vehicle until the thermostat opens.
12. Once the vehicle has cooled, recheck the coolant level.

3.8L Engine

1. Before servicing the vehicle, refer to the Precautions Section.

2. Relieve fuel system pressure.
3. Drain the cooling system.
4. Remove or disconnect the following:
- Negative battery cable

- Intake manifold
- Timing belt
- Front exhaust pipe
- Water outlet pipe assembly and O-ring
- Heater hose connections
- Water passage assembly and gasket
- Water pipe and O-ring

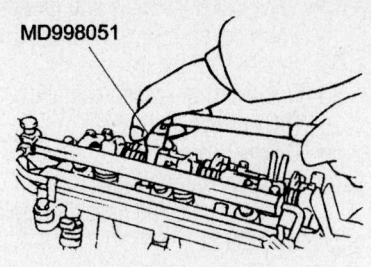

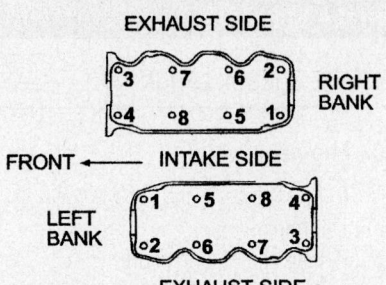

Cylinder head bolt removal sequence—3.8L engine

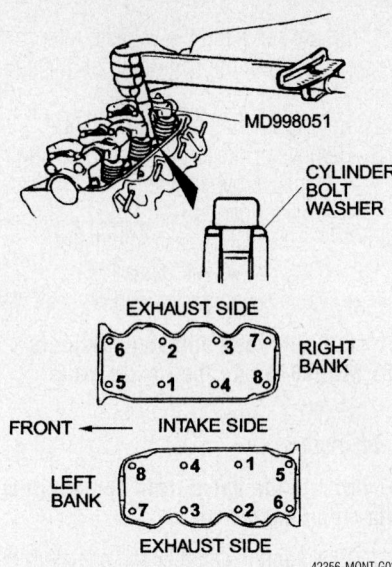

Cylinder head bolt torque sequence—3.8L engine

- Breather hose
- Spark plug wires; tag before disconnecting
- Ignition coil
- Oxygen (O_2) sensor connector
- Engine oil dipstick assembly and O-ring
- Intake manifold plenum stay
- Rocker cover
- Camshaft Position (CMP) sensor connector

5. Loosen the cylinder head mounting bolts in 3 steps, in the sequence shown. Lift off the cylinder head assembly and remove the head gasket.

To install:

6. Thoroughly clean and dry the mating surfaces of the head and block. Check the cylinder head for cracks, damage or engine coolant leakage. Remove scale, sealing compound and carbon. Clean oil passages thoroughly. Check the head for flatness. End to end, the head should be within 0.0012 in. (0.030mm), normally with 0.008 in. (0.203mm) the maximum allowed out of true. The total thickness allowed to be removed from the head and block is 0.008 in. (0.203mm) maximum.

7. Place a new head gasket on the cylinder block with the identification marks in the front top (upward) position. Do not use sealer on the gasket.

8. Install or connect the following:
- Cylinder head on the block. Be sure the head bolt washers are installed with the chamfered edge upward (see illustration). Using 3 even

steps, torque the head bolts in sequence, to 77–83 ft. lbs. (105–113 Nm) with a torque wrench and Special Tool No. MD998501.
- Rocker cover
- Intake manifold plenum stay
- Engine oil dipstick assembly and O-ring
- O_2 sensor connector
- Ignition coil
- Spark plug wires
- Breather hose
- Water pipe and O-ring
- Water passage assembly and gasket
- Heater hose connections
- Water outlet pipe assembly and O-ring
- Front exhaust pipe
- Timing belt

- Intake manifold
- Negative battery cable

9. Change the engine oil and oil filter.
10. Refill the system with coolant.
11. Run the vehicle until the thermostat opens.
12. Once the vehicle has cooled, recheck the coolant level.

Rocker Arms/Shafts

REMOVAL & INSTALLATION

3.0L Engine

1. Before servicing the vehicle, refer to the Precautions Section.
2. Remove or disconnect the following:
- Negative battery cable

✻✻ CAUTION

Work must be started after 90 seconds from the time the ignition switch is turned to the LOCK position and the negative battery cable is disconnected.

- Valve cover
- Auto lash adjuster retainers SST MD998443 on the rocker arms
- Rocker arms, rocker shafts and bearing caps, as an assembly

To install:

3. Inspect the bearing journals on the camshaft and the cylinder head.
4. Lubricate the camshaft journals and camshaft with clean engine oil.
5. Install the rocker arms, rocker arm shaft and the rocker shaft spring as follows:
   a. Temporarily tighten the rocker shaft with the bolts so that the intake valve rocker arms do not push on the valves.
   b. Insert the rocker shaft spring from

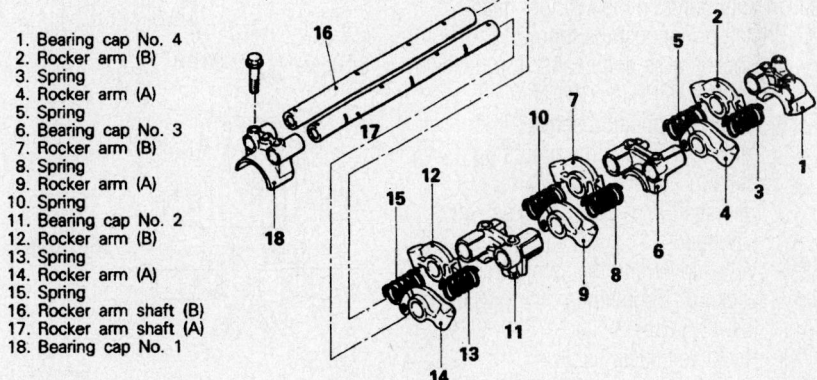

1. Bearing cap No. 4
2. Rocker arm (B)
3. Spring
4. Rocker arm (A)
5. Spring
6. Bearing cap No. 3
7. Rocker arm (B)
8. Spring
9. Rocker arm (A)
10. Spring
11. Bearing cap No. 2
12. Rocker arm (B)
13. Spring
14. Rocker arm (A)
15. Spring
16. Rocker arm shaft (B)
17. Rocker arm shaft (A)
18. Bearing cap No. 1

Rocker arms and related components—3.0L engine

above and mount it at right angles to the plug guide.

c. Before installing the exhaust rocker arms and the rocker arm shaft, mount the rocker shaft spring.

d. Remove tool SST MD998443 used to hold the lash adjuster in position.

e. Check to ensure that the flat side of the rocker shaft is perpendicular to the cylinder head, and facing the valves.

f. Gradually tighten the bearing caps in 2 or 3 steps. In the final step, tighten to 23 ft. lbs. (31 Nm).

6. Install or connect the following:
- Valve cover and new gasket. Torque the bolt to 2–3 ft. lbs. (3–4 Nm).
- Negative battery cable

7. Start the engine and check for leaks and proper operation.

3.5L Engine

1. Before servicing the vehicle, refer to the Precautions Section.
2. Relieve the fuel system pressure.
3. Remove or disconnect the following:
- Negative battery cable
- Valve cover and the semi-circular packing.
- Crankshaft Position (CKP) sensor, matchmark for reassembly
- Camshaft Position (CMP) sensor, if equipped

➡ Install auto lash adjuster retainers SST MD998443 on the rocker arms

- Rocker arms and shafts
- Lash adjusters

4. Check the camshaft journals for wear or damage. Check the cam lobes for damage. Also, check the cylinder head oil holes for clogging.

To install:

➡ Lubricate the valve train components with clean engine oil.

5. Bleed and install the lash adjusters in their original bores in the cylinder head.
6. Install or connect the following:
- Rocker arms and shafts. Torque the bolts to 23 ft. lbs. (31 Nm).
- Camshaft position sensor, if removed. Torque the mounting bolts to 78 inch lbs. (9 Nm).
- Camshaft Position (CMP) sensor, if equipped
- Valve cover and the semi-circular packing. Torque the bolts to 2.5 ft. lbs. (3.5 Nm).
- Negative battery cable

7. Run vehicle and check for leaks.

3.8L Engine

1. Before servicing the vehicle, refer to the precautions in the beginning of this section.
2. Relieve the fuel system pressure.
3. Remove or disconnect the following:
- Negative battery cable
- Oil filler cap
- Positive Crankcase Ventilation (PCV) valve and gasket
- Valve cover, gasket and oil seal(s)

➡ Install auto lash adjuster retainers SST MD998443 on the rocker arms

- Rocker arms and shafts

To install:

➡ Lubricate the valve train components with clean engine oil.

4. Bleed and install the lash adjusters to in their original bores in the cylinder head.
5. Install or connect the following:
- Rocker arms and shafts. Torque the bolts to 21–25 ft. lbs. (28–34 Nm).

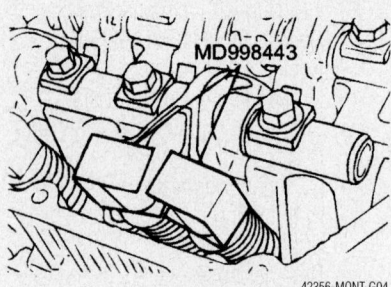

42356-MONT-G04

Install the special tool to prevent the lash adjusters from falling to the floor during rocker arm removal—3.8L engine

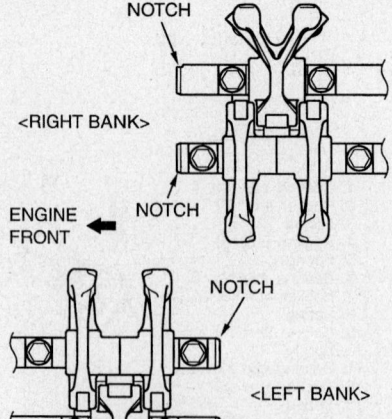

09482_MONT_G0008

Rocker arm installation identification points

- Valve cover oil seals and gasket
- Valve cover. Torque the bolts to 27–35 inch lbs.
- Positive Crankcase Ventilation (PCV) valve and gasket
- Oil filler cap
- Negative battery cable

6. Start the engine and check for leaks. Correct as required.

Intake Manifold

REMOVAL & INSTALLATION

3.0L Engine

1. Before servicing the vehicle, refer to the Precautions Section.
2. Relieve the fuel pressure.

✳✳ CAUTION

The fuel injection system remains under pressure after the engine has been OFF. Properly relieve fuel pressure before disconnecting any fuel lines. Failure to do so may result in fire or personal injury.

3. Drain the engine coolant.
4. Remove or disconnect the following:
- Negative battery cable

✳✳ CAUTION

Work must be started after 90 seconds from the time the ignition switch is turned to the LOCK position and the negative battery cable is disconnected.

- Air intake hose from the throttle body
- Positive Crankcase Ventilation (PCV) hose
- Exhaust Gas Recirculation (EGR) valve
- Manifold Differential Pressure (MDP) sensor
- Vacuum hoses from the throttle body and air intake plenum
- Accelerator cable and the throttle control cable
- Coolant hoses
- Engine oil filler neck bracket from the air intake plenum
- EGR tube from the air intake plenum
- Plenum brackets
- Air intake plenum assembly from the intake manifold and remove. Note the position of the mounting bolts as they are removed.

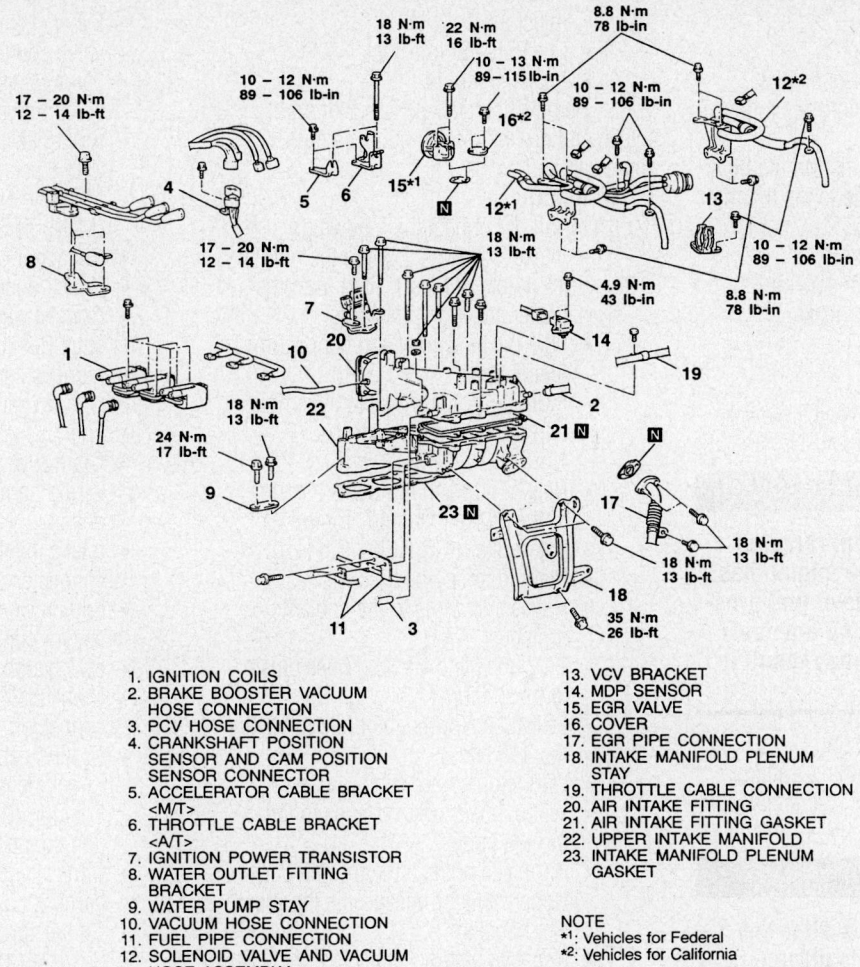

1. IGNITION COILS
2. BRAKE BOOSTER VACUUM HOSE CONNECTION
3. PCV HOSE CONNECTION
4. CRANKSHAFT POSITION SENSOR AND CAM POSITION SENSOR CONNECTOR
5. ACCELERATOR CABLE BRACKET <M/T>
6. THROTTLE CABLE BRACKET <A/T>
7. IGNITION POWER TRANSISTOR
8. WATER OUTLET FITTING BRACKET
9. WATER PUMP STAY
10. VACUUM HOSE CONNECTION
11. FUEL PIPE CONNECTION
12. SOLENOID VALVE AND VACUUM HOSE ASSEMBLY

13. VCV BRACKET
14. MDP SENSOR
15. EGR VALVE
16. COVER
17. EGR PIPE CONNECTION
18. INTAKE MANIFOLD PLENUM STAY
19. THROTTLE CABLE CONNECTION
20. AIR INTAKE FITTING
21. AIR INTAKE FITTING GASKET
22. UPPER INTAKE MANIFOLD
23. INTAKE MANIFOLD PLENUM GASKET

NOTE
*1: Vehicles for Federal
*2: Vehicles for California

7924UG41

Upper intake manifold and related components—3.0L engine

- Fuel hose from the fuel rail
- Fuel return line and vacuum hose from the fuel pressure regulator
- Electrical connectors from the injectors
- Fuel rail and injectors
- Intake manifold

5. Remove the gaskets and thoroughly clean and dry the mating surfaces of the manifold and heads.

To install:

6. Install or connect the following:
- Intake manifold. Torque the nuts to 16 ft. lbs. (21 Nm) start from the center and working outward.

7. Connect the hoses and connect the wires to the coolant switches.

- Fuel rail assembly and connect the fuel hoses
- New gasket and the air intake plenum to the intake manifold. Torque the nuts/bolts to 13 ft. lbs. (17 Nm).
- Plenum brackets

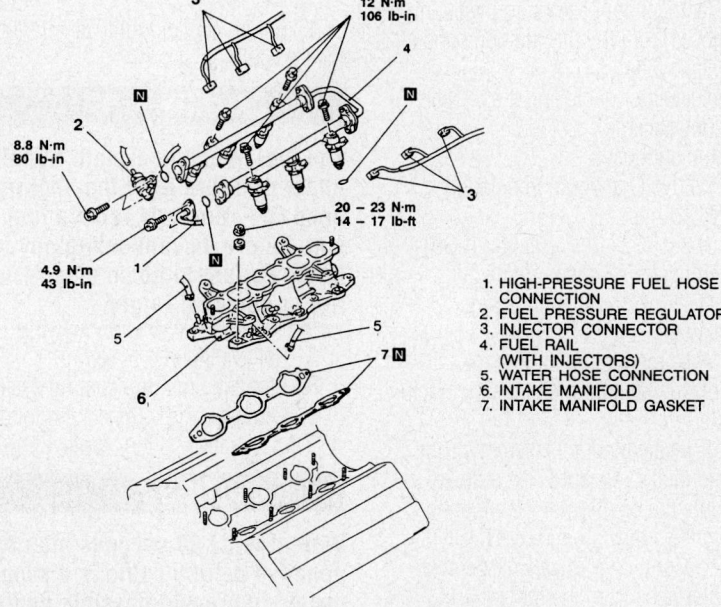

1. HIGH-PRESSURE FUEL HOSE CONNECTION
2. FUEL PRESSURE REGULATOR
3. INJECTOR CONNECTOR
4. FUEL RAIL (WITH INJECTORS)
5. WATER HOSE CONNECTION
6. INTAKE MANIFOLD
7. INTAKE MANIFOLD GASKET

7924UG42

Lower intake manifold and related components—3.0L engine

- PCV hose and vacuum hose cluster to the plenum
- EGR tube
- EGR temperature sensor wire
- Wires, hoses and linkages to the throttle body
- Air intake hose to the throttle body
- Upper radiator hose to the thermostat housing
- Negative battery cable

8. Refill the radiator with coolant.
9. Check fuel system for leaks.

3.5L Engine

1. Before servicing the vehicle, refer to the Precautions Section.

✳✳ CAUTION

The fuel injection system remains under pressure after the engine has been OFF. Properly relieve fuel pressure before disconnecting any fuel lines. Failure to do so may result in fire or personal injury.

2. Relieve the fuel pressure.
3. Partially drain the cooling system.
4. Remove or disconnect the following:
   - Negative battery cable

✳✳ CAUTION

Wait at least 90 seconds after the negative battery cable is disconnected to prevent possible deployment of the air bag.

- Air intake hose from the throttle body
- Electrical connectors and vacuum hoses from the throttle body and air intake plenum
- Accelerator cable and the throttle control cable
- Coolant hoses
- Positive Crankcase Ventilation (PCV) hose
- Exhaust Gas Recirculation (EGR) temperature sensor connector
- EGR tube from the air intake plenum
- Intake manifold plenum cover
- Intake manifold plenum stay brackets
- Air intake plenum assembly from the intake manifold and remove. Note the position of the mounting bolts as they are removed
- Induction control valve assembly
- Fuel hose from the fuel rail
- Fuel return line and vacuum hose from the fuel pressure regulator

- Electrical connectors from the injectors
- Fuel rail and injectors
- Intake manifold

5. Remove the gaskets and thoroughly clean and dry the mating surfaces of the manifold and heads.

To install:

6. Install or connect the following:
   - Intake manifold. Tighten the nuts to 16 ft. lbs. (21 Nm). Start from the center and work outward.
   - Fuel rail assembly and connect the fuel hoses
   - Induction control valve assembly and tighten to 72 inch lbs. (9 Nm).
   - New gasket and air intake plenum to the intake manifold. Torque the nuts/bolts to 13 ft. lbs. (18 Nm).
   - Plenum to engine brackets
   - Hoses and wires to the coolant switches
   - PCV hose and vacuum hose cluster to the plenum
   - EGR tube. Torque the bolts to 13 ft. lbs. (18 Nm).
   - EGR temperature sensor wire
   - Wires, hoses and linkages to the throttle body
   - Air intake hose to the throttle body
   - Upper radiator hose to the thermostat housing
   - Negative battery cable

7. Refill the radiator with coolant.
8. Check the system for fuel leaks.
9. Set all adjustments to specifications.

3.8L Engine

1. Before servicing the vehicle, refer to the Precautions Section.

✳✳ CAUTION

The fuel injection system remains under pressure after the engine has been OFF. Properly relieve fuel pressure before disconnecting any fuel lines. Failure to do so may result in fire or personal injury.

2. Relieve the fuel pressure.
3. Partially drain the cooling system.
4. Remove or disconnect the following:
   - Negative battery cable

✳✳ CAUTION

Wait at least 90 seconds after the negative battery cable is disconnected to prevent possible deployment of the air bag.

- Throttle body
- Exhaust Gas Recirculation (EGR) valve connector
- Evaporative emission (EVAP) purge solenoid valve connector
- Right bank Heated Oxygen Sensor (HO2S) connector connection
- Manifold Differential Pressure (MDP) sensor connector
- Capacitor connector
- Knock Sensor (KS) connector
- Control wiring harness and Camshaft Position (CMP) sensor wiring harness combination connector
- Ground cable
- Fuel injector connector
- Control wiring harness and injector wiring harness combination connector
- Intake manifold tuning solenoid connector
- Engine Coolant Temperature (ECT) sensor connector
- ECT gauge unit connector
- Crankshaft Position (CKP) sensor connector
- Ground cable
- Knock Sensor (KS) and Camshaft Position (CMP) sensor combination connector
- Control wiring harness and injector harness combination connector
- Connector bracket
- Positive Crankcase Ventilation (PCV) hose connection
- Fuel pipe
- Vacuum hose connection
- Water outlet fitting bracket
- EGR pipe
- EGR pipe and gasket
- Intake manifold plenum stay
- Right bank HO2S connector
- Fuel pipe clip
- Intake manifold plenum
- Intake manifold plenum gasket
- Manifold Differential Pressure (MDP) sensor and O-ring
- Solenoid valve and vacuum hose assembly
- Capacitor
- EGR valve and gasket
- Purge hose
- EVAP purge solenoid valve
- Intake manifold tuning valve assembly
- Intake manifold tuning valve gaskets P and S
- High pressure fuel hose connection and O-ring
- Fuel pressure regulator and O-ring
- Fuel injector connectors
- Fuel rail (with injectors attached)

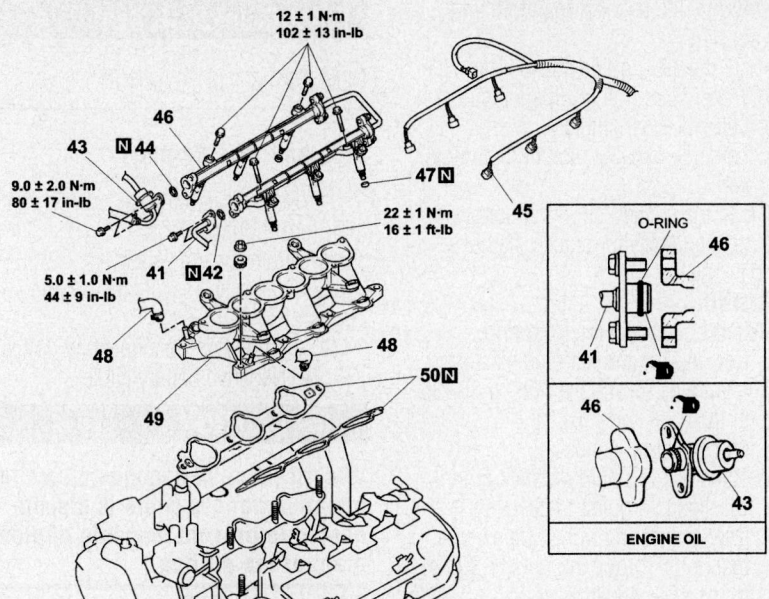

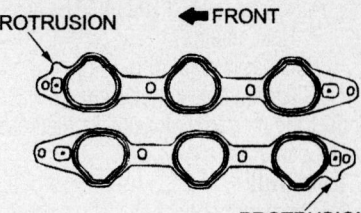

Intake manifold gasket positioning—3.8L engine

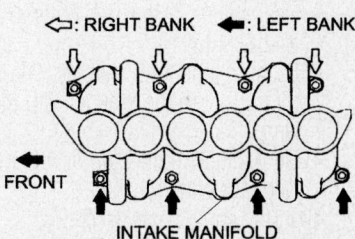

Intake manifold retaining nut locations—3.8L engine

Proper installation of the intake manifold plenum gasket—3.8L engine

41. FUEL HIGH-PRESSURE HOSE CONNECTION
42. O-RING
43. FUEL PRESSURE REGULATOR
44. O-RING
45. INJECTOR CONNECTOR
46. FUEL RAIL (WITH INJECTORS)
47. INSULATORS
48. WATER HOSE CONNECTION
49. INTAKE MANIFOLD
50. INTAKE MANIFOLD GASKET

42356-MONT-G06

Intake manifold and related components—3.8L engine

- Insulators
- Water hose connection
- Intake manifold retainers, manifold and gasket. Thoroughly clean and dry the mating surfaces of the manifold and heads.

To install:

5. Install or connect the following:
   - New intake manifold gasket. Make sure the gaskets are installed with the protrusions as shown in the illustration.
   - Intake manifold
6. Install the intake manifold nuts and torque to specification.
   - Water hose connection
   - Insulators
   - Fuel rail and injector assembly
   - Fuel injector connectors
   - Fuel pressure regulator and O-ring
   - High pressure fuel hose connection and O-ring
   - Intake manifold tuning valve gaskets P and S
   - Intake manifold tuning valve assembly
   - EVAP purge solenoid valve
   - Purge hose
   - EGR valve and gasket
   - Capacitor
   - Solenoid valve and vacuum hose assembly
   - Manifold Differential Pressure (MDP) sensor and O-ring
   - Intake manifold plenum gasket
   - Intake manifold plenum
   - Fuel pipe clip
   - Right bank HO2S connector
   - Intake manifold plenum stay
   - EGR pipe and gasket
   - EGR pipe
   - Water outlet fitting bracket
   - Vacuum hose connection
   - Fuel pipe
   - PCV hose connection
   - Connector bracket
   - Control wiring harness and injector harness combination connector
   - KS and CMP sensor combination connector
   - Ground cable
   - CKP sensor connector
   - ECT gauge unit connector
   - ECT sensor connector
   - Intake manifold tuning solenoid connector
   - Control wiring harness and injector wiring harness combination connector
   - Fuel injector connector
   - Ground cable
   - Control wiring harness and CMP sensor wiring harness combination connector
   - KS connector
   - Capacitor connector
   - MDP sensor connector
   - Right bank HO2S connector
   - EVAP purge solenoid valve connector
   - EGR valve connector
   - Throttle body
   - Negative battery cable
7. Refill the radiator with coolant.
8. Check the system for fuel leaks.
9. Set all adjustments to specifications.

Exhaust Manifold

REMOVAL & INSTALLATION

3.0L and 3.5L Engines

1. Remove or disconnect the following:
   - Negative battery cable
   - Exhaust pipe from the exhaust manifolds

- Oil dipstick, guide and O-ring
- Heat shields
- Exhaust manifolds

2. Clean the gasket mounting surfaces. Inspect the manifolds for cracks, flatness and/or damage.

To install:

3. Install or connect the following:
- New gasket and exhaust manifold. Torque the nuts to 33 ft. lbs. (44 Nm) on 3.0L engines and 22 ft. lbs. (29 Nm) on 3.5L engines.
- Heat shield, Torque the bolts to 10 ft. lbs. (14 Nm).
- Exhaust pipe to the exhaust manifolds. Torque the nuts to 35ft. (49 Nm).
- Oil dipstick, guide and new O-ring
- Negative battery cable

4. Start the engine and check for exhaust leaks.

3.8L Engine

1. Remove or disconnect the following:
- Negative battery cable
- Front exhaust pipe from the exhaust manifolds
- Air cleaner assembly
- Battery and battery tray
- Exhaust Gas Recirculation (EGR) pipe and gasket
- Right side heat shield

- Right side exhaust manifold and gasket
- Oil dipstick, guide and O-ring
- Transmission fluid dipstick guide
- Left side heat shield
- Left side exhaust manifold and gasket

2. Clean the gasket mounting surfaces. Inspect the manifolds for cracks, flatness and/or damage.

To install:

3. Install or connect the following:
- New gasket and left side exhaust manifold. Torque the nuts to 30–36 ft. lbs. (39–49 Nm).
- Left side heat shield
- Transmission fluid dipstick guide
- Oil dipstick, guide and O-ring
- New gasket and right side exhaust manifold. Torque the nuts to 30–36 ft. lbs. (39–49 Nm).
- Right side heat shield
- EGR gasket and pipe
- Battery tray and battery
- Air cleaner assembly
- Front exhaust pipe to the exhaust manifolds. Torque the nuts to 35ft. (49 Nm).
- Oil dipstick, guide and new O-ring
- Negative battery cable

4. Start the engine and check for exhaust leaks.

Front Crankshaft Seal

REMOVAL & INSTALLATION

3.0L and 3.5L Engines

1. Before servicing the vehicle, refer to the Precautions Section.
2. Drain the crankcase.
3. Drain and recycle the engine coolant.
4. Remove or disconnect the following:
- Negative battery cable

☀☀ CAUTION

Wait at least 90 seconds after the negative battery cable is disconnected to prevent possible deployment of the air bag.

- Cooling fan
- Accessory drive belts
- Alternator
- Engine undercover, if equipped
- Power steering oil pump assembly
- Air conditioner compressor and bracket, if equipped
- Timing indicator bracket
- Accessory mount assembly
- Crankshaft pulley
- Timing belt covers and the timing belt
- Crankshaft sprocket

5. Cut out a portion in the crankshaft oil seal lip and pry out the oil seal with a flat prying tool, being careful not to damage the crankshaft.

To install:

6. Coat the lip of the new seal with oil and install the seal using the proper seal driver.

7. Install or connect the following:
- Crankshaft sprocket and the timing belt
- Timing belt covers
- Crankshaft pulley. Torque the bolt to 134 ft. lbs. (181Nm).
- Accessory mount Assembly. Torque the bolts to 33 ft. lbs. (44 Nm).
- Timing indicator bracket. Torque the bolts to 97 inch lbs. (11 Nm).
- Air conditioner compressor and bracket, if equipped
- Power steering oil pump assembly
- Engine undercover, if equipped
- Alternator
- Accessory drive belts
- Cooling fan
- Negative battery cable

8. Refill the crankcase.
9. Refill the cooling system.

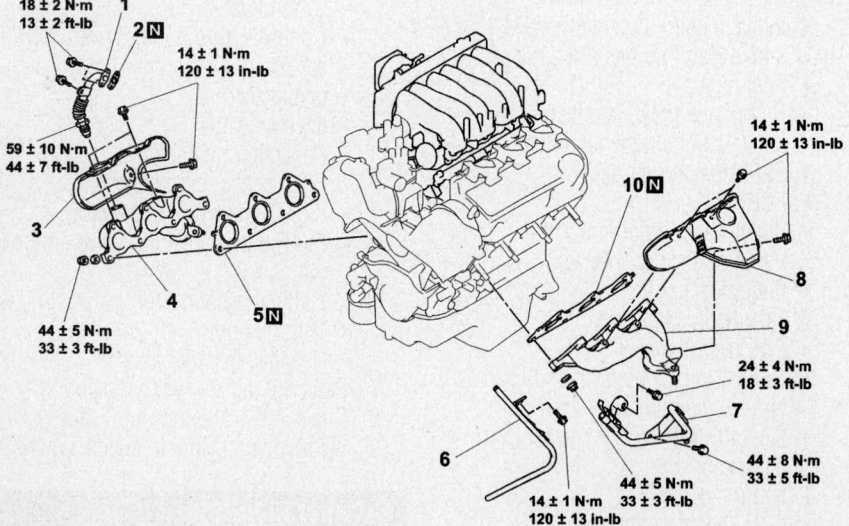

18 ± 2 N·m
13 ± 2 ft-lb   1
2 N

14 ± 1 N·m
120 ± 13 in-lb

59 ± 10 N·m
44 ± 7 ft-lb

3

4   5 N

44 ± 5 N·m
33 ± 3 ft-lb

14 ± 1 N·m
120 ± 13 in-lb

10 N

8

9

24 ± 4 N·m
18 ± 3 ft-lb

7

44 ± 8 N·m
33 ± 5 ft-lb

6

44 ± 5 N·m
33 ± 3 ft-lb
14 ± 1 N·m
120 ± 13 in-lb

1. EGR PIPE
2. EGR PIPE GASKET
3. HEAT PROTECTOR <RH>
4. EXHAUST MANIFOLD <RH>
5. EXHAUST MANIFOLD GASKET <RH>
6. ENGINE OIL DIPSTICK GUIDE
7. TRANSMISSION FLUID DIPSTICK GUIDE
8. HEAT PROTECTOR <LH>
9. EXHAUST MANIFOLD <LH>
10. EXHAUST MANIFOLD GASKET <LH>

42356-MONT-G10

Exhaust manifolds and related components—3.8L engine

10. Start the engine and check for proper operation.

3.8L Engine

1. Before servicing the vehicle, refer to the Precautions Section.
2. Drain the crankcase.
3. Drain the engine coolant.
4. Remove or disconnect the following:
  • Negative battery cable

✳✳ CAUTION

Wait at least 90 seconds after the negative battery cable is disconnected to prevent possible deployment of the air bag.

  • Timing belt
  • Crankshaft sprocket
  • Crankshaft Position (CKP) sensor
  • Crankshaft sensing blade
  • Crankshaft spacer and key
  • Front oil seal

To install:
  • Front oil seal. Apply oil to the seal, and install using Crankshaft Front Oil Seal Installer tool no. MD998717.
  • Crankshaft key and spacer.
  • Crankshaft sensing blade
  • CKP sensor

➡**To be sure the crankshaft pulley bolt does not loosen, make sure the clean the mating areas of the crankshaft, spacer, sensing blade and sprocket.**

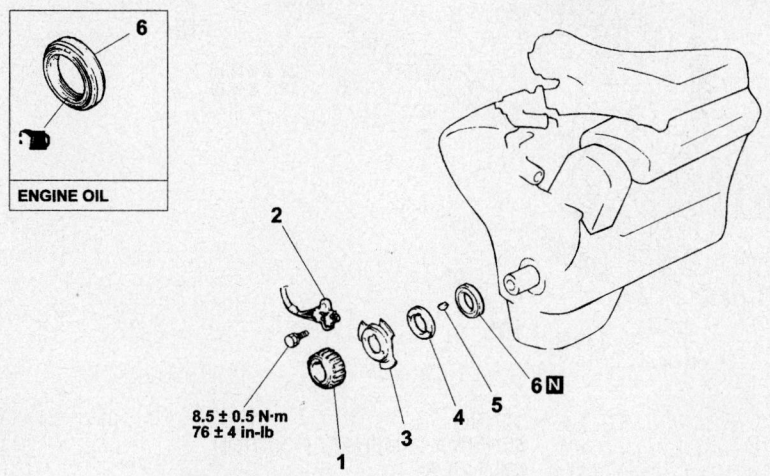

8.5 ± 0.5 N·m
76 ± 4 in-lb

1. CRANKSHAFT SPROCKET	4. CRANKSHAFT SPACER
2. CRANKSHAFT POSITION SENSOR	5. KEY
3. CRANKSHAFT SENSING BLADE	6. CRANKSHAFT FRONT OIL SEAL

42356-MONT-G11

Crankshaft front oil seal and related components—3.8L engine

  • Crankshaft sprocket
  • Timing belt

Camshaft and Valve Lifters

REMOVAL & INSTALLATION

3.0L and 3.5L Engines

1. Before servicing the vehicle, refer to the Precautions Section.
2. Relieve the fuel system pressure.
3. Drain the engine oil and coolant.
4. Remove or disconnect the following:
  • Negative battery cable

✳✳ CAUTION

Work must be started after 90 seconds from the time the ignition switch is turned to the LOCK position and the negative battery cable is disconnected.

  • Intake manifold plenum
  • Valve cover
  • Timing belt
  • Sprocket from the camshaft
5. Install auto lash adjuster retainers SST MD998443 on the rocker arms.
6. Remove or disconnect the following:
  • Distributor and the distributor extension, if equipped
  • Rocker arms, rocker shafts and bearing caps, as an assembly
  • Thrust cage and O-ring
  • Camshaft from the cylinder head

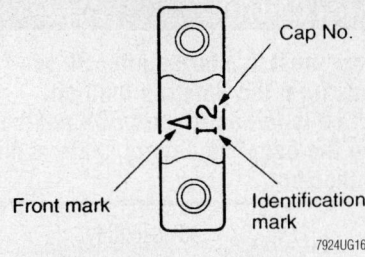

Camshaft bearing cap identification marks—3.5L engine

7. Inspect the bearing journals on the camshaft and the cylinder head.
To install:
8. Lubricate the camshaft journals and camshaft with clean engine oil
9. Install or connect the following:
  • Camshaft in the cylinder head.
  • Thrust cage and O-ring. Torque the bolts to 109 inch lbs. (12 Nm).
  • Rocker arms, rocker arm shaft and the rocker shaft spring.
10. Temporarily tighten the rocker shaft with the bolts positioned so that the intake valve rocker arms do not push the valves.
11. Install or connect the following:
  • Rocker shaft spring from above and mount it at right angles to the plug guide.
12. Before installing the exhaust rocker arms and the rocker arm shaft, mount the rocker shaft spring.
13. Remove the SST used to hold the lash adjuster in position.
14. Check to ensure that the flat side of the rocker shaft is perpendicular to the cylinder head, and facing the valves.
15. Gradually tighten the bearing caps in 2 or 3 steps. In the final step tighten to 23 ft. lbs. (31 Nm).
16. Install or connect the following:
  • Distributor, if removed
  • Sprockets. Torque the bolts to 65 ft. lbs. (88 Nm).
  • Timing belt and timing belt cover
  • Valve cover. Torque the bolts to 26 inch lbs. (3.4 Nm).
  • Intake manifold plenum
  • Negative battery cable
17. Start the engine and check for leaks and proper operation.
18. Refill the coolant and crankcase.

3.8L Engine

1. Before servicing the vehicle, refer to the Precautions Section.
2. Relieve the fuel system pressure.
3. Remove or disconnect the following:
  • Negative battery cable

✳✳ CAUTION

Work must be started after 90 seconds from the time the ignition switch is turned to the LOCK position and the negative battery cable is disconnected.

- Cylinder head assembly
- Camshaft sprocket, using special tools M998715 and MB990767

➡**Install auto lash adjuster retainers SST MD998443 on the rocker arms**

- Intake rocker arm, shaft and lash adjuster assembly. Loosen the rocker arm assembly mounting bolt and remove the assembly with the bolt still attached.
- Exhaust rocker arm, shaft and lash adjuster assembly. Loosen the rocker arm assembly mounting bolt and remove the assembly with the bolt still attached.
- Camshaft Position (CMP) sensor support and O-ring
- Sensing CMP cylinder
- Camshaft

To install:

- Camshaft
- Sensing CMP cylinder
- CMP sensor support and O-ring
- Exhaust rocker arm, shaft and lash adjuster assembly
- Intake rocker arm, shaft and lash adjuster assembly
- Camshaft sprocket
- Cylinder head assembly
- Negative battery cable

4. Run vehicle and check for leaks.

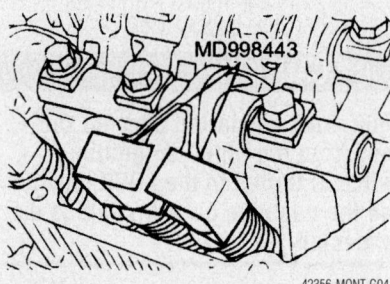

42356-MONT-G04

Install the special tool to prevent the lash adjusters from falling to the floor during rocker arm removal—3.8L engine

INSPECTION

1. Before servicing the vehicle, refer to the Precautions Section.
2. Remove the camshaft from the engine.
3. Check the camshaft bearing journals for damage and binding.
4. If the journals are binding, check the cylinder head for damage.
5. Check the cylinder head for clogged oil holes.
6. As required, check the tooth surface of the distributor drive gear teeth of the camshaft. Replace the camshaft if wear is evident.
7. Check the camshaft surface for abnormal wear and damage. Replace the camshaft, as required.
8. Measure the cam height and replace the camshaft if not within specification.
9. If equipped with a 3.0L engine, the intake camshaft standard specification is 1.485 inches. Minimum limit is 1.465 inches. The exhaust camshaft standard specification is 1.462 inches. Minimum limit is 1.443.

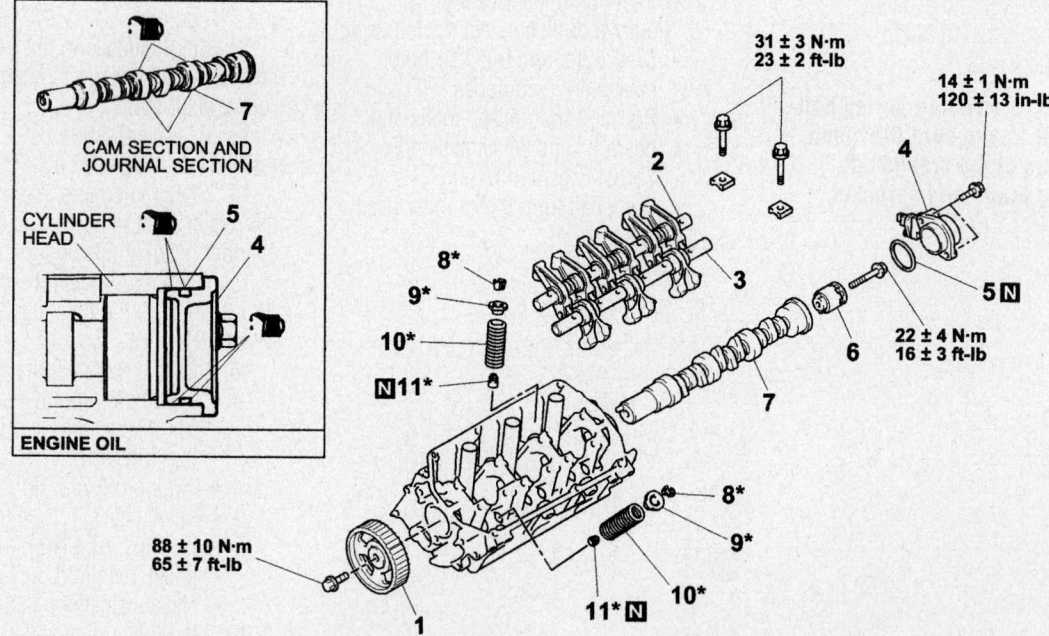

1. CAMSHAFT SPROCKET
2. ROCKER ARM, SHAFT AND LASH ADJUSTER ASSEMBLY (INTAKE SIDE)
3. ROCKER ARM, SHAFT AND LASH ADJUSTER ASSEMBLY (EXHAUST SIDE)
4. CAMSHAFT POSITION SENSOR SUPPORT

5. O-RING
6. SENSING CAMSHAFT POSITION CYLINDER
7. CAMSHAFT
8. VALVE SPRING RETAINER LOCKS
9. VALVE SPRING RETAINERS
10. VALVE SPRINGS
11. VALVE STEM SEALS

42356-MONT-G12

Camshaft and related components—3.8L engine

10. If equipped with a 3.5L engine, the intake camshaft standard specification is 1.472 inches. Minimum limit is 1.452 inches. The exhaust camshaft standard specification is 1.452 inches. Minimum limit is 1.443 inches.

11. If equipped with a 3.8L engine, the intake camshaft standard specification is 1.485 inches. Minimum limit is 1.462 inches. The exhaust camshaft standard specification is 1.465 inches. Minimum limit is 1.443 inches.

Valve Lash

ADJUSTMENT

1. Before servicing the vehicle, refer to the Precautions Section.

2. Check the engine oil. Change or adjust level as required.

3. Idle the engine for one to three minutes, to allow it to warm up.

4. Repeat the operation as indicated in the illustration. At a no load condition check for abnormal noise.

➡**Usually the abnormal noise is eliminated after repetition of the operation ten to thirty times. If no change is observed, suspect that the abnormal noise is due to other factors.**

5. After elimination of the abnormal noise, repeat the operation (in the illustration) five more times.

6. Run the engine at idle for one to three minutes to be sure the abnormal noise has been eliminated.

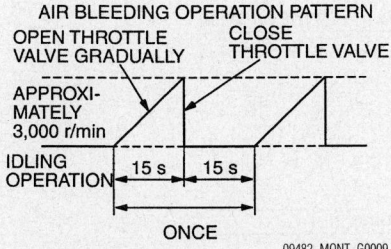

Air bleeding operation pattern

Oil Pan

REMOVAL & INSTALLATION

3.0L Engine

1. Before servicing the vehicle, refer to the Precautions Section.

2. Drain the engine oil.

3. Remove or disconnect the following:

- Negative battery
- Engine under cover
- Alternator and belt
- Stabilizer bar
- Front exhaust pipe
- Actuator assembly and heat protector
- Oil dipstick
- Crossmember assembly
- Automatic transmission oil dipstick assembly
- Exhaust pipe support bracket
- Transmission stay
- Oil pan, lower
- Oil screen and baffle plate
- Oil pan upper

To install:

4. Before installing, thoroughly clean the oil pan and cylinder block mating surfaces.

5. Apply liquid gasket around the surface of the oil pan.

➡**Assemble the oil pan to the cylinder block within 15 minutes after applying the liquid gasket.**

6. Install or connect the following:

- Oil pan upper. Torque the bolts to 53 inch lbs. (6.0 Nm).
- Oil screen and baffle plate. Torque the bolts to 14 ft. lbs. (19 Nm).
- Oil pan, lower. Torque the bolts to 53 inch lbs. (6.0 Nm).
- Transmission stay. Torque the bolts to 26 ft. lbs. (35 Nm).
- Exhaust pipe support bracket. Torque the bolts to 35 ft. lbs. (49 Nm).
- Automatic transmission oil dipstick assembly. Torque the bolts to 33 ft. lbs. (44 Nm).
- Crossmember assembly. Torque the bolts to 80 ft. lbs. (108 Nm).
- Oil dipstick. Torque the bolts to 35 ft. lbs. (48 Nm).
- Actuator assembly and heat protector
- Front exhaust pipe
- Stabilizer bar
- Alternator and belt
- Engine under cover
- Negative battery

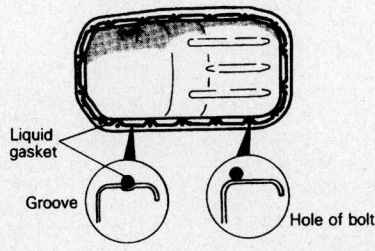

Apply a bead of sealant around the oil pan flange as shown—all engines are similar

3.5L Engine

1. Before servicing the vehicle, refer to the Precautions Section.

2. Drain the engine oil.

3. Remove or disconnect the following:

- Negative battery cable
- Skid plate and the engine undercover
- Front exhaust pipe, if necessary
- Catalytic converter
- Lower oil pan
- Front differential carrier
- Cover
- Oil dipstick
- Oil pan upper
- Oil screen

To install:

4. Before installing, thoroughly clean the oil pan and cylinder block mating surfaces.

5. Apply liquid gasket around the surface of the oil pan.

➡**Assemble the oil pan to the cylinder block within 15 minutes after applying the liquid gasket.**

6. Install or connect the following:

- Oil screen. Torque the bolts to 13 ft. lbs. (19 Nm).
- Oil pan upper. Torque the bolts to 48 inch lbs. (6 Nm).
- Oil dipstick. Torque the bolts to 39 inch lbs. (4.8 Nm).
- Cover. Torque the bolts to 84–108 inch lbs. (10–12 Nm).
- Front differential carrier
- Lower oil pan. Torque the bolts to 84–108 inch lbs. (10–12 Nm).
- Catalytic converter
- Front exhaust pipe, if necessary. Torque the bolts to 35 ft. lbs. (49 Nm).
- Skid plate and the engine undercover
- Negative battery cable

3.8L Engine

1. Before servicing the vehicle, refer to the Precautions Section.

2. Drain the engine oil. Drain the transmission fluid.

3. Remove or disconnect the following:

- Negative battery cable

✳✳ CAUTION

Work must be started after 90 seconds from the time the ignition switch is turned to the LOCK position and the negative battery cable is disconnected.

- Skid plate and the engine under-cover
- Starter motor
- Right and left halfshaft connections
- Front differential No. 2 crossmember
- Drain plug, gasket and cover
- Transmission fluid dipstick assembly and O-ring
- Engine oil dipstick and O-ring
- Oil pan retainers and oil pan. Screw

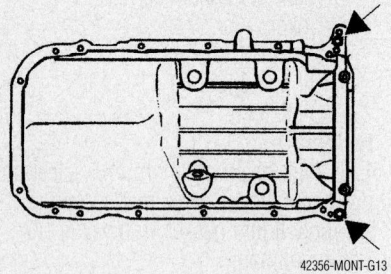

Screw the M10 bolts that hold the oil pan to the transmission assembly into the bolt holes shown by arrows—3.8L engine

the M10 bolts holding the oil pan to the transmission assembly into the bolt holes shown in the illustration to remove the pan.

- Oil screen
- Oil pan gasket

To install:

4. Before installing, thoroughly clean the oil pan and cylinder block mating surfaces.

5. Apply liquid gasket around the surface of the oil pan.

➥**Assemble the oil pan to the cylinder block within 30 minutes after applying the liquid gasket.**

6. Install or connect the following:
- Oil screen. Torque the bolts to 12–16 ft. lbs. (16–22 Nm).
- Oil pan. Torque the bolts, in sequence, to specifications shown the illustration. The bolt holes for bolts 13 and 14 are cut away on the transmission side. Be sure you do not insert the bolts at an angle.
- Engine oil dipstick and O-ring

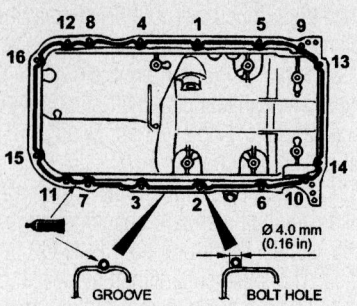

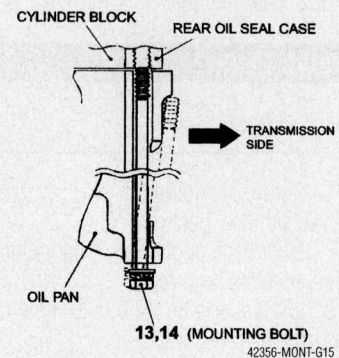

Oil pan bolt torque sequence—3.8L engine

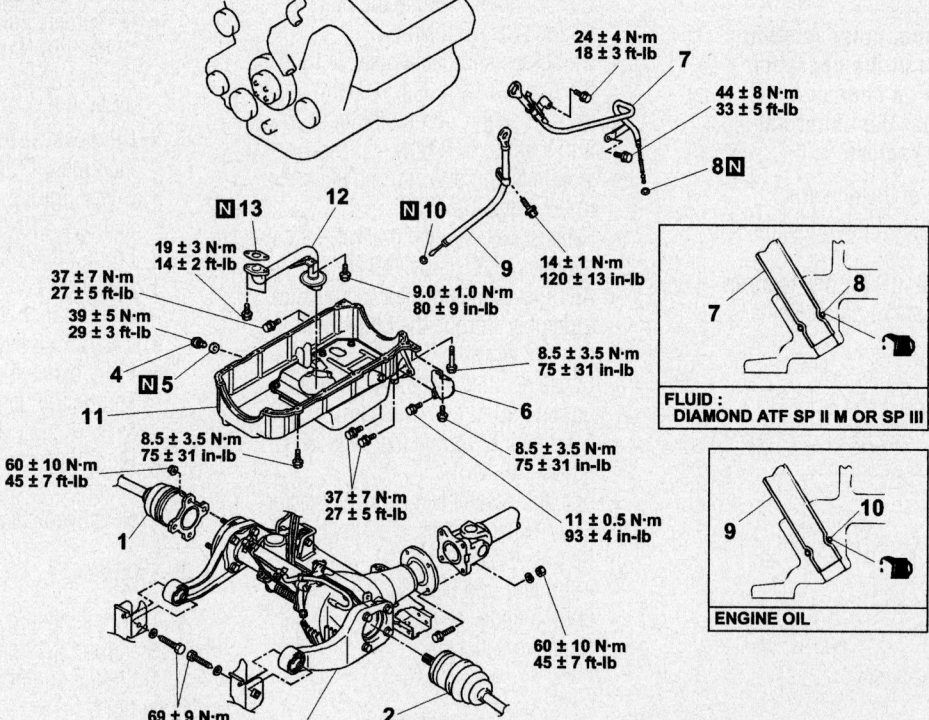

1. DRIVE SHAFT (RH) CONNECTION
2. DRIVE SHAFT (LH) CONNECTION
3. FRONT DIFFERENTIAL NUMBER 2 CROSSMEMBER ASSEMBLY
4. DRAIN PLUG
5. DRAIN PLUG GASKET
6. COVER
7. TRANSMISSION FLUID DIPSTICK ASSEMBLY
8. O-RING
9. ENGINE OIL DIPSTICK ASSEMBLY
10. O-RING
11. OIL PAN
12. OIL SCREEN
13. GASKET

Oil pan and related components—3.8L engine

- Transmission fluid dipstick assembly and O-ring
- Drain plug, gasket and cover
- Front differential No. 2 crossmember
- Right and left halfshaft connections
- Skid plate and the engine undercover
- Negative battery cable

Oil Pump

REMOVAL & INSTALLATION

1. Before servicing the vehicle, refer to the Precautions Section.
2. Drain the engine oil.
3. Remove or disconnect the following:
   - Negative battery cable

- Timing belt
- Oil pressure switch
- Oil dipstick
- Oil pan(s)
- Oil baffle and screen
- Oil pump mounting bolts and the pump from the front of the engine

➡ **Note the position of each oil pump case retaining bolts to facilitate installation. The bolts are of different length.**

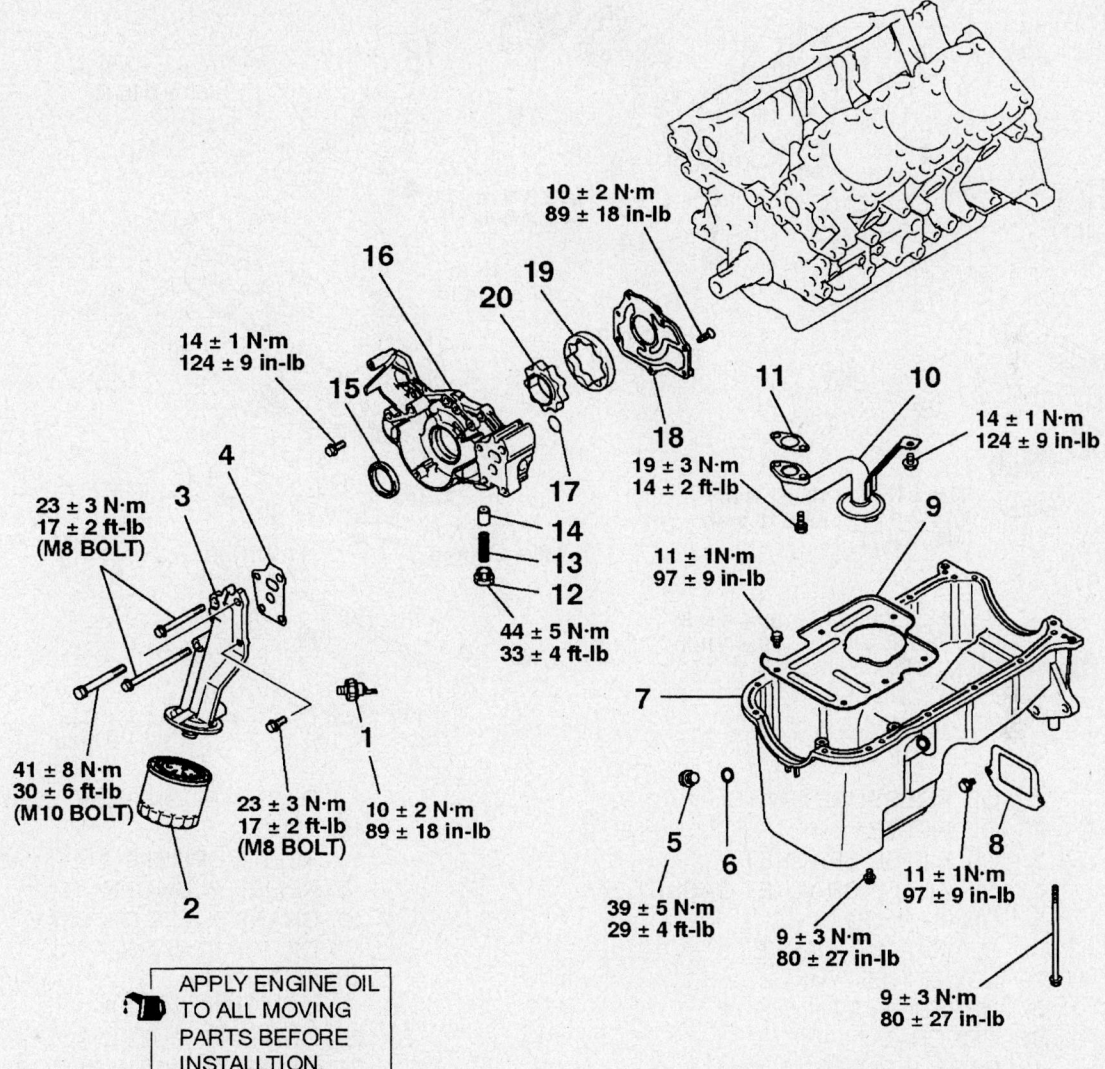

APPLY ENGINE OIL TO ALL MOVING PARTS BEFORE INSTALLTION.

1. OIL PRESSURE SWITCH
2. OIL FILTER
3. OIL FILTER BRACKET
4. OIL FILTER BRACKET GASKET
5. DRAIN PLUG
6. DRAIN PLUG GASKET
7. OIL PAN
8. COVER
9. BAFFLE PLATE
10. OIL SCREEN
11. OIL SCREEN GASKET
12. PLUG
13. RELIEF SPRING
14. RELIEF PLUNGER
15. CRANKSHAFT OIL SEAL
16. OIL PUMP CASE
17. O-RING
18. OIL PUMP COVER
19. OIL PUMP OUTER ROTOR
20. OIL PUMP INNER ROTOR

Oil pump and related components—3.0L engine

09482_MONT_G0016

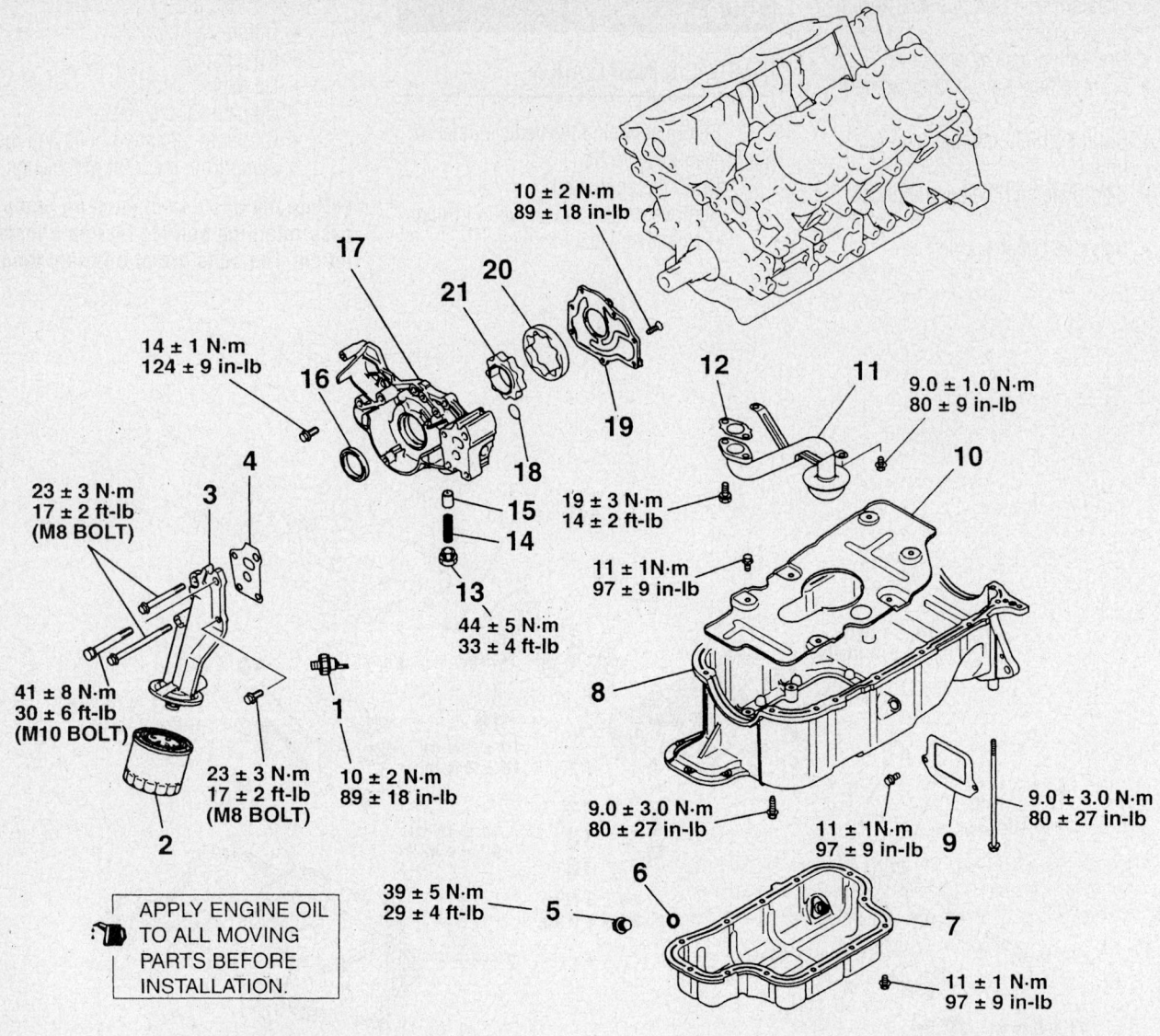

10 ± 2 N·m
89 ± 18 in-lb

14 ± 1 N·m
124 ± 9 in-lb

23 ± 3 N·m
17 ± 2 ft-lb
(M8 BOLT)

41 ± 8 N·m
30 ± 6 ft-lb
(M10 BOLT)

23 ± 3 N·m
17 ± 2 ft-lb
(M8 BOLT)

10 ± 2 N·m
89 ± 18 in-lb

44 ± 5 N·m
33 ± 4 ft-lb

9.0 ± 1.0 N·m
80 ± 9 in-lb

19 ± 3 N·m
14 ± 2 ft-lb

11 ± 1N·m
97 ± 9 in-lb

9.0 ± 3.0 N·m
80 ± 27 in-lb

11 ± 1N·m
97 ± 9 in-lb

9.0 ± 3.0 N·m
80 ± 27 in-lb

39 ± 5 N·m
29 ± 4 ft-lb

11 ± 1 N·m
97 ± 9 in-lb

APPLY ENGINE OIL TO ALL MOVING PARTS BEFORE INSTALLATION.

1. OIL PRESSURE SWITCH
2. OIL FILTER
3. OIL FILTER BRACKET
4. OIL FILTER BRACKET GASKET
5. DRAIN PLUG
6. DRAIN PLUG GASKET
7. LOWER OIL PAN
8. UPPER OIL PAN
9. COVER
10. BAFFLE PLATE
11. OIL SCREEN
12. OIL SCREEN GASKET
13. PLUG
14. RELIEF SPRING
15. RELIEF PLUNGER
16. CRANKSHAFT OIL SEAL
17. OIL PUMP CASE
18. O-RING
19. OIL PUMP COVER
20. OIL PUMP OUTER ROTOR
21. OIL PUMP INNER ROTOR

09482_MONT_G0010

Oil pump and related components—3.5L engine

To install:

4. Clean the gasket mounting surfaces of the pump and engine block.

5. Prime the pump by pouring fresh oil into the inlet and turning the rotors or by packing pump with petroleum jelly. Using a new gasket, install the oil pump on the engine and tighten all bolts to 10 ft. lbs. (14 Nm).

6. Clean out the oil pick-up or replace as required. Replace the oil pick-up gasket ring and install the pick-up to the pump.

7. Install or connect the following:
• Oil filter and the bracket
• Oil baffle and screen
• Oil pan(s)

• Oil pressure switch
• Timing belt
• Dipstick
• Negative battery cable

8. Refill the engine with the proper amount of oil.

9. Start the engine and check for proper oil pressure. Check for leaks.

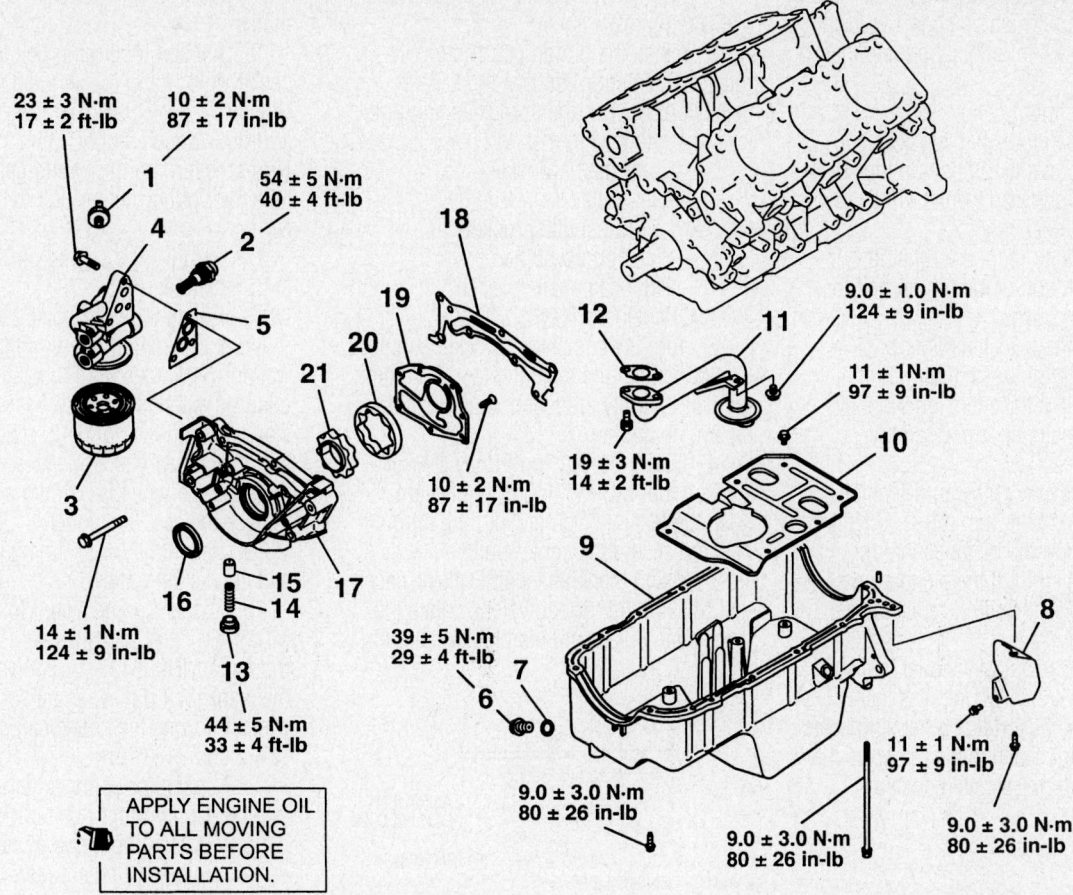

23 ± 3 N·m
17 ± 2 ft-lb

10 ± 2 N·m
87 ± 17 in-lb

54 ± 5 N·m
40 ± 4 ft-lb

9.0 ± 1.0 N·m
124 ± 9 in-lb

11 ± 1N·m
97 ± 9 in-lb

19 ± 3 N·m
14 ± 2 ft-lb

10 ± 2 N·m
87 ± 17 in-lb

14 ± 1 N·m
124 ± 9 in-lb

44 ± 5 N·m
33 ± 4 ft-lb

39 ± 5 N·m
29 ± 4 ft-lb

9.0 ± 3.0 N·m
80 ± 26 in-lb

9.0 ± 3.0 N·m
80 ± 26 in-lb

11 ± 1 N·m
97 ± 9 in-lb

9.0 ± 3.0 N·m
80 ± 26 in-lb

APPLY ENGINE OIL
TO ALL MOVING
PARTS BEFORE
INSTALLATION.

1. ENGINE OIL PRESSURE SWITCH
2. OIL COOLER BY-PASS VALVE
3. OIL FILTER
4. OIL FILTER BRACKET
5. OIL FILTER BRACKET GASKET
6. DRAIN PLUG
7. DRAIN PLUG GASKET
8. COVER
9. OIL PAN
10. BAFFLE PLATE
11. OIL SCREEN

12. OIL SCREEN GASKET
13. RELIEF PLUG
14. RELIEF SPRING
15. RELIEF PLUNGER
16. CRANKSHAFT OIL SEAL
17. OIL PUMP CASE
18. OIL PUMP CASE GASKET
19. OIL PUMP COVER
20. OIL PUMP OUTER ROTOR
21. OIL PUMP INNER ROTOR

09482_MONT_G0019

Oil pump and related components—3.8L engine

Timing Belt

REMOVAL & INSTALLATION

3.0L and 3.5L Engines

1. Before servicing the vehicle, refer to the Precautions Section.
2. Drain the engine coolant.
3. Remove or disconnect the following:
   - Negative battery cable
   - Upper radiator hose
   - Cooling fan shroud assembly
   - Cooling fan-to-clutch bolts and the fan

   - Cooling fan clutch-to-water pump nuts and the clutch assembly
   - Drive belts for the alternator, power steering pump and air conditioning compressor
   - Electrical connectors from the alternator
   - Alternator-to-engine bolts and the alternator bracket-to-engine bolts
   - Alternator and bracket from the engine
   - Power steering pump cover
   - Power steering pump-to-engine bolts and move the pump aside with the hoses and electrical connector attached

   - Air conditioning compressor-to-bracket bolts and move the compressor aside with the lines and electrical connector attached
   - Air conditioning compressor bracket-to-engine bolts and the bracket
   - Timing indicator bracket (near crankshaft pulley) bolts and the bracket
   - Accessory mount assembly-to-engine bolts and the mount assembly
   - Upper timing belt cover assembly
4. Using the End Yoke Holder tool MD990767 and 2 Crankshaft Pulley Holder

Pin tools MD998715, or equivalent to hold the crankshaft pulley, and a socket wrench, remove the crankshaft pulley bolt and the pulley.

- Lower timing belt cover

5. Rotate the crankshaft clockwise to align the timing marks to position the No. 1 cylinder at the Top Dead Center (TDC) of its compression stroke.

6. Use chalk to mark the rotating (clockwise) direction of the timing belt for reinstallation purposes.

7. Loosen the auto-tensioner pulley center bolt and remove the timing bolt.

8. Remove the auto-tensioner pulley and the auto-tensioner arm assembly.

To install:

9. Press the end of the auto-tensioner inward with 72–145 ft. lbs. (98–196 Nm) of force and measure the distance that the pushrod is pushed in. If the standard distance is not 0.04 in. (1mm), replace the auto-tensioner.

10. Position the auto-tensioner in a soft-jawed vise and SLOWLY compress the pushrod until the pushrod and housing holes align; then, install a setting pin to secure the auto-tensioner in the retracted position.

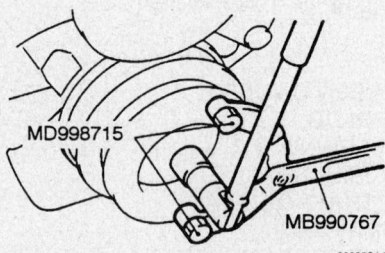

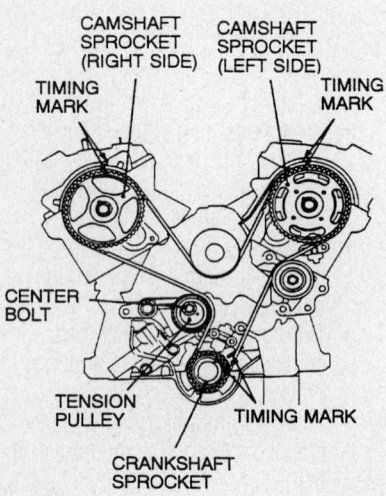

Crankshaft pulley bolt removal—3.0L and 3.5L engines

Timing belt alignment marks—3.0L and 3.5L engines

11. Align the camshaft and crankshaft TDC timing marks.

12. Install the timing belt (noting its rotational direction) so that there is no deflection between the sprockets and pulleys in the following manner:

- Crankshaft sprocket
- Idler pulley
- Left camshaft sprocket
- Water pump pulley
- Right camshaft sprocket
- Tension pulley

13. Turn the camshaft sprocket counter-clockwise until the tension side of the timing belt is firmly stretched, then, recheck the timing marks.

14. Using the Tension Pulley Socket Wrench tool MD998767, or equivalent, push the tensioner pulley into the timing belt and secure the center bolt.

15. Using the Crankshaft Pulley Spacer tool MD998769, or equivalent, rotate the crankshaft ¼ turn counterclockwise, then,

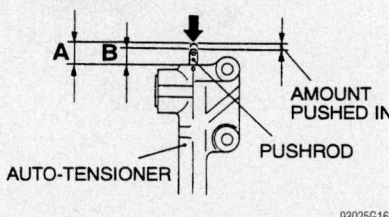

Auto-tensioner movement—3.0L, 3.5L and 3.8L engines

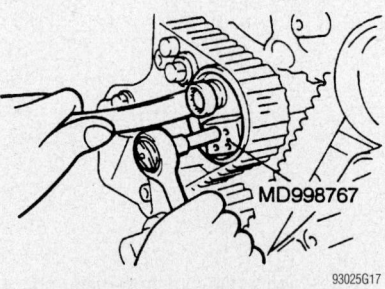

Timing belt tensioner pulley adjustment—3.0L, 3.5L and 3.8L engines

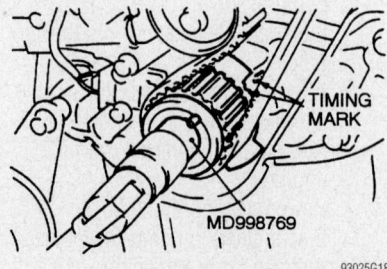

Crankshaft spacer tool installation—3.0L, 3.5L and 3.8L engines

turn it again clockwise to align the timing marks.

16. Loosen the timing belt tensioner center bolt. Using the Tension Pulley Socket Wrench tool MD998767, or equivalent, and a torque wrench, apply 39 inch lbs. (4.4 Nm) pressure on the timing belt. Torque the tensioner pulley center bolt to 35 ft. lbs. (48 Nm).

17. Remove the setting pin from the auto-tensioner.

18. Rotate the crankshaft 2 complete revolutions and realign the timing marks. Then, wait for 5 minutes until the auto-tensioner pushrod extends to its standard value. If the standard value is not 0.15–0.20 in. (3.8–5.0mm), repeat the adjustment procedure. If the standard value is still not achieved, replace the auto-tensioner.

19. Install the lower timing belt cover and crankshaft pulley.

20. Using the End Yoke Holder tool MD990767 and 2 Crankshaft Pulley Holder Pin tools MD998715, or equivalent to hold the crankshaft pulley, and a socket torque wrench, torque the crankshaft pulley bolt to 134 ft. lbs. (181 Nm).

- Install or connect the following:
- Upper timing belt cover assembly
- Remaining items by reversing the removal procedures
- Negative battery cable

21. Refill the cooling system.

3.8L Engine

1. Before servicing the vehicle, refer to the Precautions Section.

2. Drain the engine coolant.

3. Remove or disconnect the following:

- Negative battery cable
- Skid plate and under cover
- Battery and battery tray
- Air cleaner assembly
- Radiator shroud cover
- Drive belt
- Cooling fan and pulley
- Drive belt auto tensioner
- Accessory mount stay
- Power steering pump. Unbolt and position aside; do not disconnect the fluid lines.
- A/C compressor. Unbolt and position aside; do not disconnect the refrigerant lines.
- Compressor bracket
- Cooling fan bracket assembly
- Accessory mount assembly
- Timing belt upper cover
- Crankshaft pulley, using special tools MB991800 and MB991802
- Timing belt indicator bracket

- Timing belt lower cover
- Auto tensioner

4. Turn the crankshaft clockwise to align the timing marks and set the No. 1 cylinder at Top Dead Center (TDC). If you are reusing the timing belt, mark the flat side of the belt with an arrow showing the clockwise direction.

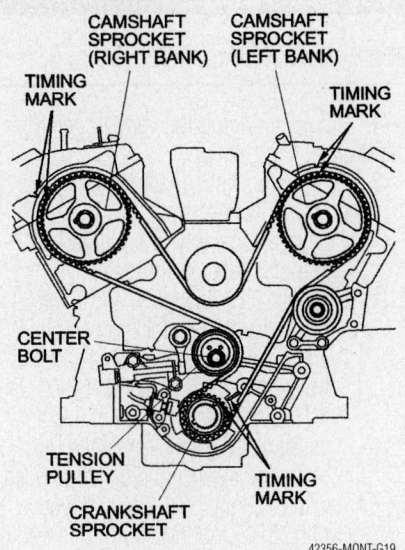

Timing belt alignment marks—3.8L engine

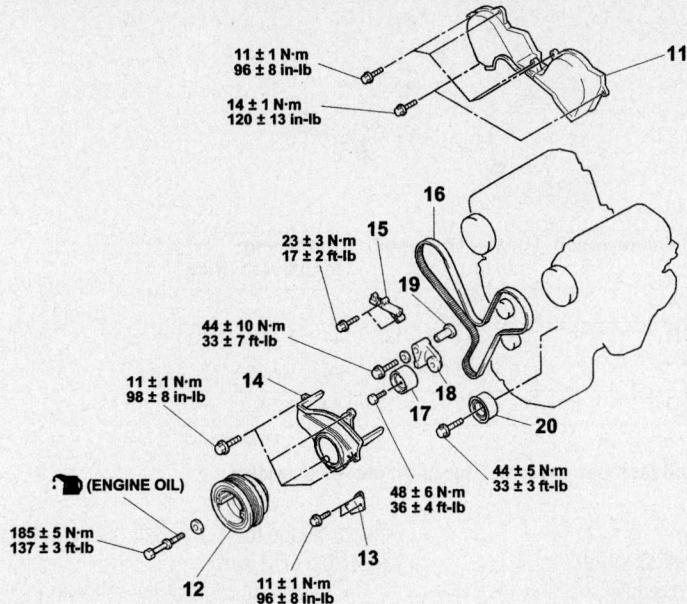

11. TIMING BELT UPPER COVER ASSEMBLY
12. CRANKSHAFT PULLEY
13. TIMING BELT INDICATOR BRACKET
14. TIMING BELT LOWER COVER ASSEMBLY
15. AUTO-TENSIONER
16. TIMING BELT
17. TENSION PULLEY
18. TENSIONER ARM ASSEMBLY
19. SHAFT
20. IDLER PULLEY

Timing belt and related components—3.8L engine

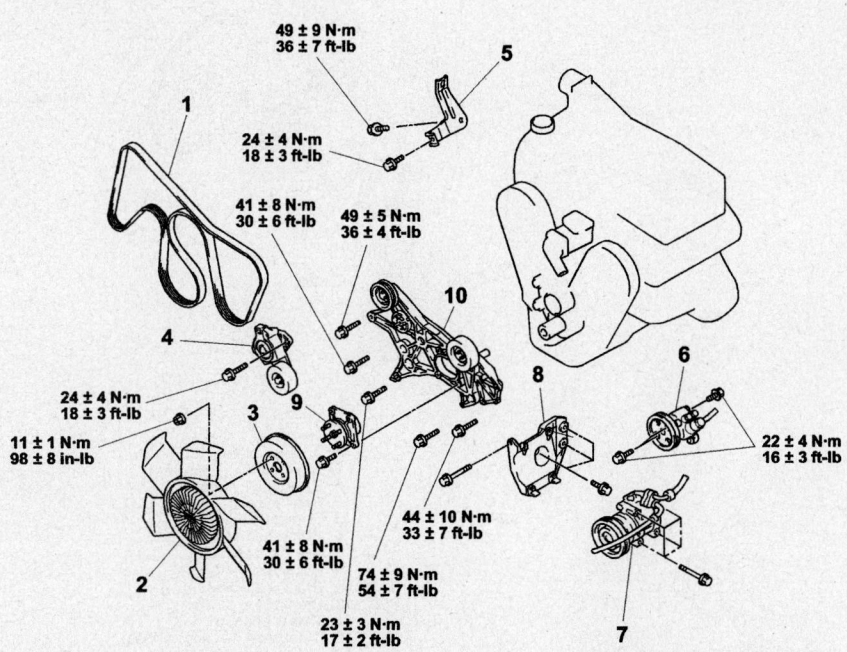

1. DRIVE BELT
2. COOLING FAN
3. COOLING FAN PULLEY
4. DRIVE BELT AUTO TENSIONER
5. ACCESSORY MOUNT STAY
6. POWER STEERING OIL PUMP ASSEMBLY
7. A/C COMPRESSOR ASSEMBLY
8. COMPRESSOR BRACKET
9. COOLING FAN BRACKET ASSEMBLY
10. ACCESSORY MOUNT ASSEMBLY

Accessory mount and related components—3.8L engine

5. Loosen the center bolt of the tension pulley, and then remove the timing belt.
6. Remove or disconnect the following:
- Tension pulley
- Tensioner arm assembly
- Shaft
- Idler pulley

To install:
7. Install or connect the following:
- Idler pulley
- Shaft
- Tensioner arm assembly
- Tension pulley

8. Press the end of the auto-tensioner inward with 72–145 ft. lbs. (98–196 Nm) of force and measure the distance that the pushrod is pushed in. If the standard distance is not 0.04 in. (1mm), replace the auto-tensioner.

9. Position the auto-tensioner in a soft-jawed vise and SLOWLY compress the pushrod until the pushrod and housing holes align; then, install a setting pin to secure the auto-tensioner in the retracted position.

10. Align the camshaft and crankshaft TDC timing marks.

11. Install the timing belt (noting its rotational direction) so that there is no deflection between the sprockets and pulleys in the following manner:
- Crankshaft sprocket

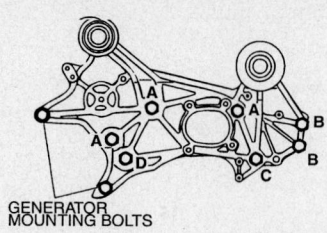

Bolt (symbol)	Diameter × length mm (in)	Tightening torque N·m (ft-lb)
A	10 × 100 (0.4 × 3.9)	41 ± 8 (30 ± 6)
B	10 × 30 (0.4 × 1.2)	41 ± 8 (30 ± 6)
C	10 × 100 (0.4 × 3.9)	44 ± 10 (33 ± 7)
D	12 × 100 (0.5 × 3.9)	74 ± 9 (54 ± 7)

09482_MONT_G0011

Accessory mount bolt locations and torque specifications—3.8L engine

- Idler pulley
- Left camshaft sprocket
- Water pump pulley
- Right camshaft sprocket
- Tension pulley

12. Turn the camshaft sprocket counter-clockwise until the tension side of the timing belt is firmly stretched, then, recheck the timing marks.

13. Using the Tension Pulley Socket Wrench tool MD998767, or equivalent, push the tensioner pulley into the timing belt and secure the center bolt.

14. Using the Crankshaft Pulley Spacer tool MD998769, or equivalent, rotate the crankshaft ¼ turn counterclockwise, then, turn it again clockwise to align the timing marks.

15. Loosen the timing belt tensioner center bolt. Using the Tension Pulley Socket Wrench tool MD998767, or equivalent, and a torque wrench, apply 39 inch lbs. (4.4 Nm) pressure on the timing belt. Torque the tensioner pulley center bolt to 35 ft. lbs. (48 Nm).

16. Remove the setting pin from the auto-tensioner.

17. Rotate the crankshaft 2 complete revolutions and realign the timing marks. Then, wait for 5 minutes until the auto-tensioner pushrod extends to its standard value. If the standard value is not 0.15–0.20 in. (3.8–5.0mm) for 2003 engines and 0.19–0.22 in. (4.8–5.5mm) for 2004–2006 engines, repeat the adjustment procedure. If the standard value is still not achieved, replace the auto-tensioner.

18. Install the lower timing belt cover and crankshaft pulley.

19. Using the End Yoke Holder tool MD990767 and 2 Crankshaft Pulley Holder Pin tools MD998715, or equivalent to hold the crankshaft pulley, and a socket torque wrench, torque the crankshaft pulley bolt to 134 ft. lbs. (181 Nm).

- Install or connect the following:
- Timing belt upper cover
- Accessory mount assembly
- Cooling fan bracket assembly
- Compressor bracket
- A/C compressor
- Power steering pump
- Accessory mount stay
- Drive belt auto tensioner

- Cooling fan pulley and fan
- Drive belt
- Radiator shroud cover
- Air cleaner assembly
- Battery tray and battery
- Under cover and skid plate
- Negative battery cable

20. Refill the cooling system.

Rear Main Seal

REMOVAL & INSTALLATION

1. Before servicing the vehicle, refer to the Precautions Section.

2. Remove or disconnect the following:
- Transmission
- Clutch assembly, if equipped
- Transfer case, if equipped
- Flywheel or driveplate and adapter plate, matchmark for reassembly. If equipped with 3.0L engine, use the Mitsubishi tools (MB990767-01 and MIT308239) to hold the crankshaft and flywheel stationary while loosening the flywheel bolts. If equipped with 3.5L and 3.8L engines, use Mitsubishi tool (MD998781) to hold the flywheel in position.

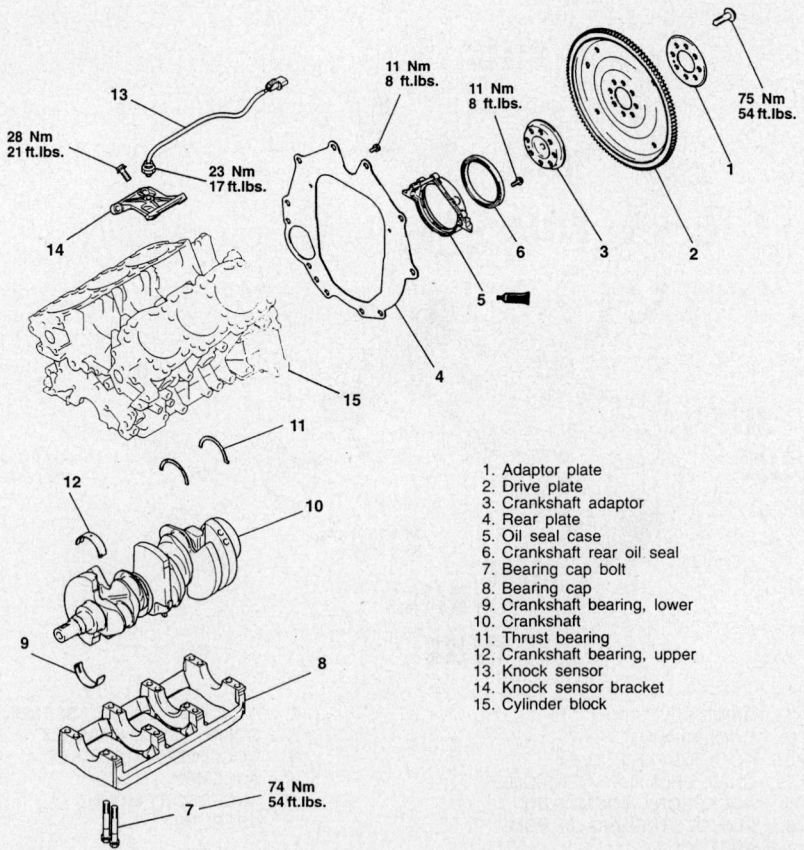

1. Adaptor plate
2. Drive plate
3. Crankshaft adaptor
4. Rear plate
5. Oil seal case
6. Crankshaft rear oil seal
7. Bearing cap bolt
8. Bearing cap
9. Crankshaft bearing, lower
10. Crankshaft
11. Thrust bearing
12. Crankshaft bearing, upper
13. Knock sensor
14. Knock sensor bracket
15. Cylinder block

7924UG22

Crankshaft rear main seal and related components—3.5L engine

3. Remove the rear oil seal as follows:

  a. Cut out a portion in the crankshaft oil seal lip.

  b. Cover the tip of a small prytool with a cloth and apply it to the cutout in the oil seal to pry the oil seal out.

✳✳ CAUTION

Take care not to damage the crankshaft and oil seal case.

To install:

4. Inspect the sealing surface at the rear of the crankshaft. If a deep groove is worn into the surface, the crankshaft will have to be replaced. Coat the sealing lip of the seal with fresh, clean engine oil. Press the new seal into the case with a seal installing tool. The seal must be pressed in squarely until it bottoms in the case. It is necessary to use the proper tool (MD998718-01) to fit the seal into place.

5. Install or connect the following:

- Rear plate
- Transmission mounting plate
- Flywheel or drive plate and adapter
- Transmission and related components as necessary

Piston and Ring

POSITION

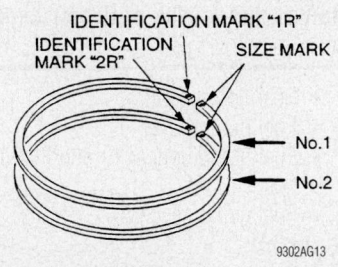

Piston ring identification

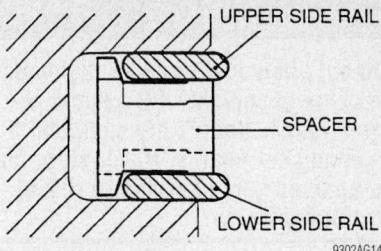

Oil ring identification

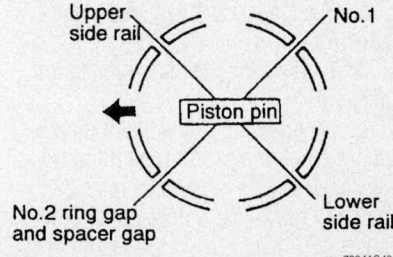

Piston ring end-gap spacing

FUEL SYSTEM

Fuel System Service Precautions

Safety is the most important factor when performing not only fuel system maintenance but any type of maintenance. Failure to conduct maintenance and repairs in a safe manner may result in serious personal injury or death. Maintenance and testing of the vehicle's fuel system components can be accomplished safely and effectively by adhering to the following rules and guidelines.

- To avoid the possibility of fire and personal injury, always disconnect the negative battery cable unless the repair or test procedure requires that battery voltage be applied.
- Always relieve the fuel system pressure prior to disconnecting any fuel system component (injector, fuel rail, pressure regulator, etc.), fitting or fuel line connection. Exercise extreme caution when relieving fuel system pressure, to avoid exposing your skin, face and eyes to fuel spray. Please be advised that fuel under pressure may penetrate the skin or any part of the body that it contacts.
- Always place a shop towel or cloth around the fitting or connection prior to loosening to absorb any excess fuel due to spillage. Ensure that all fuel spillage (should it occur) is quickly removed from engine surfaces. Ensure that all fuel soaked cloths or towels are deposited into a suitable waste container.
- Always keep a dry chemical (Class B) fire extinguisher near the work area.

- Do not allow fuel spray or fuel vapors to come into contact with a spark or open flame.
- Always use a back-up wrench when loosening and tightening fuel line connection fittings. This will prevent unnecessary stress and torsion to fuel line piping. Always follow the proper torque specifications.
- Always replace worn fuel fitting O-rings with new. Do not substitute fuel hose where fuel pipe is installed.

Fuel System Pressure

RELIEVING

✳✳ CAUTION

The fuel system is under constant pressure, even with the engine off. This pressure must be relieved before disconnecting any fuel system component, fitting or fuel line connection. Failure to do so may result in personal injury.

Montero

1. Turn the ignition switch to **LOCK**.
2. Fold down the second seat.
3. Remove the upper and lower service hole cover and packing.
4. Disconnect the fuel pump module connector.
5. Start the engine.

6. After the engine stalls, turn the ignition switch **OFF** and reconnect the fuel pump connector.
7. Disconnect the negative battery cable.

Montero Sport

1. Disconnect the fuel pump electrical connector, located at the rear side of the fuel tank.
2. Start the engine.
3. After the engine stalls, turn the ignition switch **OFF** and reconnect the fuel pump connector.
4. Disconnect the negative battery cable.

Fuel Filter

REMOVAL & INSTALLATION

Montero

The manufacturer does not provide a procedure or maintenance interval for replacing the fuel filter.

Montero Sport

✳✳ CAUTION

The fuel injection system remains under pressure after the engine has been OFF. Properly relieve fuel pressure before disconnecting any fuel lines. Failure to do so may result in fire or personal injury.

✳✳ CAUTION

Do not allow fuel spray or fuel vapors to come in contact with a spark or open flame. Keep a dry chemical fire extinguisher nearby. Never store fuel in an open container due to risk of fire or explosion.

1. Relieve the fuel system pressure.
2. Before servicing the vehicle, refer to the Precautions Section.
3. Disconnect the negative battery cable.
4. Remove the fuel filter protector if equipped.
5. Using a back-up wrench disconnect the fuel line(s) from the filter. If the filter uses a push-on type connector, press the retainer to release the connection.
6. Remove the filter from the mounting bracket.

To install:
7. Position the filter to the mounting bracket in the proper direction.
8. Connect the fuel lines to the filter. Use a back-up wrench to hold the fuel filter. Torque the banjo bolt(s) to 18–25 ft. lbs. (25–35 Nm) or the line fitting to 27 ft. lbs. (36 Nm).
9. Install the fuel filter protector if equipped.
10. Connect the negative battery cable.
11. Start the engine and check for leaks.

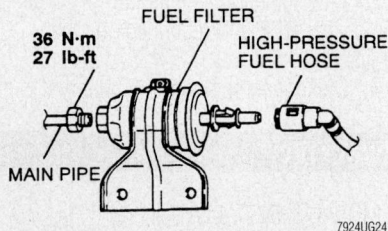

Fuel filter removal—Montero Sport

Fuel Pump

REMOVAL & INSTALLATION

Montero

➡The manufacturer recommends draining of the fuel tank.

1. Before servicing the vehicle, refer to the Precautions Section.
2. Relieve the fuel system pressure.
3. Remove or disconnect the following:
   • Negative battery cable

✳✳ CAUTION

The fuel injection system remains under pressure after the engine has been OFF. Properly relieve fuel pressure before disconnecting any fuel lines. Failure to do so may result in fire or personal injury. Do not allow fuel spray or fuel vapors to come in contact with a spark or open flame. Keep a dry chemical fire extinguisher nearby. Never store fuel in an open container due to risk of fire or explosion.

   • Rear floor carpeting
   • Fuel pump cover
   • Fuel pump connector and the fuel hoses
   • Fuel pump assembly

To install:
4. Install or connect the following:
   • Fuel pump assembly into the fuel tank. Torque the nuts to 18–26 inch lbs.
   • Fuel lines and the fuel pump connector

   • Fuel pump cover
   • Rear floor carpeting.
   • Negative battery cable
5. Refill the fuel tank, if drained
6. Start the engine and check for leaks and proper operation.

Montero Sport

1. Before servicing the vehicle, refer to the Precautions Section.
2. Properly relieve the fuel system pressure.
3. Remove the fuel tank drain plug and drain the fuel from the tank.

✳✳ CAUTION

The fuel injection system remains under pressure after the engine has been OFF. Properly relieve fuel pressure before disconnecting any fuel lines. Failure to do so may result in fire or personal injury.

4. Remove or disconnect the following:
   • Negative battery cable

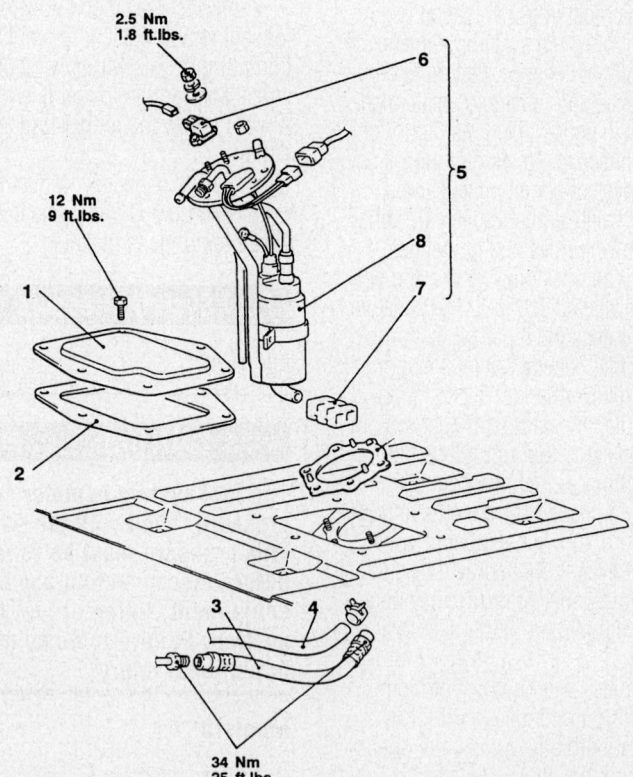

1. Floor cover
2. Packing
3. High-pressure fuel hose
4. Fuel return hose connection
5. Fuel pump and filter assembly
6. Fuel tank differential pressure sensor
7. Filter
8. Fuel pump assembly

Fuel pump and related components—Montero

- Fuel tank protector, if equipped
- Fuel tank from the vehicle
- Fuel pump retaining screws and the pump from the tank

To install:

5. Clean the seal area of the tank.
6. Install or connect the following:
   - New gasket
   - Fuel pump in the same position as originally installed.
   - Fuel pump retaining screws, Torque the nuts to 22 inch lbs. (2.5 Nm).
   - Fuel tank. Torque the bolts to 20 ft. lbs. (27 Nm).
   - Fuel tank drain plug and the fuel tank protector, if equipped
   - Negative battery cable
7. Refill the fuel tank and install the cap.
8. Check fuel system for leaks.

Fuel Rail

REMOVAL & INSTALLATION

3.0L and 3.5L Engines

1. Before servicing the vehicle, refer to the Precautions Section.
2. Properly relieve the fuel system pressure.
3. Remove or disconnect the following:
   - Air cleaner assembly, ducts and air intake hose
   - Intake manifold plenum
   - Fuel return line
   - Pressure regulator, vacuum line and O-ring
   - High pressure fuel hose and O-ring
   - Injector connector
   - Fuel pipe and O-rings
   - Fuel rails and insulators

To install:

4. Install or connect the following:
   - Fuel rail and insulators. Torque the bolts to 106 inch lbs. (12 Nm).
   - Injector connector
   - High pressure fuel hose and O-ring. Torque the bolts to 43 inch lbs. (4.9 Nm).
   - Pressure regulator, vacuum line and O-ring. Torque the bolts to 78 inch lbs. (8.8 Nm).
   - Fuel return line
   - Intake manifold plenum

- Air cleaner assembly, ducts and air intake hose

3.8L Engine

1. Before servicing the vehicle, refer to the Precautions Section.
2. Properly relieve the fuel system pressure. Disconnect the negative battery cable.
3. Remove or disconnect the following:
   - Air cleaner assembly, ducts and air intake hose
   - Intake manifold plenum
   - Ignition coil connectors and coils
   - Injector connectors
   - High pressure fuel hose and O-ring
   - Blow-by hose connection
   - Fuel rail, injector and damper assembly, as necessary. Only earlier production vehicles will have a fuel damper.
   - Insulators from the fuel rail

To install:

4. Install or connect the following:
   - Fuel rail and insulators. Torque the bolts to 102 inch lbs. (12 Nm).
   - Injector connectors
   - Blow-by hose connection
   - High pressure fuel hose and O-ring. Torque the bolts to 44 inch lbs. (5.0 Nm).
   - Ignition coils and connectors
   - Intake manifold plenum
   - Air cleaner assembly, ducts and air intake hose
5. Connect the negative battery cable.

Fuel Injectors

REMOVAL & INSTALLATION

1. Before servicing the vehicle, refer to the Precautions Section.

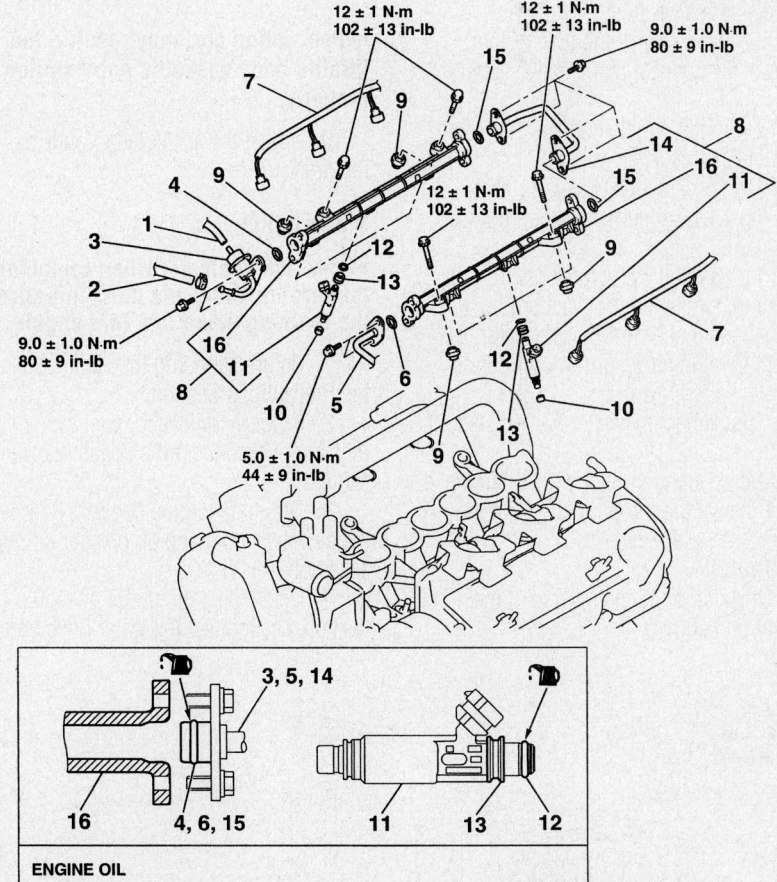

1.	VACUUM HOSE	6.	O-RING
2.	FUEL RETURN HOSE CONNECTION	7.	INJECTOR CONNECTORS
3.	FUEL PRESSURE REGULATOR	8.	FUEL INJECTORS, FUEL PIPE AND FUEL RAILS ASSEMBLY
4.	O-RING	9.	INSULATORS
5.	FUEL HIGH-PRESSURE HOSE CONNECTION	10.	INSULATORS
		11.	INJECTORS

09482_MONT_G0012

Fuel rail and related components—3.8L engine

2. Properly relieve the fuel system pressure. Disconnect the negative battery cable.

3. Remove the fuel rail.

4. Remove the insulator from the fuel injector. Remove the fuel injector from the fuel rail. Discard the O-ring.

To install:

5. Installation is the reverse of the removal procedure.

6. Be sure to use new gaskets and O-rings, as required.

7. Connect the negative battery cable.

8. Run the engine and check for leaks, correct as required.

9. Adjust all fluid levels, as required.

Throttle Body

REMOVAL & INSTALLATION

3.0L and 3.5L Engines

1. Before servicing the vehicle, refer to the Precautions Section.

2. Properly relieve the fuel system pressure. Disconnect the negative battery cable.

3. Drain the engine coolant. Remove the air cleaner assembly.

4. Disconnect the accelerator cable connection. Disconnect the vacuum hose connection.

5. If equipped with cruise control, disconnect the cruise control vacuum hose.

6. Disconnect the TPS connector. Disconnect the idle air control motor connector.

7. Disconnect the water hose connection.

8. Remove the throttle body retaining bolts. Remove the throttle body from the engine. Discard the gasket.

To install:

9. Installation is the reverse of the removal procedure.

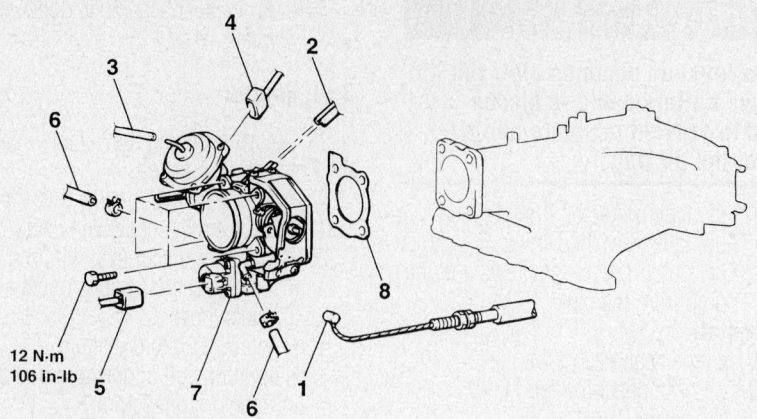

1. ACCELERATOR CABLE CONNECTION
2. VACUUM HOSE CONNECTION
3. VACUUM HOSE CONNECTION <VEHICLES WITH AUTO-CRUISE CONTROL>
4. THROTTLE POSITION SENSOR CONNECTOR
5. IDLE AIR CONTROL MOTOR CONNECTOR
6. WATER HOSE CONNECTION
7. THROTTLE BODY
8. THROTTLE BODY GASKET

12 N·m
106 in-lb

09482_MONT_G0013

Throttle body and related components—Montero Sport with 3.5L engine

➡**Poor idling etc. may result if the throttle body gasket is not installed properly.**

10. Adjust the accelerator cable, as required.

3.8L Engine

➡**On 2006 vehicles, when replacing the throttle body, you must initialize the learning value in a MFI engine.**

1. Before servicing the vehicle, refer to the Precautions Section.

2. Properly relieve the fuel system pressure. Disconnect the negative battery cable.

3. Drain the engine coolant. Remove the air intake duct and air cleaner housing assembly.

4. Disconnect the purge hose clip connection. Disconnect the water hose connection. Disconnect the TPS sensor connector. Remove the purge hose bracket.

5. Remove the throttle body retaining bolts. Remove the throttle body from the engine. Discard the gasket.

To install:

➡**If equipped, do not loosen the retaining screws for the resin cover of the throttle body assembly. If the screws are loosened, the sensor incorporated in the resin cover becomes misaligned and the throttle body can not work properly.**

6. Align the recess on the intake manifold plenum with the projection of the throttle body gasket. Install the gasket.

➡**Poor idling etc. may result if the throttle body gasket is not installed properly.**

7. Install the throttle body to the engine. Torque the retaining bolts to 18–24 ft. lbs. (24–32 Nm).

8. Continue the installation in the reverse order of the removal procedure.

9. Connect the negative battery cable.

10. On 2006 vehicles, complete the vehicle initialization procedure.

➡**To complete the initialization procedure the following tools are needed. MB991958 scan tool, MB991824 VCI, MB991827 MUT III USB cable, MB991910 MUT III Main harness "A".**

11. Connect the scan tool to the data link connector. To prevent damage to the scan tool be sure that the ignition switch is

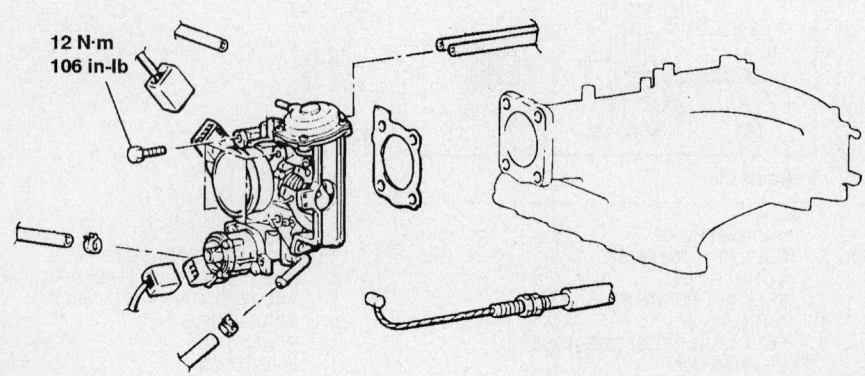

12 N·m
106 in-lb

09482_MONT_G0017

Throttle body and related components—Montero Sport with 3.0L engine

15. Initialize the learning value.

16. After initialization complete the idle learning procedure.

➡This procedure must be performed when the PCM is replaced, or when the learning value is initialized, as the idling is not stabilized because the learning value in the MFI engine is not completed.

17. Start the engine. Allow the coolant temperature to reach 176 degrees or more.

18. Stop the engine and place the ignition switch in the **LOCK** position.

19. After ten seconds, restart the engine.

20. For ten minutes carry out the idling procedure below to confirm that the engine has normal idling.

- Position the transaxle selector lever in the **P** range
- The engine fan is not to be operated
- The engine coolant temperature should be 176 degrees or more

➡If the engine stalls during idling, check the throttle valve of the throttle body for dirt. Correct and perform the procedure again.

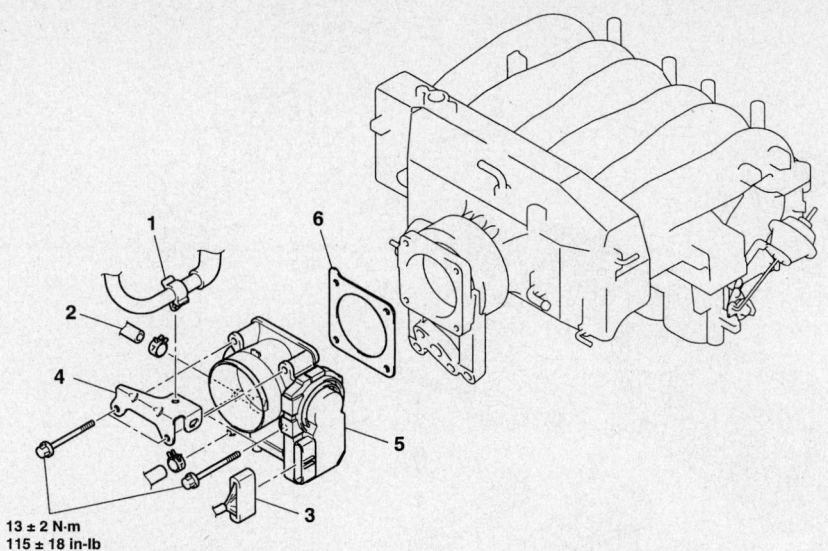

1. PURGE HOSE CLIP CONNECTION
2. WATER HOSES CONNECTION
3. THROTTLE POSITION SENSOR CONNECTOR
4. PURGE HOSE BRACKET
5. THROTTLE BODY
6. THROTTLE BODY GASKET

13 ± 2 N·m
115 ± 18 in-lb

09482_MONT_G0014

Throttle body and related components—Montero with 3.5L and 3.8L engines

in the **LOCK** position before connecting the scan tool.

12. Turn the ignition switch to the **ON** position.

13. Select "check mode" from the menu screen.

14. Select "erase memory" from the menu screen.

DRIVE TRAIN

Transmission

REMOVAL & INSTALLATION

Montero

1. Before servicing the vehicle, refer to the Precautions Section.

2. Disconnect the negative battery cable. Remove the skid plate.

3. Drain the transmission fluid. Drain the transfer case fluid.

4. Shift the transfer case lever in the 2H position.

5. Matchmark the front driveshaft by making mating marks on the differential companion flange and the flange yoke.

6. Matchmark the rear driveshaft by making mating marks on the differential companion flange and the outer race of the ball joint assembly.

7. Remove the front and rear driveshafts.

➡Be careful not to bend the joint portion when removing the rear driveshaft, as damage to the joint boot will result. The driveshaft is made of CFRP, if a

tear, chip or distortion is found on the driveshaft, replace it. Do not shock, drop or scratch the driveshaft. Once removed wrap the driveshaft with a rubber sheet of 0.039 inch thick material or shaft protector part number MR534564 so that the driveshaft is protected.

8. Disconnect the front exhaust pipe.

9. Remove the radiator shroud lower cover.

10. Remove the transmission fluid dipstick and guide assembly.

11. Disconnect the transmission control cable connection. Remove the cover.

12. Remove the heater hose connection retaining bracket bolts.

13. Remove the torque converter and drive plate connection bolts.

14. Disconnect the starter electrical connections. Remove the starter retaining bolts. Remove the starter from the vehicle. Remove the starter cover.

15. Remove the oil pan connection bolts. Remove the battery cable connection.

16. Remove the fluid cooler tube connection.

17. Remove the dynamic damper. Remove the tension wire bracket.

18. Support the transmission assembly using a transmission jack. Remove the transmission mount center member assembly.

19. Remove the transmission mount insulator assembly.

20. Disconnect the transmission harness connector connection.

21. Remove the transmission retaining bolts.

➡The size of the mounting bolts are different, note the location for proper reinstallation

22. Remove the transmission/transfer case from the vehicle. Remove the ground wire.

To install:

23. Lift the transmission and transfer case assembly into position, using a transmission jack.

24. Install the transmission retaining bolts.

➡Be sure to install the bolts in the same holes as they were removed from.

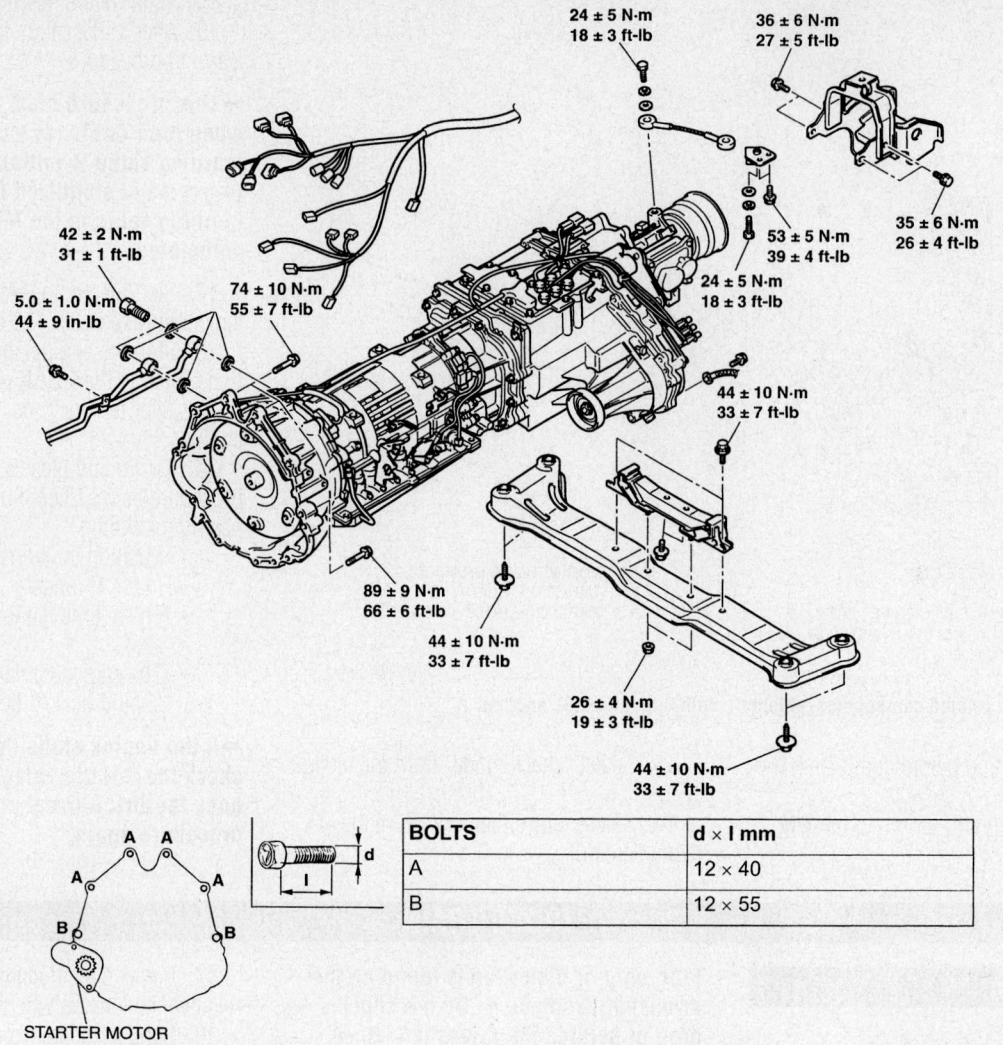

24 ± 5 N·m / 18 ± 3 ft-lb	36 ± 6 N·m / 27 ± 5 ft-lb
42 ± 2 N·m / 31 ± 1 ft-lb	35 ± 6 N·m / 26 ± 4 ft-lb
5.0 ± 1.0 N·m / 44 ± 9 in-lb	53 ± 5 N·m / 39 ± 4 ft-lb
74 ± 10 N·m / 55 ± 7 ft-lb	24 ± 5 N·m / 18 ± 3 ft-lb
89 ± 9 N·m / 66 ± 6 ft-lb	44 ± 10 N·m / 33 ± 7 ft-lb
44 ± 10 N·m / 33 ± 7 ft-lb	
26 ± 4 N·m / 19 ± 3 ft-lb	
44 ± 10 N·m / 33 ± 7 ft-lb	

STARTER MOTOR

BOLTS	d × l mm
A	12 × 40
B	12 × 55

09482_MONT_G0015

Automatic transmission and transfer case—Montero

25. Fill the transmission and transfer case with the proper grade and type fluids.

Montero Sport

1. Before servicing the vehicle, refer to the Precautions Section.
2. Disconnect the negative battery cable. Remove the skid plate.
3. Drain the transmission fluid.
4. Matchmark the driveshaft by making mating marks on the differential companion flange and the outer race of the ball joint assembly. Remove the driveshaft.
5. On 4WD vehicles, matchmark the front driveshaft by making mating marks on the differential companion flange and the flange yoke. Remove the driveshaft.

➡Be careful not to bend the joint portion when removing the rear driveshaft, as damage to the joint boot will result. Once removed wrap the driveshaft so that it protected.

6. Disconnect the catalytic converter and the front exhaust pipe.
7. Remove the transmission fluid dipstick guide. Remove the dust shield cover.
8. On 2WD vehicles, disconnect the transmission control cable connection.
9. Disconnect the ground cable. Disconnect the speed sensor connection. Disconnect the transmission range selector connection.
10. On 4WD vehicles, disconnect the high/low detection switch connection.
11. Disconnect the solenoid valve connection. Disconnect the output shaft speed sensor connection. Disconnect the input shaft speed sensor connection.
12. Disconnect and plug the transmission fluid lines. Remove the starter motor cover. Disconnect the starter electrical connections. Remove the starter retaining bolts. Remove the starter from the engine.
13. Remove the torque converter to flex plate retaining bolts.

14. Support the transmission assembly, using a transmission jack.
15. Remove the number two crossmember assembly.
16. On 2WD vehicles, remove the air breather pipe assembly. On 4WD vehicles, remove the rear engine mount bracket.
17. Remove the transmission assembly to engine retaining bolts.

➡The size of the mounting bolts are different, note the location for proper reinstallation

18. Carefully remove the transmission assembly from the vehicle.
   To install:
19. Lift the transmission assembly into position, using a transmission jack.
20. Install the transmission assembly retaining bolts.

➡Be sure to install the bolts in the same holes as they were removed from.

21. Fill the transmission with the proper grade and type fluid.

22. Fill the transfer case with the proper grade and type fluid.

23. Check and adjust the selector lever operation, as required.

Transfer Case Assembly

REMOVAL & INSTALLATION

The transfer case is removed from the vehicle along with the transmission. Refer to the Transmission Removal and Installation procedure for information.

Halfshaft

REMOVAL & INSTALLATION

Montero

FRONT

1. Before servicing the vehicle, refer to the Precautions Section.

2. Raise and support the vehicle safely. Remove the tire and wheel assembly.

3. Remove the under cover. Remove the skid plate.

4. Disconnect the shock absorber and stabilizer bar lower link.

5. Remove the front brake caliper and position it to the side.

6. Disconnect the knuckle and tie rod end connection.

7. Remove the dust cap, cotter pin, locknut and washer.

8. Disconnect the brake hose and VSS clamp bracket.

9. Remove the wheel speed sensor.

10. Remove the upper control arm ball joint connection.

11. Remove the halfshaft retaining bolts.

➡ **When pulling the halfshaft out from the differential carrier, be careful that the spline part of the halfshaft does not damage the oil seal.**

12. On the left side halfshaft, remove the circlip.

To install:

13. Installation is the reverse of the removal procedure.

REAR

1. Before servicing the vehicle, refer to the Precautions Section.

2. Raise and support the vehicle safely. Remove the tire and wheel assembly.

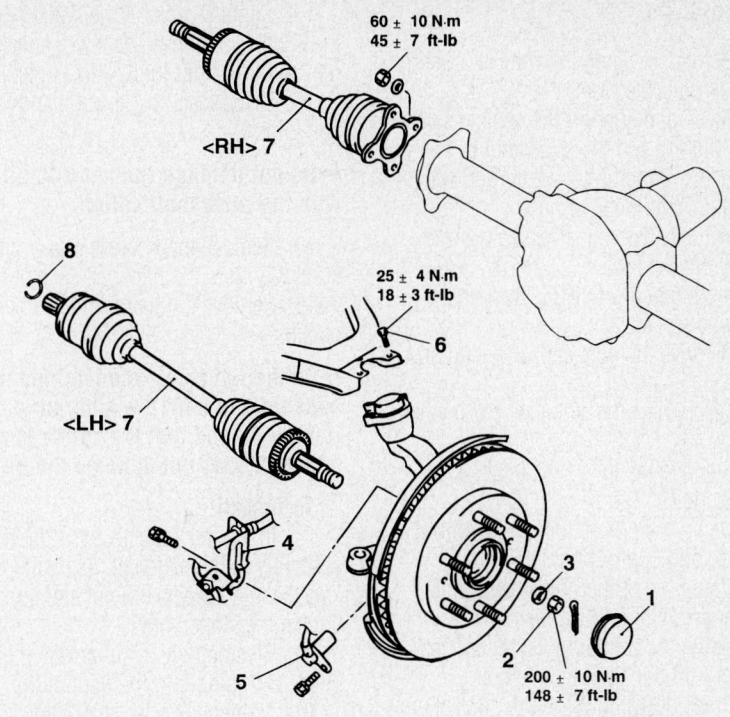

1. HUB CAP
2. CASTLE NUT
3. WASHER
4. BRAKE HOSE AND VEHICLE SPEED SENSOR CLAMP BRACKET
5. WHEEL SPEED SENSOR
6. UPPER ARM AND UPPER ARM BALL JOINT CONNECTION
7. DRIVE SHAFT
8. CIRCLIP <LH>

09482_MONT_G0036

Front halfshaft and related components—Montero

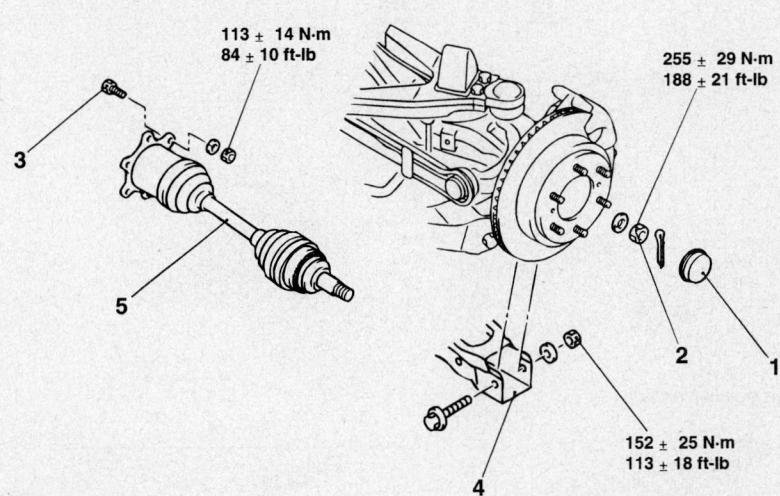

1. HUB CAP
2. DRIVE SHAFT NUT
3. COMPANION SHAFT AND DRIVE SHAFT CONNECTION
4. KNUCKLE AND LOWER ARM CONNECTION
5. REAR DRIVE SHAFT ASSEMBLY

09482_MONT_G0044

Rear halfshaft and related components—Montero

3. Remove the dust cap, cotter pin and halfshaft nut.

4. Remove the companion shaft and halfshaft connection bolts.

5. Support the lower control arm with a floor jack. Compress the coil spring in order to remove the retaining bolts. Remove the knuckle and lower control arm retaining bolts.

6. Remove the halfshaft from the vehicle.

To install:

7. Installation is the reverse of the removal procedure.

Montero Sport

1. Before servicing the vehicle, refer to the Precautions Section.

2. Raise and support the vehicle safely. Remove the tire and wheel assembly.

3. Remove the under cover. Remove the skid plate. Drain the gear oil.

4. Remove the caliper assembly and position it to the side.

5. If equipped with ABS, remove the front speed sensor.

6. Remove the dust cap, snapring and shim.

7. Disconnect the stabilizer bar connection.

8. Remove the lower shock absorber mounting bolt.

9. Using the proper tools, disconnect the tie rod end connection.

10. Using the proper tools, disconnect the lower arm ball joint connection.

11. Using the proper tools, disconnect the upper arm ball joint connection.

12. Press down on the lower control arm and remove the upper knuckle towards you.

➡**Pull the DOJ side of the halfahsft assembly out slightly from the front differential carrier.**

13. Slightly back off the halfshaft from the knuckle. Remove the lower knuckle retaining nut from the lower arm ball joint.

14. Disconnect the knuckle and lower ball joint.

➡**Do not damage the knuckle oil seals with the driveshaft spline.**

15. Remove the knuckle and front hub assembly from the halfshaft assembly.

16. Remove the halfshaft assembly and circlip.

➡**On the left side, when pulling the halfshaft out from the differential carrier, be careful that the spline part of the halfshaft does not damage the oil seal.**

To install:

17. Insert the knuckle and front hub assembly to the halfshaft. Assemble the knuckle and lower ball joint and temporarily tighten the slotted nut.

18. Press up the lower arm and lock the upper ball joint onto the upper arm.

19. Tighten the lower ball joint mounting nuts to 87–131 ft. lbs.

20. Push the halfshaft in by hand toward the knuckle until they touch.

21. Measure the clearance between the drive flange and the spacer, specification should be 0.02–0.03 inch.

22. If the end play is not within specification, adjust by selecting a shim that will bring the end play within specification.

➡**Shims are available from 0.01 inch thick to 0.02 inch thick, in increments of 0.004 inches, and from 0.04 inch to 0.07 inch thick in increments of 0.01 inches.**

23. Continue the installation in the reverse order of the removal procedure.

24. Be sure to correct and adjust the gear oil level, as necessary.

CV-Joints

OVERHAUL

Montero

1. Before servicing the vehicle, refer to the Precautions Section.

2. Raise and support the vehicle safely. Remove the tire and wheel assembly.

3. Remove the halfshaft from the vehicle.

➡**Never disassemble the UJ assembly except when replacing the UJ boot.**

4. Remove the small and large DOJ boot bands.

5. Remove the circlip. Remove the DOJ outer race. Remove the snapring.

6. Remove the inner race, cage and ball assembly.

7. Remove the DOJ boot.

8. Remove the small and large UJ boot bands, boot and assembly.

9. On the left side, remove the circlip. Remove the dust cover.

To install:

10. Assembly is the reverse of the disassembly procedure.

11. When installing the DOJ boot, position the outer race so that the distance between the boot bands is 3.1–3.3 inch. Remove part of the DOJ outer race to release the air pressure inside the boot.

Montero Sport

1. Before servicing the vehicle, refer to the Precautions Section.

2. Remove or disconnect the following:
- Front wheel
- Halfshaft
- Small and larger band
- Circlip
- Double Offset Joint (DOJ) outer race
- Dust cover
- Circlip

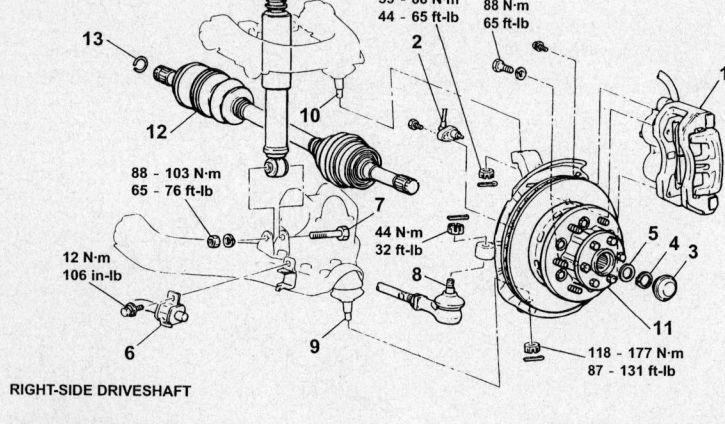

LEFT-SIDE DRIVESHAFT

59 - 88 N·m
44 - 65 ft-lb

88 N·m
65 ft-lb

88 - 103 N·m
65 - 76 ft-lb

12 N·m
106 in-lb

44 N·m
32 ft-lb

118 - 177 N·m
87 - 131 ft-lb

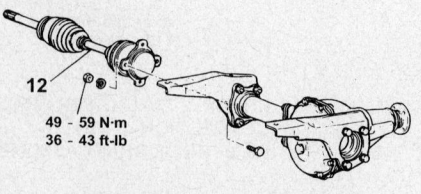

RIGHT-SIDE DRIVESHAFT

49 - 59 N·m
36 - 43 ft-lb

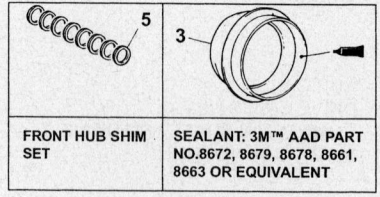

AC004699 AB

FRONT HUB SHIM SET	SEALANT: 3M™ AAD PART NO.8672, 8679, 8678, 8661, 8663 OR EQUIVALENT

1. CALIPER ASSEMBLY
2. FRONT SPEED SENSOR <VEHICLES WITH ABS>
3. HUB CAP
4. SNAP RING
5. SHIM
6. STABILIZER BAR CONNECTION
7. SHOCK ABSORBER LOWER MOUNTING BOLT
8. TIE ROD END CONNECTION
9. LOWER ARM BALL JOINT CONNECTION
10. UPPER ARM BALL JOINT CONNECTION
11. KNUCKLE AND FRONT HUB ASSEMBLY
12. DRIVESHAFT ASSEMBLY
13. CIRCLIP

09482_MONT_G0037

Front halfshaft and related components—Montero Sport

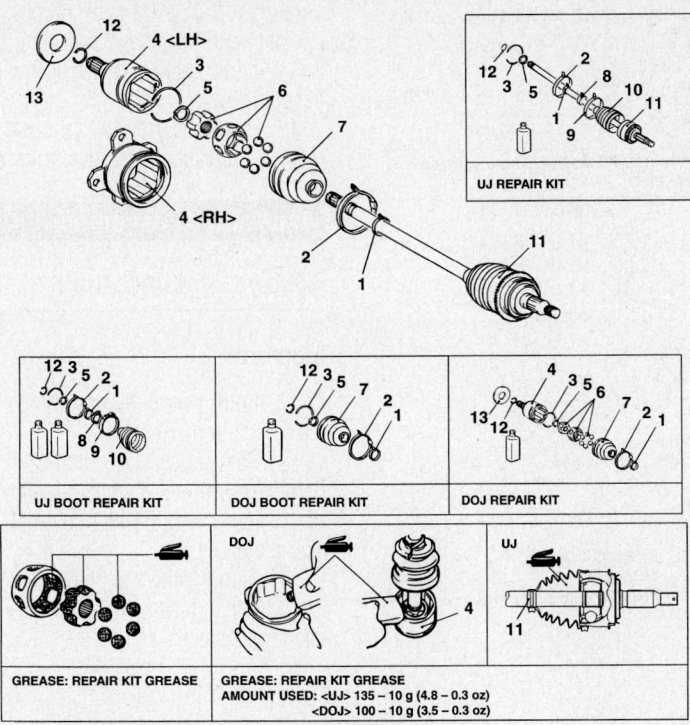

1. DOJ BOOT BAND (SMALL)
2. DOJ BOOT BAND (LARGE)
3. CIRCLIP
4. DOJ OUTER RACE
5. SNAP RING
6. INNER RACE, CAGE AND BALL ASSEMBLY
7. DOJ BOOT
8. UJ BOOT BAND (SMALL)
9. UJ BOOT BAND (LARGE)
10. UJ BOOT
11. UJ ASSEMBLY
12. CIRCLIP
13. DUST COVER

09482_MONT_G0038

CV-Joint and related components—Montero

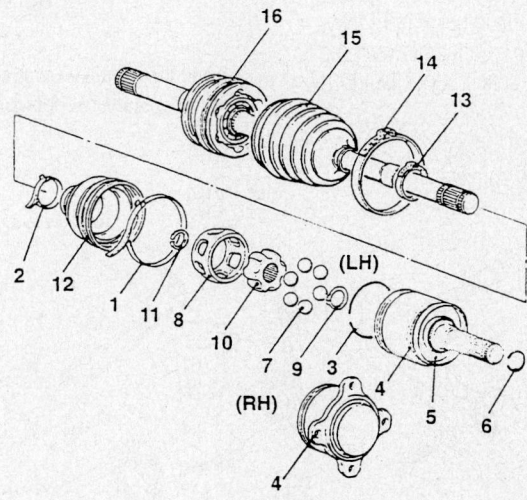

1. D.O.J. boot band (large)
2. D.O.J. boot band (small)
3. Circlip
4. D.O.J. outer race
5. Dust cover
6. Circlip
7. Ball
8. D.O.J. cage
9. Snap ring
10. D.O.J. inner race
11. Circlip
12. D.O.J. boot
13. B.J. boot band (small)
14. B.J. boot band (large)
15. B.J. boot
16. B.J. assembly

9308UG06

CV-Joint and related components—Montero Sport

- Balls from the cage
- Cage from the inner race. Turn the cage so that the projections of the inner race align with the recesses of the cage.
- Snapring from the shaft
- DOJ inner race
- Slide the boot off
- Birfield Joint (BJ) small and larger bands
- BJ boot
- BJ assembly

To install:

3. Check the shaft and splines for damage or wear. Inspect the cage, race and balls for any sign of corrosion, wear, cracking or damage. Clean all the parts thoroughly and air dry them completely before installation. Any remaining cleaning solvent can dissolve the lubricating grease.

4. Tool MB991561 can be used to crimp the bands in place.

5. Install or connect the following:
- BJ assembly
- BJ boot, slid the small end of the boot until only one shaft groove can be seen.
- BJ small band, crimp the band. Fill the BJ boot with 4.6 oz (130 g) of grease.
- BJ larger band, crimp the band
- DOJ small band and boot, fill the boot with grease
- DOJ cage onto the halfshaft so that the smaller diameter side is installed first.
- Circlip
- DOJ inner race and new snapring, apply grease to the inner race
- Balls into the cage, grease to the ball areas of the cage and race
- Outer race, fill the outer race about ⅓ full of grease.
- Dust cover
- Circlip
- Large boot band, release the air from the boot then crimp
- Halfshaft

Front Axle Hub

REMOVAL & INSTALLATION

Montero Sport

1. Before servicing the vehicle, refer to the Precautions Section.

2. Raise and support the vehicle safely. Remove the tire and wheel assembly.

3. Remove the brake caliper assembly and position it to the side, with wire.

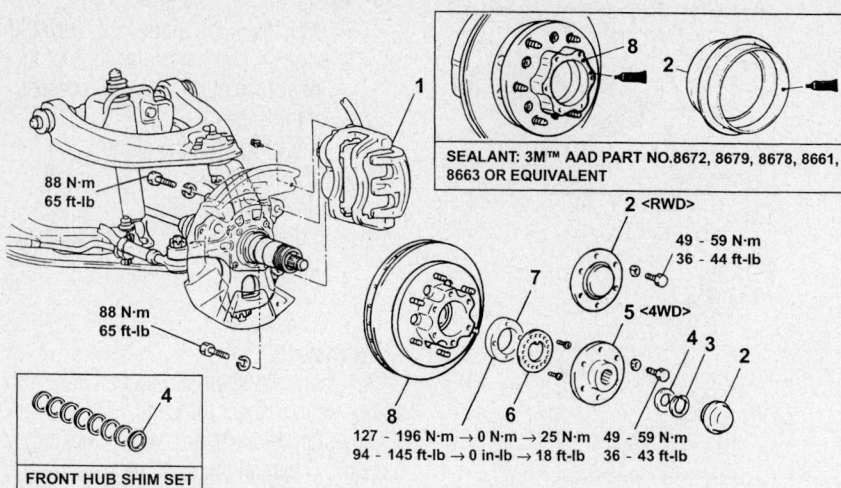

SEALANT: 3M™ AAD PART NO.8672, 8679, 8678, 8661, 8663 OR EQUIVALENT

2 <RWD>
49 - 59 N·m
36 - 44 ft-lb

5 <4WD>
49 - 59 N·m
36 - 43 ft-lb

127 - 196 N·m → 0 N·m → 25 N·m
94 - 145 ft-lb → 0 in-lb → 18 ft-lb

FRONT HUB SHIM SET

1. CALIPER ASSEMBLY
2. HUB CAP
• DRIVE SHAFT END PLAY
   ADJUSTMENT <4WD>
3. SNAP RING
4. SHIM
5. DRIVE FLANGE
• HUB ROTARY SLIDING
   RESISTANCE AND WHEEL
   ADJUSTMENT <4WD>

6. SPRING WASHER
7. JAM NUT
8. FRONT HUB ASSEMBLY

09482_MONT_G0047

Front axle hub and related components—Montero Sport

4. Remove the dust cap, snapring and shim. Remove the drive flange.

5. Remove the spring washer. Remove the jamnut, using tool MB990954, or equivalent.

6. Remove the front hub assembly from the vehicle.

To install:

7. Install the front hub assembly.

8. On 4WD vehicles, use a spring scale to measure the hub rotary sliding resistance. Specification should be 2.7–11.5 inch lbs. If not within specification remove the lock washer. If the resistance is lower than specification, use tool MB9909954 to tighten the jamnut. If the resistance is higher than specification, use MB9909954 to tighten the jamnut. Install a dial indicator gauge and move the hub in the axial direction. Measure the wheel bearing movement. Specification should be 0.002 inch or less. If not within specification, remove the spring washer and use special tool MB990954 to tighten the jamnut. If adjustment is not possible, disassemble the hub and inspect each component.

9. Install the lock washer.

➡**If the hole position is not aligned with the jam nut, move it within a range of not more than a 20 degree angle, until the holes are aligned.**

10. Install the drive flange, shim and snapring.

11. On 4WD vehicles, push the halfshaft in by hand towards the knuckle until they touch. Measure the clearance between the drive flange and the spacer. Specification should be 0.02-0.03 inch. If not within specification select and adjusting shim that will bring the halfshaft end play within to specification.

12. Continue the installation in the reverse order of the removal procedure.

Inner Shaft Assembly

REMOVAL & INSTALLATION

Montero

1. Before servicing the vehicle, refer to the Precautions Section.

2. Raise and support the vehicle safely. Remove the tire and wheel assembly.

3. Remove the under cover. Remove the skid plate.

4. Remove the right side halfshaft.

5. Loosen the differential mounting insulator bolts.

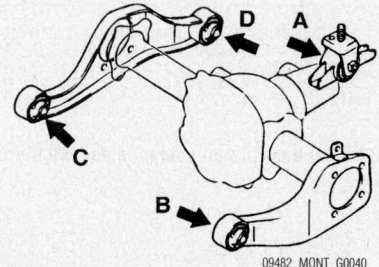

09482_MONT_G0040

Front differential mounting insulator bolt torque sequence—Montero

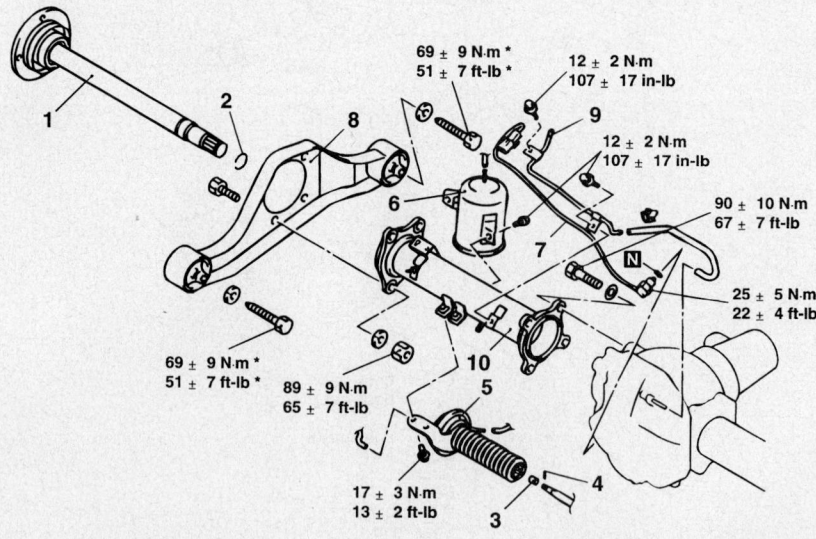

69 ± 9 N·m *
51 ± 7 ft-lb *

12 ± 2 N·m
107 ± 17 in-lb

12 ± 2 N·m
107 ± 17 in-lb

90 ± 10 N·m
67 ± 7 ft-lb

25 ± 5 N·m
22 ± 4 ft-lb

69 ± 9 N·m *
51 ± 7 ft-lb *

89 ± 9 N·m
65 ± 7 ft-lb

17 ± 3 N·m
13 ± 2 ft-lb

1. INNER SHAFT
2. CLIP
3. COLLAR
4. PIN
5. ACTUATOR ASSEMBLY
6. VACUUM TANK ASSEMBLY

7. FREE WHEEL ENGAGE SWITCH
   ASSEMBLY
8. DIFFERENTIAL MOUNTING
   BRACKET <RH>
9. BREATHER PIPE
10. HOUSING TUBE

09482_MONT_G0039

Front inner shaft assembly and related components—Montero

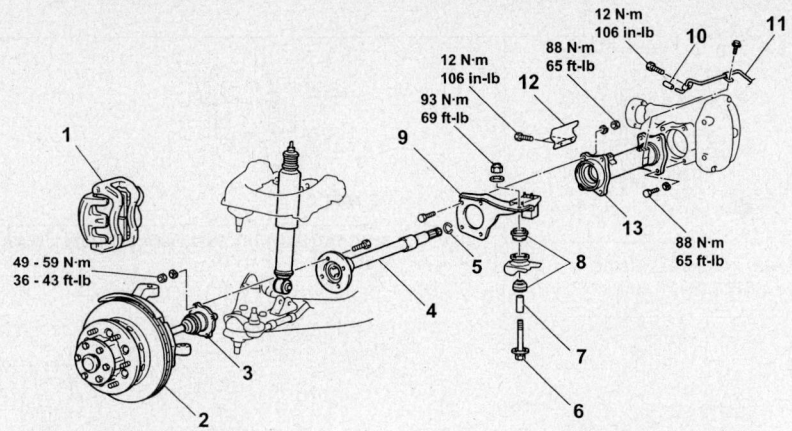

1. CALIPER ASSEMBLY
2. KNUCKLE AND HUB ASSEMBLY
3. DRIVESHAFT <RH>
4. INNER SHAFT
5. CIRCLIP
6. PIN
7. SPACER
8. DIFFERENTIAL MOUNTING CUSHION
9. DIFFERENTIAL MOUNTING BRACKET <RH>
10. BREATHER HOSE
11. BREATHER PIPE
12. HEAT PROTECTOR
13. HOUSING TUBE ASSEMBLY

09482_MONT_G0041

Front inner shaft assembly and related components—Montero Sport

6. Remove the inner shaft. Remove the circlip.

➡ **When pulling the inner shaft from the front differential carrier with tools MB9909906 and MB990211, be careful that the spline part of the inner shaft does not damage the oil seal.**

7. Remove the collar, pin and actuator assembly.
8. Remove the vacuum tank assembly.
9. Remove the free wheel engage switch assembly.
10. Properly support the differential assembly, using a floor jack.
11. Remove the right side differential mounting bracket.
12. Remove the breather pipe. Remove the housing tube.

To install:

13. Installation is the reverse of the removal procedure.
14. When installing the differential mounting insulator bolts lower the vehicle to the ground after temporarily tightening the bolts to prevent the bushing from twisting. Then fully tighten the bolts in the unladen condition to 44–58 ft. lbs. and in the proper sequence.

Montero Sport

1. Before servicing the vehicle, refer to the Precautions Section.
2. Raise and support the vehicle safely. Remove the tire and wheel assembly.
3. Remove the under cover. Drain the gear oil.

4. Remove the brake caliper and position it to the side, using wire.
5. Remove the knuckle and hub assembly.
6. Remove the halfshaft assembly.
7. Remove the inner shaft. Remove the circlip.

➡ **When pulling the inner shaft from the front differential carrier with tools MB9909906 and MB990211, be careful**

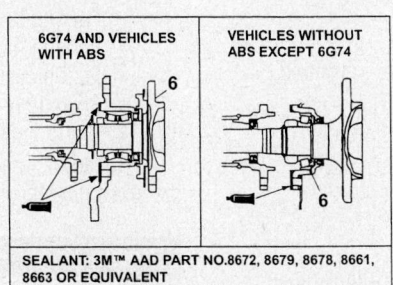

SEALANT: 3M™ AAD PART NO.8672, 8679, 8678, 8661, 8663 OR EQUIVALENT

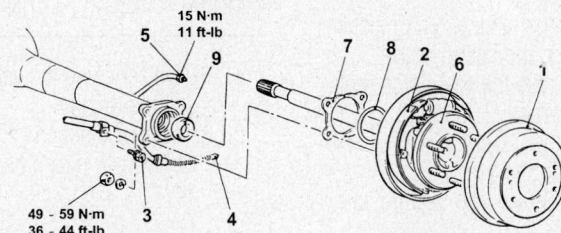

1. BRAKE DRUM
2. SHOE AND LINING ASSEMBLY
3. PARKING BRAKE CABLE, SPEED SENSOR CABLE <VEHICLES WITH ABS> ATTACHMENT BOLT
4. PARKING BRAKE CABLE CONNECTION
5. BRAKE TUBE
6. AXLE SHAFT ASSEMBLY
7. SHIM <VEHICLES WITHOUT ABS EXCEPT 6G74>
8. O-RING
9. OIL SEAL

09482_MONT_G0042

Rear axle shaft and related components—Montero Sport with drum brakes

that the spline part of the inner shaft does not damage the oil seal.

8. Remove the circlip, pin and spacer.
9. Properly support the differential assembly using a floor jack.
10. Remove the differential mounting cushion. Remove the right side differential mounting bracket.
11. Remove the breather hose. Remove the breather pipe. Remove the heat protector.
12. Remove the housing tube assembly.

To install:

13. Installation is the reverse of the removal procedure.
14. When installing the inner shaft, drive it into position using tools MB990906 and MB990211.

➡ **Be careful not to damage the lip of the dust seal and/or the oil seal.**

15. Fill the front differential with the proper grade and type gear oil.

Rear Axle Shaft , Bearing and Seal

REMOVAL & INSTALLATION

Montero Sport

DRUM BRAKES

1. Before servicing the vehicle, refer to the Precautions Section.

2. Raise and support the vehicle safely. Remove the tire and wheel assembly.

3. Drain the brake fluid.

4. Remove the brake drum. Remove the brake shoes.

5. Remove the parking brake cable retaining bolt.

6. If equipped with ABS remove the speed sensor cable.

7. Disconnect the parking brake cable connection.

8. Disconnect the brake line hose.

9. Using tool MB990211 and MB990241, pull the rear axle shaft from the axle shaft housing.

10. Using tools MB990211 and MB990212 remove the oil seal.

To install:

11. Installation is the reverse of the removal procedure.

12. Use tools MB990930 and MB990938 to drive the new oil seal into position.

13. Be sure to bleed the brakes and adjust the fluid level, as required.

DISC BRAKES

1. Before servicing the vehicle, refer to the Precautions Section.

2. Remove or disconnect the following:
- Drain brake fluid
- Wheel and tire assembly
- Brake line
- Rear brake assembly
- Parking brake cable and assembly
- Axle shaft
- Snapring
- Retainer
- Bearing inner race, inner and outer
- Oil seal
- Bearing case
- O-ring
- Oil seal

To install:

3. Install or connect the following:
- Bearing case
- Bearing inner race, outer. Press the bearing into the bearing case.
- Oil seal. Press the seal using tools MB990932 and MB990938.
- Bearing inner race, inner. Press the bearing into the bearing case.
- Axle shaft, Place into bearing case

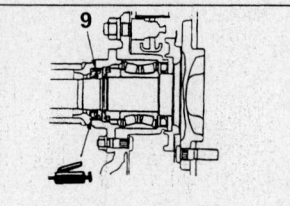

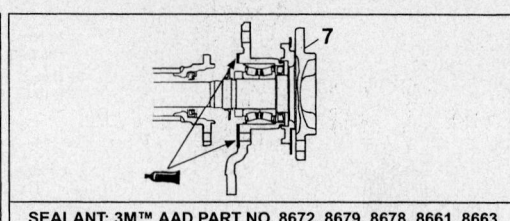

APPLY TO ENTIRE INSIDE DIAMETER OF OIL SEAL LIP

SEALANT: 3M™ AAD PART NO. 8672, 8679, 8678, 8661, 8663 OR EQUIVALENT

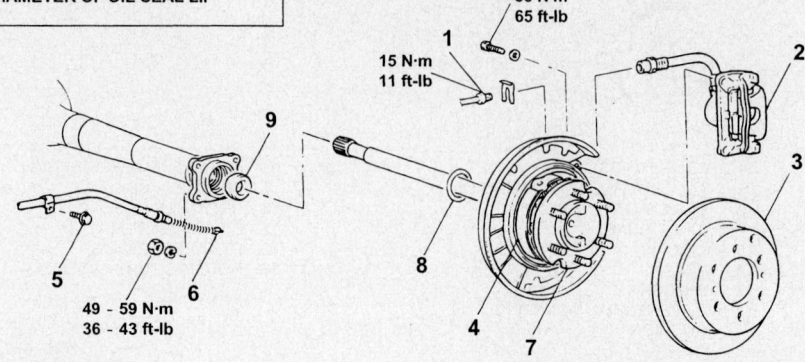

88 N·m
65 ft-lb

15 N·m
11 ft-lb

49 - 59 N·m
36 - 43 ft-lb

1. BRAKE TUBE
2. CALIPER ASSEMBLY
3. BRAKE DISC
4. PARKING BRAKE SHOE
5. PARKING BRAKE CABLE AND SPEED SENSOR <VEHICLES WITH ABS> ATTACHING BOLT
6. PARKING BRAKE CABLE
7. AXLE SHAFT ASSEMBLY
8. O-RING
9. OIL SEAL

09482_MONT_G0043

Rear axle shaft and related components—Montero Sport with disc brakes

- Retainer, press onto shaft
- Snapring
- Oil seal and O-ring into axle shaft
4. Axle shaft assembly into axle.
  - Parking brake cable end
  - Parking brake cable attaching bolt
  - Rear brake assembly
  - Brake line
  - Wheel and tire assembly
5. Bleed the brakes. Check and adjust the brake fluid level

Pinion Seal

REMOVAL & INSTALLATION

1. Before servicing the vehicle, refer to the Precautions Section.

2. Remove or disconnect the following:
- Driveshaft, matchmark for reassembly

- Pinion nut and washer using a suitable pinion flange holding tool
- Companion flange from the drive pinion

3. Pry the pinion seal out of the differential carrier.

To install:

4. Clean and inspect the sealing surface of the housing.

5. Install or connect the following:
- New seal into the housing until the flange on the seal is flush with the carrier. Using a seal driver.

6. With the seal installed, the pinion bearing preload must be set.
- Pinion nut (a new self-locking pinion nut must be used).
- Driveshaft, align the matchmarks

7. Check the level of the differential lubricant when finished.

STEERING

Air Bag

✳✳ CAUTION

These vehicles are equipped with an air bag system. The system must be disabled before performing service on or around system components, steering column, instrument panel components, wiring and sensors. Failure to follow safety and disabling procedures could result in accidental air bag deployment, possible personal injury and unnecessary system repairs.

PRECAUTIONS

Several precautions must be observed when handling the inflator module to avoid accidental deployment and possible personal injury.

- Never carry the inflator module by the wires or connector on the underside of the module.
- When carrying a live inflator module, hold securely with both hands, and ensure that the bag and trim cover are pointed away.
- Place the inflator module on a bench or other surface with the bag and trim cover facing up.
- With the inflator module on the bench, never place anything on or close to the module that may be thrown in the event of an accidental deployment.

DISARMING

To avoid personal injury when working on vehicles equipped with an air bag, the negative battery cable must be disconnected and at least 90 seconds must elapse before working on the system. Failure to do so may result in deployment of the air bag.

Power Steering Gear

REMOVAL & INSTALLATION

Montero

➡The vehicle is equipped with a supplemental restraint system (SRS), turn the front wheel to the straight ahead position and remove the ignition key to prevent the steering wheel from turning. Failure to do so may damage the

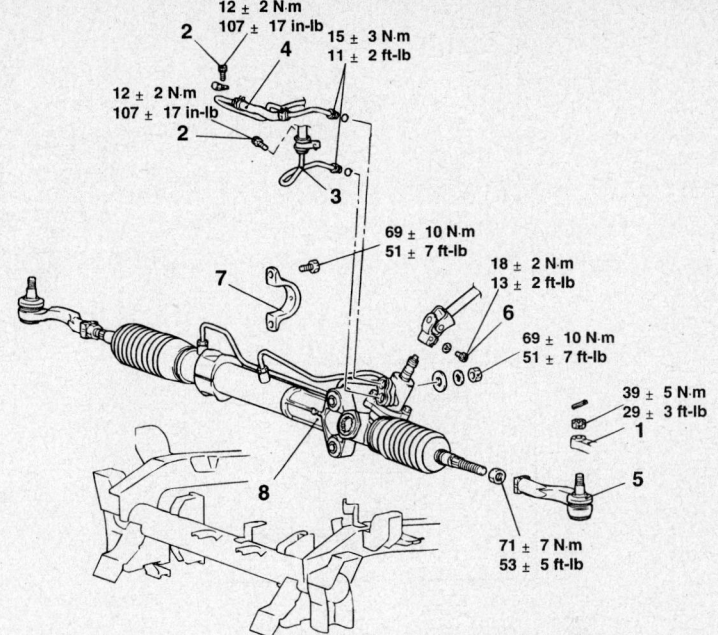

1. TIE ROD END AND KNUCKLE CONNECTION
2. BOLT
3. PRESSURE HOSE ASSEMBLY
4. RETURN TUBE
5. TIE ROD END (LH)
- INSTALLATION OF DIFFERENTIAL MOUNT BRACKET (LH)

6. STEERING SHAFT ASSEMBLY AND GEAR BOX CONNECTING BOLT
7. GEAR BOX CLAMP
8. GEAR BOX ASSEMBLY

09482_MONT_G0020

Power steering gear and related components—Montero

SRS clock spring and render the SRS system inoperative, risking serious injury.

1. Disconnect the negative battery cable.

✳✳ CAUTION

Wait at least 90 seconds after the negative battery cable is disconnected to prevent possible deployment of the air bag.

2. Drain the power steering fluid.
3. Remove or disconnect the following:
   - Engine under cover
   - Tie rod ends
   - Steering hoses
   - Left side differential mount bracket
   - Intermediate shaft-to-gear box bolt
   - Gear mounting clamps and gear
4. Installation is the reverse of removal. Observe the following torques:
   - Mounting clamp bolts: 51 ft. lbs. (69Nm)
   - Tie rod end ball stud nuts: 29 ft. lbs. (39Nm)
   - Intermediate shaft pinch bolt: 13 ft. lbs. (18Nm)

Montero Sport

➡The vehicle is equipped with a supplemental restraint system (SRS), turn the front wheel to the straight ahead position and remove the ignition key to prevent the steering wheel from turning. Failure to do so may damage the SRS clock spring and render the SRS system inoperative, risking serious injury.

1. Disconnect the negative battery cable.

✳✳ CAUTION

Wait at least 90 seconds after the negative battery cable is disconnected to prevent possible deployment of the air bag.

2. Drain the power steering fluid.
3. Remove or disconnect the following:
   - Pinch bolt securing the steering shaft to the steering gear
   - Pitman arm from the relay rod
   - Fluid lines from the steering gear
   - Mounting bolts securing the gear to the frame rail and steering gear

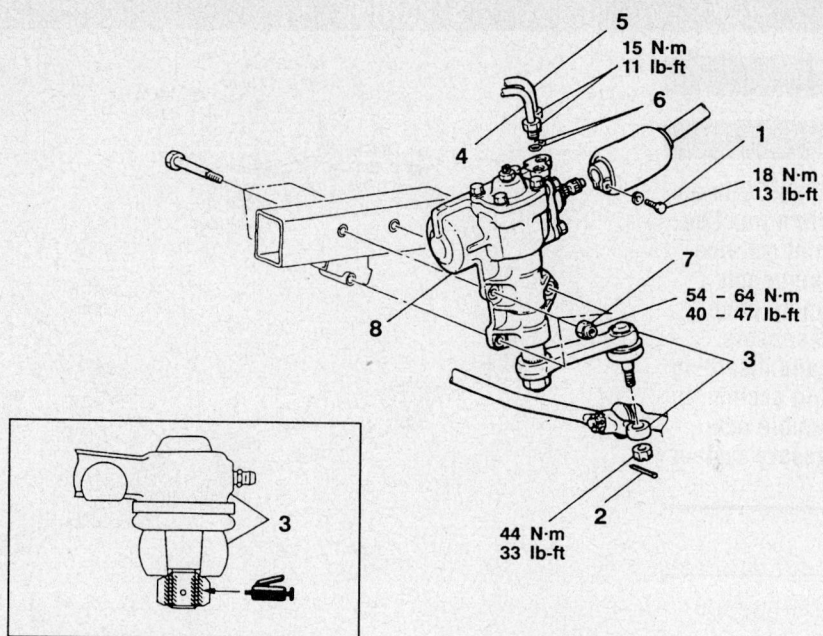

1. CONNECTING BOLT FOR STEERING GEAR BOX AND STEERING SHAFT
2. COTTER PIN
3. CONNECTION FOR PITMAN ARM AND RELAY ROD
4. PRESSURE TUBE
5. RETURN TUBE
6. O-RING
7. SELF-LOCKING NUT
8. POWER STEERING GEAR BOX

7924UG33

Power steering gear and related components—Montero Sport

To install:

4. Install or connect the following:
- Steering gear on the frame rail. Torque the nuts to 40–47 ft. lbs. (54–64 Nm).
- Fluid lines to the steering gear use a new O-rings. Torque the fittings to 11 ft. lbs. (15 Nm).
- Relay rod on the Pitman arm. Torque the nut to 33 ft. lbs. (44 Nm).
- Steering shaft on the steering gear.

Torque the bolt to 13 ft. lbs. (18 Nm).
- Negative battery cable

5. Refill and bleed the power steering system.

FRONT SUSPENSION

Shock Absorber

REMOVAL & INSTALLATION

Montero

1. Before servicing the vehicle, refer to the Precautions Section.
2. Remove the upper control arm assembly.
3. Remove the battery and the battery tray. Remove the radiator condenser tank. Remove the air cleaner assembly.
4. Remove the upper shock absorber mounting cap. Remove the mounting nuts.
5. Remove the lower shock absorber mounting bolt and nut.
6. Remove the shock absorber from the vehicle.

To install:

7. Install the shock absorber. Temporarily tighten the lower nut and bolt, but do not torque to specification at this time.

➡ **To prevent the bushings from breakage, the lower shock absorber nut and bolt should not be torqued to specification until the vehicle is on the ground in the unladen position.**

8. Torque the lower nut and bolt to 108–130 ft. lbs. (148–176 Nm)
9. Torque the upper nuts to 28–36 ft. lbs. (39–49 Nm).
10. Continue the installation in the reverse order of the removal procedure.
11. Test drive the vehicle and check the alignment.
12. Check and adjust all fluid levels, as required.

Montero Sport

1. Before servicing the vehicle, refer to the Precautions Section.
2. Remove the upper shock mounting nut, washer and bushing.
3. Raise and support the vehicle safely.
4. Remove the lower mounting bolts.
5. Remove the shock absorber from the vehicle.

To install:

➡ **If the shock absorber has a white paint mark on the lower end, be sure the mark faces the outside of the vehicle when installed.**

6. Install the shock absorber. Torque the lower nut to 65–76 ft. lbs. (88–103 Nm) and the upper nut to 11 ft. lbs. (15 Nm).
7. Test drive the vehicle and check the alignment.

Coil Spring

REMOVAL & INSTALLATION

Montero

1. Before servicing the vehicle, refer to the Precautions Section.
2. Remove or disconnect the following:
   - Shock absorber
3. Compress the coil spring until there is a clearance on both ends
4. Remove or disconnect the following:
   - Self locking nut
   - Seat
   - Collar
   - Bushing
   - Bracket
   - Upper pad
   - Cup
   - Helper rubber
   - Spring
   - Lower pad
5. Installation is the reverse of removal. Torque the center nut to 17 ft. lbs. (22 Nm).

Stabilizer Bar

REMOVAL & INSTALLATION

Montero

1. Before servicing the vehicle, refer to the Precautions Section.
2. Raise and support the vehicle safely.
3. Remove the under cover.
4. Remove the left and right stabilizer link assemblies.
5. Remove the left and right stabilizer bar retaining bolts. Remove the bushings.

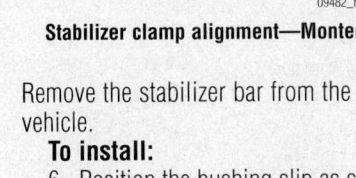

Stabilizer clamp alignment—Montero

Remove the stabilizer bar from the vehicle.

To install:

6. Position the bushing slip as shown in the illustration. Position the stabilizer clamps as shown in the illustration.
7. Install the stabilizer bar. Tighten the mounting bolts to 72–87 ft. lbs. (34–54 Nm).
8. Continue the installation in the reverse order of the removal procedure.

Montero Sport

1. Before servicing the vehicle, refer to the Precautions Section.
2. Raise and support the vehicle safely.
3. Remove the under cover.
4. Remove the outside left and right stabilizer bracket bushing retaining screws. Remove the brackets. Remove the bushings.
5. Remove the inside left and right stabilizer bracket bushing retaining bolts. Remove the brackets. Remove the bushings.
6. Remove the stabilizer bar from the vehicle.
7. Remove the stabilizer link assembly.

To install:

8. Installation is the reverse of the removal procedure.
9. Tighten the stabilizer link so that in the tightened position the threaded rod of the assembly protrudes about 0.2–0.3 inch.

Torsion Bar

REMOVAL & INSTALLATION

Montero Sport

1. Before servicing the vehicle, refer to the Precautions Section.
2. Remove the under cover and skid plate, if equipped.
3. Support the lower arm with a jack.
4. Remove or disconnect the following:
   - Heat protector, right side only
   - Bump stopper
   - Anchor adjustment nut and arm assembly

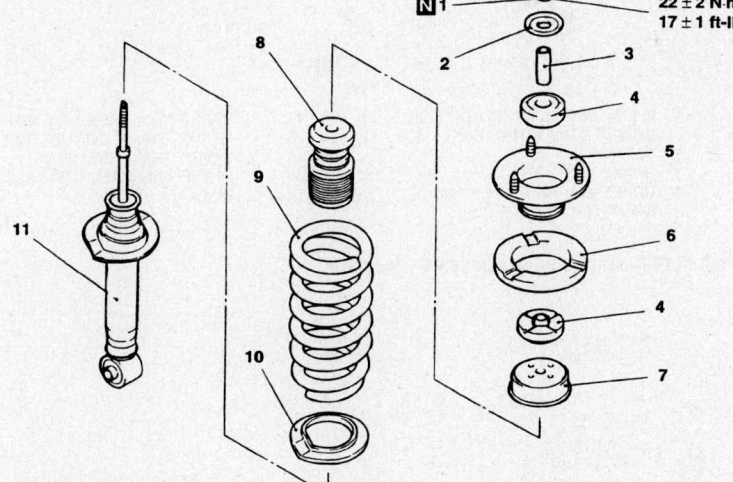

1. SELF-LOCKING NUT
2. SEAT
3. COLLAR
4. UPPER BUSHING
5. SPRING BRACKET ASSEMBLY
6. SPRING UPPER PAD
7. CUP ASSEMBLY
8. HELPER RUBBER
9. COIL SPRING
10. SPRING LOWER PAD
11. SHOCK ABSORBER ASSEMBLY

Front shock absorber and coil spring—Montero

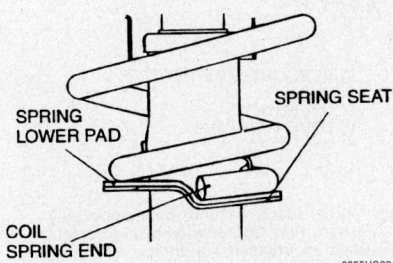

Coil spring installation locating point—Montero

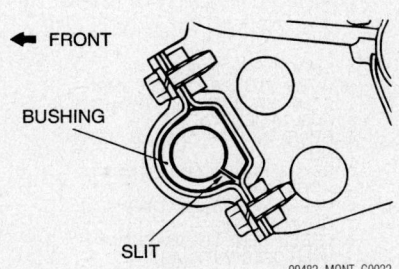

Stabilizer bushing alignment—Montero

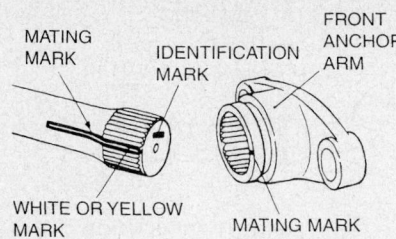

MATING MARK — IDENTIFICATION MARK — FRONT ANCHOR ARM

WHITE OR YELLOW MARK — MATING MARK

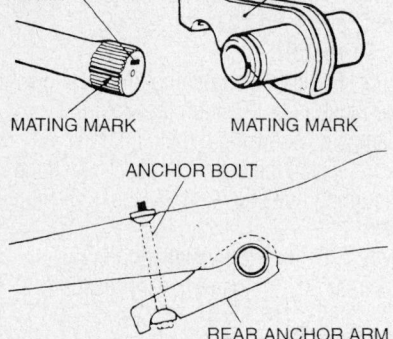

IDENTIFICATION MARK — REAR ANCHOR ARM ASSEMBLY

MATING MARK — MATING MARK

ANCHOR BOLT

REAR ANCHOR ARM

09482_MONT_G0021

Torsion bar identification and alignment points—Montero Sport

- Anchor collar
- Torsion bar
- Dust covers
- Heat covers, right side only

To install:

5. Check the identification marks at the end of the left and right torsion bars (R= right side, L= left side).

6. When installing the torsion bar, align the white mark on the serrated section of the torsion bar with the mating mark on the anchor arm.

7. Mount the anchor bolt (see illustration) and install the rear anchor arm adjusting nut.

8. Install or connect the following:
- Heat covers, right side only
- Dust covers
- Anchor collar
- Anchor adjustment nut and arm assembly.
- Heat protector, right side only

Upper Control Arm

REMOVAL & INSTALLATION

Montero

1. Before servicing the vehicle, refer to the Precautions Section.

2. Raise and support the vehicle safely. Remove the tire and wheel assembly.

⚠️ **CAUTION**
*: Indicates parts which should be temporarily tightened, and then fully tightened with the vehicle on the ground in an unladen condition.

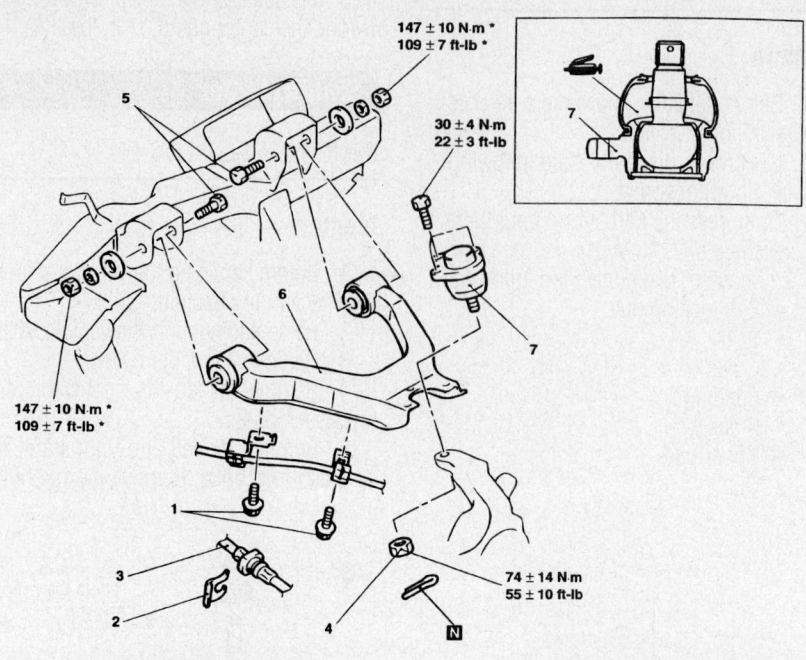

147 ± 10 N·m *
109 ± 7 ft-lb *

30 ± 4 N·m
22 ± 3 ft-lb

147 ± 10 N·m *
109 ± 7 ft-lb *

74 ± 14 N·m
55 ± 10 ft-lb

1. FRONT WHEEL SPEED SENSOR BRACKET MOUNTING BOLT
2. CLIP
3. BRAKE HOSE
4. UPPER ARM ASSEMBLY AND KNUCKLE CONNECTION
5. UPPER ARM ASSEMBLY AND FRONT FRAME CONNECTION
6. UPPER ARM ASSEMBLY
7. UPPER ARM BALL JOINT ASSEMBLY

9355UG03

Upper control arm and related components—Montero

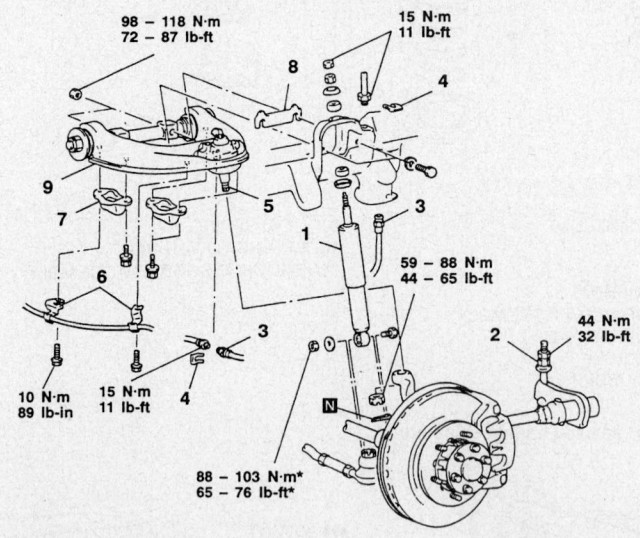

98 – 118 N·m
72 – 87 lb-ft

15 N·m
11 lb-ft

59 – 88 N·m
44 – 65 lb-ft

44 N·m
32 lb-ft

10 N·m
89 lb-in

15 N·m
11 lb-ft

88 – 103 N·m*
65 – 76 lb-ft*

1. SHOCK ABSORBER
● BUMP STOPPER AND BUMP STOPPER BRACKET CLEARANCE ADJUSTMENT
2. REAR ANCHOR ARM ADJUSTING NUT
3. BRAKE HOSE CONNECTION
4. HOSE CLIP
5. UPPER ARM BALL JOINT CONNECTION
6. SPEED SENSOR BRACKET <VEHICLES WITH ABS>
7. REBOUND STOPPER
8. SHIMS
9. UPPER ARM
10. UPPER ARM BALL JOINT ASSEMBLY

Caution
*: Indicates parts which should be temporarily tightened, and then fully tightened with the vehicle on the ground in an unladen condition.

7924UG34

Upper control arm and related components—Montero Sport

3. Properly support the lower control arm with a jack.

4. Remove the wheel speed sensor bracket mounting bolt.

5. Remove the brake hose clip. Remove the brake hose.

6. Using the proper tools, separate the upper arm ball joint and steering knuckle connection.

7. Remove the upper arm assembly and front frame retaining bolts and nuts.

8. Remove the upper control arm from the vehicle.

9. As required, and using the proper tools remove the ball joint from the upper control arm.

To install:

10. Installation is the reverse of the removal procedure.

11. Check the front end alignment.

Montero Sport

1. Before servicing the vehicle, refer to the Precautions Section.

2. Drain the brake fluid.

3. Raise and support the vehicle safely. Remove the tire and wheel assembly.

4. Properly support the lower control arm with a jack.

5. Remove the shock absorber.

6. Loosen the anchor arm bolt of the torsion bar, all the way.

➡**Be sure that when the anchor arm adjusting nut is loosened to use a jack to support the lower arm.**

7. Disconnect the brake line hose fitting. Remove the brake line hose clamp.

8. Using the proper tools, separate the upper arm ball joint and steering knuckle connection.

9. If equipped with ABS, remove the speed sensor retaining bracket. Remove the rebound stopper. Remove the shims.

10. Remove the upper control arm retaining bolts. Remove the upper control arm from the vehicle.

11. As required, and using the proper tools remove the ball joint from the upper control arm.

To install:

12. Installation is the reverse of the removal procedure.

13. Install the upper control arm so that the word "OUT" on the upper arm shaft is facing toward the outside of the vehicle.

14. Bleed the brake system. Correct and adjust the brake fluid level, as required.

15. Check the front end alignment.

Lower Control Arm

REMOVAL & INSTALLATION

Montero

1. Before servicing the vehicle, refer to the Precautions Section.

2. Raise and support the vehicle safely. Remove the tire and wheel assembly.

3. Remove the under cover. Remove the skid plate.

4. Disconnect the shock absorber and stabilizer link at the lower control arm assembly.

5. Remove the front brake assembly.

6. Disconnect the steering knuckle and tie rod end connection, using the proper tools.

7. Remove the dust cap, castle nut and washer. Remove the brake hose and vehicle speed sensor clamp bracket. Remove the wheel speed sensor.

8. Disconnect the upper control arm ball joint connection, using the proper tools.

9. Remove the halfshaft.

➡**When pulling the halfshaft out of the differential carrier, be careful that the spline part of the halfshaft does not damage the oil seal.**

10. Remove the lower ball joint cotter pin and retaining nut.

11. Using the proper tools disconnect the lower arm ball joint and knuckle assembly connection.

12. Using the proper tools, disconnect the tie rod end and knuckle assembly connection.

13. Remove the lower control arm ball joint retaining bolts. Remove the ball joint.

14. Remove the hub and knuckle assembly.

15. Disconnect the shock absorber and lower arm assembly connection. Remove the bump stopper.

16. Disconnect the lower arm assembly and stabilizer link connection.

17. Make an alignment mark on the bracket and eccentric cam bolt. Remove the lower control arm mounting bolts. Remove the lower control arm from the vehicle.

To install:

18. Installation is the reverse of the removal procedure.

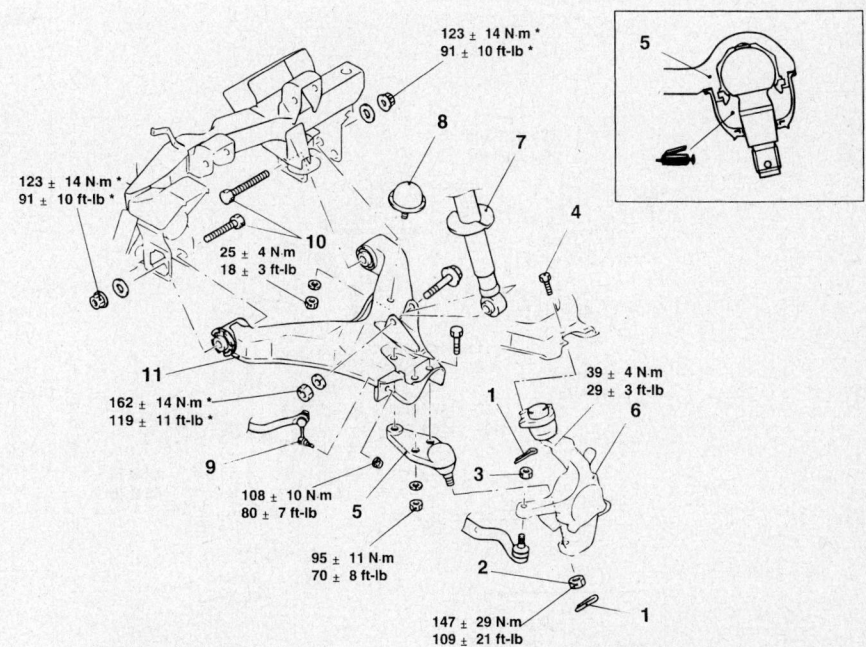

1. COTTER PIN
2. LOWER ARM BALL JOINT AND KNUCKLE ASSEMBLY CONNECTION
3. TIE ROD END AND KNUCKLE ASSEMBLY CONNECTION
4. UPPER ARM AND UPPER ARM BALL JOINT CONNECTION
5. LOWER ARM BALL JOINT
6. HUB AND KNUCKLE ASSEMBLY
7. SHOCK ABSORBER AND LOWER ARM ASSEMBLY CONNECTION
8. BUMP STOPPER
9. LOWER ARM ASSEMBLY AND STABILIZER LINK CONNECTION
10. LOWER ARM MOUNTING BOLT
11. LOWER ARM ASSEMBLY

09482_MONT_G0024

Lower control arm and related components—Montero

19. Torque all retaining bolts to specifications.

➡ **Bolt torques are indicated in the accompanying illustration. To prevent the bushings from breakage, the parts indicated with an asterisk (*) should be temporarily tightened, and then fully tightened with the vehicle on the ground in the unladen condition.**

20. Check the front end alignment.

Montero Sport

1. Before servicing the vehicle, refer to the Precautions Section.
2. Raise and support the vehicle safely. Remove the tire and wheel assembly.
3. Remove the under cover. Remove the skid plate.
4. Remove the torsion bar. Remove the heat cover, right side. Remove the dust covers.
5. Disconnect the lower arm ball joint connection. Disconnect the stabilizer bar connection. Remove the bump stopper.
6. Remove the lower arm shaft and front anchor arm.

7. Remove the lower control arm retaining bolts. Remove the lower control arm from the vehicle.
8. Remove the lower control arm ball joint assembly. Remove the stopper bolt.

To install:

9. Installation is the reverse of the removal procedure.
10. Torque all retaining bolts to specifications.

➡ **Bolt torques are indicated in the accompanying illustration. To prevent the bushings from breakage, the parts indicated with an asterisk (*) should be temporarily tightened, and then fully tightened with the vehicle on the ground in the unladen condition.**

11. Check the front end alignment.

CONTROL ARM BUSHING REPLACEMENT

1. Before servicing the vehicle, refer to the Precautions Section.
2. Raise and support the vehicle safely. Remove the tire and wheel assembly.
3. Remove the lower control arm and place in a vise.

4. Using a bushing removal tool remove the bushing from its mounting

To install:

5. Position the bushing with the larger end facing the front of the vehicle.

➡ **Coat the bushing with a soap solution and take care not to twist.**

6. Using the bushing installation tool, press the bushing into the bracket.
7. Install the lower control arm.
8. Install the tire and wheel assembly. Lower the vehicle.

Wheel Bearings

ADJUSTMENT

Montero

The bearings are integral with the hub. No adjustment is possible.

Montero Sport

1. With the caliper removed, check the rotational starting torque. Rotational torque should be 2.7–11.5 inch lbs. (0.3–1.3 Nm).

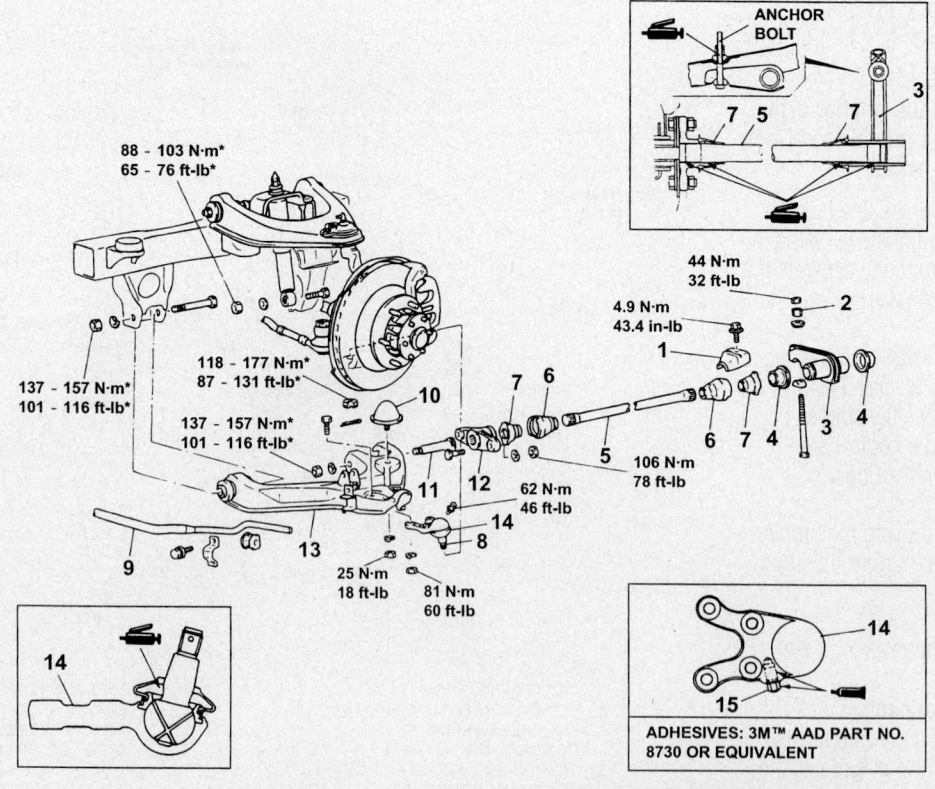

1. HEAT PROTECTOR (RIGHT SIDE ONLY)
2. ANCHOR ARM ASSEMBLY ADJUSTING NUT
3. REAR ANCHOR ARM ASSEMBLY
4. ANCHOR COLLAR
5. TORSION BAR
6. HEAT COVER (RIGHT SIDE ONLY)

09482_MONT_G0025

Lower control arm and related components—Montero Sport

Rotational torque can be adjusted by tightening or loosening the adjusting nut.

2. Check the hub axial. Endplay should not exceed 0.002 inch (0.05mm). If adjusting nut tightening does not bring the axial play within specifications, the bearings must be replaced.

3. Check hub endplay. Endplay should be 0.02–0.03 inch (0.4–0.7mm). Shims are available to adjust endplay.

4. Install the hub assembly. Tighten the nut to 94–145 ft. lbs. (127–196 Nm). Loosen it completely. Tighten the nut to 18 ft. lbs. (25Nm), then back it off 30 degrees.

REMOVAL & INSTALLATION

Montero

The wheel bearings are not replaceable. If defective, the hub/bearing assembly must be replaced.

1. Before servicing the vehicle, refer to the Precautions Section.

2. Remove or disconnect the following:
   - Tire and wheel assembly
   - Hub cover
   - Nut
   - Washer
   - Brake hose
   - Speed sensor
   - Caliper
   - Rotor
   - Dust cover
   - Tie rod end
   - Upper and lower arms from the knuckle
   - Rotor shield
   - Hub/knuckle assembly

3. Mount the assembly in a vise. Install tool MB990998, or equivalent on the hub. On 2002–2003 vehicles tighten the nut to 167–209 ft. lbs. (226–284 Nm) and on 2004–2006 vehicles tighten the nut to 141–155 ft. lbs. (190–210 Nm). Check the rotation starting torque. Torque should be 15.48 inch lbs. (1.75 Nm). Wheel bearing backlash should be 0.

4. If the hub is to be replaced, remove the hub-to-knuckle bolts.

5. Installation is the reverse of removal. Observe the following torques:
   - Hub-to-knuckle bolts: 65 ft. lbs. (88 Nm)

6. On 2002–2003 vehicles tighten the hub nut to 167–209 ft. lbs. (226–284 Nm) and on 2004–2006 vehicles tighten the hub nut to 141–155 ft. lbs. (190–210 Nm).

Montero Sport

1. Before servicing the vehicle, refer to the Precautions Section.

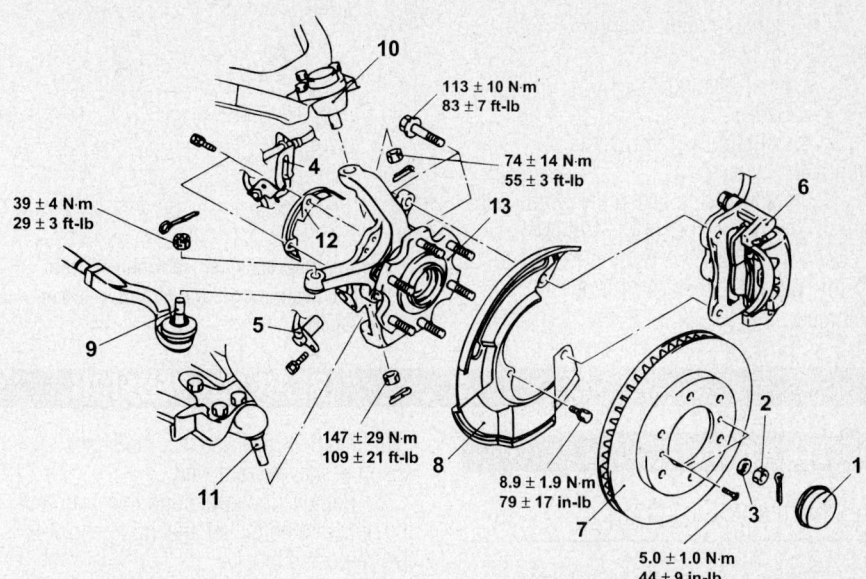

1. HUB CAP
2. CASTLE NUT
3. WASHER
4. BRAKE HOSE AND VEHICLE SPEED SENSOR CLAMP BRACKET
5. WHEEL SPEED SENSOR
6. DISC BRAKE ASSEMBLY
7. BRAKE DISC
8. DUST COVER
9. TIE ROD END, HUB AND KNUCKLE ASSEMBLY CONNECTION
10. UPPER ARM, HUB AND KNUCKLE ASSEMBLY CONNECTION
11. LOWER ARM, HUB AND KNUCKLE ASSEMBLY CONNECTION
12. ROTOR PROTECTOR
13. HUB AND KNUCKLE ASSEMBLY

09482_MONT_G0026

Front hub and bearing assembly—Montero

2. Remove or disconnect the following:
   - Wheel
   - Caliper
   - Hub cover
   - Snapring
   - Shim
   - Drive flange
   - Spring washer

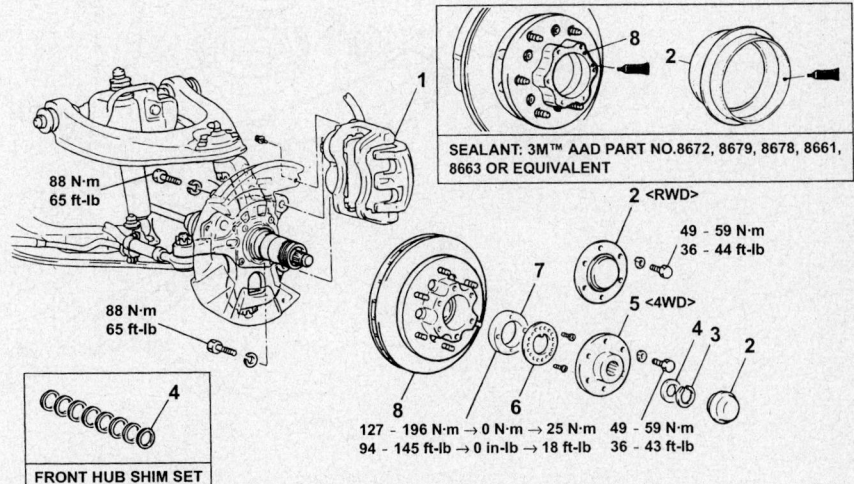

1. CALIPER ASSEMBLY
2. HUB CAP
3. SNAP RING
4. SHIM
5. DRIVE FLANGE
6. SPRING WASHER
7. JAM NUT
8. FRONT HUB ASSEMBLY

09482_MONT_G0027

Front hub and bearing assembly—Montero Sport

- Nut
- Hub and outer bearing
- Oil seal
- Inner bearing
- Races

3. Installation is the reverse of removal.

4. Install the hub assembly. Tighten the nut to 94–145 ft. lbs. (127–196 Nm). Loosen it completely. Tighten the nut to 18 ft. lbs. (25 Nm), then back it off 30 degrees.

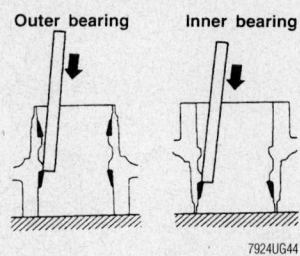

The bearing races can be removed from the hub using a drift and hammer—Montero Sport

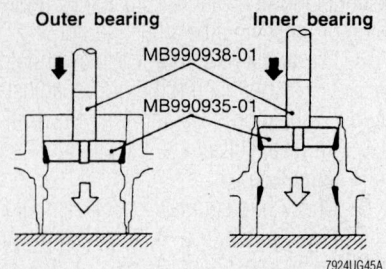

Install the new races into the hub using the proper size driver—Montero Sport

REAR SUSPENSION

Shock Absorber

REMOVAL & INSTALLATION

1. Before servicing the vehicle, refer to the Precautions Section.

2. Raise and support the vehicle safely.

3. Support the rear axle assembly with a hydraulic floor jack, so that the shock absorber may be removed.

4. Remove the upper and lower mounting nuts and bolts that attach the shock to the frame and bracket.

5. Remove the shock absorber from the vehicle.

To install:

6. Installation is the reverse of the removal procedure.

7. Torque all retaining bolts to specifications.

➡ **Bolt torques are indicated in the illustration. To prevent the bushings from breakage, the parts indicated with an asterisk (*) should be temporarily tightened, and then fully tightened with the vehicle on the ground in the unladen condition.**

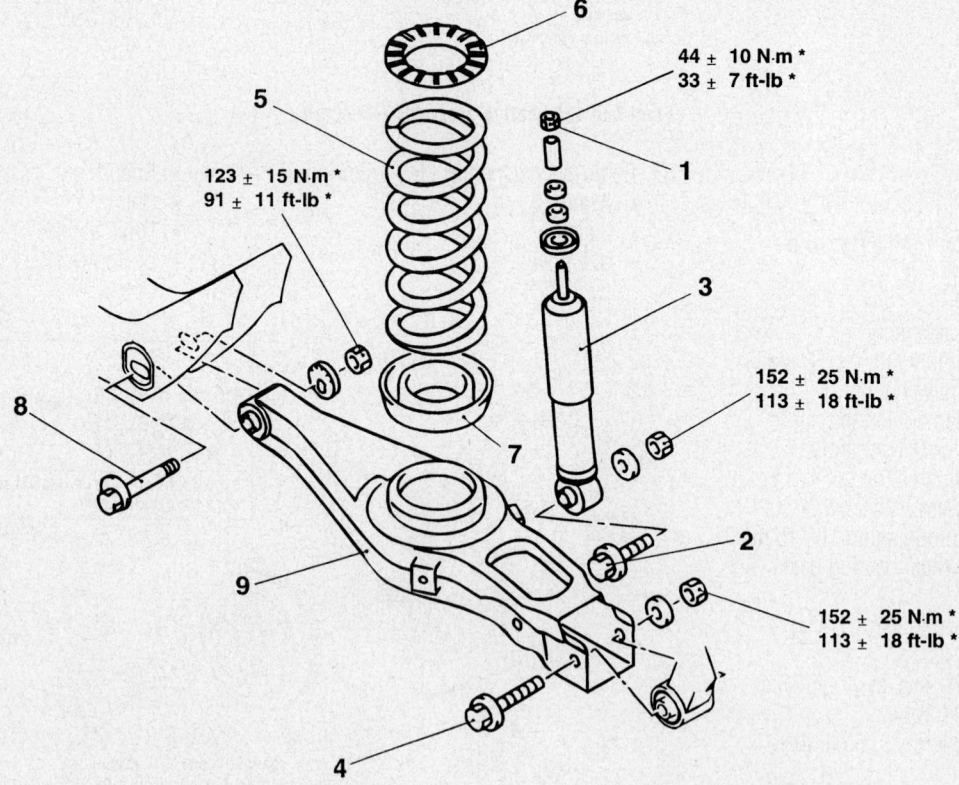

1. SHOCK ABSORBER MOUNTING NUT
2. SHOCK ABSORBER MOUNTING BOLT
3. SHOCK ABSORBER
4. LOWER ARM MOUNTING BOLT
5. COIL SPRING
6. SPRING UPPER PAD
7. SPRING LOWER PAD
8. BOLT ASSEMBLY (CAMBER ADJUSTING BOLT)
9. LOWER ARM ASSEMBLY

Shock absorber and related torque specifications—Montero

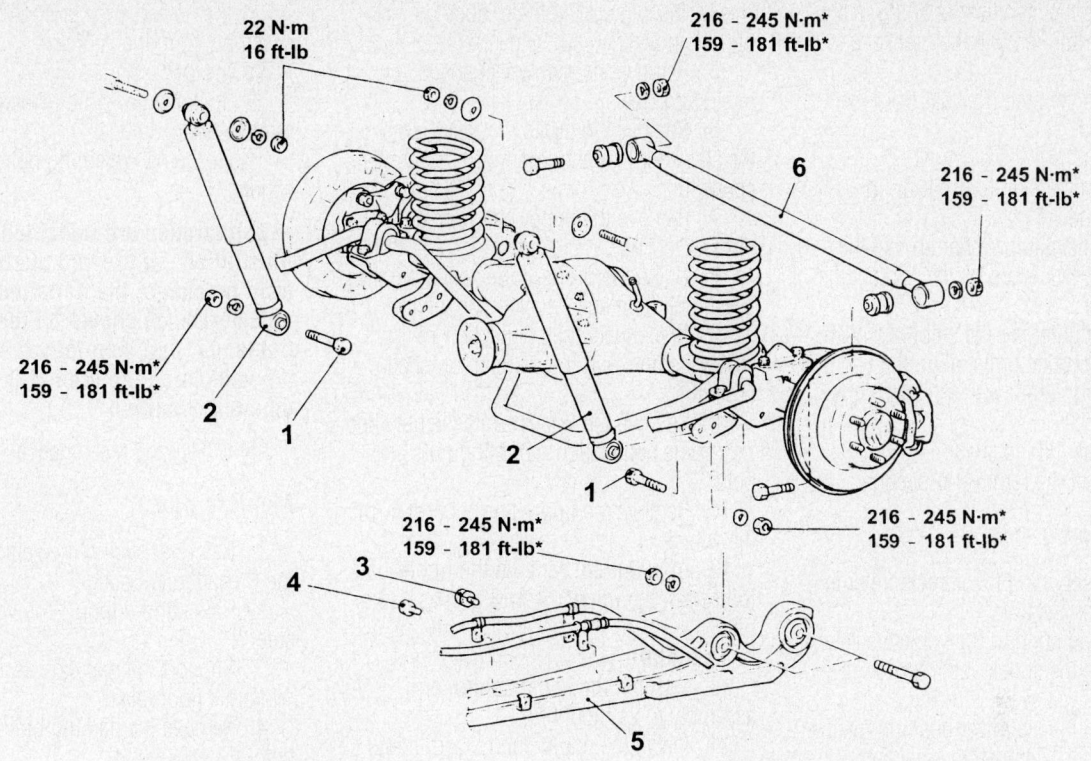

22 N·m
16 ft-lb

216 - 245 N·m*
159 - 181 ft-lb*

216 - 245 N·m*
159 - 181 ft-lb*

216 - 245 N·m*
159 - 181 ft-lb*

216 - 245 N·m*
159 - 181 ft-lb*

216 - 245 N·m*
159 - 181 ft-lb*

1. SHOCK ABSORBER MOUNTING
   BOLT
2. SHOCK ABSORBER
3. PARKING BRAKE CABLE
   ATTACHING BOLT (LH)
4. SPEED SENSOR ATTACHING
   BOLT (LH) <VEHICLES WITH
   ABS>
5. LOWER ARM (LH)
6. LATERAL ROD

09482_MONT_G0029

Shock absorber and related torque specifications—Montero Sport

8. On Montero, check the rear wheel alignment.

Coil Spring

REMOVAL & INSTALLATION

Montero

1. Before servicing the vehicle, refer to the Precautions Section.
2. Raise and support the vehicle safely.
3. Support the rear axle assembly with a hydraulic floor jack, so that the shock absorber may be removed.
4. Disconnect the lower shock absorber retaining bolt.
5. Position a jack under the lower arm assembly. Compress the coil spring to remove the lower arm retaining bolt.
6. Lower the floor jack and remove the coil spring from the vehicle.
7. Remove the spring upper and lower pads.

To install:
8. Installation is the reverse of the removal procedure.

9. Torque all retaining bolts to specifications.

➡Bolt torques are indicated in the illustration. To prevent the bushings from breakage, the parts indicated with an asterisk (*) should be temporarily tightened, and then fully tightened with the vehicle on the ground in the unladen condition.

10. Check the rear wheel alignment.

Montero Sport

1. Before servicing the vehicle, refer to the Precautions Section.
2. Support the weight of the axle.
3. Remove or disconnect the following:
   • Breather hose
   • Parking brake cable attaching bolt
   • ABS speed sensor attaching bolt
   • Brake hose connection
   • Lower shock mounting bolt
   • Bolt that attaches the lateral rod to the body
   • Stabilizer bar
4. Lower the axle and remove the coil spring and seat

To install:
5. Installation is the reverse of the removal procedure.
6. Torque all retaining bolts to specifications.

➡Bolt torques are indicated in the illustration. To prevent the bushings from breakage, the parts indicated with an asterisk (*) should be temporarily tightened, and then fully tightened with the vehicle on the ground in the unladen condition.

7. Be sure to bleed the brakes. Fill the master cylinder with the proper grade and type brake fluid.

Stabilizer Bar

REMOVAL & INSTALLATION

Montero

1. Before servicing the vehicle, refer to the Precautions Section.
2. Raise and support the vehicle safely.
3. Remove the stabilizer link assembly.

4. Remove the stabilizer bushing retaining bracket bolts. Remove the bracket and bushing.

5. Remove the stabilizer bar from the vehicle.

To install:

6. Position the stabilizer bar on its mounting in the vehicle.

7. Install the stabilizer bar so that the identification mark faces the left side of the vehicle.

8. Before tightening the stabilizer bushing retaining bracket bolts, align the end of the identification mark with the end of the bushing.

9. Continue the installation in the reverse order of the removal procedure.

Montero Sport

1. Before servicing the vehicle, refer to the Precautions Section.

2. Raise and support the vehicle safely.

3. Remove the shock absorber lower retaining bolt.

4. Remove the stabilizer bushing retaining bracket bolts. Remove the bracket and bushing.

5. Remove the stabilizer bar mounting bolt and nut. Remove the joint cup, rubber bushing and collar.

6. Remove the stabilizer bar from the vehicle.

To install:

7. Position the stabilizer bar on its mounting in the vehicle.

8. Temporarily tighten the lower shock absorber retaining bolt. Torque to specification once the vehicle is lowered to the ground.

9. Continue the installation in the reverse order of the removal procedure.

➡**Bolt torques are indicated in the illustration. To prevent the bushings from breakage, the parts indicated with an asterisk (*) should be temporarily tightened, and then fully tightened with the vehicle on the ground in the unladen condition.**

Upper Control Arm

REMOVAL & INSTALLATION

Montero

1. Before servicing the vehicle, refer to the Precautions Section.

2. Drain some brake fluid from the master cylinder, as required.

3. Raise and support the vehicle safely.

4. Disconnect the brake hose connection. Remove the clip.

5. Remove the rear wheel speed sensor retaining screws.

6. Remove the bump stopper. Remove the stabilizer link and upper arm assembly connection.

7. Remove the upper ball joint cotter pin and locknut.

8. Using the proper tools, separate the upper ball joint and knuckle connection.

9. Disconnect the connection of the DOJ assembly and the companion shaft assembly.

10. Remove the upper arm assembly and rear frame connection mounting nuts and bolts.

11. Remove the upper control arm from the vehicle.

12. As required, remove the upper ball joint from the upper control arm assembly.

To install:

13. Position the upper control arm assembly to its mounting.

14. Install the upper control arm retaining nuts and bolts.

15. After installing the upper arm, tighten the connecting bolt of the DOJ assembly and the companion shaft assembly to 74–94 ft. lbs.

16. Continue the installation in the reverse order of the removal procedure.

17. Check and adjust the rear wheel alignment.

18. Bleed the brake system. Fill the master cylinder with the proper grade and type brake fluid.

Lower Control Arm

REMOVAL & INSTALLATION

Montero

1. Before servicing the vehicle, refer to the Precautions Section.

2. Raise and support the vehicle safely.

3. Support the rear axle assembly with a hydraulic floor jack.

4. Disconnect the lower shock absorber retaining bolt.

5. Position a jack under the lower arm assembly. Compress the coil spring to remove the lower arm retaining bolt. Lower the floor jack and remove the coil spring. Remove the spring upper and lower pads.

6. Make mating marks to the bracket and the camber adjusting bolt. Remove the camber adjusting bolt.

7. Remove the lower control arm assembly from the vehicle.

To install:

8. Installation is the reverse of the removal procedure.

9. Torque all retaining bolts to specifications.

➡**Bolt torques are indicated in the illustration. To prevent the bushings from breakage, the parts indicated with an asterisk (*) should be temporarily tightened, and then fully tightened with the vehicle on the ground in the unladen condition.**

10. Check the rear wheel alignment.

Montero Sport

1. Before servicing the vehicle, refer to the Precautions Section.

2. Raise and support the vehicle safely.

3. Support the rear axle assembly with a hydraulic floor jack.

4. Remove the parking brake attaching bolt.

5. If equipped with ABS, remove the speed sensor attaching bolt.

6. Remove the lower control arm retaining bolts. Remove the lower control arm from the vehicle.

To install:

7. Installation is the reverse of the removal procedure.

8. Torque all retaining bolts to specifications.

➡**Bolt torques are indicated in the illustration. To prevent the bushings from breakage, the parts indicated with an asterisk (*) should be temporarily tightened, and then fully tightened with the vehicle on the ground in the unladen condition.**

CONTROL ARM BUSHING REPLACEMENT

1. Before servicing the vehicle, refer to the Precautions Section.

2. Remove the wheel.

3. Remove the lower control arm.

4. On Montero, use tool MB990881 to press out the bushing.

5. On Montero Sport, use tools MB990881, MB99097 and MB991318 to press out the bushing.

To install:

6. Position the bushing with the larger end facing the front of the vehicle.

7. On Montero use tool MB990881 to press the bushing into the bracket.

8. On Montero Sport, use tools

MB990881, MB99097 and MB991318 to press the bushing into the bracket.

9. Install the lower control arm.

10. Install the wheel.

Wheel Bearings

ADJUSTMENT

Montero

The rear wheel bearings are not adjustable. If the bearings are noisy or become loose, they must be replaced.

REMOVAL & INSTALLATION

Montero

1. Before servicing the vehicle, refer to the Precautions Section.

2. Raise and support the vehicle safely.

3. Remove the tire and wheel assembly.

4. Remove the dust cap. Remove the cotter pin. Remove the half shaft locknut.

5. Remove the caliper assembly and position it to the side.

6. Remove the rotor.

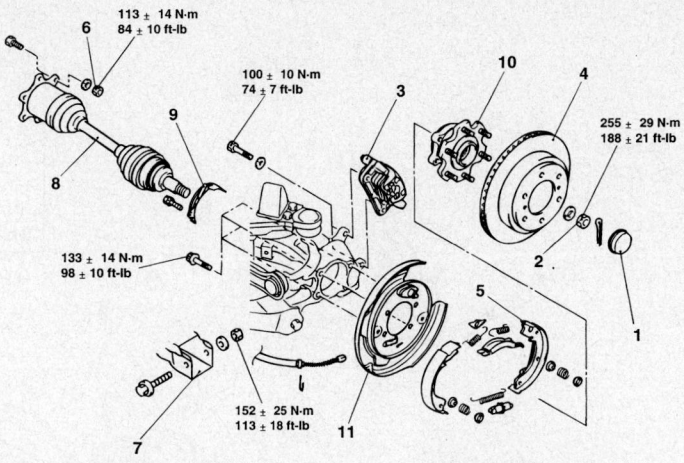

1. HUB CAP
2. DRIVE SHAFT NUT
3. DISC BRAKE ASSEMBLY
4. BRAKE DISC
5. SHOE AND LINING ASSEMBLY
6. COMPANION SHAFT AND DRIVE SHAFT CONNECTION
7. LOWER ARM AND KNUCKLE CONNECTION
8. REAR DRIVE SHAFT ASSEMBLY
9. ABS ROTOR PROTECTOR
10. REAR HUB ASSEMBLY
11. BACKING PLATE

09482_MONT_G0030

Rear hub and bearing assembly—Montero

7. Remove the parking brake shoe and lining assembly.

8. Remove the companion shaft and halfshaft retaining bolts.

9. Remove the lower arm and knuckle connection.

10. Remove the halfshaft assembly. Remove the ABS rotor protector.

11. Remove the rear hub and bearing assembly.

To install:

12. Installation is the reverse of the removal procedure.

13. Adjust the parking brake lever stroke, as required.

BRAKES

Brake Caliper

REMOVAL AND INSTALLATION

Front

1. Before servicing the vehicle, refer to the Precautions Section.

2. Raise and safely support the vehicle.

3. Remove or disconnect the following:
   - Wheel and tire assembly
   - Brake hose from the caliper brake line and remove the retaining clip
   - Caliper guide pin bolts
   - Caliper, by lifting it from the caliper support

To install:

4. Make sure the disc brake pad shims and clips are properly positioned.

5. Position the caliper over the rotor so the caliper engages the adapter correctly

6. Install or connect the following:
   - Mounting pins
   - Brake hose to the caliper brake line
   - Retaining clip

7. Bleed the brake system.
   - Wheel and tire assembly

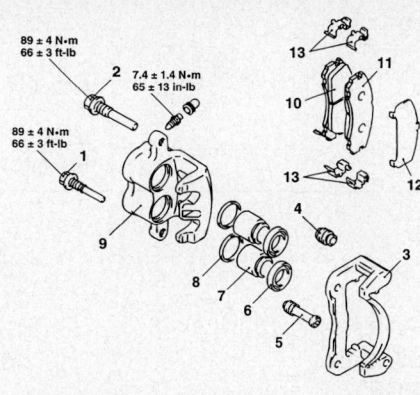

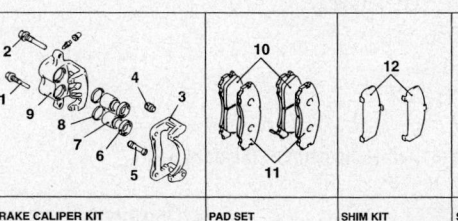

1. GUIDE PIN LOCK BOLT
2. GUIDE PIN
3. CALIPER SUPPORT, PAD, CLIP AND SHIM ASSEMBLY
4. BOOT
5. BUSHING
6. PISTON BOOT
7. PISTON
8. PISTON SEAL
9. CALIPER BODY
10. PAD AND WEAR INDICATOR ASSEMBLY
11. PAD ASSEMBLY
12. SHIM
13. CLIP

BRAKE CALIPER KIT   PAD SET   SHIM KIT   SEAL AND BOOT KIT

GREASE

CLIP KIT

09482_MONT_G0031

Front disc brake caliper and related components—Montero

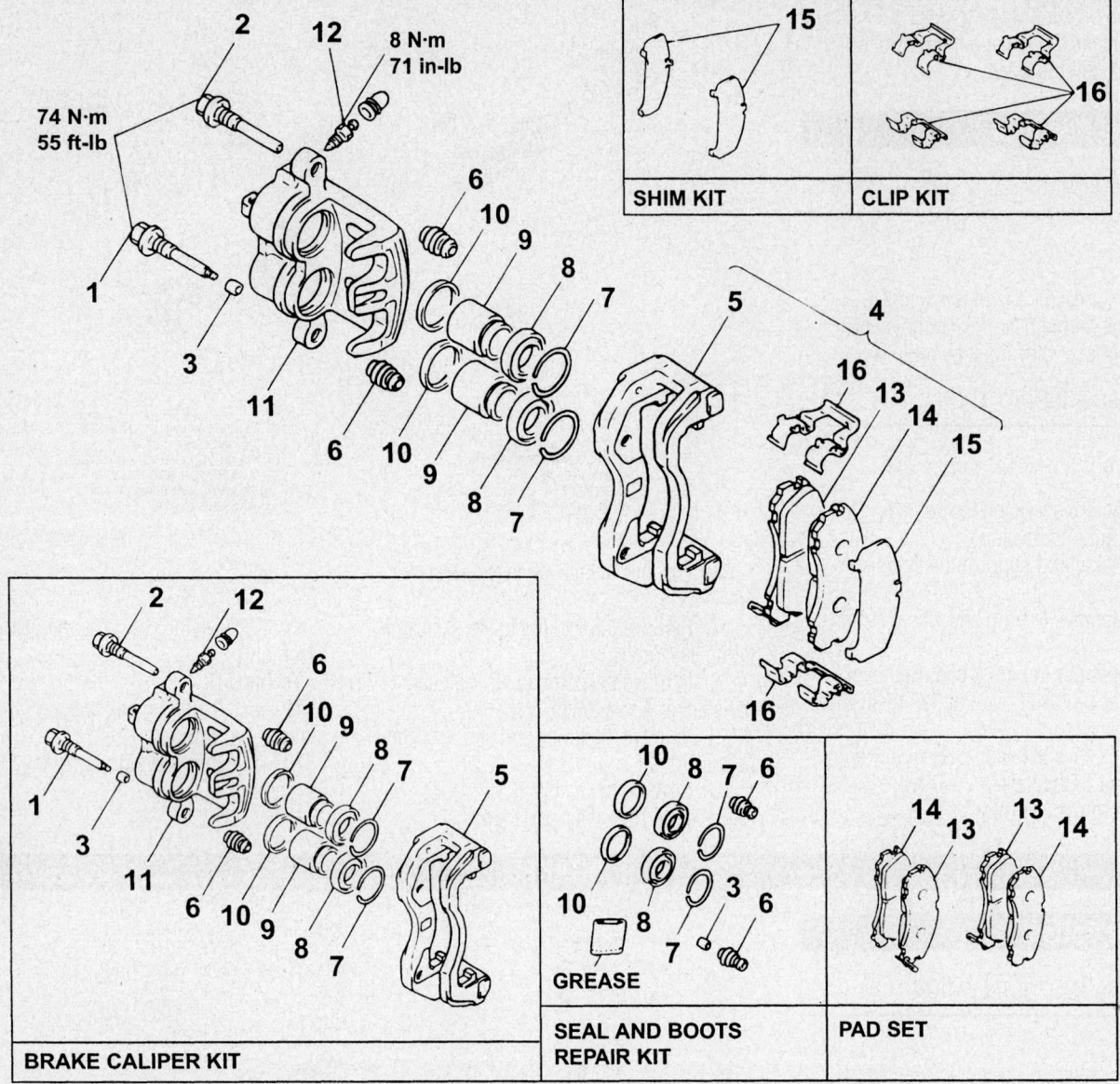

Front disc brake caliper and related components—Montero Sport

1. LOCK PIN
2. GUIDE PIN
3. BUSHING
4. CALIPER SUPPORT, PAD, CLIP AND SHIM ASSEMBLY
5. CALIPER SUPPORT
6. PIN BOOT
7. BOOT RING
8. PISTON BOOT
9. PISTON
10. PISTON SEAL
11. CALIPER BODY
12. BLEEDER SCREW
13. PAD AND WEAR INDICATOR ASSEMBLY
14. PAD ASSEMBLY
15. OUTER SHIM
16. CLIP

09482_MONT_G0032

Rear

1. Before servicing the vehicle, refer to the Precautions Section.

2. Raise and safely support the vehicle.

3. Remove or disconnect the following:
   • Wheel and tire assembly
   • Brake hose from the caliper brake line and remove the retaining clip
   • Caliper guide pin bolts
   • Caliper by lifting it from the caliper support

To install:

4. Make sure the disc brake pad shims and clips are properly positioned.

5. Position the caliper over the rotor so the caliper engages the adapter correctly.

6. Install or connect the following:
   • Mounting pins
   • Brake hose to the caliper brake line
   • Retaining clip

7. Bleed the brake system.
   • Wheel and tire assembly

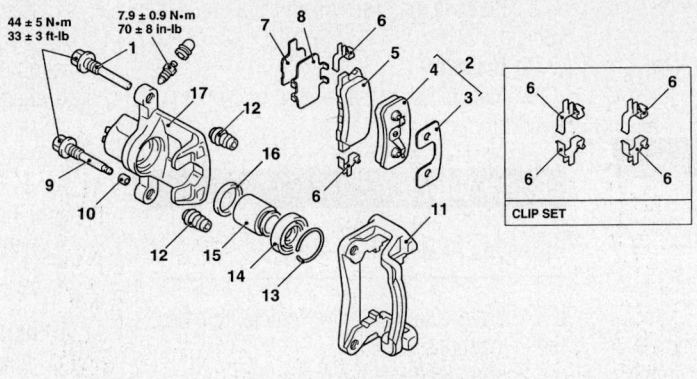

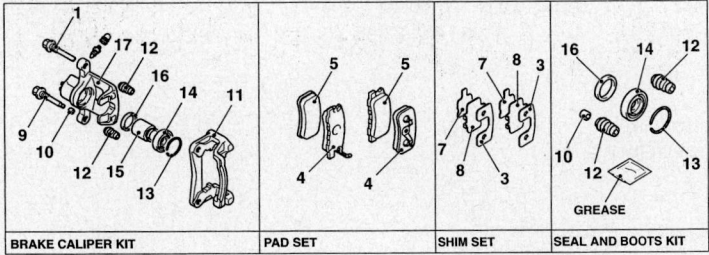

1. GUIDE PIN
2. PAD AND CLIP ASSEMBLY AND SHIM
3. SHIM
4. PAD AND CLIP ASSEMBLY
5. PAD ASSEMBLY
6. PAD CLIP
7. SHIM B
8. SHIM A
9. LOCK PIN
10. BUSHING
11. CALIPER SUPPORT
12. PIN BOOT
13. BOOT RING
14. PISTON BOOT
15. PISTON
16. PISTON SEAL
17. CALIPER BODY

09482_MONT_G0033

Rear disc brake caliper and related components—Montero

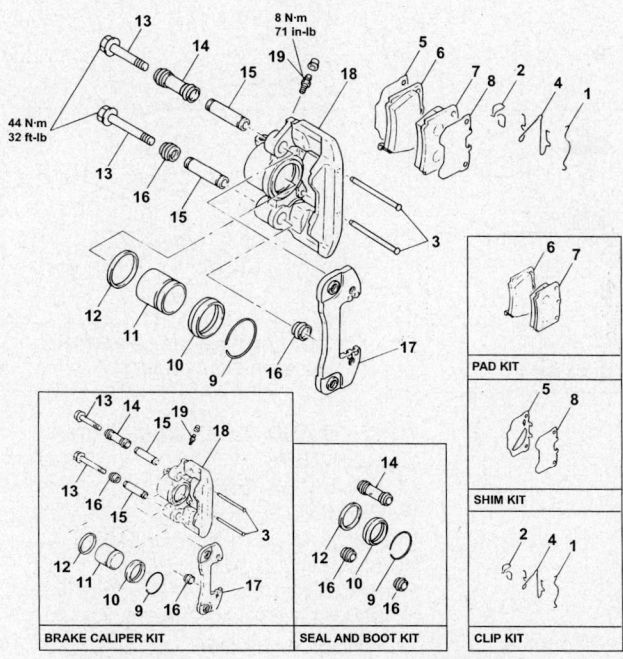

1. CLIP
2. K-SPRING
3. PAD PIN
4. SPRING
5. INNER SHIM
6. PAD AND WEAR INDICATOR
   ASSEMBLY
7. PAD ASSEMBLY
8. OUTER SHIM
9. RETAINING RING
10. PISTON BOOT
11. PISTON
12. PISTON SEAL
13. SLEEVE BOLT
14. BUSHING
15. SLEEVE
16. PIN BOOT
17. INNER CALIPER
18. TORQUE PLATE
19. BLEEDER SCREW

09482_MONT_G0034

Rear disc brake caliper and related components—Montero Sport

Disc Brake Pads

REMOVAL AND INSTALLATION

1. Before servicing the vehicle, refer to the Precautions Section.

2. Remove ½ of the brake fluid from the master cylinder.

3. Raise and safely support the vehicle.

4. Remove or disconnect the following:
- Wheel and tire assembly
- Lower caliper guide pin bolt
- Caliper from the caliper support
- Disc brake pads, shims, and the clips from the caliper support

To install:

5. Clean the exposed portion of the caliper piston, then press the piston back into the caliper bore using the old inner brake pad and a C-clamp.

6. Install or connect the following:
- Disc brake pads, shims, and the clips. Make sure the shims and clips are properly positioned.
- Caliper over the rotor so the caliper engages the adapter correctly
- Mounting pin(s) and the rear caliper.
- Wheel and tire assembly and lower the vehicle

7. Apply the brake pedal several times until a firm pedal is obtained. Check the fluid level in the master cylinder and add fluid, as necessary.

Brake Drums

REMOVAL AND INSTALLATION

1. Before servicing the vehicle, refer to the Precautions Section.
2. Raise and safely support the vehicle.
3. Remove the tire and wheel assembly.
4. loosen the parking brake cable adjusting nut.

5. Remove the brake drum from the vehicle.

To install:

6. Installation is the reverse of the removal procedure.

Brake Shoes

REMOVAL AND INSTALLATION

1. Before servicing the vehicle, refer to the Precautions Section.
2. Remove ½ of the brake fluid from the master cylinder.
3. Raise and safely support the vehicle.

4. Remove the tire and wheel assembly.
5. loosen the parking brake cable adjusting nut.
6. Remove the brake drum.
7. Remove the shoe to lever spring, adjuster lever and auto adjuster assembly.
8. Remove the retainer spring. Remove the shoe hold down cup, spring, and cup.
9. Remove the shoe to shoe spring.
10. Remove the shoe and lining assembly.

To install:

11. Installation is the reverse of the removal procedure.
12. Be sure to check and adjust the brake fluid level, as required.

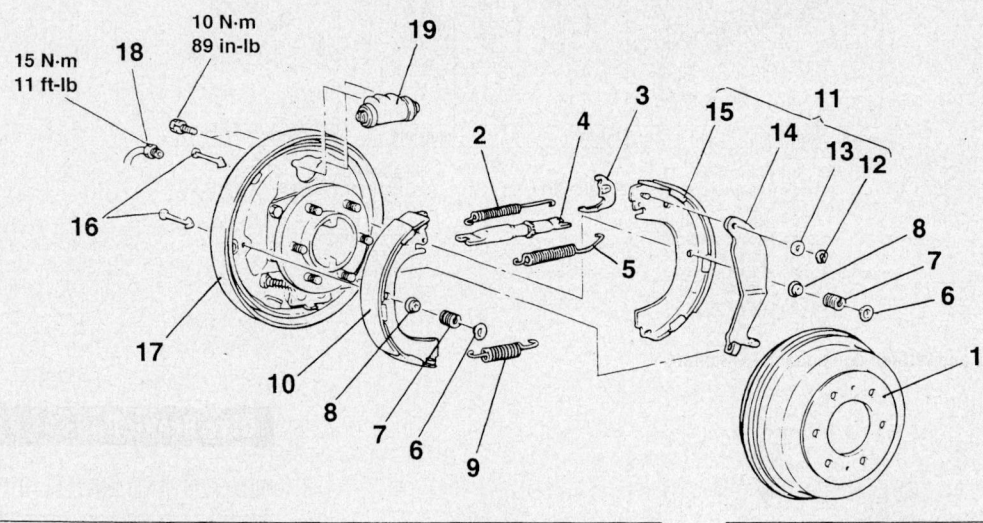

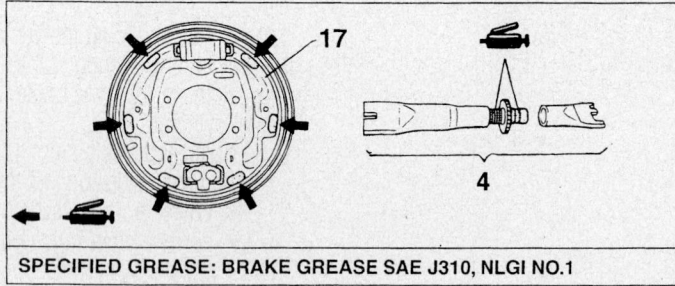

SPECIFIED GREASE: BRAKE GREASE SAE J310, NLGI NO.1

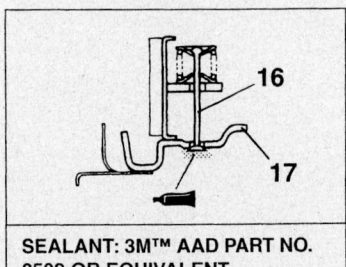

SEALANT: 3M™ AAD PART NO. 8509 OR EQUIVALENT

1. BRAKE DRUM
2. SHOE-TO-LEVER SPRING
3. ADJUSTER LEVER
4. AUTO ADJUSTER ASSEMBLY
5. RETAINER SPRING
6. SHOE HOLD-DOWN CUP
7. SHOE HOLD-DOWN SPRING
8. SHOE HOLD-DOWN CUP
9. SHOE-TO-SHOE SPRING
10. SHOE AND LINING ASSEMBLY
11. SHOE AND LEVER ASSEMBLY
12. RETAINER
13. WAVE WASHER
14. PARKING LEVER
15. SHOE AND LINING ASSEMBLY
16. SHOE HOLD-DOWN PIN
17. BACKING PLATE
18. BRAKE TUBE CONNECTION
19. WHEEL CYLINDER

09482_MONT_G0035

Rear drum brake and related components—Montero Sport

MITSUBISHI

14

Outlander

SPECIFICATIONS AND MAINTENANCE CHARTS

ENGINE AND VEHICLE IDENTIFICATION

Engine								Model Year	
Code ①	Liters (cc)	Cu. In	Cyl.	Fuel Sys.	Type	Eng. Mfg.		Code ②	Year
4G69/G	2.4 (2378)	143	4	MFI	SOHC	Mitsubishi		3	2003
4G69/F	2.4 (2378)	143	4	MFI	SOHC	Mitsubishi		4	2004
								5	2005
								6	2006

MFI: Multiport fuel injection

SOHC: Single overhead camshaft

① Engine ID / 8th digit of the VIN

② 10th digit of the VIN

09482_OUTL_C0001

GENERAL ENGINE SPECIFICATIONS

Year	Model	Engine Displacement Liters	Engine ID/VIN	Net Horsepower @ rpm	Net Torque @ rpm (ft. lbs.)	Bore x Stroke (in.)	Compression Ratio	Oil Pressure @ rpm
2003	Outlander	2.4	4G69/G	160@5750	162@4000	3.43x3.94	9.5:1	①
2004	Outlander	2.4	4G69/G	160@5750	162@4000	3.43x3.94	9.5:1	①
2005	Outlander	2.4	4G69/F	160@5750	162@4000	3.43x3.94	9.5:1	①
2006	Outlander	2.4	4G69/F	160@5750	162@4000	3.43x3.94	9.5:1	①

① 11.4 psi or more at curb idle speed

09482_OUTL_C0002

ENGINE TUNE-UP SPECIFICATIONS

Year	Engine Displacement Liters	Engine ID/VIN	Spark Plugs Gap (in.)	Ignition Timing (deg.) MT	Ignition Timing (deg.) AT	Fuel Pump (psi)	Idle Speed (rpm) MT	Idle Speed (rpm) AT	Valve Clearance In.	Valve Clearance Ex.
2003	2.4	4G69/G	0.028-0.031	—	2-8B	38	—	600-800	①	②
2004	2.4	4G69/G	0.028-0.031	—	2-8B	38	—	600-800	①	②
2005	2.4	4G69/F	0.028-0.031	2-8B	2-8B	38	580-780	580-780	①	②
2006	2.4	4G69/F	0.028-0.031	2-8B	2-8B	38	580-780	580-780	①	②

NOTE: The Vehicle Emission Control Information label often reflects specification changes made during production. The label figures must be used if they differ from those in this chart.

B: Before top dead center

① Engine cold: 0.004
  Engine hot: 0.008

② Engine cold: 0.008
  Engine hot: 0.012

09482_OUTL_C0003

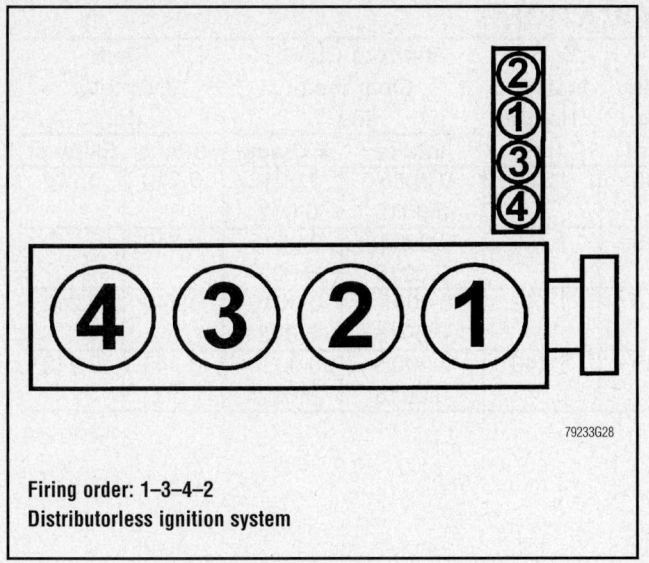

Firing order: 1–3–4–2
Distributorless ignition system

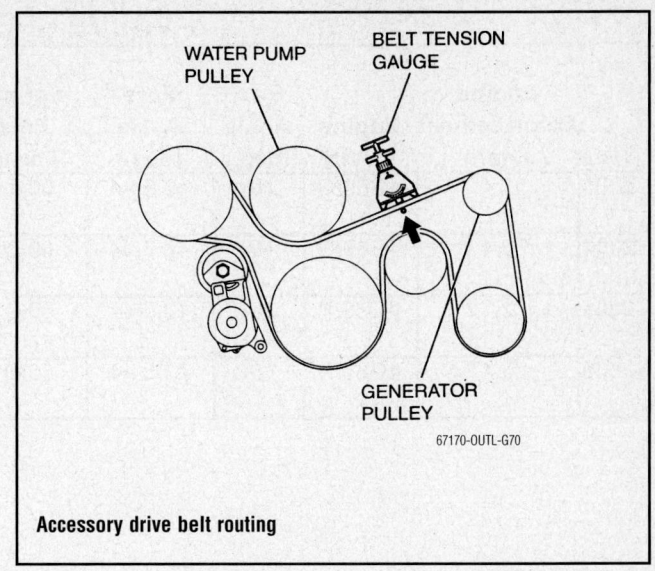

Accessory drive belt routing

CAPACITIES

Year	Model	Engine Displacement Liters	Engine ID/VIN	Engine Oil with Filter	Transmission (pts.) 5-Spd	Transmission (pts.) Auto.	Transfer Case (pts.)	Drive Axle Front (pts.)	Drive Axle Rear (pts.)	Fuel Tank (gal.)	Cooling System (qts.)
2003	Outlander	2.4	4G69/G	4.5	—	①	1.12	—	1.16	15.7	7.4
2004	Outlander	2.4	4G69/G	4.5	—	①	1.12	—	1.16	15.7	7.4
2005	Outlander	2.4	4G69/F	4.5	②	①	1.12	—	1.16	15.7	7.4
2006	Outlander	2.4	4G69/F	4.5	②	①	1.12	—	1.16	15.7	7.4

NOTE: All capacities are approximate. Add fluid gradually and ensure a proper fluid level is obtained.

① FWD transaxle: 16.2 pts.
  AWD tranaxle: 17.2 pts.
② FWD transaxle: 4.6 pts.
  AWD tranaxle: 4.9 pts.

VALVE SPECIFICATIONS

Year	Engine Displacement Liters	Engine ID/VIN	Seat Angle (deg.)	Face Angle (deg.)	Spring Test Pressure (lbs. @ in.)	Spring Installed Height (in.)	Stem-to-Guide Clearance (in.)		Stem Diameter (in.)	
							Intake	Exhaust	Intake	Exhaust
2003	2.4	4G69/G	NA	43.5-44	60@1.740	1.740	0.0008-0.0016	0.0016-0.0024	0.240	0.240
2004	2.4	4G69/G	NA	43.5-44	60@1.740	1.740	0.0008-0.0016	0.0016-0.0024	0.240	0.240
2005	2.4	4G69/F	NA	43.5-44	60@1.740	1.740	0.0008-0.0016	0.0016-0.0024	0.240	0.240
2006	2.4	4G69/F	NA	43.5-44	60@1.740	1.740	0.0008-0.0016	0.0016-0.0024	0.240	0.240

09482_OUTL_C0005

CAMSHAFT SPECIFICATIONS CHART
All measurements are given in inches.

Year	Engine Displ. Liters	Engine VIN	Journal Dia.	Brg. Oil Clearance	Shaft End-play	Runout	Lobe Height	
							Intake	Exhaust
2003	2.4	4G69/G	1.8000	NS	NS	NS	①	②
2004	2.4	4G69/G	1.8000	NS	NS	NS	③	④
2005	2.4	4G69/F	1.8000	NS	NS	NS	⑤	④
2006	2.4	4G69/F	1.8000	NS	NS	NS	⑤	④

NS - Not specified by manufacturer

① Standard value: 1.472
  Minimum value: 1.452

② Standard value: 1.450
  Minimum value: 1.430

③ Standard value low speed cam "A": 1.332
  Minimum value low speed cam "A": 1.313
  Standard value low speed cam "B": 1.471
  Minimum value low speed cam "B": 1.451
  Standard value high speed cam: 1.465
  Minimum value high speed cam: 1.445

④ Standard value: 1.491
  Minimum value: 1.471

⑤ Standard value low speed cam "A" and "B": 1.475
  Minimum value low speed cam "A" and "B": 1.455
  Standard value high speed cam: 1.465
  Minimum value high speed cam: 1.445

09482_OUTL_C0006

CRANKSHAFT AND CONNECTING ROD SPECIFICATIONS

All measurements are given in inches.

Year	Engine Displacement Liters	Engine ID/VIN	Crankshaft				Connecting Rod		
			Main Brg. Journal Dia.	Main Brg. Oil Clearance	Shaft End-play	Thrust on No.	Journal Diameter	Oil Clearance	Side Clearance
2003	2.4	4G69/G	2.240	0.0008-0.0015	0.0020-0.0090	3	NA	NA	0.0040-0.0090
2004	2.4	4G69/G	2.240	0.0008-0.0015	0.0020-0.0090	3	NA	NA	0.0040-0.0090
2005	2.4	4G69/F	2.240	0.0008-0.0015	0.0020-0.0090	3	NA	NA	0.0040-0.0090
2006	2.4	4G69/F	2.240	0.0008-0.0015	0.0020-0.0090	3	NA	NA	0.0040-0.0090

NA - Not Available

09482_OUTL_C0007

PISTON AND RING SPECIFICATIONS

All measurements are given in inches.

Year	Engine Displacement Liters	Engine ID/VIN	Piston Clearance	Ring Gap			Ring Side Clearance		
				Top Compression	Bottom Compression	Oil Control	Top Compression	Bottom Compression	Oil Control
2003	2.4	4G69/G	0.0008-0.0019	0.0060-0.0120	0.0110-0.0170	0.0040-0.0160	0.0012-0.0028	0.0008-0.0023	NA
2004	2.4	4G69/G	0.0008-0.0019	0.0060-0.0120	0.0110-0.0170	0.0040-0.0160	0.0012-0.0028	0.0008-0.0023	NA
2005	2.4	4G69/F	0.0008-0.0019	0.0060-0.0120	0.0110-0.0170	0.0040-0.0160	0.0012-0.0028	0.0008-0.0023	NA
2006	2.4	4G69/F	0.0008-0.0019	0.0060-0.0120	0.0110-0.0170	0.0040-0.0160	0.0012-0.0028	0.0008-0.0023	NA

NA - Not Available

09482_OUTL_C0008

TORQUE SPECIFICATIONS
All readings in ft. lbs.

Year	Engine Displacement Liters	Engine ID/VIN	Cylinder Head Bolts	Main Bearing Bolts	Rod Bearing Bolts	Crankshaft Damper Bolts	Flywheel Bolts	Manifold Intake	Manifold Exhaust	Spark Plugs	Oil Pan Drain Plug
2003	2.4	4G69/G	①	18 ②	15 ②	123	98	③	④	18	26-31
2004	2.4	4G69/G	①	18 ②	15 ②	123	98	③	④	18	26-31
2005	2.4	4G69/F	①	18 ②	15 ②	123	98	③	⑤	18	26-31
2006	2.4	4G69/F	①	18 ②	15 ②	123	98	③	⑤	18	26-31

① Step 1: Tighten all bolts to 58 ft. lbs.
Step 2: Loosen all bolts to 0 ft. lbs.
Step 3: Tighten all bolts to 15 ft. lbs.
Step 4: Tighten all bolts 90 degrees.
Step 5: Tighten all bolts an additional 90 degrees.

② Torque to specification plus
an additional 90 degrees.

③ Bolt: 18 ft. lbs.
Nut: 15 ft. lbs.

④ Bracket (bolt & washer assembly) bolt: 26 ft. lbs.
Bracket (flange) bolt: 27 ft. lbs.
Nut: 36 ft. lbs.

⑤ Cover bolt, except California: 124 inch lbs.
Cover bolt, California: 22 ft. lbs.
Exhaust bracket bolt, California: 24 ft. lbs.
Nut: 36 ft. lbs.

09482_OUTL_C0009

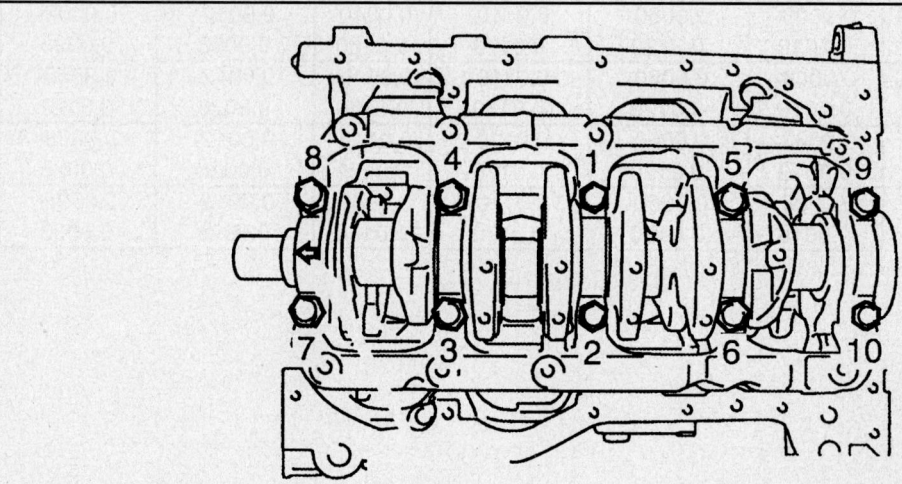

09482_OUTL_G0001

Main bearing torque sequence

WHEEL ALIGNMENT

Year	Model		Caster Range (+/-Deg.)	Caster Preferred Setting (Deg.)	Camber Range (+/-Deg.)	Camber Preferred Setting (Deg.)	Toe-in (in.)
2003	Outlander	F	0.30	+3.10	0.30	0	0.04 +/- 0.09
		R	—	—	0.30	-0.40	0.12 +/- 0.08
2004	Outlander	F	0.30	+3.10	0.30	0	0.04 +/- 0.09
		R	—	—	0.30	-0.40	0.12 +/- 0.08
2005	Outlander	F	0.30	+3.10	0.30	0	0.04 +/- 0.09
		R	—	—	0.30	-0.40	0.12 +/- 0.08
2006	Outlander	F	0.30	+3.10	0.30	0	0.04 +/- 0.09
		R	—	—	0.30	-0.40	0.12 +/- 0.08

09482_OUTL_C0010

TIRE, WHEEL AND BALL JOINT SPECIFICATIONS

Year	Model	OEM Tires Standard	OEM Tires Optional	Tire Pressures (psi) Front	Tire Pressures (psi) Rear	Wheel Size	Ball Joint Inspection ①	Lug Nut (ft. lbs.)
2003	Outlander	P225/60R16	None	32	30	6-JJ	U: 4-26 in. L: 0-35 in.	66-80
2004	Outlander	P225/60R16	None	32	30	6-JJ	U: 4-26 in. L: 0-35 in.	66-80
2005	Outlander	P225/60R16	P215/55R17	32	30	②	U: 4-26 in. L: 0-35 in.	66-80
2006	Outlander	P225/60R16	P215/55R17	32	30	②	U: 4-26 in. L: 0-35 in.	66-80

OEM: Original Equipment Manufacturer

PSI: Pounds Per Square Inch. If specification differs from the one located on driver's door, use the driver's door specification

L: Lower

U: Upper

① Torque required in inch lbs. to rotate ball joint when removed from the knuckle

② Standard: 6-JJ. Optional: 6.5JJ

09482_OUTL_C0011

BRAKE SPECIFICATIONS

All measurements in inches unless noted

Year	Model		Brake Disc Original Thickness	Brake Disc Minimum Thickness	Brake Disc Maximum Runout	Brake Drum Diameter Original Inside Diameter	Max. Wear Limit	Maximum Machine Diameter	Minimum Lining Thickness Front	Minimum Lining Thickness Rear	Brake Caliper Bracket Bolts (ft. lbs.)	Brake Caliper Mounting Bolts (ft. lbs.)
2003	Outlander	F	1.020	0.960	0.0015	—	—	—	0.080	—	28	74
		R	—	—	—	9.000	—	9.080	—	0.040	—	—
2004	Outlander	F	1.020	0.960	0.0015	—	—	—	0.080	—	28	74
		R	—	—	—	9.000	—	9.080	—	0.040	—	—
2005	Outlander	F	0.940	0.880	0.0015	—	—	—	0.080	—	55	74
		R	0.390	0.330	0.0019	—	—	—	—	0.080	33	33
2006	Outlander	F	0.940	0.880	0.0015	—	—	—	0.080	—	55	74
		R	0.390	0.330	0.0019	—	—	—	—	0.080	33	33

NA: Not Available

F: Front

R: Rear

09482_OUTL_C0012

SCHEDULED MAINTENANCE INTERVALS
Mitsubishi—Outlander

TO BE SERVICED	TYPE OF SERVICE	VEHICLE MILEAGE INTERVAL (x1000)													
		7.5	15	22.5	30	37.5	45	52.5	60	67.5	75	82.5	90	97.5	102.5
Engine oil & filter	R	✓	✓	✓	✓	✓	✓	✓	✓	✓	✓	✓	✓	✓	✓
Air cleaner element	R				✓				✓				✓		
Automatic transaxle fluid	S/I		✓		✓		✓		✓		✓		✓		
Manual transaxle fluid	S/I		✓		✓		✓		✓		✓		✓		
Brake hoses	S/I		✓		✓		✓		✓		✓		✓		
Disc brake pads	S/I		✓		✓		✓		✓		✓		✓		
Driveshaft boots	S/I		✓		✓		✓		✓		✓		✓		
Valve clearance	S/I				✓				✓				✓		
Engine coolant (2003)	R				✓				✓				✓		
Engine coolant (2004-2006)	R								✓				✓		
Spark plugs (standard)	R				✓				✓				✓		
Spark plugs (platinum)	R								✓				✓		
Spark plugs (iridium)	R														✓
Ball joints & steering linkage seals	S/I				✓				✓				✓		
Drive belt(s)	S/I				✓				✓				✓		
Exhaust system	S/I				✓				✓				✓		
Fuel hoses	S/I				✓				✓				✓		
Transfer case fluid	S/I				✓				✓				✓		
Transfer case fluid	R								✓						
Rear drum brake linings & rear wheel cylinders	S/I				✓				✓				✓		
Ignition cables	R								✓						
Timing belt(s)	R								✓						
EVAP system (except canister)	S/I								✓						
Fuel system (tank, pipe line, connection & fuel tank filler tube cap)	S/I								✓						
Tires (rotate)	S/I	✓	✓	✓	✓	✓	✓	✓	✓	✓	✓	✓	✓	✓	✓

R: Replace   S/I: Service or Inspect

FREQUENT OPERATION MAINTENANCE (SEVERE SERVICE)

If a vehicle is operated under any of the following conditions it is considered severe service:

- Extremely dusty areas.

- 50% or more of the vehicle operation is in 32°C (90°F) or higher temperatures, or constant operation in temperatures below 0°C (32°F).

- Prolonged idling (vehicle operation in stop and go traffic).

- Frequent short running periods (engine does not warm to normal operating temperatures).

- Police, taxi, delivery usage or trailer towing usage.

Oil & oil filter: change every 3750 miles.

Disc brake pads: service or inspect every 6000 miles.

Rear drum brake linings and rear wheel cylinders: service or inspect every 15,000 miles

Air filter element: service or inspect every 15,000 miles.

Automatic transaxle fluid & filter: replace every 15,000 miles.

09482_OUTL_C0013

PRECAUTIONS

Before servicing any vehicle, please be sure to read all of the following precautions, which deal with personal safety, prevention of component damage, and important points to take into consideration when servicing a motor vehicle:

• Never open, service or drain the radiator or cooling system when the engine is hot; serious burns can occur from the steam and hot coolant.

• Observe all applicable safety precautions when working around fuel. Whenever servicing the fuel system, always work in a well-ventilated area. Do not allow fuel spray or vapors to come in contact with a spark, open flame, or excessive heat (a hot drop light, for example). Keep a dry chemical fire extinguisher near the work area. Always keep fuel in a container specifically designed for fuel storage; also, always properly seal fuel containers to avoid the possibility of fire or explosion. Refer to the additional fuel system precautions in this section.

• Fuel injection systems often remain pressurized, even after the engine has been turned **OFF**. The fuel system pressure must be relieved before disconnecting any fuel lines. Failure to do so may result in fire and/or personal injury.

• Brake fluid often contains polyglycol ethers and polyglycols. Avoid contact with the eyes and wash your hands thoroughly after handling brake fluid. If you do get brake fluid in your eyes, flush your eyes with clean, running water for 15 minutes. If

eye irritation persists, or if you have taken brake fluid internally, IMMEDIATELY seek medical assistance.

• The EPA warns that prolonged contact with used engine oil may cause a number of skin disorders, including cancer. You should make every effort to minimize your exposure to used engine oil. Protective gloves should be worn when changing oil. Wash your hands and any other exposed skin areas as soon as possible after exposure to used engine oil. Soap and water, or waterless hand cleaner should be used.

• All new vehicles are now equipped with an air bag system. The system must be disabled before performing service on or around system components, steering column, instrument panel components, wiring and sensors. Failure to follow safety and disabling procedures could result in accidental air bag deployment, possible personal injury, and unnecessary system repairs.

• Always wear safety goggles when working with, or around, the air bag system. When carrying a non-deployed air bag, be sure the bag and trim cover are pointed away from your body. When placing a non-deployed air bag on a work surface, always face the bag and trim cover upward, away from the surface. This will reduce the motion of the module if it is accidentally deployed. Refer to the additional air bag system precautions later in this section.

• Clean, high quality brake fluid from a sealed container is essential to the safe and

proper operation of the brake system. You should always buy the correct type of brake fluid for your vehicle. If the brake fluid becomes contaminated, completely flush the system with new fluid. Never reuse any brake fluid. Any brake fluid that is removed from the system should be discarded. Also, do not allow any brake fluid to come in contact with a painted surface; it will damage the paint.

• Never operate the engine without the proper amount and type of engine oil; doing so will result in severe engine damage.

• Timing belt maintenance is extremely important. Many models utilize an interference-type, non-freewheeling engine. If the timing belt breaks, the valves in the cylinder head may strike the pistons, causing potentially serious (also time-consuming and expensive) engine damage. Refer to the maintenance interval charts in the front of this section for the recommended replacement interval for the timing belt, and to the timing belt procedure for belt replacement and inspection.

• Disconnecting the negative battery cable on some vehicles may interfere with the functions of the on-board computer system(s) and may require the computer to undergo a relearning process once the negative battery cable is reconnected.

• When servicing drum brakes, only disassemble and assemble one side at a time, leaving the remaining side intact for reference.

ENGINE REPAIR

➡ **Disconnecting the negative battery cable on some vehicles may interfere with the functions of the on board computer systems and may require the computer to undergo a relearning process, once the negative battery cable is reconnected.**

Alternator

REMOVAL & INSTALLATION

1. Before servicing the vehicle, refer to the Precautions Section.
2. Remove or disconnect the following:
   • Negative battery cable
   • Engine undercover
   • Drive belt
   • A/C compressor connector and connector clamp

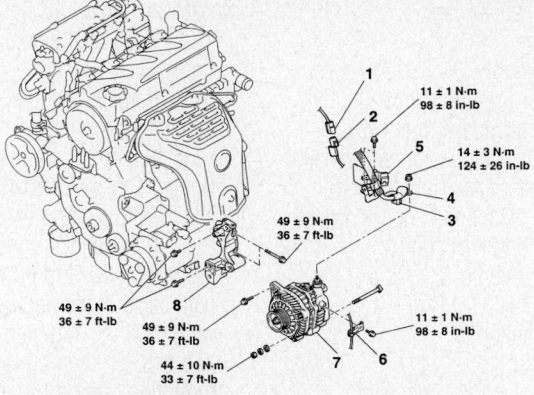

1. A/C COMPRESSOR ASSEMBLY CONNECTOR
2. A/C COMPRESSOR ASSEMBLY CONNECTOR CLAMP
3. GENERATOR CONNECTOR
4. GENERATOR TERMINAL
5. CONNECTOR BRACKET
6. HARNESS BRACKET
7. GENERATOR
8. GENERATOR MOUNTING BRACKET

67170-0UTL-G01

Alternator and related components

- Alternator connector and terminal
- Alternator connector bracket
- Harness bracket
- Alternator, lifting it up out of the vehicle

3. If necessary, remove the lower timing belt cover and alternator mounting bracket.

To install:

4. If removed, install the alternator mounting bracket and lower timing belt cover.

5. Install or connect the following:

- Alternator, lifting it up out of the vehicle
- Harness bracket
- Alternator connector bracket
- Alternator terminal and connector
- A/C compressor connector clamp and connector
- Drive belt
- Engine undercover
- Negative battery cable

Ignition Timing

ADJUSTMENT

The ignition timing is controlled by the Electronic Control Module (ECM) and is not adjustable. However it can be inspected using a scan tool.

Engine Assembly

REMOVAL & INSTALLATION

1. Before servicing the vehicle, refer to the Precautions Section.
2. Relieve fuel system pressure.
3. Remove or disconnect the following:
- Negative battery cable
- Undercover
- Hood assembly
- Air cleaner assembly and all adjoining air intake duct work
4. Drain the engine coolant, engine oil, transaxle and transfer case (if equipped).
5. Remove or disconnect the following:
- Battery and battery tray
- Accelerator cable
- Radiator
- Front exhaust pipe
- Drive belt
6. Detach the electrical connectors and/or vacuum lines from the following components, as necessary:
- Control wiring harness
- Battery wiring harness
- Evaporative emission (EVAP) vacuum connection

- Brake booster vacuum hose connection
- Air Conditioning (A/C) compressor
- Power steering oil pressure switch
- Crank angle sensor
- Manifold differential pressure sensor
- Exhaust Gas Recirculation (EGR) solenoid valve
- Ignition coil
- Fuel injectors
- Throttle Position (TP) sensor
- Idle Air Control (IAC) motor
- Engine Coolant Temperature (ECT) sensor
- Camshaft Position (CMP) sensor
- Knock Sensor (KS)
- ECT gauge unit
- Heated Oxygen Sensor (HO_2S)
- Starter and alternator
- Oil pressure switch

7. Remove or disconnect the following:
- Power steering pump and brace. Position the assembly aside, but do NOT disconnect the fluid line.
- A/C compressor, but do NOT disconnect the lines

➡**Matchmark the installed position of the radiator hoses before disconnecting them.**

- Upper and lower radiator hoses
- Heater and purge hoses
- Fuel lines. Discard the O-rings

8. Pre-tighten the two bolts on the vehicle to assemble the radiator support upper insulator to set Mitsubishi Special tool no. MB991928 or MB991895. See the accompanying figure.
- Transaxle assembly
- Ground cable connection
- Self-locking nuts

9. Remove the engine front mounting bracket as follows:

a. Support the engine with a suitable floor jack.

b. Remove the special tools that were installed for transaxle removal.

c. Support the engine with Mitsubishi Special tools MB991454 and MB991527 attached to a chain block or engine hoist.

d. Place a jack under the oil pan with a block of wood in between to protect the pan. Jack up the engine to take the weight off the engine mount insulator and bracket, then remove the front mounting bracket mounting nuts and bolts, then remove the bracket.

10. Make sure that all cables, hoses and harnesses are disconnected from the engine, then use the engine hose to slowly lift the engine up and out of the engine compartment.

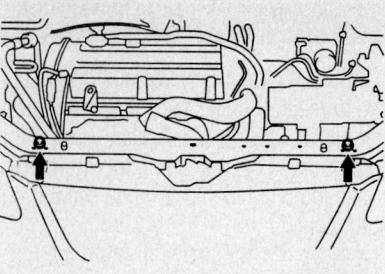

67170-OUTL-G02

Pre-tighten the two bolts on the vehicle to assemble the radiator support upper insulator for the engine hanger (MB991928 or MB991895)

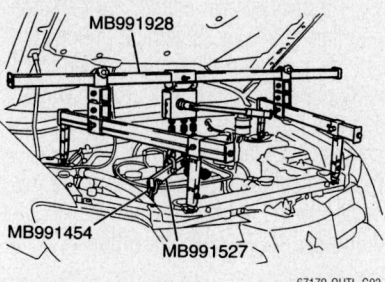

67170-OUTL-G03

View of the tool MB991928 installed, necessary to support the engine when the transaxle is removed

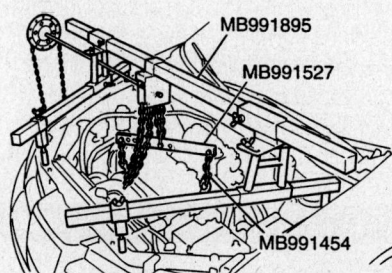

67170-OUTL-G04

View of the tool MB991895 installed, necessary to support the engine when the transaxle is removed

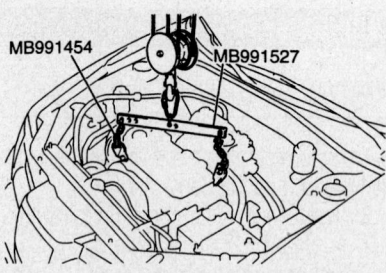

67170-OUTL-G05

View of the special tools needed to support the engine during mount removal

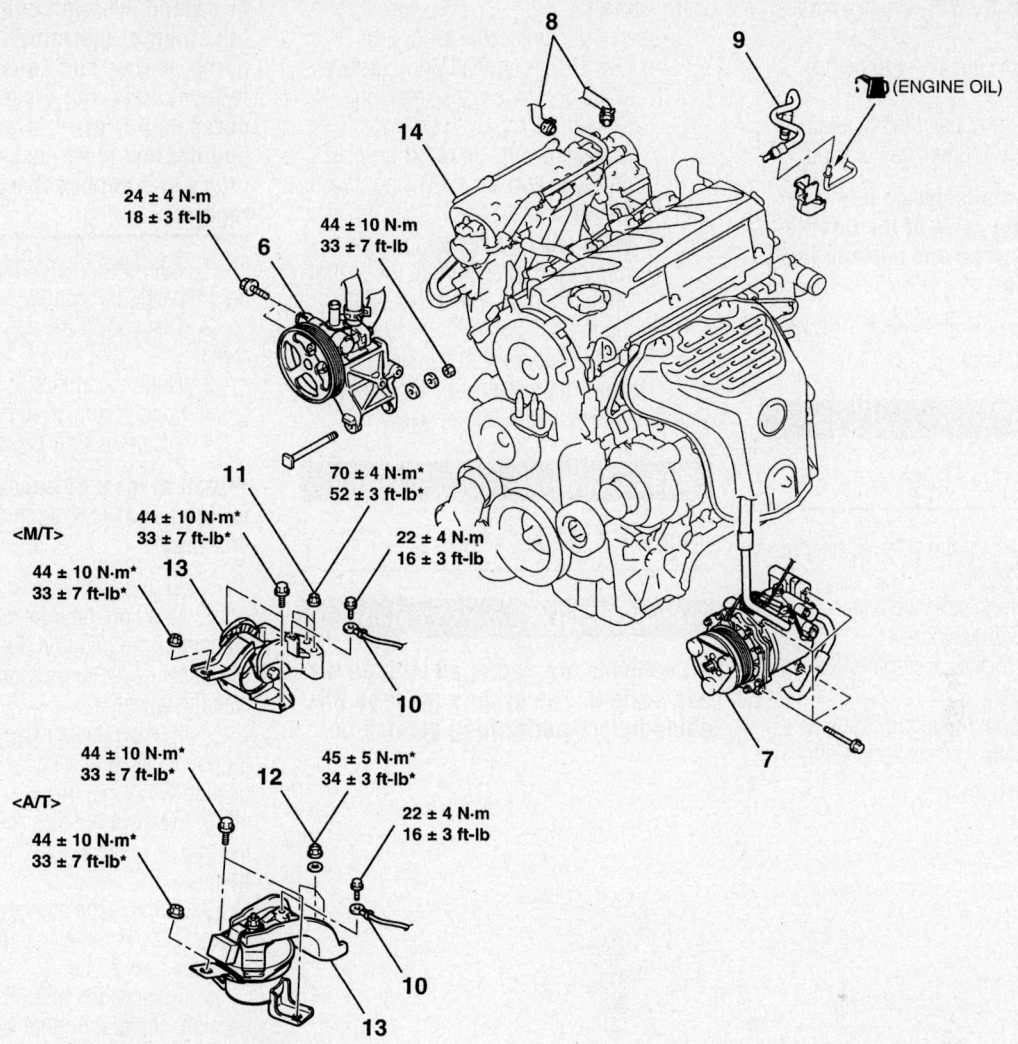

8

9

🛢 (ENGINE OIL)

14

24 ± 4 N·m
18 ± 3 ft-lb

6

44 ± 10 N·m
33 ± 7 ft-lb

11

70 ± 4 N·m*
52 ± 3 ft-lb*

<M/T>

44 ± 10 N·m*
33 ± 7 ft-lb*

22 ± 4 N·m
16 ± 3 ft-lb

44 ± 10 N·m*
33 ± 7 ft-lb*

13

10

7

44 ± 10 N·m*
33 ± 7 ft-lb*

12

45 ± 5 N·m*
34 ± 3 ft-lb*

<A/T>

44 ± 10 N·m*
33 ± 7 ft-lb*

22 ± 4 N·m
16 ± 3 ft-lb

10

13

6. POWER STEERING OIL PUMP
   AND BRACKET ASSEMBLY
7. A/C COMPRESSOR AND
   CLUTCH ASSEMBLY
8. HEATER WATER HOSES
   CONNECTION
9. FUEL HIGH-PRESSURE HOSE
   CONNECTION

10. GROUND CABLE CONNECTION
11. SELF-LOCKING NUTS <M/T>
12. SELF-LOCKING NUTS <A/T>
13. ENGINE MOUNTING INSULATOR
14. ENGINE ASSEMBLY

09482_OUTL_G0002

Engine mounts and related components

To install:

11. Installation is the reverse of the removal procedure.

➡**Bolt torques are indicated in the accompanying illustration. The parts indicated with an asterisk (*) should be temporarily tightened, and then fully tightened with the engine weight applied on the body of the vehicle.**

12. Fill the coolant system and engine crankcase.

13. Connect the negative battery cable.

14. On 2006 vehicles, complete the vehicle initialization procedure.

➡**To complete the initialization procedure the following tools are needed. MB991958 scan tool, MB991824 VCI, MB991827 MUT III USB cable, MB991910 MUT III Main harness "A"**

15. Connect the scan tool to the data link connector. To prevent damage to the scan tool be sure that the ignition switch is in the LOCK position before connecting the scan tool.

16. Turn the ignition switch to the ON position.

17. Select "check mode" from the menu screen.

18. Select "erase memory" from the menu screen.

19. Initialize the learning value.

20. On 2006 vehicles, after initialization complete the idle learning procedure.

21. Start the engine. Allow the coolant temperature to reach 176 degrees or more.

22. Stop the engine and place the ignition switch in the LOCK position.

23. After ten seconds, restart the engine.

24. For ten minutes carry out the idling procedure below to confirm that the engine has normal idling.

• Position the transaxle selector lever in the "P" range, for automatic

transaxle and in neutral for manual transaxle
- The engine fan is not to be operated
- The engine coolant temperature should be 176 degrees or more

➡️**If the engine stalls during idling, check the throttle valve of the throttle body for dirt. Correct and perform the procedure again.**

25. Run the engine and check for leaks.
26. Install the hood.

Water Pump

REMOVAL & INSTALLATION

1. Before servicing the vehicle, refer to the Precautions Section.
2. Disconnect the negative battery cable.
3. Drain the engine coolant.
4. Remove or disconnect the following:
   - Timing belt
   - Water pump mounting bolts
   - Water pump, gasket and O-ring

To install:
5. Install or connect the following:
   - New O-ring on the O-ring grooved at the tip of the water inlet pipe. Coat the O-ring with water or coolant. Do not allow oil or other grease to contact the O-ring. Insert the water inlet pipe.
   - Water pump to the engine block, with new gasket. Torque the mounting bolts to 10 ft. lbs. (14 Nm)
   - Timing belt
6. Refill the engine with coolant.
7. Connect the negative battery cable, start the engine and check for leaks.

Heater Core

REMOVAL & INSTALLATION

※※ CAUTION

All vehicles are equipped with an air bag system. The system must be disabled before performing service on or around system components, steering column, instrument panel components, wiring and sensors. Failure to follow safety and disabling procedures could result in accidental air bag deployment, possible personal injury and unnecessary system repairs.

1. Before servicing the vehicle, refer to the Precautions Section.
2. Disconnect the negative battery cable.
3. Drain the cooling system. Discharge the air conditioning system.
4. Disarm the air bag system.

➡️**Wait at least 60 seconds before performing and work once the system is disarmed.**

5. Remove the floor console assembly.
6. Position the front wheels in the straight ahead position. Remove the steering wheel cover, located underneath the steering wheel.
7. Remove the air bag module by sliding the clock spring connector and disconnecting it.
8. Remove the steering wheel, using the proper removal tool. Remove the sub harness.
9. Remove the steering column lower cover. Remove the clock spring and store it in a clean dry place.
10. Remove the instrument panel clock garnish, center air outlet, clock, driver's side garnish and meter bezel.
11. Remove the combination meter. Remove the hood lock release handle. Remove the instrument panel parcel box.
12. Disconnect the right side harness and junction block connection.
13. Remove the instrument lower panel, left side air conditioning outlet and instrument lower panel bracket.
14. Remove the instrument panel driver's side bracket and knee absorber.
15. If not previously removed, remove the steering column cover.
16. Remove the air mix door cable, blow vent changeover damper cable connection, heater control switch assembly and air conditioning switch. Remove the knob.
17. Remove the center panel. Remove the manual air conditioner control panel, if equipped.
18. Remove the radio and box assembly. Remove the radio bracket. Remove the center lower panel cover.
19. Remove the accessory socket, air bag cutoff switch indicator, seat belt indicator and center lower box.

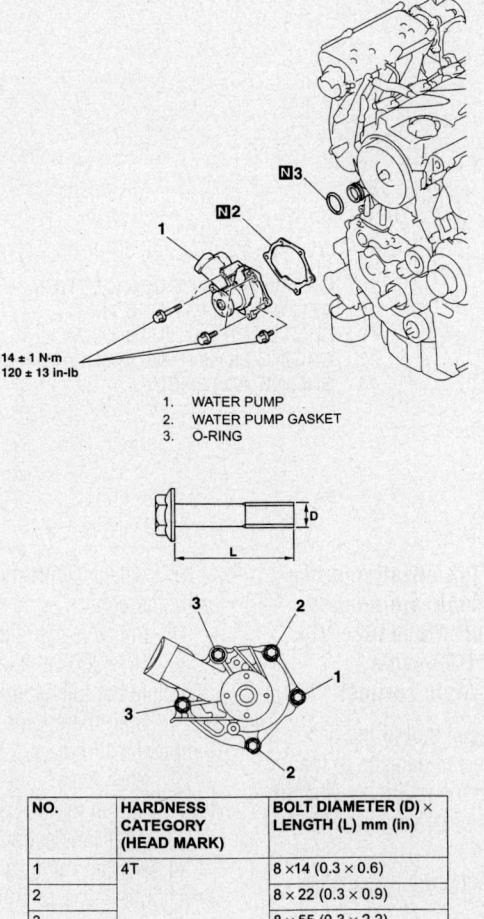

14 ± 1 N·m
120 ± 13 in-lb

1. WATER PUMP
2. WATER PUMP GASKET
3. O-RING

NO.	HARDNESS CATEGORY (HEAD MARK)	BOLT DIAMETER (D) × LENGTH (L) mm (in)
1	4T	8 × 14 (0.3 × 0.6)
2		8 × 22 (0.3 × 0.9)
3		8 × 55 (0.3 × 2.2)

61710-OUTL-G07

Water pump and bolt torque sequence

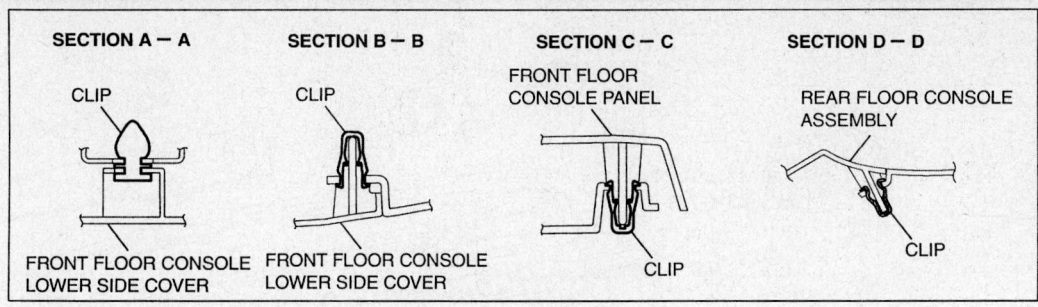

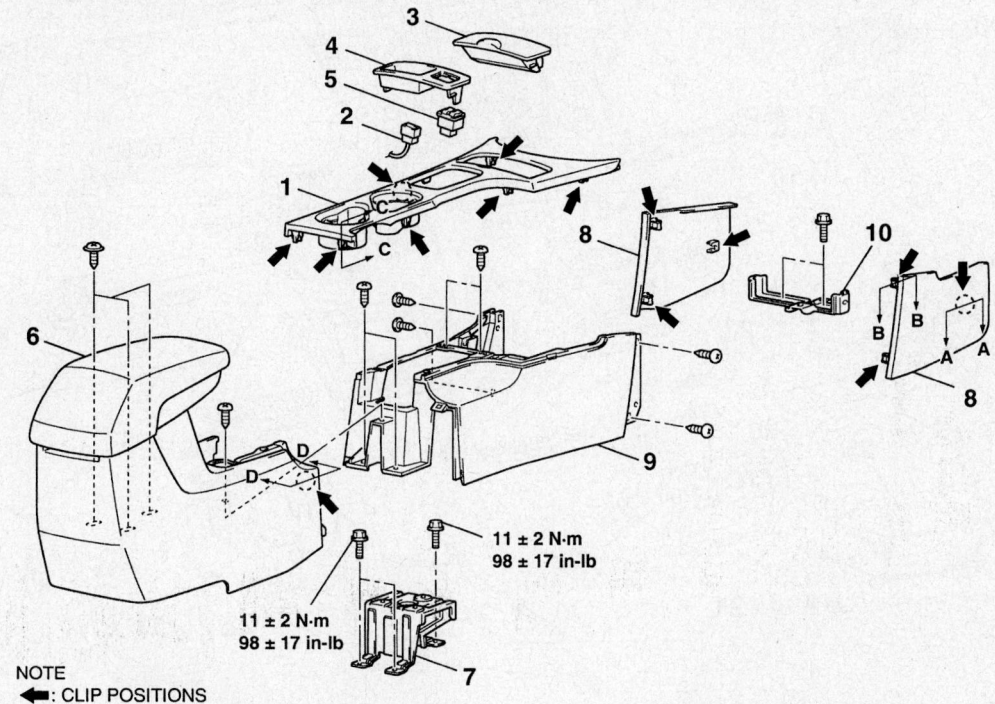

NOTE

◀ : CLIP POSITIONS

1. FRONT FLOOR CONSOLE PANEL
2. HARNESS CONNECTOR
   <VEHICLE WITH HEATED SEAT>
3. TRAY ASSEMBLY <VEHICLE
   WITHOUT HEATED SEAT>
4. HEATED SEAT SWITCH TRAY
   <VEHICLE WITH HEATED SEAT>
5. HEATED SEAT SWITCH
   <VEHICLE WITH HEATED SEAT>

6. REAR FLOOR CONSOLE
   ASSEMBLY
7. REAR FLOOR CONSOLE
   BRACKET
8. FRONT FLOOR CONSOLE
   LOWER SIDE COVER
9. FRONT FLOOR CONSOLE
10. FRONT FLOOR CONSOLE
    BRACKET

09482_OUTL_G0003

Floor console and related components

20. Remove the center lower cover, back bone spacer, glove box stopper and glove box.

21. Remove the passenger's side knee absorber. Remove the passenger's side connector block. Remove the instrument panel side cover.

22. Disconnect the DIN cable clamp, if equipped with amplifier, roof harness connection, ground cable, instrument panel and body harness connector, steering shaft harness connector and clamp, brake light switch harness connector, SRS-ECU harness connector and steering lock cable clamp.

23. Remove the steering shaft cover installation bolt.

➡The tilt lever should be in the lock position before the steering column shaft assembly is removed, and should remain in the locked position until it is reinstalled. If the steering column shaft assembly is removed with the tilt lever released, or the tilt lever is released after removal the assembly cannot be reinstalled correctly. If the assembly is installed incorrectly, the collision energy absorbing mechanism may be damaged.

24. Ensure that the tilt lever is in the lock position. Remove the steering shaft mounting bolts. Remove the assembly from the vehicle.

25. Remove the instrument panel assembly from the vehicle.

➡Do not subject the SRS-ECU to any shocks when removing or installing the instrument panel.

26. Remove the intake duct, right side foot duct, joint duct and center duct.

27. Remove the heater hose connection, flexible suction hose connection, liquid pipe connection and O-ring and left side foot duct.

28. Remove the heater case from the vehicle.

29. Disassemble the heater unit as necessary for access to components.

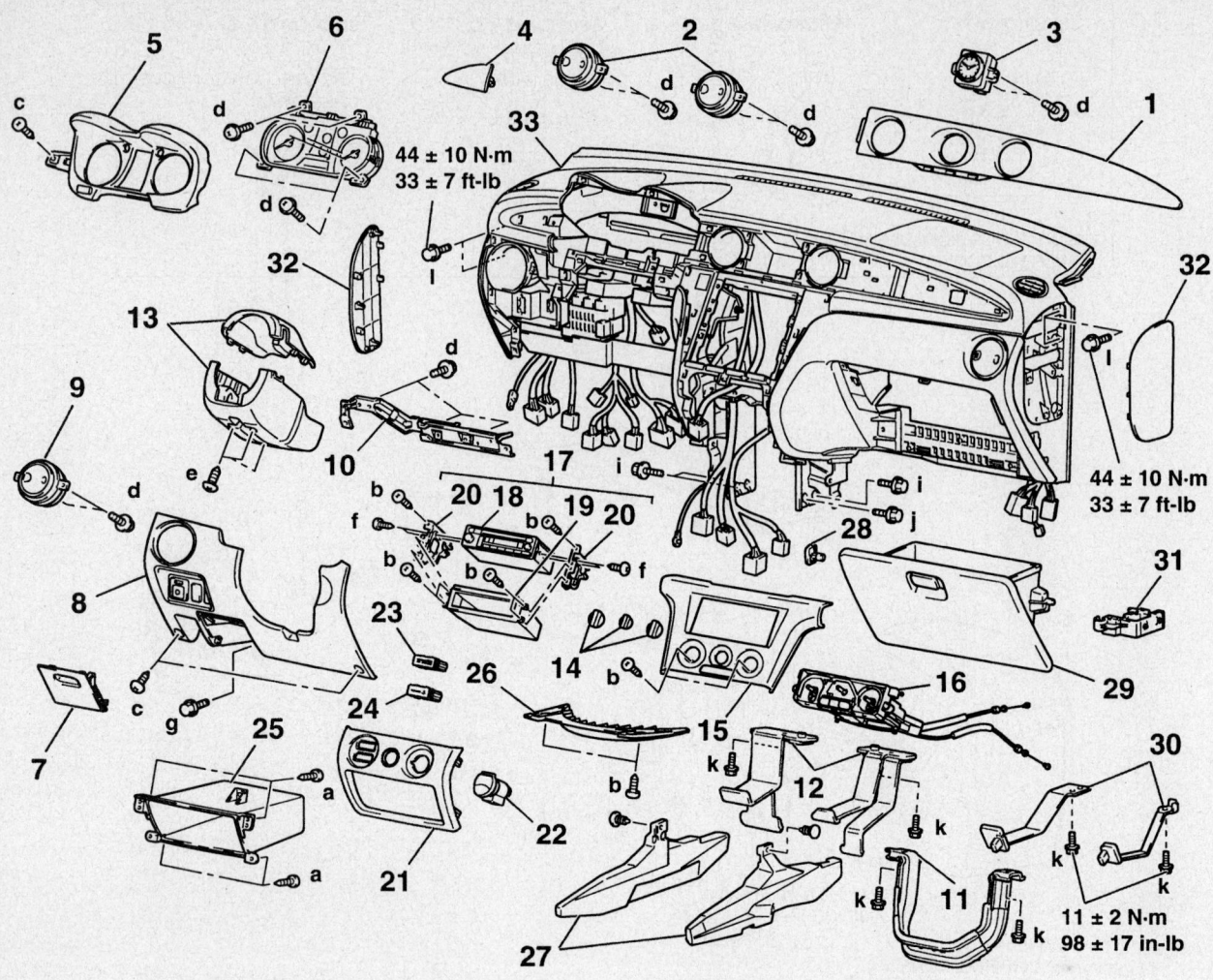

44 ± 10 N·m
33 ± 7 ft-lb

44 ± 10 N·m
33 ± 7 ft-lb

11 ± 2 N·m
98 ± 17 in-lb

1. INSTRUMENT PANEL CLOCK GARNISH
2. INSTRUMENT PANEL AIR OUTLET (CENTER)
3. CLOCK
4. INSTRUMENT PANEL DRIVER'S SIDE GARNISH
5. METER BEZEL
6. COMBINATION METER
7. INSTRUMENT PANEL PARCEL BOX
8. INSTRUMENT LOWER PANEL
9. INSTRUMENT PANEL AIR OUTLET (LH)
10. INSTRUMENT LOWER PANEL BRACKET
11. INSTRUMENT PANEL DRIVER SIDE BRACKET
12. KNEE ABSORBER (DRIVER'S SIDE)
13. COLUMN COVER
14. KNOB
15. CENTER PANEL

16. MANUAL AIR CONDITIONER CONTROL PANEL
17. AUDIO AND BOX ASSEMBLY
18. AUDIO
19. BOX
20. RADIO BRACKET
21. CENTER LOWER PANEL
22. ACCESSORY SOCKET
23. AIR BAG CUT-OFF SWITCH INDICATOR
24. SEAT BELT INDICATOR
25. CENTER LOWER BOX
26. CENTER LOWER COVER
27. BACK BONE SPACER
28. GLOVE BOX STOPPER
29. GLOVE BOX
30. KNEE ABSORBER (PASSENGER'S SIDE)
31. CONNECTOR BLOCK
32. INSTRUMENT PANEL SIDE COVER
33. INSTRUMENT PANEL ASSEMBLY

09482_OUTL_G0004

Instrument panel and related components

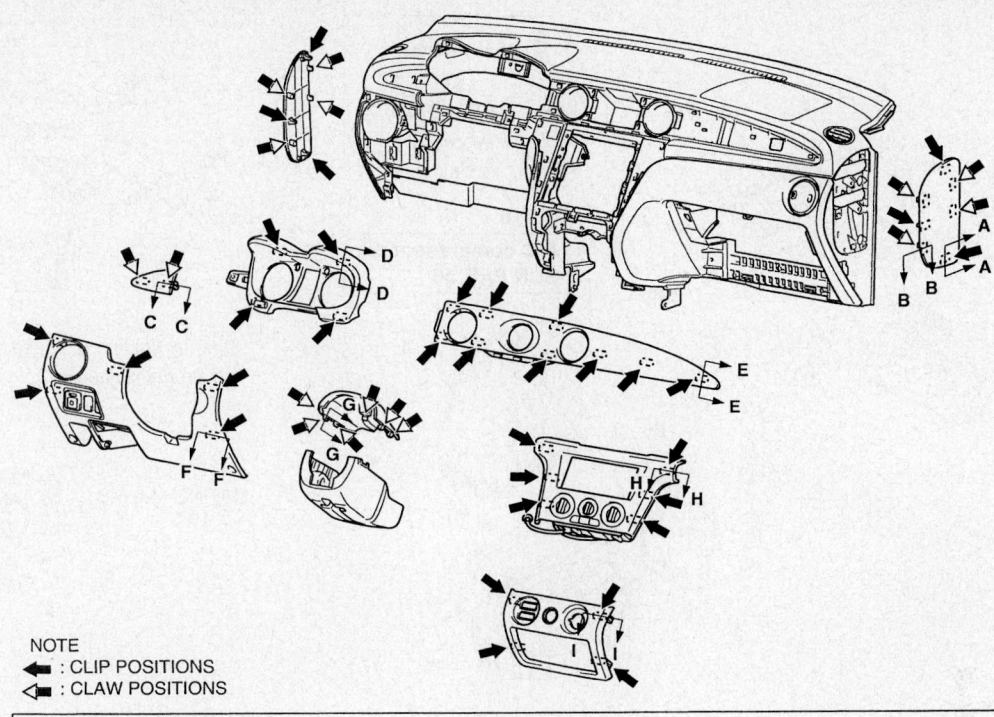

NOTE
⬅ : CLIP POSITIONS
◁ : CLAW POSITIONS

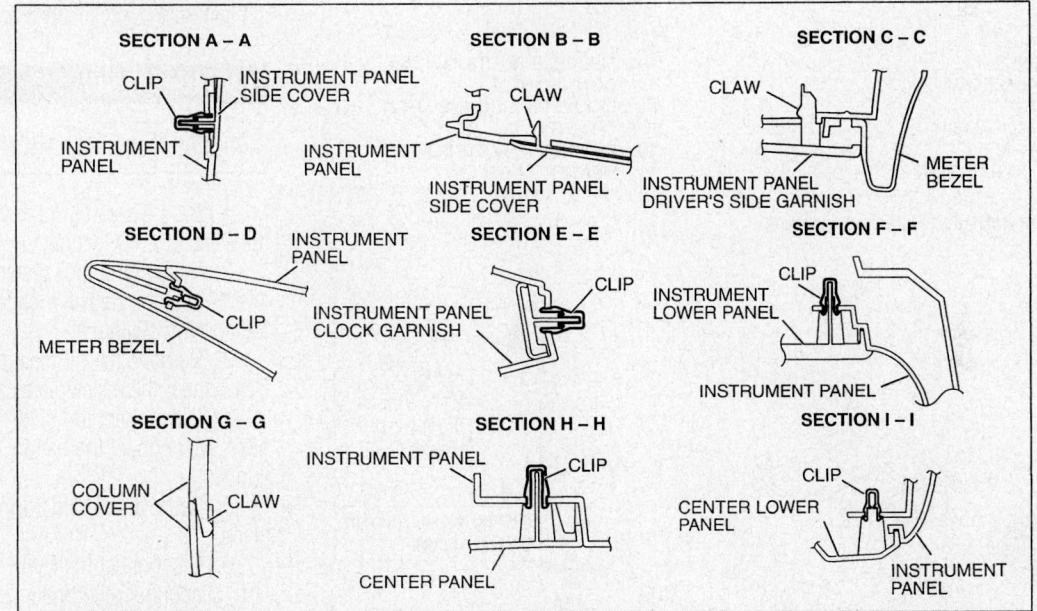

SECTION A – A
CLIP
INSTRUMENT PANEL
SIDE COVER
INSTRUMENT
PANEL

SECTION B – B
CLAW
INSTRUMENT
PANEL
INSTRUMENT PANEL
SIDE COVER

SECTION C – C
CLAW
INSTRUMENT PANEL
DRIVER'S SIDE GARNISH
METER
BEZEL

SECTION D – D INSTRUMENT
PANEL
METER BEZEL
CLIP

SECTION E – E
INSTRUMENT PANEL
CLOCK GARNISH
CLIP

SECTION F – F
CLIP
INSTRUMENT
LOWER PANEL
INSTRUMENT PANEL

SECTION G – G
COLUMN
COVER
CLAW

SECTION H – H
INSTRUMENT PANEL
CLIP
CENTER PANEL

SECTION I – I
CLIP
CENTER LOWER
PANEL
INSTRUMENT
PANEL

09482_OUTL_G0005

Instrument panel retention clip locations

To install:

30. Installation is the reverse of the removal procedure.

➡**When installing the heater/evaporator unit do not allow it to bump against the SRS-ECU or any of its components.**

31. Ensure that the tilt lever is in the lock position. Install the steering column shaft.

➡**Do not release the tilt lever until the steering column shaft has been installed. Torque the retaining bolts 80–124 inch. lbs.**

➡**When installing the clock spring, ensure that the mating marks are properly aligned. If not the steering wheel may not rotate completely during a turn, or the flat cable in the clock spring could be damaged. This would prevent normal SRS operation and possibly cause serious injury to the driver.**

32. Turn the front wheels to the straight ahead position. Then install the clock spring to the column switch.

33. To align the mating marks of the

clock spring, turn the clock spring counterclockwise fully. Then turn it back approximately 3-3/4 turns counterclockwise to align the mating marks.

➡**When installing the steering wheel, be sure that the harness of the clock spring does not become caught or tangled.**

34. Check and adjust all fluid levels. Properly recharge the air conditioning system.

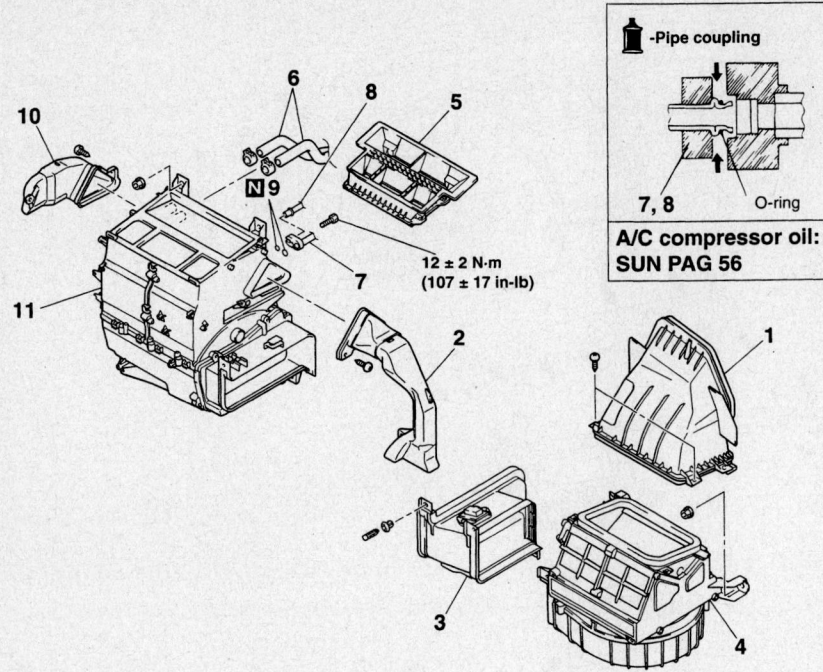

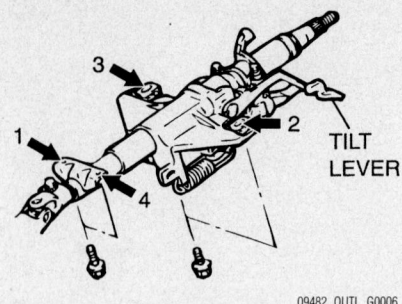

Steering column retaining bolt locations and torque sequence

12 ± 2 N·m
(107 ± 17 in-lb)

-Pipe coupling

7, 8   O-ring

A/C compressor oil: SUN PAG 56

TILT LEVER

09482_OUTL_G0006

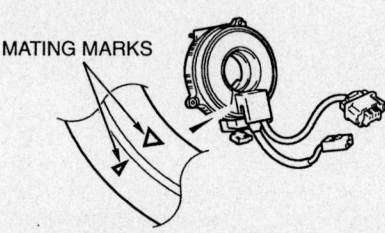

MATING MARKS

09482_OUTL_G0007

Clock spring alignment marks

1. INTAKE DUCT
2. RIGHT-HAND FOOT DUCT
3. JOINT DUCT
4. BLOWER ASSEMBLY
5. CENTER DUCT
6. HEATER HOSE CONNECTION
7. FLEXIBLE SUCTION HOSE CONNECTION
8. LIQUID PIPE CONNECTION
9. O-RING
10. LEFT-HAND FOOT DUCT
11. HEATER CASE

67170-OUTL-G08

Heater/evaporator unit and related components

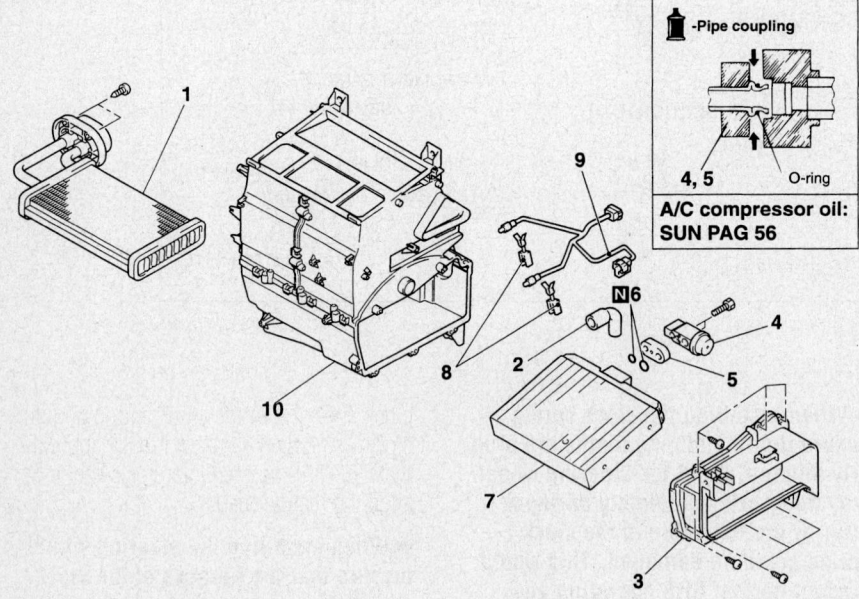

-Pipe coupling

4, 5   O-ring

A/C compressor oil: SUN PAG 56

1. HEATER CORE
2. DRAIN HOSE
3. EVAPORATOR COVER
4. EXPANSION VALVE
5. JOINT
6. O-RING
7. EVAPORATOR
8. AIR THERMO SENSOR CLIP
9. AIR THERMO SENSOR
10. HEATER CASE

67170-OUTL-G09

Heater and evaporator core locations

Cylinder Head

REMOVAL & INSTALLATION

1. Before servicing the vehicle, refer to the Precautions Section.

2. Relieve the fuel system pressure. Disconnect the negative battery cable. Drain the cooling system.

3. Remove the air cleaner assembly. Disconnect the accelerator cable.

4. Disconnect the control wiring harness. Disconnect the battery wiring harness connector.

5. Remove the radiator hose lower clamp.

6. Disconnect the EVAP hose connection. Disconnect the brake booster vacuum hose, pressure hose clamp and knock sensor connector.

7. Remove the engine oil dipstick and dipstick guide. Remove the knock sensor connector connection. Remove the battery wiring harness connection. Remove the intake manifold stay.

8. Remove the exhaust manifold. Remove the timing belt upper cover.

9. Turn the crankshaft clockwise, align the timing marks on the camshaft sprocket to position number one piston to TDC of its compression stroke. Never rotate the crankshaft in the counterclockwise direction.

10. Remove the timing belt under cover rubber plug, using tool MD998738. Screw

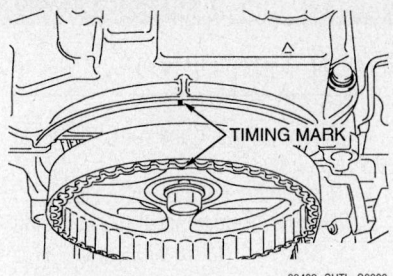

Timing mark alignment

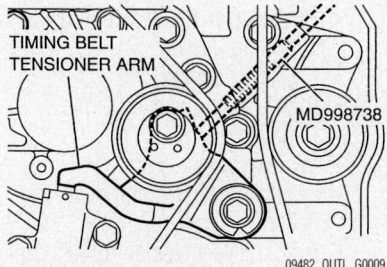

Tool MD998738 installation

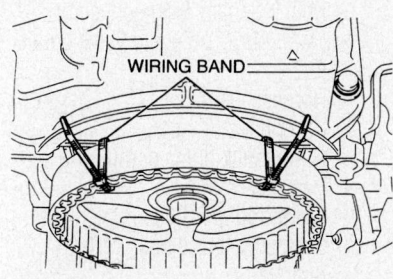

Securing the camshaft sprocket

the tool in until it contacts the timing belt tensioner arm.

11. Secure the camshaft and valve timing belt, with wire to prevent slippage between the camshaft sprocket and valve timing belt. Remove the camshaft sprocket. Do not turn the crankshaft after the camshaft sprocket is removed.

12. Remove the upper radiator hose connection and water cooler hose connection. Disconnect the high pressure fuel hose connection.

13. Remove the valve cover.

14. Loosen all cylinder head bolts in two or three steps and in the proper bolt removal sequence. Remove the cylinder head from the engine. Discard the cylinder head gasket.

➡️If the cylinder head bolts cannot be pulled out due to the washer being trapped in the valve spring, raise the bolt slightly, then remove it while holding it with a magnet.

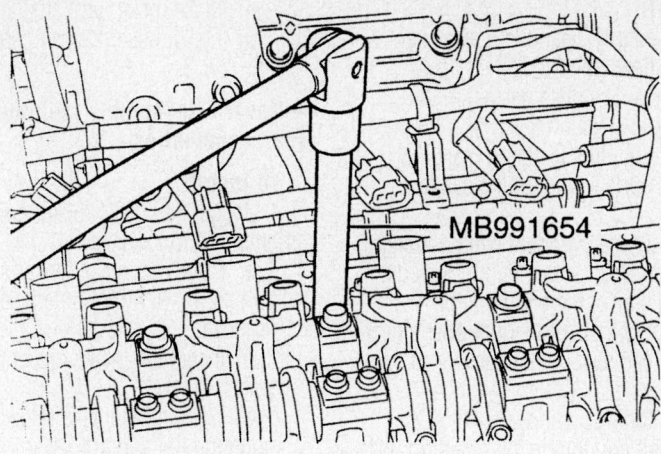

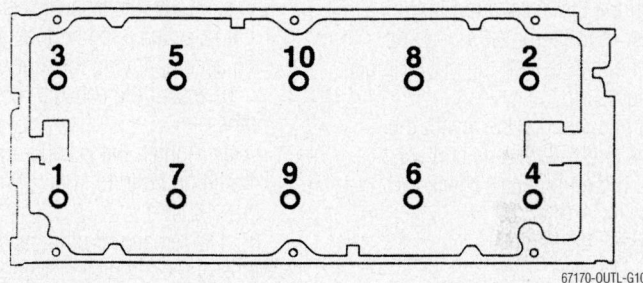

Cylinder head bolt removal sequence

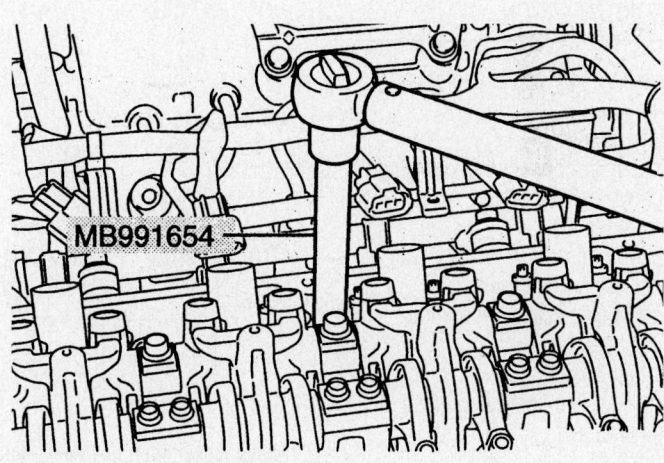

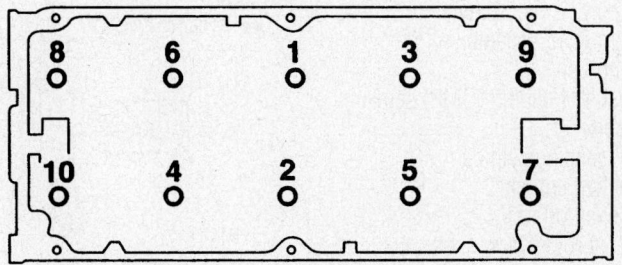

Cylinder head bolt torque sequence

To install:

15. Thoroughly clean the mating surfaces of the head and block.

16. Place a new head gasket on the cylinder block. Position the gasket on the block so that the identification mark "381" is at the top surface and on the exhaust side.

17. Inspect the cylinder head bolt length prior to installation. If the length exceeds 3.91 in. (99.4mm), the bolt must be replaced. Apply a small amount of engine oil to the thread section and washer of the bolt.

18. Position the cylinder head on the engine. Install and torque the retaining bolts to specification and in the proper sequence.

19. Torque the camshaft bolt to 58—72 ft. lbs. (80—98 Nm)

20. Continue the installation in the reverse order of the removal procedure.

21. Fill the engine with the proper grade and type engine oil.

22. Fill the cooling system with the proper grade and type engine coolant.

23. Start the engine and check for leaks.

24. Roadtest the vehicle.

Rocker Arms/Shafts

REMOVAL & INSTALLATION

1. Before servicing the vehicle, refer to the Precautions Section.
2. Drain the engine oil.
3. Remove or disconnect the following:
   - Negative battery cable
   - Air cleaner
   - Ignition coils
   - Upper timing belt cover
   - Positive Crankcase Ventilation (PCV) hose
   - Breather hose
   - Control wiring harness connection
   - Engine hanger
   - Oil control valve
   - Oil pressure switch
   - Rocker arm (valve) cover and gasket
   - Oil seals, if necessary
   - Accumulator assembly
   - Timing belt
   - Camshaft Position (CMP) sensor support
   - CMP sensing cylinder
   - Camshaft sprocket
   - Camshaft oil seal
   - Exhaust rocker arm shaft caps
   - Exhaust rocker arm and shaft assembly
   - Intake rocker arm shaft caps
   - Intake rocker arm and shaft assembly

➡ **Never disassemble the rocker arm shaft assemblies.**

To install:

4. Install the intake rocker arm and shaft assembly as follows:
   a. Place the intake rocker shaft so that the 0.22 in. (5.5mm) hole faces the cylinder head.
   b. Install the intake rocker arm shaft caps.
   c. Tighten the intake rocker shaft mounting bolts to 106–124 inch lbs. (12–14 Nm).

5. Install the exhaust rocker arm and shaft assembly as follows:
   a. Place the exhaust rocker shaft so that its notch is positioned as shown in the accompanying illustration.
   b. Install the exhaust rocker arm shaft caps.
   c. Tighten the exhaust rocker shaft mounting bolts to 106–124 inch lbs. (12–14 Nm).

6. The remainder of installation is the reverse of the removal procedure.

7. Fill the crankcase with oil and connect the negative battery cable.

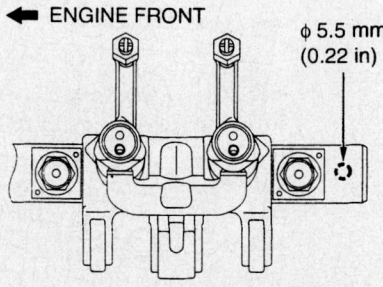

Intake rocker shaft position—hole must face the cylinder head

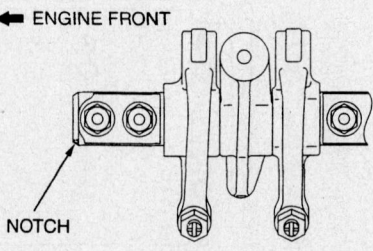

Exhaust rocker shaft position

Intake Manifold

REMOVAL & INSTALLATION

1. Before servicing the vehicle, refer to the Precautions Section.
2. Relieve the fuel system pressure.
3. Remove or disconnect the following:
   - Battery negative cable and drain the cooling system
   - Air cleaner assembly
   - Throttle body
   - Fuel rail assembly
   - Exhaust Gas Recirculation (EGR) valve and gasket
   - Capacitor connector
   - Brake booster vacuum hose
   - Capacitor
   - Vacuum pipe
   - Brake booster vacuum hose connection
   - Evaporative emission (EVAP) canister vacuum hose
   - Manifold Absolute Pressure (MAP) sensor
   - Knock Sensor (KS) connector
   - EVAP purge solenoid valve connector
   - Harness clamp
   - KS connector clamp
   - Oil dipstick guide and O-ring
   - Harness clamp
   - Lower radiator hose. Matchmark the clamp to the hose before removal.
   - Heater hose
   - Engine Coolant Temperature (ECT) gauge unit connector
   - Thermostat housing and O-ring
   - KS harness clamp
   - MAP sensor and harness clamp
   - Intake manifold stay
   - Intake manifold bolt and manifold.
   - Intake manifold gasket and discard

To install:

4. Clean all gasket material from the cylinder head intake mounting surface and intake manifold assembly.

5. Install the intake manifold, using a new gasket. Torque the manifold in a criss-cross pattern, starting from the inside and working outwards to specification.

6. Install or connect the following:
   - Intake manifold stay. Torque the bolt to 21–25 ft. lbs. (29–33 Nm)
   - MAP sensor and harness clamp
   - KS harness clamp
   - Thermostat housing and O-ring. Clean the gasket mating surfaces, then apply a 3mm bead of 3M™

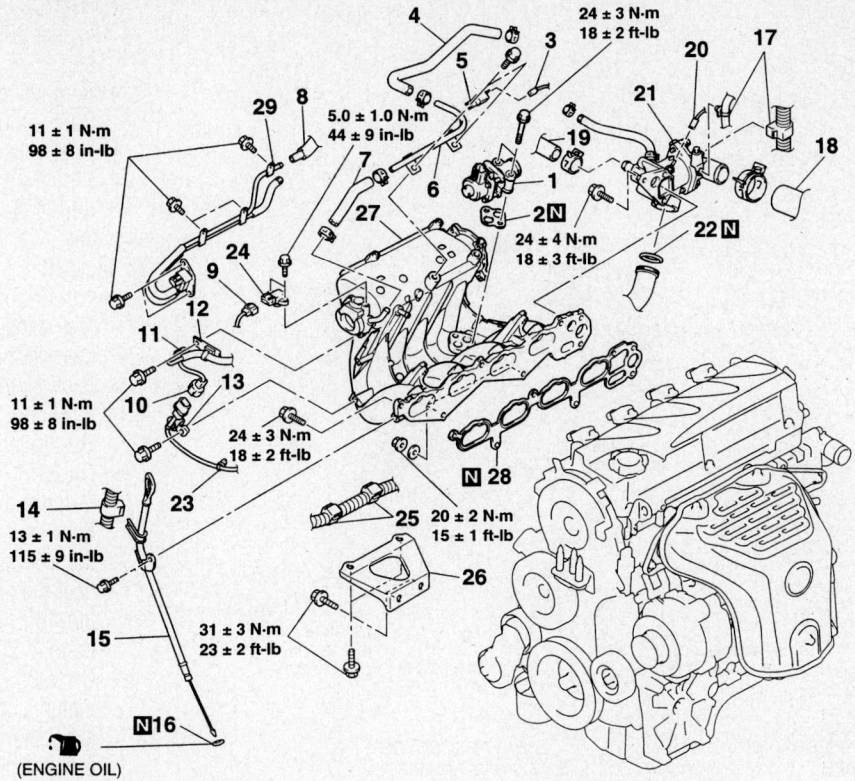

11 ± 1 N·m
98 ± 8 in-lb

5.0 ± 1.0 N·m
44 ± 9 in-lb

24 ± 3 N·m
18 ± 2 ft-lb

24 ± 4 N·m
18 ± 3 ft-lb

11 ± 1 N·m
98 ± 8 in-lb

24 ± 3 N·m
18 ± 2 ft-lb

13 ± 1 N·m
115 ± 9 in-lb

20 ± 2 N·m
15 ± 1 ft-lb

31 ± 3 N·m
23 ± 2 ft-lb

(ENGINE OIL)

1. EXHAUST GAS RECIRCULATION VALVE
2. EXHAUST GAS RECIRCULATION VALVE GASKET
3. CAPACITOR CONNECTOR
4. BRAKE BOOSTER VACUUM HOSE
5. CAPACITOR
6. BRAKE BOOSTER VACUUM PIPE
7. BRAKE BOOSTER VACUUM HOSE
8. EVAPORATIVE EMISSION CANISTER VACUUM HOSE CONNECTION
9. MANIFOLD ABSOLUTE PRESSURE SENSOR CONNECTOR
10. KNOCK SENSOR CONNECTOR
11. EVAPORATIVE EMISSION PURGE SOLENOID VALVE CONNECTOR
12. HARNESS CRAMP
13. KNOCK SENSOR CONNECTOR CRAMP
14. HARNESS CRAMP
15. OIL DIPSTICK GUIDE

16. O-RING
17. HARNESS CRAMP
18. RADIATOR LOWER HOSE
19. HEATER HOSE CONNECTION
20. ENGINE COOLANT TEMPERATURE GAUGE UNIT CONNECTOR
21. THERMOSTAT HOUSING ASSEMBLY
22. O-RING
23. KNOCK SENSOR HARNESS CRAMP
24. MANIFOLD ABSOLUTE PRESSURE SENSOR
25. HARNESS CRAMP
26. INTAKE MANIFOLD STAY
27. INTAKE MANIFOLD
28. INTAKE MANIFOLD GASKET
29. EVAPORATIVE EMISSION PURGE SOLENOID VALVE, EVAPORATIVE EMISSION VACUUM HOSE AND PIPE ASSEMBLY

67170-OUTL-G16

Intake manifold and related components

ADD part No. 8672, 3M™ ADD part No. 8679/8678 or equivalent.
- ECT gauge unit connector
- Heater hose
- Lower radiator hose. Matchmark the clamp to the hose before removal.
- Harness clamp
- Oil dipstick guide and O-ring
- KS connector clamp
- Harness clamp
- EVAP purge solenoid valve connector
- KS connector

- MAP sensor
- EVAP canister vacuum hose
- Brake booster vacuum hose connection
- Vacuum pipe
- Capacitor
- Brake booster vacuum hose
- Capacitor connector
- EGR valve and gasket
- Fuel rail assembly
- Throttle body
- Air cleaner assembly
7. Fill the system with coolant.
8. Connect the negative battery cable.

Exhaust Manifold

REMOVAL & INSTALLATION

1. Before servicing the vehicle, refer to the Precautions Section.
2. Remove or disconnect the following:
   - Negative battery cable
   - Engine under cover, 2005–2006 vehicles
   - Front exhaust pipe bracket
   - Front exhaust pipe from the exhaust manifold

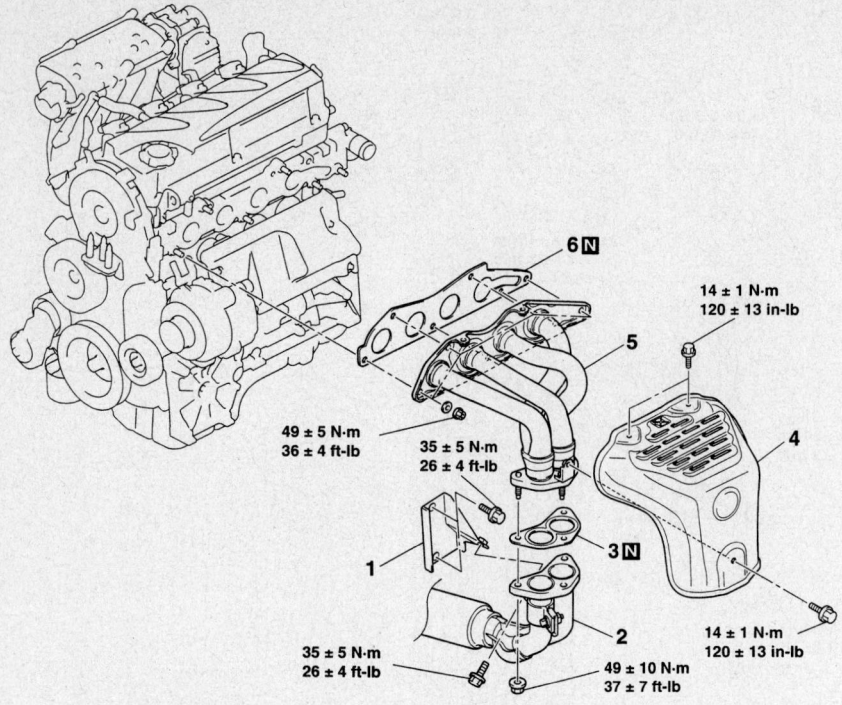

1. FRONT EXHAUST PIPE BRACKET
2. FRONT EXHAUST PIPE
   CONNECTION
3. GASKET
4. HEAT PROTECTOR
5. EXHAUST MANIFOLD
6. EXHAUST MANIFOLD GASKET

67170-OUTL-G17

Exhaust manifold and related components—except 2005–2006 California vehicles

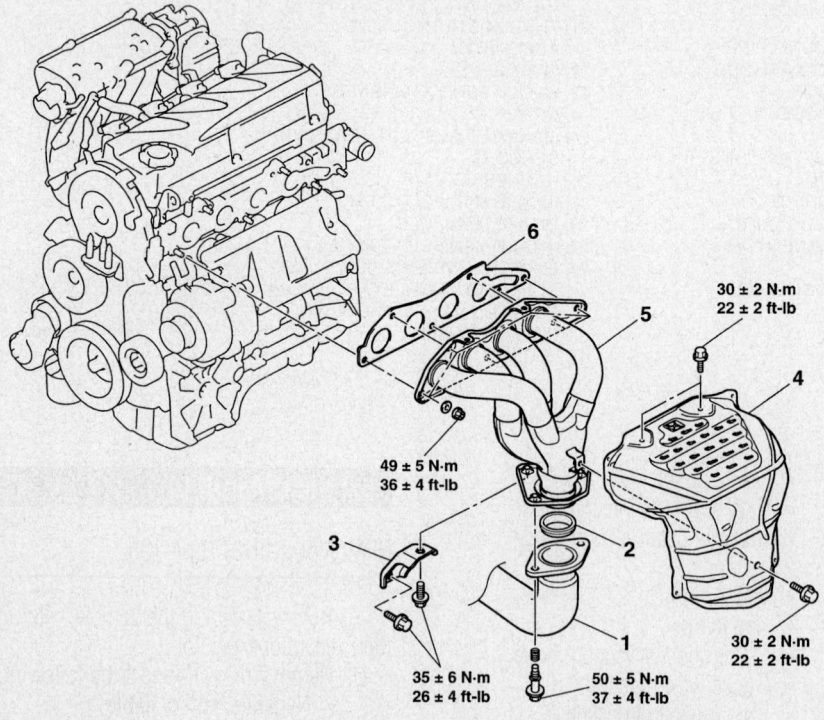

1. FRONT EXHAUST PIPE
   CONNECTION
2. SEAL RING
3. EXHAUST MANIFOLD BRACKET
4. HEAT PROTECTOR
5. EXHAUST MANIFOLD
6. EXHAUST MANIFOLD GASKET

09482_OUTL_G0010

Exhaust manifold and related components—2005–2006 California vehicles

- Remove and discard the gasket, except 2005–2006 California vehicles
- Remove and discard the seal ring, 2005–2006 California vehicles
- Heat shield
- Mounting nuts, the exhaust manifold, and the exhaust manifold gasket

To install:

3. Install or connect the following:
- New exhaust manifold gasket to the cylinder head
- Exhaust manifold. Torque the mounting nuts to specification.
- Heat shield and tighten the bolts
- On 2005–2006 vehicles, except California install a new gasket between the exhaust manifold and the front exhaust pipe and reconnect the pipe. Torque the nuts to 30–44 ft. lbs. (39–59 Nm) and the bolts to 22–30 ft. lbs. (30–40 Nm).
- On 2005–2006 California vehicles install a new seal ring between the exhaust manifold and the front exhaust pipe and reconnect the pipe. Torque the bolts to 30–44 ft. lbs.
- Front exhaust pipe bracket and tighten the retainers
- Negative battery cable

4. Start the engine and check for any exhaust leaks.

Front Crankshaft Seal

REMOVAL & INSTALLATION

1. Before servicing the vehicle, refer to the Precautions Section.

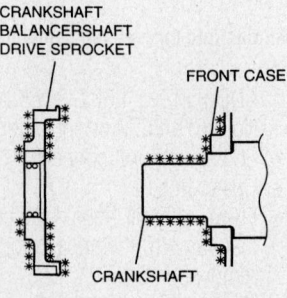

67170-OUTL-G18

Crankshaft balancer shaft drive sprocket installation

2. Remove or disconnect the following:
- Negative battery cable
- Timing belt
- Crankshaft sprocket
- Crankshaft key

3. Carefully pry the oil seal out of the front case. Be careful not to damage the oil seal bore or the crankshaft sealing surface.

To install:

4. Apply clean engine oil to the oil seal lip. Using a seal driver, install the oil seal.

5. Install or connect the following:
- Crankshaft key
- Crankshaft sprocket. Clean and

degrease the sprocket and install as shown in the accompanying figure.
- Timing belt
- Negative battery cable

Camshaft and Valve Lifters

REMOVAL & INSTALLATION

1. Before servicing the vehicle, refer to the Precautions Section.
2. Drain the engine oil.
3. Remove or disconnect the following:
- Negative battery cable

- Air cleaner
- Ignition coils
- Upper timing belt cover
- Positive Crankcase Ventilation (PCV) hose
- Breather hose
- Control wiring harness connection
- Engine hanger
- Oil control valve
- Oil pressure switch
- Rocker arm (valve) cover and gasket
- Oil seals, if necessary
- Accumulator assembly

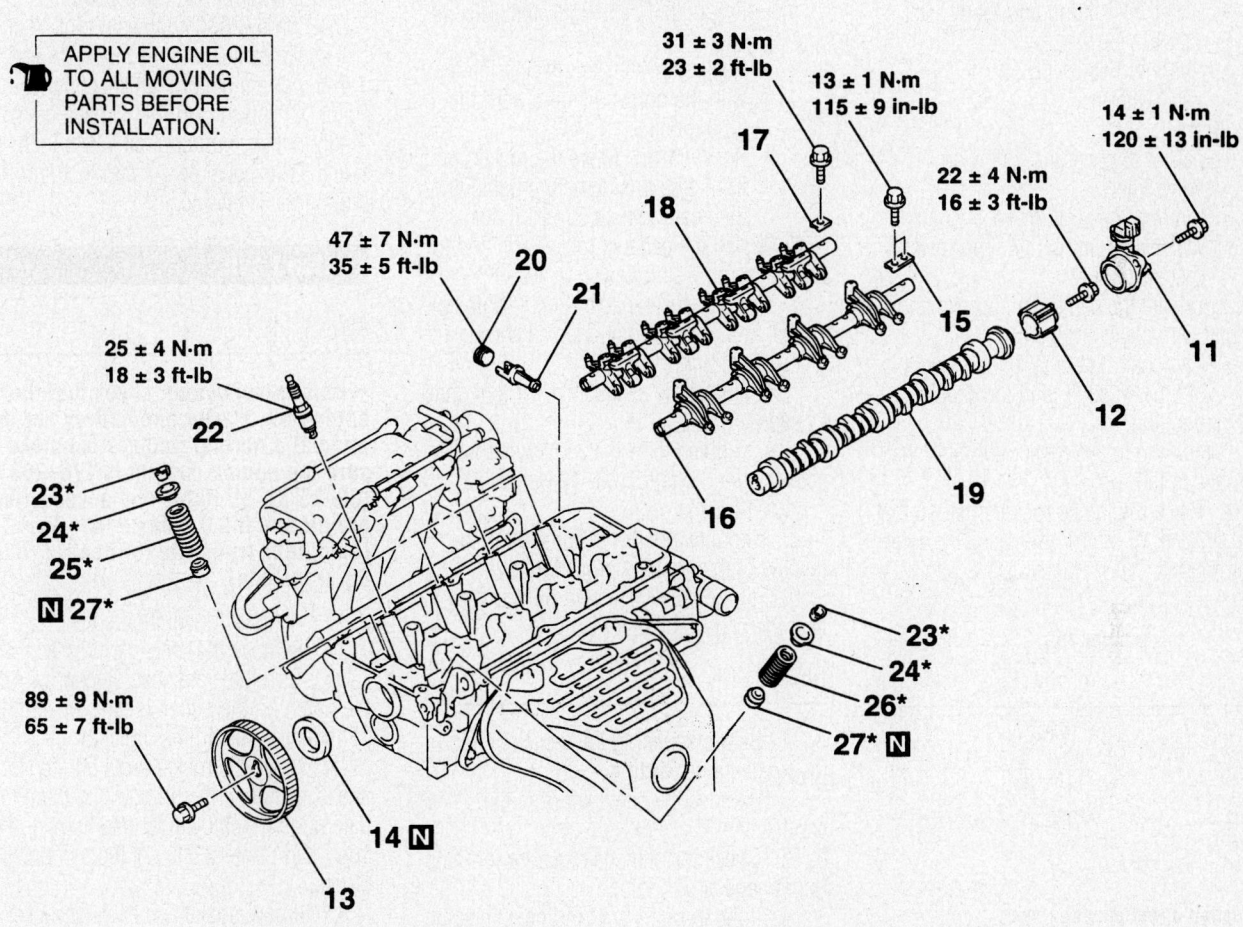

APPLY ENGINE OIL TO ALL MOVING PARTS BEFORE INSTALLATION.

31 ± 3 N·m
23 ± 2 ft-lb

13 ± 1 N·m
115 ± 9 in-lb

14 ± 1 N·m
120 ± 13 in-lb

22 ± 4 N·m
16 ± 3 ft-lb

47 ± 7 N·m
35 ± 5 ft-lb

25 ± 4 N·m
18 ± 3 ft-lb

89 ± 9 N·m
65 ± 7 ft-lb

11. CAMSHAFT POSITION SENSOR SUPPORT
12. CAMSHAFT POSITION SENSING CYLINDER
13. CAMSHAFT SPROCKET
14. CAMSHAFT OIL SEAL
15. EXHAUST ROCKER ARM SHAFT CAPS
16. EXHAUST ROCKER ARM AND SHAFT ASSEMBLY
17. INLET ROCKER ARM SHAFT CAPS

18. INLET ROCKER ARM AND SHAFT ASSEMBLY
19. CAMSHAFT
20. CYLINDER HEAD PLUG
21. OIL CONTROL VALVE FILTER
22. SPARK PLUGS
23. VALVE SPRING RETAINER LOCKS
24. VALVE SPRING RETAINERS
25. INLET VALVE SPRINGS
26. EXHAUST VALVE SPRINGS
27. VALVE STEM SEALS

67170-OUTL-G13

Camshaft and related components

- Timing belt
- Camshaft Position (CMP) sensor support
- CMP sensing cylinder
- Camshaft sprocket
- Camshaft oil seal
- Exhaust rocker arm shaft caps
- Exhaust rocker arm and shaft assembly
- Intake rocker arm shaft caps
- Intake rocker arm and shaft assembly
- Camshaft. Inspect the bearing journals on the camshaft and the cylinder head.
- Water inlet fitting and thermostat case
- Cylinder head plug
- Valve lifters

To install:

4. Install or connect the following:
- Valve lifters
- Cylinder head plug
- Water inlet fitting and thermostat case assembly

5. Lubricate the camshaft journals and camshaft with clean engine oil
- Camshaft. Make sure the dowel pin of the camshaft is in the proper position.

6. Install the intake rocker arm and shaft assembly as follows:
a. Place the intake rocker shaft so that the 0.22 in. (5.5mm) hole faces the cylinder head.

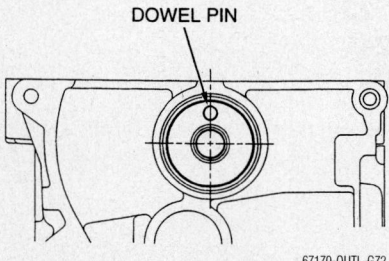

DOWEL PIN

67170-OUTL-G72

Camshaft dowel pin alignment

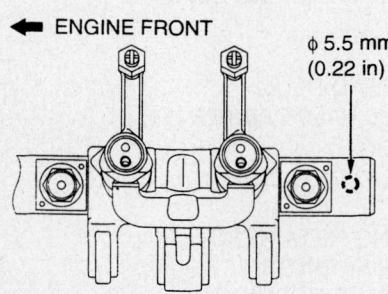

← ENGINE FRONT

φ 5.5 mm
(0.22 in)

67170-OUTL-G14

Intake rocker shaft position—hole must face the cylinder head

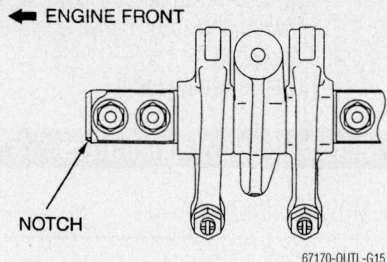

← ENGINE FRONT

NOTCH

67170-OUTL-G15

Exhaust rocker shaft position

b. Install the intake rocker arm shaft caps.

c. Tighten the intake rocker shaft mounting bolts to 21–25 ft. lbs. (28–34 Nm).
- Rocker arm shafts, placing the end with the notched side toward the timing belt
- Rocker arms. Move the rocker arms from side to side before tightening the shaft bolts to 21–25 ft. lbs.

7. Install the exhaust rocker arm and shaft assembly as follows:
a. Place the exhaust rocker shaft so that its notch is positioned as shown in the illustration.

b. Install the exhaust rocker arm shaft caps.

c. Tighten the exhaust rocker shaft mounting bolts to 106–124 inch lbs. (12–14 Nm).

8. The remainder of installation is the reverse of the removal procedure.

9. Fill the crankcase with oil and connect the negative battery cable.

INSPECTION

1. Before servicing the vehicle, refer to the Precautions Section.

2. Remove the camshaft from the engine.

3. Check the camshaft bearing journals for damage and binding.

4. If the journals are binding, check the cylinder head for damage.

5. Check the cylinder head for clogged oil holes.

6. Check the camshaft surface for abnormal wear and damage. Replace the camshaft, as required.

7. Measure the cam height and replace the camshaft if not within specification.

8. 2003 engine: Intake camshaft 1.472 inch, minimum limit 1.452 inch. Exhaust camshaft 1.450 inch, minimum limit 1.430 inch.

9. 2004 engine: Point "A" intake cam (low speed cam) "A" 1.332 inch, minimum limit 1.313 inch. Point "B" intake cam (low

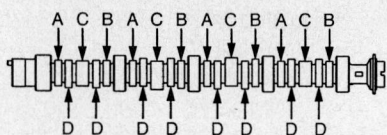

A C B A C B A C B A C B A C B

D D D D D D D D

09482_OUTL_G0011

Camshaft measurement point locations—2004–2006 engines

speed cam) "B" 1.471 inch, minimum limit 1.451 inch. Point "C" intake cam (high speed cam) "C" 1.465 inch, minimum limit 1.445 inch. Point "D" exhaust cam 1.491 inch, minimum limit 1.471 inch.

10. 2005–2006 engines: Point "A and B" intake cam (low speed cam) "A and "B" 1.475 inch, minimum limit 1.455 inch. Point "C" intake cam (high speed cam) "C" 1.465 inch, minimum limit 1.445 inch. Point "D" exhaust cam 1.465 inch, minimum limit 1.445 inch.

Valve Lash

ADJUSTMENT

➡Before inspection, check that the engine oil, starter and battery are operating at a normal range. Also make sure the engine coolant is 176–203°F (80–95°C), all lights and accessories are off and the transaxle is in Park (automatic transaxle) or neutral (manual transaxle).

1. Remove the ignition coils.
2. Remove the rocker arm (valve) cover.
3. Turn the crankshaft clockwise until the notch on the pulley is lines up with the "T" mark on the timing indicator.
4. Move the rocker arms on the No. 1 and 4 cylinders up and down by hand to determine which cylinder had its piston at Top Dead Center (TDC) of the compression stroke.
5. If both intake and exhaust rocker arms have a valve lash, the piston in the cylinder corresponding to these rocker arms is at TDC.
6. Valve clearance inspection and adjustment can be performed on the rocker arms denoted by the white arrows in the figure, when the No. 1 cylinder piston is at TDC and on rocker denoted by the black arrows in the figure when the No. 4 cylinder piston is at TDC.
7. Measure the valve clearance. If the clearance is not within specification, loosen the rocker arm locknut and adjust the clearance using a thickness gauge while turning the adjusting screw.

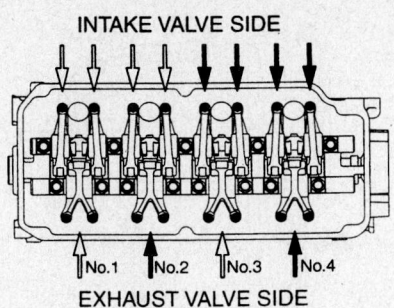

INTAKE VALVE SIDE

No.1 No.2 No.3 No.4

EXHAUST VALVE SIDE

67170-OUTL-G73

Valve lash adjusting points

8. While holding the adjusting screw with a screwdriver to prevent it from turning, tighten the locknut to 6.5 ft. lbs. (9 Nm).

9. Turn the crankshaft a full rotation (360°) to line up the notch on the crankshaft pulley with the "T" mark on the timing indicator.

10. Repeat the adjustment steps on the other valves.

11. Install the rocker cover.

12. Install the ignition coils.

Oil Pan

REMOVAL & INSTALLATION

FWD Vehicles

1. Before servicing the vehicle, refer to the Precautions Section.

2. Remove the negative battery cable.

3. Remove the engine undercover.

4. Drain the engine oil. Install the oil pan drain plug with a new gasket and tighten to 29 ft. lbs. (39 Nm).

5. Remove or disconnect the following:
- Front exhaust pipe
- Torque converter/bell housing inspection cover
- Bolts attaching the oil pan to the cylinder block

6. Remove the oil pan assembly by using a hammer to tap Mitsubishi Special tool no. MD998727 into the range shown in the illustration of the oil pan and engine block. Slide the tool sideways to separate the oil pan from the block.

To install:

7. Clean the sealing surface on the oil pan and engine block. Apply a continuous 4mm bead of sealant to the oil pan. The oil pan MUST be installed within 15 minutes of applying the sealant.

8. Install or connect the following:
- Oil pan to the cylinder block and tighten the bolts to 80 inch lbs. (9 Nm)
- Torque converter/bell housing inspection cover. Torque the bottom bolts to 19 ft. lbs. (26 Nm) and the flange bolts to 80 inch lbs. (9 Nm). Refer to accompanying illustration.
- Front exhaust pipe
- Engine undercover

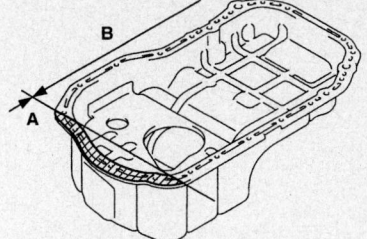

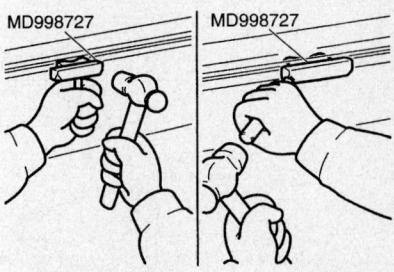

MD998727   MD998727

09482_OUTL_G0012

Use the special tool between the oil pan and block in area B only. Do NOT insert the pan separator tool into area A—FWD vehicles

❋❋ WARNING

Wait at least 1 hour after tightening the oil pan bolts to refill the crankcase and start the engine.

9. Refill the engine with oil. Connect the negative battery cable. Start the engine and check for leaks.

AWD Vehicles

1. Before servicing the vehicle, refer to the Precautions Section.

2. Remove the negative battery cable.

3. Remove the engine undercover.

4. Drain the engine oil. Install the oil pan drain plug with a new gasket and tighten to 29 ft. lbs. (39 Nm).

5. Remove or disconnect the following:
- Front exhaust pipe
- Center member
- Transaxle housing front lower cover stay
- Torque converter housing front lower (inspection) cover

➡ **There are 2 sizes of oil pan bolts used. Note the locations of the bolts as you remove them.**

- Lower oil pan mounting bolts. Place a piece of wood at the rear of the lower pan and strike it with a hammer to remove the lower pan.
- Upper oil pan mounting bolts. Screw bolts into the bolt holes (A) shown in the illustration, then lift and remove the upper oil pan.

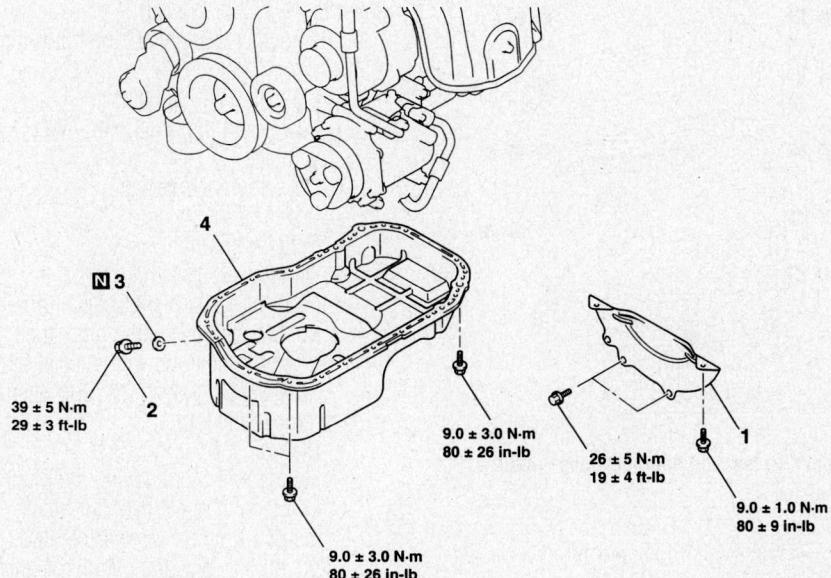

39 ± 5 N·m
29 ± 3 ft-lb

9.0 ± 3.0 N·m
80 ± 26 in-lb

9.0 ± 3.0 N·m
80 ± 26 in-lb

26 ± 5 N·m
19 ± 4 ft-lb

9.0 ± 1.0 N·m
80 ± 9 in-lb

1. TORQUE CONVERTER HOUSING FRONT LOWER COVER
2. ENGINE OIL PAN DRAIN PLUG
3. ENGINE OIL PAN DRAIN PLUG GASKET
4. ENGINE OIL PAN

67170-OUTL-G21

Oil pan and related components—FWD vehicles

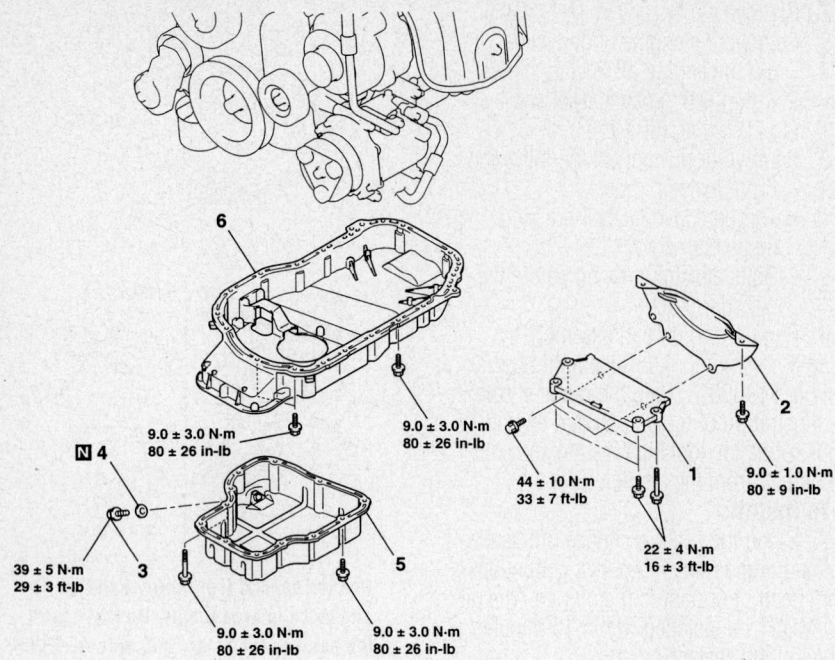

9.0 ± 3.0 N·m
80 ± 26 in-lb

9.0 ± 3.0 N·m
80 ± 26 in-lb

9.0 ± 1.0 N·m
80 ± 9 in-lb

44 ± 10 N·m
33 ± 7 ft-lb

22 ± 4 N·m
16 ± 3 ft-lb

39 ± 5 N·m
29 ± 3 ft-lb

9.0 ± 3.0 N·m
80 ± 26 in-lb

9.0 ± 3.0 N·m
80 ± 26 in-lb

1. TRANSAXLE HOUSING FRONT LOWER COVER STAY
2. TORQUE CONVERTER HOUSING FRONT LOWER COVER
3. ENGINE OIL PAN DRAIN PLUG
4. ENGINE OIL PAN DRAIN PLUG GASKET
5. ENGINE LOWER OIL PAN
6. ENGINE UPPER OIL PAN

67170-OUTL-G23

Upper and lower oil pans and related components—AWD vehicles

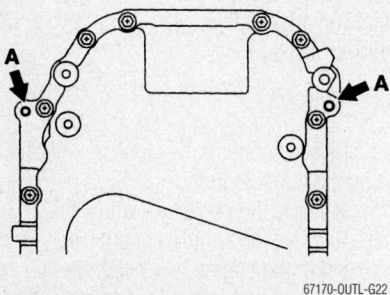

67170-OUTL-G22

To remove the upper oil pan, screw bolts into the bolt holes A, then lift and remove the upper oil pan—AWD vehicles

To install:

6. Clean the sealing surface on the oil pan and engine block. Apply a continuous 4mm bead of sealant to the oil pan. The oil pan MUST be installed within 15 minutes of applying the sealant.

7. Install or connect the following:
- Upper oil pan to the block and tighten the bolts to 80 inch lbs. (9 Nm)
- Lower oil pan to the block and tighten the bolts to 80 inch lbs. (9 Nm)
- Torque converter lower (inspection) cover. Torque the bolts to 80 inch lbs. (9 Nm).

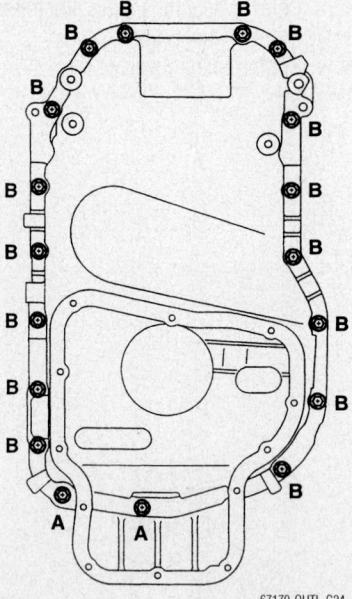

67170-OUTL-G24

Upper oil pan bolt hole locations—AWD vehicles

- Transaxle housing front lower cover stay. Tighten the bolts as shown in the accompanying illustration.
- Center member
- Front exhaust pipe
- Engine undercover

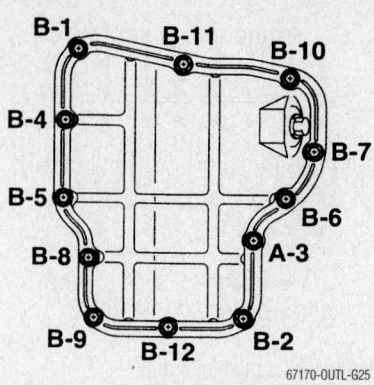

67170-OUTL-G25

Lower oil pan bolt hole locations—AWD vehicles

※※ WARNING

Wait at least 1 hour after tightening the oil pan bolts to refill the crankcase and start the engine.

8. Refill the engine with oil. Connect the negative battery cable. Start the engine and check for leaks.

Oil Pump

REMOVAL & INSTALLATION

1. Before servicing the vehicle, refer to the Precautions Section.
2. Remove the negative battery cable.
3. Remove the engine undercover.
4. Drain the engine oil. Install the oil pan drain plug with a new gasket and tighten to 29 ft. lbs. (39 Nm).
5. Remove or disconnect the following:
- Timing belt
- Oil pan, FWD vehicles
- Lower and upper oil pan, AWD vehicles
- Oil screen and gasket
- Relief spring
- Relief plunger
- Oil filter bracket and gasket
- Plug. Fit Mitsubishi Special tool no. MD998162 on the plug, then hold it in position with special tool MD998783, loosen the plug, then remove the tools.
- O-ring
- Flange bolt by removing the plug on the side of the cylinder block. Insert a Phillips screwdriver into the hold to lock the counterbalance shaft, then loosen the flange bolt.
- Front case and gasket
- Oil pump cover
- Oil pump driven gear
- Oil pump drive gear

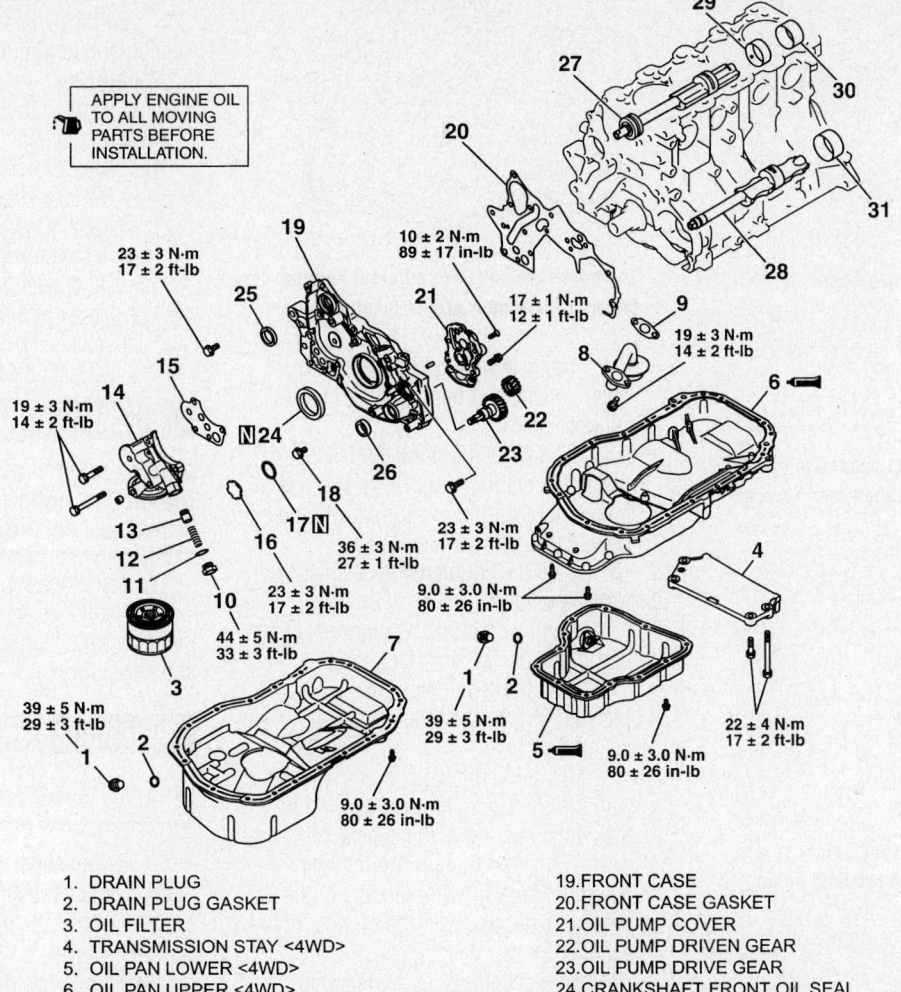

APPLY ENGINE OIL TO ALL MOVING PARTS BEFORE INSTALLATION.

23 ± 3 N·m
17 ± 2 ft-lb

10 ± 2 N·m
89 ± 17 in-lb

17 ± 1 N·m
12 ± 1 ft-lb

19 ± 3 N·m
14 ± 2 ft-lb

19 ± 3 N·m
14 ± 2 ft-lb

23 ± 3 N·m
17 ± 2 ft-lb

9.0 ± 3.0 N·m
80 ± 26 in-lb

36 ± 3 N·m
27 ± 1 ft-lb

23 ± 3 N·m
17 ± 2 ft-lb

44 ± 5 N·m
33 ± 3 ft-lb

22 ± 4 N·m
17 ± 2 ft-lb

39 ± 5 N·m
29 ± 3 ft-lb

39 ± 5 N·m
29 ± 3 ft-lb

9.0 ± 3.0 N·m
80 ± 26 in-lb

9.0 ± 3.0 N·m
80 ± 26 in-lb

1. DRAIN PLUG
2. DRAIN PLUG GASKET
3. OIL FILTER
4. TRANSMISSION STAY <4WD>
5. OIL PAN LOWER <4WD>
6. OIL PAN UPPER <4WD>
7. OIL PAN <2WD>
8. OIL SCREEN
9. OIL SCREEN GASKET
10. RELIEF PLUG
11. GASKET
12. RELIEF SPRING
13. RELIEF PLUNGER
14. OIL FILTER BRACKET
15. OIL FILTER BRACKET GASKET
16. PLUG
17. O-RING
18. FLANGE BOLT

19. FRONT CASE
20. FRONT CASE GASKET
21. OIL PUMP COVER
22. OIL PUMP DRIVEN GEAR
23. OIL PUMP DRIVE GEAR
24. CRANKSHAFT FRONT OIL SEAL
25. OIL PUMP OIL SEAL
26. COUNTERBALANCE SHAFT OIL SEAL
27. COUNTERBALANCE SHAFT, LEFT
28. COUNTERBALANCE SHAFT, RIGHT
29. COUNTERBALANCE SHAFT, FRONT BEARING
30. COUNTERBALANCE SHAFT, REAR BEARING, RIGHT
31. COUNTERBALANCE SHAFT, REAR BEARING, LEFT

09482_OUTL_G0013

Oil pump and related components

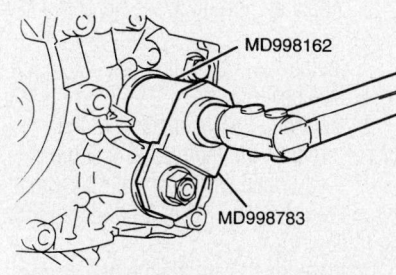

MD998162

MD998783

67170-OUTL-G31

To remove or install the plug, fit special tool MD998162 on the plug and hold it in place with tool MD998783

- Crankshaft front oil seal
- Oil pump oil seal
- Left and right counterbalance shafts
- Front counterbalance shaft bearing, using Mitsubishi Special tool no. MD998371
- Rear left and right counterbalance shaft bearings using Mitsubishi Special tool nos. MD991603 and MD998372

To install:

6. Install the left counterbalance shaft rear bearing as follows:

a. Install Mitsubishi Special tool no. MD991603 to the cylinder block.

b. Apply engine oil to the rear bearing outer surface and bearing hole in the cylinder block.

c. Use Mitsubishi Special tool MD998705 to install the rear bearing. The left rear bearing has no oil holes.

7. Install the right counterbalance shaft rear bearing as follows:

a. Install the guide pin of special tool MD998705 in the threaded hole of the cylinder block.

b. Align the ratchet ball of special tool MD998705 with the oil hold in the rear bearing to install the bearing of the special tool MD998705.

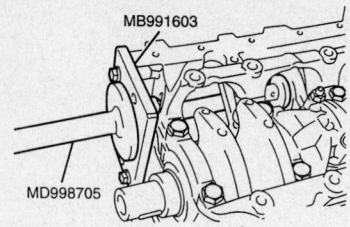

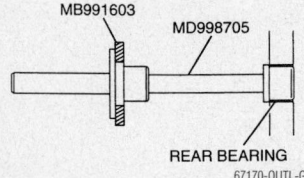

View of the special tools needed to install the left counterbalance shaft rear bearing

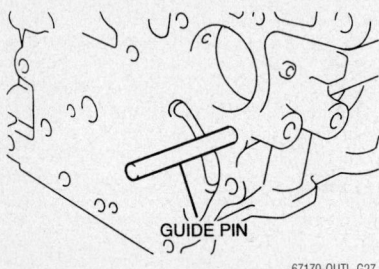

Install the guide pin of special tool MD998705 into the threaded hold of the cylinder block

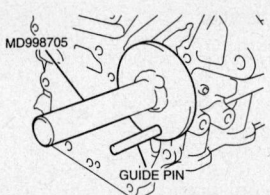

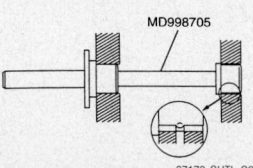

Use special tool MD998705 to install the rear bearing. Make sure the oil hole of the bearing is aligned with the oil hole of the cylinder block

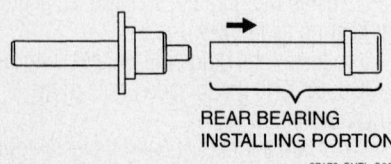

Remove the rear bearing installing part from special tool MD998705

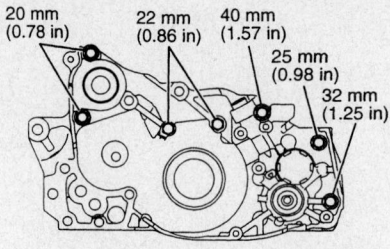

The front case bolts are different lengths. Install bolts in their proper location

    c. Apply engine oil to the bearing outer surface and bearing hole in the cylinder block.

    d. Use special tool MD998705 to install the rear bearing. Make sure the oil hole of the bearing is aligned with the oil hole of the cylinder block.

  8. Install the counterbalance shaft front bearing as follows:

    a. Remove the rear bearing installing part from special tool MD998705.

    b. Install the guide pin of special tool MD998705 into the threaded hole of the cylinder block.

    c. Align the ratchet ball of the tool with the oil hole in the rear bearing to install the bearing of the special tool.

    d. Using the tool, install the rear bearing. Make sure the oil hole of the bearing is aligned with the oil hole of the cylinder block.

  9. Use a suitable socket wrench to install the counterbalance shaft oil seal into the front case.

  10. Install the oil pump oil seal, using a suitable socket to press the seal into place.

  11. Use special tool MD998375 to install the front oil seal into the front case.

  12. Install the oil pump gears into the front case and make sure the alignment marks line up.

  13. Install the oil pump case as follows:

    a. Place special tool MD998285 on the front end of the crankshaft, then apply a thin coating of engine oil to the outer surface of the special tool.

    b. Apply engine oil to the lip of the crankshaft front oil seal.

    c. Install the front case. Be careful not to damage the oil seal.

    d. Install the bolts, noting their different lengths and tighten to 17 ft. lbs. (23 Nm).

  14. Install the flange bolt, using a screwdriver to lock the counterbalance shaft and tighten to 27 ft. lbs. (36 Nm). Pull out the screwdriver and screw the plug back in.

  15. Install the plug as follows:

    a. Install a new O-ring to the groove of the front case.

    b. Install the plug to the front case.

    c. Fit special tool MD998162 on the plug and hold it in place with tool MD998783.

    d. Tighten the plug to 17 ft. lbs. (23 Nm). Remove the special tools.

  16. Install or connect the following:

- Oil filter bracket and gasket
- Relief plunger
- Relief spring
- Oil screen and gasket
- Upper and lower oil pans, AWD vehicles
- Oil pan, FWD vehicles

✷✷ WARNING

Wait at least 1 hour after tightening the oil pan bolts to refill the crankcase and start the engine.

  17. Install the timing belt.

  18. Refill the engine with oil. Connect the negative battery cable. Start the engine and check for leaks.

Rear Main Seal

REMOVAL & INSTALLATION

  1. Before servicing the vehicle, refer to the Precautions Section.

  2. Remove the transaxle from the vehicle.

  3. If equipped with automatic transaxle, remove the driveplate bolts. Use special tool MD998781 to hold the driveplate while loosening the bolts. Remove the driveplate adapter plate and driveplate.

  4. If equipped with manual transaxle, remove the flywheel bolts. Remove the flywheel adapter plate, flywheel and flywheel adapter plate.

  5. Remove the crankshaft bushing. Carefully pry the seal out of the oil seal case without damaging the sealing surface of the crankshaft.

To install:

  6. Apply engine oil to the lip of the new seal.

  7. Install the seal in the case using the proper size seal driver.

  8. Continue the installation in the reverse order of the removal procedure.

  9. If equipped with automatic transaxle, torque the driveplate retaining bolts to 98 ft. lbs. (132 Nm).

  10. If equipped with manual transaxle, torque the flywheel retaining bolts to 98 ft. lbs. (132 Nm).

  11. Install the transaxle assembly.

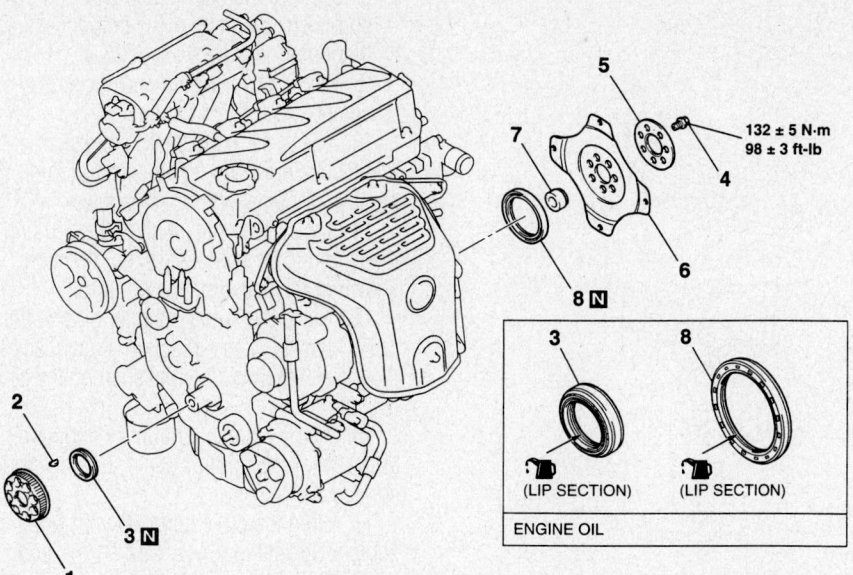

132 ± 5 N·m
98 ± 3 ft-lb

ENGINE OIL

1. CRANKSHAFT BALANCER
   SHAFT DRIVE SPROCKET
2. CRANK SHAFT KEY
3. CRANKSHAFT FRONT OIL SEAL
4. A/T DRIVE PLATE BOLTS

5. A/T DRIVE PLATE ADAPTER
   PLATE
6. A/T DRIVE PLATE
7. CRANKSHAFT BUSH
8. CRANKSHAFT REAR OIL SEAL

67170-OUTL-G19

Crankshaft rear main seal and related components—automatic transaxle

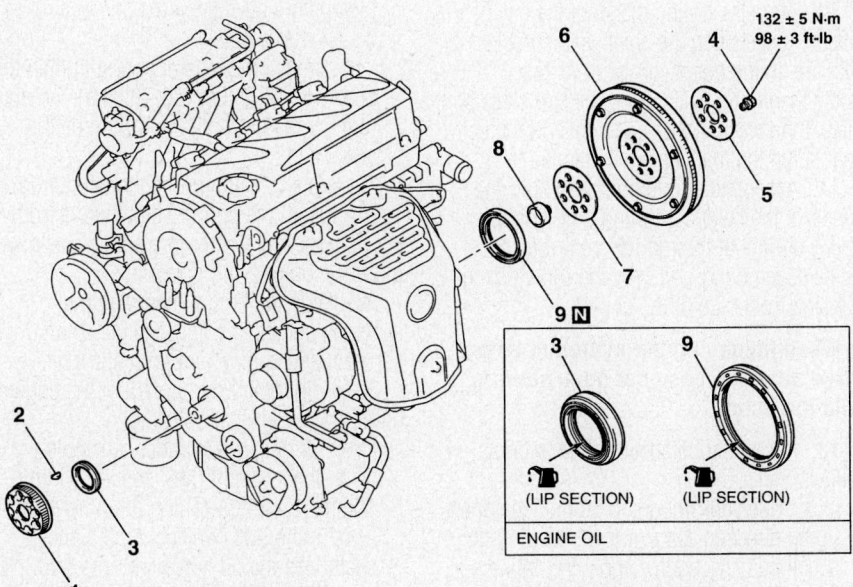

132 ± 5 N·m
98 ± 3 ft-lb

ENGINE OIL

1. CRANKSHAFT BALANCER
   SHAFT DRIVE SPROCKET
2. CRANKSHAFT KEY
3. CRANKSHAFT FRONT OIL SEAL
4. FLYWHEEL BOLTS

5. FLYWHEEL ADAPTER PLATE
6. FLYWHEEL ASSEMBLY
7. FLYWHEEL ADAPTER PLATE
8. CRANKSHAFT BUSH
9. CRANKSHAFT REAR OIL SEAL

09482_OUTL_G0014

Crankshaft rear main seal and related components—manual transaxle

Timing Belt

REMOVAL & INSTALLATION

1. Before servicing the vehicle, refer to the Precautions Section.

2. Remove or disconnect the following:
   - Engine undercover
   - Drive belt
   - Crankshaft damper pulley
   - Control wiring harness connection
   - Battery wiring harness connection
   - Connector
   - Engine front mounting bracket
   - Harness bracket
   - Upper timing belt cover
   - Water pump pulley
   - Idler pulley
   - Auto-tensioner
   - Lower timing belt cover

3. Position the engine so that the No. 1 piston is at Top Dead Center (TDC).

➡**If the timing belts are going to be reused, mark the direction of rotation on the belt. This will ensure the belt is reinstalled in same direction, extending belt life.**

4. To loosen the timing (outer) belt tensioner, remove the rubber plug, then install Mitsubishi Special tool MD998738 to the slot and screw inward until the it contacts the timing belt tensioner arm.

5. Gradually screw in the tool, then align the timing belt tensioner adjuster rod set hold A with the timing belt tensioner adjusted cylinder set hole B.

6. Insert a wire or pin in the set hole aligned.

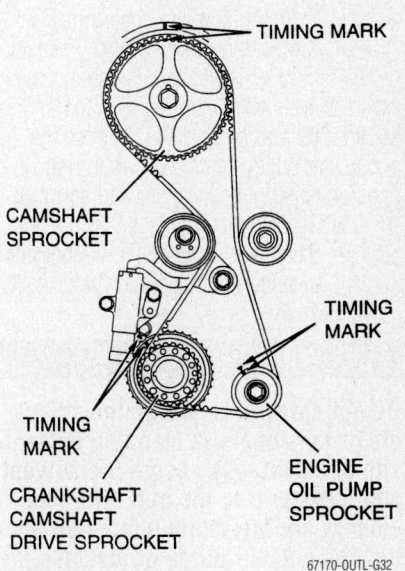

TIMING MARK

CAMSHAFT
SPROCKET

TIMING
MARK

TIMING
MARK

CRANKSHAFT
CAMSHAFT
DRIVE SPROCKET

ENGINE
OIL PUMP
SPROCKET

67170-OUTL-G32

Timing mark alignment

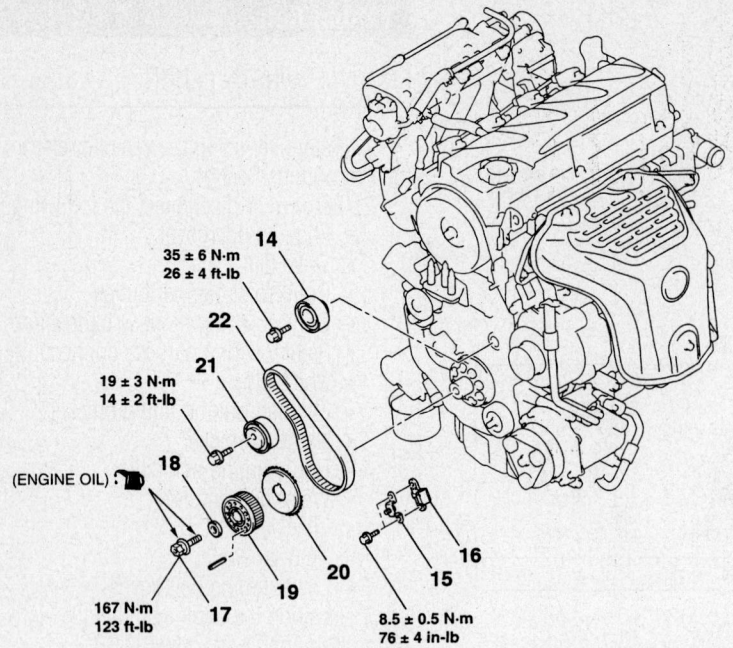

35 ± 6 N·m
26 ± 4 ft-lb — **14**

22

19 ± 3 N·m
14 ± 2 ft-lb — **21**

(ENGINE OIL) — **18**

167 N·m
123 ft-lb — **17**

20

19

15

16

8.5 ± 0.5 N·m
76 ± 4 in-lb

67170-OUTL-G40

14. TIMING BELT IDLER PULLEY
15. TIMING BELT LOWER COVER BRACKET
16. CRANKSHAFT POSITION SENSOR
17. CRANKSHAFT PULLEY CENTER BOLT
18. CRANKSHAFT PULLEY WASHER
19. CRANKSHAFT CAMSHAFT DRIVE SPROCKET
20. CRANKSHAFT ANGLE SENSING BLADE
21. BALANCER TIMING BELT TENSIONER
22. BALANCER TIMING BELT

Outer timing belt and related components

7. Loosen the timing belt tensioner pulley bolts and remove the outer timing belt.

8. Remove or disconnect the following:
• Timing belt tensioner pulley
• Timing belt tensioner arm
• Timing belt tensioner adjuster
• Timing belt idler pulley
• Timing belt lower cover bracket
• Crankshaft Position (CKP) sensor

9. Hold the crankshaft/camshaft drive sprocket with Mitsubishi Special tools MD991367 and MD991385. Loosen the crankshaft pulley center bolt, then remove the crankshaft pulley washer and sprocket.
• Crankshaft angle sensing blade
• Balancer timing belt tensioner and balancer timing (inner) belt.

To install:

10. Align the timing marks of the balancer shaft sprockets and the crankshaft sprocket with the timing marks on the front case. Route the timing belt around the sprockets so there is no slack in the upper span of the belt and the timing marks are still aligned.

11. Assemble and temporarily hold the center of the balancer timing belt tensioner pulley so it is at the top left from the center of the assembling bolt, and the pulley flange is at the front side of the engine.

➡ **When tightening the mounting bolts, make sure the tensioner does not turn with the bolts.**

12. Adjust the balancer belt tension as follows:
a. Use your fingers to lift the tensioner in the direction of the arrow shown in the accompanying figure. Apply 22–30 inch lbs. (2.6–3.4 Nm) to the timing belt so the belt is taut without any looseness.
b. Tighten the assembling bolt to 12–16 ft. lbs. (16–22 Nm). Then, fix the timing belt tensioner.
c. Turn the crankshaft clockwise 2 revolutions to set the No. 1 cylinder to TDC of its compression stroke and check that the timing marks are still aligned.

d. Apply about 22 lbs. (100 N) of pressure to the center area between the sprockets, then check the belt deflection. It should measure 0.20–0.27 in. (5–7mm). If not within specifications, adjust the belt tension again.

13. Clean the crankshaft, crankshaft angle sensing blade, drive sprocket and crankshaft pulley washer, as shown.

14. Install the crankshaft angle sending blade and drive sprocket in the direction shown in the accompanying figure.

15. Place the larger chamfer side of the pulley washer in the direction shown, and then assemble on the crankshaft pulley center bolt.

16. Apply a small amount of engine oil to the crankshaft pulley center bolt bearing surface and screw.

17. Hold the drive sprocket with Mitsubishi Special tool nos. MB991367 and MD991385, then torque the center bolt to 123 ft. lbs. (167 Nm).

18. Install or connect the following:
• CKP sensor
• Timing belt lower cover bracket
• Timing belt idler pulley

19. Install the timing belt tensioner arm as follows:
a. If the adjuster rod is fully extended, use a vise to slowly compress the timing belt adjuster rod and align set hole "A". Insert a pin or wire into the holes. If installing a new tensioner, it will already be set with a pin.
b. Install the tensioner and tighten the bolts 15–19 ft. lbs. (20–26 Nm). Do not remove the pin until the timing belt is tensioned.

20. Install the timing belt tensioner arm.

21. Install the timing belt tensioner pulley. Temporarily tighten the pulley as shown in the accompanying figure.

22. Align the timing marks on the camshaft sprocket, crankshaft camshaft drive sprocket and engine oil pump sprocket.

23. Adjust the timing mark of the engine oil pump sprocket as follows:
a. Unplug the cylinder block plug.
b. Insert a bolt (M6, section width 10mm, 45mm long) into the plug hole.
c. If the bolt contacts the balancer shaft, turn the oil sprocket 1 rotation.
d. Re-adjust the timing mark, and then check to make sure the bolt fits. Do not remove the bolt until the timing belt is assembled.

24. Install the timing belt as follows:
a. Place the belt on the timing belt tensioner pulley and cam/crank drive sprocket, then support it with your left hand so it does not slide.

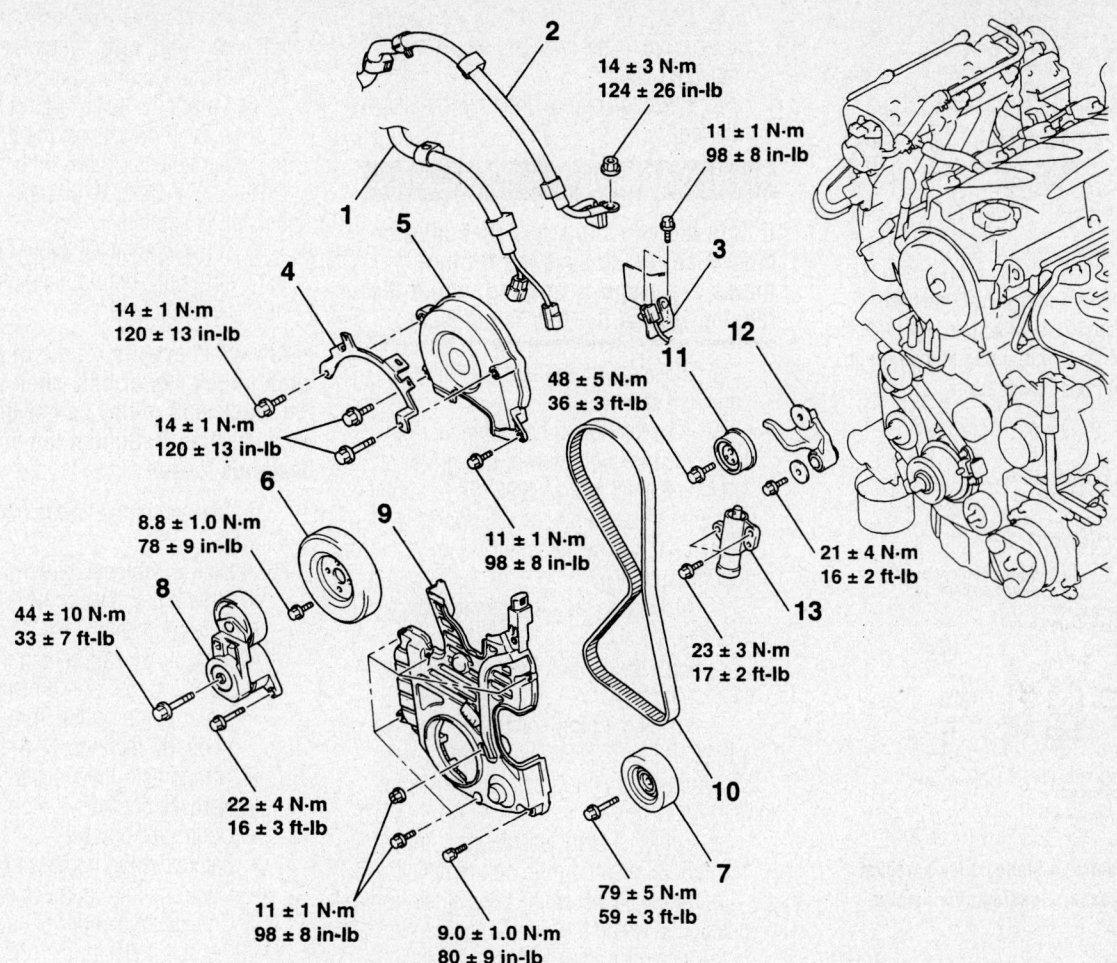

14 ± 3 N·m
124 ± 26 in-lb

11 ± 1 N·m
98 ± 8 in-lb

14 ± 1 N·m
120 ± 13 in-lb

14 ± 1 N·m
120 ± 13 in-lb

48 ± 5 N·m
36 ± 3 ft-lb

11 ± 1 N·m
98 ± 8 in-lb

8.8 ± 1.0 N·m
78 ± 9 in-lb

44 ± 10 N·m
33 ± 7 ft-lb

21 ± 4 N·m
16 ± 2 ft-lb

23 ± 3 N·m
17 ± 2 ft-lb

22 ± 4 N·m
16 ± 3 ft-lb

11 ± 1 N·m
98 ± 8 in-lb

9.0 ± 1.0 N·m
80 ± 9 in-lb

79 ± 5 N·m
59 ± 3 ft-lb

1. CONTROL WIRING HARNESS CONNECTION
2. BATTERY WIRING HARNESS CONNECTION
3. CONNECTOR BRACKET
4. HARNESS BRACKET
5. TIMING BELT UPPER COVER
6. WATER PUMP PULLEY
7. IDLER PULLEY

8. AUTO-TENSIONER
9. TIMING BELT LOWER COVER
10. VALVE TIMING BELT
11. TIMING BELT TENSIONER PULLEY
12. TIMING BELT TENSIONER ARM
13. TIMING BELT TENSIONER ADJUSTER

67170-0UTL-G41

Inner timing belt and related components

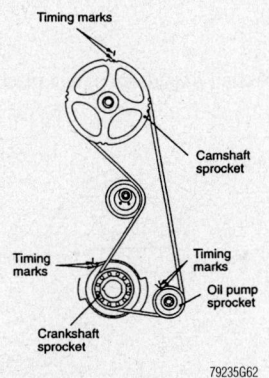

Timing marks

Camshaft sprocket

Timing marks

Timing marks

Oil pump sprocket

Crankshaft sprocket

79235G62

Proper alignment of the timing belt sprocket marks for belt service

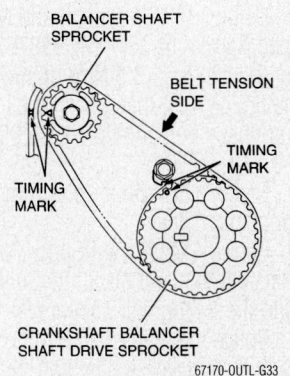

BALANCER SHAFT SPROCKET

BELT TENSION SIDE

TIMING MARK

TIMING MARK

CRANKSHAFT BALANCER SHAFT DRIVE SPROCKET

67170-0UTL-G33

Timing belt "B" (balancer belt) installation mark alignment

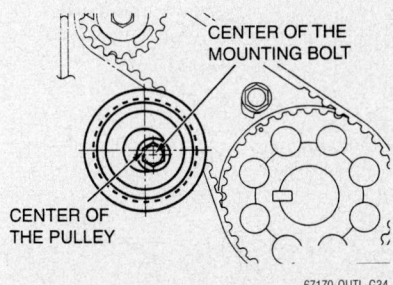

CENTER OF THE MOUNTING BOLT

CENTER OF THE PULLEY

67170-0UTL-G34

Assemble and temporarily hold the center of the balancer belt tensioner pulley so it's at the top left from the center of the assembling bolt, and the pulley flange is at the front side of the engine

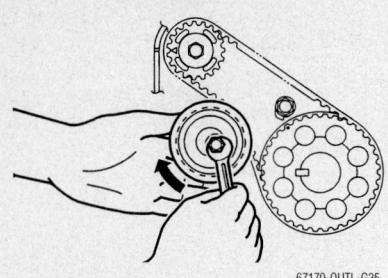

67170-OUTL-G35

Lift the tensioner in the direction of the arrow, then apply 22–30 inch lbs. (2.6–3.4 Nm) of pressure to the timing belt so the it is taut without any looseness

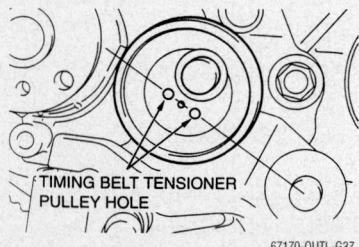

○ : CLEAN
✳ : CLEAN AND DEGREASE
● : APPLY ENGINE OIL

CRANKSHAFT PULLEY CENTER BOLT
CRANKSHAFT PULLEY WASHER
CRANKSHAFT ANGLE SENSING BLADE
CRANKSHAFT CAMSHAFT DRIVE SPROCKET
CRANKSHAFT
◄ ENGINE FRONT

67170-OUTL-G36

Crankshaft, sensing blade, drive sprocket and pulley washer cleaning and installation direction

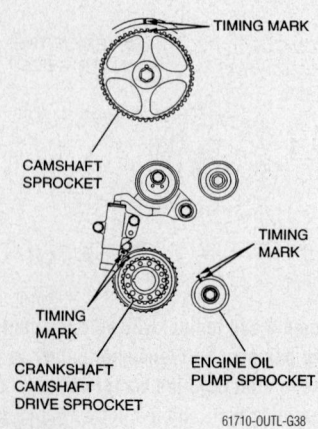

TIMING BELT TENSIONER PULLEY HOLE

67170-OUTL-G37

Temporary alignment of the timing belt tensioner pulley

TIMING MARK
CAMSHAFT SPROCKET
TIMING MARK
TIMING MARK
CRANKSHAFT CAMSHAFT DRIVE SPROCKET
ENGINE OIL PUMP SPROCKET

61710-OUTL-G38

Timing mark and sprocket alignment

b. Place the timing belt on the engine oil pump sprocket while pulling it with your right hand.

c. Place the belt on the timing belt idler pulley.

✳✳ WARNING

Rotate the camshaft sprocket counterclockwise. Make sure the timing marks are properly aligned, while the tension side of the belt is correct.

d. Place the timing belt on the camshaft sprocket.

e. Turn the timing belt tensioner pulley in the direction shown, using Mitsubishi special tool MD998767 to apply tension to the timing belt. Temporarily tighten and fix the tensioner pulley mounting bolt.

f. Make sure the timing marks and still aligned.

g. Remove the bolt from the cylinder block plug hole.

h. Install the plug and tighten to 21–25 ft. lbs. (27–33 Nm).

25. Adjust the timing belt tension as follows:

a. Set Mitsubishi Special tool no. MD998738, used during belt removal.

b. Gradually screw the special tool to a position in which the wire or pin inserted in the tensioner adjuster moves slightly.

c. Rotate the crankshaft counterclockwise ¼-turn.

d. Rotate the crankshaft clockwise, aligning each timing mark to set the No. 1 cylinder to TDC.

e. Loosen the tensioner pulley mounting bolt.

✳✳ WARNING

When tightening the mounting bolt, make sure the pulley does not rotate with the bolt.

f. Using a torque wrench and Mitsubishi Special tool no. MD998767, apply 31 inch lbs. (3.5 Nm) of pressure to the timing belt, then install the tensioner pulley bolt and tighten to 33–39 ft. lbs. (43–53 Nm).

g. Remove the wire or pin from the tensioner adjuster.

h. Remove Mitsubishi Special tool no. MD998738, and install the rubber plug to the timing belt undercover.

i. Rotate the crankshaft 2 full rotations in the clockwise direction, then wait 15 minutes to continue the procedure.

j. Insert the wire or pin removed in

Step G, and make sure it can be pulled out with a light load. When it can be lightly removed, the timing belt is tensioned properly. If so, remove the pin.

k. If the projection of the timing belt tensioner adjuster rod is within 0.15–0.17 inch. (3.8–4.5mm), proper tension is applied.

l. If the wire or pin cannot be removed easily, the timing belt tension must be readjusted.

➡Always check the torque of the crankshaft pulley center bolt when turning the crankshaft pulley center bolt counterclockwise. Retighten the bolt if it becomes loose.

m. Make sure the timing marks are aligned.

26. Install or connect the following:
- Timing belt lower cover
- Auto-tensioner
- Idler pulley and tighten the bolt to 56–62 ft. lbs. (74–84 Nm)
- Water pump pulley. Torque the bolt to 69–87 inch lbs. (7.8–9.8 Nm).
- Timing belt upper cover
- Harness bracket
- Connector bracket
- Battery wiring harness connection
- Control wiring harness connection
- Crankshaft damper pulley. Torque the bolt to 18 ft. lbs. (25 Nm).
- Engine undercover

Piston and Ring

POSITIONING

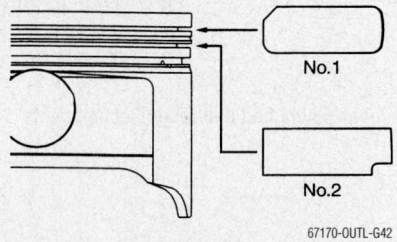

No.1
No.2

67170-OUTL-G42

Compression ring identification marks

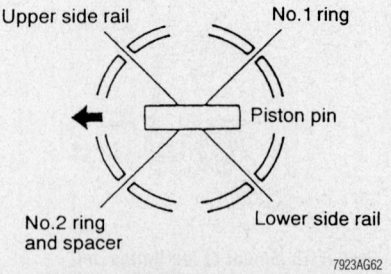

Upper side rail
No.1 ring
Piston pin
No.2 ring and spacer
Lower side rail

7923AG62

Piston ring end-gap spacing

FUEL SYSTEM

Fuel System Service Precautions

Safety is the most important factor when performing not only fuel system maintenance but any type of maintenance. Failure to conduct maintenance and repairs in a safe manner may result in serious personal injury or death. Maintenance and testing of the vehicle's fuel system components can be accomplished safely and effectively by adhering to the following rules and guidelines.

• To avoid the possibility of fire and personal injury, always disconnect the negative battery cable unless the repair or test procedure requires that battery voltage be applied.

• Always relieve the fuel system pressure prior to disconnecting any fuel system component (injector, fuel rail, pressure regulator, etc.), fitting or fuel line connection. Exercise extreme caution whenever relieving fuel system pressure, to avoid exposing skin, face and eyes to fuel spray. Please be advised that fuel under pressure may penetrate the skin or any part of the body that it contacts.

• Always place a shop towel or cloth around the fitting or connection prior to loosening to absorb any excess fuel due to spillage. Ensure that all fuel spillage (should it occur) is quickly removed from engine surfaces. Ensure that all fuel soaked cloths or towels are deposited into a suitable waste container.

• Always keep a dry chemical (Class B) fire extinguisher near the work area.

• Do not allow fuel spray or fuel vapors to come into contact with a spark or open flame.

• Always use a back-up wrench when loosening and tightening fuel line connection fittings. This will prevent unnecessary stress and torsion to fuel line piping. Always follow the proper torque specifications.

• Always replace worn fuel fitting O-rings with new. Do not substitute fuel hose or equivalent, where fuel pipe is installed.

Fuel System Pressure

RELIEVING

1. Before servicing the vehicle, refer to the Precautions Section.
2. Turn the ignition to the **OFF** position.
3. Loosen the fuel filler cap to release fuel tank pressure.

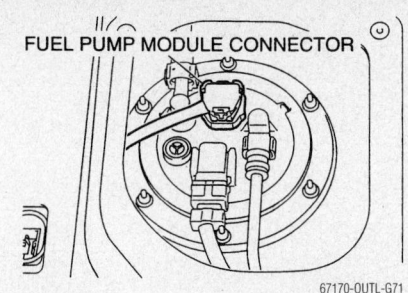

FUEL PUMP MODULE CONNECTOR

67170-OUTL-G71

Fuel pump module connector location

4. Remove the rear seat assembly, then remove the protector and disconnect the fuel pump module connector.
5. Start the vehicle and allow it to run until it stalls from lack of fuel. Turn the key to the **OFF** position.
6. Disconnect the negative battery cable, then reconnect the fuel pump connector. Install the rear seat assembly.

Fuel Pump

REMOVAL & INSTALLATION

1. Before servicing the vehicle, refer to the Precautions Section.
2. Properly relieve the fuel system pressure.
3. Disconnect the negative battery cable.
4. Remove or disconnect the following:
   • Rear seat cushion
   • Retainer screws and service hole cover
   • Harness electrical connector(s)
   • Fuel lines
   • Fuel pump module mounting nuts and plate

✳✳ WARNING

Be careful not to damage the module unit and float, when removing the fuel pump module from the fuel tank.

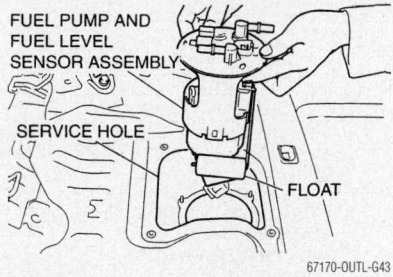

FUEL PUMP AND FUEL LEVEL SENSOR ASSEMBLY

SERVICE HOLE

FLOAT

67170-OUTL-G43

Fuel pump module and related components

• Fuel pump module (FWD) or Fuel pump and fuel level sensor (AWD) from the service hole.
5. Disassemble the fuel pump module as necessary.
6. Installation is the reverse of the removal procedure. Torque the fuel pump module place nuts to 22 inch lbs. (2.5 Nm).

Fuel Rail

REMOVAL & INSTALLATION

1. Before servicing the vehicle, refer to the Precautions Section.
2. Relieve the fuel system pressure.
3. Disconnect the negative battery cable.
4. Wrap the connection with a shop towel and disconnect the high pressure fuel line at the fuel rail.

✳✳ CAUTION

Observe all applicable safety precautions when working around fuel. Whenever servicing the fuel system, always work in a well ventilated area. Do not allow fuel spray or vapors to come in contact with a spark or open flame. Keep a dry chemical fire extinguisher near the work area. Always keep fuel in a container specifically designed for fuel storage; also, always properly seal fuel containers to avoid the possibility of fire or explosion.

5. Remove or disconnect the following:
   • Air cleaner
   • Positive Crankcase Ventilation (PCV) hose
   • Capacitor connector
   • Ignition coil connectors
   • Electrical connector from each injector, label for reference
   • Exhaust Gas Recirculation (EGR) valve connector
   • Rocker arm (valve) cover bracket bolts
   • High pressure fuel hose
   • Bolt(s) holding the fuel rail to the manifold. Carefully lift the rail up and remove it with the injectors attached. Take great care not to drop an injector. Place the rail and injectors in a safe location on the workbench; protect the tips of the injectors from dirt and/or impact.
   • Injector insulators from the intake manifold, and discard

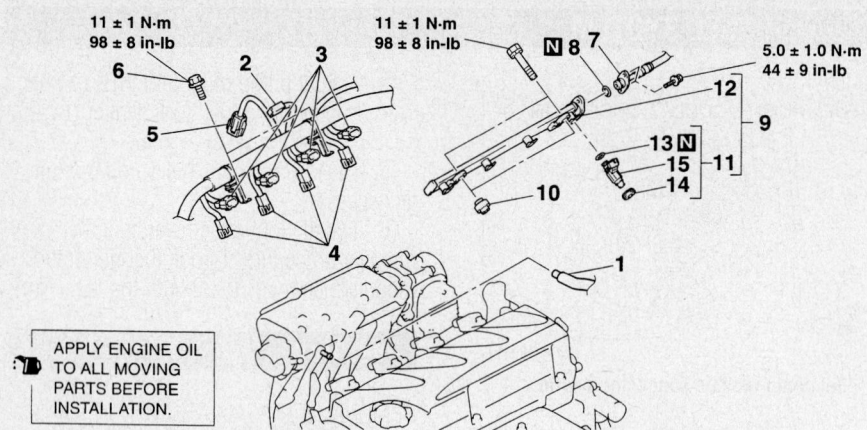

11 ± 1 N·m
98 ± 8 in-lb

11 ± 1 N·m
98 ± 8 in-lb

5.0 ± 1.0 N·m
44 ± 9 in-lb

APPLY ENGINE OIL TO ALL MOVING PARTS BEFORE INSTALLATION.

1. PCV HOSE CONNECTION
2. CAPACITOR CONNECTOR
3. IGNITION COIL CONNECTORS
4. INJECTOR CONNECTORS
5. EGR VALVE CONNECTOR
6. ROCKER COVER BRACKET INSTALLATION BOLTS
7. FUEL HIGH-PRESSURE HOSE CONNECTION
8. O-RING
9. FUEL RAIL AND INJECTOR ASSEMBLY
10. INSULATORS
11. INJECTOR ASSEMBLY
12. FUEL RAIL
13. O-RING
14. GROMMETS
15. INJECTORS

67170-OUTL-G44

Fuel injector rail and related components

To install:
6. Install the fuel rail.

➡Make certain that each injector fits correctly into its port and that the rubber insulators for the fuel rail mounts are in position.

7. Install or connect the following:
- Engine oil to a new O-ring and install on the high pressure fuel line
- High pressure fuel line to the fuel rail. Tighten to 44 inch lbs. (5 Nm).
- Rocker arm (valve) cover bracket bolts
- EGR valve connector
- Fuel injector connectors
- Ignition coil connectors
- Capacitor connector
- PCV hose connector
- Negative battery cable

8. Pressurize the fuel system and inspect all connections for leaks.

Fuel Injector

REMOVAL & INSTALLATION

1. Before servicing the vehicle, refer to the Precautions Section.
2. Relieve the fuel system pressure.
3. Disconnect the negative battery cable.
4. Wrap the connection with a shop towel and disconnect the high pressure fuel line at the fuel rail.

✳✳ CAUTION

Observe all applicable safety precautions when working around fuel. Whenever servicing the fuel system, always work in a well ventilated area. Do not allow fuel spray or vapors to come in contact with a spark or open flame. Keep a dry chemical fire extinguisher near the work area. Always keep fuel in a con-

tainer specifically designed for fuel storage; also, always properly seal fuel containers to avoid the possibility of fire or explosion.

5. Remove the fuel rail assembly.
6. Remove the injectors from the fuel rail by pulling gently in a straight outward motion. Make certain the grommet and O-ring come off with the injector.

To install:
7. Install a new insulator in each injector port in the manifold.
8. Remove the old grommet and O-ring from each injector. Install a new grommet and O-ring; coat the O-ring lightly with clean, thin oil.
9. Install the injector into the fuel rail, constantly turning the injector left and right during installation. When fully installed, the injector should still turn freely in the rail. If it does not, remove the injector and inspect the O-ring for deformation or damage.
10. Continue the installation in the reverse order of the removal procedure.
11. Pressurize the fuel system and inspect all connections for leaks.

Throttle Body

REMOVAL & INSTALLATION

➡On 2006 vehicles, when replacing the throttle body, you must initialize the learning value in a MFI engine.

1. Before servicing the vehicle, refer to the Precautions Section.

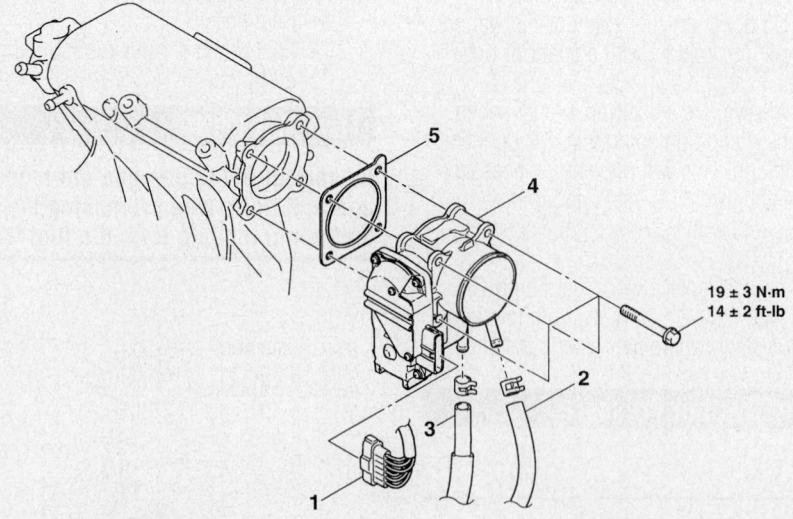

19 ± 3 N·m
14 ± 2 ft-lb

1. THROTTLE POSITION SENSOR CONNECTOR
2. WATER RETURN HOSE CONNECTION
3. WATER FEED HOSE CONNECTION
4. THROTTLE BODY ASSEMBLY
5. THROTTLE BODY GASKET

09482_OUTL_G0015

Throttle body and related components

2. Properly relieve the fuel system pressure. Disconnect the negative battery cable.

3. Drain the engine coolant. Remove the air cleaner assembly.

4. Disconnect the throttle position sensor connector. Disconnect the water hose connection. Disconnect the water feed hose connection.

5. Remove the throttle body retaining bolts. Remove the throttle body from the engine. Discard the gasket.

To install:

➡️**Do not loosen the retaining screws for the resin cover of the throttle body assembly. If the screws are loosened, the sensor incorporated in the resin cover becomes misaligned and the throttle body can not work properly.**

6. Align the recess on the intake manifold plenum with the projection of the throttle body gasket. Install the gasket.

➡️**Poor idling etc. may result if the throttle body gasket is not installed properly.**

7. Install the throttle body to the engine. Torque the retaining bolts to 18–24 ft. lbs. (24–32 Nm).

8. Continue the installation in the reverse order of the removal procedure.

9. Connect the negative battery cable.

10. On 2005–2006 vehicles, turn the ignition **ON** and then **OFF**, and keep it off for at least ten seconds.

11. On 2006 vehicles, complete the vehicle initialization procedure.

➡️**To complete the initialization procedure the following tools are needed. MB991958 scan tool, MB991824 VCI, MB991827 MUT III USB cable, MB991910 MUT III Main harness "A".**

12. Connect the scan tool to the data link connector. To prevent damage to the scan tool be sure that the ignition switch is in the **LOCK** position before connecting the scan tool.

13. Turn the ignition switch to the **ON** position.

14. Select "check mode" from the menu screen.

15. Select "erase memory" from the menu screen.

16. Initialize the learning value.

17. After initialization complete the idle learning procedure.

➡️**This procedure must be performed when the PCM is replaced, or when the learning value is initialized, as the idling is not stabilized because the learning value in the MFI engine is not completed.**

18. Start the engine. Allow the coolant temperature to reach 176 degrees or more.

19. Stop the engine and place the ignition switch in the **LOCK** position.

20. After ten seconds, restart the engine.

21. For ten minutes carry out the idling procedure below to confirm that the engine has normal idling.

- Position the transaxle selector lever in the **P** range
- The engine fan is not to be operated
- The engine coolant temperature should be 176 degrees or more

➡️**If the engine stalls during idling, check the throttle valve of the throttle body for dirt. Correct and perform the procedure again.**

DRIVE TRAIN

Transaxle

REMOVAL & INSTALLATION

Automatic

1. Before servicing the vehicle, refer to the Precautions Section.

2. Install or connect the following:
- Negative battery cable
- Front and side engine undercovers

3. Drain the transaxle fluid. Drain the transfer case fluid, if equipped.
- Front exhaust pipe
- Propeller shaft
- Air cleaner assembly
- Battery and battery tray
- Hood
- Transaxle control cable
- A/T control solenoid valve assembly connector
- Transmission range switch connector
- Input and output shaft speed sensor connectors
- A/T fluid cooler hose from the radiator. Plug the hose.
- Starter mounting bolt
- Loosen the upper engine-to-transaxle bolts. Do not remove the bolts yet.

9. A/T FLUID COOLER HOSE (A/T FLUID COOLER SIDE)
10. TRANSAXLE STAY <AWD>
11. BELL HOUSING COVER
12. DRIVE PLATE BOLTS
- CENTERMEMBER AND FRONT ROLL STOPPER ASSEMBLY (REFER TO GROUP 32, ENGINE ROLL STOPPER AND CENTERMEMBER P.32-6).
- TRANSFER ASSEMBLY <AWD> (REFER TO P.23A-376).

13. TRANSAXLE MOUNT
14. TRANSAXLE MOUNT STOPPER
15. TRANSAXLE LOWER CONNECTING BOLTS
16. TRANSAXLE ASSEMBLY

Automatic transaxle and related components

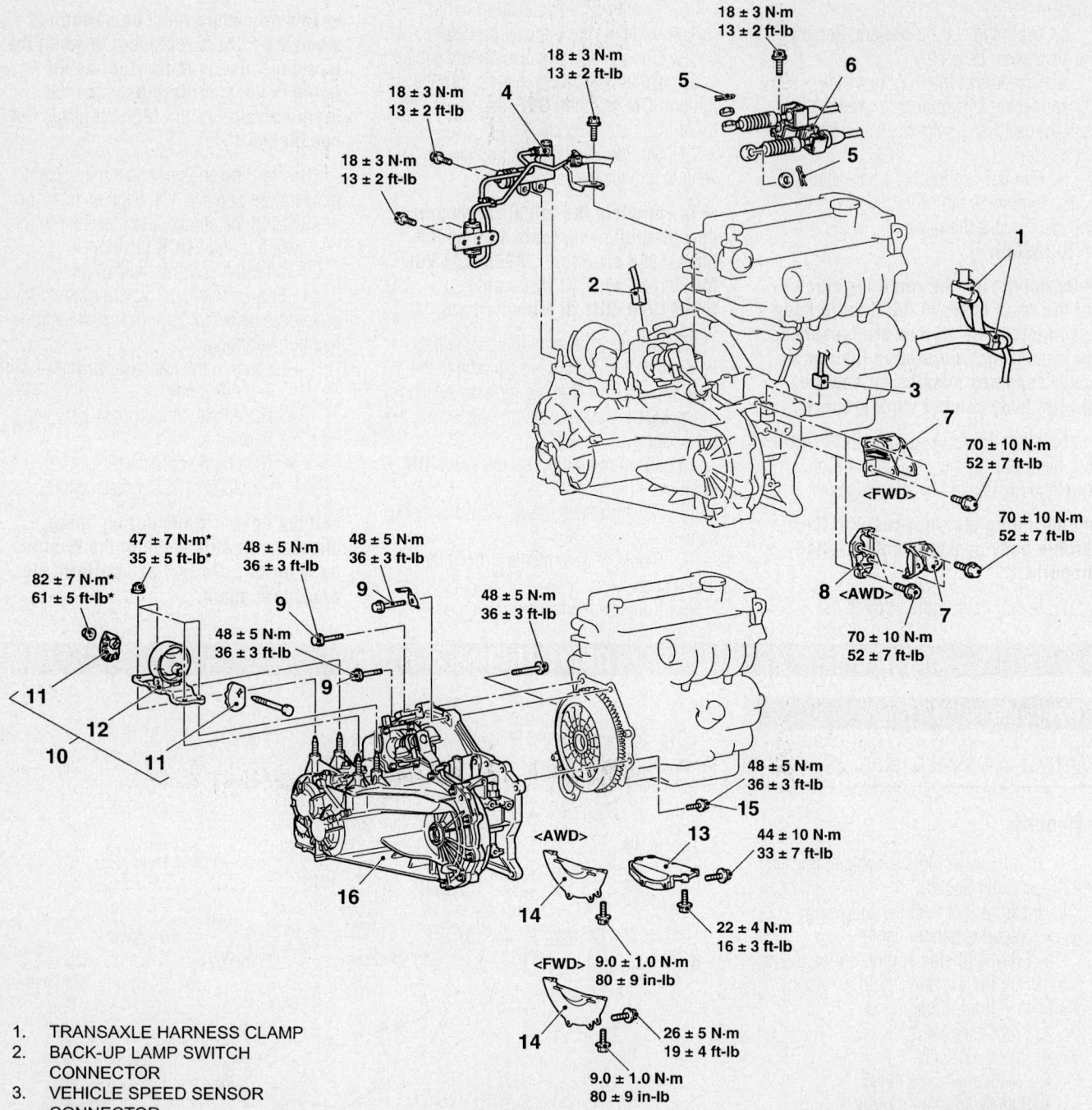

1. TRANSAXLE HARNESS CLAMP
2. BACK-UP LAMP SWITCH CONNECTOR
3. VEHICLE SPEED SENSOR CONNECTOR
4. CLUTCH RELEASE CYLINDER AND CLUTCH OIL PIPE
5. SNAP PIN
6. CABLE BRACKET AND CABLE ASSEMBLY (TRANSAXLE SIDE)
7. REAR ROLL MOUNT BRACKET
8. REAR ROLL MOUNT ADAPTER BRACKET <AWD>

9. TRANSAXLE ASSEMBLY UPPER PART BOLTS
10. TRANSAXLE MOUNTING INSULATOR ASSEMBLY
11. TRANSAXLE MOUNTING INSULATOR STOPPER
12. TRANSAXLE MOUNTING INSULATOR

13. TRANSAXLE HOUSING FRONT LOWER COVER STAY <AWD>
14. FLYWHEEL HOUSING FRONT LOWER COVER
15. TRANSAXLE ASSEMBLY LOWER PART COUPLING BOLTS
16. TRANSAXLE ASSEMBLY

Manual transaxle and related components

09482_OUTL_G0016

- A/T fluid cooler hose from the cooler. Plug the hose.
- Halfshaft and inner shaft, FWD vehicles
- Halfshaft and output shaft assembly, AWD vehicles
- Transaxle stay
- Bell housing cover
- Drive plate bolts. Turn the crankshaft to access the bolts. Push the torque converter into the transaxle side and make sure the torque converter does not stay on the engine side.
- Center member and front roll stopper assembly
- Transfer case, AWD vehicles
- Rear roll stopper, FWD vehicles
- Roll rod bracket, AWD vehicles

4. Carefully support the transaxle with a jack, slightly raise the transaxle, then remove the transaxle mount.
- Transaxle mount stopper

5. Install a suitable engine support assembly, then raise and safely support the vehicle.
- Transaxle to engine bolts and lower the transaxle from the vehicle

6. Support the transaxle with a suitable jack.
- Lower transaxle-to-engine mounting bolts
- Transaxle from the vehicle

To install:

7. Installation is the reverse of the removal procedure, noting the following:
- Transaxle-to-engine lower mounting bolts: 34–40 ft. lbs. (46–54 Nm)
- Driveplate bolts: 34–40 ft. lbs. (46–54 Nm)
- Transaxle mount bracket nuts: 36 ft. lbs. (49 Nm)
- Transaxle mount stopper nuts: 56–66 ft. lbs. (75–89 Nm)
- Transaxle-to-engine upper mounting bolts: 34–40 ft. lbs. (46–54 Nm)

8. Fill the transaxle using the proper grade and type transaxle fluid.

9. If AWD, fill the transfer case to the correct level.

10. Check and adjust the front wheel alignment.

11. Check the speedometer and gear selector for proper operation.

12. Start the engine and check for leaks.

Manual

1. Before servicing the vehicle, refer to the Precautions Section.

2. Disconnect the negative battery cable. Remove the air cleaner assembly.

3. Remove the front under cover. Remove the battery and battery tray.

4. Remove the halfshaft. Remove the output shaft, if equipped with AWD.

5. Drain the transaxle fluid. Drain the transfer case fluid, if equipped.

6. Disconnect the front exhaust pipe.

7. Disconnect the transaxle harness clamp, back up lamp switch and VSS connector.

8. Remove the clutch release cylinder and clutch oil pipe.

9. Remove the snapring, cable bracket and cable assembly (transaxle side).

10. If equipped with AWD, remove the transfer case.

11. Remove the rear roll mount bracket.

12. If equipped with AWD, remove the rear roll mount adapter bracket.

13. Install the engine/transaxle supporting tools.

14. Remove the upper transaxle mounting bolts. As necessary, remove the starter.

15. Properly support the transaxle assembly. Remove the transaxle mounting insulator assembly (including the insulator stopper).

16. With the engine/transaxle supporting tools still installed, remove the transaxle housing front lower cover stay, if equipped with AWD.

17. Remove the flywheel housing front lower cover.

18. Remove the lower transaxle mount-

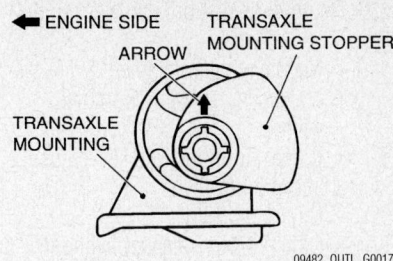

09482_OUTL_G0017

Insulator stopper installation locating point—Manual transaxle

ing bolts. Carefully remove the transaxle from the vehicle, using the proper removal equipment.

To install:

19. Installation is the reverse of the removal procedure, noting the following:

20. Install the transaxle mounting insulator stopper so that the arrow points upward, as indicated in the illustration.

21. Fill the transaxle using the proper grade and type transaxle fluid.

22. If equipped with AWD, fill the transfer case to the correct level.

23. Check and adjust the front wheel alignment.

24. Start the engine and check for leaks.

Clutch

REMOVAL & INSTALLATION

1. Before servicing the vehicle, refer to the Precautions Section.

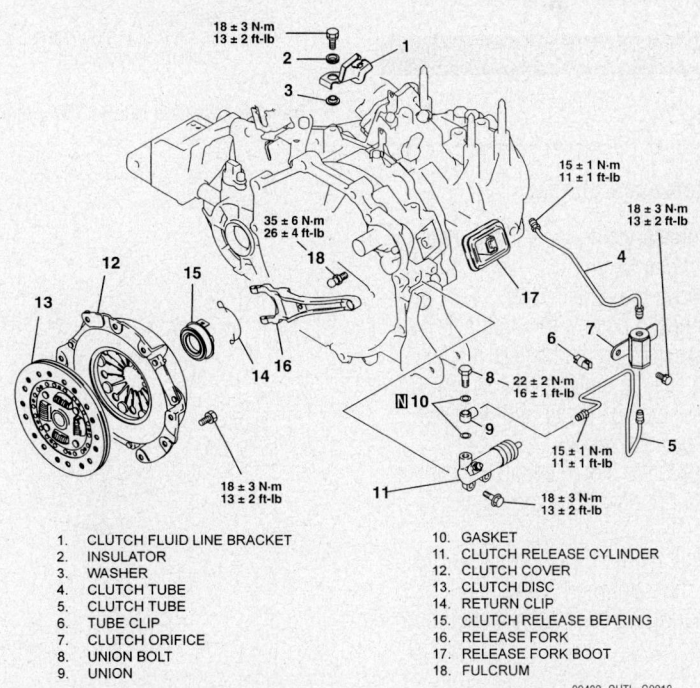

1. CLUTCH FLUID LINE BRACKET	10. GASKET
2. INSULATOR	11. CLUTCH RELEASE CYLINDER
3. WASHER	12. CLUTCH COVER
4. CLUTCH TUBE	13. CLUTCH DISC
5. CLUTCH TUBE	14. RETURN CLIP
6. TUBE CLIP	15. CLUTCH RELEASE BEARING
7. CLUTCH ORIFICE	16. RELEASE FORK
8. UNION BOLT	17. RELEASE FORK BOOT
9. UNION	18. FULCRUM

09482_OUTL_G0018

Clutch assembly and related components

2. Remove or disconnect the following:
- Negative battery cable
- Transaxle assembly from the vehicle
- Clutch fluid line bracket, insulator and washer
- Clutch fluid line
- Clutch slave (release) cylinder
- Release bearing
- Clutch cover (pressure plate) attaching bolts, cover plate and clutch disc. If the pressure plate is to be reused, loosen the bolts in a diagonal pattern, 1 or 2 turns at a time. This will prevent warping the clutch cover assembly.

3. Carefully inspect the condition of the clutch components and replace any worn or damaged parts.

To install:

4. Inspect the flywheel for heat damage or cracks. Resurface or replace the flywheel as required.

5. Using the proper alignment tool, install the clutch disc to the flywheel. Install the clutch cover (pressure plate) assembly. Install the retainer bolts and tighten a little at a time, in a diagonal sequence. Tighten them to a final torque of 11–15 ft. lbs. Remove the aligning tool.

6. Install or connect the following:
- Boot
- Clutch line, washer, insulator and bracket
- Transaxle assembly

7. Check for proper clutch operation.

Hydraulic Clutch System

BLEEDING

Clutch Release Cylinder

1. Before servicing the vehicle, refer to the Precautions Section.

2. Remove the under cover.

3. Connect a hose with a bottle to the bleeder screw. Open the bleeder nipple.

4. Depress the clutch pedal slowly. Open the bleeder screw to let air and fluid out. Close the bleeder screw.

5. Release the clutch pedal. Repeat the procedure until only fluid and no air is expelled.

6. Check the fluid reservoir level to ensure it stays between "MAX" and "MIN" throughout the clutch bleeding process.

Transfer Case Assembly

REMOVAL & INSTALLATION

1. Before servicing the vehicle, refer to the Precautions Section.

2. Drain the transaxle fluid and transfer case fluid.

3. Remove or disconnect the following:
- Engine undercover(s)
- Front exhaust pipe
- Propeller shaft
- Center member
- Air guide
- Dust seal guard, manual transaxle
- Halfshaft and output shaft
- Rear roll stopper bolt and nut
- Transfer case mounting bolts. Use a suitable tool to slide the transaxle to the front of the vehicle to make a suitable opening between the transaxle and crossmember. Pull the transfer case out of the opening.

To install:

4. Install or connect the following:
- Transfer case to the transaxle. Torque the bolts to 44–58 ft. lbs. (60–78 Nm).
- Rear roll stopper bolt and nut hand-tight. After the full weight of the vehicle is on the ground, torque the nut to 34–44 ft. lbs. (45–59 Nm).
- Halfshaft and output shaft
- Air guide
- Dust seal guard, manual transaxle
- Center member
- Propeller shaft
- Front exhaust pipe
- Engine undercover(s)

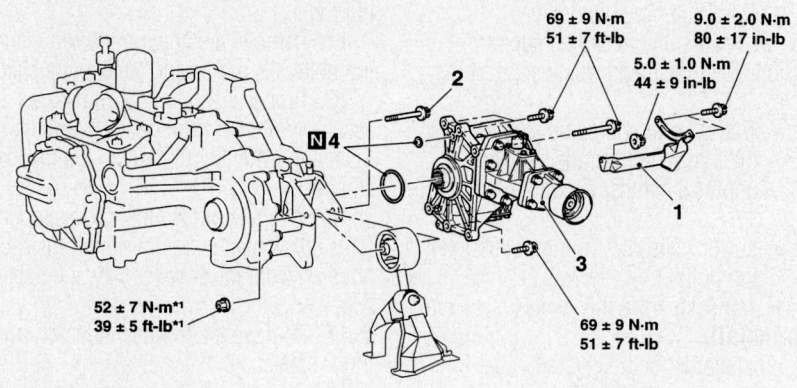

1. AIR GUIDE
2. REAR ROLL STOPPER CONNECTING BOLT
3. TRANSFER ASSEMBLY
4. O RING

67170-OUTL-G46

Transfer case and related components—vehicles with automatic transaxle

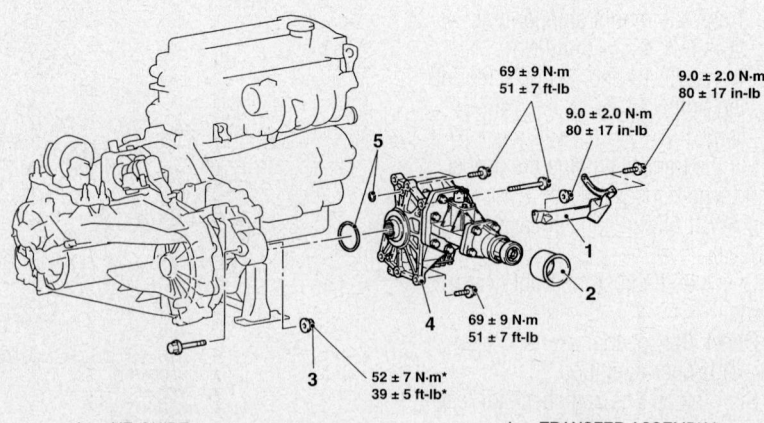

1. AIR GUIDE
2. DUST SEAL GUARD
3. REAR ROLL STOPPER NUT
4. TRANSFER ASSEMBLY
5. O-RING

09482_OUTL_G0019

Transfer case and related components—vehicles with manual transaxle

5. Fill the transaxle and the transfer case to the correct level.

Halfshaft

REMOVAL & INSTALLATION

Front

➡**If the vehicle is equipped with ABS, do not strike the ABS rotors installed to the outer BL outer race of the halfshaft against other parts when removing or installing the halfshaft. Otherwise the ABS rotors will be damaged.**

1. Before servicing the vehicle, refer to the Precautions Section.
2. Raise the vehicle and support it safely.
3. Drain the transaxle fluid.
4. Drain the transfer case, if equipped with AWD.
5. Remove the wheel and tire assembly.
6. Disconnect the front exhaust pipe.
7. Remove the cotter pin, halfshaft nut and washer.
8. If equipped with Anti-Lock Brake (ABS), remove the front wheel speed sensor and harness bracket.
9. Remove or disconnect the following:
   • Brake hose bracket
   • Self-locking nut
   • Stabilizer rubber insulator
   • Stabilizer link assembly
   • Lower control arm ball joint and tie rod end from the steering knuckle

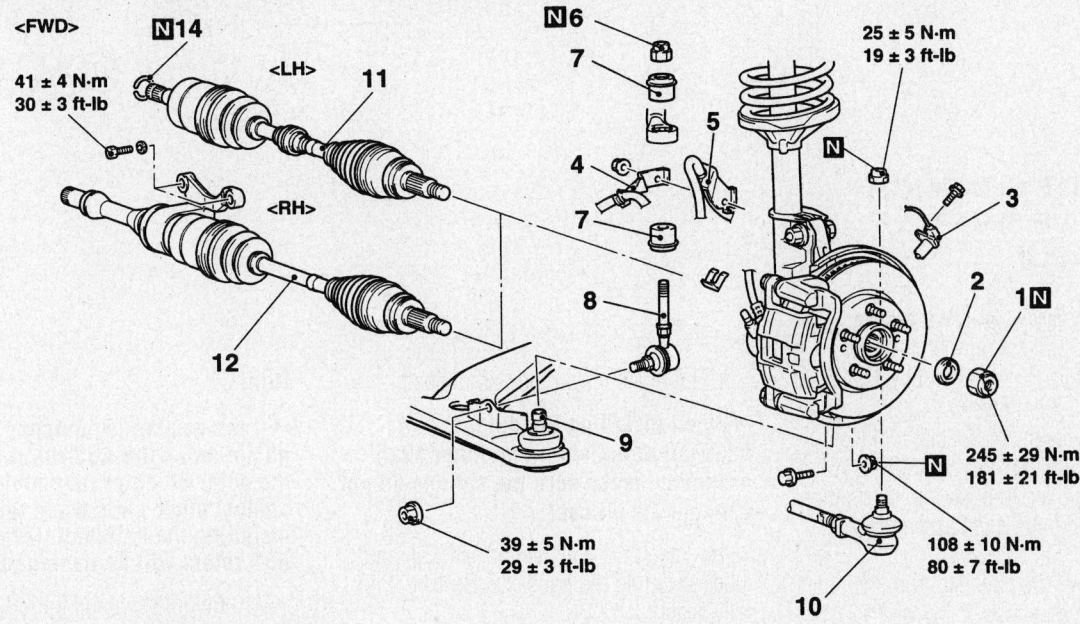

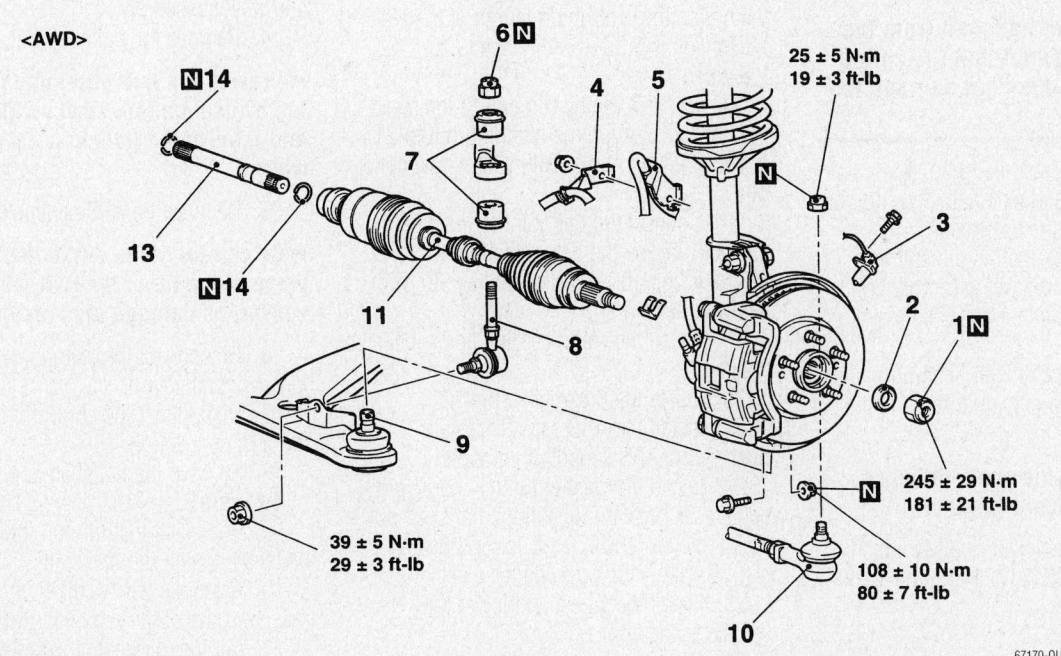

Front halfshaft and related components

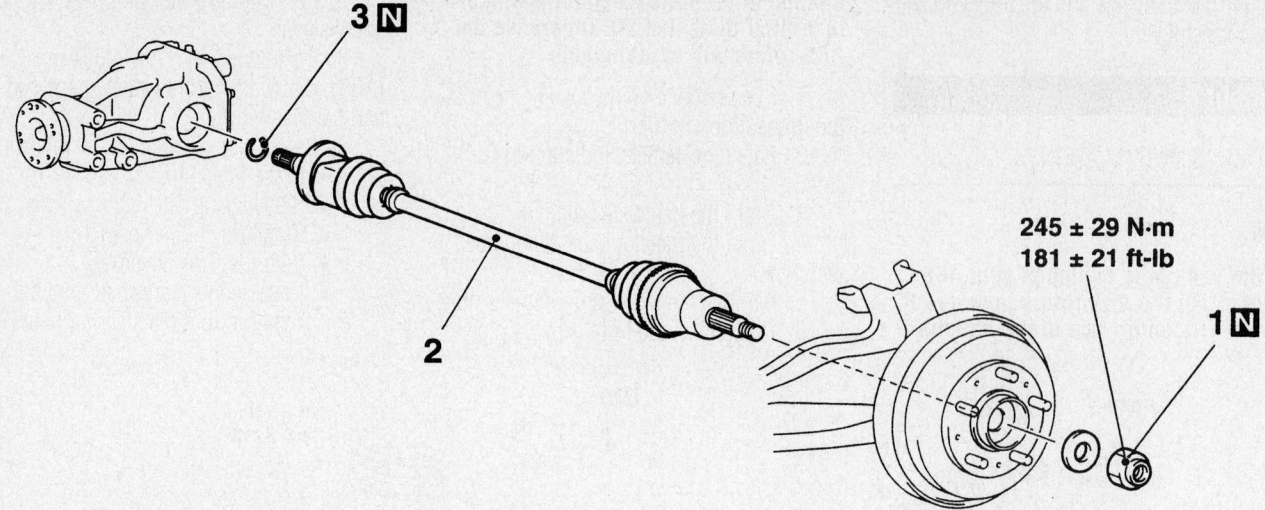

3 N

245 ± 29 N·m
181 ± 21 ft-lb

1 N

2

1. DRIVE SHAFT NUT
2. DRIVE SHAFT
3. CIRCLIP

67170-OUTL-G48

Rear halfshaft assembly—AWD vehicle

10. To remove the halfshaft or halfshaft and inner shaft as follows:

a. Use Mitsubishi Special tool Nos. MB990241, MB991354 and MB990767, or suitable puller, to push the halfshaft or halfshaft and inner shaft assembly from the hub.

b. Remove the halfshaft from the hub by pulling the bottom of the rotor toward you.

❋❋ WARNING

When pulling the halfshaft from the transaxle, be careful that the spline of the halfshaft does not damage the oil seal.

c. Insert a prybar between the transaxle case and halfshaft, then pry the halfshaft out of the transaxle.

d. If the inner shaft is difficult to remove, tap the bracket assembly lightly with a plastic hammer, then remove the inner shaft.

e. Cover the halfshaft opening in the transaxle case to prevent foreign debris from entering.

➡**Do not pull on the shaft; doing so damages the inboard joint.**

11. For AWD vehicles, Use Mitsubishi Special tool No. MB991721 to remove the output shaft.

To install:

12. Replace the circlips on the ends of the halfshafts.

13. If AWD, install the output shaft.

➡**When installing the output shaft, halfshaft or halfshaft and inner shaft assembly, make sure the splines do not damage the oil seal.**

14. Insert the halfshaft or halfshaft and inner shaft into the transaxle. Be sure it is fully seated.

15. Push out on the knuckle assembly and install the halfshaft through the hub.

16. Install the self-locking nut. Tighten the nut so the protruding length of the stabilizer link is 0.35–0.39 inches (9.0–9.8mm).

17. Connect the tie rod end to the steering knuckle. Torque the retaining nut to 21 ft. lbs. (29 Nm) and secure with a new cotter pin.

18. Connect the ball joint to the steering knuckle. Torque the new retaining nut to 43–52 ft. lbs. (60–72 Nm) and secure with a new cotter pin.

19. Install the stabilizer link and rubber insulator.

20. Install the brake hose bracket.

21. Install the ABS sensor harness bracket and ABS sensor, if equipped.

22. Install the washer so the chamfered edge faces outward. Install the halfshaft nut and tighten to 160–201 ft. lbs. (226–274 Nm). Secure with a new cotter pin.

23. Install the wheel and lower the vehicle to the floor.

24. Fill the transaxle and transfer case, if equipped, with fluid.

Rear

➡**If the vehicle is equipped with ABS, do not strike the ABS rotors installed to the outer BL outer race of the halfshaft against other parts when removing or installing the halfshaft. Otherwise the ABS rotors will be damaged.**

1. Before servicing the vehicle, refer to the Precautions Section.

2. Raise and safely support the vehicle.

3. Drain the gear oil. Remove the tire and wheel assembly.

4. Remove the halfshaft nut and washer.

➡**Prevent the hub assembly from turning by using a tool such as MB990767 and remove the halfshaft nut and washer.**

5. Remove the ABS sensor, if equipped.

➡**Be careful not to strike the pole piece at the tip of the rear wheel speed sensor, as damage may result.**

6. Disconnect the lower arm and trailing arm connection.

7. Disconnect the control link and trailing arm connection.

8. Remove the halfshaft and circlip.

To install:

9. Replace the circlip on the end of the halfshaft. Install the halfshaft.

10. Continue the installation in the reverse order of the removal procedure.

11. Fill the differential with the proper grade and type fluid.

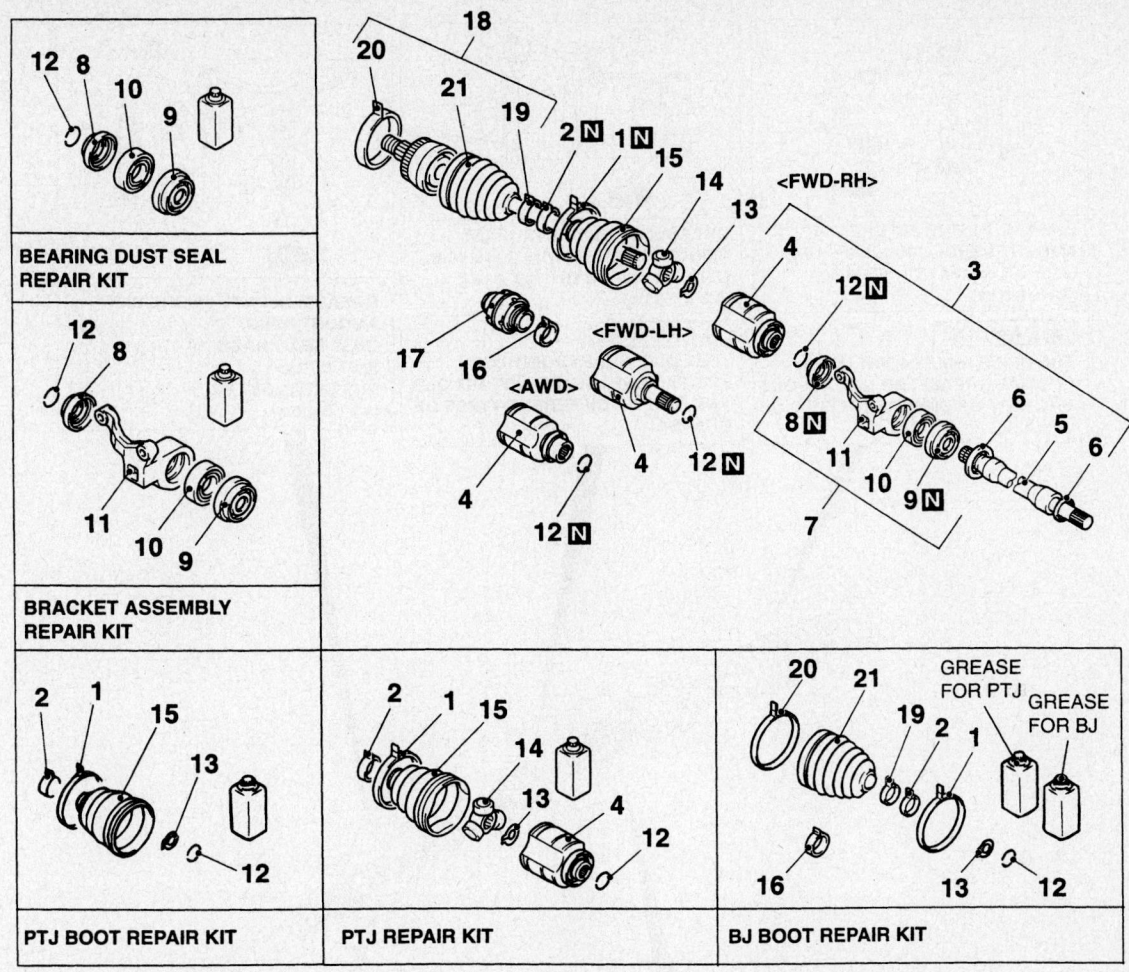

BEARING DUST SEAL
REPAIR KIT

BRACKET ASSEMBLY
REPAIR KIT

PTJ BOOT REPAIR KIT

PTJ REPAIR KIT

BJ BOOT REPAIR KIT

1. PTJ BOOT BAND (LARGE)
2. PTJ BOOT BAND (SMALL)
3. PTJ CASE AND INNER SHAFT
   ASSEMBLY <FWD-RH>
4. PTJ CASE
5. INNER SHAFT <FWD-RH>
6. DUST COVER <FWD-RH>
7. BRACKET ASSEMBLY
   <FWD-RH>
8. DUST SEAL OUTER <FWD-RH>
9. DUST SEAL INNER <FWD-RH>
10. CENTER BEARING <FWD-RH>

11. CENTER BEARING BRACKET
    <FWD-RH>
12. CIRCLIP
13. SNAP RING
14. SPIDER ASSEMBLY
15. PTJ BOOT
16. DAMPER BAND
17. DYNAMIC DAMPER
18. BJ ASSEMBLY
19. BJ BOOT BAND (SMALL)
20. BJ BOOT BAND (LARGE)
21. BJ BOOT

67170-OUTL-G49

Front halfshaft assembly exploded view

12. Check and adjust the rear alignment, as required.

CV-JOINT OVERHAUL

Front

➡ **If the vehicle is equipped with ABS, do not strike the ABS rotors installed to the outer BL outer race of the halfshaft against other parts when removing or installing the halfshaft. Otherwise the ABS rotors will be damaged.**

1. Before servicing the vehicle, refer to the Precautions Section.
2. Raise and safely support the vehicle.
3. Remove or disconnect the following:
   - Halfshaft assembly from the vehicle
   - Large Pillow Tripod Joint (PTJ) boot band
   - Small PTJ boot band
   - PTJ case and inner shaft assembly, right side FWD vehicles
   - PTJ case
   - Inner shaft and dust cover, right side FWD vehicles
   - Bracket assembly, right side FWD vehicles
   - Outer and inner dust seals, right side FWD vehicles
   - Center bearing and bracket, right side FWD vehicles
   - Circlip
   - Snapring
   - Spider assembly
   - PTJ boot
   - Damper band
   - Dynamic Damper
   - Birfield Joint (BJ) assembly

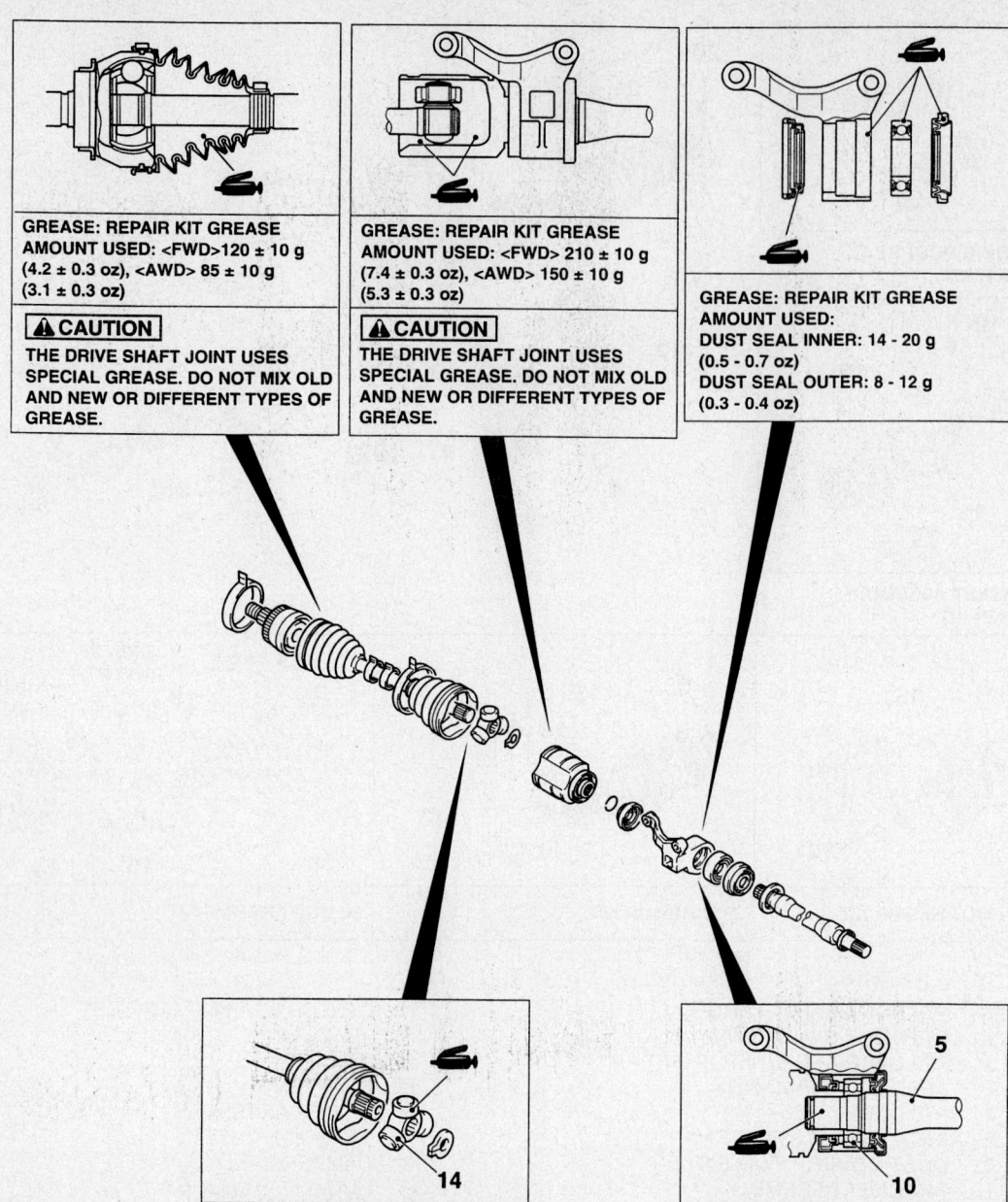

GREASE: REPAIR KIT GREASE AMOUNT USED: <FWD>120 ± 10 g (4.2 ± 0.3 oz), <AWD> 85 ± 10 g (3.1 ± 0.3 oz)

⚠ CAUTION

THE DRIVE SHAFT JOINT USES SPECIAL GREASE. DO NOT MIX OLD AND NEW OR DIFFERENT TYPES OF GREASE.

GREASE: REPAIR KIT GREASE AMOUNT USED: <FWD> 210 ± 10 g (7.4 ± 0.3 oz), <AWD> 150 ± 10 g (5.3 ± 0.3 oz)

⚠ CAUTION

THE DRIVE SHAFT JOINT USES SPECIAL GREASE. DO NOT MIX OLD AND NEW OR DIFFERENT TYPES OF GREASE.

GREASE: REPAIR KIT GREASE AMOUNT USED: DUST SEAL INNER: 14 - 20 g (0.5 - 0.7 oz) DUST SEAL OUTER: 8 - 12 g (0.3 - 0.4 oz)

GREASE: REPAIR KIT GREASE

GREASE: REPAIR KIT GREASE

67170-OUTL-G74

Front halfshaft assembly lubrication points and grease specifications

- Small and large BJ boot bands
- BJ boot

To install:

➡ Refer to the illustration for the lubrication points and grease specifications.

- BJ boot
- Large and small BJ boot bands
- BJ assembly
- Dynamic Damper
- Damper band
- PTJ boot
- Spider assembly.
- Snapring

- Circlip
- Center bearing and bracket, right side FWD vehicles
- Outer and inner dust seals, right side FWD vehicles
- Bracket assembly, right side FWD vehicles
- Inner shaft and dust cover, right side FWD vehicles
- PTJ case
- PTJ case and inner shaft assembly, right side FWD vehicles
- Small PTJ boot band
- Large PTJ boot band
- Halfshaft assembly into the vehicle

Rear

➡ If the vehicle is equipped with ABS, do not strike the ABS rotors installed to the outer BL outer race of the halfshaft against other parts when removing or installing the halfshaft. Otherwise the ABS rotors will be damaged.

1. Before servicing the vehicle, refer to the Precautions Section.
2. Raise and safely support the vehicle.
3. Remove or disconnect the following:
   - Halfshaft assembly from the vehicle
   - Large and small Tripod Joint (TJ) boot bands

- TJ case
- Circlip
- Snapring
- Spider assembly
- TJ boot
- Large and small Birfield Joint (BJ) boot bands

- BJ boot
- BJ assembly

To install:
- BJ assembly
- BJ boot
- Small and large Birfield Joint (BJ) boot bands

- TJ boot
- Spider assembly
- Snapring
- Circlip
- TJ case
- Small and large TJ boot bands
- Halfshaft assembly to the vehicle

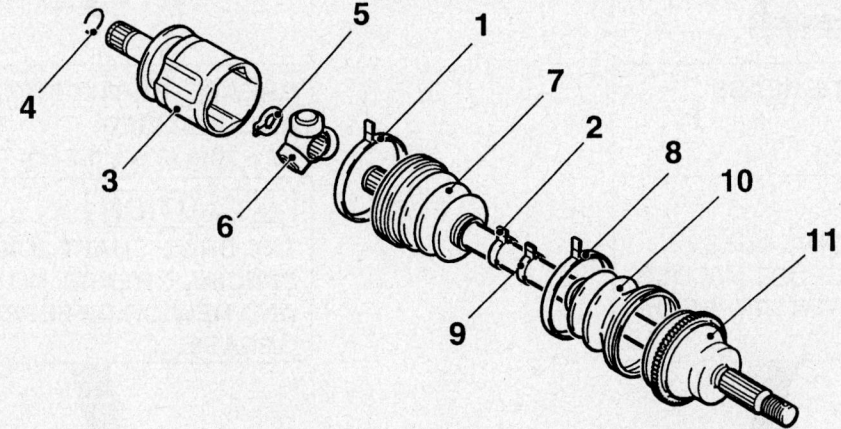

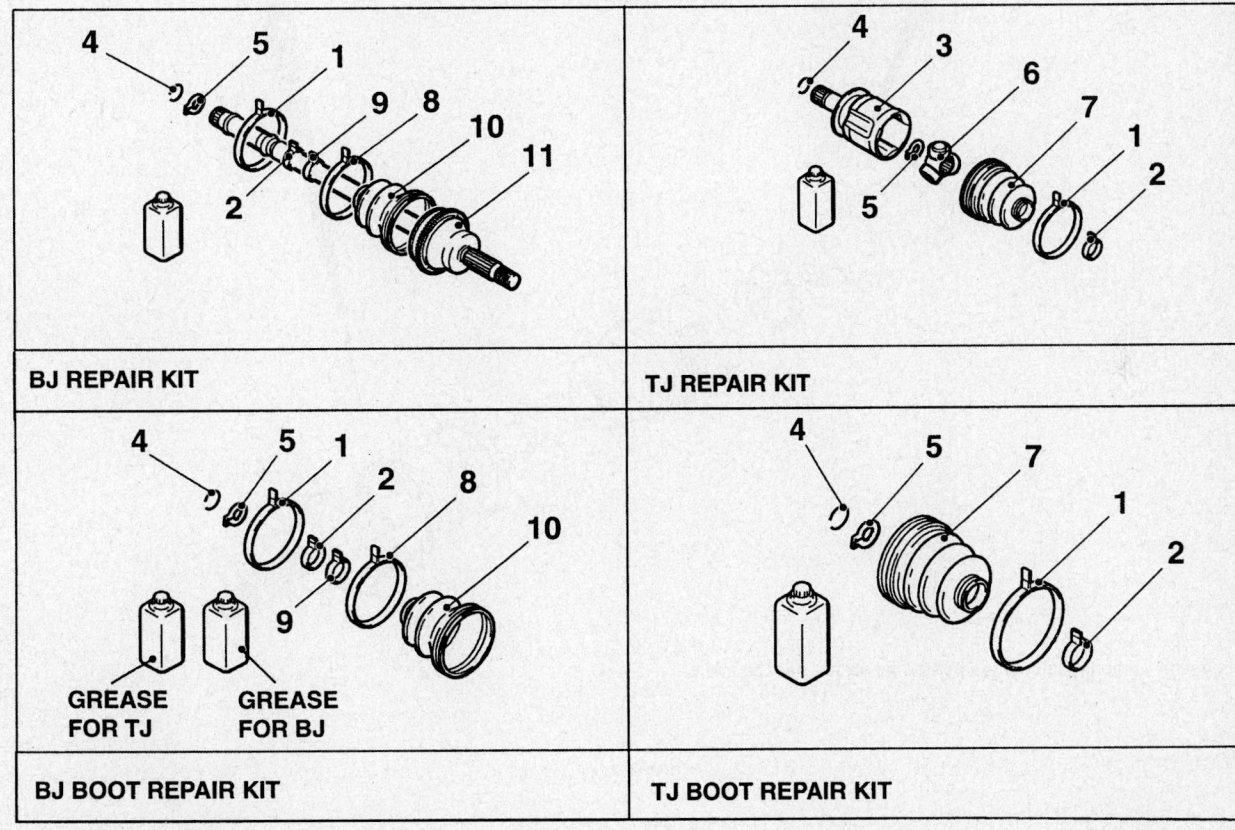

BJ REPAIR KIT

TJ REPAIR KIT

GREASE FOR TJ   GREASE FOR BJ

BJ BOOT REPAIR KIT

TJ BOOT REPAIR KIT

1. TJ BOOT BAND (LARGE)
2. TJ BOOT BAND (SMALL)
3. TJ CASE
4. CIRCLIP
5. SNAP RING
6. SPIDER ASSEMBLY

7. TJ BOOT
8. BJ BOOT BAND (LARGE)
9. BJ BOOT BAND (SMALL)
10. BJ BOOT
13. BJ ASSEMBLY

09482_OUTL_G0020

Rear halfshaft assembly exploded view

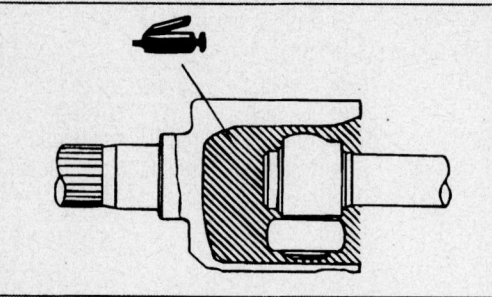

GREASE: REPAIR KIT GREASE
AMOUNT USED:
110 ± 10 g (3.9 ± 0.3 oz)

⚠ **CAUTION**

THE DRIVE SHAFT JOINT USES
SPECIAL GREASE. DO NOT MIX OLD
AND NEW OR DIFFERENT TYPES OF
GREASE.

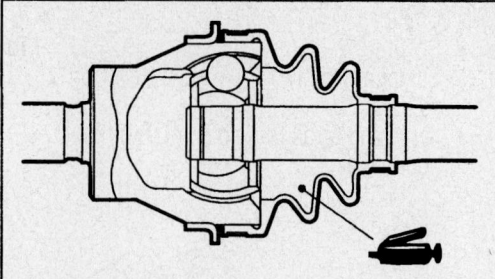

GREASE: REPAIR KIT GREASE
AMOUNT USED:
75 ± 10 g (2.6 ± 0.3 oz)

⚠ **CAUTION**

THE DRIVE SHAFT JOINT USES
SPECIAL GREASE. DO NOT MIX OLD
AND NEW OR DIFFERENT TYPES OF
GREASE.

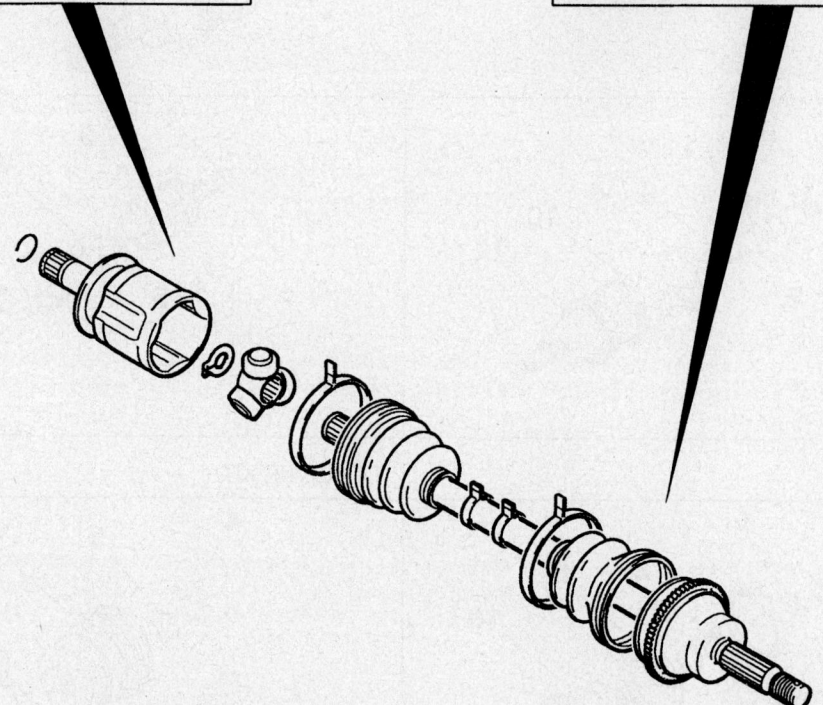

09482_OUTL_G0021

Rear halfshaft assembly lubrication points and grease specifications

STEERING

Air Bag

※※ CAUTION

All vehicles are equipped with an air bag system. The system must be disabled before performing service on or around system components, steering column, instrument panel components, wiring and sensors. Failure to follow safety and disabling procedures could result in accidental air bag deployment, possible personal injury and unnecessary system repairs.

PRECAUTIONS

Several precautions must be observed when handling the inflator module to avoid accidental deployment and possible personal injury.

• Never carry the inflator module by the wires or connector on the underside of the module.

• When carrying a live inflator module, hold securely with both hands, and ensure that the bag and trim cover are pointed away.

• Place the inflator module on a bench or other surface with the bag and trim cover facing up.

• With the inflator module on the bench, never place anything on or close to the module which may be thrown in the event of an accidental deployment.

DISARMING

1. Before servicing the vehicle, refer to the Precautions Section.
2. Position the front wheels in the straight-ahead position and place the key in the **LOCK** position. Remove the key from the ignition lock cylinder.
3. Disconnect the negative battery cable and insulate the cable end with high-quality electrical tape or similar non-conductive wrapping.

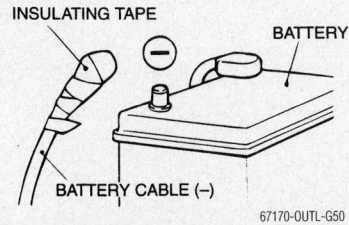

INSULATING TAPE

BATTERY

BATTERY CABLE (−)

67170-OUTL-G50

Insulate the negative battery cable to prevent accidental deployment of the air bag

4. Wait at least 1 minute before working on the vehicle. The air bag system is designed to retain enough voltage to deploy the air bag for a short period of time after the battery has been disconnected.

REARMING

1. Connect the negative battery cable, turn the ignition switch to the **ON** position and check the Supplemental Restraint System (SRS) warning light for proper operation.

Power Steering Gear

REMOVAL & INSTALLATION

➡**Prior to removal of the steering gear box, center the front wheels and remove the ignition key. Failure to do so may damage the Supplemental Restraint System (SRS) clock spring and render SRS system inoperative.**

1. Before servicing the vehicle, refer to the Precautions Section.
2. Drain the power steering system.
3. Remove or disconnect the following:
   • Battery negative cable. Insulate the cable end with high-quality electrical tape or similar non-conductive wrapping Wait at least 1 minute before proceeding.
   • Heated Oxygen (HO$_2$) sensor and remove the front exhaust pipe
4. Properly support the engine.
   • Roll stopper mounting bolts and the 4 center member installation bolts. Remove the center member.
   • Center member
5. Remove the steering wheel, as follows:
   a. Pry the cover from the steering wheel.
   b. Detach the connector, and retainers, as necessary then remove the air bag module assembly
   c. Remove the steering wheel nut, then use a puller to remove the steering wheel.
   d. Detach the sub-harness.
6. Remove the clock spring and put aside in a safe place.
7. Remove the stabilizer link and lower control arm assembly.
8. Remove or disconnect the following:
   • Shaft cover
   • Steering column shaft assembly and steering gear
   • Tie rod end from the steering knuckle, using a suitable puller

• Rear roll stopper bolt
• Fluid return hose and return tube. Plug the hoses. Remove and discard the O-ring.
• Pressure hose connection and hose gasket

9. Use a transaxle jack to support the crossmember, then remove the crossmember mounting nuts and bolts. Lower the crossmember with the rear roll stopper, stabilizer bar, return tube and steering gear.
10. Remove the following from the crossmember.
   • Rear roll stopper
   • Joint cover grommet
   • Steering extension
   • Return tube
   • Crossmember
   • Steering gear

To install:
11. Install or connect the following:
   • Steering gear into the vehicle. Tighten the bolts to 45–59 ft. lbs. (60–80 Nm)
   • Crossmember to the vehicle with the return tube, joint cover grommet, steering extension and rear roll stopper. Tighten the crossmember assembly bolts to 30–44 ft. lbs. (39–59 Nm) and the nuts to 116–130 ft. lbs. (158–176 Nm).
   • Pressure hose with a new gasket. Tighten the bolt to 37–47 ft. lbs. (50–64 Nm).
   • Fluid return tube with a new O-ring
   • Return hose connection
   • Rear roll stopper connecting bolt
   • Tie rod end to the steering knuckle
   • Steering gear connecting bolt and steering column shaft assembly
   • Shaft cover
   • Stabilizer and lower control arm assembly
   • Center member and roll stopper mounting bolts
   • Front exhaust pipe and HO2 sensor, if removed
   • Clock spring
   • Steering wheel. Tighten the nut to 24–36 ft. lbs. (33–49 Nm).
   • Attach the sub-harness
   • Air bag module, retainer and electrical connector
   • Cover to the steering wheel
   • Wheels and connect the negative battery cable
12. Refill the reservoir and bleed the system.
13. Perform a front end alignment.

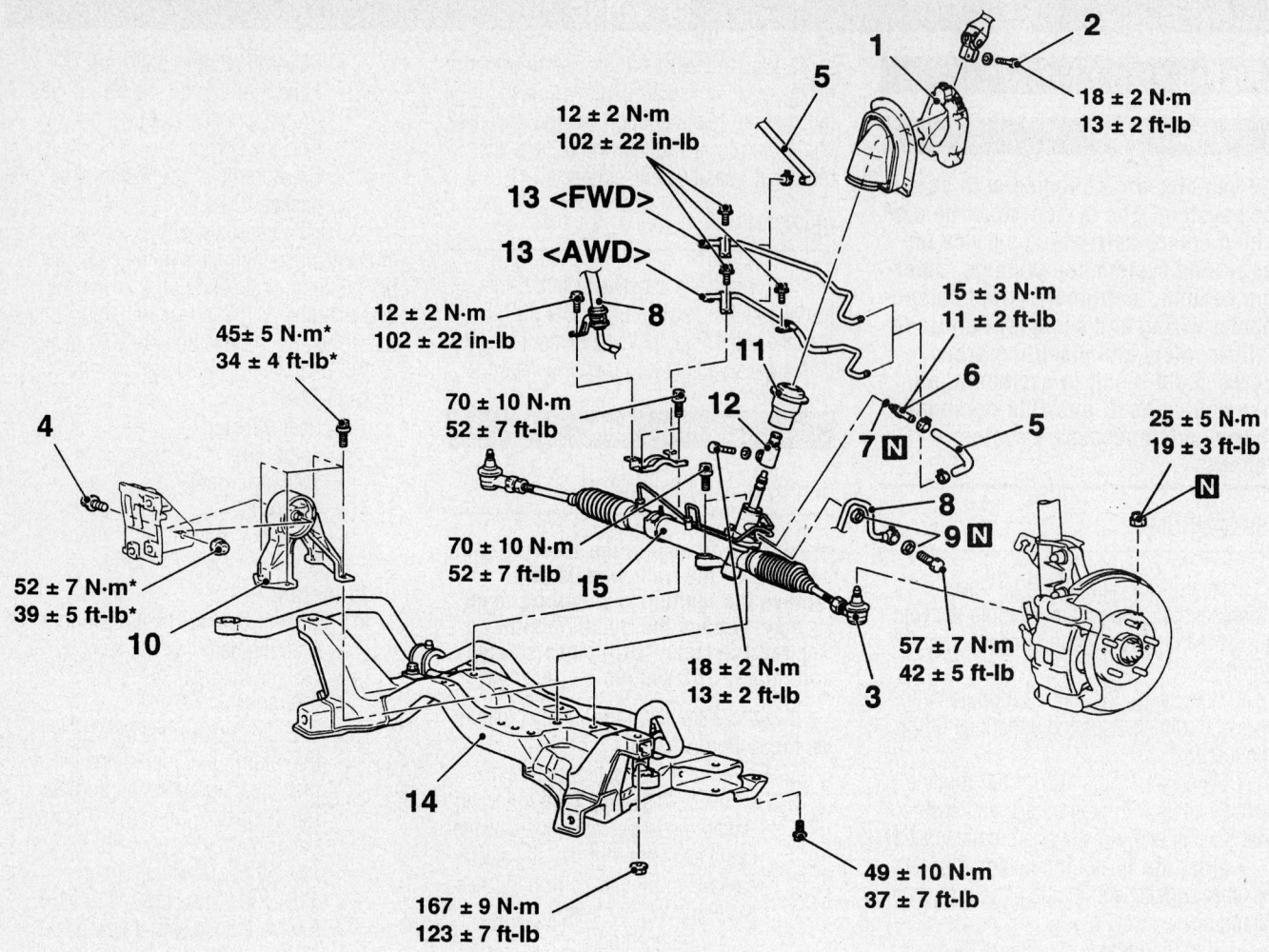

1. SHAFT COVER
2. STEERING COLUMN SHAFT ASSEMBLY AND STEERING GEAR CONNECTING BOLT
3. TIE ROD END AND KNUCKLE CONNECTION
4. REAR ROLL STOPPER CONNECTING BOLT
5. RETURN HOSE CONNECTION
6. RETURN TUBE
7. O-RING
8. PRESSURE HOSE CONNECTION
9. GASKET
10. REAR ROLL STOPPER
11. JOINT COVER GROMMET
12. STEERING EXTENSION
13. RETURN TUBE
14. CROSSMEMBER
15. STEERING GEAR

67170-OUTL-G51

Power steering gear and related components

FRONT SUSPENSION

Strut

REMOVAL & INSTALLATION

→If the vehicle is equipped with ABS, be careful when handling the pole piece at the tip of the speed sensor so as not to damage it by striking against other parts.

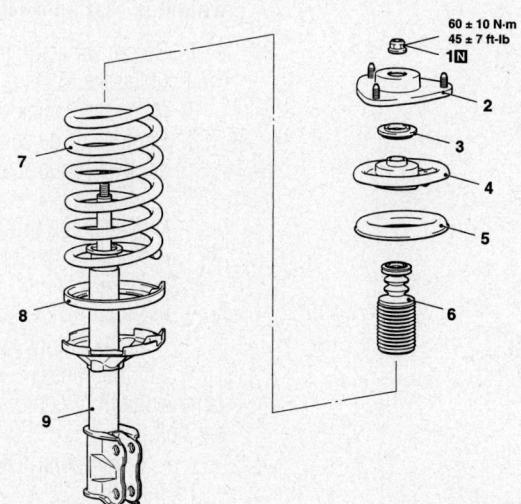

1. FRONT ABS SENSOR HARNESS BRACKET <VEHICLES WITH ABS>
2. BRAKE HOSE BRACKET
3. KNUCKLE CONNECTION

4. STRUT MOUNTING NUT
5. INSULATOR CLIP
6. STIFFENER PLATE
7. STRUT ASSEMBLY

67170-OUTL-G52

Front strut assembly and related components

1. Before servicing the vehicle, refer to the Precautions Section.
2. Disconnect the negative battery cable.
3. Remove the windshield washer fluid tank.
4. Raise and safely support vehicle. Remove the tire and wheel assembly.
5. If equipped with ABS, disconnect the front speed sensor harness bracket from the strut.

6. Remove the brake hose bracket.
7. Support the lower arm using floor jack. Remove the lower strut to knuckle bolts.

→**Before removing the top bolts, make matchmarks on the body and the strut insulator for proper reassembly.**

8. Remove the strut upper mounting nuts, insulator clip and stiffener plate.
9. Remove the strut assembly from the vehicle.
   To install:
10. Install or connect the following:
    - Strut to the vehicle
    - Stiffener plate
    - Insulator clip
    - Top mounting nuts and tighten to 30–36 ft. lbs. (39–49 Nm)
11. Position the strut on the knuckle and install the mounting bolts. While holding the head of the lower mounting bolt, tighten the nuts to 116–130 ft. lbs. (158–175 Nm).
12. Install or connect the following:
    - Brake hose bracket and the ABS harness brake, if equipped
    - Wheel and tire assembly
    - Windshield washer fluid reservoir
13. Perform a front end alignment.

OVERHAUL

1. Before servicing the vehicle, refer to the Precautions Section.
2. Remove the strut assembly from the vehicle.
3. Compress the coil spring using a spring compressor until the spring just comes away from one of the seats.
4. Remove or disconnect the following:
   - Center nut from the strut. Do NOT use an impact wrench to remove the center nut.
   - Strut insulator assembly
   - Bearing
   - Upper spring seat and upper spring pad
   - Rubber bumper
   - Coil spring
   - Lower spring pad
   - Strut assembly
   To install:
5. Install or connect the following:
   - Strut assembly
   - Lower spring pad
   - Compressed spring on the strut assembly
   - Rubber bumper

1. SELF-LOCKING NUT
2. STRUT INSULATOR ASSEMBLY
3. BEARING
4. UPPER SPRING SEAT
5. UPPER SPRING PAD
6. BUMP RUBBER
7. COIL SPRING

8. LOWER SPRING PAD
9. STRUT ASSEMBLY

67170-OUTL-G53

Front strut assembly exploded view

- Upper spring pad and spring seat
- Bearing
- Strut insulator
- Nut and tighten it to 38–52 ft. lbs. (50–70 Nm)
- Strut to the vehicle

Stabilizer Bar

REMOVAL & INSTALLATION

➡**Prior to removal of the stabilizer bar, center the front wheels and remove the ignition key. Failure to do so may damage the Supplemental Restraint System (SRS) clock spring and render SRS system inoperative.**

1. Before servicing the vehicle, refer to the Precautions Section.
2. Disarm the air bag system. Disconnect the negative battery cable.
3. Raise and support the vehicle safely. Remove the side under cover and the center under cover.
4. Be sure the front wheels in the straight ahead position. Remove the steering wheel cover, located underneath the steering wheel.

5. Remove the air bag module by sliding the clock spring connector and disconnecting it.
6. Remove the steering wheel, using the proper removal tool. Remove the sub harness.
7. Remove the steering column lower cover. Remove the clock spring and store it in a clean dry place.
8. Remove the center member. Disconnect the front exhaust pipe.
9. Disconnect the power steering hose clamps. Remove the roll stopper connecting bolt.
10. Remove the steering shaft cover. Remove the steering gear and joint connecting bolt.
11. Use a transaxle jack to hold the crossmember, and then remove the crossmember mounting nuts and bolts

➡**Lower the crossmember until the fixture, bushings, stabilizer links and stabilizer bar can be removed. Be careful not to lower the crossmember excessively, otherwise the power steering return hose bracket may deform.**

12. Remove the stabilizer bar bracket retaining bolts. Remove the bushings.

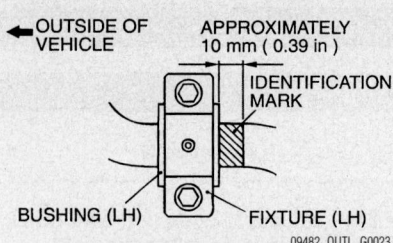

Front stabilizer bushing identification

13. Remove the stabilizer link locknut, rubber, link, rubber and bolt.
14. Remove the stabilizer bar from the vehicle.

To install:
15. Installation is the reverse of the removal procedure.
16. Align the stabilizer bar identification mark with the right end of the bushing.
17. Torque all retaining bolts to specifications.

➡**Bolt torques are indicated in the accompanying illustration. Bolt torques indicated with an asterisk (*) should be temporarily tightened, and then fully tightened with the vehicle on the ground in the unladen condition.**

18. Check the front end alignment.

Lower Control Arm

REMOVAL & INSTALLATION

➡**The suspension components should not be fully tightened until the vehicle's weight is resting on its wheels.**

1. Before servicing the vehicle, refer to the Precautions Section.
2. Raise the vehicle and support safely.
3. Remove or disconnect the following:
- Lower control arm-to-knuckle bolt and nut
- Wheel and tire assembly
- Separate the lower ball joint from its mounting
- Stabilizer bar self-locking nut, rubber bushings, and stabilizer link
4. Lift the transaxle with a jack, and then remove the front control arm-to-crossmember bolt.
- Lower control arm

To install:
5. Install or connect the following:
- Lower control arm into the vehicle
- Lower ball joint to its mounting
- Lower control arm-to-crossmember bolts. Torque the bottom bolt to 123 ft. lbs. (167 Nm) and the side bolt snug until the vehicle is lowered.

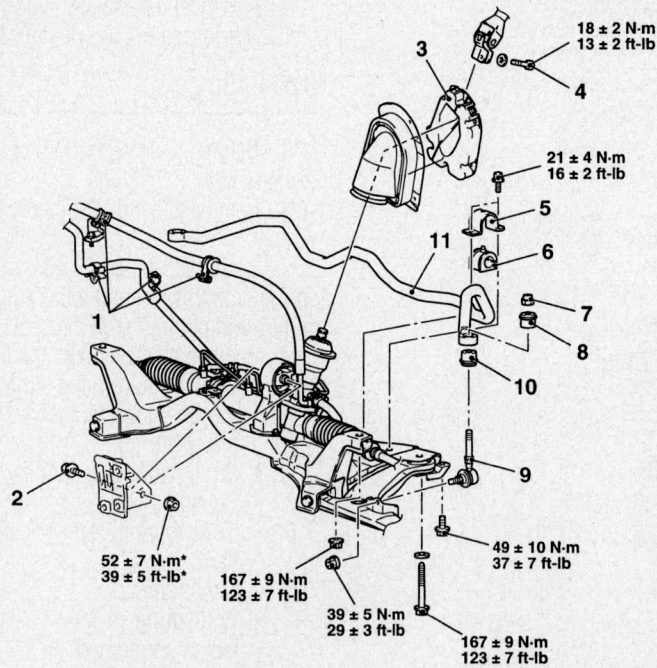

No.	Component	No.	Component
1.	POWER STEERING HOSE CLAMP	6.	BUSHING
2.	REAR ROLL STOPPER CONNECTING BOLT	7.	SELF-LOCKING NUT
3.	STEERING SHAFT COVER	8.	STABILIZER RUBBER
4.	STEERING GEAR AND JOINT CONNECTING BOLT	9.	STABILIZER LINK
5.	FIXTURE	10.	STABILIZER RUBBER
		11.	STABILIZER BAR

09482_OUTL_G0022

Front stabilizer and related components

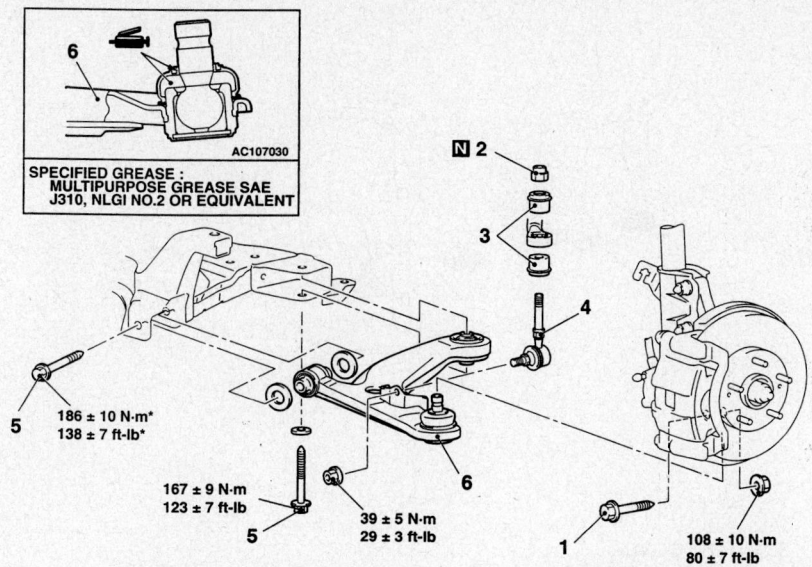

SPECIFIED GREASE :
MULTIPURPOSE GREASE SAE
J310, NLGI NO.2 OR EQUIVALENT

186 ± 10 N·m*
138 ± 7 ft-lb*

167 ± 9 N·m
123 ± 7 ft-lb

39 ± 5 N·m
29 ± 3 ft-lb

108 ± 10 N·m
80 ± 7 ft-lb

1. LOWER ARM AND KNUCKLE
   CONNECTION
2. SELF-LOCKING NUT
3. STABILIZER RUBBER

4. STABILIZER LINK ASSEMBLY
5. LOWER ARM AND CROSSMEMBER
   CONNECTION
6. LOWER ARM ASSEMBLY

67170-OUTL-G55

Front lower control arm and related components

- Lower control arm-to-steering knuckle bolt. Torque to 80 ft. lbs. (108 Nm).
- Stabilizer link, bushings and new self-locking nut. Tighten the self-locking nut until the stabilizer link threads protrude 0.35–0.39 inch (9.0–9.8mm).

6. Lower the vehicle, install the wheels, then with the weight of the vehicle on the wheels, torque the side control arm-to-crossmember bolt to 137 ft. lbs. (186 Nm).

7. Check and adjust the front wheel alignment.

CONTROL ARM BUSHING REPLACEMENT

1. Before servicing the vehicle, refer to the Precautions Section.
2. Raise and support the vehicle safely. Remove the tire and wheel assembly.
3. Remove the lower control arm and place in a vise.
4. Using a bushing removal tool remove the bushing from its mounting

To install:

➡ Coat the bushing with a soap solution and take care not to twist.

5. Using the bushing installation tool, press the bushing into the bracket.
6. Install the lower control arm.
7. Install the tire and wheel assembly. Lower the vehicle.

Wheel Bearings

ADJUSTMENT

The front wheel bearings are not adjustable. If the bearings are noisy or become loose, they must be replaced.

REMOVAL & INSTALLATION

➡ If the vehicle is equipped with ABS, do not strike the ABS rotors installed to the outer BL outer race of the halfshaft against other parts when removing or installing the halfshaft. Otherwise the ABS rotors will be damaged.

1. Before servicing the vehicle, refer to the Precautions Section.
2. Raise and safely support the vehicle.
3. Remove or disconnect the following:
   - Front wheel and tire assembly
   - Wheel Speed Sensor (WSS), vehicles with Anti-Lock Brake (ABS)
   - Caliper and suspend it out of the way with wire or string
   - Brake rotor
   - Axle nut and washer, using special tool MB990767 to hold the hub secure while removing the nut. Discard the nut.
   - Stabilizer rubber insulator
   - Stabilizer link from the lower control arm
   - Lower ball joint from the steering knuckle
   - Tie rod end from the steering knuckle. Do not remove the nut from the tie rod end. Loosen the nut and use special tool MB991113 or MB990635 to avoid damaging the threads.
   - Halfshaft from the hub and knuckle using the proper puller

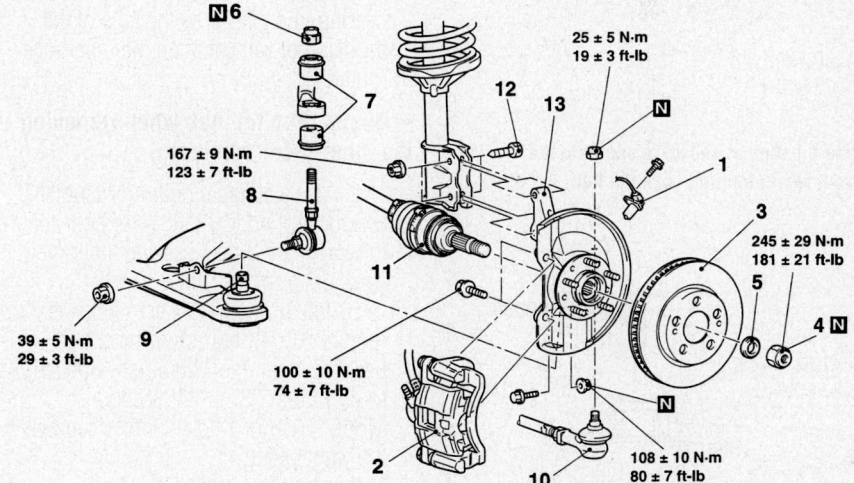

25 ± 5 N·m
19 ± 3 ft-lb

167 ± 9 N·m
123 ± 7 ft-lb

39 ± 5 N·m
29 ± 3 ft-lb

100 ± 10 N·m
74 ± 7 ft-lb

245 ± 29 N·m
181 ± 21 ft-lb

108 ± 10 N·m
80 ± 7 ft-lb

1. FRONT ABS SENSOR <VEHICLES WITH ABS>
2. CALIPER ASSEMBLY
3. BRAKE DISC
4. DRIVE SHAFT NUT
5. WASHER
6. SELF-LOCKING NUT
7. STABILIZER RUBBER
8. STABILIZER LINK ASSEMBLY

9. CONNECTION FOR LOWER ARM BALL JOINT
10. CONNECTION FOR TIE ROD END
11. DRIVE SHAFT
12. FRONT STRUT TO HUB AND KNUCKLE MOUNTING BOLT AND NUT
13. HUB AND KNUCKLE

67170-OUTL-G57

Front hub and related components

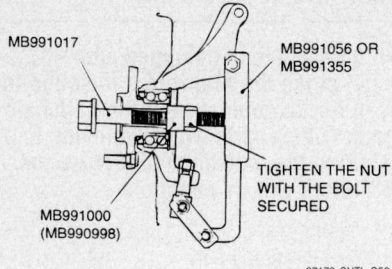

MB991017

MB991056 OR
MB991355

MB991000
(MB990998)

TIGHTEN THE NUT
WITH THE BOLT
SECURED

67170-OUTL-G58

Use the special tools to pull the front hub from the knuckle

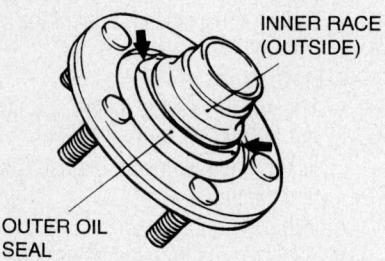

INNER RACE
(OUTSIDE)

OUTER OIL
SEAL

67170-OUTL-G59

Crush the oil seal on the front wheel bearing in the 2 places shown (arrows) so the tabs of the special tool will catch on the inner race

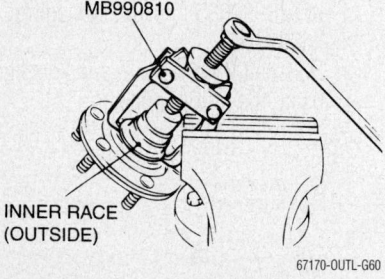

MB990810

INNER RACE
(OUTSIDE)

67170-OUTL-G60

Use a puller as shown to separate the front wheel bearing from the hub

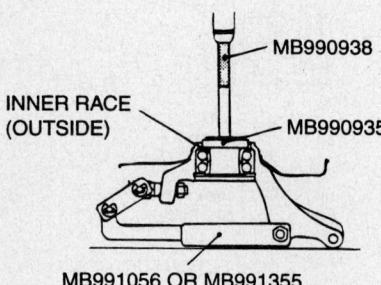

MB990938

INNER RACE
(OUTSIDE)

MB990935

MB991056 OR MB991355

67170-OUTL-G61

Place the inner race that was removed from the hub on the wheel bearing, and then use the special tools to remove the wheel bearing

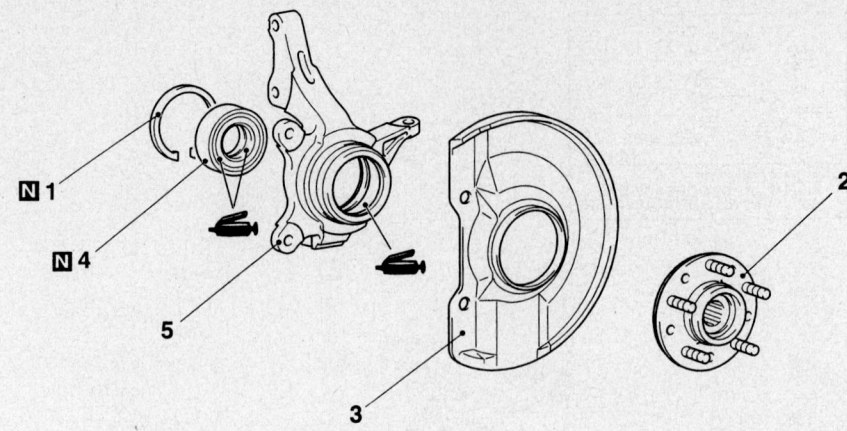

N 1

N 4

5

3

2

67170-OUTL-G64

1. SNAP RING
2. HUB
3. DUST COVER
4. WHEEL BEARING
5. KNUCKLE

Front bearing and related components

- Front strut-to-hub and knuckle mounting bolt and nut
- Hub and knuckle from the vehicle

4. Disassemble the hub and knuckle as follows:

a. Remove the snap ring.

b. Use Mitsbisishi Special tools MB991017, MB991056 or MB991355, MB991000 (MB990998) to pull the hub out of the knuckle.

c. Remove the dust cover.

d. Crush the oil seal on the wheel bearing in 2 places so the tabs of the special tool will catch on the inner race (outside).

➡**Do not drop the hub when removing the inner race (outside).**

e. Use a suitable puller (Mitsubishi Special tool no MB990810 or equivalent) to remove the wheel bearing inner race (outside) from the hub.

f. Install the inner race that was removed from the hub to the wheel bearing, then use Mitsubishi Special tools MB990935, 990938 and MB9911056 or MB991355 to remove the wheel bearing.

To install:

5. Assemble the hub and knuckle as follows:

a. Fill the wheel bearing with multipurpose grease. Apply a thin coat of grease to the knuckle and bearing mating surfaces.

✳✳ WARNING

To avoid damaging the wheel bearing, press the outer race when pressing in the wheel bearing.

b. Use Mitsubishi Special tools MB990883 and MB990890 to press in the bearing.

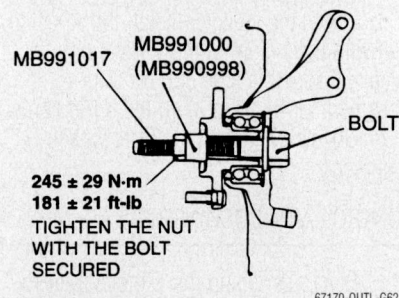

MB991017

MB991000
(MB990998)

BOLT

245 ± 29 N·m
181 ± 21 ft-lb
TIGHTEN THE NUT
WITH THE BOLT
SECURED

67170-OUTL-G62

Tighten Mitsubishi Special tools to 160–202 ft. lbs. (216–274 Nm) to press the hub into the knuckle

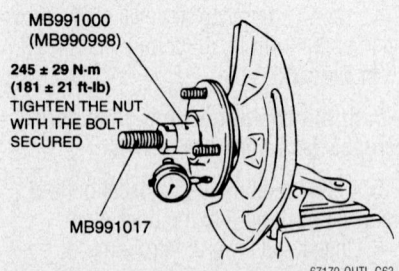

MB991000
(MB990998)

245 ± 29 N·m
(181 ± 21 ft-lb)
TIGHTEN THE NUT
WITH THE BOLT
SECURED

MB991017

67170-OUTL-G63

Front wheel bearing end play check

c. Install the snap ring.
d. Install the dust cover.
e. Install the hub. Tighten Mitsubishi Special tools MB991000 (MB990998) and MB991017 to 160–202 ft. lbs. (216–274 Nm), then press the hub into the knuckle.
f. Rotate the hub to seat the bearing.
g. Measure the hub starting torque using Mitsubishi Special tools MB990326 and MB990685. The starting torque must be within 16 inch lbs. (1.8 Nm), and the hub rotation must be smooth.

6. Check the wheel bearing play as follows:
a. Measure the wheel bearing play. If not within the limit range of 0.002 inch (0.05mm), while the nut is tightened to 160–202 ft. lbs. (216–274 Nm), the assembly has probably been incorrectly installed. You must replace the bearing and re-install.
7. Install or connect the following:
- Hub and knuckle assembly
- Front strut-to-hub bolt and nut. Torque to 123 ft. lbs. (167 Nm).
- Halfshaft

- Tie rod end. Torque the nut to 19 ft. lbs. (25 Nm).
- Lower control arm ball joint. Tighten the nut to 19 ft. lbs. (25 Nm).
- Stabilizer shaft
- New axle nut and washer and tighten to 181 ft. lbs. (245 Nm).
- Brake rotor and caliper. Torque the caliper bolts to 74 ft. lbs. (100 Nm).
- Front WSS, if equipped
- Front wheel and tire assembly
8. Lower the vehicle

REAR SUSPENSION

Shock Absorber

REMOVAL & INSTALLATION

1. Before servicing the vehicle, refer to the Precautions Section.
2. Remove or disconnect the following:
- Stabilizer link connection
3. Support the lower control arm with a jack.
- Lower control arm and trailing arm bolt
- Upper shock absorber mounting nut
- Shock absorber-to-lower control arm attaching bolt
- Shock absorber assembly

To install:
4. Position the shock absorber into the vehicle. Install the spring seat stepped section so that it points toward the rear side of the vehicle.
5. Install or connect the following:
- Shock absorber-to-lower control arm bolt and nut. Tighten the nut to 59–81 ft. lbs. (80–110 Nm).
- Upper shock mounting nut and tighten to 30–36 ft. lbs. (39–49 Nm)
- Lower control arm and trailing arm
- Stabilizer link. Tighten the nuts to 26–32 ft. lbs. (34–44 Nm).

➡**Bolt torques are indicated in the accompanying illustration. Bolt torques indicated with an asterisk (*) should be temporarily tightened, and then fully tightened with the vehicle on the ground in the unladen condition.**

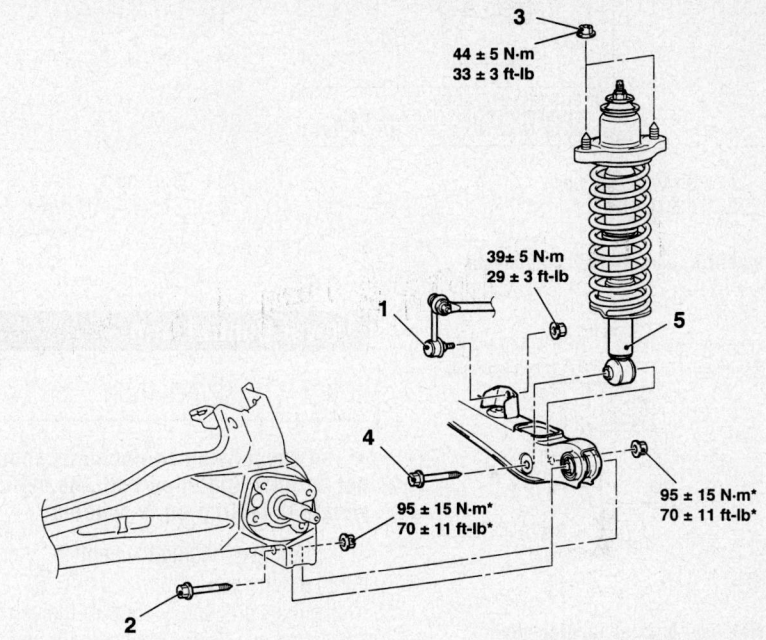

1.	STABILIZER LINK CONNECTION
2.	LOWER ARM AND TRAILING ARM CONNECTION
3.	SHOCK ABSORBER MOUNTING NUT
4.	SHOCK ABSORBER AND LOWER ARM CONNECTING BOLT
5.	SHOCK ABSORBER ASSEMBLY

67170-OUTL-G54

Rear shock absorber and related components

Stabilizer Bar

REMOVAL & INSTALLATION

1. Before servicing the vehicle, refer to the Precautions Section.
2. Raise and support the vehicle safely.
3. Remove the stabilizer link assembly.
4. Remove the stabilizer bushing retaining bracket bolts. Remove the bracket and bushing.

5. Remove the stabilizer bar from the vehicle.
To install:
6. Position the stabilizer bar on its mounting in the vehicle.
7. Install the stabilizer bar so that the stoppers come inboard of the bushings.
8. Continue the installation in the reverse order of the removal procedure.

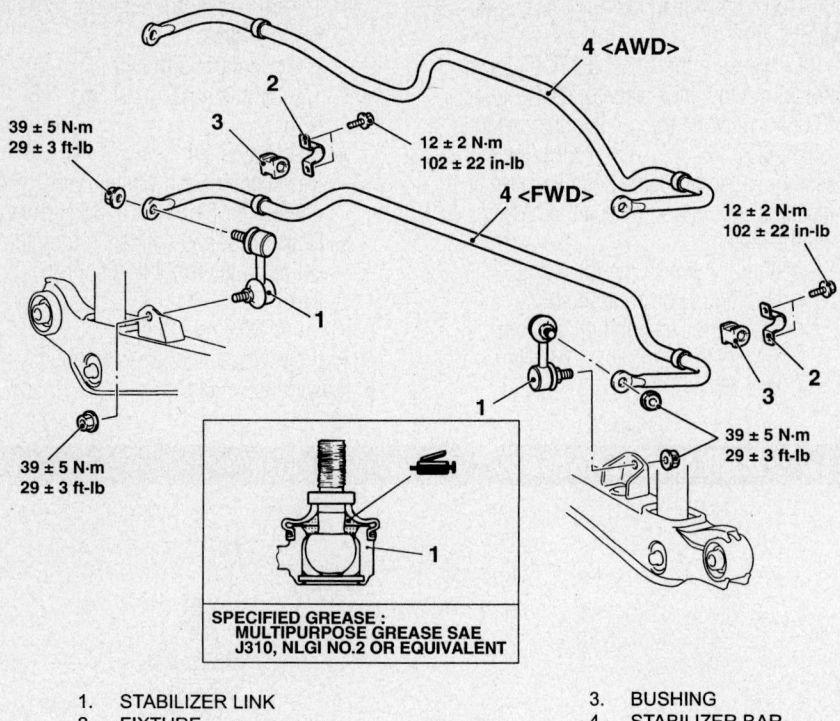

**SPECIFIED GREASE :
MULTIPURPOSE GREASE SAE
J310, NLGI NO.2 OR EQUIVALENT**

1. STABILIZER LINK
2. FIXTURE
3. BUSHING
4. STABILIZER BAR

09482_OUTL_G0024

Rear stabilizer and related components

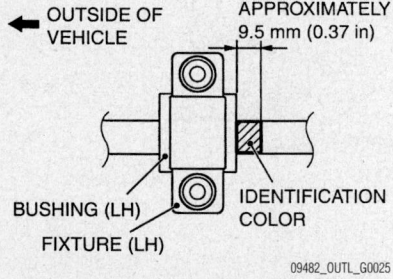

← OUTSIDE OF VEHICLE

APPROXIMATELY 9.5 mm (0.37 in)

BUSHING (LH)
FIXTURE (LH)
IDENTIFICATION COLOR

09482_OUTL_G0025

Rear stabilizer bushing identification

Upper Control Arm

REMOVAL & INSTALLATION

➡**The suspension components should not be fully tightened until the vehicle's weight is resting on its wheels.**

1. Before servicing the vehicle, refer to the Precautions Section.
2. Raise and safely support the vehicle. Support the trailing arm assembly.
3. Remove the upper arm mounting bolts. Remove the upper arm from the vehicle.

To install:

4. Installation is the reverse of the removal procedure.
5. Be sure to install the upper arm so that the identification mark ("A") faces inside the vehicle.

- Mounting bolt and upper control arm
- Stabilizer link nut and separate the link from the lower control arm

4. Support the lower control arm with a jack, remove the retaining bolts and separate the lower control arm from the trailing arm.

- Lower shock absorber-to-lower control arm nut and bolt
- Lower control arm

To install:

5. Install or connect the following:

- Lower control arm and secure the retainers hand-tight only at this time
- Shock absorber to the lower control arm. Secure with the nut and bolt and tighten until just snug
- Lower control arm to the trailing arm. Secure the retainers hand-tight.
- Stabilizer link and secure with the nut. Tighten to 26–35 ft. lbs. (34–44 Nm).
- Upper arm. Install the arm so the "A" mark faces inside of the vehicle. Tighten the retainers snug. Final tightening will occur when the vehicle's weight is on the suspension.
- Control link, so that its identification mark faces the front-outside of the vehicle
- Rear wheel and tire assembly and lower the vehicle.

6. At this time, torque the components to the following specifications:

- Lower control arm-to-trailing arm retainers: 59–81 ft. lbs. (80–110 Nm)
- Lower shock absorber nut: 59–81 ft. lbs. (80–110 Nm)

Lower Control Arm

REMOVAL & INSTALLATION

➡**The suspension components should not be fully tightened until the vehicle's weight is resting on its wheels.**

1. Before servicing the vehicle, refer to the Precautions Section.
2. Raise and safely support the vehicle.
3. Remove or disconnect the following:

- Control link, after matchmarking its installed position

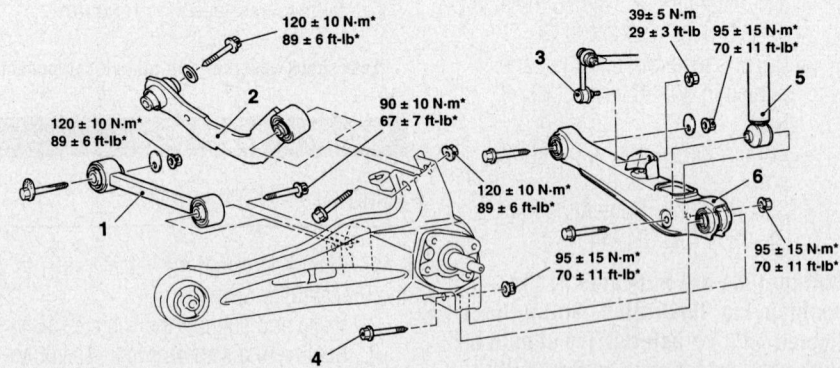

1. CONTROL LINK
2. UPPER ARM
3. STABILIZER LINK CONNECTION
4. LOWER ARM AND TRAILING ARM CONNECTION
5. SHOCK ABSORBER CONNECTION
6. LOWER ARM

67170-OUTL-G56

Rear upper control arm and related components

- Upper control arm retainers: 83–95 ft. lbs. (110–130 Nm)
- Control link retainers: 83–95 ft. lbs. (110–130 Nm)

➡Bolt torques are indicated in the accompanying illustration. Bolt torques indicated with an asterisk (*) should be temporarily tightened, and then fully tightened with the vehicle on the ground in the unladen condition.

7. Check and adjust the rear wheel alignment.

CONTROL ARM BUSHING REPLACEMENT

1. Before servicing the vehicle, refer to the Precautions Section.
2. Raise and support the vehicle safely. Remove the tire and wheel assembly.
3. Remove the lower control arm and place in a vise.
4. Using a bushing removal tool remove the bushing from its mounting
To install:

➡Coat the bushing with a soap solution and take care not to twist.

5. Using the bushing installation tool, press the bushing into the bracket.
6. Install the lower control arm.
7. Install the tire and wheel assembly. Lower the vehicle.

Wheel Bearings

ADJUSTMENT

The rear wheel bearings are not adjustable. If the bearings are noisy or become loose, they must be replaced.

REMOVAL & INSTALLATION

FWD Vehicles

DRUM BRAKES

➡If equipped with ABS, care must be used not to scratch or damage the teeth of the ABS rotor. The ABS rotor must never be dropped. If the teeth of the ABS rotor are chipped, the ABS rotor will not be able to accurately detect the wheel rotation speed and the system will not function normally.

➡The rear hub assembly should not be disassembled. When removing the rear hub assembly, the wheel bearing inner race may be left at the spindle side. In this case, always replace the rear hub assembly; otherwise the hub will dam-

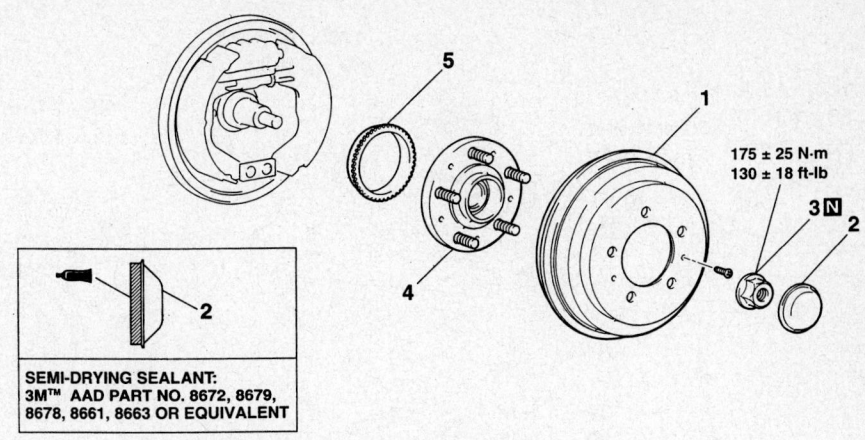

SEMI-DRYING SEALANT:
3M™ AAD PART NO. 8672, 8679, 8678, 8661, 8663 OR EQUIVALENT

1. REAR DRUM
2. HUB CAP
3. SELF-LOCKING NUT
4. REAR HUB ASSEMBLY
5. ABS ROTOR <VEHICLES WITH ABS>

67170-0UTL-G65

FWD rear wheel hub and bearing—drum brakes

age the seal, causing oil leaks and excessive play.

1. Before servicing the vehicle, refer to the Precautions Section.
2. Raise and safely support the vehicle.
3. Remove or disconnect the following:
   - Rear wheel
   - Brake drum
   - Dust cap and self-locking nut
   - Rear hub assembly
   - ABS rotor, if equipped, by pressing it out of the rear hub
To install:
4. Install or connect the following:
   - ABS rotor, if equipped, and press it into the rear hub

- Rear hub assembly using a new flange nut. Torque the flange nut to 130 ft. lbs. (180 Nm).
- Dust cap
- Brake drum
- Rear wheel assembly and lower the vehicle to the floor

DISC BRAKES

➡If equipped with ABS, care must be used not to scratch or damage the teeth of the ABS rotor. The ABS rotor must never be dropped. If the teeth of the ABS rotor are chipped, the ABS rotor will not be able to accurately detect the wheel rotation speed and the system will not function normally.

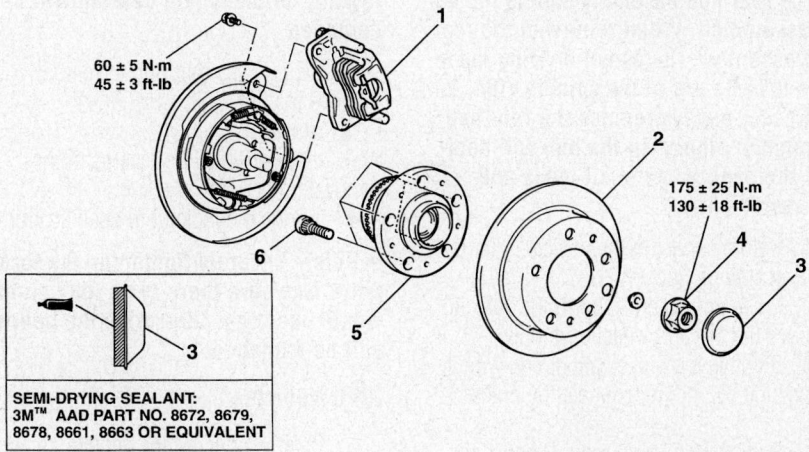

SEMI-DRYING SEALANT:
3M™ AAD PART NO. 8672, 8679, 8678, 8661, 8663 OR EQUIVALENT

1. CALIPER ASSEMBLY
2. BRAKE DISC
3. HUB CAP
4. SELF-LOCKING NUT
5. REAR HUB ASSEMBLY
6. HUB BOLT

09482_0UTL_G0026

FWD rear wheel hub and bearing—disc brakes

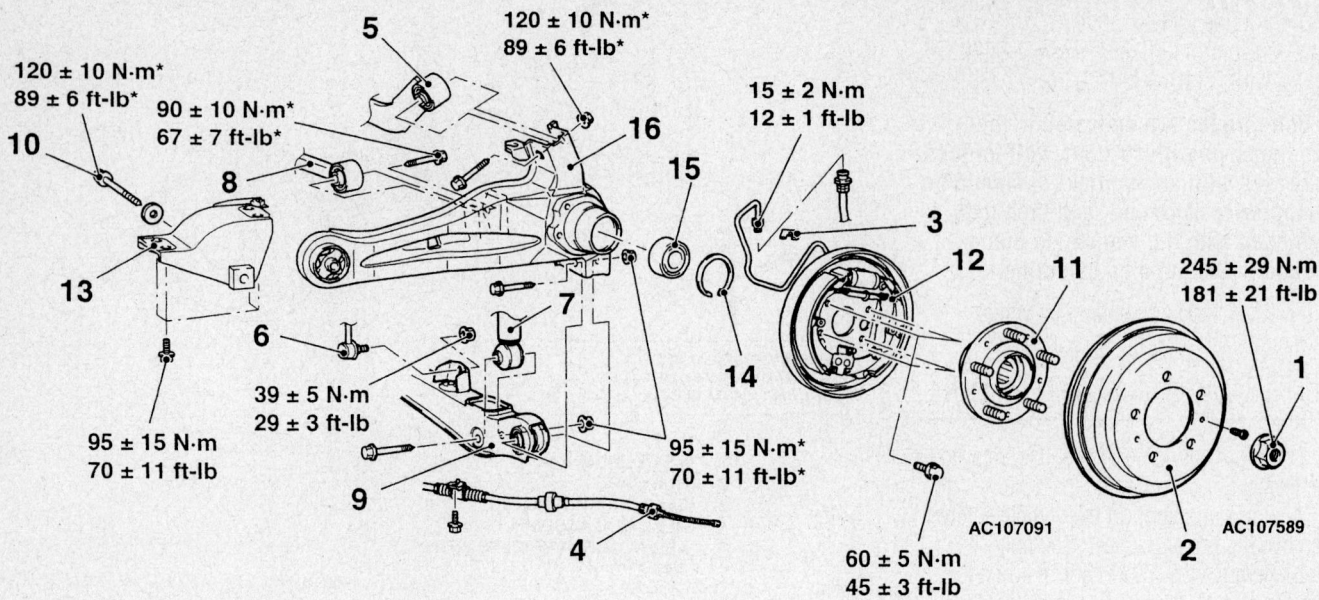

120 ± 10 N·m*
89 ± 6 ft-lb*

90 ± 10 N·m*
67 ± 7 ft-lb*

120 ± 10 N·m*
89 ± 6 ft-lb*

15 ± 2 N·m
12 ± 1 ft-lb

245 ± 29 N·m
181 ± 21 ft-lb

39 ± 5 N·m
29 ± 3 ft-lb

95 ± 15 N·m
70 ± 11 ft-lb

95 ± 15 N·m*
70 ± 11 ft-lb*

60 ± 5 N·m
45 ± 3 ft-lb

AC107091

AC107589

09482_OUTL_G0027

1. DRIVE SHAFT NUT
2. REAR DRUM
3. BRAKE HOSE AND TRAILING ARM CONNECTION
4. PARKING BRAKE CABLE CONNECTION
5. UPPER ARM AND TRAILING ARM CONNECTION
6. LOWER ARM AND STABILIZER LINK CONNECTION
7. LOWER ARM AND SHOCK ABSORBER CONNECTION

8. CONTROL LINK AND TRAILING ARM CONNECTION
9. LOWER ARM AND TRAILING ARM CONNECTION
10. TRAILING ARM AND TRAILING ARM BRACKET CONNECTING BOLT
11. REAR HUB ASSEMBLY
12. REAR BRAKE ASSEMBLY
13. TRAILING ARM BRACKET
14. SNAP RING
15. WHEEL BEARING
16. TRAILING ARM

AWD rear wheel hub and bearing—drum brakes

➡The rear hub assembly should not be disassembled. When removing the rear hub assembly, the wheel bearing inner race may be left at the spindle side. In this case, always replace the rear hub assembly; otherwise the hub will damage the seal, causing oil leaks and excessive play.

1. Before servicing the vehicle, refer to the Precautions Section.
2. Raise and safely support the vehicle. Remove the tire and wheel assembly.
3. Remove the brake caliper and wire it to the side. Do not disconnect the brake line.
4. Remove the brake rotor. Remove the dust cap. Remove the locknut.

➡Do not apply the vehicle weight to the wheel bearing while loosening the

locknut, or the wheel bearing will be damaged.

5. Remove the rear hub assembly. Remove the hub bolt.
   To install:
6. Installation is the reverse of the removal procedure.
7. Torque the locknut to specification.

➡Before securely tightening the locknut, make sure there is no load on the wheel bearings. Otherwise the bearing will be damaged.

AWD Vehicles

1. Before servicing the vehicle, refer to the Precautions Section.
2. Raise and safely support the vehicle. Remove the tire and wheel assembly.
3. Remove the brake caliper and wire it

to the side. Do not disconnect the brake line.
4. Remove the brake rotor. Remove the locknut.
5. On disc brakes, use special tools MB990211, MB990244 and MB991354, to pull out the rear hub assembly.
6. On drum brakes, use special tools MB990211 and MB990241, to pull out the rear hub assembly.
   To install:
7. Installation is the reverse of the removal procedure.
8. Torque the locknut to specification.

➡Bolt torques are indicated in the accompanying illustration. Bolt torques indicated with an asterisk (*) should be temporarily tightened, and then fully tightened with the vehicle on the ground in the unladen condition.

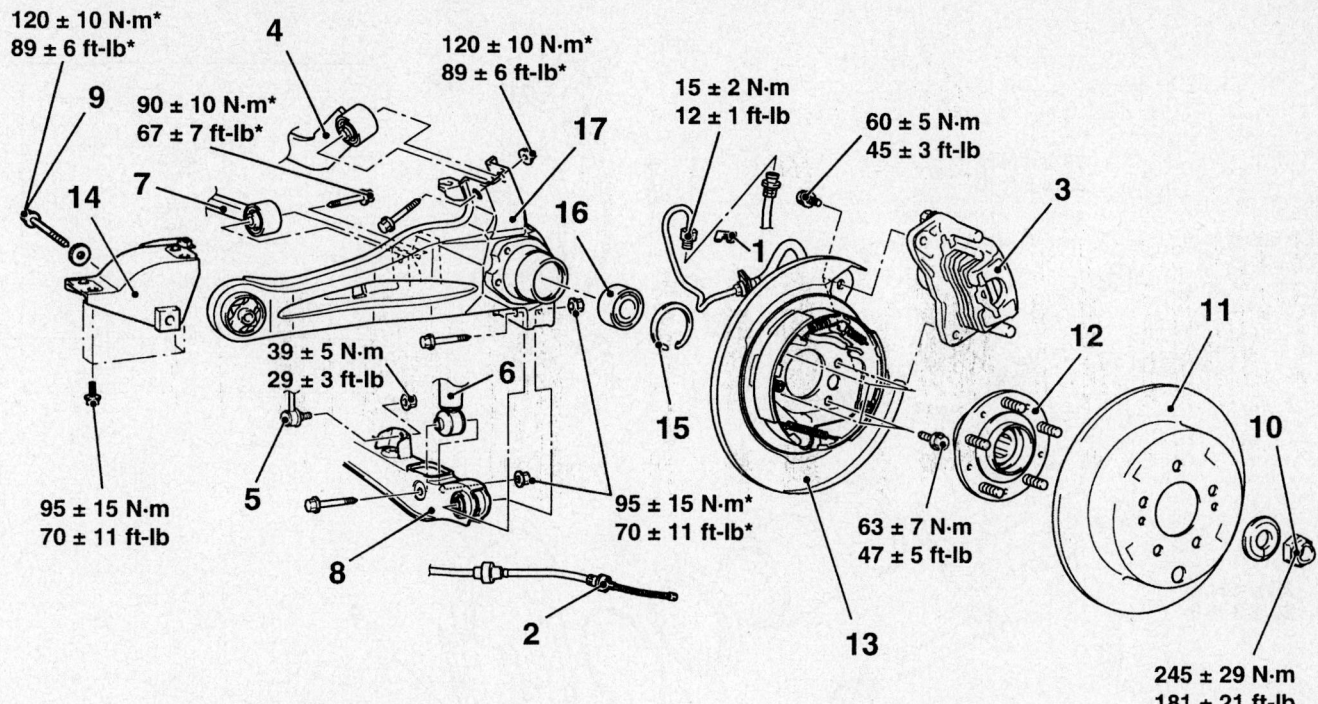

120 ± 10 N·m*
89 ± 6 ft-lb*

90 ± 10 N·m*
67 ± 7 ft-lb*

120 ± 10 N·m*
89 ± 6 ft-lb*

15 ± 2 N·m
12 ± 1 ft-lb

60 ± 5 N·m
45 ± 3 ft-lb

39 ± 5 N·m
29 ± 3 ft-lb

95 ± 15 N·m
70 ± 11 ft-lb

95 ± 15 N·m*
70 ± 11 ft-lb*

63 ± 7 N·m
47 ± 5 ft-lb

245 ± 29 N·m
181 ± 21 ft-lb

1. BRAKE HOSE AND TRAILING ARM CONNECTION
2. PARKING BRAKE CABLE CONNECTION
3. REAR BRAKE CALIPER ASSEMBLY
4. UPPER ARM AND TRAILING ARM CONNECTION
5. LOWER ARM AND STABILIZER LINK CONNECTION
6. LOWER ARM AND SHOCK ABSORBER CONNECTION
7. CONTROL LINK AND TRAILING ARM CONNECTION
8. LOWER ARM AND TRAILING ARM CONNECTION
9. TRAILING ARM AND TRAILING ARM BRACKET CONNECTING BOLT
10. DRIVESHAFT NUT
11. BRAKE DISC DRIVESHAFT CONNECTION
12. REAR HUB ASSEMBLY
13. REAR PARKING BRAKE ASSEMBLY
14. TRAILING ARM BRACKET
15. SNAP RING
16. WHEEL BEARING
17. TRAILING ARM

09482_OUTL_G0028

AWD rear wheel hub and bearing—disc brakes

BRAKES

Brake Caliper

REMOVAL & INSTALLATION

1. Before servicing the vehicle, refer to the Precautions Section.
2. Drain brake fluid from the master cylinder. Raise and support the vehicle safely.
3. Remove or disconnect the following:

- Wheels
- Brake hose. Discard the gaskets.
- Caliper mounting bolts
- Caliper assembly

To install
4. Install or connect the following:

- Brake caliper into position. Tighten the bolts to specification.
- Brake hose with new gaskets.
- Bleed the brake system.
- Wheels

Disc Brake Pads

REMOVAL & INSTALLATION

1. Before servicing the vehicle, refer to the Precautions Section.
2. Raise and support the vehicle safely.
3. Remove or disconnect the following:

- Some of the brake fluid from the master cylinder reservoir
- Wheels

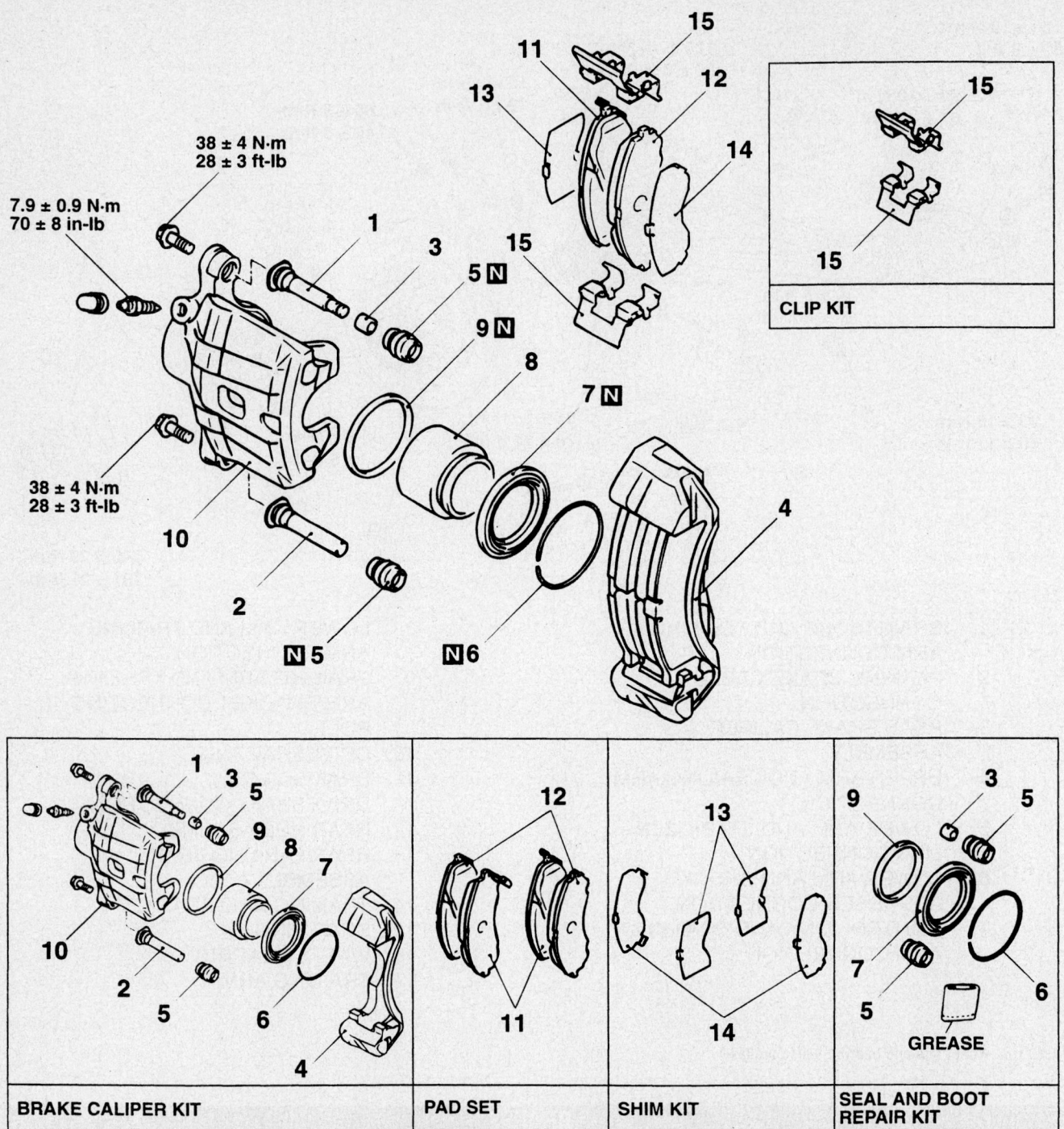

38 ± 4 N·m
28 ± 3 ft-lb

7.9 ± 0.9 N·m
70 ± 8 in-lb

38 ± 4 N·m
28 ± 3 ft-lb

CLIP KIT

BRAKE CALIPER KIT

PAD SET

SHIM KIT

SEAL AND BOOT
REPAIR KIT

GREASE

1. GUIDE PIN
2. LOCK PIN
3. BUSHING
4. CALIPER SUPPORT (INCLUDING PAD, CLIP, AND SHIM)
5. PIN BOOT
6. BOOT RING
7. PISTON BOOT
8. PISTON
9. PISTON SEAL
10. CALIPER BODY
11. PAD AND WEAR INDICATOR ASSEMBLY
12. PAD ASSEMBLY
13. INNER SHIM
14. OUTER SHIM
15. CLIP

67170-OUTL-G67

Front brake caliper and related components—2003–2004 vehicles

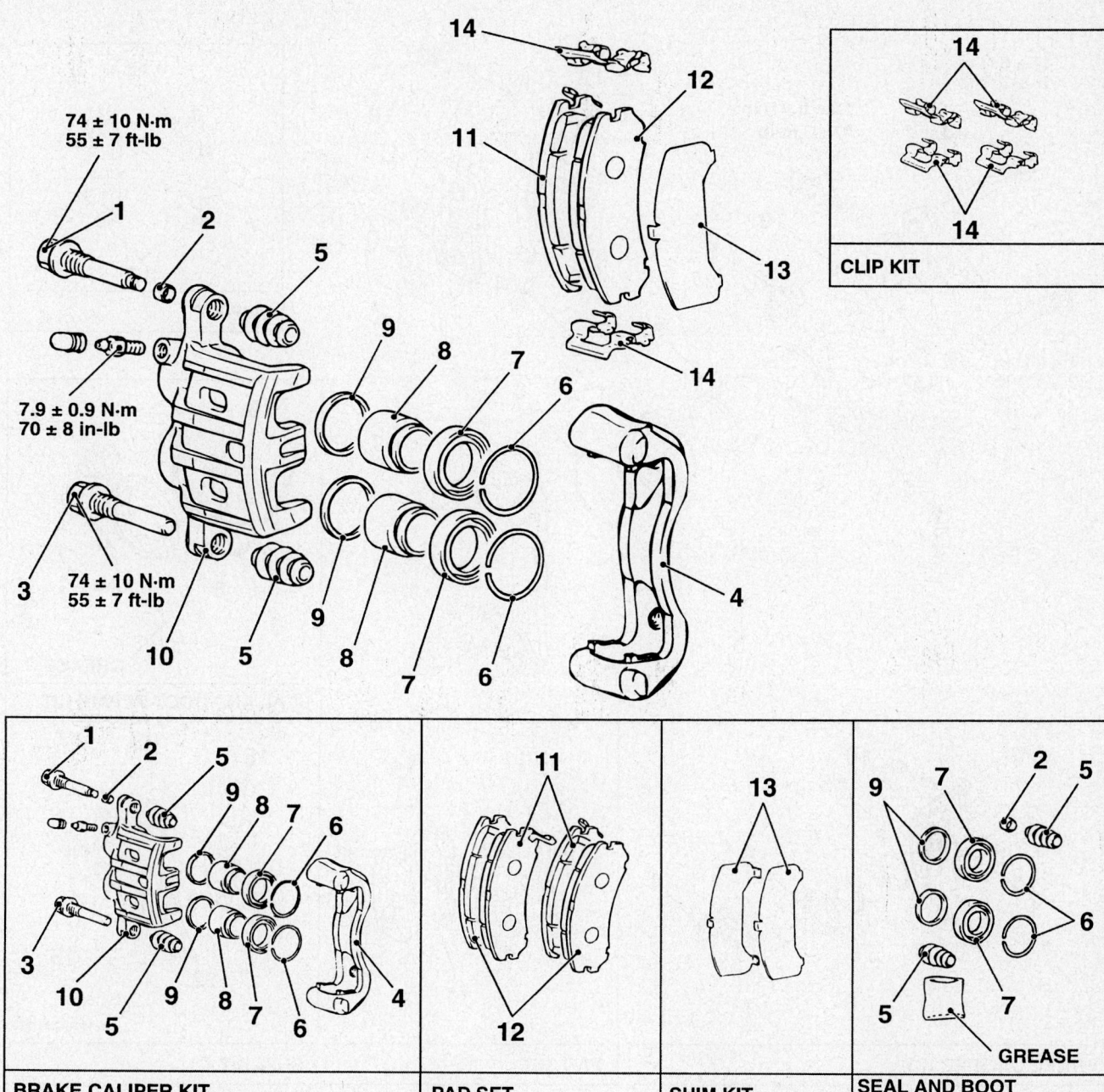

74 ± 10 N·m
55 ± 7 ft-lb

7.9 ± 0.9 N·m
70 ± 8 in-lb

74 ± 10 N·m
55 ± 7 ft-lb

CLIP KIT

BRAKE CALIPER KIT

PAD SET

SHIM KIT

SEAL AND BOOT
REPAIR KIT

GREASE

1. LOCK PIN
2. BUSHING
3. GUIDE PIN
4. CALIPER SUPPORT (INCLUDING
   PAD, CLIP, AND SHIM)
5. PIN BOOT
6. BOOT RING
7. PISTON BOOT

8. PISTON
9. PISTON SEAL
10. CALIPER BODY
11. PAD AND WEAR INDICATOR
    ASSEMBLY
12. PAD ASSEMBLY
13. OUTER SHIM
14. CLIP

09482_OUTL_G0029

Front brake caliper and related components—2005–2006 vehicles

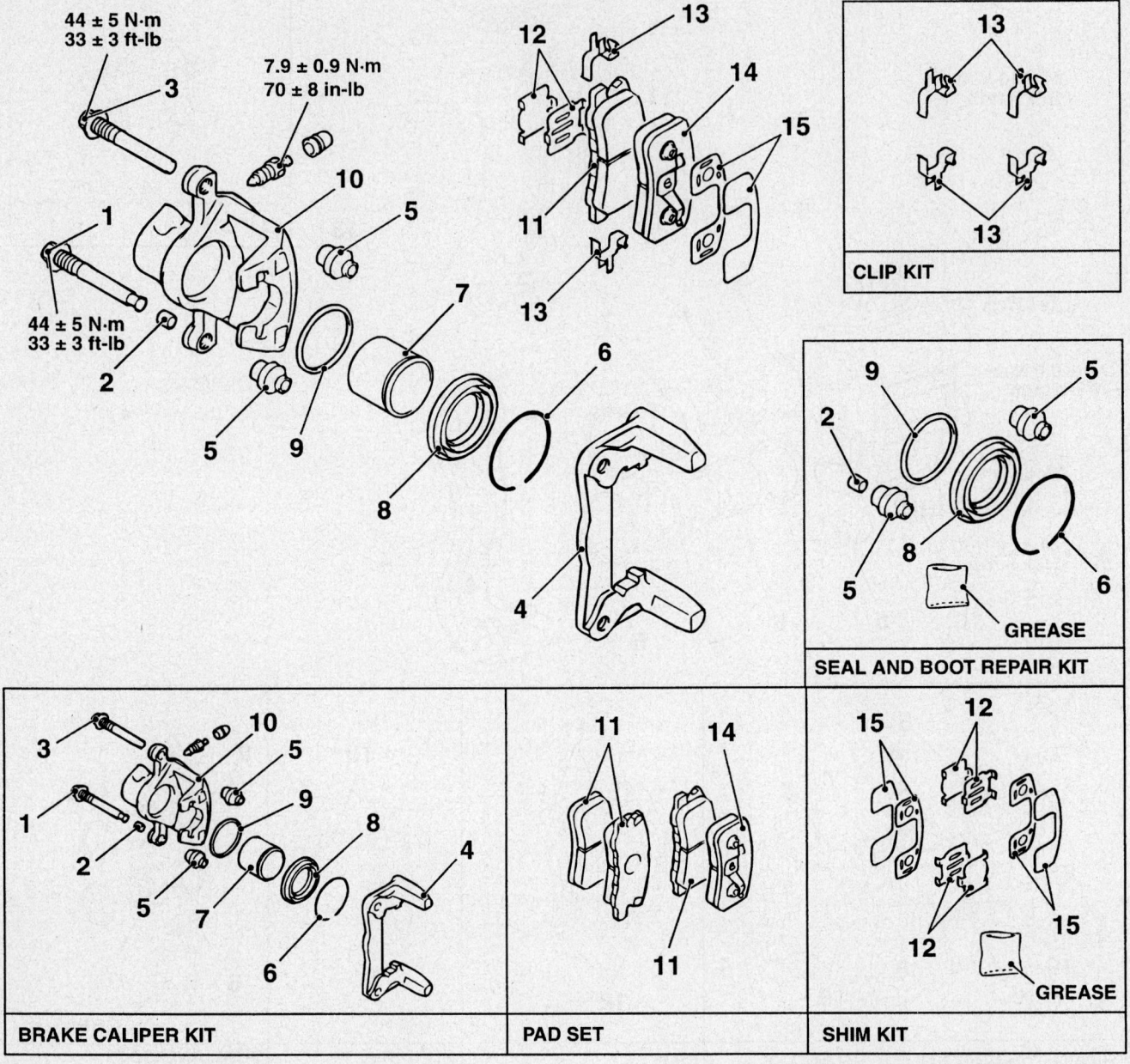

44 ± 5 N·m
33 ± 3 ft-lb

7.9 ± 0.9 N·m
70 ± 8 in-lb

44 ± 5 N·m
33 ± 3 ft-lb

CLIP KIT

SEAL AND BOOT REPAIR KIT

GREASE

BRAKE CALIPER KIT

PAD SET

SHIM KIT

GREASE

1. LOCK PIN
2. BUSHING
3. GUIDE PIN
4. CALIPER SUPPORT
5. PIN BOOT
6. BOOT RING
7. PISTON
8. PISTON BOOT

9. PISTON SEAL
10. CALIPER BODY
11. PAD AND WEAR INDICATOR
    ASSEMBLY
12. INNER SHIM
13. CLIP
14. PAD ASSEMBLY
15. OUTER SHIM

09482_OUTL_G0030

Rear brake caliper and related components—2005–2006 vehicles

- Caliper guide and lock pins and lift the caliper assembly from the caliper support.

➡**On some vehicles, the front caliper can be flipped up by leaving the upper pin in place and using it as a pivot point.**

- Brake pads, spring clip and shims

To install:

4. Compress pistons back into the caliper bore.

5. Lubricate slide points and install the brake pads, shims and spring clips.

6. Install the caliper.

7. Lubricate and install the caliper guide and lock pins in their original positions.

8. Check and adjust the brake fluid level, as required.

9. Install the wheels.

Brake Drums

REMOVAL & INSTALLATION

1. Before servicing the vehicle, refer to the Precautions Section.

2. Raise and support the vehicle safely.

3. Remove or disconnect the following:
- Wheels
- Drum

To install:
- Drum
- Wheels

Brake Shoes

REMOVAL & INSTALLATION

1. Before servicing the vehicle, refer to the Precautions Section.

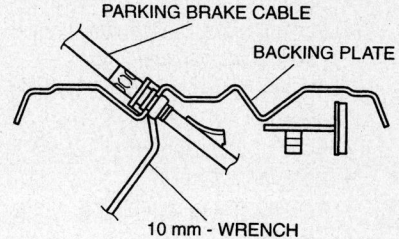

67170-OUTL-G68

Disconnecting the parking brake cable

2. Remove some of the brake fluid from the master cylinder reservoir.

3. Raise and support the vehicle safely.

4. Loosen the parking brake cable adjusting nut.

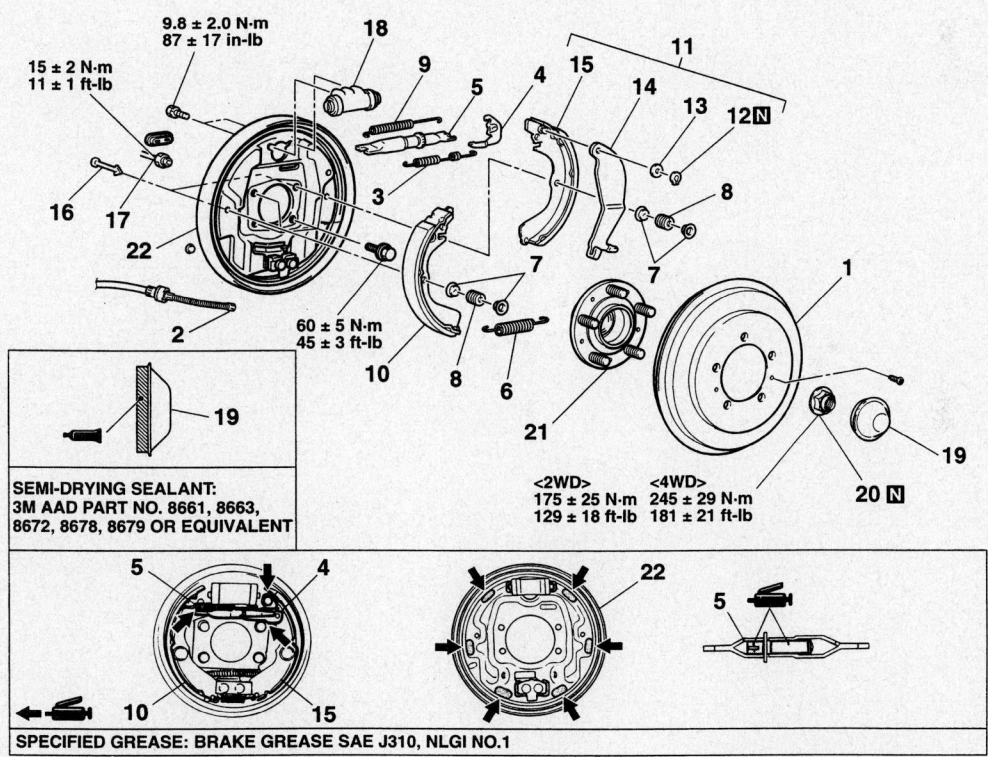

SEMI-DRYING SEALANT:
3M AAD PART NO. 8661, 8663, 8672, 8678, 8679 OR EQUIVALENT

SPECIFIED GREASE: BRAKE GREASE SAE J310, NLGI NO.1

1. BRAKE DRUM
2. PARKING BRAKE CABLE CONNECTION
3. SHOE-TO-SHOE SPRING
4. ADJUSTER LEVER
5. AUTO ADJUSTER ASSEMBLY
6. RETAINER SPRING
7. SHOE HOLD-DOWN CUP
8. SHOE HOLD-DOWN SPRING
9. SHOE-TO-LEVER SPRING
10. SHOE AND LINING ASSEMBLY
11. SHOE AND LEVER ASSEMBLY
12. RETAINER
13. WAVE WASHER
14. PARKING LEVER
15. SHOE AND LINING ASSEMBLY
16. SHOE HOLD-DOWN PIN
17. BRAKE PIPE CONNECTION
18. WHEEL CYLINDER ASSEMBLY
19. HUB CAP (2WD)
20. LOCK NUT
21. REAR HUB ASSEMBLY
22. BACKING PLATE

67170-OUTL-G69

Rear drum brake and related components

5. Remove or disconnect the following:
- Wheels
- Brake drum
- Parking brake cable connection. Use a 10mm wrench to help remove the cable from the backing plate.
- Shoe-to-shoe spring
- Adjuster assembly
- Auto adjuster assembly
- Retainer spring
- Shoe hold-down cup and spring
- Shoe-to-lever spring
- Shoe and lining assembly
- Shoe and lever assembly
- Shoe from the lever by inserting a prytool to open the retainer joint, then remove the retainer

To install

6. Lubricate the backing plate bosses, anchor pin, and parking brake actuating mechanism with lithium-based grease.

7. Install or connect the following:
- Lever to the shoe and secure with the retainer
- Shoe and lever assembly
- Shoe and lining assembly
- Shoe-to-lever spring
- Shoe hold-down spring
- Shoe hold-down cup
- Retainer spring
- Auto adjuster assembly
- Adjuster lever
- Shoe-to-shoe spring
- Parking brake cable

8. Pre-adjust the shoes so the drum slides on with a light drag and install brake drum.
- Wheel bearing dust cap and adjust the rear brake shoes

9. Tighten the parking brake cable adjusting nut.

10. Check and adjust the brake fluid level, as required.

SPECIFICATIONS AND MAINTENANCE CHARTS

ENGINE AND VEHICLE IDENTIFICATION

	Engine						Model Year	
Code ①	Liters (cc)	Cu. In.	Cyl.	Fuel Sys.	Engine Type	Eng. Mfg.	Code ②	Year
K	3.7 (3701)	226	6	SEFI	SOHC	Chrysler	6	2006
N	4.7 (4701)	287	8	SEFI	SOHC	Chrysler		

① 8th position of VIN
② 10th position of VIN

09482_RAID_C0001

GENERAL ENGINE SPECIFICATIONS

Year	Engine Displacement Liters	Engine VIN	Net Horsepower @ rpm	Net Torque @ rpm (ft. lbs.)	Bore x Stroke (in.)	Compression Ratio	Oil Pressure @ rpm
2006	3.7	K	211@5200	236@4000	3.66x3.40	9.6:1	25-110@3000
	4.7	N	235@4800	295@3200	3.66x3.40	9.0:1	35-105@3000

09482_RAID_C0002

GASOLINE ENGINE TUNE-UP SPECIFICATIONS

Year	Engine Displacement Liters	Engine VIN	Spark Plug Gap (in.)	Ignition Timing (deg.)	Fuel Pump (psi)	Idle Speed (rpm)	Valve Clearance Intake	Valve Clearance Exhaust
2006	3.7	K	0.042	①	44-54	②	HYD	HYD
	4.7	N	0.040	①	47-51	②	HYD	HYD

NOTE: The Vehicle Emission Control Information (VECI) label often reflects specification changes made during production.
The label figures must be used if they differ from those in this chart.

HYD: Hydraulic

① Ignition timing is controlled by the PCM and is not adjustable.

② Idle speed is controlled by the PCM and is not adjustable

09482_RAID_C0003

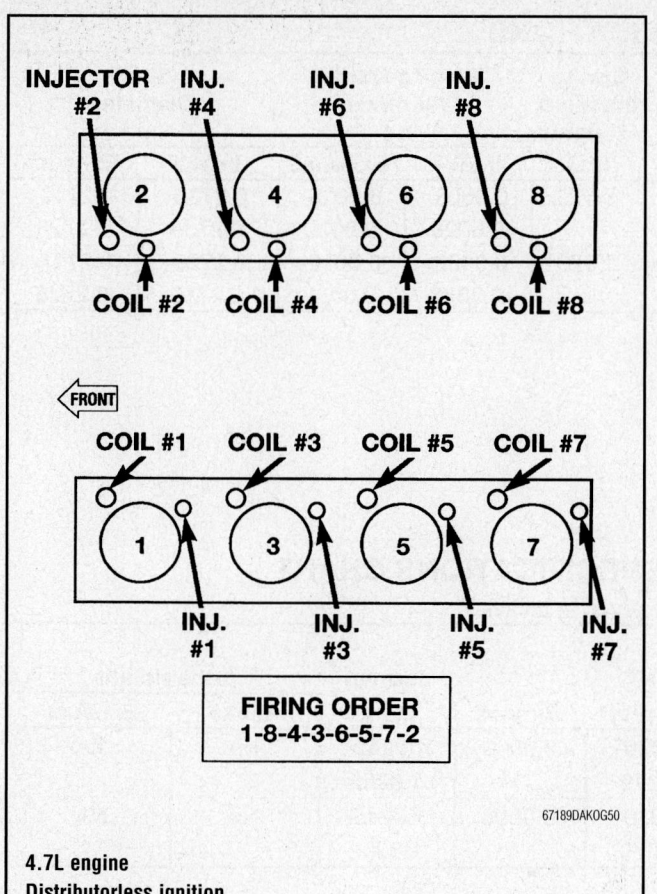

INJECTOR #2   INJ. #4   INJ. #6   INJ. #8

COIL #2   COIL #4   COIL #6   COIL #8

◄FRONT

COIL #1   COIL #3   COIL #5   COIL #7

INJ. #1   INJ. #3   INJ. #5   INJ. #7

FIRING ORDER
1-8-4-3-6-5-7-2

67189DAK0G50

4.7L engine
Distributorless ignition

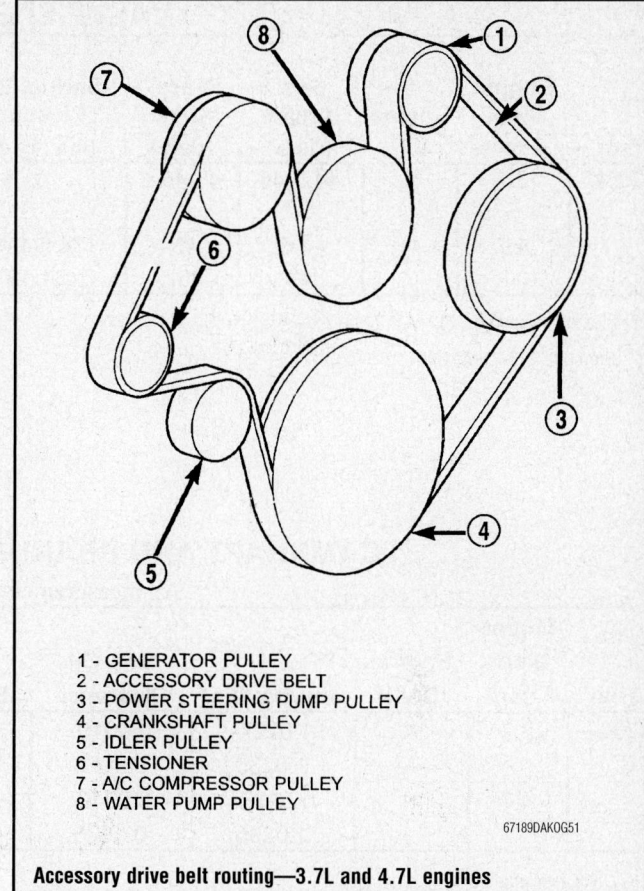

1 - GENERATOR PULLEY
2 - ACCESSORY DRIVE BELT
3 - POWER STEERING PUMP PULLEY
4 - CRANKSHAFT PULLEY
5 - IDLER PULLEY
6 - TENSIONER
7 - A/C COMPRESSOR PULLEY
8 - WATER PUMP PULLEY

67189DAK0G51

Accessory drive belt routing—3.7L and 4.7L engines

CAPACITIES

Year	Engine Displ. Liters	Engine VIN	Oil with Filter (qts.)	Transmission (pts.) Manual	Transmission (pts.) Auto.	Transfer Case (pts.)	Drive Axle Front (pts.)	Drive Axle Rear (pts.)	Fuel Tank (gal.)	Cooling System (qts.)
2006	3.7	K	5.0	4.65	①	②	3.5	③	25.0	16.2
	4.7	N	6.0	4.65	①	②	3.5	③	25.0	16.2

NOTE: All capacities are approximate. Add fluid gradually and check to be sure a proper fluid level is obtained.

① 42RLE: 8.0 pts.
  545RFE 2wd: 11.0 pts.
  545RFE 4wd: 13.0 pts.

② NV233: 2.5 pts.
  NV244: 2.85 pts.

③ The following values include 0.25 pt. of friction
  modifier for LSD axles.
  8.25 in. axle: 4.4 pts.
  9.25 in. axle: 4.9 pts.

09482_RAID_C0004

VALVE SPECIFICATIONS

Year	Engine Displ. Liters	Engine VIN	Seat Angle (deg.)	Face Angle (deg.)	Spring Test Pressure (lbs. @ in.)	Spring Installed Height (in.)	Stem-to-Guide Clearance (in.)		Stem Diameter (in.)	
							Intake	Exhaust	Intake	Exhaust
2006	3.7	K	44.5-45	45-45.5	①	1.579	0.0008-0.0028	0.0019-0.0039	0.2729-0.2739	0.2717-0.2728
	4.7	N	44.5-45	45-45.5	174.5-195.6 @1.137	1.579	0.0008-0.0028	0.0019-0.0039	0.2729-0.2739	0.2717-0.2728

① Intake: 213-234@1.107

Exhaust: 215-219@1.067

09482_RAID_C0005

CAMSHAFT AND BEARING SPECIFICATIONS CHART

All measurements are given in inches.

Year	Engine Displ. Liters	Engine ID/VIN	Journal Dia.	Brg. Oil Clearance	Shaft End-play	Runout	Journal Bore	Lobe Height	
								Intake	Exhaust
2006	3.7	K	1.0227-1.0235	0.0010-0.0026	0.0030-0.0079	0.0008	1.0245-1.0252	NA	NA
	4.7	N	1.0227-1.0235	0.0010-0.0026	0.0030-0.0079	0.0008	1.0245-1.0252	NA	NA

NA: Not Available

09482_RAID_C0006

CRANKSHAFT AND CONNECTING ROD SPECIFICATIONS

All measurements are given in inches.

Year	Engine Displ. Liters	Engine VIN	Crankshaft				Connecting Rod		
			Main Brg. Journal Dia.	Main Brg. Oil Clearance	Shaft End-play	Thrust on No.	Journal Diameter	Oil Clearance	Side Clearance
2006	3.7	K	2.4996-2.5005	0.0008-0.0018	0.0021-0.0112	2	2.2792-2.2798	0.0002-0.0011	0.0040-0.0138
	4.7	N	2.4996-2.5005	0.0002-0.0018	0.0021-0.0112	2	2.2793-2.2798	0.0002-0.0011	0.0040-0.0138

09482_RAID_C0007

PISTON AND RING SPECIFICATIONS

All measurements are given in inches.

Year	Engine Displ. Liters	Engine VIN	Piston Clearance	Ring Gap			Ring Side Clearance		
				Top Comp.	Bottom Comp.	Oil Control	Top Comp.	Bottom Comp.	Oil Control
2006	3.7	K	0.0014	0.0079-0.0142	0.0146-0.0249	0.0100-0.0300	0.0020-0.0037	0.0016-0.0031	0.0007-0.0091
	4.7	N	0.0014	0.0146-0.0249	0.0146-0.0249	0.0099-0.0300	0.0020-0.0037	0.0016-0.0031	0.0175-0.0185

09482_RAID_C0008

TORQUE SPECIFICATIONS

All readings in ft. lbs.

Year	Engine Displ. Liters	Engine VIN	Cylinder Head Bolts	Main Bearing Bolts	Rod Bearing Bolts	Crankshaft Damper Bolts	Flywheel Bolts	Manifold		Spark Plugs	Oil Pan Drain Plug
								Intake	Exhaust		
2006	3.7	K	①	②	④	130	70	9	18	20	25
	4.7	N	①	③	④	130	45	⑤	18	20	25

① See text

② Bed plate bolt sequence. Refer to illustration

    Step 1: Hand tighten bolts 1D,1G and 1F until the bedplate contacts the block.

    Step 2: Tighten bolts 1A - 1J to 54 N·m (40 ft. lbs.)

    Step 3: Tighten bolts 1 - 8 to 7 N·m (5 ft. lbs.)

    Step 4: Turn bolts 1 - 8 an additional 90°.

    Step 5: Tighten bolts A - E 27 N·m (20 ft. lbs.).

③ Bed plate bolt sequence. Refer to illustration

    Step 1: Bolts A-L to 40 ft. lbs.

    Step 2: Bolts 1-10 25 inch lbs.

    Step 3: Bolts 1-10 plus 90 degrees

    Step 4: Bolts A1-A6 20 ft. lbs.

④ 20 ft. lbs. plus 90 degrees

⑤ See illustration in text section

    Step 1: 1-4 to 72 inch lbs. in 12 inch lb. Increments

    Step 2: bolts 5-12: 72 inch lbs.

    Step 3: Check that all bolts are at 72 inch lbs.

    Step 4: All bolts, in sequence, to 12 ft. lbs.

    Step 5: Check that all bolts are at 12 ft. lbs.

09482_RAID_C0009

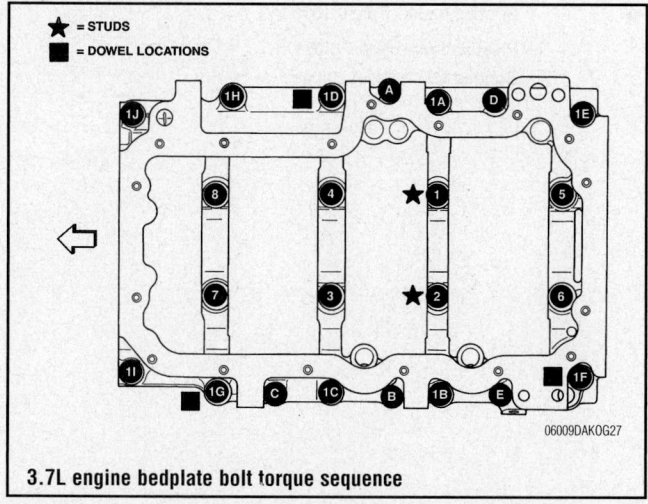

3.7L engine bedplate bolt torque sequence

06009DAKOG27

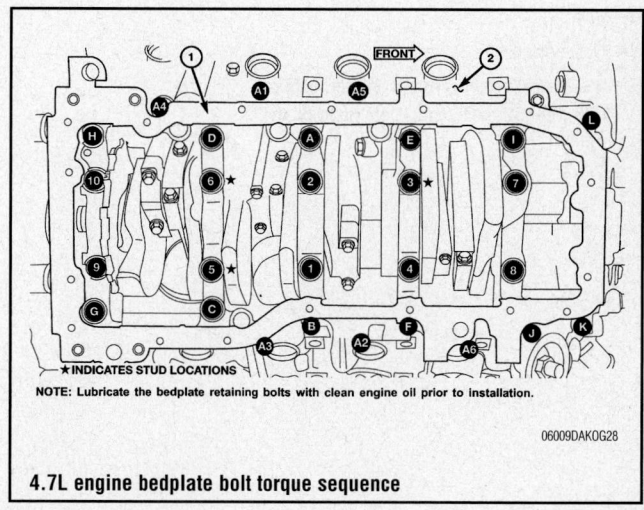

4.7L engine bedplate bolt torque sequence

06009DAKOG28

WHEEL ALIGNMENT

Year	Model	Wheel Base (in.)		Caster Range (+/-Deg.)	Caster Preferred Setting (Deg.)	Camber Range (+/-Deg.)	Camber Preferred Setting (Deg.)	Toe-in (Deg.)
2006	All	131	Left	0.05	+3.50	0.50	-0.25	0.20+/-0.10
			Right	0.05	+3.70	0.50	-0.25	0.20+/-0.10

09482_RAID_C0010

TIRE, WHEEL AND BALL JOINT SPECIFICATIONS

Year	Model	OEM Tires Standard	OEM Tires Optional	Tire Pressures (psi) Front	Tire Pressures (psi) Rear	Wheel Size	Ball Joint Inspection	Lug Nut Torque ft. lbs.
2006	LS	P245/70R16	P255/65R16	①	①	NA		
			P265/70R16	①	①	NA		
	DuroCross 2WD	P265/70R16	-	①	①	NA	0.020 in.	135
	DuroCross 4WD	LT265/70R16	-	①	①	NA		
	XLS	P265/65R17		①	①	NA		

OEM: Original Equipment Manufacturer

PSI: Pounds Per Square Inch

① See the tire placard on the vehicle

09482_RAID_C0011

BRAKE SPECIFICATIONS
All measurements in inches unless noted

Year		Brake Disc Original Thickness	Brake Disc Minimum Thickness	Brake Disc Maximum Run-out	Brake Drum Original Inside Diameter	Brake Drum Max. Wear Limit	Brake Drum Maximum Machine Diameter	Minimum Lining Thickness Front	Minimum Lining Thickness Rear	Brake Caliper Bracket Bolts (ft. lbs.)	Brake Caliper Mounting Bolts (ft. lbs.)
2006	F	1.100	1.039	0.0010	-	-	-	②	-	130	26
	R	-	-	-	11.50	①	11.693	-	③	-	-

F: Front

R: Rear

NA: Not Available

① Maximum allowable drum diameter, either from wear or machining, is stamped on the drum.

② Riveted brake pads: 0.0625 in.
  Bonded brake pads: 0.1875 in.

③ Riveted brake shoes: 0.031 in.
  Bonded brake shoes: 0.0625 in.

09482_RAID_C0012

SCHEDULED MAINTENANCE INTERVALS
2006 Mistubishi Raider

TO BE SERVICED	TYPE OF SERVICE	VEHICLE MILEAGE INTERVAL (x1000)													
		6	12	18	24	30	36	42	48	54	60	66	72	78	84
Engine coolant	R	Replace every 60 months or 102,000 miles, whichever comes first													
Accessory drive belt ①	S/I										✓				
Engine oil & filter	R	✓	✓	✓	✓	✓	✓	✓	✓	✓	✓	✓	✓	✓	✓
Tires	Rotate	✓	✓	✓	✓	✓	✓	✓	✓	✓	✓	✓	✓	✓	✓
PCV valve ①	S/I										✓				
Brake linings	S/I			✓			✓			✓			✓		
Air cleaner element	S/I					✓					✓				
Spark plugs	R					✓					✓				
Transfer case fluid level ②	I					✓					✓				

R: Replace   S/I: Service or Inspect   L: Lubricate   Adj: Adjust

① Replace if necessary.

② Replace every 120,000 miles

FREQUENT OPERATION MAINTENANCE (SEVERE SERVICE)

If a vehicle is operated under any of the following conditions it is considered severe service:

- Extremely dusty areas.
- 50%or more of the vehicle operation is in 32°C (90°F) or higher temperatures, or constant operation in temperatures below 0°C (32°F).
- Prolonged idling (vehicle operation in stop and go traffic.
- Frequent short running periods (engine does not warm to normal operating temperatures).
- Police, taxi, delivery usage or trailer towing usage.

Oil & oil filter change: change every 3000 miles.

Air filter/air pump air filter: change every 24,000 miles.

Engine coolant level, hoses & clamps: check every 6,000 miles.

Exhaust system: check every 6000 miles.

Drive belts: check every 18,000 miles; replace every 24,000 miles.

Crankcase inlet air filter (6 & 8 cyl.): clean every 24,000 miles.

Oxygen sensor: replace every 82,500 miles.

Automatic transmission fluid, filter & bands: change & adjust every 12,000 miles.

Steering linkage: lubricate every 6000 miles.

Rear axle fluid: change every 15,000 miles.

09482_RAID_C0013

ENGINE REPAIR

➡ **Disconnecting the negative battery cable on some vehicles may interfere with the functions of the on board computer system. The computer may undergo a relearning process once the negative battery cable is reconnected.**

Distributor

The 3.7L and 4.7L engines are equipped with a Distributorless Ignition System (DIS).

Alternator

REMOVAL & INSTALLATION

1. Before servicing the vehicle, refer to the Precautions Section.
2. Remove or disconnect the following:
   - Negative battery cable
   - Accessory drive belt
   - Alternator harness connectors
   - Mounting bolts and alternator

➡ **There are 1 vertical and 2 horizontal bolts.**

To install:

3. Before servicing the vehicle, refer to the Precautions Section.

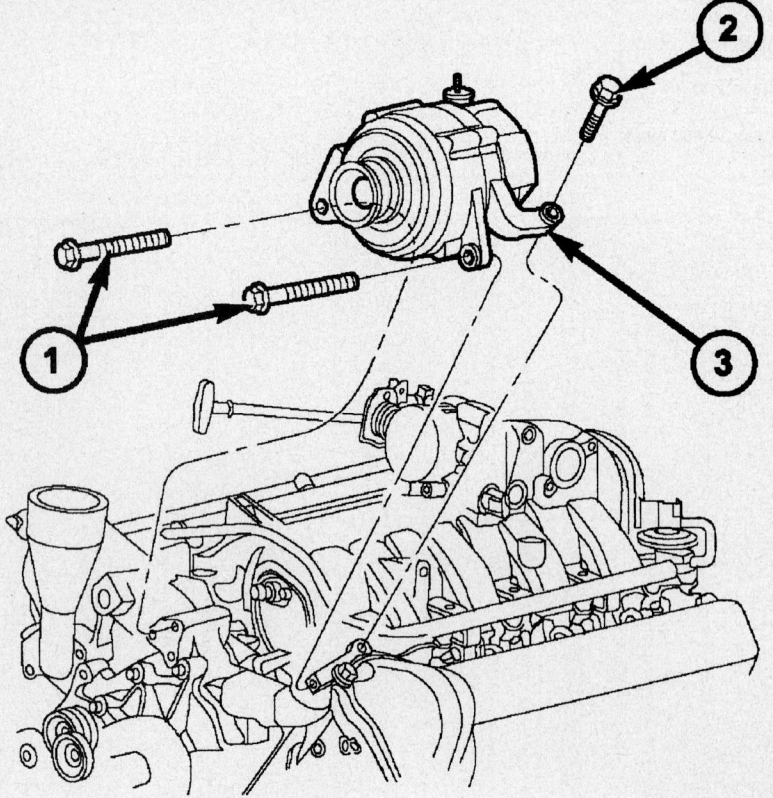

Alternator mounting—3.7L and 4.7L engines

06009DAK0G01

4. Install the alternator and tighten the bolts to the following specifications:
   - Vertical bolt and long horizontal bolt: 40 ft. lbs. (55 Nm)
   - Short horizontal bolt: 55 ft. lbs. (74 Nm).
5. Install or connect the following:
   - Alternator harness connectors
   - Accessory drive belt
   - Negative battery cable

Ignition Timing

ADJUSTMENT

The ignition timing is controlled by the Powertrain Control Module (PCM) and is not adjustable.

Engine Assembly

REMOVAL & INSTALLATION

3.7L Engine

1. Before servicing the vehicle, refer to the Precautions Section.
2. Discharge the A/C system.

3. Drain the cooling system.
4. Release the fuel rail pressure.
5. Remove or disconnect the following:
   - Battery
   - Air intake assembly
   - Upper fan shroud
   - Accessory drive belt
   - Viscous fan assembly
6. Remove the A/C compressor and position out of the way.
7. Remove the generator and secure away from engine.
8. Remove the power steering pump with lines attached and secure away from engine.

➡ **Do not remove the phenolic pulley from the P/S pump. It is not required for P/S pump removal.**

9. Remove or disconnect the following:
   - Heater hoses
   - Upper radiator hose
   - Lower radiator hose
   - Transmission oil cooler lines at the radiator
   - Radiator/cooling module assembly
   - Engine to body ground straps at the left side of cowl
10. Disconnect the engine wiring harness at the following points:
    - Intake air temperature (IAT) sensor
    - Fuel Injectors
    - Throttle Position (TPS) Switch
    - Idle Air Control (IAC) Motor
    - Engine Oil Pressure Switch
    - Engine Coolant Temperature (ECT) Sensor
    - Manifold Absolute Pressure MAP) Sensor
    - Camshaft Position (CMP) Sensor
    - Coil Over Plugs
    - Crankshaft Position Sensor
11. Remove the coil over plugs.
12. Remove fuel rail and secure away from engine.

➡ **It is not necessary to release the quick connect fitting from the fuel supply line for engine removal.**

13. Remove or disconnect the following:
    - PCV hose
    - Breather hoses
    - Vacuum hose for the power brake booster
    - Knock sensors
    - Engine oil dipstick tube
    - Intake manifold
14. Install the Special Tool 8247 Engine Lifting Fixture, using original fasteners from the intake manifold and fuel rail.

15. Remove or disconnect the following:
   - Oxygen sensor wiring
   - Crankshaft position sensor
   - Engine block heater power cable, if equipped

16. Disconnect the front driveshaft at the front differential and secure out of the way.

17. Remove or disconnect the following:
   - Pinion bracket
   - Starter
   - Two ground straps from the lower left hand side and one ground strap from the lower right hand side of the engine
   - Structural cover between engine and transmission
   - Exhaust crossover pipe from exhaust manifolds
   - Torque converter bolts, if equipped

➠Matchmark the bolts for reassembly in the original position.

18. Support the transmission with a suitable jack.

19. Connect a suitable engine hoist to the engine lift plate.

20. Remove the bellhousing-to-engine mounting bolts.

21. Remove the left and right engine mount bolts.

22. Remove the engine from the vehicle.

To install:

23. Position the engine into the vehicle.

24. Install or connect the following:
   - Bellhousing-to-engine mounting bolts. Tighten to 30 ft. lbs. (41 Nm).
   - Engine mount bolts: 2WD vehicles to 70 ft. lbs. (95 Nm); 4WD vehicles to 75 ft. lbs. (102 Nm).

25. Remove the jack from under the transmission.

26. Remove the Engine Lifting Fixture Tool.

27. Install or connect the following:
   - Torque converter bolts in their original positions
   - Starter
   - Crankshaft position sensor
   - Engine block heater power cable, if equipped
   - Structural cover
   - Two ground straps from the lower left hand side and one ground strap from the lower right hand side of the engine
   - Pinion bracket
   - Exhaust pipe to the crossover
   - Oxygen sensor wiring

28. Remove the Engine Lifting Fixture.

29. Install or connect the following:
   - Knock sensors

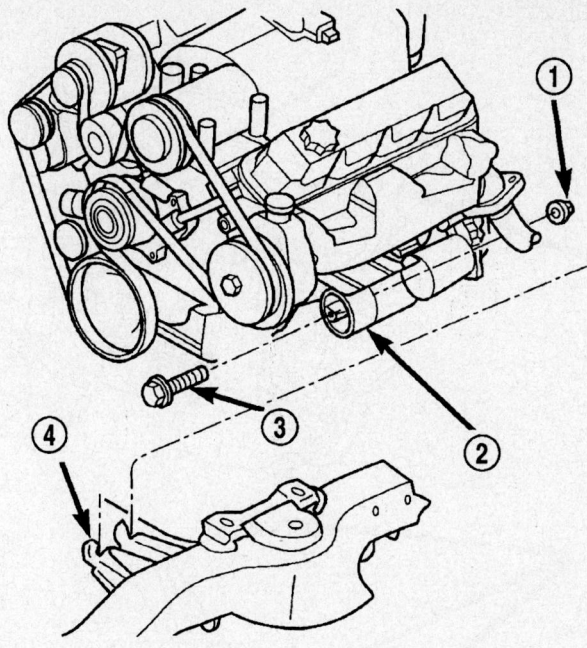

1 - LOCKNUT AND WASHER
2 - ENGINE MOUNT/INSULATOR
3 - THROUGH BOLT
4 - FRAME

67189DAK0G49

Engine mount through-bolt removal—3.7L engine w/2wd

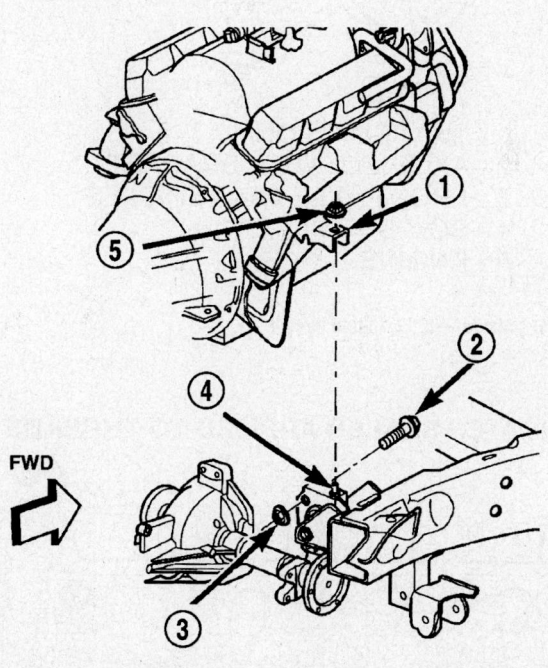

1 - ENGINE MOUNT BRACKET (2)
2 - THROUGH BOLT (2)
3 - LOCKNUT AND WASHER (2)
4 - ENGINE ISOLATOR TO ENGINE MOUNT BRACKET STUD (2)
5 - LOCKNUT (2)

67189DAK0G47

Engine mount through-bolt removal—3.7L engine w/4wd

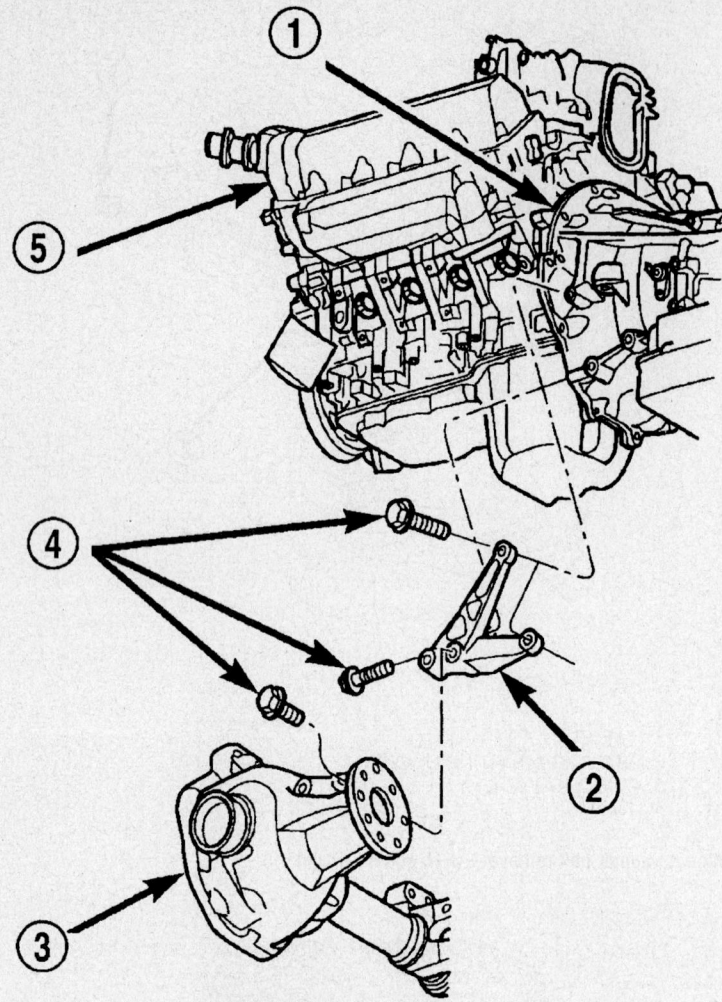

1 - TRANSMISSION
2 - AXLE ISOLATOR BRACKET
3 - FRONT AXLE 4X4 VEHICLES
4 - BOLTS
5 - ENGINE

67189DAK0G48

Axle isolator bracket removal—3.7L engine w/4wd

♦ **INDICATES SEALER APPLIED TO THREADS**

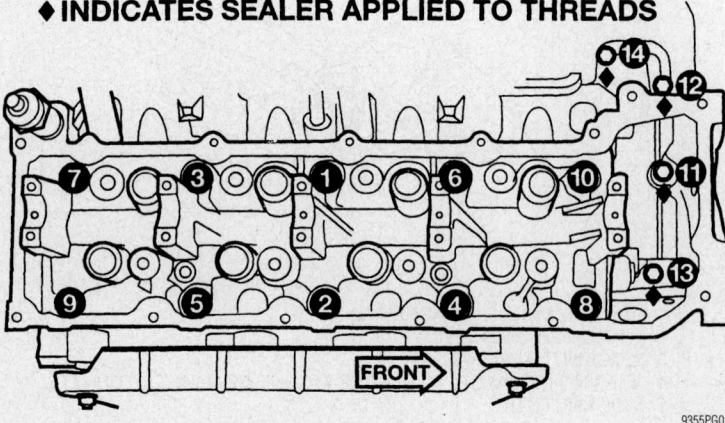

9355PG01

Tighten the structural cover bolts in this order

- Engine to body ground straps at the left side of cowl
- Intake manifold
- Engine oil dipstick tube
- Vacuum hose for the power brake booster
- Breather hoses
- PCV hose
- Coil over plugs
- Fuel rail

30. Connect the engine harness at the following points:
- Intake Air Temperature (IAT) Sensor
- Idle Air Control (IAC) Motor
- Fuel Injectors
- Throttle Position (TPS) Switch
- Engine Oil Pressure Switch
- Engine Coolant Temperature (ECT) Sensor
- Manifold Absolute Pressure (MAP) Sensor
- Camshaft Position (CMP) Sensor
- Coil Over Plugs
- Crankshaft position sensor

31. Install or connect the following:
- Radiator/cooling module assembly
- Lower radiator hose
- Upper radiator hose
- Throttle cables
- Heater hoses
- Power steering pump
- Alternator
- A/C compressor
- Accessory drive belt
- Viscous fan assembly
- Upper fan shroud
- Radiator core support bracket
- Air intake assembly
- Battery cables

32. Fill the engine with oil to the correct level.

33. Fill the cooling system to the correct level.

34. Start the engine and check for leaks.

Water Pump

REMOVAL & INSTALLATION

1. Before servicing the vehicle, refer to the Precautions Section.

2. Drain the cooling system.

3. Remove or disconnect the following:
- Negative battery cable
- Fan and clutch assembly from the pump
- Fan shroud mounting screws
- Fan shroud and fan assembly from the vehicle

➡**If you're reusing the fan clutch, keep it upright to avoid silicone fluid loss.**

★ **INDICATES STUD LOCATIONS**

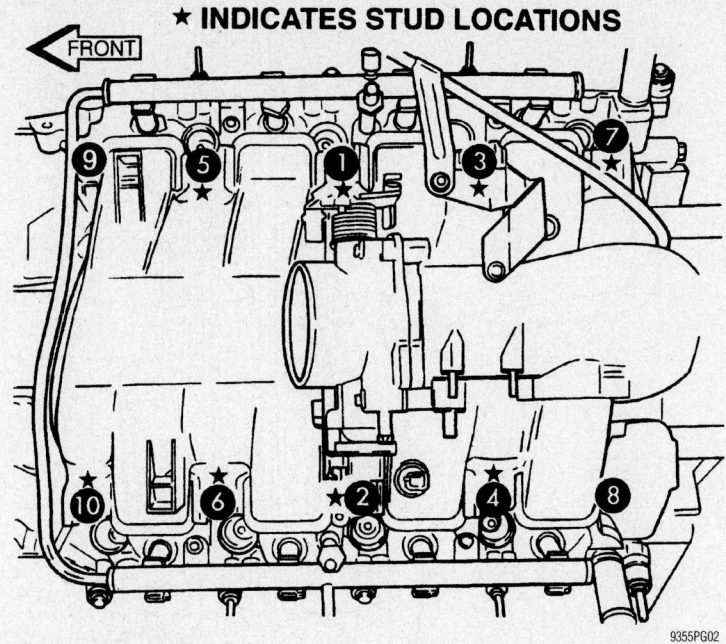

9355PG02

Water pump tightening sequence

- Accessory drive belt
- Upper radiator hose
- Water pump

4. Installation is the reverse of removal. Using a new gasket, tighten the water pump bolts, in sequence, to 43 ft. lbs. (58 Nm).

Heater Core

REMOVAL & INSTALLATION

1. Before servicing the vehicle, refer to the Precautions Section.
2. Recover the refrigerant from the A/C system.
3. Drain the cooling system.
4. Remove or disconnect the following:
- Negative battery cable
- Shift boot, if equipped with mini floor console
- Rubber liners in the center console to reveal the screws
- Bottom tray in the center console, if equipped
- Screws that screw the floor console
- Console assembly wiring harness, if equipped
- Floor console
- Outer front seat belt retractor
- Drivers side door sill retaining tabs
- Drivers side door sill trim cover
- Driver's side instrument panel end cap
- Parking brake release from the release rod
- Steering column lower cover

5. Position the steering wheel with the wheel straight ahead.
6. Remove or disconnect the following:

- Steering column tilt lever
- Steering column shrouds
- Brake switch electrical connector
- Steering column wiring harnesses
- Shift cable from the column shift actuator and bracket
- SKIM module and electrical connector
- Upper steering shaft coupler bolt from the column

7. Separate the steering shaft from the coupler.
8. Remove the steering column mountings to remove the column assembly.
9. Remove or disconnect the following:

- Upper mounting bolts from the pedal support
- Left and right A-pillar trim panels
- Defroster grille
- Defroster electrical connectors
- Five upper fence line bolts
- HVAC electrical connector
- Left A-pillar mounting bolt
- ABS module electrical connector
- Center wiring harness from the center support
- Lower center support bolts
- Left side support bolts
- Passenger door sill trim cover
- Passenger side end cap
- Antenna cable under the glove box
- Ground cables under the glove box area

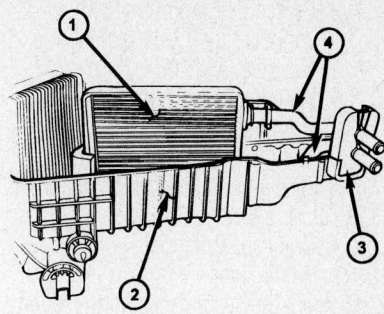

1. Heater Core
2. HVAC Housing
3. Rubber Flange
4. Heater Core Tubes

09482_RAID_G0001

Exploded view of the heater core assembly

- Antenna amplifier connector, if equipped
- Right side support bolts

10. With assistance, remove the instrument panel assembly.
11. Remove the HVAC housing mounting bolts from the passenger compartment side.
12. Disconnect the following from the HVAC housing:

- A/C lines
- Heater hoses
- Fresh air inlet screen

13. Pull the HVAC housing assembly rearward so the mounting studs and drain tube clear the dash panel to remove.
14. Lift the heater core from the lower half of the HVAC housing.
15. Installation is the reverse of removal.

Cylinder Head

REMOVAL & INSTALLATION

3.7L Engine

1. Before servicing the vehicle, refer to the Precautions Section.
2. Drain the cooling system.
3. Properly relieve the fuel system pressure.
4. Remove or disconnect the following:

- Negative battery cable
- Air intake assembly
- Exhaust pipe at the exhaust manifold
- Intake manifold
- Master cylinder and booster assembly
- Fuel injector harnesses
- Left side breather tube
- Cylinder head cover
- Engine cooling fan and shroud
- Accessory drive belt

- Power steering pump without disconnect the lines and hang securely, left side

5. Rotate the crankshaft so that the crankshaft timing mark aligns with the Top Dead Center (TDC) mark on the front cover, and the **V6** marks on the camshaft sprockets are at 12 o'clock as shown.
- Crankshaft damper
- Front cover

6. Lock the secondary timing chain to the idler sprocket with Timing Chain Locking tool 8429.

7. Matchmark the secondary timing chain one link on each side of the **V6** mark to the camshaft sprocket.

8. Remove or disconnect the following:
- Secondary timing chain tensioner
- Cylinder head access plug
- Secondary timing chain guide
- Camshaft sprocket
- Cylinder head and gasket

➡**The cylinder head is retained by twelve bolts. Four of the bolts are smaller and are at the front of the head.**

To install:

9. Clean the cylinder head surfaces of old gasket material.

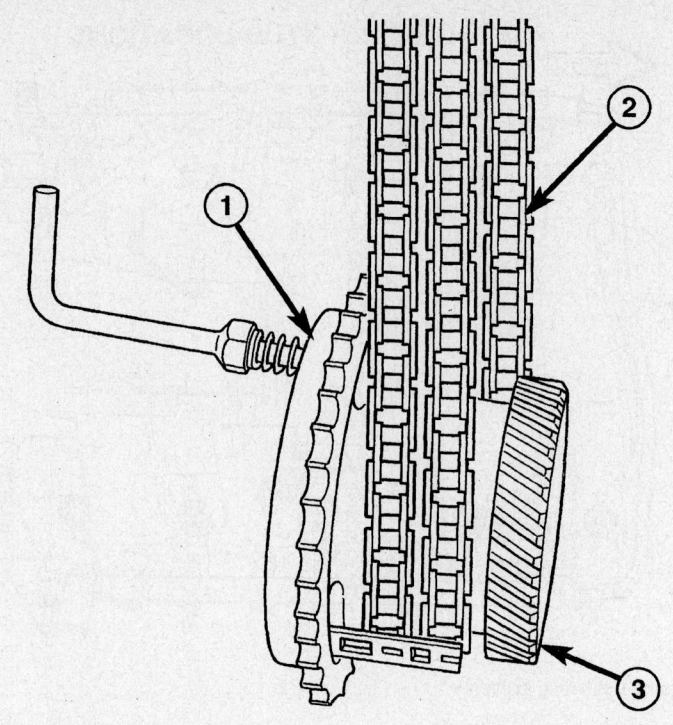

1 - SPECIAL TOOL 8429

2 - CAMSHAFT CHAIN

3 - CRANKSHAFT TIMING GEAR

9355PG05

Camshaft locking tool—3.7L

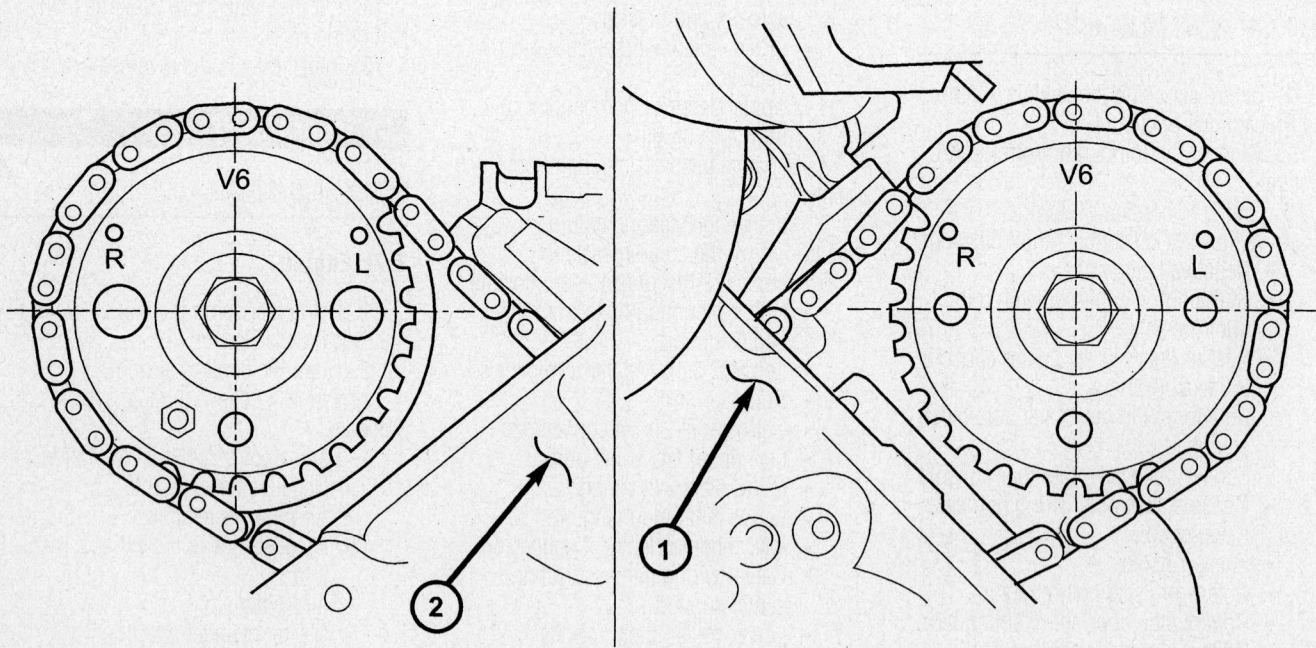

1 - LEFT CYLINDER HEAD
2 - RIGHT CYLINDER HEAD

9355PG04

Camshaft sprocket timing marks—3.7L

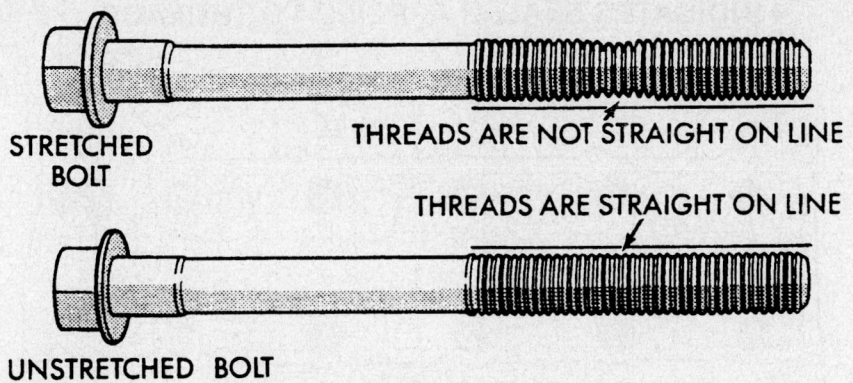

STRETCHED BOLT

THREADS ARE NOT STRAIGHT ON LINE

THREADS ARE STRAIGHT ON LINE

UNSTRETCHED BOLT

9302PG10

Examine the head bolts for signs of stretching—3.7L engine

❋❋ WARNING

Use a plaster or wooden scraper only.

10. Check the cylinder head bolts for signs of stretching and replace as necessary.

11. Lubricate the threads of the 11mm bolts with clean engine oil.

12. Coat the threads of the 8mm bolts with Mopar® Lock and Seal Adhesive.

13. Install the cylinder heads. Use new gaskets and tighten the bolts, in sequence, as follows:

  a. Step 1: Bolts 1–8 to 20 ft. lbs. (27 Nm)

  b. Step 2: Bolts 1–8 verify torque without loosening

  c. Step 3: Bolts 9–12 to 10 ft. lbs. (14 Nm)

  d. Step 4: Bolts 1–8 plus ¼ (90 degree) turn

  e. Step 5: Bolts 1–8 plus ¼ (90 degree) turn again

  f. Step 6: Bolts 9–12 to 19 ft. lbs. (26 Nm)

14. Position the secondary chain onto the camshaft sprocket, making sure one marked chain link is on either side of the V6 mark on the gear.

15. Using Special Tool 8428 Camshaft Wrench, position the sprocket onto the camshaft. Tighten retaining bolt to 90 ft. lbs. (122 Nm).

16. Install or connect the following:
  • Secondary timing chain guide
  • Cylinder head access plug
  • Secondary timing chain tensioner.

17. Remove the Timing Chain Locking tool.

18. Install or connect the following:
  • Front cover
  • Crankshaft damper. Torque the bolt to 130 ft. lbs. (175 Nm).
  • Power steering pump, left side only
  • Accessory drive belt
  • Engine cooling fan and shroud
  • Cylinder head cover. Tighten to 105 inch lbs. (12 Nm).
  • Left side breather tube

LEFT BANK

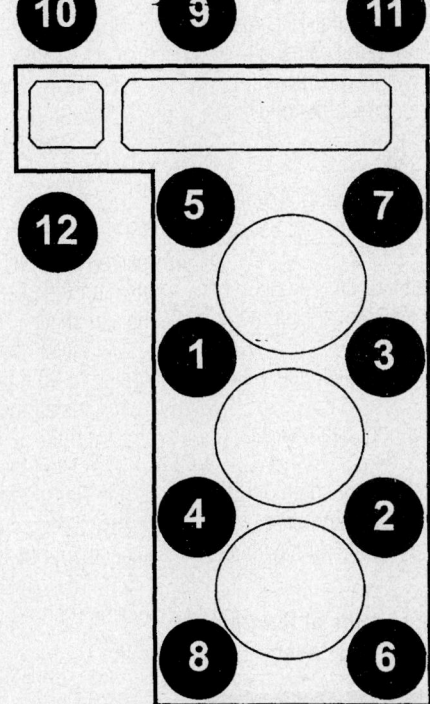

RIGHT BANK

9355PG03

Cylinder head bolt torque sequence—3.7L

- Fuel injector harnesses
- Intake manifold
- Exhaust pipe to the exhaust manifold
- Air intake assembly
- Negative battery cable

19. Fill and bleed the cooling system.

20. Start the engine, check for leaks and repair if necessary.

4.7L Engine

LEFT SIDE

1. Before servicing the vehicle, refer to the Precautions Section.

2. Drain the cooling system.

3. Remove or disconnect the following:
- Negative battery cable
- Air intake assembly
- Exhaust pipe at the exhaust manifold
- Intake manifold
- Master cylinder and booster assembly
- Injector wiring harnesses
- Left side breather tube
- Cylinder head cover
- Fan shroud and fan assembly
- Accessory drive belt
- Power steering pump without disconnecting and hang securely

4. Rotate the crankshaft until the damper mark is aligned with the TDC mark. Verify that the V8 mark on the camshaft sprocket is at the 12 o'clock position.

5. Remove or disconnect the following:
- Crankshaft damper
- Front cover

6. Lock the secondary timing chains to the idler sprocket with Special Tool 8515, or equivalent.

7. Mark the secondary timing chain, on link on either side of the V8 mark on the cam sprocket.

8. Remove the left side secondary chain tensioner.

9. Remove the cylinder head access plug.

10. Remove the secondary chain guide.

11. Remove the camshaft sprocket.

12. Remove the head bolts and cylinder head.

➡There are 4 smaller bolts at the front of the head. Don't overlook these.

✲✲ WARNING

Don't lay the head on its sealing surface. Due to the design of the head gasket, any distortion to the head sealing surface will result in leaks.

◆ INDICATES SEALER APPLIED TO THREADS

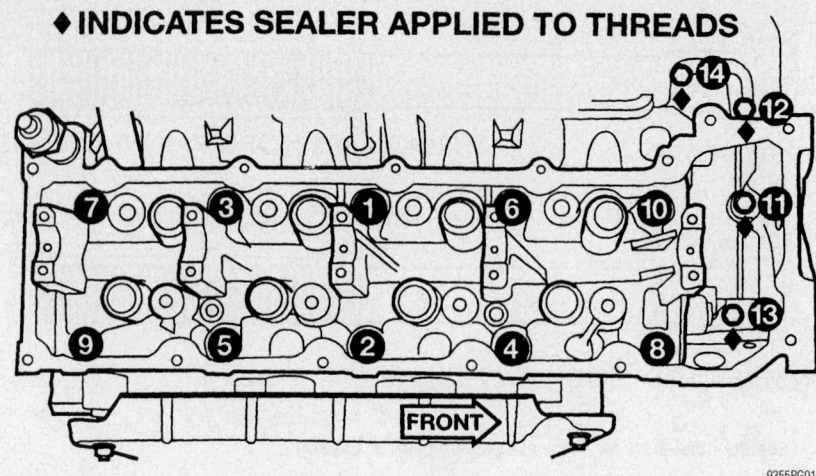

Cylinder head tightening sequence—4.7L

To install:

➡**Check the head bolts. If any necking is observed, replace the bolt.**

13. Lubricate the threads of the 11mm bolts with clean engine oil.

14. Coat the threads of the 8mm bolts with Mopar® Lock and Seal Adhesive.

15. Install the cylinder head with a new gasket. Tighten the bolts as follows:

a. Step 1: Bolts 1-10 to 15 ft. lbs. (20 Nm)

b. Step 2: Bolts 1-10 to 35 ft. lbs. (47 Nm)

c. Step 3: Bolts 11-14 to 18 ft. lbs. (25 Nm)

d. Step 4: Bolts 1-10 plus 90 degrees

e. Step 5: Bolts 11-14 to 22 ft. lbs. (30 Nm)

16. Position the secondary chain onto the camshaft sprocket, making sure one marked chain link is on either side of the V8 mark on the gear and position the gear onto the camshaft.

17. Install the camshaft sprocket and tighten to 90 ft. lbs. (122 Nm).

18. Install the left side secondary timing chain guide.

19. Install the cylinder head access plug.

20. Reset and install the left side secondary timing chain tensioner and remove Special Tool 8515.

21. The remainder of installation is the reverse order of removal. Note the following torques:

a. Crankshaft damper to 130 ft. lbs. (175 Nm).

b. Cylinder head cover bolts to 105 inch lbs. (12 Nm).

22. Fill the cooling system to the correct level.

23. Start the engine and check for leaks.

RIGHT SIDE

1. Before servicing the vehicle, refer to the Precautions Section.

2. Drain the cooling system.

3. Remove or disconnect the following:
- Negative battery cable
- Air intake assembly
- Exhaust pipe at the exhaust manifold
- Intake manifold
- Accessory drive belt
- A/C compressor, and hang securely aside
- Heater hoses
- Fuel injector and ignition coil connectors
- PCV hose
- Fuel injector wiring harnesses
- Ignition coil wiring harness
- Right side breather tube
- Cylinder head cover
- Fan shroud
- Oil filler housing from the cylinder head

4. Rotate the crankshaft until the damper mark is aligned with the TDC mark. Verify that the V8 mark on the camshaft sprocket is at the 12 o'clock position.

5. Remove or disconnect the following:
- Vibration damper
- Timing chain cover

6. Lock the secondary timing chains to the idler sprocket with Special Tool 8515, or equivalent.

7. Mark the secondary timing chain, on link on either side of the V8 mark on the cam sprocket.

8. Remove the right side secondary chain tensioner.

9. Remove the cylinder head access plug.

10. Remove the secondary chain guide.

11. Remove the camshaft sprocket.

❄ WARNING

Do not pry on the target wheel for any reason!

12. Remove the head bolts and cylinder head.

➡️**There are 4 smaller bolts at the front of the head. Don't overlook these.**

❄ WARNING

Don't lay the head on its sealing surface. Due to the design of the head gasket, any distortion to the head sealing surface will result in leaks.

To install:

➡️**Check the head bolts. If any necking is observed, replace the bolt.**

13. Lubricate the threads of the 11mm bolts with clean engine oil.

14. Coat the threads of the 8mm bolts with Mopar® Lock and Seal Adhesive.

15. Install the cylinder head with a new gasket. Tighten the bolts as follows:

    a. Step 1: Bolts 1-10 to 15 ft. lbs. (20 Nm)

    b. Step 2: Bolts 1-10 to 35 ft. lbs. (47 Nm)

    c. Step 3: Bolts 11-14 to 18 ft. lbs. (25 Nm)

    d. Step 4: Bolts 1-10 plus 90 degrees

    e. Step 5: Bolts 11-14 to 22 ft. lbs. (30 Nm)

16. Position the secondary chain onto the camshaft sprocket, making sure one marked chain link is on either side of the V8 mark on the sprocket and position the sprocket onto the camshaft.

17. Tighten the camshaft sprocket bolt to 90 ft. lbs. (122 Nm).

18. Install the right side secondary chain guide.

19. Install the right side cylinder head access plug.

20. Reset and install the right side secondary chain tensioner and remove Special Tool 8515.

21. The remainder of the installation is the reverse order of removal. Note the following torques:

    a. Crankshaft damper to 130 ft. lbs. (175 Nm)

    b. Cylinder head cover bolts to 105 inch lbs. (12 Nm)

    c. Radiator upper shroud to 95 inch lbs. (11 Nm)

22. Refill the cooling system to the correct level.

23. Start the engine and check for leaks.

Rocker Arms/Shafts

REMOVAL & INSTALLATION

3.7L and 4.7L Engines

1. Before servicing the vehicle, refer to the Precautions Section.
2. Remove or disconnect the following:
   - Negative battery cable
   - Cylinder head cover
3. Rotate the crankshaft so that the piston of the cylinder to be serviced is at Bottom Dead Center (BDC) and both valves are closed.
4. Use special tool 8516 to depress the valve and remove the rocker arm.
5. Repeat for each rocker arm to be serviced.

➡️**Keep valvetrain components in order for reassembly.**

To install:

6. Rotate the crankshaft so that the piston of the cylinder to be serviced is at BDC.
7. Compress the valve spring and install each rocker arm in its original position.
8. Repeat for each rocker arm to be installed.
9. Install or connect the following:
   - Cylinder head cover
   - Negative battery cable

Intake Manifold

REMOVAL & INSTALLATION

3.7L Engine

1. Before servicing the vehicle, refer to the Precautions Section.
2. Drain the cooling system.
3. Relieve the fuel system pressure.
4. Remove or disconnect the following:
   - Negative battery cable
   - Air cleaner assembly
   - Accelerator cable
   - Cruise control cable
5. Disconnect the electrical connectors for:

    a. Manifold Absolute Pressure (MAP) sensor

    b. Intake Air Temperature (IAT) sensor

    c. Throttle Position (TP) sensor

    d. Idle Air Control (IAC) valve

6. Remove or disconnect the following:
   - Positive Crankcase Ventilation (PCV) valve and hose
   - Canister purge vacuum line
   - Brake booster vacuum line
   - Cruise control servo hose
   - Accessory drive belt
   - Alternator electrical connections
   - Engine ground straps
   - Ignition coil towers

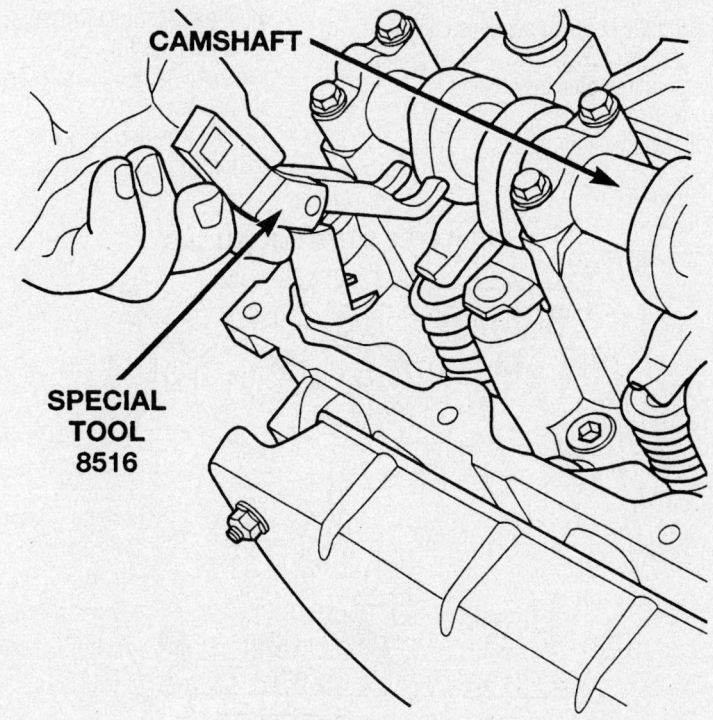

CAMSHAFT

SPECIAL TOOL 8516

Rocker arm service

9302PG13

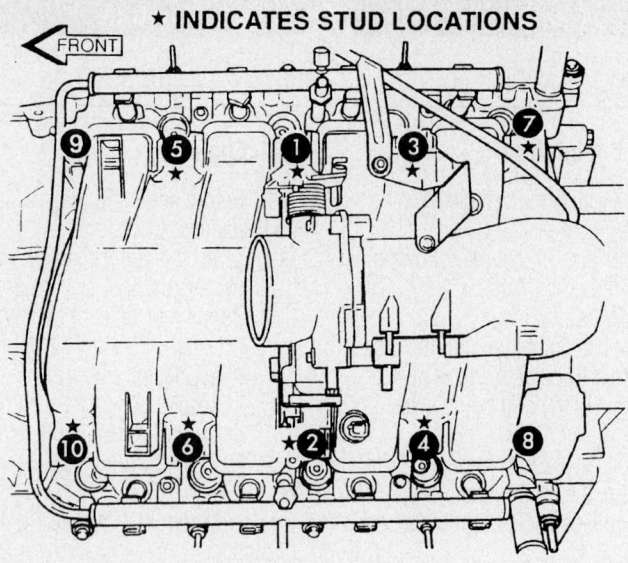

★ INDICATES STUD LOCATIONS

FRONT

9302PG14

Intake manifold torque sequence—3.7L engine

- Oil dipstick tube
- Fuel line
- Fuel rail
- Throttle body assembly
- Heater hoses
- Coolant temperature sensor connector
- Intake manifold. Remove the fasteners in reverse of the tightening sequence.

To install:

7. Install the intake manifold using new gaskets. Tighten the bolts, in sequence, to 105 inch lbs. (12 Nm).

8. The remainder of the installation is the reverse order of removal.

9. Start the engine and check for leaks.

4.7L Engine

1. Before servicing the vehicle, refer to the Precautions Section.

2. Drain the cooling system.

3. Relieve the fuel system pressure.

4. Remove or disconnect the following:
   - Negative battery cable
   - Wiper module
   - Air intake assembly
   - Accelerator cable
   - Cruise control cable
   - Manifold Absolute Pressure (MAP) sensor connector
   - Intake Air Temperature (IAT) sensor connector
   - Throttle Position (TP) sensor connector

- Idle Air Control (IAC) valve connector
- Engine Coolant Temperature (ECT) sensor
- Positive Crankcase Ventilation (PCV) valve and hose
- Alternator electrical connector
- A/C compressor connectors
- Canister purge vacuum line
- Brake booster vacuum line
- Cruise control servo hose
- Accessory drive belt
- Engine ground straps
- Ignition coil towers
- Oil dipstick tube
- Fuel supply rail
- Throttle body assembly and mounting bracket
- Heater hoses
- Coolant temperature sensor
- Intake manifold. Remove the fasteners in reverse of the tightening sequence.

To install:

5. Install the intake manifold using new gaskets. Tighten the bolts in sequence to 105 inch lbs. (12 Nm).

6. The remainder of installation is the reverse order of removal.

7. Fill the cooling system.

8. Start the engine and check for leaks.

Exhaust Manifold

REMOVAL & INSTALLATION

3.7L Engines

1. Before servicing the vehicle, refer to the Precautions Section.

2. Remove or disconnect the following:

- Negative battery cable
- Exhaust manifold heat shields
- Exhaust Gas Recirculation (EGR) tube
- Exhaust Y-pipe
- Exhaust manifolds

To install:

➡If the exhaust manifold studs came out with the nuts when removing the exhaust manifolds, replace them with new studs.

3. Install or connect the following:
   - Exhaust manifolds. Torque the fasteners to 20 ft. lbs. (27 Nm), starting with the center nuts and work out to the ends.
   - Exhaust Y-pipe
   - EGR tube

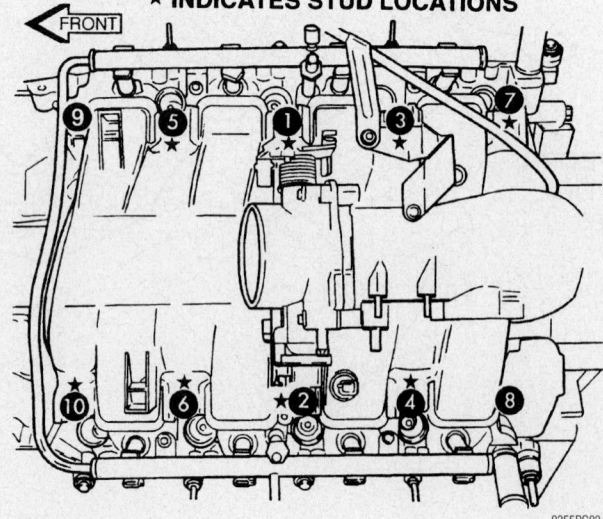

★ INDICATES STUD LOCATIONS

FRONT

9355PG02

Intake manifold torque sequence—4.7L

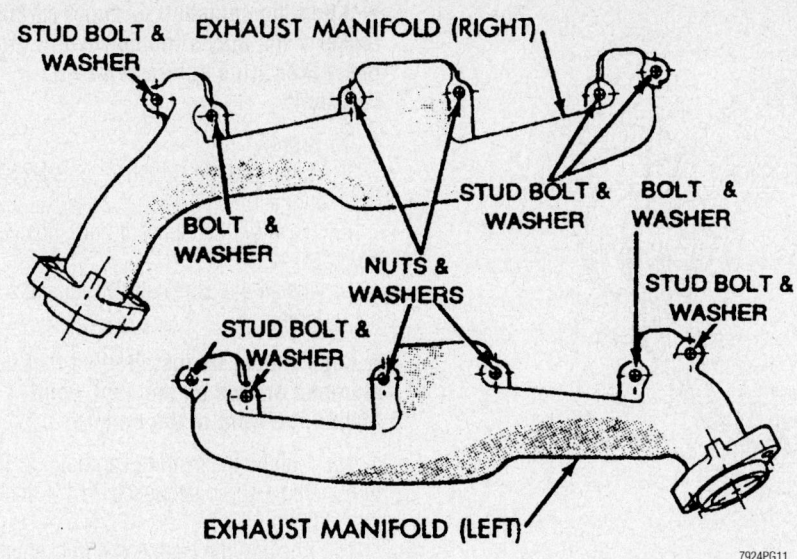

Exhaust manifold fastener locations—3.7L engines

- Exhaust manifold heat shields
- Negative battery cable

4. Start the engine, check for leaks and repair if necessary.

4.7L Engine

RIGHT

1. Before servicing the vehicle, refer to the Precautions Section.
2. Drain the cooling system.
3. Remove or disconnect the following:
   - Negative battery cable

- Air intake assembly
- Accessory drive belt
- A/C compressor
- A/C accumulator support bracket
- Heater hoses
- Exhaust manifold heat shields
- Exhaust down pipe from the manifold
- Starter motor
- Exhaust manifolds

To install:

4. Install or connect the following:
   - Exhaust manifolds, using new gas-

kets. Tighten the bolts to 18 ft. lbs. (25 Nm), starting with the inner bolts and work out to the ends.
- Manifold heat shield. Tighten to 72 inch lbs. (8 Nm) and then back off 45 degrees.
- Starter motor
- Exhaust down pipe to the manifold
- Heater hoses
- A/C accumulator bracket
- A/C compressor
- Accessory drive belt
- Air intake assembly
- Battery

5. Fill the cooling system.
6. Start the engine and check for leaks.

Camshaft

REMOVAL & INSTALLATION

3.7L Engines

1. Before servicing the vehicle, refer to the Precautions Section.
2. Disconnect the negative battery cable.
3. Remove the cylinder head cover.
4. Set the No. 1 cylinder to Top Dead Center (TDC). The camshaft sprocket alignment mark should be at the 12 o'clock position.

➡ **Keep all valvetrain components in order for assembly.**

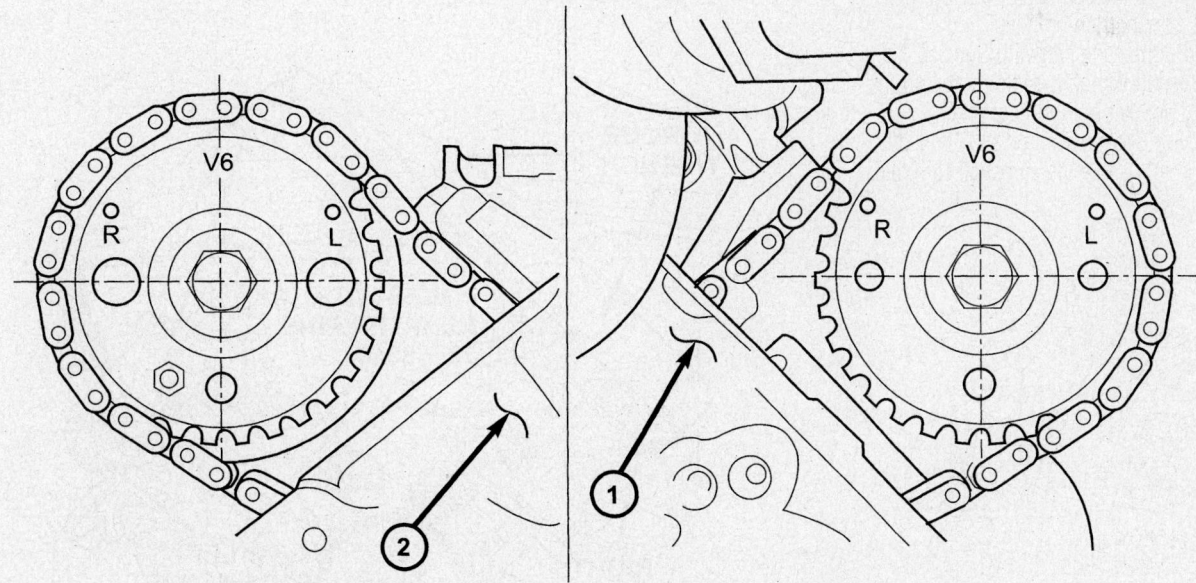

1 - LEFT CYLINDER HEAD
2 - RIGHT CYLINDER HEAD

Camshaft sprocket timing mark—3.7L Engine

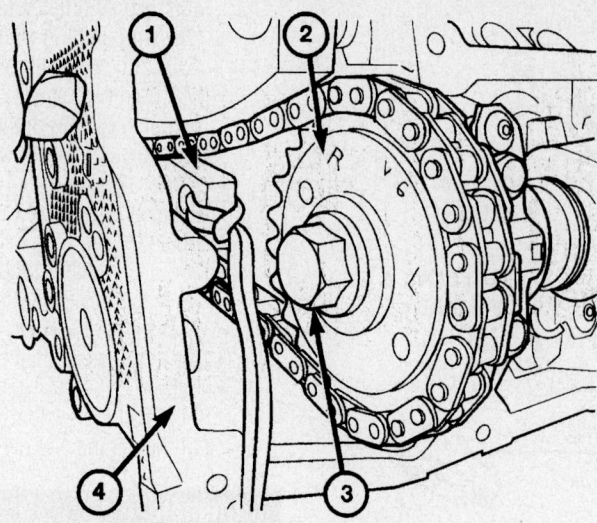

1 - SPECIAL TOOL 8379
2 - CAMSHAFT SPROCKET
3 - CAMSHAFT SPROCKET BOLT
4 - CYLINDER HEAD

09482_RAID_G0002

Chain Tensioner Retaining Wedge installation

5. Loosen, but DO NOT remove the camshaft sprocket retaining bolt. Leave the bolt snug against the sprocket.

6. Position Special Tool 8379 timing chain wedge between the timing chain strands, tap the tool to securely wedge the timing chain against the tensioner arm and guide.

7. Remove the camshaft position sensor, right camshaft only.

8. Hold the camshaft with Special Tool 8428 Camshaft Wrench, while removing the camshaft sprocket bolt and sprocket.

9. Using Special Tool 8428 Camshaft Wrench, gently allow the camshaft to rotate

5° clockwise until the camshaft is in the neutral position (no valve load).

10. Starting at the outside working inward, loosen the camshaft bearing cap retaining bolts ½ turn at a time. Repeat until the load is off the bearing caps.

11. Remove the camshaft bearing caps and camshaft.

➡When the camshaft is removed the rocker arms may slide downward. Mark the rocker arms before removing camshaft.

To install:

12. Lubricate the camshaft journals with clean engine oil

13. Position the camshaft into the cylinder head.

14. Install the camshaft bearing caps and hand tighten.

➡Caps should be installed so that the stamped arrows on the caps point toward the front of the engine.

15. Tighten the bearing cap bolts in ½ turn increments, in sequence, to 100 inch lbs. (11 Nm).

16. Position the camshaft sprocket into the timing chain aligning the alignment mark between the two marked chain links (Two links marked during removal).

17. Using Tool 8428 Camshaft Wrench, rotate the camshaft until the camshaft sprocket dowel is aligned with the slot in the camshaft sprocket. Install the sprocket onto the camshaft.

✷✷ CAUTION

Remove any excess oil from the camshaft sprocket bolt. Failure to do so can cause bolt over-torque.

18. Remove excess oil from bolt, then install the camshaft sprocket retaining bolt and hand tighten.

1 - Camshaft hole
2 - Special Tool 8428

09482_RAID_G0003

Hold the camshaft with Camshaft Wrench to remove

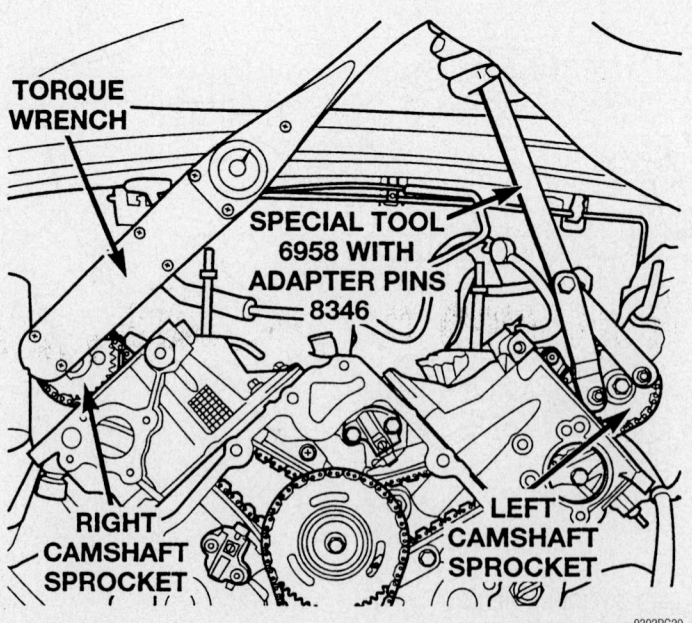

9302PG20

Hold the left camshaft sprocket with a spanner wrench while removing or installing the camshaft sprocket bolts—3.7L engine

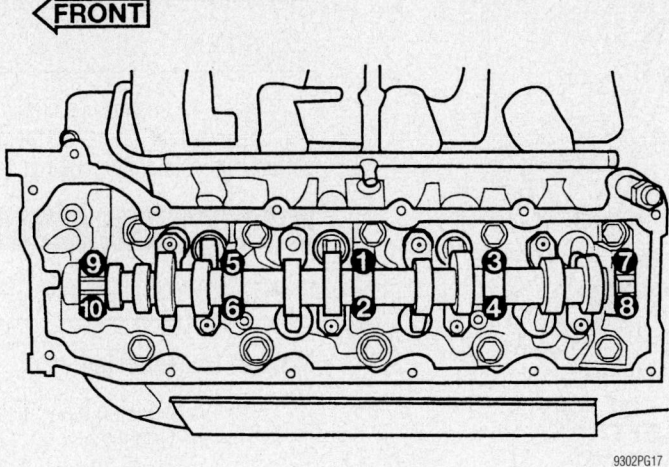

Camshaft bearing cap bolt tightening sequence—3.7L engine

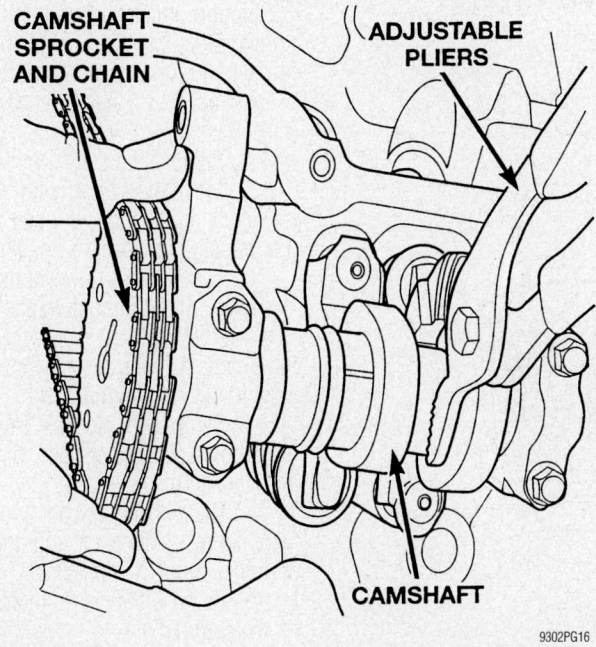

Turn the camshaft with pliers, if needed, to align the dowel in the sprocket—3.7L engine

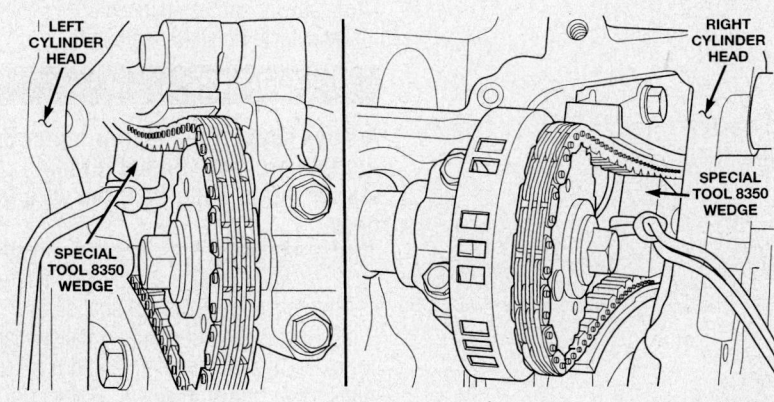

Chain Tensioner Retaining Wedges—4.7L engine

19. Remove Special Tool 8379 timing chain wedge.

20. Using Special Tool 6958 spanner wrench with adapter pins 8346, torque the camshaft sprocket retaining bolt to 90 ft. lbs. (122 Nm).

21. Install or connect the following:
  • Camshaft position sensor, right side only
  • Cylinder head cover

22. Start the engine and check for leaks.

4.7L Engine

1. Before servicing the vehicle, refer to the Precautions Section.

2. Remove or disconnect the following:
  • Negative battery cable
  • Cylinder head covers
  • Rocker arms
  • Hydraulic lash adjusters

➡**Keep all valvetrain components in order for assembly.**

3. Set the engine at Top Dead Center (TDC) of the compression stroke for the No. 1 cylinder.

4. Install Timing Chain Wedge 8350 to retain the chain tensioners.

5. Matchmark the timing chains to the camshaft sprockets.

6. Install Camshaft Holding Tool 6958 and Adapter Pins 8346 to the left camshaft sprocket.

7. Remove or disconnect the following:
  • Right camshaft timing sprocket and target wheel
  • Left camshaft sprocket
  • Camshaft bearing caps, by reversing the tightening sequence
  • Camshafts

To install:

8. Install or connect the following:
  • Camshafts. Tighten the bearing cap bolts in ½ turn increments, in sequence, to 100 inch lbs. (11 Nm).
  • Target wheel to the right camshaft
  • Camshaft timing sprockets and chains, by aligning the matchmarks

9. Remove the tensioner wedges and tighten the camshaft sprocket bolts to 90 ft. lbs. (122 Nm).

10. Install or connect the following:
  • Hydraulic lash adjusters in their original locations
  • Rocker arms in their original locations
  • Cylinder head covers
  • Negative battery cable

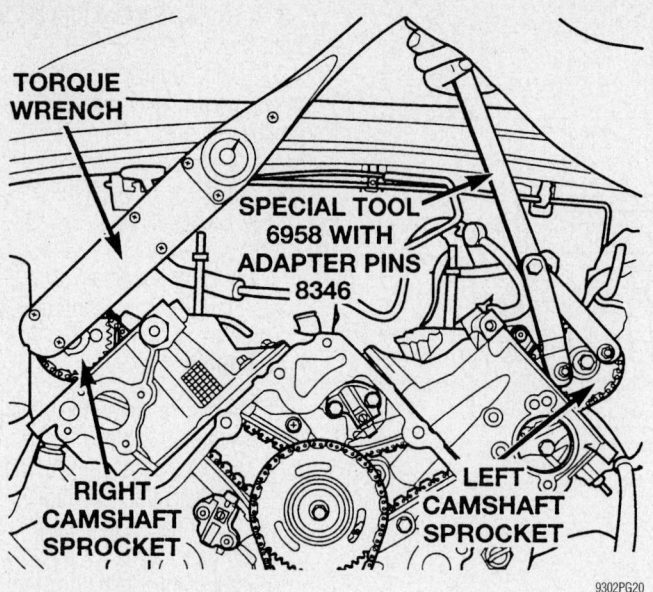

Hold the left camshaft sprocket with a spanner wrench while removing or installing the camshaft sprocket bolts—4.7L engine

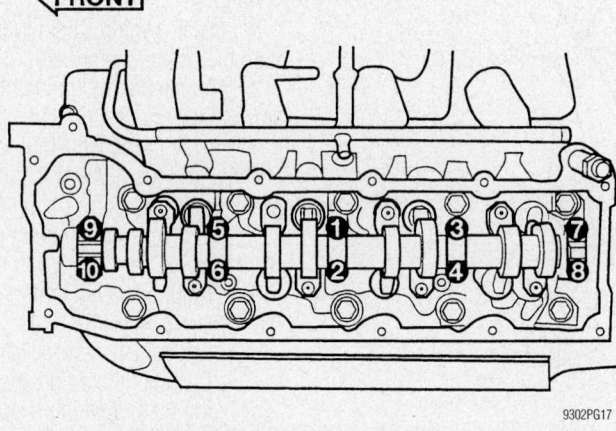

Camshaft bearing cap bolt tightening sequence—4.7L engine

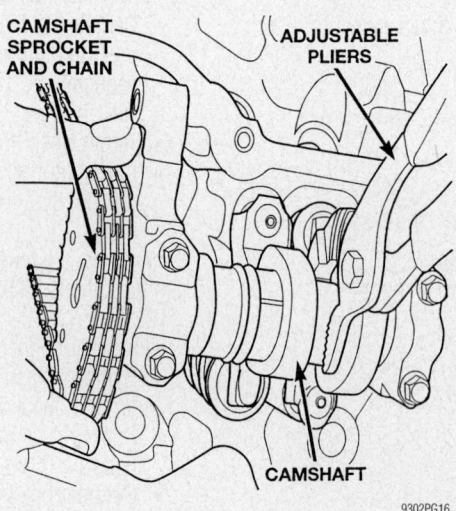

Turn the camshaft with pliers, if needed, to align the dowel in the sprocket—4.7L engine

Valve Lash

ADJUSTMENT

All gasoline engines covered in this section use hydraulic lifters. No maintenance or periodic adjustment is required.

Starter Motor

REMOVAL & INSTALLATION

Manual Transmission

1. Before servicing the vehicle, refer to the Precautions Section.
2. Disconnect and isolate negative battery cable.
3. If equipped with 4WD:
   a. Remove the bracket bolts for the support bracket between the front axle and side of the transmission.
   b. Pry the support bracket slightly to gain access to the starter lower mounting bolt.
4. Remove the starter mounting bolts.
5. Move the starter motor towards front of vehicle far enough for nose of starter pinion housing to clear the housing.
6. Tilt the nose downwards and lower starter motor far enough to access and remove the nut that secures the battery positive cable wire harness connector eyelet to solenoid battery terminal stud.
7. Remove the positive cable wire harness connector eyelet from solenoid battery terminal stud.
8. Disconnect the battery positive cable wire harness connector from the solenoid terminal connector receptacle.
9. Remove the starter motor.
To install:
10. Position starter motor to transmission housing.
11. Connect the battery cable solenoid terminal wire harness connector to connector receptacle on starter solenoid.

✳✳ CAUTION

Always support the starter motor during this process. Do not let the starter motor hang from the wire harness.

12. Install the battery cable eyelet terminal onto solenoid B (+) terminal stud.
13. Install the nut securing battery cable eyelet terminal to starter solenoid B (+) terminal stud. Tighten nut to 13.6 Nm (120 inch lbs.).

14. Position the starter motor over stud on transmission housing.

15. Loosely install the washers, bolt, and nut to starter. Tighten bolt and nut to 67.8 Nm (50 ft. lbs.).

16. Connect the negative battery cable.

Automatic Transmission

1. Before servicing the vehicle, refer to the Precautions Section.

2. Disconnect and isolate negative battery cable.

3. If equipped with 4WD:

a. Remove the bracket bolts for the support bracket between the front axle and side of the transmission.

b. Pry the support bracket slightly to gain access to the starter lower mounting bolt.

4. Remove mounting bolts (rearward facing) securing starter motor to the transmission housing.

❋❋ CAUTION

Always support starter motor during this process. Do not let starter motor hang from wire harness.

5. Lower starter motor from front of transmission housing far enough to access and remove nut securing battery positive cable eyelet terminal to the starter solenoid B (+) terminal stud.

6. Remove the battery cable eyelet terminal from solenoid B (+) terminal stud.

7. Disconnect the battery cable solenoid terminal wire harness connector from receptacle on starter solenoid.

8. Remove the starter motor.

To install:

9. Position the starter motor to transmission housing.

10. Connect battery cable solenoid terminal wire harness connector to connector receptacle on starter solenoid.

11. Install the battery cable eyelet terminal onto solenoid B (+) terminal stud.

12. Install and tighten nut securing battery cable eyelet terminal to starter solenoid B (+) terminal stud. Tighten nut to 13.6 Nm (120 inch lbs.).

13. Position starter motor to transmission housing and loosely install two bolts/washers.

14. Tighten bolts to 67.8 Nm (50 ft. lbs.).

15. Connect negative battery cable.

Oil Pan

REMOVAL & INSTALLATION

3.7L Engine

1. Before servicing the vehicle, refer to the Precautions Section.

2. Drain the engine oil.

3. Disconnect the negative battery cable.

4. Install engine support fixture 8354. Do not raise engine at this time.

5. Loosen both left and right side engine mount through bolts. Do not remove bolts.

6. Remove the structural dust cover, if equipped.

7. Remove the front crossmember.

8. Raise engine to provide clearance to remove oil pan.

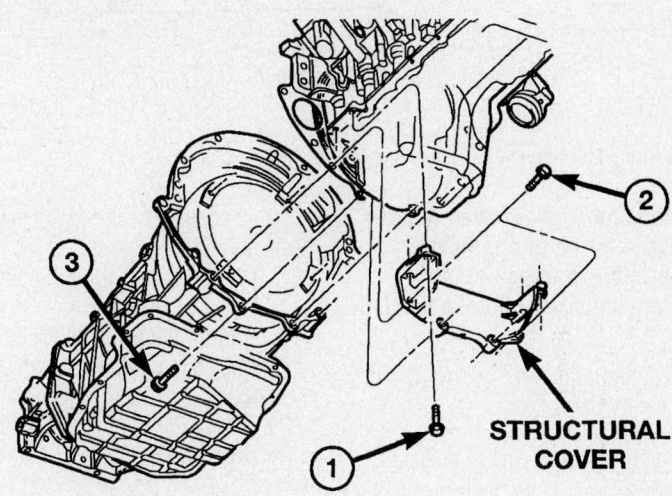

1 - BOLT
2 - BOLT
3 - BOLT

Structural cover torque sequence—3.7L and 4.7L engines

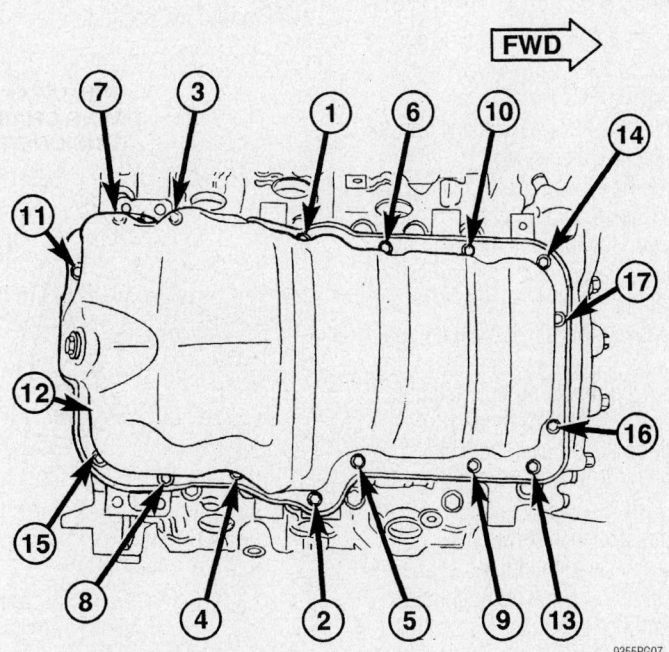

Oil pan bolt torque sequence—3.7L Engine

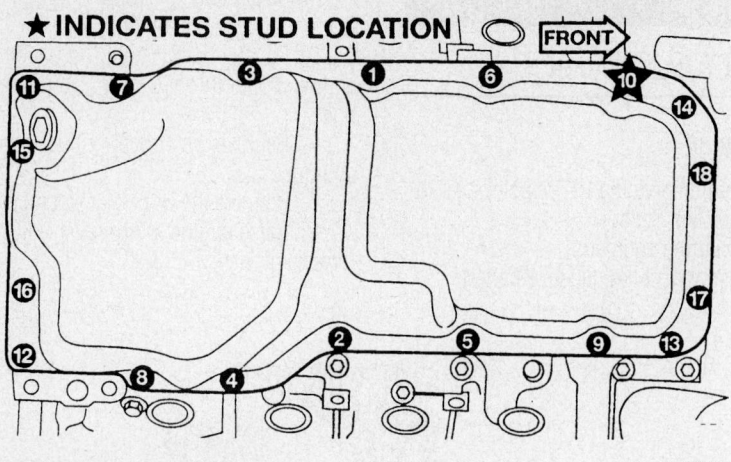

★ **INDICATES STUD LOCATION**

FRONT

Oil pan mounting bolt tightening sequence—4.7L engine

9302PG21

➡ **Raise the engine just enough to provide clearance for oil pan removal. Check for proper clearance at fan shroud to fan and cowl to intake manifold.**

9. If equipped with 4WD, remove or disconnect the following:
  • Pinion bracket
  • Front driveshaft at the front axle
  • Front axle mounting bolts
  • Lower the front axle using a suitable jack.
10. Remove the oil pan mounting bolts and oil pan.

➡ **Do not pry on oil pan or oil pan gasket. Gasket is integral to engine windage tray and does not come out with oil pan.**

11. Unbolt oil pump pickup tube and remove tube.
To install:
12. Inspect the integral windage tray and gasket and replace as needed.
13. Clean the oil pan gasket mating surface of the bedplate and oil pan.
14. Position the integrated oil pan gasket/windage tray assembly.
15. Install the oil pickup tube
16. If removed, install stud at position No. 9.
17. Install the mounting bolt and nuts. Tighten nuts to 28 Nm (20 ft. lbs.).
18. Position the oil pan and install the mounting bolts. Tighten the mounting bolts to 15 Nm (11 ft. lbs.) in the sequence shown.
19. Lower the engine into mounts.
20. Install both the left and right side engine mount through bolts. Tighten the nuts to 68 Nm (50 ft. lbs.).
21. Remove the lifting device.

22. If equipped with 4WD, install or connect the following:
  • Front axle mounting bolts
  • Pinion bracket
  • Front driveshaft to front axle
23. Install structural dust cover, if equipped.
24. Install the front crossmember.
25. Refill the engine oil to the correct level.
26. Reconnect the negative battery cable.
27. Start engine and check for leaks.

Oil Pump

REMOVAL & INSTALLATION

1. Before servicing the vehicle, refer to the Precautions Section.

2. Drain the engine oil.
3. Remove or disconnect the following:
  • Oil Pan
  • Timing chain cover
  • Timing chains and tensioners
  • Oil pump
4. Installation is the reverse of removal. Torque the pump bolts, in sequence, to 21 ft. lbs. (28 Nm).
5. Refill the engine oil to the correct level.

Rear Main Seal

REMOVAL & INSTALLATION

1. Before servicing the vehicle, refer to the Precautions Section.
2. Remove or disconnect the following:
  • Transmission
  • Flexplate

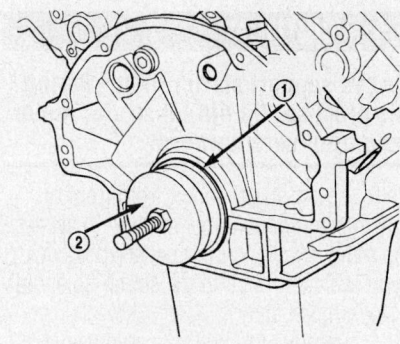

09482_RAID_G0005

Removing the rear main seal (1) with special tool 8506 (2)

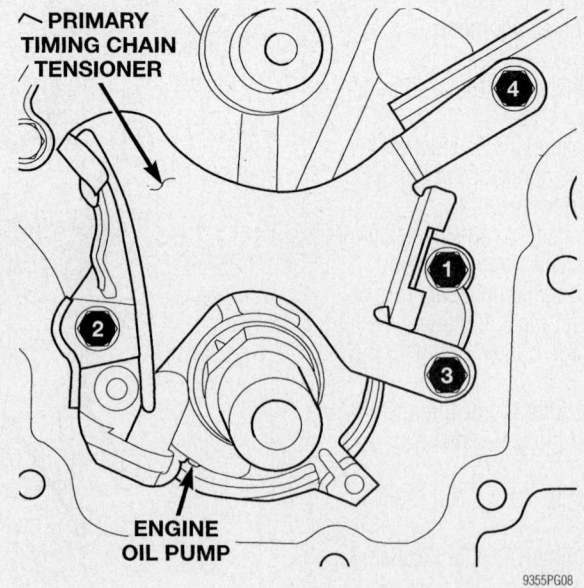

PRIMARY TIMING CHAIN TENSIONER

ENGINE OIL PUMP

9355PG08

Oil pump bolt torque sequence

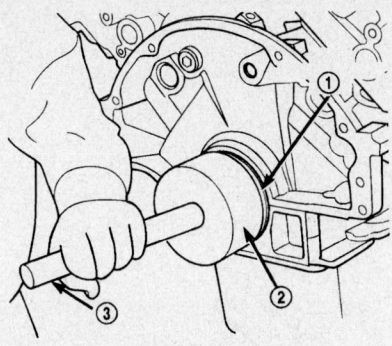

Installing rear main seal using Special Tool 8349 (2) and C-4171 (3)

09482_RAID_G0004

3. Thread Oil Seal Remover 8506 into the rear main seal as far as possible and remove the rear main seal.

To install:

4. Install or connect the following:
- Seal Guide 8349-2 onto the crankshaft
- Rear main seal on the seal guide
- Rear main seal, using the Crankshaft Rear Oil Seal Installer 8349 and Driver Handle C-4171; tap it into place until the installer is flush with the cylinder block
- Flexplate. Tighten the bolts to 45 ft. lbs. (60 Nm).
- Transmission

5. Start the engine and check for leaks.

Timing Chain, Sprockets, Front Cover and Seal

REMOVAL & INSTALLATION

3.7L Engines

1. Before servicing the vehicle, refer to the Precautions Section.
2. Drain the cooling system.
3. Remove or disconnect the following:
- Negative battery cable
- Cylinder head covers
- Radiator fan

4. Rotate the crankshaft so that the crankshaft timing mark aligns with the Top Dead Center (TDC) mark on the front cover, and the **V6** marks on the camshaft sprockets are at 12 o'clock.
- Power steering pump
- Access plugs from the cylinder heads
- Oil fill housing
- Crankshaft damper

5. Compress the primary timing chain tensioner and install a lockpin.
6. Remove the secondary timing chain tensioners.
7. Hold the left camshaft with adjustable pliers and remove the sprocket and chain. Rotate the **left** camshaft 5 degrees **clockwise** to the neutral position.
8. Hold the right camshaft with

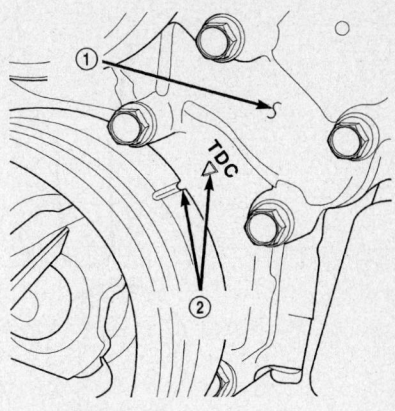

1 - TIMING CHAIN COVER
2 - CRANKSHAFT TIMING MARKS

9355PG10

Crankshaft timing marks—3.7L

adjustable pliers and remove the camshaft sprocket. Rotate the **right** camshaft 45 degrees **counterclockwise** to the neutral position.

9. Remove the primary timing chain and sprockets.

To install:

10. Use a small prytool to hold the ratchet pawl and compress the secondary timing chain tensioners in a vise and install locking pins.

➡**The black bolts fasten the guide to the engine block and the silver bolts fasten the guide to the cylinder head.**

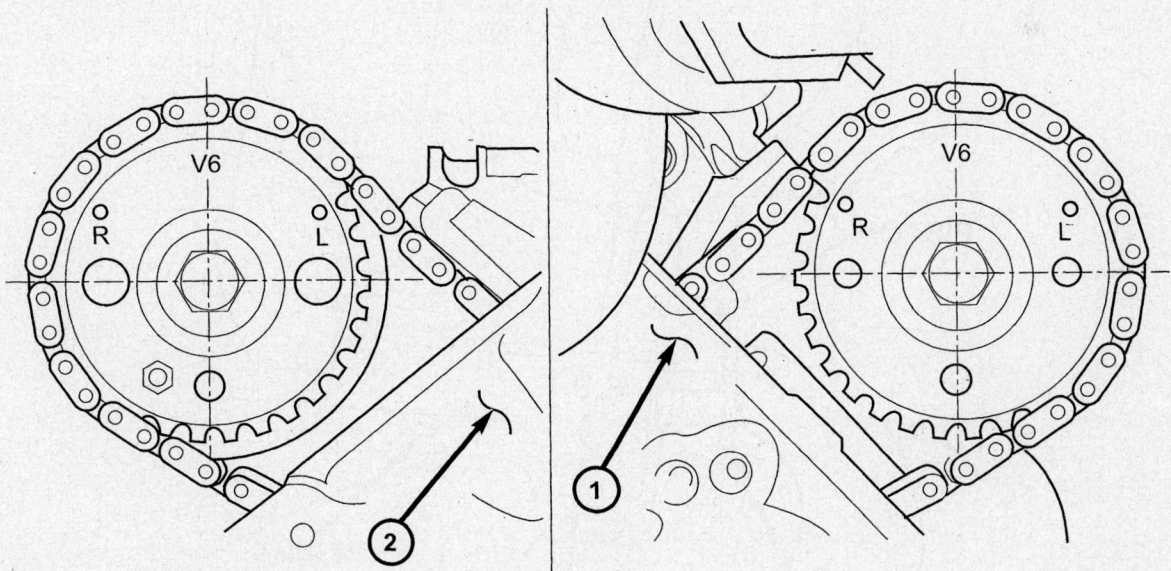

1 - LEFT CYLINDER HEAD
2 - RIGHT CYLINDER HEAD

9355PG09

Camshaft sprocket timing marks—3.7L

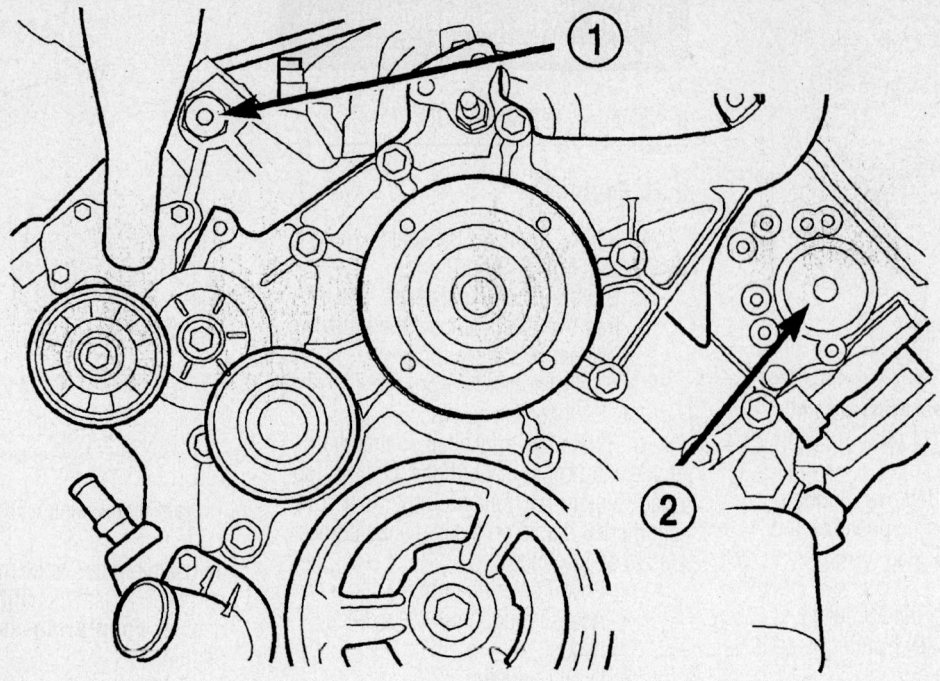

1 - RIGHT CYLINDER HEAD ACCESS PLUG
2 - LEFT CYLINDER HEAD ACCESS PLUG

9355PG11

Cylinder head access plugs—3.7L

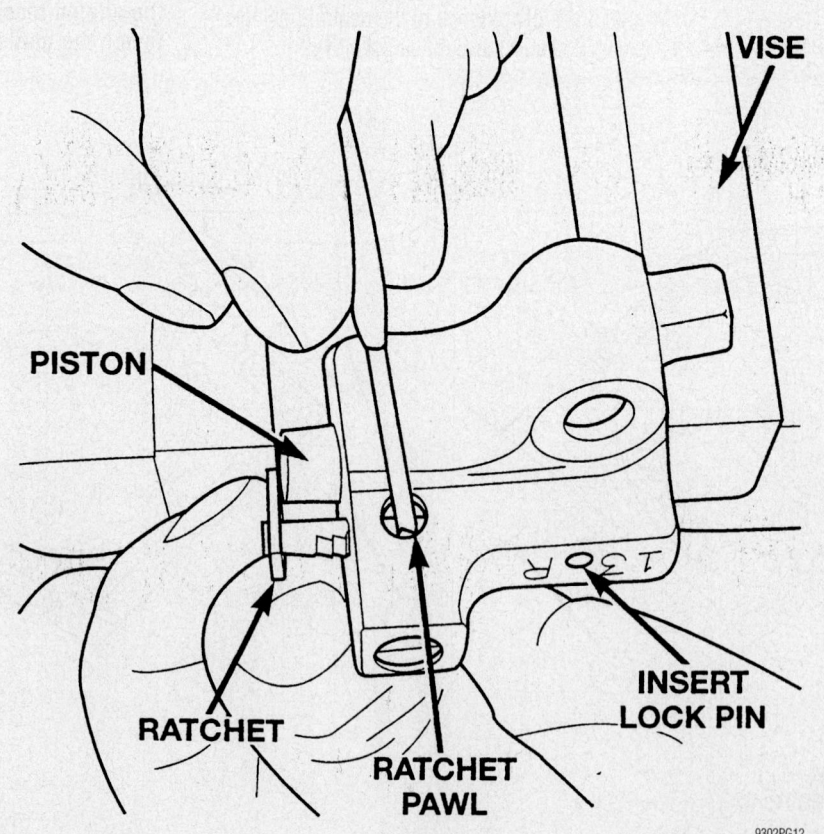

Secondary timing chain tensioner preparation—3.7L engine

9302PG12

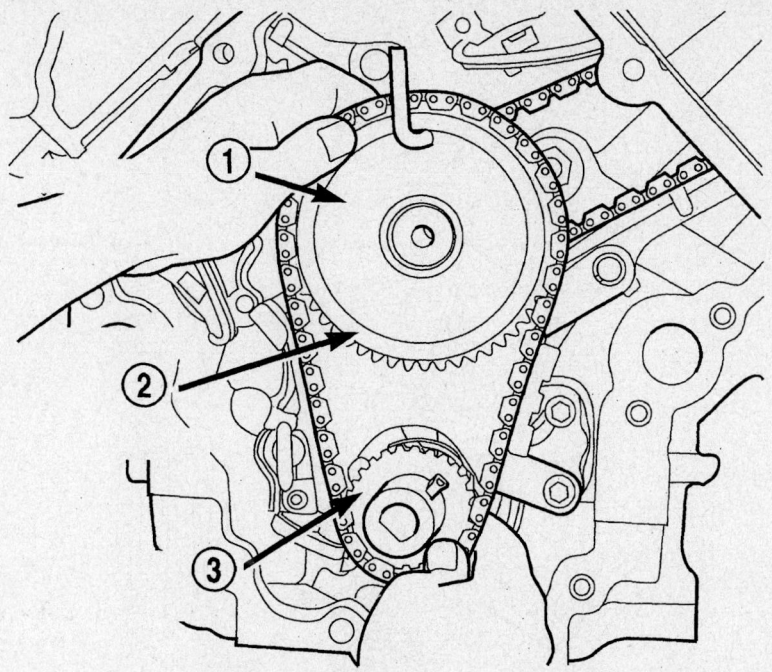

1 - SPECIAL TOOL 8429
2 - PRIMARY CHAIN IDLER SPROCKET
3 - CRANKSHAFT SPROCKET

9355PG12

Installing the idler gear and timing chains—3.7L

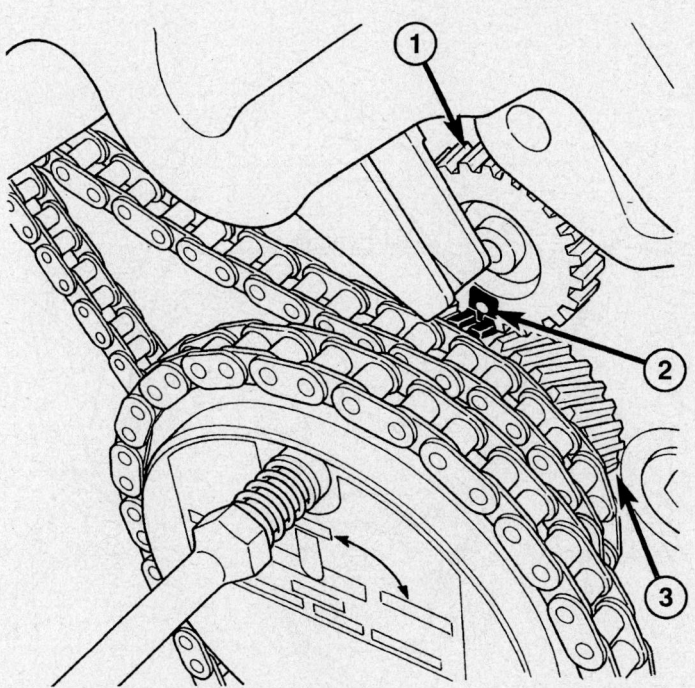

1 - COUNTERBALANCE SHAFT

2 - TIMING MARKS

3 - IDLER SPROCKET

9355PG13

Counterbalance shaft timing marks—3.7L

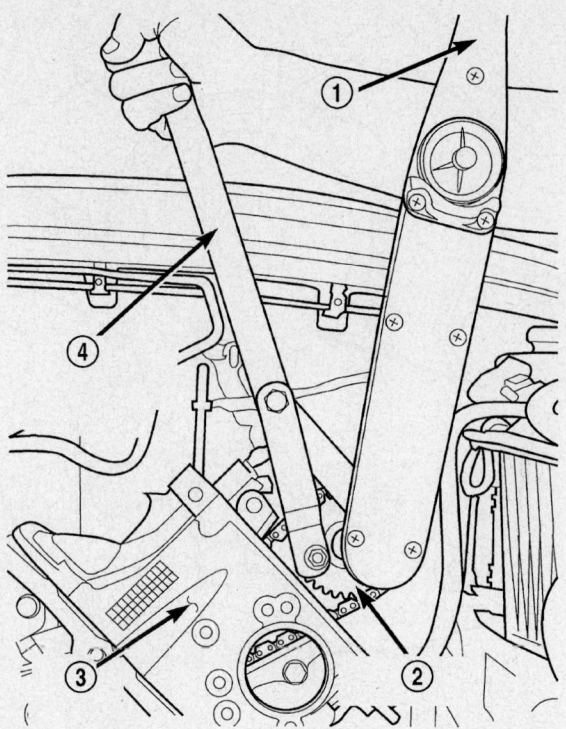

1 - TORQUE WRENCH
2 - CAMSHAFT SPROCKET
3 - LEFT CYLINDER HEAD
4 - SPECIAL TOOL 6958 SPANNER WITH ADAPTER PINS 8346

9355PG14

Tightening the left side camshaft sprocket—3.7L

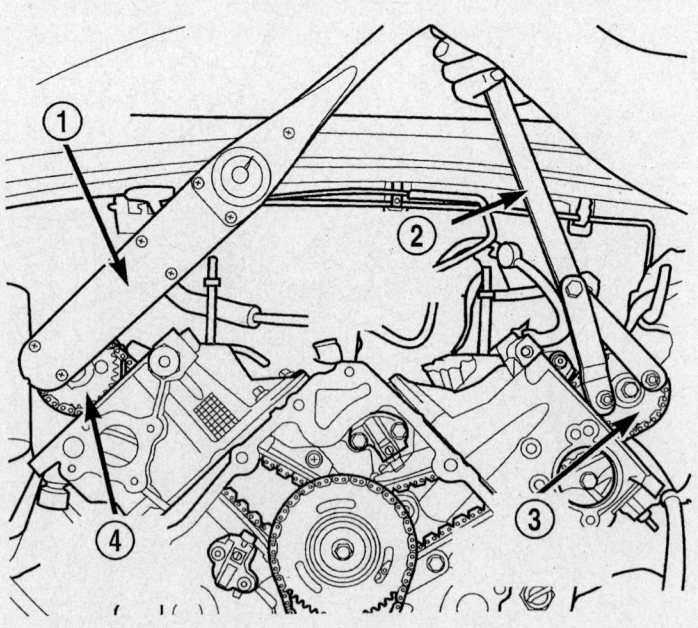

1 - TORQUE WRENCH
2 - SPECIAL TOOL 6958 WITH ADAPTER PINS 8346
3 - LEFT CAMSHAFT SPROCKET
4 - RIGHT CAMSHAFT SPROCKET

9355PG15

Tightening the right side camshaft sprocket—3.7L

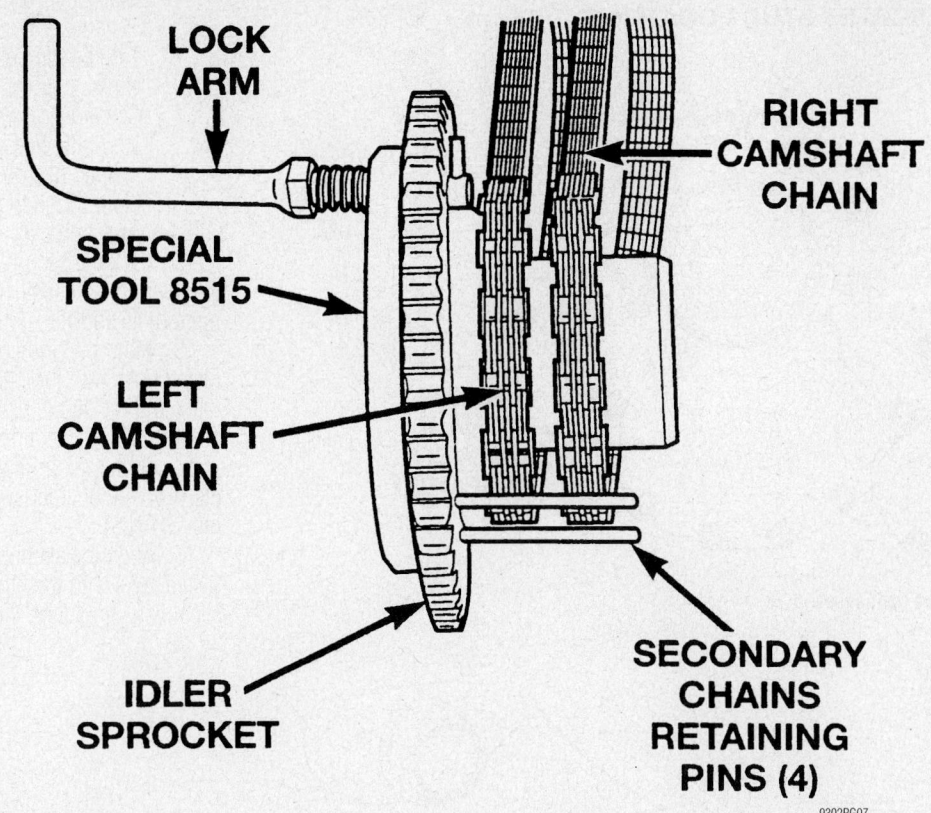

Use the Timing Chain Locking tool to lock the timing chains on the idler gear—3.7L engine

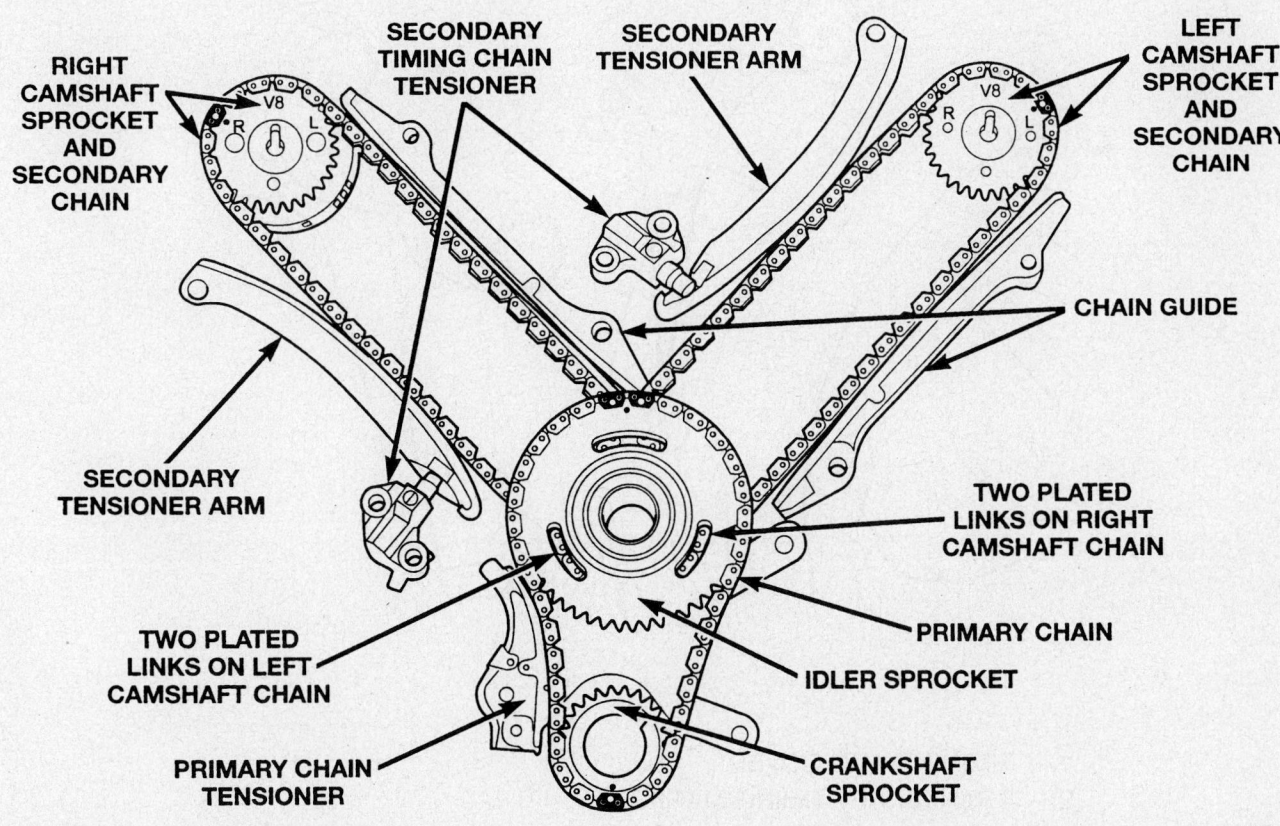

Timing chain system and alignment marks—3.7L engine

★ **INDICATES STUD LOCATIONS**

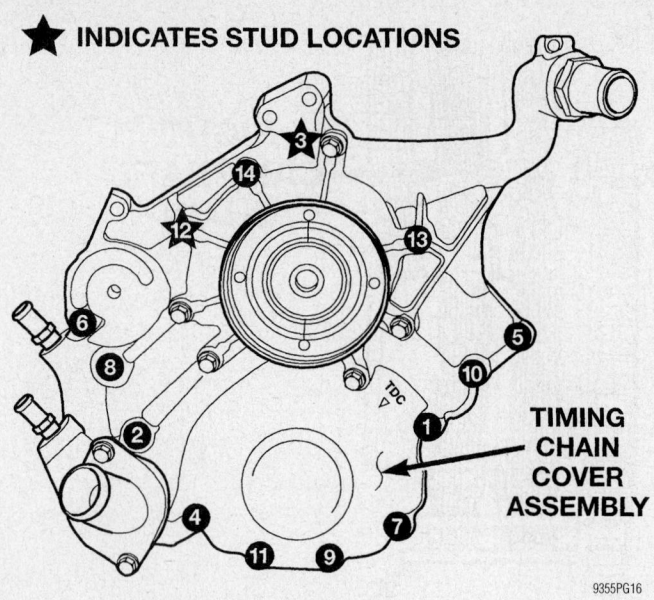

Timing cover bolt torque sequence—3.7L

9355PG16

11. Install or connect the following:
- Secondary timing chain guides. Tighten the bolts to 21 ft. lbs. (28 Nm).
- Secondary timing chains to the idler sprocket so that the double plated links on each chain are visible through the slots in the primary idler sprocket

12. Lock the secondary timing chains to the idler sprocket with Timing Chain Locking tool as shown.

13. Align the primary chain double plated links with the idler sprocket timing mark and the single plated link with the crankshaft sprocket timing mark.

14. Install the primary chain and sprockets. Tighten the idler sprocket bolt to 25 ft. lbs. (34 Nm).

15. Align the secondary chain single plated links with the timing marks on the secondary sprockets. Align the dot at the **L**

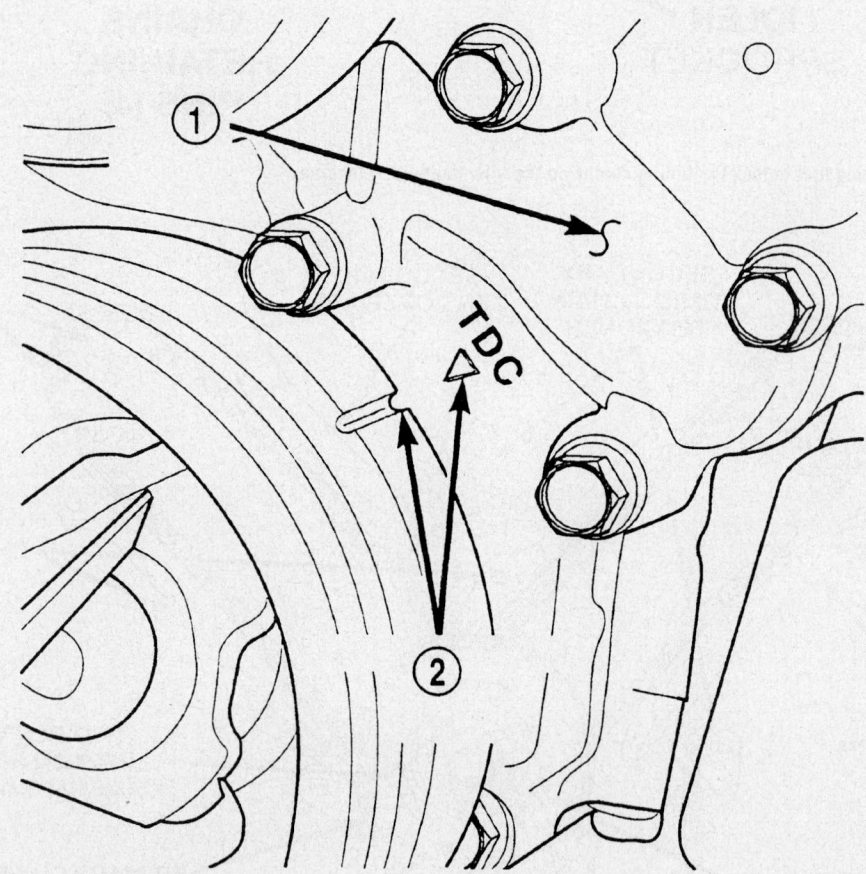

1 – TIMING CHAIN COVER
2 – CRANKSHAFT TIMING MARKS

9308PG04

Crankshaft timing marks—4.7L engine

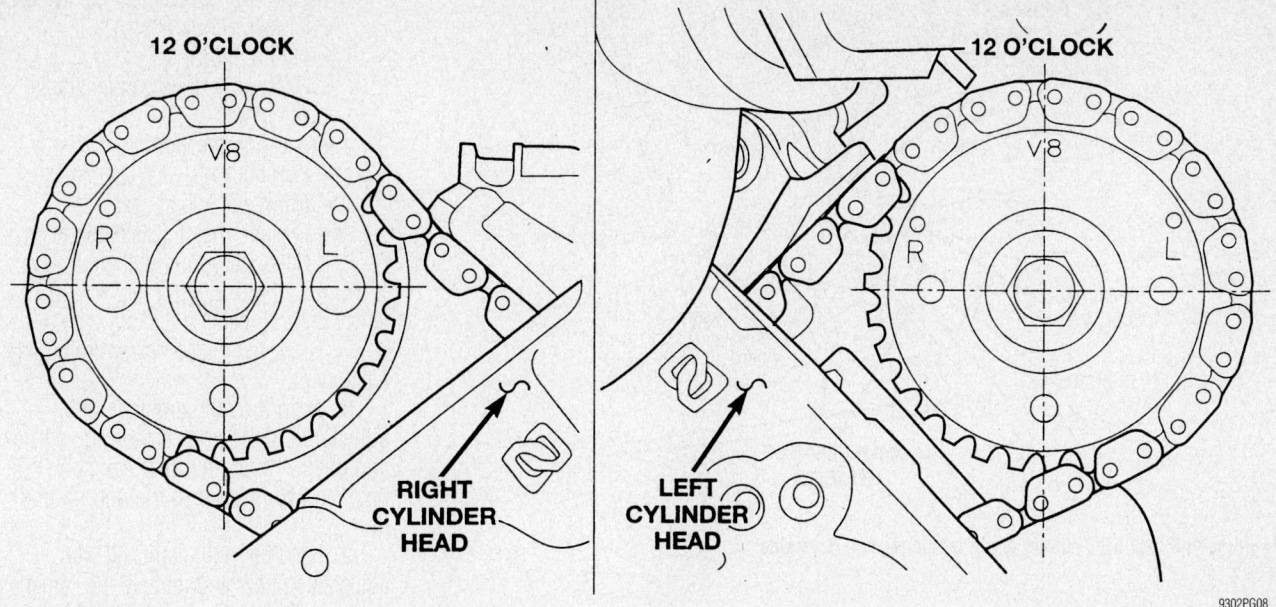

Camshaft positioning—4.7L engine

mark on the left sprocket with the plated link on the left chain and the dot at the **R** mark on the right sprocket with the plated link on the right chain.

16. Rotate the camshafts back from the neutral position and install the camshaft sprockets.

17. Remove the secondary chain locking tool.

18. Remove the primary and secondary timing chain tensioner locking pins.

19. Hold the camshaft sprockets with a spanner wrench and tighten the retaining bolts to 90 ft. lbs. (122 Nm).

20. Install or connect the following:
- Front cover. Tighten the bolts, in sequence, to 40 ft. lbs. (54 Nm).
- Front crankshaft seal
- Cylinder head access plugs
- A/C compressor
- Alternator
- Accessory drive belt tensioner. Tighten the bolt to 40 ft. lbs. (54 Nm).
- Oil fill housing
- Crankshaft damper. Tighten the bolt to 130 ft. lbs. (175 Nm).
- Power steering pump
- Lower radiator hose
- Heater hoses
- Accessory drive belt
- Engine cooling fan and shroud
- Camshaft Position (CMP) sensor
- Valve covers
- Negative battery cable

21. Fill and bleed the cooling system.

22. Start the engine, check for leaks and repair if necessary.

4.7L Engine

1. Before servicing the vehicle, refer to the Precautions Section.

2. Drain the cooling system.

3. Remove or disconnect the following:
- Negative battery cable
- Valve covers
- Camshaft Position (CMP) sensor
- Engine cooling fan and shroud
- Accessory drive belt
- Heater hoses
- Lower radiator hose
- Power steering pump

4. Rotate the crankshaft so that the crankshaft timing mark aligns with the Top Dead Center (TDC) mark on the front cover, and the **V8** marks on the camshaft sprockets are at 12 o'clock.

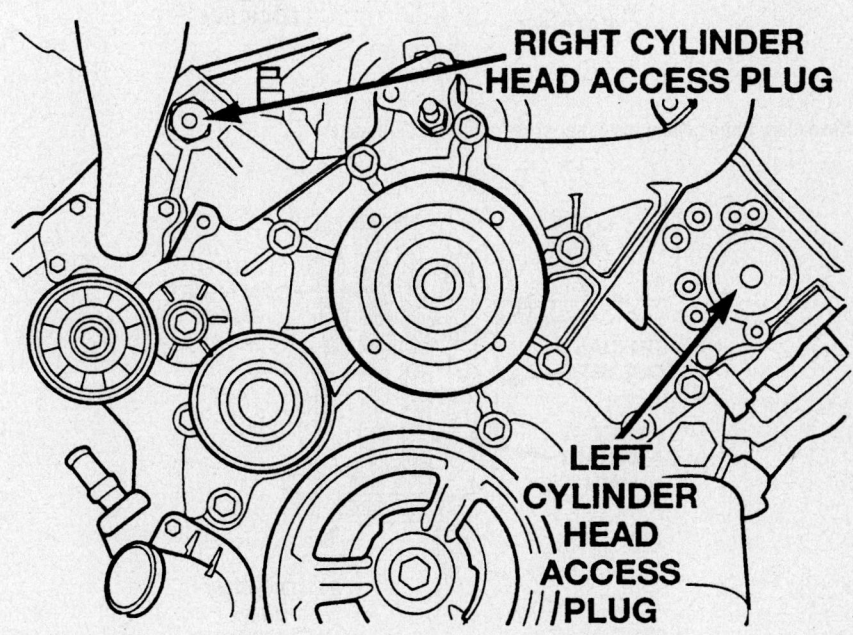

Cylinder head access plug locations—4.7L engine

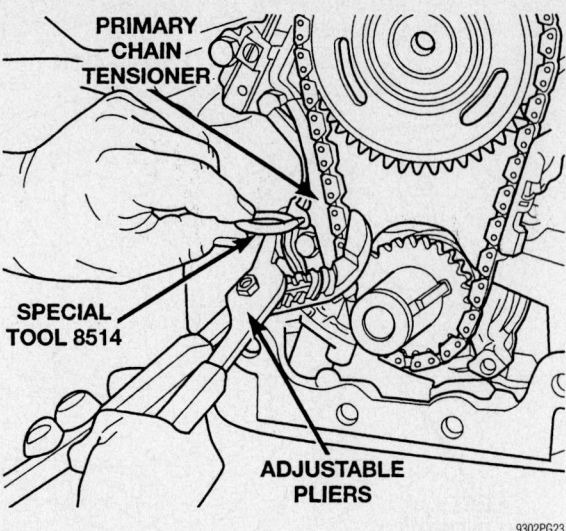

Compress and lock the primary chain tensioner—4.7L engine

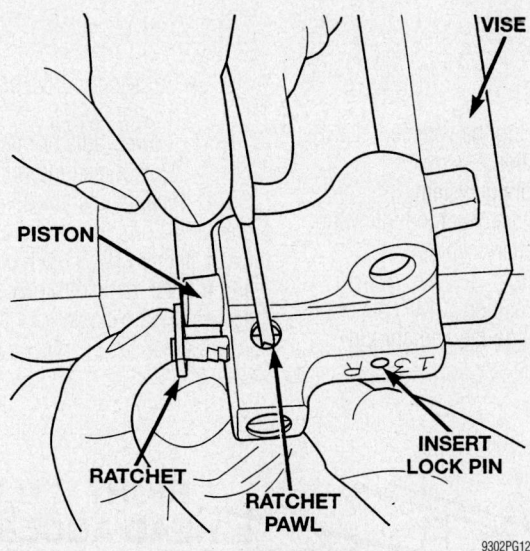

Secondary timing chain tensioner preparation—4.7L engine

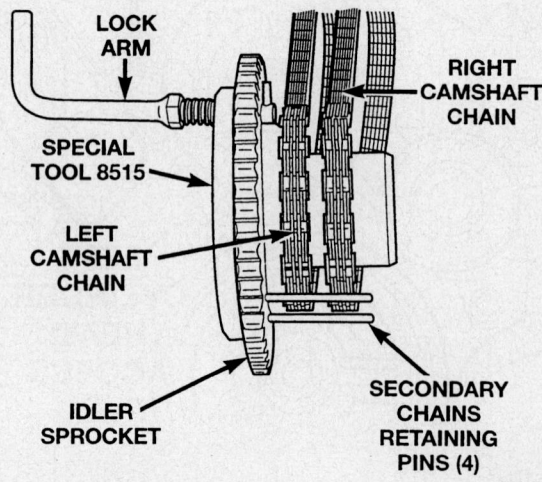

Use the Timing Chain Locking tool to lock the timing chains on the idler gear—4.7L engine

5. Remove or disconnect the following:
   - Crankshaft damper
   - Oil fill housing
   - Accessory drive belt tensioner
   - Alternator
   - A/C compressor
   - Front cover
   - Front crankshaft seal
   - Cylinder head access plugs
   - Secondary timing chain guides

6. Compress the primary timing chain tensioner and install a lockpin.

7. Remove the secondary timing chain tensioners.

8. Hold the left camshaft with adjustable pliers and remove the sprocket and chain. Rotate the **left** camshaft 15 degrees **clockwise** to the neutral position.

9. Hold the right camshaft with adjustable pliers and remove the camshaft sprocket. Rotate the **right** camshaft 45 degrees **counterclockwise** to the neutral position.

10. Remove the primary timing chain and sprockets.

To install:

11. Use a small prytool to hold the ratchet pawl and compress the secondary timing chain tensioners in a vise and install locking pins.

➡ **The black bolts fasten the guide to the engine block and the silver bolts fasten the guide to the cylinder head.**

12. Install or connect the following:
   - Secondary timing chain guides. Tighten the bolts to 21 ft. lbs. (28 Nm).
   - Secondary timing chains to the idler sprocket so that the double plated links on each chain are visible through the slots in the primary idler sprocket

13. Lock the secondary timing chains to the idler sprocket with Timing Chain Locking tool 8515 as shown.

14. Align the primary chain double plated links with the idler sprocket timing mark and the single plated link with the crankshaft sprocket timing mark.

15. Install the primary chain and sprockets. Tighten the idler sprocket bolt to 25 ft. lbs. (34 Nm).

16. Align the secondary chain single plated links with the timing marks on the secondary sprockets. Align the dot at the **L** mark on the left sprocket with the plated link on the left chain and the dot at the **R** mark on the right sprocket with the plated link on the right chain.

17. Rotate the camshafts back from the

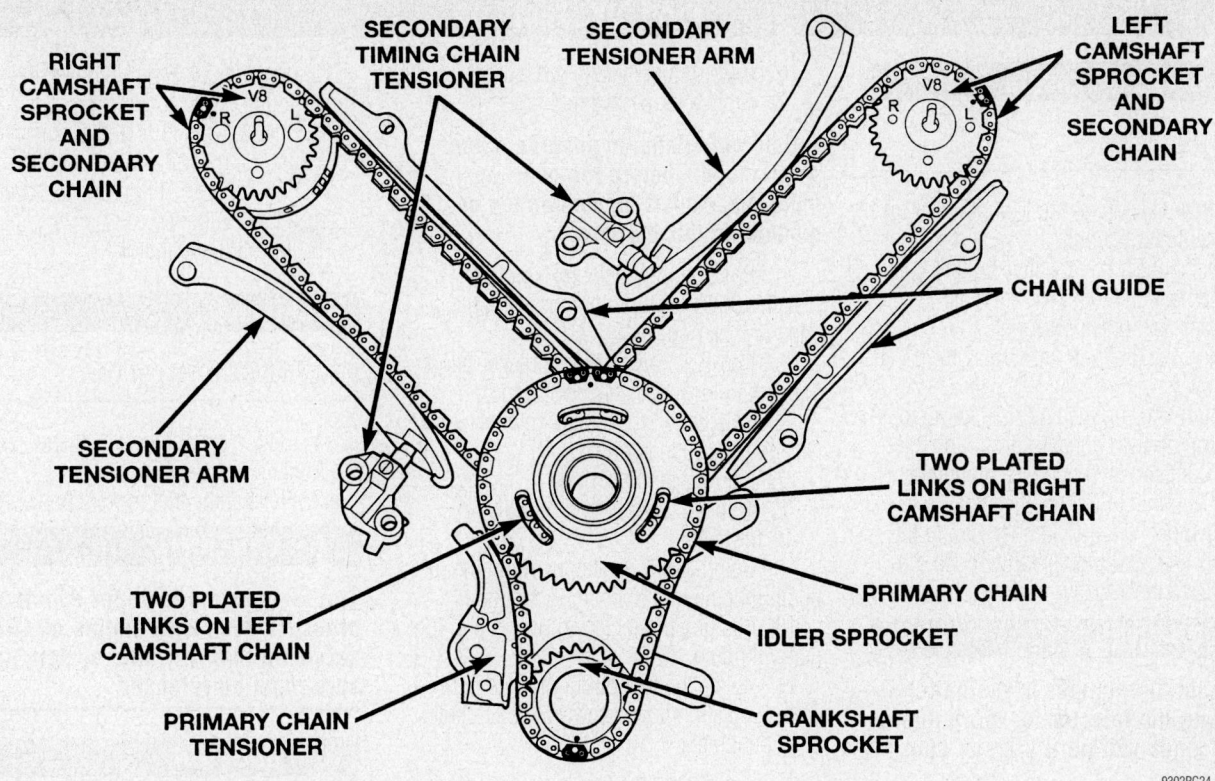

Timing chain system and alignment marks—4.7L engine

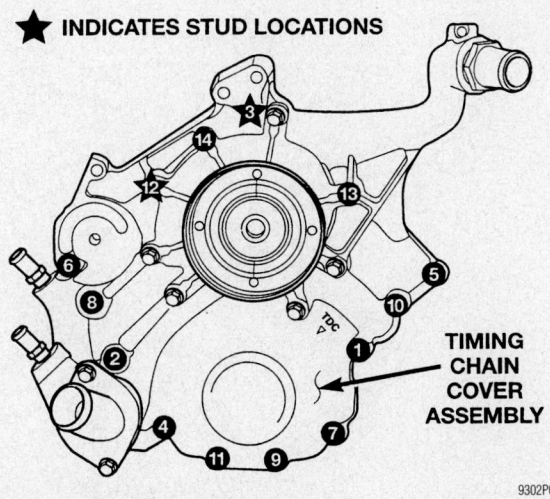

★ INDICATES STUD LOCATIONS

Timing chain cover bolt torque sequence—4.7L engine

neutral position and install the camshaft sprockets.

18. Remove the secondary chain locking tool.

19. Remove the primary and secondary timing chain tensioner locking pins.

20. Hold the camshaft sprockets with a spanner wrench and tighten the retaining bolts to 90 ft. lbs. (122 Nm).

21. Install or connect the following:
  • Front cover. Tighten the bolts, in sequence, to 40 ft. lbs. (54 Nm).

  • Front crankshaft seal
  • Cylinder head access plugs
  • A/C compressor
  • Alternator
  • Accessory drive belt tensioner. Tighten the bolt to 40 ft. lbs. (54 Nm).
  • Oil fill housing
  • Crankshaft damper. Tighten the bolt to 130 ft. lbs. (175 Nm).
  • Power steering pump
  • Lower radiator hose

  • Heater hoses
  • Accessory drive belt
  • Engine cooling fan and shroud
  • Camshaft Position (CMP) sensor
  • Valve covers
  • Negative battery cable
22. Fill the cooling system.
23. Start the engine and check for leaks.

Piston and Ring

POSITIONING

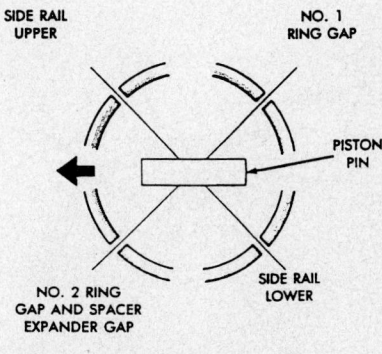

Piston ring end-gap spacing. Position raised "F" on piston towards front of engine—3.7L and 4.7L engines

FUEL SYSTEM

Fuel System Pressure

RELIEVING

1. Before servicing the vehicle, refer to the Precautions Section.
2. Disconnect the negative battery cable.
3. Remove the fuel tank filler cap to release any fuel tank pressure.
4. Remove the fuel pump relay from the PDC.
5. Start and run the engine until it stops.
6. Unplug the connector from any injector and connect a jumper wire from either injector terminal to the positive battery terminal. Connect another jumper wire to the other terminal and momentarily touch the other end to the negative battery terminal.

⚙ WARNING

Just touch the jumper to the battery. Powering the injector for more than a few seconds will permanently damage it.

7. Place a rag below the quick-disconnect coupling at the fuel rail and disconnect it.

Fuel Filter

REMOVAL & INSTALLATION

The fuel filter mounts inside the fuel pump module and is a non-serviceable part.

Fuel Pump

REMOVAL & INSTALLATION

1. Before servicing the vehicle, refer to the Precautions Section.

2. Release fuel system pressure.
3. Drain and remove fuel tank.

➡ **Note the rotational position of fuel pump module before removal. An indexing arrow is located on top of module for this purpose.**

4. Position the Locking Remover/Installer 9340 Tool into notches on outside edge of lock ring.
5. Install a ½ inch drive breaker bar to Locking Remover/Installer 9340 tool.
6. Rotate the breaker bar counter-clockwise to remove lock ring.
7. Remove the lock ring.
8. Remove fuel pump module
To install:
9. Using a new gasket, position the fuel pump module into opening in fuel tank.
10. Position the lock ring over top of fuel pump module.
11. Rotate the fuel pump module until embossed alignment arrow points to center alignment mark.

➡ **Be sure the fuel fitting on top of pump module is pointed to drivers side of vehicle.**

12. Install Lock Ring Remover/Installer Tool 9340 to the lock ring.
13. Install ½ inch drive breaker into Lock Ring Remover/Installer Tool 9340
14. Tighten lock Ring until all seven notches have engaged.
15. Install the fuel tank.

Fuel Injector

REMOVAL & INSTALLATION

1. Before servicing the vehicle, refer to the Precautions Section.
2. Relieve the fuel system pressure.

✳✳ CAUTION

The fuel system is under constant pressure even with engine off. Before servicing fuel rail, fuel system pressure must be released.

✳✳ CAUTION

The left and right fuel rails are replaced as an assembly. Do not

1 - MOUNTING BOLTS (4)
2 - QUICK-CONNECT FITTING
3 - FUEL RAIL
4 - INJ. #1
5 - INJ. #3
6 - INJ. #5
7 - INJ. #2
8 - INJ. #4
9 - INJ. #6
10 - CONNECTOR TUBE

67189DAKOG42

Fuel rail components—3.7L engine

06009DAKOG03

Using Special Tool 9340 to remove the lock ring

attempt to separate rail halves at connector tube. Due to design of tube, it does not use any clamps. Never attempt to install a clamping device of any kind to tube. When removing fuel rail assembly for any reason, be careful not to bend or kink tube.

3. Remove or disconnect the following:
- Fuel tank filler tube cap
- Negative battery cable
- Air intake assembly
- Fuel line latch clip and fuel line at fuel rail
- Vacuum lines at throttle body
- Fuel injector electrical connectors
- Electrical connectors at throttle body sensors
- Ignition coils
- Fuel rail mounting bolts
- Fuel rail with injectors attached

4. Disconnect the clip that retains the fuel injector to fuel rail to remove the injector.

To install:

➡ **Apply a small amount of clean engine oil to each fuel injector o-ring. This will help in fuel rail installation.**

5. Install the injectors and injector clips to fuel rail.

6. Position fuel rail/fuel injector assembly to machined injector openings in cylinder head.

7. Guide each injector into cylinder head.

✳✳ CAUTION

Be careful not to tear injector o-rings.

8. Push right side of fuel rail down until fuel injectors have bottomed on cylinder head shoulder.

9. Push left fuel rail down until injectors have bottomed on cylinder head shoulder.

10. Install or connect the following:
- Fuel rail mounting bolts and tighten to 27 Nm (20 ft. lbs.).
- Ignition coils
- Electrical connectors to throttle body
- Electrical connectors to the fuel injectors.
- Vacuum lines to throttle body
- Fuel supply hose to the fuel rail.
- Air intake assembly
- Negative battery cable

11. Start engine and check for leaks.

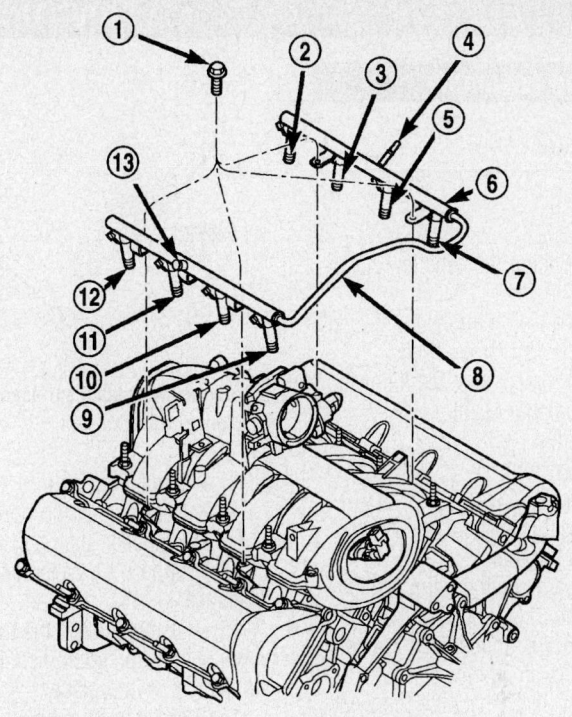

1 - MOUNTING BOLTS (4)
2 - INJ.#7
3 - INJ.#5
4 - QUICK-CONNECT FITTING
5 - INJ.#3

67189DAKOG44

Fuel rail components—4.7L engine

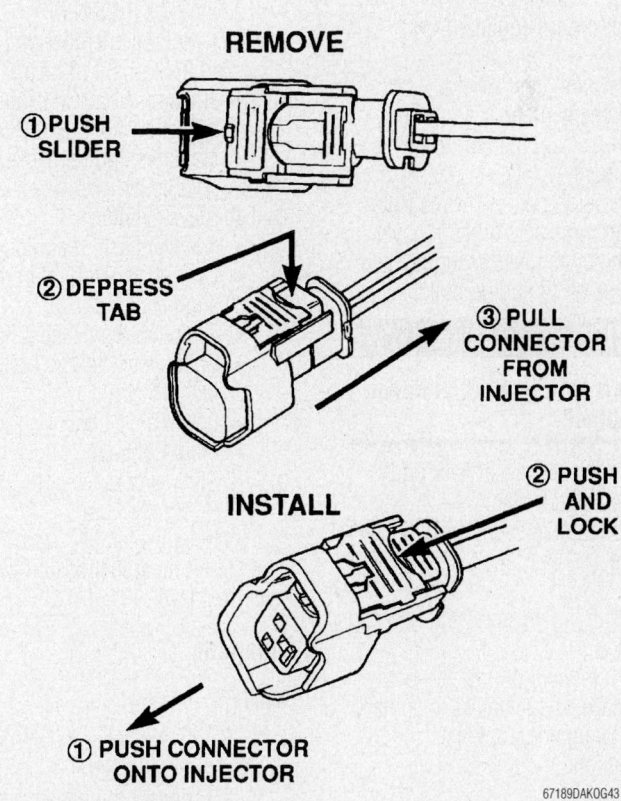

REMOVE

① PUSH SLIDER

② DEPRESS TAB

③ PULL CONNECTOR FROM INJECTOR

INSTALL

② PUSH AND LOCK

① PUSH CONNECTOR ONTO INJECTOR

67189DAKOG43

Fuel injector connector removal

DRIVE TRAIN

Transmission Assembly

REMOVAL & INSTALLATION

Manual

1. Before servicing the vehicle, refer to the Precautions Section.
2. Shift transmission into Neutral.
3. Drain the transmission fluid.
4. Remove or disconnect the following:
   - Negative battery cable
   - Shift knob
   - Shift lever boot
   - Shift lever extension
   - Lower shift lever boot assembly from the floorpan
   - Shift tower from the transmission
   - Skid plate, if equipped with 4WD
5. Matchmark the driveshaft companion flanges for installation reference.
6. Remove or disconnect the following:
   - Driveshafts
   - Y-pipe from the exhaust manifolds.
   - Backup light switch connector
   - Clutch slave cylinder splash shield, if equipped.
   - Clutch slave cylinder
7. Support the transmission with a suitable jack.
8. Remove or disconnect the following:
   - Transfer case if equipped
   - Starter
   - Transmission dust shield
   - Rear crossmember
   - Bolts/nuts from the rear transmission mount
   - Transmission harness wires from clips on transmission shift cover
9. Lower the transmission slightly and remove transmission to engine bolts.

✳✳ CAUTION

Do not remove structural dust cover from engine block.

10. Slide transmission rearward until input shaft clears clutch disc.
11. Lower the transmission and remove from the vehicle.

To install:

12. Clean the transmission front housing mounting surface
13. Apply a light coat of high temperature bearing grease or equivalent to contact surfaces of following components
    - Release fork
    - Ball stud
    - Release bearing slide surface

09482_RAID_G0006

Grease the contact surfaces as shown—Manual transmission

   - Input shaft splines
   - Release bearing bore
   - Propeller shaft slip yoke
14. Support the transmission with a suitable jack.
15. Raise and align the transmission input shaft with clutch disc, then slide transmission into place.
16. Verify the front housing is fully seated.
17. Install or connect the following:
    - Transmission to engine bolts without washers and tighten to 30 ft. lbs. (41 Nm)
    - Transmission bolts with washers and tighten to 50 ft. lbs. (68 Nm)
    - Dust shield and tighten bolts to 40 inch lbs. (4.5 Nm)
    - Rear crossmember and tighten nuts to 75 ft. lbs. (102 Nm)
    - Transmission rear mount bolts and tighten to 50 ft. lbs. (68 Nm)
    - Transmission harnesses to clips on shift cover
    - Slave cylinder
    - Transfer case, if equipped
    - Driveshafts with reference marks aligned
    - Y-pipe to the exhaust manifolds
    - Shift tower, tighten bolts to 88 inch lbs. (10 Nm).
    - Lower shift boot
    - Floor console
    - Shift lever extension
    - Upper shift boot
    - Negative battery cable
18. Refill the transmission with fluid to the correct level.

Automatic

545RFE

1. Before servicing the vehicle, refer to the Precautions Section.
2. Disconnect the negative battery cable.
3. Remove the skid plates, if equipped.
4. Mark the driveshaft and axle companion flanges for assembly alignment.
5. Remove or disconnect the following:
   - Rear driveshaft
   - Front driveshaft, if equipped
   - Engine to transmission collar
   - Exhaust support bracket from the rear of the transmission
   - Any necessary exhaust components required for clearance
   - Starter motor
6. Rotate the crankshaft in clockwise direction until converter bolts are accessible. Then remove bolts one at a time.
7. Remove or disconnect the following:
   - Output speed sensor connector
   - Input speed sensor connector
   - Transmission solenoid/TRS assembly connector
   - Line pressure sensor connector
   - Gearshift cable from transmission manual valve lever
   - Transmission fluid cooler lines
   - Transmission vent hose from the transmission
8. Support rear of engine with safety stand or jack.
9. Raise the transmission slightly with service jack to relieve load on crossmember and supports.
10. Remove or disconnect the following:
    - Bolts securing rear support and cushion to transmission and crossmember
    - Bolts attaching crossmember to frame and remove crossmember
    - Transfer case, if equipped
    - All remaining converter housing bolts
11. Carefully work transmission and torque converter assembly rearward off engine block dowels.
12. Hold torque converter in place during transmission removal.
13. Lower transmission and remove the transmission assembly.

To install:

➡ **If a replacement transmission is being installed, transfer any components necessary, such as the manual shift lever and shift cable bracket, from the original transmission onto the replacement transmission.**

14. Raise the transmission and align the torque converter with the drive plate and transmission converter housing with the engine block.

15. Move transmission forward and align the converter housing with engine block dowels.

16. Carefully work transmission forward and over engine block dowels until converter hub is seated in crankshaft.

✱✱ CAUTION

Verify that no wires, or the transmission vent hose, have become trapped between the engine block and the transmission.

17. Use two mounting bolts to attach the transmission to the engine.
18. Install or connect the following:
- Remaining torque converter housing to engine bolts. Tighten to 50 ft. lbs. (68 Nm).
- Transfer case, if equipped
- Rear transmission crossmember. Tighten crossmember to frame bolts to 50 ft. lbs. (68 Nm).
- Rear support to transmission. Tighten the bolts to 35 ft. lbs. (47 Nm).
19. Lower transmission onto crossmember and install bolts attaching transmission mount to crossmember. Tighten clevis bracket to crossmember bolts to 35 ft. lbs. (47 Nm). Tighten the clevis bracket to rear support bolt to 50 ft. lbs. (68 Nm).
20. Install or connect the following:
- Gearshift cable to transmission
- Wires to solenoid and pressure switch assembly connector, input and output speed sensors, and line pressure sensor.

➡**Be sure the transmission harnesses are properly routed.**

- Torque converter-to-driveplate bolts. Tighten to 22 ft. lbs. (31 Nm).
- Starter motor and cooler line bracket
- Cooler lines to transmission
- Transmission fill tube
- Exhaust components
- Engine collar onto the transmission and the engine. Tighten the bolts to 40 ft. lbs. (54 Nm).
- Rear driveshaft
- Front driveshaft, if equipped
- Skidplates, if equipped
- Negative battery cable
21. Adjust the gearshift cable as necessary.
22. Refill the transmission with fluid to the correct level.

42RLE

1. Before servicing the vehicle, refer to the Precautions Section.
2. Mark the driveshaft and axle companion flanges for assembly alignment.
3. Remove or disconnect the following:
- Negative battery cable
- Skid plates, if equipped
- Rear driveshaft
- Front driveshaft, if equipped
- Input and output speed sensors
- Transfer case shift motor and mode sensor assembly
- Variable line pressure connector from the transmission, if equipped
- Transmission range sensor
- Wires from the solenoid/pressure switch assembly
- Bolts holding the exhaust crossover pipe to the pre-catalytic converter pipe flanges
- Bolts holding the exhaust crossover pipe to the catalytic converter flange.
- Starter motor
- Engine to transmission collar
4. Rotate the crankshaft in a clockwise direction until converter bolts are accessible. Remove the bolts one at a time.
5. Remove or disconnect the following:
- Transmission vent hose
- Transfer case, if equipped
6. Support the rear of engine with a suitable jack.
7. Raise the transmission slightly with a suitable jack to relieve the load on the crossmember and supports.
8. Remove the bolts securing rear support and cushion to transmission and crossmember.
9. Remove the bolts attaching the crossmember to the frame and remove the crossmember.
10. Disconnect the transmission fluid cooler lines
11. Remove all remaining converter housing bolts.
12. Carefully work the transmission assembly rearward off engine block dowels.

➡**Hold the torque converter in place during transmission removal.**

13. Lower the transmission assembly and remove from the vehicle.
To install:

➡**If a replacement transmission is being installed, transfer any components necessary, such as the manual shift lever and shift cable bracket, from the original transmission onto the replacement transmission.**

14. Raise the transmission and align the torque converter with the drive plate and transmission converter housing with the engine block.
15. Move the transmission forward and align the converter housing with engine block dowels.
16. Carefully work the transmission forward and over engine block dowels until converter hub is seated in crankshaft.

➡**Verify that no wires, or the transmission vent hose, have become trapped between the engine block and the transmission.**

17. Install two mounting bolts to attach the transmission to the engine.
18. Install or connect the following:
- Remaining torque converter housing to engine bolts. Tighten to 50 ft. lbs. (68 Nm)
- Transfer case, if equipped
- Rear transmission crossmember. Tighten the crossmember to frame bolts to 50 ft. lbs. (68 Nm)
- Rear support to transmission. Tighten the bolts to 35 ft. lbs. (47 Nm)
19. Lower the transmission assembly onto the crossmember and install bolts attaching transmission mount to crossmember. Tighten clevis bracket to crossmember bolts to 35 ft. lbs. (47 Nm). Tighten the clevis bracket to rear support bolt to 50 ft. lbs. (68 Nm)
20. Install or connect the following:
- Gearshift cable to support bracket and transmission manual lever.
- Input and output speed sensor and the transmission range sensor.
- Variable line pressure connector, if equipped
- Wires to the solenoid/pressure switch assembly
- Torque converter-to-driveplate bolts. Tighten to 65 ft. lbs. (88 Nm).
- Starter motor and cooler line bracket
- Cooler lines to the transmission
- Transmission fill tube
- Exhaust components
- Rear driveshaft
- Front driveshaft, if equipped
- Skidplates, if equipped
- Negative battery cable
21. Adjust gearshift cable if necessary.
22. Refill the transmission with fluid to the correct level.

Clutch

REMOVAL & INSTALLATION

1. Before servicing the vehicle, refer to the Precautions Section.

2. Remove the transmission and clutch housing as assembly.

➡**If pressure plate is being removed for access to another component, mark position of pressure plate cover on flywheel with small punch marks.**

3. Loosen pressure plate cover bolts evenly and in rotation to relieve spring tension. Loosen bolts a few threads at a time to avoid warping cover.

4. Remove cover bolts, pressure plate and clutch disc.

To install:

➡**Clean flywheel surface with solvent. Scuff sand the surface with 120/180 grit emery cloth to remove minor scratches and glazing.**

5. Check new clutch disc for runout and free operation on input shaft splines.

6. Lubricate crankshaft pilot bearing with a NLGI—2 rated grease.

7. Position clutch disc with pressure plate on the flywheel.

8. Insert alignment tool or spare input shaft through clutch disc and into pilot bearing.

9. Verify that the disc hub is positioned correctly. The raised portion of the hub faces away from the flywheel.

10. Install the cover bolts finger tight.

11. Tighten cover bolts evenly (and in rotation) a few threads at a time. Cover bolts must be tightened evenly and to specified torque to avoid distorting cover.

12. Tighten the cover bolts as follows:

   a. 5/16 in. bolts to 23 Nm (17 ft. lbs.).

   b. 3/8 in. bolts to 41 Nm (30 ft. lbs.).

13. Apply light coat of high temperature bearing grease to splines of transmission input shaft and to release bearing slide surface of front bearing retainer.

✳✳ WARNING

Do not over-lubricate shaft splines. This could result in grease contamination of disc.

14. Install the transmission assembly.

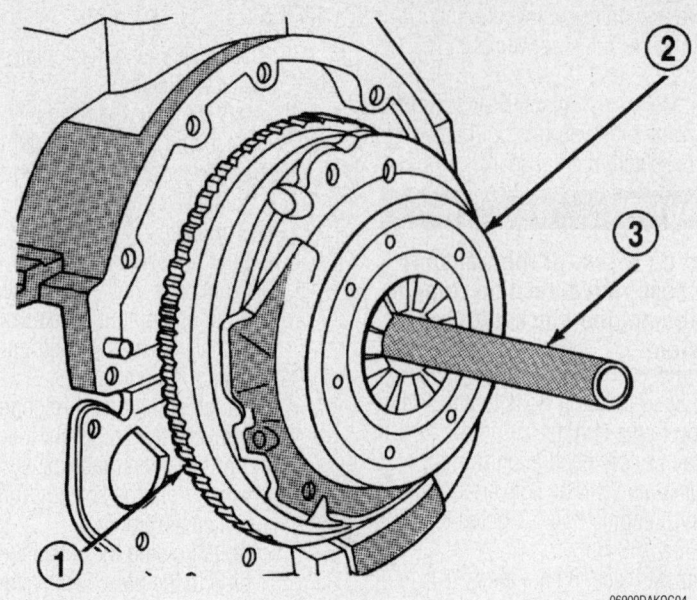

Clutch installation. (1) flywheel, (2) pressure plate, (3) alignment tool

06009DAKOG04

Hydraulic Clutch System

BLEEDING

The system is self-bleeding. Press the clutch pedal repeatedly to release air from the fluid. The air will be vented from the reservoir.

Transfer Case Assembly

REMOVAL & INSTALLATION

1. Before servicing the vehicle, refer to the Precautions Section.

2. Drain the transfer case fluid.

3. Shift the transfer case into 2WD.

4. Mark the front and rear propeller shafts for alignment reference.

5. Support the transmission with a suitable jack stand.

6. Remove the rear crossmember and skid plate, if equipped.

7. Disconnect the front and rear driveshafts at the transfer case.

8. Disconnect the transfer case shift motor and mode sensor wire connectors.

9. Disconnect the transfer case vent hose.

10. Support the transfer case with a suitable transmission jack and secure the transfer case to the jack with chains.

11. Remove the nuts attaching transfer case to the transmission.

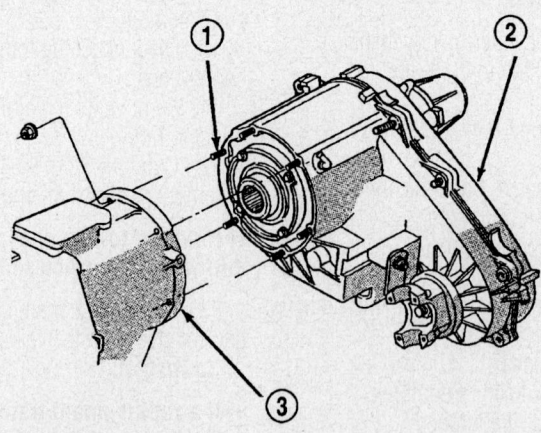

1 - MOUNTING STUDS
2 - TRANSFER CASE
3 - TRANSMISSION

67189DAKOG41

Typical transfer case mounting

12. Pull the transfer case and jack rearward to disengage the transfer case.

13. Remove the transfer case assembly.

To install:

14. Align the transfer case and transmission shafts and install the transfer case onto the transmission.

15. Install and the transfer case attaching nuts. Tighten to 20–25 ft. lbs. (27–34 Nm).

16. Connect the vent hose.

17. Connect the shift motor and mode sensor wiring connectors. Secure the wire harness to clips on the transfer case.

18. Align and connect the driveshafts.

19. Install the rear crossmember and skid plate, if equipped. Tighten the crossmember bolts to 30 ft. lbs. (41 Nm).

20. Refill the transfer case with fluid to the correct level.

21. Verify proper transfer case shift operation.

Halfshaft

REMOVAL & INSTALLATION

Front

1. Before servicing the vehicle, refer to the Precautions Section.

2. Remove or disconnect the following:

- Front wheel
- Skid plate, if equipped
- Hub nut and washer
- Brake caliper and rotor
- ABS wheel speed sensor if equipped
- Hub bearing bolts from the knuckle
- Hub bearing and brake shield from the knuckle

3. Support the half shaft at the CV joint housings.

4. Position two pry bars behind the inner CV housing and disengage the CV joint from the axle.

5. Remove the half shaft from the vehicle.

To install:

6. Apply a light coating of wheel bearing grease on the axle splines.

7. Insert the half shaft stub through the steering knuckle and onto the axle. Verify the shaft snapring engages with the groove on the inside of the joint housing.

8. Clean the hub bearing bore and hub bearing mating surface of all foreign materials. Apply a light coating of grease to all mating surfaces.

9. Install the hub bearing onto the axle half shaft and steering knuckle.

10. Install the hub bearing bolts and tighten to 120 ft. lbs. (163 Nm).

11. Install the ABS wheel speed sensor, if equipped.

12. Install brake rotor and caliper.

13. Apply the brakes and tighten hub nut to 185 ft. lbs. (251 Nm).

14. Install the skid plate, if equipped.

15. Install the wheel and tire assembly.

CV-Joints

OVERHAUL

Outer Joint

2005

1. Before servicing the vehicle, refer to the Precautions Section.

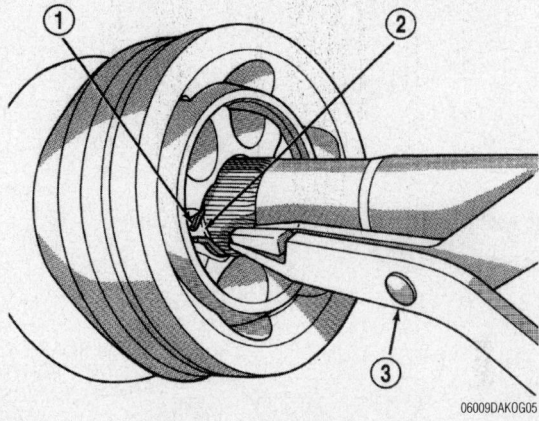

06009DAKOG05

Removing the snapring (1) from the shaft (2) with snapring pliers (3)—Outer Joint

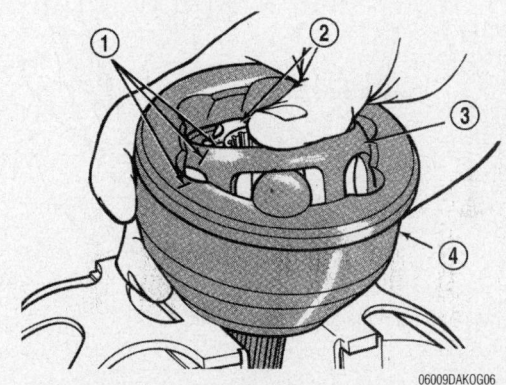

06009DAKOG06

Make alignment marks (1) on the inner race/hub (2) and cage (3)—Outer Joint

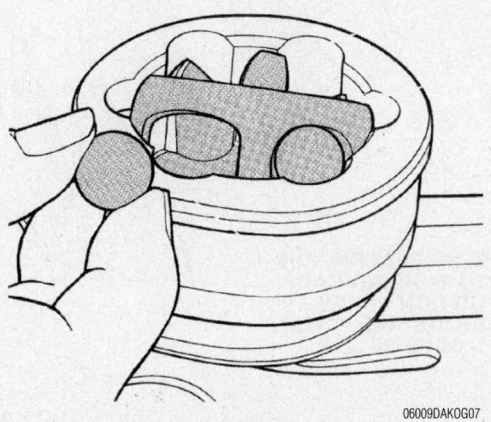

06009DAKOG07

Removing the balls from the bearing cage—Outer Joint

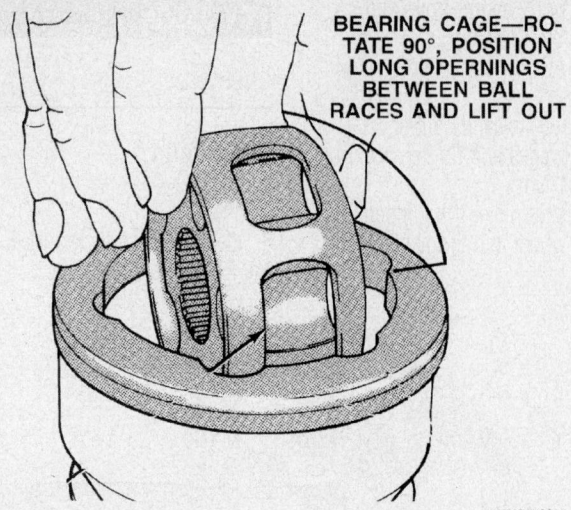

BEARING CAGE—RO-
TATE 90°, POSITION
LONG OPERNINGS
BETWEEN BALL
RACES AND LIFT OUT

06009DAKOG08

Removing the cage and inner race from the housing—Outer Joint

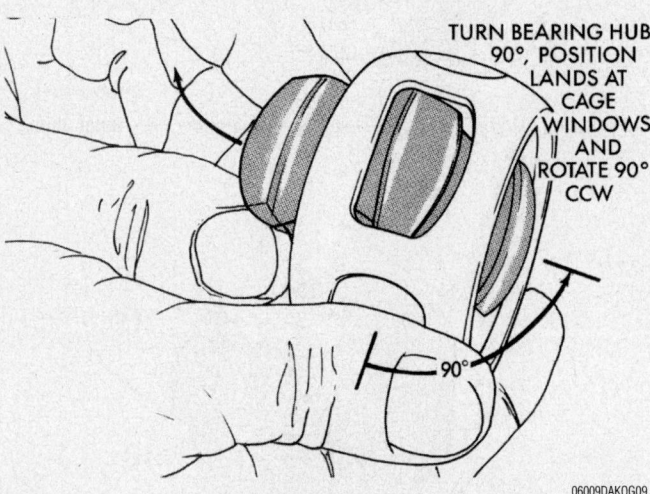

TURN BEARING HUB
90°, POSITION
LANDS AT
CAGE
WINDOWS
AND
ROTATE 90°
CCW

90°

06009DAKOG09

Removing the inner race/hub from the cage—Outer Joint

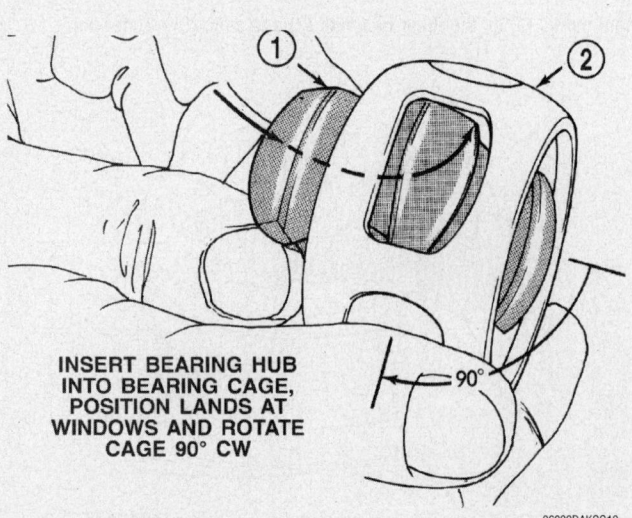

① ②

INSERT BEARING HUB
INTO BEARING CAGE,
POSITION LANDS AT
WINDOWS AND ROTATE
CAGE 90° CW

90°

06009DAKOG10

Assembling the inner race cage and housing—Outer Joint

2. Remove the halfshaft and secure it in vise.

Take care not to damage the CV-joint housing or halfshaft.

3. Remove the boot clamps with a cut-off wheel or grinder.
4. Slide the boot down the shaft.
5. Remove any excess lubricant to expose the CV-joint snapring.
6. Spread snapring and slide the joint off the shaft.
7. Slide the boot off the shaft and discard old boot.
8. Matchmark the inner race/hub, bearing cage and housing with dabs of paint.
9. Clamp the CV-joint in a vertical position in the vise.
10. Press down one side of the bearing cage to gain access to the ball at the opposite side.

➡If joint is tight, use a hammer and brass drift to loosen the bearing hub. Do not contact the bearing cage with the drift.

11. Remove ball from the bearing cage.
12. Repeat step above until all six balls are removed from the bearing cage.
13. Lift cage and inner race upward and out from the housing.
14. Turn inner race 90° in the cage and rotate the inner race/hub out of the cage.

To install:

15. Apply a light coat of grease to the CV-joint components before assembling them.
16. Align the inner race, cage and housing according to the alignment reference marks.
17. Insert the inner race into the cage and rotate race into the cage.
18. Rotate the inner race/hub in the cage.
19. Insert the cage into the housing.
20. Rotate the cage 90° into the housing.
21. Apply lubricant included with replacement boot/joint to the ball races. Spread lubricant equally among all the races.
22. Tilt inner race/hub and cage and install the balls.
23. Place new clamps onto new boot and slide boot onto the shaft to its original position.
24. Apply the remaining lubricant to the CV-joint and boot.

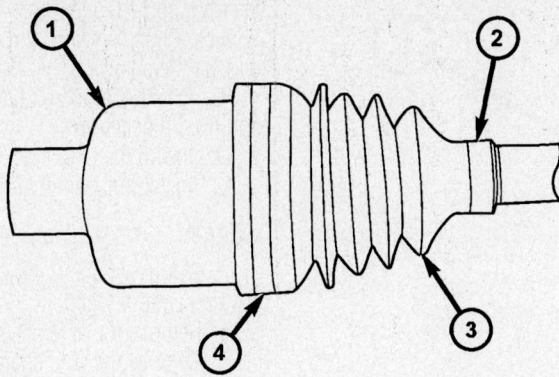

1 - C/V HOUSING
2 - CLAMP
3 - BOOT
4 - CLAMP

67189DAK0G33

Inner Joint boot clamp locations

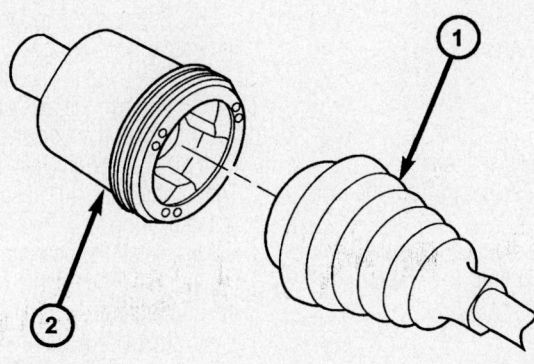

1 - BOOT
2 - HOUSING

67189DAK0G35

Remove the housing from the halfshaft—Inner Joint

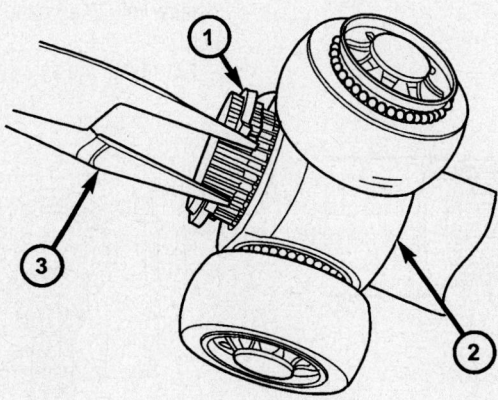

1 - SNAP RING
2 - TRIPOD
3 - PLIERS

67189DAK0G34

Removing the snapring—Inner Joint

25. Push the joint onto the shaft until the snapring seats in the groove. Pull on the joint to verify the span ring has engaged.

26. Position the boot on the joint in its original position. Ensure the boot is not twisted and remove any excess air.

27. Secure both boot clamps with Clamp Installer C-4975A, or equivalent. Place the tool on the clamp bridge and tighten until the jaws of the tool are closed.

Inner Joint

1. Before servicing the vehicle, refer to the Precautions Section.

2. Secure the halfshaft in a vise.

3. Remove the boot clamps with a cut-off wheel or grinder.

✳✳ WARNING

Do not damage the CV housing or half shaft with the cut-off wheel or grinder.

4. Remove the housing from the half shaft and slide the boot down shaft.

5. Remove the housing bushing from the housing.

6. Remove the tripod snapring.

7. Remove the tripod and boot from the halfshaft.

8. Clean and inspect the CV components for excessive wear and damage. Replace the tripod as a unit only if necessary.

To install:

9. Slide a new boot down the halfshaft.

10. Install the tripod and tripod snapring on the halfshaft.

11. Pack the grease supplied with the joint/boot into the housing and boot.

12. Coat the tripod with the supplied grease.

13. Install a new bushing onto the housing.

14. Insert the tripod and shaft in the housing.

15. Position the boot on the joint in its original position.

➡**Verify the boot is not twisted and remove any excess air.**

16. Secure both boot clamps with Clamp Installer C-4975A, or equivalent. Place the tool on the clamp bridge and tighten the tool until the jaws of the tool are closed.

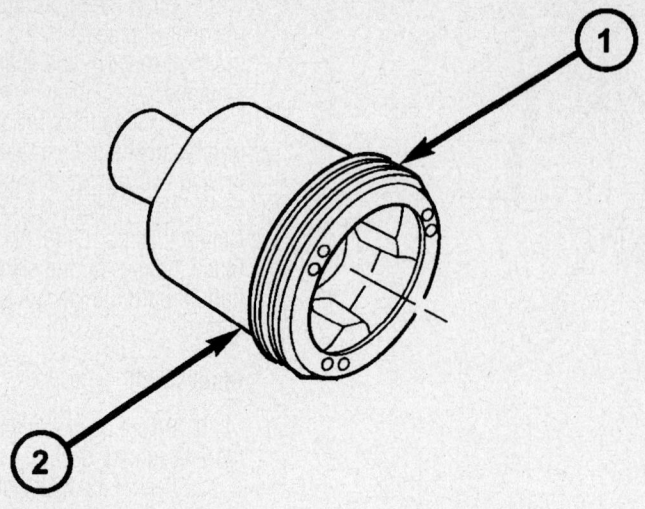

1 - HOUSING
2 - BUSHING

67189DAKOG37

Install a new bushing onto the joint housing—Inner Joint

Front Axle Shaft, Bearing and Seal

REMOVAL & INSTALLATION

AXLE SHAFTS

1. Before servicing the vehicle, refer to the Precautions Section.
2. Place the transmission in neutral.
3. Remove the half shaft from vehicle.
4. Remove skid plate, if equipped.

5. Clean the axle seal area.
6. Remove the snapring from the axle shaft.
7. Remove the axle with Special Tool 8420A and slide hammer C-3752, or equivalent.

To install:

➡**Use care to prevent shaft splines from damaging axle shaft seal lip.**

8. Lubricate the bearing bore and seal lip with gear lubricant.
9. Insert the axle shaft through seal,

bearing, and engage it into the side gear splines. Push firmly on the axle shaft to engage the snapring.
10. Check the differential fluid level and add fluid if necessary.
11. Install the skid plate, if necessary.
12. Install the half shaft.

SEALS

1. Before servicing the vehicle, refer to the Precautions Section.
2. Remove the axle shaft.
3. Remove the axle shaft seal with a small prybar.

To install:

4. Wipe the axle shaft tube bore clean.
5. Install a new axle shaft seal with installer 8402 and handle C-4171, or equivalent.
6. Install the axle shaft and half shaft.

BEARINGS

1. Before servicing the vehicle, refer to the Precautions Section.
2. Remove the axle shaft.
3. Remove the axle shaft seal.
4. Install the axle shaft bearing remover C-4660-A, or equivalent, in the bearing. Then tighten the nut to spread the remover in the bearing.
5. Install the bearing remove cup, bearing and nut. Then tighten the nut to draw the bearing out.
6. Inspect the axle shaft tube bore for roughness and burrs. Remove as necessary.

To install:

7. Wipe the axle shaft tube bore clean.
8. Install the axle shaft bearing with installer 5063 and handle C-4171, or equivalent.
9. Install a new axle shaft seal with installer 8402 and handle C-4171, or equivalent.
10. Install the axle shaft and half shaft.

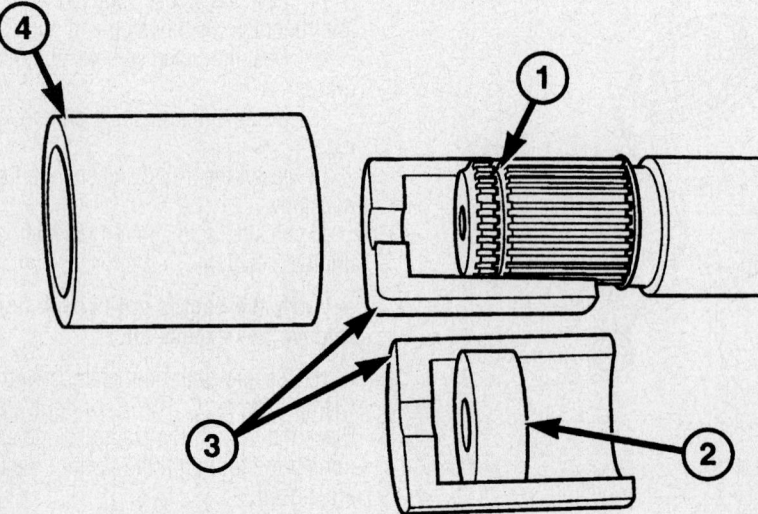

06009DAKOG11

Assembling the Axle Shaft Removal Tool 8420A

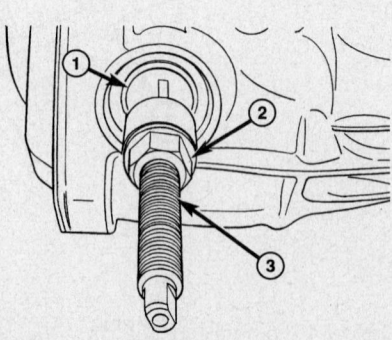

06009DAKOG12

Removing the axle shaft bearing (1) using remover C-4660-A (3) and nut (2)

Rear Axle Shaft, Bearing and Seal

REMOVAL & INSTALLATION

Axle Shaft

1. Before servicing the vehicle, refer to the Precautions Section.
2. Remove the rear brake components.
3. Remove the differential cover and drain the fluid.
4. Rotate the differential case to access the pinion mate shaft lock screw. Remove the lock screw and pinion mate shaft from the differential case.
5. Push the axle shaft inward then remove the C-clip from the axle shaft.
6. Remove the axle shaft.

To install:

7. Lubricate the bearing bore and seal lip with gear lubricant. Insert the axle shaft through the seal, bearing and engage it into side gear splines.

➡**Use care to prevent the shaft splines from damaging the axle shaft seal.**

8. Insert the C-clip in end of axle shaft. Push the axle shaft outward to seat the C-clip in side gear.
9. Insert the pinion shaft into the differential case, through the thrust washers and differential pinions.
10. Align the hole in the pinion mate shaft with the hole in the differential case and install the lock screw with Loctite® on the threads. Tighten the lock screw to 8 ft. lbs. (11 Nm).
11. Install the differential cover and tighten the bolts in a criss-cross pattern to 30 ft. lbs. (41 Nm).
12. Fill the differential with gear lubricant to the bottom of the fill plug hole.
13. Install the fill hole plug.
14. Install the rear brake components.

Axle Shaft Seal

1. Before servicing the vehicle, refer to the Precautions Section.
2. Remove the axle shaft.
3. Remove the axle shaft seal from the end of the axle tube with a seal pick.

To install:

4. Wipe the axle tube bore clean. Remove any old sealer or burrs from the tube.
5. Coat the lip of the new seal with axle lubricant for protection prior to installing the axle shaft.
6. Install the new axle seal with Installer

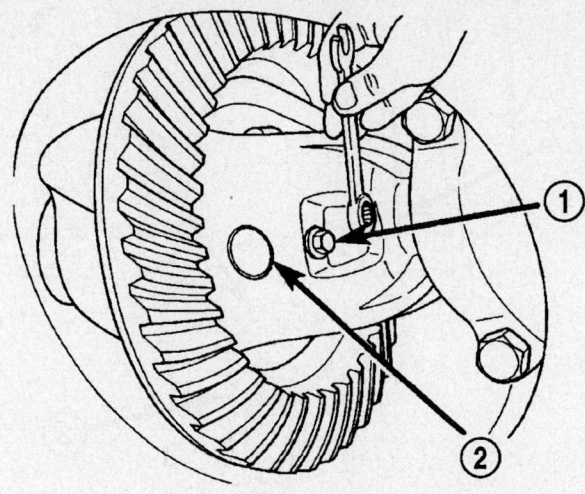

1 - LOCK SCREW
2 - PINION MATE SHAFT

67189DAKOG29

Remove the pinion mate shaft and lock screw—Rear axle

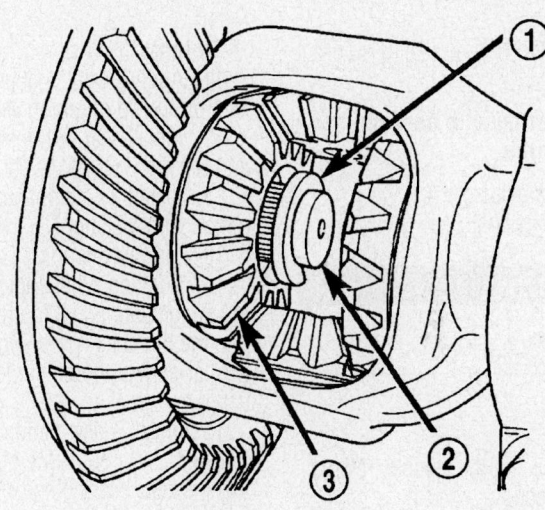

1 - C-LOCK
2 - AXLE SHAFT
3 - SIDE GEAR

67189DAKOG30

Remove the axle shaft C-clip—Rear axle

C-4198 and Handle C-4171, or equivalent. When the tool contacts the axle tube, the seal is installed to the correct depth.

7. Install the axle shaft.

Axle Bearings

1. Before servicing the vehicle, refer to the Precautions Section.
2. Remove the axle shaft.
3. Remove the axle shaft seal from the axle tube with a seal pick.

➡**The seal and bearing can be removed at the same time with the bearing removal tool.**

4. Remove the axle shaft bearing with Bearing Removal Tool Set 6310 and Adapter Foot 6310-9, or equivalent.

To install:

5. Wipe the axle tube bore clean. Remove any old sealer or burrs from the tube.
6. Install the axle shaft bearing with

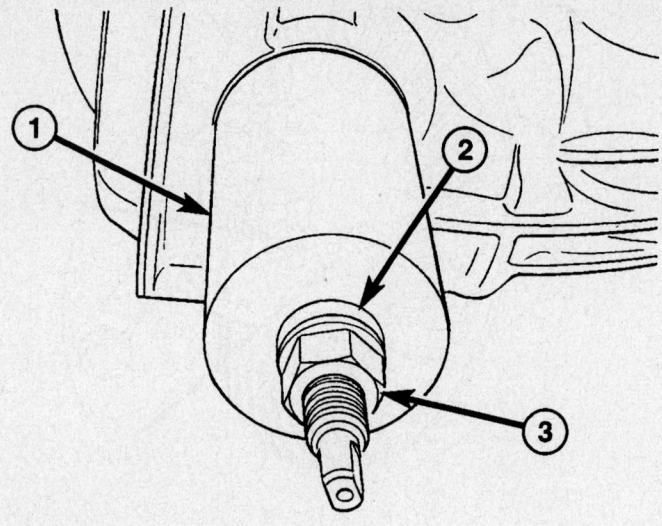

1 - REMOVER CUP
2 - BEARING
3 - NUT

67189DAK0G32

Bearing Removal Tool Set 6310—Rear Axle

Installer C-4198 and Handle C-4171, or equivalent.

➡️ **Install the bearing with part number against the installer.**

7. Install a new axle seal.
8. Install the axle shaft

Pinion Seal

REMOVAL & INSTALLATION

Front Axle

1. Before servicing the vehicle, refer to the Precautions Section.
2. Remove both half shafts.
3. Matchmark the front driveshaft and pinion companion flange for installation reference, if equipped.
4. Remove the front driveshaft, if equipped
5. Rotate the pinion gear three or four times and verify pinion rotates smoothly.
6. Record the pinion rotating torque with an inch pound torque wrench, for installation reference.
7. Position Holder 6719A or equivalent against the companion flange and install four bolts and washers into the threaded holes and tighten the bolts.
8. Remove the pinion nut.
9. Remove the companion flange with Puller C-452, or equivalent.
10. Remove pinion seal with a seal pick.

To install:
11. Apply a light coating of gear lubricant on the lip of pinion seal.
12. Install seal with Installer C-3972-A and Handle C-4171, or equivalent,
13. Install the companion flange onto the pinion with Installer C-3718 and Holder 6719A, or equivalent.
14. Position holder against the companion flange and install four bolts and washers into the threaded holes. Tighten the bolt and washer so that the holder is held to the flange.
15. Install a new pinion nut onto the pinion shaft and tighten the pinion nut until there is zero bearing end-play.

➡️ **Do not exceed the minimum tightening torque when installing the companion flange at this point. Damage to the collapsible spacer or bearings may result.**

16. Tighten the nut to 200 ft. lbs. (271 Nm).

➡️ **Never loosen pinion nut to decrease pinion bearing rotating torque and never exceed specified preload torque. If preload torque or rotating torque is exceeded a new collapsible spacer must be installed.**

17. Record the pinion rotating torque using a torque wrench. The rotating torque should be equal to the reading recorded during removal plus 5 inch lbs. (0.56 Nm).

18. If the rotating torque is low, tighten the pinion nut in 5 ft. lbs. (6.8 Nm) increments until the proper rotating torque is achieved.

➡️ **If the maximum tightening torque is reached prior to reaching the required rotating torque, the collapsible spacer may have been damaged. Replace the collapsible spacer.**

19. Install the driveshaft with the reference marks aligned.
20. Install half shafts.

Rear Axle

1. Before servicing the vehicle, refer to the Precautions Section.
2. Matchmark the universal joint, companion flange and pinion shaft for installation reference.
3. Disconnect the rear driveshaft.
4. Remove companion flange bolts and secure the shaft in an upright position to prevent damage to the rear universal joint.
5. Remove brake drums to prevent any drag.
6. Rotate companion flange three or four times and verify flange rotates smoothly.
7. Measure rotating torque of the pinion with an inch pound torque wrench and record the reading for installation reference.
8. Install bolts into two of the threaded holes in the companion flange 180° apart.
9. Position Holder 6719 or equivalent against the companion flange and install a bolt and washer into one of the remaining threaded holes. Tighten the bolts so the Holder 6719 or equivalent is held to the flange.
10. Remove the pinion nut and washer.
11. Remove companion flange with Remover C-452, or equivalent.
12. Remove pinion seal with a pry tool or slidehammer mounted screw.
To install:
13. Apply a light coating of gear lubricant on the lip of pinion seal.
14. Install a new pinion seal with Installer C-4076-B and Handle C-4735, or equivalent.
15. Install companion flange on the end of the shaft with the reference marks aligned.
16. Install bolts into two of the threaded holes in the companion flange 180° apart.
17. Position Holder 6719, or equivalent, against the companion flange and install a bolt and washer into one of the remaining

threaded holes. Tighten the bolts so Holder 6719 is held to the flange.

18. Install companion flange on pinion shaft with Installer C-3718 and Holder 6719, or equivalent.

19. Install the pinion washer and a new pinion nut. The convex side of the washer must face outward.

➡**Do not exceed the minimum tightening torque when installing the companion flange retaining nut at this point. Damage to collapsible spacer or bearings may result.**

20. Hold companion flange with Holder 6719 and tighten the pinion nut to 210 ft. lbs. (285 Nm). Rotate pinion several revolutions to ensure the bearing rollers are seated.

21. Rotate the pinion flange with an inch pound torque wrench. Rotating torque should be equal to the reading recorded during removal plus 5 inch lbs. (0.56 Nm).

➡**Never loosen pinion nut to decrease pinion bearing rotating torque and never exceed specified preload torque. If rotating torque is exceeded, a new collapsible spacer must be installed.**

22. If rotating torque is too low, use Holder 6719 to hold the companion flange and tighten pinion nut in 5 ft. lbs. (6.8 Nm) increments until proper rotating torque is achieved.

➡**The seal replacement is unacceptable if final pinion nut torque is less than 210 ft. lbs. (285 Nm).**

➡**The bearing rotating torque should be constant during a complete revolution of the pinion. If the rotating torque varies, this indicates a binding condition.**

23. Install driveshaft with the installation reference marks aligned.

24. Tighten companion flange bolts to 80 ft. lbs. (108 Nm).

25. Install the rear brake components.

26. Check the differential housing lubricant level.

STEERING AND SUSPENSION

Air Bag

DISARMING

1. Disconnect and isolate the negative battery cable. Wait 2 minutes for the system capacitor to discharge before performing any service.

2. When repairs are completed, connect the negative battery cable.

Rack and Pinion Steering Gear

REMOVAL & INSTALLATION

1. Before servicing the vehicle, refer to the Precautions Section.

2. Siphon out as much power steering fluid as possible from the pump.

3. Lock the steering wheel.

4. Remove or disconnect the following:
- Front wheels
- Nuts from the tie rod ends
- Tie rod ends from the knuckles
- Steering gear pinch bolt
- Lower steering coupling from the steering gear

5. Turn the steering gear to the full right position.

➡**Protect the end of hoses to prevent contamination to the system and damage to the O-rings.**

6. Remove the power steering lines from the gear.

7. Remove the steering gear mounting bolts and nuts.

8. Tip the steering gear assembly forward to allow clearance and move to the right then tip the gear downward on the left side to remove from the vehicle.

To install:

➡**Before installing gear inspect bushings and replace if worn or damaged.**

9. Install the steering gear assembly to the vehicle and tighten mounting nuts and bolts to 190 ft. lbs. (258 Nm).

10. Install the power steering lines to the steering gear. Tighten the pressure hose to 23 ft. lbs. (31 Nm) and tighten the return hose to 27 ft. lbs. (37 Nm).

11. Slide the shaft coupler onto the steering gear. Install a new bolt and tighten to 42 ft. lbs. (57 Nm).

12. Clean the tie rod end studs and knuckle tapers.

13. Install the tie rod ends into the steering knuckles and tighten the nuts to 60 ft. lbs. (81 Nm).

14. Install the front wheels.

15. Refill the power steering system with fluid to the correct level.

16. Check the wheel alignment and adjust, as necessary.

Front Shock Absorber

REMOVAL & INSTALLATION

1. Before servicing the vehicle, refer to the Precautions Section.

2. Remove the front wheel.

3. Support the lower control arm outboard end.

4. Remove the upper shock nuts.

5. Remove the stabilizer link lower nut and then separate the stabilizer link from the lower control arm to gain access to the lower shock nut.

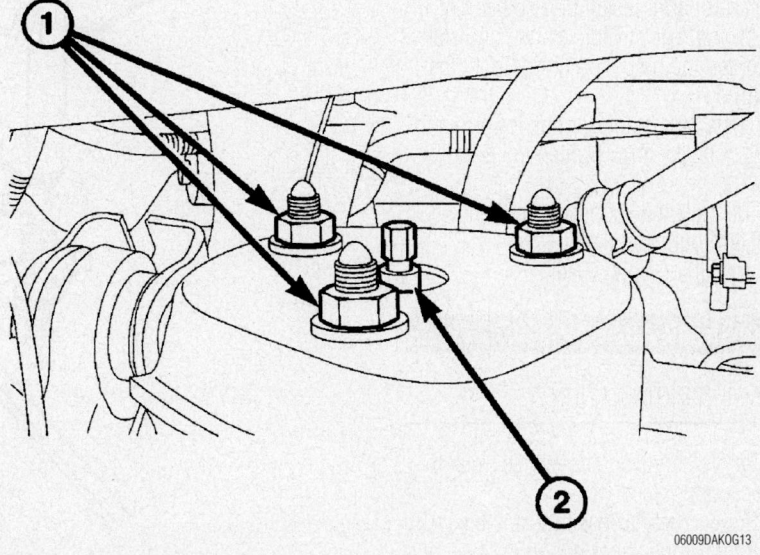

06009DAK0G13

Remove the upper shock nuts (1)

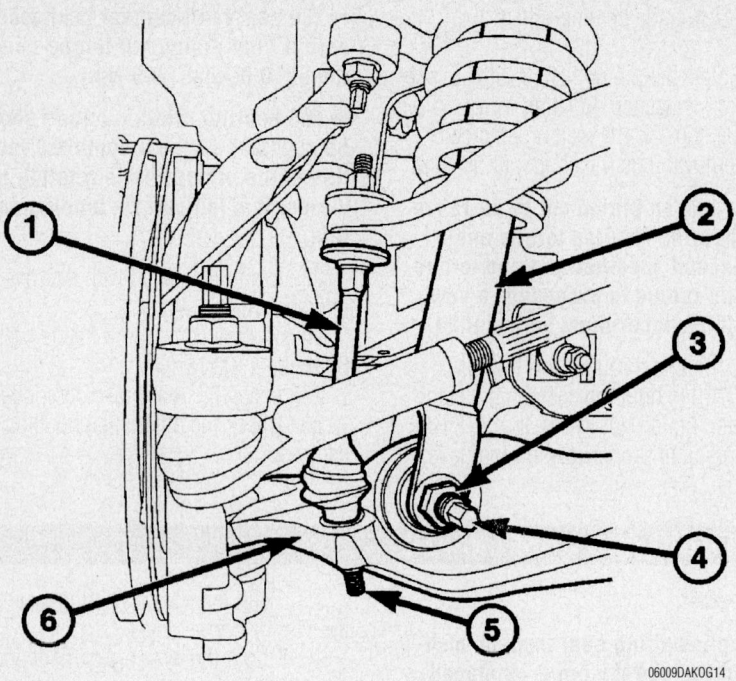

Stabilizer link (1), strut (2), nut (3), lower strut bolt (4), link lower nut (5), lower control arm (6)

06009DAKOG14

6. Remove the lower shock bolt and nut.

7. Remove the shock.

To install:

➡**All suspension components should be tightened with the weight of the vehicle on them (curb height).**

8. Install the upper part of the shock into the frame bracket.

9. Install the upper nuts. Tighten to 45 ft. lbs. (61 Nm).

10. Install the lower part of the shock into the lower control arm shock bushing.

11. Position shock module clevis to lower control arm. Install bolt so head of bolt is facing rear of vehicle and hand start nut. Tighten the bolt and nut to 155 ft. lbs. (210 Nm).

12. Install the stabilizer link lower nut to the lower control arm. Tighten to 75 ft. lbs. (102 Nm).

13. Remove the support from the lower control arm outboard end.

14. Install the front wheel.

Rear Shock Absorbers

REMOVAL & INSTALLATION

1. Before servicing the vehicle, refer to the Precautions Section.

2. Support rear axle with a suitable jack.

3. Remove the shock absorber lower nut and bolt from the axle bracket.

4. Remove the shock absorber upper nut and bolt from the frame bracket and remove the shock absorber.

To install:

5. Install the shock absorber and upper mounting bolt and nut. Tighten the nut to 70 ft. lbs. (95 Nm).

6. Install the shock absorber into the axle bracket. Install the bolt and nut and tighten the nut to 70 ft. lbs. (95 Nm).

Coil Spring

REMOVAL & INSTALLATION

1. Before servicing the vehicle, refer to the Precautions Section.

2. Remove the shock assembly.

3. Install the shock assembly in the Branick 7200T spring removal/installation tool or equivalent press.

4. Compress the spring.

5. Position Special Tool 9362 Wrench on the shock shaft retaining nut. Insert an 8 mm socket though Wrench onto hex located on end of shock shaft. While holding shock shaft from turning, remove nut from shock shaft using Wrench.

6. Remove the upper shock nut.

7. Remove the shock.

8. Remove the shock upper mounting plate.

9. Remove and inspect the upper and lower spring isolators.

To install:

10. Install the lower isolator.

11. Install the upper isolator.

12. Position the shock into the coil spring.

13. Install the upper shock mounting plate.

14. Install Wrench (on end of a torque wrench), Special Tool 9362, on shock shaft retaining nut. Next, insert 8 mm socket though Wrench onto hex located on end of shock shaft. While holding shock shaft from turning, tighten nut using Wrench to 33 ft. lbs. (45 Nm).

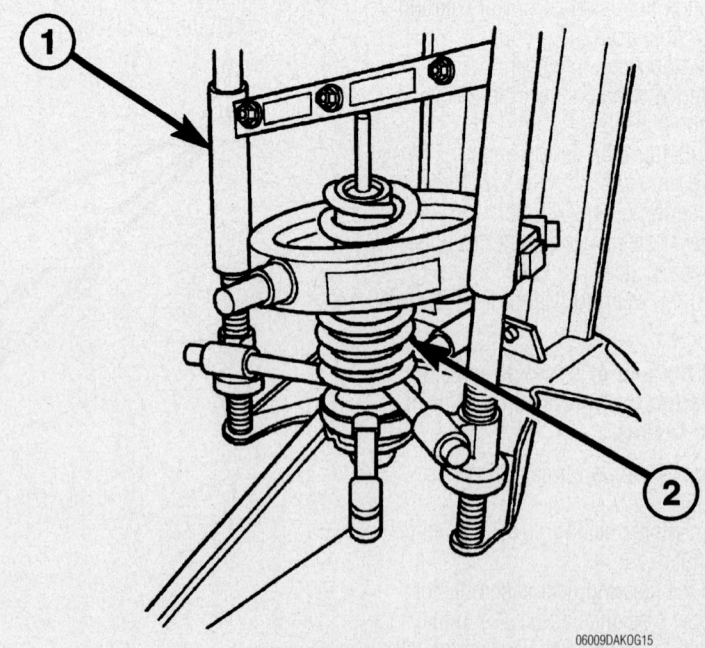

06009DAKOG15

Strut assembly (2) mounted in the compressor (1)

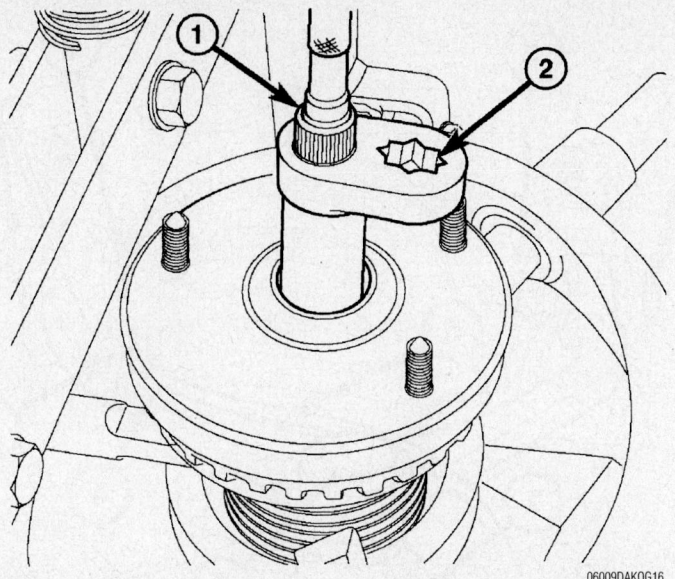

06009DAKOG16

Position Special Tool 9362 (2), or equivalent, on shock shaft retaining nut (1)

15. Install the shock upper mounting nut.

16. Decompress the spring.

17. Remove the shock assembly from the spring compressor tool.

18. Install the shock assembly.

Leaf Spring

REMOVAL & INSTALLATION

1. Before servicing the vehicle, refer to the Precautions Section.

✳✳ CAUTION

The rear of the vehicle must be lifted only with a jack or hoist. The lift must be placed under the frame rail crossmember located aft of the rear axle. Use care to avoid bending the side rail flange.

2. Raise the vehicle at the frame.

3. Use a hydraulic jack to relieve the axle weight.

4. Remove the rear wheel.

5. Remove the nuts, the U-bolts and spring plate from the axle.

6. Loosen and remove the bolt and then remove the flag nut through the access hole in the bracket from the spring front eye.

7. Remove the nut and bolt that attaches the spring shackle to the rear frame bracket.

8. Remove the spring from the vehicle.

9. Remove the shackle from the spring.

To install:

10. Install the spring shackle on the spring finger tight.

11. Position the spring on the rear axle pad. Make sure the spring center bolt is inserted in the pad locating hole.

12. Align front spring eye with the bolt hole in the front frame bracket. Install the spring eye bolt and flag nut through the access hole in the frame and tighten the bolt finger-tight.

13. Align spring shackle eye with the bolt hole in the rear frame bracket. Install the bolt and nut and tighten the spring shackle eye nut finger-tight.

14. Install the U-bolts, spring plate and nuts.

15. Tighten the U-bolt nuts to 110 ft. lbs. (149 Nm).

16. Install the rear wheel.

17. Remove the support stands from under the frame rails. Lower the vehicle until the springs are supporting the weight of the vehicle.

18. Tighten the spring eye pivot bolt and flag nut to 125 ft. lbs. (170 Nm).

19. Tighten the upper shackle bolt and nut and the lower shackle bolt and nut to 125 ft. lbs. (170 Nm).

Upper Ball Joint

REMOVAL & INSTALLATION

The Raider models utilize an upper control arm with an integral ball joint. If the ball joint is damaged or worn, the upper control arm must be replaced.

Lower Ball Joint

REMOVAL & INSTALLATION

1. Before servicing the vehicle, refer to the Precautions Section.

2. Remove the front wheel.

3. Remove the brake caliper and rotor.

4. Remove the outer tie rod retaining nut from the knuckle.

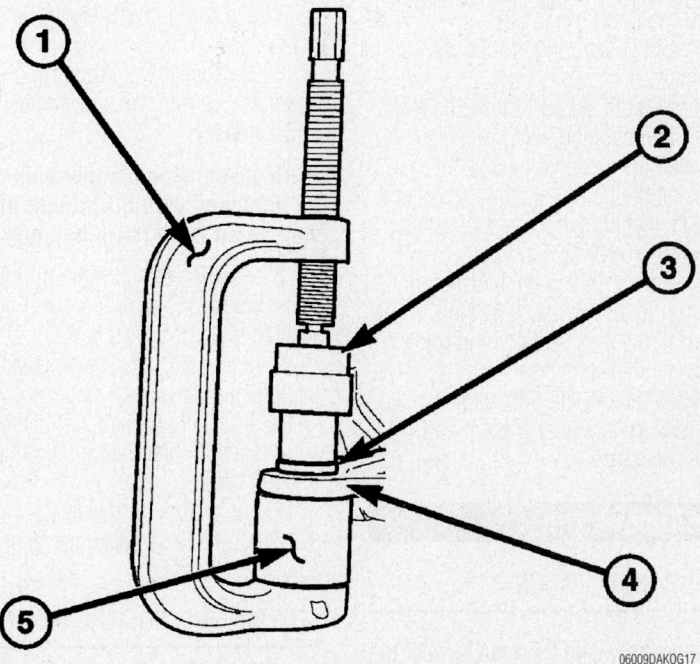

06009DAKOG17

Install the ball joint (3) into the control arm and press in using special tools C-4212-F (1), 8441-4 (2) and 9654-1 (4).

5. Separate the tie rod from the steering knuckle using Special Tool C-3894-A.

6. Remove the upper ball joint nut, then separate the upper ball joint from the knuckle using special tool 8677, or equivalent.

7. Remove the lower ball joint nut, then separate the lower ball joint from the steering knuckle using special tool 8677.

8. Remove the steering knuckle.

9. If equipped with 4WD, move the halfshaft to the side and support the halfshaft out of the way.

10. Remove the snapring from the ball joint flange.

➡Extreme pressure lubrication must be used on the threaded portions of the tool. This will increase the longevity of the tool and insure proper operation during the removal and installation process.

11. Press the ball joint (3) from the lower control arm (4) using special tools C-4212-F (Press) (1), 8445-3 (Driver) (2) and 9604 (Receiver) (5).

To install:

➡Extreme pressure lubrication must be used on the threaded portions of the tool. This will increase the longevity of the tool and insure proper operation during the removal and installation process.

12. Install the ball joint (3) into the control arm and press in using special tools C-4212-F (press) (1), 8441-4 (Receiver) (2) and 9654-1 (Driver) (4).

13. Install the snapring around the ball joint flange.

14. Remove the support for the halfshaft and install into position, 4WD models only.

15. Install the steering knuckle.

16. Install the tie rod end into the steering knuckle, then install the retaining nut and tighten to 75 Nm (55 ft. lbs.).

17. Install and tighten the halfshaft nut (if equipped) to 185 ft. lbs. (251 Nm).

18. Install the brake caliper and rotor.

19. Install the front wheel.

20. Check the vehicle ride height.

21. Check the wheel alignment and adjust, as necessary.

Upper Control Arm

REMOVAL & INSTALLATION

1. Before servicing the vehicle, refer to the Precautions Section.

2. Raise and support vehicle.

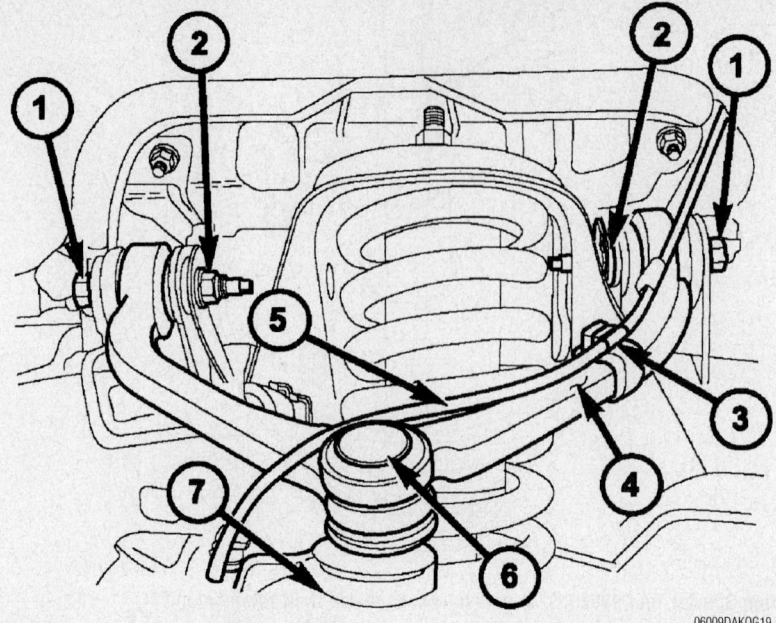

Upper control arm mounting, 2005 models. (1) bolts, (2) nuts, (3) brackets, (4) control arm, (5) ABS wheel speed wire, (6) ball joint, (7) knuckle

3. Remove front wheel.

4. Remove the nut from upper ball joint.

5. Separate the upper ball joint from the steering knuckle with a ball joint puller.

※ WARNING

When installing the tool to separate the ball joint, be careful not to damage the ball joint seal.

6. Remove the wheel speed sensor wire from the retaining brackets to the upper control arm.

7. Remove the control arm pivot bolts and nuts and remove control arm.

To install:

➡All suspension components should be tightened with the weight of the vehicle on them (curb height).

8. Position the control arm into the frame brackets. Install bolts and nuts. Tighten to 102 Nm (75 ft. lbs.).

9. Reposition the wheel speed wire into the retaining brackets.

10. Insert the ball joint in steering knuckle and tighten ball joint nut to 70 ft. lbs. (95 Nm).

11. Install the front wheel.

12. Check the wheel alignment and adjust, as necessary.

CONTROL ARM BUSHING REPLACEMENT

The control arm bushings are serviced with the control arm as an assembly.

Lower Control Arm

REMOVAL & INSTALLATION

1. Before servicing the vehicle, refer to the Precautions Section.

2. Remove the front wheel.

3. Remove the brake caliper assembly and rotor.

4. Disconnect the wheel speed sensor at the wheel well.

5. Remove tie rod end jam nut.

6. Disconnect the tie rod from the knuckle using special tool C-3894-A, or equivalent.

7. Remove the front halfshaft nut, if equipped with 4WD.

8. Remove the upper ball joint nut. Separate the upper ball joint from the steering knuckle with a remover.

9. Remove the lower ball joint nut. Separate the lower ball joint from the steering knuckle with a remover.

10. Remove or disconnect the following:

- Steering knuckle
- Stabilizer bar link
- Shock absorber lower bolt and nut
- Lower control arm bolts, nuts and washers
- Lower control arm

To install:

➡All suspension components should be tightened with the weight of the vehicle on them (curb height).

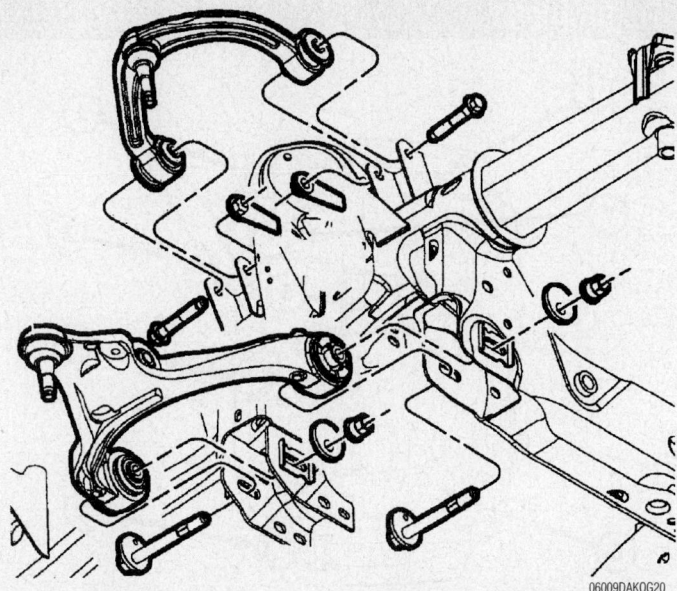

Lower and upper control arm installation

11. Position the lower control arm at the frame rail brackets. Install the pivot bolts washers and nuts. Tighten the nuts finger-tight.

❊❊ CAUTION

The ball joint stud taper must be CLEAN and DRY before installing the knuckle. Clean the stud taper with mineral spirits to remove dirt and grease.

12. Install the steering knuckle.

13. Insert the lower ball joint into the steering knuckle. Install and tighten the retaining nut to 95 ft. lbs. (129 Nm).

14. Install shock absorber lower bolt and nut. Tighten to 155 ft. lbs. (210 Nm).

15. Install the front halfshaft nut, if equipped with 4WD.

16. Insert the upper ball joint into the steering knuckle. Install and tighten the retaining nut to 70 ft. lbs. (95 Nm).

17. Install the stabilizer bar link and tighten to 75 ft. lbs. (102 Nm).

18. Tighten the lower control arm pivot nut and bolts to 105 ft. lbs. (142 Nm).

19. Insert the outer tie rod end into the steering knuckle. Install and tighten the retaining nut to 55 ft. lbs. (75 Nm).

20. Install the brake caliper and rotor.

21. Install the front wheel.

22. Check the wheel alignment and adjust, as necessary.

CONTROL ARM BUSHING REPLACEMENT

The control arm bushings are serviced with the control arm as an assembly.

Wheel Bearings

ADJUSTMENT

The hub/bearing assembly is not adjustable.

REMOVAL & INSTALLATION

1. Before servicing the vehicle, refer to the Precautions Section.

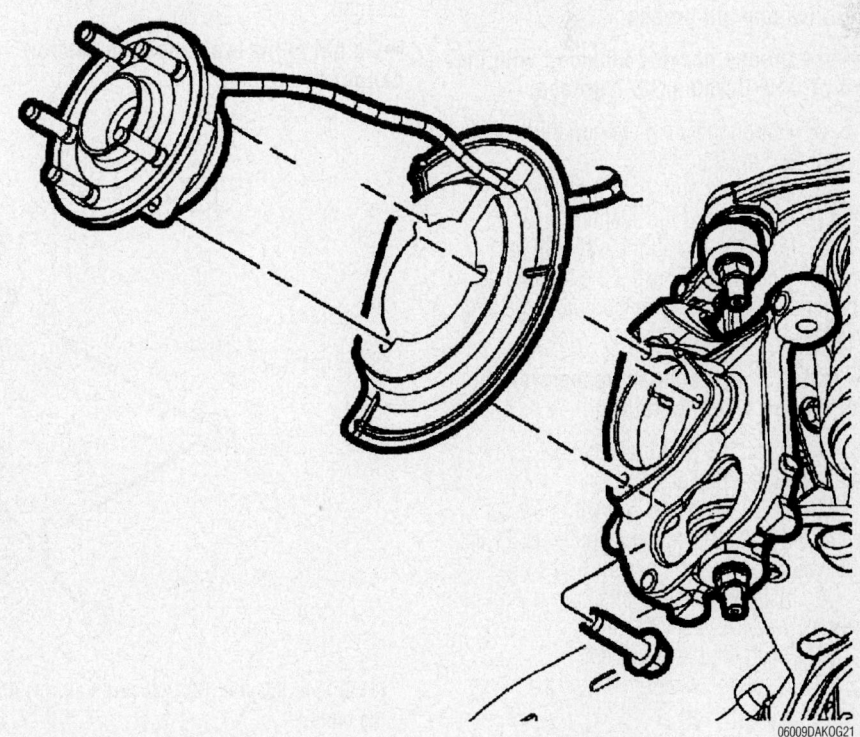

Front hub installation

2. Remove the wheel.

3. Remove the brake caliper and rotor.

4. Remove the ABS wheel speed sensor if equipped.

5. Remove the halfshaft nut, if equipped with 4WD.

❊❊ WARNING

Do not strike the knuckle with a hammer to remove the tie rod end or the ball joint. Damage to the steering knuckle will occur.

6. Pull down on the steering knuckle to separate the halfshaft from the hub/bearing (4WD).

7. Remove the three hub/bearing mounting bolts from the steering knuckle.

8. Slide the hub/bearing out of the steering knuckle.

9. Remove the brake dust shield.

To install:

10. Install the brake dust shield.

11. Install the hub/bearing into the steering knuckle and tighten the bolts to 120 ft. lbs. (163 Nm).

12. Install the brake rotor and caliper.

13. Install the ABS wheel speed sensor if equipped.

14. Install the halfshaft nut, if equipped with 4WD. Tighten to 185 ft. lbs. (251 Nm).

15. Install the wheel.

BRAKES

Brake Caliper

REMOVAL & INSTALLATION

1. Before servicing the vehicle, refer to the Precautions Section.

> **⚡ CAUTION**
>
> **Never allow the disc brake caliper to hang from the brake hose. Damage to the brake hose with result. Provide a suitable support to hang the caliper securely.**

2. Remove the wheel.
3. Compress the disc brake caliper.
4. Remove the banjo bolt and discard the copper washers.
5. Remove the caliper slide pin bolts.
6. Remove the disc brake caliper from the caliper adapter.
7. Remove the caliper slide pins from the adapter.

To install:

> **⚡ WARNING**
>
> **Petroleum based grease should not be used on any of the rubber components of the caliper, Use only Non-Petroleum based grease.**

➡ **Clean slide pin bores thoroughly to remove any old grease.**

➡ **Use grease packets included with the kit or Dow Corning-807T grease.**

8. Thoroughly coat the new slide pins on all working surfaces.
9. Install the boot onto the slide pin and then insert into the adapter.
10. Push the pin all the way into the adapter and carefully expel the trapped air by gently pushing on the boot near the slide pin head.

➡ **Install a new copper washers on the banjo bolt when installing**

11. Install the disc brake caliper to the brake caliper adapter.
12. Install the banjo bolt with new copper washers to the caliper. Tighten to 21 ft. lbs. (28 Nm).
13. Install the caliper slide pin bolts. Tighten to 24 ft. lbs. (32 Nm).
14. Bleed the brake system.
15. Install the wheel.

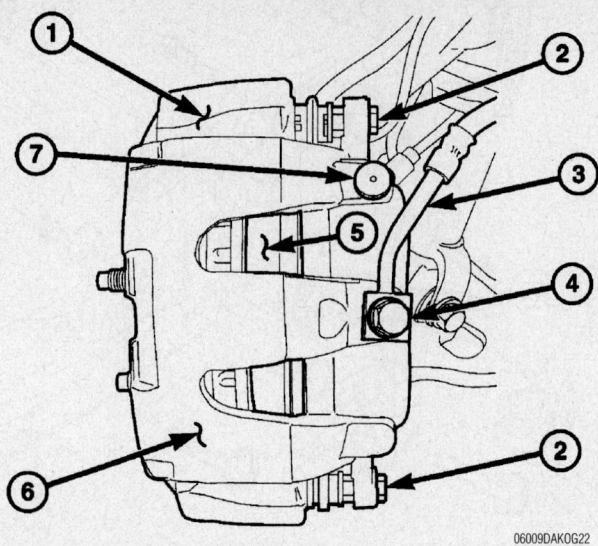

06009DAKOG22

(1) Caliper mounting adaptor, (2) Caliper mounting bolts, (3) Brake line, (4) Banjo bolt, (5) Pad, (6) Caliper

Disc Brake Pads

REMOVAL & INSTALLATION

1. Before servicing the vehicle, refer to the Precautions Section.
2. Remove the wheel.
3. Compress the caliper.
4. Remove the caliper slide pin bolts.
5. Remove the caliper from the caliper adapter.

➡ **Do not allow brake hose to support caliper assembly.**

6. Remove the inboard brake pad from the caliper adapter.
7. Remove the outboard brake pad from the caliper adapter.
8. Remove the anti-rattle clips from the pad.

To install:

9. Bottom pistons in caliper bore with C-clamp. Place an old brake shoe between a C-clamp and caliper piston.
10. Clean the caliper mounting adapter.
11. Install new anti-rattle clips to the brake pads.
12. Install the inboard brake pad in the adapter.

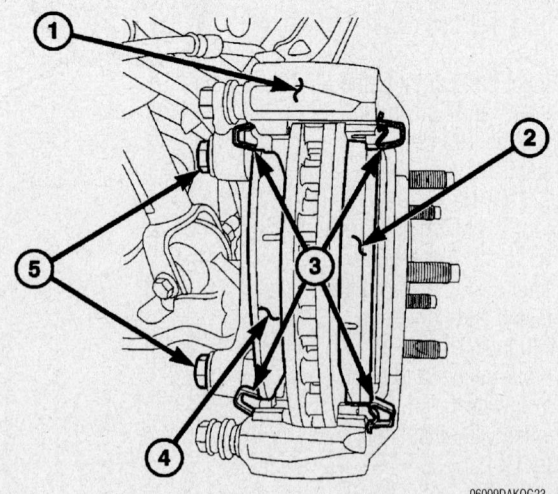

06009DAKOG23

(1) Caliper adapter, (2) Outboard pad, (3) Anti-rattle clips, (4) Inboard pad, (5) Caliper mounting bolts

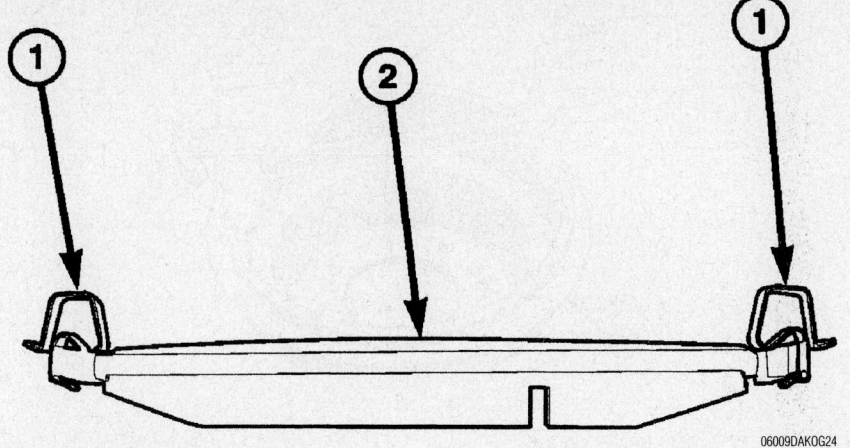

Anti-rattle clips (1) installed on the pad (2)

13. Install the outboard brake pad in the adapter.

14. Install the caliper over rotor. Then, push the caliper onto the adapter.

15. Install the caliper slide pin bolts.

16. Install the wheel.

17. Apply brakes several times to seat caliper pistons and brake shoes and obtain firm pedal.

18. Top off master cylinder fluid level if required.

Brake Drums

REMOVAL & INSTALLATION

1. Before servicing the vehicle, refer to the Precautions Section.

2. Remove the axle shaft nuts, washers and cones. If the cones do not readily release, rap the axle shaft sharply in the center.

3. Remove the axle shaft.

4. Remove the outer hub nut.

5. Straighten the lockwasher tab and remove it along with the inner nut and bearing.

6. Carefully remove the drum.

To install:

7. Position the drum on the axle housing.

8. Install the bearing and inner nut. While rotating the wheel and tire, tighten the adjusting nut until a slight drag is felt.

9. Back off the adjusting nut ⅙ turn so that the wheel rotates freely without excessive end-play.

10. Install the lockrings and nut. Place a new gasket on the hub and install the axle shaft, cones, lockwashers and nuts.

11. Install the wheel and tire.

12. Road-test the vehicle.

Brake Shoes

REMOVAL & INSTALLATION

1. Before servicing the vehicle, refer to the Precautions Section.

2. Remove the wheel.

3. Remove the clip nuts securing brake drum to wheel studs.

4. Remove the drum. If drum is difficult to remove, remove rear plug from access hole in support plate. Back-off self adjusting by inserting a thin screwdriver into access hole and push lever away from adjuster star wheel. Then insert an adjuster tool into brake adjusting hole rotate adjuster star wheel to retract brake shoes.

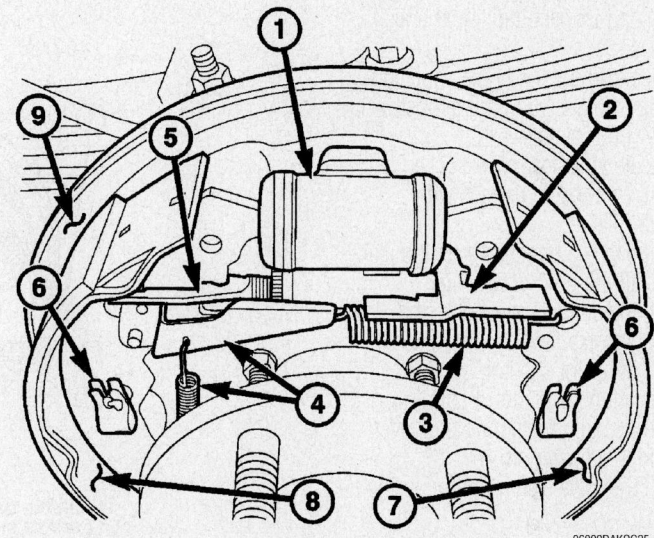

Rear brake parts, 2005 models. (1) wheel cylinder, (2) parking brake lever, (3) return spring, (4 & 5) adjuster spring and lever, (6) hold-down clips, (7 & 8) brake shoes, (9) backing plate

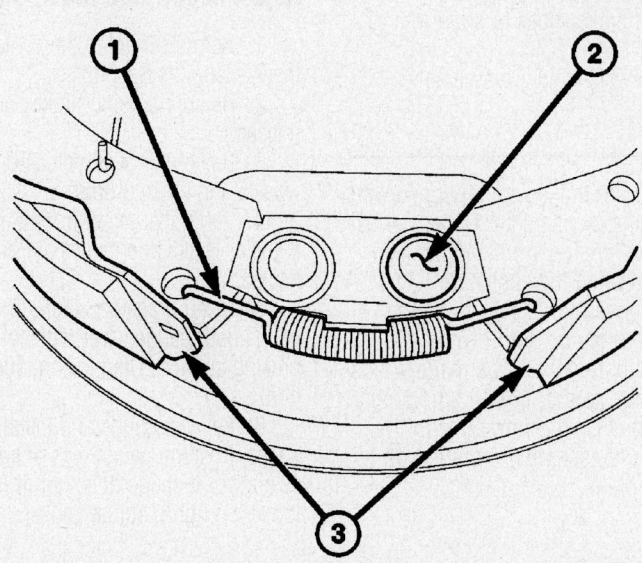

Lower return spring (1), cam (2) and brake shoes, (3)—2005 models

5. Vacuum brake components to remove brake lining dust.

6. Remove shoe return spring with brake spring pliers tool.

7. Remove adjuster spring and lever. Disengage lever from spring by sliding lever forward to clear pivot and work lever out from under spring.

8. Disengage and remove shoe return spring from brake shoes.

9. Remove the brake shoe hold down clips.

10. Remove the rear brake shoe from support plate.

11. Remove the front brake shoe from support plate.

12. Remove the park brake lever from the brake shoe.

To install:

13. Clean and inspect individual brake components.

14. Lubricate where the brake shoe contacts the support plate with high temperature grease or Lubriplate®.

15. Lubricate adjuster screw socket, nut, button and screw thread surfaces with grease or Lubriplate®.

16. Install parking brake lever to the rear shoe and install the hold down clip.

17. Install the adjuster strut onto the shoes and park brake lever.

18. Install the front shoe on support plate, and install the hold down clip.

19. Install the adjuster spring and lever in the slot in the adjuster strut.

20. Install the lower return spring to the shoes.

21. Verify adjuster operation. Pull both shoes outward to move the adjuster lever to rotate the star wheel. Be sure adjuster lever properly engages star wheel teeth.

22. Adjust brake shoes to drum with a brake gauge.

23. Install wheel and tire assembly.

ADJUSTMENTS

The rear drum brakes are equipped with a self-adjusting mechanism. Under normal circumstances, the only time adjustment is required is when the shoes are replaced, removed for access to other parts, or when one or both drums are replaced. Adjustment can be made with a standard brake gauge or with adjusting tool. Adjustment is performed with the complete brake assembly installed on the backing plate.

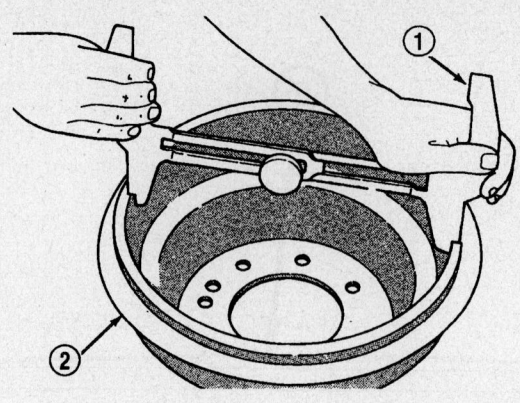

1 - BRAKE GAUGE
2 - BRAKE DRUM

67189DAKOG03

Adjusting gauge on the drum

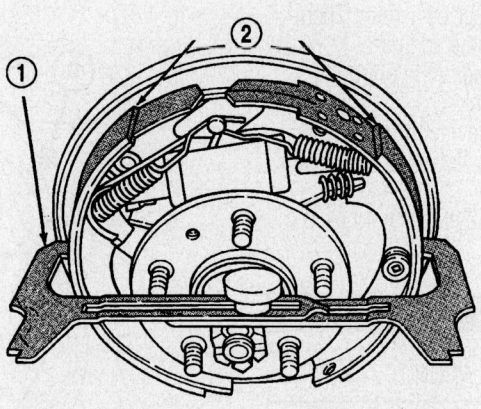

1 - BRAKE GAUGE
2 - BRAKE SHOES

67189DAKOG01

Adjustment with a brake adjusting gauge

Adjustment with a Brake Gauge

1. Before servicing the vehicle, refer to the Precautions Section.

2. Be sure parking brakes are fully released.

3. Raise rear of vehicle and remove wheels and brake drums.

4. Verify that left and right automatic adjuster levers and cables are properly connected.

5. Insert brake gauge in drum. Expand gauge until gauge inner legs contact drum braking surface. Then lock gauge in position.

6. Reverse gauge and install it on brake shoes. Position gauge legs at shoe centers as shown. If gauge does not fit (too loose/too tight), adjust shoes.

7. Pull shoe adjuster lever away from adjuster screw star wheel.

8. Turn adjuster screw star wheel (by hand) to expand or retract brake shoes. Continue adjustment until gauge outside legs are light drag-fit on shoes.

9. Install brake drums and wheels and lower vehicle.

10. Drive vehicle and make one forward stop followed by one reverse stop. Repeat procedure 8-10 times to operate automatic adjusters and equalize adjustment.

➡**Bring vehicle to complete standstill at each stop. Incomplete, rolling stops will not activate automatic adjusters.**

NISSAN

Altima • Sentra

SPECIFICATIONS AND MAINTENANCE CHARTS

ENGINE AND VEHICLE IDENTIFICATION

Engine							Model Year	
Code ①	Liters (cc)	Cu. In.	Cyl.	Fuel Sys.	Engine Type	Eng. Mfg.	Code ②	Year
QG18DE	1.8 (1769)	108	4	MFI	DOHC	Nissan	2	2002
QR25DE	2.5 (2488)	152	4	MFI	DOHC	Nissan	3	2003
VQ35DE	3.5 (3498)	213	6	MFI	DOHC	Nissan	4	2004
							5	2005
							6	2006

MFI: Multi-port Fuel Injection

DOHC: Double Overhead Camshaft

① The Engine Code is stamped on the engine block near the starter.

② 10th position of the Vehicle Identification Number (VIN)

09482_ALTI_C0001

GENERAL ENGINE SPECIFICATIONS

Year	Model	Engine Displacement Liters	Engine Series (ID/VIN)	Net Horsepower @ rpm	Net Torque @ rpm (ft. lbs.)	Bore x Stroke (in.)	Compression Ratio	Oil Pressure @ rpm
2002	Altima	2.5	QR25DE	150@5600	154@5600	3.50X3.94	9.5:1	60@3000
	Altima	3.5	VQ35DE	260@6000	260@4800	3.76X3.20	10.3:1	43@2000
	Sentra	1.8	QG18DE	126@6000	129@2400	3.15x3.46	9.5:1	43@2000
	Sentra	2.5	QR25DE	150@5600	154@5600	3.50x3.94	9.5:1	60@3000
2003	Altima	2.5	QR25DE	150@5600	154@5600	3.50X3.94	9.5:1	60@3000
	Altima	3.5	VQ35DE	260@6000	260@4800	3.76X3.20	10.3:1	43@2000
	Sentra	1.8	QG18DE	126@6000	129@2400	3.15x3.46	9.5:1	43@2000
	Sentra	2.5	QR25DE	150@5600	154@5600	3.50x3.94	9.5:1	60@3000
2004	Altima	2.5	QR25DE	150@5600	154@5600	3.50X3.94	9.5:1	60@3000
	Altima	3.5	VQ35DE	260@6000	260@4800	3.76X3.20	10.3:1	43@2000
	Sentra	1.8	QG18DE	126@6000	129@2400	3.15x3.46	9.5:1	43@2000
	Sentra	2.5	QR25DE	150@5600	154@5600	3.50x3.94	9.5:1	60@3000
2005	Altima	2.5	QR25DE	150@5600	154@5600	3.50x3.94	9.5:1	60@3000
	Altima	3.5	VQ35DE	260@6000	260@4800	3.76X3.20	10.3:1	43@2000
	Sentra	1.8	QG18DE	126@6000	129@2400	3.15x3.46	9.5:1	43@2000
	Sentra	2.5	QR25DE	150@5600	154@5600	3.50x3.94	9.5:1	60@3000
2006	Altima	2.5	QR25DE	175@6000	180@4000	3.50X3.94	9.5:1	60@3000
	Altima	3.5	VQ35DE	260@6000	260@4800	3.76X3.20	10.3:1	43@2000
	Sentra	1.8	QG18DE	126@6000	129@2400	3.15x3.46	9.5:1	43@2000
	Sentra	2.5	QR25DE	150@5600	154@5600	3.50x3.94	9.5:1	60@3000

09482_ALTI_C0002

ENGINE TUNE-UP SPECIFICATIONS

Year	Engine Displacement Liters	Engine ID/VIN	Spark Plug Gap (in.)	Ignition Timing (deg.) MT	Ignition Timing (deg.) AT	Fuel Pump (psi) ①	Idle Speed (rpm) MT	Idle Speed (rpm) AT ②	Valve Clearance Intake ③	Valve Clearance Exhaust ③
2002	1.8	QG18DE	0.043	9B	9B	51	600-700	750-850	0.013-0.016	0.015-0.018
	2.5	QR25DE	0.043	15B	15B	33	650-750	650-750	0.013-0.016	0.013-0.016
	3.5	VQ35DE	0.043	—	15B	34	650-750	650-750	0.012-0.016	0.012-0.017
2003	1.8	QG18DE	0.043	7B	18B	51	600-700	750-850	0.013-0.016	0.015-0.018
	2.5	QR25DE	0.043	15B	15B	33	650-750	650-750	0.013-0.016	0.013-0.016
	3.5	VQ35DE	0.043	—	15B	51	650-750	650-750	0.012-0.016	0.012-0.017
2004	1.8	QG18DE	0.043	7B	18B	51	600-700	750-850	0.013-0.016	0.015-0.018
	2.5	QR25DE	0.043	15B	15B	33	650-750	650-750	0.013-0.016	0.013-0.016
	3.5	VQ35DE	0.043	—	15B	51	650-750	650-750	0.012-0.016	0.012-0.017
2005	1.8	QG18DE	0.043	7B	18B	51	600-700	750-850	0.013-0.016	0.015-0.018
	2.5	QR25DE	0.043	15B	15B	33	650-750	650-750	0.013-0.016	0.013-0.016
	3.5	VQ35DE	0.043	—	15B	51	575-675	625-725	0.012-0.016	0.012-0.017
2006	1.8	QG18DE	0.043	7B	18B	51	600-700	750-850	0.013-0.016	0.015-0.018
	2.5	QR25DE	0.043	15B	15B	33	650-750	650-750	0.013-0.016	0.013-0.016
	3.5	VQ35DE	0.043	—	15B	51	575-675	625-725	0.012-0.016	0.012-0.017

NOTE: The Vehicle Emission Control Information label often reflects specification changes made during production.

The label figures must be used if they differ from those in this chart.

B: Before top dead center

① System pressure at idle with vacuum hose connected; should increase to 43 psi when disconnected

② Automatic transmission in neutral

③ Engine warm

09482_ALTI_C0003

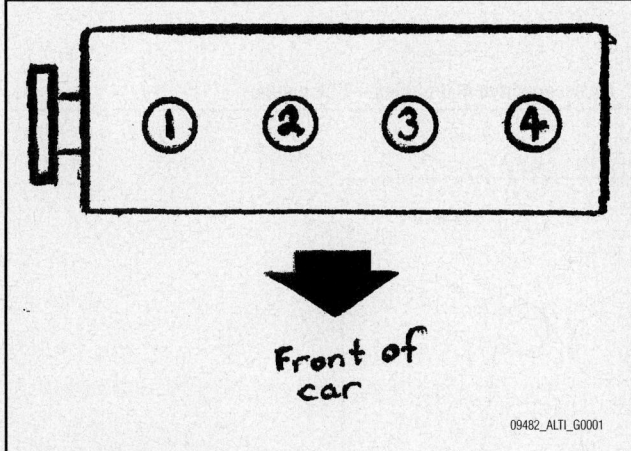

1.8L (QG18DE) Engine
Firing order: 1–3–4–2
Distributorless ignition system (one coil on each cylinder)

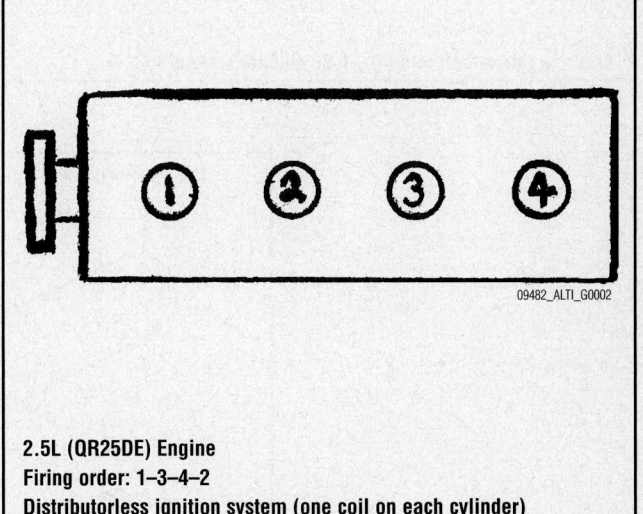

2.5L (QR25DE) Engine
Firing order: 1–3–4–2
Distributorless ignition system (one coil on each cylinder)

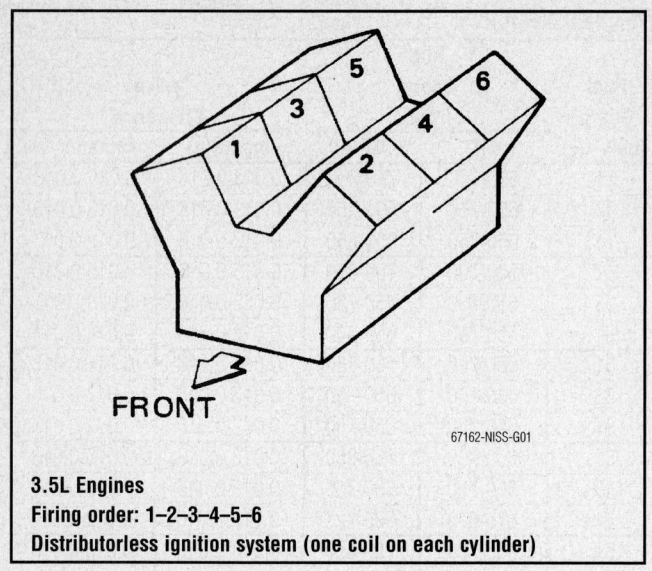

3.5L Engines
Firing order: 1–2–3–4–5–6
Distributorless ignition system (one coil on each cylinder)

67162-NISS-G01

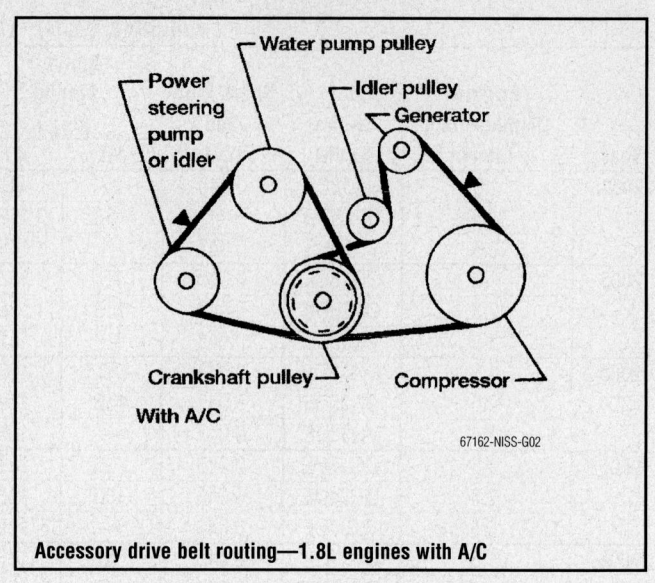

With A/C

67162-NISS-G02

Accessory drive belt routing—1.8L engines with A/C

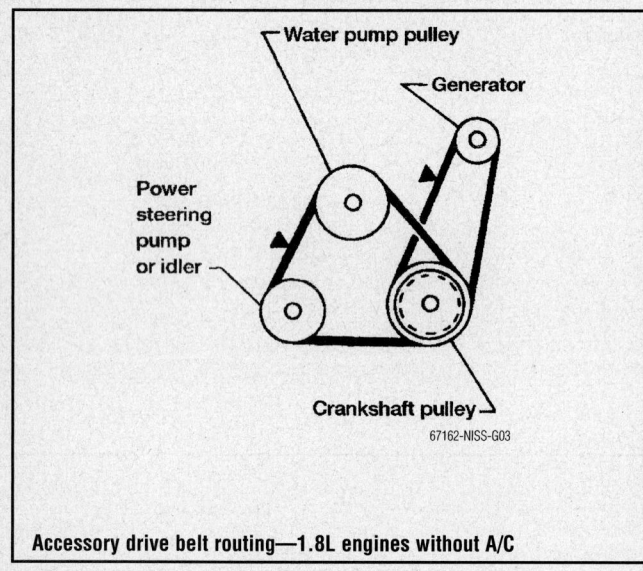

67162-NISS-G03

Accessory drive belt routing—1.8L engines without A/C

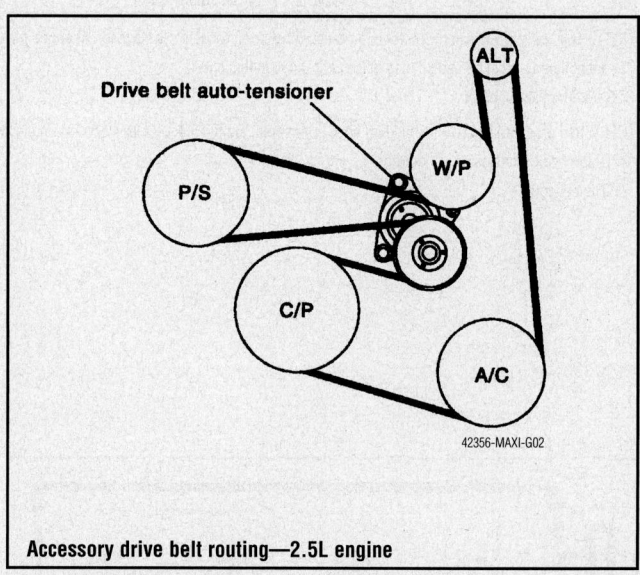

42356-MAXI-G02

Accessory drive belt routing—2.5L engine

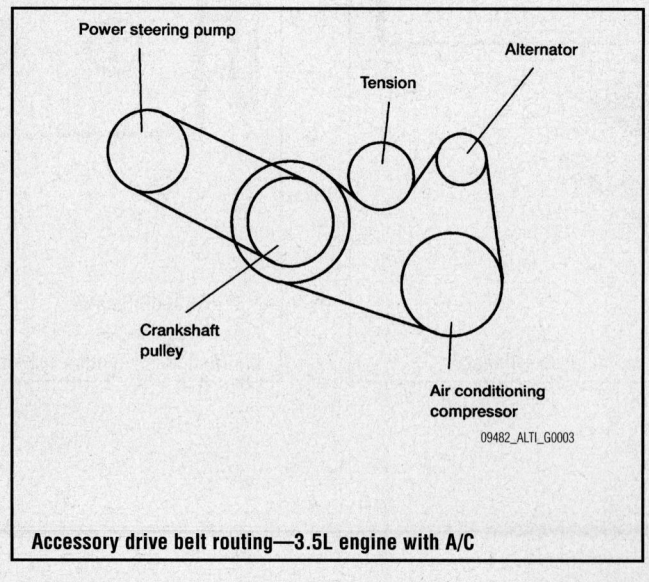

09482_ALTI_G0003

Accessory drive belt routing—3.5L engine with A/C

CAPACITIES

Year	Model	Engine ID/VIN	Engine Displacement Liters	Engine Oil with Filter (qts.)	Transmission (pts.)		Drive Axle Rear (pts.)	Fuel Tank (gal.)	Cooling System (qts.)
					5-Spd	Auto.			
2002	Altima	QR25DE	2.5	4.5	6.8	19.5	—	15.9	8.2
	Altima	VQ35DE	3.5	4.3	6.8	19.0	—	15.9	9.3
	Sentra	QG18DE	1.8	3.5	①	14.8	—	13.2	②
	Sentra	QR25DE	2.5	3.4	7.5	14.8	—	13.0	6.0
2003	Altima	QR25DE	2.5	4.5	6.8	19.5	—	15.9	8.2
	Altima	VQ35DE	3.5	4.3	6.8	19.0	—	15.9	9.3
	Sentra	QG18DE	1.8	3.5	①	14.8	—	13.2	②
	Sentra	QR25DE	2.5	4.2	4.9	18	—	13.2	6.4
2004	Altima	QR25DE	2.5	4.5	4.9	19.5	—	20.0	8
	Altima	VQ35DE	3.5	4.3	4.9	19.5	—	20.0	8.5
	Sentra	QG18DE	1.8	2.8	6.3	14.3	—	13.2	12.5
	Sentra	QR25DE	2.5	4.2	4.9	18	—	13.2	6.4
2005	Altima	QR25DE	2.5	4.5	4.9	19.5	—	20.0	8
	Altima	VQ35DE	3.5	4.3	4.9	19.5	—	20.0	8.5
	Sentra	QG18DE	1.8	2.8	6.3	14.3	—	13.2	12.5
	Sentra	QR25DE	2.5	4.2	4.9	18	—	13.2	6.4
2006	Altima	QR25DE	2.5	4.5	4.9	19.5	—	20.0	8
	Altima	VQ35DE	3.5	4.3	4.9	19.5	—	20.0	8.5
	Sentra	QG18DE	1.8	2.8	6.3	14.3	—	13.2	12.5
	Sentra	QR25DE	2.5	4.2	4.9	18	—	13.2	6.4

NOTE: All capacities are approximate. Add fluid gradually and check to be sure a proper fluid level is obtained.

① RS5F31A: 6.5 pts.
RS5F32V: 8.0 pts.

② GA16DE with MT: 5.5 qts.
GA16DE with AT: 6.0 qts.

09482_ALTI_C0004

VALVE SPECIFICATIONS

Year	Engine ID/VIN	Engine Displacement Liters	Seat Angle (deg.)	Face Angle (deg.)	Spring Test Pressure (lbs. @ in.)	Spring Installed Height (in.)	Stem-to-Guide Clearance (in.)		Stem Diameter (in.)	
							Intake	Exhaust	Intake	Exhaust
2002	SR20DE	1.8	45	45.25-45.75	137@1.181	NA	0.0008-0.0021	0.0016-0.0029	0.2348-0.2354	0.2341-0.2346
	QR25DE	2.5	45.15-45.45	NA	34-39@1.39	NA	0.0009-0.0013	0.010-0.013	0.2348-0.2354	0.2344-0.2350
	VQ35DE	3.5	45.15-45.45	NA	91.5-103.2@1.094	1.796	0.010-0.013	0.011-0.015	0.2348-0.2354	0.2344-0.2350
2003	QG18DE	1.8	44.53-45.07	NA	83@0.931	NA	0.0008-0.0020	0.0016-0.0028	0.2152-0.2157	0.2144-0.2150
	QR25DE	2.5	45.15-45.45	NA	34-39@1.39	NA	0.0009-0.0013	0.010-0.013	0.2348-0.2354	0.2344-0.2350
	VQ35DE	3.5	45.15-45.45	NA	91.5-103.2@1.094	NA	0.010-0.013	0.011-0.015	0.2348-0.2354	0.2344-0.2350
2004	QG18DE	1.8	44.53-45.07	NA	83@0.931	NA	0.0008-0.0020	0.0016-0.0028	0.2152-0.2157	0.2144-0.2150
	QR25DE	2.5	45.15-45.45	NA	34-39@1.39	NA	0.0008-0.0021	0.0012-0.0025	0.2348-0.2354	0.2344-0.2350
	VQ35DE	3.5	45.15-45.45	NA	91.5-103.2@1.094	NA	0.010-0.013	0.011-0.015	0.2348-0.2354	0.2344-0.2350
2005	QG18DE	1.8	44.53-45.07	NA	83@0.931	NA	0.0008-0.0020	0.0016-0.0028	0.2152-0.2157	0.2144-0.2150
	QR25DE	2.5	45.15-45.45	NA	34-39@1.39	NA	0.0008-0.0021	0.0012-0.0025	0.2348-0.2354	0.2344-0.2350
	VQ35DE	3.5	45.15-45.45	NA	91.5-103.2@1.094	NA	0.010-0.013	0.011-0.015	0.2348-0.2354	0.2344-0.2350
2006	QG18DE	1.8	44.53-45.07	NA	83@0.931	NA	0.0008-0.0020	0.0016-0.0028	0.2152-0.2157	0.2144-0.2150
	QR25DE	2.5	45.15-45.45	NA	34-39@1.39	NA	0.0008-0.0021	0.0012-0.0025	0.2348-0.2354	0.2344-0.2350
	VQ35DE	3.5	45.15-45.45	NA	91.5-103.2@1.094	NA	0.010-0.013	0.011-0.015	0.2348-0.2354	0.2344-0.2350

NA: Not Available

09482_ALTI_C0005

CRANKSHAFT AND CONNECTING ROD SPECIFICATIONS
All measurements are given in inches.

| Year | Engine Displacement Liters | Engine ID/VIN | Crankshaft | | | | Connecting Rod | | |
			Main Brg. Journal Dia.	Main Brg. Oil Clearance	Shaft End-play	Thrust on No.	Journal Diameter	Oil Clearance	Side Clearance
2002	1.8	QG18DE	1.9668-1.9671	0.0007-0.0017	0.0024-0.0089	3	1.6929-1.6934	0.0006-0.0015	0.0079-0.0185
	2.5	QR25DE	2.1636-2.1645	①	0.0039-0.0102	3	1.8898-1.8903	0.0004-0.0014	0.0079-0.0138
	3.5	VQ35DE	2.3603-2.3612	0.0014-0.0021	0.0039-0.0098	3	1.7704-1.7706	0.0013-0.0023	0.0079-0.0138
2003	1.8	QG18DE	1.9668-1.9671	0.0007-0.0017	0.0024-0.0089	3	1.6929-1.6934	0.0006-0.0015	0.0079-0.0185
	2.5	QR25DE	2.1636-2.1645	①	0.0039-0.0102	3	1.8898-1.8903	0.0004-0.0014	0.0079-0.0138
	3.5	VQ35DE	2.3603-2.3612	0.0014-0.0021	0.0039-0.0098	3	1.7704-1.7706	0.0013-0.0023	0.0079-0.0138
2004	1.8	QG18DE	1.9668-1.9671	0.0007-0.0017	0.0024-0.0089	3	1.6929-1.6934	0.0006-0.0015	0.0079-0.0185
	2.5	QR25DE	2.1636-2.1645	①	0.0039-0.0102	3	1.8898-1.8903	0.0011-0.0018	0.0079-0.0138
	3.5	VQ35DE	2.3603-2.3612	0.0014-0.0021	0.0039-0.0098	3	1.7704-1.7706	0.0013-0.0023	0.0079-0.0138
2005	1.8	QG18DE	1.9668-1.9671	0.0007-0.0017	0.0024-0.0089	3	1.6929-1.6934	0.0006-0.0015	0.0079-0.0185
	2.5	QR25DE	2.1636-2.1645	①	0.0039-0.0102	3	1.8898-1.8903	0.0011-0.0018	0.0079-0.0138
	3.5	VQ35DE	2.3603-2.3612	0.0014-0.0021	0.0039-0.0098	3	1.7704-1.7706	0.0013-0.0023	0.0079-0.0138
2006	1.8	QG18DE	1.9668-1.9671	0.0007-0.0017	0.0024-0.0089	3	1.6929-1.6934	0.0006-0.0015	0.0079-0.0185
	2.5	QR25DE	2.1636-2.1645	①	0.0039-0.0102	3	1.8898-1.8903	0.0011-0.0018	0.0079-0.0138
	3.5	VQ35DE	2.3603-2.3612	0.0014-0.0018	0.0039-0.0098	3	1.7704-1.7706	0.0013-0.0023	0.0079-0.0138

① Nos. 1, 3 and 5: 0.0005-0.0009 in.
  Nos. 2 and 4: 0.0007-0.0011

09482_ALTI_C0006

PISTON AND RING SPECIFICATIONS
All measurements are given in inches.

Year	Engine Displacement Liters	Engine ID/VIN	Piston Clearance	Ring Gap			Ring Side Clearance		
				Top Compression	Bottom Compression	Oil Control	Top Compression	Bottom Compression	Oil Control
2002	1.8	QG18DE	0.0010-0.0018	0.0079-0.0154	0.0126-0.0220	0.0079-0.0272	0.0018-0.0031	0.0012-0.0028	0.0026-0.0053
	2.5	QR25DE	0.0004-0.0012	0.0083-0.0122	0.0126-0.0185	0.0079-0.0236	0.0018-0.0031	0.0012-0.0028	0.0026-0.0053
	3.5	VQ35DE	0.0004-0.0012	0.0091-0.0130	0.0130-0.0189	0.0079-0.0197	0.0018-0.0031	0.0012-0.0028	0.0026-0.0053
2003	1.8	QG18DE	0.0010-0.0018	0.0079-0.0154	0.0126-0.0220	0.0079-0.0272	0.0018-0.0031	0.0012-0.0028	0.0026-0.0053
	2.5	QR25DE	0.0004-0.0012	0.0083-0.0122	0.0126-0.0185	0.0079-0.0236	0.0018-0.0031	0.0012-0.0028	0.0026-0.0053
	3.5	VQ35DE	0.0004-0.0012	0.0091-0.0130	0.0130-0.0189	0.0079-0.0197	0.0018-0.0031	0.0012-0.0028	0.0026-0.0053
2004	1.8	QG18DE	0.0010-0.0018	0.0079-0.0154	0.0126-0.0220	0.0079-0.0272	0.0018-0.0031	0.0012-0.0028	0.0026-0.0053
	2.5	QR25DE	0.0004-0.0012	0.0083-0.0122	0.0126-0.0185	0.0079-0.0236	0.0018-0.0031	0.0012-0.0028	0.0026-0.0053
	3.5	VQ35DE	0.0004-0.0012	0.0091-0.0130	0.0130-0.0189	0.0079-0.0197	0.0018-0.0031	0.0012-0.0028	0.0026-0.0053
2005	1.8	QG18DE	0.0010-0.0018	0.0079-0.0154	0.0126-0.0220	0.0079-0.0272	0.0018-0.0031	0.0012-0.0028	0.0026-0.0053
	2.5	QR25DE	0.0004-0.0012	0.0083-0.0122	0.0126-0.0185	0.0079-0.0236	0.0018-0.0031	0.0012-0.0028	0.0026-0.0053
	3.5	VQ35DE	0.0004-0.0012	0.0091-0.0130	0.0130-0.0189	0.0079-0.0197	0.0018-0.0031	0.0012-0.0028	0.0026-0.0053
2006	1.8	QG18DE	0.0010-0.0018	0.0079-0.0154	0.0126-0.0220	0.0079-0.0272	0.0018-0.0031	0.0012-0.0028	0.0026-0.0053
	2.5	QR25DE	0.0004-0.0012	0.0083-0.0122	0.0126-0.0185	0.0079-0.0236	0.0018-0.0031	0.0012-0.0028	0.0026-0.0053
	3.5	VQ35DE	0.0004-0.0012	0.0091-0.0130	0.0130-0.0189	0.0079-0.0197	0.0018-0.0031	0.0012-0.0028	0.0026-0.0053

09482_ALTI_C0007

TORQUE SPECIFICATIONS

All readings in ft. lbs.

Year	Engine Displacement Liters (cc)	Engine ID/VIN	Cylinder Head Bolts	Main Bearing Bolts	Rod Bearing Bolts	Crankshaft Damper Bolts	Flywheel Bolts	Manifold		Spark Plugs	Oil Drain Plug
								Intake	Exhaust		
2002	1.8 (1769)	QG18DE	①	34-38	②	98-112	③	13-15	20-22	18	25
	2.5 (2488)	QR25DE	⑬	⑭	⑮	⑪	76-83	13-15	29-32	18	25
	3.5 (3498)	VQ35DE	⑯	⑰	⑱	⑪	61-69	⑲	21-24	18	25
2003	1.8 (1769)	QG18DE	①	34-38	②	98-112	③	13-15	20-22	18	25
	2.5 (2488)	QR25DE	⑬	⑭	⑮	⑪	76-83	13-15	29-32	18	25
	3.5 (3498)	VQ35DE	⑯	⑰	⑱	⑪	61-69	⑲	21-24	18	25
2004	1.8 (1769)	QG18DE	①	34-38	②	98-112	③	13-15	20-22	18	25
	2.5 (2488)	QR25DE	⑬	⑭	⑮	⑪	76-83	13-15	29-32	18	25
	3.5 (3498)	VQ35DE	⑯	⑰	⑱	⑪	61-69	⑲	21-24	18	25
2005	1.8 (1769)	QG18DE	①	34-38	②	98-112	③	13-15	20-22	18	25
	2.5 (2488)	QR25DE	⑬	⑭	⑮	⑪	76-83	13-15	29-32	18	25
	3.5 (3498)	VQ35DE	⑯	⑰	⑱	⑪	61-69	⑲	21-24	18	25
2006	1.8 (1769)	QG18DE	①	34-38	②	98-112	③	13-15	20-22	18	25
	2.5 (2488)	QR25DE	⑬	⑭	⑮	⑪	76-83	13-15	29-32	18	25
	3.5 (3498)	VQ35DE	⑯	⑰	⑱	⑪	61-69	⑲	21-24	18	25

① Bolt Nos. 1-10:
Step 1: 22 ft. lbs.
Step 2: 43 ft. lbs.
Step 3: Loosen completely then retorque to 22 ft. lbs.
Step 4: 43 ft. lbs. or an additional 50-55 degrees
Bolt Nos. 11-14: Torque last, to 74 inch lbs.

② Step 1: 12 ft. lbs.
Step 2: 19 ft. lbs. or an additional 35-40 degrees

③ Manual transmission: 61-69 ft. lbs.
Automatic transmission: 69-76 ft. lbs.

④ Step 1: 29 ft. lbs.
Step 2: 58 ft. lbs.
Step 3: Loosen completely then retorque to 30 ft. lbs.
Step 4: Turn each bolt, in sequence,
an additional 90-100 degrees
Step 5: Repeat Step 4

⑤ Step 1: 20-24 ft. lbs.
Step 2: 75-80 degrees
Step 3: Loosen completely and retorque to 24-28 ft. lbs.
Step 4: 45-50 degree turn

⑥ 12 ft. lbs. plus an additional 60-65 degrees

⑦ Step 1: 22 ft. lbs.
Step 2: 58 ft. lbs.
Step 3: Loosen completely then retorque to 22 ft. lbs.
Step 4: 58 ft. lbs. or an additional 80-85 degrees

⑧ Step 1: 29-36 ft. lbs.
Step 2: Plus 60-65 degrees

⑨ Step 1: 3.6-7.2 ft. lbs.
Step 2: 20-23 ft. lbs.

⑩ Step 1: 29 ft. lbs.
Step 2: 90 ft. lbs.
Step 3: Loosen completely and retorque to 25-33 ft. lbs.
Step 4: Plus 90 ft. lbs. or 70 degrees
Step 5: Tighten two bolts marked with an "X" to 7-9 ft. lbs.

⑪ Step 1: 29-36 ft. lbs.
Step 2: 60-66 degrees

⑫ Step 1: 10-12 ft. lbs.
Step 2: 43-48 ft. lbs. or an additional 60-65 degrees

⑬ Step 1: 72 ft. lbs.
Step 2: Loosen completely, then retorque to 26-32 ft. lbs.
Step 3: Turn each bolt, in sequence, an additional 75-80 degrees
Step 4: Turn each bolt, in sequence, an additional 75-80 degrees

⑭ Bolt Nos. 1-10:
Step 1: 27-31 ft. lbs.
Step 2: Torque an additional 60-65 degrees
Bolt Nos. 11-14: Torque last, to 17-20 ft. lbs.

⑮ Step 1: 14-15 ft. lbs.
Step 2: 85-95 degrees

⑯ Step 1: 72 ft. lbs.
Step 2: Loosen bolts completely
Step 3: 26-32 ft. lbs.
Step 4: Tighten an additional 90-95 degrees
Step 5: Repeat Step 4

⑰ Step 1: Shift crankshaft to align the bearing beam
Step 2: Tighten all bolts to 24-28 ft. lbs.
Step 3: Tighten an additional 90-95 degrees

⑱ Step 1: Tighten to 15 ft. lbs.
Step 2: Tighten an additional 90-95 degrees

⑲ Step 1: Tighten to 4-7 ft. lbs.
Step 2: Tighten to 20-23 ft. lbs.

09482_ALTI_C0008

WHEEL ALIGNMENT

Year	Model		Caster Range (+/-Deg.)	Caster Preferred Setting (Deg.)	Camber Range (+/-Deg.)	Camber Preferred Setting (Deg.)	Toe-in (in.)
2002	Altima	F	0.75	+2.66	0.75	-0.10	0.04 +/- 0.04
		R	—	—	0.75	-1.25	0.08 +/- 0.04
	Sentra	F	0.75	+1.42	0.75	-0.58	0.08 +/- 0.08
		R	—	—	0.75	+1.00	0.04 +/- 0.15
2003	Altima	F	0.75	+2.66	0.75	-0.10	0.04 +/- 0.04
		R	—	—	0.75	-1.25	0.08 +/- 0.04
	Sentra	F	0.75	+1.42	0.75	-0.58	0.08 +/- 0.08
		R	—	—	0.75	+1.00	0.04 +/- 0.15
2004	Altima ③	F	0.75	+2.83	0.75	-0.25	0.04 +/- 0.04
		R	—	—	0.30	-1.25	0.15 +/- 0.06
	Altima ④	F	0.75	+2.83	0.75	-0.34	0.04 +/- 0.04
		R	—	—	0.30	-0.25	0.16 +/- 0.06
	Sentra ⑤	F	0.75	+1.60	0.75	-0.42	0.08 +/- 0.04
		R	—	—	0.75	-1.00	0.04 +/- 0.16
	Sentra ⑥	F	0.75	+1.72	0.75	-0.45	0.08 +/- 0.04
		R	—	—	0.75	-1.00	0.04 +/- 0.16
2005	Altima ③	F	0.75	+2.83	0.75	-0.25	0.04 +/- 0.04
		R	—	—	0.30	-1.25	0.15 +/- 0.06
	Altima ④	F	0.75	+2.83	0.75	-0.34	0.04 +/- 0.04
		R	—	—	0.30	-0.25	0.16 +/- 0.06
	Sentra ⑤	F	0.75	+1.60	0.75	-0.42	0.08 +/- 0.04
		R	—	—	0.75	-1.00	0.04 +/- 0.16
	Sentra ⑥	F	0.75	+1.72	0.75	-0.45	0.08 +/- 0.04
		R	—	—	0.75	-1.00	0.04 +/- 0.16
2006	Altima ③	F	0.75	+2.83	0.75	-0.25	0.04 +/- 0.04
		R	—	—	0.30	-1.25	0.15 +/- 0.06
	Altima ④	F	0.75	+2.83	0.75	-0.34	0.04 +/- 0.04
		R	—	—	0.30	-0.25	0.16 +/- 0.06
	Sentra ⑤	F	0.75	+1.60	0.75	-0.42	0.08 +/- 0.04
		R	—	—	0.75	-1.00	0.04 +/- 0.16
	Sentra ⑥	F	0.75	+1.72	0.75	-0.45	0.08 +/- 0.04
		R	—	—	0.75	-1.00	0.04 +/- 0.16

① With P225/55R16, P215/55R16 tires
② With P205/65R15 tires
③ With 2.5L engine
④ With 3.5L engine
⑤ With 1.8L engine
⑥ With 2.5L engine
⑦ 0.04 +/- 0.03 with 17 inch tire
   0.07 +/- 0.03 with 18 inch tire

09482_ALTI_C0009

TIRE, WHEEL AND BALL JOINT SPECIFICATIONS

Year	Model	OEM Tires		Tire Pressures (psi)		Wheel Size	Lug Nut Torque Ft. Lbs.
		Standard	Optional	Front	Rear		
2002	Altima	P195/65R15	P205/60R15	30	30	6-JJ	80
	Sentra, base	P155/80R13	None	26	26	5-J	80
	Sentra XE	P175/70R13	None	26	26	5-J	80
	Sentra GXE	P175/65R14	None	26	26	5.5-JJ	80
	Sentra GLE	P175/65R14	None	26	26	5.5-JJ	80
	Sentra SE	P195/55R15	None	30	30	6-JJ	80
2003	Altima	P195/65R15	P205/60R15	30	30	6-JJ	80
	Sentra, base	P155/80R13	None	26	26	5-J	80
	Sentra XE	P175/70R13	None	26	26	5-J	80
	Sentra GXE	P175/65R14	None	26	26	5.5-JJ	80
	Sentra GLE	P175/65R14	None	26	26	5.5-JJ	80
	Sentra SE	P195/55R15	None	30	30	6-JJ	80
2004	Altima	P205/65TR16	P215/55R17	30	30	6.5-JJ/7-JJ	80
	Sentra	P195/60HR15	P195/55HR16	33	30	6-JJ	80
	Sentra SER-V	P215/45ZR17	None	33	33	7-JJ	80
2005	Altima	P215/60TR16	None	29	29	6.5-JJ	80
	Altima 3.5 SE	P215/55VR17	P215/55HR17	33	30	7-JJ	80
	Altima SE-R	P225/45YR18	None	35	35	8-JJ	80
	Sentra	P195/60HR15	P195/55HR16	33	30	6-JJ	80
	Sentra SER-V	P215/45ZR17	None	33	33	7-JJ	80
2006	Altima	P215/60TR16	None	29	29	6.5-JJ	80
	Altima 3.5 SE	P215/55VR17	P215/55HR17	33	30	7-JJ	80
	Altima SE-R	P225/45YR18	None	35	35	8-JJ	80
	Sentra	P195/60HR15	P195/55HR16	33	30	6-JJ	80
	Sentra SER-V	P215/45ZR17	None	33	33	7-JJ	80

OEM: Original Equipment Manufacturer

PSI: Pounds Per Square Inch

09482_ALTI_C0010

BRAKE SPECIFICATIONS
All measurements in inches unless noted

Year	Model		Brake Disc Original Thickness	Brake Disc Minimum Thickness	Brake Disc Maximum Run-out	Brake Drum Diameter Original Inside Diameter	Brake Drum Diameter Max. Wear Limit	Brake Drum Diameter Maximum Machine Diameter	Minimum Lining Thickness Front	Minimum Lining Thickness Rear	Brake Caliper Bracket Bolts (ft. lbs.)	Brake Caliper Mounting Bolts (ft. lbs.)
2002	Altima	F	1.020	0.866	0.003	—	—	—	0.079	—	53-72	16-23
		R	0.350	0.310	0.003	—	—	—	—	0.059	—	—
	Sentra	F	0.710	0.630	0.003	—	—	—	0.079	—	40-47	12-14
		R	0.280	0.236	0.003	7.09	7.13	7.13	—	0.059	—	—
2003	Altima	F	1.020	0.866	0.003	—	—	—	0.079	—	53-72	16-23
		R	0.350	0.310	0.003	—	—	—	—	0.059	—	—
	Sentra	F	0.710	0.630	0.003	—	—	—	0.079	—	40-47	12-14
		R	0.280	0.236	0.003	7.09	7.13	7.13	—	0.059	—	—
2004	Altima	F	1.020	0.866	0.003	—	—	—	0.079	—	53-72	16-23
		R	0.350	0.310	0.003	—	—	—	—	0.059	28-38	23-30
	Sentra 1.8L	F	0.870	0.790	0.003	—	—	—	0.079	—	53-72	16-23
		R	—	—	—	8.05	—	—	—	0.059	—	—
	Sentra 2.5L	F	1.181	1.118	0.002	—	—	—	0.079	—	112	—
		R	0.350	0.315	0.003	—	—	—	—	0.059	28-38	16-23
2005	Altima	F	1.020	0.866	0.003	—	—	—	0.079	—	53-72	16-23
		R	0.350	0.310	0.003	—	—	—	—	0.059	28-38	23-30
	Sentra 1.8L	F	0.870	0.790	0.003	—	—	—	0.079	—	53-72	16-23
		R	—	—	—	8.05	—	—	—	0.059	—	—
	Sentra 2.5L	F	1.181	1.118	0.002	—	—	—	0.079	—	112	—
		R	0.350	0.315	0.003	—	—	—	—	0.059	28-38	16-23
2006	Altima	F	1.020	0.866	0.003	—	—	—	0.079	—	53-72	16-23
		R	0.350	0.310	0.003	—	—	—	—	0.059	28-38	23-30
	Sentra 1.8L	F	0.870	0.790	0.003	—	—	—	0.079	—	53-72	16-23
		R	—	—	—	8.05	—	—	—	0.059	—	—
	Sentra 2.5L	F	1.181	1.118	0.002	—	—	—	0.079	—	112	—
		R	0.350	0.315	0.003	—	—	—	—	0.059	28-38	16-23

NA: Not Available

① Front brake model CLZ25VD, rear brake model AD14VE.

② Front brake model OPB27VA, rear brake model OPB13VB.

09482_ALTI_C0011

SCHEDULED MAINTENANCE INTERVALS
Nissan—Altima & Sentra

TO BE SERVICED	TYPE OF SERVICE	VEHICLE MILEAGE INTERVAL (x1000)												
		7.5	15	22.5	30	37.5	45	52.5	60	67.5	75	82.5	90	97.5
Engine oil & filter	R	✓	✓	✓	✓	✓	✓	✓	✓	✓	✓	✓	✓	✓
Brake lines & cables	S/I		✓		✓		✓		✓		✓		✓	
Brake pads, discs, drums & linings	S/I		✓		✓		✓		✓		✓		✓	
Driveshaft boots	S/I		✓		✓		✓		✓		✓		✓	
Exhaust system	S/I				✓				✓				✓	
Transaxle fluid	S/I		✓		✓		✓		✓		✓		✓	
Air cleaner filter	R				✓				✓				✓	
Spark plugs (except platinum)	R				✓				✓				✓	
Spark plugs (platinum tip)	R								✓					
Idle RPM (Sentra)	S/I				✓				✓				✓	
Steering gear & linkage, axle & suspension parts	S/I				✓				✓				✓	
Engine coolant	R								✓					
Timing belt	R								✓					
Drive belts	S/I								✓					
Fuel lines	S/I								✓					
Vapor lines	S/I								✓					

R: Replace   S/I: Service or Inspect

FREQUENT OPERATION MAINTENANCE (SEVERE SERVICE)

If a vehicle is operated under any of the following conditions it is considered severe service:

- Extremely dusty areas.
- 50% or more of the vehicle operation is in 32°C (90°F) or higher temperatures, or constant operation in temperatures below 0°C (32°F).
- Prolonged idling (vehicle operation in stop and go traffic).
- Frequent short running periods (engine does not warm to normal operating temperatures).
- Police, taxi, delivery usage or trailer towing usage.

Oil & oil filter: change every 3750 miles.

Brake pads & discs: service or inspect every 7500 miles.

Driveshaft boots: service or inspect every 7500 miles.

Exhaust system: service or inspect every 7500 miles.

Steering gear & linkage, axle & suspension parts: service or inspect every 7500 miles.

Steering linkage ball joints & front suspension ball joints: service or inspect every 7500 miles.

Air cleaner filter: service or inspect every 15,000 miles.

09482_ALTI_C0012

ENGINE REPAIR

Distributor

REMOVAL

The Nissan 1.8L, 2.5L and 3.5L engines are equipped with a Distributorless Ignition System (DIS).

Ignition Timing

ADJUSTMENT

➡ **The ignition timing is not adjustable. If not within specifications, further diagnostic inspectn is required. The following procedure is for viewing the ignition timing setting.**

Visually check the air cleaner, intake hoses, ducts, Exhaust Gas Recirculation (EGR) valve operation and electrical connections prior to the adjustment of the ignition timing. Correct or repair any problem as required. Be sure to inspect the throttle valve and Throttle Position (TP) sensor for proper operation.

1. Before servicing the vehicle, refer to the Precautions Section.
2. Locate the timing marks on the crankshaft pulley and the front of the engine.
3. Clean the timing marks.
4. The ignition timing specifications are as follows:
   - 4–14 degrees BTDC for the 1.8L engine (2002 model only)
   - 13–23 degrees BTDC for the 1.8L engine (2003–2006 models with A/T)
   - 2–12 degrees BTDC for the 1.8L engine (2003–2006 models with M/T)
   - 10–20 degrees Before Top Dead Center (BTDC) for 2.5L and 3.5L engines
5. Using chalk or white paint, color the mark on the crankshaft pulley and the mark on the scale, that will indicate the correct timing when aligned with the notch on the crankshaft pulley.
6. Attach a tachometer to the engine.
7. Attach a timing light to the engine to number 1 cylinder ignition wire.
8. Turn all electrical equipment and accessories **OFF**.
9. Check to be sure all of the wires clear the fan, then, start the engine and allow it to reach normal operating temperatures.

10. Block the front wheels and set the parking brake. Shift the transmission into **NEUTRAL** for manual transmission and automatic transmissions. Do not stand in front of the vehicle when making adjustments.
11. Perform the following procedures:
    a. Race the engine at 2000 rpm for about 2 minutes under a no-load condition; be sure all of the accessories are turned **OFF**.
    b. Perform on board engine diagnostics and repair any fault code.
    c. Race the engine at 2000 rpm for about 2 minutes under a no-load condition.
    d. Turn the engine **OFF** and disconnect the TP sensor.
    e. Start and race the engine 2–3 times under no-load, then run the engine at idle speed.
12. Aim the timing light at the timing marks. If the marks on the pulley and the engine are aligned when the light flashes, the timing is correct. Turn the engine **OFF** and remove the tachometer and the timing light. If the marks are not in alignment, proceed with the following steps.
13. Turn the engine **OFF**.
14. Check the Camshaft Position (CMP) sensor (PHASE), Crankshaft Position (CKP) sensor (REF) and CKP sensor (POS). Replace if necessary.
15. If the ignition timing is still not correct, substitute a known good Electronic Control Module (ECM).

➡ **The ECM may be the cause of the problem but this is rarely the case.**

16. Turn the engine **OFF** and remove the tachometer and the timing light.

Alternator

REMOVAL

1.8L Engine

1. Before servicing the vehicle, refer to the Precautions Section.
2. Remove or disconnect the following:
   - Negative battery cable
   - 2 lead wires and connector from the alternator
   - Drive belt adjusting bolt, loosen only
   - Drive belt
   - Alternator

2.5L Engine

1. Before servicing the vehicle, refer to the Precautions Section.
2. Remove or disconnect the following:
   - Negative battery cable
   - Engine cover
   - Engine undercover
   - Alternator drive belt
   - Alternator mounting bolts
   - Alternator from the engine

3.5L Engines

1. Before servicing the vehicle, refer to the Precautions Section.
2. Remove or disconnect the following:
   - Negative battery cable
   - Right side engine undercover and side inspection cover
   - Radiator
   - Drive belt
   - Alternator and A/C compressor harness connectors
   - Upper and lower alternator bolts
   - Alternator

INSTALLATION

1.8L Engine

1. Install or connect the following:
   - Alternator and retaining bolts loosely
   - Belt and connect the wiring
2. Adjust the drive belt.
3. Torque the retaining bolts to 25 ft. lbs. (34 Nm).
4. Connect the negative battery cable.

2.5L Engines

1. Install or connect the following:
   - Alternator and torque the upper bolt to 18–23 ft. lbs. (26–31 Nm) and the lower bolt to 33–38 ft. lbs. (44–52 Nm)
   - Drive belt. Properly tension the belt
   - Engine undercover
   - Engine cover
   - Negative battery cable

3.5L Engine

Install or connect the following:
- Alternator
- Upper and lower alternator bolts.
Tighten the upper bolt to 12–15 ft. lbs. (16–20 Nm) and the lower bolt to 32–38 ft. lbs. (44–52 Nm).

- Alternator and A/C compressor harness connectors
- Drive belt
- Radiator
- Right side engine undercover and side inspection cover
- Negative battery cable

Engine Assembly

REMOVAL & INSTALLATION

Sentra

➡**The engine and transaxle are removed as one unit from the underside of the vehicle.**

1. Before servicing the vehicle, refer to the Precautions Section.
2. Relieve the fuel system pressure.
3. Drain the coolant from the radiator and the engine block.
4. Drain the engine oil.
5. Remove or disconnect the following:
   - Negative and positive battery cables
   - Battery and tray from the vehicle
   - Both front wheels
   - Engine undercovers and the engine side covers
   - Air cleaner assembly and air duct
   - Vacuum hoses. Make sure to note the locations prior to disconnection them.
   - Heater hoses from the engine
   - Automatic transmission cooler hoses from the transaxle, if equipped
   - Power steering hoses
   - Fuel hoses from the engine
   - Harness and wiring connections. Make sure to note the locations prior to disconnecting them.
   - Throttle cable and the cruise control cable
   - Control cable, if equipped with an automatic transmission
   - Cooling fans, radiator and the recovery tank
   - Front halfshafts from the vehicle
   - Front exhaust pipe
   - Starter motor and intake manifold support brackets
   - Engine drive belts
   - Alternator and adjusting brackets
   - Power steering pump and A/C compressor. It is not necessary to disconnect the lines.
6. Position a transmission jack under the transaxle and support the engine with engine slinger.

- Center crossmember
- Front stabilizer bar, if necessary
- Engine mounting bolts from both sides of the engine
7. Slowly lower the jacking devices and remove the engine and transaxle from the vehicle.

To install:
8. Install or connect the following:
   - Engine and transaxle assembly
   - Mounting bolts to both sides of the engine and torque the bolts to 44 ft. lbs. (60 Nm)
9. For vehicles with manual transaxles, adjust the height of the mounting bracket (buffer rod). The distance between the 2 through-bolts should be 2.13–2.20 in. (54–56mm).
10. Install or connect the following:
    - Center crossmember and torque the bolts to 40 ft. lbs. (54 Nm)
11. Remove the engine support jacks and engine slinger.
12. Install or connect the following:
    - A/C compressor and power steering pump
    - Alternator and brackets
    - Starter motor and intake manifold support bracket
    - Front exhaust pipe
    - Drive belts
    - Both front halfshafts
    - Radiator, cooling fans and recovery tank
    - Control cable, automatic transmissions only
    - Throttle and cruise control cables, if equipped
    - Wiring harness and electrical connections
    - Power steering hoses and fuel line
    - Transmission cooler lines, if equipped
    - Vacuum hoses
    - Air cleaner assembly
    - Engine side and under covers
    - Both front wheels
    - Battery tray and battery
    - Both battery cables
13. Fill the engine with clean oil.
14. Fill the cooling system.
15. Start the engine and check for leaks. Make all the necessary adjustments.

Altima

➡**The engine and transaxle must be removed as a single unit. The engine and transaxle are removed from under the vehicle.**

1. Before servicing the vehicle, refer to the Precautions Section.

2. Release fuel system pressure.
3. Drain the cooling system.
4. Drain the engine oil.
5. Remove or disconnect the following:
   - Battery cables and the battery tray
   - Air cleaner assembly
   - Both front wheels
   - Engine under cover and engine hood
   - Cooler lines from the radiator, if equipped with an automatic transaxle
   - Upper and lower hoses from the radiator
   - Radiator assembly
   - Heater hoses from the engine
   - Throttle cable and cruise control cable, if equipped
   - Fuel feed and return hoses
   - All the necessary vacuum hoses and electrical connectors. Label all wires and hoses before disconnecting them.
   - Wiring from starter motor
   - Slave cylinder from the transaxle, if equipped. It is not necessary to disconnect the hydraulic hose.
   - Engine drive belts. Be sure to mark belts for reinstallation.
   - Alternator, A/C compressor and the power steering pump
   - Both halfshafts from the transaxle and support the engine with slinger and support the transaxle with proper jack
   - Left and right engine mounting through-bolts
   - Crossmember
   - Front and rear engine mounts
6. Lower the transaxle and engine assembly from the vehicle.

➡**The engine and transaxle assembly should be removed through the bottom of the vehicle. Do not attempt to remove the assembly from above.**

To install:
7. Raise the transaxle and engine assembly to the vehicle.
8. Install or connect the following:
   - Front and rear engine mounts. Torque the mounting bolts to 55 ft. lbs. (75 Nm).
   - Crossmember and torque the bolts to 57–72 ft. lbs. (77–98 Nm)
   - Left and right engine mounting through bolts. Torque the bolts to 72 ft. lbs. (98 Nm).
9. Remove the engine and transaxle support jacks.
   - Both halfshafts

- Power steering pump, A/C compressor and alternator
- Slave cylinder, if equipped
- Starter motor
- Drive belts
- Vacuum hoses and electrical connectors
- Fuel feed and return lines
- Throttle and cruise control cables, if equipped
- Radiator
- Heater and radiator hoses
- Cooler lines, if equipped
- Engine side and under covers
- Both front wheels
- Air cleaner assembly
- Battery tray and battery
- Both battery cables
- Hood
10. Fill the cooling system.
11. Fill the engine with clean oil.
12. Start the vehicle, check for leaks and repair if necessary.

Water Pump

REMOVAL & INSTALLATION

1.8L Engine

1. Before servicing the vehicle, refer to the Precautions Section.
2. Drain the cooling system.
3. Remove or disconnect the following:

- Negative battery cable
- Cylinder head front mounting bracket and loosen the water pump pulley bolts
- Engine drive belts
- Water pump pulley
- Coolant hoses from the water inlet and thermostat housing
- Water pump and thermostat housing

4. Remove all traces of gasket material from sealing surfaces.

To install:

5. Apply a continuous bead of liquid sealer to the sealing surface of the thermostat housing. The sealant should be 0.079–0.118 in. (2–3mm) diameter.
6. Install or connect the following:

- Water pump. Torque the bolts to 56–73 inch lbs. (7–8 Nm).
- Pulley to the water pump and tighten the mounting bolts to 56–73 inch lbs. (7–8 Nm)
- Coolant hoses to the thermostat housing
- Drive belts and adjust as needed

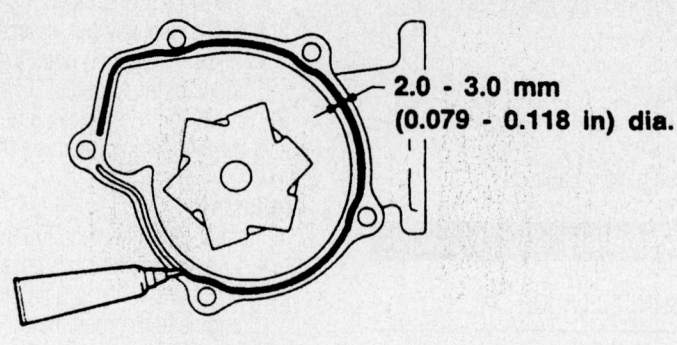

2.0 - 3.0 mm (0.079 - 0.118 in) dia.

7923QG06

Apply RTV sealant to the water pump sealing surface as shown— 1.8L engines

- Cylinder head front mounting bracket
- Negative battery cable

7. Fill the cooling system
8. Start the engine, check for leaks and repair if necessary.

2.5L Engine

1. Before servicing the vehicle, refer to the Precautions Section.
2. Drain the cooling system.
3. Disconnect the negative battery cable
4. On Altima remove or disconnect the following:

- Upper and lower engine covers
- Coolant reservoir
- Power Distribution Module (PDM) and move aside
- Passenger front wheel
- Engine ground

5. Remove or disconnect the following:

- Drive belt
- Radiator hose
- Alternator
- Water pump

To install:

6. Be sure all gasket surfaces are clean and properly apply a continuous bead of silicone sealer to the pump.
7. Install or connect the following:

- Water pump and torque the bolts to 16–20 ft. lbs. (21–28 Nm)

8. Reverse the removal procedure to complete installation.
9. Connect the negative battery cable.
10. Fill the cooling system.
11. Start the engine, check for leaks and repair if necessary.

3.5L Engine

1. Before servicing the vehicle, refer to the Precautions Section.
2. Drain the cooling system.
3. Position a jack under the oil pan for support. Be sure to place a block of wood on the jack for protection to the engine parts.
4. Remove or disconnect the following:

- Negative battery cable
- Right side engine mount and bracket
- Drive belts and the idler pulley bracket
- Chain tensioner cover and the water pump cover

5. Push the timing chain tensioner sleeve and apply a stopper pin so it does not return.

- Timing chain tensioner assembly
- 3 bolts that secure the water pump

6. Rotate the crankshaft 20 degrees counterclockwise to provide timing chain slack.
7. Put M8 bolts in 2 M8 threaded holes of the water pump.
8. Tighten each bolt by turning alternately ½ turn until they reach the timing chain rear case. Be sure to turn each bolt ½ turn at a time to prevent damage.
9. Lift up the water pump and remove it.
10. When removing the water pump, do not allow the water pump gear to hit the timing chain.
11. Remove and discard the O-rings from the water pump.
12. Clean all traces of liquid gasket from the water pump and covers.

To install:

13. Install or connect the following:

- Water pump using new O-rings to the engine block. Torque the 3 water pump mounting bolts evenly to 75–95 inch lbs. (8.5–10.7 Nm).

14. Rotate the crankshaft pulley to its original position by turning it 20 degrees clockwise.

- Timing chain tensioner and torque the bolts to 62–82 inch lbs. (7–9.3 Nm)

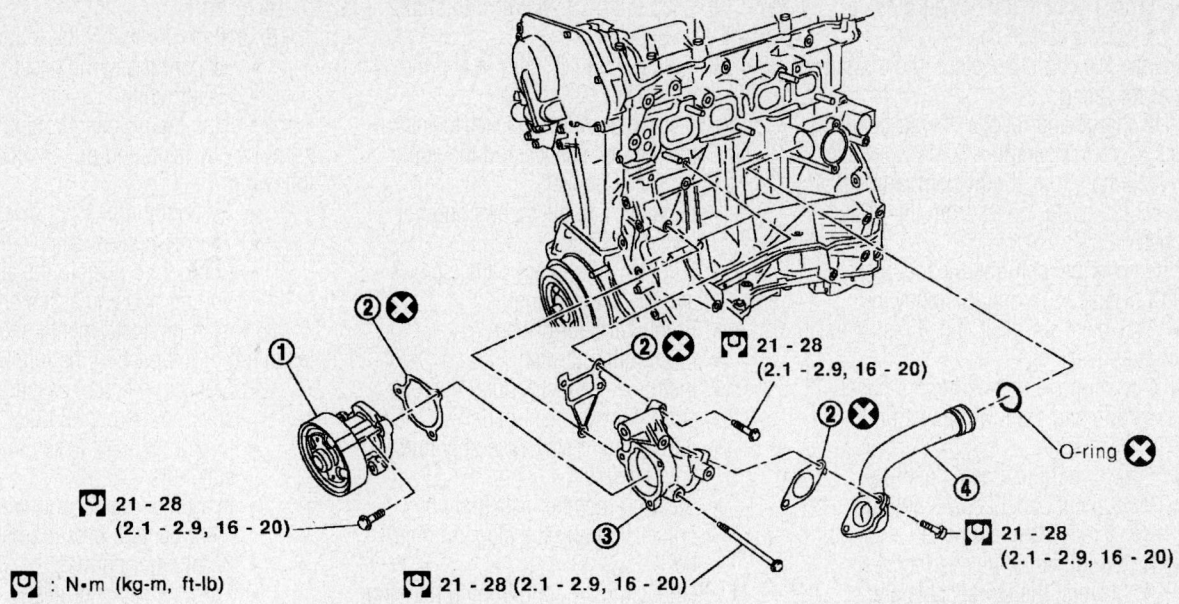

1. Water pump
4. Water pipe
2. Gasket
3. Water pump housing

67162-NISS-G07

Exploded view of the water pump assembly–2.5L engine

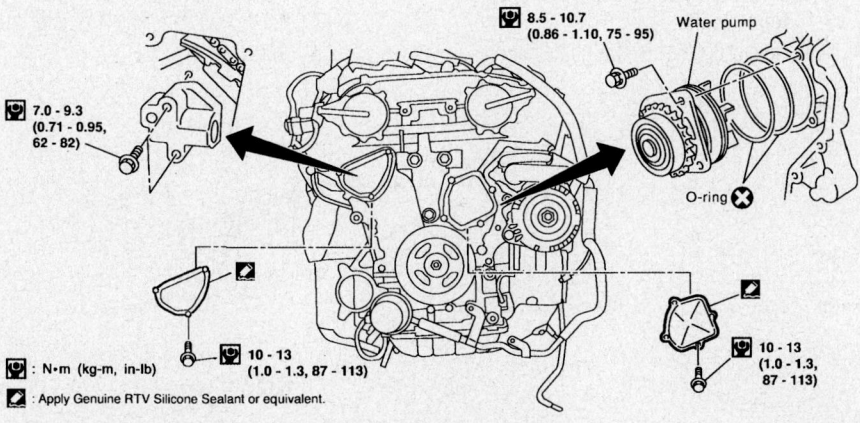

09482_ALTI_G0004

Water pump and timing cover assembly—3.5L engine

15. Remove the stopper pin from the timing chain tensioner.

16. Apply a continuous 0.091–0.130 in. (2.3–3.3mm) bead of liquid sealant to the mating surfaces of the timing chain tensioner and water pump covers.

- Timing chain tensioner and water pump covers to the engine block. Torque the bolts to 87–113 inch lbs. (10–13 Nm).
- Drive belts and the idler pulley bracket
- Right side engine mounting bracket and the engine mount
- Negative battery cable

17. Remove the jack from under the engine and install the drain plugs to the cylinder block.

18. Fill the cooling system.

19. Start the engine, check for leaks and repair if necessary.

Heater Core

REMOVAL & INSTALLATION

Altima

1. Before servicing the vehicle, refer to the Precautions Section.

2. Position the steering wheel in the straight-ahead position.

3. Turn the ignition switch OFF.

4. Disconnect the negative (-) battery cable; then, the positive (+) battery cable.

➡**Wait for a least 3 minutes after disconnecting the battery cables for the charge in the air bag circuit to dissipate before working on the air bag module(s).**

5. Remove the driver's side SRS and steering wheel by removing or disconnecting the following:

- Lower lid from the steering wheel and disconnect the driver's air bag module connector
- Left and right side lids from the steering wheel
- Special bolts from both side of the steering wheel using a tamper resistant Torx® wrench (T50)
- Air bag module and store it face up
- Horn's electrical connector and remove the steering wheel nut
- Steering wheel from the steering column using a suitable puller

6. Remove the passenger's side SRS by removing or disconnecting the following:

- Glove box door and the glove box
- Front passenger's air bag module connector
- 2 special bolts using a tamper resistant Torx® wrench (T50)
- 4 passenger's air bag-to-instrument panel nuts

- Front passenger's air bag module and store it face up.

7. Drain the cooling system into a clean container for reuse.

8. Discharge and recover the air conditioning system refrigerant.

9. Working in the engine compartment, disconnect the heater hoses from the heater core tubes.

10. Remove the instrument panel by removing or disconnecting the following:

- Kick plate and dash side finisher on the driver's side
- 2 lower panel-to-instrument panel screws and the lower panel on the driver's side
- 2 lower reinforcement panel-to-instrument panel screws and the lower reinforcement panel
- 6 steering column cover screws, the covers, the spiral cable and combination switch
- 2 cluster lid "A" screws and the cluster lid "A"
- 3 combination meter screws, disconnect the electrical harness connector and remove the combination meter
- Switch panel
- Instrument panel lower covers

- Snap out the transmission shifter finisher (boot)
- 4 cluster lid "C" screws and the cluster lid "C"
- 4 audio and deck pocket-to-instrument panel screws and the audio and deck pocket
- 5 center console screws and the center console
- 2 center instrument panel screws and the center panel
- Front defroster grilles
- Front pillar garnish
- Instrument panel 3 nuts/4 screws and the instrument panel
- 8 instrument stay assembly nuts and the stay
- Steering member assembly 5 nuts/1 bolt and the steering member

11. Remove the air conditioning housing assembly by removing or disconnecting the following:

- Refrigerant lines from the air conditioning housing assembly
- Thermo control amp
- Air conditioning housing assembly

12. Remove or disconnect the following:
- Heater unit
- Heater core from the heater unit

To install:

13. Install or connect the following:
- Heater core to the heater unit
- Heater unit

14. Install the air conditioning housing assembly by installing or connecting the following:

- Air conditioning housing assembly
- Thermo control amp
- Refrigerant lines to the air conditioning housing assembly

15. Install the instrument panel by installing or connecting the following:

- Steering member assembly and the steering member 5 nuts/1 bolt
- Instrument stay assembly and the 8 stay nuts
- Instrument panel and the instrument panel 3 nuts/4 screws
- Front pillar garnish
- Front defroster grilles
- Center instrument panel and the 2 center panel screws
- Center console and the 5 center console screws
- Audio and deck pocket and the 4 audio and deck pocket-to-instrument panel screws
- Cluster lid "C" and the 4 cluster lid "C" screws

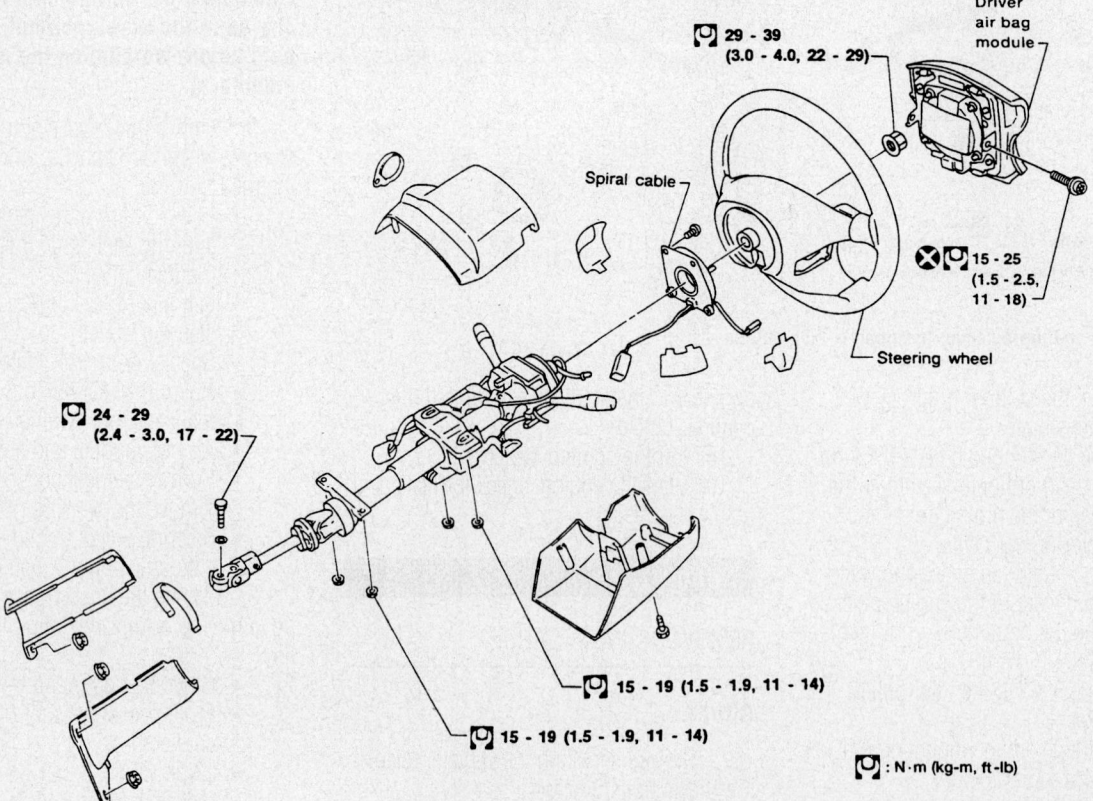

29 - 39 (3.0 - 4.0, 22 - 29)

Driver air bag module

Spiral cable

15 - 25 (1.5 - 2.5, 11 - 18)

Steering wheel

24 - 29 (2.4 - 3.0, 17 - 22)

15 - 19 (1.5 - 1.9, 11 - 14)

15 - 19 (1.5 - 1.9, 11 - 14)

: N·m (kg-m, ft-lb)

93112GD9

Exploded view of the steering wheel and air bag module—Altima

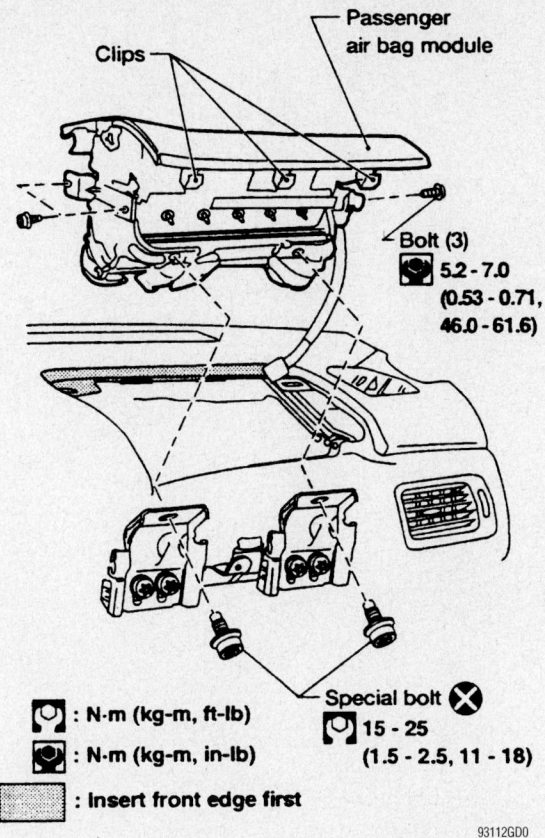

Clips

Passenger air bag module

Bolt (3)

🔧 5.2 - 7.0
(0.53 - 0.71,
46.0 - 61.6)

🔧 : N·m (kg-m, ft-lb)

🔧 : N·m (kg-m, in-lb)

▭ : Insert front edge first

Special bolt ⊗
🔧 15 - 25
(1.5 - 2.5, 11 - 18)

93112GD0

Exploded view of the passenger's side air bag module—Altima

- Snap in the transmission shifter finisher (boot)
- Instrument panel lower covers
- Switch panel
- Combination meter, connect the electrical harness connector and install the 3 combination meter screws
- Cluster lid "A" and the 2 cluster lid "A" screw
- Combination switch, the spiral cable, the covers and the 6 steering column cover screws
- Lower reinforcement panel and the 2 lower reinforcement panel-to-instrument panel screws
- Lower panel and the 2 lower panel-to-instrument panel screws, on the driver's side
- Kick plate and dash side finisher, on the driver's side

16. Working in the engine compartment, connect the heater hoses to the heater core tubes.

17. Install the passenger's side SRS by installing or connecting the following:
- Front passenger's air bag module
- 4 passenger's air bag-to-instrument panel nuts
- 2 new special bolts and torque

using a tamper resistant Torx® wrench (T50) to 11–18 ft. lbs. (15–25 Nm)
- Front passenger's air bag module connector
- Glove box door and the glove box

18. Install the driver's side SRS and steering wheel by installing or connecting the following:
- Steering wheel to the steering column
- Steering wheel nut and torque the nut to 22–29 ft. lbs.
- Horn's electrical connector
- Air bag module
- New special bolts to both sides of the steering wheel and torque the bolts using a tamper resistant Torx® wrench (T50) to 11–18 ft. lbs. (15–25 Nm).
- Both the left and right side lids to the steering wheel
- Lower lid to the steering wheel and connect the driver's air bag module connector

19. Refill the cooling system.

20. Connect the positive (+) battery cable; then, the negative (-) battery cable.

21. Evacuate, charge and leak test the air conditioning system refrigerant.

22. Operate the engine to normal operating temperatures; then, check the climate control operation and check for leaks.

Sentra

1. Before servicing the vehicle, refer to the Precautions Section.

2. Position the steering wheel in the straight-ahead position.

3. Turn the ignition switch OFF.

4. Disconnect the negative (-) battery cable; then, the positive (+) battery cable.

➡ **Wait for a least 3 minutes after disconnecting the battery cables for the charge in the air bag circuit to dissipate before working on the air bag module(s).**

5. Remove the driver's side SRS and steering wheel by performing the following procedure:
- Remove the lower lid from the steering wheel and disconnect the driver's air bag module connector.
- Remove both the left and right side lids from the steering wheel.
- Using a tamper resistant Torx® wrench (T50), remove the special bolts from both side of the steering wheel.
- Carefully, remove the air bag module and store it face up.
- Disconnect the horn's electrical connector and remove the steering wheel nut.
- Using a steering wheel puller, press the steering wheel from the steering column.

6. Remove the passenger's side SRS by performing the following procedure:
- Remove the glove box door and the glove box.
- Disconnect the front passenger's air bag module connector.
- Using a tamper resistant Torx® wrench (T50), remove the 2 special bolts.
- Remove the 4 passenger's air bag-to-instrument panel nuts.
- Carefully, remove the front passenger's air bag module and store it face up.

7. Drain the cooling system into a clean container for reuse.

8. Discharge and recover the air conditioning system refrigerant.

9. Working in the engine compartment, disconnect the heater hoses from the heater core tubes.

10. Remove the instrument panel by performing the following procedures:

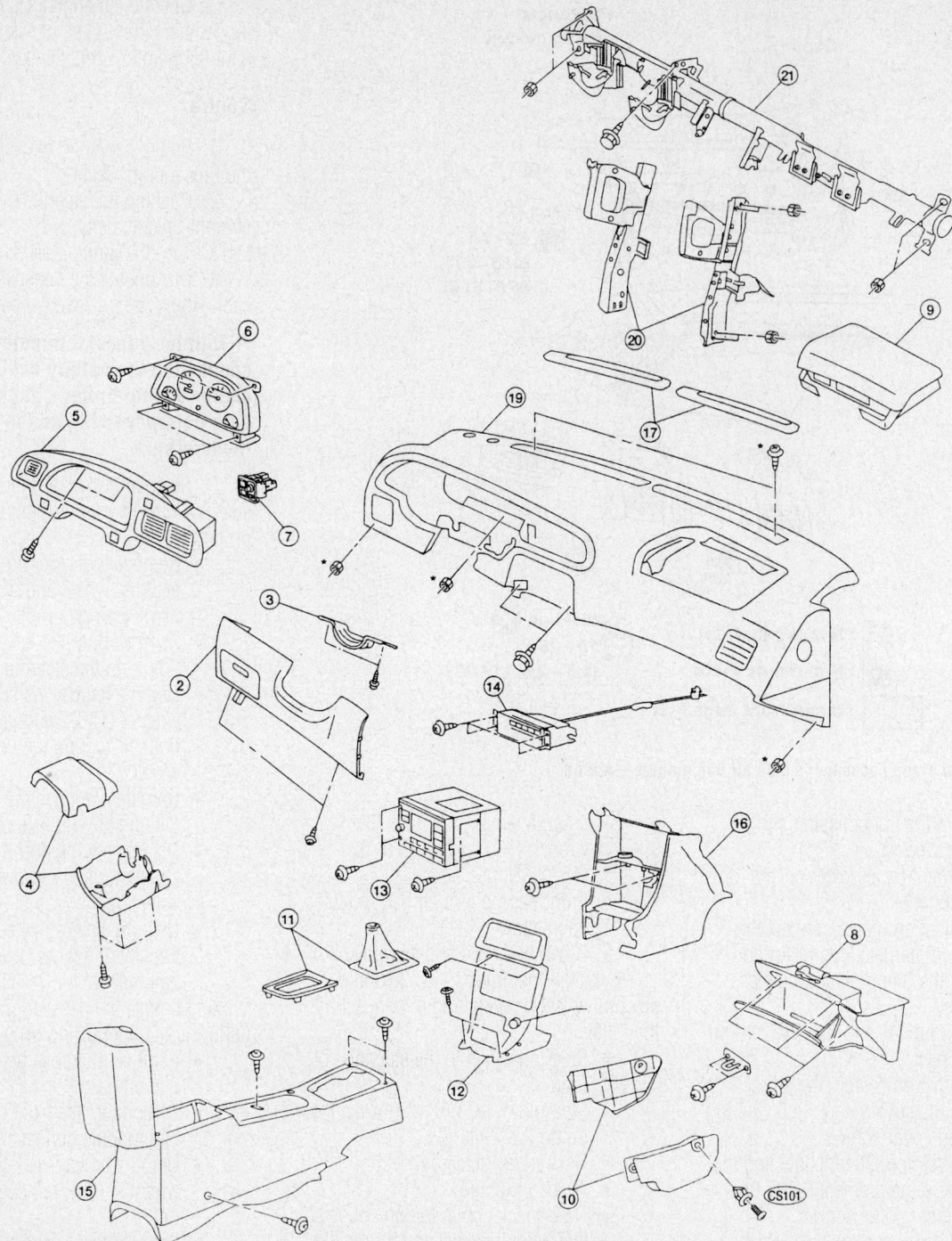

*: Instrument panel assembly mounting bolts, screws and nuts.

1. Remove kick plate and dash side finisher on driver side
2. Instrument lower panel on driver side
3. Dash lower reinforcement panel
4. Steering column covers, spiral cable and combination switch
5. Cluster lid A
6. Combination meter
7. Switch panel
8. Glove box assembly
9. Remove passenger side air bag moldule
10. Instrument lower covers
11. A/T finisher or M/T boot

12. Cluster lid C
13. Audio and deck pocket
14. A/C & heater control
15. Center console assembly
16. Instrument center panel
17. Front defroster grilles
18. Front pillar garnish
19. Instrument panel assembly
20. Instrument stay assemblies, if necesary
21. Steering member assembly, if necessary

93112GE1

Exploded view of the instrument panel assembly—Altima

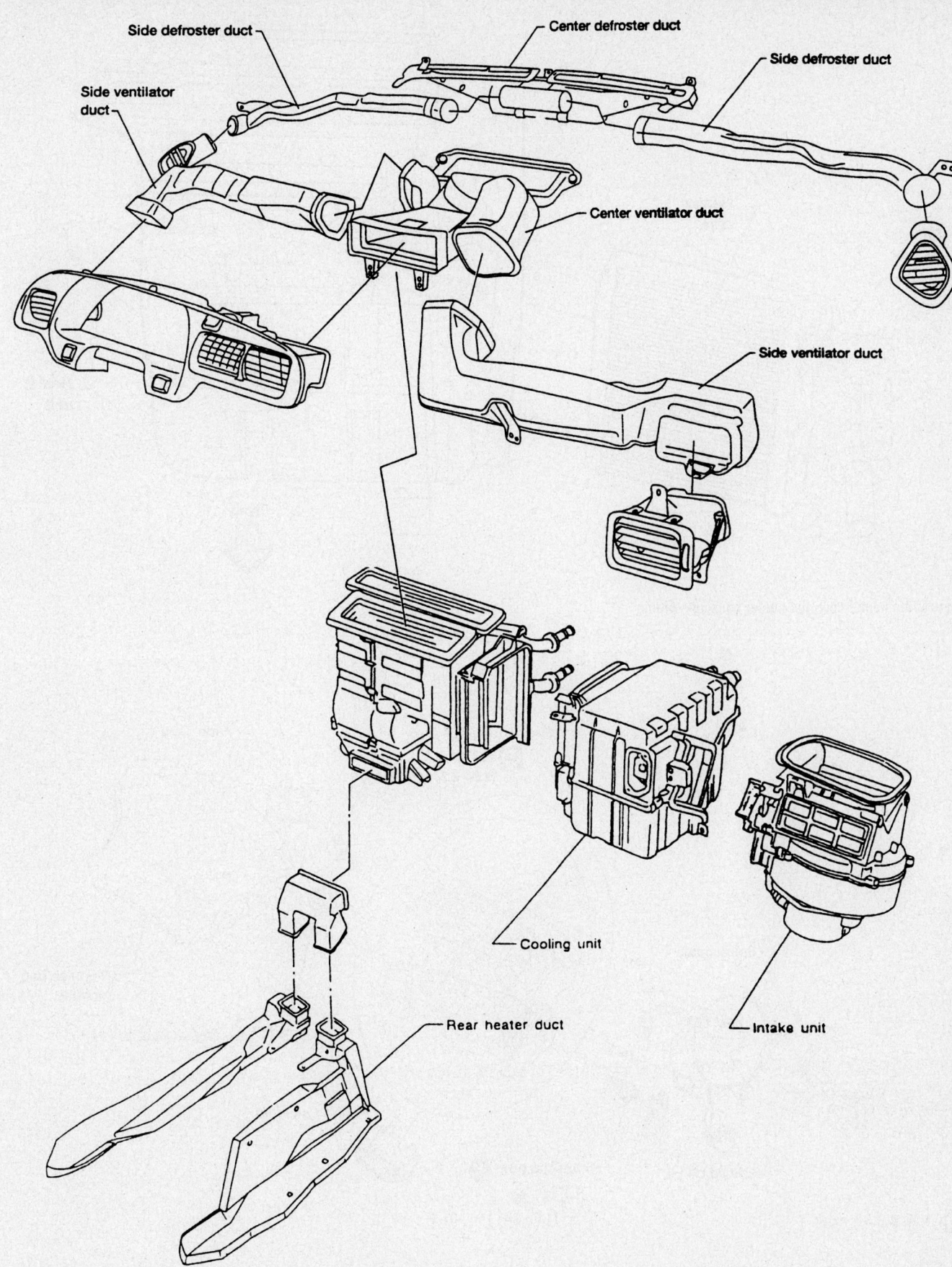

Side defroster duct

Center defroster duct

Side defroster duct

Side ventilator duct

Center ventilator duct

Side ventilator duct

Cooling unit

Rear heater duct

Intake unit

93112GE2

Exploded view of the heater housing assembly and related components—Altima

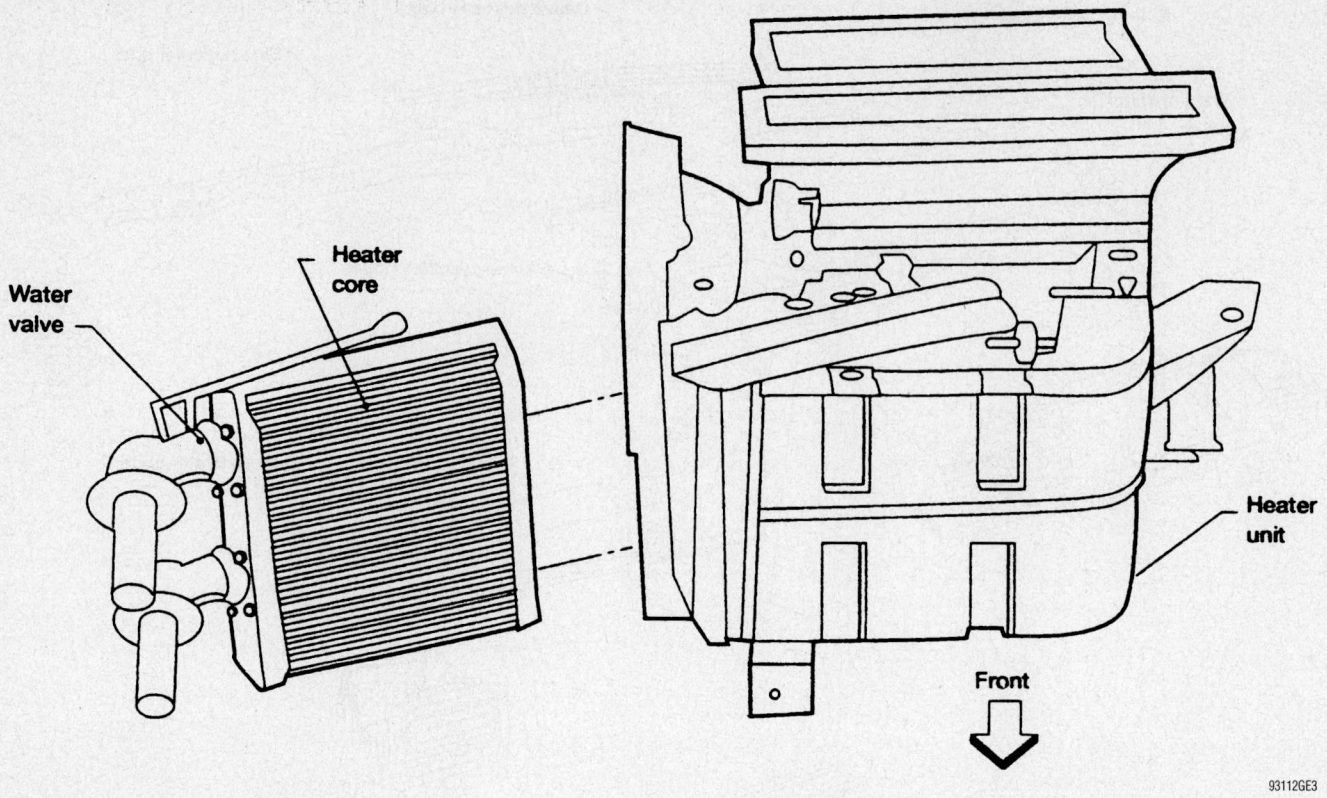

Water valve

Heater core

Heater unit

Front

93112GE3

View of the heater core and heater housing—Altima

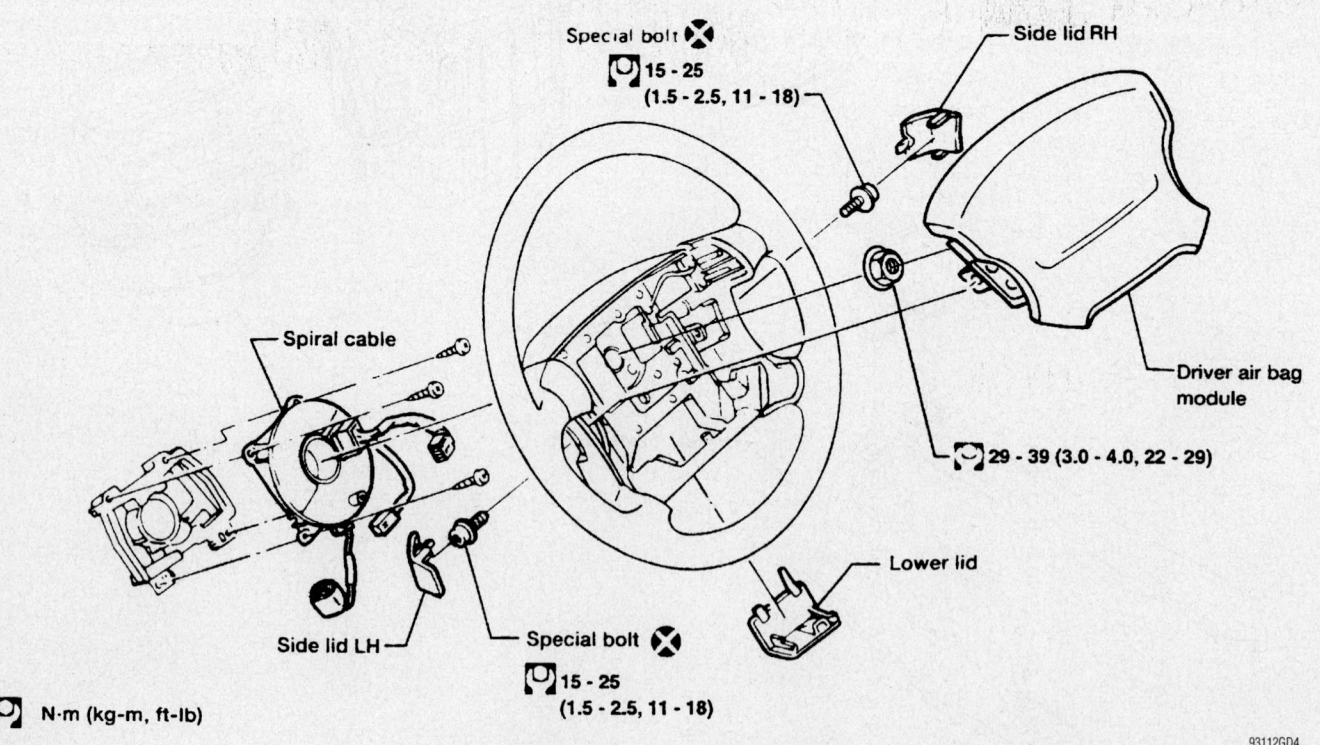

Special bolt ⊗
🔧 15 - 25
(1.5 - 2.5, 11 - 18)

Side lid RH

Spiral cable

Driver air bag module

🔧 29 - 39 (3.0 - 4.0, 22 - 29)

Lower lid

Side lid LH

Special bolt ⊗
🔧 15 - 25
(1.5 - 2.5, 11 - 18)

🔧 N·m (kg-m, ft-lb)

93112GD4

Exploded view of the steering wheel and air bag module—Sentra

- On the driver's side, remove the 2 lower panel-to-instrument panel screws and the lower panel.
- Remove the 2 lower reinforcement panel-to-instrument panel screws and the lower reinforcement panel.
- Remove the 6 steering column cover screws, the cover and combination switch.
- Remove the 2 cluster lid "A" screws and the cluster lid "A".
- Remove the 3 combination meter screws, disconnect the electrical harness connector and remove the combination meter.
- Remove the ashtray.
- Remove the cluster lid "C" mask, screw and the cluster lid "C".
- Remove the 8 audio and air conditioning control assembly-to-instrument panel screws, the electrical connectors and the audio and air conditioning control assembly.
- Remove the transmission shifter finisher.
- Remove the rear console mask, the 4 screws and the rear console.
- Remove the 4 front console screws and the front console.
- Remove the front pillar garnish.
- Remove the lower dash side garnish.
- Remove the instrument panel mask.
- Remove the instrument panel-to-chassis nuts/bolts and the instrument panel.

11. Remove the air conditioning housing assembly by performing the following procedure:
- Disconnect the refrigerant lines from the air conditioning housing assembly.
- Disconnect the thermo control amp.
- Remove the air conditioning housing assembly.

12. Remove the heater unit.
13. Remove the heater core from the heater unit.

To install:
14. Install or connect the following:
- Heater core to the heater unit
- Heater unit

15. Install the air conditioning housing assembly by performing the following procedure:
- Install the air conditioning housing assembly.
- Connect the thermo control amp.
- Connect the refrigerant lines to the air conditioning housing assembly.

16. Install the instrument panel by installing or connecting the following:

- Instrument panel and the instrument panel-to-chassis nuts/bolts
- Instrument panel mask
- Lower dash side garnish
- Front pillar garnish
- Front console and the 4 front console screws
- Rear console, the 4 screws and the rear console mask
- Transmission shifter finisher
- Air conditioning control assembly, the electrical connectors and the audio and the 8 audio and air conditioning control assembly-to-instrument panel screws.
- Cluster lid "C", the screw and the cluster lid "C" mask
- Ashtray
- Combination meter, connect the electrical harness connector and install the 3 combination meter screws
- Cluster lid "A" and the 2 cluster lid "A" screws
- Combination switch, the cover and the 6 steering column cover screws
- Lower reinforcement panel and the 2 lower reinforcement panel-to-instrument panel screws
- Lower panel and the 2 lower panel-to-instrument panel screws, on the driver's side

17. Working in the engine compartment, connect the heater hoses to the heater core tubes.

18. Install the passenger's side SRS by installing or connecting the following:
- Front passenger's air bag module

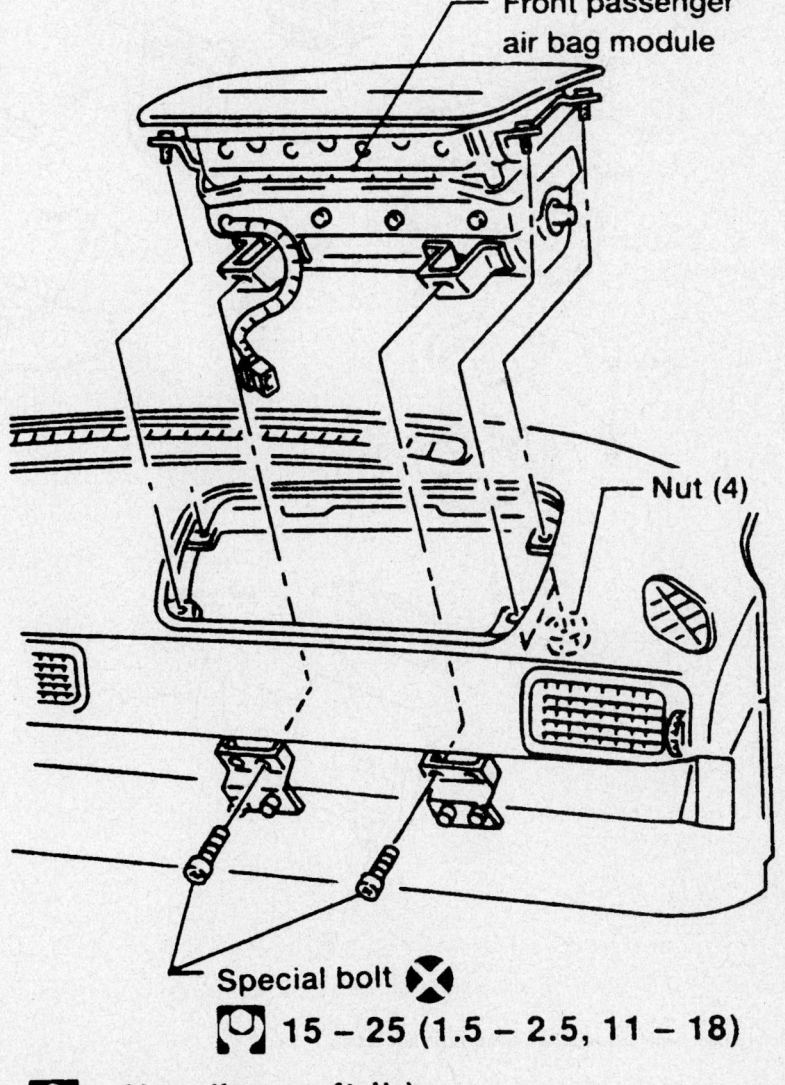

Front passenger air bag module

Nut (4)

Special bolt ⊗
15 – 25 (1.5 – 2.5, 11 – 18)

: N·m (kg-m, ft-lb)

93112GD5

Exploded view of the passenger's side air bag module—Sentra

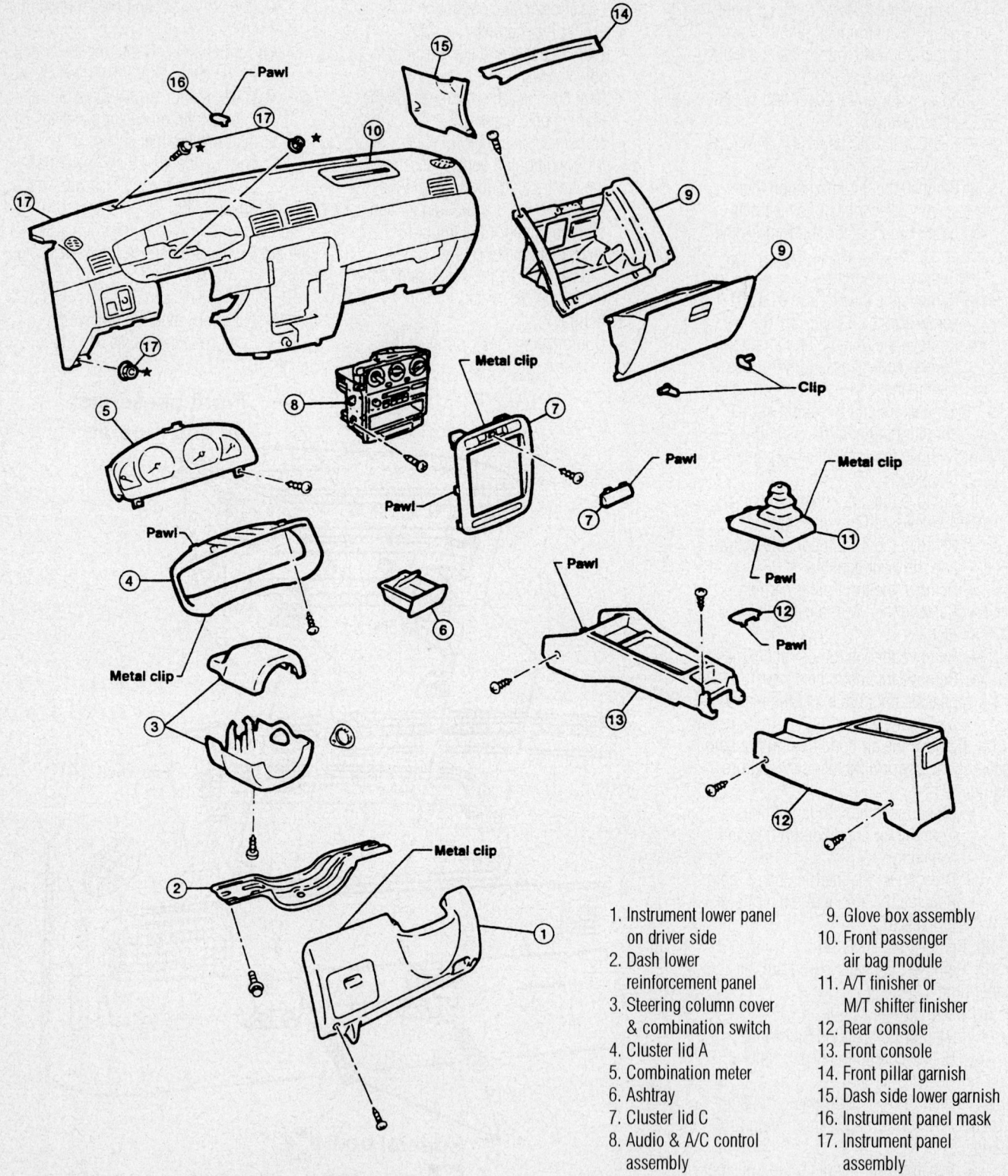

1. Instrument lower panel on driver side
2. Dash lower reinforcement panel
3. Steering column cover & combination switch
4. Cluster lid A
5. Combination meter
6. Ashtray
7. Cluster lid C
8. Audio & A/C control assembly
9. Glove box assembly
10. Front passenger air bag module
11. A/T finisher or M/T shifter finisher
12. Rear console
13. Front console
14. Front pillar garnish
15. Dash side lower garnish
16. Instrument panel mask
17. Instrument panel assembly

★ : Instrument panel assembly mounting bolts and nuts.

93112GD6

Exploded view of the instrument panel assembly—Sentra

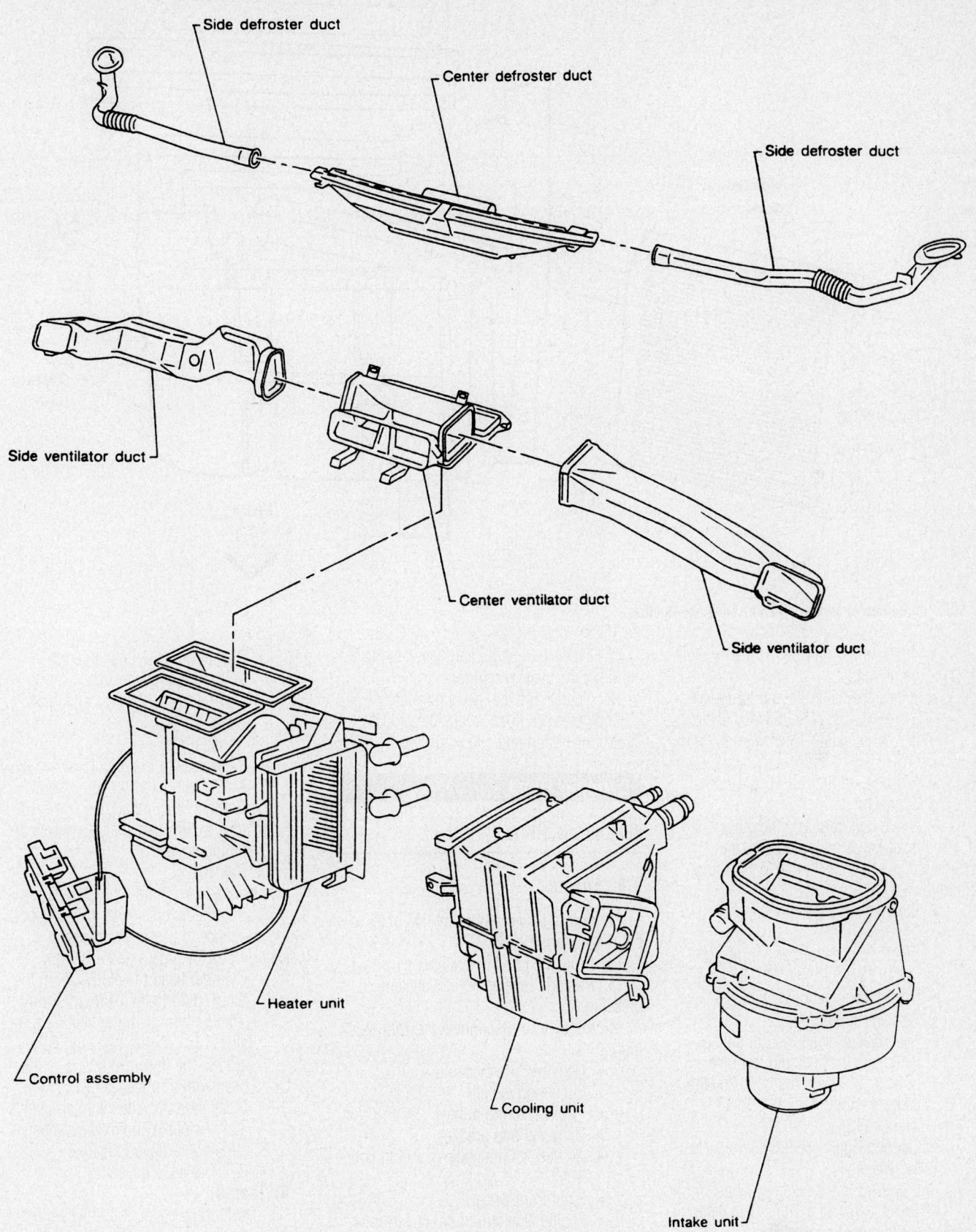

Side defroster duct

Center defroster duct

Side defroster duct

Side ventilator duct

Center ventilator duct

Side ventilator duct

Heater unit

Control assembly

Cooling unit

Intake unit

93112GD7

Exploded view of the heater housing assembly and related components—Sentra

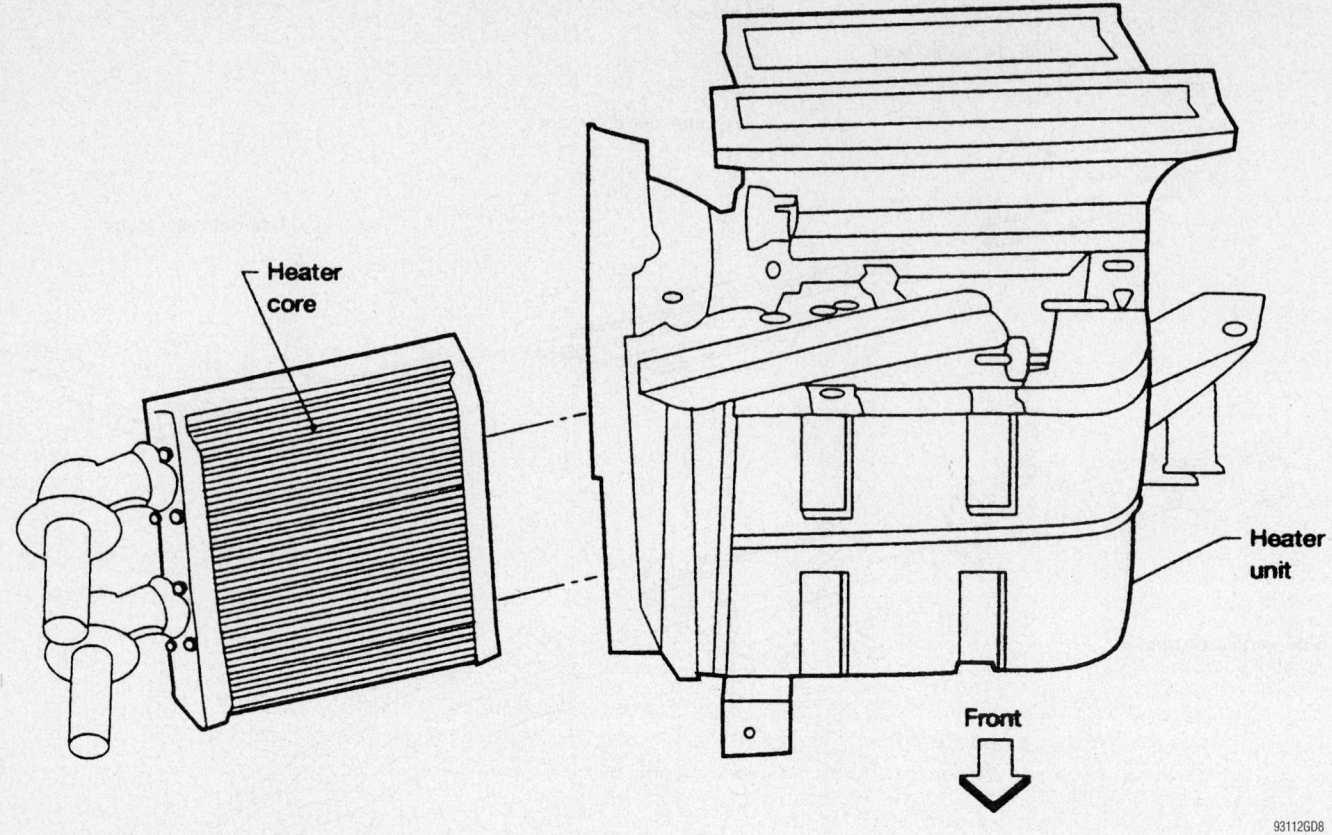

93112GD8

View of the heater core and heater housing—Sentra

- 4 passenger's air bag-to-instrument panel nuts
- 2 new special bolts and torque to 11–18 ft. lbs. (15–25 Nm), using a tamper resistant Torx® wrench (T50)
- Front passenger's air bag module connector
- Glove box door and the glove box

19. Install the driver's side SRS and steering wheel by installing or connecting the following:

- Steering wheel to the steering column
- Steering wheel nut and torque the nut to 22–29 ft. lbs. (30–39 Nm)
- Horn's electrical connector
- Air bag module
- New special bolts to both side of the steering wheel and torque the bolts to 11–18 ft. lbs. (15–25 Nm), using a tamper resistant Torx® wrench (T50)
- Left and right side lids to the steering wheel
- Lower lid to the steering wheel
- Driver's air bag module connector.

20. Refill the cooling system.
21. Connect the positive (+) battery cable; then, the negative (-) battery cable.

22. Evacuate, charge and leak test the air conditioning system refrigerant.
23. Operate the engine to normal operating temperatures; then, check the climate control operation and check for leaks.

Cylinder Head

REMOVAL & INSTALLATION

1.8L Engine

1. Before servicing the vehicle, refer to the Precautions Section.
2. Drain the cooling system.
3. Properly relieve the fuel system pressure.
4. Remove or disconnect the following:

- Negative battery cable
- Engine drive belts
- Power steering pulley
- Oil pump and bracket
- Air duct to the intake manifold collector
- Right front wheel
- Engine side and under covers
- Front exhaust tube
- Cylinder head front mounting bracket

- Rocker cover by loosening the bolts in numerical order
- Distributor, plug wires and spark plugs
- Spark plugs
- Intake manifold support and set the No. 1 cylinder at the Top Dead Center (TDC) position
- Idler pulley, camshaft sprockets and timing chains
- Camshafts

5. Loosen the cylinder head bolts in 2–3 steps in the reverse order of the tightening sequence to prevent warpage or cracking of the cylinder head assembly.

- Cylinder head (carefully), from the block, pulling the head up evenly from both ends. If the head seems stuck, do not pry it off. Tap lightly around the lower perimeter of the head with a rubber mallet to help break the seal. The cylinder head and the intake and exhaust manifolds are removed together.
- Cylinder head gasket(s)

To install:

6. Thoroughly clean both the cylinder block and head mating surfaces. Avoid scratching either surface.
7. Coat the threads and the seating sur-

face of the head bolts with clean engine oil. Install the cylinder head assembly (always replace the head gasket). Install head bolts (with washers) in their proper locations.

8. For 1.8L engines, tighten the bolts in sequence, as follows:

    a. Step 1: Bolts 1–10 to 22 ft. lbs. (29 Nm).

    b. Step 2: Bolts 1–10 to 43 ft. lbs. (59 Nm).

    c. Step 3: Loosen bolts 1–10 completely.

    d. Step 4: Bolts 1–10 to 22 ft. lbs. (29 Nm).

    e. Step 5: Bolts 1–10 plus 50–55 degrees.

    f. Step 6: Bolts 11–14 to 74 inch lbs. (8 Nm).

9. Install or connect the following:
- Camshafts

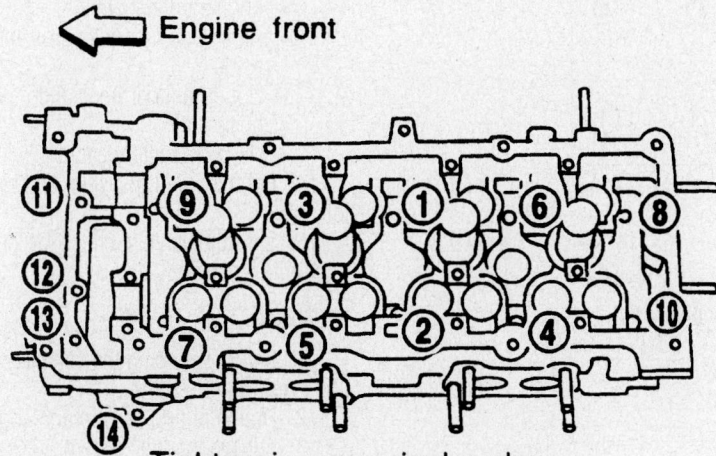

Engine front

Tighten in numerical order.

9347QG01

Cylinder head torque sequence—1.8L engine

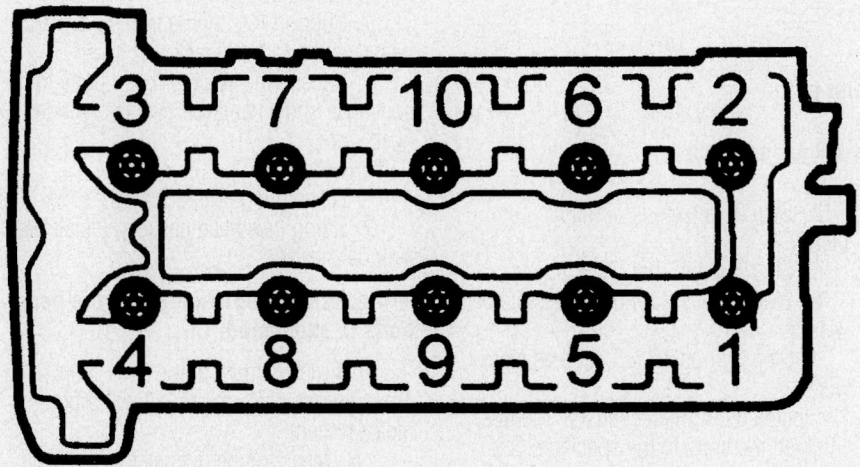

9347UG03

Tighten the rocker cover bolts in sequence—1.8L engines

- Idler pulley, camshaft sprockets and timing chains
- Intake manifold support
- Distributor
- Spark plugs and wires
- Distributor cap
- Rocker arm cover and torque the bolts to 34 inch lbs. (4 Nm)
- Cylinder head front mounting bracket
- Front exhaust tube
- Engine side and under covers
- Right front wheel
- Air duct to the intake manifold collector
- Oil pump and bracket
- Power steering pulley
- Drive belts
- Negative battery cable

10. Fill the cooling system.

11. Start the vehicle, check for leaks and repair if necessary.

2.5L Engine

ALTIMA

1. Before servicing the vehicle, refer to the Precautions Section.

2. Drain cooling system and engine oil.

3. Relieve the fuel system pressure.

4. Remove or disconnect the following:
- Negative battery cable
- Upper radiator hose and heater hose
- Right side fuse/relay box and move aside
- Engine undercover
- Resonator and air cleaner case
- Engine top cover
- All electrical and vacuum lines at the head
- Fuse hose quick connector at fuel tube side
- Engine harness and power steering hose bracket
- Timing chain
- Drive belt tensioner
- Camshafts
- Spark plugs
- Exhaust manifold

5. Support the engine from above and below with a suitable jack and hoist.

6. Remove or disconnect the following:
- Power steering pump and reservoir and wire aside
- Auxiliary drive belts
- A/C compressor and wire aside
- Upper sway bar links
- Front and rear engine mount through bolts
- Lower ball joints
- 2 steering gear housing mounting bolts
- Front suspension member bolts and front suspension member

7. Loosen the cylinder head bolts in the sequence shown and remove the cylinder head.

To install:

8. Clean the gasket surfaces.

9. Using new head gaskets, install the cylinder heads.

➡**If possible, replacement of the head bolts is suggested.**

10. If replacement of the head bolts is not possible, perform the following bolt measurement:

    a. Measure the diameter of the head bolt 0.43 in. (11mm) from the bottom of the bolt.

    b. Measure the diameter of the head

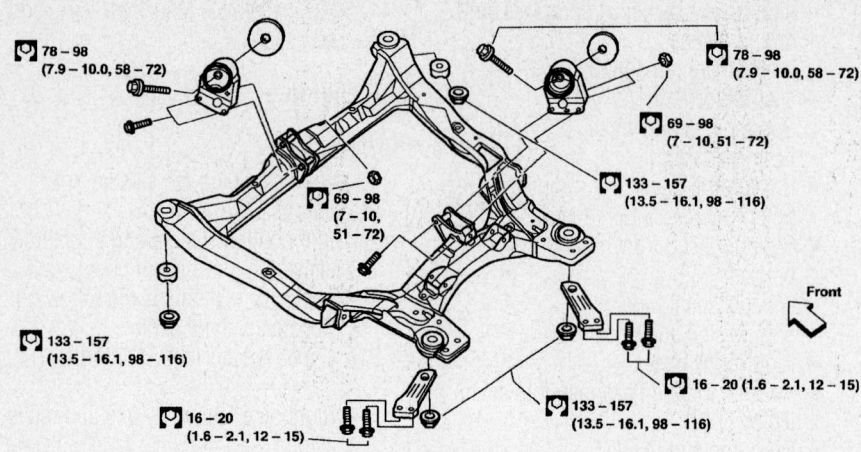

Removing the front suspension member—Altima

67162-NISS-G12

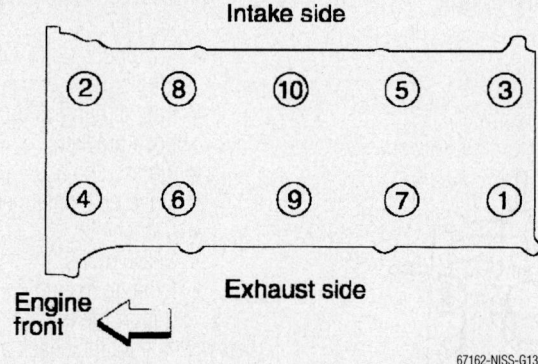

Cylinder head bolt removal sequence—2.5L engine–Altima and Sentra

67162-NISS-G13

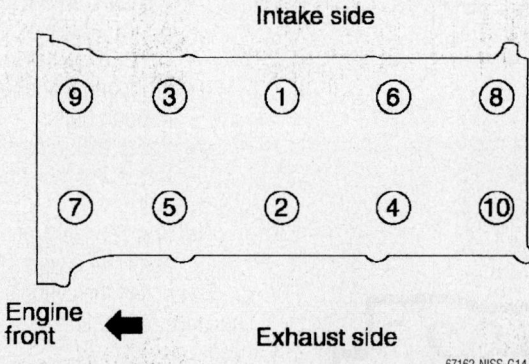

Cylinder head bolt tightening sequence—2.5L engine–Altima and Sentra

67162-NISS-G14

bolt 2.17 in. (55mm) from the bottom of the bolt.

  c. Whenever the size difference between the 2 measurements exceeds 0.0091 in. (0.23mm) the head bolts must be replaced.

11. Apply clean engine oil to the bolt threads and seating surfaces and tighten the cylinder head bolts in sequence as follows:

  a. Step 1: 72 ft. lbs. (98 Nm).

  b. Step 2: Loosen all the bolts completely.

  c. Step 3: 29 ft. lbs. (33 Nm).

  d. Step 4: Plus 75–80 degrees clockwise.

  e. Step 5: Plus 75–80 degrees clockwise.

12. Install the front suspension member and tighten the bolts to the specifications shown in the illustration.

13. Install or connect the following:

- 2 steering gear housing mounting bolts
- Lower ball joints
- Front and rear engine mount through bolts
- Upper sway bar links
- A/C compressor
- Auxiliary drive belts
- Power steering pump and reservoir
- Exhaust manifold
- Spark plugs
- Camshafts
- Drive belt tensioner
- Timing chain
- Engine harness and power steering hose bracket
- Fuse hose quick connector at fuel tube side
- All electrical and vacuum lines at the head
- Engine top cover
- Resonator and air cleaner case
- Engine undercover
- Right side fuse/relay box and move aside
- Upper radiator hose and heater hose
- Negative battery cable

14. Fill the cooling system and engine oil

15. Start the vehicle, check for leaks and repair if necessary.

SENTRA

1. Before servicing the vehicle, refer to the Precautions Section.

2. Drain cooling system and engine oil.

3. Relieve the fuel system pressure.

4. Remove or disconnect the following:

- Negative battery cable
- Strut tower cross brace
- Timing chain
- Camshafts
- Exhaust manifold

5. Support the engine from above and below with a suitable jack and hoist.

6. Loosen the cylinder head bolts in the sequence shown and remove the cylinder head.

To install:

7. Clean the gasket surfaces.

8. Using new head gaskets, install the cylinder heads.

➥**If possible, replacement of the head bolts is suggested.**

9. If replacement of the head bolts is not possible, perform the following bolt measurement:

  a. Measure the diameter of the head bolt 0.47 in. (12mm) from the bottom of the bolt.

b. Measure the diameter of the head bolt 2.17 in. (55mm) from the bottom of the bolt.

c. Whenever the size difference between the 2 measurements exceeds 0.0091 in. (0.23mm) the head bolts must be replaced.

10. Apply clean engine oil to the bolt threads and seating surfaces and tighten the cylinder head bolts in sequence as follows:

a. Step 1: 72 ft. lbs. (98 Nm).

b. Step 2: Loosen all the bolts completely.

c. Step 3: 29 ft. lbs. (33 Nm).

d. Step 4: Plus 75–80 degrees clockwise.

e. Step 5: Plus 75–80 degrees clockwise.

11. Install or connect the following:
- Exhaust manifold
- Camshafts
- Timing chain
- Strut tower cross brace
- Negative battery cable

12. Fill the cooling system and engine oil

13. Start the vehicle, check for leaks and repair if necessary.

3.5L Engine

ALTIMA

➡ You must remove the engine from the vehicle in order to remove the cylinder head, for this procedure.

1. Before servicing the vehicle, refer to the Precautions Section.

2. Relieve the fuel system pressure.

3. Drain the engine oil.

4. Drain the cooling system.

➡ Before detaching any hoses or connectors, note the locations for reassembly.

5. Remove or disconnect the following:
- Negative battery cable
- Engine assembly
- Exhaust manifold

6. Place the engine on a suitable workstand.
- Oil pan
- Timing chain
- Intake manifold
- Water outlet
- Rear timing chain case bolts, in the sequence shown
- Rear timing chain case
- O-rings from the cylinder head and block
- Intake valve timing control solenoid valves

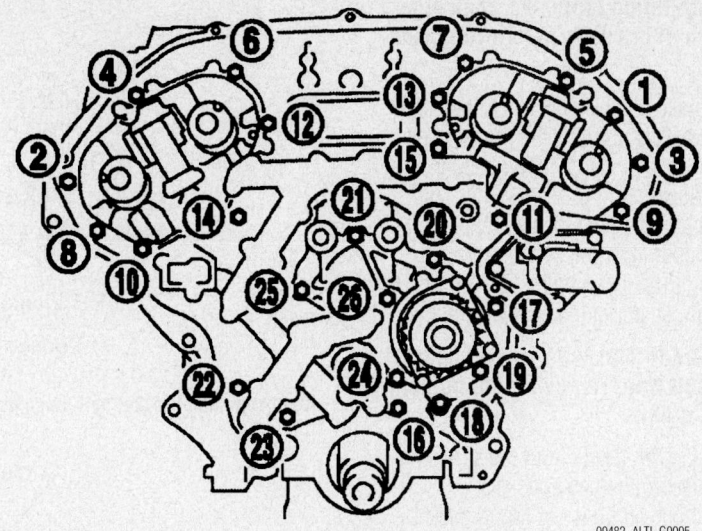

09482_ALTI_G0005

Loosen the rear timing chain case bolts in sequence—3.5L engine

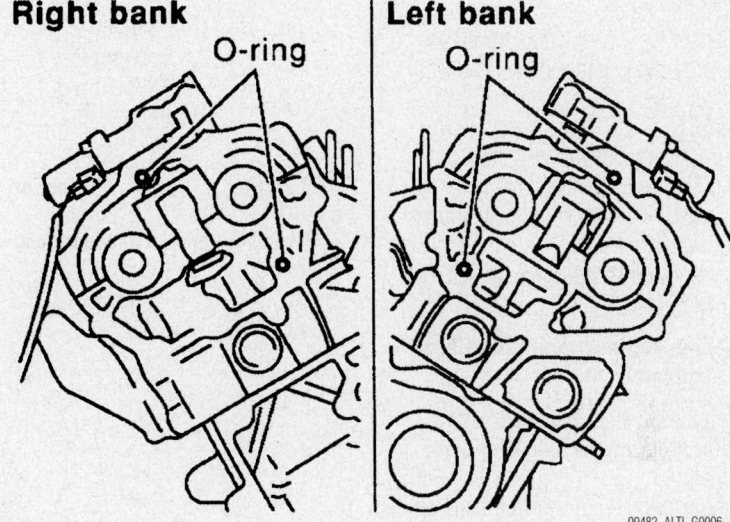

09482_ALTI_G0006

Remove the O-rings from the cylinder head—3.5L engine

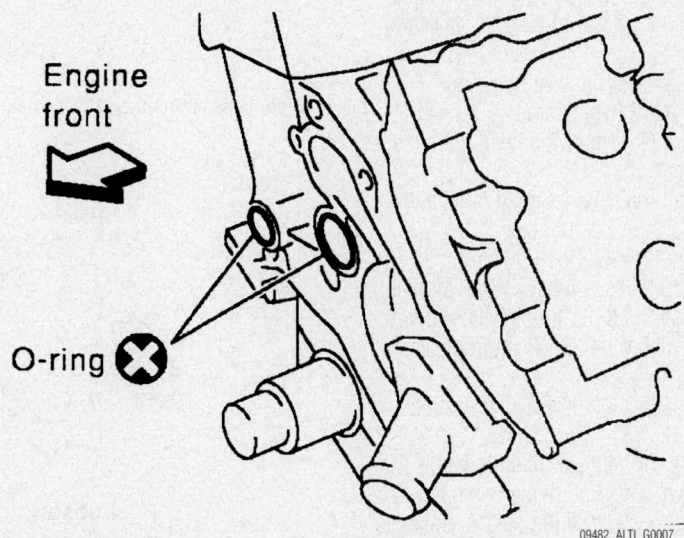

09482_ALTI_G0007

Remove the O-rings from the engine block—3.5L engine

➥**For installation purposes, matchmark the camshaft brackets before removing them.**

- Intake and exhaust camshafts and brackets. Loosen the bracket bolts in several steps, in the sequence shown.
- Right and left side cam chain tensioner from the cylinder head
- Cylinder head bolts. Loosen in several steps, in the sequence shown.

➥**A warped or cracked cylinder head could result from removing the bolts in incorrect order.**

- Cylinder heads from the vehicle
- Discard the head gaskets

7. Remove all traces of liquid gasket from the timing chain case and from the water pump covers.

8. Remove all traces of liquid gasket from the engine block.

9. Inspect the timing chain for excessive wear or damage and replace as necessary.

To install:

10. Turn the crankshaft until the No. 1 piston is a Top Dead Center (TDC) on compression stroke. The crankshaft key should face toward the right bank.

11. Using new head gaskets, install the cylinder heads.

➥**If possible, replacement of the head bolts is suggested.**

12. If replacement of the head bolts is not possible, perform the following bolt measurement:

   a. Measure the diameter of the head bolt 0.43 in. (11mm) from the bottom of the bolt.

   b. Measure the diameter of the head bolt 1.89 in. (48mm) from the bottom of the bolt.

   c. Whenever the size difference between the 2 measurements exceeds 0.0043 in. (0.11mm) the head bolts must be replaced.

13. Install the cylinder head bolts and torque in sequence as follows:

   a. Step 1: 72 ft. lbs. (98 Nm).

   b. Step 2: Completely loosen all bolts.

   c. Step 3: 26–32 ft. lbs. (34–44 Nm).

   d. Step 4: plus 90–95 degrees clockwise.

   e. Step 5: plus 90–95 degrees clockwise.

14. Install or connect the following:

- Camshafts and related components
- Intake valve timing control solenoid valves

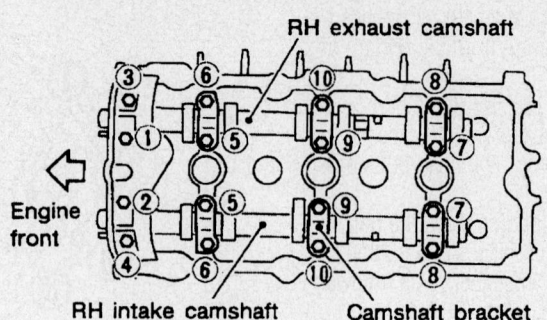

Loosen in numerical order.

09482_ALTI_G0008

Right camshaft bracket bolt loosening sequence—3.5L engine

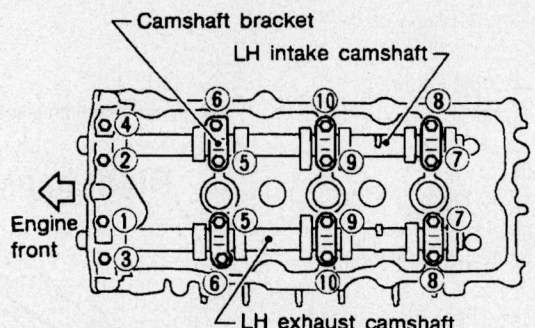

Loosen in numerical order.

09482_ALTI_G0009

Left camshaft bracket bolt loosening sequence—3.5L engine

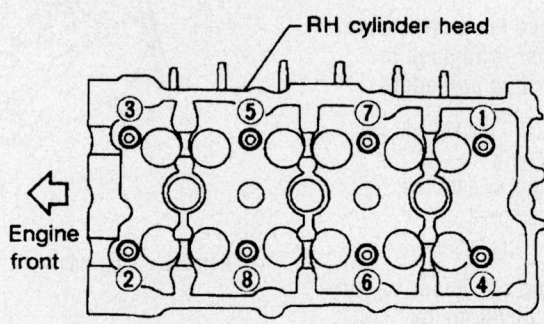

Loosen in numerical order.

09482_ALTI_G0010

Right cylinder head bolt loosening sequence—3.5L engine

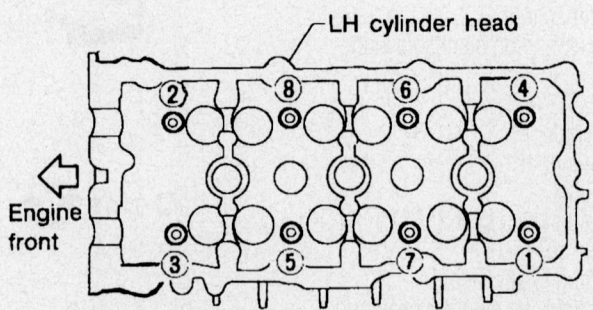

Loosen in numerical order.

09482_ALTI_G0011

Left cylinder head bolt loosening sequence—3.5L engine

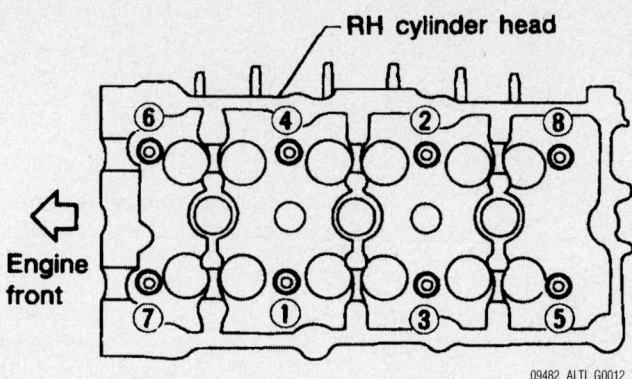

Right cylinder head bolt torque sequence—3.5L engine

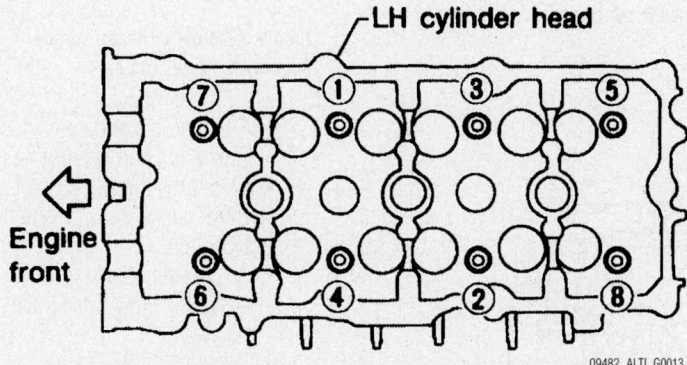

Left cylinder head bolt torque sequence—3.5L engine

- New O-rings to the front of the engine block and cylinder head

15. Apply sealant to the hatched portion of the of the rear timing chain case.

16. Align the rear timing chain case with the dowel pins and install onto the cylinder heads and engine block.

17. Torque the rear timing chain case mounting bolts in sequence to 105–121 inch lbs. (11.8–13.7 Nm).

18. Install or connect the following:
- Water outlet
- Intake manifold
- Timing chain
- Oil pan
- Exhaust manifold
- Engine assembly into the vehicle
- Negative battery cable

19. Fill the cooling system.

20. Fill the engine with clean oil.

21. Start the vehicle, check for leaks and repair if necessary.

Rocker Arms

REMOVAL & INSTALLATION

The 1.8L, 2.5L and 3.5L engines do not utilize rocker arms. The valves are actuated directly by the camshafts.

Intake Manifold

REMOVAL & INSTALLATION

1.8L Engine

1. Before servicing the vehicle, refer to the Precautions Section.
2. Relieve the fuel system pressure.
3. Drain the cooling system.
4. Remove or disconnect the following:
- Negative battery cable
- Air cleaner assembly
- Throttle linkage, electrical connections and vacuum lines from the throttle body
- Intake manifold collector support brackets

5. The throttle body can be removed from the manifold at this point or can be removed as an assembly with the intake manifold.

- Bolts holding the upper portion of the intake to the lower portion. Remove the bolts in reverse order of the tightening sequence.
- Upper portion of the intake
- Fuel injector wiring harness connectors and the vacuum line from the fuel pressure regulator

- Fuel hoses from the fuel rail assembly
- Bolts that secure the fuel rail to the intake
- Injectors with the fuel rail assembly
- Intake manifold retaining bolts in the proper sequence and separate the manifold from the cylinder head. Remove the bolts in reverse order of the tightening sequence.
- Intake manifold gasket and clean all the gasket contact surfaces thoroughly with a gasket scraper and suitable solvent. All traces of old gasket material must be removed to ensure proper sealing.
- Inspect the intake manifold for cracks. Using a metal straightedge, check the surface of the intake manifold for warpage.

To install:

6. Install or connect the following:
- New intake manifold gasket onto the cylinder head and position the lower intake manifold over the mounting studs and onto the gasket.
- Intake manifold and torque the bolts to 13–15 ft. lbs. (18–21 Nm) in sequence
- Injectors with the fuel rail assembly. Be sure to install the fuel rail insulators. Torque the bolts in 2 steps to 13–15 ft. lbs. (18–21 Nm).
- Fuel injector wiring harness connectors and the vacuum line to the fuel pressure regulator
- Fuel hoses to the fuel rail assembly using new hose clamps
- Intake manifold collector using a new gasket. Torque the bolts to 13–15 ft. lbs. (18–21 Nm) in sequence.
- Throttle body or throttle chamber, if removed. Torque the bolts in a crisscross pattern to 13–16 ft. lbs. (18–22 Nm).

➡Be sure to properly position the throttle body gasket with the cut out facing down.

- Intake manifold collector support brackets
- Throttle linkage, electrical connections and vacuum lines
- Air cleaner
- Negative battery cable

7. Fill the cooling system to the proper level.

8. Start the engine, check for leaks and repair if necessary.

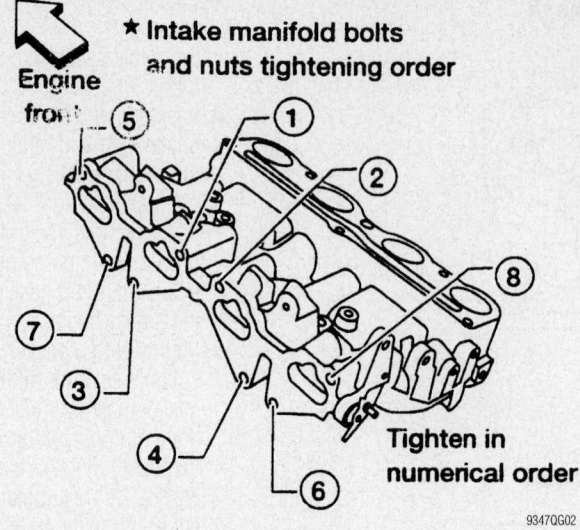

★ Intake manifold bolts and nuts tightening order

Engine front

Tighten in numerical order

9347QG02

Lower intake manifold torque sequence—1.8L engine

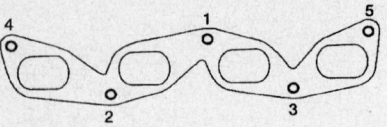

67162-NISS-G16

Intake manifold torque sequence—2.5L engine

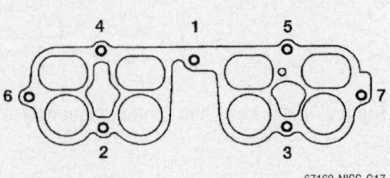

67162-NISS-G17

Intake manifold collector torque sequence—2.5L engine

- Fuel line quick connector
- Brake booster vacuum hose
- Electronic throttle control actuator
- EVAP canister purge solenoid
- PCV hose
- Air cleaner case and duct
- Mass Air Flow (MAF) sensor connector
- Negative battery cable

7. Fill the cooling system to the proper level.

8. Start the vehicle, check for leaks and repair if necessary.

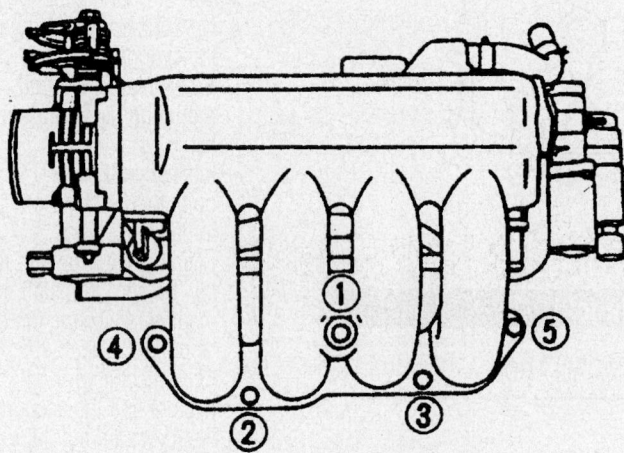

Tighten in numerical order.

7923QG17

Upper intake manifold torque sequence—1.8L engine

2.5L Engine

1. Before servicing the vehicle, refer to the Precautions Section.
2. Relieve the fuel system pressure.
3. Drain the cooling system.
4. Remove or disconnect the following:
- Negative battery cable
- Mass Air Flow (MAF) sensor connector
- Air cleaner case and duct
- PCV hose
- EVAP canister purge solenoid
- Electronic throttle control actuator
- Brake booster vacuum hose
- Fuel line quick connector
- Intake manifold collector harness and vacuum hose
- Intake manifold collector-to-intake manifold bolts/nuts in the reverse

sequence of the tightening procedure and separate the intake manifold from the intake manifold collector
- Intake manifold in the reverse sequence of the tightening procedure.

5. Using a putty knife, clean the gasket mounting surfaces. Check the intake manifold/collector for cracks and warpage.

To install:

6. Install or connect the following:
- Intake manifold with new gaskets and torque the bolts, in sequence, to 13–15 ft. lbs. (18–22 Nm)
- Intake manifold collector using new gaskets and torque the bolts in sequence to 13–15 ft. lbs. (18–22 Nm)
- Intake manifold collector harness and vacuum hose

3.5L Engine

1. Before servicing the vehicle, refer to the Precautions Section.
2. Drain the cooling system.
3. Release the fuel system pressure.
4. Disconnect the negative battery cable.
5. Remove or disconnect the following:
- Throttle body coolant hoses
- Electrical connectors from the Throttle Position (TP) sensor
- Hoses from the throttle body, the Exhaust Gas Recirculation (EGR) valve, intake manifold collector, Idle Air Control (IAC) valve and the fuel pressure regulator
- Canister purge hose and blow-by hose
- EGR guide tube
- Accelerator cable from the throttle body
- Intake manifold collector support brackets
- Right side electrical connectors from the ignition coils
- Electrical connector from the crank angle sensor and the power transistor, if necessary

- Intake manifold collector-to-intake manifold bolts/nuts and the intake manifold collector

6. Remove the fuel injector assembly by performing the following procedures:

a. Detach the electrical connectors from the fuel injectors.

b. Disconnect the fuel lines from the fuel injector assembly.

c. Remove the fuel rail-to-cylinder head bolts.

d. Remove the fuel rail assembly from the engine.

7. On all models, remove or disconnect the following:

- Intake manifold bolts/nuts in the reverse of the installation sequence
- Intake manifold from the engine and discard the gaskets

8. Clean all gasket mounting surfaces.

To install:

9. Using new gaskets, install the intake manifold to the engine.

10. Torque the bolts in sequence as follows:

a. Step 1: 44–86 inch lbs. (5–10 Nm).

b. Step 2: 20–23 ft. lbs. (26–31 Nm).

11. Install or connect the following:

12. Install the fuel injector assembly by performing the following procedures:

a. Install the fuel rail assembly to the engine.

b. Install the fuel rail-to-cylinder head bolts and tighten the bolts to 15–20 ft. lbs. (21–26 Nm) in 2 progressive steps.

c. Connect the fuel lines to the fuel injector assembly.

d. Connect the electrical connectors to the fuel injectors.

13. Install the intake manifold collector. Torque the bolts to 13–15 ft. lbs. (18–21 Nm).

14. Install or connect the following:

- Crank angle sensor and transmitter electrical connectors
- Right side ignition coil electrical connectors
- Intake manifold collector support brackets
- Accelerator cable to the throttle body
- EGR guide tube
- Canister purge and blow by hoses
- Throttle body, EGR valve and intake manifold collector hoses
- IAC valve and fuel pressure regulator hoses

- TP sensor electrical connector
- Throttle body coolant hose

15. Connect the negative battery cable.

16. Fill the cooling system.

17. Start the vehicle, check for leaks and repair if necessary.

Exhaust Manifold

REMOVAL & INSTALLATION

1.8L Engine

1. Before servicing the vehicle, refer to the Precautions Section.

2. Remove or disconnect the following:

- Negative battery cable
- Engine undercovers
- Air cleaner or collector assembly
- Heat shields from the manifold and front exhaust pipe
- Front exhaust pipe from the exhaust manifold
- Temperature sensors, Oxygen (O_2) sensors and air induction pipes from the manifold
- Manifold support brackets
- Exhaust manifold attaching nuts and the manifold from the block. Discard the exhaust manifold gaskets.

3. Clean the gasket surfaces and check the manifold for cracks and warpage.

To install:

4. Install the exhaust manifold with a new gasket.

5. Tighten the mounting nuts with washers in sequence to 20–22 ft. lbs. (26–29 Nm).

6. Install or connect the following:

- Temperature sensors, O_2 sensors and air induction pipes
- Manifold support brackets
- Exhaust pipe to the manifold using a new gasket. Torque the nuts to 21–25 ft. lbs. (28–33 Nm).
- Heat shields
- Air cleaner or collector assembly
- Engine undercovers
- Negative battery cable

7. Start the engine and check for exhaust leaks.

2.5L Engine

1. Before servicing the vehicle, refer to the Precautions Section.

2. Remove or disconnect the following:

- Negative battery cable
- Engine undercover
- Air/fuel ratio sensor connector

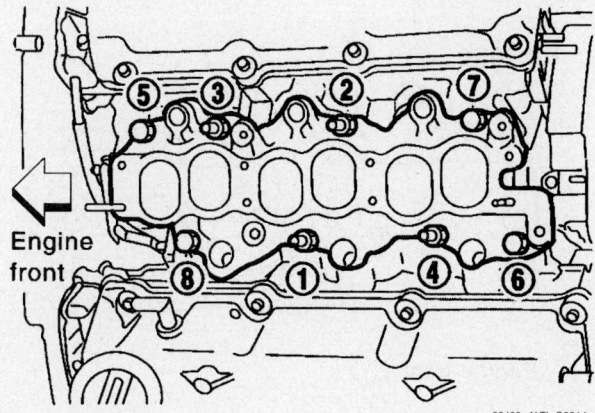

09482_ALTI_G0014

Intake manifold torque sequence—3.5L engine

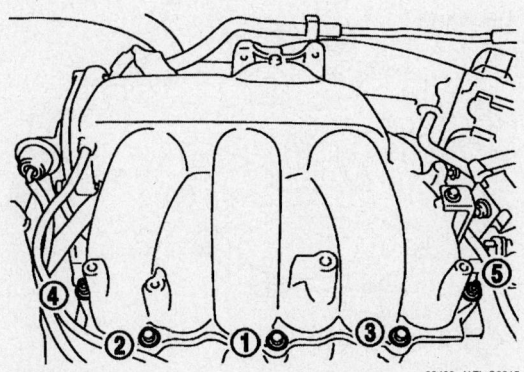

09482_ALTI_G0015

Upper intake manifold collector torque sequence—3.5L engine

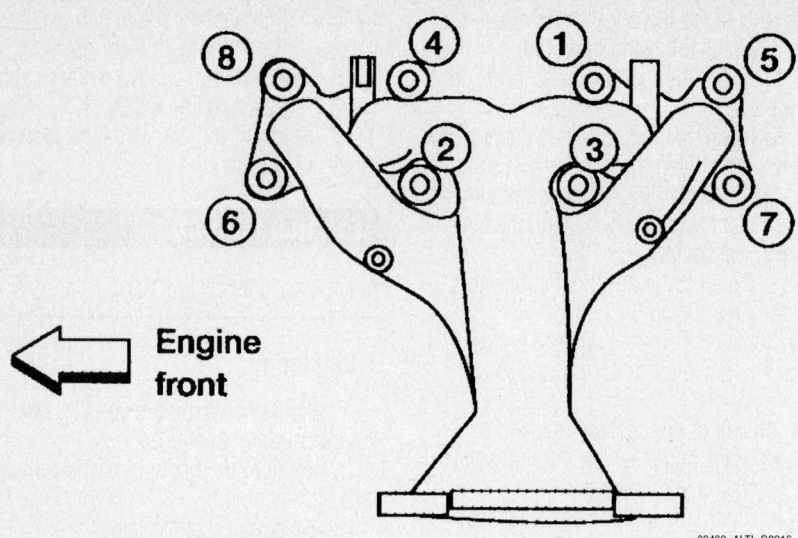

Exhaust manifold bolt tightening sequence—1.8L engine

• Oxygen (O2) sensor electrical connector
• Air/fuel ratio sensor
• Oxygen (O2) sensor
• Upper and lower exhaust manifold covers
• Exhaust manifold/catalyst assembly in reverse of the tightening sequence

To install:
3. Clean all gasket mounting surfaces and install new gaskets.
4. Install or connect the following:
• Exhaust manifold/catalyst and torque the nuts to 29–32 ft. lbs. (40–44 Nm)

➥After tightening, go back and retighten nuts numbers 1 and 3 to the specification.

• Air/fuel ratio sensor after coating the threads with anti-sieze
• Oxygen (O2) sensor after coating the threads with anti-sieze
• Oxygen (O2) sensor electrical connector
• Air/fuel ratio sensor connector
• Engine undercover

• Negative battery cable
5. Start the engine and check for exhaust leaks.

3..5L Engine

1. Before servicing the vehicle, refer to the Precautions Section.
2. Remove or disconnect the following:

• Negative battery cable
• Exhaust manifolds from the exhaust pipes
• Protective covers from the manifolds
• Heated Oxygen Sensor (HO2S) from the manifold, if equipped
• Exhaust manifold-to-engine mounting nuts
• Manifolds from the engine and discard the gaskets

To install:
3. Clean all gasket mounting surfaces. Install new gaskets.
4. Install or connect the following:
• Exhaust manifold and torque the nuts in steps to 21–24 ft. lbs. (28–33 Nm)
• Protective shields and torque the bolts in steps to 46–57 inch lbs. (5–7 Nm)
• Exhaust manifolds to the exhaust pipes and torque the nuts to 32–37 ft. lbs. (43–50 Nm)
• HO2S sensor to the manifold and torque the fastener to 30–44 ft. lbs. (40–60 Nm), if equipped
• Negative battery cable
5. Start the engine and check for exhaust leaks.

RH bank

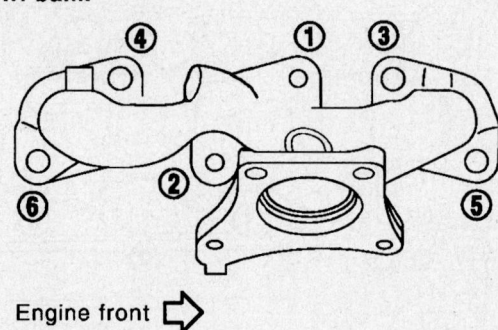

Right exhaust manifold bolt tightening sequence—3.5L engine

LH bank

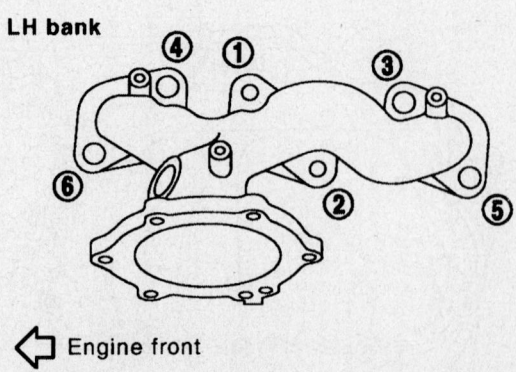

Left exhaust manifold bolt tightening sequence—3.5L engine

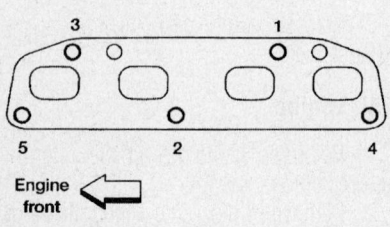

Exhaust manifold bolt tightening sequence—2.5L engine

Front Crankshaft Seal

REMOVAL & INSTALLATION

➡The front crankshaft seal procedure is applicable to timing belt-equipped engines only. For the front seal on engines equipped with timing chains, refer to the timing chain, sprockets, front cover and seal procedure later in this section.

Camshaft and Valve Lifters

REMOVAL & INSTALLATION

1.8L Engine

1. Before servicing the vehicle, refer to the Precautions Section.
2. Drain the cooling system.
3. Relieve the fuel system pressure.
4. Remove or disconnect the following:
   • Negative battery cable
   • All engine drive belts
   • Exhaust pipe from the exhaust manifold
   • Power steering pulley and pump with the mounting bracket
   • Cylinder head cover
   • Distributor assembly
   • Timing chain tensioners and camshaft sprocket

➡Before the camshafts are removed from the cylinder head, note the positioning of the pins at the end of the camshafts for reassembly purposes.

   • Camshaft bearing caps in sequence
   • Camshafts from the cylinder head
   • Idler sprocket bolt. These parts should be reassembled in their original position.
   • Shims from the tops of the lifters. Be sure to note the position of each shim.
   • Valve lifters from the bores in the cylinder head. Note the positioning of the lifters for reassembly.
5. Measure the diameter of the lifters. The diameter should be 1.1795–1.1801 in. (29.960–29.975mm).
6. Measure the diameter of the lifter bores. The diameter should be 1.1811–1.1819 in. (30.000–30.021mm).
7. Clearance between the lifter and bore should be 0.0010–0.0024 in. (0.025–0.061mm).
   To install:
8. Install or connect the following:
   • Lifters and shims to the cylinder

head in the proper locations as noted during removal

➡The exhaust and intake camshafts are marked with identification stamps. (E for exhaust and I for intake).

   • Camshafts to the cylinder head and position the intake camshaft knock pin at the 9 o'clock position and the exhaust camshaft at the 12 o'clock position

9. Install the camshaft bearing caps and tighten the mounting bolts as follows:
   a. Bolts 11 through 15, then bolts 1 through 10: 18 inch lbs. (2 Nm).
   b. Bolts 1 through 15: 53 inch lbs. (6 Nm).
   c. Bolts 1 through 14: 87–105 inch lbs. (10–12 Nm).
   d. Bolt 15: 56–73 inch lbs. (7–8 Nm).

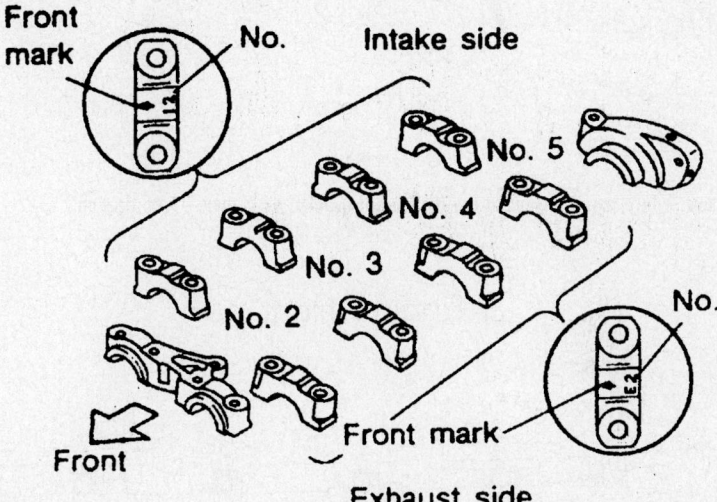

Be sure to install the camshaft bearing caps in their original positions—1.8L engine

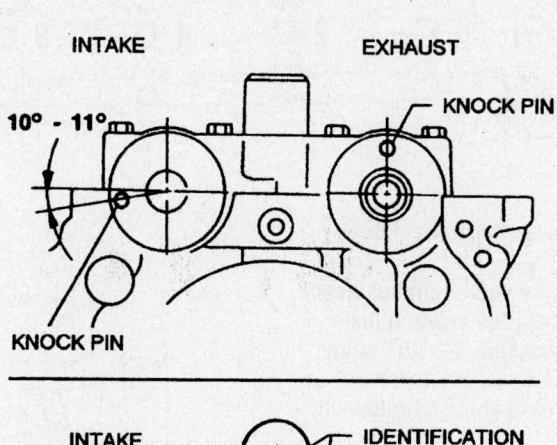

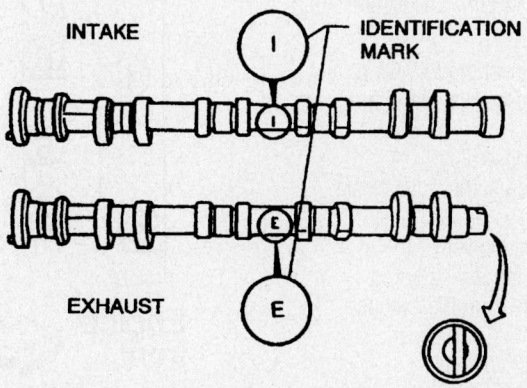

Positioning and identification of the camshafts—1.8L engine

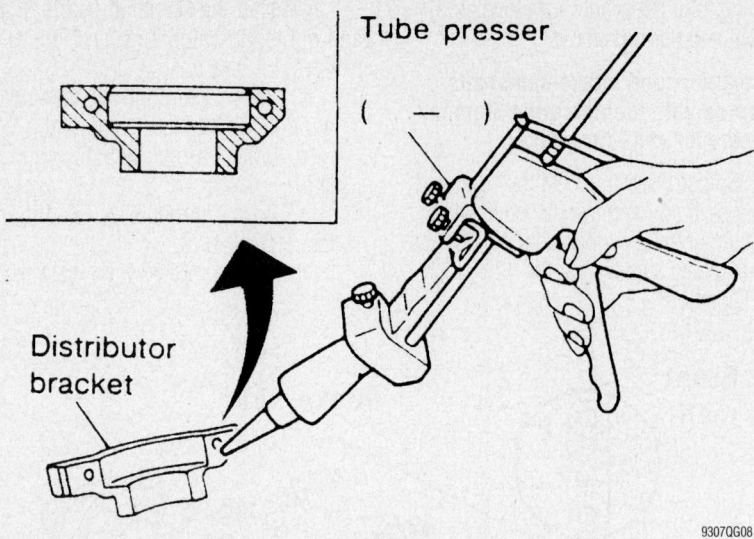

If removed, apply liquid gasket to the distributor bracket as shown—1.8L engine

9307QG08

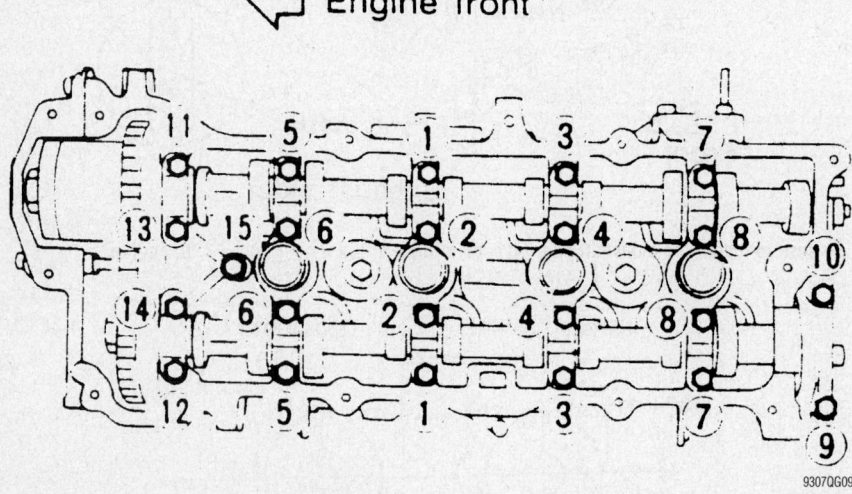

⇐ Engine front

Camshaft bolt torque sequence—1.8L engine

9307QG09

➡ If any part of the valvetrain has been has been replaced, the valve adjustment must be checked. DO NOT adjust the valves or rotate the camshafts at this point. Internal engine damage will result.

10. Install or connect the following:
- Camshaft sprockets with timing chains
- Distributor assembly

11. Check and adjust the valve clearance.
- Cylinder head cover
- Power steering pulley pump
- Exhaust pipe from the exhaust manifold
- All engine drive belts
- Negative battery cable

12. Fill the cooling system.

13. Start the vehicle, check for leaks and repair if necessary.

2.5L Engine

1. Before servicing the vehicle, refer to the Precautions Section.
2. Relieve the fuel system pressure.
3. Drain coolant from the engine and radiator.
4. Remove or disconnect the following:
- Negative battery cable
- Engine undercover
- Right front wheel
- Valve cover
- Drive belt
- Coolant reservoir
- Variable timing control solenoid connector
- Camshaft position sensor
- Intake valve timing control cover in reverse of the tightening sequence
- Set the No. 1 piston at Top Dead Center (TDC) on its compression stroke

5. Check that the mating marks on the camshaft sprockets are lined up with the Yellow links on the timing chain. If not, rotate the crankshaft one revolution until the links line up.
6. Remove the timing chain guide out between the camshaft sprockets through the front cover.
7. Push in the plunger on the timing chain tensioner, then insert a pin to hold the tensioner retracted.
8. Remove the timing chain tensioner .
9. While holding the flat of the camshaft with an open end wrench, loosen and remove the camshaft sprockets.

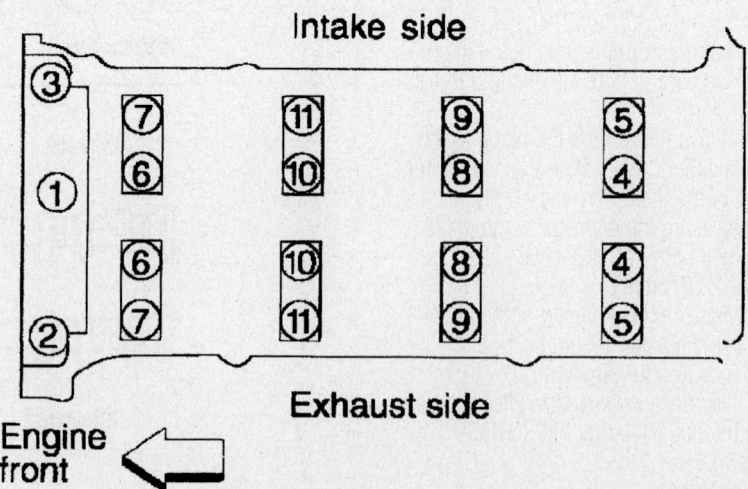

Camshaft bearing cap removal sequence—2.5L engines

67162-NISS-G24

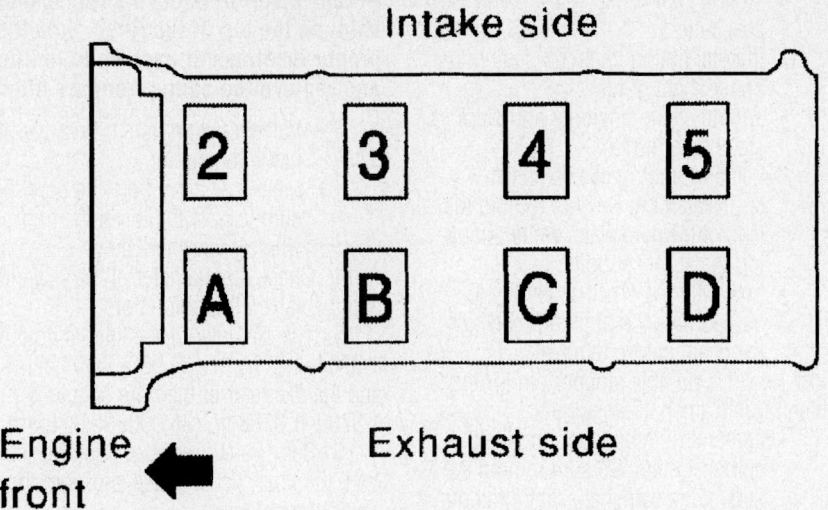

Identifiying camshaft bearing cap installation marks—2.5L engines

67162-NISS-G25

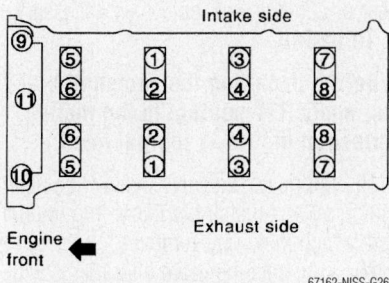

Camshaft bearing cap tightening sequence—2.5L engines

67162-NISS-G26

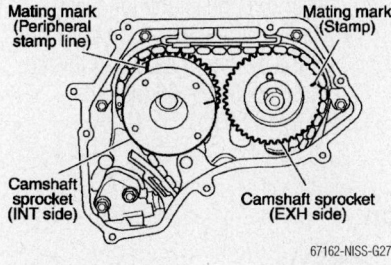

Camshaft sprocket alignment marks—2.5L engines

67162-NISS-G27

10. Loosen the camshaft brackets in the order shown .

11. Remove the camshafts.

12. Remove the valve lifters.

To install:

➡**When installing the valve components, apply a coat of clean engine oil to the component.**

13. Install or connect the following:
- Lifters into the lifter bores from which they were removed

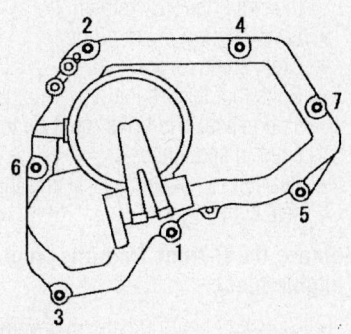

Intake valve timing cover tightening sequence—2.5L engines

67162-NISS-G28

- Valve shims to the lifters from which they came

14. Install the camshafts so the dowel pin on the intake camshaft is placed at the three o'clock position and the exhaust camshaft pin is placed at the twelve o'clock position.

15. Install the camshaft caps with the identifying marks placed in the correct position as shown.

16. Place a bead of sealant of the No. 1 bearing cap bottom edge.

17. Place a bead of sealant on the back side of the front cover where the No. 1 bearing cap lines up.

18. Apply sealant to the bolt hole on the front cover.

19. Install the No. 1 bearing cap so the sealant joints line up.

20. Tighten the camshaft bearing caps bolts in the proper sequence as follows:

    a. Step 1 bolts 9–11: 17 inch lbs. (2 Nm).

    b. Step 2 bolts 1–8: 17 inch lbs. (2 Nm).

    c. Step 3 bolts 1–11: 52 inch lbs. (6 Nm).

    d. Step 4 bolts 1–11: 80–104 inch lbs. (9–12 Nm).

21. Install the camshaft sprockets so the mating marks are aligned as shown.

22. Install or connect the following:
- Timing chain tensioner and remove the pin
- Intake valve timing control cover and tighten the bolts in the sequence shown.
- Check and adjust the valve clearance as necessary.
- Timing chain guide between the camshaft sprockets through the front cover
- Camshaft position sensor
- Variable timing control solenoid connector
- Coolant reservoir
- Drive belt
- Valve cover
- Right front wheel
- Negative battery cable
- Engine undercover

23. Fill the cooling system.

24. Start the vehicle, check for leaks and repair if necessary.

3.5L Engine

1. Before servicing the vehicle, refer to the Precautions Section.

2. Relieve the fuel system pressure.

3. Drain the engine oil.

4. Drain the cooling system.

5. Remove or disconnect the following:
- Negative battery cable
- Left side rocker cover ornament

➡**Before detaching any hoses or connectors, note the locations for reassembly.**

- Air duct to intake manifold hose, collector hose, blow-by hose and vacuum hoses
- Fuel hoses and detach the harness connections
- Canister purge hoses
- Water hoses from the cylinder head and intake manifold
- All 6 ignition coils from the spark plugs
- Spark plugs
- Bolts that secure the Exhaust Gas Recirculation (EGR) tube
- EGR tube
- Intake manifold collector supports and the collector
- Bolts that secure the fuel tube and the fuel tube

- Bolts that secure the intake manifold to the engine block and the manifold. Loosen the bolts in the reverse sequence of the tightening procedure.
- Left-hand and right-hand rocker covers from the cylinder head
- Engine undercovers
- Right front wheel and engine side covers
- Drive belts and idler pulley
- Power steering oil pump belt and the power steering oil pump assembly
- Camshaft Position (CMP) sensor (PHASE) and Crankshaft Position (CKP) sensors (REF)/(POS)

6. Set the No. 1 piston to Top Dead Center (TDC) of compression stroke by rotating the crankshaft.

- Ring gear cover access plate. Loosen the crankshaft pulley bolt while securing the ring gear so the crankshaft cannot rotate
- Crankshaft pulley, using a suitable puller
- Air conditioning compressor and bracket
- Front exhaust pipe and its support

7. Hang the engine at the right and left side engine slingers with a suitable hoist.

8. Support the transaxle with jack.

- Right side engine mounting, mounting bracket and nuts
- Center crossmember assembly
- Steel (lower) oil pan bolts in the reverse of the installation sequence

9. Insert a seal cutter between the steel and aluminum oil pan

10. Tapping the cutter with a hammer, slide it around the entire edge of the oil pan. Be careful not to damage the aluminum mating surface of the upper oil pan.

- Steel oil pan and the oil strainer
- Aluminum (upper) oil pan bolts in the reverse of the installation sequence
- Transaxle bolts that secure the oil pan

11. Insert a seal cutter between the aluminum oil pan and the engine block.

12. Tapping the cutter with a hammer, slide it around the entire edge of the oil pan. Be careful not to damage the mating surfaces of the oil pan or engine block.

- Oil pan from the vehicle
- Water pump cover and the bolts that secure the front timing chain case cover

- Timing chain case cover, using the seal cutter
- Internal timing chain guide and the upper chain guide
- Timing chain tensioner and slack side chain guide
- Left and right intake camshaft sprockets first. Be sure to hold the flats of the camshafts while removing the sprocket bolts.
- Lower timing chain assembly. Be sure to note the aligning marks of the chain before removal.

13. Insert a suitable stopper pin for the left and right camshaft tensioners.

- Left and right exhaust camshaft sprocket bolts. Be sure to hold the flats of the camshafts while removing the sprocket bolts.
- Upper timing chain assembly. Be sure to note the aligning marks of the chain before removal.
- Lower timing chain guide
- Crankshaft sprocket
- Bolts that secure the rear timing chain case. The bolts must be loosened in sequence.
- Rear timing case cover, using the seal cutter

➡**Remove the O-rings from the front of the engine block.**

- Camshaft bearing caps in several steps. The bearing caps MUST be loosened in sequence.

➡**Keep all bearing caps and camshafts in proper order for installation.**

- Left-hand and right-hand camshaft tensioners from the cylinder head
- Camshafts from the cylinder heads

➡**The valve lifters have a replaceable shim on the top of the lifter. Note the proper locations of each shim to lifter and remove the shims from the lifters.**

- Valve adjusting shim from the lifter, using a magnet
- Lifter assembly from the bore. Be sure to note the locations from where each lifter came.

14. Check the diameter of the valve lifter and the valve lifter guide bore.

15. The diameter of the lifter should be 1.3764–1.3770 in. (34.960–34.975mm) and the diameter of the bore should be 1.3780–1.3788 in. (35.000–35.021mm).

16. Remove all traces of liquid gasket from the timing chain case and from the water pump covers.

17. Remove all traces of liquid gasket from the engine block.

18. Inspect the camshafts for excessive wear or damage and replace as necessary.

To install:

➡**Before installing the camshaft brackets, apply RTV sealant to the mating surface of the No. 1 journal head.**

19. Lubricate the valve lifters with clean engine oil and install the lifters into the bore from which they were removed.

20. Lubricate the valve lifter shims with clean engine oil and install the shims into the lifter from which they were removed.

21. Turn the crankshaft clockwise until the No. 1 piston is set 240 degrees before TDC on compression stroke.

22. Install or connect the following:

- Camshaft tensioners on both sides of the cylinder heads and torque the bolts to 75–96 inch lbs. (8.4–10.8 Nm)

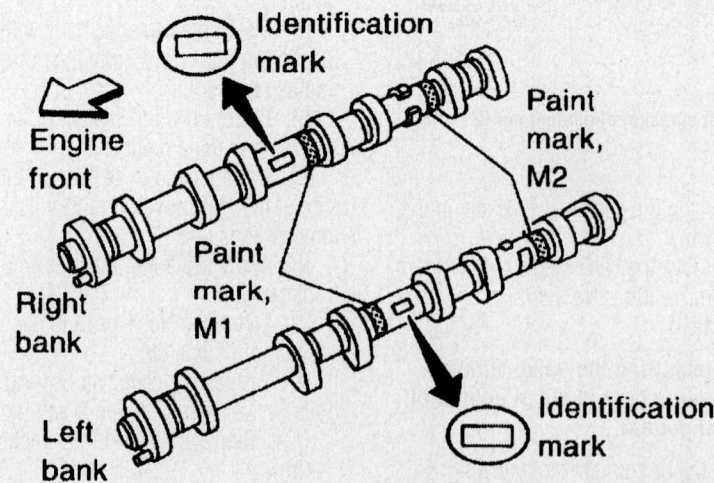

Camshaft identification marks—3.5L engines

09482_ALTI_G0019

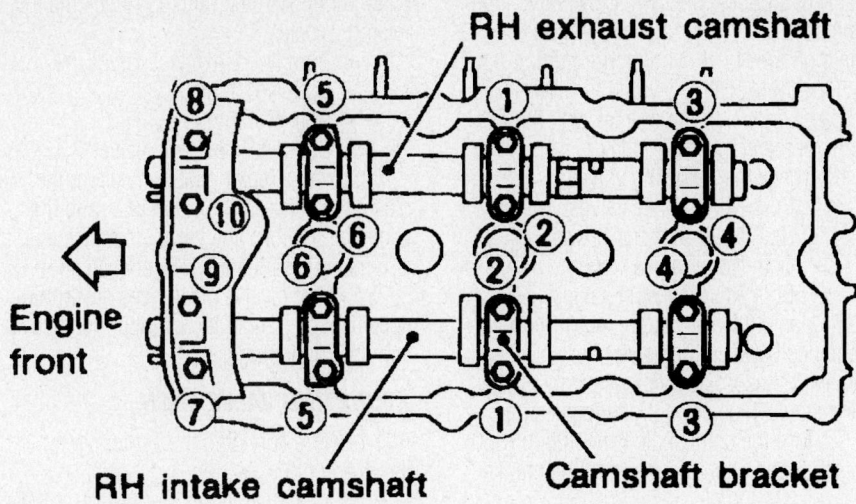

Right cylinder head camshaft bearing cap tightening sequence—3.5L engines

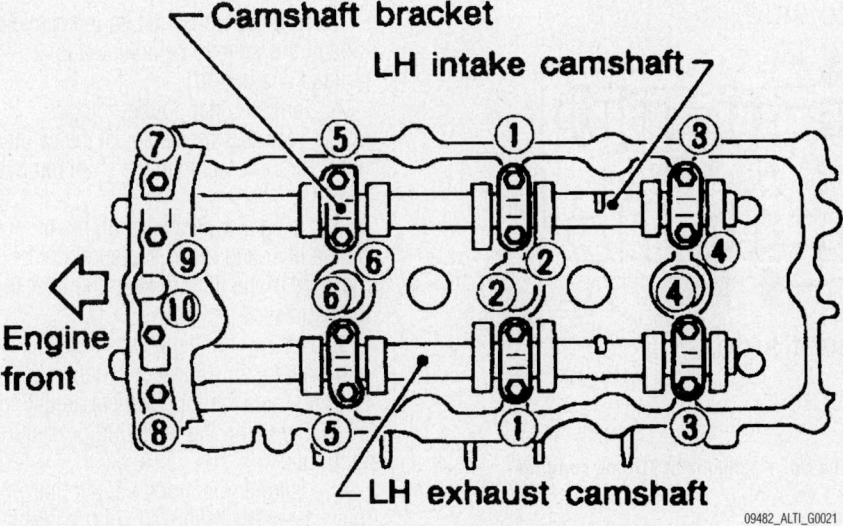

Left cylinder head camshaft bearing cap tightening sequence—3.5L engines

➡The camshafts can be identified by the paint marks on the camshaft. The left cylinder head camshafts have a YELLOW paint mark and the right cylinder head camshafts have a WHITE paint mark.

- Exhaust and intake camshafts and install the bearing caps. Before installing the No. 1 bearing cap, apply liquid gasket to the corners of the cap.

➡When installing the camshafts, position the camshaft keys at the 12 o'clock position in respect to the cylinder head angle.

23. Torque the camshaft bearing caps as follows:

    a. Bolts No. 7–10: 17 inch lbs. (2 Nm).

    b. Bolts No. 1–6: 17 inch lbs. (2 Nm).

    c. Bolts No. 1–10: 52 inch lbs. (6 Nm).

    d. Bolts No. 1–10: 81–104 inch lbs. (9–11 Nm).

24. Install new O-rings to the front of the engine block.

25. Apply sealant to the hatched portion of the of the rear timing chain case.

26. Align the rear timing chain case with the dowel pins and install onto the cylinder heads and engine block.

27. Torque the rear timing chain case mounting bolts in sequence to 105–121 inch lbs. (11.8–13.7 Nm).

28. Install the crankshaft sprocket with the mating mark facing out.

29. Rotate the crankshaft clockwise and position the crankshaft to TDC of compres-

sion stroke and align the dowels of the camshaft sprockets to the 12 o'clock position in respect to the cylinder head.

30. Install the lower chain guide on the dowel pin with the front mark on the guide facing upward.

31. On a workbench, align the marks on the intake and exhaust camshaft sprockets with the marks of the chain.

32. Put the exhaust camshaft sprockets onto the dowel pin and torque the bolts to 88–95 ft. lbs. (119–128 Nm). Be sure to secure the camshafts while tightening the bolts.

33. Install or connect the following:

- Timing chains, sprockets and related components
- Transaxle bolts that secure the oil pan
- Oil pan strainer and torque the bolts to 12–14 ft. lbs. (16–19 Nm)

34. Apply a 0.177–0.217 in. (4.5–5.5mm) continuous bead of liquid gasket to the lower oil pan mating surface and install the oil pan. Torque the bolts in sequence to 57–66 inch lbs. (6.4–7.5 Nm).

- Center crossmember assembly
- Right side engine mounting bracket and mount assembly

35. Remove the engine slinger assembly.

- Front exhaust pipe and its support
- Air conditioning compressor and bracket
- Crankshaft pulley to the crankshaft and install the mounting bolt. Torque the bolt to 14–22 ft. lbs. (20–29 Nm). Torque the crankshaft bolt an additional 60–66 degrees clockwise. This is about the angle from one hexagon bolt head corner to another
- CMP sensor, PHASE and CKP sensors
- Power steering pump
- Idler pulley and all belts
- Engine side and under covers
- Right front wheel
- Intake manifold
- Rocker covers
- Fuel tube
- Intake manifold support and collector
- EGR tube
- Spark plugs and ignition coils
- Coolant hoses
- Canister purge hoses
- Fuel feed and return lines
- Vacuum hoses
- Negative battery cable

36. Fill the cooling system.

37. Fill the engine with clean oil.

38. Start the vehicle, check for leaks and repair if necessary.

Valve Lash

ADJUSTMENT

1.8L and 2.5L Engines

CHECKING VALVE LASH

1. Before servicing the vehicle, refer to the Precautions Section.

2. Run the engine until it reaches normal operating temperature and shut if off.

3. Remove the cylinder head cover and all the spark plugs.

4. Set the No. 1 cylinder at Top Dead Center (TDC) on its compression stroke. Align the pointer with the TDC mark on the crankshaft pulley. Check that the valve lifters on the No. 1 cylinder are loose and valve lifters on the No. 4 cylinder are tight. If not, turn the crankshaft 1 revolution (360 degrees) and align the pointer with the TDC mark on the crankshaft pulley.

5. Check the following valves:
   - Both No. 1 intake valves
   - Both No. 1 exhaust valves
   - Both No. 2 intake valves
   - Both No. 3 exhaust valves

6. Using a feeler gauge, measure the clearance between the valve lifter and the camshaft. Record any valve clearance measurements which are out of specification.

7. Turn the crankshaft 1 revolution (360 degrees) and align the mark on the crankshaft pulley with the pointer. Check the following valves:
   - Both No. 2 exhaust valves
   - Both No. 3 intake valves
   - Both No. 4 intake valves
   - Both No. 4 exhaust valves

8. Using a feeler gauge, measure the clearance between the valve lifter and the camshaft. Record any valve clearance measurements which are out of specification.

9. If all the valve clearances are within specification, install the cylinder head cover and the spark plugs.

ADJUSTING VALVE LASH

1. Before servicing the vehicle, refer to the Precautions Section.

2. If an adjustment is necessary, adjust the valve clearance while engine is cold by removing the adjusting shim. The adjusting shim can be removed by using the following procedures:

   a. Turn the crankshaft so the camshaft lobe of the valve to be adjusted is pointed straight up.

   b. Turn the lifter so the notch is pointed towards the center of the cylinder head; this will facilitate the shim removal process.

   c. Using a depressor tool, push down on the lifter and insert a keeper tool on the edge of the lifter to keep the lifter in the depressed position.

   d. Remove the depressor tool and remove the shim with a magnet.

3. Determine the replacement adjusting shim size by using the following procedures and formula:

   a. Using a micrometer determine thickness of the removed shim.

   b. Calculate the thickness of a new adjusting shim so valve clearance is within the specified values.

   c. R = thickness of the removed shim.

   d. N = thickness of the new shim.

   e. M = measured valve clearance.

   f. 1.8L engine: Intake shim determination formula: $N = R + (M - 0.0146$ in. or 0.37mm$)$

   g. 1.8L engine: Exhaust shim determination formula: $N = R + (M - 0.0157$ in. or 0.40mm$)$

Shims are available in different sizes from 0.0772–0.1055 in. (1.96–2.68mm) in increments of 0.0008 in. (0.02mm). The thickness is stamped on the shim; this side is always installed facing down. Select new shims with thickness as close as possible to calculated valve and install it in the lifter.

4. Install the new shim onto the lifter.

5. Depress the lifter and remove the keeper tool. Remove the depressor tool and

No. 1 cylinder at TDC of its compression stroke

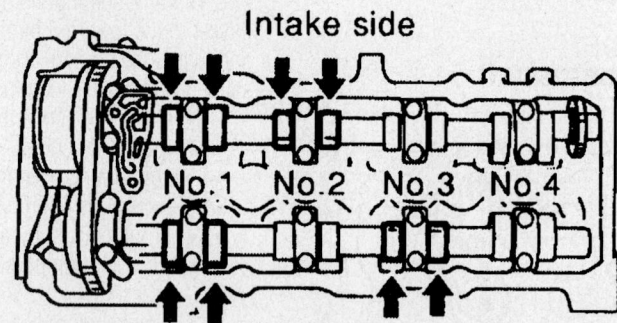

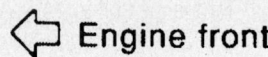

09482_ALTI_G0022

Measure the clearance of the valves indicated when the No. 1 piston is at TDC on compression—1.8L and 2.4L engines

No. 4 cylinder at TDC of its compression stroke

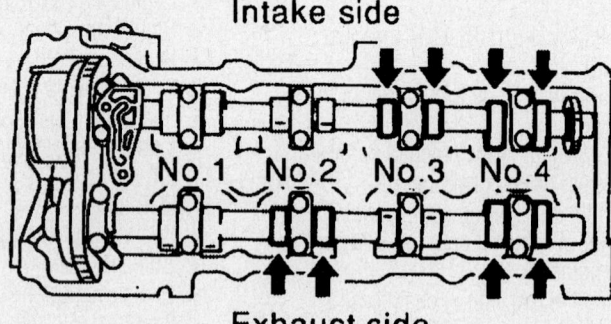

09482_ALTI_G0023

Measure the clearance of the valves indicated when the No. 4 piston is at TDC on compression—1.8L and 2.4L engines

recheck the valve clearance. Repeat this procedure for any other valves requiring adjustment.

6. Install the cylinder head cover and spark plugs when all valve adjustments are finished.

3.5L Engine

➡**Check and adjust the valve clearances while the engine is cold and not running.**

CHECKING VALVE LASH

1. Before servicing the vehicle, refer to the Precautions Section.

2. Remove or disconnect the following:
   • Intake manifold collector
   • Left and right rocker covers
   • Spark plugs

3. Set the No. 1 cylinder at Top Dead Center (TDC) on its compression stroke. Align the pointer with the TDC mark on the crankshaft pulley. Check that the valve lifters on the No. 1 cylinder are loose and valve lifters on the No. 4 cylinder are tight. If not, turn the crankshaft 1 revolution (360 degrees) and align the pointer with the TDC mark on the crankshaft pulley.

4. Check the following valves:
   • Both No. 1 intake valves
   • Both No. 2 exhaust valves
   • Both No. 3 exhaust valves
   • Both No. 6 intake valves

5. Using a feeler gauge, measure the clearance between the valve lifter and the camshaft. Record any valve clearance measurements that are out of specification.
   • Intake valve clearance (cold) is 0.010–0.013 in. (0.26–0.34mm) and exhaust valve clearance (cold) is 0.011–0.015 in. (0.29–0.37mm).
   • Intake valve clearance (warm) is 0.012–0.016 in. (0.30–0.41mm) and exhaust valve clearance (warm) is 0.012–0.017 in. (0.30–0.43mm).

6. Turn the crankshaft 240 degrees and set the No. 3 cylinder to TDC of its compression stroke.

7. Check the following valves:
   • Both No. 2 intake valves
   • Both No. 3 intake valves
   • Both No. 4 exhaust valves
   • Both No. 5 exhaust valves

8. Using a feeler gauge, measure the clearance between the valve lifter and the camshaft. Record any valve clearance measurements that are out of specification.
   • Intake valve clearance (cold) is 0.010–0.013 in. (0.26–0.34mm) and exhaust valve clearance (cold) is 0.011–0.015 in. (0.29–0.37mm).

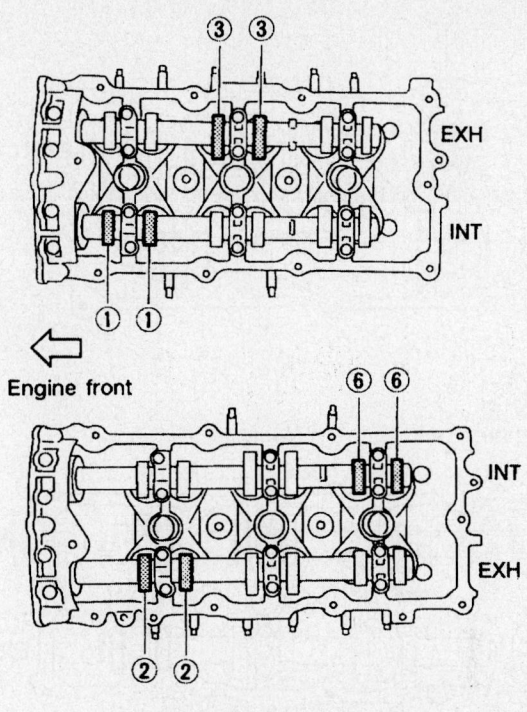

Measure the valves indicated while the No. 1 piston is at TDC on the compression stroke—3.5L engine

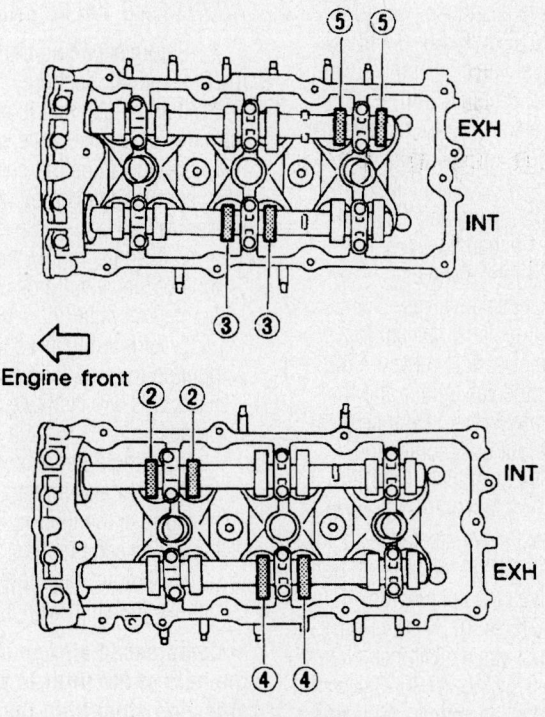

Measure the valves indicated while the No. 3 piston is at TDC on the compression stroke—3.5L engine

RH cylinder head

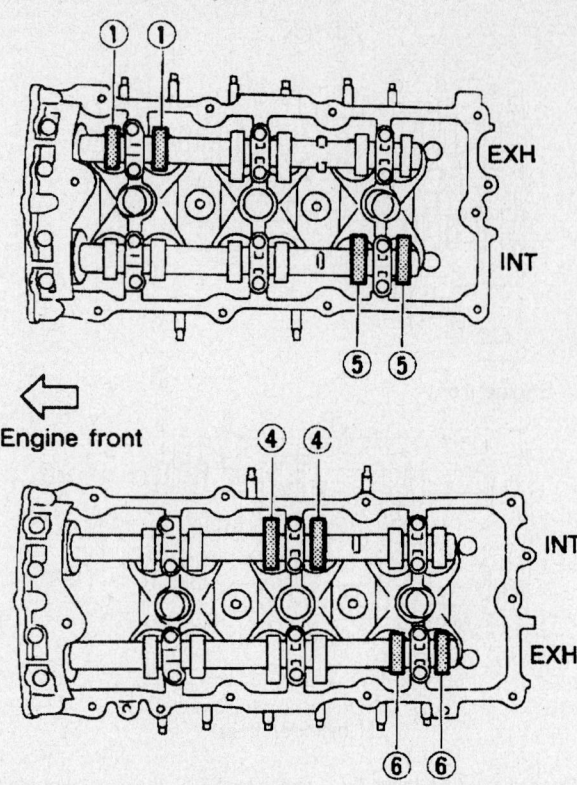

Engine front

LH cylinder head

09482_ALTI_G0026

Measure the valves indicated while the No. 5 piston is at TDC on compression—3.5L engine

- Intake valve clearance (warm) is 0.012–0.016 in. (0.30–0.41mm) and exhaust valve clearance (warm) is 0.012–0.017 in. (0.30–0.43mm).

9. Turn the crankshaft 240 degrees and set the No. 5 cylinder to TDC of its compression stroke.

10. Check the following valves:
- Both No. 1 exhaust valves
- Both No. 4 intake valves
- Both No. 5 intake valves
- Both No. 6 exhaust valves

11. Using a feeler gauge, measure the clearance between the valve lifter and the camshaft. Record any valve clearance measurements that are out of specification.
- Intake valve clearance (cold) is 0.010–0.013 in. (0.26–0.34mm) and exhaust valve clearance (cold) is 0.011–0.015 in. (0.29–0.37mm).
- Intake valve clearance (warm) is 0.012–0.016 in. (0.30–0.41mm) and exhaust valve clearance (warm) is 0.012–0.017 in. (0.30–0.43mm).

12. If all the valve clearances are within specification, install the cylinder head cover, spark plugs and the intake manifold collector.

ADJUSTING VALVE LASH

1. Before servicing the vehicle, refer to the Precautions Section.

2. If an adjustment is necessary, adjust the valve clearance while engine is cold by removing the adjusting shim. The adjusting shim can be removed by using the following procedures:

a. Turn the crankshaft so the camshaft lobe of the valve to be adjusted is pointed straight up.

b. Turn the lifter so the notch is pointed towards the center of the cylinder head; this will facilitate the shim removal process.

c. Using a depressor tool, push down on the lifter and insert a keeper tool on the edge of the lifter to keep the lifter in the depressed position.

d. Remove the depressor tool and remove the shim with a magnet.

➡**Compressed air can be blown into the hole of the lifter to separate the adjusting shim from the lifter.**

3. Determine the replacement adjusting shim size by using the following procedures and formula:

a. Using a micrometer determine thickness of the removed shim.

b. Calculate the thickness of a new adjusting shim so valve clearance is within the specified values.

c. R = thickness of the removed shim.

d. N = thickness of the new shim.

e. M = measured valve clearance.

- Intake shim determination formula: $N = R + (M - 0.0118 \text{ in. or } 0.30\text{mm})$
- Exhaust shim determination formula: $N = R + (M - 0.0130 \text{ in. or } 0.33\text{mm})$

4. Shims are available in 64 sizes from 0.0913–0.1161 in. (2.32–2.95mm) in steps of 0.004 in. (0.01mm). The thickness is stamped on the shim; this side is always installed facing down. Select new shims with thickness as close as possible to calculated valve and install it in the lifter.

5. Install the new shim onto the lifter.

6. Depress the lifter and remove the keeper tool. Remove the depressor tool and recheck the valve clearance. Repeat this procedure for any other valves requiring adjustment.

7. When all valve adjustments are finished, install the cylinder head cover, spark plugs and the intake manifold collector.

Starter Motor

REMOVAL & INSTALLATION

1.8L Engine

1. Before servicing the vehicle, refer to the Precautions Section.

2. Remove or disconnect the following:
- Negative battery cable
- Wiring at the starter
- Bolts attaching the starter to the engine
- Starter

To install:

3. Install or connect the following:
- Starter and torque the bolts to 70 inch lbs. (8 Nm)
- Starter electrical connections
- Negative battery cable

2.5L Engines

1. Remove or disconnect the following:
- Negative battery cable
- Air cleaner cover and duct
- Harness bracket
- Wiring at the starter
- Starter

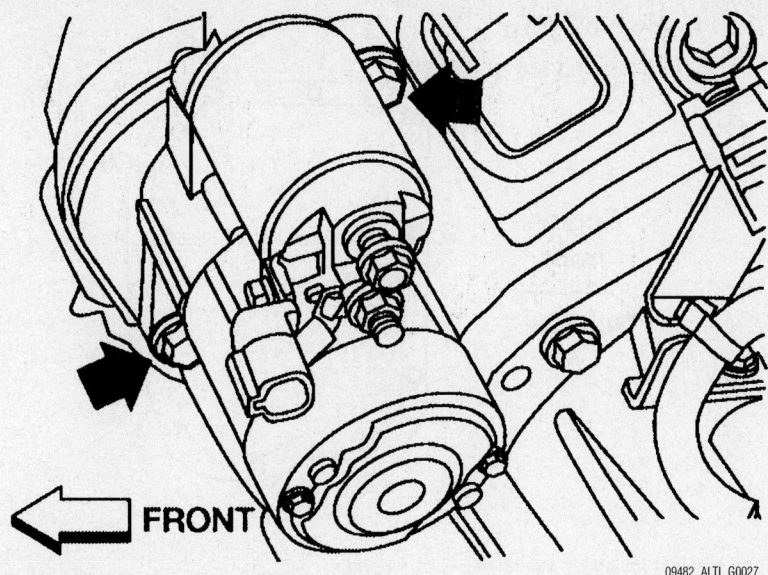

Starter location and mounting detail—3.5L engine

09482_ALTI_G0027

To install:

2. Install or connect the following:
   - Starter and torque the bolts to 60 inch lbs. (7 Nm)
   - Starter electrical connectors
   - Harness bracket
   - Air inlet tube
   - Negative battery cable

3.5L Engine

1. Remove or disconnect the following:
   - Negative battery cable
   - Air cleaner case (upper) and air cleaner to electric throttle control actuator tube.
   - Harness protector from the harness
   - Starter wiring at the starter
   - Starter-to-engine bolts
   - Starter from the vehicle

To install:

2. Install or connect the following:
   - Starter and torque the mounting bolts to 37–45 ft. lbs. (49–62 Nm).
   - Starter wiring
   - Harness protector
   - Air duct
   - Negative battery cable

Oil Pan

REMOVAL & INSTALLATION

1.8L Engine

1. Before servicing the vehicle, refer to the Precautions Section.
2. Drain the engine oil.
3. Remove or disconnect the following:

- Negative battery cable
- Engine undercovers
- Front exhaust tube and properly support the transaxle assembly
- Center crossmember
- Support brackets from the sides of the oil pan
- Rear cover plate, models equipped with a automatic transaxle
- Oil pan mounting bolts

4. Using an oil pan seal cutter, separate the oil pan from the engine.

❋❋ WARNING

Do not drive the seal cutter into the oil pump or rear oil seal retainer portion, for the aluminum mating surfaces will be damaged. Do not use a prybar to remove the oil pan; the flange will be deformed.

5. Clean all the sealing surfaces.

To install:

6. Apply sealant to the rear oil seal retainer.
7. Apply a 0.128–0.177 in. (3.5–4.5mm) continuous bead of liquid gasket to the oil pan mating surface.
8. Install or connect the following:
   - Oil pan and torque bolts, in sequence, to 56–73 inch lbs. (6.3–8.3 Nm)
   - Rear cover plate, models equipped with a automatic transaxle
   - Oil pan support brackets
   - Center crossmember
   - Front exhaust tube
   - Engine undercovers
   - Oil pan plug, using a new gasket and tighten the plug to 21–28 ft. lbs. (7–8 Nm)
   - Negative battery cable
9. After 30 minutes of gasket curing time, refill the oil pan with the specified quantity of clean oil.
10. Start the vehicle, check for leaks and repair if necessary.

2.5L Engine

ALTIMA

1. Before servicing the vehicle, refer to the Precautions Section.
2. Drain the engine oil.
3. Remove or disconnect the following:

   - Negative battery cable
   - Engine undercover
   - Front exhaust pipe
   - Power steering hose bracket from the collector
4. Support the engine from above and below with a suitable jack and hoist.
5. Remove or disconnect the following:

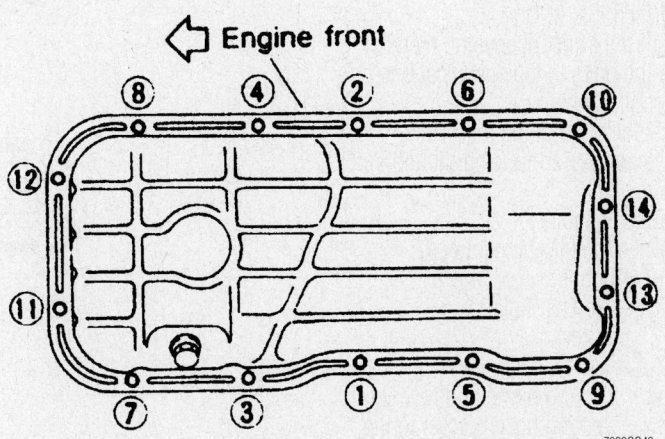

Tighten the oil pan bolts in the correct sequence to prevent oil leakage—1.8L engines

7923QG42

- Power steering pump and reservoir and wire aside
- Auxiliary drive belts
- A/C compressor and wire aside
- Upper sway bar links
- Front and rear engine mount through bolts
- Lower ball joints
- 2 steering gear housing mounting bolts
- Front suspension member bolts and front suspension member
- Remove the lower oil pan bolts in the sequence shown

6. Install a seal cutter between the steel oil pan and the aluminum oil pan

7. Tapping the cutter with a hammer, slide it around the entire edge of the oil pan. Take care not to damage the aluminum oil pan.

8. Remove or disconnect the following:
- Steel oil pan
- Oil pickup screen
- Rear plate cover and 4 engine-to-transmission bolts
- Remove the upper oil pan bolts in the sequence shown

9. Install a seal cutter between the block and the upper oil pan

10. Remove the upper oil pan.

To install:

11. Carefully scrape the old gasket material away from the pan and cylinder block mounting surfaces, then apply a continuous 3.5–4.5mm wide bead of liquid gasket around the oil pan. Install the pan within 5 minutes or else this step will have to be repeated.

12. Install or connect the following:
- Upper oil pan and torque the bolts in sequence to 16 ft. lbs. (22 Nm)
- Rear plate cover and 4 engine-to-transmission bolts
- Oil pickup screen
- Lower oil pan bolts in sequence to 61 inch lbs. (7 Nm).

13. Install the front suspension member and tighten the bolts to the specifications shown in the illustration.

14. Install or connect the following:
- 2 steering gear housing mounting bolts
- Lower ball joints
- Front and rear engine mount through bolts
- Upper sway bar links
- A/C compressor
- Auxiliary drive belts
- Power steering pump and reservoir
- Power steering hose bracket to the collector
- Front exhaust pipe

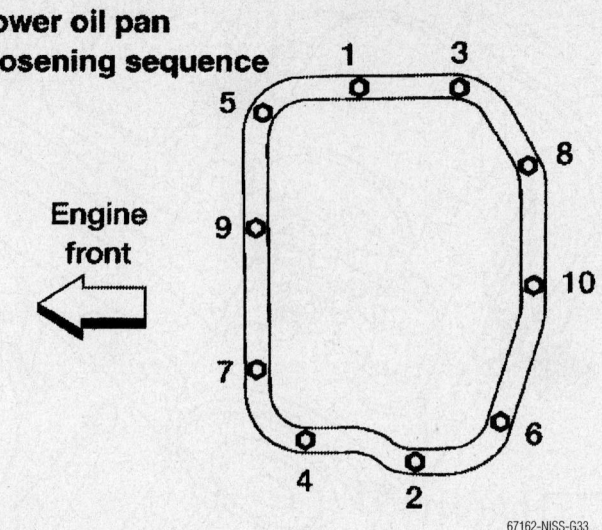

Lower oil pan bolt loosening sequence—2.5L engine

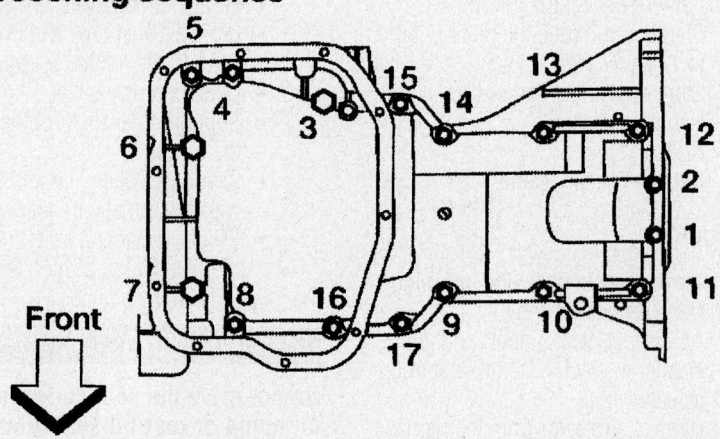

Upper oil pan bolt loosening sequence—2.5L engine

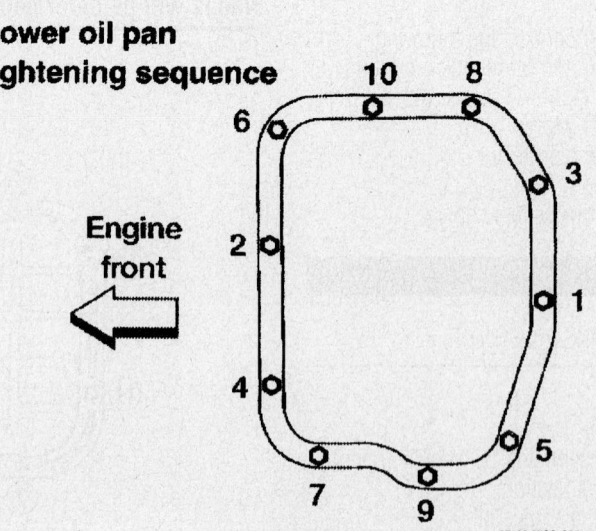

Lower oil pan bolt tightening sequence—2.5L engine

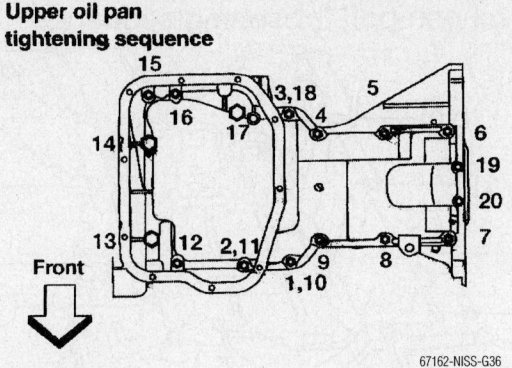

Upper oil pan tightening sequence

Front

67162-NISS-G36

Upper oil pan bolt tightening sequence—2.5L engine

- Engine undercover
- Negative battery cable

15. Wait 30 minutes before refilling the crankcase with clean oil.

16. Fill the engine with clean oil.

17. Start the vehicle, checks for leaks and repair if necessary.

SENTRA

1. Before servicing the vehicle, refer to the Precautions Section.

2. Drain the engine oil.

3. Remove or disconnect the following:
   - Negative battery cable
   - Engine undercover
   - Front exhaust pipe

4. Support the engine from above and below with a suitable jack and hoist.

5. Remove or disconnect the following:

- Front and rear engine mount through bolts
- Center crossmember
- A/C compressor and wire aside
- Remove the lower oil pan bolts in the sequence shown

6. Install a seal cutter between the steel oil pan and the aluminum oil pan

7. Tapping the cutter with a hammer, slide it around the entire edge of the oil pan. Take care not to damage the aluminum oil pan.

8. Remove or disconnect the following:
   - Steel oil pan
   - Oil pickup screen
   - Rear plate cover and 4 engine-to-transmission bolts
   - Remove the upper oil pan bolts in the sequence shown

9. Install a seal cutter between the block and the upper oil pan

10. Remove the upper oil pan.

To install:

11. Carefully scrape the old gasket material away from the pan and cylinder block mounting surfaces, then apply a continuous 3.5–4.5mm wide bead of liquid gasket around the oil pan. Install the pan within 5 minutes or else this step will have to be repeated.

12. Install or connect the following:
   - Upper oil pan and torque the bolts in sequence to 16 ft. lbs. (22 Nm)
   - Rear plate cover and 4 engine-to-transmission bolts
   - Oil pickup screen
   - Lower oil pan bolts in sequence to 61 inch lbs. (7 Nm).
   - A/C compressor
   - Center crossmember
   - Front and rear engine mount through bolts
   - Front exhaust pipe
   - Engine undercover
   - Negative battery cable

13. Wait 30 minutes before refilling the crankcase with clean oil.

14. Fill the engine with clean oil.

15. Start the vehicle, checks for leaks and repair if necessary.

3.5L Engine

1. Before servicing the vehicle, refer to the Precautions Section.

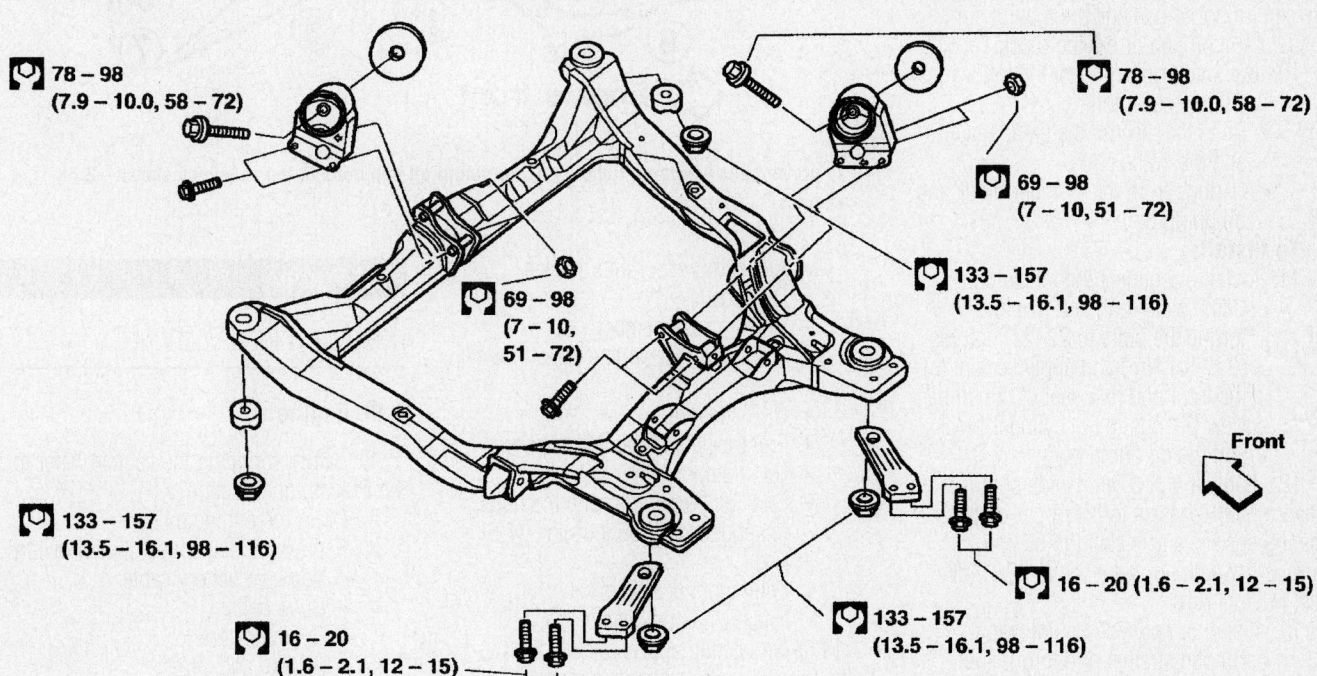

78 – 98
(7.9 – 10.0, 58 – 72)

78 – 98
(7.9 – 10.0, 58 – 72)

69 – 98
(7 – 10, 51 – 72)

69 – 98
(7 – 10, 51 – 72)

133 – 157
(13.5 – 16.1, 98 – 116)

133 – 157
(13.5 – 16.1, 98 – 116)

Front

16 – 20 (1.6 – 2.1, 12 – 15)

133 – 157
(13.5 – 16.1, 98 – 116)

16 – 20
(1.6 – 2.1, 12 – 15)

Exploded view of the front suspension member—Altima

67162-NISS-G12

2. Drain the engine oil
3. Remove or disconnect the following:
- Negative battery cable
- Engine undercovers
- Steel (lower) oil pan bolts in the reverse of the installation sequence

4. Insert a seal cutter between the steel and aluminum oil pan.

5. Tapping the cutter with a hammer, slide it around the entire edge of the oil pan. Be careful not to damage the aluminum mating surface of the upper oil pan.
- Steel oil pan and the oil strainer
- Front exhaust pipe and its support

6. Hang the engine at the right and left side engine slingers with a suitable hoist.

7. Position a suitable jack under the transaxle.
- Crankshaft Position (CKP) sensors (REFERENCE and POSITION) from the oil pan
- Front and rear engine mounting nuts and bolts
- Center crossmember assembly
- Engine drive belts
- A/C compressor and mounting bracket
- Rear cover plate and the lower transaxle bolts
- Aluminum (upper) oil pan bolts in the reverse of the installation sequence

8. Insert a seal cutter between the aluminum oil pan and the engine block.

9. Tapping the cutter with a hammer, slide it around the entire edge of the oil pan. Be careful not to damage the mating surfaces of the oil pan or engine block.

10. Remove or disconnect the following:
- Oil pan assembly
- Bolts that secure the baffle plate and the baffle plate
- O-rings from the cylinder block and oil pump body

To install:

11. Install or connect the following:
- Baffle plate to the oil pan and torque the bolts to 22–27 inch lbs. (2.5–3.1 Nm) and apply sealant to the front and rear seal of the oil pan
- New O-rings to the cylinder block and the oil pump body

12. Apply a 4.5–5.5mm wide continuous bead of liquid gasket to the upper oil pan mating surface and install the oil pan. Torque the bolts in sequence to 12–14 ft. lbs. (16–19 Nm).

13. Install or connect the following:
- Oil pan strainer and torque the bolts to 12–14 ft. lbs. (16–19 Nm)
- Rear cover plate and lower transaxle bolts

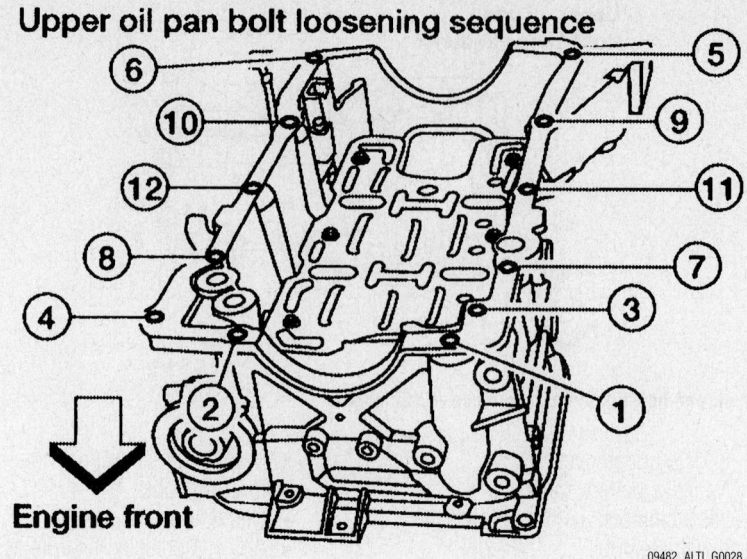

Upper oil pan bolt loosening sequence

Engine front

09482_ALTI_G0028

Aluminum (upper) oil pan bolt loosening sequence—3.5L engine

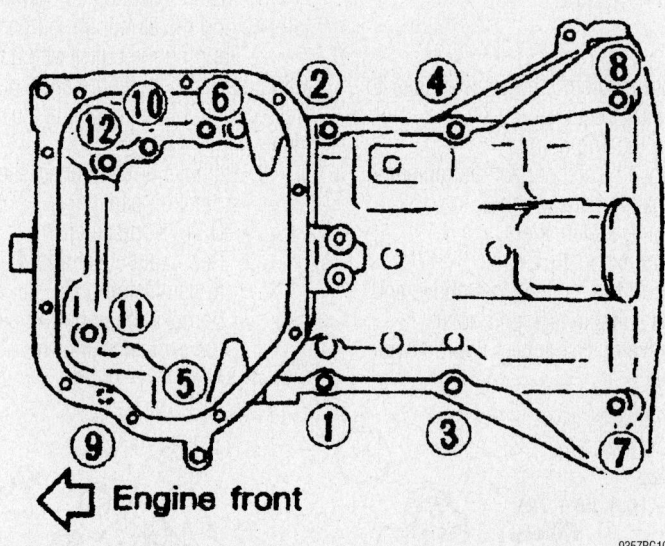

Engine front

9357RG10

To prevent pan warpage, tighten the aluminum oil pan bolts in the sequence shown—3.5L engines

- A/C compressor and bracket
- Drive belts
- Center crossmember
- Front and rear engine mount hardware
- CKP sensors
- Front exhaust tube and support
- Oil strainer
- Steel oil pan and torque the bolts, in sequence, to 66 inch lbs. (7.5 Nm)
- Engine under covers
- Negative battery cable

14. After waiting approximately 30 minutes, fill the engine with clean oil.

15. Start the vehicle, check for leaks and repair if necessary.

Oil Pump

REMOVAL & INSTALLATION

1.8L Engine

1. Before servicing the vehicle, refer to the Precautions Section.

2. Drain the engine oil

3. Remove or disconnect the following:
- Negative battery cable
- Drive belts
- Cylinder head
- Oil pan and strainer
- Engine front cover
- Oil pump from the front cover

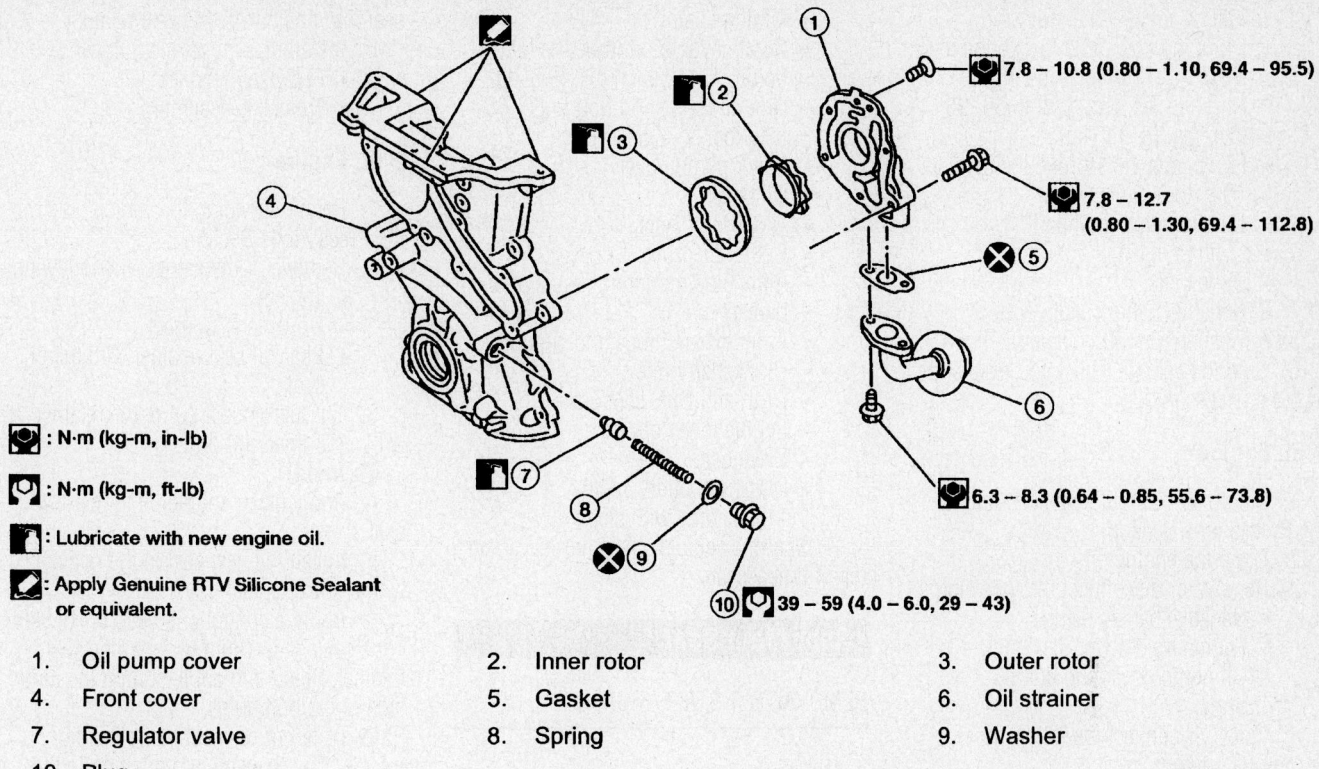

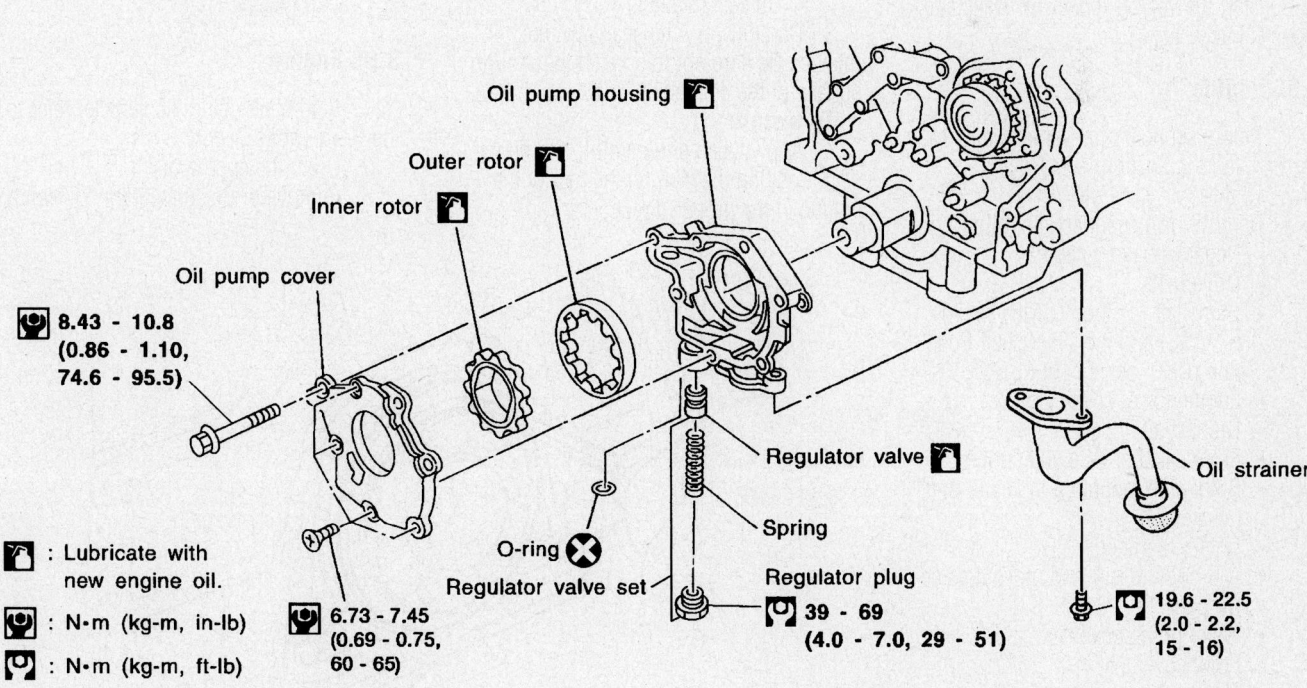

: N·m (kg-m, in-lb)

: N·m (kg-m, ft-lb)

: Lubricate with new engine oil.

: Apply Genuine RTV Silicone Sealant or equivalent.

1. Oil pump cover
2. Inner rotor
3. Outer rotor
4. Front cover
5. Gasket
6. Oil strainer
7. Regulator valve
8. Spring
9. Washer
10. Plug

09482_ALTI_G0029

Exploded view of the oil pump assembly–1.8L engine

7.8 – 10.8 (0.80 – 1.10, 69.4 – 95.5)

7.8 – 12.7 (0.80 – 1.30, 69.4 – 112.8)

6.3 – 8.3 (0.64 – 0.85, 55.6 – 73.8)

39 – 59 (4.0 – 6.0, 29 – 43)

Oil pump housing

Outer rotor

Inner rotor

Oil pump cover

8.43 - 10.8 (0.86 - 1.10, 74.6 - 95.5)

Regulator valve

Oil strainer

O-ring

Spring

Regulator valve set

Regulator plug

: Lubricate with new engine oil.

: N·m (kg-m, in-lb)

: N·m (kg-m, ft-lb)

6.73 - 7.45 (0.69 - 0.75, 60 - 65)

39 - 69 (4.0 - 7.0, 29 - 51)

19.6 - 22.5 (2.0 - 2.2, 15 - 16)

09482_ALTI_G0030

Exploded view of the oil pump assembly–3.5L engine

To install:

4. Install or connect the following:
- Oil pump cover and torque the long bolt to 69–113 inch lbs. (8–13 Nm) and the short bolt to 69–96 inch lbs. (8–11 Nm)
- Front cover and torque the bolts to 43 ft. lbs. (58 Nm)
- Oil strainer and oil pan
- Cylinder head
- Drive belts
- Negative battery cable

5. Fill the engine with clean oil.

6. Start the vehicle, check for leaks and repair if necessary.

2.5L Engine

1. Before servicing the vehicle, refer to the Precautions Section.
2. Drain the engine oil
3. Remove or disconnect the following:
- Negative battery cable
- Engine front cover
- Oil pump cover and pump

To install:

4. Install or connect the following:
- Oil pump
- Oil pump cover and torque the bolts to 61 inch lbs. (7 Nm) and shorter bolt to 69 inch lbs. (8 Nm)
- Front cover
- Negative battery cable

5. Fill the engine with clean oil.

6. Start the vehicle, check for leaks and repair if necessary

3.5L Engine

1. Before servicing the vehicle, refer to the Precautions Section.
2. Drain the engine oil
3. Remove or disconnect the following:
- Negative battery cable
- Drive belts
- Camshaft Position (CMP) sensor (PHASE) and the Crankshaft Position (CKP) sensor (REF)/(POS)
- Engine lower covers
- Crankshaft pulley
- Front exhaust tube and support
- Right side mounting insulator and bracket
- Center member
- A/C compressor and move it aside
- Oil pans
- Water pump cover
- Front cover
- Timing chain
- Oil pump assembly

4. Clean all mating surfaces.

To install:

5. Install or connect the following:

- Oil pump
- Timing chain
- Front cover and torque the long bolt to 75–96 inch lbs. (8.5–11 Nm) and the short bolt to 60–65 inch lbs. (7–7.5 Nm)
- Water pump cover
- Oil pans
- A/C compressor
- Center member
- Right side mounting insulator and bracket
- Front exhaust tube and support
- Crankshaft pulley
- CMP and CKP sensors
- Engine lower covers
- Drive belts
- Negative battery cable

6. Fill the engine with clean oil

7. Start the vehicle, check for leaks and repair if necessary.

Rear Main Seal

REMOVAL & INSTALLATION

1.8L Engine

1. Before servicing the vehicle, refer to the Precautions Section.
2. Remove or disconnect the following:
- Transaxle
- Driveplate/flywheel
- Oil seal retainer

3. Carefully pry the seal from the retainer. Be sure not to scratch the sealing surface of the crankshaft or oil seal bore.

To install:

4. Apply clean engine oil to the new seal. Position the seal on the rear of the engine in the proper direction.

5. Using a suitable seal driver, tap the seal into position in the seal retainer.

6. Install or connect the following:
- Flywheel/flexplate
- Transaxle assembly

2.5L Engine

1. Before servicing the vehicle, refer to the Precautions Section.
2. Remove or disconnect the following:
- Transaxle
- Driveplate/flywheel
- Rear oil seal retainer with the oil seal

3. Tap the oil seal out of the retainer with a hammer and drift.

To install:

4. Apply clean engine oil to the new seal.

5. Install the new seal in the retainer with a suitable seal driver.

6. Apply a continuous bead of RTV silicone sealant, 2–3mm wide, to the seal retainer. Be sure to apply around the inner side of the bolt holes.

7. Install or connect the following:
- Tap the seal into position in the seal retainer, using a suitable seal driver
- Oil seal retainer and torque the bolts to 56–66 inch lbs. (6.5–7.5 Nm)
- Driveplate/flywheel
- Transaxle

3.5L Engine

1. Before servicing the vehicle, refer to the Precautions Section.
2. Drain the engine oil.
3. Remove or disconnect the following:

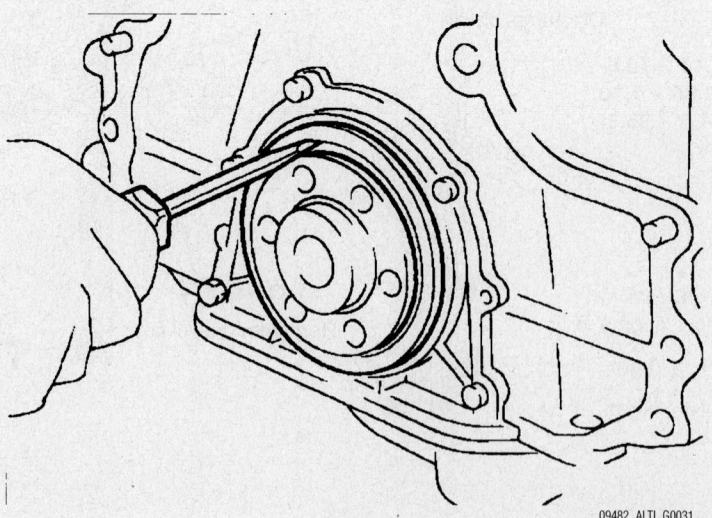

Carefully pry the rear main seal out of the retainer on the rear of the engine—1.8L engine

09482_ALTI_G0031

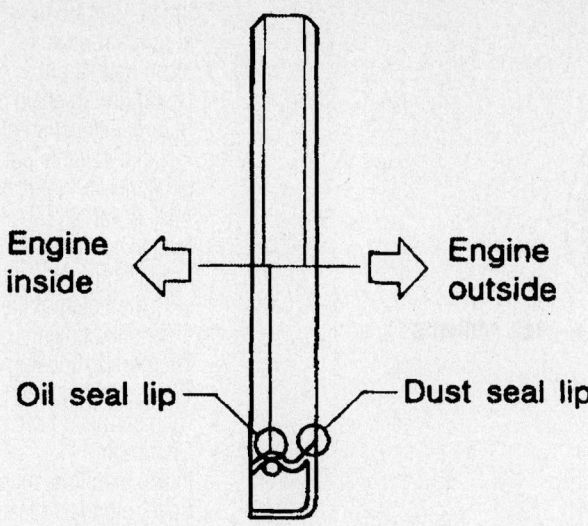

Engine inside ⇐ ⇒ Engine outside

Oil seal lip ——— ——— Dust seal lip

09482_ALTI_G0032

Be sure to install the seal in the correct orientation—1.8L engine

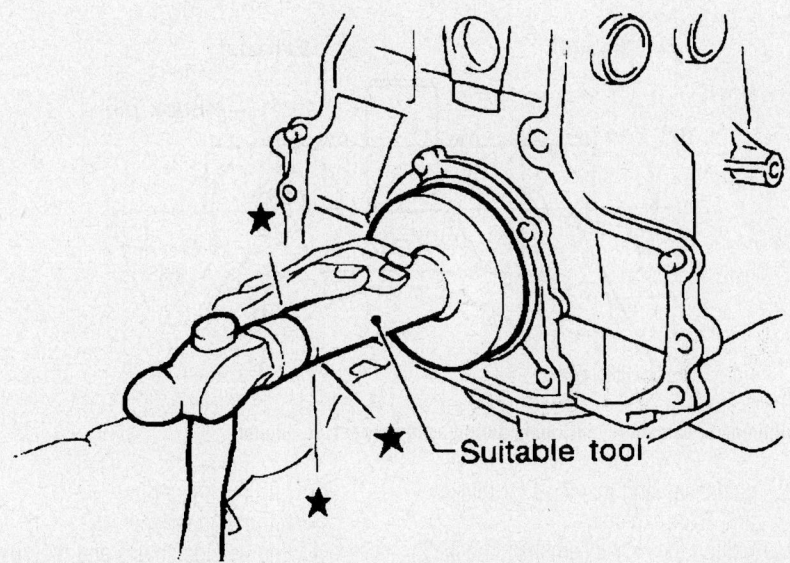

★ Suitable tool

09482_ALTI_G0033

Install the rear main seal using a suitable driver—1.8L engine

- Transaxle
- Driveplate/flywheel
- Oil pan
- Oil seal retainer

4. Tap the oil seal out of the retainer with a hammer and drift.

5. Clean all mating surfaces of any residual liquid gasket.

To install:

6. Install or connect the following:
- New seal into the retainer
- Oil seal retainer
- Oil pan
- Driveplate/flywheel
- Transaxle

7. Fill the engine with clean oil.

8. Start the vehicle, check for leaks and repair if necessary.

Timing Chain, Sprockets, Front Cover and Seal

REMOVAL & INSTALLATION

1.8L Engine

1. Before servicing the vehicle, refer to the Precautions Section.

2. Relieve the fuel system pressure.

3. Drain the cooling system.

4. Remove or disconnect the following:
- Negative battery cable
- Upper radiator hose
- Engine drive belts
- Power steering pulley and the pump with bracket

- Air duct from the intake manifold collector
- Right front wheel and engine side covers
- Engine undercovers
- Front exhaust pipe
- Cylinder head front mounting bracket
- Cylinder head cover from the engine
- Rocker cover
- Distributor cap
- Spark plugs
- Intake manifold support and set the No. 1 piston at the Top Dead Center (TDC) compression stroke
- Distributor
- Cylinder head front cover
- Water pump pulley
- Thermostat housing
- Lower timing chain tensioner
- Upper timing chain tensioner and slack side timing chain guide
- Idler sprocket bolt
- Camshaft sprocket bolts and the sprockets from the camshafts. Be sure to mark the sprockets for proper reinstallation.
- Camshaft mounting caps by loosening the bolts in 2 or 3 steps
- Camshafts from the engine
- Idler sprocket bolt
- Cylinder head with the manifolds
- Idler sprocket shaft from the rear side
- Upper timing chain and support the engine assembly
- Center crossmember
- Oil pan and strainer assembly
- Crankshaft pulley
- Engine front mount and bracket
- Bolts that secure the front timing cover and the cover from the engine. Once the timing chain cover is removed, drive out the old oil seal.
- Idler sprocket and the lower timing chain
- Oil pump drive spacer and the crankshaft sprocket
- Timing chain guide

To install:

5. Drive a new oil seal into the front cover. Lubricate the oil seal lip with clean engine oil.

6. Confirm that No. 1 piston is set at Top Dead Center (TDC) on compression stroke.

7. Install or connect the following:
- Crankshaft sprocket with the marks of the sprocket facing the front of the engine

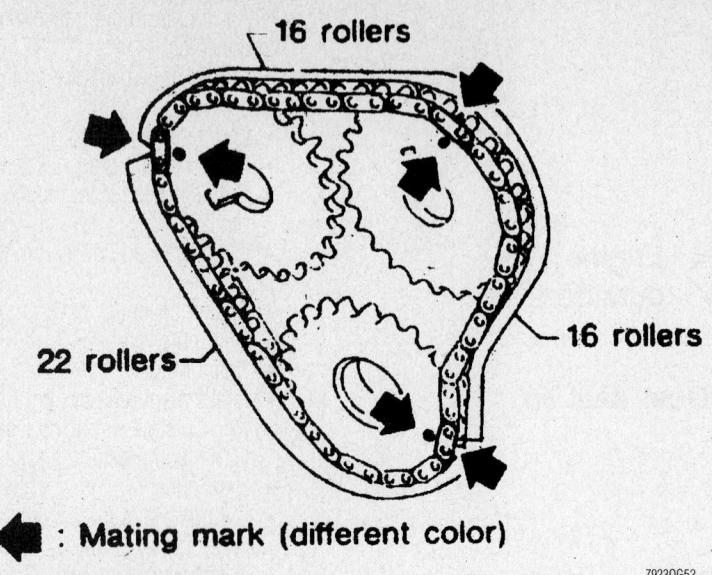

: Mating mark (different color)

16 rollers

16 rollers

22 rollers

79230G52

Be sure to align the camshaft sprockets with the timing chain—1.8L engine

- Oil pump drive spacer and the chain guide
- Lower timing chain. Set the chain by aligning its mating mark with the one on the crankshaft sprocket. Be sure the sprocket's mating mark faces the front of the engine.

➡**The number of links between the alignment marks are the same for the left and the right side.**

- Crankshaft sprocket and the lower timing chain. Set the timing chain by aligning its mating mark with the one on the crankshaft sprocket. Be sure sprocket's mating mark faces engine front.
- Front cover assembly, using liquid gasket
- Engine front mounting bracket and the engine mount
- Oil strainer, oil pan assembly and the crankshaft pulley
- Center crossmember

8. Set the idler sprocket by aligning the mating mark on the larger sprocket with the silver mating mark on the lower timing chain.

- Upper timing chain and set it by aligning the mating mark on the smaller sprocket with the silver mating marks on the upper timing chain. Be sure sprocket marks face engine front.
- Idler sprocket shaft to the rear side
- Cylinder head assembly
- Idler sprocket bolt. Be sure to lubricate the bolt with clean engine oil.
- Exhaust and intake camshafts. The

the bolts to 86 ft. lbs. (117 Nm). Be sure to lubricate the bolts with clean engine oil.
- Upper timing chain tensioner. Before installation of the tensioner, install a suitable pin to hold the tensioner in the relaxed position. After installing the chain tensioner, remove the pin.
- Lower timing chain tensioner. Be sure the notch of the gasket is positioned down.
- Thermostat housing
- Water pump pulley
- Cylinder head front cover
- Distributor
- Intake manifold support
- Spark plugs and leads
- Distributor cap
- Rocker cover
- Cylinder head cover

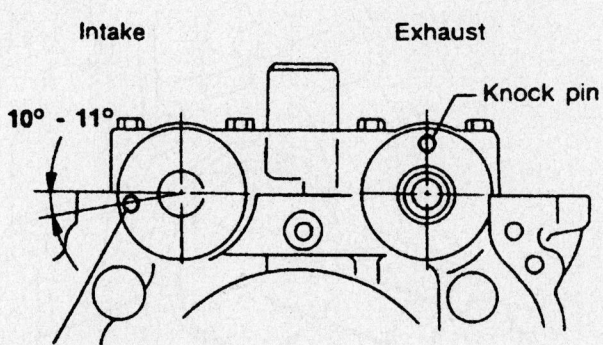

Positioning of camshaft knock pins during assembly—1.8L engine

Intake

Exhaust

10° - 11°

Knock pin

Knock pin

79230G53

camshafts and marked **I** for intake and **E** for exhaust.

9. Position the intake camshaft knock pin at the 9 o'clock position and the exhaust camshaft knock pin at the 12 o'clock position.

10. Install or connect the following:
- Camshaft bearing caps and distributor bracket. Apply liquid sealant to the distributor bracket.

11. Torque the mounting bolts in sequence as follows:
  a. Bolts 11–15, then bolts 1–10: 18 inch lbs. (2.0 Nm).
  b. Bolts 1–15: 52 inch lbs. (6 Nm).
  c. Bolts 1–14: 104 inch lbs. (12 Nm).
  d. Bolt 15: 73 inch lbs. (8 Nm).

12. Install or connect the following:
- Camshaft sprockets with timing chain. Set the camshaft sprockets by aligning the mating marks of the timing chain with the marks on the camshaft sprockets.
- Camshaft sprocket bolts. Torque

- Front exhaust pipe
- Engine under covers
- Engine side covers and the right front wheel
- Air duct to the intake manifold collector
- Power steering pulley and oil pump
- Drive belts
- Upper radiator hose
- Negative battery cable

13. Fill the cooling system.

14. Start the vehicle, check for leaks and repair if necessary.

2.5L Engine

1. Before servicing the vehicle, refer to the Precautions Section.

2. Drain the engine oil.

3. Drain the cooling system.

4. Relieve the fuel system pressure.

5. Disconnect the negative battery cable.

6. Support the engine and transaxle with suitable tools.

**Upper oil pan
tightening sequence**

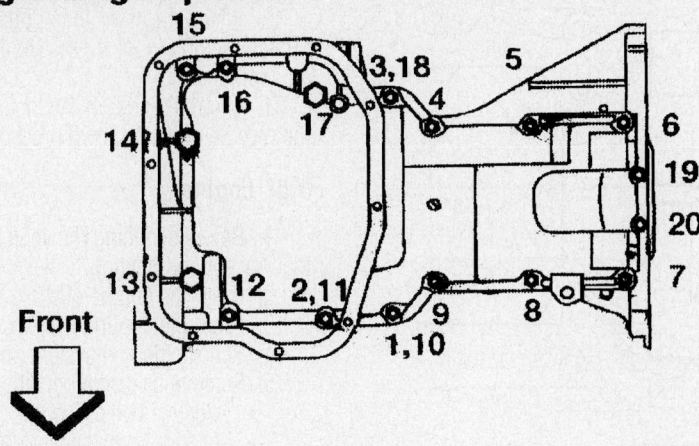

Front

67162-NISS-G36

Exploded view of the timing chain assembly—2.5L engine

7. On Altima, remove or disconnect the following:
- Engine under cover
- Upper and lower oil pans
- Alternator
- Engine cover
- Variable timing control solenoid conector
- Engine ground
- Coolant reservoir tank
- Right side fuse and relay box and move aside
- Right engine mount and bracket

8. On Sentra, remove or disconnect the following:
- Air cleaner and duct
- Spark plugs
- Valve cover
- Coolant reservoir tank
- Auxiliary drive belt tensioner
- Alternator
- Strut tower cross brace
- A/C compressor and wire aside
- Power steering pump and reservoir and wire aside
- Upper and lower oil pans

9. On all models, remove or disconnect the following:
- Intake valve timing control cover in reverse of the tightening sequence
- Timing chain guide between the camshaft sprockets and through the front cover
- Rotate the crankshaft clockwise and set the No. 1 piston to TDC of the compression stroke
- Crankshaft pulley using a suitable puller
- Front timing chain cover bolts in reverse order of the tightening sequence
- Front timing chain case
- Insert a suitable stopper pin in the timing chain tensioner
- Chain tensioner

10. While holding the camshaft with an open end wrench on the camshaft flat, remove the camshaft sprocket bolts and remove the camshaft sprockets.

11. Remove the timing chain slack guide, tension guide, timing chain and oil pump drive spacer.

12. Lift up on the timing chain tensioner lever for the balancer and release the ratchet.

13. Push the tensioner sleeve in and hold.

14. Insert a stopper pin to secure the tensioner sleeve

15. Remove the balancer tensioner.

16. Remove the timing chain for the balancer unit.

17. Remove the balancer unit using the sequence shown.

To install:

18. If replacement of the balancer bolts is not possible, perform the following bolt measurement:

a. Measure the diameter of the head bolt 0.39 in. (10mm) from the bottom of the bolt.

67162-NISS-G37

Timing chain cover bolt removal sequence shown—2.5L engine

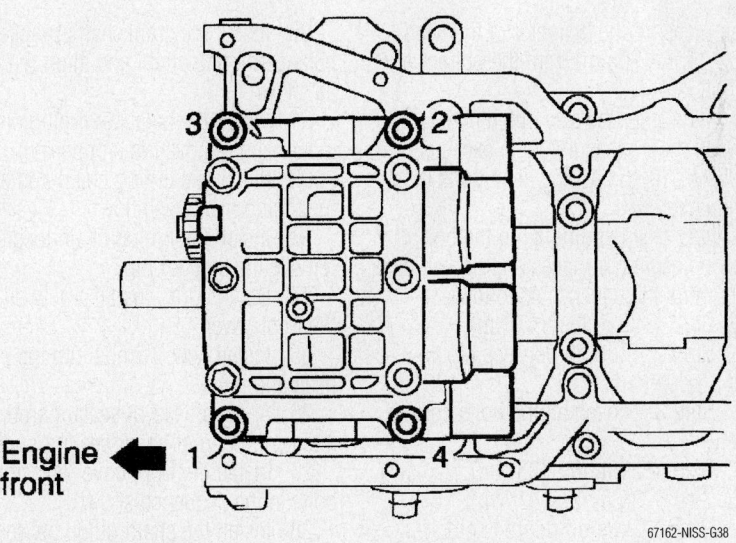

Engine front ← 1

67162-NISS-G38

Balancer unit mounting bolts removal sequence shown—2.5L engine

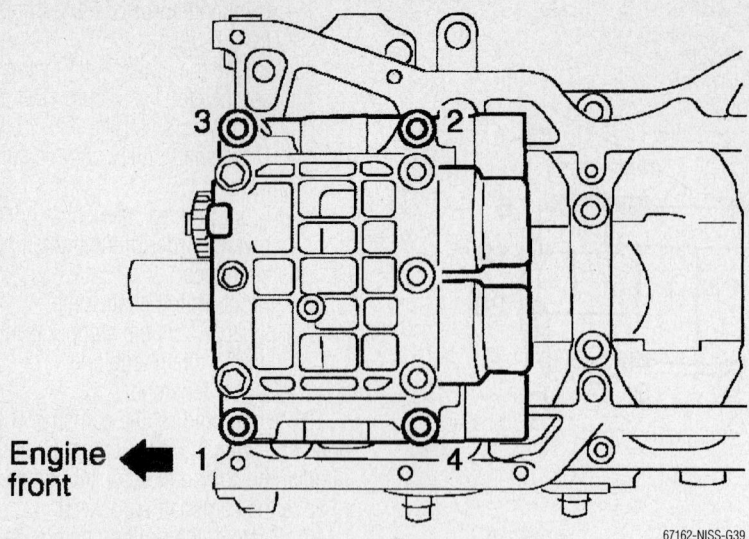

Balancer unit mounting bolts tightening sequence shown—2.5L engine

67162-NISS-G39

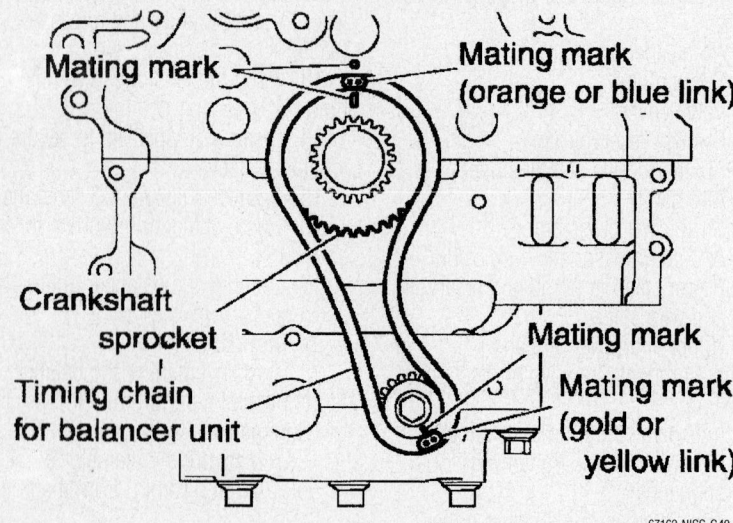

Balancer timing chain alignment marks—2.5L engine

67162-NISS-G40

b. Measure the diameter of the head bolt 2.75 in. (70mm) from the bottom of the bolt.

c. Whenever the size difference between the 2 measurements exceeds 0.0059 in. (0.15mm) the head bolts must be replaced.

19. Apply clean engine oil to the bolt threads and seating surfaces and tighten the balancer bolts in sequence as follows:

    a. Step 1: 35 ft. lbs. (47 Nm).

    b. Step 2: Plus 90 degrees clockwise

    c. Step 3: Loosen all the bolts completely.

    d. Step 4: 29 ft. lbs. (33 Nm).

    e. Step 5: 35 ft. lbs. (47 Nm).

    f. Step 6: Plus 90 degrees clockwise

20. Install the crankshaft sprocket and balancer timing chain and align the marks as shown.

21. Install the balancer timing chain tensioner and remove the stopper pin.

22. Install the timing chain and and align the timing marks as shown.

23. Install the timing chain tensioner and remove the stopper pin.

24. Using adrift, install a new oil seal in the front cover.

25. Install new O-rings into the cylinder head and block.

26. Apply a bead of sealant around the inside of the front cover sealing surface.

27. Install the front cover and tighten the bolts in the sequence shown.

28. Install the chain guide between camshaft sprockets.

29. Install the intake valve timing control cover in the sequence shown and tighten the bolts to 10 ft. lbs. (13 Nm).

30. Install the crankshaft pulley and tighten the bolt to 31 ft. lbs. (42 Nm), plus an additional 60°.

31. The remainder of the installation is the reverse of the removal procedure.

3.5L Engine

1. Before servicing the vehicle, refer to the Precautions Section.

2. Drain the engine oil.

3. Drain the cooling system.

4. Relieve the fuel system pressure.

5. Remove or disconnect the following:

- Negative battery cable
- Left side rocker cover ornament

➡ **Before detaching any hoses or connectors, note the locations for reassembly.**

- Air duct to intake manifold hose, collector hose, blow-by hose and vacuum hoses
- Fuel hoses and detach the harness connections
- Canister purge hoses
- Water hoses from the cylinder head and intake manifold
- All 6 ignition coils from the spark plugs
- Spark plugs
- Bolts that secure the Exhaust Gas Recirculation (EGR) tube and remove the tube
- Intake manifold collector supports and the collector
- Fuel tube assembly
- Intake manifold. Loosen the bolts in the reverse sequence of the tightening procedure.
- Left-hand and right-hand intake valve timing control solenoid valves
- Left-hand and right-hand rocker covers from the cylinder head
- Engine undercovers
- Right front wheel and the engine side covers
- Drive belts and the idler pulley
- Power steering oil pump belt and the power steering oil pump assembly
- Camshaft Position (CMP) sensor (PHASE) and Crankshaft Position (CKP) sensors (REF)/(POS)

6. Set the No. 1 piston to Top Dead Center (TDC) of compression stroke by rotating the crankshaft.

7. Loosen the crankshaft pulley bolt

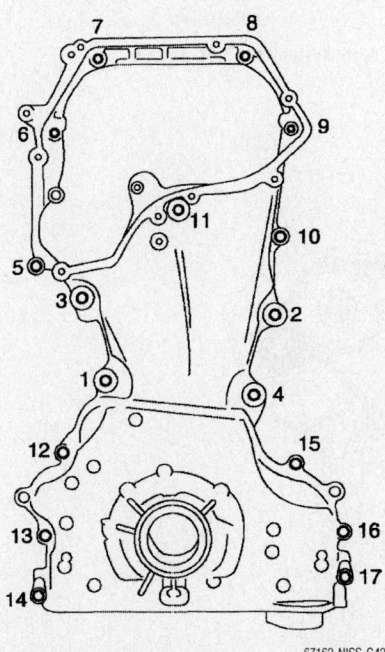

Mating mark (peripheral stamp line)

Gold or yellow link

Timing chain

Chain guide

Gold or yellow link

Mating mark (stamp)

Camshaft sprocket (INT side)

Camshaft sprocket (EXH side)

Chain tensioner

Tension guide

Slack guide

Mating mark (lug)

Crankshaft key

Crankshaft sprocket

Mating mark (stamp)

Orange or blue link

Orange or blue link

Mating mark (stamp)

Timing chain (for balancer unit)

Chain tensioner

Mating mark (stamp)

Balancer unit sprocket component

Gold or yellow link

67162-NISS-G41

Timing chain alignment marks—2.5L engine

Timing chain cover bolt removal sequence shown—2.5L engine

67162-NISS-G42

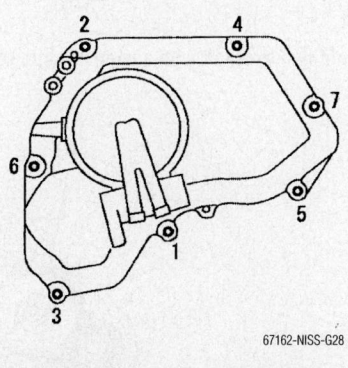

67162-NISS-G28

Intake valve timing control cover tightening sequence shown—2.5L engine

reverse order shown in the accompanying figure. In the cover, the shaft is engaged with the center hole of the intake camshaft sprocket. Remove it straight out until the engagement comes off.
- A/C compressor and bracket
- Front exhaust pipe and its support

8. Hang the engine at the right and left side engine slingers with a suitable hoist.

9. Support the transaxle with jack.
- Right side engine mounting and bracket
- Center crossmember assembly
- Upper and lower oil pans
- Water pump cover
- Bolts that secure the front timing chain case, in sequence
- Timing chain case cover using a seal cutter
- Internal timing chain guide and the upper chain guide
- Timing chain tensioner and slack side chain guide
- Left and right intake camshaft sprockets first. Be sure to hold the flats of the camshafts while removing the sprocket bolts.
- Lower timing chain assembly. Be

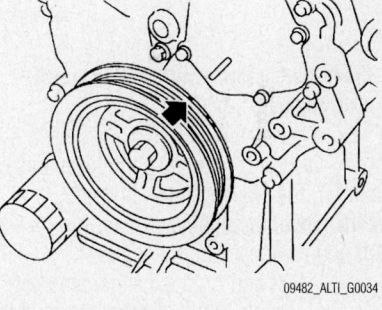

09482_ALTI_G0034

Set the No. 1 piston to Top Dead Center (TDC)—3.5L engine

while securing the ring gear so the crankshaft cannot rotate.
- Ring gear cover access plate

➡**Use care not to damage the ring gear teeth.**

- Crankshaft pulley using a suitable puller
- Intake valve timing control valve cover. Loosen the bolts in the

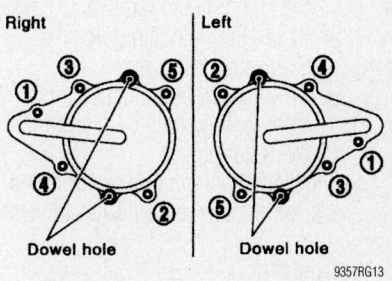

9357RG13

Loosen the intake valve timing control valve cover bolts in the reverse of the order shown—3.5L engine

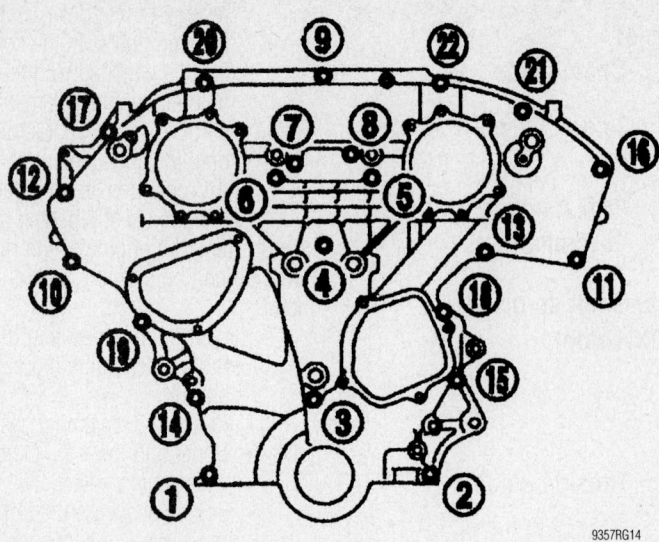

Remove the front timing chain case mounting bolts in the reverse of the sequence shown—3.5L engines

compression stroke and align the dowels of the camshaft sprockets to the 12 o'clock position in respect to the cylinder head.

- Lower timing chain guide. The front mark on the guide should face upwards.

14. On a work bench, align the marks on the intake and exhaust camshaft sprockets with the marks of the chain.

- Exhaust camshaft sprockets onto the dowel pin and torque the mounting bolts to 88–95 ft. lbs. (119–128 Nm). Be sure to secure the camshafts while tightening the bolts.
- Timing chains and sprockets to the intake camshafts. Be sure to align the timing chain and sprocket mating marks.
- Left and right camshaft tensioner stopper pins

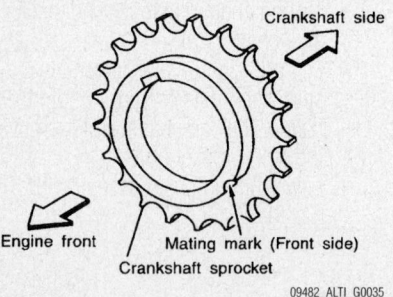

Crankshaft sprocket with mating marks—3.5L engine

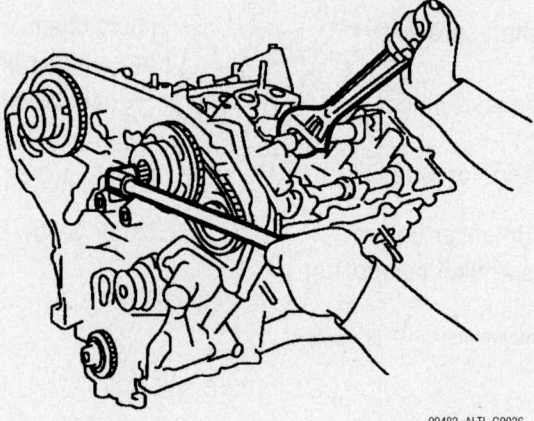

Hold the camshaft with a wrench while removing the sprocket bolts—3.5L engine

sure to note the aligning marks of the chain before removal.

10. Insert a suitable stopper pin for the left and right camshaft tensioners.

- Left and right exhaust camshaft sprocket bolts. Be sure to hold the flats of the camshafts while removing the sprocket bolts.
- Upper timing chain assembly. Be sure to note the aligning marks of the chain before removal.
- Lower timing chain guide
- Crankshaft sprocket
- All traces of liquid gasket from the front timing chain case and from the water pump

11. Inspect the timing chain for excessive wear or damage and replace as necessary.

To install:

12. Install or connect the following:

- Crankshaft sprocket with the mating mark facing out

13. Position the crankshaft to TDC of

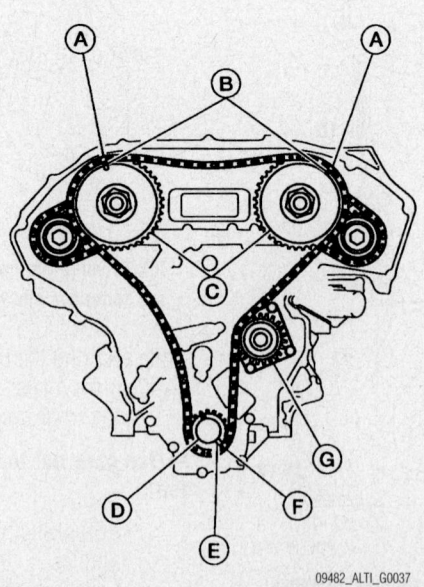

Timing chain alignment marks—3.5L engine

Front timing chain case

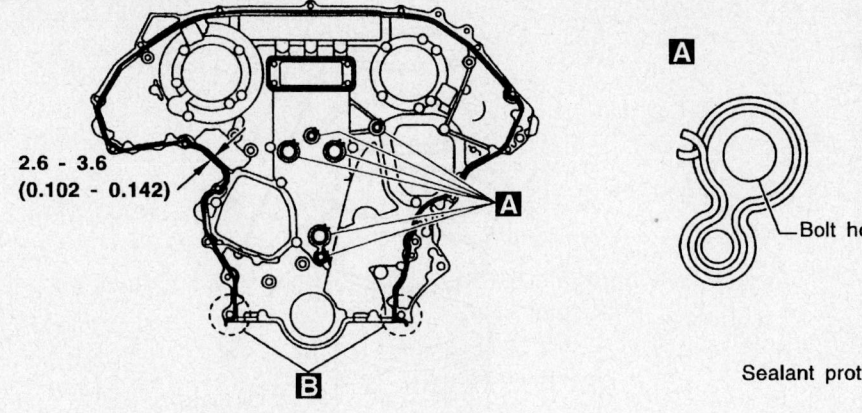

2.6 - 3.6
(0.102 - 0.142)

A   A   B

Bolt hole—   —Bolt hole   Protrusion

Sealant protrusion away from bolt hole

09482_ALTI_G0038

Application of liquid gasket to the front timing case—3.5L engine

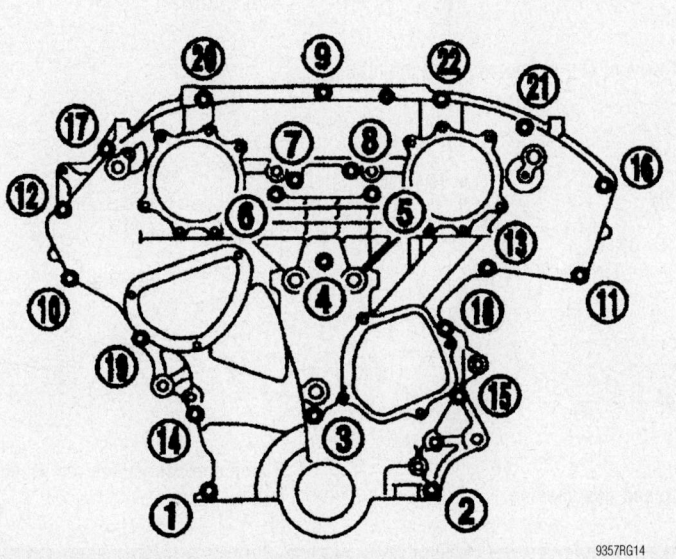

9357RG14

Tighten the front timing chain case bolts according to the sequence shown—3.5L engines

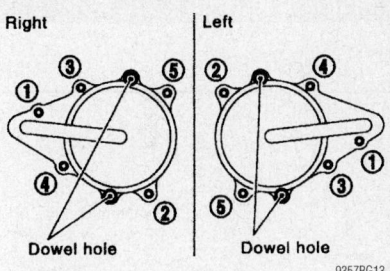

Right | Left

Dowel hole   Dowel hole

9357RG13

Tighten the intake valve timing control valve cover bolts in sequence—3.5L engine

➡**Leave the bolts unattended for 30 minutes or more after tightening. This will allow the liquid gasket to cure sufficiently.**

22. Apply a 0.091–0.130 in. (2.3–3.3mm) continuous bead of liquid gasket to the water pump cover and install the cover. Torque the bolts to 84–108 inch lbs. (10–13 Nm).

23. Install or connect the following:
- Oil pans
- Center crossmember
- Right side engine mount and bracket
- Front exhaust pipe and remove the transaxle support
- A/C compressor and bracket
- Crankshaft pulley
- Ring gear access cover plate
- CMP sensor and CKP sensors
- Power steering pump
- Idler pulley and drive belts
- Engine side cover and right front wheel
- Engine under covers
- Intake valve timing control sole-

15. Align the mating mark on the crankshaft with the matchmark (gold link) on the lower timing chain.

16. Install the lower timing chain to the water pump sprocket.

17. Working counterclockwise, install the lower timing chain camshaft sprockets. Be sure to align the sprocket marks with the blue links of the timing chain during installation.

18. Install or connect the following:
- Intake sprocket and torque the bolts to 88–95 ft. lbs. (119–128 Nm). Be sure to secure the camshafts while tightening the bolts.
- Internal timing chain guide, upper timing chain guide, lower timing

chain tensioner and slack side timing chain guide

19. Torque the tensioner mounting bolt to 75–96 inch lbs. (8.4–10.8 Nm) and the guide bolts to 108–168 inch lbs. (13–19 Nm).

20. Apply a 0.102–0.142 in. (2.6–3.6mm) continuous bead of liquid gasket to all necessary areas as shown on the front timing cover.
- Timing cover evenly and gently. Be sure to align the dowel pin holes.

21. Torque the mounting bolts in sequence as follows:
   a. Bolts No. 1 and 2: 19–23 ft. lbs. (26–31 Nm).
   b. Bolts No. 3–20: 105–121 inch lbs. (11.8–13.7 Nm).

noid valves with new covers. Tighten the cover bolts in the proper sequence.

- Rocker covers
- Intake manifold
- Fuel tube assembly
- Intake manifold collector and support
- EGR tube
- Spark plugs and ignition coils
- Coolant hoses
- Fuel hoses
- Air duct assembly and hoses
- Left side rocker cover ornament
- Negative battery cable

24. Fill the cooling system.
25. Fill the engine with clean oil.
26. Start the vehicle, check for leaks and repair if necessary.

Piston and Ring

POSITIONING

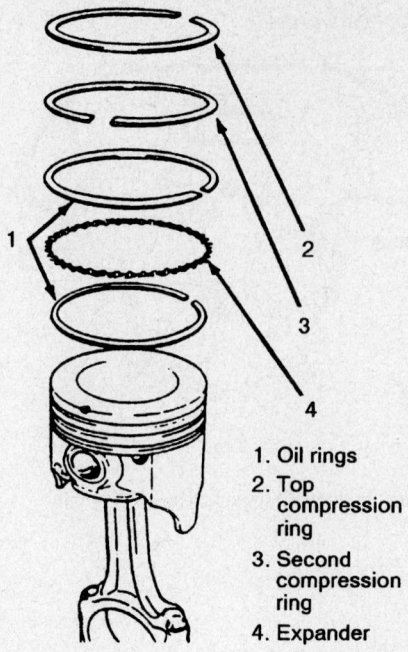

1. Oil rings
2. Top compression ring
3. Second compression ring
4. Expander

7923AG76

Exploded view of common piston ring mounting

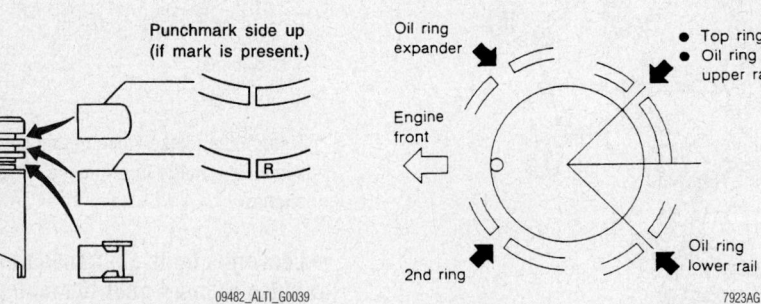

Punchmark side up (if mark is present.)

09482_ALTI_G0039

Piston ring positioning—3.5L engine

Oil ring expander

Engine front

2nd ring

• Top ring
• Oil ring upper rail

Oil ring lower rail

7923AG73

Piston ring end-gap spacing

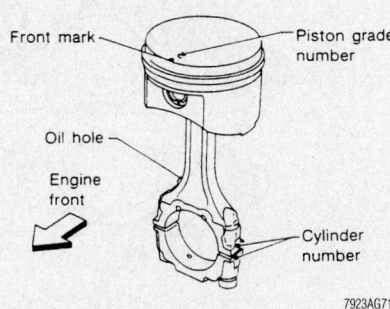

Front mark

Piston grade number

Oil hole

Engine front

Cylinder number

7923AG71

Piston and connecting rod assembly positioning

FUEL SYSTEM

Fuel System Service Precautions

Safety is the most important factor when performing not only fuel system maintenance but any type of maintenance. Failure to conduct maintenance and repairs in a safe manner may result in serious personal injury or death. Maintenance and testing of the vehicle's fuel system components can be accomplished safely and effectively by adhering to the following rules and guidelines.

- To avoid the possibility of fire and personal injury, always disconnect the negative battery cable unless the repair or test procedure requires that battery voltage be applied.
- Always relieve the fuel system pressure prior to disconnecting any fuel system component (injector, fuel rail, pressure regulator,

etc.), fitting or fuel line connection. Exercise extreme caution whenever relieving fuel system pressure, to avoid exposing skin, face and eyes to fuel spray. Please be advised that fuel under pressure may penetrate the skin or any part of the body that it contacts.

- Always place a shop towel or cloth around the fitting or connection prior to loosening to absorb any excess fuel due to spillage. Ensure that all fuel spillage (should it occur) is quickly removed from engine surfaces. Ensure that all fuel soaked cloths or towels are deposited into a suitable waste container.
- Always keep a dry chemical (Class B) fire extinguisher near the work area.
- Do not allow fuel spray or fuel vapors to come into contact with a spark or open flame.

- Always use a back-up wrench when loosening and tightening fuel line connection fittings. This will prevent unnecessary stress and torsion to fuel line piping. Always follow the proper torque specifications.
- Always replace worn fuel fitting O-rings with new. Do not substitute fuel hose where fuel pipe is installed.

Fuel System Pressure

RELIEVING

The fuel pump fuse is located in the dash fuse box or in the engine compartment fuse box. Check the lid of the fuse box for exact location.

1. Before servicing the vehicle, refer to the Precautions Section.

2. Remove the fuel pump fuse.
3. Start the engine.
4. Start the engine and run until the engine stalls.
5. After the engine stalls, try to restart the engine; if the engine will not start, the fuel pressure has been released.
6. Turn the ignition switch **OFF**. Reinstall the fuel pump fuse into the fuse block.

➡**Do not crank the engine or turn the ignition switch ON after the fuel pump fuse has been reinstalled, or the fuel pressure will be re-established.**

Fuel Filter

REMOVAL & INSTALLATION

Altima

➡**On Altima models, the fuel filter and fuel pump are replaced as an assembly. See the fuel pump procedure to remove the fuel filter on these models.**

Sentra

1. Before servicing the vehicle, refer to the Precautions Section.

2. Properly relieve fuel system pressure.
3. Remove or disconnect the following:
- Negative battery cable
- Fuel pump assembly from fuel tank
- Fuel filter from fuel level sensor unit by gently releasing the tabs

✳✳ WARNING

Apply mating marks on fuel filter and fuel level sensor unit forproper alignment for installation.

- Fuel pressure regulator from the fuel filter by pulling off the retaining clip

To install:
4. Install or connect the following:
- New filter and secure the filter into the fuel level sensor unit
- Fuel level sensor unit into the fuel tank
- Negative battery cable
5. Start the vehicle, check for leaks and repair if necessary.
6. Install the inspection cover and the rear seat

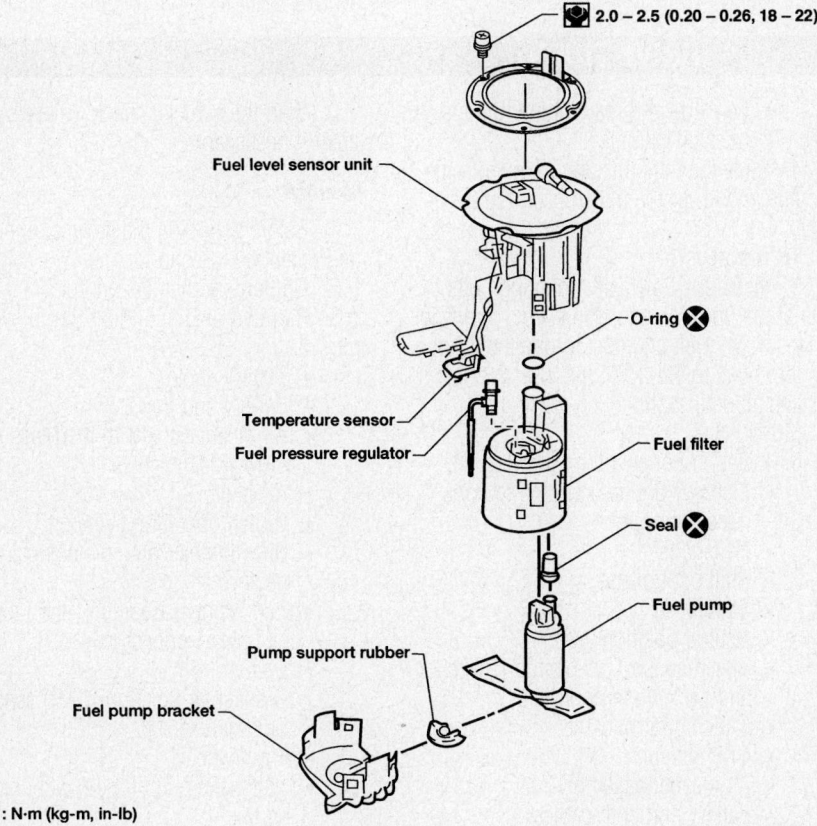

Exploded view of the fuel level sensor unit and filter assembly—Sentra shown

2.0 – 2.5 (0.20 – 0.26, 18 – 22)

Fuel level sensor unit
O-ring
Temperature sensor
Fuel pressure regulator
Fuel filter
Seal
Fuel pump
Pump support rubber
Fuel pump bracket

: N·m (kg-m, in-lb)

09482_ALTI_G0040

Fuel Pump

REMOVAL & INSTALLATION

Sentra

The fuel pump is located in the fuel tank on all vehicles. In-tank fuel pumps are accessible by lifting up the rear seat to gain access to the inspection cover.
1. Before servicing the vehicle, refer to the Precautions Section.
2. Relieve the fuel system pressure.
3. Remove or disconnect the following:
- Negative battery cable
- Rear seat from the vehicle
- Inspection cover that is located under the rear seat
- Inlet and outlet fuel lines from the fuel pump assembly
- Fuel pump and gauge wiring connections
- 6 mounting bolts that secure the fuel pump assembly to the top of the fuel tank
4. Raise up the fuel pump assembly and detach the fuel tubes and connector.
- Fuel gauge assembly
- Fuel pump with the fuel chamber
5. Pull up the front of the fuel pump chamber and slide the chamber forward.
- Fuel pump from the chamber
- O-ring seal or gasket

To install:
6. Install or connect the following:
- Fuel pump to the fuel pump chamber and slide chamber rearward
- Fuel pump with the fuel pump chamber
- Fuel gauge assembly using a new O-ring
- Fuel tubes and connector. Use new hoses and clamps.
- 6 mounting bolts to the top of the fuel gauge unit and torque the bolts to 18–22 inch lbs. (2–2.5 Nm)
- Fuel pump and gauge wiring connections
- New inlet and outlet fuel lines to the fuel pump assembly
- Negative battery cable
7. Start the vehicle, check for leaks and repair if necessary.
- Inspection cover and the rear seat

Altima

1. Before servicing the vehicle, refer to the Precautions Section.
2. Relieve the pressure from the fuel system.
3. Remove or disconnect the following:

- Negative battery cable
- Rear seat and the access cover
- Fuel pump electrical connector
- Fuel lines from the fuel pump assembly
- Locking ring
- Fuel gauge assembly
- Fuel tube and connector from the fuel gauge

➡**When the fuel sending unit needs to be removed, pull the tab upwards. The tab is located on the sending unit, opposite the end of the float. After the tab is pulled, the sending unit will lift straight out of the tank bracket.**

- Fuel pump by pinching the 2 locking tabs together. Lift the fuel pump assembly straight upward and out of fuel tank.
- O-ring and discard

4. Place a clean rag in the hole to keep out dirt.

To install:

5. Remove the rag
6. Install or connect the following:
- New O-ring and fuel pump
- Electrical connection and fuel tube to the fuel gauge sending unit
- Fuel sending unit into the tank

➡**Verify that the mark on the fuel tank and the components are aligned when installing the pump and fuel gauge sending unit.**

7. If equipped with a threaded lock ring torque the ring to 22–26 ft. lbs. (30–35 Nm). If equipped with a lock ring held by six retaining screws, torque the screws to 18–22 inch lbs. (2–2.5 Nm).
- Fuel lines and fuel pump electrical connector. Always install new clamps on the fuel lines.
- Negative battery cable

8. Start the engine, check for leaks and repair if necessary.
- Fuel pump access cover
- Rear seat

Fuel Injector

REMOVAL & INSTALLATION

1. Before servicing the vehicle, refer to the Precautions Section.
2. Relieve the fuel system pressure
3. Remove or disconnect the following:
- Negative battery cable
- Intake manifold collector
- Vacuum hose from the pressure regulator

- Fuel hoses from the rail
- Injector electrical connectors
- Fuel rail bolts
- Injector rail assembly with injectors from the intake manifold
- Injector from the rail by pushing on the injector tail piece
- Discard injector O-rings

To install:

4. Clean the injector tail piece and lubricate new O-rings with a smear of clean engine oil.
5. Install or connect the following:
- New O-rings
- Injector to the fuel rail
- Fuel rail and the injectors as an assembly to the intake manifold

6. Install the fuel rail bolts and tighten in 2 steps as follows:
   a. Step 1: Bolts to 84–96 inch lbs. (9–10 Nm).
   b. Step 2: Bolts to 15–20 ft. lbs. (21–26 Nm).
7. Install or connect the following:
- Injector electrical connectors
- Vacuum hose to the pressure regulator
- Fuel hoses to the rail
- All remaining components in the reverse order of removal

DRIVE TRAIN

Transaxle Assembly

REMOVAL & INSTALLATION

Manual

SENTRA

1. Before servicing the vehicle, refer to the Precautions Section.
2. Drain the fluid from the transaxle.
3. Remove or disconnect the following:
- Both battery cables
- Battery and bracket from the vehicle
- Crankshaft Position (CKP) sensor from the transaxle
- Air cleaner assembly
- All electrical connectors from the transaxles
- Control cable from the transaxle
- Speed sensor, OD position switch and Back-up lamp switch
- Neutral position switch
- Starter
- Shift control rod
- Halfshafts and properly support the transmission

- Left side and rear engine to transmission mounts

4. Slide the transmission away from the engine and lower the transmission assembly.

To install:

5. Install the transaxle mounting bolts in the proper location as noted during removal.
   a. On 1.8L engines: torque the 2 bottom bolts to 12–15 ft. lbs. (16–21 Nm) and all other bolts to 22–30 ft. lbs. (30–40 Nm).
6. Install or connect the following:
- Left side and rear engine to transmission mounts
- Halfshafts
- Shift control rod
- Starter
- Neutral position switch
- Speed sensor, OD position switch and back-up lamp switch
- Clutch control cable
- CKP sensor
- Air cleaner assembly
- Battery and both cables

7. Fill the transmission with clean oil to the proper level.

8. Start the vehicle, check for leaks and repair if necessary.

ALTIMA

1. Before servicing the vehicle, refer to the Precautions Section.
2. Drain the transmission fluid.
3. Remove or disconnect the following:
- Battery cables
- Battery and tray
- Air cleaner box with the Mass Air Flow (MAF) sensor
- Air duct
- Clutch operating cylinder
- Speedometer pinion electrical connectors
- Park/Neutral Position (PNP) switch electrical connectors
- Starter
- Crankshaft Position (CKP) sensor
- Shift control rod
- Front wheels
- Halfshafts and properly support the engine
- Rear and left side engine mounts
- Transaxle assembly

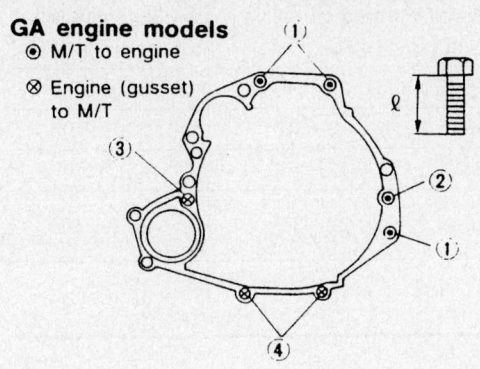

GA engine models
◉ M/T to engine
⊗ Engine (gusset) to M/T

Bolt No.	Tightening torque N·m (kg-m, ft-lb)	"ℓ" mm (in)
①	30 - 40 (3.1 - 4.1, 22 - 30)	70 (2.76)
②	30 - 40 (3.1 - 4.1, 22 - 30)	85 (3.35)
③	30 - 40 (3.1 - 4.1, 22 - 30)	30 (1.18)
④	16 - 21 (1.6 - 2.1, 12 - 15)	25 (0.98)
Front gusset to engine	30 - 40 (3.1 - 4.1, 22 - 30)	20 (0.79)
Rear gusset to engine	16 - 21 (1.6 - 2.1, 12 - 15)	16 (0.63)

9307QG13

Bolt locations and torque specifications—1.8L engine with manual transaxle

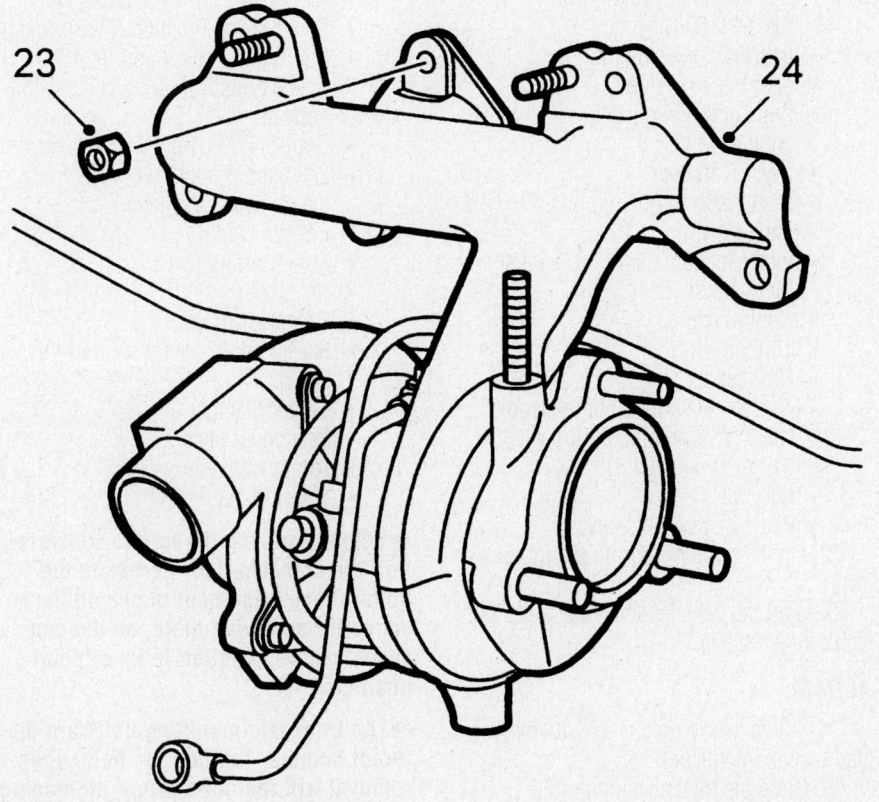

9347UG11

Bolt locations and torque specifications for the Altima with a manual transaxle

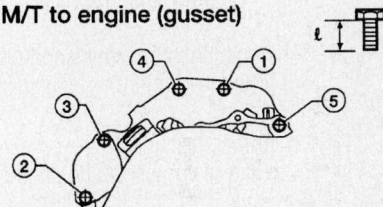

M/T to engine (gusset)

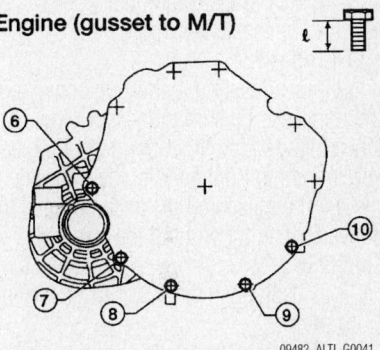

Engine (gusset to M/T)

09482_ALTI_G0041

Manual transaxle bolt torque specifications and locations—Altima with 3.5L engine

To install:

4. Install the transaxle assembly into the vehicle.

5. Torque the 4 lower mounting bolts to 22–30 ft. lbs. (30–40 Nm) and all remaining bolts to 29–36 ft. lbs. (39–49 Nm).

6. Install or connect the following:
   • Rear and left side engine mounts
   • Halfshafts and remove the engine support
   • Both front wheels
   • Shift control rod
   • CKP sensor
   • Starter and torque the bolts to 30 ft. lbs. (41 Nm)
   • PNP switch electrical connectors
   • Speedometer pinion connectors
   • Clutch operating cylinder
   • Air duct and air cleaner assembly
   • Battery tray and battery
   • Both battery cables

7. Fill the transmission with clean fluid.

8. Start the vehicle, check for leaks and repair if necessary.

Automatic

SENTRA

1. Before servicing the vehicle, refer to the Precautions Section.

2. Drain the fluid from the transaxle.

3. Remove or disconnect the following:
   • Both battery cables
   • Battery and bracket from the vehicle
   • Crankshaft Position (CKP) sensor from the transaxle
   • Air cleaner assembly
   • Torque converter clutch solenoid valve electrical connector
   • Inhibitor switch and Vehicle Speed Sensor (VSS) electrical connectors
   • Throttle wire from the engine side
   • Control cable

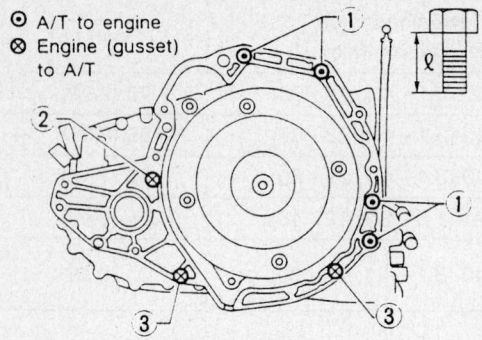

⊙ A/T to engine
⊗ Engine (gusset) to A/T

Bolt No.	Tightening torque N·m (kg-m, ft-lb)	Bolt length "ℓ" mm (in)
①	30 - 40 (3.1 - 4.1, 22 - 30)	50 (1.97)
②	30 - 40 (3.1 - 4.1, 22 - 30)	30 (1.18)
③	16 - 21 (1.6 - 2.1, 12 - 15)	25 (0.98)
Front gusset to engine	30 - 40 (3.1 - 4.1, 22 - 30)	20 (0.79)
Rear gusset to engine	16 - 21 (1.6 - 2.1, 12 - 15)	16 (0.63)

9307QG11

Bolt locations and torque specifications—1.8L engine with automatic transaxle

- Oil cooler hoses
- Halfshafts
- Intake manifold support bracket
- Starter
- Upper engine to transmission bolts and properly support the transmission
- Center member
- Front and rear gussets and engine rear plate
- Rear transaxle to engine bracket
- Rear transaxle mount
- Transaxle assembly

To install:

When connecting the torque converter to the transaxle, be sure to measure the distance between the mounting lug of the converter and the front edge of the transaxle.

4. The measured distance between the converter and the front of the transaxle should be:

   a. 1.8L engines: 0.831 in. (21.1mm) or more.

5. Raise the transaxle and install to engine drive plate.

6. Install the transaxle mounting bolts in the proper location as noted during removal.

7. On 1.8L engines torque the 2 bottom bolts to 12–15 ft. lbs. (16–21 Nm) and all other bolts to 22–30 ft. lbs. (30–40 Nm).

8. Install or connect the following:

- Torque converter to the drive plate and torque the bolts to 33–43 ft. lbs. (44–59 Nm)
- Rear transmission mount
- Rear transmission bracket
- Front and rear gussets and the rear engine plate
- Center member
- Starter and torque the bolts to 31 ft. lbs. (42 Nm)
- Intake manifold support bracket
- Both half shafts
- Control cable
- Throttle wire
- CKP sensor
- Torque converter clutch solenoid valve, VSS and inhibitor switch electrical connectors
- Air duct
- Battery and both cables

9. Fill the transmission to the proper level.

10. Start the vehicle, check for leaks and repair if necessary.

ALTIMA

1. Before servicing the vehicle, refer to the Precautions Section.

2. Drain the transmission fluid.

3. Remove or disconnect the following:

- Battery cables
- Battery and tray
- Air cleaner and resonator
- Park/Neutral Position (PNP) switch
- Revolution sensor and Vehicle Speed Sensor (VSS) electrical connectors
- Crankshaft Position (CKP) sensor
- Left hand mounting bracket from the transaxle and body
- Control cable
- Both front wheels
- Halfshafts
- Oil cooler pipes
- Starter and properly support the engine
- Center member
- Rear cover plate
- Torque converter
- Transaxle assembly

➡ **When removing the torque converter, turn the crankshaft for access to the bolts. Place alignment marks on the converter and drive plate, so the converter can be installed in its original position.**

➡ **The transaxle mounting bolts are different lengths. Tagging the bolts upon removal will facilitate proper tightening during installation.**

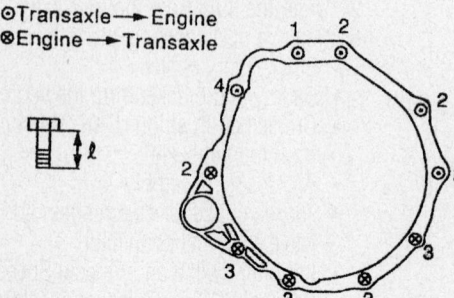

⊙ Transaxle → Engine
⊗ Engine → Transaxle

Bolt No.	Tightening torque N-m (kg-m, ft-lb)	ℓ mm (in)
1	70 - 79 (7.1 - 8.1, 52 - 58)	65 (2.56)
2	70 - 79 (7.1 - 8.1, 52 - 58)	52 (2.05)
3	70 - 79 (7.1 - 8.1, 52 - 58)	40 (1.57)
4	78 - 98 (7.9 - 10.0, 58 - 72)	124 (4.88)

09482_ALTi_G0042

Automatic transaxle bolt torque specifications and locations—2002–2004 Altima

Bolt No.	Tightening torque N·m (kg-m, ft-lb)	ℓ mm (in)
1	70 - 79 (7.2 - 8.0, 52 - 58)	55 (2.17)
2	41.2 - 52.0 (4.2 - 5.3, 31 - 38)	40 (1.57)
3	70 - 79 (7.2 - 8.0, 52 - 58)	55 (2.17)

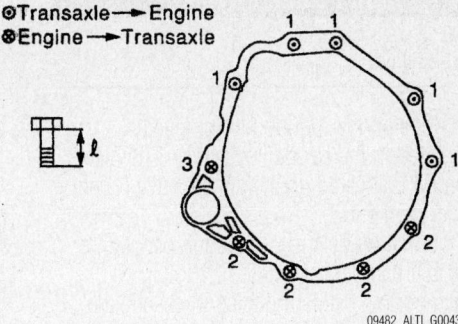

09482_ALTI_G0043

Automatic transaxle bolt torque specifications and locations—2005–2006 Altima with RE5F22A transaxle

Bolt No.	1	2	3	4	5	6
Number of bolts	4	1	1	2	2	1
Bolt length "ℓ" mm (in)	49 (1.93)	40 (1.57)	45 (1.77)	40 (1.57)	30 (1.18)	45 (1.77)
Tightening torque N·m (kg-m, ft-lb)	75 (7.7, 55)	35 (3.3, 26)	75 (7.7, 55)	43 (4.4, 32)		35 (3.6, 26)

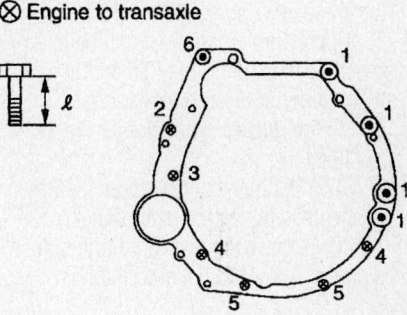

09482_ALTI_G0044

Automatic transaxle bolt torque specifications and locations—2005–2006 Altima with RE4F04B transaxle

To install:

➡**When installing the torque converter to the transaxle, measure the depth of the converter to ensure proper installation.**

4. Using a straight edge across the mounting flange, measure the depth of the converter. The measurement is to the bolt mounting flange of the converter.

5. The depth measurement of the converter should be 0.75 in. (19mm) or more.

➡**The transaxle mounting bolts are different lengths and require special torque specifications. Use care when installing and tightening these bolts.**

6. Install the transaxle assembly into the vehicle.

7. Refer to the diagram for the automatic transaxle mounting bolt torque specifications.

8. Torque the bolts holding the converter to the flexplate to 33–43 ft. lbs. (44–59 Nm).

9. Install or connect the following:
- Rear cover plate
- Center member
- Starter and remove the engine support
- Oil cooler pipes

- Halfshafts and both front wheels
- Control cable
- Left hand mounting bracket
- CKP sensor
- Revolution and VSS sensor electrical connectors
- PNP switch
- Air cleaner and resonator
- Battery, tray and both cables

10. Fill the transaxle with the proper type and amount of fluid.

11. Start the vehicle, check for leaks and repair if necessary.

Clutch

REMOVAL & INSTALLATION

1. Before servicing the vehicle, refer to the Precautions Section.

2. Remove or disconnect the following:
- Transmission/transaxle assembly

3. Insert a clutch disc centering tool into the clutch disc hub for support.
- Pressure plate bolts evenly in reverse order of the tightening sequence, a little at a time to prevent distortion
- Clutch assembly

- Throw-out bearing from the clutch lever

To install:

4. Apply a light coating of chassis lube to the clutch disc spleens, input shaft and pilot bearing. Use a disc centering tool to aid installation.

5. Install or connect the following:
- Disc and pressure plate

6. Torque the pressure plate bolts in a crisscross pattern and in several steps to 16–22 ft. lbs. (20–26 Nm).

7. Install or connect the following:
- New throw-out bearing in the clutch release lever. Remove the clutch disc centering tool.
- Transaxle into the vehicle. If the mating surfaces will not come together, do not force the units together. Remove the transaxle and recheck that the disc is centered.

➡**DO NOT draw the transaxle to the engine with the bolts. This may damage the clutch and/or transaxle. Also, be careful not to move the throw-out bearing when installing the transaxle.**

8. After the transaxle is installed, connect the clutch cable and check operation before complete reassembly.

9. Adjust the clutch pedal as necessary.

Hydraulic Clutch System

BLEEDING

Bleeding is required to remove air trapped in the hydraulic system. The bleed screw is located on the clutch slave (operating) cylinder.

Some models are also equipped with a clutch damper mechanism. The clutch damper mechanism is bled in exactly the same manner as the operating cylinder. It should be bled along with the operating cylinder.

1. Before servicing the vehicle, refer to the Precautions Section.
2. Remove the bleed screw dust cap.
3. Attach a transparent vinyl tube to the bleed screw, immersing the free end in a clean container of clean brake fluid.
4. Fill the master cylinder with the proper fluid.
5. Open the bleed screw about ¾ turn.
6. Depress the clutch pedal quickly. Hold it down. Have an assistant tighten the bleed screw. Allow the pedal to return slowly.

7. Repeat the above procedure until no more air bubbles are seen in the fluid container.
8. Remove the bleed tube.
9. Replace the dust cap and refill the master cylinder.
10. Bleed the clutch damper, if equipped.

Halfshaft

REMOVAL & INSTALLATION

Sentra

➡ **The halfshafts will require a special tool for the spline alignment of the halfshaft end into the transaxle case. Do not perform this procedure without access to this tool. The Kent Moore tool Number is J-34296 and J-34297**

1. Before servicing the vehicle, refer to the Precautions Section.
2. Raise the front of the vehicle and support it on jackstands, then remove the wheel and the tire assembly.
3. Remove or disconnect the following:

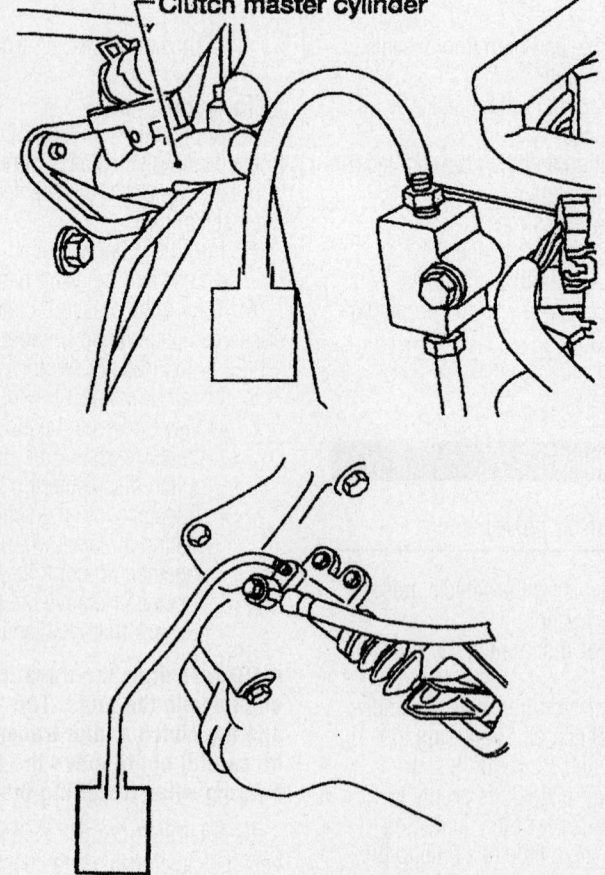

Clutch system bleeding points—Altima

09482_ALTI_G0045

- Wheel
- Hub nut using a bar to hold the wheel from turning
- Clip and separate the brake hose from the strut
- Caliper assembly and support it with a wire. Do not allow the caliper to hang from the brake hose.
- Bolts that secure the strut to the steering knuckle

➡ **Cover the halfshaft boots with shop towels to protect them during removal of the shaft.**

- Halfshaft from the knuckle by lightly tapping it with a hammer. If it is hard to remove, use a puller.

4. Remove the halfshaft from the transaxle as follows:
   a. Models without support bearing: Pry the halfshaft from the transaxle.
   b. Models with support bearing: Remove the support bearing bolts and pull the halfshaft from transaxle.

➡ **When removing the halfshaft from the transaxle, do not pull on the halfshaft. The halfshaft will separate at the sliding joint (damaging the boot). Use a small prybar to remove it from the transaxle. Be sure to replace the oil seal in the transaxle.**

5. Remove the halfshaft from the vehicle.

To install:

6. Use a new circlip on the halfshaft and install a new oil seal to the transaxle.

➡ **When installing the halfshaft into the transaxle, use a oil seal protector tool to protect the oil seal from damage.**

7. Install or connect the following:
- Halfshaft assembly into the transaxle

➡ **After installation of the halfshaft, try to pull the flange out by hand. If it pulls out, the circular clip is not locked into the transaxle.**

- Support bearing bracket and torque the bolts to 19–26 ft. lbs. (25–35 Nm)

8. Lubricate the splines of the halfshaft and insert the shaft through the steering knuckle.
9. Align the steering knuckle with the lower strut mount. Torque the bolts to 68–82 ft. lbs. (92–111 Nm).
- Disc brake caliper and the brake hose to the strut with the clip

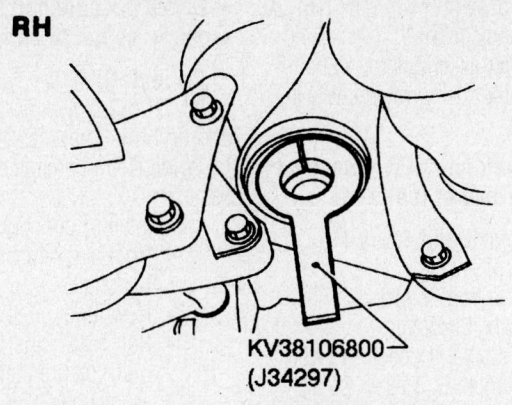

RH

KV38106800
(J34297)

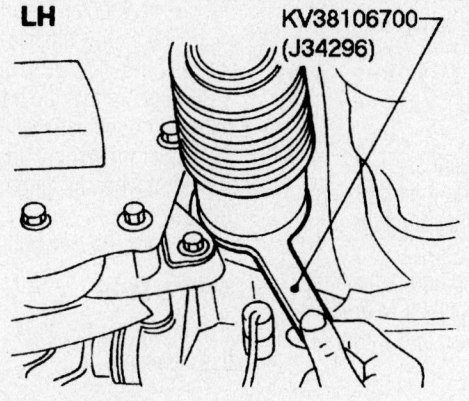

LH

KV38106700
(J34296)

7923QG75

Halfshaft installation tools—Sentra

- Washer and hub nut to the halfshaft and torque the nut to 145–202 ft. lbs. (197–274 Nm)
- Adjusting cap and a new cotter pin in drive axle
- Wheel and tire assembly and lower the vehicle

10. Road test the vehicle for proper operation.

Altima

1. Before servicing the vehicle, refer to the Precautions Section.

2. Raise and safely support the vehicle with the front wheels hanging freely.

3. Remove or disconnect the following:
  - Front wheels from the vehicle

➡**The brake caliper does not need to be disconnected from the knuckle.**

- Cotter pin from the castellated nut on the wheel hub
- Wheel bearing locknut

➡**Cover the CV-joint boots with a shop towel or waste cloth so not to damage them when removing the halfshaft.**

- Cotter pin and castle nut from the lower ball joint
4. Strike the knuckle with a hammer

and pull down on the transverse link to separate the lower ball joint from the knuckle.
  - Tie rod end from the steering knuckle
  - Halfshaft from the steering knuckle by tapping it with a block of wood and a mallet

5. Using a prybar, reach through the engine crossmember and carefully pry the right inner CV-joint from the transaxle.

6. If equipped with manual transaxle, carefully pry the left inner CV-joint from the transaxle.

7. If equipped with automatic transaxle, insert a long tool into the opening for the right halfshaft and strike the tool to with a hammer.

8. Remove the left halfshaft from the transaxle.

To install:

➡**Whenever the halfshafts are removed, the axle seals should be replaced.**

9. When installing the shafts into the transaxle, use a new oil seal and install an alignment tool along the inner circumference of the oil seal.

10. Install or connect the following:
  - Halfshaft into the transaxle, align

the serration's and remove the alignment tool

11. Push the halfshaft, then press-fit the circular clip on the shaft into the clip groove on the side gear.

➡**After insertion, attempt to pull the flange out of the side joint to be sure the circular clip is properly seated in the side gear and will not come out.**

- Halfshaft into the steering knuckle
- Lower ball joint and tie rod end and torque the lower ball joint-to-control arm nuts to 52–64 ft. lbs. (71–86 Nm) and the tie rod end-to-steering knuckle nut to 22–29 ft. lbs. (29–39 Nm)
- New cotter pins to the castle nuts
- Axle nut and torque the locknut to 174–231 ft. lbs. (235–314 Nm)
- New cotter pin on the wheel hub and install the wheel
- Front wheels to the vehicle

12. Road test the vehicle for proper operation.

13. Check the transaxle fluid level and top off as necessary.

CV-Joints

OVERHAUL

Sentra

TRANSAXLE SIDE

1. Before servicing the vehicle, refer to the Precautions Section.

2. Disassemble the joint as follows:
  a. Remove the boot bands.
  b. Matchmark the slide joint housing and inner race before separating the assembly.
  c. Matchmark the spider assembly and drive shaft.
  d. Remove the snapring.
  e. Remove the spider assembly.
  f. Remove the boot.

➡**Cover the halfshaft serrations with tape, so as not to damage the boot.**

To install:

3. Assemble the joint as follows:
  a. Thoroughly clean all parts in solvent and dry with compressed air. Check parts for evidence of damage and replace as necessary.
  b. Install the boot and new small boot band on the halfshaft.

4. Install the spider assembly. Confirm that the matchmarks are aligned.

5. Install a new outer snapring.

6. Pack the CV joint with 5.5–5.8 ounces (155–165g) of grease.

  a. Install the slide joint housing.

7. Set the boot so that it does not swell or deform when its length is 4–4.07 in. (101.5–103.5mm).

8. Lock the new boot bands securely.

WHEEL SIDE

1. Before servicing the vehicle, refer to the Precautions Section.

The joint on the wheel side cannot be disassembled.

2. Prior to separating the joint assembly, matchmark the halfshaft and joint assembly.

3. Separate the joint using a slide hammer.

4. Remove the boot bands.

To assemble:

5. Thoroughly clean all parts in solvent and dry with compressed air. Check parts for evidence of damage and replace as necessary.

➡**Cover the halfshaft serrations with tape, so as not to damage the boot.**

6. Install the boot and small boot band on the halfshaft.

7. Set the joint assembly onto the halfshaft and align the matchmarks.

8. Attach the joint assembly to the halfshaft by lightly tapping the serrated end with a plastic hammer.

➡**Using a metal hammer may damage the threads on the end of the joint.**

9. Pack the CV joint with 4.6–4.41 ounces (115–125g) of grease.

10. Ensure that the boot is properly installed on the halfshaft groove.

11. Set the boot so that it does not swell or deform when its length is 3.78–3.86 in. (96–98mm).

12. Lock the new boot bands securely.

Altima

TRANSAXLE SIDE

1. Remove the boot bands.

2. Matchmark the slide joint housing and inner race, prior to separating the joint assembly.

3. Pry off the snapring and remove the ball cage, inner race and balls as a unit.

4. Remove the snapring and withdraw the boot.

➡**Cover the halfshaft serrations with tape, so as not to damage the boot.**

To install:

5. Thoroughly clean all parts in solvent and dry with compressed air. Check parts for evidence of damage and replace as necessary.

6. Install or connect the following:
  • Boot and new boot band on the halfshaft
  • New inner snapring
  • Ball cage, inner race and balls as a unit. Ensure that the matchmarks are aligned.
  • New outer snapring

7. Pack the halfshaft with 5.0–6.0 ounces (165–175 g) of grease.

8. Ensure that the boot is properly installed on the halfshaft groove.

9. Set the boot so that it does not swell or deform when its length is 3.82–3.90 in. (97–99mm).

10. Lock the new boot bands securely.

WHEEL SIDE

The joint on the wheel side cannot be disassembled.

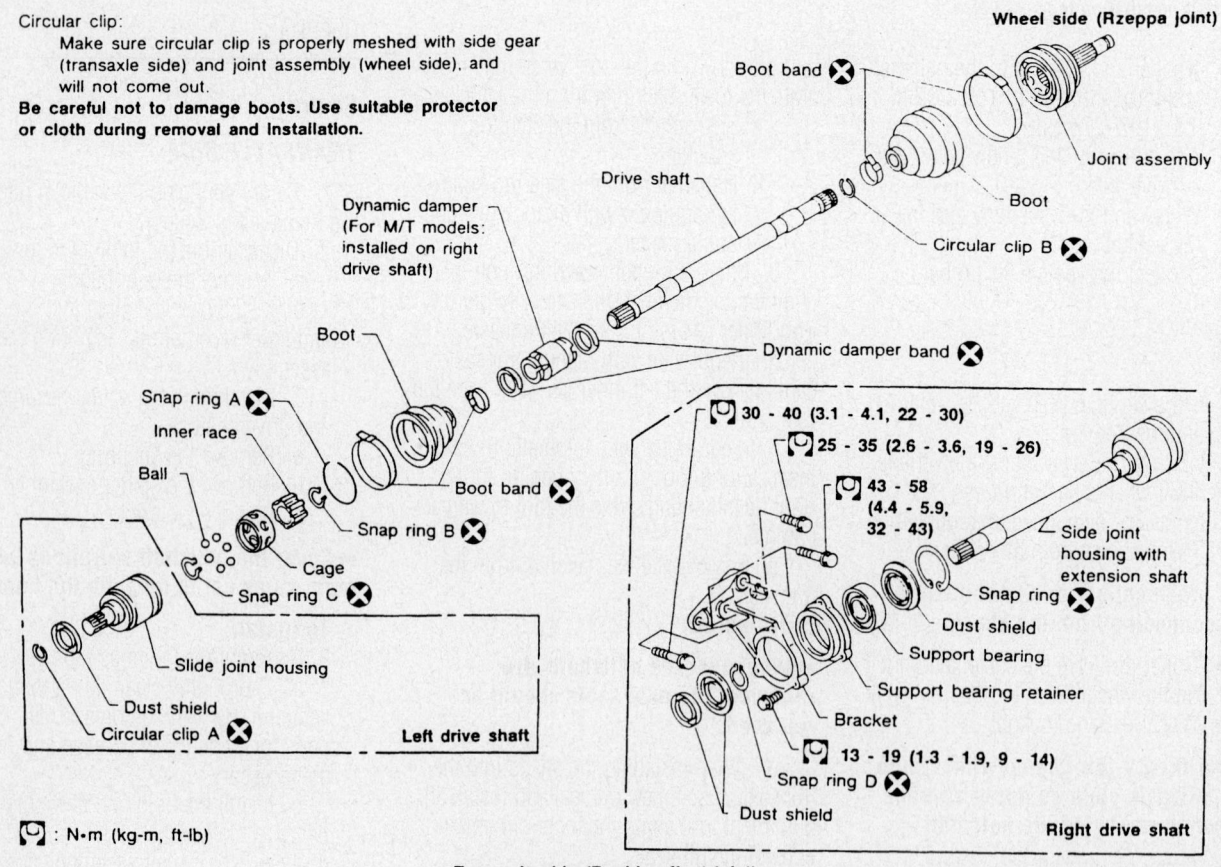

Circular clip:
  Make sure circular clip is properly meshed with side gear (transaxle side) and joint assembly (wheel side), and will not come out.
Be careful not to damage boots. Use suitable protector or cloth during removal and installation.

Wheel side (Rzeppa joint)
Boot band ⊗
Joint assembly
Boot
Circular clip B ⊗
Drive shaft
Dynamic damper (For M/T models: installed on right drive shaft)
Dynamic damper band ⊗
Boot
Snap ring A ⊗
Inner race
Ball
Boot band ⊗
Snap ring B ⊗
Cage
Snap ring C ⊗
Slide joint housing
Dust shield
Circular clip A ⊗
Left drive shaft

30 - 40 (3.1 - 4.1, 22 - 30)
25 - 35 (2.6 - 3.6, 19 - 26)
43 - 58 (4.4 - 5.9, 32 - 43)
Side joint housing with extension shaft
Snap ring ⊗
Dust shield
Support bearing
Support bearing retainer
Bracket
13 - 19 (1.3 - 1.9, 9 - 14)
Snap ring D ⊗
Dust shield
Right drive shaft

⬚ : N•m (kg-m, ft-lb)

Transaxle side (Double offset joint)

89617G09

Exploded view of the halfshafts and related components

The inner CV joint uses a large C-clip to retain the ball and cage assembly in the outer housing

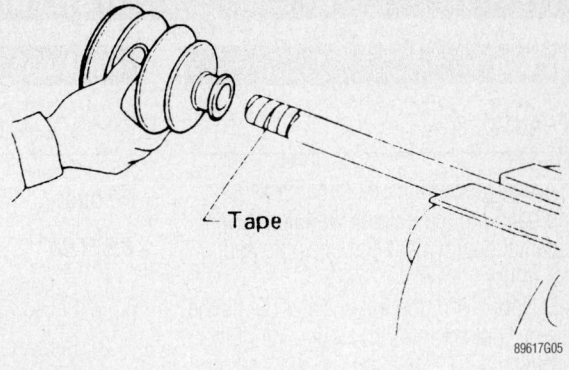

Use vinyl tape and wrap the end of the shaft to protect the boot during installation

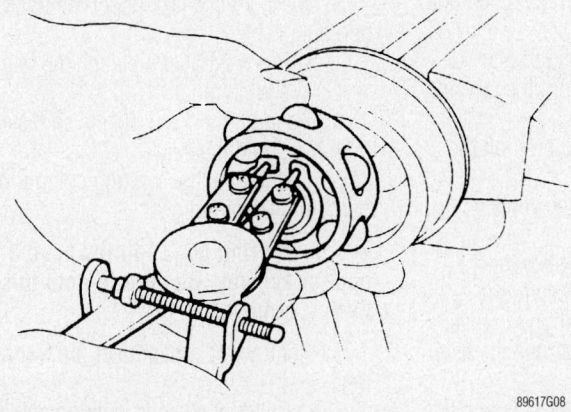

After the outer housing is removed, the ball and cage assembly can slide from the shaft by removing the C-clip

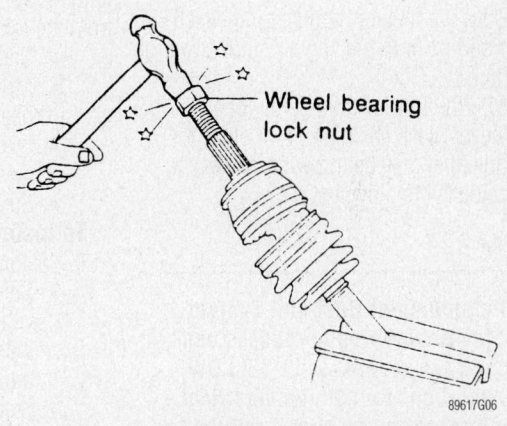

Use an old nut to protect the threads when tapping the outer CV joint onto the shaft

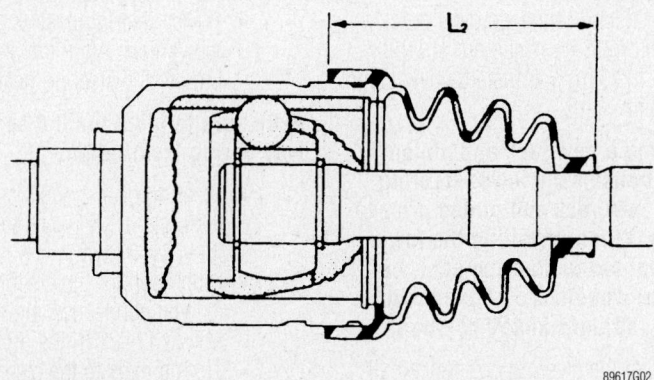

Make sure to properly position the boot before tightening the boot clamps

1. Prior to separating the joint assembly, matchmark the halfshaft and joint assembly.

2. Separate the joint using a slide hammer.

3. Remove the boot bands.

To assemble:

4. Thoroughly clean all parts in solvent and dry with compressed air. Check parts for evidence of damage and replace as necessary.

➡ **Cover the halfshaft serrations with tape, so as not to damage the boot.**

5. Install the boot and small boot band on the halfshaft.

6. Set the joint assembly onto the halfshaft and align the matchmarks.

7. Attach the joint assembly to the halfshaft by lightly tapping the serrated end with a plastic hammer.

➡ **Using a metal hammer may damage the threads on the end of the joint.**

8. Pack the halfshaft with 3.5–4.0 ounces (100–115 g) of grease.

9. Ensure that the boot is properly installed on the halfshaft groove.

10. Set the boot so that it does not swell or deform when its length is 3.327–3.406 in. (84.5–86.5mm).

11. Lock the new boot bands securely.

STEERING & SUSPENSION

Air Bag

PRECAUTIONS

Several precautions must be observed when handling the inflator module to avoid accidental deployment and possible personal injury.

1. Never carry the inflator module by the wires or connector on the underside of the module.

2. When carrying a live inflator module, hold securely with both hands and ensure that the bag and trim cover are pointed away.

3. Place the inflator module on a bench or other surface with the bag and trim cover facing up.

4. With the inflator module on the bench, never place anything on or close to the module that may be thrown in the event of an accidental deployment.

DISARMING

➡ **All Supplemental Restraint System (SRS) electrical wiring harnesses and connectors are covered with YELLOW outer insulation. Do not use electrical test equipment on any circuit related to the SRS (air bag) sensors. When installing SRS components, always install with the arrow marks facing the front of the vehicle.**

To disarm the SRS system turn the ignition switch to **OFF** position. Then, disconnect the both battery cables starting with the negative cable first and wait at least 10 minutes after the cables are disconnected. Be sure to insulate the battery terminal ends.

REARMING

To arm the Supplemental Restraint System (SRS) system turn the ignition switch to **OFF** position. Connect the both battery cables starting with the positive cable first.

➡ **The SRS or air bag system is equipped with a self-diagnostic operation. After turning the ignition key to the ON or START position, the AIR BAG warning lamp will illuminate for 7 seconds. After 7 seconds, the AIR BAG lamp will extinguish if no malfunction is detected. If the AIR BAG lamp does not extinguish after 7 seconds, check the SRS self-diagnostic system for a malfunction.**

Rack and Pinion Steering Gear

REMOVAL & INSTALLATION

Manual

SENTRA

1. Before servicing the vehicle, refer to the Precautions Section.

2. Remove or disconnect the following:
   - Front wheels
   - Both tie rod ends from the steering knuckles

3. Matchmark the steering column shaft to the lower joint and remove the pinch bolt from the joint.
   - Steering gear mounting bolts
   - Mounting clamps from the steering gear
   - Steering gear by sliding it off the steering shaft
   - Steering gear from the vehicle

To install:

4. Install or connect the following:
   - Steering gear assembly to the vehicle. Be sure to align the matchmarks of the rack with the marks on the steering shaft.
   - Steering gear mounting clamps and torque the bolts to 58 ft. lbs. (78 Nm)
   - Lower joint-to-steering column pinch bolt and torque the bolt to 22 ft. lbs. (29 Nm)
   - Tie rod end to the steering knuckle and torque the castle nut to 29 ft. lbs. (39 Nm) and install a new cotter pin

➡ **If installing a new rack and pinion assembly, transfer the lower steering joint to the new rack and pinion prior to installation. When installing the lower steering joint to the steering gear, be sure that the wheels are aligned with the vehicle (straight-ahead position).**

5. To center the steering gear, turn it all the way to the lock position on one side. Now, count the number of turns it takes to get to the opposite side lock position. Turn the steering gear ½ the number of turns towards the original starting position. The steering rack should now be centered. When connecting the steering joint to the steering column shaft, be sure to align the matchmarks made during disassembly.

6. Install the front wheels

7. Check the vehicle's alignment.

Power

SENTRA

1. Before servicing the vehicle, refer to the Precautions Section.

2. Remove or disconnect the following:
   - Low pressure hose clamp
   - Low pressure hose at the steering gear. Be sure to use a pan to catch the fluid.
   - Flare nut and the high pressure tube at the steering gear, then drain the fluid from the gear
   - Tie rod ends from the steering knuckle

3. Place a floor jack under the transaxle and support it.
   - Front exhaust pipe and the rear engine mount

4. Position the front wheels so they are pointing straight ahead.

5. Matchmark the steering column lower joint to the steering gear.

➡ **The steering gear splines have a flat spot or keyway. Be sure to note this during removal.**

6. Remove or disconnect the following:
   - Bolt that secures steering column lower joint
   - Bolts, steering gear unit and the linkage

To install:

7. Install or connect the following:
   - Power steering gear assembly to the vehicle. Align the steering column to the steering gear.

➡ **Be sure to align the flat spot or keyway during installation.**

   - Steering gear mounts and torque the bolts in sequence to 54–72 ft. lbs. (73–97 Nm)
   - Pinch bolt for the steering column-to-gear connection and torque the bolt to 17–22 ft. lbs. (24–29 Nm)
   - Tie rod ends to the steering knuckle and torque the nut to 22–29 ft. lbs. (29–39 Nm). Tighten the tie rod mounting nut further so the groves in the nut align with first cotter pin hole. Install a new cotter pin.
   - Power steering low pressure hose and torque the fitting to 20–29 ft. lbs. (27–39 Nm)
   - Power steering high pressure and torque the fitting to 11–18 ft. lbs. (15–25 Nm)

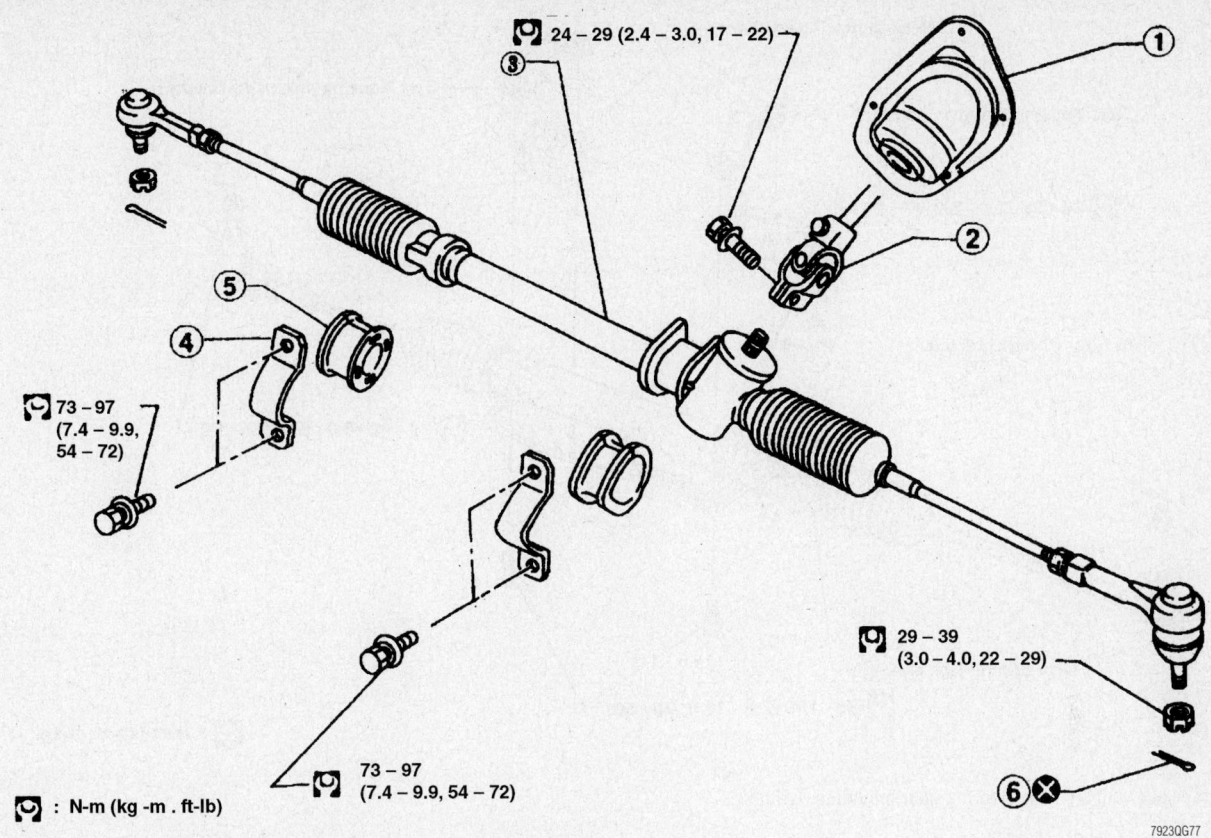

24 – 29 (2.4 – 3.0, 17 – 22)

73 – 97
(7.4 – 9.9,
54 – 72)

29 – 39
(3.0 – 4.0, 22 – 29)

73 – 97
(7.4 – 9.9, 54 – 72)

: N-m (kg -m . ft-lb)

7923QG77

Exploded view of the manual rack and pinion steering gear mounting—Sentra

- Rear engine mount and remove the floor jack
- Front exhaust pipe assembly using new gaskets

8. Fill the power steering system and start the engine.

9. Check the wheel alignment.

ALTIMA

1. Before servicing the vehicle, refer to the Precautions Section.

2. Disconnect the negative battery cable and disarm the air bag.

3. Remove or disconnect the following:
- Bolt securing the lower steering column shaft to the power steering gear assembly. Be sure to match-mark the shaft from the steering gear to the steering column joint for correct installation.
- Hoses from the power steering gear and plug the hoses to prevent leakage

- Cotter pins and castle nuts from the tie rod ends
- Tie rod ends from the steering knuckle, using a ball joint separating tool
- Front exhaust pipe mounting nuts and bolts
- Front exhaust pipe from the vehicle
- Control cable or linkage from the transmission and position it out of the way, if necessary
- Power steering gear mounting bolts or nuts
- Steering gear from the vehicle. Use care when separating the steering column joint.

4. Inspect the steering gear mount bushings and replace as necessary.

To install:

5. Align the steering column-to-steering gear matchmark and install the steering gear to the vehicle. Be sure to properly install the mounting bushings and hand-tighten the mounting nuts or bolts.

➡When installing the lower steering joint to the steering gear, be sure that the wheels are aligned straight and the steering joint slot is aligned.

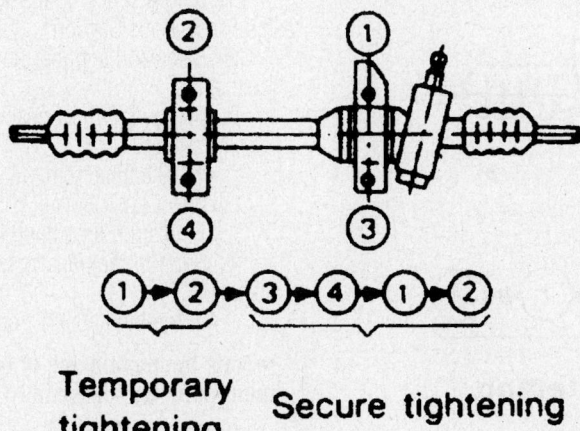

Temporary tightening   Secure tightening

7923QG78

Tighten the power steering gear mounting bolts according to the sequence shown—Sentra

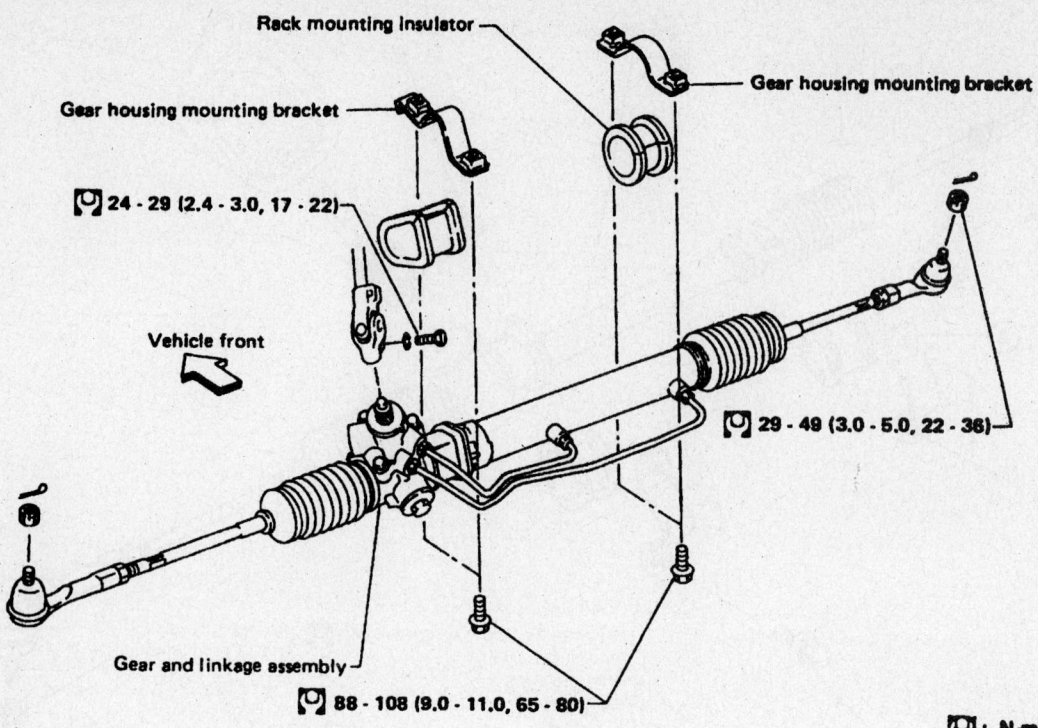

Rack mounting insulator

Gear housing mounting bracket

Gear housing mounting bracket

[○] 24 - 29 (2.4 - 3.0, 17 - 22)

Vehicle front

[○] 29 - 49 (3.0 - 5.0, 22 - 36)

Gear and linkage assembly

[○] 88 - 108 (9.0 - 11.0, 65 - 80)

[○] : N·m (kg-m, ft-lb)

79230G79

Exploded view of the power steering gear mounting—Altima

6. Torque the steering gear mounts to 54–72 ft. lbs. (73–97 Nm) in the sequence illustrated.

7. Install or connect the following:
- Pinch bolt securing the lower steering column shaft to the power steering gear assembly and torque the pinch bolt to 17–22 ft. lbs. (24–29 Nm)
- Tie rod end to steering knuckle and torque the castle nut to

22–29 ft. lbs. (29–39 Nm). Tighten the castle nut further to align the slot in the castle nut with the cotter pin hole and install a new cotter pin.
- Control cable or linkage to the transmission, if removed
- Front exhaust pipe assembly, using new gaskets
- Power steering hoses to the steering gear

8. Start the engine and fill the power steering reservoir.

9. Perform a front end alignment.

10. Connect the negative battery cable.

11. If equipped, enable the air bag system.

Strut and Coil Spring

REMOVAL & INSTALLATION

Front

SENTRA

1. Before servicing the vehicle, refer to the Precautions Section.

2. Raise and support the vehicle on jackstands.

3. Remove or disconnect the following:
- Wheel
- Brake tube from the strut
- Anti-Lock Brake (ABS) wiring from the strut, if equipped

4. Support the transverse link with a jackstand.
- Steering knuckle from the strut

➡ Note the positioning of the strut alignment mark for reassembly purposes.

5. Support the strut and remove the 3 upper attaching nuts. Remove the strut from the vehicle.

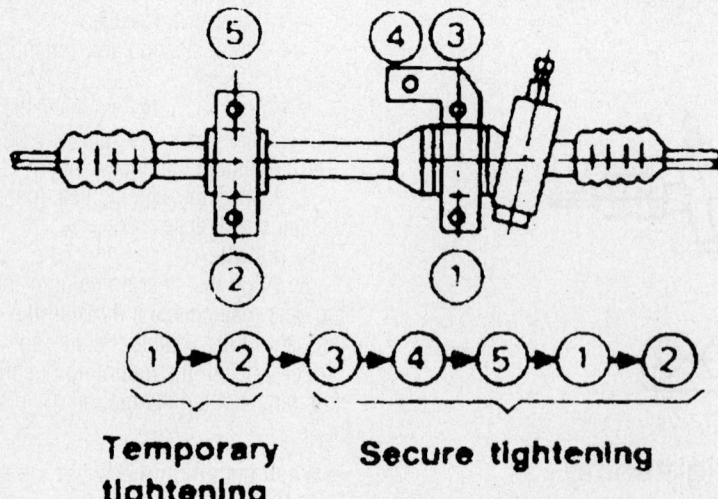

Temporary tightening

Secure tightening

79230G80

Tighten the mounting bolts using the illustrated procedure—Altima

❋❋ CAUTION

Never loosen the center spring retaining nut until the coil spring is compressed, or serious injury or vehicle damage may occur.

6. Place the strut assembly in a vise with a holding tool or in a spring compressor.

7. Loosen the piston rod locknut.

8. Compress the spring with the spring compressor, then remove the piston rod locknut.

➡ **Before removing the strut from the coil spring, note the positioning of the strut in relationship to the coil spring for reassembly.**

9. Remove or disconnect the following:

- Strut mounting insulator bracket, strut mounting bearing, upper spring seat and the upper spring rubber seat
- Strut, leaving the coil spring compressed
- Piston boot and rebound bumper from the strut

To install:

10. Install or connect the following:

- Rebound bumper and the boot to the strut piston
- Strut into the coil spring, be sure the strut and spring are properly positioned
- Upper spring rubber seat, upper spring seat, strut mounting bearing and the strut mounting insulator bracket. Be sure that the cutout on the upper spring seat is facing the outside of the vehicle.

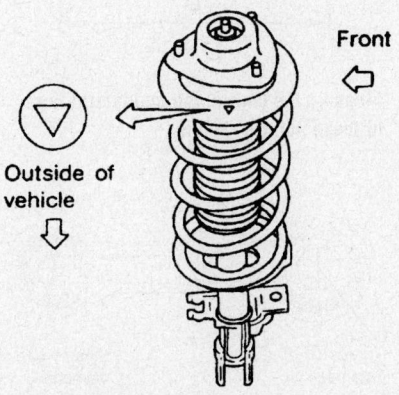

During assembly, be sure to point the alignment mark toward the outside of the vehicle—Sentra

- Piston rod locknut. Remove the tool and torque the piston rod locknut to 43–54 ft. lbs. (59–74 Nm).

➡ **When installing the strut, be sure to position the alignment mark toward the outside of the vehicle.**

- Strut to the vehicle
- 3 upper attaching nuts and torque the nuts to 18–22 ft. lbs. (25–29 Nm)
- Steering knuckle to the strut and torque the bolts to 68–82 ft. lbs. (92–111 Nm)
- Brake tube to the strut and the ABS wiring to the strut, if it was removed

11. Bleed the brake system and install the wheel.

12. Perform a front end alignment.

ALTIMA

1. Before servicing the vehicle, refer to the Precautions Section.

2. Raise and support the vehicle on jackstands.

3. Remove or disconnect the following:

- Wheel
- Brake tube from the strut
- Anti-Lock Brake (ABS) wiring from the strut, if equipped with ABS

4. Support the transverse link with a jackstand.

- Steering knuckle from the strut and properly support the strut
- 3 upper attaching nuts
- Strut from the vehicle

❋❋ WARNING

Never loosen the center spring retaining nut until the coil spring is compressed, or serious injury or vehicle damage may occur.

To install:

➡ **When installing the strut, be sure to position the alignment mark toward the outside of the vehicle.**

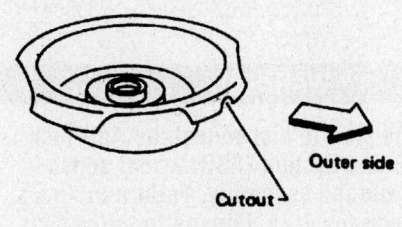

Position the alignment mark toward the outside of the vehicle—Altima

5. Install or connect the following:

- Strut to the vehicle
- 3 upper attaching nuts and torque the nuts to 29–40 ft. lbs. (39–54 Nm)
- Steering knuckle to the strut and torque the nuts to 123–137 ft. lbs. (167–186 Nm)
- Brake tube to the strut and the ABS wiring to the strut, if removed

6. Bleed the brake system and install the wheel.

7. Lower the vehicle and perform a front end alignment.

Rear

SENTRA

1. Before servicing the vehicle, refer to the Precautions Section.

2. Remove or disconnect the following:

- Rear wheel
- Trim panel from the trunk to gain access to the upper mounting nuts of the strut
- Protective cap from the upper portion of the strut

3. Position a floor jack under the rear axle for support.

➡ **Note and mark the positioning of the upper strut plate to the vehicle body.**

❋❋ CAUTION

Never remove the center strut nut until the strut is removed from the vehicle and the spring is safely compressed.

4. Remove or disconnect the following:

- Lower strut mounting through-bolt
- 2 upper strut mounting nuts and the strut from the vehicle

5. Place the strut assembly in a vise with a holding tool or in a spring compressor.

6. Loosen the piston rod locknut.

7. Compress the spring with the spring compressor, then remove the piston rod locknut.

➡ **Before removing the strut from the coil spring, note the positioning of the strut in relationship to the coil spring for reassembly.**

8. Remove or disconnect the following:

- Strut mounting insulator bracket, strut mounting bearing, upper spring seat and the upper spring rubber seat
- Strut, leaving the coil spring compressed

- Piston boot and rebound bumper from the strut

To install:

9. Install or connect the following:
- Rebound bumper and the boot to the strut piston
- Strut into the coil spring, be sure the strut and spring are properly positioned
- Upper spring rubber seat, upper spring seat, strut mounting bearing and the strut mounting insulator bracket. Be sure that the cutout on the upper spring seat is facing the outside of the vehicle.
- Piston rod locknut and torque the locknut to 13–17 ft. lbs. (18–24 Nm)

10. Remove the spring compressor from the coil spring.
- Strut and torque the 2 upper mounting nuts to 12–14 ft. lbs. (16–19 Nm)
- Upper mount protective cap
- Through-bolt to the lower mount of the strut and torque the bolt to 72–87 ft. lbs. (98–118 Nm)
- Trunk trim panel
- Rear wheel

11. Lower the vehicle and perform an alignment.

ALTIMA

1. Before servicing the vehicle, refer to the Precautions Section.
2. Remove or disconnect the following:
- Rear wheels from the vehicle and support the rear axle with a jack
- Strut lower mounting through-bolts

➡**Be sure to note the position the strut upper plate to the vehicle for reinstallation purposes.**

- 2 nuts from the top of the strut
- Strut as an assembly

❋❋ CAUTION

Do not remove the center locknut from the strut assembly until the strut is safely compressed.

3. Compress the strut coil spring with a spring compressor.
- Strut assembly center locknut

➡**Before removing the strut from the coil spring, note the positioning of the strut in relationship to the coil spring for reassembly.**

- Strut leaving the coil spring compressed

➡**Mark the coil spring position to the strut assembly for reinstallation purposes.**

4. To remove the spring from the strut assembly, perform the following steps:
  a. Compress the coil spring with the proper compressor tool.
  b. Remove the center retaining nut holding strut mounting insulator.
  c. Slowly decompress the coil spring.
  d. Remove the strut mounting insulator.
  e. Remove coil spring.

To install:

5. Install or connect the following:
- Coil spring onto the strut assembly. Be sure to align the matchmarks made during the removal procedure.
- Strut mounting insulator and compress the coil spring assembly

➡**It will be necessary to use a new locknut for the center retaining nut of the coil spring.**

- Center retaining nut and torque to 43–58 ft. lbs. (59–78 Nm). Be sure the spring is seated properly on the strut and in the mounting insulator.

6. Slowly remove the spring compressor tool.
- Strut assembly and torque the upper nuts to 31–40 ft. lbs. (42–54 Nm)
- Lower strut through-bolt and torque to 123–137 ft. lbs. (167–186 Nm)

➡**Be sure to hold the through-bolt and tighten the nuts.**

7. Install the wheels, lower the vehicle and perform a front end alignment.

Torsion Bars

REMOVAL & INSTALLATION

Sentra

1. Before servicing the vehicle, refer to the Precautions Section.
2. Loosen the lug nuts.
3. Raise and safely support the vehicle
4. Remove or disconnect the following:
- Wheels

❋❋ WARNING

Be sure to disconnect the Anti-lock Brake System (ABS) wheel sensor from the assembly. Failure to do so may result in damage to the sensor wire and the sensor becoming inoperative.

- Brake calipers and suspend them with a piece of wire. Do not let them hang by the hose.

5. Using a transmission jack, raise the torsion beam a little, then remove the suspension mounting bolts.
6. Lower the jack and remove the suspension assembly.
7. Remove the lateral link and control rod.
8. Inspect the torsion beam and control rod for cracks, wear and deformation. The length of the lateral link and control rod is as follows:
- A—8.15–8.19 in. (207–208mm)
- B—15.51–15.55 in. (394–395mm)
- C—23.66–23.74 in. (601–603mm)
- D—4.17–4.25 in. (106–108mm)

To install:

9. When installing the control rod, connect the bushing with the smaller inner diameter to the lateral link. Install the lateral link and the control rod on the torsion beam. Place the lateral link with the arrow topside.
10. Place the lateral link and control rod horizontally against the beam and tighten the bolts. Refer to the illustration.
11. Secure the torsion beam to the vehicle. Make sure the lateral link is horizontal, then tighten the link to the chassis.

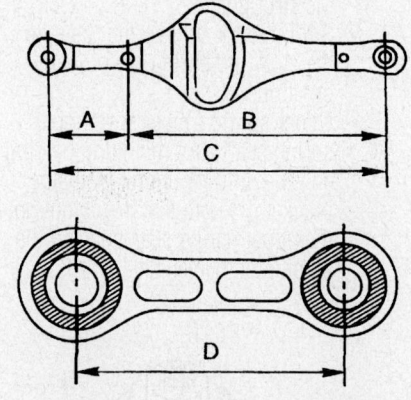

09482_ALTI_G0046

Measure the control rod and lateral links at these points—Sentra

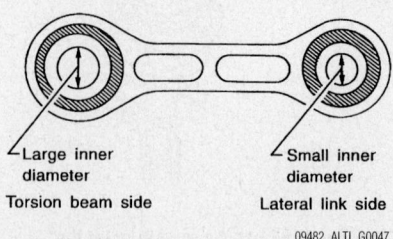

Large inner diameter
Torsion beam side

Small inner diameter
Lateral link side

09482_ALTI_G0047

Be sure to install the control rod correctly—Sentra

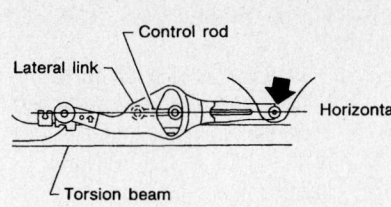

09482_ALTI_G0048

The lateral link must be in the horizontal position when tightening the bolts—Sentra

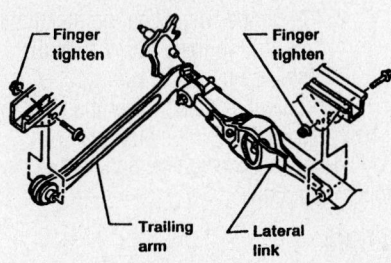

09482_ALTI_G0049

Tighten the torsion beam-to-chassis bolts with the suspension unloaded—Sentra

12. Attach the struts to the torsion beam and tighten the fasteners.

13. Tighten the torsion beam-to-chassis bolts.

14. Install the calipers, ABS sensor and wheels. Lower the vehicle to the ground.

Lower Ball Joint

REMOVAL & INSTALLATION

The ball joint is an integral part of the lower control arm. If the ball joint is defective the control arm must be replaced.

Lower Control Arm (Transverse Link)

REMOVAL & INSTALLATION

Sentra

1. Before servicing the vehicle, refer to the Precautions Section.

2. Remove or disconnect the following:
   - Front wheels
   - Disc brake caliper from the steering knuckle

❋❋ WARNING

DO NOT allow the disc brake caliper to hang from the brake hose. Support the disc caliper with safety wire.

- Cotter pin and loosen the wheel bearing locknut
- Cotter pin and the castle nut from the tie rod ball joint. Separate the tie rod with a suitable puller.
- 2 bolts that secure the lower portion of the strut to the steering knuckle

3. Using a plastic or rubber mallet, tap on the loosened wheel bearing locknut to loosen the halfshaft in the knuckle. Remove the locknut and remove the halfshaft from the steering knuckle. Be sure to cover the CV-joints with a shop rag.

➡**Support the halfshaft assembly with wire. Do not allow the halfshaft to hang by the inner joint.**

- Nut that secures the stabilizer link to the lower control arm
- Link from the control arm. Note the positioning of the washers and spacers for reassembly.
- Cotter pin and castle nut from the lower ball joint
- Lower ball joint from the knuckle
- Knuckle from the vehicle
- Mounting nuts/bolts that secure the lower control arm to the frame
- Control arm from the vehicle

To install:

➡**Final tightening of all suspension components should take place with the weight of the vehicle on the wheels.**

4. Install the lower control arm assembly and torque mounting bolts/nuts as follows:

   a. Through bolt and nut: 76–90 ft. lbs. (103–123 Nm).

   b. 2 saddle bracket mounting bolts: 58–72 ft. lbs. (78–98 Nm).

5. Install or connect the following:
   - Steering knuckle to the lower ball joint and torque the castle nut to 43–54 ft. lbs. (59–74 Nm). Install a new cotter pin.
   - Stabilizer link to the lower control arm and torque the nut to 12–16 ft. lbs. (16–22 Nm)
   - Halfshaft through the wheel bearing
   - Wheel bearing locknut. Do not torque the locknut at this time.
   - Steering knuckle to the strut and torque the bolts to 68–82 ft. lbs. (92–111 Nm)
   - Tie rod end and torque the castle nut to 22–29 ft. lbs. (29–39 Nm). Install a new cotter pin.
   - Disc brake caliper to the steering knuckle

6. Tighten the halfshaft mounting nut (hub nut) and torque the nut to 145–202 ft. lbs. (196–274 Nm). It may be necessary to have an assistant hold the brake pedal while tightening the locknut. Install the adjusting cap and a new cotter pin.

7. Install the front wheels, lower the vehicle and perform a front end alignment.

Altima

➡**The lower ball joint is integral with the lower control arm (transverse link). They are removed and replaced as an assembly.**

1. Before servicing the vehicle, refer to the Precautions Section.

2. Remove or disconnect the following:
   - Front wheels
   - Stabilizer bar. The bar is removed by unfastening the nut that hold the bar to the transverse link gusset plate.

➡**Take note of position of marks on clamp face and stabilizer bar for reassembling.**

   - Lower ball joint to knuckle cotter pin and nut
   - Ball joint stud from knuckle using the proper tool
   - Transverse link mounting bolts and nuts
   - Link

To install:

3. Install or connect the following:
   - Transverse link with mounting bolts and torque nuts and bolts to 87–108 ft. lbs. (118–147 Nm)

➡**The final tightening of suspension components must be done with wheels on the ground and vehicle at curb weight.**

   - Lower ball joint to the knuckle and torque the nut to 52–64 ft. lbs. (71–86 Nm) and install a new cotter pin
   - Stabilizer bar link to the transverse link and torque the nuts to 30–35 ft. lbs. (41–47 Nm)

4. Install wheels and safely lower vehicle to ground.

5. Check the front end alignment.

CONTROL ARM BUSHING REPLACEMENT

The bushing is an integral part of the lower control arm. If the bushing is defective the control arm must be replaced.

Wheel Bearings

ADJUSTMENT

Front

➡ **Whenever the hub or bearing assemblies are removed, the wheel bearing must be replaced. Never reuse the old bearing assembly.**

The wheel bearings are sealed and are not adjustable. If defective, replacement is the only option.

Rear

If the wheel hub bearing assembly is removed, it must be replaced.

➡ **The wheel hub bearing assembly is not repairable; it must be replaced when defective.**

1. Before servicing the vehicle, refer to the Precautions Section.
2. Torque the wheel bearing locknut to 138–188 ft. lbs. (187–255 Nm).
3. Verify that the wheel bearings operate smoothly.
4. Install a new cotter pin into the spindle to hold the wheel bearing locknut.
5. Install a dial indicator to the rear wheel hub bearing assembly and check the axial end-play; it should be less than 0.0020 in. (0.05mm).
6. Install the grease cap.
7. If the axial end-play exceeds specifications, the wheel bearing must be replaced.

REMOVAL & INSTALLATION

Front

SENTRA

➡ **Whenever the hub or bearing assembly is removed, the wheel bearing assembly must be replaced. Never reuse the old bearing assembly.**

1. Before servicing the vehicle, refer to the Precautions Section.
2. Remove or disconnect the following:
   - Front wheel
   - Wheel bearing/axle shaft locknut while depressing the brake pedal
   - Brake caliper and support it with a piece of wire. It is not necessary to disconnect the brake line from the caliper.
   - Anti-Lock Brake System (ABS) sensor from the steering knuckle

➡ **Do not depress the brake pedal or twist the brake line.**
   - Tie rod end
   - Halfshaft from the knuckle by slightly tapping with a soft hammer. Position the axle shaft nut on the threads of the shaft to protect them when lightly tapping.
   - Lower ball joint nut and separate
   - 2 strut-to-knuckle retaining bolts and separate
   - Steering knuckle from the vehicle
3. Place the assembly in a vise. Drive the hub with the inner race from the knuckle with a suitable tool. Remove the inner and outer grease seals.
   - Bearing inner race and outer grease seal from the hub
   - Snapring and press the bearing outer race to remove the bearing from the steering knuckle

To install:
4. Press a new wheel bearing into the knuckle assembly not exceeding 3.3 tons (2994 kg) pressure.
5. Install or connect the following:
   - Snapring and pack the grease seal lips with chassis grease
   - Inner and outer grease seals
6. Press the wheel hub into the knuckle not exceeding 3.3 tons (2994 kg) pressure.
7. Check bearing operation and by applying 3.9–5.5 tons (3538–4990 kg) pressure to the hub assembly. Spin the hub several times in both directions.
8. Be sure the bearings rotate freely. If

the bearings do not rotate freely, replace the bearings.
9. Install or connect the following:
   - Knuckle and wheel hub assembly
   - Lower ball joint and torque the nut to 43–54 ft. lbs. (59–74 Nm). Install a new cotter pin.
   - Strut and torque the bolts to 68–82 ft. lbs. (92–118 Nm)
   - Tie rod end and tighten the nut to 22–29 ft. lbs. (29–39 Nm). Install a new cotter pin.
   - Disc brake caliper
   - Torque the wheel bearing locknut to 145–203 ft. lbs. (196–275 Nm). Install a new cotter pin.
   - Front wheels and lower the vehicle
10. Check the vehicle's alignment.
11. Road test the vehicle and verify proper operation.

ALTIMA

➡ **Whenever the hub or bearing assembly is removed, the wheel bearing assembly must be replaced. Never reuse the old bearing assembly.**

1. Before servicing the vehicle, refer to the Precautions Section.
2. Remove or disconnect the following:
   - Knuckle assembly from the vehicle
   - Hub with the inner race from the steering knuckle, using a shop press and a suitable tool
   - Bearing inner race from the hub, using a shop press and a suitable tool
   - Outer grease seal

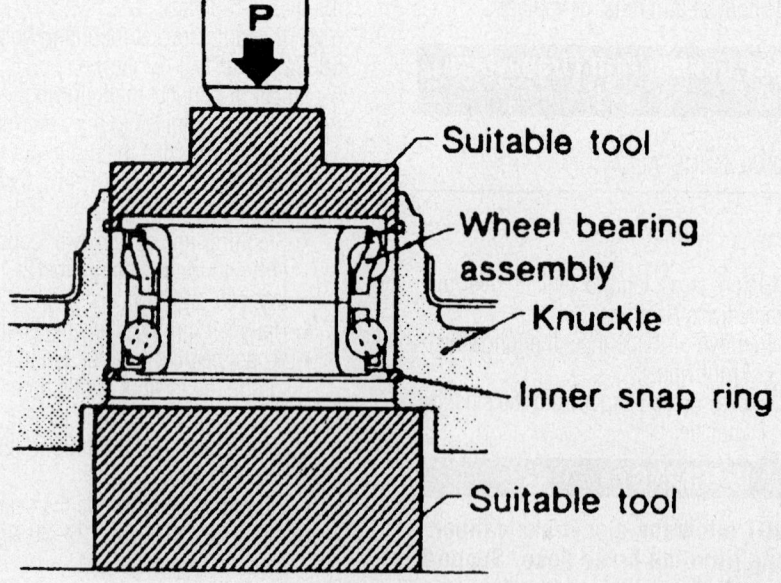

Typical method of installing the wheel bearing

79230G85

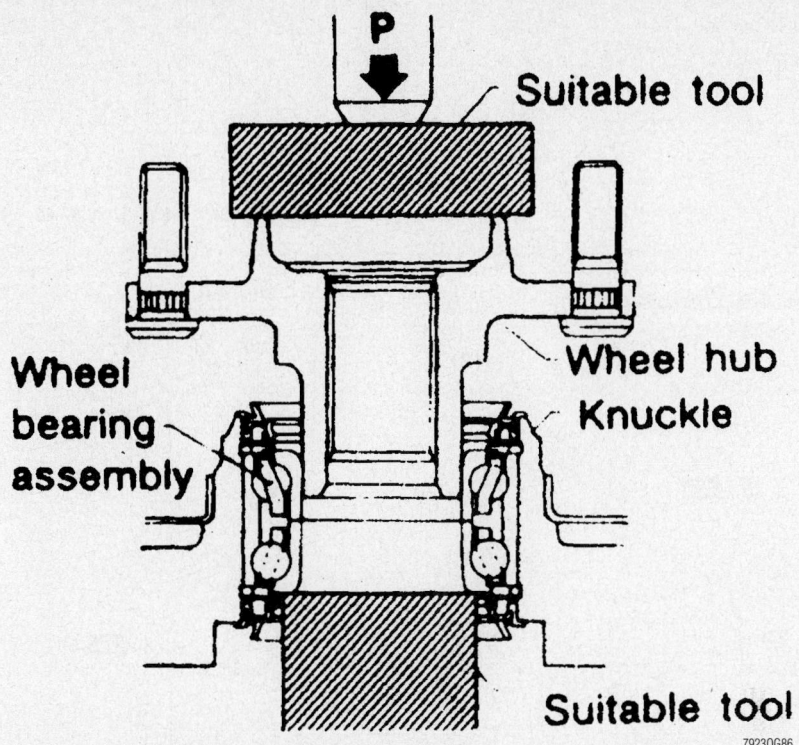

Use a press to install the hub into the knuckle assembly

- Inner grease seal from the steering knuckle, using a prybar
- Inner and outer snaprings from the steering knuckle, using snapring pliers
- Sealed bearing assembly from the steering knuckle, using a shop press and a suitable tool

3. Inspect the hub, steering knuckle and snaprings for cracks and/or wear; if necessary, replace the damaged part(s).

To install:

4. Install or connect the following:
- Inner snapring in the steering knuckle groove

- New wheel bearing assembly into the steering knuckle, using a shop press and a suitable tool, until it seats, using a maximum pressure of 3 tons (2722 kg)
- Outer snapring

5. Pack the new grease seal lips with multi-purpose grease.

- New outer grease seal into the steering knuckle, using a shop press and a suitable tool
- Hub into the steering knuckle, using a shop press and a suitable tool, until it seats, using a maximum pressure of 5.5 tons (4990 kg); be careful not to damage the grease seal

6. To check the bearing operation, perform the following procedures:
   a. Increase the press pressure to 3.5–5.0 tons (3175–4536 kg).
   b. Spin the steering knuckle, several turns, in both directions.
   c. Be sure the wheel bearings operate smoothly.

7. If the wheel bearings do not operate smoothly, replace the wheel bearing assembly.

8. Install the knuckle assembly.

9. Install the halfshaft into the hub. Torque the locknut to 174–231 ft. lbs. (235–314 Nm).

10. Install the wheel assembly and lower the vehicle.

11. Road test the vehicle and verify proper operation.

BRAKES

Brake Caliper

REMOVAL & INSTALLATION

FRONT

1. Before servicing the vehicle, refer to the Precautions Section.

2. Remove or disconnect the following:
- Front wheels
- Brake fluid hose
- Pin bolts
- Caliper assembly from the vehicle

To install:

3. Use a large C-clamp to press the caliper piston back into the caliper.

4. Install or connect the following:
- New pads, new shims and pad retainers

- Brake caliper and torque the pin bolts to 23 ft. lbs. (31 Nm)
- Brake line to the caliper, using new copper washers, and torque the connecting bolt to 12–14 ft. lbs. (17–20 Nm)
- Wheels

5. Bleed the brake system and top off the master cylinder as necessary.

Rear

1. Before servicing the vehicle, refer to the Precautions Section.

2. Remove or disconnect the following:
- Rear wheels
- Parking brake cable and the lock spring
- Brake fluid hose from the caliper
- Caliper pin bolts and remove the caliper

To install:

3. Turn the piston clockwise back into the caliper body. Remove some brake fluid from the master cylinder, if necessary. Take care not to damage the piston boot.

4. Coat the pad contact area on the mounting support with a silicone based grease.

5. Install or connect the following:
- New pads, shims and the pad springs
- Caliper body into position and torque the caliper pin bolts to 16–23 ft. lbs. (22–31 Nm)
- Brake fluid hose, using new copper washers, and tighten the flare nut to 12–14 ft. lbs. (17–20 Nm)
- Lock spring and the parking brake cable

6. Bleed the brake system and top off the master cylinder as necessary.

7. Replace the wheels.

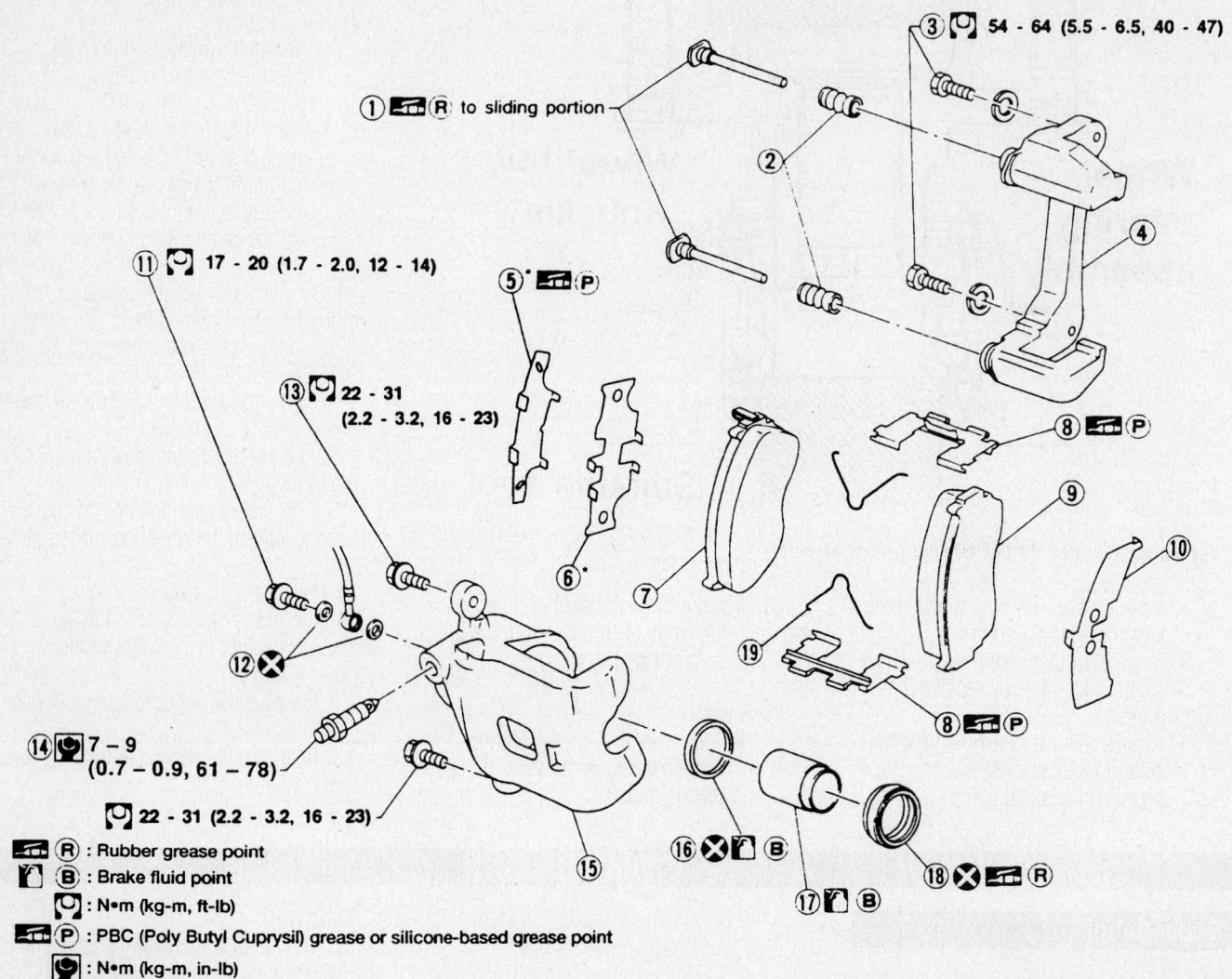

① ⟨R⟩ to sliding portion

③ ⟨⟩ 54 - 64 (5.5 - 6.5, 40 - 47)

②

④

⑪ ⟨⟩ 17 - 20 (1.7 - 2.0, 12 - 14)

⑤ ⟨P⟩

⑬ ⟨⟩ 22 - 31
(2.2 - 3.2, 16 - 23)

⑧ ⟨P⟩

⑨

⑩

⑥*

⑦

⑲

⑫ ⊗

⑧ ⟨P⟩

⑭ ⟨⟩ 7 - 9
(0.7 - 0.9, 61 - 78)

⑯ ⊗ ⟨B⟩

⑱ ⊗ ⟨R⟩

⟨⟩ 22 - 31 (2.2 - 3.2, 16 - 23)

⑰ ⟨B⟩

⑮

⟨R⟩ : Rubber grease point
⟨B⟩ : Brake fluid point
⟨⟩ : N•m (kg-m, ft-lb)
⟨P⟩ : PBC (Poly Butyl Cuprysil) grease or silicone-based grease point
⟨⟩ : N•m (kg-m, in-lb)
* : If equipped

①	Main pin	⑧	Pad retainer	⑭	Bleed valve
②	Pin boot	⑨	Outer pad	⑮	Cylinder body
③	Torque member fixing bolt	⑩	Outer shim	⑯	Piston seal
④	Torque member	⑪	Connecting bolt	⑰	Piston
⑤	Shim cover*	⑫	Copper washer	⑱	Piston boot
⑥	Inner shim*	⑬	Main pin bolt	⑲	Pad return spring
⑦	Inner pad				

93016G51

Front brake caliper—Sentra

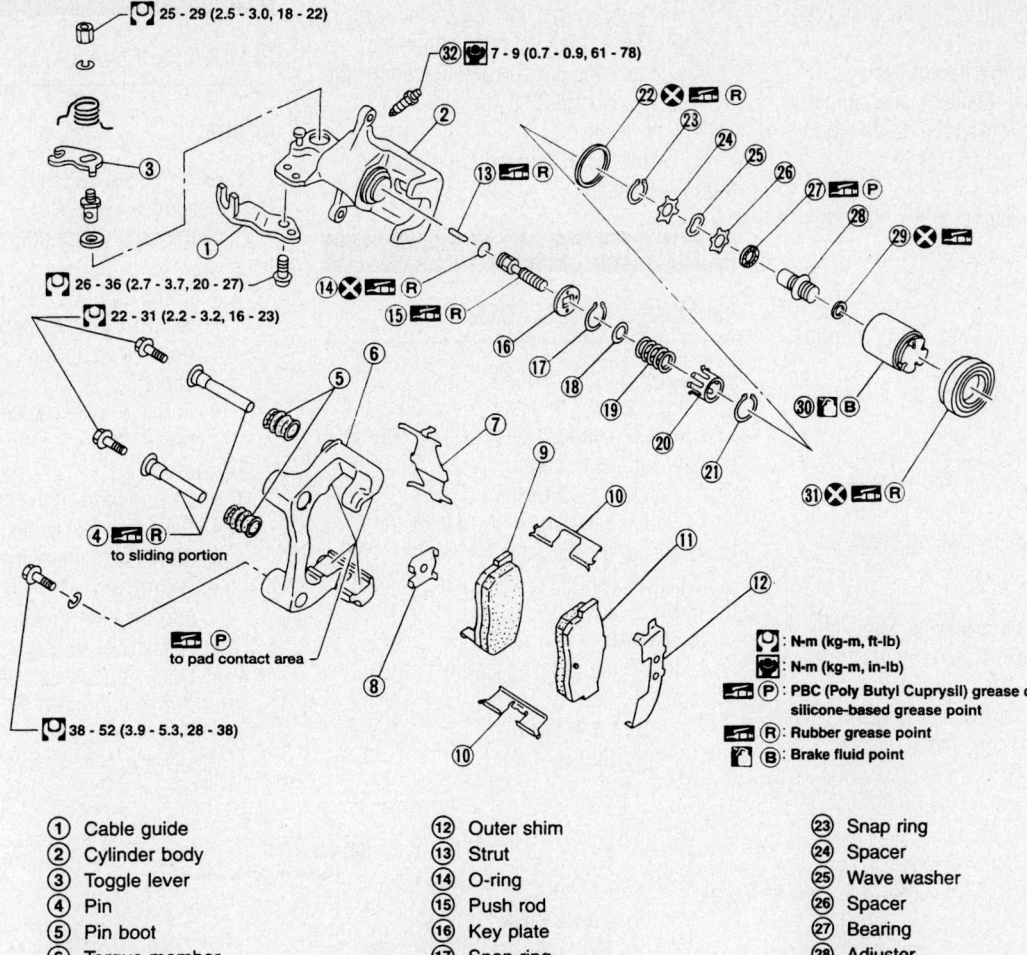

25 - 29 (2.5 - 3.0, 18 - 22)

32 ⚙ 7 - 9 (0.7 - 0.9, 61 - 78)

26 - 36 (2.7 - 3.7, 20 - 27)

22 - 31 (2.2 - 3.2, 16 - 23)

to sliding portion

(P) to pad contact area

38 - 52 (3.9 - 5.3, 28 - 38)

🔧 : N·m (kg-m, ft-lb)
⚙ : N·m (kg-m, in-lb)
(P) : PBC (Poly Butyl Cuprysil) grease or silicone-based grease point
(R) : Rubber grease point
(B) : Brake fluid point

① Cable guide
② Cylinder body
③ Toggle lever
④ Pin
⑤ Pin boot
⑥ Torque member
⑦ Retainer
⑧ Inner shim
⑨ Inner pad
⑩ Pad retainer
⑪ Outer pad
⑫ Outer shim
⑬ Strut
⑭ O-ring
⑮ Push rod
⑯ Key plate
⑰ Snap ring
⑱ Seat
⑲ Spring
⑳ Spring cover
㉑ Snap ring
㉒ Piston seal
㉓ Snap ring
㉔ Spacer
㉕ Wave washer
㉖ Spacer
㉗ Bearing
㉘ Adjuster
㉙ Cup
㉚ Piston
㉛ Piston boot
㉜ Air bleeder

93016G52

Rear disc brakes—Sentra shown

Disc Brake Pads

REMOVAL & INSTALLATION

Sentra

FRONT

1. Before servicing the vehicle, refer to the Precautions Section.
2. Remove or disconnect the following:
   - Wheels
   - Bottom guide pin from the caliper and swing the caliper cylinder body up
   - Brake pad retainers and the pads
To install:
3. Compress the piston of the disc brake caliper.
4. Install or connect the following:
   - Brake pads, shims, and retainers
   - Caliper assembly. Torque the guide pin to 23–30 ft. lbs. (31–41 Nm).
   - Wheels
5. Check the master cylinder and add fluid if necessary.

REAR

1. Before servicing the vehicle, refer to the Precautions Section.
2. Remove or disconnect the following:
   - Wheels
   - Parking brake cable bracket bolt
   - Pin bolts and lift off the caliper body
   - Pad springs and the pads and shims
To install:
3. Turn the piston clockwise back into the caliper body. Take care not to damage the piston boot.

4. Coat the pad contact area on the mounting support with a silicone based grease.
5. Install or connect the following:
   - Pads, shims and retainer springs
   - Caliper body into position in the mounting support and tighten the pin bolts to 28–38 ft. lbs. (38–52 Nm)
   - Wheels and bleed the system if necessary.

Altima

FRONT

1. Before servicing the vehicle, refer to the Precautions Section.
2. Remove or disconnect the following:
   - Wheels
   - Bottom guide pin from the caliper and swing the caliper cylinder body up
   - Brake pad retainers and the pads

To install:

3. Compress the piston of the disc brake caliper.

4. Install or connect the following:
- Brake pads, retainers, and caliper assembly. Torque the guide pin to 16–23 ft. lbs. (22–31 Nm).
- Wheels

5. Check the master cylinder and add fluid if necessary.

REAR

1. Before servicing the vehicle, refer to the Precautions Section.

2. Remove or disconnect the following:
- Rear wheels
- Parking brake cable bracket bolt
- Pin bolts and lift off the caliper body
- Pad springs by pulling them out
- Pads and shims

To install:

3. Turn the piston clockwise back into the caliper body. Take care not to damage the piston boot.

4. Coat the pad contact area on the mounting support with a silicone based grease.

5. Install or connect the following:
- Pads, shims, and the pad springs
- Caliper body into position in the mounting support and tighten the pin bolts to 16–23 ft. lbs. (22–31 Nm)
- Wheels

6. Check the master cylinder and add fluid if necessary.

Brake Drums

REMOVAL & INSTALLATION

Sentra

1. Before servicing the vehicle, refer to the Precautions Section.

2. Remove the wheels.

3. Remove the brake drum from the brake shoes. If necessary, two 8mm x 1.25 bolts can be used to press the drum from the hub.

To install:

4. Install the drum assembly to the vehicle.

5. Install the wheels.

6. Adjust the rear brakes.

Brake Shoes

REMOVAL & INSTALLATION

Sentra

1. Before servicing the vehicle, refer to the Precautions Section.

2. Remove or disconnect the following:
- Brake drum
- Return springs, adjuster assembly, hold-down springs, and brake shoes
- Parking brake cable from the toggle lever

To install:

3. Install or connect the following:
- Parking brake cable
- Shoes with the hold-down springs
- Return springs, by hooking them into the new shoes
- Adjuster assembly
- Drums and wheels. Adjust the brakes and bleed the hydraulic system, if necessary.

4. Check the parking brake adjustment.

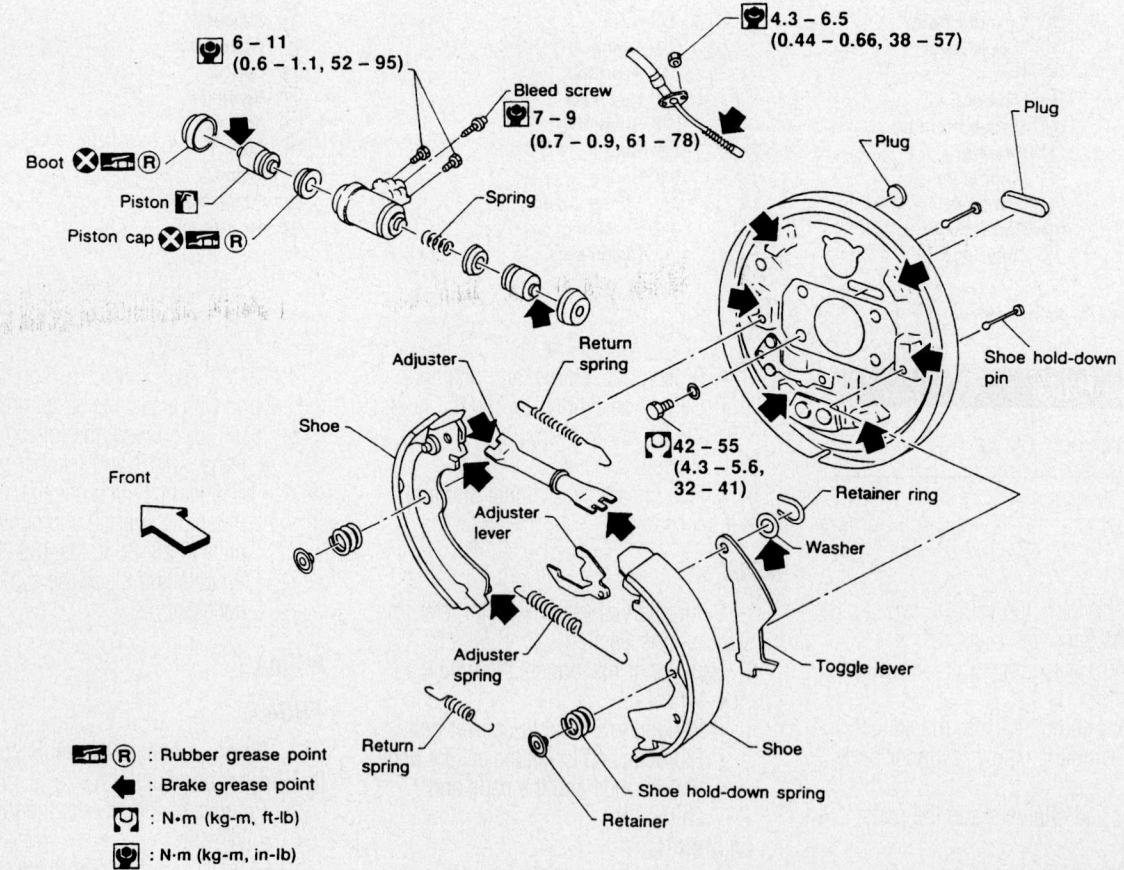

Exploded view of the rear drum brakes—Sentra

93016G53

SPECIFICATIONS AND MAINTENANCE CHARTS

ENGINE AND VEHICLE IDENTIFICATION

		Engine						Model Year	
Code ①	Liters (cc)	Cu. In.	Cyl.	Fuel Sys.	Engine Type	Eng. Mfg.	Code ②		Year
VQ35DE	3.5 (3498)	213	6	MFI	DOHC	Nissan	2		2002
							3		2003
							4		2004
							5		2005
							6		2006

MFI: Multi-port Fuel Injection

DOHC: Double Overhead Camshaft

① The Engine Code is stamped on the engine block near the starter.

② 10th position of the Vehicle Identification Number (VIN)

09482_ZMAX_C0001

GENERAL ENGINE SPECIFICATIONS

Year	Model	Engine Displacement Liters	Engine Series (ID/VIN)	Net Horsepower @ rpm	Net Torque @ rpm (ft. lbs.)	Bore x Stroke (in.)	Com-pression Ratio	Oil Pressure @ rpm
2002	Maxima	3.5	VQ35DE	260@6000	260@4800	3.76X3.20	10.3:1	43@2000
2003	Maxima	3.5	VQ35DE	260@6000	260@4800	3.76X3.20	10.3:1	43@2000
	350Z	3.5	VQ35DE	260@6000	260@4800	3.76X3.20	10.3:1	43@2000
2004	Maxima	3.5	VQ35DE	260@6000	260@4800	3.76X3.20	10.3:1	43@2000
	350Z	3.5	VQ35DE	260@6000	260@4800	3.76X3.20	10.3:1	43@2000
2005	Maxima	3.5	VQ35DE	260@6000	260@4800	3.76X3.20	10.0:1	43@2000
	350Z	3.5	VQ35DE	260@6000	260@4800	3.76X3.20	10.3:1	43@2000
2006	Maxima	3.5	VQ35DE	260@6000	260@4800	3.76X3.20	10.0:1	43@2000
	350Z	3.5	VQ35DE	260@6000	260@4800	3.76X3.20	10.3:1	43@2000

09482_ZMAX_C0002

ENGINE TUNE-UP SPECIFICATIONS

Year	Engine Displacement Liters	Engine ID/VIN	Spark Plug Gap (in.)	Ignition Timing (deg.) MT	AT	Fuel Pump (psi) ①	Idle Speed (rpm) MT	AT ②	Valve Clearance (in.) Intake ③	Exhaust ③
2002	3.5	VQ35DE	0.043	15B	15B	34	575-675	625-725	0.010-0.013	0.011-0.015
2003	3.5	VQ35DE	0.043	15B	15B	51	④	⑤	0.010-0.013	0.011-0.015
2004	3.5	VQ35DE	0.043	15B	15B	51	④	⑤	0.010-0.013	0.011-0.015
2005	3.5	VQ35DE	0.043	15B	15B	51	④	⑤	0.010-0.013	0.011-0.015
2006	3.5	VQ35DE	0.043	15B	15B	51	④	⑤	0.010-0.013	0.011-0.015

NOTE: The Vehicle Emission Control Information label often reflects specification changes made during production.

The label figures must be used if they differ from those in this chart.

B: Before top dead center

① System pressure at idle with vacuum hose connected; should increase to 43 psi when disconnected

② Automatic transmission in neutral

③ Engine cold

④ Maxima: 575-675 rpm; 350Z: 600-700 rpm

⑤ Maxima: 625-725 rpm; 350Z: 600-700 rpm

09482_ZMAX_C0003

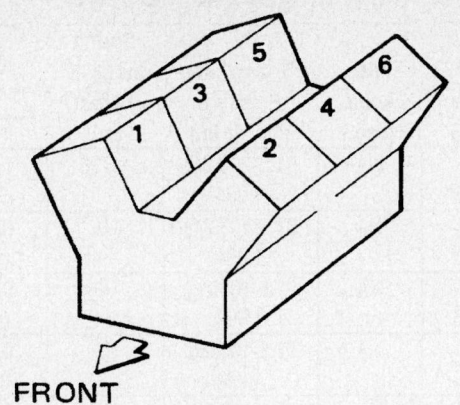

FRONT

67162-NISS-G01

3.5L Engines
Firing order: 1–2–3–4–5–6
Distributorless ignition system (one coil on each cylinder)

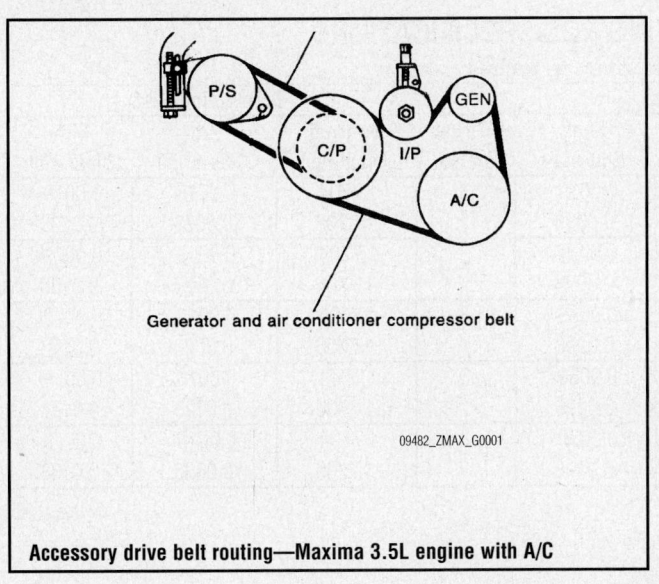

Generator and air conditioner compressor belt

09482_ZMAX_G0001

Accessory drive belt routing—Maxima 3.5L engine with A/C

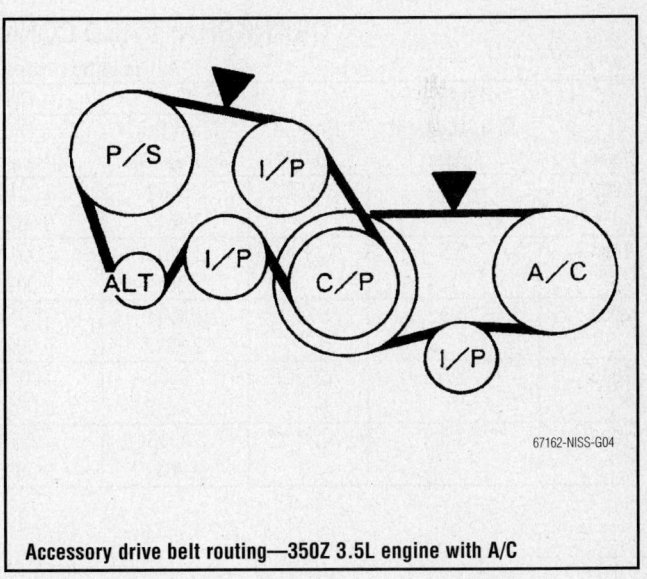

67162-NISS-G04

Accessory drive belt routing—350Z 3.5L engine with A/C

CAPACITIES

Year	Model	Engine ID/VIN	Engine Displacement Liters	Engine Oil with Filter (qts.)	Transmission (pts.) 5-Spd	Auto.	Drive Axle Rear (pts.)	Fuel Tank (gal.)	Cooling System (qts.)
2002	Maxima	VQ35DE	3.5	4.25	4.8	18.0	—	18.5	8.2
2003	Maxima	VQ35DE	3.5	4.25	4.8	18.0	—	18.5	8.2
	350Z	VQ35DE	3.5	5.0	6.1	21.2	3.0	20.0	9.25
2004	Maxima	VQ35DE	3.5	4.25	4.3	①	—	20.0	7.8
	350Z	VQ35DE	3.5	5.0	6.1	21.2	3.0	20.0	9.25
2005	Maxima	VQ35DE	3.5	4.5	4.5	15.5	—	20.0	9.0
	350Z	VQ35DE	3.5	5.0	6.1	21.2	3.0	20.0	9.25
2006	Maxima	VQ35DE	3.5	4.5	4.5	15.5	—	20.0	9.0
	350Z	VQ35DE	3.5	5.0	6.1	21.2	3.0	20.0	9.25

NOTE: All capacities are approximate. Add fluid gradually and check to be sure a proper fluid level is obtained.

① 4-speed A/T: 19.5 pts.; 5-speed A/T: 15.5 pts.

VALVE SPECIFICATIONS

Year	Engine ID/VIN	Engine Displacement Liters	Seat Angle (deg.)	Face Angle (deg.)	Spring Test Pressure (lbs. @ in.)	Spring Installed Height (in.)	Stem-to-Guide Clearance (in.)		Stem Diameter (in.)	
							Intake	Exhaust	Intake	Exhaust
2002	VQ35DE	3.5	45.15-45.45	NA	91.5-103.2@1.094	1.796	0.010-0.013	0.011-0.015	0.2348-0.2354	0.2344-0.2350
2003	VQ35DE	3.5	45.15-45.45	NA	91.5-103.2@1.094	NA	0.010-0.013	0.011-0.015	0.2348-0.2354	0.2344-0.2350
2004	VQ35DE	3.5	45.15-45.45	NA	91.5-103.2@1.094	NA	0.010-0.013	0.011-0.015	0.2348-0.2354	0.2344-0.2350
2005	VQ35DE	3.5	45.15-45.45	NA	91.5-103.2@1.094	NA	0.010-0.013	0.011-0.015	0.2348-0.2354	0.2344-0.2350
2006	VQ35DE	3.5	45.15-45.45	NA	91.5-103.2@1.094	NA	0.010-0.013	0.011-0.015	0.2348-0.2354	0.2344-0.2350

NA: Not Available

09482_ZMAX_C0005

CRANKSHAFT AND CONNECTING ROD SPECIFICATIONS

All measurements are given in inches.

Year	Engine Displacement Liters	Engine ID/VIN	Crankshaft				Connecting Rod		
			Main Brg. Journal Dia.	Main Brg. Oil Clearance	Shaft End-play	Thrust on No.	Journal Diameter	Oil Clearance	Side Clearance
2002	3.5	VQ35DE	2.3603-2.3612	0.0014-0.0021	0.0039-0.0098	3	1.7704-1.7706	0.0013-0.0023	0.0079-0.0138
2003	3.5	VQ35DE	2.3603-2.3612	0.0014-0.0021	0.0039-0.0098	3	1.7704-1.7706	0.0013-0.0023	0.0079-0.0138
2004	3.5	VQ35DE	2.3603-2.3612	0.0014-0.0021	0.0039-0.0098	3	1.7704-1.7706	0.0013-0.0023	0.0079-0.0138
2005	3.5	VQ35DE	2.3603-2.3612	0.0014-0.0021	0.0039-0.0098	3	1.7704-1.7706	0.0013-0.0023	0.0079-0.0138
2006	3.5	VQ35DE	2.3603-2.3612	0.0014-0.0018	0.0039-0.0098	3	1.7704-1.7706	0.0013-0.0023	0.0079-0.0138

09482_ZMAX_C0006

PISTON AND RING SPECIFICATIONS

All measurements are given in inches.

Year	Engine Displacement Liters	Engine ID/VIN	Piston Clearance	Ring Gap			Ring Side Clearance		
				Top Compression	Bottom Compression	Oil Control	Top Compression	Bottom Compression	Oil Control
2002	3.5	VQ35DE	0.0004-0.0012	0.0091-0.0130	0.0130-0.0189	0.0079-0.0197	0.0018-0.0031	0.0012-0.0028	0.0026-0.0053
2003	3.5	VQ35DE	0.0004-0.0012	0.0091-0.0130	0.0130-0.0189	0.0079-0.0197	0.0018-0.0031	0.0012-0.0028	0.0026-0.0053
2004	3.5	VQ35DE	0.0004-0.0012	0.0091-0.0130	0.0130-0.0189	0.0079-0.0197	0.0018-0.0031	0.0012-0.0028	0.0026-0.0053
2005	3.5	VQ35DE	0.0004-0.0012	0.0091-0.0130	0.0130-0.0189	0.0079-0.0197	0.0018-0.0031	0.0012-0.0028	0.0026-0.0053
2006	3.5	VQ35DE	0.0004-0.0012	0.0091-0.0130	0.0130-0.0189	0.0079-0.0197	0.0018-0.0031	0.0012-0.0028	0.0026-0.0053

09482_ZMAX_C0007

TORQUE SPECIFICATIONS

All readings in ft. lbs.

Year	Engine Displacement Liters (cc)	Engine ID/VIN	Cylinder Head Bolts	Main Bearing Bolts	Rod Bearing Bolts	Crankshaft Damper Bolts	Flywheel Bolts	Manifold Intake	Manifold Exhaust	Spark Plugs	Oil Drain Plug
2002	3.5 (3498)	VQ35DE	①	②	③	④	61-69	⑤	21-24	18	25
2003	3.5 (3498)	VQ35DE	①	②	③	④	61-69	⑤	⑥	18	25
2004	3.5 (3498)	VQ35DE	①	②	③	④	61-69	⑤	⑥	18	25
2005	3.5 (3498)	VQ35DE	①	②	③	④	61-69	⑤	⑥	18	25
2006	3.5 (3498)	VQ35DE	①	②	③	④	61-69	⑤	⑥	18	25

① Step 1: 72 ft. lbs.
Step 2: Loosen bolts completely
Step 3: 26-32 ft. lbs.
Step 4: Tighten an additional 90-95 degrees
Step 5: Repeat Step 4

② Step 1: Shift crankshaft to align the bearing beam
Step 2: Tighten all bolts to 24-28 ft. lbs.
Step 3: Tighten an additional 90-95 degrees

③ Step 1: Tighten to 15 ft. lbs.
Step 2: Tighten an additional 90-95 degrees

④ Step 1: 29-36 ft. lbs.
Step 2: 60-66 degrees

⑤ Step 1: Tighten to 4-7 ft. lbs.
Step 2: Tighten to 20-23 ft. lbs.
Step 3: Tighten, again, to 20-23 ft. lbs.

⑥ Maxima: 21-24 ft. lbs.
350Z: 22 ft. lbs.

09482_ZMAX_C0008

WHEEL ALIGNMENT

Year	Model		Caster Range (+/-Deg.)	Caster Preferred Setting (Deg.)	Camber Range (+/-Deg.)	Camber Preferred Setting (Deg.)	Toe-in (in.)
2002	Maxima ①	F	0.75	+2.75	0.75	-0.25	0.04 +/- 0.04
		R	—	—	0.75	-1.00	0.04 +/- 0.04
	Maxima ②	F	0.75	+2.75	0.75	-0.33	0.04 +/- 0.04
		R	—	—	0.75	-1.00	0.04 +/- 0.04
2003	Maxima ①	F	0.75	+2.75	0.75	-0.25	0.04 +/- 0.04
		R	—	—	0.75	-1.00	0.04 +/- 0.04
	Maxima ②	F	0.75	+2.75	0.75	-0.33	0.04 +/- 0.04
		R	—	—	0.75	-1.00	0.04 +/- 0.04
	350Z	F	0.75	+8.17	0.75	-0.58	0.04 +/- 0.04
		R	—	—	0.50	+1.58	③
2004	Maxima	F	0.75	+2.83	0.75	-0.25	0.02 +/- 0.04
		R	—	—	0.50	-0.67	0.16 +/- 0.06
	350Z	F	0.75	+8.17	0.75	-0.58	0.04 +/- 0.04
		R	—	—	0.50	+1.58	③
2005	Maxima	F	0.75	+2.83	0.75	-0.25	0.02 +/- 0.04
		R	—	—	0.50	-0.67	0.16 +/- 0.06
	350Z	F	0.75	+8.17	0.75	-0.58	0.04 +/- 0.04
		R	—	—	0.50	+1.58	⑦
2006	Maxima	F	0.75	+2.83	0.75	-0.25	0.02 +/- 0.04
		R	—	—	0.50	-0.67	0.16 +/- 0.06
	350Z	F	0.75	+8.17	0.75	-0.58	0.04 +/- 0.04
		R	—	—	0.50	+1.58	③

① With P225/55R16, P215/55R16 tires

② With P205/65R15 tires

③ 0.04 +/- 0.03 with 17 inch tire
0.07 +/- 0.03 with 18 inch tire

09482_ZMAX_C0009

TIRE, WHEEL AND BALL JOINT SPECIFICATIONS

Year	Model	OEM Tires Standard	OEM Tires Optional	Tire Pressures (psi) Front	Tire Pressures (psi) Rear	Wheel Size	Lug Nut Torque Ft. Lbs.
2002	Maxima GLE	P215/55R16	None	29	29	6.5-JJ	80
	Maxima GXE	P205/65SR15	None	29	29	6JJ	80
	Maxima SE	P215/55R16	P225/50R17	29	29	6.5J/7J	80
2003	Maxima GLE	P215/55R16	None	29	29	6.5-JJ	80
	Maxima GXE	P205/65SR15	None	29	29	6JJ	80
	Maxima SE	P215/55R16	P225/50R17	29	29	6.5J/7J	80
	350Z Front	225/50R17	225/45R18	26	26	7.5-JJ/8-JJ	80
	350Z Rear	235/50R17	245/45R18	26	26	8-JJ/8.5-JJ	80
2004	Maxima	P225/55VR17	245/45VR18	29	29	7-JJ/7.5-JJ	80
	350Z Front	225/50R17	225/45R18	26	26	7.5-JJ/8-JJ	80
	350Z Rear	235/50R17	245/45R18	26	26	8-JJ/8.5-JJ	80
2005	Maxima	P225/55VR17	245/45VR18	29	29	7-JJ/7.5-JJ	80
	350Z Front	225/50R17	225/45R18	26	26	7.5-JJ/8-JJ	80
	350Z Rear	235/50R17	245/45R18	26	26	8-JJ/8.5-JJ	80
2006	Maxima	P225/55VR17	245/45VR18	29	29	7-JJ/7.5-JJ	80
	350Z Front	225/50R17	225/45R18	26	26	7.5-JJ/8-JJ	80
	350Z Rear	235/50R17	245/45R18	26	26	8-JJ/8.5-JJ	80

OEM: Original Equipment Manufacturer

PSI: Pounds Per Square Inch

09482_ZMAX_C0010

BRAKE SPECIFICATIONS
All measurements in inches unless noted

Year	Model		Original Thickness	Minimum Thickness	Maximum Run-out	Front	Rear	Bracket Bolts (ft. lbs.)	Mounting Bolts (ft. lbs.)
2002	Maxima	F	0.940	0.866	0.003	0.079	—	53-72	23
		R	0.350	0.315	0.003	—	0.059	—	—
2003	Maxima	F	0.940	0.866	0.003	0.079	—	53-72	23
		R	0.350	0.315	0.003	—	0.059	—	—
	350Z ①	F	0.945	0.866	0.002	0.079	—	113-114	17-22
		R	0.630	0.551	0.004	—	0.079	53-71	13-14
	350Z ②	F	1.181	1.118	0.002	0.079	—	—	112
		R	0.866	0.795	0.002	—	0.079	53-71	—
2004	Maxima	F	1.100	0.1.02	0.003	0.079	—	53-72	23
		R	0.350	0.315	0.002	—	0.059	101-129	53-71
	350Z ①	F	0.945	0.866	0.002	0.079	—	113-114	17-22
		R	0.630	0.551	0.004	—	0.079	53-71	13-14
	350Z ②	F	1.181	1.118	0.002	0.079	—	—	112
		R	0.866	0.795	0.002	—	0.079	53-71	—
2005	Maxima	F	1.100	0.1.02	0.003	0.079	—	53-72	23
		R	0.350	0.315	0.002	—	0.059	101-129	53-71
	350Z ①	F	0.945	0.866	0.002	0.079	—	113-114	17-22
		R	0.630	0.551	0.004	—	0.079	53-71	13-14
	350Z ②	F	1.181	1.118	0.002	0.079	—	—	112
		R	0.866	0.795	0.002	—	0.079	53-71	—
2006	Maxima	F	1.100	0.1.02	0.003	0.079	—	53-72	23
		R	0.350	0.315	0.002	—	0.059	101-129	53-71
	350Z ①	F	0.945	0.866	0.002	0.079	—	113-114	17-22
		R	0.630	0.551	0.004	—	0.079	53-71	13-14
	350Z ②	F	1.181	1.118	0.002	0.079	—	—	112
		R	0.866	0.795	0.002	—	0.079	53-71	—
	350Z ③	F	1.102	1.024	0.001	0.079	—	—	113
		R	0.630	0.551	0.002	—	0.079	53-71	—

NA: Not Available

① Front brake model CLZ25VD, rear brake model AD14VE.

② Front brake model OPB27VA/Brembo, rear brake model OPB13VB/Brembo.

③ Except CLZ25VD/OPB27VA/Brembo caliper.

09482_ZMAX_C0011

SCHEDULED MAINTENANCE INTERVALS
Nissan—Maxima & 350Z

TO BE SERVICED	TYPE OF	VEHICLE MILEAGE INTERVAL (x1000)												
		7.5	15	22.5	30	37.5	45	52.5	60	67.5	75	82.5	90	97.5
Engine oil & filter	R	✓	✓	✓	✓	✓	✓	✓	✓	✓	✓	✓	✓	✓
Brake lines & cables	S/I		✓		✓		✓		✓		✓		✓	
Brake pads & discs	S/I		✓		✓		✓		✓		✓		✓	
Driveshaft boots	S/I		✓		✓		✓		✓		✓		✓	
Exhaust system	S/I				✓				✓				✓	
Transmission or transaxle fluid	S/I		✓		✓		✓		✓		✓		✓	
Air cleaner filter	R				✓				✓				✓	
Spark plugs (except platinum)	R				✓				✓				✓	
Spark plugs (platinum tip)	R								✓					
Steering gear & linkage, axle & suspension parts	S/I				✓				✓				✓	
Engine coolant	R								✓					
Drive belts	S/I								✓					
Fuel lines	S/I								✓					
Vapor lines	S/I								✓					

R: Replace   S/I: Service or Inspect

FREQUENT OPERATION MAINTENANCE (SEVERE SERVICE)

If a vehicle is operated under any of the following conditions it is considered severe service:

- Extremely dusty areas.

- 50% or more of the vehicle operation is in 32°C (90°F) or higher temperatures, or constant operation in temperatures below 0°C (32°F).

- Prolonged idling (vehicle operation in stop and go traffic).

- Frequent short running periods (engine does not warm to normal operating temperatures).

- Police, taxi, delivery usage or trailer towing usage.

Oil & oil filter: change every 3750 miles.

Brake pads & discs: service or inspect every 7500 miles.

Driveshaft boots: service or inspect every 7500 miles.

Exhaust system: service or inspect every 7500 miles.

Steering gear & linkage, axle & suspension parts: service or inspect every 7500 miles.

Steering linkage ball joints & front suspension ball joints: service or inspect every 7500 miles.

Air cleaner filter: service or inspect every 15,000 miles.

09482_ZMAX_C0012

ENGINE REPAIR

Distributor

The Nissan 3.5L engine is equipped with a Distributorless Ignition System (DIS).

Alternator

REMOVAL

1. Before servicing the vehicle, refer to the Precautions Section.
2. Remove or disconnect the following:
- Negative battery cable
- Right side engine undercover and side inspection cover
- Radiator on Maxima
- Radiator fan on 350Z
- Drive belt
- Alternator and A/C compressor harness connectors
- Upper and lower alternator bolts
- Alternator

INSTALLATION

Install or connect the following:
- Alternator
- Upper and lower alternator bolts. On Maxima tighten the upper bolt to 12–15 ft. lbs. (16–20 Nm) and the lower bolt to 32–38 ft. lbs. (44–52 Nm). On 350Z, tighten the upper bolt to 48 ft. lbs. (65 Nm) and the lower bolts to 21 ft. lbs. (28 Nm).
- Alternator and A/C compressor harness connectors
- Drive belt
- Radiator on Maxima
- Radiator fan on 350Z
- Right side engine undercover and side inspection cover
- Negative battery cable

Ignition Timing

ADJUSTMENT

➡The ignition timing is not adjustable. If not within specifications, further diagnostic inspection is required. The following procedure is for viewing the ignition timing setting.

Visually check the air cleaner, intake hoses, ducts, Exhaust Gas Recirculation (EGR) valve operation and electrical connections prior to the adjustment of the ignition timing. Correct or repair any problem as required. Be sure to inspect the throttle valve and Throttle Position (TP) sensor for proper operation.

1. Before servicing the vehicle, refer to the Precautions Section.
2. Locate the timing marks on the crankshaft pulley and the front of the engine.
3. Clean the timing marks.

➡The ignition timing specification is 15 +/- 5 degrees Before Top Dead Center (BTDC).

4. Using chalk or white paint, color the mark on the crankshaft pulley and the mark on the scale, that will indicate the correct timing when aligned with the notch on the crankshaft pulley.
5. Attach a tachometer to the engine.
6. Attach a timing light to the engine to number 1 cylinder ignition wire.
7. Turn all electrical equipment and accessories **OFF**.
8. Check to be sure all of the wires clear the fan, then, start the engine and allow it to reach normal operating temperatures.
9. Block the front wheels and set the parking brake. Shift the transmission into **NEUTRAL** for manual transmission and automatic transmissions. Do not stand in front of the vehicle when making adjustments.
10. Perform the following procedures:
    a. Race the engine at 2000 rpm for about 2 minutes under a no-load condition; be sure all of the accessories are turned **OFF**.
    b. Perform on board engine diagnostics and repair any fault code.
    c. Race the engine at 2000 rpm for about 2 minutes under a no-load condition.
    d. Turn the engine **OFF** and disconnect the TP sensor.
    e. Start and race the engine 2–3 times under no-load, then run the engine at idle speed.

➡The ignition timing specification is 15 +/- 5 degrees BTDC.

11. Aim the timing light at the timing marks. If the marks on the pulley and the engine are aligned when the light flashes, the timing is correct. Turn the engine **OFF** and remove the tachometer and the timing light. If the marks are not in alignment, proceed with the following steps.
12. Turn the engine **OFF**.
13. Check the Camshaft Position (CMP) sensor (PHASE), Crankshaft Position (CKP) sensor (REF) and CKP sensor (POS). Replace if necessary.
14. If the ignition timing is still not correct, substitute a known good Electronic Control Module (ECM).

➡The ECM may be the cause of the problem but this is rarely the case.

15. Turn the engine **OFF** and remove the tachometer and the timing light.

Engine Assembly

REMOVAL & INSTALLATION

Maxima

It is recommended the engine and transaxle be removed as a single unit. If need be, the units may be separated after removal.

➡The engine and transaxle assembly must be removed from the underside of the vehicle.

1. Before servicing the vehicle, refer to the Precautions Section.
2. Release the fuel system pressure.
3. Drain the cooling system.
4. Drain the engine oil.
5. Drain the automatic transaxle, if equipped.
6. Remove or disconnect the following:
- Negative battery cable
- Hood
- Engine under cover
- All vacuum hoses, fuel lines, wires and connectors; tag before disconnecting
7. If necessary, remove the wiper motor and linkage as follows:
- Operate wiper motor one full cycle, then turn off
- Remove the wiper arm nut covers, then the nuts
- Remove the wiper arms
- Remove fender covers
- Remove cowl top clips, then partially lift cowl top for access
- Disconnect windshield washer tube
- Remove cowl top seal and weatherstrip seal
- Remove cowl top cover
- Disconnect wiper motor connector
- Remove bracket and wiper motor assembly
8. Remove or disconnect the following:
- Front exhaust pipe from the manifold

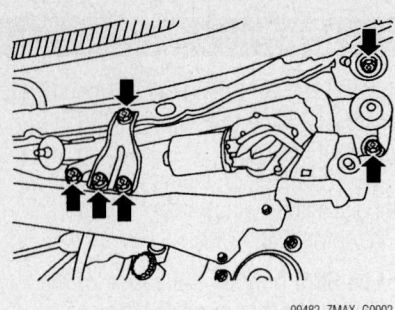

09482_ZMAX_G0002

**Removal of motor and bracket assembly—
Maxima**

- Ball joints from the steering knuckle
- Halfshafts
- Radiator and fans
- Drive belts
- Alternator
- A/C compressor. Position it aside with the lines attached. Do NOT disconnect the refrigerant lines.
- Power steering pump and position aside with the lines attached. Do NOT disconnect the fluid lines.

9. Place a suitable jack under the transaxle. Install engine slingers and a suitable engine hoist. Raise the engine for access to the left side engine mount.

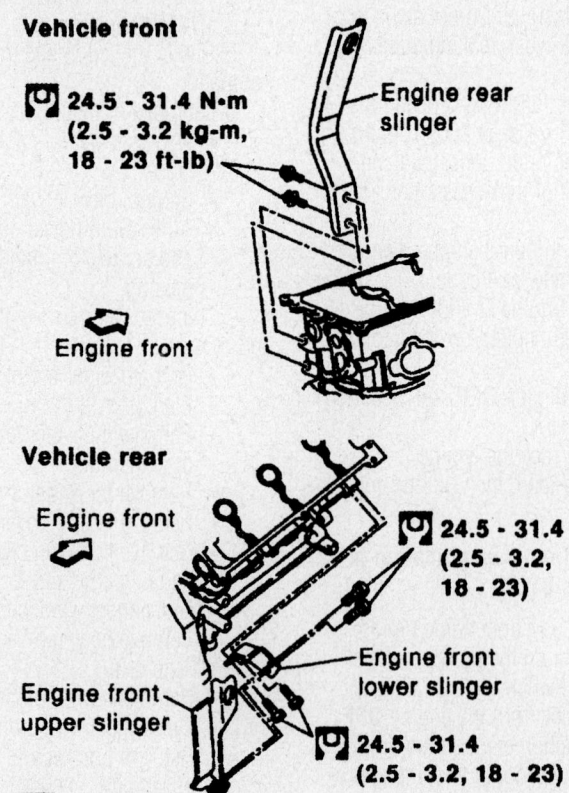

Vehicle front

⬛ 24.5 - 31.4 N•m
(2.5 - 3.2 kg-m,
18 - 23 ft-lb)

Engine rear slinger

Engine front

Vehicle rear

Engine front

Engine front upper slinger

Engine front lower slinger

⬛ 24.5 - 31.4
(2.5 - 3.2,
18 - 23)

⬛ 24.5 - 31.4
(2.5 - 3.2, 18 - 23)

⬛ : N•m (kg-m, ft-lb)

9357RG01

Installation of engine slingers to lift the engine

- Left side engine mount
- Control and support rods from the transaxle, manual transaxle only
- Control cable from the transaxle, automatic transaxle only
- Right side engine mount
- Center member, then carefully and slowly lower the transmission jack

10. Lower the engine/transaxle assembly onto an engine stand.

➡ **When lowering the engine out, guide it carefully to avoid hitting any other components.**

To install:

11. Installation is the reverse of the removal procedure, noting the following points:

a. If equipped with electronically controlled engine mounts, install them to the specifications shown in the accompanying figure.

b. Make sure to connect all vacuum hoses, lines, and electrical connectors as tagged during removal.

c. Fill the cooling system.

d. Fill the engine with clean oil.

e. Start the vehicle, check for leaks and repair if necessary.

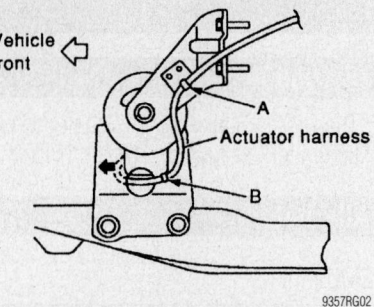

Vehicle front

A

Actuator harness

B

9357RG02

For electronically controlled engine mounts, the proper length from A to B is 6.69 in. (170mm)

350Z

It is recommended the engine and transaxle be removed as a single unit. If need be, the units may be separated after removal.

➡ **The engine and transaxle assembly must be removed from the underside of the vehicle.**

1. Before servicing the vehicle, refer to the Precautions Section.
2. Release the fuel system pressure.
3. Drain the cooling system.
4. Drain the engine oil.
5. Drain the automatic transaxle, if equipped.
6. Discharge and recover the A/C refrigerant
7. Remove or disconnect the following:
   - Negative battery cable
   - Hood
   - Strut tower bar
   - Engine under cover
   - Wiper arms and cowl top
   - All vacuum hoses, fuel lines, wires and connectors; tag before disconnecting
   - Front wheels
   - Air cleaner case and duct
   - Cooling fan, reservoir and hoses
   - Heater hoses
   - Battery ground at cylinder head
   - Battery positive cable harness
   - A/C lines from compressor
   - 2 body ground cables
   - Fuel feed and EVAP hoses
   - Power steering pump and lines and wire to engine
8. From inside the passenger side of the vehicle remove the following:
   - Kick panel
   - Dash side finish panel
   - Lower instrument panel cover
   - ECM and TCM harness connectors
9. Pull the connectors out of the pas-

senger side into the engine compartment and secure them to the engine.

10. Remove or disconnect the following:
- Front exhaust pipe from the manifold
- Steering column lower shaft
- Propeller shaft
- Shift lever and clutch slave cylinder on man. trans. models
- Automatic transmission control rod
- Upper rear oil pan plate
- Front stabilizer bar
- Steering outer socket from steering knuckle
- Front transverse link
- Place a suitable jack under the front suspension member and transmission

- Rear engine crossmember bolts
- Front suspension member bolts and nuts, then carefully and slowly lower the jack

➡**When lowering the engine out, guide it carefully to avoid hitting any other components.**

To install:

11. Installation is the reverse of the removal procedure, noting the following points:

a. Tighten the engine mounts, brackets and mounting members to the specified torque as shown in the illustration. When tightening the engine brackets, tighten the upper bolts first.

b. Tighten the rear engine mounting brackets in the sequence shown.

c. Tighten the suspension member bolts to 80–93 ft. lbs. (108–127 Nm) on 2003–2005 models or 65 ft. lbs. (87.5 Nm) on 2006 models.

d. Tighten the strut tower bar bolts 23–25 ft. lbs. (30–34 Nm) on 2003–2005 models or 33 ft. lbs. (45 Nm) on 2006 models.

e. Make sure to connect all vacuum hoses, lines, and electrical connectors as tagged during removal.

f. Fill the cooling system.

g. Fill the engine with clean oil.

h. Start the vehicle, check for leaks and repair if necessary.

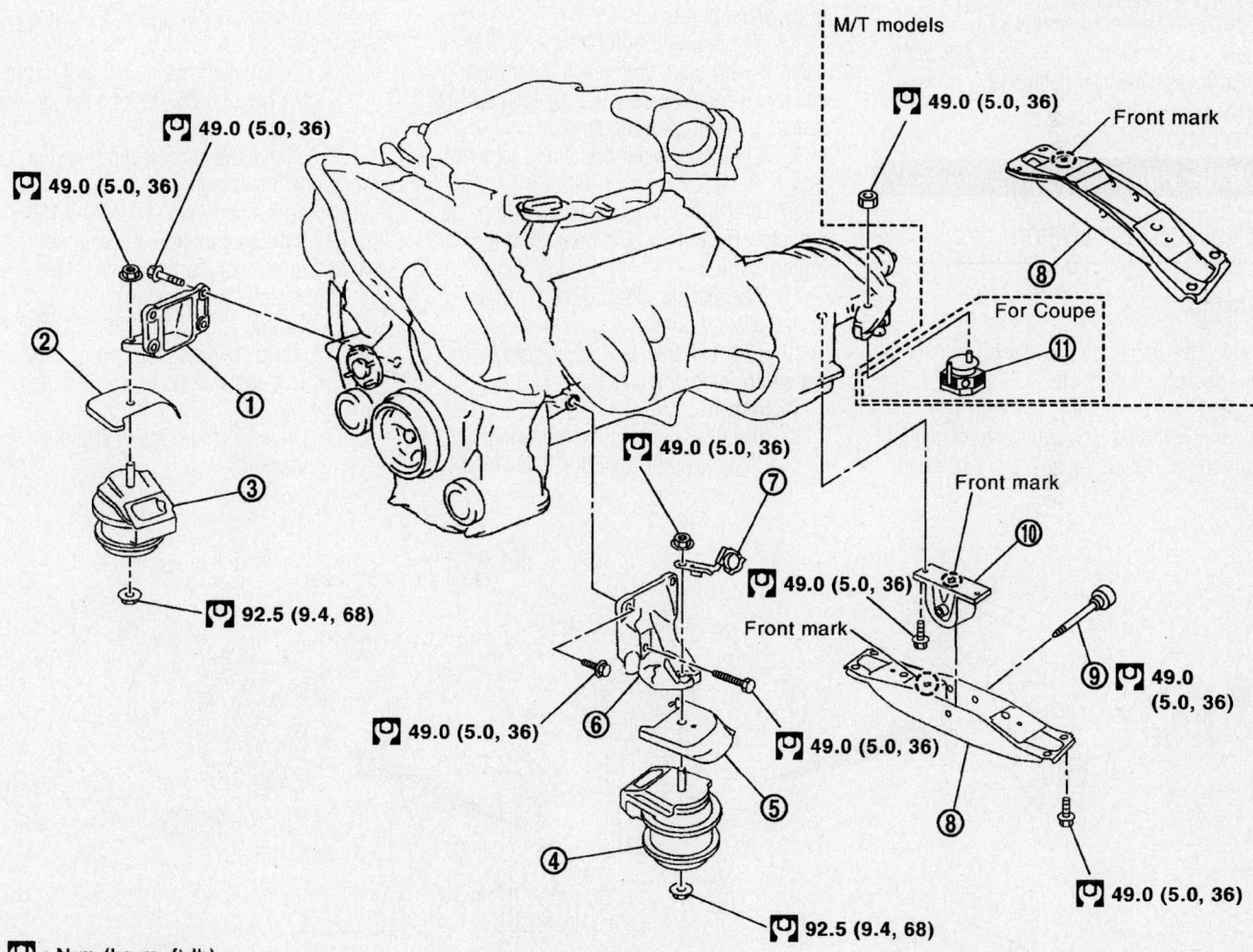

1. Engine mounting bracket (RH)
2. Heat insulator (RH)
3. Engine mounting Insulator (RH)
4. Engine mounting insulator (LH)
5. Heat insulator (LH)
6. Engine mounting bracket (LH)
7. Harness bracket
8. Rear engine mounting member
9. Mass damper
10. Engine mounting insulator (rear)
11. Dynamic damper

Exploded view of the engine mounting assemblies—350Z

67162-NISS-G05

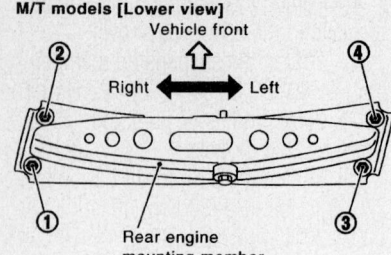

M/T models [Lower view]

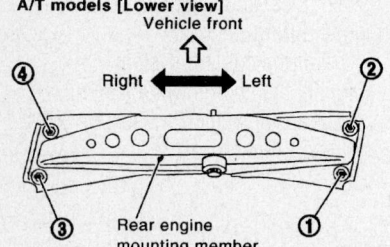

A/T models [Lower view]

67162-NISS-G06

Rear engine member tightening sequence—350Z

Water Pump

REMOVAL & INSTALLATION

Maxima

1. Before servicing the vehicle, refer to the Precautions Section.
2. Drain the cooling system.
3. Position a jack under the oil pan for support. Be sure to place a block of wood on the jack for protection to the engine parts.
4. Remove or disconnect the following:
   - Negative battery cable
   - Right side engine mount and bracket
   - Drive belts and the idler pulley bracket
   - Chain tensioner cover and the water pump cover
5. Push the timing chain tensioner sleeve and apply a stopper pin so it does not return.
   - Timing chain tensioner assembly
   - 3 bolts that secure the water pump
6. Rotate the crankshaft 20 degrees counterclockwise to provide timing chain slack.
7. Put M8 bolts in 2 M8 threaded holes of the water pump.
8. Tighten each bolt by turning alternately ½ turn until they reach the timing chain rear case. Be sure to turn each bolt ½ turn at a time to prevent damage.
9. Lift up the water pump and remove it.
10. When removing the water pump, do not allow the water pump gear to hit the timing chain.
11. Remove and discard the O-rings from the water pump.
12. Clean all traces of liquid gasket from the water pump and covers.

To install:

13. Install or connect the following:
    - Water pump using new O-rings to

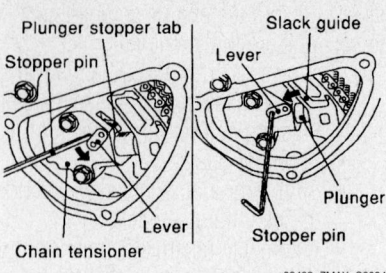

09482_ZMAX_G0004

Apply a stopper pin so the timing chain tensioner sleeve does not return—3.5L engine–Maxima

the engine block. Torque the 3 water pump mounting bolts evenly to 75–95 inch lbs. (8.5–10.7 Nm).

14. Rotate the crankshaft pulley to its original position by turning it 20 degrees clockwise.
    - Timing chain tensioner and torque the bolts to 75–89 inch lbs. (9–10 Nm)
15. Remove the stopper pin from the timing chain tensioner.
16. Apply a continuous 0.091–0.130 in. (2.3–3.3mm) bead of liquid sealant to the mating surfaces of the timing chain tensioner and water pump covers.
    - Timing chain tensioner and water pump covers to the engine block. Torque the bolts to 87–113 inch lbs. (10–13 Nm).
    - Drive belts and the idler pulley bracket

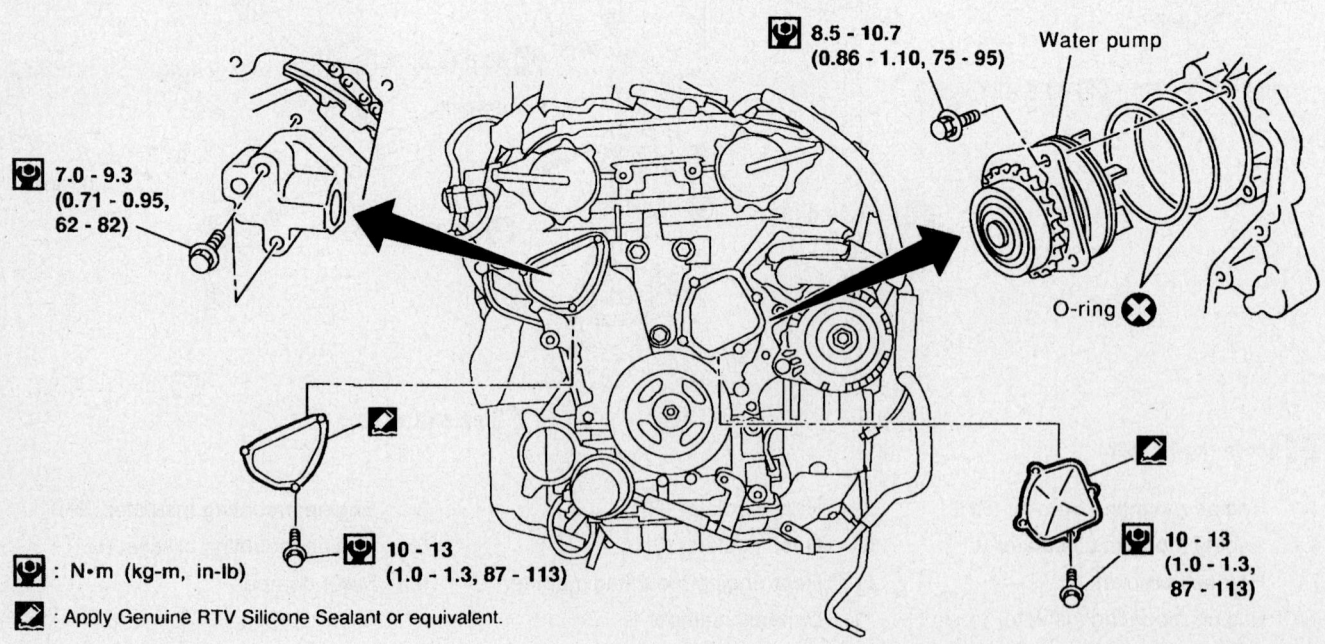

09482_ZMAX_G0003

Water pump and timing cover assembly—3.5L engine–Maxima

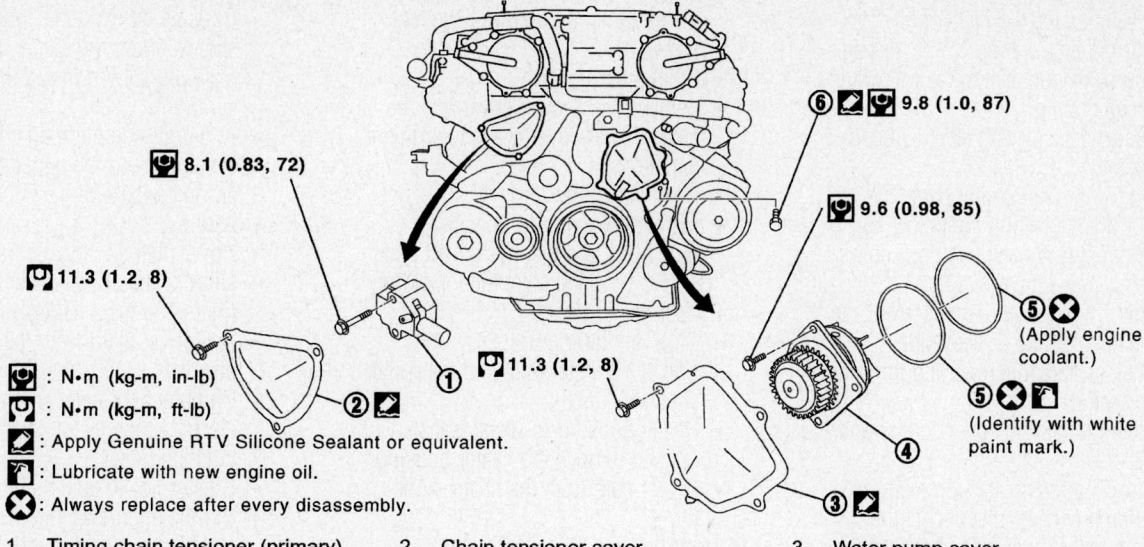

6 ⚆✏️⚒ 9.8 (1.0, 87)

⚒ 8.1 (0.83, 72)

9.6 (0.98, 85)

⚒ 11.3 (1.2, 8)

5 ✖️
(Apply engine coolant.)

⚒ : N•m (kg-m, in-lb)
⚆ : N•m (kg-m, ft-lb)
✏️ : Apply Genuine RTV Silicone Sealant or equivalent.
⚒ : Lubricate with new engine oil.
✖️ : Always replace after every disassembly.

⚆ 11.3 (1.2, 8)

5 ✖️ ⚒
(Identify with white paint mark.)

1. Timing chain tensioner (primary)
2. Chain tensioner cover
3. Water pump cover
4. Water pump
5. O- ring
6. Water drain plug (front)

67162-NISS-G08

Exploded view of water pump mounting—3.5L engine–350Z

- Right side engine mounting bracket and the engine mount
- Negative battery cable

17. Remove the jack from under the engine and install the drain plugs to the cylinder block.

18. Fill the cooling system.

19. Start the engine, check for leaks and repair if necessary.

350Z

1. Before servicing the vehicle, refer to the Precautions Section.

2. Drain the cooling system.

3. Remove or disconnect the following:
- Negative battery cable
- Accessory drive belts
- Radiator hoses
- Cooling fan
- Water drain plug on water pump side of block
- Timing chain tensioner cover
- Water pump cover
- Primary timing chain tensioner
- Water pump mounting bolts

4. Turn the crankshaft pulley counter-clockwise until the timing chain slack on the water pump pulley is at maximum.

5. Place M8 bolts in the upper and lower M8 threaded holes of the water pump.

6. Tighten each bolt by turning alternately ½ turn until they reach the timing chain rear case. Be sure to turn each bolt ½ turn at a time to prevent damage.

7. Lift up the water pump and remove it.

8. When removing the water pump, do not allow the water pump gear to hit the timing chain.

9. Remove and discard the O-rings from the water pump.

10. Clean all traces of liquid gasket from the water pump and covers.

To install:

11. Install the water pump using new O-rings to the engine block. Lubricate the inner O-ring with clean engine oil and the outer O-ring with engine coolant. Ensure the water pump sprocket and timing chain are engaged. Torque the 3 water pump mounting bolts evenly to 85 inch lbs. (10 Nm).

12. Rotate the crankshaft pulley clockwise so the timing chain on the tensioner side is loose.

13. Install the primary timing chain tensioner.

14. Apply a continuous 0.091–0.130 in. (2.3–3.3mm) bead of liquid sealant to the mating surfaces of the timing chain tensioner and water pump covers.

15. Install the timing chain tensioner and water pump covers to the engine block. Torque the bolts to 97 inch lbs. (11 Nm).

16. Install or connect the following:
- Water drain plug
- Cooling fan
- Radiator hoses
- Accessory drive belts
- Negative battery cable

17. Fill the cooling system.

18. Start the engine, check for leaks and repair if necessary.

Heater Core

REMOVAL & INSTALLATION

Maxima

2002–2003 MODELS

1. Before servicing the vehicle, refer to the Precautions Section.

2. Disconnect the negative battery terminal.

✳✳ CAUTION

After disconnecting the negative battery cable, wait for at least 3 minutes for the SRS modules to deplete its energy.

3. Drain the cooling system into a clean container for reuse.

4. Remove the air bag module and steering wheel by removing or disconnecting the following:
- Place the front wheels in the straight-ahead position
- Lower lid and disconnect the air bag electrical connector at the bottom of the steering wheel
- Side lids from both sides of the steering wheel
- Torx® bolts using a Torx® wrench T50 from both side of the steering wheel; then, discard the bolts
- Air bag module from the steering wheel
- Steering wheel nut

- Steering wheel from the steering column using a suitable puller

5. Disarm the passenger's side air bag by removing or disconnecting the following:
- Glove box lid
- Passenger's air bag electrical connector

6. Remove the instrument panel by removing or disconnecting the following:
- Upper and lower glove box screws and remove the glove box
- Lower instrument panel screws and the panel at the driver's side
- Knee protector screws and the knee protector
- Steering column cover screws and the cover
- Combination switch-to-steering column screws, disconnect the electrical connector and the combination switch
- Cluster lid "A" screws and the lid
- Combination meter screws, disconnect the electrical connectors and the combination meter
- Center ventilator with the switch panel using a suitable prytool
- Cover plate (automatic transmission) or the shifter cover plate (manual transmission)
- Ashtray
- Upper and lower audio/air conditioning control unit assembly screws and the assembly
- Console box screws and the con-

sole box (under the shifter cover plate); be sure to remove the rear screws
- Front pillar garnish
- Left and right lower cover and the center lower cover at the instrument panel dash
- Defroster grille
- Instrument panel-to-chassis nuts/bolts and the instrument panel

7. Remove or disconnect the following:
- Rear heater ducts
- Side ventilator ducts
- Center defroster duct and the center ventilator duct
- Heater housing-to-chassis fasteners and remove the heater housing
- Heater core from the heater housing

To install:

8. Install or connect the following:
- Heater core to the heater housing
- Heater housing and the heater housing-to-chassis fasteners
- Center ventilator duct and the center defroster duct
- Side ventilator ducts
- Rear heater ducts

9. Install the instrument panel by installing or connecting the following:
- Instrument panel and the instrument panel-to-chassis nuts/bolts
- Defroster grille
- Left and right lower cover and the center lower cover

- Front pillar garnish
- Console box and the console box screws under the shifter cover plate; be sure to install the rear screws
- Audio/air conditioning control unit assembly and the upper and lower assembly screws
- Ashtray
- Cover plate (automatic transmission) or the shifter cover plate (manual transmission)
- Center ventilator with the switch panel
- Combination meter, connect the electrical connectors and the combination meter screws
- Cluster lid "A" and the lid screws
- Combination switch, connect the electrical connector and the combination switch-to-steering column screws
- Steering column cover and the cover screws
- Knee protector and the knee protector screws
- Lower instrument panel and the panel screws on the driver's side
- Glove box and the upper and lower glove box screws

10. Arm the passenger's side air bag by installing or connecting the following:
- Passenger's air bag electrical connector
- Glove box lid

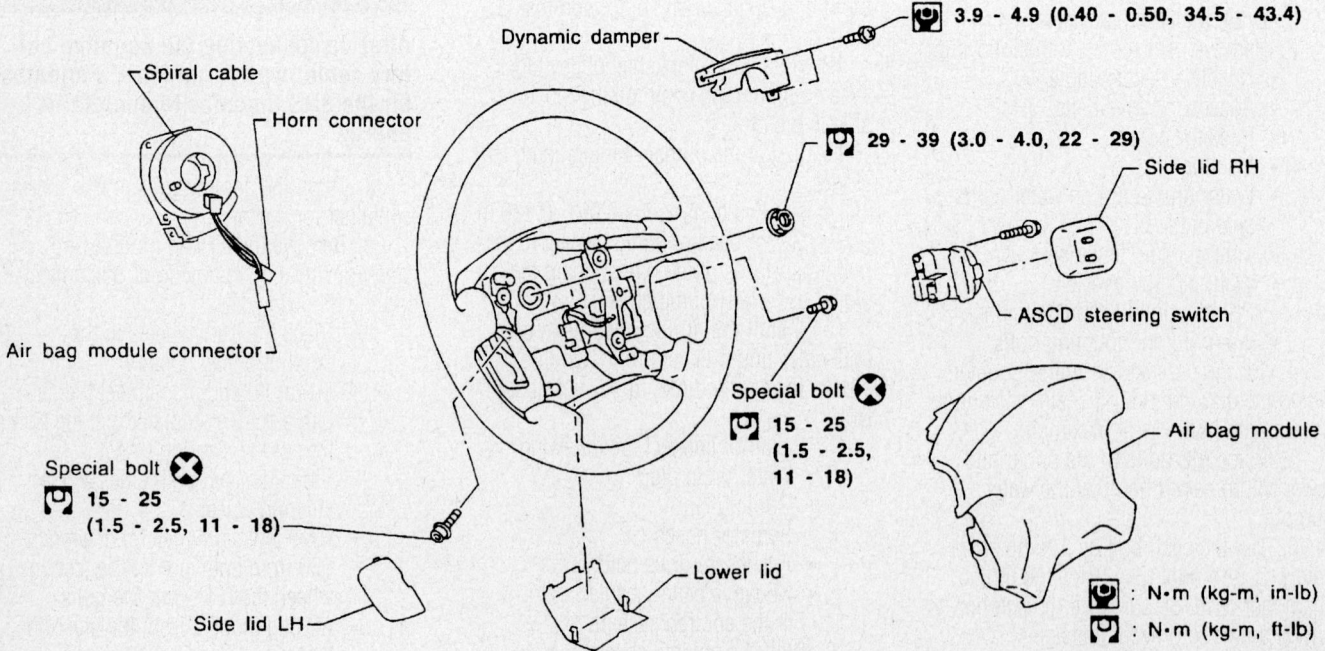

Exploded view of the air bag module and steering wheel—Maxima

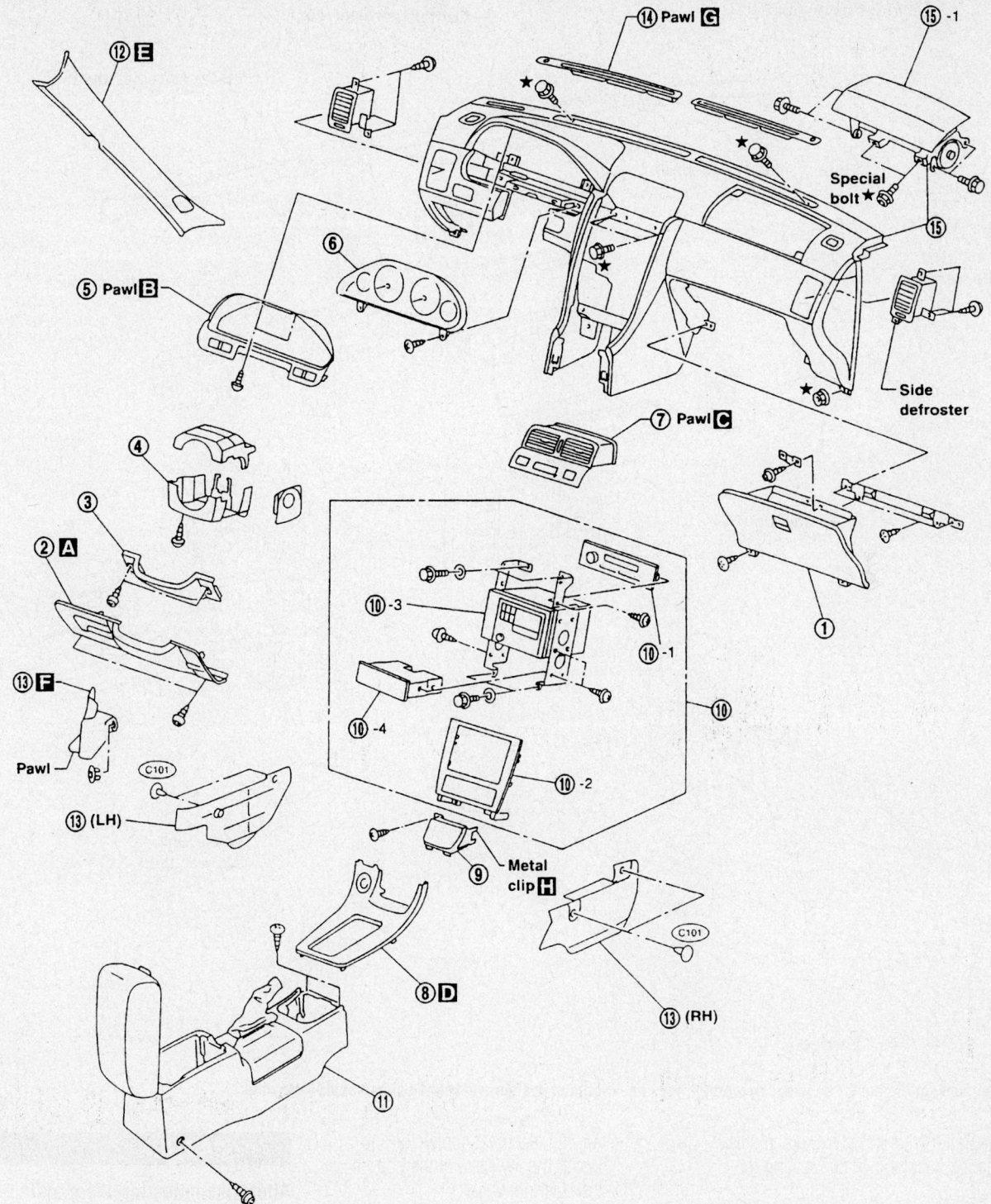

★: Instrument panel assembly mounting bolts and nuts

1. Glove box assembly
2. Instrument lower panel on driver side
3. Knee protector assembly
4. Steering column cover & combination switch
5. Cluster lid A
6. Combination meter
7. Center ventilator with switch panel
8. A/T shifter cover plate or M/T shifter cover plate
9. Ashtray
10. Audio & A/C control unit assembly
11. Console box
12. Front pillar garnish
13. Instrument dash: lower cover and center lower cover on LH, RH
14. Defroster grille
15. Instrument panel assembly
15. -1 Passenger air bag module

93112GK2

Exploded view of the instrument panel, console and related components—Maxima

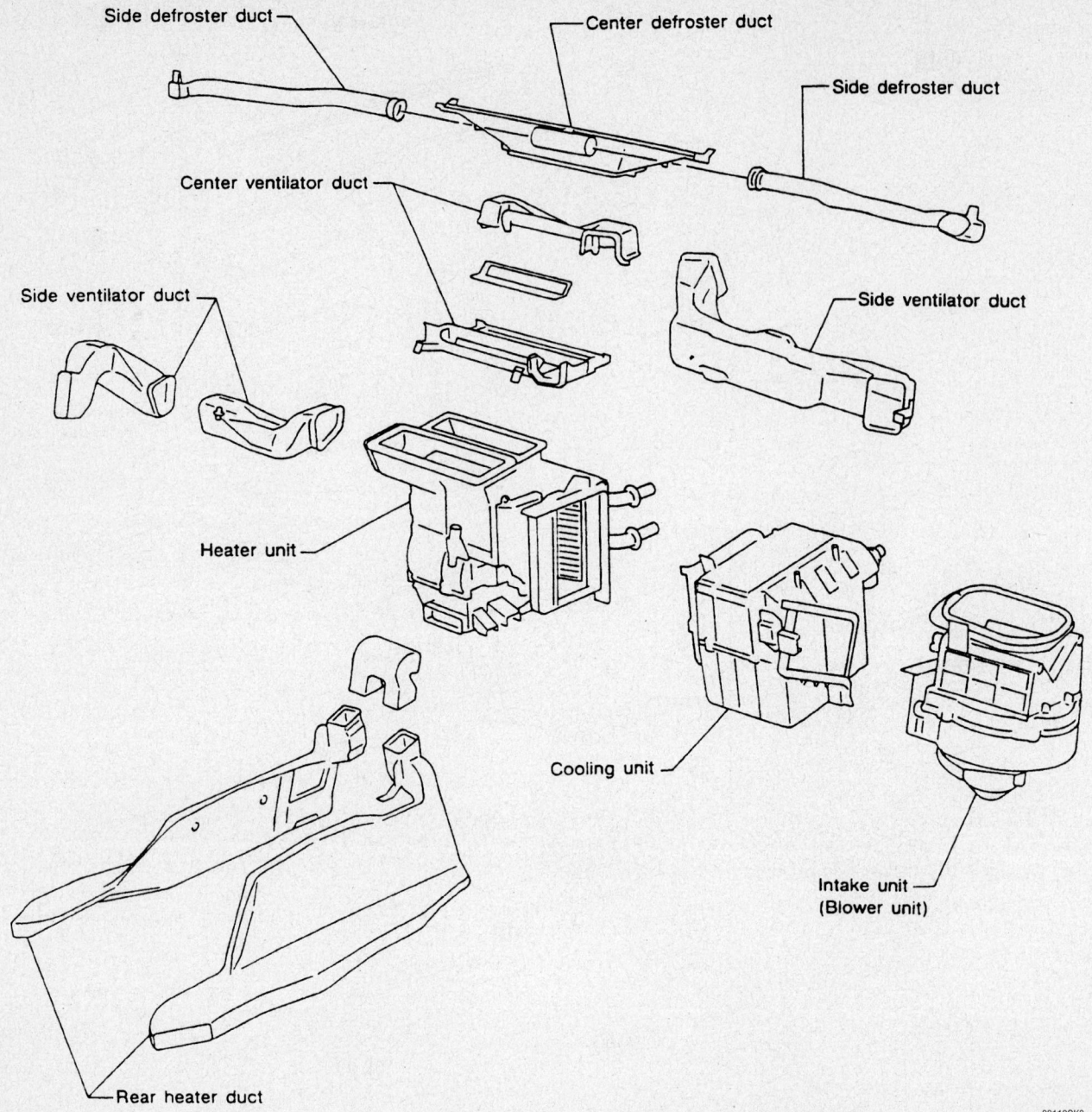

Side defroster duct

Center defroster duct

Side defroster duct

Center ventilator duct

Side ventilator duct

Side ventilator duct

Heater unit

Cooling unit

Intake unit
(Blower unit)

Rear heater duct

93112GK3

Exploded view of the heater housing, evaporator housing, ventilator system and related components—Maxima

11. Install the air bag module and steering wheel by installing or connecting the following:

- Steering wheel to the steering column
- Steering wheel nut and torque it to 22–29 ft. lbs. (29–39 Nm)
- Air bag module to the steering wheel
- Torque the new Torx® bolts (using a Torx® wrench T50), at both side of the steering wheel to 11–18 ft. lbs. (15–25 Nm).
- Side lids to both sides of the steering wheel

- Air bag electrical connector and install the lower lid at the bottom of the steering wheel

12. Refill the cooling system.
13. Connect the negative battery terminal.
14. Operate the engine to normal operating temperatures; then, check the climate control operation and check for leaks.

2004–2006 MODELS

1. Before servicing the vehicle, refer to the Precautions Section.
2. Disconnect the negative battery terminal.
3. Disconnect battery positive terminal.

⁂ CAUTION

After disconnecting the negative battery cable, wait for at least 3 minutes for the SRS modules to deplete its energy.

4. Discharge the refrigerant from the A/C system.
5. Drain the coolant from the cooling system.
6. Disconnect the heater hoses from the heater core pipes.
7. Disconnect the refrigerant lines from the evaporator.

8. Remove the instrument panel by removing or disconnecting the following:
- Fuse block cover
- Lower driver instrument panel screws
- Lower driver instrument panel
- Aspirator tube and in vehicle temperature sensor
- Electrical harness connectors
- Right side instrument lower cover
- Glove box pins and remove glove box door
- Glove box housing screws
- Trunk cancel switch harness
- Glove box lamp and harness from glove box housing
- Glove box housing
- Both rear seat cushions and both rear seatbacks
- Screws and pass through assembly
- Clips and right lower side cover and left lower side cover
- Screws, disconnect harness and remove rear console assembly
- A/T or M/T finisher
- Cluster lid C screws
- Pull cluster lid C towards rear of vehicle to release clips
- Cluster lid C electrical connectors
- Cluster lid D screws
- Pull cluster lid D toward rear of vehicle to release clips
- Cluster lid D electrical connectors

✻✻ WARNING

To avoid damage, eject map DVD-ROM disk before removing the NAVI control unit.

✻✻ WARNING

Cover NAVI control unit with a cloth and avoid contact of NAVI control unit with brackets that may cause scratches or damage to NAVI control unit.

- Front center console
- Control device
- Pull up parking brake lever and remove parking brake lever finisher
- Move front seats forward and remove front center console assembly
- NAVI control unit mounting screws (if equipped)
- NAVI control unit connectors
- NAVI control unit
- Center stack mounting screws and center stack
- Center stack electrical harness connectors and GPS antenna

- Security light flasher/passenger air bag off indicator
- GPS antenna
- Right side instrument mask, sun sensor
- Left side instrument mask, optical sensor
- Partially remove and place the front door welts aside
- Left side instrument panel finisher
- Right side instrument panel finisher
- Left front pillar garnish and right front pillar garnish
- Kicking plate
- Right lower dash side trim and left lower dash side trim
- Passenger air bag to steering member assembly bolt
- Passenger side air bag electrical connector
- Tilt motor and telescopic motor electrical connector (if equipped)
- Steering lock escutcheon
- Set the front wheels in the straight-ahead position.

✻✻ CAUTION

When servicing the SRS, do not work from directly in front of air bag module.

✻✻ WARNING

Do not attempt to repair or replace damaged direct-connect SRS component connectors. If a driver air bag direct-connect harness connector is damaged, the spiral cable must be replaced.

- Bolt covers
- Left and right side bolts
- Lift the driver air bag module from the steering wheel
- The air bag harness connectors and driver air bag module

✻✻ WARNING

Always place air bag module with pad side facing upward. Do not insert any foreign objects (screwdriver, etc.) into air bag module or harness connectors. Do not disassemble air bag module. Do not use old bolts after removal; replace with new bolts. Do not expose the air bag module to temperatures exceeding 90 degrees C (194 degrees F). Replace the air bag module if it has been dropped or sustained an impact. Do not allow oil, grease or water to come in contact with the air bag module.

- Steering wheel center nut
- Steering wheel using a steering wheel removal tool
- Place a piece of tape across the spiral cable so it will not be rotated out of position
- Combination meter
- Left side instrument panel screws to allow the ignition switch to clear the instrument panel during removal
- Disconnect the following:
a. Telescopic sensor, if equipped
b. Tilt sensor, if equipped
c. Headlamp switch
d. Combination switch
e. Spiral cable
f. Key in reminder
g. Immobilizer
h. Illumination lamp
i. Ignition switch
j. Column harness clips, position aside
- Shaft lower cover
- Pinch bolt.
- Steering column nuts and steering column
- Instrument panel

9. Remove the ECM.

10. Disconnect the blower motor, intake door motor, and fan control amp. connector.

11. Disconnect the main harness from the top of the blower unit.

12. Remove the two bolts and one screw from the blower unit as shown.

13. Remove the blower unit.

14. Disconnect the mode door motor and the air mix door motor connectors.

15. Remove the heater and cooling unit.

16. Remove the heater core pipe support screws and then remove the heater core pipe support.

17. Remove the air mix door motor (passenger side).

18. Remove the heater core cover screws and then remove the heater core cover.

19. Remove the heater core.

To install:

20. Installation is the reverse of the removal procedure, noting the following points:

✻✻ CAUTION

When installing the steering column, finger-tighten all of the lower bracket and joint retaining bolt; then tighten them to specification. Do not apply undue stress to the steering column.

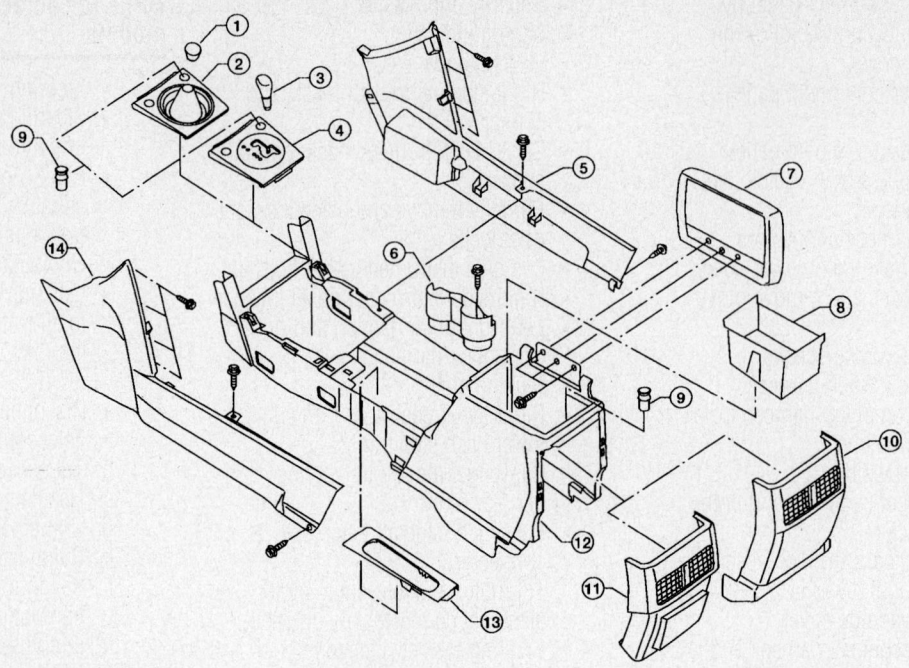

1. M/T shift knob	2. M/t console finisher	3. A/T shift knob
4. A/T console finisher	5. Console cover, RH	6. Cup holder insert
7. Console lid assembly	8. Console rear pocket	9. Power socket assembly
10. Rear finisher assembly, 5 seat model	11. Rear finisher assembly, 4 seat model	12. Front center console
13. Parking brake lever finisher	14. Console cover, LH	

09482_ZMAX_G0005

Exploded view of the center console assembly—2004–2006 Maxima

a. For the steering column, be sure to align slit of the coupling joint with projection on dust cover. Insert the joint until both surfaces make contact. Torque the joint retaining bolt to 32 ft. lbs. (44 Nm). Torque the lower bracket bolts to 13 ft. lbs. (17 Nm).

b. After installation, turn steering wheel to make sure it moves smoothly. Ensure the number of turns are the same from the straight-forward position to left and right locks. Be sure that the steering wheel is in a neutral position when driving straight ahead.

c. Align spiral cable correctly when installing steering wheel. Make sure that the spiral cable is in the neutral position. The neutral position is detected by turning left 2.5 revolutions from the right end position and ending with the knob at the top.

21. Torque the steering wheel center nut to 25 ft. lbs. (34 Nm).

❈❈ WARNING

The spiral cable may snap due to steering operation if the cable is installed in an improper position.

22. Make sure to connect all electrical connectors including the following:

a. Tilt motor and telescopic motor electrical connector, if equipped.
b. Telescopic sensor, if equipped
c. Tilt sensor, if equipped
d. Headlamp switch
e. Combination switch
f. Spiral cable
g. Key in reminder
h. Immobilizer
i. Illumination lamp
j. Ignition switch
k. Column harness clips, position aside
l. Make sure that any/all vacuum lines and hoses are reconnected.
23. Recharge the A/C system.
24. Fill the cooling system.
25. Reconnect battery positive terminal first; the negative battery terminal second.

350Z

1. Before servicing the vehicle, refer to the Precautions Section.
2. Discharge the A/C system using approved recycling equipment.
3. Drain the cooling system.
4. Remove or disconnect the following:

• Hood ledge cover
• Both wiper arms
• Cowl rubber seal
• Cowl top cover and washer hose
• Evaporator lines from the firewall and cap openings
• Electronic throttle control assembly
• Heater hoses
• Kick panels on both sides
• Foot rests
• Passenger side lower instrument panel cover
• Instrument panel side finish panels on both sides
• Cluster Lid C
• Data link connector
• Hood lock cable
• Steering column lower cover
• 4 bolts and combination meter
5. Position the steering wheel in the straight-ahead position.
6. Turn the ignition switch OFF.
7. Disconnect the negative (−) battery cable; then, the positive (+) battery cable.

➡**Wait for a least 3 minutes after disconnecting the battery cables for the charge in the air bag circuit to dissipate before working on the air bag module(s).**

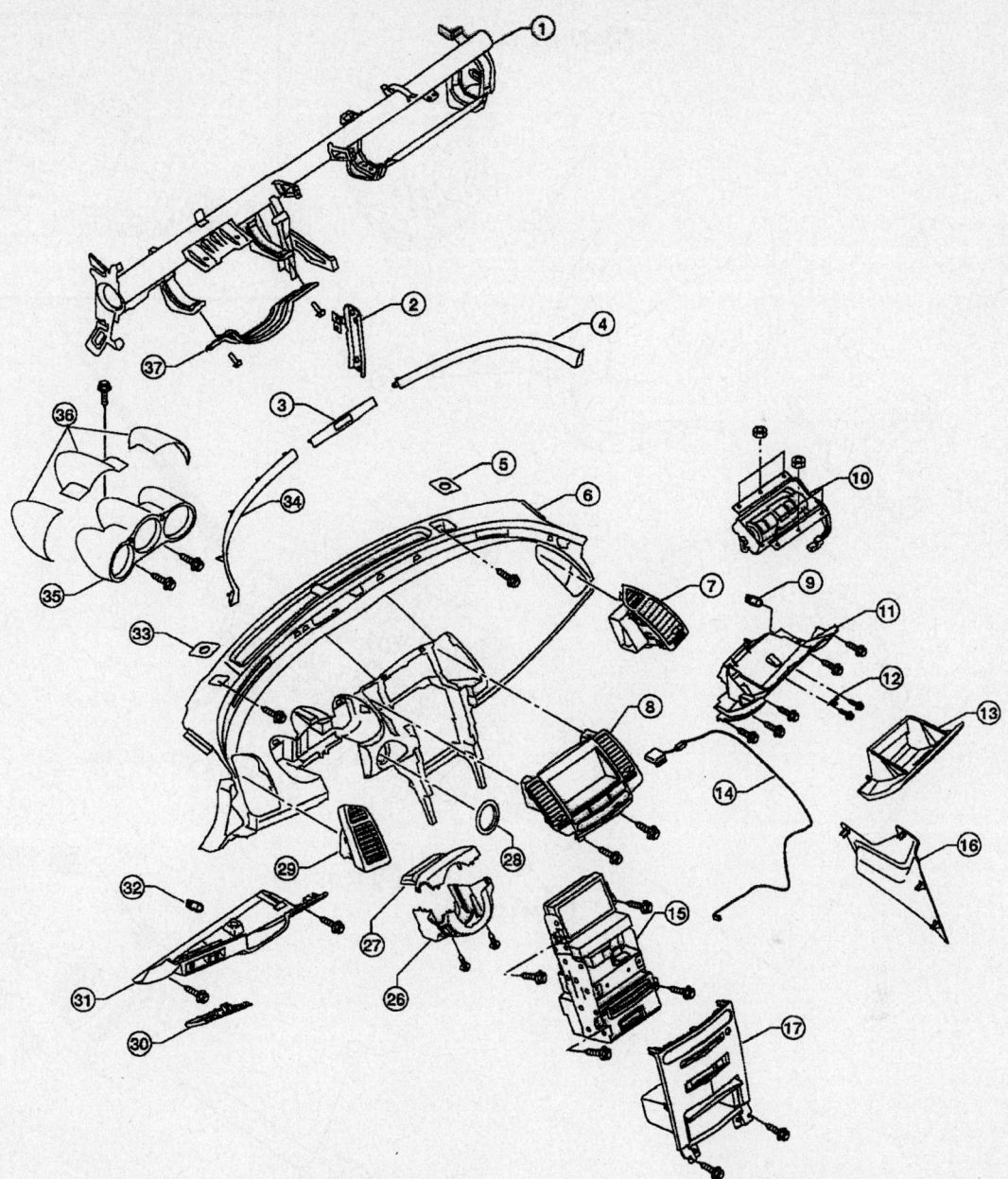

1.	Steering member assembly	2.	Instrument stay, driver	3.
				Security light finisher/Passenger air bag off indicator
4.	Instrument panel finisher, RH	5.	Instrument mask RH, sun sensor	6.	Instrument panel
7.	Side ventilator assembly, RH	8.	Cluster lid D	9.	Glove box bulb
10.	Front passenger air bag module	11.	Instrument passenger lower panel	12.	Glove box striker
13.	Glove box assembly	14.	GPS antenna assembly	15.	Center stack
16.	Instrument lower cover, RH	17.	Cluster lid C	18.	M/T shift knob
19.	M/T finisher	20.	A/T shift knob	21.	A/T finisher
22.	Center console assembly	23.	Parking brake lever finisher	24.	Rear upper console assembly
25.	Rear console assembly	26.	Steering column cover, lower	27.	Steering column cover, upper
28.	Steering lock escutcheon	29.	Side ventilator assembly, LH	30.	Fuse block cover
31.	Lower driver instrument panel	32.	Lower driver instrument panel bulb	33.	Instrument mask LH, optical sensor
34.	Instrument panel finisher, LH	35.	Combination meter assembly	36.	Combination meter covers
37.	Lower knee protector, LH				

Exploded view of the instrument panel assembly—2004–2006 Maxima

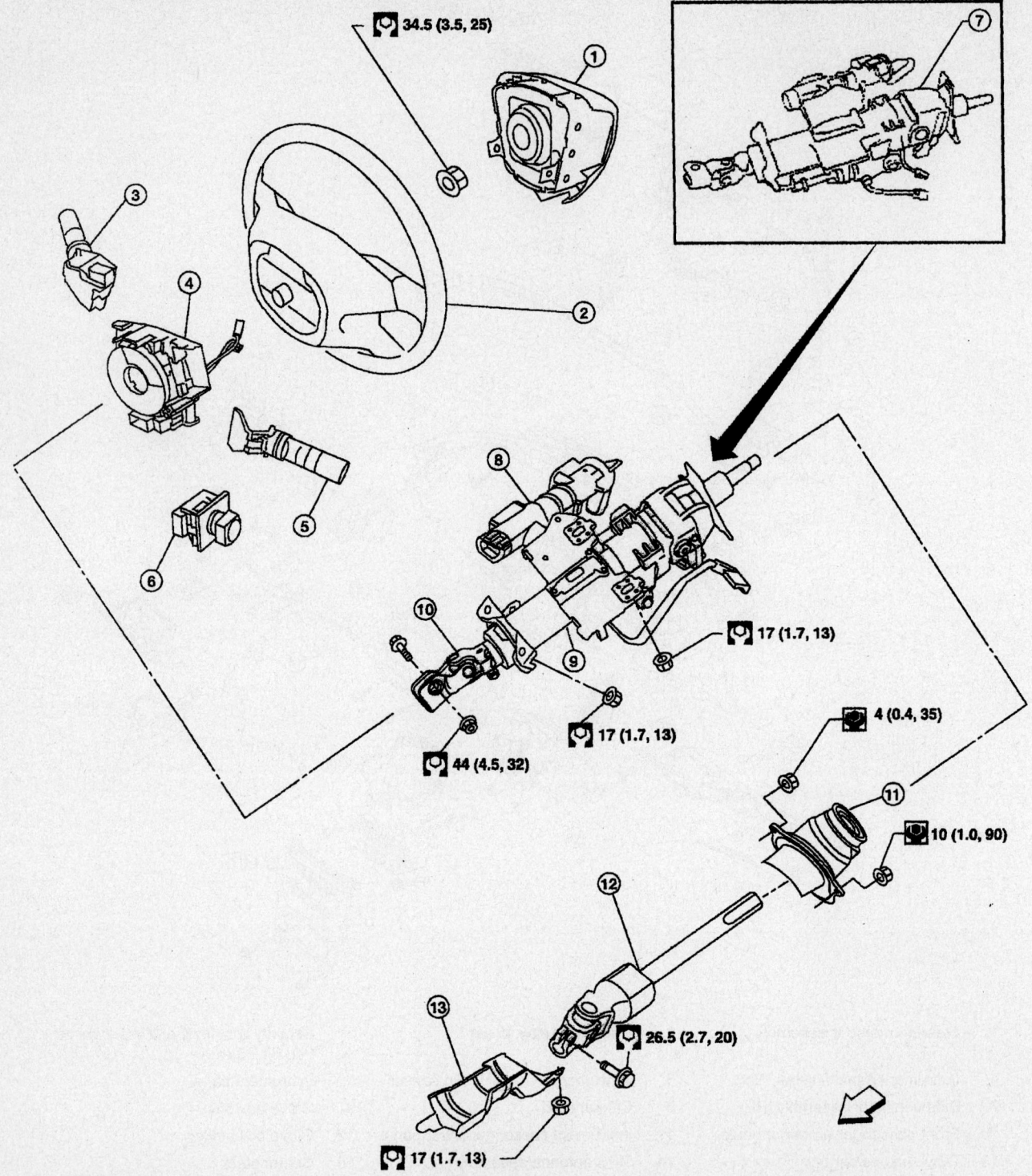

1. Driver air bag module
2. Steering wheel
3. Head lamp switch
4. Spiral cable
5. Combination switch
6. ADP steering switch
7. Steering column (electric tilt/tele-scopic type)
8. Ignition switch
9. Steering column (manual tilt/tele-scope type)
10. Upper joint
11. Hole cover
12. Lower Joint and shaft assembly
13. Shaft lower cover
⇐ Front

09482_ZMAX_G0007

Exploded view of the steering wheel and steering column assemblies—2004–2006 Maxima

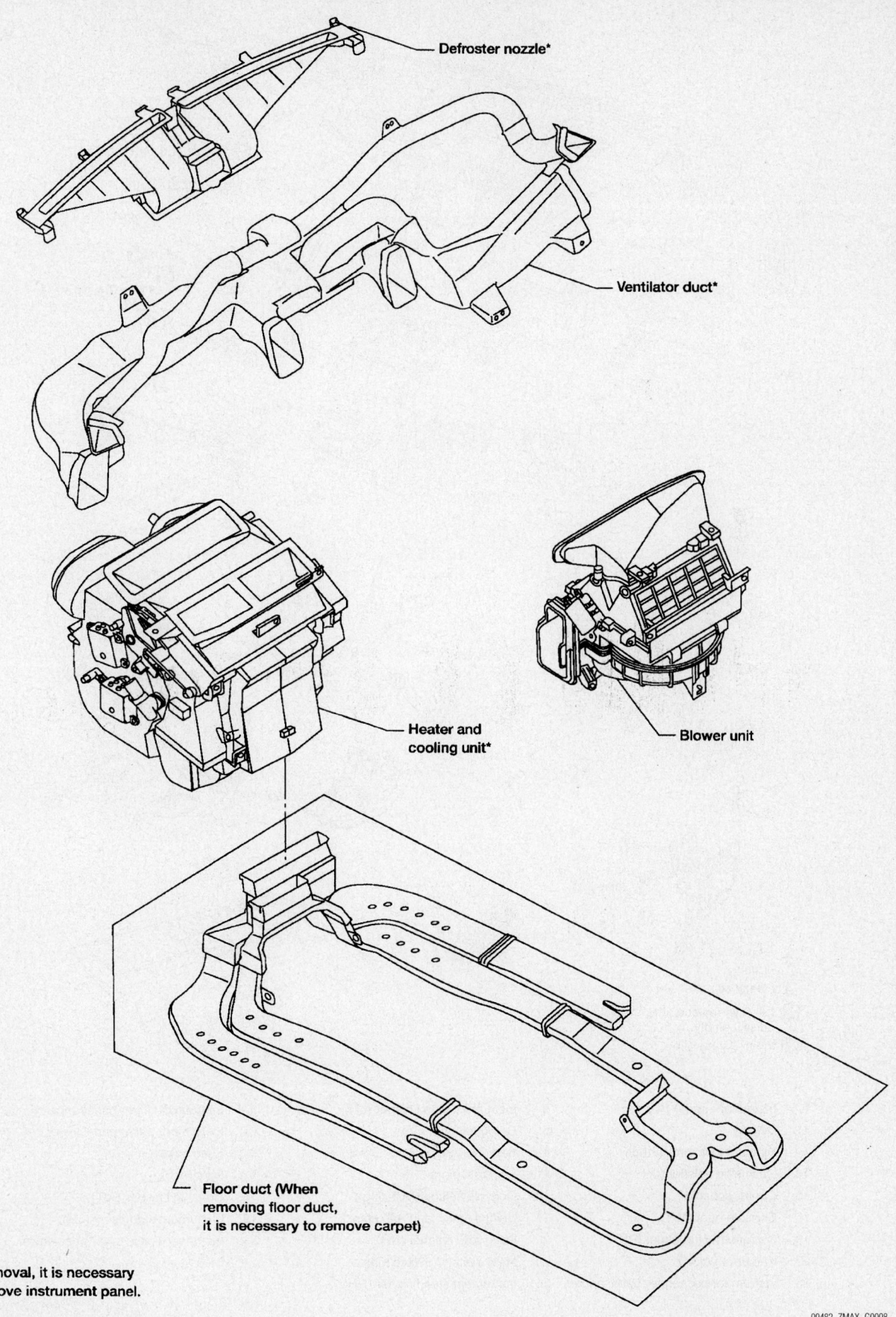

Defroster nozzle*

Ventilator duct*

Heater and
cooling unit*

Blower unit

Floor duct (When
removing floor duct,
it is necessary to remove carpet)

* For removal, it is necessary
to remove instrument panel.

09482_ZMAX_G0008

The heater and cooling unit with related components—2004–2006 Maxima

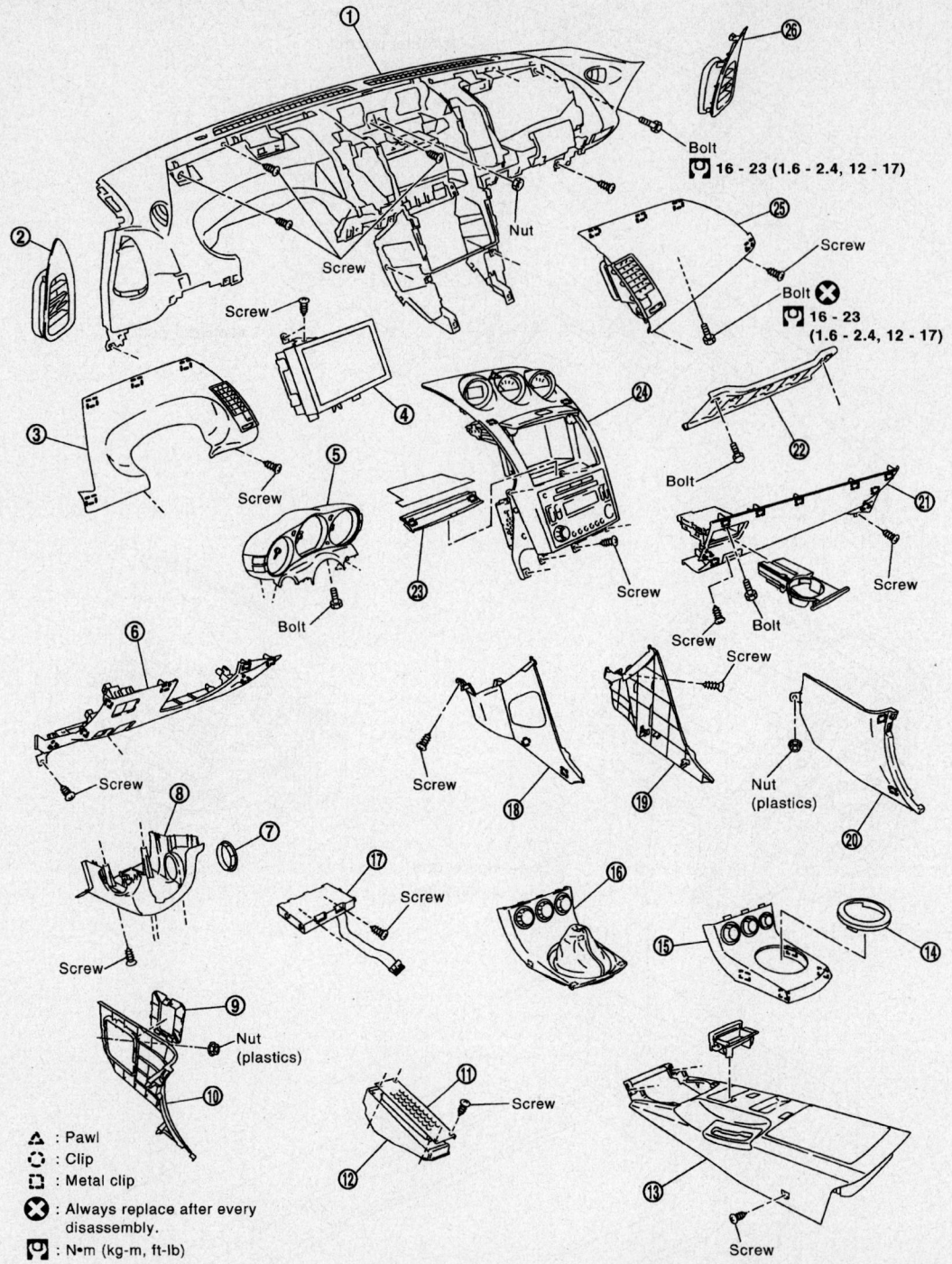

Bolt
⟐ 16 - 23 (1.6 - 2.4, 12 - 17)

Bolt ✖
⟐ 16 - 23
(1.6 - 2.4, 12 - 17)

△ : Pawl
◌ : Clip
▢ : Metal clip
✖ : Always replace after every disassembly.
⟐ : N•m (kg-m, ft-lb)

1.	Instrument panel	2.	Instrument side finisher (LH)	3.	Instrument driver panel upper
4.	Display	5.	Combination meter	6.	Instrument driver panel lower
7.	Steering lock escutcheon	8.	Steering column lower cover	9.	Fuse cover
10.	Dash side finisher (LH)	11.	Foot rest plate	12.	Foot rest
13.	Center console	14.	Console finisher (A/T ring)	15.	Console finisher (A/T)
16.	Console boot (M/T)	17.	Unified meter and A/C amp	18.	Instrument side panel (LH)
19.	Instrument side panel (RH)	20.	Dash side finisher (RH)	21.	Instrument passenger panel lower
22.	Knee protector	23.	NAVI switch / Switch mask	24.	Cluster lid C
25.	Instrument passenger panel upper	26.	Instrument side finisher (RH)		

67162-NISS-G10

Exploded view of the instrument panel assembly—350Z

8. Remove the driver's side SRS and steering wheel by performing the following procedure:
- Remove both the left and right side lids from the steering wheel.
- Using a tamper resistant Torx® wrench (T50), remove the special bolts from both sides of the steering wheel.
- Steering wheel switch sub-harness connector
- Air bag harness connector
- Carefully remove the air bag module and store it face up.

9. Remove or disconnect the following:
- Steering wheel
- Steering column upper cover
- Spiral cable connector
- Combination switch
- Automatic transmission console finisher panel, if equipped
- Manual transmission shift knob, if equipped
- Center console
- Cup holder
- Passenger side lower instrument panel cover
- Instrument panel side cover
- Navigation switch cover panel and switch connector
- Audio cluster lid
- Audio unit and meter assembly
- Display unit
- Garnish panels and side finishers
- Passenger air bag connector
- Passenger air bag bolt and passenger air bag
- ECM and bracket

- Intake door motor and blower motor connectors
- 2 screws, 1 bolt and the blower unit
- Left and right instrument panel stays
- Defroster and ventilation ducts
- Heating–A/C unit
- Heater pipe cover, support and grommet
- Slide the heater core out of the heating–A/C unit

To install:

10. Install or connect the following:
- Heater core to the heater–A/C unit
- Heater pipe cover, support and grommet
- Heater–A/C unit and tighten the bolts to 61 inch lbs. (7 Nm)
- Defroster and ventilation ducts
- Left and right instrument panel stays
- 2 screws, 1 bolt and the blower unit
- Intake door motor and blower motor connectors
- ECM and bracket
- Passenger air bag and tighten the bolt to 15–21 ft. lbs. (20–29 Nm)
- Passenger air bag connector
- Garnish panels and side finishers
- Display unit
- Audio unit and meter assembly
- Audio cluster lid
- Navigation switch cover panel and switch connector
- Instrument panel side cover
- Passenger side lower instrument panel cover

- Cup holder
- Center console
- Manual transmission shift knob, if equipped
- Automatic transmission console finisher panel, if equipped
- Combination switch
- Spiral cable connector
- Steering column upper cover
- Steering wheel and tighten the bolt to 22–28 ft. lbs. (30–39 Nm)
- Driver air bag module and tighten the bolts to 83 inch lbs. (9.4 Nm)
- Air bag harness connector
- Steering wheel switch sub-harness connector
- Both the left and right side lids to steering wheel
- 4 bolts and combination meter
- Steering column lower cover
- Hood lock cable
- Data link connector
- Cluster lid C
- Instrument panel side finish panels on both sides
- Passenger side lower instrument panel cover
- Foot rests
- Kick panels on both sides
- Heater hoses
- Electronic throttle control assembly
- Evaporator lines to the firewall using new O-rings
- Cowl top cover and washer hose
- Cowl rubber seal
- Both wiper arms
- Hood ledge cover

11. Refill the cooling system.

12. Connect the positive (+) battery cable; then, the negative (−) battery cable.

13. Evacuate, charge and leak test the air conditioning system refrigerant.

14. Operate the engine to normal operating temperatures; then, check the climate control operation and check for leaks.

Cylinder Head

REMOVAL & INSTALLATION

Maxima

➡You must remove the engine from the vehicle in order to remove the cylinder head, for this procedure.

1. Before servicing the vehicle, refer to the Precautions Section.
2. Relieve the fuel system pressure.
3. Drain the engine oil.
4. Drain the cooling system.

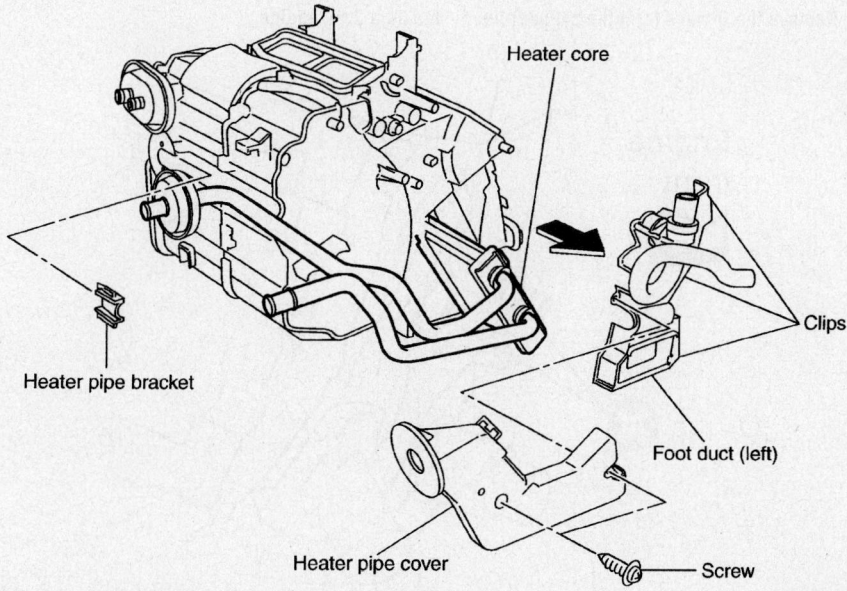

Heater core

Heater pipe bracket

Clips

Foot duct (left)

Heater pipe cover

Screw

67162-NISS-G11

View of the heater core and heater housing—350Z

➡**Before detaching any hoses or connectors, note the locations for reassembly.**

5. Remove or disconnect the following:
   - Negative battery cable
   - Engine assembly
   - Exhaust manifold
6. Place the engine on a suitable workstand.
   - Oil pan
   - Timing chain
   - Intake manifold
   - Water outlet
   - Rear timing chain case bolts, in the sequence shown
   - Rear timing chain case
   - O-rings from the cylinder head and block
   - Intake valve timing control solenoid valves

➡**For installation purposes, matchmark the camshaft brackets before removing them.**

   - Intake and exhaust camshafts and brackets. Loosen the bracket bolts in several steps, in the sequence shown.
   - Right and left side cam chain tensioner from the cylinder head
   - Cylinder head bolts. Loosen in several steps, in the sequence shown.

➡**A warped or cracked cylinder head could result from removing the bolts in incorrect order.**

   - Cylinder heads from the vehicle
   - Discard the head gaskets
7. Remove all traces of liquid gasket from the timing chain case and from the water pump covers.
8. Remove all traces of liquid gasket from the engine block.
9. Inspect the timing chain for excessive wear or damage and replace as necessary.

To install:

10. Turn the crankshaft until the No. 1 piston is a Top Dead Center (TDC) on compression stroke. The crankshaft key should face toward the right bank.
11. Using new head gaskets, install the cylinder heads.

➡**If possible, replacement of the head bolts is suggested.**

12. If replacement of the head bolts is not possible, perform the following bolt measurement:
   a. Measure the diameter of the head bolt 0.43 in. (11mm) from the bottom of the bolt.

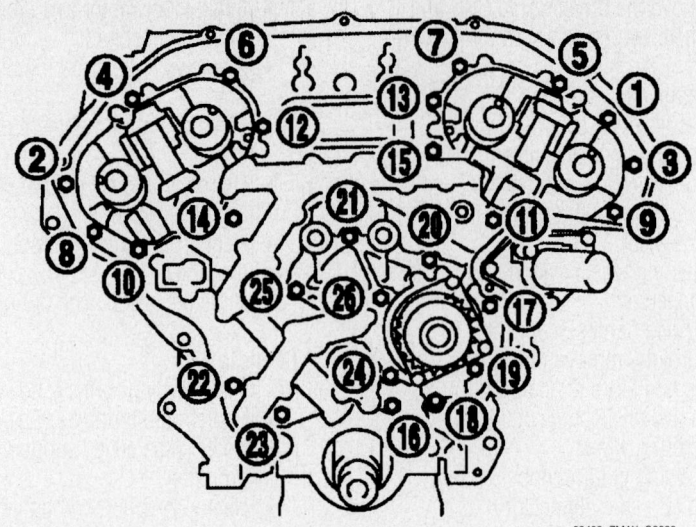

09482_ZMAX_G0009

Loosen the rear timing chain case bolts in sequence—Maxima 3.5L engine

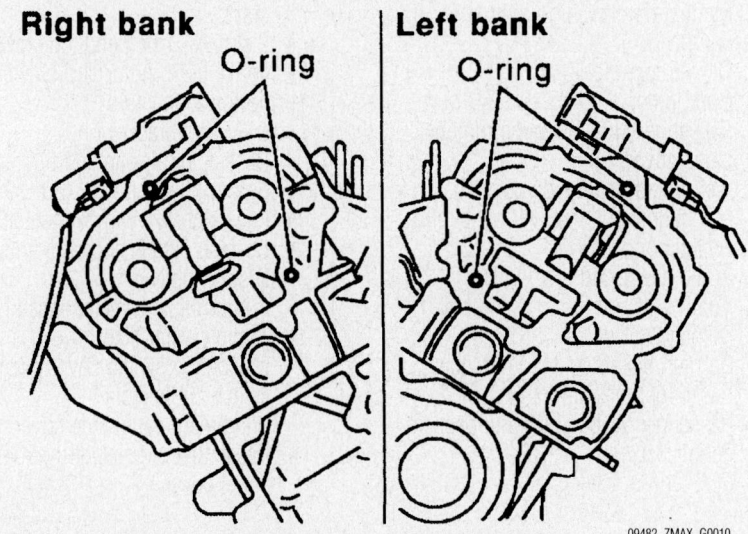

09482_ZMAX_G0010

Remove the O-rings from the cylinder head—Maxima 3.5L engine

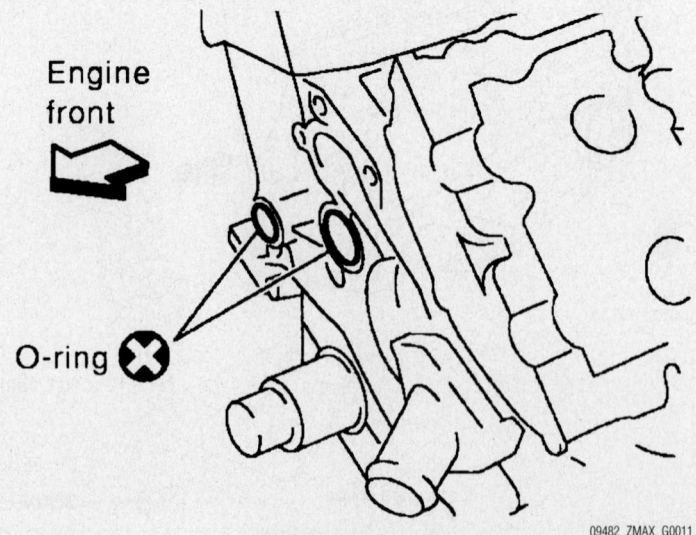

09482_ZMAX_G0011

Remove the O-rings from the engine block—Maxima 3.5L engine

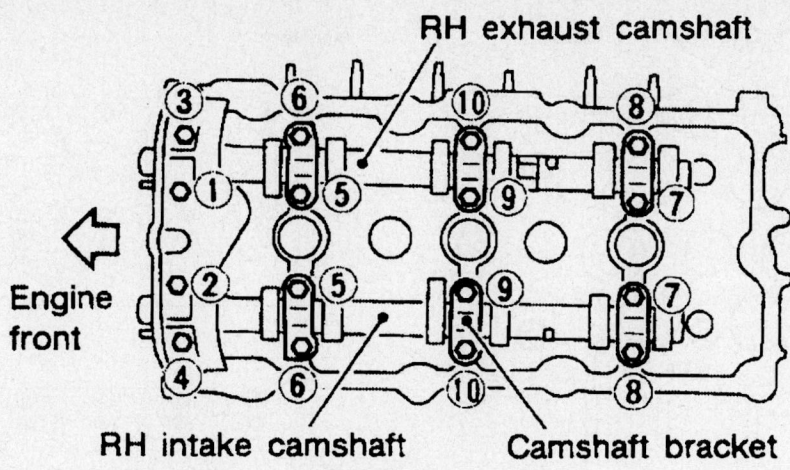

Right camshaft bracket bolt loosening sequence—Maxima 3.5L engine

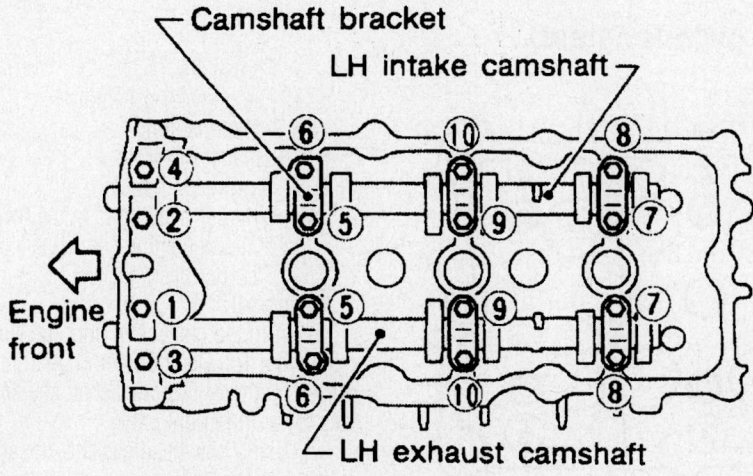

Left camshaft bracket bolt loosening sequence—Maxima 3.5L engine

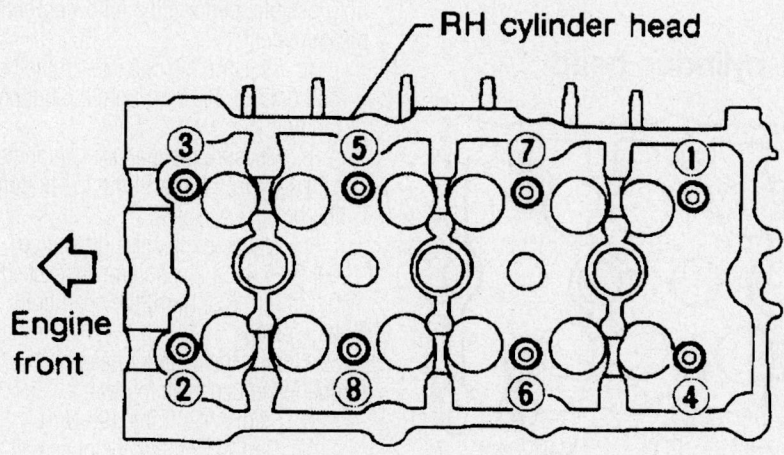

Right cylinder head bolt loosening sequence—Maxima 3.5L engine

b. Measure the diameter of the head bolt 1.89 in. (48mm) from the bottom of the bolt.

c. Whenever the size difference between the 2 measurements exceeds 0.0043 in. (0.11mm) the head bolts must be replaced.

13. Install the cylinder head bolts and torque in sequence as follows:

a. Step 1: 72 ft. lbs. (98 Nm).

b. Step 2: Completely loosen all bolts.

c. Step 3: 26–32 ft. lbs. (34–44 Nm).

d. Step 4: plus 90–95 degrees clockwise.

e. Step 5: plus 90–95 degrees clockwise.

14. Install or connect the following:
- Camshafts and related components
- Intake valve timing control solenoid valves
- New O-rings to the front of the engine block and cylinder head

15. Apply sealant to the hatched portion of the of the rear timing chain case.

16. Align the rear timing chain case with the dowel pins and install onto the cylinder heads and engine block.

17. Torque the rear timing chain case mounting bolts in sequence to 105–121 inch lbs. (11.8–13.7 Nm).

18. Install or connect the following:
- Water outlet
- Intake manifold
- Timing chain
- Oil pan
- Exhaust manifold
- Engine assembly into the vehicle
- Negative battery cable

19. Fill the cooling system.

20. Fill the engine with clean oil.

21. Start the vehicle, check for leaks and repair if necessary.

350Z

➡You must remove the engine from the vehicle in order to remove the cylinder head, for this procedure.

1. Before servicing the vehicle, refer to the Precautions Section.

2. Relieve the fuel system pressure.

3. Drain the engine oil.

4. Drain the cooling system.

➡Before detaching any hoses or connectors, note the locations for reassembly.

5. Remove or disconnect the following:
- Negative battery cable
- Engine assembly
- Timing chain

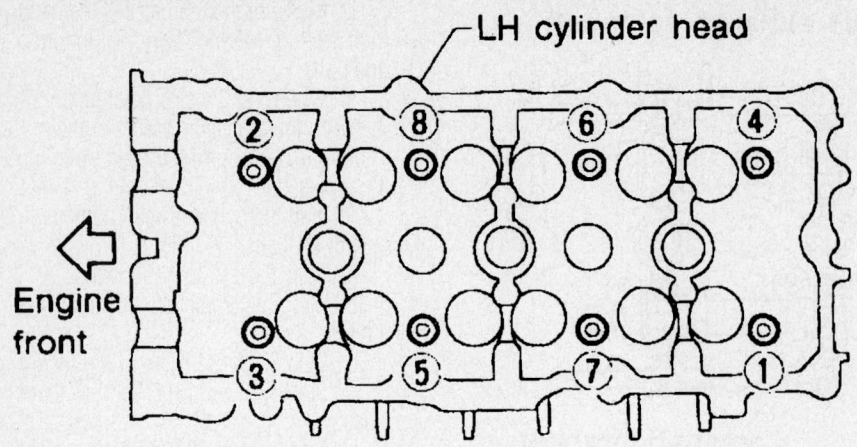

Loosen in numerical order.

09482_ZMAX_G0015

Left cylinder head bolt loosening sequence—Maxima 3.5L engine

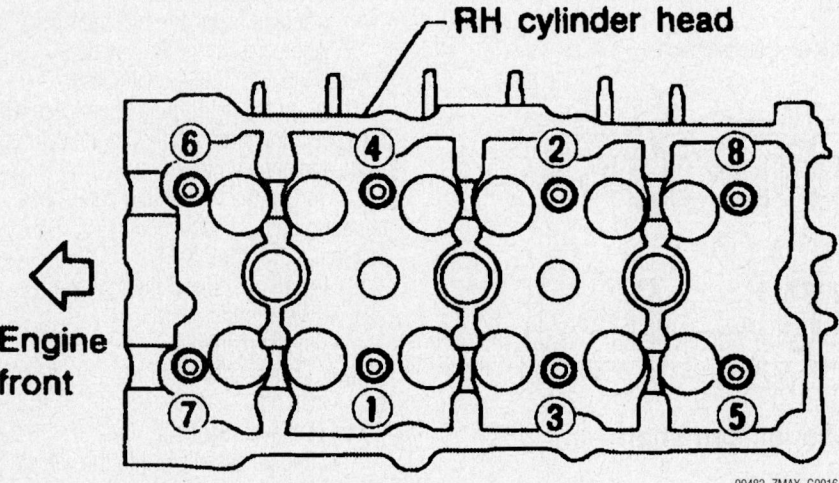

09482_ZMAX_G0016

Right cylinder head bolt torque sequence—Maxima 3.5L engine

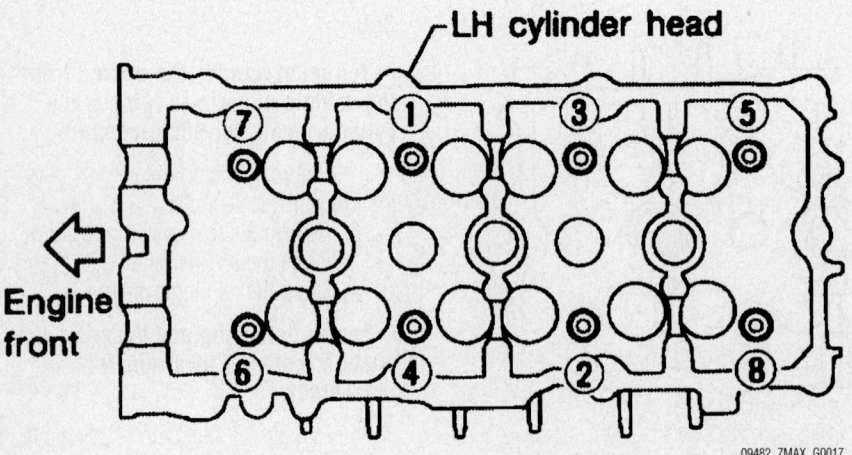

09482_ZMAX_G0017

Left cylinder head bolt torque sequence—Maxima 3.5L engine

Right bank

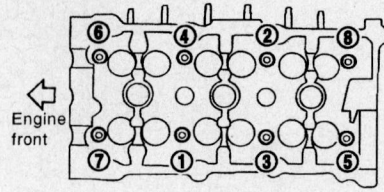

Left bank

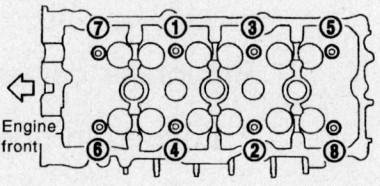

67162-NISS-G15

Cylinder head bolt torque sequence—350Z 3.5L engine

- Camshafts
- Fuel injector assembly
- Intake manifold
- Exhaust manifold
- Thermostat housing
- Cylinder head bolts in the reverse of the tightening sequence
- Cylinder head gaskets

To install:

6. Turn the crankshaft until the No. 1 piston is a Top Dead Center (TDC) on compression stroke. The crankshaft key should face toward the right bank.

7. Using new head gaskets, install the cylinder heads.

➡**If possible, replacement of the head bolts is suggested.**

8. If replacement of the head bolts is not possible, perform the following bolt measurement:

  a. Measure the diameter of the head bolt 0.43 in. (11mm) from the bottom of the bolt.

  b. Measure the diameter of the head bolt 1.89 in. (48mm) from the bottom of the bolt.

  c. Whenever the size difference between the 2 measurements exceeds 0.0043 in. (0.11mm) the head bolts must be replaced.

9. Install the cylinder head bolts and torque in sequence as follows:

  a. Step 1: 72 ft. lbs. (98 Nm).

  b. Step 2: Completely loosen all bolts.

  c. Step 3: 26–32 ft. lbs. (34–44 Nm).

  d. Step 4: plus 90–95 degrees clockwise.

e. Step 5: plus 90–95 degrees clockwise.

10. Install or connect the following:
- Thermostat housing
- Exhaust manifold
- Intake manifold
- Fuel injector assembly
- Camshafts
- Timing chain
- Engine assembly
- Negative battery cable

11. Fill the cooling system.

12. Fill the engine with clean oil.

13. Start the vehicle, check for leaks and repair if necessary.

Rocker Arms

REMOVAL & INSTALLATION

The Nissan 3.5L engine does not utilize rocker arms. The valves are actuated directly by the camshafts.

Intake Manifold

REMOVAL & INSTALLATION

1. Before servicing the vehicle, refer to the Precautions Section.

2. Drain the cooling system.

3. Release the fuel system pressure.

4. Disconnect the negative battery cable.

5. On 350Z remove or disconnect the following:
- Strut tower bar
- Engine cover
- Air cleaner case and duct
- Electronic throttle control actuator bolt in the sequence shown
- Fuel injector and fuel tube assembly
- Vacuum and water hoses from intake manifold collector
- EVAP solenoid valve bracket
- Upper intake manifold collector bolts in reverse of the tightening sequence
- PCV hose
- Lower intake manifold collector bolts in reverse of the tightening sequence

6. On Maxima remove or disconnect the following:
- Throttle body coolant hoses
- Electrical connectors from the Throttle Position (TP) sensor
- Hoses from the throttle body, the Exhaust Gas Recirculation (EGR) valve, intake manifold collector, Idle

Air Control (IAC) valve and the fuel pressure regulator
- Canister purge hose and blow-by hose
- EGR guide tube
- Accelerator cable from the throttle body
- Intake manifold collector support brackets
- Right side electrical connectors from the ignition coils
- Electrical connector from the crank angle sensor and the power transistor, if necessary

- Intake manifold collector-to-intake manifold bolts/nuts and the intake manifold collector

7. Remove the fuel injector assembly by performing the following procedures:

a. Detach the electrical connectors from the fuel injectors.

b. Disconnect the fuel lines from the fuel injector assembly.

c. Remove the fuel rail-to-cylinder head bolts.

d. Remove the fuel rail assembly from the engine.

8. Remove or disconnect the following:

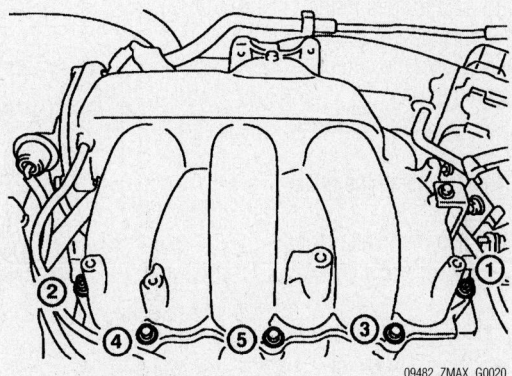

Upper intake manifold collector loosening sequence—Maxima 3.5L engine

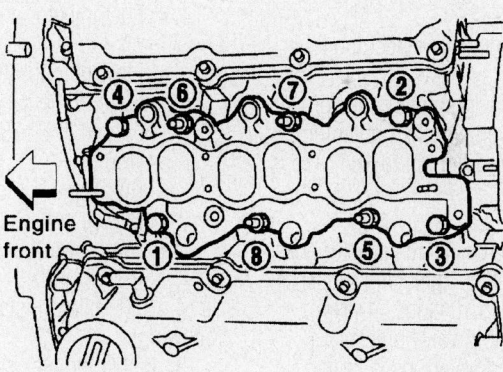

Intake manifold bolt loosening sequence—3.5L engines

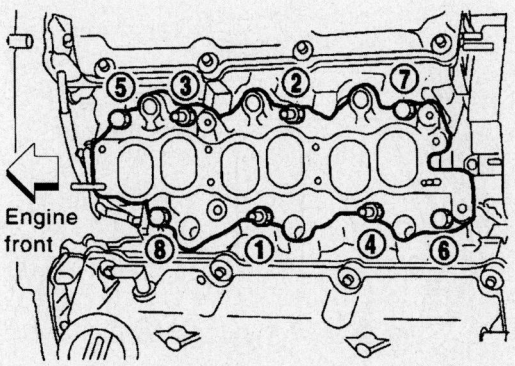

Intake manifold torque sequence—3.5L engines

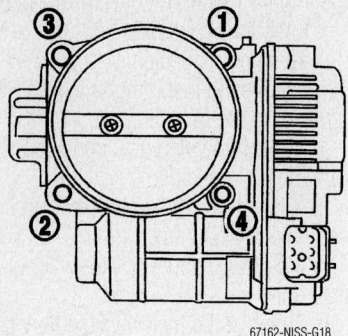

67162-NISS-G18

Electronic throttle actuator removal and installation sequence—350Z 3.5L engine

- Intake manifold bolts/nuts in the reverse of the installation sequence
- Intake manifold from the engine and discard the gaskets

9. Clean all gasket mounting surfaces.

To install:

10. Using new gaskets, install the intake manifold to the engine.

11. If necessary, tighten the intake manifold stud bolts to 87–104 inch lbs. (10–12 Nm).

12. Torque the intake manifold bolts in sequence as follows:

   a. Step 1: 4–7 ft. lbs. (5–10 Nm).

   b. Step 2: 20–23 ft. lbs. (26–31 Nm).

   c. Step 3: Tighten again to 20–23 ft. lbs. (26–31 Nm).

13. On 350Z install or connect the following:

- Lower intake manifold collector bolts in reverse of the tightening sequence and tighten to 10 ft. lbs. (14 Nm).
- PCV hose
- Upper intake manifold collector bolts in reverse of the tightening sequence and tighten to 10 ft. lbs. (14 Nm).
- EVAP solenoid valve bracket
- Vacuum and water hoses from intake manifold collector

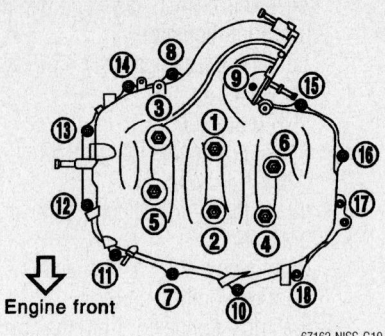

67162-NISS-G19

Upper intake manifold collector tightening sequence—350Z 3.5L engine

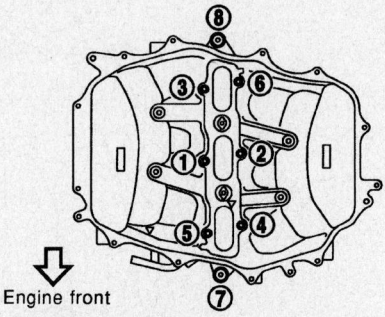

67162-NISS-G20

Lower intake manifold collector tightening sequence—350Z 3.5L engine

- Fuel injector and fuel tube assembly
- Electronic throttle control actuator bolts in the sequence shown and tighten to 64–85 inch lbs. (7–10 Nm)
- Air cleaner case and duct
- Engine cover
- Strut tower bar and tighten to 24 ft. lbs. (32 Nm).

14. On Maxima install or connect the following:

15. Install the fuel injector assembly by performing the following procedures:

   a. Install the fuel rail assembly to the engine.

   b. Install the fuel rail-to-cylinder head bolts and tighten the bolts to 15–20 ft. lbs. (21–26 Nm) in 2 progressive steps.

   c. Connect the fuel lines to the fuel injector assembly.

   d. Connect the electrical connectors to the fuel injectors.

   e. Install the intake manifold collector. Torque the fasteners to 13–15 ft. lbs. (17.5–21.5 Nm).

16. Install or connect the following:

- Crank angle sensor and transmitter electrical connectors
- Right side ignition coil electrical connectors
- Intake manifold collector support brackets
- Accelerator cable to the throttle body
- EGR guide tube
- Canister purge and blow by hoses
- Throttle body, EGR valve and intake manifold collector hoses
- IAC valve and fuel pressure regulator hoses
- TP sensor electrical connector
- Throttle body coolant hose

17. On all models, connect the negative battery cable.

18. Fill the cooling system.

19. Start the vehicle, check for leaks and repair if necessary.

Exhaust Manifold

REMOVAL & INSTALLATION

Maxima

1. Before servicing the vehicle, refer to the Precautions Section.

2. Remove or disconnect the following:

- Negative battery cable
- Exhaust manifolds from the exhaust pipes
- Protective covers from the manifolds
- Heated Oxygen Sensor (HO2S) from the manifold, if equipped
- Exhaust manifold-to-engine mounting nuts
- Manifolds from the engine and discard the gaskets

To install:

3. Clean all gasket mounting surfaces. Install new gaskets.

4. Install or connect the following:

- Exhaust manifold and torque the nuts in steps to 21–24 ft. lbs. (29–33 Nm)

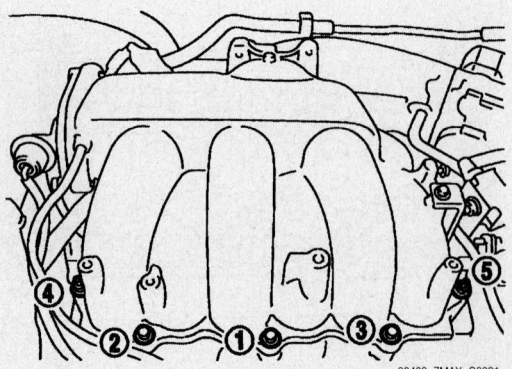

09482_ZMAX_G0021

Upper intake manifold collector tightening sequence—Maxima 3.5L engine

- Protective shields and torque the bolts in steps to 46–57 inch lbs. (5–7 Nm)
- Exhaust manifolds to the exhaust pipes and torque the nuts to 32–37 ft. lbs. (43–50 Nm)
- HO2S sensor to the manifold and torque the fastener to 30–44 ft. lbs. (40–60 Nm), if equipped
- Negative battery cable

5. Start the engine and check for exhaust leaks.

350Z

1. Before servicing the vehicle, refer to the Precautions Section.
2. Drain the engine coolant
3. Remove or disconnect the following:
- Negative battery cable
- Strut tower bar
- Engine cover
- Air cleaner case and duct

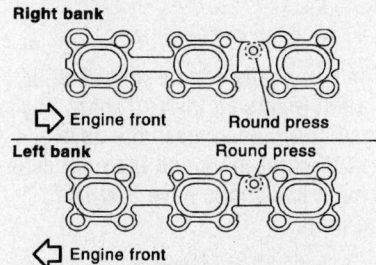

Right bank

Engine front **Round press**

Left bank **Round press**

Engine front

67162-NISS-G22

Exhaust manifold gasket identification—350Z 3.5L engine

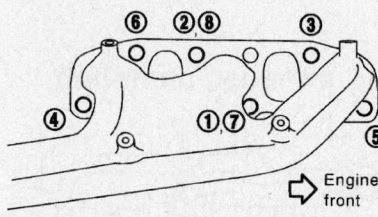

Right bank

Engine front

Left bank

Engine front

67162-NISS-G23

Exhaust manifold bolt tightening sequence—350Z 3.5L engine

- Heated oxygen sensor No. 2 connectors and sensors
- Exhaust mounting bracket between transmission and catalytic converters
- Catalytic converters
- Heated oxygen sensor No. 1 connectors and sensors
- Water and heater pipe on both sides
- Heat shield
- Exhaust manifold bolts in reverse of the tightening sequence
- Manifold gaskets

To install:

4. Clean all gasket mounting surfaces. Install new gaskets noting the correct placement.
5. Installation is the reverse of the removal procedure noting the following:
- Exhaust manifold and torque the nuts in sequence to 22 ft. lbs. (30 Nm)
- Heat shields and torque the bolts in steps to 51 inch lbs. (6 Nm)
- Exhaust manifolds to the exhaust pipes and torque the nuts to 46 ft. lbs. (63 Nm)
- Oxygen sensors and torque the fastener to 33 ft. lbs. (45 Nm)

Camshaft and Valve Lifters

REMOVAL & INSTALLATION

Maxima

1. Before servicing the vehicle, refer to the Precautions Section.
2. Relieve the fuel system pressure.
3. Drain the engine oil.
4. Drain the cooling system.
5. Remove or disconnect the following:
- Negative battery cable
- Left side rocker cover ornament

➡️**Before detaching any hoses or connectors, note the locations for reassembly.**

- Air duct to intake manifold hose, collector hose, blow-by hose and vacuum hoses
- Fuel hoses and detach the harness connections
- Canister purge hoses
- Water hoses from the cylinder head and intake manifold
- All 6 ignition coils from the spark plugs
- Spark plugs
- Bolts that secure the Exhaust Gas Recirculation (EGR) tube

- EGR tube
- Intake manifold collector supports and the collector
- Bolts that secure the fuel tube and the fuel tube
- Bolts that secure the intake manifold to the engine block and the manifold. Loosen the bolts in the reverse sequence of the tightening procedure.
- Left-hand and right-hand rocker covers from the cylinder head
- Engine undercovers
- Right front wheel and engine side covers
- Drive belts and idler pulley
- Power steering oil pump belt and the power steering oil pump assembly
- Camshaft Position (CMP) sensor (PHASE) and Crankshaft Position (CKP) sensors (REF)/(POS)

6. Set the No. 1 piston to Top Dead Center (TDC) of compression stroke by rotating the crankshaft.
- Ring gear cover access plate. Loosen the crankshaft pulley bolt while securing the ring gear so the crankshaft cannot rotate
- Crankshaft pulley, using a suitable puller
- Air conditioning compressor and bracket
- Front exhaust pipe and its support

7. Hang the engine at the right and left side engine slingers with a suitable hoist.
8. Support the transaxle with jack.
- Right side engine mounting, mounting bracket and nuts
- Center crossmember assembly
- Steel (lower) oil pan bolts in the reverse of the installation sequence

9. Insert a seal cutter between the steel and aluminum oil pan
10. Tapping the cutter with a hammer, slide it around the entire edge of the oil pan. Be careful not to damage the aluminum mating surface of the upper oil pan.
- Steel oil pan and the oil strainer
- Aluminum (upper) oil pan bolts in the reverse of the installation sequence
- Transaxle bolts that secure the oil pan

11. Insert a seal cutter between the aluminum oil pan and the engine block.
12. Tapping the cutter with a hammer, slide it around the entire edge of the oil pan. Be careful not to damage the mating surfaces of the oil pan or engine block.
- Oil pan from the vehicle
- Water pump cover and the bolts

that secure the front timing chain case cover
- Timing chain case cover, using the seal cutter
- Internal timing chain guide and the upper chain guide
- Timing chain tensioner and slack side chain guide
- Left and right intake camshaft sprockets first. Be sure to hold the flats of the camshafts while removing the sprocket bolts.
- Lower timing chain assembly. Be sure to note the aligning marks of the chain before removal.

13. Insert a suitable stopper pin for the left and right camshaft tensioners.
- Left and right exhaust camshaft sprocket bolts. Be sure to hold the flats of the camshafts while removing the sprocket bolts.
- Upper timing chain assembly. Be sure to note the aligning marks of the chain before removal.
- Lower timing chain guide
- Crankshaft sprocket
- Bolts that secure the rear timing chain case. The bolts must be loosened in sequence.
- Rear timing case cover, using the seal cutter

➡Remove the O-rings from the front of the engine block.

- Camshaft bearing caps in several steps. The bearing caps MUST be loosened in sequence.

➡Keep all bearing caps and camshafts in proper order for installation.

- Left-hand and right-hand camshaft tensioners from the cylinder head
- Camshafts from the cylinder heads

➡The valve lifters have a replaceable shim on the top of the lifter. Note the proper locations of each shim to lifter and remove the shims from the lifters.

- Valve adjusting shim from the lifter, using a magnet
- Lifter assembly from the bore. Be sure to note the locations from where each lifter came.

14. Check the diameter of the valve lifter and the valve lifter guide bore.
15. The diameter of the lifter should be 1.3764–1.3770 in. (34.960–34.975mm) and the diameter of the bore should be 1.3780–1.3788 in. (35.000–35.021mm).
16. Remove all traces of liquid gasket from the timing chain case and from the water pump covers.

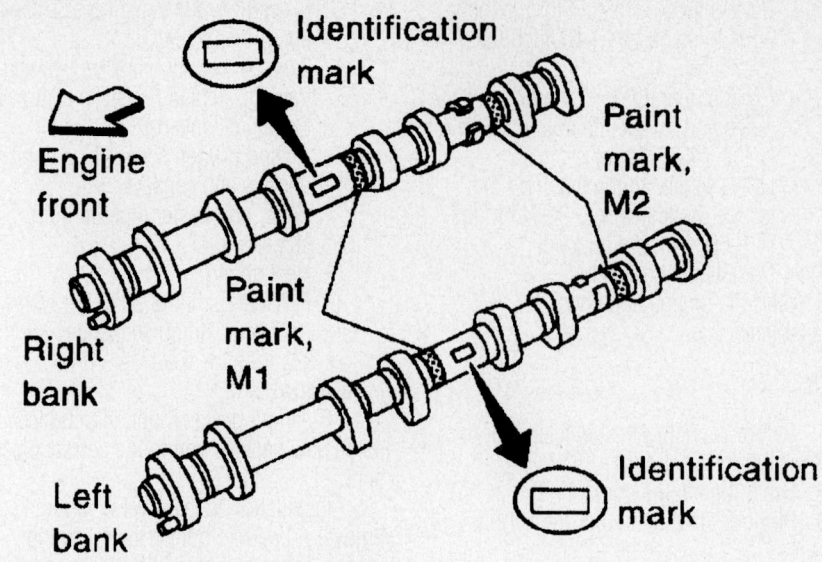

Camshaft identification marks—3.5L engines

17. Remove all traces of liquid gasket from the engine block.
18. Inspect the camshafts for excessive wear or damage and replace as necessary.

To install:

➡Before installing the camshaft brackets, apply RTV sealant to the mating surface of the No. 1 journal head.

19. Lubricate the valve lifters with clean engine oil and install the lifters into the bore from which they were removed.
20. Lubricate the valve lifter shims with clean engine oil and install the shims into the lifter from which they were removed.
21. Turn the crankshaft clockwise until the No. 1 piston is set 240 degrees before TDC on compression stroke.

22. Install or connect the following:
- Camshaft tensioners on both sides of the cylinder heads and torque the bolts to 75–96 inch lbs. (8.4–10.8 Nm)

➡The camshafts can be identified by the paint marks on the camshaft. The left cylinder head camshafts have a YELLOW paint mark and the right cylinder head camshafts have a WHITE paint mark.

- Exhaust and intake camshafts and install the bearing caps. Before installing the No. 1 bearing cap, apply liquid gasket to the corners of the cap.

➡When installing the camshafts, position the camshaft keys at the 12 o'clock

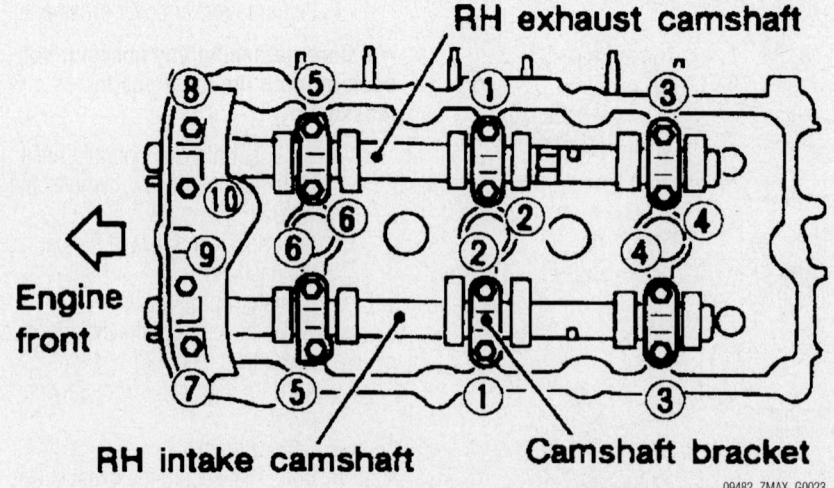

Right cylinder head camshaft bearing cap tightening sequence—3.5L engine

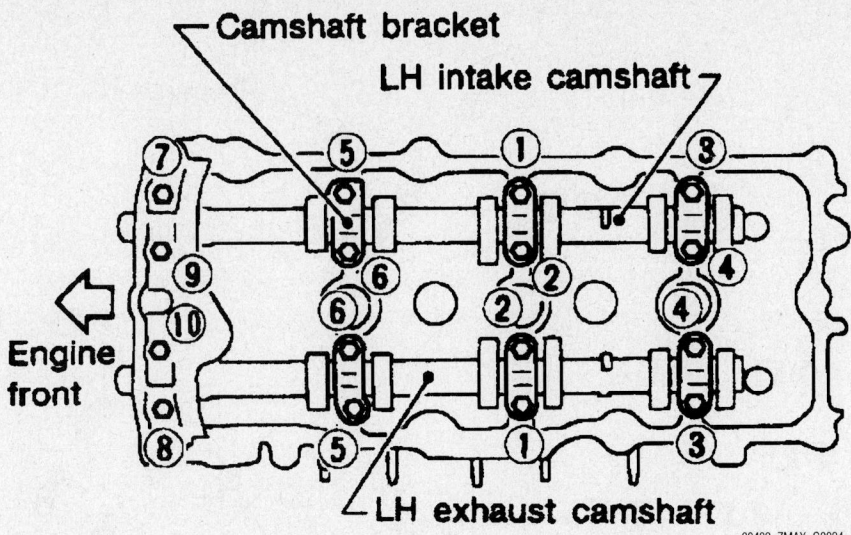

Left cylinder head camshaft bearing cap tightening sequence—3.5L engine

09482_ZMAX_G0024

position in respect to the cylinder head angle.

23. Torque the camshaft bearing caps as follows:

    a. Bolts No. 7–10: 17 inch lbs. (2 Nm).

    b. Bolts No. 1–6: 17 inch lbs. (2 Nm).

    c. Bolts No. 1–10: 52 inch lbs. (6 Nm).

    d. Bolts No. 1–10: 81–104 inch lbs. (9–11 Nm).

24. Install new O-rings to the front of the engine block.

25. Apply sealant to the hatched portion of the of the rear timing chain case.

26. Align the rear timing chain case with the dowel pins and install onto the cylinder heads and engine block.

27. Torque the rear timing chain case mounting bolts in sequence to 105–121 inch lbs. (11.8–13.7 Nm).

28. Install the crankshaft sprocket with the mating mark facing out.

29. Rotate the crankshaft clockwise and position the crankshaft to TDC of compression stroke and align the dowels of the camshaft sprockets to the 12 o'clock position in respect to the cylinder head.

30. Install the lower chain guide on the dowel pin with the front mark on the guide facing upward.

31. On a workbench, align the marks on the intake and exhaust camshaft sprockets with the marks of the chain.

32. Put the exhaust camshaft sprockets onto the dowel pin and torque the bolts to 88–95 ft. lbs. (119–128 Nm). Be sure to secure the camshafts while tightening the bolts.

33. Install or connect the following:

- Timing chains, sprockets and related components
- Transaxle bolts that secure the oil pan
- Oil pan strainer and torque the bolts to 12–14 ft. lbs. (16–19 Nm)

34. Apply a 0.177–0.217 in. (4.5–5.5mm) continuous bead of liquid gasket to the lower oil pan mating surface and install the oil pan. Torque the bolts in sequence to 57–66 inch lbs. (6.4–7.5 Nm).

- Center crossmember assembly
- Right side engine mounting bracket and mount assembly

35. Remove the engine slinger assembly.

- Front exhaust pipe and its support
- Air conditioning compressor and bracket
- Crankshaft pulley to the crankshaft and install the mounting bolt. Torque the bolt to 14–22 ft. lbs. (20–29 Nm). Torque the crankshaft bolt an additional 60–66 degrees clockwise. This is about the angle from one hexagon bolt head corner to another
- CMP sensor, PHASE and CKP sensors
- Power steering pump
- Idler pulley and all belts
- Engine side and under covers
- Right front wheel
- Intake manifold
- Rocker covers
- Fuel tube
- Intake manifold support and collector
- EGR tube
- Spark plugs and ignition coils
- Coolant hoses

- Canister purge hoses
- Fuel feed and return lines
- Vacuum hoses
- Negative battery cable

36. Fill the cooling system.

37. Fill the engine with clean oil.

38. Start the vehicle, check for leaks and repair if necessary.

350Z

1. Before servicing the vehicle, refer to the Precautions Section.

2. Relieve the fuel system pressure.

3. Drain the engine oil.

4. Drain the cooling system.

5. Disconnect the negative battery cable.

6. Remove the timing chain case, camshaft sprockets, timing chain and rear timing chain case.

7. Remove or disconnect the following:

- Camshaft position sensors (PHASE) from the back of the cylinder heads
- Intake valve timing control solenoid valves and discard the gaskets
- Camshaft bearing caps in the reverse of the tightening sequence
- Camshafts
- Valve lifters
- Secondary timing chain tensioners

To install:

➡**Before installing the camshaft brackets, apply RTV sealant to the mating surface of the No. 1 journal head.**

8. Lubricate the valve lifters with clean engine oil and install the lifters into the bore from which they were removed.

9. Lubricate the valve lifter shims with clean engine oil and install the shims into the lifter from which they were removed.

10. Ensure the crankshaft is set to TDC for the No. 1 cylinder.

11. Install the camshaft tensioners using new O-rings on both sides of the cylinder heads. The sliding part faces downward on the right head and upward on the left head. Torque the bolts to 75 inch lbs. (8.4 Nm).

➡**The camshafts can be identified by the paint marks on the camshaft. The intake camshafts have a PINK paint mark and the exhaust camshafts have a ORANGE paint mark.**

- Install the camshafts so the large and small pin holes are located on the front face of the camshafts at 180° intervals.

12. Install the bearing caps aligning the stamp marks on the caps as shown.

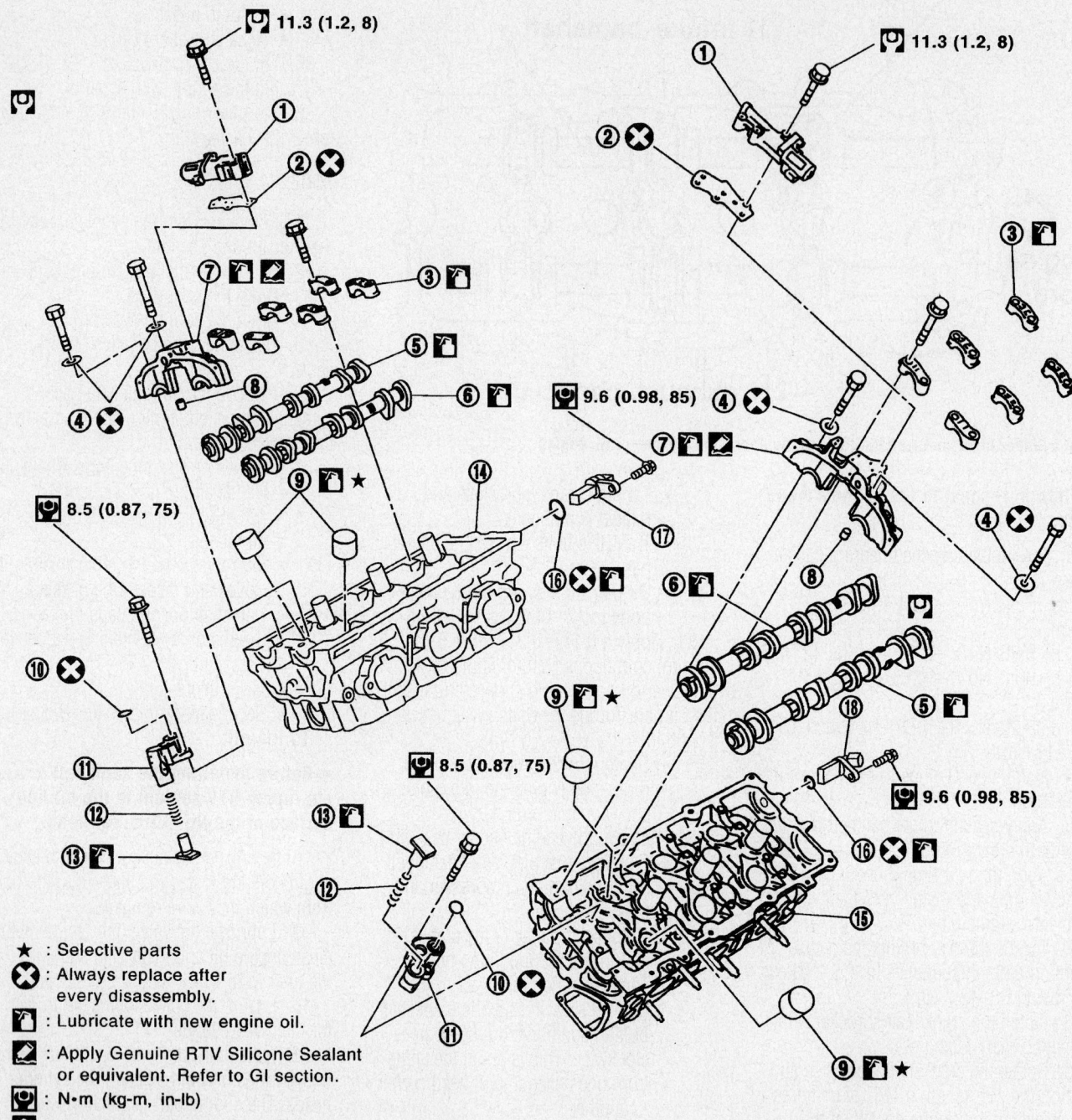

★ : Selective parts

⊗ : Always replace after every disassembly.

🔧 : Lubricate with new engine oil.

✎ : Apply Genuine RTV Silicone Sealant or equivalent. Refer to GI section.

⚙ : N•m (kg-m, in-lb)

🔩 : N•m (kg-m, ft-lb)

1. Intake valve timing control solenoid valve	2. Gasket	3. Camshaft bracket (No. 2 to No. 4)
4. Seal washer	5. Camshaft (EXH)	6. Camshaft (INT)
7. Camshaft bracket (No. 1)	8. Dowel pin	9. Valve lifter
10. O-ring	11. Timing chain tensioner (Secondary)	12. Spring
13. Plunger	14. Cylinder head (right bank)	15. Cylinder head (left bank)
16. O-ring	17. Camshaft position sensor (PHASE) (right bank)	18. Camshaft position sensor (PHASE) (left bank)

67162-NISS-G29

Exploded view of camshaft assemblies—350Z 3.5L engine

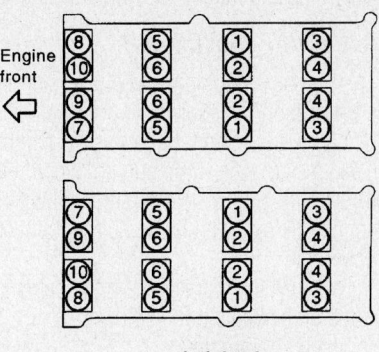

Camshaft bearing cap tightening sequence—350Z 3.5L engine

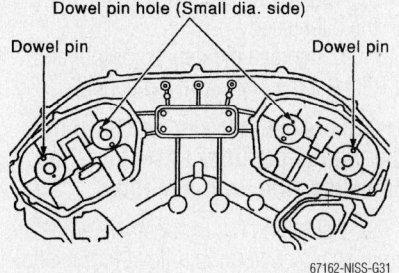

Camshaft dowel pin installation location—350Z 3.5L engine

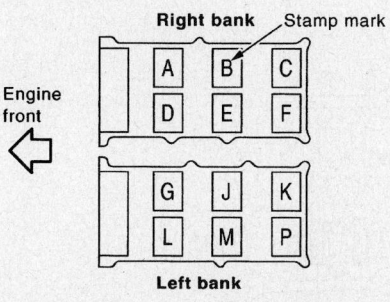

Camshaft bearing cap stamp positioning—350Z 3.5L engine

13. Torque the camshaft bearing caps as follows:

    a. Bolts No. 7–10: 17 inch lbs. (2 Nm).

    b. Bolts No. 1–6: 17 inch lbs. (2 Nm).

    c. Bolts No. 1–10: 52 inch lbs. (6 Nm).

    d. Bolts No. 1–6: 81–104 inch lbs. (9–11 Nm).

    e. Bolts No. 7–10: 74–91 inch lbs. (8.3–10.3 Nm).

14. Check and adjust the valve clearance.

15. Install or connect the following:

- Intake valve timing control solenoid valves using new gaskets and tighten the bolts to 8 ft. lbs. (11.3 Nm)
- Camshaft position sensors (PHASE) and tighten the bolts to 85 inch lbs. (9.6 Nm)
- Timing chain case, camshaft sprockets, timing chain and rear timing chain case
- Negative battery cable

16. Fill the cooling system.

17. Fill the engine with clean oil.

18. Start the vehicle, check for leaks and repair if necessary.

Valve Lash

ADJUSTMENT

➡**Check and adjust the valve clearances while the engine is cold and not running.**

CHECKING VALVE LASH

1. Before servicing the vehicle, refer to the Precautions Section.

2. Remove or disconnect the following:
- Intake manifold collector
- Left and right rocker covers
- Spark plugs

3. Set the No. 1 cylinder at Top Dead Center (TDC) on its compression stroke. Align the pointer with the TDC mark on the crankshaft pulley. Check that the valve lifters on the No. 1 cylinder are loose and valve lifters on the No. 4 cylinder are tight. If not, turn the crankshaft 1 revolution (360 degrees) and align the pointer with the TDC mark on the crankshaft pulley.

4. Check the following valves:
- Both No. 1 intake valves
- Both No. 2 exhaust valves
- Both No. 3 exhaust valves
- Both No. 6 intake valves

5. Using a feeler gauge, measure the clearance between the valve lifter and the camshaft. Record any valve clearance measurements that are out of specification. Intake valve clearance (cold) is 0.010–0.013 in. (0.26–0.34mm) and exhaust valve clearance (cold) is 0.011–0.015 in. (0.29–0.37mm).

6. Turn the crankshaft 240 degrees and set the No. 3 cylinder to TDC of its compression stroke.

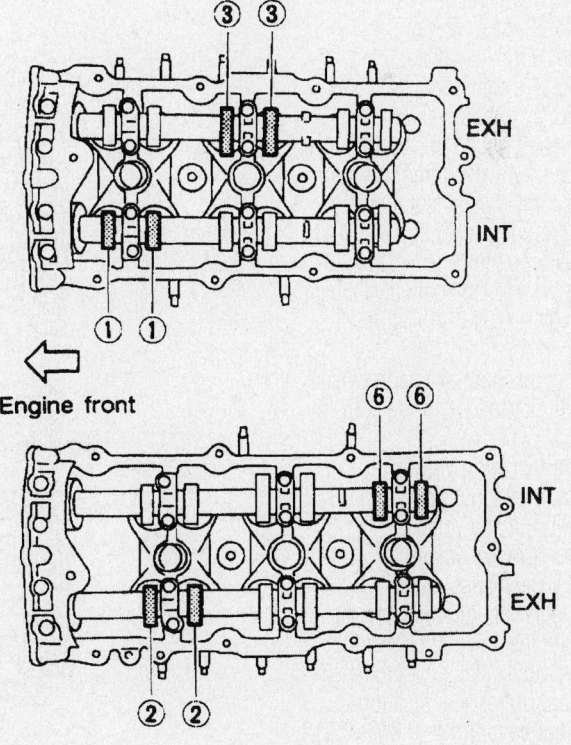

Measure the valves indicated while the No. 1 piston is at TDC on the compression stroke—3.5L engine

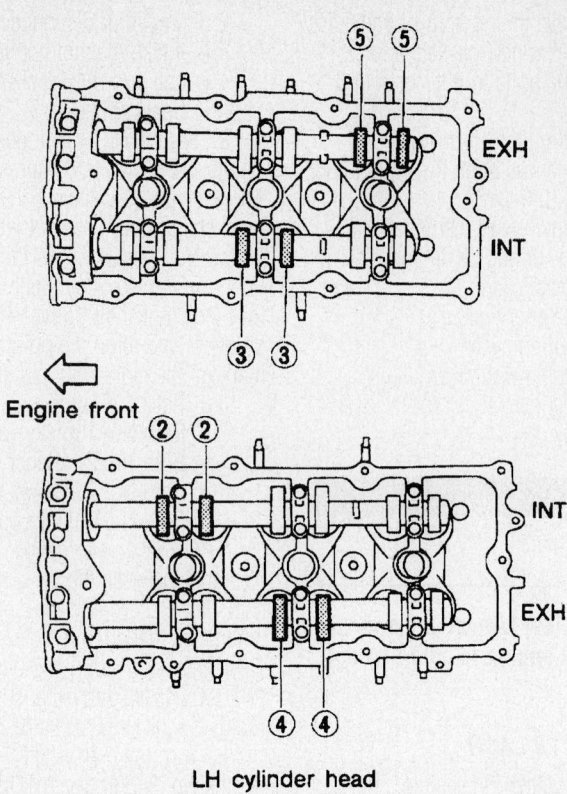

Engine front

LH cylinder head

09482_ZMAX_G0026

Measure the valves indicated while the No. 3 piston is at TDC on the compression stroke—3.5L engine

7. Check the following valves:
- Both No. 2 intake valves
- Both No. 3 intake valves
- Both No. 4 exhaust valves
- Both No. 5 exhaust valves

8. Using a feeler gauge, measure the clearance between the valve lifter and the camshaft. Record any valve clearance measurements that are out of specification. Intake valve clearance (cold) is 0.010–0.013 in. (0.26–0.34mm) and exhaust valve clearance (cold) is 0.011–0.015 in. (0.29–0.37mm).

9. Turn the crankshaft 240 degrees and set the No. 5 cylinder to TDC of its compression stroke.

10. Check the following valves:
- Both No. 1 exhaust valves
- Both No. 4 intake valves
- Both No. 5 intake valves
- Both No. 6 exhaust valves

11. Using a feeler gauge, measure the clearance between the valve lifter and the camshaft. Record any valve clearance measurements that are out of specification. Intake valve clearance (cold) is 0.010–0.013 in. (0.26–0.34mm) and exhaust valve clearance (cold) is 0.011–0.015 in. (0.29–0.37mm).

12. If all the valve clearances are within

specification, install the cylinder head cover, spark plugs and the intake manifold collector.

ADJUSTING VALVE LASH

1. Before servicing the vehicle, refer to the Precautions Section.

2. If an adjustment is necessary, adjust the valve clearance while engine is cold by removing the adjusting shim. The adjusting shim can be removed by using the following procedures:

   a. Turn the crankshaft so the camshaft lobe of the valve to be adjusted is pointed straight up.

   b. Turn the lifter so the notch is pointed towards the center of the cylinder head; this will facilitate the shim removal process.

   c. Using a depressor tool, push down on the lifter and insert a keeper tool on the edge of the lifter to keep the lifter in the depressed position.

   d. Remove the depressor tool and remove the shim with a magnet.

➡**Compressed air can be blown into the hole of the lifter to separate the adjusting shim from the lifter.**

3. Determine the replacement adjusting shim size by using the following procedures and formula:

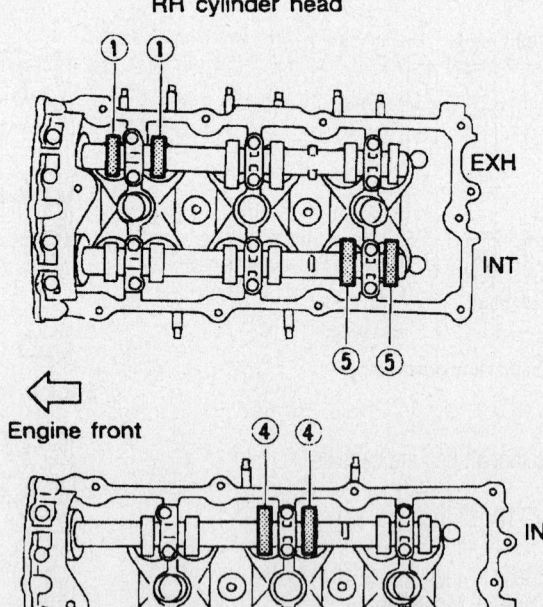

RH cylinder head

Engine front

LH cylinder head

09482_ZMAX_G0027

Measure the valves indicated while the No. 5 piston is at TDC on compression—3.5L engine

a. Using a micrometer determine thickness of the removed shim.

b. Calculate the thickness of a new adjusting shim so valve clearance is within the specified values.

c. R = thickness of the removed shim.

d. N = thickness of the new shim.

e. M = measured valve clearance.

- Intake shim determination formula: N = R + (M−0.0118 in. or 0.30mm)

- Exhaust shim determination formula: N = R + (M−0.0130 in. or 0.33mm)

4. Shims are available in 64 sizes from 0.0913–0.1161 in. (2.32–2.95mm) in steps of 0.004 in. (0.01mm). The thickness is stamped on the shim; this side is always installed facing down. Select new shims with thickness as close as possible to calculated valve and install it in the lifter.

5. Install the new shim onto the lifter.

6. Depress the lifter and remove the keeper tool. Remove the depressor tool and recheck the valve clearance. Repeat this procedure for any other valves requiring adjustment.

7. When all valve adjustments are finished, install the cylinder head cover, spark plugs and the intake manifold collector.

Starter Motor

REMOVAL & INSTALLATION

1. Remove or disconnect the following:
- Negative battery cable

- Air duct
- Harness protector from the harness
- Starter wiring at the starter
- Starter-to-engine bolts
- Starter from the vehicle

To install:

2. Install or connect the following:
- Starter and torque the long bolt to 57–72 ft. lbs. (77–98 Nm) on Maxima and 41 ft. lbs. (55 Nm) on 350Z. torque the short bolt to 22–30 ft. lbs. (30–41 Nm)
- Starter wiring
- Harness protector
- Air duct
- Negative battery cable

Oil Pan

REMOVAL & INSTALLATION

1. Before servicing the vehicle, refer to the Precautions Section.

2. Drain the engine oil.

3. Remove or disconnect the following:
- Negative battery cable
- Engine undercovers
- Steel (lower) oil pan bolts in the reverse of the installation sequence

4. Insert a seal cutter between the steel and aluminum oil pan.

5. Tapping the cutter with a hammer, slide it around the entire edge of the oil pan. Be careful not to damage the aluminum mating surface of the upper oil pan.
- Steel oil pan and the oil strainer
- Front exhaust pipe and its support

6. Hang the engine at the right and left side engine slingers with a suitable hoist.

7. Position a suitable jack under the transaxle.
- Crankshaft Position (CKP) sensors (REFERENCE and POSITION) from the oil pan
- Front and rear engine mounting nuts and bolts
- Center crossmember assembly
- Engine drive belts
- A/C compressor and mounting bracket
- Rear cover plate and the lower transaxle bolts
- Aluminum (upper) oil pan bolts in the reverse of the installation sequence

8. Insert a seal cutter between the aluminum oil pan and the engine block.

9. Tapping the cutter with a hammer, slide it around the entire edge of the oil pan. Be careful not to damage the mating surfaces of the oil pan or engine block.

10. Remove or disconnect the following:
- Oil pan assembly
- Bolts that secure the baffle plate and the baffle plate
- O-rings from the cylinder block and oil pump body

To install:

11. Install or connect the following:
- Baffle plate to the oil pan and torque the bolts to 22–27 inch lbs. (2.5–3.1 Nm) and apply sealant to the front and rear seal of the oil pan
- New O-rings to the cylinder block and the oil pump body

12. Apply a 4.5–5.5mm wide continuous bead of liquid gasket to the upper oil pan mating surface and install the oil pan. Torque the bolts in sequence to 12–14 ft. lbs. (16–19 Nm).

13. Install or connect the following:
- Oil pan strainer and torque the bolts to 12–14 ft. lbs. (16–19 Nm)
- Rear cover plate and lower transaxle bolts
- A/C compressor and bracket
- Drive belts
- Center crossmember
- Front and rear engine mount hardware
- CKP sensors
- Front exhaust tube and support
- Oil strainer
- Steel oil pan and torque the bolts, in sequence, to 66 inch lbs. (7.5 Nm)
- Engine under covers
- Negative battery cable

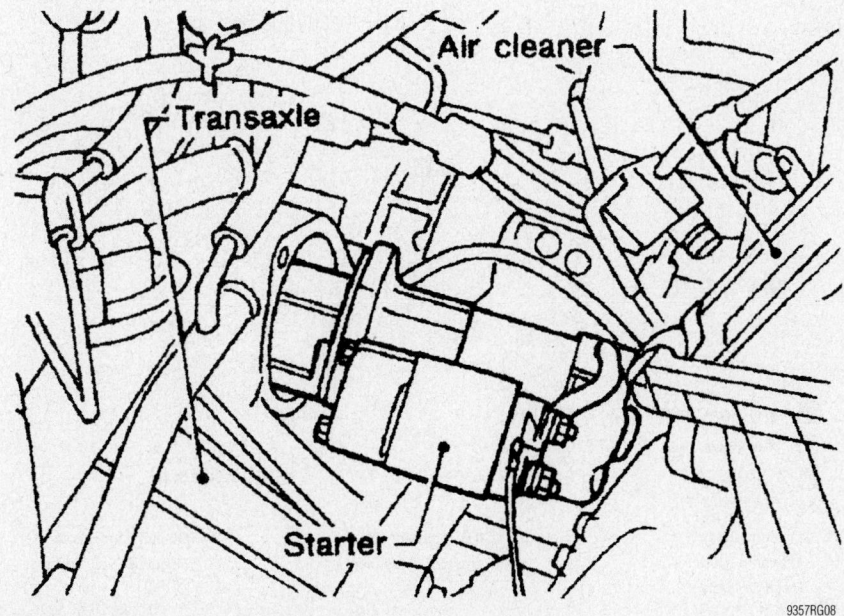

Starter location and mounting detail—3.5L Maxima

9357RG08

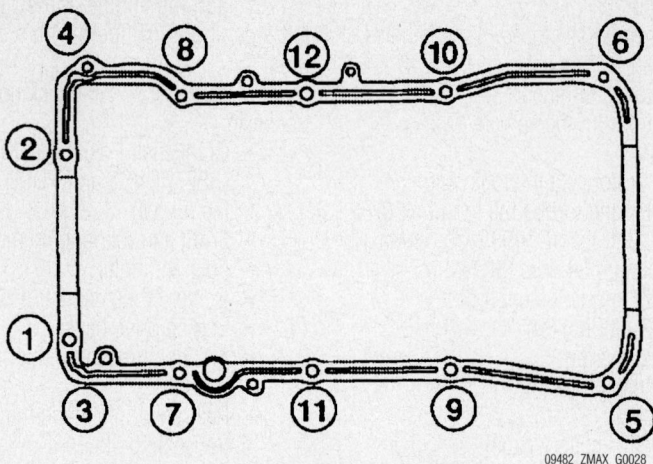

09482_ZMAX_G0028

Aluminum (upper) oil pan bolt loosening sequence—3.5L engine

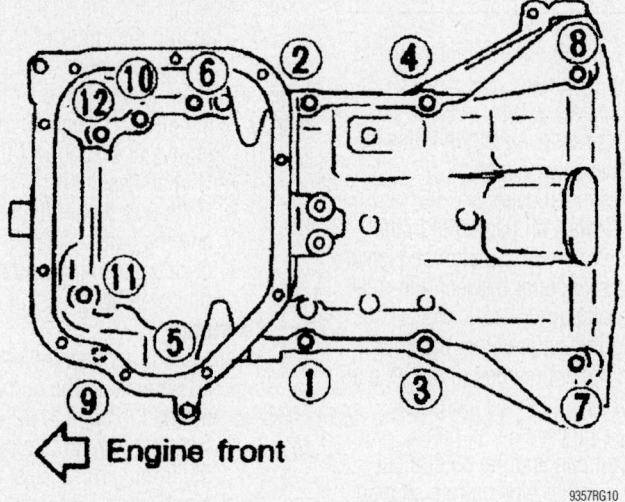

⬅ Engine front

9357RG10

To prevent pan warpage, tighten the aluminum oil pan bolts in the sequence shown—3.5L engine

14. After waiting approximately 30 minutes, fill the engine with clean oil.

15. Start the vehicle, check for leaks and repair if necessary.

Oil Pump

REMOVAL & INSTALLATION

1. Before servicing the vehicle, refer to the Precautions Section.
2. Drain the engine oil.
3. Remove or disconnect the following:
   - Negative battery cable
   - Drive belts
   - Camshaft Position (CMP) sensor (PHASE) and the Crankshaft Position (CKP) sensor (REF)/(POS)
   - Engine lower covers
   - Crankshaft pulley

- Front exhaust tube and support
- Right side mounting insulator and bracket
- Center member
- A/C compressor and move it aside
- Oil pans
- Water pump cover
- Front cover
- Timing chain
- Oil pump assembly

4. Clean all mating surfaces.

To install:

5. Install or connect the following:
   - Oil pump
   - Timing chain
   - Front cover and torque the long bolt to 62 inch lbs. (7 Nm) and the short bolt to 61 inch lbs. (6.5 Nm)
   - Water pump cover
   - Oil pans
   - A/C compressor
   - Center member
   - Right side mounting insulator and bracket
   - Front exhaust tube and support
   - Crankshaft pulley
   - CMP and CKP sensors
   - Engine lower covers
   - Drive belts
   - Negative battery cable

6. Fill the engine with clean oil.
7. Start the vehicle, check for leaks and repair if necessary.

Rear Main Seal

REMOVAL & INSTALLATION

1. Before servicing the vehicle, refer to the Precautions Section.

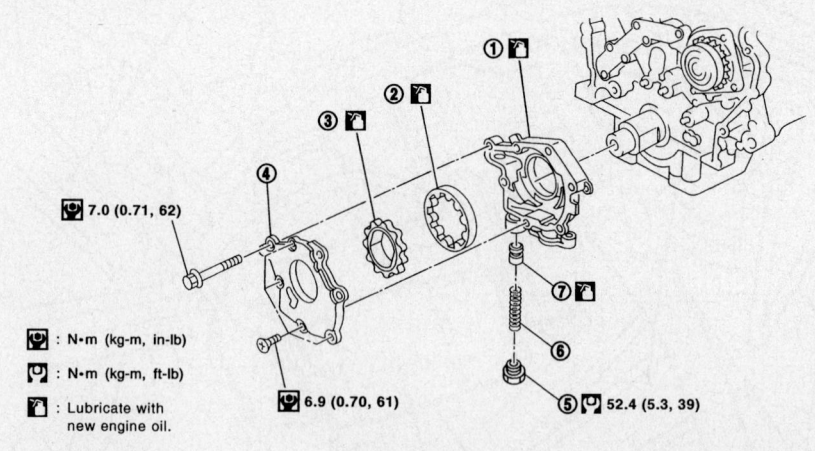

7.0 (0.71, 62)

: N·m (kg-m, in-lb)

: N·m (kg-m, ft-lb)

: Lubricate with new engine oil.

6.9 (0.70, 61)

52.4 (5.3, 39)

1. Oil pump body
2. Oil pump outer rotor
3. Oil pump inner rotor
4. Oil pump cover
5. Regulator valve plug
6. Regulator valve spring
7. Regulator valve

09482_ZMAX_G0029

Exploded view of the oil pump assembly–3.5L engine

2. Drain the engine oil.

3. Remove or disconnect the following:
- Transaxle
- Driveplate/flywheel
- Oil pan
- Oil seal retainer

4. Tap the oil seal out of the retainer with a hammer and drift.

5. Clean all mating surfaces of any residual liquid gasket.

To install:

6. Install or connect the following:
- New seal into the retainer
- Oil seal retainer
- Oil pan
- Driveplate/flywheel
- Transaxle

7. Fill the engine with clean oil.

8. Start the vehicle, check for leaks and repair if necessary.

Timing Chain, Sprockets, Front Cover and Seal

REMOVAL & INSTALLATION

Maxima

1. Before servicing the vehicle, refer to the Precautions Section.

2. Drain the engine oil.

3. Drain the cooling system.

4. Relieve the fuel system pressure.

5. Remove or disconnect the following:
- Negative battery cable
- Left side rocker cover ornament

➡**Before detaching any hoses or connectors, note the locations for reassembly.**

- Air duct to intake manifold hose, collector hose, blow-by hose and vacuum hoses
- Fuel hoses and detach the harness connections
- Canister purge hoses
- Water hoses from the cylinder head and intake manifold
- All 6 ignition coils from the spark plugs
- Spark plugs
- Bolts that secure the Exhaust Gas Recirculation (EGR) tube and remove the tube
- Intake manifold collector supports and the collector
- Fuel tube assembly
- Intake manifold. Loosen the bolts in the reverse sequence of the tightening procedure.
- Left-hand and right-hand intake

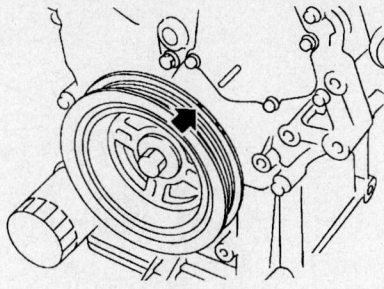

09482_ZMAX_G0030

Set the No. 1 piston to Top Dead Center (TDC)—3.5L engine

valve timing control solenoid valves
- Left-hand and right-hand rocker covers from the cylinder head
- Engine undercovers
- Right front wheel and the engine side covers
- Drive belts and the idler pulley
- Power steering oil pump belt and the power steering oil pump assembly
- Camshaft Position (CMP) sensor (PHASE) and Crankshaft Position (CKP) sensors (REF)/(POS)

6. Set the No. 1 piston to Top Dead Center (TDC) of compression stroke by rotating the crankshaft.

7. Loosen the crankshaft pulley bolt while securing the ring gear so the crankshaft cannot rotate.
- Ring gear cover access plate

➡**Use care not to damage the ring gear teeth.**

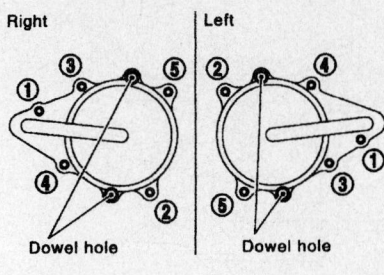

9357RG13

Loosen the intake valve timing control valve cover bolts in the reverse of the order shown—3.5L engine

- Crankshaft pulley using a suitable puller
- Intake valve timing control valve cover. Loosen the bolts in the reverse order shown in the accompanying figure. In the cover, the shaft is engaged with the center hole of the intake camshaft sprocket. Remove it straight out until the engagement comes off.
- A/C compressor and bracket
- Front exhaust pipe and its support

8. Hang the engine at the right and left side engine slingers with a suitable hoist.

9. Support the transaxle with jack.
- Right side engine mounting and bracket
- Center crossmember assembly
- Upper and lower oil pans
- Water pump cover
- Bolts that secure the front timing chain case, in sequence

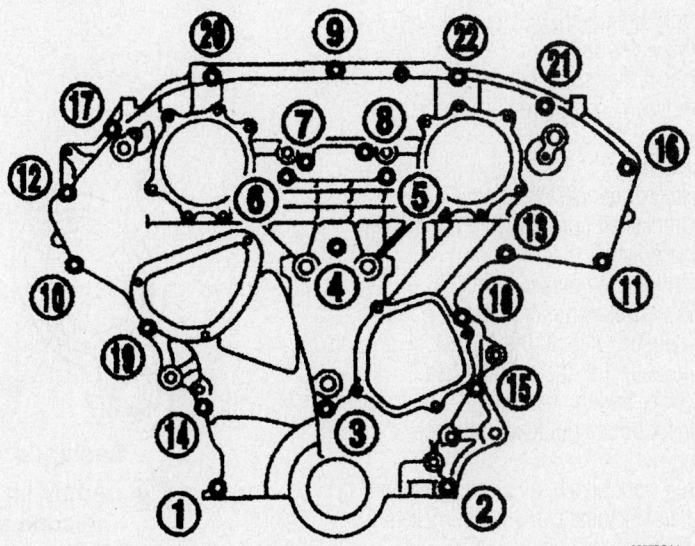

9357RG14

Remove the front timing chain case mounting bolts in the reverse of the sequence shown—3.5L engines

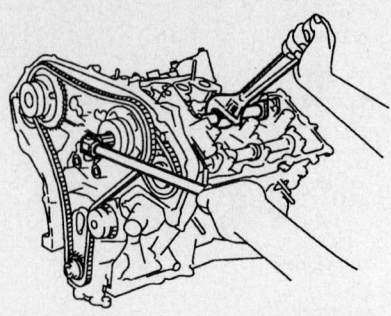

09482_ZMAX_G0031

Hold the camshaft with a wrench while removing the sprocket bolts—3.5L engine

- Timing chain case cover using a seal cutter
- Internal timing chain guide and the upper chain guide
- Timing chain tensioner and slack side chain guide
- Left and right intake camshaft sprockets first. Be sure to hold the flats of the camshafts while removing the sprocket bolts.
- Lower timing chain assembly. Be sure to note the aligning marks of the chain before removal.

10. Insert a suitable stopper pin for the left and right camshaft tensioners.

- Left and right exhaust camshaft sprocket bolts. Be sure to hold the flats of the camshafts while removing the sprocket bolts.
- Upper timing chain assembly. Be sure to note the aligning marks of the chain before removal.
- Lower timing chain guide
- Crankshaft sprocket
- All traces of liquid gasket from the front timing chain case and from the water pump

11. Inspect the timing chain for excessive wear or damage and replace as necessary.

To install:

12. Install or connect the following:

- Crankshaft sprocket with the mating mark facing out

13. Position the crankshaft to TDC of compression stroke and align the dowels of the camshaft sprockets to the 12 o'clock position in respect to the cylinder head.

- Lower timing chain guide. The front mark on the guide should face upwards.

14. On a work bench, align the marks on the intake and exhaust camshaft sprockets with the marks of the chain.

- Exhaust camshaft sprockets onto the dowel pin and torque the

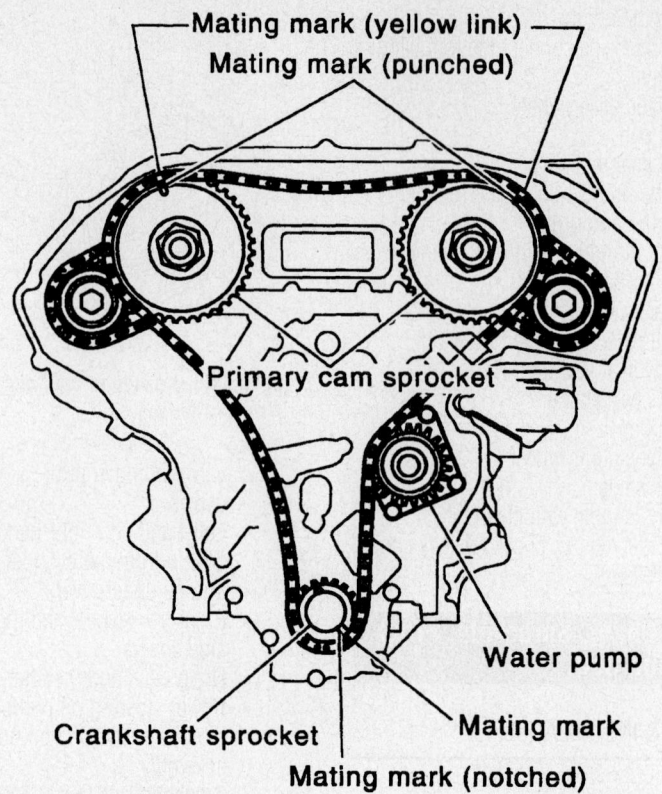

09482_ZMAX_G0032

Timing chain alignment marks—3.5L engine

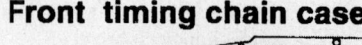

Front timing chain case

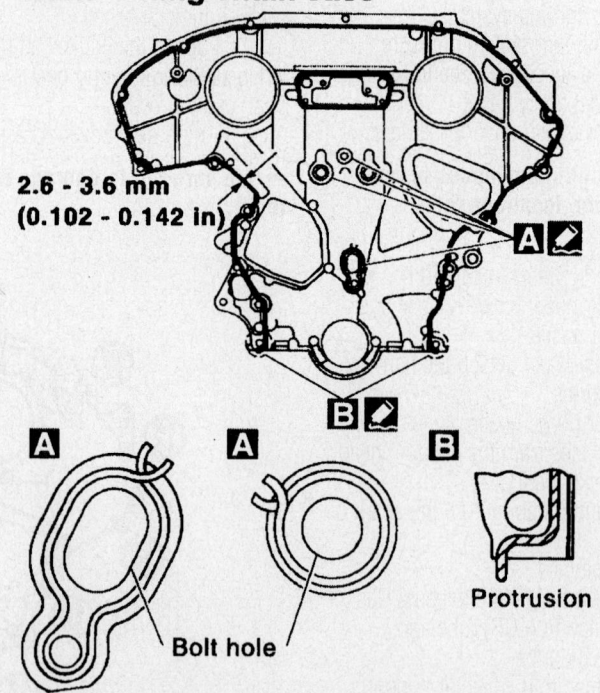

Sealant protrusion away from bolt hole

🔧 : Apply liquid gasket. (Use Genuine RTV silicone sealant or equivalent. Refer to GI section.)

09482_ZMAX_G0033

Application of liquid gasket to the front timing case—3.5L engine

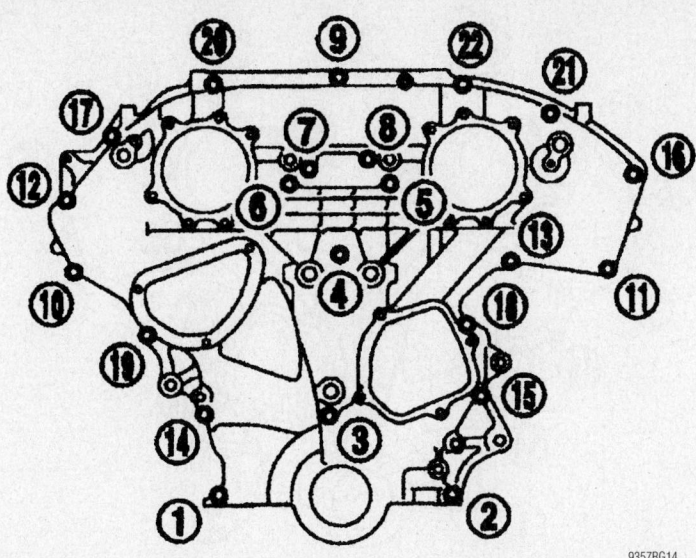

Tighten the front timing chain case bolts according to the sequence shown—3.5L engines

mounting bolts to 88–95 ft. lbs. (119–128 Nm). Be sure to secure the camshafts while tightening the bolts.
- Timing chains and sprockets to the intake camshafts. Be sure to align the timing chain and sprocket mating marks.
- Left and right camshaft tensioner stopper pins

15. Align the mating mark on the crankshaft with the matchmark (gold link) on the lower timing chain.

16. Install the lower timing chain to the water pump sprocket.

17. Working counterclockwise, install the lower timing chain camshaft sprockets. Be sure to align the sprocket marks with the blue links of the timing chain during installation.

18. Install or connect the following:
- Intake sprocket and torque the bolts to 88–95 ft. lbs. (119–128 Nm). Be sure to secure the camshafts while tightening the bolts.
- Internal timing chain guide, upper timing chain guide, lower timing chain tensioner and slack side timing chain guide.

19. Torque the tensioner mounting bolt to 75–96 inch lbs. (8.4–10.8 Nm) and the guide bolts to 108–168 inch lbs. (13–19 Nm).

20. Apply a 0.102–0.142 in. (2.6–3.6mm) continuous bead of liquid gasket to all necessary areas as shown on the front timing cover.
- Timing cover evenly and gently. Be sure to align the dowel pin holes.

21. Torque the mounting bolts in sequence as follows:
   a. Bolts No. 1 and 2: 19–23 ft. lbs. (26–31 Nm).
   b. Bolts No. 3–20: 105–121 inch lbs. (11.8–13.7 Nm).

➡ Leave the bolts unattended for 30 minutes or more after tightening. This will allow the liquid gasket to cure sufficiently.

22. Apply a 0.091–0.130 in. (2.3–3.3mm) continuous bead of liquid gasket to the water pump cover and install the cover. Torque the bolts to 84–108 inch lbs. (10–13 Nm).

23. Install or connect the following:
- Oil pans
- Center crossmember
- Right side engine mount and bracket
- Front exhaust pipe and remove the transaxle support
- A/C compressor and bracket

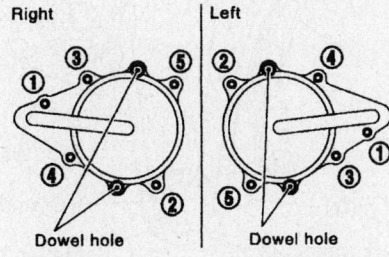

Tighten the intake valve timing control valve cover bolts in sequence—3.5L engine

- Crankshaft pulley
- Ring gear access cover plate
- CMP sensor and CKP sensors
- Power steering pump
- Idler pulley and drive belts
- Engine side cover and right front wheel
- Engine under covers
- Intake valve timing control solenoid valves with new covers. Tighten the cover bolts in the proper sequence.
- Rocker covers
- Intake manifold
- Fuel tube assembly
- Intake manifold collector and support
- EGR tube
- Spark plugs and ignition coils
- Coolant hoses
- Fuel hoses
- Air duct assembly and hoses
- Left side rocker cover ornament
- Negative battery cable

24. Fill the cooling system.
25. Fill the engine with clean oil.
26. Start the vehicle, check for leaks and repair if necessary.

350Z

1. Before servicing the vehicle, refer to the Precautions Section.
2. Drain the engine oil.
3. Drain the cooling system.
4. Relieve the fuel system pressure.
5. Remove or disconnect the following:
- Negative battery cable
- Engine cover
- Air cleaner case assembly
- Engine harnesses from timing chain case
- Upper and lower intake manifold collectors
- Cooling fan
- Drive belts
- A/C compressor and wire aside
- Power steering pump and wire aside
- Power steering pump bracket
- Alternator
- Water bypass hose, clamp and idler pulley bracket from timing chain case
- Intake valve timing control valve covers. Loosen the bolts in the reverse order shown in the accompanying figure. In the cover, the shaft is engaged with the center hole of the intake camshaft sprocket. Remove it straight out until the engagement comes off.

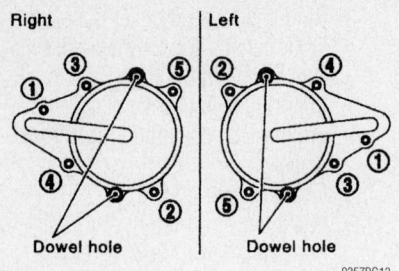

Right | Left

Dowel hole | Dowel hole

9357RG13

Loosen the intake valve timing control valve cover bolts in the reverse of the order shown—350Z 3.5L engine

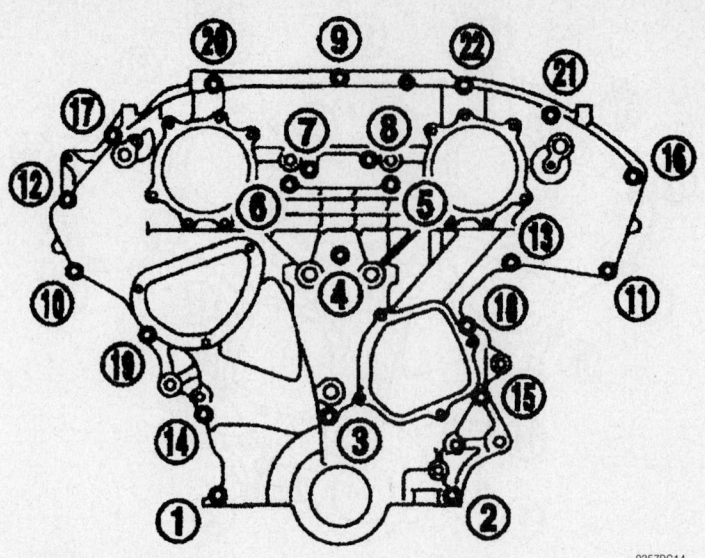

9357RG14

Remove the front timing chain case mounting bolts in the reverse of the sequence shown—350Z 3.5L engine

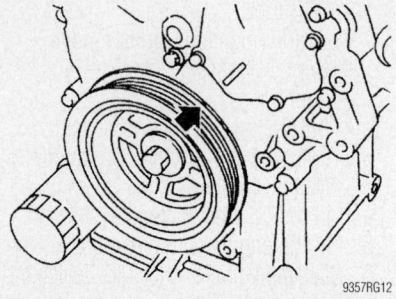

9357RG12

Set the No. 1 piston to Top Dead Center (TDC)—350Z 3.5L engine

- O-ring from timing chain case on both sides
- Both valve covers
- Rotate the crankshaft clockwise and set the No. 1 piston to TDC of the compression stroke
6. Make sure the intake and exhaust camshaft lobes are facing toward the inside of the cylinder head
7. Remove the starter and lock the flywheel through the starter mounting hole.
8. Loosen the crankshaft pulley bolt.
- Crankshaft pulley using a suitable puller

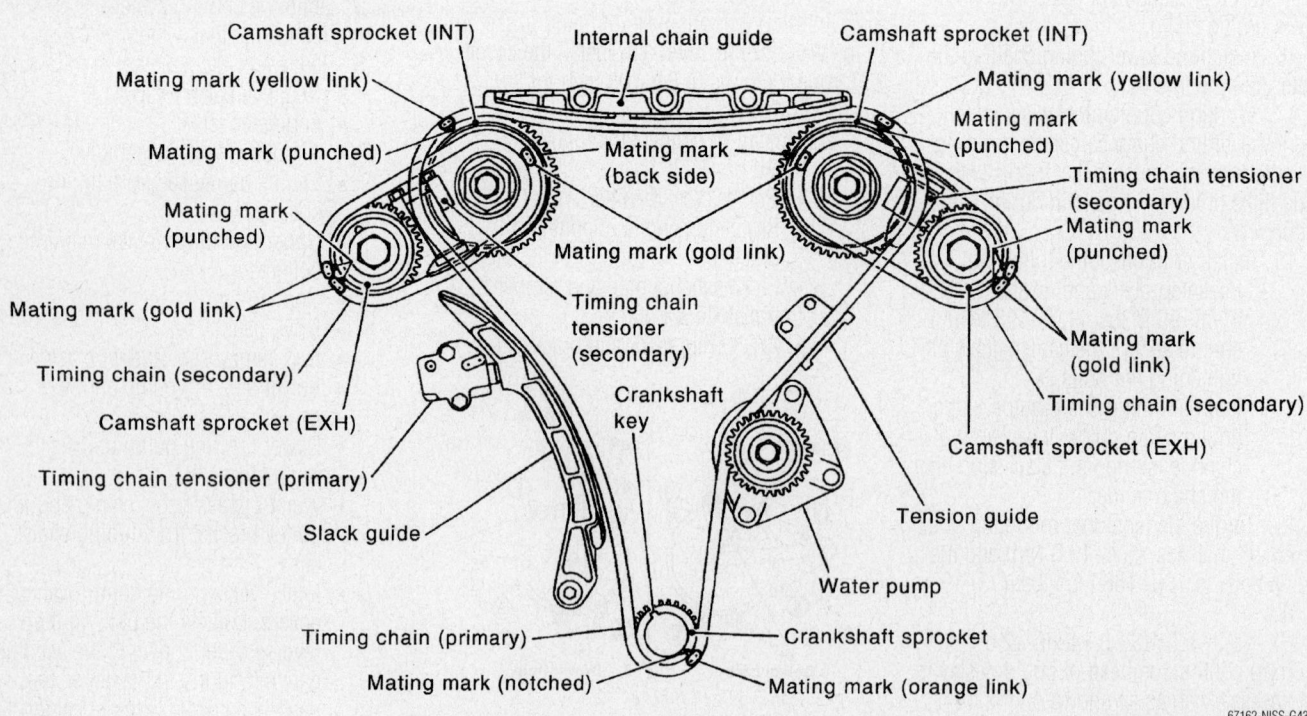

67162-NISS-G43

Timing chain alignment marks—350Z 3.5L engine

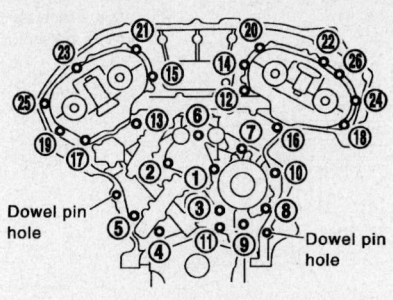

Rear timing chain case tightening sequence—350Z 3.5L engine

- Upper and lower oil pans
- Front timing chain cover bolts in reverse order of the tightening sequence
- Front timing chain case
- O-rings from rear timing chain cover
- Water pump and chain tensioner cover from rear cover
- Pry out front oil seal
- Insert a suitable stopper pin for the left and right primary camshaft tensioners
- Primary chain tensioners
- Timing chain guide, slack guide and tension guide

- Primary timing chain and crank-shaft sprocket
- Insert a suitable stopper pin for the left and right secondary camshaft tensioners
- Secondary chain tensioners
- Left and right intake camshaft sprocket bolts first, then exhaust sprocket bolts. Be sure to hold the flats of the camshafts while removing the sprocket bolts.
- Lower timing chain assembly with camshaft sprockets. Be sure to note the aligning marks of the chain before removal.

9. Remove the rear timing chain case bolts in the reverse order of the tightening sequence.

10. Remove the O-rings from the cylinder heads and block.
- All traces of liquid gasket from the front and timing chain case and from the water pump

11. Inspect the timing chain for excessive wear or damage and replace as necessary.

To install:
12. Install or connect the following:
- New O-rings to the cylinder heads and block

- Apply a bead of sealant to the back side of the rear timing chain case
- Install the rear timing chain case and tighten the bolts in sequence to 10 ft. lbs. (14 Nm). After tightening, retighten them again to 10 ft. lbs. (14 Nm).

13. Position the crankshaft to TDC of compression stroke and align the dowels of the camshaft sprockets to the 12 o'clock position in respect to the cylinder head.

14. Install the secondary timing chain and camshaft sprockets aligning the timing marks and timing chain links as shown.

15. Tighten the camshaft sprocket bolts to 76 ft. lbs. (103 Nm).

16. Remove the pins from the secondary timing chain tensioners.

17. Install the primary timing chain tension guide.

18. Install the crankshaft sprocket with the mating marks on the front side.

19. Install the primary timing chain and so the mating marks are aligned as shown.

20. Install the internal chain guide and slack guide.

Example: Right bank (Rear view)

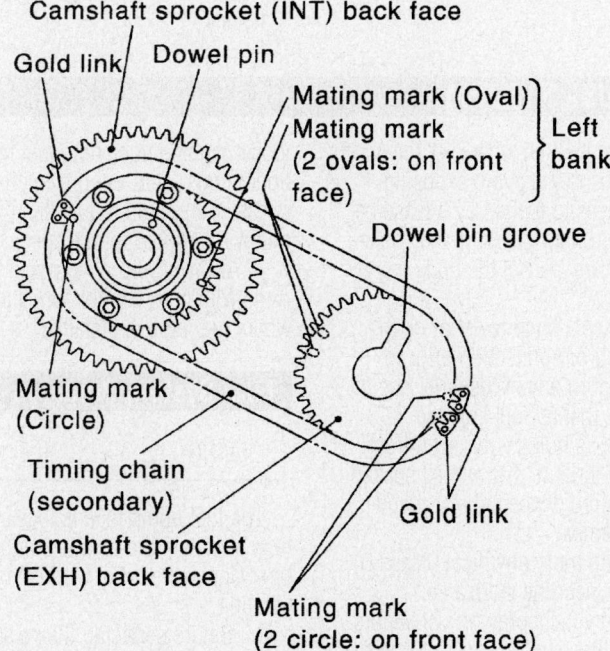

Secondary timing chain alignment marks—350Z 3.5L engine

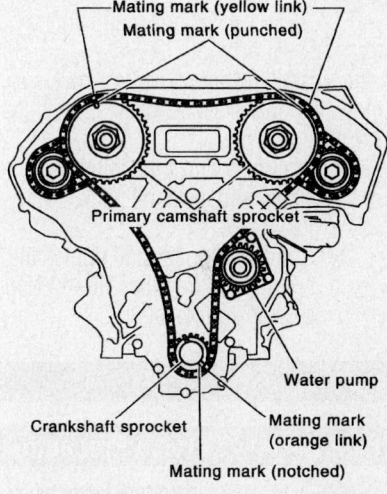

Primary timing chain alignment marks— 350Z 3.5L engine

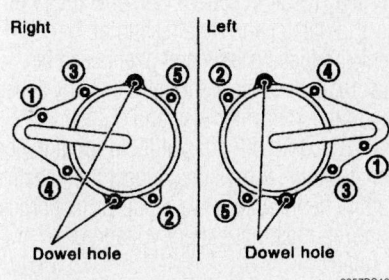

Tighten the intake valve timing control valve cover bolts in sequence—3.5L engine

21. Install the primary timing chain tensioner and tighten the bolt to 72 inch lbs. (8 Nm). Remove the stopper pin

22. Double check that the mating marks on the timing chain and sprockets are in the correct locations.

23. Install new O-rings on the timing chain case.

24. Apply clean engine oil to the front oil seal and dust seal lips.

25. Use a drift and press fit the oil seal into the timing chain case.

26. Apply liquid gasket to the water pump and chain tensioner cover openings, then install the covers.

27. Apply liquid gasket to the back side of the timing chain case cover.

28. Install the dowel pin on the rear timing chain case into the dowel pin of the front timing chain case.

29. Install the front timing chain case and tighten the bolts in sequence to 19–23 ft. lbs. (26–31 Nm) for M8 bolts and 9–10 ft. lbs. (12–14 Nm) for M6 bolts. After tightening, retighten them again to the same specification.

30. Install seal rings in the timing control cover shaft grooves and apply liquid gasket to the covers.

31. Install new O-rings in the timing chain case oil holes.

32. Install the timing control covers and tighten the bolt in sequence to 72 inch lbs. (8 Nm).

33. Install or connect the following:
- Upper and lower oil pans
- Valve covers
- Crankshaft pulley and tighten the bolt to 29–36 ft. lbs. (40–49 Nm), plus an additional 60°.

34. The remainder of the installation is the reverse of the removal procedure.

35. Fill the cooling system.

36. Fill the engine with clean oil.

37. Start the vehicle, check for leaks and repair if necessary.

Piston and Ring

POSITIONING

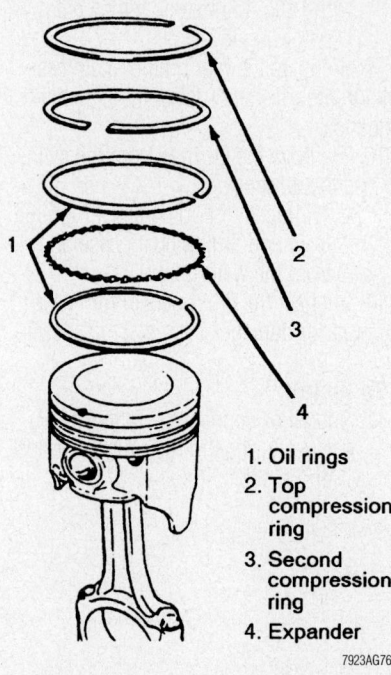

1. Oil rings
2. Top compression ring
3. Second compression ring
4. Expander

7923AG76

Exploded view of common piston ring mounting

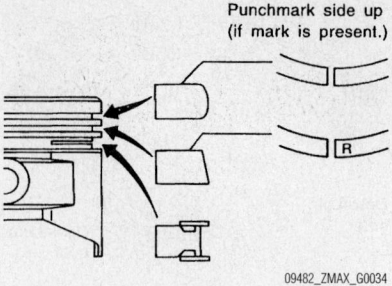

Punchmark side up (if mark is present.)

09482_ZMAX_G0034

Piston ring positioning—3.5L engine

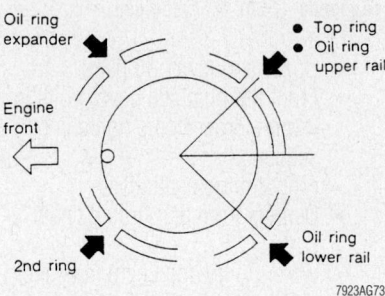

Oil ring expander

Top ring
Oil ring upper rail

Engine front

Oil ring lower rail

2nd ring

7923AG73

Piston ring end-gap spacing

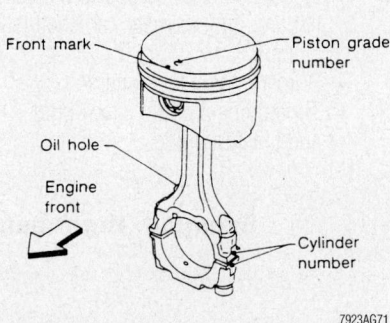

Front mark

Piston grade number

Oil hole

Engine front

Cylinder number

7923AG71

Piston and connecting rod assembly positioning

FUEL SYSTEM

Fuel System Service Precautions

Safety is the most important factor when performing not only fuel system maintenance but any type of maintenance. Failure to conduct maintenance and repairs in a safe manner may result in serious personal injury or death. Maintenance and testing of the vehicle's fuel system components can be accomplished safely and effectively by adhering to the following rules and guidelines.

- To avoid the possibility of fire and personal injury, always disconnect the negative battery cable unless the repair or test procedure requires that battery voltage be applied.

- Always relieve the fuel system pressure prior to disconnecting any fuel system component (injector, fuel rail, pressure regulator, etc.), fitting or fuel line connection.

Exercise extreme caution whenever relieving fuel system pressure, to avoid exposing skin, face and eyes to fuel spray. Please be advised that fuel under pressure may penetrate the skin or any part of the body that it contacts.

- Always place a shop towel or cloth around the fitting or connection prior to loosening to absorb any excess fuel due to spillage. Ensure that all fuel spillage (should it occur) is quickly removed from engine surfaces. Ensure that all fuel soaked cloths or towels are deposited into a suitable waste container.

- Always keep a dry chemical (Class B) fire extinguisher near the work area.

- Do not allow fuel spray or fuel vapors to come into contact with a spark or open flame.

- Always use a back-up wrench when loosening and tightening fuel line connection fittings. This will prevent unnecessary stress and torsion to fuel line piping. Always follow the proper torque specifications.

- Always replace worn fuel fitting O-rings with new. Do not substitute fuel hose where fuel pipe is installed.

Fuel System Pressure

RELIEVING

The fuel pump fuse is located in the dash fuse box or in the engine compartment fuse box. Check the lid of the fuse box for exact location.

1. Before servicing the vehicle, refer to the Precautions Section.

2. Remove the fuel pump fuse.

3. Start the engine.

4. Start the engine and run until the engine stalls.

5. After the engine stalls, try to restart the engine; if the engine will not start, the fuel pressure has been released.

6. Turn the ignition switch **OFF**. Reinstall the fuel pump fuse into the fuse block.

➡**Do not crank the engine or turn the ignition switch ON after the fuel pump fuse has been reinstalled, or the fuel pressure will be re-established.**

Fuel Filter

REMOVAL & INSTALLATION

➡**On 350Z models, the fuel filter and fuel pump are replaced as an assembly. See the fuel pump procedure to remove the fuel filter on these models.**

Maxima

1. Before servicing the vehicle, refer to the Precautions Section.

2. Properly relieve fuel system pressure.

3. Remove or disconnect the following:
   - Negative battery cable
   - Rear seat bottom
   - Inspection hole cover
   - Electrical and quick connectors
   - Six screws
   - Fuel level sensor unit and fuel pump assembly
   - Flange and snap fit portion of the fuel pump
   - Fuel tank temperature sensor harness

- Fuel level sensor flange
- Fuel pump connector
- Quick connectors from the fuel level sensor
- Fuel level sensor from the chamber
- Fuel filter from the chamber

To install:

4. Install or connect the following:
   - Fuel filter to the chamber
   - Fuel level sensor to the chamber
   - Quick connectors to the fuel level sensor
   - Fuel pump connector
   - Fuel level sensor flange
   - Fuel tank temperature sensor harness
   - Fuel pump assembly to the fuel tank
   - Screws and electrical connectors
   - Quick connectors
   - Negative battery cable

5. Start the vehicle, check for leaks and repair if necessary.
   - Inspection hole cover
   - Rear seat bottom

Fuel Pump

REMOVAL & INSTALLATION

Maxima

1. Before servicing the vehicle, refer to the Precautions Section.

2. Relieve the fuel system pressure

3. Remove or disconnect the following:
   - Negative battery cable

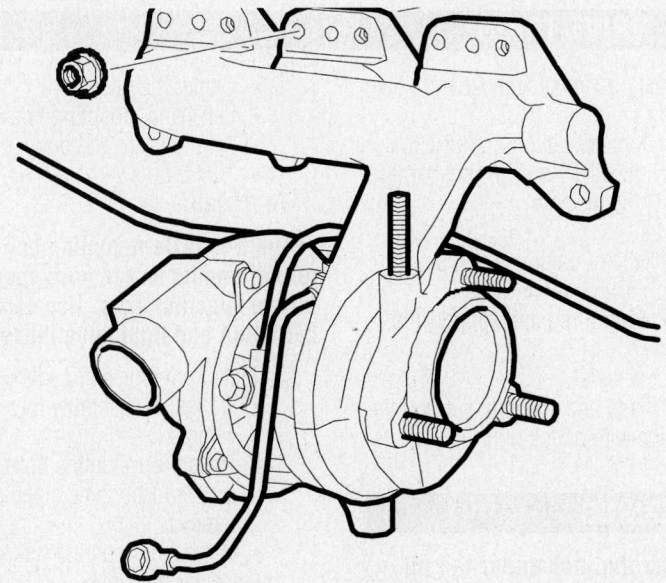

Remove the fuel filter from the fuel chamber—Maxima

- Rear seat or open the access panel in the trunk
- Fuel gauge electrical connector and pump electrical connector
- Fuel outlet and the return hoses
- Fuel tank, if necessary

4. On some models you need to remove the fuel pump assembly-to-fuel tank bolts and lift the fuel pump assembly from the fuel tank.

5. On other models you need to remove the locking ring and raise the fuel pump from the tank. Disconnect the feed tube while raising the pump.

6. Discard the O-ring. Plug the fuel tank opening with a clean rag to prevent dirt from entering the system.

➡**When removing or installing the fuel pump assembly, be careful not to damage or deform it and always install a new O-ring.**

To install:

7. Remove the rag

8. Install or connect the following:
   - Fuel pump assembly into the fuel tank using a new O-ring
   - Fuel pump assembly-to-fuel tank bolts and torque the bolts to 17–22 inch lbs. (2.0–2.5 Nm)
   - Locking ring assembly and tighten
   - Fuel tank assembly, if removed
   - Fuel lines and the electrical connectors. Always use new clamps when reconnecting fuel line hoses.

➡**When installing the upper plate, be sure to align the mark with the center marks on the fuel tank.**

   - Negative battery cable

9. Start the engine, check for fuel leaks and repair if necessary.

10. Install the fuel pump access cover.

➡**On some models, the Check Engine Light will stay ON after installation is completed. The memory code in the control unit must be erased. This code is stored for an open fuel pump circuit, this is caused when the fuel pressure is released. To erase the code, disconnect the battery cable for 10 seconds, then reconnect after installation of fuel pump.**

350Z

1. Before servicing the vehicle, refer to the Precautions Section.

➡**If the fuel tank is more than three quarter full, some fuel will have to be drained before removing the fuel pump/fuel filter assembly.**

2. Relieve the fuel system pressure
3. Remove or disconnect the following:
   - Negative battery cable
   - Rear floor box
   - Fuel pump inspection cover
   - Harness connector and fuel feed tube
   - Fuel feed tube quick connector
   - Retainer
   - Raise the fuel pump assembly and remove the fuel hose connector
   - Remove the fuel pump assembly
   - Reverse the removal procedure to install

Fuel Injector

REMOVAL & INSTALLATION

Maxima

1. Before servicing the vehicle, refer to the Precautions Section.
2. Relieve the fuel system pressure
3. Remove or disconnect the following:
   - Negative battery cable
   - Intake manifold collector
   - Vacuum hose from the pressure regulator
   - Fuel hoses from the rail
   - Injector electrical connectors
   - Fuel rail bolts
4. To remove the fuel injector from the fuel rail, expand and remove the clips securing the injectors and press the fuel injector out from the fuel rail. Discard the O-rings.

To install:
5. Apply a thin coat of engine oil to the new O-rings, install them on the

injectors, then press the injector into the fuel rail.
6. Install or connect the following:
   - New injector retaining clips
   - New injector gaskets onto the manifold
   - Fuel rail assembly to the engine
   - Fuel rail-to-cylinder head bolts and torque the bolts to 84–96 inch lbs. (9.3–10.8 Nm). Then tighten them again to 16–19 ft. lbs. (21–26 Nm).
   - Fuel lines to the rail assembly
   - Vacuum hose to the fuel pressure regulator
   - Electrical connectors to the fuel injectors
   - Intake manifold collector
   - Negative battery cable.
7. Start the engine and check for leaks.

350Z

1. Before servicing the vehicle, refer to the Precautions Section.
2. Relieve the fuel system pressure
3. Remove or disconnect the following:
   - Negative battery cable
   - Engine cover
   - Fuel feed hose and damper
   - Under vehicle fuel line quick connectors
   - Upper and lower intake manifold collectors
   - Fuel injector harness connectors
   - Fuel rail mounting bolts in reverse of the tightening sequence
   - Intake manifold spacers
4. To remove the fuel injector from the fuel rail, expand and remove the clips secur-

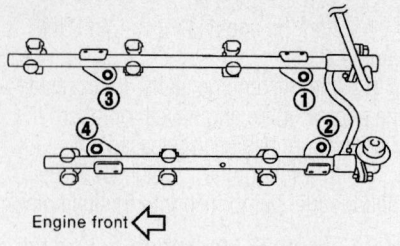

Engine front ⇐

67162-NISS-G47

Fuel rail tightening sequence–350Z

ing the injectors and press the fuel injector out from the fuel rail. Discard the O-rings.

To install:

➡ **Upper and lower O-rings are different. The Blue rings go on the fuel tube side and the Brown rings go on the nozzle side.**

5. Apply a thin coat of engine oil to the new O-rings, install them on the injectors, then press the injector into the fuel rail.
6. Install or connect the following:
   - New injector retaining clips
   - Intake manifold spacers
   - Fuel rail mounting bolts in sequence and tighten to 8 ft. lbs. (11 Nm), then to 16–19 ft. lbs. (21–27 Nm).
   - Fuel injector harness connectors
   - Upper and lower intake manifold collectors
   - Under vehicle fuel line quick connectors
   - Fuel feed hose and damper
   - Engine cover
   - Negative battery cable
7. Start the engine and check for leaks.

DRIVE TRAIN

Transaxle Assembly

REMOVAL & INSTALLATION

Maxima

MANUAL

1. Before servicing the vehicle, refer to the Precautions Section.
2. Drain the fluid from the transaxle.
3. Remove or disconnect the following:
   - Battery cables
   - Battery and the battery tray
   - Air cleaner and Mass Air Flow (MAF) sensor
   - Air breather hose
   - Control cable and cable mounting bracket

   - Clutch operating cylinder and hose clamps
   - Speedometer pinion, Park/Neutral Position (PNP) and ground harness switch connectors
   - Starter
   - Crankshaft Position (CKP) sensor (POS) from the transaxle
   - Shift control rod and support rod bracket
   - Front wheels
   - Halfshafts and properly support the engine with a jack under the oil pan

✳ WARNING

Do not place the jack under the oil pan drain plug.

   - Center member
   - Left-hand mounting bracket from the transaxle and body
   - Transaxle

To install:

➡ **The transaxle mounting bolts are different lengths and require special torque specifications. Use care when installing and tightening these bolts.**

4. Install or connect the following:
   - Transaxle assembly into the vehicle
   - Transaxle mounting bolts in the proper location as noted during removal
   - Left hand mounting bracket
   - Center member
   - Halfshafts and front wheels

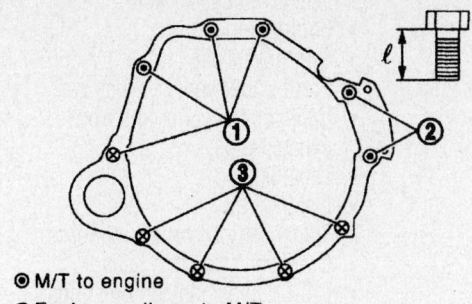

⊙ M/T to engine
⊗ Engine or oil pan to M/T

Bolt No.	Tightening torque N·m (kg-m, ft-lb)	"ℓ" mm (in)
1	69.6 - 79.4 (7.1 - 8.0, 52 - 58)	52 (2.05)
2	69.6 - 79.4 (7.1 - 8.0, 52 - 58)	113 (4.45)
3	36 - 47 (3.7 - 4.7, 27 - 34)	40 (1.57)

9357RG15

Manual transaxle bolt torque specifications and locations—Maxima with 3.5L engine

- Shift control rod and support rod bracket
- Starter
- CKP sensor
- Speedometer pinion, PNP switch and ground harness switch connectors
- Clutch operating cylinder and hose clamp
- Control cable and cable mounting bracket
- Air breather hose
- Air cleaner and MAF sensor
- Battery and tray
- Battery cables

5. Fill the transaxle with clean fluid.
6. Start the vehicle, check for leaks and repair if necessary.

AUTOMATIC

1. Before servicing the vehicle, refer to the Precautions Section.
2. Drain the transaxle fluid.
3. Remove or disconnect the following:
- Battery cables
- Battery and tray
- Air cleaner and resonator
- Park/Neutral Position (PNP) switch
- Revolution sensor and Vehicle Speed Sensor (VSS) electrical connectors
- Crankshaft Position (CKP) sensor

- Left hand mounting bracket from the transaxle and body
- Control cable
- Both front wheels
- Halfshafts
- Oil cooler pipes
- Starter and properly support the engine
- Center member
- Rear cover plate
- Torque converter
- Transaxle assembly

➡**When removing the torque converter, turn the crankshaft for access to the bolts. Place alignment marks on the converter and drive plate, so the converter can be installed in its original position.**

➡**The transaxle mounting bolts are different lengths. Tagging the bolts upon removal will facilitate proper tightening during installation.**

To install:

➡**When installing the torque converter to the transaxle, measure the depth of the converter to ensure proper installation.**

4. Using a straight edge across the mounting flange, measure the depth of the converter. The measurement is to the bolt mounting flange of the converter.

5. The depth measurement of the converter should be 0.75 in. (19mm) or more.

➡**The transaxle mounting bolts are different lengths and require special torque specifications. Use care when installing and tightening these bolts.**

6. Install the transaxle assembly into the vehicle.
7. Refer to the diagram for the automatic transaxle mounting bolt torque specifications.
8. Torque the bolts holding the converter to the flexplate to 33–43 ft. lbs. (44–59 Nm).
9. Install or connect the following:
- Rear cover plate
- Center member
- Starter and remove the engine support
- Oil cooler pipes
- Halfshafts and both front wheels
- Control cable
- Left hand mounting bracket
- CKP sensor
- Revolution and VSS sensor electrical connectors
- PNP switch
- Air cleaner and resonator
- Battery, tray and both cables

10. Fill the transaxle with the proper type and amount of fluid.

⊙ A/T to engine
⊗ Engine to A/T

Bolt No.	Tightening torque N·m (kg-m, ft-lb)	ℓ mm (in)
1	69.6 - 79.4 (7.1 - 8.0, 52 - 58)	65 (2.56)
2	69.6 - 79.4 (7.1 - 8.0, 52 - 58)	52 (2.05)
3	69.6 - 79.4 (7.1 - 8.0, 52 - 58)	40 (1.57)

9357RG16

Automatic transaxle bolt torque specifications and locations—Maxima with 3.5L engine

11. Start the vehicle, check for leaks and repair if necessary.

Transmission Assembly

REMOVAL & INSTALLATION

350Z

MANUAL

1. Before servicing the vehicle, refer to the Precautions Section.
2. Drain the fluid from the transmission.
3. Remove or disconnect the following:
   - Negative battery cable
   - Engine undercover
   - Strut tower bar
   - Front under vehicle cross bar
   - Catalytic converter and front exhaust pipe
   - Drive shaft
   - Shift lever assembly from the control rod
   - Shifter console and boot
   - Shift lever from shift housing
   - Clutch slave cylinder
   - Crankshaft position sensor
   - Back-up light and neutral safety switch

- Wire harnesses from transmission
- Starter
- Transmission cover plate

4. Place a transmission jack under the transmission.
5. Remove the rear engine crossmember.
6. Remove the transmission mounting bolts and the transmission.

To install:

➡ The transmission mounting bolts are different lengths and require special torque specifications. Use care when installing and tightening these bolts.

7. Installation is the reverse of the removal procedure noting the following:
   a. Tighten the rear engine crossmember bolts to 36 ft. lbs. (49 Nm).
8. Fill the transmission with clean fluid.
9. Start the vehicle, check for leaks and repair if necessary.

AUTOMATIC

1. Before servicing the vehicle, refer to the Precautions Section.
2. Drain the fluid from the transmission.
3. Remove or disconnect the following:

- Negative battery cable
- Engine undercover
- Strut tower bar
- Front under vehicle cross bar
- Catalytic converter and front exhaust pipe
- Drive shaft
- Control rod
- Transmission harness connector
- Crankshaft position sensor
- Cooler lines and cap openings
- Air breather hose
- Starter
- Torque converter dust cover
- Torque converter bolts

4. Place a transmission jack under the transmission.
5. Remove the rear engine crossmember.
6. Remove the transmission mounting bolts and the transmission.

To install:

➡ The transmission mounting bolts are different lengths and require special torque specifications. Use care when installing and tightening these bolts.

7. Installation is the reverse of the removal procedure noting the following:
   a. Tighten the rear engine crossmember bolts to 36 ft. lbs. (49 Nm).

Bolt No.	1	2	3	4	5
Quantity	1	5	2	2	2
"ℓ" mm (in)	55 (2.17)	65 (2.56)	50 (1.97)	35 (1.38)	65 (2.56)
Tightening torque N·m (kg-m, ft-lb)	75 (7.7, 55)		55.4 (5.7, 41)	46.6 (4.8, 34)	55.4 (5.7, 41)

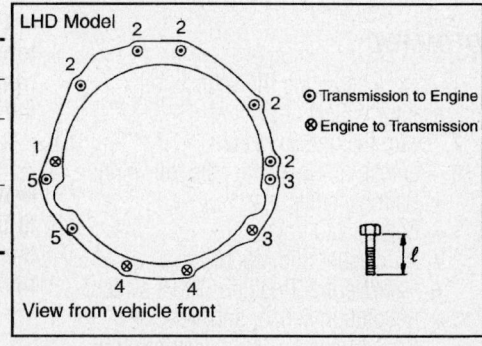

LHD Model
⊙ Transmission to Engine
⊗ Engine to Transmission
View from vehicle front

67162-NISS-G48

Manual transaxle bolt torque specifications and locations—350Z

Bolt No.	1	2	3	4
Number of bolts	1	5	2	2
Bolt length "ℓ" mm (in)	55 (2.17)	65 (2.56)	56 (2.20)	35 (1.38)
Tightening torque N·m (kg-m, ft-lb)	70 - 80 (7.2 - 8.1, 52 - 59)		49.0 - 61.8 (5.0 - 6.3, 37 - 45)	41.2 - 52.0 (4.2 - 5.3, 31 - 38)

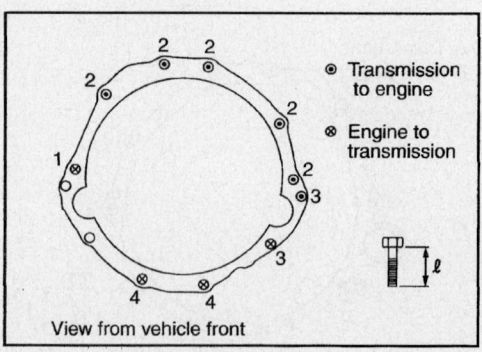

⊙ Transmission to engine
⊗ Engine to transmission
View from vehicle front

67162-NISS-G49

Automatic transaxle bolt torque specifications and locations—350Z

b. Tighten the torque converter bolts to 33—42 ft. lbs. (44–58 Nm).

8. Fill the transmission with clean fluid.

9. Start the vehicle, check for leaks and repair if necessary.

Clutch

REMOVAL & INSTALLATION

1. Before servicing the vehicle, refer to the Precautions Section.

2. Remove or disconnect the following:

- Transmission/transaxle assembly

3. Insert a clutch disc centering tool into the clutch disc hub for support.

- Pressure plate bolts evenly in reverse order of the tightening sequence, a little at a time to prevent distortion
- Clutch assembly
- Throw-out bearing from the clutch lever

To install:

4. Apply a light coating of chassis lube to the clutch disc spleens, input shaft and pilot bearing. Use a disc centering tool to aid installation.

5. Install or connect the following:

- Disc and pressure plate

6. On Maxima, torque the pressure plate bolts in a crisscross pattern in the following 2 steps:

a. Step 1: 7–14 ft. lbs. (10–20 Nm).

b. Step 2: 25–33 ft. lbs. (34–44 Nm).

7. On 350Z, torque the pressure plate bolts in a crisscross pattern in the following 2 steps:

a. Step 1: 11 ft. lbs. (115 Nm).

b. Step 2: 29 ft. lbs. (340 Nm).

8. Install or connect the following:

- New throw-out bearing in the clutch release lever. Remove the clutch disc centering tool.
- Transaxle into the vehicle. If the mating surfaces will not come together, do not force the units together. Remove the transaxle and recheck that the disc is centered.

➡**DO NOT draw the transaxle to the engine with the bolts. This may damage the clutch and/or transaxle. Also, be careful not to move the throw-out bearing when installing the transaxle.**

9. After the transaxle is installed, connect the clutch cable and check operation before complete reassembly.

10. Adjust the clutch pedal as necessary.

Hydraulic Clutch System

BLEEDING

Bleeding is required to remove air trapped in the hydraulic system. The bleed screw is located on the clutch slave (operating) cylinder.

Some models are also equipped with a clutch damper mechanism. The clutch damper mechanism is bled in exactly the same manner as the operating cylinder. It should be bled along with the operating cylinder.

1. Before servicing the vehicle, refer to the Precautions Section.

2. Remove the bleed screw dust cap.

3. Attach a transparent vinyl tube to the bleed screw, immersing the free end in a clean container of clean brake fluid.

4. Fill the master cylinder with the proper fluid.

5. Open the bleed screw about ¾ turn.

6. Depress the clutch pedal quickly. Hold it down. Have an assistant tighten the bleed screw. Allow the pedal to return slowly.

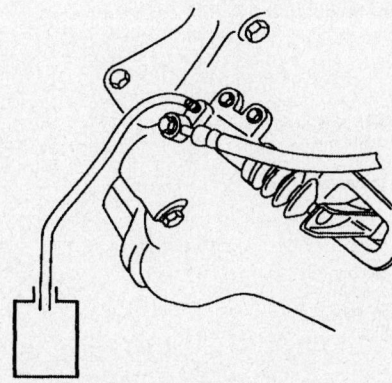

09482_ZMAX_G0035

Clutch system bleeding at the clutch operating cylinder—Maxima

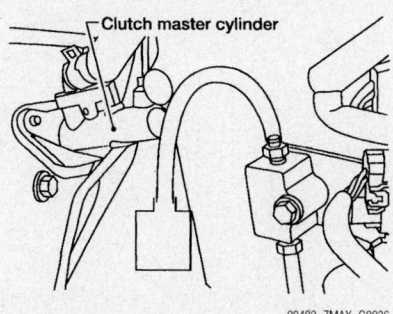

09482_ZMAX_G0036

Clutch system bleeding from the air bleeder valve on the bleed connector—Maxima

7. Repeat the above procedure until no more air bubbles are seen in the fluid container.

8. Remove the bleed tube.

9. Replace the dust cap and refill the master cylinder.

10. Bleed the clutch damper, if equipped.

Halfshaft

REMOVAL & INSTALLATION

Maxima

1. Before servicing the vehicle, refer to the Precautions Section.

2. Raise and support the front of the vehicle safely and remove the wheels.

3. Remove or disconnect the following:

- Anti-Lock Brake (ABS) wheel sensor and move it out of the way
- Brake hose from the strut
- Wheel bearing locknut
- Bolts attaching the steering knuckle to the strut. Matchmark the bolts before removal.

➡**Cover axle boots with waste cloth so as not to damage them when removing halfshaft.**

- Halfshaft from the knuckle by slightly tapping it
- Bolts attaching the support bearing to the support bearing bracket
- Halfshaft from the transaxle with a flat-bladed tool, if equipped with a manual transaxle

4. If equipped with a automatic transaxle perform the following:

a. Remove the right halfshaft from the vehicle.

b. Insert a flat-bladed tool into the transaxle where the right halfshaft was, place the end of the tool on the halfshaft, then, drive the left shaft from the pinion side gear.

5. Remove or disconnect the following:

- Support bearing bolts and the halfshaft from the vehicle
- Discard the circlip on the end of the halfshaft
- Seal from the transaxle

To install:

6. Install or connect the following:

- New seal into the transaxle and install a halfshaft alignment tool into the transaxle seal
- New circlip to the halfshaft, then insert the halfshaft into the transaxle

7. With the serration's aligned remove the alignment tool.

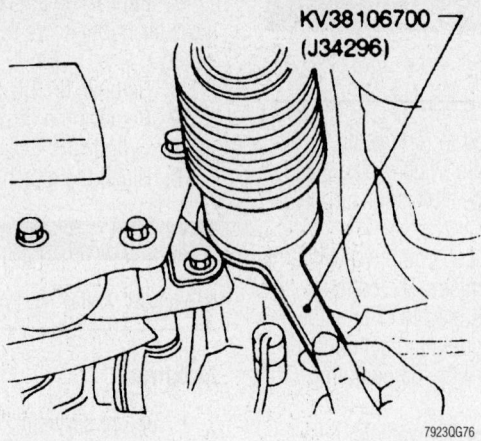

Left halfshaft alignment tool—Maxima

8. Push the halfshaft fully into the transaxle to seat the circlip. Try to pull the halfshaft from the transaxle by hand to verify that the circlip is properly seated.

- Support bearing and torque the bolts to 10–14 ft. lbs. (13–19 Nm)
- Halfshaft into the steering knuckle and install the hub locknut, do not tighten the hub nut
- Steering knuckle to the strut
- Strut mounting bolts to the matchmarks and torque the bolts to 103–117 ft. lbs. (140–159 Nm)

- Brake hose to the strut
- ABS wheel sensor and torque the bolt to 13–17 ft. lbs. (18–24 Nm)
- Front wheels and torque hub locknut to 174–231 ft. lbs. (235–314 Nm)

9. Check and/or adjust the wheel alignment as necessary.

350Z

1. Before servicing the vehicle, refer to the Precautions Section.

2. Raise and support the rear of the vehicle safely and remove the wheels.

3. Remove or disconnect the following:
- Cotter pin and axle nut
- Stabilizer bar connecting rod

4. Remove the nuts and bolts between the side flange and the drive shaft.

5. Use a suitable puller and remove the drive shaft from the axle.

To install:

6. Install or connect the following:
- New seal into the transaxle and install halfshaft
- Tighten the side flange bolts to 47–58 ft. lbs. (63–79 Nm)
- Stabilizer bar connecting rod
- Axle nut and new cotter pin. Tighten the axle nut to 152–202 ft. lbs. (206–274 Nm)
- Rear wheels

CV-Joints

OVERHAUL

Maxima

TRANSAXLE SIDE

1. Remove the boot bands.
2. Matchmark the slide joint housing

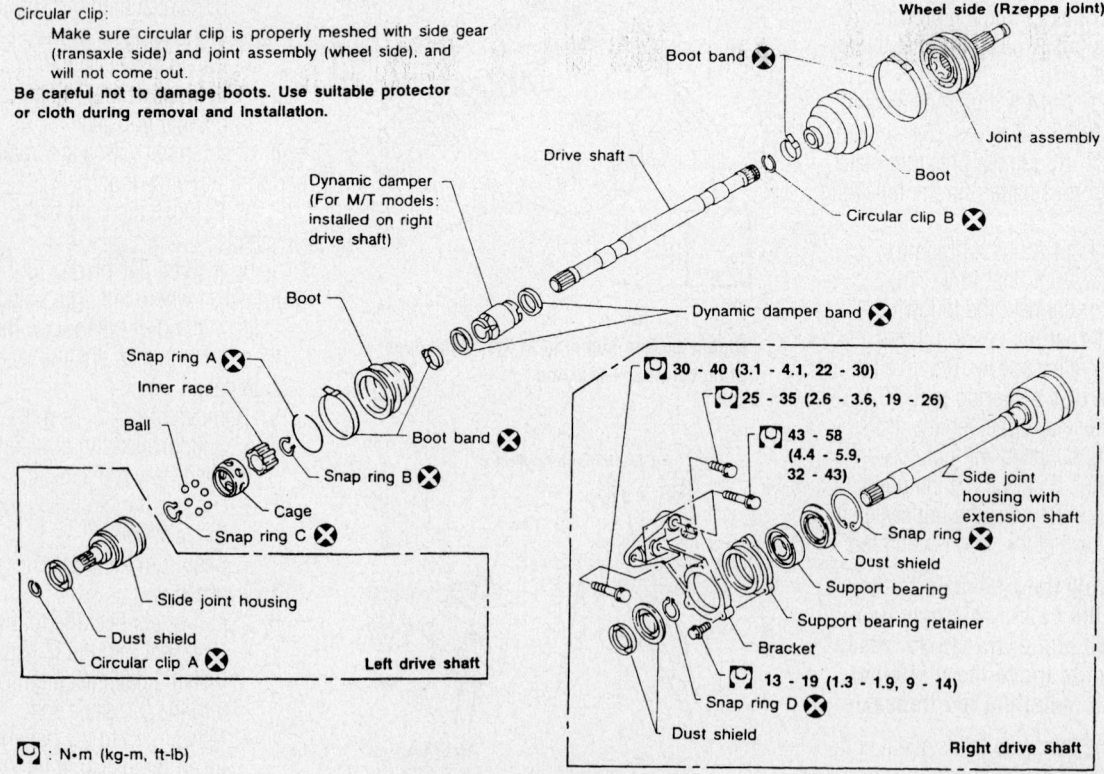

Exploded view of the halfshafts and related components

and inner race, prior to separating the joint assembly.

3. Pry off the snapring and remove the ball cage, inner race and balls as a unit.

4. Remove the snapring and withdraw the boot.

To install:

➡**Cover the halfshaft serrations with tape, so as not to damage the boot.**

5. Thoroughly clean all parts in solvent and dry with compressed air. Check parts for evidence of damage and replace as necessary.

6. Install or connect the following:
- Boot and new boot band on the halfshaft
- New inner snapring
- Ball cage, inner race and balls as a unit. Confirm that the matchmarks are aligned.
- New outer snapring

7. Pack the CV joint with 5.0–6.0 ounces (165–175 g) of grease.

8. Ensure that the boot is properly installed on the halfshaft groove.

9. Set the boot so that it does not swell or deform when its length is 3.86 in. (98mm).

10. Lock the new boot bands securely.

WHEEL SIDE

The joint on the wheel side cannot be disassembled.

1. Prior to separating the joint assembly, matchmark the halfshaft and joint assembly.

2. Separate the joint using a slide hammer.

3. Remove the boot bands and the boot.

To install:

4. Thoroughly clean all parts in solvent and dry with compressed air. Check parts for evidence of damage and replace as necessary.

➡**Cover the halfshaft serrations with tape, so as not to damage the boot.**

5. Install the boot and small boot band on the halfshaft.

6. Set the joint assembly onto the halfshaft and align the matchmarks.

7. Attach the joint assembly to the halfshaft by lightly tapping the serrated end with a plastic hammer.

➡**Using a metal hammer may damage the threads on the end of the joint.**

8. Pack the CV joint with 4.76–5.11 ounces (135–145 g) of grease.

9. Ensure that the boot is properly installed on the halfshaft groove.

10. Set the boot so that it does not swell or deform when its length is 3.82 in. (97mm).

11. Lock the new boot bands securely.

89617G07

The inner CV joint uses a large C-clip to retain the ball and cage assembly in the outer housing

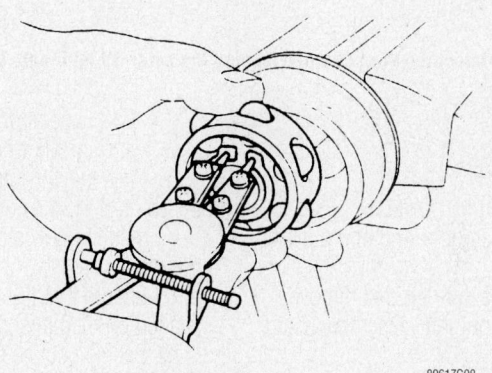

89617G08

After the outer housing is removed, the ball and cage assembly can slide from the shaft by removing the C-clip

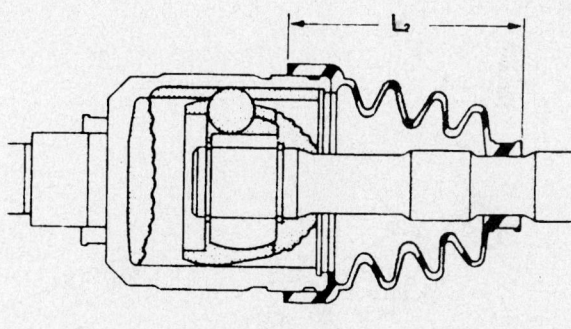

89617G02

Make sure to properly position the boot before tightening the boot clamps

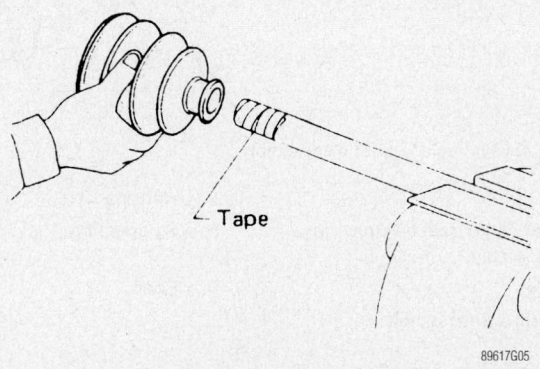

Tape

89617G05

Use vinyl tape and wrap the end of the shaft to protect the boot during installation

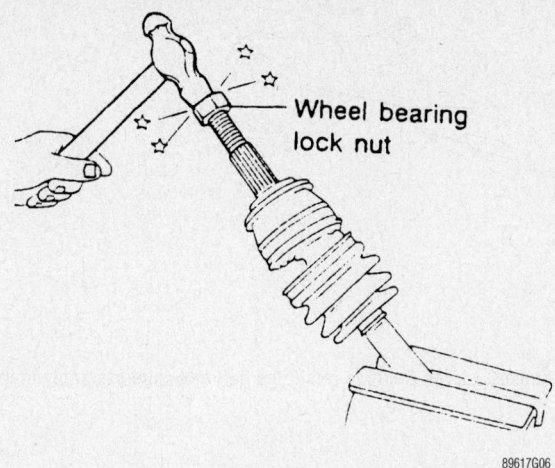

89617G06

Use an old nut to protect the threads when tapping the outer CV joint onto the shaft

350Z

FINAL DRIVE SIDE

1. Remove the boot bands.
2. Remove the stopper ring and pull out of the housing.
3. Remove the snap ring and remove the ball cage and inner race assembly using a puller.
4. Remove the boot from the shaft.

To assemble:

5. Install or connect the following:

- Boot and new boot band on the halfshaft

- Ball cage and inner race as a unit and install a new snap ring.

6. Pack the CV joint with 4.4–4.7 ounces (124–134 g) of grease.
7. Install a new stopper ring in the housing.
8. Ensure that the boot is properly installed on the halfshaft groove.
9. Set the boot so that it does not swell or deform when its length is 3.70 in. (94mm).
10. Lock the new boot bands securely.

WHEEL SIDE

The joint on the wheel side cannot be disassembled.

1. Prior to separating the joint assembly, matchmark the halfshaft and joint assembly.
2. Separate the joint using a slide hammer.
3. Remove the boot bands.

To assemble:

4. Thoroughly clean all parts in solvent and dry with compressed air. Check parts for evidence of damage and replace as necessary.

➡**Cover the halfshaft serrations with tape, so as not to damage the boot.**

5. Install the boot and small boot band on the halfshaft.
6. Set the joint assembly onto the halfshaft and align the matchmarks.
7. Attach the joint assembly to the halfshaft by lightly tapping the serrated end with a plastic hammer.

➡**Using a metal hammer may damage the threads on the end of the joint.**

8. Pack the halfshaft with 3.0–3.4 ounces (86–96 g) of grease.
9. Ensure that the boot is properly installed on the halfshaft groove.
10. Set the boot so that it does not swell or deform when its length is 3.82 in. (0.97mm).
11. Lock the new boot bands securely.

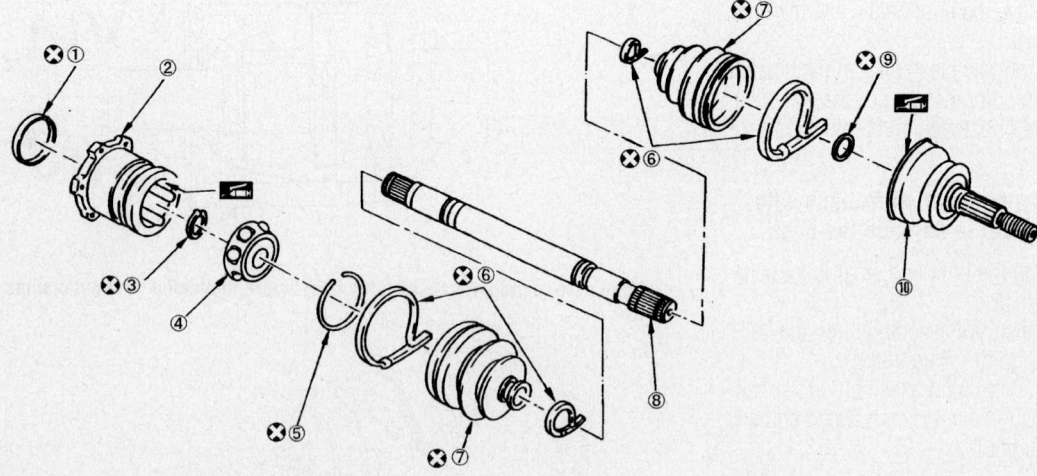

⊗ : Always replace after disassembly

1. Plug	2. Housing	3. Snap ring
4. Ball cage/Steel ball/Inner race assembly	5. Stopper ring	6. Boot band
7. Boot	8. Shaft	9. Circular clip
10. Joint sub-assembly		

67162-NISS-G50

Exploded view of the rear halfshaft—350Z

STEERING

Air Bag

PRECAUTIONS

Several precautions must be observed when handling the inflator module to avoid accidental deployment and possible personal injury.

1. Never carry the inflator module by the wires or connector on the underside of the module.

2. When carrying a live inflator module, hold securely with both hands and ensure that the bag and trim cover are pointed away.

3. Place the inflator module on a bench or other surface with the bag and trim cover facing up.

4. With the inflator module on the bench, never place anything on or close to the module that may be thrown in the event of an accidental deployment.

DISARMING

➡**All Supplemental Restraint System (SRS) electrical wiring harnesses and connectors are covered with YELLOW outer insulation. Do not use electrical test equipment on any circuit related to the SRS (air bag) sensors. When installing SRS components, always install with the arrow marks facing the front of the vehicle.**

To disarm the SRS system turn the ignition switch to **OFF** position. Then, disconnect the both battery cables starting with the negative cable first and wait at least 10 minutes after the cables are disconnected. Be sure to insulate the battery terminal ends.

REARMING

To arm the Supplemental Restraint System (SRS) system turn the ignition switch to **OFF** position. Connect the both battery cables starting with the positive cable first.

➡**The SRS or air bag system is equipped with a self-diagnostic operation. After turning the ignition key to the ON or START position, the AIR BAG warning lamp will illuminate for 7 seconds. After 7 seconds, the AIR BAG lamp will extinguish if no malfunction is detected. If the AIR BAG lamp does not extinguish after 7 seconds, check the SRS self-diagnostic system for a malfunction.**

Rack and Pinion Steering Gear

REMOVAL & INSTALLATION

Maxima

1. Before servicing the vehicle, refer to the Precautions Section.

2. Disconnect both battery cables and wait at least 10 minutes after the battery cables are disconnected. This will disarm the air bag system so the steering wheel can be removed.

3. Point the front tires straight ahead and lock the steering in this position.

❊❊ WARNING

Do not turn the steering wheel or column with the lower joint removed from the steering column or the spiral cable may be damaged.

4. Remove the steering wheel.

➡**The steering wheel must be removed before disconnecting the steering column lower joint to avoid damaging the Supplemental Restraint System (SRS) spiral cable.**

5. Raise and support the vehicle safely and remove the front wheels.

6. Remove or disconnect the following:

- Tie rod ends from the steering knuckles
- Front exhaust tube and properly support the engine
- Bolts attaching the engine mounts to the engine mounting center member
- Engine mounting center member and rear engine mount
- Front stabilizer bar from the vehicle
- Nuts attaching the hole cover to the bulkhead

7. Move the hole cover aside and disconnect the lower joint from the rack and pinion. Matchmark the pinion shaft and the pinion housing to record the steering neutral position.

- Power steering fluid pipes from the rack and pinion
- Bolts attaching the mounting brackets
- Rack and pinion from the vehicle

To install:

8. Position the rack and pinion in the vehicle and install the mounting brackets. Torque the mounting nuts and bolts in the proper sequence to 54–72 ft. lbs. (73–97 Nm).

9. Install or connect the following:

- New O-rings to the power steering fluid pipes and connect them to the rack and pinion. Torque the low pressure line 20–29 ft. lbs. (27–39 Nm) and the high pressure line to 11–18 ft. lbs. (15–25 Nm).

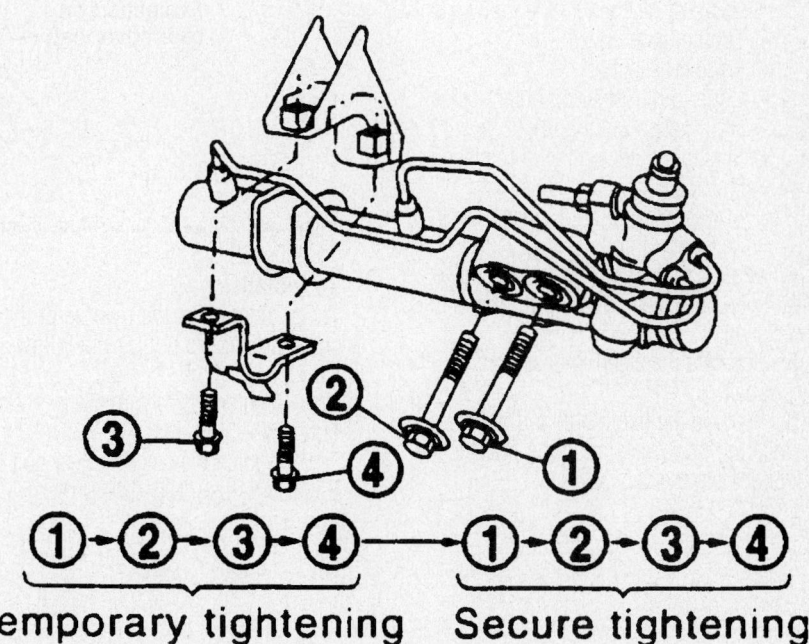

Temporary tightening   **Secure tightening**

9307QG21

Tighten the mounting bolts using the illustrated procedure—Maxima

10. Align the lower steering joint to the pinion shaft and install the joint onto the pinion shaft. Torque the bolt to 17–22 ft. lbs. (24–29 Nm).

- Hole cover and torque the nuts to 36–43 inch lbs. (4–5 Nm)
- Front stabilizer
- Engine mounting center member and torque the bolts to 57–72 ft. lbs. (77–98 Nm)
- Engine mounts to the center member. Torque the bolts to 57–72 ft. lbs. (77–98 Nm). Remove the support from the engine.
- Remaining components in the reverse order of removal

11. Torque the tie rod end nuts to 22–29 ft. lbs. (29–39 Nm), then install a new cotter pin.

12. Fill the power steering reservoir with fluid and bleed the air from the power steering system.

13. Check the vehicle front end alignment and adjust as necessary.

350Z

1. Before servicing the vehicle, refer to the Precautions Section.

※※ WARNING

Do not turn the steering wheel or column with the lower joint removed from the steering column or the spiral cable may be damaged.

2. Raise and support the vehicle safely and remove the front wheels.

3. Remove or disconnect the following:
- Engine undercover
- Front crossbar
- Cotter pin at steering outer socket, then loosen mounting nut
- Separate outer socket from steering knuckle
- Power steering fluid pipes from the rack and pinion

4. Loosen upper yoke bolt and remove lower yoke bolt, then slide lower joint onto lower shaft.

5. Separate the steering gear from the lower shaft.

6. Remove the fixing bolt and remove the steering gear, rack mounting bracket and insulator from the vehicle.

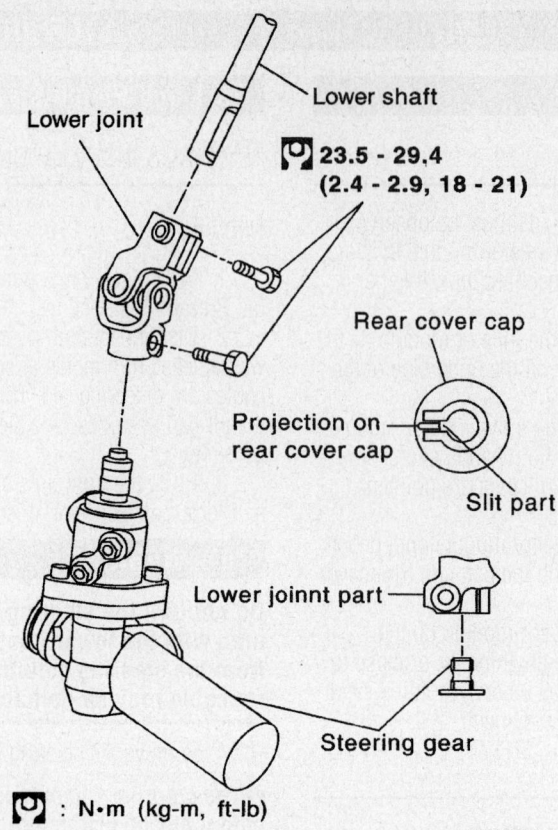

: N•m (kg-m, ft-lb)

67162-NISS-G51

Separating steering shaft from steering gear—350Z

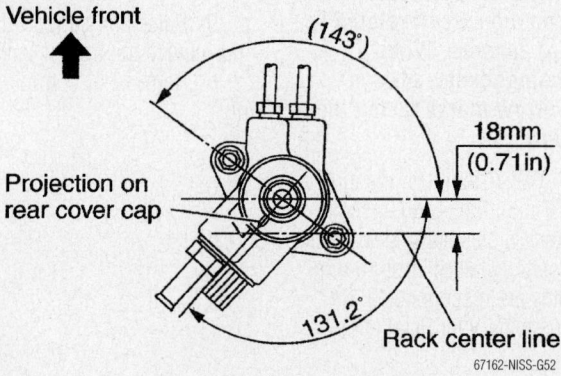

Proper orientation of the steering lower joint-to-gear—350Z

67162-NISS-G52

To install:

7. Installation is the reverse of the removal procedure noting the following:

a. Tighten the rack mounting bracket bolt to 46–56 ft. lbs. (62–76 Nm).

b. Tighten the steering gear bolt to 89–103 ft. lbs. (120–140 Nm).

c. Tighten the outer socket nut to 22–28 ft. lbs. (30–40 Nm).

d. Orientate the steering column yoke to the steering gear to the specifications shown.

e. Tighten the steering column lower joint bolts to gear bolt to 18–21 ft. lbs. (24–34 Nm).

FRONT SUSPENSION

Strut and Coil Spring

REMOVAL & INSTALLATION

Maxima

1. Before servicing the vehicle, refer to the Precautions Section.
2. Raise and safely support the vehicle.

3. Remove or disconnect the following:
- Wheel
4. Matchmark the position of the strut-to-steering knuckle location.
- Brake hose from the strut
- Anti-Lock Brake (ABS) wheel sensor and move it out of the way
- Bolts attaching the steering knuckle to the strut. Matchmark the bolts before removal.

5. Open the hood and remove the strut attaching nuts while holding the strut.

✲✲ CAUTION

Do not remove the center locknut from the strut assembly until the strut is safely compressed.

- Strut from the vehicle
6. Place the strut assembly in a vise

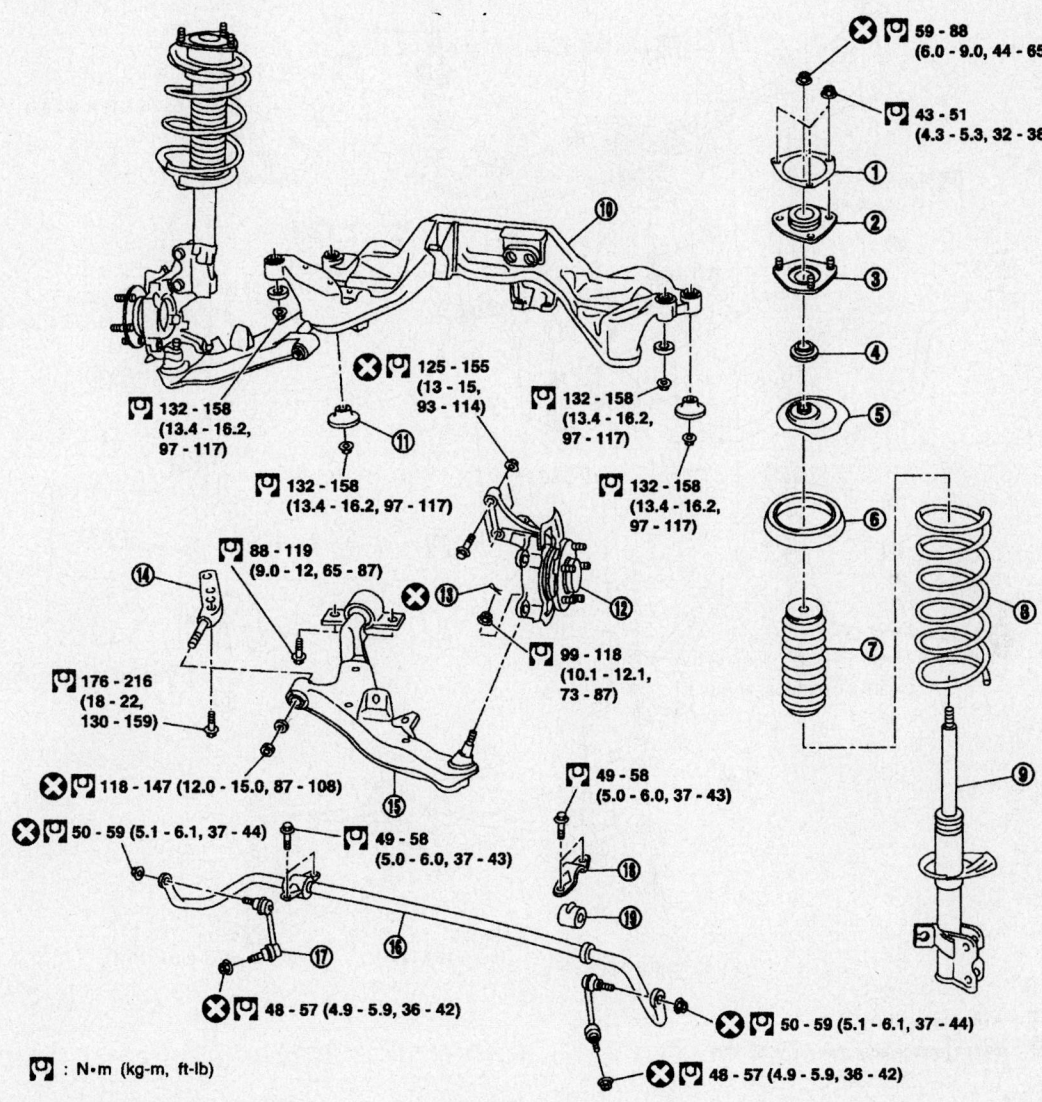

1. Strut spacer
2. Strut mount insulator
3. Strut mount bracket
4. Strut mount bearing
5. Spring upper seat
6. Spring rubber seat
7. Bound bumper rubber

8. Coil spring
9. Shock absorber
10. Suspension member
11. Rebound stopper
12. Wheel hub and steering knuckle
13. Cotter pin

14. Bush link pin
15. Transverse link
16. Stabilizer
17. Connecting rod
18. Stabilizer clamp
19. Bushing

Exploded view of the front suspension—2002 Maxima shown

9357RG17

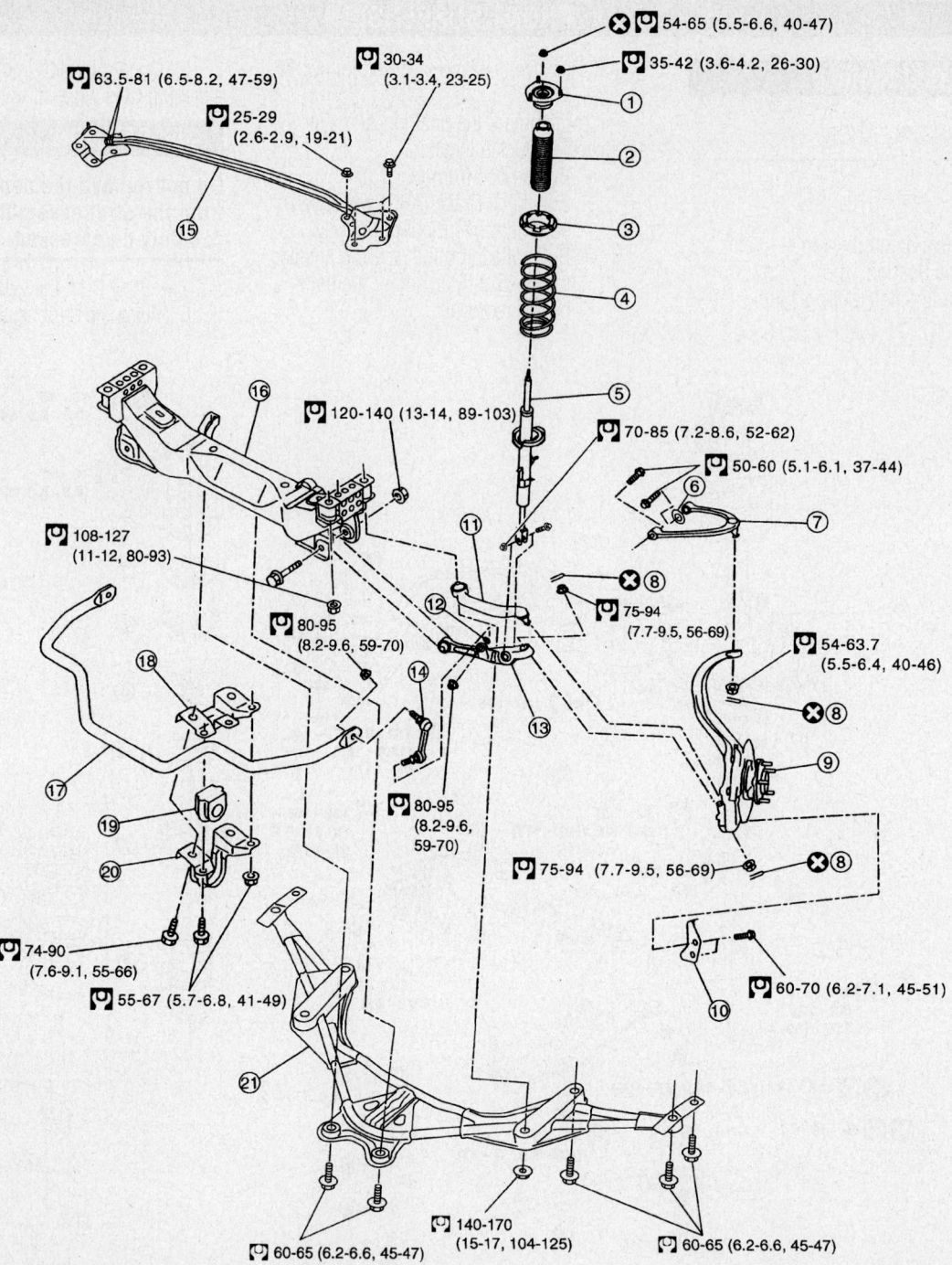

54-65 (5.5-6.6, 40-47)

35-42 (3.6-4.2, 26-30)

63.5-81 (6.5-8.2, 47-59)

30-34 (3.1-3.4, 23-25)

25-29 (2.6-2.9, 19-21)

120-140 (13-14, 89-103)

70-85 (7.2-8.6, 52-62)

50-60 (5.1-6.1, 37-44)

108-127 (11-12, 80-93)

75-94 (7.7-9.5, 56-69)

80-95 (8.2-9.6, 59-70)

54-63.7 (5.5-6.4, 40-46)

80-95 (8.2-9.6, 59-70)

75-94 (7.7-9.5, 56-69)

74-90 (7.6-9.1, 55-66)

60-70 (6.2-7.1, 45-51)

55-67 (5.7-6.8, 41-49)

60-65 (6.2-6.6, 45-47)

140-170 (15-17, 104-125)

60-65 (6.2-6.6, 45-47)

: N•m (kg-m, ft-lb)

: Always replace after every disassembly

1.	Mounting insulator	2.	Bound bumper	3.	Spring upper seat
4.	Coil spring	5.	Shock absorber	6.	Stopper rubber
7.	Upper link	8.	Cotter pin	9.	Front axle
10.	Steering stopper bracket	11.	Compression rod	12.	Washer
13.	Transverse link	14.	Stabilizer connecting rod	15.	Tower bar
16.	Front suspension member	17.	Stabilizer bar	18.	Stabilizer clamp bracket
19.	Stabilizer bushing	20.	Stabilizer clamp	21.	Front cross bar

67162-NISS-G53

Exploded view of the front suspension—350Z

with a holding tool or in a spring compressor.

7. Loosen the piston rod locknut.

❋❋ CAUTION

Do not remove the piston rod locknut, the spring is under tension and can cause serious personal injury.

8. Compress the spring with the spring compressor, then remove the piston rod locknut.

➡**Before removing the strut from the coil spring, note the positioning of the strut in relationship to the coil spring for reassembly.**

9. Remove or disconnect the following:
- Strut mounting insulator bracket, strut mounting bearing, upper spring seat and the upper spring rubber seat
- Strut, leaving the coil spring compressed
- Piston boot and rebound bumper from the strut

To install:

10. Install or connect the following:
- Rebound bumper and the boot to the strut piston
- Strut into the coil spring, be sure the strut and spring are properly positioned
- Upper spring rubber seat, upper spring seat, strut mounting bearing and the strut mounting insulator bracket. Be sure that the cutout on the upper spring seat is facing the outside of the vehicle.
- Piston rod locknut. Remove the tool and torque the piston rod locknut to 44–65 ft. lbs. (59–88 Nm).
- Strut into the strut tower
- New attaching nuts and torque to 32–38 ft. lbs. (43–51 Nm)
- Bolts attaching the steering knuckle to the strut and align the matchmarks and torque to 103–117 ft. lbs. (140–159 Nm)
- ABS wheel sensor and torque to 13–17 ft. lbs. (18–24 Nm)
- Brake hose to the strut
- Front wheels

11. Lower the vehicle.

12. Check and/or adjust the wheel alignment as necessary.

350Z

1. Before servicing the vehicle, refer to the Precautions Section.

2. Raise and safely support the vehicle.

3. Remove or disconnect the following:
- Wheel
- Engine undercover
- Anti-Lock Brake (ABS) wheel sensor and move it out of the way
- Brake hose from the strut
- Strut-to-transverse link nut and bolt

4. Open the hood and remove the strut attaching nuts while holding the strut.

❋❋ CAUTION

Do not remove the center locknut from the strut assembly until the strut is safely compressed.

- Strut from the vehicle

5. Place the strut assembly in a vise with a holding tool or in a spring compressor.

6. Loosen the piston rod locknut.

❋❋ CAUTION

Do not remove the piston rod locknut, the spring is under tension and can cause serious personal injury.

7. Compress the spring with the spring compressor, then remove the piston rod locknut

➡**Before removing the strut from the coil spring, note the positioning of the strut in relationship to the coil spring for reassembly.**

8. Remove or disconnect the following:
- Strut mounting insulator bracket, strut mounting bearing, upper spring seat and the upper spring rubber seat
- Strut, leaving the coil spring compressed
- Piston boot and rebound bumper from the strut

To install:

9. Install or connect the following:
- Rebound bumper and the boot to the strut piston
- Strut into the coil spring, be sure the strut and spring are properly positioned
- Upper spring rubber seat, upper spring seat, strut mounting bearing and the strut mounting insulator bracket. Be sure that the cutout on the upper spring seat is facing the outside of the vehicle.
- Piston rod locknut. Remove the tool and torque the piston rod locknut to 40–47 ft. lbs. (54–65 Nm).
- Strut into the strut tower
- New attaching nuts and torque to 26–30 ft. lbs. (35–42 Nm)

- Bolts attaching the strut to the transverse link and torque to 52–62 ft. lbs. (70–80 Nm)
- ABS wheel sensor
- Brake hose to the strut
- Engine undercover
- Front wheels

10. Lower the vehicle.

11. Check and/or adjust the wheel alignment as necessary.

Lower Control Arm (Transverse Link)

REMOVAL & INSTALLATION

Maxima

2002–2003 MODELS

1. Before servicing the vehicle, refer to the Precautions Section.

2. Remove or disconnect the following:
- Front wheels
- Anti-Lock Brake (ABS) wheel sensor and move it out of the way
- Wheel bearing locknut
- Tie rod from the steering knuckle
- Bolts attaching the strut to the steering knuckle. Matchmark the bolts before removal.
- Halfshaft from the steering knuckle by lightly tapping the end of the shaft
- Steering knuckle and the lower ball joint
- Stabilizer bar from the lower control arm
- Bolts attaching the link bushing pin to the chassis
- Nut attaching the link to the control arm and the link, if necessary
- Bolts attaching the compression rod bushing clamp
- Lower control arm/traverse link

To install:

3. Install or connect the following:
- Lower control arm and the compression rod bushing clamp into the vehicle
- Link bushing pin, if removed from the control arm

4. Tighten all bolts and nuts until they are snug enough to support the weight of the vehicle but not fully tight, the bolts should be torqued to specification with the vehicle on the floor.

➡**Always use a new nut when installing the ball joint to the control arm.**

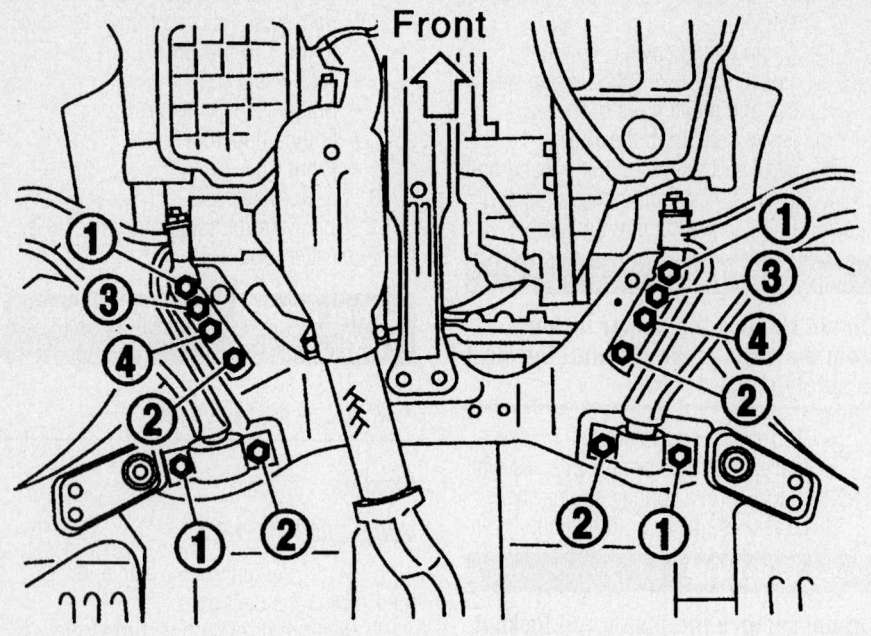

Front

09482_ZMAX_G0042

Bolt tightening sequence for the lower control arms—2002–2003 Maxima

- Steering knuckle to the strut and to the halfshaft
- Strut mounting bolts and torque the bolts to 103–117 ft. lbs. (140–159 Nm)
- Tie rod ball joint and torque the nut to 22–29 ft. lbs. (29–39 Nm)
- Wheel bearing locknut
- ABS wheel sensor and torque the bolt to 13–17 ft. lbs. (18–24 Nm)
- Front wheels, lower the vehicle and torque hub locknut to 174–231 ft. lbs. (235–314 Nm)

5. Torque the bolts attaching the compression rod bushing clamp and the link bushing pin, in the proper sequence to 87–108 ft. lbs. (118–147 Nm).

6. If the link bushing pin was removed from the control arm torque the attaching nut to 87–108 ft. lbs. (118–147 Nm).

7. Torque the sway bar attaching nut to 30–35 ft. lbs. (41–47 Nm).

8. Check the vehicle alignment.

2004–2006 MODELS

1. Before servicing the vehicle, refer to the Precautions Section.

2. Remove the front wheel(s).

3. Remove the lower ball joint cotter pin and remove the lower ball joint nut.

➡**Discard the cotter pin and use a new cotter pin for installation.**

4. Disconnect the lower ball joint from the steering knuckle using a ball joint separator tool.

5. Remove the member stay pin nut and two bolts and remove the member stay pin.

6. Remove the two transverse link pivot bolts.

7. Remove the transverse link bolt and remove the transverse link from the front suspension member.

8. Check the transverse link for damage, cracks or deformation. Replace it if necessary.

9. Check the bushing for damage, cracks and deformation. Replace the transverse link if necessary.

To install:

10. Installation is the reverse of the removal procedure, noting the following points:

 a. Tighten the transverse link mounting bolt to 77 ft. lbs. (105 Nm).

 b. Tighten the transverse link pivot bolts to 103 ft. lbs. (105 Nm).

 c. Tighten the ball joint nut to 58 ft. lbs. (78 Nm).

➡**During installation, the final tightening must be done with the vehicle in unloaded condition and the tires on the ground.**

 d. Install front wheel(s).

11. After installation, check the wheel alignment.

CONTROL ARM BUSHING REPLACEMENT

The bushing is an integral part of the lower control arm. If the bushing is defective the control arm must be replaced.

Lower Ball Joint

REMOVAL & INSTALLATION

The ball joint is an integral part of the lower control arm. If the ball joint is defective the control arm must be replaced.

Wheel Bearings

ADJUSTMENT

➡**Whenever the hub or bearing assemblies are removed, the wheel bearing must be replaced. Never reuse the old bearing assembly.**

The wheel bearings are sealed and are not adjustable. If defective, replacement is the only option.

REMOVAL & INSTALLATION

Maxima

➡**Whenever the hub or bearing assembly is removed, the wheel bearing assembly must be replaced. Never reuse the old bearing assembly.**

1. Before servicing the vehicle, refer to the Precautions Section.

2. Remove or disconnect the following:

- Knuckle assembly from the vehicle
- Hub with the inner race from the steering knuckle, using a shop press and a suitable tool
- Bearing inner race from the hub, using a shop press and a suitable tool
- Outer grease seal
- Inner grease seal from the steering knuckle, using a prybar
- Inner and outer snaprings from the steering knuckle, using snapring pliers
- Sealed bearing assembly from the steering knuckle, using a shop press and a suitable tool

3. Inspect the hub, steering knuckle and snaprings for cracks and/or wear; if necessary, replace the damaged part(s).

To install:

4. Install or connect the following:

- Inner snapring in the steering knuckle groove
- New wheel bearing assembly into the steering knuckle, using a shop press and a suitable tool, until it seats, using a maximum pressure of 3 tons (2722 kg)
- Outer snapring

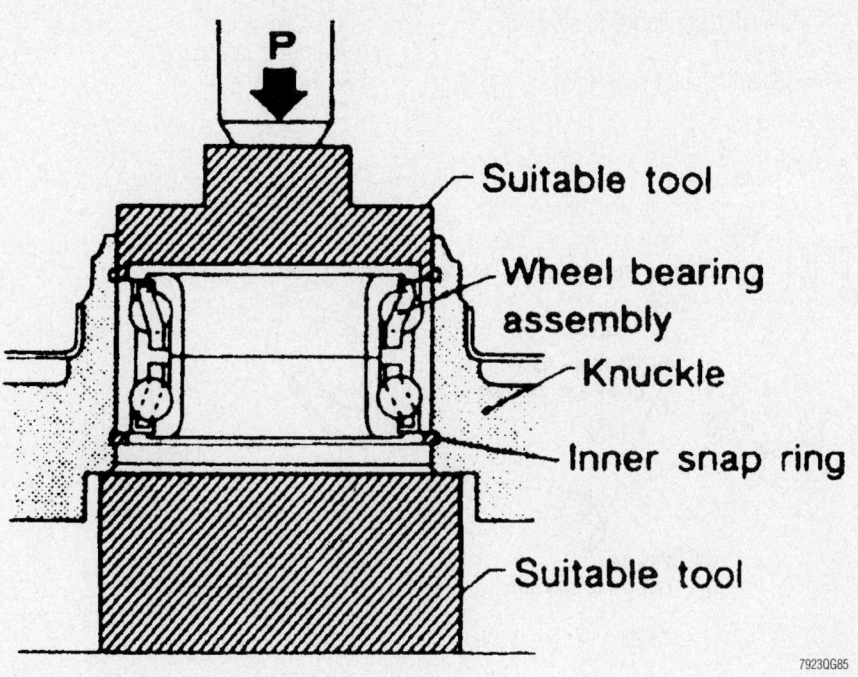

Typical method of installing the wheel bearing

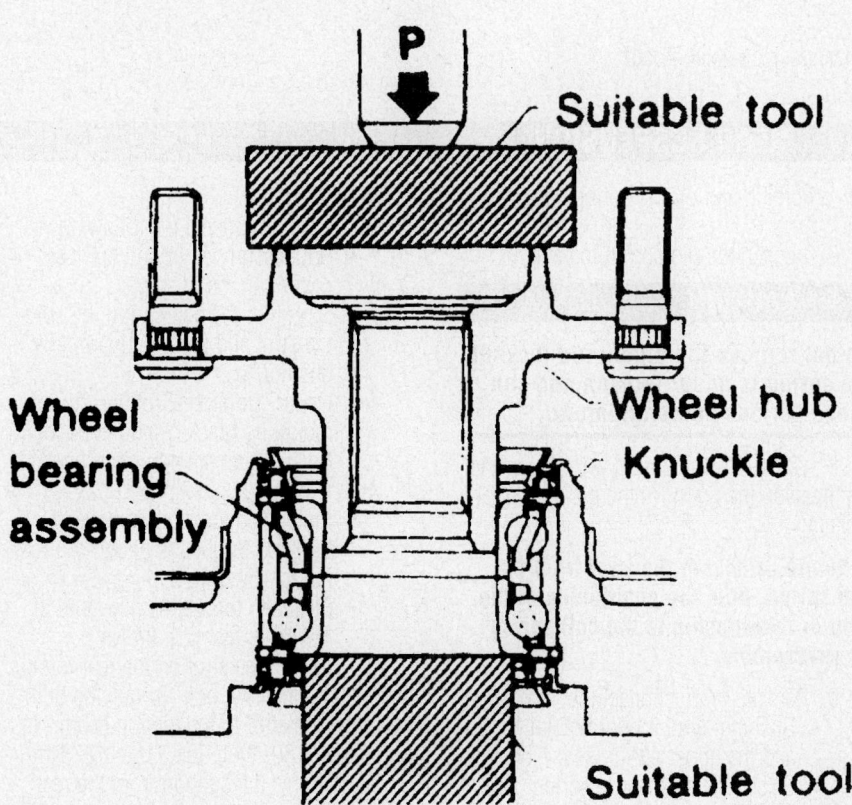

Use a press to install the hub into the knuckle assembly

5. Pack the new grease seal lips with multi-purpose grease.
 • New outer grease seal into the steering knuckle, using a shop press and a suitable tool
 • Hub into the steering knuckle, using a shop press and a suitable tool, until it seats, using a maximum pressure of 5.5 tons (4990 kg); be careful not to damage the grease seal

6. To check the bearing operation, perform the following procedures:
  a. Increase the press pressure to 3.5–5.0 tons (3175–4536 kg).
  b. Spin the steering knuckle, several turns, in both directions.
  c. Be sure the wheel bearings operate smoothly.

7. If the wheel bearings do not operate smoothly, replace the wheel bearing assembly.

8. Install the knuckle assembly.

9. Install the halfshaft into the hub. Torque the locknut to 174–231 ft. lbs. (235–314 Nm).

10. Install the wheel assembly and lower the vehicle.

11. Road test the vehicle and verify proper operation.

350Z

➡If the wheel bearing is damaged, the steering knuckle and bearing must be replaced as an assembly.

1. Before servicing the vehicle, refer to the Precautions Section.

2. Remove or disconnect the following:

 • Front wheel
 • Engine undercover
 • Brake caliper and wire aside
 • Brake rotor
 • ABS sensor
 • Brake hose bracket
 • Loosen steering outer socket nut
 • Separate outer socket from steering knuckle
 • Upper link from knuckle
 • Transverse link from knuckle
 • Compression rod from knuckle
 • Loosen steering knuckle nut
 • Knuckle and hub assembly from the vehicle
 • Separate the wheel hub from the knuckle

To install:

3. Install the wheel hub to the knuckle.

4. Install or connect the following:
 • Knuckle and hub assembly
 • Tighten the steering knuckle/hub nut to 58–72 ft. lbs. (79–98 Nm)

- Compression rod to knuckle and tighten the nut to 56–69 ft. lbs. (75–94 Nm)
- Transverse link to knuckle and tighten the nut to 56–69 ft. lbs. (75–94 Nm)

- Upper link to knuckle and tighten the nut to 40–46 ft. lbs. (54–64 Nm)
- Install outer socket to steering knuckle
- Steering outer socket nut

- Brake hose bracket
- ABS sensor
- Brake rotor
- Brake caliper
- Engine undercover
- Front wheel

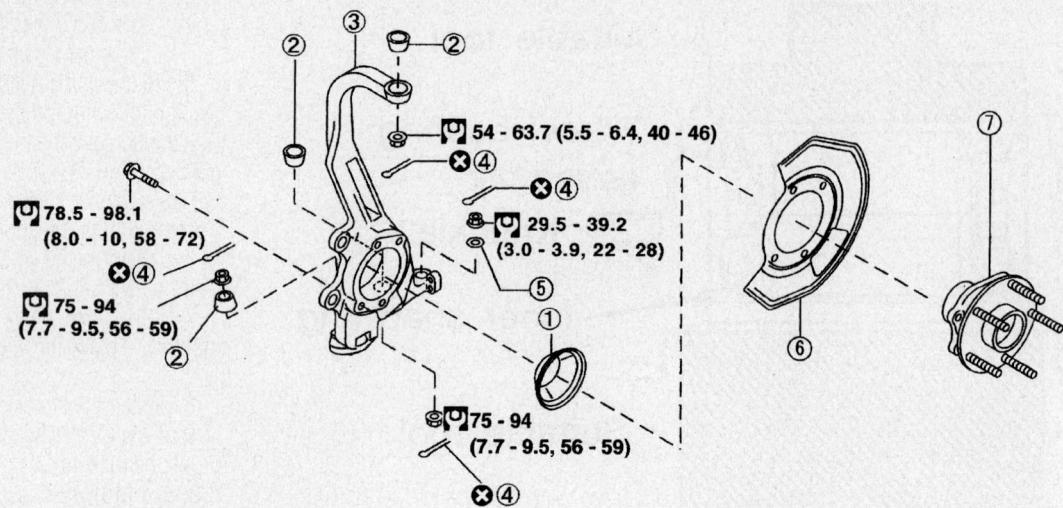

⊡ : N•m (kg-m, ft-lb)

✖ : Always replace after disassembly

1. Hub cap
4. Cotter pin
7. Wheel hub and bearing assembly

2. Ball seat
5. Washer

3. Steering knuckle
6. Splash guard

67162-NISS-G55

Exploded view of the front steering knuckle and wheel bearing assembly—350Z

REAR SUSPENSION

Strut and Coil Spring

REMOVAL & INSTALLATION

2002–2003 Maxima

1. Before servicing the vehicle, refer to the Precautions Section.
2. Remove the rear wheels.
3. Support the rear torsion beam assembly with a jack.
4. Open the trunk and remove the 2 nuts attaching the strut to the vehicle.

✳✳ CAUTION

Do not remove the center locknut from the strut assembly until the strut is safely compressed.

5. Remove the bolt attaching the strut to the rear torsion beam assembly and remove the strut.
6. Place the strut assembly in a vise

with a holding tool or in a spring compressor.
7. Loosen the piston rod locknut.

✳✳ CAUTION

Do not remove the piston rod locknut, the spring is under tension and can cause serious personal injury.

8. Compress the spring with the spring compressor, then remove the piston rod locknut.

➡ **Before removing the strut from the coil spring, note the positioning of the strut in relationship to the coil spring for reassembly.**

9. Remove or disconnect the following:
- Bushing, strut mounting bracket and the upper spring seat rubber
- Strut, leaving the coil spring compressed
- Bushing, bound bumper cover and the bound bumper

To install:
10. Install or connect the following:
- Bound bumper, bound bumper cover and the bushing
- Strut into the coil spring, be sure the strut and spring are properly positioned
- Upper spring seat rubber, strut mounting bracket and the bushing. Be sure that the mounting bracket is properly positioned.
- Piston rod locknut. Remove the tool and torque the piston rod locknut to 15–18 ft. lbs. (20–24 Nm).
- Strut and torque the new nuts to 18–25 ft. lbs. (25–34 Nm)
11. Position the strut on the rear torsion beam and install the bolt. Torque the bolt attaching the strut to the torsion beam assembly to 80–94 ft. lbs. (108–127 Nm).
12. Remove the support from the rear torsion beam.
13. Install the rear wheels and lower the vehicle.

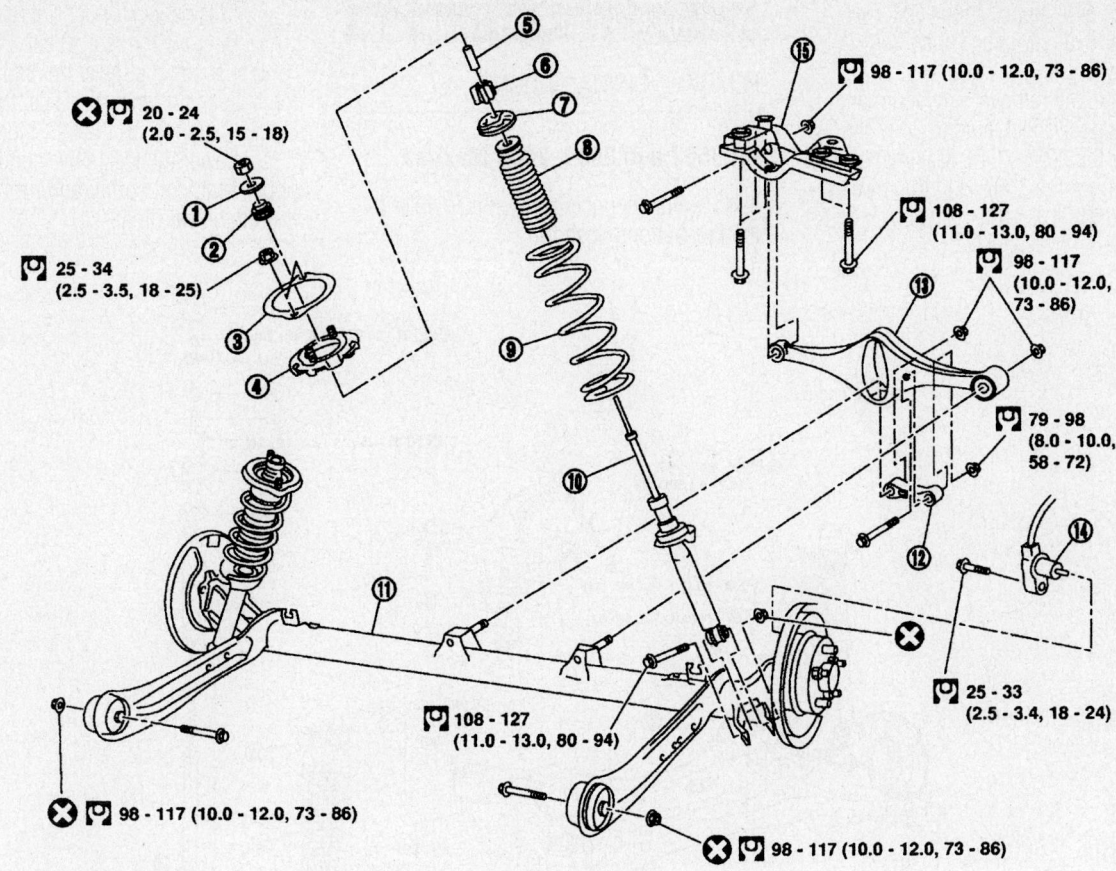

Exploded view of the rear suspension—2002 Maxima shown

1. Washer
2. Bushing
3. Shock absorber mounting seal
4. Shock absorber mounting bracket
5. Distance tube
6. Bushing
7. Bound bumper cover
8. Bound bumper
9. Coil spring
10. Shock absorber
11. Torsion beam
12. Control rod
13. Lateral link
14. ABS sensor
15. Suspension member

9357RG18

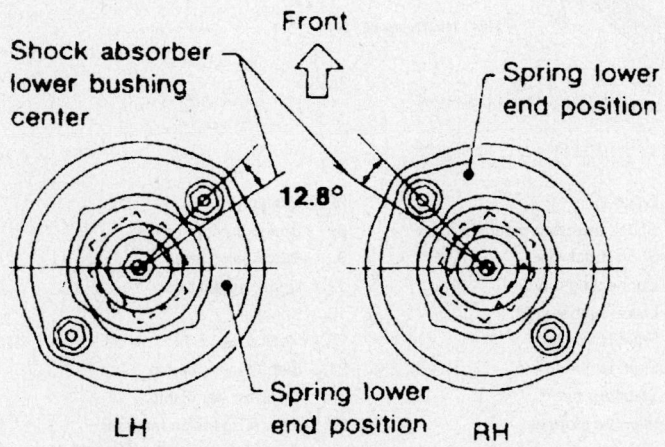

Positioning of the strut mounting brackets—Maxima

7923QG84

14. Check the vehicle's alignment and adjust as necessary.

Shock Absorber

REMOVAL & INSTALLATION

All 350Z and 2004–2006 Maxima

1. Before servicing the vehicle, refer to the Precautions Section.
2. Set a transmission jack on the rear lower link to remove the lower shock absorber nut and bolt.
3. Remove the rear wheel.
4. Remove the transmission jack from rear lower link.

5. Remove the upper shock absorber nuts and remove rear shock absorber.

To install:

6. Reverse the removal procedure and tighten the upper shock nuts to 20 ft. lbs. (24 Nm) for the 2004–2006 Maxima; 21 ft. lbs. (28 Nm) for the 350Z, and the lower bolt to 74–88 ft. lbs. (100–120 Nm).

Coil Spring and Rear Lower Link

REMOVAL & INSTALLATION

All 350Z and 2004–2006 Maxima

1. Before servicing the vehicle, refer to the Precautions Section.

2. Place a jack under the rear lower link.
3. Remove the rear wheel.
4. Loosen the lower link nut and bolt on the suspension member side, then remove the bolt and nut on the axle side.
5. Lower the jack slowly and remove the upper seat, coil spring and rubber sheet from the lower link.

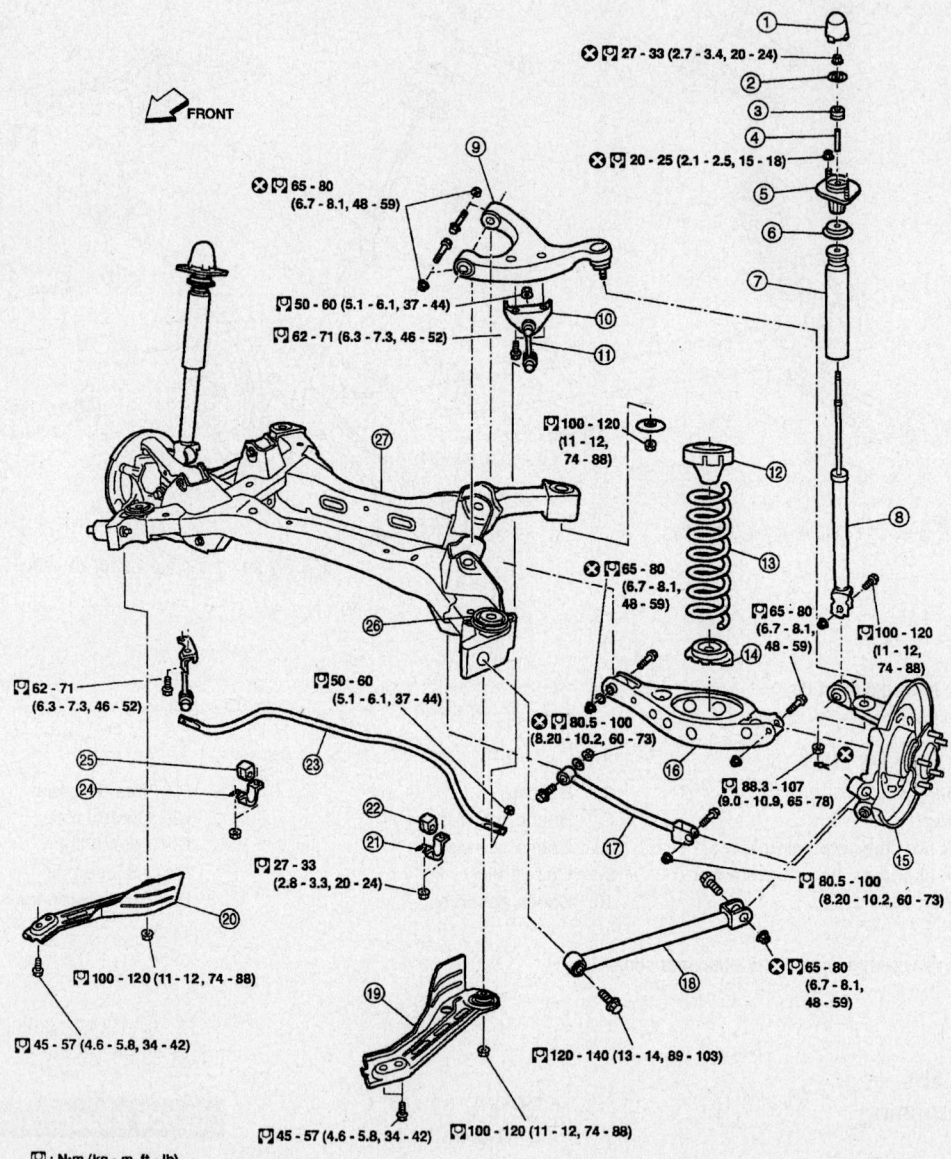

: N·m (kg - m, ft - lb)
: Always replace after every disassembly.

1. Cap	2. Washer	3. Bushing
4. Distance tube	5. Shock absorber mount bracket	6. Bound bumper cover
7. Bound bumper	8. Shock absorber	9. Suspension arm
10. Connecting rod mount bracket	11. Connecting rod	12. Upper rubber seat
13. Coil spring	14. Lower rubber seat	15. Knuckle
16. Rear lower link	17. Front lower link	18. Radius rod
19. Member stay	20. Member stay	21. Stabilizer bar clamp
22. Bushing	23. Stabilizer bar	24. Stabilizer bar clamp
25. Bushing	26. Member stopper	27. Rear suspension member

09482_ZMAX_G0037

Exploded view of the rear suspension—2004–2006 Maxima

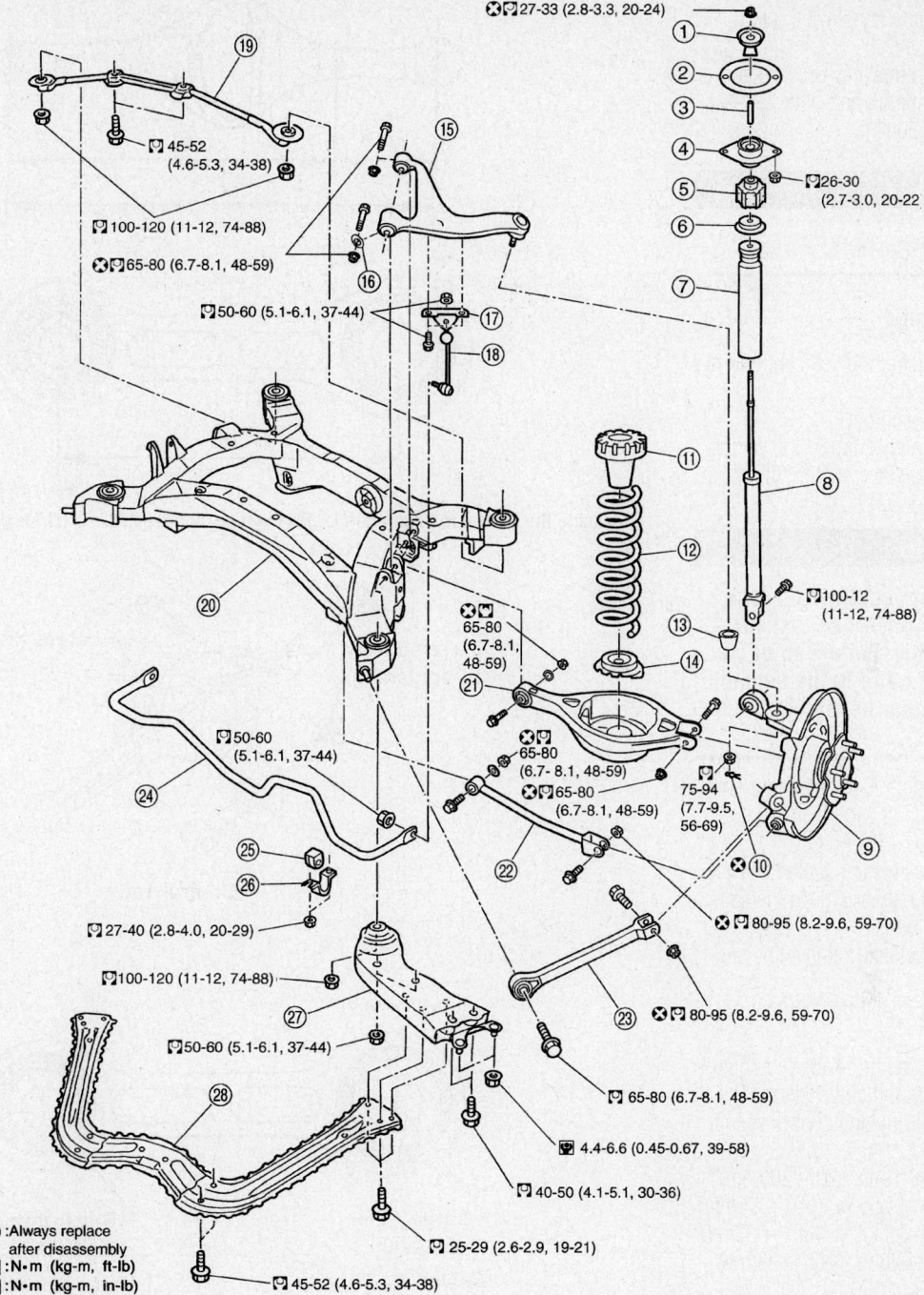

⊗ :Always replace
after disassembly
🔧:N•m (kg-m, ft-lb)
🔧:N•m (kg-m, in-lb)

1. Bushing	2. Mounting seal	3. Distance tube
4. Mounting seal bracket	5. Bushing	6. Bound bumper cover
7. Bound bumper	8. Shock absorber	9. Axle
10. Cotter pin	11. Upper seat	12. Coil spring
13. Ball seat	14. Rubber seat	15. Suspension arm
16. Stopper rubber	17. Stabilizer connecting rod mounting bracket	18. Stabilizer connecting rod
19. Rear pin stay	20. Rear suspension member	21. Rear lower link
22. Front lower link	23. Radius rod	24. Stabilizer bar
25. Stabilizer bushing	26. Stabilizer clamp	27. Member stay
28. Tunnel stay		

67162-NISS-G54

Exploded view of the rear suspension—350Z

6. Remove the lower link nut and bolt on the axle side to remove the lower link.

To install:

7. Reverse the removal procedure and tighten the lower link nut and bolts to 48–59 ft. lbs. (65–80 Nm).

Torsion Beam

REMOVAL & INSTALLATION

2002–2003 Maxima

1. Before servicing the vehicle, refer to the Precautions Section.
2. Loosen the lug nuts.
3. Raise and safely support the vehicle
4. Remove or disconnect the following:
   • Wheels

※※ WARNING

Be sure to disconnect the Anti-lock Brake System (ABS) wheel sensor from the assembly. Failure to do so may result in damage to the sensor wire and the sensor becoming inoperative.

   • Brake calipers and suspend them with a piece of wire. Do not let them hang by the hose.
5. Using a transmission jack, raise the torsion beam a little, then remove the suspension mounting bolts.
6. Lower the jack and remove the suspension assembly.
7. Remove the lateral link and control rod.
8. Inspect the torsion beam and control rod for cracks, wear and deformation. The length of the lateral link and control rod is as follows:
   • A—8.15–8.19 in. (207–208mm)
   • B—15.51–15.55 in. (394–395mm)
   • C—23.66–23.74 in. (601–603mm)
   • D—4.17–4.25 in. (106–108mm)

To install:

9. When installing the control rod, connect the bushing with the smaller inner diameter to the lateral link. Install the lateral link and the control rod on the torsion beam. Place the lateral link with the arrow topside.
10. Place the lateral link and control rod horizontally against the beam and tighten the bolts. Refer to the illustration.
11. Secure the torsion beam to the vehicle. Make sure the lateral link is horizontal, then tighten the link to the chassis.
12. Attach the struts to the torsion beam and tighten the fasteners.

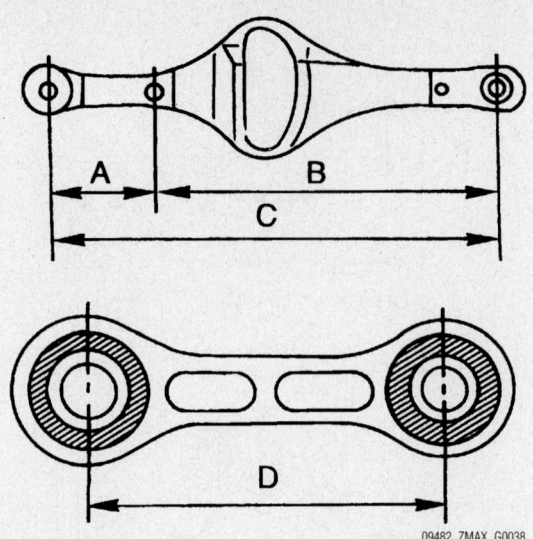

Measure the control rod and lateral links at these points—2002–2003 Maxima

09482_ZMAX_G0038

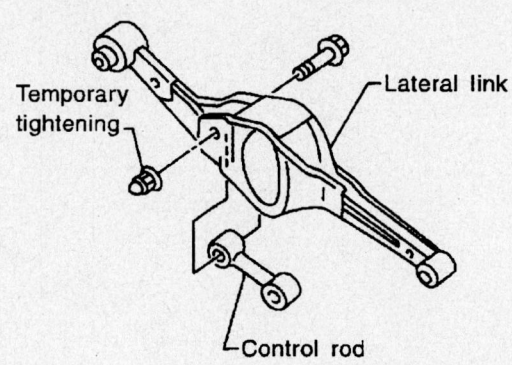

Temporary tightening — Lateral link — Control rod

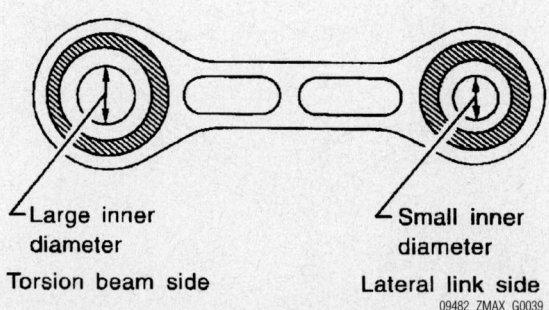

Large inner diameter
Torsion beam side

Small inner diameter
Lateral link side

09482_ZMAX_G0039

Be sure to install the control rod correctly—2002–2003 Maxima

13. Tighten the torsion beam-to-chassis bolts unloaded.
14. Install the calipers, ABS sensor and wheels. Lower the vehicle to the ground.

Wheel Bearings

ADJUSTMENT

If the wheel hub bearing assembly is removed, it must be replaced.

➡️**The wheel hub bearing assembly is not repairable; it must be replaced when defective.**

1. Before servicing the vehicle, refer to the Precautions Section.
2. Torque the wheel bearing locknut to 138–188 ft. lbs. (187–255 Nm).
3. Verify that the wheel bearings operate smoothly.
4. Install a new cotter pin into the spindle to hold the wheel bearing locknut.

5. Install a dial indicator to the rear wheel hub bearing assembly and check the axial end-play; it should be less than 0.0020 in. (0.05mm).

6. Install the grease cap.

7. If the axial end-play exceeds specifications, the wheel bearing must be replaced.

REMOVAL & INSTALLATION

If the wheel hub bearing assembly is removed, it must be replaced.

➡**If the vehicle is equipped with Anti-Lock Brake (ABS), the sensor must be removed to protect the sensor and its wiring.**

1. Before servicing the vehicle, refer to the Precautions Section.

2. Raise and safely support the vehicle. Remove the rear wheel(s).

3. If equipped with disc brakes, remove or disconnect the following:

- Brake caliper and hang it by a piece of wire
- Brake caliper support
- Disc brake pads
- Brake rotor

4. If equipped with drum brakes, remove or disconnect the following:

- Brake drum
- Brake shoe assembly, if necessary
- Grease cap

5. Remove the cotter pin, wheel bearing locknut, washer and the wheel hub bearing assembly. A slide hammer may be needed to remove the hub bearing assembly.

➡**The wheel hub bearing assembly is not repairable; it must be replaced when defective.**

To install:

➡**If the vehicle is equipped with ABS, the sensor ring must be removed and installed on the new hub.**

6. Apply oil to the threaded portion of the spindle and both sides of the plain washer.

7. Install the wheel hub bearing assembly, the washer and the wheel bearing locknut. Torque the wheel bearing locknut to 138–188 ft. lbs. (187–255 Nm).

8. Verify that the wheel bearings operate smoothly.

9. Install or connect the following:

- New cotter pin into the spindle to hold the wheel bearing locknut

10. Install a dial micrometer to the rear wheel hub bearing assembly and check the axial end-play. It should be less than 0.0020 in. (0.05mm).

- Grease cap
- ABS sensor and its wiring, if removed
- Brake assembly and the wheels

FRONT DISC BRAKES

Brake Caliper

REMOVAL & INSTALLATION

Maxima

1. Before servicing the vehicle, refer to the Precautions Section.

2. Remove or disconnect the following:

- Front wheels
- Brake fluid hose
- Pin bolts
- Caliper assembly from the vehicle

To install:

3. Use a large C-clamp to press the caliper piston back into the caliper.

4. Install or connect the following:

- New pads, new shims and pad retainers
- Brake caliper and torque the pin bolts to 23 ft. lbs. (31 Nm)
- Brake line to the caliper, using new copper washers, and torque the connecting bolt to 12–14 ft. lbs. (17–20 Nm).
- Wheels

5. Bleed the brake system and top off the master cylinder as necessary.

350Z with CLZ25VD Caliper

1. Before servicing the vehicle, refer to the Precautions Section.

2. Remove or disconnect the following:

- Front wheels

- Brake fluid hose
- Sliding pin bolts
- Caliper assembly from the vehicle

To install:

3. Use a large C-clamp to press the caliper piston back into the caliper.

4. Install or connect the following:

- New pads, new shims and pad retainers
- Brake caliper and torque the sliding pin bolts to 17–22 ft. lbs. (22–31 Nm).
- Brake line to the caliper, using new copper washers, and torque the connecting bolt to 12–14 ft. lbs. (17–20 Nm).
- Wheels

5. Bleed the brake system and top off the master cylinder as necessary.

350Z with OPB27VA/Brembo Caliper

1. Before servicing the vehicle, refer to the Precautions Section.

2. Remove or disconnect the following:

- Front wheels
- Caliper bolts
- Caliper assembly from the vehicle

To install:

3. Use a large C-clamp to press the caliper piston back into the caliper.

4. Install or connect the following:

- New pads, new shims and pad retainers
- Brake caliper and torque the mounting bolts to 112 ft. lbs. (152 Nm).
- Wheels

5. Bleed the brake system and top off the master cylinder as necessary.

350Z Except CLZ25VD/OPB27VA/ Brembo Caliper (2006 Model Only)

1. Before servicing the vehicle, refer to the Precautions Section.

2. Remove tires from vehicle.

3. Fasten disc rotor using wheel nut.

4. Drain brake fluid gradually (from bleed valve while depressing brake pedal).

5. Remove union bolt, and then remove brake hose from caliper assembly.

6. Remove torque member mounting bolts (from torque member), and remove caliper assembly (from vehicle with a power tool).

✳✳ WARNING

Do not drop brake pads.

7. Remove disc rotor.

➡**Put matching marks on both disc rotor and wheel hub when removing disc rotor.**

To install:

8. Install disc rotor.

➡**Align the matching marks of disc rotor and wheel hub, which were marked at the time of removal when reusing disc rotor.**

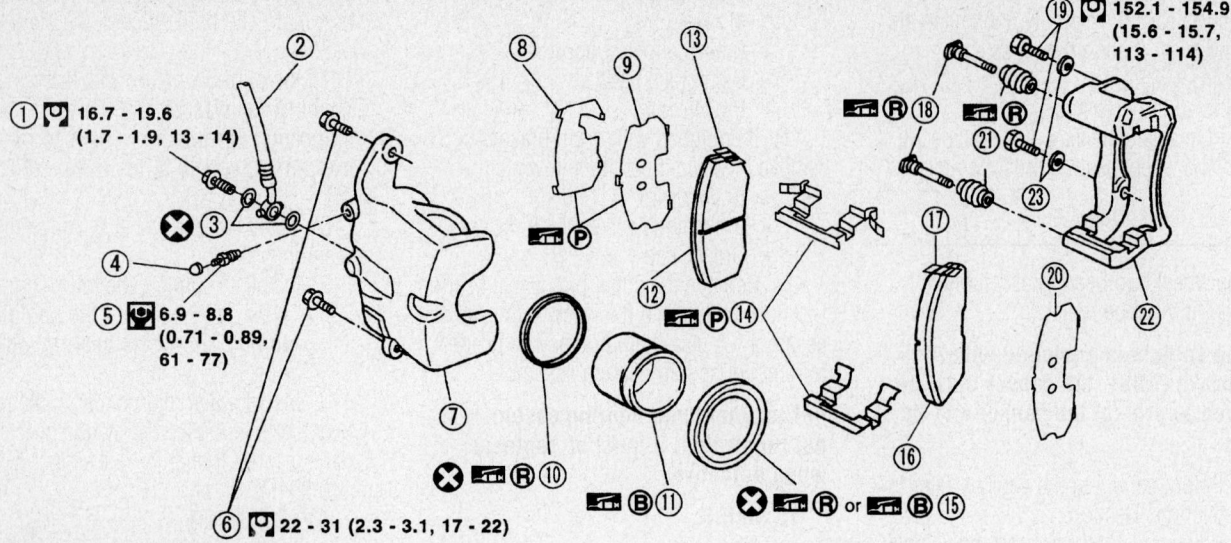

- ① 16.7 - 19.6 (1.7 - 1.9, 13 - 14)
- ⑤ 6.9 - 8.8 (0.71 - 0.89, 61 - 77)
- ⑥ 22 - 31 (2.3 - 3.1, 17 - 22)
- ⑲ 152.1 - 154.9 (15.6 - 15.7, 113 - 114)

- ℗ : PBC (Poly Butyl Cuprysil) grease or silicone-based grease
- ℝ : Rubber Gress
- Ⓑ : Brake fluid
- N•m (kg-m, ft-lb)
- N•m (kg-m, in-lb)
- ✕ : Always replace after every disassembly.

1.	Union bolt	2.	Brake hose	3.	Copper washer
4.	Cap	5.	Bleed valve	6.	Sliding pin bolt
7.	Cylinder body	8.	Inner shim cover	9.	Inner shim
10.	Piston seal	11.	Piston	12.	Inner pad
13.	Pad wear sensor	14.	Pad retainer	15.	Piston boot
16.	Outer pad	17.	Pad wear sensor	18.	Sliding pin bolt
19.	Torque member bolts	20.	Outer shim	21.	Slide pin boot
22.	Torque member	23.	Washer		

09482_ZMAX_G0040

Front brake caliper assembly—350Z with CLZ25VD caliper

9. Install caliper assembly to vehicle, and tighten torque member mounting bolts to 113 ft. lbs. (154 Nm).

➡ **Before installing torque member to vehicle, wipe oil and grease on washer seats on steering knuckle and mounting surface of torque member.**

10. Install a projection of brake hose metal fitting by aligning with protrusions on cylinder body, and tighten union bolt to 13 ft. lbs. (18 Nm).

✳✳ WARNING

Refill with new brake fluid "DOT 3". Never reuse drained brake fluid.

11. Bleed the brake system and top off the master cylinder as necessary.
12. Install tires to vehicle.

Disc Brake Pads

REMOVAL & INSTALLATION

Maxima

1. Before servicing the vehicle, refer to the Precautions Section.
2. Remove or disconnect the following:
   - Wheels
   - Bottom guide pin from the caliper and swing the caliper cylinder body up
   - Brake pad retainers and the pads

To install:
3. Compress the piston of the disc brake caliper.
4. Install or connect the following:
   - Brake pads, retainers, and caliper assembly. Torque the guide pin to 16–23 ft. lbs. (22–31 Nm).
   - Wheels
5. Check the master cylinder and add fluid if necessary.

350Z

1. Before servicing the vehicle, refer to the Precautions Section.
2. Remove or disconnect the following:
   - Front wheels
   - Sliding pin bolts
   - Rotate the caliper up and remove the brake pads

To install:
3. Install or connect the following:
   - New pads, new shims and pad retainers
   - Brake caliper
   - Wheels
4. Bleed the brake system and top off the master cylinder as necessary.

REAR DISC BRAKES

Brake Caliper

REMOVAL & INSTALLATION

Maxima

1. Before servicing the vehicle, refer to the Precautions Section.
2. Remove or disconnect the following:

- Rear wheels
- Parking brake cable and the lock spring
- Brake fluid hose from the caliper
- Caliper pin bolts and remove the caliper

To install:

3. Turn the piston clockwise back into the caliper body. Remove some brake fluid from the master cylinder, if necessary. Take care not to damage the piston boot.

4. Coat the pad contact area on the mounting support with a silicone based grease.

5. Install or connect the following:

- New pads, shims and the pad springs
- Caliper body into position and

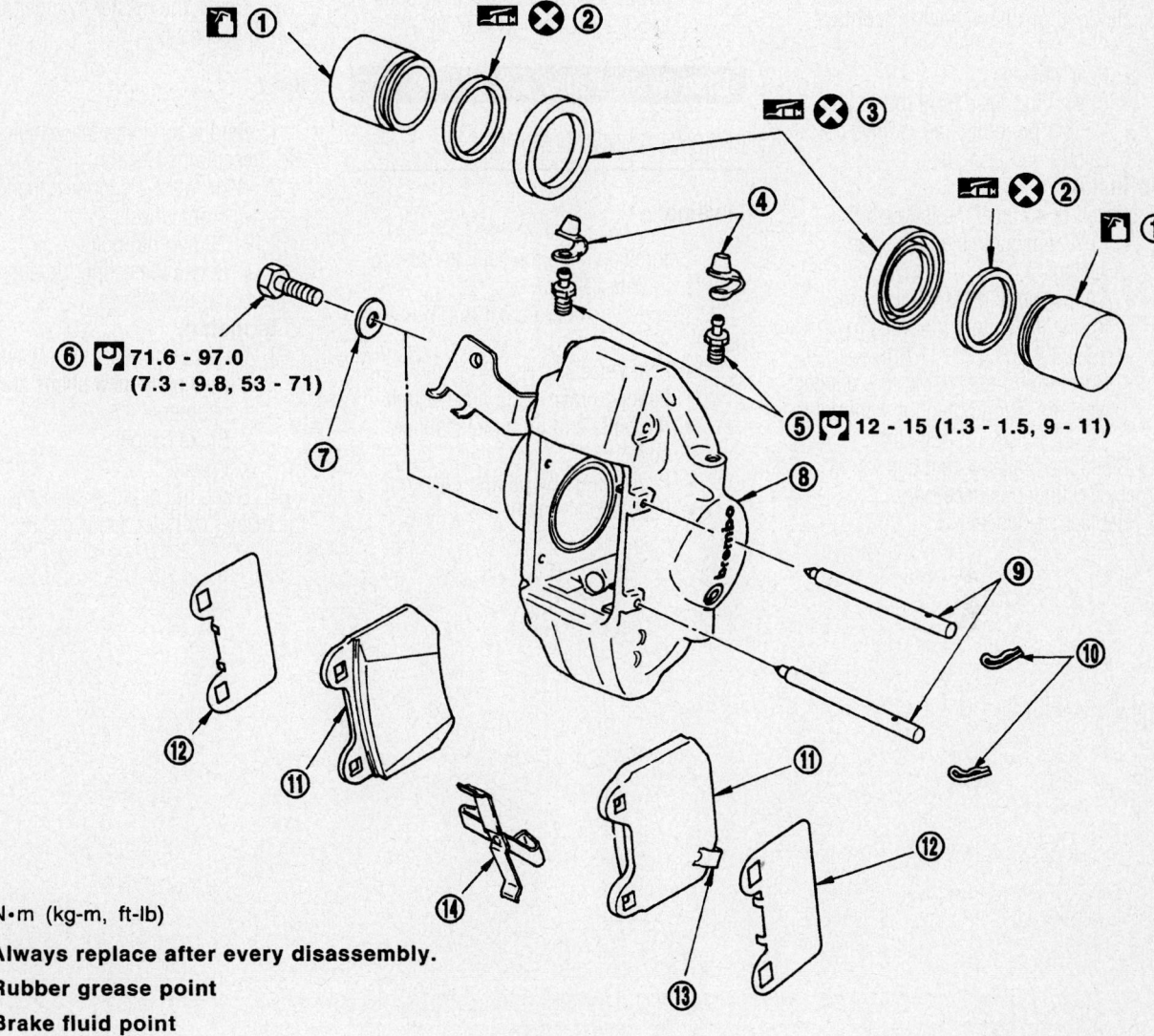

⑥ 🔧 71.6 - 97.0 (7.3 - 9.8, 53 - 71)

⑤ 🔧 12 - 15 (1.3 - 1.5, 9 - 11)

🔧 : N•m (kg-m, ft-lb)

❌ : **Always replace after every disassembly.**

⬛ : **Rubber grease point**

🛢 : **Brake fluid point**

1.	Piston	2.	Piston seal	3.	Piston boot
4.	Cap	5.	Bleed valve	6.	Bolt
7.	Washer	8.	Caliper	9.	Pad pins
10.	Clips	11.	Brake pad	12.	Shim cover
13.	Pad wear sensor	14.	Cross spring		

09482_ZMAX_G0041

Rear brake caliper assembly—350Z with OPB13VB/Brembo caliper

torque the caliper pin bolts to 16–23 ft. lbs. (22–31 Nm)
- Brake fluid hose, using new copper washers, and tighten the flare nut to 12–14 ft. lbs. (17–20 Nm)
- Lock spring and the parking brake cable

6. Bleed the brake system and top off the master cylinder as necessary.

7. Replace the wheels.

350Z with AD14VE Caliper

1. Before servicing the vehicle, refer to the Precautions Section.

2. Remove or disconnect the following:

- Rear wheels
- Brake fluid hose from the caliper
- Caliper pin bolts and remove the caliper

To install:

3. Install or connect the following:

- New pads, shims and the pad springs
- Caliper body into position and torque the caliper pin bolts to 16–23 ft. lbs. (22–31 Nm)
- Brake fluid hose, using new copper washers, and tighten the flare nut to 12–14 ft. lbs. (17–20 Nm)

4. Bleed the brake system and top off the master cylinder as necessary.

5. Replace the wheels.

350Z with OPB13VB/Brembo Caliper

1. Before servicing the vehicle, refer to the Precautions Section.

2. Remove or disconnect the following:

- Rear wheels
- Caliper pin bolts and remove the caliper

To install:

3. Install or connect the following:

- New pads, shims and the pad springs
- Caliper body into position and torque the caliper mounting bolts to 53–71 ft. lbs. (72–97 Nm)

Disc Brake Pads

REMOVAL & INSTALLATION

Maxima

1. Before servicing the vehicle, refer to the Precautions Section.

2. Remove or disconnect the following:

- Rear wheels
- Parking brake cable bracket bolt
- Pin bolts and lift off the caliper body
- Pad springs by pulling them out
- Pads and shims

To install:

3. Turn the piston clockwise back into the caliper body. Take care not to damage the piston boot.

4. Coat the pad contact area on the mounting support with a silicone based grease.

5. Install or connect the following:

- Pads, shims, and the pad springs
- Caliper body into position in the mounting support and tighten the pin bolts to 16–23 ft. lbs. (22–31 Nm)
- Wheels

6. Check the master cylinder and add fluid if necessary.

350Z

1. Before servicing the vehicle, refer to the Precautions Section.

2. Remove or disconnect the following:

- Front wheels
- Sliding pin bolts
- Rotate the caliper up and remove the brake pads

To install:

3. Install or connect the following:

- New pads, new shims and pad retainers
- Brake caliper
- Wheels

4. Bleed the brake system and top off the master cylinder as necessary.

NISSAN

Frontier • Xterra

18

SPECIFICATIONS AND MAINTENANCE CHARTS

ENGINE AND VEHICLE IDENTIFICATION

ID/Code ①	Liters (cc)	Cu. In.	Cyl.	Fuel Sys.	Engine Type	Eng. Mfg.	Code ②	Year
KA24DE/D	2.4 (2389)	146	4	MFI	DOHC	Nissan	2	2002
QR25DE/B	2.5 (2488)	152	4	MFI	DOHC	Nissan	3	2003
VG33E/E	3.3 (3277)	199	6	MFI	SOHC	Nissan	4	2004
VG33ER/M	3.3 (3277)	199	6	MFI	SOHC	Nissan	5	2005
VQ40DE/A	4.0 (3954)	241	6	MFI	DOHC	Nissan	6	2006

Engine columns span ID/Code through Eng. Mfg.; Model Year spans Code and Year.

MFI: Multi-port Fuel Injection

SOHC: Single Overhead Camshaft

DOHC: Double Overhead Camshafts

① 4th digit of the Vehicle Identification Number (VIN)

② 10th digit of the Vehicle Identification Number (VIN)

09482_FRON_C0001

GENERAL ENGINE SPECIFICATIONS

Year	Model	Engine Displacement Liters	Engine ID	Net Horsepower @ rpm	Net Torque @ rpm (ft. lbs.)	Bore x Stroke (in.)	Compression Ratio	Oil Pressure @ rpm
2002	Frontier	2.4	KA24DE	143@5200	154@4000	3.50x3.78	9.2:1	60-70@3000
	Frontier	3.3	VG33E	170@4800	200@2800	3.60X3.27	8.9:1	60-65@2000
	Frontier	3.3	VG33ER	210@4800	231@2800	3.60X3.27	8.3:1	60-65@2000
	Xterra	2.4	KA24DE	143@5200	154@4000	3.50X3.78	9.2:1	60-70@3000
	Xterra	3.3	VG33E	170@4800	200@2800	3.60X3.27	8.9:1	60-65@2000
	Xterra	3.3	VG33ER	210@4800	231@2800	3.60X3.27	8.3:1	60-65@2000
2003	Frontier	2.4	KA24DE	143@5200	154@4000	3.50x3.78	9.2:1	60-70@3000
	Frontier	3.3	VG33E	170@4800	200@2800	3.60X3.27	8.9:1	60-65@2000
	Frontier	3.3	VG33ER	210@4800	246@2800	3.60X3.27	8.3:1	60-65@2000
	Xterra	2.4	KA24DE	143@5200	154@4000	3.50X3.78	9.2:1	60-70@3000
	Xterra	3.3	VG33E	170@4800	200@2800	3.60X3.27	8.9:1	60-65@2000
	Xterra	3.3	VG33ER	210@4800	231@2800	3.60X3.27	8.3:1	60-65@2000
2004	Frontier	2.4	KA24DE	143@5200	154@4000	3.50x3.78	9.2:1	60-70@3000
	Frontier	3.3	VG33E	170@4800	200@2800	3.60X3.27	8.9:1	60-65@2000
	Frontier	3.3	VG33ER	210@4800	246@2800	3.60X3.27	8.3:1	60-65@2000
	Xterra	2.4	KA24DE	143@5200	154@4000	3.50X3.78	9.2:1	60-70@3000
	Xterra	3.3	VG33E	170@4800	200@2800	3.60X3.27	8.9:1	60-65@2000
	Xterra	3.3	VG33ER	210@4800	231@2800	3.60X3.27	8.3:1	60-65@2000
2005	Frontier	2.5	QR25DE	154@5200	173@4400	3.50X3.94	9.5:1	43@2000
	Frontier	4.0	VQ40DE	265@5600	284@4000	3.76X3.62	9.7:1	43@2000
	Xterra	4.0	VQ40DE	265@5600	284@4000	3.76X3.62	9.7:1	43@2000
2006	Frontier	2.5	QR25DE	154@5200	173@4400	3.50X3.94	9.5:1	43@2000
	Frontier	4.0	VQ40DE	265@5600	284@4000	3.76X3.62	9.7:1	43@2000
	Xterra	4.0	VQ40DE	265@5600	284@4000	3.76X3.62	9.7:1	43@2000

MFI: Multi-port Fuel Injection

09482_FRON_C0002

ENGINE TUNE-UP SPECIFICATIONS

Year	Engine Displ. Liters	Engine ID	Spark Plug Gap (in.)	Ignition Timing (deg.)		Fuel Pump (psi)	Idle Speed (rpm)		Valve Clearance (in.)	
				MT	AT		MT	AT ②	In.	Ex.
2002	2.4	KA24DE	0.039-0.043	18-22B	18-22B	34 ①	750-850	750-850	0.012-0.015	0.013-0.016
	3.3	VG33E	0.039-0.043	13-17B	13-17B	34 ①	700-800	700-800	HYD	HYD
	3.3	VG33ER	0.043	10B	10B	35 ①	700-800	700-800	HYD	HYD
2003	2.4	KA24DE	0.039-0.043	18-22B	18-22B	34 ①	750-850	750-850	0.012-0.015	0.013-0.016
	3.3	VG33E	0.039-0.043	13-17B	13-17B	34 ①	700-800	700-800	HYD	HYD
	3.3	VG33ER	0.043	10B	10B	35 ①	700-800	700-800	HYD	HYD
2004	2.4	KA24DE	0.043	18-22B	18-22B	34 ①	750-850	750-850	0.012-0.015	0.013-0.016
	3.3	VG33E	0.043	13-17B	13-17B	34 ①	700-800	700-800	HYD	HYD
	3.3	VG33ER	0.043	10B	10B	35 ①	700-800	700-800	HYD	HYD
2005	2.5	QR25DE	0.043	10-20B	10-20B	51 ③	575-675	650-750	④	⑤
	4.0	VQ40DE	0.043	10-20B	10-20B	51 ③	575-675	650-750	⑥	⑦
2006	2.5	QR25DE	0.043	10-20B	10-20B	51 ③	575-675	650-750	④	⑤
	4.0	VQ40DE	0.043	10-20B	10-20B	51 ③	575-675	650-750	⑥	⑦

NOTE: The Vehicle Emission Control Information label often reflects specification changes made during production. The label figures must be used if they differ from those in this chart.

B: Before top dead center

HYD: Hydraulic

① System pressure at idle with vacuum hose connected
   Should increase to 43 psi when disconnected

② Automatic transmission in Neutral

③ At idle

④ 0.009-0.013 cold
   0.012-0.016 hot

⑤ 0.010-0.013 cold
   0.012-0.017 hot

⑥ 0.009-0.013 cold
   0.012-0.016 hot

⑦ 0.011-0.015 cold
   0.012-0.017 hot

09482_FRON_C0003

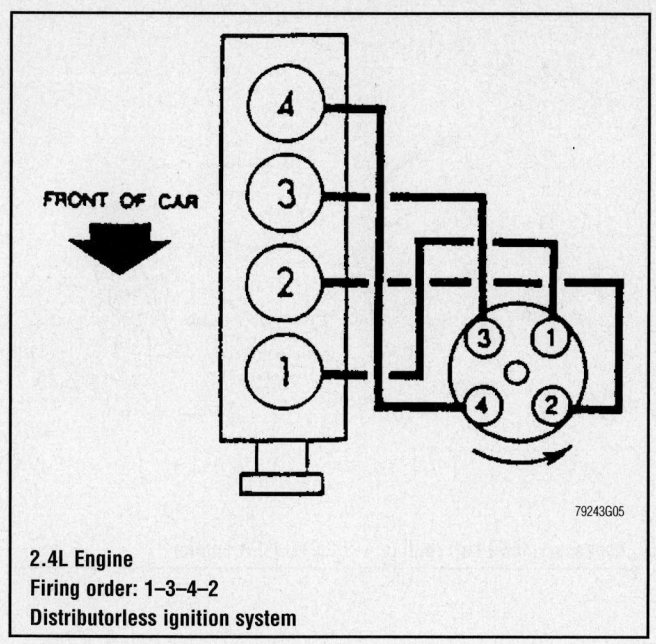

2.4L Engine
Firing order: 1–3–4–2
Distributorless ignition system

79243G05

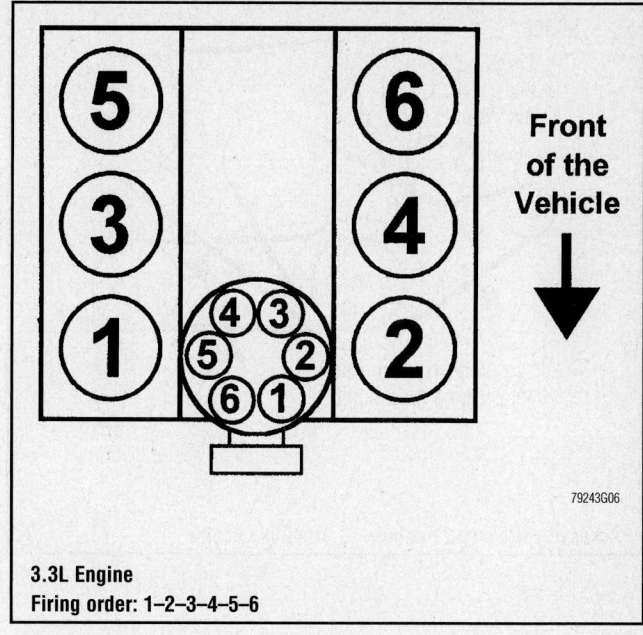

Front of the Vehicle

3.3L Engine
Firing order: 1–2–3–4–5–6

79243G06

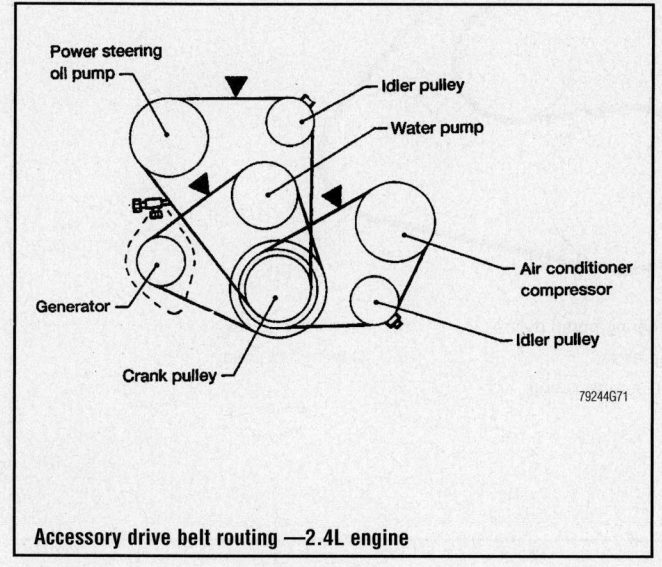

Accessory drive belt routing —2.4L engine

79244G71

Power steering oil pump
Idler pulley
Water pump
Air conditioner compressor
Idler pulley
Generator
Crank pulley

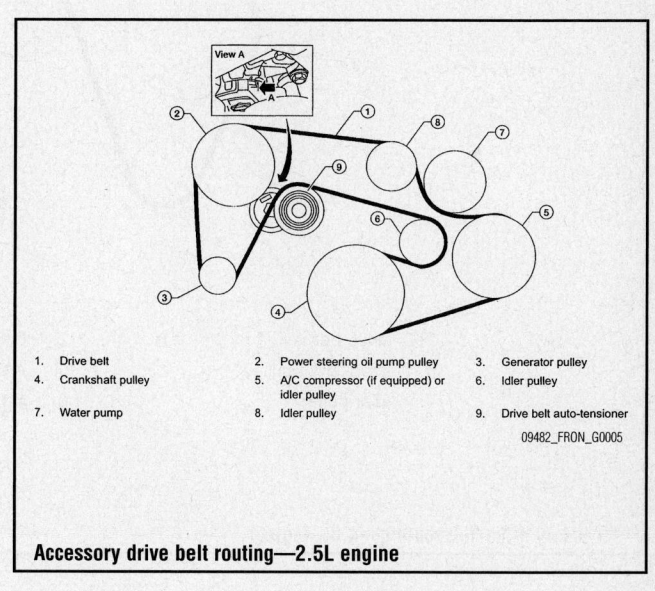

Accessory drive belt routing—2.5L engine

1.	Drive belt	2.	Power steering oil pump pulley	3.	Generator pulley
4.	Crankshaft pulley	5.	A/C compressor (if equipped) or idler pulley	6.	Idler pulley
7.	Water pump	8.	Idler pulley	9.	Drive belt auto-tensioner

09482_FRON_G0005

View A

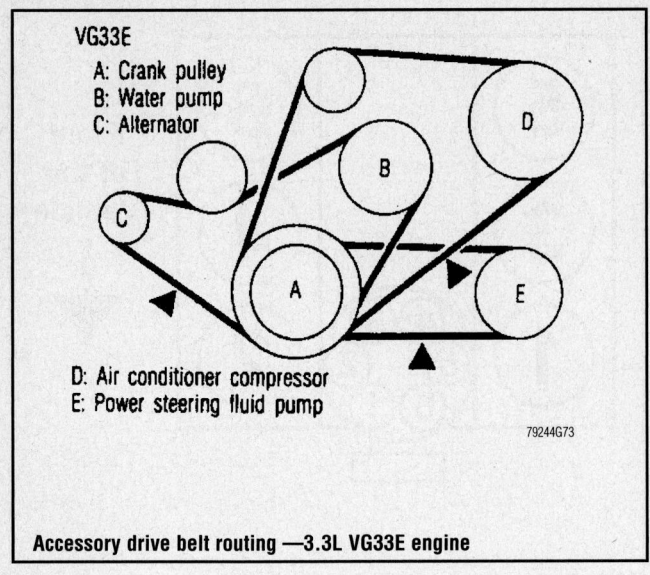

Accessory drive belt routing —3.3L VG33E engine

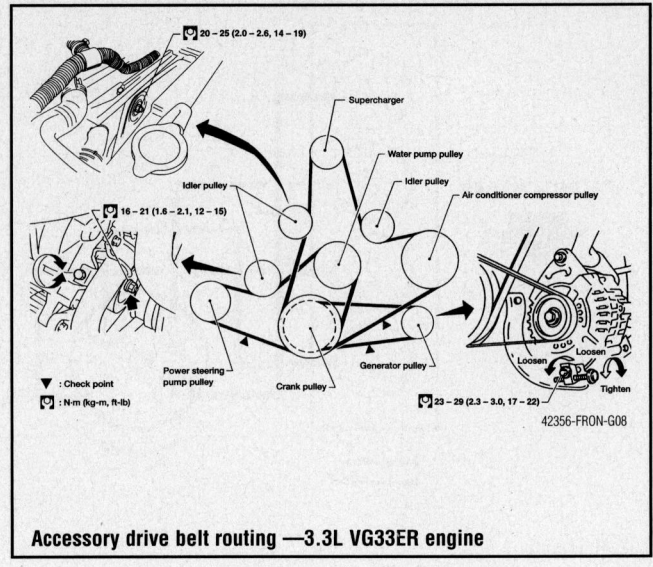

Accessory drive belt routing —3.3L VG33ER engine

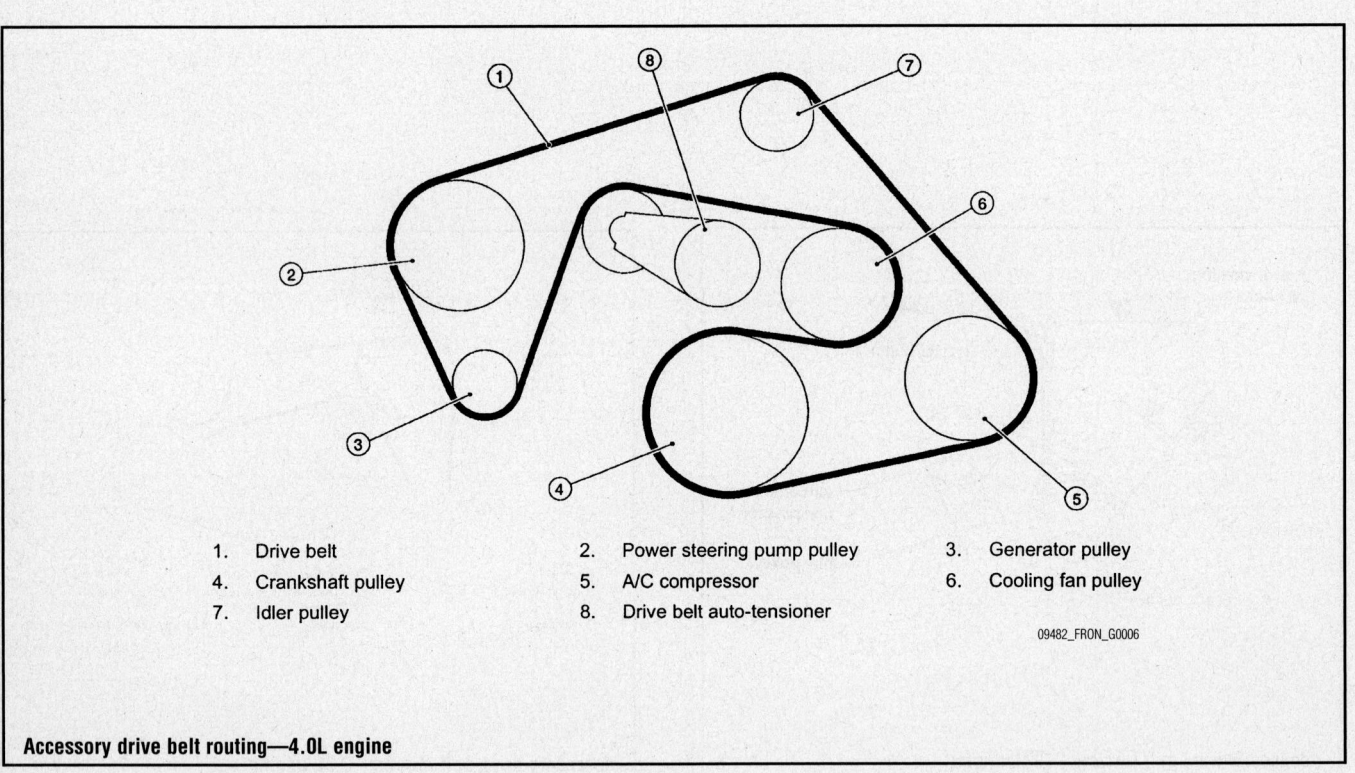

1. Drive belt
2. Power steering pump pulley
3. Generator pulley
4. Crankshaft pulley
5. A/C compressor
6. Cooling fan pulley
7. Idler pulley
8. Drive belt auto-tensioner

09482_FRON_G0006

Accessory drive belt routing—4.0L engine

CAPACITIES

Year	Model	Engine Displacement Liters	Engine ID	Engine Oil with Filter (qts.)	Transmission (pts.)		Transfer Case (pts.)	Drive Axle		Fuel Tank (gal.)	Cooling System (qts.)
					5-Spd	Auto.		Front (pts.)	Rear (pts.)		
2002	Frontier	2.4	KA24DE	3.50	4.25	—	—	—	2.4	15.9	7.75
	Frontier	3.3	VG33E	3.50	①	②	4.8	5.8	3.12	19.4	11.6
	Xterra	2.4	KA24DE	3.75	4.25	—	—	—	3.1	19.4	7.75
	Xterra	3.3	VG33E	3.50	①	②	4.8	3.1	5.9	19.4	11.6
	Xterra	3.3	VG33ER	3.50	①	②	4.8	3.1	5.9	19.4	11.6
2003	Frontier	2.4	KA24DE	3.50	4.25	—	—	—	2.4	15.9	7.75
	Frontier	3.3	VG33E	3.50	①	②	4.8	5.8	3.12	19.4	11.6
	Frontier	3.3	VG33ER	3.50	①	②	4.8	5.8	3.12	19.4	11.6
	Xterra	2.4	KA24DE	3.75	4.25	—	—	—	3.1	19.4	7.75
	Xterra	3.3	VG33E	3.50	①	②	4.8	3.1	5.9	19.4	11.6
	Xterra	3.3	VG33ER	3.50	①	②	4.8	3.1	5.9	19.4	11.6
2004	Frontier	2.4	KA24DE	3.75	4.25	16.75	—	—	2.4	15.9	7.75
	Frontier	3.3	VG33E	3.50	①	②	4.8	3.75	5.8	19.4	11.6
	Frontier	3.3	VG33ER	3.50	①	②	4.8	3.75	5.8	19.4	11.6
	Xterra	2.4	KA24DE	3.75	4.25	16.75	—	—	5.8	19.4	7.75
	Xterra	3.3	VG33E	3.50	①	②	4.8	3.75	5.9	19.4	11.6
2005	Frontier	2.5	QR25DE	5.15	6.15	21.50	—	—	3.3	21.2	10.0
	Frontier	4.0	VQ40DE	5.15	③	21.50	2.1	1.75	④	21.2	11.0
	Xterra	4.0	VQ40DE	3.50	③	21.50	2.1	1.75	④	21.2	11.0
2006	Frontier	2.5	QR25DE	5.15	6.15	21.50	—	—	3.3	21.2	10.0
	Frontier	4.0	VQ40DE	5.15	③	21.50	2.1	1.75	④	21.2	11.0
	Xterra	4.0	VQ40DE	3.50	③	21.50	2.1	1.75	④	21.2	11.0

NOTE: All capacities are approximate. Add fluid gradually and check to be sure a proper fluid level is obtained.

① 2WD: 5.8; 4WD: 10.75

② 2WD: 17.5; 4WD: 18 pts.

③ 2WD: 8.3; 4WD: 8.9

④ C200: 3.3; M226: 4.25

09482_FRON_C0004

VALVE SPECIFICATIONS

Year	Engine Displacement Liters	Engine ID	Seat Angle (deg.)	Face Angle (deg.)	Spring Test Pressure (lbs. @ in.)	Spring Installed Height (in.)	Stem-to-Guide Clearance (in.)		Stem Diameter (in.)	
							Intake	Exhaust	Intake	Exhaust
2002	2.4	KA24DE	45	45.5	93.9@1.15	NA	0.0008-0.0021	0.0016-0.0029	0.2742-0.2748	0.2734-0.2740
	3.3	VG33E	45	45.25-46.75	①	NA	0.0008-0.0021	0.0016-0.0029	0.2742-0.2748	0.3135-0.3138
	3.3	VG33ER	45	45.25-46.75	①	NA	0.0008-0.0021	0.0012-0.0019	0.2742-0.2748	0.3135-0.3138
2003	2.4	KA24DE	45	45.5	93.9@1.15	NA	0.0008-0.0021	0.0016-0.0029	0.2742-0.2748	0.2734-0.2740
	3.3	VG33E	45	45.25-46.75	①	NA	0.0008-0.0021	0.0016-0.0029	0.2742-0.2748	0.3135-0.3138
	3.3	VG33ER	45	45.25-46.75	①	NA	0.0008-0.0021	0.0012-0.0019	0.2742-0.2748	0.3135-0.3138
2004	2.4	KA24DE	45	45.5	93.9@1.15	NA	0.0008-0.0021	0.0016-0.0029	0.2742-0.2748	0.2734-0.2740
	3.3	VG33E	45	45.25-46.75	①	NA	0.0008-0.0021	0.0016-0.0029	0.2742-0.2748	0.3135-0.3138
	3.3	VG33ER	45	45.25-46.75	①	NA	0.0008-0.0021	0.0012-0.0019	0.2742-0.2748	0.3135-0.3138
2005	2.5	QR25DE	②	NA	NA	1.390	0.0008-0.0021	0.0012-0.0025	0.2348-0.2354	0.2344-0.2350
	4.0	VQ40DE	②	NA	③	1.853	0.0008-0.0021	0.0012-0.0025	0.2348-0.2354	0.2344-0.2350
2006	2.5	QR25DE	②	NA	NA	1.390	0.0008-0.0021	0.0012-0.0025	0.2348-0.2354	0.2344-0.2350
	4.0	VQ40DE	②	NA	③	1.853	0.0008-0.0021	0.0012-0.0025	0.2348-0.2354	0.2344-0.2350

NA: Not Available

① Inner: 57.3 @ 0.984
Outer: 117.7 @ 1.181

② 45 degrees 45'

③ Installation: 37–42@1.457
Valve open: 84–95@1.071

09482_FRON_C0005

CAMSHAFT SPECIFICATIONS CHART

All measurements are given in inches.

Year	Engine Displ. Liters	Engine VIN	Journal Dia.	Brg. Oil Clearance	Shaft End-play	Runout	Lobe Height	
							Intake	Exhaust
2002	2.4	KA24E	1.0998-1.1006	0.0018-0.0035	0.0028-0.0058	0.0008	1.6440-1.6510	1.6460-1.6540
	3.3	VG33E	①	0.0024-0.0041	0.0012-0.0024	0.0016	1.5332-1.5407	1.5332-1.5407
	3.3	VG33ER	①	0.0024-0.0041	0.0012-0.0024	0.0016	1.5332-1.5407	1.5332-1.5407
2003	2.4	KA24E	1.0998-1.1006	0.0018-0.0035	0.0028-0.0058	0.0008	1.6440-1.6510	1.6460-1.6540
	3.3	VG33E	①	0.0024-0.0041	0.0012-0.0024	0.0016	1.5332-1.5407	1.5332-1.5407
	3.3	VG33ER	①	0.0024-0.0041	0.0012-0.0024	0.0016	1.5332-1.5407	1.5332-1.5407
2004	2.4	KA24E	1.0998-1.1006	0.0018-0.0035	0.0028-0.0058	0.0008	1.6440-1.6510	1.6460-1.6540
	3.3	VG33E	①	0.0024-0.0041	0.0012-0.0024	0.0016	1.5332-1.5407	1.5332-1.5407
	3.3	VG33ER	①	0.0024-0.0041	0.0012-0.0024	0.0016	1.5332-1.5407	1.5332-1.5407
2005	2.5	QR25DE	②	0.0018-0.0034	0.0045-0.0074	0.0008	1.7722-1.7797	1.7313-1.7388
	4.0	VQ40DE	③	④	0.0045-0.0074	0.0010	1.7900-1.7921	1.7746-1.7821
2006	2.5	QR25DE	②	0.0018-0.0034	0.0045-0.0074	0.0008	1.7722-1.7797	1.7313-1.7388
	4.0	VQ40DE	③	④	0.0045-0.0074	0.0010	1.7900-1.7921	1.7746-1.7821

NS - Not specified by manufacturer

① right camshaft: 1.8472-1.8480, except rear journal. 1.6701-1.6709 rear journal.
   left camshaft: 1.8472-1.8480, except rear journal and front jurnal. 1.6701-1.6709, rear journal. 1.8866-1.8874, front journal.

② No.1: 1.0998-1.1006. No's. 2, 3, 4, 5: 0.9226-0.9234

③ No.1: 1.0211-1.0218. No's. 2, 3, 4: 0.9230-0.9238

④ No.1: 1.0018-0.0034. No's. 2, 3, 4: 0.0014-0.0030

CRANKSHAFT AND CONNECTING ROD SPECIFICATIONS

All measurements are given in inches.

Year	Engine Displacement Liters	Engine ID	Crankshaft				Connecting Rod		
			Main Brg. Journal Dia.	Main Brg. Oil Clearance	Shaft End-play	Thrust on No.	Journal Diameter	Oil Clearance	Side Clearance
2002	2.4	KA24DE	2.3609-2.3612	0.0008-0.0019	0.0020-0.0071	3	1.9672-1.9675	0.0004-0.0014	0.0080-0.0160
	3.3	VG33E	2.4790-2.4793	0.0011-0.0022	0.0020-0.0067	4	1.9967-1.9675	0.0006-0.0021	0.0079-0.0138
	3.3	VG33ER	①	②	0.0020-0.0067	4	1.9667-1.9675	0.0009-0.0025	0.0079-0.0138
2003	2.4	KA24DE	2.3609-2.3612	0.0008-0.0019	0.0020-0.0071	3	1.9672-1.9675	0.0004-0.0014	0.0080-0.0160
	3.3	VG33E	2.4790-2.4793	0.0011-0.0022	0.0020-0.0067	4	1.9967-1.9675	0.0006-0.0021	0.0079-0.0138
	3.3	VG33ER	①	②	0.0020-0.0067	4	1.9667-1.9675	0.0009-0.0025	0.0079-0.0138
2004	2.4	KA24DE	2.3609-2.3612	0.0008-0.0019	0.0020-0.0071	3	1.9672-1.9675	0.0004-0.0014	0.0080-0.0160
	3.3	VG33E	2.4790-2.4793	0.0011-0.0022	0.0020-0.0067	4	1.9967-1.9675	0.0006-0.0021	0.0079-0.0138
	3.3	VG33ER	①	②	0.0020-0.0067	4	1.9667-1.9675	0.0009-0.0025	0.0079-0.0138
2005	2.5	QR25DE	③	④	0.0039-0.0102	NA	NA	0.0015-0.0022	0.0079-0.0138
	4.0	VQ40DE	⑤	0.0014-0.0018	0.0039-0.0098	NA	NA	0.0013-0.0023	0.0079-0.0138
2006	2.5	QR25DE	③	④	0.0039-0.0102	NA	NA	0.0015-0.0022	0.0079-0.0138
	4.0	VQ40DE	⑤	0.0014-0.0018	0.0039-0.0098	NA	NA	0.0013-0.0023	0.0079-0.0138

NA: Not Available

① Except No. 1

    Grade 0: 2.4790-2.4793

    Grade 1: 2.4787-2,4790

    Grade 2: 2.4784-2.4787

    No. 1

    Grade 3: 2.4683-2.4793

    Grade 4: 2.4789-2.4791

    Grade 5: 2.4786-2.4789

    Grade 6: 2.4784-2.4786

② No. 1: 0.0012-0.0019

    No's. 2, 3, 4: 0.0015-0.0026

③ There are 24 different grades, ranging from A (2.1645) to 7 (2.1636)

④ No. 1, 3, 5: 0.0011-0.0017

    No's. 2, 4: 0.0016-0.0022

⑤ There are 24 different grades, ranging from A (2.7549) to 7 (2.7540)

PISTON AND RING SPECIFICATIONS

All measurements are given in inches.

Year	Engine Displacement Liters	Engine ID	Piston Clearance	Ring Gap			Ring Side Clearance		
				Top Comp.	Bottom Comp.	Oil Control	Top Comp.	Bottom Comp.	Oil Control
2002	2.4	KA24E	0.0008-0.0016	0.011-0.021	0.018-0.027	0.008-0.027	0.0016-0.0031	0.0012-0.0028	0.0026-0.0053
	3.3	VG33E	①	0.0083-0.0157	0.0197-0.0272	0.0079-0.0272	0.0009-0.0030	0.0012-0.0028	0.0006-0.0073
	3.3	VG33ER	②	0.0083-0.0122	0.0197-0.0236	0.0079-0.0236	0.0016-0.0031	0.0012-0.0028	0.0006-0.0073
2003	2.4	KA24E	0.0008-0.0016	0.011-0.021	0.018-0.027	0.008-0.027	0.0016-0.0031	0.0012-0.0028	0.0026-0.0053
	3.3	VG33E	①	0.0083-0.0157	0.0197-0.0272	0.0079-0.0272	0.0009-0.0030	0.0012-0.0028	0.0006-0.0073
	3.3	VG33ER	②	0.0083-0.0122	0.0197-0.0236	0.0079-0.0236	0.0016-0.0031	0.0012-0.0028	0.0006-0.0073
2004	2.4	KA24E	0.0008-0.0016	0.011-0.021	0.018-0.027	0.008-0.027	0.0016-0.0031	0.0012-0.0028	0.0026-0.0053
	3.3	VG33E	①	0.0083-0.0157	0.0197-0.0272	0.0079-0.0272	0.0009-0.0030	0.0012-0.0028	0.0006-0.0073
	3.3	VG33ER	②	0.0083-0.0122	0.0197-0.0236	0.0079-0.0236	0.0016-0.0031	0.0012-0.0028	0.0006-0.0073
2005	2.5	QR25DE	0.0004-0.0012	0.0083-0.0122	0.0146-0.0205	0.0079-0.0236	0.0016-0.0031	0.0012-0.0028	0.0026-0.0053
	4.0	VQ40DE	0.0004-0.0012	0.0091-0.0130	0.0130-0.0189	0.0079-0.0197	0.0018-0.0031	0.0012-0.0028	0.0026-0.0053
2006	2.5	QR25DE	0.0004-0.0012	0.0083-0.0122	0.0146-0.0205	0.0079-0.0236	0.0016-0.0031	0.0012-0.0028	0.0026-0.0053
	4.0	VQ40DE	0.0004-0.0012	0.0091-0.0130	0.0130-0.0189	0.0079-0.0197	0.0018-0.0031	0.0012-0.0028	0.0026-0.0053

① Except cylinders 3 and 4: 0.0010 - 0.0018 in.
   Cylinders 3 and 4: 0.0006 - 0.0010 in.

② Cylinders 3, 4: 0.0006-0.0010 in.
   Cylinders 1, 2, 5, 6: 0.0010-0.0018 in.

09482_FRON_C0008

TORQUE SPECIFICATIONS
All readings in ft. lbs.

Year	Engine Displacement Liters	Engine ID	Cylinder Head Bolts	Main Bearing Bolts	Rod Bearing Bolts	Crankshaft Damper Bolts	Flywheel Bolts	Manifold Intake	Manifold Exhaust	Spark Plugs	Oil Pan Drain Plug
2002	2.4	KA24DE	①	34-41	②	105-112	105-112	12-14	27-35	14-22	22-29
	3.3	VG33E	③	67-74	②	141-156	61-69	③	21-25	14-22	22-29
	3.3	VG33ER	③	67-74	④	141-156	61-69	③	21-25	14-22	22-29
2003	2.4	KA24DE	①	34-41	②	105-112	105-112	12-14	27-35	14-22	22-29
	3.3	VG33E	③	67-74	②	141-156	61-69	③	21-25	14-22	22-29
	3.3	VG33ER	③	67-74	④	141-156	61-69	③	21-25	14-22	22-29
2004	2.4	KA24DE	①	34-41	②	105-112	105-112	12-14	27-35	14-22	22-29
	3.3	VG33E	③	67-74	②	141-156	61-69	③	21-25	14-22	22-29
	3.3	VG33ER	③	67-74	④	141-156	61-69	③	21-25	14-22	22-29
2005	2.5	QR25DE	⑤	⑥	⑦	⑧	80	⑨	⑩	14-22	25
	4.0	VQ40DE	⑪	⑫	14	⑬	65	⑭	⑮	14-22	25
2006	2.5	QR25DE	⑤	⑥	⑦	⑧	80	⑨	⑩	14-22	25
	4.0	VQ40DE	⑪	⑫	14	⑬	65	⑭ ⑮	⑯	14-22	25

① Step 1: 22 ft. lbs.
Step 2: 59 ft. lbs.
Step 3: Loosen completely then retorque to 22 ft. lbs.
Step 4: 18-25 ft. lbs.
Step 5: Plus 86-91 degrees

② 10-12 ft. lbs. plus 60-65 degrees or 28-33 ft. lbs.

③ The cylinder heads and the lower intake manifold are installed together
Step 1: Tighten the cylinder head bolts to 22 ft. lbs.
Step 2: Tighten the cylinder head bolts to 43 ft. lbs.
Step 3: Loosen the cylinder head bolts completely
Step 4: Tighten the cylinder head bolts to 84 inch lbs.
Step 5: Tighten the intake manifold fasteners to 35 inch lbs.
Step 6: Tighten the intake manifold fasteners to 13 ft. lbs.
Step 7: Tighten the intake manifold fasteners to 12-14 ft. lbs.
Step 8: Loosen all intake manifold fasteners completely
Step 9: Tighten the cylinder head bolts to 22 ft. lbs.
Step 10: Tighten the cylinder head bolts 60-65 degrees
Step 11: Tighten the cylinder head sub-bolts to 80-105 inch lbs.
Step 12: Tighten the intake manifold fasteners to 35 inch lbs.
Step 13: Tighten the intake manifold fasteners to 78 inch lbs.
Step 14: Tighten the intake manifold fasteners to 70-84 inch lbs.

④ 10-12 ft. lbs. +60-65 degrees

⑤ Step 1: 72 ft. lbs.
Step 2: loosen completely to 0 ft. lbs.
Step 3: 29 ft. lbs.
Step 4: Plus 75 degrees clockwise
Step 5: Plus 75 degrees clockwise

⑥ Step 1: bolts 11-22 19 ft. lbs.
Step 2: bolts 1-10 29 ft. lbs.
Step 3: bolts 1-10 Plus 60-65 degrees

⑦ Step 1: 20 ft. lbs.
Step 2: loosen to 0 ft. lbs.
Step 3: 14 ft. lbs.
Step 4: Plus 85-95 degrees

⑧ Step 1: 31 ft. lbs.
Step 2: Plus 60 degrees

⑨ 83 inch lbs.

⑩ Stud bolt: 11 ft. lbs.
Nuts: 31 ft. lbs.

⑪ Step 1: 72 ft. lbs.
Step 2: loosen completely to 0 ft. lbs.
Step 3: 29 ft. lbs.
Step 4: Plus 90 degrees clockwise
Step 5: Plus 90 degrees clockwise

⑫ Bolts: 17-24 (M8) 16 ft. lbs.
Install rear main seal
Bolts: 1-16 (M10) 26 ft. lbs.
Bolts: 1-16 (M10) Plus 90 degrees clockwise

⑬ Step 1: 33 ft. lbs.
Step 2: Plus 84-90 degrees clockwise

⑭ Intake manifold collector:
Bolts and nuts: 8 ft. lbs.
Stud bolts: 61 inch lbs.

⑮ Intake manifold:
Bolts and nuts: 5 ft. lbs., than 21 ft. lbs.
Studs: 8 ft. lbs.

⑯ Stud bolts: 11 ft. lbs.
Nuts: 22 ft. lbs.

09482_FRON_C0009

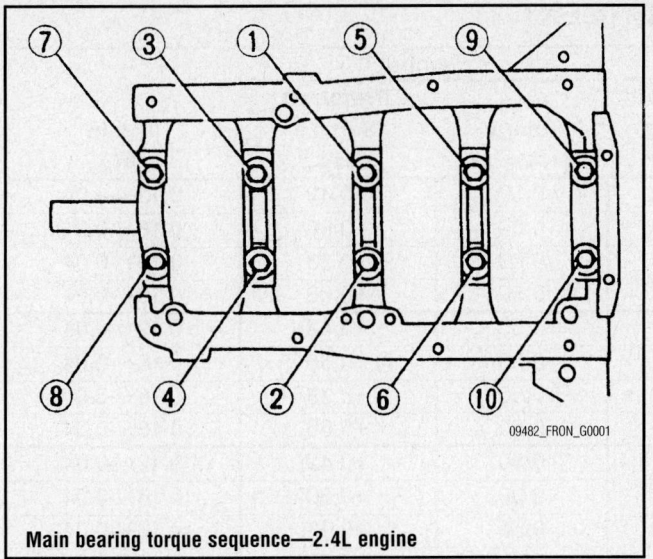

Main bearing torque sequence—2.4L engine

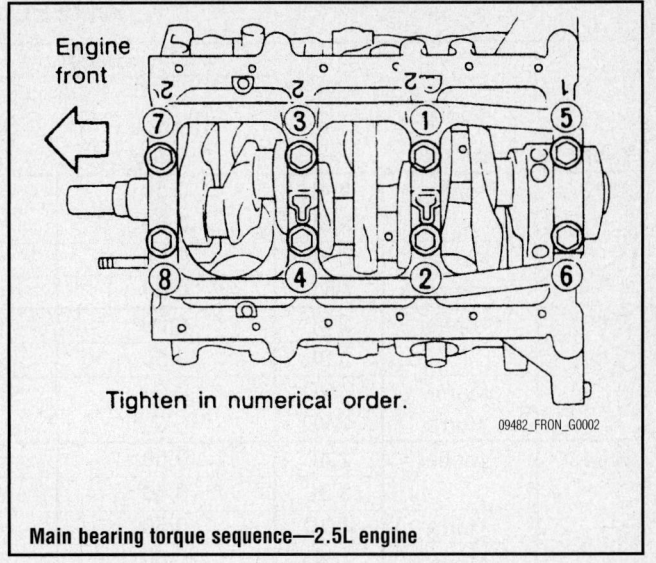

Tighten in numerical order.

Main bearing torque sequence—2.5L engine

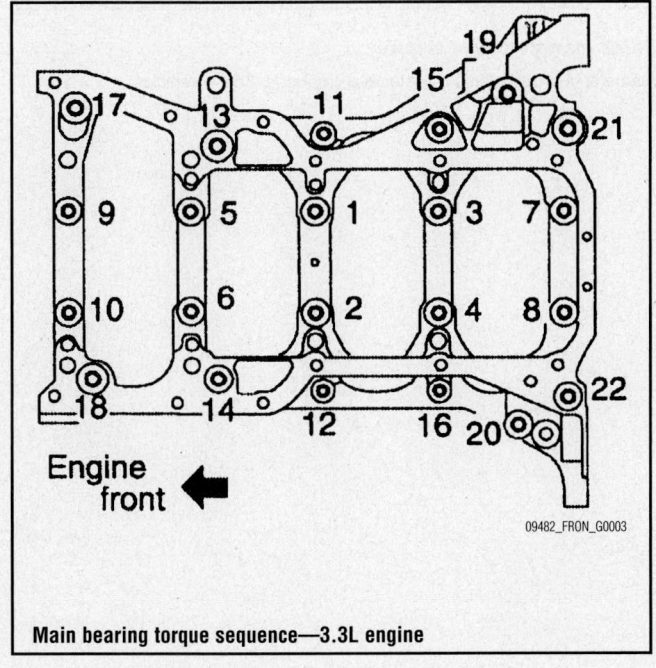

Engine front

Main bearing torque sequence—3.3L engine

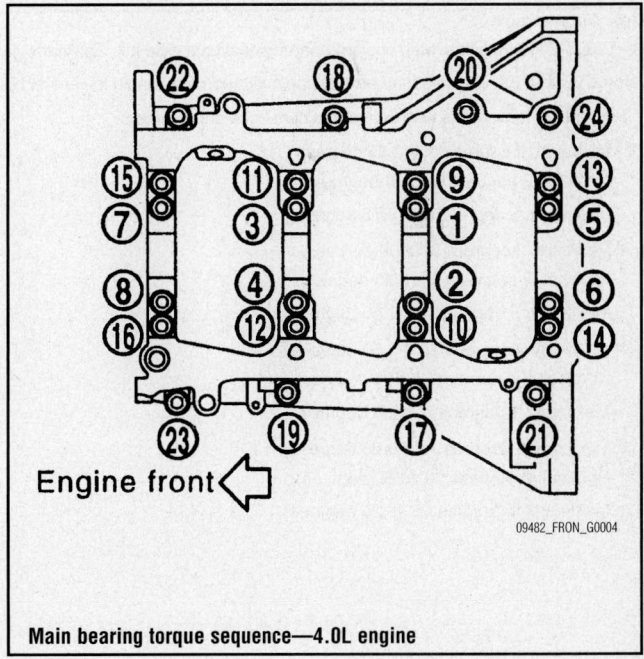

Engine front

Main bearing torque sequence—4.0L engine

WHEEL ALIGNMENT

Year	Model		Caster Range (+/-Deg.)	Caster Preferred Setting (Deg.)	Camber Range (+/-Deg.)	Camber Preferred Setting (Deg.)	Toe-in (in.)
2002	Frontier	2.4L	0.50	+0.60	0.50	+0.42	0.12+/-0.04
		3.3L	0.50	+2.17	0.50	+0.60	0.16+/-0.04
	Xterra	2WD	0.50	+2.57	0.50	+0.33	0.16+/-0.04
	Xterra	4WD	0.50	+2.10	0.50	+0.60	0.16+/-0.04
2003	Frontier	2.4L	0.50	+0.60	0.50	+0.42	0.12+/-0.04
		3.3L	0.50	+2.17	0.50	+0.60	0.16+/-0.04
	Xterra	2WD	0.50	+2.57	0.50	+0.33	0.16+/-0.04
	Xterra	4WD	0.50	+2.10	0.50	+0.60	0.16+/-0.04
2004	Frontier	2.4L	0.50	+0.60	0.50	+0.42	0.12+/-0.04
		3.3L	0.50	+2.17	0.50	+0.60	0.16+/-0.04
	Xterra	2WD	0.50	+2.57	0.50	+0.33	0.16+/-0.04
	Xterra	4WD	0.50	+2.10	0.50	+0.60	0.16+/-0.04
2005	Frontier	2WD	①	①	③	③	NA
		4WD	②	②	④	④	NA
	Xterra	2WD	①	①	③	③	NA
	Xterra	4WD	②	②	④	④	NA
2006	Frontier	2WD	①	①	③	③	NA
		4WD	②	②	④	④	NA
	Xterra	2WD	①	①	③	③	NA
	Xterra	4WD	②	②	④	④	NA

NA: Not Available

On 2005-2006 vehicles, fuel, coolant and engine oil must be full. Spare tire, jack, hand tools and mats must be in place.

Some 2005-2006 vehicles may be equipped with non adjustable lower link bolts and washers. In order to adjust caster and camber on these vehicles,
first replace these bolts with adjustable cam bolts and washers.

① Minimum: 2 degrees 15' (2.25 degrees)
  Nominal: 3 degrees 0' (3.00 degrees)
  Maximum: 3 degrees 45' (3.75 degrees)

② Minimum: 2 degrees 0' (2.00 degrees)
  Nominal: 2 degrees 45' (2.75 degrees)
  Maximum: 3 degrees 30' (3.50 degrees)

③ Minimum: 0 degrees 30' (-0.50 degrees)
  Nominal: 0 degrees 15' (0.25 degrees)
  Maximum: 1 degrees 0' (1.00 degrees)

④ Minimum: 0 degrees 15' (-0.25 degrees)
  Nominal: 0 degrees 30' (0.50 degrees)
  Maximum: 1 degrees 15' (1.25 degrees)

09482_FRON_C0010

TIRE, WHEEL AND BALL JOINT SPECIFICATIONS

Year	Model	OEM Tires Standard	OEM Tires Optional	Tire Pressures (psi) Front	Tire Pressures (psi) Rear	Wheel Size	Ball Joint Inspection	Lugnut Torque (ft. lbs.)
2002	Frontier 2wd 4-Cyl.	P225/70R15	None	②	②	NA	U: 0.020 in. L: ①	87-108
	Frontier 4wd 6-Cyl.	P255/65R16	None	②	②	NA	U: 0.020 in. L: ①	87-108
	Frontier 2wd SC 6-Cyl.	P265/55R17	None	②	②	NA	U: 0.020 in. L: ①	87-108
	Frontier 4wd SC 6-Cyl	P265/65R17	None	②	②	NA	U: 0.020 in. L: ①	87-108
	Frontier 4wd XE 6-cyl.	P265/70R15	None	②	②	NA	U: 0.020 in. L: ①	87-108
	Frontier 6-Cyl. XE Desert Runner	P265/70R15	None	②	②	NA	U: 0.020 in. L: ①	87-108
	Frontier 6-Cyl. SE Desert Runner	P255/65R16	None	②	②	NA	U: 0.020 in. L: ①	87-108
	Frontier SE V6	P265/70R16	None	②	②	NA	U: 0.020 in. L: ①	87-108
	Frontier 2wd SC V6 Crew Cab	P265/55R17	None	②	②	NA	U: 0.020 in. L: ①	87-108
	Frontier 4wd SC V6 Crew Cab	P265/65R17	None	②	②	NA	U: 0.020 in. L: ①	87-108
	Xterra SE, SE S/C, and XE S/C	P265/70R16	None	②	②	7JJ	③	87-108
	Xterra XE, XE V6	P265/70R15	None	②	②	7JJ	③	87-108
2003	Frontier 2wd 4-Cyl.	P225/70R15	None	②	②	NA	U: 0.020 in. L: ①	87-108
	Frontier 4wd 6-Cyl.	P255/65R16	None	②	②	NA	U: 0.020 in. L: ①	87-108
	Frontier 2wd SC 6-Cyl.	P265/55R17	None	②	②	NA	U: 0.020 in. L: ①	87-108
	Frontier 4wd SC 6-Cyl	P265/65R17	None	②	②	NA	U: 0.020 in. L: ①	87-108
	Frontier 4wd XE 6-cyl.	P265/70R15	None	②	②	NA	U: 0.020 in. L: ①	87-108
	Frontier 6-Cyl. XE Desert Runner	P265/70R15	None	②	②	NA	U: 0.020 in. L: ①	87-108

09482_FRON_C0011

TIRE, WHEEL AND BALL JOINT SPECIFICATIONS

Year	Model	OEM Tires Standard	OEM Tires Optional	Tire Pressures (psi) Front	Tire Pressures (psi) Rear	Wheel Size	Ball Joint Inspection	Lugnut Torque (ft. lbs.)
2003 cont.	Frontier 6-Cyl. SE Desert Runner	P255/65R16	None	②	②	NA	U: 0.020 in. L: ①	87-108
	Frontier SE V6	P265/70R16	None	②	②	NA	U: 0.020 in. L: ①	87-108
	Frontier 2wd SC V6 Crew Cab	P265/55R17	None	②	②	NA	U: 0.020 in. L: ①	87-108
	Frontier 4wd SC V6 Crew Cab	P265/65R17	None	②	②	NA	U: 0.020 in. L: ①	87-108
	Xterra SE, SE S/C, and XE S/C	P265/70R16	None	②	②	7JJ	③	87-108
	Xterra XE, XE V6	P265/70R15	None	②	②	7JJ	③	87-108
2004	Frontier 2wd 4-Cyl.	P225/70R15	None	②	②	NA	U: 0.020 in. L: ①	87-108
	Frontier 4wd 6-Cyl.	P255/65R16	None	②	②	NA	U: 0.020 in. L: ①	87-108
	Frontier 2wd SC 6-Cyl.	P265/55R17	None	②	②	NA	U: 0.020 in. L: ①	87-108
	Frontier 4wd SC 6-Cyl	P265/65R17	None	②	②	NA	U: 0.020 in. L: ①	87-108
	Frontier 4wd XE 6-cyl.	P265/70R15	None	②	②	NA	U: 0.020 in. L: ①	87-108
	Frontier 6-Cyl. XE Desert Runner	P265/70R15	None	②	②	NA	U: 0.020 in. L: ①	87-108
	Frontier 6-Cyl. SE Desert Runner	P255/65R16	None	②	②	NA	U: 0.020 in. L: ①	87-108
	Frontier SE V6	P265/70R16	None	②	②	NA	U: 0.020 in. L: ①	87-108
	Frontier 2wd SC V6 Crew Cab	P265/55R17	None	②	②	NA	U: 0.020 in. L: ①	87-108
	Frontier 4wd SC V6 Crew Cab	P265/65R17	None	②	②	NA	U: 0.020 in. L: ①	87-108
	Xterra SE, SE S/C, and XE S/C	P265/70R16	None	②	②	7JJ	③	87-108
	Xterra XE, XE V6	P265/70R15	None	②	②	7JJ	③	87-108
2005	Frontier XE	P235/75R15	None	②	②	7J	U: 0.020 in. L: ①	98
	Frontier SE	P265/70R16	None	②	②	7J	U: 0.020 in. L: ①	98
	Frontier off road	P265/75R16	None	②	②	7J	U: 0.020 in. L: ①	98
	Frontier LE	P265/65R17	None	②	②	7.5J	U: 0.020 in. L: ①	98
	Xterra S	P265/70R16	None	②	②	7J	U: 0.020 in. L: ①	98
	Xterra S-O/R	P265/75R16	None	②	②	7J	U: 0.020 in. L: ①	98
	Xterra SE	P255/65R17	None	②	②	7.5J	U: 0.020 in.	98

TIRE, WHEEL AND BALL JOINT SPECIFICATIONS

| Year | Model | OEM Tires | | Tire Pressures (psi) | | Wheel Size | Ball Joint Inspection | Lugnut Torque (ft. lbs.) |
		Standard	Optional	Front	Rear			
2006	Frontier XE	P235/75R15	None	②	②	7J	U: 0.020 in. L: ①	98
	Frontier SE	P265/70R16	None	②	②	7J	U: 0.020 in. L: ①	98
	Frontier NISMO off road	P265/75R16	None	②	②	7J	U: 0.020 in. L: ①	98
	Frontier LE	P265/65R17	None	②	②	7.5J	U: 0.020 in. L: ①	98
	Xterra S	P265/70R16	None	②	②	7J	U: 0.020 in. L: ①	98
	Xterra S-O/R	P265/75R16	None	②	②	7J	U: 0.020 in. L: ①	98
	Xterra SE	P255/65R17	None	②	②	7.5J	U: 0.020 in. L: ①	98

OEM: Original Equipment Manufacturer

PSI: Pounds Per Square Inch

L: Lower

U: Upper

① Replace if any measurable movement is found.

② See placard on vehicle

③ Axial play

Upper: 0

Lower: 0.008 in.

09482_FRON_C0013

BRAKE SPECIFICATIONS
All measurements in inches unless noted

| Year | Model | Brake Disc | | | Brake Drum Diameter | | | Minimum Lining Thickness | | Brake Caliper | |
		Original Thickness	Minimum Thickness	Maximum Runout	Original Inside Diameter	Max. Wear Limit	Maximum Machine Diameter	Front	Rear	Bracket Bolts (ft. lbs.)	Mounting Bolts (ft. lbs.)
2002	Frontier	①	②	0.003	③	NA	④	0.079	0.059	⑦	17-22
	Xterra	1.100	1.024	0.003	11.61	NA	11.67	0.079	0.059	101-130	17-22
2003	Frontier	NA	⑤	0.003	③	NA	⑥	0.079	0.059	⑦	17-22
	Xterra	1.100	1.024	0.003	11.61	NA	11.67	0.079	0.059	101-130	17-22
2004	Frontier	NA	⑤	0.003	③	NA	⑥	0.079	0.059	⑦	17-22
	Xterra	1.100	1.024	0.003	11.61	NA	11.67	0.079	0.059	101-130	17-22
2005	Frontier	⑧	⑨	0.002	⑩	⑪	—	0.079	0.079	⑫	⑬
	Xterra	1.102	1.024	0.002	⑭	⑪	—	0.079	0.079	⑫	⑬
2006	Frontier	⑧	⑨	0.002	⑩	⑪	—	0.079	0.079	⑫	⑬
	Xterra	1.102	1.024	0.002	⑭	⑪	—	0.079	0.079	⑫	⑬

NA: Information not available

① 2WD: 0.870
4WD: 1.020

② 2WD: 0.787
4WD: 0.945

③ 2WD: 10.20
4WD: 11.60

④ 2WD: 10.30
4WD: 11.67

⑤ 4-cyl.: 0.945
6-cyl.: 1.024

⑥ 4-cyl.: 10.30
6-cyl.: 11.67

⑦ 4-cyl.: 53-72 ft. lbs.
6-cyl.: 101-130 ft. lbs.

⑧ 4-cyl.: 0.710
6-cyl.: 1.100

⑨ 4-cyl.: 0.630
6-cyl.: 1.024

⑩ rear disc brakes: 0.710

⑪ rear disc brakes: 0.630

⑫ front: 136 ft. lbs.
rear: 76 ft. lbs.

⑬ front: 32 ft. lbs.
rear: 24 ft. lbs.

⑭ rear disc brakes: 0.709

09482_FRON_C0014

SCHEDULED MAINTENANCE INTERVALS
Nissan Frontier and Xterra

TO BE SERVICED	TYPE OF SERVICE	VEHICLE MILEAGE INTERVAL (x1000)												
		7.5	15	22.5	30	37.5	45	52.5	60	67.5	75	82.5	90	97.5
Engine oil & filter	R	✓	✓	✓	✓	✓	✓	✓	✓	✓	✓	✓	✓	✓
Brake lines & cables	S/I		✓		✓		✓		✓		✓		✓	
Brake pads, discs, drums & linings	S/I		✓		✓		✓		✓		✓		✓	
Driveshaft boots & propeller shaft	S/I				✓				✓				✓	
Front wheel bearings (4x2) 2002-2004	S/I				✓				✓				✓	
Front wheel bearings (4x4) 2002-2004	S/I				✓				✓				✓	
Automatic & manual transmission, transfer & differential gear oil ①	S/I		✓		✓		✓		✓		✓		✓	
Air cleaner filter	R				✓				✓				✓	
Engine coolant 2002-2004	R				✓				✓				✓	
Engine coolant 2005-2006	R								✓				✓	
PCV filter (KA24E)	R				✓				✓				✓	
Spark plugs (except platinum)	R				✓				✓				✓	
Spark plugs (platinum)	R	replace every 100,000 miles												
Drive belt(s)	S/I				✓				✓				✓	
Exhaust system	S/I				✓				✓				✓	
Fuel lines	S/I				✓				✓				✓	
Steering gear (box) & linkage, axle & suspension parts	S/I				✓				✓				✓	
Vapor lines	S/I	every 5,000 miles												
Tires (rotate)	S/I													
Timing belt ②	R													

R: Replace   S/I: Service or Inspect

① Differential (w/limited-slip differential) oil: replace oil every 30,000 miles, 2002-2004 vehicles.

② Timing belt: replace at 105,000 miles.

FREQUENT OPERATION MAINTENANCE (SEVERE SERVICE)

If a vehicle is operated under any of the following conditions it is considered severe service:

- Extremely dusty areas.

- 50% or more of the vehicle operation is in 32°C (90°F) or higher temperatures, or constant operation in temperatures below 0°C (32°F).

- Prolonged idling (vehicle operation in stop and go traffic).

- Frequent short running periods (engine does not warm to normal operating temperatures).

- Police, taxi, delivery usage or trailer towing usage.

Oil & oil filter: replace every 3750 miles.

Brake pads, discs, drums & linings: service or inspect every 7500 miles.

Driveshaft boots & propeller shaft: service or inspect every 7500 miles.

Exhaust system: service or inspect every 7500 miles.

Steering gear (box) & linkage, (steering damper-4x4), axle & suspension parts: service or inspect every 7500 miles.

Steering linkage ball joints & front suspension ball joints: service or inspect every 7500 miles.

ENGINE REPAIR

➡Disconnecting the negative battery cable on some vehicles may interfere with the functions of the on board computer system. The computer may undergo a relearning process once the negative battery cable is reconnected.

Relearning Procedures

ACCELERATOR PEDAL RELEASED POSITION LEARNING

Adjustment

➡Accelerator pedal released position learning is an operation to learn the fully released position of the accelerator pedal by monitoring the accelerator pedal position sensor output signal. It must be performed each time the harness connector of the accelerator pedal position sensor or the ECM is disconnected.

1. Make sure that the accelerator pedal is in the fully released position.
2. Turn the ignition switch ON. Wait at least 2 seconds.
3. Turn the ignition switch OFF. Wait at least 10 seconds.
4. Turn the ignition switch ON. Wait at least 2 seconds.
5. Turn the ignition switch OFF. Wait at least 10 seconds.

THROTTLE VALVE CLOSED POSITION LEARNING

Adjustment

➡Throttle valve closed position learning is an operation to learn the fully closed position of the throttle valve by monitoring the throttle valve position sensor output signal. It must be performed each time the harness connector of the electrical throttle valve control actuator or ECM is disconnected.

1. Make sure that the accelerator pedal is in the fully released position.
2. Turn the ignition switch ON.
3. Turn the ignition switch OFF. Wait at least 10 seconds.
4. Make sure that the throttle valve moves during the 10 seconds, by confirming the operating sound.

IDLE AIR VOLUME LEARNING

➡Idle air volume learning is an operation to learn the idle air volume that keeps the engine within a specific range. It must be performed each time the electronic throttle control actuator or ECM is replaced, or if the idle speed and ignition timing is out of specification.

Pre-Adjustment

➡Before performing the idle air volume learning procedure, be sure the following conditions are satisfied. Learning will be cancelled if any of the following conditions are missed for even a moment.

1. Be sure battery voltage is at least 12.9 volts at idle.
2. Be sure engine coolant temperature is at least 158–212 degrees.
3. Be sure PNP switch in ON.
4. Be sure electric load switch is OFF (air conditioning, headlights, rear defogger etc.).

➡On vehicles equipped with daytime running lights, if the parking brake is applied before the engine is started the headlights will not turn on.

5. Be sure the steering wheel is in the straight ahead position.
6. Be sure the vehicle is stopped.
7. Be sure the transmission is warmed up.

➡If using CONSULT-II, drive the vehicle until the "fluid temp SE" in the data monitor mode of the CVT system indicates less than 0.0 volt. For vehicles without CONSULT-II, drive the vehicle for 10 minutes.

Adjustment

WITH CONSULT-II

1. Perform accelerator pedal released position learning procedure.
2. Perform throttle valve closed position learning procedure.
3. Perform pre-adjustment procedure.
4. Start and run the engine until normal operating temperature is reached.
5. Select "idle air vol learn" in work support mode.
6. Touch "start" and wait 20 seconds.
7. Be sure that "CMPLT" is displayed on the screen. If not displayed, idle air volume learning will not be carried out successfully.
8. Rev the engine two or three times and make sure that the idle speed and ignition timing are within specification.

WITHOUT CONSULT-II

➡It is better to count the time accurately, using a clock. It is impossible to switch the diagnostic mode when the accelerator pedal position sensor circuit has a malfunction.

1. Perform accelerator pedal released position learning procedure.
2. Perform throttle valve closed position learning procedure.
3. Perform pre-adjustment procedure.
4. Start and run the engine until normal operating temperature is reached.
5. Turn the ignition switch to the OFF position. Wait at least 10 seconds.
6. Confirm that the accelerator pedal is fully released. Turn the ignition switch ON and wait 3 seconds.
7. Fully depress the accelerator, fully release the accelerator pedal.

➡Repeat the above step 5 times, within 5 seconds.

8. Wait 7 seconds. Fully depress the accelerator pedal and keep it depressed for about 20 seconds, until the MIL stops blinking and turned ON.
9. Fully depress the accelerator pedal within 3 seconds after the MIL turned ON.
10. Start the engine and let it idle.
11. Wait 20 seconds.
12. Rev the engine two or three times and make sure that the idle speed and ignition timing are within specification.

STEERING ANGLE SENSOR NEUTRAL POSITION

➡After removing/installing or replacing steering and suspension components which effect wheel alignment or after adjusting wheel alignment, or the steering angle sensor or the ABS actuator electrical unit be sure to adjust the neutral position of the steering angle sensor before running the vehicle.

1. Position the steering wheel in the straight ahead position.
2. Drive the vehicle at 10 mph for more than 10 minutes.
3. When this procedure is complete the

SLP indicator lamp and the VDC OFF indicator lamp will turn off.

Distributor

REMOVAL

1. Before servicing the vehicle, refer to the Precautions Section.
2. Remove or disconnect the following:
   - Negative battery cable
   - Distributor cap
   - Distributor wiring harness connector
3. Matchmark the rotor to the distributor housing and the distributor housing to the cylinder head.
4. Remove the distributor.

INSTALLATION

Timing Not Disturbed

1. Install or connect the following:
   - Distributor by aligning the matchmarks made during removal
   - Distributor wiring harness connector
   - Distributor cap
   - Negative battery cable
2. Check the ignition timing and adjust, as necessary.

Timing Disturbed

2.4L ENGINE

1. Set the engine to Top Dead Center (TDC) of the compression stroke for the No. 1 cylinder.
2. Install the distributor so that the distributor shaft engages the oil pump driveshaft.
3. Check that the distributor rotor is aligned, as shown.

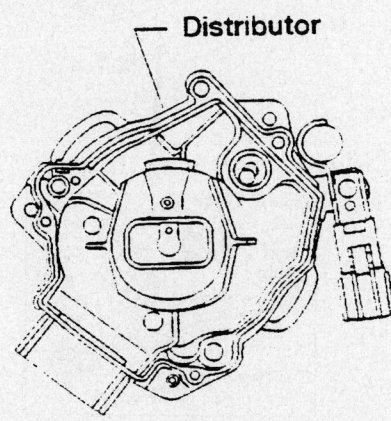

Distributor rotor alignment with the engine at Top Dead Center (TDC)—2.4L engine

9308VG01

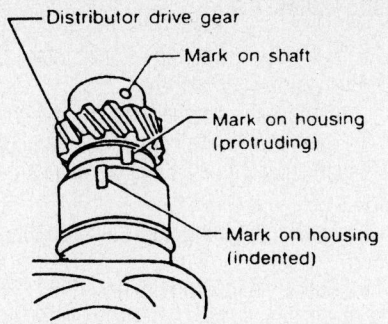

Distributor shaft alignment—3.3L engine

7924VG28

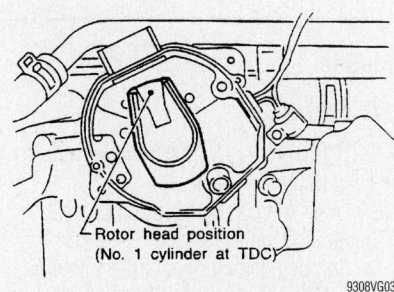

Distributor rotor alignment—3.3L engine

9308VG03

4. Install or connect the following:
   - Distributor cap
   - Distributor harness connector
5. Check the ignition timing and adjust, as necessary.

3.3L ENGINE

1. Set the engine to Top Dead Center (TDC) of the compression stroke for the No. 1 cylinder.
2. Align the index mark on the distributor shaft with the protrusion on the distributor housing.
3. Install the distributor and check that the distributor rotor is aligned.
4. Install or connect the following:
   - Distributor cap
   - Distributor harness connector
5. Check the ignition timing and adjust, as necessary.

Alternator

REMOVAL & INSTALLATION

2.4L Engine

1. Before servicing the vehicle, refer to the Precautions Section.
2. Remove or disconnect the following:
   - Negative battery cable
   - Engine under cover
   - Right splash shield
   - Alternator harness connectors
   - Alternator belt
   - Alternator

To install:

3. Install or connect the following:
   - Alternator
   - Alternator belt. Tighten the adjustment bolt to 12–14 ft. lbs. (16–19 Nm) and the pivot bolt to 32–38 ft. lbs. (44–52 Nm).
   - Alternator harness connectors
   - Right splash shield
   - Engine under cover
   - Negative battery cable

2.5L Engine

1. Before servicing the vehicle, refer to the Precautions Section.
2. Disconnect the negative battery cable.
3. Remove the fan shroud. Remove the drive belt.
4. Disconnect the alternator harness electrical connectors.
5. Remove the alternator mounting nut. Remove the upper alternator mounting bolt.
6. Remove the alternator from the vehicle.

To install:

7. Installation is the reverse of the removal procedure.
8. Be sure that the alternator spacer is in place on the lower mounting stud.

3.3L Engine

1. Before servicing the vehicle, refer to the Precautions Section.
2. Remove or disconnect the following:
   - Negative battery cable
   - Alternator harness connectors
   - Engine under cover
   - Alternator belt
   - Alternator

To install:

3. Install or connect the following:
   - Alternator
   - Alternator belt. Tighten the adjustment bolt to 12–14 ft. lbs. (16–19 Nm) and the pivot bolts to 16–22 ft. lbs. (22–30 Nm).
   - Engine under cover
   - Alternator harness connectors
   - Negative battery cable

4.0L Engine

1. Before servicing the vehicle, refer to the Precautions Section.
2. Disconnect the negative battery cable.
3. Remove the fan shroud, on Frontier.
4. Remove the drive belt.
5. Disconnect the alternator harness electrical connectors.

6. Remove the alternator mounting nut. Remove the upper alternator mounting bolt.

7. Remove the alternator from the vehicle.

To install:

8. Installation is the reverse of the removal procedure.

Ignition Timing

ADJUSTMENT

2002–2004

➡**Ignition timing is set with the engine at operating temperature, transmission in Neutral and all electrical accessories OFF.**

1. Before servicing the vehicle, refer to the Precautions Section.

2. Attach a timing light to the No. 1 spark plug wire.

3. Start the engine and allow it to reach normal operating temperature.

4. Check that the idle speed is less than 1000 rpm.

5. Run the engine at 2000 rpm for 2 minutes.

6. Rev the engine to 3000 rpm 2–3 times and allow it to idle for 1 minute.

7. Check for the presence of Diagnostic Trouble Codes (DTC) and service as necessary.

8. Run the engine at 2000 rpm for 2 minutes.

9. Stop the engine and disconnect the Throttle Position (TP) sensor.

10. Start the engine and rev it to 3000 rpm 2–3 times and allow it to idle.

11. Set the base timing to 8–12 degrees Before Top Dead Center (BTDC).

12. Tighten the distributor lockbolt to 83–113 inch lbs. (9–13 Nm).

13. Set the base idle speed to 700–800 rpm.

14. Stop the engine and connect the TP sensor.

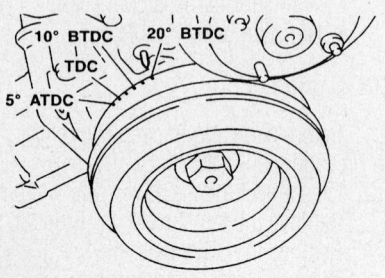

Typical timing indicator

7924VG04

2005–2006

1. Before servicing the vehicle, refer to the Precautions Section.

2. Remove the number one ignition coil.

3. Connect the number one ignition coil and spark plug with a suitable high tension wire.

4. Attach the timing light clamp to the wire.

5. Check the ignition timing.

Engine Assembly

REMOVAL & INSTALLATION

Frontier

2.4L ENGINE

1. Before servicing the vehicle, refer to the Precautions Section.

2. Drain the cooling system.

3. Relieve the fuel system pressure.

4. Remove or disconnect the following:
- Negative battery cable
- Hood
- Air cleaner assembly
- Idle Air Control (IAC) valve and solenoid connectors
- Throttle Position (TP) sensor and switch connectors
- Engine Coolant Temperature (ECT) sensor connector
- Manifold Absolute Pressure (MAP) sensor connector and vacuum line
- Evaporative Emissions (EVAP) canister purge valve connector and vacuum line

- Mass Air Flow (MAF) sensor connector
- Brake booster vacuum line
- Fuel lines
- Exhaust Gas Recirculation (EGR) temperature sensor connector
- Throttle cable
- Accessory drive belts
- Radiator and hoses
- Heater hoses
- Exhaust manifold heat shield
- Heated Oxygen (HO2S) sensor connectors
- Exhaust front pipe
- A/C compressor, if equipped
- Power steering pump, if equipped
- Crankshaft Position (CKP) sensor
- Starter motor
- Transmission
- Left and right engine mounts
- Engine

To install:

5. Install or connect the following:
- Engine. Tighten the engine mount nuts to 30–38 ft. lbs. (41–52 Nm).
- Transmission
- Starter motor
- CKP sensor
- Power steering pump, if equipped
- A/C compressor, if equipped
- Exhaust front pipe
- HO2S sensor connectors
- Exhaust manifold heat shield
- Heater hoses
- Radiator and hoses
- Accessory drive belts
- Throttle cable
- EGR temperature sensor connector

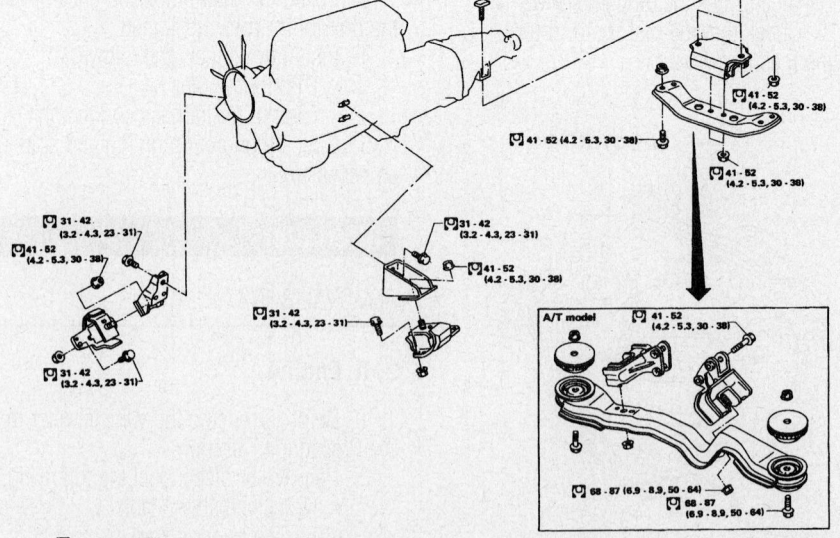

Engine mounts and related components—2.4L engine

7924VG05

- Fuel lines
- Brake booster vacuum line
- MAF sensor connector
- EVAP canister purge valve connector and vacuum line
- MAP sensor connector and vacuum line
- ECT sensor connector
- TP sensor and switch connectors
- IAC valve and solenoid connectors
- Air cleaner assembly
- Hood
- Negative battery cable

6. Fill the cooling system.
7. Start the engine and check for leaks.

2.5L ENGINE

➡**Be sure to disarm the SRS system, prior to working on the vehicle. Turn the ignition switch OFF, disconnect both battery cables and wait at least three minutes before starting any work.**

1. Before servicing the vehicle, refer to the Precautions Section.
2. Properly release the fuel system pressure. Disconnect the negative battery cable.
3. Drain the radiator. Drain the engine oil. Drain the automatic transmission fluid, if equipped.
4. Matchmark and remove the hood.
5. Remove the air duct and the air cleaner assembly.
6. Disconnect the vacuum hose between the vehicle and the engine and position it to the side.
7. Remove the coolant hoses. Remove the radiator.
8. Remove the drive belts. Remove the cooling fan.
9. Disconnect the engine electrical harness from the engine side and position it to the side.
10. Disconnect the engine harness ground wires.
11. Disconnect the power steering reservoir tank and position it to the side.
12. Remove the power steering pump from the engine and position it to the side.
13. Remove the air conditioning compressor retaining bolts. Position the air conditioning compressor to the side.
14. Disconnect the brake booster vacuum line. Disconnect the EVAP line.
15. Disconnect the fuel line hoses at the engine side connection.
16. Disconnect and plug the heater hoses at the cowl.
17. Remove the automatic transmission oil level indicator stick and tube assembly, if equipped.

18. Remove the three way catalyst.
19. Properly support the engine using an engine support tool.
20. Remove the transmission.
21. Connect a suitable engine lifting fixture and remove the engine from the vehicle.

➡**Before lifting the engine check to be sure that all necessary electrical connections, vacuum lines and grounding wires have been disconnected, so as not to interfere with the engine removal. Also, check that all mounting bolts have been removed before lifting the engine from the vehicle.**

➡**Be careful not to damage the drive plate. Avoid deforming and damaging the signal plate teeth, on the drive plate. If the drive plate is removed from the engine, position it with the signal plate surface facing other than downward. Keep magnetic materials away from the signal plate.**

To install:

22. Installation is the reverse of the removal procedure.
23. Be sure to fill the engine with the proper grade and type engine oil and engine coolant.
24. Be sure to fill the automatic transmission with the proper grade and type transmission fluid, if equipped.

25. Start the engine and check for leaks, correct as required.

3.3L ENGINE

1. Before servicing the vehicle, refer to the Precautions Section.
2. Drain the cooling system.
3. Relieve the fuel system pressure.
4. Recover the A/C refrigerant, if equipped.
5. Remove or disconnect the following:

- Negative battery cable
- Hood
- Air cleaner assembly
- Idle Air Control (IAC) valve and solenoid connectors
- Throttle Position (TP) sensor and switch connectors
- Engine Coolant Temperature (ECT) sensor connector
- Manifold Absolute Pressure (MAP) sensor connector and vacuum line
- Evaporative Emissions (EVAP) canister purge valve connector and vacuum line
- Mass Air Flow (MAF) sensor connector
- Brake booster vacuum line
- Fuel lines
- Exhaust Gas Recirculation (EGR) temperature sensor connector
- Throttle cable
- Accessory drive belts
- Cooling fan and shroud

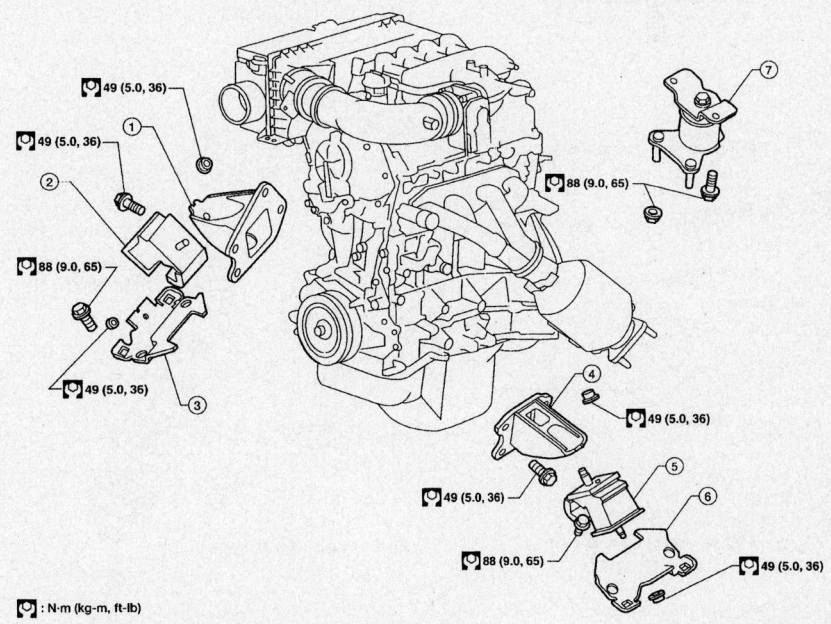

: N·m (kg-m, ft-lb)

1. RH engine mounting bracket (upper)
2. RH engine mounting insulator
3. RH engine mounting bracket (lower)
4. LH engine mounting bracket (upper)
5. LH engine mounting insulator
6. LH engine mounting bracket (lower)
7. Rear engine mounting insulator

09482_FRON_G0007

Engine mounts and related components—2.5L engine

- Radiator and hoses
- Engine under cover
- A/C compressor manifold
- Power steering pump
- Heated Oxygen (HO2S) sensor connectors
- Exhaust front pipes
- Crankshaft Position (CKP) sensor
- Starter motor
- Transmission
- Left and right engine mounts
- Engine

➡When removing the engine mounts, do not loosen the 4 mount cover nuts. The mount is fluid filled and will not function if the fluid leaks out.

To install:
6. Install or connect the following:
- Engine. Tighten the engine mount nuts to 43–58 ft. lbs. (59–78 Nm).
- Transmission
- Starter motor
- CKP sensor
- Exhaust front pipes
- HO2S sensor connectors
- Power steering pump
- A/C compressor manifold
- Engine under cover
- Radiator and hoses
- Cooling fan and shroud
- Accessory drive belts
- Throttle cable
- EGR temperature sensor connector
- Fuel lines
- Brake booster vacuum line

- MAF sensor connector
- EVAP canister purge valve connector and vacuum line
- MAP sensor connector and vacuum line
- ECT sensor connector
- TP sensor and switch connectors
- IAC valve and solenoid connectors
- Air cleaner assembly
- Hood
- Negative battery cable
7. Fill the cooling system.
8. Recharge the A/C system, if equipped.
9. Start the engine and check for leaks.

4.0L ENGINE

➡Be sure to disarm the SRS system, prior to working on the vehicle. Turn the ignition switch OFF, disconnect both battery cables and wait at least three minutes before starting any work.

1. Before servicing the vehicle, refer to the Precautions Section.
2. Properly release the fuel system pressure. Disconnect the negative battery cable.
3. Drain the radiator. Drain the engine oil. Drain the automatic transmission fluid, if equipped.
4. Matchmark and remove the hood.
5. Remove the engine cover. Remove the air duct and the air cleaner assembly.
6. Disconnect the vacuum hose between the vehicle and the engine and position it to the side.

7. Remove the coolant hoses. Remove the radiator.
8. Remove the drive belts. Remove the cooling fan.
9. Disconnect the engine electrical harness from the engine side and position it to the side.
10. Disconnect the engine harness ground wires.
11. Disconnect the power steering reservoir tank and position it to the side.
12. Remove the power steering pump from the engine and position it to the side.
13. Remove the air conditioning compressor retaining bolts. Position the air conditioning compressor to the side.
14. Disconnect the brake booster vacuum line. Disconnect the EVAP line.
15. Disconnect the fuel line hoses at the engine side connection.
16. Disconnect and plug the heater hoses at the cowl.
17. Remove the automatic transmission oil level indicator stick and tube assembly, if equipped.
18. If equipped with 4WD, remove the final drive assembly.
19. Remove the three way catalyst.
20. Properly support the engine using an engine support tool.
21. Remove the transmission.
22. Connect a suitable engine lifting fixture and remove the engine from the vehicle.

➡Before lifting the engine check to be sure that all necessary electrical connections, vacuum lines and grounding wires have been disconnected, so as not to interfere with the engine removal. Also, check that all mounting bolts have been removed before lifting the engine from the vehicle.

➡Be careful not to damage the drive plate. Avoid deforming and damaging the signal plate teeth, on the drive plate. If the drive plate is removed from the engine, position it with the signal plate surface facing other than downward. Keep magnetic materials away from the signal plate.

To install:
23. Installation is the reverse of the removal procedure.
24. Be sure to fill the engine with the proper grade and type engine oil and engine coolant.
25. Be sure to fill the automatic transmission with the proper grade and type transmission fluid, if equipped.
26. Start the engine and check for leaks, correct as required.

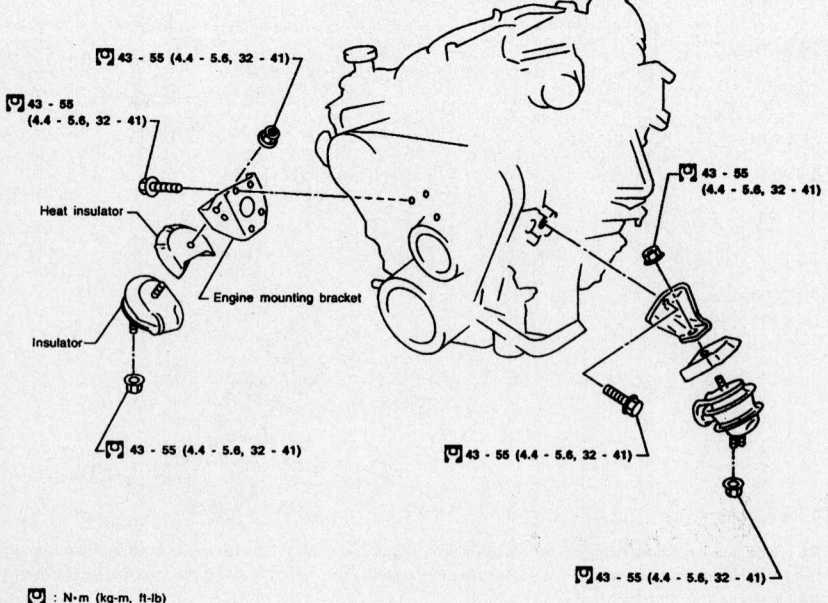

43 - 55 (4.4 - 5.6, 32 - 41)

43 - 55 (4.4 - 5.6, 32 - 41)

Heat insulator

Engine mounting bracket

Insulator

43 - 55 (4.4 - 5.6, 32 - 41)

43 - 55 (4.4 - 5.6, 32 - 41)

43 - 55 (4.4 - 5.6, 32 - 41)

43 - 55 (4.4 - 5.6, 32 - 41)

: N·m (kg-m, ft-lb)

7924VG11

Engine mounts and related components—Frontier with 3.3L engine

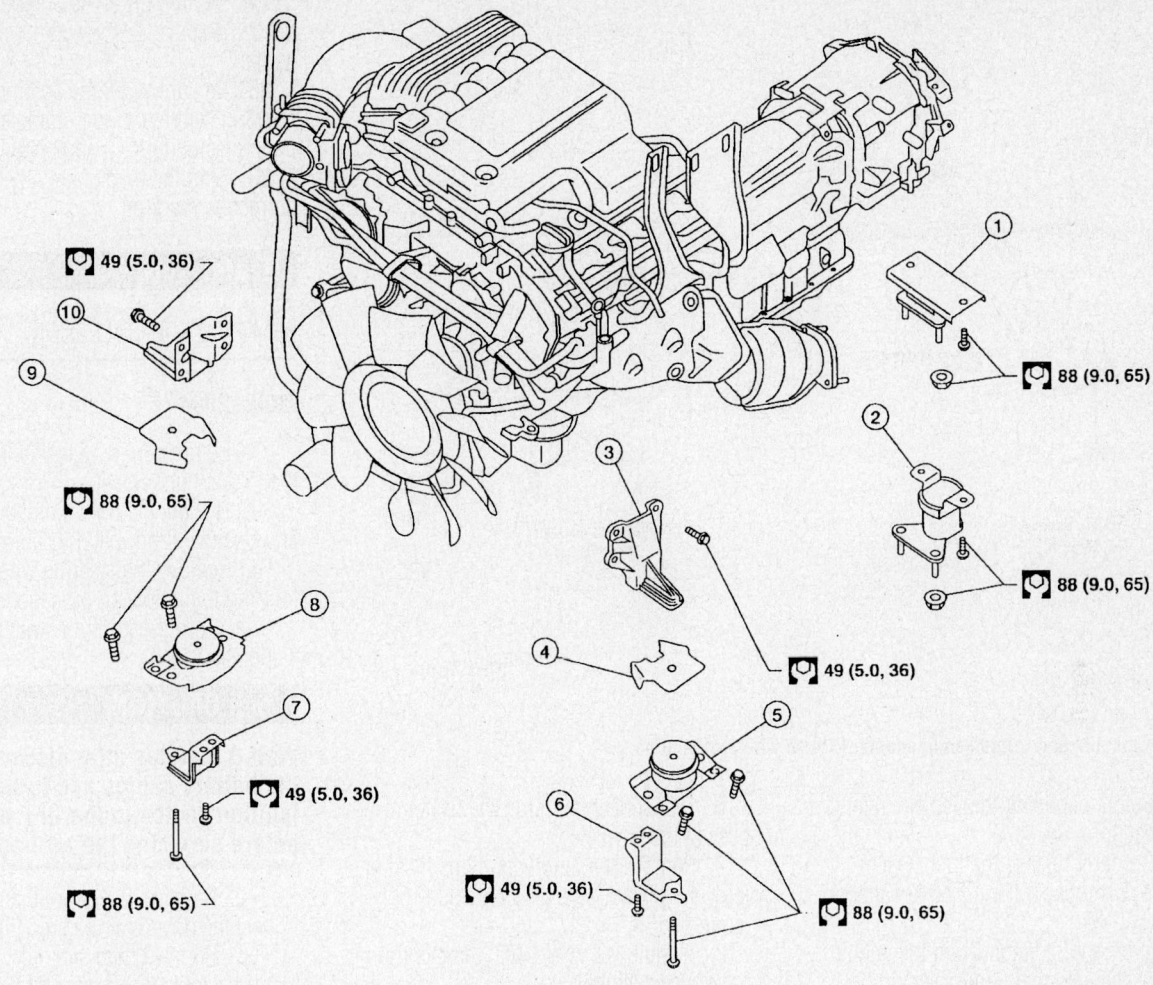

49 (5.0, 36)

88 (9.0, 65)

88 (9.0, 65)

88 (9.0, 65)

49 (5.0, 36)

88 (9.0, 65)

49 (5.0, 36)

49 (5.0, 36)

88 (9.0, 65)

: N·m (kg-m, ft-lb)

1. Rear engine mounting insulator 4x4
2. Rear engine mounting insulator 4x2
3. LH engine mounting bracket (upper)
4. LH heat shield plate
5. LH engine mounting insulator
6. LH engine mounting bracket (lower)
7. RH engine mounting bracket (lower)
8. RH engine mounting insulator (upper)
9. RH heat shield plate
10. RH engine mounting bracket (upper)

09482_FRON_G0008

Engine mounts and related components—4.0L engine

Xterra

2002–2004

➡**Do not loosen front engine mounting insulator cover securing bolts. When cover is removed, damper oil flows out and mounting insulator will not function.**

1. Remove engine undercover and hood.
2. Drain coolant from cylinder block and radiator.
3. Remove vacuum hoses, fuel tubes, wires, harnesses and connectors.
4. Before disconnecting fuel hose, release fuel pressure from fuel line.
5. Remove radiator with shroud and cooling fan.
6. Remove drive belts.
7. Discharge refrigerant.
8. Remove A/C compressor manifold.
9. Remove power steering oil pump from engine.
10. Remove front exhaust tubes.
11. Remove transmission from vehicle.
12. Install engine slingers. Tighten the slinger bolts to 15–20 ft. lbs. (20–26 Nm).
13. Hoist engine with engine slingers and remove engine mounting nuts from both sides.
14. Lift and remove engine from vehicle.
To install:
15. Installation is the reverse of removal. See the accompanying illustration for installation torques.

2005–2006

➡**Be sure to disarm the SRS system, prior to working on the vehicle. Turn the ignition switch OFF, disconnect both battery cables and wait at least three minutes before starting any work.**

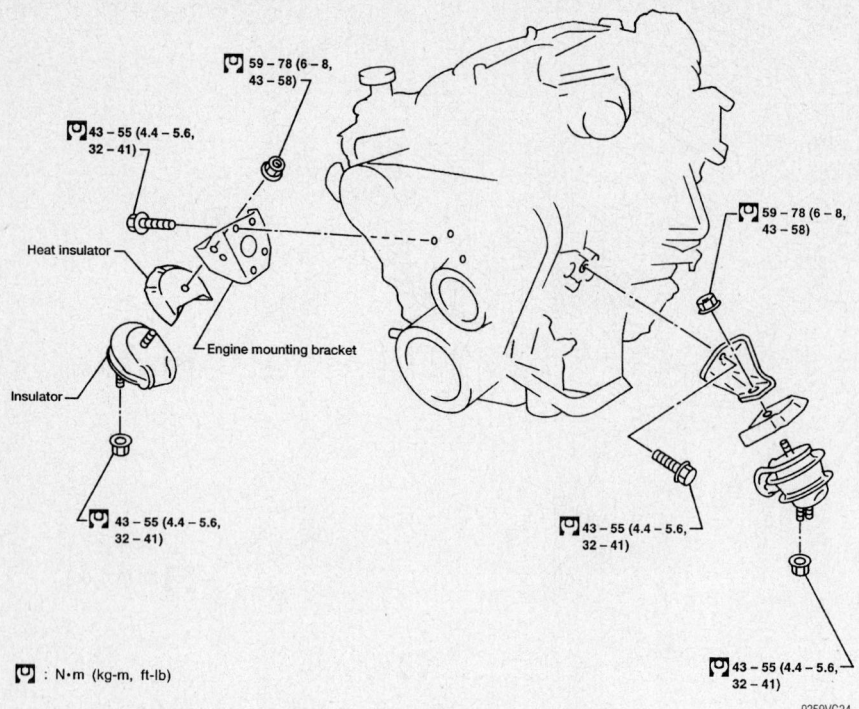

: N•m (kg-m, ft-lb)

9359VG34

Engine mounts and related components—Xterra with 3.3L engine

1. Before servicing the vehicle, refer to the Precautions Section.

2. Properly release the fuel system pressure. Disconnect the negative battery cable.

3. Drain the radiator. Drain the engine oil. Drain the automatic transmission fluid, if equipped.

4. Matchmark and remove the hood.

5. Remove the engine cover. Remove the air duct and the air cleaner assembly.

6. Disconnect the vacuum hose between the vehicle and the engine and position it to the side.

7. Remove the coolant hoses. Remove the radiator.

8. Remove the drive belts. Remove the cooling fan.

9. Disconnect the engine electrical harness from the engine side and position it to the side.

10. Disconnect the engine harness ground wires.

11. Disconnect the power steering reservoir tank and position it to the side.

12. Remove the power steering pump from the engine and position it to the side.

13. Remove the air conditioning compressor retaining bolts. Position the air conditioning compressor to the side.

14. Disconnect the brake booster vacuum line. Disconnect the EVAP line.

15. Disconnect the fuel line hoses at the engine side connection.

16. Disconnect and plug the heater hoses at the cowl.

17. Remove the automatic transmission oil level indicator stick and tube assembly, if equipped.

18. If equipped with 4WD, remove the final drive assembly.

19. Remove the three way catalyst.

20. Properly support the engine using an engine support tool.

21. Remove the transmission.

22. Connect a suitable engine lifting fixture and remove the engine from the vehicle.

➡ Before lifting the engine check to be sure that all necessary electrical connections, vacuum lines and grounding wires have been disconnected, so as not to interfere with the engine removal. Also, check that all mounting bolts have been removed before lifting the engine from the vehicle.

➡ Be careful not to damage the drive plate. Avoid deforming and damaging the signal plate teeth, on the drive plate. If the drive plate is removed from the engine, position it with the signal plate surface facing other than downward. Keep magnetic materials away from the signal plate.

To install:

23. Installation is the reverse of the removal procedure.

24. Be sure to fill the engine with the proper grade and type engine oil and engine coolant.

25. Be sure to fill the automatic transmission with the proper grade and type transmission fluid, if equipped.

26. Start the engine and check for leaks, correct as required.

Heater Core

REMOVAL & INSTALLATION

2002–2004

1. Before servicing the vehicle, refer to the Precautions Section.

2. Disconnect both the negative (1st) and positive (2nd) battery cables.

3. Remove the steering wheel by performing the following procedure:

  a. Turn the ignition switch to the OFF position.

✳✳ CAUTION

Wait 3 minutes after disconnecting the battery cables and turning the ignition switch to the OFF position before servicing the air bag system.

  b. Remove the lower lid and disconnect the driver's air bag module connector.

  c. Remove both side lids.

  d. Using the Tamper Resistant Torx® Wrench size T50, remove the special bolts from both sides of the steering wheel and discard them.

  e. Remove the SRS module from the steering wheel.

✳✳ CAUTION

Always store the SRS module face up.

  f. Position the steering wheel in the straight-ahead position.

  g. Disconnect the horn connector and remove the steering wheel nut.

  h. Using a steering wheel puller, press the steering wheel from the steering column.

4. Remove the passenger's side air bag by disconnecting or removing the following items:

  a. Turn the ignition switch to the OFF position.

✳✳ CAUTION

Wait 3 minutes after disconnecting the battery cables and turning the ignition switch to the OFF position before servicing the air bag system.

b. Open the glove box door.

c. Working inside the glove box, open the lower instrument panel lid.

d. Remove the passenger's air bag module connector clip from the lid.

e. Disconnect the passenger's SRS module connector.

f. Remove the glove box and the lower passenger's side instrument panel.

g. Using the Tamper Resistant Torx® Wrench size T50, remove the SRS module-to-instrument panel special bolts and discard them.

h. Remove the 4 SRS module-to-instrument panel mounting nuts.

i. Release the SRS module-to-instrument panel clips and remove the SRS module.

✳✳ CAUTION

Always store the SRS module face up.

5. Drain the cooling system into a clean container for reuse.

6. Working inside the engine compartment, disconnect the 2 heater hoses from the heater core.

7. Discharge and recover the air conditioning system refrigerant.

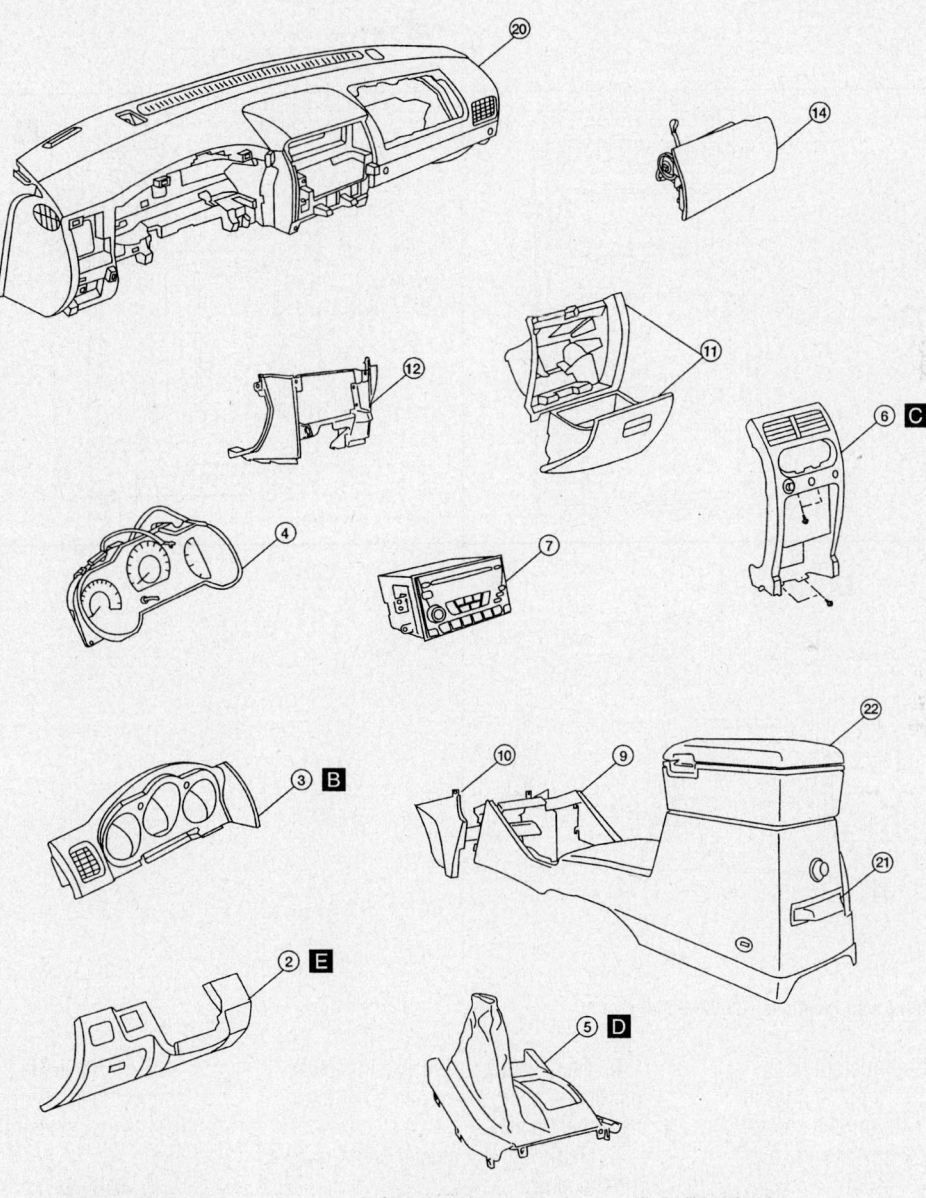

1 Steering column cover	9 Center console	17 A-pillar assist grips
2 Instrument lower panel driver side	10 Lower center instrument cover	18 A-pillar trim panels
3 Cluster lid A	11 Glove box striker	19 Body side welt
4 Combination meter	12 Glove box assembly	20 Instrument panel assembly
5 Remove shift bezel	13 Lower center instrument panel	21 Cup holder assembly (if equipped)
6 Remove cluster lid C	14 Passenger side air bag module	22 Armrest assembly
7 Audio unit	15 Front door kicking plates	
8 A/C & heater control	16 Dash side lower finishers	

09482_FRON_G0014

Instrument panel and related components—2002–2004

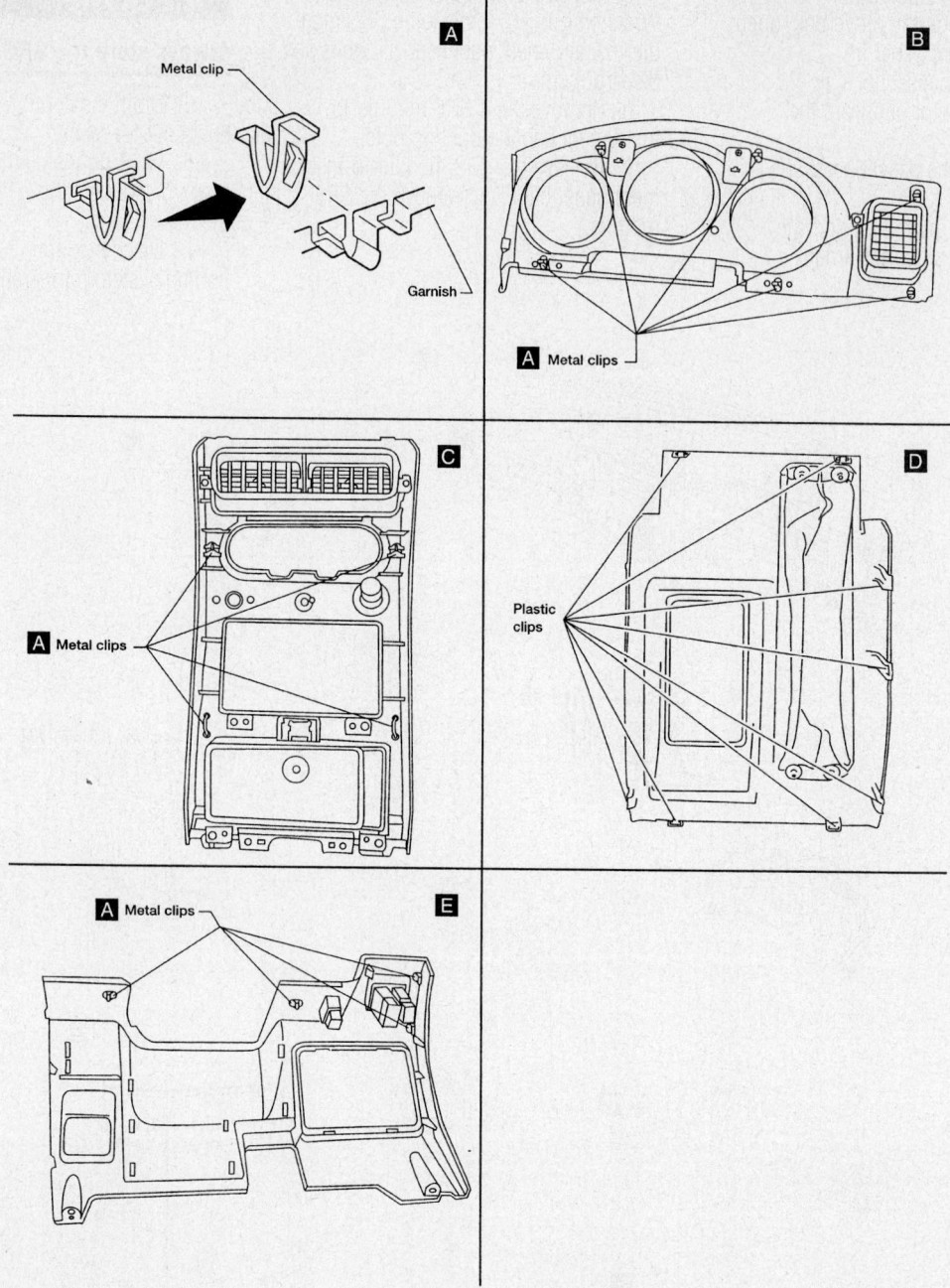

09482_FRON_G0015

Instrument panel retaining clip locations—2002–2004

8. Disconnect both refrigerant lines from the evaporator core. Plug the lines to prevent moisture from entering the system.

9. Remove the glove box and the mating trim.

10. Disconnect the thermal amp connector.

11. Remove the air conditioning housing assembly from the vehicle.

12. Remove the instrument panel assembly by performing the following procedure:

   a. Remove the 4 steering column cover screws; then, separate and remove the steering column covers.

   b. Remove the 2 driver's side lower instrument panel screws and the lower instrument panel.

   c. Remove the 4 cluster cover screws and the cluster cover.

   d. Remove the 6 combination meter screws; then, disconnect the combination meter electrical connector and remove the meter.

   e. Remove the 2 glove box screws and the glove box.

   f. Remove the 2 instrument stay cover screws; then, disconnect the electrical harness connectors and remove the stay cover.

   g. Remove the 2 cluster lid "C" screws; then, disconnect the electrical harness connectors and remove the cluster lid "C".

   h. Remove the 4 audio and deck pocket screws; then, disconnect the electrical harness connectors and remove the audio and deck pocket.

   i. Disconnect the ASCD main switch connector.

   j. Remove the 2 meter cover screws; then, disconnect the electrical harness connectors and remove the meter cover.

   k. Remove the 4 air conditioning-

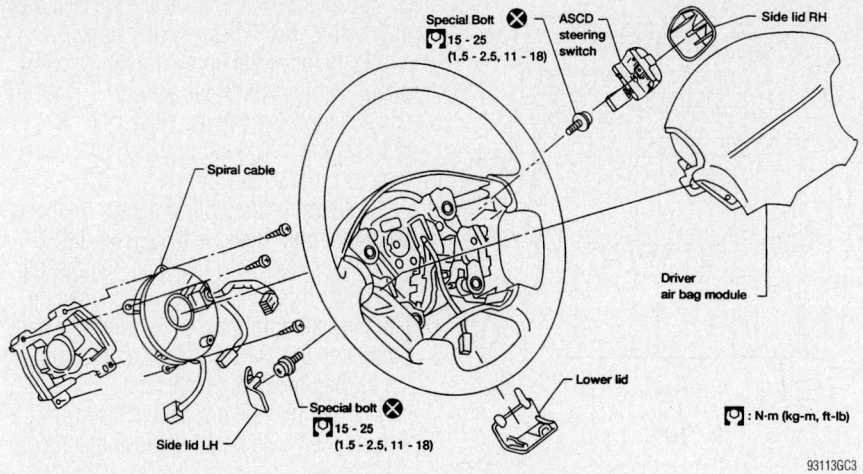

Steering wheel and SRS module and related components—2002 Frontier

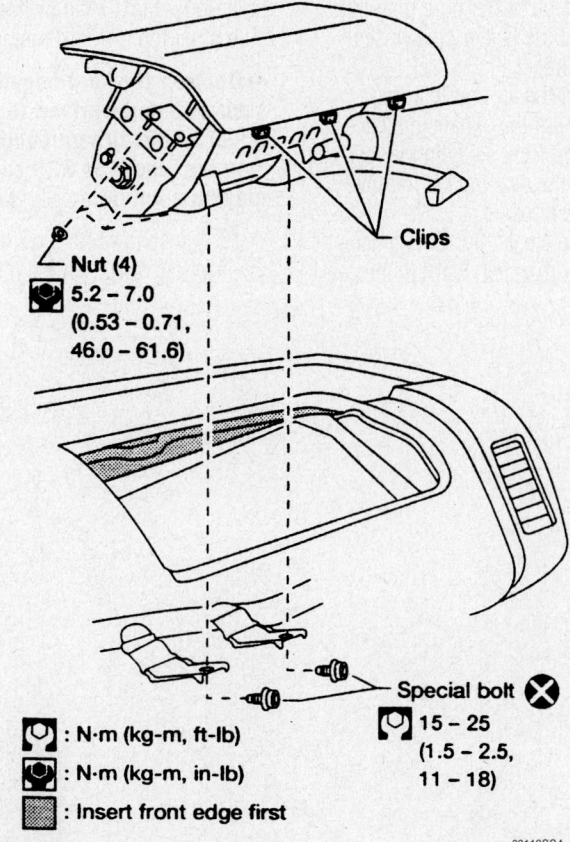

Passenger's side SRS module and related components—2002 Frontier

heater control screws; then, disconnect the control cables and remove the air conditioning-heater control.

l. Remove the front pillar garnish.

m. Remove the 3 instrument panel assembly nuts and 2 bolts; then, remove the instrument panel.

13. Remove the heater housing assembly.

14. Remove the heater core from the heater housing assembly.

To install:

15. Install the heater core to the heater housing assembly.

16. Install the heater housing assembly.

17. Install the instrument panel assembly by performing the following procedure:

a. Install the instrument panel and the 3 instrument panel assembly nuts and 2 bolts.

b. Install the front pillar garnish.

c. Install the air conditioning-heater control, connect the control cables and install the 4 air conditioning/heater control screws.

d. Install the meter cover, connect the electrical harness connectors and install the 2 meter cover screws.

e. Connect the ASCD main switch connector.

f. Install the audio and deck pocket, connect the electrical harness connectors and install the 4 audio and deck pocket screws.

g. Install the cluster lid "C", connect the electrical harness connectors and install the 2 cluster lid "C" screws.

h. Install the stay cover, connect the electrical harness connectors and the 2 instrument stay cover screws.

i. Install the glove box and the 2 glove box screws.

j. Install the meter, connect the combination meter electrical connector and install the 6 combination meter screws.

k. Install the cluster cover and the 4 cluster cover screws.

l. Install the lower instrument panel and the 2 driver's side lower instrument panel screws.

m. Install the steering column covers and the 4 steering column cover screws.

18. Install the air conditioning housing assembly to the vehicle.

19. Connect the thermal amp connector.

20. Install the glove box and the mating trim.

21. Connect both refrigerant lines to the evaporator core.

22. Inside the engine compartment, connect the heater hoses to the heater core.

23. Refill the cooling system.

24. Install the passenger's side air bag by performing the following procedure:

a. Install the SRS module and secure the SRS module-to-instrument panel clips.

b. Install the 4 SRS module-to-instrument panel mounting nuts.

c. Using the Tamper Resistant Torx® Wrench size T50, install the new special SRS module-to-instrument panel bolts.

d. Install the lower passenger's side instrument panel and the glove box.

e. Connect the passenger's SRS module connector.

f. Install the passenger's air bag module connector clip to the lid.

g. Inside the glove box, close the lower instrument panel lid.

h. Close the glove box door.

25. Install the steering wheel by performing the following procedure:

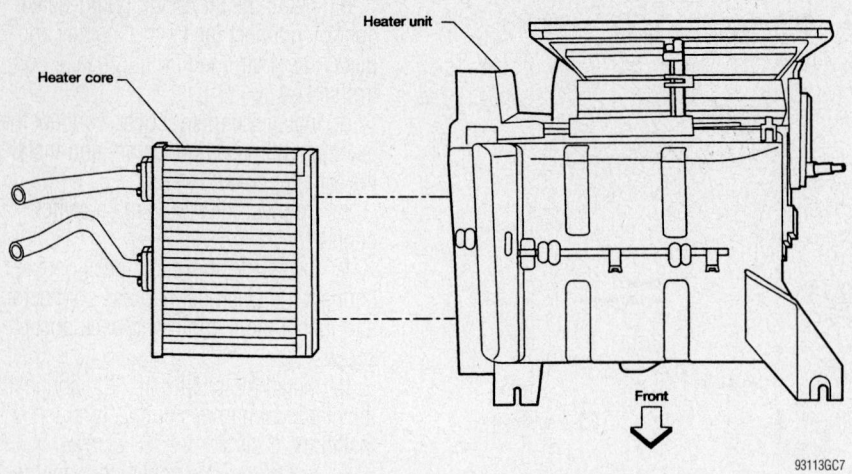

Heater core and related components—2002–2004

a. Align the spiral cable pin guide and install the steering wheel by pulling the spiral cable connectors through it.

b. Connect the horn connector and connect the spiral cable by aligning the pawls in the steering wheel.

c. Install the steering wheel nut and torque it to 22–29 ft. lbs. (29–39 Nm).

d. Install the SRS module to the steering wheel.

e. Using the Tamper Resistant Torx® Wrench size T50, install the new special bolts to both sides of the steering wheel.

f. Install the lower lid and disconnect the driver's air bag module connector.

g. Install both side lids.

h. Rotate the steering wheel fully right and left to make sure that the spiral cable is set in the neutral position.

26. Connect both the positive (1st) and negative (2nd) battery cables.

27. Evacuate and charge the air conditioning system refrigerant.

28. Run the engine to normal operating temperatures; then, check the climate control operation and check for leaks.

2005–2006

➡Be sure to disarm the SRS system, prior to working on the vehicle. Turn the ignition switch OFF, disconnect both battery cables and wait at least three minutes before starting any work.

1. Before servicing the vehicle, refer to the Precautions Section.

2. Position the front wheels in the straight ahead direction.

3. Disconnect the negative battery cable. Disconnect the positive battery cable.

4. Drain the cooling system.

5. Properly discharge the air conditioning system.

6. If equipped with the 4.0L engine, remove the right side heater core pipe nuts.

7. Disconnect the heater core hoses from the heater core.

8. Disconnect the air conditioning refrigerant lines from the expansion valve.

9. Position the front seats in the rearmost position on the seat tracks. Remove the lower instrument panel.

10. Remove the cluster lid "C". Remove the transmission trim panel. Remove the cen-ter console screws. Disconnect the electrical connectors. Remove the center console.

11. Remove the upper front pillar trim panel. Remove the steering lock escutcheon. Remove the cluster lid "A". Remove the combination meter. Disconnect the electrical connections.

12. Remove the optical sensor. Remove the audio unit. Remove the cluster lid "D".

13. Remove the glove box. Remove the two bolts, through the glove box opening, retaining the front passenger's side air bag module to the steering member. Disconnect the air bag module connectors.

14. Remove the instrument stay right side and left side bolts. Remove the instrument panel.

15. Remove the two front floor ducts.

16. To remove the driver's side air bag module, locate the retaining clip access hole under the steering wheel. Insert a suitable blunt tool (4mm-6mm in size)

➡Do not use sharp edged objects, such as a screwdriver, to release the driver's side airbag module from the steering wheel as SRS components may be unintentionally damaged.

17. Press upward, toward the center of the steering wheel, on the retaining clip

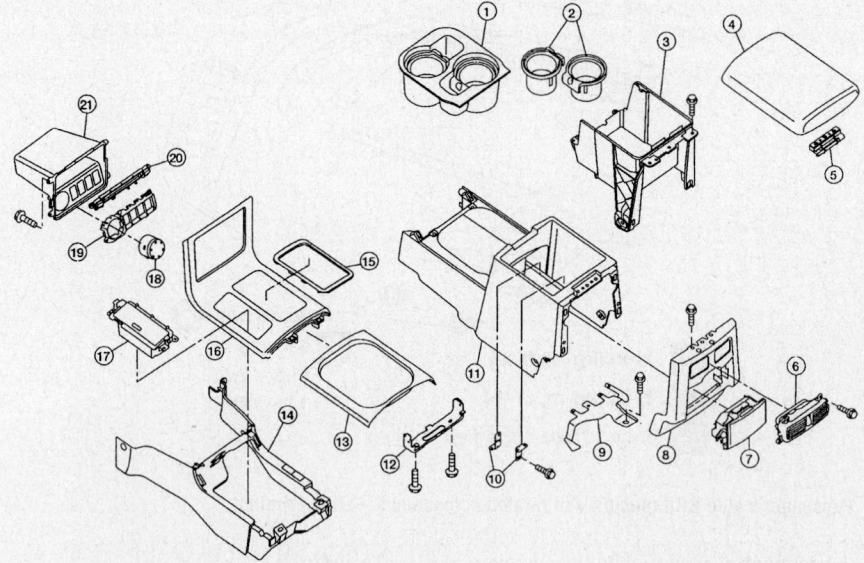

1.	Cup holder assembly	2.	Cup holder insert	3.	Center console bin
4.	Center console lid	5.	Hinge	6.	Ventilator console grille
7.	Rear cup holder assembly	8.	Rear finisher assembly	9.	Wire harness bracket
10.	Bracket DVD	11.	Center console rear base	12.	Bracket
13.	Cup holder finisher	14.	Center console front base	15.	A/T finisher bezel
16.	A/T finisher	17.	Ash tray	18.	Switch assembly
19.	Switch finisher	20.	CD changer door	21.	Console bin

09482_FRON_G0011

Center console and related components—2005–2006

until the air bag module is released from the steering wheel.

18. Lift the air bag module from the steering wheel. Disconnect the electrical connectors. Remove the air bag module.

19. Disconnect the steering wheel switches. Remove the steering wheel center nut. Using a steering wheel removal tool, remove the steering wheel.

20. Remove the steering column upper and lower covers. Disconnect the wiper and washer switch connector. While pressing the tabs, pull the wiper and washer switch away from the spiral cable to remove it.

21. Disconnect the light and turn signal switch connector. While pressing the tabs, pull the light and turn signal switch toward the driver's door to remove it.

22. Remove the screws. While pressing the tab, pull the spiral cable away from the steering column assembly. Disconnect the electrical connectors.

➡**With the steering linkage disconnected, the spiral cable may snap by turning the steering wheel beyond the limited number of turns. The spiral cable can be turned counterclockwise about 2.5 turns from the neutral position.**

23. Remove the lower knee protector.

24. Remove the locknut and bolt from the upper joint and then separate the upper joint from the upper shaft.

25. Remove the three nuts and bolt from the steering column and then remove the steering column assembly from the steering member.

26. Remove the hole cover seal and clamp. Remove the hole cover nuts, remove the hole cover from the dash panel.

27. Remove the bolt from the lower joint of the lower joint shaft and remove the lower joint shaft from the vehicle.

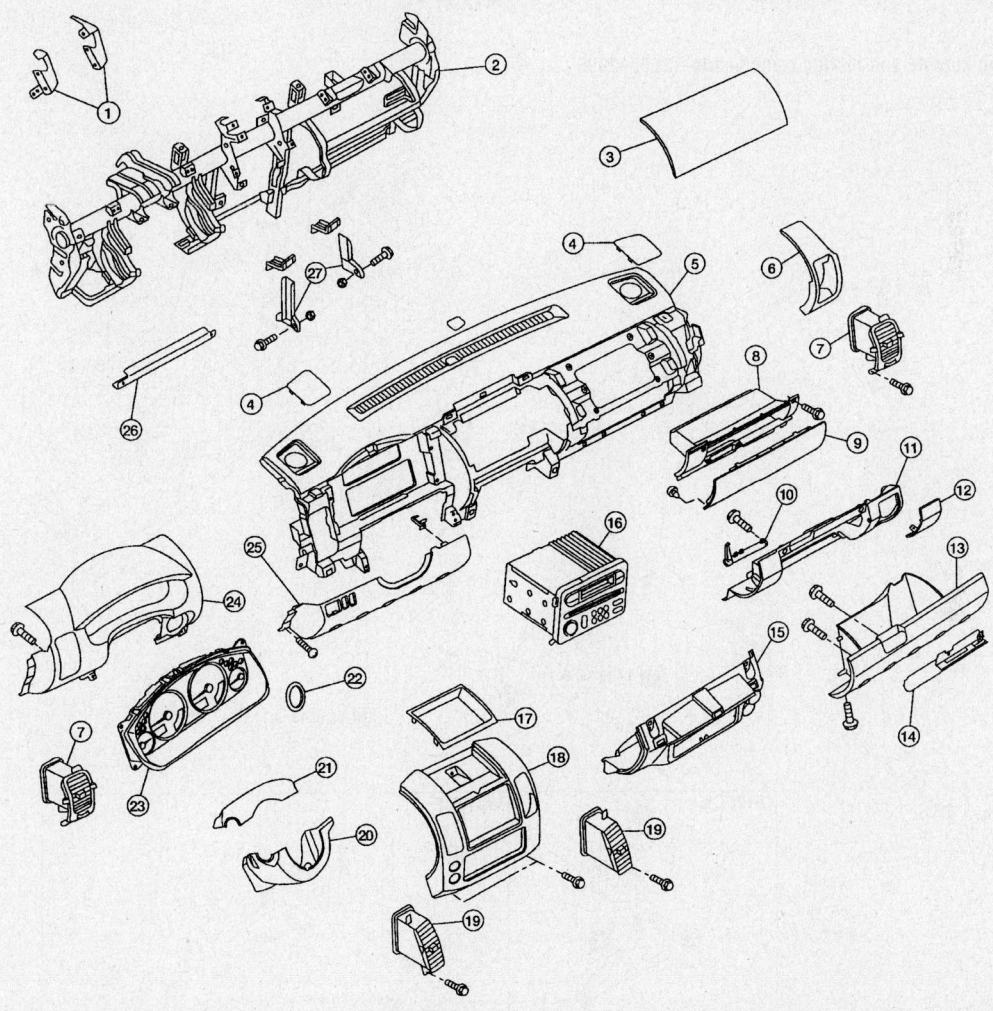

1.	Display unit bracket RH/LH	2.	Steering member assembly	3.	Passenger air bag module cover
4.	Speaker grille RH/LH	5.	Instrument panel and pad assembly	6.	Instrument side finisher
7.	Side ventilator assembly RH/LH	8.	Upper glove box bin	9.	Upper glove box door
10.	Lower glove box damper assembly	11.	Lower instrument panel RH	12.	Fuse block cover
13.	Lower glove box assembly	14.	Lower glove box latch assembly	15.	Cluster lid D
16.	Audio unit	17.	Storage tray	18.	Cluster lid C
19.	Center ventilator assembly RH/LH	20.	Steering column cover lower	21.	Steering column cover upper
22.	Steering lock escutcheon	23.	Combination meter	24.	Cluster lid A
25.	Lower instrument panel LH	26.	Knee protector brace	27.	Instrument stay RH/LH

09482_FRON_G0009

Instrument panel and related components—2005-2006

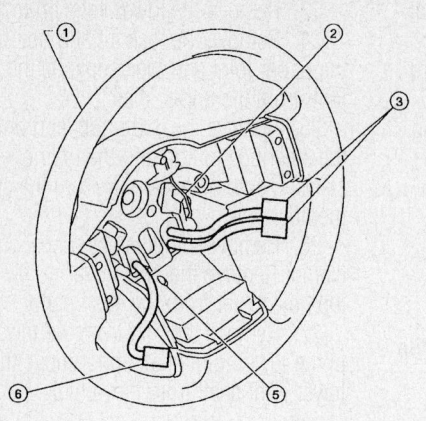

1. Steering wheel
2. Retaining clip
3. Driver air bag module connectors
4. Driver air bag module
5. Retaining clip access hole
6. Horn connector

09482_FRON_G0010

Driver's side air bag module and related components—2005–2006

28. Disconnect the instrument panel wire harness at the right and left in-line connector brackets, and the fuse block (SMJ) electrical connectors.

29. Remove the covers and than remove the three steering member bolts from each side to disconnect the steering member from the vehicle body.

30. Remove the heater/evaporator case assembly with it attached to the steering member from the vehicle.

31. Separate the steering member from the heater/evaporator unit.

32. Remove the heater cover retaining screws. Remove the cover.

33. Remove the heater core and the evaporator pipe bracket. Remove the heater core.

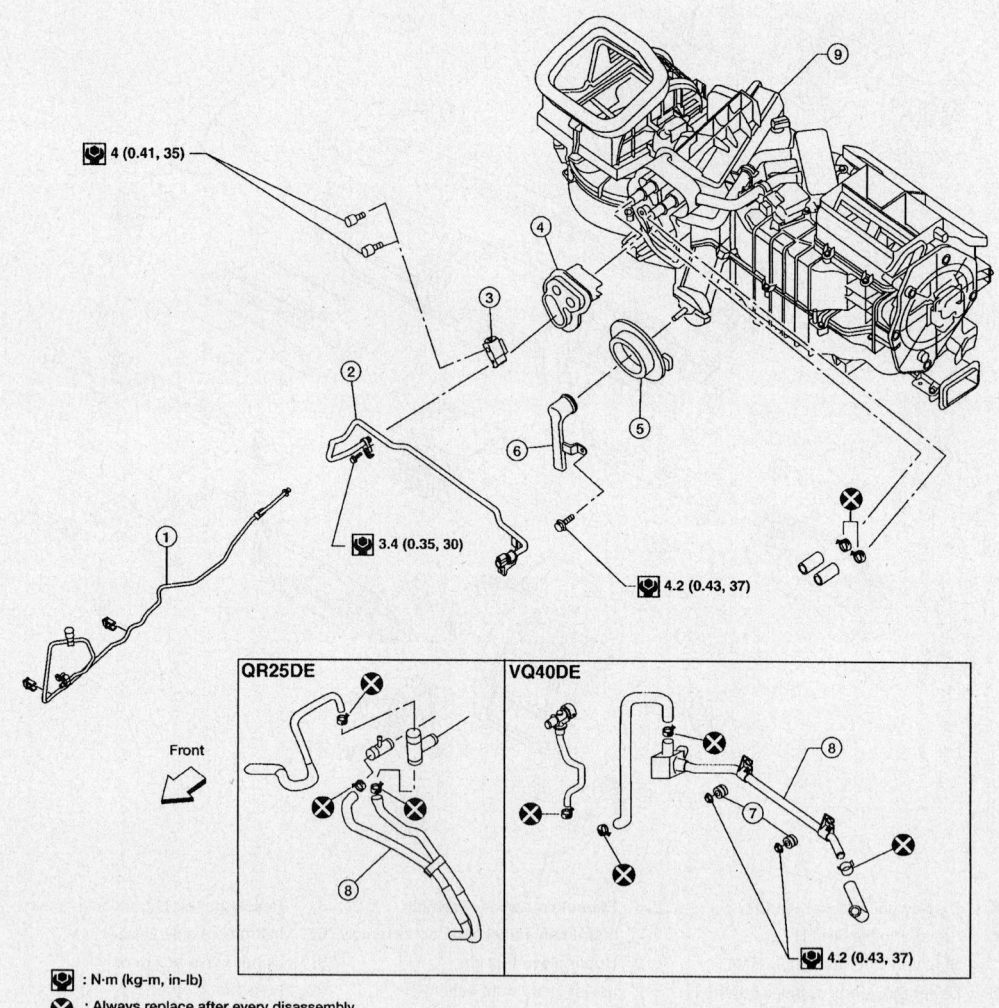

QR25DE

VQ40DE

Front

[icon] : N·m (kg-m, in-lb)

[icon] : Always replace after every disassembly.

1. High-pressure A/C pipe
2. Low-pressure A/C pipe
3. Expansion valve
4. Heater core and evaporator pipes grommet
5. A/C drain hose grommet
6. A/C drain hose
7. Heater core pipe mounts
8. Heater core pipes
9. Heater and cooling unit assembly

09482_FRON_G0012

Heater/evaporator core and related components—2005–2006

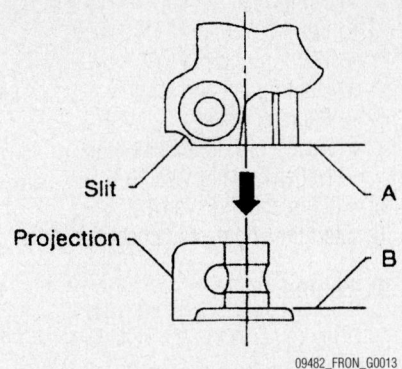

Lower joint installation—2005–2006

To install:

34. Installation is the reverse of the removal procedure.

➡ **If the in-cabin microfilters are contaminated with coolant, replace them.**

35. Be sure to use new steering column retaining bolts and pinch bolt, as required.

➡ **When installing the steering column, finger tighten all of the lower bracket and joint bolts and then tighten them to specification. Do not apply undue stress to the steering column.**

36. With the wheels in the straight ahead position align the slit of the lower joint with the projection on the dust cover. Insert the joint until surface "A" contacts surface "B"

37. Be sure to align the spiral cable correctly when installing the steering wheel. Make sure that the cable is in the neutral position. The neutral position is detected by turning left 2.6 revolutions from the right end position and ending with the locating pin at the top.

38. To adjust the steering angle sensor neutral position, position the steering wheel in the straight ahead position and rive the vehicle at 10 mph or more for ten minutes. When the procedure is complete, the SLP

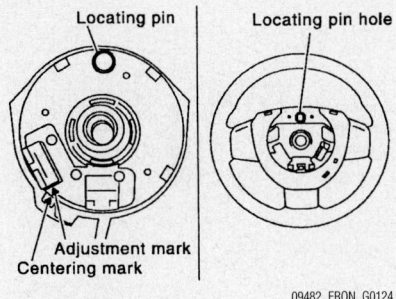

Spiral cable installation and locating point—2005–2006

indicator lamp and the VDC OFF indicator lamp will turn off.

39. Be sure to fill the cooling system with the proper grade and type coolant.

40. Be sure to recharge the air conditioning system.

41. Check and adjust the front end alignment, as necessary.

Water Pump

REMOVAL & INSTALLATION

2.4L Engine

1. Before servicing the vehicle, refer to the Precautions Section.

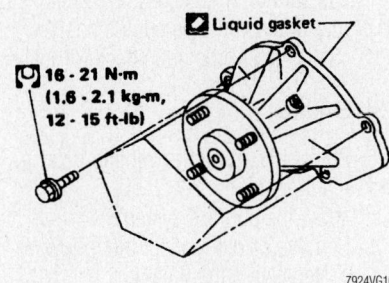

Water pump—2.4L engine

2. Drain the cooling system.
3. Remove or disconnect the following:

- Negative battery cable
- Accessory drive belts
- Cooling fan
- Water pump

To install:

4. Install or connect the following:

- Water pump. Apply sealant and tighten the bolts to 12–15 ft. lbs. (16–21 Nm).
- Cooling fan
- Accessory drive belts
- Negative battery cable

Diameter of liquid gasket:
2.0 - 3.0 mm (0.079 - 0.118 in)

Liquid gasket application—2.4L engine

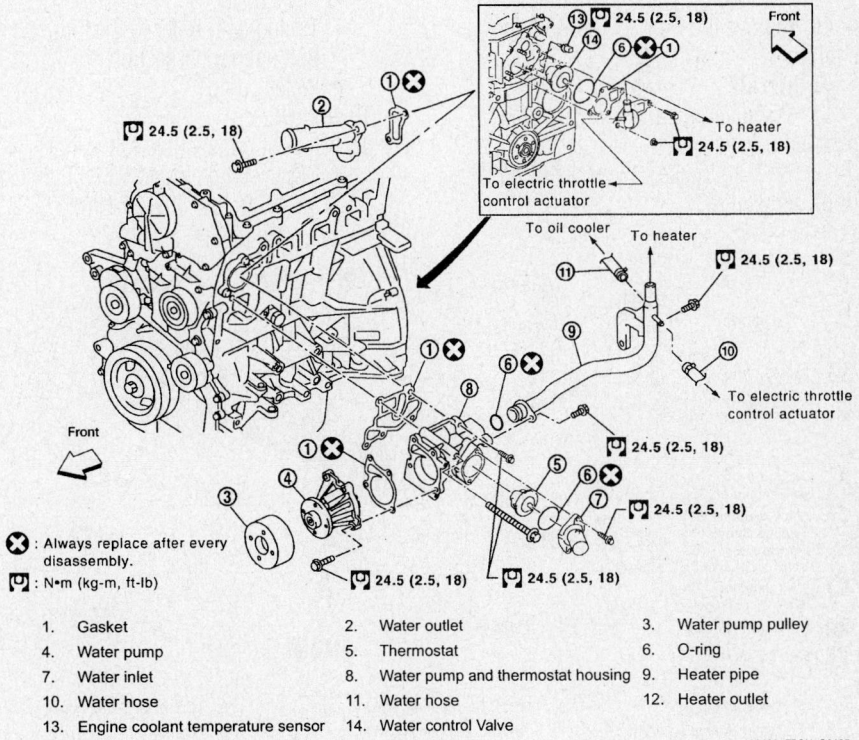

❌ : Always replace after every disassembly.

🔧 : N•m (kg-m, ft-lb)

1. Gasket
2. Water outlet
3. Water pump pulley
4. Water pump
5. Thermostat
6. O-ring
7. Water inlet
8. Water pump and thermostat housing
9. Heater pipe
10. Water hose
11. Water hose
12. Heater outlet
13. Engine coolant temperature sensor
14. Water control Valve

Water pump and related components—2.5L engine

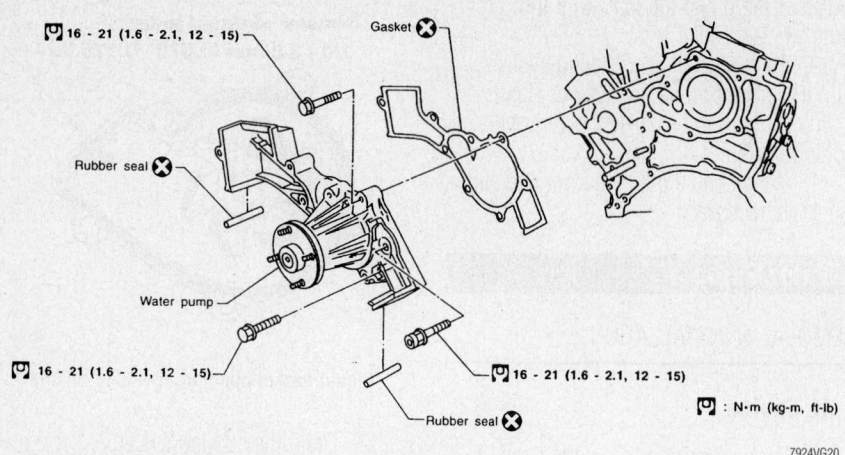

Water pump and related components—3.3L engine

5. Fill the cooling system.
6. Start the engine and check for leaks.

2.5L Engine

1. Before servicing the vehicle, refer to the Precautions Section.
2. Disconnect the negative battery cable. Drain the cooling system.
3. Remove the air duct. Remove the drive belt.
4. Remove the upper and lower radiator hoses.
5. Remove the cooling fan and the water pump pulley.
6. Remove the water pump retaining bolts. Remove the water pump from the engine.

To install:

7. Installation is the reverse of the removal procedure.
8. Be sure to use new gaskets and O-rings, as required.

9. Be sure to fill the cooling system with the proper grade and type engine coolant.
10. Start the engine and check for leaks.

3.3L Engine

1. Before servicing the vehicle, refer to the Precautions Section.
2. Drain the cooling system.
3. Remove or disconnect the following:
   • Negative battery cable
   • Accessory drive belts
   • Radiator hoses
   • Cooling fan and shroud
   • Water pump pulley
   • Front cover
   • Timing belt. Refer to the Timing Belt unit repair section.
   • Water pump

To install:

4. Install or connect the following:
   • Water pump. Tighten the bolts to 12–15 ft. lbs. (16–21 Nm).

• Timing belt
• Front cover
• Water pump pulley
• Cooling fan and shroud
• Radiator hoses
• Accessory drive belts
• Negative battery cable
5. Fill the cooling system.
6. Start the engine and check for leaks.

4.0L Engine

1. Before servicing the vehicle, refer to the Precautions Section.
2. Disconnect the negative battery cable. Drain the cooling system.
3. Remove the undercover. Remove the drive belts.
4. Remove the radiator upper and lower hoses. Remove the cooling fan.
5. Remove the chain tensioner cover and water pump cover from the front timing case, using tool KV10111100 (J-37228) or equivalent.
6. To remove the timing chain tensioner (primary), loosen the clip of the timing chain tensioner (primary) and release the plunger stopper. Insert the plunger into the tensioner body by pressing the slack guide. Keep the slack guide pressed and hold the plunger in by pushing the stopper pin through the tensioner body hole and plunger groove. Turn the crankshaft pulley clockwise so that the timing chain on the timing chain tensioner (primary) side is loose. Remove the bolts and remove the timing chain tensioner (primary).

➡**Be careful not to drop the bolts inside the timing chain case.**

7. Remove the three water pump retaining bolts. Secure a gap between the water pump gear and the timing chain, by turning the crankshaft pulley counterclockwise until timing chain looseness on the water pump sprocket becomes maximum.
8. Screw M8 bolts approximately 1.97 inch in length into the water pumps upper

▢ : Lubricate with new engine oil.
▢ : Apply Genuine RTV Silicone Sealant or equivalent. Refer to GI section.
✖ : Always replace after every disassembly.
▢ : N•m (kg-m, in-lb)
▢ : N•m (kg-m, ft-lb)

1. Water pump
2. Timing chain tensioner (primary)
3. Chain tensioner cover
4. Water drain plug (front)
5. Water pump cover
6. O-ring
7. O-ring

Water pump and related components—4.0L engine

09482_FRON_G0016

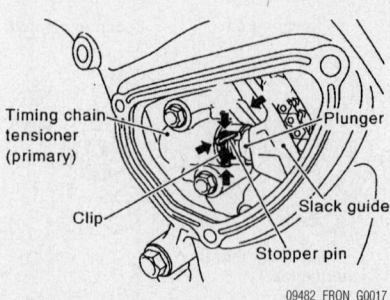

Chain tensioner (primary) cover removal—4.0L engine

09482_FRON_G0017

and lower bolt holes until they reach the timing chain case.

9. Alternately tighten each bolt for a half turn and pull out the water pump.

➡**Pull the pump straight out while preventing the vane from contacting the socket in the installation area. Remove the pump without causing the sprocket to contact the timing chain.**

10. Remove the M8 bolts. Remove and discard the O-rings.

To install:

11. Installation is the reverse of the removal procedure.

12. Be sure to use new gaskets and O-rings, as required.

13. When installing the water pump make sure that the timing chain and water pump sprocket are engaged. Tighten the bolts alternately and evenly to specification.

14. Before installing the chain tensioner cover and the water pump cover be sure to apply a continuous bead of sealant to the mating surfaces of the covers.

➡**Do not allow the sealant to set for more than five minutes before installing the covers.**

15. Be sure to fill the cooling system with the proper grade and type engine coolant.

16. Start the engine and check for leaks.

17. Let the engine idle for about three minutes than rev it up to 3,000 rpm's under a no load condition to purge air from the high pressure chamber of the chain tensioner. The engine may produce a rattling noise. This indicates that air still remains in the chamber and is not a matter of concern.

Cylinder Head

REMOVAL & INSTALLATION

2.4L Engine

1. Before servicing the vehicle, refer to the Precautions Section.

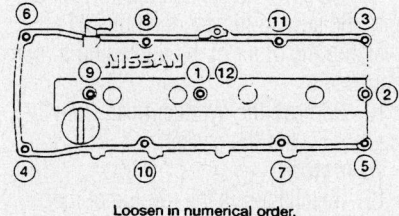

Loosen in numerical order.

9308VG06

Valve cover bolt loosening sequence—2.4L engine

2. Drain the cooling system.
3. Relieve the fuel system pressure.
4. Remove or disconnect the following:
   - Negative battery cable
   - Air cleaner assembly
   - Spark plug wires
   - Radiator hoses
   - Accessory drive belts
   - Fuel lines
   - Intake manifold
   - Exhaust manifold
   - Valve cover. Remove the bolts in the sequence shown.
   - Camshaft sprocket cover
   - Camshaft sprockets and upper timing chain
5. Wedge the lower timing chain in place to prevent the chain tensioner from expanding.
6. Remove or disconnect the following:
   - Timing chain idler sprocket
   - Camshafts
   - Cylinder head. Loosen the bolts in several passes and in sequence as shown.

To install:

7. Install the cylinder head with a new gasket. Tighten the bolts in sequence as follows:
   a. Step 1: 22 ft. lbs. (30 Nm)
   b. Step 2: 59 ft. lbs. (79 Nm)
   c. Step 3: Loosen all bolts completely
   d. Step 4: 18–25 ft. lbs. (25–34 Nm)
   e. Step 5: Plus 86–91 degrees
8. Install or connect the following:

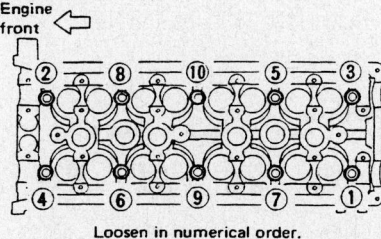

Loosen in numerical order.

9308VG04

Cylinder head bolt loosening sequence—2.4L engine

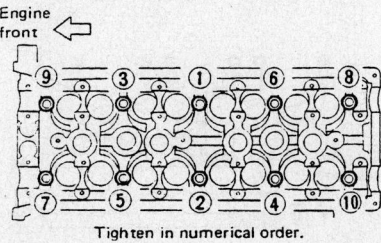

Tighten in numerical order.

9308VG05

Cylinder head bolt torque sequence—2.4L engine

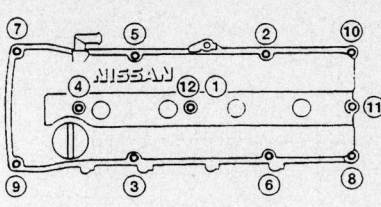

Tighten in numerical order.

9308VG07

Valve cover bolt torque sequence—2.4L engine

   - Camshafts
   - Timing chain idler sprocket and lower timing chain. Remove the wedge and tighten the bolt to 48–61 ft. lbs. (66–83 Nm).
   - Camshaft sprockets and upper timing chain. Tighten the bolts to 123–130 ft. lbs. (167–177 Nm).
   - Camshaft sprocket cover
   - Valve cover. Tighten the bolts in sequence to 69–95 ft. lbs. (8–11 Nm).
   - Exhaust manifold
   - Intake manifold
   - Fuel lines
   - Accessory drive belts
   - Radiator hoses
   - Spark plug wires
   - Air cleaner assembly
   - Negative battery cable
9. Fill the cooling system.
10. Start the engine and check for leaks.

2.5L Engine

1. Before servicing the vehicle, refer to the Precautions Section.

2. Properly relieve the fuel system pressure.

3. Disconnect the negative battery cable. Drain the cooling system. Drain the engine oil.

4. Remove the intake manifold and fuel tube assembly.

5. Remove the fuel injector and fuel tube assembly.

6. Remove the exhaust manifold and the three way catalyst.

7. Remove the water outlet. Remove the heater outlet.

8. Remove the front cover and the timing chain.

9. Remove the camshafts.

10. Remove the cylinder head retaining bolts. Be sure to remove the bolts by reversing the order of the tightening torque sequence.

11. Remove the cylinder head from the engine. Discard the gasket.

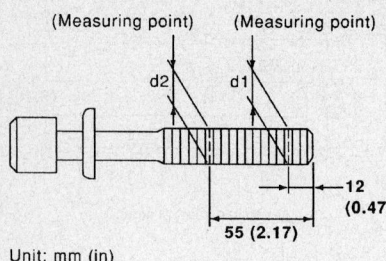

Cylinder head bolt measurement—2.5L engine

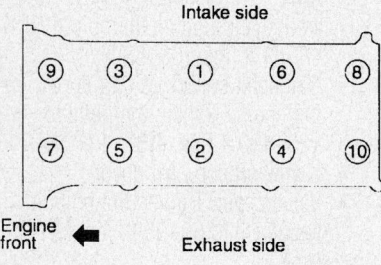

Cylinder head bolt torque sequence—2.5L engine

To install:

12. Installation is the reverse of the removal procedure.

13. Be sure to inspect the cylinder head bolts. Replace as required.

➡**Head bolts are tightened by plastic zone tightening method. Whenever the size difference between "d1" and "d2" exceeds the limit, replace the bolt. "d1"-"d2" limit is 0.0091. If reduction of the outer diameter appears in a position other than "d2", use it the "d2" point.**

14. Install the new cylinder head gasket. Torque the cylinder head bolts to specification and in the proper sequence.

15. Be sure to fill the cooling system with the proper grade and type engine coolant.

16. Be sure to fill the engine with the proper grade and type motor oil.

17. Start the engine and check for leaks.

3.3L Engine

1. Before servicing the vehicle, refer to the Precautions Section.

2. Drain the cooling system.

3. Relieve the fuel system pressure.

4. Remove or disconnect the following:
   • Negative battery cable
   • Accessory drive belts
   • Front cover

• Timing belt. Refer to the Timing Belt unit repair section.
• Upper intake manifold
• Lower intake manifold
• Camshaft sprockets
• Rear timing cover
• Distributor
• Exhaust front pipes
• A/C compressor
• Alternator
• Power steering pump
• Accessory brackets
• Valve covers. Loosen the bolts in several passes and in sequence.
• Cylinder heads with the exhaust manifolds attached. Loosen the bolts in several passes and in sequence.

➡**The cylinder head bolts vary in length. Note the bolt locations for assembly.**

To install:

5. Install the cylinder heads and the lower intake manifold at the same time. Tighten the bolts in sequence as follows:

   a. Step 1: Tighten the cylinder head bolts to 22 ft. lbs. (29 Nm)

   b. Step 2: Tighten the cylinder head bolts to 43 ft. lbs. (59 Nm)

   c. Step 3: Loosen all cylinder head bolts completely

   d. Step 4: Tighten the cylinder head bolts to 84 inch lbs. (10 Nm)

   e. Step 5: Tighten the intake manifold fasteners to 35 inch lbs. (4 Nm)

   f. Step 6: Tighten the intake manifold fasteners to 13 ft. lbs. (18 Nm)

Cylinder head bolt torque sequence—3.3L engine

   g. Step 7: Tighten the intake manifold fasteners to 12–14 ft. lbs. (16–20 Nm)

   h. Step 8: Loosen all intake fasteners completely

   i. Step 9: Tighten the cylinder head bolts to 22 ft. lbs. (29 Nm)

   j. Step 10: Tighten the cylinder head bolts 60–65 degrees **OR** tighten to 40–47 ft. lbs. (54–64 Nm)

   k. Step 11: Tighten the cylinder head sub-bolts to 80–105 inch lbs. (9–12 Nm)

   l. Step 12: Tighten the intake manifold fasteners to 35 inch lbs. (4 Nm)

   m. Step 13: Tighten the intake manifold fasteners to 78 inch lbs. (9 Nm)

   n. Step 14: Tighten the intake manifold fasteners to 70–84 inch lbs. (6–7 Nm)

6. Install or connect the following:
   • Valve covers
   • Accessory brackets
   • Power steering pump
   • Alternator
   • A/C compressor
   • Exhaust front pipes
   • Distributor
   • Rear timing cover
   • Camshaft sprockets
   • Upper intake manifold
   • Timing belt
   • Front cover
   • Accessory drive belts
   • Negative battery cable

7. Fill the cooling system.

8. Start the engine and check for leaks.

4.0L Engine

1. Before servicing the vehicle, refer to the Precautions Section.

2. Properly relieve the fuel system pressure.

3. Disconnect the negative battery cable. Drain the cooling system.

4. Remove the camshaft.

5. Remove the intake manifold.

6. Remove the exhaust manifold.

7. Remove the water inlet and thermostat assembly.

8. Remove the water outlet, water pipe and heater pipe.

9. Remove the cylinder head retaining bolts. Be sure to remove the bolts by reversing the order of the tightening torque sequence.

10. Remove the cylinder head from the engine. Discard the gasket.

To install:

11. Installation is the reverse of the removal procedure.

12. Be sure to inspect the cylinder head bolts. Replace as required.

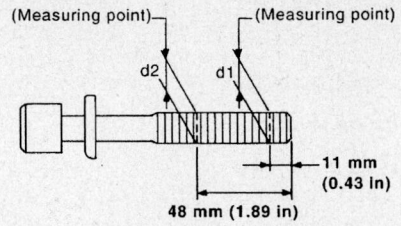

Cylinder head bolt measurement—4.0L engine

➡Head bolts are tightened by plastic zone tightening method. Whenever the size difference between "d1" and "d2" exceeds the limit, replace the bolt. "d1"-"d2" limit is 0.0043. If reduction of the outer diameter appears in a position other than "d2", use it the "d2" point.

13. Install the new cylinder head gasket. Turn the crankshaft until the number one piston is at TDC.

➡The crankshaft key should line up with the right bank center line, see illustration.

14. Torque the cylinder head bolts to specification and in the proper sequence.

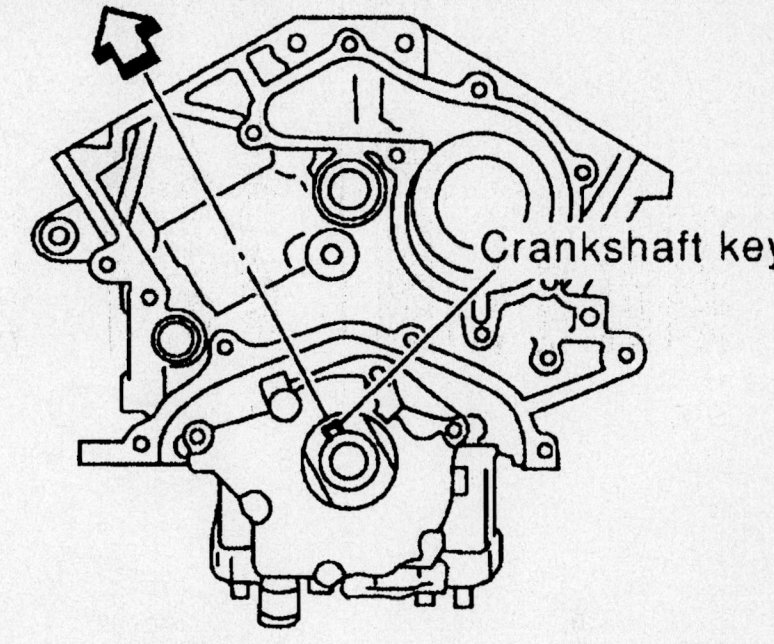

Cylinder head and crankshaft key alignment—4.0L engine

Right bank

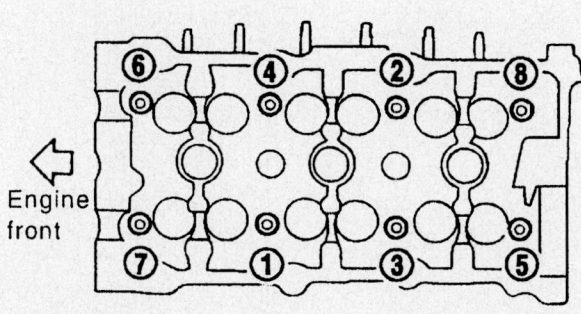

Left bank

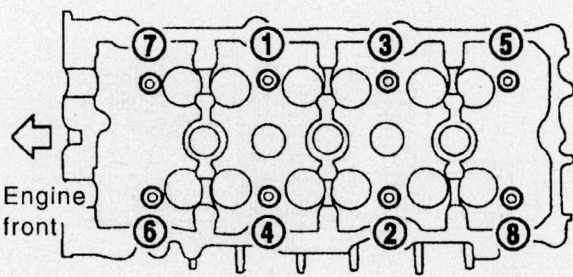

Cylinder head bolt torque sequence—4.0L engine

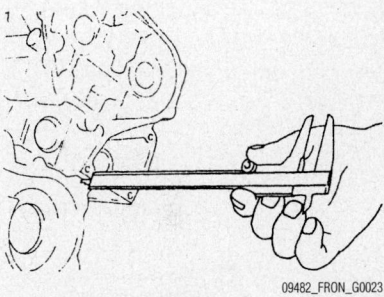

Cylinder head to cylinder block installation measurement—4.0L engine

15. Measure the distance between the front end faces of the cylinder block and the cylinder head on both the left and right banks. If the measured value is not within specification reinstall the cylinder head. Specification is 0.555–0.587 inch.

16. Be sure to fill the cooling system with the proper grade and type engine coolant.

17. Start the engine and check for leaks.

Supercharger

REMOVAL & INSTALLATION

3.3L Engine

1. Before servicing the vehicle, refer to the Precautions Section.

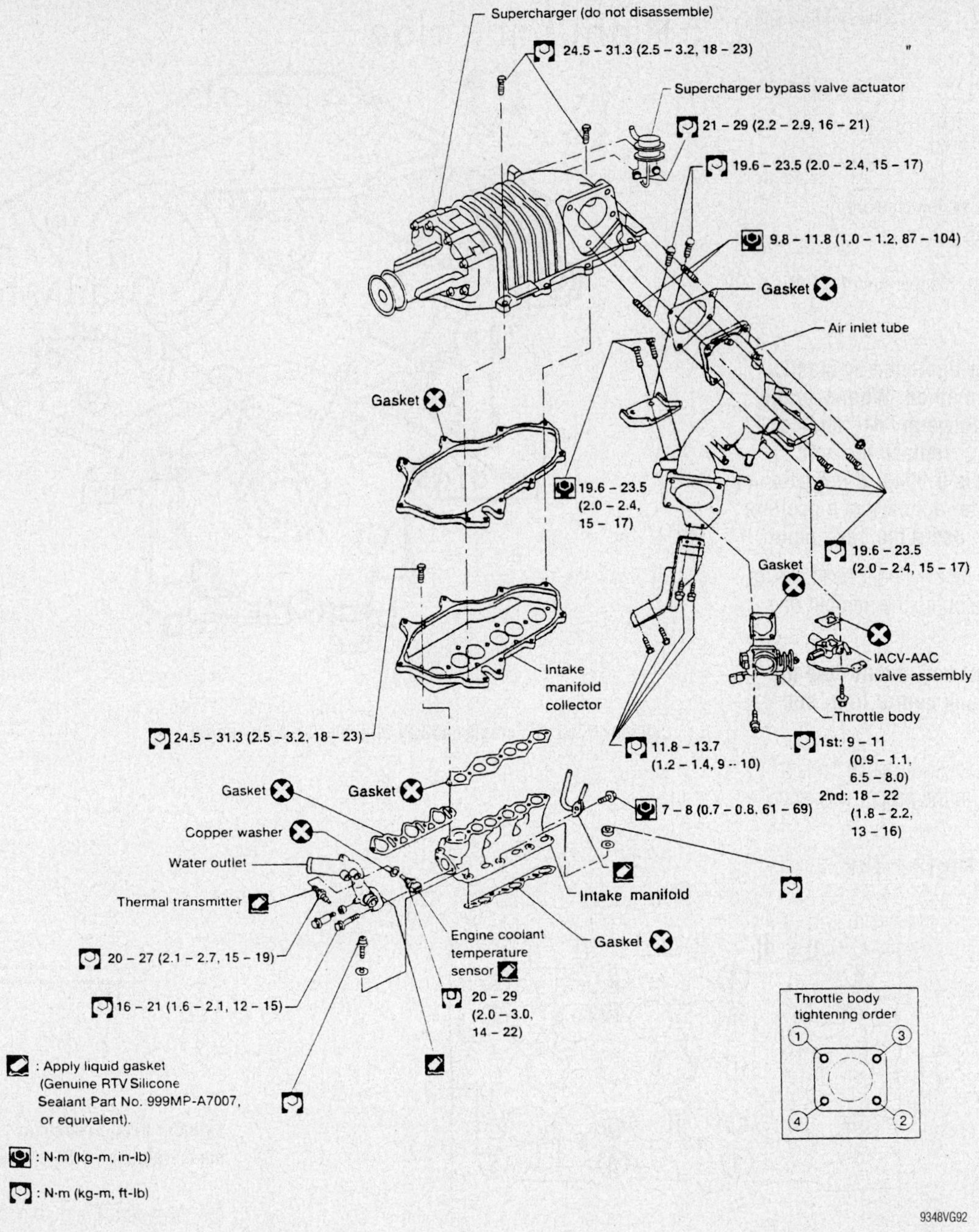

Supercharger (do not disassemble)

24.5 – 31.3 (2.5 – 3.2, 18 – 23)

Supercharger bypass valve actuator

21 – 29 (2.2 – 2.9, 16 – 21)

19.6 – 23.5 (2.0 – 2.4, 15 – 17)

9.8 – 11.8 (1.0 – 1.2, 87 – 104)

Gasket

Air inlet tube

Gasket

19.6 – 23.5 (2.0 – 2.4, 15 – 17)

19.6 – 23.5 (2.0 – 2.4, 15 – 17)

Gasket

IACV-AAC valve assembly

Throttle body

24.5 – 31.3 (2.5 – 3.2, 18 – 23)

Intake manifold collector

11.8 – 13.7 (1.2 – 1.4, 9 – 10)

1st: 9 – 11 (0.9 – 1.1, 6.5 – 8.0)
2nd: 18 – 22 (1.8 – 2.2, 13 – 16)

Gasket   Gasket

7 – 8 (0.7 – 0.8, 61 – 69)

Copper washer

Water outlet

Intake manifold

Thermal transmitter

Gasket

20 – 27 (2.1 – 2.7, 15 – 19)

Engine coolant temperature sensor

16 – 21 (1.6 – 2.1, 12 – 15)

20 – 29 (2.0 – 3.0, 14 – 22)

: Apply liquid gasket (Genuine RTV Silicone Sealant Part No. 999MP-A7007, or equivalent).

: N·m (kg-m, in-lb)

: N·m (kg-m, ft-lb)

Throttle body tightening order

```
1        3
    
4        2
```

9348VG92

Supercharger and related components—3.3L engine

2. Drain the coolant.
3. Remove or disconnect the following:

- Negative battery cable
- Accelerator cable
- ASCD cable at the throttle body
- Air inlet duct
- PCV hoses
- Resonator hose
- Supercharger pulley cover
- Supercharger drive belt
- Air inlet tube supports
- Air inlet tube
- EVAP vacuum hose
- Brake booster hose

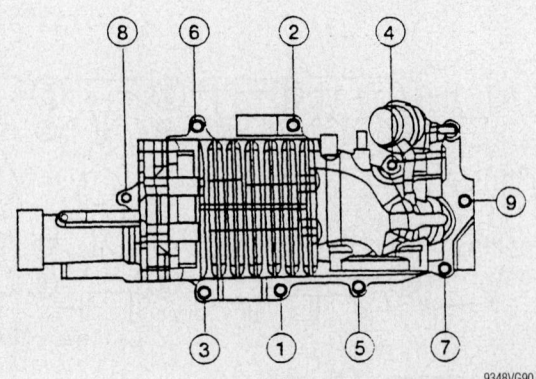

9348VG90

Supercharger bolt torque sequence. Loosen in reverse order—3.3L engine

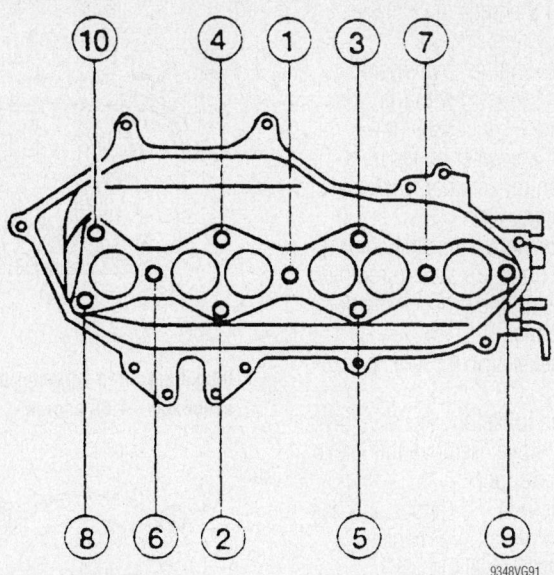

Supercharged intake manifold collector bolt torque sequence. Loosen in reverse order—3.3L engine

- All remaining hoses and wires in the way of removal
- Intake manifold collector
- Heater hoses
- Supercharger

To install:

4. Installation is the reverse of removal. Observe the following torques:
- Supercharger mounting bolts: 18–23 ft. lbs. (24–31 Nm).
- Air inlet tube-to-supercharger: 15–17 ft. lbs. (20–24 Nm).

Rocker Arms/Shafts

REMOVAL & INSTALLATION

3.3L Engine

1. Before servicing the vehicle, refer to the Precautions Section.

2. Remove or disconnect the following:

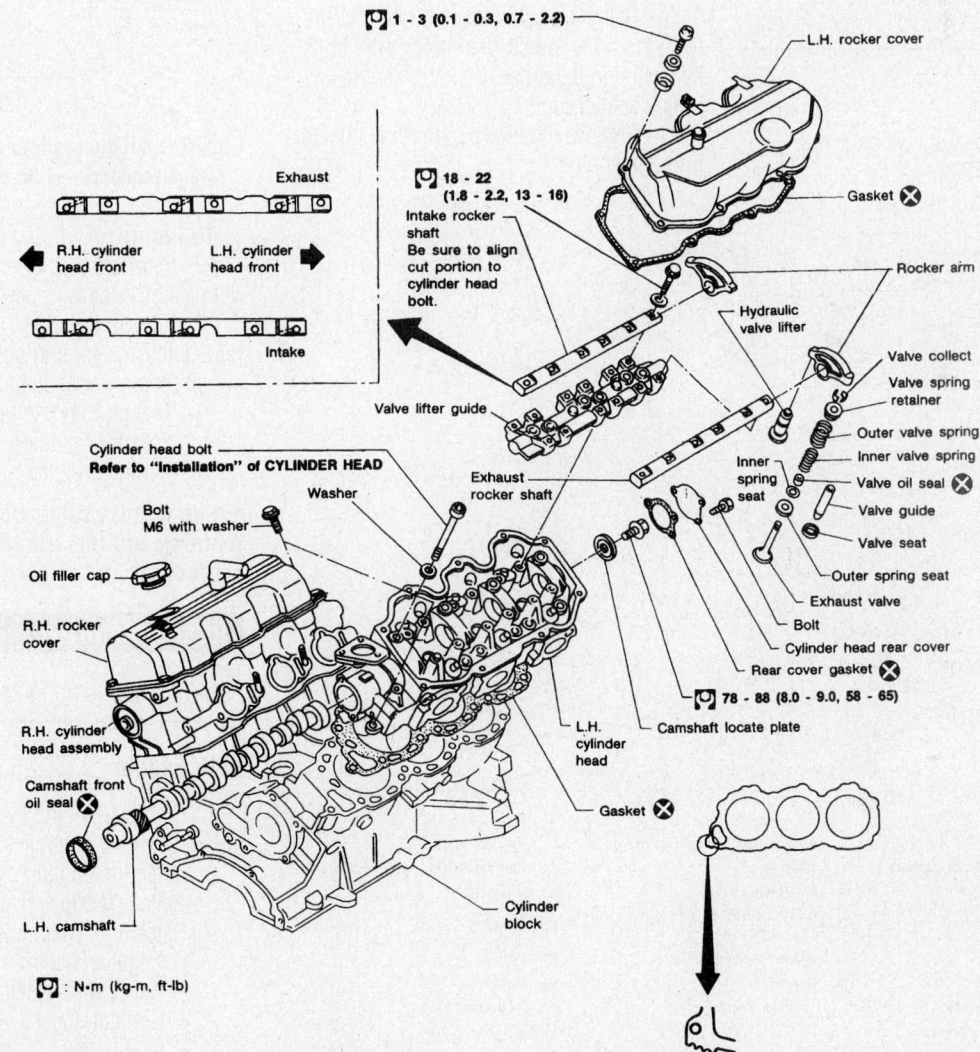

Rocker arms/shafts and related components—3.3L engine

- Negative battery cable
- Supercharger or upper intake manifold
- Valve covers
- Rocker arm and shaft assemblies
- Rocker arms from the shafts

➡️Keep all valve train components in order for assembly.

To install:

3. Lubricate all contact points with clean engine oil and assemble the rocker arms to the shafts in their original positions.

4. Install or connect the following:
- Rocker arm and shaft assemblies. Tighten the bolts to 13–16 ft. lbs. (18–22 Nm).
- Valve covers
- Upper intake manifold
- Negative battery cable

5. Start the engine and check for leaks.

Intake Manifold Collector

REMOVAL & INSTALLATION

4.0L Engine

1. Before servicing the vehicle, refer to the Precautions Section.

2. Disconnect the negative battery cable. Drain the cooling system.

3. Remove the engine cover. Remove the air cleaner case (upper) with the mass air flow sensor and air duct assembly.

4. Disconnect the water hoses from the electric throttle control actuator. Disconnect the harness connector.

5. Remove the electric throttle control actuator retaining bolts. Be sure to remove the bolts by reversing the order of the tightening torque sequence.

6. Remove the electric throttle control actuator.

7. Remove the brake booster vacuum hose and the PCV hose. Remove the intake manifold collector support.

8. Disconnect the EVAP hoses and harness connector from the EVAP canister purge volume control solenoid valve. Remove the EVAP canister purge volume control solenoid valve.

9. Remove the VIAS control solenoid valve and vacuum tank.

10. Remove the intake manifold collector retaining bolts. Be sure to remove the bolts by reversing the order of the tightening torque sequence.

11. Remove the intake manifold collector from the engine.

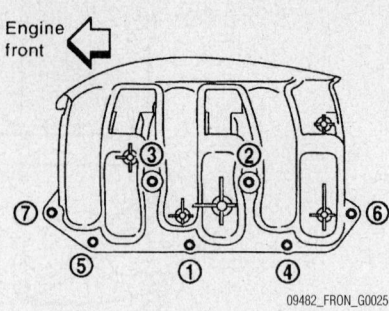

Intake manifold collector bolt torque sequence—4.0L engine

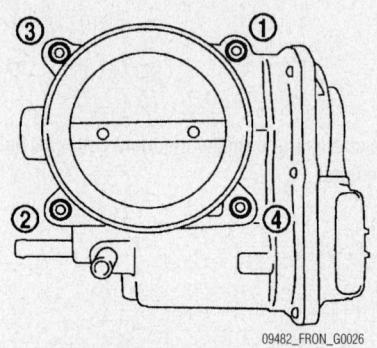

Electric throttle control actuator bolt torque sequence—4.0L engine

To install:

12. Installation is the reverse of the removal procedure.

13. Be sure to tighten the intake manifold collector retaining bolts to specification and in the proper sequence.

14. Be sure to tighten the electric throttle control actuator retaining bolts to specification and in the proper sequence.

➡️See throttle valve closed position learning and idle air volume learning procedures, for relearning information.

Intake Manifold

REMOVAL & INSTALLATION

2.4L Engine

1. Before servicing the vehicle, refer to the Precautions Section.

2. Drain the cooling system.

3. Relieve the fuel system pressure.

4. Remove or disconnect the following:
- Negative battery cable
- Air cleaner assembly
- Coolant hoses
- Fuel lines
- Accelerator cable

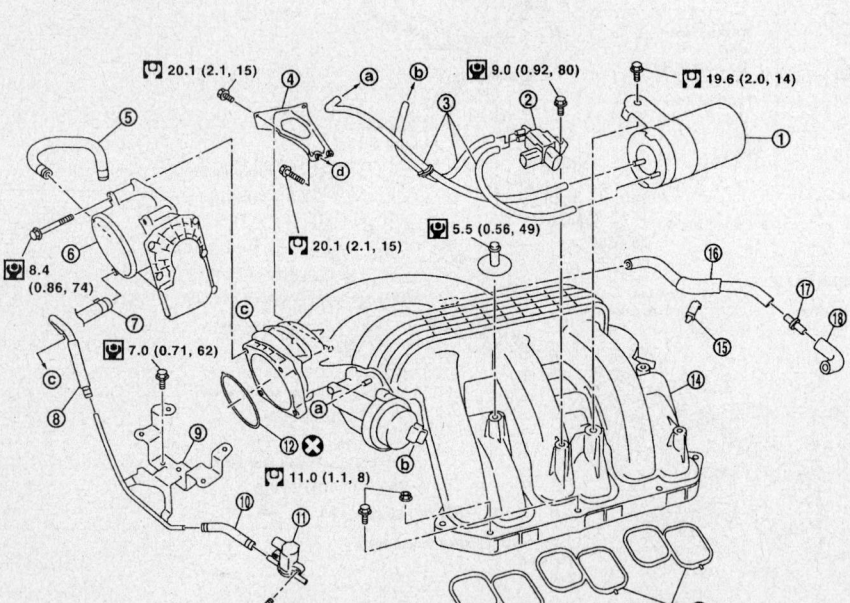

1. Vacuum tank	2. VIAS control solenoid valve	3. Vacuum hose
4. Intake manifold collector support	5. Water hose	6. Electric throttle control actuator
7. Water hose	8. EVAP hose	9. Bracket
10. EVAP hose	11. EVAP canister purge volume control solenoid valve	12. Gasket
13. Gasket	14. Intake manifold collector	15. Clip
16. PCV hose	17. Connector	18. PCV hose
a. To intake manifold collector	b. To power valve	c. To throttle body
d. To cylinder head (RH bank)		

Intake manifold collector and related components—4.0L engine

- Cruise control cable, if equipped
- Positive Crankcase Ventilation (PCV) valve and hose
- Exhaust Gas Recirculation (EGR) tube
- EGR temperature sensor connector
- Idle Air Control (IAC) valve and solenoid connectors
- Throttle Position (TP) sensor and switch connectors
- Engine Coolant Temperature (ECT) sensor connector
- Manifold Absolute Pressure (MAP) sensor connector and vacuum line
- Evaporative Emissions (EVAP) canister purge valve vacuum line
- Brake booster vacuum line
- Fuel injector connectors
- Intake manifold bracket
- Intake manifold. Loosen the fasteners in reverse of the torque sequence.

To install:

5. Install or connect the following:
   - Intake manifold. Tighten the bolts to 12–14 ft. lbs. (16–19 Nm).
   - Intake manifold bracket. Tighten the bolts to 24–28 ft. lbs. (32–38 Nm).
   - Fuel injector connectors
   - Brake booster vacuum line
   - EVAP canister purge valve vacuum line
   - MAP sensor connector and vacuum line
   - ECT sensor connector
   - TP sensor and switch connectors
   - IAC valve and solenoid connectors
   - EGR temperature sensor connector
   - EGR tube
   - PCV valve and hose
   - Cruise control cable, if equipped
   - Accelerator cable
   - Fuel lines
   - Coolant hoses
   - Air cleaner assembly

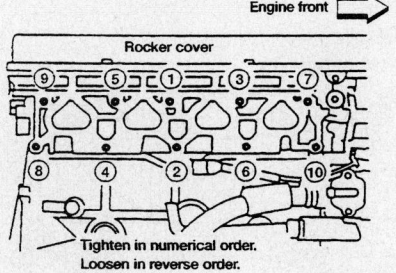

Intake manifold bolt torque sequence—2.4L engine

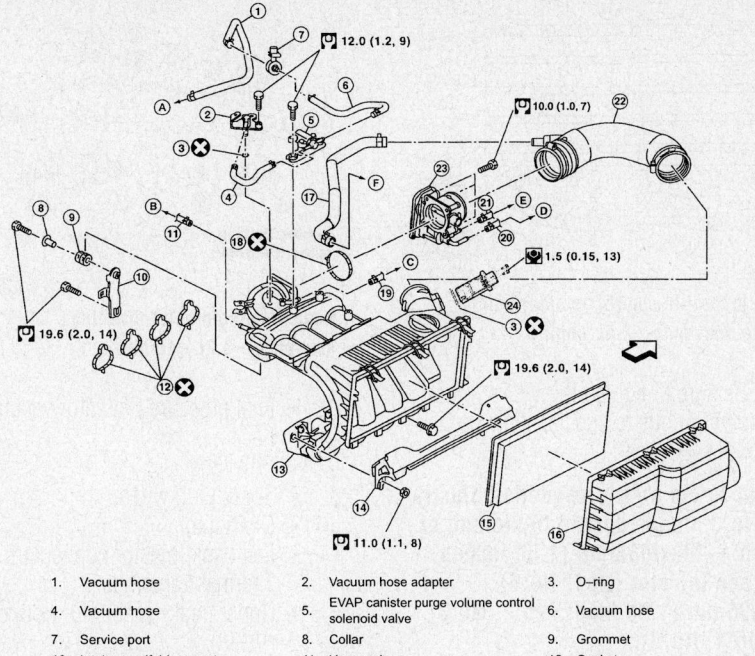

1. Vacuum hose	2. Vacuum hose adapter	3. O–ring
4. Vacuum hose	5. EVAP canister purge volume control solenoid valve	6. Vacuum hose
7. Service port	8. Collar	9. Grommet
10. Intake manifold support	11. Vacuum hose	12. Gasket
13. Intake manifold	14. Fuel tube protector	15. Air cleaner
16. Air cleaner case	17. PCV hose	18. Gasket
19. PCV hose	20. Water hose	21. Water hose
22. Air duct	23. Electric throttle control actuator	24. Mass air flow sensor
A. To vacuum pipe (EVAP canister)	B. To brake booster	C. To PCV valve
D. To heater outlet	E. To heater pipe	F. To rocker cover
→ Engine front		

Intake manifold and related components—2.5L Engine

- Negative battery cable
6. Fill the cooling system.
7. Start the engine and check for leaks.

2.5L Engine

1. Before servicing the vehicle, refer to the Precautions Section.
2. Properly relieve the fuel system pressure.
3. Disconnect the negative battery cable. Drain the cooling system.
4. Remove the air cleaner case, air cleaner and air duct.
5. Disconnect the water hoses from the electric throttle control actuator.
6. Remove the mass air flow sensor from the intake manifold. Remove the quick connector cap and disconnect the quick connector at the engine side.
7. Remove the electric throttle control actuator retaining bolts. Be sure to remove the bolts by reversing the order of the tightening torque sequence.
8. Remove the electric throttle control actuator and gasket.
9. Disconnect the harness, vacuum hoses and PCV hoses from the intake manifold and position them to the side.
10. Remove the intake manifold retaining bolts. Be sure to remove the bolts by

reversing the order of the tightening torque sequence.

11. Remove the intake manifold, fuel tube protector and gasket from the engine.

12. As necessary, remove the EVAP canister purge volume solenoid valve and vacuum hose adapter from the intake manifold.

13. Disconnect the sub frame harness from the fuel injectors. Remove the fuel tube and fuel injector assembly from the intake manifold, if required.

To install:

14. Installation is the reverse of the removal procedure.

15. Be sure to use new gaskets.

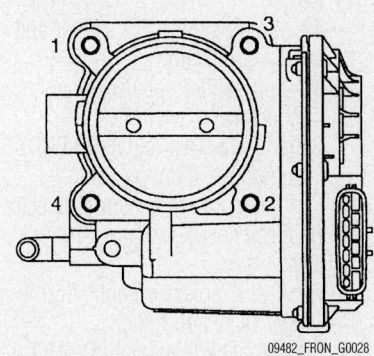

Intake manifold bolt torque sequence—2.5L engine

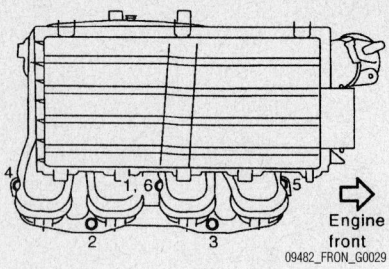

Electric throttle control actuator bolt torque sequence—2.5L engine

16. Be sure to tighten the intake manifold retaining bolts to specification and in the proper sequence.

➡️ **Refer to the torque sequence illustration, No.6 means double tightening of bolt No.1. M8xM38mm (1.50 inches) are green in color (No1, No.6). M8xM35mm (1.38 inch) (No.2, No.3). Nut (No.4, No.5).**

17. Be sure to tighten the electric throttle control actuator retaining bolts to specification and in the proper sequence.

➡️ **See throttle valve closed position learning and idle air volume learning procedures, for relearning information.**

18. Be sure to fill the cooling system with the proper grade and type engine coolant.

19. Start the engine and check for leaks.

3.3L Engine

1. Before servicing the vehicle, refer to the Precautions Section.
2. Drain the cooling system.
3. Relieve the fuel system pressure.
4. Remove or disconnect the following:
   - Negative battery cable
   - Air intake duct
   - Accelerator cable
   - Cruise control cable
   - Idle Air Control (IAC) valve connector
   - Throttle Position (TP) sensor and switch connectors
   - Ignition coil and power transistor connectors
   - Exhaust Gas Recirculation (EGR) Solenoid valve connector
   - EGR temperature sensor connector
   - Radiator hoses
   - Heater hoses
   - Positive Crankcase Ventilation (PCV) valve and hose
   - Evaporative Emissions (EVAP) canister vacuum and purge hoses
   - Brake booster vacuum hose

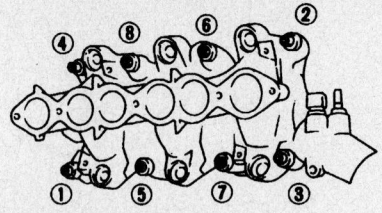

Loosen bolts in numerical order.

Intake manifold bolt loosening sequence—3.3L engine

   - Fuel pressure regulator vacuum hose
   - EGR tube
   - Spark plug wires
   - Distributor
   - Left bank injector connectors
   - Thermal transmitter
   - Upper intake manifold ground cable (VG33)
   - Supercharger (VG33ER)
   - Breather pipe
   - Upper intake manifold (VG33)
   - Intake manifold collector (VG33ER)
   - Fuel lines
   - Right bank injector connectors
   - Fuel supply manifold
   - Engine Coolant Temperature (ECT) sensor connector
   - Lower intake manifold. Loosen the fasteners in the sequence shown.

To install:

5. Install the lower intake manifold with a new gasket.
6. Tighten the fasteners in sequence as follows:
   a. Step 1: 35 inch lbs. (4 Nm)
   b. Step 2: 78 inch lbs. (9 Nm)
   c. Step 3: 70–84 inch lbs. (8–10 Nm)
7. Install or connect the following:
   - ECT sensor connector
   - Fuel supply manifold
   - Right bank injector connectors
   - Fuel lines
   - Upper intake manifold

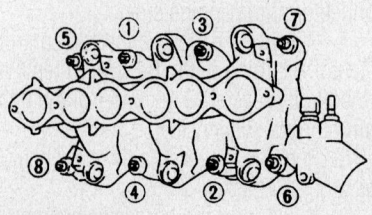

Tighten bolts in numerical order.

Intake manifold bolt torque sequence— 3.3L engine

   - Breather pipe
   - Upper intake manifold ground cable
   - Thermal transmitter
   - Left bank injector connectors
   - Distributor
   - Spark plug wires
   - EGR tube
   - Fuel pressure regulator vacuum hose
   - Brake booster vacuum hose
   - EVAP canister vacuum and purge hoses
   - PCV valve and hose
   - Heater hoses
   - Radiator hoses
   - EGR temperature sensor connector
   - EGR Solenoid valve connector
   - Ignition coil and power transistor connectors
   - TP sensor and switch connectors
   - IAC valve connector
   - Cruise control cable
   - Accelerator cable
   - Air intake duct
   - Negative battery cable

8. Fill the cooling system.
9. Start the engine and check for leaks.

4.0L Engine

1. Before servicing the vehicle, refer to the Precautions Section.
2. Properly relieve the fuel system pressure.
3. Disconnect the negative battery cable. Drain the cooling system.
4. Remove the intake manifold collector.
5. Remove the fuel tube and fuel injector assembly.
6. Remove the intake manifold retaining bolts. Be sure to remove the bolts by reversing the order of the tightening torque sequence.
7. Remove the intake manifold from the engine.

To install:

8. Installation is the reverse of the removal procedure.

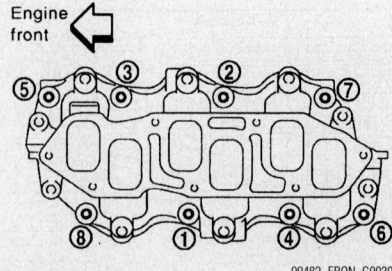

Intake manifold bolt torque sequence— 4.0L engine

9. Be sure to use new gaskets.

10. Be sure to tighten the intake manifold retaining bolts to specification and in the proper sequence in two or more steps.

Exhaust Manifold

REMOVAL & INSTALLATION

2.4L Engine

1. Before servicing the vehicle, refer to the Precautions Section.

2. Remove or disconnect the following:
- Negative battery cable
- Heated Oxygen (HO2S) sensor connector
- Exhaust manifold heat shield
- Exhaust Gas Recirculation (EGR) tube
- Exhaust front pipe
- Exhaust manifold. Loosen the nuts in reverse of the torque sequence.

To install:

3. Install or connect the following:
- Exhaust manifold. Tighten the nuts in sequence to 28–35 ft. lbs. (37–48 Nm).
- Exhaust front pipe. Tighten the fasteners to 32–37 ft. lbs. (43–50 Nm).
- EGR tube. Tighten the flange fittings to 29–36 ft. lbs. (39–49 Nm).
- Exhaust manifold heat shield. Tighten the bolts to 45–57 inch lbs. (5–7 Nm).
- HO2S sensor connector
- Negative battery cable

4. Start the engine and check for leaks.

2.5L Engine

1. Before servicing the vehicle, refer to the Precautions Section.

2. Properly relieve the fuel system pressure.

3. Disconnect the negative battery cable.

4. Remove the quick connector cap and disconnect the quick connector at the engine side.

5. Remove the air duct and PCV hose.

6. Remove the electric throttle control actuator retaining bolts. Be sure to remove the bolts by reversing the order of the tightening torque sequence.

7. Remove the electric throttle control actuator and gasket.

8. Disconnect the harness connector of the air fuel ratio sensor and the harness from the bracket and middle clamp. Remove the air fuel ratio sensor using tool J-44626, or equivalent.

➡**Be careful not to damage the air fuel ratio sensor. Discard the sensor if it has been dropped from a height of more than 19.7 inches on to a hard surface.**

9. Remove the front exhaust tube. Remove the exhaust manifold cover.

10. Remove the bracket between the exhaust manifold three way catalyst assembly and the transmission assembly.

11. Remove the exhaust manifold retaining bolts. Be sure to remove the bolts by reversing the order of the tightening torque sequence.

12. Remove the exhaust manifold from the engine. Discard the gasket.

To install:

13. Installation is the reverse of the removal procedure.

14. Be sure to use new gaskets.

15. Be sure to tighten the exhaust manifold retaining bolts to specification and in the proper sequence.

➡**Before installing a new air fuel sensor apply anti seize lubricant to the threads. Do not over torque the sensor, doing so may cause damage to the sensor resulting in the MIL light coming on.**

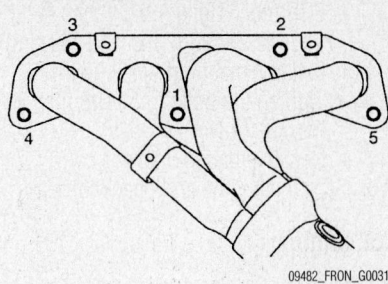

Exhaust manifold bolt torque sequence—2.5L engine

➡**See throttle valve closed position learning and idle air volume learning procedures, for relearning information.**

3.3L Engine

1. Before servicing the vehicle, refer to the Precautions Section.

2. Remove or disconnect the following:
- Negative battery cable
- Exhaust manifold heat shields
- Exhaust Gas Recirculation (EGR) tube
- Heated Oxygen (HO2S) sensor connectors
- Exhaust front pipes
- Exhaust manifolds with catalytic converters attached. Loosen the nuts in the reverse of the torque sequence.

To install:

3. Install or connect the following:
- Exhaust manifolds with catalytic converters attached. Tighten the nuts in sequence to 21–25 ft. lbs. (28–33 Nm).
- Exhaust front pipes. Tighten the bolts to 21–25 ft. lbs. (28–33 Nm).
- Heated Oxygen (HO2S) sensor connectors

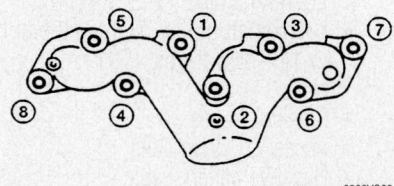

Exhaust manifold bolt torque sequence—2.4L engine

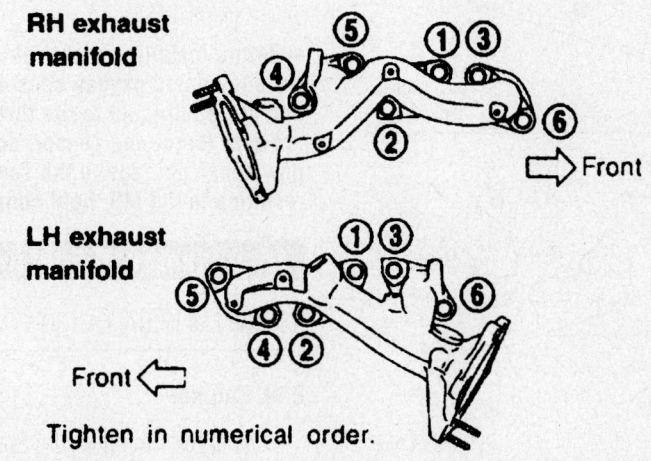

Exhaust manifold bolt torque sequence—3.3L engine

- EGR tube. Tighten the flange fittings to 29–36 ft. lbs. (39–49 Nm).
- Exhaust manifold heat shields. Tighten the bolts to 84–96 inch lbs. (9–11 Nm)
- Negative battery cable

4. Start the engine and check for leaks.

4.0L Engine

LEFT SIDE

1. Before servicing the vehicle, refer to the Precautions Section.
2. Disconnect the negative battery cable.
3. Remove the engine under cover.
4. Disconnect the harness connector and remove the heated oxygen sensor.

➡ **Be careful not to damage the air fuel ratio sensor. Discard the sensor if it has been dropped from a height of more than 19.7 inches on to a hard surface.**

5. Remove the center exhaust tube, main muffler and left front exhaust tube.
6. Remove the exhaust manifold cover.
7. Disconnect the harness connector of the air fuel ratio sensor and the harness from the bracket and middle clamp. Remove the air fuel ratio sensor using tool J-44626, or equivalent.

➡ **Be careful not to damage the air fuel ratio sensor. Discard the sensor if it has been dropped from a height of more than 19.7 inches on to a hard surface.**

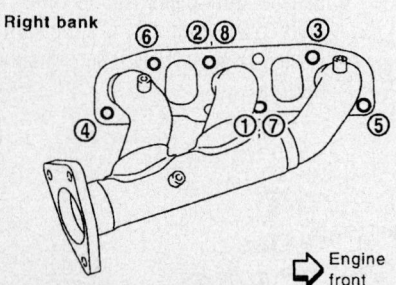

Right bank

Engine front ⇨

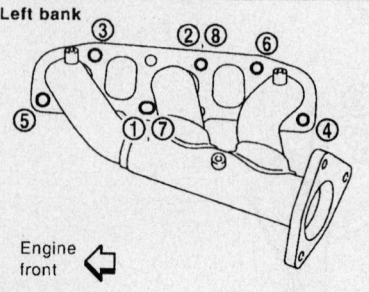

Left bank

Engine front ⇦

09482_FRON_G0032

Exhaust manifold bolt torque sequence— 4.0L engine

8. Remove the three way catalyst.
9. Remove the exhaust manifold retaining bolts. Be sure to remove the bolts by reversing the order of the tightening torque sequence.
10. Remove the exhaust manifold from the engine. Discard the gasket.

To install:

11. Installation is the reverse of the removal procedure.
12. Be sure to use new gaskets.
13. Be sure to tighten the exhaust manifold retaining bolts to specification and in the proper sequence.

➡ **Before installing a new air fuel sensor and heated oxygen sensor apply anti seize lubricant to the threads. Do not over torque the sensor, doing so may cause damage to the sensor resulting in the MIL light coming on.**

RIGHT SIDE

1. Before servicing the vehicle, refer to the Precautions Section.
2. Disconnect the negative battery cable.
3. Remove the engine from the vehicle. Position the assembly in a suitable holding fixture.
4. Remove the exhaust manifold retaining bolts. Be sure to remove the bolts by reversing the order of the tightening torque sequence.

➡ **Disregard the numerical order of No.7 and No.8 in the removal process.**

5. Discard the gaskets.

To install:

6. Installation is the reverse of the removal procedure.
7. Be sure to use new gaskets.
8. Be sure to tighten the exhaust manifold retaining bolts to specification and in the proper sequence.

➡ **Before installing a new air fuel sensor and heated oxygen sensor apply anti seize lubricant to the threads. Do not over torque the sensor, doing so may cause damage to the sensor resulting in the MIL light coming on.**

Front Crankshaft Seal

REMOVAL & INSTALLATION

2.4L Engine

Refer to the Timing Chain, Sprockets, Front Cover and Seal procedure in this section.

2.5L Engine

1. Before servicing the vehicle, refer to the Precautions Section.
2. Disconnect the negative battery cable.
3. Remove the engine undercover.
4. Remove the fan shroud. Remove the cooling fan.
5. Remove the drive belts.
6. Hold the crankshaft pulley with a suitable tool. Loosen the crankshaft pulley retaining bolt.
7. Pull the pulley out about 0.39 inch. Remove the crankshaft pulley bolt.
8. Attach a pulley puller in the M6 thread hole on the crankshaft pulley. Remove the crankshaft pulley.
9. Using a seal removal tool, remove the oil seal from its mounting.

➡ **Be careful not to damage the front cover and/or the crankshaft.**

To install:

10. Installation is the reverse order of the removal procedure.
11. Press fit the seal until it is flush with the front end surface of the front cover, using the proper tools.

3.3L Engine

1. Before servicing the vehicle, refer to the Precautions Section.
2. Drain the cooling system.
3. Remove or disconnect the following:

- Negative battery cable
- Accessory drive belts
- Radiator hoses
- Crankshaft pulley
- Front cover
- Timing belt. Refer to the Timing Belt unit repair section.
- Crankshaft timing sprocket
- Front crankshaft seal

To install:

4. Install or connect the following:

- Front crankshaft seal flush with the oil pump housing
- Crankshaft timing sprocket
- Timing belt
- Front cover. Tighten the bolts to 26–43 inch lbs. (3–5 Nm).
- Crankshaft pulley. Tighten the bolt to 141–156 ft. lbs. (191–211 Nm).
- Radiator hoses
- Accessory drive belts
- Negative battery cable

5. Fill the cooling system.
6. Start the engine and check for leaks.

4.0L Engine

1. Before servicing the vehicle, refer to the Precautions Section.
2. Disconnect the negative battery cable.
3. Remove the engine undercover.
4. Remove the cooling fan.
5. Remove the drive belts.
6. Remove the starter. Position tool KV10117700 (J-44716) or equivalent.
7. Loosen the crankshaft pulley retaining bolt and locate the bolt seating surface, which is about 0.39 inch from its original position.

➡**Do not remove the crankshaft pulley bolt. Keep the loosened pulley bolt in place to protect the removed crankshaft pulley from dropping.**

8. Pull the pulley with both hands and remove it from its mounting. Remove the bolt and pulley from the engine.
9. Using a seal removal tool, remove the oil seal from its mounting.

➡**Be careful not to damage the front cover and/or the crankshaft.**

To install:

10. Installation is the reverse order of the removal procedure.

11. Press fit until the height of the front oil seal is level with the mounting surface, using the proper tools.

Camshaft and Valve Lifters

REMOVAL & INSTALLATION

2.4L Engine

1. Before servicing the vehicle, refer to the Precautions Section.
2. Remove or disconnect the following:
   - Negative battery cable
   - Air cleaner assembly

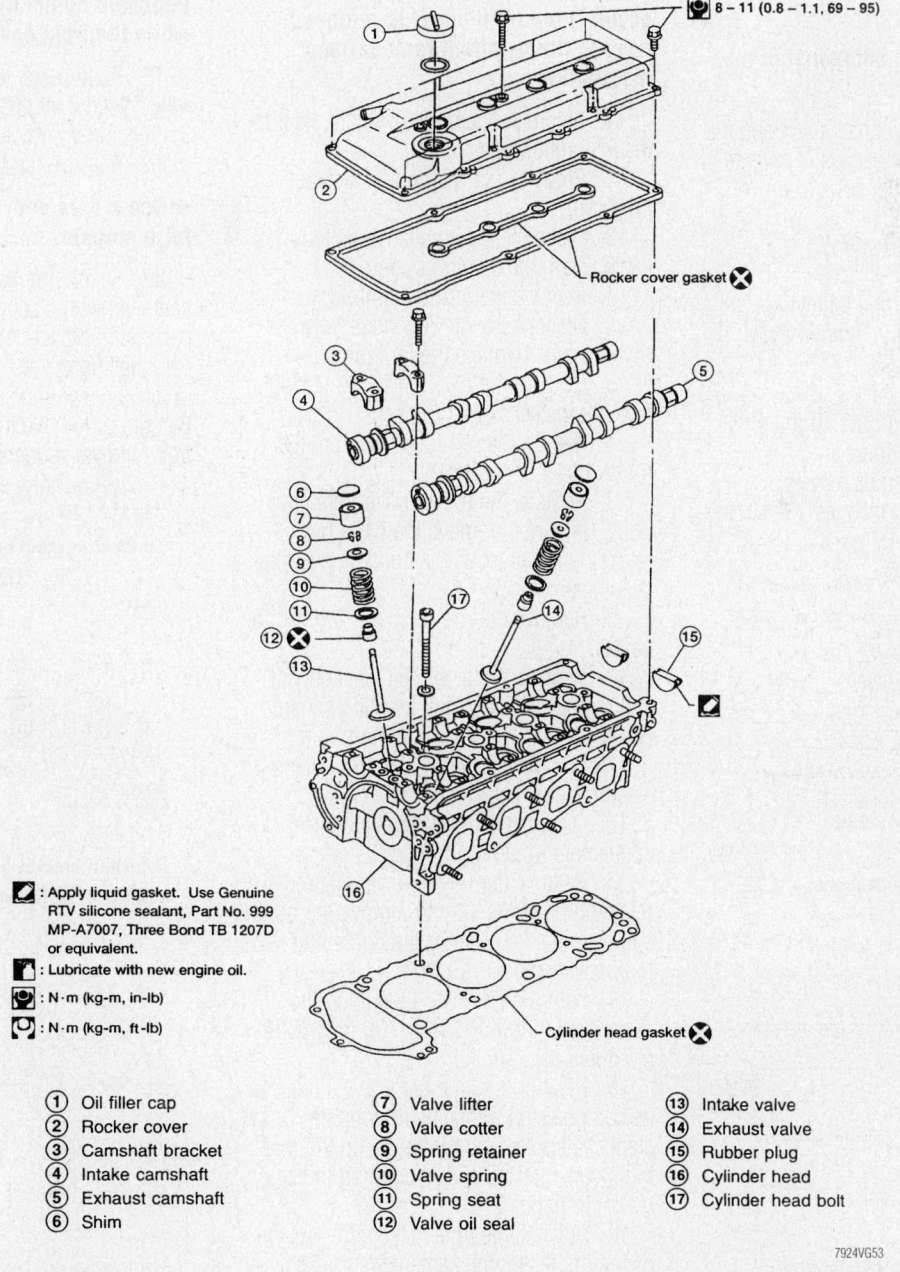

8 – 11 (0.8 – 1.1, 69 – 95)

Rocker cover gasket ⊗

Cylinder head gasket ⊗

🔧 : Apply liquid gasket. Use Genuine RTV silicone sealant, Part No. 999 MP-A7007, Three Bond TB 1207D or equivalent.
🔧 : Lubricate with new engine oil.
⊙ : N·m (kg-m, in-lb)
⊙ : N·m (kg-m, ft-lb)

① Oil filler cap	⑦ Valve lifter	⑬ Intake valve
② Rocker cover	⑧ Valve cotter	⑭ Exhaust valve
③ Camshaft bracket	⑨ Spring retainer	⑮ Rubber plug
④ Intake camshaft	⑩ Valve spring	⑯ Cylinder head
⑤ Exhaust camshaft	⑪ Spring seat	⑰ Cylinder head bolt
⑥ Shim	⑫ Valve oil seal	

7924VG53

Camshafts and related components—2.4L engine

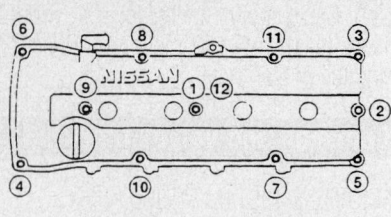

Loosen in numerical order.

9308VG06

Valve cover bolt loosening sequence—2.4L engine

- Spark plug wires
- Valve cover. Remove the bolts in the sequence shown.
- Camshaft sprocket cover
- Camshaft sprockets and upper timing chain

➡ Keep all valvetrain components in order for assembly.

- Camshaft bearing caps. Loosen the bolts in several passes in reverse of the torque sequence.
- Camshafts
- Valve lifters and shims

To install:

3. Install or connect the following:
- Valve lifters and shims in their original positions
- Camshafts

4. Install the bearing caps. Tighten the bolts in sequence as follows:
   a. Step 1: 17 inch lbs. (2 Nm)
   b. Step 2: 80–104 inch lbs. (9–12 Nm)

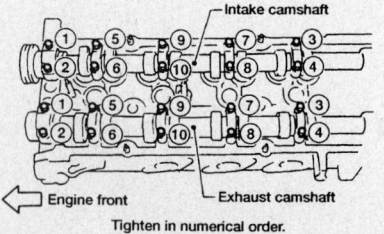

Tighten in numerical order.
Loosen in reverse order.

7924VG51

Bearing cap bolt torque sequence—2.4L engine

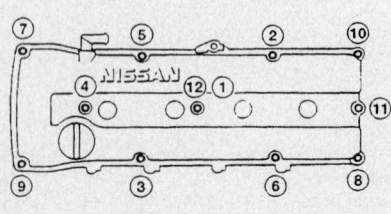

Tighten in numerical order.

9308VG07

Valve cover bolt torque sequence—2.4L engine

5. Install or connect the following:
- Camshaft sprockets and upper timing chain. Tighten the sprocket bolts to 123–130 ft. lbs. (167–177 Nm).
- Camshaft sprocket cover
- Valve cover. Tighten the bolts in sequence to 69–95 inch lbs. (8–11 Nm).
- Spark plug wires
- Air cleaner assembly
- Negative battery cable

2.5L Engine

➡ The procedure below describes removal and installation of the camshaft without removing the front cover. If the front cover is removed refer to timing chain removal and installation.

1. Before servicing the vehicle, refer to the Precautions Section.
2. Properly relieve the fuel system pressure.
3. Disconnect the negative battery cable. Drain the cooling system.
4. Remove the intake manifold.
5. Disconnect the PCV hose from the rocker cover. Remove the ignition coil.
6. Remove the PCV valve and O-ring from the rocker cover, if necessary.
7. Remove the oil filler cap from the rocker cover, if necessary.
8. Remove the rocker cover retaining bolts. Be sure to remove the bolts by reversing the order of the tightening torque sequence.
9. Remove the rocker cover. Discard the gasket.
10. Remove the drive belt. Disconnect and remove the camshaft position sensor (PHASE).
11. Disconnect the IVT control solenoid electrical connector.
12. Disconnect the ground electrical connectors from the front cover.
13. Remove the IVT control solenoid retaining bolts. Be sure to remove the bolts by reversing the order of the tightening torque sequence.
14. Remove the cover by cutting the sealant using tool KV10111100 (J-37228) or equivalent.
15. Position the number one cylinder on its compression stroke by rotating the crankshaft pulley clockwise. Align the mating marks for TDC with the timing indicator on the front cover.
16. At the same time make sure that the mating marks on the camshaft sprockets are lined up with the yellow links in the timing

chain. If not rotate the crankshaft one more turn to line up the mating marks to the yellow links.
17. Pull the timing chain guide out between the camshaft sprockets through the front cover.
18. Line up the mating marks on the camshaft sprockets with the yellow links in the timing chain and paint an indelible mating mark on the sprocket and timing chain link plate.

➡ Do not rotate the crankshaft or the camshaft while the timing chain is removed.

➡ Chain tension holding work is not necessary. Crankshaft sprocket and timing chain do not disconnect structurally while the front cover is attached.

19. Push in the tensioner plunger and hold. Insert a stopper pin into the hole on the tensioner body to hold the chain tensioner. Remove the timing chain tensioner.

➡ Use a wire with 0.02 inch diameter for a stopper pin.

20. Secure the hexagonal part of the camshaft with a suitable tool. Loosen the camshaft sprocket bolts and remove the camshaft sprockets.
21. Loosen the camshaft bracket bolts. Be sure to remove the bolts by following the bolt removal sequence.

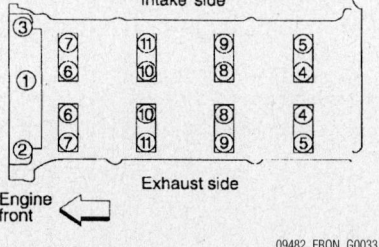

Camshaft bracket bolts loosening sequence

09482_FRON_G0033

Camshaft bracket bolt loosening sequence—2.5L engine

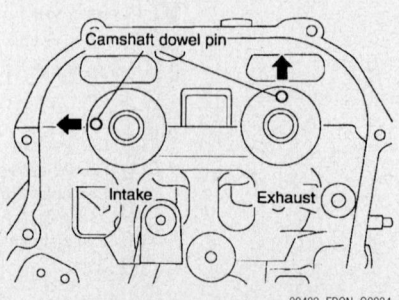

09482_FRON_G0034

Camshaft positioning—2.5L engine

22. Remove the camshafts and brackets from the engine.

23. Remove the number one camshaft bracket by tapping lightly with a rubber mallet. Note the positions for installation.

24. As necessary, remove the valve lifters. Be sure to keep them in the proper order for installation.

To install:

25. Inspect the camshafts, replace as required.

26. Install the camshafts so that the camshaft dowel pins on the front side are positioned as indicated in the illustration.

27. Remove any foreign material from the camshaft bracket backside and from the cylinder head face.

28. Install the camshaft brackets (No.2–No.5) aligning the identification marks on the upper surface as indicated in the illustration.

➡**Install so that the identification mark can be correctly read when viewed from the exhaust side.**

29. To install camshaft bracket NO.1, apply liquid gasket to the bracket.

➡**After installation be sure to wipe excessive gasket material from part "A", as indicated in the illustration. Be sure to use genuine RTV silicone sealant or equivalent.**

30. Apply liquid gasket to camshaft bracket No.1 contact surface on the front cover back-

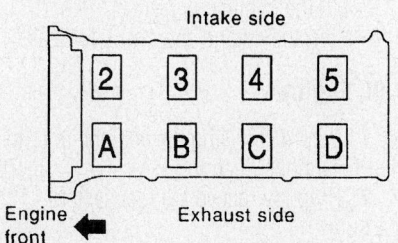

Camshaft bracket identification—2.5L engine

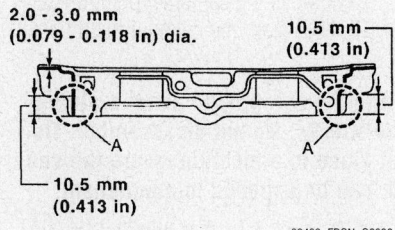

Camshaft bracket No.1 sealant application point "A"—2.5L engine

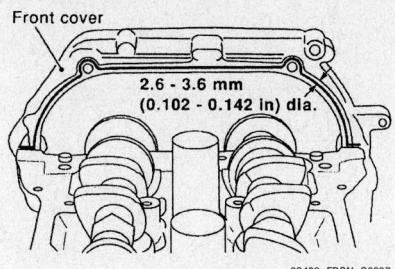

Camshaft bracket No.1 sealant application outside bolt hole—2.5L engine

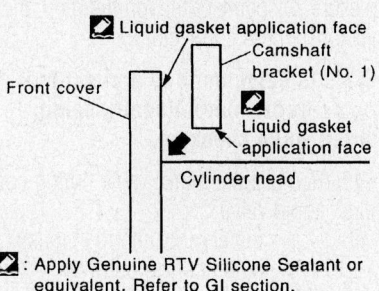

Camshaft bracket No.1 sealant application locating points—2.5L engine

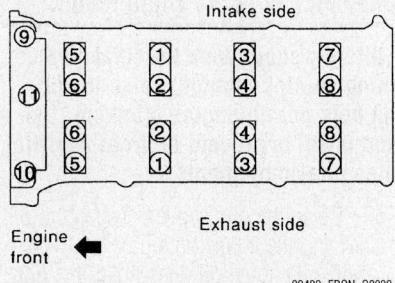

Camshaft bracket bolt torque sequence—2.5L engine

side. Apply liquid gasket to the outside bolt hole on the front cover. Be sure to use genuine RTV silicone sealant or equivalent.

31. Locate camshaft bracket No.1 near installation position and install it without disturbing the liquid gasket applied to the surfaces. Be sure to use genuine RTV silicone sealant or equivalent.

32. Tighten the camshaft bracket bolts in the proper sequence and to specification.

Bolts 9–11 to 17 inch lbs.
Bolts 1–8 to 17 inch lbs.
Bolts 1–11 to 52 inch lbs.
Bolts 1–11 to 92 inch lbs.

➡**After tightening the bolts be sure to wipe off any excessive liquid gasket. Be sure to use genuine RTV silicone sealant or equivalent.**

33. Install the camshaft position sensor. Install the camshaft sprockets.

➡**Install them by aligning the mating marks on each camshaft sprocket with the paint marks on the timing chain link plates, which were made during removal.**

✳✳ CAUTION

Aligned mating marks could slip. Therefore, after matching them, hold the timing chain in place by hand. Before and after installing the chain tensioner, make sure again that the mating marks have not slipped.

34. Install the chain tensioner. After installation, pull the stopper pin off completely, and make sure that the chain tensioner plunger is released.

➡**Before installation of the chain tensioner, it is possible to rematch the marks on the timing chain with new ones on each sprocket.**

35. Install the chain guide. Install oil rings to the camshaft sprocket (INT) insertion points on backside of intake valve timing control cover. Install the O-ring to the front cover.

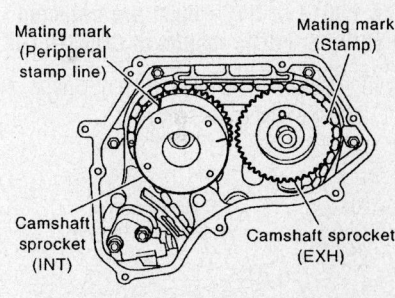

Camshaft alignment mating marks—2.5L engine

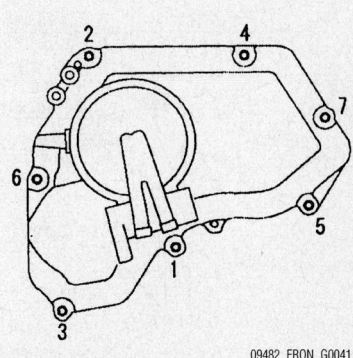

Intake valve timing control valve cover bolt torque sequence—2.5L engine

36. Apply a 0.083–0.122 inch diameter bead of liquid gasket to the intake valve timing control cover. Be sure to use genuine RTV silicone sealant or equivalent.

37. Install the cover. Tighten the bolts in the proper sequence. Connect the ground cables and install the harness clip.

38. Check and adjust the valve clearance, as required.

→**If hydraulic pressure inside the timing chain tensioner drops after removal/ installation, slack in the guide may generate a pounding noise during and just after engine start. This is normal the noise will stop after hydraulic pressure rises.**

39. Continue the installation in the reverse order of the removal procedure.

40. Apply liquid gasket, be sure to use genuine RTV silicone sealant or equivalent, to the positions shown in the illustration. Refer to figure "a" to apply liquid gasket to joint part of camshaft bracket No.1 and cylinder head. Refer to figure "b" to apply liquid gasket in 90 degrees to figure "b".

41. Install the rocker cover. Torque the retaining bolts to 18 inch lbs and than to 73 inch lbs, in the proper sequence.

42. Inspect the camshaft sprocket (INT) oil groove.

→**Perform this inspection only when DTC P0011 or DTC P0021 are detected in self diagnostic results of CONSULT-II.**

43. Be sure the engine is cold. Check and adjust oil level, as required.

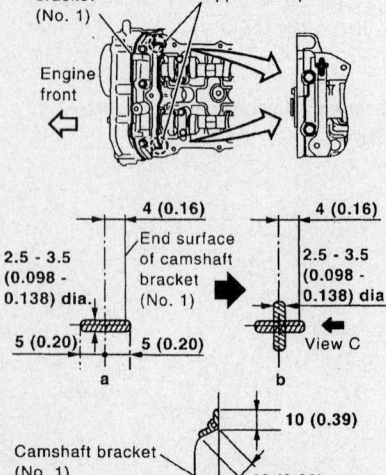

Rocker cover sealant application locating points—2.5L engine

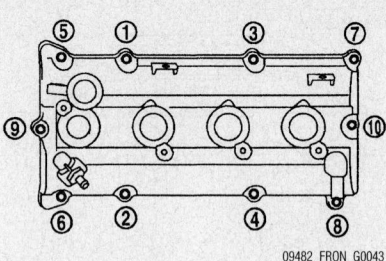

09482_FRON_G0043

Rocker cover bolt torque sequence—2.5L engine

44. Properly release the fuel system pressure. Disconnect the ignition coil and injector harness connectors.

→**This is being done to prevent the engine from unintentionally being started while checking.**

45. Remove the intake valve timing control solenoid valve.

46. Crank the engine, and then make sure that engine oil comes out from camshaft bracket (No.1) oil hole.

✳✳ WARNING

Be careful not to touch rotating parts, (drive belt, idler pulley, crankshaft pulley etc) as injury could result.

→**Oil may squirt from the intake valve timing control solenoid valve installation hole during engine cranking. Use a shop towel to prevent oil from squirting on engine components.**

47. Clean the oil groove between the oil strainer and the intake timing control solenoid valve if engine oil does not come out from camshaft bracket (No.1) oil hole.

48. Remove the components between the intake valve timing control solenoid valve and the camshaft sprocket (INT). Check each oil groove for clogging.

49. After inspection install any removed components.

3.3L Engine

1. Before servicing the vehicle, refer to the Precautions Section.

2. Drain the cooling system.

3. Remove or disconnect the following:
   - Negative battery cable
   - Upper intake manifold
   - Valve covers

→**Keep all valvetrain components in order for assembly.**

   - Rocker arm and shaft assemblies
   - Valve lifter guide and valve lifters. Attach a wire to the top of the lifters

so that they will not drop from the lifter guide.
   - Radiator
   - Accessory drive belts
   - Front cover
   - Timing belt. Refer to the Timing Belt unit repair section.
   - Camshaft sprockets
   - Camshaft seals
   - Rear timing cover
   - Distributor
   - Cylinder head rear covers
   - Camshaft locating plates
   - Camshafts

To install:

4. Install or connect the following:
   - Camshafts
   - Camshaft locating plates. Tighten the bolts to 58–65 ft. lbs. (78–88 Nm).
   - Cylinder head rear covers
   - Distributor
   - Rear timing cover
   - Camshaft seals
   - Camshaft sprockets. Tighten the bolts to 58–65 ft. lbs. (78–88 Nm).
   - Timing belt
   - Front cover
   - Accessory drive belts
   - Radiator
   - Valve lifter guide and valve lifters
   - Rocker arm and shaft assemblies. Tighten the bolts to 13–16 ft. lbs. (18–22 Nm).
   - Valve covers
   - Upper intake manifold
   - Negative battery cable

5. Fill the cooling system.

6. Start the engine and check for leaks.

4.0L Engine

1. Before servicing the vehicle, refer to the Precautions Section.

2. Properly relieve the fuel system pressure.

3. Disconnect the negative battery cable. Remove the engine cover.

4. Remove the front timing chain case, camshaft sprocket, timing chain and rear timing chain case.

5. Remove the camshaft position sensor (PHASE) from the cylinder head back side.

→**Handle carefully to avoid dropping and shocks. Do not disassemble. Do not place in a location where the sensor can be exposed to magnetism.**

6. Remove the intake manifold collector.

7. Separate the engine harness and

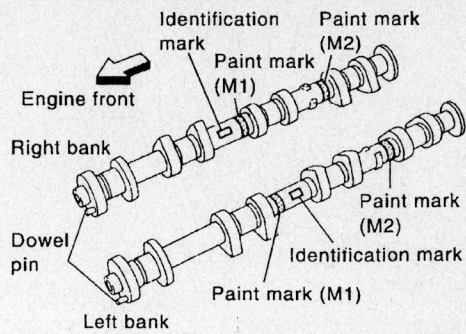

Bank	INT/EXH	Dowel pin	Paint marks		Identification mark
			M1	M2	
RH	INT	No	Green	No	RE
	EXH	Yes	No	White	RE
LH	INT	No	Green	No	LH
	EXH	Yes	No	White	LH

09482_FRON_G0044

Camshaft identification—4.0L engine

remove their brackets from the rocker covers. Remove the harness bracket from the cylinder head, if necessary.

8. Remove the ignition coil. Remove the PCV hoses. Remove the oil filler cap, if necessary.

9. Loosen the rocker cover retaining bolts, in the reverse order of the tightening sequence.

10. Remove the rocker covers from the engine.

11. Remove the intake valve timing control solenoid valves. Discard the gaskets.

12. Mark the camshaft brackets and bolts for reinstallation. Remove the camshaft bracket bolts. Be sure to remove the bolts by reversing the order of the tightening torque sequence and in several steps.

13. Remove the camshafts.

14. If required, remove the valve lifters. Identify them for reinstallation in their original locations.

15. Remove the timing chain tensioner (secondary) from the cylinder head. Remove the timing chain tensioner (secondary) with its stopper pin attached.

To install:

16. Inspect the camshafts, replace as required.

17. Install the timing chain tensioners (secondary) on both sides of the cylinder head. Be sure to use new O-rings.

➡**Install the tensioner with its stopper pin attached. Install the tensioner with the sliding part facing downward on the right cylinder head and with the sliding part facing upward on the left cylinder head.**

18. Install the valve lifters, in their original bores.

19. Install the camshafts, with the dowel pin attached to its front end face on the exhaust side.

➡**Follow the identification marks for proper placement and direction.**

20. Install the camshaft so that the dowel pin hole and dowel pin on the front end face are positioned as shown in the illustration (No.1 piston at TDC on its compression stroke).

➡**Large and small pin holes are located on the front end face of the camshaft (INT), at intervals of 180 degrees. Face small diameter side pin hole upward (in cylinder head upper face direction).**

➡**Though the camshaft does not stop at the portion as shown, for placement of the cam nose, it is generally accepted that the camshaft is placed for the same direction as shown.**

21. Install the camshaft brackets in the same position that they were removed.

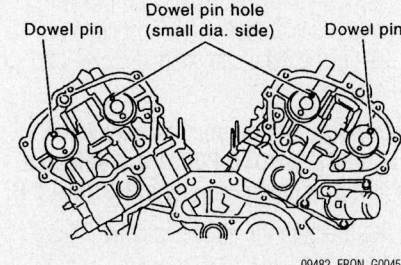

Camshaft dowel pin positioning—4.0L engine

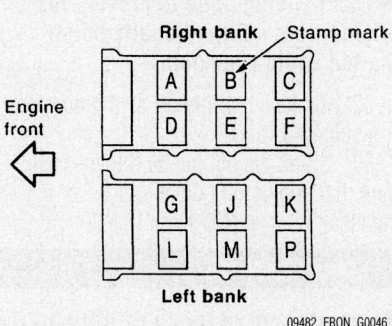

Camshaft bearing cap identification—4.0L engine

Install brackets No.2–No.4 aligning the stamp marks as indicated in the illustration.

➡**There are no identification marks indicating left or right for camshaft bracket No.1.**

Camshaft bracket (No. 1)

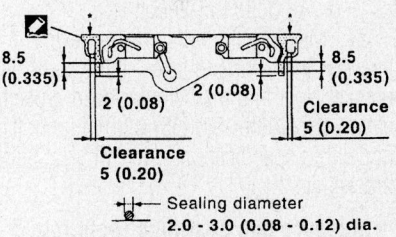

* : Remove the protruding liquid gasket from front face. (Remove the hardened liquid gasket from surface only.)

▨ : Apply Genuine RTV Silicone Sealant or equivalent. Refer to GI section.

Unit: mm (in)

09482_FRON_G0047

Camshaft sealant application and location—4.0L engine

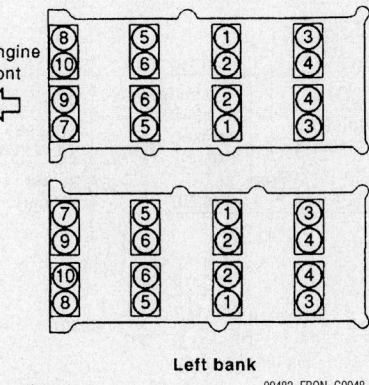

Camshaft bearing bracket bolt torque sequence—4.0L engine

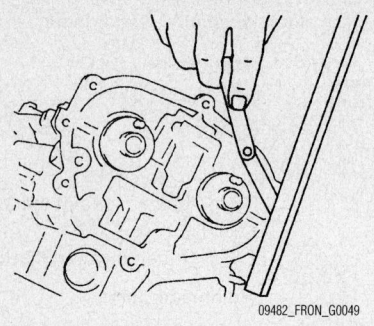

Camshaft bracket and cylinder head measurement—4.0L engine

22. Apply liquid gasket to the mating surfaces of camshaft bracket No.1 as shown in the illustration on both the left and right cylinder heads. Be sure to use genuine RTV sealant, or equivalent.

23. Tighten the camshaft bracket bolts in the proper sequence and to specification.

   a. Bolts 7–10 to 17 inch lbs.

   b. Bolts 1–6 to 17 inch lbs.

   c. All bolts to 52 inch lbs.

   d. All bolts to 92 inch lbs.

24. Measure the difference in levels between the front end faces of the camshaft bracket No.1 and the cylinder head. Specification should be -0.0055–0.0055 inch. If not within specification, reinstall camshaft bracket No.1.

➡**Measure two positions (both intake and exhaust side) for a single bank.**

25. Check and adjust valve clearance, as required.

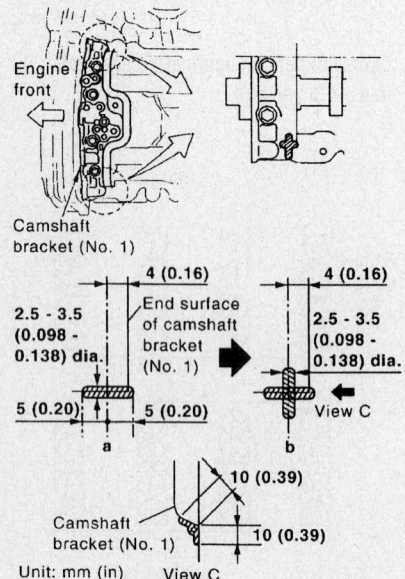

Rocker cover sealant application locating points—4.0L engine

Right bank

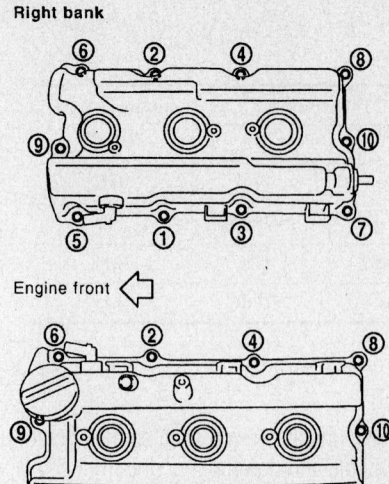

Engine front ⇦

Left bank

Rocker cover bolt torque sequence—4.0L engine

26. Apply liquid gasket, be sure to use genuine RTV silicone sealant or equivalent, to the positions shown in the illustration. Refer to figure "a" to apply liquid gasket to joint part of camshaft bracket No.1 and cylinder head. Refer to figure "b" to apply liquid gasket to the figure "a" squarely.

27. Install the rocker cover. Torque the retaining bolts to 17 inch lbs and than to 74 inch lbs, in the proper sequence.

28. Continue the installation in the reverse order of the removal procedure.

29. Inspect the camshaft sprocket (INT) oil groove.

➡**Perform this inspection only when DTC P0011 or DTC P0021 are detected in self diagnostic results of CONSULT-II.**

30. Be sure the engine is cold. Check and adjust oil level, as required.

31. Properly release the fuel system pressure. Disconnect the ignition coil and injector harness connectors.

➡**This is being done to prevent the engine from unintentionally being started while checking.**

32. Remove the intake valve timing control solenoid valve.

33. Crank the engine, and then make sure that engine oil comes out from camshaft bracket (No.1) oil hole.

✳✳ WARNING

Be careful not to touch rotating parts, (drive belt, idler pulley, crankshaft pulley etc) as injury could result.

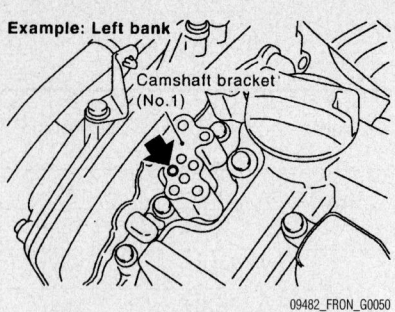

Camshaft bracket (No1) oil hole location

➡**Oil may squirt from the intake valve timing control solenoid valve installation hole during engine cranking. Use a shop towel to prevent oil from squirting on engine components.**

34. Clean the oil groove between the oil strainer and the intake timing control solenoid valve if engine oil does not come out from camshaft bracket (No.1) oil hole.

35. Remove the components between the intake valve timing control solenoid valve and the camshaft sprocket (INT). Check each oil groove for clogging.

36. After inspection install any removed components.

INSPECTION

1. Before servicing the vehicle, refer to the Precautions Section.

2. Remove the camshaft from the engine.

3. Check the camshaft bearing journals for damage and binding.

4. If the journals are binding, check the cylinder head for damage.

5. Check the cylinder head for clogged oil holes.

6. Check the camshaft surface for abnormal wear and damage. Replace the camshaft, as required.

7. Measure the camshaft lobe surface and replace the camshaft if not within specification.

8. Measure the camshaft journal diameter and replace the camshaft if not within specification.

9. Measure the camshaft run out and replace the camshaft if not within specification.

Valve Lash

ADJUSTMENT

2.4L Engine

➡**Measure valve clearance with the engine warm.**

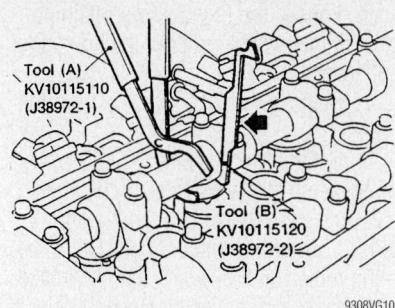

Valve adjustment tools (A) and (B)—2.4L engine

1. Before servicing the vehicle, refer to the Precautions Section.
2. Remove the valve cover.
3. Set the engine to the top of the compression stroke with the valves closed for the cylinder to be measured.
4. Check the valve clearance. The valve clearance specifications are as follows:
- Intake: 0.012–0.015 in. (0.31–0.39mm)
- Exhaust: 0.013–0.016 in. (0.33–0.41mm)

5. If adjustment is necessary, compress the valve spring with Tool **A** and insert Tool **B** to hold the valve in the open position as shown.
6. Replace the shims as necessary to achieve the correct valve clearance.
7. Repeat for each valve to be adjusted.

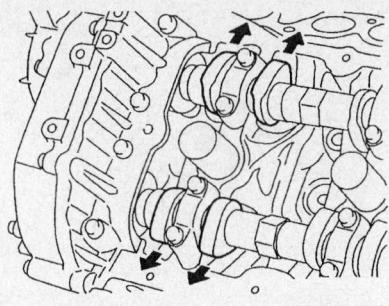

No.1 cylinder at TDC (compression stroke)—2.5L engine

2.5L Engine

1. Before servicing the vehicle, refer to the Precautions Section.
2. Disconnect the negative battery cable. Drain the cooling system.
3. Remove the intake manifold.
4. Disconnect the PCV hose from the rocker cover. Remove the ignition coil.
5. Remove the PCV valve and O-ring from the rocker cover, if necessary.
6. Remove the oil filler cap from the rocker cover, if necessary.
7. Remove the rocker cover retaining bolts. Be sure to remove the bolts by reversing the order of the tightening torque sequence.
8. Remove the rocker cover. Discard the gasket.

9. Remove the undercover. Remove the lower radiator shroud.
10. Set the No.1 cylinder at TDC of its compression stroke by rotating the crankshaft pulley clockwise to align the TDC mark to the timing indicator on the front cover. At the same time make sure that both the intake and exhaust cam noses of the NO.1 cylinder face outward, as indicated by the arrows in the illustration. If not, rotate the crankshaft in the clockwise direction 360 degrees.
11. Use a feeler gauge and measure the clearance between the valve lifter and the camshaft.
12. With the No.1 piston at TDC, refer to the illustration and measure the valve clearances at the locations marked with an "X". The "X" locations are indicated in the illustration with an arrow.
13. Rotate the crankshaft pulley clockwise 360 degrees and align the TDC mark to the timing indicator on the front cover.
14. With the No.4 piston at TDC, refer to the illustration and measure the valve clearances at the locations marked with an "X". The "X" locations are indicated in the illustration with an arrow.
15. If measurements are not within specification, proceed to the next step.
16. Remove the camshaft. Remove the valve lifters that are not within specification.
17. Measure the center thickness of the removed lifters, using a micrometer.

No. 1 cylinder compression TDC

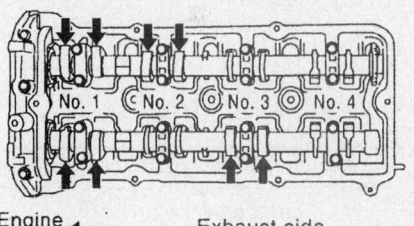

Valve adjustment measurement No.1 cylinder at TDC (compression stroke)—2.5L engine

Measuring position		No. 1 CYL.	No. 2 CYL.	No. 3 CYL.	No. 4 CYL.
No. 1 cylinder at compression TDC	INT	×	×		
	EXH	×		×	

No. 4 cylinder compression TDC

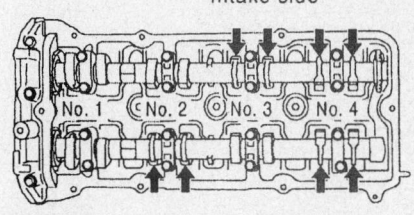

Valve adjustment measurement No.4 cylinder at TDC (compression stroke)—2.5L engine

Measuring position		No. 1 CYL.	No. 2 CYL.	No. 3 CYL.	No. 4 CYL.
No. 4 cylinder at compression TDC	INT			×	×
	EXH		×		×

18. Use the equation (t=t1+(C1-C2) to calculate valve lifter thickness for replacement.

➡ t= valve lifter thickness to be replaced. t1= removed valve lifter thickness. C1= measured valve clearance. C2= standard valve clearance.

19. Thickness of the new valve lifter can be identified by the stamp mark on the reverse side (inside the cylinder). The stamp mark "696" indicates 6.96 mm (0.2740 inch) thickness.

➡ Available thickness of a valve lifter ranges from 6.96–7.46 mm (0.2740–0.2937 inch) in steps of 0.02 mm (0.0008 inch). There are 26 different sizes.

20. Install the selected valve lifters.
21. Install the camshaft.
22. Manually rotate the crankshaft pulley in the clockwise direction a few rotations.
23. Check the valve clearance and be sure it is within specification.
24. When installing the rocker cover, apply liquid gasket, be sure to use genuine RTV silicone sealant or equivalent, to the positions shown in the illustration. Refer to figure "a" to apply liquid gasket to joint part of camshaft bracket No.1 and cylinder head. Refer to figure "b" to apply liquid gasket in 90 degrees to figure "b".
25. Install the rocker cover. Torque the retaining bolts to 18 inch lbs and than to 73 inch lbs, in the proper sequence.

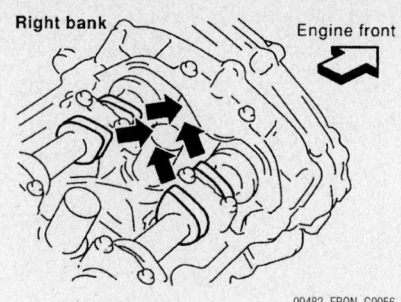

09482_FRON_G0056

No.1 cylinder at TDC (compression stroke)—4.0L engine

26. Continue the installation in the reverse of the removal procedure.

3.3L Engine

These engines are equipped with hydraulic valve lifters that do not require periodic adjustment.

4.0L Engine

1. Before servicing the vehicle, refer to the Precautions Section.
2. Disconnect the negative battery cable. Remove the engine under cover.
3. Remove the intake manifold collector.
4. Separate the engine harness and remove their brackets from the rocker covers. Remove the harness bracket from the cylinder head, if necessary.
5. Remove the ignition coil. Remove the PCV hoses. Remove the oil filler cap, if necessary.

6. Loosen the rocker cover retaining bolts, in the reverse order of the tightening sequence.
7. Remove the rocker covers from the engine.
8. Set the No.1 cylinder at TDC of its compression stroke by rotating the crankshaft pulley clockwise to align the timing mark (grooved line without color) with the timing indicator. Make sure that the intake and exhaust cam noses on No.1 cylinder (engine front side on right bank) are in alignment as shown in the illustration. If not, rotate the crankshaft in the clockwise direction 360 degrees.
9. Use a feeler gauge and measure the clearance between the valve lifter and the camshaft.
10. With the No.1 piston at TDC, refer to the illustration and measure the valve clearances at the locations marked with an "X". The "X" locations are indicated in the illustration with an arrow.
11. Rotate the crankshaft pulley clockwise 240 degrees (when viewed from the engine front) to align No.3 cylinder at TDC on the compression stroke.

➡ The crankshaft pulley bolt flange has a stamped line every 60 degrees, which can be used as a guide to rotation angle.

12. With the No.3 piston at TDC, refer to the illustration and measure the valve clearances at the locations marked with an "X". The "X" locations are indicated in the illustration with an arrow.

Measuring position (right bank)		No. 1 CYL.	No. 3 CYL.	No. 5 CYL.
No. 1 cylinder at compression TDC	EXH		×	
	INT	×		
Measuring position (left bank)		No. 2 CYL.	No. 4 CYL.	No. 6 CYL.
No. 1 cylinder at compression TDC	INT			×
	EXH	×		

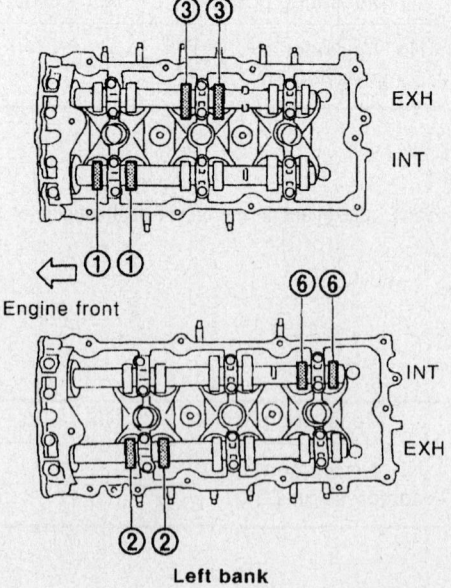

09482_FRON_G0057

Valve adjustment measurement No.1 cylinder at TDC (compression stroke)—4.0L engine

13. Rotate the crankshaft pulley clockwise 240 degrees (when viewed from the engine front) to align No.5 cylinder at TDC on the compression stroke.

➡ **The crankshaft pulley bolt flange has a stamped line every 60 degrees, which can be used as a guide to rotation angle.**

14. With the No.5 piston at TDC, refer to the illustration and measure the valve clearances at the locations marked with an "X". The "X" locations are indicated in the illustration with an arrow.

15. If measurements are not within specification, proceed to the next step.

16. Remove the camshaft. Remove the valve lifters that are not within specification.

17. Measure the center thickness of the removed lifters, using a micrometer.

18. Use the equation (t=t1+(C1-C2) to calculate valve lifter thickness for replacement.

➡ **t= valve lifter thickness to be replaced. t1= removed valve lifter thickness. C1= measured valve clearance. C2= standard valve clearance.**

19. Intake valve lifter thickness of the new valve lifter can be identified by the stamp mark on the reverse side (inside the cylinder). The stamp mark "788U" indicates 7.88 mm (0.3102 inch) thickness.

➡ **Available thickness of a valve lifter ranges from 7.88–8.40 mm (0.3102–0.3307 inch) in steps of 0.02 mm (0.0008 inch). There are 27 different sizes.**

20. Exhaust valve lifter thickness of the new valve lifter can be identified by the stamp mark on the reverse side (inside the

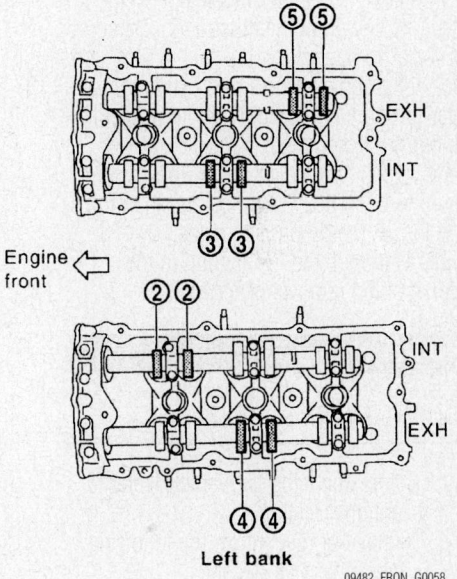

Measuring position (right bank)		No. 1 CYL.	No. 3 CYL.	No. 5 CYL.
No. 3 cylinder at compression TDC	EXH			×
	INT		×	
Measuring position (left bank)		No. 2 CYL.	No. 4 CYL.	No. 6 CYL.
No. 3 cylinder at compression TDC	INT	×		
	EXH		×	

Valve adjustment measurement No.3 cylinder at TDC (compression stroke)—4.0L engine

09482_FRON_G0058

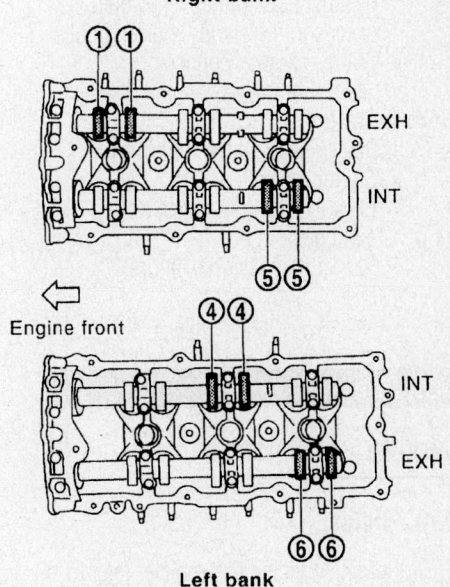

Measuring position (right bank)		No. 1 CYL.	No. 3 CYL.	No. 5 CYL.
No. 5 cylinder at compression TDC	EXH	×		
	INT			×
Measuring position (left bank)		No. 2 CYL.	No. 4 CYL.	No. 6 CYL.
No. 5 cylinder at compression TDC	INT		×	
	EXH			×

Valve adjustment measurement No.5 cylinder at TDC (compression stroke)—4.0L engine

09482_FRON_G0059

cylinder). The stamp mark "N788" indicates 7.88 mm (0.3102 inch) thickness.

➡ **Available thickness of a valve lifter ranges from 7.88–8.36 mm (0.3102–0.3291 inch) in steps of 0.02 mm (0.0008 inch). There are 25 different sizes.**

21. Install the selected valve lifters.
22. Install the camshaft.
23. Manually rotate the crankshaft pulley in the clockwise direction a few rotations.
24. Check the valve clearance and be sure it is within specification.
25. When installing the rocker cover, apply liquid gasket, be sure to use genuine RTV silicone sealant or equivalent, to the positions shown in the illustration. Refer to figure "a" to apply liquid gasket to joint part of camshaft bracket No.1 and cylinder head. Refer to figure "b" to apply liquid gasket to the figure "a" squarely.
26. Install the rocker cover. Torque the retaining bolts to 17 inch lbs and than to 74 inch lbs, in the proper sequence.
27. Continue the installation in the reverse of the removal procedure.

Starter Motor

REMOVAL & INSTALLATION

1. Before servicing the vehicle, refer to the Precautions Section.
2. Remove or disconnect the following:
   - Negative battery cable
   - Engine under cover
   - On 2005–2006 Frontier and Xterra with 4.0L engine, remove the exhaust manifold cover to gain access to the starter retaining bolts
   - Starter harness connectors
   - Starter motor

To install:
3. Install or connect the following:
   - Starter motor
   - Exhaust manifold cover, as equipped
   - Starter harness connectors
   - Engine under cover
   - Negative battery cable

Oil Pan

REMOVAL & INSTALLATION

2.4L Engine

1. Before servicing the vehicle, refer to the Precautions Section.
2. Drain the engine oil.

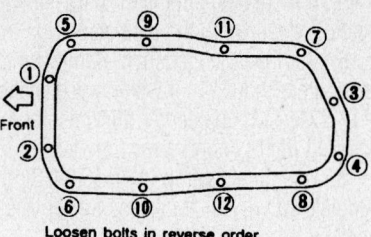

Oil pan bolt removal sequence—2.4L engine

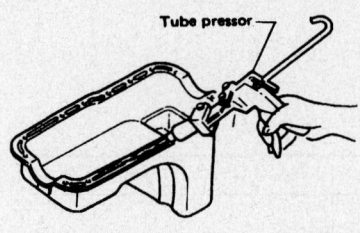

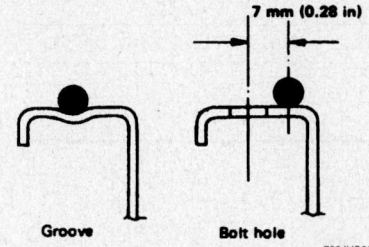

Oil pan sealant application—2.4L engine

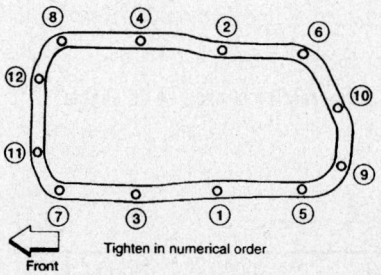

Oil pan bolt installation sequence—2.4L engine

3. Remove or disconnect the following:
   - Negative battery cable
   - Engine under cover
   - Stabilizer bar
   - Oil pan. Loosen the bolts in the sequence shown.

To install:
4. Apply a continuous bead of sealant 0.138–0.177 in. (3.5–4.5mm) to the oil pan mating surface.
5. Install or connect the following:
   - Oil pan. Tighten the bolts in sequence to 60–72 inch lbs. (7–8 Nm).

- Stabilizer bar. Tighten the bracket bolts to 38–45 ft. lbs. (51–61 Nm) and the link nuts to 12–16 ft. lbs. (16–22 Nm).
- Engine under cover
- Negative battery cable

➡ **Wait 30 minutes after installation of the oil pan to allow the sealant to cure before adding oil.**

6. Fill the crankcase to the correct level.
7. Start the engine and check for leaks.

2.5L Engine

1. Before servicing the vehicle, refer to the Precautions Section.
2. Disconnect the negative battery cable.
3. Remove the engine under cover. Drain the engine oil.
4. If equipped with automatic transmission, remove the fluid cooler tube.
5. Loosen the oil pan retaining bolts, in the reverse order of the installation sequence.
6. Insert a seal cutter tool between the oil pan and the cylinder block, and slide it by tapping on the side of the tool with a hammer.
7. Remove the oil pan from the engine.

To install:
8. Be sure to clean all the oil gasket material from both the oil pan and the cylinder block surfaces, using the proper tools.
9. Apply a continuous bead of sealant 0.138–0.177 in. (3.5–4.5mm) to the oil pan mating surface.
10. Install the oil pan to the cylinder block. This must be done within 5 minutes after applying the liquid gasket.
11. Torque the bolts to specification and in the proper sequence.
12. Continue the installation in the reverse order of the removal procedure.

➡ **Wait 30 minutes after installation of the oil pan to allow the sealant to cure before adding oil.**

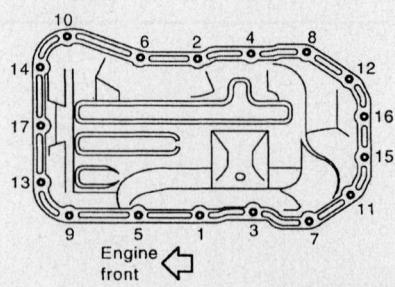

Oil pan bolt torque sequence—2.5L engine

13. Fill the crankcase to the correct level.
14. Start the engine and check for leaks.

3.3L Engine

2WD

1. Before servicing the vehicle, refer to the Precautions Section.
2. Drain the engine oil.
3. Remove or disconnect the following:
- Negative battery cable
- Engine under cover
- Stabilizer bar
- Front crossmember
- Starter motor
- Transmission mount
- Left and right motor mounts
- Power steering gear

4. Raise and support the engine for clearance.
5. Remove or disconnect the following:
- Oil pan bolts in the sequence
- Oil pan

To install:

6. Apply a continuous bead of sealant 0.138–0.177 in. (3.5–4.5mm) to the oil pan mating surface.
7. Install or connect the following:
- Oil pan. Tighten the bolts in reverse of the removal sequence to 62 inch lbs. (7 Nm).
- Power steering gear
- Left and right motor mounts
- Transmission mount
- Starter motor
- Front crossmember
- Stabilizer bar
- Engine under cover
- Negative battery cable

➡**Wait 30 minutes after installation of the oil pan to allow the sealant to cure before adding oil.**

8. Fill the crankcase to the correct level.
9. Start the engine and check for leaks.

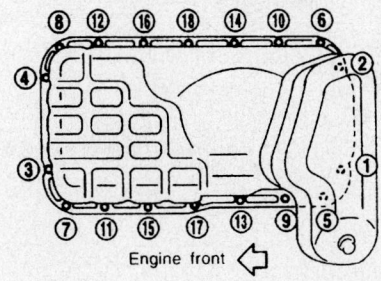

Oil pan bolt removal sequence—3.3L engine

4WD

1. Before servicing the vehicle, refer to the Precautions Section.
2. Drain the engine oil.
3. Remove or disconnect the following:
- Negative battery cable
- Engine under cover
- Stabilizer bar brackets
- Front driveshaft
- Axle halfshafts
- Front suspension crossmember
- Front differential and mounting bracket
- Starter motor
- Transmission mount
- Left and right motor mounts
- Power steering gear
- Relay rod

4. Raise and support the engine for clearance.
5. Remove or disconnect the following:
- Oil pan bolts in the sequence
- Oil pan

To install:

6. Apply a continuous bead of sealant 0.138–0.177 in. (3.5–4.5mm) to the oil pan mating surface.
7. Install or connect the following:
- Oil pan. Tighten the bolts in reverse of the removal sequence to 62 inch lbs. (7 Nm).
- Relay rod
- Power steering gear
- Left and right motor mounts
- Transmission mount
- Starter motor
- Front differential and mounting bracket
- Front suspension crossmember
- Axle halfshafts
- Front driveshaft
- Stabilizer bar brackets
- Engine under cover
- Negative battery cable

➡**Wait 30 minutes after installation of the oil pan to allow the sealant to cure before adding oil.**

8. Fill the crankcase to the correct level.
9. Start the engine and check for leaks.

4.0L Engine

LOWER

1. Before servicing the vehicle, refer to the Precautions Section.
2. Disconnect the negative battery cable.
3. Remove the engine under cover. Drain the engine oil.

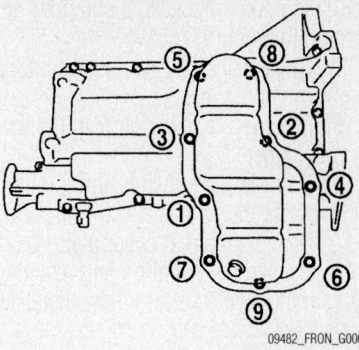

Lower oil pan bolt torque sequence—4.0L engine

4. Loosen the oil pan retaining bolts, in the reverse order of the installation sequence.
5. Insert a seal cutter tool between the oil pan and the cylinder block, and slide it by tapping on the side of the tool with a hammer.
6. Remove the oil pan from the engine.

To install:

7. Be sure to clean all the oil gasket material from both the oil pan and the cylinder block surfaces, using the proper tools.
8. Apply a continuous bead of sealant 0.138–0.177 in. (3.5–4.5mm) to the oil pan mating surface. Be sure to use genuine RTV sealant, or equivalent.
9. Install the oil pan to the cylinder block. This must be done within 5 minutes after applying the liquid gasket.
10. Torque the bolts to specification and in the proper sequence.
11. Continue the installation in the reverse order of the removal procedure.

➡**Wait 30 minutes after installation of the oil pan to allow the sealant to cure before adding oil.**

12. Fill the crankcase to the correct level.
13. Start the engine and check for leaks.

UPPER

1. Before servicing the vehicle, refer to the Precautions Section.
2. Disconnect the negative battery cable.
3. Remove the air duct. Remove the engine under cover.
4. Drain the engine oil. Drain the engine coolant.
5. Remove the final drive, if equipped with 4WD.
6. Disconnect the steering gear lower shaft joint bolt and steering gear nuts and bolts, position the assembly out of the way.
7. Remove the starter.
8. Disconnect the automatic transmis-

sion fluid cooler brackets, if equipped and position them out of the way.

9. Remove the oil filter, as necessary. Remove the oil cooler.

10. Remove the lower oil pan. Remove the oil strainer.

11. Remove the transmission joint bolts which pierce the oil pan.

12. Remove the rear cover plate.

13. Loosen the upper oil pan retaining bolts, in the reverse order of the installation sequence.

14. Insert a seal cutter tool between the oil pan and the cylinder block, and slide it by tapping on the side of the tool with a hammer.

15. Remove the oil pan from the engine. Remove the O-rings from the bottom lower cylinder block and oil pump.

To install:

16. Be sure to clean all the oil gasket material from both the oil pan and the cylinder block surfaces, using the proper tools.

17. Install new O-rings on the bottom lower cylinder block and oil pump.

18. Apply a continuous bead of sealant 0.138–0.177 in. (3.5–4.5mm) to the lower cylinder block mating surfaces of the upper oil pan. Be sure to use genuine RTV sealant, or equivalent.

➡**For bolt holes marked with a solid black triangle, apply liquid gasket outside the hole. Apply a bead of sealant 0.177–0.217 inch diameter to area "A".**

19. Install the upper oil pan. This must be done within 5 minutes after applying the liquid gasket.

20. Torque the bolts to specification and in the proper sequence. There are two types of bolts M8X100mm (3.97 inch) bolts 7, 11, 12, 13 and M8X25mm (0.98 inch) except 7, 11, 12 and 13.

21. Tighten the transmission joint bolts.

22. Install the oil strainer to the upper oil pan.

23. Continue the installation in the reverse order of the removal procedure.

➡**Wait 30 minutes after installation of the oil pan to allow the sealant to cure before adding oil.**

24. Fill the crankcase to the correct level.

25. Start the engine and check for leaks.

Oil Pump

REMOVAL & INSTALLATION

2.4L Engine

1. Before servicing the vehicle, refer to the Precautions Section.

2. Set the engine to Top Dead Center (TDC) of the compression stroke for the No. 1 cylinder.

3. Remove or disconnect the following:

- Distributor cap
- Distributor
- Engine under cover
- Stabilizer bar
- Oil pump and drive spindle

To install:

4. Fill the pump housing with engine oil, then align the punch mark on the spindle with the hole in the oil pump as shown.

5. Install or connect the following:

- Oil pump and drive spindle. Tighten the mounting bolts to 96–132 inch lbs. (11–15 Nm).
- Stabilizer bar
- Engine under cover
- Distributor
- Distributor cap

6. Start the engine and check for leaks.

7. Check the ignition timing and adjust, as necessary.

3.3L Engine

1. Before servicing the vehicle, refer to the Precautions Section.

2. Drain the engine oil.

3. Drain the cooling system.

4. Remove or disconnect the following:

- Negative battery cable
- Accessory drive belts
- Radiator hoses
- Crankshaft pulley
- Front cover
- Timing belt. Refer to the Timing Belt unit repair section.
- Crankshaft timing sprocket

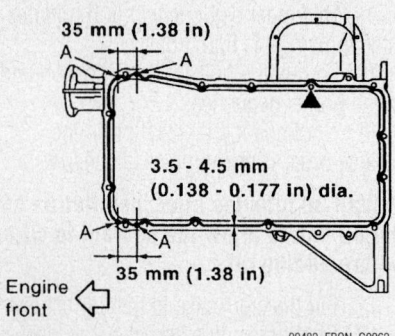

35 mm (1.38 in)

A   A

3.5 - 4.5 mm (0.138 - 0.177 in) dia.

A   A

35 mm (1.38 in)

Engine front

09482_FRON_G0062

Upper oil pan sealant application—4.0L engine

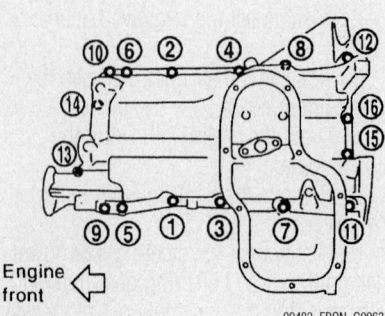

⑩ ⑥ ② ④ ⑧ ⑫
⑭   ⑯
⑬   ⑮
⑨ ⑤ ① ③ ⑦ ⑪

Engine front

09482_FRON_G0063

Upper oil pan bolt torque sequence—4.0L engine

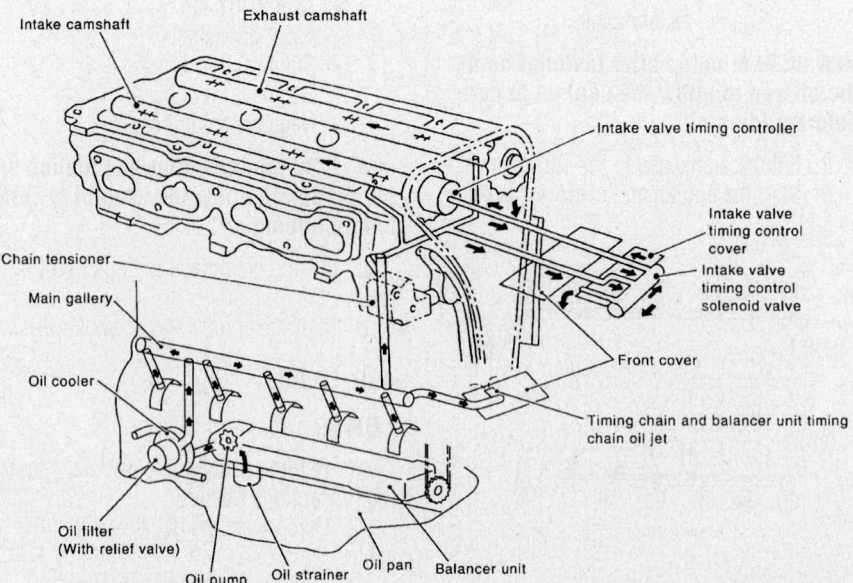

Intake camshaft — Exhaust camshaft
Chain tensioner
Main gallery
Oil cooler
Intake valve timing controller
Intake valve timing control cover
Intake valve timing control solenoid valve
Front cover
Timing chain and balancer unit timing chain oil jet
Oil filter (With relief valve)
Oil pump   Oil strainer   Oil pan   Balancer unit

09482_FRON_G0064

Oil pump and related components—2.5L engine

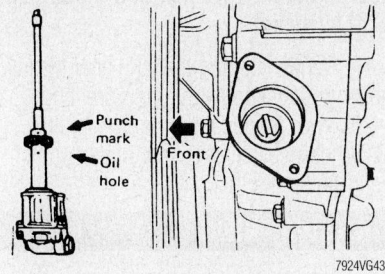

Align the punch mark with the oil hole before oil pump installation—2.4L engine

- Oil pan
- Oil pump pickup tube
- Oil pump

To install:

5. Install or connect the following:
- Oil pump. Tighten the large bolts to 16–22 ft. lbs. (22–29 Nm) and the small bolts to 55–74 inch lbs. (6–8 Nm).
- Oil pump pickup tube. Tighten the flange bolts to 12 ft. lbs. (16 Nm) and the bracket bolt to 55–74 inch lbs. (6–8 Nm).
- Oil pan
- Crankshaft timing sprocket
- Timing belt
- Front cover
- Crankshaft pulley
- Radiator hoses
- Accessory drive belts
- Negative battery cable

6. Fill the cooling system.

7. Fill the crankcase to the correct level.

8. Start the engine and check for leaks.

4.0L Engine

1. Before servicing the vehicle, refer to the Precautions Section.

2. Disconnect the negative battery cable. Drain the engine oil. Drain the engine coolant.

3. Remove the lower oil pan.

4. Remove the upper oil pan.

5. Remove the front timing chain case and timing chain (primary).

6. Remove the oil pump from the engine

To install:

7. Installation is the reverse of the removal procedure.

➡**Wait 30 minutes after installation of the oil pan to allow the sealant to cure before adding oil.**

8. Fill the crankcase to the correct level.

9. Start the engine and check for leaks.

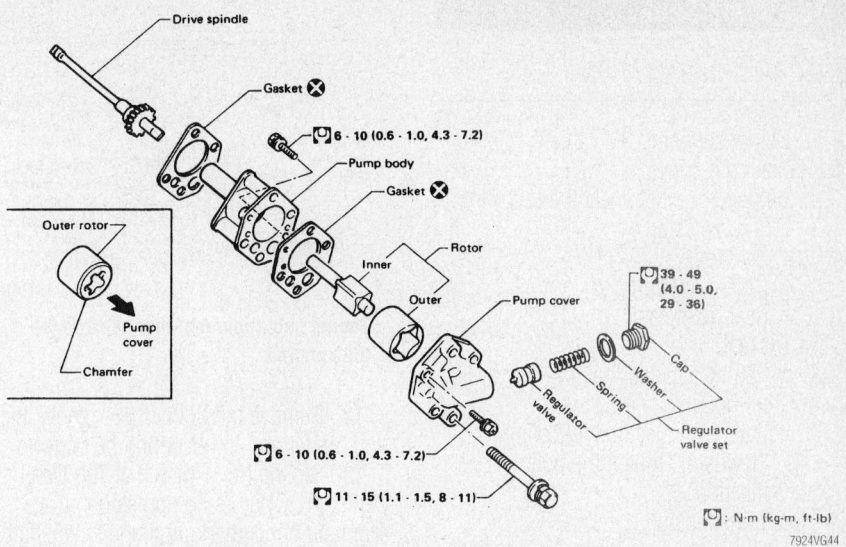

Oil pump exploded view—2.4L engine

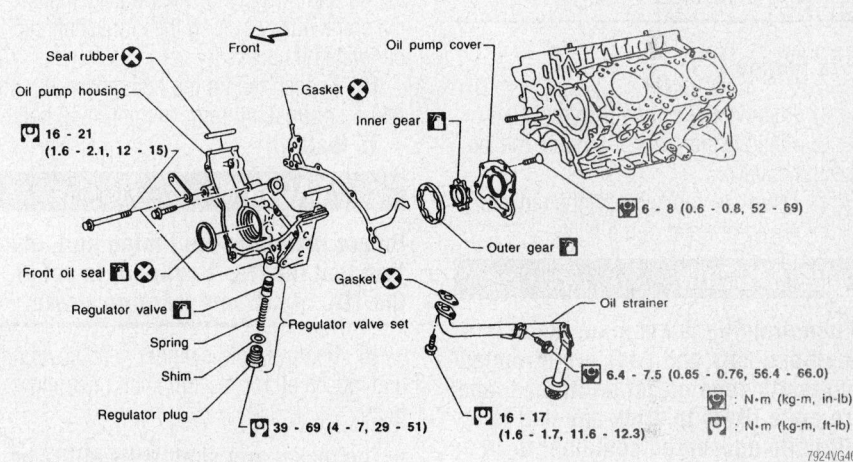

Oil pump and related components—3.3L engine

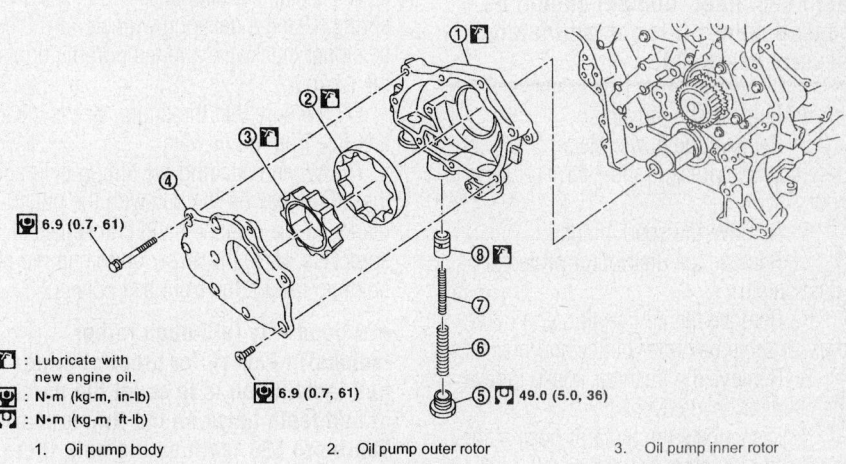

1. Oil pump body	2. Oil pump outer rotor	3. Oil pump inner rotor
4. Oil pump cover	5. Regulator valve plug	6. Regulator valve spring
7. Regulator valve spring	8. Regulator valve	

Oil pump and related components—4.0L engine

Rear Main Seal

REMOVAL & INSTALLATION

1. Before servicing the vehicle, refer to the Precautions Section.
2. Remove or disconnect the following:
   - Transmission
   - Flywheel
   - Clutch, if equipped
   - Rear main seal

To install:

3. Install the seal so that it is flush with the retainer housing.
4. Install or connect the following:
   - Flywheel. Tighten the bolts to specification.
   - Transmission

Timing Belt

REMOVAL & INSTALLATION

3.3L Engine

1. Remove the engine undercover.
2. Remove the radiator shroud, the fan and the pulleys.
3. Drain the coolant from the radiator and remove the water pump hose.

✷✷ CAUTION

When draining the coolant, keep in mind that cats and dogs are attracted by the ethylene glycol antifreeze, and are quite likely to drink any that is left in an uncovered container or in puddles on the ground. This will prove fatal in sufficient quantity. Always drain the coolant into a sealable container. Coolant should be reused unless it is contaminated or several years old.

4. Remove the radiator.
5. Remove the power steering, air conditioning compressor and alternator drive belts.
6. Remove the spark plugs.
7. Remove the distributor protector (dust shield).
8. Remove the air conditioning compressor drive belt idler pulley and bracket.
9. Remove the fresh air intake tube at the cylinder head cover.
10. Disconnect the radiator hose at the thermostat housing.
11. Remove the crankshaft pulley bolt, then pull off the pulley with a suitable puller.

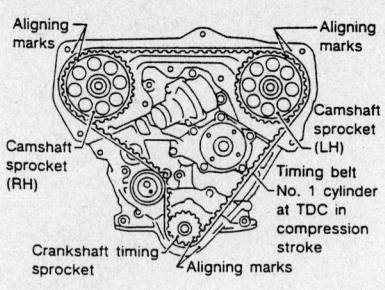

Timing belt alignment mark locations—3.3L engine

12. Remove the bolts, then remove the front upper and lower timing belt covers.
13. Set the No. 1 piston at Top Dead Center (TDC) of its compression stroke. Align the punchmark on the left camshaft sprocket with the punchmark on the timing belt upper rear cover. Align the punchmark on the crankshaft sprocket with the notch on the oil pump housing. Temporarily install the crank pulley bolt so the crankshaft can be rotated if necessary.
14. Loosen the timing belt tensioner and return spring, then remove the timing belt.

To install:

✷✷ CAUTION

Before installing the timing belt, confirm that the No. 1 cylinder is set at the TDC of the compression stroke.

15. Remove both cylinder head covers and loosen all rocker arm shaft retaining bolts.

➡**The rocker arm shaft bolts MUST be loosened so that the correct belt tension can be obtained.**

16. Install the tensioner and the return spring. Using a hexagon wrench, turn the tensioner clockwise and temporarily tighten the locknut.
17. Be sure that the timing belt is clean and free from oil or water.
18. When installing the timing belt, align the white lines on the belt with the punchmarks on the camshaft and crankshaft sprockets. Have the arrow on the timing belt pointing toward the front belt covers.

➡**A good way (although rather tedious!) to check for proper timing belt installation is to count the number of belt teeth between the timing marks. There are 133 teeth on the belt; there should be 40 teeth between the timing marks on the left and right side camshaft sprockets, and 43 teeth between the timing marks on the left**

side camshaft sprocket and the crankshaft sprocket.

19. While keeping the tensioner steady, loosen the locknut with a hex wrench.
20. Turn the tensioner approximately 70–80 degrees clockwise with the wrench, then tighten the locknut.

✷✷ WARNING

If any binding is felt when adjusting the timing belt tension by turning the crankshaft, STOP turning the engine, because the pistons may be hitting the valves.

21. Turn the crankshaft in a clockwise direction several times, then **slowly** set the No. 1 piston to TDC of the compression stroke.
22. Apply 22 lbs. (10 kg) of pressure (push it in!) to the center span of the timing belt between the right side camshaft sprocket and the tensioner pulley, then loosen the tensioner locknut.
23. Using a 0.0138 in. (0.35mm) thick feeler gauge (the actual width of the blade **must** be ½ in. or 13mm!), turn the crankshaft clockwise (**slowly!**). The timing belt should move approximately 2½ teeth. Tighten the tensioner locknut, turn the crankshaft slightly and remove the feeler gauge.
24. Slowly rotate the crankshaft clockwise several more times, then set the No. 1 piston to TDC of the compression stroke.
25. Position the 2 timing covers on the block, then tighten the mounting bolts to 24 ft. lbs. (35 Nm).
26. Press the crankshaft pulley onto the shaft, then tighten the bolt to 90–98 ft. lbs. (123–132 Nm).
27. Connect the radiator hose to the thermostat housing.
28. Reconnect the fresh air intake tube at the cylinder head cover.
29. Install the air conditioning compressor drive belt idler pulley and bracket.
30. Install the distributor protector (dust shield).
31. Install the spark plugs.
32. Install the power steering, air conditioning compressor and alternator drive belts.
33. Install the radiator.
34. Reconnect the water pump hose and fill the engine with coolant. Install the fan shroud and pulleys.
35. Install the engine undercover.
36. Start the engine and check for any leaks.

Timing Chain, Sprockets, Front Cover and Seal

REMOVAL & INSTALLATION

2.4L Engine

1. Before servicing the vehicle, refer to the Precautions Section.
2. Drain the cooling system.
3. Drain the engine oil.
4. Set the engine to Top Dead Center (TDC) of the compression stroke for the No. 1 cylinder.
5. Remove or disconnect the following:

- Negative battery cable
- Air cleaner assembly
- Spark plug wires
- Cooling fan and shroud
- Distributor
- Valve cover
- Accessory drive belts
- Power steering pump and brackets
- A/C compressor and bracket
- Idler pulleys
- Water pump pulley
- Crankshaft pulley
- Front crankshaft seal
- Oil pump and drive spindle
- Oil pan
- Upper timing cover
- Lower timing cover
- Upper timing chain tensioner
- Upper timing chain and camshaft

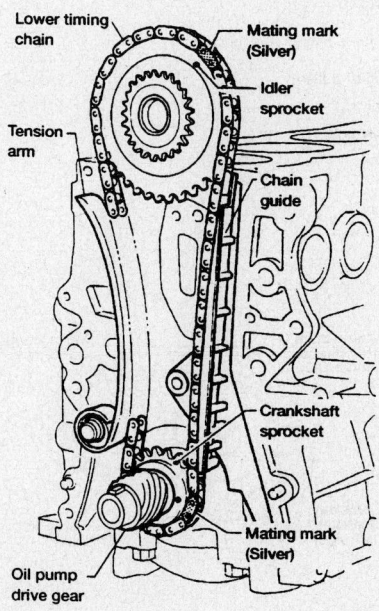

Lower timing chain alignment—2.4L engine

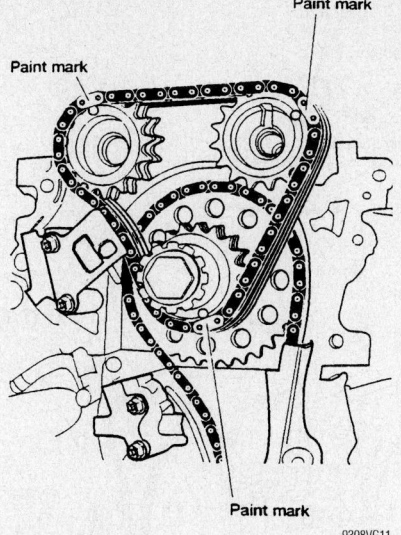

Upper timing chain alignment—2.4L engine

sprockets. Matchmark the timing chain to the sprockets.
- Lower timing chain tensioner
- Lower timing chain and idler sprocket. Matchmark the timing chain to the sprockets.

To install:

6. Install or connect the following:
- Lower timing chain and idler sprocket with the timing marks aligned as shown. Tighten the idler sprocket bolt to 48–61 ft. lbs. (66–83 Nm).
- Lower timing chain tensioner. Tighten the bolts to 56–66 inch lbs. (6.5–7.5 Nm).
- Upper timing chain and camshaft sprockets with the timing marks aligned as shown. Tighten the camshaft sprocket bolts to 123–130 ft. lbs. (167–177 Nm).
- Upper timing chain tensioner. Tighten the bolts to 56–66 inch lbs. (6.5–7.5 Nm).
- Lower timing cover. Tighten the large bolts to 12–14 ft. lbs. (16–19 Nm) and the small bolts to 56–66 inch lbs. (6.5–7.5 Nm).
- Upper timing cover. Tighten the large bolts to 12–14 ft. lbs. (16–19 Nm) and the small bolts to 56–66 inch lbs. (6.5–7.5 Nm).
- Oil pan
- Oil pump and drive spindle
- Front crankshaft seal
- Crankshaft pulley. Tighten the bolt to 105–112 ft. lbs. (142–152 Nm).
- Water pump pulley
- Idler pulleys

- A/C compressor and bracket
- Power steering pump and brackets
- Accessory drive belts
- Valve cover
- Distributor
- Cooling fan and shroud
- Spark plug wires
- Air cleaner assembly
- Negative battery cable
7. Fill the cooling system.
8. Fill the crankcase to the correct level.
9. Start the engine and check for leaks.
10. Check the ignition timing and adjust, as necessary.

2.5L Engine

1. Before servicing the vehicle, refer to the Precautions Section.
2. Properly relieve the fuel system pressure.
3. Disconnect the negative battery cable.
4. Remove the air cleaner and the air duct assembly.
5. Remove the spark plugs.
6. Remove the intake manifold.
7. Disconnect the PCV hose from the rocker cover. Remove the ignition coil.
8. Remove the PCV valve and O-ring from the rocker cover, if necessary.
9. Remove the oil filler cap from the rocker cover, if necessary.
10. Remove the rocker cover retaining bolts. Be sure to remove the bolts by reversing the order of the tightening torque sequence.
11. Remove the rocker cover. Discard the gasket.
12. Remove the coolant reservoir tank. Remove the auxiliary drive belt auto-tensioner.
13. Remove the alternator. Remove the strut tower brace.
14. Remove the air conditioning compressor and position it to the side. Do not disconnect the refrigerant lines.
15. Remove the power steering pump and reservoir tank; position the assembly to the side. Do not disconnect the fluid lines.
16. Remove the upper and lower oil pan. Remove the strainer.
17. Remove the IVT control cover bolts. Be sure to remove the bolts by reversing the order of the tightening torque sequence.
18. Remove the cover by cutting the sealant using tool KV10111100 (J-37228) or equivalent.
19. Position the number one cylinder on its compression stroke by rotating the crankshaft pulley clockwise. Align the mat-

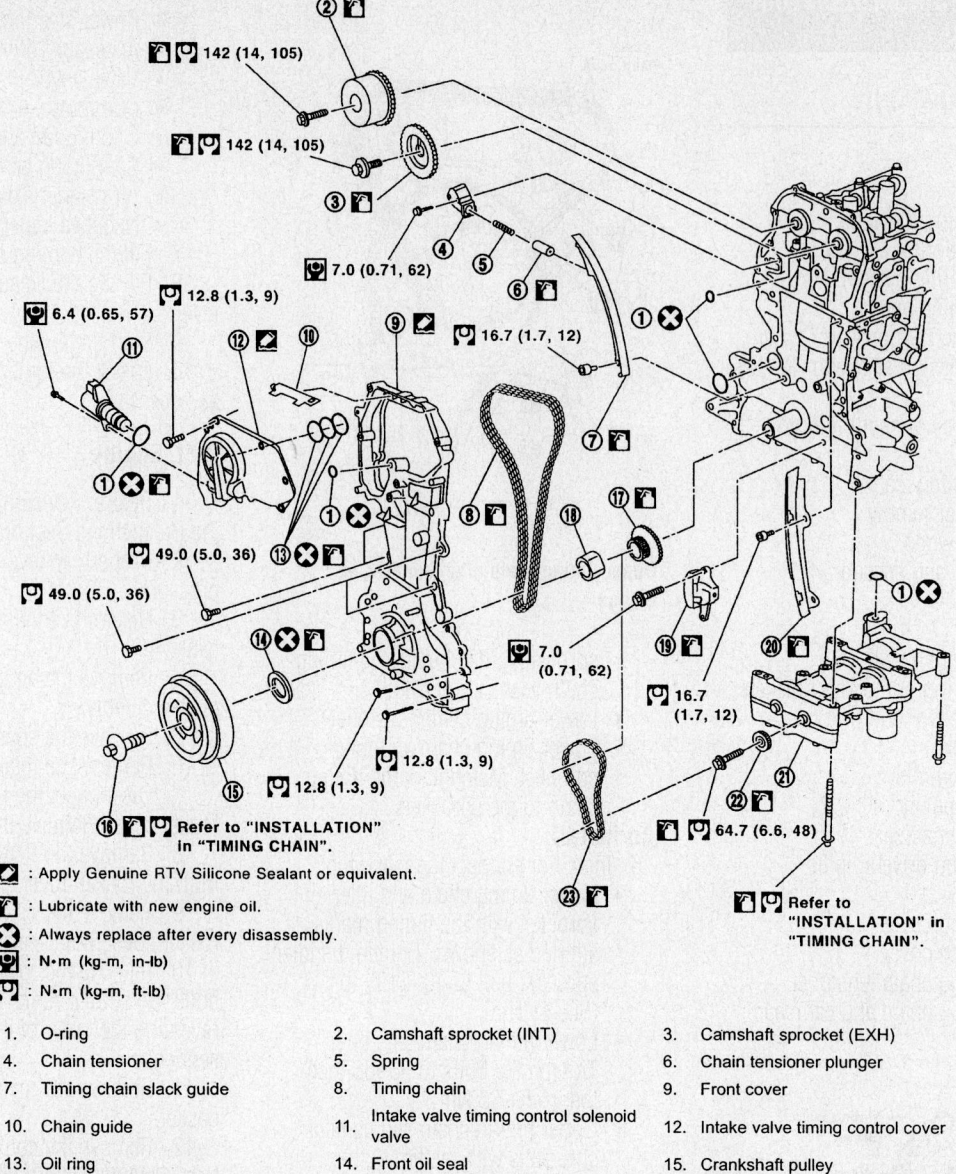

142 (14, 105)

142 (14, 105)

7.0 (0.71, 62)

12.8 (1.3, 9)

6.4 (0.65, 57)

16.7 (1.7, 12)

49.0 (5.0, 36)

49.0 (5.0, 36)

7.0 (0.71, 62)

16.7 (1.7, 12)

12.8 (1.3, 9)

12.8 (1.3, 9)

64.7 (6.6, 48)

Refer to "INSTALLATION" in "TIMING CHAIN".

Refer to "INSTALLATION" in "TIMING CHAIN".

: Apply Genuine RTV Silicone Sealant or equivalent.

: Lubricate with new engine oil.

: Always replace after every disassembly.

: N•m (kg-m, in-lb)

: N•m (kg-m, ft-lb)

1. O-ring	2. Camshaft sprocket (INT)	3. Camshaft sprocket (EXH)
4. Chain tensioner	5. Spring	6. Chain tensioner plunger
7. Timing chain slack guide	8. Timing chain	9. Front cover
10. Chain guide	11. Intake valve timing control solenoid valve	12. Intake valve timing control cover
13. Oil ring	14. Front oil seal	15. Crankshaft pulley
16. Crankshaft pulley bolt	17. Crankshaft sprocket	18. Spacer
19. Balancer unit timing chain tensioner	20. Timing chain tension guide	21. Balancer unit
22. Balancer unit sprocket	23. Balancer unit timing chain	

09482_FRON_G0066

Timing chain and related components—2.5L engine

ing marks for TDC with the timing indicator on the front cover.

20. At the same time make sure that the mating marks on the camshaft sprockets are lined up as indicated in the illustration. If not rotate the crankshaft one more turn to line up the mating marks.

21. Hold the crankshaft pulley with a suitable tool. Loosen the crankshaft pulley retaining bolt.

22. Pull the pulley out about 0.39 inch. Remove the crankshaft pulley bolt.

23. Attach a pulley puller in the M6 thread hole on the crankshaft pulley. Remove the crankshaft pulley.

➡Be careful not to damage the front cover and/or the crankshaft.

24. Loosen the front cover retaining bolts, in the order indicated in the bolt loosening sequence illustration. Remove the front cover. Be careful not to damage the mating surfaces.

25. Using a seal removal tool, remove the oil seal, as required.

26. To remove the timing chain, push in on the chain tensioner plunger. Insert a stopper pin into the hole on the chain tensioner body to secure the chain tensioner plunger. Remove the chain tensioner.

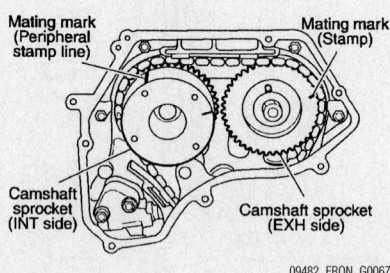

Mating mark (Peripheral stamp line)

Mating mark (Stamp)

Camshaft sprocket (INT side)

Camshaft sprocket (EXH side)

09482_FRON_G0067

Timing chain alignment marks—2.5L engine

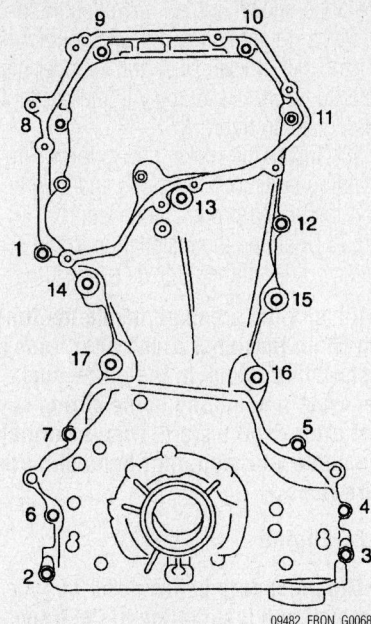

Front cover bolt removal sequence—2.5L engine

Remove the chain. Do not rotate the crankshaft with the chain removed.

➡**Use a 0.02 inch (approximate) metal pin as a stopper pin.**

27. Remove the camshaft sprockets.

28. Remove the timing chain slack guide, timing chain tensioner guide and spacer.

29. Remove the balancer unit timing chain tensioner by lifting the lever up and releasing the ratchet claw for return proof. Push the tensioner sleeve in and hold it. Matching the hole on the lever with the one on the body, insert a stopper pin to secure the tensioner sleeve. Remove the balancer unit timing chain tensioner.

➡**Use a 0.04 inch (approximate) metal pin as a stopper pin.**

30. Secure the hexagonal portion of the balancer shaft using a suitable tool. Loosen the balancer unit sprocket bolt.

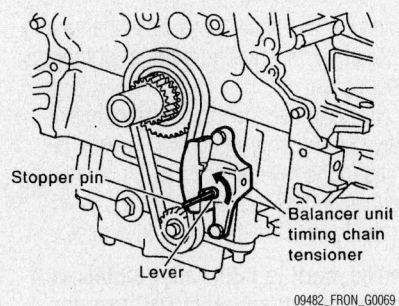

Balance unit stopper pin installation—2.5L engine

31. Remove the balancer unit timing chain, balancer unit sprocket and crankshaft sprocket.

➡**When removing the balancer unit timing chain, remove the crankshaft sprocket and balancer unit sprocket at the same time.**

32. Loosen the balancer unit mounting bolts, in the order of the tightening sequence. Remove the balancer unit. Do not disassemble the balancer unit. Bolts one and four use a E14 torx head socket.

To install:

33. Check the chain for cracks and excessive wear, replace as required.

34. Measure the balancer unit bolt outer diameters ("d1" and "d2") at two positions, as shown in the illustration. If reduction appears in the "A" range, regard it as "d2". Specification is as follows: ("d1"—"d2"): 0.0059 inch. If it exceeds the specification (large difference in dimensions) replace the balancer unit bolt with a new one.

35. Measure the balancer bolt unit length. If it exceeds the specification replace

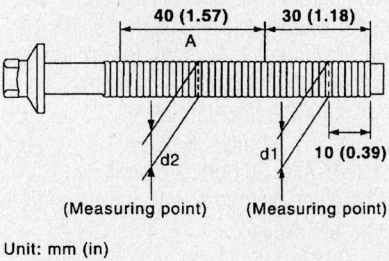

Unit: mm (in)

Balance unit bolt measurement—2.5L engine

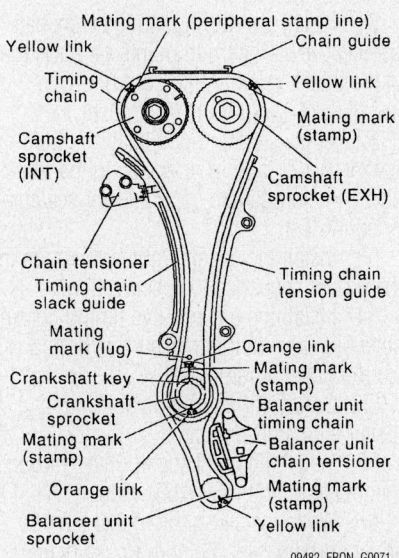

Timing chain alignment—2.5L engine

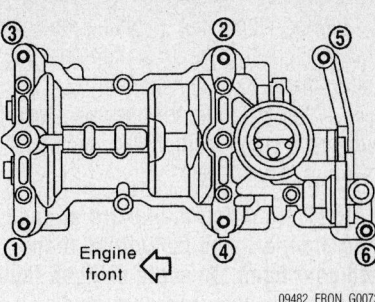

Balance unit bolt torque sequence—2.5L engine

the balancer unit bolt with a new one. Specification is 6.974 inch.

36. Make sure that the crankshaft key is pointing straight up. Install the O-ring to the balancer unit.

37. Install the balancer unit. Torque the bolts to specification and in the proper sequence using tool KV10112100 (BT8653-A) or equivalent.

  a. Step 1: bolts 1–4 to 35 ft. lbs.

  b. Step 2: bolts 1–4 100 degrees clockwise

  c. Step 3: bolts 1–4 loosen in the reverse order of the tightening sequence to zero

  d. Step 4: bolts 1–4 to 35 ft. lbs.

  e. Step 5: bolts 1–4 100 degrees clockwise

  f. Step 6: bolts 5–6 to 22 ft. lbs.

➡**Check the tightening angle using a tool or a protractor. Do not make a judgment by visual check alone.**

38. Install the crankshaft sprocket, balancer unit sprocket and balancer timing chain.

39. Make sure that the crankshaft sprocket is positioned with the mating marks on the cylinder block and crankshaft sprocket meeting at the top.

40. Install it by aligning the mating marks on each sprocket and balancer unit timing chain.

41. Secure the hexagonal portion of the balancer shaft using a suitable tool. Tighten the balancer unit sprocket bolt to specification.

➡**Install the crankshaft sprocket, balancer unit sprocket and balancer unit timing chain at the same time.**

42. Install the balancer unit timing chain tensioner.

➡**After installation, make sure that the mating marks have not slipped. Remove the stopper pin and release the tensioner sleeve.**

43. Align the mating marks on each sprocket and timing chain. Install the timing chain and related parts.

44. Before and after installing the chain tensioner, check again to be sure that the mating marks have not slipped.

45. After installing the chain tensioner, remove the stopper pin. Make sure that the tensioner moves freely.

➡ **After the mating marks are aligned, keep them aligned by holding them with your hand. To avoid skipped teeth, do not rotate the crankshaft and camshaft until the cover is installed.**

➡ **Before installing the chain tensioner, it is possible to change the position of the mating mark on the timing chain for that on each sprocket for alignment.**

46. Install the front cover oil seal. Install O-rings to the cylinder head and the cylinder block,

47. Apply a continuous bead of liquid gasket to the front cover. Be sure to use genuine RTV sealant, or equivalent.

➡ **Sealant application instructions differ depending on position, refer to the illustration for positioning. Detail "A", cross over the start of the application and the end. Detail "B", apply liquid gasket outside of the bolt holes. For all bolt holes other than "B", apply to the inside. Detail "C", between here only, apply a bead of sealant 0.177–0.217 inch diameter.**

48. Make sure that the mating marks of the chain and each sprocket are still aligned.

49. Install the front cover. Torque the

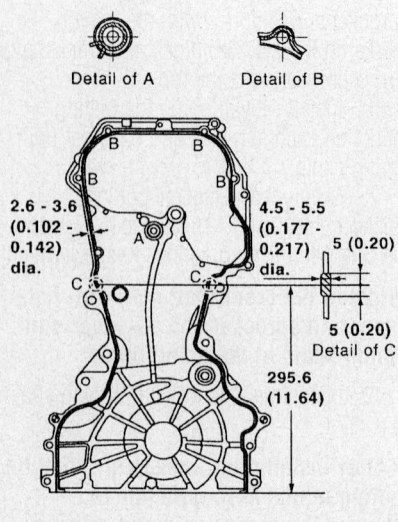

Unit: mm (in)

09482_FRON_G0073

Front cover sealant application with respect to positioning—2.5L engine

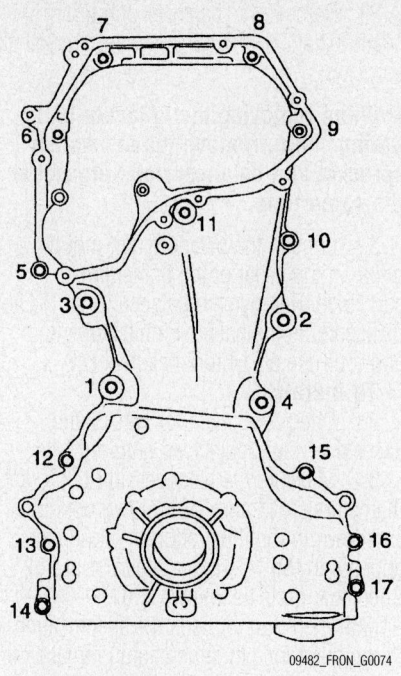

09482_FRON_G0074

Front cover bolt torque sequence—2.5L engine

retaining bolts to specification and in the proper sequence. Bolt position 5, 10, 14 and 17: 45mm (1.77 inch). Except the above (except 1 to 4): 20mm (0.79 inch).

   a. M6 bolts: 9 ft. lbs.

   b. M10 bolts: 36 ft. lbs.

   c. After all bolts are tightened, retighten them to specification and in the proper sequence

➡ **Be sure to wipe off any excess liquid gasket leaking to the surface for installing the oil pan.**

50. Install the chain guide between the camshaft sprockets.

51. Install the oil rings to the camshaft sprocket (INT) insertion points on backside of the intake valve timing control cover. Install the O-ring to the front cover.

52. Apply a continuous bead of liquid gasket, 0.122 inch in diameter, to the front cover. Be sure to use genuine RTV sealant, or equivalent.

53. Install the cover. Tighten the bolts in the proper sequence to specification.

54. Install the intake valve timing control solenoid valve to the intake valve timing control cover, if removed.

55. Connect the ground cables, and install the harness clip.

56. Install the crankshaft pulley. Torque the retaining bolt to specification.

57. When installing the rocker cover, apply liquid gasket, be sure to use genuine RTV silicone sealant or equivalent, to the

positions shown in the illustration. Refer to figure "a" to apply liquid gasket to joint part of camshaft bracket No.1 and cylinder head. Refer to figure "b" to apply liquid gasket in 90 degrees to figure "b".

58. Install the rocker cover. Torque the retaining bolts to 18 inch lbs and than to 73 inch lbs, in the proper sequence.

59. Continue the installation in the reverse order of the removal procedure.

➡ **If hydraulic pressure inside the timing chain tensioner drops after removal/installation, slack in the guide may generate a pounding noise during and just after engine start. This is normal the noise will stop after hydraulic pressure rises.**

4.0L Engine

➡ **The procedure below describes the removal and installation of the front timing case and timing chain related parts and rear timing chain case, when the upper oil pan needs to be removed or installed. When only the timing chain (primary) is being removed it is not necessary to remove the rocker covers.**

1. Before servicing the vehicle, refer to the Precautions Section.

2. Properly relieve the fuel system pressure.

3. Disconnect the negative battery cable.

4. Remove the engine cover. Drain the engine oil. Drain the engine coolant.

5. Remove the upper and lower oil pans.

6. Remove the radiator cooling fan assembly. Remove the drive belts.

7. Separate the engine wiring harnesses by removing their brackets from the front timing chain case.

8. Remove the power steering pump from the bracket with the fluid hoses attached. Position the assembly to the side. Do not disconnect the hoses. Remove the bracket.

9. Remove the alternator. Remove the water bypass hose, water hose clamp and idler pulley bracket from the front timing chain case.

10. Remove the left and right intake valve timing control covers. Loosen the bolts in the reverse order of the tightening sequence. Use tool KV10111100 (J-37228) or equivalent to cut the liquid gasket seal.

➡ **The shaft is internally jointed with the camshaft sprocket (INT) center hole. When removing, keep it horizontal until it is completely disconnected.**

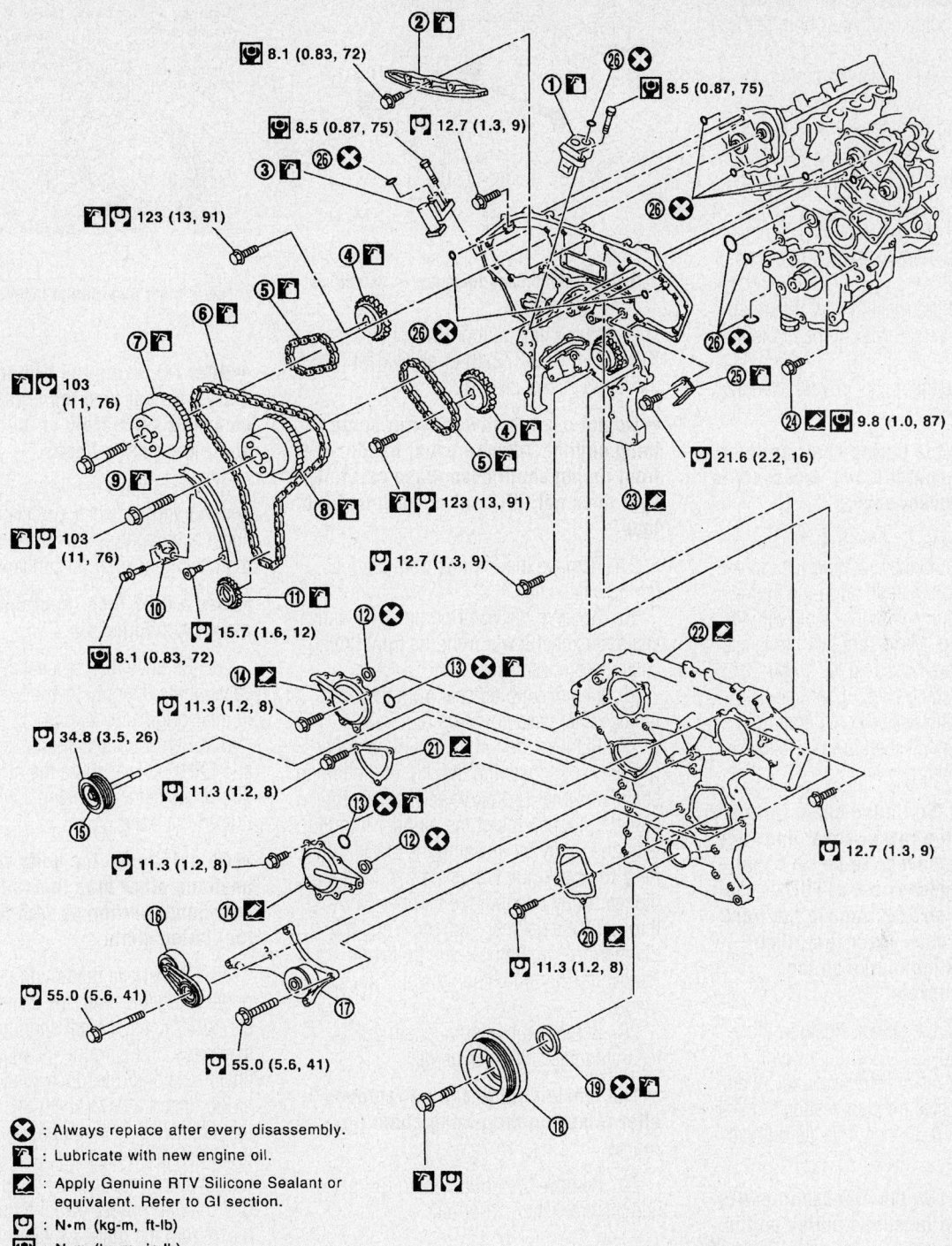

8.1 (0.83, 72)

8.5 (0.87, 75)   12.7 (1.3, 9)

8.5 (0.87, 75)

123 (13, 91)

103 (11, 76)

103 (11, 76)

8.1 (0.83, 72)

15.7 (1.6, 12)

11.3 (1.2, 8)

34.8 (3.5, 26)

11.3 (1.2, 8)

11.3 (1.2, 8)

55.0 (5.6, 41)

55.0 (5.6, 41)

123 (13, 91)

12.7 (1.3, 9)

9.8 (1.0, 87)

21.6 (2.2, 16)

12.7 (1.3, 9)

11.3 (1.2, 8)

⊗ : Always replace after every disassembly.

◨ : Lubricate with new engine oil.

◩ : Apply Genuine RTV Silicone Sealant or equivalent. Refer to GI section.

◻ : N•m (kg-m, ft-lb)

◼ : N•m (kg-m, in-lb)

1. Timing chain tensioner (secondary) (left bank)	2. Internal chain guide	3. Timing chain tensioner (secondary) (right bank)
4. Camshaft sprocket (EXH)	5. Timing chain (secondary)	6. Timing chain (primary)
7. Camshaft sprocket (INT)	8. Camshaft sprocket (INT)	9. Slack guide
10. Timing chain tensioner (primary)	11. Crankshaft sprocket	12. Collared O-ring
13. O-ring	14. Intake valve timing control cover	15. Idler pulley
16. Drive belt auto tensioner	17. Cooling fan bracket	18. Crankshaft pulley
19. Front oil seal	20. Water pump cover	21. Chain tensioner cover
22. Front timing chain case	23. Rear timing chain case	24. Water drain plug (front)
25. Tension guide	26. O-ring	

Timing chain and related components—4.0L engine

11. Remove the collared O-rings from the front timing chain case on both the left and right side.

12. Remove the intake manifold collector.

13. Separate the engine harness and remove their brackets from the rocker covers. Remove the harness bracket from the cylinder head, if necessary.

14. Remove the ignition coil. Remove the PCV hoses. Remove the oil filler cap, if necessary.

15. Loosen the rocker cover retaining bolts, in the reverse order of the tightening sequence.

16. Remove the rocker covers from the engine.

➡When only the timing chain (primary) is being removed it is not necessary to remove the rocker covers.

17. Set the No.1 cylinder at TDC of its compression stroke by rotating the crankshaft pulley clockwise to align the timing mark (grooved line without color) with the timing indicator. Make sure that the intake and exhaust cam noses on No.1 cylinder (engine front side on right bank) are in alignment as shown in the illustration. If not, rotate the crankshaft in the clockwise direction 360 degrees.

➡When only the timing chain (primary) is removed, the rocker cover does not need to be removed. To be sure that the No.1 cylinder is set at TDC on the compression stroke, remove the front timing chain case cover first, then check the mating marks on the camshaft sprockets.

18. Remove the starter. Position tool KV10117700 (J-44716) or equivalent.

19. Loosen the crankshaft pulley retaining bolt and locate the bolt seating surface, which is about 0.39 inch from its original position.

➡Do not remove the crankshaft pulley bolt. Keep the loosened pulley bolt in place to protect the removed crankshaft pulley from dropping.

20. Pull the pulley with both hands and remove it from its mounting. Remove the bolt and pulley from the engine.

21. Loosen and remove the two bolts of the upper oil pan.

22. Loosen the front timing chain cover retaining bolts in the reverse order of the tightening sequence.

23. Insert a suitable tool in the notch at the top of the front timing chain case and pry off the case by moving the tool as

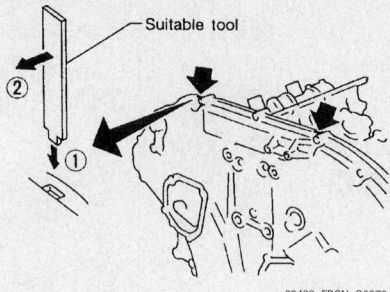

Tool installation and location—4.0L engine

shown in the illustration. Use tool KV10111100 (J-37228) or equivalent to cut the liquid gasket seal.

➡Do not use a screwdriver or something similar. After removal handle the front timing chain cover case carefully so it does not tilt, cant or warp under a load.

24. Remove the O-rings from the rear timing chain case.

25. Remove the water pump cover and chain tensioner cover from the front timing chain case cover, as required.

26. Remove the oil seal from the front timing chain case cover, as required.

27. Remove the timing chain tensioner (primary) by loosening the clip of the timing chain tensioner (primary) and release the plunger stopper. Insert the plunger into the tensioner body by pressing the slack guide. Keep the slack guide pressed and hold the plunger in by pushing the stopper pin through the tensioner body hole and the plunger groove. Remove the bolts and remove the timing chain tensioner (primary).

28. Remove the internal chain guide, tension guide and slack guide.

➡The tension guide can be removed after removing the timing chain (primary).

29. Remove the timing chain (primary) and the crankshaft sprocket.

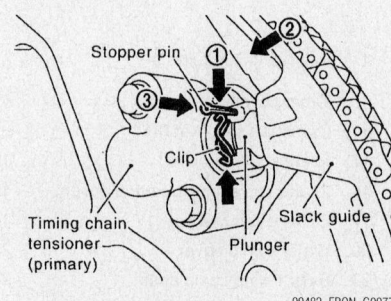

Timing chain tensioner (primary)—4.0L engine

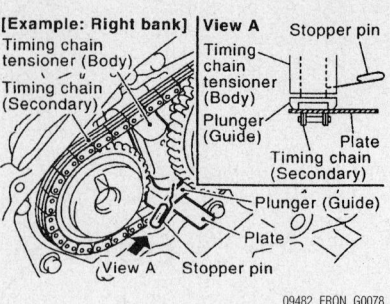

Resin plate installation location—4.0L engine

➡After removing the timing chain (primary), do not turn the crankshaft and camshaft separately or the valves will strike the piston heads.

30. To remove the timing chain (secondary) and camshaft sprockets, attach a suitable stopper pin to the right and left timing chain tensioner (secondary).

➡Use a 0.02 inch (approximate) metal pin as a stopper pin.

31. Remove the camshafts. Remove the valve lifters. Identify them for reinstallation in their original locations.

32. Remove the camshaft sprocket (INT and EXH) bolts. Secure the hexagonal portion of the camshaft using a wrench to loosen the bolts.

➡Do not loosen the bolts with securing anything other than the camshaft hexagonal portion or with tensioning the timing chain.

33. To remove the timing chain (secondary) together with the camshaft sprockets, turn the crankshaft slightly to secure slackness of the timing chain on the timing chain tensioner (secondary) side.

34. Insert a 0.020 inch thick metal or resin plate between the timing chain and timing chain plunger (guide). Remove the timing chain (secondary) together with the camshaft sprockets with the timing chain loose from the guide groove.

✳✳ CAUTION

Be careful of the plunger coming off when removing the timing chain (secondary). This is because the plunger of the timing chain tensioner (secondary) moves during operation, leading to coming off its fixed stopper pin.

➡The camshaft sprocket (INT) is a one piece integrated design sprocket for the timing chain (primary) and for the tim-

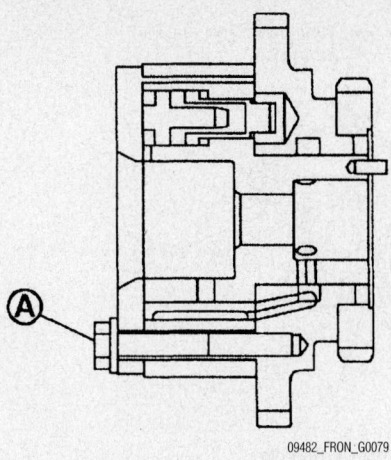

09482_FRON_G0079

Camshaft sprocket bolt location—4.0L engine

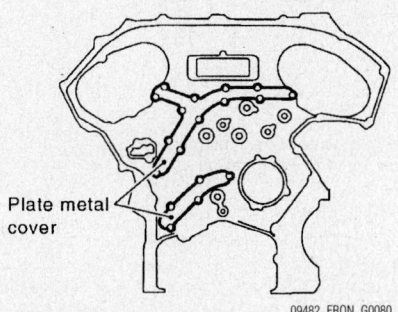

Plate metal cover

09482_FRON_G0080

Metal cover plate location on rear timing case cover—4.0L engine

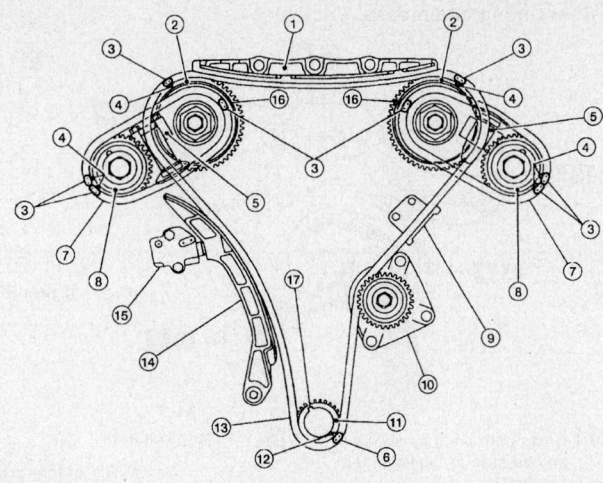

1.	Internal chain guide	2.	Camshaft sprocket (intake)	3.	Mating mark (copper link)
4.	Mating mark (punched)	5.	Secondary timing chain tensioner	6.	Mating mark (yellow link)
7.	Secondary timing chain	8.	Camshaft sprocket (exhaust)	9.	Tensioner guide
10.	Water pump	11.	Crankshaft sprocket	12.	Mating mark (notched)
13.	Primary timing chain	14.	Slack guide	15.	Primary timing chain tensioner
16.	Mating mark (back side)	17.	Crankshaft key		

09482_FRON_G0081

Timing chain alignment—4.0L engine

ing chain (secondary). When handling the sprocket avoid shock to the sprocket. Do not disassemble or loosen bolt "A", as shown in the illustration.

35. Remove the rear timing chain case cover bolts, in the reverse order of the tightening sequence. Using the proper tool cut the liquid gasket sealant seal. Remove the cover.

➡Do not remove the metal cover of the oil passage. After removal handle the case carefully so it does not tilt, cant or warp under a load.

36. Remove the O-rings from the cylinder head and No.1 camshaft bracket. Remove the O-rings from the cylinder block.

37. If necessary, remove the timing chain tensioners (secondary) from the cylinder head by first removing the No.1 camshaft bracket. Remove the timing chain tensioners (secondary) with the stopper pin attached.

To install:

38. Check the chain for cracks and excessive wear, replace as required.

39. Be sure to remove all old gasket material from bolts and bolt holes.

40. If removed install the timing chain tensioners (secondary) to the cylinder head.

41. Install camshaft brackets No.1.

42. To install the rear timing chain case cover, first install new O-rings to the cylinder block, Install new O-rings to the cylinder head and camshaft bracket No.1.

43. Apply liquid gasket sealant to the rear timing chain case back side, as shown in the illustration. Be sure to use genuine RTV sealant, or equivalent.

➡For "A" in the figure, completely wipe out excessive liquid gasket extended on a portion touching at engine coolant. Apply liquid gasket on the installation position of the water pump and cylinder head very completely.

44. Align the rear timing case with dowel pins (right and left) on the cylinder block. Install the rear timing chain case. Make sure that the O-rings stay in place during installation to the cylinder block, cylinder head and camshaft bracket No.1.

45. Tighten the bolts to specification and in the proper sequence.

   a. Bolt length: 0.79 inch. Bolt position: 1,2,3,6,7,8,9,and 10.

   b. Bolt length: 0.63 inch. Bolt position: except 1,2,3,6,7,8,9,and 10.

   c. Torque bolts to 9 ft. lbs.

   d. After all bolts are tightened,

retighten them to specification and in the proper sequence

➡Be sure to wipe off any excess liquid gasket leaking to the surface for installing the oil pan.

46. After installing the rear timing case, check the surface height deference between the rear timing chain case and the lower cylinder block. Specification should be -0.0094–0.0055 inch. If not within specification, repeat the installation procedure.

47. Install the water pump, using new O-rings.

48. Make sure that the dowel pin hole, dowel pin of camshaft and crankshaft key are located with number one piston at TDC on the compression stroke.

➡Though the camshaft does not stop at the position, as shown in the illustration, for placement of the cam nose it is generally accepted that the camshaft is placed for the same direction as the illustration. Camshaft dowel pin hole (intake side): at the cylinder head upper face side in each bank. Camshaft dowel pin hole (exhaust side): at the cylinder head upper face side in each bank. Crankshaft key: at the cylinder head side of the right bank. Hole on the small diameter side must be used for the intake side dowel pin hole.

49. To install the timing chains (secondary) and camshaft sprockets, push the

Rear timing chain case: Back side

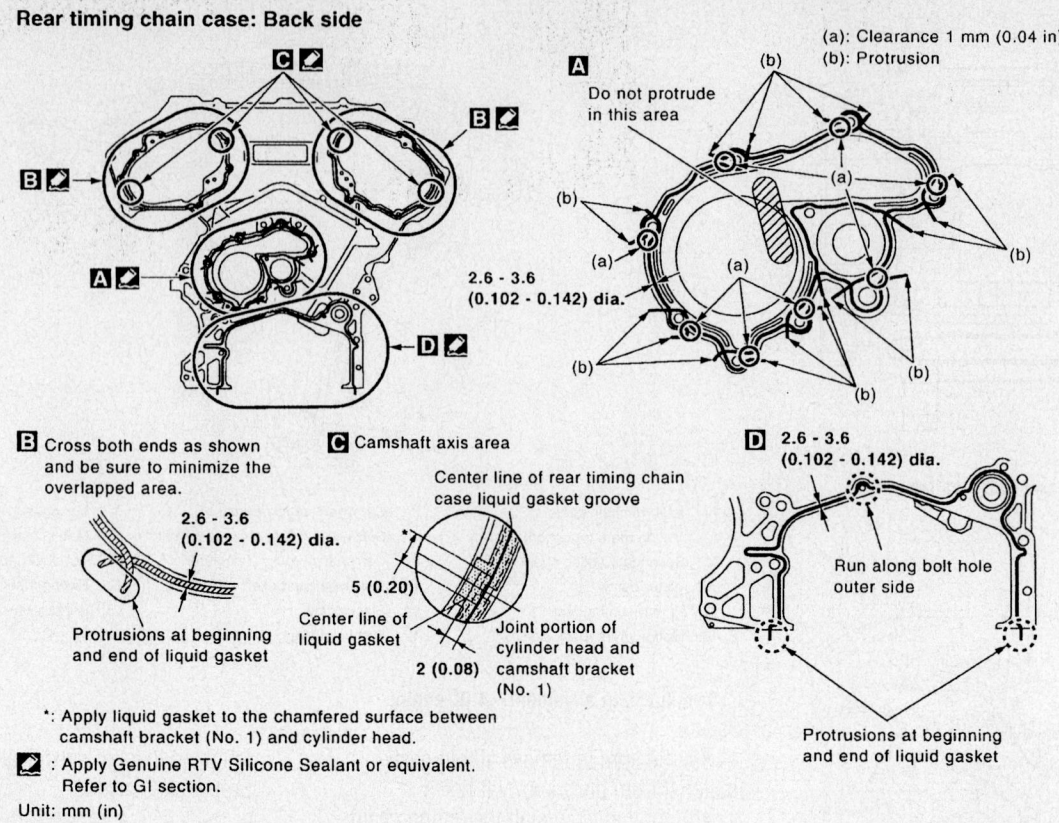

(a): Clearance 1 mm (0.04 in)
(b): Protrusion

A Do not protrude in this area

2.6 - 3.6 (0.102 - 0.142) dia.

B Cross both ends as shown and be sure to minimize the overlapped area.

2.6 - 3.6 (0.102 - 0.142) dia.

Protrusions at beginning and end of liquid gasket

C Camshaft axis area

Center line of rear timing chain case liquid gasket groove

5 (0.20)

Center line of liquid gasket

Joint portion of cylinder head and camshaft bracket (No. 1)

2 (0.08)

D 2.6 - 3.6 (0.102 - 0.142) dia.

Run along bolt hole outer side

Protrusions at beginning and end of liquid gasket

*: Apply liquid gasket to the chamfered surface between camshaft bracket (No. 1) and cylinder head.

✐ : Apply Genuine RTV Silicone Sealant or equivalent. Refer to GI section.

Unit: mm (in)

09482_FRON_G0082

Rear timing chain cover sealant application—4.0L engine

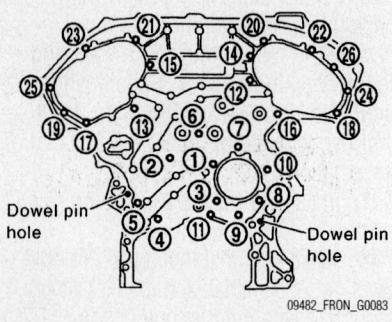

Dowel pin hole

Dowel pin hole

09482_FRON_G0083

Rear timing chain cover bolt torque sequence—4.0L engine

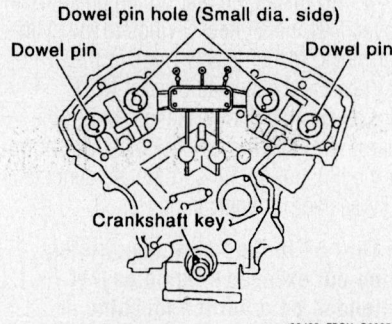

Dowel pin hole (Small dia. side)

Dowel pin

Dowel pin

Crankshaft key

09482_FRON_G0085

Dowel pin and crankshaft key alignment—4.0L engine

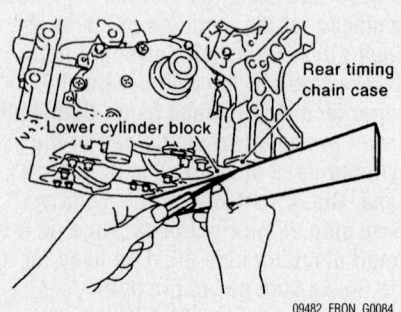

Rear timing chain case

Lower cylinder block

09482_FRON_G0084

Checking surface height—4.0L engine

plunger of the timing chain tensioner (secondary) and keep it pressed in with the stopper pin.

➡ **Mating surfaces between the timing chain and sprockets slip easily. Confirm all mating mark positions repeatedly during the installation process.**

50. Install the timing chains (secondary) and camshaft sprockets (INT and EXH).

51. Align the mating marks on the timing chain (secondary) cooper color link, with the ones on the camshaft sprockets (INT and EXH) punched and install them.

➡ **Mating marks for the camshaft sprocket (INT) are on the back side of the camshaft sprocket (secondary). There are two types of mating marks, circle and oval. They should be used for the right and the left banks, respectively. Right bank: circle type. Left bank: oval type.**

52. Align the dowel pin and pin hole on the camshafts with the groove and the dowel pin on the sprockets, and install them.

53. On the exhaust side, align the pin hole on the small diameter side of the camshaft front end with the dowel pin on the back side of the camshaft sprocket, and install them.

54. On the exhaust side, align the dowel pin on the camshaft front end with the pin groove on the camshaft sprocket, and install them.

➡ **In case that the positions of each mating mark and each dowel pin will not fit on the mating marks, make a fine adjustment to the position holding the hexagonal portion on the camshaft with a wrench, or equivalent.**

➡ **Bolts for the camshaft sprockets must be tightened. Tightening them by** hand is enough to prevent the dislocation of the dowel pins. It may be difficult to visually check the dislocation of mating marks during and after installation. To make the matching easier, make a mating mark on the top of the sprocket teeth and its extended line in advance with paint.

55. After confirming that the mating marks are aligned, tighten the camshaft sprocket bolts.

56. Pull the stopper pins out from the timing chain tensioners (secondary). Install the tension guide.

57. To install the timing chain (primary), install the crankshaft sprocket. Be sure that the mating marks on the crankshaft sprocket, face the front of the engine.

58. Install the timing chain (primary).

➡ Install the timing chain (primary) so that the mating mark punched (B) on the camshaft sprocket is aligned with the copper link (A) on the timing chain, while the mating mark notched (E) on

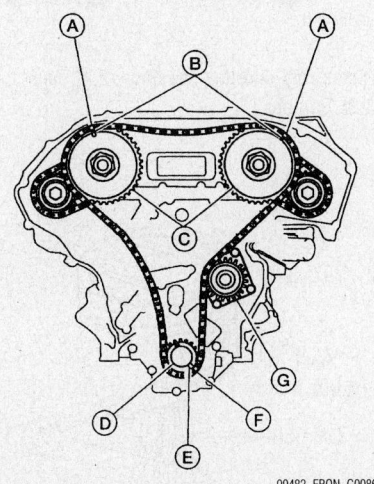

Timing chain (primary) alignment—4.0L engine

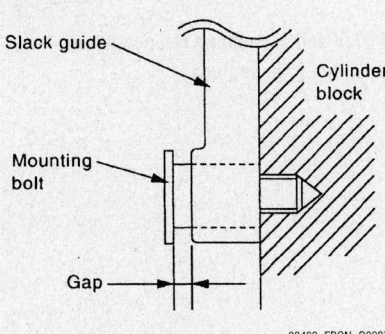

Slack guide mounting bolt gap—4.0L engine

the crankshaft sprocket (D) is aligned with the yellow link (F) on the timing chain, as shown in the illustration. If it is difficult to align mating marks (A) with (B) and (E) with (F) of the timing chain (primary) with each sprocket, gradually turn the camshaft using a wrench on the hexagonal portion to align it with the timing marks. During alignment be careful to prevent dislocation of the mating marks alignments of the timing chains (secondary). Note (G) indicates the water pump.

59. Install the internal chain guide, slack guide and timing chain tensioner (primary).

➡ Do not over tighten the slack guide bolts. It is normal for a gap to exist under the bolt seats when the bolts are tightened to specification.

60. When installing the timing chain tensioner (primary), push in the plunger and keep it pressed in with the stopper pin. Remove any dirt on the surfaces. After installation, pull out the stopper pin by pressing the slack guide.

61. Make sure, again, that the mating marks on the camshaft sprockets and timing chain have not slipped out of alignment. Install new O-rings on the rear timing chain case.

62. Install a new front seal in the front timing chain case cover.

63. Install the water pump cover and chain tensioner cover to the front timing chain case cover. Apply a continuous bead of liquid gasket (0.091–0.130 inch diameter) to the front timing chain case cover before installing the water pump cover and chain tensioner cover. Be sure to use genuine RTV sealant, or equivalent.

64. Before installing the front timing chain case cover apply a continuous bead of liquid gasket (0.102–0.142 inch in diameter) to the front timing chain case back side, as shown in the illustration. Be sure to use genuine RTV sealant, or equivalent.

65. Install new O-rings on the rear timing chain case. To assemble the front timing chain case cover, fit the lower end of the front timing chain case tightly onto the top face of the oil pan (upper). From the fitting point, make entire front timing chain case contact rear timing chain case completely.

➡ Since the front timing chain case cover is offset for difference of holt holes, tighten the bolts temporarily while holding the front timing chain case cover from the front and the top. Now insert a dowel pin while holding the front timing chain case cover from the front and the top.

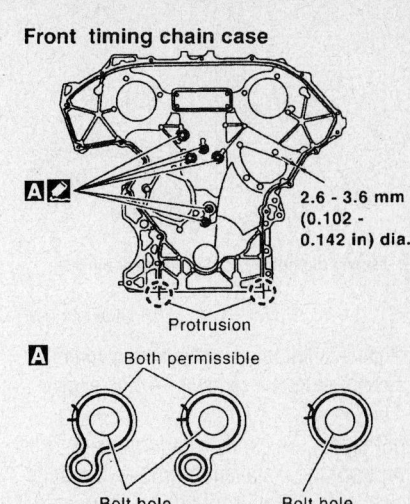

Front timing chain case

2.6 - 3.6 mm (0.102 - 0.142 in) dia.

Protrusion

Ⓐ Both permissible

Bolt hole   Bolt hole

Liquid gasket protrusion away from bolt hole

🔷 : Apply Genuine RTV silicone sealant or equivalent. Refer to GI section.

Front timing chain cover sealant application—4.0L engine

66. Once the cover is installed, torque the retaining bolts to specification and in the proper sequence. There are four different types of bolts.

a. Bolt diameter: 0.39 inch. Bolt position: 1–5. Torque to 41 ft. lbs.

b. Bolt diameter: 0.24 inch. Bolt position: 6–25. Torque to 9 ft. lbs.

c. After all bolts are tightened, retighten them to specification and in the proper sequence

67. Install the two bolts in the oil pan (upper). Torque to 16 ft. lbs.

68. Install new seal rings in the shaft grooves of the right and left intake valve timing control covers.

69. Apply a continuous bead of liquid gasket (0.083–0.122 inch in diameter) to the covers. Be sure to use genuine RTV sealant, or equivalent.

70. Install new collared O-rings in the front timing chain case oil hole (left and

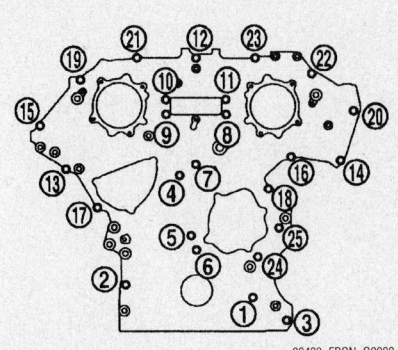

Front timing chain cover bolt torque sequence—4.0L engine

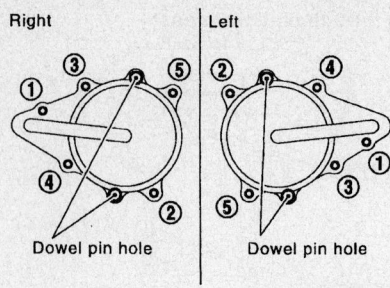

09482_FRON_G0090

Right and left intake valve timing control cover bolt torque sequence—4.0L engine

right sides). Be careful not to move the seal ring from the installation groove, align the dowel pins on the front timing chain case with the holes to install the intake valve timing control covers.

71. Tighten the bolts in sequence and to specification.

72. Install the crankshaft pulley. Torque to specification

73. Install the upper and lower oil pans.

74. Install the intake manifold collector.

75. Before installing the rocker cover, apply liquid gasket, be sure to use genuine RTV silicone sealant or equivalent, to the positions shown in the illustration. Refer to figure "a" to apply liquid gasket to joint part of camshaft bracket No.1 and cylinder head. Refer to figure "b" to apply liquid gasket to the figure "a" squarely.

76. Install the rocker cover. Torque the retaining bolts to 17 inch lbs and than to 74 inch lbs, in the proper sequence.

77. Continue the installation in the reverse order of the removal procedure.

➥**If hydraulic pressure inside the timing chain tensioner drops after removal/installation, slack in the guide may generate a pounding noise during and just after engine start. This is normal the noise will stop after hydraulic pressure rises.**

Piston and Ring

POSITIONING

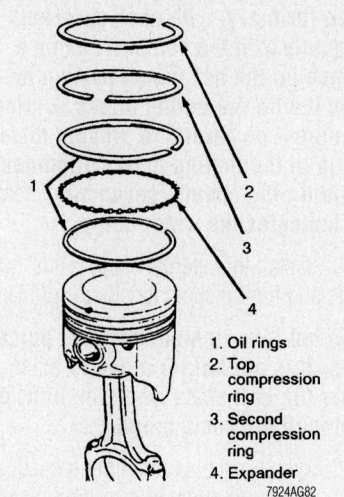

1. Oil rings
2. Top compression ring
3. Second compression ring
4. Expander

7924AG82

Piston ring positioning—2.4L engine

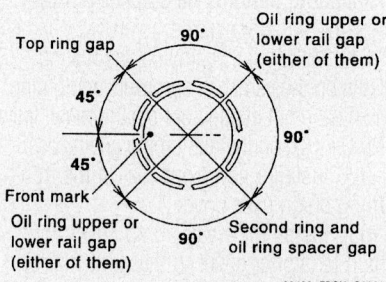

09482_FRON_G0091

Piston ring positioning—2.5L engine

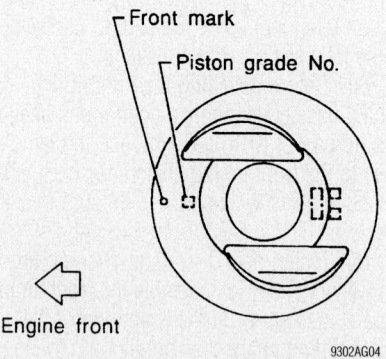

9302AG04

Piston ring positioning—3.3L engine

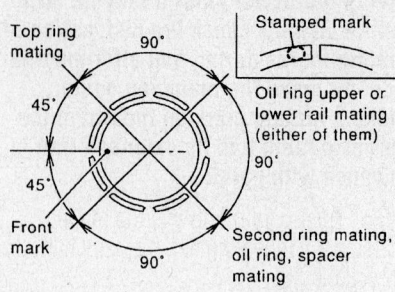

09482_FRON_G0092

Piston ring positioning—4.0L engine

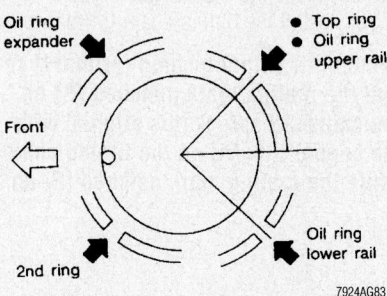

7924AG83

Piston ring end-gap spacing—2.4L and 3.3L engines

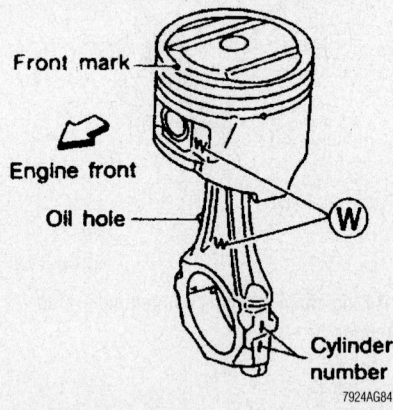

7924AG84

Piston and connecting rod positioning—2.4L and 3.3L engines

FUEL SYSTEM

Fuel System Service Precautions

Safety is the most important factor when performing not only fuel system maintenance but any type of maintenance. Failure to conduct maintenance and repairs in a safe manner may result in serious personal injury or death. Maintenance and testing of the vehicle's fuel system components can be accomplished safely and effectively by adhering to the following rules and guidelines.

• To avoid the possibility of fire and personal injury, always disconnect the negative battery cable unless the repair or test procedure requires that battery voltage be applied.

• Always relieve the fuel system pressure prior to disconnecting any fuel system component (injector, fuel rail, pressure regulator, etc.), fitting or fuel line connection. Exercise extreme caution whenever relieving fuel system pressure, to avoid exposing skin, face and eyes to fuel spray. Please be advised that fuel under pressure may penetrate the skin or any part of the body that it contacts.

• Always place a shop towel or cloth around the fitting or connection prior to loosening to absorb any excess fuel due to spillage. Ensure that all fuel spillage (should it occur) is quickly removed from engine surfaces. Ensure that all fuel soaked cloths or towels are deposited into a suitable waste container.

• Always keep a dry chemical (Class B) fire extinguisher near the work area.

• Do not allow fuel spray or fuel vapors to come into contact with a spark or open flame.

• Always use a back-up wrench when loosening and tightening fuel line connection fittings. This will prevent unnecessary stress and torsion to fuel line piping. Always follow the proper torque specifications.

• Always replace worn fuel fitting O-rings with new. Do not substitute fuel hose or equivalent, where fuel pipe is installed.

Fuel System Pressure

RELIEVING

1. Before servicing the vehicle, refer to the Precautions Section.
2. Remove the fuel pump fuse from the panel.
3. Start the engine and allow it to run until it stalls. Crank the engine for a few seconds to relieve additional fuel pressure.
4. Disconnect the negative battery cable.
5. When repairs are complete, replace the fuel pump fuse and connect the negative battery cable.

Fuel Filter

REMOVAL & INSTALLATION

➡**On 2002–2004 vehicles the fuel filter is located under the vehicle near the fuel tank.**

1. Before servicing the vehicle, refer to the Precautions Section.
2. Properly relieve the fuel system pressure.
3. Disconnect the negative battery cable.
4. Remove or disconnect the following:

   • Fuel filter shield, if equipped
   • Fuel lines
   • Fuel filter from the bracket

To install:
5. Install or connect the following:
   • Fuel filter to the bracket
   • Fuel lines
   • Fuel filter shield, if equipped
   • Negative battery cable.
6. Start the engine and check for leaks.

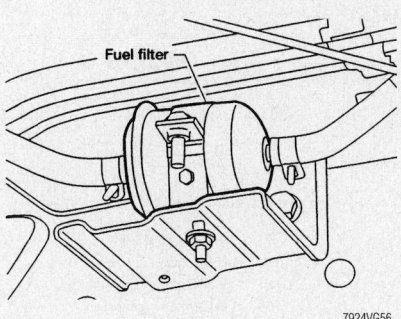

Fuel filter

Fuel filter

7924VG56

Typical fuel filter locations—2002–2004 vehicles

Fuel Pump

REMOVAL & INSTALLATION

Frontier

2002–2004

1. Before servicing the vehicle, refer to the Precautions Section.
2. Relieve the fuel system pressure.
3. Drain the fuel tank.
4. Remove or disconnect the following:
   • Negative battery cable
   • Fuel pump module harness connectors
   • Filler hose shield
   • Fuel pressure and return lines
   • Filler hose
   • Vent hose
   • Evaporative Emissions (EVAP) hose
   • Fuel tank skid plate
   • Fuel tank
   • Fuel level sender
   • Fuel pump

To install:
5. Install or connect the following:
   • Fuel pump
   • Fuel level sender. Tighten the screws to 17–23 inch lbs. (2.0–2.5 Nm).
   • Fuel tank. Tighten the bolts to 27–36 ft. lbs. (37–49 Nm).
   • Fuel tank skid plate. Tighten the bolts to 27–36 ft. lbs. (37–49 Nm).
   • EVAP hose
   • Vent hose
   • Filler hose
   • Fuel pressure and return lines
   • Filler hose shield
   • Fuel pump module harness connectors
   • Negative battery cable
6. Fill the fuel tank.
7. Start the engine and check for leaks.

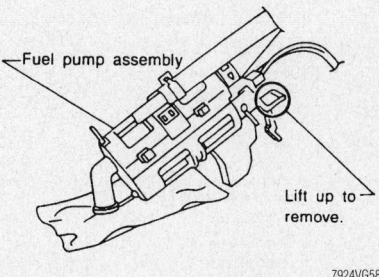

Fuel pump assembly

Lift up to remove.

7924VG58

Remove the fuel pump with bracket while lifting the pawl of the pump bracket upward—2002–2004 vehicles

2005–2006

➡️**Be sure to check the fuel gauge indicator. Make sure that it reads less than FULL. If not drain some fuel until the gauge reads less than FULL.**

1. Before servicing the vehicle, refer to the Precautions Section.
2. Properly relieve the fuel system pressure.
3. Disconnect the negative battery cable.
4. Remove the fuel filler cap. Remove the left rear tire and wheel assembly.
5. Remove the fuel tank shield.
6. Properly support the fuel tank. Remove the fuel tank retaining straps.
7. Lower the fuel tank to gain access to the top of the fuel pump assembly.

8. Disconnect the fuel pump assembly electrical connector, EVAP hose and the fuel feed hose from the molded clip in the side of the fuel tank.
9. Disconnect the quick connector.
10. Lower the fuel tank and remove it from the vehicle. Remove the fuel pump assembly lockring.
11. Disconnect the EVAP hose from the molded clip in the top of the fuel tank.
12. Remove the fuel pump assembly. Discard the O-ring.

To install:
13. Installation is the reverse of the removal procedure.
14. Be sure to use a new O-ring upon installation.
15. Turn the ignition switch ON, but do not start the engine. Check the fuel lines

and hose connections for leaks while applying fuel pressure to the system.
16. Start the engine and check for fuel leaks, correct as required.

Xterra

2002–2004

1. Before servicing the vehicle, refer to the Precautions Section.
2. Relieve the fuel system pressure.
3. Remove the rear seat cushion and the access panel.
4. Remove or disconnect the following:
   • Negative battery cable
   • Fuel pump electrical connectors
5. Matchmark the installed position of the fuel line quick connect fittings, then disconnect the fittings by holding the sides of the connector, push in the tabs and pull out the tube inserted in the retainer.

➡️**The tube can be removed when the tabs are completely pushed in. Do NOT use any tools to remove the quick connector.**

   • Six screws
   • Fuel level sensor retainer and fuel level sensor
   • Fuel pump with the bracket, while lifting the pawl of the fuel bracket upward
   • Fuel level sensor

To install:
6. Installation is the reverse of the removal procedure.
7. Start the engine and check for leaks.

2005–2006

➡️**Be sure to check the fuel gauge indicator. Make sure that it reads less than FULL. If not drain some fuel until the gauge reads less than FULL.**

1. Before servicing the vehicle, refer to the Precautions Section.
2. Properly relieve the fuel system pressure.
3. Disconnect the negative battery cable.
4. Remove the fuel filler cap. Remove the left rear tire and wheel assembly.
5. Disconnect the lower fuel filler hose from the fuel tank, the EVAP hose and the vent pipe quick connector.

➡️**Disconnect the fuel feed hose from the molder clip in the side of the fuel tank.**

6. Remove the four tank shield retaining bolts. Remove the tank shield.
7. Remove the driveshaft.

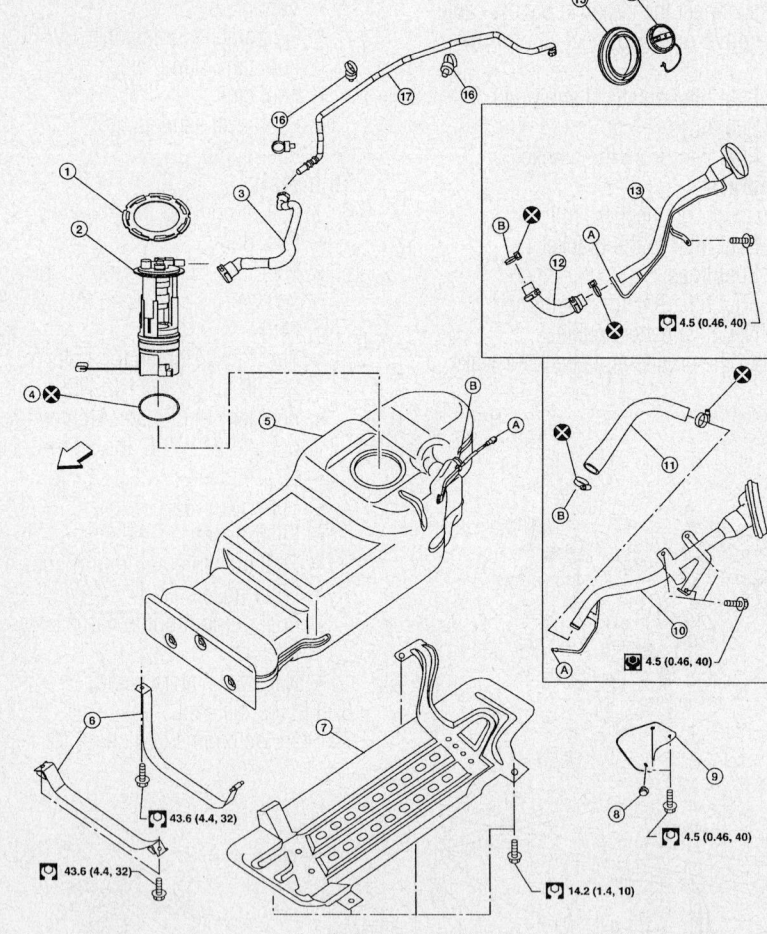

1.	Lock ring	2.	Fuel level sensor, fuel filter, and fuel pump assembly	3.	EVAP hose
4.	Fuel level sensor, fuel filter, and fuel pump assembly O-ring	5.	Fuel tank	6.	Fuel tank straps
7.	Fuel tank shield	8.	Clip	9.	Fuel filler pipe shield
10.	Fuel filler pipe	11.	Fuel filler hose	12.	Fuel filler hose
13.	Fuel filler pipe	14.	Fuel filler cap	15.	Fuel filler pipe grommet
16.	Clamp	17.	EVAP canister hose	A.	Fuel filler hose connection
B.	Fuel filler hose vent connection	⇐	Front		

09482_FRON_G0093

Fuel pump and related components—2005–2006 Frontier

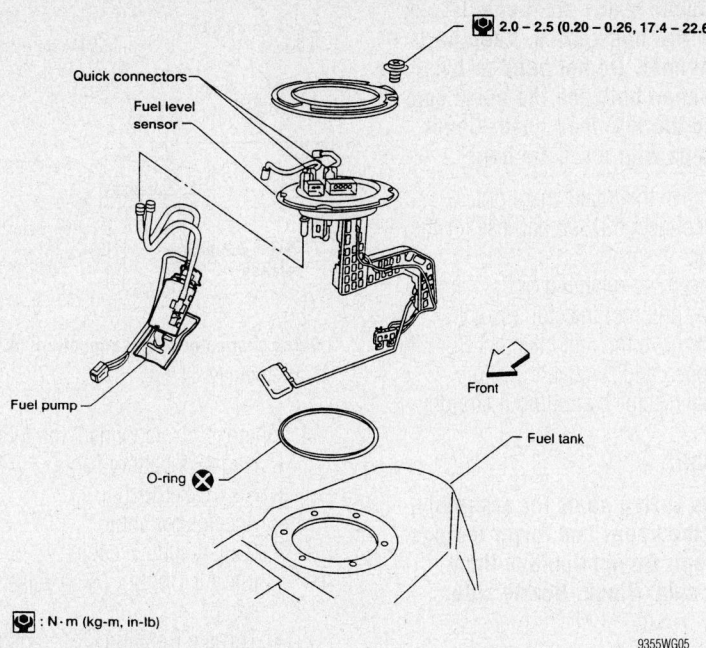

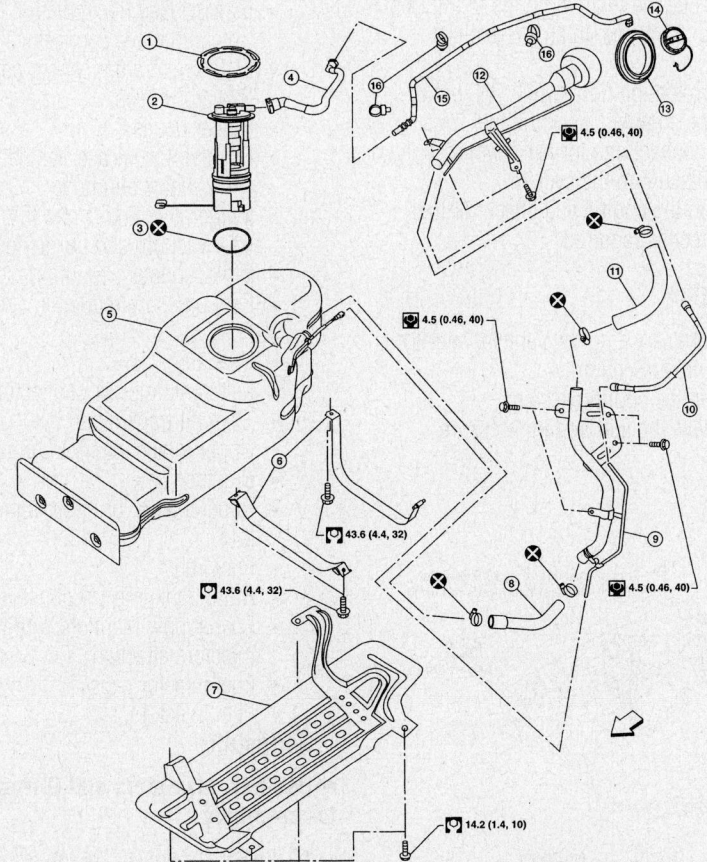

Fuel pump and related components—2002–2004 Xterra

9355WG05

| : N·m (kg-m, in-lb) |

Fuel pump and related components—2005–2006 Xterra

1.	Lock ring	2.	Fuel level sensor, fuel filter, and fuel pump assembly	3.	Fuel level sensor, fuel filter, and fuel pump assembly O-ring
4.	EVAP hose	5.	Fuel tank	6.	Fuel tank straps
7.	Fuel tank shield	8.	Lower fuel filler hose	9.	Fuel filler pipe and vent pipe
10.	Vent hose	11.	Upper fuel filler hose	12.	Fuel filler pipe and cup
13.	Fuel filler hose grommet	14.	Fuel filler cap	15.	EVAP Canister hose
16.	Clip	⇐	Front		

09482_FRON_G0094

8. Properly support the fuel tank. Remove the three fuel tank retaining strap bolts. Remove the fuel tank straps.

9. Lower the fuel tank to gain access to the top of the fuel pump assembly.

➡**Be careful not to lower the tank too much as you do not want to damage the fuel feed hose and the fuel pump assembly.**

10. Disconnect the fuel pump assembly electrical connector, and the fuel feed hose.

11. Disconnect the quick connector.

12. Lower the fuel tank and remove it from the vehicle. Disconnect the EVAP hose from the fuel pump and remove the EVAP hose from the molded clip in the top of the fuel tank.

13. Remove the fuel pump assembly lockring. Remove the fuel pump assembly. Discard the O-ring.

To install:

14. Installation is the reverse of the removal procedure.

15. Be sure to use a new O-ring upon installation.

16. Turn the ignition switch ON, but do not start the engine. Check the fuel lines and hose connections for leaks while applying fuel pressure to the system.

17. Start the engine and check for fuel leaks, correct as required.

Fuel Injectors

REMOVAL & INSTALLATION

2.4L Engine

1. Before servicing the vehicle, refer to the Precautions Section.

2. Relieve the fuel system pressure.

3. Remove or disconnect the following:
   - Negative battery cable
   - Air cleaner assembly
   - Fuel lines
   - Fuel pressure regulator vacuum line
   - Fuel injector connectors
   - Fuel supply manifold with the injectors attached
   - Fuel injector caps
   - Fuel injectors

To install:

➡**Use new insulators and O-ring seals for assembly.**

4. Install or connect the following:
   - Fuel injectors
   - Fuel injector caps. Tighten the screws to 26–34 inch lbs. (3–4 Nm).

- Fuel supply manifold with the injectors attached. Tighten the bolts to 96–132 inch lbs. (11–15 Nm).
- Fuel injector connectors
- Fuel pressure regulator vacuum line
- Fuel lines
- Air cleaner assembly
- Negative battery cable

5. Start the engine and check for leaks.

2.5L Engine

1. Before servicing the vehicle, refer to the Precautions Section.

2. Properly relieve the fuel system pressure.

3. Disconnect the negative battery cable. Remove the fuel filler cap.

4. Remove the quick connector cap (engine side). With the sleeve side of the quick connector release facing the quick connector, install the quick connector release on to the tube. Insert the quick connector release into the quick connector until the sleeve contacts and goes no further. Hold the quick connector release in that position.

➡**Disconnect the quick connector using tool J-45488, or equivalent, not by picking out the retainer tabs. Inserting the quick connector hard will not disconnect the quick connector. Hold the quick connector release where it contacts and goes no further.**

5. Draw and pull out the quick connector straight from the fuel tube. Grasp the quick connector holding "A" in the illustration. Do not pull with lateral force applied and the O-ring inside the quick connector could be damaged.

➡**Have a cloth ready, as fuel will leak out. Avoid fire and sparks. Keep parts away from heat. Do not bend or twist the connection between the quick connector and the fuel feed hose. Cover the openings with a plastic bag.**

6. Remove the intake manifold.

7. Disconnect the sub harness for the fuel injector.

8. Loosen the retaining bolts. Remove the fuel tube and fuel injector assembly.

9. To remove the fuel injectors from the fuel tube, open and remove the clip. Remove the injector by pulling it straight out.

To install:

➡**Use new O-ring seals for assembly. Note that the upper and lower O-rings are different. Do not confuse them. Fuel tube side: Black. Nozzle side: Green.**

10. Installation is the reverse of the removal procedure.

11. When installing the fuel feed tube be sure to torque the retaining bolts to 9 ft. lbs and then to 21 ft. lbs. in an alternating order.

12. Turn the ignition switch ON, but do not start the engine. Check the fuel lines and hose connections for leaks while applying fuel pressure to the system.

13. Start the engine and check for fuel leaks, correct as required.

3.3L Engine

1. Before servicing the vehicle, refer to the Precautions Section.

2. Drain the cooling system.

3. Relieve the fuel system pressure.

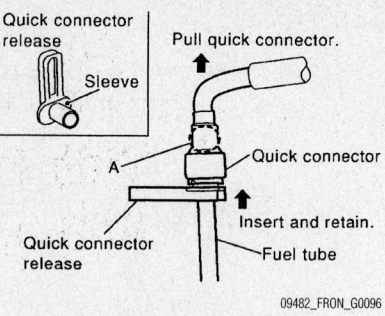

Quick connector release location "A"—2.5L engine

4. Remove or disconnect the following:
- Negative battery cable
- Air intake duct
- Accelerator cable
- Cruise control cable
- Idle Air Control (IAC) valve connector
- Throttle Position (TP) sensor and switch connectors
- Ignition coil and power transistor connectors
- Exhaust Gas Recirculation (EGR) Solenoid valve connector
- EGR temperature sensor connector
- Radiator hoses
- Heater hoses
- Positive Crankcase Ventilation (PCV) valve and hose
- Evaporative Emissions (EVAP) canister vacuum and purge hoses
- Brake booster vacuum hose
- Fuel pressure regulator vacuum hose
- EGR tube
- Left bank injector connectors
- Thermal transmitter
- Upper intake manifold ground cable
- Breather pipe
- Supercharger or upper intake manifold
- Fuel lines
- Right bank injector connectors
- Fuel supply manifold with the injectors attached
- Fuel injector caps
- Fuel injectors

To install:

➡**Use new insulators and O-ring seals for assembly.**

5. Install or connect the following:
- Fuel injectors
- Fuel injector caps. Tighten the screws to 26–34 inch lbs. (3–4 Nm).
- Fuel supply manifold with the injectors attached. Tighten the bolts to 96–132 inch lbs. (11–15 Nm).

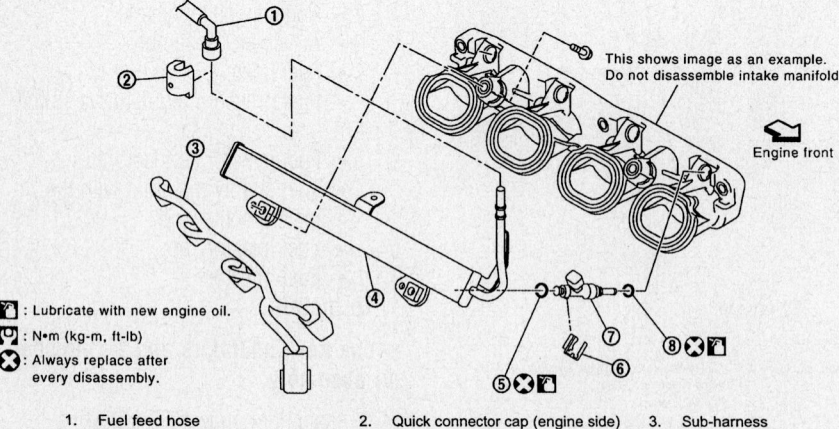

- : Lubricate with new engine oil.
- : N•m (kg-m, ft-lb)
- : Always replace after every disassembly.

1.	Fuel feed hose	2.	Quick connector cap (engine side)	3.	Sub-harness
4.	Fuel tube	5.	O-ring (black)	6.	Clip
7.	Fuel injector	8.	O-ring (green)		

Fuel injector tube and related components—2.5L engine

09482_FRON_G0095

- Right bank injector connectors
- Fuel lines
- Upper intake manifold
- Breather pipe
- Upper intake manifold ground cable
- Thermal transmitter
- Left bank injector connectors
- EGR tube
- Fuel pressure regulator vacuum hose
- Brake booster vacuum hose
- EVAP canister vacuum and purge hoses
- PCV valve and hose
- Heater hoses
- Radiator hoses
- EGR temperature sensor connector
- EGR Solenoid valve connector
- Ignition coil and power transistor connectors
- TP sensor and switch connectors
- IAC valve connector
- Cruise control cable
- Accelerator cable
- Air intake duct
- Negative battery cable

6. Fill the cooling system.
7. Start the engine and check for leaks.

4.0L Engine

1. Before servicing the vehicle, refer to the Precautions Section.
2. Properly relieve the fuel system pressure.
3. Disconnect the negative battery cable. Remove the fuel filler cap.
4. Remove the intake manifold collector.
5. Remove the quick connector cap (engine side). With the sleeve side of the quick connector release facing the quick connector, install the quick connector release on to the tube. Insert the quick connector release into the quick connector until the sleeve contacts and goes no further. Hold the quick connector release in that position.

➡**Disconnect the quick connector using tool J-45488, or equivalent, not by picking out the retainer tabs. Inserting the quick connector hard will not disconnect the quick connector. Hold the quick connector release where it contacts and goes no further.**

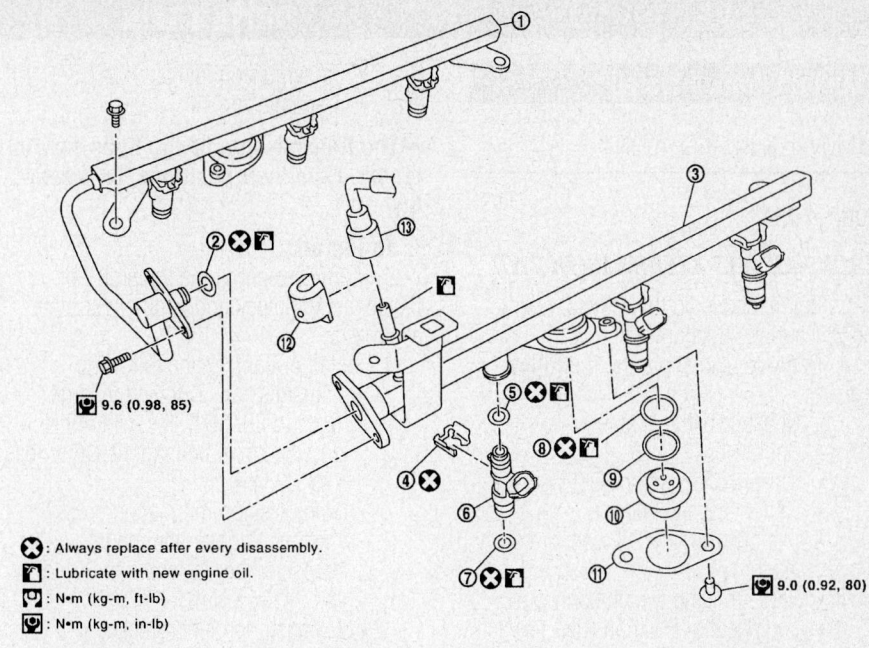

9.6 (0.98, 85)

⊗ : Always replace after every disassembly.
▨ : Lubricate with new engine oil.
⬛ : N•m (kg-m, ft-lb)
⬛ : N•m (kg-m, in-lb)

9.0 (0.92, 80)

1.	Fuel tube (RH)	2.	O-ring	3.	Fuel tube (LH)
4.	Clip	5.	O-ring (blue)	6.	Fuel injector
7.	O-ring (brown)	8.	O-ring	9.	Spacer
10.	Fuel damper	11.	Fuel damper cap	12.	Quick connector cap
13.	Fuel feed hose				

09482_FRON_G0097

Fuel injector tube and related components—4.0L engine

6. Draw and pull out the quick connector straight from the fuel tube. Grasp the quick connector holding "A" in the illustration. Do not pull with lateral force applied and the O-ring inside the quick connector could be damaged.

➡**Have a cloth ready, as fuel will leak out. Avoid fire and sparks. Keep parts away from heat. Do not bend or twist the connection between the quick connector and the fuel feed hose. Cover the openings with a plastic bag.**

7. Remove the PCV hose between the rocker covers.
8. Disconnect the harness for the fuel injector.
9. Loosen the retaining bolts. Remove the fuel tube and fuel injector assembly. Remove the bolts which connect the left and right fuel tubes.
10. To remove the fuel injectors from the fuel tube, open and remove the clip. Remove the injector by pulling it straight out.

11. Disconnect the right fuel tube from the left fuel tube. Loosen the bolts, to remove the fuel damper cap and fuel damper, if necessary.

To install:

➡**Use new O-ring seals for assembly. Note that the upper and lower O-rings are different. Do not confuse them. Fuel tube side: Blue. Nozzle side: Brown.**

12. Installation is the reverse of the removal procedure.
13. When installing the fuel feed tube be sure to torque the retaining bolts to 7 ft. lbs and then to 16 ft. lbs. in an alternating order.
14. Turn the ignition switch ON, but do not start the engine. Check the fuel lines and hose connections for leaks while applying fuel pressure to the system.
15. Start the engine and check for fuel leaks, correct as required.

DRIVE TRAIN

Manual Transmission

REMOVAL & INSTALLATION

2002–2004

FRONTIER AND XTERRA WITH 2WD

1. Before servicing the vehicle, refer to the Precautions Section.

2. Remove or disconnect the following:

- Negative battery cable
- Shift lever
- Crankshaft Position (CKP) sensor
- Clutch slave cylinder
- Vehicle Speed (VSS) sensor connector
- Back-up lamp switch connector
- Park/Neutral Position (PNP) switch connector
- Rear Heated Oxygen (HO2S) sensor connector
- Starter motor
- Driveshaft
- Exhaust mounting bracket
- Transmission mount and crossmember. Support the transmission.

- Transmission flange bolts
- Transmission

→ The transmission flange bolts vary in length. Note their positions for assembly.

To install:

3. Apply sealant to the transmission flange, engine block and engine rear plate as shown.

4. Install or connect the following:

- Transmission. Tighten the large bolts to 29–36 ft. lbs. (39–49 Nm) and the small bolts to 12–16 ft. lbs. (16–22 Nm).
- Transmission mount and crossmember. Tighten the mount and crossmember fasteners to 30–38 ft. lbs. (41–52 Nm).
- Exhaust mounting bracket
- Driveshaft
- Starter motor
- HO2S sensor connector
- PNP switch connector
- Back-up lamp switch connector
- VSS sensor connector
- Clutch slave cylinder
- CKP sensor

- Shift lever
- Negative battery cable

FRONTIER WITH 4WD

1. Before servicing the vehicle, refer to the Precautions Section.

2. Remove or disconnect the following:

- Negative battery cable
- Shift lever
- Transfer case select lever
- Crankshaft Position (CKP) sensor
- Clutch slave cylinder
- Vehicle Speed (VSS) sensor connector
- Back-up lamp switch connector
- Park/Neutral Position (PNP) switch connector
- Rear Heated Oxygen (HO2S) sensor connector
- Starter motor
- Front and rear driveshafts
- Exhaust front pipes
- Exhaust center pipe
- Torsion bars and mounts
- Rear torsion bar cross mount
- Transmission mount and crossmember. Support the transmission.

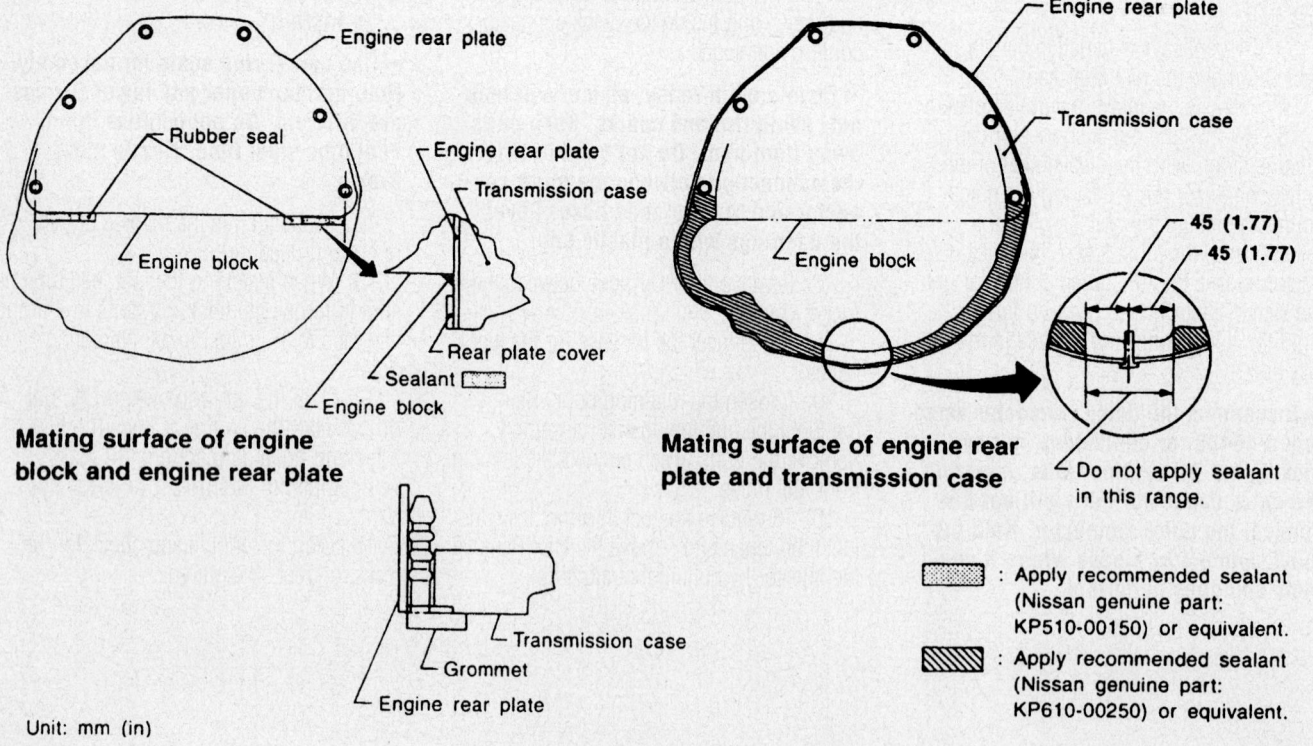

Mating surface of engine block and engine rear plate

Mating surface of engine rear plate and transmission case

45 (1.77)
45 (1.77)

Do not apply sealant in this range.

: Apply recommended sealant (Nissan genuine part: KP510-00150) or equivalent.

: Apply recommended sealant (Nissan genuine part: KP610-00250) or equivalent.

Unit: mm (in)

7924VG61

Apply sealant to the indicated areas between the engine block, transmission and engine rear plate—4WD vehicles

Bolt No.	1	2	3
Quantity	4	1	6
Bolt length " ℓ " mm (in)	60 (2.36)		65 (2.56)
Tightening torque N·m (kg-m, ft-lb)	34.3 (3.5, 25)		75 (7.7, 55)

⦿ Transmission to Engine
⦻ Transmisstion to Dust Cover

View from vehicle rear

09482_FRON_G0098

Five speed transmission bolt tightening sequence—2005–2006

- Transmission flange bolts
- Transmission

➥ **The transmission flange bolts vary in length. Note their positions for assembly.**

To install:

3. Apply sealant to the transmission flange, engine block, and engine rear plate as shown.

4. Install or connect the following:
- Transmission. Tighten the large bolts to 29–36 ft. lbs. (39–49 Nm) and the small bolts to 22–29 ft. lbs. (29–39 Nm).
- Transmission mount and cross-member. Tighten the mount and crossmember fasteners to 30–38 ft. lbs. (41–52 Nm).
- Rear torsion bar cross mount
- Torsion bars and mounts
- Exhaust center pipe
- Exhaust front pipes
- Front and rear driveshafts
- Starter motor
- HO_2S sensor connector
- PNP switch connector
- Back-up lamp switch connector
- VSS sensor connector
- Clutch slave cylinder
- CKP sensor
- Transfer case select lever
- Shift lever
- Negative battery cable

XTERRA WITH 4WD

1. Before servicing the vehicle, refer to the Precautions Section.

2. Remove or disconnect the following:
- Negative battery cable
- Shift lever
- Transfer case select lever
- Crankshaft Position (CKP) sensor
- Clutch slave cylinder
- Vehicle Speed (VSS) sensor connector
- Back-up lamp switch connector

- Park/Neutral Position (PNP) switch connector
- Rear Heated Oxygen (HO_2S) sensor connector
- Starter motor
- Front and rear driveshafts
- Exhaust front pipes
- Exhaust center pipe
- Torsion bars and mounts
- Rear torsion bar cross mount
- Transmission mount and cross-member. Support the transmission.
- Transmission flange bolts
- Transmission

➥ **The transmission flange bolts vary in length. Note their positions for assembly.**

To install:

3. Apply sealant to the transmission flange, engine block, and engine rear plate as shown.

4. Install or connect the following:
- Transmission. Tighten the large bolts to 29–36 ft. lbs. (39–49 Nm) and the small bolts to 22–29 ft. lbs. (29–39 Nm).
- Transmission mount and cross-member. Tighten the mount and crossmember fasteners to 30–38 ft. lbs. (41–52 Nm).
- Rear torsion bar cross mount
- Torsion bars and mounts
- Exhaust center pipe
- Exhaust front pipes
- Front and rear driveshafts
- Starter motor
- HO_2S sensor connector
- PNP switch connector
- Back-up lamp switch connector
- VSS sensor connector
- Clutch slave cylinder
- CKP sensor
- Transfer case select lever
- Shift lever
- Negative battery cable

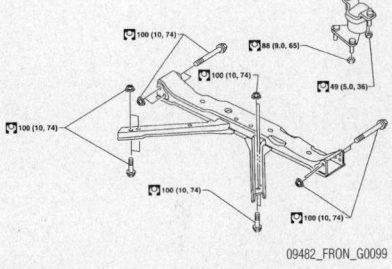

09482_FRON_G0099

Five speed transmission crossmember bolt tightening specifications—2005–2006

2005–2006

FIVE SPEED

1. Before servicing the vehicle, refer to the Precautions Section.

2. Disconnect the negative battery cable.

3. Raise and support the vehicle safely. Drain the transmission fluid.

4. Remove the shift lever assembly. Remove the rear driveshaft. Remove the gusset.

5. Disconnect the heated oxygen sensor connector and remove the wire harness from the transmission.

6. Disconnect the back up light switch and the PNP switch connectors.

7. Remove the clutch slave cylinder from the transmission. Remove the starter.

8. Support the transmission using a suitable jack.

9. Remove the transmission dust cover. Remove the transmission to engine retaining bolts.

10. Remove the nuts securing the insulator to the crossmember. Remove the crossmember retaining bolts. Remove the crossmember.

➥ **Be sure that the transmission is properly supported.**

11. Remove the air breather hose and the breather tube.

Quantity	10
Bolt length " ℓ " mm (in)	65 (2.56)
Tightening torque N·m (kg-m, ft-lb)	75 (7.7, 55)

View from vehicle rear

09482_FRON_G0100

Six speed transmission bolt tightening sequence—2005–2006

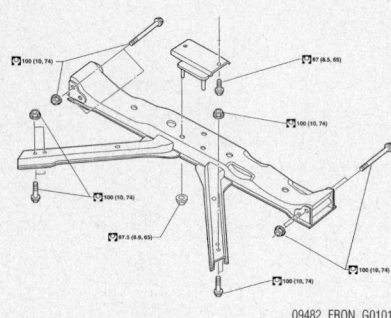

09482_FRON_G0101

Six speed transmission crossmember bolt tightening specifications—2005–2006

12. Separate the transmission from the engine and remove it from the vehicle.

To install:

13. Installation is the reverse of the removal procedure.

14. Torque the transmission to engine retaining bolts to specification and in the proper sequence.

15. Start the engine and check for leaks, correct as required.

16. Roadtest the vehicle and check for vibrations, correct as required.

SIX SPEED

1. Before servicing the vehicle, refer to the Precautions Section.

2. Disconnect the negative battery cable.

3. Raise and support the vehicle safely. Drain the transmission fluid.

4. Remove the shift lever assembly. Remove the left fender protector.

5. Remove the crankshaft position sensor (POS) from the transmission. Be careful not to damage the sensor edge.

6. Remove the undercovers. Remove the front crossmember. Remove the starter.

7. If equipped with 4WD, remove the front and rear driveshafts.

8. If equipped with 2WD, remove the rear driveshaft.

9. Remove the left and right front

exhaust tubes. Remove the clutch slave cylinder from the transmission.

10. Support the transmission using a suitable jack.

11. Remove the nuts securing the insulator to the crossmember. Remove the crossmember retaining bolts. Remove the crossmember.

➡**Be sure that the transmission is properly supported.**

12. Tilt the transmission slightly to gain clearance between the body and the transmission, and then remove the air breather hose and the breather tube.

13. Disconnect the back up light electrical connector and the PNP switch connector.

14. If equipped with 4WD, disconnect the 4LO switch connector, wait detection switch connector, ATP switch connector and the transfer control device connector.

15. Remove the wiring harness from the retainers.

16. Remove the transmission to engine retaining bolts. Separate the transmission from the engine and remove it from the vehicle.

To install:

17. Installation is the reverse of the removal procedure.

18. Torque the transmission to engine retaining bolts to specification and in the proper sequence.

Automatic Transmission

REMOVAL & INSTALLATION

2002–2004

2WD

1. Before servicing the vehicle, refer to the Precautions Section.

2. Remove or disconnect the following:

- Negative battery cable
- Crankshaft Position (CKP) sensor
- Exhaust front pipes
- Exhaust rear pipes
- Transmission dipstick tube
- Transmission oil cooler lines
- Driveshaft
- Shift cable
- Transmission control harness connectors
- Vehicle Speed (VSS) sensor connector
- Starter motor
- Torque converter
- Transmission mount and crossmember. Support the transmission.
- Transmission flange bolts
- Transmission

➡**The transmission flange bolts vary in length. Note their positions for assembly.**

To install:

3. Install or connect the following:

- Transmission. Tighten the large bolts to 29–36 ft. lbs. (39–49 Nm) and the small bolts to 22–29 ft. lbs. (29–39 Nm).
- Transmission mount and crossmember. Tighten the mount and crossmember fasteners to 30–38 ft. lbs. (41–52 Nm).
- Torque converter. Tighten the bolts to 33–43 ft. lbs. (44–59 Nm).
- Starter motor
- VSS sensor connector
- Transmission control harness connectors
- Shift cable
- Driveshaft
- Transmission oil cooler lines
- Transmission dipstick tube
- Exhaust rear pipes
- Exhaust front pipes
- CKP sensor
- Negative battery cable

4WD

1. Before servicing the vehicle, refer to the Precautions Section.
2. Remove or disconnect the following:
   - Negative battery cable
   - Crankshaft Position (CKP) sensor
   - Exhaust front pipes
   - Exhaust rear pipes
   - Transmission dipstick tube
   - Transmission oil cooler lines
   - Front and rear driveshafts
   - Transfer case linkage
   - Shift cable
   - Transmission control harness connectors
   - Vehicle Speed (VSS) sensor connector
   - Starter motor
   - Torque converter
   - Transmission mount and crossmember. Support the transmission.
   - Transmission flange bolts
   - Transmission

➡**The transmission flange bolts vary in length. Note their positions for assembly.**

To install:

3. Install or connect the following:
4. Install or connect the following:
   - Transmission. Tighten the large bolts to 29–36 ft. lbs. (39–49 Nm) and the small bolts to 22–29 ft. lbs. (29–39 Nm).
   - Transmission mount and crossmember. Tighten the mount and crossmember fasteners to 30–38 ft. lbs. (41–52 Nm).
   - Torque converter. Tighten the bolts to 33–43 ft. lbs. (44–59 Nm).
   - Starter motor
   - VSS sensor connector
   - Transmission control harness connectors
   - Shift cable
   - Transfer case linkage
   - Front and rear driveshafts

- Transmission oil cooler lines
- Transmission dipstick tube
- Exhaust rear pipes
- Exhaust front pipes
- CKP sensor
- Negative battery cable

2005–2006

2WD

➡**Before removing the transmission remove the crankshaft position sensor (POS) from the transmission assembly.**

1. Before servicing the vehicle, refer to the Precautions Section.
2. Disconnect the negative battery cable.
3. Raise and support the vehicle safely. Drain the transmission fluid.
4. Remove the transmission fluid indicator. Remove the left fender protector.
5. Remove the crankshaft position sensor (POS) from the transmission. Do not disassemble it. Do not place it in an area affected by magnetism.
6. Remove the engine under covers. Remove the front crossmember. Remove the starter. Remove the rear driveshaft.
7. If equipped with a 4.0L engine, remove the left and right exhaust tubes.
8. Remove the selector control cable and bracket from the transmission. Disconnect and plug the cooler lines.
9. Remove the dust cover from the torque converter housing. Remove the torque converter to flex plate retaining bolts. There are four of them.

➡**Always rotate the crankshaft in the clockwise direction as viewed from the front of the engine.**

10. Support the transmission using a suitable jack.
11. Remove the insulator retaining nuts. Remove the crossmember retaining bolts. Remove the crossmember.

➡**Be sure that the transmission is properly supported.**

12. Tilt the transmission slightly to gain clearance between the body and the transmission, and then remove the air breather hose and the breather tube.
13. Disconnect the transmission harness connector. Remove the wiring harness from the retainers. Remove the fluid indicator pipe.
14. Remove the transmission to engine retaining bolts. Remove the transmission from the vehicle.

To install:

15. Installation is the reverse of the removal procedure.
16. After installing the torque converter to the transmission, check dimension "A" to be sure it is within specification. Specification is 0.98 inch or more.
17. Torque the transmission to engine retaining bolts to specification and in the proper sequence. Torque bolts to 55 ft. lbs if equipped with 4.0L engine. If equipped with 2.5L engine, see illustration.

4WD

➡**Before removing the transmission remove the crankshaft position sensor (POS) from the transmission assembly.**

1. Before servicing the vehicle, refer to the Precautions Section.

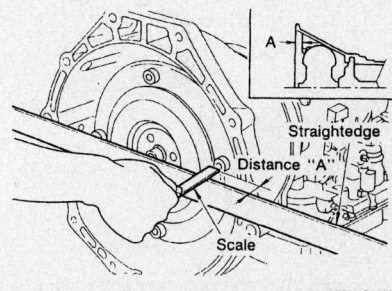

09482_FRON_G0102

Torque converter installation measurement "A"—2005–2006

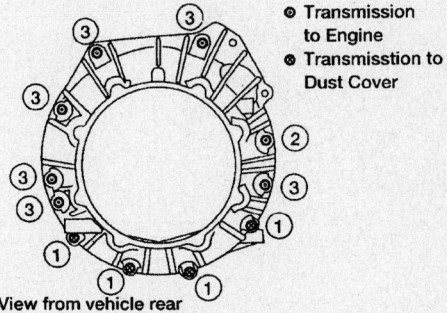

View from vehicle rear

09482_FRON_G0103

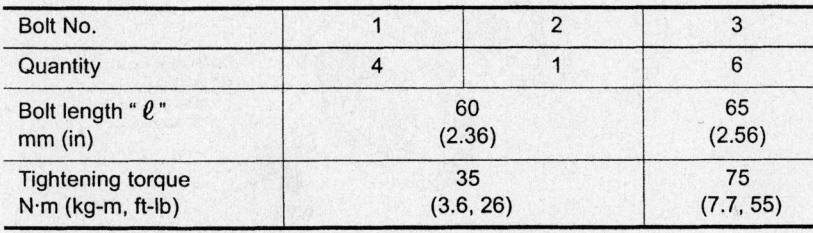

Bolt No.	1	2	3
Quantity	4	1	6
Bolt length "ℓ" mm (in)		60 (2.36)	65 (2.56)
Tightening torque N·m (kg-m, ft-lb)		35 (3.6, 26)	75 (7.7, 55)

2.5L engine transmission bolt tightening sequence—2005–2006

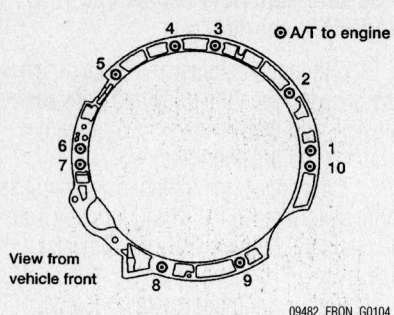

4.0L engine transmission bolt tightening sequence—2005–2006

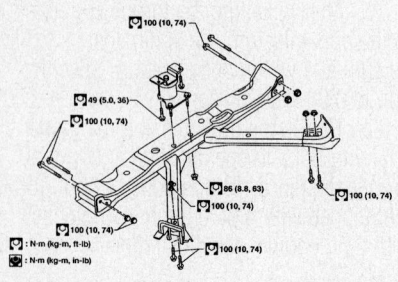

2.5L crossmember bolt tightening specifications—2005–2006

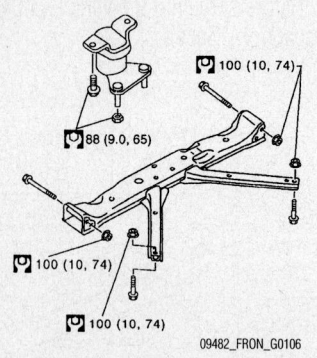

4.0L with 2WD crossmember bolt tightening specifications—2005–2006

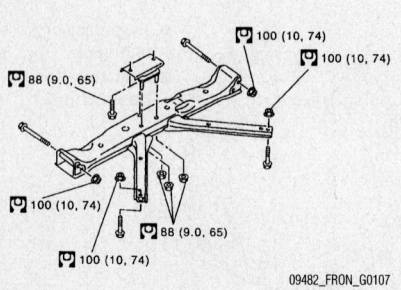

4.0L with 4WD crossmember bolt tightening specifications—2005–2006

2. Disconnect the negative battery cable.

3. Raise and support the vehicle safely. Drain the transmission fluid.

4. Remove the transmission fluid indicator. Remove the left fender protector.

5. Remove the crankshaft position sensor (POS) from the transmission. Do not disassemble it. Do not place it in an area affected by magnetism.

6. On frontier, remove the air dam.

7. Remove the engine under covers. Remove the front crossmember. Remove the starter. Remove the rear driveshaft.

8. Remove the left and right exhaust tubes.

9. Remove the selector control cable and bracket from the transmission. Disconnect and plug the cooler lines.

10. Remove the dust cover from the torque converter housing. Remove the torque converter to flex plate retaining bolts. There are four of them.

➡**Always rotate the crankshaft in the clockwise direction as viewed from the front of the engine.**

11. Support the transmission using a suitable jack.

12. Remove the insulator retaining nuts. Remove the crossmember retaining bolts. Remove the crossmember.

➡**Be sure that the transmission is properly supported.**

13. Tilt the transmission slightly to gain clearance between the body and the transmission, and then remove the air breather hose and the breather tube.

14. Disconnect the transmission harness connector. Disconnect the 4LO switch connector, wait detection switch connector, ATP switch connector and the transfer control device connector.

15. Remove the wiring harness from the retainers. Remove the fluid indicator pipe.

16. Remove the transmission to engine retaining bolts. Remove the transmission from the vehicle, with the transfer case attached.

17. As required, remove the transfer case from the transmission.

To install:

18. Installation is the reverse of the removal procedure.

19. After installing the torque converter to the transmission, check dimension "A" to be sure it is within specification. Specification is 0.98 inch or more.

20. Torque the transmission to engine retaining bolts to specification and in the proper sequence. Torque bolts to 55 ft. lbs.

Clutch

REMOVAL & INSTALLATION

1. Before servicing the vehicle, refer to the Precautions Section.

2. Remove or disconnect the following:
   - Negative battery cable
   - Transmission
   - Pressure plate. Loosen the bolts evenly in ½ turn steps.
   - Clutch disc

To install:

3. Install or connect the following:
   - Clutch disc and pressure plate.

➡**On 2002–2004 vehicles, tighten the pressure plate bolts evenly in ½ turns to 16–22 ft. lbs. (22–29 Nm). On 2005–2006 vehicles with 2.5L engine tighten the pressure plate bolts in sequence to 11 ft. lbs. and then to 29 ft. lbs. On 2005–2006 vehicles with 4.0L engine tighten the pressure plate bolts in sequence to 11 ft. lbs. and then to 19 ft. lbs.**

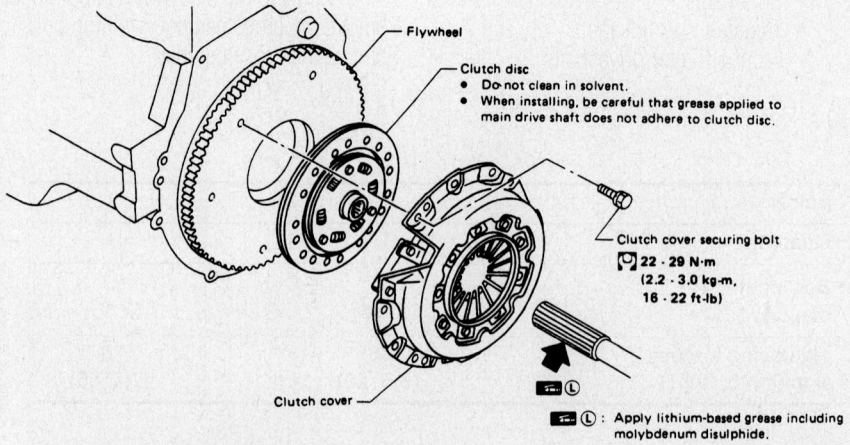

Clutch and related components

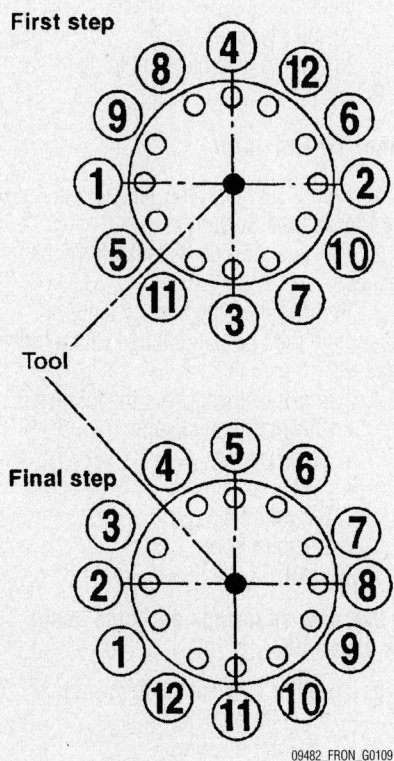

Clutch/pressure plate bolt tightening sequence—2.5L engine

09482_FRON_G0108

Clutch/pressure plate bolt tightening sequence—4.0L engine

09482_FRON_G0109

- Transmission
- Negative battery cable

ADJUSTMENT

1. Before servicing the vehicle, refer to the Precautions Section.

2. Check to see that the clevis pin floats freely in the bore of the clutch pedal. It should not be bound by the clevis or the clutch pedal.

3. If it is not free check that the pedal stopper bolt or ASCD clutch switch is not applying pressure to the clutch pedal causing the clevis pin to bind.

4. To adjust, loosen the lock nut and turn the pedal stopper bolt or ASCD clutch switch. Tighten the lock nut.

5. Verify that the clevis pin floats in the bore of the clutch pedal. It should not be bound by the clutch pedal.

6. If the clevis pin is still not free, remove it and check for deformation or damage. Replace as required.

7. Check the clutch pedal stroke for free range of movement.

8. With the clevis pin removed, manually move the clutch pedal up and down to determine if it moves freely. If not, replace the assembly.

9. Adjust the clearance "C" while fully depressing the clutch pedal (with the clutch interlock switch). Specification should be 0.004–0.039 inch.

10. Check the clutch hydraulic system for problems. Repair and replace components, as required.

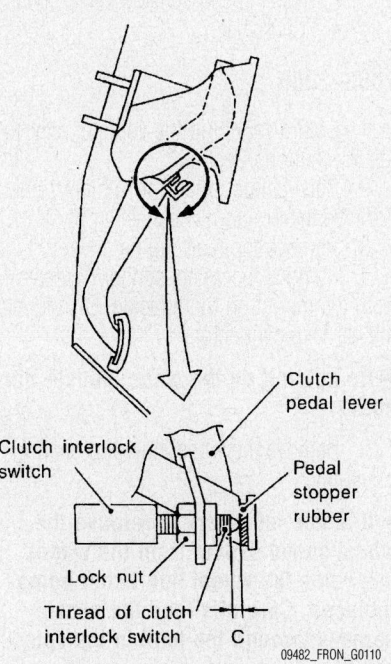

09482_FRON_G0110

Clutch pedal adjustment

11. If repair was necessary, bleed the system.

Hydraulic Clutch System

BLEEDING

1. Before servicing the vehicle, refer to the Precautions Section.

2. Fill the system with the proper grade and type fluid.

3. Have an assistant pump the clutch pedal slowly several times and hold it depressed.

4. Open the slave cylinder bleeder screw and allow air to escape.

5. Close the bleeder screw before releasing the clutch pedal.

6. Repeat until all air is purged from the clutch hydraulic system.

7. Refill the reservoir to the full mark.

Transfer Case Assembly

REMOVAL & INSTALLATION

2002–2004

1. Before servicing the vehicle, refer to the Precautions Section.

2. Remove or disconnect the following:

- Negative battery cable
- Front and rear driveshafts
- Torsion bars and mounts
- Rear torsion bar crossmember
- Exhaust front pipes
- Exhaust rear pipes
- Vehicle Speed (VSS) sensor connector
- Transfer case shift linkage
- Transfer case neutral switch connector
- 4WD switch connector
- Vent hose
- Transfer case flange bolts
- Transfer case

To install:

3. Install or connect the following:

- Transfer case. Tighten the flange bolts to 23–30 ft. lbs. (31–41 Nm).
- Vent hose
- 4WD switch connector
- Transfer case neutral switch connector
- Transfer case shift linkage
- VSS sensor connector
- Exhaust rear pipes
- Exhaust front pipes
- Rear torsion bar crossmember
- Torsion bars and mounts

- Front and rear driveshafts
- Negative battery cable

2005–2006

1. Before servicing the vehicle, refer to the Precautions Section.

2. Disconnect the negative battery cable.

3. Switch the 4WD switch to 2WD. Set the transfer case to 2WD.

4. Raise and support the vehicle safely. Drain the transfer case fluid fluid.

5. Remove the transmission undercover.

6. Remove the center exhaust tube and main muffler.

7. Remove the front and rear driveshafts.

8. Remove the transmission bolts. Properly support the transmission and transfer case assembly, using a suitable jack.

9. Remove the transmission crossmember.

➡**Support the transmission and transfer case using two suitable jacks while removing the transmission crossmember.**

10. Disconnect the ATP electrical connector, the 4LO switch connector, the wait detection switch and the transfer control device.

11. Disconnect each air breather hose from the transfer control device and the breather tube.

12. Remove the transfer case to transmission retaining bolts.

➡**Support the transmission and transfer case, using a suitable jack.**

13. Remove the transfer case from the vehicle.

To install:

14. Installation is the reverse of the removal procedure.

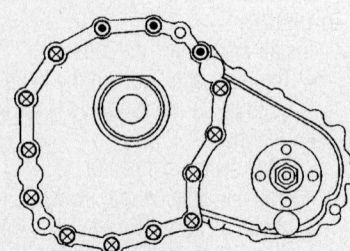

⊙ : Transfer ➡ Transmission
⊗ : Transmission ➡ Transfer

09482_FRON_G0111

Transfer case bolt tightening sequence—2005–2006

15. Tighten the transfer case to transmission retaining bolts to specification and in the proper sequence. Specification is 27 ft. lbs.

16. Start the engine and check for leaks, correct as required.

Halfshaft

REMOVAL & INSTALLATION

2002–2004

1. Before servicing the vehicle, refer to the Precautions Section.

2. Remove or disconnect the following:

- Front wheel
- Wheel speed sensor, if equipped
- Locking hub or drive flange
- Snapring
- Spindle washer
- Thrust washer
- Inner CV-joint bolts
- Axle halfshaft. Separate the stub shaft from the spindle by tapping with a plastic hammer.

To install:

3. Install or connect the following:

- Axle halfshaft. Guide the stub shaft into the spindle and tighten the inner CV-joint bolts to 25–33 ft. lbs. (34–44 Nm).
- Thrust washer
- Spindle washer
- Snapring
- Locking hub or drive flange
- Wheel speed sensor, if equipped
- Front wheel

2005–2006

1. Before servicing the vehicle, refer to the Precautions Section.

2. Raise and support the vehicle safely. Remove the tire and wheel assembly.

3. Remove the rear engine cover.

4. Remove the wheel sensor harness from the mount on the knuckle. Disconnect the harness connector.

➡**Do not pull on the wheel sensor harness.**

5. Remove the wheel hub and bearing assembly.

➡**It is not necessary to remove the wheel speed sensor from the wheel hub when the wheel hub is not being replaced. Carefully feed the sensor harness through the hole in the splash shield.**

6. Separate the upper link ball joint stud from the steering knuckle using tool ST29020001 (J-24319-01) or equivalent.

7. Remove the halfshaft assembly from the vehicle by prying the halfshaft from the front final drive using the proper tool.

To install:

8. Installation is the reverse of the removal procedure.

9. Be sure to use a new differential side oil seal.

CV-Joints

OVERHAUL

Outer CV-Joint

1. Before servicing the vehicle, refer to the Precautions Section.

2. Remove the axle halfshaft from the vehicle.

3. Remove the CV-joint boot clamps and push the boot away from the joint.

4. Remove the CV-joint from the axle shaft by tapping it with a brass hammer.

To install:

➡**Use new circlips and boot clamps for assembly.**

5. Install the CV-joint to the axle shaft by tapping it with a brass hammer.

6. Pack the joint with grease.

7. Install the boot clamps.

8. Install the axle halfshaft to the vehicle.

Inner Tri-Pot Joint

1. Before servicing the vehicle, refer to the Precautions Section.

2. Remove the axle halfshaft from the vehicle.

3. Remove the plug seal by tapping around the joint housing flange with a brass hammer.

4. Remove or disconnect the following:

- CV-joint boot clamps
- Snapring
- Spider assembly
- CV-joint housing
- CV-joint boot

To install:

➡**Use new snaprings and plug seals for assembly.**

5. Install or connect the following:

- CV-joint boot
- CV-joint housing
- Spider assembly
- Snapring. Pack the joint with grease.

- CV-joint boot clamps
- Plug seal

6. Install the axle halfshaft to the vehicle.

Spindle Bearings

REMOVAL, PACKING & INSTALLATION

1. Before servicing the vehicle, refer to the Precautions Section.
2. Remove or disconnect the following:
   - Front wheel
   - Locking hub or drive flange
   - Brake caliper and support
   - Wheel speed sensor, if equipped
   - Axle halfshaft
   - Outer tie rod ends
   - Upper ball joint or steering knuckle bracket bolts
   - Lower ball joint
   - Steering knuckle
   - Inner seal
   - Thrust washer
   - Spindle bearing

To install:

3. Install or connect the following:
   - Spindle bearing. Coat the bearing with multi-purpose grease.
   - Thrust washer
   - Inner seal
   - Steering knuckle
   - Lower ball joint
   - Upper ball joint or steering knuckle bracket bolts
   - Outer tie rod ends
   - Axle halfshaft
   - Wheel speed sensor, if equipped
   - Brake caliper and support
   - Locking hub or drive flange
   - Front wheel

Rear Axle Shaft, Bearing and Seal

REMOVAL & INSTALLATION

1. Before servicing the vehicle, refer to the Precautions Section.
2. Remove or disconnect the following:
   - Rear wheel and tire assembly
   - Wheel speed sensor
   - Brake drum or rotor
   - Brake shoes or caliper assembly
   - Parking brake cable
   - Brake fluid line
   - Bearing cage and backing plate bolts
   - Axle shaft assembly
   - Axle seal

   - Wheel speed sensor rotor, if equipped
   - Lockwasher
   - Bearing locknut
   - Flat washer
   - Wheel bearing
   - Wheel bearing cage grease seal

To install:

➡️**Use new lockwashers, seals and bearings for assembly.**

3. Install or connect the following:
   - Wheel bearing cage grease seal
   - Wheel bearing
   - Flat washer
   - Bearing locknut
   - Lockwasher
   - Wheel speed sensor rotor, if equipped
   - Axle seal
   - Axle shaft assembly
   - Bearing cage and backing plate bolts
   - Brake fluid line
   - Parking brake cable
   - Brake shoes or caliper assembly
   - Brake drum or rotor
   - Wheel speed sensor, if equipped
   - Rear wheel and tire assembly

4. Bleed the rear brakes and check the rear axle lubricant level.

Pinion Seal

Front

2002–2004

1. Before servicing the vehicle, refer to the Precautions Section.
2. Remove or disconnect the following:
   - Driveshaft
   - Front wheels
   - Front brake calipers

➡️**The front brake calipers must be removed so that there is no additional drag when measuring pinion bearing preload.**

3. Use an inch lb. torque wrench and measure the amount of torque required to maintain pinion rotation through several revolutions.
4. Remove or disconnect the following:
   - Pinion flange
   - Oil seal

To install:

5. Install or connect the following:
   - Pinion seal
   - Pinion flange
6. Rotate the pinion flange occasionally

while tightening the flange nut to make sure the pinion bearings seat correctly.

7. Take frequent bearing preload torque readings. Tighten the flange nut to achieve the preload torque readings originally recorded. Do not exceed 137–180 ft. lbs. (186–245 Nm) torque when tightening the pinion flange nut.

❋❋ CAUTION

If the bearing preload can not be achieved at the specified torque, remove the pinion bearing and install a new adjustment spacer.

8. Install or connect the following:
   - Front brake calipers
   - Front wheels
   - Driveshaft. Tighten the fasteners to 29–33 ft. lbs. (39–44 Nm).
9. Fill the differential with gear lubricant and check for leaks.

2005–2006

1. Before servicing the vehicle, refer to the Precautions Section.
2. Raise and support the vehicle safely. Remove the tire and wheel assemblies.
3. Remove the brake calipers and position them to the side. Do not disconnect the fluid lines. Do not press the brake pedal with the calipers removed.
4. Remove the ABS sensor harness from its mount on the knuckle.
5. Support the lower link, using a suitable jack. Separate the upper ball joint stud from the steering knuckle using tool ST2902001 (J-24319-01) or equivalent.

➡️**Support the lower link using a suitable jack.**

6. Remove the rear engine cover.
7. Remove the right and left halfshafts from the front final drive, using the proper tool.
8. Disconnect the front driveshaft from the front final drive. Reposition it out of the way, using wire.
9. Measure the drive pinion bearing preload with tool ST3127S00 (J-25765-A) or equivalent. Record the measurement.
10. Put a mating mark on the end of the drive pinion in line with the mating mark "B" on the companion flange.
11. Remove the drive pinion lock nut using tool KV38108300, or equivalent. Discard the lock nut. Do not reuse it.
12. Using the proper tool, remove the companion flange. Using tool KV381054S0 (J-34286) or equivalent, remove the front oil seal.

To install:

13. Installation is the reverse of the removal procedure.

14. Measure the drive pinion bearing preload with the resistance tool. Specification should be equal to the measurement taken during the removal procedure plus an additional 5 inch lb. If the specification is low, tighten the new drive pinion lock nut in 5 ft. lb increments until the drive pinion preload is met.

➡ **Never loosen the drive pinion nut to decrease drive pinion bearing preload. Do not exceed specified preload. If preload torque is exceeded a new collapsible spacer must be installed. If maximum torque is reached prior to reaching the required preload, the collapsible spacer may have been damaged. Replace the collapsible spacer.**

15. Drive pinion lock nut torque is 138–216 ft. lbs.

16. Continue the installation in the reverse order of the removal procedure.

Rear

2002–2004 2WD

1. Before servicing the vehicle, refer to the Precautions Section.
2. Remove or disconnect the following:
   • Driveshaft
   • Rear wheels
   • Brake drums

➡ **The rear brake drums must be removed so that there is no additional drag when measuring pinion bearing preload.**

3. Use an inch lb. torque wrench and measure the amount of torque required to maintain pinion rotation through several revolutions.
4. Remove or disconnect the following:
   • Pinion flange
   • Wheel speed sensor and rotor, if equipped
   • Oil seal
   • Pinion bearing
   • Collapsible spacer

To install:

➡ **Use a new collapsible spacer and wheel speed sensor rotor for assembly.**

5. Install or connect the following:
   • Collapsible spacer
   • Pinion bearing
   • Pinion seal
   • Pinion flange

6. Rotate the pinion flange occasionally while tightening the flange nut to make sure the pinion bearings seat correctly.

7. Take frequent bearing preload torque readings. Tighten the flange nut to achieve the preload torque readings originally recorded. Do not exceed 137–180 ft. lbs. (186–245 Nm) torque when tightening the pinion flange nut.

✷✷ CAUTION

Never loosen the pinion nut to reduce bearing preload. If it is necessary to reduce bearing preload, install a new collapsible spacer.

8. Install or connect the following:
   • Brake drums
   • Rear wheels
   • Driveshaft. Tighten the fasteners to 58–65 ft. lbs. (78–88 Nm).
9. Fill the differential with gear lubricant and check for leaks.

2002–2004 4WD

1. Before servicing the vehicle, refer to the Precautions Section.
2. Remove or disconnect the following:
   • Driveshaft
   • Rear wheels
   • Brake drums

➡ **The rear brake drums must be removed so that there is no additional drag when measuring pinion bearing preload.**

3. Use an inch lb. torque wrench and measure the amount of torque required to maintain pinion rotation through several revolutions.
4. Remove or disconnect the following:
   • Pinion flange
   • Oil seal

To install:
5. Install or connect the following:
   • Pinion seal
   • Pinion flange

6. Rotate the pinion flange occasionally while tightening the flange nut to make sure the pinion bearings seat correctly.

7. Take frequent bearing preload torque readings. Tighten the flange nut to achieve the preload torque readings originally recorded. Do not exceed 137–180 ft. lbs. (186–245 Nm) torque when tightening the pinion flange nut.

✷✷ CAUTION

If the bearing preload can not be achieved at the specified torque,

remove the pinion bearing and install a new adjustment spacer.

8. Install or connect the following:
   • Brake drums
   • Rear wheels
   • Driveshaft. Tighten the fasteners to 58–65 ft. lbs. (78–88 Nm).
9. Fill the differential with gear lubricant and check for leaks.

2005–2006 C200 AXLE

1. Before servicing the vehicle, refer to the Precautions Section.
2. Raise and support the vehicle safely.
3. Remove the driveshaft.
4. Put a mating mark on the end of the drive pinion. The mating mark should be in line with the mating mark "A" on the companion flange.
5. Remove the drive pinion lock nut using tool KV38108300, or equivalent.
6. Using the proper tool, remove the companion flange. Using tool KV381054S0 (J-34286) or equivalent, remove the front oil seal.

To install:
7. Installation is the reverse of the removal procedure.

2005–2006 M226 AXLE

1. Before servicing the vehicle, refer to the Precautions Section.
2. Raise and support the vehicle safely. Remove the tire and wheel assemblies.
3. Remove the driveshaft.
4. Remove the brake calipers and rotors.
5. Rotate the pinion three or four times using tool ST3127S00 (J-25765-A) or equivalent. Record the rotating torque.
6. Remove the drive pinion lock nut.
7. Put a mating mark on the end of the drive pinion. The mating mark should be in line with the mating mark "B" on the companion flange.
8. Remove the drive pinion lock nut using tool KV38108300, or equivalent.
9. Using the proper tool, remove the companion flange. Using tool KV381054S0 (J-34286) or equivalent, remove the front oil seal.

To install:
10. Installation is the reverse of the removal procedure.
11. Measure the drive pinion bearing preload with the resistance tool. Specification should be equal to the measurement taken during the removal procedure plus an additional 5 inch lb. If the specification is

low, tighten the new drive pinion lock nut in 5 ft. lb increments until the drive pinion preload is met.

➡**Never loosen the drive pinion nut to decrease drive pinion bearing preload. Do not exceed specified preload. If preload torque is exceeded a new collapsible spacer must be**

installed. **If maximum torque is reached prior to reaching the required preload, the collapsible spacer may have been damaged. Replace the collapsible spacer.**

12. Turn the drive pinion in both directions several times to set the bearing rollers.

13. Check the preload with tool ST3127S000 (J-25765-A) or equivalent. Total preload, with oil seal, for gear ratio 3.538 should be 22–40 inch lbs., for gear ratio 3.692 should be 21–39 inch lbs

14. Continue the installation in the reverse order of the removal procedure.

STEERING

Air Bag

✳✳ CAUTION

The SRS system must be disarmed before performing service on, or around, system components, the steering column, instrument panel components, wiring and sensors. Failure to follow the safety precautions and the disarming procedure could result in accidental air bag deployment, possible injury and unnecessary system repairs.

PRECAUTIONS

Several precautions must be observed when handling the inflator module to avoid accidental deployment and possible personal injury.

• Never carry the inflator module by the wires or connector on the underside of the module.

• When carrying a live inflator module, hold securely with both hands, and ensure that the bag and trim cover are pointed away.

• Place the inflator module on a bench or other surface with the bag and trim cover facing up.

• With the inflator module on the bench, never place anything on or close to the module which may be thrown in the event of an accidental deployment.

DISARMING

To disarm the **SRS** system turn the ignition switch to the **OFF** position. Then, disconnect both battery cables starting with the negative cable first and wait at least 3 minutes after the cables are disconnected.

To rearm the **SRS** system, turn the ignition switch to the **OFF** position. Connect both battery cables starting with the positive cable first.

Power Steering Gear

REMOVAL & INSTALLATION

2002–2004

1. Before servicing the vehicle, refer to the Precautions Section.
2. Disarm the SRS system.
3. Disconnect the negative battery cable.
4. Remove or disconnect the following:

- Pitman arm
- Steering column intermediate shaft
- Power steering hoses
- Steering gear

To install:
5. Install or connect the following:
- Steering gear. Tighten the bolts to 62–71 ft. lbs. (84–96 Nm).
- Power steering hoses. Tighten the banjo fittings to 29–38 ft. lbs. (39–51 Nm).
- Steering column intermediate shaft. Tighten the pinch bolt to 17–22 ft. lbs. (24–29 Nm).
- Pitman arm. Tighten the nut to 102-130 ft. lbs. (138-176 Nm) (4-cyl.) or 174–195 ft. lbs. (235–265 Nm) (6-cyl.).
6. Check the wheel alignment and adjust, as necessary.

2005–2006

➡**The spiral cable may snap due to steering operation if the steering column is separated from the steering gear assembly. Be sure to secure the steering wheel to avoid turning.**

1. Before servicing the vehicle, refer to the Precautions Section.
2. Position the front wheels in the straight ahead position.
3. Disarm the SRS system.
4. Disconnect the negative battery cable.
5. Drain the power steering fluid.
6. Raise and support the vehicle safely. Remove the tire and wheel assemblies.
7. Remove the undercover.
8. If equipped with 4WD, remove the final drive, then support the halfshafts, using wire.
9. Remove the stabilizer bar brackets, and position the stabilizer bar aside.
10. Remove and discard the cotter pins at the steering outer sockets. Loosen the outer socket locknuts.
11. Remove the steering gear outer sockets from the steering knuckles, using tool HT72520000 (J-25730-A) or equivalent.
12. Disconnect and plug the power steering fluid lines at the steering gear.

- Plug openings of gear housing, and securely locate hose connectors at a position higher than oil pump and cover with rag.
- Be extremely careful to prevent entry of foreign matter into hoses through connectors.
- When installing gear arm, align four grooves of gear arm serrations with four projections of sector shaft serrations, and install and tighten lock washer and nut.

⟳ : N·m (kg-m, ft-lb)

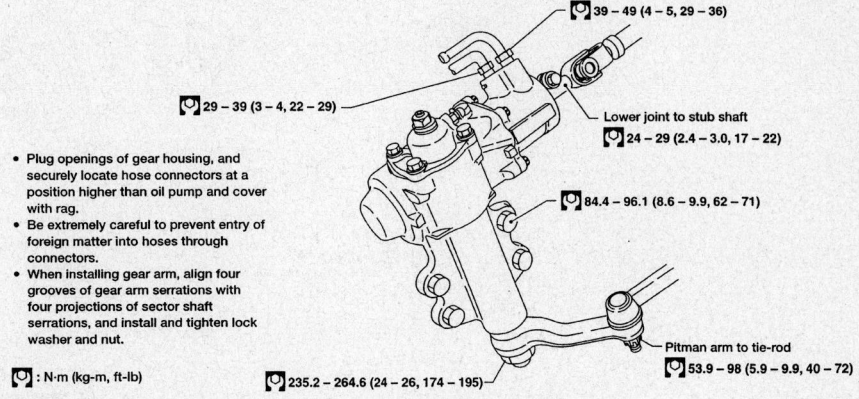

39 – 49 (4 – 5, 29 – 36)
29 – 39 (3 – 4, 22 – 29)
Lower joint to stub shaft
24 – 29 (2.4 – 3.0, 17 – 22)
84.4 – 96.1 (8.6 – 9.9, 62 – 71)
Pitman arm to tie-rod
53.9 – 98 (5.9 – 9.9, 40 – 72)
235.2 – 264.6 (24 – 26, 174 – 195)

09482_FRON_G0112

Power steering gear and related components—2002–2004

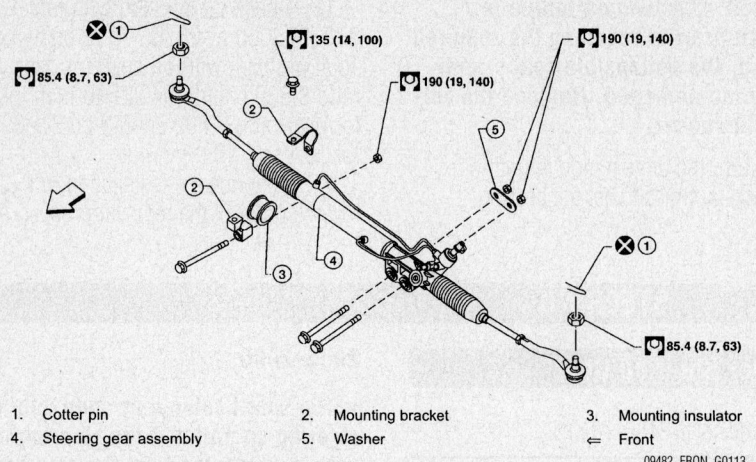

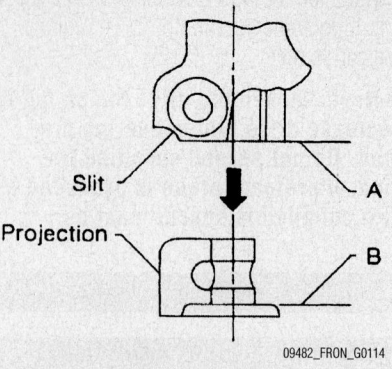

1. Cotter pin
2. Mounting bracket
3. Mounting insulator
4. Steering gear assembly
5. Washer
⇐ Front

09482_FRON_G0113

Power steering gear lower joint installation alignment—2005–2006

09482_FRON_G0114

Power steering gear and related components—2005–2006

13. Remove the bolt from the lower joint of the lower joint assembly. Separate the lower joint from the steering gear assembly. Be careful not to damage the lower joint.

14. Remove the steering gear retaining nuts and bolts. Remove the steering gear from the vehicle.

To install:

15. With the steering wheel in the straight ahead position, align the slit of the lower joint with the projection on the dust cover. Insert the joint until both surfaces contact each other.

16. Continue the installation in the reverse order of the removal procedure.

17. Check and adjust the front alignment, as required.

18. Bleed the power steering system.

19. Fill the power steering pump with the proper grade and type fluid.

➡**After removing/installing or replacing steering and suspension components**

which effect wheel alignment or after adjusting wheel alignment, or the steering angle sensor or the ABS actuator electrical unit be sure to adjust the neutral position of the steering angle sensor before running the vehicle.

20. Position the steering wheel in the straight ahead position.

21. Drive the vehicle at 10 mph for more than 10 minutes.

22. When this procedure is complete the SLP indicator lamp and the VDC OFF indicator lamp will turn off.

FRONT SUSPENSION

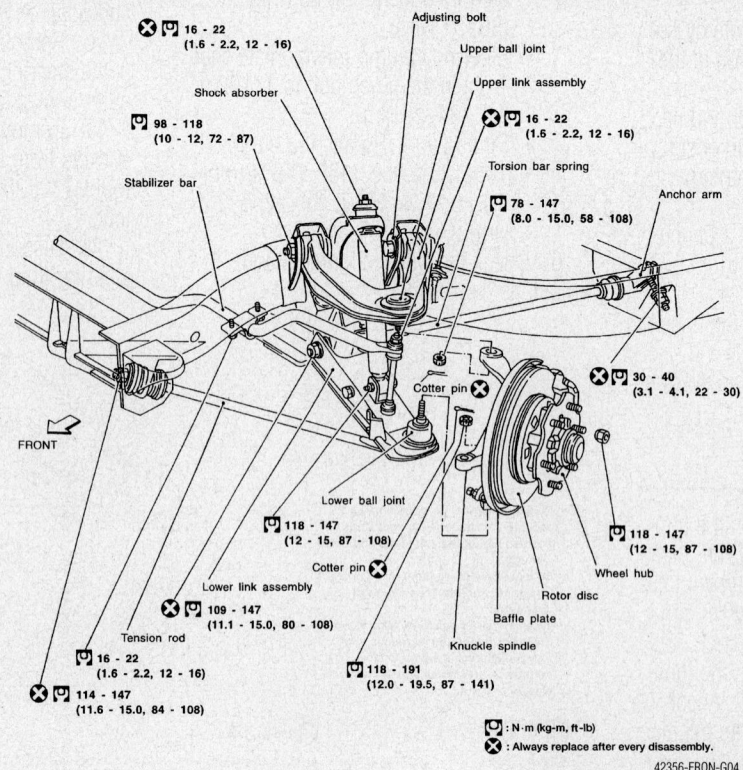

⊗ : Always replace after every disassembly.

42356-FRON-G04

Front suspension and related components—2WD 4-cyl. 2003–2004 Frontier

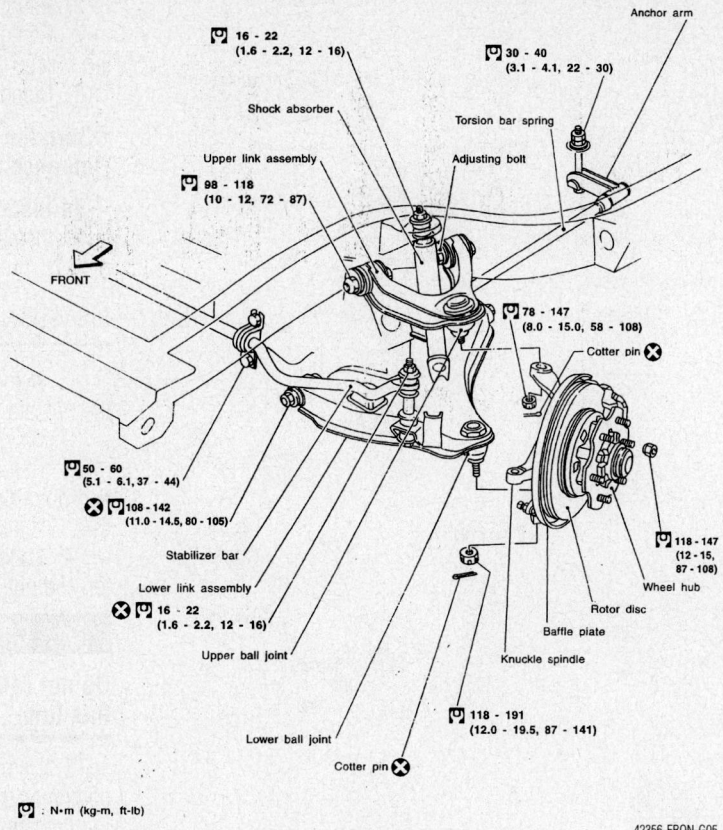

Front suspension and related components—2WD 6-cyl. 2003–2004 Frontier and Xterra

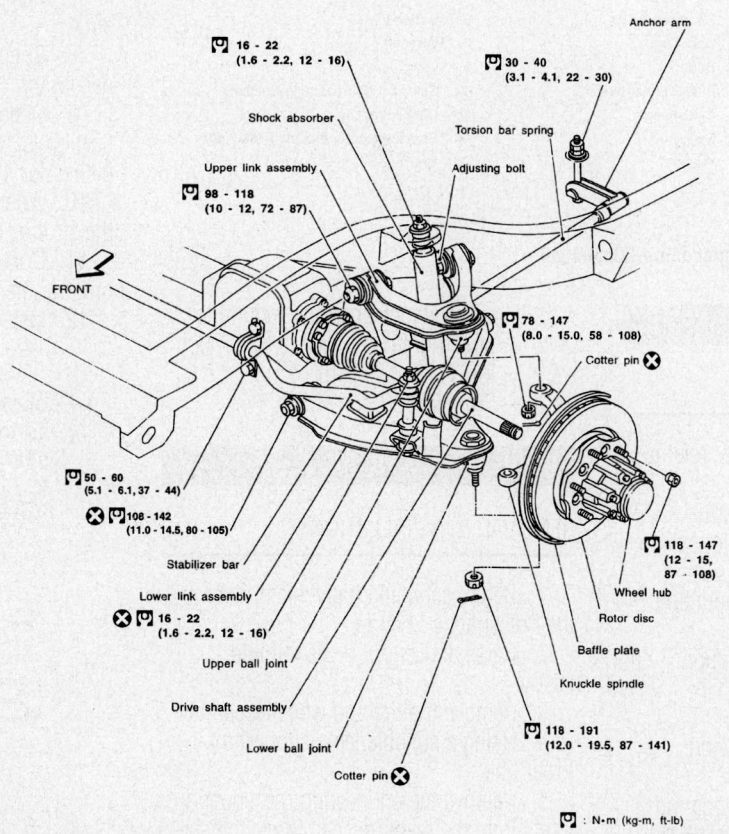

Front suspension and related components—4WD 2003–2004 Frontier and Xterra

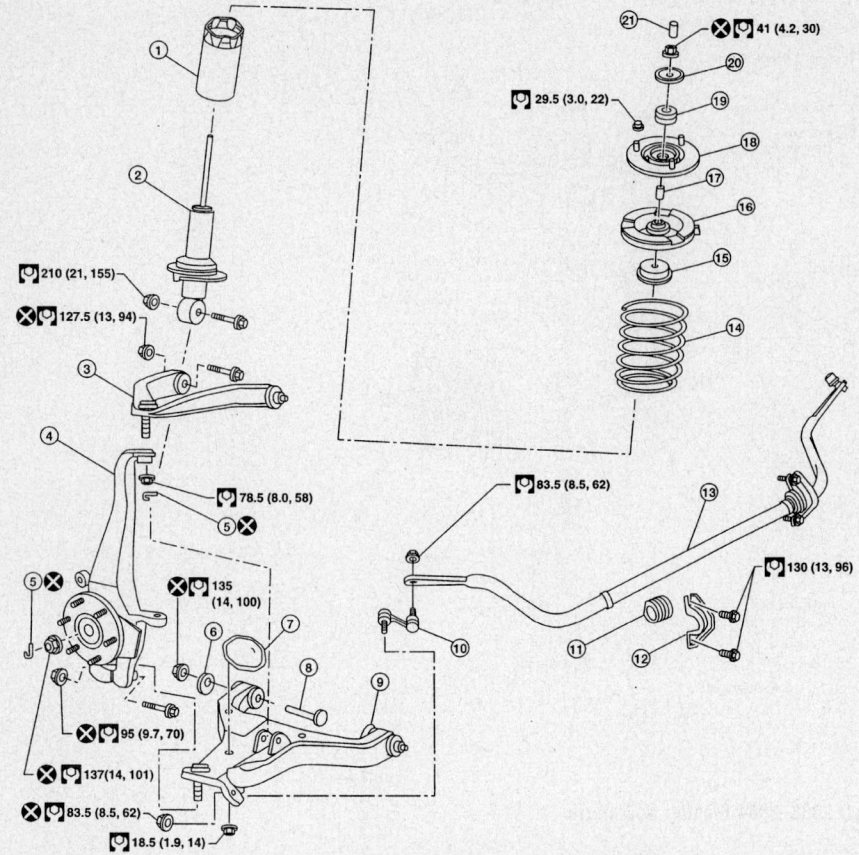

#		#		#	
1.	Dust cover	2.	Shock absorber	3.	Upper link
4.	Steering knuckle	5.	Cotter pin	6.	Washer
7.	Jounce bumper	8.	Bolt	9.	Lower link
10.	Connecting rod	11.	Stabilizer bar bushing	12.	Stabilizer bar mounting bracket
13.	Stabilizer bar	14.	Coil spring	15.	Dust cover cap
16.	Upper spring seat	17.	Spacer	18.	Shock absorber mounting insulator
19.	Spacer	20.	Washer	21.	Cap

09482_FRON_G0115

Front suspension and related components—2005–2006

Shock Absorber

REMOVAL & INSTALLATION

1. Before servicing the vehicle, refer to the Precautions Section.
2. Support the lower control arm.
3. Remove or disconnect the following:
   • Front wheel
   • Lower shock absorber mounting bolt
   • Upper shock absorber mounting nut
   • Shock absorber

To install:

4. Install or connect the following:
   • Shock absorber
   • Upper shock absorber mounting nut. Tighten the nut to 12–16 ft. lbs. (16–22 Nm).

• Lower shock absorber mounting bolt. Tighten the bolt to 87–106 ft. lbs. (118–147 Nm).
• Front wheel

Strut

REMOVAL & INSTALLATION

1. Before servicing the vehicle, refer to the Precautions Section.
2. Raise and support the vehicle safely.
3. Remove the tire and wheel assembly.
4. Using a suitable jack, support the lower link.
5. Remove the connecting rod upper joints from the stabilizer bar. Move the stabilizer out of the way.

6. Remove the strut lower bolt and nut.
7. Remove the three upper strut mounting nuts.
8. Remove the strut from the vehicle.

➡ **Turn the steering knuckle to gain clearance for removal.**

To install:

9. Installation is the reverse of the removal procedure.

DISASSEMBLY & ASSEMBLY

1. Before servicing the vehicle, refer to the Precautions Section.
2. Raise and support the vehicle safely.
3. Remove the strut from the vehicle and position it in a suitable holding fixture.
4. Loosen, but do not remove, the piston rod nut.

✳✳ CAUTION

Do not remove the piston rod nut at this time.

5. Using a spring compressor tool, compress the spring until the strut mounting insulator can be turned by hand.
6. Remove the piston rod and locknut. Discard the locknut. Do not reuse it.
7. Remove the components from the strut.
8. Check and replace components, as necessary.
9. Check the free spring height. Specification is 13.6 inches for 2WD and 14.0 inches for 4WD.
10. When installing the coil spring on the strut be sure it is positioned properly.
11. Continue the assembly in the reverse order of the disassembly.
12. Install the strut.

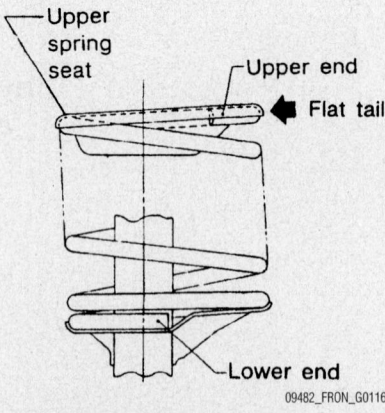

09482_FRON_G0116

Front strut coil spring positioning— 2005–2006

Torsion Bar

REMOVAL & INSTALLATION

1. Before servicing the vehicle, refer to the Precautions Section.
2. Matchmark the torsion bar to the control arm mount and the anchor arm.
3. Measure the adjustment bolt protrusion as shown and note the length (L) for assembly.
4. Loosen the adjustment bolt so that all tension is released.
5. Remove the torsion bar mount from the control arm and remove the torsion bar.

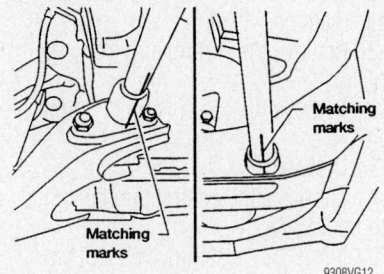

Torsion bar matchmarks

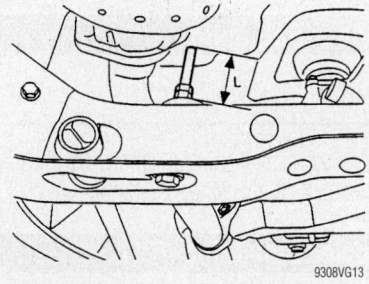

Adjustment bolt measurement (L)

To install:
6. Align the matchmarks and install the torsion bar. Tighten the large mount nut to 66–87 ft. lbs. (89–118 Nm) and the small nut to 33–44 ft. lbs. (45–60 Nm).
7. Tighten the adjustment bolt to achieve the measurement (L) noted earlier. Tighten the locknut to 22–30 ft. lbs. (30–40 Nm).
8. If a new torsion bar is being installed, set length (L) to 2.68 inches.

Stabilizer Bar

REMOVAL & INSTALLATION

2002–2004

1. Before servicing the vehicle, refer to the Precautions Section.

2. Raise and support the vehicle safely.
3. Remove the stabilizer bar bracket bolts.
4. Remove the stabilizer bar connecting bolts.
5. Remove the stabilizer bar from the vehicle.

To install:
6. Installation is the reverse of the removal procedure.

2005–2006

1. Before servicing the vehicle, refer to the Precautions Section.
2. Raise and support the vehicle safely.
3. On Xterra, remove the front valance center.
4. Remove the engine undercover.
5. Remove the connecting rod nuts.
6. Loosen the top bolts for the stabilizer bar mounting brackets. Remove the lower bolts from the mounting brackets.
7. Remove the stabilizer bar from the vehicle.
8. Remove the bushings from the stabilizer bar.

To install:
9. Installation is the reverse of the removal procedure.

Upper Control Arm

REMOVAL & INSTALLATION

2002

1. Before servicing the vehicle, refer to the Precautions Section.
2. Raise and support the vehicle safely.
3. Support the lower control arm.
4. Remove or disconnect the following:
   - Front wheel
   - Shock absorber
   - Upper ball joint
   - Control arm mounting bolts
   - Upper control arm

To install:
5. Install or connect the following:
   - Upper control arm. Tighten the mounting bolts to 72–87 ft. lbs. (98–118 Nm).
   - Upper ball joint. Tighten the nut to 58–108 ft. lbs. (78–147 Nm).
   - Shock absorber
   - Front wheel
6. Check the wheel alignment and adjust, as necessary.

2003–2004

1. Before servicing the vehicle, refer to the Precautions Section.

2. Raise and support the vehicle safely.
3. Remove shock absorber.
4. Separate upper ball joint stud from knuckle spindle.

✳✳ CAUTION

Support lower link with jack.

5. Put matching marks on adjusting bolts and remove adjusting bolts.

To install:
6. While aligning the adjusting bolts with the matching marks, install the upper link. If a new upper link or any other suspension part is installed, align the matching mark with the slit as indicated in the accompanying figure, then install the upper link.
7. Install shock absorber.
8. Tighten adjusting bolts under unladen condition (fuel, radiator coolant, and engine oil full; with spare tire, jack, hand tools, and mats in designated positions) with tires on ground. See the accompanying illustrations for the proper torques.
9. After installing, check wheel alignment. Adjust if necessary.

2005–2006

1. Before servicing the vehicle, refer to the Precautions Section.
2. Raise and support the vehicle safely.
3. Remove the tire and wheel assembly.
4. Using a suitable jack, support the lower control arm.
5. If working on the left side, remove the bolt from the lower joint of the lower joint shaft, then reposition the lower joint

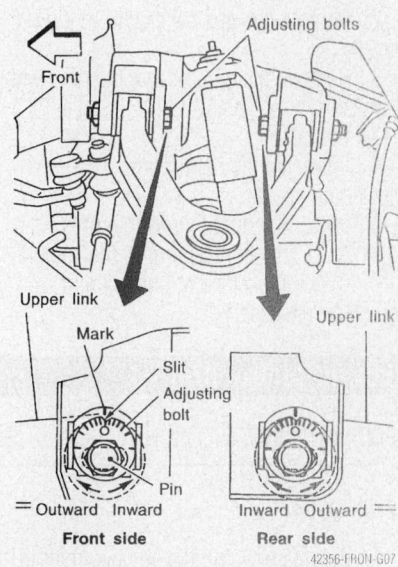

Adjusting bolt alignment—2003– Frontier and Xterra

shaft out of the way. Do not damage the lower joint.

6. Remove the cotter pin and nut from the upper control arm ball joint.

7. Separate the upper control arm ball joint stud from the steering knuckle, using tool ST29020001 (J-24319-01) or equivalent.

8. Remove the upper control arm retaining bolts and nuts.

9. Remove the upper control arm from the vehicle.

To install:

10. Installation is the reverse of the removal procedure.

11. Be sure to replace all wearable components, as required.

12. Check and adjust the front alignment, as required.

➡ **After removing/installing or replacing steering and suspension components which effect wheel alignment or after adjusting wheel alignment, or the steering angle sensor or the ABS actuator electrical unit be sure to adjust the neutral position of the steering angle sensor before running the vehicle.**

13. Position the steering wheel in the straight ahead position.

14. Drive the vehicle at 10 mph for more than 10 minutes.

15. When this procedure is complete the SLP indicator lamp and the VDC OFF indicator lamp will turn off.

CONTROL ARM BUSHING REPLACEMENT

1. Before servicing the vehicle, refer to the Precautions Section.

2. Remove the control arm from the vehicle.

3. Remove the control arm bushing with a press.

To install:

4. Lubricate the control arm bushings with liquid soap.

5. Install the bushings with a press.

6. Install the control arm to the vehicle.

7. Check the wheel alignment and adjust, as necessary.

Lower Control Arm

REMOVAL & INSTALLATION

2002–2004

1. Before servicing the vehicle, refer to the Precautions Section.

2. Raise and support the vehicle safely.

3. Remove or disconnect the following:
- Front wheel
- Torsion bar
- Shock absorber
- Stabilizer bar link
- Axle halfshaft, if equipped
- Lower ball joint
- Control arm mounting bolts
- Lower control arm

To install:

4. Install or connect the following:
- Lower control arm. Tighten the mount bolts to 80–105 ft. lbs. (108–142 Nm).
- Lower ball joint. Tighten the nut to 87–141 ft. lbs. (118–191 Nm).
- Axle halfshaft, if equipped
- Stabilizer bar link
- Shock absorber
- Torsion bar
- Front wheel

5. Check the wheel alignment and adjust, as necessary.

2005–2006

1. Before servicing the vehicle, refer to the Precautions Section.

2. Raise and support the vehicle safely.

3. Remove the tire and wheel assembly.

4. Remove the lower strut bolt.

5. Remove the stabilizer bar connecting rod lower nut. Separate the connecting rod from the lower link.

6. If equipped with 4WD, remove the halfshaft.

7. Remove the pinch bolt from the steering knuckle. Separate the lower control arm ball joint stud from the steering knuckle, using the proper tool.

8. Remove the lower control arm adjusting bolts and nuts. Lower the control arm and remove it from the vehicle.

9. Remove the jounce bumper from the lower control arm.

To install:

10. Installation is the reverse of the removal procedure.

11. Be sure to replace all wearable components, as required.

➡ **Some vehicles may be equipped with straight (non-adjustable) lower control arm bolts and washers. In order to adjust camber and caster on these vehicles, first replace the lower control arm bolts and washers with adjustable (cam) bolts and washers.**

12. Check and adjust the front alignment, as required.

➡ **After removing/installing or replacing steering and suspension components**

which effect wheel alignment or after adjusting wheel alignment, or the steering angle sensor or the ABS actuator electrical unit be sure to adjust the neutral position of the steering angle sensor before running the vehicle.

13. Position the steering wheel in the straight ahead position.

14. Drive the vehicle at 10 mph for more than 10 minutes.

15. When this procedure is complete the SLP indicator lamp and the VDC OFF indicator lamp will turn off.

CONTROL ARM BUSHING REPLACEMENT

1. Before servicing the vehicle, refer to the Precautions Section.

2. Remove the control arm from the vehicle.

3. Remove the control arm bushing with a press.

To install:

4. Lubricate the control arm bushings with liquid soap.

5. Install the bushings with a press.

6. Install the control arm to the vehicle.

7. Check the wheel alignment and adjust, as necessary.

Upper Ball Joint

REMOVAL & INSTALLATION

The upper ball joint is serviced with the upper control arm as an assembly.

Lower Ball Joint

REMOVAL & INSTALLATION

The lower ball joint is serviced with the lower control arm as an assembly.

Wheel Bearings

ADJUSTMENT

2002–2004

2WD

➡ **Use a new split pin for assembly.**

1. Before servicing the vehicle, refer to the Precautions Section.

2. Remove or disconnect the following:
- Dust cap
- Split pin
- Spindle nut cap

3. Tighten the spindle nut to 25–29 ft. lbs. (34–39 Nm).

4. Spin the hub several times to fully seat the bearings.

5. Retighten the spindle nut to 25–29 ft. lbs. (34–39 Nm).

6. Loosen the spindle nut 45–60 degrees and install the spindle nut cap and split pin.

7. Install the dust cap.

4WD

1. Before servicing the vehicle, refer to the Precautions Section.

2. Remove or disconnect the following:

- Locking hub or driveplate
- Snapring
- Spindle washer
- Thrust washer
- Lockwasher

3. Tighten the wheel bearing locknut to 58–72 ft. lbs. (78–98 Nm).

4. Loosen the locknut fully.

5. Tighten the wheel bearing locknut to 4–13 inch lbs. (0.5–1.5 Nm).

6. Spin the hub several times to fully seat the bearings.

7. Retighten the wheel bearing locknut to 4–13 inch lbs. (0.5–1.5 Nm).

8. Install or connect the following:

- Lockwasher. Tighten the retaining screw to 10–16 inch lbs. (1–2 Nm).
- Thrust washer
- Spindle washer
- Snapring
- Locking hub or driveplate

2005–2006

1. Before servicing the vehicle, refer to the Precautions Section.

2. Raise and support the vehicle safely.

3. Remove the tire and wheel assembly.

4. Move the wheel hub in the axial direction by hand. Make sure that there is no looseness of the wheel bearing. Axial end play should be 0.002 inch or less.

5. Rotate the wheel hub and make sure there is no unusual noise or other irregular conditions. Replace the hub and wheel bearing assembly, as necessary.

REMOVAL & INSTALLATION

2002–2004

2WD

1. Before servicing the vehicle, refer to the Precautions Section.

2. Raise and support the vehicle safely.

3. Remove or disconnect the following:

- Front wheel
- Brake caliper and support
- Dust cap
- Split pin
- Spindle nut cap
- Spindle nut
- Bearing washer
- Outer bearing
- Hub and brake rotor assembly
- Inner grease seal
- Inner wheel bearing

To install:

4. Install or connect the following:

- Inner wheel bearing
- Inner grease seal
- Hub and brake rotor assembly
- Outer bearing
- Bearing washer
- Spindle nut. Adjust the wheel bearings.
- Spindle nut cap
- Split pin
- Dust cap
- Brake caliper and support
- Front wheel

4WD

1. Before servicing the vehicle, refer to the Precautions Section.

2. Raise and support the vehicle safely.

3. Remove or disconnect the following:

- Front wheel
- Brake caliper and support
- Locking hub or driveplate
- Snapring
- Spindle washer
- Thrust washer
- Lockwasher
- Wheel bearing locknut
- Outer bearing
- Hub and brake rotor assembly
- Inner grease seal
- Inner wheel bearing

To install:

4. Install or connect the following:

- Inner wheel bearing
- Inner wheel bearing
- Inner grease seal
- Hub and brake rotor assembly
- Outer bearing
- Wheel bearing locknut. Adjust the wheel bearings.
- Lockwasher
- Thrust washer
- Spindle washer
- Snapring
- Locking hub or driveplate
- Brake caliper and support
- Front wheel

2005–2006

1. Before servicing the vehicle, refer to the Precautions Section.

2. Raise and support the vehicle safely.

3. Remove the tire and wheel assembly.

4. Remove the caliper and position it to the side with wire. Do not disconnect the brake fluid line.

➡**Do not press the brake pedal while the brake caliper is removed.**

5. Matchmark the brake rotor and the wheel hub. Remove the brake rotor.

6. Remove the cotter pin. Remove the lock nut.

7. Remove the halfshaft from the wheel hub and bearing assembly.

8. Remove the wheel sensor from the hub and bearing assembly. Do not pull on the wheel sensor harness.

9. Remove the wheel hub and bearing assembly bolts.

10. Remove the splash guard. Remove the wheel hub and bearing assembly from the steering knuckle.

➡**Carefully remove the wheel sensor and harness through the hole in the splash guard.**

To install:

11. Inspect the wheel sensor O-ring, replace the wheel speed sensor assembly, as required.

12. Installation is the reverse of the removal procedure.

13. Be sure to use new bolts when installing the wheel hub and bearing assembly.

REAR SUSPENSION

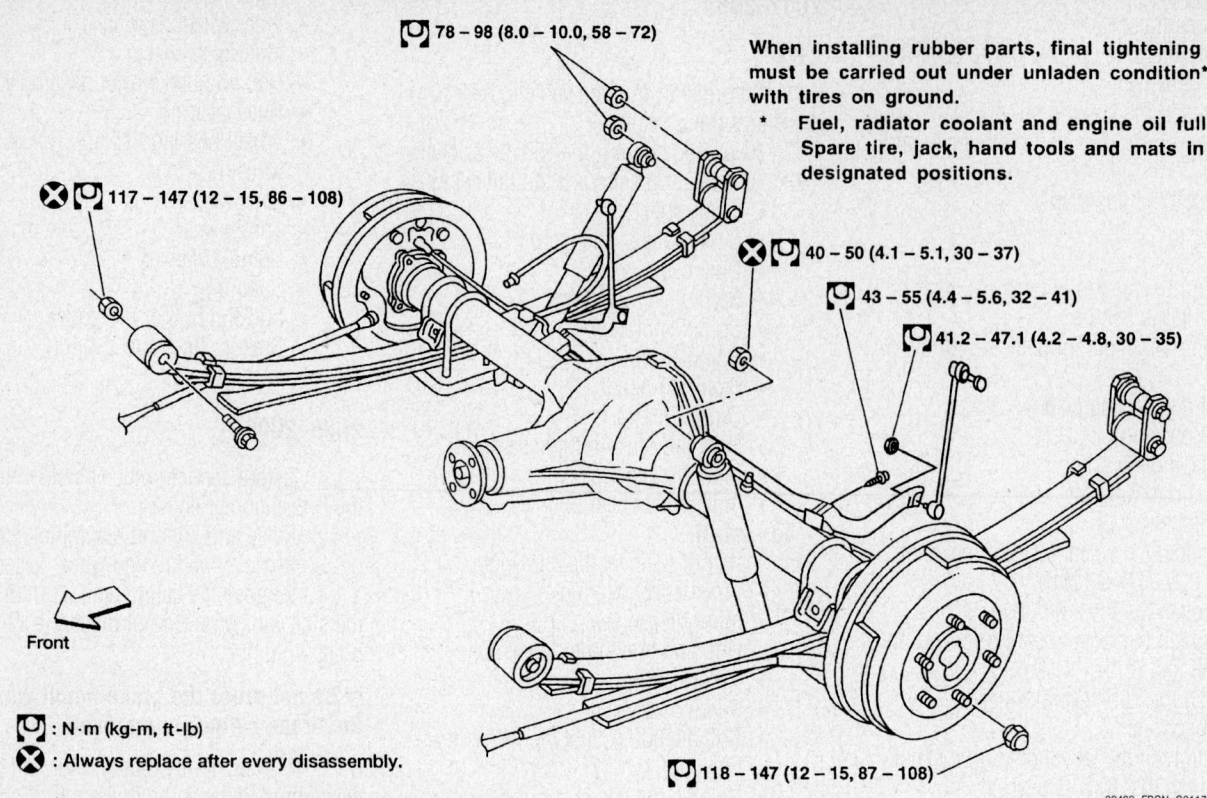

78 – 98 (8.0 – 10.0, 58 – 72)

When installing rubber parts, final tightening must be carried out under unladen condition* with tires on ground.

* Fuel, radiator coolant and engine oil full. Spare tire, jack, hand tools and mats in designated positions.

117 – 147 (12 – 15, 86 – 108)

40 – 50 (4.1 – 5.1, 30 – 37)

43 – 55 (4.4 – 5.6, 32 – 41)

41.2 – 47.1 (4.2 – 4.8, 30 – 35)

Front

: N·m (kg-m, ft-lb)

: Always replace after every disassembly.

118 – 147 (12 – 15, 87 – 108)

09482_FRON_G0117

Rear suspension and related components—2WD 4-cyl. 2002–2004

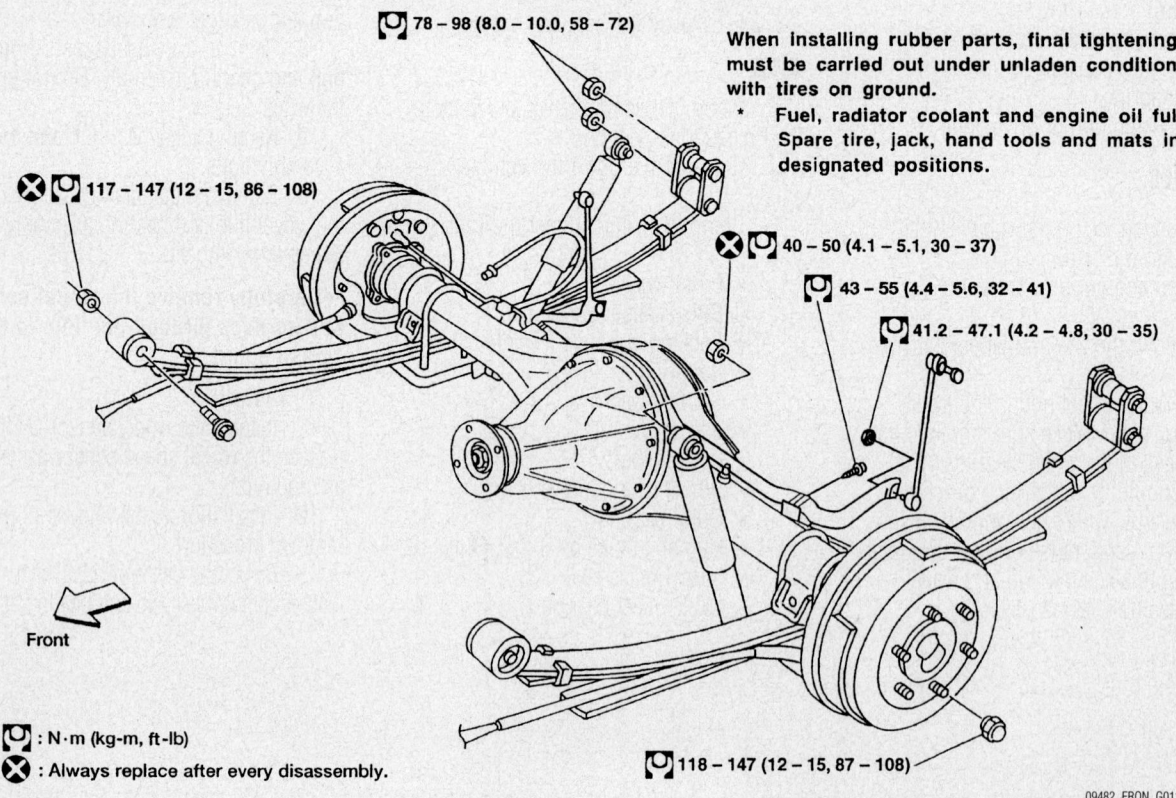

78 – 98 (8.0 – 10.0, 58 – 72)

When installing rubber parts, final tightening must be carried out under unladen condition* with tires on ground.

* Fuel, radiator coolant and engine oil full. Spare tire, jack, hand tools and mats in designated positions.

117 – 147 (12 – 15, 86 – 108)

40 – 50 (4.1 – 5.1, 30 – 37)

43 – 55 (4.4 – 5.6, 32 – 41)

41.2 – 47.1 (4.2 – 4.8, 30 – 35)

Front

: N·m (kg-m, ft-lb)

: Always replace after every disassembly.

118 – 147 (12 – 15, 87 – 108)

09482_FRON_G0118

Rear suspension and related components—2WD 6-cyl. 2002–2004

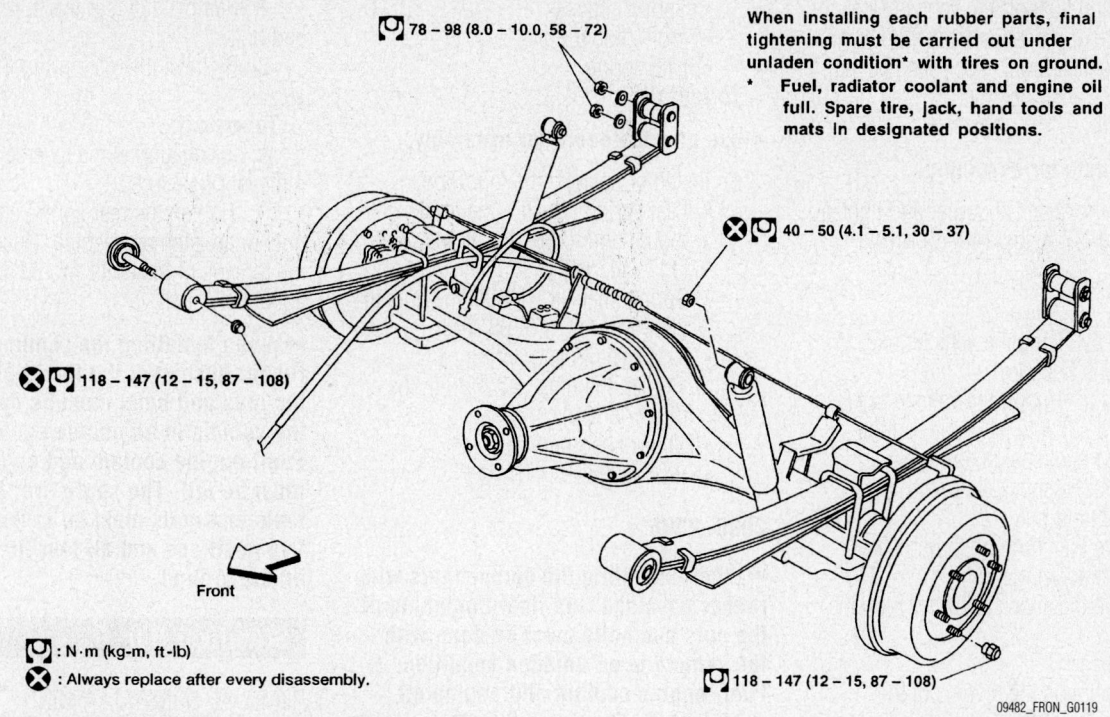

78 – 98 (8.0 – 10.0, 58 – 72)

When installing each rubber parts, final tightening must be carried out under unladen condition* with tires on ground.
* Fuel, radiator coolant and engine oil full. Spare tire, jack, hand tools and mats in designated positions.

✕ 🔧 40 – 50 (4.1 – 5.1, 30 – 37)

✕ 🔧 118 – 147 (12 – 15, 87 – 108)

Front

🔧 : N·m (kg-m, ft-lb)
✕ : Always replace after every disassembly.

🔧 118 – 147 (12 – 15, 87 – 108)

09482_FRON_G0119

Rear suspension and related components—4WD 2002–2004

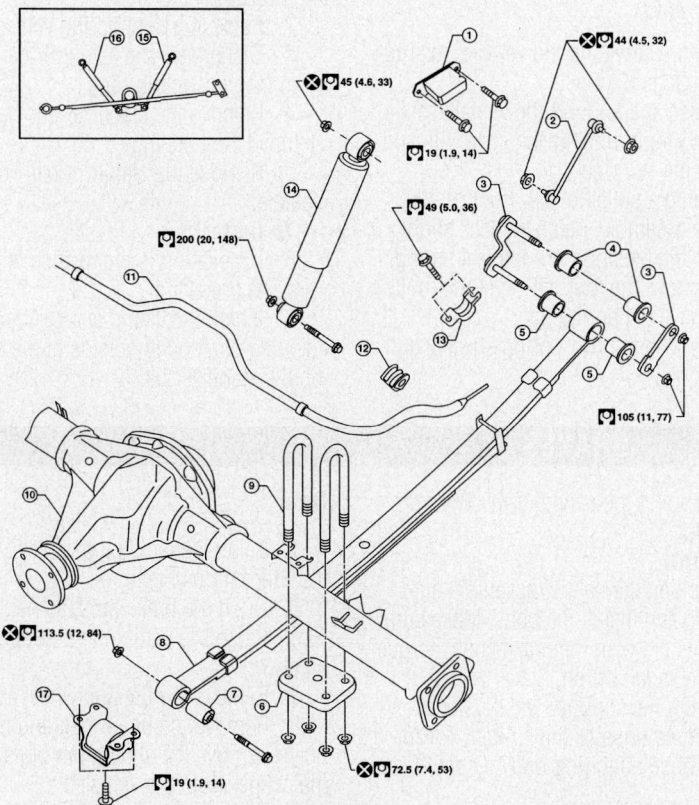

✕ 🔧 45 (4.6, 33)

1

✕ 🔧 44 (4.5, 32)

🔧 19 (1.9, 14)

2

🔧 49 (5.0, 36)

3

🔧 200 (20, 148)

4

3

5

🔧 105 (11, 77)

11

12

13

✕ 🔧 113.5 (12, 84)

10

9

8

7

6

5

17

7

6

🔧 19 (1.9, 14)

✕ 🔧 72.5 (7.4, 53)

1. Bumper	2. Connecting rod	3. Rear spring shackle
4. Rear spring shackle bushing	5. Rear spring bushing (rear)	6. Rear spring pad
7. Rear spring bushing (front)	8. Rear leaf spring	9. Rear spring clip U-bolts
10. Rear final drive	11. Stabilizer bar	12. Bushing
13. Stabilizer bar clamp	14. Shock absorber	15. Shock absorber (left side)
16. Shock absorber (right side)	17. Dynamic damper	

09482_FRON_G0120

Rear suspension and related components—2005–2006

Shock Absorber

REMOVAL & INSTALLATION

2002–2004

FRONTIER

1. Before servicing the vehicle, refer to the Precautions Section.
2. Raise and support the vehicle safely.
3. Support the rear axle.
4. Remove or disconnect the following:
   • Lower shock absorber bolt
   • Upper shock absorber bolt
   • Shock absorber

To install:

➡**Use new fasteners for assembly.**

5. Install the shock absorber and tighten the bolts to 49–65 ft. lbs. (67–88 Nm).
6. Before servicing the vehicle, refer to the Precautions Section.
7. Remove or disconnect the following:
   • Upper and lower shock absorber nuts
   • Shock absorber

To install:

➡**Use new nuts for assembly.**

8. Install the shock absorber and tighten the nuts to 30–37 ft. lbs. (40–50 Nm).

XTERRA

1. Before servicing the vehicle, refer to the Precautions Section.

2. Raise and support the vehicle safely.
3. Remove or disconnect the following:
- Upper and lower shock absorber nuts
- Shock absorber

To install:

➡**Use new nuts for assembly.**

4. Install the shock absorber and tighten the nuts to 30–37 ft. lbs. (40–50 Nm).

2005–2006

1. Before servicing the vehicle, refer to the Precautions Section.
2. Raise and support the vehicle safely.
3. Using a suitable jack, support the final drive and suspension assembly.
4. Remove the shock absorber upper mounting bolt and nut.
5. Remove the shock absorber lower mounting bolt and nut.
6. Remove the shock absorber from the vehicle.
To install:
7. Installation is the reverse of the removal procedure.

Leaf Springs

REMOVAL & INSTALLATION

2002–2004

1. Before servicing the vehicle, refer to the Precautions Section.
2. Support the vehicle at the frame.
3. Support the axle with a floor jack.
4. Remove or disconnect the following:
- Rear wheels
- Shock absorbers
- Axle U-bolts and spring pad

- Spring shackle
- Front mount bolt
- Leaf spring

To install:

➡**Use new fasteners for assembly.**

5. Install or connect the following:
- Leaf spring. Tighten the front mount bolt to 86–108 ft. lbs. (117–147 Nm).
- Spring shackle. Tighten the nuts to 58–72 ft. lbs. (78–98 Nm).
- Axle U-bolts and spring pad. Tighten the nuts to 72–80 ft. lbs. (98–108 Nm).
- Shock absorbers
- Rear wheels

2005–2006

➡**When installing the components with rubber bushings, the final tightening of the nuts and bolts must be done with the vehicle in an unladen condition. Fuel, engine coolant and engine oil must be full. The spare tire, jack, hand tools and mats must be in their respective positions and all four tires must be on the ground.**

1. Before servicing the vehicle, refer to the Precautions Section.
2. Raise and support the vehicle safely.
3. On Xterra, remove the spare tire and the right side tail pipe.
4. Using a suitable jack, support the final drive assembly. Raise the jack slightly to remove the tension from the leaf spring.
5. Remove the four spring clip U-bolts nuts. Remove the spring pad.
6. Remove the rear spring shackle and bushings.

7. Remove the rear leaf spring front nut and bolt.
8. Remove the rear spring from the vehicle.
To install:
9. Installation is the reverse of the removal procedure.
10. Tighten the rear spring clip U-bolts nuts in an alternate fashion. Once complete the threads of the bolts should be exposed about 0.12 inch.

➡**When installing the components with rubber bushings, the final tightening of the nuts and bolts must be done with the vehicle in an unladen condition. Fuel, engine coolant and engine oil must be full. The spare tire, jack, hand tools and mats must be in their respective positions and all four tires must be on the ground.**

Stabilizer Bar

REMOVAL & INSTALLATION

1. Before servicing the vehicle, refer to the Precautions Section.
2. Raise and support the vehicle safely.
3. Disconnect the stabilizer bar ends from the connecting rods.
4. Remove the stabilizer bar clamps. Remove the bushings.
5. Remove the stabilizer bar from the vehicle.
To install:
6. Installation is the reverse of the removal procedure.
7. Install the clamp and bushings so they are positioned outside of the crimp ring on the stabilizer bar.

BRAKES

Brake Caliper

REMOVAL & INSTALLATION

Front

1. Before servicing the vehicle, refer to the Precautions Section.
2. Drain the brake fluid, as necessary.
3. Raise the vehicle and support safely.
4. Remove the tire and wheel assembly.
5. Remove the bolt attaching the brake hose to the caliper. Plug the brake hose to prevent brake fluid loss.
6. Remove the caliper support mount-

ing bolts and lift the caliper assembly from the knuckle.
To install
7. Position the caliper assembly onto the knuckle and install the bolts. Make sure the rotor fits between the brake pads. Torque the bolts to specification.
8. Using new copper washers, connect the brake hose to the caliper. Torque the brake hose attaching bolt to specification.
9. Bleed the brake system.
10. Apply the brake pedal and inspect the system. Ensure proper operation and no leakage.
11. Install tire and wheel assembly. Lower the vehicle and roadtest.

Rear

1. Before servicing the vehicle, refer to the Precautions Section.
2. Drain the brake fluid, as necessary.
3. Raise the vehicle and support safely.
4. Remove the tire and wheel assembly.
5. Remove the union bolt and brake hose. Remove the sliding pin bolts. Remove the caliper from the vehicle.
To install
6. Installation is the reverse of the removal procedure.
7. Bleed the brake system.
8. Apply the brake pedal and inspect the system. Ensure proper operation and no leakage.

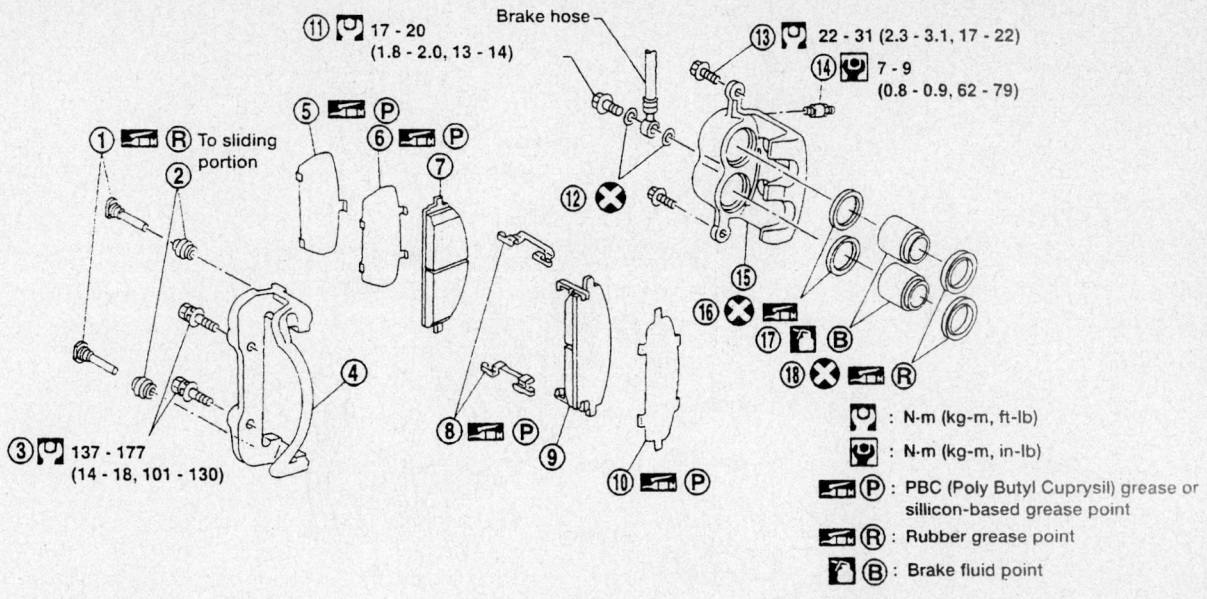

1. Main pin
2. Pin boot
3. Torque member fixing bolt
4. Torque member
5. Shim cover (if so equipped)
6. Inner shim
7. Inner pad
8. Pad retainer
9. Outer pad
10. Outer shim
11. Connecting bolt
12. Copper washer
13. Main pin bolt
14. Bleed valve
15. Cylinder body
16. Piston seal
17. Piston
18. Piston boot

9348VG95

Front disc brake and related components—2002

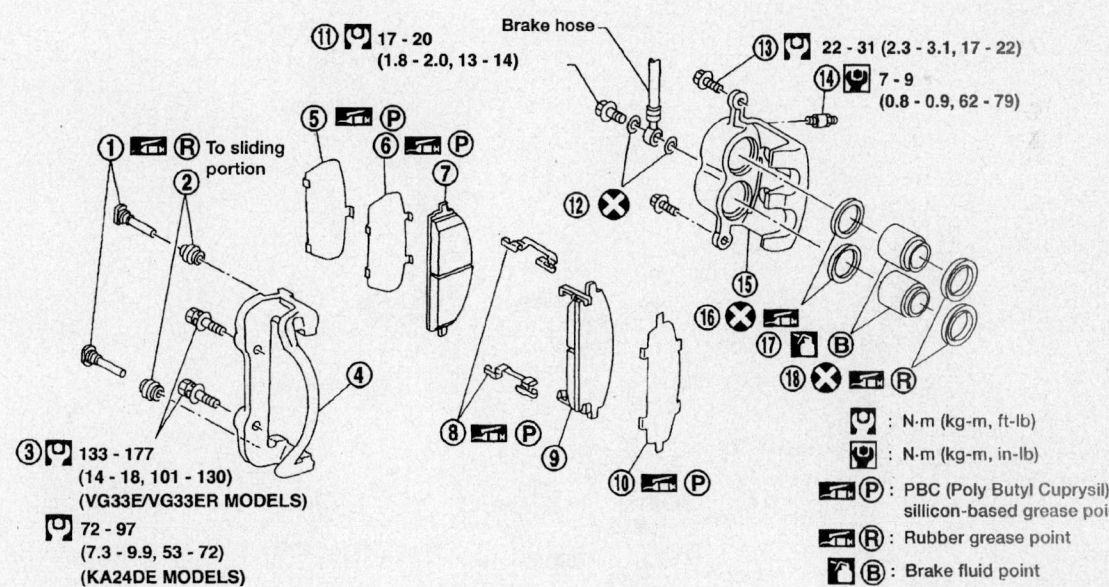

1. Main pin
2. Pin boot
3. Torque member fixing bolt
4. Torque member
5. Shim cover (if equipped)
6. Inner shim
7. Inner pad
8. Pad retainer
9. Outer pad
10. Outer shim
11. Connecting bolt
12. Copper washer
13. Main pin bolt
14. Bleed valve
15. Cylinder body
16. Piston seal
17. Piston
18. Piston boot

42356-FRON-G01

Front disc brake and related components—2003–2004

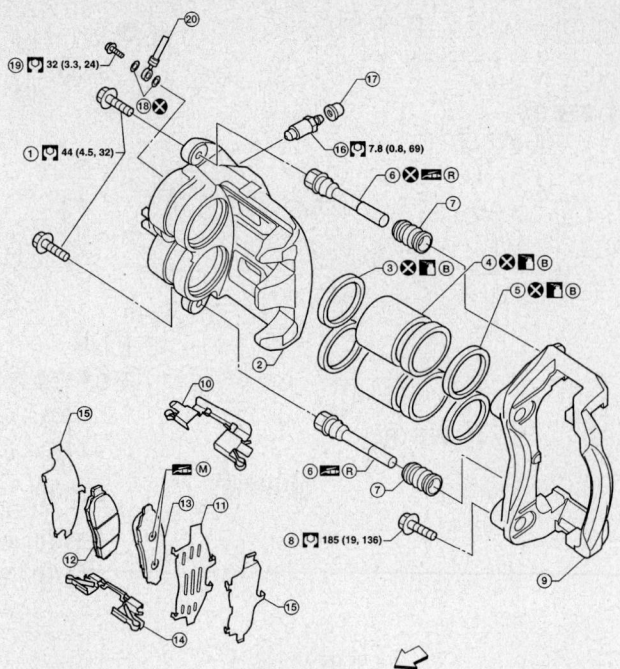

1. Sliding pin bolt
2. Cylinder body
3. Piston seal
4. Piston
5. Piston boot
6. Sliding pin
7. Sliding pin boot
8. Torque member bolt
9. Torque member
10. Pad retainer
11. Inner shim
12. Inner brake pad
13. Outer brake pad
14. Pad retainer
15. Outer shim
16. Bleed valve
17. Cap
18. Copper washers
19. Union bolt
20. Brake hose

09482_FRON_G0121

Front disc brake and related components—2005–2006

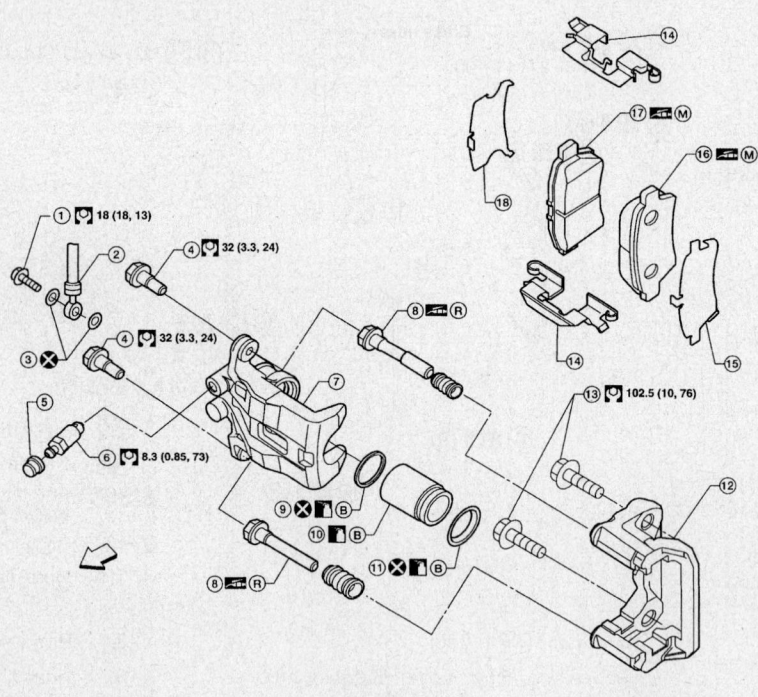

1. Union bolt
2. Brake hose
3. Copper washers
4. Sliding pin bolt
5. Cap
6. Bleed valve
7. Cylinder body
8. Sliding pin
9. Piston seal
10. Piston
11. Piston boot
12. Torque member
13. Torque member bolt
14. Pad retainer
15. Outer shim
16. Outer brake pad
17. Inner brake pad
18. Inner shim

09482_FRON_G0122

Rear disc brake and related components—Frontier 2005–2006

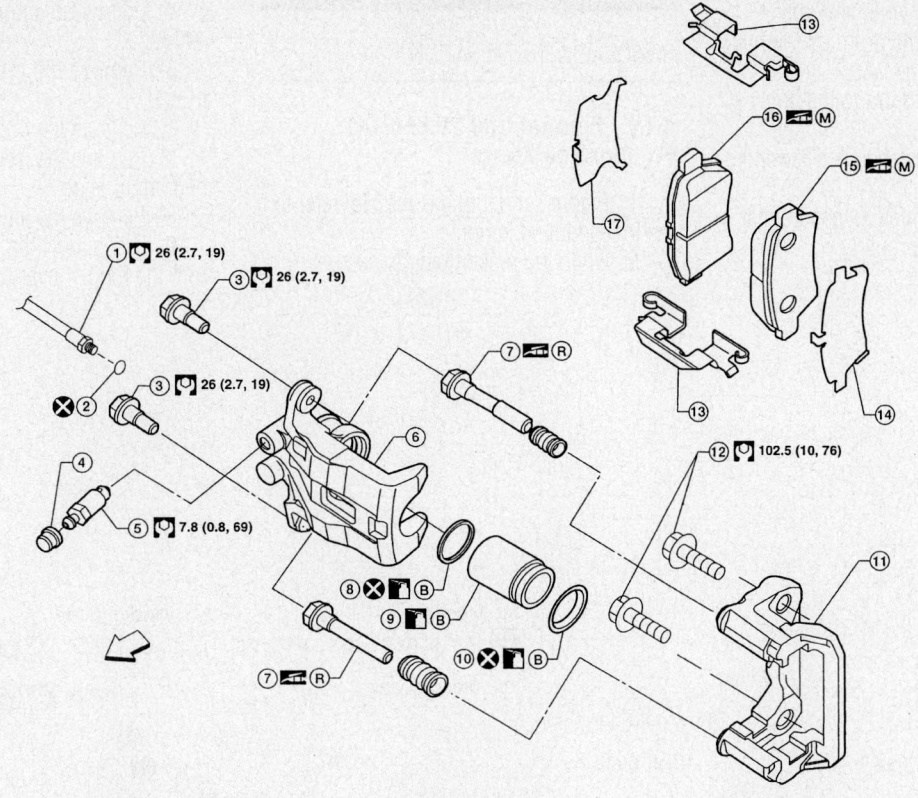

1.	Brake hose	2.	Copper washer	3.	Sliding pin bolt
4.	Cap	5.	Bleed valve	6.	Cylinder body
7.	Sliding pin	8.	Piston seal	9.	Piston
10.	Piston boot	11.	Torque member	12.	Torque member bolt
13.	Pad retainer	14.	Outer shim	15.	Outer brake pad
16.	Inner brake pad	17.	Inner shim	⇐:	Front

09482_FRON_G0123

Rear disc brake and related components—Xterra 2005–2006

9. Install tire and wheel assembly. Lower the vehicle and roadtest.

Disc Brake Pads

REMOVAL & INSTALLATION

Front

1. Before servicing the vehicle, refer to the Precautions Section.
2. Drain the brake fluid, as necessary.
3. Raise the vehicle and support safely.
4. Remove the bottom pin from the caliper and swing the caliper cylinder body upward; support the caliper with a wire.
5. Remove the brake pad retainers, shims and the pads.
   To install:
6. Compress the piston of the disc brake caliper.
7. Install the brake pads and caliper assembly.

Rear

1. Before servicing the vehicle, refer to the Precautions Section.
2. Drain the brake fluid, as necessary.
3. Raise the vehicle and support safely.
4. Remove the tire and wheel assembly.
5. Remove the top bolt from the caliper.
6. Swing the caliper open and remove the pads.
   To install:
7. Compress the piston of the disc brake caliper.
8. Install the brake pads and caliper assembly.

Brake Drums

REMOVAL & INSTALLATION

1. Before servicing the vehicle, refer to the Precautions Section.
2. Drain the brake fluid, as necessary.
3. Release the parking brake.

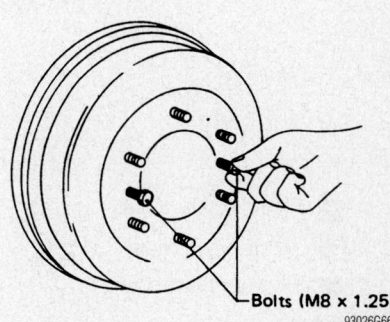

Bolts (M8 x 1.25)
93026G66

Install and tighten 2 bolts to remove a stubborn brake drum

4. Raise the vehicle and support safely.
5. Remove the tire and wheel assembly.
6. Pull the brake drum from the hub. If difficult to remove try the following:
   a. Strike the face of the drum with a plastic or rubber mallet. This will break free any rust that may develop between the drum and the hub.

b. Install 2, M8x1.25mm bolts into the holes in the drum and gradually tighten them to pull the drum off the hub.

To install:

7. Install the brake drum to the hub.

8. Install the wheel.

9. Remove the jackstands and lower the vehicle.

10. Road-test the vehicle to ensure that the brakes are working properly.

Brake Shoes

REMOVAL & INSTALLATION

4-Cyl. Frontier and 2002 6-Cyl. Frontier and Xterra

1. Before servicing the vehicle, refer to the Precautions Section.

2. Drain the brake fluid, as necessary.

3. Release the parking brake.

4. Safely raise and support the vehicle.

5. Remove the rear wheel and drum.

6. Remove the hold-down pin retainers.

7. Remove the leading shoe and then the trailing shoe.

8. Remove the adjuster.

9. Disconnect the parking brake cable from the toggle lever on the rear shoe.

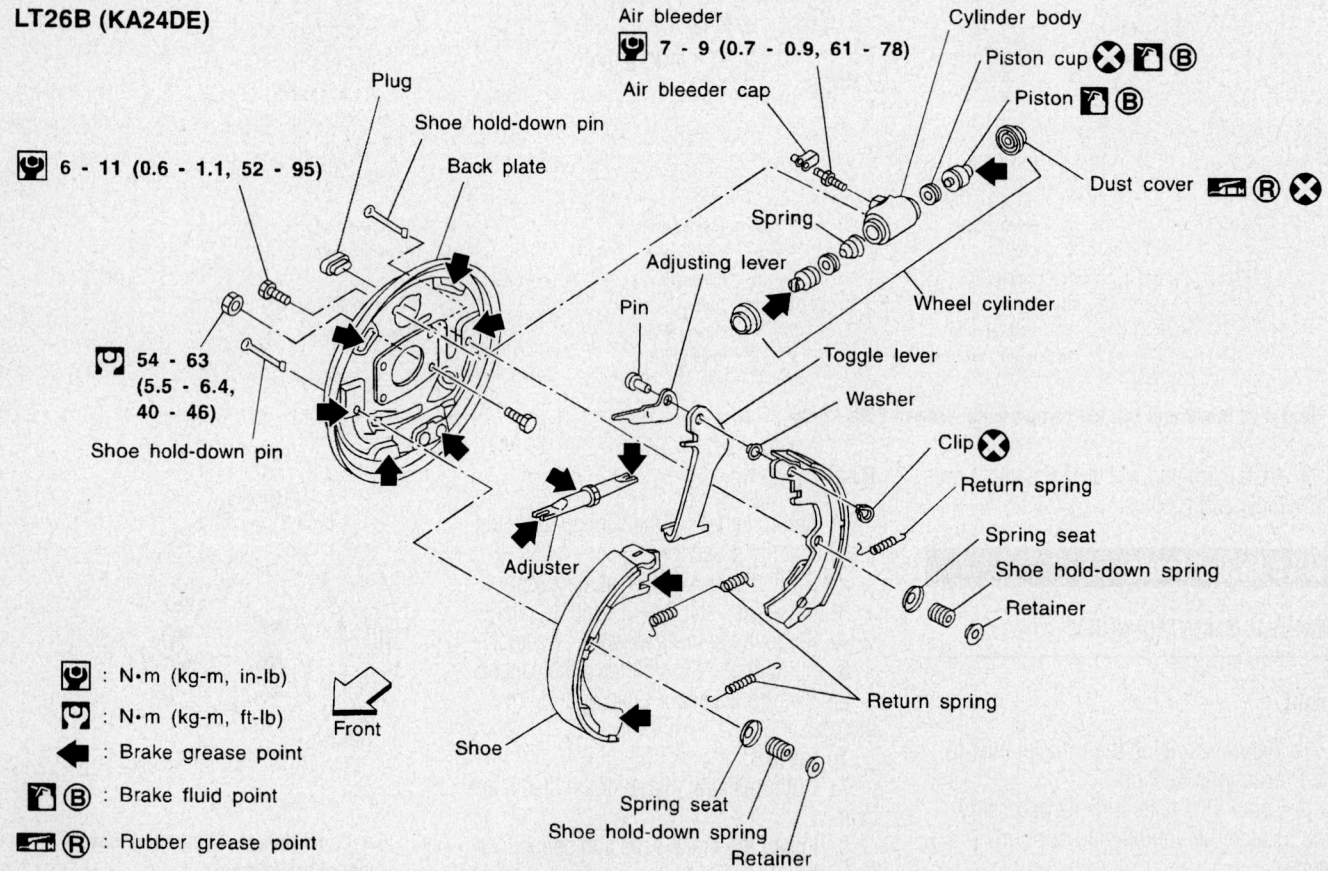

LT26B (KA24DE)

Plug
Shoe hold-down pin
Back plate
6 - 11 (0.6 - 1.1, 52 - 95)
54 - 63 (5.5 - 6.4, 40 - 46)
Shoe hold-down pin

Air bleeder
7 - 9 (0.7 - 0.9, 61 - 78)
Air bleeder cap
Spring
Adjusting lever
Pin

Cylinder body
Piston cup
Piston
Dust cover
Wheel cylinder
Toggle lever
Washer
Clip
Return spring
Spring seat
Shoe hold-down spring
Retainer

Adjuster
Return spring
Shoe
Spring seat
Shoe hold-down spring
Retainer

: N•m (kg-m, in-lb)
: N•m (kg-m, ft-lb)
Front
: Brake grease point
B : Brake fluid point
R : Rubber grease point

9348VG96

Rear drum brake assembly and related components—4-cyl. Frontier

LT30A (VG33E and VG33ER)

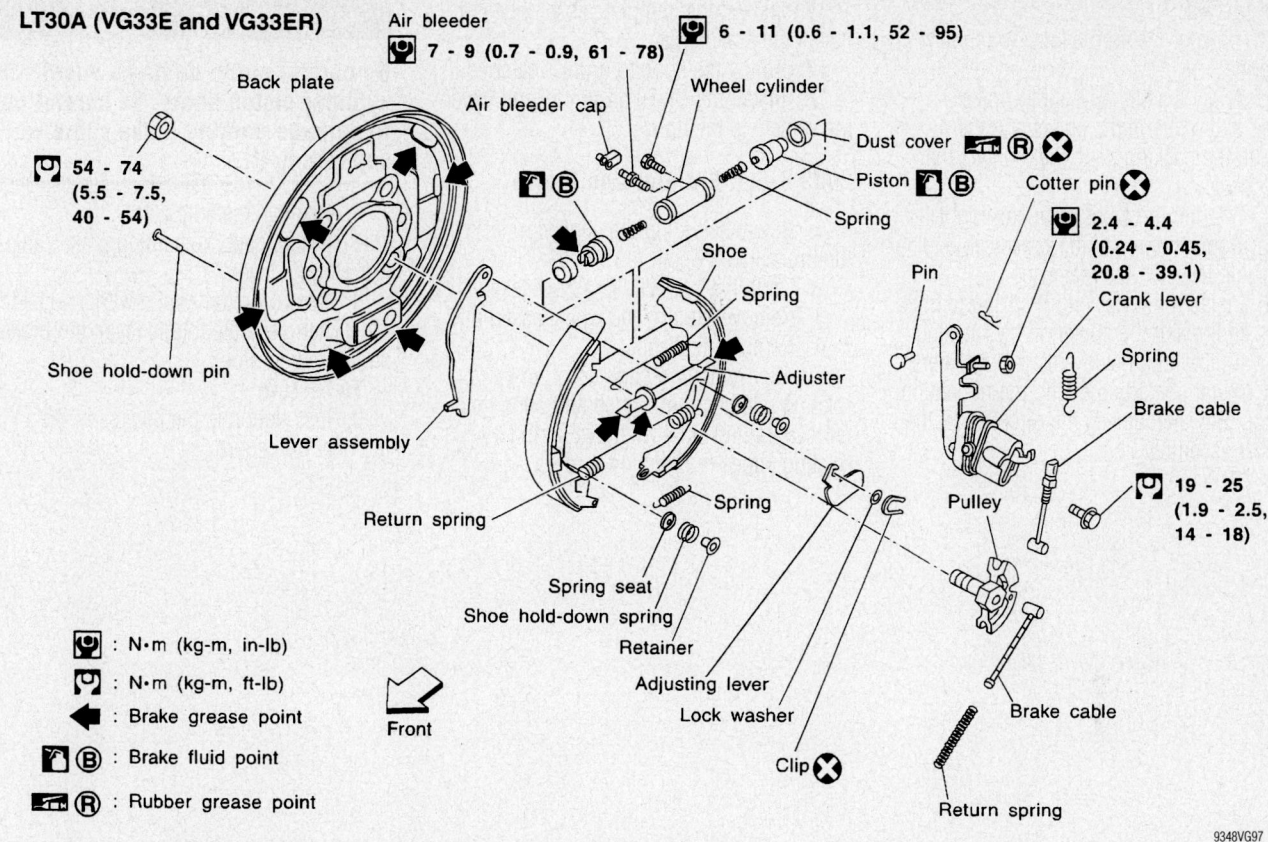

Rear drum brake assembly and related components—2002 6-cyl. Frontier and Xterra

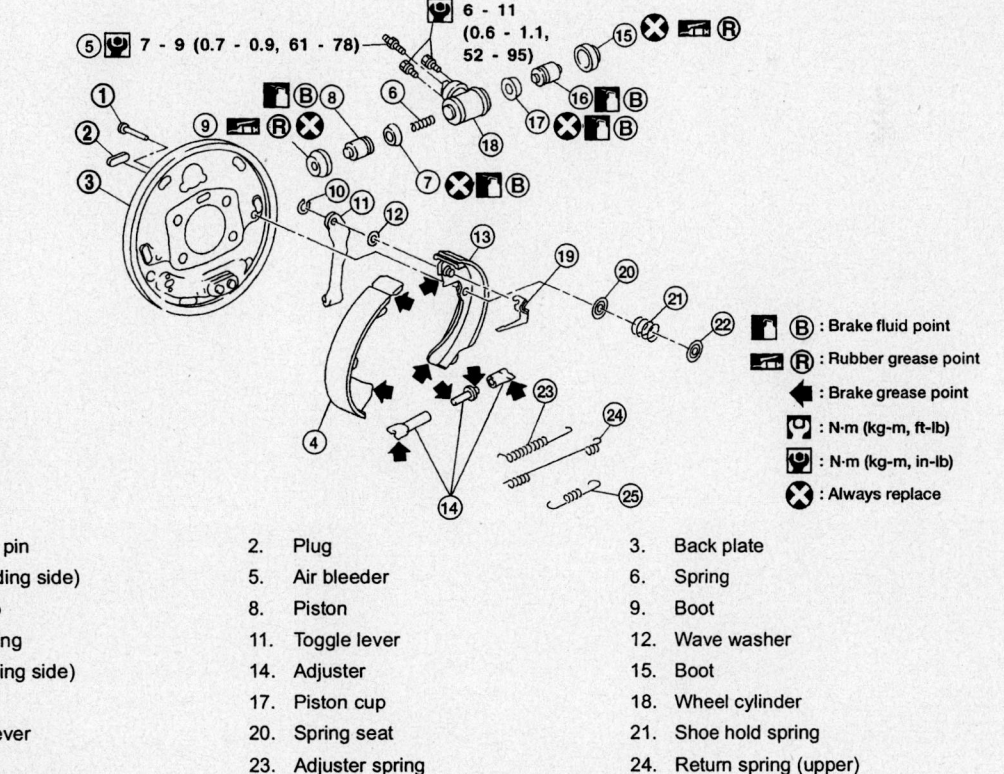

1. Shoe hold pin	2. Plug	3. Back plate
4. Shoe (leading side)	5. Air bleeder	6. Spring
7. Piston cup	8. Piston	9. Boot
10. Retainer ring	11. Toggle lever	12. Wave washer
13. Shoe (trailing side)	14. Adjuster	15. Boot
16. Piston	17. Piston cup	18. Wheel cylinder
19. Adjuster lever	20. Spring seat	21. Shoe hold spring
22. Retainer	23. Adjuster spring	24. Return spring (upper)
25. Return spring (lower)		

Rear drum brake assembly and related components—2003 6-cyl. Frontier and Xterra

To install:

10. Transfer the toggle lever to the new rear shoe.

11. Apply a small amount of brake grease to the tips of the shoes and the 6 pads on the backing plate that contact the brake shoe.

12. Shorten the adjuster by turning it.

13. Connect the parking brake cable to the toggle lever on the rear shoe.

14. Install the lower return spring to both shoes and install the shoes on the backing plate with the hold down pins and retainers.

15. Install the adjuster and the remaining springs. Pay attention to the direction of the adjuster assembly.

16. Inspect the complete assembly and install the brake drum.

17. Adjust the shoe to drum clearance.

18. Install the wheel assembly and lower the vehicle to the floor.

2003 6-Cyl. Frontier and Xterra

1. Before servicing the vehicle, refer to the Precautions Section.

2. Drain the brake fluid, as necessary.

3. Remove the tire and wheel assembly.

4. Remove the drum.

5. After removing shoe hold pin by rotating retainer, remove leading shoe then remove trailing shoe. Remove spring by rotating shoes in direction arrow.

⁂ WARNING

Be careful not to damage wheel cylinder piston boots. Be careful not to damage parking brake cable when separating it.

6. Remove the adjuster.

7. Disconnect the parking brake cable from toggle lever.

8. Remove retainer clip with a suitable tool. Then separate toggle lever and brake shoe (trailing side).

To install:

9. Installation is the reverse of the removal procedure.

SPECIFICATIONS AND MAINTENANCE CHARTS

ENGINE AND VEHICLE IDENTIFICATION

Engine								Model Year	
Code	Liters (cc)	Cu. In.	Cyl.	Fuel Sys.	Engine	Eng. Mfg.		Code	Year
VQ35DE	3.5 (3498)	213	6	MFI	DOHC	Nissan		3	2003
MFI: Multi-port Fuel Injection								4	2004
DOHC: Double Overhead Camshaft								5	2004
								6	2006

09482_MURA_C0001

GENERAL ENGINE SPECIFICATIONS

Year	Model	Engine Displacement Liters	Engine ID	Net Horsepower @ rpm	Net Torque @ rpm (ft. lbs.)	Bore x Stroke (in.)	Com-pression Ratio	Oil Pressure @ rpm
2003	Murano	3.5	VQ35DE	240@6000	265@3200	3.76X3.20	10.0:1	43@2000
2004	Murano	3.5	VQ35DE	240@6000	265@3200	3.76X3.20	10.0:1	43@2000
2005	Murano	3.5	VQ35DE	240@6000	265@3200	3.76X3.20	10.0:1	43@2000
2006	Murano	3.5	VQ35DE	240@6000	265@3200	3.76X3.20	10.0:1	43@2000

09482_MURA_C0002

ENGINE TUNE-UP SPECIFICATIONS

Year	Engine Displacement Liters	Engine ID	Spark Plug Gap (in.)	Ignition Timing (deg.)	Fuel Pump (psi)	Idle Speed RPM	Valve Clearance (in.) In.	Valve Clearance (in.) Ex.
2003	3.5	VQ35DE	0.043	15B	51 ①	600-700	③	④
2004	3.5	VQ35DE	0.043	15B	51 ①	600-700	③	④
2005	3.5	VQ35DE	0.043	②	51 ①	600-700	③	④
2006	3.5	VQ35DE	0.043	②	51 ①	600-700	③	④

NOTE: The Vehicle Emission Control Information label often reflects specification changes made during production. The label figures must be used if they differ from those in this chart.

B: Before top dead center

① At idle

② 10-20B

③ 0.010-0.013 cold
0.012-0.016 hot

④ 0.011-0.015 cold
0.012-0.017 hot

09482_MURA_C0003

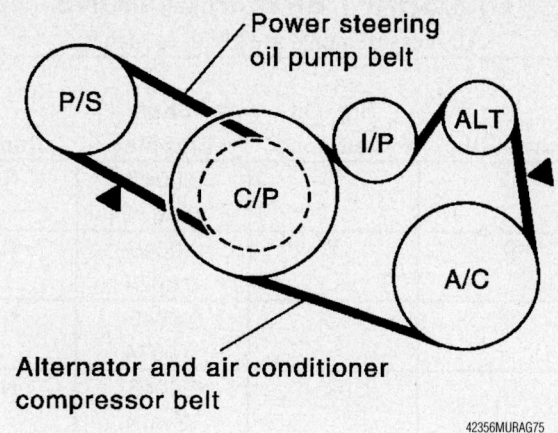

Power steering
oil pump belt

Alternator and air conditioner
compressor belt

42356MURAG75

Accessory drive belt routing

CAPACITIES

Year	Model	Engine Displacement Liters	Engine ID	Engine Oil with Filter (qts.)	Transmission (pts.)	Transfer Case (pts.)	Drive Axle Front (pts.)	Drive Axle Rear (pts.)	Fuel Tank (gal.)	Cooling System (qts.)
2003	Murano	3.5	VQ35DE	4.25	21.5	0.63	—	1.5	21.6	9.75
2004	Murano	3.5	VQ35DE	4.25	21.5	0.63	—	1.1	21.6	9.75
2005	Murano	3.5	VQ35DE	4.25	21.5	0.63	—	1.1	21.6	10.0
2006	Murano	3.5	VQ35DE	4.25	21.5	0.63	—	1.1	21.6	10.0

NOTE: All capacities are approximate. Add fluid gradually and check to be sure a proper fluid level is obtained.

09482_MURA_C0004

VALVE SPECIFICATIONS

Year	Engine Displacement Liters	Engine ID	Seat Angle (deg.)	Face Angle (deg.)	Spring Test Pressure (lbs. @ in.)	Spring Installed Height (in.)	Stem-to-Guide Clearance (in.) Intake	Stem-to-Guide Clearance (in.) Exhaust	Stem Diameter (in.) Intake	Stem Diameter (in.) Exhaust
2003	3.5	VQ35DE	45.15-45.45	45	45.4@1.457	1.457	0.0008-0.0021	0.0012-0.0025	0.2348-0.2354	0.2344-0.2350
2004	3.5	VQ35DE	45.15-45.45	45	45.4@1.457	1.457	0.0008-0.0021	0.0012-0.0025	0.2348-0.2354	0.2344-0.2350
2005	3.5	VQ35DE	45.15-45.45	45	45.4@1.457	1.457	0.0008-0.0021	0.0012-0.0025	0.2348-0.2354	0.2344-0.2350
2006	3.5	VQ35DE	45.15-45.45	45	45.4@1.457	1.457	0.0008-0.0021	0.0012-0.0025	0.2348-0.2354	0.2344-0.2350

09482_MURA_C0005

CAMSHAFT SPECIFICATIONS

All measurements are given in inches.

Year	Engine Displ. Liters	Engine VIN	Journal Dia.	Brg. Oil Clearance	Shaft End-play	Runout	Lobe Height Intake	Lobe Height Exhaust
2003	3.5	VQ35DE	①	②	0.0045-0.0074	NS	1.7663-1.7738	1.7663-1.7738
2004	3.5	VQ35DE	①	②	0.0045-0.0074	NS	1.7663-1.7738	1.7663-1.7738
2005	3.5	VQ35DE	①	②	0.0045-0.0074	NS	1.7663-1.7738	1.7663-1.7738
2006	3.5	VQ35DE	①	②	0.0045-0.0074	NS	1.7663-1.7738	1.7663-1.7738

NS - Not specified by manufacturer

① No.1: 1.0211-1.0218
No.2, No.3, No.4: 0.9230-0.9238

② No.1: 1.0018-1.0034
No.2, No.3, No.4: 0.0014-0.0030

09482_MURA_C0006

CRANKSHAFT AND CONNECTING ROD SPECIFICATIONS

All measurements are given in inches.

Year	Engine Displacement Liters	Engine ID	Crankshaft Main Brg. Journal Dia.	Crankshaft Main Brg. Oil Clearance	Crankshaft Shaft End-play	Crankshaft Thrust on No.	Connecting Rod Journal Diameter	Connecting Rod Oil Clearance	Connecting Rod Side Clearance
2003	3.5	VQ35DE	①	0.0014-0.0018	0.0118	4	②	0.0013-0.0023	0.0079-0.0138
2004	3.5	VQ35DE	①	0.0014-0.0018	0.0118	4	②	0.0013-0.0023	0.0079-0.0138
2005	3.5	VQ35DE	①	0.0014-0.0018	0.0118	4	②	0.0013-0.0023	0.0079-0.0138
2006	3.5	VQ35DE	①	0.0014-0.0018	0.0118	4	②	0.0013-0.0023	0.0079-0.0138

① There are 24 different grades, ranging from A (2.3612) to 7 (2.3603)

② Grade 0: 0.0591-0.0592. Identification color: black
Grade 1: 0.0592-0.0593. Identification color: brown
Grade 2: 0.0593-0.0594. Identification color: green

09482_MURA_C0007

PISTON AND RING SPECIFICATIONS
All measurements are given in inches.

Year	Engine Displacement Liters	Engine ID	Piston Clearance	Ring Gap			Ring Side Clearance		
				Top Comp.	Bottom Comp.	Oil Control	Top Comp.	Bottom Comp.	Oil Control
2003	3.5	VQ35DE	0.0004-0.0012	0.0091-0.0130	0.0130-0.0189	0.0079-0.0197	0.0018-0.0031	0.0012-0.0028	0.0026-0.0053
2004	3.5	VQ35DE	0.0004-0.0012	0.0091-0.0130	0.0130-0.0189	0.0079-0.0197	0.0018-0.0031	0.0012-0.0028	0.0026-0.0053
2005	3.5	VQ35DE	0.0004-0.0012	0.0091-0.0130	0.0130-0.0189	0.0079-0.0197	0.0018-0.0031	0.0012-0.0028	0.0026-0.0053
2006	3.5	VQ35DE	0.0004-0.0012	0.0091-0.0130	0.0130-0.0189	0.0079-0.0197	0.0018-0.0031	0.0012-0.0028	0.0026-0.0053

09482_MURA_C0008

TORQUE SPECIFICATIONS
All readings in ft. lbs.

Year	Engine Displacement Liters	Engine ID	Cylinder Head Bolts	Main Bearing Bolts	Rod Bearing Bolts	Crankshaft Damper Bolts	Driveplate Bolts	Manifold		Spark Plugs	Oil Pan Drain Plug
								Intake	Exhaust		
2003	3.5	VQ35DE	①	②	③	④	61-69	⑤	21-24	14-22	25
2004	3.5	VQ35DE	①	②	③	④	61-69	⑤	21-24	14-22	25
2005	3.5	VQ35DE	⑥	⑦	③	⑧	65	⑤	21-24	18	25
2006	3.5	VQ35DE	⑥	⑦	③	⑧	65	⑤	21-24	18	25

① Step 1: 72 ft. lbs.
   Step 2: Loosen all bolts completely
   Step 3: 25-33 ft. lbs.
   Step 4: +90 degrees
   Step 5: +90 degrees

② Step 1: 24-28 ft. lbs.
   Step 2: +90 degrees

③ Step 1: 15 ft. lbs.
   Step 2: +90 degrees

④ 29-36 ft. lbs. +60-66 degrees

⑤ Step 1: 5 ft. lbs
   Step 2: 20-23 ft. lbs.

⑥ Step 1: 72 ft. lbs.
   Step 2: Loosen all bolts completely
   Step 3: 29 ft. lbs.
   Step 4: +90 degrees
   Step 5: +90 degrees

⑦ Step 1: 26 ft. lbs.
   Step 2: +90 degrees

⑧ 33 ft. lbs. +60 degrees

09482_MURA_C0009

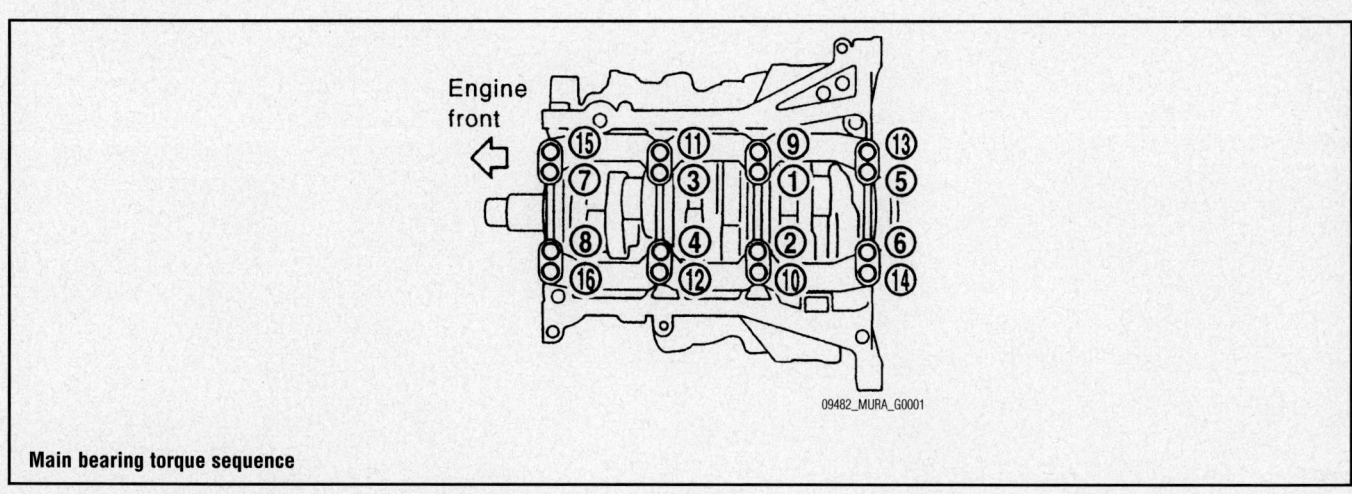

09482_MURA_G0001

Main bearing torque sequence

WHEEL ALIGNMENT

Year	Model		Caster Range (+/-Deg.)	Caster Preferred Setting (Deg.)	Camber Range (+/-Deg.)	Camber Preferred Setting (Deg.)	Toe-in (in.)	Kingpin Inclination (Deg.)
2003	Murano	F	0.75	+2.58	0.25	-0.33	0.02+/-0.04	14.33
		R	—	—	①	①	②	—
2004	Murano	F	0.75	+2.58	0.25	-0.33	0.02+/-0.04	14.33
		R	—	—	①	①	②	—
2005	Murano	F	0.75	+2.58	0.25	-0.33	0.02+/-0.04	14.33
		R	—	—	①	①	②	—
2006	Murano	F	0.75	+2.58	0.25	-0.33	0.02+/-0.04	14.33
		R	—	—	①	①	②	—

① minimum, except SE: -1 degree 16'

minimum, SE: -1 degree 18'

nominal, except SE: -0 degree 46'

nominal, SE: -0 degree 48'

② minimum, except SE: 0.055"'

minimum, SE: 0.067"

nominal, except SE: 0.126"

nominal, SE: 0.138"

maximum, except SE: 0.197"

maximum, SE: 0.209"

09482_MURA_C0010

TIRE, WHEEL AND BALL JOINT SPECIFICATIONS

Year	Model	OEM Tires Standard	OEM Tires Optional	Tire Pressures (psi) Front	Tire Pressures (psi) Rear	Wheel Size	Ball Joint Inspection	Lugnut Torque (ft. lbs.)
2003	Murano	P235/65SR18	none	33	33	7.5JJ	①	90
2004	Murano	P235/65SR18	none	33	33	7.5JJ	①	90
2005	Murano	P235/65TR18	none	33	33	7.5JJ	①	73-90
2006	Murano	P235/65TR18	none	33	33	7.5JJ	①	73-90

OEM: Original Equipment Manufacturer

PSI: Pounds Per Square Inch

NA: Not available

① Rotating torque: 5-30 inch lbs.

09482_MURA_C0011

BRAKE SPECIFICATIONS
All measurements in inches unless noted

Year	Model		Brake Disc Original Thickness	Brake Disc Minimum Thickness	Brake Disc Maximum Runout	Minimum Lining Thickness	Brake Caliper Bracket Bolts (ft. lbs.)	Brake Caliper Mounting Bolts (ft. lbs.)
2003	Murano	F	1.102	1.024	0.0016	0.079	101-129	17-22
		R	0.630	0.551	0.0020	0.079	53-71	28-35
2004	Murano	F	1.102	1.024	0.0016	0.079	101-129	17-22
		R	0.630	0.551	0.0020	0.079	53-71	28-35
2005	Murano	F	1.102	1.024	0.0016	0.079	101-129	17-22
		R	0.630	0.551	0.0020	0.079	53-71	28-35
2006	Murano	F	1.102	1.024	0.0016	0.079	101-129	17-22
		R	0.630	0.551	0.0020	0.079	53-71	28-35

F: Front
R: Rear

09482_MURA_C0012

SCHEDULED MAINTENANCE INTERVALS
Nissan—Murano

TO BE SERVICED	TYPE OF SERVICE	7.5	15	22.5	30	37.5	45	52.5	60
Engine oil & filter	R	✓	✓	✓	✓	✓	✓	✓	✓
Brake lines & cables	S/I		✓		✓		✓		✓
Brake pads, discs	I		✓		✓		✓		✓
Driveshaft boots & propeller shaft	L/I		✓		✓		✓		✓
CVT, transfer case and differential fluid	I		✓		✓		✓		
Air cleaner filter	R				✓				✓
Drive belt(s) ①	S/I								
Engine coolant ②	R								✓
Spark plugs	R			Platinum plugs, every 100,000 miles					
Cabin air filter	R		✓		✓		✓		✓
Exhaust system	I			✓					✓
Fuel lines	S/I				✓				✓
Steering gear, linkage, axle & suspension parts	I		✓		✓		✓		✓
Tires (rotate)	S/I			every 5,000-6,000 miles					
Vapor lines	S/I				✓				✓

R: Replace   S/I: Service or Inspect   L: Lubricate

① First a 60,000, then every 15,000 miles

② After 60,000, replace every 30,000

FREQUENT OPERATION MAINTENANCE (SEVERE SERVICE)

If a vehicle is operated under any of the following conditions it is considered severe service:

- Extremely dusty areas.

- 50% or more of the vehicle operation is in 32°C (90°F) or higher temperatures, or constant temperatures below 0°C (32°F).

- Prolonged idling (vehicle operation in stop and go traffic).

- Frequent short running periods (engine does not warm to normal operating temperatures).

- Police, taxi, delivery usage or trailer towing usage.

Oil & oil filter: replace every 3750 miles.

Brake pads, discs, drums & linings: service or inspect every 7500 miles.

Driveshaft boots & propeller shaft: service or inspect every 7500 miles.

Exhaust system: service or inspect every 7500 miles.

Steering gear (box) & linkage, (steering damper-4x4), axle & suspension parts: service or inspect every 7500 miles.

Steering linkage ball joints & front suspension ball joints: service or inspect every 7500 miles.

09482_MURA_C0013

ENGINE REPAIR

➡Disconnecting the negative battery cable on some vehicles may interfere with the functions of the on board computer system. The computer may undergo a relearning process once the negative battery cable is reconnected.

Relearning Procedures

ACCELERATOR PEDAL RELEASED POSITION LEARNING

Adjustment

➡Accelerator pedal released position learning is an operation to learn the fully released position of the accelerator pedal by monitoring the accelerator pedal position sensor output signal. It must be performed each time the harness connector of the accelerator pedal position sensor or the ECM is disconnected.

1. Make sure that the accelerator pedal is in the fully released position.
2. Turn the ignition switch ON. Wait at least 2 seconds.
3. Turn the ignition switch OFF. Wait at least 10 seconds.
4. Turn the ignition switch ON. Wait at least 2 seconds.
5. Turn the ignition switch OFF. Wait at least 10 seconds.

THROTTLE VALVE CLOSED POSITION LEARNING

Adjustment

➡Throttle valve closed position learning is an operation to learn the fully closed position of the throttle valve by monitoring the throttle valve position sensor output signal. It must be performed each time the harness connector of the electrical throttle valve control actuator or ECM is disconnected.

1. Make sure that the accelerator pedal is in the fully released position.
2. Turn the ignition switch ON.
3. Turn the ignition switch OFF. Wait at least 10 seconds.
4. Make sure that the throttle valve moves during the 10 seconds, by confirming the operating sound.

IDLE AIR VOLUME LEARNING

➡Idle air volume learning is an operation to learn the idle air volume that keeps the engine within a specific range. It must be performed each time the electronic throttle control actuator or ECM is replaced, or if the idle speed and ignition timing is out of specification.

Pre-Adjustment

➡Before performing the idle air volume learning procedure, be sure the following conditions are satisfied. Learning will be cancelled if any of the following conditions are missed for even a moment.

1. Be sure battery voltage is at least 12.9 volts at idle.
2. Be sure engine coolant temperature is at least 158–212 degrees.
3. Be sure PNP switch in ON.
4. Be sure electric load switch is OFF (air conditioning, headlights, rear defogger etc.).

➡On vehicles equipped with daytime running lights, if the parking brake is applied before the engine is started the headlights will not turn on.

5. Be sure the steering wheel is in the straight ahead position.
6. Be sure the vehicle is stopped.
7. Be sure the transmission is warmed up.

➡For vehicles with CONSULT-II, drive the vehicle until the "fluid temp SE" in the data monitor mode of the CVT system indicates less than 0.0 volt. For vehicles without CONSULT-II, drive the vehicle for 10 minutes.

Adjustment

WITH CONSULT-II

1. Perform accelerator pedal released position learning procedure.
2. Perform throttle valve closed position learning procedure.
3. Perform pre-adjustment procedure.
4. Start and run the engine until normal operating temperature is reached.
5. Select "idle air vol learn" in work support mode.
6. Touch "start" and wait 20 seconds.
7. Be sure that "CMPLT" is displayed on the screen. If not displayed, idle air volume learning will not be carried out successfully.
8. Rev the engine two or three times and make sure that the idle speed and ignition timing are within specification.

WITHOUT CONSULT-II

➡It is better to count the time accurately, using a clock. It is impossible to switch the diagnostic mode when the accelerator pedal position sensor circuit has a malfunction.

1. Perform accelerator pedal released position learning procedure.
2. Perform throttle valve closed position learning procedure.
3. Perform pre-adjustment procedure.
4. Start and run the engine until normal operating temperature is reached.
5. Turn the ignition switch to the OFF position. Wait at least 10 seconds.
6. Confirm that the accelerator pedal is fully released. Turn the ignition switch ON and wait 3 seconds.
7. Fully depress the accelerator, fully release the accelerator pedal.

➡Repeat the above step 5 times, within 5 seconds.

8. Wait 7 seconds. Fully depress the accelerator pedal and keep it depressed for about 20 seconds, until the MIL stops blinking and turned ON.
9. Fully depress the accelerator pedal within 3 seconds after the MIL turned ON.
10. Start the engine and let it idle.
11. Wait 20 seconds.
12. Rev the engine two or three times and make sure that the idle speed and ignition timing are within specification.

Alternator

REMOVAL & INSTALLATOIN

1. Before servicing the vehicle, refer to the Precautions Section.
2. Drain the cooling system.
3. Remove or disconnect the following:
   • Negative battery cable
   • Alternator harness connectors
   • Engine right side under cover
   • Radiator
   • Remove alternator and air conditioner compressor belt.
   • Idler pulley
   • Alternator

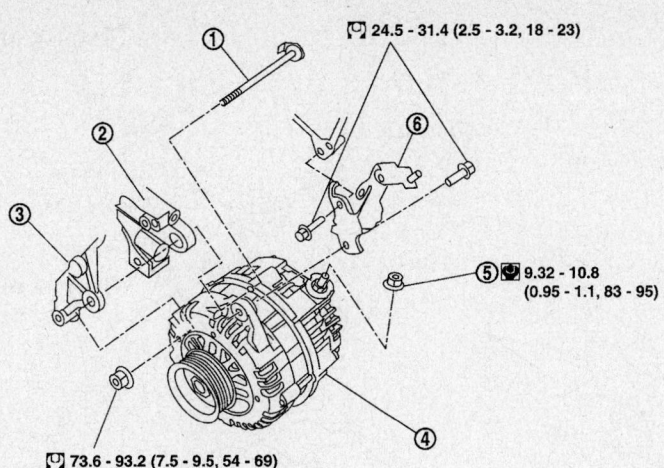

⚙ 24.5 - 31.4 (2.5 - 3.2, 18 - 23)

⚙ 9.32 - 10.8 (0.95 - 1.1, 83 - 95)

⚙ 73.6 - 93.2 (7.5 - 9.5, 54 - 69)

🔧 : N·m (kg-m, in-lb)
⚙ : N·m (kg-m, ft-lb)

| 1. | Through bolt | 2. | Cylinder block | 3. | Timing chain case |
| 4. | Alternator | 5. | B terminal nut | 6. | Alternator bracket |

42356-MURA-G01

Alternator and related components

To install:
4. Install or connect the following:
- Alternator
- Idler pulley
- Alternator belt. Tighten the through-bolts to 18–23 ft. lbs. (25–31 Nm).
- Engine under cover
- Radiator
- Alternator harness connectors
- Negative battery cable
5. Refill the cooling system, using the proper grade and type coolant.

Ignition Timing

ADJUSTMENT

Ignition timing is not adjustable, however it can be checked.
1. Before servicing the vehicle, refer to the Precautions Section.
2. Remove the number one ignition coil.
3. Connect the number one ignition coil and spark plug with a suitable high tension wire.
4. Attach the timing light clamp to the wire.
5. Check the ignition timing.

Engine Assembly

REMOVAL & INSTALLATION

1. Before servicing the vehicle, refer to the Precautions Section.
2. Properly release the fuel pressure.
3. Disconnect the negative battery cable.
4. Matchmark and remove the hood.
5. Remove the engine cover, and the splash guards.

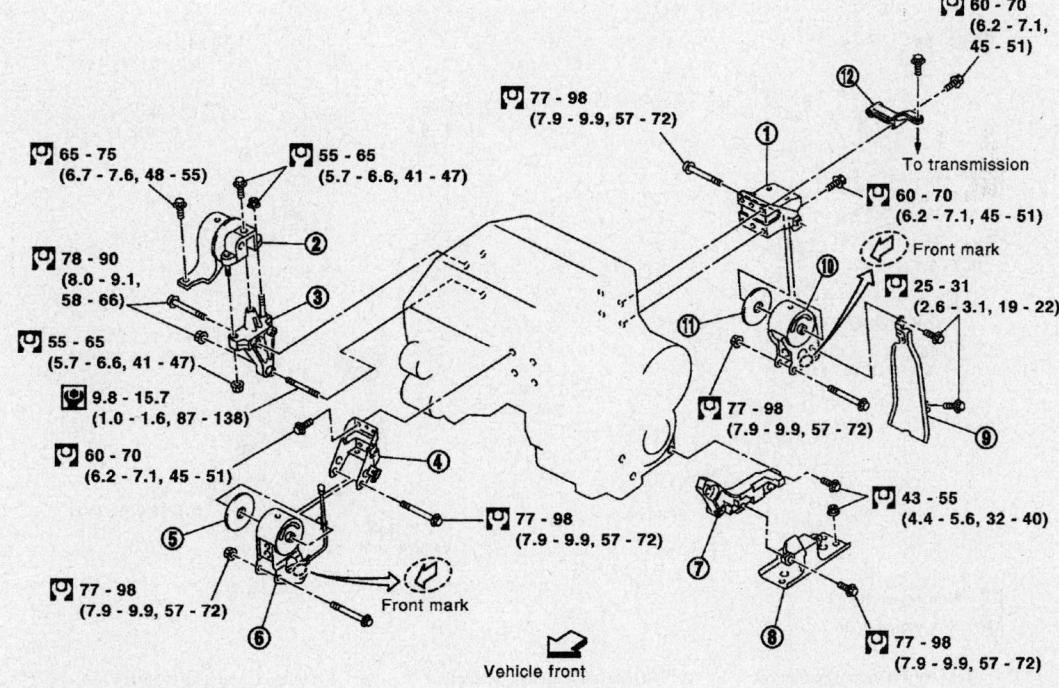

🔧 : N·m (kg-m, ft-lb)
⚙ : N·m (kg-m, in-lb)

| 1. | Rear engine mounting bracket | 2. | RH engine mounting insulator | 3. | RH engine mounting bracket |
| 4. | Front engine mounting bracket | 5. | Stopper | 6. | Front engine mounting insulator |

42356-MURA-G02A

Engine mounting and related components—2003 2WD

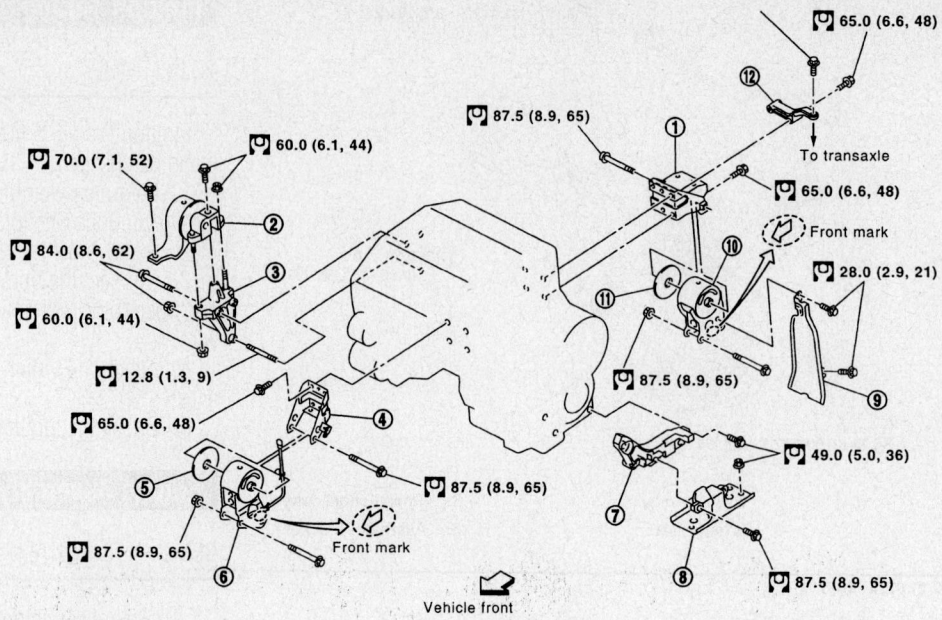

Engine mounting and related components—2004–2006 2WD

: N•m (kg-m, ft-lb)

1. Rear engine mounting bracket
2. RH engine mounting insulator
3. RH engine mounting bracket
4. Front engine mounting bracket
5. Stopper
6. Front engine mounting insulator
7. LH engine mounting bracket
8. LH engine mounting insulator
9. Engine mounting air guide
10. Rear engine mounting insulator
11. Stopper
12. Bracket

09482_MURA_G0002

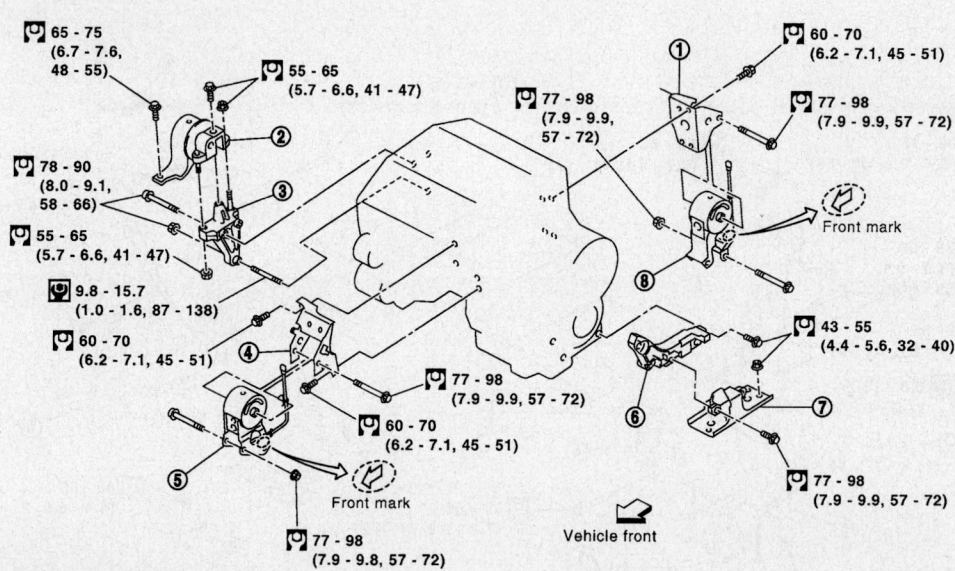

: N•m (kg-m, ft-lb)

: N•m (kg-m, in-lb)

1. Rear engine mounting bracket
2. RH engine mounting insulator
3. RH engine mounting bracket
4. Front engine mounting bracket
5. Front engine mounting insulator
6. LH engine mounting bracket
7. LH engine mounting insulator
8. Rear engine mounting insulator

42356-MURA-G03A

Engine mounting and related components—2003 AWD

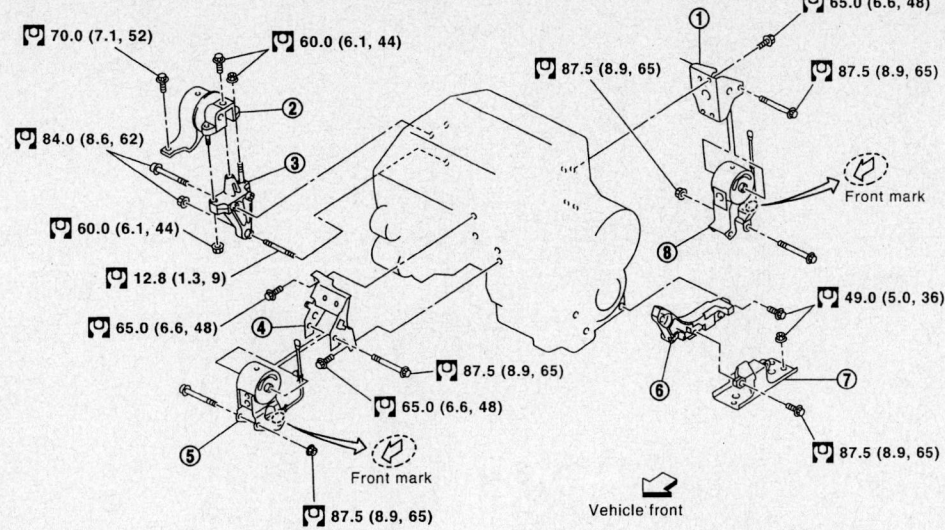

70.0 (7.1, 52)

60.0 (6.1, 44)

65.0 (6.6, 48)

87.5 (8.9, 65)

87.5 (8.9, 65)

84.0 (8.6, 62)

60.0 (6.1, 44)

Front mark

12.8 (1.3, 9)

49.0 (5.0, 36)

65.0 (6.6, 48)

87.5 (8.9, 65)

65.0 (6.6, 48)

87.5 (8.9, 65)

Front mark

Vehicle front

87.5 (8.9, 65)

: N•m (kg-m, ft-lb)

1. Rear engine mounting bracket
2. RH engine mounting insulator
3. RH engine mounting bracket
4. Front engine mounting bracket
5. Front engine mounting insulator
6. LH engine mounting bracket
7. LH engine mounting insulator
8. Rear engine mounting insulator

09482_MURA_G0003

Engine mounting and related components—2004–2006 AWD

6. Drain engine coolant.
7. Remove or disconnect the following:
- Battery and tray
- Air inlet duct
- Air duct and air cleaner case (upper) assembly with mass air flow sensor
- Power brake booster vacuum hose
- Drive belts
- Radiator assembly, coolant reservoir, and system hoses
- RH windshield wiper arm and right font cowl top cover
- Engine room harness from the ECM side
- Heater hoses
- Wheel and tires
- A/C compressor with piping connected, and temporarily secure it aside
- Fuel hose quick connector at vehicle piping side
- Transaxle shift control cable.
- Starter motor
- Front exhaust tube
- Reservoir tank for the power steering from engine compartment bracket and position it aside.
- Power steering gear from steering lower joint

- Steering outer socket from steering knuckle
- Stabilizer connecting rod
- Propeller shaft (AWD vehicles)
- Left and right front halfshafts
- Power steering piping from power steering oil cooler

8. Position a manual lift table caddy under the engine and transaxle assembly.
9. Remove the right engine mounting insulator
10. Remove mounting bolt between transverse link and front suspension member
11. Carefully lower the engine, transaxle, transfer case (AWD vehicles) and front suspension member assembly with the manual lift table caddy, avoiding interference with the vehicle body.

✳✳ WARNING

Before and during this procedure, always check if any harnesses are left connected. Avoid any damage to, or any oil or grease smearing or spills onto the engine mounting insulators.

12. Remove the crankshaft position sensor (POS).

13. Disconnect front suspension mounting nuts and bolts to remove engine, transaxle, transfer case (AWD vehicles) and front suspension member assembly as a unit.
14. Separate the engine, transaxle and transfer case (AWD vehicles) assembly and front suspension member.

To install:

15. Installation is the reverse order of removal. See the accompanying illustrations for the proper torque values.

➡**On 2005–2006 vehicles, see accelerator pedal released position learning, throttle valve closed position learning and idle air volume learning procedures.**

Heater Core

REMOVAL & INSTALLATION

➡**Be sure to disarm the SRS system, prior to working on the vehicle. Turn the ignition switch OFF, disconnect both battery cables and wait at least three minutes before starting any work.**

1. Before servicing the vehicle, refer to the Precautions Section.

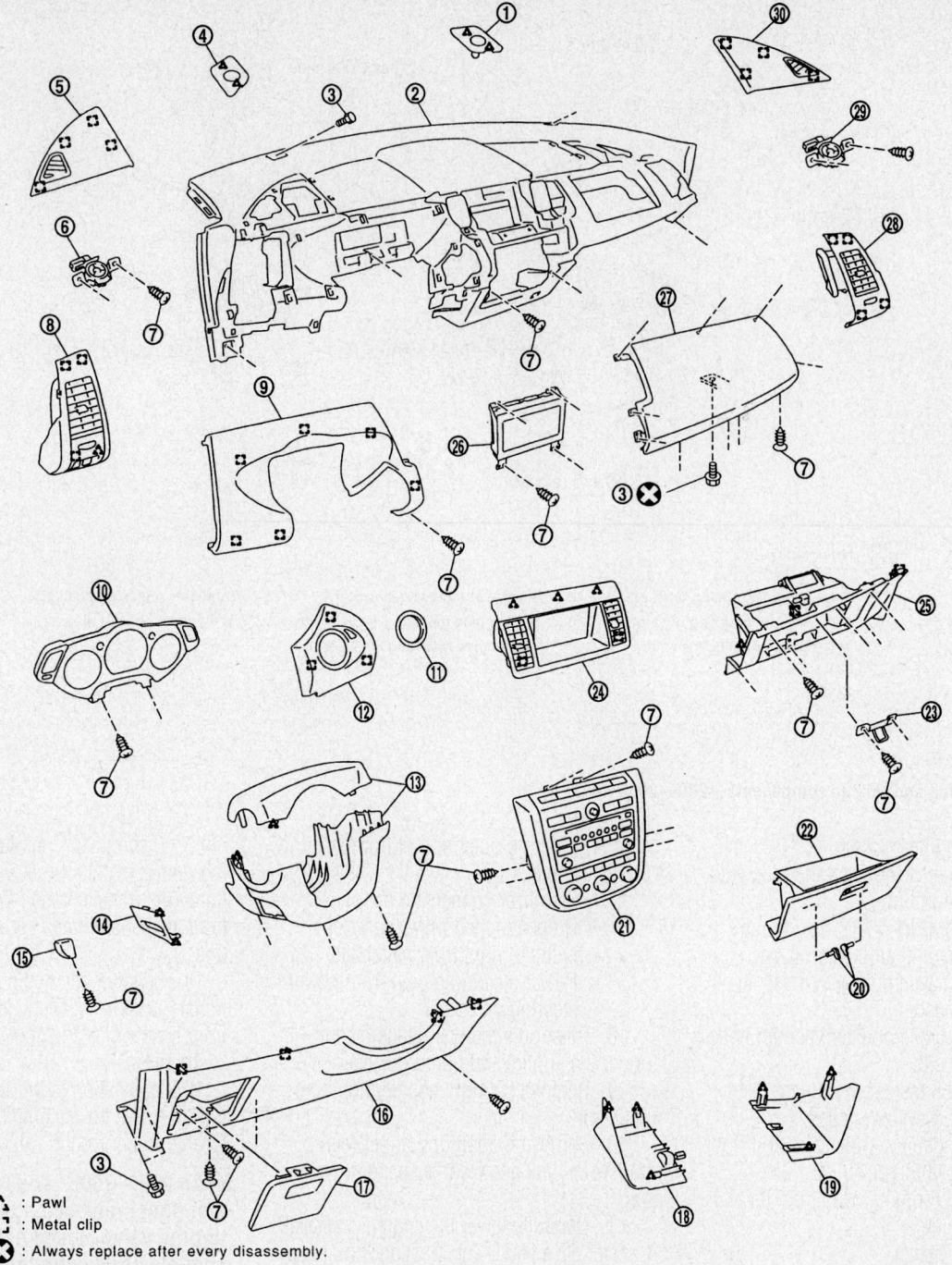

△ : Pawl

⎡⎤
⎣⎦ : Metal clip

⊗ : Always replace after every disassembly.

1. Instrument mask (RH; Sunload. sensor)	2. Instrument panel	3. Bolt
4. Instrument mask (LH; Auto light sensor)	5. Instrument side finisher (LH)	6. Tweeter (LH)
7. Screw	8. Side ventilator assembly (LH)	9. Instrument driver upper panel
10. Combination meter	11. Steering lock escutcheon	12. Ignition key finisher
13. Steering column cover	14. Tilt lever mask	15. Tilt lever knob
16. Instrument driver lower panel	17. Fuse lid	18. Instrument stay cover (LH)
19. Instrument stay cover (RH)	20. Glove box pin	21. Cluster lid C
22. Glove box assembly	23. Glove box striker	24. Center ventilator
25. Instrument passenger lower panel	26. Display	27. Instrument passenger upper panel
28. Side ventilator assembly (RH)	29. Tweeter (RH)	30. Instrument side finisher (RH)

09482_MURA_G0004

Instrument panel and related components

2. Discharge the air conditioning system.

3. Disconnect both battery cables.

4. Drain coolant from cooling system.

5. Remove both right/left wiper arms.

6. Remove cowl top seal rubber.

7. Remove clips from cowl top cover (right) and remove cowl top cover (right).

8. Remove clips from cowl top cover (left) and remove cowl top cover (left)

9. Remove washer nozzles and hose from cowl top cover.

10. Remove cowl top cover.

11. Disconnect evaporator-side one-touch joints.

  a. Install a disconnector tool (High-pressure side: 92530-89908, Low pressure side: 92530-89916) on A/C piping.

  b. Slide a disconnector toward vehicle front until it clicks.

  c. Slide A/C piping toward vehicle front and disconnect it.

❈❈ WARNING

Seal connection opening of piping with a cap or vinyl tape to avoid exposure to atmosphere.

12. Disconnect two heater hoses from heater core.

13. Remove fuse lid.

14. Remove instrument driver lower panel screws.

15. Remove data link connector.

16. Pull to disengage metal clip by removing panel in horizontal direction.

17. Disconnect in-vehicle sensor and each electrical parts.

18. Remove bolts, and remove hood lock opener.

19. Remove tilt lever knob screws.

20. Remove the knob by picking it up and pulling it out. Using a remover, ply and remove tilt lever mask.

21. Remove steering column cover screws.

22. Disengage the tab, and remove steering column cover. After removing combination meter screws, remove harness connector.

23. Using a remover, pry and remove side ventilator assembly (left/right).

24. Disconnect aiming switch harness connector and VDC switch harness connector only left side.

25. Insert a remover into lower space of instrument side finisher (left/right) and remove by lifting.

26. Remove screws and glove box striker, disconnect connectors, and remove instrument passenger lower panel assembly.

27. Detach the damper from glove box right side.

28. Remove glove box pins, and remove glove box.

29. Insert a remover into lower space of center ventilator and remove by lifting.

30. Insert a remover into upper space of center ventilator and the upper clip is removed.

31. Remove screws. Disconnect harness connector, and remove tweeter with right and left part.

32. Insert a remover into front space of instrument stay cover (left/right) and detach.

33. Disconnect the left side harness connector only.

34. Remove cluster lid screws.

35. Disconnect A/C and AV harness connectors, and remove cluster lid C.

36. Remove display unit screws.

37. Disconnect harness connector and remove display.

38. Using a remover and disengage the ignition key finisher metal clips

39. Disconnect harness connector.

40. Using a remover, pry and remove instrument mask.

41. Disconnect harness connector.

42. Remove screw. Disengage the metal clip and remove instrument driver upper panel.

43. Remove the bolt and screws and remove the front passenger air bag module.

44. Disconnect metal clips, then remove instrument passenger upper panel.

45. Disconnect harness connector.

46. Pull to inside of vehicle, disconnect metal clips and remove front pillar garnish.

47. Remove bolts and screws, and remove instrument panel from passenger door opening portion.

48. Tweeter and sensor harness clip are removed from the duct.

49. Remove instrument panel assembly.

50. Remove ECM with bracket attached.

51. Remove nuts (2), then bolts (2) and screw (1), then remove blower unit.

➡**Move blower unit to the right and remove locating pin (1) and joint. Then remove blower unit downward.**

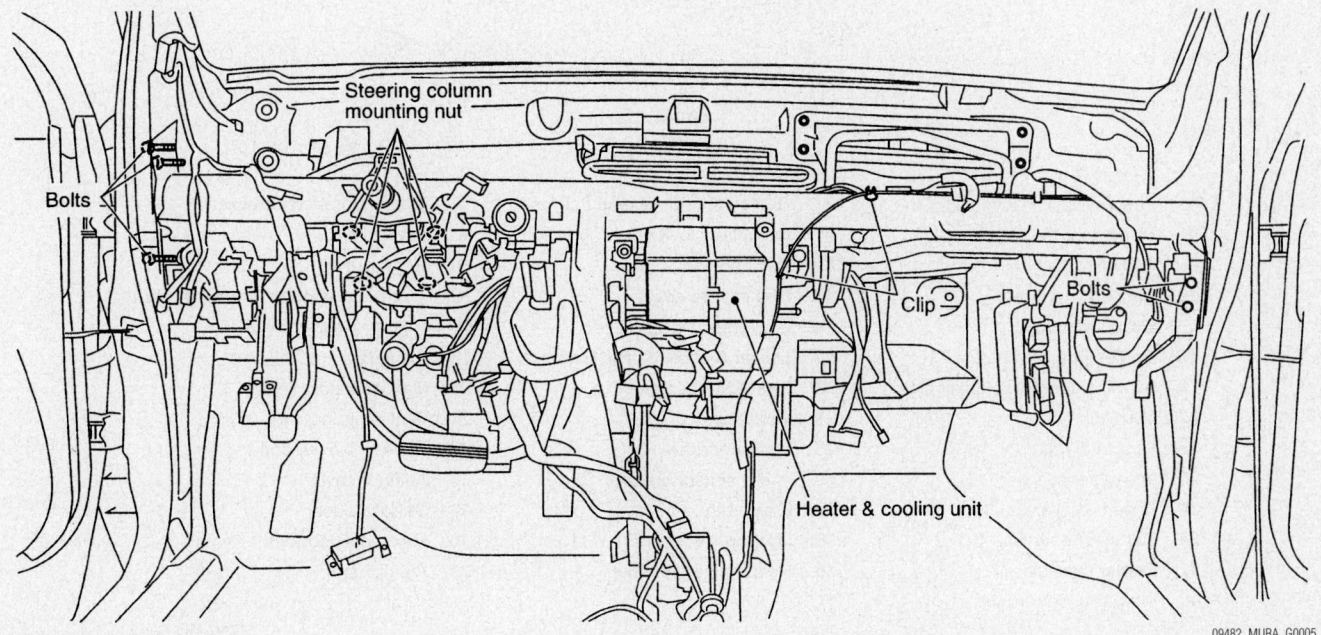

Steering member location

09482_MURA_G0005

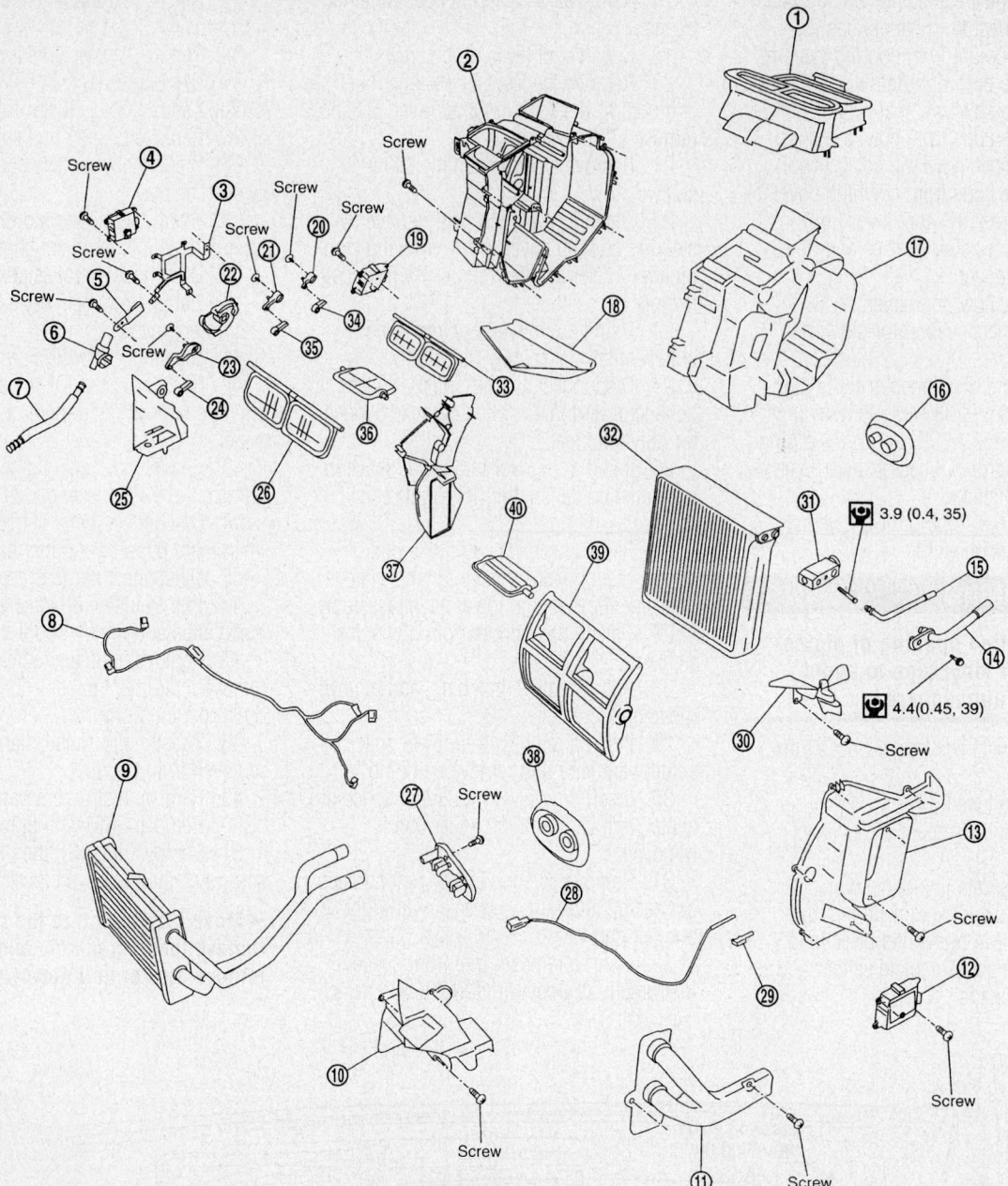

3.9 (0.4, 35)

4.4 (0.45, 39)

1. Adaptor duct
2. Heater & cooling case (left)
3. Mode door motor bracket
4. Mode door motor
5. Instrument lower cover bracket
6. Aspirator
7. Aspirator duct
8. Sub harness
9. Heater core
10. Foot duct (right)
11. Heater core cover
12. Air mix door motor (passenger side)
13. Evaporator cover
14. Low-pressure pipe 2
15. High-pressure pipe 2
16. Cooler pipe grommet
17. Heater & cooling case (right)
18. Insulator
19. Air mix door motor (driver side)
20. Defroster door lever
21. Max. cool door lever
22. Main link
23. Ventilator door lever
24. Ventilator door link
25. Foot duct (left)
26. Ventilator door
27. Heater pipe support
28. Intake sensor
29. Intake sensor bracket
30. Adaptor cover
31. Expansion valve
32. Evaporator
33. Defroster door
34. Defroster door link
35. Max. cool door link
36. Max. cool door (left)
37. Center case
38. Heater pipe grommet
39. Air mix door
40. Max. cool door (right)

09482_MURA_G0006

Heater/evaporator core and related components

52. Disconnect intake door motor connector and blower fan motor connector.

53. Remove harness clips (2) from blower unit.

54. Remove blower unit.

55. Remove clips from vehicle harness from steering member.

56. Remove instrument stays (driver side and passenger side).

57. Remove rear ventilator duct1 and front floor duct.

58. Remove mounting screws from heater & cooling unit.

59. Remove the steering member, and then remove heater & cooling unit.

60. Remove foot duct (right).

61. Remove heater core cover.

62. Remove heater pipe support and heater pipe grommet.

63. Slide heater core to passenger side.

To install:

64. Installation is the reverse of removal. Note the following points:

- Replace O-rings for A/C piping with new ones, coated with compressor oil.
- Connection point for female-side piping is thin. So, when inserting male-side piping, take care not to deform female-side piping. Slowly insert in axial direction.
- Insert one-touch joint connection point securely until it clicks.
- After piping has been connected, pull male-side piping by hand to check that piping does not come off.

- When recharging refrigerant, check for leaks.

Water Pump

REMOVAL & INSTALLATION

1. Before servicing the vehicle, refer to the Precautions Section.

2. Disconnect the negative battery cable.

3. Remove drive belts.

4. Remove undercover.

5. Drain engine coolant from radiator.

6. Remove water drain plug on water pump side of cylinder block.

7. Support lower oil pan bottom with a transmission jack.

8. Remove right engine mounting insulator and mounting bracket.

9. Remove idler pulley bracket.

10. Remove chain tensioner cover and water pump cover.

11. Remove the chain tensioner assembly in the following procedure.

   a. Pull the lever down and release the plunger stopper tab.

   b. Insert the stopper pin into the tensioner body hole to hold the lever and keep the stopper tab released.

   c. Insert the plunger into the tensioner body by pressing the timing chain slack guide.

   d. Keep the slack guide pressed and hold the plunger in by pushing the stopper pin deeper through the lever and into the tensioner body hole.

   e. Turn crankshaft pulley approximately 20 degrees clockwise so that the timing chain on the chain tensioner side is loose.

12. Remove chain tensioner.

13. Remove the 3 water pump bolts. Secure a gap between water pump gear and timing chain, by turning crankshaft pulley approximately 20 degrees counterclockwise.

14. Screw M8 bolts, pitch: 1.25mm (0.049in) length: Approx. 50 mm (1.97in) into water pump's upper and lower mounting-bolt holes until they reach timing chain case. Then, alternately tighten each bolt for a half turn, and pull out water pump.

15. Pull straight out while preventing vane from contacting socket in installation area. Remove water pump without causing sprocket to contact timing chain.

16. Remove M8 bolts and O-rings from water pump.

To install:

17. Install new O-rings to water pump.

18. Apply engine oil and engine coolant to the O-rings as shown. Locate the O-ring with white paint mark to engine front side.

19. Install the water pump.

➡**Do not allow cylinder block to nip the O-rings when installing the water pump.**

20. Check that timing chain and water pump sprocket are engaged.

21. Insert water pump by tightening mounting bolts alternately and evenly.

22. Remove dust and foreign material

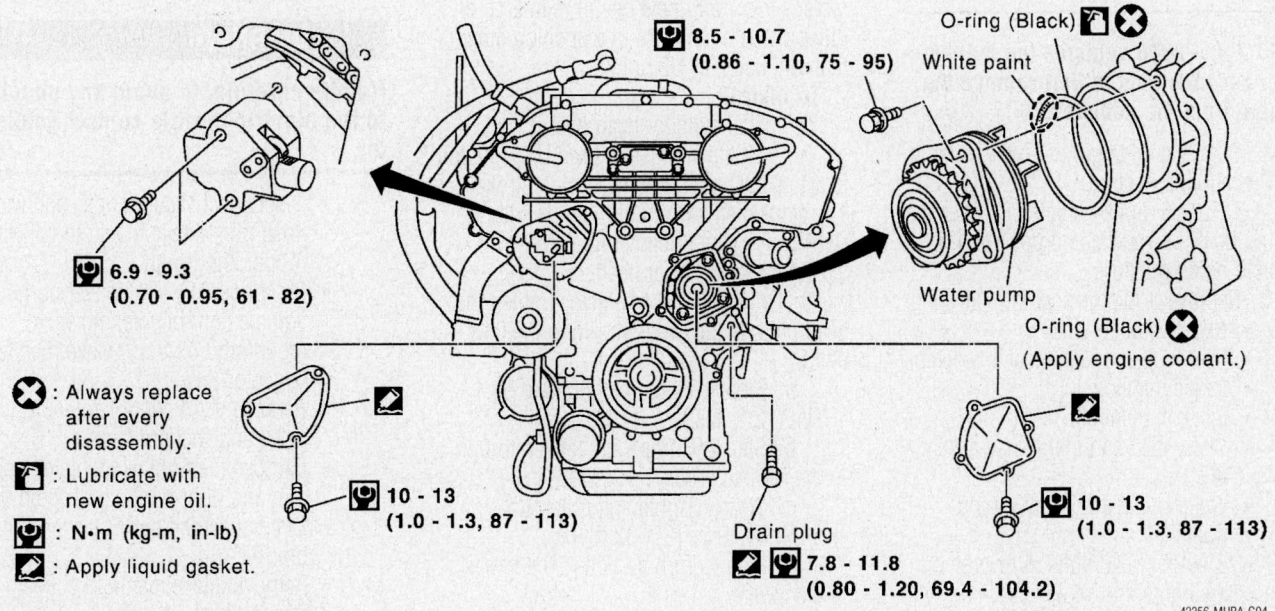

Water pump and related components

completely from backside of chain tensioner and from installation area of rear timing chain case.

23. Turn the crankshaft pulley clockwise so that the timing chain on the timing chain tensioner side is loose.

➡ **When installing the timing chain tensioner, engine oil should be applied to the oil hole and tensioner.**

24. Install the timing chain tensioner.
25. Remove the stopper pin.
26. Install chain tensioner cover and water pump cover.

   a. Before installing, remove all traces of liquid gasket from mating surface of water pump cover and chain tensioner cover using a scraper. Also remove traces of liquid gasket from the mating surface of the front cover.

   b. Apply a continuous bead of liquid gasket, to mating surface of chain tensioner cover and water pump cover. Use RTV silicon sealant or equivalent

27. Install water drain plug on water pump side of cylinder block.
28. Installation is in the reverse order of removal for remaining parts.
29. After starting engine, let idle for three minutes, then rev engine up to 3,000 rpm under no load to purge air from the high-pressure chamber of the chain tensioner. The engine may produce a rattling noise. This indicates that air still remains in the chamber and is not a matter of concern.

Cylinder Head

REMOVAL & INSTALLATION

➡ **On 2005–2006 vehicles the manufacturer recommends to first remove the engine from the vehicle.**

1. Before servicing the vehicle, refer to the Precautions Section.
2. On 2005–2006 vehicles, remove the engine from the vehicle and position it in a suitable holding fixture.
3. Remove or disconnect the following:
   - Negative battery cable
   - Fuel tube and fuel injector assembly
   - Intake manifold
   - Exhaust manifold
   - Water inlet and thermostat assembly
   - Water outlet and water piping
   - Camshaft
   - Cylinder head bolts in reverse of the tightening sequence
4. Inspect the bolts. Cylinder head bolts are tightened by plastic zone tighten-

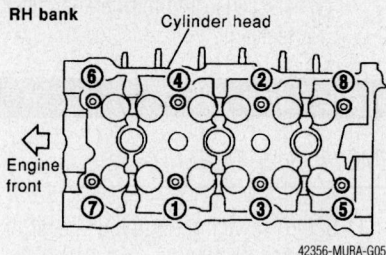

Cylinder head bolt torque sequence—right side

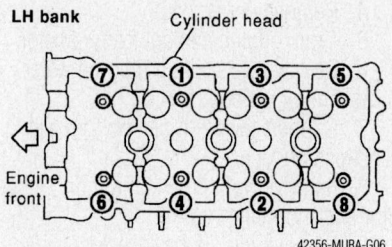

Cylinder head bolt torque sequence—left side

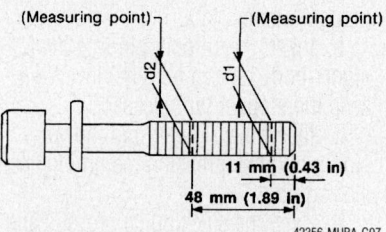

Cylinder head bolt measurement

ing method. Whenever the size difference between d1 and d2 exceeds the limit, replace them with new one. If reduction of outer diameter appears in a position other than d2, use it as d2 point.

To install:

5. Install cylinder head gasket.
6. Turn the crankshaft until No. 1 piston is set at TDC on the compression stroke. The crankshaft key should line up with the right bank cylinder center line as shown.
7. Install cylinder head.
8. On 2003–2004 vehicles tighten the head bolts in the order shown in the illustration.

   a. Step 1: Tighten all bolts to 98.1 Nm (72 ft. lbs.).
   b. Step 2: Completely loosen to 0 in the reverse order.
   c. Step 3: Tighten all bolts to 34.3–44.1 Nm (25–33 ft. lbs.).
   d. Step 4: Turn all bolts 90 degrees clockwise.
   e. Step 5: Turn all bolts an additional 90 degrees clockwise.

9. On 2005–2006 vehicles tighten the head bolts in the order shown in the illustration.

   a. Step 1: Tighten all bolts to 98.1 Nm (72 ft. lbs.).
   b. Step 2: Completely loosen to 0 in the reverse order.
   c. Step 3: Tighten all bolts to 39.2 (29 ft. lbs.).
   d. Step 4: Turn all bolts 90 degrees clockwise.
   e. Step 5: Turn all bolts an additional 90 degrees clockwise.

10. After installing cylinder head, measure distance between front end faces of cylinder block and cylinder head (left and right banks). If measurement is outside the specified range, reinstall cylinder head.

11. The remainder of installation is the reverse of removal.

Intake Manifold

REMOVAL & INSTALLATION

Upper and Lower Intake Manifold Collectors

1. Before servicing the vehicle, refer to the Precautions Section.
2. Remove or disconnect the following:
   - Negative battery cable
   - Remove engine cover
   - Drain engine coolant
   - Remove air duct
   - Remove electric throttle control actuator. Loosen bolts in the reverse order of that shown in the illustration.

✳✳ WARNING

Handle carefully to avoid any shock to the electric throttle control actuator.

   - Disconnect vacuum hose and water hose from intake manifold collector (upper and lower).
   - Disconnect EVAP canister purge volume control solenoid valve mounting bolt from intake manifold collector (lower).
   - Remove VIAS control solenoid valve and vacuum tank
   - Remove the right windshield wiper arm and right front cowl top cover
   - Disconnect the power steering hose bracket
   - Remove intake manifold collector support bracket
   - Remove PCV hose

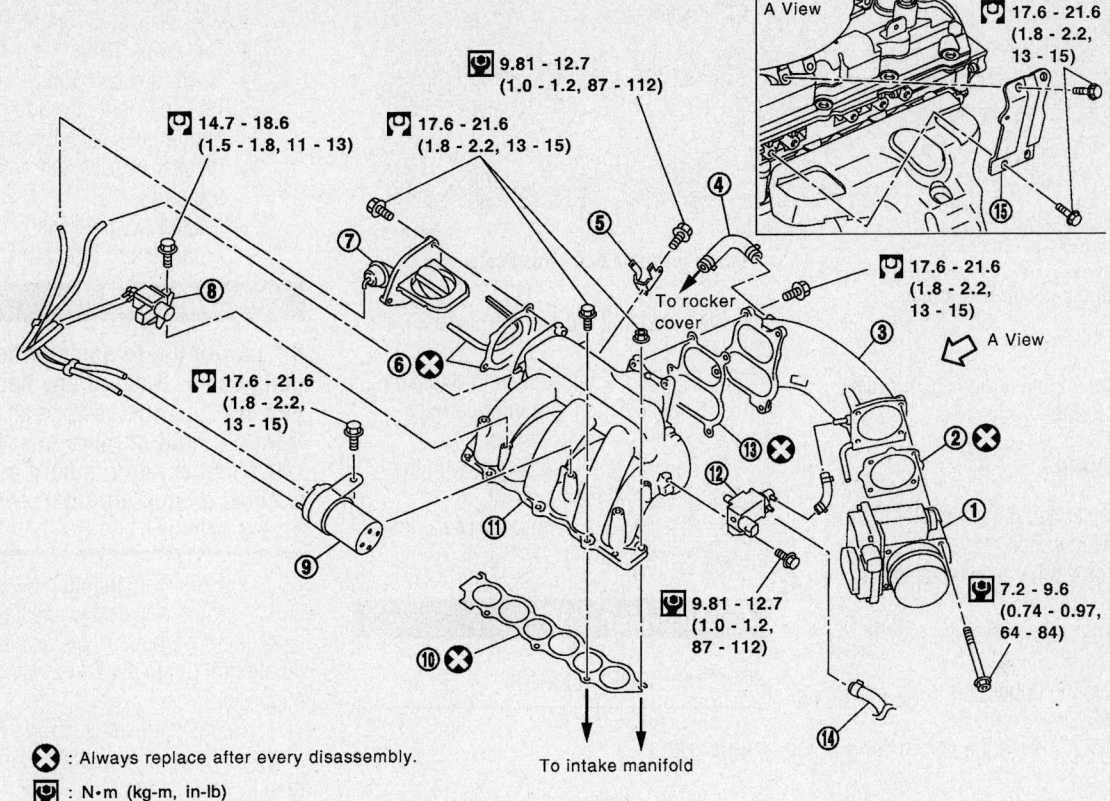

9.81 - 12.7
(1.0 - 1.2, 87 - 112)

14.7 - 18.6
(1.5 - 1.8, 11 - 13)

17.6 - 21.6
(1.8 - 2.2, 13 - 15)

17.6 - 21.6
(1.8 - 2.2,
13 - 15)

A View

17.6 - 21.6
(1.8 - 2.2,
13 - 15)

To rocker cover

A View

17.6 - 21.6
(1.8 - 2.2,
13 - 15)

9.81 - 12.7
(1.0 - 1.2,
87 - 112)

7.2 - 9.6
(0.74 - 0.97,
64 - 84)

To intake manifold

❌ : Always replace after every disassembly.

🔧 : N•m (kg-m, in-lb)

🔧 : N•m (kg-m, ft-lb)

1. Electric throttle control actuator
2. Gasket
3. Intake manifold collector (upper)
4. PCV hose
5. Harness bracket
6. Gasket
7. Power valve
8. VIAS control solenoid valve
9. Vacuum tank
10. Gasket
11. Intake manifold collector (lower)
12. EVAP canister purge volume control solenoid valve
13. Gasket
14. EVAP hose
15. Support bracket

42356-MURA-G08

Upper and lower intake manifold collectors and related components

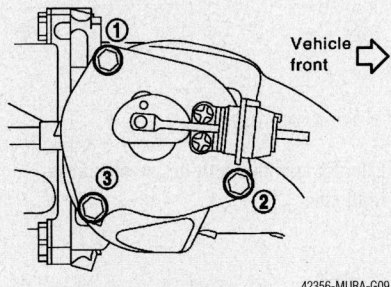

Vehicle front

42356-MURA-G09

Power valve installation

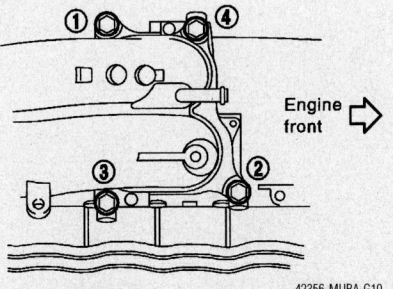

Engine front

42356-MURA-G10

Upper collector bolt torque sequence

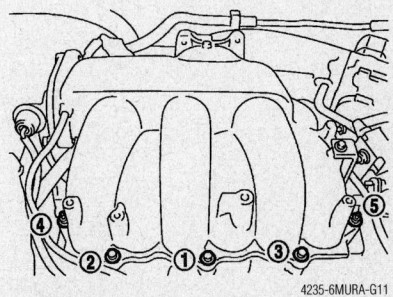

4235-6MURA-G11

Lower collector bolt torque sequence

- Loosen bolts in reverse order of illustration, and remove intake manifold collector (upper and lower) assembly
- Loosen bolts in reverse order of illustration to remove intake manifold collector (upper)
- Remove power valve in reverse order of illustration

To install:
3. Using straightedge and feeler gauge, inspect the surface distortion of intake manifold collector (lower). If it exceeds the limit (0.004 inch), replace the intake manifold collector.
4. Installation is the reverse of the removal procedure.
5. Tighten all mounting bolts and nuts

to specification in two or more steps in numerical order shown in illustration.
6. Tighten the power valve bolts in the sequence shown in the illustration.
7. Tighten the upper and lower intake manifold collector bolts in the proper sequence and to specification.
8. Be sure to fill the engine with the proper grade and type coolant.

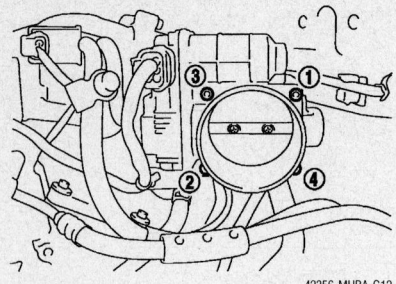

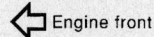

42356-MURA-G12

Throttle control actuator bolt torque sequence

9. Start the engine and check for leaks, correct as required.

Intake Manifold

1. Before servicing the vehicle, refer to the Precautions Section.

2. Disconnect the negative battery cable.

3. Properly relieve the fuel system pressure.

4. Remove the upper and lower intake manifold collectors.

5. Remove the fuel tube and fuel injector assembly.

6. Loosen bolts and nuts in reverse order of illustration to remove intake manifold assembly from the engine.

To install:

7. Using straightedge and feeler gauge, inspect the surface distortion of each surface on the intake manifold. If it exceeds the limit (0.004 inch), replace the intake manifold.

8. Installation is the reverse of the removal procedure.

9. If the intake manifold stud bolts were

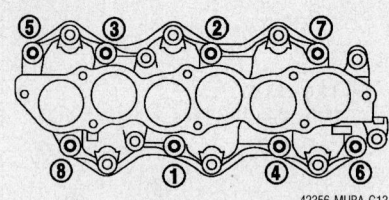

⬅ Engine front

42356-MURA-G13

Intake manifold bolt torque sequence

removed, install them and torque to 8 inch lbs.

10. Torque the intake manifold retaining bolts to specification and in the proper sequence.

11. Be sure to fill the engine with the proper grade and type coolant.

12. Start the engine and check for leaks, correct as required.

Exhaust Manifold

REMOVAL & INSTALLATION

2003–2004

1. Before servicing the vehicle, refer to the Precautions Section.

2. Drain the cooling system.

3. Remove or disconnect the following:
- Negative battery cable
- Exhaust front tube
- Rear engine mount insulator (2WD vehicles), when right exhaust manifold and three way catalyst is removed
- Right windshield wiper arm and right front cowl top cover, when

right exhaust manifold and three way catalyst is removed
- Both wiper arms
- Cowl top seal rubber
- Left and right cowl top cover clips and remove cowl top covers
- Washer nozzles and hose from cowl top cover
- Heated oxygen sensor 1 and 2 on both left and right bank.

✳✳ WARNING

Be careful not to damage heated oxygen sensor. Discard any heated oxygen sensor which has been dropped from a height of more than 0.5 m (19.7 inches) onto a hard surface such as a concrete floor; replace with a new sensor.

- Exhaust manifold covers and the three way catalyst heat shields

4. Remove bolts in the reverse order of illustration to remove three way catalyst supports.

5. Remove the left and right three way catalysts by loosening the bolts first and then removing the nuts.

6. Loosen the nuts in the reverse order of the torque sequence and remove the exhaust manifolds from the engine.

To install:

7. Use a straightedge and feeler gauge to check the flatness of the exhaust manifold

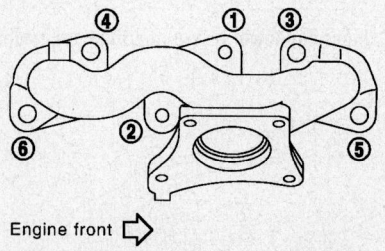

RH bank

Engine front ⮕

42356-MURA-G15

Exhaust manifold bolt torque sequence— right side

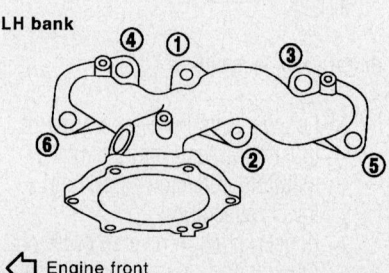

LH bank

⬅ Engine front

42356-MURA-G16

Exhaust manifold bolt torque sequence— left side

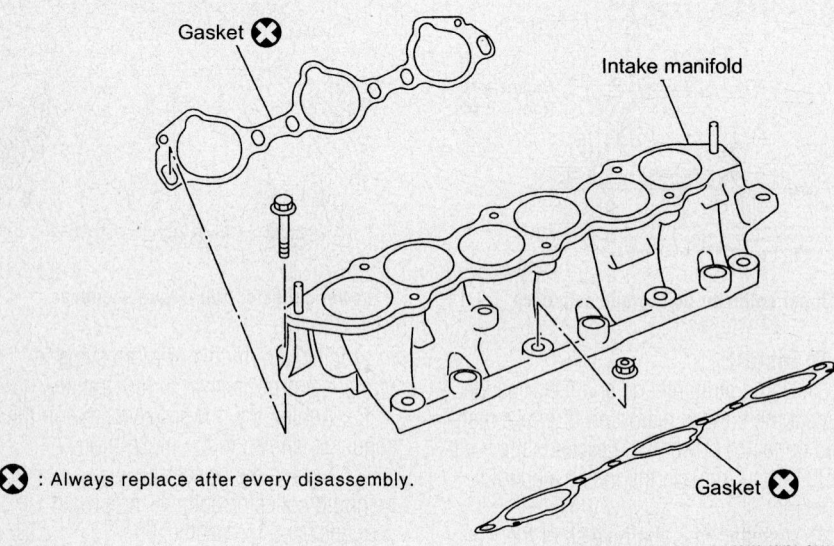

❌ : Always replace after every disassembly.

09482_MURA_G0007

Intake manifold and related components

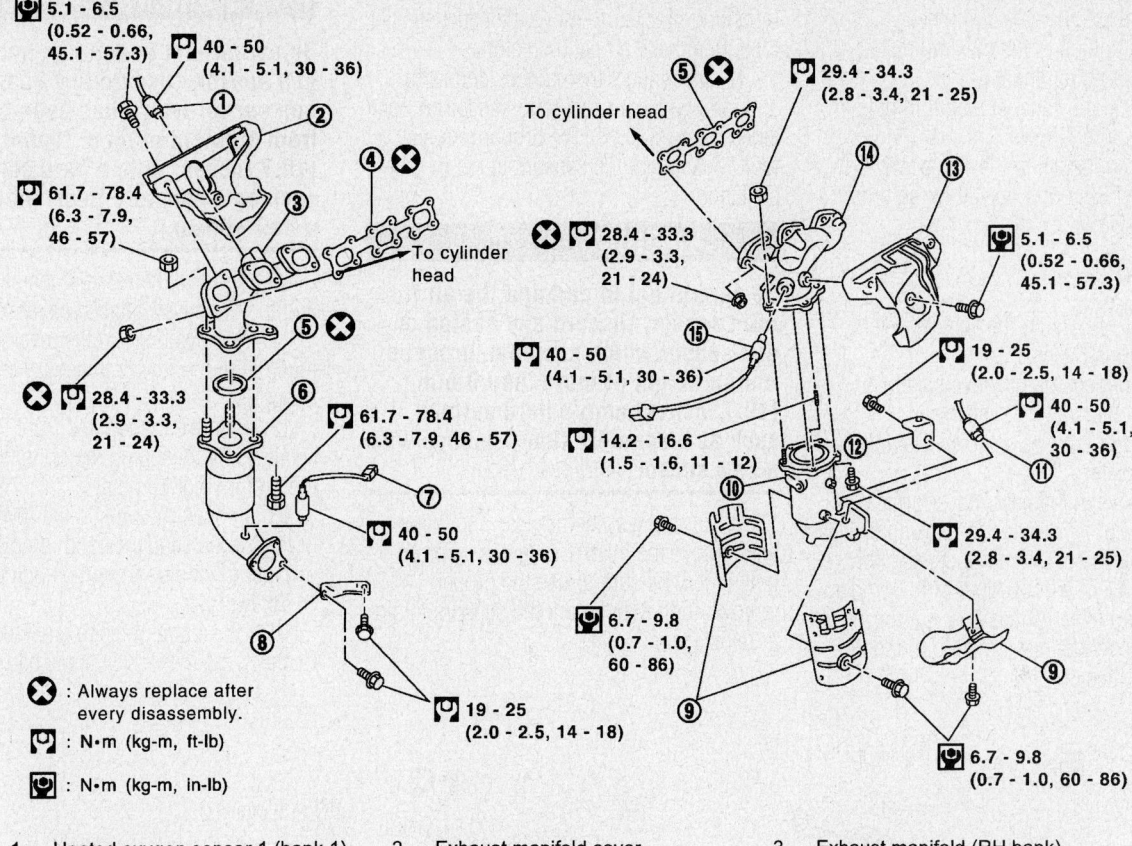

- 5.1 - 6.5 (0.52 - 0.66, 45.1 - 57.3)
- 40 - 50 (4.1 - 5.1, 30 - 36)
- 29.4 - 34.3 (2.8 - 3.4, 21 - 25)
- To cylinder head
- 61.7 - 78.4 (6.3 - 7.9, 46 - 57)
- To cylinder head
- 5.1 - 6.5 (0.52 - 0.66, 45.1 - 57.3)
- 28.4 - 33.3 (2.9 - 3.3, 21 - 24)
- 28.4 - 33.3 (2.9 - 3.3, 21 - 24)
- 40 - 50 (4.1 - 5.1, 30 - 36)
- 19 - 25 (2.0 - 2.5, 14 - 18)
- 61.7 - 78.4 (6.3 - 7.9, 46 - 57)
- 14.2 - 16.6 (1.5 - 1.6, 11 - 12)
- 40 - 50 (4.1 - 5.1, 30 - 36)
- 40 - 50 (4.1 - 5.1, 30 - 36)
- 29.4 - 34.3 (2.8 - 3.4, 21 - 25)
- 6.7 - 9.8 (0.7 - 1.0, 60 - 86)
- 19 - 25 (2.0 - 2.5, 14 - 18)
- 6.7 - 9.8 (0.7 - 1.0, 60 - 86)

- ❌ : Always replace after every disassembly.
- : N•m (kg-m, ft-lb)
- : N•m (kg-m, in-lb)

1. Heated oxygen sensor 1 (bank 1)
2. Exhaust manifold cover
3. Exhaust manifold (RH bank)
4. Gasket
5. Gasket
6. Three way catalyst (manifold) (RH bank)
7. Heated oxygen sensor 2 (bank 1)
8. Support (RH)
9. Three way catalyst heat shield
10. Three way catalyst (manifold) (LH bank)
11. Heated oxygen sensor 2 (bank 2)
12. Support (LH)
13. Exhaust manifold cover
14. Exhaust manifold (LH bank)
15. Heated oxygen sensor 1 (bank 2)

42356-MURA-G14

Exhaust manifold and related components—2003–2004

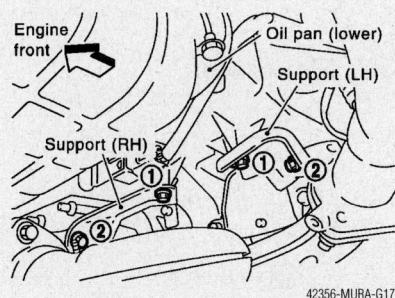

42356-MURA-G17

TWC support bolt torque sequence

mating surfaces. If it exceeds the limit (0.012 inch), replace the exhaust manifold.

8. Installation is in the reverse of removal. Check the accompanying illustrations for the proper torque sequences and specification. Note the following:

- When installing the heated oxygen sensor, tighten to the middle of specified torque range, because the

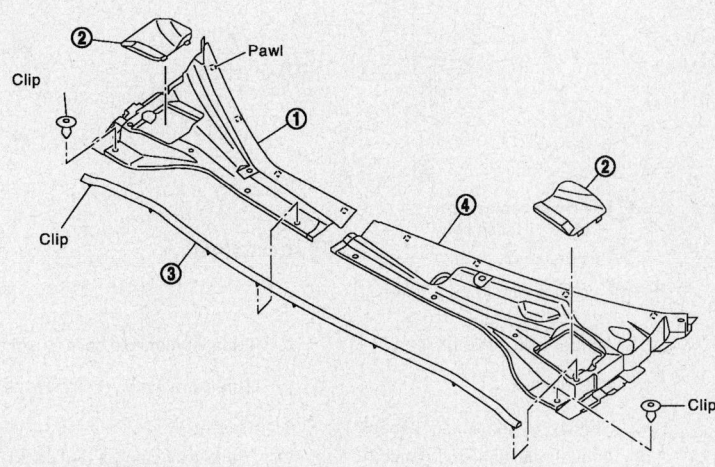

1. Cowl top cover (right)
2. Cap (left / right)
3. Cowl top seal rubber
4. Cowl top cover (left)

42356-MURA-G18

Extension cowl top and related components

length of the torque wrench may increase the actual tightness. Do not tighten to the maximum specified torque range.

- Install the exhaust manifold nuts in the order shown.
- When installing the Three Way Catalyst Supports, install in the order shown.

2005–2006

1. Before servicing the vehicle, refer to the Precautions Section.

2. Disconnect the negative battery cable. Drain the cooling system.

3. Remove the engine cover. Remove the under cover.

4. Remove the radiator cover grilles, air duct (inlet) and air cleaner cases (upper) with air flow sensor and air duct assembly.

5. Remove the front wiper arm. Remove the upper and lower extension cowl top.

6. Remove the intake manifold upper and lower collectors.

7. Remove the exhaust front tube mounting bracket. Remove the exhaust front tube. Remove the heat insulator.

8. Disconnect the harness connector and remove the air fuel ratio sensor on both sides. Be sure to use the proper removal tool. Matchmark the sensors to aid in reinstallation.

※※ WARNING

Be careful not to damage the air flow ratio sensor. Discard any heated oxygen sensor which has been dropped from a height of more than 0.5 m (19.7 inches) onto a hard surface such as a concrete floor; replace with a new sensor.

9. Disconnect the harness connector and remove the heated oxygen sensor on both sides. Be sure to use the proper removal tool. Matchmark the sensors to aid in reinstallation.

※※ WARNING

Be careful not to damage heated oxygen sensor. Discard any heated oxygen sensor which has been dropped from a height of more than 0.5 m (19.7 inches) onto a hard surface such as a concrete floor; replace with a new sensor.

10. Remove the exhaust manifold covers and the three way catalyst covers.

11. Remove bolts in the reverse order of illustration to remove three way catalyst supports.

12. Remove the left and right three way catalysts by loosening the bolts first and then removing the nuts.

13. Loosen the nuts in the reverse order of the torque sequence and remove the exhaust manifolds from the engine.

To install:

14. Use a straightedge and feeler gauge to check the flatness of the exhaust manifold

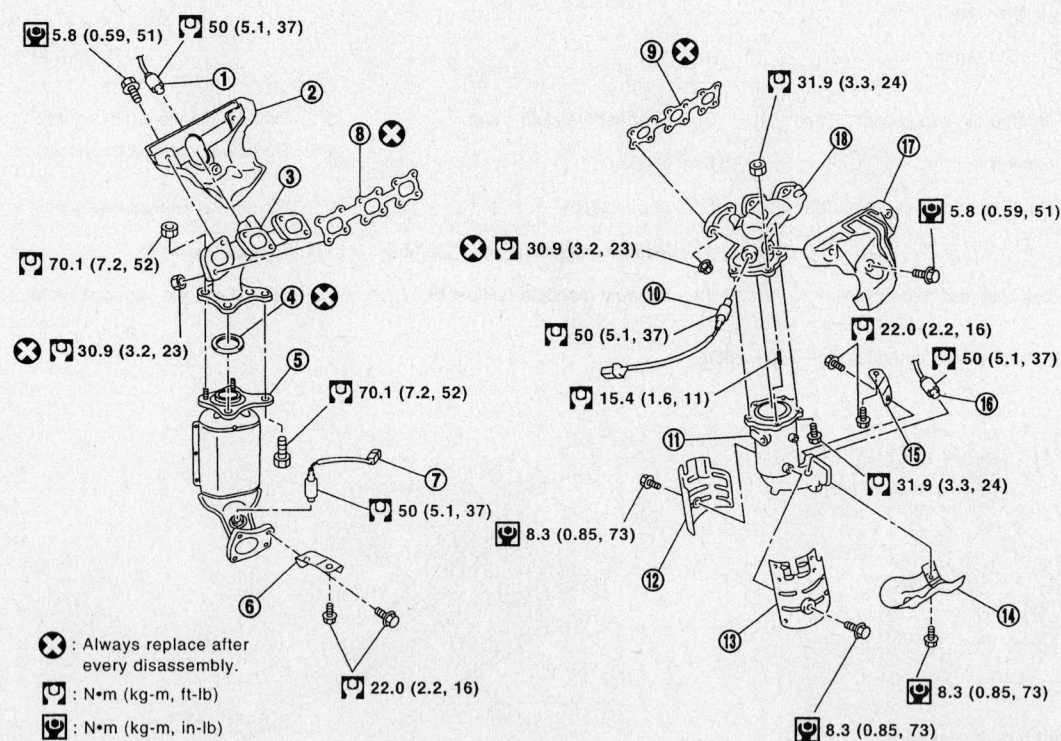

- ⊗ : Always replace after every disassembly.
- 🔧 : N•m (kg-m, ft-lb)
- 🔧 : N•m (kg-m, in-lb)

1.	Air fuel ratio sensor 1 (bank 1)	2.	Exhaust manifold cover (right bank)	3.	Exhaust manifold (right bank)
4.	Ring gasket	5.	Three way catalyst (right bank)	6.	Three way catalyst support (right bank)
7.	Heated oxygen sensor 2 (bank 1)	8.	Gasket	9.	Gasket
10.	Air fuel ratio sensor 1 (bank 2)	11.	Three way catalyst (left bank)	12.	Three way catalyst cover
13.	Three way catalyst cover	14.	Three way catalyst cover	15.	Three way catalyst support (left bank)
16.	Heated oxygen sensor 2 (bank 2)	17.	Exhaust manifold cover (left bank)	18.	Exhaust manifold (left bank)

09482_MURA_G0008

Exhaust manifold and related components—2005–2006

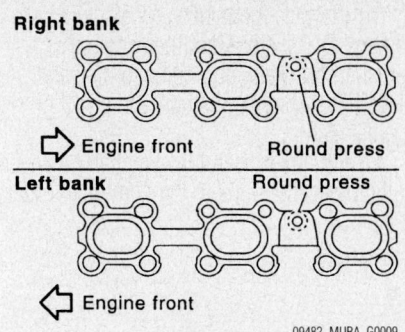

Right bank

⟹ Engine front   Round press

Left bank   Round press

⟸ Engine front

09482_MURA_G0009

Exhaust manifold gasket positioning

mating surfaces. If it exceeds the limit (0.012 inch), replace the exhaust manifold.

15. Using new gaskets, install the exhaust manifold on the engine. Torque the retaining bolts to specification and in the proper sequence.

16. Install the air fuel ratio sensors and the heated oxygen sensors in their proper positions. The glass tube color of the air fuel ratio sensor is Black. The glass tube color of the heated oxygen sensor is White.

➡**Before installing these sensors, clean the exhaust system threads using thread cleaner and tool J43897-18 or tool J43897-12. Apply anti-seize lubricant.**

17. Install the sensors. See illustration for proper torque specifications.

➡**Do not over torque these sensors. Doing so may cause damage to them, resulting in the ML light coming on.**

18. Continue the installation in the reverse order of the removal procedure.

Camshaft and Valve Lifters

REMOVAL

➡**On 2005–2006 vehicles the manufacturer recommends to first remove the engine from the vehicle.**

1. Before servicing the vehicle, refer to the Precautions Section.
2. Disconnect the negative battery cable. Drain the cooling system.
3. On 2005–2006 vehicles, remove the engine from the vehicle and position it in a suitable holding fixture.

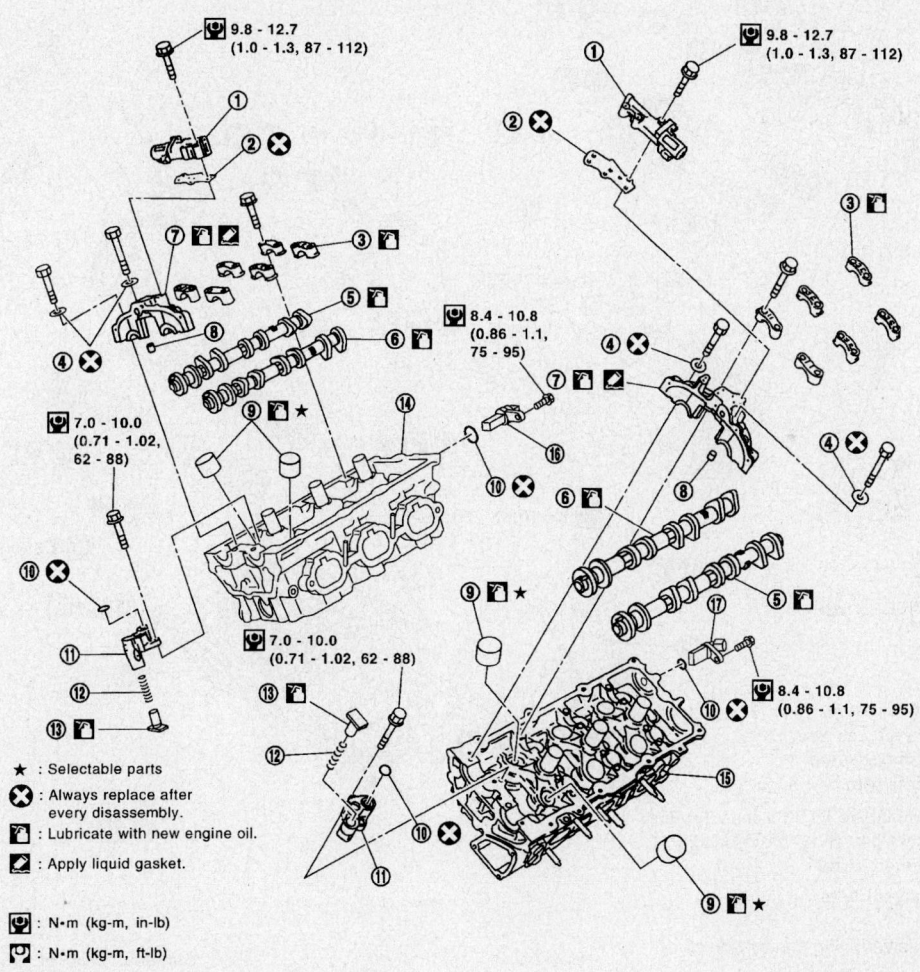

★ : Selectable parts

❌ : Always replace after every disassembly.

: Lubricate with new engine oil.

: Apply liquid gasket.

: N·m (kg-m, in-lb)

: N·m (kg-m, ft-lb)

1. Intake valve timing control solenoid valve	2. Gasket	3. Camshaft bracket (No.2 to No.4)
4. Seal washer	5. Camshaft (EXH)	6. Camshaft (INT)
7. Camshaft bracket (No.1)	8. Dowel pin	9. Valve lifter
10. O-ring	11. Chain tensioner	12. Spring
13. Plunger	14. Cylinder head (RH bank)	15. Cylinder head (LH bank)
16. Camshaft position sensor (PHASE) (RH bank)	17. Camshaft position sensor (PHASE) (LH bank)	

Camshafts and related components—2003–2004

42356-MURA-G19

4. Remove front timing chain case, camshaft sprocket, timing chain and rear timing chain case.

5. If necessary, remove camshaft position sensor (PHASE) (right and left banks) from cylinder head back side.

➡**Handle carefully to avoid dropping and shocks. Do not disassemble. Do not allow metal powder to adhere to magnetic part at sensor tip. Do not place sensors in a location where they are exposed to magnetism.**

6. Remove the intake valve timing control solenoid valves. Discard the intake valve timing control solenoid valve gaskets and use new gaskets for installation.

7. Remove the intake and exhaust camshaft brackets. Mark the camshafts, camshaft brackets, and bolts so they are placed in the same position and direction for installation.

8. Equally loosen the camshaft bracket bolts in several steps in the numerical order shown.

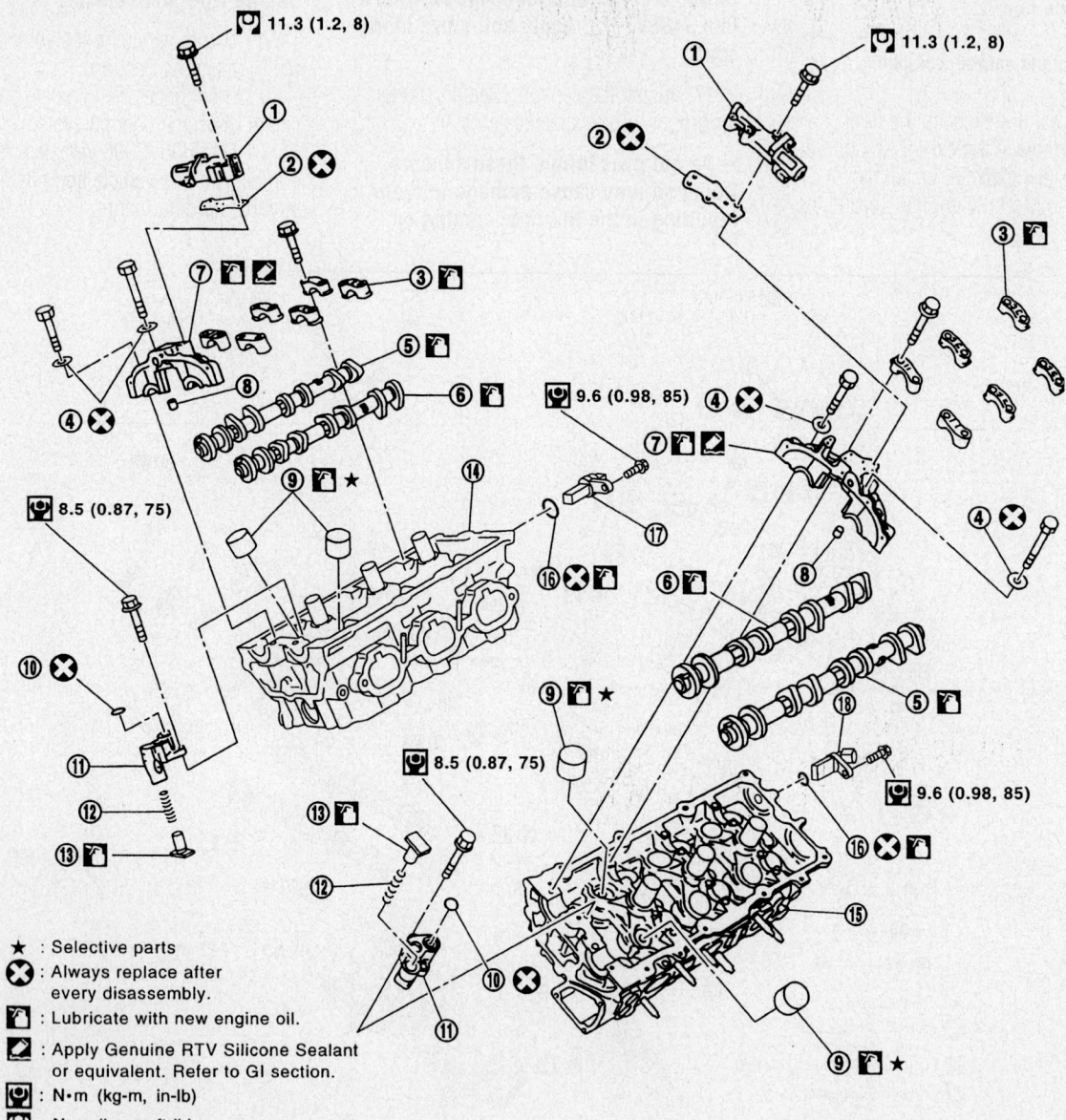

★ : Selective parts

⊗ : Always replace after every disassembly.

🛢 : Lubricate with new engine oil.

🖌 : Apply Genuine RTV Silicone Sealant or equivalent. Refer to GI section.

⚙ : N·m (kg-m, in-lb)

⚙ : N·m (kg-m, ft-lb)

1.	Intake valve timing control solenoid valve	2.	Gasket	3.	Camshaft bracket (No. 2 to No. 4)
4.	Seal washer	5.	Camshaft (EXH)	6.	Camshaft (INT)
7.	Camshaft bracket (No. 1)	8.	Dowel pin	9.	Valve lifter
10.	O-ring	11.	Timing chain tensioner (Secondary)	12.	Spring
13.	Plunger	14.	Cylinder head (right bank)	15.	Cylinder head (left bank)
16.	O-ring	17.	Camshaft position sensor (PHASE) (right bank)	18.	Camshaft position sensor (PHASE) (left bank)

09482_MURA_G0010

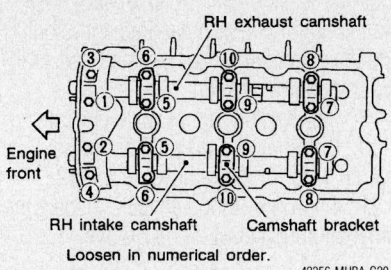

Right side camshaft bolt loosening sequence—2003–2004

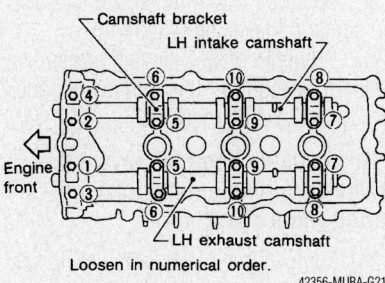

Left side camshaft bolt loosening sequence—2003–2004

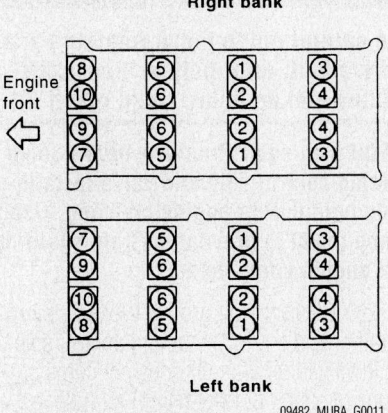

Camshaft bolt loosening sequence—2005–2006

9. Remove the camshaft.

10. Remove valve lifter. Identify installation positions, and store them without mixing them up.

11. Remove secondary timing chain tensioner from cylinder head with its stopper pin attached.

To install:

12. Install secondary chain tensioners on both sides of cylinder head.

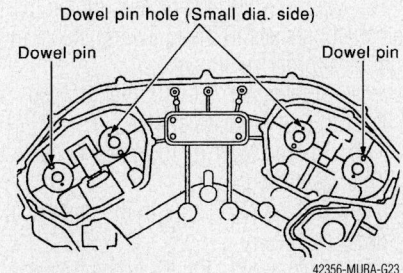

Dowel pin orientation

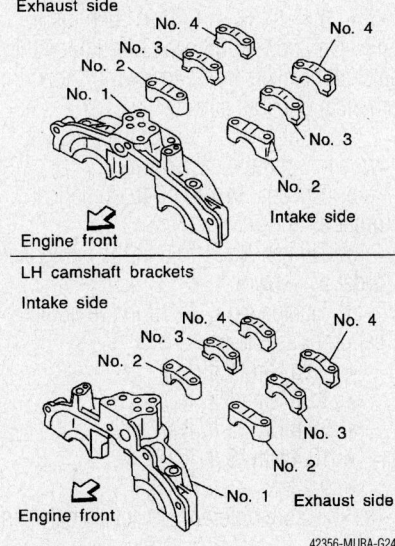

Camshaft bracket identification

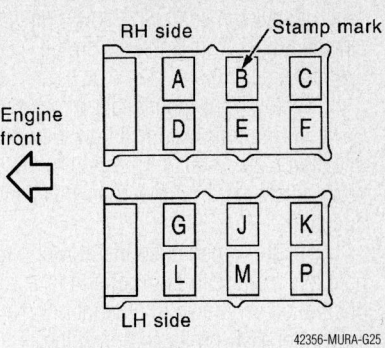

Align the stamp marks as shown

Chain tensioner location and installation

a. Install chain tensioner with its stopper pin attached.

b. Install tensioner with sliding part facing downward on right side cylinder head, and with sliding part facing upward on left side cylinder head.

c. Install O-ring.

13. Install valve lifters in their original position.

Valve lifter outer diameter (Intake and exhaust)

• 33.977–33.987mm (1.3377–1.3381 inch)

Standard (Intake and exhaust)

• 34.000–34.016 mm (1.3386–1.3392 inch)

Standard (Intake and exhaust)

• 0.013–0.039 mm (0.0005–0.0015 in)

14. Install camshafts.

Bank	INT/EXH	Dowel pin	Paint marks		ID mark
			M1	M2	
RH	INT	No	Pink	No	RE
	EXH	Yes	No	Orange	RE
LH	INT	No	Pink	No	LH
	EXH	Yes	No	Orange	LH

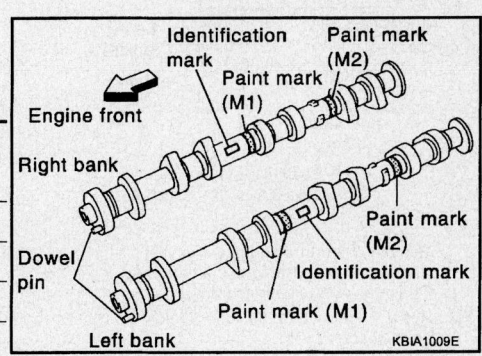

Camshaft identification

a. Install camshaft with dowel pin attached to its front end face on the exhaust side.

b. Follow your identification marks made during removal, or follow the identification marks that are present on the new camshafts for proper placement and direction.

c. Install camshaft so that dowel pin hole and dowel pin on front end face are positioned as shown in illustration. (No. 1 cylinder TDC on its compression stroke)

➡ **Large and small pin holes are located on front end face of intake camshaft, at intervals of 180 degrees. Face small diameter side pin hole upward (in cylinder head upper face direction).**

15. Install camshaft brackets.

a. Remove foreign material completely from camshaft bracket backside and from cylinder head installation face.

b. Install camshaft bracket in original position and direction as shown in illustration.

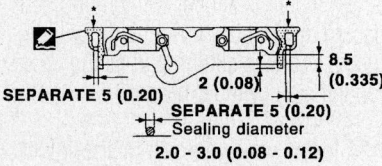

SEPARATE 5 (0.20)
SEPARATE 5 (0.20)
2 (0.08)
8.5 (0.335)
Sealing diameter
2.0 - 3.0 (0.08 - 0.12)

* : Remove the protruding sealant from front face. (Remove the hardended sealant from surface only.)

◪ : Apply liquid gasket

Unit: mm (in)

42356-MURA-G26

RTV sealer application—2003–2004

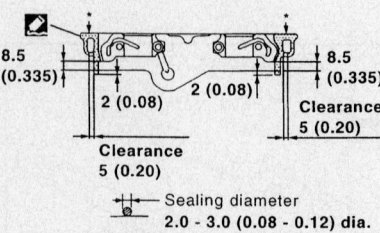

8.5 (0.335)
8.5 (0.335)
2 (0.08)
2 (0.08)
Clearance 5 (0.20)
Clearance 5 (0.20)
Sealing diameter
2.0 - 3.0 (0.08 - 0.12) dia.

* : Remove the protruding liquid gasket from front face. (Remove the hardened liquid gasket from surface only.)

◪ : Apply Genuine RTV Silicone Sealant or equivalent.

Unit: mm (in)

09482_MURA_G0013

RTV sealer application—2005–2006

c. Install No.2 to 4 camshaft brackets aligning the stamp marks as shown.

➡ **There are no identification marks indicating left and right for No. 1 camshaft bracket.**

d. Apply sealant to mating surface of No.1 camshaft bracket as shown on right and left banks.

➡ **Use RTV silicone sealant or equivalent.**

16. On 2003–2004 vehicles, tighten the camshaft brackets in the following steps, in numerical order as shown.

a. Tighten No. 7 to 10, then tighten No.1 to 6 in order as shown.

b. Tighten No.1 to 10 in numerical order as shown.

c. Tighten No. 1 to 6 in the numerical order as shown.

d. Tighten No. 7 to 10 in the numerical order as shown.

- 1.96 Nm (17 inch lbs.)
- 5.88 Nm (52 inch lbs.)
- 9.02–11.8 Nm (80–104 inch lbs.)
- 8.3–10.3 Nm (74–91 inch lbs.)

17. On 2005–2006 vehicles, tighten the camshaft brackets in the following steps, in numerical order as shown.

a. Tighten No. 7 to 10, then tighten No.1 to 6 in order as shown.

b. Tighten No.1 to 10 in numerical order as shown.

c. Tighten No. 1 to 6 in the numerical order as shown.

d. Tighten No. 7 to 10 in the numerical order as shown.

- 1.96 Nm (1 ft. lb.)
- 1.96 Nm (1 ft. lb.)
- 5.88 Nm (4 ft. lbs.)
- 10.4 Nm (8 ft. lbs.)
- 9.3 Nm

18. Measure difference in levels between

RH bank

Engine front
8 5 1 3
10 6 2 4
9 6 2 4
7 5 1 3

7 5 1 3
9 6 2 4
10 6 2 4
8 5 1 3

LH bank

42356-MURA-G27

Camshaft bolt torque sequence

front end faces of No. 1 camshaft bracket and cylinder head. If measurement is outside the specified range, reinstall camshaft and camshaft bracket.

19. Inspect and adjust valve clearance.

20. Continue the installation in the reverse order of the removal procedure.

21. On 2005–2006 vehicles, inspect the camshaft sprocket (INT) oil groove.

➡ **Perform this inspection only when DTC P0011 or DTC P0021 are detected in self diagnostic results of CONSULT-II.**

22. Be sure the engine is cold. Check and adjust oil level, as required.

23. Properly release the fuel system pressure. Disconnect the ignition coil and injector harness connectors.

➡ **This is being done to prevent the engine from unintentionally being started while checking.**

24. Remove the intake valve timing control solenoid valve.

25. Crank the engine, and then make sure that engine oil comes out from camshaft bracket (No.1) oil hole.

✵ WARNING

Be careful not to touch rotating parts, (drive belt, idler pulley, crankshaft pulley etc) as injury could result.

➡ **Oil may squirt from the intake valve timing control solenoid valve installation hole during engine cranking. Use a shop towel to prevent oil from squirting on engine components.**

26. Clean the oil groove between the oil strainer and the intake timing control solenoid valve if engine oil does not come out from camshaft bracket (No.1) oil hole.

27. Remove the components between the intake valve timing control solenoid valve and the camshaft sprocket (INT). Check each oil groove for clogging.

28. After inspection install any removed components.

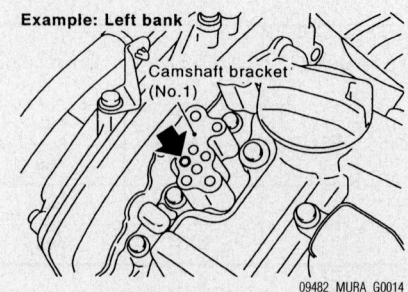

Example: Left bank

Camshaft bracket (No.1)

09482_MURA_G0014

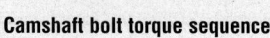

Camshaft bracket (No1) oil hole location

INSPECTION

Camshaft Runout

1. Put V block on precise flat bed, and support No. 2 and No. 4 journal of camshaft.

2. Set dial gauge vertically to No. 3 journal.

3. Turn camshaft to one direction with hands, and measure camshaft runout on dial gauge. (Total indicator reading)

4. If it exceeds the limit, replace camshaft.

Camshaft Cam Height

1. Measure camshaft cam height. Limit: 0.05 mm (0.0020 in). Standard cam height (intake and exhaust): 44.865–45.055 mm (1.7663–1.7738 in)

2. If wear is beyond the limit, replace camshaft.

Camshaft Journal Clearance

OUTER DIAMETER OF CAMSHAFT JOURNAL

1. Measure outer diameter of camshaft journal.

Cam wear limit: 0.2 mm (0.008 inch)
Standard outer diameter:
- No. 1: 25.935–25.955 mm (1.0211–1.0218 in)
- No. 2, 3, 4: 23.445–23.465 mm (0.9230–0.9238 in)

INNER DIAMETER OF CAMSHAFT BRACKET

1. Tighten camshaft bracket bolt with specified torque.

2. Using inside micrometer, measure inner diameter "A" of camshaft bracket.

3. Specification is as follows:
- No. 1: 26.000–26.021 mm (1.0236–1.0244 in)
- No. 2, 3, 4: 23.500–23.521 mm (0.9252–0.9260 in)

CALCULATION OF CAMSHAFT JOURNAL CLEARANCE

Journal clearance = inner diameter of camshaft bracket—outer diameter of camshaft journal. When outside the limit, replace either or both camshaft and cylinder head.

➡**Inner diameter of camshaft bracket is manufactured together with cylinder head. Replace the whole cylinder head assembly.**

1. Specification is as follows:
- No. 1: 0.045–0.0034 mm (0.0018–0.0034 in)
- No. 2, 3, 4: 0.035–0.076 mm (0.0014–0.0030 in)
- Limit: 0.15 mm (0.0059 in)

Camshaft End Play

1. Install dial gauge in thrust direction on front end of camshaft. Measure end play of dial gauge when camshaft is moved forward and backward.

2. Standard specification should be 0.115–0.188 mm (0.0045–0.0074 in). Limit specification should be 0.24 mm (0.0094 in).

3. When out of the limit, replace with new camshaft and measure again.

4. When out of the limit again, replace with new cylinder head.

Valve Lash

CHECKING

Perform inspection as follows after removal, installation or replacement of camshaft or valve-related parts, or if there is unusual engine conditions regarding valve clearance.

1. Remove right and left rocker covers.

2. Measure valve clearance as below:

a. Set No.1 cylinder at TDC of its compression stroke. Align crankshaft pulley timing mark (grooved line without color) with timing indicator. Check that No. 1 cylinder intake and exhaust cam nose is facing in direction shown in illustration. If not, rotate crankshaft pulley 360 degrees clockwise (when viewed from front).

Measuring position (RH bank)		No.1 CYL.	No.3 CYL.	No.5 CYL.
No.1 cylinder at TDC	EXH		×	
	INT	×		
Measuring position (LH bank)		No.2 CYL.	No.4 CYL.	No.6 CYL.
No.1 cylinder at TDC	INT			×
	EXH	×		

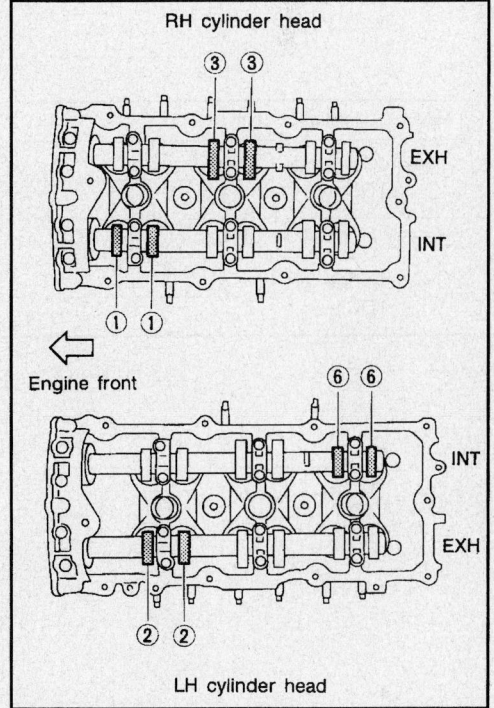

Valve clearance inspection—No.1 TDC

b. Using a feeler gauge, measure valve clearance. Standard: -0.14 to 0.14 mm (-0.0055 to 0.0055 in)

➡️ **If inspection was carried out with cold engine, check that values with fully warmed up engine are still within specifications.**

c. Rotate crankshaft by 240 degrees clockwise (when viewed from front) to align No. 3 cylinder at TDC of its compression stroke.

➡️ **Crankshaft pulley mounting bolt flange has a stamped line every 60 degrees. They can be used as a guide to rotation angle.**

d. Turn crankshaft pulley clockwise by 240 degrees from the position of No. 5 cylinder at compression TDC.

3. For measurements that are outside the specified range, perform adjustment below.

Measuring position (RH bank)		No.1 CYL.	No.3 CYL.	No.5 CYL.
No.3 cylinder at TDC	EXH			×
	INT		×	
Measuring position (LH bank)		No.2 CYL.	No.4 CYL.	No.6 CYL.
No.3 cylinder at TDC	INT	×		
	EXH		×	

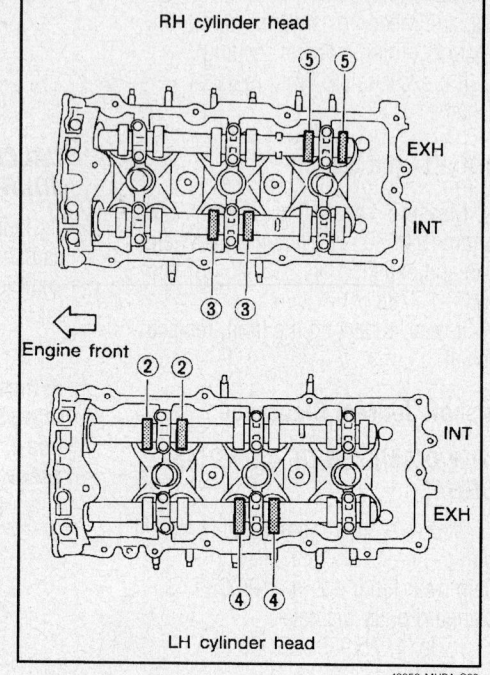

42356-MURA-G29

Valve clearance inspection—No.3 TDC

Measuring position (RH bank)		No.1 CYL.	No.3 CYL.	No.5 CYL.
No.5 cylinder at TDC	EXH	×		
	INT			×
Measuring position (LH bank)		No.2 CYL.	No.4 CYL.	No.6 CYL.
No.5 cylinder at TDC	INT		×	
	EXH			×

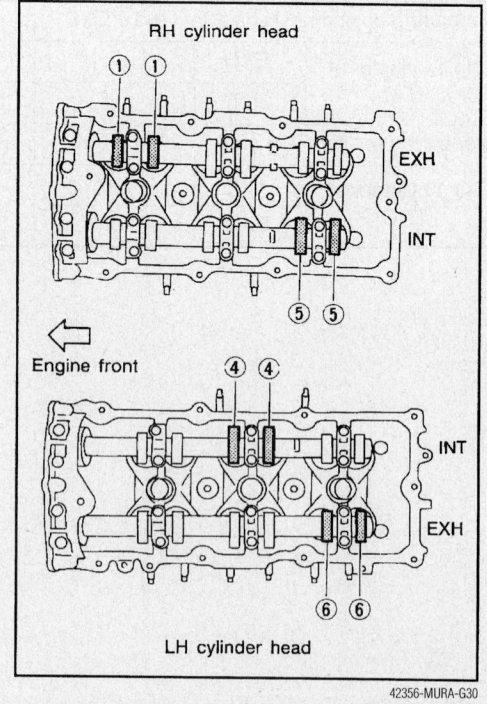

42356-MURA-G30

Valve clearance inspection—No.5 TDC

ADJUSTMENT

Perform adjustment depending on selected head thickness of valve lifter. The specified valve lifter thickness is the dimension at normal temperatures. Ignore dimensional differences caused by temperature. Use the specifications for hot engine condition to adjust.

1. Remove camshaft.
2. Remove the valve lifters at the locations that are outside the standard.
3. Measure the center thickness of the removed valve lifters with a micrometer.
4. Use the equation below to calculate valve lifter thickness for replacement.
Valve lifter thickness calculation:
- Thickness of replacement valve lifter = $t1 + (C1 - C2)$
- $t1$ = Thickness of removed valve lifter
- $C1$ = Measured valve clearance
- $C2$ = Standard valve clearance:

Thickness of a new valve lifter can be identified by stamp marks on the reverse side (inside the cylinder). Stamp mark 788U or 788R indicates 7.88 mm (0.3102 in) in thickness.

➠**2 types of stamp marks are used for parallel setting and for manufacturer identification.**

Available thickness of valve lifter: 27 sizes with range 7.88 to 8.40 mm (0.3102 to 0.3307 in) in steps of 0.02 mm (0.0008 in) (when manufactured at factory).

5. Install the selected valve lifter.
6. Install camshaft.
7. Manually turn crankshaft pulley a few turns.
8. Check that valve clearances for cold engine are within specifications by referring to the specified values.
9. After completing the repair, check valve clearances again with the specifications for warmed engine. Make sure the values are within specifications.
Valve clearance:
- Intake: 0.30 mm (0.012 in)
- Exhaust: 0.33 mm (0.013 in)

Starter Motor

REMOVAL & INSTALLATION

1. Before servicing the vehicle, refer to the Precautions Section.
2. Remove or disconnect the following:
- Battery
- Air intake duct
- Battery bracket

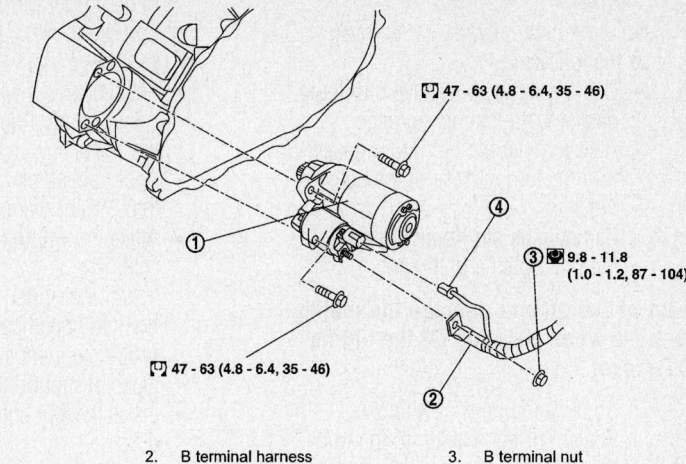

🔧 47 - 63 (4.8 - 6.4, 35 - 46)

🔧 9.8 - 11.8 (1.0 - 1.2, 87 - 104)

🔧 47 - 63 (4.8 - 6.4, 35 - 46)

🔩 : N·m (kg-m, in-lb)
🔧 : N·m (kg-m, ft-lb)

1. Starter motor
4. S connector
2. B terminal harness
3. B terminal nut

42356-MURA-G32

Starter and related components

- Transaxle dipstick tube on AWD vehicles. Be sure to drain the transaxle fluid.
- S connector
- B terminal nut
- Starter motor mounting bolts
- Starter motor to the direction of upper side the vehicle
3. Install in the reverse order of removal. Observe the following toques:
- B terminal nut: 9.8–11.8 Nm (87–104 inch lbs.)
- Starter motor mounting bolt: 47–63 Nm (35–46 ft. lbs.)
- Battery bracket mounting bolt: 14–20 Nm (10–15 ft. lbs.)

Oil Pan

REMOVAL & INSTALLATION

2WD

1. Before servicing the vehicle, refer to the Precautions Section.

➠**When removing the upper oil pan from the engine, first remove the crankshaft position sensor (POS). Be careful not to damage sensor edges or signal plate teeth.**

2. Remove engine cover.
3. Remove right splash guard.
4. Remove the front right road wheel and tire.
5. Drain engine oil.
6. Drain engine coolant.
7. Remove oil filter.
8. Remove oil cooler and water pipes.
9. Remove all drive belts.
10. Remove A/C compressor with pip-
ing connected, and temporarily secure it aside.
11. Remove exhaust front tube.
12. Remove the heated oxygen sensor 2 (bank 2) and remove the three way catalyst (manifold) (bank 2) from the exhaust manifold.
13. Loosen lower oil pan bolts in reverse order of the installation sequence.
14. Insert a seal cutter (special service tool) between the lower oil pan and the upper oil pan.

➠**Be careful not to damage the mating surface. Do not insert a screwdriver; this will damage the mating surfaces.**

15. Slide seal cutter (special service tool) by tapping on the side of the tool with a hammer.
16. Remove lower oil pan.
17. Remove oil strainer.
18. Remove the oil pressure switch.
19. If not already removed, remove crankshaft position sensor (POS).

➠**Handle carefully to avoid dropping and shocks. Do not disassemble. Do not allow metal powder to adhere to magnetic part at sensor tip. Do not place sensors in a location where they are exposed to magnetism.**

20. Remove the four engine-to-transaxle bolts.
21. Remove upper oil pan. Loosen bolts in reverse order of the installation sequence.
22. Insert an appropriate size tool into the notch of the upper oil pan shown (1). Pry off the upper oil pan by moving the tool up and down shown (2).
23. Remove O-rings from the bottom of the cylinder block and oil pump body.
24. Remove oil pan gasket.

To install:

Installation is the reverse of removal. Note the following:

- Use a scraper to remove old liquid gasket from mating surfaces
- Also remove the old liquid gasket from mating surface of the cylinder block
- Remove the old liquid gasket from the bolt holes and threads

➡ **Do not scratch or damage the mating surfaces when cleaning off the old liquid gasket.**

- Apply Genuine RTV Silicone Sealant or equivalent, to the front timing chain case gasket and the rear oil seal retainer gasket shown
- To install, align protrusion of oil pan gasket with notches of front timing chain case and rear oil seal retainer
- Install oil pan gasket with smaller arc to front timing chain case side
- Install new O-rings on the cylinder block and oil pump side
- Apply a continuous bead of sealant to the cylinder block mating surface

of the upper oil pan to a limited portion shown. Use RTV silicone sealant or equivalent

- For bolt holes with marks (5 locations), apply liquid gasket outside the holes
- Apply a bead of 4.5 to 5.5 mm (0.177 to 0.217 in) diameter to area "A"
- Attaching within 5 minutes after coating
- Install the upper oil pan. Torque the bolts to specification and in the proper sequence. There are two types of mounting bolts.
- Install the four engine-to-transaxle bolts
- Install oil strainer to oil pump
- Use a scraper to remove old liquid gasket from mating surfaces. Also remove old liquid gasket from mating surface of upper oil pan.
- Apply a continuous bead of sealant to the lower oil pan. Use RTV silicone sealant. Be sure the sealant is 4.5–5.5 mm (0.177–0.217 inch) wide. Attach within 5 minutes after coating.

- Install lower oil pan. Torque the bolts to specification and in the proper sequence.
- Install oil pan drain plug. Refer to illustration for pan washer installation.
- Install in the reverse order of removal after this step
- Wait at least 30 minutes after oil pan is installed, before adding engine oil
- Before starting engine, check the levels of engine coolant, engine oil and working fluid. If less than required quantity, fill to the specified level.
- Use procedure below to check for fuel leakage
- Turn ignition switch ON (with engine stopped). With fuel pressure applied to fuel piping, check for fuel leakage at connection points.
- Start engine. With engine speed increased, check again for fuel leakage at connection points.
- Run engine to check for unusual noise and vibration

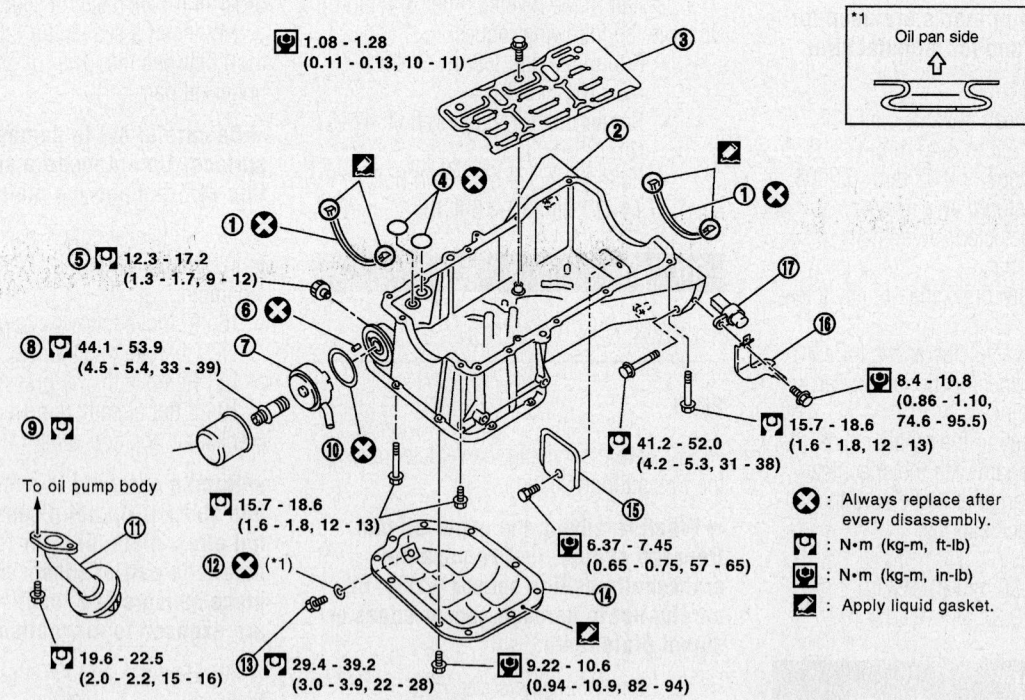

1. Gasket
2. Upper oil pan
3. Baffle plate
4. O-ring
5. Oil pressure switch
6. Relief valve
7. Oil cooler
8. Oil cooler connector
9. Oil filter
10. Gasket
11. Oil strainer
12. Gasket
13. Drain plug
14. Lower oil pan
15. Rear cover plate
16. Heated oxygen sensor (bank 2) harness clamp (2WD models)
17. Crankshaft position sensor (POS)

42356-MURA-G33

Oil pan and related components—2003–2004

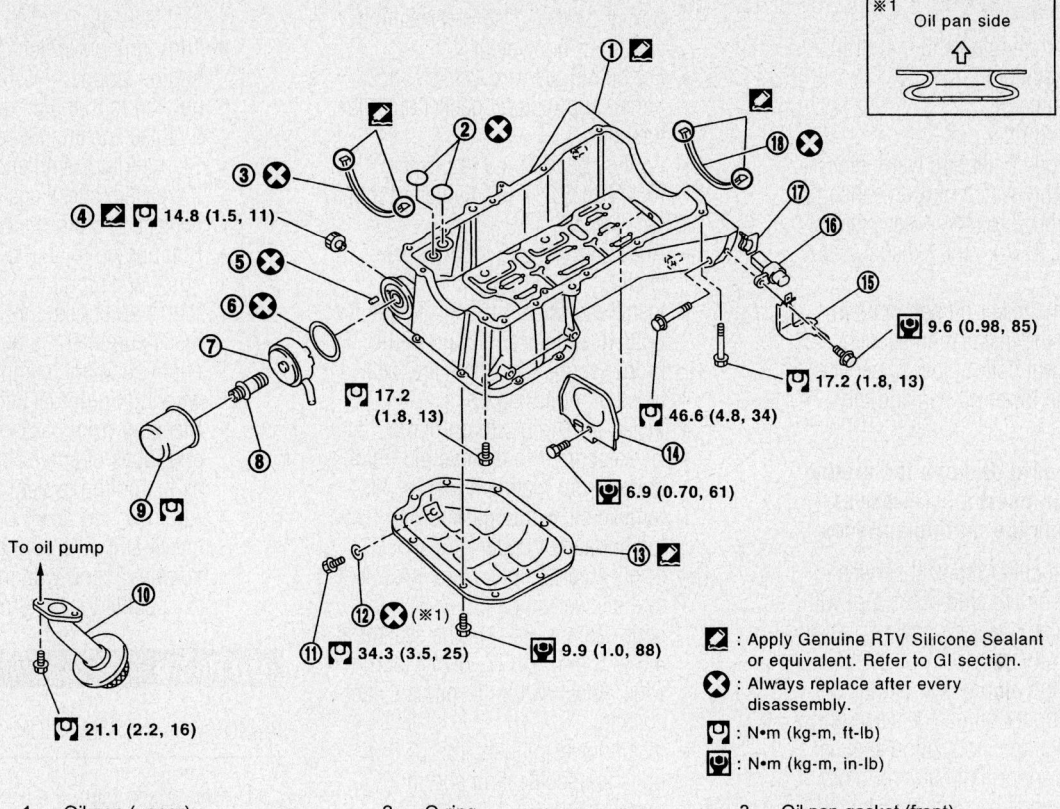

Oil pan and related components—2005–2006

1. Oil pan (upper)
4. Oil pressure switch
7. Oil cooler
10. Oil strainer
13. Oil pan (lower)
16. Crankshaft position sensor (POS)

2. O-ring
5. Relief valve
8. Connector bolt
11. Drain plug
14. Rear plate cover
17. Seal rubber

3. Oil pan gasket (front)
6. O-ring
9. Oil filter
12. Drain plug washer
15. Harness bracket (2WD models)
18. Oil pan gasket (rear)

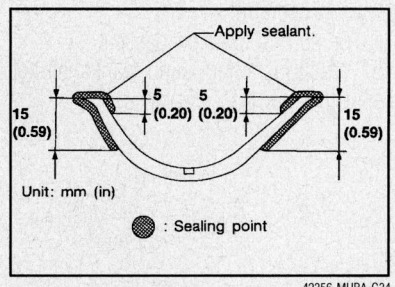

RTV sealer application at the timing case

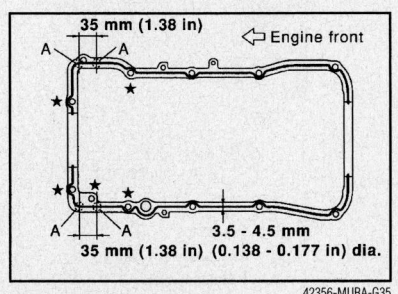

RTV sealer application on the pan

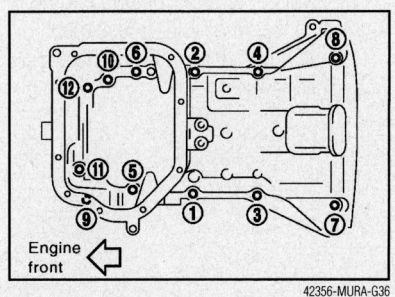

Upper oil pan bolt torque sequence

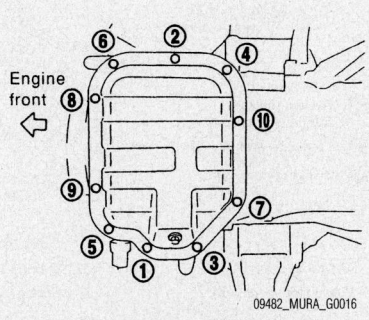

Lower oil pan bolt torque sequence

• Warm up engine thoroughly to make sure there is no leakage of engine coolant, engine oil and working fluid, fuel and exhaust gas
• Bleed air from passages in pipes and tubes of applicable lines, such as in cooling system
• After cooling down engine, again check amounts of engine coolant, engine oil and working fluid. Refill to specified level, if necessary

AWD

1. Before servicing the vehicle, refer to the Precautions Section.

→When removing the upper oil pan from the engine, first remove the crankshaft position sensor (POS). Be careful not to damage sensor edges or signal plate teeth.

2. Remove engine assembly from vehicle, and separate front suspension

member, transaxle and transfer case assembly from engine.

3. Remove the engine and mount it onto the engine stand.

4. Drain engine oil.

5. Remove oil filter.

6. Remove oil cooler and water pipes.

7. Remove the heated oxygen sensor 2 (bank 2) and remove the three way catalyst (manifold) (bank 2) from the exhaust manifold.

8. Loosen the lower oil pan bolts in reverse order of the installation sequence.

9. Insert a seal cutter (special service tool) between the lower oil pan and the upper oil pan.

➡ **Be careful not to damage the mating surface. Do not insert a screwdriver, this will damage the mating surfaces.**

10. Slide seal cutter (special service tool) by tapping on the side of the tool with a hammer. Remove lower oil pan.

11. Remove oil strainer.

12. Remove the oil pressure switch.

13. Remove upper oil pan. Loosen bolts in reverse order of the installation sequence.

14. Insert an appropriate size tool into the notch of the upper oil pan shown (1). Pry off the upper oil pan by moving the tool up and down shown (2).

15. Remove O-rings from the bottom of the cylinder block and oil pump body.

16. Remove oil pan gasket.

To install:
Installation is the reverse of removal. Note the following:

- Use a scraper to remove old liquid gasket from mating surfaces
- Also remove the old liquid gasket from mating surface of the cylinder block
- Remove the old liquid gasket from the bolt holes and threads

➡ **Do not scratch or damage the mating surfaces when cleaning off the old liquid gasket.**

- Apply Genuine RTV Silicone Sealant or equivalent, to the front timing chain case gasket and the rear oil seal retainer gasket shown
- To install, align protrusion of oil pan gasket with notches of front timing chain case and rear oil seal retainer
- Install oil pan gasket with smaller arc to front timing chain case side
- Install new O-rings on the cylinder block and oil pump side
- Apply a continuous bead of sealant to the cylinder block mating surface

of the upper oil pan to a limited portion shown. Use RTV silicone sealant or equivalent.

- For bolt holes with marks (5 locations), apply liquid gasket outside the holes
- Apply a bead of 4.5 to 5.5 mm (0.177 to 0.217 in) diameter to area "A"
- Attaching within 5 minutes after coating
- Install the upper oil pan. Torque the bolts to specification and in the proper sequence. There are two types of mounting bolts.
- Install oil strainer to oil pump
- Use a scraper to remove old liquid gasket from mating surfaces. Also remove old liquid gasket from mating surface of upper oil pan
- Apply a continuous bead of sealant to the lower oil pan. Use RTV silicone sealant. Be sure the sealant is 4.5–5.5 mm (0.177–0.217 inch) wide. Attach within 5 minutes after coating
- Install lower oil pan. Torque the bolts to specification and in the proper sequence.
- Install oil pan drain plug. Refer to illustration for pan washer installation.
- Install in the reverse order of removal
- Wait at least 30 minutes after oil pan is installed, before adding oil
- Before starting engine, check the levels of engine coolant, engine oil and working fluid. If less than required quantity, fill to the specified level

- Use procedure below to check for fuel leakage
- Turn ignition switch ON (with engine stopped). With fuel pressure applied to fuel piping, check for fuel leakage at connection points.
- Start engine. With engine speed increased, check again for fuel leakage at connection points.
- Run engine to check for unusual noise and vibration
- Warm up engine thoroughly to make sure there is no leakage of engine coolant, engine oil and working fluid, fuel and exhaust gas
- Bleed air from passages in pipes and tubes of applicable lines, such as in cooling system
- After cooling down engine, again check amounts of engine coolant, engine oil and working fluid. Refill to specified level, if necessary

Oil Pump

REMOVAL & INSTALLATION

1. Before servicing the vehicle, refer to the Precautions Section.

2. Remove the upper and lower oil pans.

3. Remove the oil strainer.

4. Remove front timing chain case and timing chain (primary).

5. Remove oil pump assembly.

To install:

6. Installation is the reverse of the removal procedure.

7. Be sure to use new gaskets.

8. When installing, align crankshaft flat faces with inner rotor flat faces.

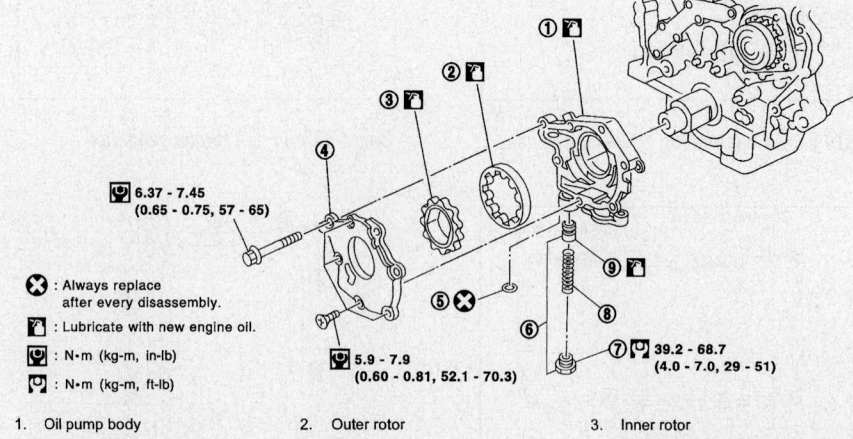

6.37 - 7.45 (0.65 - 0.75, 57 - 65)

❌ : Always replace after every disassembly.

🛢 : Lubricate with new engine oil.

⊡ : N·m (kg-m, in-lb)

⊡ : N·m (kg-m, ft-lb)

5.9 - 7.9 (0.60 - 0.81, 52.1 - 70.3)

39.2 - 68.7 (4.0 - 7.0, 29 - 51)

1. Oil pump body	2. Outer rotor	3. Inner rotor
4. Oil pump cover	5. O-ring	6. Regulator valve set
7. Regulator valve plug	8. Spring	9. Regulator valve

Oil pump and related components

42356-MURA-G37

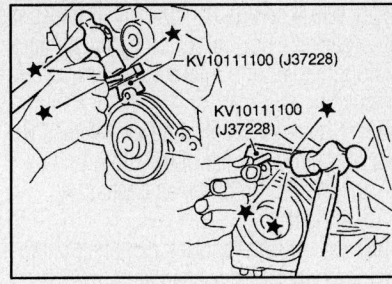

KV10111100 (J37228)

KV10111100 (J37228)

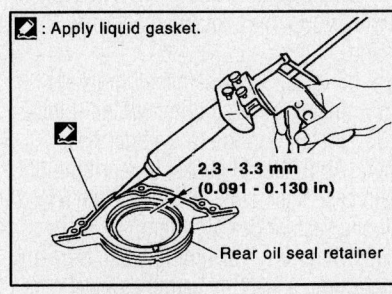

🖉 : Apply liquid gasket.

🖉

2.3 - 3.3 mm
(0.091 - 0.130 in)

Rear oil seal retainer

42356-MURA-G38

Applying sealer to the engine rear main seal

Rear Main Seal

REMOVAL & INSTALLATION

1. Before servicing the vehicle, refer to the Precautions Section.

2. Remove engine from vehicle, and separate front suspension member, transaxle and transfer case (AWD vehicles) assembly from engine.

3. Remove drive plate.

4. Remove upper oil pan.

5. Use a seal cutter (special service tool) to cut away liquid gasket and remove rear oil seal retainer.

To install:

6. Remove old liquid gasket from mating surface of cylinder block and oil pan using scraper.

7. Apply liquid gasket to rear oil seal retainer as shown in the illustration. Use RTV silicone sealant. Assembly should be done within 5 minutes after coating.

8. Install rear oil seal retainer to cylinder block.

9. Perform the remaining steps in the reverse order of removal.

Timing Chain, Sprockets, Front Cover and Seal

REMOVAL & INSTALLATION

1. Before servicing the vehicle, refer to the Precautions Section.

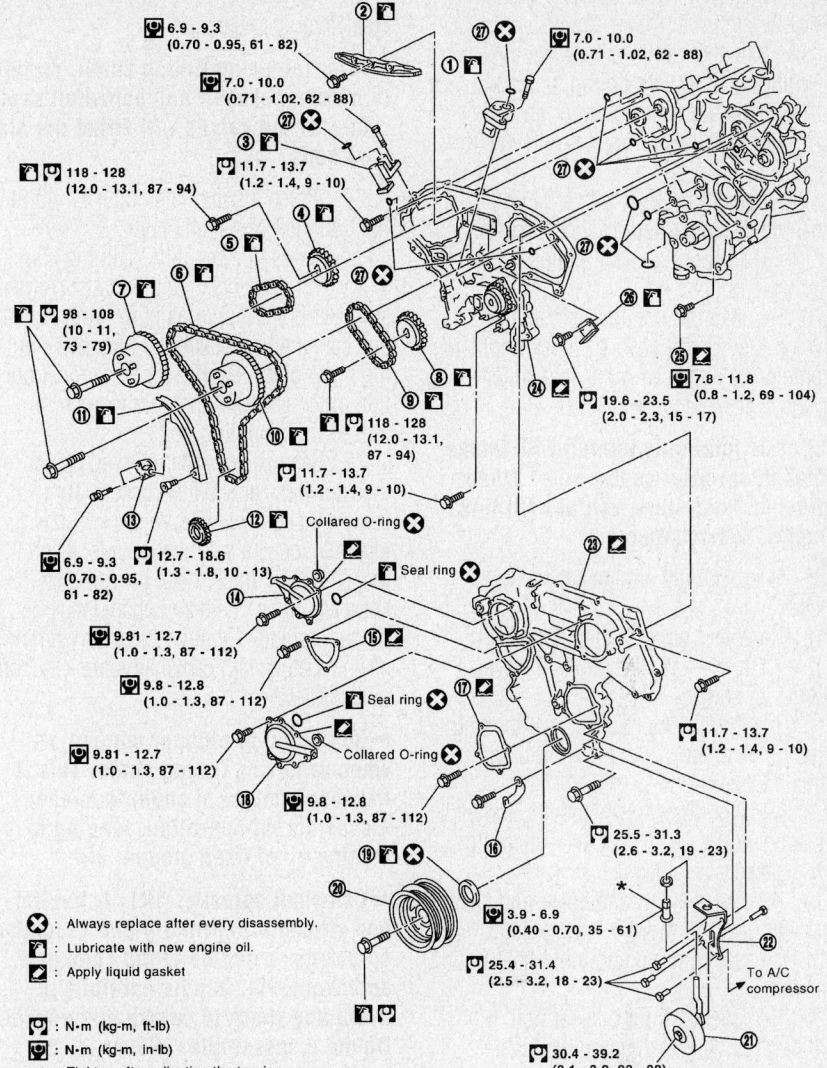

6.9 - 9.3
(0.70 - 0.95, 61 - 82)

7.0 - 10.0
(0.71 - 1.02, 62 - 88)

118 - 128
(12.0 - 13.1, 87 - 94)

98 - 108
(10 - 11, 73 - 79)

7.0 - 10.0
(0.71 - 1.02, 62 - 88)

11.7 - 13.7
(1.2 - 1.4, 9 - 10)

7.8 - 11.8
(0.8 - 1.2, 69 - 104)

19.6 - 23.5
(2.0 - 2.3, 15 - 17)

118 - 128
(12.0 - 13.1, 87 - 94)

11.7 - 13.7
(1.2 - 1.4, 9 - 10)

6.9 - 9.3
(0.70 - 0.95, 61 - 82)

12.7 - 18.6
(1.3 - 1.8, 10 - 13)

Collared O-ring

Seal ring

9.81 - 12.7
(1.0 - 1.3, 87 - 112)

9.8 - 12.8
(1.0 - 1.3, 87 - 112)

Seal ring

11.7 - 13.7
(1.2 - 1.4, 9 - 10)

9.81 - 12.7
(1.0 - 1.3, 87 - 112)

Collared O-ring

9.8 - 12.8
(1.0 - 1.3, 87 - 112)

25.5 - 31.3
(2.6 - 3.2, 19 - 23)

3.9 - 6.9
(0.40 - 0.70, 35 - 61)

25.4 - 31.4
(2.5 - 3.2, 18 - 23)

To A/C
compressor

30.4 - 39.2
(3.1 - 3.9, 23 - 28)

❌ : Always replace after every disassembly.
🛢 : Lubricate with new engine oil.
🖉 : Apply liquid gasket

🔧 : N·m (kg-m, ft-lb)
🔧 : N·m (kg-m, in-lb)
★ : Tighten after adjusting the tension.

1. Timing chain tensioner (secondary)	2. Internal chain guide	3. Timing chain tensioner (secondary)
4. Camshaft sprocket (EXH)	5. Timing chain (secondary)	6. Timing chain (primary)
7. Camshaft sprocket (INT)	8. Camshaft sprocket (EXH)	9. Timing chain (secondary)
10. Camshaft sprocket (INT)	11. Slack guide	12. Crankshaft sprocket
13. Timing chain tensioner (primary)	14. Intake valve timing control cover	15. Chain tensioner cover
16. Water hose clamp	17. Water pump cover	18. Intake valve timing control cover
19. Front oil seal	20. Crankshaft pulley	21. Idler pulley
22. Idler pulley bracket	23. Front timing chain case	24. Rear timing chain case
25. Water drain plug	26. Tension guide	27. O-ring

42356-MURA-G39

Timing chain and related components

2. Remove engine assembly from vehicle, and separate front suspension member, transaxle and transfer case (AWD vehicles) assembly from engine.

3. Drain the engine oil.

4. Remove the engine harnesses.

5. Remove the water hoses.

6. Remove the EVAP canister purge volume control solenoid valve.

7. Remove the drive belts and idler pulley bracket.

8. Remove the power steering oil pump assembly.

9. Remove the alternator.

10. Remove the upper and lower intake manifold collectors.

11. Remove the right and left rocker covers.

12. Remove the crankshaft position sensor (POS).

➡ **Handle carefully to avoid dropping and shocks. Do not disassemble. Do not allow metal powder to adhere to magnetic part at sensor tip. Do not place sensors in a location where they are exposed to magnetism.**

13. Obtain compression TDC of No.1 cylinder as follows. Rotate crankshaft pulley clockwise to align timing mark (grooved line without color) with timing indicator.

14. Remove lower and upper oil pans.

15. Remove the crankshaft pulley as follows:

 a. Lock crankshaft with a hammer handle or similar tool to loosen bolts.

 b. Remove crankshaft pulley with a suitable puller.

16. Remove the right and left intake valve timing control covers. Loosen bolts in reverse order as shown. Use seal cutter to cut liquid gasket for removal.

➡**Shaft is internally jointed with intake camshaft sprocket center hole. When removing, keep it horizontal until it is completely disconnected.**

17. Remove right engine mounting bracket.

18. Remove front timing chain case.

 a. Loosen mounting bolts in reverse order as shown.

 b. Insert the appropriate size tool into the notch at the top of the front timing chain case as shown (1).

 c. Pry off the case by moving the tool as shown (2). Use seal cutter to cut liquid gasket for removal.

19. Remove water pump cover and chain tensioner cover from front timing chain case. Use seal cutter to cut liquid gasket for removal.

20. Remove the front oil seal from the front timing chain case using a suitable tool. Use screwdriver for removal. Exercise care not to damage front timing chain case.

21. Remove internal chain guide, timing chain tensioner, tension guide and slack guide.

22. Remove timing chain tensioner as follows:

 a. Pull lever down and release plunger stopper tab. Plunger stopper tab can be pushed up to release (coaxial structure with lever).

 b. Insert stopper pin into tensioner body hole to hold lever, and keep the tab released.

➡**An Allen wrench [2.5 mm (0.098 in)] is used for a stopper pin as an example.**

 c. Insert plunger into tensioner body by pressing the slack guide.

 d. Keep the slack guide pressed and hold it by pushing the stopper pin through the lever hole and body hole.

 e. Remove the mounting bolts and remove the timing chain tensioner.

23. Remove timing chain (primary) and crankshaft sprocket.

➡**After removing timing chain, do not turn the crankshaft and camshaft separately, or the valves will strike the piston heads.**

24. Attach a suitable stopper pin to the right and left camshaft chain tensioners (for secondary timing chains).

25. Remove intake and exhaust camshaft sprocket bolts. Apply paint to timing chain and camshaft sprockets for alignment during installation. Secure the hexagonal portion of the camshaft using a wrench to loosen the mounting bolts.

26. Remove secondary timing chain together with camshaft sprockets. Turn camshaft slightly to secure slackness of timing chain on chain tensioner side. Insert 0.5 mm (0.020 in) thick metal or resin plate between timing chain and chain tensioner plunger (guide). Remove secondary timing chain together with camshaft sprockets from guide groove.

➡**Be careful of plunger coming off when removing timing chain. This is because plunger of chain tensioner moves during operation, leading to coming off of fixed stopper pin.**

➡**Camshaft sprocket (INT) is two-for-one structure of primary and secondary sprockets. When handling camshaft sprocket (INT), handle carefully to avoid any shock to camshaft sprocket. Do not disassemble. (Never loosen bolts "A" and "B" as shown).**

27. Remove chain tension guide.

28. Remove the water pump.

29. Remove rear timing chain case as follows:

 a. Loosen and remove mounting bolts in reverse order as shown.

 b. Cut the sealant using a seal cutter and remove rear timing chain case.

➡**Do not remove plate metal cover of oil passage. After removing chain case, do not apply any load which affects flatness.**

30. Remove right and left camshaft chain tensioners from cylinder head as follows if necessary.

 a. Remove No.1 camshaft brackets.

 b. Remove secondary chain tensioners with stopper pin attached.

31. Use a scraper to remove all traces of liquid gasket from front and rear timing chain cases, and opposite mating surfaces. Remove old liquid gasket from the bolt hole and thread.

32. Use a scraper to remove all traces of liquid gasket from water pump cover, chain tensioner cover and intake valve timing control covers.

33. Check for cracks and any excessive wear at the roller links of the timing chain. Replace the timing chain as necessary.

To install:

34. Install right and left camshaft chain tensioners to cylinder head as follows if removed.

 a. Install secondary chain tensioners with stopper pin attached and new O-ring.

 b. Install No.1 camshaft brackets.

35. Install O-rings onto cylinder block.

36. Install O-rings to cylinder head.

37. Apply liquid gasket to rear timing chain case back side as shown. Use RTV silicone sealant or equivalent.

38. Align the rear timing chain case and water pump assembly with the dowel pins (right and left) on the cylinder block and install the case. Make sure the O-rings stay in place during installation to cylinder block and cylinder head.

 a. Tighten the mounting bolts in the numerical order as shown. There are two bolt lengths used.

 • Bolt length: Bolt position 20 mm (0.79 inch)

 • 1, 2, 3, 6, 7, 8, 9, 10 16 mm (0.63 in): Except the above 11.7–13.7 Nm (9–10 ft. lbs.)

 Standard Rear timing chain case to cylinder block: -0.24 to 0.14 mm (-0.0094 to 0.0055 inch)

 b. After all bolts are temporarily tightened, retighten them to the specification in the numerical order as shown. If the RTV protrudes, wipe it off immediately.

39. After installing rear timing chain case, check surface height difference between following parts on oil pan mounting surface. If not within standard, repeat above installation procedure.

40. Install chain tension guide.

41. Position the crankshaft so No. 1 piston is set at TDC on the compression stroke. Make sure that the dowel pin hole,

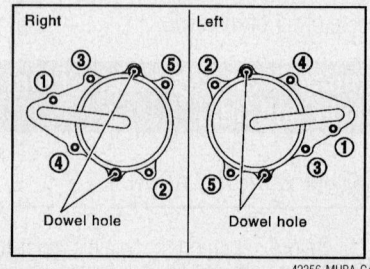

Timing control cover bolt torque sequence

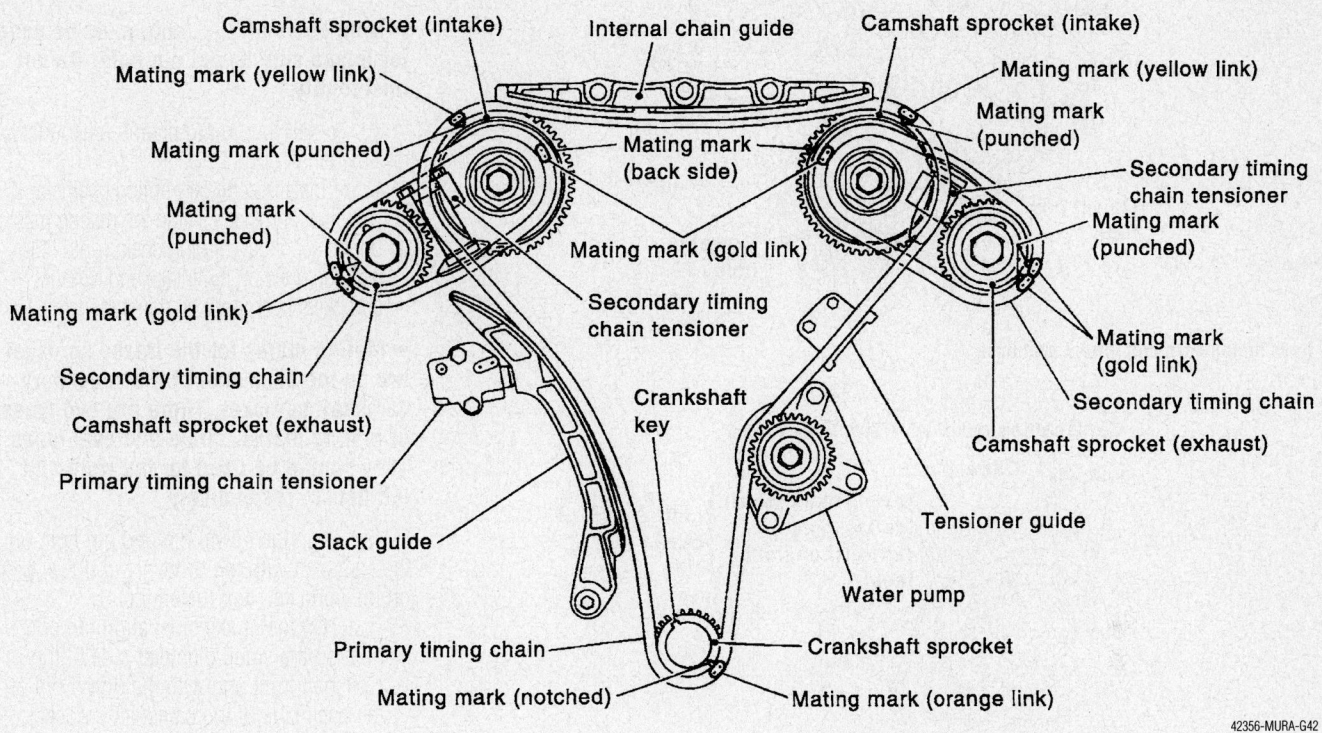

Camshaft sprocket (intake) — Internal chain guide — Camshaft sprocket (intake)

Mating mark (yellow link) — Mating mark (yellow link)

Mating mark (punched) — Mating mark (back side) — Mating mark (punched)

Mating mark (punched) — Secondary timing chain tensioner

Mating mark (gold link) — Mating mark (punched)

Mating mark (gold link) — Secondary timing chain tensioner — Mating mark (gold link)

Secondary timing chain — Crankshaft key — Secondary timing chain

Camshaft sprocket (exhaust) — Camshaft sprocket (exhaust)

Primary timing chain tensioner — Tensioner guide

Slack guide — Water pump

Primary timing chain — Crankshaft sprocket

Mating mark (notched) — Mating mark (orange link)

42356-MURA-G42

Timing chain mating marks

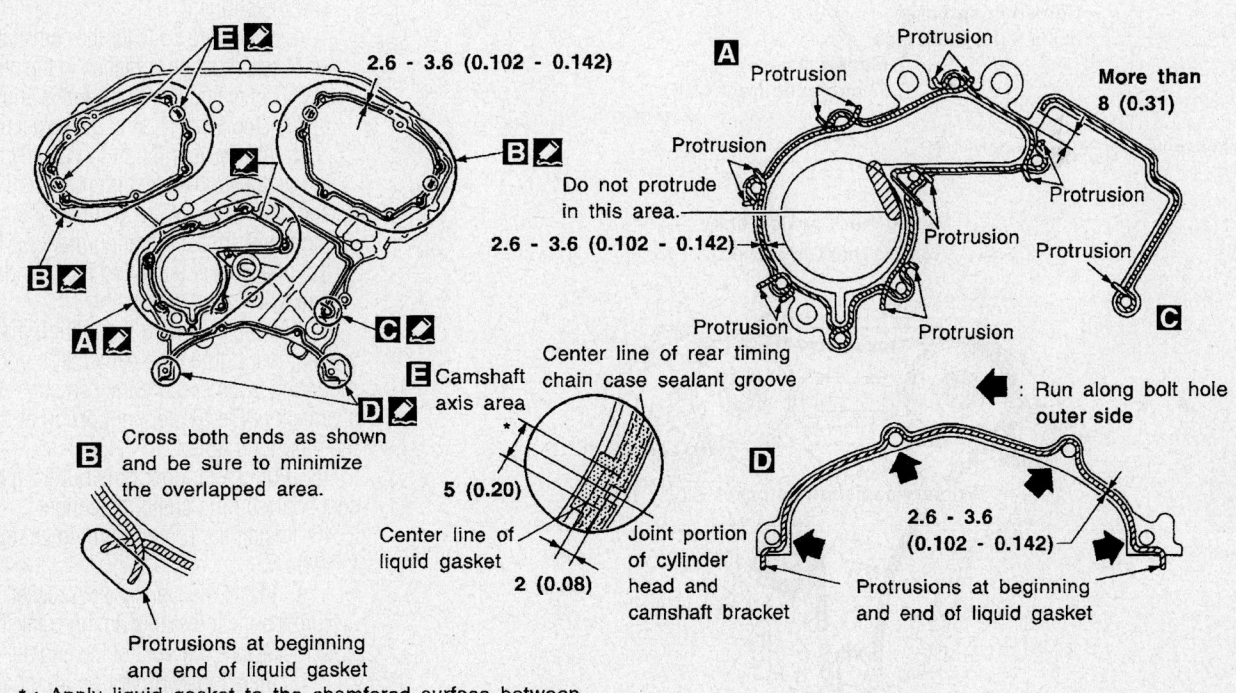

2.6 - 3.6 (0.102 - 0.142)

More than 8 (0.31)

Protrusion

Do not protrude in this area.

2.6 - 3.6 (0.102 - 0.142)

E Camshaft axis area

Center line of rear timing chain case sealant groove

Cross both ends as shown and be sure to minimize the overlapped area.

5 (0.20)

Center line of liquid gasket

2 (0.08)

Joint portion of cylinder head and camshaft bracket

: Run along bolt hole outer side

2.6 - 3.6 (0.102 - 0.142)

Protrusions at beginning and end of liquid gasket

Protrusions at beginning and end of liquid gasket

* : Apply liquid gasket to the chamfered surface between camshaft bracket and cylinder head.

: Apply liquid gasket.

Unit: mm (in)

42356-MURA-G43

Rear timing case sealer application

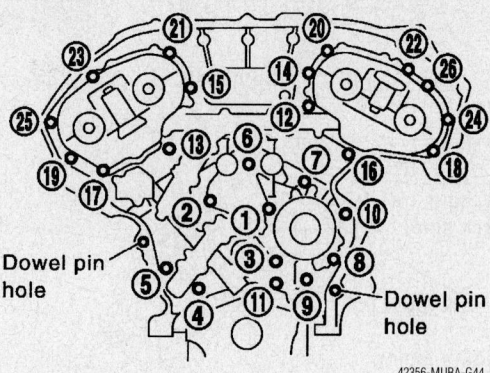

Dowel pin hole

Dowel pin hole

42356-MURA-G44

Rear timing case bolt torque sequence

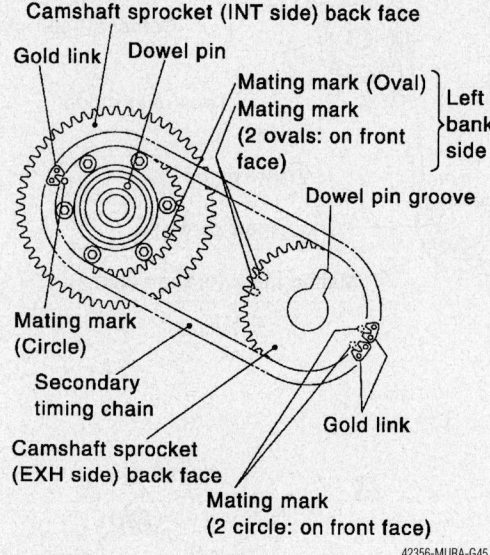

Camshaft sprocket (INT side) back face

Gold link   Dowel pin

Mating mark (Oval)   Left bank side
Mating mark (2 ovals: on front face)

Dowel pin groove

Mating mark (Circle)

Secondary timing chain

Gold link

Camshaft sprocket (EXH side) back face

Mating mark (2 circle: on front face)

42356-MURA-G45

Secondary timing chain alignment

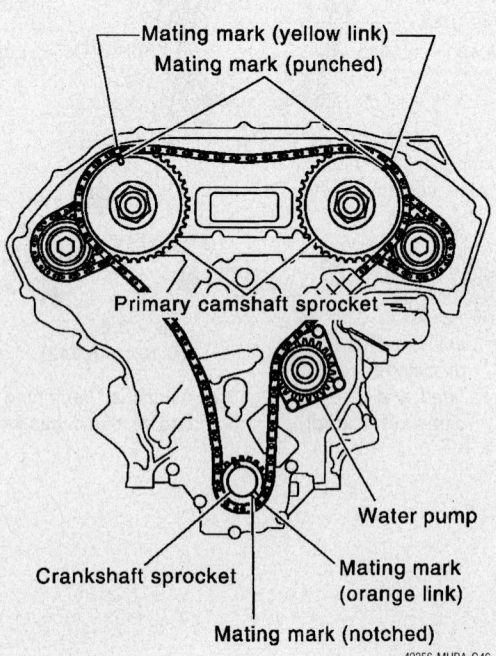

Mating mark (yellow link)
Mating mark (punched)

Primary camshaft sprocket

Water pump

Crankshaft sprocket

Mating mark (orange link)

Mating mark (notched)

42356-MURA-G46

Primary timing chain alignment

dowel pin and crankshaft key are located as shown.

➡ **Hole on small dia. side must be used for intake side dowel pin hole. Do not misidentify.**

42. Install the timing chains (secondary) and camshaft sprockets.

a. Install secondary timing chains and camshaft sprockets. Align the mating marks on the secondary timing chain (gold link) with the ones on the intake and exhaust sprockets (stamped), and install them.

➡ **Mating marks for the intake sprocket are on the back side of the secondary camshaft sprocket. There are two types of mating marks, circle and oval types. They should be used for the right and left banks, respectively.**

43. Align the dowel pin and pin hole on the camshaft with the groove and dowel pin on the sprocket, and install them.

a. On the intake side, align the pin hole on the small diameter side of the camshaft front end with the dowel pin on the back side of the camshaft sprocket, and install them.

b. On the exhaust side, align the dowel pin on the camshaft front end with the pin groove on the camshaft sprocket, and install them.

c. Mounting bolts for the camshaft sprockets must be tightened in the next step. Tightening them by hand is enough to prevent the dislocation of the dowel pins.

d. It may be difficult to visually check the dislocation of mating marks during and after installation. To make the matching easier, make a mating mark on the top of sprocket teeth and its extended line in advance with paint.

44. After confirming the mating marks are aligned, tighten the camshaft sprocket mounting bolts. Secure the camshaft using a wrench at the hexagonal portion to tighten the mounting bolts.

45. Pull the stopper pins out from the secondary timing chain tensioners.

46. Install the primary timing chain as follows:

a. Install the crankshaft sprocket. Make sure the mating marks on the crankshaft sprocket face the front of the engine.

b. Install the primary timing chain. Install primary timing chain so the mating mark (punched) on camshaft sprocket is aligned with the yellow link on the timing chain, while the mating mark (notched) on the crankshaft sprocket is aligned with the orange one on the timing chain, as shown.

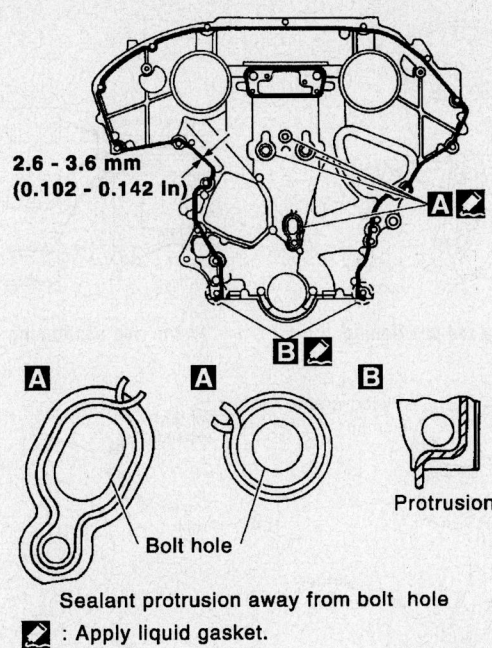

2.6 - 3.6 mm
(0.102 - 0.142 In)

Protrusion

Bolt hole

Sealant protrusion away from bolt hole

: Apply liquid gasket.

42356-MURA-G47

Sealer application on front case

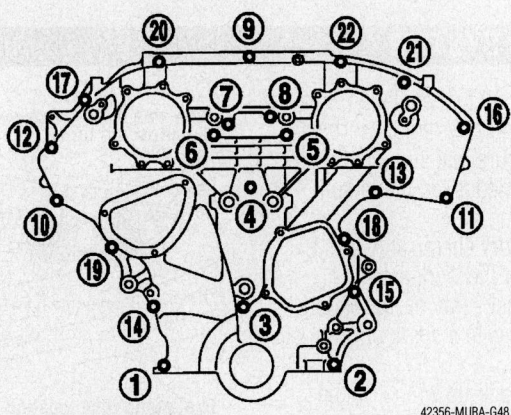

42356-MURA-G48

Front timing case bolt torque sequence

c. When it is difficult to align mating marks of the primary timing chain with each sprocket, gradually turn the camshaft using a wrench on the hexagonal portion to align it with the mating marks.

d. During alignment, be careful to prevent dislocation of mating mark alignments of the secondary timing chains.

47. Install the internal chain guide and tension guide.

48. Install slack guide. Do not over tighten the slack guide mounting bolts. It is normal for a gap to exist under the bolt seats when the mounting bolts are tightened to specification.

49. Install chain tensioner for slack guide. When installing the chain tensioner,
push in the sleeve and keep it pressed in with the stopper pin. Remove any dirt and foreign materials completely from the back and the mounting surfaces of the chain tensioner. After installation, pull out the stopper pin by pressing the slack guide.

50. Reconfirm that the mating marks on the sprockets and the timing chain have not slipped out of alignment.

51. Install new O-rings on the rear timing chain case.

52. Install the front oil seal on the front timing chain case. Apply new engine oil to the oil seal edges. Install it so that each seal lip is oriented as shown in illustration. Using a suitable drift, press-fit oil seal until it becomes flush with timing chain case end face. Make sure the garter spring is in position and seal lip is not inverted.

53. Install the water pump cover and the chain tensioner cover to front cover. Apply RTV silicone sealant.

54. Install front timing chain case as follows:

a. Apply liquid gasket to front timing chain case back side as shown.

b. Install dowel pin on the rear timing chain case into dowel pin hole on front timing chain case.

c. Tighten bolts to the specified torque in order shown in the illustration.
- 8 mm (0.31 in) dia. bolts 1, 2: 25.5–31.3 Nm (19–23 ft. lbs.)
- 6 mm (0.24 in) dia. bolts Except the above: 11.7–13.7 Nm (9–10 ft. lbs.)

d. After tightening, retighten them to specified torque in numerical order shown in illustration.

e. After tightening, retighten them to specified torque in numerical order shown in illustration.

55. After installing the front timing chain case, check the surface height difference between the following parts on the oil pan mounting surface. If not within specification, repeat the installation procedure.

56. Install right and left intake valve timing control covers as follows:

a. Install seal rings in shaft grooves.

b. Apply liquid gasket to the intake valve timing control covers. Use RTV Silicone Sealant.

c. Install collared O-ring in front cover oil hole (left and right sides).

d. Being careful not to move the seal ring from the installation groove, align the dowel pins on the chain case with the holes to install the intake valve timing control covers.

e. Tighten bolts in the numerical order as shown.

57. Install right and left rocker covers.

58. Install crankshaft pulley as follows:

a. Fix crankshaft using a hammer shaft or an equivalent tool.

b. Install crankshaft pulley, taking care not to damage front oil seal. When press-fitting crankshaft pulley with a plastic hammer, tap on its center portion (not circumference).

c. Tighten bolt to 39.2 to 49.0 Nm (29 to 36 ft. lbs.).

d. Put a paint mark on crankshaft pulley aligning with angle mark on crankshaft pulley bolt. Then, further retighten bolt by 60 to 65 degrees.

59. Rotate crankshaft pulley in normal direction (clockwise when viewed from front) to confirm it turns smoothly.

60. For the following operations, perform steps in the reverse order of removal.

➡If hydraulic pressure inside chain tensioner drops after removal/installation, slack in the guide may generate a pounding noise during and just after engine start. However, this is not unusual. Noise will stop after hydraulic pressure rises.

Piston and Ring

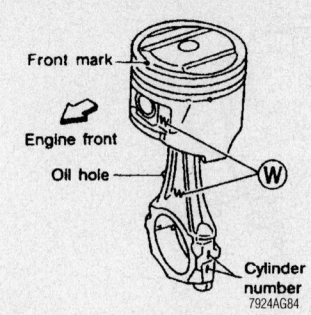

Piston and connecting rod positioning

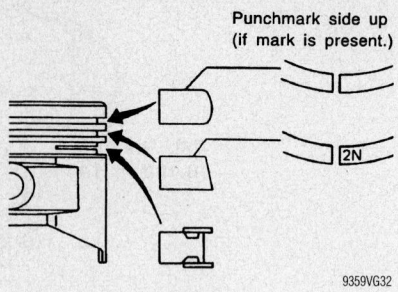

Piston ring positioning

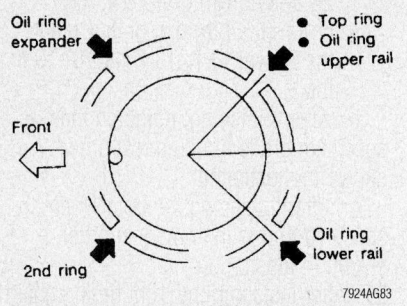

Piston ring end-gap spacing

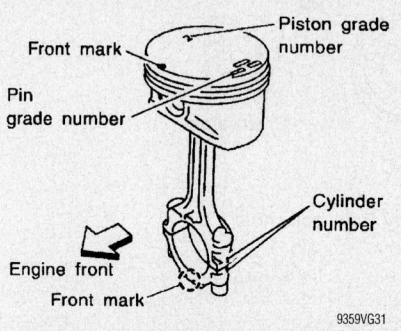

Piston and connecting rod positioning

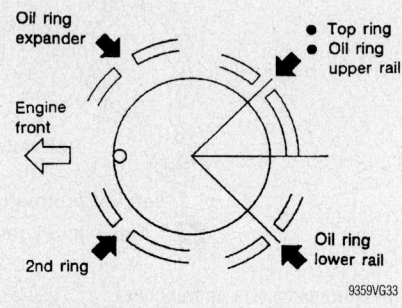

Piston ring positioning

FUEL SYSTEM

POSITIONING

Fuel System Service Precautions

Safety is the most important factor when performing not only fuel system maintenance but any type of maintenance. Failure to conduct maintenance and repairs in a safe manner may result in serious personal injury or death. Maintenance and testing of the vehicle's fuel system components can be accomplished safely and effectively by adhering to the following rules and guidelines.

• To avoid the possibility of fire and personal injury, always disconnect the negative battery cable unless the repair or test procedure requires that battery voltage be applied.

• Always relieve the fuel system pressure prior to disconnecting any fuel system component (injector, fuel rail, pressure regulator, etc.), fitting or fuel line connection. Exercise extreme caution whenever relieving fuel system pressure, to avoid exposing skin, face and eyes to fuel spray. Please be advised that fuel under pressure may penetrate the skin or any part of the body that it contacts.

• Always place a shop towel or cloth around the fitting or connection prior to loosening to absorb any excess fuel due to spillage. Ensure that all fuel spillage (should it occur) is quickly removed from engine surfaces. Ensure that all fuel soaked cloths or towels are deposited into a suitable waste container.

• Always keep a dry chemical (Class B) fire extinguisher near the work area.

• Do not allow fuel spray or fuel vapors to come into contact with a spark or open flame.

• Always use a back-up wrench when loosening and tightening fuel line connection fittings. This will prevent unnecessary stress and torsion to fuel line piping. Always follow the proper torque specifications.

• Always replace worn fuel fitting O-rings with new. Do not substitute fuel hose or equivalent, where fuel pipe is installed.

Fuel System Pressure

RELIEVING

1. Before servicing the vehicle, refer to the Precautions Section.
2. Remove the fuel pump fuse from the panel.
3. Start the engine and allow it to run until it stalls. Crank the engine for a few seconds to relieve additional fuel pressure.
4. Turn the ignition switch OFF.
5. Disconnect the negative battery cable.

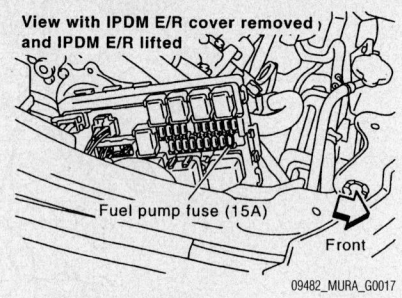

Fuel pump fuse location

6. When repairs are complete, replace the fuel pump fuse and connect the negative battery cable.

Fuel Pump

REMOVAL & INSTALLATION

1. Before servicing the vehicle, refer to the Precautions Section.
2. Relieve the fuel system pressure.
3. Open the fuel filler lid.
4. Open the filler cap and release the pressure inside the fuel tank.
5. Remove or disconnect the following:
   • Rear seat cushion trim and pad bolts, then lift up rear seat cushion.
   • Inspection hole cover for main and sub fuel level sensor unit by turning clips counterclockwise by 90°.

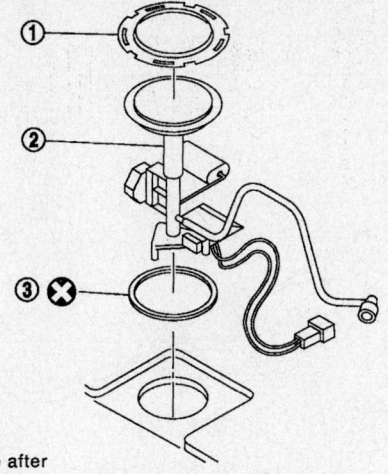

Left side

⊗ : Always replace after every disassembly.

1. Retainer
2. Sub fuel level sensor unit
3. O-ring
4. Main fuel level sensor unit, fuel filter and fuel pump assembly

42356-MURA-G49

Fuel pump and related components

- Harness connector and quick connectors for EVAP/Vent line hose and fuel feed tube. Disconnect EVAP/Vent line hose connector (push in tubs and pull out).
- Remove the retainer for main and sub fuel level sensor unit with fuel tank lock ring wrench (SST) by turning counterclockwise.
- Remove main fuel level sensor unit, fuel filter and fuel pump assembly, and sub fuel level sensor unit. Raise the main fuel level sensor unit, fuel filter and fuel pump assembly, and disconnect the fuel hose connector (push in tabs and pull out) and sub fuel level sensor unit harness connector. Raise and release the sub fuel level sensor unit to remove.

To install:

6. Installation is the reverse of removal. Note the following:

- Connect fuel hose connector (push in until it stops) and sub fuel level sensor unit harness connector.
- Align the direction mark on main and sub fuel level sensor unit with that on fuel tank as shown in the illustration.
- Install the inspection hole cover with front mark (arrow) facing front of the vehicle (Both for right and left). Lock the clips by turning clockwise.

7. Connect the quick connector as follows.

- Check the connection for damage or any foreign materials.
- Align the connector with the tube, then insert the connector straight into the tube until a click is heard.
- After connecting, make sure that the connection is secure by following method.
- Pull the tube and the connector to make sure they are securely connected. Visually confirm that the two retainer tabs are connected to the connector.

8. Turn ignition switch ON (with engine stopped), then check connections for leaks by applying fuel pressure to fuel piping.

9. Start the engine and let it idle and make sure there are no fuel leaks at the fuel system connections.

Fuel Injectors

REMOVAL & INSTALLATION

1. Before servicing the vehicle, refer to the Precautions Section.
2. Remove the engine cover.
3. Properly release the fuel pressure.
4. Remove the right windshield wiper arm and the right cowl top cover.
5. Remove radiator cover grille, air duct (inlet), air cleaner case, air duct assembly and mass air flow sensor.

6. On 2003 vehicles, disconnect electric throttle control actuator and engine coolant hoses. Disconnect vacuum hose, all fuel injector electrical connectors, and PCV hose. Remove the vacuum tank from intake manifold collector (lower). Disconnect the power steering hose bracket.

7. Remove the intake manifold upper and lower collectors.

➡**The intake manifold collector (upper) should be moved aside with water hoses connected.**

8. Remove fuel feed hose (with damper) from fuel tube.

9. Disconnect fuel feed hose (with damper) quick connector at vehicle piping side. When separating fuel feed hose and centralized under-floor piping connection, disconnect quick connector with the following procedure.

   a. Remove quick connector cap from quick connector.
   b. Disconnect quick connector from centralized under-floor piping.

10. Remove harness connector from fuel injector.

11. Loosen the mounting bolts in the same sequence as the installation sequence. Remove fuel tube and fuel injector assembly.

12. Remove fuel injector from fuel tube with following procedure.

   a. Open and remove clip.
   b. Remove fuel injector from the fuel tube by pulling straight.

13. Remove fuel damper from fuel tube.

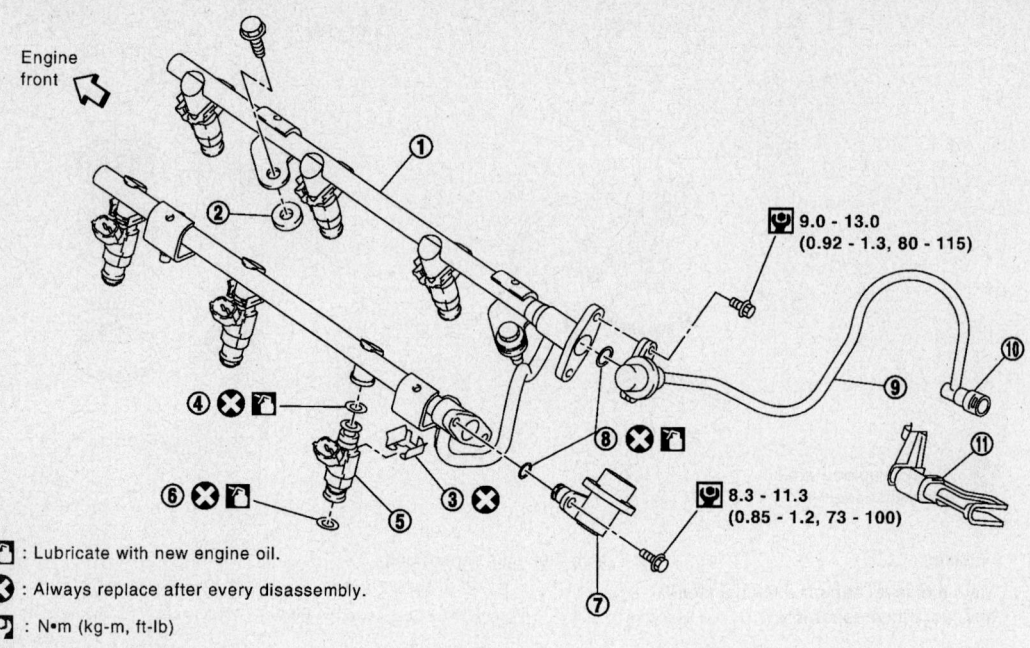

Engine front

9.0 - 13.0
(0.92 - 1.3, 80 - 115)

8.3 - 11.3
(0.85 - 1.2, 73 - 100)

🔧 : Lubricate with new engine oil.

❌ : Always replace after every disassembly.

🔧 : N•m (kg-m, ft-lb)

🔧 : N•m (kg-m, in-lb)

1. Fuel tube	2. Insulator	3. Clip
4. O-ring (black)	5. Fuel injector	6. O-ring (green)
7. Fuel damper	8. O-ring	9. Fuel feed hose (with damper)
10. Quick connector	11. Quick connector cap	

42356-MURA-G50

Fuel injector rail and related components

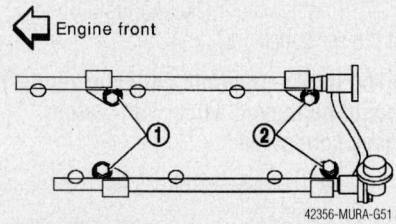

Engine front

42356-MURA-G51

**Fuel rail bolt tightening sequence—
2003–2004**

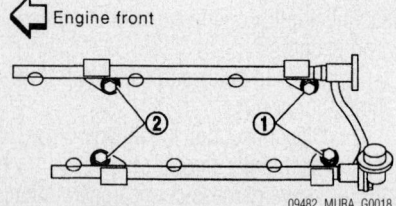

Engine front

09482_MURA_G0018

**Fuel rail bolt tightening sequence—
2005–2006**

To install:

14. Install fuel damper. Insert fuel damper straight into fuel tube. Tighten mounting bolts evenly in turn. After tightening mounting bolts, make sure that there is no gap between flange and fuel tube. When handling O-rings, be careful of the following:

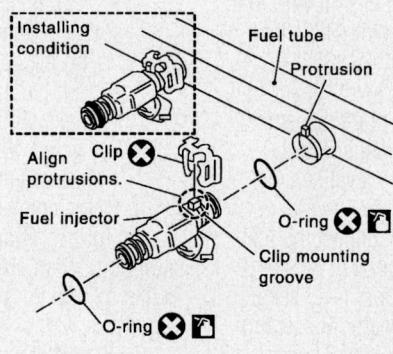

Installing condition

Fuel tube

Protrusion

Align protrusions.

Clip ❌

Fuel injector

Clip mounting groove

O-ring ❌ 🔧

O-ring ❌ 🔧

❌ : Always replace after every disassembly.

🔧 : Lubricate with new engine oil.

42356-MURA-G52

Fuel injector and related components

- Handle O-ring with bare hands. Never wear gloves.
- Lubricate O-ring with new engine oil.
- Do not clean O-ring with solvent.
- Make sure that O-ring and its mating part are free of foreign material.
- When installing O-ring, be careful not to scratch it with tool or fingernails. Also be careful not to twist or stretch O-ring. If O-ring was stretched while it was being attached, do not insert it quickly into fuel tube.

15. Install O-rings on the fuel injector. Upper and lower O-ring are different.

16. Install fuel injector to fuel tube with the following procedure.

a. Insert clip into clip mounting groove on fuel injector. Insert clip so that lug "A" of fuel injector matches notch "A" of the clip.

➡**Do not reuse clip. Replace it with a new one. Be careful to keep clip from interfering with O-ring. If interference occurs, replace O-ring.**

b. Insert fuel injector into fuel tube with clip attached. Insert it while matching it to the axial center. Insert fuel injector so that lug "B" of fuel tube matches notch "B" of the clip. Make sure that fuel tube flange is securely fixed in flange groove on clip.

c. Make sure that installation is complete by checking that fuel injector does not rotate or come off.

17. Tighten mounting bolts in two steps (first to 7 ft. lbs and then to 17 ft. lbs.) in the proper sequence.

18. Connect fuel injector harness.

19. Install intake manifold collector (upper and lower).

20. Connect quick connector between fuel feed hose (with damper) and centralized under-floor piping connection with the following procedure:

a. Check the connection for damage and foreign materials.

b. Align the quick connector with the tube, then insert the connector straight into the tube until a click is heard.

c. After connecting the quick connector, use the following method to make sure it is full connected. Visually confirm that the two retainer tabs are connected to the connector. Pull the tube and the connector to make sure they are securely connected.

d. Install quick connector cap to quick connector connection. Install quick connector cap with arrow on surface facing in direction of quick connector.

➡**If cap cannot be installed smoothly, quick connector may have not been installed correctly. Check connection again.**

21. The remainder of installation is the reverse of removal.

22. Turn ignition switch ON (with engine stopped). With fuel pressure applied to fuel piping, check for fuel leakage at connection points.

23. Start engine. With engine speed increased, check again for fuel leakage at connection points.

DRIVE TRAIN

Transaxle

REMOVAL & INSTALLATION

The engine and transaxle are removed as an assembly from the vehicle.

1. Before servicing the vehicle, refer to the Precautions Section.

2. Disconnect the negative battery cable. Remove the engine under cover.

3. Remove the air guide. Remove the exhaust front tube.

4. Remove dust cover from converter housing part.

5. Turn crankshaft clockwise and remove the four tightening nuts for drive plate and torque converter.

6. Remove the four lower transaxle

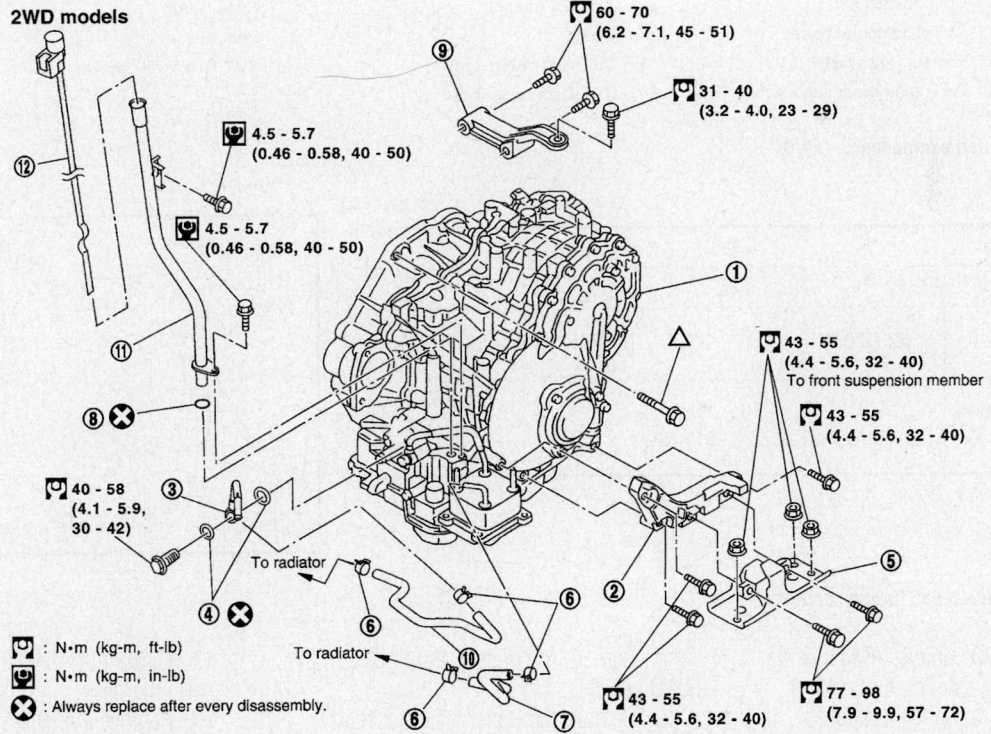

2WD models

60 - 70 (6.2 - 7.1, 45 - 51)
31 - 40 (3.2 - 4.0, 23 - 29)
4.5 - 5.7 (0.46 - 0.58, 40 - 50)
4.5 - 5.7 (0.46 - 0.58, 40 - 50)
43 - 55 (4.4 - 5.6, 32 - 40) To front suspension member
43 - 55 (4.4 - 5.6, 32 - 40)
40 - 58 (4.1 - 5.9, 30 - 42)
To radiator
To radiator
43 - 55 (4.4 - 5.6, 32 - 40)
77 - 98 (7.9 - 9.9, 57 - 72)

: N•m (kg-m, ft-lb)
: N•m (kg-m, in-lb)
⊗ : Always replace after every disassembly.

1.	Transaxle assembly	2.	LH engine mounting bracket	3.	Fluid cooler tube
4.	Copper washer	5.	LH engine mounting insulator	6.	Hose clamp
7.	CVT fluid cooler hose	8.	O-ring	9.	Rear gusset
10.	CVT fluid cooler hose	11.	CVT fluid charging pipe	12.	CVT fluid level gauge

42356-MURA-G53

Transaxle and related components—2WD

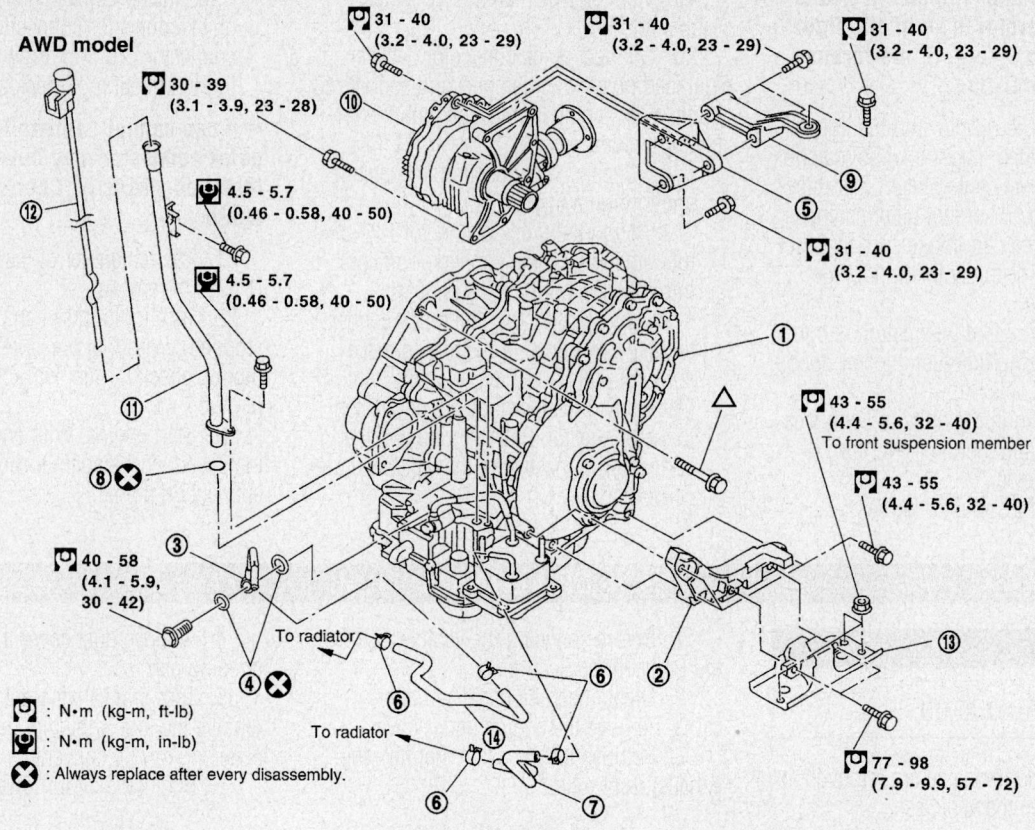

AWD model

- 31 - 40 (3.2 - 4.0, 23 - 29)
- 31 - 40 (3.2 - 4.0, 23 - 29)
- 31 - 40 (3.2 - 4.0, 23 - 29)
- 30 - 39 (3.1 - 3.9, 23 - 28)
- 4.5 - 5.7 (0.46 - 0.58, 40 - 50)
- 4.5 - 5.7 (0.46 - 0.58, 40 - 50)
- 31 - 40 (3.2 - 4.0, 23 - 29)
- 43 - 55 (4.4 - 5.6, 32 - 40) To front suspension member
- 43 - 55 (4.4 - 5.6, 32 - 40)
- 40 - 58 (4.1 - 5.9, 30 - 42)
- 77 - 98 (7.9 - 9.9, 57 - 72)

To radiator
To radiator

⬛ : N•m (kg-m, ft-lb)
⬛ : N•m (kg-m, in-lb)
⊗ : Always replace after every disassembly.

1. Transaxle assembly
2. LH engine mounting bracket
3. Fluid cooler tube
4. Copper washer
5. Transfer gusset
6. Hose clamp
7. CVT fluid cooler hose
8. O-ring
9. Rear gusset
10. Transfer assembly
11. CVT fluid charging pipe
12. CVT fluid level gauge
13. LH engine mounting insulator
14. CVT fluid cooler hose

42356-MURA-G54

Transaxle and related components—AWD

Bolt No.	1	2	3	4
Number of bolts	1	2	2	4
Bolt length "ℓ" mm (in)	52 (2.05)	36 (1.42)	105 (4.13)	35 (1.38)
Tightening torque N-m (kg-m, ft-lb)	70 - 79 (7.1 - 8.0, 51 - 58)			42-52 (4.3 - 5.3, 31 - 38)

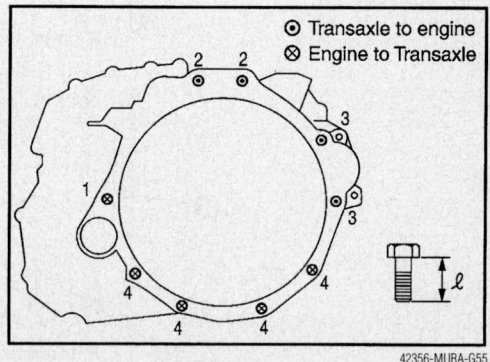

⊙ Transaxle to engine
⊗ Engine to Transaxle

42356-MURA-G55

Transaxle bolt location and torque sequence

retaining bolts, 2WD vehicles. Remove the six lower transaxle retaining bolts, AWD vehicles. See illustration for location.

7. Remove the engine and transaxle from the vehicle, as an assembly.

8. Remove the halfshaft.

➡ Be sure to replace the new differential side oil seal every removal of halfshaft.

9. Remove the transfer case gusset. (AWD vehicles)

10. Remove the transfer case assembly.

➡ Be sure to replace the new differential side oil seal (converter housing side only) whenever the transfer case is removed.

11. Remove the filler pipe. Discard the O-ring.

12. Disconnect the harness connector and wire harness.

13. Remove the POS sensor, from engine assembly.

14. Remove the starter.

15. Remove CVT fluid cooler valve assembly. (With CVT fluid cooler tube assembly and heater hose).

16. Install the slinger to transaxle assembly.

: Bolt (4)

09482_MURA_G0019

Transaxle bolt removal location—2WD

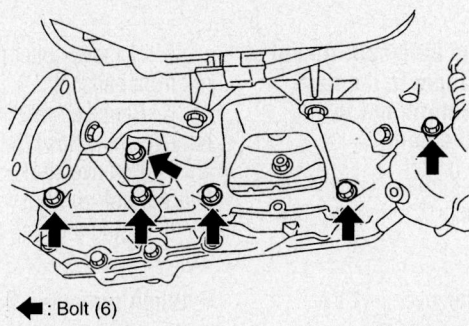

: Bolt (6)

09482_MURA_G0020

Transaxle bolt removal location—AWD

17. Remove the rear gusset.

18. Remove the left engine mounting bracket and the left engine mounting insulator.

19. Remove the front suspension member from the engine/transaxle assembly.

20. Remove the transaxle assembly bolts.

21. Separate the transaxle from the engine with a hoist. Secure torque converter to prevent it from dropping.

➡️ **After installing a torque converter to a transaxle, be sure to check dimension A to ensure it is within the reference value limit. Specification should be 0.55 inch or more.**

To install:

22. Installation is the reverse of the removal. Note the following:

- Screw and set the locator into the stud bolts for the torque converter locate
- Rotate the torque converter to allow the locator to go down
- Rotate the drive plate so that the hole of the drive plate locator faces down
- Installing transaxle assembly from engine assembly with a hoist
- When installing fluid cooler tube to transaxle assembly, transaxle assembly the part with the tube aligned with the rib
- When installing CVT fluid cooler

valve assembly to the engine, torque the bolts to: 28–32 Nm (21–23 ft. lbs.)

- Align the positions of tightening nuts for drive plate with those of the torque converter, and temporar-

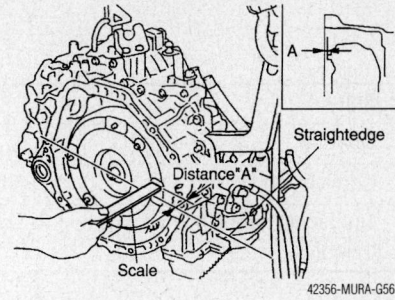

42356-MURA-G56

Torque converter installation measurement

ily tighten the nuts. Then, tighten the nuts to the specified torque.

- Install POS sensor
- After completing installation, check for fluid leakage, fluid level, and the positions of CVT
- When replacing the CVT assembly, erase EEP ROM in TCM

Transfer Case Assembly

REMOVAL & INSTALLATION

The engine and transaxle assembly must first be removed from the vehicle before the transfer case can be removed.

1. Before servicing the vehicle, refer to the Precautions Section.

2. Remove the engine and transaxle assembly, from the vehicle.

3. Remove gusset mounting bolts, and

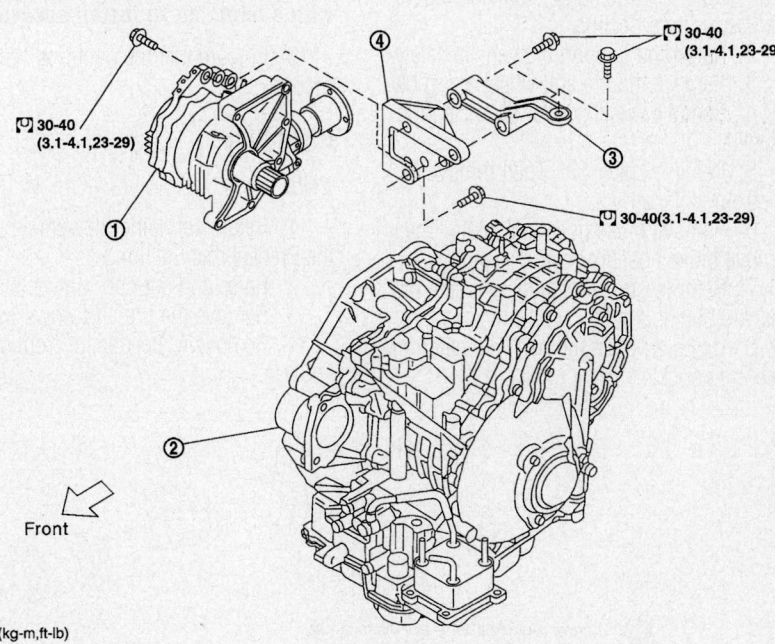

30-40
(3.1-4.1,23-29)

30-40
(3.1-4.1,23-29)

30-40(3.1-4.1,23-29)

Front

: N·m(kg-m,ft-lb)

1. Transfer assembly
4. Transfer gusset
2. Transaxle assembly
3. Rear gusset

42356-MURA-G57

Transfer case and related components

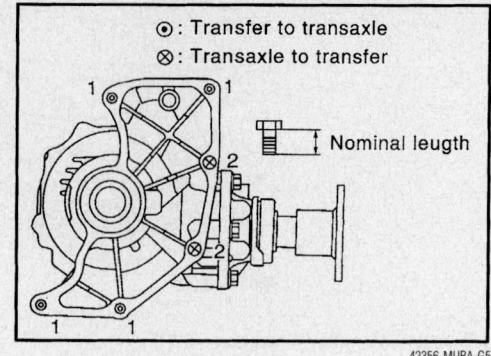

Bolt No.	1	2
Quantity	4	2
Nominal length mm (in)	65 (2.56)	40 (1.57)
Tightening torque [N·m (kg·m, ft.-lb.)]	29.4 - 39.2 (3.0 - 3.9, 22 - 28)	

⊙ : Transfer to transaxle
⊗ : Transaxle to transfer

Nominal leugth

42356-MURA-G58

Transfer case bolt location and torque sequence

then remove gusset from engine and transaxle.

4. Remove transfer case mounting bolts and separate transfer case from transaxle.

➡**After removing transfer case from transaxle, be sure to replace differential side oil seal of the transaxle side with new one.**

To install:

5. Installation is the reverse of the removal procedure.

Front Halfshaft

REMOVAL & INSTALLATION

Left Side

1. Before servicing the vehicle, refer to the Precautions Section.
2. Raise and support the vehicle safely.
3. Remove the tire and wheel assembly.
4. Remove wheel sensor from steering knuckle.
5. Remove cotter pin. Then remove lock nut from halfshaft.
6. Remove brake hose lock plate. Then remove brake hose from strut assembly.
7. Remove strut assembly and steering knuckle bolt and nut.
8. Using a puller, remove halfshaft from steering knuckle.

➡**When removing the halfshaft, do not apply an excessive angle to the halfshaft joint. Also be careful not to excessively extend the joint.**

9. Remove halfshaft from the transaxle, using special tool KV40107500, or equivalent.

To install:

10. Installation is the reverse of the removal procedure.
11. Be sure to replace the differential side oil seal.
12. Be sure to replace all non reusable components with new ones.

➡**In order to prevent damage to the transaxle side oil seal, first install an oil seal protector tool onto the oil seal before inserting the halfshaft. Slide the halfshaft into the slide joint and tap with a hammer, to install securely.**

13. Be sure that the circlip is secured properly.

Right Side

2WD

1. Before servicing the vehicle, refer to the Precautions Section.
2. Raise and support the vehicle safely.
3. Remove the tire and wheel assembly.
4. Remove wheel sensor from steering knuckle.

5. Remove cotter pin. Then remove lock nut from halfshaft.
6. Remove brake hose lock plate. Then remove brake hose from strut assembly.
7. Remove strut assembly and steering knuckle bolt and nut.
8. Using a puller, remove halfshaft from axle.

➡**When removing the halfshaft, do not apply an excessive angle to the halfshaft joint. Also be careful not to excessively extend the joint.**

9. Remove support bearing bolts, and pull halfshaft from transaxle. Pry off halfshaft from transaxle.

To install:

10. Installation is the reverse of the removal procedure.
11. Be sure to replace the differential side oil seal.
12. Be sure to replace all non reusable components with new ones.

AWD

1. Before servicing the vehicle, refer to the Precautions Section.
2. Raise and support the vehicle safely.
3. Remove the tire and wheel assembly.
4. Remove wheel sensor from steering knuckle.

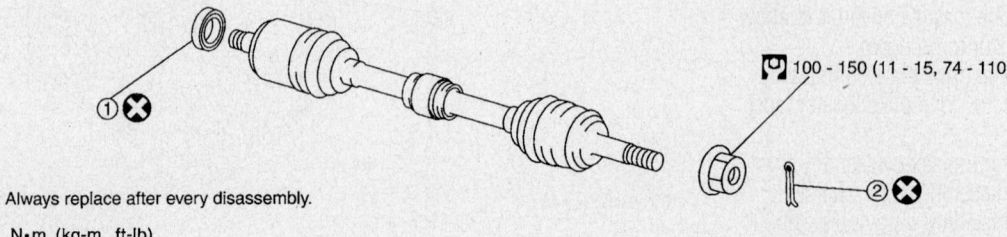

100 - 150 (11 - 15, 74 - 110)

⊗ : Always replace after every disassembly.

🔧 : N·m (kg-m, ft-lb)

1.　Dust shield
2.　Cotter pin

42356-MURA-G59

Front halfshaft—left side

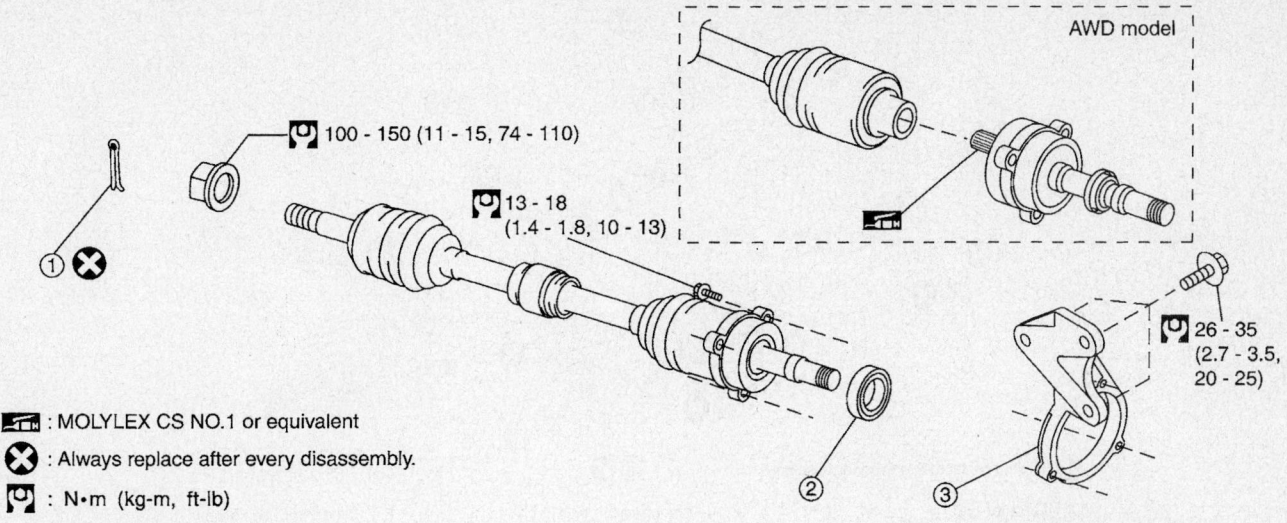

: MOLYLEX CS NO.1 or equivalent

: Always replace after every disassembly.

: N•m (kg-m, ft-lb)

1. Cotter pin

2. Dust shield

3. Support bearing bracket

42356-MURA-G60

Front halfshaft—right side

5. Remove brake hose lock plate. Then remove brake hose from strut assembly.
6. Remove cotter pin. Then remove lock nut from halfshaft.
7. Remove strut assembly and steering knuckle bolt and nut.
8. Using a puller, remove halfshaft from the axle.

➡**When removing the halfshaft, do not apply an excessive angle to the half-shaft joint. Also be careful not to excessively extend the joint.**

9. Remove halfshaft from the transaxle, using special tool KV40107500, or equivalent.
    To install:
10. Installation is the reverse of the removal procedure.
11. Be sure to replace the differential side oil seal.
12. Be sure to replace all non reusable components with new ones.

Rear Halfshaft

REMOVAL & INSTALLATION

1. Before servicing the vehicle, refer to the Precautions Section.
2. Raise and support the vehicle safely.
3. Remove the tire and wheel assembly.
4. Remove wheel sensor from axle. Do not pull on the wheel sensor harness.
5. Remove cotter pin. Then remove lock nut from halfshaft.
6. Remove parking cable and parking brake shoe from back plate.
7. Remove wheel hub and bearing assembly bolts, then remove wheel hub and bearing assembly from axle.
8. Use a wheel wrench or other tool to remove halfshaft from final drive.
    To install:
9. Installation is the reverse of the removal procedure.

10. Be sure to replace all non reusable components with new ones.

➡**In order to prevent damage to the final drive oil seal, first install an oil seal protector tool onto the oil seal before inserting the halfshaft. Slide the halfshaft into the slide joint and tap with a hammer, to install securely.**

11. Be sure that the circlip is secured properly.

Front Halfshaft CV-Joints

OVERHAUL

Outer CV-Joint

1. Before servicing the vehicle, refer to the Precautions Section.
2. Remove the axle halfshaft from the vehicle.

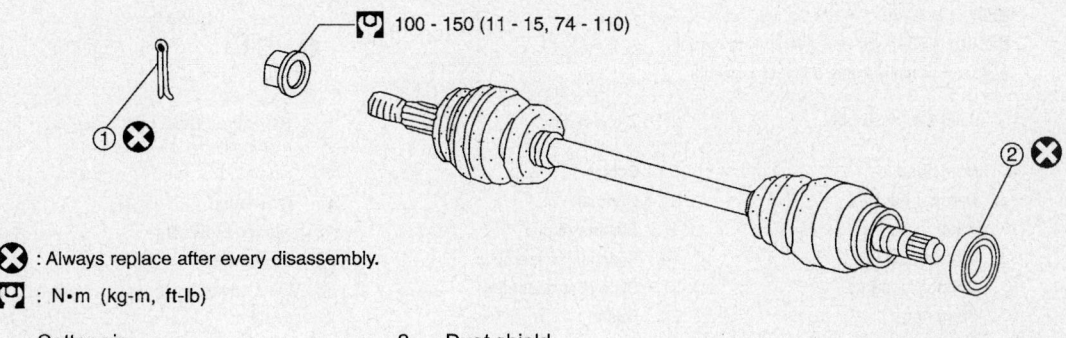

: Always replace after every disassembly.

: N•m (kg-m, ft-lb)

1. Cotter pin

2. Dust shield

42356-MURA-G63

Rear halfshaft and related components

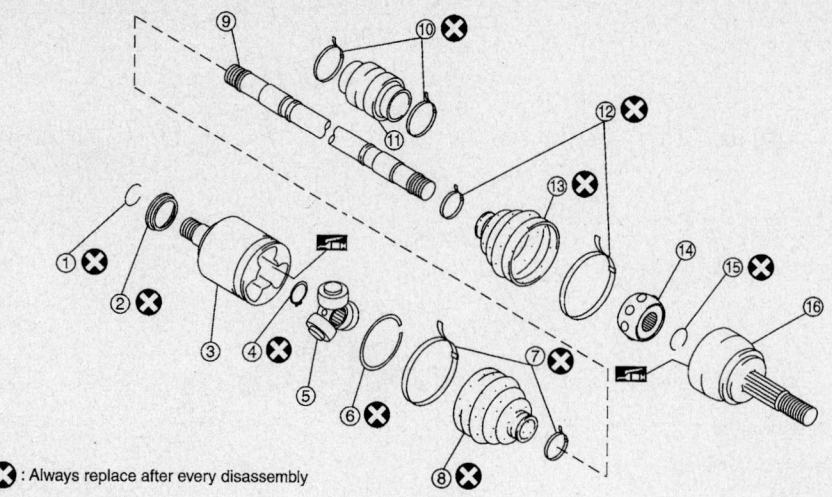

❌ : Always replace after every disassembly

1. Circular clip
2. Dust cover
3. Slide joint assembly
4. Snap ring
5. Spider assembly
6. Stopper ring
7. Boot band
8. Boot
9. Shaft
10. Damper band
11. Damper
12. Boot band
13. Boot
14. Ball cage / Steel ball / Inner race assembly
15. Circular clip
16. Joint sub-assembly

42356-MURA-G61

Front halfshaft exploded view—left side

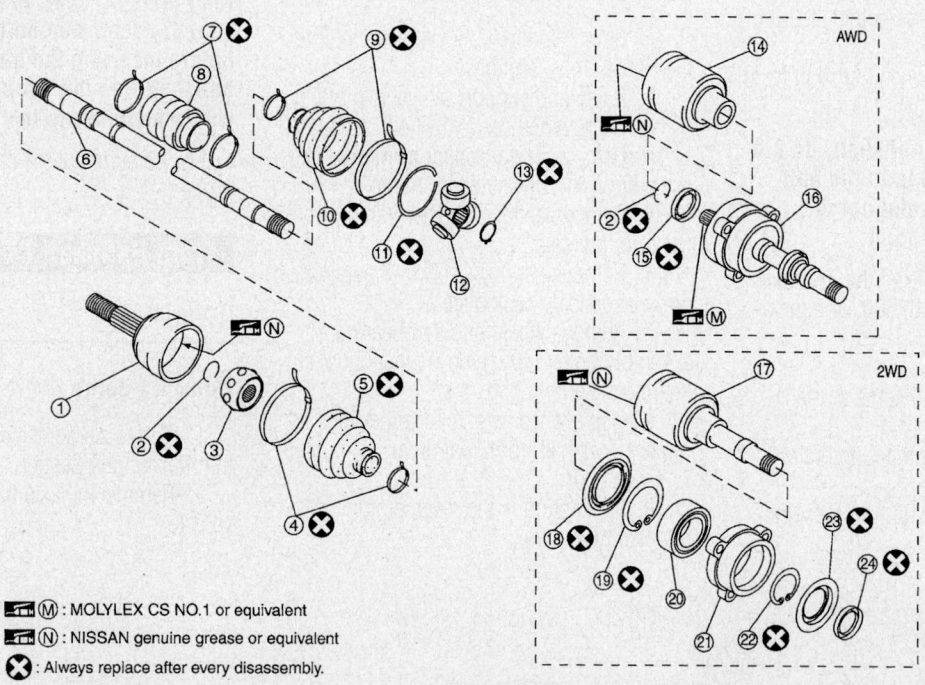

⊞Ⓜ : MOLYLEX CS NO.1 or equivalent

⊞Ⓝ : NISSAN genuine grease or equivalent

❌ : Always replace after every disassembly.

1. Joint sub-assembly
2. Circular clip
3. Ball cage / Steel ball / Inner race assembly
4. Boot band
5. Boot
6. Shaft
7. Damper band
8. Damper
9. Boot band
10. Boot
11. Stopper ring
12. Spider assembly
13. Circular clip
14. Slide joint assembly
15. Dust cover
16. Support bearing
17. Slide joint assembly
18. Dust cover
19. Snap ring
20. Bearing
21. Bracket
22. Snap ring
23. Dust cover
24. Dust cover

42356-MURA-G62

Front halfshaft exploded view—right side

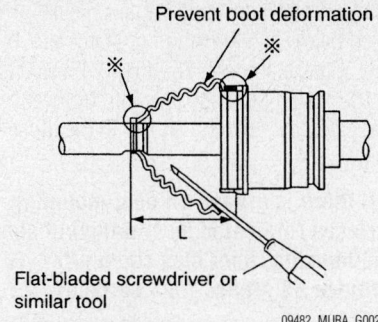

Front halfshaft outer cv-joint boot installation measurement

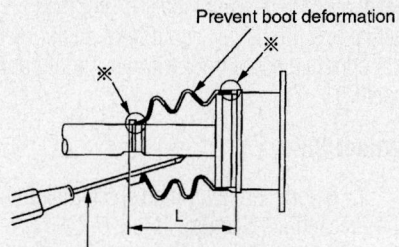

Boot installation measurement

3. Remove the CV-joint boot clamps and push the boot away from the joint.

4. Remove the CV-joint from the axle shaft by tapping it with a brass hammer.

To install:

→**Use new circlips and boot clamps for assembly.**

5. Install the CV-joint to the axle shaft by tapping it with a brass hammer.

6. Pack the joint with grease. Approximately 5.11–5.82 ounces.

7. Install the boot clamps.

→**Install the boot securely into the grooves (indicated by the * marks, in the illustration). Remove all grease from the boot mounting surfaces.**

8. Make sure that boot installation length "L" is 5.05 inch, as indicated in the illustration. Bleed excessive air from the boot to prevent deformation.

9. Install the axle halfshaft to the vehicle.

Inner Tri-Pot Joint

1. Before servicing the vehicle, refer to the Precautions Section.

2. Remove the axle halfshaft from the vehicle.

3. Remove the plug seal by tapping around the joint housing flange with a brass hammer.

4. Remove or disconnect the following:
- CV-joint boot clamps
- Snapring
- Spider assembly
- CV-joint housing
- CV-joint boot

To install:

→**Use new snaprings and plug seals for assembly.**

5. Install or connect the following:
- CV-joint boot
- CV-joint housing
- Spider assembly
- Snapring. Pack the joint with grease. Approximately 8.11–8.82 ounces.
- CV-joint boot clamps
- Plug seal

→**Install the boot securely into the grooves (indicated by the * marks, in the illustration). Remove all grease from the boot mounting surfaces.**

6. Make sure that boot installation length "L" is 4.02 inch, as indicated in the illustration. Bleed excessive air from the boot to prevent deformation.

7. Install the axle halfshaft to the vehicle.

Rear Halfshaft CV-Joints

OVERHAUL

Final Drive Side

1. Before servicing the vehicle, refer to the Precautions Section.

2. Remove the axle halfshaft from the vehicle.

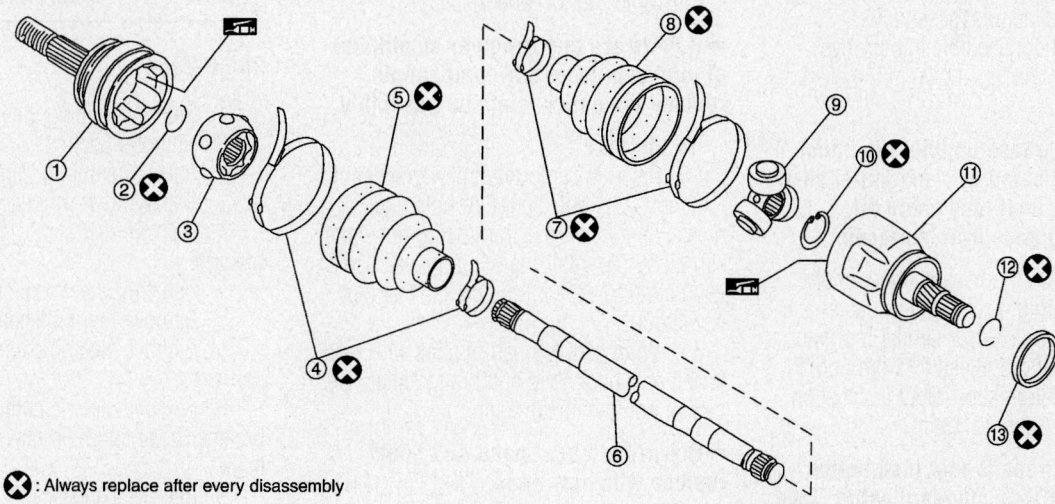

❌ : Always replace after every disassembly

1. Joint sub-assembly
2. Circular clip
3. Ball cage/Steel ball/Inner race assembly
4. Boot band (Wheel side)
5. Boot (Wheel side)
6. Shaft
7. Boot band (Final drive side)
8. Boot (Final drive side)
9. Spider assembly
10. Snap ring
11. Slide joint assembly
12. Circular clip
13. Dust shield

Rear halfshaft exploded view

3. Press shaft in a vise.
4. Remove boot band.

➡**When retaining shaft in a vice, always use copper or aluminum plates between vise and shaft.**

5. Put matching marks on slide joint assembly and shaft before separating slide joint assembly.
6. Put matching marks on spider assembly and shaft.
7. Remove snap ring, then remove spider assembly from shaft.
8. Remove boot from shaft.
9. Remove old grease on slide joint assembly with paper towels.
10. Remove circular clip and dust shield from slide joint assembly.

To install:
11. Install new boot and new small boot band on shaft.

➡**Cover shaft serration with tape to prevent damage to boot during installation.**

12. Remove protective tape wound around serrated part of shaft.
13. Install spider assembly securely, making sure the matching marks which were made during disassembly are properly aligned.
14. Install new snap ring.
15. Insert the amount of new grease listed below into housing from large end of boot. Grease amount: 85–95 g (3.0–3.35 oz)
16. Install slide joint assembly.
17. Install boot securely into grooves (indicated by * marks) shown in the illustration.

➡**If there is grease on boot mounting surfaces (indicated by * marks) of shaft and housing, boot may come off. Remove all grease from surfaces.**

18. Make sure boot installation length "L" is the length indicated below. Insert a flat-bladed screwdriver or similar tool into smaller side of boot. Bleed air from boot to prevent boot deformation. Boot installation length "L": 79.6 mm (3.13 in)

➡**Boot may break if boot installation length is less than standard value. Take care not to touch the tip of screwdriver to inside of boot.**

19. Secure big and small ends of boot with new boot bands as shown in the illustration.

➡**Discard old boot bands; replace with new ones.**

20. After installing housing and shaft, rotate boot to check whether or not the actual position is correct. If boot position is not correct, secure boot with new boot band again.

Wheel Side

1. Before servicing the vehicle, refer to the Precautions Section.
2. Remove the axle halfshaft from the vehicle.
3. Place shaft in a vise.

➡**When retaining shaft in a vise, always use copper or aluminum plates between vise and shaft.**

4. Remove boot bands. Then remove boot from joint sub-assembly.
5. Screw a halfshaft puller (suitable tool) 30 mm (1.18 in) or more into threaded part of joint sub-assembly. Pull joint sub-assembly out of shaft.

➡**If joint sub-assembly cannot be removed after five or more unsuccessful attempts, replace the entire halfshaft assembly. Align sliding hammer and halfshaft and remove them by pulling directly.**

6. Remove boot from shaft.
7. Remove circular clip from shaft.
8. While rotating ball cage, remove old grease on joint sub-assembly with paper towels.
9. Replace halfshaft if there is any runout, cracking, or other damage.

➡**If there are any irregular conditions of joint sub-assembly components, replace the entire joint sub-assembly.**

To install:
10. Insert the amount of new grease into joint sub-assembly serration hole until grease begins to ooze from ball groove and serration hole. After inserting grease, use a shop cloth to wipe off old grease that has oozed out.
11. Wind serrated part of shaft with tape. Install new boot band and boot to shaft. Be careful not to damage boot.

➡**Discard old boot band and boot; replace with new ones.**

12. Remove protective tape wound around serrated part of shaft.
13. Install new circular clip to shaft. At this time, circular clip must fit securely into shaft groove. Attach nut to joint sub-assembly. Use a wooden hammer to press-fit.

➡**Discard old circular clip; replace with new one.**

14. Insert the amount of new grease listed below into housing from large end of boot. Grease amount: 75–85 g (2.65–3.0 oz)
15. Install boot securely into grooves (indicated by * marks) shown in the illustration.

➡**If there is grease on boot mounting surfaces (indicated by * marks) of shaft and housing, boot may come off. Remove all grease from surfaces.**

16. Make sure boot installation length "L" is the length indicated below. Insert a flat-bladed screwdriver or similar tool into smaller side of boot. Bleed air from boot to prevent boot deformation. Boot installation length "L": 67.7mm (2.67 in)

➡**Boot may break if boot installation length is less than standard value. Be careful that screwdriver tip does not contact inside surface of boot.**

17. Secure big and small ends of boot with new boot bands as shown in the illustration.

➡**Discard old boot bands; replace with new ones.**

18. Check installation status of boot. Rotate joint to make sure boot is securely in place. If not, reinstall using a new boot band.

Rear Final Drive

REMOVAL & INSTALLATION

2003–2004

1. Before servicing the vehicle, refer to the Precautions Section.
2. Raise and support the vehicle safely.
3. Remove rear propeller shaft.
4. Remove rear stabilizer mounting bracket.
5. Remove wheel sensor.
6. Remove rear halfshaft.
7. Remove electric controlled coupling connector.
8. Remove electric controlled coupling breather hose and rear final drive breather hose.
9. Remove canister.
10. Set Transmission Jack to rear final drive assembly, and then remove nuts from rear suspension member.

➡**Do not place a transmission jack on the rear cover (aluminum case).**

11. Remove bolt and nut from final drive mount bracket, and then remove rear final drive assembly from vehicle.

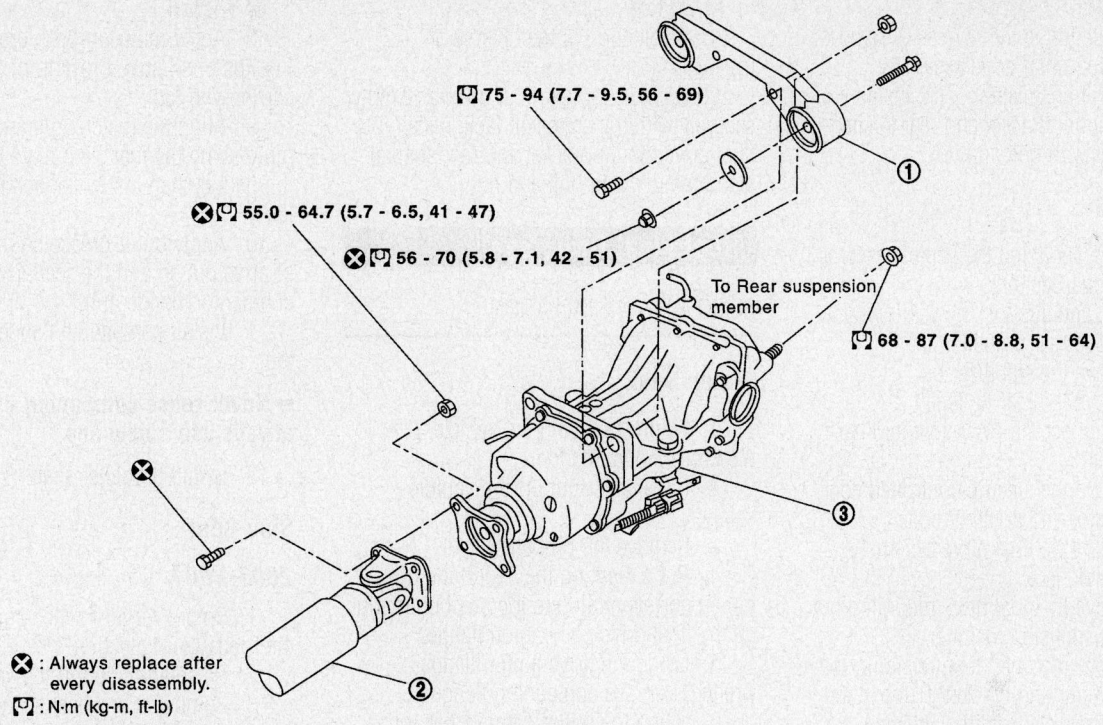

⊗🔧 55.0 - 64.7 (5.7 - 6.5, 41 - 47)

⊗🔧 56 - 70 (5.8 - 7.1, 42 - 51)

🔧 75 - 94 (7.7 - 9.5, 56 - 69)

To Rear suspension member

🔧 68 - 87 (7.0 - 8.8, 51 - 64)

⊗ : Always replace after every disassembly.

🔧 : N·m (kg-m, ft-lb)

1. Final drive mount bracket
2. Rear propeller shaft
3. Rear final drive assembly

42356-MURA-G66

Rear final drive and related components—2003–2004

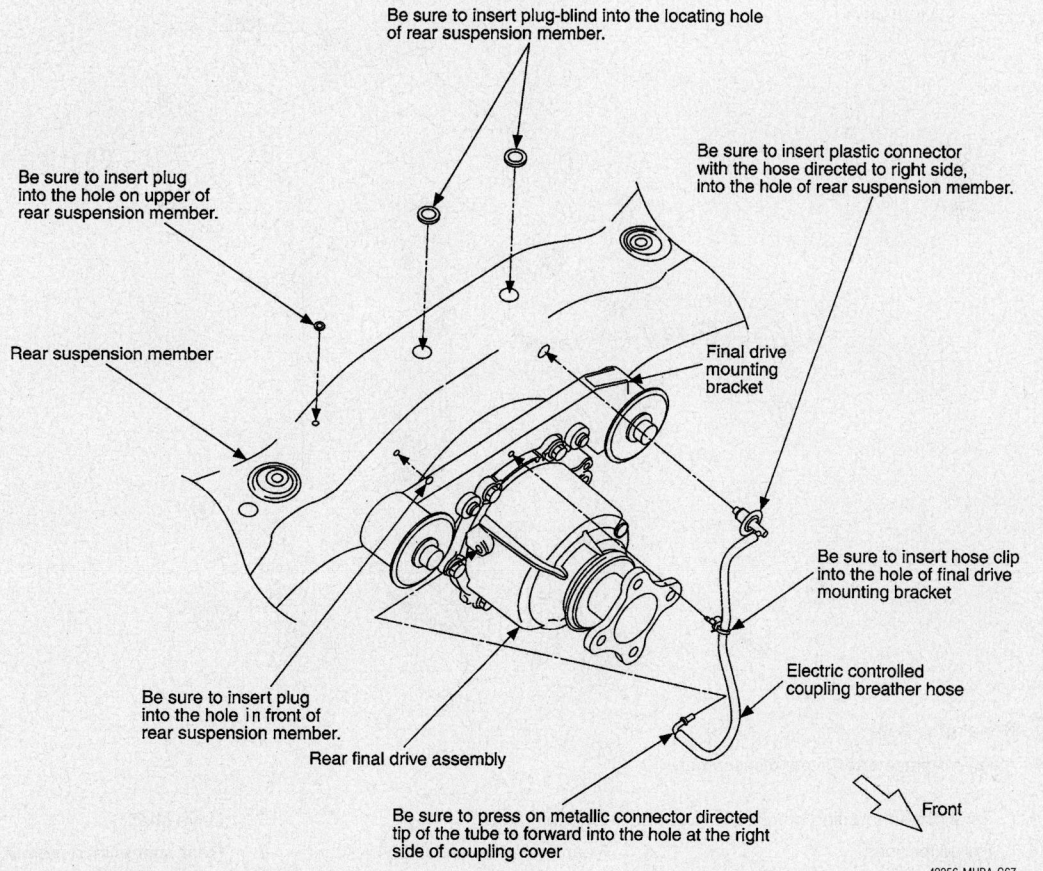

Be sure to insert plug-blind into the locating hole of rear suspension member.

Be sure to insert plastic connector with the hose directed to right side, into the hole of rear suspension member.

Be sure to insert plug into the hole on upper of rear suspension member.

Rear suspension member

Final drive mounting bracket

Be sure to insert hose clip into the hole of final drive mounting bracket

Electric controlled coupling breather hose

Be sure to insert plug into the hole in front of rear suspension member.

Rear final drive assembly

Be sure to press on metallic connector directed tip of the tube to forward into the hole at the right side of coupling cover

Front

42356-MURA-G67

Electronically controlled coupling breather hose and related components

To install:

12. Installation is the reverse of removal. Supporting rear final drive assembly securely with transmission jack, install it to final drive mount bracket and rear suspension member with bolt and nut.

2005–2006

1. Before servicing the vehicle, refer to the Precautions Section.
2. Raise and support the vehicle safely.
3. Remove rear propeller shaft.
4. Remove the stabilizer bar.
5. Remove the rear halfshaft.
6. Disconnect the AWD solenoid harness connector.
7. Remove the electric controlled coupling connector and breather hose.
8. Support the final drive assembly, using a suitable jack.
9. Remove the final drive mounting nut at the rear suspension member.
10. Remove the final drive retaining bolts at the mounting bracket. Remove the final drive assembly from the vehicle.

➡**Secure the final drive assembly in a suitable holding fixture as you are removing it from the vehicle.**

To install:

11. Installation is the reverse of removal.
12. Supporting rear final drive assembly securely with transmission jack, install it to final drive mount bracket and rear suspension member with bolt and nut.

Rear Final Drive Seals

REMOVAL & INSTALLATION

Pinion Seal

1. Before servicing the vehicle, refer to the Precautions Section.
2. Raise and support the vehicle safely.
3. Remove the propeller shaft.
4. Put a mark on the end of the drive pinion corresponding to the position mark on the final drive companion flange.
5. Using the drive pinion flange wrench, Remove companion flange nut.
6. Using the puller, remove the companion flange.
7. Using the side bearing outer race puller, remove front oil seal.

To install:

8. Apply multi-purpose grease to sealing lips of oil seal. Press front oil seal into carrier with tool.
9. Align the matching mark of drive pinion with the matching mark of companion flange, then install the companion flange.
10. Apply oil or grease on the screw part of drive pinion and the seating surface of companion flange nut.
11. Install companion flange nut with tool.

➡**Never reuse companion flange nut, always use a new one.**

12. Install propeller shaft.

Side Seal

2003–2004

1. Before servicing the vehicle, refer to the Precautions Section.
2. Raise and support the vehicle safely.
3. Remove rear wheel sensor.
4. Remove rear axle assembly.
5. Remove rear halfshaft.
6. Using flat tip screwdriver, remove side oil seal.

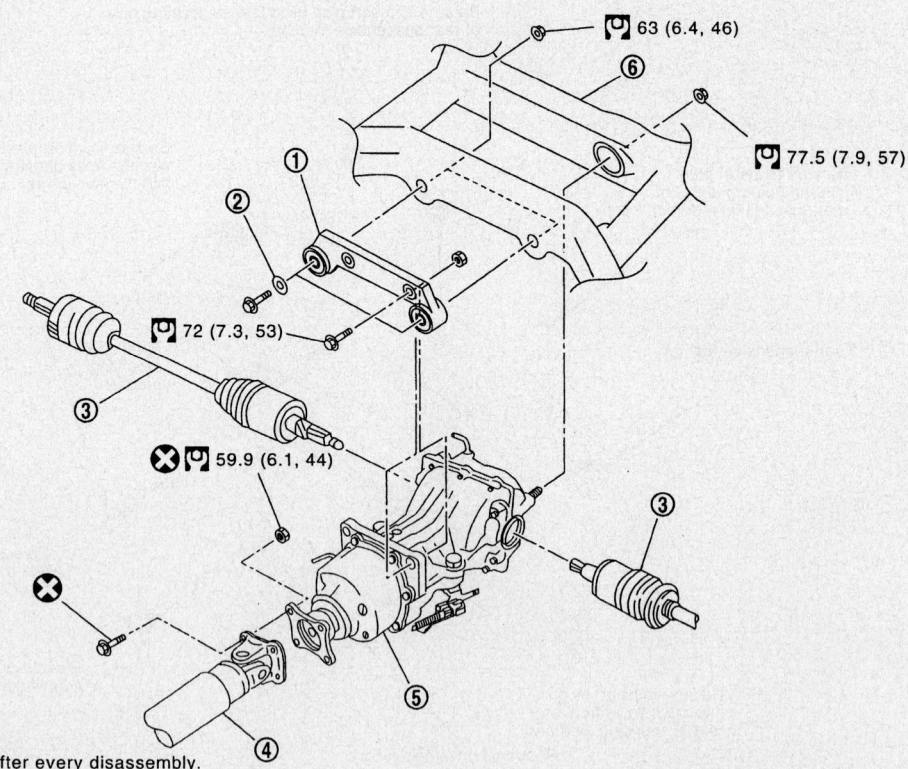

63 (6.4, 46)
77.5 (7.9, 57)
72 (7.3, 53)
59.9 (6.1, 44)

🔧 : N•m (kg-m, ft-lb)
❌ : Always replace after every disassembly.

| 1. | Final drive mounting bracket | 2. | Washer | 3. | Drive shaft |
| 4. | Propeller shaft | 5. | Rear final drive assembly | 6. | Rear suspension member |

09482_MURA_G0022

Rear final drive and related components—2005–2006

To install:

7. Apply multi-purpose grease to sealing lips of oil seal.

8. Using the drift, press-fit oil seal so that its surface comes face to face with the end surface of the case.

9. Install rear halfshaft.

10. Install rear axle assembly.

2005–2006

1. Before servicing the vehicle, refer to the Precautions Section.

2. Raise and support the vehicle safely.

3. Remove the rear halfshaft.

4. Using the proper tool, remove the side oil seal.

➡Be careful not to damage the gear carrier and cover.

To install:

5. Apply multi-purpose grease to sealing lips of oil seal.

6. Install the seal until it is flush with the case end. Use a drift and tap with a hammer.

7. Install the halfshaft.

STEERING

Air Bag

✳✳ CAUTION

The SRS system must be disarmed before performing service on, or around, system components, the steering column, instrument panel components, wiring and sensors. Failure to follow the safety precautions and the disarming procedure could result in accidental air bag deployment, possible injury and unnecessary system repairs.

PRECAUTIONS

Several precautions must be observed when handling the inflator module to avoid accidental deployment and possible personal injury.

• Never carry the inflator module by the wires or connector on the underside of the module.

• When carrying a live inflator module, hold securely with both hands, and ensure that the bag and trim cover are pointed away.

• Place the inflator module on a bench or other surface with the bag and trim cover facing up.

• With the inflator module on the bench, never place anything on or close to the module which may be thrown in the event of an accidental deployment.

DISARMING

To disarm the **SRS** system turn the ignition switch to the **OFF** position. Then, disconnect both battery cables starting with the negative cable first and wait at least 3 minutes after the cables are disconnected.

To rearm the **SRS** system, turn the ignition switch to the **OFF** position. Connect both battery cables starting with the positive cable first.

Rack and Pinion Steering Gear

REMOVAL & INSTALLATION

➡Be sure to disarm the SRS system, prior to working on the vehicle. Turn the ignition switch OFF, disconnect both battery cables and wait at least three minutes before starting any work.

2WD

1. Before servicing the vehicle, refer to the Precautions Section.

2. Disarm the SRS system. Disconnect the negative battery cable.

3. Set wheels in the straight ahead position.

4. Raise and support the vehicle safely.

5. Remove locknut and bolt, then separate lower joint from upper joint.

6. Remove the tires and wheels.

7. Confirm that the slit on lower joint fits with the projection on rear cover cap,

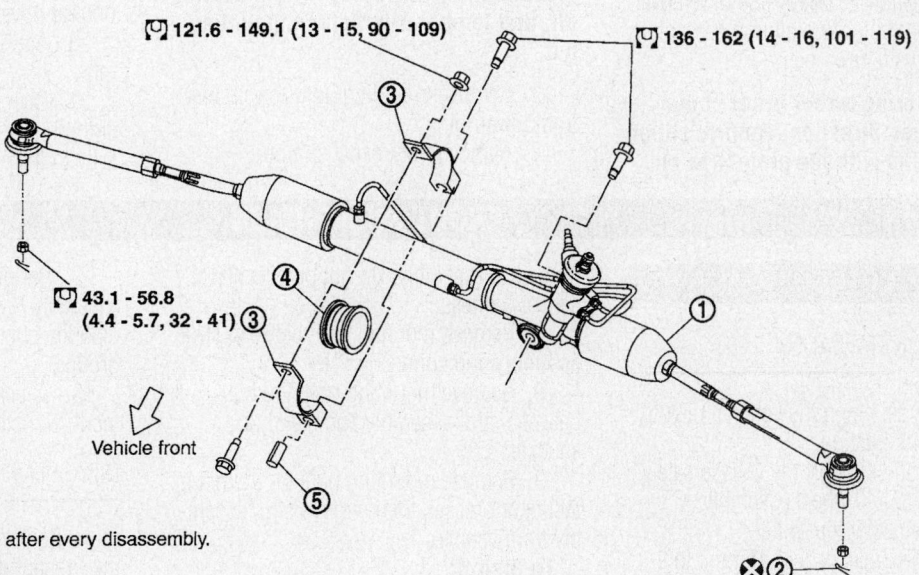

⊗ : Always replace after every disassembly.
🔧 : N·m(kg-m,ft-lb)

1. Steering gear assembly
2. Cotter pin
3. Rack mounting bracket
4. Rack mounting insulator
5. Sleeve

Steering gear and related components

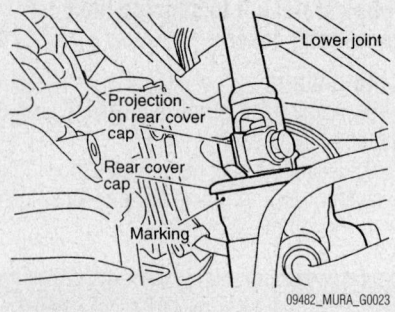

Steering gear pinch bolt location and alignment

also the matchmark on steering gear assembly fits with the projection on rear cover cap.

8. Remove cotter pin at steering knuckle, and then loosen mounting nut.

9. Use a ball joint remover to remove steering outer socket from steering knuckle. Be careful not to damage ball joint boot.

➡ **To prevent damage to threads and to prevent ball joint remover from coming off, temporarily tighten mounting nut.**

10. Remove oil pipes (high pressure side and low pressure side) from steering gear assembly, then drain fluid from pipes.

11. Remove mounting bolt (lower side) from lower joint.

12. Remove mounting bolts and nut from steering gear assembly, and then remove steering gear assembly, rack mounting bracket, rack mounting insulator and sleeve from vehicle.

To install:

13. Installation is the reverse of removal.

14. Be sure to replace all non reusable components with new ones.

➡ **When steering wheel is set in the straight ahead direction, confirm slit of lower joint fits with the projection on**

rear cover cap, also the matchmarks on steering gear assembly fit with the projection on rear cover cap.

15. After installation, bleed air from piping.

16. Check if steering wheel turns smoothly when it is turned several times fully to the end of the left and right.

AWD

1. Before servicing the vehicle, refer to the Precautions Section.

2. Disarm the SRS system. Disconnect the negative battery cable.

3. Set wheels in the straight ahead position.

4. Raise and support the vehicle safely.

5. Remove locknut and bolt, then separate lower joint from upper joint.

6. Remove tires and wheels.

7. Remove the undercover.

8. Confirm slit of lower joint fits with the projection on rear cover cap, furthermore marking position on steering gear assembly nearly fits with the projection on rear cover cap.

9. Remove oil pipes (high pressure side and low pressure side) from steering gear assembly, then drain fluid from pipes.

10. Remove cotter pin at steering knuckle, then loosen mounting nut.

11. Use a ball joint remover to remove steering outer socket from steering knuckle. Be careful not to damage ball joint boot.

➡ **To prevent damage to threads and to prevent ball joint remover from coming off, and temporarily tighten mounting nut.**

12. Remove mounting bolt (lower side) from lower joint.

13. Remove front exhaust tube.

14. Remove rear propeller shaft.

15. Remove mounting nuts on lower position from stabilizer connecting rod.

16. Remove mounting bolts from stabilizer clamp and hang stabilizer on vehicle.

17. Remove steering hydraulic piping bracket from front suspension member.

18. Disconnect electrical rear engine mounting actuator harness connector.

19. Set jack under engine and front suspension member.

20. Remove mounting bolts from rear engine mounting insulator.

21. Loosen mounting nuts of front suspension member (front side).

22. Remove mounting bolts from member stay (body side), then loosen mounting nuts of member stay (front suspension member side).

23. Move jack down slowly (front suspension member side) to remove rear engine mounting insulator from engine and front suspension member.

24. Remove mounting bolts and nut from steering gear assembly, and then remove steering gear assembly, rack mounting bracket, rack mounting insulator and sleeve from vehicle.

To install:

25. Installation is the reverse of removal.

26. Be sure to replace all non reusable components with new ones.

27. When steering wheel is set in the straight ahead direction, confirm slit of lower joint fits with the projection on rear cover cap, also the matchmarks on steering gear assembly nearly fits with the projection on rear cover cap.

28. After installation, bleed air from piping.

29. Check if steering wheel turns smoothly when it is turned several times fully to the end of the left and right.

FRONT SUSPENSION

Strut

REMOVAL & INSTALLATION

1. Before servicing the vehicle, refer to the Precautions Section.

2. Raise and support the vehicle safely. Remove the tire and wheel assembly.

3. Remove cowl top grille.

4. Remove brake caliper. Hang it in a place where it will not interfere with work.

5. Remove lock plate from brake hose from strut assembly.

6. Remove harness from wheel sensor

from strut assembly. Do not pull on wheel sensor harness.

7. Remove mounting nut between strut assembly and connecting rod.

8. Remove mounting bolt and nut between strut assembly and steering knuckle.

9. Remove mounting nuts on mounting insulator bracket, then remove strut assembly from vehicle.

To install:

10. Installation is the reverse of the removal procedure.

11. Be sure to replace all non reusable components with new ones.

12. Perform final tightening of the strut assembly lower side (rubber bushing) under unladen conditions with tires on level ground.

13. Check and adjust the front end alignment, as necessary.

DISASSEMBLY

1. Before servicing the vehicle, refer to the Precautions Section.

2. Remove the strut assembly.

3. Compress the coil spring and remove the piston rod nut.

4. Remove or disconnect the following:

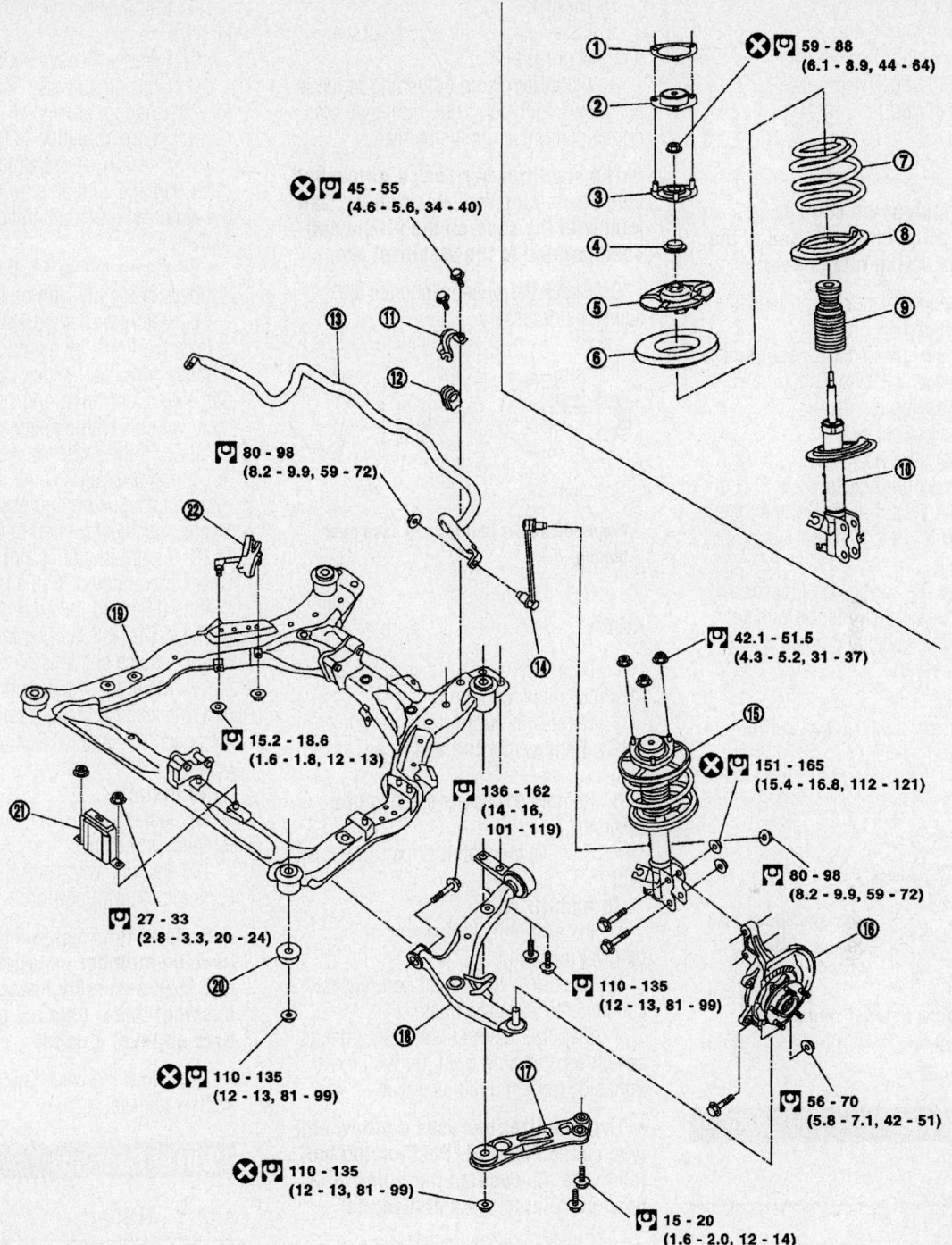

59 - 88
(6.1 - 8.9, 44 - 64)

45 - 55
(4.6 - 5.6, 34 - 40)

80 - 98
(8.2 - 9.9, 59 - 72)

42.1 - 51.5
(4.3 - 5.2, 31 - 37)

15.2 - 18.6
(1.6 - 1.8, 12 - 13)

151 - 165
(15.4 - 16.8, 112 - 121)

136 - 162
(14 - 16,
101 - 119)

80 - 98
(8.2 - 9.9, 59 - 72)

27 - 33
(2.8 - 3.3, 20 - 24)

110 - 135
(12 - 13, 81 - 99)

56 - 70
(5.8 - 7.1, 42 - 51)

110 - 135
(12 - 13, 81 - 99)

110 - 135
(12 - 13, 81 - 99)

15 - 20
(1.6 - 2.0, 12 - 14)

: N•m (kg-m, ft-lb)

: Always replace after every disassembly.

1.	Upper mounting plate	2.	Mounting insulator	3.	Mounting insulator bracket
4.	Mounting bearing	5.	Spring upper seat	6.	Spring upper rubber seat
7.	Coil spring	8.	Spring lower rubber seat	9.	Bound bumper
10.	Strut	11.	Stabilizer clamp	12.	Stabilizer bushing
13.	Stabilizer	14.	Connecting rod	15.	Strut assembly
16.	Front axle	17.	Member stay	18.	Transverse link
19.	Front suspension member	20.	Rebound stopper	21.	Damper assembly
22.	Air guide				

42356-MURA-G69

Front suspension and related components

- Upper strut mount
- Strut mount bracket
- Upper strut bearing
- Spring upper seat
- Coil spring

ASSEMBLY

➡**Face the side of the coil spring downward. Align the lower end of the spring to the spring rubber seat.**

1. Be sure to replace all non reusable components with new ones.
2. Install the spring compressor tool. Install or connect the following:
- Coil spring
- Spring upper seat
- Upper strut bearing
- Strut mount bracket
- Upper strut mount. Tighten the piston rod nut to 43–58 ft. lbs. (59–78 Nm).

3. Remove the spring compressor and install the strut assembly to the vehicle.
4. Check the wheel alignment and adjust, as necessary.

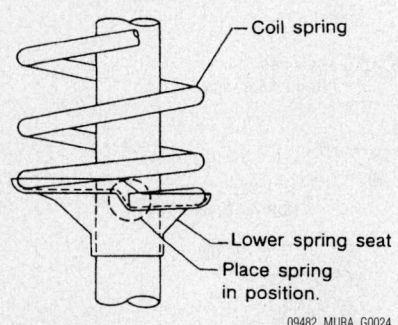

Front coil spring to lower spring seat alignment

Stabilizer Bar

REMOVAL & INSTALLATION

2WD

1. Before servicing the vehicle, refer to the Precautions Section.
2. Raise and support the vehicle safely. Remove the tire and wheel assembly.
3. Remove the power steering gear assembly.
4. Remove the stabilizer connecting rod lower nut.
5. Separate the stabilizer bar and stabilizer connecting rod, using the proper tools.
6. Remove the stabilizer clamp mounting bolts. Remove the stabilizer bar from the vehicle.

To install:
7. Installation is the reverse of the removal procedure.
8. Stabilizer clamp tightening order, is as follows: Left side front, right side rear, right side front and left side rear.

➡**The stabilizer bar uses a pillow ball type connecting rod. Position the ball joint with the case on the pillow ball head parallel to the stabilizer bar.**

9. Check the wheel alignment and adjust, as necessary.

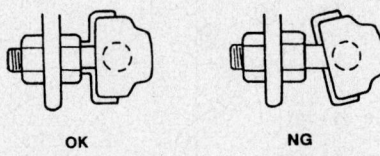

Proper stabilizer ball joint to case positioning

AWD

1. Before servicing the vehicle, refer to the Precautions Section.
2. Raise and support the vehicle safely. Remove the tire and wheel assembly.
3. Remove the power steering gear assembly.
4. Remove the stabilizer bar from the vehicle.

To install:
5. Installation is the reverse of the removal procedure.
6. Be sure to replace all non reusable components with new ones.
7. Stabilizer clamp tightening order, is as follows: Left side front, right side rear, right side front and left side rear.

➡**The stabilizer bar uses a pillow ball type connecting rod. Position the ball joint with the case on the pillow ball head parallel to the stabilizer bar.**

8. Check the wheel alignment and adjust, as necessary.

Transverse Link

REMOVAL & INSTALLATION

1. Before servicing the vehicle, refer to the Precautions Section.
2. Raise and support the vehicle safely. Remove the tire and wheel assembly.
3. Remove mounting bolt between transverse link and front suspension member.

4. Remove transverse link from steering knuckle.
5. Remove transverse link from vehicle.
6. Check transverse link and bushing for deformation, cracks, or damage. If any non-standard condition is found, replace it.
7. Check boot of ball joint for cracks or other damage, and also for grease leakage. If any non-standard condition is found, replace it.
8. Manually move ball stud to confirm it moves smoothly with no binding.
9. Before measurement, move ball joint at least ten times by hand to check for smooth movement. Hook spring scale at ball stud. Confirm spring scale measurement value is within specifications (3.08–20.5 lbs.) when ball stud begins moving. If it is outside the specified range, replace suspension arm assembly. Swing torque specification is 0.5–0.30 inch lbs.
10. Attach mounting nut to ball stud. Check that rotating torque is within specifications 0.5–0.30 inch lbs.) with a preload gauge (SST). If it is outside the specified range, replace suspension arm assembly.
11. Move tip of ball joint in axial direction to check for looseness. If it is outside the specified range (0.0 inch), replace suspension arm assembly.

To install:
12. Installation is the reverse of the removal procedure.
13. Be sure to replace all non reusable components with new ones.

➡**Perform final tightening of front suspension member installation position and strut assembly lower side (rubber bushing) under unladen conditions with tires on level ground.**

14. Check the wheel alignment and adjust, as necessary.

Wheel Bearings

ADJUSTMENT

The front wheel bearings are part of a unitized hub and are not adjustable. Move the wheel hub in the axial direction by hand. Make sure that there is no looseness of the wheel bearing. Axial end play is 0.002 inch or less.

REMOVAL & INSTALLATION

1. Before servicing the vehicle, refer to the Precautions Section.
2. Raise and support the vehicle safely. Remove the tire and wheel assembly.

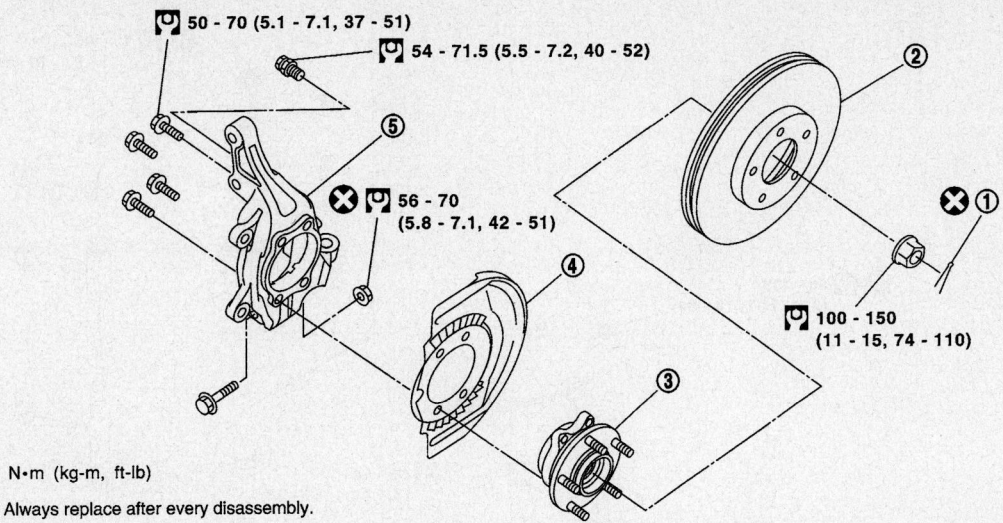

- ☐ 50 - 70 (5.1 - 7.1, 37 - 51)
- ☐ 54 - 71.5 (5.5 - 7.2, 40 - 52)
- ☒ ☐ 56 - 70 (5.8 - 7.1, 42 - 51)
- ☐ 100 - 150 (11 - 15, 74 - 110)

☐ : N·m (kg-m, ft-lb)

☒ : Always replace after every disassembly.

1. Cotter pin
2. Disc rotor
3. Wheel hub and bearing assembly
4. Splash guard
5. Steering knuckle

42356-MURA-G71

Front wheel bearing and related components

3. Remove the brake caliper. Hang it in a place where it will not interfere with work. Avoid depressing brake pedal while brake caliper is removed.

4. Put alignment marks on disc rotor and wheel hub and bearing assembly, then remove disc rotor.

5. Remove wheel sensor from steering knuckle. Do not pull on the wheel sensor harness.

6. Remove cotter pin, and then remove lock nut from halfshaft.

7. Remove steering outer socket and cotter pin at steering knuckle, then loosen mounting nut.

8. Use a ball joint remover (SST) to remove steering outer socket from steering knuckle. Be careful not to damage ball joint boot.

➡To prevent damage to threads and to prevent ball joint remover (SST) from coming off suddenly, temporarily tighten mounting nut.

9. Using a puller (suitable tool), remove wheel hub and bearing assembly from halfshaft.

10. Remove wheel hub and bearing assembly bolt.

11. Remove splash guard and wheel hub and bearing assembly from steering knuckle.

12. Remove strut assembly and steering knuckle bolts and nuts.

13. Remove transverse link and steering knuckle bolt and nut.

14. Remove steering knuckle from vehicle.

To install:

15. Installation is the reverse of the removal procedure.

16. Check for deformity, cracks and damage on each parts, replace if necessary.

17. Check for boot breakage, axial looseness, and torque of transverse link ball joint.

18. Be sure to replace all non reusable components with new ones.

19. Check the wheel alignment and adjust, as necessary.

REAR SUSPENSION

Shock Absorber

REMOVAL & INSTALLATION

1. Before servicing the vehicle, refer to the Precautions Section.

2. Raise and support the vehicle safely. Remove the tire and wheel assembly.

3. Remove bolt in lower side of shock absorber assembly.

4. Remove mounting seal bracket nuts from shock absorber upper side and remove shock absorber assembly from vehicle.

To install:

5. Installation is the reverse of the removal procedure.

6. Be sure to replace all non reusable components with new ones.

➡Perform final tightening of the shock absorber lower side (rubber bushing) under unladen conditions with tires on level ground.

Coil Spring

REMOVAL & INSTALLATION

1. Before servicing the vehicle, refer to the Precautions Section.

2. Raise and support the vehicle safely. Remove the tire and wheel assembly.

3. Set jack under rear lower link.

4. Loosen bolt and nut between rear lower link and suspension member, and then remove bolt and nut between rear axle and rear lower link.

5. Slowly lower jack, then remove upper seat, coil spring and rubber seat from rear lower link.

6. Remove bolt and nut between rear suspension member and rear lower link.

To install:

7. Installation is the reverse of the removal procedure.

8. Be sure to replace all non reusable components with new ones.

➡Perform final tightening of the rear suspension member and axle installation position (rubber bushing) under unladen conditions with tires on level ground.

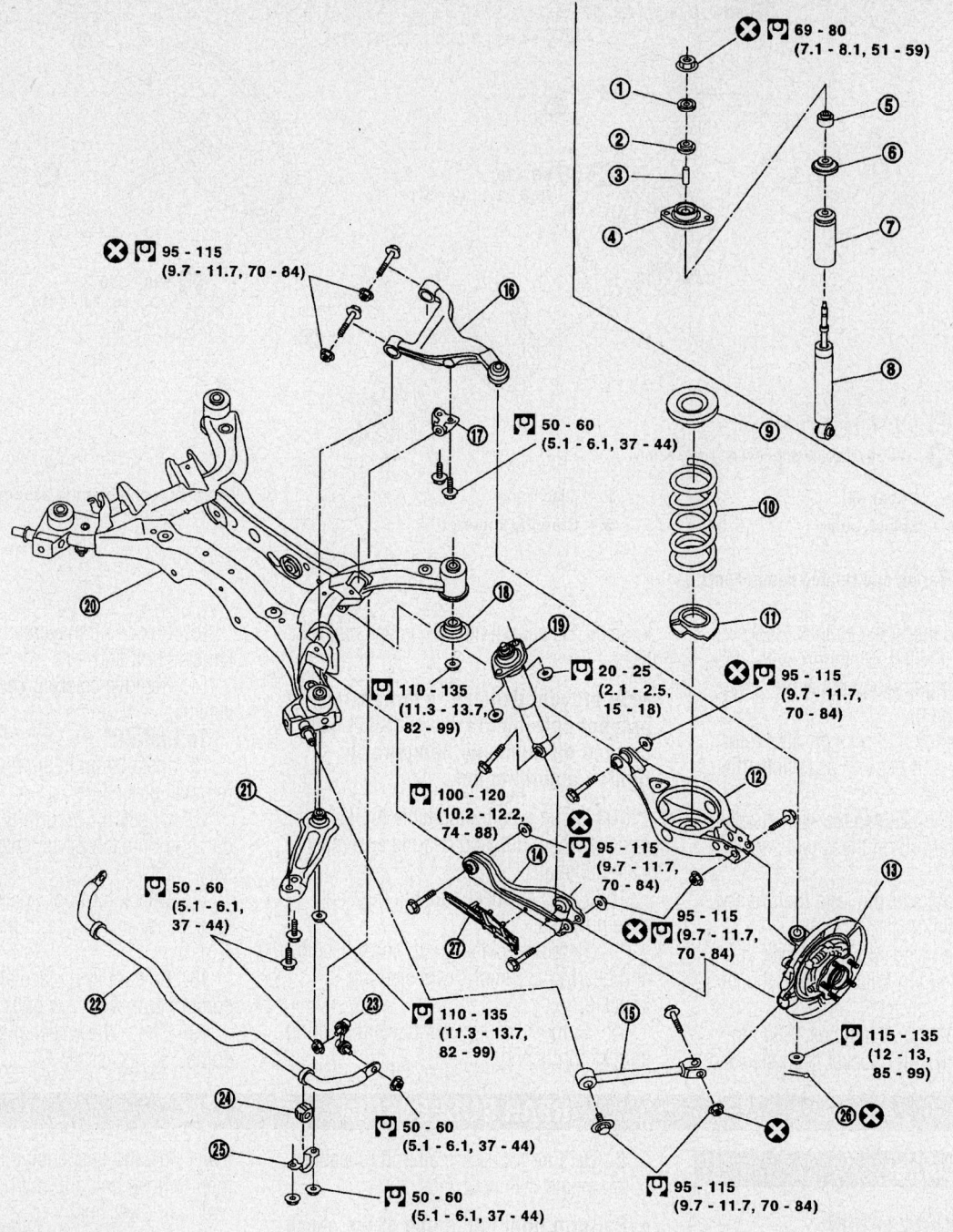

❌ 🔧 : N•m (kg-m, ft-lb)

❌ : Always replace after every disassembly.

1.	Outer washer	2.	Bushing A	3.	Distance tube
4.	Mounting seal bracket	5.	Bushing B	6.	Bound bumper cover
7.	Bound bumper	8.	Shock absorber	9.	Upper seat
10.	Coil spring	11.	Rubber seat	12.	Rear lower link
13.	Axle	14.	Front lower link	15.	Radius rod
16.	Suspension arm	17.	Stabilizer connecting rod mount bracket	18.	Rebound stopper
19.	Shock absorber assembly	20.	Rear suspension member	21.	Member stay
22.	Stabilizer bar	23.	Stabilizer connecting rod	24.	Stabilizer bushing
25.	Stabilizer clamp	26.	Cotter pin	27.	Front lower link protector

42356-MURA-G70

Rear suspension and related components

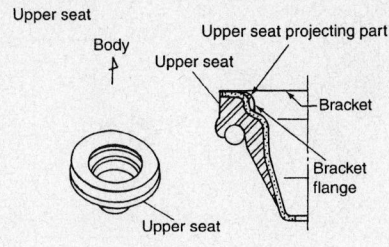

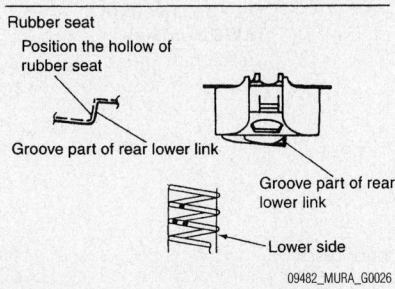

Rear coil spring rubber seat alignment

➡**Insert bracket tabs (3) and the inside protrusion on upper seat into each other beforehand as shown in the illustration. Match up rubber seat indentions and rear lower link grooves and attach.**

Stabilizer Bar

REMOVAL & INSTALLATION

1. Before servicing the vehicle, refer to the Precautions Section.
2. Raise and support the vehicle safely. Remove the tire and wheel assembly.
3. Remove the lower side retaining nut on the stabilizer connecting rod. Remove the stabilizer connecting rod from the stabilizer bar.
4. Remove the retaining nut on the stabilizer clamp and remove the stabilizer from the vehicle.
    To install:
5. Installation is the reverse of the removal procedure.

➡**The stabilizer bar uses a pillow ball type connecting rod. Position the ball joint with the case on the pillow ball head parallel to the stabilizer bar.**

6. When the bushing and the clamp are installed to the stabilizer bar, position the bushing and clamp inside of the side slip prevention clamp.
7. Check the wheel alignment and adjust, as necessary.

Suspension Arm

REMOVAL & INSTALLATION

1. Before servicing the vehicle, refer to the Precautions Section.
2. Raise and support the vehicle safely. Remove the tire and wheel assembly.
3. Remove the coil spring.
4. Remove the wheel sensor and sensor harness from the axle and suspension arm.
5. Remove the stabilizer connecting rod mounting bracket from the suspension arm.
6. Position a jack under the front lower link.
7. Remove the fuel filler tube retaining bolt, left side only.
8. Remove the retaining nuts and bolts between the suspension arm and the rear suspension member.
9. Remove the cotter pin from the suspension arm ball joint. Loosen the nut.
10. Use a ball joint remover tool and remove the suspension arm from the axle. Be careful not to damage the ball joint boot.
11. Remove the suspension arm from the vehicle.
    To install:
12. Installation is the reverse of the removal procedure.
13. Be sure to replace all non reusable components with new ones.

➡**Perform final tightening of the rear suspension member installation position (rubber bushing) under unladen conditions with tires on level ground.**

14. Check and adjust the rear alignment, as necessary.

Radius Arm

REMOVAL & INSTALLATION

1. Before servicing the vehicle, refer to the Precautions Section.
2. Raise and support the vehicle safely. Remove the tire and wheel assembly.
3. Remove the coil spring.
4. Remove the wheel sensor and sensor harness from the axle and suspension arm.
5. Remove the retaining bolt in the lower side of the shock absorber.
6. Remove the retaining bolt and nut in the axle side of the front lower link.
7. Loosen the retaining bolt and nut of the front lower link in the side of the suspension member.
8. Remove the retaining bolt and nut in the axle side of the radius rod.

9. Remove the retaining bolt in the rear suspension member side of the radius rod. Remove the radius rod from the vehicle.
    To install:
10. Installation is the reverse of the removal procedure.
11. Be sure to replace all non reusable components with new ones.

➡**Perform final tightening of the rear suspension member and axle installation position (rubber bushing) under unladen conditions with tires on level ground.**

12. Check and adjust the rear alignment, as necessary.

Front Lower Link

REMOVAL & INSTALLATION

1. Before servicing the vehicle, refer to the Precautions Section.
2. Raise and support the vehicle safely. Remove the tire and wheel assembly.
3. Remove the coil spring.
4. Remove the wheel sensor and sensor harness from the axle and suspension arm.
5. Remove the retaining bolt in the lower side of the shock absorber.
6. Remove the stabilizer bushing and clamp from the suspension member.
7. Remove the retaining bolt and nut between the front lower link and the suspension member.
8. Remove the retaining bolt and nut between the front lower link and the axle.
9. Remove the front lower link from the vehicle.
    To install:
10. Installation is the reverse of the removal procedure.
11. Be sure to replace all non reusable components with new ones.

➡**Perform final tightening of the rear suspension member and axle installation position under unladen conditions with tires on level ground.**

12. Check and adjust the rear alignment, as necessary.

Wheel Bearings

ADJUSTMENT

The front wheel bearings are part of a unitized hub and are not adjustable. Move the wheel hub in the axial direction by hand. Make sure that there is no looseness of the

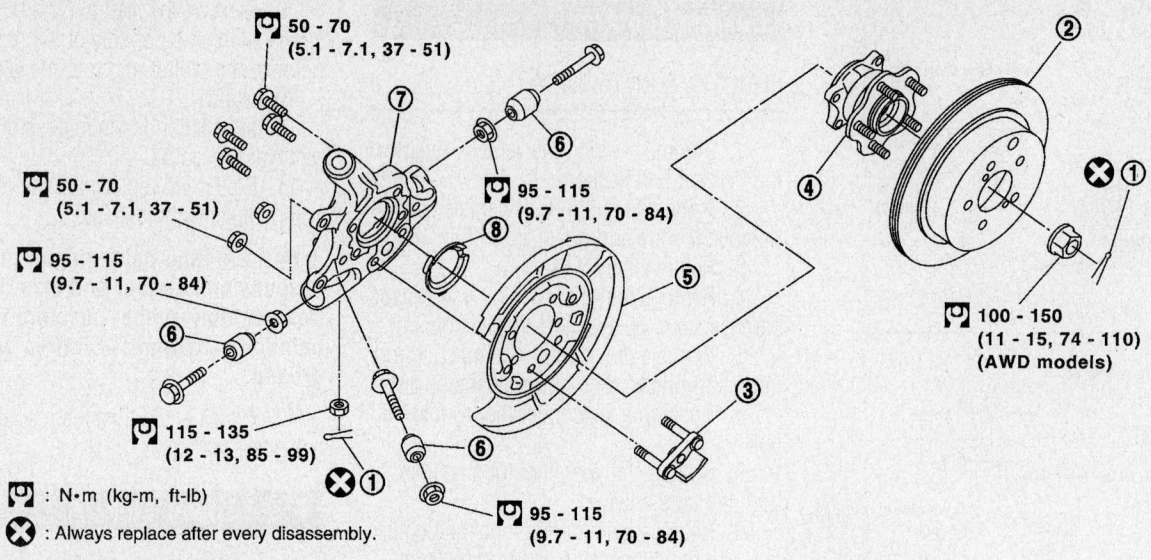

50 - 70
(5.1 - 7.1, 37 - 51)

50 - 70
(5.1 - 7.1, 37 - 51)

95 - 115
(9.7 - 11, 70 - 84)

95 - 115
(9.7 - 11, 70 - 84)

115 - 135
(12 - 13, 85 - 99)

95 - 115
(9.7 - 11, 70 - 84)

100 - 150
(11 - 15, 74 - 110)
(AWD models)

: N·m (kg-m, ft-lb)

: Always replace after every disassembly.

1. Cotter pin (AWD models)	2. Disc rotor	3. Anchor block
4. Wheel hub and bearing assembly	5. Back plate	6. Bushing
7. Axle	8. Axle cap (2WD models) Dust shield (AWD models)	

42356-MURA-G72

Rear wheel bearings and related components

wheel bearing. Axial end play is 0.002 inch or less.

REMOVAL & INSTALLATION

1. Before servicing the vehicle, refer to the Precautions Section.
2. Raise and support the vehicle safely. Remove the tire and wheel assembly.
3. Remove brake caliper. Hang it in a place where it will not interfere with work.
4. Put alignment marks on disc rotor and wheel hub and bearing assembly, then remove disc rotor.
5. Remove wheel sensor from axle. Do not pull on the wheel speed sensor harness.
6. Remove parking cable and parking brake shoe from back plate.
7. Remove cotter pin. Then remove lock nut from halfshaft. (AWD vehicles)
8. Using a puller (suitable tool), remove wheel hub and bearing assembly from halfshaft. (AWD vehicles)

9. Remove wheel hub and bearing assembly from axle.
10. Loosen bolts and nuts of front lower link, radius rod and rear lower link in side of suspension member.
11. Remove shock absorber bolt (lower), front lower link bolt and nut (axle-side) while supporting rear lower link with jack.
12. Remove bolt and nut in axle side of rear lower link. Then remove coil spring.
13. Remove bolt and nut in axle side of radius rod.
14. Remove suspension arm and cotter pin at axle, then loosen mounting nut.
15. Use a ball joint remover (suitable tool) to remove suspension arm from axle. Be careful not to damage ball joint boot.

➡ **To prevent damage to threads and to prevent ball joint remover (suitable tool) from coming off suddenly, and temporarily tighten mounting nut.**

16. Remove axle from vehicle.
17. Remove nuts from anchor block, then remove anchor block and back plate from axle.
18. Remove axle cap (2WD) or dust shield (AWD) from the axle.
To install:
19. Check for deformity, cracks and damage on each parts, replace if necessary.
20. Check for boot breakage, axial looseness, and torque of suspension arm ball joint.
21. Installation is the reverse of the removal procedure.
22. Be sure to replace all non reusable components with new ones.

➡ **Perform final tightening of installation position of suspension links (rubber bushing) under unladen conditions with tires on level ground.**

23. Check and adjust the rear alignment, as necessary.

BRAKES

Brake Caliper

REMOVAL AND INSTALLATION

Front

1. Before servicing the vehicle, refer to the Precautions Section.
2. Raise and support the vehicle

safely. Remove the tire and wheel assembly.
3. Drain brake fluid.
4. Remove union bolts and torque member bolts, and remove brake caliper assembly.
5. Remove disc rotor.
To install:
6. Install disc rotor.

7. Install caliper assembly to the vehicle, and tighten bolts to the specified torque shown in the illustration.
8. Install brake hose to the brake caliper assembly, and tighten union bolts to the specified torque shown in the illustration.

➡ **Do not reuse the copper washer for union bolts. Attach the brake hose to the brake hose mounting boss.**

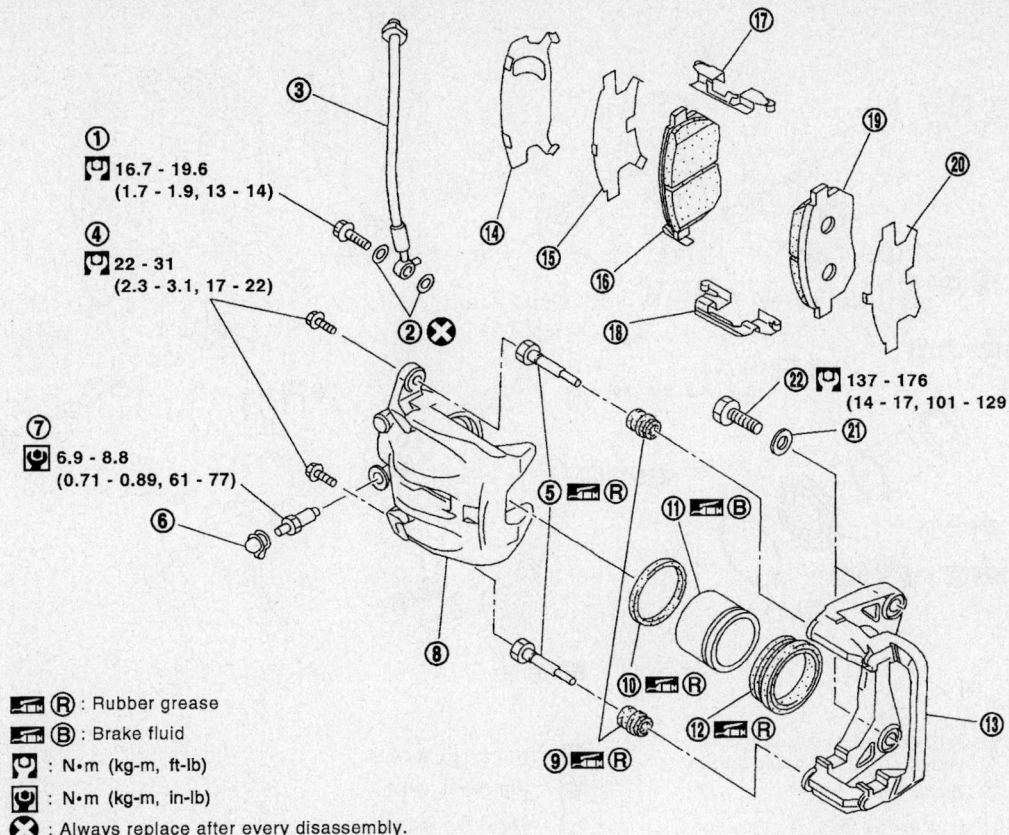

①
[torque] 16.7 - 19.6
(1.7 - 1.9, 13 - 14)

④
[torque] 22 - 31
(2.3 - 3.1, 17 - 22)

⑦
[torque] 6.9 - 8.8
(0.71 - 0.89, 61 - 77)

㉒ [torque] 137 - 176
(14 - 17, 101 - 129)

🔳Ⓡ : Rubber grease
🔳Ⓑ : Brake fluid
[symbol] : N•m (kg-m, ft-lb)
[symbol] : N•m (kg-m, in-lb)
✖ : Always replace after every disassembly.

1. Union bolt	2. Copper washer	3. Brake hose
4. Sliding pin bolt	5. Sliding pin	6. Cap
7. Bleed valve	8. Cylinder body	9. Sliding pin boot
10. Piston seal	11. Piston	12. Piston boot
13. Torque member	14. Inner shim cover	15. Inner shim
16. Inner pad	17. Pad retainer (Upper)	18. Pad retainer (Lower)
19. Outer pad	20. Outer shim	21. Washer
22. Torque member fixing bolt		

42356-MURA-G73

Front disc brake caliper and related components

9. Refill new brake fluid and bleed air.
10. Install the wheels.

Rear

1. Before servicing the vehicle, refer to the Precautions Section.
2. Raise and support the vehicle safely. Remove the tire and wheel assembly.
3. Drain brake fluid.
4. Remove union bolts and torque member bolts, and remove brake caliper assembly.
To install:
5. Install disc rotor.
6. Install caliper assembly to the vehicle, and tighten bolts to the specified torque shown in the illustration.
7. Install brake hose to caliper assembly and tighten union bolts to the specified torque shown in the illustration.

➡**Do not reuse the copper washer for union bolts. Attach brake hose to the brake hose mounting boss.**

8. Refill new brake fluid and bleed air. Refer
9. Install the tires to the vehicle.

Disc Brake Pads

REMOVAL & INSTALLATION

Front

1. Before servicing the vehicle, refer to the Precautions Section.
2. Raise and support the vehicle safely. Remove the tire and wheel assembly.
3. Remove sliding pin bolt (top).

4. Suspend cylinder body with a wire, and remove pads, pad retainers, shims from torque member.
To install:
5. Apply PBC (Poly Butyl Cuprysil) grease or silicon-based grease to the rear of the pad and to both sides of the shim, and attach the inner shim and shim cover to the inner pad, and the outer shim and outer shim cover to the outer pad.
6. Attach the pad retainer and pad to the torque member.
7. Push the piston in so that the pad is attached and attach the cylinder body to the torque member.
8. Install the sliding pin bolt (top) and tighten to the specified torque.
9. Check brake for drag.
10. Install the tires.

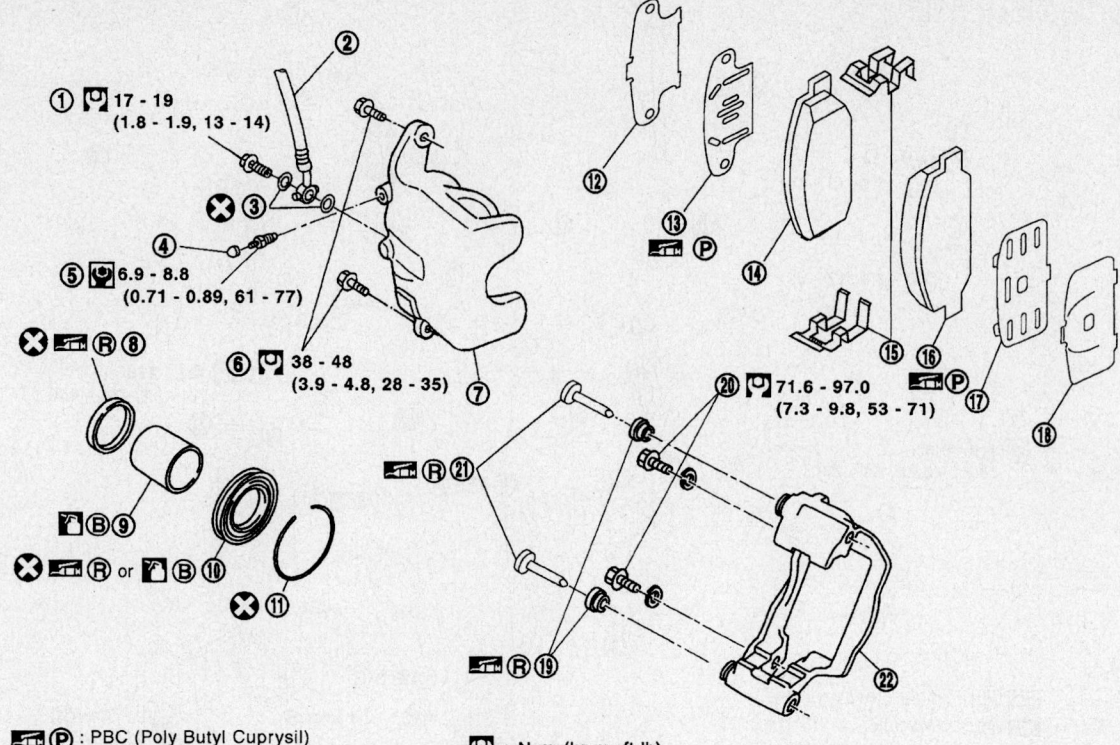

① 🔧 17 - 19
(1.8 - 1.9, 13 - 14)

⑤ 🔧 6.9 - 8.8
(0.71 - 0.89, 61 - 77)

⑥ 🔧 38 - 48
(3.9 - 4.8, 28 - 35)

⑳ 🔧 71.6 - 97.0
(7.3 - 9.8, 53 - 71)

🔧(P) : PBC (Poly Butyl Cuprysil)
grease or silicone-based grease.

🔧(B) : Brake fluid.

🔧(R) : Rubber grease.

🔧 : N•m (kg-m, ft-lb)

🔧 : N•m (kg-m, in-lb)

❌ : Always replace after every disassembly.

1. Union bolt	2. Brake hose	3. Copper washer
4. Cap	5. Bleed valve	6. Sliding pin bolt
7. Cylinder body	8. Piston seal	9. Piston
10. Piston boot	11. Retaining ring	12. Inner shim cover
13. Inner shim	14. Inner pad	15. Pad retainer
16. Outer pad	17. Outer shim	18. Outer shim cover
19. Slide pin boot	20. Torque member fixing bolt	21. Sliding pin
22. Torque member		

42356-MURA-G74

Rear disc brake caliper and related components

Rear

1. Before servicing the vehicle, refer to the Precautions Section.

2. Raise and support the vehicle safely. Remove the tire and wheel assembly.

3. Remove sliding pin bolt (top).

4. Suspend cylinder body with a wire, and remove pads, pad retainers, shims from torque member.

To install:

5. Apply PBC (Poly Butyl Cuprysil) grease or silicon based grease to the rear of the pad and to both sides of the shim, and attach the inner shim and shim cover to the inner pad, and the outer shim and outer shim cover to the outer pad.

6. Attach the pad retainer and pad to the torque member.

7. Push the piston in so that the pad is attached and attach the cylinder body to the torque member.

8. Install the sliding pin bolt (one on top) and tighten to the specified torque.

9. Check brake for drag.

10. Install the tires.

NISSAN

Quest

20

SPECIFICATION AND MAINTENANCE CHARTS

VEHICLE AND ENGINE IDENTIFICATION CHART

Engine Code								Model Year	
Code	Liters (cc)	Cu. In.	Cyl.	Fuel Sys.	Engine Type	Eng. Mfg.		Code ①	Year
T	3.3 (3275)	200	6	SEFI	SOHC	Nissan		2	2002
VQ35DE	3.5 (3498)	213	6	MFI	DOHC	Nissan		3	2003
								4	2004
								5	2005
								6	2006

MFI: Multi-port Fuel Injection

SEFI: Sequential Multi-port Fuel Injection

09482_NIQU_C0001

GENERAL ENGINE SPECIFICATIONS

Year	Engine Displacement Liters (VIN)	Net Horsepower @ rpm	Net Torque @ rpm (ft. lbs.)	Bore x Stroke (in.)	Com-pression Ratio	Oil Pressure @ rpm
2001	3.3 (T)	195@4500	190@3800	3.60x3.27	8.9:1	60-65@2000
2002	3.3 (T)	195@4500	190@3800	3.60x3.27	8.9:1	60-65@2000
2004	3.5 (VQ35DE)	240@5800	242@4400	3.76X3.20	10.3:1	43@2000
2005-06	3.5 (VQ35DE)	240@5800	242@4400	3.76X3.20	10.3:1	43@2000

MFI: Multiport fuel injection

SEFI: Sequential Multi-port Fuel Injection

09482_NIQU_C0002

ENGINE TUNE-UP SPECIFICATIONS

Year	Engine Displacement Liters (VIN)	Spark Plug Gap (in.)	Ignition Timing (deg.) MT	AT	Fuel Pump (psi)	Idle Speed (rpm) MT	AT ②	Valve Clearance In.	Ex.
2001	3.3 (T)	0.043	—	13-17B	34 ①	—	650-750	HYD	HYD
2002	3.3 (T)	0.043	—	13-17B	34 ①	—	650-750	HYD	HYD
2004	3.5 (VQ35DE)	0.043	—	10-20B	51 ③	—	650-750	HYD	HYD
2005-06	3.5 (VQ35DE)	0.043	—	10-20B	51 ③	—	650-750	HYD	HYD

NOTE: The Vehicle Emission Control Information label must be used if they differ from those in this chart.

B: Before top dead center

HYD: Hydraulic

① System pressure at idle with vacuum hose connected should increase to 43 psi when disconnected

② Transmission in Neutral

③ System pressure at idle

09482_NIQU_C0003

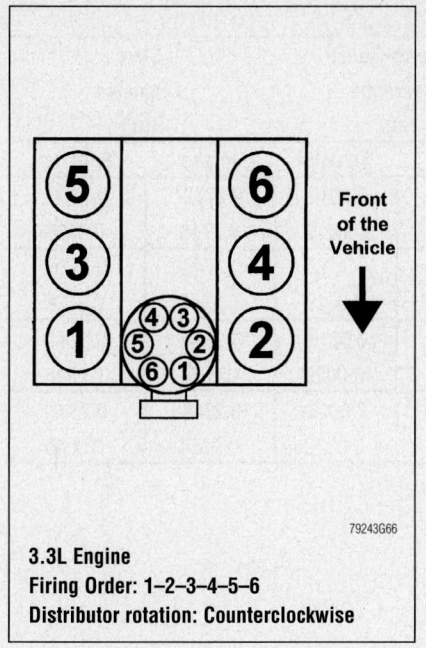

3.3L Engine
Firing Order: 1–2–3–4–5–6
Distributor rotation: Counterclockwise

79243G66

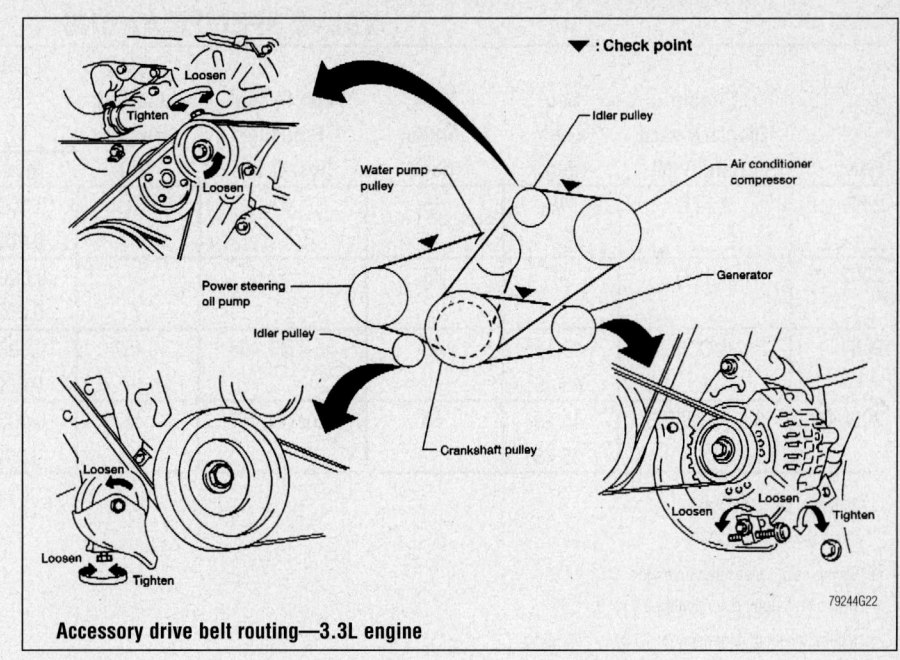

Accessory drive belt routing—3.3L engine

79244G22

CAPACITIES

Year	Model	Engine Displacement Liters (VIN)	Engine Oil with Filter (qts.)	Transmission (pts.) 4-Spd	5-Spd	Auto.	Drive Axle Front (pts.)	Rear (pts.)	Fuel Tank (gal.)	Cooling System (qts.) ②
2001	Quest	3.3 (T)	4.0	—	—	20.0	①	—	20	11.25
2002	Quest	3.3 (T)	4.0	—	—	20.0	①	—	20	11.25
2004	Quest	3.5 (VQ35DE)	3.78	—	—	③	—	—	20	10.50
2005-06	Quest	3.5 (VQ35DE)	3.78	—	—	③	—	—	20	10.50

NOTE: All capacities are approximate. Add fluid gradually and check to be sure a proper fluid level is obtained.

① Included in transaxle capacity

② Includes reservoir tank.

③ 4 speed: 9 qts.
   5 speed: 7 7/8 qts.

09482_NIQU_C0004

VALVE SPECIFICATIONS

Year	Engine Displacement Liters (VIN)	Seat Angle (deg.)	Face Angle (deg.)	Spring Test Pressure (lbs. @ in.)	Spring Installed Height (in.)	Stem-to-Guide Clearance (in.)		Stem Diameter (in.)	
						Intake	Exhaust	Intake	Exhaust
2001	3.3 (T)	45	45	①	②	0.0008-0.0021	0.0012-0.0019	0.2742-0.2748	0.3136-0.3138
2002	3.3 (T)	45	45	①	②	0.0008-0.0021	0.0012-0.0019	0.2742-0.2748	0.3136-0.3138
2004	3.5 (VQ35DE)	45.15-45.45	45	45.4@1.457	1.457	0.0008-0.0021	0.0016-0.0029	0.2348-0.2354	0.2341-0.2346
2005-06	3.5 (VQ35DE)	45.15-45.45	45	45.4@1.457	1.457	0.0008-0.0021	0.0016-0.0029	0.2348-0.2354	0.2341-0.2346

① Outer spring: 118@1.81

Inner spring: 57.3@0.984

② Spring height measured unloaded

Minimum length. outer spring: 2.016

Minimum length. inner spring: 1.736

09482_NIQU_C0005

CRANKSHAFT AND CONNECTING ROD SPECIFICATIONS
All measurements are given in inches.

Year	Engine Displacement Liters (VIN)	Crankshaft				Connecting Rod		
		Main Brg. Journal Dia.	Main Brg. Oil Clearance	Shaft End-play	Thrust on No.	Journal Diameter	Oil Clearance	Side Clearance
2001	3.3 (T)	2.4791-2.4793	0.0011-0.0022	0.0020-0.0067	3	1.9667-1.9675	0.0006-0.0021	0.0079-0.00138
2002	3.3 (T)	2.4791-2.4793	0.0011-0.0022	0.0020-0.0067	3	1.9667-1.9675	0.0006-0.0021	0.0079-0.00138
2004	3.5 (VQ35DE)	①	0.0014-0.0018	0.0118	4	②	0.0013-0.0023	0.0079-0.0138
2005-06	3.5 (VQ35DE)	①	0.0014-0.0018	0.0118	4	②	0.0013-0.0023	0.0079-0.0138

① There are 24 different grades, ranging from A (2.3612) to 7 (2.3603)

② Grade 0: 0.0591-0.0592

Grade 1: 0.0592-0.0593

Grade 2: 0.0593-0.0594

09482_NIQU_C0006

PISTON AND RING SPECIFICATIONS
All measurements are given in inches.

Year	Engine Displacement Liters (VIN)	Piston Clearance	Ring Gap			Ring Side Clearance		
			Top Compression	Bottom Compression	Oil Control	Top Compression	Bottom Compression	Oil Control
2001	3.3 (T)	①	0.0083-0.0122	0.0197-0.0236	0.0079-0.0236	0.0016-0.0031	0.0012-0.0028	0.0006-0.0073
2002	3.3 (T)	①	0.0083-0.0122	0.0197-0.0236	0.0079-0.0236	0.0016-0.0031	0.0012-0.0028	0.0006-0.0073
2004	3.5 (VQ35DE)	0.0004-0.0012	0.0091-0.0130	0.0130-0.0189	0.0079-0.0197	0.0018-0.0031	0.0012-0.0028	0.0026-0.0053
2005-06	3.5 (VQ35DE)	0.0004-0.0012	0.0091-0.0130	0.0130-0.0189	0.0079-0.0197	0.0018-0.0031	0.0012-0.0028	0.0026-0.0053

① Journals 1, 2 and 6: 0.0010 - 0.0018 in.

Journals 3 and 4: 0.0006 - 0.0010 in.

Journal 5: 0.0012 - 0.0016 in.

09482_NIQU_C0007

TORQUE SPECIFICATIONS
All readings in ft. lbs.

Year	Engine Displacement Liters (VIN)	Cylinder Head Bolts	Main Bearing Bolts	Rod Bearing Bolts	Crankshaft Damper Bolts	Flywheel Bolts	Manifold		Spark Plugs	Oil Pan Drain Plug
							Intake	Exhaust		
2001	3.3 (T)	①	②	②	141-156	61-69	①	13-16	14-22	24
2002	3.3 (T)	①	②	②	141-156	61-69	①	13-16	14-22	24
2004	3.5 (VQ35DE)	③	④	⑤	⑥	61-69	14	23	18	25
2005-06	3.5 (VQ35DE)	③	④	⑤	⑥	61-69	14	23	18	25

① Intake manifold and cylinder heads are installed at the same time.

Step 1: cylinder head bolts to 22 ft. lbs.

Step 2: cylinder head bolts to 43 ft. lbs.

Step 3: Loosen all bolts completely

Step 4: cylinder head bolts to 7 ft. lbs.

Step 5: Intake manifold bolts to 2.9 (4Nm) ft. lbs.

Step 6: Intake manifold bolts to 13 ft. lbs.

Step 7: Intake manifold bolts to 14 ft. lbs.

Step 8: Loosen all maifold bolts completely

Step 9: Cylinder head bolts to 26 inch lbs.

Step 10: cylinder head bolts to 47 ft. lbs. Or, an additional 65 degrees

Step 11: cylinder head sub-bolts to 104 inch lbs.

Step 12: Intake manifold bolts to 36 inch lbs.

Step 13: Intake manifold bolts to 78 inch lbs.

Step 14: Intake manifold bolts to 84 inch lbs.

② Step 1: 34-37 ft. lbs.

Step 2: 67-74 ft. lbs.

③ Step 1: 72 ft. lbs.

Step 2: Loosen all bolts completely

Step 3: 29 ft. lbs.

Step 4: +90 degrees

Step 5: +90 degrees

④ Step 1: 26 ft. lbs.

Step 2: +90-95 degrees

⑤ Step 1: 15 ft. lbs.

Step 2: +90-95 degrees

⑥ 32 ft. lbs. +60-65 degrees

09482_NIQU_C0008

WHEEL ALIGNMENT

Year	Model		Caster Range (Deg.)	Caster Preferred Setting (Deg.)	Camber Range (Deg.)	Camber Preferred Setting (Deg.)	Toe-in (in.)
2001	Quest	F	0.75	2.75	0.75	-0.25	0.04 +/- 0.04
		R	—	—	0.75	-1.00	0.04 +/- 0.16
2002	Quest	F	0.75	2.75	0.75	-0.25	0.04 +/- 0.04
		R	—	—	0.75	-1.00	0.04 +/- 0.16
2004	Quest ① ②	F	③	③	④	④	0.02 +/- 0.04
		R	—	—	⑤	⑤	0.02 +/- 0.16
2005-06	Quest ① ②	F	③	③	④	④	0.02 +/- 0.04
		R	—	—	⑤	⑤	0.02 +/- 0.16

① Vehicle unladen

② Specifications are decimal degrees

③ Minimum: 2.17
Nominal: 2.92
Maximum: 3.67
Left and right difference 0.75 or less

④ Minimum: - 0.92
Nominal: - 0.17
Maximum: 0.58
Left and right difference 0.75 or less

⑤ Minimum: - 0.24
Nominal: - 0.26
Maximum: - 0.76

09482_NIQU_C0009

TIRE, WHEEL AND BALL JOINT SPECIFICATIONS

Year	Model	OEM Tires Standard	OEM Tires Optional	Tire Pressures (psi) Front	Tire Pressures (psi) Rear	Wheel Size	Ball Joint Inspection	Lug Nut
2001	Quest	P215/70R15	P225/60R16	30	30	Std: 5.5-JJ Opt: 6.5-JJ	①	80
2002	Quest	P215/70R15	P225/60R16	30	30	Std: 5.5-JJ Opt: 6.5-JJ	①	80
2004	Quest	P225/65R16	P225/60R17	35	35	NA	①	83
2005-06	Quest	P225/65R16	P225/60R17	35	35	NA	①	83

NA: Not Available

OEM: Original Equipment Manufacturer

PSI: Pounds Per Square Inch

① Replace if any measurable movement is found.

09482_NIQU_C0010

BRAKE SPECIFICATIONS
All measurements in inches unless noted

Year	Model		Brake Disc Original Thickness	Brake Disc Minimum Thickness	Brake Disc Maximum Runout	Brake Drum Diameter Original Inside Diameter	Brake Drum Diameter Max, Wear Limit	Brake Drum Diameter Maximum Machine Diameter	Minimum Lining Thickness	Brake Caliper Bracket-to-Hub Bolt (ft. lbs.)	Brake Caliper Mounting Pin or Bolt (ft. lbs.)
2001	Quest	F	1.002	0.940	0.0028	—	—	—	0.079	—	18-25
		R	—	—	—	9.84	9.90	9.86	0.079	—	—
2002	Quest	F	1.002	0.940	0.0028	—	—	—	0.079	—	18-25
		R	—	—	—	9.84	9.90	9.86	0.079	—	—
2004	Quest	F	1.100	1.02	0.0006	—	—	—	0.079	115	20
		R	0.630	0.551	0.0006	—	—	—	0.079	62	32
2005-06	Quest	F	1.100	1.02	0.0006	—	—	—	0.079	115	20
		R	0.630	0.551	0.0006	—	—	—	0.079	62	32

NOTE: Due to changes made during production, refer to the manufacturer's specifications if they differ from those in this chart

F: Front

R: Rear

09482_NIQU_C0011

SCHEDULED MAINTENANCE INTERVALS
2001—02 NISSAN QUEST

TO BE SERVICED	TYPE OF SERVICE	VEHICLE MILEAGE INTERVAL (x1000)												
		5	10	15	20	25	30	35	40	45	50	55	60	65
Engine oil & filter	R	✓	✓	✓	✓	✓	✓	✓	✓	✓	✓	✓	✓	✓
Rotate tires	S/I	✓		✓		✓		✓		✓		✓		✓
Engine coolant strength hoses & clamps	S/I			✓			✓			✓			✓	
Air cleaner filter	R						✓						✓	
Automatic transmission fluid & filter	R						✓						✓	
Engine coolant ①	R						✓						✓	
PCV valve	R												✓	
Spark plugs ②	R													
Drive belts ③	S/I			✓			✓			✓			✓	
Timing belt ④	S/I													
Exhaust system & heat shields	S/I			✓			✓			✓			✓	
Drive shaft boots	S/I			✓			✓			✓		✓		
Front & rear brake components	S/I	✓	✓	✓	✓	✓	✓	✓	✓	✓	✓	✓	✓	✓

R: Replace   S/I: Service or Inspect

① Engine coolant: change every 30,000 miles or 36 months.

② Replace every 105,000 miles

③ Inspect every 15, 000 miles or 12 months and replace every 60,000 miles or 48 months.

④ Replace every 105,000 miles

FREQUENT OPERATION MAINTENANCE (SEVERE SERVICE)

If a vehicle is operated under any of the following conditions it is considered severe service:

- Extremely dusty areas.

- 50% or more of the vehicle operation is in 32°C (90°F) or higher temperatures, or constant operation in temperatures below 0°C (32°F).

- Prolonged idling (vehicle operation in stop and go traffic).

- Frequent short running periods (engine does not warm to normal operating temperatures).

- Police, taxi, delivery usage or trailer towing usage.

Engine oil & filter: replace every 3000 miles.

Rotate tires initially at 6000 miles and every 9000 miles thereafter.

Air cleaner filter: change every 15,000 miles.

Engine coolant strength, hoses & clamps: check every 15,000 miles.

Exhaust system: check every 15,000 miles.

Automatic transmission fluid & filter: change every 21,000 miles.

09482_NIQU_C0012

SCHEDULED MAINTENANCE INTERVALS
2004—06 NISSAN QUEST

TO BE SERVICED	TYPE OF SERVICE	VEHICLE MILEAGE INTERVAL (x1000)															
		3.75	7.5	11.3	15	18.8	22.5	26.3	30	33.8	37.5	41.3	45	48.8	52.5	56.3	60
Engine oil & filter	R	✓	✓	✓	✓	✓	✓	✓	✓	✓	✓	✓	✓	✓	✓	✓	✓
Brake lines & cables	S/I				✓				✓				✓				✓
Brake pads and rotors	S/I		✓		✓		✓		✓		✓		✓		✓		✓
Driveshaft boots & propeller shaft	L/I		✓		✓		✓		✓		✓		✓		✓		✓
Transaxle fluid ①	I																✓
Air cleaner filter ②	R																✓
Drive belt(s) ③	S/I																✓
Engine coolant ④	R																✓
Spark plugs	R	platinum tipped plugs every 105,000 miles															
Cabin air filter	R				✓				✓				✓				✓
Exhaust system	I		✓		✓		✓		✓		✓		✓		✓		✓
Fuel lines	S/I								✓								✓
Steering gear, linkage, axle & suspension parts	I		✓		✓		✓		✓		✓		✓		✓		✓
Valve clearance ⑤																	
Vapor lines	S/I																✓

R: Replace   S/I: Service or Inspect   L: Lubricate

① Replace first a 60,000 or 48 months, inspect every 15,000 miles

② If operating in dusty conditions more frequent maintenance may be required

③ Replace first a 60,000 or 48 months, inspect every 15,000 miles, replace if damaged during any inspection

④ Replace first after 60,000 or 48 months, replace every 30,000 or 24 months

⑤ If valve noise increases inspect clearance

FREQUENT OPERATION MAINTENANCE (SEVERE SERVICE)

If a vehicle is operated under any of the following conditions it is considered severe service:

- Extremely dusty areas.

- 50% or more of the vehicle operation is in 32°C (90°F) or higher temperatures, or constant operation in temperatures below 0°C (32°F).

- Prolonged idling (vehicle operation in stop and go traffic).

- Frequent short running periods (engine does not warm to normal operating temperatures).

- Police, taxi, delivery usage or trailer towing usage.

Oil & oil filter: replace every 3750 miles.

Brake pads, discs, drums & linings: service or inspect every 7500 miles.

Driveshaft boots & propeller shaft: service or inspect every 7500 miles.

Exhaust system: service or inspect every 7500 miles.

Steering gear (box) & linkage, (steering damper-4x4), axle & suspension parts: service or inspect every 7500 miles.

Steering linkage ball joints & front suspension ball joints: service or inspect every 7500 miles.

09482_NIQU_C0013

ENGINE REPAIR

Distributor

REMOVAL

This applies to the 3.3L engine only.
1. Before servicing the vehicle, refer to the precautions section.
2. Remove or disconnect the following:
  • Negative battery cable
  • Distributor cap
  • Distributor wiring harness connector
3. Matchmark the rotor to the distributor housing and the distributor housing to the cylinder head.
4. Remove the distributor hold-down bolt and the distributor.

INSTALLATION

Timing Not Disturbed

1. Install or connect the following:
  • Distributor and align the matchmarks made during removal. Tighten the hold-down bolt to 10–12 ft. lbs. (14–17 Nm).
  • Distributor wiring harness connector
  • Distributor cap
  • Negative battery cable
2. Check the ignition timing and adjust, as necessary.

Timing Disturbed

1. Set the engine to Top Dead Center (TDC) of the compression stroke for the No. 1 cylinder.
2. Align the index mark on the distributor shaft with the protrusion on the distributor housing.
3. Install the distributor and check that the distributor rotor is aligned.

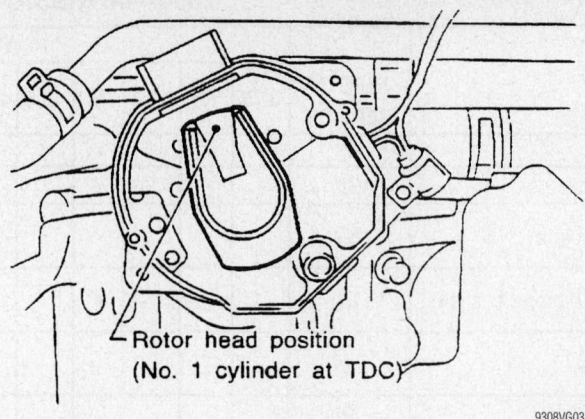

Rotor head position
(No. 1 cylinder at TDC)

9308VG03

Distributor rotor alignment

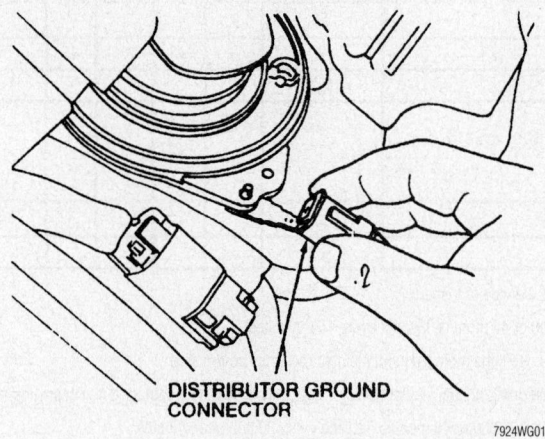

DISTRIBUTOR GROUND CONNECTOR

7924WG01

Disengage the distributor ground connector when removing the distributor

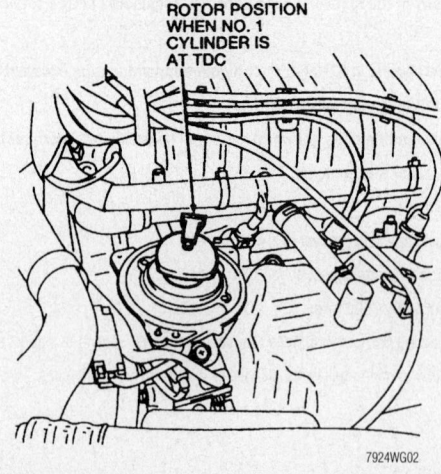

ROTOR POSITION WHEN NO. 1 CYLINDER IS AT TDC

7924WG02

Note the position of the rotor when the No. 1 piston is at TDC on the compression stroke

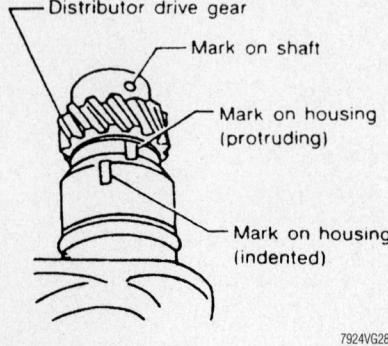

Distributor drive gear
Mark on shaft
Mark on housing (protruding)
Mark on housing (indented)

7924VG28

Distributor shaft alignment

4. Install or connect the following:
  • Distributor. Tighten the hold-down bolt to 10–12 ft. lbs. (14–17 Nm).
  • Distributor cap
  • Distributor harness connector
5. Check the ignition timing and adjust, as necessary.

Alternator

REMOVAL

3.3L Engine

1. Before servicing the vehicle, refer to the precautions section.
2. Remove or disconnect the following:
   - Negative battery cable
   - Idler adjusting bolt, loosen
   - A/C belt
   - Engine undercover
   - Alternator electrical connectors and bracket
   - Alternator mounting bolts
   - Alternator belt
   - Alternator

3.5L Engine

1. Before servicing the vehicle, refer to the precautions section.
2. Disconnect the negative battery terminal.
3. Remove radiator.
4. Remove the drive belt.
5. Remove idler pulley.
6. Remove the alternator adjustable top mount.
7. Remove the alternator lower bolt and nut.
8. Disconnect the alternator harness connectors.
9. Remove the alternator upper bolt.
10. Remove the alternator.

INSTALLATION

3.3L Engine

1. Before servicing the vehicle, refer to the precautions section.
2. Install the components in the reverse order of removal. Tighten the fasteners to the following specifications:
   a. Alternator mounting bolts to 16–22 ft. lbs. (22–29 Nm).

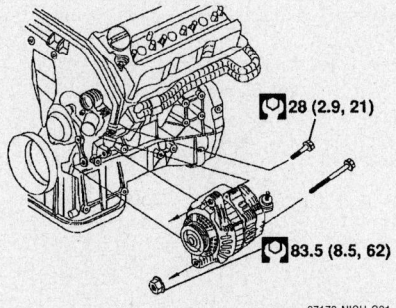

28 (2.9, 21)

83.5 (8.5, 62)

67170-NIQU-G01

Alternator exploded view–3.5L engine

b. Alternator bracket bolt to 12–15 ft. lbs. (16–20 Nm).

3.5L Engine

1. Before servicing the vehicle, refer to the precautions section.
2. Install the alternator.
3. Install the alternator upper bolt. Tighten to 21 ft. lbs. (28 Nm).
4. Connect the alternator harness connectors.
5. Install the alternator lower bolt and nut. Tighten to 62 ft. lbs. (83 Nm).
6. Install the alternator adjustable top mount.
7. Install idler pulley.
8. Install the drive belt.
9. Install radiator.
10. Connect the negative battery terminal.

Ignition Timing

ADJUSTMENT

3.3L Engine

1. Before servicing the vehicle, refer to the precautions section.
2. Check for trouble codes and make necessary repairs if needed.
3. Apply the parking brake and be sure that the vehicle is in PARK.
4. Start and run the engine until it reaches normal operating temperature.
5. Run the engine at about 2000 rpm for 2 minutes under no-load.
6. Turn off all electrical loads.
7. Disconnect the Throttle Position (TP) sensor electrical connector.
8. Be sure the engine speed is 700–800 rpm.

9. Rev the engine 2 or 3 times to 2,000–3,000 rpm and return the engine to idle speed.
10. Connect a timing light to the No. 1 cylinder spark plug wire at the distributor end and check the ignition timing. Be sure that the timing pointer is pointing to the 15° BTDC mark on the crankshaft pulley.

➡**Each notch on the crankshaft pulley represents 5°.**

11. If the timing is not within the specification, loosen the distributor mounting bolt and adjust the distributor until the timing is at the proper specification.
12. Tighten the distributor mounting bolt to 10–12 ft. lbs., (14–17 Nm).
13. Stop the engine and connect the TP sensor.

Engine Assembly

REMOVAL & INSTALLATION

3.3L Engine

1. Before servicing the vehicle, refer to the precautions section.
2. Properly relieve the fuel system pressure.
3. Drain the coolant and crankcase.
4. Remove or disconnect the following:
   - Negative battery cable
   - Front wheels
   - All vacuum hoses, fuel lines, wires, harnesses and connectors that would interfere with engine removal
   - Exhaust tube
   - Ball joints
   - Drive shafts
5. Recover the refrigerant from the A/C system

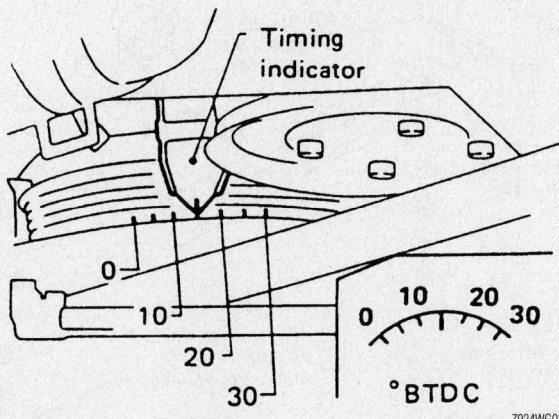

7924WG03

Adjust the timing so the pointer on the engine indicates 15° before top dead center (3 notches from TDC) on the crankshaft pulley.

- A/C compressor manifold
- Power steering pump

6. Support the engine using a suitable lift.

- Left hand engine mount bolts

- Right hand engine mount
- Rear A/C refrigerant line bracket, if equipped
- Crossmember

7. Lower the engine and transaxle assembly and remove it from the vehicle.

To install:

8. Installation is the reverse of removal. Refer to the accompanying engine mounting

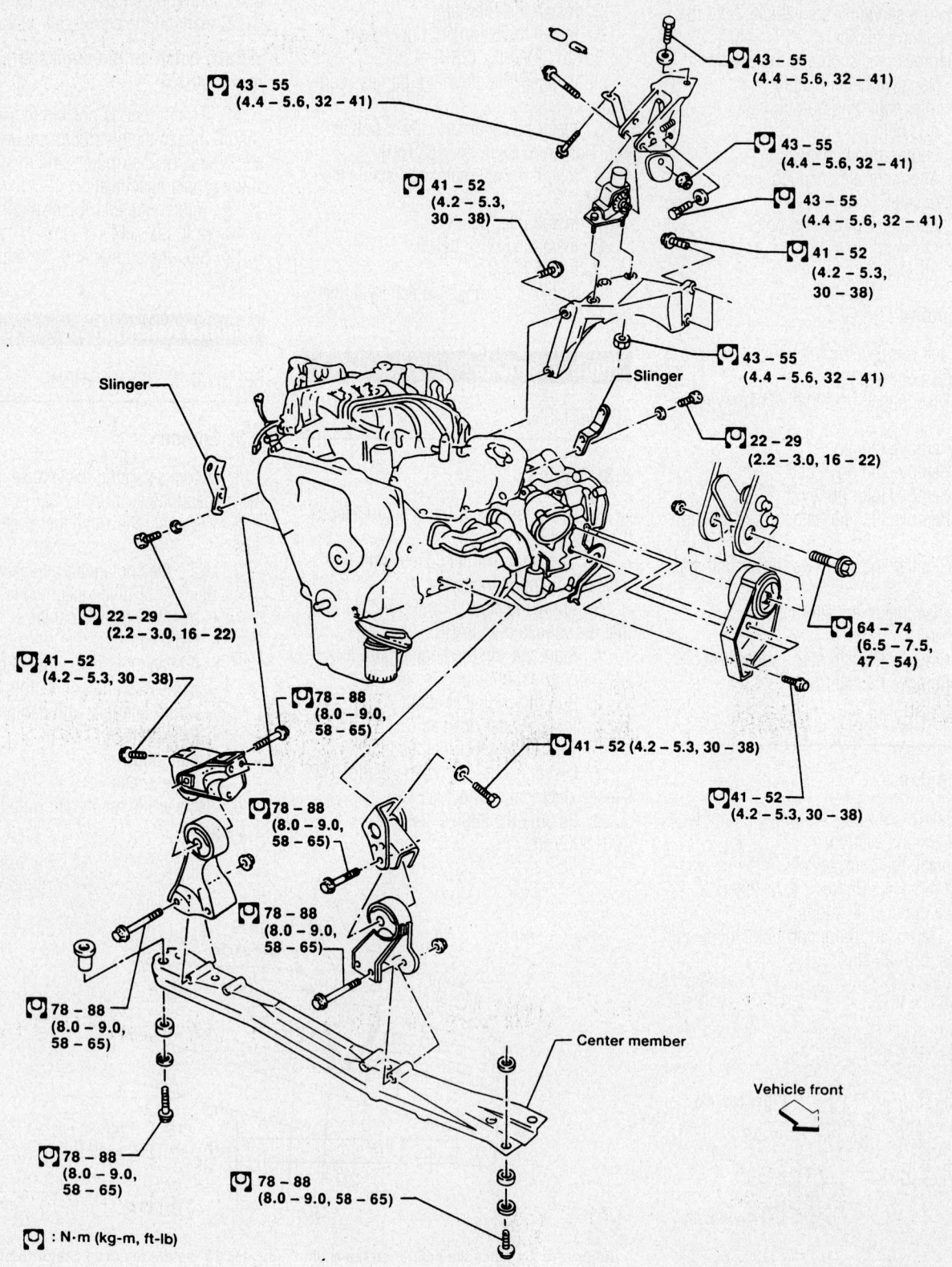

43 – 55
(4.4 – 5.6, 32 – 41)

43 – 55
(4.4 – 5.6, 32 – 41)

41 – 52
(4.2 – 5.3, 30 – 38)

43 – 55
(4.4 – 5.6, 32 – 41)

43 – 55
(4.4 – 5.6, 32 – 41)

41 – 52
(4.2 – 5.3, 30 – 38)

43 – 55
(4.4 – 5.6, 32 – 41)

Slinger

Slinger

22 – 29
(2.2 – 3.0, 16 – 22)

22 – 29
(2.2 – 3.0, 16 – 22)

41 – 52
(4.2 – 5.3, 30 – 38)

64 – 74
(6.5 – 7.5, 47 – 54)

78 – 88
(8.0 – 9.0, 58 – 65)

41 – 52 (4.2 – 5.3, 30 – 38)

78 – 88
(8.0 – 9.0, 58 – 65)

41 – 52
(4.2 – 5.3, 30 – 38)

78 – 88
(8.0 – 9.0, 58 – 65)

78 – 88
(8.0 – 9.0, 58 – 65)

Center member

Vehicle front

78 – 88
(8.0 – 9.0, 58 – 65)

78 – 88
(8.0 – 9.0, 58 – 65)

: N·m (kg-m, ft-lb)

9302WG01

Engine mounting components and specifications—3.3L engines

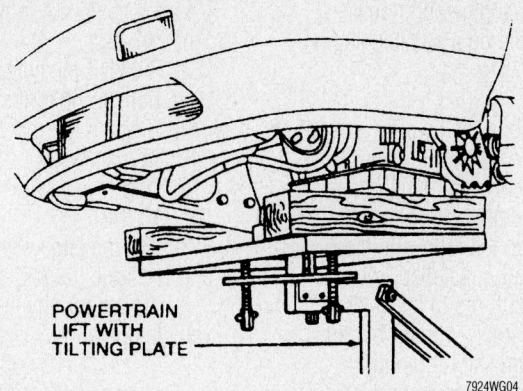

POWERTRAIN
LIFT WITH
TILTING PLATE

7924WG04

Carefully lower the engine/transaxle assembly from the vehicle.

illustration for all necessary torque specifications.

3.5L Engine

1. Before servicing the vehicle, refer to the precautions section.
2. Disconnect the battery cables.
3. Drain the engine oil, coolant and transmission fluid.
4. Remove the cowl top extension.
5. Disconnect the engine harness from the Powertrain Control Module (PCM) and the two connections at the right hand strut tower.
6. Disconnect the engine harness ground connections.
7. Disconnect the coolant valve and position the valve aside.
8. Disconnect the Mass Air Flow (MAF) sensor connector.
9. Remove the fresh air intake tube and air cleaner–to–electric throttle control actuator tube attached to the air cleaner lid.
10. Remove the lower air cleaner case.
11. Remove the engine cover.
12. Remove the battery and tray.
13. Relieve the fuel system pressure.
14. Disconnect fuel hose quick connection at vehicle piping side as follows:
    a. Remove the connector cap from the fuel hose.
    b. Squeeze the two tabs and pull the fuel hose from the fuel line.

➡**If the connector and the tube are stuck together, push and pull several times until they start to move, then disconnect them by pulling.**

⁂ **CAUTION**

The tube can be removed when the tabs are completely depressed. Do not twist it more than necessary.

➡**Do not use any tools to remove the quick connector. Keep the resin tube away from heat. Be especially careful when welding near the tube.? Prevent acid liquids such as battery electrolyte, etc. from getting on the resin tube. Do not bend or twist the tube during removal or installation. Do not remove the remaining retainer on the tube. When the tube is replaced, also replace the retainer with a new one. To keep the connecting portion clean and to avoid damage and foreign materials entering, cover the ends of the fuel tubes with plastic bags or something similar.**

15. Remove the radiator assembly, coolant reservoir, and hoses.
16. Disconnect the power brake booster vacuum hose from the back of the intake manifold collector.
17. Disconnect the EVAP canister purge volume control solenoid valve hose.
18. Disconnect heater hoses at the water outlet and heater pipe.
19. Disconnect the two fusible link connectors at the positive battery terminal.
20. Disconnect two engine harness connectors below the MAF sensor attached to the shock tower.
21. Disconnect the harness retainers and position the engine harness aside.
22. Remove the ground cable and ground wire from transaxle.
23. Disconnect the transaxle shift controls.
24. Remove the drive belts.
25. Remove the front exhaust tube and hanger.
26. Remove the front drive shafts.
27. Remove the lower ball joint pinch bolt, then separate the transverse link from the steering knuckle.
28. Remove the power steering line bracket from the front suspension member.

29. Remove the mounting bolts on the lower side of the steering gear.
30. Disconnect the front engine mount electrical connector.
31. Disconnect the connecting rod from the front.
32. Disconnect the power steering line brackets from the rear engine mount insulator and rear of the lower intake manifold collector.
33. Remove engine oil cooler pipe bolts.
34. Discharge and recover the A/C refrigerant.
35. Remove A/C low and high–pressure flexible hoses.
36. Remove the A/C compressor.
37. Disconnect the transaxle breather hose.
38. Disconnect the power steering pressure switch.
39. Disconnect the harness retainer from power steering oil pump bracket.
40. Remove the power steering adjusting bar and power steering pump, without disconnecting the lines from the engine and position and secure it aside.
41. Remove the Crankshaft Position (CKP) sensor.
42. Remove the rear cover plate and bolts securing the torque converter–to–drive plate.
43. Position a transmission jack under the engine/transaxle assembly.
44. On 4 speed transmission equipped vehicles, remove the left hand transaxle mount through bolt.
45. Remove the right hand engine mount insulator nuts and bolt.
46. Remove the front suspension member and engine/transaxle assembly as follows:
    a. Remove the right and left hand member pin stay bolts.
    b. Remove the front suspension member nuts and cups and carefully lower the front suspension member and engine/transaxle assembly avoiding interference with the vehicle body.

⁂ **CAUTION**

Make sure to disconnect electrically controlled engine mounting insulator harness clips from the front suspension member prior to removal. Before and during this procedure, always check if any harnesses are left connected. Avoid any damage to, or any oil/grease smearing or spills onto the engine mounting insulators.

47. Remove the starter motor.

48. Disconnect the electrical connectors, harness retainers and remove the harnesses.

49. Disconnect the transmission cooler hoses and remove the cooler.

50. Remove the front and rear engine mount through bolts.

51. On 5 speed transmission equipped models, remove the left hand transaxle mount bolts.

52. Raise the engine/transaxle and remove the front suspension member.

53. Remove the transmission cooler valve from the engine with the hoses attached.

54. Separate the engine and transaxle and mount the engine on a suitable engine stand.

To install:

55. Installation is in the reverse order of removal. Tighten the bolts to the specifications shown in the engine mounting and front suspension member exploded view illustrations and note the following steps:

   a. With converter installed, rotate crankshaft several turns to check that transaxle rotates freely without binding. Tighten the transmission–to–engine bolts for the 4 or 5 speed transmission vehicles as shown in the accompanying illustrations.

   b. Tighten the converter–to–drive plate bolts to 40 ft. lbs. (54 Nm).

   c. Tighten the rear cover plate bolt to 61 inch lbs. (7 Nm).

   d. Tighten the CKP sensor to 85 inch lbs. (10 Nm).

   e. Fill the engine with oil, and coolant and recharge the A/C system.

   f. Check for any leaks and for proper vehicle operation.

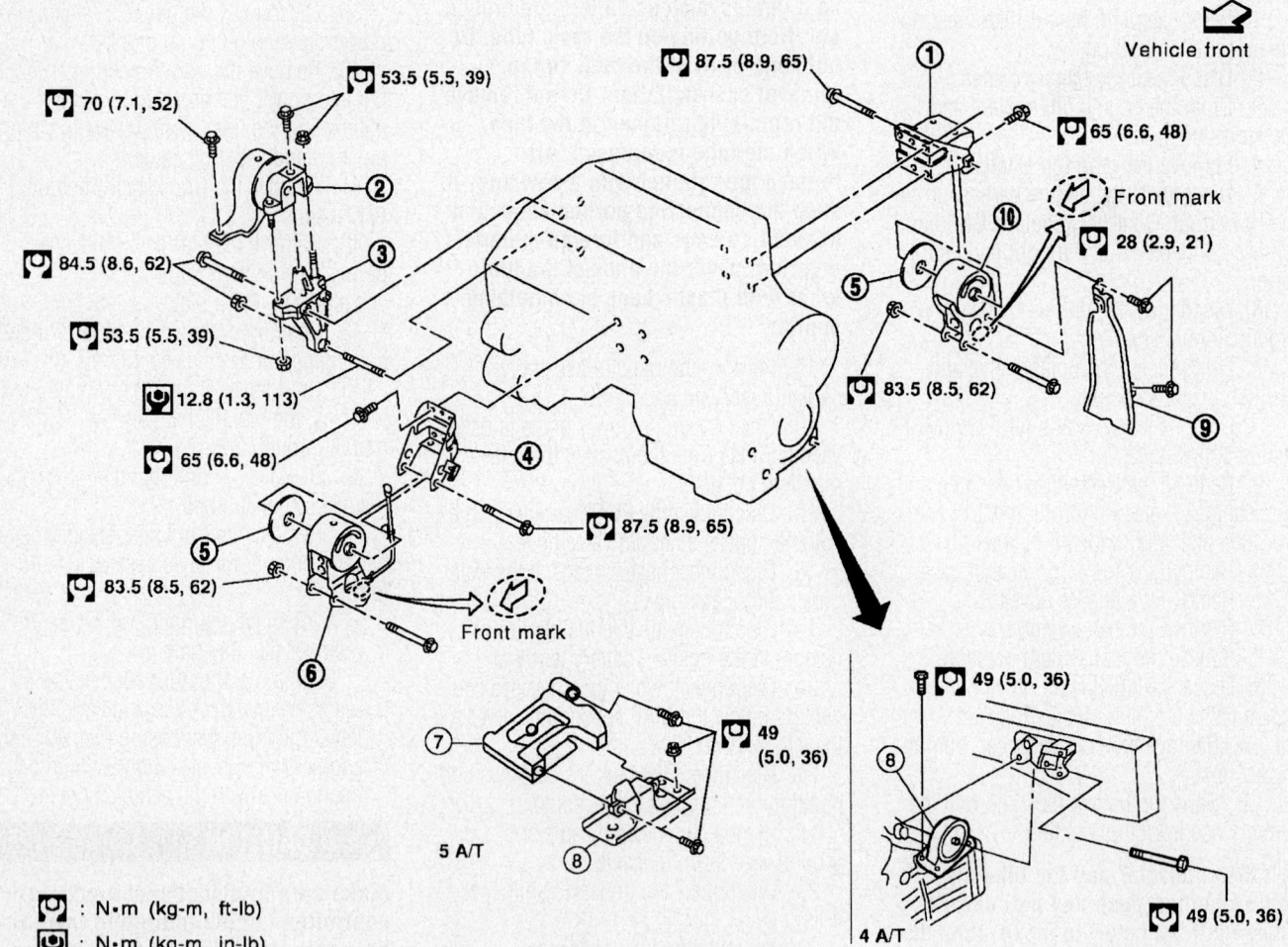

: N•m (kg-m, ft-lb)

: N•m (kg-m, in-lb)

1. Rear engine mounting bracket
2. RH engine mounting insulator
3. RH engine mount
4. Front engine mount
5. Stopper
6. Front engine mounting insulator
7. LH transaxle mount
8. LH transaxle mounting insulator
9. Air guide
10. Rear engine mounting insulator

Exploded view of the engine mounting–3.5L engine

67170-NIQU-G73

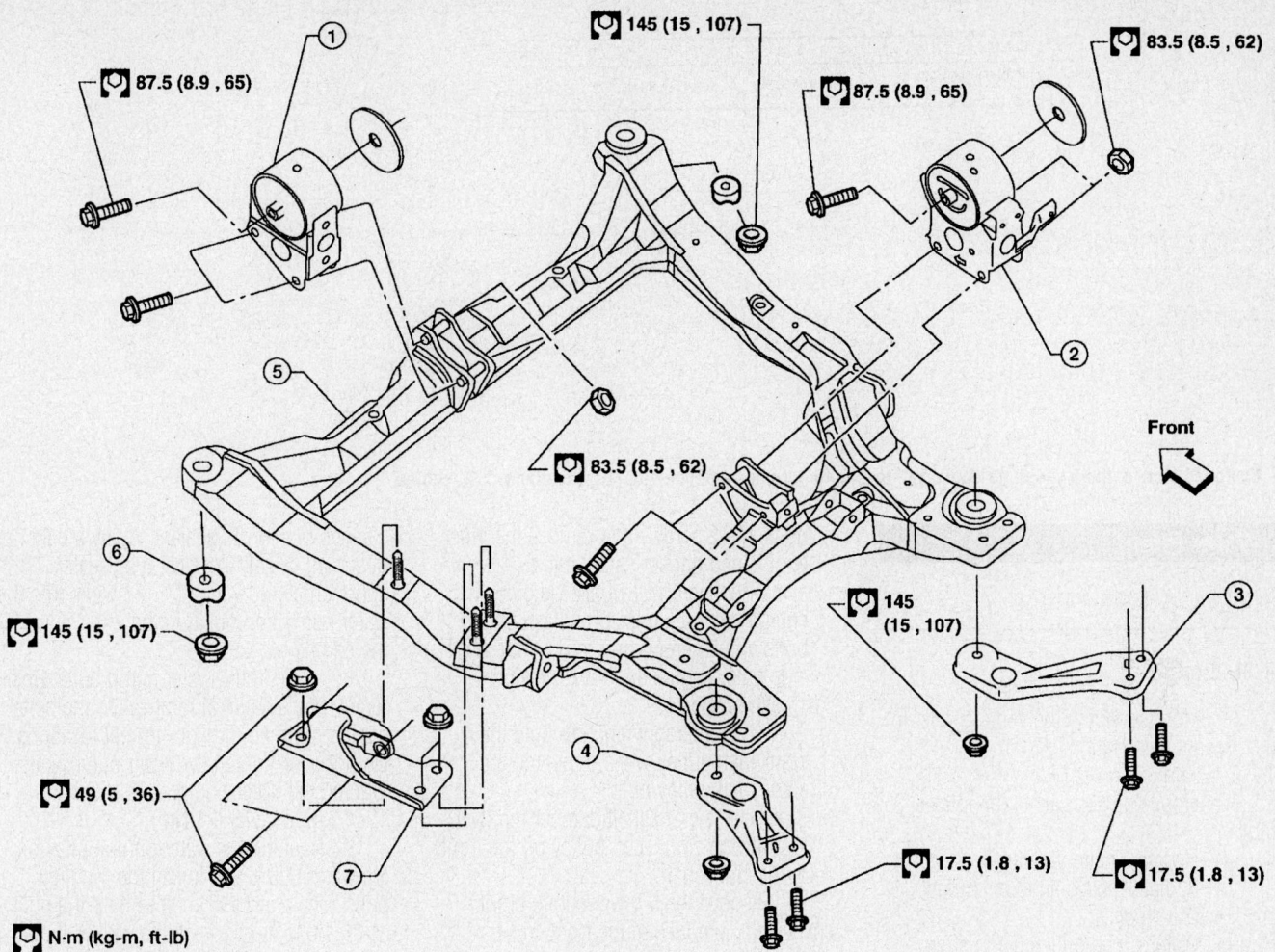

1. Front engine mount
2. Rear engine mount
3. Member pin stay, RH
4. Member pin stay, LH
5. Front suspension member
6. Cup
7. LH transaxle mounting insulator (5 A/T)

67170-NIQU-G02

Exploded view of the front suspension member mounting–3.5L engine

Bolt No.	1	2	3	4	5	6	7	8	9
Tightening torque N·m (kg-m, ft-lb)		74.5 (7.6, 55)				41.5 (4.2, 31)			

4 A/T bolt tightening sequence

LH view

RH view

67170-NIQU-G74

Exploded view of the 4 speed transmission–to–engine torque sequence and specifications–3.5L engine

Bolt No.	1	2	3	4	5	6	7	8	9
Tightening torque N·m (kg-m, ft-lb)	74.5 (7.6, 55)					41.5 (4.2, 31)			

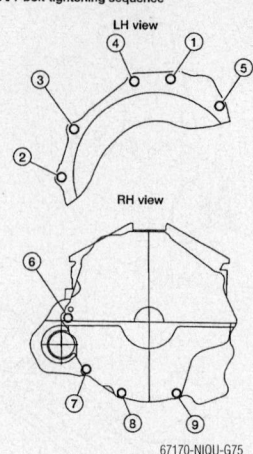

5 A/T bolt tightening sequence

67170-NIQU-G75

Exploded view of the 5 speed transmission–to–engine torque sequence and specifications–3.5L engine

Water Pump

REMOVAL & INSTALLATION

3.3L Engine

1. Before servicing the vehicle, refer to the precautions section.
2. Drain the coolant.
3. Remove or disconnect the following:

- Negative battery cable
- Radiator hoses and fan shroud
- Drive belts
- Water pump pulley using strap wrench 303–D055–(D85L–6000–A) to hold the pulley while removing the bolts

4. Remove the crankshaft pulley using the following procedure:

   a. Raise and safely support the vehicle.

   b. Remove the 5 right side inner engine and transmission splash shield bolts and 2 screws and remove the inner engine and transmission shield.

   c. Remove the 4 right side outer engine and transmission splash shield bolts and 2 screws and remove the right side outer engine and transmission splash shields.

   d. Use a strap wrench to hold the crankshaft pulley while removing the crankshaft pulley bolt.

   e. Use a crankshaft damper remover to draw the crankshaft pulley off the front of the crankshaft.

5. Remove the 5 lower engine front cover bolts and take of the front cover.

6. Remove the 6 water pump bolts. Make note of the locations of the bolts since one should be a stud/bolt and must be returned to its original location. Remove the water pump.

To install:

7. Clean all parts well. The bolt threads should be cleaned of any old sealer or corrosion. Be sure the mating surfaces between the water pump and the engine block are cleaned of any old sealant. Apply a continuous bead of gasket maker type sealer approximately ⅛ inch (3mm) wide onto the water pump and position the water pump on the engine block.

8. Install the 6 water pump bolts. Refer to any notes made at removal so the bolts can be returned to their original locations. Do not over-tighten the water pump bolts. Tighten the water pump bolts evenly to 12–15 ft. lbs. (16–21 Nm).

9. Position the water pump pulley on the water pump and install the 4 pulley bolts. Use a strap wrench to hold the pulley as the bolts are tightened to 12–15 ft. lbs. (16–21 Nm).

10. Install the front engine cover and the 5 lower front cover bolts. Tighten to 27–44 inch lbs. (3–5 Nm).

11. Install the crankshaft pulley using the following procedure:

    a. Install the crankshaft pulley and pulley bolt.

    b. Hold the pulley with a strap wrench. Tighten the crankshaft pulley bolt to 90–98 ft. lbs. (123–132 Nm).

    c. Install the inner and outer engine and transmission splash shields.

12. Install the drive belts.
13. Connect the negative battery cable.
14. Refill the cooling system.
15. Start the engine and check for leaks.

3.5L Engine

1. Before servicing the vehicle, refer to the precautions section.
2. Remove drive belts.
3. Remove undercover.
4. Drain engine coolant from radiator.
5. Remove water drain plug on water pump side of cylinder block.
6. Support lower oil pan bottom with a transmission jack.

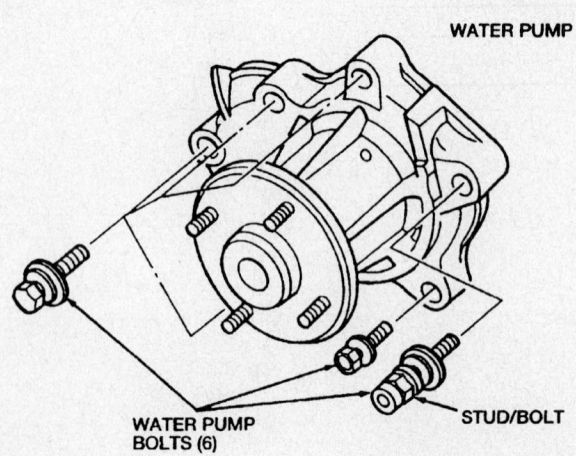

WATER PUMP

WATER PUMP BOLTS (6)

STUD/BOLT

7924WG05

Water pump mounting. Note the location of the stud/bolt—3.3L engine

7. Remove right engine mounting insulator and mounting bracket.

8. Remove idler pulley bracket.

9. Remove chain tensioner cover and water pump cover.

10. Remove the chain tensioner assembly in the following procedure.

  a. Pull the lever down and release the plunger stopper tab.

  b. Insert the stopper pin into the tensioner body hole to hold the lever and keep the stopper tab released.

  c. Insert the plunger into the tensioner body by pressing the timing chain slack guide.

  d. Keep the slack guide pressed and hold the plunger in by pushing the stopper pin deeper through the lever and into the tensioner body hole.

  e. Turn crankshaft pulley approximately 20 degrees clockwise so that the timing chain on the chain tensioner side is loose.

11. Remove chain tensioner.

12. Remove the 3 water pump bolts. Secure a gap between water pump gear and timing chain, by turning crankshaft pulley approximately 20 degrees counterclockwise.

13. Screw M8 bolts, pitch: 1.25mm (0.049in) length: Approx. 50 mm (1.97in) into water pump's upper and lower mounting-bolt holes until they reach timing chain case. Then, alternately tighten each bolt for a half turn, and pull out water pump.

14. Pull straight out while preventing the vane from contacting socket in the installation area. Remove the water pump without the sprocket contacting the timing chain.

15. Remove the M8 bolts and O-rings from water pump.

To install:

16. Install new O-rings to water pump.

17. Apply engine oil and engine coolant to the O-rings as shown. Locate the O-ring with white paint mark to engine front side.

18. Install the water pump.

➡**Do not allow cylinder block to nip the O-rings when installing the water pump.**

19. Check that timing chain and water pump sprocket are engaged.

20. Insert the water pump by tightening the mounting bolts alternately and evenly.

21. Remove the dust and foreign material completely from the backside of the chain tensioner and from the installation area of the rear timing chain case.

22. Turn the crankshaft pulley clockwise so that the timing chain on the timing chain tensioner side is loose.

➡**When installing the timing chain tensioner, engine oil should be applied to the oil hole and tensioner.**

23. Install the timing chain tensioner.

24. Remove the stopper pin.

25. Install chain tensioner cover and water pump cover.

  a. Before installing, remove all traces of liquid gasket from mating surface of water pump cover and chain tensioner cover using a scraper. Also remove traces of liquid gasket from the mating surface of the front cover.

  b. Apply a continuous bead of liquid gasket, to the mating surface of the chain tensioner cover and the water pump cover. Use RTV silicon sealant or equivalent

26. Install the water drain plug on the water pump side of the cylinder block.

27. Installation is in the reverse order of removal for any remaining parts.

28. After starting the engine, let it idle for three minutes, then rev the engine up to 3,000 rpm under no load to purge the air from the high-pressure chamber of the chain tensioner. The engine may produce a rattling noise. This indicates that air still remains in the chamber and is not a matter of concern.

Heater Core

REMOVAL & INSTALLATION

Front System

2001–02 MODELS

1. Before servicing the vehicle, refer to the precautions section.

2. Disconnect the negative battery cable.

3. Drain the cooling system into a clean container for reuse.

4. Remove or disconnect the following:
  • Heater hoses at the bulkhead and plug

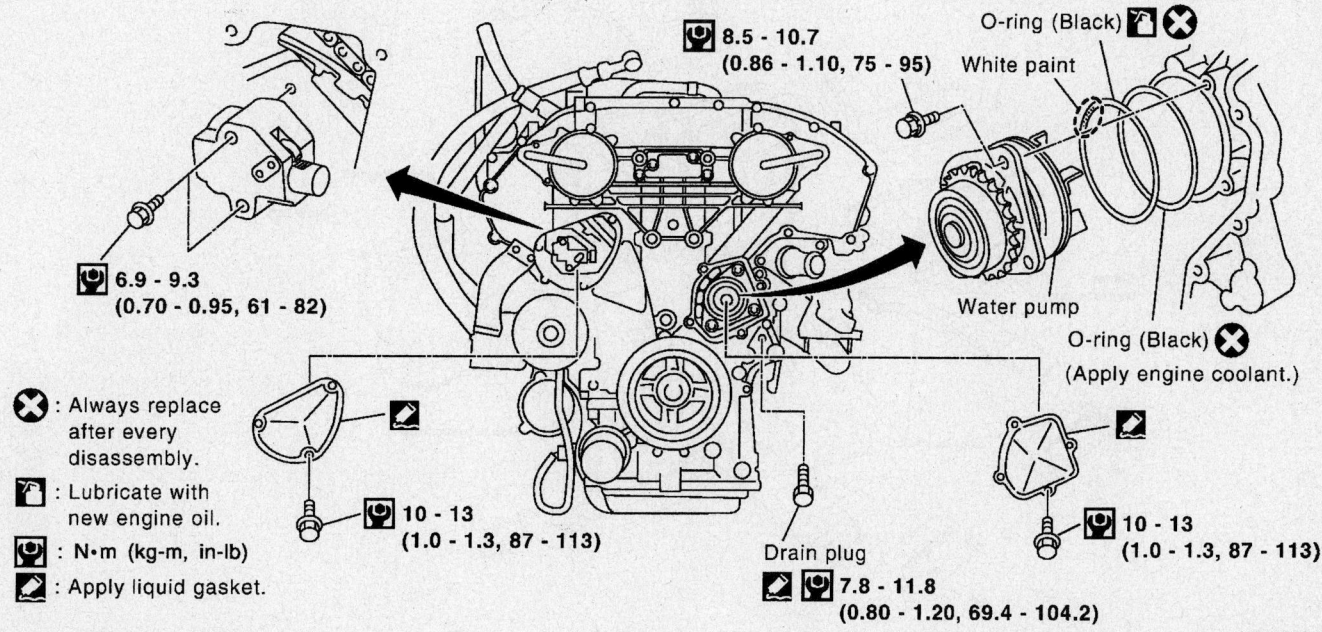

8.5 - 10.7
(0.86 - 1.10, 75 - 95)

O-ring (Black)
White paint

Water pump

O-ring (Black)
(Apply engine coolant.)

6.9 - 9.3
(0.70 - 0.95, 61 - 82)

❌ : Always replace after every disassembly.

🔧 : Lubricate with new engine oil.

🔧 : N•m (kg-m, in-lb)

📐 : Apply liquid gasket.

10 - 13
(1.0 - 1.3, 87 - 113)

10 - 13
(1.0 - 1.3, 87 - 113)

Drain plug

7.8 - 11.8
(0.80 - 1.20, 69.4 - 104.2)

42356-MURA-G04

Water pump mounting—3.5L engine

- Storage bin, then both side covers by the bin and the footlamp, if equipped
- Control console bezel (1 screw in the center), then the ashtray assembly
- Climate control console screws, pull the console rearward and detach the electrical connectors
- 4 radio assembly screws and take the radio out of the vehicle
- Floor duct and the right and left knee reinforcement plates
- ABS control module.

5. The speed control module, keyless entry module (if equipped) and the passive restraint (air bag) module are all located behind the center console and can be removed after detaching the respective connectors and removing the retaining nuts or screws.

❄❄ WARNING

The control modules are very sensitive to static electricity and can be damaged if exposed to static or stray electrical impulses.

6. Remove or disconnect the following:
- Center air duct
- 2 ground wire bolts, the U-bracket and the 2 console brackets
- Glove box and lamp
- Accelerator pedal and pedal stop
- Floor air duct
- Temperature blend sir door actuator and mode door actuator by unfastening the attaching bracket bolts

and detaching the electrical connections
- Center distribution duct
- 4 evaporator/blower assembly screws, the 4 heater assembly screws and the heater assembly
- Heater pipe plate from the assembly
- Heater core retainer, disengage the shut-off valve control rod
- Heater core from the assembly

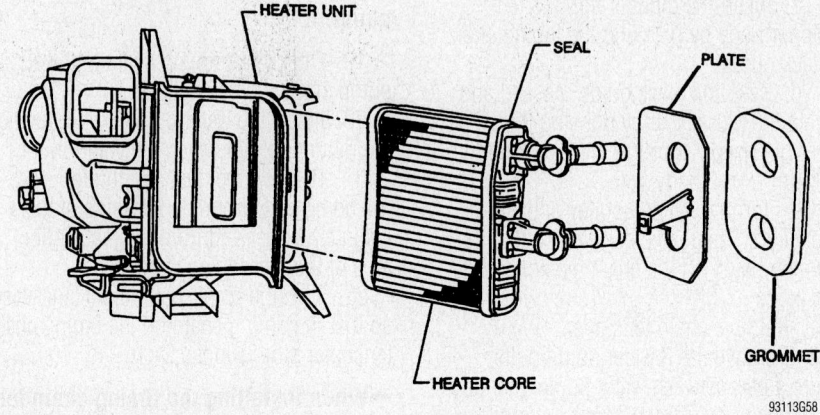

View the front heater core and heater housing assembly

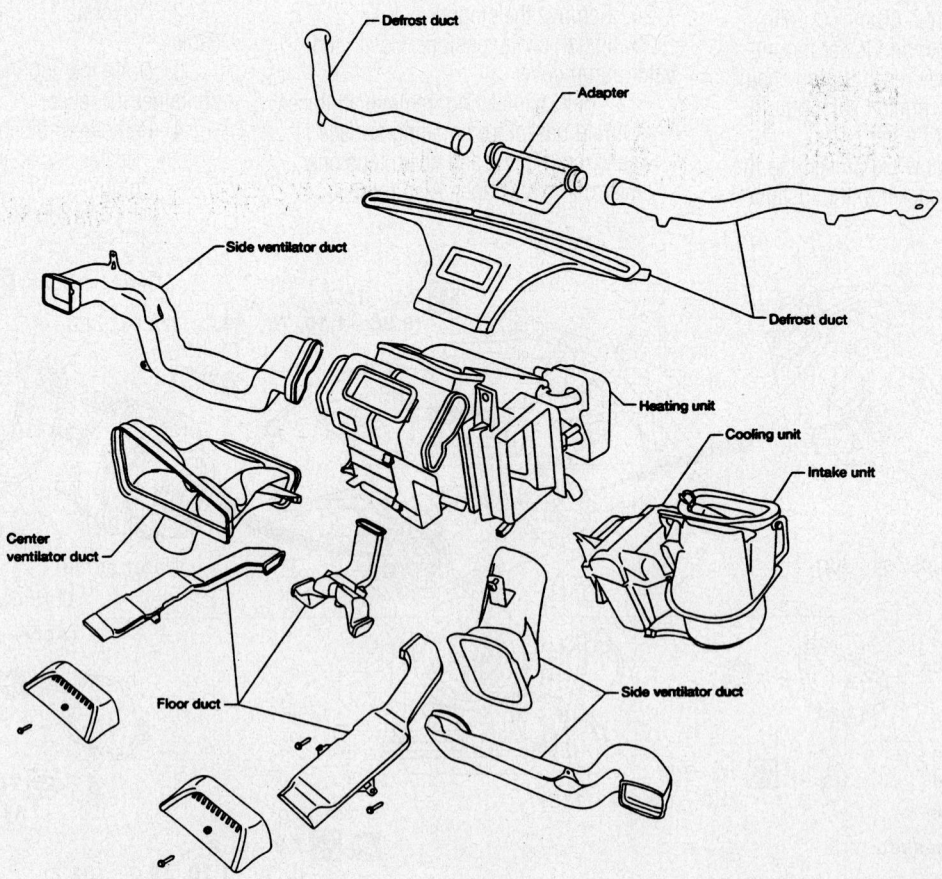

Exploded view the front heater/air conditioning assembly

To install:

7. Install or connect the following:
    *   Heater core to the case, the retainer and pipe plate
    *   Heater assembly in the vehicle and attach the 4 retaining screws
    *   Center distribution duct, the blend air and mode door actuators
    *   Floor air duct
    *   Accelerator stop and pedal
    *   Glove box and lamp, then the center console and U-brackets
    *   Center air duct, the passive restraint, the keyless entry, the speed control and the ABS modules, as removed
    *   Remaining center console components
    *   Heater hoses to the heater core
8. Refill the cooling system.
9. Connect the negative battery cable.
10. Run the engine to normal operating temperatures; then, check the climate control operation and check for leaks.

2004–06 MODELS

1. Before servicing the vehicle, refer to the precautions section.

2. Discharge the refrigerant from the A/C system.
3. Drain the engine cooling system.
4. Remove the cowl top extension as follows:
    a. Remove the wiper arms.
    b. Release the clips under cowl top cover and clips on hood–ledge.
    c. Remove the seal by releasing the ends from tabs the on the cowl top cover and releasing the plastic clips.
    d. Remove the weatherstrip.
    e. Remove the cowl top cover.
    f. Remove the cowl top seal.
    g. Remove the windshield washer nozzles and the hoses from cowl top cover.
    h. Remove the heater pump from cowl top extension.
    i. Disconnect the clip attaching the coolant control valve hose to the cowl top extension.
    j. Remove cowl top extension.
5. Remove the exhaust system.
6. Disconnect the front heater hoses from the front heater core.
7. Disconnect the high/low pressure pipe from the front expansion valve.

8. Move the two front seats to the rear–most position on the seat track.
9. Remove the instrument panel and console panel as follows:
    a. Remove the center console lower cover.
    b. Slide the selector knob cover downwards to reveal the selector knob latch.
    c. Gently pry the selector knob latch outward to release it, then lift the selector knob up to remove it.
    d. Disconnect the harness connectors and remove the cluster lid C.
    e. Disconnect the harness connectors and remove the left hand instrument lower panel.
    f. Remove the storage tray.
    g. Disconnect the glove box light and remove the right hand instrument lower panel.
    h. Remove the glove box housing.
    i. Disconnect the harness connectors and remove the center console.
    j. Remove left hand center console side finisher. Wrap the tip of flat-bladed screwdriver with a cloth when removing the clips from the finishers. When

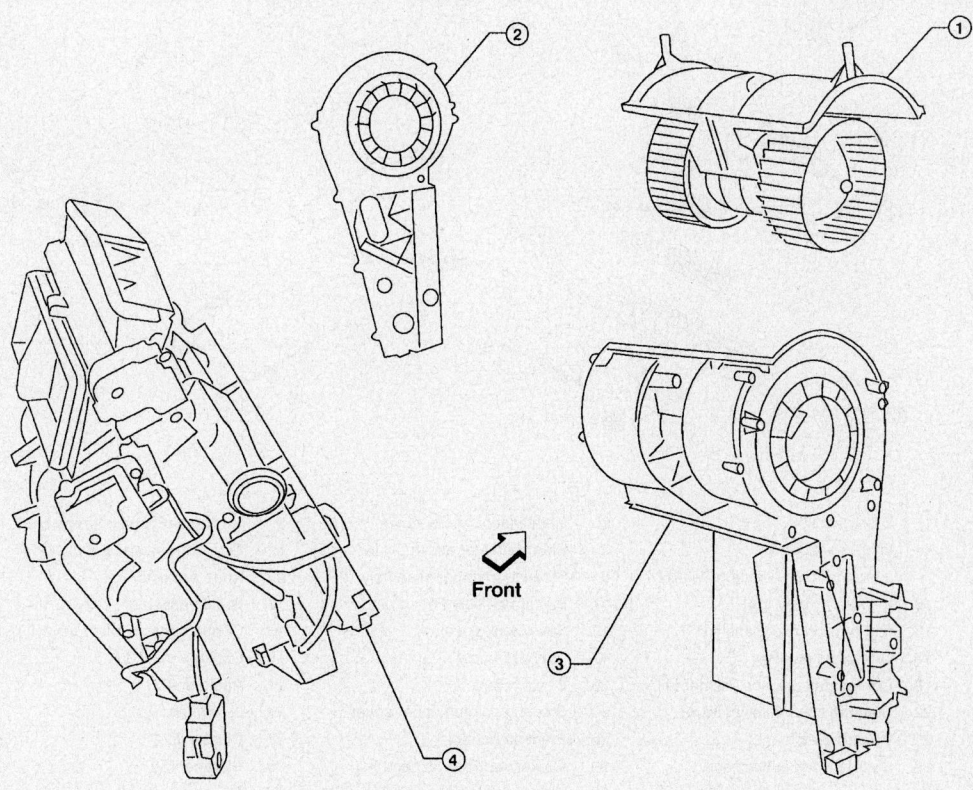

1. Front blower motor
2. Blower motor side cover
3. Blower motor case
4. Heater core and evaporator case

Front

67170-NIQU-G85

Exploded view of the front heater core assembly and related components—2004–06 models

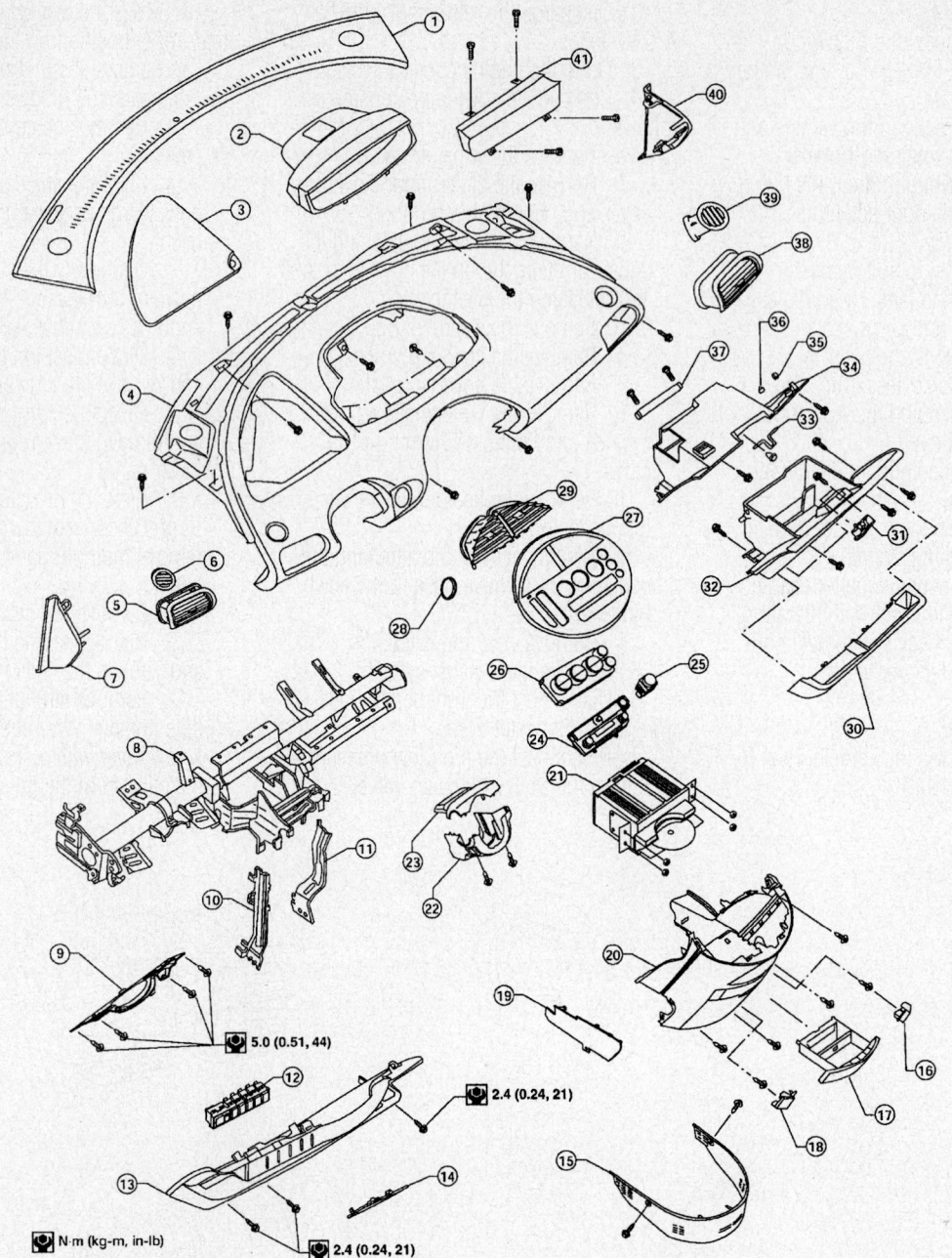

5.0 (0.51, 44)

2.4 (0.24, 21)

N·m (kg-m, in-lb)

2.4 (0.24, 21)

1. Defrost grille	2. Combination meter cover	3. Instrument panel storage bin
4. Instrument panel	5. Side ventilator assembly LH	6. Side defroster assembly LH
7. Instrument panel side cover LH	8. Steering member assembly	9. Knee protector
10. Instrument stay LH	11. Instrument stay RH	12. Switch assembly
13. Instrument lower panel LH	14. Fuse block cover	15. Center console lower cover
16. Console mask RH	17. Tray assembly	18. Console mask LH
19. Center console side finisher LH	20. Center console	21. Audio unit
22. Steering column lower cover	23. Steering column upper cover	24. AV switch
25. Hazard switch	26. Front air control	27. Cluster lid C
28. Steering lock escutcheon	29. Center ventilator assembly	30. Storage tray
31. Glove box latch assembly	32. Instrument lower panel RH	33. Glove box striker
34. Glove box housing	35. Glove box lamp receptacle	36. Glove box lamp
37. Glove box damper	38. Side ventilator assembly RH	39. Side defroster assembly RH
40. Instrument panel side cover RH	41. Combination meter	

Exploded view of the instrument panel components—2004–06 models

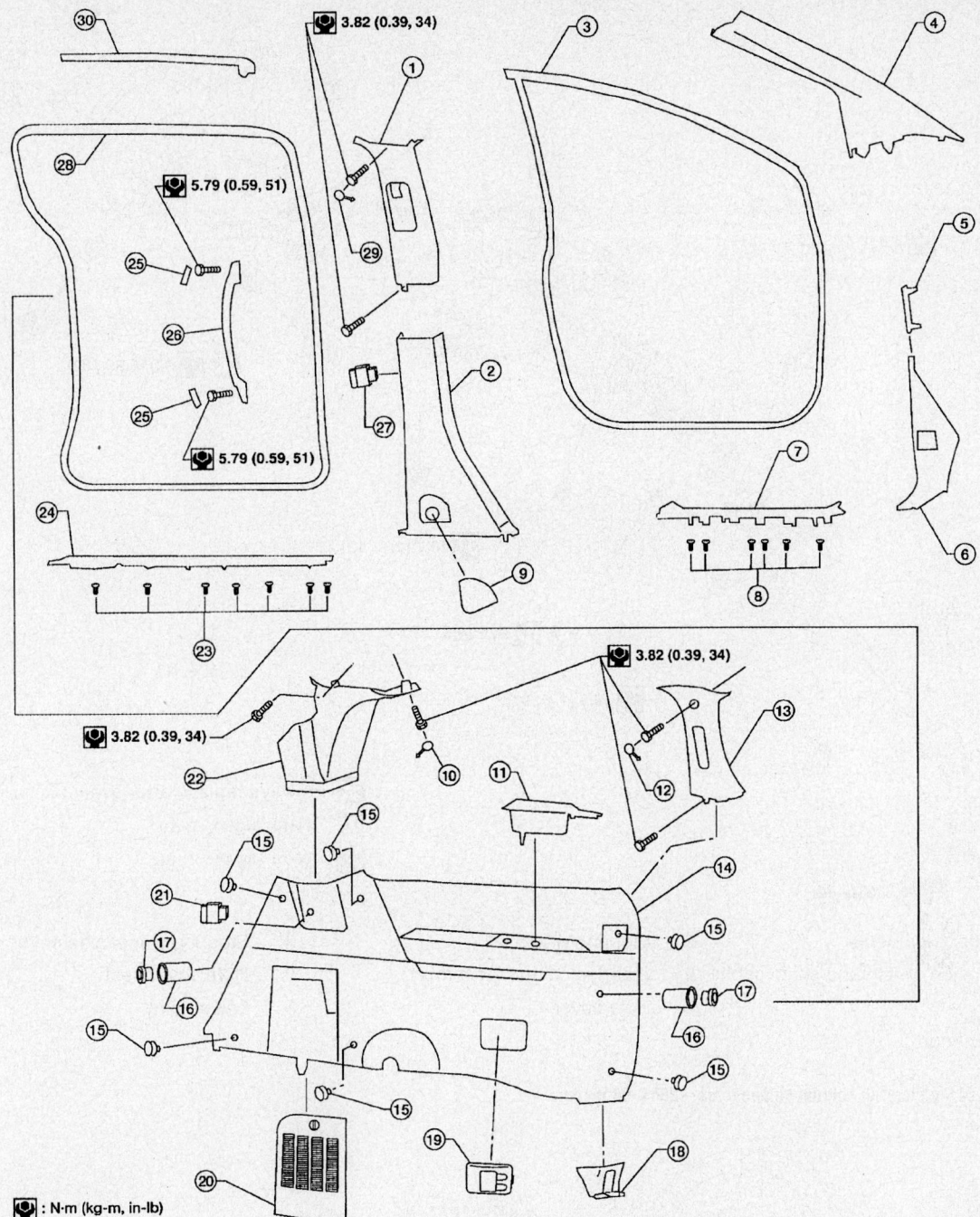

3.82 (0.39, 34)

5.79 (0.59, 51)

5.79 (0.59, 51)

3.82 (0.39, 34)

3.82 (0.39, 34)

: N·m (kg-m, in-lb)

1.	Center pillar upper finisher LH	2.	Center pillar lower finisher LH	3.	Front door welt
4.	Front pillar finisher LH	5.	Upper dash side finisher LH	6.	Lower dash side finisher LH
7.	Front kicking plate	8.	Trim clips	9.	Center pillar lower escutcheon LH
10.	Rear finisher cover	11.	Luggage side cup holder LH	12.	Center finisher cover
13.	Rear center pillar finisher LH	14.	Rear lower finisher assembly LH	15.	Cargo net hooks
16.	Power point	17.	Power point cap	18.	Rear kick escutcheon LH
19.	Luggage side lower escutch-eon LH	20.	Luggage side panel LH	21.	Power sliding door switch assembly
22.	Rear pillar upper finisher LH	23.	Trim clips	24.	Rear kicking plate
25.	Assist grip end caps	26.	Center pillar assist grip	27.	Back door open/close switch
28.	Sliding door welt	29.	Center pillar upper cover	30.	Sliding door drip weatherstrip LH

67170-NIQU-G87

Exploded view of the left hand center console side finisher components—2004–06 models

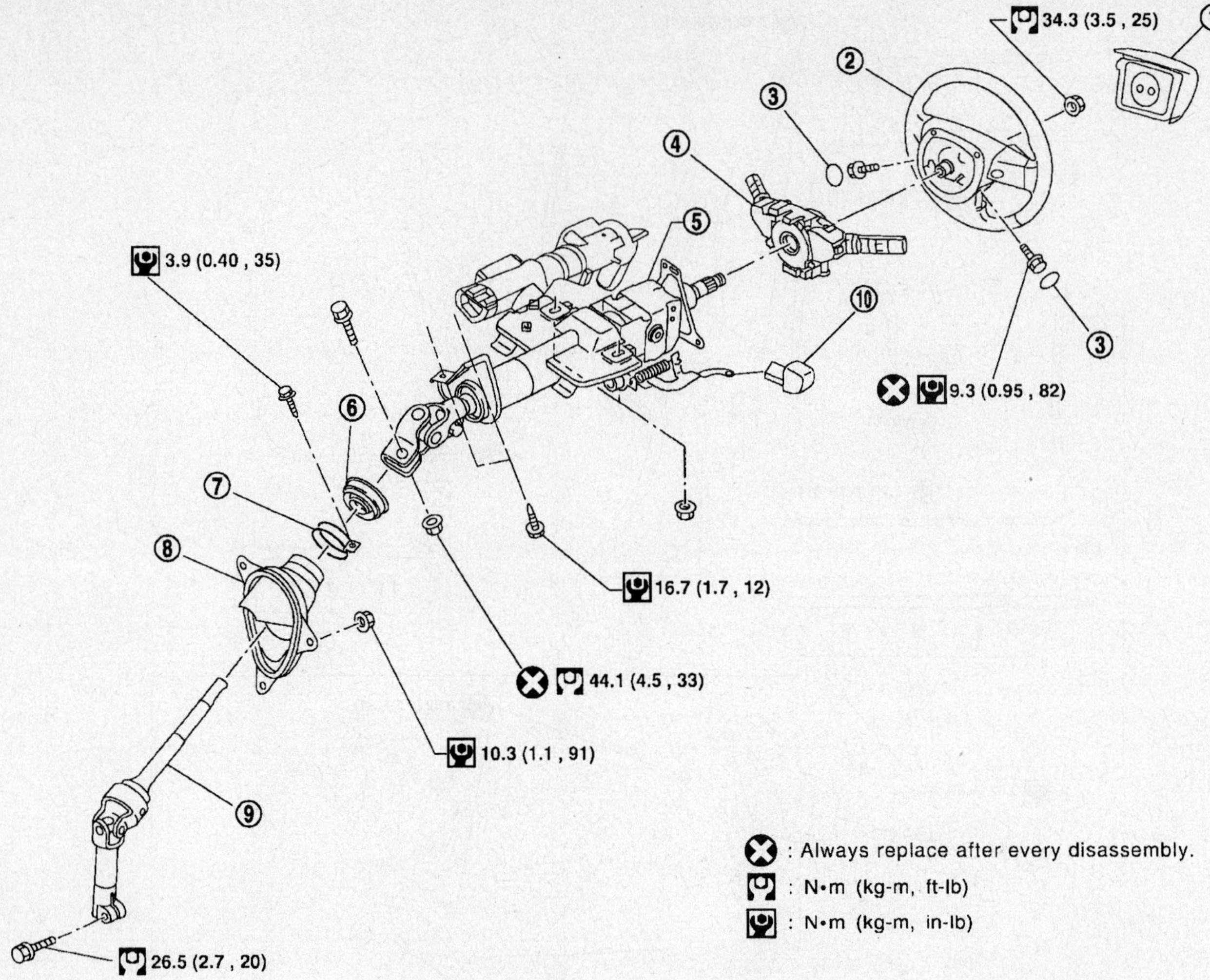

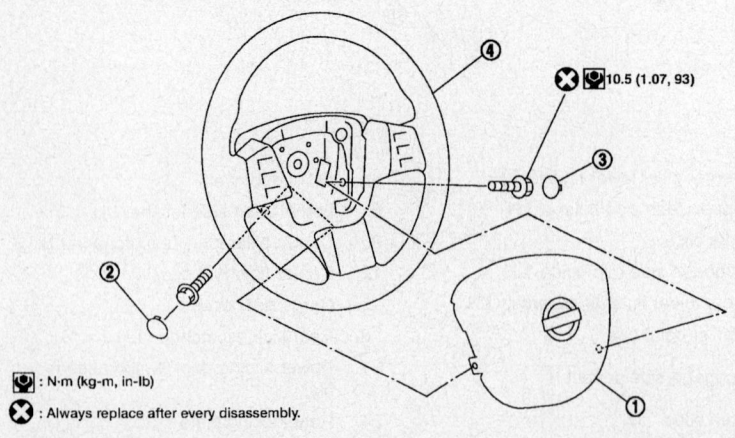

🔩 34.3 (3.5 , 25)

🔩 3.9 (0.40 , 35)

🔩 16.7 (1.7 , 12)

❌ 🔩 44.1 (4.5 , 33)

🔩 10.3 (1.1 , 91)

🔩 26.5 (2.7 , 20)

❌ 🔩 9.3 (0.95 , 82)

❌ : Always replace after every disassembly.

🔩 : N•m (kg-m, ft-lb)

🔩 : N•m (kg-m, in-lb)

1. Driver air bag module
2. Steering wheel
3. Steering wheel side cover
4. Combination switch and spiral cable
5. Steering column assembly
6. Hole cover seal
7. Clamp
8. Hole cover
9. Lower joint
10. Tilt lever knob

67170-NIQU-G88

Exploded view of the steering column components—2004–06 models

❌ 🔩 10.5 (1.07 , 93)

🔩 : N•m (kg-m, in-lb)

❌ : Always replace after every disassembly.

1. Driver air bag module
2. Side lid LH
3. Side lid RH
4. Steering wheel

67170-NIQU-G98

Remove the side lids from the retainers retaining the drivers side air bag—2004–06 models

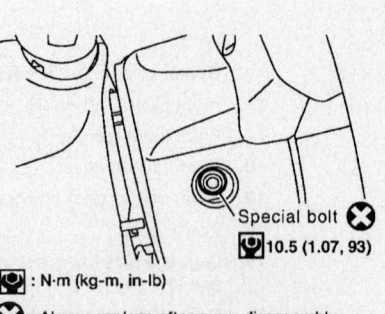

Special bolt ❌

🔩 10.5 (1.07 , 93)

🔩 : N•m (kg-m, in-lb)

❌ : Always replace after every disassembly.

67170-NIQU-G99

Remove the left and right side air bag bolts—2004–06 models

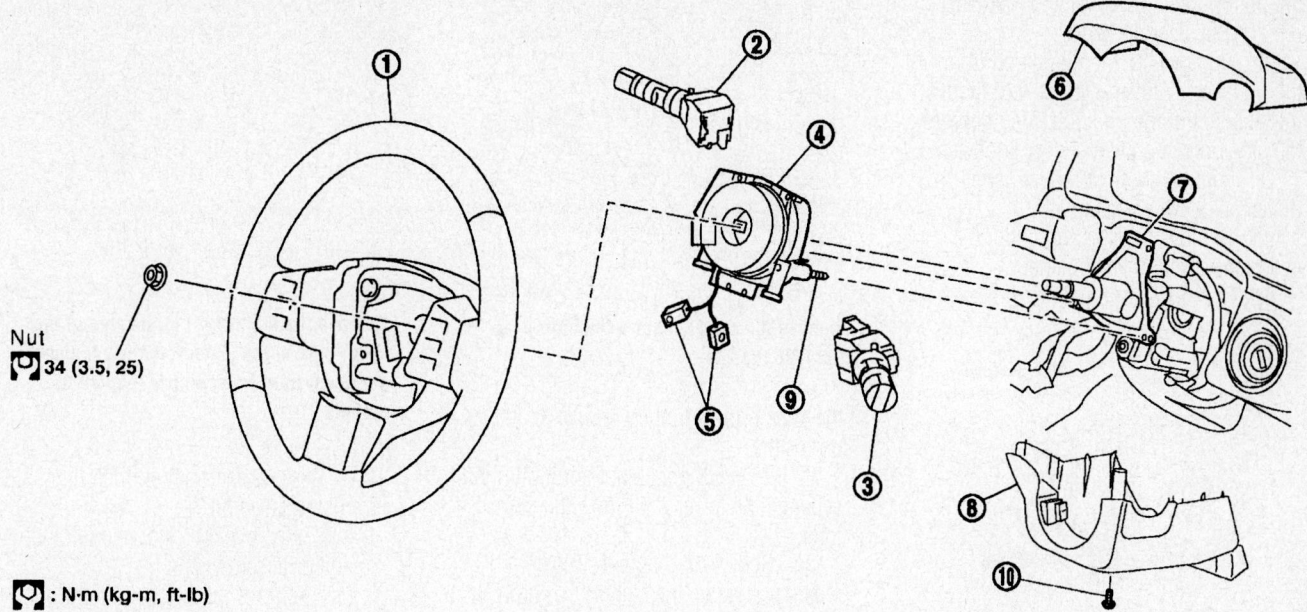

Nut
⬡ 34 (3.5, 25)

⬡ : N·m (kg-m, ft-lb)

1. Steering wheel
4. Spiral cable
7. Column assembly
10. Screw

2. Lighting and turn signal switch
5. Driver air bag module connector
8. Column cover lower

3. Wiper and washer switch
6. Column cover upper
9. Screw (Do not remove)

67170-NIQU-G92

Exploded view of the steering wheel, spiral cable and air bag components—2004–06 models

removing or installing the body side welts, do not allow butyl seal to come in contact with the pillar finisher.

k. Disconnect the harness connectors and remove the combination meter cover and the combination meter.

l. Remove the instrument panel storage bin.

m. Remove the knee protector.

n. Remove the upper and lower steering column covers.

✳✳ CAUTION

Before servicing the air bag system, turn ignition switch OFF, disconnect both battery cables and wait at least 3 minutes. When servicing the SRS, do not work from directly in front of air bag module.

o. Set the front wheels in the straight-ahead position.

p. Remove the side lids from the retainers retaining the driver's side air bag.

q. Remove the left and right side bolts.

r. Lift the driver's side air bag from the steering wheel.

➡**When Disconnecting or connecting the air bag harness or horn connectors, always pull up to release the black locking tab prior to removing the connector from the air bag component. Always push down to lock the black locking tab after installing the connector to the air bag component. When locked, the black locking tab is level with the connector housing.**

s. Disconnect the air bag harness and horn connectors, then remove the driver air bag module.

t. Remove the steering wheel center nut.

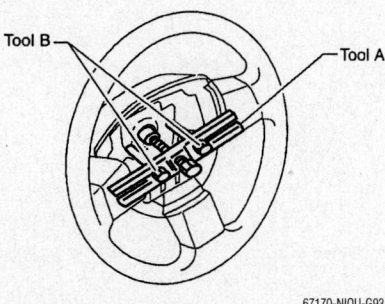

67170-NIQU-G93

Remove the steering wheel using puller tools J–1859–A and J–42578—2004–06 models

u. Remove the steering wheel using puller tools J–1859–A and J–42578.

v. Remove the column cover upper and lower.

w. Remove wiper washer switch connector, then pinch the tabs at the wiper and washer switch base and slide the switch away from the steering column.

x. While pressing tabs, pull the light and turn signal switch toward the driver door and disconnect from the base.

y. Remove the screws, release the tab, and remove the spiral cable.

✳✳ CAUTION

Do not disassemble spiral cable. Do not apply lubricant to the spiral cable.

z. Remove the spiral cable connectors.

✳✳ CAUTION

With the steering linkage disconnected, the spiral cable may snap by turning the steering wheel beyond the limited number of turns. The spiral cable can be turned counterclockwise about 2.5 turns from the right end position.

aa. Inspect the steering wheel near the puller holes for damage. If damaged, replace the steering wheel.

bb. Remove the tilt lever knob from the tilt lever by inserting a suitable tool into the slot of the tilt knob, then depress the tab and withdraw the lever knob.

cc. Remove the instrument panel driver's side lower panel.

dd. Remove the steering column cover and ignition key finisher.

ee. Remove the mounting screws of

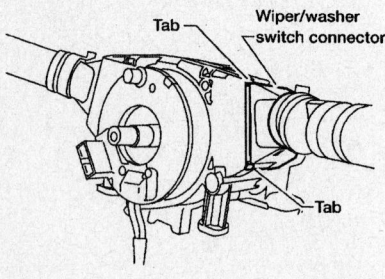

67170-NIQU-G94

Remove wiper washer switch connector, pinch the tabs at the wiper and washer switch base and slide the switch away from the column—2004–06 models

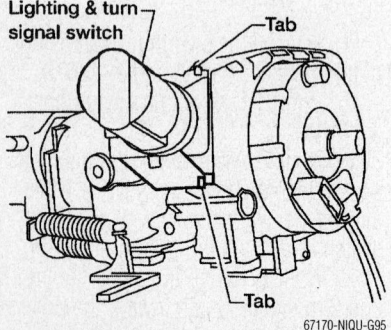

67170-NIQU-G95

While pressing tabs, pull the light and turn signal switch toward the driver door and disconnect from the base—2004–06 models

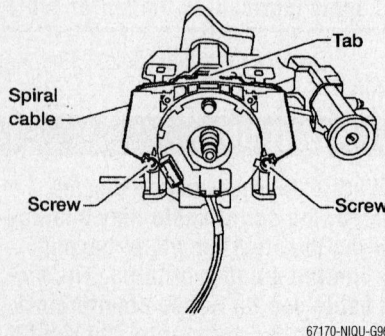

67170-NIQU-G96

Remove the screws, release the tab, and remove the spiral cable—2004–06 models

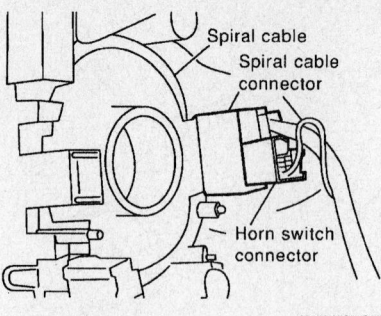

67170-NIQU-G97

Remove the spiral cable connectors—2004–06 models

the knee protector, then remove the knee protector.

ff. Remove the lock nut and bolt, then separate the lower joint from the upper joint.

gg. Remove the nuts from steering member, remove the steering column assembly from the steering member.

hh. Remove the hole cover seal and clamp.

ii. Remove the nuts, then remove the hole cover from the dash panel.

jj. Remove the mounting bolt located at the lower side of the lower joint and remove the lower joint from vehicle.

kk. Remove the defrost grille.

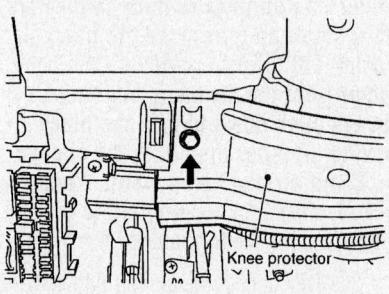

67170-NIQU-G89

Remove the knee protector—2004–06 models

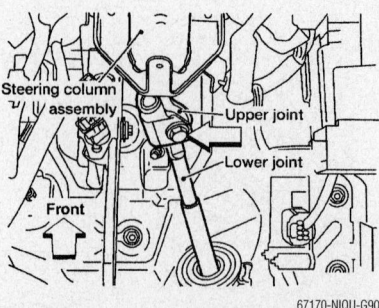

67170-NIQU-G90

Remove the lock nut and bolt, then separate the lower joint from the upper joint—2004–06 models

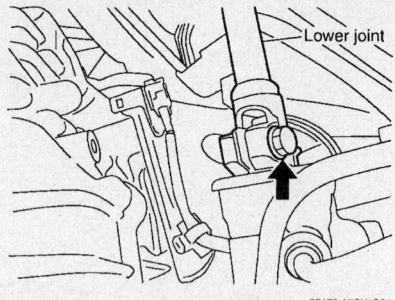

67170-NIQU-G91

Remove the mounting bolt located at the lower side of the lower joint and remove the lower joint from vehicle—2004–06 models

ll. Remove the right and left hand side defroster assemblies.

mm. Remove the right and left side ventilator assemblies.

nn. Remove the right and left front pillar finishers.

oo. Remove the instrument panel.

10. Disconnect the instrument panel wire harness at the right and left hand in–line connector brackets, and the fuse block electrical connectors.

11. Disconnect the steering member from each side of the vehicle body.

12. Remove the front heater and cooling unit assembly with it attached to the steering member, from the vehicle.

✳✳ CAUTION

Use care not to damage the seats and interior trim panels when removing the front heater and cooling unit assembly with it attached to the steering member.

13. Remove the front heater and cooling unit assembly from the steering member.

14. Remove the blower motor side cover.

15. Remove the front blower motor.

16. Remove heater core and evaporator case bottom cover.

17. Remove the blower motor case.

18. Remove the front heater core.

➡ **If the in-cabin microfilters are contaminated from coolant leaking from the heater core, replace the microfilters before installing the new front heater core.**

To install:

19. Installation is the reverse of removal, refer to the component exploded views for component locations and torque specifications.

20. Replace the O-ring of the low-pres-

sure flexible hose and high-pressure flexible hose with a new one, and apply compressor oil to it when installing it.

21. Fill the engine cooling system.

22. Recharge the A/C system and check for leaks.

Rear Auxiliary System

2001–02 MODELS

➡**The rear heater/air conditioning assembly must be removed as a complete unit in order to remove the heater core and/or evaporator core.**

1. Before servicing the vehicle, refer to the precautions section.

2. Disconnect the negative battery cable.

3. Drain the cooling system into a clean container for reuse.

4. Discharge and recover the air conditioning system refrigerant.

5. Remove or disconnect the following:

- Heater hoses at the bulkhead and plug
- Center seats
- 2 left half seat belt lower anchor bolts
- Left rear cargo net retainers, if equipped
- Lift gate scuff plate and the 3 screws from the left rear quarter trim panel. Gently pry the rear seat remote control (if equipped) from the trim panel. Disconnect the remote control wiring connector and remove the rear radio control panel. Pull the top of the trim panel away from the body.
- Rear climate control panel wiring, if equipped
- Left front lap belt guide from the left quarter trim panel and pass the belt through the trim panel
- Trim panel from the vehicle
- Upper duct from the assembly (6 screws)
- Blower motor and resistor wiring
- Temperature blend and vent door actuator connectors

6. Raise and safely support the vehicle. Use the spring lock coupling tool to disconnect and plug the refrigerant line connections from beneath the vehicle.

7. Lower the vehicle.

8. Remove or disconnect the following:

- 4 heater/air conditioning assembly bolts and the assembly from the vehicle

- Heater core and/or evaporator core from the assembly

To install:

9. Install or connect the following:

- Heater core and/or evaporator core into the assembly and the 4 retaining bolts

10. Raise and safely support the vehicle.

11. Using new O-rings, reconnect the refrigerant lines to the evaporator.

12. Lower the vehicle.

13. Install or connect the following:

- All wiring connectors
- Upper air duct with the 6 screws
- Trim panel and pass the lap seat belt through the panel slot
- Rear climate control panel
- Rear radio and rear remote control
- Remaining trim panel and components

14. Refill the cooling system.

15. Connect the negative battery cable.

16. Evacuate and charge the air conditioning system.

17. Run the engine to normal operating temperatures; then, check the climate control operation and check for leaks.

2004–06 MODELS

1. Before servicing the vehicle, refer to the precautions section.

2. Partially drain the engine cooling system.

3. Remove the rear right hand interior trim panel.

4. Disconnect the rear heater core hoses from the rear heater core.

5. Remove the rear heater core bracket.

6. Remove the rear heater core.

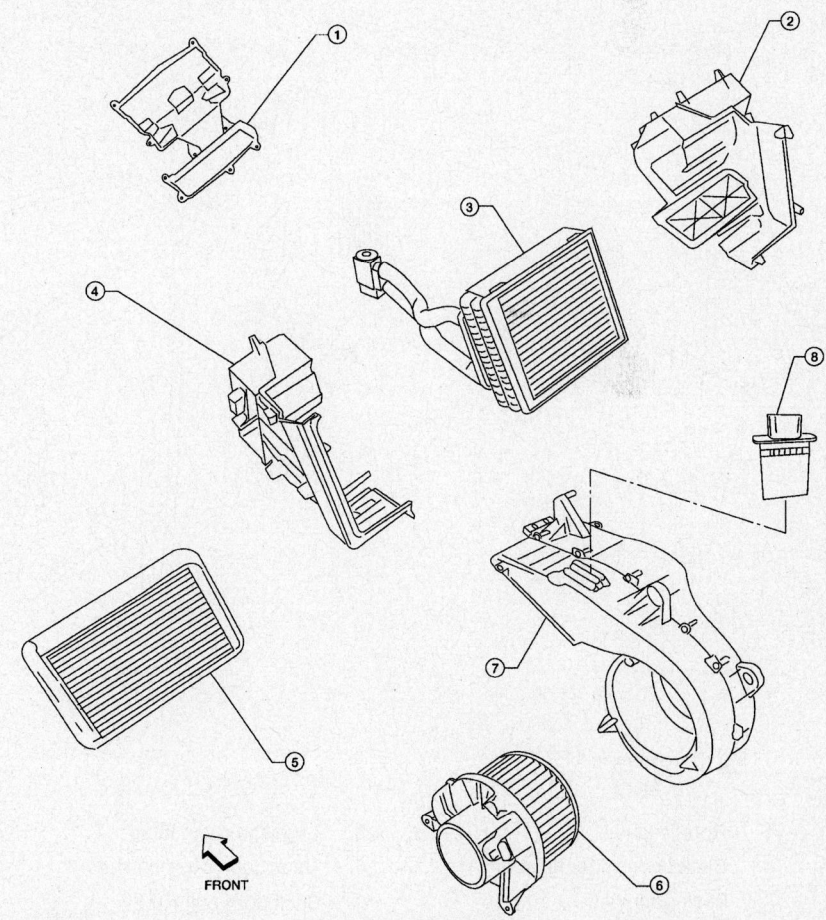

1.	Front cover	2.	Evaporator and heater core case	3.	Evaporator
4.	Side cover	5.	Heater core	6.	Rear blower motor
7.	Blower motor case	8.	Rear blower motor resistor		

67170-NIQU-GAA

Exploded view of the rear heater core assembly—2004–06 models

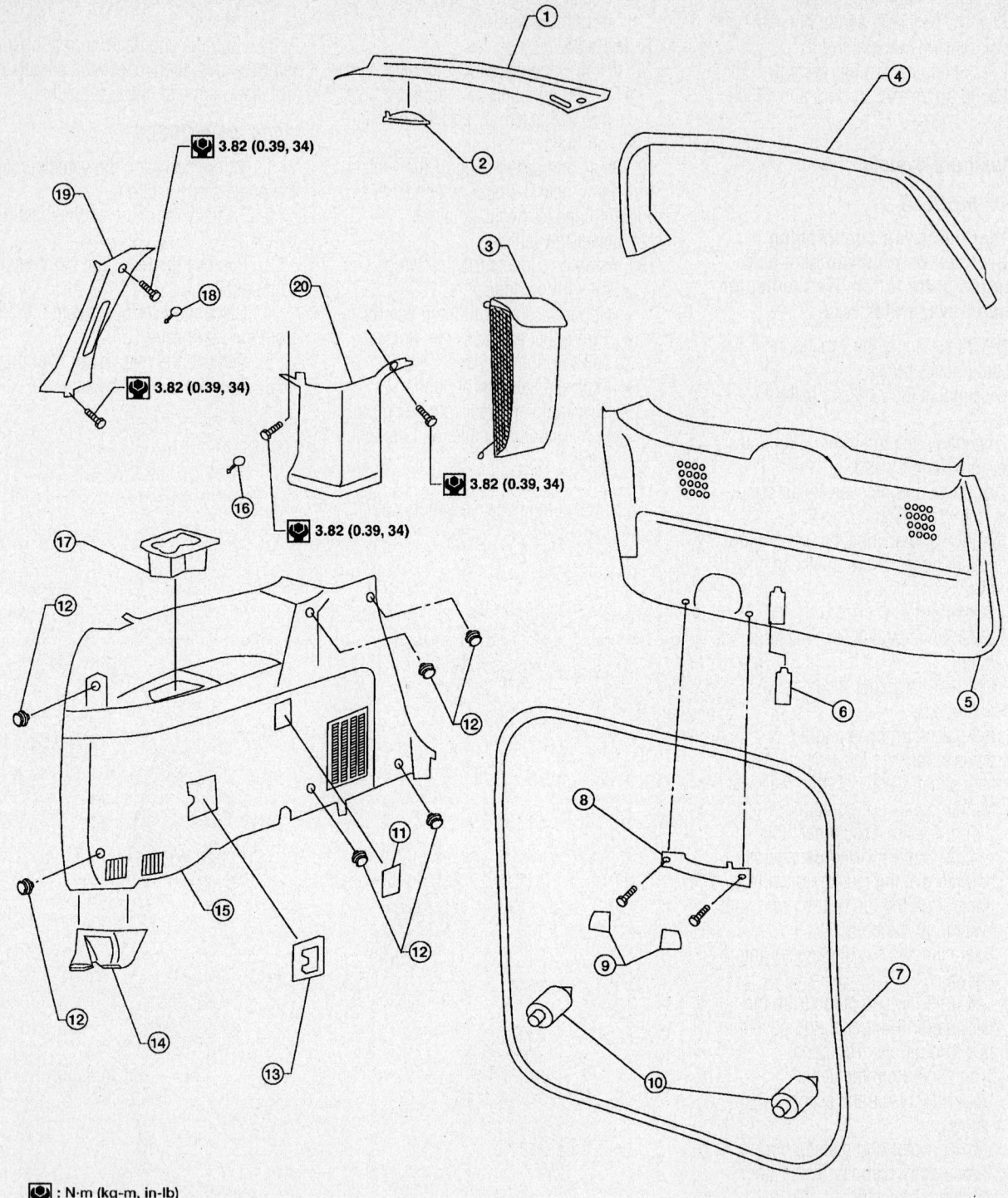

3.82 (0.39, 34)

3.82 (0.39, 34)

3.82 (0.39, 34)

3.82 (0.39, 34)

: N·m (kg-m, in-lb)

1.	Roof finisher	2.	Luggage room lamp	3.	Cargo net
4.	Back door upper finisher	5.	Back door lower finisher	6.	Back door mask
7.	Back door welt	8.	Back door pull handle	9.	Pull handle covers RH/LH
10.	Back door bumpers	11.	Gas leak check lid	12.	Cargo net hooks
13.	Luggage side lower escutcheon RH	14.	Rear kick escutcheon RH	15.	Rear lower finisher assembly RH
16.	Rear pillar cover RH	17.	Cup holder luggage side RH	18.	Center pillar cover RH
19.	Rear center pillar finisher RH	20.	Rear pillar upper finisher RH		

67170-NIQU-GAB

Exploded view of the rear trim components—2004–06 models

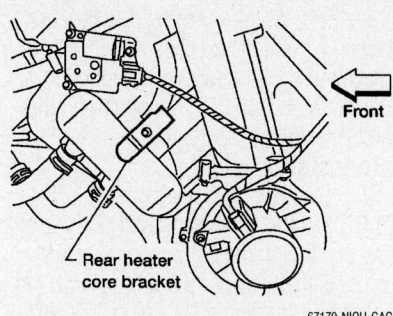

Remove the rear heater core bracket—
2004–06 models

67170-NIQU-GAC

Cylinder Head

REMOVAL & INSTALLATION

3.3L Engine

The factory specifies that the cylinder head bolts ARE NOT to be reused. Obtain the proper replacement parts before beginning this procedure. Check carefully that all bolts are removed before attempting to remove a cylinder head. A tab, part of the head, contains 1 lightly tightened head bolt that is external to the valve cover. Do not overlook this "hidden" bolt or the head will be damaged.

1. Before servicing the vehicle, refer to the precautions section.
2. Properly relieve the fuel system pressure.
3. Drain the coolant.
4. Remove or disconnect the following:
   - Negative battery cable
   - Air intake tube
   - Timing belt
   - Upper intake manifold (plenum)
   - Fuel feed and return hoses from the fuel rail
   - Fuel injector's electrical connections
   - Fuel rail and injectors as an assembly
   - Intake manifold (lower)
   - Camshaft sprockets
   - Rear timing belt cover
   - Distributor
   - Harness clamp from the right hand rocker cover
   - Exhaust tube from the left hand manifold
   - Left hand exhaust manifold from the right hand exhaust manifold
   - Left hand manifold-to-bracket bolt
   - A/C compressor, alternator and their brackets
   - Rocker covers

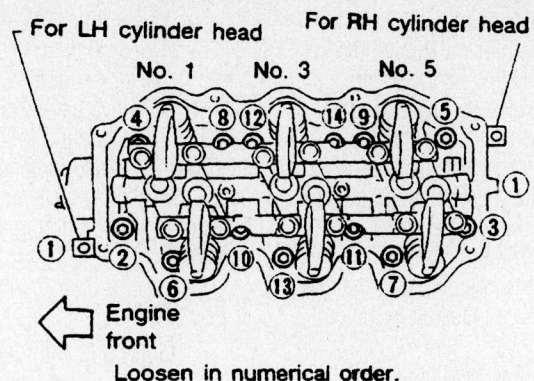

Loosen in numerical order.

9348WG08

Remove the cylinder head bolts in the sequence shown—3.3L engines

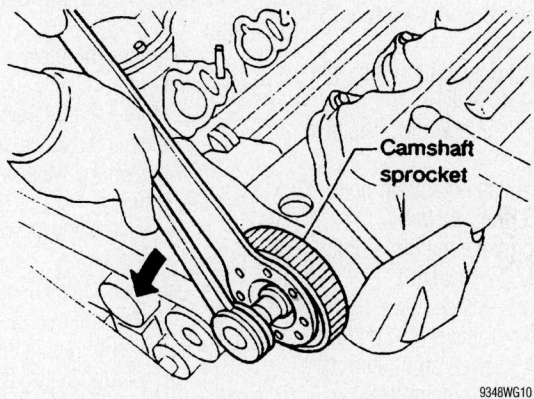

9348WG10

Hold the camshaft sprocket while removing the sprocket retaining bolt—3.3L engines

- Cylinder head bolts in the sequence illustrated using several passes
- Cylinder head with the exhaust manifold and gasket. Discard the gasket.
- Exhaust manifold from the head

To install:
5. Clean all parts well.
6. Inspect the cylinder head for damage, cracks and leakage of water and oil. If necessary, replace the head. Check the head gasket surface for burrs and nicks. If the head is cracked, it must be replaced.
7. Install the exhaust manifold on the cylinder head.

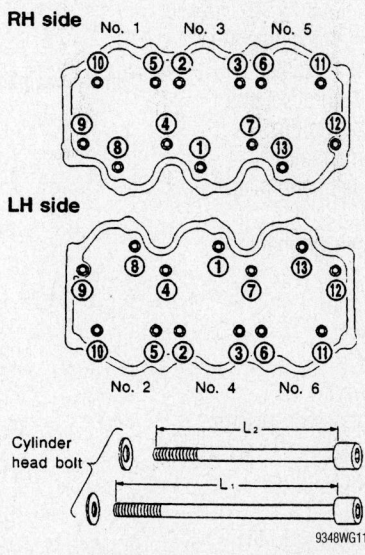

9348WG11

Tighten the cylinder head bolts in sequence as shown—3.3L engines

8. Position a new head gasket and the cylinder head on the block. Examine the head bolt washers. Note that the washers have a chamfer or bevel on one side. The beveled side should face "up" when installed. Examine the new replacement head bolts. There are different lengths. The head bolts in positions 4, 7, 9 and 12 are 5.00 inches (127mm) long and the rest are 4.17 inches (106mm) long. Be sure the new cylinder head bolts are installed in the correct positions.

9. Tighten the new head bolts in the following sequence:

  a. First pass: cylinder head bolts to 22 ft. lbs. (29 Nm).

  b. Second pass: cylinder head bolts to 43 ft. lbs. (59 Nm).

  c. Third pass: Loosen all of the cylinder head bolts completely.

  d. Fourth pass: cylinder head bolts to 7 ft. lbs. (10 Nm).

  e. Fifth pass: intake manifold bolts and nuts to 2.9 ft. lbs. (4 Nm).

  f. Sixth pass: intake manifold bolts and nuts to 13 ft. lbs. (18 Nm).

  g. Seventh pass: intake manifold bolts and nuts to 12–14 ft. lbs. (16–20 Nm).

  h. Eight pass: Loosen all of the intake manifold bolts and nuts completely.

  i. Ninth pass: cylinder head bolts to 22 ft. lbs. (29 Nm).

  j. Tenth pass: cylinder head bolts to 40–47 ft. lbs. (54–64 Nm).

  k. Eleventh pass: cylinder head sub-bolts to 6.7–8.7 ft. lbs. (9–12 Nm).

  l. Twelfth pass: intake manifold bolts and nuts to 2.9 ft. lbs. (4 Nm).

  m. Thirteenth pass: intake manifold bolts and nuts to 6.5 ft. lbs. (9 Nm).

  n. Fourteenth pass: intake manifold bolts and nuts to 6–7 ft. lbs. (8–10 Nm).

- Rocker covers
- A/C compressor, alternator brackets
- A/C compressor and alternator
- Left hand manifold-to-bracket bolt
- Left hand exhaust manifold to the right hand exhaust manifold
- Exhaust tube to the left hand manifold
- Harness clamp to the right hand rocker cover
- Distributor
- Rear timing belt cover
- Camshaft sprockets
- Intake manifold (lower)
- Fuel rail and injectors as an assembly
- Fuel injector's electrical connections
- Fuel feed and return hoses to the fuel rail

- Upper intake manifold (plenum)
- Timing belt
- Air intake tube
- Negative battery cable

10. Fill the cooling system. An oil and filter change is recommended.

11. Start the vehicle and check for leaks. Check the ignition timing and adjust as required.

3.5L Engine

1. Before servicing the vehicle, refer to the precautions section.

2. Before servicing the vehicle, refer to the precautions section.

3. Disconnect the negative battery cable.

4. Remove the fuel tube and fuel injector assembly.

5. Remove the intake manifold.

6. Remove the exhaust manifold.

7. Remove the water inlet and thermostat assembly.

8. Remove the water outlet and water piping.

9. Remove the camshaft.

10. Remove the cylinder head bolts in reverse of the tightening sequence.

11. Inspect the bolts. Cylinder head bolts are tightened by plastic zone tightening method. Whenever the size difference between d1 and d2 exceeds the limit, replace them with new one. If reduction of outer diameter appears in a position other than d2, use it as d2 point.

To install:

12. Install cylinder head gasket.

13. Turn the crankshaft until No. 1 piston is set at TDC on the compression stroke. The crankshaft key should line up with the right bank cylinder center line as shown.

14. Install cylinder head. Tighten the head bolts in the order shown in illustration.

  a. Step 1: Tighten all bolts to 72 ft. lbs. (98 Nm).

  b. Step 2: Completely loosen to 0 in the reverse order.

  c. Step 3: Tighten all bolts to 26–32 ft. lbs. (34.3–44.1 Nm).

  d. Step 4: Turn all bolts 90 degrees clockwise.

  e. Step 5: Turn all bolts an additional 90 degrees clockwise.

15. After installing cylinder head, measure distance between front end faces of cylinder block and cylinder head (left and right banks). If measurement is outside the specified range, reinstall cylinder head.

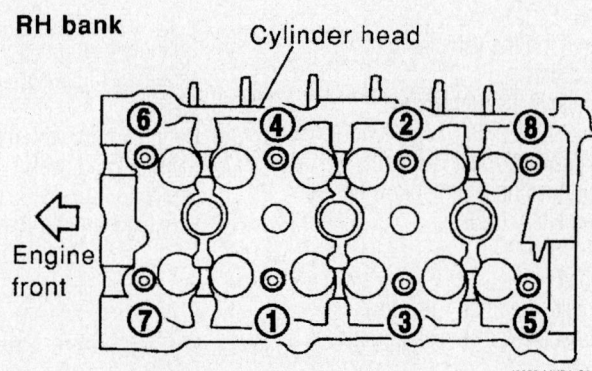

42356-MURA-G05

Cylinder head bolt torque sequence—3.5L engine–right side

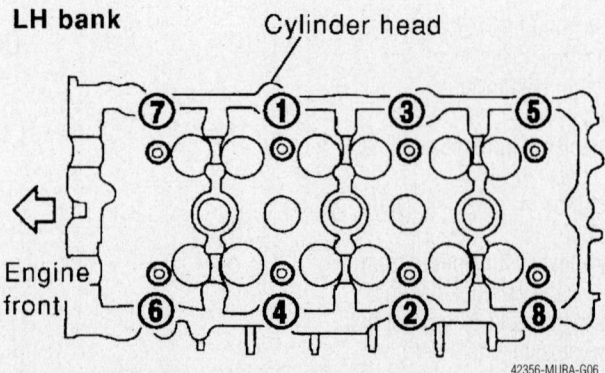

42356-MURA-G06

Cylinder head bolt torque sequence—3.5L engine–left side

Cylinder head bolt

(Measuring point)   (Measuring point)

d2   d1

11 mm (0.43 in)

48 mm (1.89 in)

42356-MURA-G07

Cylinder head bolt measurement—3.5L engine

16. The remainder of installation is the reverse of removal.

Rocker Arms/Shafts

REMOVAL & INSTALLATION

3.3L Engine

1. Before servicing the vehicle, refer to the precautions section.
2. Remove or disconnect the following:
   - Negative battery cable
   - Upper intake manifold

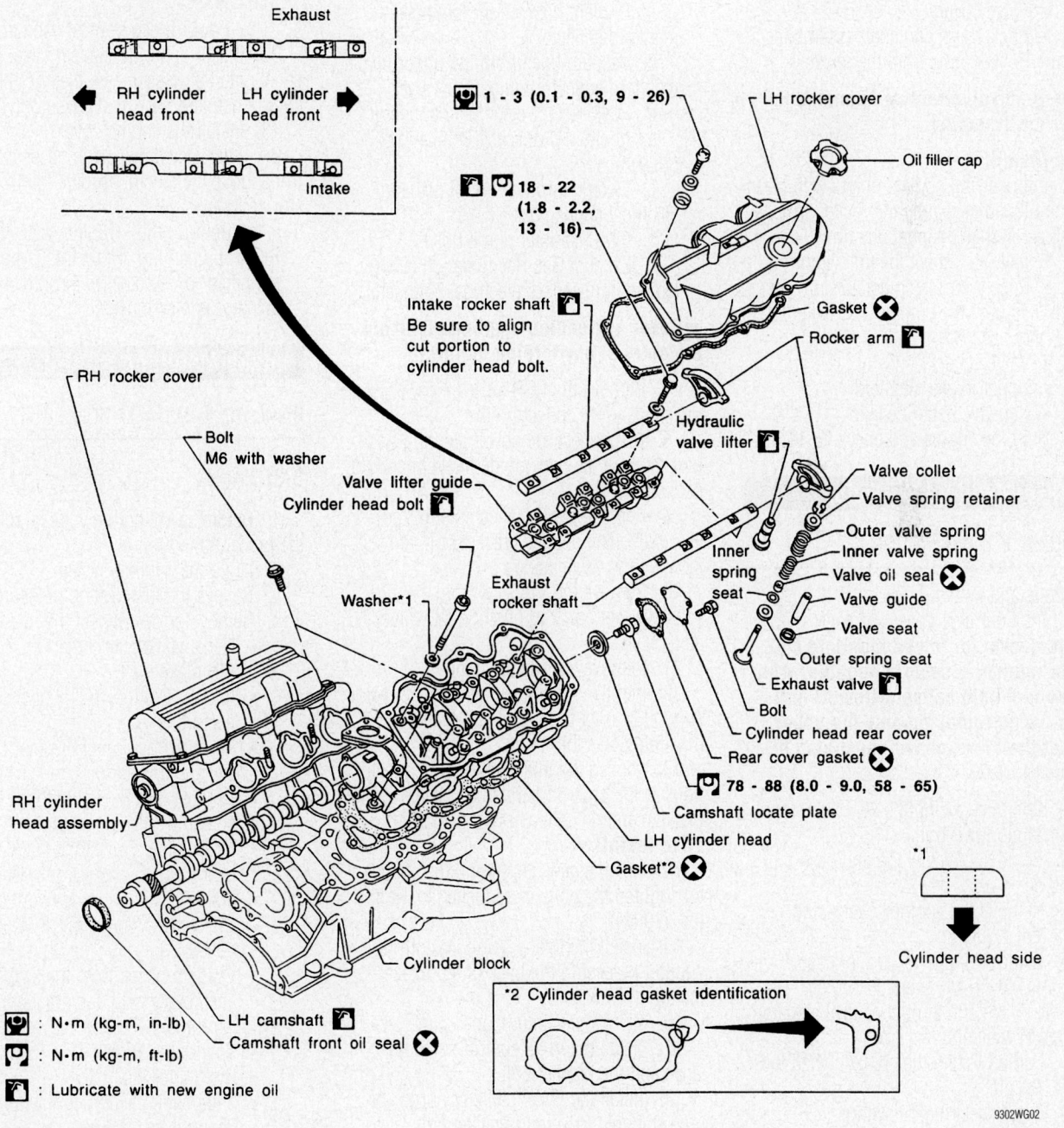

Rocker arm and shaft components—3.3L engine

9302WG02

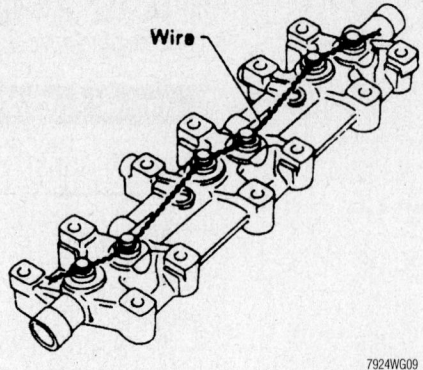

7924WG09

Wire the lifters on top of the guide so they won't fall out when the guide is removed from the head—3.3L engines

- Valve covers
- Rocker arm and shaft assemblies
- Rocker arms from the shafts

➥**Keep all valvetrain components in order for assembly.**

To install:

3. Lubricate all contact points with clean engine oil and assemble the rocker arms to the shafts in their original positions.

4. Install or connect the following:
- Rocker arm and shaft assemblies. Tighten the bolts to 13–16 ft. lbs. (18–22 Nm).
- Valve covers
- Upper intake manifold
- Negative battery cable

5. Start the engine and check for leaks.

Intake Manifold Collector

REMOVAL & INSTALLATION

3.5L Engine

➥**The gasket for intake manifold collector (upper) is secured together with intake manifold collector(lower) bolt. Thus, when replacing only the upper gasket the lower gasket must also be replaced.**

1. Before servicing the vehicle, refer to the precautions section.

2. Remove the cowl top and cowl top extension.

3. Remove the engine cover using power tool.

4. Remove upper air cleaner case, Mass Air Flow (MAF) sensor, and air cleaner–to–electric throttle control actuator tube as an assembly.

5. Partially drain the coolant when the engine is cool.

6. Disconnect the following:
   a. Power brake booster vacuum hose.
   b. Coolant hoses from the intake manifold collector
   c. Vacuum lines from the upper intake manifold collector and power valve.
   d. Fuel injector electrical connectors.
   e. Positive Crankcase Ventilation (PCV) hose.
   f. Electric throttle control actuator electrical connectors.
   g. EVAP canister purge hose.
   h. Exhaust Gas Recirculation (EGR) temperature sensor electrical connector.

➥**Cover any engine openings to avoid the entry of any foreign material.**

7. Remove the EGR tube–to–lower intake manifold collector nuts.

8. Disconnect the power steering hose bracket from the back of the intake manifold collector.

9. Remove the EVAP canister purge volume solenoid valve bracket bolt and position the valve aside.

10. Remove the VIAS control solenoid valve bracket bolt and position the valve aside.

11. Remove the vacuum tank.

12. Remove the intake manifold collector support bracket from the back of the intake manifold collector.

13. Loosen the intake manifold collector bolts in the order illustrated and remove the intake manifold collector and gasket.

To install:

14. Install a new gasket and the collector. Tighten the bolts in sequence to 14 ft. lbs. (19 Nm).

15. Install the intake manifold collector support bracket to the back of the intake manifold collector.

16. Install the vacuum tank.

17. Install the VIAS control solenoid valve and the bracket bolt.

18. Install the EVAP canister purge volume solenoid valve and bracket bolt.

19. Connect the power steering hose bracket to the back of the intake manifold collector.

20. Install the EGR tube–to–lower intake manifold collector nuts.

21. Connect the following:
   a. EGR temperature sensor electrical connector
   b. EVAP canister purge hose.
   c. Electric throttle control actuator electrical connectors.
   d. PCV hose.
   e. Fuel injector electrical connectors.
   f. Vacuum lines to the upper intake manifold collector and power valve.
   g. Coolant hoses to the intake manifold collector.
   h. Power brake booster vacuum hose.

22. Install upper air cleaner case, MAF sensor, and air cleaner–to–electric throttle control actuator tube as an assembly.

23. Install the engine cover.

24. After installation, it is necessary to re-calibrate the electric throttle control actuator as follows:
   a. Perform the "Throttle Valve Closed Position Learning" when the harness connector of the electric throttle control actuator is disconnected.

Intake Manifold

REMOVAL & INSTALLATION

3.3L Engine

1. Before servicing the vehicle, refer to the precautions section.

2. Drain the cooling system.

3. Relieve the fuel system pressure.

4. Remove or disconnect the following:
- Negative battery cable
- Air intake duct
- Idle Air Control (IAC) valve connectors
- Throttle Position (TP) sensor and switch connectors
- Exhaust Gas Recirculation (EGR) solenoid valve connector
- Evaporative Emissions (EVAP) canister vacuum and purge hoses
- Water, heater and Positive Crankcase Ventilation (PCV) valve hoses
- Vacuum hoses from the EVAP canister, brake cylinder, pressure regulator and EGR tube
- Spark plug wires
- Distributor cap
- 3 left bank injector connectors
- Thermal transmitter

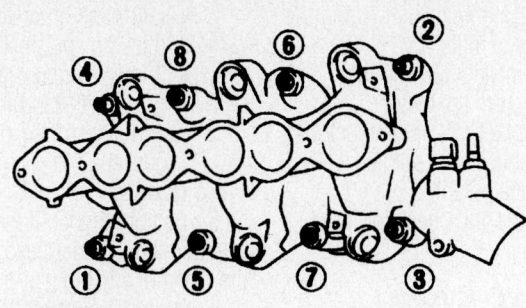

Loosen bolts in
numerical order.

7924VG32

Intake manifold loosening sequence—3.3L engine

- Ground harness
- Breather pipe
- Upper manifold
- Fuel feed and return lines from the fuel rail
- Right injector harness connectors
- Fuel rail and injectors
- Coolant temperature switch harness connector
- Water hose from the thermostat
- Lower manifold bolts in the sequence illustrated.
- Manifold gasket and discard

To install:
5. Install the lower intake manifold with a new gasket.
    a. Step 1: 35 inch lbs. (4 Nm)
    b. Step 2: 78 inch lbs. (9 Nm)
    c. Step 3: 70–84 inch lbs. (8–10 Nm)
6. Install or connect the following:
- ECT sensor connector
- Fuel supply manifold
- Right bank injector connectors
- Fuel lines
- Upper intake manifold
- Breather pipe
- Upper intake manifold ground cable
- Thermal transmitter
- Left bank injector connectors
- Distributor

- Spark plug wires
- EGR tube
- Fuel pressure regulator vacuum hose
- Brake booster vacuum hose
- EVAP canister vacuum and purge hoses
- PCV valve and hose
- Heater hoses
- Radiator hoses
- EGR temperature sensor connector
- EGR solenoid valve connector
- Ignition coil and power transistor connectors
- TP sensor and switch connectors
- IAC valve connector
- Cruise control cable
- Accelerator cable
- Air intake duct
- Negative battery cable
7. Fill the cooling system.
8. Start the engine and check for leaks.

3.5L Engine

1. Before servicing the vehicle, refer to the precautions section.
2. Relieve the fuel system pressure.
3. Remove the intake manifold collector.
4. Remove the fuel rail with the fuel injectors as an assembly.

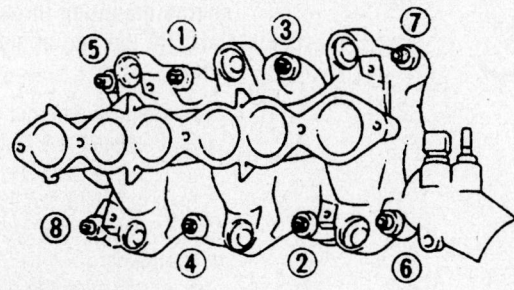

Tighten bolts in
numerical order.

7924VG33

Intake manifold tightening sequence—3.3L engine

5. Loosen the intake manifold nuts and bolts in the sequence illustrated.
6. Remove the intake manifold.
To install:
7. Install the intake manifold. Tighten the intake manifold nuts and bolts in the sequence illustrated in the following steps:
    a. Stud bolts: 96 inch lbs. (10 Nm).
    b. Manifold bolts first pass: 65 inch lbs. (7 Nm).
    c. Manifold bolts second pass: 21 ft. lbs. (29 Nm).
8. Install the fuel rail with the fuel injectors as an assembly.
9. Install the intake manifold collector.

Exhaust Manifold

REMOVAL & INSTALLATION

3.3L Engine

REAR (RIGHT-HAND) EXHAUST MANIFOLD

1. Before servicing the vehicle, refer to the precautions section.
2. Remove or disconnect the following:
- Negative battery cable
- Radiator overflow hose from the radiator
- Radiator coolant-recovery reservoir off of the bracket
- Reservoir
- Air cleaner intake tube and the engine air intake resonator
- 6 rear (right-hand) exhaust manifold crossover tube heat-shield bolts and the heat shields
- 2 nuts and the 1 bolt securing the rear (right-hand) exhaust manifold tube to the front (left-hand) exhaust manifold. Discard the gasket.
- Transmission fluid level indicator tube heat shield
3. Disengage the following electrical connectors:
- Idle switch
- Throttle Position (TP) sensor
- Exhaust Gas Recirculation (EGR) control solenoid
4. Raise and safely support the vehicle.
5. Remove or disconnect the following:
- EGR valve-to-back-pressure transducer valve tube nut and position it out of the way
- 2 EGR valve-to-exhaust manifold tube nuts and tube
- 6 rear exhaust manifold nuts in the reverse order of the tightening sequence
6. Safely lower the vehicle, remove the

exhaust manifold and discard the exhaust manifold gasket.

To install:

7. Raise and safely support the vehicle.

8. Be sure that both the exhaust manifold and the cylinder head mating surfaces are clean of any old gasket material.

9. Install or connect the following:
- Rear (right-hand) exhaust manifold gasket onto the exhaust manifold mounting studs

10. Lower the vehicle safely.
- Rear (right-hand) exhaust manifold onto the studs
- 6 rear (right-hand) exhaust manifold nuts. Tighten the nuts in sequence to 13–16 ft. lbs. (18–22 Nm).
- EGR valve-to-exhaust manifold tube and tube nuts
- EGR valve-to-back-pressure transducer valve tube nut

11. Lower the vehicle carefully.

12. Reconnect the following electrical connectors:
- EGR solenoid
- TP sensor
- Idle switch
- Transmission fluid level indicator tube heat shield
- New gasket between the front (left-hand) exhaust manifold and the rear exhaust manifold crossover tube
- 2 nuts and the 1 bolt securing the rear (right-hand) exhaust manifold crossover tube to the front (left-hand) exhaust manifold. Tighten the rear exhaust manifold crossover tube-to-front (left-hand) exhaust manifold nuts and bolt.
- Rear (right-hand) exhaust manifold crossover tube heat shield with the 6 mounting bolts

- Rear (right-hand) exhaust manifold crossover tube bolts
- Air cleaner intake tube and the engine air intake resonator
- Radiator coolant recovery reservoir
- Radiator overflow hose to the radiator
- Negative battery cable

13. Start the engine and check for leaks and proper operation.

FRONT (LEFT-HAND) EXHAUST MANIFOLD

1. Before servicing the vehicle, refer to the precautions section.

2. Remove or disconnect the following:
- Negative battery cable and wait at least 90 seconds before performing any work. This allows time for the SRS or air bag system to deplete its back up energy supply.
- 2 nuts and the 1 bolt securing the front (left-hand) exhaust manifold to the rear (right-hand) exhaust manifold crossover tube. Discard the gasket.

3. Remove the transmission fluid level indicator tube heat shield.
- 6 front (left-hand) exhaust manifold nuts in 2 steps in the reverse order of the tightening sequence. Do not remove the 3 lower front (left-hand) exhaust manifold nuts.
- Front (left-hand) exhaust manifold-to-mounting bracket bolt

4. Raise and safely support the vehicle.
- Heated Oxygen Sensor (HO2S) electrical connector
- 3 front (left-hand) exhaust manifold-to-inlet pipe nuts
- Exhaust system flex tube bracket bolt
- Left-hand inner engine and transmission splash shield bolts and screws

- Left-hand inner engine and transmission splash shield
- 3 lower exhaust manifold nuts
- Front (left-hand) exhaust manifold and discard the exhaust manifold gasket

To install:

5. Be sure that both the exhaust manifold and the cylinder head mating surfaces are clean of any old gasket material.

6. Install or connect the following:
- New front exhaust manifold gasket in place
- Front (left-hand) exhaust manifold
- 3 lower exhaust manifold mounting nuts. Do not tighten the nuts at this time.
- Left-hand inner engine and transmission splash shield with their mounting bolts and screws
- Exhaust system flex tube bracket bolt
- 3 exhaust manifold-to-exhaust inlet pipe nuts
- HO2S electrical connector

7. Lower the vehicle.
- Front (left-hand) exhaust manifold-to-mounting bracket bolt
- 3 upper exhaust manifold mounting bolts and tighten all 6 exhaust manifold mounting bolts in sequence to 13–16 ft. lbs. (18–22 Nm)
- Transmission fluid level indicator tube heat shield
- 2 nuts and the 1 bolt securing the front (left-hand) exhaust manifold to the rear (right-hand) exhaust manifold crossover tube
- Negative battery cable

8. Start the engine, check for leaks and road test for proper operation.

3.5L Engine

1. Before servicing the vehicle, refer to the precautions section.

➡**When removing the front and rear engine mounting through bolts and nuts, lift the engine up slightly for safety.**

2. Remove the front wheels.

3. Remove the engine undercover.

4. Remove the wheel splash shields.

5. If removing the left hand exhaust manifold, remove the radiator and cooling fan assembly.

6. If removing the right hand exhaust manifold, remove the front suspension member as follows:

   a. Remove the left hand transaxle

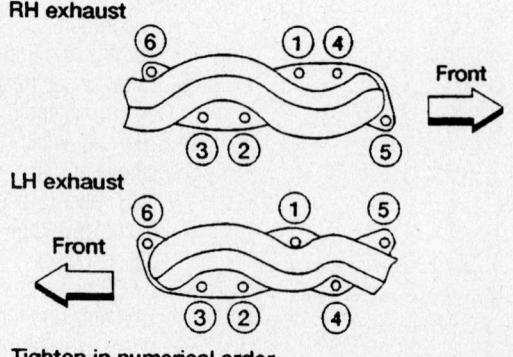

RH exhaust

LH exhaust

Tighten in numerical order.

9348WG12

To avoid warping the exhaust manifolds, use this sequence when loosening the bolts—3.3L engine

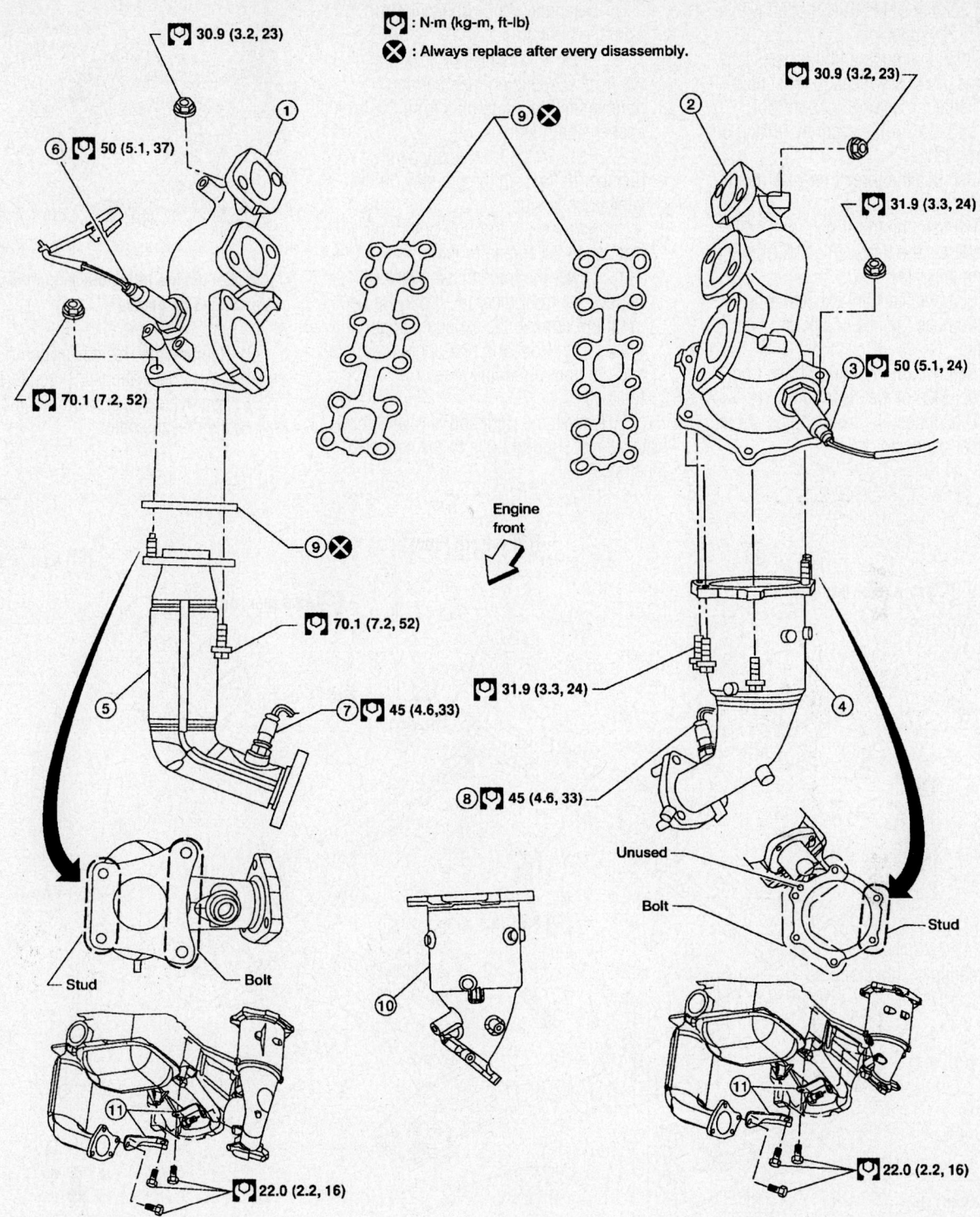

1. Exhaust manifold (RH bank)
2. Exhaust manifold (LH bank)
3. Air fuel ratio (A/F) sensor 1 (bank
4. Three way catalyst (manifold) (bank 2) (4 A/T only)
5. Three way catalyst (manifold) (bank 1)
6. Air fuel ratio (A/F) sensor 1 (bank
7. Heated oxygen sensor 2 (bank 1)
8. Heated oxygen sensor 2 (bank 2) (4 A/T only)
9. Gasket
10. Three way catalyst (manifold) (bank 2) (5 A/T only)
11. Three way catalyst supports

67170-NIQU-G04

Exploded view of the exhaust manifold and catalyst assembly–3.5L engine

mount insulator nuts, if equipped with a 5 speed transmission.

b. Attach an engine lifting bracket to the transaxle at the location illustrated.

c. Install an engine support tool. Make sure the tool is securely resting on the hood ledge.

d. Remove the three transaxle mount insulator nuts.

e. Remove the lower ball joint bolt and separate the transverse link from the steering knuckle.

f. Remove the front exhaust tube.

g. Remove the power steering line bracket.

h. Remove the mounting bolts from the lower side of the steering gear.

i. Disconnect the front engine mount electrical connector.

j. Disconnect the connecting rod from the front strut.

k. Place a transmission jack under the front suspension member and remove the mounting nuts from the front suspension member.

l. Remove the front suspension member bolts from the pin stay on the vehicle body side.

m. Remove the through bolts from the front and rear engine mounts.

n. Lower the transmission jack to remove the front member. It may be necessary to remove the exhaust hanger bracket, the front and rear engine mounts and the transverse link to enable removal.

7. Remove the right and left hand three way catalyst support bolts in the sequence illustrated.

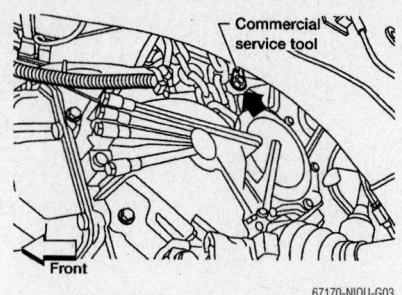

Attach a engine lifting bracket to the tranaxle–3.5L engine

8. Remove the Heated Oxygen (HO2S) sensors, Air Fuel Ratio (AFR) sensors.

9. Remove exhaust manifold and three way catalyst heat shields.

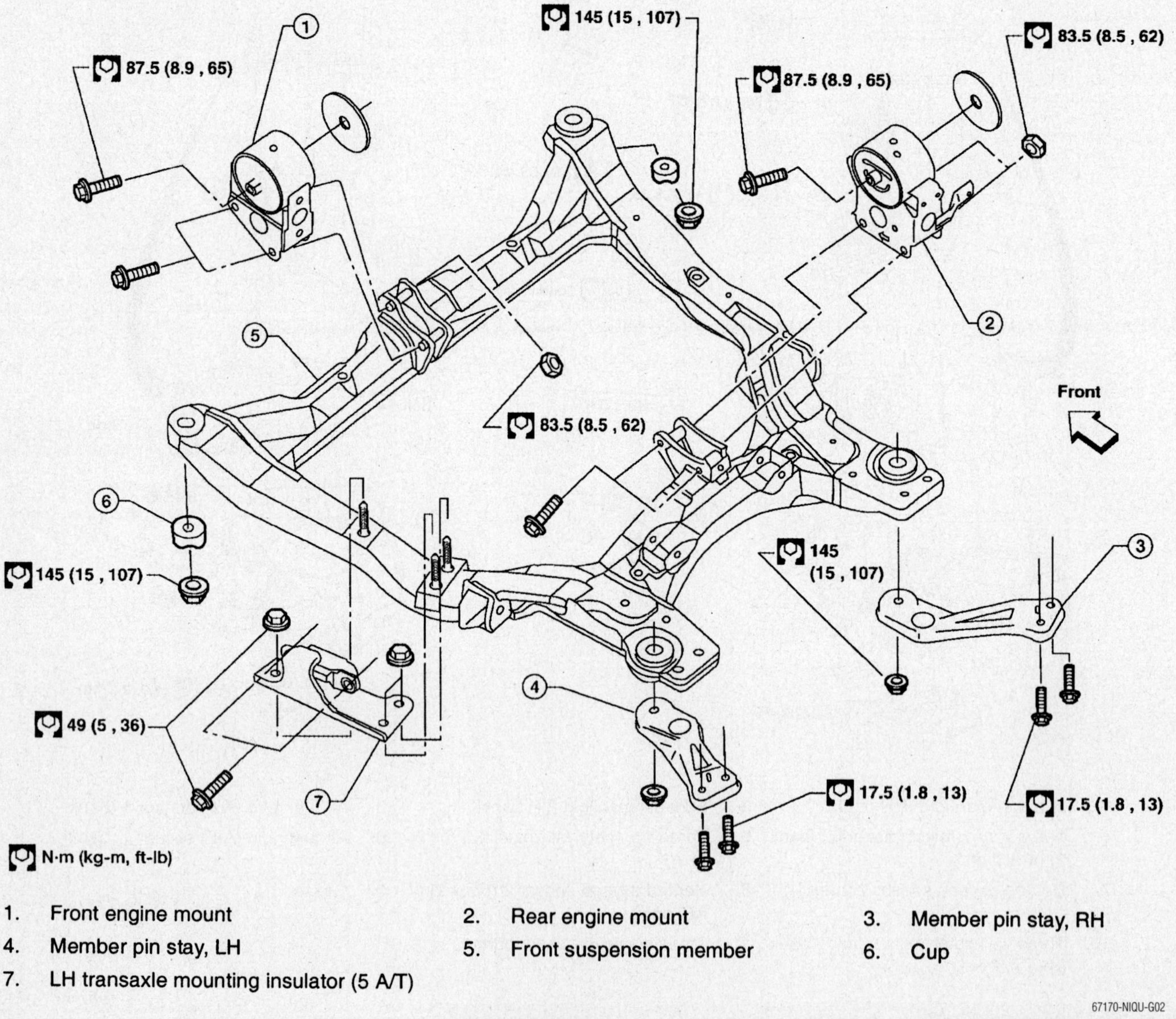

87.5 (8.9 , 65)

145 (15 , 107)

83.5 (8.5 , 62)

87.5 (8.9 , 65)

Front

83.5 (8.5 , 62)

145 (15 , 107)

145 (15 , 107)

49 (5 , 36)

17.5 (1.8 , 13)

17.5 (1.8 , 13)

N·m (kg-m, ft-lb)

1. Front engine mount
4. Member pin stay, LH
7. LH transaxle mounting insulator (5 A/T)

2. Rear engine mount
5. Front suspension member

3. Member pin stay, RH
6. Cup

Exploded view of the front suspension member mounting–3.5L engine

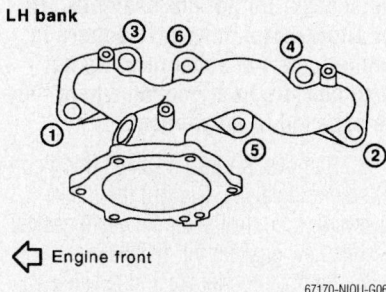

Remove the right hand manifold nuts in this sequence–3.5L engine

Remove the left hand manifold nuts in this sequence–3.5L engine

10. Remove the three way catalyst by loosening the bolts first and then removing the nuts and through bolts.

11. Remove the exhaust manifolds by loosening the nuts in the sequence illustrated.

To install:

12. Install the manifolds and tighten the nuts in sequence to 23 ft. lbs. (30 Nm).

13. Install the three way catalyst by tightening the nuts first and then the bolts and tighten to 16 ft. lbs. (22 Nm) in the sequence illustrated.

14. Install the exhaust manifold and three way catalyst heat shields.

15. Install the Heated Oxygen (HO2S) sensors and Air Fuel Ratio (AFR) sensors. Tighten to the specifications shown in the exploded view of the exhaust manifold assembly illustration.

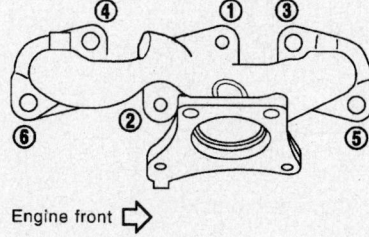

Install the right hand manifold nuts in this sequence–3.5L engine

Install the left hand manifold nuts in this sequence–3.5L engine

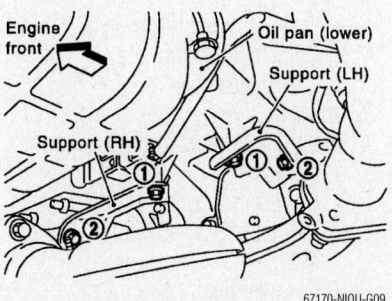

Install the three way catalyst by tightening the nuts first and then the bolts in the sequence shown–3.5L engine

16. Install the front support member in the reverse order of removal and tighten the retainers to the specifications shown in the front suspension member mounting exploded view illustration.

17. Install any remaining components in the reverse order of removal.

Starter

REMOVAL & INSTALLATION

3.3L Engine

1. Before servicing the vehicle, refer to the precautions section.

2. Remove or disconnect the following:
   - Battery negative cable
   - Air cleaner
   - Nut attaching the positive cable to the starter
   - Positive cable from the starter
   - S-terminal connector
   - 2 starter bolts and the starter

To install:

3. Installation is the reverse of removal.

4. Tighten the starter bolts to 17–19 ft. lbs. (23–26 Nm) and the nut that attaches the positive battery cable to the starter to 87–104 inch lbs. (10–12 Nm).

3.5L Engine

4 SPEED MODELS

1. Before servicing the vehicle, refer to the precautions section.

2. Disconnect the negative battery terminal.

3. Remove the upper air cleaner case and the air cleaner to electric throttle control actuator tube.

4. Remove the harness protector from the starter harness.

5. Disconnect the starter harness connectors.

6. Remove the two starter mounting bolts.

7. Remove the starter.

To install:

8. Installation is the reverse of removal. Tighten the upper bolt to 41 ft. lbs. (55 Nm) and the lower bolt to 55 ft. lbs. (74 Nm).

5 SPEED MODELS

1. Before servicing the vehicle, refer to the precautions section.

2. Disconnect the negative battery terminal.

3. Remove the starter insulator.

4. Remove the harness protector from the starter harness.

5. Disconnect the starter harness connectors.

6. Remove the two starter mounting bolts.

7. Remove the starter.

To install:

8. Installation is the reverse of removal. Tighten the bolts to 41 ft. lbs. (55 Nm).

Front Crankshaft Seal

REMOVAL & INSTALLATION

3.3L Engine

1. Before servicing the vehicle, refer to the precautions section.

2. Remove or disconnect the following:
   - Negative battery cable
   - Drive belts
   - Radiator hoses
   - Crankshaft pulley
   - Front cover
   - Timing belt
   - Crankshaft timing sprocket
   - Crankshaft seal using a suitable prytool

To install:

3. Install or connect the following:
   - Crankshaft seal using a driver and a hammer until its flush with the housing

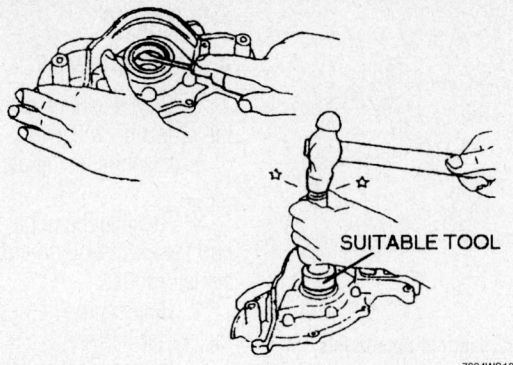

7924WG13

Removal and installation of the front crankshaft seal—3.3L engines

- Crankshaft timing sprocket
- Timing belt
- Front cover and tighten the bolts to 26–43 inch lbs. (3–5 Nm)
- Crankshaft pulley and tighten the bolt to 141–156 ft. lbs. (191–211 Nm)
- Radiator hoses
- Drive belts
- Negative battery cable

4. Fill the cooling system, start the vehicle and check for leaks.

3.3L Engine

Refer to the front timing cover case removal and installation for seal removal.

Camshaft And Valve Lifters

REMOVAL & INSTALLATION

3.3L Engine

1. Before servicing the vehicle, refer to the precautions section.
2. Drain the cooling system.
3. Remove or disconnect the following:
   - Negative battery cable
   - Upper intake manifold
   - Valve covers

➡Keep all valvetrain components in order for assembly.

- Rocker arm and shaft assemblies
- Valve lifter guide and valve lifters. Attach a wire to the top of the lifters so that they will not drop from the lifter guide.
- Radiator
- Accessory drive belts
- Front cover
- Timing belt
- Camshaft sprockets
- Camshaft seals
- Rear timing cover
- Distributor

- Cylinder head rear covers
- Camshaft locating plates
- Camshafts

To install:
4. Install or connect the following:
   - Camshafts
   - Camshaft locating plates. Tighten the bolts to 58–65 ft. lbs. (78–88 Nm).
   - Cylinder head rear covers
   - Distributor
   - Rear timing cover
   - Camshaft seals
   - Camshaft sprockets. Tighten the bolts to 58–65 ft. lbs. (78–88 Nm).
   - Timing belt
   - Front cover
   - Accessory drive belts
   - Radiator
   - Valve lifter guide and valve lifters
   - Rocker arm and shaft assemblies. Tighten the bolts to 13–16 ft. lbs. (18–22 Nm).
   - Valve covers
   - Upper intake manifold
   - Negative battery cable
5. Fill the cooling system.
6. Start the engine and check for leaks.

3.5L Engine

1. Before servicing the vehicle, refer to the precautions section.
2. Before servicing the vehicle, refer to the precautions section.
3. Drain the cooling system.
4. Remove the front timing chain case, camshaft sprocket, timing chain and the rear timing chain case.
5. If necessary, remove both the Camshaft Position (CMP) sensors from cylinder head back side.

➡**Handle carefully to avoid dropping and shocks. Do not disassemble. Do not allow metal powder to adhere to magnetic part at sensor tip. Do not place sensors in a location where they are exposed to magnetism.**

6. Remove the intake valve timing control solenoid valves. Discard the intake valve timing control solenoid valve gaskets and use new gaskets for installation.
7. Remove the intake and exhaust camshaft brackets. Mark the camshafts, camshaft brackets, and bolts so they are placed in the same position and direction for installation.
8. Equally loosen the camshaft bracket bolts in several steps in the numerical order shown.
9. Remove the camshaft.
10. Remove the valve lifter. Identify installation positions, and store them without mixing them up.
11. Remove the secondary timing chain tensioner from cylinder head with its stopper pin attached.
12. Inspect the Camshaft Runout as follows:
    a. Put V block on precise flat bed, and support No. 2 and No. 4 journal of camshaft.
    b. Set dial gauge vertically to No. 3 journal.

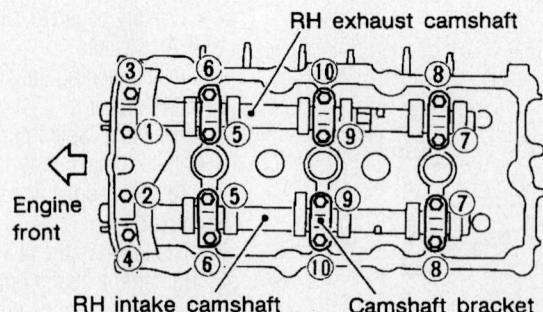

RH exhaust camshaft

Engine front

RH intake camshaft   Camshaft bracket

Loosen in numerical order.

42356-MURA-G20

Camshaft bracket bolt loosening sequence—3.5L right side

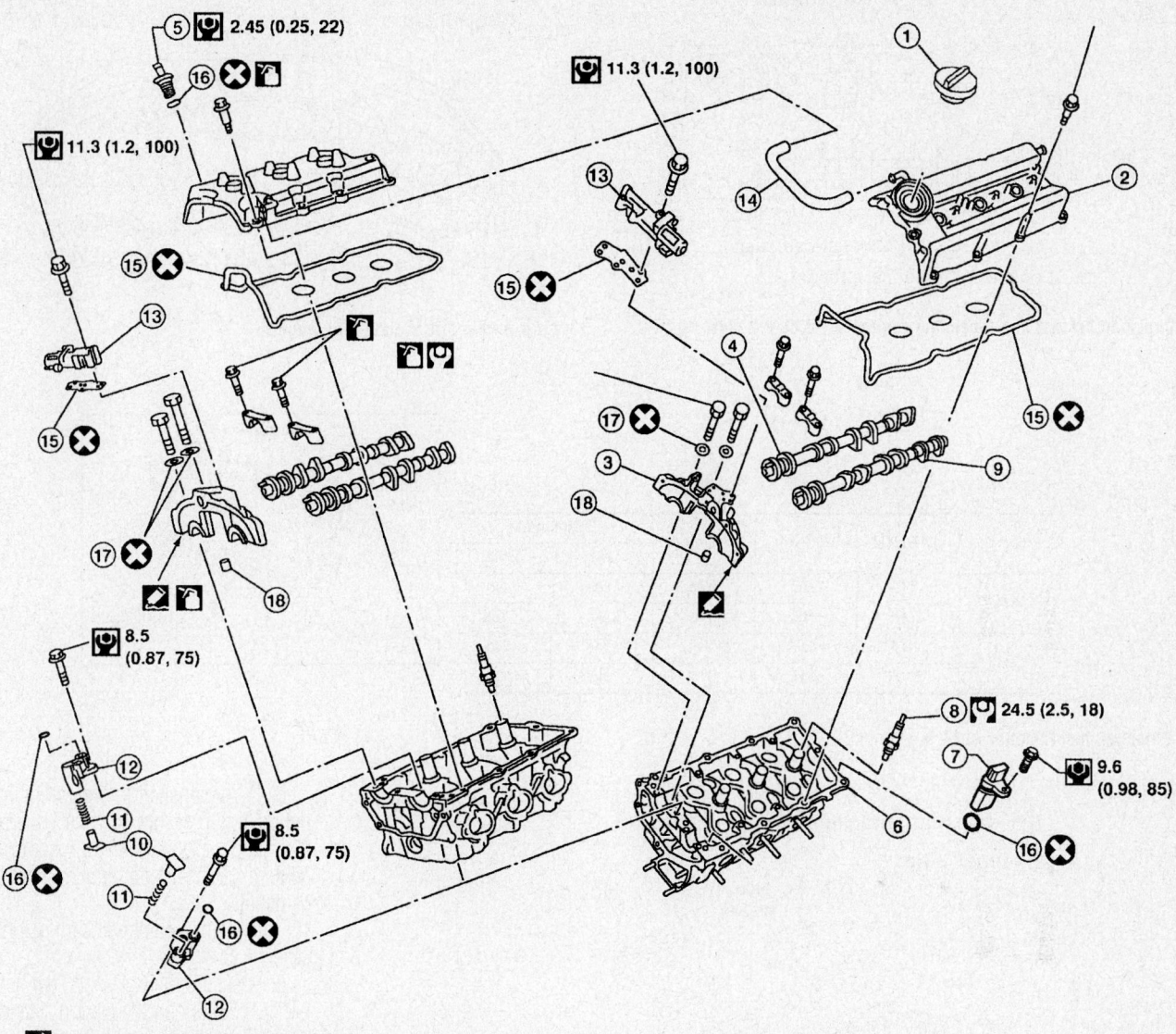

: Apply Genuine Silicone RTV Sealant or equivalent.
  Refer to GI Section.

: Lubricate with engine oil

: N·m (kg-m, ft-lb)

: N·m (kg-m, in-lb)

: Always replace after every disassembly.

1. Oil filler cap	2. Rocker cover (LH)	3. Camshaft bracket (LH)
4. Camshaft (INT)	5. PCV valve	6. Cylinder head (LH)
7. Camshaft position sensor (PHASE)	8. Spark plug	9. Camshaft (EXH)
10. Tensioner sleeve	11. Tensioner spring	12. Camshaft chain tensioner
13. IVT control solenoid valve	14. PCV hose	15. Gasket
16. O-ring	17. Seal washer	18. Dowel pin

CAUTION:
Apply new engine oil to parts marked in illustration before installation.

67170-NIQU-G68

Exploded view of the camshaft and related components–3.5L engine

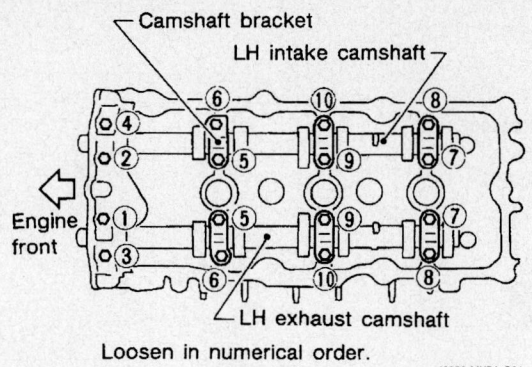

Camshaft bracket bolt loosening sequence—3.5L left side

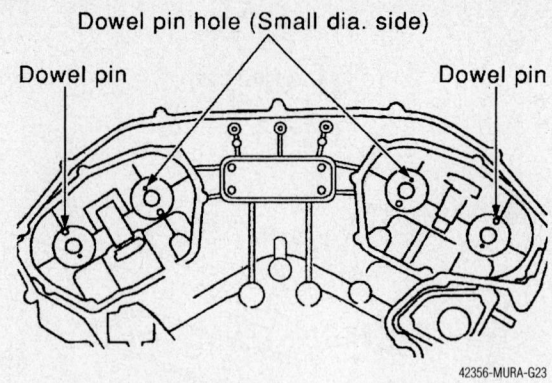

Dowel pin orientation–3.5L engine

Bank	INT/EXH	ID mark	Drill mark	Paint marks	
				M1	M2
RH	INT	RE	Yes	Yes	No
	EXH	RE	No	No	Yes
LH	INT	LH	Yes	Yes	No
	EXH	LH	No	No	Yes

Camshaft identification–3.5L engine

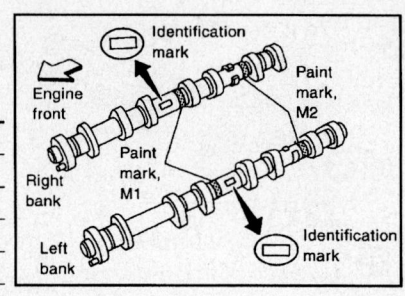

RH camshaft brackets

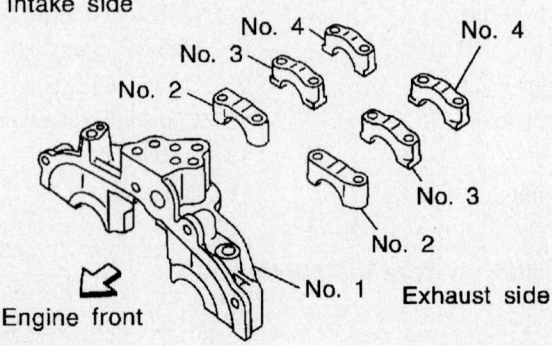

Camshaft bracket identification–3.5L engine

c. Turn camshaft to one direction with hands, and measure camshaft runout on dial gauge. (Total indicator reading)

d. If it exceeds the limit, replace camshaft.

13. Inspect the camshaft cam height as follows:

a. Measure the camshaft cam height. Limit: 0.0020 inch (0.05 mm). Standard cam height (intake and exhaust): 1.7663–1.7738 inch (44.865–45.055 mm)

b. If wear is beyond the limit, replace camshaft.

14. Inspect the camshaft journal clearance of the outer diameter of the camshaft journal as follows:

a. Measure outer diameter of camshaft journal.

Cam wear limit: 0.008 inch (0.2 mm)
Standard outer diameter:
- No. 1: 1.0211–1.0218 inch (25.935–25.955 mm)
- No. 2, 3, 4: 0.9230–0.9238 inch (23.445–23.465 mm)

15. Inspect the camshaft journal clearance of the inner diameter of the camshaft bracket as follows:

a. Tighten camshaft bracket bolt with specified torque.

b. Using inside micrometer, measure inner diameter "A" of camshaft bracket.

16. Calculate the camshaft journal clearance as follows:

Journal clearance = inner diameter of camshaft bracket—outer diameter of camshaft journal. When outside the limit, replace either or both camshaft and cylinder head.

➡**Inner diameter of camshaft bracket is manufactured together with cylinder head. Replace the whole cylinder head assembly.**

17. Inspect the camshaft end play as follows:

a. Install dial gauge in thrust direction on front end of camshaft. Measure end play of dial gauge when camshaft is moved forward and backward.

b. When out of the limit, replace with new camshaft and measure again.

c. When out of the limit again, replace with new cylinder head.

To install:

18. Install secondary chain tensioners on both sides of cylinder head.

a. Install chain tensioner with its stopper pin attached.

b. Install tensioner with sliding part facing downward on right side cylinder head, and with sliding part facing upward on left side cylinder head.

c. Install O-ring as shown.

19. Install valve lifters in their original position.

Valve lifter outer diameter (Intake and exhaust)

• 1.3377–1.3381 inch (33.977–33.987 mm)

Standard (Intake and exhaust)

• 1.3386–1.3392 inch (34.000–34.016 mm)

Standard (Intake and exhaust)

• 0.0005–0.0015 inch (0.013–0.039 mm)

20. Install camshafts.

a. Install camshaft with dowel pin attached to its front end face on the exhaust side.

b. Follow your identification marks made during removal, or follow the identification marks that are present on the new camshafts for proper placement and direction.

c. Install camshaft so that dowel pin hole and dowel pin on front end face are positioned as shown in illustration. (No. 1 cylinder TDC on its compression stroke)

➡**Large- and small-pin holes are located on front end face of intake camshaft, at intervals of 180 degrees. Face small dia. side pin hole upward (in cylinder head upper face direction).**

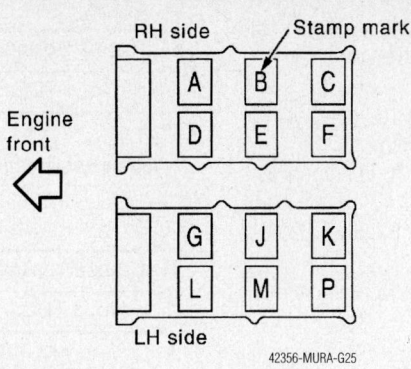

RH side   Stamp mark

Engine front

42356-MURA-G25

LH side

Align the stamp marks as shown

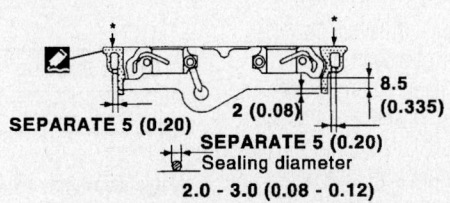

8.5 (0.335)

SEPARATE 5 (0.20)   2 (0.08)

SEPARATE 5 (0.20)

Sealing diameter

2.0 - 3.0 (0.08 - 0.12)

* : Remove the protruding sealant from front face. (Remove the hardended sealant from surface only.)

◪ : Apply liquid gasket

Unit: mm (in)

42356-MURA-G26

RTV sealer application

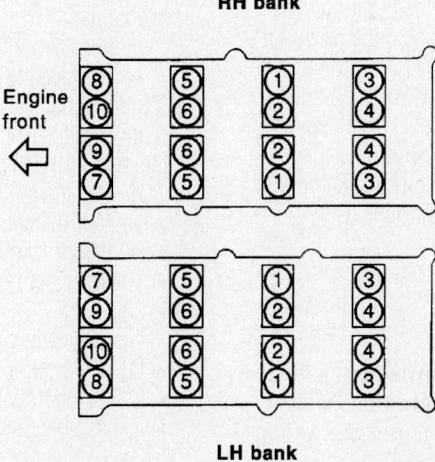

RH bank

Engine front

LH bank

42356-MURA-G27

Camshaft bracket bolt torque sequence

21. Install camshaft brackets.

a. Remove foreign material completely from camshaft bracket backside and from cylinder head installation face.

b. Install camshaft bracket in original position and direction as shown in illustration.

c. Install No.2 to 4 camshaft brackets aligning the stamp marks as shown.

➡**There are no identification marks indicating left and right for No. 1 camshaft bracket.**

d. Apply sealant to mating surface of No.1 camshaft bracket as shown on right and left banks.

➡**Use RTV silicone sealant or equivalent.**

22. Tighten the camshaft brackets in the following steps, in numerical order as shown.

    a. Tighten No. 7 to 10, then tighten No.1 to 6 in order as shown.

    b. Tighten No.1 to 10 in numerical order as shown.

    c. Tighten No. 1 to 6 in the numerical order as shown.

    d. Tighten No. 7 to 10 in the numerical order as shown.

- 17 inch lbs. (1.96 Nm)
- 52 inch lbs. (5.88 Nm)
- 80–104 inch lbs. (9.02–11.8 Nm)
- 74–91 inch lbs. (8.3–10.3 Nm)

23. Measure difference in levels between front end faces of No. 1 camshaft bracket and cylinder head. If measurement is outside the specified range, re-install camshaft and camshaft bracket.

24. Inspect and adjust valve clearance.

25. Install in the reverse order of removal after this step.

26. Inspect the valve clearance as follows:

Perform inspection as follows after removal, installation or replacement of camshaft or valve-related parts, or if there is unusual engine conditions regarding valve clearance.

    a. Remove right and left rocker covers.

    b. Measure valve clearance as below:

- Set No.1 cylinder at TDC of its compression stroke. Align crankshaft pulley timing mark (grooved line without color) with timing indicator. Check that No. 1 cylinder intake and exhaust cam nose is facing in direction shown in illustration. If not, rotate crankshaft pulley 360 degrees clockwise (when viewed from front).
- Using a feeler gauge, measure valve clearance. Standard: -0.0055–0.0055 inch (-0.14 to 0.14 mm)

➡ **If inspection was carried out with cold engine, check that values with fully warmed up engine are still within specifications.**

- Rotate crankshaft by 240 degrees clockwise (when viewed from front) to align No. 3 cylinder at TDC of its compression stroke.

➡ **Crankshaft pulley mounting bolt flange has a stamped line every 60 degrees. They can be used as a guide to rotation angle.**

- Turn crankshaft pulley clockwise by 240 degrees from the position of

Crank Position	Valve No. 1	Valve No. 2	Valve No. 3	Valve No. 6
No. 1 TDC	Intake	Exhaust	Exhaust	Intake

67170-NIQU-G70

Valve clearance inspection—No.1 TDC on 3.5L engine

Crank Position	Valve No. 2	Valve No. 3	Valve No. 4	Valve No. 5
No. 3 TDC	Intake	Intake	Exhaust	Exhaust

67170-NIQU-G71

Valve clearance inspection—No.3 TDC on 3.5L engine

Crank Position	Valve No. 1	Valve No. 4	Valve No. 5	Valve No. 6
No. 5 TDC	Exhaust	Intake	Intake	Exhaust

67170-NIQU-G72

Valve clearance inspection—No.5 TDC on 3.5L engine

No. 5 cylinder at compression TDC.

    c. For measurements that are outside the specified range, perform adjustment below.

27. Adjust the clearance as follows:

Perform adjustment depending on selected head thickness of valve lifter. The specified valve lifter thickness is the dimension at normal temperatures. Ignore dimensional differences caused by temperature. Use the specifications for hot engine condition to adjust.

    a. Remove camshaft.

    b. Remove the valve lifters at the locations that are outside the standard.

    c. Measure the center thickness of the removed valve lifters with a micrometer.

    d. Use the equation below to calculate valve lifter thickness for replacement. Valve lifter thickness calculation:

- Thickness of replacement valve lifter = $t1 + (C1 - C2)$
- $t1$ = Thickness of removed valve lifter
- $C1$ = Measured valve clearance
- $C2$ = Standard valve clearance:

Thickness of a new valve lifter can be identified by stamp marks on the reverse side (inside the cylinder). Stamp mark 788U or 788R indicates 0.3102 inch (7.88 mm) in thickness.

➡ **2 types of stamp marks are used for parallel setting and for manufacturer identification.**

Available thickness of valve lifter: 27 sizes with range 0.3102–0.3307 inch (7.88–8.40 mm) in steps of 0.0008 inch (0.02 mm) (when manufactured at factory).

    a. Install the selected valve lifter.

    b. Install camshaft.

    c. Manually turn crankshaft pulley a few turns.

    d. Check that valve clearances for cold engine are within specifications by referring to the specified values.

    e. After completing the repair, check valve clearances again with the specifications for warmed engine. Make sure the values are within specifications.

Valve clearance:

- Intake: 0.012 inch (0.30 mm)
- Exhaust: 0.013 inch (0.33 mm)

Valve clearance:

Unit: mm (in)

	Cold	Hot * (reference data)
Intake	0.26 - 0.34 (0.010 - 0.013)	0.304 - 0.416 (0.012 - 0.016)
Exhaust	0.29 - 0.37 (0.011 - 0.015)	0.308 - 0.432 (0.012 - 0.016)

*: Approximately 80°C (176°F)

42356-MURA-G31

Valve clearance specifications–3.5L engines

Valve Lash

ADJUSTMENT

The engines covered in this section use hydraulic valve lifters that automatically adjust the valve lash. No periodic adjustment is needed.

Oil Pan

REMOVAL & INSTALLATION

3.3L Engine

1. Before servicing the vehicle, refer to the precautions section.
2. Drain the engine oil.
3. Remove or disconnect the following:
- Negative battery cable
- Front engine mount (support) insulator through-bolt
- Rear engine mount (support) through-bolt
- 2 rear refrigerant/heater pipe hold down bracket bolts
- 4 crossmember (also called a transverse member) bolts, and remove the crossmember.
- Exhaust inlet pipe
- 4 rear transaxle-to-engine brace bolts and the 5 front transaxle-to-engine brace bolts
- Front transaxle-to-engine brace
- Low oil level sensor electrical connector
- 18 oil pan bolts in the reverse order of the tightening sequence, working from the outside, towards the center bolts.
- Oil pan and discard the seals

To install:

4. Clean all parts well. Be sure that all old sealing material is removed from the oil pan and engine mating surfaces.
5. Position new oil pan seals. Apply Loctite® Ultra Gray 599 Silicone Sealer, or equivalent, to the ends of the oil pan seals.
6. Apply a bead of Loctite® Ultra Gray 599 Silicone Sealer or equivalent to the oil pan gasket rail inboard of the bolt holes.
7. Install or connect the following:
- Oil pan on the engine block. Tighten the 18 oil pan bolts in sequence, working from the inside, towards the outer bolts. Do not over-tighten. Tighten to 62–70 inch lbs. (7–8 Nm).
- Low oil level sensor electrical connector.
- Front and rear transaxle braces.

Tighten all bolts to 22–30 ft. lbs. (30–40 Nm).
- Exhaust inlet pipe
- Crossmember and tighten the bolts to 58–65 ft. lbs. (78–88 Nm).
- Both engine support through-bolts and tighten to 58–65 ft. lbs. (78–88 Nm).

8. Remove the support jack from under the crankshaft pulley.
9. Lower the vehicle.
10. Fill the engine with the specified engine oil to the required level.
11. Connect the negative battery cable. Start the engine and check for leaks.

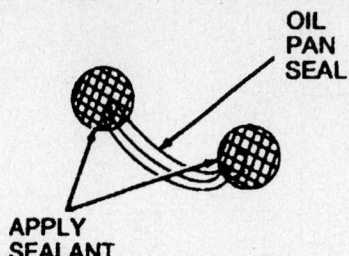

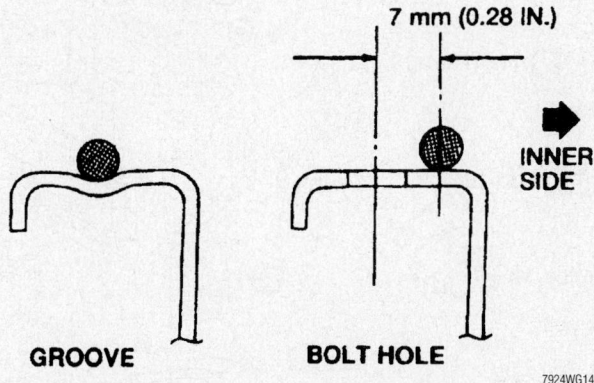

Apply RTV silicone sealer to the seal ends and to the oil pan gasket rail

TIGHTENING SEQUENCE

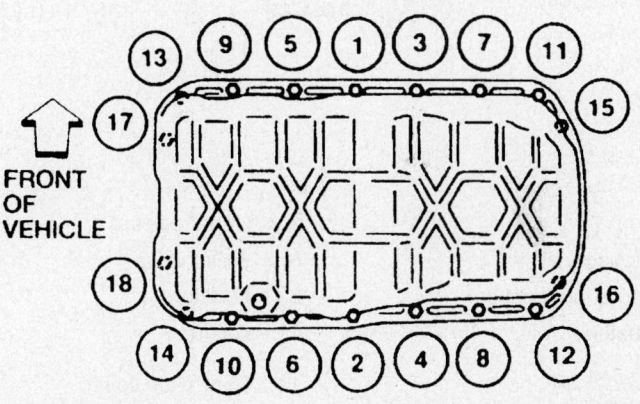

Tighten the 18 oil pan bolts in sequence, working from the inside, towards the outer bolts

3.5L Engine

➡When removing the front and rear engine mounting through bolts and nuts, lift the engine up slightly for safety.

✳✳ CAUTION

When removing the upper oil pan from the engine, first remove the Crankshaft Position (CKP) sensor.

1. Before servicing the vehicle, refer to the precautions section.
2. Remove the front right hand wheel.
3. Drain engine oil and coolant.

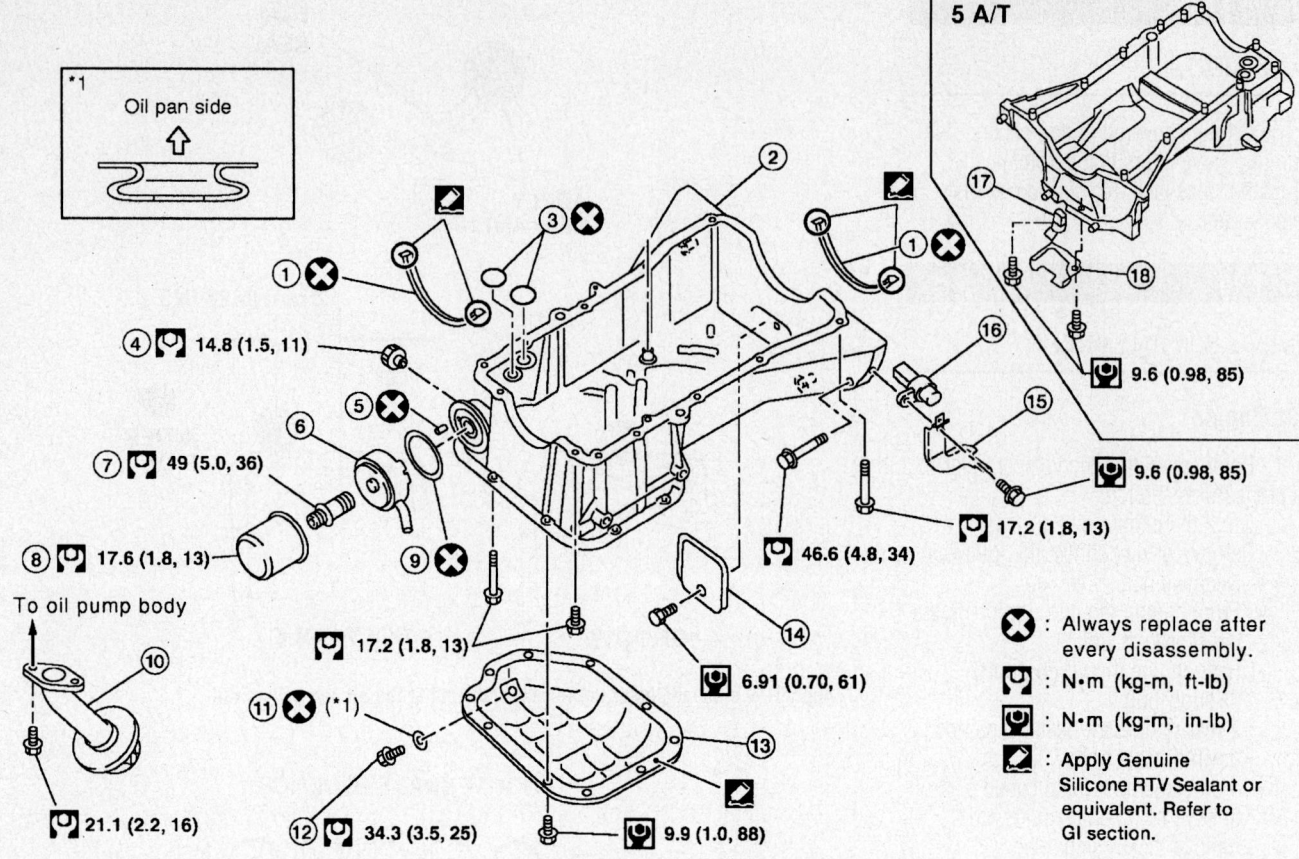

Exploded view of the oil pan assembly–3.5L engine

1. Gasket
2. Upper oil pan
3. O-ring
4. Oil pressure switch
5. Relief valve
6. Oil cooler
7. Oil cooler connection
8. Oil filter
9. Gasket
10. Oil strainer
11. Gasket
12. Drain plug
13. Lower oil pan
14. Rear plate cover
15. Heated oxygen sensor (bank 2) harness clamp (4 A/T only)
16. Crankshaft position sensor (POS) (4 A/T)
17. Crankshaft position sensor (POS) (5 A/T)
18. Crankshaft position sensor (POS) (5 A/T) shield

67170-NIQU-G10

4. Remove the oil dipstick.
5. Remove the engine undercover.
6. Remove the right hand inner fender splash shield.
7. Remove the A/C drive belt.
8. Remove the front exhaust tube.
9. Remove the coolant pipe bolts.
10. Discharge and recover the A/C refrigerant.
11. Remove the A/C compressor.
12. Disconnect the coolant lines from the engine oil cooler and plug the lines them to prevent coolant loss.

13. Remove the oil filter and engine oil cooler from the upper oil pan.
14. Remove the oil pressure switch, and the CKP sensor from the upper oil pan.
15. Remove the front driveshafts.
16. Remove the front suspension member as follows:
   a. Remove the left hand transaxle mount insulator nuts, if equipped with a 5 speed transmission.
   b. Attach a engine lifting bracket to the tranaxle at the location illustrated.

c. Install an engine support tool. Make sure the tool is securely resting on the hood ledge.
d. Remove the three transaxle mount insulator nuts.
e. Remove the lower ball joint bolt and separate the transverse link from the steering knuckle.
f. Remove the front exhaust tube.
g. Remove the power steering line bracket.
h. Remove the mounting bolts from the lower side of the steering gear.

i. Disconnect the front engine mount electrical connector.

j. Disconnect the connecting rod from the front strut.

k. Place a transmission jack under the front suspension member and remove the mounting nuts from the front suspension member.

l. Remove the front suspension member bolts from the pin stay on the vehicle body side.

m. Remove the through bolts from the front and rear engine mounts.

n. Lower the transmission jack to remove the front member. It may be necessary to remove the exhaust hanger bracket, the front and rear engine mounts and the transverse link to enable removal.

17. Disconnect the Heated Oxygen (HO2S) sensors and Air Flow Ratio (AFR) sensors and remove the two three way catalysts from the exhaust manifolds.

18. Remove the rear plate cover from the upper oil pan.

19. Loosen the lower oil pan bolts in the sequence illustrated.

20. Remove the lower oil pan by inserting a suitable prytool between the lower and the upper pans.

➡Do not insert a screwdriver, this will damage the mating surfaces. Slide the Tool by tapping its side with a hammer to remove the lower oil pan from the upper oil pan.

21. Remove the four upper oil pan–to–transaxle bolts.

22. Remove O2S sensor wire retainer bracket.

23. Remove the upper oil pan by loosening the bolts in the sequence illustrated.

24. Insert an appropriate size tool into the notch of the upper oil pan as shown in the accompanying illustration.

25. Pry off the upper oil pan by moving the tool up and down as illustrated.

26. If re-installing the original oil pan, remove the old sealant from the mating surfaces using a scraper.

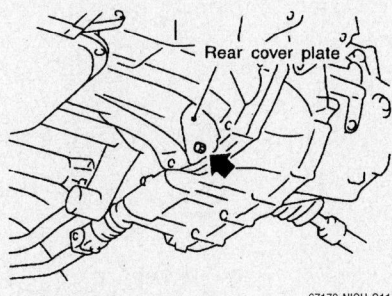

Location of the rear plate–3.5L engine

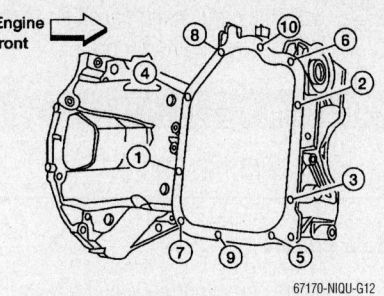

Loosen the lower oil pan bolts in this sequence–3.5L engine

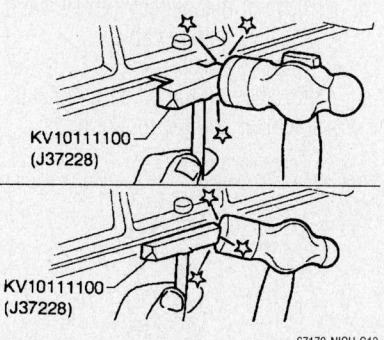

Separate the lower pan from the upper pan using a suitable tool and a hammer–3.5L engine

To install:

❉❉ CAUTION

Wait at least 30 minutes before refilling the engine with oil.

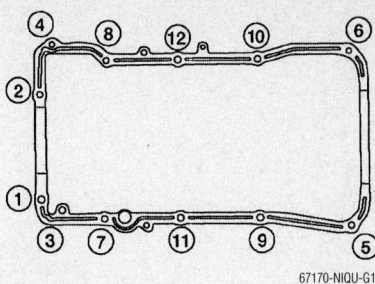

Loosen the upper oil pan bolts in this sequence–3.5L engine

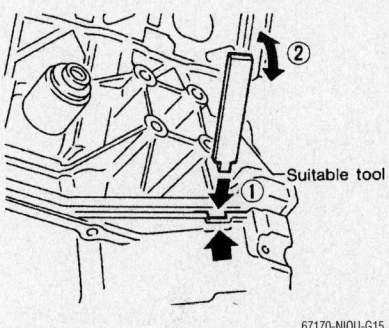

Pry off the upper oil pan by moving the tool up and down–3.5L engine

27. Apply a 0.15–0.197 inch (4–5mm) bead of sealant to the cylinder block mating surface of the upper oil pan as illustrated. Attach the pan within 5 minutes of applying the sealant.

28. Install new O-rings on the cylinder block and oil pump body.

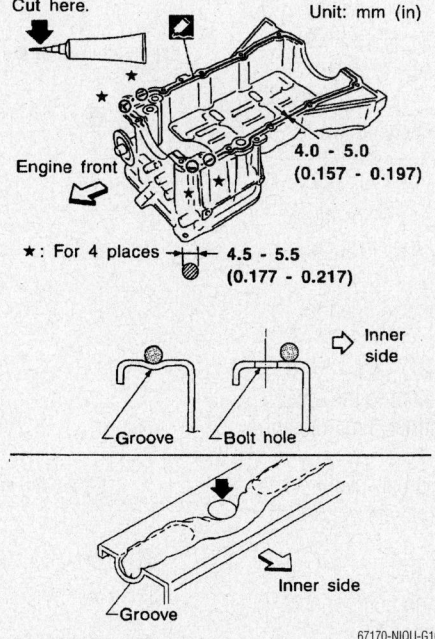

Apply a 0.15–0.197 inch (4–5mm) bead of sealant to the cylinder block mating surface of the upper oil pan–3.5L engine

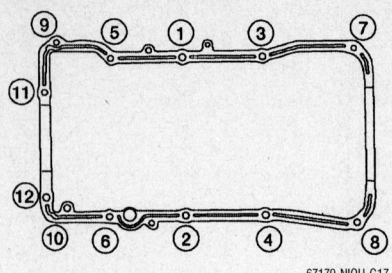

Tighten the upper oil pan bolts in this sequence—3.5L engine

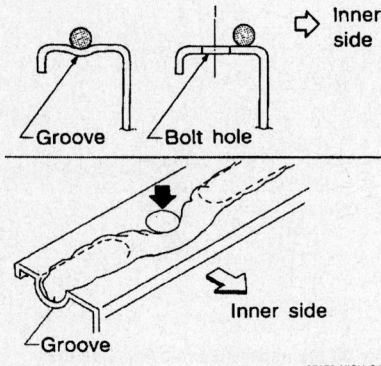

Apply a 0.177–0.217 inch (4.5–5.5mm) bead of sealant to the lower oil pan—3.5L engine

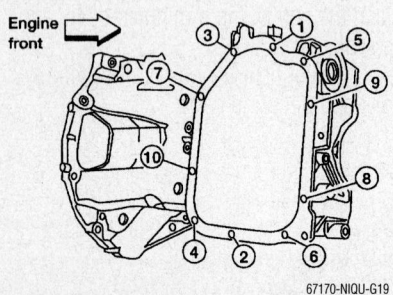

Tighten the lower oil pan bolts in this sequence—3.5L engine

29. Install the upper oil pan. Tighten upper oil pan bolts in the order illustrated to 13 ft. lbs. (17 Nm).

30. Install the four upper oil pan to transaxle bolts.

31. Apply a 0.177–0.217 inch (4.5–5.5mm) bead of sealant to the lower oil pan. Attach the pan within 5 minutes of applying the sealant.

32. Install the lower oil pan. Tighten the lower oil pan bolts in sequence to 88 inch lbs. (10 Nm).

33. Install rear plate cover.

34. Install the remaining components is in the reverse order of removal.

35. Wait at least 30 minutes before refilling the engine with oil.

36. Start the engine and check for leaks.

37. Inspect the engine oil level.

Oil Pump

REMOVAL & INSTALLATION

3.3L Engine

1. Before servicing the vehicle, refer to the precautions section.

2. Drain the engine oil.

3. Drain the cooling system.

4. Remove or disconnect the following:
   • Negative battery cable
   • Oil pan

5. After removing the oil pan, reinstall the crossmember and the mount bolts.

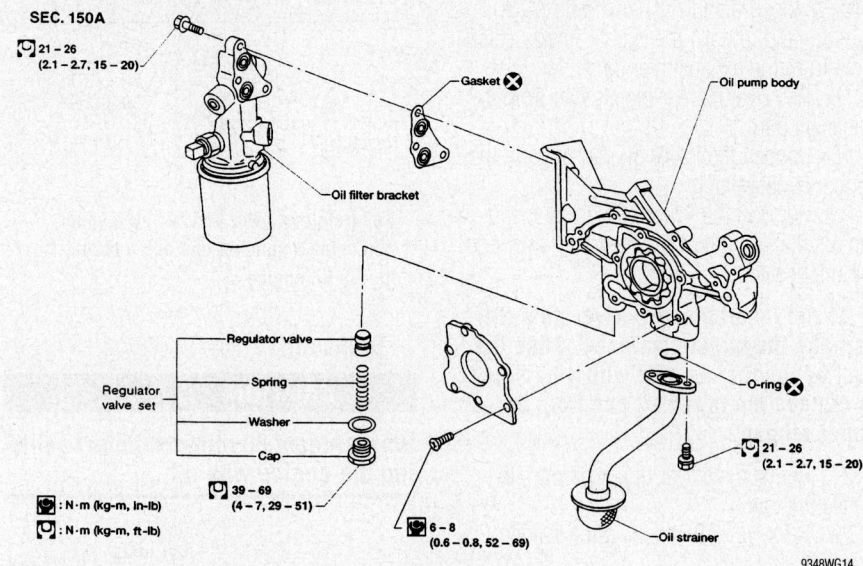

Exploded view of the oil pump assembly—3.3L engine

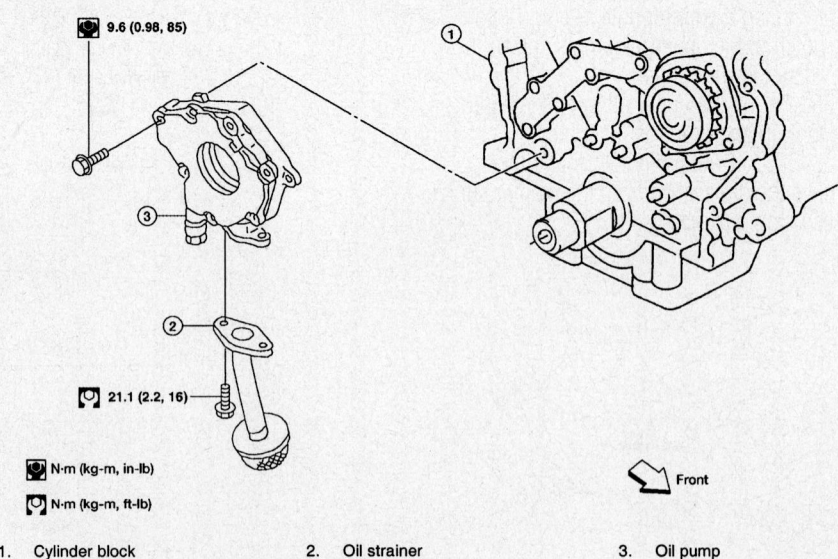

1. Cylinder block   2. Oil strainer   3. Oil pump

Oil pump exploded view—3.5L engine

• Timing belt
• Crankshaft sprocket and timing belt plate
• Oil pump

To install:

6. Install or connect the following:
   • Oil pump-to-body bolts to 52–69 inch lbs. (6–8 Nm)
   • Timing belt plate and tighten the bolts to 52–69 inch lbs. (6–8 Nm)
   • Crankshaft sprocket
   • Timing belt
   • Oil pan
   • Negative battery cable

7. Fill the cooling system.

8. Fill the crankcase to the correct level.

9. Start the engine and check for leaks.

3.5L Engine

1. Before servicing the vehicle, refer to the precautions section.
2. If equipped with a 5 speed transmission, remove the engine/transmission and front suspension member.
3. Remove front timing chain case and timing chain (primary).
4. Remove oil pan and oil strainer.
5. Remove oil pump assembly.
6. Installation is the reverse of removal. When installing, align crankshaft flat faces with inner rotor flat faces. Tighten the pump bolts to 85 inch lbs. (10 Nm) and the strainer bolts to 16 ft. lbs. (21 Nm).

Rear Main Seal

REMOVAL & INSTALLATION

3.3L Engine

1. Before servicing the vehicle, refer to the precautions section.
2. Disconnect the negative battery cable.
3. Remove the transaxle from the vehicle.
4. Remove the flexplate from the crankshaft.
5. Remove the rear oil seal retainer.

❋❋ WARNING

Do not scratch the seal bore of the oil seal retainer when removing the oil seal.

6. Remove the oil seal from the seal retainer.
To install:
7. Apply clean engine oil to the lip and outer surface of the new seal to aid during installation.
8. Install the seal in the retainer using a suitable seal driver.

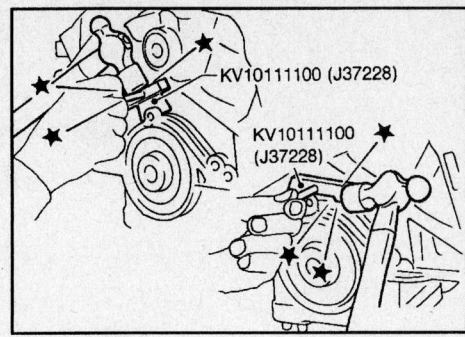

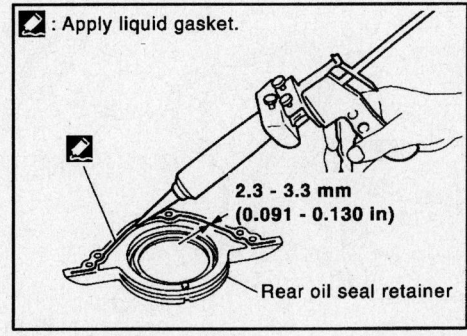

Applying sealer to the rear main seal

9. Using a new gasket install the retainer on the engine. Tighten the bolts to 52–61 inch lbs. (6–7 Nm).
10. Install the flexplate. Tighten the bolts to 61–69 ft. lbs. (83–93 Nm).
11. Install the transaxle and remaining components.

3.5L Engine

1. Before servicing the vehicle, refer to the precautions section.
2. Remove the drive plate.
3. Remove the engine from the vehicle.
4. Remove the upper oil pan.
5. Use a seal cutter (special service tool) to cut away the liquid gasket and remove the rear oil seal retainer.

To install:
6. Remove the old liquid gasket from the mating surface of the cylinder block and the oil pan using a scraper.
7. Apply liquid gasket to the rear oil seal retainer as shown in the illustration. Use RTV silicone sealant. Assembly should be done within 5 minutes after coating.
8. Install the rear oil seal retainer to the cylinder block.
9. Perform the remaining steps in the reverse order of removal.

Front Timing Cover

REMOVAL & INSTALLATION

3.5L Engine

1. Before servicing the vehicle, refer to the precautions section.

➡**This section describes procedures for removal/installation procedure of the front timing chain case and timing chain related parts without removing the upper oil pan from the vehicle. If the upper oil needs to be removed or installed, or when rear timing chain case is removed or installed, remove both oil pans first. Then remove front timing chain case and timing chain related parts.**

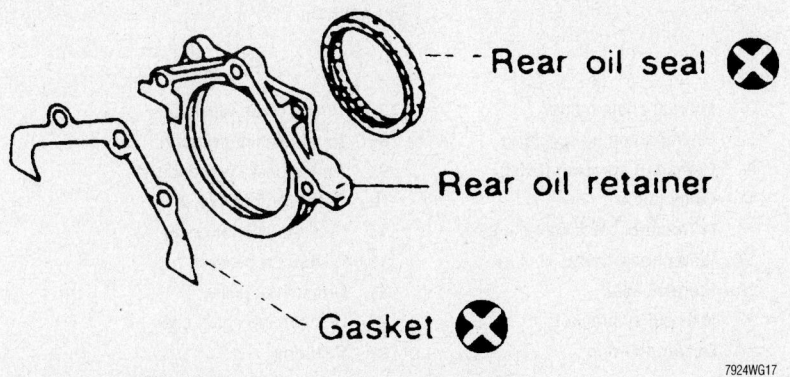

Exploded view of the oil seal, retainer and gasket—3.3L engine

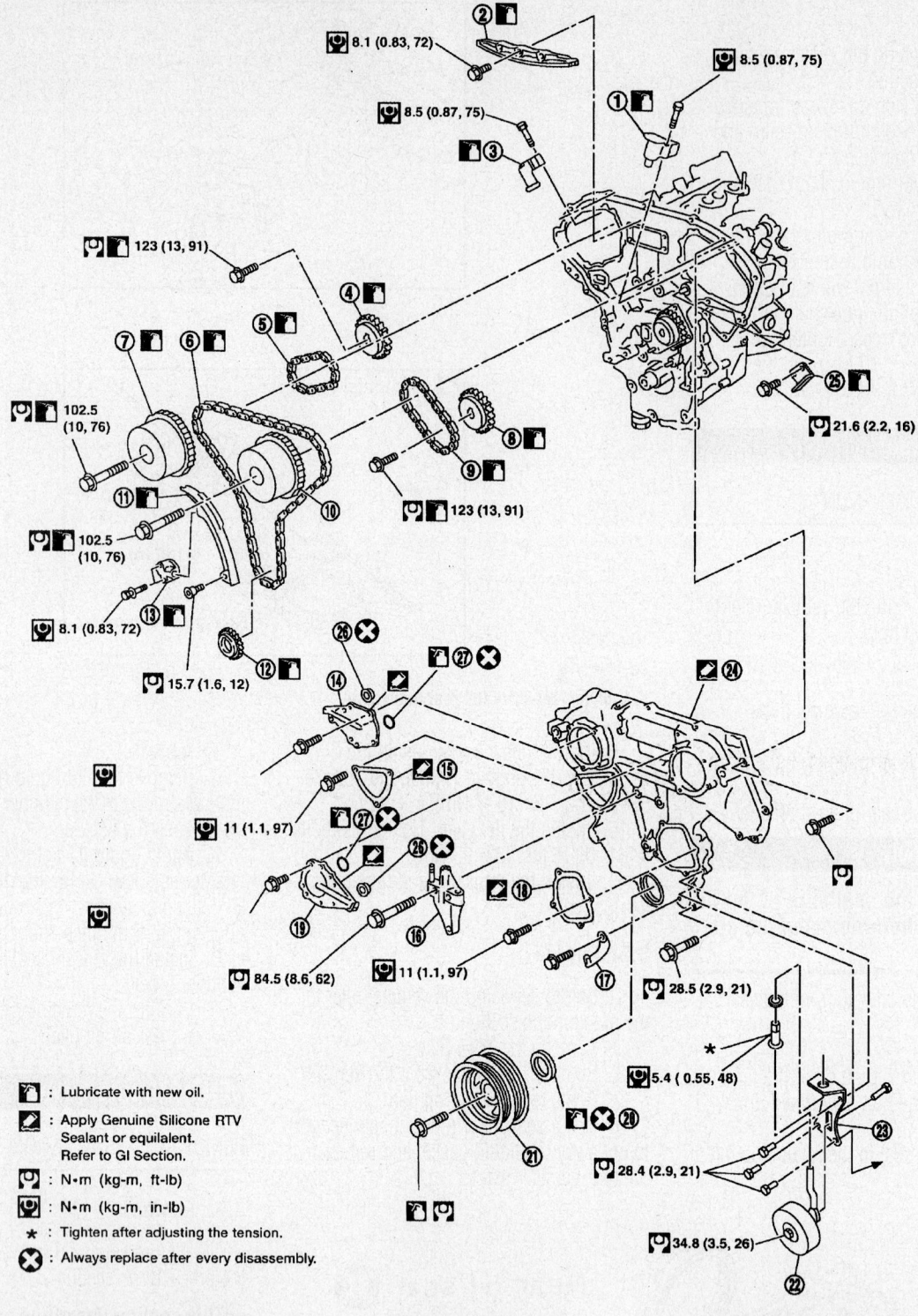

8.1 (0.83, 72)

8.5 (0.87, 75)

8.5 (0.87, 75)

③

②

①

123 (13, 91)

④ ⑤ ⑥ ⑦

102.5 (10, 76)

⑧

⑨

123 (13, 91)

⑪ ⑩

102.5 (10, 76)

⑬

8.1 (0.83, 72)

15.7 (1.6, 12)

⑫

⑭

㉖

㉗

24

25

21.6 (2.2, 16)

15

11 (1.1, 97)

㉗

26

18

84.5 (8.6, 62)

11 (1.1, 97)

19

16

17

28.5 (2.9, 21)

5.4 (0.55, 48)

20

21

23

28.4 (2.9, 21)

34.8 (3.5, 26)

22

*

: Lubricate with new oil.

: Apply Genuine Silicone RTV Sealant or equilalent. Refer to GI Section.

: N•m (kg-m, ft-lb)

: N•m (kg-m, in-lb)

★ : Tighten after adjusting the tension.

Ⓧ : Always replace after every disassembly.

1. Timing chain tensioner	2. Internal chain guide	3. Timing chain tensioner
4. Camshaft sprocket (EXH)	5. Timing chain (secondary)	6. Timing chain (primary)
7. Camshaft sprocket (INT)	8. Camshaft sprocket (EXH)	9. Timing chain (secondary)
10. Camshaft sprocket (INT)	11. Slack guide	12. Crankshaft sprocket
13. Timing chain tensioner	14. IVT control valve cover - right	15. Chain tensioner cover
16. RH engine mounting bracket	17. Water hose clamp	18. Water pump cover
19. IVT control valve cover - left	20. Front oil seal	21. Crankshaft pulley
22. Idler pulley	23. Idler pulley bracket	24. Front timing chain case
25. Timing tension guide	26. Collared O-ring	27. Seal ring

67170-NIQU-G48

Exploded view of the timing chain, cover and related components–3.5L engine

2. Disconnect the negative battery cable.

3. Drain the engine cooling system and engine oil.

4. Remove the engine cover.

5. Remove the upper air cleaner case, Mass Air Flow (MAF) sensor and air cleaner–to–electric throttle control actuator tube.

6. Remove the engine coolant reservoir.

7. Remove the cowl top and cowl top extension.

8. Remove the IPDM E/R and position aside. Remove the bracket.

9. Remove the front right hand wheel.

10. Remove the engine undercover.

11. Remove the right hand inner splash shield.

12. Remove the drive belts and idler pulley.

13. Recover the A/C refrigerant and remove the A/C compressor.

14. Remove engine oil cooler pipe bolts.

15. Remove the power steering oil pump and reservoir tank with lines attached and position them aside.

16. Remove the lower oil pan.

17. Remove the alternator.

18. Disconnect the engine harness and position aside.

19. Remove the A/C low-pressure flexible hose.

20. Support the engine using a suitable support device and remove the right hand engine mount insulator, mount and bracket.

21. Remove the chain tensioner cover and water pump cover.

22. Remove the left and right IVT control covers. Loosen the IVT control cover bolts in the order illustrated.

➡**The shaft in the cover is inserted into the center hole of the intake camshaft sprocket. Remove the cover by pulling straight out until the cover disengages from the camshaft sprocket.**

23. Remove the starter.

24. Remove the intake manifold collector.

25. Remove the ignition coils. Make sure to note the locations for installation purposes.

26. Remove the spark plugs.

27. Remove the engine oil dipstick.

28. If removing the secondary timing chains, remove the rocker covers as follows:

   a. Remove the engine cover.

   b. Remove side engine covers.

   c. If removing right hand rocker cover, disconnect the Mass Air Flow (MAF) sensor electrical connector and remove the air cleaner–to–electric throttle control actuator tube and air cleaner lid.

   d. If removing right hand rocker cover, remove the following:

   e. Front cowl panel.

   f. Windshield wiper arms and motor assembly.

   g. Intake manifold collector.

   h. If removing left hand rocker cover, disconnect the Air Fuel Ratio (AFR) sensor.

   i. Remove the ignition coils.

   j. Position engine harness aside.

   k. Disconnect Positive Crankcase Ventilation (PCV) hose.

   l. Remove the dipstick.

   m. Remove the rocker cover bolts in the sequence illustrated.

29. Remove the IVT control solenoid valves and discard the gaskets.

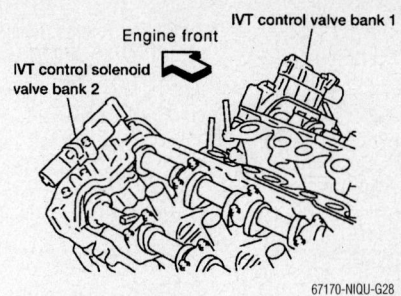

Remove the IVT control solenoid valves–3.5L engine

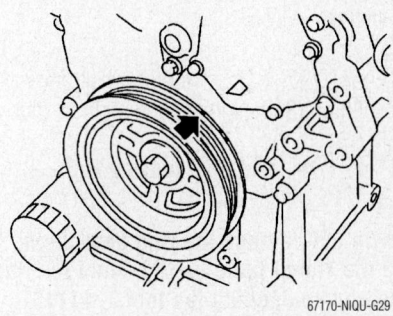

Position the engine at Top Dead Center (TDC) of No. 1 cylinder–3.5L engine

30. Position the engine at Top Dead Center (TDC) of No. 1 cylinder as follows:

   a. Rotate crankshaft pulley clockwise to align timing mark (grooved line without color) with timing indicator.

   b. Check that intake and exhaust camshaft lobes on No. 1 cylinder on the right bank of engine are located as illus-

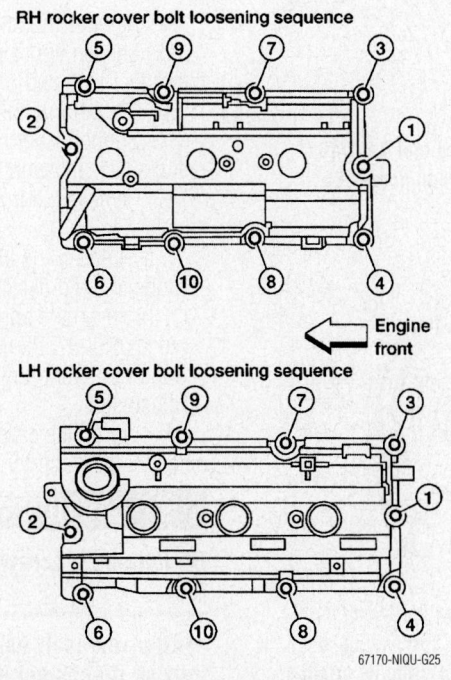

Loosen the rocker cover bolts in this sequence–3.5L engine

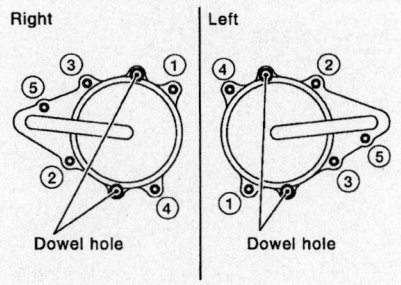

Loosen the IVT control cover bolts in this sequence–3.5L engine

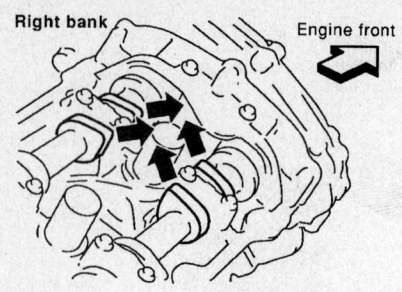

67170-NIQU-G30

Check that intake and exhaust camshaft lobes on No. 1 cylinder on the right bank of engine are located as shown—3.5L engine

trated. If not, turn the crankshaft one revolution (360 degrees) and align as illustrated.

31. Lock the ring gear using tool J–44716 attached to the starter bolt hole.

➡**Do not damage the ring gear teeth, or the signal plate teeth behind the ring gear, when setting the tool J-44716.**

32. Remove the crankshaft pulley as follows:

a. Loosen crankshaft pulley bolt using tool KV10109300 and locate bolt

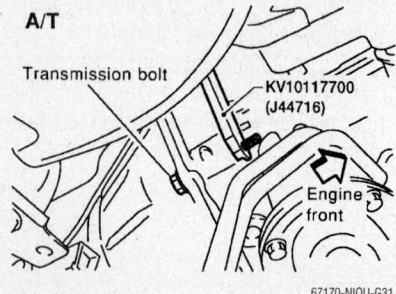

67170-NIQU-G31

Lock the ring gear using tool J–44716 attached to the starter bolt hole–3.5L engine

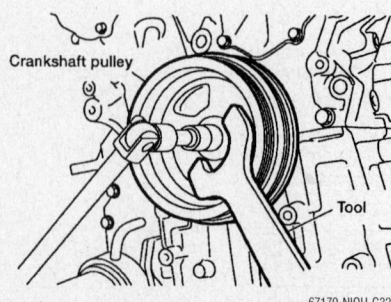

67170-NIQU-G32

Loosen crankshaft pulley bolt using tool KV10109300 and locate bolt seating surface at 0.39 inch (10mm) from its original position–3.5L engine

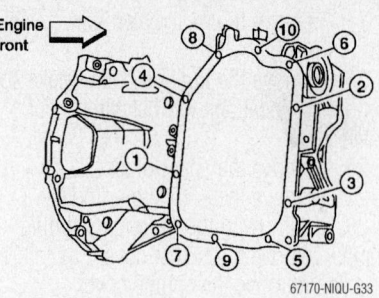

67170-NIQU-G33

Loosen lower oil pan front bolts in this order–3.5L engine

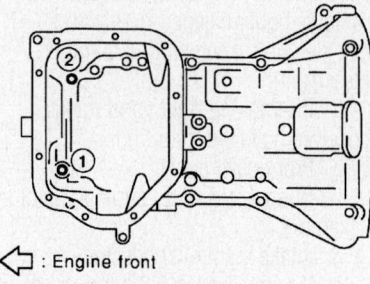

67170-NIQU-G34

Loosen upper oil pan front bolts in this order–3.5L engine

seating surface at 0.39 inch (10 mm) from its original position.

33. Position a pulley puller at recess hole of crankshaft pulley to remove crankshaft pulley.

✴✴ CAUTION

Do not use a puller claw on crankshaft pulley periphery.

34. Remove the lower oil pan.
35. Loosen upper oil pan front bolts in the order illustrated.
36. Temporarily install lower oil pan.
37. Support front of engine under oil pan using a jack and wood block.
38. Remove the front timing chain case as follows:

a. Loosen the front timing chain case bolts in the order illustrated.

b. Insert the appropriate size tool such as J–37228 into the notch at the top of the front timing chain case as illustrated.

39. Pry off the case by moving the suitable tool back and forward.

✴✴ CAUTION

Do not use a screwdriver or similar tool.

➡**After removal, handle the cover carefully so it does not bend, or warp under a load.**

40. Remove the water pump cover and chain tensioner cover from the front timing chain case using Tool J–37228.

✴✴ CAUTION

Do not insert a screwdriver, this will damage the mating surfaces.

41. Remove the front oil seal from the front timing chain case using a suitable tool being careful not to damage the front cover.

42. If necessary, remove timing chain and related parts.

43. Use a scraper to remove all of the old sealant from the front timing chain mat-

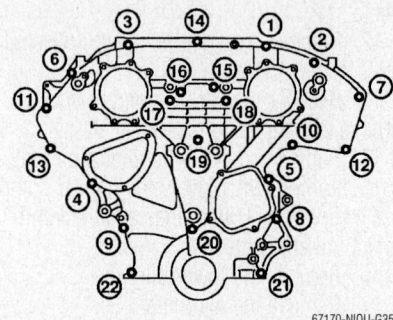

67170-NIQU-G35

Loosen the front timing chain case bolts in this order–3.5L engine

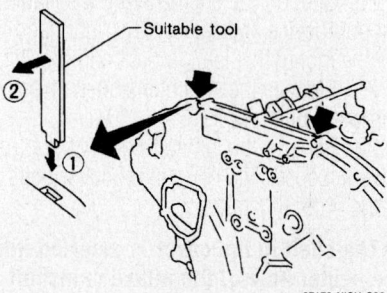

67170-NIQU-G36

Insert the appropriate size tool such as J–37228 into the notch at the top of the front timing chain case–3.5L engine

67170-NIQU-G37

Remove the front oil seal from the front timing chain case using a suitable tool–3.5L engine

ing surfaces. Be careful not to damage the mating surfaces while cleaning.

To install:

44. Install the timing chain and related parts, if removed.

45. Install the left and right dowel pins into the front timing chain case up to a point close to taper in order to shorten protrusion length.

46. Install the front oil seal on the front timing chain case. Apply new engine oil to the oil seal edges. Install it so that each seal lip is oriented as illustrated.

47. Make sure the garter spring is in position and seal lip is not inverted.

48. Apply a 0.102–0.142 inch (2.6–3.6mm) bead of RTV sealant to front timing chain case as illustrated. Make sure to wipe off the protruding sealant prior to installation.

49. Install the dowel pin on the rear timing chain case into the dowel pin hole in the front timing chain case.

50. Apply a 0.138–0.77 inch (3.5–4.5mm) bead of RTV sealant to the top surface of the upper oil pan as illustrated.

51. Install the front timing chain case as follows:

    a. Install the lower end of the front

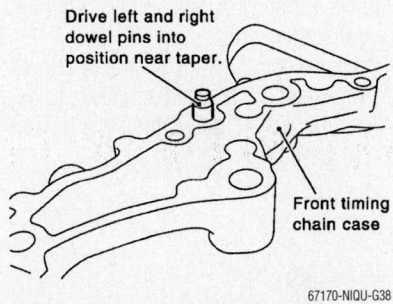

Install the left and right dowel pins into the front timing chain case up to a point close to taper in order to shorten protrusion length–3.5L engine

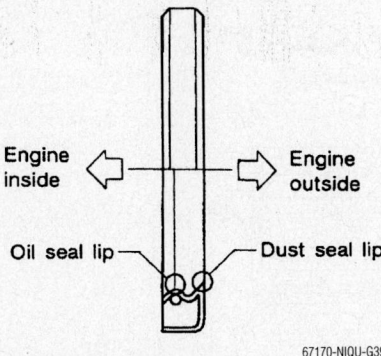

Install the front oil seal on the front timing chain case–3.5L engine

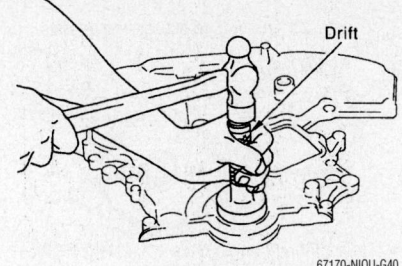

Make sure the garter spring is in position and seal lip is not inverted when installing the front oil seal–3.5L engine

timing chain case tightly onto the top surface of the upper oil pan. Make sure that the oil pan gasket is in place.

    b. While pressing the front timing chain case from its front and top as illustrated, hammer the dowel pin until the outer end becomes flush with the surface.

52. Loosely install the front timing chain case bolts, refer to the illustration for bolt location.

53. Tighten the front timing chain case bolts in the order illustrated. Bolt positions in the illustration and torques are as follows:

    a. Bolt location 1 and 2 with a diameter of 0.31 inch (8 mm) are torqued to 21 ft. lbs. (28 Nm).

    b. Bolt locations for bolts 3 through 22 with a diameter of 0.24 inch (6 mm) are torqued to 9 ft. lbs. (13 Nm).

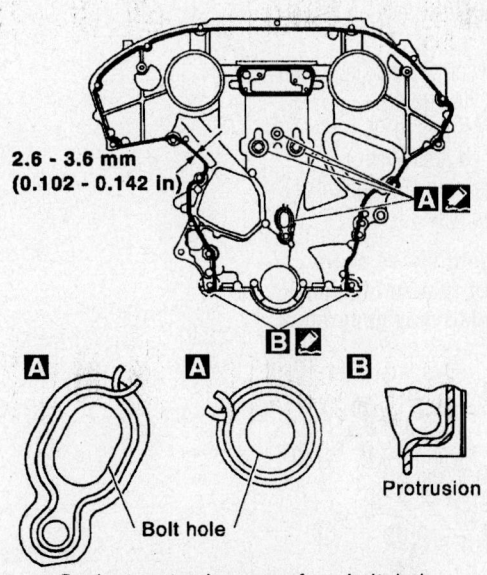

Apply a 0.102–0.142 inch (2.6–3.6mm) bead of RTV sealant to front timing chain case–3.5L engine

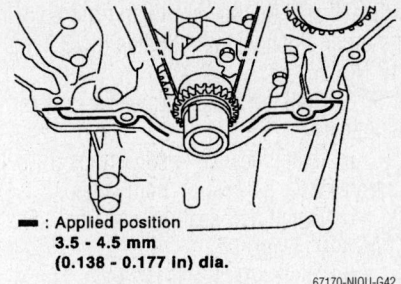

Apply a 0.138–0.77 inch (3.5–4.5mm) bead of RTV sealant to top surface of the upper oil pan–3.5L engine

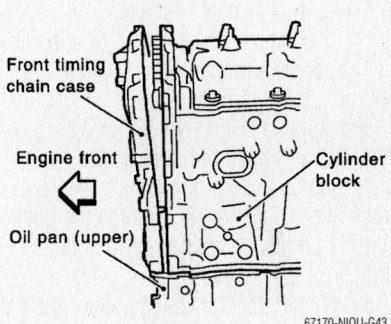

Install lower end of front timing chain case tightly onto top surface of the upper oil pan–3.5L engine

54. Install the upper oil pan front bolts in the order illustrated and tighten to 13 ft. lbs. (17 Nm).

55. Install the IVT control valve covers as follows:

    a. Install new collared O-rings in front cover oil hole on both sides.

    b. Install new seal rings on the IVT control covers.

    c. Apply a 0.083–0.122 inch (2.1–3.1mm) bead of RTV sealant to the IVT control covers as illustrated.

56. Being careful not to move the seal ring from the installation groove, align the dowel pins on the chain case with the holes to install the IVT control covers.

57. Tighten the intake valve timing control cover bolts in the order illustrated to 100 inch lbs. (11 Nm).

58. Apply a 0.091–0.130 inch (2.3–3.3mm) bead of liquid gasket to the water pump cover and the chain tensioner cover and install covers. Tighten the bolts on the pump and cover to 97 inch lbs. (11 Nm).

59. Install the crankshaft pulley, lubricate the pulley bolt with clean engine oil and tighten the bolt in two steps:

    a. Step 1: 32 ft. lbs. (44 Nm).

    b. Step 2: 60–65 degrees clockwise.

60. Rotate crankshaft pulley in a clockwise direction as viewed from front of the engine to confirm it turns smoothly.

61. Install the rocker covers in the reverse order of removal, make sure to perform the following:

62. Apply sealant to the areas on the front corners.

63. Tighten the rocker cover bolts in two steps in the sequence illustrated:

    a. Step 1: 17 inch lbs. (1.96 Nm).

    b. Step 2: 74 inch lbs. (8 Nm).

64. Install the remaining components is in reverse order of removal.

➡ **If hydraulic pressure inside chain tensioner drops after removal/installation, slack in the guide may generate a**

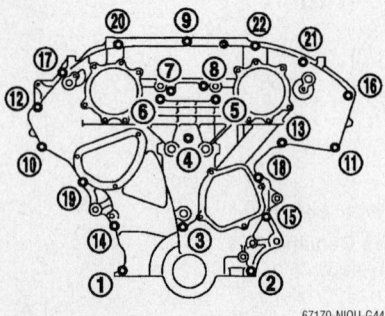

Tighten the front timing chain case bolts in the order shown–3.5L engine

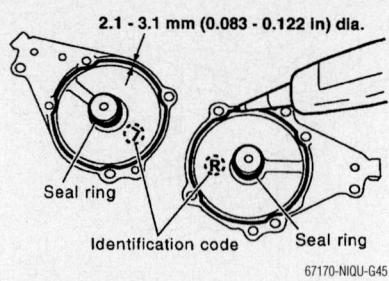

Apply a 0.083–0.122 inch (2.1–3.1mm) bead of RTV sealant to the IVT control covers–3.5L engine

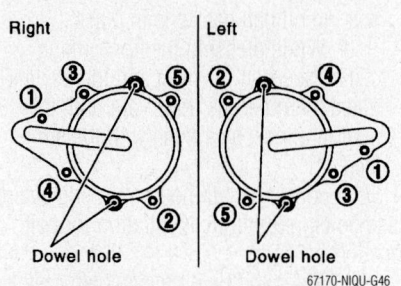

Tighten the intake valve timing control cover bolts in this order–3.5L engine

pounding noise during and just after engine start. This is normal. Noise will stop after hydraulic pressure rises.

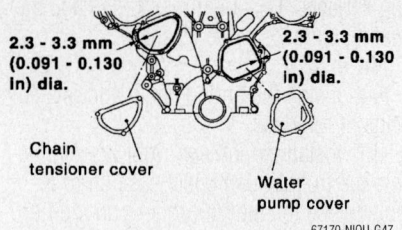

Apply a 0.091–0.130 inch (2.3–3.3mm) bead of liquid gasket to the water pump cover and the chain tensioner cover–3.5L engine

Timing Belt

REMOVAL & INSTALLATION

3.3L Engines

On this vehicle, right side refers to the "rear" components (near the firewall) and left side refers to the "front" components (near the radiator).

1. If the timing belt is to be removed, it is good practice to turn the crankshaft until the engine is at Top Dead Center (TDC) of the No. 1 cylinder, compression stroke (firing position), before beginning work. This should align all timing marks and serve as a reference for all work that follows. After verifying that the engine is at TDC for the No. 1 cylinder, do not crank the engine or allow the crankshaft or camshaft sprockets to be turned otherwise engine timing will be lost.

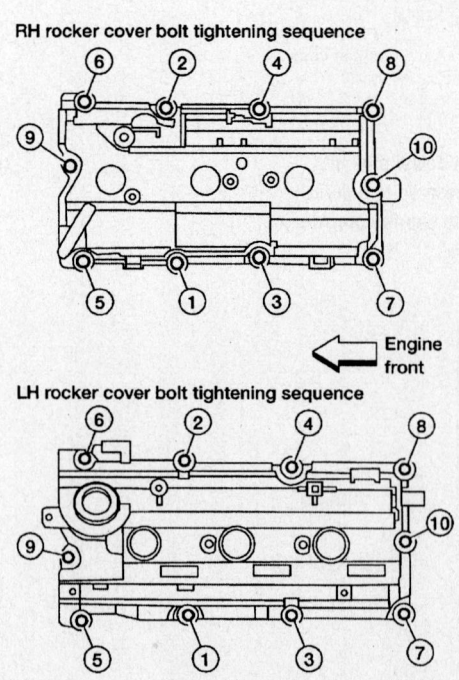

Tighten the rocker cover bolts in this sequence–3.5L engine

2. Before servicing the vehicle, refer to the precautions section.

3. Drain the cooling system.

4. Remove or disconnect the following:

- Negative battery cable
- Alternator drive belt, water pump and power steering pump belt and the air conditioning compressor belt, if equipped
- 3 air conditioning compressor drive belt idler pulley bolts and the idler pulley, if equipped with air conditioning
- Upper radiator hose bracket bolt
- Upper hose with the bracket from the vehicle
- Water bypass hose from between the thermostat housing and the lower water hose connection
- Main wiring harness from the upper engine front cover
- 8 upper engine front cover bolts and the upper cover
- Right side front wheel and tire assembly
- 4 right side engine and transmission splash shield bolts and 2 screws, and right side outer engine and transaxle splash shield

5. Use a strap wrench to hold the water pump pulley. Remove the 4 pulley bolts, and the water pump pulley.

6. Use a strap wrench to hold the crankshaft pulley. Remove the center pulley bolt, and the crankshaft pulley using a harmonic balancer (damper) puller to draw the pulley from the front of the crankshaft.

- 5 lower engine front cover bolts, then remove the lower engine front cover

7. Be sure that the timing marks between the crankshaft sprocket and the oil pump housing align.

8. If the timing belt is to be reused, mark an arrow on the belt indicating the direction of rotation. The directional arrow is necessary to ensure that the timing belt, if it to be reused, can reinstalled in the same direction.

9. Loosen the timing belt tensioner nut and slip the timing belt off of the sprockets.

10. If necessary, the camshaft sprockets can be removed. A special spanner tool is designed to hold the sprocket to keep it from turning while the center bolt is being loosened. Use care if using substitutes.

➡ **The sprockets are not interchangeable.**

11. If necessary, the crankshaft sprocket can be removed. The outer timing belt guide (looks like a large washer) and the crank-

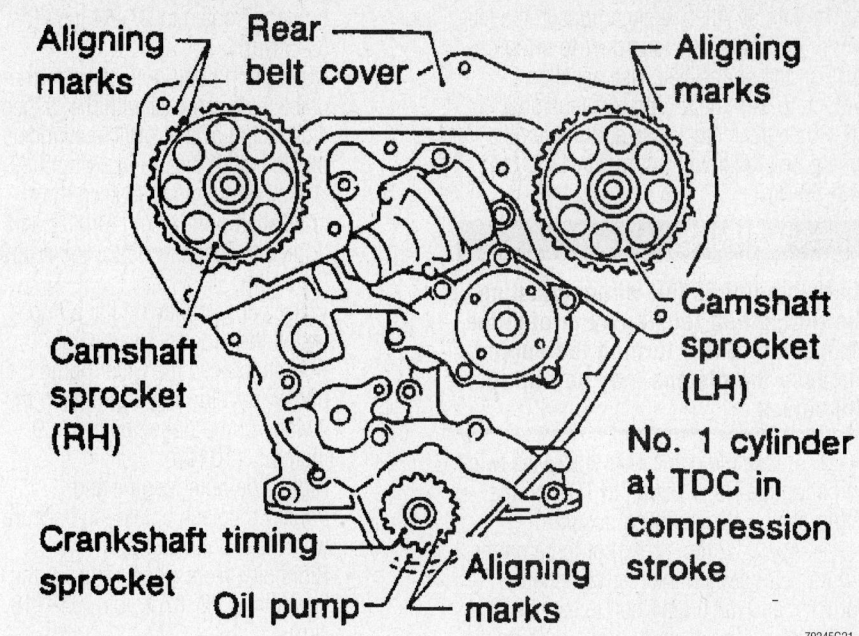

Use a shop rag to clean the alignment marks for the timing belt— 3.3L engines

shaft sprocket simply pull off the front of the crankshaft.

➡ **Be careful, there are 2 crankshaft keys. Use care not to loose them.**

To install:

12. Clean all parts well. If removed, inspect the crankshaft sprocket for warping or abnormal wear. Check the sprocket teeth for wear, deformation, chipping or other damage. Replace as necessary. Clean the sprocket mounting surface to ease installation. Install the key. Slip the sprocket onto the crankshaft. Tap it in place with a suitably-sized socket.

13. If removed, inspect the camshaft sprockets for damage and wear. Replace as required. The sprockets should be marked **L3** to designate the front, or left side camshaft and **R3** to designate the rear, or right side camshaft. Use care to install the sprockets properly. A special spanner tool is designed to hold the sprocket to keep it from turning while the center bolt is being tightened. Use care if using a substitute. Tighten the camshaft sprocket center bolts to 61 ft. lbs. (83 Nm). Verify that the timing marks on the camshaft sprockets and the timing marks on the rear cover (called the seal plate) are aligned.

14. Use an Allen wrench to turn the timing belt tensioner clockwise until the belt tensioner spring is fully extended. Temporarily tighten the tensioner nut to 32–43 ft. lbs. (43–58 Nm).

15. If a new timing belt is to be installed, look for a printed arrow on the belt. Be sure

the arrow is pointing away from the engine. If the original timing belt is to be reused, be sure that the directional arrow that was marked at disassembly is facing the correct direction.

16. A new Original Equipment Manufacture (OEM) timing belt should have 3 white timing marks on it that indicate the correct timing positions of the camshafts and the crankshaft. These marks are to help ensure that the engine is properly timed. When the engine is properly timed, each white timing mark on the timing belt will be aligned with the corresponding camshaft and crankshaft timing mark on the sprocket. Because the white timing marks are not evenly spaced, the technician needs to use care in installing the belt. There should be 40 timing belt teeth between the timing marks on the front and rear camshaft sprockets and 43 teeth between the timing mark on the front camshaft sprocket and the timing mark on the crankshaft sprocket.

17. Verify that the camshaft timing marks are aligned with the timing marks on the rear cover (seal plate) and that the crankshaft sprocket timing mark is aligned with the timing mark on the oil pump housing.

18. Install the timing belt starting at the crankshaft sprocket and moving around the camshaft sprockets following a counterclockwise path. Do not allow any slack in the timing belt between the sprockets. After all of the timing marks are aligned with the timing belt installed, slip the timing belt onto the belt tensioner.

19. While holding the timing belt ten-

sioner with an Allen wrench, loosen the tensioner nut. Allow the tensioner to put pressure on the timing belt. Use an Allen wrench to turn the timing belt tensioner 70–80 degrees clockwise and tighten the timing belt tensioner nut to 32–43 ft. lbs. (43–58 Nm).

✴✴ WARNING

If any binding is felt when adjusting the timing belt tension by turning the crankshaft, STOP turning the engine, because the pistons may be hitting the valves.

20. Rotate the crankshaft clockwise twice and align the No. 1 piston to TDC on the compression stroke (firing position).

21. Apply 22 lbs. (10kg) of force on the timing belt between the rear camshaft sprocket and the timing belt tensioner. An assistant may be needed. While holding the timing belt tensioner steady with an Allen wrench, loosen the timing belt tensioner nut. Remove the Allen wrench and adjust the timing belt tensioner using the following procedure:

   a. Install a 0.0138 in. (0.35mm) thick and 0.500 in. (12.7mm) wide feeler gauge where the timing belt just starts to go around the tensioner (approximately the 4 o'clock position, looking at the tensioner).

   b. Turn the crankshaft sprocket clockwise, which should force the feeler gauge between the timing belt and the tensioner, up to a position on the tensioner of about 1 o'clock.

   c. Tighten the timing belt tensioner nut to 61 ft. lbs. (83 Nm).

   d. Turn the crankshaft clockwise to rotate the feeler gauge out from between the timing belt tensioner and the timing belt.

22. Rotate the crankshaft clockwise twice, and once again align the No. 1 piston to TDC on the compression stroke (firing position).

23. Apply 22 lbs. (10kg) of force on the timing belt between the front and rear camshaft sprockets. Measure the amount of belt deflection. Belt deflection should be between 0.51–0.59 in. (13–15mm). If belt deflection is out of specification, repeat Steps 29 through 33. If the timing belt deflection cannot be adjusted into specification, the timing belt will have to be replaced.

24. Install or connect the following:
- Lower engine front cover and the 5 lower cover bolts. Do not over

tighten. Tighten to 27–44 inch lbs. (3–5 Nm).
- Outer timing belt guide next to the crankshaft sprocket with the dished side facing away from the cylinder block. Install the crankshaft pulley. Use a strap wrench to keep the crankshaft pulley from turning and tighten the center bolt to 148 ft. lbs. (201 Nm).
- Water pump pulley on the pump. Install the 4 bolts. Use a strap wrench to keep the water pump pulley from turning and tighten the 4 water pump pulley bolts to 89 inch lbs. (10 Nm).
- Right side outer engine and transaxle splash shield, and secure with the 4 bolts and 2 screws
- Right side front wheel. Tighten the lug nuts to 72–87 ft. lbs. (98–118 Nm).
- Upper engine timing belt front cover, and tighten the 8 bolts to 27–44 inch lbs. (3–5 Nm)
- Main wiring harness on the upper engine front cover
- Water bypass hose between the thermostat housing and water connection
- Upper radiator hose between the radiator and the water hose connection. Secure the hoses with clamps. Install the upper radiator hose bracket. Tighten the bracket bolt to 34–58 ft. lbs. (46–65 Nm).
- Air conditioning compressor drive belt idler pulley and install the 3 bolts. Tighten to 15 ft. lbs. (21 Nm), if equipped
- Alternator drive belt, the water pump and power steering pump drive belt and the air conditioning compressor drive belt, if equipped
- Negative battery cable

25. Fill the cooling system.

26. Start the engine and allow it to warm to operating temperature. Check and adjust the ignition timing. Road test to verify correct engine operation.

Timing Chain, Sprockets And Rear Case

REMOVAL & INSTALLATION

3.5L Engines

4–SPEED ENGINE

1. Before servicing the vehicle, refer to the precautions section.

2. Disconnect the battery cables.
3. Drain the engine oil and coolant.
4. Remove the engine cover.
5. Remove the upper air cleaner case, Mass Air Flow (MAF) sensor and air cleaner to electric throttle control actuator tube.
6. Remove the battery and tray.
7. Disconnect the heater pump and position aside.
8. Remove the cowl top and cowl top extension.
9. Disconnect the engine room harness from the Powertrain Control Module (PCM) and the two connections at the right hand strut tower.
10. Disconnect the engine harness ground connections.
11. Remove the radiator assembly, coolant reservoir, and all system hoses.
12. Remove the idler pulley and bracket.
13. Remove the upper and lower oil pans.
14. Remove the alternator.
15. Disconnect the engine harness and position it aside.
16. Support the engine using a suitable support device and remove the right hand engine mount insulator, mount and bracket.
17. Remove the timing chain tensioner cover.
18. Remove both the IVT control covers, remove the bolts in the sequence illustrated.

➡**The shaft in the cover is inserted into the center hole of the intake camshaft sprocket. Remove the cover by pulling straight out until the cover disengages from the camshaft sprocket.**

19. Remove the starter.
20. Remove the intake manifold collector.
21. Remove the rocker covers as follows:
   a. Remove the engine cover.
   b. Remove side engine covers.
22. If removing right hand rocker cover, remove the following:
- Front cowl panel

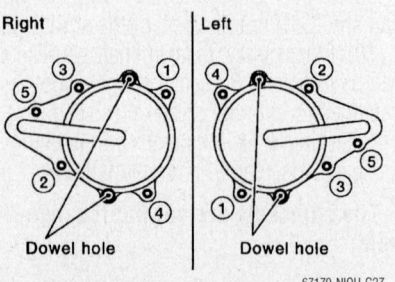

Loosen the IVT control cover bolts in this sequence–3.5L engine

67170-NIQU-G27

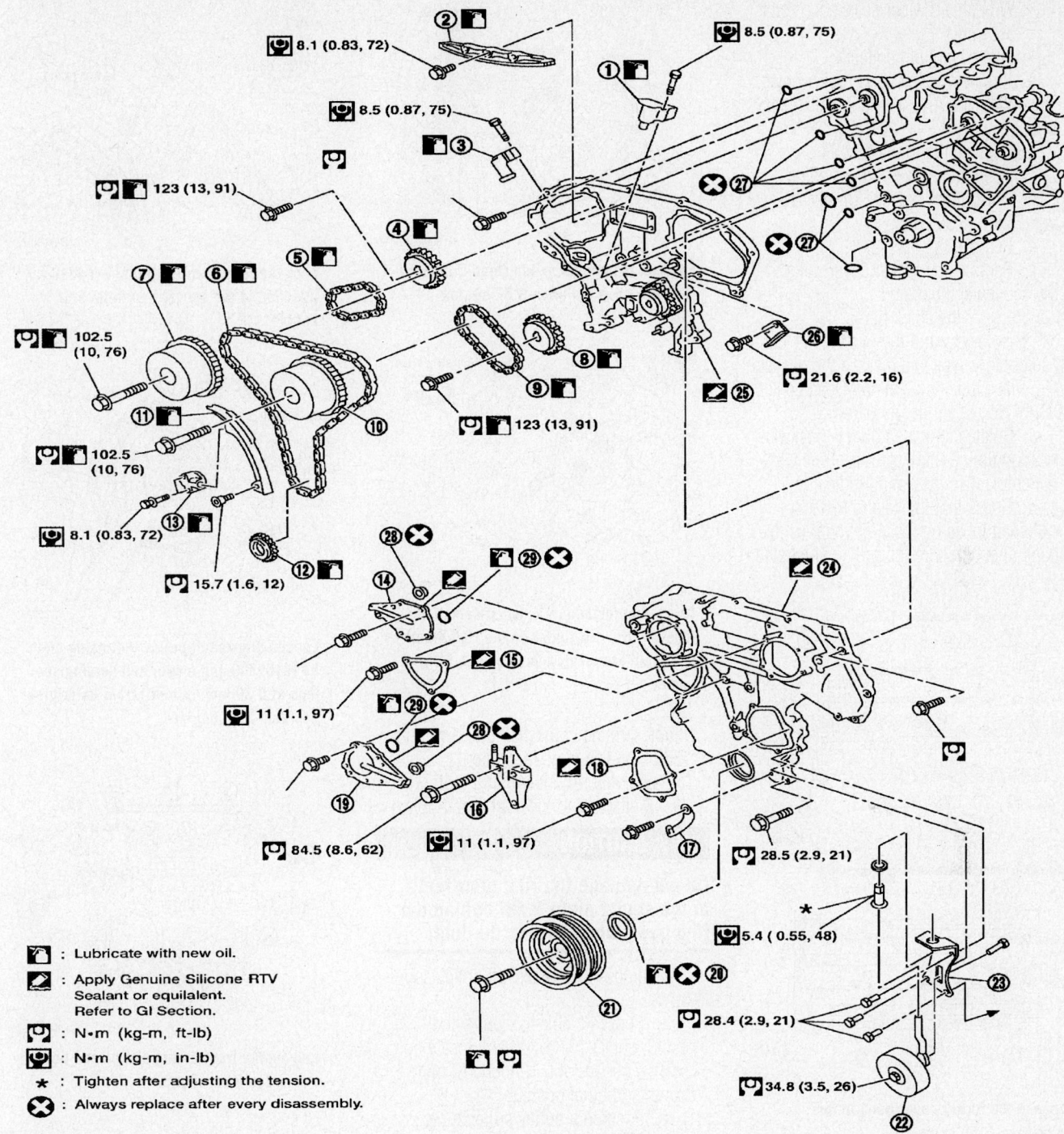

: Lubricate with new oil.

: Apply Genuine Silicone RTV Sealant or equilalent. Refer to GI Section.

: N•m (kg-m, ft-lb)

: N•m (kg-m, in-lb)

* : Tighten after adjusting the tension.

: Always replace after every disassembly.

1.	Timing chain tensioner	2.	Internal chain guide	3.	Timing chain tensioner
4.	Camshaft sprocket (EXH)	5.	Timing chain (secondary)	6.	Timing chain (primary)
7.	Camshaft sprocket (INT)	8.	Camshaft sprocket (EXH)	9.	Timing chain (secondary)
10.	Camshaft sprocket (INT)	11.	Slack guide	12.	Crankshaft sprocket
13.	Timing chain tensioner	14.	IVT control valve cover - right	15.	Chain tensioner cover
16.	RH engine mounting bracket	17.	Water hose clamp	18.	Water pump cover
19.	IVT control valve cover - left	20.	Front oil seal	21.	Crankshaft pulley
22.	Idler pulley	23.	Idler pulley bracket	24.	Front timing chain case
25.	Rear timing chain case	26.	Timing chain tension guide	27.	O-ring
28.	Collared O-ring	29.	Seal ring		

Exploded view of the timing chain and related components—3.5L engine

- Windshield wiper arms and motor assembly
- Intake manifold collector.

a. If removing left hand rocker cover, disconnect the Air Fuel Ratio (AFR) sensor.

b. Remove the ignition coils.

c. Position engine harness aside.

d. Disconnect Positive Crankcase Ventilation (PCV) hose.

e. Remove the dipstick.

f. Remove the rocker cover bolts in the sequence illustrated.

23. Remove the spark plugs.

24. Disconnect and remove the IVT control solenoid valves and discard the gaskets

25. Place the engine at Top Dead center (TDC) of No. 1 cylinder as follows:

a. Rotate crankshaft pulley clockwise to align timing mark (grooved line without color) with the timing indicator.

b. Check that intake and exhaust camshaft lobes on No. 1 cylinder (right bank of engine) are located as illustrated.

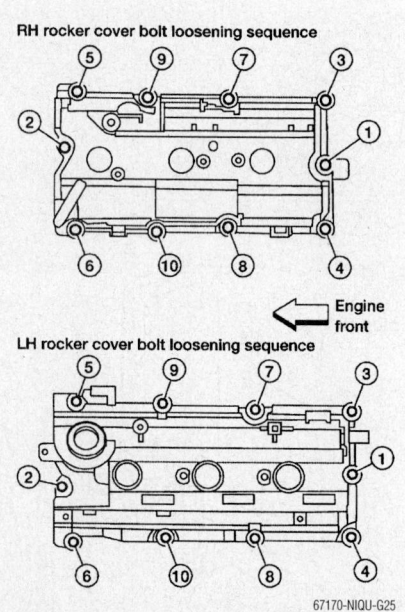

RH rocker cover bolt loosening sequence

LH rocker cover bolt loosening sequence

67170-NIQU-G25

Loosen the rocker cover bolts in this sequence–3.5L engine

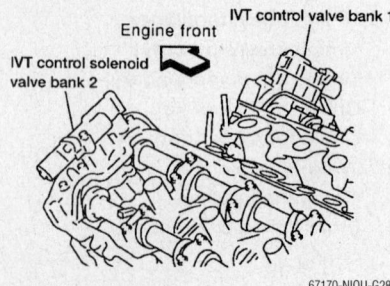

67170-NIQU-G28

Remove the IVT control solenoid valves–3.5L engine

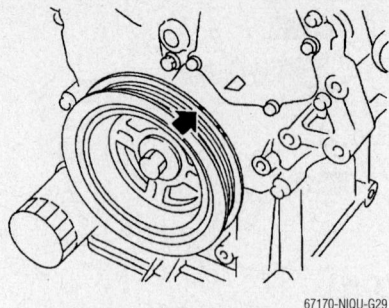

67170-NIQU-G29

Position the engine at Top Dead Center (TDC) of No. 1 cylinder–3.5L engine

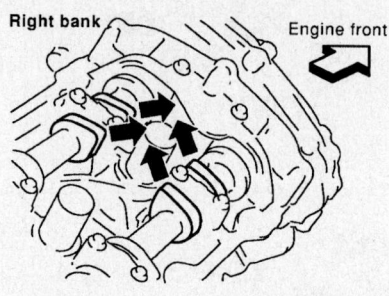

67170-NIQU-G30

Check that intake and exhaust camshaft lobes on No. 1 cylinder on the right bank of engine are located as shown–3.5L engine

If not, turn the crankshaft one revolution (360°) and align as illustrated.

26. Lock the ring gear using tool J–44716 attached to the starter bolt hole.

✱✱ CAUTION

Do not damage the ring gear teeth, or the signal plate teeth behind the ring gear, when setting the tool.

27. Remove the crankshaft pulley as follows:

a. Loosen crankshaft pulley bolt using tool KV10109300 and locate bolt seating surface at 0.39 inch (10 mm) from its original position.

b. Position a pulley puller at recess hole of the crankshaft pulley to remove the crankshaft pulley.

✱✱ CAUTION

Do not use a puller claw on the crankshaft pulley.

28. Remove the front timing chain case after loosening the front timing chain case bolts in the order illustrated.

29. Remove the internal chain guide.

30. Remove the timing chain tensioner and slack guide as follows:

a. Put matchmarks on the timing

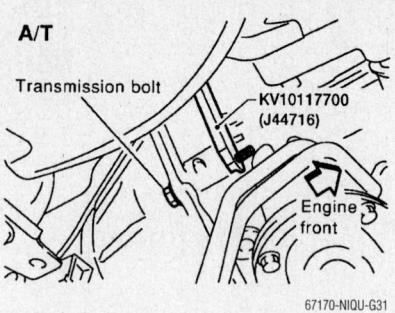

67170-NIQU-G31

Lock the ring gear using tool J–44716 attached to the starter bolt hole–3.5L engine

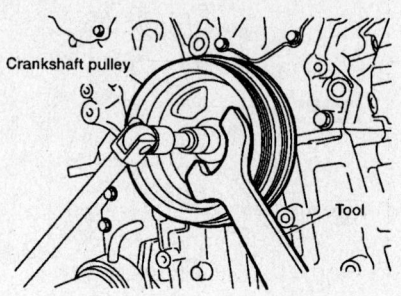

67170-NIQU-G32

Loosen crankshaft pulley bolt using tool KV10109300 and locate bolt seating surface at 0.39 inch (10mm) from its original position–3.5L engine

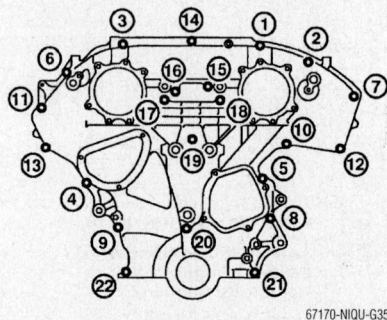

67170-NIQU-G35

Loosen the front timing chain case bolts in this order–3.5L engine

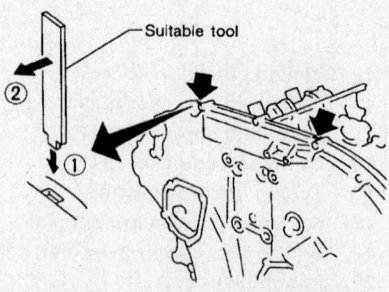

67170-NIQU-G36

Insert the appropriate size tool such as J–37228 into the notch at the top of the front timing chain case–3.5L engine

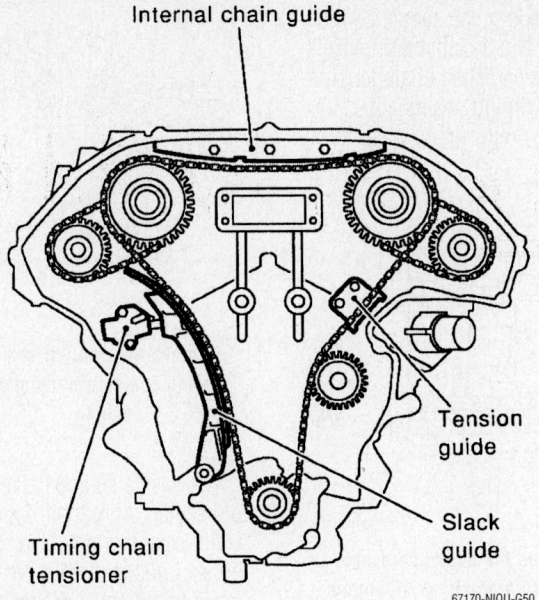

Location of the internal chain guide–3.5L engine

67170-NIQU-G50

chain and sprockets to indicate the correct position of the components to aid during installation.

b. Pull the lever down and release the plunger stopper tab. The plunger stopper tab can be pushed up to release (coaxial structure with lever).

c. Insert a 0.098 inch (2.5mm) Allen wrench into the tensioner body hole to hold the lever, and keep the tab released.

d. Insert the plunger into tensioner body by pressing the slack side chain guide.

e. Keep the slack side chain guide pressed and hold it by pushing a stopper pin such as a 0.098 inch (2.5mm) Allen wrench through the lever hole and body hole.

f. Remove the timing chain tensioner bolts and the timing chain tensioner.

g. Remove the slack guide bolt and the guide.

31. Remove primary timing chain and crankshaft sprocket.

✳✳ CAUTION

After removing timing chain, do not turn the crankshaft and camshaft separately, or the valves will strike the pistons.

32. Attach a suitable stopper pin such as a 0.098 inch (2.5mm) Allen wrench to the right and left camshaft chain tensioners, for secondary timing chains.

33. Remove the intake and exhaust camshaft sprocket bolts. Put matchmarks on the timing chain and camshaft sprockets to aid alignment during installation.

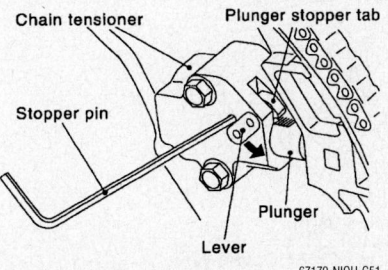

Insert a 0.098 inch (2.5mm) Allen wrench into the tensioner body hole to hold the chain tensioner lever–3.5L engine

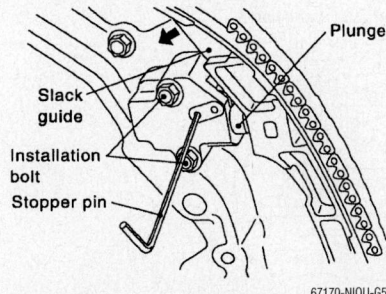

Keep the slack side chain guide pressed and hold it by pushing a stopper pin–3.5L engine

34. Use a wrench on the hexagonal portion of the camshaft and loosen the bolts as illustrated.

35. Remove the secondary timing chains with camshaft sprockets as follows:

a. Rotate camshaft slightly, and slacken timing chain of timing chain tensioner side.

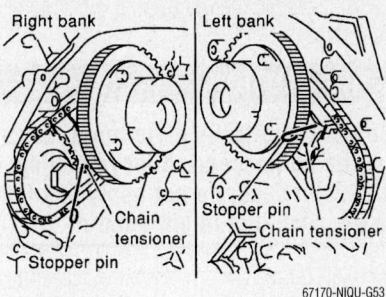

67170-NIQU-G53

Attach a suitable stopper pin such as a 0.098 inch (2.5mm) Allen wrench to the right and left camshaft chain tensioners–3.5L engine

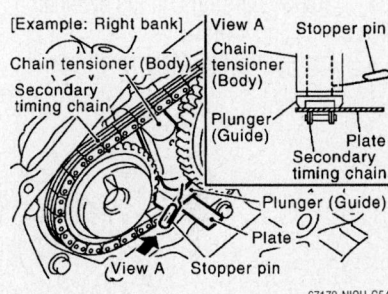

67170-NIQU-G54

Insert a 0.020 inch (0.5 mm) metal or resin plate into the guide between timing chain and chain tensioner plunger–3.5L engine

b. Insert a 0.020 inch (0.5 mm) metal or resin plate into the guide between timing chain and chain tensioner plunger.

36. Remove the camshaft sprocket and secondary timing chain with the timing chain removed from guide groove.

➡**The intake camshaft sprocket is two-for-one structure of primary and secondary sprockets. Handle the intake sprockets as an assembly.**

✳✳ CAUTION

The chain tensioner plunger can move while stopper pin is inserted in tensioner. The plunger can come out of tensioner when timing chain is removed. Use caution during removal.

➡**Do not disassemble the intake sprockets (never loosen bolts A and B shown in the accompanying illustration).**

37. Remove the timing chain tension guide.

38. Remove the rear timing chain case as follows:

☀☀ CAUTION

Do not remove the plate metal cover for the oil passage. After removing the chain case, do not apply any load to the case that might bend it.

a. Loosen and remove the rear timing chain case bolts in the order illustrated.

b. Cut the sealant using tool J–37228 and remove the rear timing chain case.

39. Disconnect the inlet coolant hose.

40. Remove the inlet coolant housing, gasket and thermostat.

41. Remove O–rings on the cylinder head and cylinder block.

42. Loosen the No. 1 camshaft bracket

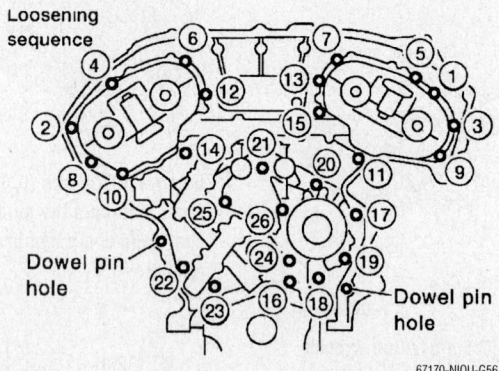

Remove the rear timing chain case bolts in the order illustrated–3.5L engine

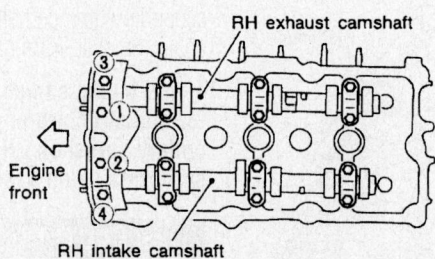

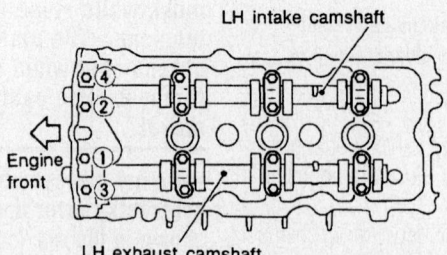

Loosen the No. 1 camshaft bracket bolts in several steps in the sequence illustrated–3.5L engine

bolts in several steps in the sequence illustrated and remove No. 1 camshaft brackets.

43. Remove the camshaft chain tensioners, for secondary timing chains.

44. If necessary, remove the water pump.

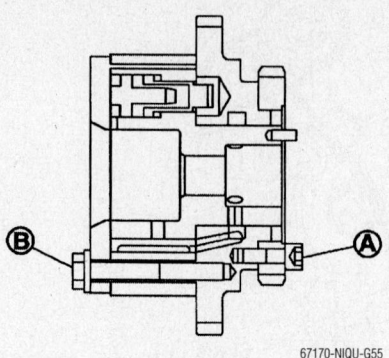

Do not disassemble the intake sprockets (never loosen bolts A and B)–3.5L engine

Remove the front oil seal from the front timing chain case using a suitable tool–3.5L engine

45. Use a scraper to remove all of the old sealant from the front and rear timing chain case mating surfaces and the bolt holes, bolts, camshaft No. 1 bracket and water pump cover. Being careful not to damage the mating surfaces.

46. Remove the front oil seal from the front timing chain case using a suitable tool.

47. Check for cracks and any excessive wear at the roller links of the timing chain. Replace the timing chain as necessary.

To install:

48. Install the camshaft chain tensioners, for secondary timing chain Tighten the bolts to 75 inch lbs. (9 Nm).

49. Before installing the No. 1 camshaft bracket, apply sealant to mating surface and wipe off any excess sealant before installation.

50. Tighten the No. 1 camshaft bracket in three steps, in the order illustrated as follows:

a. Step 1: 17 inch lbs. (2 Nm).

b. Step 2: 52 inch lbs. (6 Nm).

c. Step 3: 92 inch lbs. (10 Nm).

51. Install the thermostat, gasket and coolant inlet housing. Tighten the bolts to 87 inch lbs. (10 Nm).

52. Install rear timing chain case as follows:

a. Install new O–rings on cylinder block.

b. Install new O–rings on cylinder head.

c. Install the water pump, if removed.

53. Apply silicone sealant to rear timing chain case as illustrated, clean off the protruding sealant prior to installation.

a. Align the rear timing chain case with the dowel pins on the cylinder block and install the case. Make sure the O–rings stay in place during installation.

b. Tighten the rear timing chain case bolts in the order illustrated.

c. Install the 0.79 inch (20 mm) bolts

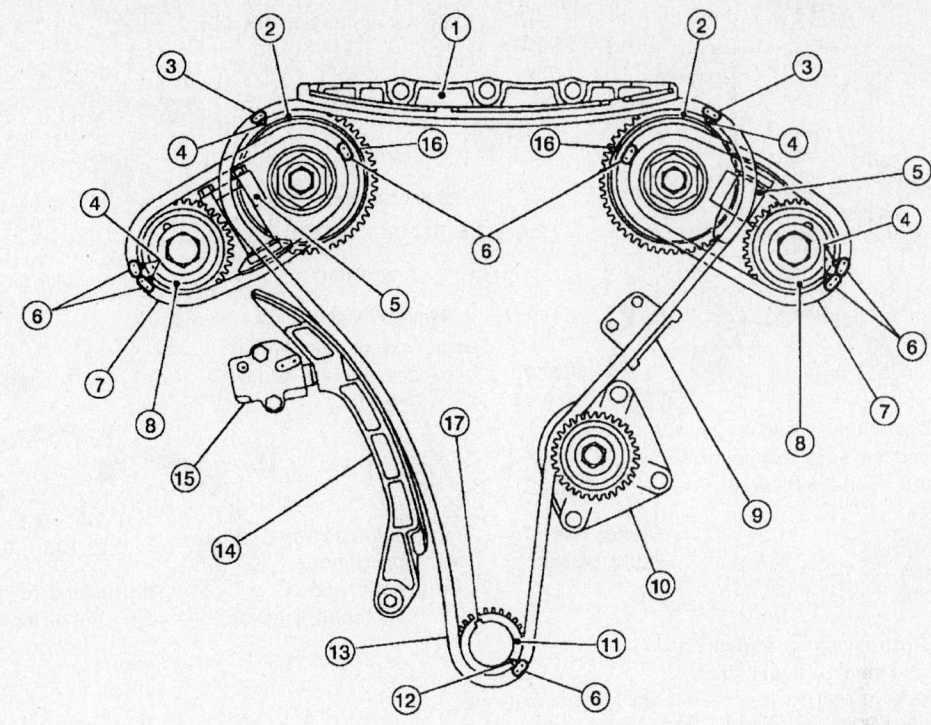

1.	Internal chain guide	2.	Camshaft sprocket (intake)	3.	Mating mark (copper link)
4.	Mating mark (punched)	5.	Secondary timing chain tensioner	6.	Mating mark (gold link)
7.	Secondary timing chain	8.	Camshaft sprocket (exhaust)	9.	Tensioner guide
10.	Water pump	11.	Crankshaft sprocket	12.	Mating mark (notched)
13.	Primary timing chain	14.	Slack guide	15.	Primary timing chain tensioner
16.	Mating mark (back side)	17.	Crankshaft key		

67170-NIQU-G58

Make sure all timing marks are aligned as shown when the chain is installed–3.5L engine

in locations 1, 2, 3, 6, 7, 8, 9 and 10. Tighten to 9 ft. lbs. (13 Nm). Install the 0.63 inch (16 mm) bolts in locations 4, 5, 11 through 26 and tighten to 9 ft. lbs. (13 Nm).

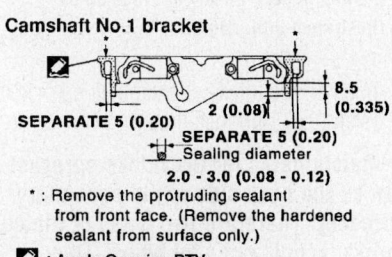

Camshaft No.1 bracket

8.5 (0.335)

SEPARATE 5 (0.20)   2 (0.08)

SEPARATE 5 (0.20)
Sealing diameter
2.0 - 3.0 (0.08 - 0.12)

* : Remove the protruding sealant from front face. (Remove the hardened sealant from surface only.)

: Apply Genuine RTV Silicone Sealant or equivalent. Refer to GI section.

67170-NIQU-G59

Before installing the No. 1 camshaft bracket, apply sealant to mating surface and wipe off any excess sealant–3.5L engine

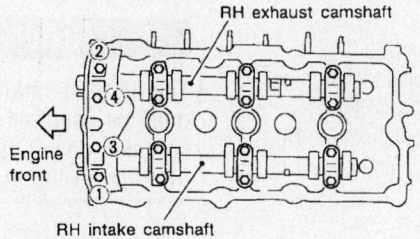

RH exhaust camshaft

Engine front

RH intake camshaft

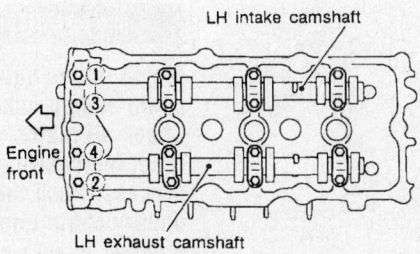

LH intake camshaft

Engine front

LH exhaust camshaft

Loosen in numerical order.

67170-NIQU-G60

Camshaft bolt tightening sequence–3.5L engine

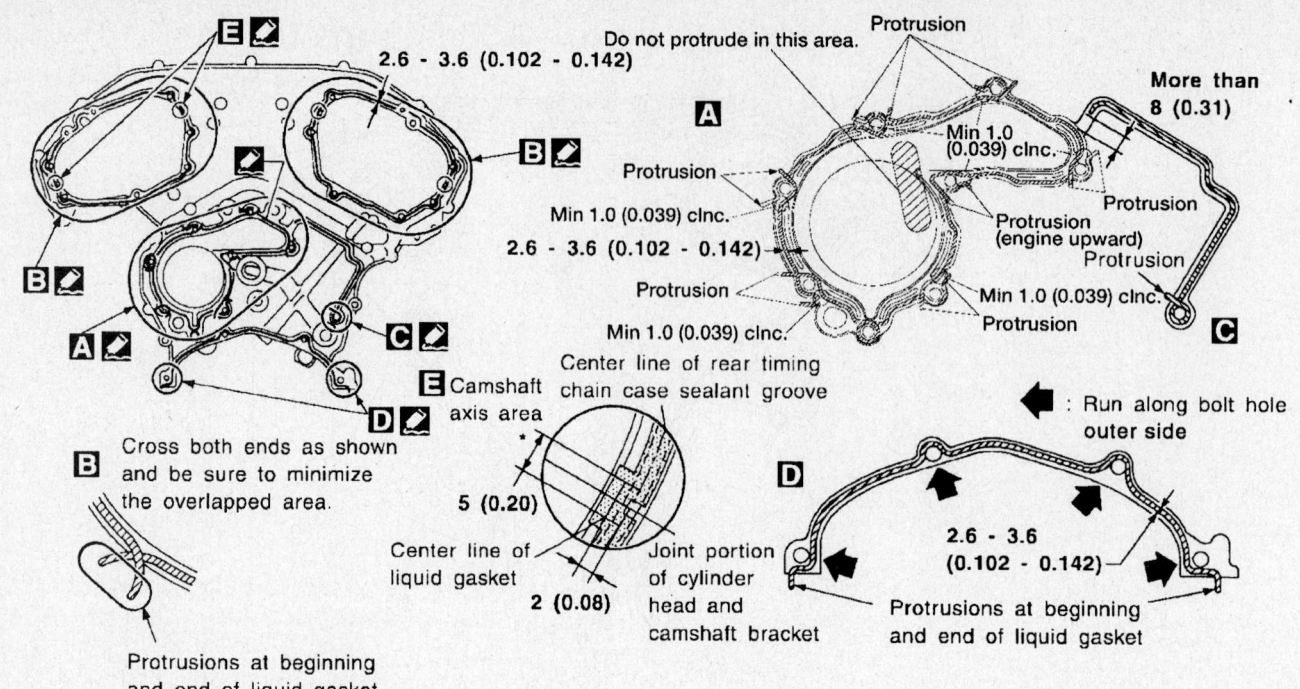

2.6 - 3.6 (0.102 - 0.142)

Do not protrude in this area.

Protrusion

More than 8 (0.31)

A

Protrusion

Min 1.0 (0.039) clnc.

Min 1.0 (0.039) clnc.

2.6 - 3.6 (0.102 - 0.142)

Protrusion (engine upward)

Protrusion

Protrusion

Min 1.0 (0.039) clnc.

Protrusion

Min 1.0 (0.039) clnc.

C

E Camshaft axis area

Center line of rear timing chain case sealant groove

: Run along bolt hole outer side

5 (0.20)

Center line of liquid gasket

2 (0.08)

Joint portion of cylinder head and camshaft bracket

D

B Cross both ends as shown and be sure to minimize the overlapped area.

2.6 - 3.6 (0.102 - 0.142)

Protrusions at beginning and end of liquid gasket

Protrusions at beginning and end of liquid gasket

* : Apply liquid gasket to the chamfered surface between camshaft bracket and cylinder head.

: Apply liquid gasket. Use Genuine RTV silicone sealant or equivalent. Refer to GI section.

Unit: mm (in)

67170-NIQU-G61

Apply silicone sealant to rear timing chain case as illustrated–3.5L engine

d. After all bolts are initially tightened, retighten them to the specification in the order illustrated.

e. After installing rear timing chain case, check surface height difference between the rear timing chain case to cylinder block. The measurement should be 0.0094–0.0055 inch (0.24–0.14 mm). If not within specification, repeat the cover installation procedure.

54. Install the timing chain tension guide. Tighten the bolts to 9 ft. lbs. (13 Nm).

55. Position the crankshaft so the No. 1 piston is set at TDC on the compression stroke.

56. Make sure that the dowel pin hole, dowel pin and crankshaft key are located as illustrated. The camshaft dowel pin hole (intake side): at cylinder head upper face side in each bank. The camshaft dowel pin (exhaust side): at cylinder head upper face side in each bank. The crankshaft key: at cylinder head side of right hand bank.

✳✳ CAUTION

The hole on small diameter side must be used for the intake camshaft sprocket dowel pin. Do not misidentify (ignore the big diameter side).

57. Install the secondary timing chains and camshaft sprockets as follows:

✳✳ CAUTION

Match marks between the timing chain and sprockets can slip easily. Check all match mark positions repeatedly during the installation process. Push the sleeve of the secondary chain tensioner and keep it pressed in with a stopper pin.

a. Align the match marks on the secondary timing chain (gold link) with the

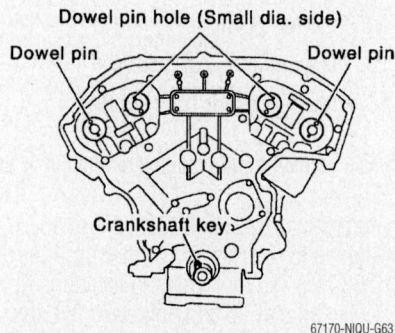

Dowel pin hole (Small dia. side)

Dowel pin

Dowel pin

Crankshaft key

67170-NIQU-G63

Make sure that the dowel pin hole, dowel pin and crankshaft key are located as illustrated–3.5L engine

ones on the intake and exhaust sprockets (stamped), and install them.

➡Match marks for the intake sprocket are on the back side of the secondary sprocket. There are two types of match marks, round and oval types. They should be used for the right hand and left hand banks, respectively. The right hand banks use round type and the left hand bank use oval type.

b. Align the dowel pin and pin hole on the camshaft with the groove and

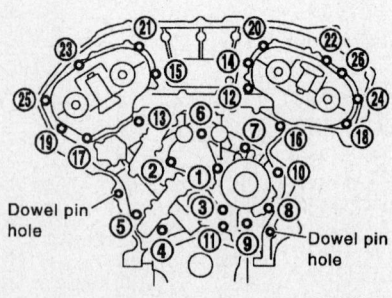

Dowel pin hole

Dowel pin hole

67170-NIQU-G62

Rear timing chain case torque sequence–3.5L engine

Example: Right bank side (Rear view)

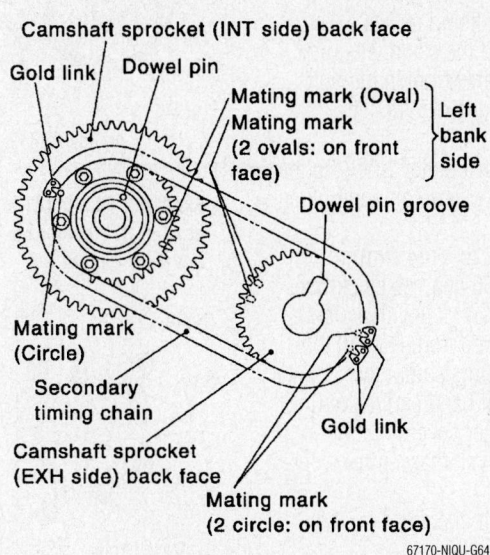

Match marks for the intake sprocket are on the back side of the secondary sprocket. There are two types of match marks, round and oval types–3.5L engine

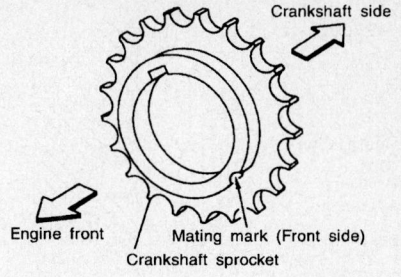

Make sure the mating marks on the crankshaft sprocket face the front of the engine–3.5L engine

dowel pin on the sprocket, and install them.

c. On the intake side, align the pin hole on the small diameter side of the camshaft front end with the dowel pin on the back side of the camshaft sprocket, and install them.

d. On the exhaust side, align the dowel pin on the camshaft front end with the pin groove on the camshaft sprocket, and install them.

e. Tighten the camshaft sprocket bolts by hand to prevent the dislocation of the dowel pins.

58. Tighten the timing chain tension guide bolts to 16 ft. lbs. (21 Nm).

➥It may be difficult to visually check the dislocation of mating marks during and after installation. To make the matching easier, make a mating mark on the sprocket teeth in advance with paint.

59. After confirming the mating marks are aligned, tighten the camshaft sprocket bolts to 76 ft. lbs. (102 Nm).

60. Remove the stopper pins out from the timing chain tensioners, on secondary timing chains.

61. Install the crankshaft sprocket on the crankshaft. Make sure the mating marks on the crankshaft sprocket face the front of the engine.

62. Install the primary timing chain as follows:

a. Install the primary timing chain so

the mating mark (punched) on the camshaft sprocket is aligned with the copper link on the timing chain, while the mating mark (notched) on the crankshaft sprocket is aligned with the gold link on the timing chain, as illustrated.

➥When it is difficult to align mating marks of the primary timing chain with each sprocket, gradually turn the camshaft using a wrench on the hexagonal portion to align it with the mating

marks. During alignment, be careful to prevent dislocation of mating mark alignments of the secondary timing chains.

63. Install the internal chain guide. Tighten the bolts to 72 inch lbs. (8 Nm).

64. Install the slack guide and tighten to 12 ft. lbs. (15 Nm). Do not over–tighten the slack guide installation bolt. It is normal for a gap to exist under the bolt seats when the bolt is tightened to specification.

65. Install the timing chain tensioner for the slack guide as follows:

a. When installing the chain tensioner, push in the sleeve and keep it pressed in with the stopper pin.

b. Remove any dirt and foreign materials completely from the back and the mounting surfaces of the chain tensioner.

c. Tighten the bolts to 72 inch lbs. (8 Nm).

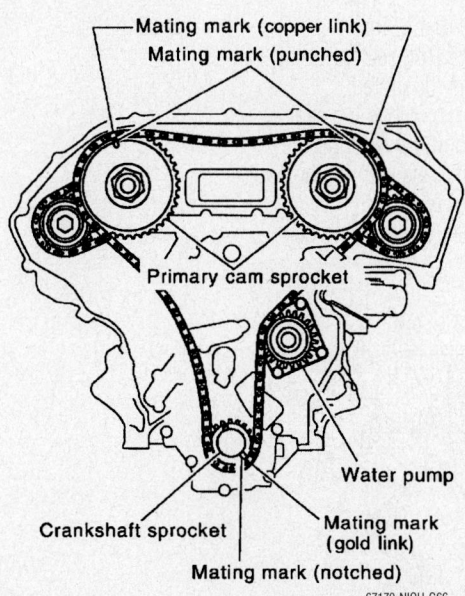

Install the primary timing chain so the match mark (punched) on the cam sprocket is aligned with the copper link on the timing chain, while the mating mark (notched) on the crank sprocket is aligned with the gold link on the timing chain–3.5L engine

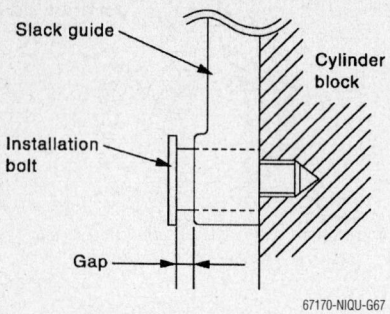

67170-NIQU-G67

It is normal for a gap to exist under the bolt seats when the internal chain guide bolt is tightened –3.5L engine

d. After installation, pull out the stopper pin by pressing the slack guide.

66. Confirm that the match marks on the sprockets and the timing chain have not slipped out of alignment.

67. Install new O–rings on the rear timing chain case.

68. Install the front oil seal on the front timing chain case using a suitable tool. Apply clean engine oil to the oil seal edges. Install the seal so that each seal lip is oriented as illustrated. Using a suitable drift, press-fit oil seal until it becomes flush with timing chain case end face. Make sure the garter spring in the oil seal is in position and seal lip is not inverted.

69. Install the front timing cover case.

70. Install IVT control valve covers as follows:

   a. Install new collared O–rings in front cover oil hole on both sides.

   b. Install new seal rings on the IVT control covers.

   c. Apply a 0.083–0.122 inch (2.1–3.1mm) bead of RTV sealant to the IVT control covers as illustrated.

71. Being careful not to move the seal ring from the installation groove, align the dowel pins on the chain case with the holes to install the IVT control covers.

72. Tighten the intake valve timing control cover bolts in the order illustrated to 100 inch lbs. (11 Nm).

73. Apply a 0.091–0.130 inch (2.3–3.3mm) bead of liquid gasket to the water pump cover and the chain tensioner cover and install covers. Tighten the bolts on the pump and cover to 97 inch lbs. (11 Nm).

74. Install crankshaft pulley lubricate the pulley bolt with clean engine oil and tighten the bolt in two steps:

   a. Step 1: 32 ft. lbs. (44 Nm).

   b. Step 2: 60–65 degrees clockwise.

75. Rotate crankshaft pulley in a clockwise direction as viewed from front of the engine to confirm it turns smoothly.

76. Install the right hand engine mount.

77. Install the rocker covers in the reverse order of removal, make sure to perform the following:

78. Apply sealant to the areas on the front corners.

79. Tighten the rocker cover bolts in two steps in the sequence illustrated:

   a. Step 1: 17 inch lbs. (1.96 Nm).

   b. Step 2: 74 inch lbs. (8 Nm).

80. Install the remaining components is in reverse order of removal.

➡ **If hydraulic pressure inside chain tensioner drops after removal/installation, slack in the guide may generate a pounding noise during and just after engine start. This is normal. Noise will stop after hydraulic pressure rises.**

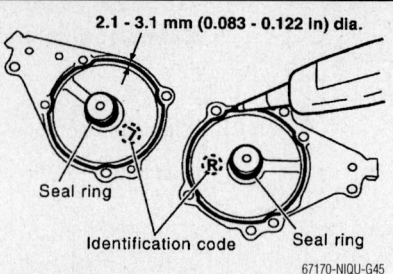

2.1 - 3.1 mm (0.083 - 0.122 In) dia.

Seal ring / Identification code / Seal ring

67170-NIQU-G45

Apply a 0.083–0.122 inch (2.1–3.1mm) bead of RTV sealant to the IVT control covers–3.5L engine

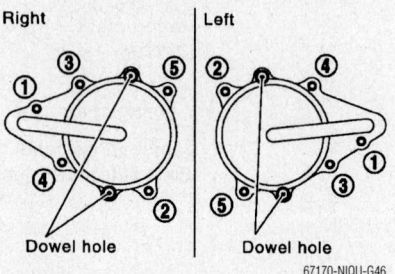

Right   Left

Dowel hole   Dowel hole

67170-NIQU-G46

Tighten the intake valve timing control cover bolts in this order–3.5L engine

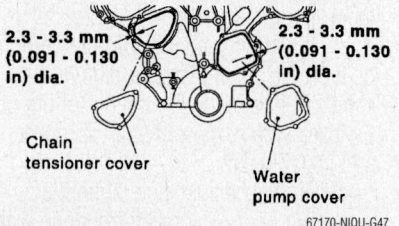

2.3 - 3.3 mm (0.091 - 0.130 In) dia.

2.3 - 3.3 mm (0.091 - 0.130 In) dia.

Chain tensioner cover   Water pump cover

67170-NIQU-G47

Apply a 0.091–0.130 inch (2.3–3.3mm) bead of liquid gasket to the water pump cover and the chain tensioner cover–3.5L engine

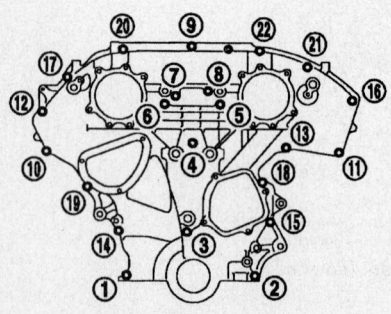

67170-NIQU-G44

Tighten the front timing chain case bolts in the order shown–3.5L engine

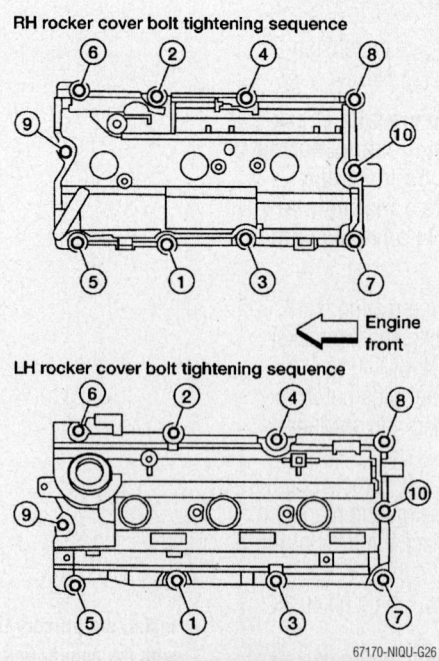

RH rocker cover bolt tightening sequence

← Engine front

LH rocker cover bolt tightening sequence

67170-NIQU-G26

Tighten the rocker cover bolts in this sequence–3.5L engine

5–SPEED MODELS

1. Before servicing the vehicle, refer to the precautions section.

2. Remove the engine from the vehicle.

3. Remove the idler pulley and bracket.

4. Remove the upper and lower oil pans.

5. Remove the alternator.

6. Remove the timing chain tensioner cover.

7. Remove both the IVT control covers, remove the bolts in the sequence illustrated.

➡ **The shaft in the cover is inserted into the center hole of the intake camshaft sprocket. Remove the cover by pulling straight out until the cover disengages from the camshaft sprocket.**

8. Remove the starter.

9. Remove the intake manifold collector.

10. Remove the rocker covers as follows:

a. Remove the engine cover.

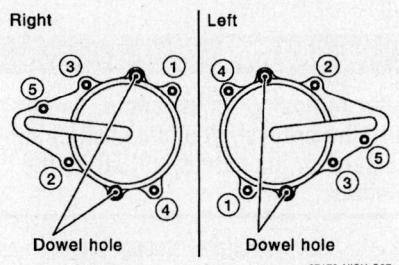

Loosen the IVT control cover bolts in this sequence–3.5L engine

b. Remove side engine covers.

11. If removing right hand rocker cover, remove the following:

• Front cowl panel

• Windshield wiper arms and motor assembly

• Intake manifold collector.

a. If removing left hand rocker cover, disconnect the Air Fuel Ratio (AFR) sensor.

b. Remove the ignition coils.

c. Position engine harness aside.

d. Disconnect Positive Crankcase Ventilation (PCV) hose.

e. Remove the dipstick.

f. Remove the rocker cover bolts in the sequence illustrated.

12. Remove the spark plugs.

13. Disconnect and remove the IVT control solenoid valves and discard the gaskets

14. Place the engine at Top Dead center (TDC) of No. 1 cylinder as follows:

a. Rotate crankshaft pulley clockwise to align timing mark (grooved line without color) with the timing indicator.

b. Check that intake and exhaust camshaft lobes on No. 1 cylinder (right bank of engine) are located as illustrated. If not, turn the crankshaft one revolution (360 degrees) and align as illustrated.

15. Lock the ring gear using tool J–44716 attached to the starter bolt hole.

❋❋ CAUTION

Do not damage the ring gear teeth, or the signal plate teeth behind the ring gear, when setting the tool.

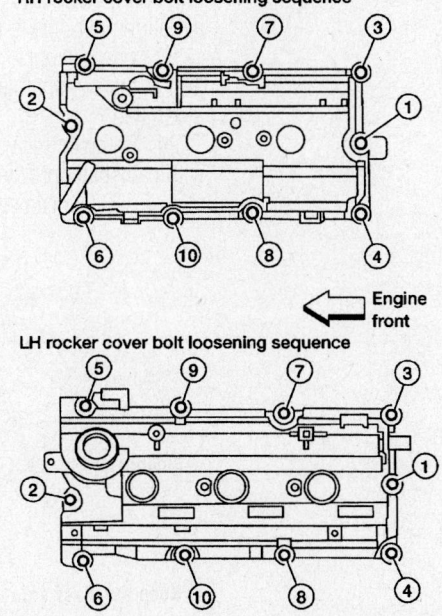

RH rocker cover bolt loosening sequence

LH rocker cover bolt loosening sequence

Loosen the rocker cover bolts in this sequence–3.5L engine

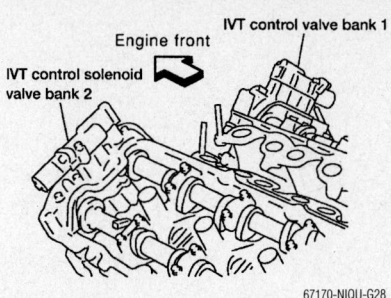

Remove the IVT control solenoid valves–3.5L engine

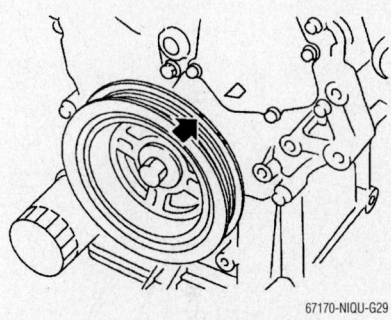

Position the engine at Top Dead Center (TDC) of No. 1 cylinder–3.5L engine

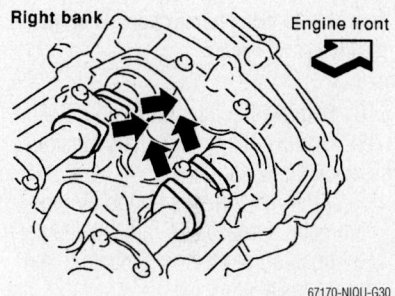

Check that intake and exhaust camshaft lobes on No. 1 cylinder on the right bank of engine are located as shown–3.5L engine

16. Remove the crankshaft pulley as follows:

a. Loosen crankshaft pulley bolt using tool KV10109300 and locate bolt seating surface at 0.39 inch (10 mm) from its original position.

b. Position a pulley puller at recess hole of crankshaft pulley to remove crankshaft pulley.

❋❋ CAUTION

Do not use a puller claw on crankshaft pulley.

17. Remove the front timing chain case after loosening the front timing chain case bolts in the order illustrated.

A/T

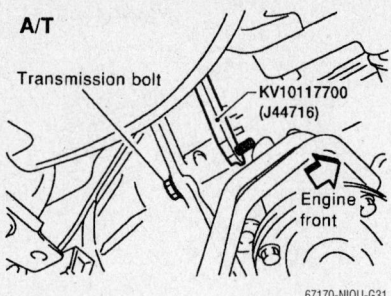

67170-NIQU-G31

Lock the ring gear using tool J–44716 attached to the starter bolt hole–3.5L engine

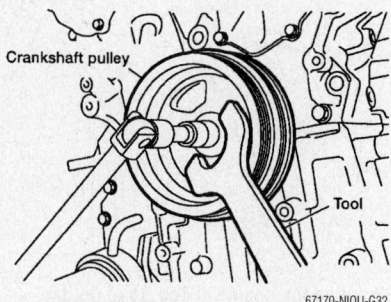

67170-NIQU-G32

Loosen crankshaft pulley bolt using tool KV10109300 and locate bolt seating surface at 0.39 inch (10mm) from its original position–3.5L engine

18. Remove the internal chain guide.
19. Remove the timing chain tensioner and slack guide as follows:

    a. Put matchmarks on the timing chain and sprockets to indicate the correct position of the components to aid during installation.

    b. Pull the lever down and release the

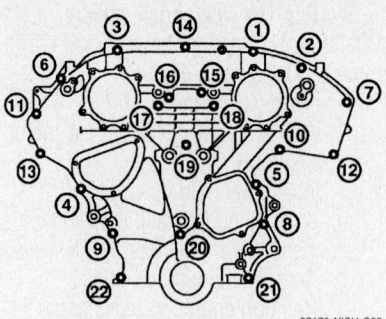

67170-NIQU-G35

Loosen the front timing chain case bolts in this order–3.5L engine

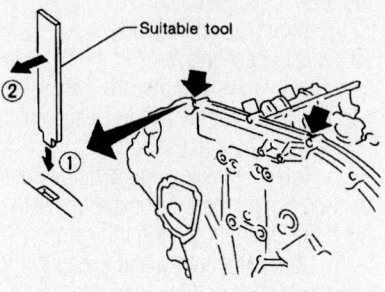

67170-NIQU-G36

Insert the appropriate size tool such as J–37228 into the notch at the top of the front timing chain case–3.5L engine

plunger stopper tab. The plunger stopper tab can be pushed up to release (coaxial structure with lever).

    c. Insert a 0.098 inch (2.5mm) Allen wrench into the tensioner body hole to hold the lever, and keep the tab released.

    d. Insert the plunger into tensioner body by pressing the slack side chain guide.

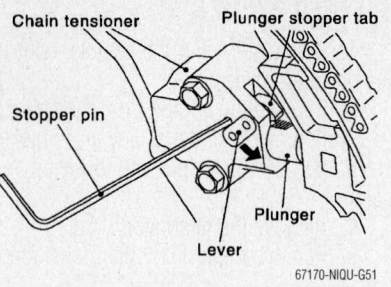

67170-NIQU-G51

Insert a 0.098 inch (2.5mm) Allen wrench into the tensioner body hole to hold the chain tensioner lever–3.5L engine

    e. Keep the slack side chain guide pressed and hold it by pushing a stopper pin such as a 0.098 inch (2.5mm) Allen wrench through the lever hole and body hole.

    f. Remove the timing chain tensioner bolts and the timing chain tensioner.

    g. Remove the slack guide bolt and the guide.

20. Remove primary timing chain and crankshaft sprocket.

✷✷ CAUTION

After removing timing chain, do not turn the crankshaft and camshaft separately, or the valves will strike the pistons.

21. Attach a suitable stopper pin such as a 0.098 inch (2.5mm) Allen wrench to the right and left camshaft chain tensioners, for secondary timing chains.

22. Remove the intake and exhaust camshaft sprocket bolts. Put matchmarks on the timing chain and camshaft sprockets to aid alignment during installation.

23. Use a wrench on the hexagonal portion of the camshaft and loosen the bolts as illustrated.

24. Remove the secondary timing chains with camshaft sprockets as follows:

    a. Rotate camshaft slightly, and

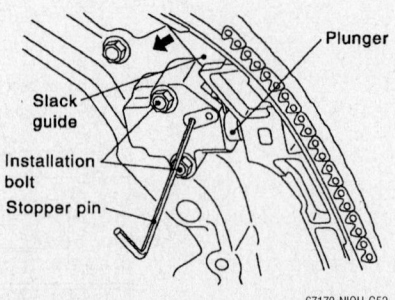

67170-NIQU-G52

Keep the slack side chain guide pressed and hold it by pushing a stopper pin–3.5L engine

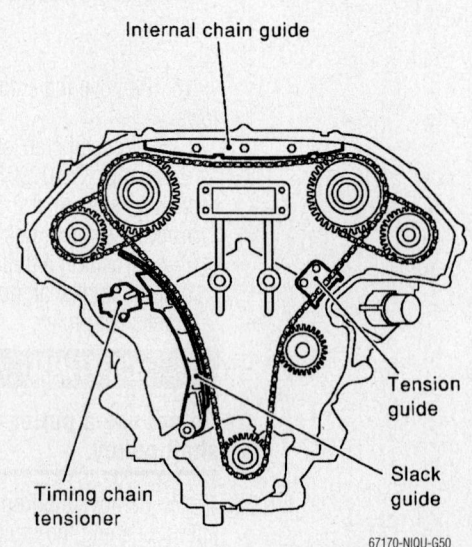

67170-NIQU-G50

Location of the internal chain guide–3.5L engine

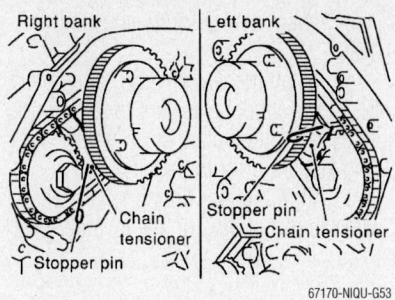

Attach a suitable stopper pin such as a 0.098 inch (2.5mm) Allen wrench to the right and left camshaft chain tensioners–3.5L engine

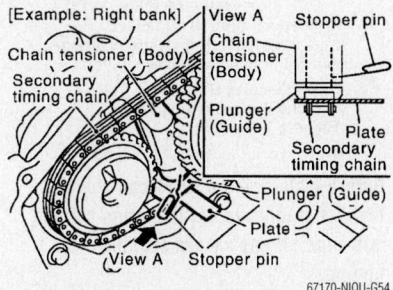

Insert a 0.020 inch (0.5 mm) metal or resin plate into the guide between timing chain and chain tensioner plunger–3.5L engine

slacken timing chain of timing chain tensioner side.

b. Insert a 0.020 inch (0.5 mm) metal or resin plate into the guide between timing chain and chain tensioner plunger.

25. Remove the camshaft sprocket and secondary timing chain with the timing chain removed from guide groove.

➡ The intake camshaft sprocket is two-for-one structure of primary and secondary sprockets. Handle the intake sprockets as an assembly.

✳✳ CAUTION

The chain tensioner plunger can move while stopper pin is inserted in tensioner. The plunger can come out of tensioner when timing chain is removed. Use caution during removal.

➡ Do not disassemble the intake sprockets (never loosen bolts A and B shown in the accompanying illustration).

26. Remove the timing chain tension guide.

27. Remove the rear timing chain case as follows:

✳✳ CAUTION

Do not remove the plate metal cover for the oil passage. After removing the chain case, do not apply any load to the case that might bend it.

a. Loosen and remove the rear timing chain case bolts in the order illustrated.
b. Cut the sealant using tool J–37228 and remove the rear timing chain case.
28. Disconnect the inlet coolant hose.
29. Remove the inlet coolant housing, gasket and thermostat.
30. Remove the O–rings on the cylinder head and cylinder block.

31. Loosen the No. 1 camshaft bracket bolts in several steps in the sequence illustrated and remove No. 1 camshaft brackets.

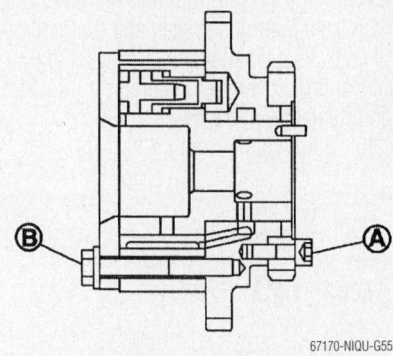

Do not disassemble the intake sprockets (never loosen bolts A and B)–3.5L engine

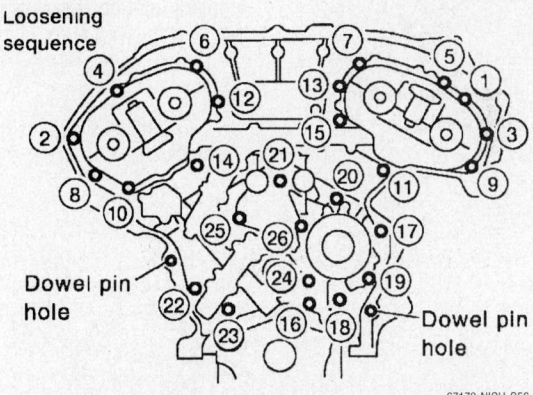

Remove the rear timing chain case bolts in the order illustrated–3.5L engine

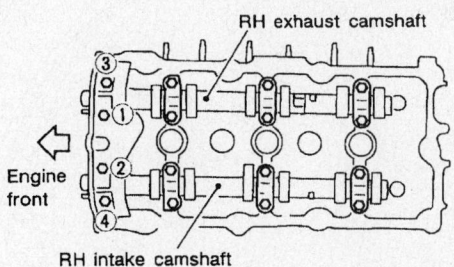

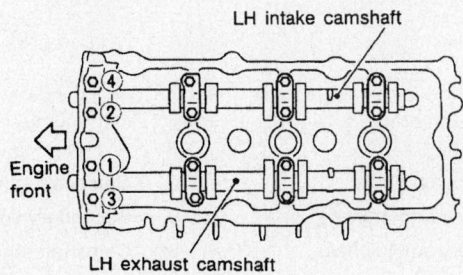

Loosen the No. 1 camshaft bracket bolts in several steps in the sequence illustrated–3.5L engine

32. Remove the camshaft chain tensioners, for secondary timing chains.

33. If necessary, remove the water pump.

34. Use a scraper to remove all of the old sealant from the front and rear timing chain case mating surfaces and the bolt holes, bolts, camshaft No. 1 bracket and water pump cover. Being careful not to damage the mating surfaces.

35. Remove the front oil seal from the

67170-NIQU-G37

Remove the front oil seal from the front timing chain case using a suitable tool–3.5L engine

front timing chain case using a suitable tool.

36. Check for cracks and any excessive wear at the roller links of the timing chain. Replace the timing chain as necessary.

To install:

37. Install the camshaft chain tensioners, for secondary timing chain Tighten the bolts to 75 inch lbs. (9 Nm).

38. Before installing the No. 1 camshaft bracket, apply sealant to mating surface and wipe off any excess sealant before installation.

39. Tighten the No. 1 camshaft bracket in three steps, in the order illustrated as follows:

   a Step 1: 17 inch lbs. (2 Nm).

   b. Step 2: 52 inch lbs. (6 Nm).

   c. Step 3: 92 inch lbs. (10 Nm).

40. Install the thermostat, gasket and coolant inlet housing. Tighten the bolts to 87 inch lbs. (10 Nm).

41. Install rear timing chain case as follows:

   a. Install new O–rings on cylinder block.

b. Install new O–rings on cylinder head.

c. Install the water pump, if removed.

42. Apply silicone sealant to rear timing chain case as illustrated, clean off the protruding sealant prior to installation.

   a. Align the rear timing chain case with the dowel pins on the cylinder block

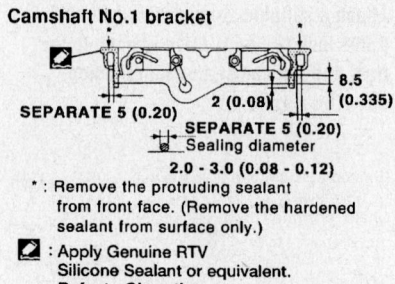

67170-NIQU-G59

Before installing the No. 1 camshaft bracket, apply sealant to mating surface and wipe off any excess sealant–3.5L engine

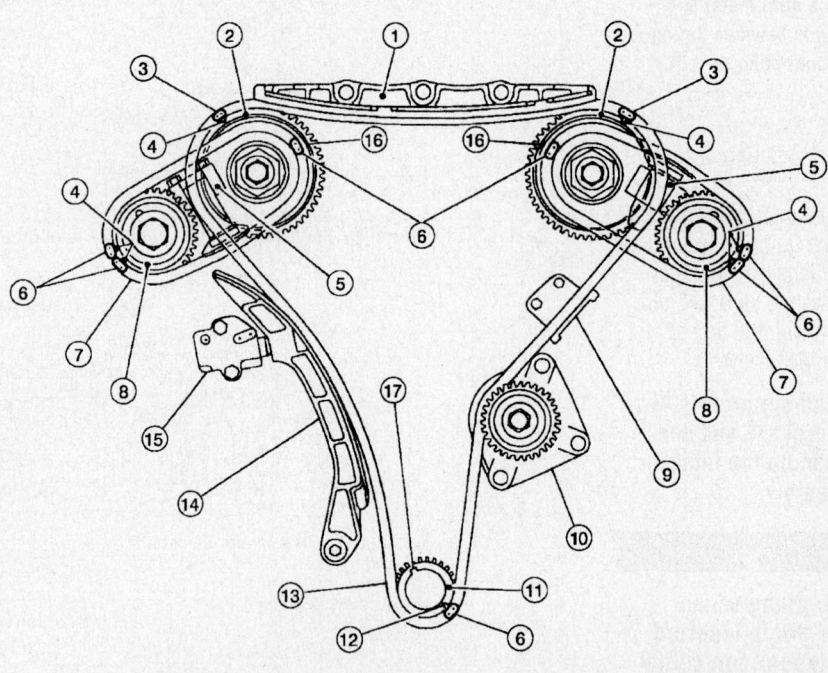

1. Internal chain guide	2. Camshaft sprocket (intake)	3. Mating mark (copper link)
4. Mating mark (punched)	5. Secondary timing chain tensioner	6. Mating mark (gold link)
7. Secondary timing chain	8. Camshaft sprocket (exhaust)	9. Tensioner guide
10. Water pump	11. Crankshaft sprocket	12. Mating mark (notched)
13. Primary timing chain	14. Slack guide	15. Primary timing chain tensioner
16. Mating mark (back side)	17. Crankshaft key	

67170-NIQU-G58

Make sure all timing marks are aligned as shown when the chain is installed–3.5L engine

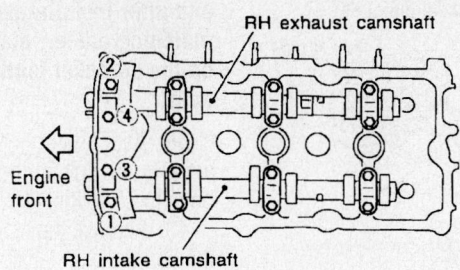

RH exhaust camshaft

RH intake camshaft

Engine front

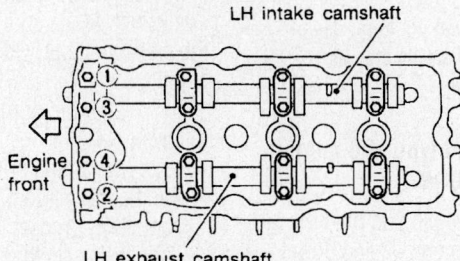

LH intake camshaft

LH exhaust camshaft

Engine front

Loosen in numerical order.

67170-NIQU-G60

Camshaft bolt tightening sequence–3.5L engine

and install the case. Make sure the O–rings stay in place during installation.

b. Tighten the rear timing chain case bolts in the order illustrated.

c. Install the 0.79 inch (20 mm) bolts in locations 1, 2, 3, 6, 7, 8, 9 and 10. Tighten to 9 ft. lbs. (13 Nm). Install the 0.63 inch (16 mm) bolts in locations 4, 5, 11 through 26 and tighten to 9 ft. lbs. (13 Nm).

d. After all the bolts are initially tightened, retighten them to the specification in the order illustrated.

e. After installing the rear timing chain case, check the surface height difference between the rear timing chain case to cylinder block. The measurement should be 0.0094–0.0055 inch (0.24–0.14 mm). If not within specification, repeat the cover installation procedure.

43. Install the timing chain tension guide. Tighten the bolts to 9 ft. lbs. (13 Nm).

44. Position the crankshaft so the No. 1 piston is set at TDC on the compression stroke.

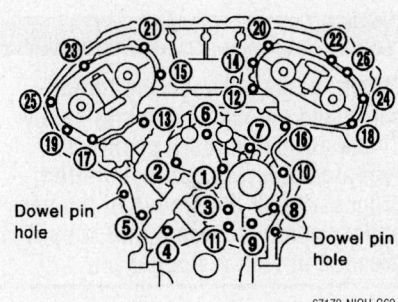

Dowel pin hole

Dowel pin hole

67170-NIQU-G62

Rear timing chain case torque sequence–3.5L engine

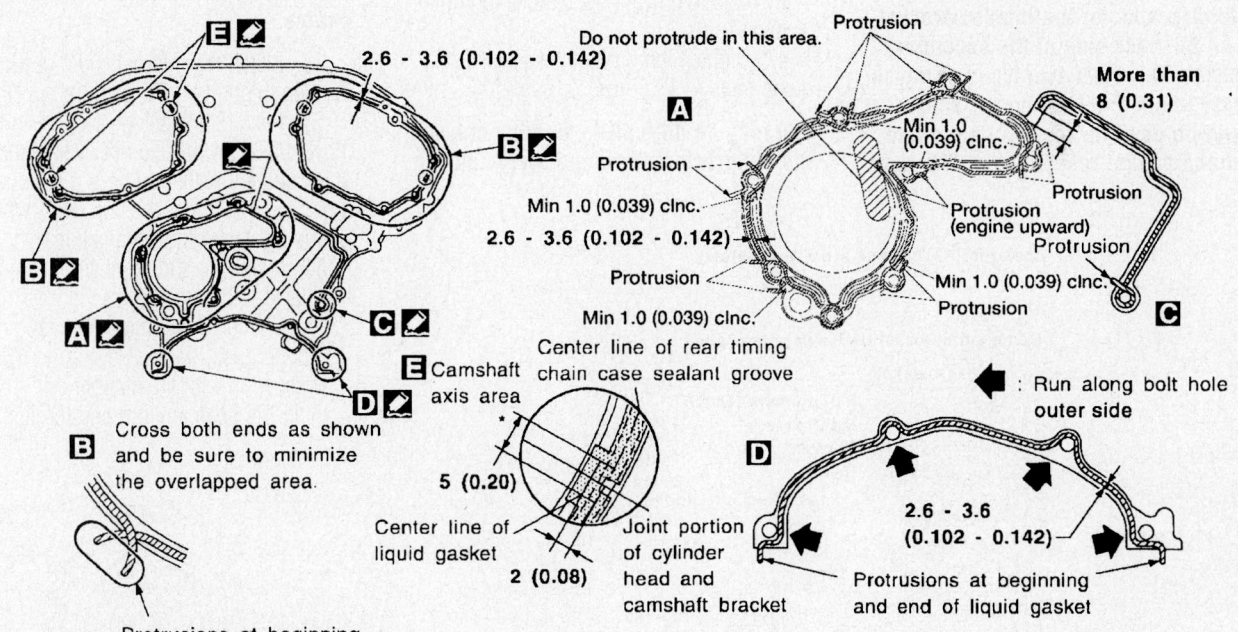

2.6 - 3.6 (0.102 - 0.142)

Do not protrude in this area.

Protrusion

More than 8 (0.31)

Protrusion

Min 1.0 (0.039) clnc.

2.6 - 3.6 (0.102 - 0.142)

Protrusion

Min 1.0 (0.039) clnc.

Protrusion (engine upward)

Protrusion

Protrusion

Min 1.0 (0.039) clnc.

Protrusion

E Camshaft axis area

Center line of rear timing chain case sealant groove

5 (0.20)

Center line of liquid gasket

2 (0.08)

Joint portion of cylinder head and camshaft bracket

B Cross both ends as shown and be sure to minimize the overlapped area.

Protrusions at beginning and end of liquid gasket

: Run along bolt hole outer side

2.6 - 3.6 (0.102 - 0.142)

Protrusions at beginning and end of liquid gasket

* : Apply liquid gasket to the chamfered surface between camshaft bracket and cylinder head.

: Apply liquid gasket. Use Genuine RTV silicone sealant or equivalent. Refer to GI section.

Unit: mm (in)

67170-NIQU-G61

Apply silicone sealant to rear timing chain case as illustrated–3.5L engine

45. Make sure that the dowel pin hole, dowel pin and crankshaft key are located as illustrated. The camshaft dowel pin hole (intake side): at cylinder head upper face side in each bank. The camshaft dowel pin (exhaust side): at cylinder head upper face side in each bank. The crankshaft key: at cylinder head side of right hand bank.

✳✳ CAUTION

The hole on small diameter side must be used for the intake camshaft sprocket dowel pin. Do not misidentify (ignore the big diameter side).

46. Install the secondary timing chains and camshaft sprockets as follows:

✳✳ CAUTION

Match marks between the timing chain and sprockets can slip easily. Check all match mark positions repeatedly during the installation process. Push the sleeve of the secondary chain tensioner and keep it pressed in with a stopper pin.

a. Align the match marks on the secondary timing chain (gold link) with the ones on the intake and exhaust sprockets (stamped), and install them.

➡Match marks for the intake sprocket are on the back side of the secondary sprocket. There are two types of match marks, round and oval types. They should be used for the right hand and left hand banks, respectively. The right

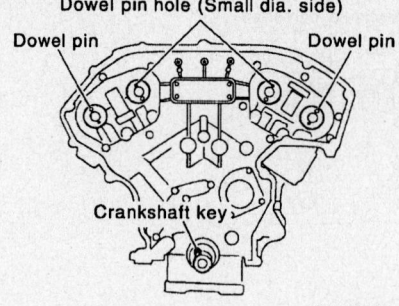

Dowel pin hole (Small dia. side)
Dowel pin   Dowel pin
Crankshaft key

67170-NIQU-G63

Make sure that the dowel pin hole, dowel pin and crankshaft key are located as illustrated—3.5L engine

hand banks use round type and the left hand bank use oval type.

b. Align the dowel pin and pin hole on the camshaft with the groove and dowel pin on the sprocket, and install them.

c. On the intake side, align the pin hole on the small diameter side of the camshaft front end with the dowel pin on the back side of the camshaft sprocket, and install them.

d. On the exhaust side, align the dowel pin on the camshaft front end with the pin groove on the camshaft sprocket, and install them.

e. Tighten the camshaft sprocket bolts by hand to prevent the dislocation of the dowel pins.

47. Tighten the timing chain tension guide bolts to 16 ft. lbs. (21 Nm).

➡It may be difficult to visually check the dislocation of mating marks during

and after installation. To make the matching easier, make a mating mark on the sprocket teeth in advance with paint.

48. After confirming the mating marks are aligned, tighten the camshaft sprocket bolts to 76 ft. lbs. (102 Nm).

49. Remove the stopper pins out from the timing chain tensioners, on secondary timing chains.

50. Install the crankshaft sprocket on the crankshaft. Make sure the mating marks on the crankshaft sprocket face the front of the engine.

51. Install the primary timing chain as follows:

a. Install the primary timing chain so the mating mark (punched) on the camshaft sprocket is aligned with the copper link on the timing chain, while the mating mark (notched) on the crankshaft sprocket is aligned with the gold link on the timing chain, as illustrated.

➡When it is difficult to align mating marks of the primary timing chain with each sprocket, gradually turn the camshaft using a wrench on the hexagonal portion to align it with the mating marks. During alignment, be careful to prevent dislocation of mating mark alignments of the secondary timing chains.

52. Install the internal chain guide. Tighten the bolts to 72 inch lbs. (8 Nm).

53. Install the slack guide and tighten to 12 ft. lbs. (15 Nm). Do not over–tighten the slack guide installation bolt. It is normal for a gap to exist under the bolt seats when the bolt is tightened to specification.

54. Install the timing chain tensioner for the slack guide as follows:

a. When installing the chain tensioner, push in the sleeve and keep it pressed in with the stopper pin.

b. Remove any dirt and foreign mate-

Example: Right bank side (Rear view)

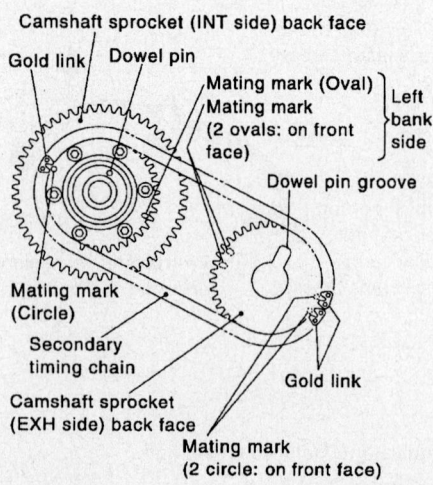

Camshaft sprocket (INT side) back face
Gold link   Dowel pin
Mating mark (Oval)   } Left
Mating mark   } bank
(2 ovals: on front   } side
face)
Dowel pin groove
Mating mark (Circle)
Secondary timing chain
Gold link
Camshaft sprocket (EXH side) back face
Mating mark (2 circle: on front face)

67170-NIQU-G64

Match marks for the intake sprocket are on the back side of the secondary sprocket. There are two types of match marks, round and oval types—3.5L engine

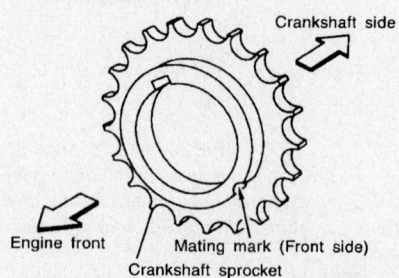

Crankshaft side
Engine front   Mating mark (Front side)
Crankshaft sprocket

67170-NIQU-G65

Make sure the mating marks on the crankshaft sprocket face the front of the engine—3.5L engine

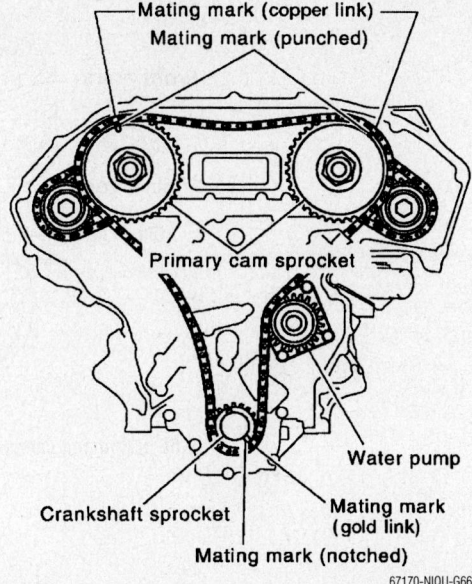

67170-NIQU-G66

Install the primary timing chain so the match mark (punched) on the cam sprocket is aligned with the copper link on the timing chain, while the mating mark (notched) on the crank sprocket is aligned with the gold link on the timing chain–3.5L engine

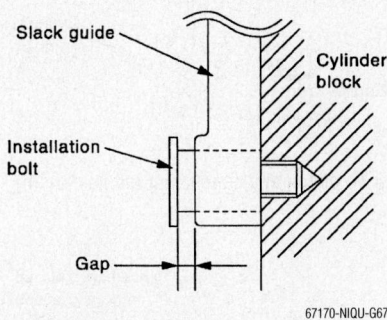

67170-NIQU-G67

It is normal for a gap to exist under the bolt seats when the internal chain guide bolt is tightened –3.5L engine

rials completely from the back and the mounting surfaces of the chain tensioner.

   c. Tighten the bolts to 72 inch lbs. (8 Nm).

   d. After installation, pull out the stopper pin by pressing the slack guide.

55. Confirm that the match marks on the sprockets and the timing chain have not slipped out of alignment.

56. Install new O–rings on the rear timing chain case.

57. Install the front oil seal on the front timing chain case using a suitable tool. Apply clean engine oil to the oil seal edges. Install the seal so that each seal lip is oriented as illustrated. Using a suitable drift, press-fit oil seal until it becomes flush with timing chain case end face. Make sure the garter spring in the oil seal is in position and seal lip is not inverted.

58. Install the front timing cover case.

59. Install IVT control valve covers as follows:

   a. Install new collared O–rings in front cover oil hole on both sides.

   b. Install new seal rings on the IVT control covers.

   c. Apply a 0.083–0.122 inch (2.1–3.1mm) bead of RTV sealant to the IVT control covers as illustrated.

60. Being careful not to move the seal ring from the installation groove, align the dowel pins on the chain case with the holes to install the IVT control covers.

61. Tighten the intake valve timing control cover bolts in the order illustrated to 100 inch lbs. (11 Nm).

62. Apply a 0.091–0.130 inch (2.3–3.3mm) bead of liquid gasket to the water pump cover and the chain tensioner cover and install covers. Tighten the bolts on the pump and cover to 97 inch lbs. (11 Nm).

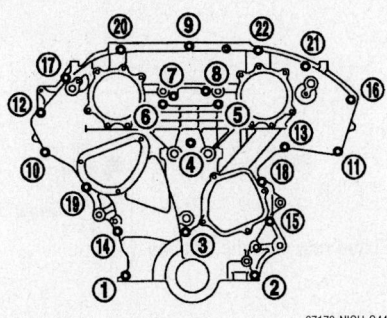

67170-NIQU-G44

Tighten the front timing chain case bolts in the order shown–3.5L engine

63. Install crankshaft pulley lubricate the pulley bolt with clean engine oil and tighten the bolt in two steps:

   a. Step 1: 32 ft. lbs. (44 Nm).

   b. Step 2: 60–65 degrees clockwise.

64. Rotate crankshaft pulley in a clockwise direction as viewed from front of the engine to confirm it turns smoothly.

65. Install the rocker covers in the reverse order of removal, make sure to perform the following:

66. Apply sealant to the areas on the front corners.

67. Tighten the rocker cover bolts in two steps in the sequence illustrated:

   a. Step 1: 17 inch lbs. (1.96 Nm).

   b. Step 2: 74 inch lbs. (8 Nm).

68. Install the remaining components is in reverse order of removal.

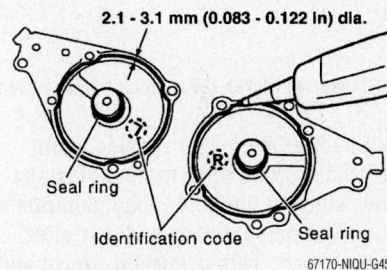

67170-NIQU-G45

Apply a 0.083–0.122 inch (2.1–3.1mm) bead of RTV sealant to the IVT control covers–3.5L engine

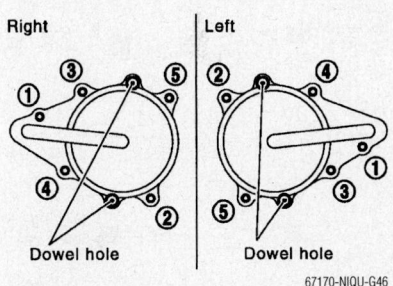

67170-NIQU-G46

Tighten the intake valve timing control cover bolts in this order–3.5L engine

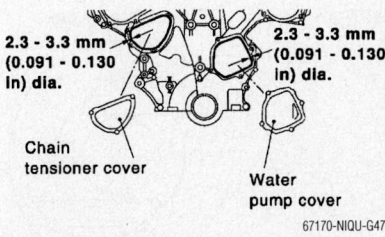

67170-NIQU-G47

Apply a 0.091–0.130 inch (2.3–3.3mm) bead of liquid gasket to the water pump cover and the chain tensioner cover–3.5L engine

RH rocker cover bolt tightening sequence

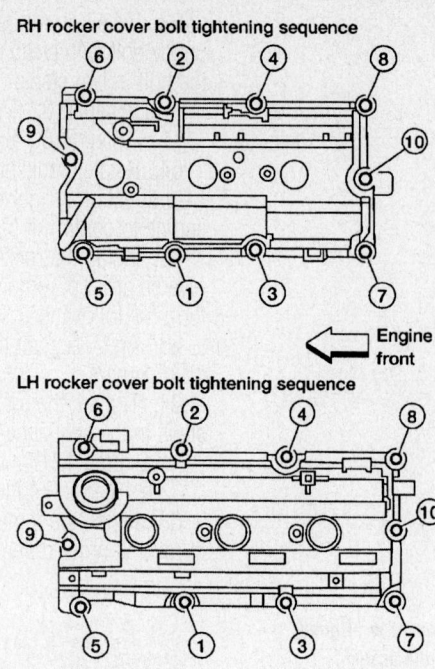

Engine front

LH rocker cover bolt tightening sequence

67170-NIQU-G26

Tighten the rocker cover bolts in this sequence–3.5L engine

→ **If hydraulic pressure inside chain tensioner drops after removal/installation, slack in the guide may generate a pounding noise during and just after engine start. This is normal. Noise will stop after hydraulic pressure rises.**

Piston and Ring Positioning

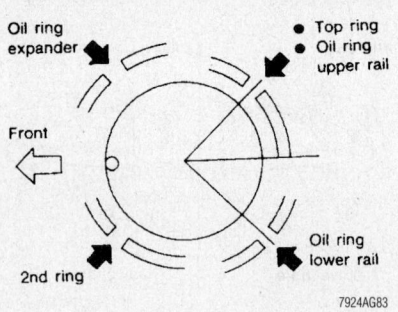

3.3L engines piston ring end-gap spacing

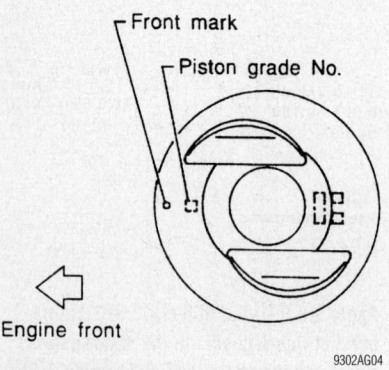

3.3L engine piston positioning

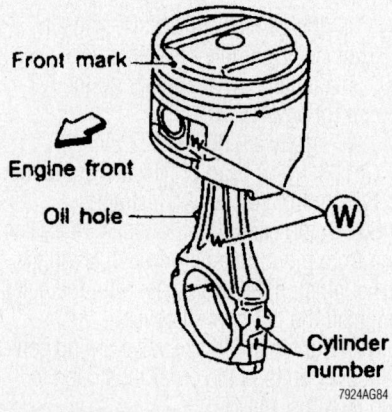

3.3L engines piston and connecting rod assembly positioning

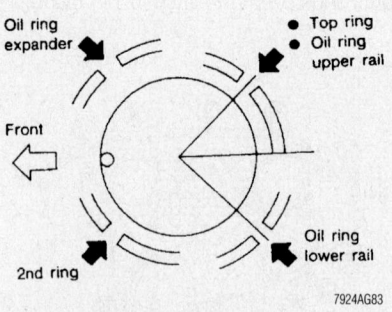

Piston ring end-gap spacing

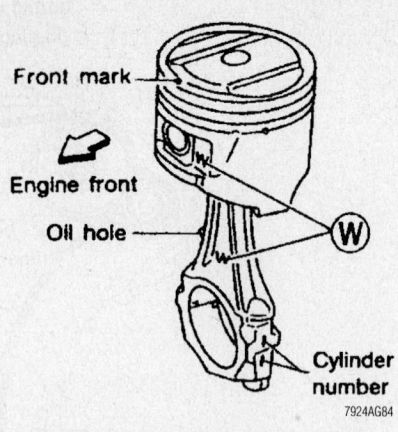

3.5L piston and connecting rod positioning

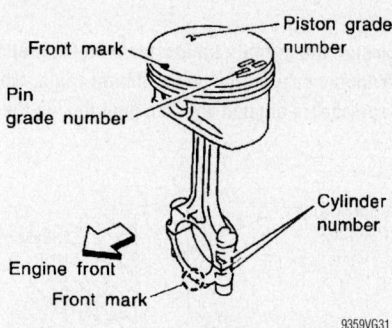

3.5L piston and connecting rod positioning

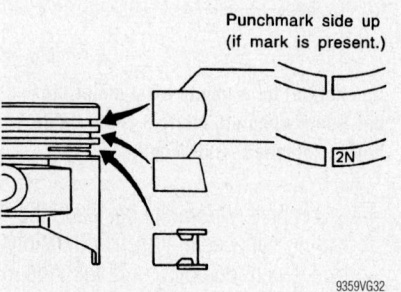

3.5L piston ring positioning

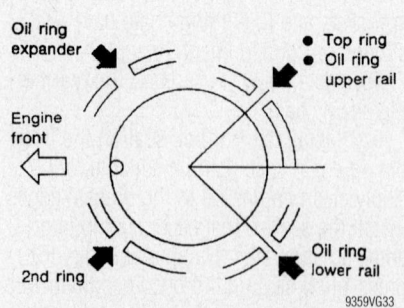

3.5L piston ring positioning

FUEL SYSTEM

Fuel System Service Precautions

Safety is the most important factor when performing not only fuel system maintenance but any type of maintenance. Failure to conduct maintenance and repairs in a safe manner may result in serious personal injury or death. Maintenance and testing of the vehicle's fuel system components can be accomplished safely and effectively by adhering to the following rules and guidelines.

• To avoid the possibility of fire and personal injury, always disconnect the negative battery cable unless the repair or test procedure requires that battery voltage be applied.

• Always relieve the fuel system pressure prior to disconnecting any fuel system component (injector, fuel rail, pressure regulator, etc.), fitting or fuel line connection. Exercise extreme caution whenever relieving fuel system pressure, to avoid exposing skin, face and eyes to fuel spray. Please be advised that fuel under pressure may penetrate the skin or any part of the body that it contacts.

• Always place a shop towel or cloth around the fitting or connection prior to loosening to absorb any excess fuel due to spillage. Ensure that all fuel spillage (should it occur) is quickly removed from engine surfaces. Ensure that all fuel soaked cloths or towels are deposited into a suitable waste container.

• Always keep a dry chemical (Class B) fire extinguisher near the work area.

• Do not allow fuel spray or fuel vapors to come into contact with a spark or open flame.

• Always use a back-up wrench when loosening and tightening fuel line connection fittings. This will prevent unnecessary stress and torsion to fuel line piping. Always follow the proper torque specifications.

• Always replace worn fuel fitting O-rings with new. Do not substitute fuel hose or equivalent, where fuel pipe is installed.

Fuel System Pressure

RELIEVING

1. Before servicing the vehicle, refer to the precautions section.
2. Remove the left side engine compartment relay panel cover.
3. Locate and remove the fuel pump relay from the relay panel.

4. Start the engine.
5. Allow the engine to run until it stalls from fuel starvation. After the engine stalls, crank the engine over 2 more times to ensure all pressure has been released.
6. Turn the ignition switch to the **OFF** position and install the fuel pump relay.
7. Most service work that follows fuel pressure relief also requires that the negative battery cable (ground) be disconnected before service work begins. This also prevents accidental fuel pump energizing that could pressurize the system.

Fuel Filter

REMOVAL & INSTALLATION

3.3L Engine

IN-LINE—EXCEPT CALIFORNIA

1. Before servicing the vehicle, refer to the precautions section.
2. Relieve the fuel system pressure using the recommended procedure.
3. Disconnect the negative battery cable.
4. Raise and safely support the vehicle.
5. Remove the fuel hose clamps.
6. Disconnect and plug the hoses to prevent leakage.
7. Remove the fuel filter from the bracket.

To install:

8. Install the fuel filter into the bracket with the arrow facing up, in the direction of the fuel travel to the engine.
9. Reconnect the fuel hoses.
10. Install and tighten the hose clamps. Verify that the clamps are properly tightened. System operating pressure is approximately 36 psi (248 kPa) and fuel will leak is connections are not properly made.
11. Lower the vehicle.
12. Reconnect the negative battery cable.
13. Check for leaks.

IN-LINE—CALIFORNIA

1. Before servicing the vehicle, refer to the precautions section.
2. Relieve the fuel system pressure using the recommended procedure.
3. Disconnect the negative battery cable.
4. Raise and safely support the vehicle.
5. Remove the filter splash shield bolts.
6. Disconnect the lines from each end of the filter and plug the hoses to prevent leakage.

7. Loosen the filter bracket nuts.
8. Remove the fuel filter from the bracket.

To install:

9. Install the fuel filter into the bracket with the arrow facing forward. Tighten the bracket bolts to 44 inch lbs. (5 Nm).
10. Reconnect the fuel hoses.
11. Lower the vehicle.
12. Reconnect the negative battery cable.
13. Check for leaks.

Fuel Pump

REMOVAL & INSTALLATION

3.3L Engine

1. Before servicing the vehicle, refer to the precautions section.
2. Properly relieve the fuel system pressure.
3. Disconnect the negative battery cable.
4. Raise and safely support the vehicle.
5. Remove the fuel tank as follows:
   a. Drain the fuel from the tank.
   b. Remove the filler protector.
   c. Disconnect the filler tube.
   d. Detach any electrical connectors related to the fuel pump and fuel level sending unit.
   e. Detach the fuel line quick connectors.
   f. Safely support the fuel tank.
   g. Remove the tank mounting straps, then lower the tank out of the vehicle.
6. Remove the 6 fuel pump bolts.
7. Lift the fuel pump out of the fuel tank. Use care. The fuel level sensor and fuel pump and bracket must be tipped to remove it from the fuel tank. Do not lift the fuel sensor and pump assembly straight out of the fuel tank or damage to the level sensor may occur.
8. Remove the 2 bolts attaching the level sensor to the fuel pump.
9. Remove the fuel pump level sensor and the gasket.
10. Discard the gasket.
11. Remove the fuel pump from the bracket.

To install:

12. Position the fuel level sensor on the fuel pump and bracket and install the 2 bolts.
13. Install a new level sensor gasket. Carefully install the level sensor and pump assembly.

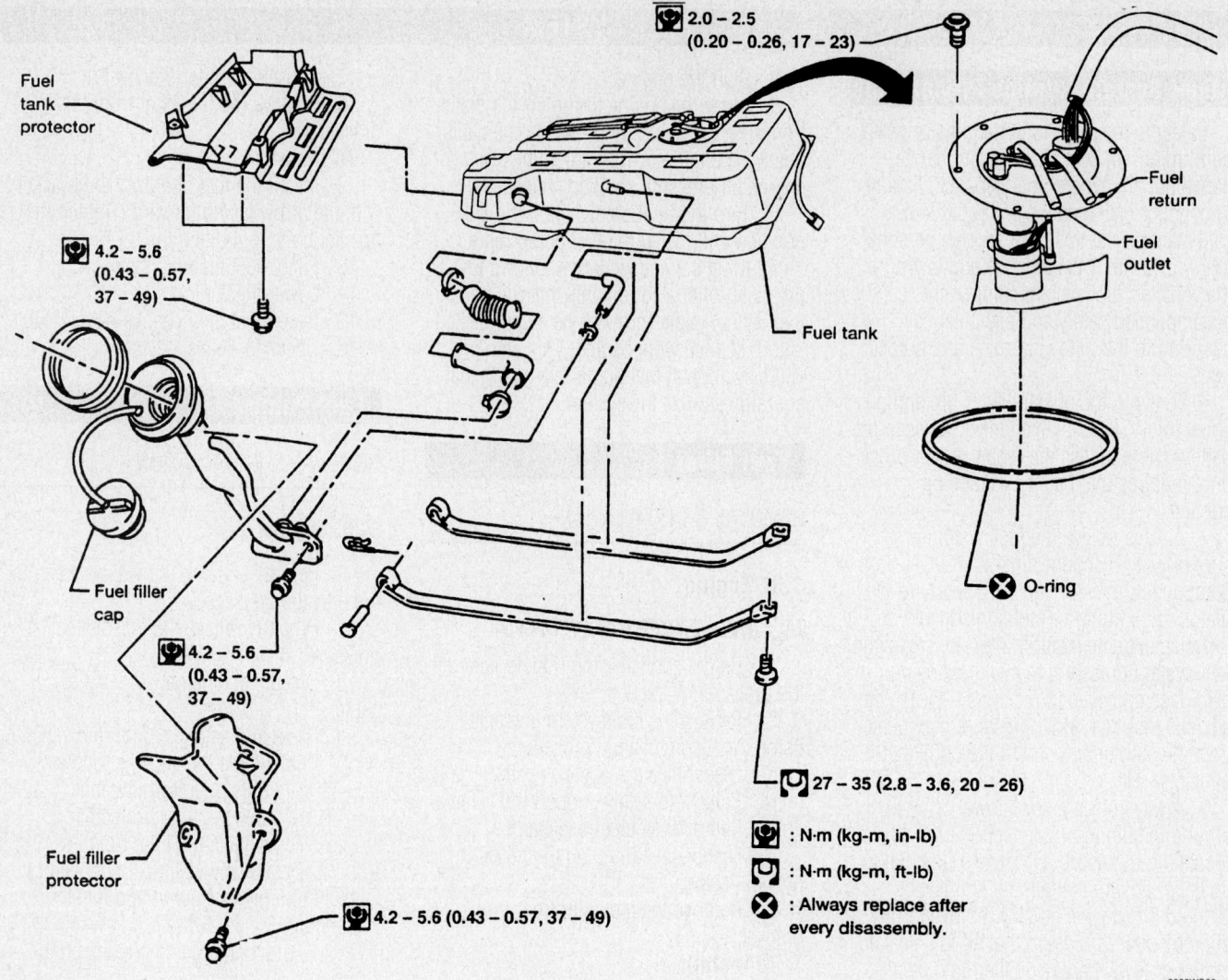

Fuel tank protector

☒ 2.0 – 2.5 (0.20 – 0.26, 17 – 23)

☒ 4.2 – 5.6 (0.43 – 0.57, 37 – 49)

Fuel return

Fuel outlet

Fuel tank

Fuel filler cap

☒ 4.2 – 5.6 (0.43 – 0.57, 37 – 49)

Fuel filler protector

☒ 4.2 – 5.6 (0.43 – 0.57, 37 – 49)

⊗ O-ring

[☒] 27 – 35 (2.8 – 3.6, 20 – 26)

☒ : N·m (kg-m, in-lb)

[☒] : N·m (kg-m, ft-lb)

⊗ : Always replace after every disassembly.

9302WG03

Fuel tank and related components

14. Install the 6 fuel pump bolts. Do not over-tighten the bolts. Tighten the bolts to just 17–23 inch lbs. (2–3 Nm).

15. Install the fuel tank in the reverse order of removal, be sure to tighten the tank mounting straps to 20–26 ft. lbs. (27–35 Nm).

16. Lower the vehicle. Refill the fuel tank as required.

17. Connect the negative battery cable. Verify that the fuel pump relay has been properly installed. Start the engine and check for proper operation.

3.5L Engine

1. Before servicing the vehicle, refer to the precautions section.

2. Disconnect the negative battery cable.

3. Drain the fuel tank.

4. Open the fuel door and unscrew the fuel filler cap to release the pressure inside the fuel tank.

5. Release the fuel system pressure

6. Remove the center exhaust tube, with mufflers attached.

7. Disconnect the parking brake cables from the equalizer, then disconnect the three parking brake cable mounting brackets on each cable and position the cables out of the way.

8. Remove the fuel tank protector.

9. Disconnect the fuel filler hose, recirculation hose and EVAP canister hose at the fuel tank.

10. Disconnect the fuel tank mounting straps while supporting the fuel tank.

11. Lower the fuel tank to access the top of the fuel level sensor unit, fuel filter, and fuel pump assembly.

12. Disconnect the fuel level sensor unit, fuel filter, and fuel pump assembly electrical connector, and the fuel feed hose from the fuel level sensor unit, fuel filter, and fuel pump assembly.

13. Remove the fuel tank.

14. Remove the lock ring using a socket drive handle and tool J–16214.

15. Remove the fuel level sensor, fuel filter, and fuel pump assembly from the fuel tank.

❋❋ CAUTION

Do not bend the float arm during removal.

16. Make sure the fuel level sensor, fuel filter, and fuel pump is free from defects and foreign materials.

To install:

17. Install the fuel level sensor, fuel filter, and fuel pump assembly with the fuel feed hose facing the front of the vehicle.

18. Turn the lock ring until the lock ring is fully rotated into the fuel tank lock tabs as illustrated.

19. Install the fuel tank in the reverse order of removal.

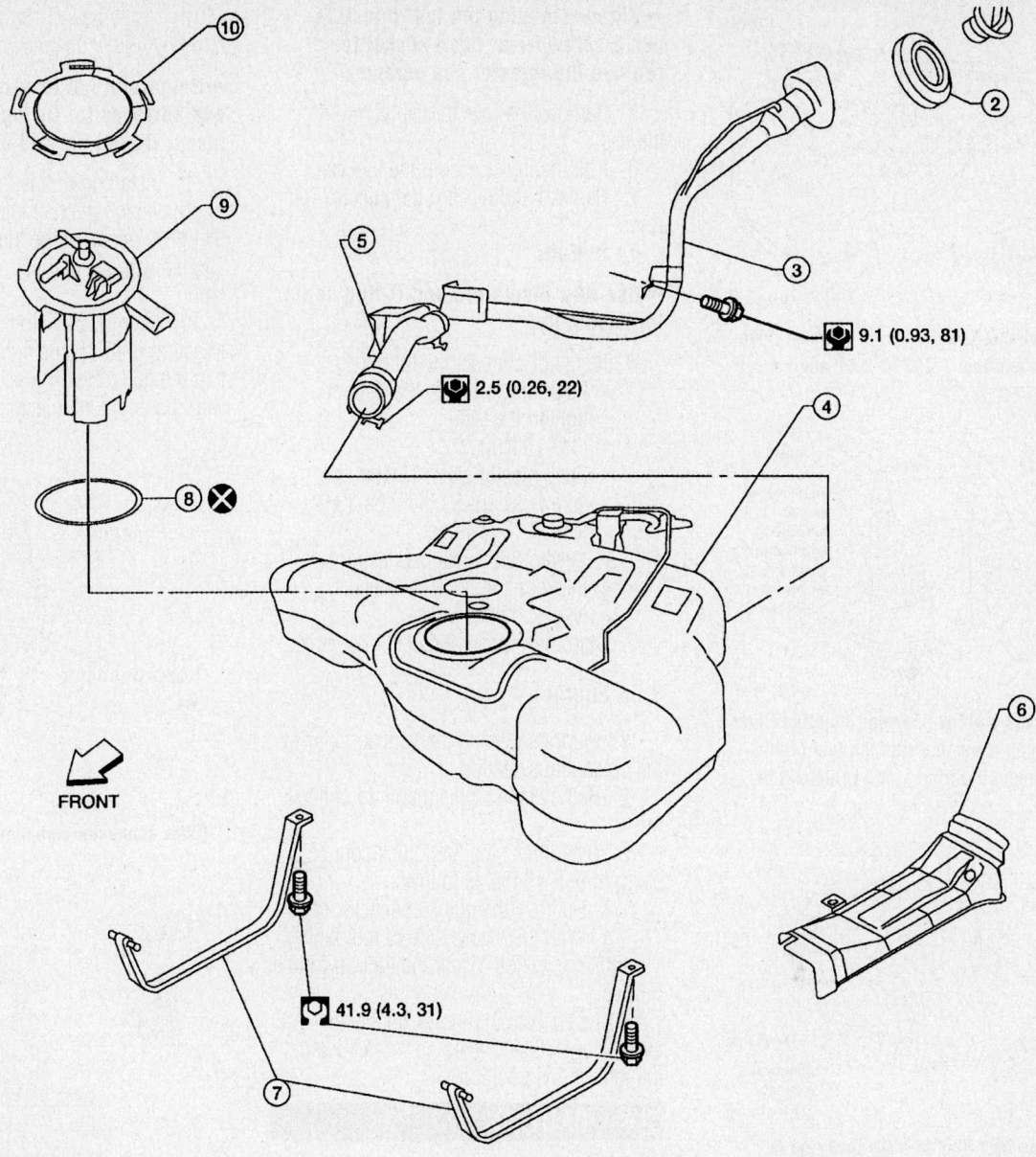

FRONT

9.1 (0.93, 81)

2.5 (0.26, 22)

41.9 (4.3, 31)

N·m (kg-m, in-lb)

N·m (kg-m, ft-lb)

Always replace after every disassembly

1. Fuel filler cap	2. Grommet	3. Fuel filler tube
4. Fuel tank	5. Fuel filler hose	6. Fuel tank protector
7. Fuel tank mounting straps	8. O-ring	9. Fuel level sensor unit, fuel filter, and fuel pump assembly
10. Lock ring		

67170-NIQU-G77

Exploded view of the fuel tank and pump assembly–3.5L engine

20. Before tightening the fuel tank mounting straps, temporarily install the filler hose, recirculation hose, and signal hose. Tighten the straps to 31 ft. lbs. (41 Nm).

21. Connect the quick connector as follows:

a. Check the connection for damage or any foreign materials.

b. Align the connector with the tube, then insert the connector straight into the tube until a click is heard.

c. After the tube is connected, make sure the connection is secure by pulling

on the tube and the connector to make sure they are securely connected.

22. Turn the ignition switch to ON with the engine OFF to check the connections for fuel leaks with the electric fuel pump applying fuel pressure to the fuel lines.

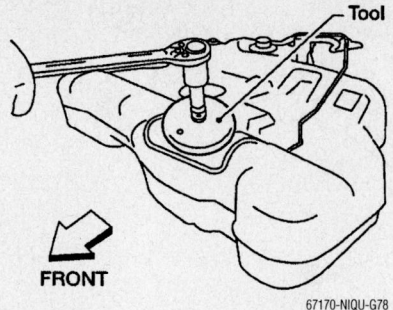

Remove the lock ring using a socket drive handle and tool J–16214–3.5L engine

67170-NIQU-G78

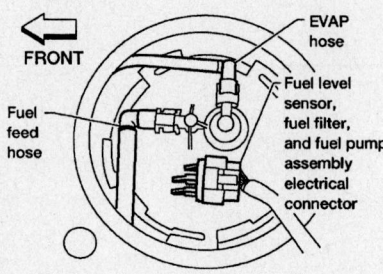

Install the fuel level sensor, fuel filter, and fuel pump assembly with the fuel feed hose facing the front of the vehicle–3.5L engine

67170-NIQU-G79

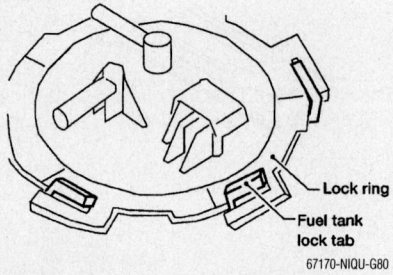

Turn the lock ring until the lock ring is fully rotated into the fuel tank lock tabs–3.5L engine

67170-NIQU-G80

23. Start the engine and let it idle to check that there are no fuel leaks at the fuel system tube and hose connections.

Fuel Injector

REMOVAL AND INSTALLATION

3.3L Engine

1. Before servicing the vehicle, refer to the precautions section.
2. Disconnect the negative battery cable.
3. If removing a rear injector, remove the upper intake manifold.
4. Disengage the injector electrical connection.

➡ When removing the fuel injectors, use a screwdriver head socket to remove the injector cap screws.

5. Remove the injector cap screws and the cap.
6. Pull the injector from the fuel rail.
7. Remove and discard the injector O-rings.

To install:

➡ Use new insulators and O-ring seals for assembly.

8. Install or connect the following:
   • Fuel injectors with the rail and tighten the fasteners to 8–11 ft. lbs. (11–15 Nm)
   • Fuel injector caps. Tighten the screws to 26–33 inch lbs. (3–4 Nm).
   • Injector electrical connections.
   • Intake manifold, if removed.
   • Negative battery cable.
9. Start the vehicle and check for leaks.

3.5L Engine

1. Before servicing the vehicle, refer to the precautions section.
2. Remove the intake manifold collector.
3. Disconnect the fuel quick connector using tool J–45488 as follows:
   a. Remove the quick connector cap.
   b. With the sleeve side of tool facing quick connector, install the tool on to fuel tube.
4. Insert Tool into quick connector until sleeve contacts and goes no further. Hold the tool in that position.

✳ CAUTION

Inserting the tool hard will not disconnect quick connector. Hold tool where it contacts and will go no further.

   a. Pull the quick connector straight out from the fuel tube. Pull the quick connector holding it at the (A) position, as illustrated.

➡ Do not pull with lateral force applied as the O-ring inside the quick connector may be damaged.

5. Remove the fuel rail with the fuel injectors attached, from the intake manifold.
6. Remove the fuel injector O-rings and use new O-rings for installation.

To install:

7. Install the fuel rails with fuel injectors attached.

8. Install new O-rings. Lubricate the O-rings by lightly coating with new engine oil.

➡ Be careful not to damage the O-rings and surfaces for O-ring sealing surfaces. Do not expand or twist O-rings.

9. Install new clips, position the clips in grooves on the fuel injectors. Make sure that protrusions of the fuel injectors are aligned with the cutouts of clips after installation.
10. After properly inserting the fuel injectors onto the fuel tube assembly, check that the fuel tube protrusions are engaged with those of fuel injectors, and the flanges

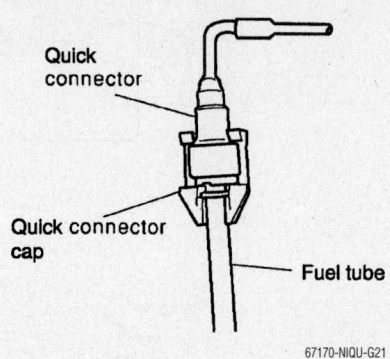

67170-NIQU-G21

Quick connector components–3.5L engine

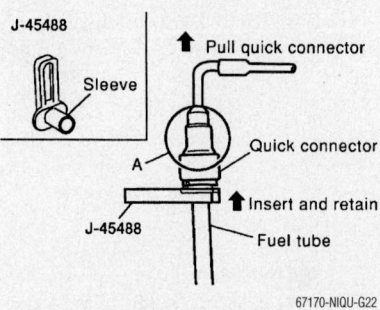

67170-NIQU-G22

Use tool J–45488 to disconnect the quick connector–3.5L engine

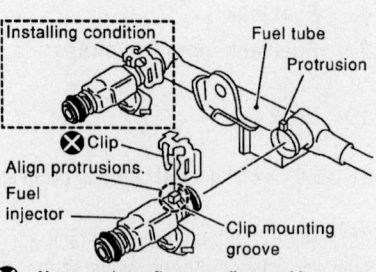

67170-NIQU-G23

Position clips in grooves on the fuel injectors. Make sure that protrusions of fuel injectors are aligned with cutouts of clips after installation–3.5L engine

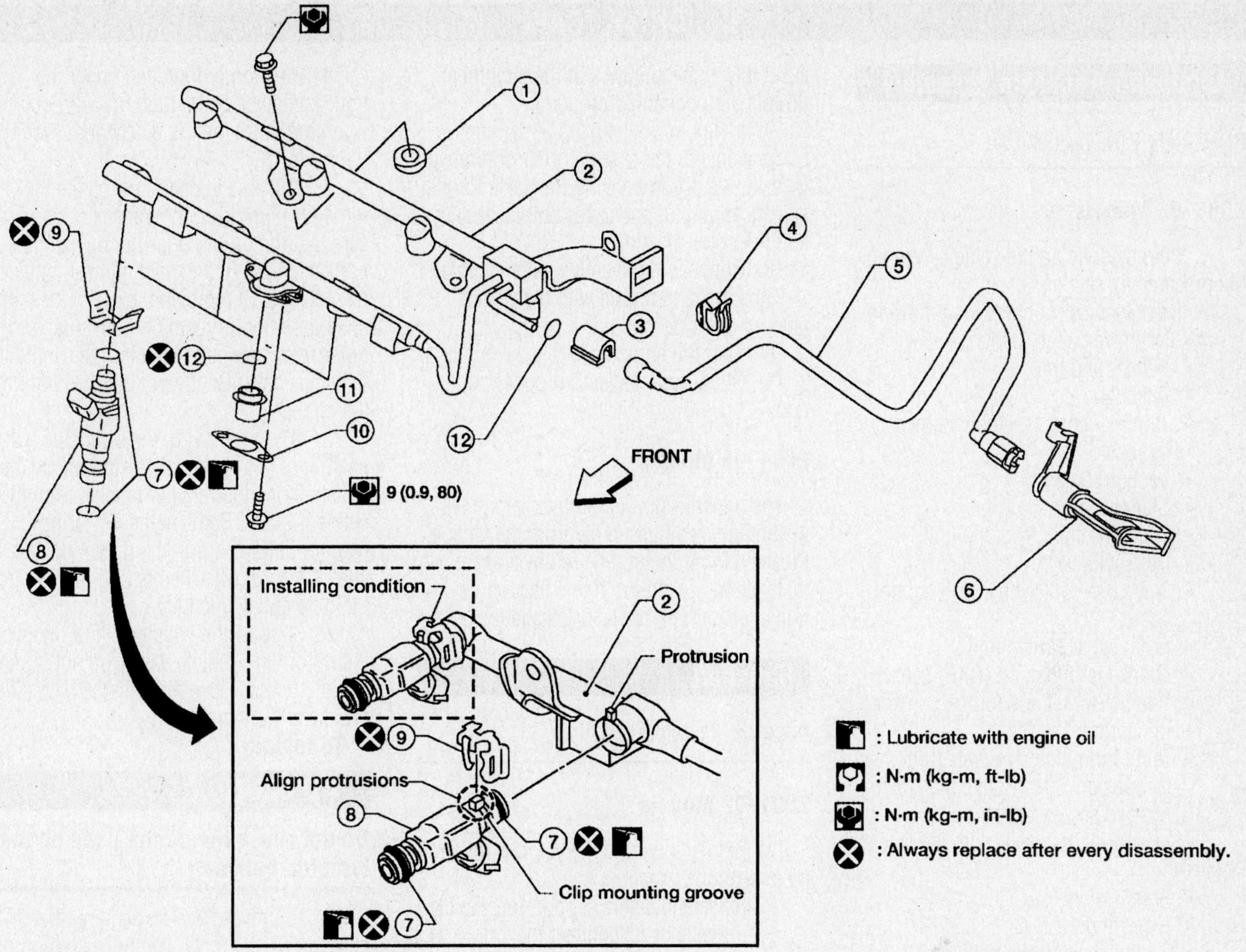

FRONT

9 (0.9, 80)

Installing condition

Protrusion

Align protrusions

Clip mounting groove

: Lubricate with engine oil

: N·m (kg-m, ft-lb)

: N·m (kg-m, in-lb)

: Always replace after every disassembly.

1. Insulator	2. Fuel tube assembly	3. Connector cap
4. Clip	5. Fuel hose	6. Connector cap
7. O-ring	8. Fuel injector	9. Clip
10. Fuel damper retainer	11. Fuel damper	12. O-ring

67170-NIQU-G20

Exploded view of the fuel rail and injector assembly–3.5L engine

of the fuel tube assembly are fully engaged with the clips.

11. Tighten the fuel tube assembly bolts in the sequence illustrated, in two steps:
    a. Step 1: 89 inch lbs. (10 Nm).
    b. Step 2: 16 ft. lbs. (22 Nm).

12. Install the quick connector as follows:
    a. Make sure no dirt or foreign objects are around the tube and quick connector to avoid damage.
    b. Align the center to insert the quick connector straight onto the fuel tube.
    c. Insert the fuel tube until a click is heard.

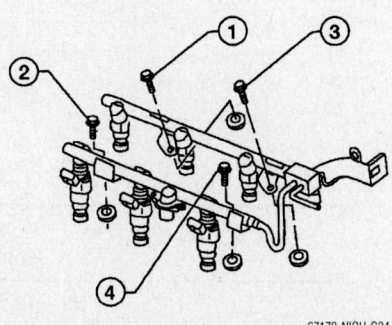

67170-NIQU-G24

Tighten the fuel tube assembly bolts in this sequence–3.5L engine

13. Install the remaining components is in the reverse of removal.

14. Make sure there is no fuel leakage at connections as follows:
    a. Apply fuel pressure to fuel lines by turning ignition switch **ON** with the engine **OFF** and check for fuel leaks at connections.
    b. Start the engine and rev it up and check for fuel leaks at connections. Use mirrors for checking on connections out of the direct line of sight.

DRIVE TRAIN

Automatic Transaxle Assembly

REMOVAL & INSTALLATION

2001–02 Models

1. Before servicing the vehicle, refer to the precautions section.
2. Remove or disconnect the following:
   - Negative battery cable
   - Battery and tray
   - Resonator
   - Terminal cord assembly harness connector
   - Vacuum lines
   - Starter motor
   - Transaxle fluid from the unit
   - Halfshafts
   - Transaxle cooler hose and control cable
   - Front exhaust manifold
   - Crankshaft Position (CKP) sensor
   - Engine gusset and torque converter undercover
   - Bolts from the drive plate from the torque converter. Rotate the crankshaft to access all the bolts.
3. Support the transaxle with a suitable jack.
   - Front mount
   - Rear mount
   - Bolts attaching the transaxle to the engine
4. Carefully separate the transaxle assembly from the engine assembly. Lower the assembly from the vehicle.

To install:

5. Be sure that the transaxle is secured firmly to the transaxle jack.
6. Carefully raise the transaxle into the vehicle and align the transaxle to the engine assembly, making sure that the alignment dowels are positioned properly.
7. Install or connect the remaining components in the reverse order of removal. Refer to the accompanying transaxle torque specification illustration for bolt locations and their specifications.
8. Connect the negative battery cable.
9. Fill the transaxle with the correct amount and type of fluid.
10. Start the engine.
11. Check for leaks and proper operation.

2004–06 Models

The transmission is removed along with the engine as an assembly from the vehicle. Please refer to the engine removal and installation procedure in this section for transmission removal and installation

Halfshaft

REMOVAL & INSTALLATION

2001–02 Models

1. Before servicing the vehicle, refer to the precautions section.
2. Raise and safely support the vehicle.
3. Remove or disconnect the following:
   - Wheel
   - Fender splash shield
   - Cotter pin, nut retainer, and the hub retainer washers from the front hub assembly
   - Lower ball joint from the knuckle
   - Sway bar from the lower control arm at the sway bar link nut
   - Halfshaft and CV-joint from the wheel hub

4. Position a drain pan under the transaxle since some fluid may run out when the inner joint is disengaged from the transaxle.
5. A prybar is used to separate the inner CV-joint from the transaxle. Use great care that the prybar does not damage the transaxle case, differential oil seal, outer race or boot. If removing the left side halfshaft, position prybars on both sides of the outer race, between the outer race and the transaxle case. Gently pry outward to unseat the circlip.
6. When removing the right side halfshaft, it is not be necessary to remove the halfshaft bearing retainer bracket from the cylinder block. Remove the 3 bearing retainer bolts and pull the right side halfshaft CV-joint with the bearing retainer from the differential side gear.
7. Support the halfshafts and remove them from the vehicle. Use care not to damage the boots. Place the halfshafts on a flat, protected work area.

To install:

❊❊ CAUTION

Do not reuse the circlip used on the left side halfshaft.

8. To prevent over-expanding the circlip, install the circlip carefully, starting one end in the shaft groove, then working the circlip over the CV-joint splined end. Always use a new circlip. No circlip is used on the right side halfshaft.
9. Inspect the CV-joint boots. If service is required, replace the CV-joint boots.
10. Inspect the differential oil seals. If damaged, the factory recommends using a hook-type puller and slide hammer arrange-

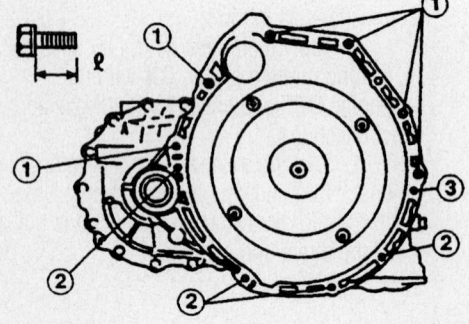

Bolt No.	Tightening torque N·m (kg-m, ft-lb)	ℓ mm (in)
1	39 - 49 (4.0 - 5.0, 29 - 36)	60 (2.36)
2	30 - 40 (3.1 - 4.1, 22 - 30)	25 (0.98)
3*	30 - 40 (3.1 - 4.1, 22 - 30)	25 (0.98)

*: TORX bolt

9348WG17

Transaxle torque specification and locations—3.3L engine

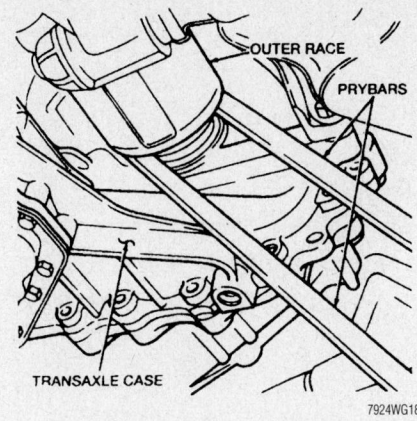

Removing the left side halfshaft by gently prying with 2 prybars to unseat the circlip

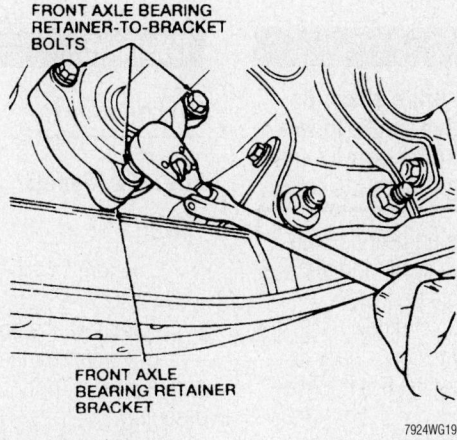

Right side halfshaft bearing retainer bracket

ment to remove the seals. A seal driver is used to install the replacement differential oil seals.

11. If installing the left side halfshaft and CV-joint assembly, position the CV-joint so the splines are aligned with the differential side gear splines, then push the halfshaft joint into the differential case. As the circlip locks into the differential side gear groove, a click will be felt.

12. If installing the right side halfshaft and CV-joint assembly, simply push the

CV-joint into the differential side gear. Position the bearing retainer onto the bearing retainer bracket that should still be on the cylinder block. Install the 3 bolts and tighten to 8–14 ft. lbs. (13–19 Nm).

13. Install or connect the following:
- Halfshaft
- Lower ball joint and tighten the lower ball joint stud nut to 52–63 ft. lbs. (71–86 Nm). Secure the nut with a new cotter pin.
- Sway bar link to the lower control

arm and tighten the link nut to 12–16 ft. lbs. (16–22 Nm).
- Wheel outer bearing retainer, washer and axle nut. Tighten the hub nut to 174–231 ft. lbs. (235–314 Nm). Install the nut retainer and secure with a new cotter pin.
- Splash shield
- Wheel. Tighten the lug nuts to 72–87 ft. lbs. (98–118 Nm).

14. Lower the vehicle.
15. Check the transaxle fluid level.
16. Road test the vehicle to verify correct operation and no noise or vibration.

2004–06 Models

LEFT SIDE

1. Before servicing the vehicle, refer to the precautions section.
2. Remove the wheel.
3. Remove the ABS sensor from the steering knuckle.
4. Remove the cotter pin, then remove the lock nut from the drive shaft.
5. Remove the brake hose lock plate, then remove the brake hose from the strut.
6. Remove the lower ball joint pinch bolt, then separate the lower ball joint from the steering
knuckle.
7. Using a puller, remove the drive shaft from the wheel hub and bearing assembly.

✳✳ CAUTION

When removing the drive shaft, do not apply an excessive angle to the drive shaft joint. Also be careful not to excessively extend the slide joint.

8. Remove the drive shaft from the transaxle using a suitable tool and drive shaft puller KV40107500.
9. Move the joint up and down, left and right, and in an axial direction. Check for

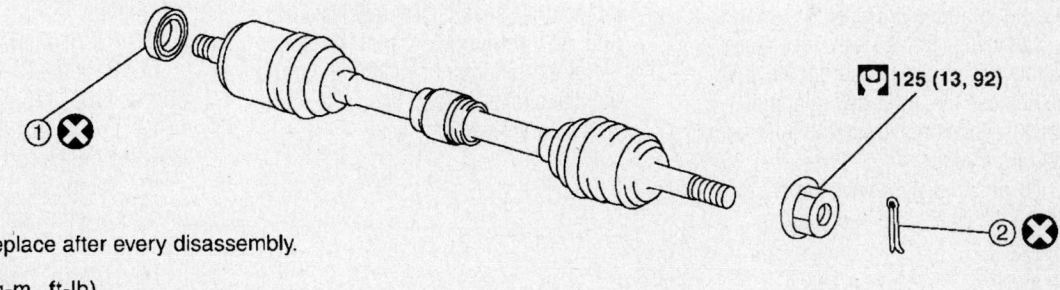

❌ : Always replace after every disassembly.

🔧 : N·m (kg-m, ft-lb)

Exploded view of the left front halfshaft—2004–06 models

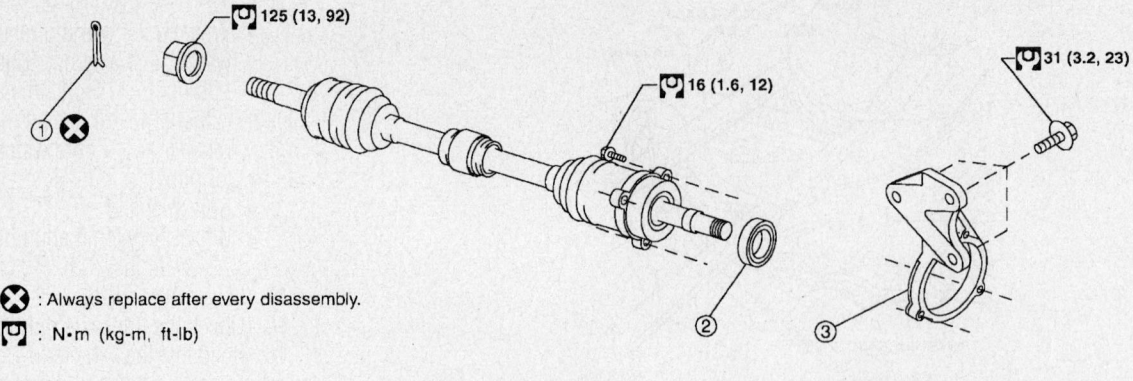

125 (13, 92)

16 (1.6, 12)

31 (3.2, 23)

⊗ : Always replace after every disassembly.

⊡ : N·m (kg-m, ft-lb)

1. Cotter pin
2. Differential side oil seal
3. Support bearing bracket

67170-NIQU-G82

Exploded view of the right front halfshaft—2004–06 models

any rough movement or significant loose-ness.

10. Check the boot for cracks or other damage, and for grease leakage. If dam-aged, disassemble the drive shaft to verify damage, and repair or replace as necessary.

To install:

11. Installation is in the reverse order of removal. Tighten the axle nut to 92 ft. lbs. (125 Nm).

12. Please note the following:

a. Be sure to replace the differential side oil seal with a new one every time drive shaft is removed on 4 speed transaxle models.

b. Install new circlip on the drive shaft in the circular clip groove on the transaxle side. Make sure the new circlip on the drive shaft is securely fastened. After its insertion, try to pull the flange out of the slide joint by hand. If it pulls out, the circlip is not properly meshed with the transaxle side gear.

RIGHT SIDE

1. Before servicing the vehicle, refer to the precautions section.

2. Remove the wheel.

3. Remove the ABS sensor from the steering knuckle.

4. Remove the cotter pin, then remove the lock nut from the drive shaft.

5. Remove the brake hose lock plate, then remove the brake hose from the strut.

6. Remove the lower ball joint pinch bolt, then separate the lower ball joint from the steering knuckle.

7. Using a puller, remove the drive shaft from the wheel hub and bearing assembly.

※※ CAUTION

When removing the drive shaft, do not apply an excessive angle to the drive shaft joint. Also be careful not to excessively extend the slide joint.

8. Remove the support bearing bolts.

9. Remove the drive shaft from the transaxle using a suitable prytool.

10. Move the joint up and down, left and right, and in an axial direction. Check for any rough movement or significant loose-ness.

11. Check the boot for cracks or other damage, and for grease leakage. If dam-aged, disassemble drive shaft to verify dam-age, and repair or replace as necessary.

To install:

12. Installation is in the reverse order of removal. Tighten the axle nut to 92 ft. lbs. (125 Nm) and the support bearing bolts to 23 ft. lbs. (31 Nm).

13. Please note the following:

a. Be sure to replace the differential side oil seal with a new one every time drive shaft is removed on 4 speed transaxle models.

b. Install new circlip on the drive shaft in the circular clip groove on transaxle side. Make sure the new circlip on the drive shaft is securely fastened. After its insertion, try to pull the flange out of the slide joint by hand. If it pulls out, the circlip is not properly meshed with the transaxle side gear.

CV-Joint

OVERHAUL

2001–02 Models

INNER

1. Before servicing the vehicle, refer to the precautions section.

2. Remove the boot bands.

3. Matchmark the slide joint housing and inner race, prior to separating the joint assembly.

4. Pry off the snapring and remove the ball cage, inner race and balls as a unit.

5. Remove the snapring and withdraw the boot.

To install:

➡ **Cover the halfshaft serrations with tape, so as not to damage the boot.**

6. Thoroughly clean all parts in solvent and dry with compressed air. Check parts for evidence of damage and replace as nec-essary.

7. Install the boot and new boot band on the halfshaft.

8. Install a new inner snapring.

9. Install the ball cage, inner race and balls as a unit. Confirm that the matchmarks are aligned.

10. Install a new outer snapring.

11. Pack the CV-joint with 5.0–6.0 ounces (165–175 g) of grease.

12. Ensure that the boot is properly installed on the halfshaft groove.

13. Set the boot so that it does not swell or deform when its length is 3.86 in. (98mm).

14. Lock the new boot bands securely.

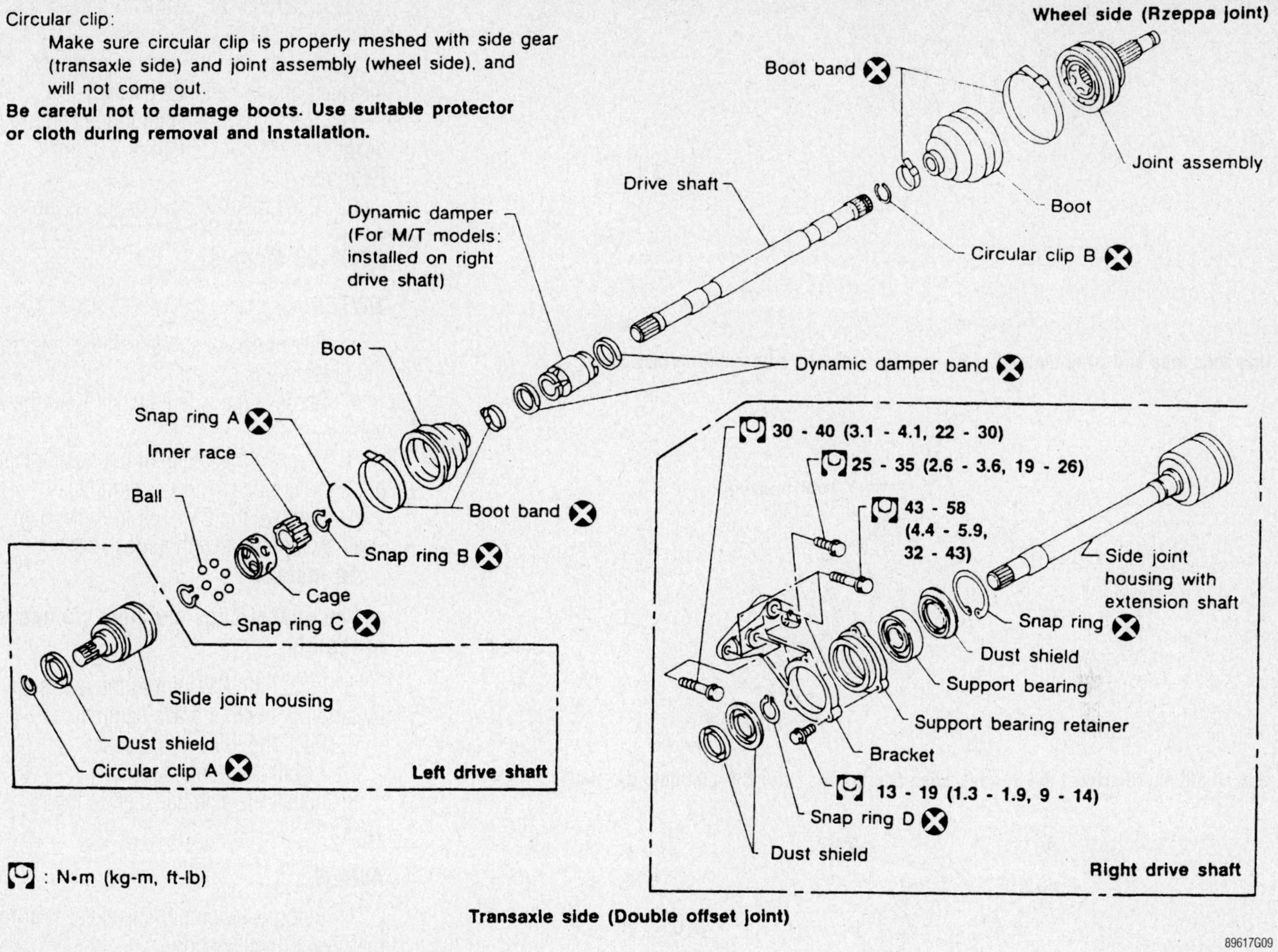

Circular clip:
Make sure circular clip is properly meshed with side gear (transaxle side) and joint assembly (wheel side), and will not come out.
Be careful not to damage boots. Use suitable protector or cloth during removal and installation.

Wheel side (Rzeppa joint)

Boot band ⊗

Joint assembly

Boot

Drive shaft

Circular clip B ⊗

Dynamic damper (For M/T models: installed on right drive shaft)

Dynamic damper band ⊗

Boot

Snap ring A ⊗

Inner race

Ball

Boot band ⊗

Snap ring B ⊗

Cage

Snap ring C ⊗

Slide joint housing

Dust shield

Circular clip A ⊗

Left drive shaft

⊙ 30 - 40 (3.1 - 4.1, 22 - 30)

⊙ 25 - 35 (2.6 - 3.6, 19 - 26)

⊙ 43 - 58 (4.4 - 5.9, 32 - 43)

Side joint housing with extension shaft

Snap ring ⊗

Dust shield

Support bearing

Support bearing retainer

Bracket

⊙ 13 - 19 (1.3 - 1.9, 9 - 14)

Snap ring D ⊗

Dust shield

Right drive shaft

⊙ : N•m (kg-m, ft-lb)

Transaxle side (Double offset joint)

89617G09

Exploded view of the halfshafts and related components

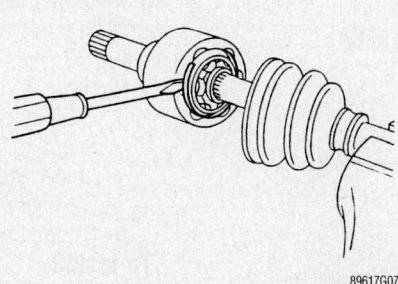

89617G07

The inner CV-joint uses a large C-clip to retain the ball and cage assembly in the outer housing

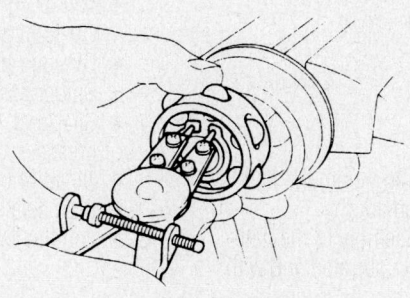

89617G08

After the outer housing is removed, the ball and cage assembly can slide from the shaft by removing the C-clip

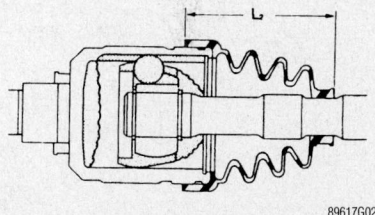

89617G02

Make sure to properly position the boot before tightening the boot clamps

OUTER

1. Before servicing the vehicle, refer to the precautions section.
  The joint on the wheel side cannot be disassembled.

2. Prior to separating the joint assembly, matchmark the halfshaft and joint assembly.

3. Separate the joint using a slide hammer.

4. Remove the boot bands and the boot.

To install:

5. Thoroughly clean all parts in solvent and dry with compressed air. Check parts

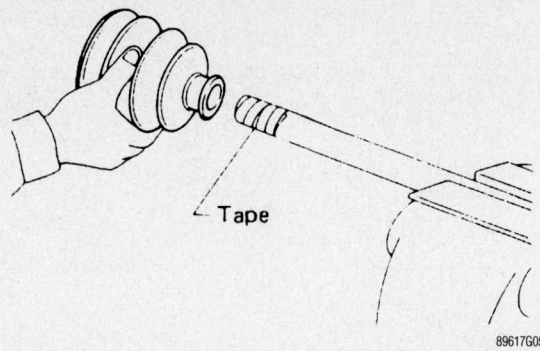

Use vinyl tape and wrap the end of the shaft to protect the boot during installation

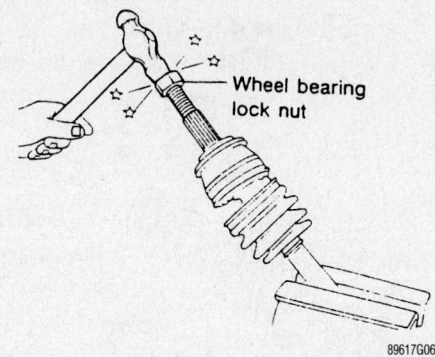

Wheel bearing lock nut

Use an old nut to protect the threads when tapping the outer CV-joint onto the shaft

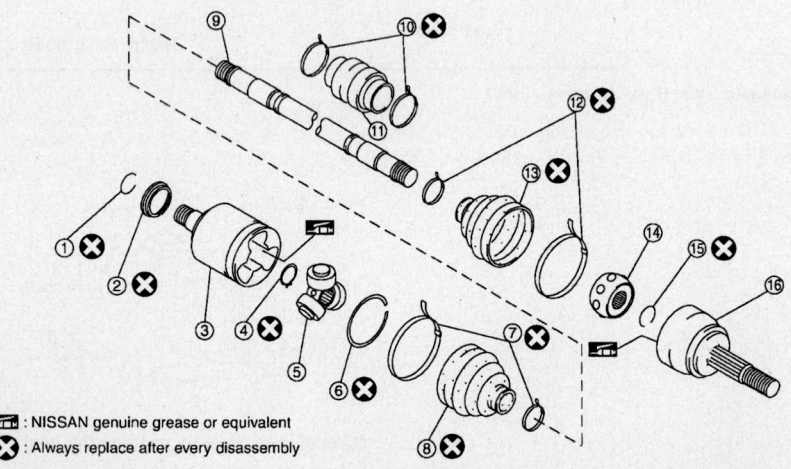

🔧 : NISSAN genuine grease or equivalent

❌ : Always replace after every disassembly

1. Circlip
2. Differential side oil seal
3. Slide joint housing
4. Snap ring
5. Spider assembly
6. Stopper ring
7. Boot band
8. Boot
9. Shaft
10. Damper band (5 A/T)
11. Damper (5 A/T)
12. Boot band
13. Boot
14. Ball cage / Steel ball / Inner race assembly
15. Circlip
16. Joint sub-assembly

67170-NIQU-G83

Left side front halfshaft components exploded view—2004–06 models

for evidence of damage and replace as necessary.

➡**Cover the halfshaft serrations with tape, so as not to damage the boot.**

6. Install the boot and small boot band on the halfshaft.

7. Set the joint assembly onto the halfshaft and align the matchmarks.

8. Attach the joint assembly to the halfshaft by lightly tapping the serrated end with a plastic hammer.

➡**Using a metal hammer may damage the threads on the end of the joint.**

9. Pack the CV-joint with 4.76–5.11 ounces (135–145 g) of grease.

10. Ensure that the boot is properly installed on the halfshaft groove.

11. Set the boot so that it does not swell or deform when its length is 3.82 in. (97mm).

12. Lock the new boot bands securely.

2004–06 Models

OUTER

1. Before servicing the vehicle, refer to the precautions section.

2. Remove the axle halfshaft from the vehicle.

3. Remove the CV-joint boot clamps and push the boot away from the joint.

4. Remove the CV-joint from the axle shaft by tapping it with a brass hammer.

To install:

➡**Use new circlips and boot clamps for assembly.**

5. Install the CV-joint to the axle shaft by tapping it with a brass hammer.

6. Pack the joint with grease.

7. Install the boot clamps.

8. Install the axle halfshaft to the vehicle.

INNER

1. Before servicing the vehicle, refer to the precautions section.

2. Remove the axle halfshaft from the vehicle.

3. Remove the plug seal by tapping around the joint housing flange with a brass hammer.

4. Remove or disconnect the following:
   - CV-joint boot clamps
   - Snapring
   - Spider assembly
   - CV-joint housing
   - CV-joint boot

To install:

➡**Use new snaprings and plug seals for assembly.**

5. Install or connect the following:
   - CV-joint boot
   - CV-joint housing
   - Spider assembly
   - Snapring. Pack the joint with grease.
   - CV-joint boot clamps
   - Plug seal

6. Install the axle halfshaft to the vehicle.

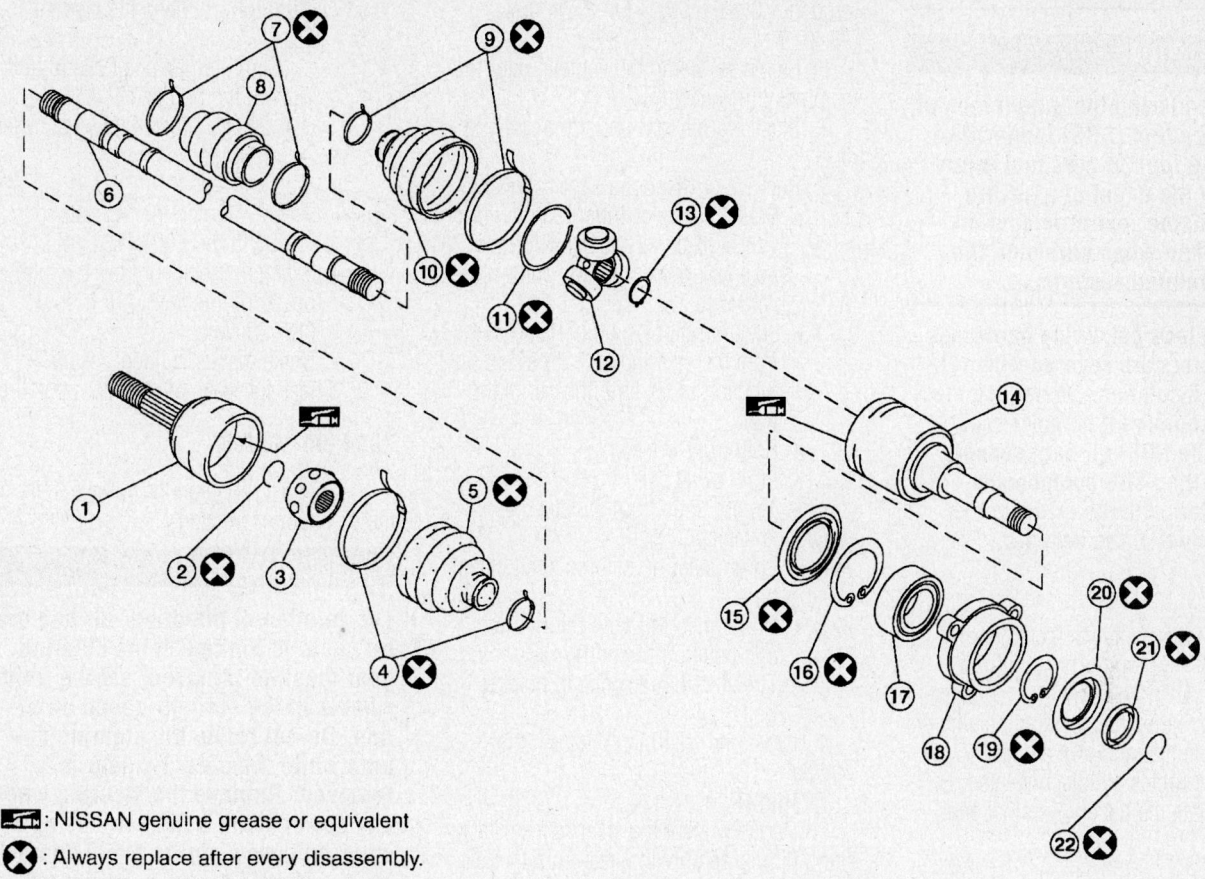

: NISSAN genuine grease or equivalent

: Always replace after every disassembly.

1. Joint sub-assembly	2. Circlip	3. Ball cage / Steel ball / Inner race assembly
4. Boot bands	5. Boot	6. Shaft
7. Damper bands (5 A/T)	8. Damper (5 A/T)	9. Boot band
10. Boot	11. Stopper ring	12 Spider assembly
13. Snap ring	14. Slide joint housing	15. Dust cover
16. Snap ring	17. Bearing	18. Bracket
19. Snap ring	20. Dust cover	21. Differential side oil seal
22. Circlip		

67170-NIQU-G84

Right side front halfshaft components exploded view—2004–06 models

STEERING AND SUSPENSION

Air Bag

PRECAUTIONS

Several precautions must be observed when handling the inflator module to avoid accidental deployment and possible personal injury.

• Never carry the inflator module by the wires or connector on the underside of the module.

• When carrying a live inflator module, hold securely with both hands, and ensure that the bag and trim cover are pointed away.

• Place the inflator module on a bench or other surface with the bag and trim cover facing up.

• With the inflator module on the bench, never place anything on or close to the module which may be thrown in the event of an accidental deployment.

DISARMING

☆☆ CAUTION

To avoid rendering the Supplemental Restraint System (SRS) inoperative, which could lead to personal injury or death in the event of a severe frontal collision, extreme caution must be taken when servicing the electrical related systems.

➡️**All SRS electrical wiring harnesses and connectors are covered with YELLOW outer insulation. Do not use electrical test equipment on any circuit related to the SRS (air bag) sensors. When installing SRS components, always install with the arrow marks facing the front of the vehicle.**

Disarming

To disarm the Supplemental Restraint System (SRS) system turn the ignition switch to the **OFF** position. Then, disconnect the both battery cables starting with the negative cable first and wait at least 10 minutes after the cables are disconnected. Be sure to insulate the battery terminal ends.

Arming

To arm the Supplemental Restraint System (SRS) system turn the ignition switch to **OFF** position. Connect the both battery cables starting with the positive cable first.

➡️**The SRS or air bag system is equipped with a self-diagnostic operation. After turning the ignition key to the ON or START position, the AIR BAG warning lamp will illuminate for 7 seconds. After 7 seconds, the AIR BAG lamp will extinguish if no malfunction is detected. If the AIR BAG lamp does not extinguish after 7 seconds, check the SRS self-diagnostic system for a malfunction.**

Power Rack and Pinion

REMOVAL & INSTALLATION

2001–02 Models

The power steering gear is held in position by 2 steering gear brackets and insulators. Note that the housing may move slightly when the steering wheel is turned. If the housing moves more than 0.080 inch (2mm), replace the steering gear insulators. If one or both of the brackets move, check

the torque of the bracket bolts. The correct torque for these bolts is 54–72 ft. lbs. (73–97 Nm).

1. Before servicing the vehicle, refer to the precautions section.
2. Place a drain pan under the steering rack.
3. Remove or disconnect the following:
   - Brake master cylinder remote reservoir bracket screws. Position the reservoir out of the way and secure with wire.
   - Junction block/high pressure line from the steering rack. Position the junction block and line out of the way.
   - Both front wheels
   - Front sway bar
   - Tie rod ends from the steering knuckles
   - Lower steering column shaft clamp bolt
   - Power steering fluid return hose and position out of the way.
   - The steering rack clamp bracket bolts
4. Lower the steering rack from the vehicle.

   To install:
5. Carefully slide the steering gear rack and pinion assembly in place from the left side of the vehicle. Position the input shaft so it is just below the lower steering column shaft clamp.
6. Raise the steering gear until the plastic aligning tab on the input shaft enters the clamp bolt gap on the lower column shaft. Do not install the clamp bolt yet.
7. Examine the steering gear brackets. They should be marked UP with arrows pointing to one end of the bracket. Be sure the brackets are installed correctly. Tighten the steering gear bracket bolts to 54–72 ft. lbs. (73–97Nm) in sequence, working counterclockwise from the number 1 bolt (upper right side).

8. Install or connect the following:
   - Fluid return line to the steering gear
   - Steering column shaft clamp bolt. Tighten the bolt to 17–22 ft lbs. (24–29 Nm). Install the dust cover.
   - Tie rod ends
   - Stabilizer bar
   - Wheel. Tighten the lug nuts to 72–87 ft. lbs. (98–118 Nm).
   - Junction block. Tighten the high-pressure line to 11–18 ft. lbs. (15–25 Nm).
   - Brake master cylinder reservoir
9. Check for leaks and proper operation.

2004–06 Models

1. Before servicing the vehicle, refer to the precautions section.

☆☆ CAUTION

The rotation of the driver air bag spiral cable is limited. If the steering gear must be removed, set the front wheels in the straight-ahead direction. Do not rotate the steering column while the steering gear is removed. Remove the steering wheel and spiral cable before removing the steering lower joint to avoid damaging the supplemental restraint system spiral cable.

☆☆ CAUTION

Before servicing the air bag system, turn ignition switch OFF, disconnect both battery cables and wait at least 3 minutes. When servicing the SRS, do not work from directly in front of the air bag module.

2. Remove the steering wheel and spiral gear as follows:
   a. Set the front wheels in the straight-ahead position.

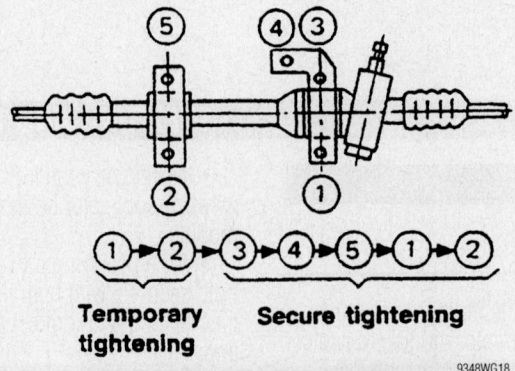

Temporary tightening   Secure tightening

9348WG18

Tighten the power steering rack mounting bolts in the sequence shown

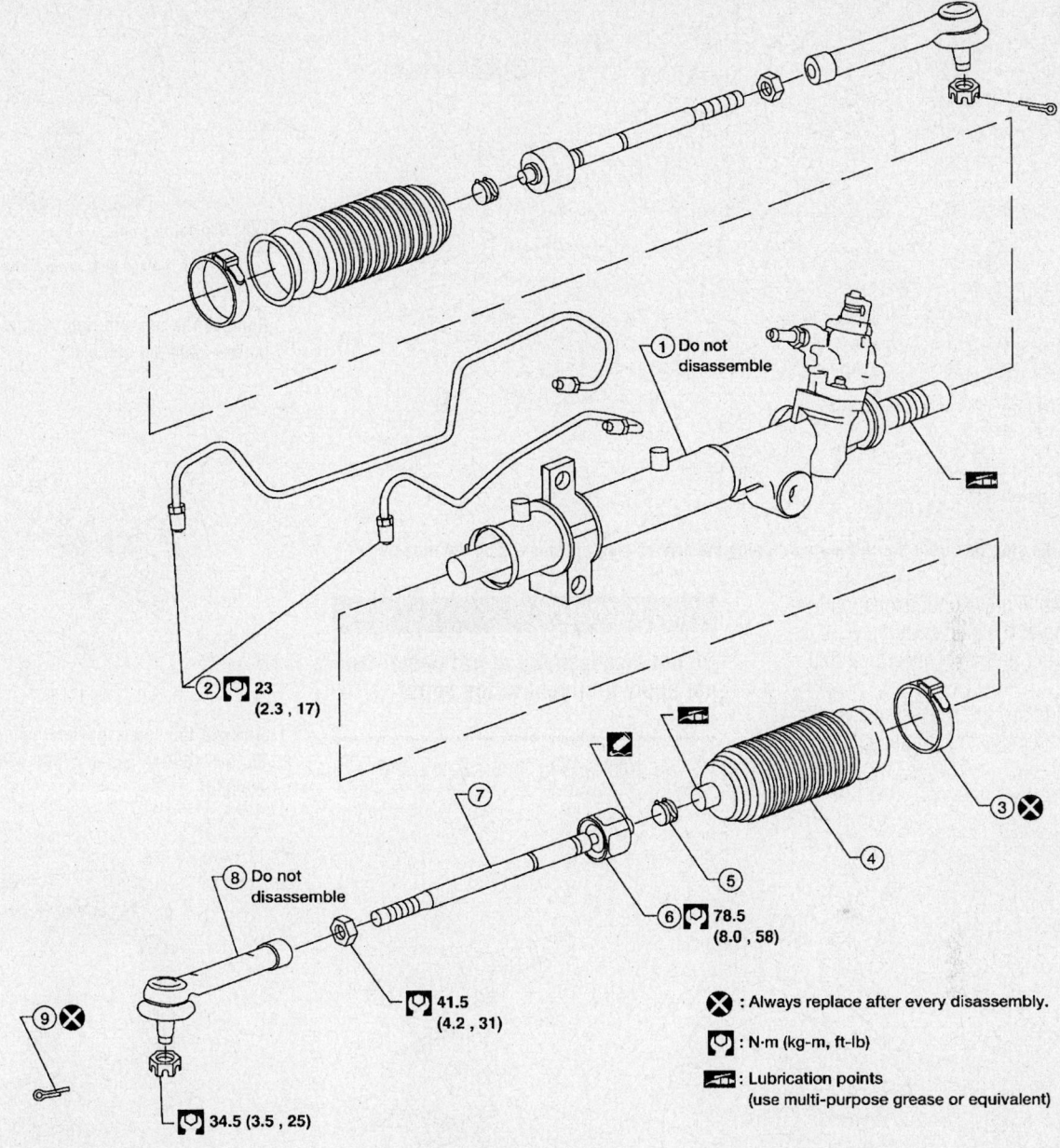

1. Steering gear
2. Gear housing fluid tubes
3. Boot clamp
4. Dust boot
5. Boot band
6. Tie-rod inner socket
7. Inner tie-rod
8. Outer tie-rod
9. Cotter pin

⊗ : Always replace after every disassembly.

N·m (kg-m, ft-lb)

: Lubrication points
(use multi-purpose grease or equivalent)

67170-NIQU-GAD

Exploded view of the steering gear mounting with torque specifications—2004–06 models

b. Remove the side lids from the retainers retaining the driver's side air bag.

c. Remove the left and right side bolts.

d. Lift the driver's side air bag from the steering wheel.

➡**When Disconnecting or connecting the air bag harness or horn connectors, always pull up to release the black**

locking tab prior to removing the connector from the air bag component. **Always push down to lock the black locking tab after installing the connector to the air bag component. When locked, the black locking tab is level with the connector housing.**

e. Disconnect the air bag harness and horn connectors, then remove the driver air bag module.

f. Remove the steering wheel center nut.

g. Remove the steering wheel using puller tools J–1859–A and J–42578.

h. Remove the cupper and lower column covers.

i. Remove the wiper washer switch connector, then pinch the tabs at the wiper and washer switch base and slide the switch away from the steering column.

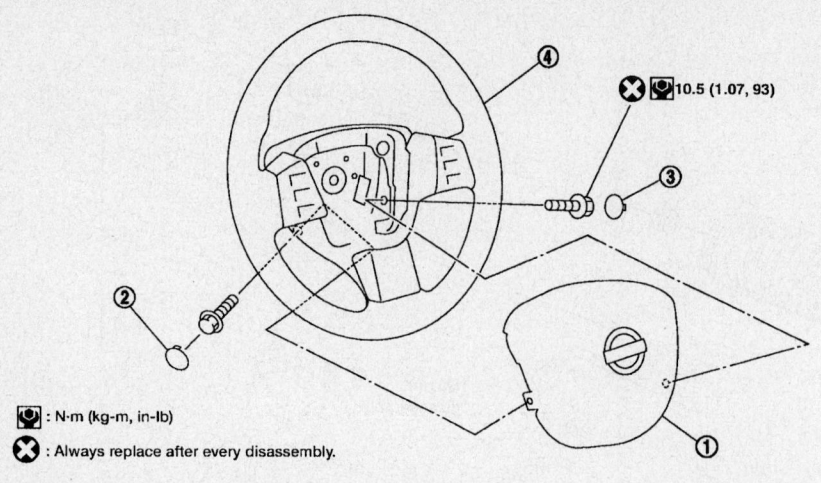

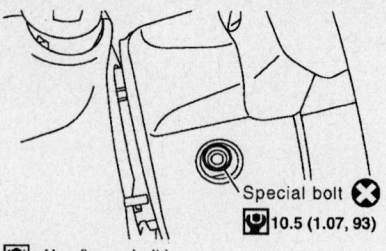

: N·m (kg-m, in-lb)

: Always replace after every disassembly.

Remove the left and right side air bag bolts—2004–06 models

1. Driver air bag module
2. Side lid LH
3. Side lid RH
4. Steering wheel

67170-NIQU-G98

Remove the side lids from the retainers retaining the drivers side air bag—2004–06 models

j. While pressing the tabs, pull the light and turn signal switch toward the driver door and disconnect from the base.

k. Remove the screws, release the tab, and remove the spiral cable.

❊❊ CAUTION

Do not disassemble spiral cable. Do not apply lubricant to the spiral cable.

l. Remove the spiral cable connectors.

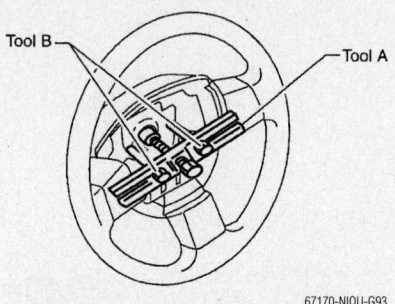

Remove the steering wheel using puller tools J–1859–A and J–42578—2004–06 models

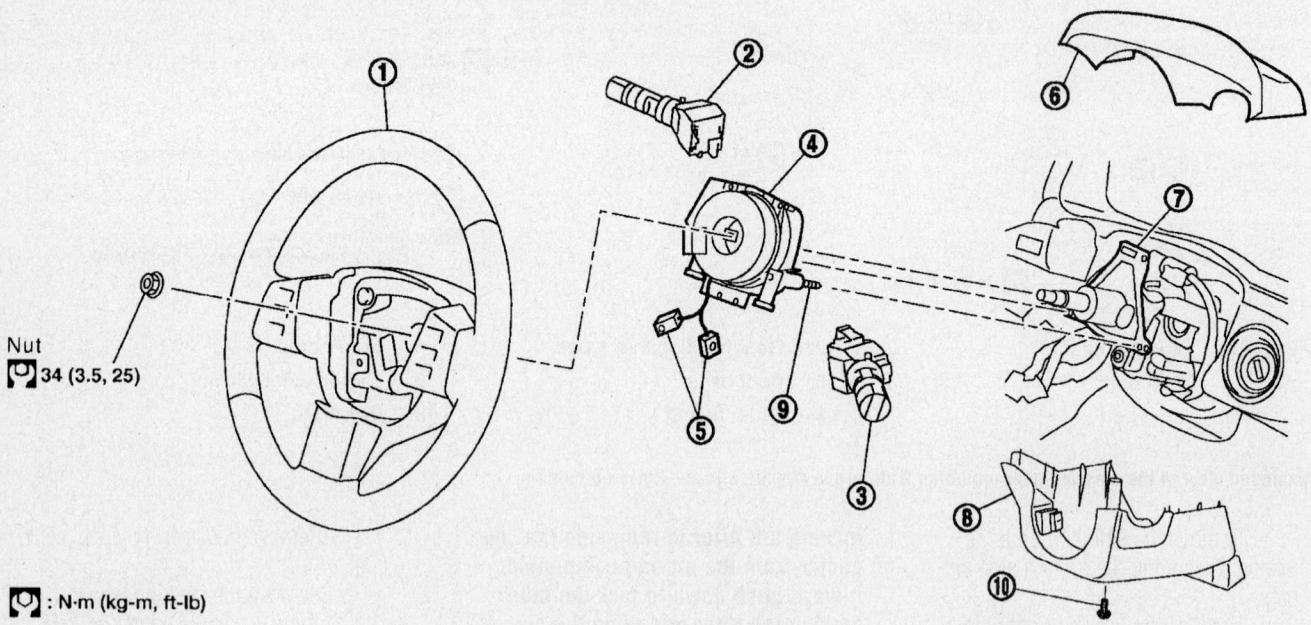

Nut
34 (3.5, 25)

: N·m (kg-m, ft-lb)

1. Steering wheel
4. Spiral cable
7. Column assembly
10. Screw

2. Lighting and turn signal switch
5. Driver air bag module connector
8. Column cover lower

3. Wiper and washer switch
6. Column cover upper
9. Screw (Do not remove)

67170-NIQU-G92

Exploded view of the steering wheel, spiral cable and air bag components—2004–06 models

✳ CAUTION

With the steering linkage disconnected, the spiral cable may snap by turning the steering wheel beyond the limited number of turns. The spiral cable can be turned counterclockwise about 2.5 turns from the right end position.

m. Inspect the steering wheel near the puller holes for damage. If damaged, replace the steering wheel.

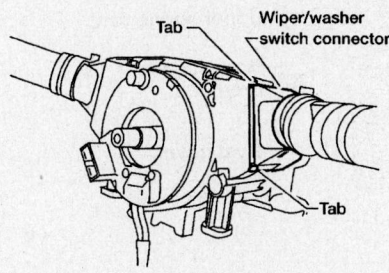

67170-NIQU-G94

Remove wiper washer switch connector, pinch the tabs at the wiper and washer switch base and slide the switch away from the column—2004–06 models

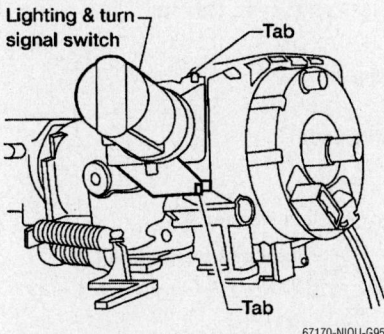

67170-NIQU-G95

While pressing tabs, pull the light and turn signal switch toward the driver door and disconnect from the base—2004–06 models

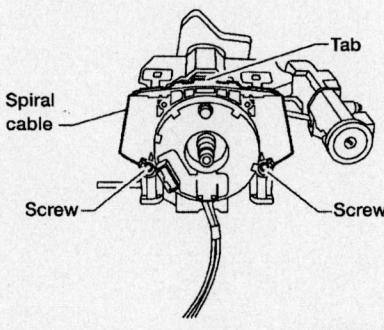

67170-NIQU-G96

Remove the screws, release the tab, and remove the spiral cable—2004–06 models

3. Remove the two front wheels.
4. Remove the cotter pins and nuts, then disconnect the outer tie-rod ends from the knuckle.
5. Disconnect the outer stabilizer bar ends from the connecting rods.
6. Remove the front stabilizer bar bracket rear bolts and loosen the front bolt.
7. Remove the lower joint pinch bolt.
8. Drain the power steering fluid.
9. Disconnect the power steering high and low pressure lines from the steering gear.
10. Position the stabilizer bar up and out of the way.
11. Remove the two gear housing mounting bolts. Do not remove the gear

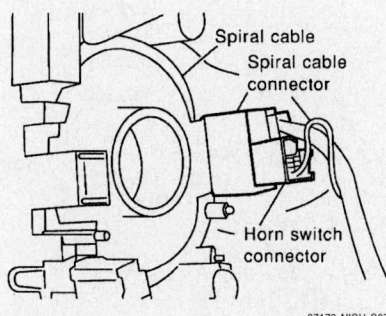

67170-NIQU-G97

Remove the spiral cable connectors—2004–06 models

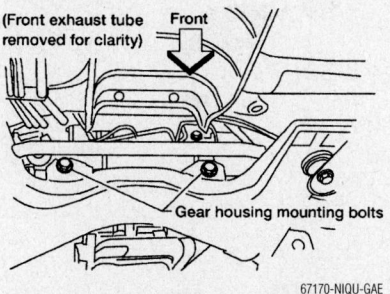

67170-NIQU-GAE

Location of the steering gear housing mounting bolts—2004–06 models

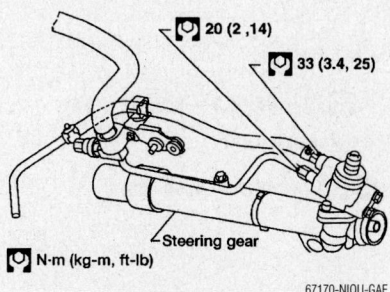

67170-NIQU-GAF

Use the specified tightening torque when installing the high pressure and low-pressure hose connections—2004–06 models

housing mounting bracket from the gear housing.
12. Remove the power steering gear and linkage assembly.
To install:
13. Installation is in the reverse order of removal, please note the following:
a. Use the specified tightening torque when installing the high pressure and low-pressure hose connections.

✳ CAUTION

Excessive tightening will damage threads of connection or O-ring.

b. The O-ring in low-pressure hose connector is larger than that in high-pressure connector. Take care to install the proper O-ring.
c. Initially, tighten the nut on the tie-rod outer socket and knuckle arm to the specification shown in the exploded view illustration of 25 ft. lbs. (34 Nm). Then tighten further to align nut groove with the first pin hole so that the cotter pin can be installed. The tightening torque must not exceed 36 ft. lbs. (49 Nm).
14. Refill the power steering system and bleed after installation.

MacPherson Strut

REMOVAL & INSTALLATION

2001–02 Models

1. Before servicing the vehicle, refer to the precautions section.
2. Disconnect the negative battery cable.
3. Matchmark the front strut upper mounting bracket and the chassis strut tower.
4. Raise and safely support the vehicle.
5. Remove the front wheel.
6. If equipped, remove the 2 front brake anti-lock sensor cable bracket bolts and position the anti-lock sensor cable out of the way.
7. Detach the brake tube from the strut.
8. Support the control arm.
9. Matchmark the knuckle to the strut so it can installed in the same position. This is important for the camber angle of the front wheel.
10. Remove the strut-to-steering knuckle bolts.
11. Support the strut and remove the 3 upper strut-to-chassis nuts. Remove the strut from the vehicle.

When installing rubber parts, final tightening must be carried out under unladen condition* with tires on ground.
*: Fuel, radiator coolant and engine oil full. Spare tire, jack, hand tools and mats in designated positions.

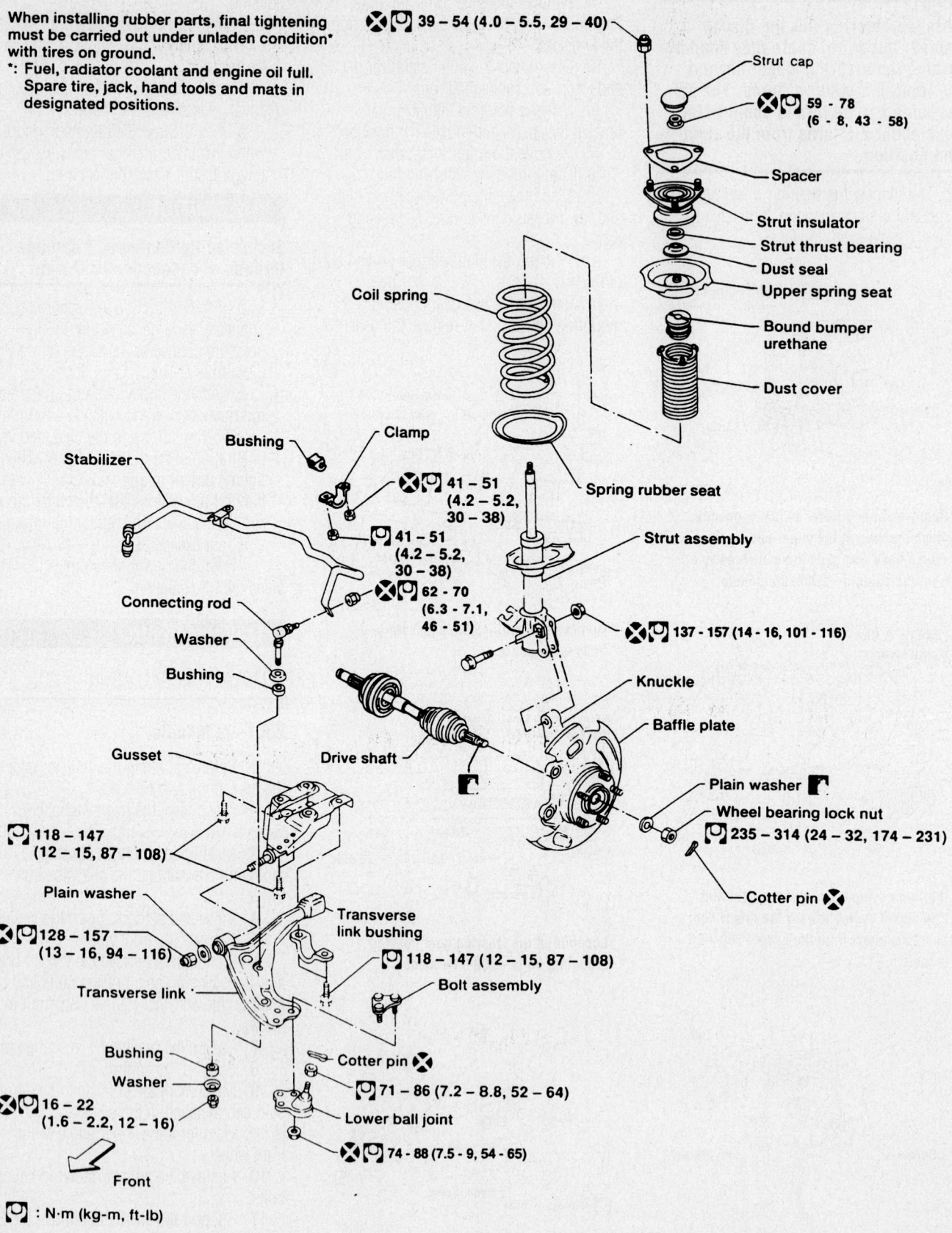

⊗ 39 – 54 (4.0 – 5.5, 29 – 40)

Strut cap

⊗ 59 – 78 (6 - 8, 43 - 58)

Spacer

Strut insulator

Strut thrust bearing

Dust seal

Upper spring seat

Bound bumper urethane

Dust cover

Coil spring

Spring rubber seat

Bushing

Clamp

⊗ 41 – 51 (4.2 – 5.2, 30 – 38)

41 – 51 (4.2 – 5.2, 30 – 38)

Stabilizer

⊗ 62 - 70 (6.3 - 7.1, 46 - 51)

Strut assembly

⊗ 137 - 157 (14 - 16, 101 - 116)

Connecting rod

Washer

Bushing

Knuckle

Baffle plate

Drive shaft

Gusset

Plain washer

Wheel bearing lock nut

⊗ 235 - 314 (24 - 32, 174 - 231)

118 – 147 (12 – 15, 87 – 108)

Plain washer

Cotter pin ⊗

⊗ 128 – 157 (13 – 16, 94 – 116)

Transverse link bushing

118 – 147 (12 – 15, 87 – 108)

Bolt assembly

Transverse link

Bushing

Cotter pin ⊗

Washer

71 – 86 (7.2 – 8.8, 52 – 64)

⊗ 16 – 22 (1.6 – 2.2, 12 – 16)

Lower ball joint

⊗ 74 - 88 (7.5 - 9, 54 - 65)

Front

: N·m (kg-m, ft-lb)

9302WG04

Coil spring and strut assembly—2001–02 models

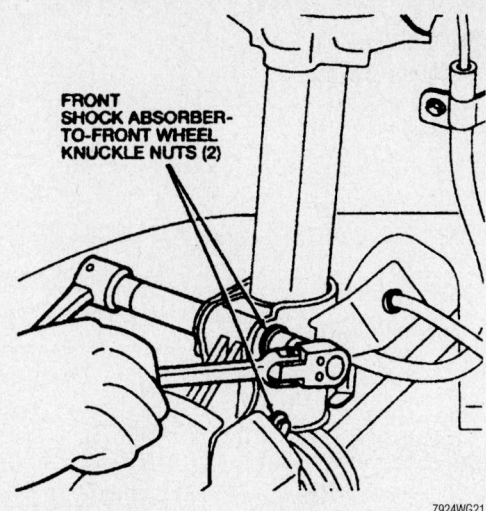

FRONT
SHOCK ABSORBER-
TO-FRONT WHEEL
KNUCKLE NUTS (2)

7924WG21

The strut is attached to the knuckle with 2 large bolts—2001–02 models

✳✳ WARNING

Never loosen the strut center nut until the spring is compressed or serious injury or vehicle damage may occur.

12. Place the strut and coil spring assembly in a suitable vise and remove the strut nut cover.

13. Slightly loosen, but **do not** remove the front strut nut.

If desired, use the following steps to remove the coil spring from the strut.

14. Using an approved coil spring compressor, compress the coil spring.

15. Remove the strut assembly top nut.

16. Remove the following components from the strut assembly:

- Upper mounting bracket
- Strut bearing
- The bearing seat.
- Upper coil spring seat and dust boot
- Coil spring

17. Slowly release the tension of the coil spring compressor and remove the coil spring from the compressor tool.

18. Remove the coil spring insulator and slide the jounce bumper off of the strut assembly.

To install:

19. Slide the jounce bumper onto the strut assembly and install the coil spring insulator.

20. Carefully compress the coil spring with an approved coil spring compressor.

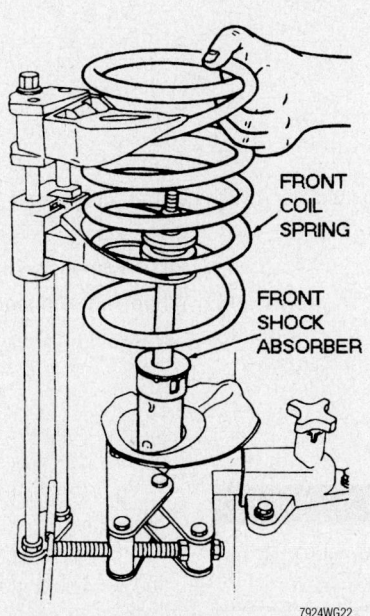

FRONT
COIL
SPRING

FRONT
SHOCK
ABSORBER

7924WG22

Compress the coil spring in an approved spring compressor—2001–02 models

21. Reinstall the following components to the strut assembly:
- Coil spring

➡**Install the coil spring to the strut assembly with the end of the spring in the lower coil spring seat indentation.**

- Upper coil spring seat and dust boot
- Bearing seat and the bearing
- Upper mounting bracket

22. Install and tighten the strut assembly nut and tighten the nut to 43–58 ft. lbs. (59–78 Nm).

23. Install the strut assembly onto the vehicle and tighten the following:
- Strut-to-body nuts: 29–40 ft. lbs. (39–54 Nm)
- Strut-to-knuckle bolts: 101–116 ft. lbs. (137–157 Nm)

24. Reattach the brake tube to the strut assembly.

25. Install and tighten the 2 front brake anti-lock sensor cable bracket bolts.

26. Reinstall the tire and wheel assembly.

27. Connect the negative battery cable and the adjustable strut electrical connectors, if equipped.

28. Check and/or adjust the wheel alignment.

2004–06 Models

1. Before servicing the vehicle, refer to the precautions section.

2. Remove the wheels.

3. Remove the cowl top extension as follows:

a. Remove the wiper arms.

b. Release the clips under cowl top cover and clips on hood–ledge.

c. Remove the seal by releasing the ends from tabs the on the cowl top cover and releasing the plastic clips.

d. Remove the weatherstrip.

e. Remove the cowl top cover.

f. Remove the cowl top seal.

g. Remove the windshield washer nozzles and the hoses from cowl top cover.

h. Remove the heater pump from cowl top extension.

i. Disconnect the clip attaching the coolant control valve hose to the cowl top extension.

j. Remove cowl top extension.

4. Disconnect the ABS sensor wire and front brake hose from the brackets on the front strut.

5. Disconnect the connecting rod upper link.

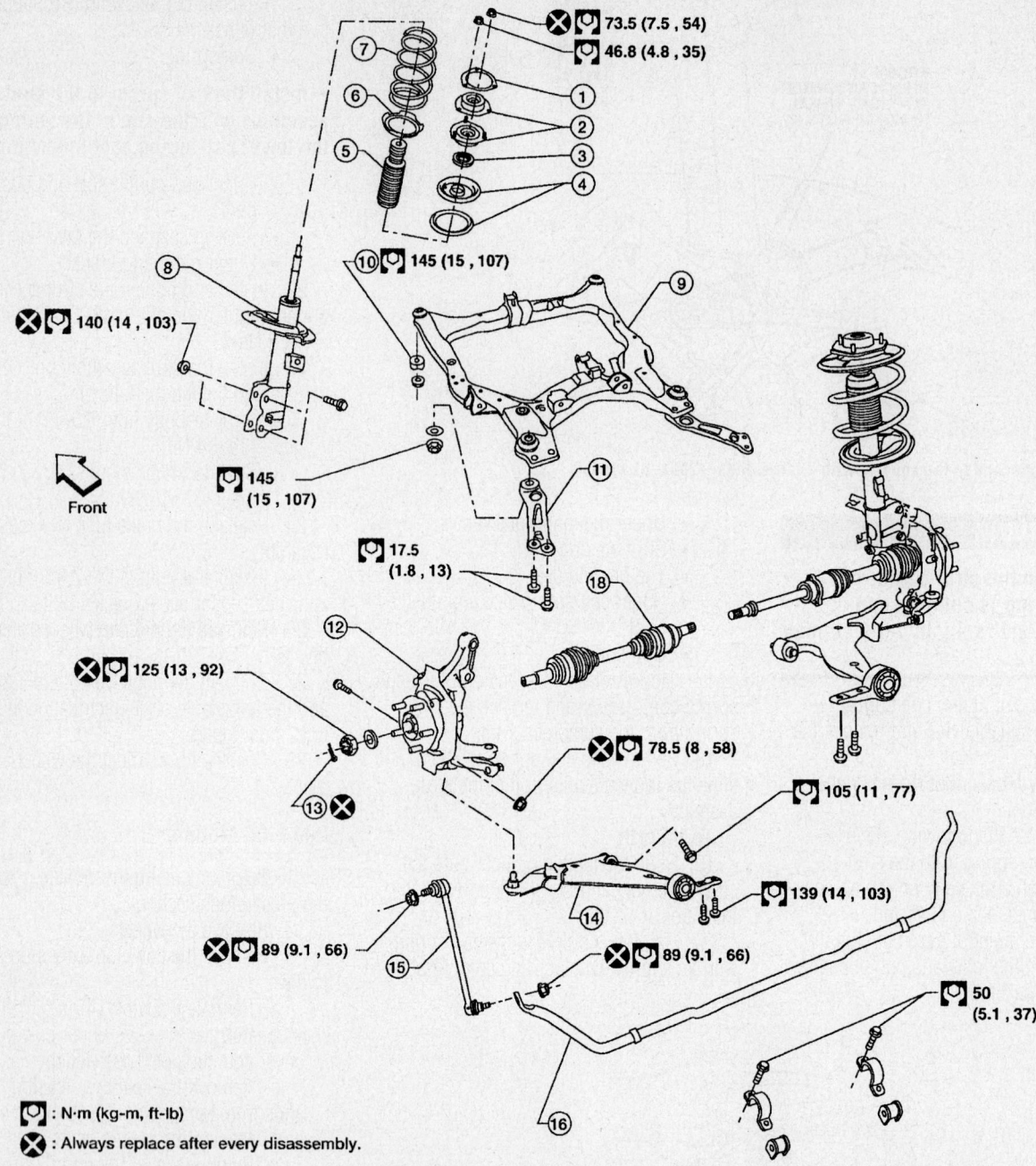

73.5 (7.5 , 54)
46.8 (4.8 , 35)
145 (15 , 107)
140 (14 , 103)
145 (15 , 107)
17.5 (1.8 , 13)
125 (13 , 92)
78.5 (8 , 58)
105 (11 , 77)
139 (14 , 103)
89 (9.1 , 66)
89 (9.1 , 66)
50 (5.1 , 37)

Front

: N·m (kg-m, ft-lb)

: Always replace after every disassembly.

1.	Gasket	2.	Shock absorber mounting insulator	3.	Upper rubber seat
4.	Shock absorber bushing	5.	Dust cover	6.	Coil spring
7.	Lower rubber seat	8.	Shock absorber	9.	Front suspension member
10.	Cup	11.	Member pin stay	12.	Wheel hub and steering knuckle assembly
13.	Cotter pin	14.	Transverse link	15.	Connecting rod
16.	Stabilizer bar				

67170-NIQU-GAG

Exploded view of the front suspension assembly—2004–06 models

6. Support the wheel hub and steering knuckle assembly with mechanics wire.

7. Remove the lower shock absorber bolts and nuts.

8. Remove the three upper strut mounting nuts.

✳✳ CAUTION

Do not remove the piston rod lock nut on vehicle.

9. Remove the coil spring and strut assembly.

10. Place the assembly in a vise, then loosen, but do not remove the piston rod lock nut.

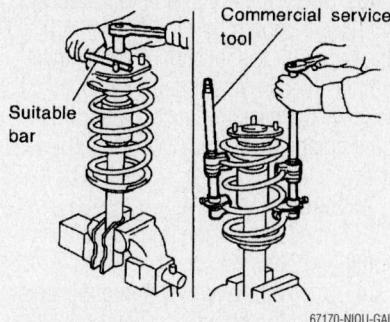

67170-NIQU-GAI

Place the strut assembly in a vise, then loosen, but not removing the piston rod lock nut as illustrated—2004–06 models

➡**Do not remove piston rod lock nut at this time.**

11. Compress the spring using a spring compressor until the shock absorber mounting insulator can be turned by hand.

✳✳ CAUTION

Make sure that the pawls of the two spring compressors are firmly hooked on the spring. The spring compressors must be tightened alternately and evenly so as not to tilt the spring.

12. Remove the piston rod lock nut and spring.
13. Check the cemented rubber-to-metal portion for separation or cracks. Check the rubber parts for deterioration and replace if necessary.
14. Check the thrust bearing parts for abnormal noise or excessive rattle in axial direction and replace if necessary.
15. Check the spring for cracks, deformation or other damage and replace if necessary.

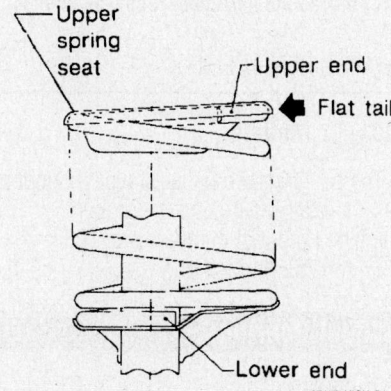

67170-NIQU-GAJ

When installing the coil spring on strut, it must be positioned as illustrated—2004–06 models

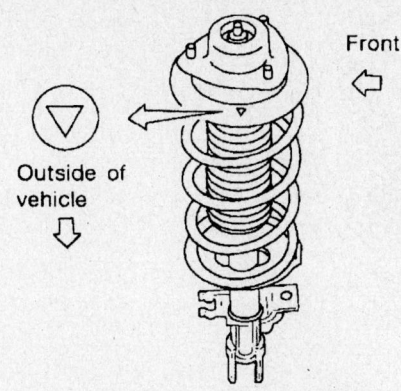

67170-NIQU-GAK

Install the upper spring seat with alignment mark facing the outer side of vehicle, in line with strut-to-knuckle attachment points—2004–06 models

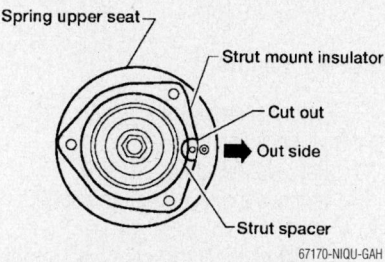

67170-NIQU-GAH

Make sure the front strut spacer is positioned as illustrated—2004–06 models

16. Check the free spring height. The SE model height is 13.39 inch (340 mm) and the SL height is 13.78 inch (350 mm). Replace if not as specified.

To install:

17. When installing the coil spring on strut, it must be positioned as illustrated.
18. Install the upper spring seat with alignment mark facing the outer side of vehicle, in line with strut-to-knuckle attachment points.
19. Install and compress the spring.
20. Tighten the piston rod lock nut to 54 ft. lbs. (73 Nm).
21. Installation of the assembly is in the reverse order of removal. Make sure the front strut spacer is positioned as illustrated. Tighten the upper bolts to 35 ft. lbs. (46 Nm) and the lower bolts to 103 ft. lbs. (140 Nm).
22. Check that the front wheel alignment

Shock Absorber

REMOVAL & INSTALLATION

1. Before servicing the vehicle, refer to the precautions section.

2. Raise and safely support the vehicle.
3. Support the rear axle and slightly lower the vehicle enough to lessen tension on the shock absorber.
4. Remove the lower shock absorber retaining nut and washer.
5. Disconnect the lower end of the shock absorber from the mounting stud.
6. Remove the shock absorber upper end retaining nut and washer.
7. Remove the shock absorber from the vehicle.

To install:

8. Install the shock absorber onto the upper and lower mounting studs of the vehicle.
9. Install the washers and retaining nuts. Tighten the upper and lower retaining nuts to 22–30 ft. lbs. (30–41 Nm) on 2001–02 models or 22 ft. lbs. (30 Nm) for the upper bolts and 47 ft. lbs. (64 Nm) on the lower bolts.
10. Lower the vehicle.

Lower Ball Joints

REMOVAL & INSTALLATION

2001–02 Models

To check if ball joint replacement is required, raise and safely support the vehicle clear of the floor and try to rock the wheel up and down. If any play is felt, have an assistant rock the wheel while observing the front suspension lower arm ball joint at the bottom of the steering knuckle. If any movement is seen, the ball joint should be replaced. If not, any wheel play indicates wheel bearing wear.

1. Before servicing the vehicle, refer to the precautions section.
2. Raise and safely support the vehicle.
3. Remove the tire and wheel.
4. Remove and discard the ball joint cotter pin. Loosen the ball joint attaching nut from the steering knuckle. Because of tight clearance, the nut likely cannot be removed until the ball joint stud is loosened and lowered slightly.
5. Strike the front knuckle with a hammer while pulling down on the lower control arm. There should now be enough clearance to allow removal of the ball joint stud nut. Separate the ball joint from the steering knuckle.
6. Remove the 3 bolts attaching the ball joint to the control arm.
7. Remove the ball joint from the control arm.

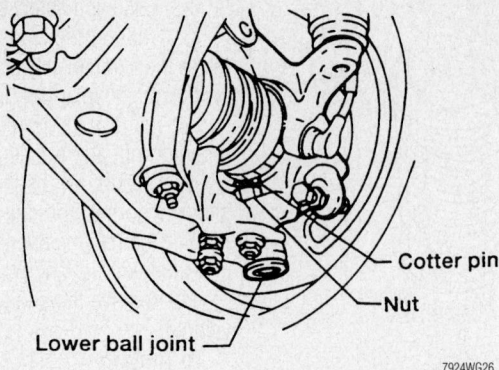

Loosen the nut on the lower ball joint stud—2001–02 models

To install:

8. Install the ball joint to the control arm and install the attaching bolts.

9. Tighten the bolts to 54–65 ft. lbs. (74–88 Nm).

10. Install the ball joint into the steering knuckle, just enough to get the nut started on the stud. Then, push the ball joint stud fully in place. Tighten the nut to 52–63 ft. lbs. (71–86 Nm). Secure the nut with a new cotter pin.

11. Install the tire and wheel.

12. Lower the vehicle.

13. A front end alignment check is recommended.

2004–06 Models

The ball joint on 2004–06 models is an integral part of the lower control arm (transverse link) assembly. If the ball joint is found to be defective, replace the transverse link.

Lower Control Arm

REMOVAL & INSTALLATION

2001–02 Models

1. Before servicing the vehicle, refer to the precautions section.

2. Remove the wheel.

3. Disconnect the ball joint.

4. Disconnect the stabilizer bar from the control arm.

5. Remove the 2 rear arm bolts and the mounting bracket.

6. Remove the lower arm nut.

7. Pull the rear of the arm down and gently pry the arm forward and off the gusset.

8. Installation is the reverse of removal. Observe the following torques:

- Stabilizer bar-to-lower arm: 12–16 ft. lbs. (16–22 Nm)
- Lower arm rear bolts: 87–108 ft. lbs. (118–147 Nm)
- Lower arm nuts: 94–115 ft. lbs. (128–156 Nm)
- Ball stud nut: 56–80 ft. lbs. (76–109 Nm)

2004–06 Models

1. Before servicing the vehicle, refer to the precautions section.

2. Remove the wheel.

3. Remove the lower ball joint pinch bolt.

4. Separate the transverse link from the steering knuckle assembly.

5. Remove the two transverse link pivot bolts.

6. Remove the transverse link from the front suspension member.

7. Check the transverse link for damage, cracks or any deformities and replace as necessary.

8. Check the bushing for damage, cracks or any deformities. Replace the transverse link if necessary.

9. Check the ball joint for excessive play. Replace the transverse link assembly if the lower ball joint stud is worn, the ball joint is hard to swing, the ball joint play in axial directions or end play is excessive.

➡Before checking the axial forces and end play, turn the lower ball joint at least 10 revolutions so that the ball joint is properly broken in.

10. The swinging force (measuring from the cotter pin hole of the ball stud) is 1.8–12.3 ft. lbs. (7.8–54.9 N). Refer to the illustration for the location of the swinging force measurement point identified by the letter **A**.

11. The turning force is 4.3–30.4 inch lbs. (0.49–3.43 Nm). Refer to the illustration for the location of the turning force measurement point identified by the letter **B**.

12. Check the transverse link vertical end play, it should be zero. Refer to the illustration for the location of the vertical end play measurement point identified by the letter **C**. Check the dust cover for damage. Replace it and the cover clamp if necessary.

To install:

13. Hand tighten the transverse link mounting bolts.

14. Once all components are installed final tighten the transverse link bolts with the vehicle at curb weight and the tires on the ground. Tighten the ball joint pinch bolt to 58 ft. lbs. (78 Nm), the two outer link bolts to 103 ft. lbs. (139 Nm) and the single inner bolt to 77 ft. lbs. (105 Nm). Refer the exploded view of the front suspension assembly illustration for bolt locations.

15. Installation of the remaining components is in the reverse order of removal.

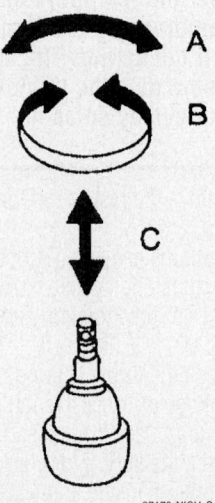

Ball joint swinging force (A), turning force (B) and transverse link vertical end play (C) measuring locations—2004–06 models

BUSHING REPLACEMENT

2001–02 Models

The bushings are press-fit types. Support the arm in a press, using the proper adapters. Ford tool numbers are: T93P-5493-A, T75L-1165-B and -DA.

Wheel Bearings

ADJUSTMENT

The wheel bearings are not adjustable. If the bearings become loose or make noise, they must be replaced using the following procedure.

REMOVAL & INSTALLATION

Front

2001–02 MODELS

1. Before servicing the vehicle, refer to the precautions section.
2. Raise and safely support the vehicle.
3. Remove the wheel and tire.

4. Remove the brake caliper assembly. DO NOT disconnect the brake hose. Hang the caliper on a piece of wire from a near by support such as the strut.
5. Remove the brake rotor.
6. Remove and discard the cotter pin from the end of the outboard CV-joint stub shaft. Remove the hub nut retainer, washer and the hub nut. There should be another washer under the hub nut that acts as a front wheel bearing outer bearing retainer.
7. Disengage the lower ball joint stud

from the steering knuckle using the following procedure.

a. Remove and discard the cotter pin from the front lower ball joint.
b. Loosen the lower ball joint nut until it contacts the front halfshaft joint.
c. Strike the front knuckle with a hammer while pulling down on the lower control arm until the ball joint stud separates from the knuckle.
d. Remove the ball joint nut.

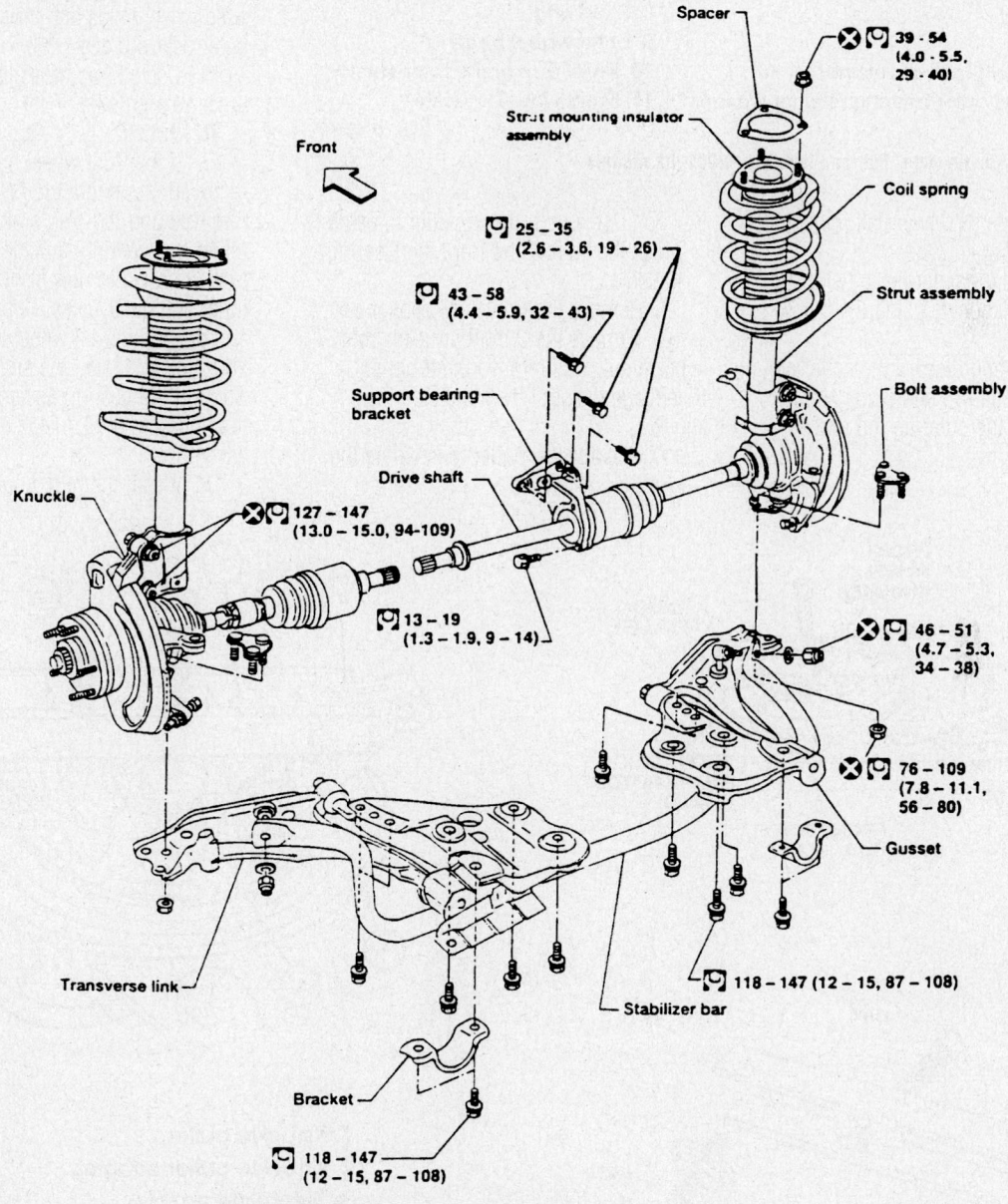

Exploded view of the front suspension and drive axles—2001–02 models

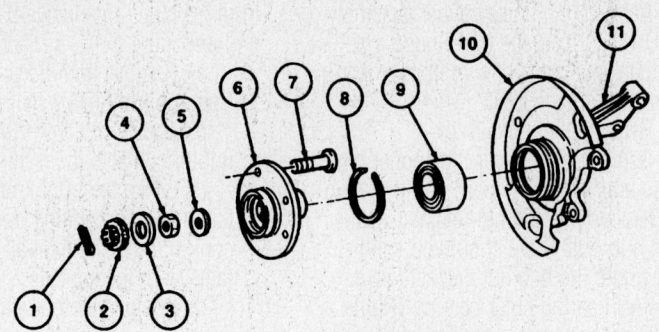

1. Cotter pin
2. Nut retainer
3. Insulator
4. Front axle wheel hub retainer
5. Front wheel outer bearing retainer washer
6. Wheel hub
7. Wheel hub bolt
8. Snap ring
9. Front wheel bearing
10. Front disc brake rotor shield
11. Front wheel knuckle

7924WG23

Exploded view of the knuckle, hub and bearing—2001–02 models

e. Disengage the lower ball joint stud from the steering knuckle.

8. Disengage the outer tie rod end stud from the steering knuckle using the following procedure.

   a. Remove and discard the cotter pin from the outer tie rod end stud.

   b. Remove the outer tie rod end retaining nut.

   c. Use a tie rod end puller to carefully press the tie rod end from the steering knuckle.

9. Remove the front ABS sensor bolt.

10. Remove the 2 front strut-to-front knuckle nuts and remove the 2 bolts. Disengage the strut from the steering knuckle.

11. Use a 2-jaw puller to separate the front halfshaft outboard CV-joint stub shaft from the knuckle/bearing assembly.

12. Remove the front wheel hub, knuckle and wheel bearing assembly from the vehicle.

13. If the knuckle is being replaced with a service part, change over the steering stop bolt and jam nut from the old knuckle to the replacement part.

14. To remove the front wheel bearing, jig up a puller to bear against the front wheel bearing inner race and pull the race from the hub/knuckle assembly.

15. Use a shop press to press out damaged wheel studs and also to press out the outer bearing race.

16. Use a shop press to press out the inner bearing race.

To install:

17. If the front wheel bearings were removed, assemble the ABS sensing ring, if removed and the disc brake dust shield under the steering knuckle. Use a shop press to push in new front wheel bearing inner and outer races. Support the knuckle and press the front wheel bearing into the knuckle and install the snapring retainer. Support the bearing assemblies and press the hub onto the knuckle and wheel bearing assembly.

18. Install the hub, knuckle and bearings

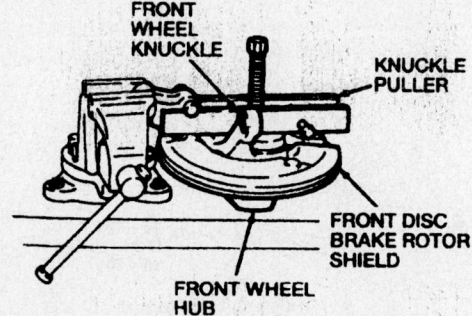

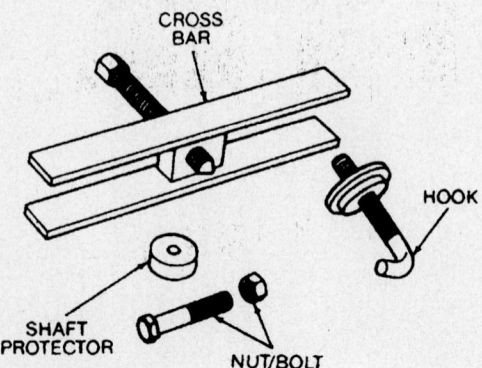

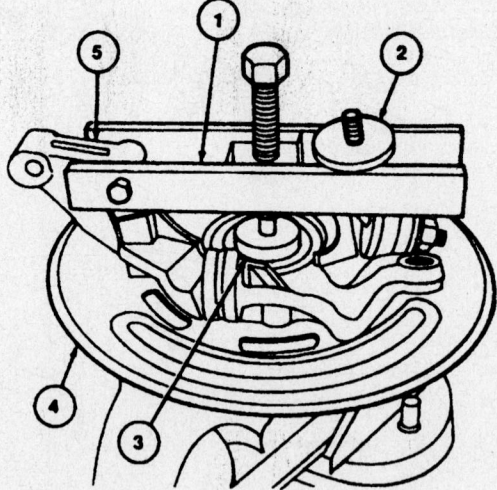

1. Knuckle puller
2. Knuckle puller adapter
3. Step plate adapter
4. Front disc brake rotor shield
5. Front wheel knuckle

7924WG27

Example of a puller set up to bear against the front wheel bearing inner race—2001–02 models

as an assembly. Position the assembly on the halfshaft outer CV-joint stub axle end. Guide the knuckle into the front strut and install the 2 knuckle-to-strut bolts and nuts. Tighten the nuts to 83–91 ft. lbs. (113–123 Nm).

19. Install the ABS sensor bolt. Do not over-tighten. Tighten to just 16–21 inch lbs. (1.8–2.4 Nm).

20. Install the outer tie rod end to the steering knuckle. Tighten the nut to 22–29 ft. lbs. (29–39 Nm). If the cotter pin holes do not align, tighten the nut slightly until they do. Never loosen the nut to align the holes. Secure the nut with a new cotter pin.

21. Start the lower ball joint stud to the steering knuckle and partially install the nut, then push the ball joint stud fully in place. Tighten the ball joint stud nut to 52–63 ft. lbs. (71–86 Nm). Secure the nut with a new cotter pin.

22. Install the front wheel outer bearing retaining washer and the hub retainer nut. Tighten to 174–231 ft. lbs. (235–314 Nm). Install the nut retainer, insulator and a new cotter pin.

23. Install the front brake rotor and install the disc brake caliper.

24. If removed, install the steering stop bolt.

25. Install the tire and wheel assembly. Tighten the lug nuts to 72–87 ft. lbs. (98 to 118 Nm).

26. Lower the vehicle. Pump the brake pedal slowly to seat the front brake pads. Do not move the vehicle until a firm pedal is obtained.

27. A front end alignment is recommended.

2004–06 Models

1. Before servicing the vehicle, refer to the precautions section.

2. Remove the wheel.

3. Remove the brake caliper, leaving the line attached and suspend the caliper using a piece of wire.

4. Place alignment marks on the brake rotor, wheel hub and bearing assembly, then remove the rotor.

5. Remove the ABS sensor from the steering knuckle.

6. Remove the cotter pin and lock nut from the drive shaft.

7. Remove the steering outer socket cotter pin at the steering knuckle, then loosen the mounting nut.

8. Disconnect the steering outer socket from the steering knuckle using tool J25730–A. Be careful not to damage ball joint boot.

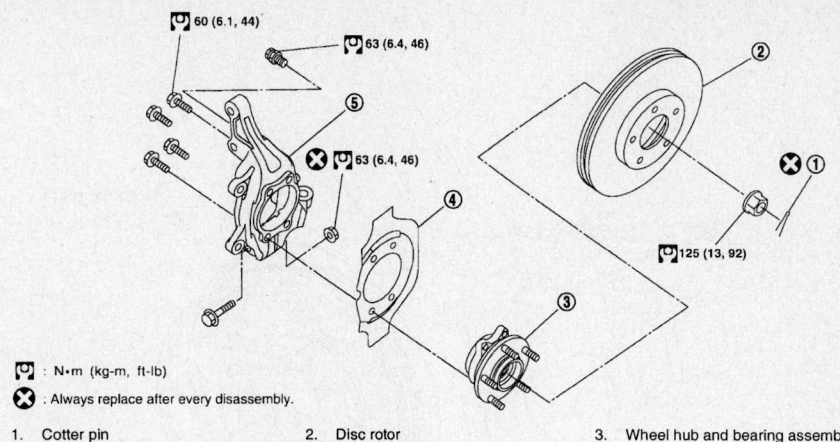

⌷ 60 (6.1, 44)   ⌷ 63 (6.4, 46)   ②
✕ ⌷ 63 (6.4, 46)   ④   ✕ ①
⌷ 125 (13, 92)   ③

⌷ : N•m (kg-m, ft-lb)
✕ : Always replace after every disassembly.

1. Cotter pin	2. Disc rotor	3. Wheel hub and bearing assembly
4. Splash guard	5. Steering knuckle	

67170-NIQU-GAM

Exploded view of the front wheel hub and knuckle components—2004–06 models

✳✳ CAUTION

To prevent damage to the threads and to prevent the tool from coming off suddenly, temporarily tighten mounting nut.

9. Remove the transverse link and steering knuckle pinch bolt and nut.

10. Remove the wheel hub and bearing assembly from drive shaft using a suitable puller.

➡**When removing the wheel hub and bearing assembly, do not apply an excessive angle to drive shaft joint. Be careful not to excessively extend the slide joint and support the driveshaft when removing.**

11. Remove the wheel hub and bearing assembly bolts.

12. Remove the splash guard and the wheel hub and bearing assembly from steering knuckle.

To install:

13. Installation is in the reverse order of removal. Tighten the bolts to the specifications shown in the exploded view of the front wheel hub and knuckle illustration. Also please note the following:

a. Replace the differential side oil seal with a new one every time driveshaft is removed on 4 speed transmission models

b. When installing the wheel hub and bearing assembly to steering knuckle, align cutout in toner ring cover with ABS sensor mounting hole in steering knuckle.

c. When installing the rotor on the wheel hub and bearing assembly, align the marks made prior to removal.

d. A front end alignment is recommended.

Rear

2001–02 MODELS

1. Before servicing the vehicle, refer to the precautions section.

2. Raise and safely support the vehicle.

3. Remove the rear wheel(s).

4. Remove the brake drum.

5. Remove the grease cap for the hub.

6. Remove and discard the cotter pin.

7. Remove the wheel bearing nut and washer.

8. Remove the rear wheel hub and bearing assembly.

To install:

9. Install the rear wheel hub and bearing assembly onto the vehicle.

10. Install the rear wheel bearing washer and nut and tighten the bearing nut to 159–210 ft. lbs. (216–284 Nm). Install a new cotter pin.

11. Install the wheel hub grease cap. Install the brake drum.

12. Install the rear wheel(s) and lug nuts. Tighten the lug nuts, in a star sequence, to 72–87 ft. lbs. (98–118 Nm).

13. Lower the vehicle.

2004–06 Models

1. Before servicing the vehicle, refer to the precautions section.

2. Remove the rear wheel.

3. Remove the brake caliper assembly without disconnecting the hydraulic line and suspend the caliper aside using wire.

4. Release the parking brake and remove the rotor.

5. Remove the rear ABS sensor, then position it aside using wire.

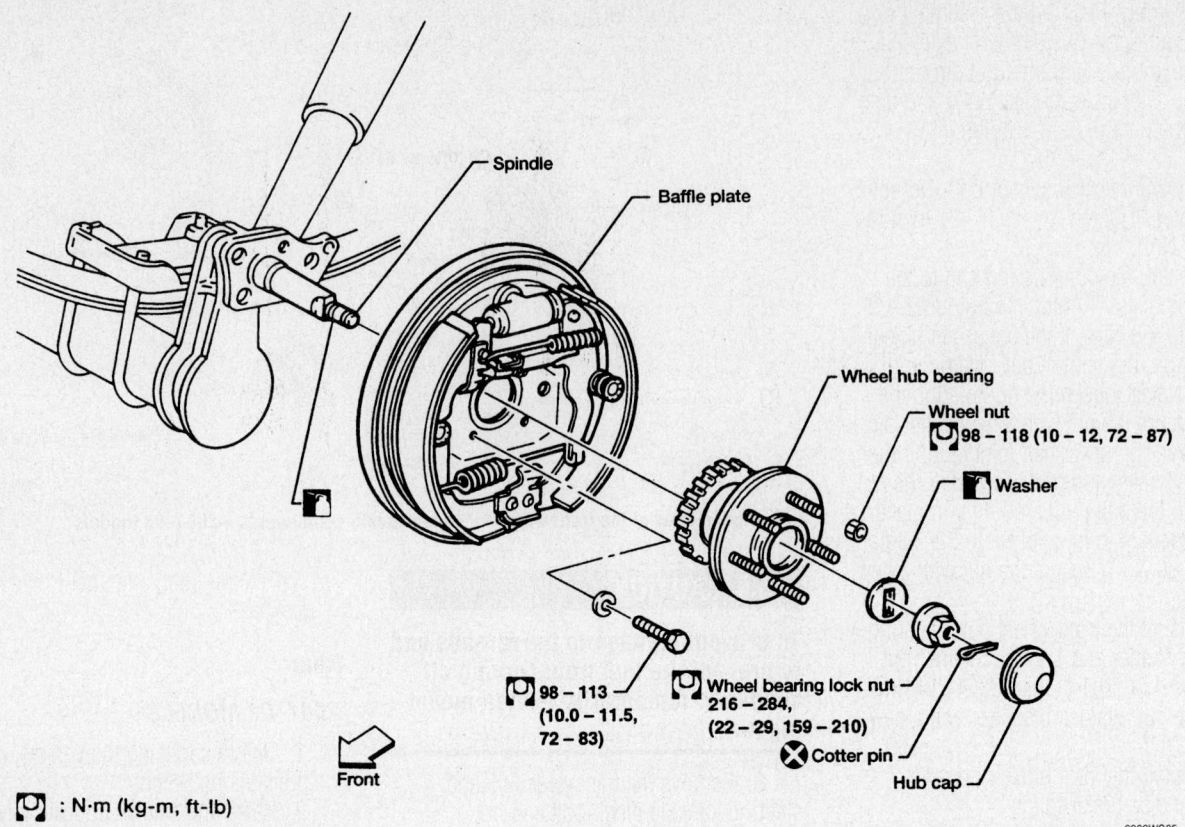

☐ : N·m (kg-m, ft-lb)

9302WG05

Rear hub assembly—2001–02 models

6. Remove the wheel hub assembly from the knuckle.

7. Check for cracks, and damage on the wheel hub assembly and replace as necessary.

To install:

8. Installation is in the reverse order of removal. Tighten the wheel hub bolts to 44 ft. lbs. (60 Nm).

9. Check that the wheel bearing operates smoothly.

10. Check that the wheel hub bearing axial end play is within 0.004 inch (0.1mm) or less using a dial indicator.

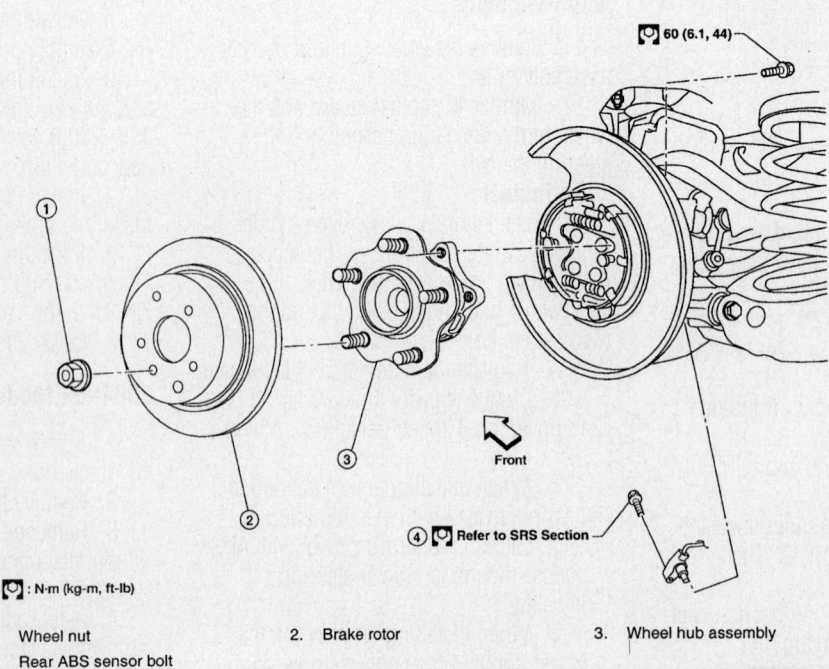

☐ : N·m (kg-m, ft-lb)

1. Wheel nut
2. Brake rotor
3. Wheel hub assembly
4. Rear ABS sensor bolt

Exploded view of the rear wheel hub assembly—2004–06 models

67170-NIQU-GAN

BRAKES

Brake Caliper

REMOVAL & INSTALLATION

2001–02 Models

The front disc brake caliper slides on 2 stainless steel locating pins. The front disc brakes use a conventional pin slider-type front disc brake caliper with a 10.875 inch (27.6cm) front disc rotor. The front disc brake caliper is attached to the front suspension with 2 Torx® head brake caliper bolts. Rubber insulators isolate the stainless steel locating pins from direct contact with the front disc brake caliper. The front disc brake calipers must be removed to replace the front brake pads.

1. Before servicing the vehicle, refer to the precautions section.
2. Remove or disconnect the following:
   • Wheel and tire

➡**If the brake caliper is being removed for brake pad replacement only, DO NOT disconnect the brake hose.**

   • 2 caliper pin bolts. Most applications will require a Torx® T-40 bit to remove the 2 brake caliper bolts.
3. If the brake caliper is being removed just for brake service, with the brake hose still attached to the caliper, use a length of wire to support the caliper from the front shock absorber. Do not let the caliper hang by the brake hose. If the caliper is being completely removed from the vehicle for overhaul, use care not to drip brake fluid on the paint.

➡**If both calipers are being completely removed from the vehicle at the same time, mark them Left and Right so the calipers can be reinstalled to their original locations. The reason for this is that the bleeder screws must be positioned on the top of the front disc brake caliper when installed on the vehicle.**

To install:

4. Clean all parts well. Use a C-clamp and a used brake pad to push the caliper piston fully in the piston bore. Inspect the caliper pins and clean any dirt and debris.
5. Install the caliper onto the rotor. Make sure the inboard and outboard brake pads are properly positioned.
6. Lubricate the stainless steel locating pins with a Silicone Dielectric Compound such as Ford DZAZ-19A331-A or equivalent silicone grease. Install the 2 caliper pin bolts and torque to 18–25 ft. lbs. (24–34 Nm).
7. If disconnected, install the brake hose using a new replacement copper washer, install the banjo bolt and torque to 12–14 ft. lbs. (17–20 Nm).
8. If the brake hose had been disconnected, bleed the brake system.
9. Install the wheel and tire.
10. Torque the lug nuts to 72–87 ft. lbs. (98–118 Nm).
11. Check the master cylinder reservoir and add fresh DOT 3 brake fluid as required.
12. Lower the vehicle. Pump the brake pedal slowly until a firm brake pedal is obtained, indicating that the brake pads are properly seated, before attempting to move the vehicle. Road-test and check for proper brake operation.

2004–06 Models

1. Before servicing the vehicle, refer to the precautions section.
2. Remove the wheel.
3. Drain the brake fluid.
4. Remove brake hose bolt and caliper bolts, then remove caliper assembly.

To install:

5. Install caliper assembly, and tighten bolts to 115 ft. lbs. (156 Nm) on the front caliper or 62 ft. lbs. (84 Nm) on the rear caliper.

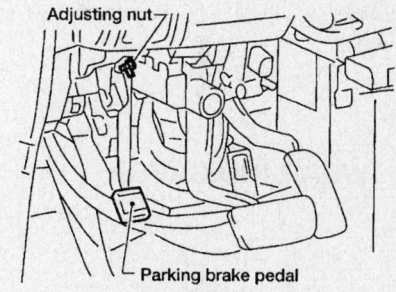

67170-NIQU-GAO

Location of the brake pedal adjusting nut—2004–06 models

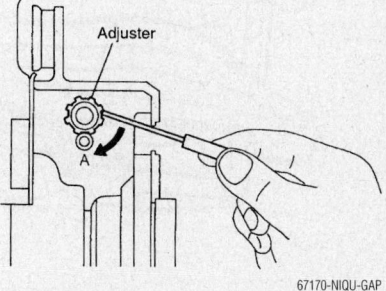

67170-NIQU-GAP

Location of the parking brake adjuster—2004–06 models

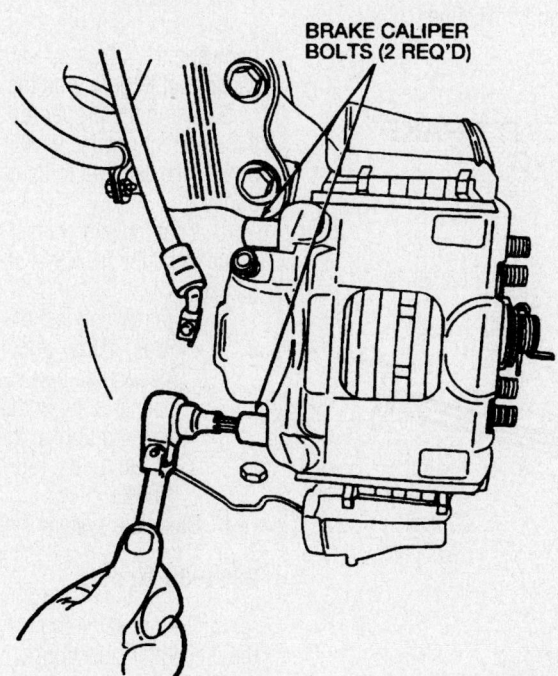

93026G10

Caliper pin bolt removal—2001–02 models

6. Install brake hose to the caliper assembly using a new copper washer, and tighten the bolt to 13 ft. lbs. (18 Nm).

7. Refill the system with new brake fluid and bleed the brakes.

8. If the rear caliper was removed, adjust the parking brake as follows:

a. Insert a deep socket wrench to rotate adjusting nut and loosen the cable sufficiently. Then, return the pedal to the free height.

b. Using the lug nuts, secure the rotor to the hub and prevent it from tilting.

c. Remove the adjusting hole plug installed on the rotor. Using a suitable tool, rotate the adjuster in direction **A** as illustrated, until the rotor is locked. After locking, turn the adjuster in the opposite direction by 5 or 6 notches.

d. Rotate the rotor to make sure there is no drag and install the adjusting hole plug.

e. Pump the brake pedal 10 or more times with a force of 66 lbs. (294 N).

f. Rotate the adjusting nut with a deep socket to adjust the pedal stroke. The stroke should be 5 to 6 notches

➡ **Do not reuse the adjusting nut after removing it.**

g. When parking brake pedal is operated at specified force, make sure the force is 44 lbs. (196 N).

h. With the pedal completely returned, make sure there is no drag on the rear brake.

9. Install the wheel.

Disc Brake Pads

REMOVAL & INSTALLATION

Front

2001–02 MODELS

1. Before servicing the vehicle, refer to the precautions section.

2. Remove or disconnect the following:

- Wheels
- Bottom guide pin from the caliper and swing the caliper cylinder body upward; support the caliper with a wire
- Brake pad retainers and the pads

To install:

3. Compress the piston of the disc brake caliper.

4. Install or connect the following:
- Brake pads and caliper assembly. Torque the guide pin to 23–30 ft. lbs. (31–41 Nm).
- Wheels

5. Apply the brakes a few times to seat the pads. Check the master cylinder and add fluid if necessary. Bleed the brakes, if necessary.

2004–06 Models

1. Before servicing the vehicle, refer to the precautions section.

2. Remove the wheels.

3. Remove the bottom guide pin from the caliper and swing the caliper cylinder body upward; support the caliper with a wire.

4. Remove the brake pad retainers and the pads.

To install:

5. Compress the piston of the disc brake caliper.

6. Install the retainers and brake pads.

7. Install the caliper assembly. Torque the guide pin to 20 ft. lbs. (26 Nm).

8. Install the wheels.

9. Apply the brakes a few times to seat the pads. Check the master cylinder and add fluid if necessary. Bleed the brakes, if necessary.

Rear

2001–02 MODELS

1. Before servicing the vehicle, refer to the precautions section.

➡ **Do not press the piston into the bore as performed on the front disc brakes. Due to the parking brake mechanism, the caliper piston must be turned into the bore using a special tool.**

2. Remove or disconnect the following:

- Rear wheels

3. Release the parking brake.
- Parking brake cable bracket bolt
- Pin bolts and lift off the caliper body
- Pad springs
- Pads and shims

To install:

4. Clean the piston end of the caliper body and the area around the pin holes. Be careful not to get oil on the rotor.

5. Using the proper tool, carefully turn the piston clockwise back into the caliper body. Take care not to damage the piston boot.

6. Coat the pad contact area on the mounting support with a silicone based grease.

7. Install or connect the following:
- Pads, shims, and the pad springs. Always use new shims.
- Caliper body in the mounting support and tighten the pin bolts to 28–38 ft. lbs. (38–52 Nm)
- Wheels

8. Bleed the system if necessary.

2004–06 Models

1. Before servicing the vehicle, refer to the precautions section.

2. Remove the wheels.

3. Remove the bottom guide pin from

WEAR INDICATOR

OUTBOARD FRONT BRAKE SHOE AND LINING

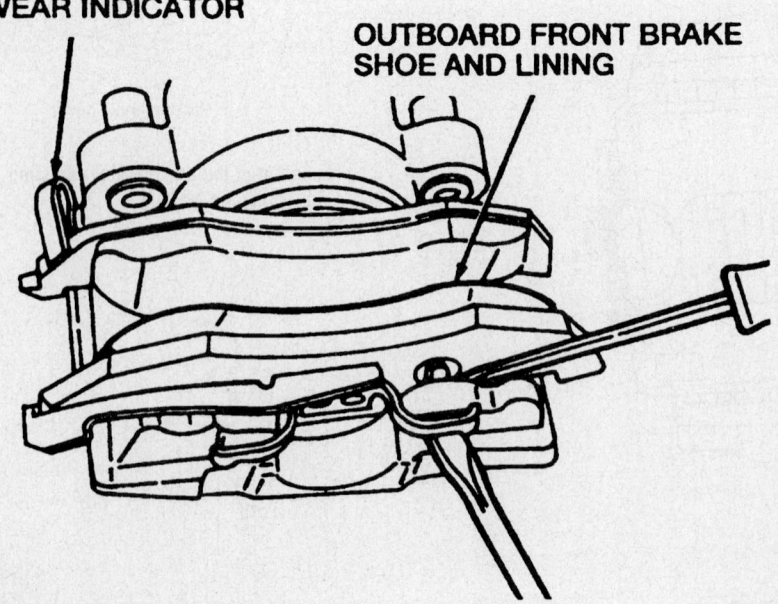

93026G33

Replacing the disc brake pads—2001–02 models

the caliper and swing the caliper cylinder body upward; support the caliper with a wire.

4. Remove the brake pad retainers, shims and the pads.

To install:

5. Compress the piston of the disc brake caliper.

6. Coat the pad contact area on the mounting support with a silicone based grease.

7. Install the retainers and brake pads.

8. Install the caliper assembly. Torque the guide pin to 32 ft. lbs. (43 Nm).

9. Install the wheels.

10. Apply the brakes a few times to seat the pads. Check the master cylinder and add fluid if necessary. Bleed the brakes, if necessary.

Brake Drums

REMOVAL & INSTALLATION

2001–02 Models

1. Before servicing the vehicle, refer to the precautions section.

The rear drum brakes used on these vehicles are conventional expanding shoe-type with the brake shoe lining applied to the inside of the rotating drum. An incremental brake adjuster screw is designed to actuate whenever sufficient wear occurs.

2. Remove or disconnect the following:
   • Wheel and tire
   • Brake drum by pulling it from the wheel studs

3. If necessary for brake drum removal, pry off the access hole plug from the access hole. Insert a screwdriver and a brake adjustment tool. Press the screwdriver

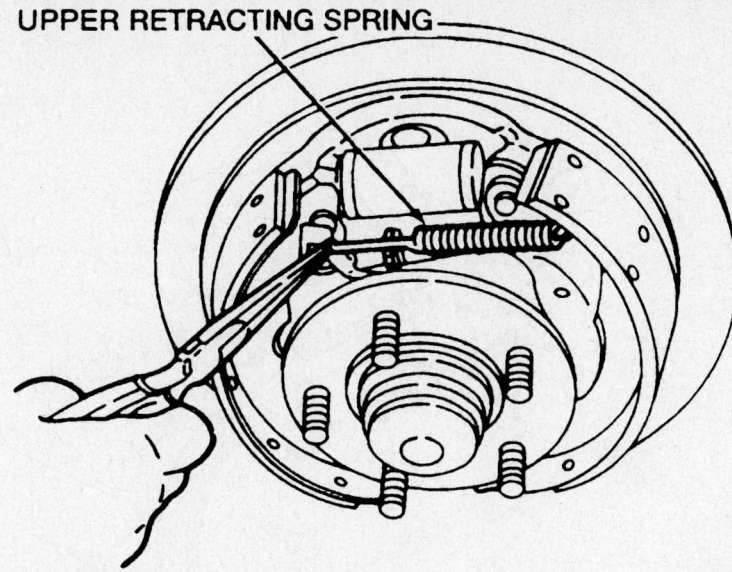

UPPER RETRACTING SPRING

93026G30

Remove the upper retracting spring—2001–02 models

against the adjusting lever to disengage it from the adjuster. Loosen the adjuster using the brake adjusting tool.

To install:

4. Clean all parts well. It is good practice to inspect the wheel cylinder for leaks anytime the brake drum is removed. If a new replacement brake drum is being installed, inspect it for a protective coating on the machined inside braking surface. Remove any coating with suitable solvent.

5. Install or connect the following:
   • Brake drum onto the wheel studs

6. In most all cases, manual brake adjustment IS NOT recommended. Adjustment is performed by driving the vehicle and applying the brakes.
   • Tire and wheel and torque the fasteners to 72–87 ft. lbs. (98–118 Nm)

7. Adjust the rear brake shoes by sharply applying the brakes several times while driving the vehicle alternately forwards and backwards. Check the brake operation by making several stops while driving forward.

Brake Shoes

REMOVAL & INSTALLATION

2001–02 Models

1. Before servicing the vehicle, refer to the precautions section.

The rear drum brakes use an internal rear wheel cylinder with expanding shoes and lining that are applied against a rotating brake drum. An incremental brake adjuster screw is actuated whenever sufficient wear occurs. Brake adjustment takes place in forward or reverse braking but not with parking brake application.

2. Remove or disconnect the following:
   • Wheels
   • Brake drum
   • Parking brake rear cable and conduit from the parking brake lever
   • 2 brake shoe hold-down springs and the 2 brake shoe hold-down pins
   • Upper retracting spring
   • Lower retracting spring
   • Brake adjuster screw
   • Rear brake shoes and linings from the brake backing plate
   • Parking brake lever clip and washer
   • Parking brake lever from the secondary brake shoe and lining

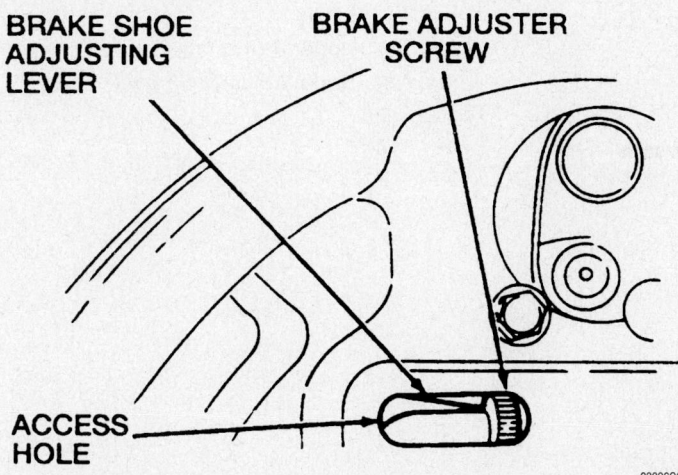

BRAKE SHOE ADJUSTING LEVER

BRAKE ADJUSTER SCREW

ACCESS HOLE

93026G28

Brake shoe adjustment may need to be loosened to remove the brake drum—2001–02 models

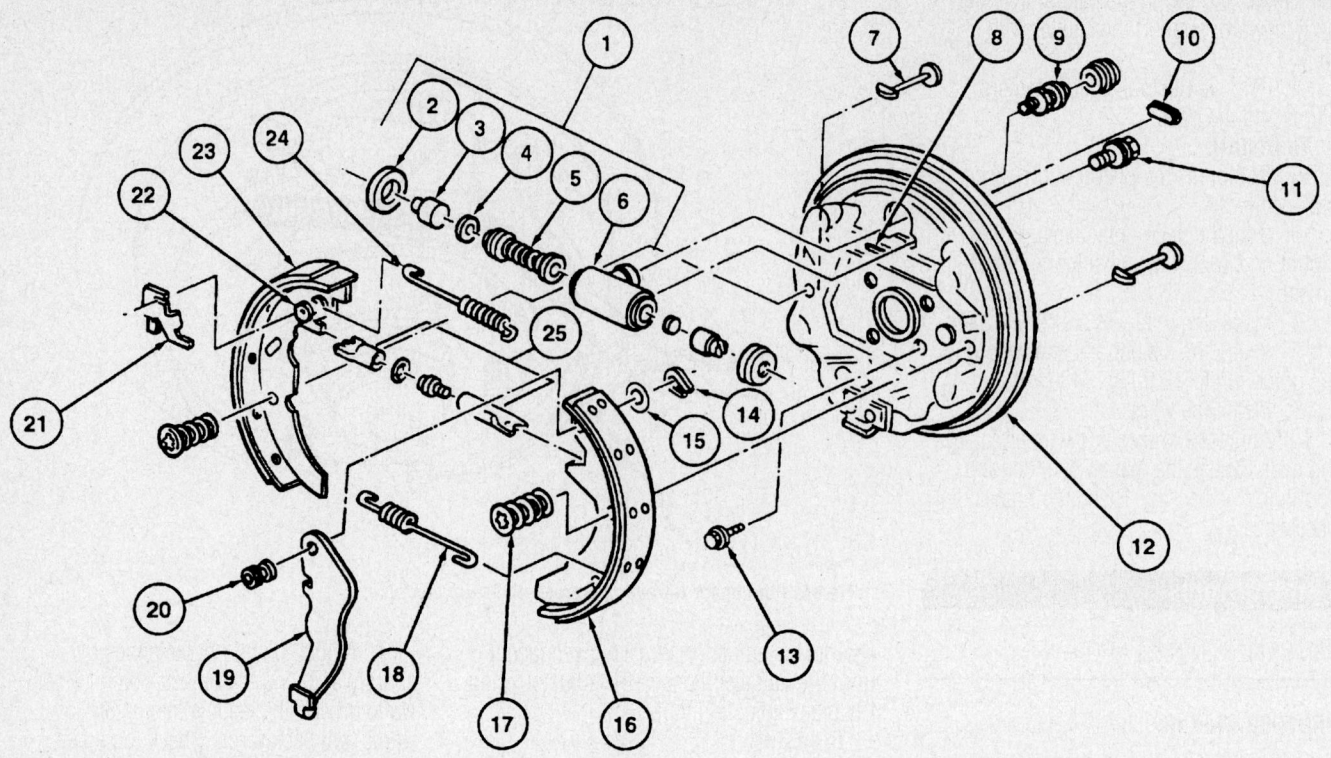

1 Rear Wheel Cylinder
2 Dust Boot (2 Req'd)
3 Wheel Cylinder Piston (2 Req'd)
4 Cup (2 Req'd)
5 Wheel Cylinder Piston Cup Spring
6 Wheel Cylinder Housing
7 Brake Shoe Hold-Down Pin (2 Req'd)
8 Access Hole
9 Rear Brake Bleeder Screw
10 Access Hole Plug
11 Rear Wheel Cylinder Bolt (2 Req'd)
12 Rear Brake Backing Plate

13 Rear Brake Backing Plate Bolts (4 Req'd)
14 Parking Brake Lever Clip
15 Spring Washer
16 Secondary Brake Shoe and Lining
17 Brake Shoe Hold-Down Spring
18 Lower Retracting Spring
19 Parking Brake Lever
20 Parking Brake Lever Pin
21 Brake Shoe Adjusting Lever
22 Adjuster Lever Pin
23 Primary Brake Shoe and Lining
24 Upper Retracting Spring
29 Brake Adjuster Screw

Rear drum brake assembly and related components—2001–02 models

93026G29

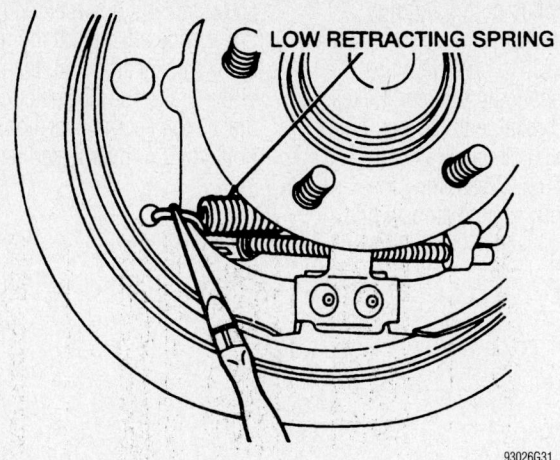

LOW RETRACTING SPRING

93026G31

Remove the lower retracting spring—2001–02 models

To install:

3. Clean all parts well.

4. Inspect the wheel cylinder for signs of leaking. Service as required.

5. Inspect the retracting springs for heat damage, bends or damage to the coils or shank or loss of tension. A good retracting spring will make a full thud when dropped on a concrete floor. A heat-damaged retracting spring that has lost tension will make a distinctive ringing sound when dropped on a concrete floor.

6. Check the brake backing plate for signs of scoring. The shoe contact points must be smooth and have a light coating of lithium grease. Verify that the brake lining thickness is between 0.059–0.232 in. (1.5–5.9mm). Failure to replace worn rear brake shoes will result in a scored drum.

7. Inspect the brake drum for scratches, scoring, bell mouth and out-of-round conditions. Remove minor scores on a brake drum with sandpaper. Do not refinish brake drums to remove scoring marks. A brake drum surface that is highly polished can cause the brakes to lock up. Remove polished surfaces with sandpaper or refinish the brake drum. Refinish a brake drum that is out-of-round enough to cause vehicle vibration or noise when braking. Remove only enough surface metal to true-up the brake drum. Brake drum maximum inside diameter is shown on each drum. If the maximum inside diameter shown on the brake drum is exceeded through wear or refinishing, replace the brake drum. After a brake drum is refinished, wipe the refinished surface with a cloth soaked in clean denatured alcohol. If one brake drum is refinished, the brake drum on the opposite side of the vehicle should also be refinished to the same diameter. The standard inner brake drum diameter is 9.840 inches (250.0mm). Replace the brake drum if worn beyond 9.900 inches (251.5mm).

8. Install or connect the following:
- Parking brake lever to the secondary brake shoe and lining with a new parking brake lever clip
- Secondary (rear) shoe on the backing plate and the brake shoe hold-down spring and pin
- Primary (front) shoe on the backing plate and the brake shoe hold-down spring and pin
- Parking brake rear cable and conduit to the parking brake lever
- Lower retracting spring to the rear brake shoes

9. Apply a light coat of high-quality grease to the threaded areas of the adjuster nut and adjuster socket. Turn the adjuster nut all the way down on the brake adjuster screw, then loosen the adjuster ½ turn. Install the adjuster screw in the slots on the rear brake shoes. The wider slot on the socket must fit in the slot on the primary (front) brake shoe. The slot on the adjuster nut end must fit into the slots in the secondary (rear) brake shoe and parking brake lever.

10. Install or connect the following:
- Brake shoe adjusting lever on the adjuster lever pin
- Upper retracting spring in the slot on the secondary shoe and in the slot on the brake shoe adjusting lever. The brake shoe adjusting

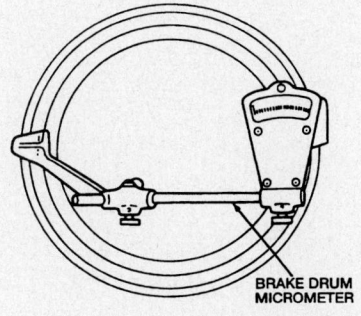

BRAKE DRUM MICROMETER

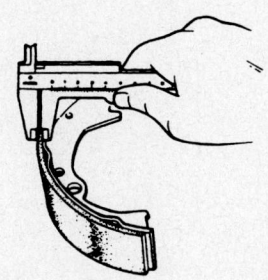

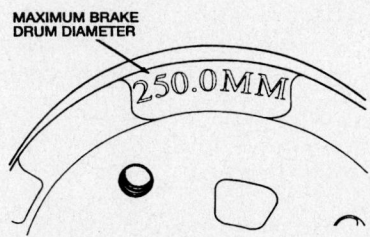

MAXIMUM BRAKE DRUM DIAMETER

250.0MM

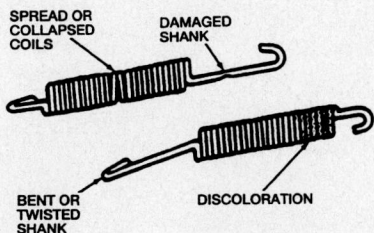

SPREAD OR COLLAPSED COILS

DAMAGED SHANK

BENT OR TWISTED SHANK

DISCOLORATION

93026G32

These size checks should be made and the retracting springs' condition checked. Replace questionnable parts—2001–02 models

lever should contact the brake adjuster screw.
- Brake drum
- Tire and wheel and torque the fasteners to 72–87 ft. lbs. (98–118 Nm).

➡**In most all cases, manual brake adjustment IS NOT recommended. Adjustment is performed by driving the vehicle and applying the brakes.**

11. The rear brakes do not require adjustment when being serviced to obtain a firm brake pedal feel. To achieve a firm brake pedal after servicing the rear brakes, sharply apply the brake pedal several times while driving the vehicle alternately forwards and backwards. Check the brake operation by making several stops while driving forward. The self-adjusting mechanism will sufficiently adjust the rear brake shoes without any manual tightening at the brake shoe adjuster. If the rear brake shoes are manually adjusted, the additional action of the brake shoe adjuster can cause the brakes to become over-tightened and result in binding or overheated rear brakes.

NISSAN

Titan

21

SPECIFICATIONS AND MAINTENANCE CHARTS

ENGINE AND VEHICLE IDENTIFICATION

Engine							Model Year	
Code ①	Liters (cc)	Cu. In.	Cyl.	Fuel Sys.	Engine	Eng. Mfg.	Code ②	Year
VK56DE	5.6 (5552)	338.8	8	MFI	DOHC	Nissan	4	2004
							5	2005
							6	2006

MFI: Multi-port Fuel Injection

DOHC: Double Overhead Camshafts

② 10th digit of the Vehicle Identification Number (VIN)

09482_TITAN_C0001

GENERAL ENGINE SPECIFICATIONS

Year	Model	Engine Displacement Liters	Engine ID	Net Horsepower @ rpm	Net Torque @ rpm (ft. lbs.)	Bore x Stroke (in.)	Com-pression Ratio	Oil Pressure @ rpm
2004	Titan	5.6	VK56DE	305@4900	379@3600	3.86X3.62	9.8:1	43@2000
2005	Titan	5.6	VK56DE	305@4900	379@3600	3.86X3.62	9.8:1	43@2000
2006	Titan	5.6	VK56DE	305@4900	379@3600	3.86X3.62	9.8:1	43@2000

09482_TITAN_C0002

ENGINE TUNE-UP SPECIFICATIONS

Year	Engine Displacement Liters	Engine ID	Spark Plug Gap (in.)	Ignition Timing	Fuel Pump (psi) ①	Idle Speed ②	Valve Clearance (in.) In.	Valve Clearance (in.) Ex.
2004	5.6	VK56DE	0.043	15B	51	650	0.010-0.013	0.011-0.015
2005	5.6	VK56DE	0.043	15B	51	650	0.010-0.013	0.011-0.015
2006	5.6	VK56DE	0.043	15B	51	650	0.010-0.013	0.011-0.015

NOTE: The Vehicle Emission Control Information label often reflects specification changes made during production. The label figures must be used if they differ from those in this chart.

B: Before top dead center

① System pressure at idle

② Automatic transmission in Neutral

09482_TITAN_C0003

CAPACITIES

Year	Model	Engine Displacement Liters	Engine ID	Engine Oil with Filter (qts.)	Transmission (pts.)	Transfer Case (pts.)	Drive Axle Front (pts.)	Rear (pts.)	Fuel Tank (gal.)	Cooling System (qts.)
2004	Titan	5.6	VK56DE	6.2	22.5	6.25	3.375	4.25	28.0	13
2005	Titan	5.6	VK56DE	6.2	22.5	6.25	3.375	4.25	28.0	13
2006	Titan	5.6	VK56DE	6.2	22.5	6.25	3.375	4.25	28.0	13

NOTE: All capacities are approximate. Add fluid gradually and check to be sure a proper fluid level is obtained.

09482_TITAN_C0004

VALVE SPECIFICATIONS

Year	Engine Displacement Liters	Engine ID	Seat Angle (deg.)	Face Angle (deg.)	Spring Test Pressure (lbs. @ in.)	Spring Installed Height (in.)	Stem-to-Guide Clearance (in.) Intake	Exhaust	Stem Diameter (in.) Intake	Exhaust
2004	5.6	VK56DE	45.15-45.45	45	37.0@1.457	1.9913	0.0008-0.0021	0.0012-0.0025	0.2348-0.2354	0.2344-0.2350
2005	5.6	VK56DE	45.15-45.45	45	37.0@1.457	1.9913	0.0008-0.0021	0.0012-0.0025	0.2348-0.2354	0.2344-0.2350
2006	5.6	VK56DE	45.15-45.45	45	37.0@1.457	1.9913	0.0008-0.0021	0.0012-0.0025	0.2348-0.2354	0.2344-0.2350

09482_TITAN_C0005

CAMSHAFT AND BEARING SPECIFICATIONS CHART

All measurements are given in inches.

Year	Engine Displ. Liters	Engine ID/VIN	Journal Dia.	Brg. Oil Clearance	Shaft End-play	Runout	Journal Bore	Lobe Height Intake	Exhaust
2004	5.6	VK56DE	1.0218-1.0224	0.0012-0.0027	0.0045-0.0074	0.0008	1.0236-1.0244	1.7506-1.7581	1.7506-1.7581
2005	5.6	VK56DE	1.0218-1.0224	0.0012-0.0027	0.0045-0.0074	0.0008	1.0236-1.0244	1.7506-1.7581	1.7506-1.7581
2006	5.6	VK56DE	1.0218-1.0224	0.0012-0.0027	0.0045-0.0074	0.0008	1.0236-1.0244	1.7506-1.7581	1.7506-1.7581

09482_TITAN_C0006

CRANKSHAFT AND CONNECTING ROD SPECIFICATIONS

All measurements are given in inches.

Year	Engine Displ. Liters	Engine ID	Crankshaft Main Brg. Journal Dia.	Main Brg. Oil Clearance	Shaft End-play	Thrust on No.	Connecting Rod Journal Diameter	Oil Clearance	Side Clearance
2004	5.6	VK56DE	①	②	0.0118	3	③	0.0008-0.0015	0.0079-0.0157
2005	5.6	VK56DE	①	②	0.0118	3	③	0.0008-0.0015	0.0079-0.0157
2006	5.6	VK56DE	①	②	0.0118	3	③	0.0008-0.0015	0.0079-0.0157

① There are 24 different grades, ranging from G (2.5183) to 9 (2.5173)

② No. 1 and 5: 0.00004-0.0004

No. 2, 3 and 4: 0.0003-0.0007

③ Grade 0: 2.2441-2.2441

Grade 1: 2.2441-2.2442

Grade 2: 2.2442-2.2442

Grade 3: 2.2442-2.2443

Grade 4: 2.2443-2.2443

Grade 5: 2.2443-2.2443

Grade 6: 2.2443-2.2444

Grade 7: 2.2444-2.2444

Grade 8: 2.2444-2.2444

Grade 9: 2.2444-2.2445

Grade A: 2.2445-2.2445

Grade B: 2.2445-2.22446

Grade C: 2.2446-2.2446

09482_TITAN_C0007

PISTON AND RING SPECIFICATIONS

All measurements are given in inches.

Year	Engine Displacement Liters	Engine ID	Piston Clearance	Ring Gap Top Comp.	Bottom Comp.	Oil Control	Ring Side Clearance Top Comp.	Bottom Comp.	Oil Control
2004	5.6	VK56DE	0.0004-0.0012	0.0091-0.0130	0.0091-0.0130	0.0079-0.0236	0.0014-0.0033	0.0012-0.0028	0.0006-0.0073
2005	5.6	VK56DE	0.0004-0.0012	0.0091-0.0130	0.0091-0.0130	0.0079-0.0236	0.0014-0.0033	0.0012-0.0028	0.0006-0.0073
2006	5.6	VK56DE	0.0004-0.0012	0.0091-0.0130	0.0091-0.0130	0.0079-0.0236	0.0014-0.0033	0.0012-0.0028	0.0006-0.0073

09482_TITAN_C0008

TORQUE SPECIFICATIONS
All readings in ft. lbs.

Year	Engine Displacement Liters	Engine ID	Cylinder Head Bolts	Main Bearing Bolts	Rod Bearing Bolts	Crankshaft Damper Bolts	Flywheel Bolts	Manifold Intake	Manifold Exhaust	Spark Plugs	Oil Pan Drain Plug
2004	5.6	VK56DE	①	②	③	④	65	73	21	18	25
2005	5.6	VK56DE	①	②	③	④	65	73	21	18	25
2006	5.6	VK56DE	①	②	③	④	65	73	21	18	25

① Step 1: 72 ft. lbs

   Step 2: Loosen all bolts completely

   Step 3: 33 ft. lbs.

   Step 4: +60 degrees

   Step 5: +60 degrees

② Step 1: Main Bolts to 29 ft. lbs.

   Step 2: Sub-bolts to 22 ft. lbs.

   Step 3: Main Bolts +40 degrees

   Step 4: Sub-Bolts +30 degrees

   Step 5: Side Bolts to 36 ft. lbs.

③ Step 1: 11 ft. lbs.

   Step 2: +90 degrees

④ Step 1: 65 ft. lbs.

   Step 2: +90 degrees

09482_TITAN_C0009

WHEEL ALIGNMENT

Year	Model	Caster Range (+/-Deg.)	Caster Preferred Setting (Deg.)	Camber Range (+/-Deg.)	Camber Preferred Setting (Deg.)	Toe-in (in.)
2004	Titan ①	0.75	②	0.75	③	0.11+/-0.05
2005	Titan ①	0.75	②	0.75	③	0.11+/-0.05
2006	Titan ①	0.75	②	0.75	③	0.11+/-0.05

① Assumes P245/70R17 tire

② 4x2: +3.27
   4x4: +2.37

③ 4x2: -0.12
   4x4: +0.43

09482_TITAN_C0010

TIRE, WHEEL AND BALL JOINT SPECIFICATIONS

Year	Model	OEM Tires Standard	OEM Tires Optional	Tire Pressures (psi) Front	Tire Pressures (psi) Rear	Wheel Size	Ball Joint Inspection	Lugnut Torque (ft. lbs.)
2004	Titan	P245/70R17	①	35	35	17	②	98
2005	Titan	P245/70R17	①	35	35	17	②	98
2006	Titan	P245/70R17	①	35	35	17	②	98

OEM: Original Equipment Manufacturer

PSI: Pounds Per Square Inch

① P285/70R17

   P265/70R18

② Axial play

   Upper: 0

09482_TITAN_C0011

BRAKE SPECIFICATIONS

All measurements in inches unless noted

| Year | Model | | Brake Disc | | | Minimum Lining Thickness | Brake Caliper | |
			Original Thickness	Minimum Thickness	Maximum Runout		Bracket Bolts (ft. lbs.)	Mounting Bolts (ft. lbs.)
2004	Titan	F	1.024	0.965	0.0016	0.039	155	32
		R	0.551	0.472	0.002	0.039	NA	24
2005	Titan	F	1.024	0.965	0.0016	0.039	155	32
		R	0.551	0.472	0.002	0.039	NA	24
2006	Titan	F	1.024	0.965	0.0016	0.039	155	32
		R	0.551	0.472	0.002	0.039	NA	24

NA: Not applicable

09482_TITAN_C0012

SCHEDULED MAINTENANCE INTERVALS
Nissan - Titan

TO BE SERVICED	TYPE OF SERVICE	7.5	15	22.5	30	37.5	45	52.5	60
Engine oil & filter	R	✓	✓	✓	✓	✓	✓	✓	✓
Brake lines & cables	S/I		✓		✓		✓		✓
Brake pads and rotors	I		✓		✓		✓		✓
Driveshaft boots & propeller shaft (4x4)	L/I		✓		✓		✓		✓
Transmission, transfer & differential gear oil	I		✓		✓		✓		✓
Air cleaner filter	R				✓				✓
Engine coolant ①	R								✓
Spark plugs (Platinum)	R				Replace every 105,000 miles				
Drive belt(s) ②	S/I								✓
Cabin air filter	R		✓		✓		✓		✓
Exhaust system	I				✓				✓
Fuel lines	S/I				✓				✓
Fuel Filter ③									
Steering gear (box) & linkage, axle & suspension parts	I				✓				✓
Vapor lines	S/I				✓				✓

R: Replace   S/I: Service or Inspect   L: Lubricate

① Coolant: After 60,000 miles, inspect every 30,000 miles.

② Drive Belts: After 60,000 miles, inspect every 15,000 miles. Replace belts if damaged.

③ Fuel Filter: Maintenance free item.

FREQUENT OPERATION MAINTENANCE (SEVERE SERVICE)

If a vehicle is operated under any of the following conditions it is considered severe service:

- Extremely dusty areas.

- Rough, muddy, or salt spread roads.

- 50% or more of the vehicle constant operation is in 32°C (90°F) or higher temperatures, or temperatures below 0°C (32°F).

- Prolonged idling (vehicle operation in stop and go traffic).

- Frequent short running periods (engine does not warm to normal operating temperatures).

- Police, taxi, delivery usage or trailer towing usage.

Oil & oil filter: replace every 3750 miles.

Brake pads, discs, drums & linings: service or inspect every 7500 miles.

Driveshaft boots & propeller shaft: service or inspect every 7500 miles.

Exhaust system: service or inspect every 7500 miles.

Steering gear (box) & linkage, (steering damper-4x4), axle & suspension parts: service or inspect every 7500 miles.

Steering linkage ball joints & front suspension ball joints: service or inspect every 7500 miles.

09482_TITAN_C0013

ENGINE REPAIR

Distributor

REMOVAL & INSTALLATION

These models use a Distributorless Ignition System (DIS) controlled by the Powertrain Control Module (PCM).

Alternator

REMOVAL & INSTALLATION

1. Before servicing the vehicle, refer to the Precautions Section.
2. Remove or disconnect the following:
   - Negative battery cable
   - Fan shroud
   - Drive belt
   - Lower alternator bracket
   - Alternator upper bolt
   - Alternator harness connectors
   - Alternator

To install:
3. Install or connect the following:
   - Alternator
   - Alternator harness connectors
   - Upper bolt, tighten to 48 ft. lbs. (65 Nm)
   - Lower bracket, tighten to 16 ft. lbs (22 Nm)
   - Drive belt
   - Fan shroud
   - Negative battery cable

Ignition Timing

ADJUSTMENT

Ignition timing is controlled by the ECM and manual adjustment is not possible.

Engine Assembly

REMOVAL & INSTALLATION

1. Before servicing the vehicle, refer to the Precautions Section.
2. Drain the cooling system.
3. Partially drain the automatic transmission fluid.
4. Relieve the fuel system pressure.
5. Remove or disconnect the following:
   - Hood
   - Cowl extension
   - Engine cover
   - Air intake assembly
   - Vacuum hose between vehicle and engine
   - Radiator hoses
   - Radiator
   - Drive belts
   - Engine fan
   - Wiring harness
   - ECM
   - Power steering reservoir tank and oil pump
   - A/C compressor
   - Brake booster vacuum line
   - EVAP line
   - Fuel hose
   - Heater hoses
   - Transmission dipstick assembly
   - Front final drive assembly (4WD models only)
   - Exhaust manifolds
6. Install engine slings onto the left and right cylinder heads and tighten to 33 ft. lbs. (45 Nm).
7. Attach an engine hoist to slings and lift engine out of the vehicle

To install:
8. Lower engine into the vehicle and remove the engine slings.
9. Install or connect the following:
   - Exhaust manifolds
   - Front final drive assembly (4WD models only)
   - Transmission dipstick assembly
   - Heater hoses
   - Fuel hose
   - EVAP line
   - Brake booster vacuum line
   - A/C compressor
   - Power steering oil pump and reservoir tank
   - ECM
   - Wiring harnesses
   - Engine fan
   - Drive belts
   - Radiator
   - Radiator hoses

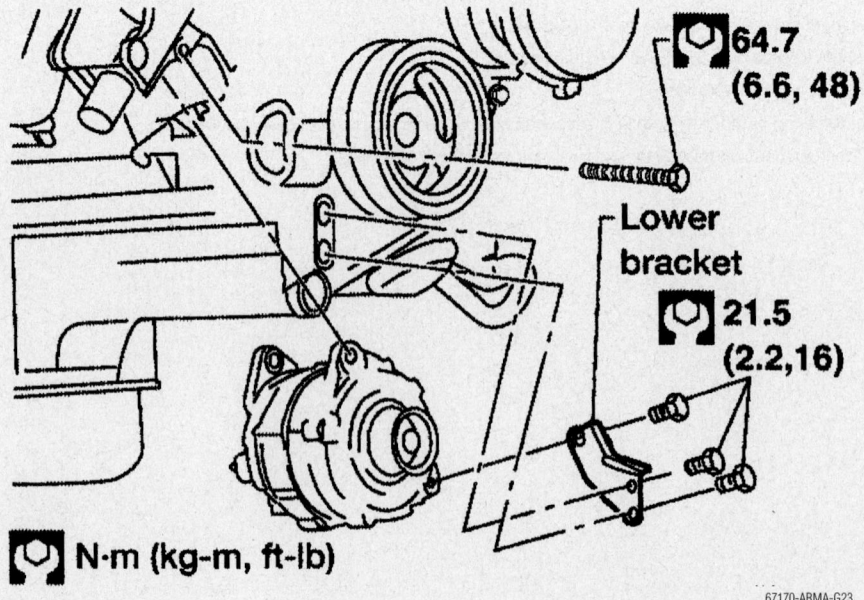

64.7
(6.6, 48)

Lower bracket
21.5
(2.2,16)

N·m (kg-m, ft-lb)

67170-ARMA-G23

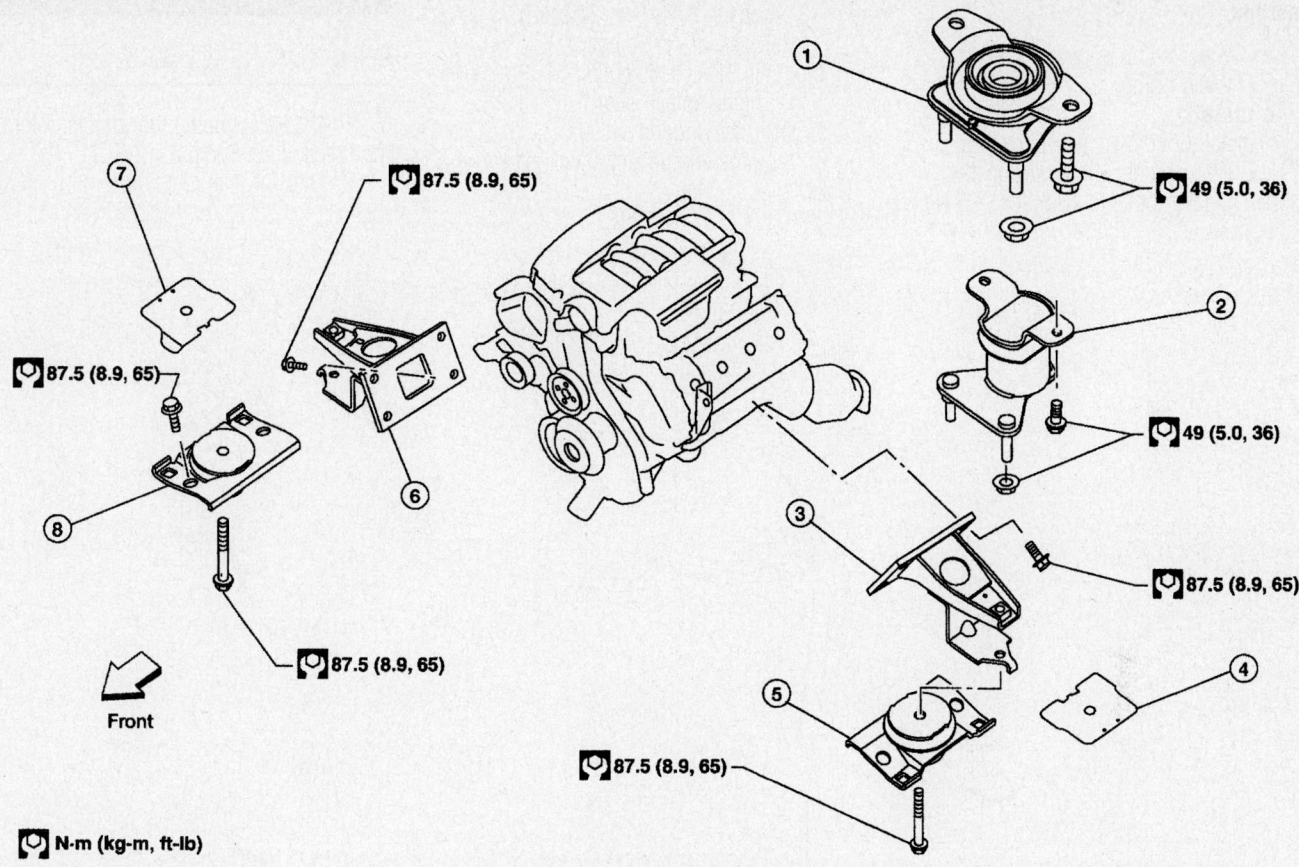

⊡ **N·m (kg-m, ft-lb)**

1. Rear engine mounting insulator 4x4
2. Rear engine mounting insulator 4x2
3. LH engine mounting bracket
4. LH Heat shield plate
5. LH engine mounting insulator
6. RH engine mounting bracket
7. RH Heat shield plate
8. RH engine mounting insulator

67170-ARMA-G24

Engine mounts

- Vacuum hose between engine and vehicle
- Air intake assembly
- Engine cover
- Cowl extension
- Hood

10. Refill the automatic transmission fluid.
11. Refill the cooling system.
12. Start the engine and check for leaks.

Water Pump

REMOVAL & INSTALLATION

1. Before servicing the vehicle, refer to the Precautions Section.
2. Drain the cooling system.
3. Remove or disconnect the following:
   - Engine splash guard
   - Air intake assembly
   - Accessory drive belt

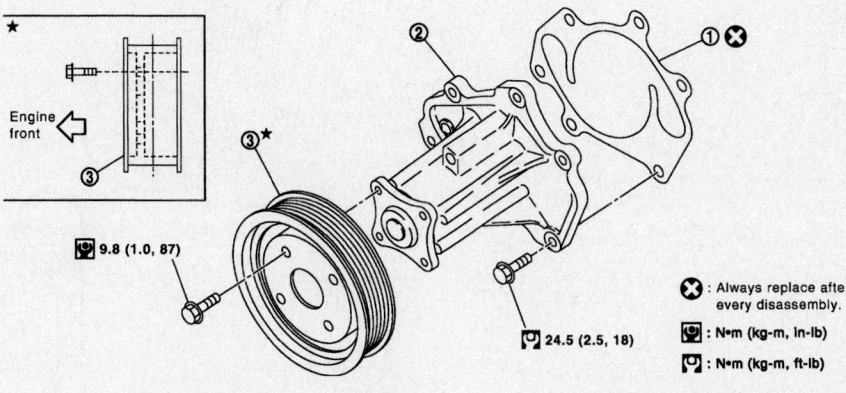

1. Gasket
2. Water pump
3. Water pump pulley

❌ : Always replace after every disassembly.
⊡ : N·m (kg-m, in-lb)
⊡ : N·m (kg-m, ft-lb)

Water pump mounting

67170-ARMA-G25

→**Leave tensioner pulley in its fixed position.**

- Water pump pulley
- Water pump

To install:

4. Install or connect the following:
- Water pump with a new gasket. Tighten bolts to 18 ft. lbs. (25 Nm).

- Water pump pulley and tighten bolts to 87 in. lbs. (10 Nm).
- Accessory drive belt
- Air intake assembly
- Engine splash guard

5. Refill the cooling system.
6. Start the engine and check for leaks.

Heater Core

REMOVAL & INSTALLATION

1. Before servicing the vehicle, refer to the Precautions Section.
2. Discharge the A/C system.

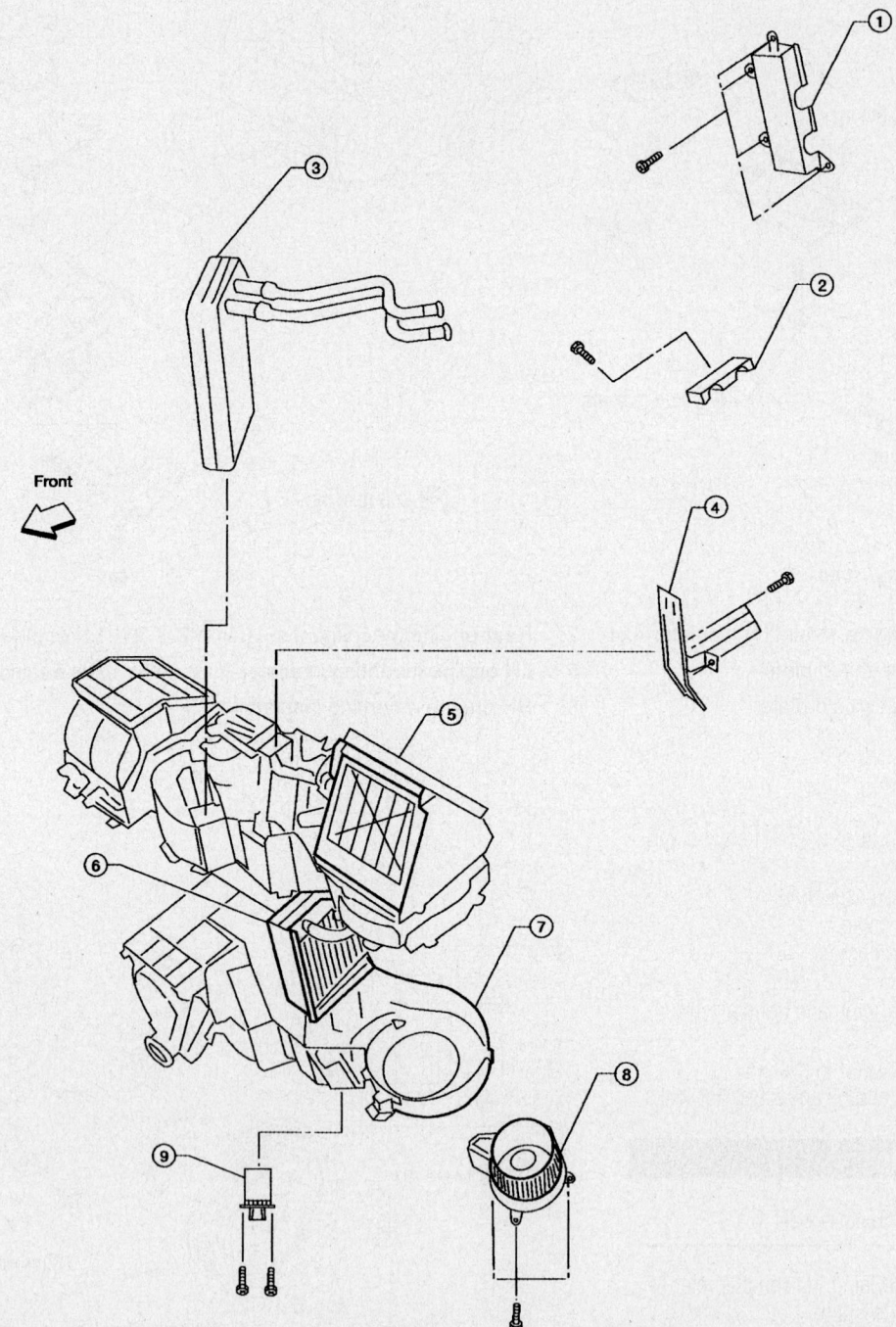

Front

1.	Heater core cover	2.	Heater core pipe bracket	3.	Heater core
4.	Upper bracket	5.	Upper heater and cooling unit case	6.	A/C evaporator
7.	Lower heater and cooling unit case	8.	Blower motor	9.	Variable blower control

67170-ARMA-G26

Front heater/AC assembly

3. Drain the cooling system.
4. Remove or disconnect the following:
   - Right and left wiper arms
   - Cowl top seal
   - Left and right-hand cowl top covers
   - Cowl top extension brackets
   - Wiper motor and connecting rod linkage
   - Windshield washer tube
   - A/C low pressure pipe bracket from the cowl top extension
   - Drain tube from each side of the cowl top extension
   - Cowl top extension
   - All necessary exhaust system components
   - Front heater hoses from the heater core
   - High/Low pressure pipes
   - Center console
   - Steering column
   - Gauge cluster
   - Defroster grille
   - Left-hand side ventilator assembly

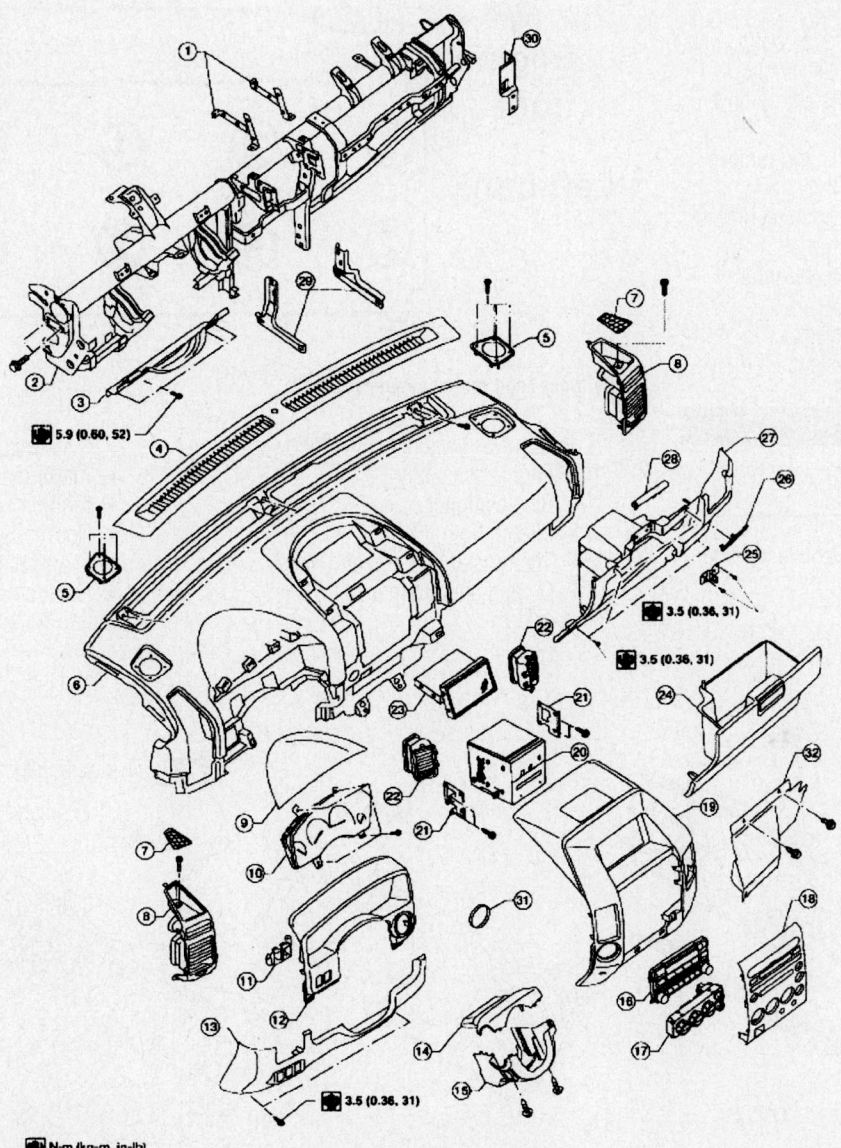

N-m (kg-m, in-lb)

1. Display unit bracket RH/LH	2. Steering member assembly	3. Lower knee protector
4. Defroster grille	5. Speaker grille RH/LH	6. Instrument panel and pad assembly
7. Deck pocket mat RH/LH	8. Side ventilator assembly RH/LH	9. Meter cover
10. Combination meter	11. Switch assembly	12. Cluster lid A
13. Lower instrument panel LH	14. Upper steering column cover	15. Lower steering column cover
16. Audio display switch assembly	17. Front air control	18. Cluster lid C
19. Cluster lid D	20. Audio unit	21. Radio Bracket RH/LH
22. Center ventilator assembly RH/LH	23. Display assembly	24. Glove box
25. Glove box lid striker	26. Fuse block cover	27. Lower instrument panel RH
28. Glove box damper	29. Instrument stay RH/LH	30. Instrument side bracket
31. Key cylinder escutcheon	32. Lower instrument panel RH	

09482_TITAN_G0001

Exploded view of the Instrument Panel

- Right-hand assist grip and windshield trim
- Passenger air bag module
- Instrument and console panels
- Instrument panel wiring harness
- Steering member from each side of the vehicle body

5. Remove the heater assembly from the vehicle, attached to the steering gear assembly.

6. Remove the heating and cooling unit assembly from the steering assembly.

7. Remove the four bolts and remove the upper bracket.

8. Remove the four bolts and remove the heater core cover.

9. Remove the heater core pipe bracket.

10. Remove the heater core.

11. Installation is the reverse order of removal.

12. Refill the cooling system.

13. Recharge the A/C system.

Cylinder Head

REMOVAL & INSTALLATION

1. Before servicing the vehicle, refer to the Precautions Section.

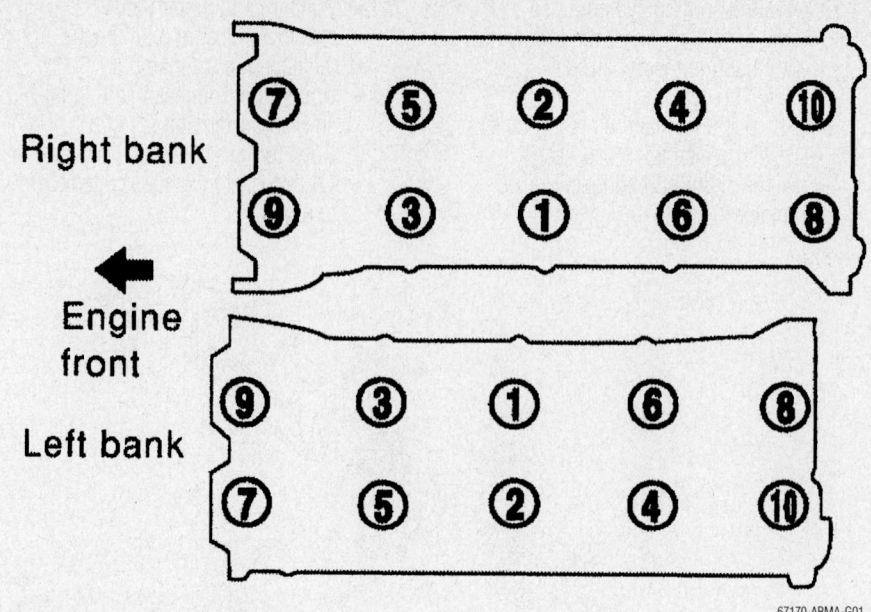

67170-ARMA-G01

Cylinder head torque sequence

2. Remove or disconnect the following:
- Engine assembly
- Belt tensioner
- Idler pulley
- Thermostat housing and hose
- Oil pan and strainer
- Fuel tube and injector assembly
- Intake manifold
- Ignition coil
- Rocker cover
- Crankshaft pulley
- Front engine cover

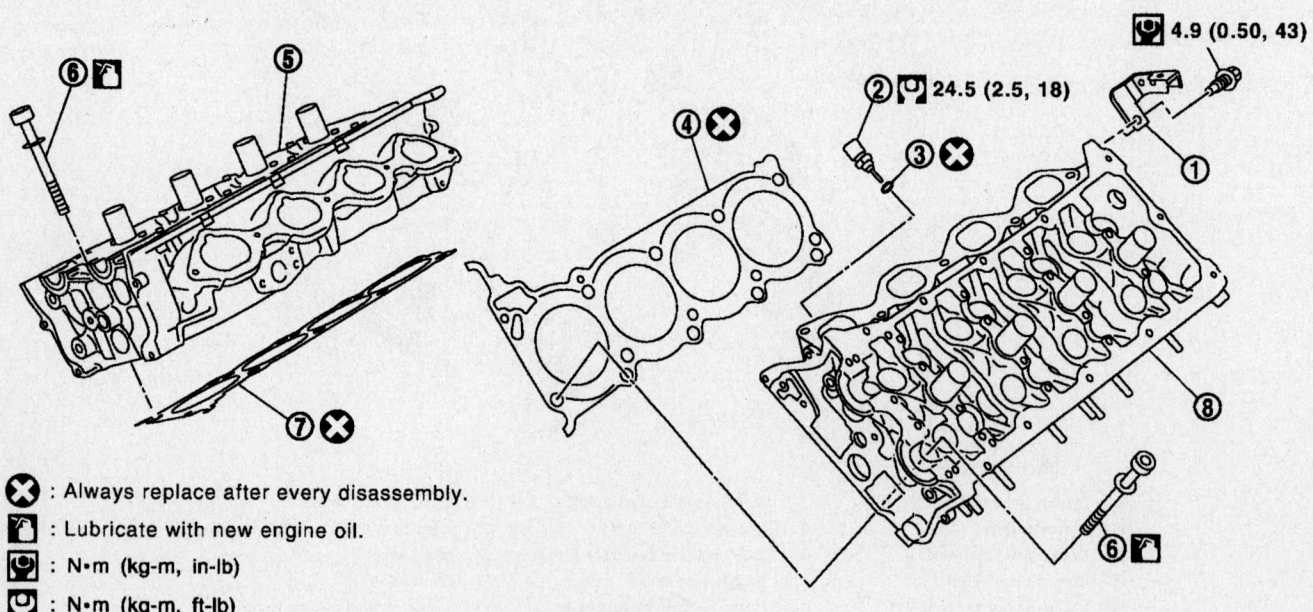

⊗ : Always replace after every disassembly.

🛢 : Lubricate with new engine oil.

🔧 : N•m (kg-m, in-lb)

🔧 : N•m (kg-m, ft-lb)

1.	Harness bracket	2.	Engine coolant temperature sensor	3.	Washer
4.	Cylinder head gasket (left bank)	5.	Cylinder head (right bank)	6.	Cylinder head bolt
7.	Cylinder head gasket (right bank)	8.	Cylinder head (left bank)		

67170-ARMA-G28

- Oil pump
- Timing chain
- Camshaft sprockets
- Camshafts
- Cylinder head, removing bolts in reverse order shown in figure

3. Install the cylinder head with a new gasket. Tighten the bolts in sequence as follows:
   a. Step 1: 72 ft. lbs. (98 Nm)
   b. Step 2: Loosen all bolts completely
   c. Step 3: 33 ft. lbs. (44 Nm)
   d. Step 4: Plus 60 degrees
   e. Step 5: Plus 60 degrees
4. Install or connect the following:
   - Camshaft
   - Camshaft sprockets
   - Timing chain
   - Oil pump
   - Front engine cover
   - Crankshaft pulley
   - Rocker cover

- Ignition coil
- Intake manifold
- Fuel tube and injector assembly
- Oil pain and strainer
- Thermostat housing and hose
- Idler pulley
- Belt tensioner
- Engine assembly

5. Start the engine and check for leaks

Intake Manifold

REMOVAL & INSTALLATION

1. Before servicing the vehicle, refer to the Precautions Section.
2. Drain the cooling system.
3. Relieve the fuel system pressure.
4. Remove or disconnect the following:
   - Engine cover
   - Air intake assembly

- Fuel tube quick connector using special tool J-45488
- Wiring harnesses and brackets from manifold
- Vacuum hoses
- PCV hose and tube
- Electric throttle control actuator, loosening bolts diagonally
- Fuel injectors
- Fuel tube assembly
- Intake manifold, removing bolts in reverse order shown in figure

To install:

5. Install the intake manifold with new gaskets. Tighten the bolts in order as shown.
6. Install or connect the following:
   - Fuel tube assembly
   - Fuel injectors
   - Electronic throttle control actuator, tightening the bolts in several steps
   - PCV hose

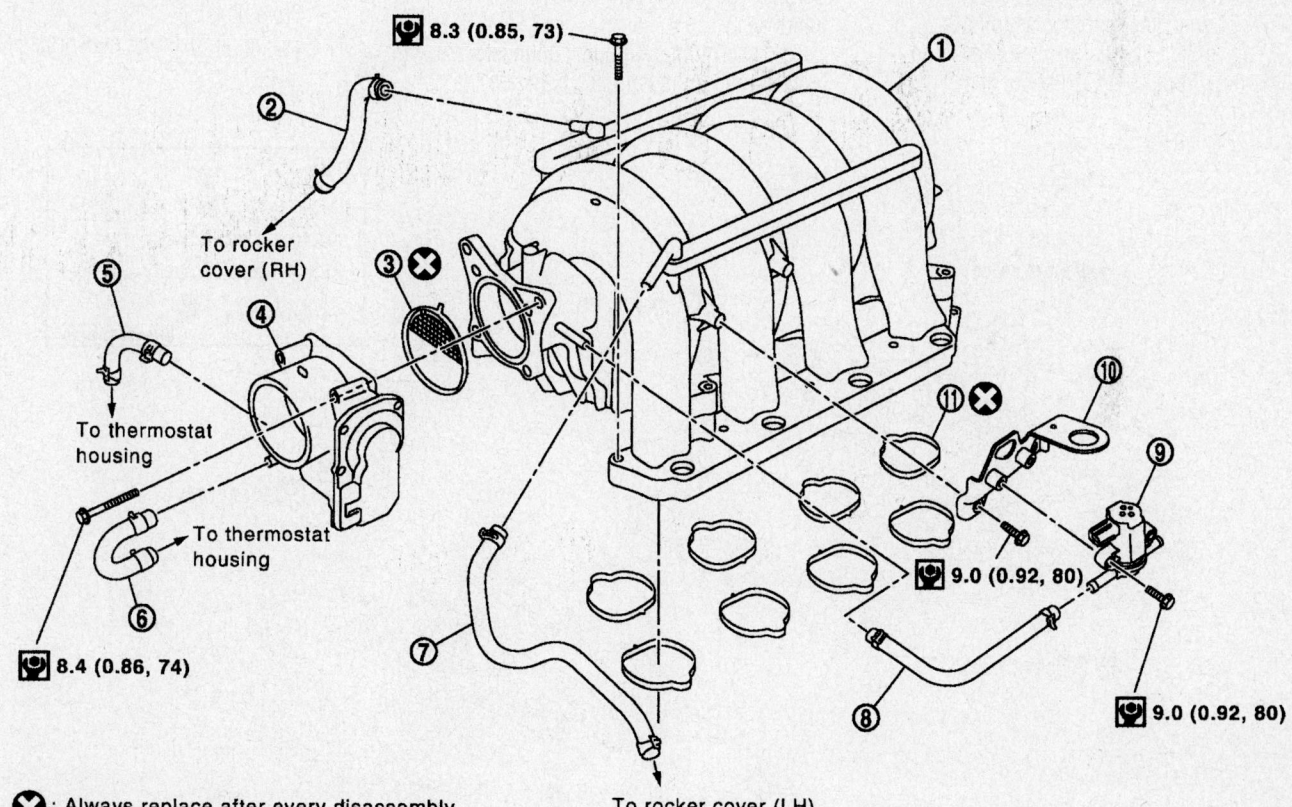

❌: Always replace after every disassembly.

🔧: N•m (kg-m, in-lb)

1.	Intake manifold	2.	PCV hose	3.	Gasket
4.	Electric throttle control actuator	5.	Water hose	6.	Water hose
7.	PCV hose	8.	EVAP hose	9.	EVAP canister purge control solenoid valve
10.	Bracket	11.	Gasket		

67170-ARMA-G29

Intake manifold and related parts

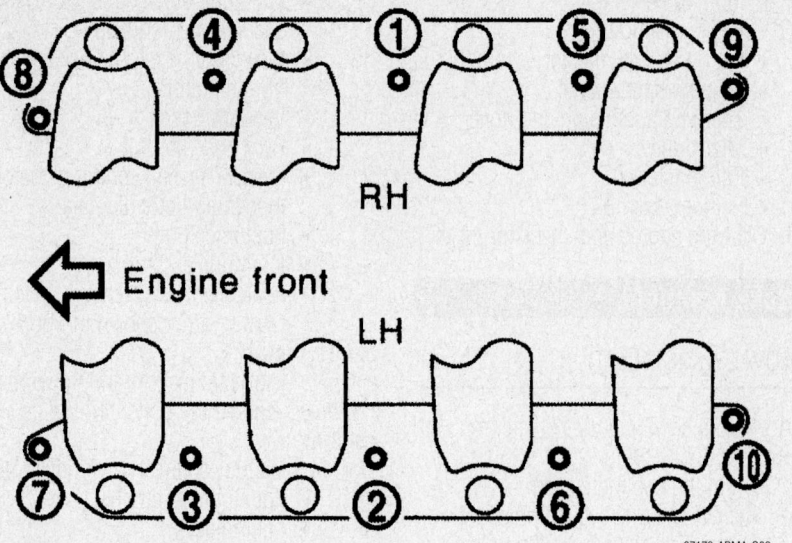

Intake manifold torque sequence

- Vacuum hoses
- Wiring harnesses

7. Connect the fuel tube as follows:
   a. Apply a thin layer of engine oil on the tube from tip end to spool end.

   b. Insert tube into quick connector past the white identification mark

   c. Insert tube into quick connector until top spool is completely inside the connector and 2nd level spool is exposed right below the connector.

   d. Pull slightly on the quick connector to ensure it is fully engaged.

   e. Install quick connector cap on quick connector joint.

8. Install or connect the following:
   - Air intake assembly
   - Engine cover
9. Refill the cooling system.
10. Start engine and check for leaks.

Exhaust Manifold

REMOVAL & INSTALLATION

1. Before servicing the vehicle, refer to the Precautions Section.
2. Drain the cooling system.
3. Remove or disconnect the following:
   - Air intake assembly
   - Engine splash guard
   - Radiator and radiator hoses
   - Drive belts
4. Remove air fuel ratio sensors as follows:

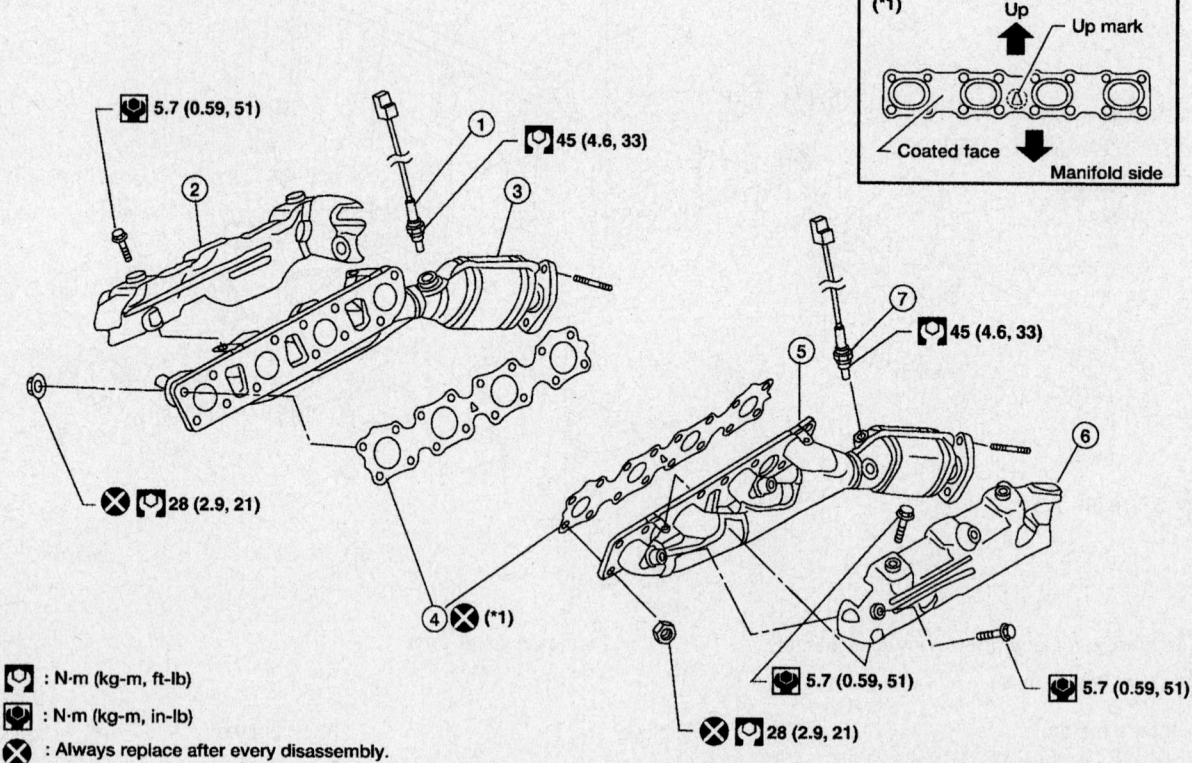

1. Air fuel ratio (A/F) sensor 1 (bank 2)
2. Exhaust manifold cover (right bank)
3. Exhaust manifold (right bank)
4. Gaskets
5. Exhaust manifold (left bank)
6. Exhaust manifold cover (left bank)
7. Air fuel ratio (A/F) sensor 1 (bank 1)

Exhaust manifolds and related parts

67170-ARMA-G30

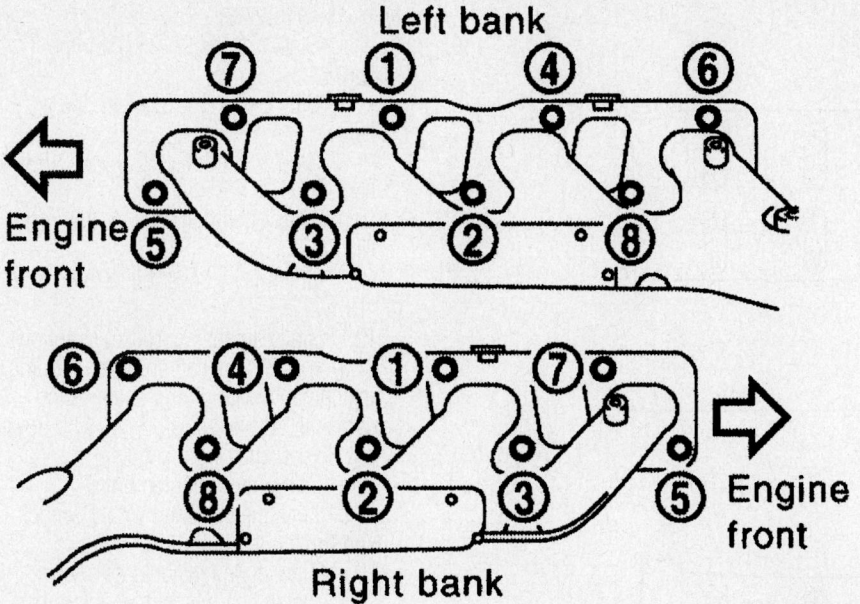

Exhaust manifold torque sequence

a. Remove engine cover.

b. Remove wiring harness from each sensor

c. Remove sensors, using special tool J-38356

5. Remove front cross bar

6. Remove left exhaust manifold as follows:

a. Remove the exhaust front tube.

b. Remove the exhaust manifold cover.

c. Loosen nuts in reverse order shown in figure.

d. Remove studs from position 2, 4, 6, and 8 and remove manifold.

7. Remove right exhaust manifold as follows:

a. Remove the exhaust front tube.

b. Remove the oil level gauge guide.

c. Remove the exhaust manifold cover.

d. Loosen nuts in reverse order shown in figure.

e. Remove studs from position 2, 4, 6, and 8 and remove manifold.

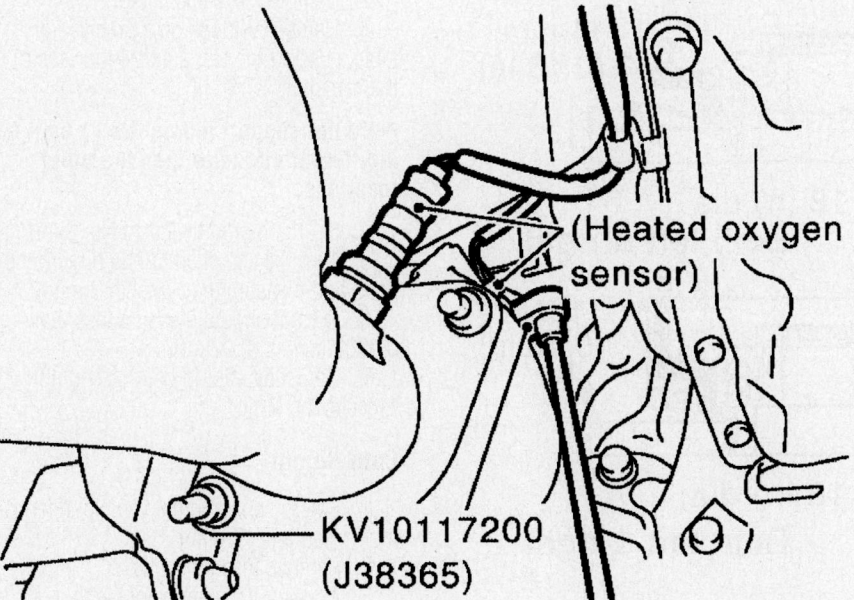

Removing Air-Fuel ratio sensors

To install:

8. Install or connect the following:

- Exhaust manifold gasket with triangle mark facing up and coated (gray) face toward exhaust manifold.
- Exhaust manifold, tightening the nuts as shown in figure
- Exhaust manifold cover
- Oil level gauge guide (right side only)
- Exhaust front tube
- Front cross bar
- Air fuel ratio sensors, with anti-seize lubricant
- Engine cover
- Drive belts
- Radiator and radiator hoses
- Engine splash guard
- Air intake assembly

9. Refill the cooling system

10. Start engine and check for leaks.

Camshaft and Valve Lifters

REMOVAL & INSTALLATION

1. Before servicing the vehicle, refer to the Precautions Section.

2. Remove rocker cover.

3. Obtain compression Top Dead Center (TDC) of No. 1 cylinder.

4. Remove timing chain case cover.

5. Matchmark the timing chain, aligning with the camshaft sprocket marks.

6. Remove chain tensioner from left bank as follows:

a. Squeeze end clips and push plunger into tensioner body.

b. Secure plunger using stopper pin.

c. Remove chain tensioner.

7. Remove chain tensioner from right bank as follows:

a. Remove chain tensioner cover using special tool J-37228.

b. Squeeze end clips and push plunger into tensioner body.

c. Secure plunger using stopper pin.

d. Remove chain tensioner.

8. With camshaft locked with a wrench, loosen bolts to remove camshaft sprocket.

9. Remove front cover bolts.

10. Remove camshaft brackets, removing bolts in reverse order shown in figure.

11. Remove camshaft.

12. Remove valve lifters.

To install:

13. Install valve lifters.

14. Install camshaft, refer to table for correct placement.

15. Install camshaft brackets as follows:

Right bank

Exhaust

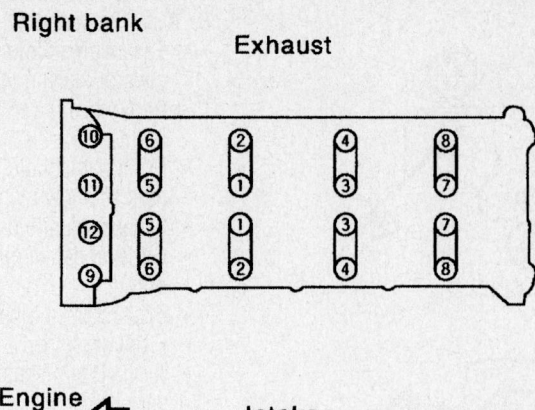

Engine
front ← —— Intake ——————

Left bank

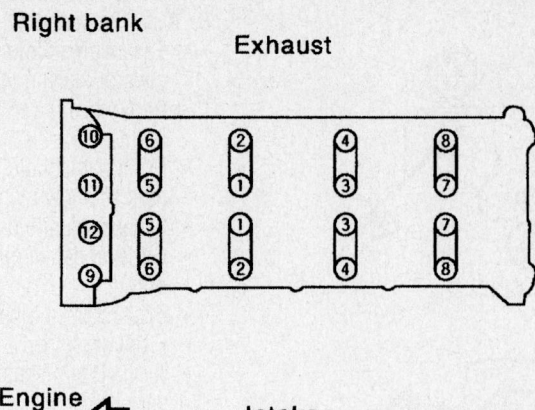

Exhaust

67170-ARMA-G04

Camshaft torque sequence

a. Refer to location mark on upper surface of bracket.

b. Installation mark should be correctly read when viewed from intake side.

16. Install camshaft bracket #1 as follows:

a. Apply liquid gasket to bracket and backside of front cover as shown in figure.

b. Carefully position and mount camshaft bracket #1.

c. Temporarily tighten front cover bolts

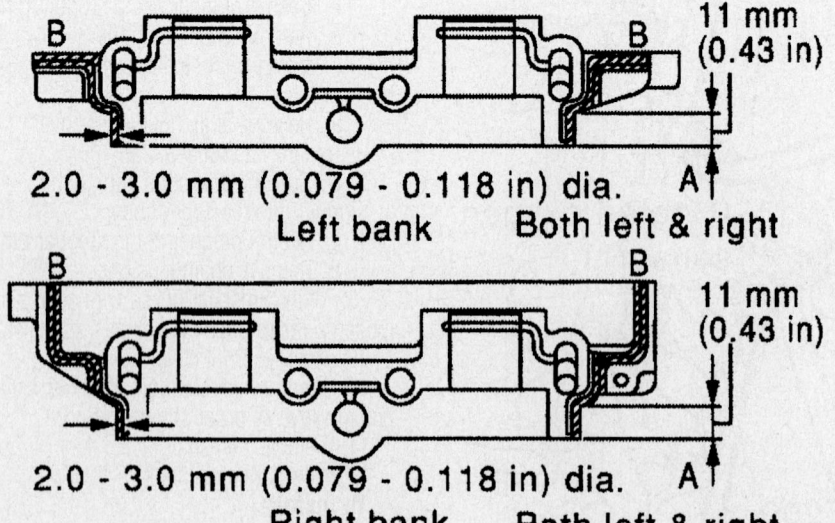

11 mm (0.43 in)

2.0 - 3.0 mm (0.079 - 0.118 in) dia.
Left bank **Both left & right**

11 mm (0.43 in)

2.0 - 3.0 mm (0.079 - 0.118 in) dia.
Right bank **Both left & right**

67170-ARMA-G05

Gasket application for camshaft bracket

17. Tighten fixing bolts for camshaft brackets as follows:

a. Step 1: Bolts 9-12: 17 in. lbs. (1.9 Nm)

b. Step 2: Bolts 1-8: 17 in. lbs. (1.9 Nm)

c. Step 3: All bolts: 52 in. lbs. (5.9 Nm)

d. Step 4: All bolts: 92 in. lbs. (10 Nm)

18. Tighten front cover bolts to 8 ft. lbs. (11 Nm)

19. Install camshaft sprocket as follows:

a. Install camshaft sprocket aligning matchmarks with timing chain. Align camshaft sprocket key groove with dowel pin on camshaft front edge.

b. Temporarily tighten bolts.

c. Lock the camshaft with a wrench and tighten the bolts.

20. Install chain tensioner as shown:

a. Install chain tensioner, compress plunger and hold with stopper pin.

b. Tighten chain tensioner bolts to 61 in. lbs. (7 Nm)

c. Remove stopper pin, release plunger and apply tension to timing chain.

d. Install chain tensioner front cover (Right-hand bank only) and tighten bolts to 80 in. lbs. (9 Nm).

21. Install timing chain cover.

22. Install rocker cover.

INSPECTION

Runout

1. Before servicing the vehicle, refer to the Precautions Section.

2. Remove the camshafts.

3. Using a V-block on a precise flat table, support the No. 2 and 4 journals of the camshaft.

➡**Do not support journal No. 1 as it has a different diameter than the other locations.**

4. Set the dial indicator to No. 3 journal.

5. Turn the camshaft to one direction by hand and measure the camshaft runout.

6. Runout should measure less than 0.0008 inches (0.02 mm).

7. Camshaft should be replaced if it exceeds the limit.

Cam Height

1. Before servicing the vehicle, refer to the Precautions Section.

2. Remove the camshafts.

3. Measure the cam height with a micrometer.

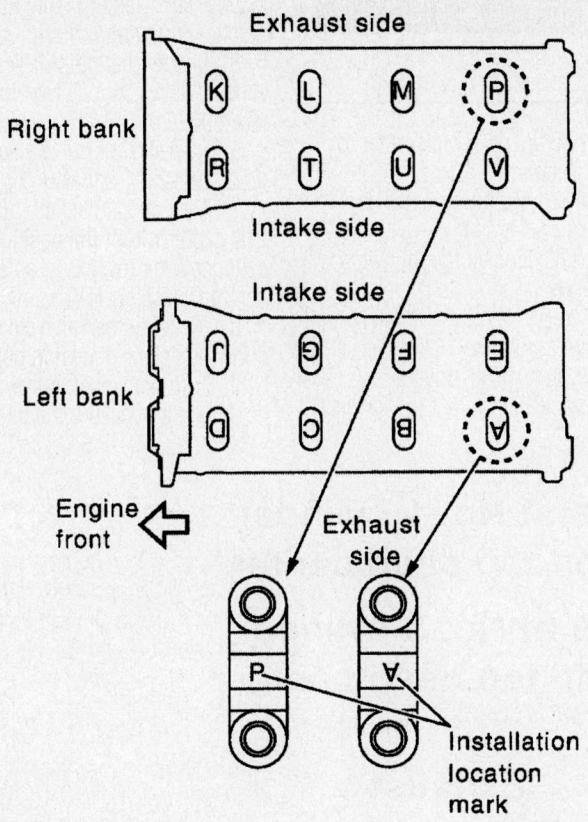

Camshaft bracket installation markings

4. The intake and exhaust camshaft should measure between 1.7506–1.7581 inches (44.465–44.655 mm).

5. Camshaft should be replaced if it exceeds the limit.

Journal Oil Clearance

1. Before servicing the vehicle, refer to the Precautions Section.

2. Remove the camshafts.

3. Measure the outer diameter of camshaft journal with micrometer and record the result.

4. Reinstall the camshaft bearing caps in accordance to the installation procedure.

5. Measure the inner diameter of the camshaft bracket ("A") with a bore gauge and record the result.

6. Subtract the camshaft journal diameter from the camshaft bracket inner diameter. The difference should measure between 0.0012–0.0027 inches (0.030–0.068 mm).

7. The camshaft or camshaft bracket should be replaced if it exceeds the limit.

➡ **The camshaft bracket cannot be replaced as an individual part, because it is machined together with the cylinder head. The entire cylinder head assembly must be replaced.**

End Play

1. Before servicing the vehicle, refer to the Precautions Section.

2. Install a dial indicator in the thrust direction on the front end of the camshaft.

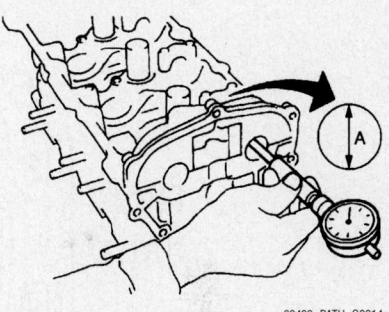

Measuring the journal bore for oil clearance

R: For right bank
L: For left bank

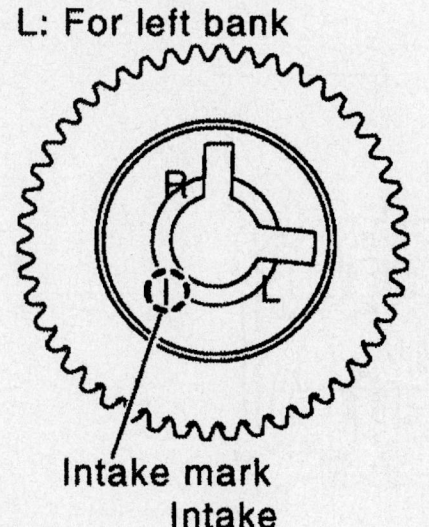

Intake mark
Intake

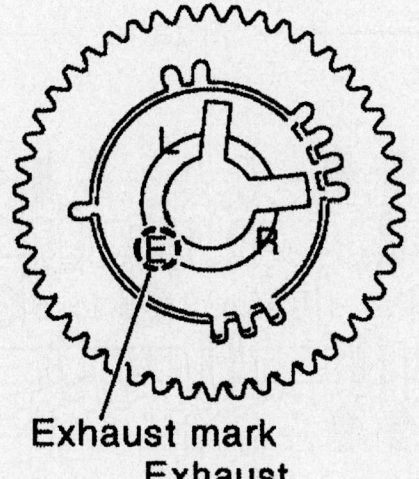

Exhaust mark
Exhaust

67170-ARMA-G16

Camshaft installation markings

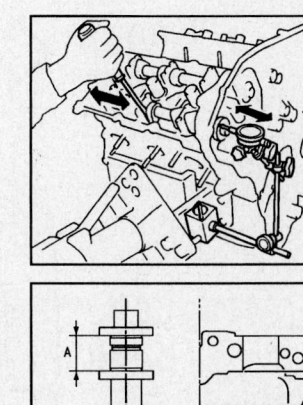

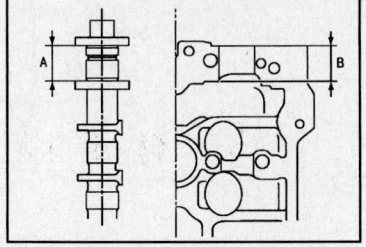

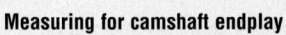

09482_PATH_G0015

Measuring for camshaft endplay

Measure the end play of the dial indicator when the camshaft is moved back and forth. The dial indicator should measure between 0.0045–0.0074 inches (0.115–0.188 mm).

3. Measure the No. 1 journal as shown. The distance ("A") should be 1.2008–1.2027 inches (30.500–30.548 mm).

4. Measure the No. 1 journal bearing as shown. The distance ("B") should be 1.1953–1.1963 inches (30.360–30.385 mm).

5. Replace either the camshaft or cylinder head assembly if the measurement is exceeded.

Valve Lash

INSPECTION

1. Before servicing the vehicle, refer to the Precautions Section.
2. Run engine to operating temperature.
3. Remove or disconnect the following:
    *   Engine cover
    *   Battery cover
    *   Air intake assembly
    *   Left and right rocker covers
4. Turn the crankshaft pulley clockwise to Top Dead Center (TDC) identification notch with timing indicator.
5. Ensure that both the intake and exhaust cam noses of the No. 1 cylinder face outside.
6. Measure the valve clearances at locations marked 'x' shown in figure.
7. Turn the crankshaft pulley clockwise 270 degrees from the position of No. 1 cylinder compression to obtain No. 3 cylinder compression TDC.
8. Measure the valve clearances at locations marked 'x' shown in next figure.
9. Turn crankshaft pulley clockwise 90 degrees and measure the intake and exhaust

↑ : Measurable at No. 1 cylinder compression top dead center

⇧ : Measurable at No. 3 cylinder compression top dead center

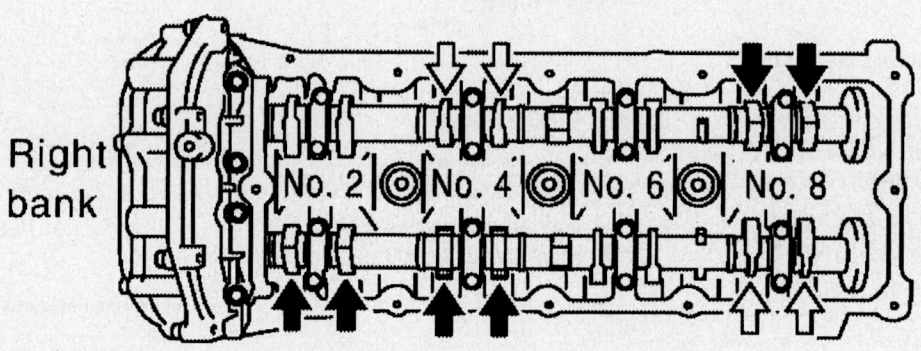

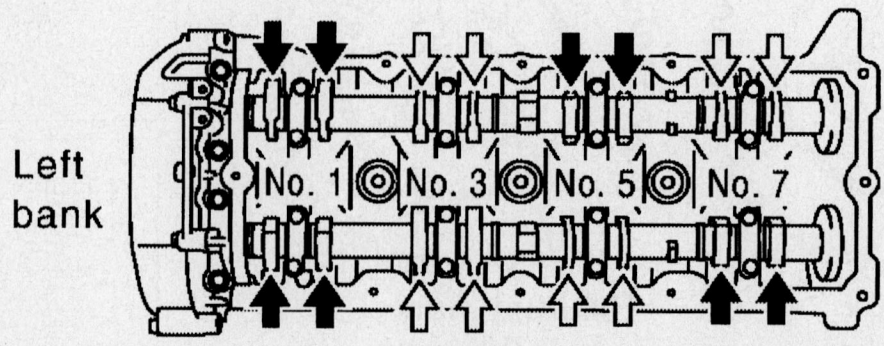

67170-ARMA-G06

Locations to measure clearance with No. 1 cylinder at TDC

↑ (solid) : Measurable at No. 1 cylinder compression top dead center

⇧ (outline) : Measurable at No. 3 cylinder compression top dead center

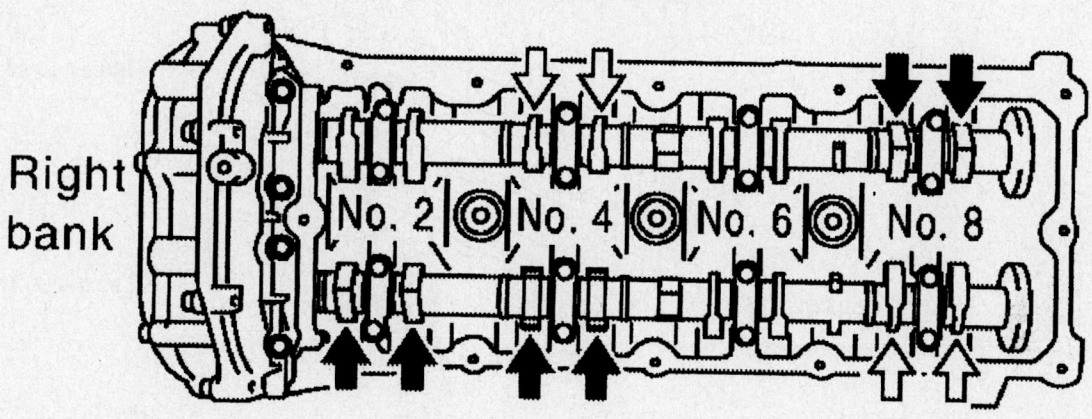

Exhaust

Right bank

No. 2   No. 4   No. 6   No. 8

Engine front ⇐ — · — Intake — · — · —

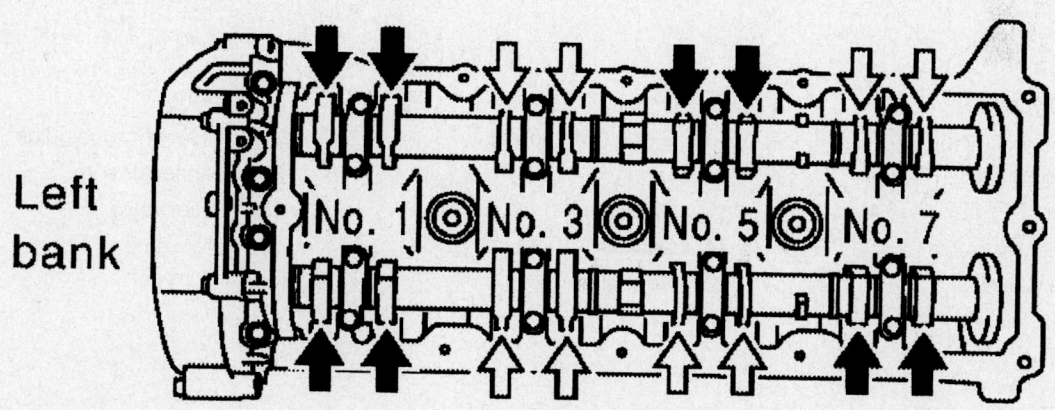

Left bank

No. 1   No. 3   No. 5   No. 7

Exhaust

67170-ARMA-G07

Locations to measure clearance with No. 3 cylinder at TDC

valve clearance of No. 6 cylinder and exhaust valve clearance of No. 2 cylinder.

ADJUSTMENT

1. Remove camshaft and valve lifter(s) out of specification.

2. Install replacement valve lifter(s).
3. Install the camshaft.
4. Manually turn the crankshaft pulley several turns.
5. Recheck valve clearances with engine at operating temperature.

Oil Pan

REMOVAL & INSTALLATION

1. Before servicing the vehicle, refer to the Precautions Section.

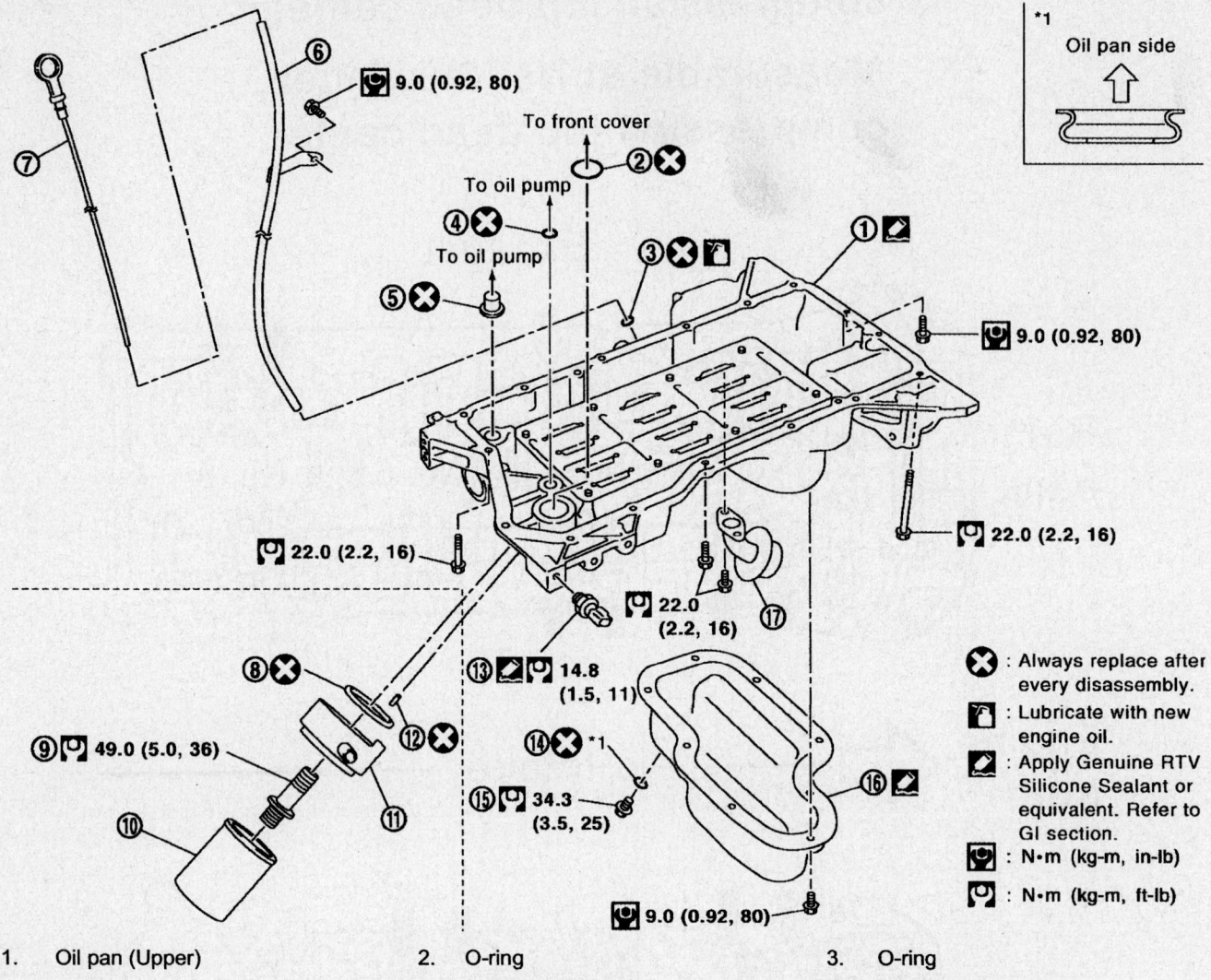

*1 Oil pan side

🔧 9.0 (0.92, 80)

To front cover

To oil pump

To oil pump

🔧 9.0 (0.92, 80)

🔧 22.0 (2.2, 16)

🔧 22.0 (2.2, 16)

🔧 22.0 (2.2, 16)

🔧 14.8 (1.5, 11)

🔧 49.0 (5.0, 36)

🔧 34.3 (3.5, 25)

*1

🔧 9.0 (0.92, 80)

❌ : Always replace after every disassembly.

🛢 : Lubricate with new engine oil.

✏ : Apply Genuine RTV Silicone Sealant or equivalent. Refer to GI section.

🔧 : N·m (kg-m, in-lb)

🔧 : N·m (kg-m, ft-lb)

1. Oil pan (Upper)	2. O-ring	3. O-ring
4. O-ring	5. O-ring (with collar)	6. Oil level gauge guide
7. Oil level gauge	8. O-ring	9. Connector bolt
10. Oil filter	11. Oil cooler	12. Relief valve
13. Oil pressure switch	14. Gasket	15. Drain plug
16. Oil pan (Lower)	17. Oil strainer	

67170-ARMA-G31

Oil pan and related parts

2. Remove engine assembly.

3. Remove lower oil pan, loosening bolts in reverse order shown in figure.

4. Remove oil strainer from upper oil pan.

5. Gently pry and remove upper oil pan from engine block.
To install:

6. Apply liquid gasket to upper oil pan mating surfaces.

7. Install new O-rings to oil pump and front cover side.

8. Tighten upper oil pan bolts in following numerical order:
   a. No. 15, 16

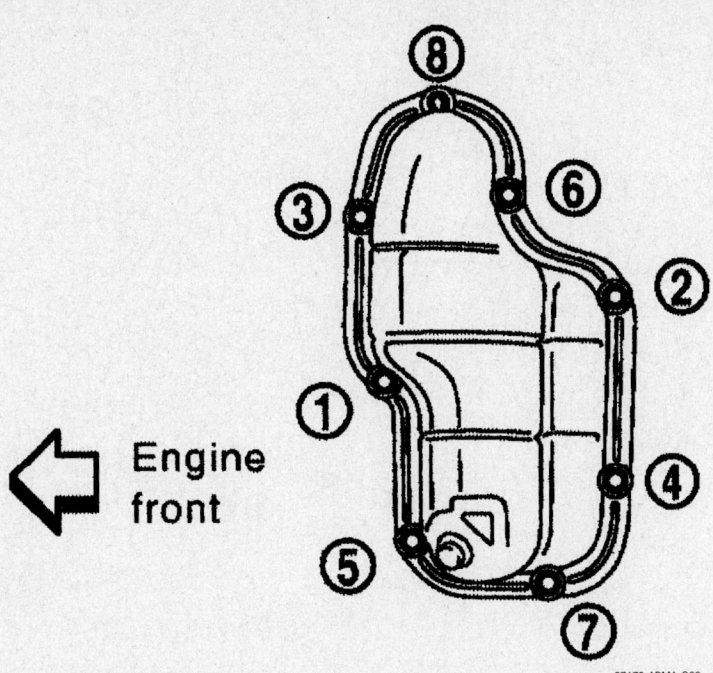

Lower oil pan torque sequence

67170-ARMA-G08

b. No. 1, 3, 5, 7, 11, 13
c. No. 2, 4, 6, 8, 10, 14
d. No. 9, 12
9. Install or connect the following:
   • Rear plate cover
   • Oil strainer to upper oil pan
   • Lower oil pan, tightening bolts in order shown in figure

Oil Pump

REMOVAL & INSTALLATION

1. Before servicing the vehicle, refer to the Precautions Section.
2. Remove or disconnect the following:
   • Timing chain cover
   • Oil pump drive spacer
   • Oil pump
To install:
3. Install or connect the following:
   • Oil pump
   • Oil pump drive spacer
   • Timing chain cover

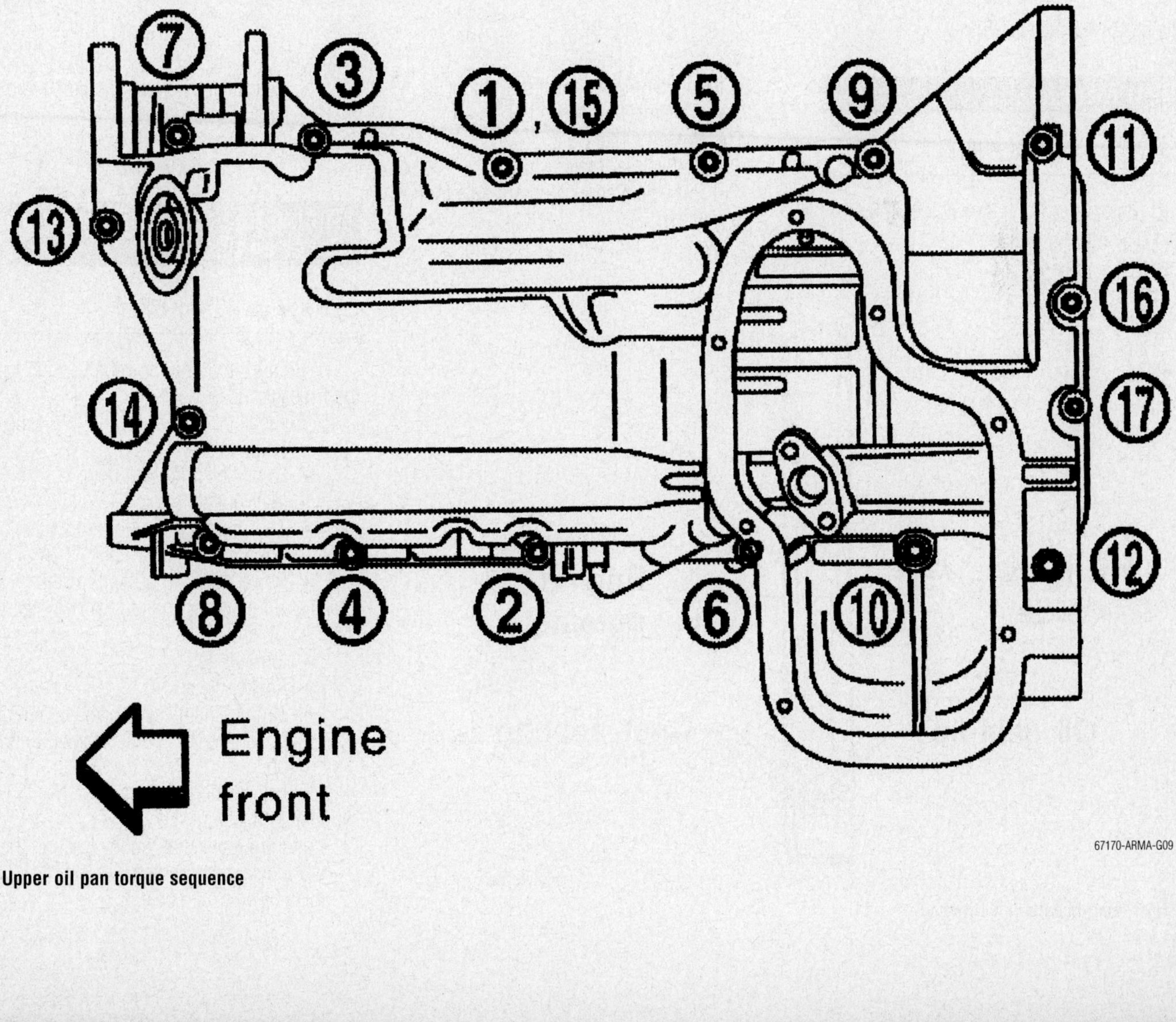

Upper oil pan torque sequence

67170-ARMA-G09

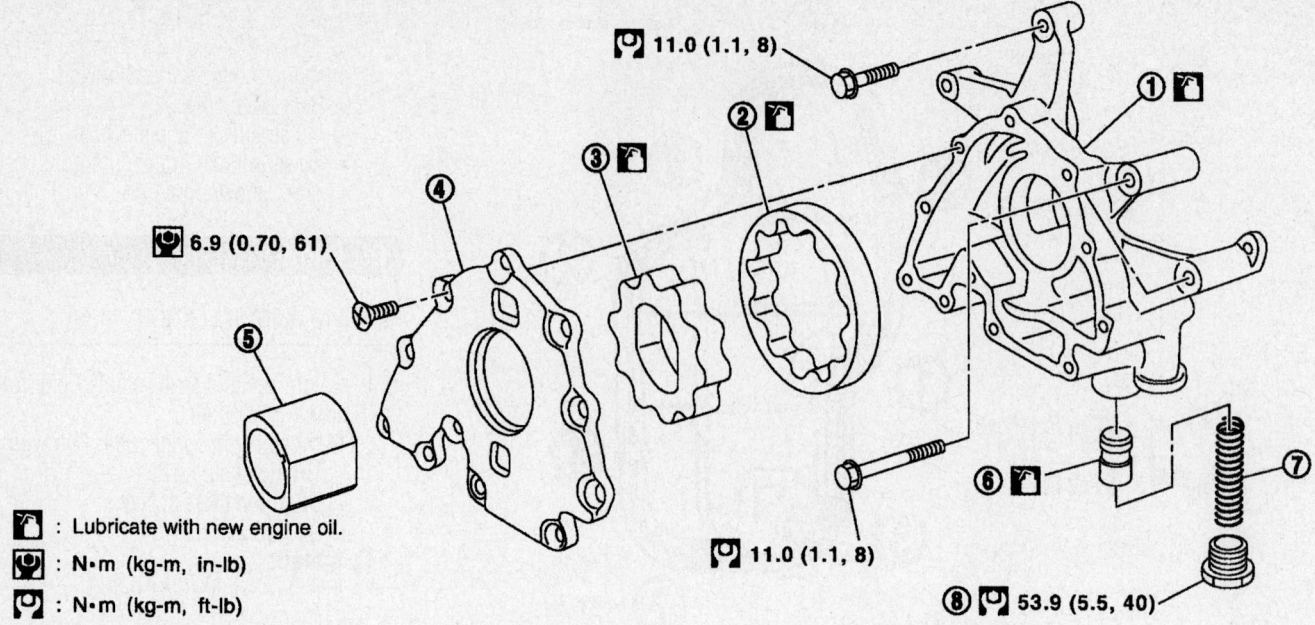

⬚ : Lubricate with new engine oil.

⬚ : N·m (kg-m, in-lb)

⬚ : N·m (kg-m, ft-lb)

1. Oil pump body
4. Oil pump cover
7. Regulator spring

2. Outer rotor
5. Oil pump drive spacer
8. Regulator plug

3. Inner rotor
6. Regulator valve

67170-ARMA-G32

Oil pump exploded view

Rear Main Seal

REMOVAL & INSTALLATION

1. Before servicing the vehicle, refer to the Precautions Section.

2. Remove or disconnect the following:
   - Transmission assembly
   - Pressure plate
   - Engine rear plate
   - Rear main seal using suitable tool

To install:
3. Install or connect the following:

   - Rear main seal using suitable tool
   - Engine rear plate
   - Pressure plate
   - Transmission assembly

Timing Chain, Sprockets, Front Cover and Seal

REMOVAL & INSTALLATION

1. Before servicing the vehicle, refer to the Precautions Section.
2. Remove or disconnect the following:
   - Engine assembly
   - Drive belt auto tensioner
   - Idler pulley
   - Thermostat housing and water hose
   - Power steering pump bracket
   - Oil pan (upper and lower)
   - Oil strainer
   - Ignition coil
   - Rocker cover
   - Timing chain case cover, loosening bolts in reverse order shown in figure
3. Obtain compression TDC of No. 1 cylinder as follows:
   a. Turn crankshaft pulley to align the TDC identification notch with timing indicator on front cover.

67170-ARMA-G33

Proper seal installation direction

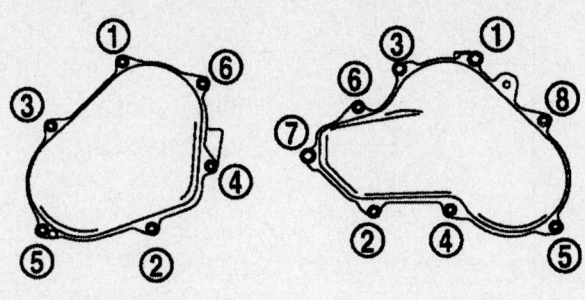

Right bank   Left bank

67170-ARMA-G10

Timing case cover torque sequence

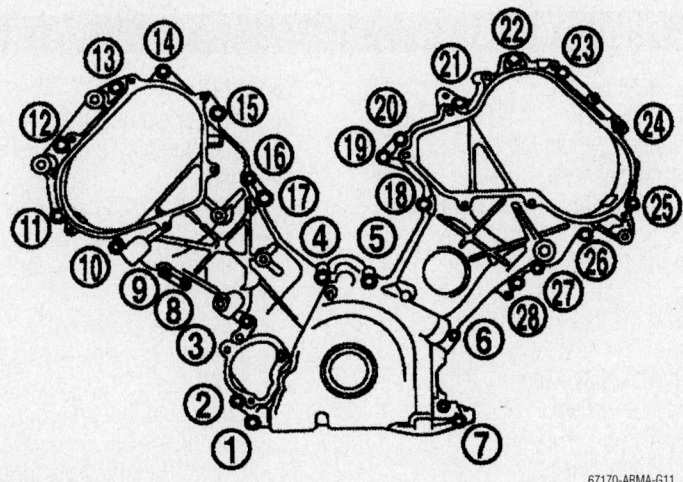

67170-ARMA-G11

Front cover torque sequence

b. Ensure intake and exhaust cam lobes of No. 1 cylinder point outside.

4. Remove or disconnect the following:
   • Crankshaft pulley from crankshaft using a suitable puller
   • Front cover, loosening bolts in reverse order shown in figure
   • Front oil seal
   • Oil pump drive spacer
   • Oil pump
   • Timing chain tensioner
   • Chain tension guide and slack guide
   • Timing chain
   • Camshaft sprocket

To install:

5. Ensure that the crankshaft key and dowel pin of each camshaft are facing the same direction.

6. Install or connect the following:
   • Camshaft sprockets
   • Timing chain
   • Chain tension guide and slack guide
   • Oil pump
   • Oil pump drive spacer
   • Front oil seal, using suitable tool
   • Front cover, using new O-rings and tighten bolts in order shown in figure
   • Chain case cover, and tighten bolts in order shown in figure
   • Crankshaft pulley and tighten bolt

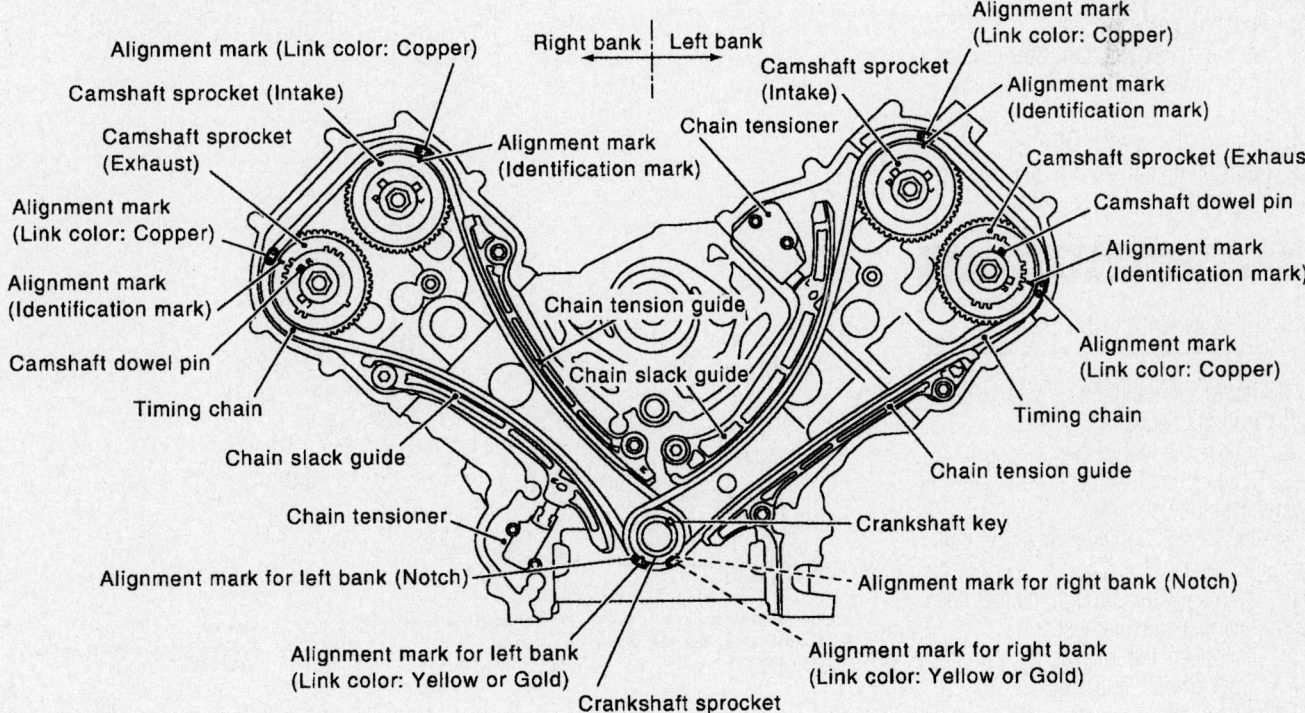

67170-ARMA-G35

Timing mark alignment

to 69 ft. lbs. (93 Nm) plus 90 degrees
- Ignition coil
- Oil strainer
- Lower and upper oil pan
- Power steering pump bracket
- Thermostat housing and water hose
- Idler pulley
- Drive belt auto tensioner
- Engine assembly

Piston and Ring

POSITIONING

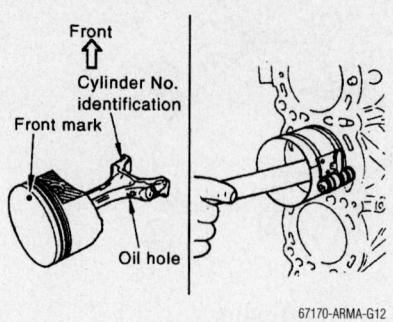

67170-ARMA-G12

Piston and rod positioning and identification

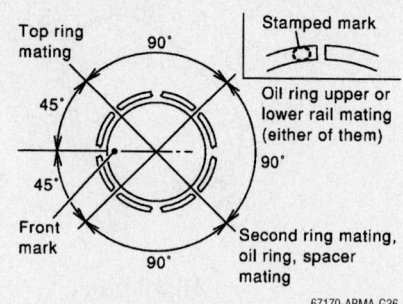

67170-ARMA-G36

Piston ring installation

FUEL SYSTEM

Relieving Fuel System Pressure

With CONSULT-II

1. Turn ignition switch **ON**.
2. Perform "FUEL PRESSURE RELEASE" in "WORK SUPPORT" mode with CONSULT-II.
3. Start engine.
4. After engine stalls, turn over the engine two or three times to release all fuel pressure.
5. Turn ignition switch **OFF**.

Without CONSULT-II

1. Remove fuel pump fuse located in IPDM E/R.
2. Start engine.
3. After engine stalls, turn over engine two or three times to release all fuel pressure.
4. Turn ignition switch **OFF**.
5. Reinstall fuel pump fuse after servicing fuel system.

Fuel Filter and Fuel Pump

REMOVAL & INSTALLATION

1. Before servicing the vehicle, refer to the Precautions Section.
2. Relieve the fuel system pressure.
3. Remove fuel filler cap to release pressure from inside tank.
4. Disconnect fuel filler hose from fuel filler pipe.
5. Drain fuel tank through the fuel filler opening using a suitable hose.
6. Disconnect the following:
   - Fuel pump line protector
   - EVAP hose
- Fuel level sensor
- Fuel filter
- Fuel pump wiring harness
- Fuel supply hose

7. Using a suitable jack to support the fuel tank, remove the strap bolts and remove the fuel tank from the vehicle.

8. Remove the lock ring using special tool J-46536.

9. Remove the following:
   - Fuel level sensor
   - Fuel filter
   - Fuel pump assembly

To install:

10. Install or connect the following:
    - Fuel pump assembly, using new O-ring
    - Fuel filter, using new filter
    - Fuel level sensor, using new sensor
    - Fuel pump assembly lock ring
    - Fuel tank
    - Fuel supply hose
    - Fuel pump wiring harness
    - EVAP hose
    - Fuel pump line protector
    - Fuel filler pipe

11. Start engine and check for leaks.

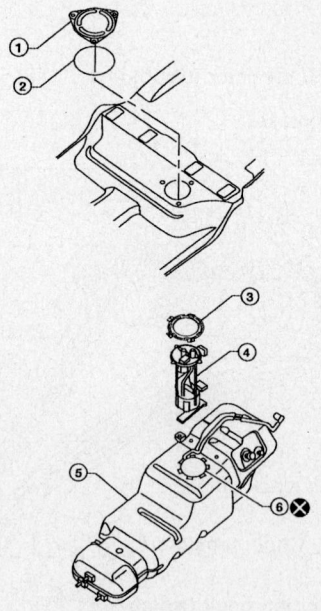

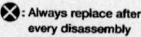: Always replace after every disassembly

1. Inspection hole cover
4. Fuel level sensor, fuel filter, and fuel pump assembly
2. Inspection hole cover O-ring
5. Fuel tank
3. Lock ring
6. Fuel level sensor, fuel filter, and fuel pump assembly O-ring

67170-ARMA-G37

Fuel pump and related parts

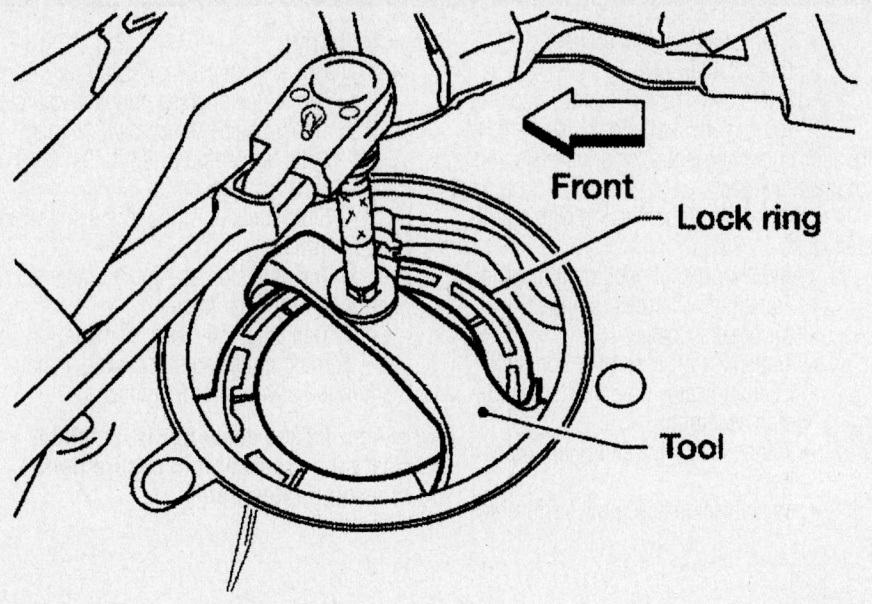

Front
Lock ring
Tool

67170-ARMA-G17

Removing fuel assembly lock ring

Fuel Injectors

REMOVAL & INSTALLATION

1. Before servicing the vehicle, refer to the Precautions Section.
2. Remove engine cover
3. Relieve fuel system pressure.
4. Remove or disconnect the following:
   - Negative battery cable
   - Fuel injector harness connectors
   - Fuel hose assembly from right and left fuel rails
   - Fuel injectors with fuel rail as an assembly
   - Fuel injector from fuel rail

To install:
5. Install or connect the following:
   - New clip onto the fuel injector
   - Fuel injector to fuel rail
   - Fuel injectors and fuel rail as an assembly to the intake manifold
   - Fuel hose assembly
   - Fuel injector harness connectors
   - Negative battery cable
   - Engine cover
6. Start engine and check for leaks.

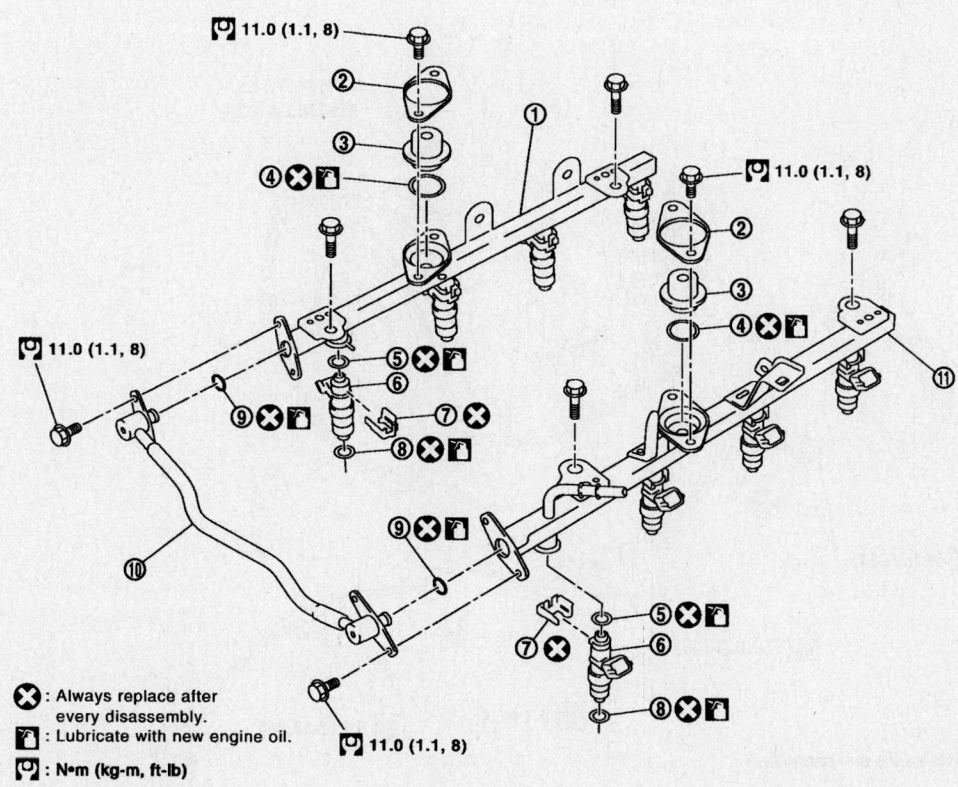

⊗ : Always replace after every disassembly.
🛢 : Lubricate with new engine oil.
🔧 : N•m (kg-m, ft-lb)

1. Fuel tube (right bank)	2. Cap	3. Fuel damper
4. O-ring	5. O-ring (Blue)	6. Fuel injector
7. Clip	8. O-ring (Brown)	9. O-ring
10. Fuel hose assembly	11. Fuel tube (left bank)	

67170-ARMA-G38

Fuel injectors and related parts

DRIVE TRAIN

Transmission Assembly

REMOVAL & INSTALLATION

2-Wheel Drive

1. Before servicing the vehicle, refer to the Precautions Section.
2. Remove or disconnect the following:
   - Negative battery cable
   - Engine cover
   - Transmission fluid indicator gauge
   - Engine splash guard
   - Exhaust front pipe
   - Center muffler
   - Rear drive shaft
   - Transmission control cable
   - Crankshaft position sensor
   - Fluid cooler tube
   - Dust cover from converter housing
3. Turning crankshaft clockwise, remove the four tightening bolts for drive plate and torque converter
4. Support the transmission with a suitable jack.
5. Remove or disconnect the following:
   - Transmission cross member
   - Air breather hose
   - Transmission assembly connector
   - Fluid indicator tube from transmission assembly
   - Transmission assembly to engine bolts
   - Transmission assembly from vehicle

To install:

6. Install or connect the following:
   - Transmission assembly into vehicle
   - Transmission assembly to engine bolts tightening to 83 ft. lbs. (113 Nm)
   - Fluid indicator tube to transmission assembly
   - Transmission assembly connector
   - Air breather hose
   - Transmission cross member
7. Turning crankshaft clockwise, install the torque converter to drive plate.

➡**After torque converter is installed, rotate the crankshaft to ensure transmission rotates freely.**

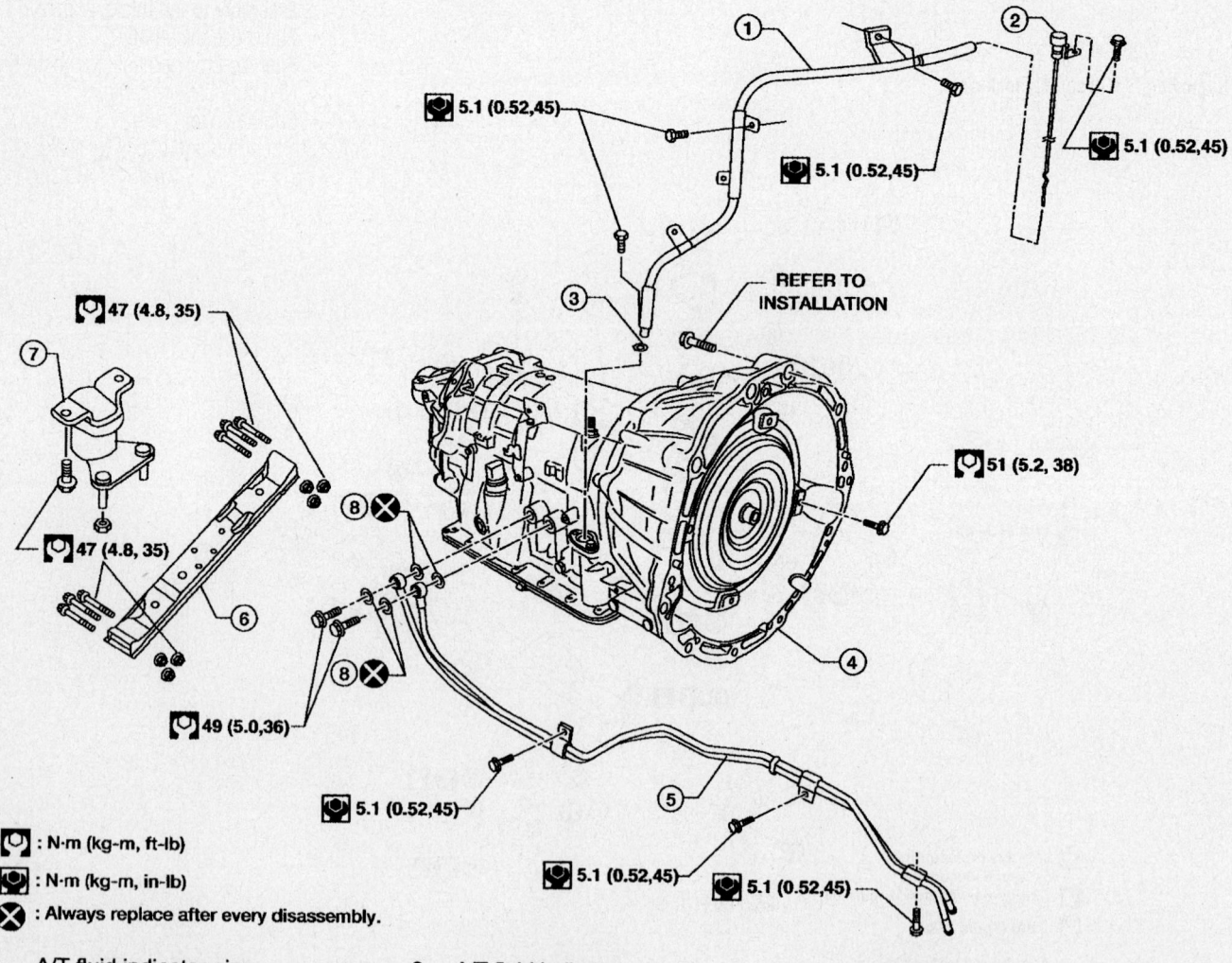

🔧 : N·m (kg-m, ft-lb)

🔧 : N·m (kg-m, in-lb)

✖ : Always replace after every disassembly.

1. A/T fluid indicator pipe
2. A/T fluid indicator
3. O-ring
4. Transmission assembly
5. A/T fluid cooler tube
6. A/T crossmember
7. Insulator
8. Copper washers

Transmission and related parts—2-wheel drive

8. Install or connect the following:
- Dust cover for converter housing
- Fluid cooler tube
- Crankshaft position sensor
- Transmission control cable
- Rear drive shaft
- Center muffler
- Exhaust front pipe
- Engine splash guard
- Transmission fluid indicator gauge
- Engine cover
- Negative battery cable

9. Start engine and check for leaks.

4-Wheel Drive

1. Before servicing the vehicle, refer to the Precautions Section.
2. Remove or disconnect the following:

- Negative battery cable
- Engine cover
- Transmission fluid indicator gauge
- Engine splash guard
- Exhaust front pipe
- Center muffler
- Drive shaft
- Transmission control cable
- Crankshaft position sensor
- Fluid cooler tube
- Dust housing for torque converter

3. Turning the crankshaft clockwise, remove the four tightening bolts for drive plate and torque converter.
4. Support the transmission assembly with a suitable jack.
5. Remove transmission cross member.
6. Tilt the transmission slightly to keep

clearance between the body and the transmission assembly, then disconnect the air breather hose.

7. Remove or disconnect the following:

- Transmission assembly connector and transfer case connector
- Fluid indicator pipe
- Transmission assembly to engine bolts
- Transmission assembly, with transfer case attached, from vehicle
- Transmission assembly from transfer case

To install:

8. Install or connect the following:
- Transfer case to transmission assembly

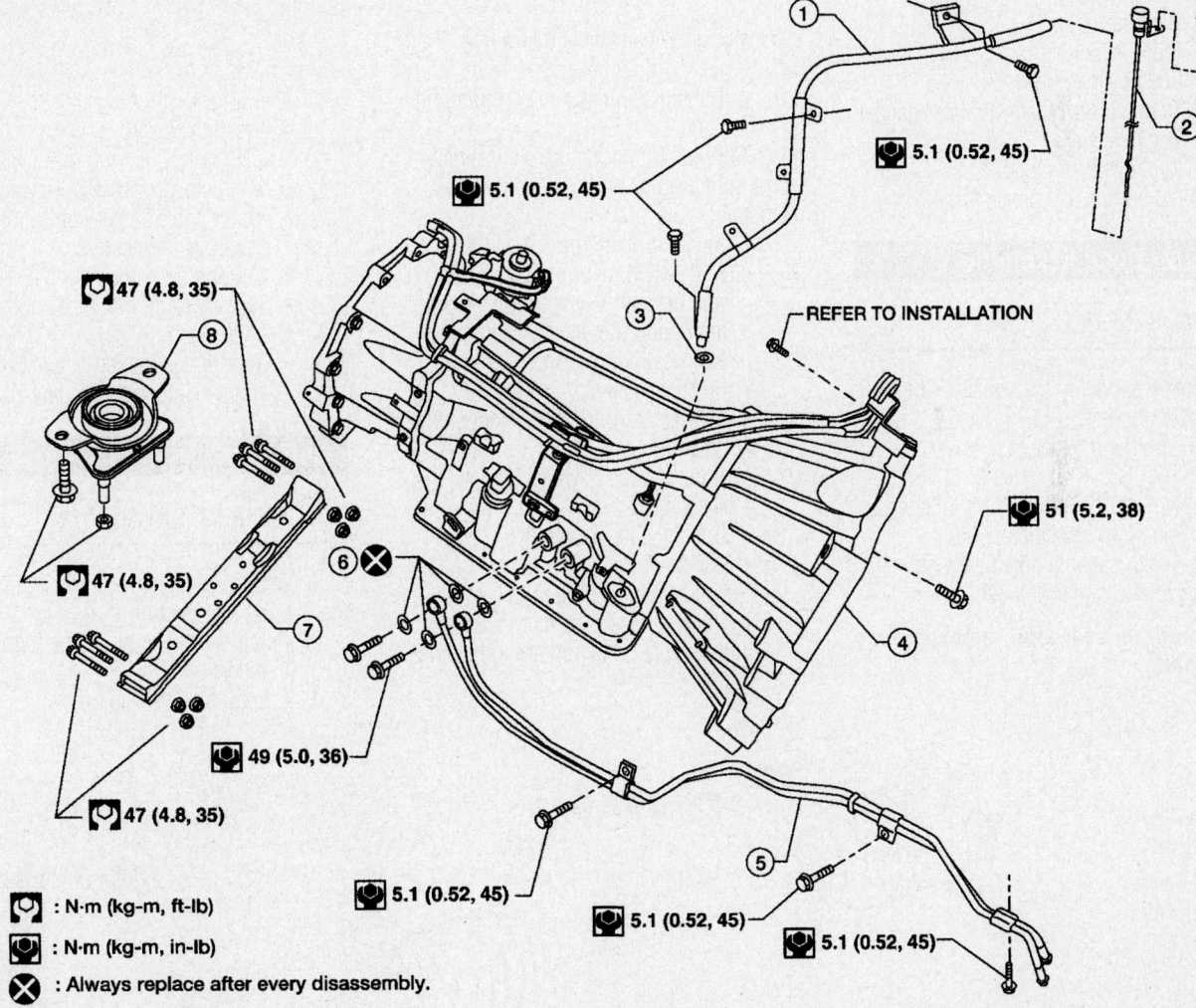

47 (4.8, 35)

47 (4.8, 35)

5.1 (0.52, 45)

5.1 (0.52, 45)

5.1 (0.52, 45)

5.1 (0.52, 45)

5.1 (0.52, 45)

49 (5.0, 36)

47 (4.8, 35)

51 (5.2, 38)

REFER TO INSTALLATION

: N·m (kg-m, ft-lb)

: N·m (kg-m, in-lb)

: Always replace after every disassembly.

1. A/T fluid indicator pipe	2. A/T fluid indicator	3. O-ring	
4. Transmission assembly	5. Fluid cooler tube	6. Copper washer	
7. A/T crossmember	8. Insulator		

67170-ARMA-G40

Transmission and related parts—with 4-wheel drive

- Transmission assembly into vehicle
- Transmission assembly to engine bolts tightening to 83 ft. lbs. (113 Nm)

9. With the transmission slightly tilted to allow clearance between body and transmission, connect the air breather hose.

10. Install the transmission cross member.

11. Turning crankshaft clockwise, install the torque converter to drive plate.

➡**After torque converter is installed, rotate the crankshaft to ensure transmission rotates freely.**

12. Install or connect the following:
- Dust housing for torque converter
- Fluid cooler tube
- Crankshaft position sensor
- Transmission control cable
- Drive shaft
- Center muffler
- Front exhaust pipe
- Engine splash guard
- Transmission fluid indicator gauge
- Engine cover
- Negative battery cable

13. Start engine and check for leaks.

Transfer Case Assembly

REMOVAL & INSTALLATION

1. Before servicing the vehicle, refer to the Precautions Section.

2. Ensure the transfer case is set to 2WD.

3. Remove or disconnect the following:
- Transmission splash guard
- Center exhaust pipe and muffler
- Front and rear drive shafts

➡**Plug rear oil seal after removing rear drive shaft.**

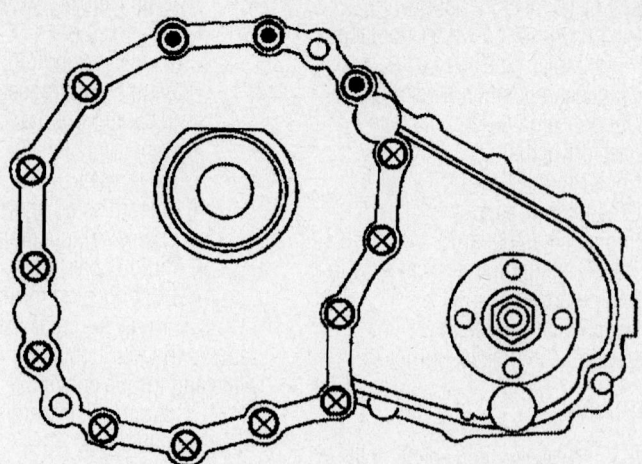

◉ : Transfer → Automatic transmission
⊗ : Automatic transmission → Transfer

67170-ARMA-G41

Transfer case mounting bolt locations

- Transmission assembly mounting bolts

4. Support the transmission assembly with a suitable jack and remove the crossmember.

5. Remove or disconnect the following:
- ATP switch, neutral 4LO switch, wait detection switch, transfer motor and transfer control device electrical connectors
- Breather hoses
- Shift actuator from the extension housing
- Transfer case to transmission assembly bolts
- Transfer case assembly

To install:

6. Install or connect the following:
- Transfer case to transmission assembly bolts tightening to 26 ft. lbs. (36 Nm)

- Shift actuator
- Breather hoses
- ATP switch, neutral 4LO switch, wait detection switch, transfer motor and transfer control device electrical connectors
- Support crossmember
- Transmission mounting bolts
- Drive shafts
- Muffler and center exhaust pipe
- Transmission splash guard

Halfshaft

REMOVAL & INSTALLATION

1. Before servicing the vehicle, refer to the Precautions Section.

2. Remove or disconnect the following:
- Wheel
- Engine splash guard

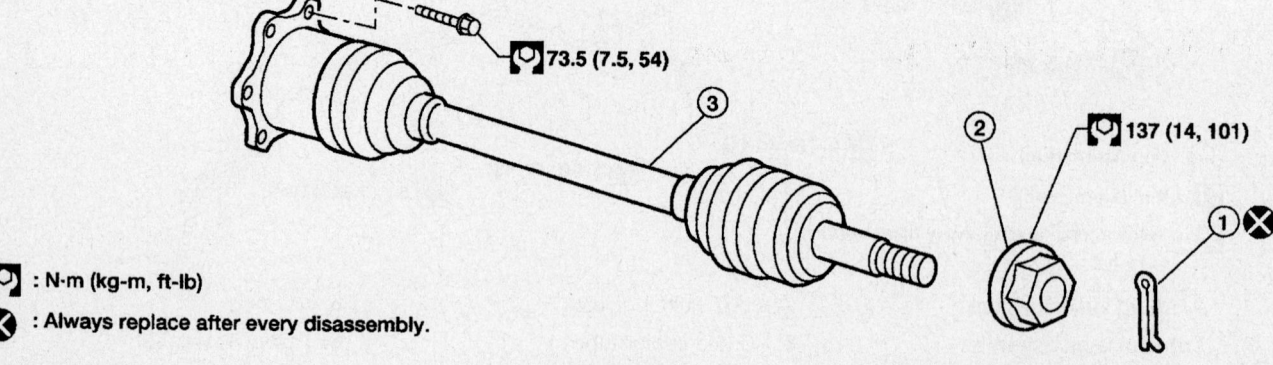

🔧 73.5 (7.5, 54)

🔧 137 (14, 101)

🔧 : N·m (kg-m, ft-lb)

⊗ : Always replace after every disassembly.

1. Cotter pin
2. Drive shaft nut
3. Drive shaft

67170-ARMA-G42

Front halfshaft

- ABS sensor harness on knuckle
- Brake caliper and suspend it aside
- Coil spring and shock absorber assembly

3. Separate upper ball joint stud from steering knuckle using special tool J-24319-01.

4. Remove or disconnect the following:
- Cotter pin and half shaft nut
- Half shaft from front differential
- Half shaft from hub and bearing assembly

To install:

5. Install or connect the following:
- Half shaft into hub
- Halt shaft into front differential
- Half shaft nut and tighten to 101 ft. lbs. and replace cotter pin
- Upper ball joint to steering knuckle
- Coil spring and shock absorber assembly
- Brake caliper
- ABS sensor
- Engine splash guard
- Wheel

CV-Joints

OVERHAUL

Inner

1. Before servicing the vehicle, refer to the Precautions Section.

2. Remove the halfshaft from the vehicle.

3. Mount halfshaft in a vise.

4. Remove the dust boot bands.

5. Remove the stopper ring with a flat-bladed screwdriver or suitable tool.

6. Remove the snap ring.

7. Disassemble the cage, ball and inner race assembly and dust boot for cleaning and inspection.

To install:

➡ **Discard old dust boot, dust boot bands and snap ring and use new ones for assembly.**

8. Wrap the serrated part of the half-shaft with tape.

9. Install new dust boot and band onto halfshaft.

10. Remove tape from serrated part of halfshaft.

11. Install the cage, ball and inner race assembly.

12. Install new snap ring.

13. Insert 4.50-5.3 oz of genuine NISSAN grease or equivalent onto the housing and install onto halfshaft.

14. Install the stopper ring onto the housing.

15. Install the dust boot into the grooves on joint sub-assembly.

16. Secure the big and small ends of the dust boot using new boot bands.

Outer

1. Before servicing the vehicle, refer to the Precautions Section.

2. Remove the halfshaft from the vehicle.

3. Mount halfshaft in a vise.

4. Remove the dust boot bands and dust boot from joint sub-assembly.

5. Insert a suitable puller into the threaded part of the halfshaft. Pull the joint sub-assembly off of the halfshaft as shown in figure.

6. Remove dust boot and circlip from halfshaft for cleaning and inspection.

To install:

➡ **Discard old dust boot, dust boot bands and circlip and use new ones for assembly.**

7. Insert genuine NISSAN grease or equivalent into the joint sub-assembly until grease oozes from the ball groove and serration hole.

8. Wrap the serrated part of the half-shaft with tape.

9. Install new dust boot and band onto halfshaft.

10. Remove tape from serrated part of the halfshaft.

11. Press-fit the new circlip to the half-shaft.

12. Insert 5.1-5.8 oz of genuine NISSAN grease or equivalent into the joint sub-assembly and large end of boot.

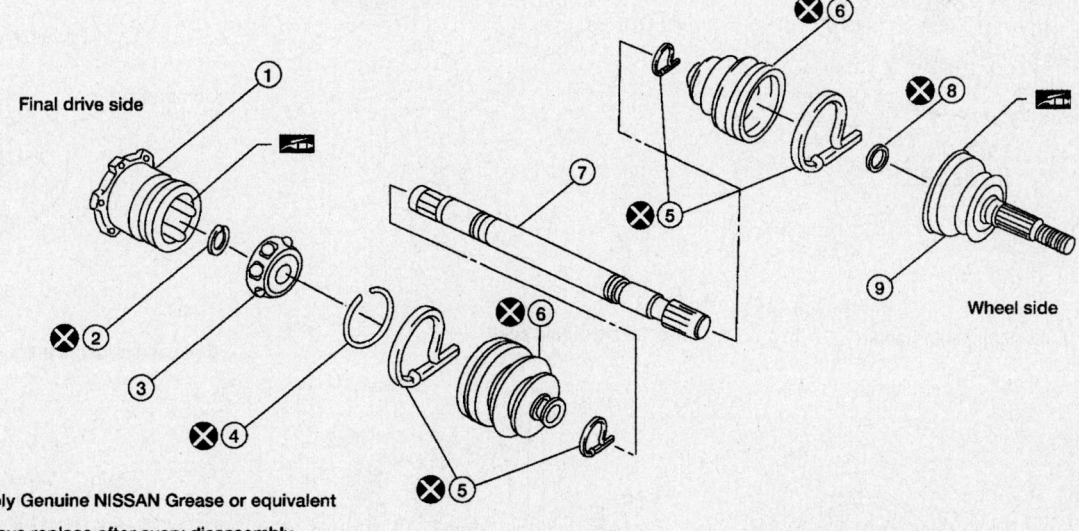

Final drive side

Wheel side

⚙ : Apply Genuine NISSAN Grease or equivalent

✕ : Always replace after every disassembly.

1.	Housing	2.	Snap ring	3.	Ball cage, steel ball, iiner race assembly
4.	Stopper ring	5.	Boot band	6.	Boot
7.	Shaft	8.	Circlip	9.	Joint sub-assembly

67170-ARMA-G43

Front halfshaft—exploded view

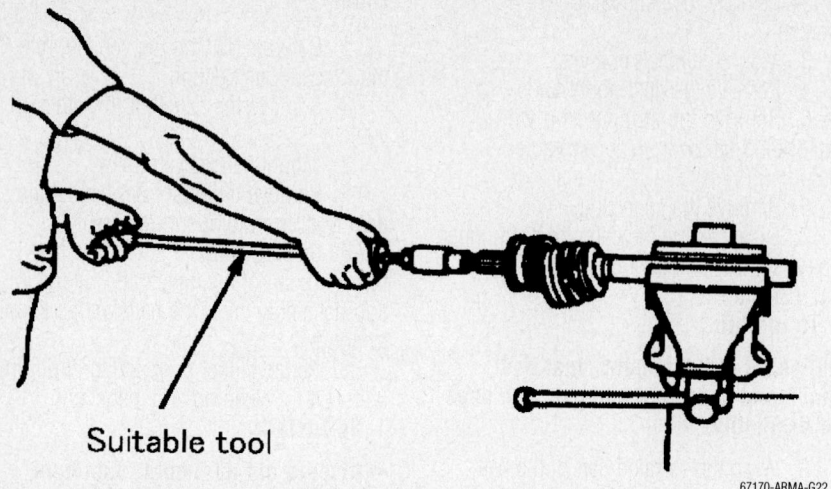

Using a suitable puller to remove joint sub-assembly.

13. Install the dust boot into the grooves on the joint sub-assembly.

14. Secure the big and small ends of the dust boot using new boot bands.

Front Differential Pinion Seal

REMOVAL & INSTALLATION

1. Before servicing the vehicle, refer to the Precautions Section.

2. Remove or disconnect the following:
   • Front drive shaft

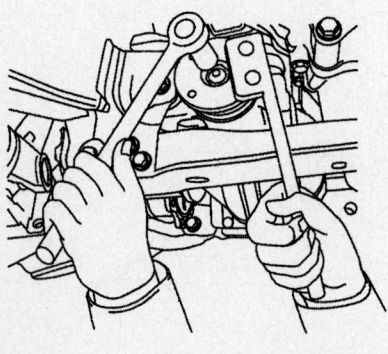

Removing the companion flange

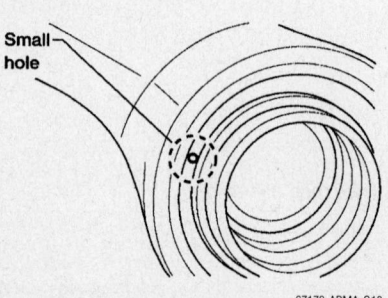

Small hole in casing

• Halfshafts

3. Measure and record the pinion bearing preload using special tool J-25765-A.

4. Loosen the pinion nut while holding the companion flange using special tool J-44195.

5. Remove the companion flange using a suitable tool.

6. Using a punch or drill, place a small hole in the case.

7. Remove the seal using special tool SP8P.

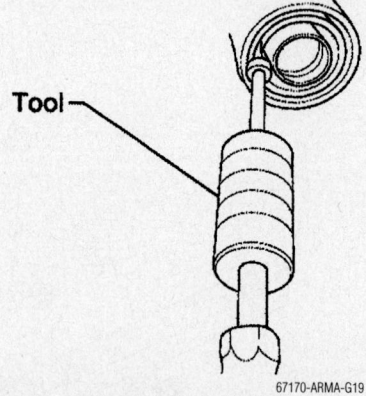

Removing pinion seal

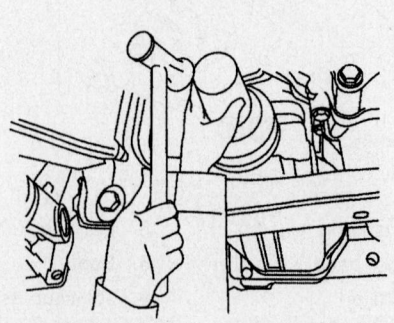

Seal installation

To install:

8. Press front seal into carrier using a suitable tool.

9. Install companion flange and new pinion nut. Tighten pinion nut until there is no end play and until recorded pinion bearing preload is met plus an additional 5 inch lbs. (0.5 Nm).

10. Install or connect the following:
   • Halfshafts
   • Front drive shaft

Rear Differential Pinion Seal

REMOVAL & INSTALLATION

1. Before servicing the vehicle, refer to the Precautions Section.

2. Remove the rear drive shaft.

3. Remove brake calipers and rotors.

4. Measure and record the total preload.

5. Matchmark the drive pinion to position 'B' on the companion flange.

6. Remove the drive pinion nut using suitable tool.

7. Remove the companion flange using suitable tool.

8. Remove the rear pinion seal using special tool J-34286.

To install:

9. Press the rear pinion seal into the carrier using suitable tool.

10. Align the matchmark on the companion flange to the drive pinion and install the companion flange.

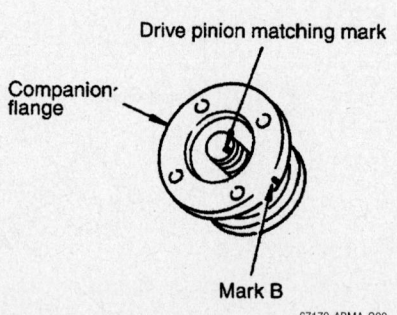

Companion flange marking

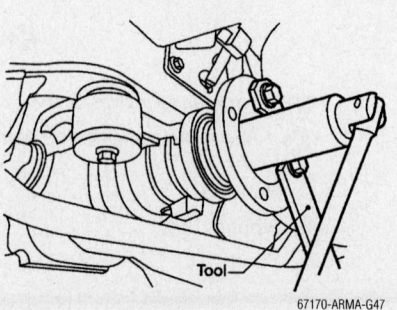

Loosening the flange nut

11. Lubricate the drive pinion threads and seating surfaces of the drive pinion nut with grease.

12. Using a new drive pinion nut, tighten to 124-274 ft. lbs. (167-372 Nm).

➡ **Final torque is determined when adjusting total preload using special tool J-25765-A.**

13. Install brake calipers and rotors.
14. Install rear drive shaft.

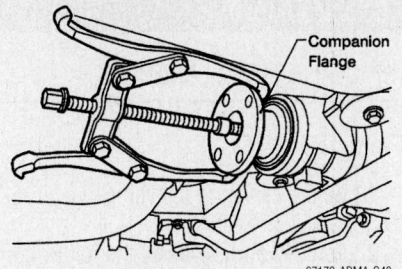

Removing the companion flange

67170-ARMA-G48

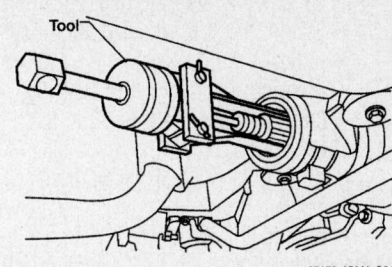

Removing the pinion seal

67170-ARMA-G21

STEERING AND SUSPENSION

Air Bag

DISARMING THE SYSTEM

1. Before servicing the vehicle, refer to the Precautions Section.
2. Disconnect both battery cables.
3. Wait at least 3 minutes before working on the vehicle. The air bag system is designed to retain enough power to deploy the air bag for a short time after the battery has been disconnected.
4. After repairs are complete, connect the negative battery cable. Turn the ignition switch to the **ON** position and check the air bag warning light blinks for proper operation.

Power Steering Gear

REMOVAL & INSTALLATION

1. Before servicing the vehicle, refer to the Precautions Section.
2. Ensure the wheels are in the straight-ahead position.
3. Remove or disconnect the following:
   • Wheels
   • Engine splash guard
4. On 4-wheel drive models only, remove front final drive and support the drive shafts.
5. Remove cotter pin at steering outer socket and loosen mounting nut.
6. Remove steering outer socket from

steering knuckle using special tool J-25730-A.

7. On 2-wheel drive models only, remove stabilizer bar mounting bolts and secure the stabilizer bar.
8. Remove or disconnect the following:
   • Oil pipes from steering gear assembly
   • Lower joint mounting bolt from lower shaft
   • Mounting bolts and nuts from steering gear assembly
   • Steering gear assembly

To install:

9. Install or connect the following:
   • Steering gear assembly, tighten nuts to 140 ft. lbs. (190 Nm)
   • Lower joint mounting bolt

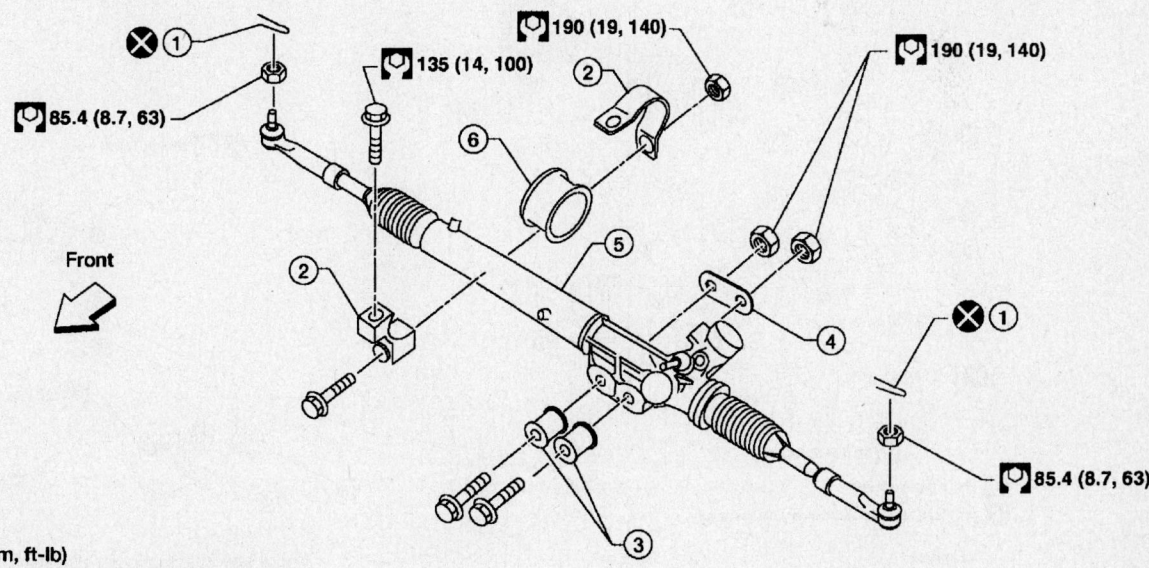

⊙ : N•m (kg-m, ft-lb)

✕ : Always replace after every disassembly.

1.	Cotter pin	2.	Mounting bracket	3.	Bushing
4.	Washer	5.	Steering gear assembly	6.	Mounting insulator

67170_TITAN_G99

Steering gear and related parts

- Oil pipes to steering gear assembly
- Stabilizer bar, 2 wheel-drive models only
- Steering outer socket to steering knuckle, tighten nut to 63 ft. lbs. (86 Nm)
- Front final drive, 4-wheel drive models only
- Engine splash guard
- Wheels

10. Check the wheel alignment and adjust as necessary.

Shock Absorber

REMOVAL & INSTALLATION

Front

1. Before servicing the vehicle, refer to the Precautions Section.
2. Remove or disconnect the following:
   - Wheel
   - Lower shock absorber bolt
   - Upper shock absorber bolts

- Coil spring and shock absorber assembly
3. Secure the shock absorber in a vice and loosen (without removing) the piston rod lock nut.
4. Install a spring compressor and tighten until the shock absorber mounting insulator can be turned by hand.
5. Remove piston rod lock nut and remove shock absorber.

To install:

6. Install upper mounting insulator in line with the lower shock absorber mount

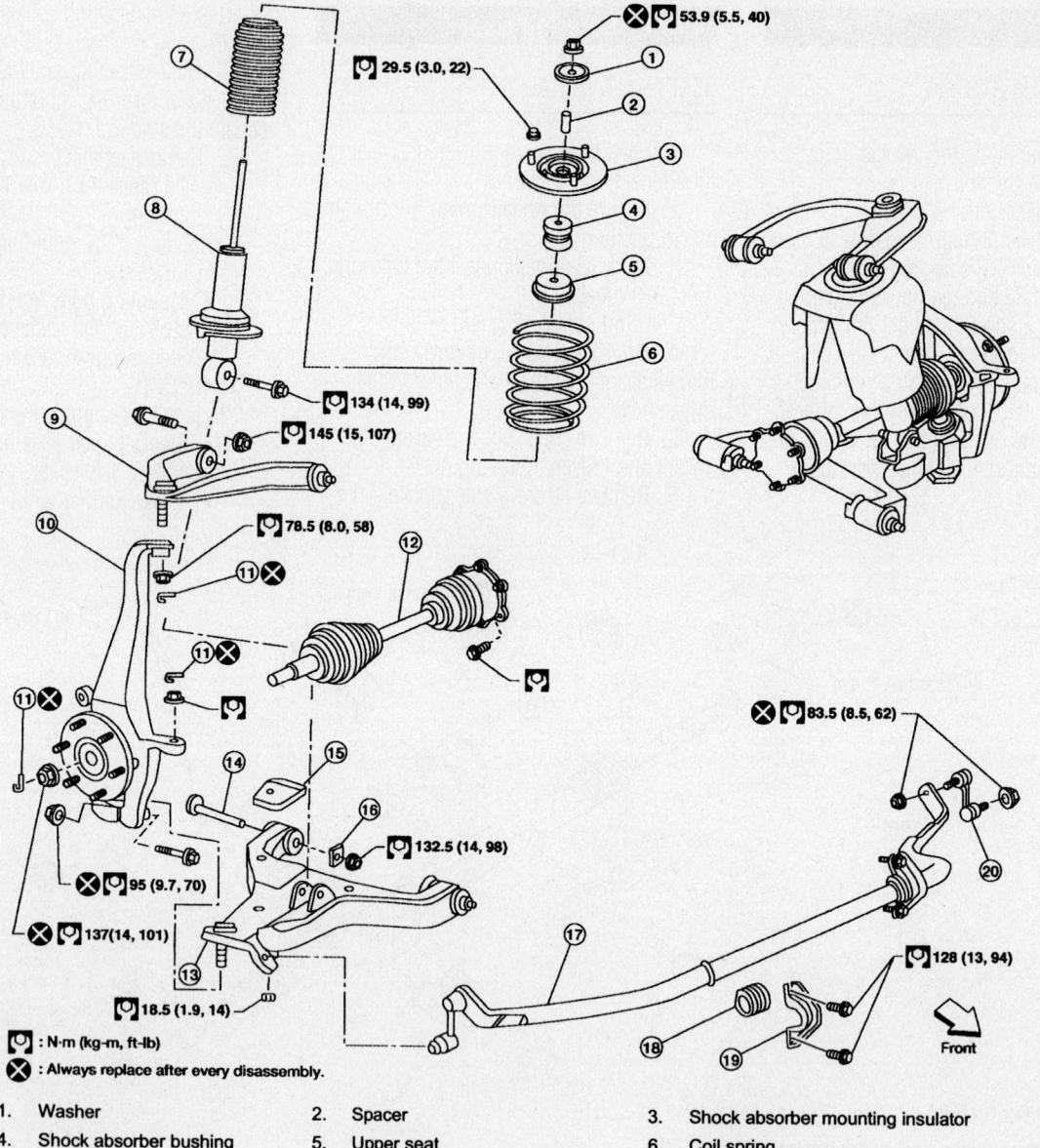

: N·m (kg-m, ft-lb)

: Always replace after every disassembly.

1.	Washer	2.	Spacer	3.	Shock absorber mounting insulator
4.	Shock absorber bushing	5.	Upper seat	6.	Coil spring
7.	Dust cover	8.	Shock absorber	9.	Upper link
10.	Steering knuckle	11.	Cotter pin	12.	Drive shaft
13.	Lower link	14.	Cam bolt	15.	Jounce bumper
16.	Cam washer	17.	Stabilizer bar	18.	Stabilizer bar bushing
19.	Stabilizer bar mounting bracket	20.	Connecting rod		

Front suspension

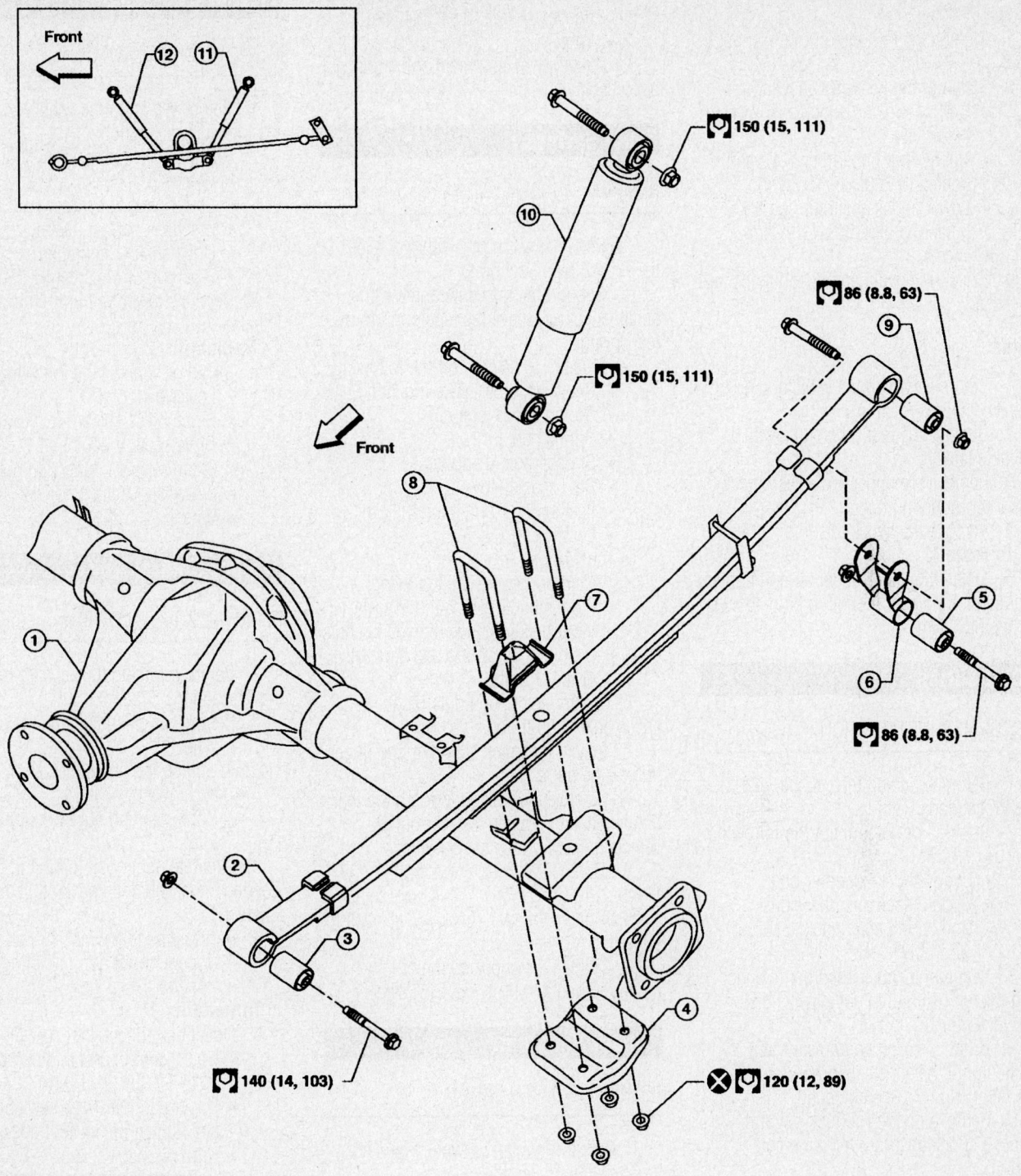

 : N·m (kg-m, ft-lb)

 : Always replace after every disassembly.

1. Rear final drive
2. Rear leaf spring
3. Rear spring bushing (front)
4. Rear spring pad
5. Rear spring shackle bushing
6. Rear spring shackle
7. Bumper
8. Rear spring clip U-bolts
9. Rear spring bushing (rear)
10. Shock absorber
11. Shock absorber (left side)
12. Shock absorber (right side)

67170_TITAN_G97

Rear suspension

and step in shock absorber lower seat as shown in figure.

7. Tighten the new piston rod lock nut to 40 ft. lbs. (54 Nm).

8. Install or connect the following:
- Coil spring and shock absorber assembly
- Upper shock absorber bolts and tighten to 22 ft. lbs (30 Nm)
- Lower shock absorber bolt and tighten to 99 ft. lbs. (134 Nm)
- Wheel

9. Check wheel alignment and adjust as necessary.

Rear

1. Before servicing the vehicle, refer to the Precautions Section.

2. Support the rear differential with a suitable jack.

3. Remove the upper and lower shock absorber mounting bolts.

4. Remove the shock absorber.

To install:

5. Install the shock absorber and tighten the upper and lower mounting bolts to 111 ft. lbs. (150 Nm).

Coil Spring

REMOVAL & INSTALLATION

1. Before servicing the vehicle, refer to the Precautions Section.

2. Remove or disconnect the following:
- Wheel
- Lower shock absorber bolt
- Upper shock absorber bolts
- Coil spring and shock absorber assembly

3. Secure the shock absorber in a vice and loosen (without removing) the piston rod lock nut.

4. Install a spring compressor and tighten until the shock absorber mounting insulator can be turned by hand.

5. Remove piston rod lock nut and remove shock absorber from the coil spring.

To install:

6. Install upper mounting insulator in line with the lower shock absorber mount and step in shock absorber lower seat as shown in figure.

7. Tighten the new piston rod lock nut to 40 ft. lbs. (54 Nm).

8. Install or connect the following:
- Coil spring and shock absorber assembly
- Upper shock absorber bolts and tighten to 22 ft. lbs (30 Nm)

- Lower shock absorber bolt and tighten to 99 ft. lbs. (134 Nm)
- Wheel

9. Check wheel alignment and adjust as necessary.

Leaf Springs

REMOVAL & INSTALLATION

1. Before servicing the vehicle, refer to the Precautions Section.

2. Support the rear differential with a suitable jack to relieve the tension from the leaf spring.

3. Remove or disconnect the following:
- Shock absorber lower mounting bolt
- Spring clip U-bolt nuts
- Spring pad
- Storage box, if equipped
- Rear shackle lower bolt
- Leaf spring front mounting bolt
- Leaf spring

To install:

4. Install or connect the following:
- Front mounting bolt and shackle lower bolt and finger tighten the nuts
- U-bolts, rear spring pad and nuts or the U-bolts

5. Tighten the U-bolt nuts diagonally and evenly to 89 ft. lbs. (120 Nm).

6. Install the shock absorber and finger tighten the nuts.

7. Remove the jack supporting the rear differential and bounce the rear of the vehicle to stabilize the suspension.

8. Tighten the front mount bolt to 103 ft. lbs. (140 Nm).

9. Tighten the rear shackle lower bolt to 63 ft. lbs. (86 Nm).

10. Tighten the shock absorber lower mounting bolt to 111 ft. lbs. (150 Nm).

Upper Ball Joint

REMOVAL & INSTALLATION

1. Before servicing the vehicle, refer to the Precautions Section.

2. Remove or disconnect the following:
- Wheel
- Cotter pin and nut from upper ball joint

3. Separate upper ball joint from steering knuckle using special tool J-24319-01

To install:

4. Install or connect the following:
- Upper ball joint
- New cotter pin and tighten nut to 58 ft. lbs. (79 Nm)
- Wheel

Lower Ball Joint

REMOVAL & INSTALLATION

1. Before servicing the vehicle, refer to the Precautions Section.

2. Remove or disconnect the following:
- Wheel
- Lower shock absorber bolt
- Stabilizer bar connecting rod
- Drive shaft, if equipped
- Pinch bolt from steering knuckle

3. Separate lower ball joint from steering knuckle

To install:

4. Install or connect the following:
- Lower ball joint
- Pinch bolt to steering knuckle
- Drive shaft, if equipped
- Stabilizer bar connecting rod
- Lower shock absorber bolt
- Wheel

Upper Control Arm

REMOVAL & INSTALLATION

1. Before servicing the vehicle, refer to the Precautions Section.

2. Remove or disconnect the following:
- Wheel
- Coil spring and shock absorber assembly
- Cotter pin and nut from upper ball joint

3. Separate upper ball joint stud from steering knuckle using special tool J-24319-01.

4. Remove the following:
- Upper control arm mounting bolts
- Upper arm

To install:

5. Install or connect the following:
- Upper control arm and tighten bolts to 107 ft. lbs. (145 Nm)
- Upper ball joint with new cotter pin and tighten nut to 58 ft. lbs. (79 Nm)
- Coil spring and shock absorber assembly
- Wheel

Lower Control Arm

REMOVAL & INSTALLATION

1. Before servicing the vehicle, refer to the Precautions Section.

2. Remove or disconnect the following:
- Wheel

- Lower shock absorber bolt
- Stabilizer bar connecting rod
- Drive shaft, if equipped
- Pinch bolt from steering knuckle

3. Separate lower ball joint from steering knuckle.
4. Remove the following:
- Lower control arm adjusting bolts
- Lower control arm

To install:

5. Install or connect the following:
- Lower control arm and tighten adjusting bolts to 98 ft. lbs. (133 Nm)
- Lower ball joint
- Pinch bolt
- Drive shaft, if equipped
- Stabilizer bar connected rod
- Lower shock absorber bolt
- Wheel

Wheel Bearings

REMOVAL & INSTALLATION

Front

1. Before servicing the vehicle, refer to the Precautions Section.
2. Remove or disconnect the following:
- Wheel
- Engine splash guard

- Brake caliper without disconnecting the hydraulic lines, and reposition aside with wire

3. Matchmark the brake rotor to the wheel hub and remove the brake rotor.
4. Remove or disconnect the following:
- Cotter pin and lock nut from drive shaft
- Drive shaft from wheel hub and bearing assembly
- ABS sensor
- Wheel hub and bearing assembly bolts
- Wheel hub and bearing assembly

To install:

5. Install or connect the following:
- Wheel hub and bearing assembly, using new bolts and tighten to 155 ft. lbs. (210 Nm)
- ABS sensor
- Drive shaft to wheel hub and bearing assembly
- Cotter pin and lock nut and tighten to 101 ft. lbs. (137 Nm)
- Brake rotor
- Brake caliper
- Engine splash guard
- Wheel

Rear

1. Before servicing the vehicle, refer to the Precautions Section.

2. Remove or disconnect the following:
- ABS sensor
- Rear brake rotor
- Parking brake assembly
- Four axle shaft bearing cage nuts and lock washers
- Axle shaft assembly using special tool J-25604-01 and J-25840-A
- Snap ring from axle shaft

3. Remove the bearing ring retainer by drilling 3/4 of thickness of the ring and using a hammer and chisel to break the ring free.
4. Remove the axle shaft bearing from the axle shaft using special tool 205-D002.
5. Remove and discard the axle oil seal.
6. Remove the wheel bearing assembly.

To install:

7. Install or connect the following:
- Wheel bearing assembly
- New axle oil seal
- Axle shaft bearing on axle shaft
- Bearing ring retainer onto the axle shaft
- Snap ring
- Axle shaft assembly into the axle shaft housing
- Axle shaft bearing cage lock washers and nuts and tighten to 87 ft. lbs. (118 Nm)
- Parking brake assembly
- Rear brake rotor
- ABS sensor

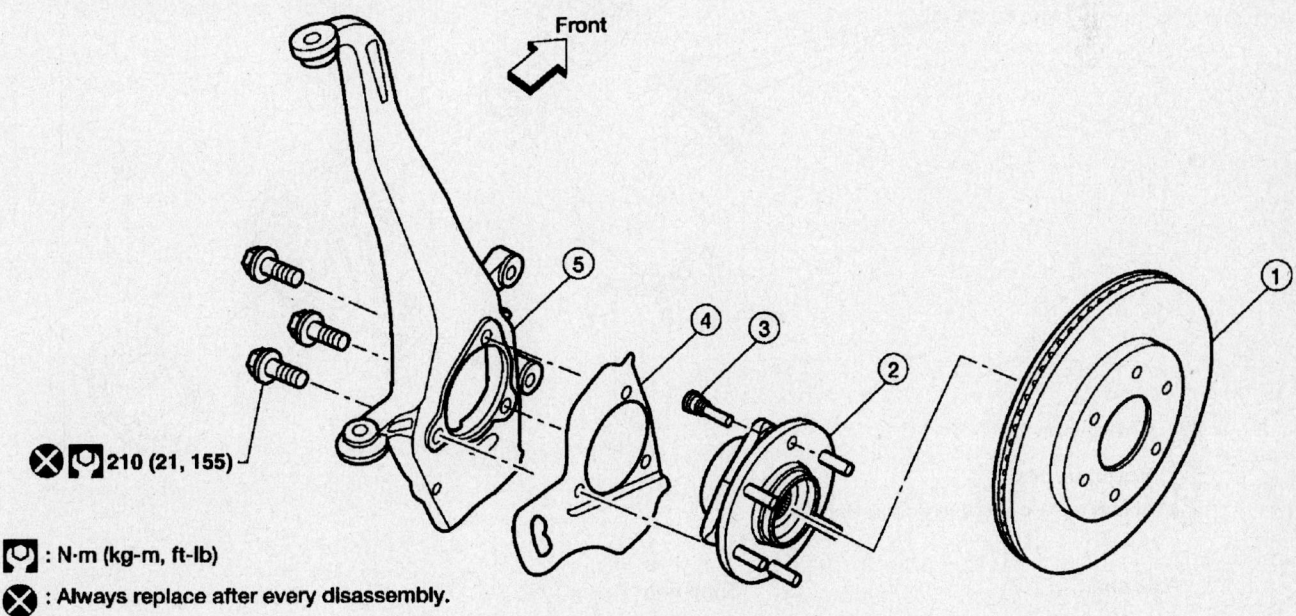

210 (21, 155)

: N·m (kg-m, ft-lb)

: Always replace after every disassembly.

1. Disc rotor
4. Splash guard
2. Wheel hub and bearing assembly
5. Steering knuckle
3. Wheel stud

Front hub

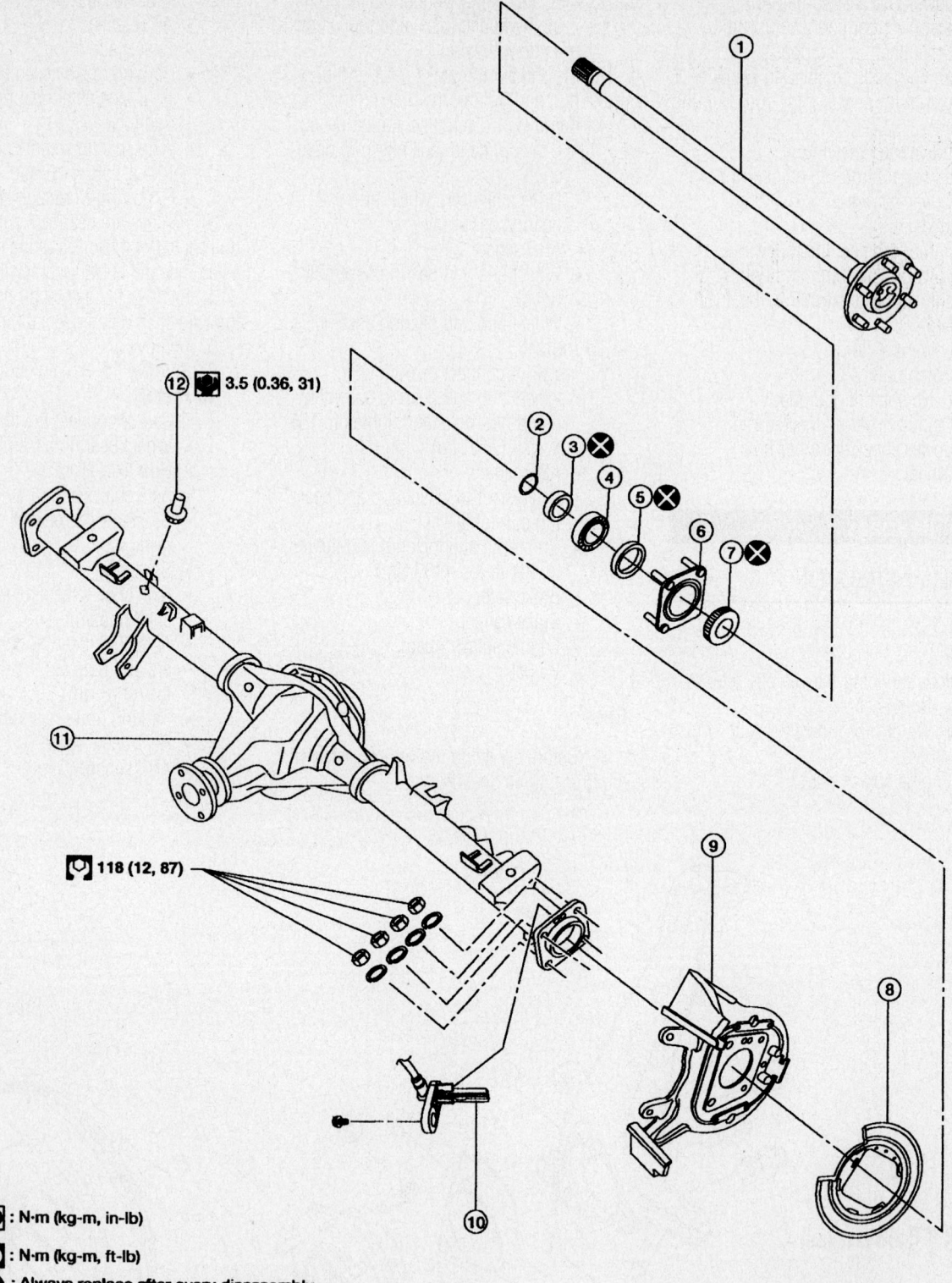

: N·m (kg-m, in-lb)

: N·m (kg-m, ft-lb)

: Always replace after every disassembly.

1.	Axle shaft	2.	Snap ring	3.	Bearing ring retainer
4.	Axle shaft bearing and cup	5.	Axle oil seal	6.	Axle shaft bearing cage
7.	ABS sensor rotor	8.	Back plate	9.	Torque member
10.	ABS sensor	11.	Rear final drive	12.	Breather

67170_TITAN_G95

Rear axle shaft and related parts

BRAKES

Brake Caliper

REMOVAL & INSTALLATION

Front

1. Before servicing the vehicle, refer to the Precautions Section.
2. Drain brake fluid as necessary.
3. Remove or disconnect the following:
   - Wheel
   - Union bolt
   - Caliper-to-torque member slide

pins, or remove the caliper and torque member as an assembly.
   - Brake caliper

To install:
4. Install or connect the following:
   - Brake caliper, tighten torque member bolts to 155 ft. lbs. (210 Nm); the caliper slide pins to 32 ft. lbs. (44 Nm)
   - Union bolt and tighten to 13 ft. lbs. (18 Nm)
5. Fill the master cylinder and bleed the brake system.
6. Install the wheels.

Rear

1. Before servicing the vehicle, refer to the Precautions Section.
2. Remove brake hose and mounting bolts.
3. Remove brake caliper assembly.

To install:
4. Install caliper assembly and tighten mounting bolts to 24 ft. lbs. (32 Nm).
5. Install brake hose and tighten to 13 ft. lbs. (18 Nm).

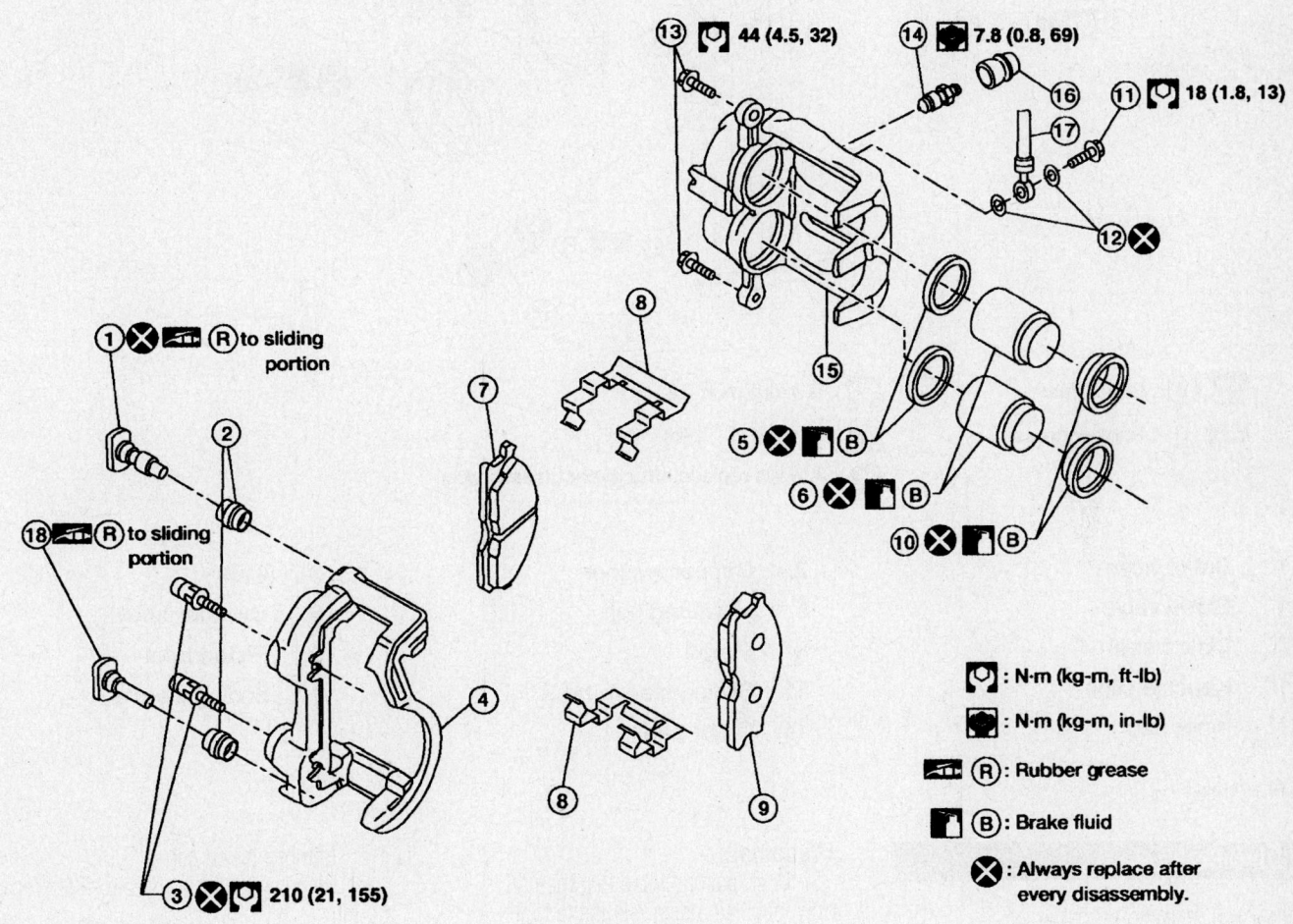

- $\boxed{}$: N·m (kg-m, ft-lb)
- $\boxed{}$: N·m (kg-m, in-lb)
- (R) : Rubber grease
- (B) : Brake fluid
- ⊗ : Always replace after every disassembly.

1. Upper sliding pin	2. Sliding pin boot	3. Torque member bolt
4. Torque member	5. Piston seal	6. Piston
7. Inner pad	8. Pad retainer	9. Outer pad
10. Piston boot	11. Union bolt	12. Copper washer
13. Sliding pin bolt	14. Bleed valve	15. Cylinder body
16. Cap	17. Brake hose	18. Lower sliding pin

67170_TITAN_G94

Front disc brake

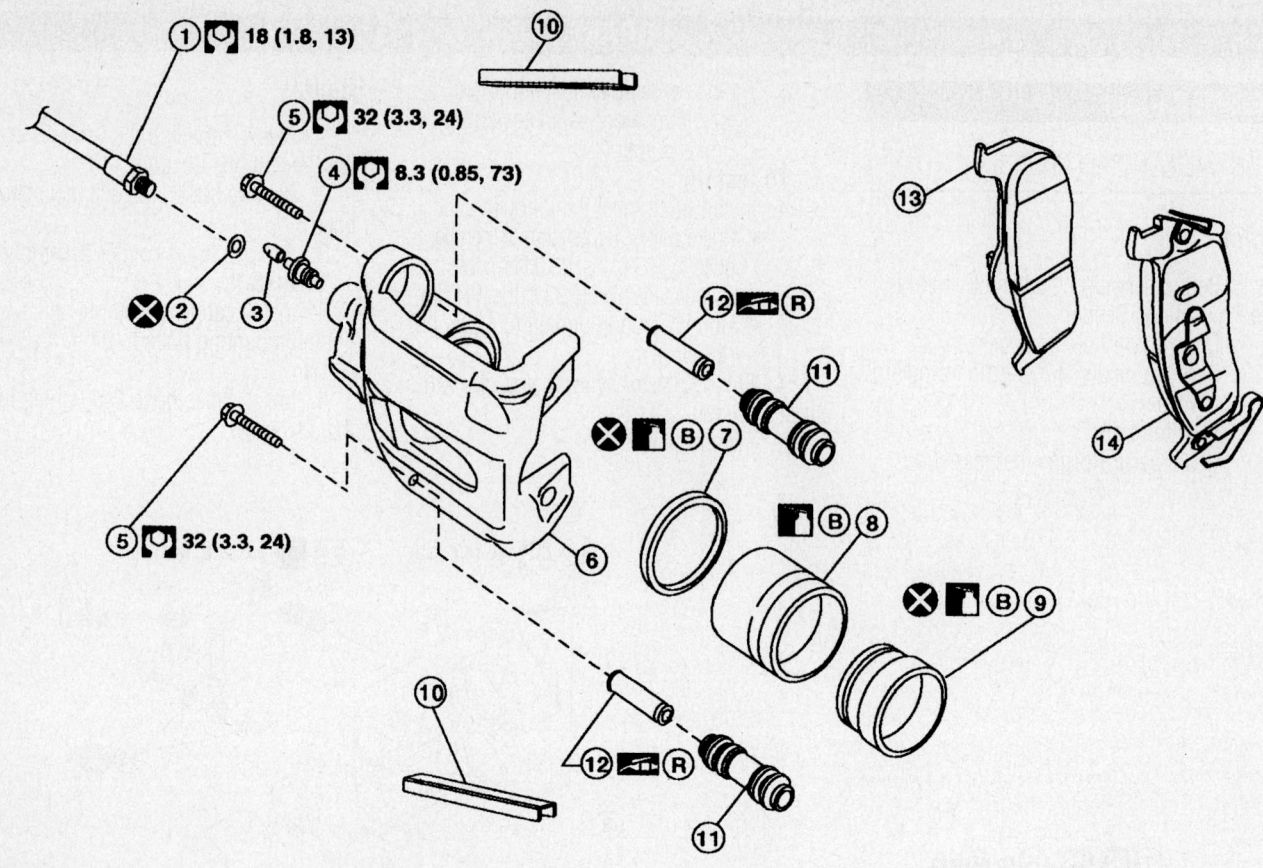

① 🔧 18 (1.8, 13)
⑤ 🔧 32 (3.3, 24)
④ 🔧 8.3 (0.85, 73)
⑤ 🔧 32 (3.3, 24)

🛢️ (B) : Brake fluid   🔧 : N·m (kg-m, ft-lb)

🛠️ (R) : Rubber grease   🔧 : N·m (kg-m, in-lb)

✕ : Always replace after every disassembly.

1. Brake hose	2. Copper washer	3. Cap
4. Bleed valve	5. Mounting bolt	6. Cylinder body
7. Piston seal	8. Piston	9. Piston boot
10. Knuckle slide	11. Sliding sleeve boot	12. Sliding sleeve
13. Inner pad	14. Outer pad	

Rear brakes

67170_TITAN_G93

Disc Brake Pads

REMOVAL & INSTALLATION

Front

1. Before servicing the vehicle, refer to the Precautions Section.
2. Remove the wheel.
3. Remove lower sliding pin bolt.
4. Suspend brake caliper with a remove and remove brake pad and shim from torque member.

To install:
5. Push pistons in so that the pad is firmly installed, using a suitable tool.
6. Mount the brake caliper to torque member.
7. Attach pad retainer to torque member.
8. Lubricate lower sliding pin bolt with a thin layer of silicone grease and install. Torque to 32 ft. lbs. (44 Nm).
9. Install the wheel.

Rear

1. Before servicing the vehicle, refer to the Precautions Section.
2. Remove the wheel.
3. Remove mounting bolt from the top mount.
4. Swing brake caliper open and remove the brake pads.

To install:
5. Push pistons in so that the pad is firmly installed, using a suitable tool.
6. Install pads to the brake caliper.
7. Install top mounting bolt and tighten to 24 ft. lbs. (32 Nm).
8. Install the wheel.

Commonly Used Abbreviations

2
2WD	Two Wheel Drive

4
4WD	Four Wheel Drive

A
A/C	Air Conditioning
ABDC	After Bottom Dead Center
ABS	Anti-lock Brakes
AC	Alternating Current
ACL	Air cleaner
ACT	Air Charge Temperature
AIR	Secondary Air Injection
ALCL	Assembly Line Communications Link
ALDL	Assembly Line Diagnostic Link
AT	Automatic Transaxle/Transmission
ATDC	After Top Dead Center
ATF	Automatic Transmission Fluid
ATS	Air Temperature Sensor
AWD	All Wheel Drive

B
BAP	Barometric Absolute Pressure
BARO	Barometric Pressure
BBDC	Before Bottom Dead Center
BCM	Body Control Module
BDC	Bottom Dead Center
BPT	Backpressure Transducer
BTDC	Before Top Dead Center
BVSV	Bimetallic Vacuum Switching Valve

C
CAC	Charge Air Cooler
CARB	California Air Resources Board
CAT	Catalytic Converter
CCC	Computer Command Control
CCCC	Computer Controlled Catalytic Converter
CCCI	Computer Controlled Coil Ignition
CCD	Computer Controlled Dwell
CDI	Capacitor Discharge Ignition
CEC	Computerized Engine Control
CFI	Continuous Fuel Injection
CIS	Continuous Injection System
CIS-E	Continuous Injection System - Electronic
CKP	Crankshaft Position
CL	Closed Loop
CMP	Camshaft Position
CPP	Clutch Pedal Position
CTOX	Continuous Trap Oxidizer System
CTP	Closed Throttle Position
CVC	Constant Vacuum Control
CYL	Cylinder

D
DBC	Dual Bed Catalyst
DC	Direct Current
DFI	Direct Fuel Injection
DIS	Distributorless Ignition System
DLC	Data Link Connector
DMM	Digital Multimeter
DOHC	Double Overhead Camshaft
DRB	Diagnostic Readout Box
DTC	Diagnostic Trouble Code
DTM	Diagnostic Test Mode
DVOM	Digital Volt/Ohmmeter

E
EBCM	Electronic Brake Control Module
ECM	Engine Control Module
ECT	Engine Coolant Temperature
ECU	Engine Control Unit or Electronic Control Unit
EDIS	Electronic Distributorless Ignition System
EEC	Electronic Engine Control
EEPROM	Electrically Erasable Programmable Read Only Memory
EFE	Early Fuel Evaporation
EGR	Exhaust Gas Recirculation
EGRT	Exhaust Gas Recirculation Temperature
EGRVC	EGR Valve Control
EPROM	Erasable Programmable Read Only Memory
EVAP	Evaporative Emissions
EVP	EGR Valve Position

F
FBC	Feedback Carburetor
FEEPROM	Flash Electrically Erasable Programmable Read Only Memory
FF	Flexible Fuel
FI	Fuel Injection
FT	Fuel Trim
FWD	Front Wheel Drive

G
GND	Ground

H
HAC	High Altitude Compensation
HEGO	Heated Exhaust Gas Oxygen sensor
HEI	High Energy Ignition
HO2 Sensor	Heated Oxygen Sensor

I
IAC	Idle Air Control
IAT	Intake Air Temperature
ICM	Ignition Control Module
IFI	Indirect Fuel Injection
IFS	Inertia Fuel Shutoff
ISC	Idle Speed Control
IVSV	Idle Vacuum Switching Valve

Commonly Used Abbreviations

K

KOEO	Key On, Engine Off
KOER	Key ON, Engine Running
KS	Knock Sensor

M

MAF	Mass Air Flow
MAP	Manifold Absolute Pressure
MAT	Manifold Air Temperature
MC	Mixture Control
MDP	Manifold Differential Pressure
MFI	Multiport Fuel Injection
MIL	Malfunction Indicator Lamp or Maintenance
MST	Manifold Surface Temperature
MVZ	Manifold Vacuum Zone

N

NVRAM	Nonvolatile Random Access Memory

O

O2 Sensor	Oxygen Sensor
OBD	On-Board Diagnostic
OC	Oxidation Catalyst
OHC	Overhead Camshaft
OL	Open Loop

P

P/S	Power Steering
PAIR	Pulsed Secondary Air Injection
PCM	Powertrain Control Module
PCS	Purge Control Solenoid
PCV	Positive Crankcase Ventilation
PIP	Profile Ignition Pick-up
PNP	Park/Neutral Position
PROM	Programmable Read Only Memory
PSP	Power Steering Pressure
PTO	Power Take-Off
PTOX	Periodic Trap Oxidizer System

R

RABS	Rear Anti-lock Brake System
RAM	Random Access Memory
ROM	Read Only Memory
RPM	Revolutions Per Minute
RWAL	Rear Wheel Anti-lock Brakes
RWD	Rear Wheel Drive

S

SBC	Single Bed Converter
SBEC	Single Board Engine Controller
SC	Supercharger
SCB	Supercharger Bypass
SFI	Sequential Multiport Fuel Injection
SIR	Supplemental Inflatable Restraint
SOHC	Single Overhead Camshaft
SPL	Smoke Puff Limiter
SPOUT	Spark Output
SRI	Service Reminder Indicator
SRS	Supplemental Restraint System
SRT	System Readiness Test
SSI	Solid State Ignition
ST	Scan Tool
STO	Self-Test Output

T

TAC	Thermostatic Air Cleaner
TBI	Throttle Body Fuel Injection
TC	Turbocharger
TCC	Torque Converter Clutch
TCM	Transmission Control Module
TDC	Top Dead Center
TFI	Thick Film Ignition
TP	Throttle Position
TR Sensor	Transaxle/Transmission Range Sensor
TVV	Thermal Vacuum Valve
TWC	Three-way Catalytic Converter

V

VAF	Volume Air Flow, or Vane Air Flow
VAPS	Variable Assist Power Steering
VRV	Vacuum Regulator Valve
VSS	Vehicle Speed Sensor
VSV	Vacuum Switching Valve

W

WOT	Wide Open Throttle
WU-TWC	Warm Up Three-way Catalytic Converter

ENGLISH TO METRIC CONVERSION: TORQUE

To convert foot-pounds (ft. lbs.) to Newton-meters (Nm), multiply the number of ft. lbs. by 1.36

To convert Newton-meters (Nm) to foot-pounds (ft. lbs.), multiply the number of Nm by 0.7376

ft. lbs.	Nm	ft. lbs.	Nm	ft. lbs.	Nm	ft. lbs.	Nm
0.1	0.1	34	46.2	76	103.4	118	160.5
0.2	0.3	35	47.6	77	104.7	119	161.8
0.3	0.4	36	49.0	78	106.1	120	163.2
0.4	0.5	37	50.3	79	107.4	121	164.6
0.5	0.7	38	51.7	80	108.8	122	165.9
0.6	0.8	39	53.0	81	110.2	123	167.3
0.7	1.0	40	54.4	82	111.5	124	168.6
0.8	1.1	41	55.8	83	112.9	125	170.0
0.9	1.2	42	57.1	84	114.2	126	171.4
1	1.4	43	58.5	85	115.6	127	172.7
2	2.7	44	59.8	86	117.0	128	174.1
3	4.1	45	61.2	87	118.3	129	175.4
4	5.4	46	62.6	88	119.7	130	176.8
5	6.8	47	63.9	89	121.0	131	178.2
6	8.2	48	65.3	90	122.4	132	179.5
7	9.5	49	66.6	91	123.8	133	180.9
8	10.9	50	68.0	92	125.1	134	182.2
9	12.2	51	69.4	93	126.5	135	183.6
10	13.6	52	70.7	94	127.8	136	185.0
11	15.0	53	72.1	95	129.2	137	186.3
12	16.3	54	73.4	96	130.6	138	187.7
13	17.7	55	74.8	97	131.9	139	189.0
14	19.0	56	76.2	98	133.3	140	190.4
15	20.4	57	77.5	99	134.6	141	191.8
16	21.8	58	78.9	100	136.0	142	193.1
17	23.1	59	80.2	101	137.4	143	194.5
18	24.5	60	81.6	102	138.7	144	195.8
19	25.8	61	83.0	103	140.1	145	197.2
20	27.2	62	84.3	104	141.4	146	198.6
21	28.6	63	85.7	105	142.8	147	199.9
22	29.9	64	87.0	106	144.2	148	201.3
23	31.3	65	88.4	107	145.5	149	202.6
24	32.6	66	89.8	108	146.9	150	204.0
25	34.0	67	91.1	109	148.2	151	205.4
26	35.4	68	92.5	110	149.6	152	206.7
27	36.7	69	93.8	111	151.0	153	208.1
28	38.1	70	95.2	112	152.3	154	209.4
29	39.4	71	96.6	113	153.7	155	210.8
30	40.8	72	97.9	114	155.0	156	212.2
31	42.2	73	99.3	115	156.4	157	213.5
32	43.5	74	100.6	116	157.8	158	214.9
33	44.9	75	102.0	117	159.1	159	216.2

METRIC TO ENGLISH CONVERSION: TORQUE

To convert foot-pounds (ft. lbs.) to Newton-meters (Nm), multiply the number of ft. lbs. by 1.36
To convert Newton-meters (Nm) to foot-pounds (ft. lbs.), multiply the number of Nm by 0.7376

Nm	ft. lbs.	Nm	ft. lbs.	Nm	ft. lbs.	Nm	ft. lbs.	Nm	ft. lbs.
0.1	0.1	34	25.0	76	55.9	118	86.8	160	117.6
0.2	0.1	35	25.7	77	56.6	119	87.5	161	118.4
0.3	0.2	36	26.5	78	57.4	120	88.2	162	119.1
0.4	0.3	37	27.2	79	58.1	121	89.0	163	119.9
0.5	0.4	38	27.9	80	58.8	122	89.7	164	120.6
0.6	0.4	39	28.7	81	59.6	123	90.4	165	121.3
0.7	0.5	40	29.4	82	60.3	124	91.2	166	122.1
0.8	0.6	41	30.1	83	61.0	125	91.9	167	122.8
0.9	0.7	42	30.9	84	61.8	126	92.6	168	123.5
1	0.7	43	31.6	85	62.5	127	93.4	169	124.3
2	1.5	44	32.4	86	63.2	128	94.1	170	125.0
3	2.2	45	33.1	87	64.0	129	94.9	171	125.7
4	2.9	46	33.8	88	64.7	130	95.6	172	126.5
5	3.7	47	34.6	89	65.4	131	96.3	173	127.2
6	4.4	48	35.3	90	66.2	132	97.1	174	127.9
7	5.1	49	36.0	91	66.9	133	97.8	175	128.7
8	5.9	50	36.8	92	67.6	134	98.5	176	129.4
9	6.6	51	37.5	93	68.4	135	99.3	177	130.1
10	7.4	52	38.2	94	69.1	136	100.0	178	130.9
11	8.1	53	39.0	95	69.9	137	100.7	179	131.6
12	8.8	54	39.7	96	70.6	138	101.5	180	132.4
13	9.6	55	40.4	97	71.3	139	102.2	181	133.1
14	10.3	56	41.2	98	72.1	140	102.9	182	133.8
15	11.0	57	41.9	99	72.8	141	103.7	183	134.6
16	11.8	58	42.6	100	73.5	142	104.4	184	135.3
17	12.5	59	43.4	101	74.3	143	105.1	185	136.0
18	13.2	60	44.1	102	75.0	144	105.9	186	136.8
19	14.0	61	44.9	103	75.7	145	106.6	187	137.5
20	14.7	62	45.6	104	76.5	146	107.4	188	138.2
21	15.4	63	46.3	105	77.2	147	108.1	189	139.0
22	16.2	64	47.1	106	77.9	148	108.8	190	139.7
23	16.9	65	47.8	107	78.7	149	109.6	191	140.4
24	17.6	66	48.5	108	79.4	150	110.3	192	141.2
25	18.4	67	49.3	109	80.1	151	111.0	193	141.9
26	19.1	68	50.0	110	80.9	152	111.8	194	142.6
27	19.9	69	50.7	111	81.6	153	112.5	195	143.4
28	20.6	70	51.5	112	82.4	154	113.2	196	144.1
29	21.3	71	52.2	113	83.1	155	114.0	197	144.9
30	22.1	72	52.9	114	83.8	156	114.7	198	145.6
31	22.8	73	53.7	115	84.6	157	115.4	199	146.3
32	23.5	74	54.4	116	85.3	158	116.2	200	147.1
33	24.3	75	55.1	117	86.0	159	116.9	201	147.8

ENGLISH/METRIC CONVERSION: TEMPERATURE

To convert Fahrenheit (F°) to Celsius (C°), take F° temperature and subtract 32, multiply the result by 5 and divide the result by 9
To convert Celsius (C°) to Fahrenheit (F°), take C° temperature and multiply it by 9, divide the result by 5 and add 32

F°	C°	F°	C°	C°	F°	C°	F°
-40	-40.0	150	65.6	-38	-36.4	46	114.8
-35	-37.2	155	68.3	-36	-32.8	48	118.4
-30	-34.4	160	71.1	-34	-29.2	50	122
-25	-31.7	165	73.9	-32	-25.6	52	125.6
-20	-28.9	170	76.7	-30	-22	54	129.2
-15	-26.1	175	79.4	-28	-18.4	56	132.8
-10	-23.3	180	82.2	-26	-14.8	58	136.4
-5	-20.6	185	85.0	-24	-11.2	60	140
0	-17.8	190	87.8	-22	-7.6	62	143.6
1	-17.2	195	90.6	-20	-4	64	147.2
2	-16.7	200	93.3	-18	-0.4	66	150.8
3	-16.1	205	96.1	-16	3.2	68	154.4
4	-15.6	210	98.9	-14	6.8	70	158
5	-15.0	212	100.0	-12	10.4	72	161.6
10	-12.2	215	101.7	-10	14	74	165.2
15	-9.4	220	104.4	-8	17.6	76	168.8
20	-6.7	225	107.2	-6	21.2	78	172.4
25	-3.9	230	110.0	-4	24.8	80	176
30	-1.1	235	112.8	-2	28.4	82	179.6
35	1.7	240	115.6	0	32	84	183.2
40	4.4	245	118.3	2	35.6	86	186.8
45	7.2	250	121.1	4	39.2	88	190.4
50	10.0	255	123.9	6	42.8	90	194
55	12.8	260	126.7	8	46.4	92	197.6
60	15.6	265	129.4	10	50	94	201.2
65	18.3	270	132.2	12	53.6	96	204.8
70	21.1	275	135.0	14	57.2	98	208.4
75	23.9	280	137.8	16	60.8	100	212
80	26.7	285	140.6	18	64.4	102	215.6
85	29.4	290	143.3	20	68	104	219.2
90	32.2	295	146.1	22	71.6	106	222.8
95	35.0	300	148.9	24	75.2	108	226.4
100	37.8	305	151.7	26	78.8	110	230
105	40.6	310	154.4	28	82.4	112	233.6
110	43.3	315	157.2	30	86	114	237.2
115	46.1	320	160.0	32	89.6	116	240.8
120	48.9	325	162.8	34	93.2	118	244.4
125	51.7	330	165.6	36	96.8	120	248
130	54.4	335	168.3	38	100.4	122	251.6
135	57.2	340	171.1	40	104	124	255.2
140	60.0	345	173.9	42	107.6	126	258.8
145	62.8	350	176.7	44	111.2	128	262.4

LENGTH CONVERSION

To convert inches (in.) to millimeters (mm), multiply the number of inches by 25.4

To convert millimeters (mm) to inches (in.), multiply the number of millimeters by 0.04

Inches	Millimeters	Inches	Millimeters	Inches	Millimeters	Inches	Millimeters
0.0001	0.00254	0.005	0.1270	0.09	2.286	4	101.6
0.0002	0.00508	0.006	0.1524	0.1	2.54	5	127.0
0.0003	0.00762	0.007	0.1778	0.2	5.08	6	152.4
0.0004	0.01016	0.008	0.2032	0.3	7.62	7	177.8
0.0005	0.01270	0.009	0.2286	0.4	10.16	8	203.2
0.0006	0.01524	0.01	0.254	0.5	12.70	9	228.6
0.0007	0.01778	0.02	0.508	0.6	15.24	10	254.0
0.0008	0.02032	0.03	0.762	0.7	17.78	11	279.4
0.0009	0.02286	0.04	1.016	0.8	20.32	12	304.8
0.001	0.0254	0.05	1.270	0.9	22.86	13	330.2
0.002	0.0508	0.06	1.524	1	25.4	14	355.6
0.003	0.0762	0.07	1.778	2	50.8	15	381.0
0.004	0.1016	0.08	2.032	3	76.2	16	406.4